RUSSIAN–ENGLISH

SCIENTIFIC AND TECHNICAL DICTIONARY

IN TWO VOLUMES

Volume 1

А–О

RUSSIAN–ENGLISH SCIENTIFIC AND TECHNICAL DICTIONARY

M. H. T. and V. L. ALFORD

Volume 1

A–O

PERGAMON PRESS

OXFORD · NEW YORK · BEIJING · FRANKFURT
SÃO PAULO · SYDNEY · TOKYO · TORONTO

U.K.	Pergamon Press, Headington Hill Hall, Oxford OX3 0BW, England
U.S.A.	Pergamon Press, Maxwell House, Fairview Park, Elmsford, New York 10523, U.S.A.
PEOPLE'S REPUBLIC OF CHINA	Pergamon Press, Room 4037, Qianmen Hotel, Beijing, People's Republic of China
FEDERAL REPUBLIC OF GERMANY	Pergamon Press, Hammerweg 6, D-6242 Kronberg, Federal Republic of Germany
BRAZIL	Pergamon Editora, Rua Eça de Queiros, 346, CEP 04011, Paraiso, São Paulo, Brazil
AUSTRALIA	Pergamon Press Australia, P.O. Box 544, Potts Point, N.S.W. 2011, Australia
JAPAN	Pergamon Press, 8th Floor, Matsuoka Central Building, 1-7-1 Nishishinjuku, Shinjuku-ku, Tokyo 160, Japan
CANADA	Pergamon Press Canada, Suite No 271, 253 College Street, Toronto, Ontario, Canada M5T 1R5

First edition 1970

Reprinted 1974; 1981; 1984, 1988

Library of Congress Catalog Card No. 73—88348

Printed in Great Britain by A. Wheaton & Co. Ltd., Exeter
ISBN 0-08-012227-2

CONTENTS

FOREWORD

THE publication of an entirely new dictionary containing more than 100,000 entries is a rare event. It is not surprising. A work of this size takes over ten man-years to write and it is seldom that any of the few people with the necessary experience can spare the time for such an undertaking.

The coverage of the dictionary is even greater than the number of entries suggests. All Russian word components, both morphemes and longer combining forms, are shown and these make it possible to deduce the meanings of many words not included in the dictionary. Thus, it has only been necessary to give examples of words based on international roots or forming part of systematized vocabularies.

The dictionary is designed to give the fullest possible coverage of all text in 94 specialized fields. For this purpose it has a completely reworked vocabulary of basic Russian. This will be invaluable to scientists learning the language, for whom ordinary words are harder to deduce than technical terms, and it will be useful to translators in establishing correct formality and permissible degrees of synonymity.

One of the most original features of the dictionary is that it has been planned to assist the latest techniques of vocabulary learning. In the past, vocabulary acquisition has been a major obstacle in achieving an ability to read extensively. Language students were expected to teach themselves by incessant reference to dictionaries. This procedure is so time-consuming and inefficient that many never succeeded in reaching a useful standard. Experiments conducted by M. H. T. Alford in the Language Centre at the University of Essex have now established means of learning large vocabularies in a reasonable amount of employed time. The section following this foreword explains how this objective can be achieved. It is recommended that anyone who uses a dictionary more than five times a day should study the instructions carefully.

Although an inadequate knowledge of grammar can be improved more quickly than an insufficient vocabulary, it is the lexicographer's duty to assume that his client's grasp of Russian grammar is weak. In consequence, the present dictionary is particularly well provided with grammatical information. Besides the standard (canonical) forms of the words, the endings of certain regular cases are provided in order to illustrate their paradigms. All irregular forms are shown with, where necessary, their own separate alphabetical entries. Consequently, foreknowledge, other than of Russian alphabetical order, is not required for tracing information. A supplementary dictionary of word endings in right-to-left alphabetical order is also provided and this will be of great assistance in determining the grammatical functions of word forms not appearing in the main body of the dictionary.

No other large-scale dictionary of scientific and technical Russian shows the stress of all words. This information is here given for the words appearing in each entry and for other inflected forms, where the stress is different. This will provide an indispensable aid to interpreters and to those learning to speak Russian. Readers will be able to acquire a large vocabulary with correct pronunciation and thus avoid prejudicing their use of the spoken language.

It is particularly appropriate that the vast amount of effort required to make this dictionary has been devoted to the scientific and technical registers of Russian. The importance of the Soviet contribution to the advancement of science and technology is already great and it continues to increase. In addition, Russian publications now provide access to the world's literature with a thoroughness and speed not always matched by English language sources.

This dictionary should play its part in promoting the use of the Russian language among people who know English. It has been designed primarily for British and American scientists and translators, the English being defined by *Webster's International Dictionary*. The needs of users whose native language is not English, have also been studied.

The dictionary does not cover belles-lettres or popular science, but there will be few serious students of contemporary affairs in the U.S.S.R. who will not find it of service.

PETER STREVENS

Professor of Applied Linguistics
and Director of the Language Centre,
University of Essex

TRANSLITERATION OF RUSSIAN AND LATIN LETTERS

THERE are 32 letters in the Russian alphabet and 26 in Latin. A one-for-one correspondence is not, therefore, possible. Numerous systems have been employed in the past. Nearly all the variations are represented among the four given below. The table does not provide explanations about some small points. In general, accents and diacritics may be omitted for simplicity. The I.S.O. system allows certain of the B.S. variations as alternatives.

For details, the British Standard can be purchased from:

> The British Standards Institution,
> British Standards House,
> 2, Park Street,
> London, W.1.

At the time of going to press, the I.S.O. system was in the form of Draft Recommendation No. 1243 for the Revision of the I.S.O. Recommendation R9-1955. The issuing body was:

> International Organization for Standardization,
> I.S.O. Central Secretariat,
> 1, rue de Varembé,
> Geneva.

TRANSLITERATION

Cyrillic character		B.S. 2979/58	I.S.O. 1967	Library of Congress	Royal Geog. Soc. U.S. Board (BGN/PCGN) 1959
А	а	a	a	a	a
Б	б	b	b	b	b
В	в	v	v	v	v
Г	г	g	g	g	g
Д	д	d	d	d	d
Е	е	e	e	e	ye, e
Ё	ё	ë	ë	e	yë, ë
Ж	ж	zh	z	zh	zh
З	з	z	z	z	z
И	и	i	i	i	i
Й	й	ĭ	j	ĭ	y
К	к	k	k	k	k
Л	л	l	l	l	l
М	м	m	m	m	m
Н	н	n	n	n	n
О	о	o	o	o	o
П	п	p	p	p	p
Р	р	r	r	r	r
С	с	s	s	s	s
Т	т	t	t	t	t
Тс	тс	t-s			
У	у	u	u	u	u
Ф	ф	f	f	f	f
Х	х	kh	h	k͡h	kh
Ц	ц	ts	c	t͡s	ts
Ч	ч	ch	č	ch	ch
Ш	ш	sh	š	sh	sh
Щ	щ	shch	šč	shch	shch
Ъ	ъ	″	″	″	″
Ы	ы	ȳ	y	ẏ	y
Ь	ь	′	′	′	′
Э	э	é	e	e	e
Ю	ю	yu	ju	i͡u	yu
Я	я	ya	ja	i͡a	ya

INSTRUCTIONS FOR LEARNING LARGE VOCABULARIES

Purpose of the Instructions

Attempts to learn large vocabularies by incessantly looking up the meanings of foreign words waste a great deal of time and often lead to failure. If the instructions given here are followed carefully and methodically, learning will be much more efficient. The dictionary can then be used in the proper way for occasional reference.

Basis of the Instructions

In 1966 the Office for Scientific and Technical Information provided the University of Essex with a grant of £22,460 for developing new methods of learning to read foreign languages. Most of the ensuing work has been devoted to the problems of how to learn large vocabularies. The instructions given below are based on this research and have been tested in numerous experiments carried out with scientists learning Russian.

Learning to Identify a Stimulus

The learning of each vocabulary item involves two stages:
(1) The identification of the foreign word or term (the stimulus).
(2) The remembering of its meaning, either in the form of the English equivalent or of associated information (the response).

Identification of a word is the process of remembering that the stimulus being encountered on the present occasion is the same as a particular stimulus encountered previously. The most effective stimulus obtained from a printed word is the idea of its sound, the auditory image. Some people have useful visual images of print as well, but these seldom persist in long term memory. Words are, therefore, normally identified by their pronunciation. For remembering, this does not have to be correct (though there are good reasons for trying to make it so). Consequently, vocabulary can be learnt without a teacher or recordings. When the latter are available, it will be found that beginners are confident of forming auditory images from print after listening to 200–300 Russian words (spaced 30 words a day with recordings lasting 6–10 minutes, followed by memorizing of the words).

Words composed of familiar parts will, on average, be identified more quickly and remembered longer than words made up of unfamiliar parts. It is thus advantageous to learn the internal structure of words.

The significant parts of words are called morphemes. In Russian these are similar to English. A few hundred of them form many thousands of structured words, such as:

UN·EVEN·NESS·ES
prefix·root·suffix·inflection

These words always have a root and may have some or none of the other components. Examples are:

ATOM·	(root only)
ATOM·IZATION·	(root·suffix·)
ATOM·IZ·ES	(root·suffix·inflection)
INTER·ATOM·IC·	(prefix·root·suffix·)

Some words have more than one root, e.g. GEO·GRAPH·Y, and there are also unstructured words, such as "but", "when", "and", etc.

The entries in this dictionary show the components of structured Russian words by dividing them with raised points. Students will find that they soon learn the morphemes if they always pay attention to these divisions. When making notes of vocabulary from the dictionary, words should be copied with the dividing points and, as will be seen later, memorizing should be carried out with reference to the components thus indicated.

Words in texts do not contain marks to show their morphemes. When these are not all recognized, a strategy for analysing the word can help. In Russian there is only a small number of prefixes. All the common ones are listed on front end paper. There is a similar number of inflections. These will be found in the Supplementary Dictionary of Russian Word Endings (see page 1409) which also contains many of the common suffixes. Reference to these two sources will often clarify the morphemes at either end of the word and, in many cases, what is left over will be the root.

If a root is not recognized, it may be new to one's experience or it may be a variant of a root already learnt. Many Russian roots have several forms and they often differ only in the last letter or two. These consonant variations follow a distinct pattern which can be found in the table of equivalent letters on front end paper. The alternation of consonants shown in this table is responsible for much so-called irregularity. This type of irregularity raises problems which can often be resolved by deduction. Thus, if one knows that г is interchangeable with ж and that мож·ет = it can, one might expect мог·ут to mean "they can".

Learning to Remember a Response

When a foreign word (or multi-word term) is identified as a particular stimulus which has occurred before, remembering has to be extended to its meaning, initially in the form of its English equivalent and later, when familiarity has been achieved, directly to the same reference background as evoked by the English.

The pairing of a stimulus with a response is often regarded as the process of linking two items. In the case of instantaneous response, this appears to be so. But when the remembering of past learning is limited and slow, it can be observed that it is two extensive events which are being associated. The recall of less important parts of the response event, such as the first letter of the English word or the general context of the last encounter, often helps to reinstate the particular information which the memory is seeking. In consequence, the efficiency of learning is considerably increased by building a network of associations for each item.

The first step is to find the meanings of the separate morphemes. These do not always indicate word meanings, but provide valuable clues. The following table illustrates this point:

A Russian word	B Morpheme indication of meaning		C Word meaning
1. в·ход·ить	IN·GO·	infinitive	to enter
2. вы·ход·ить	OUT·GO·	,,	to leave
3. на·ход·ить	ON·GO·	,,	to find
4. рас·ход·ить	APART·GO·	,,	to separate

Given knowledge about the morphemes in column B, one should be able to deduce the meanings of the words in column C, with the possible exception of word No. 3.

Information about morphemes can be found in most text-books (and the present authors have a book on Russian morphemes in preparation). It is also possible to determine morpheme meaning from the dictionary by comparing any entry (a) with its neighbours and (b) with its unprefixed form or with forms having other prefixes. The common elements of meaning can then be ascribed to the common morphemes.

After an introduction to Russian grammar, everyone has some knowledge of the significance of certain morphemes. This evidence and anything else that imagination or ingenuity can produce should be used in the following procedure for learning each Russian–English vocabulary item.

(1) Note any SIMILARITY between the Russian word and its English equivalent. Do the Russian morphemes match the meaning of the English morphemes? Do the words have any common letters? Is there an unusual or archaic English word which would more readily recall the Russian than the English form being learnt?

(2) Find any CONTRAST between the Russian word and its English equivalent. Are the morphemes contradictory to the senses of the whole words? What dissimilarities are there in the letters or in the lengths of the words?

(3) Consider what will happen when you next meet the Russian word without the English equivalent. What evidence would you have to help in the DEDUCTION of its meaning?

(4) Seek any physical EXPERIENCE you can obtain from the Russian word. Say it and its English equivalent aloud and with emphasis (the dictionary gives you the stressed syllable for every Russian word). If you re-encounter the word and cannot remember it, pronounce it or let your tongue move as if doing so. You may remember your motor reactions and then other parts of the learning event. If the Russian word describes an object which you can look at or a sensation which can be experienced (e.g. rough, smooth, hot, cold, wet, dry, etc.), seek the appropriate sensation while thinking about the Russian word.

(5) Try to remember any mental IMAGE you can associate with the Russian word. Images of formerly experienced physical stimulus may be visual, auditory, tactile, olfactory, or motor.

(6) Practise the REPETITION of the Russian and English words. Periods of repetition should be kept short and should be interspersed with other forms of learning.

(7) If, on testing, there is any difficulty in remembering a particular vocabulary item, try the INTEGRATION of all the six forms of learning listed above. Carry out each type of learning in quick succession. In any case, some variation of technique is always advantageous. It is particularly important as one gets tired. A useful reminder of this fact can be displayed on a card listing the different approaches:

SIMILARITY

CONTRAST

DEDUCTION

EXPERIENCE

IMAGE

REPETITION

INTEGRATION

INVENTION

(8) The last means of learning given on the list, INVENTION, refers to all forms of association which are linguistically useless, but assist learning. Any bizarre or striking story can be invented to link the Russian word with the English. For example, in order to remember ТОК = current (elec), one could invent the mnemonic "ТIK–ТОК mechanisms are seldom driven by electric current". Such ideas are very effective in assisting memory. But there is a limit to the amount of nonsense material that one wants to stow away in the brain. Invention should, therefore, be kept for separating confused items or overcoming other difficulties.

These eight different means of learning a Russian–English vocabulary item form a receptive phase in memorization. Not all need be employed and preferred techniques can be given priority, but some variation is important. In particular, repetition should not be used exclusively.

Whatever means are employed in the RECEPTIVE PHASE, the time spent in this kind of activity should be kept to LESS THAN HALF THE TOTAL LEARNING TIME. The rest should be employed in repeated attempts to remember and the checking of the results of each attempt. Attempted remembering of each item has to be carried out at intervals and therefore has to be considered in relation to the learning of vocabularies composed of many words.

Learning a Large Vocabulary

In order to acquire a large vocabulary reasonably quickly, it is necessary to learn 15–30 Russian words a day for 5 or 6 days a week. Using a method described below, a beginner with a good memory will learn 30 Russian words in about 30 minutes. For a University student with average memory this time increases to 40–45 minutes. If more than 30 words a day are learnt, efficiency will fall. It takes less than half the above times to learn 15 words a day and this number may be preferred, if a prolonged task and slower progress are acceptable.

The words to be memorized should be written on separate cards or pieces of paper. This is important to avoid list effects when learning. List structure aids the remembering of items in the list and they may subsequently not be recognized in other surroundings.

One side of each card is used for testing and should have only the Russian word. The other side is for the receptive phase and for checking and should have both Russian and English words, together with any other available information, such as grammatical classification or morpheme meaning.

The checking side can also have a layout giving the Russian word and, on successive lines, first letter of English, first and last letter of English, English letters in scrambled order, the English word, e.g.

ТОК

C

C T

TERNRUC

CURRENT

When an attempt to remember the meaning of the Russian word shown on the testing side fails, the card can be turned over and further attempts to remember can be made with the uncovering of the successive lines which provide increasingly detailed prompts.

When students write their own cards, the procedure contributes to learning. Copying is useful for people with below average memories, but for others it is marginally less efficient than learning directly from reading. Teachers can, therefore, help the students with better memories by preparing cards for them to use.

The cards which are to be learnt on one day are gathered into a pack. A 30-card pack is then divided roughly into three parts (a 15-card pack into two). The words in the first part are then briefly examined using the receptive phase techniques previously described. This is followed by testing whether they have been remembered. If one or more words are forgotten in testing the whole part, the cards are shuffled and testing continues until all are remembered correctly. (Remembering with the help of prompts counts as forgetting.)

The other parts of the pack are then learnt in the same way. Finally, all the parts are shuffled together and testing continues until the whole pack is remembered correctly. This concludes the day's learning. There is no need to go on learning since revision at later stages is used to eliminate the effects of forgetting. It may, however, be found convenient to study the words at odd moments before the systematic learning. This makes use of time that would otherwise be spent unprofitably. People with poor memories also benefit from several learning sessions in one day.

It is one of the commonplaces of psychology that items learnt and forgotten will be relearnt in less time than will be required to learn new material. A system of checking is, therefore, necessary to make use of past effort which would otherwise be wasted. When such a system is in operation, the initial learning need not aim at 100% retention. Owing to many chance factors, some items are more easily remembered than others. At initial learning the easy and difficult items often cannot be distinguished and, if learning aims at 100% retention, all items have to be treated as difficult. This leads to a waste of effort in learning easy vocabulary. It is more efficient to plan for a lower retention and relearn items which prove to be difficult.

The initial learning routine described in the preceding paragraphs has a typical retention pattern for 30-card packs, as follows:

Check 1 Next day—no intervening learning, 1–3 words forgotten.
Check 2 About 1 week later—5 other packs learnt meanwhile, 5–10 words forgotten.
Check 3 About 2 weeks later—10 other packs learnt meanwhile, 5–10 words forgotten.

All checks, though very brief, provide an element of fresh learning. The differences in the number forgotten are due to chance factors and not to personal memorizing ability, which has been equalized in the initial learning by the different times taken to achieve the same criterion.

The testing routine shown above has been very satisfactory for the fast (30 words a day) learning schedule. The interval between checks should be regulated by the number of words learnt in between, i.e. about 120–150. No special relearning effort is required after checks 1 and 2 unless forgetting exceeds the highest figures given. Words not remembered at the third check should be put into a new pack and relearnt when the pack contains 30–45 cards. The time taken on relearning is about half that of initial learning.

In the middle of learning a large vocabulary a daily quota of memorizing will consist of learning a new pack and checking three others. This presents no difficulties, but it is necessary to be methodical.

Cards used for learning should never be thrown away. If, for any reason, a person's Russian reading is interrupted for a long period, he will forget much of his vocabulary, but since it is precisely the forgotten material that will be learnt most quickly, the original instructional data should be used for remedial measures.

Often it is not possible for dictionary users to make notes of words after they look them up. At the same time, frequent use of a dictionary involves forgetting the meanings of many of the words which have been so laboriously found. A procedure is required to ensure that the effort expended on forgotten items is not wasted.

It is, therefore, advantageous to make a pencil mark against a dictionary entry each time it is referred to. Later, one can browse through the dictionary checking the words which were previously not known. If time permits, they can then be copied onto cards and subsequently learnt in the manner described. In any case, entries consulted twice or more, due to forgetting and not to elucidation of multiple meaning, ought to be prepared for systematic learning. It must always be borne in mind that remembering a word:

(1) Saves dictionary search time.

(2) Reduces distraction in the learning of other words.

(3) Contributes to the establishment of context which often enables the meanings of unknown words to be deduced. This not only saves further search time, but is also one of the most efficient means of learning.

(4) Increases the speed of reading and allows greater concentration on the subject matter.

INSTRUCTIONS FOR USE OF DICTIONARY

Looking up Single Words

The dictionary entries are in Russian alphabetical order (see end papers). The dictionary (canonical) forms of single words are:

Nouns	nominative singular.
Adjectives	masculine nominative singular.
Adverbs	(a) invariable
	(b) generally not given when they can be deduced from related adjectives.
Verbs	imperfective infinitive.
Participles	(a) if given, masculine nominative singular (like adjectives)
	(b) generally not given when they can be deduced from infinitives of verbs.
Gerunds	(a) invariable
	(b) generally not given when they can be deduced from infinitives of verbs.
Pronouns	masculine nominative singular.
Prepositions	invariable.
Conjunctions	invariable.

Examples

Text word	до·крит·ическ·ого
Dictionary word	до·крит·и́ческ·ий = subcritical.
Text word	когда
Dictionary word	когда́ = when.

Perfective verbs and irregular forms of all words are listed in the dictionary with cross references to entries in the cases given above.

The look-up procedure for a single word is, therefore, a simple matching process. Invariable words must match exactly, but the others may have a different inflection. Inflections are all letters coming after the last of the raised points dividing the dictionary words.

If everything matches except the letters after the second last point, the word has a different suffix. In these cases it is often possible to deduce the meaning of the text word. The Supplementary Dictionary of Russian Word Endings (see page 1409) will assist this process.

Examples

Text word	волос·ност·ь
Dictionary word	волос·н·о́й = capillary (*a !j*)

See ending ·ност·ь in the Supplementary Dictionary of Russian Word Endings
= feminine noun often denoting a quality expressed by similar adjective.
therefore волос·ност·ь = capillarity.

If a text word is not found in the dictionary, one should see if its first few letters match the prefixes (letters before first raised point) of dictionary words near the location where the absent word would have been. If they do, one should also look for an entry without a prefix or with a different prefix (see common prefixes, front end paper). In this way senses of the absent word's prefix and root may be found. This evidence, together with context, will often make it possible to deduce the meaning of the whole word.

Looking up Hyphenated Words

In Russian as in English there is considerable variation in the use of hyphens. The following search procedure should, therefore, be followed:

(1) Look up the first word and then seek the second word as a sub-entry below.

(2) If (1) produces no result, look up the hyphenated words as if they were written as one continuous word.

(3) If (1) and (2) produce no result, look up both words separately.

Looking up Multi-word Terms

Russian is also like English in the construction of multi-word terms. One of the words, usually the first noun, is dominant. Entries have to be located in the alphabetical position of these key words because:

(1) Certain of the qualifying words may be omitted for brevity and occasionally their order may vary.

(2) Three- and four-word terms can often be deduced from two-word terms linked by key words.

Examples

Text entry — continuous strip mill

Dictionary entries — mill, continuous
mill, strip

However, it is not possible to make an invariable rule that all entries will be found in the alphabetical place of their first nouns. A small number of nouns, such as у·стрóй·ств·о, у·станóв·к·а and машйн·а, are of such generality that the preceding adjective is usually the dominant word of the term. In these cases terms are located under the adjectives with a cross reference "see associated adjectives" in the noun entry. There are also a few nouns, such as кисл·от·á, which form part of so many hundreds of terms, that it would be wasteful of space to list every qualifying adjective by itself and again under кисл·от·á.

Instructions for looking up multi-word terms are, therefore:

(1) Find the first noun.

(2) Find the sub-entry with words coming after the first noun, if there are any that belong to the term.

(3) Find the sub-entry with the first and subsequent adjectives.

(4) When noun entries refer users to the associated adjectives, terms are in the alphabetical order of the nouns found there. This allows sub-entries to be traced immediately without knowledge of the gender of the term.

Interpretation of an Entry

When a word or a multi-word term is found, the information given in the entry requires to be interpreted.

The morpheme (prefix · root · suffix · inflection) structure of the Russian word is indicated by raised points. The morpheme boundaries have been adjusted in a few cases to be as consistent as possible and thus assist learning and cross reference.

Any letters after the last point are the inflection and are liable to change according to the use of the word in sentences. All the inflected forms of a word are called its paradigm. The forms of the entry words given in the dictionary were listed in the first paragraph of this section. Immediately after these words, between commas and still in heavy type, are other inflected forms illustrating part of their paradigms. These show:

Nouns	genitive singular
Adjectives	nominative feminine and neuter
Verbs	3rd person plural
	present of imperfective verbs
	future of perfective verbs.

All irregularities, including changes of stress, are also given. When inflections other than the above are shown, they are preceded by a bracketed abbreviation describing what they are. (All abbreviations are listed on back end paper).

The information about inflected forms is followed by a grammatical classification of the word, e.g. (*m*) = masculine noun, or (*perf*) = perfective verb.

Bracketed abbreviations before the English equivalent show a discipline or branch of technology where the term is in common use. Its use is not, of course, limited to these spheres. Bracketed information after the English equivalent is self-explanatory and will not be found in the list of abbreviations.

English equivalents sharing the same meaning are separated by commas. Those with different meanings are separated by semicolons.

The spelling is taken from *Webster's Third New International Dictionary*. Where British and American forms differ, the British is given before the American. Where British and American word usage differs, the words are followed by (Brit) or (U.S.). An oblique stroke (/) between two English words signifies that either can be used in the combination shown.

REFERENCE SOURCES

A VERY large number of books have been consulted in making this dictionary. Many of them are of limited interest to the reader since they are not prepared for easy reference in seeking encyclopaedic information about specific terms. The list given below is, therefore, confined to works which can be used to trace individual Russian or English words and obtain detailed information about their meaning and use.

Chambers Technical Dictionary, Ed. C. F. Tweney *et al.* W. & R. Chambers Ltd., London, 1959.

Dictionary and Encyclopaedia of Paper and Paper-making, E. J. Labarre. Swets and Zeitlinger, Amsterdam, 1952.

Dictionary of American–English Usage, Margaret Nicholson. Oxford University Press, New York, 1957.

Dictionary of Metallurgy, A. D. Merriman. Macdonald & Evans, London, 1958.

Dictionary of Named Effects and Laws, D. W. G. Ballentyne and L. E. Q. Walker. Chapman & Hall Ltd., London, 1961.

Dictionary of Photography, A. L. M. Sowerby. Iliffe & Sons Ltd., London, 1957.

Dictionary of Scientific Terms, I. F. and W. D. Henderson (7th ed. J. H. Kenneth). Oliver & Boyd, Edinburgh, 1960.

Encyclopaedia Britannica, Vols. 1–24, Encyclopaedia Britannica, Chicago, etc., 1959.

Encyclopaedia of the Iron and Steel Industry, A. K. Osborne. Technical Press Ltd., London.

Encyclopaedic Dictionary of Electronics and Nuclear Engineering, Robert I. Sarbacher. Sir Isaac Pitman & Sons Ltd., London, 1959.

Encyclopaedic Dictionary of Physics, Vols. 1–8, Ed. J. Thewlis. Pergamon Press, Oxford, 1961–1963.

Forbes' Russian Grammar, J. C. Dumbreck. Oxford University Press, 1964.

Glossary of Terms in Nuclear Science and Technology, American Society of Mechanical Engineers, New York, 1957.

High Frequency Words and Occurrence Forms in Russian Physics, A. S. Kozak, RAND Corporation, Santa Monica, Cal., 1962.

International Code of Zoological Nomenclature, International Trust for Zoological Nomenclature, London, 1961.

Kempe's Engineers Year-Book, Morgan Brothers Ltd., London.

Making, Shaping and Treating of Steel, United States Steel Corporation, Pittsburgh, 1957.

Manufacture of Iron and Steel, Vols. 1–4, G. R. Bashforth, Chapman & Hall Ltd., London, 1960–1962.

Mathematics Dictionary, Ed. James and James. D. Van Nostrand Company, Inc., Princeton etc., 1959.

McGraw-Hill Encyclopedia of Science and Technology, Vols. 1–15. McGraw-Hill Book Co., New York, 1960.

Metallurgical Dictionary, J. G. Henderson *et al.* Reinhold Publishing Corporation, New York, 1953.

Modern Building Encyclopaedia, Odhams Press Ltd., London, 1959.

Petroleum Production Engineering, L. C. Uren. McGraw-Hill Book Co. Inc., New York, etc., 1953.

Roget's Thesaurus of English Words and Phrases. Longmans Green & Co., London, etc.

Rückläufiges Wörterbuch der Russischen Sprache der Gegenwart, H. H. Bielfeldt. Akademie Verlag, Berlin, 1965.

Russian Word Count, Harry H. Josselson. Wayne University Press, Detroit, 1953.

Russian Word Count, E. Steinfeldt. Press Publishers, Moscow.

Russischen Verben (*Die*), E. Daum, W. Schenk. Veb Verlag Enzyklopädie, Leipzig, 1963.

Russisches Etymologisches Wörterbuch, Vols. 1–3, Max Vasmer. Carl Winter Universitätsverlag, Heidelberg, 1953.

Russisches rückläufiges Wörterbuch, Vols. 1–2, R. Greve and B. Kroesche. Otto Harrassowitz, Wiesbaden, 1959.

Scientific Russian, George E. Condoyannis. John Wiley & Sons Inc., New York, 1959.

Short Russian Reference Grammar, I. M. Pulkina. Foreign Languages Publishing House, Moscow, 1960.

Shorter Oxford English Dictionary, Vols. 1–2, Oxford, 1962.

Van Nostrand Chemists Dictionary, Honig *et al.* Van Nostrand, Princeton, etc., 1953.

Webster's Biographical Dictionary, G. and C. Merriam Co., Springfield, Mass.

Webster's Third New International Dictionary, Ed. Philip B. Gove. London, G. Bell & Sons Ltd.; Springfield, Mass., G. and C. Merriam Co.

Атомная энергия (краткая энциклопедия), В. С. Емельянов *et al.* Большая Советская Энциклопедия.

Большая советская энциклопедия (второе издание), 1–50, *Ed.* С. И. Вавилов. Большая Советская Энциклопедия. 1949–1957. Year Books 1960–1967.

Краткая химическая энциклопедия, 1–5. Советская Энциклопедия Москва, 1961–1967.

Краткий словарь ботанических терминов, Д. П. Викторов. Наука, Москва–Ленинград, 1964.

Краткий словарь синонимов русского языка, В. Н. Клюева. Учпедгиз, Москва, 1961.

Краткий справочник по русской грамматике, И. М. Пулькина. Учпедгиз, Москва, 1961.

Краткий справочник по химии, О. Д. Куриленко *et al.* Наукова Думка, Киев, 1965.

Краткий этимологический словарь русского языка, Н. М. Шанский *et al.* Учпедгиз, Москва, 1961.

Морской словарь, 1–2, Военное Издательство, Москва, 1959.

Морфология современного русского языка, И. Г. Голанов. Высшая Школа, Москва, 1962.

Орфографический словарь русского языка, С. Г. Бархударов *et al.* Госиздат. Иностранных и Национальных Словарей. Москва, 1963.

Петрографический словарь, Ф. Ю. Левинсон-Лессинг *et al.* Госгеолтехиздат, Москва, 1963.

Словарь американской лингвистической терминологии, Э. Хэмп. Прогресс, Москва, 1964.

Словарь иностранных слов, Издательство Советская Энциклопедия, Москва, 1964.

Словарь лингвистических терминов, О. С. Ахманова. Советская Энциклопедия, Москва, 1966.

Словарь по геологии нефти, М. Ф. Мирчинка. Гостоптехиздат, Ленинград, 1958.

Словарь русского языка, 1–4, Академия Наук СССР, Институт Языкознания, Москва, 1957–1961.

Словарь русского языка, С. И. Ожегов. Госиздат. Иностранных и Национальных Словарей, Москва, 1964.

Словарь сокращений русского языка, Д. И. Алексеев *et al.* Госиздат. Иностранных и Национальных Словарей, Москва, 1963.

Словарь ударений для работников радио и телевидения, Г. Агеенко, М. Зарва. Госиздат Иностранных и Национальных Словарей, Москва, 1960.

Справочник мер, В. А. Соколов, Л. М. Красавин. Внешторгиздат, Москва, 1956.

Справочник металлиста, 1–5, С. А. Чернавский. Машгиз, Москва, 1957.

Справочник по русской грамматике для иностранцев, Н. Г. Хромец, А. П. Вейсман. Издательство ВПШ и АОН, Москва, 1961.

Справочник по элементарной математике, М. Я. Выгодский. Физматгиз, Москва, 1962.

Справочник химика, 1–3, С. А. Зонис, Г. А. Симонов. Госхимиздат, Ленинград, 1962.

Толковый словарь математических терминов, В. А. Диткин *et al.* Издательство Просвещение, Москва, 1965.

Толковый словарь русского языка, 1–4, Д. Н. Ушаков *et al.* Москва, 1935.

Физико-химические свойства элементов, Г. В. Самсонов. Наукова Думка, Киев, 1965.

Физический энциклопедический словарь, 1–5, Советская Энциклопедия, Москва, 1960–1966.

Экспортно-импортный словарь, 1–3, Б. Т. Колпаков *et al.* Внешторгиздат, Москва, 1954.

Энциклопедический сельскохозяйственный словарь-справочник, Госиздат. Сельскохозяйственной Литературы, Москва, 1959.

Энциклопедический словарь, 1–2, Б. А. Введенский *et al.* Госиздат. Советская Энциклопедия, Москва, 1963.

ACKNOWLEDGEMENTS

AMONG experienced lexicographers we are indebted to N. S. Doniach, S. Massey and M. Lambert. Mr. Lambert made available to us a selection of basic terms from his *Russian–English Dictionary for the Rubber and Plastics Industries*, which should be consulted for detailed research in this field. C. I. Pedersen made available some vocabulary which he had collected in his work as a translator.

Professor F. Cap of Innsbruck University provided us with the Russian originals of material being abstracted for the Library of Congress. This was most valuable in enabling us to check our coverage of important vocabulary.

During the many years that this dictionary has been in preparation we have received much help from institutions and industrial companies. Often, several people in each organization have given us advice or materials and it is not possible to thank everyone by name. We would, however, like to express particular appreciation to F. Liebesny and R. Sewell, who are both well known in the field of scientific information. Our gratitude is due to:

Albright and Wilson Limited
Aluminium Development Association
Aslib
Austin Motor Company Limited
Babcock and Wilcox Limited
A. S. D. Barrett, Consultant on High Vacuum Technology
Borax Consolidated Limited
Bowater Research and Development Company Limited
Brewing Industry Research Foundation
British Aluminium Company Limited
British Cast Iron Research Association
British Ceramic Research Association
British Cotton Industry Research Association
British Glass Industry Research Association
British Oxygen Research and Development Limited
British Standards Institution
Davy and United Engineering Company Limited
Dowty Limited
Dunlop Research Centre
English Electric Company Limited
Esso Research Limited
Federation of British Industries
Fisons Limited
H. J. Heinz Company Limited

Imperial Chemical Industries Limited
Institute of Metals
Institute of Petroleum
Iron and Steel Institute
Lewis's Medical Scientific and Technical Lending Library
Loewy Engineering Company Limited
J. Lyons and Company Limited
Manchester Oil Refinery Limited
Metal Box Company Limited
Mond Nickel Company Limited
Monsanto Chemicals Limited
N. L. L. (National Lending Library for Science and Technology)
W. H. A. Robertson and Company Limited
Royal Geographic Society
United Steel Companies Limited
Whitbread and Company Limited.

Much of the work on the dictionary was carried out while we were living in Dornbirn, Vorarlberg, Austria. We are more grateful than we can say to our friends, Hans and Lottie Knapp, for putting their house at our disposal. We can best express our feelings to the numerous Austrians who made our stay so pleasant, through formal thanks to the Landeshauptmann of Vorarlberg for permitting us to spend four very happy years in that beautiful part of the world.

RUSSIAN-ENGLISH DICTIONARY А–О

RUSSIAN-ENGLISH SCIENTIFIC AND TECHNICAL DICTIONARY

by

M.H.T. and V.L. ALFORD

Errata to Vol. 1

Page	Column	Line	Read	For
14	right	6 from bottom	актива́ц·и·я	алтива́ц·и·я
88	left	5 from bottom	буты́л·к·а	бут л·к·а
126	right	7 from top	~ культу́р·а	~ култу́р·а
32	right	7 from bottom	main entry, to be moved to left margin	воз·мо́ж·н·ый
152	left	12 from bottom	~, ди·хро·йч·ая	~, ди·хро·йча·ая
213	left	2 from bottom	~, больш·а́я	~, болш·а́я
230	right	5 from bottom	гига·электрон·во́льт	гига·электорн·во́льт
237	left	25 from bottom	дви́г·а·ть, -ют, or дви́ж·ут	дви́г·а·ть, -ют, ог дви́ж·ут
241	right	16 from top	двух·ба́ль·н·ый	двух·ба́лль·н·ый
487	right	7 from bottom	кри́ч·н·ый	кри́ч·но·ый
541	right	21 from bottom	мало·алкого́ль·н·ый	мало·алкого́ль·
700	left	7 from bottom	sini	ni
762	left	7 from bottom	~ эемле·трясе́ни·я	~ эемле·тряс е́ни·я

А

a (*conj*) and, but; (*abbr*) = **ампе́р·**, (elec) A, amp, ampere

a- (*prefix to int words*) a- (not, without, lacking)

аален·ск·ий, -ая, -ое, (*a*) (geol) Aalenian

ааро́н·ов·а бород·а́ (*f*) (bot) arum, *Arum*

абажу́р·, -а, (*m*) lampshade, shade, screen

абажу́р·н·ый, -ая, -ое, (*a*) of **абажур·**

абажуро·держ·а́тел·ь, -я, (*m*) (elec lighting gallery)

абак·, -а, (*m*) abacus

аба́к·а, -и, (*f*) (arch) abacus; abaca fibre, abaca, manila, manila hemp; abacus (for calculating)

аб·аксиа́ль·н·ый, -ая, -ое, (*a*) (bot) abaxial

аб·ампе́р·, -а, (*g pl*) **аб·ампе́р·,** (*m*) (elec) abampere

абандо́н·, -а, (*m*) (ins) abandon

абаси́н·, -а, (*m*) (chem) abasin acetylcarbromal

Аббе конденс·а́тор· (*m*) (elec) Abbé condenser

аббревиату́р·а, -ы, (*f*) abbreviation (initial letters of words)

аббревиату́р·н·ый, -ая, -ое, (*a*) abbreviation (initial letters of words)

аббревиа́ци·я, -и, (*f*) abbreviation

аберра́нт·н·ый, -ая, -ое, (*a*) (biol) aberrant

аб·во́льт·, -а, (*m*) (elec) abvolt

абга́лдыр·ь, -я, (*m*) chain hook

аб·ге́нри (*n indecl*) (elec) abhenry

абгра́т-пре́сс·, -а, (*m*) (mech eng) burring machine

∼ -шта́мп·, -а, (*m*) trimming die

аб·ду́ктор·, -а, (*m*) abductor

аб·ду́кци·я, -и, (*f*) abduction

а́бел·ев·, -а, -о, (*poss adj*) (math) Abelian

∼ гру́пп·а (*f*) (math) Abelian group

а́бел·ев·ый, -ая, -ое, (*a*) Abelian

∼ под·гру́пп·а (*f*) (math) Abelian subgroup

абел·и́т·, -а, (*m*) (expl) abelite

А́бел·ь, -я, (*m*) Abel (surname)

∼ -Пе́нск·ого при·бо́р· (*m*) Abel tester (for flash point of petroleum)

А́бел·я интегра́л (*m*) (math) Abelian integral

аберра́ци·я, -и, (*f*) aberration

∼, век·ов·а́я (astron etc.) secular aberration

∼ вы́с·ш·их по·ря́д·к·ов higher aberration

∼ объекти́в·а lens aberration

∼, о·ста́т·очн·ая (opt) residual aberration

∼, по·пере́ч·н·ая lateral/transverse aberration

∼, су́т·очн·ая (astron) diurnal aberration

∼, сфер·и́ческ·ая (opt) spherical aberration

∼, хромат·и́ческ·ая (opt) chromatic aberration

∼, хромо·со́м·н·ая (gen) chromosome aberration

абза́ц·, -а, (*i*) **-ем,** (*m*) (print) indention; paragraph

абзе́тцер·, -а, (*m*) continuous bucket loader

абиет·а́т, -а, (*m*) (chem) abietate

абиет·и́нов·ый, -ая, -ое, (*a*) (chem) abietic

∼ кисл·от·а́ (*f*) abietic acid

а·био·ген·е́з, -а, (*m*) (zool) abiogenesis, spontaneous generation

а·био·троф·и́·я, -и, (*f*) (gen) abiotrophy

абисса́л·ь, -и, (*f*) (ocean) abyssal zone; abyssal fauna

абисса́ль·н·ый, -ая, -ое, (*a*) (ocean) abyssal

∼ о́бласт·ь (*f*) (ocean) abyssal benthic zone

абисси́н·ск·ий, -ая, -ое, (*a*) Abyssinian

∼ коло́д·ец· (*m*) Abyssinian well

абиссо·ли́т·ов·ый, -ая, -ое, (*a*) (geol) abyssolithic

абиссо·фо́б·н·ый, -ая, -ое, (*a*) (biol) abyssophobe

абитурие́нт·, -а, (*m*) person who has completed secondary education

аб·куло́н·, -а, (*m*) (elec) abcoulomb

аблакт·и́р·овать, -у́ют, (*imp and perf*) (hortic) graft by approach

аблакт·иро́вк·а, -и, (*g pl*) **-вок·,** (*f*) ablactation; (hortic) inarching, grafting by approach

абля́ци·я, -и, (*f*) ablation

∼, лед·нико́в·ая (geol) glacial ablation

аб·мо́ (*n indecl*) (elec) abmho

аб·о́м·, -а, *(m)* (elec) abohm

аб·норма́ль·н·ый, -ая, -ое, *(a)* abnormal

абонеме́нт·, -а, *(m)* subscription

абонеме́нт·н·ый, -ая, -ое, *(a)* subscription

∼ стол· *(m)* circulating desk (library)

∼ я́щик· *(m)* private box (at post office)

абоне́нт·, -а, *(m)* subscriber

∼, вы·зыв·а́ющ·ий caller (telephones)

абоне́нт·ск·ий, -ая, -ое, *(a)* subscriber's

аборда́ж·, -а, *(i)* **-ем,** *(m)* (naut) boarding (without consent)

аб·ориге́н·, -а, *(m)* (biol) aboriginal, aborigine

аб·ориге́н·н·ый, -ая, -ое, *(a)* aboriginal

або́рт·, -а, *(m)* (med) abortion; miscarriage

аборт·и́вн·ый, -ая, -ое, *(a)* abortive

абрад·и́р·овать, -уют, *(imp)* abrade

абрази́в·, -а, *(m)* abrasive (material)

абрази́в·н·ый, -ая, -ое, *(a)* of **абрази́в·**

∼ инструме́нт· *(m)* grinding tool

∼ об·рабо́т·к·а, жи́дк·остн·ая *(f)* liquid honing

абразио́н·н·ый, -ая, -ое, *(a)* of **абра́зи·я**

абра́зи·я, -и, *(f)* abrasion, attrition

абриахани́т·, -а, *(m)* (min) abriachanite

абрико́с·, -а, *(m)* (bot) apricot, *Prunus armeniaca*

абрико́с·н·ый, -ая, -ое, *(a)* apricot

абрико́с·ов·ый, -ая, -ое, *(a)* apricot

абри́н·, -а, *(m)* (chem) abrin

а́брис·, -а, *(m)* (surv) sketch, outline, contour

∼, пол·ев·о́й (surv) field sketch

∼, ситуацио́н·н·ый location sketch

а́брис·н·ый, -ая, -ое, *(a)* of **а́брис·**

∼ кни́ж·к·а *(f)* sketchbook

абс *(abbr)* = **абсолю́т·н·ый** absolute

абс. выс. = **абсолю́т·н·ая выс·от·а́** height above sea level, absolute height

абс. ед. = **абсолю́т·н·ая един·и́ц·а** ab-unit, absolute unit

абс. эл. магн. ед. = **абсолю́т·н·ая электро·магни́т·н·ая един·и́ц·а** absolute electromagnetic unit, abs. emu

абсароки́т·, -а, *(m)* (min) absarokite

абсе́нт·, -а, *(m)* absinthe, absinth

абси́д·а, -ы, *(f)* (arch) apse, apsis

абси́нт·, -а, *(m)* = **абсе́нт·**

абсолют·и́рован·и·е, -я, *(n)* dehydration (of alcohol)

абсолю́т·н·о *(adv)* absolutely

∼ чёрн·ое те́л·о *(n)* (phys) absolute black body

абсолю́т·н·ый, -ая, -ое, *(a)* absolute

∼ вла́ж·ност·ь *(f)* (meteor) absolute humidity

∼ един·и́ц·а *(f)* absolute unit

∼ ну́л·ь *(m)* (phys) absolute zero

∼ спи́рт· *(m)* dehydrated/absolute alcohol

абсорб·а́т·, -а, *(m)* (chem) absorbate

абсорб·е́нт·, -а, *(m)* absorbent

абсо́рб·ер·, -а, *(m)* absorber; (gas) scrubber; (chem) absorber, absorption tower

∼, барбота́ж·н·ый (chem) bubble-cap absorption tower

абсорб·и́рованн·ый, -ая, -ое, *(past part pass)* of **абсорб·и́р·овать**

∼ вещ·еств·о́ *(n)* absorbed substance, absorbate

абсорб·и́р·овать, -уют, *(imp and perf)* absorb; **-ся** *(pass)* be absorbed

абсорб·и́руем·ый, -ая, -ое, *(pres part pass)* of **абсорб·и́р·овать**; absorbable

абсорб·и́рующ·ий, -ая, -ое, *(pres part act)* of **абсорб·и́р·овать**; absorbent, absorptive, absorbing

∼ вещ·еств·о́ *(n)* absorbant, absorbent

абсорбцио́·метр·, -а, *(m)* absorptiometer

абсорбцио́н·н·ый, -ая, -ое, *(a)* absorption, absorbing, absorptive

∼ раст·е́ни·е *(n)* (nucl bot) absorber plant

∼ холод·и́льн·ая маши́н·а *(f)* absorption refrigerator

абсо́рбци·я, -и, *(f)* absorption

абстра́кт·н·ый, -ая, -ое, *(a)* *(s f)* **-т·ен, -т·на,** abstract

абсу́рд·, -а, *(m)* absurdity

абце́сс·, -а, *(m)* (med) abscess

абсци́сс·а, -ы, *(f)* (math) abscissa, X-coordinate; latitude difference

абукамал·и́т·, -а, *(m)* (min) abukamalite

абули́·я, -и, *(f)* (med) abulia

аб·фара́д·а, -ы, *(f)* (elec) abfarad

а́бштрих·, -а, *(m)* arsenic-antimony slag (lead refining)

ав, а-в *(abbr)* = **ампер·вит·о́к** (elec) A-T, a-t, ampere turns

авал·и́ст·, -а, *(m)* (com) guarantor

ава́л·ь, -я, *(m)* (com) guarantee, security

аван·гáрд·, -а, (*m*) vanguard, van

аван·кáмер·а, -ы, (*f*) (hydr) entrance flume

аван·пóрт·, -а, (*m*) outer harbour

аван·пóст·, -а, (*m*) (mil) forward position

авáнс·, -а, (*m*) (fin) advance, payment on account; (agr) advance in kind; (shipb) advance coefficient, advance per revolution (of ship's propeller)

аванс·и́ровани·е, -я, (*n*) (fin) advancing

аванс·и́р·овать, -уют, (*imp and perf*) (fin) advance; **-ся** (*pass*)

авантюри́н·, -а, (*m*) (min) aventurine quartz; aventurine glass

авар·и́йн·о-с·пас·áтельн·ая служ·б·а (*f*) (naut) salvage and rescue service

авар·и́йност·ь, -и, (*f*) accident rate; proneness to accident

авар·и́йн·ый, -ая, -ое, (*a*) accident, emergency; (mar ins) average

∼ блóк· (*m*) (nucl) safety block

∼ комáнд·а (*f*) crash crew, breakdown gang; (naut) damage control party

∼ о·станóв·к·а (*f*) scram (nucl reactor)

∼ под·пи́с·к·а (*f*) (mar ins) average bond

∼ по·сáд·к·а (*f*) (air) emergency landing

∼ сигнáл· (*m*) distress signal

авáр·и·я, -и, (*f*) accident, breakdown, damage; (naut) heavy damage; (air) crash; (mar ins) average

∼, óбщ·ая (mar ins) general average

∼, чáст·н·ая (mar ins) particular average

авги́т·, -а, (*m*) (min) augite

авени́н·, -а, (*m*) (chem) avenin

авиа- (*abbr*) = **авиациóн·н·ый** aviation, aircraft, air, aerial, aero-

авиа·бáз·а, -ы, (*f*) air base, air force base

авиа·бензи́н·, -а, (*m*) aviation gasoline

авиа·бóмб·а, -ы, (*f*) bomb, aerial bomb

∼, у·правл·я́ем·ая air-to-surface missile

авиа·ви́нт·, -а, (*m*) (air) propeller, airscrew

авиа·гарнизóн·, -а, (*m*) air force station

авиа·горизóнт·, -а, (*m*) (air) gyro/ artificial horizon

авиа·дви́г·ател·ь, -я, (*m*) *see also* **дви́г·ател·ь**; aircraft engine, aero-engine

авиа·десáнт·, -а, (*m*) (mil) airborne landing

авиа·десáнт·н·ый, -ая, -ое, (*a*) (mil) airborne-landing, airborne-troops

авиа·диспéтчер·, -а, (*m*) air traffic controller

авиа·диспéтчер·ск·ий, -ая, -ое, (*a*) air-traffic-control

авиа·за·вóд·, -а, (*m*) aircraft factory

авиа·кáмер·а, -ы, (*f*) inner tube (of aircraft tyre)

авиа·констру́ктор·, -а, (*m*) aircraft designer

авиа·лéнт·а, -ы, (*f*) (a/c) wing/fabric tape

авиа·ли́н·и·я, -и, (*f*) air route

авиа·ли́т·, -а, (*m*) (met) avialite (Cu-Al-Sn-Ni-Fe alloy)

авиáл·ь, -я, (*m*) Avial (Al-Mg-Si-Mn alloy)

авиа·мастер·ск·áя, -óй, (*f decl as adj*) aircraft repair shop

авиа·мет·стáнци·я, -и, (*f*) aviation meteorological station

авиа·мотор·и́ст·, -а, (*m*) (air) mechanic

авиа·нóс·ец·, -с·ц·а, (*i*) -с·ц·ем, (*m*) (naut) aircraft carrier

авиа·о·пры́ск·ивани·е, -я, (*n*) (agr) aerial spraying

авиа·о·пы́л·иватель·, -я, (*m*) (agr) aeroduster

авиа·под·кóрм·к·а, -и, (*f*) (agr) air-laid top dressing

авиа·пóчт·а, -ы, (*f*) air mail

авиа·почт·óв·ый, -ая, -ое, (*a*) air-mail

авиа·пу́шк·а, -и, (*f*) aircraft gun

авиа·раз·вéд·к·а, -и, (*g pl*) -·ок·, (*f*) aerial reconnaissance

авиа·съ·ём·к·а, -и, (*g pl*) -м·ок·, (*f*) aerial survey

авиа·тéхн·ик·, -а, (*m*) air rigger

авиациóн·н·ый, -ая, -ое, (*a*) aviation, aircraft

∼ кероси́н· (*m*) aviation kerosine, jet fuel

∼ ми́н·а (*f*) aircraft-laid mine

авиáци·я, -и, (*f*) aviation; aircraft (*pl*)

∼, арм·éйск·ая tactical air force

∼, воéн·н·ая air force

∼, воéн·н·о-мор·ск·áя naval air arm, naval aviation

∼, гражд·áнск·ая civil aviation

авиа́ци·я
~, **санита́р·н·ая**   air medical service
~, **фронт·ов·а́я**   tactical air force
авиа·ча́ст·ь, -и, (*nom pl*) **-и,** (*g pl*)
 -е́й, (*f*) air force unit, air unit
авиа·шко́л·а, -ы, (*f*) flying training
 school
авива́ж·н·ый со·ста́·в· (*m*) (rubber)
 finish, dressing
авие́т·к·а, -и, (*g pl*) **-т·ок·,** (*f*) light
 aircraft
ави́з·, -а, (*m*) *see* **ави́зо** (fin)
ави́зо (*n indecl*) (fin) advice note, letter
 of advice; (naut) despatch vessel
авикул·о́ид·н·ый, -ая, -ое, (*a*) avi-
 culoid
авикул·я́ри·й, -я, (*m*) avicularium,
 (*pl*) avicularia
а·витамин·о́з·, -а, (*m*) (med) avitamino-
 sis
ав/м (*abbr*) = **ампер·вит·о́к· на метр·**
 ampere-turn per metre
Авога́дро (*m indecl*) Avogadro (surname)
~ **за·ко́н·** (*m*) (chem) Avogadro's hypo-
 thesis/principle
~ **числ·о́** (*n*) (chem) Avogadro con-
 stant/number
аво́сь (*adv*) perhaps, maybe
~, **на**   at random
авра́л·, -а, (*m*) (naut) major evolution,
 drill
аври́·кул·а, -ы, (*f*) (biol) auricle, auri-
 cula
аври·пункт·у́р·а, -ы, (*f*) (med) auri-
 puncture
австрал·и́йск·ий, -ая, -ое, (*a*) Aus-
 tralian
австрал·и́т·, -а, (*m*) (min) australite
австр·и́йск·ий, -ая, -ое, (*a*) Austrian
авто-   auto-, self; (*abbr*) of **автомоби́·
 ль·н·ый** (q.v.); **автомат·и́ческ·ий**
 (q.v.)
авто́ (*n indecl*) (col) car, motor car,
 automobile
авто·ба́з·а, -ы, (*f*) (М/Т) motor depot,
 motor-transport depot
авто·баро·троп·и́·я, -и, (*f*) (meteor)
 autobarotropy
авто·бензо·за·пра́в·щик·, -а, (*m*)
 gasoline/petrol bowser
авто·бетоно·меш·алк·а, -к·и, (*g pl*)
 -лок·, (*f*) transit concrete mixer
авто·би́о·граф·, -а, (*m*) autobio-
 grapher
авто·био·граф·и́ческ·ий, -ая, -ое,
 (*a*) autobiographical

авто·био·гра́ф·и·я, -и, (*f*) autobio-
 graphy
авто·блок·иро́вк·а, -и, (*g pl*) **-вок·,**
 (*f*) (rail) automatic blocking system
авто·бре́кчи·я, -и, (*f*) (geol) auto-
 breccia
авто́·бус·, -а, (*m*) bus, motor bus,
 omnibus
авто́·бус·н·ый, -ая, -ое, (*a*) of **авто·-
 бус·**
авто·ваго́н·, -а, (*m*) (rail) rail car
авто·вакци́н·а, -ы, (*f*) (bact) auto-
 vaccine
авто·во́з·, -а, (*m*) (М/Т) straddle truck
авто·воз·вра́т·, -а, (*m*) (autom) self-
 reset
авто·га́з·ов·ый вы·ключ·а́тел·ь (*m*)
 (elec) gas-blast breaker
авто·га́м·и·я, -и, (*f*) (biol) autogamy
авто·гамма·радио́·метр·, -а, (*m*)
 mobile gamma radiometer
авто·гамма-съ·ём·к·а, -и, (*g pl*)
 -м·ок·, (*f*) automobile gamma-radiation survey
авто·гара́ж·, -а́, (*i*) **-о́м,** (*m*) garage
авто·ген·е́з·, -а, (*m*) (biol) autogenesis, autogeny
авто·генера́тор·, -а, (*m*) (elec) self-
 maintained circuit
авто·ге́н·н·ый, -ая, -ое, (*a*) autogenic,
 autogenous
~ **ре́з·к·а** (*f*) oxy-acetylene cutting
~ **с·ва́р·к·а** (*f*) fusion welding
авто·гео·синклина́л·ь, -и, (*f*) (geol)
 autogeosyncline
авто·гравю́р·а, -ы, (*f*) (print) auto-
 gravure
авто́·граф·, -а, (*m*) autograph; (phot)
 setting device
авто·гре́йдер·, -а, (*m*) (civ eng) motor
 grader
авто·дезо·у·стано́в·к·а, -и, (*g pl*)
 -вок·, (*f*) truck/mobile disinfection
 unit
авто·дета́л·ь, -и, (*f*) (М/Т) vehicle
 component/part
авто·ди́н·, -а, (*m*) (rad) autodyne,
 autoheterodyne, self-heterodyne
авто·дрезн́·а, -ы, (*f*) (rail) motor
 trolley
авто·жи́р·, -а, (*m*) (a/c) autogiro, gyroplane
авто·за·во́д·, -а, (*m*) motor works,
 automobile plant
авто·за·пра́в·щик·, -а, (*m*) (М/Т)
 bowser, refueller

авто·иммуниз·áци·я, -и, (*f*) (med) autoimmunization

авто·иониз·áци·я, -и, (*f*) autoionization

авто·кáмер·а, -и, (*f*) (M/T) inner tube (of tyre)

авто:кáр·, -а, (*m*) self-propelled trolley

авто·катá·лиз·, -а, (*m*) (chem) autocatalysis

~, магмат·и́ческ·ий (geol) magmatic autocatalysis

авто·ката·лит·и́ческ·ий, -ая, -ое, (*a*) autocatalytic

авто·клáв·, -а, (*m*) (chem) autoclave

авто·клав·и́рованн·ый, -ая, -ое, (*a*) autoclaved

авто·клáз·а, -ы, (*f*) (geol) autoclase, autoclastic fissure

авто·класт·и́ческ·ий, -ая, -ое, (*a*) (geol) autoclastic

авто·колеб·áни·е, -я, (*n*) free oscillation/vibration

авто·колеб·áтельн·ый режи́м· (*m*) oscillatory state

авто·коллим·áтор·, -а, (*m*) autocollimator

авто·коллим·ацио́н·н·ый, -ая, -ое, (*a*) (opt) autocollimation

авто·компенс·áтор·, -а, (*m*) (elec) self-balancing A.C. potentiometer

авто·консерв·и́ровани·е, -я, (*m*) (food) preservation in own CO_2

авто·коррел·я́ци·я, -и, (*f*) autocorrelation

авто·крáн·, -а, (*m*) mobile crane

автóл·, -а, (*m*) automotive engine oil

авто·лáк·, -а, (*m*) automobile/motorcar varnish

авто·лéж·нев·ая дорóг·а (*f*) logging road

авто·лесо·вóз·, -а, (*m*) straddle truck

автó·лиз·, -а, (*m*) (biol) autolysis

авто·ли́ст·, -á, (*nom pl*) **-ы́,** (*g pl*) **-óв,** (*m*) (met) automobile sheet

авто·ли́т·, -а, (*m*) (geol) endogenous inclusion, cognate inclusion, autolith

авто·ли́фт·, -а, (*m*) lift truck

авто·магистрáл·ь, -и, (*f*) motor road, motor way, superhighway

автомáт·, -а, (*m*) automatic machine, automatic device, automaton; automatic machine tool; slot machine; (elec) circuit breaker; (mil) sub-machine gun, machine carbine

~ без·о·пáс·ност·и, скор·остн·óй emergency governor (steam turbine)

автомáт

~, винто·рéз·н·ый automatic screw-cutting machine

~, воз·ду́ш·н·ый (elec) air circuit breaker

~ гаш·éни·я пóл·я (elec) automatic field-suppression circuit-breaker

~, говор·я́щ·ий automatic telephone answerer

~ давл·éни·я automatic pressure control

~ за·щи́т·ы сéт·и (elec) automatic cut-out

~, кон·éчн·ый finite automaton

~ крéн·ов (air) bank-and-climb gyro unit

~, курс·ов·óй (air nav) directional control unit

~ обрáт·н·ой мóщ·ност·и (elec) reverse-current circuit breaker

~, оптимиз·и́рующ·ий automatic optimizing device

~ пере·кóс·а cyclic pitch control (helicopter rotor control)

~ пит·áни·я automatic feed-water regulator

~ при·ём·и́стост·и (air) automatic acceleration control

~, про·дóль·н·о-по·перéч·н·ый (air nav) bank-and-climb control unit

~ -регул·я́тор· automatic governor, constant speed governor

~, рул·ев·óй (a/c) gyro-pilot, automatic steering gear

~, с·вáр·очн·ый welder, automatic welder

~ стрель·б·ы́ (mil) fire control table, fire control clock

~, тепл·ов·óй (elec) thermal circuit breaker

.~ -тóрмоз·, -а, (*m*) (a/c) wheel-brake differential unit

~ тя́г·ов·ых рас·чёт·ов automatic traction computer

~ у·правл·éни·я дáль·ност·ью (GW) range control accelerometer

~, электр·и́ческ·ий (elec) air circuit breaker

автоматиз·áци·я, -и, (*f*) automation; automatics

~, комплéкс·н·ая complex automation

автоматиз·и́р·овать, -уют, (*imp*) automate

автомáт·ик·а, -и, (*f*) automatics; automatic system

автома́т·ик·а

~, **диста́нц·ио́нн·ая** remote-control feedback automatics

~, **ме́ст·н·ая** local-control feedback automatics

~, **рефле́ктор·н·ая** feedback automatics

~, **теле·механ·и́ческ·ая** remote-control feedback automatics

автомат·и́ческ·ий, -ая, -ое, (*a*) automatic, self-operating

~ **вы·числ·и́тельн·ая маши́н·а** (*f*) automatic computer

~ **компенс·а́тор·** (*m*) (elec) self-balancing A.C. potentiometer

~ **контро́л·ь** (*m*) automatic inspection

~ **ли́н·и·я** (*f*) transfer line, automation line

~ **регул·и́ровани·е** (*n*) automatic control

~ **регул·и́ровани·е раз·ме́р·ов** automatic dimension control

~ **регул·иро́вк·а у·сил·е́ни·я** (rad) automatic gain control

~ **у·стро́й·ств·о** (*n*) (autom) automatic control system

~ **у·стро́й·ств·о, за·ви́с·им·ое** (*n*) feedback control system

~ **у·стро́й·ств·о, много·кра́т·н·ое** (*n*) repetitive-action automatic control system

~ **у·стро́й·ств·о, регул·и́рующ·ее** (*n*) automatic controller

автома́т·н·ый, -ая, -ое, (*a*) (met) free-cutting

~ **ста́л·ь** (*f*) free-cutting steel

автома́т·чик·, -а, (*m*) automatic machine operator; sub-machine gunner

авто·маши́н·а, -ы, (*f*) motor vehicle (*see also* **автомоби́л·ь**)

авто·машин·и́ст·, -а, (*m*) (autom) automatic train driver (electr)

авто·маши́н·н·ый, -ая, -ое, (*a*) of **авто·маши́н·а**

~ **мост·** (*m*) vehicular bridge

авто·механо·хор·и́·я, -и, (*f*) (bot) automechanohory

автомобиле·стро·е́ни·е, -я, (*n*) motor manufacture, motor vehicle manufacture, automobile construction

автомобил·и́ст·, -а, (*m*) motorist, driver

авто·моби́л·ь, -я, (*m*) motor vehicle; (pop) motor car, automobile

~, **груз·ов·о́й** lorry, truck

~, **легк·ов·о́й** motor car, automobile

авто·моби́л·ь

~ **-рефрижера́тор·** (*m*) refrigerator van

~ **-само·с·ва́л·** (*m*) tipping truck, dump truck

~, **санита́р·н·ый** ambulance

автомоби́ль·н·ый, -ая, -ое, (*a*) motor, automobile, mobile (pertaining to motor vehicles in general)

~ **вес·ы́** (*m pl*) weighbridge

~ **вы́ш·к·а** (*f*) tower wagon

~ **кран·** (*m*) truck-mounted crane

~ **о·пыл·ива́тел·ь** (*m*) truck-mounted crop duster

авто·моде́ль·ност·ь, -и, (*f*) (autom) similarity

авто·модул·я́ци·я, -и, (*f*) (rad) self-modulation

авто·морф·и́зм·, -а, (*m*) (math) automorphism

авто·мо́рф·н·ый, -ая, -ое, (*a*) (math) automorphic

авто·мотри́с·а, -ы, (*f*) (rail) rail car

авто·на·со́с·, -а, (*m*) fire-fighting pumping unit

авто·на·стро́й·к·а, -и, (*f*) (rad) auto-tune, automatic tuning

автоно́м·ност·ь, -и, (*f*) independence; self-sufficiency

автоно́м·н·ый, -ая, -ое, (*a*) (*s f*) **-м·ен·, -м·н·а,** autonomous, independent; self-contained; (anat) autonomic

~ **регул·и́ровани·е** (*n*) independent automatic control

авто·опера́тор·, -а, (*m*) automatic loading device (for a machine)

авто·пере·дви́ж·н·ые вес·ы́ (*m pl*) mobile scales

авто·пило́т·, -а, (*m*) (air nav) autopilot, automatic pilot; (rocket) attitude control

~, **гиро·скоп·и́ческ·ий** gyropilot

~, **цифр·ов·о́й** digital autopilot

авто·плу́г·, -а, (*nom pl*) **-а́,** (*g pl*) **-о́в,** (*m*) motor plough

авто·по·гру́з·чик·, -а, (*m*) fork truck, fork-lift truck; loader

авто·по́·езд·, -а, (*nom pl*) **-а́,** (*g pl*) **-о́в,** (*m*) (M/T) tractor-trailer unit

авто·по·и́лк·а, -и, (*g pl*) **лок·,** (*f*) (agr) automatic drinking bowl

авто·по·кры́ш·к·а, -и, (*g pl*) **-ш·ек·,** (*f*) (M/T) tyre, tyre casing

авто·потенцио́·метр·, -а, (*m*) (elec) self-balancing potentiometer

авто·при·це́п·, -а, (*m*) (M/T) trailer

авто·про·кла́д·чик·, -а, (*m*) (air nav) automatic course plotter авт·о́пси·я, -и, (*f*) (med) autopsy

а́втор·, -а, (*m*) author, writer; composer

авто·радио·гра́мм·а, -ы, (*f*) auto-radiogram, autoradiograph

авто·радио́·граф·, -а, (*m*) autoradiographer

авто·радио·гра́ф·и·я, -и, (*f*) auto-radiography

авто·регул·и́рован·и·е, -я, (*n*) = авто·регул·иро́вк·а

авто·регул·иро́вк·а, -и, (*f*) automatic control/regulation

авто·регул·и́руем·ый, -ая, -ое, (*a*) automatically controlled, self-controlled

авто·регул·я́тор·, -а, (*m*) automatic controller/regulator

авто·реду́кци·я, -и, (*f*) (surv) auto-reduction

авто·ремо́нт·н·ая маѐтер·ск·а́я (*f*) service garage, M/T repair shop

авто·рефера́т·, -а, (*m*) author's abstract

автор·из·ова́ть, -у́ют, (*imp and perf*) authorize

автор·ите́т·, -а, (*m*) authority

авто·рот·а́ци·я, -и, (*f*) autorotation

~ компре́ссор·а (a/c) windmilling

а́втор·ск·ий, -ая, -ое, (*a*) author's

~ пра́в·о (*n*) copyright

~ с·вид·е́тельств·о (*n*) inventor's certificate (often used in U.S.S.R. instead of patent)

авто·ру́ч·к·а, -и, (*g pl*) -ч·ек·, (*f*) fountain pen

авто·само·с·ва́л·, -а, (*m*) dump truck

авто·си́н·, -а, (*m*) (elec) autosyn, synchro, selsyn

авто·со́м·а, -ы, (*f*) (cyt) autosome

авто·спо́р·а, -ы, (*f*) (bot) autospore

авто·ста́ртер·, -а, (*m*) (air) Huck's starter

авто·сто·я́нк·а, -и, (*g pl*) -нок·, (*f*) (M/T) park, parking place

авто·стра́д·а, -ы, (*f*) motor road, motor way, superhighway

авто·с·це́п·к·а, -и, (*g pl*) -п·ок·, (*f*) (rail) automatic coupling, automatic coupling system

авто·теле́ж·к·а, -и, (*g pl*) -ж·ек, (*f*) self-propelled trolley

авто·ти́п·, -а, (*m*) (phot) autotype

авто·ти́п·и·я, -и, (*f*) (print, phot) autotype

авто·тка́н·ь, -и, (*f*) (text) automobile fabric

авто·то́м·и·я, -и, (*f*) (biol) autotomy, fission

авто·тра́ктор·н·ый, -ая, -ое, (*a*) motor vehicle-and-tractor, automobile-and-tractor, automotive

авто·тра́нспорт·, -а, (*m*) motor transport

~, груз·ов·о́й lorry/truck transport

авто·трансформа́тор·, -а, (*m*) (elec) auto-transformer

авто·трансформа́тор·н·ый пуск·а́·тел·ь (*m*) (elec) autotransformer starter, compensator starter

авто·тро́ф·, -а, (*m*) autotrophic plant

авто·тро́ф·н·ый, -ая, -ое, (*a*) autotrophic

авто·фаз·иро́вк·а, -и, (*f*) (elec) phase stability

авто·фрета́ж·, -а, (*m*) (gunn) auto-frettage, self-hooping

авто·хро́м·, -а, (*m*) (phot) autochrome

авто·хто́н·н·ый, -ая, -ое, (*a*) autochthonous, aboriginal, indigenous, formed where found

авто·цисте́рн·а, -ы, (*f*) road tanker, tank-truck

авто·ши́н·а, -ы, (*f*) (M/T) tire, tyre

авто·шту́рман·, -а, (*m*) (navig) ground position indicator

авто·электро́н·н·ый, -ая, -ое, (*a*) autoelectron, autoelectronic, autoelectric

~ микро·ско́п· (*m*) field-emission microscope

~ эффе́кт· (*m*) autoelectric/autoelectronic/autoelectron effect

ага́в·а, -ы, (*f*) (bot) agave

агальматоли́т·, -а, (*m*) (min) agalma-tolite, pagodite, soapstone

а·гам·и́·я, -и, (*f*) (biol) agamy

а·га́м·н·ый, -ая, -ое, (*a*) (biol) agamic, agamous

а·гамо·сперм·и́·я, -и, (*f*) (biol) agamospermy

ага́р·, -а, (*m*) (chem) agar

~ -ага́р·, -а, (*m*) (chem) agar-agar

~, мясо·пепто́н·н·ый nutrient agar

агарици́н·, -а, (*m*) (pharm) agaricin

ага́р·н·ый, -ая, -ое, (*a*) agaric; agar-agar

ага́р·ов·ый, -ая, -ое, (*a*) agaric; agar-agar

ага́т·, -а, (*m*) (min, print) agate

ага́т·н·ый, -ая, -ое, (*a*) of ага́т·

ага́т·ов·ый, -ая, -ое, (*a*) of ага́т·

~ под·ши́п·ник· (*m*) jewel bearing (instruments)

аггломера́т·, -а, (*m*) *see* **агломерат·**

аг·глютинати́в·н·ый, -ая -ое, (*a*) agglutinative

аг·глютин·а́ци·я, -и, (*f*) agglutination, clotting, clumping

аг·глютин·и́н·, -а, (*m*) (chem) agglutinin

аг·глютино·ге́н·, -а, (*m*) (med) agglutinogen

аг·градацио́н·н·ый, -ая, -ое, (*a*) (geol) aggradation

аг·града́ц·и·я, -и, (*f*) (geol) aggradation, shore deposition

аге́н·, -а, (*m*) (chem, food) agene

агениз·и́рованн·ый, -ая, -ое, (*a*) (food) agenized

аге́нт·, -а, (*m*) (*and see under qualif adj*); agent, factor, representative; agent (cause or substance); spy, secret agent

~, вос·стана́вл·ивающ·ий (chem) reducing agent

~, мор·ск·о́й shipping agent

~, регул·и́рующ·ий control agent

~, тормоз·я́щ·ий (chem, nucl) inhibitor

агент·и́ровани·е суд·о́в agency servicing (shipping)

аге́нт·ск·ий, -ая, -ое, (*a*) of **аге́нт·**

аге́нт·ств·о, -а, (*n*) agency

~, мор·ск·о́е shipping office

~, телегра́ф·н·ое news agency, telegraph agency

агент·у́р·а, -ы, (*f*) agency; secret service

агирно·примити́в·н·ая симметр·и́·я (*f*) (cryst) unsymmetrical system

агирно·центра́ль·н·ая симметр·и́·я (*f*) (cryst) pinacoidal symmetry

аги́р·н·ый, -ая, -ое, (*a*) (cryst) triclinic, agyric

~ сингóн·и·я (*f*) (cryst) triclinic system

агит- (*abbr*) of **агитацион·н·ый** propaganda

агит·а́тор·, -а, (*m*) agitator, mixer, stirrer; propagandist, agitator, canvasser

~, аэроли́фт·н·ый air-lift agitator

~ Девере Devereux agitator (ore dressing)

агитацио́н·н·ый, -ая, -ое, (*a*) agitating, stirruping; propaganda, canvassing

агит·а́ци·я, -и, (*f*) agitation, stirring; propaganda

аги́т·к·а, -и, (*g pl*) **-т·ок,** (*f*) propaganda material

агит·паро·хо́д·, -а, (*m*) political campaign vessel, propaganda ship

агит·пу́нкт·, -а, (*m*) (polit) party information office, propaganda centre/center

агломера́т·, -а, (*m*) (min, met) agglomerate; (met) sinter

агломерацио́н·н·ый, -ая, -ое, (*a*) agglomerate, agglomeration, sinter

~, маши́н·а (*f*) sinter plant

агломера́ци·я, -и, (*f*) agglomeration, sintering

агло·фа́брик·а, -и, (*f*) sinter plant

аглутина́ци·я, -и, (*f*) *see* **агглютина́ци·я**

а·гна́т·ы (*pl*) (zool) agnaths, *Agnatha*

а·гно́з·и·я, -и, (*f*) (med) agnosia

а·гон·и́ческ·ий, -ая, -ое, (*a*) agonic

~ ли́н·и·я (*f*) (geophys) agonic line

аго́н·и·я, -и, (*f*) (med) agony

агора·фо́б·и·я, -и, (*f*) (med) agoraphobia, agorophobia

аграда́ци·я, -и, (*f*) = **агграда́ци·я**

а·грануло·цит·о́з·, -а, (*m*) (med) agranulocytosis

агра́р·н·ый, -ая, -ое, (*a*) agrarian

а·гра́ф·и·я, -и, (*f*) (med) agraphia

агрега́т·, -а, (*m*) (*and see under qualif adj*); (tech) set, unit, assembly, machine, outfit, plant, accessory; (geol, min, civ eng) aggregate

~, водо·про́ч·н·ый (build) water-stable aggregate

~, кома́нд·н·ый self-contained automatic control system (aero-engine)

~, маши́н·н·ый set of machines

~, на·тёч·н·ый sinter aggregate

~, поли·кристалл·и́ческ·ий polycrystalline aggregate

~, сферо·ли́т·ов·ый (min) axialite

~, трубо·с·ва́р·очн·ый pipe-welding machine

~, трубо·про·ка́т·н·ый tube-rolling mill

~, цемент·и́ровочн·ый (oil) cementing outfit

~, электро·генера́тор·н·ый (elec) generating set

~, электро·с·ва́р·очн·ый arc-welding set

агрегат·и́ровани·е, -я, (*n*) (autom) multi-station machine system

~, паралле́ль·н·ое (*n*) (autom) parallel-operation multi-station machine system

~, паралле́ль·н·о-по·сле́д·овательн·ое parallel sequential-operation multistation machine system

агрегат·и́ровани·е
~, по·слéд·овательн·ое   sequential-operation multi-station machine
~, с·мéш·анн·ое   parallel sequential-operation multi-station machine system
агрегáт·н·ый, -ая, -ое,   (*a*) of **агрегáт·**
~ со·сто·я́ни·е   (*n*) (phys) state of aggregation
~ стан·óк·   (*m*) machine tool built out of standard units
агрегáци·я, -и,   (*f*) aggregation
агремáн·, -а,   (*m*) (dipl) agrément
агремéнт·, -а,   (*m*) (dipl) agrément
агресси́в·н·ый, -ая, -ое,   (*a*) (*s f*) **-в·ен, -в·на** agressive
агрéсси·я, -и,   (*f*) agression
агри·культу́р·а, -ы,   (*f*) (obs) agriculture, soil cultivation
агрипно·кóм·а, -ы,   (*f*) (med) agrypnocoma
агро-   (*abbr*) of **агро·нóм·ическ·ий** agronomic, agricultural, agro-
агро·био·лог·, -а,   (*m*) agrobiologist
агро·био·лóг·и·я, -и,   (*f*) agrobiology, agronomical biology, agricultural biology
агро·гóрод·, -а,   (*nom pl*) **-á,** (*g pl*) **-óв,** (*m*) agro-town, agrogorod, collective-farms settlement
агро·лесо·мелиорати́в·н·ый, -ая, -ое, (*a*) of **агролесомелиорáция**
агро·лесо·мелиорáци·я, -и,   (*f*) agricultural afforestation
агро·метеоро·лóг·и·я, -и,   (*f*) agro-meteorology
агро·нóм·, -а,   (*m*) agronomist, agriculturalist
агро·нóм·и́ческ·ий, -ая, -ое,   (*a*) agronomic, agricultural
агро·нóм·и·я, -и,   (*f*) agronomy
агросто·лóг·и·я, -и,   (*f*) (bot) agrostology
агро·тéхн·ик·а, -и,   (*f*) agricultural practices
агро·техн·и́ческ·ий, -ая, -ое,   (*a*) agricultural
агро·фито·цен·óз·, -а, (*m*) agrophytocenosis
агро·фóн·, -а,   (*m*) soil fertility
агро·хи́м·и·я, -и,   (*f*) agrochemistry, agricultural chemistry
адáжио   (*n indecl*) (print, music) adagio
ад·аксиáль·н·ый, -ая, -ое,   (*a*) adaxial
адали́н·, -а,   (*m*) (pharm) adalin
адамантин·н·ый, -ая, -ое,   (*a*) (min) adamantine

Адамáр·а, не·рáвен·ств·о   (*n*) (math) Hadamard's inequality
адамелли́т· -а,   (*m*) (geol) adamellite
адамси́т·, -а,   (*m*) (min, chem) adamsite
адапт·áци·я, -и,   (*f*) adaptation
адáпт·ер·, -а,   (*m*) adapter; (phot) adapter, film-pack adapter, holder; pickup (sound)
адапт·и́р·овать, -уют,   (*imp and perf*) adapt
адапт·и́рующ·ий, -ая, -ее,   (*pres part act*) of **адапт·и́р·овать;** adapting, adaptive
адаптó·метр·, -а,   (*m*) (opt) adaptometer
адвекти́в·н·ый, -ая, -ое,   (*a*) of **адвекци·я** advection, advective
~ тумáн·   (*m*) advection fog; sea fog
адвéкци·я, -и,   (*f*) (meteor) advection
адвенти́в·н·ый, -ая, -ое,   (*a*) adventitious, accidental, adventive
адвенти́ци·я, -и,   (*f*) (zool) adventiti,
адвокáт·, -а,   (*m*) barrister, lawyera solicitor, attorney
адвокат·у́р·а, -ы,   (*f*) legal profession, the bar
адгез·иóнн·ый, -ая, -ое,   (*a*) adhesion
адгéз·и·я, -и,   (*f*) adhesion
аддисóн·ов·а бол·éзн·ь   (*f*) (med) Addison's disease
аддити́в·н·ая, -ой,   (*f decl as adj*) (phys, math) additivity effect
аддити́в·ност·ь, -и,   (*f*) additivity
за·кóн аддити́в·ност·и   (*m*) (phys) additivity law
аддити́в·н·ый, -ая, -ое,   (*a*) (*see also* **аддити́в·н·ая** *as noun*) additive
~ свóй·ств·о   (*n*) (chem) additive property
ад·ду́кт·, -а,   (*m*) (chem) adduct
ад·ду́ктор·, -а,   (*m*) adductor
ад·ду́кци·я, -и,   (*f*) adduction
адеквáт·н·ый, -ая, -ое,   (*a*) adequate
адело·гéн·н·ый, -ая, -ое,   (*a*) adelogenic
адело·мóрф·н·ый, -ая, -ое,   (*a*) adelomorphic
аден·áз·а, -ы,   (*f*) (chem) adenase
аден·и́л·ов·ая кисл·от·á   (*f*) adenylic acid
аден·и́н, -а,   (*m*) (chem) adenine
аден·и́т·, -а,   (*m*) (med) adenitis
аден·ози́н·, -а,   (*m*) (chem) adenosin, adenine riboside
аденозин·три·фосфат·áз·а, -ы,   (*f*) (biochem) adenosine triphosphate

аден·о́ид·, -а, (*m*) (anat) adenoid

аден·о́м·а, -ы, (*f*) (med) adenoma

адено·карцин·о́м·а, -ы, (*f*) (med) adenocarcinoma

адено·то́м·и·я, -и, (*f*) (med) adenotomy

адерми́н·, -а, (*m*) (chem) pyridoxin, adermin, vitamin B₆

адиаба́т·а, -ы, (*f*) adiabatic curve, adiabatic line

∼, уда́р·н·ая (dynam) Hugoniot curve

∼, сух·а́я (meteor) dry adiabatic

адиабат·и́ческ·ий, -ая, -ое, (*a*) adiabatic

адиаба́т·н·ый, -ая, -ое, (*a*) adiabatic

∼ крив·а́я (*f*) adiabatic curve

адиагност·и́ческ·ий, -ая, -ое, (*a*) (min) adiagnostic, dubiocrystalline

адиактин·и́ческ·ий, -ая, -ое, (*a*) (phot) adiactinic

адиа́нтум·, -а, (*m*) (bot) maidenhair fern, *Adiantum*

адип·и́л·, -а, (*m*) (chem) adipyl

адип·ини́л·, -а, (*m*) adipinyl

адип·и́нов·ая кисл·от·а́ (*f*) adipic acid

адип·о́з·, -а, (*m*) (med) adiposis

администр·ати́вн·о-хозя́й·ствен·-н·ый, -ая, -ое, (*a*) administrative, internal administration; works and maintenance

администр·ати́вн·ый, -ая, -ое, (*a*) administrative, administration

администр·и́ровани·е, -я, (*n*) administration; bureaucratic administration

администр·а́ци·я, -и, (*f*) administration, administrative body

адмира́л·, -а, (*m*) Admiral

∼, ви́це- Vice-Admiral

∼, ко́нтр- Rear-Admiral

∼ фло́т·а Admiral of the Fleet

адмиралт·е́йск·ий, -ая, -ое, (*a*) Admiralty, Admiralty-pattern

адмиралт·е́йств·о, -а, (*n*) the Admiralty (outside U.S.S.R.); the Admiralty building (in Leningrad)

адмира́ль·ск·ий, -ая, -ое, (*a*) admiral's flag

адми́танс·, -а, (*m*) (elec) admittance

адон·иди́н·, -а, (*m*) (chem) adonidin

адон·изи́д·, -а, (*m*) (pharm) adoniside

адон·иле́н·, -а, (*m*) (pharm) adoniside

адон·и́н·, -а, (*m*) (biochem) adonin

адо́нис·, -а, (*m*) (bot) *Adonis vernalis*

адопт·а́ци·я, -и, (*f*) adoption

адре́м·а, -ы, (*f*) adrema, addressing machine

адрен·али́н·, -а, (*m*) (biochem) adrenalin, adrenaline

адрен·ерг·и́ческ·ий, -ая, -ое, (*a*) (physiol) adrenergic

адрено·кортико·тро́п·н·ый, -ая, -ое, (*a*) (physiol) adrenocorticotropic

адрено·стеро́н·, -а, (*m*) (biochem) adrenosterone

адрено·хро́м·, -а, (*m*) (biochem) adrenochrome

а́дрес·, -а, (*nom pl*) -а́, (*g pl*) -о́в, (*m*) address

∼ на·ча́л·а (autom) start address

∼, нул·ев·о́й (autom) zero address

∼, от·нос·и́тельн·ый (autom) relative address

∼ пере·ключ·е́ни·я (autom) reference address

по а́дрес·у care of, c/o (on letters etc.)

∼, по·дви́ж·н·ый (autom) floating address

адрес·а́нт·, -а, (*m*) sender, addresser (male)

адрес·а́нтк·а, -и, (*g pl*) -ток·, (*f*) sender, addresser (female)

адрес·а́т·, -а, (*m*) addressee (male name)

адрес·а́тк·а, -и, (*g pl*) -ток·, (*f*) addressee (female name)

адрес·а́ци·я, -и, (*f*) addressing

а́дрес·ност·ь, -и, (*f*) number of addresses (of a computer code)

а́дрес·н·ый, -ая, -ое, (*a*) address

∼ кни́г·а (*f*) directory

∼ стол· (*m*) address enquiry bureau

адрес·ова́тел·ь, -я, (*m*) addressing device/machine

адрес·ова́ть, -у́ют, (*imp and perf*) address, forward

адро́н·, -а, (*m*) (nucl) non-lepton

а́д·ск·ий, -ая, -ое, (*a*) infernal, hellish, devilish

∼ ка́мен·ь (*m*) silver nitrate, lunar caustic

адсорб·а́т·, -а, (*m*) (chem) adsorbate

адсорб·е́нт·, -а, (*m*) adsorbent

адсо́рб·ер·, -а, (*m*) adsorber

адсорб·и́рованн·ый, -ая, -ое, (*past part pass*) of **адсорб·и́р·овать**

∼ вещ·еств·о́ (*n*) adsorbed substance, adsorbate

адсорб·и́р·овать, -уют, (*imp and perf*) adsorb

адсорб·и́рующ·ий, -ая, -ее, (*pres part act*) of **адсорб·и́р·овать**; adsorptive, adsorbing, adsorbent

∼ вещ·еств·о́ (*n*) adsorbent

адсорбц·ио́нн·ый, -ая, -ое, (*a*) of **адсо́рбц·и·я**

~ колон·к·а (*f*) adsorption column

адсо́рбц·и·я, -и, (*f*) (chem) adsorption

адуля́р·, -а, (*m*) (min) adularia, glassy feldspar

адуро́л·, -а, (*m*) (phot, chem) adurol

адъю́нкт·, -а, (*m*) post-graduate at military college; assistant

адъю́нкт·а, -ы, (*f*) (math) adjunct, cofactor

~ о·пре·дел·и́тел·я (math) cofactor, signed minor

адъюста́ж·, -а, (*m*) adjustment

адъюта́нт·, -а, (*m*) (mil) aide; (naut) captain's secretary

~, ста́р·ш·ий (mil) adjutant

~, флагма́н·ск·ий flag-lieutenant

адэква́т·н·ый, -ая, -ое, (*a*) adequate

аем (*abbr*) = **а́том·н·ая един·и́ц·а ма́с·с·а** amu, atomic mass unit

ажго́н·, -а, (*m*) (pharm) omum plant, *Carum copticum*

а́жио (*n indecl*) premium (of stocks and shares)

ажиота́ж·, -а, (*i*) **-ем,** (*m*) (fin) stock-jobbing; noisy lobbying

ажит·а́ци·я, -и, (*f*) agitation

ажу́р·, -а, (*m*) keeping a day book; (text) net; (jewelry) plaiting

ажу́р·н·ый, -ая, -ое (*a*) of **ажу́р·;** open work, lattice; (arch) tracery

~ стан·о́к· (*m*) fret-saw machine, jig-saw machine

~ стро́ч·к·а (*f*) (text) hemstitch

~ ткан·ь (*f*) (text) lace-/open-work

аз·, -а́, (*nom pl*) **-ы́,** (*g pl*) **-о́в** (*m*) letter A

азале·и́н·, -а, (*m*) azalein (dye)

аза́ле·я, -и, (*f*) (bot) azalea, *Azalea*

азали́н·, -а, (*m*) (phot, chem) azaline

а́збук·а, -и, (*f*) alphabet

~ Мо́рзе (telecom) Morse code

азелайн·, -а, (*m*) (chem) azelain

азелайн·ов·ый, -ая, -ое, (*a*) azelaic

а·зео·тро́п·н·ый, -ая, -ое, (*a*) (chem) azeotropic

~ с·мес·ь (*f*) (chem) azeotropic mixture

азиатикози́д·, -а, (*m*) (pharm) asiaticoside

азиа́т·, -а, (*m*) Asian (male person)

азиа́т·ск·ий, -ая, -ое, (*a*) Asian, Asiatic

ази́д·, -а, (*m*) (chem) azide

~ кисл·от·ы́ acid azide

ази́д

~ свин·ц·а́ lead azide

азидо- (*int component*) (chem) azido-, N_3—

азими́д·, -а, (*m*) (chem) azimide, triazol

азимидо- (*int component*) (chem) azimino-

азими́н·а, -ы, (*f*) (bot) papaw, *Asimina Adans*

а́зимут·, -а, (*m*) azimuth, true bearing

~ -дви́г·ател·ь, -я, (*m*) follow-up/ azimuth motor

~, имит·и́рованн·ый simulated azimuth

~, и́ст·инн·ый true bearing

~ -квадра́нт· (*m*) (nav) azimuth circle

~ ло́паст·и (air) blade azimuth

~, обра́т·н·ый back azimuth

~, о·пре·дел·и́ть (*perf*) take/observe an azimuth

~ при·вя́з·к·и check bearing

~ про·стир·а́ни·я (geol) strike azimuth, trend azimuth

~, прям·о́й forward azimuth

~, ра́вн·ый isoazimuth

азимут·а́льн·ый, -ая, -ое, (*a*) azimuthal

~ круг· (*m*) relative bearing ring (of a compass)

~ при·во́д· (*m*) azimuth drive, (*in gen as adj*) azimuth-driven

~ у́гол· azimuthal angle

ази́н·, -а, (*m*) (chem) azine

ази́н·ов·ый, -ая, -ое, (*a*) azine

азо- (*int component*) (chem) azo-

азо·бел·о́к·, -л·к·а́, (*m*) (biochem) azoprotein

азо·бензо́л·, -а, (*m*) azobenzene (insecticide)

а́з·ов·ый, -ая, -ое, (*a*) A-shaped

азо·гру́пп·а, -ы, (*f*) (chem) azo group

азо·ими́д·, -а, (*m*) (chem) azoimide, hydrazoic acid

а·зо́й·ск·ий, -ая, -ое, (*a*) (geol) Azoic

азо·крас·и́тел·ь, -я, (*m*) (chem) azo dye

аз·окси- (*int component*) (chem) azoxy-, —N(O)N—

азокси·со·един·е́ни·е, -я, (*n*) (chem) azoxy-compound

аз·о́л·, -а, (*m*) (phot) azol

а·зона́ль·н·ый, -ая, -ое, (*a*) azonal

~ по́чв·а (*f*) azonal soil

а·зоо·сперм·и́·я, -и, (*f*) (zool) azoospermia

азо·со·един·е́ни·е, -я, (*n*) (chem) azo compound

азо·со·чет·а́ни·е, -я, (*n*) diazo reaction (Griess)

азо·сульф·ами́д·, -а, (*m*) (pharm) azosulphamide, azosulfamide

азо́т·, -а, (*m*) (chem) nitrogen, N

~ во́з·дух·а atmospheric nitrogen

~, с·вя́з·анн·ый fixed nitrogen

~, хло́р·ист·ый nitrogen trichloride

~ ядр·а́ nuclear nitrogen

азот·ем·и́·я, -и, (*f*) (med) azotaemia, azotemia

азот·иза́ци·я, -и, (*f*) = **азот·и́рова·ни·е**

азот·и́ровани·е, -я, (*n*) (met) nitriding, nitrogen hardening, nitrarding

азо́т·ист·о- (*component*) (chem) nitrite

азотисто·водо·ро́д·н·ая кисл·от·а́ (*f*) hydrazoic acid

азотисто·ка́ли·ев·ая сол·ь (*f*) potassium nitrite

азотисто·ка́льци·ев·ая сол·ь (*f*) calcium nitrite

азотисто·ки́сл·ый, -ая, -ое, (*a*) (chem) nitrite of

~ аммо́ни·й (*m*) ammonium nitrite

~ ка́ли·й (*m*) potassium nitrite

азо́т·ист·ый, -ая, -ое, (*a*) nitrous, nitrogenous, nitrogen, nitride of

~ ан·гидр·и́д· (*m*) nitrous anhydride

~ бери́ли·й (*m*) beryllium nitride

~ кисл·от·а́ (*f*) nitrous acid

~ равно·ве́с·к·е (*n*) nitrogenous equilibrium

~ селе́н· (*m*) nitrogen selenide

азо́т·н·о- (*component*) (chem) nitrate of

азотно·ами́л·ов·ый эфи́р· (*m*) amyl nitrate

азотноватисто·ки́сл·ый, -ая, -ое, (*a*) hyponitrite of

~ серебр·о́ (*n*) silver hyponitrite

азот·нова́тист·ый, -ая, -ое, (*a*) (chem) hyponitrous

~ кисл·от·а́ (*f*) hyponitrous acid

азот·нова́т·ый, -ая, -ое, (*a*) (chem) hyponitric

~ ан·гидр·и́д· (*m*) nitrogen peroxide, nitrogen tetroxide

азотно·ка́льци·ев·ая сол·ь (*f*) nitrate of lime

азотно·ки́сл·ый, -ая, -ое, (*a*) (chem) nitrate

~ ка́ли·й (*m*) potassium nitrate

~ ли́ти·й (*m*) lithium nitrate

~ мед·ь (*f*) copper nitrate

~ ура́н· (*m*) uranium nitrate

азо́т·н·ый, -ая, -ое, (*a*) (chem) nitric, nitrogen; nitrogenous

~ кисл·от·а́ (*f*) nitric acid

~ об·ме́н· (*m*) nitrogen metabolism

~ у·добр·е́ни·е (*n*) (agr) nitrogenous fertilizer

азото·ба́ктер·, -а, (*m*) (microbiol) Azotobacter

азото·бактер·и́н·, -а, (*m*) azotobacterin (fertilizer)

азото·ге́н·, -а, (*m*) (bot) azotogen

азото·лю́б·, -а, (*m*) (bot) nitrophile

азото·ме́тр·, -а, (*m*) nitrometer, azotometer

азото·со·бир·а́тел·ь, -я, (*m*) nitrogen-fixing bacterium; nitrogen-gathering plant

азото·фикс·а́ци·я, -и, (*f*) nitrogen fixation

азот·ури́·я, -и, (*f*) (med) azoturia

азотфикс·а́ци·я, -и, (*f*) = **азото·фик·с·а́ци·я**

азул·е́н·, -а, (*m*) (chem) azulene

азур·и́н·, -а, (*m*) (chem) azurine

азур·и́нов·ая кисл·от·а́ (*f*) Cleve's acid, L acid, alpha-naphtholsulphonic acid

азур·и́т·, -а, (*m*) (min) azurite

азуро·фи́ль·н·ый, -ая, -ое, (*a*) azurophilic

айв·а́, -ы́, (*f*) (bot) quince, *Cydonia*

~, обык·нове́нн·ая common quince, *Cydonia oblonga*

айкини́т·, -а, (*m*) (min) aikinite, needle ore

а́йсберг·, -а, (*m*) iceberg

~, доко·обра́з·н·ый valley iceberg

~, столо·обра́з·н·ый tabular berg

~, фи́рн·ов·ый névé iceberg

а́йсбимс·, -а, (*m*) (ocean) ice beam

айсклауз·, -а, (*m*) (mar ins) ice clause

айсолонта́йд·, -а, (*m*) (ceram) isolantite

айх·мета́лл·, -а, (*m*) Aich's metal, naval brass

акаде́м·ик·, -а, (*m*) academician

академ·и́ческ·ий, -ая, -ое, (*a*) academic, academy

~ год· (*m*) academic year (autumn-summer)

акаде́м·и·я, -и, (*f*) academy, college

~, вое́н·н·ая military college/academy

~ на·у́к· Academy of Sciences

~ худо́ж·еств· Academy of Arts

ака́д·ск·ий, -ая, -ое, (*a*) (geol) Acadian

акакатехи́н·, -а, (*m*) (chem) acacatechin

акант·, -а, (*m*) (arch, bot) acanthus

акант·и́т·, -а, (*m*) (min) acanthite, argentite, vitreous silver

аканто·бде́лл·а·, -ы, (*f*) (zool) acanthobdella, (*pl*) *Acanthobdellida*

аканто·ли́з·, -а, (*m*) (med) acantholysis

аканто·по́р·а, -ы, (*f*) (zool) acanthopore

аканто·церат·и́д·а, -ы, (*f*) (zool) acanthoceratid

аканто·цефа́л·, -а, (*m*) (zool) acanthocephal, thorn headed worm

ака́нф·, -а, (*m*) *see* **акант·**

акапри́н·, -а, (*m*) (pharm, vet) acaprine, piroplasmin

акари·а́з·, -а, (*m*) (med) acariasis

акари·ци́д·, -а, (*m*) acaricide (insecticide)

акаро·дермат·и́т·, -а, (*m*) (med) acarodermatitis

акаро·ло́г·и·я, -и, (*f*) (ent) acarology

а·ката·фа́з·и·я, -и, (*f*) (med) acataphasia

ака́ци·я, -и, (*f*) (bot) acacia, *Acacia*

∼, бе́л·ая (bot) false acacia, *Robinia pseudoacacia*

∼, жёлт·ая (bot) Caragana

∼, на·сто·я́щ·ая acacia

∼, сенега́ль·ск·ая (bot) *Acacia senegal*

аквада́г·, -а, (*m*) (chem) aquadag

аква·мари́н·, -а, (*m*) (min) aquamarine

аква·ме́тр·и·я, -и, (*f*) (chem) water determination

аква·поли·со·един·е́ни·е, -я, (*n*) (chem) aquo-compound

акваре́л·ь, -и, (*f*) (art) water colour/color

акваре́ль·н·ый, -ая, -ое, (*a*) water-colour/color

∼ кра́с·к·а (*f*) water colour (artists' paint)

аква́риум·, -а, (*m*) aquarium

аквати́нт·а, -ы, (*f*) (print) aquatint

аквато́ри·я, -и, (*f*) water area, area of water

акведу́к·, -а, (*f*) aqueduct

аквизи́тор·, -а, (*m*) buyer

акермани́т·, -а, (*m*) (min) akermanite

ак·климат·иза́ци·я, -и, (*f*) acclimatization

ак·комода́ци·я, -и, (*f*) accommodation (adjustment)

аккомпан·и́р·овать, -уют, (*imp*) accompany (music)

акко́рд·, -а, (*m*) chord (music); contract, wages contract

акко́рд·н·ый, -ая, -ое, (*a*) of **акко́рд·**

∼ о·пла́т·а труд·а́ contract wage system

аккредити́в·, -а, (*m*) (fin) letter of credit

∼, без·от·зы́в·н·ый irrevocable letter of credit

аккредит·ова́ть, -у́ют, (*imp and perf*) (fin, dipl) accredit

аккре́ци·я, -и, (*f*) accretion

аккумул·и́ровани·е, -я, (*n*) (*v n*) of **аккумул·и́р·овать**; accumulation, storing, storage

аккумул·и́рованн·ый, -ая, -ое, (*past part pass*) of **аккумул·и́р·овать**

аккумул·и́р·овать, -уют, (*imp and perf*) accumulate, store up

аккумул·яти́вн·ый, -ая, -ое, (*a*) accumulative, (geol) aggradation

аккумул·я́тор·, -а, (*m*) (elec) storage battery, accumulator (Brit); (hydr) accumulator; (nucl) storage system/ device

∼, воз·ду́ш·н·ый (hydr) air-loaded accumulator, (compr air) air receiver

∼, гидравл·и́ческ·ий hydraulic accumulator

∼, груз·ов·о́й (hydr) weight-loaded accumulator

∼ давл·е́ни·я pressure accumulator

∼ давл·е́ни·я, воз·ду́ш·н·ый (rocket) compressed air container

∼ давл·е́ни·я, га́з·ов·ый (rocket) high pressure gas container

∼ давл·е́ни·я, жидк·ост·н·ый (rocket) liquid-propellant hot-gas generator

∼ давл·е́ни·я, порох·ов·о́й (rocket) solid-propellant, hot-gas generator

∼, желе́з·о-ка́дми·ев·о-ни́кел·ев·ый (elec) nickel–iron–alkaline accumulator

∼, инерц·ио́нн·ый energy storage device, flywheel

∼, кисл·о́тн·ый (elec) lead-acid accumulator

∼, паро·воз·ду́ш·н·ый steam accumulator

∼, ста́ртер·н·ый (elec, M/T) starting battery

∼, то́пл·ивн·ый (ICE) fuel accumulator

∼, щёлоч·н·ый (elec) alkaline accumulator

∼ я́дер·н·ой энерг·и·и nuclear energy storage system

аккумуля́тор·н·ая, -ой, (*f decl as adj*) battery room, battery space, battery compartment

аккумуля́тор·н·ый, -ая, -ое, (*a*) of **аккумуля́тор·** (*see also* **аккумуля́тор·н·ая** *as noun*)

~ **батаре́·я** (*f*) (elec) storage battery

~**, то́пл·ивн·ый на·со́с·** storage-type fuel pump

аккумул·я́ци·я, -и, (*f*) *see* **аккуму-л·и́ровани·е**

аккура́т·н·ый, -ая, -ое, (*a*) (*sf*) **-т·ен·, -т·н·а,** accurate

а·клина́ль·н·ый, -ая, -ое, (*a*) aclinal

акми́т·, -а, (*m*) (geol) acmite

а·коба́льт·о́зн·ый, -ая, -ое, (*a*) co-balt-deficient

акои́н·, -а, (*m*) (pharm) acoin, quani-caine

акони́т·, -а, (*m*) (bot) aconite, *Aconitum napellus*

акони́т·ов·ый, -ая, -ое, (*a*) aconitic

~ **кисл·от·а́** (*f*) (chem) aconitic acid

а́кр·, -а, (*m*) acre

акр·иди́н·, -а, (*m*) (chem) acridine

акриди́н·ов·ый, -ая, -ое, (*a*) acridine

акр·и́л·, -а, (*m*) (chem) acryl

акр·ил·а́н·, -а, *m* (plast) acrylan

акр·ил·а́т·, -а, (*m*) (chem) acrylate

акр·и́л·ов·ый, -ая, -ое, (*a*) acrylic

~ **кисл·от·а́** (*f*) (chem) acrylic acid

~ **смол·а́** (*f*) (plast) acrylic resin

акрилово·ки́сл·ая сол·ь (*f*) acrylate

акрило·нитр·и́л·, -а, (*m*) acrylonitrile

акрило·нитри́л·ов·ый каучу́к· (*m*) acrylonitrile rubber

а·крио·ге́·н·н·ый, -ая, -ое, (*a*) acryo-genous, non-glacial

акри·флав·и́н·, -а, (*m*) (pharm) acri-flavine

акрихи́н·, -а, (*m*) (pharm) mepacrine, quinacrine

акро·дро́м·н·ый, -ая, -ое, (*a*) (bot) acrodromous

акр·оле·и́н·, -а, (*m*) (chem) acrolein

акро·мега́л·и·я, -и, (*f*) (med) acro-megaly

акро·пета́ль·н·ый, -ая, -ое, (*a*) (bot) acropetal

акро·те́ри·й, -я, (*m*) (arch) acroterion

акро·фо́б·и·я, -и, (*f*) (med) acrophobia

акс бала́нс·а (horol) balance staff

акселер·а́тор·, -а, (*m*) accelerator

акселеро́·граф·, -а, (*m*) recording accelerometer, accelerograph

акселеро́·метр·, -а, (*m*) accelerometer

~**, угл·ов·о́й** angular accelerometer

аксиа́ль·но-ве́ктор·н·ый, -ая, -ое, (*a*) axial-vector

~ **-пере·ме́н·н·ый, -ая, -ое,** (*a*) axially increasing

аксиа́ль·н·ый, -ая, -ое, (*a*) axial

~ **на·со́с·** (*m*) axial-flow pump

аксини́т·, -а, (*m*) (min) axinite

аксио́м·а, -ы, (*f*) axiom, principle, postulate

~ **бес·кон·е́чност·и** axiom of infinity

~ **вы́·бор·а** (math) axiom of choice

~ **полн·от·ы́** (math) completeness axiom, axiom of completeness

~ **про·из·во́ль·н·ого вы́·бор·а** (math) axiom of choice

~ **с·чёт·ност·и** (math) axiom of count-ability

аксиома́т·ик·а, -и, (*f*) (math) axioma-tics

~**, со·держ·а́тельн·ая** informal axio-matics

аксиомат·и́ческ·ий, -ая, -ое, (*a*) axiomatic

аксио́·метр·, -а, (*m*) steering wheel indicator

а́ксис·, -а, (*m*) axis

аксоло́тл·ь, -я (*m*) (zool) axolotl

а́ксон·, -а, (*m*) (cyt) axon, axone, neurite

аксоно·метр·и́ческ·ий, -ая, -ое, (*a*) (math) axonometric, clinographic

аксо·пла́зм·а, -ы, (*f*) (zool) axoplasm

аксо·сти́л·ь, -я, (*m*) (zool) axostyle

а́кт·, -а, (*m*) act, event; certificate, document, statement, (law) deed, act; speech-day, commencement

~ **дел·е́ни·я** (nucl) fission event

~ **за·хва́т·а** (nucl) captive event

~**, об·вин·и́тельн·ый** (law) indict-ment

~**, при·ём·н·ый** acceptance certificate

акти́в·, -а, (*m*) active membership, activist element; (fin) assets

актива́тор·, -а, (*m*) activator, activat-ing agent, promoter

~ **вулкан·иза́ци·и** (chem) vulcaniza-tion activator

~**, основ·н·о́й** (elec) dominant activator

активац·ио́нн·ый, -ая, -ое, (*a*) acti-vation

~ **анализ·** (nucl) activation analysis

алтива́ц·и·я, -и, (*f*) activation

~ **прожёктор·а** (rad) gun activation

~ **тепло·нос·и́тел·я** (nucl) coolant activation

актив·иза́тор·, -а, (*m*) activator, ener-gizer

актив·и́р·ованн·ый, -ая, -ое, (*past part pass*) of **актив·и́р·овать;** activated

~ **като́д·** (*m*) (elec) activated cathode

~ **у́гол·ь** (*m*) (chem) activated carbon

актив·и́р·овать, -уют, (*imp*) activate

актив·и́рующ·ий аге́нт· (*m*) activating agent

акти́в·ност·ь, -и, (*f*) activity; (chem) ability to be activated

~, **втор·и́чн·ая** (nucl) second-order activity

~ **ис·то́ч·ник·а** (nucl) source strength
коэффицие́нт· акти́в·ност·и (*m*) (chem) activity coefficient

~, **на·ча́ль·н·ая** initial activity

~, **не·знач·и́тельн·ая** (nucl) trace-level activity

~ **о·круж·а́ющ·ей сред·ы́** environmental activity

~, **о·с·ко́л·очн·ая** (nucl) fission-fragment activity

~, **у·де́ль·н·ая** specific activity

акти́в·н·ый, -ая, -ое, (*a*) (*s f*) **-в·ен·, -в·н·а,** active; (elec) wattful

~ **зо́н·а** (*f*) (nucl) core, active section (of reactor)

~ **на·полн·и́тел·ь** (*m*) reinforcing filler (rubber)

~ **профи́л·ь** (*m*) active profile (of gear tooth)

~ **со·ставл·я́ющ·ая** (*f*) (elec) active component

акти́н·, -а, (*m*) (biochem) actin

актин·а́льн·ый, -ая, -ое, (*a*) (zool) actinal

актин·и́д·, -а, (*m*) (chem) actinide

актин·и́д·и·я, -и, (*f*) (bot) actinidia, *Actinidia*

акти́н·иев·ый, -ая, -ое, (*a*) of **акти́н·н·и·й**

акти́н·и·й, -я, (*m*) (chem) actinium, Ac

актин·и́ческ·ий, -ая, -ое, (*a*) actinic

актин·и́чност·ь, -и, (*f*) (phys) actin·ism, actinity

~, **от·нос·и́тельн·ая** actinic coefficient

актин·и́чн·ый, -ая, -ое, (*a*) actinic

актино·дермат·и́т·, -а, (*m*) (med) actinodermatitis

актин·о́ид·, -а, (*m*) actinoid, actiniform; (nucl) actinide

актино·ли́т·, -а, (*m*) (min) actinolite, radiated stone

~, **тонко·волокн·и́ст·ый** (min) nephrite

актино·ло́г·и·я, -и, (*f*) (phys, chem) actinology

актино́·метр·, -а, (*m*) (meteor) actinometer

актино·мик·о́з·, -а, (*m*) (vet, med) actinomycosis

актино·мице́т·, -а, (*m*) (microbiol) actinomyces, (*pl*) *Actinomyces*

актино·миц·и́н·, -а, (*m*) (pharm) actinomycin

актино·миц·ин·о́л·, -а, (*m*) (pharm) actinomycinol

актино·мо́рф·н·ый, -ая, -ое, (*a*) actinomorphic, star-shaped

актин·о́н·, -а, (*m*) (chem) actinon, An

актино·ско́п·, -а, (*m*) actinoscope

актино·сте́л·а, -ы, (*f*) (bot) actinostele

актино·терап·и́·я, -и, (*f*) (med) actinotherapy

актино·ура́н·, -а, (*m*) (chem) actino-uranium

акто·миози́н·, -а, (*m*) (biochem) acto-myosin

актуа́ль·н·ый, -ая, -ое, (*a*) (*s f*) **-л·ен·, -ль·н·а,** of current interest, up-to-date

актуа́р·и·й, -я, (*m*) actuary

аку́л·а, -ы, (*f*) (zool) shark

~, **соба́ч·ья** dog fish

аку́л·ий, -ья, -ье, (*a*) shark's

акуло·по·до́б·н·ые, -ых, (*pl decl as adj*) (zool) *Cladosechic*

аку·ме́тр·, -а, (*m*) (med) acoumeter

аку·пункт·у́р·а, -ы, (*f*) (med) acupuncture

аку́ст·ик·, -а, (*m*) acoustician, specialist in acoustics; sonar operator, hydrophone operator

аку́ст·ик·а, -и, (*f*) (phys) acoustics

~, **луч·ев·а́я** geometrical acoustics, ray acoustics

акуст·и́ческ·ий, -ая, -ое, (*a*) (phys) acoustic, acoustical

акуше́р·, -а, (*m*) (med) obstetrician

акуше́р·к·а, -и, (*f*) (med) midwife

акуше́р·ск·ий, -ая, -ое, (*a*) obstetric(al)

акуше́р·ств·о, -а, (*n*) obstetrics; midwifery

акцелер·а́тор·, -а, (*m*) *see* **акселер·а́тор·**

акцеле́ро·граф·, -а, (*m*) *see* **акселеро́·граф·**

акцеле́ро·метер·, -а, (*m*) *see* **акселеро́·метр·**

акце́нт·, -а, (*m*) accent

акце́нт·и́р·овать, -уют, (*imp and perf*) accent, accentuate

акце́пт·, -а, (*m*) (fin) acceptance, bill of acceptance

акцепт·а́нт·, -а, (*m*) (fin) acceptor

акце́птор·, -а, (*m*) acceptor

~ ио́н·ов· (phys, chem) ion acceptor

~, трой·н·о́й (rad) triple acceptor

акцессо́ри·я, -и, (*f*) accessory

акциде́нт·н·ый, -ая, -ое, (*a*) accidental; (print) jobbing

~ печа́т·ь (*f*) job printing

акциде́нци·я, -и, (*f*) job printing

акци́з·, -а, (*m*) excise

акци́з·н·ый, -ая, -ое, (*a*) excise

~ с·бор· (*m*) excise duty

акционе́р·, -а, (*m*) shareholder, stockholder

акционе́р·н·ый, -ая, -ое, (*a*) of **а́кци·я**

~ капита́л· (*m*) share capital

~ о́бщ·еств·о (*n*) joint-stock company, public company

а́кци·я, -и, (*f*) (fin) share, stock; action

~, имен·н·а́я (fin) registered share

~ на предъ·яв·и́тел·я (fin) bearer share

~, прост·а́я ordinary share

алаба́м·и·й, -я, (*m*) (chem) alabamium, Ab

алабанди́н·, -а, (*m*) (min) alabandite

ала́йт·, -а, (*m*) (min) alaite

ала́мбик·, -а, (*m*) distillation flask

алани́л·, -а, (*m*) (chem) alanyl

алани́н·, -а, (*m*) (chem) alanine

алба́н·ск·ий, -ая, -ое, (*a*) Albanian

а́лгебр·а, -ы, (*f*) algebra

~, бу́л·ев·а (math) Boolean algebra

~, коммутати́в·н·ая commutative algebra

~ Ли Lie algebra

~ ло́г·ик·и Boolean algebra

алгебра·и́ческ·ий, -ая, -ое, (*a*) algebraic

алгези·ме́тр·, -а, (*m*) (med) algesimeter

алги́н·, -а, (*m*) algin (gum)

алги́н·ов·ая кисл·от·а́ (*f*) (chem) alginic acid

алгори́тм·, -а, (*m*) (math) algorithm

~ Эвкли́д·а Euclid's algorithm

алгори́фм·, -а, (*m*) *see* **алгори́тм·**

алдебара́н·и·й, -я, (*m*) (chem) aldebaranium, ytterbium

алеба́стр·, -а, (*m*) alabaster, gypsum, plaster of Paris

алевр·и́т·, -а, (*m*) (geol) aleurite, silt

~, на·го́р·н·ый upland silt

алевро·ли́т·, -а, (*m*) (geol) aleurolite, siltstone

алевро·пели́т·ов·ый, -ая, -ое, (geol) aleuropelitic

алейро́·метр·, -а, (*m*) (biochem) aleurometer

алейро́н·, -а, (*m*) (bot) aleuron, aleurone

а·ле́йк·еми́·я, -и, (*f*) (med) aleukemic, leukemia

алейро́н·ов·ый, -ая, -ое, (*a*) aleurone

~ сло́·й (*m*) (bot) aleurone layer

~ зёрн·а (*n pl*) aleurone grains

александри́т·, -а, (*m*) (min) alexandrite

а·ле́кси·я, -и, (*f*) (med) alexia

алето·пте́рис·, -а, (*m*) (pal) alethopteris

алето·ско́п·, -а, (*m*) (phot) alethoscope

алеф·, -а, (*m*) (math) aleph

а́либи (*n indecl*) (law) alibi

алида́д·а, -ы, (*f*) (surv) alidade, sight rule; vernier plate (of theodolite)

~, за·кры́·т·ая closed-sight alidade

~ квадра́нт·а (surv) index arm/bar

~ с дио́птр·ами plane-table alidade

ализар·и́н·, -а, (*m*) (chem) alizarin

~ -шарла́х·, -а, (*m*) alizarin scarlet

ализари́н·ов·ый, -ая, -ое, (*a*) alizarin, alizaric

аликво́т·н·ый, -ая, -ое, (*a*) aliquot

алимент·а́р·н·ый, -ая, -ое, (*a*) alimentary

алиме́нт·ы, (*g pl*) -ов, (*m*) (law) alimony

али́т·, -а, (*m*) (chem) alite

алит·и́ровани·е, -я, (*n*) (met) alitizing, aluminizing, calorizing

алифат·и́ческ·ий, -ая, -ое, (*a*) (chem) aliphatic

али·цикл·и́ческ·ий, -ая, -ое, (*a*) (chem) alicyclic

алкали·метр·и́ческ·ий, -ая, -ое, (*a*) (chem) alkalimetric

алкали·ме́тр·и·я, -и, (*f*) (chem) alkalimetry

алкал·о́ид·, -а (*m*) (chem) alkaloid

алкал·о́ид·н·ый, -ая, -ое, (*a*) alkaloidal

алка́н·, -а, (*m*) (chem) alkane

алканни́н·, -а, (*m*) (chem) alkannine, alkannin

алкарси́н·, -а, (*m*) (chem) alkarsine

алк·а́ть, (*pres 3rd pl*) а́лч·ут or **алк·а́·ют,** (*imp*) crave, hunger for

алк·е́н·, -а, (*m*) (chem) alkene

алк·и́л·, -а, (*m*) (chem) alkyl

алкил·а́т·, -а, (*m*) (chem) alkylate, alkylation product

алкил·бензо́л·, -а, (*m*) alkyl benzene

алкил·е́н·, -а, (*m*) alkylene, alkene

алкил·иде́н·, -а, (*m*) alkylidene

алкил·и́ровани·е, -я, (*n*) alkylation

алкил·се́р·н·ый, -ая, -ое, (*a*) alkyl-sulphuric

алкил·фосфи́н·ист·ая кисл·от·а́ (*f*) alkylphosphinic acid

алк·и́н·, -а, (*m*) (chem) alkyne, alkine

алкогол·и́з·, -а, (*m*) (chem) alcoholysis

алкогол·иза́ци·я, -и, (*f*) alcoholiza-tion; fortification (of wine)

алкого́л·ик·, -а, (*m*) alcoholic, dipso-maniac; drunkard

алкоголи·ме́тр·и·я, -и, (*f*) alcoholi-metry, alcoholometry

алкогол·и́ческ·ий, -ая, -ое, (*a*) alco-holic

алкого́л·ь·, -я, (*m*) (chem) alcohol, spirit; spirits (drink)

алко·зо́л·ь, -я, (*m*) (chem) alcosol

алк·окси́ль·н·ый, -ая, -ое, (*a*) alk-oxyl, alkoxy

алкомакс·, -а, (*m*) (met) Alcomax

аллани́т·, -а, (*m*) (min) allanite, orthite

аллант·о́ид·н·ый, -ая, -ое, (*a*) allan-toid, sausage-shaped

алланто·и́н·, -а, (*m*) (biochem) allan-toin

алланто́·ис·, -а, (*m*) (zool) allantois

аллело·мо́рф·, -а, (*m*) (biol) allelo-morph

аллело·мо́рф·н·ый, -ая, -ое, (*a*) allelo-morphic

аллемонти́т·, -а, (*m*) (min) allemon-tite, arsenical antimony

алле́н·, -а, (*m*) (chem) allene, propa-diene

алл·ерге́н·, -а, (*m*) (pharm) allergen

алл·ерги́·я, -и, (*f*) (med) allergy

алле́·я, -и, (*f*) avenue

аллига́т·, -а, (*m*) (print) alligatum, bound with

аллига́тор·н·ые но́ж·ниц·ы (*f pl*) alligator shears

алл·и́л·, -а, (*m*) (chem) allyl

~, изорода́н·ов·ый allyl-mustard oil

~ -горч·и́чн·ое ма́сл·о (*n*) allyl-mustard oil

~, сер·ни́ст·ый (chem) allyl sulphide/sulfide, oil of garlic

аллил·е́н·, -а, (*m*) (chem) allylene

алли́л·ов·ый, -ая, -ое, (*a*) allyl

~ спи́рт· (*m*) allyl alcohol

аллил·сульфо·карб·ами́д·, -а, (*m*) allyl sulphocarbamide

аллил·тио·мочеви́н·а, -ы, (*f*) allyl thiocarbamide

алл·о́з·а, -ы, (*f*) (chem) allose

алло·за́вр·, -а, (*m*) (pal, zool) allo-saurus

алл·окса́н·, -а, (*m*) (pharm, physiol) alloxan

алло·ме́р·н·ый, -ая, -ое, (*a*) (chem) allomeric

алло·мо́рф·н·ый, -ая, -ое, (*a*) (chem) allomorphic, allomorphous

алло́нж·, -а, (*i*) **-ем,** (*m*) *see* **ало́нж**

алло·оциме́н·, -а, (*m*) (chem) allo-ocimene

алло·па́т·, -а, (*m*) (med) allopath, allopathist

алло·па́т·и·я, -и, (*f*) (med) allopathy

алло·трио·морфно·зерн·и́ст·ый, -ая, -ое, (*a*) (geol) allotriomorphic-granular, xenomorphic-granular

алло·трио·мо́рф·н·ый, -ая, -ое, (*a*) (geol) allotriomorphic, xenomorphic

алло·тро́п·и·я, -и, (*f*) (chem) allo-tropy

алло·тро́п·н·ый, -ая, -ое, (*a*) allo-tropic, allotropous

алло·тро́ф·н·ый, -ая, -ое, (*a*) (bio) allotrophic

алло·фа́н·, -а, (*m*) (min) allophane

алло·хто́н·н·ый, -ая, -ое, (*a*) (geol) allochthonous

аллювиа́ль·н·ый, -ая, -ое, (*a*) (geol) alluvial

~ на·но́с·ы (*m pl*) alluvial deposits

~ фо́рм·а (*f*) (geol) alluvium

аллю́ви·й, -я, (*m*) (geol) alluvium

аллю́р·, -а, (*m*) gait (of horses etc.), (*pl*) dressage (of horses)

алма́з·, -а, (*m*) diamond

~, гран·ён·ый cut diamond

~, ме́л·к·ий spark

~, металл·и́ческ·ий (met) diamond-substitute alloy

~, не·шлиф·о́ванн·ый brait

~, о·гран·ённ·ый cut diamond

~, техн·и́ческ·ий industrial diamond

~, чёрн·ый black diamond

~, чёрн·ый техн·и́ческ·ий carbonado

~, ювели́р·н·ый gem diamond

алмаз·н·о-металл·и́ческ·ий, -ая, -ое, (*a*) metal-bonded diamond

алма́з·но-рас·то́ч·н·ый стан·о́к· (*m*) diamond boring machine

алма́з·н·ый, -ая, -ое, (*a*) of **алма́з·**; (min) adamantine

~ **бур·е́ни·е** (*n*) diamond drilling

алмазо·за·мен·и́тел·ь, -я, (*m*) diamond substitute

алманди́н·, -а, (*m*) (min) almandine, almandite

ални (*abbr*) (met) Alni (Al–Ni alloy)

алнико (*abbr*) Alnico (Al–Ni–Co–Fe alloy)

алниси (*abbr*) Alnisi (Al–Ni–Si alloy)

а́л·о (*adv*) of **а́л·ый**

алои́н·, -а, (*m*) (pharm) aloin

ало́и·н·ый, -ая, -ое, (*a*) of **ало́э**

~ **де́рев·о** (*n*) (bot) eagle wood, aloes wood

алокси́т·, -а, (*m*) (chem) aloxite

ало́нж·, -а, (*i*) **-ем,** (*m*) (comm) allonge; (chem) adapter

ало́э (*n indecl*) (bot) aloe, (med) aloes

ал·си·фе́р·, -а, (*m*) (met) Alsifer (Al–Si–Fe alloy)

алстони́т·, -а, (*m*) = **альстони́т·**

алтаи́т·, -а, (*m*) (min) altaite

алу́нд·, -а, (*m*) alundum, (chem) abrasite

алуни́т·, -а, (*m*) (min) alunite, alumstone

алуно·ге́н·, -а, (*m*) (min) alunogen, feather alum

алуч·а́, -и, (*f*) *see* **алыч·а́**

алфави́т·, -а, (*m*) alphabet

~**, в·хо́д·н·о́й** (autom) input alphabet

алфави́т·н·ый, -ая, -ое, (*a*) alphabetic, alphabetical

~ **у·каз·а́тел·ь** (*m*) (print) index

алфи́л·, -а, (*m*) (chem) alphyl

а́лч·а (*pres ger*) of **алк·а́ть** craving

а́л·ый, -ая, -ое, (*a*) (*s f*) **ал·, ал·а,** scarlet, vermilion, light-red, bright-red; (polit) red

алыч·а́, -и, (*f*) (bot) *Prunus divaricata* (kind of damson)

альб·, -а, (*m*) (geol) Albian stage

альб·е́д·о, -а, (*n*) albedo

~**, ви́д·им·ое** (astron) apparent albedo

~ **лун·ы́** (astron) albedo of the moon

~ **нейтро́н·ов** (nucl) albedo

~**, сфер·и́ческ·ое** (astron) spherical albedo

альбедо·ме́тр·, -а, (*m*) albedometer

альбер·ти́п·и·я, -и, (*f*) (print, phot) albertype

альберт·и́т·, -а, (*m*) (min) albertite, asphaltic coal

альберт·о́л·ь, -я, (*m*) (chem) albertol

альб·ин·и́зм·, -а, (*m*) (zool) albinism

альб·ин·о́с·, -а, (*m*) (zool) albino

альб·и́т·, -а, (*m*) (min) albite

альбит·иза́ци·я, -и, (*f*) (geol) albitization

альбит·и́т·, -а, (*m*) (min) albitite

альби́т·ов·ый, -ая, -ое, (*a*) (geol) albite

альбито·фи́р·, -а, (*m*) (geol) albitophyre

альб·о́м·, -а, (*m*) album, book of illustrations

альбо·мици́н·, -а, (*m*) (pharm) albomycin

альбуми́н·, -а, (*m*) (bot) albumen; (biochem) albumin

~**, яи́ч·н·ый** egg-albumin

альбумин·а́т·, -а, (*m*) (chem) albuminate

~ **серебр·а́** albuminate of silver, argentic albuminate

альбумин·о́з·н·ый, -ая, -ое, (*a*) albuminous

альбумин·о́ид·, -а, (*m*) albuminoid

альбумин·ури́·я, -и, (*f*) (med) albuminuria

альбум·о́з·а, -ы, (*f*) (biochem) albumose, proteose

альбум·о́ид·, -а, (*m*) (biochem) albumoid, albuminoid

альбу·танни́н·, -а, (*m*) (biochem) albumin tannate

альбу·ци́д·, -а, (*m*) (pharm) albucid, sulphacetamide

альве·о́л·а, -ы, (*f*) (anat) alveola, alveolus; (bot) alveolus

альве·ол·иза́ци·я, -и, (*f*) alveolation

альве·ол·я́рн·ый, -ая, -ое, (*a*) alveolar, alveolate

альви́т·, -а, (*m*) (min) alvite

альгези·ме́тр·, -а, (*m*) (med) algesimeter

альгодони́т·, -а, (*m*) (min) algodonite

альго·ла́гни·я, -и, (*f*) (med) algolagnia

альго·ло́г·и·я, -и, (*f*) (bot) algology

аль·гра́ф·и·я, -и, (*f*) (print) algraphy

аль·деги́д·, -а, (*m*) (chem) aldehyde

~**, муравь·и́н·ый** formic aldehyde

~ **-мут·а́з·а, -ы,** (*f*) (biochem) aldehyde mutase

аль·деги́д·н·ый, -ая, -ое, (*a*) aldehyde

~ **смол·а́** (*f*) aldehyde resin

альдо·би·о́н·ов·ая кисл·от·а́ (*f*) aldobionic acid

альдо·гекс·о́з·а, -ы, (*f*) (chem) aldo-hexose

альд·о́з·а, -ы, (*f*) (chem) aldose

альдо·кетéн·, -а, (*m*) (chem) aldo-ketene

альд·окса́н·, -а, (*m*) aldoxane

альд·окси́м·, -а, (*m*) aldoxime

альд·о́л·ь, -я, (*m*) aldol

альд·о́ль·н·ый, -ая, -ое, (*a*) aldehyde, aldol

~ у·плотн·éни·е (*n*) (chem) aldol/ aldehyde condensation

альдо·пент·о́з·а, -ы, (*f*) (chem) aldo-pentose

альдо·ла́з·а, -ы, (*f*) (biochem) aldo-lase

альдре·й, -я, (*m*) (elec, met) Aldrey (Al–Mg alloy)

алькализ·а́ци·я, -и, (*f*) alkalization

алькализ·и́р·овать, -уют, (*imp and perf*) alkalify

альклэд·, -а, (*m*) (met) Alclad

алько́в·, -а, (*m*) (arch) alcove

алькуси́н·, -а, (*m*) Alcusin (Al–Cu–Si alloy)

альманди́н·, -а, (*m*) (min) almandine, almandite

альмелек·, -а, (*m*) (elec, met) Almalec

альмукантара́т·, -а, (*m*) (astron, surv) almucantar, circle of altitude

альни (*abbr*) *see* **ални**

альнико (*abbr*) *see* **алнико**

альниси (*abbr*) *see* **алниси**

альпага́ (*n indecl*) = **альпака́**

альпака́ (*n indecl*) alpaca (animal, wool, textile)

альпа́ри (*n indecl*) (fin) at par, par

альп·и́йск·ий, -ая, -ое, (*a*) alpine

~ лу́г· (*m*) (bot) alpine meadow

альпини́зм·, -а, (*m*) mountaineering

альсифéр·, -а, (*m*) *see* **алсифéр·**

альстони́т·, -а, (*m*) (min) alstonite

альт·а́зимут·, -а, (*m*) (surv, astron) altazimuth

альтакс·, -а, (*m*) (chem) altax, dibenz-thiazyl disulphide (rubber accelerator)

альтернати́в·а, -ы, (*f*) alternative

альтернати́в·н·ый, -ая, -ое, (*a*) (*s f*) **-в·ен·, -в·н·а,** alternative

альтерна́тор·, -а, (*m*) (elec) alternator, synchronous A.C. generator

альтернио́н·, -а, (*m*) alternion

альтерн·и́ровани·е, -я, (*n*) (math) alternation, antisymmetrization

альти·гра́ф·, -а, (*m*) altigraph, record-ing altimeter

альти·мéтр·, -а, (*m*) altimeter

~, автома́т·ическ·ий, recording alti-meter

альти·мéтр·и́ческ·ое кольцо́ (*n*) (rad) altitude ring

альти·мéтр·и·я, -и, (*f*) altimetry

альтиту́д·а, -ы, (*f*) altitude, absolute altitude, height above sea level

а́льф·а, -ы, (*f*) alpha, α (Greek); (bot) alfa

~ -вéтв·ь, -и, (*f*) (nucl) alpha-branch

~ -желéз·о, -а, (*n*) (met) alpha iron

~ -кальци́т·, -а, (*m*) (min) elatolite

~ -ли́н·и·я, -и, (*f*) (nucl) alpha-peak

~ -лу́ч·, -и, (*nom pl*) **-й,** (*g pl*) **-éй** (*m*) (phys) alpha-ray

~ -нафт·о́л·, -а, (*m*) α-hydroxy naphthalene, α-naphthalene

~ -трав·а́, -ы́, (*f*) (bot) esparto grass, *Lygeum spartium*

~ -фа́з·а, -ы, (*f*) (met) alpha phase

~ -част·и́ц·а, -ы, (*i*) **-ей,** (*f*) (phys) alpha particle

альфа·мéтр·, -а, (*m*) alphameter

альфа·тро́н·, -а, (*m*) (air dynam) al-phatron

аль·фо́л·ь, -я, (*m*) (met) Alfol, alumi-nium foil

альфо́·метр·, -а, (*m*) fuel-mixture indicator

А́льфорд·а, антéнн·а (rad) Alford loop

альциона́ри·я, -и, (*f*) (zool) alcyo-naria

алюмэ́л·ь, -я, (*m*) (met) alumel

алюмин·а́т·, -а, (*m*) aluminate

алюми́н·иев·ый, -ая, -ое, (*a*) alumi-nium (Brit), aluminum (U.S.); (chem) aluminic

~ кисл·от·а́ (*f*) (chem) aluminic acid

~ дубл·éни·е (*n*) aluminium/alumin-um/alum tanning

алюми́н·и·й, -я, (*m*) (chem) alumi-nium (Brit), aluminum (U.S.), Al

~, фто́р·ист·ый aluminium fluoride, cryolite

~, хло́р·ист·ый aluminium chloride

алюмин·и́т·, -а, (*m*) (min) aluminite, aluminium-potassium alum

алюмино·ки́сл·ый, -ая, -ое, (*a*) alu-minate of

~ на́три·й (*m*) sodium aluminate

алюмин·о́н·, -а, (*m*) (chem) aluminon, ammonium salt of aurintricarboxylic acid

алюмино·силика́т·, -а, (*m*) alumino-silicate

алюмино·терм·и́ческ·ий, -ая, -ое, (*a*) (met) aluminothermic

алюмино·терм·и́·я, -и, (*f*) aluminothermy

алюмо·ге́л·ь, -я, (*m*) (chem) alumogel

алюмо·кальц·и́т·, -а, (*m*) (min) alumocalcite

алюмо·ли́т·, -а, (*m*) (min) bauxite

алюмо·силика́т·, -а, (*m*) = **алюмино·силика́т·**

алюмо·терм·и́ческ·ий, -ая, -ое, (*a*) *see* **алюмино·терм·и́ческ·ий**

алюмо·терм·и́·я, -и, (*f*) *see* **алюмино·терм·и́·я**

амавро́з·, -а, (*m*) (med) amaurosis

Амадори, пере·групп·иро́вк·а (*f*) (chem) Amadori arrangement

амазони́т·, -а, (*m*) (min) amazonite, green microcline, amazon stone

амальга́м·а, -ы, (*f*) (chem, min) amalgam

амальгам·а́тор·, -а, (*m*) (min) amalgamator, amalgamating device

амальгам·ацио́н·н·ый, -ая, -ое, (*a*) amalgamating, amalgamation

амальгам·а́ци·я, -и, (*f*) (met) amalgamation

~, вну́тр·енн·яя pan amalgamation

амальг·и́р·ованн·ый, -ая, -ое, (*past part pass*) amalgamated

амальг·и́р·ующ·ий, -ая, -ее, (*pres part act*) amalgamating

аманди́н·, -а, (*m*) (chem) amandin

амари́н·, -а, (*m*) (chem) amarine

амари́в·ов·ый, -ая, -ое, (*m*) amaric

амба́р·, -а, (*m*) granary, grain store, barn; warehouse, storehouse

амба́рго, (*n indecl*) embargo

амба́р·н·ый, -ая, -ое, (*a*) of **амба́р·**

амбер·о́ид·, -а, (*m*) (min, chem) amberoid, ambroid, pressed amber

амби- (*int component*) ambi- (= both)

амби·вале́нт·ност·ь, -и, (*f*) ambivalence

амби·поля́р·н·ый, -ая, -ое, (*a*) ambipolar, bipolar

амбли·гони́т·, -а, (*m*) (min) amblygonite

амбли·оп·и́·я, -и, (*f*) (med) amblyopia

амбли·сто́м·а, -ы, (*f*) (zool) amblystoma

амбо·це́птор·, -а, (*m*) (bact) amboceptor

а́мбр·а, -ы, (*f*) ambergris

амбразу́р·а, -ы, (*f*) (arch) embrasure; (mil) gun-port

а́мбр·ов·ый, -ая, -ое, (*a*) ambergris

амбро́ин·, -а, (*m*) ambroin (dielectric)

амбро́зи·я, -и, (*f*) (bot) ragweed, *Ambrosia*

амбул·ато́ри·я, -и, (*f*) (med) out-patient clinic/department

амбул·ато́рн·ый боль·н·о́й (*m*) (med) out-patient

амбуля́кр·, -а, (*m*) (zool) ambulacrum

амбушю́р·, -а, (*m*) mouth-piece (e.g. of telephone)

амёб·а, -ы, (*f*) (zool) amoeba, ameba

амёб·н·ый, -ая, -ое, (*a*) amoebic, amebic

~ дизентери́·я (*f*) (med) amoebic dysentery

амёбо·ви́д·н·ый, -ая, -ое, (*a*) amoeboid

амёбо·ци́т·, -а, (*m*) amoebocyte

америка́н·ец·, -н·ц·а, (*i*) **-н·ц·ем,** (*m*) American (man)

америка́н·к·а, -и, (*g pl*) **-н·ок·** (*f*) American (woman); (min) sand pump

америка́н·ск·ий, -ая, -ое, (*a*) American

~ лопа́т·к·а (*f*) plasterer's trowel

амери́ц·и·й, -я, (*m*) (chem) americium, Am

амети́ст·, -а, (*m*) (min) amethyst

амиа́нт·, -а, (*m*) (min) amianthus

амианти́н·, -а, (*m*) (text) amiantin

амигдал·и́н·, -а, (*m*) (chem) amygdalin

амигдал·о́ид·, -а, (*m*) amygdaloid

ами́д·, -а, (*m*) (chem) amide

~ кисл·от·ы́ acid amide

амид·и́н·, -а, (*m*) (chem) amidine

ами́до- (*int component*) (chem) amido-

амидо·ге́н·, -а, (*m*) amidogen, amido group

амид·о́л·, -а, (*m*) (chem) amidol

амин·и́р·овать, -ую́т (*imp and perf*) (chem) aminate

амидо·сульфо́н·ов·ая кисл·от·а́ (*f*) amidosulphonic acid

амикро́н·, -а, (*m*) (phys, chem) amicron

ами́л·, -а, (*m*) (chem) amyl

амил·а́з·а, -ы, (*f*) (chem) amylase

амил·ацет·а́т·, -а, (*m*) amyl acetate

амил·е́н·, -а, (*m*) (chem) amylene

амил·нитри́т·, -а, (*m*) (chem) amyl nitrite

ами́ло- (*int component*) (chem) amylo-

ами́л·ов·ый, -ая, -ое, (*a*) of **ами́л·**

~ спи́рт·, (*m*) amyl alcohol

амил·о́з·а, -ы, (*f*) (chem) amylose

амил·о́ид·, -а, (*m*) (chem) amyloid

амил·о́ид·н·ый, -ая, -ое, (*a*) amyloid

~ пере·рожд·е́ни·е (*n*) (med) amyloid degeneration

амило·класт·и́ческ·ий, -ая, -ое, (*a*) (bot) amyloclastic

амило·лит·и́ческ·ий, -ая, -ое, (*a*) (zool) amylolytic

амило·пект·и́н·, -а, (*m*) (chem) amylopectin

амило·пла́ст·, -а, (*m*) (bot) amyloplast

амил·опс·и́н·, -а, (*m*) (chem, zool) amylopsin

амило·фосфор·ил·а́з·а, -ы, (*f*) (biochem) amylophosphorylase

ами́н·, -а, (*m*) (chem) amine

~, втор·и́чн·ый secondary amine

~, перв·и́чн·ый (chem) primary amine

~, трет·и́чн·ый tertiary amine

амин·аз·и́н·, -а, (*m*) (pharm) aminazin, chlorpromazin

амин·и́р·овать, -уют, (*imp and perf*) (chem) aminate

ами́н·н·ый, -ая, -ое, (*a*) of ами́н·

ами́но- (*int component*) (chem) amino

амино·бензо́л·, -а, (*m*) aminobenzene

амино·глюта́р·ов·ая кисл·от·а́ (*f*) glutamic acid

амино·гру́пп·а, -ы, (*f*) amino-group

амино·кисл·от·а́, -ы, (*f*) (chem) amino acid

амино·нафтали́н·, -а, (*m*) naphthylamine

амино·пла́ст·, -а, (*m*) aminoplast, amino plastic

амино·про·из·во́д·н·ое, -ого (*n decl as adj*) (chem) amino derivative

амино·смол·а́, -ы́, (*f*) (plast) amino resin

амино·спи́рт·, -а, (*m*) alkamine, amino alcohol

амино·янта́р·н·ая кисл·от·а́ (*f*) (chem) aminosuccinic acid, aspartic acid

а·мио·троф·и́ческ·ий бок·ов·о́й склер·о́з·, (*m*) (med) amyotrophic lateral sclerosis

а·мио·троф·и́·я, -и, (*f*) (med) amyotrophy, amyotrophia

амир·и́н·, -а, (*m*) (chem) amyrin

амита́л·, -а, (*m*) (pharm) amytal

а·мит·о́з·, -а, (*m*) (cyt) amitosis, (*pl*) amitoses

Ами́чи, при́зм·а (*f*) (opt) Amici prism

аммиа́к·, -а, (*m*) (chem) ammonia

~, во́д·н·ый liquid ammonia, ammonium hydroxide

аммиа́к

~, жи́дк·ий liquid ammonia, ammonium hydroxide

аммиак·а́т·, -а, (*m*) (chem) ammoniate, ammine; (*pl*) ammonia/nitrogen solutions (fertilizers)

аммиачно·сели́тр·енн·ый, -ая, -ое, (*a*) ammonium nitrate

аммиа́ч·н·ый, -ая, -ое, (*a*) ammonia, ammoniac, ammoniacal

~ вод·а́ (*f*) ammoniacal liquor, ammonia solution

~ диссоциа́тор· (*m*) ammonia cracking plant

~ кре́кер· (*m*) ammonia cracking plant

~ сели́тр·а (*f*) nitrate of ammonia

~ супер·фосф·а́т· (*m*) superphosphate of ammonia

амми́н·, -а, (*m*) ammine, ammoniate

аммона́л·, -а, (*m*) (expl) ammonal

аммо́н·иев·ый, -ая, -ое, (*a*) ammonia, ammonium

аммон·иза́ци·я, -и, (*f*) (biol) ammonization

аммо́н·и·й, -я, (*m*) (chem) ammonium

~, серно·ки́сл·ый ammonium sulphate, sulphate of ammonia

~, хло́р·ист·ый (chem, min) ammonium chloride, sal ammoniac

аммоний·фосфа́т·,-а, (*m*) see аммофо́с·

аммоний·цитро·фосфа́т·, -а, (*m*) ammonicitrophosphate (fertilizer)

аммон·и́т·, -а, (*m*) (expl, geol) ammonite

~, патрон·и́рованн·ый cartridge-packed ammonite

аммони́т·ов·ый, -ая, -ое, (*a*) (expl) ammonite; (geol) ammonitic

аммо·нитро·фо́с·, -а, (*m*) ammonitrophos (fertilizer)

аммони·фик·а́ци·я, -и, (*f*) ammonization, ammonification

аммон·оид·е́·я, -и, (pal) ammonoid, ammonoidean, (*pl*) *Ammonoidea*

аммоно·ли́з·, -а, (*m*) (chem) ammonolysis

аммото́л·, -а, (*m*) (expl) ammotol

аммо·фо́с·, -а, (*m*) ammonia phosphate fertilizer, ammophos

аммо·фо́с·к·а, -и, (*f*) "Ammophoska" (complete fertilizer)

амнио́н·, -а, (*m*) (zool) amnion, (*pl*) amnions, amnia

амниот·и́ческ·ий, -ая, -ое, (*a*) (zool) amniotic

амнио́т·ы (*nom pl*), (*g pl*) -ов, (*m*) (zool) amniotes, *Amniota*

амози́т·, **-а**, (*m*) (min) amosite

амортиз·а́тор·, **-а**, (*m*) shock absorber, buffer, damper; (gunn) recoil spring

~, ма́сл·ян·о-пневмат·и́ческ·ий oleo-pneumatic shock absorber

~, ма́сл·ян·ый oil shock absorber

~, теле·скоп·и́ческ·ий telescopic/direct-acting shock absorber

~, шнур·ов·ой (air) shock cord

амортиз·ацио́н·н·ый, **-ая**, **-ое**, (*a*) of **амортиз·а́ци·я**

~ фонд· (*m*) (fin) sinking fund

амортиз·а́ци·я, **-и**, (*f*) (mech) shock absorption, damping; (fin) amortization

амортиз·и́рованн·ый, **-ая**, **-ое**, (*past part pass*) shock-proof, damped; (fin) amortized

а·мо́рф·ност·ь, **-и**, (*f*) amorphism

а·мо́рф·н·ый, **-ая**, **-ое**, (*a*), (*s f*) **-ф·ен**, **-ф·н·а**, amorphous; non-crystalline

ампело·гра́ф·и·я, **-и**, (*f*) (hortic) ampelography

ампел·о́псис·, **-а**, (*m*) (bot) ampelopsis, *Ampelopsis*

ампело·терап·и́·я, **-и**, (*f*) (med) ampelotherapy

ампе́р·, **-а**, (*g pl*) **ампе́р·**, (*m*) (elec) ampere, A (Brit), a (U.S.), amp

~ -про·вод·ни́к·, **-а**, (*m*) ampere-wire/-conductor

~ -ча́с·, **-а**, (*m*) ampere-hour

ампер·а́ж·, **-а**, (*i*) **-ем**, (*m*) (elec) amperage

ампер·вит·о́к·, **-т·к·а**, (*m*) ampere/amper turn

~ воз·бужд·е́ни·я field ampere turn

~, раз·магни́ч·ивающ·ий demagnetizing ampere-turn, back ampere-turn

ампер·вольт·ме́тр·, **-а**, (*m*) (elec) voltammeter, voltmeter-ammeter

ампер·вольт·ом·ме́тр· те́стер· (*m*) (elec) multimeter, multi-range test set

ампер·ме́тр·, **-а**, (*m*) amperemeter, ammeter

~, анте́нн·ый radiation ammeter

~ по·сто·я́нн·ого то́к·а D.C. ammeter

~, тепл·ов·ой thermal/hot-wire ammeter

~, электро·динам·и́ческ·ий (elec) moving-coil ammeter

ампер·час·ов·о́й, **-а́я**, **-о́е**, (*a*) ampere-hour

ампи́р·, **-а**, (*m*) (arch) Empire, Empire style

амплекс·о́ид·н·ый, **-ая**, **-ое**, (*a*) amplexoid

ампли·ди́н·, **-а**, (*m*) (elec) amplidyne, amplidyne generator

ампли·тро́н·, **-а**, (*m*) (rad) amplitron

амплиту́д·а, **-ы**, (*f*) amplitude, range, crest value

~, абсолю́т·н·ая range of variation

~, двой·н·а́я double/total amplitude, peak-to-peak value

~, и́мпульс·а pulse height

~, кач·е́ни·я swing

~, колеб·а́ни·я vibration amplitude, oscillation amplitude

коэффициент· амплитуд·ы (*m*) (elec) peak/crest factor

~, поро́г·ов·ая cut-off pulse height

~, при·ли́·в·а (ocean) tidal range

~, с·бро́с·а (geol) fault amplitude, total throw

~, су́т·очн·ая (ocean) daily range

амплиту́д·н·о-и́мпульсн·ая модул·я́ци·я (*f*) (elec) pulse-amplitude modulation, PAM

амплиту́д·н·о-и́мпульс·н·ая-част·о́тн·ая модул·я́ци·я (*f*) pulse-amplitude-modulation-frequency modulation, PAM-FM

амплитудо·модул·и́рованн·ый, **-ая**, **-ое**, (*a*) amplitude-modulated

амплитудо·част·о́тн·ая характер·и́стик·а (*f*) amplitude frequency characteristic

амплиту́д·н·ый, **-ая**, **-ое**, (*a*) of **амплиту́д·а**

~, вы́·бор· (*m*) (elec) amplitude discrimination

~, ис·каж·е́ни·е (*n*) amplitude distortion

~ селе́ктор· и́мпульс·ов (*m*) pulse height selection

а́мпул·а, **-ы**, (*f*) (pharm etc.) ampoule, ampulla, vial; (zool) ampulla

~ у́·ровн·я levelling glass, spirit bulb

а́мпул·к·а, **-и**, (*g pl*) **-л·ок·**, (*dim*) of **а́мпул·а**

ампут·а́ци·я, **-и**, (*f*) (med) amputation

амп-ч (*abbr*) = **ампер·ча́с·** (elec) ah, ampere-hour

АМС (*abbr*) = **автомат·и́ческ·ая меж·планет·н·ая ста́нци·я** automatic inter-planetary station

амуни́ци·я, **-и**, (*f*) (mil) kit, accoutrements

амфи́б·и·я, -и, (*f*) amphibian (air/
sea or land/sea); (zool) amphibian
автомоби́л·ь-амфи́би·я, -и, (*f*)
amphibious vehicle
амфи·бо́л·, -а, (*m*) (min) amphibole
~ -асбе́ст·, -а, (*m*) amphibole asbestos
амфи·бол·и́т·, -а, (*m*) (geol) amphi-
bolite
амфи·га́стр·и·й, -я, (*m*) (bot) amphi-
gastrium
амфи·го́н·и·я, -и, (*f*) (biol) amphi-
gony
амфи·дипл·о́ид·, -а, (*m*) (gen) amphi-
diploid
амфи·ди́ск·, -а, (*m*) (zool) amphidisc
амфи·карп·и́·я, -и, (*f*) (bot) amphi-
carpy
амфи·ми́ксис·, -а, (*m*) (gen) amphimixis, (*pl*) amphimixes
амфио́н·, -а, (*m*) (chem) zwitterion
амфи·теа́тр·, -а, (*m*) (arch) amphi-
theatre
амфи·це́ль·н·ый, -ая, -ое, (*a*) (zool)
amphicoelus
амфо·ли́т·, -а, (*m*) (chem) ampholyte
амфо·лит·о́ид·, -а, (*m*) (chem) am-
pholytoid
амфоте́р·н·ый, -ая, -ое, (*a*) (chem)
amphoteric
АН (*abbr*) = **Акаде́м·и·я На·у́к·**
Academy of Sciences
анабази́н·, -а, (*m*) (chem) anabasine
анаба́зис·, -а, (*m*) (bot) anabasis
анабат·и́ческ·ий, -ая, -ое, (*a*) (meteor) anabatic
ана·био́з·, -а, (*m*) (zool) anabiosis,
(*pl*) anabioses
ана·боли́зм·, -а, (*m*) (bot) anabolism
ана·гли́ф·, -а, (*m*) (phot) anaglyph
ана·глифо·ско́п·, -а, (*m*) (phot) ana-
glyphoscope
анадо́л·, -а, (*m*) (pharm) anadol,
alpha-prodine
ана·дро́м·н·ый, -ая, -ое, (*a*) (fish)
anadromous
ана·ка́рд·иев·ый, -ая, -ое, (*a*) of
ана·ка́рд·и·я (chem) anarcadic
ана·ка́рд·и·я, -и, (*f*) (bot) cashew,
Anacardium
ана·клин·а́льн·ый, -ая, -ое, (*a*)
anaclinal
аналагмат·и́ческ·ий, -ая, -ое, (*a*)
analagmatic
ана́·лиз·, -а, (*m*) analysis, assay
~, актив·ацио́нн·ый activation analysis

ана́·лиз
~, ба́зис·н·ый base analysis
~, бес·кон·е́чн·о ма́л·ых (math)
infinitesimal/ordinary calculus
~, вал·ов·о́й bulk/total analysis
~, вес·ов·о́й gravimetric analysis
~, гармон·и́ческ·ий (math) harmonic/
Fourier analysis
~, изото́п·н·ый isotopic/trace analysis
~, ка́честв·енн·ый qualitative analysis
~, коли́честв·енн·ый quantitative
analysis
~, люминесце́нт·н·ый fluorimetric
analysis
~, мало·групп·ов·о́й (math) few-
group analysis
~, механ·и́ческ·ий granulometric
analysis
~, нефело·метр·и́ческ·ий (chem)
nephelometric/turbidimetric analysis
~, объ·ём·н·ый volumetric analysis
~, про·би́р·н·ый assay, assaying
~ раз·ме́р·ност·и dimensional analysis
~, регресси́в·н·ый (math) regression analysis
~, реш·а́ющ·ий arbitration analysis
~, си́т·ов·ый screen analysis
~, спектр·а́льн·ый spectral analysis
~, сум·ма́рн·ый summative analysis
~, техн·и́ческ·ий commercial analysis; proximate analysis
~, фа́з·ов·ый phase-shift analysis
~ фо́рм·ы сигна́л·ов wave-form
analysis
~, чи́сл·енн·ый numerical analysis
~, шлих·ов·о́й (min) panning
~ электр·и́ческ·ой схе́м·ы electric
circuit analysis
~, элемент·а́рн·ый ultimate/elemen-
tary analysis
анализ·а́тор·, -а, (*m*) analyzer, analyser
~, амплиту́д·н·ый (elec) pulse height
analyzer, kick sorter
~, гармон·и́ческ·ий frequency analyzer
~, дифференциа́ль·н·ый differential analyzer
~, и́мпульс·н·ый pulse height analyzer, kick sorter
~ ис·каж·е́ни·й distortion analyzer
~, коррелят́ив·н·ый correlation function analyzer
~ со·ста́·в·а с·ме́с·и (ICE) mixture
analyzer

анализ·и́р·овать, -ют,   (*imp*) analyze, analyse

анали́т·ик·, -а,   (*m*) analyst

аналит·и́ческ·ий, -ая, -ое,   (*a*) analytic(al)

~ реа́кци·я   (*f*) (chem) analytical reaction

аналит·и́чность·ь, -и,   (*f*) (math) analyticity

ана́лог·, -а,   (*m*) analogue, analog

аналог·и́чн·ый, -ая, -ое,   (*a*) (*s f*) **-·и́чен·, -·и́чн·а,**   analogous, similar

~ о́рган·ы   (*m pl*) (biol) analogous organs

анало́г·и·я, -и,   (*f*) analogy

ана́лого-ци́фр·ов·о́й пре·обра́з·ова́·тел·ь, -я,   (*m*) analogue-to-digital converter, digitizer

анало́г·ов·ый, -ая, -ое,   (*a*) (autom) analogue, analog

анальби́т·, -а,   (*m*) (min) analbite

анальгез·и́·я, -и,   (*f*) (med) analgesia

анальге́т·ик·, -а,   (*m*) analgesic (drug)

анальгет·и́ческ·ий, -ая, -ое,   (*a*) (physiol) analgesic

ана́ль·н·ый, -ая, -ое,   (*a*) (zool) anal

~ плав·ни́к·,   (*m*) (fish) anal fin

анальци́м·, -а,   (*m*) (min) analcite, analcime

ана́мнез·, -а,   (*m*) (med) anamnesis, history

ана́мн·и·и   (*nom pl*), (*g pl*) **-й,** (*f*) (zool) anamniotes, *Anamniota*

анаморф·и́ровани·е, -я,   (*n*) (opt) anamorphizing

анаморф·о́з·, -а,   (*m*) (biol) anamorphosis; (opt) anamorphose

анаморфо·ско́п·, -а,   (*m*) (phot) anamorphoscope

анаморфо́т·н·ый, -ая, -ое,   (*a*) anamorphotic; (phot) anamorphote

анана́с·, -а,   (*m*) pineapple

ана·пти́х·, -а,   (*m*) (geol) anaptychus

ан·астигма́т·, -а,   (*m*) (opt) anastigmat

ана·стом·о́з·, -а,   (*m*) (biol) anastomosis

аната́з·, -а,   (*m*) (min) anatase

анатокси́н·, -а,   (*m*) (bact) anatoxin, toxoid

ана́том·, -а,   (*m*) anatomist, dissector

анатом·и́р·овать, -ют, (*imp and perf*) make a necropsy, dissect

анатом·и́ческ·ий, -ая, -ое,   (*a*) anatomic, anatomical

анато́м·и·я, -и,   (*f*) anatomy

~, микро·скоп·и́ческ·ая   histology, microscopic anatomy

~ раст·е́ни·й   vegetable anatomy, phytotomy

ана·фа́з·а, -ы,   (*f*) (cyt) anaphase

ана·филакс·и́·я, -и,   (*f*) (med) anaphylaxis

анафоре́з·, -а,   (*m*) (chem) anaphoresis

анафоре́з·н·ый, -ая, -ое,   (*a*) anaphoretic

ана·хрон·и́зм·, -а,   (*m*) anachronism

ана·хрон·и́ческ·ий, -ая, -ое,   (*a*) anachronistic

ан·аэро́б·, -а,   (*m*) (biol) anaerobe

ан·аэро·би·о́з·, -а,   (*m*) (biol) anaerobiosis

ан·аэро·би·о́нт·, -а,   (*m*) (biol) anaerobiont

ан·аэро́б·н·ый, -ая, -ое,   (*a*) (biol) anaerobic

анга́р·, -а,   (*m*) hangar, shed

ан·гармон·и́ческ·ий, -ая, -ое,   (*a*) anharmonic

ан·гармон·и́чност·ь, -и,   (*f*) (math) anharmonic/cross ratio

ан·гедра́ль·н·ый, -ая, -ое,   (*a*) anhedral

а́нгел·, -а,   (*m*) angel

~, мор·ск·о́й   angel fish, *Squatina squatina*

ан·гидр·еми́·я, -и,   (*f*) (med) anhydraemia, anhydremia

ан·гидр·и́д·, -а,   (*m*) (chem) anhydride

~ азо́т·н·ой кисл·от·ы́   nitric anhydride, nitrogen pentoxide

~ иод·нова́т·ой кисл·от·ы́   iodic anhydride, iodine pentoxide

~ кисл·от·ы́   acid anhydride

~ кремн·ёв·ой кисл·от·ы́   anhydrous silica, silicon oxide

~ мало́н·ов·ой кисл·от·ы́   malonic anhydride, carbon suboxide

~ сер·ни́ст·ой кисл·от·ы́   sulphur dioxide

ангидри́т·, -а,   (*m*) (min) anhydrite

ангидритоли́т·, -а,   (*m*) (min) anhydrite

ан·гидро·глюк·о́з·а, -ы,   (*f*) anhydroglucose

ан·гидр·о́н·, -а,   (*m*) (chem) anhydrone

анги́н·а, -ы,   (*f*) (med) angina, quinsy; tonsilitis

~ Винце́нт·а   (med) Vincent's angina, trench mouth

~, груд·н·а́я   (med) angina pectoris

ангин·а
~, **дифтер·и́йн·ая** (med) dyptheritic pharyngitis
~, **катарр·а́льн·ая** (med) catarrhal pharyngitis
~, **лакун·а́рн·ая** (med) lacunar tonsilitis
~, **Людовик·а** (med) Ludwig's angina
~, **некр·оти́ческ·ая** (med) malignant angina
~, **плён·чат·ая** (med) membranaceous croup
~, **стрепто·ко́кк·ов·ая** (med) streptococcus angina
~, **фоллик·уля́рн·ая** (med) follicular tonsilitis
анги·о́м·а, -ы, (f) (med) angioma
ангио·стом·и́·я, -и, (f) (zool) angiostomy
ангио·спе́рм·, -а, (m) (bot) angiosperm, (pl) angiosperms, *Angiospermae*
англези́т·, -а, (m) (min) anglesite, lead spar
англ·и́йск·ий, -ая, -ое, (a) English, British
~ **бол·е́зн·ь** (f) (med) rickets
~ **була́в·к·а** (f) safety pin
~ **Лло́йд·** (m) Lloyd's Registry of Shipping
~ **сол·ь** (f) (chem) epsom salts
ангоб·, -а, (m) (cer) engobe
анго́р·ск·ий крол·ик· (m) Angora rabbit
а́нгстрём·, -а, (m) (phys) Ångstrom unit, ångstrom, Å, Å.U., A.U. (= 10^{-8} cm)
андалузи́т·, -а, (m) (min) andalusite
андези́н·, -а, (m) (min) andesine
андези́т·, -а, (m) (geol) andesite
андин·ова́ни·е, -я, (n) (com) advice note
андради́т·, -а, (m) (min) andradite
андра́з·а, -ы, (f) (zool) andrase
андро·ге́н·, -а, (m) (biochem) androgen
андроме́д·а, -ы, (f) (astron) Andromeda
андро·спо́р·а, -ы, (f) (bot) androspore
андро·стер·о́н·, -а, (m) (biochem) androsterone
андро·фо́р·, -а, (m) (bot) androphore
андр·оце́·й, -я, (m) (bot) androecium
андр·оце́й·н·ый, -ая, -ое, (a) (bot) androecial
ан·евр·и́зм·, -а, (m) (med) aneurysm, aneurism

а·невр·и́н·, -а, (m) (biochem) aneurin, aneurine, thiamine, vitamin B_1
а·нейр·и́н·, -а, (m) (biochem) aneurin, thiamine, vitamin B_1
анем·и́ческ·ий, -ая, -ое, (a) (med) anaemic, anemic
анем·и́·я, -и, (f) (med) anaemia, anemia
анемо́·граф·, -а, (m) (meteor) anemograph
анемо́·метр·, -а, (m) (meteor etc.) anemometer
~, **крыль·ча́т·ый** wheel anemometer
~, **портати́в·н·ый** (a/c) air meter
~, **ча́ш·ечн·ый** cup anemometer
анемо́н·, -а, (m) = **анемо́н·а**
анемо́н·а, -ы, (f) (bot) anemone
анемо·румбо·гра́ф·, -а, (m) (meteor) anemograph
анемо·тахо́·метр·, -а, (m) air speed indicator
анемо·фил·и́·я, -и, (f) (bot) anemophily
анемо·хор·и́·я, -и, (f) (bot) anemochory
анеро́ид·, -а, (m) aneroid barometer
анестез·и́р·овать, -уют, (*imp and perf*) (med) anaesthetize, anesthetize
анестез·и́·я, -и, (f) (med) anaesthesia, anesthesia
ането́л·, -а, (m) (chem) anethol(e)
ани́д·, -а, (m) (chem) nylon-66
анизо·га́м·и·я, -и, (f) (biol) anisogamy
анизо·ко́р·и·я, -и, (f) (med) anisocoria
анизо́·метр·, -а, (m) anisometer, anisotropy meter
анизо·метр·и́ческ·ий, -ая, -ое, (a) anisometric
анизо·пара·кла́з·, -а, (m) (geol) strike fault, longitudinal fault
анизо·троп·и́ческ·ий, -ая, -ое, (a) anisotropic
анизо·тро́п·и·я, -и, (f) anisotropy
анизо·тро́п·ност·ь, -и, (f) anisotropy
анизо·тро́п·н·ый, -ая, -ое, (a) anisotropic
анизо·тро́ф·н·ый, -ая, -ое, (a) anisotrophic
ан·изо·филл·и́·я, -и, (f) (bot) anisophylly
ани́л·, -а, (m) (chem) anil
анили́д·, -а, (m) (chem) anilide
~ **бензо́й·н·ой кисл·от·ы́** benzanilide
анил·и́н·, -а, (m) (chem) aniline
~, **чёрн·ый** aniline black

анил·и́н

~, хлористо·водо·ро́д·н·ый aniline salt/hydrochloride

анили́н·ов·ый крас·и́тел·ь, -я, (*m*) aniline dye

анио́н·, -а, (*m*) (elec) anion

аниони́т·, -а, (*m*) (chem) anion exchanger, anion-exchange resin

аниони́т·ов·ый, -ая, -ое, (*a*) anion-exchange, anionite

анио́н·н·ый, -ая, -ое, (*a*) anion, anionic

анионо·об·ме́н·н·ый, -ая, -ое, (*a*) anion-exchange

ани́с·, -а, (*m*) (bot) anise, *Pimpinella Anisum*, anise apple (Russian sort)

анис·ид·и́н·, -а, (*m*) (chem) anisidine

анис·и́л·, -а, (*m*) (chem) anisyl

ани́с·овк·а, -и, (*f*) anisette (liqueur); anise apple (Russian sort)

ани́с·ов·ый, -ая, -ое, (*a*) of **ани́с·** (chem) anisic

~ альдеги́д· (*m*) (chem) anisaldehyde

а́нкер·, -а, (*m*) stay, tie, tie bolt; (horol) crutch, pallet

анкери́т·, -а, (*m*) (min) ankerite, brown spar

а́нкер·н·ый, -ая, -ое, (*a*) of **а́нкер·**

~ ка́мен·ь, -мн·я, (*m*) (build) bond-stone

~ с·вя́з·ь (*f*) connecting rod

~ с·пу́ск· (*m*) (horol) lever escapement

~ тя́г·а (*f*) (civ eng) tie-rod

анкер·о́к·, -р·к·а́, (*m*) (naut) barrico

анке́т·а, -ы, (*f*) questionnaire

анке́т·н·ый, -ая, -ое, (*a*) of **анке́т·а**

анкил·о́з·, -а, (*m*) (med) anchylosis, ankylosis

анкило·за́вр·, -а, (*m*) (pal) ankylosaur

анкило·сто́м·а, -ы, (*f*) (med) ankylostoma

анлауфра́д·, -а, (*m*) (horol) pallet wheel

аннаберги́т·, -а, (*m*) (min) annabergite, nickel bloom

анне́кс·, -а, (*m*) annex

анне́кс·и·я, -и, (*f*) (dipl) annexation

аннели́д·, -а, (*m*) (zool) annelid, (*pl*) annelids, *Annelida*

аннигил·и́р·овать, -уют, (*imp and perf*) (phys) annihilate

аннигил·я́ци·я, -и, (*f*) annihilation

~ на·лет·у́ (nucl) annihilation in flight

~, трёх·ква́нт·ов·ая (nucl) three-quantum annihilation, three-photon annihilation

аннигил·я́ци·я

~, уда́р·н·ая (nucl) impact/pick-off annihilation

эне́рг·и·я аннигил·я́ци·и (*f*) (nucl) annihilation energy

аннот·а́ци·я, -и, (*f*) annotation, precis, synopsis

аннот·и́р·овать, -уют, (*imp and perf*) annotate, summarize

аннул·и́р·овать, -уют, (*imp and perf*) annul, cancel, abrogate, abolish, nullify

аннул·я́ри·я, -и, (*nom pl*) -и, (pal, bot) annularian, (*pl*) annularians, *Annularia*

ано·ге́н·н·ый, -ая, -ое, (*a*) anogene, (soil science)

ано́д·, -а, (*m*) (elec) anode

~ воз·бужд·е́ни·я excitation anode

~, вращ·а́ющ·ийся rotating anode

~, ди́ск·ов·ый (elec) disc anode

~, до·полн·и́тельн·ый supplementary anode

~ за·жиг·а́ни·я ignition anode

~, пе́рв·ый first/focusing anode

~, пуск·ов·о́й starting anode

~, рабо́ч·ий active anode

~, раз·рез·н·о́й split anode

~, фокус·и́рующ·ий focusing anode

анод·иза́ци·я, -и, (*f*) = **анод·и́рова·ни·е**

анод·и́ровани·е, -я, (*n*) (met) anodizing, anodic oxidation treatment

аноднⵐ·механ·и́ческ·ая об·рабо́т·к·а (*f*) (mech) electromachining

ано́д·н·ый, -ая, -ое, (*a*) anode, anodic

~ бло́к· (*m*) (elec) multi-segment anode

~ о·кисл·е́ни·е (*n*) (met) anodic oxidation, anodizing

ан·одо́нт·, -а, (zool) anodon, (*pl*) anodons, *Anodonta*

ан·окси·би·о́з·, -а, (*m*) (zool) anoxybiosis

анокси́т·, -а, (*m*) (min) anauxite

аноли́т·, -а, (*m*) (elec) anolyte

аномалист·и́ческ·ий, -ая, -ое, (*a*) (astron) anomalistic

анома́л·и·я, -и, (*f*) anomaly, abnormality

анома́ль·н·ый, -ая, -ое, (*a*) anomalous, abnormal

~ диспе́рс·и·я (*f*) anomalous dispersion (light)

~ поляриз·а́ци·я (*f*) (rad) abnormal polarization

~ рас·про·стран·е́ни·е (*n*) anopropagation

аномит·, -а, (*m*) (min) anomite
анони́м·, -а, (*m*) anonymous author; anonymous work, anonymous letter
анони́м·к·а, -и, (*g pl*) **-м·ок·,** (*f*) (pop) anonymous letter
анони́м·н·ый, -ая, -ое, (*a*) (*s f*) **-м·ен·, -м·н·а,** anonymous
ано́нс·, -а, (*m*) announcement, bill, poster, notice; (cinema) trailer
аноре́кс·и·я, -и, (*f*) (med) anorexia
а·норма́ль·н·ый, -ая, -ое, (*a*) (*s f*) **-л·ен·, -ль·н·а,** abnormal
анорти́т·, -а, (*m*) (min) anorthite, indianite
ан·орто·кла́з·, -а, (*m*) (min) anorthoclase
ан·орто·фот·и́ческ·ий, -ая, -ое, (*a*) (phot) anorthophotic
ан·осм·и́·я, -и, (*f*) (med) anosmia
ано·тро́н·, -а, (*m*) (rad) anotron
ан·о́фелес·, -а, (*m*) (ent) anopheles, malarial mosquito
анса́мбл·ь, -я, (*m*) ensemble, assembly; (theat) company, cast
~, ги́ббс·ов·ый (math) Gibb's ensemble
~, канон·и́ческ·ий (math) canonical ensemble
~, нукло́н·ов (nucl) ensemble of nucleons
ансери́н·, -а, (*m*) (chem) anserin
анста́ккер·, -а, (*m*) plank unstacker
ант·, -а, (*m*) (arch) anta
антаблеме́нт·, -а, (*m*) (arch) entablature
антагон·и́зм·, -а, (*m*) antagonism; antipathy (of minerals)
антагон·и́ст·, -а, (*m*) (biol) antagonist
антаркт·и́д·а, -ы, (*f*) the Antarctic continent
анта́ркт·ик·а, -и, (*f*) the Antarctic, Antarctic regions
антаркт·и́ческ·ий, -ая, -ое, (*a*) antarctic
анте·кла́з·а, -ы, (*f*) (geol) anteclase
анте·кли́з·а, -ы, (*f*) (geol) anteclise, arch
анте·нат·а́льн·ый, -ая, -ое, (*a*) (med) antenatal
анте́нн·а, -ы, (*f*) (rad) antenna, aerial; (zool) antenna, feeler
~, авар·и́йн·ая makeshift antenna
~, азимут·а́льн·ый azimuth antenna
~, акти́в·н·ый driven antenna
~ А́льфорд·а, ра́м·очн·ая Alford loop

анте́нн·а
~, арфо·обра́з·н·ая grid antenna
~ Бе́веридж·а Beveridge antenna
~ бег·у́щ·ей волн·ы́ travelling-wave antenna
~, ве́ер·н·ая fan antenna
~, вибра́тор·н·ая dipole antenna
~, волн·ов·а́я wave antenna
~, вы·пуск·а́ем·ая trailing antenna
~, вы·пуск·н·а́я trailing antenna
~, гла́в·н·ая main antenna
~, Г-обра́з·н·ая inverted-L antenna
~, диапазо́н·н·ая wide-band antenna, multiple-tuned antenna
~, дире́ктор·н·ая director antenna, yagi antenna/array
~, дифракцио́н·н·ая slot aerial
~, дискон·и́ческ·ая discone antenna
~, ди·электр·и́ческ·ая polyrod antenna
~, ёл·очн·ая fishbone antenna
~, квадра́т·н·ая ру́пор·н·ая pyramidal horn antenna
~, колбасо·обра́з·н·ая cage/sausage antenna
~, ко·линеа́р·н·ая Franklin antenna
~, коро́б·чат·ая box antenna
~, кру́гл·ая ру́пор·н·ая conical horn antenna
~, кры́ль·ев·ая (a/c) skid-fin antenna
~, ли́нз·ов·ая electromagnetic lens
~, луч·ев·а́я beam antenna
~, много·лу́ч·ев·ая multi-wire antenna
~, навигацио́н·н·ая (naut) main antenna
~, на·кло́н·н·ая slant antenna
~, на·пра́вл·енн·ая directional/directive antenna
~, нару́ж·н·ая ра́м·очн·ая (air) external loop aerial
~, не·на·стро́·енн·ая untuned/aperiodic antenna
~, одно·на·пра́вл·енн·ая unidirectional antenna
~, от·кры́·т·ая external aerial
~, параболо́ид·н·ая cheese antenna
~, пасси́в·н·ая parasitic antenna
~, пеленга́тор·н·ая finding antenna
~, пере·да·ю́щ·ая transmitting antenna/aerial
~, пло́ск·ая flat-top antenna
~, по·ве́рх·ностн·ая flush antenna
~, под·зе́м·н·ая buried antenna
~, прямо·уго́ль·н·ая ру́пор·н·ая rectangular-horn antenna
~ радио·лока́тор·а radar dish

антéнн·а

~, рáм·очн·ая   frame antenna, loop antenna, coil antenna

~, рýпор·н·ая   horn antenna

~, сéт·очн·ая   grid antenna

~, симметр·и́чн·ая   balanced/symmetrical antenna

~, син·фáз·н·ая   broadside antenna

~, с·лóж·н·ая   (rad) array

~, стрежн·ев·áя   mast antenna

~, турникéт·н·ая   turnstile aerial

~ угл·á воз·выш·éни·я   elevation receiver antenna

~ угл·á мéст·а   elevation receiver antenna

~, угол·кóв·ая   corner reflector

~, у·длин·ённ·ая   loaded antenna

~, у·тóпл·енн·ая   suppressing antenna

~, четверт·волн·ов·áя   quarter-wave antenna

~, шлéйф-   folded dipole antenna

~, штыр·ев·áя   whip/rod antenna

~, щел·ев·áя   slot antenna

~, экран·и́рованн·ая   screened antenna

антéнн·ул·а, -ы,   (*f*) (zool) antennule

антéнн·ый, -ая, -ое,   (*a*) of **антéнн·а**

~ полотн·ó   (*n*) antenna curtain/array

~ решёт·к·а   (*f*) antenna/aerial array

~ эффéкт   (*m*) (rad) vertical component, antenna effect

антери́д·и·й, -я,   (*m*) (bot) antheridium, antherid

анте·фи́кс·, -а,   (*m*) (arch) antefix, (*pl*) antefixae

антецедéнт·ност·ь, -и,   (*f*) antecedent behaviour/behavior

антецедéнт·н·ая дол·и́н·а   (*f*) (geol) antecedent valley

анти-   (*int component*) anti-, un-

антибáт·ност·ь, -и,   (*f*) (phys) antipathy

анти·биóт·ик·, -а,   (*m*) antibiotic

анти·биот·и́ческ·ий, -ая, -ое,   (*a*) antibiotic

анти·вибр·áтор·, -а,   (*m*) vibration mount, shock absorber

анти·ви́н·н·ая кисл·от·á   (*f*) (chem) mesotartaric acid

анти·ви́рус·, -а,   (*m*) antivirus

анти·гéл·и·й, -я,   (*m*) (meteor) anthelion

анти·гéн·, -а,   (*m*) (med) antigen

анти·гéн·ност·ь, -и,   (*f*) antigen content

анти·гигиен·и́чн·ый, -ая, -ое,   (*a*) unhygienic, insanitary

антигори́т·, -а,   (*m*) (min) antigorite

антигризýтн·ое вз·ры́в·чат·ое вещ·еств·ó   (*n*) (min) permissible explosive

антидат·и́р·овать, -уют,   (*imp and perf*) post-date (a cheque, etc.)

анти·детонáтор·, -а,   (*m*) (ICE) anti-knock compound

анти·детонациóн·н·ый, -ая, -ое,   (ICE) anti-knock

~ свóй·ств·о   (*n*) anti-knock value

анти·динатрóн·н·ая сéт·к·а   (*f*) (elec) suppressor grid

антидóт·, -а,   (*m*) = **противо·я́д·и·е** (pharm) antidote

анти·катализ·áтор·, -а,   (*m*) anticatalyst, catalyst poison

анти·катóд·, -а,   (*m*) (phys) target cathode

анти́кв·а, -ы,   (*f*) (print) antique

антиквáр·, -а,   (*m*) antiquarian

анти·клинáл·ь, -и,   (*f*) (geol) anticline

анти·клинáльн·ый, -ая, -ое,   (*a*) anticline, anticlinal

~ с·клáд·к·а   (*f*) (geol) anticline

анти·клинóр·и·й, -я,   (*m*) (geol) anticlinorium

анти·когéрер·, -а,   (*m*) (rad) decoherer, anticoherer

анти·корроз·и́йн·ый, -ая, -ое,   (*a*) anticorrosive

анти·коррозиóн·н·ый, -ая, -ое,   (*a*) anticorrosive

~ вещ·еств·ó   (*n*) corrosion inhibitor

анти·логари́фм·, -а,   (*m*) (math) antilogarithm

анти·мéр·, -а,   (*m*) (zool) antimere

антимон·áт·, -а,   (*m*) (chem) antimonate, antimoniate

антимон·и́д·, -а,   (*m*) (chem) antimonide

антимóн·и·й, -я,   (*m*) (*and see* **сурьм·á**) (met) antimony, Sb

антимон·и́л·, -а,   (*m*) (chem) antimonyl

~, хлóр·ист·ый   antimonyl chloride

антимон·и́т·, -а,   (*m*) (chem, min) antimonite, stibnite

анти·на·кип·и́н·, -а,   (*m*) boiler scale preventative

анти·на·ýч·н·ый, -ая, -ое,   (*a*) wrong, reprehensible (in scientific matters), unscientific

анти·нейтри́н·о, -а,   (*n*) (nucl) antineutrino

анти·нейтрóн·, -а,   (*m*) (nucl) antineutron

анти·нóм·и·я, -и,   (*f*) antinomy

анти·нуклóн·, -а, *(m)* antinucleon
анти·об·лед·енйтел·ь, -я, (m) (air) anti-icer, de-icer, de-icing equipment
~, сéт·очн·ый ice guard
анти·о·кисл·йтел·ь, -я, *(m)* (chem) antioxidant
анти·оксид·áнт·, -а, *(m)* (chem) antioxidant
анти·паразйт·н·ый, -ая, -ое, *(a)* (rad) antiparasitic
анти·пассáт·, -а, *(m)* (meteor) anti-/counter-trade wind
анти·пéн·н·ый, -ая, -ое, *(a)* antifoam, antifroth
~ вещ·еств·ó *(n)* antifoam agent
антипертйт·, -а, *(m)* (min) antiperthite
анти·пир·йн·, -а, *(m)* (pharm) antipyrine, phenazone
анти·пóд·, -а, *(m)* antipode; (bot) antipodal cell
~ эпи·цéнтр·а (seismol) anticentre
анти·протóн·, -а, *(m)* (nucl) antiproton, negative proton
анти·рéзус-сы́ворот·к·а, -и, *(f)* (pharm) Rh antiserum
анти·рóд·, -а, *(m)* antigenus
анти·санитáр·н·ый, -ая, -ое, *(a)* insanitary, infringing public health requirements
анти·сегнето·электр·йческ·ий, -ая, -ое, *(a)* antiferroelectric
анти·сейсм·йческ·ий, -ая, -ое, *(a)* earthquake proof/resistant
анти·селен·а, -ы, *(f)* (astron) antiselene
анти·сéпт·ик·, -а, *(m)* wood preservative
анти·сéпт·ик·а, -и, *(f)* (med) antiseptics; (med) antisepsis
анти·септ·йческ·ий, -ая, -ое, *(a)* antiseptic
~ срéд·ств·о *(n)* (pharm) antiseptic
анти·симметр·йческ·ая с·вяз·ь *(f)* (autom) antisymmetrical path
анти·скóрчинг·, -а, *(m)* (chem) anti-scorching agent
анти·со·в·пад·éни·е, -я, *(n)* anti-coincidence
анти·стóкс·ов·ая лйн·и·я *(f)* (spectr) anti-Stokes line
анти·стрéсс·ов·ый, -ая, -ое, *(a)* antistress
анти·тéз·а, -ы, *(f)* antithesis, contrast
анти·тéл·о, -а, *(nom pl)* **-á,** *(g pl)* **-тел·,** *(d pl)* **-áм,** *(n)* (bact) antibody
анти·токс·йн·, -а, *(m)* (med) antitoxin

анти·тромб·йн·, -а, *(m)* (chem) antithrombin
анти·фáг·ов·ый, -ая, -ое, *(a)* (med) antiphagal, antiphage
анти·фебр·йн·, -а, *(m)* (chem) antifebrin
анти·ферро·магнет·йзм·, -а, *(m)* (phys) antiferromagnetism
анти·ферро·магнет·йк·, -а, *(m)* antiferromagnetic, antiferromagnetic material
анти·флáминг·, -а, *(m)* (ICE) backfire screen
анти·фрйз·, -а, *(m)* antifreeze
анти·фрикциóн·н·ый, -ая, -ое, *(a)* antifriction
~ с·плав· *(m)* anti-friction alloy, bearing alloy
~ с·плав·, бéл·ый white bearing alloy, Babbitt metal
анти·хлóр·, -а, *(m)* (chem) antichlor
анти·циклóн·, -а, *(m)* (meteor) anticyclone
~, вне·троп·йческ·ий extratropical anticyclone
~, выс·óтн·ый upper-level anticyclone
~, по·движ·н·óй migratory anticyclone
анти·эрмит·ов·óй, -áя, -óе, *(a)* (math) anti-Hermitian
антоксанйн·, -а, *(m)* (bot) anthoxanin
ант·óним·, -а, *(m)* (ling) antonym
анто·филлйт·, -а, *(m)* (min) anthophyllite
анто·хлóр·, -а, *(m)* (bot) anthoclore
анто·циáн·, -а, *(m)* (bot, chem) anthocyanin
антрак·нóз·, -а, *(m)* (bot) anthracnose
антрак·óз·, -а, *(m)* (med) anthracosis
~ лёгк·их (med) anthracosis, coal miner's lung
антрáкс·, -а, *(m)* (med) anthrax
антраксилóн·, -а, *(m)* (min) anthraxylon
антранйл·, -а, *(m)* (chem) anthranil
антранйл·ов·ая кисл·от·á *(f)* (chem) anthranilic acid
антранóл·, -а, *(m)* (chem) anthranol
антра·фото·скóп·, -а, *(m)* (phot) anthraphotoscope
антра·хинóн·, -а, *(m)* (chem) anthraquinone
антра·хинóн·ов·ый крас·йтел·ь *(m)* anthraquinone dye
антрац·éн·, -а, *(m)* (chem) anthracene
антрац·éн·ов·ое мáсл·о *(n)* (chem) anthracene oil
антрац·йт·, -а, *(m)* anthracite (coal)

антрац·ит·иза́ци·я, -и, (*f*) (geol) anthracitization

антреко́т·, -а, (*m*) (food) entrecôte

антресо́л·ь, -и, (*f*) (arch) entresol, mezzanine; gallery (in room)

антропо·ге́н·, -а, (*m*) (geol) Quaternary period

антропо·гене́з·, -а, (*m*) anthropogeny

антропо·ге́н·н·ый, -ая, -ое, (*a*) anthropogenic

антропо·ге́н·и·я, -и, (*f*) anthropogeny

антропо·зо́·й, -я, (*m*) Anthropozoic era

антроп·о́ид·, -а, (*m*) (zool) anthropoid

антропо·ло́г·и·я, -и, (*f*) anthropology

антропо·ме́тр·и·я, -и, (*f*) (biol) anthropometry

антропо·хор·и́·я, -и, (*f*) (bot) anthropochoria

а́нтье (*n indecl*) (math) greatest integer

анфа́с (*adv*) full face

анфила́д·а, -ы, (*f*) (arch) enfilade

анфлёра́ж·, -а, (*m*) enfleurage (perfume making)

анхи·кристалл·и́ческ·ий, -ая, -ое, (*a*) anchicrystalline

анхи·моно·минера́ль·н·ый, -ая, -ое, (*a*) anchimonomineral

анхи·те́р·и·й, -я, (*m*) (geol, zool) anchithere, *Anchitherium*

анхи·эвтект·и́ческ·ий, -ая, -ое, (*a*) anchieutectic

анцестра́ль·н·ый, -ая, -ое, (*a*) (bot) ancestral

анце́стр·ул·а, -ы, (*f*) (bot) ancestrula

анчо́ус·, -а, (*m*) (fish) anchovy, *Engraulis encrasicholus*

а́ншлиф·, -а, (*m*) (met) crystallographic specimen; (min) polished section (for microscopic examination)

а́ншпуг·, -а, (*m*) crowbar; (naut) handspike

Аньези, ло́кон· (*m*) (math) witch of Agnesi

ан·эн·цефал·и́·я, -и, (*f*) (med) anencephaly

ао́рт·а, -ы, (*f*) aorta

аорт·и́т·, -а, (*m*) (med) aortitis

апанер·, -а, (*m*) (naut) up-and-down (of anchor)

апаре́л·ь, -и, (*f*) *see* **аппаре́л·ь**

апате·ми́йд·, -а, (*m*) (geol) apatemyid

апати́т·, -а, (*m*) (min) apatite

апатри́д·, -а, (*m*) (law) stateless person

а́пекс·, -а, (*m*) (astron) apex

апелл·и́р·овать, -ую́т, (*imp and perf*) (law) appeal

апелл·яцио́н·н·ый, -ая, -ое, (*a*) appellate, appeal

~ жа́л·об·а (*f*) appeal

апелл·я́ци·я, -и, (*f*) (law) appeal

апелл·я́нт·, -а, (*m*) (law) appellant

апельси́н·, -а, (*m*) orange; orange tree

апельси́н·н·ый, -ая, -ое, (*a*) orange

а·период·и́ческ·ий, -ая, -ое, (*a*) aperiodic

~ проце́сс· (*m*) aperiodic process

~, регул·и́ровани·е (*n*) (autom) aperiodic damping control

а·период·и́чност·ь, -и, (*f*) aperiodicity

а·период·и́чн·ый, -ая, -ое, (*a*) = **а·период·и́ческ·ий**

аперту́р·а, -ы, (*f*) aperture

~, числ·ов·а́я numerical aperture

аперту́р·н·ый, -ая, -ое, (*a*) aperture; orate

апертуро·ви́д·н·ый, -ая, -ое, (*a*) oroid

апика́ль·н·ый, -ая, -ое, (*a*) apical

аплази́·я, -и, (*f*) (med) aplasia

аплана́т·, -а, (*m*) (phot) aplanatic lens, aplanat

апланат·и́ческ·ий, -ая, -ое, (*a*) (phot) aplanatic

аплано·гаме́т·а, -ы, (*f*) (biol) aplanogamete

аплано·спо́р·а, -ы, (*f*) (bot) aplanospore

аплика́д·а, -ы, (*f*) = **апплика́т·а**

апли́т·, -а, (*m*) (geol) aplite, haplite

апли́т·ов·ый, -ая, -ое, (*a*) (geol) aplitic

апло·грани́т·, -а, (*m*) (min) aplogranite, granite-aplite

аплодисме́нт·ы (*nom pl*), (*g pl*) **-ов,** (*m*) applause

а·пноэ́·, (*n indecl*) (zool) apnoea

апо·а́стр·, -а, (*m*) (astron) apastron

апо·га́м·и·я, -и, (*f*) (biol) apogamy

апо·ге́·й, -я, (*m*) (astron) apogee; summit, height, culmination

то́ч·к·а апо·ге́·я (*f*) aphelion

апо·ге́л·и·й, -я, (*m*) (astron) aphelion

апо·де́м·а, -ы, (*f*) (zool) apodeme

аполи́д·, -а, (*m*) (law) stateless person

а·поля́р·н·ый, -ая, -ое, (*a*) non-polar

апо·ми́ксис·, -а, (*m*) (bot) apomixis

апо·морф·и́н·, -а, (*m*) (chem) apomorphine

апо·невр·о́з·, -а, (*m*) (zool) aponeurosis

апо·пле́кс·и·я, -и, (*f*) (med) apoplexy

апостерио́ри (*adv*) *a posteriori*

апостерио́р·н·ый, -ая, -ое, (*a*) *a posteriori*

апо·стро́ф·, -а, (*m*) apostrophe
апо·те́ц·и·й, -я, (*m*) (bot) apothecium
апо·фе́м·а, -ы, (*f*) (math) apothem, short radius
апо·фи́г·а, -и, (*f*) (arch) apophyge
апо·фи́з·, -а, (*m*) (anat) apophysis
апо·фи́з·а, -ы, (*f*) apophysis, offshoot, outgrowth
апо·фи́з·ов·ый, -ая, -ое, (*a*) apophysal
апо·филл·и́т·, -а, (*m*) (min) apophyllite
апо·хрома́т·, -а, (*m*) (opt) apochromatic objective/lens
аппара́т·, -а, (*m*) (*and see under qualifying adjective*); apparatus, instrument, equipment; administrative machinery/organization, the apparatus of government; torpedo tube
∼ -аргуме́нт·, -а, (*m*) (autom) final control element
∼, вы·пар·н·о́й evaporator
∼, контро́ль·н·ый (autom) monitor
∼, лет·а́тельн·ый (air) aircraft; (rocket) vehicle
∼ -фу́нкци·я, -и, (*f*) (autom) input element
аппара́т·н·ая, -ой, (*f decl as adj*) instrument room, control room, operating room (telephone)
аппара́т·н·ый, -ая, -ое, (*a*) of **аппара́т·**; (*see also* **аппара́т·н·ая** *as noun*)
аппарат·у́р·а, -ы, (*f*) (*and see under qualifying adjective*); apparatus, equipment, gear, installation
аппарат·у́рн·ый, -ая, -ое, (*a*) of **аппарат·у́р·а**
∼ фа́ктор· (*m*) instrumental factor
аппаре́л·ь, -и, (*f*) (mil) ramp
аппе́ндикс·, -а, (*m*) (anat) appendix
апплика́т·а, -ы, (*f*) (math) Z-axis
апплика́тор·, -а, (*m*) applicator
апплика́ц·и·я, -и, (*f*) (text) applique
аппре́т·, -а, (*m*) (text) finish
аппрет·и́ровани·е, -я, (*n*) (text) dressing
аппрет·и́рованн·ый, -ая, -ое, (*past part pass*) of **аппрет·и́р·овать**
∼ хо́лст· (*m*) (print) buckram
аппрет·и́р·овать, -уют, (*imp and perf*) dress (cloth, hide etc.)
аппрет·у́р·а, -ы, (*f*) (text) dressing
аппрет·у́рщик·, -а, (*m*) (text) dresser
аппроксим·а́льност·ь, -и, (*f*) approximation, approximateness
аппроксим·а́ци·я, -и, (*f*) *see also* **при·бли́ж·е́ни·е;** approximation

аппроксим·и́р·овать, -уют, (*imp and perf*) approximate
аппроксим·и́руемост·ь, -и, (*f*) (math) approximability
апре́л·ь, -я, (*m*) April
априо́ри (*adv*) *a priori*
априо́р·н·ый, -ая, -ое, (*a*) *a priori*
∼ вероя́т·ност·ь (*f*) (math) *a priori* probability
апроб·а́ци·я, -и, (*f*) approval, approbation; selection of crop for seed
апси́д·а, -ы, (*f*) (arch) apse, apsis; (astron) apsis, (*pl*) apsides
ли́н·и·я апси́д· (*f*) (astron) apse line, line of apsides
апте́к·а, -и, (*f*) pharmacy, drug-store, chemist's shop; dispensary
апте́к·арск·ий вес· (*m*) apothecaries' weight
апте́к·ар·ь, -я, (*m*) pharmacist, chemist (Brit. coll)
апте́ч·к·а, -и, (*g pl*) **-ч·ек·,** (*f*) first-aid set; tool kit
апти́х·, -а, (*m*) (geol) aptychus
а́пт·ск·ий я́рус· (*m*) (geol) Aptian stage
а́р·, -а, (*m*) are (100 m^2)
араба́н·, -а, (*m*) (chem) araban
арабе́ск·, -а, (*m*) = **арабе́ск·а**
арабе́ск·а, -и, (*g pl*) **арабе́сок·,** (*f*) (arch) arabesque
араби́н·, -а, (*m*) (phot) arabin
арабин·о́з·а, -ы, (*f*) (chem) arabinose
арабито́л·, -а, (*m*) (chem) arabitol
арабо́н·ов·ая кисл·от·а́ (*f*) arabonic acid
ара́б·ск·ий, -ая, -ое, (*a*) Arabic, arabic
∼ ци́фр·а (*f*) Arabic numeral
арав·и́йск·ий, -ая, -ое, (*a*) Arab, arabic, Arabian
∼ каме́д·ь (*f*) gum arabic
Ара́в·и·я, -и, (*f*) (geog) Arabia
арагони́т·, -а, (*m*) (min) aragonite
аралки́л·, -а, (*m*) (chem) aralkyl, alphyl, alkyl phenyl
араука́р·и·я, -и, (*f*) (bot) monkey puzzle, *Araucaria*
арахи́д·н·ый, -ая, -ое, (*a*) of **ара́хис·**
ара́хис·, -а, (*m*) (bot) ground nut, earth nut, pea nut, goober, *Arachis hypogaea* L.
ара́хис·ов·ый, -ая, -ое, (*a*) of **ара́хис·**
∼ ма́сл·о (*n*) peanut butter
арби́тр·, -а, (*m*) (law) arbitrator
арбитра́ж·, -а, (*i*) **-ем,** (*m*) (law) arbitration; (fin) arbitrage, arbitration
∼, валю́т·н·ый (fin) arbitration of exchange

арбитра́ж

~, дву·сторо́н·н·ий (fin) simple arbitration of exchange

~, мно́го·сторо́н·н·ий (fin) compound arbitration of exchange

арбитра́ж·ёр·, -а, (m) (fin) arbitrageur

арбитра́ж·н·ый, -ая, -ое, (a) arbitration

~ реш·éни·е (n) arbitrament

~ суд· (m) court of arbitration, arbitration tribunal

арбу́з·, -а, (m) (bot) water melon, *Citrullus vulgaris*

арбу́з·ов·а, реа́кци·я (f) (chem) Arbusow rearrangement

арбути́н·, -а, (m) (chem) arbutin

аргент·и́т·, -а, (m) (min) argentite, silver glance

аргенто́·метр·, -а, (m) (phot) argentometer

аргилл·и́т·, -а, (m) (min) argillite, claystone

аргилл·и́т·ов·ый, -ая, -ое, (a) (min) argillaceous

аргини́н·, -а, (m) (chem) arginine

аргиро·ди́т·, -а, (m) (min) argyrodite

аргиро·пири́т·, -а, (m) (min) argyropyrite

арго́л·, -а, (m) (chem) argol

арго́н·, -а, (m) (chem) argon, Ar, A

аргоно·ду́г·ов·ая с·ва́р·к·а (f) argon-arc welding

аргуме́нт·, -а, (m) (math) argument

~ компле́кс·н·ого числ·а́ (math) argument of a complex number

~ фу́нкц·и·й (math) argument of a function

аргумент·и́р·овать, -уют, (imp and perf) argue, prove

ардоме́тр·, -а, (m) pyrometer

а́ре·а, -ы, (f) area

~, за·мо́ч·н·ая cardinal area, interarea

~, ло́ж·н·ая pseudoarea

~, с·вя́з·очн·ая ligament area

ареа́л·, -а, (m) areal, geographic range

ареа́ль·н·ый, -ая, -ое, (a) (math) areal

арекаи́н·, -а, (m) (chem) arecaine

аре́н·а, -ы, (f) arena

~, гео·морфо·лог·и́ческ·ая geomorphological terrain

аре́нд·а, -ы, (f) lease, leasehold agreement; rent

аренд·а́тор·, -а, (m) leaseholder, lessee

аре́нд·н·ый, -ая, -ое, (a) lease, leasehold; rent

~ пла́т·а (f) rent

аренд·о́ванн·ый, -ая, -ое, (past part pass) see **аренд·ова́ть;** leasehold

аренд·ова́ть, -уют, (imp and perf) lease, rent

арен·и́т·, -а, (m) (min) arenite, arenaceous rock

арео́л·а, -ы, (f) areola

ареола́р·н·ый, -ая, -ое, (a) areolar

арео́·метр·, -а, (m) (phys) areometer, hydrometer

~, моло́ч·н·ый (food) lactometer

арео·пикно́·метр·, -а, (m) areopycnometer

аре́ст·, -а, (m) (law) arrest; attachment, sequestration, seizure

арест·а́нт·, -а, (m) arrested/detained person/man

арест·ова́ть, -у́ют, (imp and perf) arrest

арест·о́выв·ать, -ают, (imp) arrest

аржан·éц·, -н·ц·а́, (i) -н·ц·о́м, (m) (bot) timothy, *Phleum*

аржанти́н·, -а, (m) (text) argentine

аржилли́т·, -а, (m) (min) argillite

ари́д·ност·ь, -и, (f) (meteor) aridity

ари́д·н·ый, -ая, -ое, (a) (meteor) arid

ари́л·, -а, (m) (chem) aryl

арил·арсо́н·ов·ая кисл·от·а́ (f) aromatic arsonic acid

арил·и́ровани·е, -я, (n) (chem) arylation

ари́ллус·, -а, (m) (bot) aril, arillus

аристо·ге́н·, -а, (m) (phot, chem) aristogen

аристо́тел·ев·, -а, -о, (possess adj) Aristotelian

аристо́тел·евск·ий, -ая, -ое, (a) Aristotelian

а·ритм·и́ческ·ий, -ая, -ое, (a) arrhythmic

а·ритм·и́·я, -и, (f) (med) arrhythmia, arhythmia

арифме́тик·, -а, (m) arithmetician

арифме́т·ик·а, -и, (f) arithmetic

арифмет·и́ческ·ий, -ая, -ое, (a) arithmetical, arithmetic

~ у·стро́й·ств·о (n) (autom) arithmetic unit

арифмо́·метр·, -а, (m) arithmometer, hand-operated adding machine

а́рк·а, -и, (g pl) аро́к·, (f) arch

~, бес·шарни́р·н·ая rigid arch

~, двух·шарни́р·н·ая two-hinged arch

~, за·по́лн·енн·ая blind arch

~, зуб·ча́т·ая multi-foil arch

а́рк·а
~, киле·ви́д·н·ая   inflected arch
~, ланцето·ви́д·н·ая   lancet arch
~, маврита́н·ск·ая   horseshoe arch
~, много·ло́паст·н·ая   multi-foil arch
~, подково·обра́з·н·ая   horseshoe arch
~, полу·кру́гл·ая   round arch
~, полу·ци́ркуль·н·ая   round arch
~, при·ро́д·н·ая   natural bridge/arch
~ пыльц·ы́   (bot) pollen arcus
~, стре́ль·чат·ая воз·вы́ш·енн·ая   lancet arch
~, трёх·ло́паст·н·ая   trefoil arch
~, трёх·центр·ов·а́я   three-centred arch
~, трёх·шарни́р·н·ая   three-hinged arch
арка́д·а, -ы,   (f) (arch) arcade
арка́н·, -а,   (m) lasso
арка́т·н·ый, -ая, -ое,   (a) harness
~ под·вя́з·ь,   (f) (text) harness
~ шнур·   (m) (text) harness cord
аркбута́н·, -а,   (m) (arch) arc-boutant, flying buttress
арквери́т·, -а,   (m) (min) arquerite
арк·косе́канс·, -а,   (m) (math) arc/ inverse cosecant, anticosecant
арк·ко́синус·, -а,   (m) (math) arc/ inverse cosine, anticosine
арк·кота́нгенс·, -а,   (m) (math) arc/ inverse cotangent, anticotangent
арко·ви́д·н·ый, -ая, -ое,   (a) arcuate, arch-like
арко·ге́н·, -а,   (m) metal-arc welding
арко́з·, -а,   (m) (geol) arkose
арк·се́канс·, -а,   (m) (math) arc inverse secant, antisecant
арк·си́нус·, -а,   (m) (math) arc/ inverse sine, antisine
арк·та́нгенс·, -а,   (m) (math) arc tangent
а́ркт·ик·а, -и,   (f) the Arctic, arctic regions
аркт·и́ческ·ий, -ая, -ое,   (a) arctic, polar
~ фронт·   (m) (meteor) arctic front
аркто·ге́·я, -и,   (f) (zool) Arctogea
аркто·карбо́н·ов·ый, -ая, -ое,   (a) (geol) Arcto-Carboniferous
аркто·тре́т·ичн·ый -ая, -ое,   (a) (geol) Arcto-Tertiary
аркту́р·, -а,   (m) (astron) Arcturus
арк·фу́нкци·я, -и,   (f) (math) inverse trigonometric function

армату́р·а, -ы,   (f) fittings, mountings, accessories; reinforcement iron/metal (for concrete)
~, ви·т·а́я   twisted bar
~ двой·н·о́го круч·е́ни·я   twin-twisted bar
~, за·бо́рт·н·ая   (shipb) sea valves/connexions
~, коте́ль·н·ая   boiler fittings/mountings
~, му́фт·ов·ая   coupled pipe fittings
~, о·свет·и́тельн·ая   light fittings
~ период·и́ческ·ого профи́л·я   twisted ribbed bar (for concrete)
~, пуч·ко́в·ая   wire-rod bundle reinforcement (of concrete)
~, рас·пре·дел·и́тельн·ая   distribution steel/rods (for concrete)
~, се́т·чат·ая   reinforcing mesh, two-way reinforcement
~, сталь·н·а́я   reinforcement steel
~, фонта́н·н·ая   (oil) Christmas tree and tubing head
армату́р·н·ый, -ая, -ое,   (a) of армату́р·а; reinforced, reinforcement, reinforcing (of concrete)
~ се́т·к·а   (f) mesh reinforcement
~ стерж·ен·ь   (m) reinforcing bar
арм·е́йск·ий, -ая, -ое,   (a) of а́рм·и·я
арм·и́ровани·е, -я,   (n) (v n) of арм·и́р·овать; reinforcement (of concrete etc.); sheathing (of electric cable)
арм·и́рованн·ый, -ая, -ое,   (past part pass) of арм·и́р·овать; armoured, armored
арм·и́р·овать, -уют,   (imp and perf) reinforce (concrete etc.); sheath (electric cables etc.)
а́рм·и·я, -и,   (f) army; armed forces
~, ка́др·ов·ая   regular army
армко-желе́з·о, -а,   (n) armco-iron
арморика́н·ск·ий, -ая, -ое,   (a) (geol) Armorican
армо·цеме́нт·н·ый, -ая, -ое,   (a) reinforced-cement
армю́р·, -а,   (m) (text) armure
А́рндт·а-Эйстерт·а, реа́кци·я   (f) (chem) Arndt–Eistert synthesis
арокси́ль·н·ый, -ая -ое,   (a) aroxyl
арома́т·, -а,   (m) aroma, fragrance
аромат·иза́ци·я, -и,   (f) (oil, chem) aromatization
аромат·и́ческ·ий, -ая, -ое,   (a) (chem) aromatic; aromatic, fragrant
~ со·един·е́ни·е   (n) (chem) aromatic compound

аромат·и́ческ·ий, -ая, -ое
~ спи́рт·,   (*m*) (chem) aromatic alcohol

аромат·и́чн·ый, -ая, -ое,   (*a*) (*s f*) -чен·, -чн·а *see* аромáт·н·ый

аромáт·н·ый, -ая, -ое   (*a*) (*s f*) -т·ен·, -т·н·а,   scented, fragrant, sweet smelling

аро·морф·óз·, -а,   (*m*) (zool) aromorphosis

áроч·н·ый, -ая, -ое,   (*a*) of áрк·а
~ мóст·   (*m*) arched bridge

аррена́л·, -а,   (*m*) (chem) arrhenal

Аррéниус·а, теóр·и·я   (*f*) (phys, chem) Arrhenius' theory

аррено·бласт·óм·а, -ы,   (*f*) (zool) arrhenoblastoma

аррено·тóк·и·я, -и,   (*f*) (zool) arrhenotoky, arrenotoky

арретúр·, -а,   (*m*) arrester, delay mechanism detent, lock, locking device; caging device (of gyro)

арретúр·овать, -уют,   (*imp and perf*) (mech) lock, arrest; cage (a gyro)

арретúр·ующ·ий, -ая, -ее,   (*pres part act*) of арретúр·овать
~ у·стрóй·ств·о (*n*) delay mechanism

аррорýт·, -а, (*m*) (bot) arrowroot, *Maranta*; arrowroot starch

арсенáл·, -а,   (*m*) arsenal, armoury

арсенáт·, -а,   (*m*) (chem) arsenate
~ ви́смут·а, вóд·н·ый   hydrous bismuth arsenate, arsenobismite
~, дерев·яни́ст·ый   wood arsenate
~ кáльци·я   calcium arsenate
~ свинц·á   lead arsenate

арсени́д·, -а,   (*m*) (chem) arsenide
~ желéз·а   arsenoferrite

арсени́т·, -а,   (*m*) (chem) arsenite

арсено-   (*int component*) arseno-

арсено·ли́т·, -а,   (*m*) (min) arsenolite

арсено·пири́т·, -а,   (*m*) (min) arsenopyrite, arsenical pyrite

арси́н·, -а,   (*m*) (chem) arsine

арт-   (*abbr of component*) = артиллер·и́йск·ий   artillery, gunnery, gun

артезиáн·ск·ий, -ая, -ое,   (*a*) artesian
~ вод·á   (*f*) artesian water
~ колóд·ец·   (*m*) artesian well
~ на·пóр·, -а,   (*m*) (geol) artesian pressure

артéл·ь, -и,   (*f*) artel' (a small collective enterprise), co-operative

артéль·щик·, -а,   (*m*) member of an artel', co-operative worker

артериáль·н·ый, -ая, -ое,   (*a*) (anat) arterial

артереó·граф·, -а,   (*m*) (med) plethysmograph

артерио·склер·óз·, -а,   (*m*) (med) arteriosclerosis

артериóл·а, -ы,   (*f*) (zool) arteriole

артéри·я, -и,   (*f*) artery

арти́зи·я, -и,   (*f*) (bot) artisia

артикуля́т·ы   (*nom pl*) (zool) articulates, *Articulata*

артикуля́ци·я, -и,   (*f*) (ling) articulation, intelligibility

артиллер·и́йск·ий, -ая, -ое,   (*a*) artillery, gunnery, ordnance
~ дéл·о   (*n*) gunnery
~ систéм·а   (*f*) gun, gun mounting, gun and mounting
~ у·станóв·к·а   (*f*) = артиллер·и́йск·ая систéм·а

артиллер·и́ст·, -а,   (*m*) (mil) artilleryman, gunner; (naut) the gunnery officer

артиллéр·и·я, -и,   (*f*) (mil) artillery; (naut) armament
~, берег·ов·áя   coast artillery
~, бес·ствóль·н·ая   rockets, rocket armament
~, мало·кали́бер·н·ая   low calibre guns/armaments
~, мор·ск·áя   naval gunnery/ordnance
~, реакти́в·н·ая   rockets, rocket armament
~, срéд·н·яя   medium calibre artillery/guns (100–180 mm)
~, ствóль·н·ая   conventional artillery/armament

арти́ст·, -а,   (*m*) artist; actor

артист·и́ческ·ий, -ая, -ое,   (*a*) artistic

артишóк·, -а,   (*m*) (bot) artichoke, globe artichoke, *Cynara scolymus*

артри́т·, -а,   (*m*) (med) arthritis

артро·пóд·, -а,   (*nom pl*) -ы, (zool) arthropod, (*pl*) arthropods, *Arthropoda*

артро·спóр·а, -ы,   (*f*) (bot) arthrospore

артро·том·и́·я, -и,   (*f*) (med) arthrotomy

арт·у·станóв·к·а, -и,   (*g pl*) -в·ок·, (*f*) gun mounting

арт·шкóл·а, -ы,   (*f*) (mil) artillery school; (naut) gunnery school

áрф·а, -ы,   (*f*) harp (cheesemaking); cribble (potato sorting)

арфведсони́т·, -а,   (*m*) (min) arfvedsonite

архай́ческ·ий, -ая, -ое,   (*a*) archaic

архе·гони·áльн·ый, -ая, -ое,   (*a*) (bot) archegoniate

архе·го́н·и·й, -я, (*m*) (bot) archegonium

архе·диктио́н·, -а, (*m*) (zool) archedictyon

архе́·й, -я, (*m*) (geol) Archean

архе́й·ск·ая э́р·а (*f*) (geol) Archean era

архео·зо́й·ск·ая э́р·а (*f*) (geol) Archaeozoic era

архео́·лог·, -а, (*m*) archaeologist, archeologist

архео·лог·и́ческ·ий, -ая, -ое, (*a*) archaeological, archeological

архео·ло́г·и·я, -и, (*f*) archaeology, archeology

архео·пте́рикс·, -а, (*m*) (pal) Archaeopteryx

архео·пте́рис·ов·ый, -ая, -ое, (*a*) archeopteric

архе·о́рнис·, -а, (*m*) (pal) archaeornis, *Archaeornis*

архе·спо́р·и·й, -я, (*m*) (cyt) archesporium

архи́в·, -а, (*m*) archives

архива́риус·, -а, (*m*) archivist

архиво́льт·, -а, (*m*) (arch) archivolt

архи·ка́рп·, -а, (*m*) (bot) archicarp

Архиме́д·а, за·ко́н· (*m*) (phys) Archimedes' principle

архиме́д·ов·, -а, -о, (*posses adj*) Archimedean

архи·мице́т·, -а, (*nom pl*) **-ы,** (bot) archimycete, (*pl*) *Archimycetes*

архипела́г·, -а, (*m*) archipelago; the Aegean islands

архи·страти·гра́ф·и·я, -и, (*f*) archistratigraphy

архи·текто́н·ик·а, -и, (*f*) architectonics

архите́ктор·, -а, (*m*) architect

архитекту́р·а, -ы, (*f*) architecture

~, корабе́ль·н·ая naval architecture, ship construction

~, суд·ов·а́я naval architecture, ship construction

архитра́в·, -а, (*m*) (arch) architrave

архо·за́вр·, -а, (*m*) (pal) archosaurus

архо·ли́т·ов·ый, -ая, -ое, (*a*) archolithic

арч·а́, -й, (*i*) **-ч·о́й,** (*f*) (bot) juniper, *Juniperus*

арши́н·, -а, (*g pl*) **арши́н·,** (*m*) (obs) arshin (= 28 in., 0·71 m)

ары́к·, -а, (*m*) irrigation ditch/canal

ары́ч·н·ый, -ая, -ое, (*a*) of **ары́к·**

арьерга́рд·, -а, (*m*) (mil) rear-guard; (naut) rear, rear screen

а́с·, -а, (*m*) ace (airman)

а-с (*abbr*) = **ампе́р-секу́нд·а** (elec) ampere-second

асбе́ст·, -а, (*m*) (min) asbestos

~, актиноли́т·ов·ый asbestiform actinolite

~, волокн·и́ст·ый fibrous asbestos, amianthus

~, паллад·и́рованн·ый palladinized asbestos

~, пере·ли́в·чат·ый chrysotile, Canadian asbestos

~, платин·и́рованн·ый platinized asbestos

~, пря́д·ен·ый asbestos thread, spun asbestos

~, рогово·об·ма́н·ков·ый hornblende asbestos

~, тигр·о́в·ый (min) tiger's eye

асбе́ст·ов·ый, -ая, -ое, (*a*) asbestos

~ пуш·о́нк·а (*f*) flaked asbestos

~ шну́р· (*m*) asbestos rope

асбо- (*abbr*) of **асбе́ст·** asbestos

асбо·картóн·, -а, (*m*) (build) asbestos board

асбола́н·, -а, (*m*) (min) asbolane, asbolite

асболи́т·, -а, (*m*) (min) asbolite, asbolane

асбопеколи́т·, -а, (*m*) asbestos-tar plastic

асбо·фане́р·а, -ы, (*f*) asbestos-cement millboard

асбо·цеме́нт·н·ый материа́л·, (*m*) asbestos-cement sheeting

асбо·ши́фер·, -а, (*m*) asbestos shingles, asbestos-cement tiles

а·сейсм·и́ческ·ий, -ая, -ое, (*a*) non-seismic, aseismic

а·секве́нт·н·ый, -ая, -ое, (*a*) asequential

а·сексуа́ль·н·ый, -ая, -ое, (*a*) asexual

а·се́пт·ик·а, -и, (*f*) (med) asepsis

а·септ·и́ческ·ий, -ая, -ое, (*a*) aseptic

а·симметр·и́ческ·ий, -ая, -ое, (*a*) asymmetric(al), unsymmetrical

~ а́том· (*m*) (chem) asymmetric atom

а·симметр·и́чн·ый, -ая, -ое, (*a*) = **а·симметр·и́ческ·ий**

а·симметр·и́·я, -и, (*f*) asymmetry, dissymmetry

~ дел·е́ни·я (nucl) fission asymmetry

асимпто́т·а, -ы, (*f*) (math) asymptote

асимптот·и́ческ·ий, -ая, -ое, (*a*) asymptotic

а·синхро́н·н·ый, -ая, -ое, (*a*) asynchronous

а·синхро́н·н·ый, -ая, -ое
~ **маши́н·а**   (*f*) (elec) asynchronous motor

асине́рги·я, -и,   (*f*) (med) asynergy

а́ск·, -а,   (*m*) (bot) ascus, sac

аскари́д·а, -ы,   (*f*) (zool) ascaride, *Ascaris*

аскарид·о́з·, -а,   (*m*) (med, vet) ascariasis, ascaridosis

аскари́д·ов·ые, -ых,   (*pl decl as adj*) (zool) *Ascaroidea*

аскарид·о́л·, -а,   (*m*) (pharm) ascaridol

аско·ге́н·н·ый, -ая, -ое, (*a*) ascogenous

аско·го́н·, -а,   (*m*) (bot) ascogon

аско·лиша́й·ник·, -а,   (*nom pl*) **-и,** (*m*) (bot) ascolichen, (*pl*) ascolichens, *Ascolichenes*

аско·мице́т·, -а,   (*m*) (bot) ascomycete

асконо́ид·н·ый, -ая, -ое, (*a*) asconoid

аскслиза́ци·я, -и, (*f*) Ascoli procedure for testing hides for anthrax

аскорб·и́нов·ая кисл·от·а́ (*f*) ascorbic acid, vitamin C

аско·спо́р·а, -ы, (*f*) (bot) ascospore

аско·хит·о́з·, -а,   (*m*) bot) aschochitosis

а·сифо́н·н·ый, -ая, -ое,   (*a*) (zool) asiphonate

аспараги́н·, -а, (*m*) (chem) asparagine

аспараги́н·ов·ая кисл·от·а́ (*f*) (chem) aspartic acid

аспара́гус·, -а, (*m*) (hortic) asparagus (decorative varieties)

аспарт·а́з·а, -ы, (*f*) (chem) aspartase

аспа́рт·ов·ая кисл·от·а́ (*f*) (chem) aspartic acid

а́спект·, -а, (*m*) (bot) aspect

асперги́лл·, -а, (*m*) (bot) Aspergillus

аспергилл·ёз·, -а, (*m*) (med) aspergillosis

а·сперм·и́·я, -и, (*f*) (zool) aspermia

а́спид·, -а, (*m*) (zool) asp; (min) slate

аспиди́н·, -а, (*m*) (pharm) aspidine

а́спид·н·ый, -ая, -ое, (*a*) slate
~ **ка́мен·ь** (*m*) slate

аспира́нт·, -а, (*m*) post-graduate student (studying for university/research post)

аспирант·у́р·а, -ы, (*f*) system of study for qualifying for university/research posts

аспира́тор·, -а, (*m*) (chem) aspirator

аспирацио́н·н·ый, -ая, -ое, (*a*) aspiration, suction

аспира́ци·я, -и, (*f*) aspiration

аспири́н·, -а, (*m*) (pharm) aspirin

ассамбле́·я, -и,   (*f*) assembly (of international organizations)

ассениз·а́тср·, -а,   (*m*) (san eng) sewageman, nightman

ассениз·ацио́н·н·ая маши́н·а   (*f*) sewage-disposal lorry

ассениз·а́ци·я, -и,   (*f*) sewage disposal

ассигн·а́ци·я, -и,   (*f*) bank note

ассигн·ова́ни·е ··я,   (*n*) (fin) appropriation, allocation, grant

ассигн·о́вк·а, -и,   (*g pl*) **-вок·,**   (*f*) (fin) draft

ассимил·яцио́н·н·ый, -ая, -ое,   (*a*) assimilative

ассимил·я́ци·я, -и,   (*f*) assimilation
~, **кра·ев·а́я** marginal assimilation

ассист·е́нт·, -а,   (*m*) assistant; lecturer (university level); assistant examiner

ассист·и́р·овать, -уют,   (*imp*) assist, carry out the duties of assistant

ассортиме́нт·, -а,   (*m*) selection, choice, range (of goods)

ассоциати́в·н·ый, -ая, -ое,   (*a*) associative
~ **за·ко́н·**   (*m*) (math) associative law

ассоциа́ци·я, -и,   (*f*) association
~ **моле́кул·** (chem) molecular association

ассоциес·, -а,   (*m*) (bot) associes

ассоци·и́р·овать, -уют, (*imp and perf*) associate

аста́зи·я, -и,   (*f*) (med) astasia

аста́т·, -а,   (*m*) (chem) astatine, At

астати́н·, -а,   (*m*) (chem) astatine, At

астат·и́ческ·ий, -ая, -ое,   (*a*) astatic; (autom) floating
~ **регул·я́тор·**   (*m*) (autom) floating controller
~ **систе́м·а**   (*f*) (elec) astatic system
~ **со·ставл·я́ющ·ая**   (*f*) (autom) floating component

астен·и́ческ·ий, -ая, -ое,   (*a*) (med) asthenic

астено·ли́т·, -а,   (*m*) (geol) asthenolith

астено·сфе́р·а, -ы,   (*f*) (geol) asthenosphere

астер·и́зм·, -а,   (*m*) asterism, stellate opalescence

астер·о́ид·, -а,   (*m*) (astron) asteroid, planetoid, minor planet

астеро·каламит·, -а,   (*m*) (min) asterocalamite

астеро·ксило́н·, -а,   (*m*) (pal) asteroxylon

астигма́т·, -а,   (*m*) (opt) astigmatic lens

астигмат·и́зм·, -а, *(m)* (opt) astigmatism

а́стм·а, -ы, *(f)* (med) asthma

астма́т·ик·, -а, *(m)* (med) asthmatic (person)

астмат·и́ческ·ий, -ая, -ое, *(a)* (med) asthmatic

асто·ген·и́·я, -и, *(f)* (bot) astogeny

А́стон·а тёмн·ое про·стра́н·ств·о (nucl) Aston dark space

асто·хи́т·ов·ый, -ая, -ое, *(a)* astochitic

а́стр·а, -ы, *(f)* (bot) aster

астрага́л·, -а, *(m)* (build) astragal; (bot) milk vetch, *Astragalus* sp.; (anat) astragalus

астра́ль·н·ый, -ая, -ое, *(a)* astral

астра́нци·я, -и, *(f)* (bot) masterwort, *Astrantia*

астро- *(int component)* astro-, astronomical

астро́·граф·, -а, *(m)* astrograph

астр·о́ид·а, -ы, *(f)* astroid

астро·ко́мпас·, -а, *(m)* (air nav) astro-compass

астроля́би·я, -и, *(f)* (surv) astrolabe

~, призм·е́нн·ая (surv) prismatic astrolabe

~ с алида́д·ой planispheric alidade

астро·ме́тр·и·я, -и, *(f)* (astron) astrometry

астро·на·вед·е́ни·е, -я, *(n)* (rocket) star tracking

астро·навигацио́н·н·ый, -ая, -ое, *(a)* astronavigation, star navigation

~ на·вед·е́ни·е *(n)* star-tracking guidance

астро·на́вт·, -а, *(m)* astronaut

астро·на́вт·ик·а, -и, astronautics

астро·но́м·, -а, *(m)* astronomer

астро·ном·и́ческ·ий, -ая, -ое, *(a)* astronomical

~ един·и́ц·а *(f)* astronomical unit

~ стекл·о́ *(n)* (a/c) astrodome

астро·но́м·и·я, -и, *(f)* astronomy

~, пол·ев·а́я (surv) field astronomy

астро·ри́з·а, -ы, *(f)* (pal) astrorhiza

астро·спектро·скоп·и́·я, -и, *(f)* astronomical spectroscopy

астро·труб·а́, -ы́, *(nom pl)* **-тру́б·ы,** *(f)* astronomical telescope

астро·фи́з·ик·, -а, *(m)* astrophysicist

астро·фи́з·ик·а, -и, *(f)* astrophysics

астро·физ·и́ческ·ий, -ая, -ое, *(a)* astrophysical

астро·фото·гра́ф·и·я, -и, *(f)* astronomical photography

астро·ци́т·, -а, *(m)* (biol) astrocyte

асфа́льт·, -а, *(m)* asphalt, asphaltum

~, вя́з·к·ий chapapote

~, мя́гк·ий pit asphalt, viscid bitumen

~, при·ро́д·н·ый natural asphalt

асфальт·е́н·, -а, *(m)* (chem) asphaltene

асфальт·и́рованн·ый, -ая, -ое, *(past part pass)* asphalted, asphalt-impregnated, asphalt-coated

асфальт·и́т·, -а, *(m)* (min) asphaltite

~, чёрн·ый impsonite

асфальто·бе·то́н·, -а, *(m)* bituminous concrete

асфа́льт·ов·ый, -ая, -ое, *(a)* asphalt, asphaltic

~ биту́м· *(m)* asphaltic bitumen (Brit), asphalt (US)

~ го́р·н·ая по·ро́д·а *(f)* rock asphalt

~ ла́к· *(m)* asphaltic varnish

~ рас·тво́р· *(m)* asphaltic mortar

асфальто·ло́м·, -а, *(m)* (civ eng) scarifier

асфальто·у·кла́д·чик·, -а, *(m)* (civ eng) asphalt spreader

асфальто·ти́п·и·я, -и, *(f)* (phot) asphalt/bitumen process

асфи́кси·я, -и, *(f)* (med) asphyxia

асхи́ст·ов·ый, -ая, -ое, *(a)* aschistic

асцензио́н·н·ый, -ая, -ое, *(a)* ascentional

асци́ди·я, -и, *(f)* (biol) ascidium; *(pl)* (zool) *Ascidia*

асци́т·, -а, *(m)* (vet) ascites, hydroperitoneum

ат. *(abbr)* = **атмосфе́р·а техн·и́ческ·ая** technical atmosphere (unit of pressure equal to 735·6 mm Hg); = **а́том·н·ый** at., atomic

ат.⁰/₀ а́том·н·ый проце́нт· at.⁰/₀, atomic percentage

ат. вес· = **а́том·н·ый вес·** at. wt., atomic weight

ат. ед. = **а́том·н·ая един·и́ц·а** au, atomic unit

ат. ед. ма́сс·а amu., atomic mass unit

ата = **абсолю́т·н·ая атмосфе́р·а** abs. atm., absolute atmosphere

атав·и́зм·, -а, *(m)* (biol) atavism

атав·исти́ческ·ий, -ая, ое, *(a)* (biol) atavistic

ата́к·а, -и, *(f)* attack, assault

у́гол· ата́к·и *(m)* (air) angle of attack/incidence

у́гол· ата́к·и, крит·и́ческ·ий (air) angle of stall

у́гол· ата́к·и, с·ры́в·н·о́й (air) angle of stall

атакам́ит·, -а,   (*m*) (min) atacamite

атак·ов́ать, -́уют, (*imp and perf*) attack, assault

атакс·́ит·, -а,   (*m*) ataxite (meteorite)

атакс·́ит·ов·ый, -ая, -ое,   (*a*) (med) ataxitic

атакс·́ическ·ий, -ая, -ое,   (*a*) ataxic

атакс·́и·я, -и,   (*f*) (med) ataxia, ataxy

атакто·ст́ел·а, -ы,   (*f*) (bot) atacto-stele

атебр́ин·, -а,   (*m*) (pharm) atebrine, quinacrine

а·тектон·́ическ·ий, -ая, -ое,   (*a*) atectonic

ателомит·́ическ·ий, -ая, -ое,   (*a*) (cyt) atelomitic

атель́е   (*n indecl*) studio; work room(s) (at tailor's, dressmaker's, shoemaker's etc.)

~ **мод·**   fashion house

атер́ом·а, -ы,   (*f*) (med) atheroma

атеромат·́ическ·ий, -ая, -ое,   (*a*) (med) atheromatous

атеромат·́оз·, -а,   (*m*) (med) atheromatosis

атеро·склер·́оз·, -а,   (*m*) (med) atherosclerosis

атет́оз·, -а,   (*m*) (med) athetosis

ати   (*abbr*) = **из·б́ыт·очн·ая атмо·сф́ер·а** psig, gauge pressure

атиз́ин·, -а,   (*m*) (pharm) atizin

атл́ант·, -а,   (*m*) (anat) atlas

атлант·́ическ·ий, -ая, -ое,   (*a*) (geog) Atlantic

~ **пор́ог·**   (*m*) (ocean) Greenland–Scotland ridge

~ **теч·́ени·е**   (*n*) (ocean) North Atlantic current

атл́антроп·, -а,   (*m*) (pal) atlanthropus

́атлас·, -а,   (*m*) atlas (maps, drawings); **атл́ас·** (text) satin

~**, основ·н·́ой**   (text) warp satin

~**, ут́оч·н·ый** (text) weft satin

атл́ас·ист·ый, -ая, -ое,   (*a*) (text etc.) satiny

́атлас·н·ый, -ая, -ое,   (*a*) (geog etc.) atlas; **атл́ас·н·ый** (text) satin

~ **д́ерев·о**   (*nom pl*) **-́ев·ья,**   (*n*) satin wood

~ **пере·плет·́ени·е**   (*n*) (text) satin weave

атл́ет·, -а,   (*m*) athlete

атл́ет·ик·а, -и,   (*f*) athletics

атм. (*abbr*) = **атмосф́ер·а** at, atmosphere

атмид́о·метр·, -а,   (*m*) atmometer, evaporimeter

атмо·ген·́ическ·ий, -ая, -ое,   (*a*) atmogenic

атм́о·лиз·, -а,   (*m*) (chem) atmolysis

атм́о·метр·, -а,   (*m*) atmometer

атмо·сф́ер·а, -ы,   (*f*) atmosphere

~**, баро·метр·́ическ·ая**   barometric pressure

~**, в́ерх·н́яя**   (meteor) upper air

~**, вос·станов·́ительн·ая**   (met) reducing atmosphere

~**, за·щ́ит·н·ая**   (met) protective atmosphere

~**, из·б́ыт·очн·ая**   gauge atmosphere

~**, контрол·́ируем·ая**   (met) controlled atmosphere

~**, метр·́ическ·ая**   technical atmosphere

~**, норм́аль·н·ая**   atmosphere, at. (unit of pressure equal to 760 mm Hg)

~**, одно·р́од·н·ая**   homogeneous atmosphere

~**, о·кисл·́ительн·ая**   (met) oxidizing atmosphere

~**, с·поќой·н·ая**   (met) tranquil atmosphere

~**, станд́арт·н·ая**   standard atmosphere (in U.S.S.R. 760·1 mm, 0°C, 980·62 cm/sec^2)

~**, техн·́ическ·ая** technical atmosphere (unit of gauge pressure = 735·6 mm Hg at 0°C or 1 kg/cm^2)

~**, физ·́ическ·ая**   atmosphere, atm., (760 mm Hg), standard atmosphere

атмосф́ер·ик·и   (*nom pl*), (*g pl*) **-ик·,** (*f*) atmospherics

атмосфер·́ическ·ий, -ая, -ое,   (*a*) atmospheric

атмосф́ер·н·ый, -ая, -ое,   (*a*) atmospheric

~ **вод·́а**   (*f*) surface storm water

~ **по·м́ех·и**   (*f pl*) (rad) interference, atmospherics

атмосферо·ст́ой·к·ий, -ая, -ое,   (*a*) weatherproof, weather-resistant

ат. н. = **́атом·н·ый н́омер·** z, atomic number

атокс́ил·, -а,   (*m*) (chem) atoxyl

атокс́ил·ов·ый, -ая, -ое,   (*a*) atoxylic

а·токс·́ическ·ий, -ая, -ое,   (*a*) atoxic

а·токс·́ичност·ь, -и   (*f*) atoxicity, lack of toxicity

ат́олл·, -а,   (*m*) (geog) atoll

ат́олл·ов·ый, -ая, -ое,   (*a*) (geog) atoll

́атом·, -а,   (*m*) (phys) atom

~ **-акц́ептор·, -а,**   (*m*) acceptor atom

~**, бомбард·́ируем·ый**   struck atom

~**, верш·́инн·ый**   apical atom

а́том
~, в·не́др·ённ·ый   interstitial atom
~, воз·бужд·ённ·ый   excited atom
~, вы́·би·т·ый   knocked-on atom
~, гор·я́ч·ий   hot atom
~ -до́нор·, -а,   (*m*) donor atom
~, доче́р·н·ый   product atom
~, мате́р·инск·ий   parent atom
~, ме́ч·ен·ый   labelled/tagged atom
~ мише́н·и   target atom
~ от·да́·ч·и   recoil atom
~ по́лн·ост·ью ио́н·изи́рованн·ый   stripped atom
~, при́·мес·н·ый   impurity atom
~, с·вя́з·анн·ый   bound atom
~, с·мещ·ённ·ый   displaced atom
~ то́пл·ивн·ого сыр·ь·я́   fertile atom
атом·а́рн·ый, -ая, -ое, (*a*) atomic
~ водо·ро́д·   (*m*) atomic hydrogen
атом·и́зм·, -а,   (*m*) = **атом·и́стик·а**
а́том·ност·ь, -и,   (*f*) (chem) atomicity
атом·и́стик·а, -и,   (*f*) atomic theory, atomism
атом·исти́ческ·ий, -ая, -ое,   (*a*) atomistic
~ тео́р·и·я   (*f*) atomism
а́том·ник·, -а,   (*m*) (pop) atomic scientist; (polit) atomic warmonger
а́томно·водо·ро́д·н·ая с·ва́р·к·а   (*f*) atomic-hydrogen welding
а́том·н·ый, -ая, -ое, (*a*) atom, atomic
~ бо́мб·а   (*f*) atom bomb
~ вес·   (*m*) atomic weight
~ компле́кс·   (*m*) aggregate of atoms
~ коэффицие́нт·   (*m*) atomic factor
~ но́мер·   (*m*) atomic number
энерг·и·я а́том·н·ой с·вя́з·и   (*f*) atomic binding energy
атомо·хо́д·, -а,   (*m*) nuclear-propelled ship, nuclear ship
а·тон·и́ческ·ий, -ая, -ое,   (*a*) (med) atonic
а·тон·и́·я, -и,   (*f*) (med) atony
атофа́н·, -а,   (*m*) (pharm) atophan, cincophen
а·трез·и́·я, -и,   (*f*) (med) atresia
атрепс·и́·я, -и,   (*f*) (med) atrepsy
атрет·и́ческ·ий, -ая, -ое, (*a*) (med) atretic, atresic
атриа́ль·н·ый, -ая, -ое,   (*a*) (zool) atrial
а́три·й, -я,   (*m*) (zool) atrium
а́триум·, -а,   (*m*) atrium
атропами́н·, -а,   (*m*) (chem) atropamine
атропи́н·, -а,   (*m*) (chem) atropine

атроф·и́р·оваться, -уются, (*imp and perf*) atrophy
атроф·и́·я, -и,   (*f*) (biol) atrophy
атроха́ль·н·ый, -ая, -ое,   (*a*) (zool) atrochal
атс (*abbr*) **автомат·и́ческ·ая теле·фо́н·н·ая ста́нци·я** telephone exchange
аттапульги́т·, -а,   (*m*) (min) attapulgite
атташе́ (*m indecl*) (dipl) attaché (of embassy)
~, вое́н·н·ый   military attaché
~, вое́н·н·о-воз·ду́ш·н·ый   air attaché
~, вое́н·н·о-мор·ск·о́й   naval attaché
~, торг·о́в·ый   commercial attaché
аттенюа́тор·, -а,   (*m*) (elec) attenuator
~, граду·и́рованн·ый   calibrated attenuator
~, по·гло́т·и́тельн·ый   dissipative attenuator
~, пре·де́л·н·ый   cut-off attenuator, non-dissipative attenuator
аттеста́т·, -а,   (*m*) certificate, diploma, pedigree; (mil) warrant, allotment
~ зре́л·ост·и   certificate of general secondary education
аттест·а́ци·я, -и,   (*f*) testimonial; confidential report; certification
аттест·о́ванн·ый, -ая, -ое, (*past part pass*) of **аттест·ова́ть**
аттест·ова́ть, -у́ют, (*imp and perf*) give a testimonial, provide a personal reference; promote, confirm (in rank)
а́ттик·, -а,   (*m*) (arch) attic
аттри́т·, -а,   (*m*) (min) attritus
аудио·гра́мм·а, -ы,   (*f*) (acous) audiogram
ауди́о·метр·, -а,   (*m*) (acous) audiometer
аудио́н·, -а,   (*m*) (rad) audion
аудито́ри·я, -и,   (*f*) auditorium, lecture hall, lecture room; audience
ауерли́т·, -а,   (*m*) (min) auerlite
ауксано́·граф·, -а,   (*m*) (bot) auxanograph
ауксано́·метр·, -а,   (*m*) (bot) auxanometer, growth-tester
аукси́н·, -а,   (*m*) (bot, chem) auxin
ауксо·спо́р·а, -ы,   (*f*) (bot) auxospore
ауксо·хро́м·, -а,   (*m*) (chem) auxochrome
аукуби́н·, -а,   (*m*) (pharm) aucubin
аукцио́н·, -а,   (*m*) auction
аукцион·и́ст·, -а,   (*m*) auctioneer
аукцио́н·н·ый, -ая, -ое, (*a*) auction
а́ур·а, -ы,   (*f*) (med) aura

аурами́н·, -а, (*m*) (chem) auramine
аура́нци·я, -и, (*f*) (phot) aurantia
аура́т·, -а, (*m*) (chem) aurate
ауре́ли·я, -и, (*f*) (zool) aurelia
аурео·миц́и́н·, -а, (*m*) (pharm) aureomycin
аур·и́д·, -а, (*m*) (chem) auride
аур·и́н·, -а (*m*) (chem) aurine
аурикуля́ри·я, -и, (*f*) (zool) auricularia
аури·пигме́нт·, -а, (*m*) (min) orpiment
аури·хлористо·водоро́д·н·ая кисл·о·т·а́ (*f*) hydrochloramic acid
аурофо́р·, -а, (*m*) (zool) aurofore
аускульт·а́ци·я, -и, (*f*) (med) auscultation, sounding
аустени́т·, -а, (*m*) (met) austenite
~, о·ста́т·очн·ый retained austenite
аустенит·иза́ци·я, -и, (*f*) (met) austenizing, austempering, causing the austenitic transformation
аустени́т·н·ый, -ая, -ое, (*a*) (met) austenitic
а́утбридинг·, -а, (*m*) (gen) outbreeding
аутент·и́ческ·ий, -ая, -ое, (*a*) = **аутент·и́чн·ый**
аутент·и́чн·ый, -ая, -ое, (*a*) authentic
ауто·ге́з·и·я, -и, (*f*) (chem) autohesion
ауто·интоксик·а́ци·я, -и, (*f*) (med) auto-intoxication
ауто·инфе́кци·я, -и, (*f*) (zool) auto-infection
ауто·ка́рп·, -а, (*m*) (bot) autocarp
ауто·ката́·лиз·, -а, (*m*) (zool) auto-catalysis
ауто·оксид·а́ци·я, -и, (*f*) (bot) auto-oxidation
аут·опс·и́·я, -и, (*f*) (anat) autopsy
ауто·те́к·а, -и, (*f*) (bot) autotheca
аутри́ггер·, -а, (*m*) bracing jack
аут·эко·ло́г·и·я, -и, (*f*) (biol) autecology
а·фа́з·и·я, -и, (*f*) (med) aphasia
афани́т·, -а, (*m*) (min) aphanite
афани́т·ов·ый, -ая, -ое, (*a*) (geol) aphanitic
афе́ли·й, -я, (*m*) (astron) aphelion
афел·и́йн·ое рас·сто·я́ни·е (astron) aphelion distance
афе́р·а, -ы, (*f*) dishonest business, swindle
афер·и́ст·, -а, (*m*) swindler
а·филл·и́·я, -и, (*f*) (bot) aphylly
а·фи́р·ов·ый, -ая, -ое, (*a*) (geol) aphyric
афи́ш·а, -и, (*i*) **-ей,** (*f*) poster, placard, bill
а·фле́б·и·я, -и, (*f*) (bot) aphlebia

а·фока́ль·н·ая систе́м·а (*f*) (opt) afocal system, afocal lens
а·фот·и́ческ·ий, -ая, -ое, (*a*) (bot) aphotic
а́фт·а, -ы, (*f*) (med) aphtha, (*pl*) aphthae
афтон·ск·ий, -ая, -ое, (*a*) (geol) Aftonian
аффе́кт·, -а, (*m*) (med) passion
а́ффикс·, -а, (*m*) (gram) affix
аффин·а́ж·, -а, (*i*) **-ем,** (*m*) refining (of rare metals)
аффин·и́ровани·е, -я, (*n*) = **аффин·а́ж·**
аффи́н·н·ый, -ая, -ое, (*a*) (math) affine
~ с·вя́з·ност·ь (*f*) affine connection
~ гео·ме́тр·и·я (math) affine geometry
~ пре·о́браз·ова́ни·е (*n*) (math) affine transformation
ахондри́т·, -а, (*m*) (geol) achondrite
ахондро·плаз·и́·я, -и, (*f*) (med) achondroplasia
ахрои́т·, -а, (*m*) (min) achroite
ахрома́т·, -а, (*m*) achromatic lens
ахромат·изи́рованн·ый, -ая, -ое, (*past part pass*); achromatized
ахромат·и́н·, -а, (*m*) (cyt) achromatin
ахромат·и́ческ·ий, -ая, -ое, (*a*) achromatic
ахро·миц́и́н·, -а, (*m*) (pharm) achromycin
а·хром·и́·я, -и, (*f*) (zool) achromia, achroma
ахтер·лю́к·, -а, (*m*) (shipb) after provision room, barrel store
ахтер·пи́к·, -а, (*m*) (shipb) after peak
ахтер·штевен·ь, -вн·я, (*m*) (shipb) stern post
аце·нафте́н·, -а, (*m*) (chem) acenaphthene, ethylene naphthene
ацета́л·ев·ый, -ая, -ое, (*a*) of **ацет·а́л·ь**
ацет·а́л·ь, -я, (*m*) (chem) acetal, ethylidene di-ethyl-ether
ацет·альдеги́д·, -а, (*m*) (chem) acetaldehyde
ацет·ами́д·, -а, (*m*) (chem) acetamide
ацет·анили́д·, -а, (*m*) (pharm) acetanilide, acetanilid
ацет·арс·о́л·, -а, (*m*) (pharm) acetarsol, acetarsone
ацет·а́т·, -а, (*m*) (chem) acetate
~ свинц·а́ lead acetate
ацета́т·н·ый, -ая, -ое, (*a*) of **ацет·а́т·**
ацетат·целлюло́з·а, -ы, (*f*) cellulose acetate, acetylcellulose

ацет·и́л·, -а, (*m*) (chem) acetyl

ацет·и́л-а́ш-кисл·от·а́, -ы, (*f*) (chem) acetyl H-acid

ацетил·гликоко́л·ь, -я, (*m*) aceturic acid, acetylglycine

ацетиле́н·, -а, (*m*) (chem) acetylene

ацетилен·и́ст·ый, -ая, -ое, (*a*) acetylide of

∼ серебр·о́ (*n*) silver acetylide

ацетиле́н·ов·ый, -ая, -ое, (*a*) acetylene

∼ генера́тор· (*m*) acetylene generator

∼ с·ва́р·к·а (*f*) oxyacetylene welding

ацетил·ен·о-кисло·ро́д·н·ое пла́м·я (*n*) (weld) neutral flame

ацетил·и́д·, -а, (*m*) acetylide

ацетил·и́ровани·е, -я, (*n*) (chem) acetylation, acetylization

ацетил·и́рующ·ий, -ая, -ее, (*pres part act*); acetylating

ацети́л·ов·ый, -ая, -ое, (*a*) of аце-т·и́л·

∼ ан·гидр·и́д· (*m*) acetaldehyde

ацетил·салици́л·ов·ая кисл·от·а́ (*f*) (chem) acetylsalicylic acid

ацетил·холи́н·, -а, (*m*) (chem) acetyl-choline

ацетил·целлюло́з·а, -ы, (*f*) (chem) cellulose acetate, acetylcellulose

ацети́ль·н·ый, -ая, -ое, (*a*) (chem) acetyl

∼ числ·о́ (*n*) (chem) acetyl value

ацет·и́н·, -а, (*m*) (chem) acetin

ацети·фик·а́ци·я, -и, (*f*) acetification, acetic fermentation

ацето·ацетанили́д·, -а, (*m*) acetoacet-anilide

ацет·ои́н·, -а, (*m*) (chem) acetoin

ацет·о́л·, -а, (*m*) (chem) acetone alcohol

ацето́·метр·, -а, (*m*) (chem) acetometer

ацет·о́н·, -а, (*m*) (chem) acetone

ацето́н·ов·ый, -ая, -се, (*a*) (chem) acetone, acetonic

∼ кисл·от·а́ (*f*) (chem) acetonic acid

∼ ма́сл·о (*n*) (chem) acetone oil

ацето́ново·кисл·ый, -ая, -ое, (*a*) acetonate of

ацето·у́ксус·н·ый, -ая, -ое, (*a*) aceto-acetic

∼ эфи́р· (*m*) aceto-acetic ester/ether, ethyl acetoacetate

ацетон·ури́·я, -и, (*f*) (med) acetonuria

ацето·фено́н·, -а, (*m*) (chem) aceto-phenone

ациди·ме́тр·и·я, -и, (*f*) (chem) acidi-metr

ациди́т·, -а, (*m*) acid rock

ацидо́з·, -а, (*m*) (med) acidosis

ацидо́·метр·, -а, (*m*) acidometer, acidimeter

ацидо·фил·и́н·, -а, (*m*) (chem) acido-philin

ацидо·фи́ль·н·ый, -ая, -ое, (*a*) (biol) acidophilous

ацидо·фил·и́·я, -и, (*f*) (biol) acido-philia

а·цикл·и́ческ·ий, -ая, -ое, (*a*) acyclic; (chem) aliphatic

∼ со·един·е́ни·е (*n*) aliphatic compound

аци́л·, -а, (*m*) (chem) acyl

ацил·и́ровани·е, -я, (*n*) (chem) acyla-tion

ач (*abbr*) = ампе́р·-ча́с· (elec) Ahr, ampere-hour

а́ш-кисл·от·а́, -ы, (*f*) (chem) H acid

аше́ль·ск·ий, -ая, -ое, (*a*) (geol) Acheulean, Acheulian

аширит·, -а, (*m*) (min) dioptase, emerald copper

а·ши́ст·ов·ый, -ая, -ое, (*a*) (min) aschistic

аэр·а́тор·, -а, (*m*) aerator

аэр·ацио́нн·ый, -ая, -ое, (*a*) of аэр·а́ци·я

∼ фона́р·ь (*m*) roof louvre ventilator

аэр·а́ци·я, -и, (*f*) natural ventilation, airing; (san eng) aeration; aeration (soil science)

аэр·енхи́м·а, -ы, (*f*) (bot) aerenchyma, aerating tissue

аэр·и́рованн·ый, -ая, -ое, (*past part pass*) aerated

аэро- (*int component*) aero-, aircraft, flying

аэро́·б·, -а, (*m*) (zool) aerobe

аэро·би·о́з·, -а, (*m*) (biol) aerobiosis

аэро́·б·н·ый, -ая, -ое, (*a*) aerobic

аэро·ваго́н·, -а, (*m*) (rail) propeller-driven railcar

аэро·визуа́ль·н·ый, -ая, -ое, (*a*) aerovisual

аэро·вокза́л·, -а, (*m*) terminal building (at airport)

аэро·гамма·радио́·метр·, -а, (*m*) (nucl) airborne gamma-radiometer

аэро·гамма·съ·ём·к·а, -и, (*g pl*) -м·ок·, (*f*) aerial gamma-radiation survey

аэро·ге́н·н·ый, -ая, -ое, (*a*) (bact) aerogenic

аэро·гео·съ·ём·к·а, -и, (*g pl*) -м·ок·, (*f*) aerial geological survey аэро·гидр·а́льн·ый, -ая, -ое, (*a*) aerohydrous

аэро·градио́·метр·, -а, (*m*) airborne gradiometer

аэро·гра́мм·а, -ы, (*f*) (meteor) adiabatic chart

аэро́·граф·, -а, (*m*) air brush, aerograph; (meteor) meteorograph

аэро·дина́м·ик·, -а, (*m*) aerodynamicist

аэро·дина́м·ик·а, -и, (*f*) aerodynamics

~, гипер·звук·ов·ы́х скор·ост·е́й hypersonics

~ идеа́ль·н·ого га́з·а perfect gas aerodynamics

~, не·стацион·а́рн·ая transient/unsteady aerodynamics

~ сверх·звук·ов·ы́х скор·ост·е́й supersonic aerodynamics

~ с·жим·а́ем·ой сред·ы́ compressible aerodynamics

аэро·динам·и́ческ·ий, -ая, -ое, (*a*) aerodynamic; (GW) air-breathing

~ компенс·а́ци·я (dynam) aerodynamic balance

~ о·ши́б·к·а (*f*) (a/c) position error (of pressure instruments)

~ труб·а́ (*f*) (dynam) wind tunnel

~ труб·а́, балло́н·н·ая (*f*) indraft wind tunnel

~ труб·а́, ва́куум·н·ая (*f*) low-pressure/-density wind tunnel

~ труб·а́, до·звук·ов·а́я (*f*) low-speed wind tunnel

~ труб·а́, кратко·вре́мен·н·ого де́й·ств·и·я blow-off/-down wind tunnel

~ труб·а́, сверх·звук·ов·а́я super-sonic wind tunnel

аэро·дро́м·, -а, (*m*) airfield, aerodrome

~, грунт·ов·о́й grass airfield

~, за·пас·н·о́й alternative airfield

~, сухо·пу́т·н·ый landplane airfield

~, трасс·ов·о́й intermediate aerodrome

аэро·дро́м·н·ый, -ая, -ое, (*a*) airfield aerodrome

~ по·лёт· (*m*) (air) local flight

аэро·зо́л·ь, -я, (*m*), or -и, (*f*) aerosol

аэро·зо́нд·, -а, (*m*) (meteor) sounding weather balloon

аэро·клу́б·, -а, (*m*) flying club

аэро·ла́к·, -а, (*m*) aircraft dope

аэро·ли́т·, -а, (*m*) (geol) aerolite

аэро·ли́фт·, -а, (*m*) air-lift pump

аэро́·лог·, -а, (*m*) (meteor) aerologist

аэро·лог·и́ческ·ий, -ая, -ое, (*a*) aerological

аэро·ло́г·и·я, -и, (*f*) (air met) aerology

аэро·магни́т·н·ый, -ая, -ое, (*a*) aeromagnetic, aerial/airborne magnetic

~ съ·ём·к·а (*f*) aeromagnetic survey, air magnetic survey

аэро·магнито́·метр·, -а, (*m*) airborne magnetometer

аэро·мая́к·, -а, (*m*) aerodrome beacon (visual)

аэро́·метр·, -а, (*m*) aerometer

аэро·меха́н·ик·а, -и, (*f*) aeromechanics

аэро́н·, -а, (*m*) aeron (Soviet anti-emetic)

аэро·навигацио́н·н·ый, -ая, -ое, (*a*) of аэро·навиг·а́ци·я

~ ого́н·ь (*m*) aircraft navigation light

аэро·навиг·а́ци·я, -и, (*f*) air navigation

аэро·на́вт·, -а, (*m*) aeronaut

аэро·на́вт·ик·а, -и, (*f*) aeronautics

аэро·но́м·и·я, -и, (*f*) aeronomy

аэро·плён·к·а, -и, (*g pl*) -н·ок·, (*f*) (phot) aerial film

аэро·о·пры́ск·иватель·ь, -я, (*m*) (agr) aerosprayer, spraying aircraft

аэро·о·пы́л·иватель·ь, -я, (*m*) (agr) aeroduster, dusting aircraft

аэро·пла́н·, -а, (*m*) airplane, aeroplane, aircraft

аэро·по́·езд·, -а, (*nom pl*) -а́, (*g pl*) -о́в, (*m*) (rail) propeller-driven rail car

аэро·по́рт·, -а, (*f*) об аэро·по́рт·е, в аэро·порт·у́, (*m*) airport

~ в·лёт·а customs airport

аэро·прое́ктор·, -а, (*m*) aerial-photograph projector

аэро·рас·пыл·и́тел·ь, -я, (*m*) aerial crop duster

аэро·са́н·и (*nom pl*), (*g pl*) -е́й, propeller-driven sledge

аэро·се́·в·, -а, (*m*) (agr) aerial crop sowing, aerosowing

аэро·сидер·и́т·, -а, (*m*) aerosiderite (meteorite)

аэро·с·ни́м·ок·, -м·к·а, (*m*) (*see also* с·ни́м·ок·) aerial photograph

~, анаглиф·и́ческ·ий (phot) aerial anaglyph

~, перспекти́в·н·ый oblique aerial photograph

~, пла́н·ов·ый vertical aerial photograph

~, трансформ·и́рованн·ый rectified aerial photograph

аэро·ста́т·, -а, (*m*) balloon
~ за·гражд·е́ни·я barrage balloon
~, змей·ков·ый kite balloon
~, на·блюд·а́тельн·ый observation balloon
~, при·вяз·н·о́й captive balloon
~, свобо́д·н·ый free balloon
аэро·ста́т·ик·а, -и, (*f*) (mech) aerostatics
аэро·съ·ём·к·а, -и, (*g pl*) **-м·ок·,** (*f*) air survey
аэро·та́ксис·, -а, (*m*) (bot) aerotaxis
аэро·та́нк·, -а, (*m*) (san eng) aeration tank
аэро·те́нк·, -а, (*m*) *see* **аэро·та́нк·**
аэро·термо́·метр·, -а, (*m*) (air) remote indicating thermometer, transmitting thermometer
аэро·термо·терап·и́·я, -и, (*f*) (med) aerothermotherapy
аэро·троп·и́зм·, -а, (*m*) (bot) aerotropism
аэро·у·пру́г·ост·ь, -и, (*f*) (mech) aeroelasticity
аэро·у·пру́г·ий, -ая, -ое, (*a*) aeroelastic
аэро·фа́г·и·я, -и, (*f*) (med) aerophagy
аэро·фи́нишер·, -а, (*m*) arrester wire (aircraft carrier)

аэро·фи́льтр·, -а, (*m*) forced-draft percolating filter (sewage disposal)
аэро·фи́т·, -а, (*m*) (bot) aerophyte, epiphyte
аэро·фло́т·, -а, (*m*) (min) aeroflot (flotation reagent); Aeroflot (Soviet State airline)
аэро·фото·аппара́т·, -а, (*m*) (air) aerial camera, camera
аэро·фото·грам·ме́тр·и·я, -и, (*f*) aerial photogrammetry
аэро·фото·за·тво́р·, -а, (*m*) aerial camera shutter
аэро·фото·монт·а́ж·, -а, (*m*) (phot) mosaic assembly
~, маршру́т·н·ый aerial photographic strip
аэро·фото·раз·ве́д·к·а, -и, (*f*) aerial photographic reconnaissance
аэро·фото·с·ни́м·ок·, -м·к·а· = **аэ-ро·с·ним·ок·,** (q.v.)
аэро·фото·съ·ём·к·а, -и, (*g pl*) **-м·ок·,** (*f*) (*see also* **аэро·с·ни́м·ок·**) aerial photographic survey; aerial photography
аэро·фото·трансформ·а́тор·, -а, (*m*) (surv) multiplex aerial photograph projector
аэс. (*abbr*) = **а́том·н·ая электро·ста́н-ци·я** (elec) atomic power station

Б

б (*abbr*) (telecom) = **бел** b, bel; (*abbr*) (meteor) = **бар** bar; (*abbr*) (nucl) = **барн** barn

Б (*abbr*) (chem) = **Боме** Bé, Beaumé

баб·а, -ы,   (*f*) hammer, ram, monkey, tup, beetle-head (of forging hammers and pile drivers); power winch; peasant woman

баббит·, -а,   (*m*) Babbitt, bearing metal

бабингтонит·, -а,   (*m*) (min) babingtonite

Бабине, компенс·áтор·   (*m*) (opt) Babinet's compensator

~, теорéм·а (*f*)(opt) Babinet's principle

бáб·к·а, -и,   (*g pl*) **бáб·ок·** old woman, grandmother; stock, block; pin (for shaping bars); stanchion; pastern (of horse's hoof); sheaf (of grain)

~, зáд·н·яя   tailstock (of lathe)

~, перéд·н·яя   headstock (of lathe)

Бабо, за·кóн·   (*m*) (phys) Babo's law

бáб·очк·а, -и,   (*g pl*) **··очек·,**   (*f*) (zool) butterfly; (naut) goosewing; bow tie (dress)

~, ноч·н·áя   moth

~, осóтов·ая   (ent) painted lady, *Vanessa cardui*

бабуúн·, -а,   (*m*) (zool) baboon, *Papio*

бавен·ск·ий за·кóн·   (*m*) (cryst) Baveno law

багáж·, -а,   (*m*) luggage, baggage

багáж·ник·, -а,   (*m*) luggage compartment; (a/c) luggage hold

багáж·н·ый, -ая, -ое,   (*a*) of **багáж·**

бáгер·, -а,   (*m*) excavator (for peat); bagger; dredger

багер·мéйстер·, -а,   (*m*) dredger master

багéт·, -а,   (*m*) (build) baguette

багóр·, -гр·á,   (*m*) hook; boat hook; (fish) gaff

бáгр·ить, -ят,   (*imp*) gaff, spear (fish etc.); **багр·úть, -я́т,**   (*imp*) redden, colour/color dark red

багр·овé·ть, -éют,   (*imp*) become/get red, redden

багр·óв·ый, -ая, -ое,   (*a*) dark/blood red

багр·я́н·ый, -ая, -ое,   (*a*) dark/blood red

багýль·ник·, -а,   (*m*) (bot) wild rosemary, *Ledum palustre*

~, болóт·н·ый   (bot) marsh tea, *Ledum palustre*

бадáн·, -а,   (*m*) (bot) bergenia, *Bergenia*

бадделеúт·, -а,   (*m*) (min) baddeleyite

бадéйк·а, -и,   (*g pl*) **-éек·,**   (*f*) (*dim*) of **бадь·я́**

бадриджáн·, -а,   (*m*) *see* **баклажáн·**

бадь·ев·óй, -ая, -ое,   (*a*) of **бадь·я́**

бадь·я́, -й,   (*g pl*) **-д·éй,**   (*f*) tub, bucket, pail

~, водо·от·лúв·н·ая   (min) dewatering bucket

~, по·рóд·н·ая   (min) mucking bucket

~, про·хóд·ческ·ая   (min) shaft bucket

~, с·пуск·áющ·аяся   (min) descending bucket

бадья́н·, -а,   (*m*) (bot) star anise, *Illicum anisatum*

бадья́н·ов·ый, -ая, -ое,   (*a*) of **бадья́н·**

бáз·а, -ы,   (*f*) base, basis; base, depot; (math) datum length, base length, base; wheelbase (motor cars etc.)

~, воéн·н·ая   military depot

~, воéн·н·о-мор·ск·áя   naval base

~ врéмен·и   time base (oscillography etc.)

~, в·ход·н·áя   receiving depot

~, за·вóз·н·ая   issuing depot

~, кито·бóй·н·ая   whale factory ship

~ колёс·   (M/T) wheelbase

~, материáль·н·ая   material resources

~, мéр·н·ая   (naut) speed course

~, нефт·ян·áя   oil installation, petroleum depot

~, плов·ýч·ая   depot ship

~, про·лёт·н·ая   (nucl) path length

~, стáрт·ов·ая   (rocket) launching base

~, сырь·ев·áя   source of raw materials

~ тóпл·ивн·ая   (nucl) fuel resources

базáль·н·ый, -ая, -ое,   (*a*) basal

~ тéл·о   (*n*) (biol) basal body/corpuscle

базáльт·, -а,   (*m*) (geol) basalt

базáльт·ов·ый, -ая, -ое,   (*a*) basaltic

базал·я́рн·ый, -ая, -ое,   (*a*) (zool) basalar

базан·, -а,   (*m*) basil (leather)

базанúт·, -а,   (*m*) (min) basanite

базáр·, -а,   (*m*) market, bazaar

Базéдов·а бол·éзн·ь   (*f*) (med) exophthalmic goitre, Basedow's disease

бази·гáм·и·я, -и,   (*f*) (bot) basigamy

базиди·а́льн·ый, -ая, -ое,   (*a*) (bot) basidial

базѝди·й, -я,   (*m*) = **базѝди·я**

базидио·мице́т·, -а,   (*m*) (bot) basidiomycete

базидио·но́с·ец·, -с·ц·а,   (*m*) (bot) basidiophore

базидио·спо́р·а, -ы,   (*f*) (bot) basidiospore

базидио·фо́р·, -а,   (*m*) (bot) basidiophore

базѝди·я, -и,   (*f*) (bot) basidium

базили́к·, -а,   (*m*) (bot) basil, *Ocimum*

базили́к·а, -и,   (*f*) (arch) basilica

базили́к·ов·ый, -ая, -ое,   (*a*) of **базили́к·**; of **базили́к·а**

базѝл·ь, -я,   (*m*) basil (leather)

баз·и́р·овать, -уют, (*imp*) base, found, ground

ба́зис·, -а,   (*m*) basis, base; base/datum line/length/level; (cryst) basal plane

~, орто·гон·а́льн·ый   (math) orthogonal basis

ба́зис·н·ый, -ая, -ое,   (*a*) of **ба́зис·**

~ волн·а́   (*f*) (nucl) base surge

~ ли́н·и·я   (*f*) base line

~ пло́ск·ост·ь   (*f*) (cryst) basal plane

~ при·бо́р·   (*m*) (surv) taping equipment

~ се́т·ь   (*f*) (surv) triangulation system

бази́т·, -а,   (*m*) basic rock

бази·фи́ль·н·ый, -ая, -ое,   (*a*) (bot) basophilic, basophile, basiphilous

ба́з·ов·ый, -ая, -ое,   (*a*) of **ба́з·а**

~ ли́н·и·я   (*f*) base line

~ на·гру́з·к·а   (*f*) (elec) basic loading

~ тра́ль·щик·   (*m*) (naut) coastal minesweeper

базо·педио́н·, -а,   (*m*) (cryst) basopedion

базо·рас·сто·я́ни·е, -я,   (*n*) base length (machining)

базо·фи́л·, -а,   (*m*) (cyt) basophil(e), basiphil(e)

базо·фил·и́·я, -и,   (*f*) (biol) basophilia

базо·фи́ль·н·ый, -ая, -ое,   (*a*) (bot) basophylic

байба́к·, -а,   (*m*) (zool) bobac, steppe marmot, *Marmota bobac* Pall.

ба́йд·а, -ы,   (*f*) baida (fishing vessel)

байда́к·, -а,   (*m*) baidak (river boat)

байда́р·к·а, -и,   (*g pl*) **-р·ок·,**   (*f*) kayak; canoe

Ба́йер·а, спо́соб·   (*m*) (met) Bayer process

ба́йк·а, -и,   (*f*) (text) baize

бай калѝт·, -а,   (*m*) (min) baikalite

байкери́т·, -а,   (*m*) (min) Baykal ozokerite

байпа́с·, -а,   (*m*) by-pass valve

байпа́с·н·ый, -ая, -ое,   (*a*) by-pass

~ регул·и́ровани·е   (*n*) by-pass governing (of turbine)

~ шѝн·а   (*f*) (elec) auxiliary bus bar

ба́к·, -а,   (*m*) tank, container, cistern; (obs) forecastle (of ship); (mil) mess tin, fanny

~ актѝв·н·ой зо́н·ы   (nucl) core vessel/tank

~, бензѝн·ов·ый   gasoline/petrol tank

~, ва́куум·н·ый   depression tank

~, вертика́ль·н·ый   (phot) deep tank

~, водо·ме́р·н·ый   water-measuring tank

~, в·стро́·енн·ый   integral tank

~ вы́·держ·к·и   holding tank

~, за·щѝт·н·ый   (nucl) shield tank

~, консо́ль·н·ый   (a/c) wing-tip tank

~, краско·на·гнет·а́тельн·ый   paint container (for sprayer)

~, мя́гк·ий   (a/c) flexible tank

~, на·гру́ж·енн·ый   pressure tank

~ -от·се́к·, -а,   (*m*) integral tank

~, пит·а́тельн·ый   supply tank

~, протект·и́рованн·ый   bulletproof tank

~, про·я́в·очн·ый   (phot) developing tank

~, рабо́ч·ий   (a/c) feeder tank

~, рас·хо́д·н·ый   (mar eng) gravity tank, day tank; (a/c) feeder tank

~, рас·шир·и́тельн·ый   expansion tank; (st eng) flash tank

~, с·бра́с·ываем·ый   (a/c) drop tank

~, с·лив·н·о́й   (hydr) receiver tank; (nucl) dump tank

~, у·равн·и́тельн·ый   surge tank

~, фикса́ж·н·ый   (phot) hypo tank, fixing tank

бакала́вр·, -а,   (*m*) bachelor (holder of degree at British university); baccalaureate (qualification at French secondary schools)

бакале́й·н·ый, -ая, -ое,   (*a*) grocery

бакале́·я, -и,   (*f*) grocery, groceries

ба́кан·, -а,   (*m*) (naut) buoy; **бака́н·, -а,**   (*m*) lake (pigment)

бака́ут·, -а,   (*m*) lignum vitae (wood)

бакбо́рт·, -а,   (naut) port side

бакел·изѝрованн·ый, -ая, -ое,   (*a*) bakelite impregnated

бакели́т·, -а,   (*m*) (plast) bakelite

бакели́т·ов·ый, лак·   (*m*) bakelite varnish

ба́кен·, -а, (*m*) beacon; buoy (in river)

~, по·сто·я́нн·ый (air) fixed beacon

баки́нск·ий, -ая, -ое, (*a*) Baku

бакла́г·а, -и, (*f*) flask

баклажа́н·, -а, (*m*) (bot) eggplant, *Solanum melongena*

бакла́н·, -а, (*m*) (orn) cormorant

ба́к·ов·ый, -ая, -ое, (*a*) of **бак·**; (*m decl as adj*) forecastle man; bowman

~ вы·ключ·а́тел·ь (*m*) (electr) bulk-oil circuit breaker

ба́кс·, -а, (*m*) (naut) forefoot

бактери·а́льн·ый, -ая, -ое, (*a*) bacterial, bacteritic

бакте́ри·и (*f pl*) *see* **бакте́ри·я** bacteria

бактери́йн·ый, -ая, -ое, (*a*) bacterial

бактери·о́з·, -а, (*m*) (bot) bacteriosis, (*pl*) bacterioses

бактерио́·лиз·, -а, (*m*) (bact) bacteriolysis

бактерио·лизи́н·, -а, (*m*) bacteriolysin

бактерио́·лог·, -а, (*m*) bacteriologist

бактерио·лог·и́ческ·ий, -ая, -ое, (*a*) bacteriological

бактерио·ло́г·и·я, -и, (*f*) bacteriology

бактерио·пурпури́н·, -а, (*m*) bacterio-purpurin (pigment)

бактерио·скоп·и́·я, -и, (*f*) bacterioscopy

бактерио·троп·и́н·, -а, (*m*) (pharm) bacteriotropin

бактери·ури́·я, -и, (*f*) (med) bacteriuria

бактерио·фа́г·, -а, (*m*) bacteriophage

бактери·ци́д·, -а, (*m*) bactericide

бактери·ци́д·н·ый, -ая, -ое, (*a*) bactericidal

бактери·эми́·я, -и, (*f*) (med) bacteriaemia, bacteriemia

бакте́ри·я, -и, (*f*) bacterium

бактер·о́ид·, -а, (*m*) bacteroid

баку́н·, -а, (*m*) bakun (type of cheap tobacco)

бакшта́г·, -а, (*m*) (naut) backstay; reach (sailing)

~, крут·о́й reach (sailing)

~, по́лн·ый broad reach (sailing)

бакшто́в·, -а, (*m*) (naut) stern boatrope

ба́л·, -а, (*p*) **о ба́л·е, на бал·у́,** (*m*) ball (entertainment)

балано·гло́сс·, -а, (*m*) (zool) Balanoglossus

балано·пости́т·, -а, (*m*) (med) balanoposthitis

балано·фо́р·, -а, (*m*) (bot) balanophore, (*pl*) balanophores, *Balanophora*

балано·фо́р·ов·ые, -ых, (*pl decl as adj*) (bot) *Balanophoracea*

бала́нс·, -а, (*m*) balance; (fin) balance sheet, balance; pulpwood

~, акти́в·н·ый торг·о́в·ый surplus, favourable trading balance

~, газ·ов·о́й gas balance

~, плат·ёжн·ый (econ) balance of payments

~, с·во́д·н·ый (fin) balance sheet

~ -стан·о́к·, -н·к·а́, (*m*) balance stand (for testing engines)

~, тепл·ов·о́й heat balance

~, цвет·ов·о́й colour/color balance

баланс·ёр·, -а, (*m*) tightrope walker; (zool) balancer, haltere

баланс·и́р·, -а, (*m*) balance, balancer, balance arm, rocker, rocking lever; walking beam; (min) balance bob; (rad) balancer

~, по·переч·н·ый (M/T) lateral axle, lateral arm

~, про·до́ль·н·ый longitudinal beam

баланс·и́рн·ый, -ая, -ое, (*a*) of **баланс·и́р·;** balanced, balance

~ дви́г·ател·ь (*m*) balanced engine/motor

~ под·ве́с·к·а (*f*) (M/T) suspension

~ тра́верс·а (*f*) lifting beam

баланс·и́ровани·е, -я, (*n*) (*v n*) *see* **баланс·и́р·овать;** compensating

баланс·и́рованн·ый, -ая, -ое, (*past part pass*) *see* **баланс·и́р·овать;** compensated

баланс·и́р·овать, -у́ют, (*imp*) balance, keep one's balance; (fin) strike a balance, balance; (mech eng) balance

баланс·иро́в·к·а, -и, (*g pl*) **-в·ок·,** (*f*) (*v n*) *see* **баланс·и́р·овать;** (a/c) trim, trimming

~, вес·ов·а́я mass balancing

~, динам·и́ческ·ая (air) dynamic trim

~, стат·и́ческ·ая static balancing

у́гол· баланс·иро́вк·и (*m*) (air) trim angle

баланс·иро́вочн·ый, -ая, -ое, (*a*) balancing

~ стан·о́к· (*m*) balance stand (for testing engines); (M/T) wheel balancing machine; (elec) balancer

бала́нс·н·ый, -ая, -ое, (*a*) balance, balancing, balanced; (elec) compensating, balanced

~ ли́н·и·я (*f*) (elec) compensating circuit

бала́нс·н·ый, -ая -ое
~ **систе́м·а** (*f*) (autom) null-seeking system
бала́нс·ов·ый, -ая, -ое, (*a*) of **бала́нс·**
~ **пре·образ·ова́тел·ь** (*m*) (elec) balanced converter
~ **древес·и́н·а** (*f*) pulpwood
балансо·ме́р·, -а, (*m*) (met) radiation balance gauge
балантиди·а́з·, -а, (*m*) (med) balantidiasis
бала́с·, -а, (*m*) (min) balas ruby
бала́т·а, -ы, (*f*) balata (a latex)
балд·а́, -ы́, (*f*) sledge hammer, hand sledge
балдахи́н·, -а, (*m*) (arch) baldachin, baldaquin
балет·ме́йстер·, -а, (*m*) (theat) ballet master
бале́т·н·ый, -ая, -ое, (*a*) (theat) ballet, balletic
бале́т·чик·, -а, (*m*) (theat) ballet dancer (male)
ба́лз·а, -ы, (*f*) balsa (wood)
балини́т·, -а, (*m*) balinite (a laminated wood)
бали́ст·ик·а, -и, (*f*) *see* **балли́ст·ик·а**
балист·и́т·, -а, (*m*) (expl) balistite
ба́л·к·а, -и, (*g pl*) **ба́л·ок·,** (*f*) beam, girder; (naut) davit; (geog) gorge, gully, ravine
~, **а́нкер·н·ая** tie-beam
~, **вре́мен·н·ая** (rocket) jury beam
~, **второ·степе́н·н·ая** joist, secondary beam
~, **выс·о́к·ая** deep beam
~, **дву·тавр·о́в·ая** joist, H-beam, I-beam
~ **жёстк·ост·и** (civ eng) horizontal stiffening girder; truss bar (of plough)
~, **за·де́л·анн·ая** fixed beam, restrained beam
~, **зе́т·ов·ая** zed section, zee beam
~, **кил·ев·а́я** (shipb) duct keel
~, **консо́ль·н·ая** cantilever beam, overhanging beam
~, **коро́б·чат·ая** box girder/beam
~, **кра́н·ов·ая** crane jib
~, **криво·лин·е́йн·ая** camber beam
~, **много·о·по́р·н·ая** continuous beam
~ **на·сти́л·а** (civ eng) joist
~, **не·раз·ре́з·н·ая** continuous beam
~, **одно·про·лёт·н·ая** single beam
~ **пере·кры́т·и·я** floor beam
~, **по·пере́ч·н·ая** (M/T) cross-member; (a/c) transverse girder

ба́л·к·а
~, **про·до́ль·н·ая** (M/T) side-member; (a/c) stringer
~, **пя́т·ов·а́я** skewback
~, **раз·ре́з·н·а́я** simply supported beam
~, **с·дво́·енн·ая** double girder
~, **сквоз·н·а́я** continuous beam
~, **со·ста́в·н·а́я** (civ eng) compound girder
~, **с·ре́з·анн·ая** coped beam
~, **тавр·о́в·ая** T-beam
~, **тонко·сте́н·н·ая** thin-web beam
~, **хвост·ов·а́я** tail boom (of a helicopter)
~, **широко·по́л·очн·ая** wide flange beam
~, **шпанго́ут·н·ая** transverse girder
балко́н·, -а, (*m*) balcony; shelf (of dry dock etc.)
балко́н·н·ый, -ая, -ое, (*a*) of **балко́н·**
ба́лл·, -а, (*m*) mark, number, point, degree; force (of wind)
~ **Бо́форт·а** (meteor) Beaufort number, wind force
~ **о́блач·ност·и** (meteor) cloud amount
~ **си́л·ы** (seismol) degree of intensity (measured on 12-degree scale)
ба́ллас·, -а, (*m*) ballas (industrial diamond)
балла́ст·, -а, (*m*) ballast
~ **то́пл·ив·а** inert constituent (of a fuel)
балласт·ёр·, -а, (*m*) (rail) ballasting machine
балласт·и́р·овать, -уют, (*imp*) ballast
балласт·иро́вк·а, -и, (*g pl*) **-в·ок·,** (*f*) ballasting
балла́ст·н·ый, -ая, -ое, (*a*) ballast
~ **сло·й** (*m*) (rail) ballast bed, ballast depth
~ **реоста́т·** (*m*) (elec) ballast resistor, thermistor
~ **систе́м·а** (*f*) (naut) ballast system
~ **со·противл·е́ни·е** (*n*) (elec) ballast
балласто·о·чи́ст·ител·ь, -я, (*m*) (rail) ballast cleaner
ба́ллер·, -а, (*m*) spindle, barrel (of capstan); head (of rudder)
балли́ст·ик·, -а, (*m*) ballistician
балли́ст·ик·а, -и, (*f*) ballistics
~ **больш·и́х энерг·и·й** high-energy ballistics
~, **вне́ш·н·яя** external ballistics
~, **вну́тр·енн·ая** internal ballistics

баллист·ик·а
~ **ка́пел·ь** droplet ballistics
~ **кон·е́чн·ых скор·ост·е́й** terminal ballistics
~ **про·ник·нове́ни·я** penetration ballistics
баллист·и́т·, -а, (*m*) (expl) ballistite
баллист·и́ческ·ий, -ая, -ое, (*a*) ballistic
~ **досто́ин·ств·о** (*n*) ballistic advantage
~ **ка́честв·а** (*pl*) (expl) ballistic properties
ба́лл·ов·ый, -ая, -ое, (*a*) of **балл·**
балло́н·, -а, (*m*) compressed gas cylinder, gas bottle; gas balloon (storage); (rocket) tank; (elec) bulb, envelope; (M/T) balloon tire, low pressure tire
  образ·ова́ни·е балло́н·а (*n*) (text) ballooning
~, **пуск·ов·о́й** starting air bottle
~, **Целля́риус·а** (chem eng) Cellarius tourill
баллоне́т·, -а, (*m*) ballonet (in airship)
балло́н·чик·, -а, (*m*) (dim) of **балло́н·**
баллот·и́р·овать, -ую́т, (*imp*) vote, ballot; **-ся** (*pass*) stand for, run for, be a candidate
ба́л·очк·а, -и, (*f*) (dim) of **ба́л·к·а**
~, **по́д·ов·ая** firing floor beam (of furnace)
ба́л·очн·ый, -ая, -ое, (*a*) of **ба́л·к·а**
~ **спо́соб·** (*m*) beam-rolling method (for channels)
балошник·, -а, (*m*) baulk (timber)
балт·и́ец·, -и́йц·а, (*i*) **-·и·ц·ем,** (*m*) Balt
балт·и́йск·ий, -ая, -ое, (*a*) Baltic
~ **щит·** (*m*) (geol) Baltic shield
балу́н·, -а, (*m*) (naut) spinnaker
~ **-кли́вер·, -а,** (*m*) (naut) balloon jib
балы́к·, -а́, (*m*) dried fish
балык·о́в·ый, -ая, -ое, (*a*) of **балы́к·**
бальза́м·, -а, (*m*) (chem) balsam
~, **кана́д·ск·ий** (chem) Canada balsam, balsam of fir
~, **перу·а́нск·ий** balsam of Peru
~, **толу·а́нск·ий** balsam of Tolu
~ **шостако́вск·ого** vinylene, polyvinylbutyl ester
бальзам·и́н·ов·ые, -ых, (*pl decl as adj*) (bot) *Balsaminaceae*
бальзам·и́рован·и·е, -я, (*n*) embalming
бальзам·и́ческ·ий, -ая, -ое, (*a*) balsamic

бальза́м·н·ый, -ая, -ое, (*a*) balsam, balsamous
бальзамо·но́с·н·ый, -ая, -ое, (*a*) balsamiferous
ба́льм·а, -ы, (*f*) (geol) rock overhang/shelter
Ба́льмер·а, се́р·и·я (spectr) Balmer series
бальнео́·лог·, -а, (*m*) (med) balneologist
бальнео·лог·и́ческ·ий, -ая, -ое, (*a*) balneological
бальнео·ло́г·и·я, -и, (*f*) (med) balneology
бальнео·терап·и́·я, -и, (*f*) (med) balneotherapy
ба́ль·ност·ь, -и, (*f*) magnitude, intensity (as measured on a scale)
ба́ль·н·ый, -ая, -ое, (*a*) of **балл·**; of **бал·**
балюстра́д·а, -ы, (*f*) (arch) balustrade
баля́син·а, -ы, (*f*) (arch) baluster
бамбу́з·а, -ы, (*f*) (bot) bamboo, *Bambusa*
бамбу́к·, -а, (*m*) (bot) bamboo, *Bambusa*
бамбу́к·ов·ый, -ая, -ое, (*a*) bamboo
~ **медве́д·ь** (*m*) (zool) giant panda, *Ailuropodus melanoleucus*
ба́ми·я, -и, (*f*) (bot) okra, gumbo, *Hibiscus esculentus*
бана́ль·н·ый, -ая, -ое, (*a*) banal, commonplace
бана́н·, -а, (*m*) (bot) banana (fruit or tree), *Musa*
бана́н·ов·ый, -ая, -ое, (*a*) of **бана́н·** (*pl*) (bot) *Musaceae*
бана́х·ов·ое про·стра́н·ств·о (*n*) (math) Banach space
банда́ж·, -а́, (*i*) **-о́м,** (*m*) (med) truss, belt; (eng) band, banding, shrouding strip, strapping; (rail etc.) tire, tyre
бандаж·и́ровани·е, -я, (*n*) banding
банда́ж·н·ый, -ая, -ое, (*a*) of **банда́ж·**
~ **вы́·ступ·** (*m*) tenon (of turbine)
~ **желе́з·о** (*n*) hoop/band iron
~ **кольц·о́** (*n*) (elec) commutator shrink ring
~ **ле́нт·а** (*f*) shrouding strip (of turbine)
~ **ста́л·ь** (*f*) (rail) tire steel
бандеро́л·ь, -и, (*f*) postal wrapper (for journals etc.); printed matter (postal classification)
ба́н·ить, -ят, (*imp*) wash, wash out, sponge out

банк·, -а, *(m)* (fin) bank

~, акционе́р·н·ый joint-stock bank

~, рас·чёт·н·ый clearing bank

~, у·чёт·н·ый discount bank

бан·к·а, -и, *(g pl)* **-н·ок·,** *(f)* can, jar, pot, tin; (hydr) bank, shoal; (naut) thwart (of boat)

~, апте́ч·н·ая (pharm) gallipot

~, бомба́ж·н·ая (food) blown can

~, за·гре́б·н·ая (naut) stroke thwart

~, крове·со́с·н·ая (med) wet cup

~, лаборато́р·н·ая phial

~, ле́йден·ск·ая (elec) Leyden jar

~, песч·а́н·ая sandbank

~, с·бо́р·н·ая built-up can

~, у́стрич·н·ая oyster bed

банкавро́ш·, -а, *(m)* (text) fly/flyer frame

~, втор·о́й то́нк·ий (text) fine roving frame

~, пере·го́н·н·ый (text) intermediate frame

~, та́з·ов·ый slubbing frame

~, то́лст·ый slubbing frame

~, то́нк·ий roving frame

банке́т·, -а, *(m)* banquet, dinner; (min) banket; (civ eng) banquette

банк·и́р·, -а, *(m)* (fin) banker

банк·но́т·, -а, *(m)* banknote

банк·но́т·а, -ы, *(f)* banknote

ба́нк·овск·ий, -ая, -ое, *(a)* (fin) bank, banking, banker's

ба́нк·ов·ый, -ая, -ое, *(a)* bank, banking

банк·ро́т·, -а, *(m)* bankrupt

банкро́т·иться, -ятся, *(imp)* become bankrupt

ба́нник·, -а, *(m)* brush, cleaner, swab (for tubes)

ба́нник-про·тир·а́ль·ник·, -а, *(m)* piasaba brush

ба́н·очк·а, -и, *(g pl)* **--очек·,** *(f)* *(dim)* of **ба́н·к·а** (q.v.); parison (glass-making)

ба́н·очн·ый, -ая, -ое, *(a)* of **ба́н·к·а**

ба́нт·, -а, *(m)* (naut) knot, bow; patch, banding (on a sail); *(pl)* shrouds

Ба́нти, бол·е́зн·ь *(f)* (med) Banti's disease

ба́н·я, -и, *(g pl)* **бань·,** *(f)* bath; Russian bath, steam bath; bath house

~, вод·ян·а́я steam bath

~, песо́ч·н·ая sand bath

баоба́б·, -а, *(m)* (bot) baobab, *Adansonia digitata*

ба́р·, -а, *(m)* bar (unit of pressure); bar (harbour, river etc.); bar (refreshments); (min) jib (of cutter)

~, за·ру́б·н·ой cutting jib

~, ни́ж·н·ий floor jib

бараба́н·, -а, *(m)* drum, reel, trommel, barrel, roll, roller, cylinder

~, би́ль·н·ый (agr) beater drum

~, вод·ян·о́й water drum (of boiler)

~ выс·от·ы́ height drum (on airborne radar)

~, гла́д·к·ий smooth drum; recording drum (recording machine); take-off drum (sound reproducing machine)

~, гравие·мо́ечно-сорт·иро́вочн·ый wet gravel-screening trommel

~, дистанц·ио́нн·ый (gunn) range drum

~, до·полн·и́тельн·ый (horol) fusee

~, жёстк·ий (shipb) ring bulkhead

~, звук·ов·о́й recording drum (sound recording machine); sound take-off drum (sound reproducing machin e)

~, зуб·н·о́й sprocket

~, зуб·ча́т·ый sprocket

~, молот·и́ль·н·ый (agr) thrashing drum

~, на·мо́т·очн·о-на·тя́ж·н·ый (met roll) power driven reels

~, нож·ев·о́й (agr) awner (of thresher); (text) porcupine cylinder (of cotton bale opener)

~, пар·ов·о́й steam drum (of boiler)

~, печа́т·ающ·ий (cinema) printing sprocket

~, при·ём·н·ый (text) take-up roller, licker-in

~, при·ним·а́ющ·ий take-up drum/sprocket

~ с ка́бел·ем (elec) cable drum

~, тарт·а́льн·ый (oil) bailing drum

~, тормоз·н·о́й brake drum

~, цёв·очн·ый spool drum

~ шпи́л·я capstan barrel

~, штифт·ов·ый (agr) peg-type drum

бараба́н·н·о-фре́зер·н·ый стан·о́к· drum-type milling machine

бараба́н·н·ый, -ая, -ое, *(a)* of **бараба́н·**

~ гро́хот· *(m)* (min) trommel

~ перепо́н·к·а *(f)* (anat) ear drum, tympanic membrane

~ по́л·ост·ь *(f)* (anat) middle ear cavity

бара́ж·, -а, *(m)* *see* **барра́ж·**

бара́к·, -а, *(m)* barrack hut

бара́н·, -а, *(m)* ram; *(pl)* rams, sheep

бара́н·ий, -ья, -ье,   (*a*) of **бара́н·**; sheepskin; mutton

бара́н·ин·а, -ы,   (*f*) mutton

бара́ч·н·ый, -ая, -ое,   (*a*) of **бара́к·**

бара́ш·ек·, -ш·к·а,   (*m*) (*dim*) of **бара́н·**; lamb, lambkin; wing nut, fly nut; (*pl*) kink (in wire); (meteor) fleecy clouds; (naut) white horses

бара́ш·ков·ый, -ая, -ое,   (*a*) of **бара́ш·ек·**; lambskin

барбари́с·, -а,   (*m*) (bot) barberry, *Berberis*

барбе́т·, -а,   (*m*) (mil) barbette

барбитур·а́т·, -а,   (*m*) barbiturate

барбиту́р·ов·ая кисл·от·а́   (*f*) (chem) barbituric acid

барбот·а́ж·, -а,   (*m*) bubbling, sparging

барбот·а́ж·н·ый, -ая, -ое,   (*a*) of **барбот·а́ж·**

~ систе́м·а   (*f*) air-lift system

барбот·ёр·, -а,   (*m*) (chem eng) bubble-type emulsifier, sparger

баръоти́н·, -а,   (*m*) (ceram) barbotine

барбот·и́ровани·е, -я,   (*n*) = **барбот·а́ж·**

барбьери́т·, -а,   (*m*) (min) barbierite

бард·а́, -ы́,   (*f*) slop (distillery waste)

~, пив·н·а́я   brewery mash

барелье́ф·, -а,   (*m*) (arch) bas-relief

баре́ттер·, -а,   (*m*) (elec) barretter

ба́рж·а, -и, (*i*) -ей, (*g pl*) барж·,   (*f*) (naut) barge

~, на·лив·н·а́я   tank barge

~, не·само·хо́д·н·ая   dumb barge

~, сухо·гру́з·н·ая   bulk-cargo barge

ба́рж·ев·ый, -ая, -ое,   (*a*) of **ба́рж·а**

барза́с·ск·ий у́гол·ь   (*m*) liptobiolithic coal

ба́р·иев·ый, -ая, -ое,   (*a*) baric, barium

ба́р·и·й, -я,   (*m*) (chem) barium, Ba

~, е́дк·ий   barium hydroxide

~, серно·ки́сл·ый   barium sulphate/sulfate

~, хло́р·ист·ый   barium chloride

~, хромо·ки́сл·ый   barium chromate

барилье́т·, -а,   (*m*) hydraulic main (of gas producer)

барио́н·, -а,   (*m*) (phys) baryon, barion

барио́н·н·ый за·ря́д·   (*m*) (nucl) barion charge

бар·и́рованн·ый, -ая, -ое,   (*past part pass*); barium-coated

бари·сфе́р·а, -ы,   (*f*) (geol) barysphere

бар·и́т·, -а,   (*m*) (min) barytes, barite, heavy spar

барит·а́ж·, -а,   (*m*) baryta coating, barium oxide coating

барит·и́ровани·е, -я,   (*n*) = **барит·а́ж·**

бари́т·ов·ый, -ая, -ое,   (*a*) of **бар·и́т·**; barium

~ вод·а́   (*f*) (chem) baryta water

~ бел·и́л·а   (*pl*) barium white

барито·кальц·и́т·, -а,   (*m*) (min) barytocalcite

бари·центр·и́ческ·ий, -ая, -ое,   (*a*) (math) baricentric

бар·и́ческ·ий, -ая, -ое,   (*a*) barometric pressure

~ клин·   (*m*) (meteor) ridge of high pressure

~ лож·би́н·а   (*f*) trough of low pressure

~ от·ро́г·   (*m*) (meteor) ridge of high pressure

~ седл·о́   (*n*) (meteor) col, saddleback

~ седл·ови́н·а   (*f*) (meteor) col, saddleback

~ систе́м·а   (*f*) (meteor) pressure system

~ тенде́нци·я   (*f*) barometric tendency

ба́рк·, -а,   (*m*) (naut) bark, barque

ба́р·к·а, -и, (*g pl*) ба́р·ок·,   (*f*) (chem) beck, bark, vat; (naut) wooden lighter

барка́з·, -а,   (*m*) (naut) launch, long boat

барка́с·, -а,   (*m*) *see* **барка́з·**; river barge

Баркга́узен·а, колеб·а́ни·я   (*n pl*) (rad) Barkhausen–Kurz oscillations

~, эффе́кт·   (*m*) Barkhausen effect (ferromagnetism)

баркенти́н·а, -ы,   (*f*) (naut) barquentine

барко́·метр·, -а,   (*m*) barkometer

Ба́рлоу, ре́й·к·а   (*f*) (surv) Barlow levelling rod

ба́рн·, -а,   (*m*) (nucl) barn (unit of area)

Ба́рнетт·а, эффе́кт·   (*m*) (phys) Barnett effect

баро·гра́мм·а, -ы,   (*f*) barograph trace, barogram; (a/c) climb curve

баро́·граф·, -а,   (*m*) (meteor) barograph

баро·ка́мер·а, -ы,   (*f*) (air) altichamber, altitude chamber

~, ва́куум·н·ая   decompression chamber

~, компрессио́н·н·ая   compression chamber

баро́кко   (*n indecl*) (arch) baroque, barocco

баро·кли́н·н·ый, -ая, -ое,   (*a*) baroclinic

баро·кли́н·ност·ь, -и,   (*f*) (meteor) baroclinic distribution

баро·ключ·а́тел·ь, -я, (*m*) barometric switch, baroswitch

баро́·метр·, -а, (*m*) barometer

~, металл·и́ческ·ий non-liquid barometer, aneroid

~, норма́ль·н·ый standard barometer

~, рту́т·н·ый mercury barometer

~, сифо́н·н·ый syphon barometer

~ Форте́н·а Fortin barometer

~, ча́ш·ечн·ый cistern barometer

баро·метр·и́ческ·ий, -ая, -ое, (*a*) barometric

баро·регул·и́рующ·ий ве́нтил·ь (*m*) barostat

баро·реле́ (*n indecl*) barometric pressure switch

баро·ско́п·, -а, (*m*) baroscope, open-end manometer

баро·ста́т·, -а, (*m*) barostat

баро·стат·и́ческ·ий, -ая, -ое, (*a*) barometric pressure

баро·термо́·граф·, -а, (*m*) (meteor) barothermograph

баро·термо́·метр·, -а, (*m*) hypsometer

баро·тро́п·н·ый, -ая, -ое, (*a*) barotropic

баро·тро́п·ност·ь, -и, (*f*) barotropic distribution

ба́р·очник·, -а, (*m*) bargee

ба́р·очн·ый, -ая, -ое, (*a*) of **ба́р·к·а;** (naut) barque; (chem) dyeing vat; **баро́ч·н·ый, -ая, -ое,** (*a*) of **баро́кко** (arch) baroque, barocco

~ сти́л·ь (*m*) (arch) baroque, barocco

барра́ж·, -а, (*i*) **-ем,** (*m*) barrage; (naut) boom defence

барраж·и́ровани·е, -я, (*n*) making a barrage; (air) standing patrol

барра́ж·н·ый, -ая, -ое, (*a*) of **барра́ж·**

барра·тро́н·, -а, (*m*) (elec) barratron

ба́ррел·ь, -я, (*m*) (oil) barrel (unit of measurement)

баррика́д·а, -ы, (*f*) barricade

ба́рс·, -а, (*m*) (zool) panther

барсу́к·, -а́, (*m*) (zool) badger, *Meles meles*

барсук·о́в·ый, -ая, -ое, (*a*) of **барсу́к·**

барсу́ч·ий, -ья, -ье, (*a*) of **барсу́к·**

Ба́рт·а реа́кци·я (*f*) (chem) Bart reaction

бархан·, -а, (*m*) (geog) barkhan, isolated sand dune

ба́рхат·, -а, (*m*) (text) velvet

~, бума́ж·н·ый (text) velveteen

бархат·и́ст·ый, -ая, -ое, (*a*) velvety; (phot) velvet

ба́рхат·к·а, -и, (*g pl*) **-т·ок·,** (*f*) velvet ribbon

ба́рхат·н·ый, -ая, -ое, (*a*) velvet, velvety

~ де́рев·о (*n*) (bot) Amur cork tree, *Phillodendron amurense*

~ руда́ (*f*) (min) cyanotrichite

бархо́ут·, -а, (naut) rubbing strip/strake

барье́р·, -а, (*m*) barrier, obstacle

~, авар·и́йн·ый crash barrier

~, га́мов·ск·ий (nucl) Gamow barrier

~, звук·ов·о́й sound/sonic barrier

~, лед·о́в·ый ice barrier

~, потенциа́ль·н·ый (elec) potential barrier

~, тепл·ов·о́й thermal/heat barrier

~, я́дер·н·ый nuclear barrier

барье́р·н·ый, -ая, -ое, (*a*) of **барье́р·**

~ се́т·к·а (*f*) (elec) barrier grid

ба́с·, -а, (*nom pl*) **-ы́,** (*g pl*) **-о́в,** (*m*) (acous) bass

бас·и́ст·ый, -ая, -ое, (*a*) bass, deep-sounding

бас·о́в·ый, -ая, -ое, (*a*) (acous) bass

басо́н·, -а, (*m*) (text) braid, trimming

басо́н·н·ая маши́н·а (*f*) (text) small-ware loom

бассе́йн·, -а, (*m*) basin; pond, pool; dock basin; swimming pool; (arch) ornamental water, pool

~, аккумул·и́рующ·ий (hydro elec) storage basin

~, бес·с·то́ч·н·ый (geog) inland/interior drainage basin

~, брызг·а́льн·ый (thermo elec) spray cooling pond, spray pond

~, водо·ла́з·н·ый diving tank

~, водо·с·бо́р·н·ый (geog) catchment basin

~ вы́·держ·к·и (nucl) cooling pond

~, газо·но́с·н·ый (oil) gas-bearing basin

~, до·стро́·ечн·ый (shipb) fitting-out basin

~, на·по́р·н·ый (hydro elec) forebay

~, океан·и́ческ·ий ocean basin

~, о́·пыт·ов·ый (shipb) testing/experimental tank, model basin

~, от·кры́·т·ый open-air swimming pool

~, при·ли́в·н·ый (civ eng) tidal basin

~ реа́ктор·а (nucl) reactor pool

~, с·пуск·ов·о́й (shipb) launching basin

бассейн

~, судо·стро·и́тельн·ый　　(shipb) fitting-out basin

~, у́голь·н·ый　　coal basin, group of coalfields

бассейн·ов·ый, -ая, -ое,　　(a) of **бассейн·**

~ реа́ктор·　　(m) (nucl) swimming-pool reactor

бассети́т·, -а,　　(m) (min) bassetite

бастио́н·, -а,　　(m) (build) bastion

басти́т·, -а,　　(m) (min) bastite

баст·ова́ни·е, -я,　　(n) (v n) see **баст·о-ва́ть;** (text) brushing

баст·ова́ть, -у́ют,　　(imp) strike, be/go on strike (industrial dispute)

ба́стр·, -а,　　(m) brown sugar

баст·у́ющ·ий, -ая, -ее, (pres part act) of **баст·ова́ть** (m decl as adj) striker

баталёр·, -а,　　(m) (naut) purser

баталья́н·, -а,　　(m) (mil) battalion

бата́н·, -а,　　(m) batten (timber); (text) slay, batten

батаре́й·к·а, -и,　　(g pl) **-е́·ек·,**　　(f) (dim) of **батаре́·я**

батаре́й·н·ый, -ая, -ое,　　(a) of **батаре́·я;** (f decl as adj) battery room

батаре́·я, -и,　　(f) battery; troop (artillery), battery (guns)

~, аккумуля́тор·н·ая　　(elec) storage battery, accumulator

~, ано́д·н·ая　　(elec) anode battery, B-battery

~, ба́ш·ечн·ая　　(mil) turret battery

~, бу́фер·н·ая　　(elec) buffer battery

~, га́з·ов·ая　　(elec) gas battery, Grove gas battery

~ ди́ск·ов　　disc gang (of harrow)

~, до·ба́в·очн·ая　　(elec) booster battery

~, ко́кс·ов·ая　　battery of coke ovens

~, кре́мн·иев·ая　　(elec) silicon solar battery

~, куб·ов·а́я　　(oil) continuous shell still

~ на·ка́л·а　　(elec) filament heating battery, A-battery

~, полу·про·вод·нико́в·ая　　(elec) silicon solar battery

~, радиа́тор·н·ая　　central-heating radiator system

~, се́т·очн·ая　　(elec) grid battery, C-battery

~, со́лн·ечн·ая　　(elec) solar battery

~, сух·а́я　　(elec) dry battery

~, термо·электр·и́ческ·ая　　thermo-electric pile

~, у·равн·и́тельн·ая　　(elec) buffer battery

бата́т·, -а,　　(m) (bot) sweet potato, Ipomea batatos

ба́тенс·, -а,　　(m) see **ба́ттенс·**

бати·а́льн·ый, -ая, -ое,　　(a) (ocean) bathyal

~ зо́н·а　　(f) (ocean) bathyal zone

бати·лимн·и́ческ·ий, -ая, -ое, (a) (zool) bathylimnetic

бат·и́л·ов·ый спирт·　　(m) (biochem) batyl alcohol

бати·метр·и́ческ·ий, -ая, -ое,　　(a) bathymetric

бати·ме́тр·и·я, -и,　　(f) (ocean) bathymetry, sounding, deep-sea sounding

бати·пелагиа́л·ь, -и,　　(f) (ocean) bathypelagic zone

бати·ска́ф·, -а,　　(m) (ocean) bathyscaphe

бати·сфе́р·а, -ы,　　(f) (ocean) bathysphere

бати́ст·, -а,　　(m) (text) cambric, lawn

бати́ст·ов·ый, -ая, -ое,　　(a) of **бати́ст·**

бати·термо·гра́ф·, -а,　　(m) (ocean) bathythermograph

бато́кс·, -а,　　(m) (shipb) buttocks; (pl) bow and buttock lines

бато·ли́т·, -а,　　(m) (geol) batholith, bathylith

бато́·метр·, -а,　　(m) (ocean) bottom sampling water bottle, bottom sampler

бато́н·, -а,　　(m) french loaf

~, шокола́д·н·ый　　chocolate bar

батопо́рт·, -а,　　(m) floating/ship caisson

бато·хро́м·, -а,　　(m) (chem) bathochrome

бато·хро́м·н·ый эффе́кт·　　(m) (spectr) bathochromic shift

батра́к·, -а,　　(m) seasonal farm labourer

батрахо·мо́рф·, -а,　　(m) (zool) batrachomorph

ба́ттенс·, -а,　　(m) batten (wood)

ба́ушингер·а, эффе́кт·　　(m) (met) Bauschinger effect

ба́фтинг·, -а,　　(m) (air) buffeting

ба́ффтинг·, -а,　　(m) see **ба́фтинг·**

бази́л·а, -ы,　　(f) boot

~, свинц·о́в·ые　　(pl) diver's lead-soled boots

бахром·а́, -ы́,　　(f) fringe; (zool) fimbria

бахро́м·к·а, -и,　　(g pl) **-м·ок·,**　　(f) (dim) of **бахром·а́**

бахромл·е́ни·е, -я,　　(n) (phot) frilling

бахро́м·чат·ый, -ая, -ое,　　(a) fringed, fimbriate, ciliate

бахтарм·а́, -ы́,　　(f) flesh side of hide

бахч·а́, -и́, (*i*) **-о́й,** (*g pl*) **-е́й,** (*f*) (agr) field of gourds
бахч·ев·о́й, -а́я, -о́е, (*a*) of **бахч·а́;** (bot) cucurbitaceous
~ культу́р·ы (*f pl*) (agr) gourd crops
баци́лл·а, -ы, (*f*) (bact) bacillus
бацилл·ем·и́·я, -и, (*f*) (med) bacillaemia, bacillemia
бацилло·нос·и́тел·ь, -я, (*m*) bacilli-carrier
бацилл·я́рн·ый, -ая, -ое, (*a*) bacillary
баци·трац·и́н·, -а, (*m*) (pharm) bacitracin
бач·ко́в·ый, -ая, -ое, (*a*) of **бач·о́к·;** (*m decl as adj*) messman
бач·о́к·, -ч·к·а́, (*m*) (*dim*) of **бак·;** cannister, can, tank; (mil) mess kettle, fanny
~, дрена́ж·н·ый (ICE) drain trap
~, конденса́т·н·ый condensate tank
~, расшир·и́тельн·ый expansion/header tank
~, с·мыв·н·о́й flushing cistern
ба́шен·к·а, -и, (*g pl*) **-н·ок·,** (*f*) (*dim*) of **ба́шн·я;** turret
ба́шен·н·ый, -ая, -ое, (*a*) of **ба́шн·я**
~ на·са́д·к·а (*f*) (chem) tower packing
~ о·хлад·и́тел·ь (*m*) cooling tower
~ стрел·о́к· (*m*) (air) turret gunner
башма́к·, -а́, (*m*) shoe, socket, chock, clip; (a/c) tail-skid shoe
~ гу́сен·иц·ы sole plate (of tracked vehicle)
~, ка́бель·н·ый (elec) cable shoe
~, колёс·н·ый (rail) wheel drag/scotch
~, по́люс·н·ый (elec) pole shoe/piece
~, скольз·я́щ·ий towing shoe
~, тормоз·н·о́й (rail) braking chock, drag shoe; (M/T) brake block
башма́ч·н·ый, -ая, -ое, (*a*) of **башма́к·**
башне·обра́з·н·ый, -ая, -ое, (*a*) turreted
ба́шн·я, -и, (*g pl*) **ба́шен·,** (*f*) *see* **коло́нн·а** (chem); tower, turret, beacon; (mil) gun turret
~, без·на·са́д·очн·ая (chem) wetted-wall tower
~, водо·на·по́р·н·ая (civ eng) water tower
~, си́лос·н·ая (agr) silo, tower silo
~, стрел·ко́в·ая gun turret
~ туш·е́ни·я quenching tower (coke oven)
~, у́голь·н·ая coal bunker

ба́шн·я
~, у·ла́вл·иваю·щ·ая (chem) entrainment tower
башта́н·, -а. (*m*) = **бахч·а́**
бд·и́тельност·ь -и, (*f*) alertness
аппара́т· бд·и́тельност·и (*m*) (rail) acknowledger
б.е. (*abbr*) = **бел·ко́в·ая един·и́ц·а** protein unit
бева·тро́н·, -а, (*m*) (nucl) bevatron accelerator
бе́г·, -а, (*m*) run, running, running race; (*pl*) horse races, trotting races
бег·а, -и, (*f*) (elec) bega (10^9 times); **бег·а́** (*nom pl*), (*g pl*) **-о́в,** (*m*) horse/trotting races
бе́г·а·ть, -ют, (*imp indet*) run, run about
бег·а́ю·щ·ий, -ая, -ее, (*pres part act*) of **бе́г·а·ть** running; (zool) cursorial
бегемо́т·, -а, (*m*) (zool) hippopotamus
~, обык·нове́нн·ый (zool) *Hippopotamus amphibius*
беге́н·ов·ый, -ая, -ое, (*a*) behen, behenic
~ кисл·от·а́ (*f*) (chem) behenic acid, n-docosoic acid
бе́г·лост·ь, -и, (*f*) fluency; dexterity
бе́г·л·ый, -ая, -ое, (*a*) fluent; superficial, cursory; runaway, fugitive; fleeting, momentary
~ огон·ь (*m*) (mil) rapid fire
бег·ов·о́й, -а́я, -о́е, (*a*) race, racing
~ доро́ж·к·а (*f*) race track/course; race (of ball or roller bearing); wheel path (tracked vehicle)
~ колц·о́ (*n*) race (of ball or roller bearing)
бег·о́м (*adv*) at the double, running; **бег·о́м·, -а,** (*m*) (elec) begohm, 10^6 ohms
бего́ни·я, -и, (*f*) (bot) begonia, *Begonia*
бег·у́н·, -а, (*m*) runner; (*pl*) edge-runner, edge-runner mill
~ -бло́к·, -а, (*m*) (naut) leading block
бег·унко́в·ый, -ая, -ое, (*a*) of **бег·уно́к·**
бег·уно́к·, -нк·а́, (*m*) (*dim*) of **бег·у́н·;** (rail) front carrying wheels, front bogie; (M/T) distributor rotor; (text) traveller
бег·у́т·, (*pres 3rd pl*) of **беж·а́ть**
бег·у́ч·ий, -ая, -ее, (*a*) running
~ такела́ж· (*m*) (naut) running rigging
бег·у́щ·ий, -ая, -ее, (*pres part act*) of **беж·а́ть** running, travelling
~ волн·а́ (*f*) (elec) travelling/traveling wave

бегхауз·, -а, (*m*) baghouse (filtering system)

бед·а́, -ы́, (*f*) misfortune, distress; poverty

бед·не́·ть, -ют, (*imp*) become poor/ impoverished; peter out, become rarer

бе́д·н·ый, -ая, -ое, (*a*) poor; lean (gas etc.); low-grade (ores)

~ с·ме́с·ь (*f*) (ICE) lean mixture

бёдр·а, (*nom pl*) of **бедр·о́**

бёдр·енн·ый, -ая, -ое, (*a*) (anat) femoral

~ ко́ст·ь (*f*) (anat) femur, thigh bone

бедр·о́, -а́, (*nom pl*) **бёдр·а,** (*g pl*) **бёдер·,** (*i pl*) **бёдр·ами,** (*n*) (anat) thigh, femur

бе́дрок·, -а, (*m*) (geol) bedrock

бе́д·стви·е, -я, (*n*) distress, need, want **сигна́л· бе́д·стви·я** (*m*) S.O.S., distress signal

~, стих·и́йн·ое (law) act of God

беж (*indecl*) beige

беж·а́вш·ий, -ая, -ее, (*past part act*) of **беж·а́ть**

беж·а́ть, (*pres 1st, 3rd sing, 3rd pl*) **бег·у́, беж·и́т, бег·у́т,** (*imp deter*) run

бе́ж·ев·ый, -ая, -ое, (*a*) beige, fawn

бе́ж·енец·, -нц·а, (*m*) fugitive

беж·и́т, (*pres 3rd sing*) of **беж·а́ть**

без (*prep + genitive*) without

~ че́тверт·и a quarter to (of time)

без- (*prefix*) = **безо-, безъ, бес-,** *particular significances*: (1) without, lacking, absence, in-, un-, non-, a-, -less, -free; (2) unsatisfactory state, inadequacy (*nouns ending in* **-ица**)

без·апертур·н·ый, -ая, -ое, (*a*) in-aperturate

без·авар·и́йн·ый, -ая, -ое, (*a*) accident-free, without a breakdown

без·азо́т·ист·ый, -ая, -ое, (*a*) nitrogen-free

без·алкого́ль·н·ый, -ая, -ое, (*a*) non-alcoholic

без·апелляцио́н·н·ый, -ая, -ое, (*a*) (*s f*) **-о́н·ен·, -о́н·н·а** (law) without right of appeal, not subject to appeal

без·бо́ж·н·ый, -ая, -ое, (*a*) atheistic

без·бра́ч·н·ый, -ая, -ое, (*a*) celibate; (biol) agamous, agamic

без·бре́ж·н·ый, -ая, -ое, (*a*) vast, boundless

без·вариа́нт·н·ый, -ая, -ое, (*a*) invariant

без·ва́тт·н·ый, -ая, -ое, (*a*) (elec) wattless, idle, reactive

без·ве́тр·енност·ь, -и, (*f*) windlessness, calmness

без·ве́тр·енн·ый, -ая, -ое, (*a*) windless, calm

без·ве́тр·и·е, -я, (*n*) (meteor) calm

без·вихр·ев·о́й, -а́я, -о́е, (*a*) eddy-free; (math) irrotational

~ ве́ктор· (*m*) (math) irrotational vector

~ движ·е́ни·е (*n*) (hydr) viscous/ streamline/steady flow

~ по́л·е (*n*) (math) non-circuital vector field, irrotational/conservative field

без·во́д·н·ый, -ая, -ое, (*a*) (*s f*) **-д·ен·, -д·н·а,** dry, arid, waterless; (chem) anhydrous

без·во́д·ь·е, -я, (*n*) aridity, drought, shortage/lack of water

без·воз·вра́т·н·ый, -ая, -ое, (*a*) (*s f*) **-т·ен·, -т·н·а,** irrevocable, irretrievable

без·воз·ду́ш·н·ый, -ая, -ое, (*a*) (*s f*) **-ш·ен·, -ш·н·а,** airless, air-free

~ пла́м·я (*n*) non-aerated flame

~ про·стра́н·ств·о (*n*) (phys) vacuum

без·возме́зд·но (*adv*) gratis, free; without compensation/indemnity

без·воло́с·ый, -ая, -ое, (*a*) hairless, glabrous

без·вре́д·н·ый, -ая, -ое, (*a*) (*s f*) **-д·е́н·, -д·н·а́,** harmless, innocuous

без·вре́мен·н·о (*adv*) prematurely

без·вре́мен·н·ый, -ая, -ое, (*a*) untimely, ill-timed

без·вы́·ход·н·ый, -ая, -ое, (*a*) (*s f*) **-д·ен·, -д·н·а,** hopeless, desperate

без·вя́з·кост·ый, -ая, -ое, (*a*) non-viscous

без·гистере́зис·н·ый, -ая, -ое, (*a*) (phys) anhysteretic

без·гла́с·и·е, -я, (*n*) (med) aphonia loss of voice

без·гла́с·н·ый, -ая, -ое, (*a*) (*s f*) **-с·ен·, -с·н·а,** (ling) mute, voiceless

без·гра́мот·ност·ь, -и, (*f*) illiteracy; ignorance

без·гран·и́чн·ый, -ая, -ое, (*a*) (*s f*) **-чен·, -чн·а,** boundless, infinite; (math) unbounded, unlimited

без·гра́н·н·ый, -ая, -ое, (*a*) anhedral

без·де́й·стви·е, -я, (*n*) inaction, inactivity; (phys) ideal state; inertia; dormancy

без·дей·ств·овать, -уют, (*imp*) be inactive, do nothing; be stopped, be idle (of machines)

без·дель·ник·, -а, (*m*) idler

без·денеж·н·ый, -ая, -ое, (*a*) impecunious; non-cash, without money
~ **рас·чёт·** (*m*) (fin) non-cash settlement, clearing

без·дефицит·н·ый, -ая, -ое, (*a*) (fin) entailing no deficit

без·де·ятельност·ь, -и, (*f*) inactivity

без·диффузион·н·ый, -ая, -ое, (*a*) diffusionless

без·дн·а, -ы, (*f*) chasm, abyss; a lot of, a mass of

без·дожд·ь·е, -я, (*n*) drought

без·до·каз·ательн·ый, -ая, -ое, (*a*), (*s f*) **-лен·, -льн·а,** (law) unsubstantiated, unproven

без·дорож·ь·е, -я, (*n*) lack of proper roads; impassable roads

без·доход·н·ый, -ая, -ое, (*a*) (*s f*) **-д·ен·, -д·н·а,** unprofitable

без·дым·н·ый, -ая, -ое, (*a*) smokeless, smoke-free
~ **порох·** (*m*) (expl) smokeless powder, cordite

без·жабер·н·ый, -ая, -ое, (*a*) (zool) abranchial, abranchiate

без·жизн·енн·ый, -ая, -ое, (*a*) (*s f*) **-н·ен·, -н·енн·а,** azoic, lifeless

без·за·род·ышев·ый, -ая, -ое, (*a*) inembryonate

без·за·кон·н·ый, -ая, -ое, (*a*) illegal, lawless; outside the law

без·за·мк·ов·ый, -ая, -ое, (*a*) (zool) inarticulate

без·за·щит·н·ый, -ая, -ое, (*a*) unprotected

без·зерн·ист·ый, -ая, -ое, (*a*) grainless

без·золь·н·ый, -ая, -ое, (*a*) ash-free, ashless

без·зуб·к·а, -и, (*g pl*) **-б·ок·,** (*f*) (zool) fresh-water mussel, *Anodontina*

без·зуб·ый, -ая, -ое, (*a*) toothless, edentate, edentous

без·индукцион·н·ый, -ая, -ое, (*a*) noninductive

без·инерцион·н·ое реле (*n*) (elec) extra high-speed relay

без·лепест·ков·ый, -ая, -ое, (*a*) (bot) apetalous

без·лепест·н·ой, -ая, -ое, (*a*) (bot) apetalous

без·лист·н·ый, -ая, -ое, (*a*) leafless, aphyllous

без·лич·н·ый, -ая, -ое, (*a*) (*s f*) **-ч·ен·, -ч·н·а,** colourless, colorless (of personality); (gram) impersonal

без·лун·н·ый, -ая, -ое, (*a*) (*s f*) **-н·ен·, -н·н·а,** moonless

без·люд·н·ый, -ая, -ое, (*a*) (*s f*) **-д·ен·, -д·н·а,** sparsely populated, lonely, unfrequented; (*attrib*) of **без·-люд·ь·е**

без·люд·ь·е, -я, (*n*) absence/lack of people, shortage of labour/labor

без·мал·а (*adv*) almost, nearly

безмен·, -а, (*m*) steelyard
~, **пруж·инн·ый** spring-balance

безмен·ом (*adv*) line-abreast, in line abreast

без·мер·н·ый, -ая, -ое, (*a*) (*s f*) **-р·ен·, -р·н·а,** immense, immeasurable

без·молв·н·ый, -ая, -ое, (*a*) (*s f*) **-в·ен·, -в·н·а,** silent, unspoken, tacit; silent, taciturn (of persons)

без·момент·н·ый, -ая, -ое, (*a*) (mech) momentless

без·мороз·н·ый, -ая, -ое, (*a*) frost-free

без·мотор·н·ый, -ая, -ое, (*a*) engineless
~ **по·лёт·** (*m*) (air) glide

без·мяк·отн·ый, -ая, -ое, (*a*) (anat) amyelinate, non-medullated

без·на·каль·н·ый, -ая, -ое, (*a*) cold-cathode, secondary emission (of vacuum tubes)
~ **тира·трон·** (*m*) cold-cathode thyratron

без·на·лич·н·ый, -ая, -ое, (*a*) (fin) clearing
~ **рас·чёт·** (*m*) (fin) non-cash settlement, clearing

без·ног·ий, -ая, -ое, (*a*) apodal, apodous; (*m decl as adj*) apod; (*pl decl as adj*) apoda

без·нос·н·ый, -ая, -ое, (*a*) noseless; spoutless

без·нос·ый, -ая, -ое, (*a*) noseless; spoutless

без·нул·ев·ой при·бор· (*m*) suppressed-zero instrument

безо (*prep + genitive*) = **без** without

безо- (*prefix*) *see* **без-**

без·о·говор·очн·ый, -ая, -ое, (*a*) unconditional, unqualified, unreserved

без·о·пас·но (*adv*) safely

без·о·па́с·ност·ь, -и, (*f*) safety, security

　коэфицие́нт· без·о·па́с·ност·и (*m*) (eng) safety factor, factor of safety

　регуля́тор· без·о·па́с·ност·и (*m*) emergency governor

~ **труд·а́** industrial safety

　у́гол· без·о·па́с·ност·и (*m*) safety angle

без·о·па́с·н·ый, -ая, -ое, (*a*) (*s· f*) **-с·ен·, -с·н·а,** safe, safety

~, **вну́тр·енн·е** inherently safe

без·ориенти́р·н·ый, -ая, -ое, (*a*) featureless, without a landmark

~ **ме́ст·ност·ь** (*f*) (air) non-rapid fixing area

без·ору́ж·н·ый, -ая, -ое, (*a*) unarmed

без·о·с·ко́л·очн·ый, -ая, -ое, (*a*) splinterproof

без·основ·а́тельн·ый, -ая, -ое, (*a*) (*s f*) **-лен·, -льн·а,** groundless, unfounded, unsubstantiated

без·о·станов·очн·ый, -ая, -ое, (*a*) continual, unceasing; (rail) non-stop

без·о́ст·ый, -ая, -ое, (*a*) (bot) awnless

без·от·ве́т·ственн·ый, -ая, -ое, (*a*) irresponsible

без·от·ка́з·н·ый, -ая, -ое, (*a*) unfailing, trouble-free, reliable

~ **рабо́т·а** (*f*) (mech eng) smooth running

без·от·ка́т·н·ый, -ая, -ое, (*a*) (gunn) recoilless

~ **ору́д·и·е** (*n*) recoilless gun

без·от·лаг·а́тельн·ый, -ая, -ое, (*a*) urgent, immediate, pressing

без·от·ме́н·н·ый, -ая, -ое, (*a*) irrevocable

без·от·нос·и́тельн·о (*a v*) irrespective of, independent of, without reference to

без·от·ры́в·н·ый, -ая, -ое, (*a*) unbroken, intact

~ **по·гран·и́чн·ый сло́·й** (*m*) (dynam) intact boundary layer

без·от·хо́д·н·ый, -ая, -ое, (*a*) waste-free

без·от·чёт·н·ый, -ая, -ое, (*a*) not accountable/answerable; instinctive

без·о·ши́б·очн·ый, -ая, -ое, (*a*) (*s f*) **-чен·, -чн·а,** faultless, correct

без·пла́мен·н·ый, -ая, -ое, (*a*) (expl) flashless

без·по·ме́х·ов·ый, -ая, -ое, (*a*) interference-free

без·рабо́т·иц·а, -ы, (*i*) **-ей,** (*f*) unemployment

без·рабо́т·н·ый, -ая, -ое, (*a*) unemployed; (*pl as noun*) the unemployed

без·радио·акти́в·н·ый, -ая, -ое, (*a*) (nucl) cold

без·раз·де́ль·н·ый, -ая, -ое, (*a*) (*s f*) **-л·ен·, -ль·н·а,** indivisible, inseparable, undivided

без·раз·ли́ч·н·ый, -ая, -ое, (*a*) (*s f*) **-ч·ен·, -ч·н·а,** indifferent, neutral; indistinguishable, uniform

~ **бе́рег·** (*m*) (geog) table-land shore

без·раз·ме́р·н·ый, -ая, -ое, (*a*) dimensionless, non-dimensional

без·ра́м·н·ый, -ая, -ое, (*a*) frameless

без·рёбер·н·ый, -ая, -ое, (*a*) ribless, ecostate

без·резьб·о́в·ое со·един·е́ни·е (*n*) flush/slip joint

без·результа́т·н·ый, -ая, -ое, (*a*) (*s f*) **-т·ен·, -т·н·а,** resultless, ineffectual, unsuccessful

~ **коло́д·ец** barren well

без·рессо́р·н·ый, -ая, -ое, (*a*) unsprung, springless

без·рефле́кс·н·ый, -ая, -ое, (*a*) non-reflex

~ **у·стро́й·ств·о** (*n*) (autom) non-feedback control system

без·ро́г·ий, -ая, -ое, (*a*) hornless, (agr) poll, pollarded

без·ру́д·н·ый, -ая, -ое, (*a*) (min) barren, ore-free

без·угле·ро́д·ист·ый, -ая, -ое, (*a*) carbon-free

без·уда́р·н·ый, -ая, -ое, (*a*) (*s f*) **-р·ен·, -р·н·а,** shockless, without impact; unstressed (phonetics)

без·упре́ч·н·ый, -ая, -ое, (*a*) (*s f*) **-ч·ен·, -ч·н·а,** irreproachable, unexceptionable

без·у·са́д·очн·ый, -ая, -ое, (*a*) non-shrink, shrink-proof, unshrinkable

без·у·сло́в·но (*adv*) undoubtedly, of course

без·у·сло́в·н·ый, -ая, -ое, (*a*) undoubted, unconditional, absolute

~ **рефле́кс·** (*m*) unconditional reflex

без·у·спе́ш·н·о (*adv*) unsuccessfully, in vain

без·у·спе́ш·н·ый, -ая, -ое, (*a*) unsuccessful, ineffectual

безъ- (*prefix*) *see* **без-, безо-**

безъ·ём·костн·ый, -ая, -ое, (*a*) non-capacitive

безъ·язы́ч·н·ый, -ая, -ое, (*a*) (*s f*) **-ч·ен·, -ч·н·а,** (zool) aglossate, aglossal

без·из·ве́ст·н·ый, -ая, -ое, (*a*) (*s f*) -т·ен·, -т·н·а, unknown, obscure

без·ыз·луч·а́тельн·ый, -ая, -ое, (*a*) (phys) radiationless, non-radiative

без·ым·я́нн·ый, -ая, -ое, (*a*) nameless, unnamed; anonymous

~ па́л·ец· (*m*) (anat) annularis, ring finger

без·ынерц·ио́нн·ый, -ая, -ое, (*a*) inertialess

бе́йдевинд·, -а, (*m*) tack (sailing)

~, крут·о́й close-hauled tack

бейле́йт·, -а, (*m*) (min) bayleyite

Бе́йльби, сло́·й (*m*) (met) Beilby layer

Бе́йльштейн·а, реа́кц·и·я (*f*) (chem) Beilstein's test

бе́йфут·, -а, (*m*) (naut) parrel band

бе́к·, -а, (*m*) back (sport)

бека́р·, -а, (*m*) natural (music)

беке́ш·а, -и, (*i*) -ей, (*f*) short winter overcoat

Бе́кке, ме́тод· (*m*) Becke method (of determining refractive index)

бе́ккерл·и́т·, -а, (*m*) (min) becquerelite

Бе́ккерел·я, луч· (*m*) (nucl) Becquerel ray

~, эффе́кт· (*m*) (phys) Becquerel effect

Бе́кман·а, пере·групп·иро́вк·а (*f*) (chem) Beckmann rearrangement

~, термо́·метр· (*m*) Beckmann thermometer

бекме́с·, -а, (*m*) concentrated grape juice

беко́н·, -а, (*m*) (food) bacon

бе́л·, -а, (*m*) (phys) bel

беле́мнит·, -а, (*m*) (pal) belemnite

белен·а́, -ы́, (*f*) (bot) henbane, *Hyoscyamus*

бел·е́ни·е, -я, (*n*) bleaching, blanching

белен·н·о́й, -а́я, -о́е, (*a*) of белен·а́

бел·ённ·ый, -ая, -ое, (*s f*) бел·ён·, бел·ен·а́ (*past part pass*) *see* бел·и́ть

бел·ён·ый, -ая, -ое, (*a*) bleached

бел·есова́т·ый, -ая, -ое, (*a*) whitish, off-white

бел·ёс·ый, -ая, -ое, (*a*) whitish, off-white

бел·е́·ть, -ют, (*imp*) whiten, become white; show white, look white; -ся show white

бе́л·и (*nom pl*), (*g pl*) -ей, (med) leucorrhoea, the whites

бел·изн·а́, -ы́, (*f*) whiteness

бел·и́л·а (*nom pl*), (*g pl*) бел·и́л· whiting, whitewash; white (pigment)

~, бари́т·ов·ые barium white (pigment)

бел·и́л·а

~, ви́смут·ов·ые pearl white, bismuth letharge

~, жемчу́ж·н·ые pearl white, bismuth trichloride

~, свинц·о́в·ые lead white

~, тита́н·ов·ые titanium white

~, ци́нк·ов·ые zinc white

бел·и́льн·ый, -ая, -ое, (*a*) of бел·и́л·а; bleaching

~ и́звест·ь (*f*) chloride of lime, bleaching powder

бел·и́льн·я, -и, (*g pl*) -лен·, (*f*) bleaching works, bleachery

бел·и́ть, -ят, (*imp*) whiten, whitewash; bleach; blanch

бе́л·ич·ий, -ья, -ье, (*a*) of бе́л·к·а squirrel; (*pl as noun*) (zool) *Sciuridae*

~ клет·к·а (*f*) (elec) squirrel cage

бе́л·к·а, -и, (*g pl*) бе́л·ок·, (*f*) squirrel; pencil drawing (original of a blueprint)

бел·к·и́ (*nom pl*) of бел·о́к· albumins; бе́л·к·и, (*g sing*) of бе́л·к·а squirrel

бел·ков·и́н·а, -ы, (*f*) albumen, egg-white

бел·ков·и́нн·ый, -ая, -ое, (*a*) *see* бел·ко́в·ый

бел·ко́в·ый, -ая, -ое, (*a*) of бел·о́к· albuminous

~ оболо́ч·к·а (*f*) (zool) sclera

белладо́нн·а, -ы, (*f*) (bot) deadly nightshade, belladonna, *Atropa belladonna*; (chem) belladonna, atropine

бело·бо́к·ий, -ая, -ое, (*a*) (*s f*) -бок·, -бок·а, white-sided/flanked

бело·бо́ч·к·а, -и, (*f*) (zool) dolphin

бел·ови́к·, -а́, (*m*) fair copy

бело·зёрн·ый, -ая, -ое, (*a*) white-grained

бел·о́к·, -л·к·а́, (*m*) white of egg, albumen; protein, albumin; white of the eye

~, прост·о́й simple protein

~, сло́ж·н·ый conjugated protein

бело·кал·и́льн·ый, -ая, -ое, (*a*) incandescent, white-hot

бело·ко́р·ый, -ая, -ое, (*a*) (bot) white-barked

бело·кро́в·и·е, -я, (*n*) (med) leukaemia, leukemia

бело·ку́р·ый, -ая, -ое, (*a*) fair-haired, blond

бело·ру́с·ск·ий, -ая, -ое, (*a*) White Russian, Belorussian

бело·ры́б·иц·а, -ы, (*i*) -ей, (*f*) (fish) white salmon, *Stenodus leucichthys*

бело·то́ч·ечн·ый, -ая, -ое, (*a*) white-dotted, white-spotted

бело·у́с·, торч·а́щ·ий (*m*) (bot) mat-grass, *Nardus stricta*
бел·о́чн·ый, -ая, -ое, (*a*) *see* бел·-ко́в·ый
бел·у́г·а, -и, (*f*) (fish) white sturgeon, beluga, *Huso huso*; white whale (*see* бел·у́х·а)
бел·у́ж·ий, -ья, -ье, (*a*) of бел·у́г·а
бел·у́х·а, -и, (*f*) (zool) white whale, beluga, *Delphinapterus leucas*
бел·у́ш·ий, -ья, -ье, (*a*) of бел·у́х·а
бе́л·ый, -ая, -ое, (*a*), (*s f*) бел, бел·а́, бел·о́, white
~ гор·я́чк·а (*f*) (med) delirium tremens
~ гриб· (*m*) (bot) *Boletus edulis*
~ жёст·ь (*f*) (met) tin plate, terne plate
~ но́ч·и (*f pl*) Northern Summer nights, Polar day
~ ору́ж·и·е (*n*) (mil) cold steel
~ с·пла́в· (*m*) white metal
~ у́гол·ь (*m*) water power
бе́л·ь, -и, (*f*) (text) gap in printed cloth; (agr) white blister, white rust
бельведе́р·, -а, (*m*) (arch) belvedere
бельг·и́йск·ий, -ая, -ое, (*a*) Belgian
бел·ь·ё, -я́, (*n*) linen, washing, under-clothing; (ceram) white wave
бель·ев·о́й, -а́я, -о́е, (*a*) of бел·ь·ё
бельм·о́, -а́, (*nom pl*) бе́льм·а, (*g pl*) бе́льм·, (*n*) (med) wall-eye
бе́ль·н·ый, -ая, -ое, (*a*) = бе́л·ый (q.v.)
бе́льтинг·, -а, (*m*) woven belting (for conveyor)
бель·эта́ж·, -а, (*i*) -ем, (*m*) first floor (second from ground); (arch) bel-étage; (theat) dress circle
бел·я́к·, -а́, (*m*) (ocean) white foam (on waves); (zool) blue hare, mountain hare, *Lepus variabilis*; frame for stretching and softening dried skins
бел·я́нк·а, -и, (*g pl*) -нок·, (*f*) (bot) spring snowflake, *Leucojum vernum*; (bot) lactarius, *Lactarius*; (ent) white butterfly
~, ре́п·н·ая (ent) cabbage white butterfly, *Pieris rapae*
белянкини́т·, -а, (*m*) (min) belyankinite
бе́м·ск·ий, -ая, -ое, (*a*) Bohemian
~ стекл·о́ (*n*) Bohemian glass
бе́нбери (*n indecl*) (chem) Banbury/internal mixer
бенз·а́л·ь, -я, (*m*) (chem) benzal, benzylidene

бенз·а́л·ь
~, хлор·и́ст·ый benzal/benzylidene chloride
бенз·альдеги́д·, -а, (*m*) (chem) benzaldehyde
бенз·ами́д·, -а, (*m*) benzamide
бенз·ами́н·, -а, (*m*) benzamine, betacaine
бенз·амо́н·, -а, (*m*) (pharm) benzamon
бенз·амили́д·, -а, (*m*) (chem) benzanilide
бенз·антре́н·, -а, (*m*) (chem) benzanthrene
бенз·антро́н·, -а, (*m*) (chem) benzanthrone
бенз·гидро́л·, -а, (*m*) benzhydrol
бе́нзел·ь, -я, (*m*) lashing (ropes etc.); (naut) seizing
~, корен·н·о́й throat seizing
~, пло́ск·ий eye seizing
~, прям·о́й round seizing
бензедри́н·, -а, (*m*) (pharm) benzedrine
бензени́л·, -а, (*m*) (chem) benzenyl
~, хлор·и́ст·ый benzotrichloride
бензиди́н·, -а, (*m*) (chem) benzidine
бенз·и́л·, -а, (*m*) (chem) benzyl, benzil, dibenzoyl, diphenylglyoxal
~, бро́м·ист·ый benzyl bromide
бензил·ами́н·, -а, (*m*) (chem) benzilamine
бензил·иде́н·, -а, (*m*) (chem) benzylidene, benzal
~, хлор·и́ст·ый benzal chloride, benzylidene chloride
бензил·и́р·овать, -у́ют, (*imp and perf*) (chem) benzylate
бензи́л·ов·ый, -ая, -ое, (*a*) of бенз·и́л·
~ спи́рт· (*m*) benzyl alcohol
бензил·хлори́д·, -а, (*m*) (chem) benzyl chloride
бензил·целлюло́з·а, -ы, (*f*) benzyl cellulose
бензи́ль·н·ый, -ая, -ое, (*a*) of бенз·и́л·
бенз·и́н·, -а, (*m*) petrol (Brit), gasoline (U.S.), motor spirit; (chem) benzine
~, авиац·ио́нн·ый aviation spirit/gasoline
~, автомоби́ль·н·ый petrol, gasoline, motor spirit/gasoline
~, га́з·ов·ый natural gasoline
~, крекинг- cracked gasoline
~, не·о·чи́щ·енн·ый sour gasoline
~, о·твержд·ённ·ый solidified gasoline
~, о·чи́щ·енн·ый sweet gasoline

бенз·и́н
~ **прям·о́й гóн·к·и** straight-run gasoline
~, **пуск·ов·о́й** (a/c) priming fuel
~, **рифóрминг··** reformed gasoline
~, **спец·иáльн·ый** special-boiling-point spirit, SBP spirit
~, **тяж·ёл·ый** heavy benzine, ligroin
~, **экстракц·иóнн·ый** solvent spirit
~, **эталóн·н·ый** standard gasoline
~, **этил·и́рованн·ый** ethylated gasoline
бензине́ (n indecl) (gas) saturated wash oil
бензи́н·к·а·, -и, (f) petrol/gasoline stove
бензино·вóд·, -а, (m) (see also **про·-вóд·**) petrol/gasoline/fuel pipe
бензино·вóз·, -а, (m) petrol/gasoline tanker/carrier (ship)
бензи́н·ов·ый, -ая, -ое, (a) of **бенз·и́н**
бензи́н·о-кисло·рóд·н·ая у·станóв··к·а (f) (weld) oxy-benzine apparatus
бензино·мéр·, -а, (m) petrol/fuel/gasoline gauge
бензино·про·вóд·, -а, (m) (see also **про·вóд·**) petrol/gasoline feed pipe, petrol/gasoline line
бензино·у·лов·и́тел·ь, -я, (m) oil trap (sewerage systems)
бензо- (abbr) of **бензи́н·ов·ый**, petrol, gasoline; (chem) (int component) benzo-, benz-
бензо·бáк·, -а, (m) (see also **бак·**) petrol, gasoline tank
бензо·вóз·, -а, (m) (М/Т) petrol/gasoline tanker, tank truck
бензо·за·прáв·щик·, -а, (m) (М/Т) petrol/gasoline bowser
бензо·и́л·, -а, (m) (chem) benzoyl
~, **хлóр·ист·ый** benzoyl chloride
бензоил·бензóйн·ая кисл·от·á (f) benzoyl benzoic acid
бензóин·, -а, (m) (chem) benzoin
бензойно·бути́л·ов·ый эфи́р· (m) butile benzoate
бензойно·ки́сл·ый, -ая, -ое, (a) benzoate of
~ **нáтр·и·й, -я,** (m) sodium benzoate
бензойно·нáтр·иев·ая сóл·ь (f) sodium benzoate
бензóйн·ый, -ая, -ое, (a) benzoic, benzoin
~ **альдеги́д·** (m) benzaldehyde
~ **кисл·от·á** (f) benzoic acid
~ **смол·á** (f) gum benzoin

бензóйн·ый, -ая, -ое
~ **у·плот·нéни·е** (n) benzoin condensation
бензойно·ки́сл·ый, -ая, -ое, (a) benzoate of
~ **нáтр·и·й** (m) sodium benzoate
бензо·крáн·, -а, (m) (see also **крáн·**) petrol/gasoline/fuel cock
бенз·óл·, -а, (m) (chem) benzene; benzol, benzole (as fuel)
~, **мотóр·н·ый** motor benzole
бензоли́н·, -а, (m) (pharm) benzolin
бензол·сульфи́н·ов·ая кисл·от·á (f) benzene sulphinic acid
бензол·сульфо·кисл·от·á, -ы, (f) benzene sulphonic acid
бензóль·н·ый, -ая, -ое, (a) of **бен-з·óл·**
~ **ядр·ó** (n) (chem) benzene ring/nucleus
бензо·мéр·, -а, (m) petrol/fuel gauge
бензо·на·лив·н·óй, -áя, -óе, (a) petrol, gasoline (of transport)
бензо·на·сóс·, -а, (m) (see also **на·сóс·**) petrol/gasoline/fuel pump
бензо·нитри́л·, -а, (m) (chem) benzonitrile
бензо·пóмп·а, -ы, (f) (see also **пóм-п·а**) petrol/gasoline/fuel pump
бензо·про·вóд·, -а, (m) (see also **про·вóд·**) petrol/gasoline/fuel pipe
бензо·прóч·н·ый, -ая, -ое, (a) benzo-fast (dyes)
бензо·раз·дáт·очн·ая колóн·к·а (f) petrol/gasoline pump (for refuelling motor transport)
бензо·рéз·, -а, (m) (weld) gasoline cutter
бензо·стóй·к·ий, -ая, -ое, (a) petrol/gasoline-resistant
бензо·фенóн·, -а, (m) (chem) benzophenone
бензо·фи́льтр·, -а, (m) (see also **фи́льтр·**) (ICE) petrol/gasoline/fuel filter
бензо·хинóн·, -а, (m) (chem) benzoquinone
бензо·хран·и́лищ·е, -а, (n) (see also **хран·и́лищ·е**) gasoline/petrol storage depot/tank
бензо·шлáнг·, -а, (m) (see also **шлáнг·**) petrol/gasoline/fuel hose
бенз·пириди́н·, -а, (m) (chem) benzopyridine, quiniline
бенз·пирéн·, -а, (m) benzpyrene
бени́т·, -а, (m) barley sugar
бенитóит·, -а, (m) (min) benitoite

беннеттит·ов·ые, -ых, (*pl decl as adj*) (pal bot) *Bennettitales*

бентонит·, -а, (*m*) (min) bentonite

бентон·н·ый, -ая, -ое, (*a*) (biol) benthic, benthonic

бентос·, -а, (*m*) (biol) benthos

бенуар·, -а, (*m*) (theat) lower boxes

Беранже (*m indecl*) Beranger (surname)

~, вес·ы (*pl*) Beranger balance

берберин·, -а, (*m*) (chem) berberine

бергамот·, -а, (*m*) (bot) bergamot, *Citrus bergamia*; (chem) bergamot

бергинизаци·я, -и, (*f*) Bergius' process, destructive hydrogeneration (of coal)

Бергман·а, труб·к·а (*f*) (elec) Bergman cable conduit

бёрд·о, -а, (*n*) (text) reed; (naut) loom

бéрег·, -а, (*nom pl*) **-á,** (*g pl*) **-óв,** (*m*) coast, shore, beach; bank (of river); **берёг·** (*past masc sing*) of **берéч·ь**

~, болóт·ист·ый boggy margin

~, мор·ск·óй coast, seashore, beach

~, на·гóр·н·ый high bank

~, об·рéз·н·ый steeply shelving shore

~, скал·ист·ый cliffs

~, с·полз·áющ·ий caving bank

берег·овичóк·, -чк·á, (*m*) (zool) winkle

берег·ов·óй, -áя, -óе, (*a*) coast, coastal, off-shore, waterside, riverside, lake-side, littoral, alongshore

~ вéтер· (*m*) off-shore wind

~ за·щит·а (*f*) coastal protection

~ крáн· (*m*) quay crane

~ лёж·ен·ь (*m*) mudsill (of bridges)

~ о·пóр·а (*f*) abutment (of bridges)

берег·óм·ый, -ая, -ое, (*pres part pass*) of **берéч·ь**

берего·у·креп·ительн·ое со·оруж·é·ни·е (*n*) coast protection works

берег·ущ·ий, -ая, -ее, (*pres part act*) of **берéч·ь**

берёг·ш·ий, -ая, -ее, (*past part act*) of **берéч·ь**

береж·ённ·ый, -ая, -ое, (*past part pass*) of **берéч·ь**

береж·ёт· (*pres 3rd sing*) of **берéч·ь**

береж·лив·ый, -ая, -ое, (*a*) careful, economical

бéреж·н·ый, -ая, -ое, (*a*), (*s f*) **-ж·ен·, -ж·н·а,** careful

берёз·а, -ы, (*f*) (bot) birch, *Betula*

березит·, -а, (*m*) (min) beresite (mica-, pyrites- and gold-bearing rock)

березит·изáци·я, -и, (*f*) (geol) berezitation

берез·няк·, -á, (*m*) birch wood/grove

берёз·ов·ик·, -а, (*m*) (bot) rough-stalked boletus, *Boletus scaber*

берёз·ов·ый, -ая, -ое, (*a*) of **берёз·а;** (*pl decl as adj*) (bot) *Betulacea*

бер·ём, (*pres 1st pl*) of **бр·ать**

берéмен·е·ть, -ют, (*imp*) become pregnant

берéмен·ност·ь, -и, (*f*) pregnancy

~ двóй·н·ей bigeminal pregnancy

берéмен·н·ый, -ая, -ое, (*a*) pregnant, gravid

бéрест·, -а, (*m*) (bot) elm, *Ulmus*

берёст·а, -ы, (*f*) birch bark

берёст·ов·ый, -ая, -ое, (*a*) of **берёст·а**

бер·ёт, (*pres 3rd sing*) of **бр·ать**

берéч·ь, (*pres 1st, 3rd sing, 3rd pl*) **берег·ý, береж·ёт, берег·ýт,** (*past masc sing*) **берёг·** (*imp*) take care of, keep, protect; **-ся** (*passive*); be careful, take care

берилл·, -а, (*m*) (min) beryl

берилл·иев·ый, -ая, -ое, (*a*) of **берилл·и·й**

берилл·изáци·я, -и, (*f*) beryllium diffusion hardening (of steel)

берилл·и·й, -я, (*m*) (chem) beryllium, Be

бериллий·диметил·, -а, (*m*) beryllium methide

бериллий·диэтил·, -а, (*m*) beryllium ethide

бериллий·орган·ическ·ий, -ая, -ое (*a*) organo-beryllium

берилл·ов·ый, -ая, -ое, (*a*) of **берилл·**

бериллонит·, -а, (*m*) (min) beryllonite

бер·ите, (*pl imper*) of **бр·ать**

беркéл·и·й, -я, (*m*) (chem) berkelium, Bk

бéркл·и·й, -я, (*m*) *see* **беркéл·и·й**

берлин·ск·ая лазýр·ь (*f*) Prussian blue, Berlin blue

бéрм·а, -ы, (*f*) (civ eng) berm, bench

Бернýлли (*m indecl*) Bernoulli (surname)

~, за·кóн· (*m*) (math) Bernoulli's law

~, у·равн·éни·е (*n*) Bernoulli's theorem

бернýлли·ев·ые числ·а (*n pl*) (math) Bernoulli's numbers

Бертело, бóмб·а (*f*) Berthelot's calorimeter

бертолéт·ов·а сóл·ь   (*f*) (chem) Berthollet's salt, potassium chlorate

бертрандúт·, -а,   (*m*) (min) bertrandite

бер·ýщ·ий, -ая, -ее, (*pres part act*) of **бр·ать**

берц·óв·ый, -ая, -ое,   (*a*) (anat) tibial

~ **кост·ь, больш·ая**   (*f*) (anat) tibia

~ **кост·ь, мáл·ая**   (*f*) (anat) tibiale

бес- (*prefix*) *see* **без-**

бес·дислокаци́он·н·ый, -ая, -ое,   (*a*) dislocation-free

бесéд·а, -ы,   (*f*) conversation, discussion

бесéд·к·а, -и, (*g pl*) **-д·ок·,**   (*f*) summer-house; (naut) bosun's chair

бес·кáмер·н·ый, -ая, -ое,   (*a*) tubeless (of tires)

~ **по·крýш·к·а**   (*f*) tubeless tire

бес·карбонáт·н·ый, -ая, -ое,   (*a*) non-calcareous

бес·кисло·рóд·н·ый, -ая, -ое,   (*a*) oxygen-free

бес·клáпан·н·ый, -ая, -ое,   (*a*) (anat) valveless, avalvular

бес·клáсс·н·ый, -ая, -ое,   (*a*) classless

бес·козы́р·к·а, -и,   (*f*) sailor's cap, peakless cap

бес·компрéссор·н·ый, -ая, -ое,   (*a*) solid-injection (diesel); (oil) natural pressure (gas lift)

бес·кон·éчн·о   (*adv*) infinitely

~ **больш·óй, -áя, -óе, -úе** (*adv and adj*) infinite, of infinite size

~ **мáл·ый, -ая, -ое,** (*adv and adj*) (math) infinitesimal

~ **у·дал·ён·ый, -ая, -ое,** (*adv and adj*) infinite, ideal (point)

бес·кон·éчност·ь, -и,   (*f*) infinity; eternity

бес·конечно·мéр·н·ый, -ая, -ое, (*a*) (math) of an infinite number of dimensions, ∞-dimensional

бес·кон·éчн·ый, -ая, -ое,   (*a*) (*s f*) **-чен·, -чн·а,** endless, infinite, nonterminating

~ **лúн·и·я**   (*f*) (math) line at infinity; (elec) infinite line

~ **мнóж·еств·о**   (*n*) (math) infinite set

~ **по·слéд·овательност·ь** (*f*) (math) infinite sequence

~ **ряд·**   (*m*) (math) infinite series

бес·контáкт·н·ый, -ая, -ое,   (*a*) non-contact, without contact, contactless; (rad) influence

бес·контрóль·н·ый, -ая, -ое,   (*a*) uncontrolled, unchecked

бес·кóрм·иц·а, -ы,   (*i*) **-ей,**   (*f*) (agr) shortage of fodder

бес·крáсн·ый, -ая, -ое,   (*a*) containing no red

бес·кры́·л·ый, -ая, -ое,   (*a*) wingless, apterous

бес·пáлуб·н·ый, -ая, -ое,   (*a*) (shipb) undecked, open

бес·пáл·ый, -ая, -ое,   (*a*) fingerless

бес·парт·úйн·ый, -ая, -ое,   (*a*) (polit) non-party, independent; (*m decl as adj*) non-party man (in U.S.S.R. not a member of the Communist party)

бес·патéнт·н·ый, -ая, -ое,   (*a*) unlicensed

бес·пере·бóй·н·ый, -ая, -ое,   (*a*) (*s f*) **-бó·ен·, -бóй·н·а,** uninterrupted; smooth (of engine running)

бес·пере·горóд·очн·ый, -ая, -ое, (*a*) (biol) non-septate

бес·пере·сáд·очн·ый, -ая, -ое,   (*a*) through (of passenger transport)

бес·плáмен·н·ый, -ая, -ое, (*a*) flameless; (expl) flashless

бес·плáт·н·ый, -ая, -ое,   (*a*) free, gratis

бес·плóд·и·е, -я,   (*n*) (biol) barrenness, sterility

бес·плóд·н·ый, -ая, -ое,   (*a*) sterile, barren, infertile; fruitless

бес·по·ворóт·н·ый, -ая, -ое,   (*a*) unalterable, irrevocable

бес·по·дóб·н·ый, -ая, -ое,   (*a*) (*s f*) **-б·ен·, -б·н·а,** incomparable, unequalled

бес·позвон·óчн·ый, -ая, -ое,   (*a*) (zool) invertebrate; (*pl decl as adj*) *Invertebrata*

бес·покó·ить, -ят, (*imp*) worry, disturb, trouble; **-ся** (*pass*) worry

бес·покóй·н·ый, -ая, -ое, (*a*) (*s f*) **-ó·ен·, -óй·н·а,** restless, turbulent

бес·по·крóв·н·ый, -ая, -ое, (*a*) naked

бес·полéз·н·ый, -ая, -ое,   (*a*) (*s f*) **-з·ен·, -з·н·а,** useless, wasteful

бес·пóл·ый, -ая, -ое,   (*a*) sexless, asexual; agamic

~ **раз·множ·éни·е** (*n*) (biol) asexual reproduction, asexual propagation, agamogenesis

бес·пóлюс·н·ый, -ая, -ое,   (*a*) (elec) pole-less

бес·пóр·ист·ый, -ая, -ое,   (*a*) non-porous, aporous

бес·пóр·ов·ый, -ая, -ое,   (*a*) non-porous, aporous

бес·поро́ч·н·ый, -ая, -ое, (*a*) (*s f*) **-ч·ен·, -ч·н·а,** spotless, without blemish

бес·по·ря́д·ок·, -д·к·а, (*m*) disorder, confusion

бес·по·ря́д·очн·ый, -ая, -ое, (*a*) (*s f*) **-чен·, -чн·а,** disorderly, confused; random, irregular; (bot) inordinate

~ об·мо́т·к·а (*f*) (elec) random winding

бес·по·са́д·очн·ый, -ая, -ое, (*a*) (air) non-stop

бес·по·те́р·н·ый, -ая, -ое, (*a*) loss-free, zero loss

бес·по́шлин·н·ый, -ая, -ое, (*a*) (com) duty-free

бес·пре·де́ль·н·ый, -ая, -ое, (*a*) (*s f*) **-л·ен·, -ль·н·а,** limitless, unbounded

бес·пре·ры́в·н·ый, -ая, -ое, (*a*) (*s f*) **-в·ен·, -в·н·а,** uninterrupted, continuous

бес·пре·ста́н·н·ый, -ая, -ое, (*a*) (*s f*) **-н·ен·, -н·н·а,** incessant, continual

бес·при́·быль·н·ый, -ая, -ое, (*a*) (*s f*) **-л·ен·, -ль·н·а,** profitless; (met cast) riserless

бес·при·ме́р·н·ый, -ая, -ое, (*a*) (*s f*) **-р·ен·, -р·н·а,** unexampled, unparalleled

бес·при́·мес·н·ый, -ая, -ое, (*a*) pure, unalloyed, uncontaminated

бес·при·стра́ст·н·ый, -ая, -ое, (*a*) (*s f*) **-т·ен·, -т·н·а,** unbiased, impartial

~ вы·бо́р·к·а (*f*) (math) unbiased sample

бес·про·ва́ль·н·ый, -ая, -ое, (*a*) through which nothing can fall

бес·про·во́д·н·ый, -ая, -ое, (*a*) unwired, non-wired

бес·про́·волоч·н·ый, -ая, -ое, (*a*) wireless

бес·проце́нт·н·ый, -ая, -ое, (*a*) (fin) interest-free

бес·пы́ль·ников·ый, -ая, -ое, (*a*) (bot) antherless

бес·са́ж·ев·ый, -ая, -ое, (*a*) unfilled (rubber)

бес·свинц·о́в·ый, -ая, -ое, (*a*) leadless

бес·с·вя́з·н·ый, -ая·, -ое, (*a*) unconnected, incoherent

бе́ссел·ев·ы фу́нкц·и·и (*f pl*) (math) Bessel functions

Бе́ссел·ь, -я, (*m*) Bessel (surname)

бес·семено·до́ль·н·ый, -ая, -ое, (*a*) (bot) acotyledonous

Бе́ссемер·, -а, (*m*) Bessemer (surname)

бессемер·ова́ни·е, -я, (*n*) (met) acid Bessemer process (iron); Bessemerizing (copper)

~, ма́л·ое side-blowing acid Bessemer process

бессеме́р·овск·ий, -ая, -ое, (*a*) (met) Bessemer, conversion

~ проце́сс· (*m*) acid Bessemer process

~ ста́л·ь (*f*) Bessemer-quality steel

~ чугу́н· (*m*) Bessemer/conversion iron

бес·сем·я́нн·ый, -ая, -ое, (*a*) (bot) seedless

бес·серд·е́чников·ый, -ая, -ое, (*a*) coreless

бес·се́р·н·ый, -ая, -ое, (*a*) sulphur-free, sweet

бес·си́л·и·е, -я, (*n*) weakness; (biol) impotence

бес·скеле́т·н·ый, -ая, -ое, (*a*) (biol) skeletonless, naked

бес·с·ко́с·н·ый, -ая, -ое, (*a*) plain, square (of plating joints)

бес·сле́д·н·о (*adv*) tracelessly, without a trace

бес·с·ли́т·ков·ая про·ка́т·к·а (*f*) (met) continuous casting

бес·с·ме́н·н·ый, -ая, -ое, (*a*) permanent, fixed

бес·с·мы́сл·енн·ый, -ая, -ое, (*a*) (*s f*) **-л·ен·, -л·енн·а,** absurd, senseless

бес·со·зна́·тельн·ый, -ая, -ое, (*a*) (*s f*) **-лен·, -льн·а,** unconscious; unintentional

бес·со·су́д·ист·ый, -ая, -ое, (*a*) (anat) avascular

бес·спо́р·н·о it is indisputable; (*adv*) undoubtedly

бес·сро́ч·н·ый, -ая, -ое, (*a*) (*s f*) **-ч·ен·, -ч·н·а,** indefinite, undefined, termless (period of time)

~ за·ём· (*m*) (fin) undated loan

бес·сте́бель·н·ый, -ая, -ое, (*a*) (bot) acauline, acaulose

бес·с·то́ч·н·ый, -ая, -ое, (*a*) drainless; (geol) inland-drainage

бес·структу́р·н·ый, -ая, -ое, (*a*) structureless, amorphous

бес·ступ·е́нчат·ый, -ая, -ое, (*a*) stepless, stageless, without stages/steps

бес·ступ·е́нчат·ый, -ая, -ое,
~ **пере·да́·ч·а** (*f*) variable speed gear/transmission
~ **регул·и́ровани·е** (*n*) (autom) stepless control
бес·суста́в·н·ый, -ая, -ое, (*a*) (zool) inarticulate
бес·с·чёт·н·ый, -ая, -ое, (*a*) innumerable, countless
бес·та́р·н·ый, -ая, -ое, (*a*) unpacked (without packing material), bulk (of cargo)
бес·тка́н·ев·ый, -ая, -ое, (*a*) non-woven, non-fabric
бес·то́пл·ивн·ый, -ая, -ое, (*a*) (nucl) non-fuel-bearing
бес·фла́нц·ев·ый, -ая, -ое, (*a*) flush, flangeless
бес·фо́н·ов·ый, -ая, -ое, (*a*) no background
бес·фо́рм·енн·ый, -ая, -ое, (*a*) (*s f*) **-м·ен·, -м·енн·а,** shapeless, amorphous
бес·хво́ст·н·ый, -ая, -ое, (*a*) tailless
~ **само·лёт·** (*m*) flying wing
бес·хво́ст·ый, -ая, -ое, (*a*) tailless; (zool) ecaudate, anurous
бес·хо́з·н·ый, -ая, -ое, (*a*) no-man's
бес·цвет·н·ый, -ая, -ое, (*a*) (*s f*) **-т·ен·, -т·н·а,** colourless, achromatic
бес·це́ль·н·ый, -ая, -ое, (*a*) (*s f*) **-л·ен·, -ль·н·а,** aimless, pointless
бес·це́н·н·ый, -ая, -ое, (*a*), (*s f*) **-н·ен·, -н·н·а,** priceless, invaluable
бес·це́нтр·ов·о-шлиф·ова́льн·ый стан·о́к· centreless/centerless grinder, centreless/centerless grinding machine
бес·це́нтр·ов·ый, -ая, -ое, (*a*) centerless, centreless
бес·челн·о́чн·ый, стан·ок· (*m*) (text) draw loom
бес·челюст·н·о́й, -а́я, -о́е, -ы́е, (*a*) (zool) agnathous, agnathic; (*pl as noun*) *Agnatha*
бес·чешу́й·н·ый, -ая, -ое, (*a*) (zool) scaleless, alepidotous, alepidote
бес·чи́сл·енн·ый, -ая, -ое, (*a*) (*s f*) **-л·ен·, -л·енн·а,** countless, innumerable; multiple
~ **мно́ж·еств·о** (*n*) (math) infinite set
бес·череп·н·ы́е, -ы́х, (*pl decl as adj*) (zool) *Acraniata*
бес·шо́в·н·ый, -ая, -ое, (*a*) seamless
~ **труб·а́** (*f*) seamless tube, solid-drawn tube

бес·шу́м·н·ый, -ая, -ое, (*a*) (*s f*) **-м·ен·, м·н·а,** silent
~ **цеп·ь** silent-chain drive
бес·щел·ев·о́й, -а́я, -о́е, (*a*) slotless, alete
бес·щелоч·н·о́й, -а́я, -о́е, (*a*) alkali-free
бе́та (*indecl*) beta, β (Greek)
бета- (*int component*) *see following hyphenated words; see also as single word*
бе́та-акти́в·н·ый, -ая, -ое, (*a*) (nucl) beta-active
бе́та-ве́тв·ь, -и, (*f*) (nucl) beta branch
бе́та-желе́з·о, -а, (*n*) (met) beta-iron (obs)
бе́та-из·луч·е́ни·е, -я, (*n*) beta radiation
бе́та-ли́н·и·я, -и, (*f*) (nucl) beta peak
бе́та-луч·и́ (*nom pl*), (*g pl*) **-е́й,** (*m*) (phys) beta rays
бе́та-рас·па́д·, -а, (*m*) (nucl) beta decay, beta disintegration
бе́та-спектро́·метр·, -а, (*m*) beta-ray spectrometer
бе́та-тольщино·ме́р·, -а, (*m*) beta-ray gauge
бе́та-ура́н·, -а, (*m*) beta-uranium
бе́та-фа́з·а, -ы, (*f*) beta phase
бетайн·, -а, (*m*) (chem) betaine, trimethyl-glycocol
бета·ме́тер·, -а, (*m*) betameter
бета·тро́н·, -а, (*m*) (nucl) betatron, betatron accelerator
~, **без·желе́з·н·ый** air-cooled betatron
~, **двух·луч·ев·о́й** dual-beam betatron
бетафи́т·, -а, (*m*) (min) betafite
бета·фу́нкц·и·я, -и, (*f*) beta function
бета·част·и́ц·а, -ы, (*f*) (phys) beta-particle
Бете, от·ветв·и́тел·ь (*m*) (rad) Bethe hole
бето́н·, -а, (*m*) concrete
~, **арм·и́рованн·ый** reinforced concrete
~, **бар·и́тов·ый** barytes concrete
~, **вибр·и́рованн·ый** vibrated concrete
~, **водо·со·держ·а́щ·ий** water-bearing concrete
~, **извест·ко́в·ый** lime concrete
~, **крупно·по́р·ист·ый** expanding-cement concrete
~, **лёгк·ий** light-aggregate concrete, lightweight concrete
~, **лит·о́й** poured concrete

бетóн

~, **монолúтный** *in situ* concrete

~, **неармúрованный** plain/mass concrete

~, **плóтный** dense concrete, air-free concrete

~, **предварúтельно-напряжённый** prestressed concrete

~, **сбóрный** precast concrete

~, **товáрный** ready-mix concrete

~, **торкрéт-** gunite, cement-gun concrete

~, **транспóрт-** transit-mixed concrete

~, **тяжёлый** heavy-aggregate concrete, heavy-weight concrete

~, **холóдный** cold-weather concrete

~, **центрифугúрованный** spun concrete

~, **шлáковый** breeze concrete

бетонúрованный, -ая, -ое, (*past part pass*) of **бетонúровать**; concrete, concreted

бетонúровать, -уют, (*imp*) concrete

бетонирóвка, -и, (*g pl*) **-вок·,** (*f*) concreting

бетонúт, -а, (*m*) concrete stone

бетóнный, -ая, -ое, (*a*) concrete

~ **завóд·** (*m*) concrete products works, precast-concrete works

~ **издéлия** (*n pl*) concrete products, precast concrete

~ **массúв·** (*m*) concrete block

~ **структýра** (*f*) (geol) mortar/murbruck structure

бетонобóйный, -ая, -ое, (*a*) anticoncrete (of missiles)

бетоновóд·, -а, (*m*) concrete delivery line

бетономешáлка, -и, (*f*) concrete mixer

бетононасóс·, -а, (*m*) concrete pumping machine

бетоноплáст·, -а, (*m*) plastic-filled concrete, plasticized concrete

бетоноотдéлочная машúна (*f*) concrete finishing machine (roadmaking)

бетонораспределúтель·, -я, (*m*) concrete spreader (roadmaking)

бетонуклáдчик·, -а, (*m*) concrete placing machine

бетулúн·, -а, (*m*) betulin, betula camphor

бетулúновая кислотá (*f*) betulinic acid

бéть красная (*f*) (naut) wind abeam

бечевá, -ы́, (*f*) cord, twine; tow rope (rivers, canals)

бечёвка, -и, (*g pl*) **-вок·,** (*f*) string, twine

бечевнúк·, -á, (*m*) towpath

Бéше мóлот· (*m*) Béche forging hammer

бéшенство, -а, (*n*) (med) hydrophobia; (vet) rabies

бéшеный, -ая, -ое, (*a*) rabid

би- (*int component*) bi-, di-

библиóграф·, -а, (*m*) bibliographer

библиографúческий, -ая, -ое, (*a*) bibliographic(al)

библиогрáфия, -и, (*f*) bibliography

библиотéка, -и, (*f*) library

библиотéкар·ь, -я, (*m*) librarian

библиотековéдение, -я, (*n*) librarianship

библиотéчный, -ая, -ое, (*a*) of **библиотéка**

бивáк·, -а, (*m*) (mil) bivouac

бивалéнт·, -а, (*m*) (cyt) bivalent

бивалéнтный, -ая, -ое, (*a*) of **бивалéнт·**

бивáчный, -ая, -ое, (*a*) of **бивáк·**

~ **лáгер·ь** (*m*) (mil) bivouac

бивéктор·, -а, (*m*) (math) bi-vector

бúвен·ь, -вня, (*m*) tusk

бивибрáтор·, -а, (*m*) (rad) double dipole

бúвиум·, -а, (*m*) (zool) bivium

бúвший, -ая, -ее, (*past part act*) of **бить**

бигармонúческий, -ая, -ое, (*a*) biharmonic

бидистиллáт·, -а, (*m*) (chem) bidistillate

бидóн·, -а, (*m*) can, container

биéние, -я, (*n*) (*v n*) *see* **бить**; beat, pulsation; wobble, play, runout

генерáтор· биéний (*m*) (rad) beating/beat-frequency oscillator

частотá биéний (*f*) (rad) beat frequency

бизáн·ь, -и, (*f*) (naut) mizzen

~, **выносн·áя** (naut) jigger sail

~ **-мáчта, -ы,** (*f*) (naut) mizzen mast

бизóн·, -а, (*m*) (zool) bison, *Bison americanus*

бикарбонáт·, -а, (*m*) (chem) bicarbonate

~ **бáрия** barium bicarbonate

~ **нáтрия** sodium bicarbonate

би·квадра́т·н·ый, -ая, -ое, (*a*) (math) fourth-degree, biquadratic

~ у·равн·е́ни·е (*n*) (math) biquadratic equation, equation of the fourth degree

би·компа́кт·, -а, (*m*) (math) bicompactum, bicompact

би·компа́кт·н·ый, -ая, -ое, (*a*) bicompact

би·кон·и́ческ·ий, -ая, -ое, (*a*) biconical, biconic

би·криста́лл·, -а, (*m*) bi-crystal

би́кфорд·ов шнур· (*m*) (expl) Bickford/safety/miner's fuse

би·латера́ль·н·ый, -ая, -ое, (*a*) bilateral

биле́т·, -а, (*m*) ticket, card; pass permit

~, обра́т·н·ый return ticket

~, от·пуск·н·о́й (mil) leave pass

~, при·глас·и́тельн·ый invitation, invitation card

~, чле́н·ск·ий membership card

билет·ёр·, -а, (*m*) ticket collector (male)

билет·ёрш·а, -и, (*i*) **-ей,** (*f*) ticket collector (female)

биле́т·н·ый, -ая, -ое, (*a*) of **биле́т·**

~ ка́сс·а (*f*) booking office; (theat) box office

били·а́рн·ый, -ая, -ое, (*a*) (med) biliary

били·верд·и́н·, -а, (*m*) (biochem) biliverdin

би·лине́ар·н·ый, -ая, -ое, (*a*) bilinear

би·ли́нз·а, -ы, (*f*) (opt) divided lens

би·лин·е́йн·ый, -ая, -ое, (*a*) bilinear

били·руб·и́н·, -а, (*m*) (biochem) bilirubin

биллио́н·, -а, (*m*) a thousand million, milliard (Brit), billion (U.S.)

би́·л·о, -а, (*n*) beater; **би́·л·о** (*past neut sing*) of **би·ть**

би́·льн·ый, -ая, -ое, (*a*) of **би́·л·о**

би·мета́лл·, -а, (*m*) duplex metal, clad metal, laminated metal, bi-metal (U.S.)

би·металл·и́зм·, -а, (*m*) (fin) bimetallism

би·металл·и́ческ·ий, -ая, -ое, (*a*) duplex-metal, clad-metal, composite, bimetallic

~ про́·вод· (*m*) (elec) composite conductor

би·молекул·я́рн·ый, -ая, -ое, (*a*) bimolecular

би́мс·, -а, (*m*) (shipb) beam

~, по·воро́т·н·ый (shipb) cant beam

~, с·тяж·н·о́й (shipb) paddle beam

би́мс

~, съ·ём·н·ый (shipb) hatch beam

~, холост·о́й (shipb) hold beam

бин·а́рн·ый, -ая, -ое, (*a*) (chem) binary; binomial

~ систе́м·а (*f*) (chem) binary system

~ с·пла́в· (*m*) binary alloy

~ у·стано́в·к·а (*f*) binary vapour engine

бин·аура́ль·н·ый, -ая, -ое, (*a*) (acous) binaural

би·нейтр·о́н·, -а, (*m*) (nucl) bineutron, dineutron

бинокле·держ·а́тел·ь, -я, (*m*) binocular holder

бин·о́кл·ь, -я, (*m*) binoculars, field glasses

~, дн·евн·о́й prism binocular

~, ноч·н·о́й night glasses

~, обык·нове́нн·ый night glasses

~, призмат·и́ческ·ий prism binocular

бин·окул·я́рн·ый, -ая, -ое, (*a*) (opt) binocular

би·но́м·, -а, (*m*) (math) binomial

~ Ньюто́н·а (math) binomial theorem

би·номи·а́льн·ый, -ая, -ое, (*a*) binomial

~ коэффицие́нт· (*m*) binomial coefficient

би·номин·а́льн·ый, -ая, -ое, (*a*) binominal

би·норма́л·ь, -и, (*f*) (math) binormal

бинт·, -а́, (*m*) bandage; band (of book)

Био́, за·ко́н· (*m*) (opt) Biot's law

Био́-Сава́р·а, за·ко́н· (*m*) (phys) Biot–Savart law

био- (*int component*) bio-, biological

био·ген·е́з·, -а, (*m*) biogenesis

био·генет·и́ческ·ий, -ая, -ое, (*a*) biogenetic

био·ге́н·н·ый, -ая, -ое, (*a*) biogenous, biogenic

био·гео·гра́ф·и·я, -и, (*f*) biogeography

био·гео·хи́м·и·я, -и, (*f*) biogeochemistry

био·ге́рм·, -а, (*m*) (geol) bioherm

био́·граф·, -а, (*m*) biographer

био·гра́ф·и·я, -и, (*f*) biography

би·о́з·а, -ы, (*f*) (chem) biose, disaccharide

био·ли́т·, -а, (*m*) biolith (mineral of biological origin), biogenic/organic rock

био́·лог·, -а, (*m*) biologist

био·лог·и́ческ·ий, -ая, -ое, (*a*) biological

~ индик·а́тор·, (*m*) (biol) biometer

био·лог·и́ческ·ий, -ая, -ое
~ **эквивале́нт· рентге́н·а** *(m)* (nucl) roentgen-equivalent man, REM, rem
био·люмин·есце́нци·я, -и, *(f)* bioluminescence
би·о́м·, -а, *(m)* (biol) biome
био·ма́сс·а, -ы, *(f)* biomass
био·ме́тр·и·я, -и, *(f)* biometry, biometrics
био·меха́н·ик·а, -и, *(f)* biomechanics
био·ном·и́ческ·ий, -ая, -ое, *(a)* (biol) bionomic
би·о́нт·, -а, *(m)* (biol) bion, biont
био·потенциа́л·, -а, *(m)* biopotential
био·препара́т·, -а, *(m)* biological product
би·опси́·я, -и, *(f)* (med) biopsy
био·рексистаз·и́ческ·ий, -ая, -ое, *(a)* biorhexistasic
би́ос·, -а, *(m)* (biochem) bios, *(pl)* bioi, bioses
био·си́нтез·, -а, *(m)* biosynthesis
био·ста́нц·и·я, -и, *(f)* biological research station
био·стимул·я́тор·, -а, *(m)* biostimulator
био·стро́м·, -а, *(m)* biostrome
био·сфе́р·а, -ы, *(f)* biosphere
био́т·а, -ы, *(f)* (biol) biota
биот·и́н·, -а, *(m)* (biochem) biotin, vitamin H
биоти́т·, -а, *(m)* (min) biotite, black mica
биот·и́ческ·ий, -ая, -ое, *(a)* biotic
био·то́п·, -а, *(m)* (biol) biotope, community habitat
био·тро́н·, -а, *(m)* (rad) biotron
био·фа́ци·я, -и, *(f)* (geol) biofacies
био·фи́з·ик·, -а, *(m)* biophysicist
био·фи́з·ик·а, -и, *(f)* biophysics
био·фи́льтр·, -а, *(m)* filter bed (sewage treatment)
~, **конта́кт·н·ый** contact bed (sewage)
~ **не·пре·ры́в·н·ого де́й·стви·я** percolating/continuous filter (sewage)
био·фо́р·, -а, *(m)* (bot) biophore
био·хи́м·и·я, -и, *(f)* biochemistry
био·хо́р·, -а, *(m)* biochore
био·цено́з·, -а, *(m)* biocoenosis, biocenosis
би·па́ч·н·ый, -ая, -ое, *(a)* (phot) bipack
би·пирами́д·н·ый, -ая, -ое, *(a)* (cryst) dipyramidal, bipyramidal
би·пла́н·, -а, *(m)* (a/c) biplane
би·поля́р·н·ый, -ая, -ое, *(a)* bipolar
би·потенциа́ль·н·ый, -ая, -ое, *(a)* bipotential

би·при́зм·, -а, *(m)* biprism
би·радика́л·, -а, *(m)* biradical
би́рж·а, -и, *(i)* **-ей,** *(f)* exchange, market; stock exchange
~, **лес·н·а́я** (paper) outdoor wood store
~ **труд·а́,** labour/labor exchange
би́рк·а, -и, *(g pl)* **-рок·,** *(f)* nameplate, tally, tag, ticket
Би́ркгоф·, -а, *(m)* Birkhoff (surname)
бирк·ова́ни·е, -я, *(n)* (agr) labelling, tagging; ringing (pigeons etc.)
Би́ркхоф·, -а, *(m)* Birkhoff (surname)
бирма́н·ск·ий, -ая, -ое, *(a)* Burmese
бирюз·а́, -ы́, *(f)* (min) turquoise
бис- *(int component)* bis-
би́сер·, -а, *(m)* bead
би́сер·н·ый, -ая, -ое, *(a)* bead
бискви́т·, -а, *(m)* (food) biscuit
би́смут·, -а, *(m) see* **ви́смут·** (chem) bismuth, Bi
бисмутини́т·, -а, *(m)* (min) bismuthinite
бисмути́т·, -а, *(m)* (min) bismutite
бис·се́ктор·, -а, *(m)* (math) bisector of the angle between two intersecting planes
бис·сектри́с·а, -ы, *(f)* bisector (of an angle)
би́ссус·, -а, *(m)* (zool) byssus
би·сульф·а́т·, -а, *(m)* (chem) bisulphate, bisulfate, acid sulphate/sulfate
~ **ба́р·и·я** barium bisulphate
би·сульф·и́д·, -а, *(m)* (chem) bisulphide, bisulfide
би·сульф·и́т·, -а, *(m)* (chem) bisulphite, bisulfite
~ **ка́л·и·я** potassium bisulphite
би·сфено́ид·, -а, *(m)* (min) bisphenoid
би́т·в·а, -ы, *(f)* battle
би́тенг·, -а, *(m)* (naut) bitt
~ **-бо́лт·, -а,** *(m)* bitt pin
би́тер·, -а, *(m)* (agr) beater
биту́м·, -а, *(m)* bitumen, asphaltic bitumen, asphalt (U.S.)
~, **жи́дк·ий** cut-back bitumen/asphalt
~, **каменно·уго́ль·н·ый** coal-tar bitumen/asphalt
~, **нефт·ян·о́й** bitumen, asphaltic bitumen, asphalt (U.S.)
~, **о·кисл·ённ·ый** blown bitumen/asphalt
~, **ото́·гн·анн·ый** straight-run bitumen/asphalt
~, **с·вя́з·анн·ый** fixed bitumen/asphalt
~, **твёрд·ый** hard bitumen/asphalt

битумин·изáци·я, -и, (*f*) (geol) bituminization; (build) impregnation with bitumen, packing with bitumen

битумин·óзн·ый, -ая, -ое, (*a*) bituminous

битум·и́ровани·е, -я, (*n*) = **битумин·изáци·я**

битýм·н·ый, -ая, -ое, (*a*) bitumen, asphalt (U.S.)

∼ **мат·** (*m*) (build) damp-proof course

битумо·вóз·, -а, (*m*) (civ eng) bitumen distributor

би́·т·ый, -ая, -ое, (*past part pass*) of **би́·ть;** beaten, broken, cracked, crushed; dressed (of poultry etc.)

би·ть, (*3rd sing pres*) **бь·ёт·,** (*3rd pl pres*) **бь·ют,** (*imp*) beat, hit, strike, fight; kill, slaughter; gush; break; flap (e.g. of drive belts); sound, sound off (bugle call), strike (of clocks); *see also* **би́·ть·ся**

бит·ь·ё, -я́, (*n*) (*v n*) *see* **би·ть**

би́·ть·ся (*imp*) (*see also* **би·ть**) fight, struggle; knock, hit; beat (pulse etc.), throb; be breakable

би·урéт·, -а, (*m*) (chem) biuret

би·урéт·ов·ая реáкци·я (*f*) (chem) biuret reaction

би·фени́л·, -а, (*m*) (chem) biphenyl, diphenyl

би·филя́р·, -а, (*m*) (elec) bifilar, bifilar winding

би·филя́р·н·ый, -ая, -ое, (*a*) (elec) bifilar, two-wire

би·фокáль·н·ый, -ая, -ое, (*a*) (opt) bifocal

би·функционáль·н·ый, -ая, -ое, (*a*) bifunctional, difunctional

би·фурк·áци·я, -и, (*f*) bifurcation

бифштéкс·, -а, (*m*) beefsteak

би·харáктер·и́стик·а, -и, (*f*) (math) bicharacteristic

би·хромáт·, -а, (*m*) (chem) bichromate

∼ **аммóн·и·я** ammonium bichromate

би́цепс·, -а, (*m*) (anat) biceps

би·цикл·и́ческ·ий, -ая, -ое, = **би··цикл·и́чн·ый**

би·цикл·и́чн·ый, -ая, -ое, (*a*) bicyclic

бициллúн·, -а, (*m*) (pharm) bicillin, benzyl penicillin

би·ч·, -á, (*i*) **-óм,** (*m*) whip, lash; (agr) beater; (biol) flagellum

биюпгун·, -а, (*m*) (bot) *Anabasis salsa*

Блáгден·а, за·кóн· (*m*) (chem) Blagden's law

блáг·о, -а, (*n*) (*sing only*) good, wellbeing; (*esp pl*) wealth

благо- (*root*) *conveys sense*: good beneficial, harmonious

благо·вол·éни·е, -я, (*n*) goodwill; kindness

благо·вóн·н·ый, -ая, -ое, (*a*) fragrant

благо·дáр·ност·ь, -и, (*f*) gratitude, thanks

благо·дар·я́ (*prep* + *dative*) due to, owing to, thanks to

благо·дéн·ствен·ый, -ая, -ое, (*a*) flourishing, thriving, prosperous

благо·дéн·стви·е, -я, (*n*) prosperity

благо·дéн·ств·овать, -уют, (*imp*) thrive, flourish, prosper

благо·звýч·и·е, -я, (*n*) harmony

благо·на·дёж·н·ый, -ая, -ое, (*a*) trustworthy

благо·по·лýч·н·о (*adv*) all right, safely; happily, well

благо·при·я́т·н·ый, -ая, -ое, (*a*) (*s f*) **-т·ен·, -т·н·а,** favourable, favorable, propitious, opportune

благо·при·я́т·ствовани·е, -я, (*n*) favour, favor

благо·при·я́т·ствуем·ый, -ая, -ое, (*pres part pass*) favoured, favored

∼ **нáц·и·я, наи·бóл·ее** (*f*) (econ) most favoured nation

благо·при·я́т·ствующ·ий, -ая, -ее, (*pres part act*) favourable, favorable, favouring, favoring

благо·раз·ýм·н·ый, -ая, -ое, (*a*) (*s f*) **-м·ен·, -м·н·а,** reasonable, sensible, prudent

благо·рóд·н·ый, -ая, -ое, (*a*) (*s f*) **-д·ен·, -д·н·а,** noble; rare (gas), noble (metal), precious (stone)

благо·с·клóн·н·ый, -ая, -ое, (*a*) (*s f*) **-н·ен·, -н·н·а,** well-disposed, favourable, not disinclined

благо·со·сто·я́ни·е, -я, (*n*) welfare, well-being, prosperity

благо·у·стрó·енн·ый, -ая, -ое, (*a*) well-built, well-laid-out, wellarranged/organized

благо·у·стрóй·ств·о, -а, (*n*) organization of public services

блáнк·, -а, (*m*) form (paper)

∼**, печáт·н·ый** (print) letterhead

блáнк·ов·ый, -ая, -ое, (*a*) of **блáнк·**

∼ **нáд·пис·ь** (*f*) (fin) blank endorsement

∼ **пере·дáт·очн·ая нáд·пис·ь** (*f*) blank/general endorsement

бланфúкс·, -а, (*m*) (chem) blanc-fix (artificial barium sulphate) бланш·и́ровани·е, -я, (*n*) blanching; fleshing (leather)

бланш·и́рова́тел·ь, -я, (*m*) blanching machine

бланш·и́р·овать, -уют, (*imp*) blanch

бласт·и́ческ·ий, -ая, -ое, (*a*) (cyt) blastic

бласто·де́рм·а, -ы, (*f*) (cyt) blastoderm

бласт·оид·е́·я, -и, (*f*) (pal) blastoid, (*pl*) *Blastoidea*

бласто·ме́р·, -а, (*m*) (zool) blastomere

бласт·по́р·, -а, (*m*) (zool) blastopore

бласто·це́л·ь, -и, (*f*) (zool) blastocoel

бласто·ци́ст·, -а, (*m*) (biol) blastocyst

бла́ст·ул·а, -ы, (*f*) (zool) blastula

бла́угаз·, -а, (*m*) blau gas

бледн·е́·ть, -ют, (*imp*) go pale, pale, lose colour/color

бледн·и́ть, -я́т, (*imp*) make pale

бледно·жёлт·ый, -ая, -ое, (*a*) pale yellow

бле́дн·ост·ь, -и, (*f*) pallor, paleness

бле́дн·ый, -ая, -ое, (*a*) (*s f*) -ден·, -дн·а́, -дн·о, pale, pallid

блейве́йс·, -а, (*m*) lead white (pigment)

блёк·л·ый, -ая, -ое, (*a*) faded

~ ме́д·н·ая руд·а́ (*f*) (min) fahlerz, grey copper ore

блёк·н·уть, -ут, (*past masc sing*) блёк·н·ул or блёк·, (*imp*) fade; (bot) wither

блёкрот·, -а, (*m*) black rot (of grapes)

блёнд·а, -ы, (*f*) (opt) shade, mask, blind

~, со́лн·ечн·ая (phot) lens hood

бле́нкер·, -а, *see* бли́нкер·

бленно·ре́·я, -и, (*f*) (med) blennorrhoea

бле́ск·, -а, (*m*) gloss, glitter, shine, brightness, lustre; (expl) flash; (min) glance

~, желе́з·н·ый (min) iron glance, haematite

~ звезд·ы́ (astron) star brightness

~, ма́т·ов·ый dull lustre

~, металл·и́ческ·ий metallic lustre

~, молибде́н·ов·ый (min) molybdenite

~, свинц·о́в·ый (min) lead glance, galenite

~, серебр·я́н·ый (min) silver glance, argentite

~, смол·ян·о́й pitch glance

~, сурьм·я́н·ый (min) antimony glance, stibnite

~, ту́ск·л·ый dull lustre

блеско·ме́р·, -а, (*m*) brightness meter

блеск·ост·ь, -и, (*f*) (opt) glare

блес·н·а́, -ы́, (*nom pl*) блёс·н·ы, (*f*) (fish) spoon bait

блес·н·у́ть·, -у́т, (*perf*) flash, give a flash

блест·е́ть, (*imp*) (*pres 3rd pl*) блест·я́т or блещ·ут, sparkle, glitter, gleam, shine

блёст·к·а, -и, (*g pl*) -т·ок·, (*f*) (met, min) spangle

блест·я́щ·ий, -ая, -ее, (*pres part act*) of блест·е́ть; brilliant, sparkling, shining, lustrous

блеф·, -а, (*m*) bluff, pretence

блефар·и́т·, -а, (*m*) (med) blepharitis

блефаро·пла́ст·, -а, (*m*) (biol) blepharoplast

блефаро·спа́зм·, -а, (*m*) (med) blepharospasm

блеф·ова́ть, -у́ют, (*imp*) bluff, pretend

блещ·ущ·ий, -ая, -ее, (*pres part act*) of блест·е́ть; = блест·я́щ·ий

бле́·яни·е, -я, (*n*) bleating

бле́·ять, -ют, (*imp*) bleat

ближ·а́йш·ий, -ая, -ее, (*a*) nearest, next, immediate; (nucl) nearest-neighbour

~ координ·а́ци·я (*f*) (nucl) nearest-neighbour coordination

бли́ж·е (*comp*) of бли́з·к·ий

ближне·вос·то́ч·н·ый, -ая, -ое, (*n*) (geog) Near East (W. Asia and N.E. Africa)

бли́ж·н·ий, -яя, -ее, (*a*) neighbouring, near; close-range, short-range

~ бо·й (*m*) (*in g used as a*), close range (of armaments)

~ конво́·й (*m*) (naut) close escort

~ по·ря́д·ок· (*m*) (phys) short-range order

~ при·вод·н·а́я ста́нци·я (air nav) inner marker beacon

близ (*prep*+*gen*) near, close to

близзард·, -а, (*m*) (meteor) blizzard

бли́з·иться, -ятся, (*imp*) draw near, approach

бли́з·к·ий, -ая, -ое, (*a*) (*s f*) -з·ок·, -з·к·а́, -з·к·о́, nearby, close (to); imminent; like, similar (to), approximately the same (as)

~ рас·сто·я́ни·е (*n*) short distance/range

~ фо́рм·а (*f*) (biol) related form

близко·дей·ствующ·ий, -ая, -ое, (*a*) short-range

близ·леж·áщ·ий, -ая, -ее, (*a*) nearby, neighbouring

близ·нéц·, -á, (*i*) **-óм,** (*nom pl*) **-ы́,** (*m*) a twin; (*pl*) (astron) Gemini

близ·нецóв·ый, -ая, -ое, (*a*) twins', twin's

близо·рýк·ий, -ая, -ое, (*a*) short-sighted, near-sighted, myopic

близо·рýк·ост·ь, -и, (*f*) short-sighted-ness, myopia

блúз·ост·ь, -и, (*f*) nearness, closeness, proximity; intimacy

эффéкт· блúз·ост·и (*m*) proximity effect (induction heating)

блúк·, -а, (*m*) highlight, sparkle, spot/speck/patch of light; spot (on mirror galvanometer)

∼, пере·дéрж·анн·ый, (phot) burnt-out highlight

блúк·ов·ый, -ая, -ое, (*a*) of **блик·**

блúн·, -á, (*m*) pancake

∼, лед·ян·óй (geol) ice cake; (ocean) pancake ice

блинд·áж·, -а, (*i*) **-óм,** (*m*) (mil) shelter, dugout

блинд·ир·овáть, -ýют, (*imp*) make shelter/dugout; protect with armour plate

блинк-компарáтор·, -а, (*m*) (astron) blink comparator

блинк-микро·скóп·, -а, (*m*) (astron) blink microscope

блúнкер·, -а, (*m*) (elec) signal relay; flag indicator (telephone exchanges)

блúн·н·ый, -ая, -ое, (*a*) pancake

блúн·чат·ый, -ая, -ое, (*a*) pancake, pancake-shaped; pancake, layered

∼ лёд· (*m*) pancake ice

блист·áтельн·ый, -ая, -ое, (*a*) (*s f*) **-лен·, -льн·а,** brilliant, splendid

блист·á·ть, -ют, (*imp*) shine; be conspicuous

∼ от·сýт·стви·ем be conspicuous by one's absence

блóк·, -а, (*m*) block; pulley, block, sheave (lifting tackle); (elec) block unit, block assembly, element, block; (nucl) slug

∼ арифмет·úческ·их дéй·стви·й arithmetic control block (of computer)

∼ в·вóд·а и вы́·вод·а кóд·ов reader and recorder (of computer magnetic drum)

∼, вз·брóш·енн·ый (geol) thrust block

блóк

∼ вы́·бор·а пáмят·и operation decoder (of computer programme control unit)

∼ вы́·дач·и дáн·н·ых (rad) data unit

∼ дáль·ност·и (rad) range unit

∼, дву·шкúв·н·ый double block (lifting gear)

∼ дел·éни·я division block (of computer)

∼.-диагрáмм·а, -ы, (*f*) block diagram

∼, диференциáль·н·ый differential hoist

∼, дóк·ов·ый (shipb) keel block

∼ за·пáзд·ывани·я delay line store (computers)

∼ из·мер·úтел·я врéмен·и (autom) timing pulse generator, clock

∼ индик·áтор·а (rad) display unit

∼, кáбель·н·ый (elec) block cable

∼ кáдр·ов·ой раз·вёрт·к·и frame scan block (in computer CRT store)

∼ -кáртер·, -а, (*m*) (ICE) cylinder block

∼ Лю́дерс·а worm and worm-wheel block

∼ менипул·я́тор·а (rad) keying unit

∼ -механ·úзм·, -а, (*m*) (rail) block

∼ не·линеáр·ност·ей non-linear function block (of computer)

∼ не·линеáр·н·ых за·вúс·имост·и non-linear function block

∼, не·по·двúж·н·ый standing block (lifting gear)

∼, одно·шкúв·н·ый single block (lifting gear)

∼ -пере·гóн·, -а, (*m*) (rail) block section, block

∼ пере·мéн·н·ых коэффициéнт·ов coefficient setter (computers)

∼ пере·множ·éни·я multiplication block (computers)

∼ пит·áни·я power unit/pack, power supply unit

∼, по·двúж·н·ой moving/travelling block (lifting gear)

∼ -полимéр·, -а, (*m*) (chem) block polymer

∼ -полимер·изáци·я, -и, (*f*) (chem) block polymerization

∼ -пóст·, -а, (*m*) (rail) block post/station, block signal box

∼ пýск·а и синхрон·изáци·и starting and synchronization block (computer programming)

блóк

~ **раз·вёрт·ок·** storage deflection block (of computer)

~ **регенер·áци·и** regeneration block (of computer)

~ **регúстр·ов прогрáмм·ы** programme register block (of computer)

~ **-сéкц·и·я, -и,**   (*f*) (shipb) block/box-section

~ **селéктор·н·ых úмпульс·ов** selector pulse block (computer)

~ **-сополимéр·, -а,**   (*m*) (chem) block copolymer

~ **-стáнц·и·я, -и, = блок-пóст·**

~ **строч·н·ой раз·вёрт·к·и** line scan block (computer)

~ **-схéм·а, -ы,**   (*f*) block diagram

~, **тáл·ев·ый** travelling/traveling block (oil derrick)

~, **тепло·вы·дел·яющ·ий**   (nucl) fuel slug/element

~, **у·множ·áющ·ий** multiplier unit (of computer)

~, **урáн·ов·ый**   (nucl) uranium slug

~ **у·скор·úтел·ей**   (rocket) booster cluster

~ **-у·чáст·ок·, -т·к·а,** (*m*) (rail) block section, block

~ **форм·ировáни·я úмпульс·ов** control pulse-shaping block (of computer)

~, **функционáль·н·ый** function block (computers)

~ **цилúндр·ов**   (ICE) cylinder block

~ **шестерён·** multiple gear

~, **шестерён·н·ый** spur-geared block

~, **шлюп·очн·ый** (naut) boat chock

блокáд·а, -ы,   (*f*) blockade

~ **сéрд·ц·а** (med) heart-block

блокгáуз·, -а,   (*m*) (mil) blockhouse

блóкинг-генер·áтор·, -а,   (*m*) (elec) blocking oscillator

блок·ировани·е, -я, (*v n*) of **блок·úр·о·вать** see also **блок·ировк·а**

блок·úрованн·ый, -ая, -ое, (*past part pass*) of **блок·úр·овать**

~ **с·чёт·** (*m*) (fin) blocked/frozen account

блок·úр·овать, -уют,   (*imp*) block; (elec) interlock; (mil) blockade

блок·úровк·а, -и, (*g pl*) -вок·,   (*f*) (*v n*) of **блок·úр·овать;** (elec) interlock, lock; (rail) block signalling/signaling

~, **автомат·úческ·ая**   (rail) automatic blocking system

~, **за·щúт·н·ая** (elec) protective interlock

блок·úрóвк·а

~, **механ·úческ·ая**   (elec) mechanical interlock

~, **не·автомат·úческ·ая**   (rail) manual block/blocking system

~, **пере·гóн·н·ая** (rail) section blocking

блок·úровочн·ый, -ая, -ое, (*a*) blocking; interlocking

~ **конденс·áтор·,** (*m*) (elec) blocking capacitor

~ **механ·úзм·,**   (*m*) interlock (safety mechanism)

блок·úрующ·ий, -ая, -ее, (*pres part act*) of **блок·úр·овать**

~ **генер·áтор·** (*m*) (elec) blocking oscillator

блокнóт·, -а,   (*m*) note book, writing pad

блóк·ов·ый, -ая, -ое, (*a*) of **блóк·**

блоко·об·рабáт·ывающ·ий агрегáт· (*m*) (print) blocking press/machine

блоко·у·клáд·чик·, -а, (*m*) erector arm (tunnel driving)

блóкшив·, -а, (*m*) (naut) hulk

бломстрандúт·, -а, (*m*) (min) blomstrandite, betafite

блох·á, -и, (*nom pl*) **блóх·и,** (*g pl*) **блох·,** (*d pl*) **-áм,** flea

Блóх·а, полос·á (*pl*) (phys) Bloch bands

блóч·н·ый, -ая, -ое, (*a*) of **блóк·**

~ **колес·ó** (*n*) (horol) great wheel

~ **модел·úрующ·ая у·стан·óвк·а** (*f*) (electr) block-construction simulator

~ **поли·стир·óл** (*m*) (chem) block polystyrene

~ **рас·по·лож·éни·е** (*n*) (phys) lumping

~ **систéм·а** (*f*) (build) modular system

блоч·óк·, -ч·к·á, (*m*) (*dim*) of **блóк·**

блош·úн·ый, -ая, -ое, (*a*) of **блох·á**

блужд·áни·е, -я, (*n*) wandering

~, **с·луч·áйн·ое** (*n*) (math) wandering

блужд·á·ть, -ют, (*imp*) wander, roam, stray

блужд·áющ·ий, -ая, -ее, (*pres part act*) of **блужд·á·ть;** wandering, travelling, roaming, stray erratic, vagrant

~ **дюн·а** (*f*) (geol) travelling/wandering dune

~ **пóч·к·а** (*f*) (med) floating kidney

~ **нéрв·** (*m*) (zool) vagus

~ **тóк·** (*m*) (elec) ground current, stray current

~ **фóрм·а** (*f*) (biol) vagrant form

блýм·, -а, (*m*) (met) bloom

блу́минг·, -а,   (*m*) (met) blooming/cogging mill

блу́мс·, -а,   (*m*) *see* блу́м·

блю́д·ечк·о, -а,   (*g pl*) -чек·, (*n*) saucer

~, мор·ск·о́е   (zool) limpet, *Patella*

блюдечко·ви́д·н·ый, -ая, -ое,   (*a*) patelliform, patelloid, saucer/pan shaped

блю́д·о, -а,   (*n*) dish

блюд·о́м·ый, -ая, -ое, (*pres part pass*) of блюс·ти́

блюд·у́щ·ий, -ая, -ее,   (*pres part act*) of блюс·ти́

блю́д·ц·е, -а,   (*i*) -ем,   (*g pl*) -д·ец·, (*n*) saucer; (biol) patella; (geol) minor depression

блюдце·обра́з·н·ый, -ая, -ое,   (*a*) *see* блюдечко·ви́д·н·ый

блюд·ш·ий, -ая, -ее,   (*past part act*) of блюс·ти́

блюм·, -а,   (*m*) = блум· bloom

блю·л,   (*past masc sing*) of блюс·ти́

блю́минг·, -а,   (*m*) *see* блу́минг·

блюс·ти́, (*pres 1st sing*) блюд·у́, (*3rd sing*) блюд·ёт, (*3rd pl*) блюд·у́т, (*past masc sing*) блю́·л, (*f sing*) блю·ла́, (*imp*) observe

бля́х·а, -и,   (*f*) identification badge/plate

бля́ш·к·а, -и,   (*g pl*) -ш·ек·,   (*f*) (*dim*) of бля́х·а

~, мото́р·н·ая   (anat) motor end plate

боб·, -а́,   (*m*) bean

~, во́лч·ий   (bot) lupin, *Lupinus*

бобёр·, (*g sing*) бобр·а́,   (*m*) beaver skin (fur)

боби́н·а, -ы,   (*f*) bobbin, reel, spool

~, по·да·ю́щ·ая   (cinema) pay-out reel

~, при·ним·а́ющ·ая   (cinema) take-up reel

бобино·ре́з·к·а,   -и,   (*f*) paper cutting and reeling machine (cable, cigarette paper etc.)

боб·ко́в·ый, -ая, -ое,   (*a*) of боб· bean, fabacious

бобо·ви́д·н·ый, -ая, -ое,   (*a*) bean-shaped, fabiform, pisiform

боб·о́в·ый,   -ая, -ое, (*a*) of боб·; (min) pisolitic, pea; (bot) leguminous, (*pl*) *Leguminosae*

~ руд·а́   (*f*) (geol) pea grit

~ структу́р·а, -ы,   (*f*) (geol) shot structure

бо́бр·, -а,   (*m*) (zool) beaver, *Castor fiber*

~, боло́т·н·ый   coypu, swamp beaver, *Myocastor coypus*

бо́бр

~, камча́т·ск·ий   sea otter, *Enhydra lutris*

бо́бр·ик·, -а,   (*m*) (text) beaver cloth

бобр·о́в·ый, -ая, -ое, (*a*) of бобр·

~ стру·я́   (*f*) (pharm) castor

боб·ы́шк·а, -и,   (*f*) boss, teat; bolster stay (of bridge); (phot) core, spool

бовени́т·, -а,   (*m*) (min) bowenite

бога́т·ств·о, -а,   (*n*) riches, wealth

~, есте́ств·енн·ые   (*pl*) natural resources

~, минера́ль·н·ые   (*pl*) mineral resources

бога́т·ый, -ая, -ое,   (*a*) rich, wealthy; abundant, prolific; high grade (ore); (chem, met) rich in

~ мета́лл·ом   with high metal content

~ с·ме́с·ь   (*f*) rich mixture

бога́ч·е   (*comp adj*) richer, wealthier

бо́гхед·, -а,   (*m*) boghead coal

бо́д·, -а,   (*m*) (telecom) band

боденбендери́т·, -а,   (*m*) (min) bodenbenderite

бодмере́й·н·ый, -ая, -ое,   (*a*) of бодмере́·я

~ до·гово́р·   (*m*) (mar ins) bottomry bond

бодмере́·я, -и,   (*f*) (mar ins) bottomry

Бодо́, теле·гра́ф·, -а,   (*m*) Baudot telegraph system

бо́др·ый, -ая, -ое,   (*a*), (*s f*) бодр·, бодр·а́, бо́др·о, brisk, cheerful

бодр·и́ть, -я́т,   (*imp*) stimulate, invigorate

бодр·я́щ·ий, -ая, -ее, (*pres part act*) of бодр·и́ть; bracing, invigorating

Бодуи́н·а, реа́кци·я   (*f*) (chem) Baudouin reaction

бое-   (*abbr of component*) = бо·ев·о́й

бо·ев·и́к·, -а,   (*m*) (expl) primer; (cinema) hit

бо·ев·о́й, -а́я, -о́е, (*a*) military; battle, fighting, combat, operational; live, armed

~ голо́в·к·а   (*f*) warhead

~ де́й·стви·я   (*n pl*) operations, fighting, action

~ кно́п·к·а   (*f*) firing button

~ компле́кт·   (*m*) (air) operational load; (nav) outfit of ammunition, (mil) unit of fire

~ концентр·а́ци·я   (*f*) lethal concentration

~ ме́ст·о   (*n*) (mil) action station

~ механ·и́зм·   (*m*) (text) picker

бо·ев·о́й, -а́я, -о́е
~ **о·травл·я́юш·ее вещ·еств·о́** (*n*) poison/war gas, C.W. agent
~ **под·гото́в·к·а** (*f*) (mil) training
~ **по·ря́д·ок·** (*m*) (mil) battle order/ formation, operational disposition
~ **пост·** (*m*) (mil) action station
~ **пут·ь** (*m*) (air) bombing run
~ **радио·акти́в·н·ое вещ·еств·о́** (*n*) radioactive war material
~ **раз·воро́т·** (*m*) (air) chandelle
~ **сигна́л·** (*m*) (mil) alarm signal
~ **сре́д·ств·а** (*n pl*) armaments
~ **уч·е́ни·е** (*n*) (mil) action training
~ **част·ь** (*f*) warhead (of missiles etc.); (mil) operational unit; department (of warship)
бое·голо́в·к·а, -и, (*g pl*) **-в·ок·,** (*f*) warhead (of missiles)
бое·за·па́с·, -а, (*m*) ammunition, outfit of ammunition
бое·за·ря́д·, -а, (*m*) warhead charge
бо·ёк·, (*g sing*) **бой·к·а́,** (*m*) face, open die (forging); face (of a hand hammer); firing pin (of a gun); striker
~, **вс·по·мог·а́тельн·ый** pane, pean, peen (of a hand hammer)
~, **основ·н·о́й** face, hammering face
бо́·ен·, (*g pl*) of **бо́й·н·я**
бое·от·се́к·, -а, (*m*) warhead, warhead compartment
бое·пит·а́ни·е, -я, (*n*) munitions supply
бое·при·па́с·ы, -ов, (*pl*) ammunition
бое·спосо́б·ност·ь, -и, (*f*) combat/ fighting/operational efficiency
бое·у·кла́д·к·а, -и, (*f*) (mil) ammunition stowage bin
бо·е́ц·, (*g sing*) **бой·ц·а́,** (*i*) **бой·ц·о́м,** (*m*) soldier, fighter, warrior
Бо́зе-част·и́ц·а, -ы, (*f*) (nucl) Bose particle
бозо́н·, -а, (*m*) (nucl) boson, Bose particle
бо·й, -я, (*nom pl*) **бо·и́,** (*g pl*) **бо·ёв,** (*m*) battle, fight, action, fighting, combat; damaged goods, breakage, scrap; (horol) striking; (text) picking
~, **бли́ж·н·ий** close combat; (*gen as adj*) close-range
~, **ве́рх·н·ий** (text) overpick
~, **встре́ч·н·ый** (mil) contact, battle, engagement
~, **да́ль·н·ий** (*gen as adj*) long-range
~, **стеко́ль·н·ый** cullet, broken glass
бо́й·к·а (*g sing*) of **бо·ёк·** (q.v.)
бойко́т·, -а, (*m*) boycott

бойкот·и́р·овать, -уют, (*imp and perf*) boycott
бо́йлер·, -а, (*m*) cylinder-tank boiler
Бо́йл·ь, -я, (*m*) Boyle (surname)
Бо́йл·я, температу́р·а (*f*) (phys) Boyle temperature
~, **то́ч·к·а** (*f*) (phys) Boyle's temperature point
~ **-Марио́тт·а, за·ко́н·** (*m*) (phys) Boyle's law
бой·ни́ц·а, -ы, (*i*) **-ей,** (*f*) (mil) loophole, embrasure
бо́й·н·я, -и, (*g pl*) **бо́·ен·,** (*f*) slaughter house, abattoir
бо́й·щик·, -а, (*m*) breaker
~ **чугу́н·а** (met) pig breaker
бок·, -а, (*nom pl*) **-á,** (*g pl*) **-о́в,** (*m*) side, flank; profile; (shipb) sheer plan; (geol) wall (of deposit)
   в бок· (*adv*) sideways
~, **вис·я́ч·ий** hanging wall
~, **леж·а́ч·ий** (geol) lower wall
~ **о́ бок·** side by side
   по бок·а́м on either/each side
бока́л·, -а, (*m*) glass, goblet
бока́л·ец·, -л·ц·а, (*m*) (bot) scyphus
бокало·ви́д·н·ый, -ая, -ое, (*a*) cup-shaped, scyphiform, ampullaceous
бока́ль·чат·ый, -ая, -ое, (*a*) cup-shaped, scyphiform, cyathiform
бока́ль·чик·, -а, (*m*) (bot) cyathium
бо́кан·ец·, -н·ц·а, (*m*) (naut) outrigger
Бо́кли, вакуум·ме́тр· (*m*) Buckley pressure gauge
боко·бо́ч·н·ый, -ая, -ое, (*a*) collateral
бок·ови́н·а, -ы, (*f*) side, side piece; side-wall (of tires etc.)
бок·ов·о́й, -а́я, -о́е, (*a*) side, lateral, flank, cross
~ **ве́тер·** (*m*) cross wind
~ **вы́·ём·к·а** (*f*) (min) side cut
~ **ка́ч·к·а** (*f*) rolling (ship motion etc.)
~ **по́л·е** (*n*) (print) margin
~ **поло́с·а** (*f*) (rad) sideband
~ **со·един·е́ни·е** (*n*) (weld) edge joint
~ **с·пу́ск·** (*m*) (shipb) broadside launching
~ **част·от·а́** (*f*) (rad) side frequency
бок·ову́шк·а, -и, (*g pl*) **-ш·ек·,** (print) marginal note
бо́к·ом (*adv*) sideways, edge-on; (naut) beam on
боко·не́рв·н·ый, -ая, -ое, (*a*) side-veined, amphineurous, (*pl as noun*) (zool) *Amphineura*

боко·пест·ичн·ый, -ая, -ое, (*a*) (bot) pleurogynous

боко·плáв·, -а, (*m*) (zool) amphipod, scud, (*pl*) amphipods, *Amphipoda*

боко·плóд·н·ый, -ая, -ое, (*a*) (bot) pleurocarpous

боко·цвéт·н·ый, -ая, -ое, (*a*) (bot) racemose

бóкс·, -а, (*m*) box; (med) isolation room; horse box (transport); boxing (sport)

~, звуко·за·глуш·áющ·ий (cinema) blimp, bungalow, sound-proofed booth

~, огн·ев·óй (mil) firing bay

~, перчáт·очн·ый (nucl) glove box

боксúт·, -а, (*m*) (min) bauxite

~, бемúт·ов·ый boehmite

~, диаспóр·ов·ый diaspore bauxite

боксúтúт·, -а, (*m*) (min) bauxitite

болвáн·, -а, (*m*) block mould, dummy

болвáн·к·а, -и, (*g pl*) **-н·ок·,** (*f*) block, block mould, dummy; (met) billet; (timber) 7-inch baulk; dummy, dummy model

бол·ев·óй, -áя, -óе, (*a*) painful

бóл·ее (*adv*) more, more than, above (of temperatures etc.)

~ вс·егó most of all

~ выс·óк·ий higher

~ или мéн·ее more or less

тем бóл·ее all the more

бол·éзненн·ый, -ая, -ое, (*a*) ailing, sickly, morbid

болезне·твóр·н·ый, -ая, -ое, (*a*) pathogenic

болезне·у·стóй·чив·ый, -ая, -ое, (*a*) disease-resistant

бол·éзн·ь, -и, (*f*) illness, disease, malady, sickness

~, англ·úйск·ая rickets

~, водо·рóд·н·ая (met) hydrogen embrittlement

~, воз·дýш·н·ая air sickness

~, выс·óтн·ая altitude sickness

~, за·рáз·н·ая infectious/contagious disease

~, кессóн·н·ая caisson disease, the bends

~, луч·ев·áя radiation sickness

~, мор·ск·áя sea sickness

~, сáхар·н·ая diabetes

~, сен·н·áя hay fever

~, с·крбі·т·ая insidious disease

бол·é·ть, -ют, (*imp*) (+ *instr*) be ill; ache, hurt; support (a contestant)

боле·утол·яющ·ий, -ая, -ое, (*a*) analgesic, analgetic

боле·утол·яющ·ий, -ая, -ое

~ срéд·ств·о (*n*) (pharm) analgesic

болúд·, -а, (*m*) (astron) bolide, fire ball

болó·метр·, -а, (*m*) (phys) bolometer

боло·метр·úческ·ий, -ая, -ое, (*a*) bolometric

болóт·ист·ый, -ая, ое, (*a*) marshy, boggy, swampy (having many marshes etc.)

~ мéст·ност·ь (*f*) marshland

болóт·н·ый, -ая, -ое, (*a*) of **болóт·о**

~ газ· (*m*) marsh gas, methane

~ лихорáд·к·а (med) malaria

болóт·о, -а, (*n*) marsh, bog, swamp, morass

~, за·сóл·енн·ое salt marsh

~, луг·ов·óе meadow moor

~ лун·ы́ (astron) palus, lunar marsh

~, сол·ён·ое salina

~, торф·ян·óе peat bog/moor

бóлт·, -á, (*m*) bolt

~, áнкер·н·ый tie rod, long screw bolt

~, вз·рыв·н·óй explosive bolt

~, глáв·н·ый king bolt

~, за·ёрш·енн·ый rag/stone/lewis bolt

~, контáкт·н·ый contact bolt

~, крюч·ков·óй hook bolt

~, на·рез·н·óй screw bolt

~, о·слáбл·енн·ый started bolt

~, пут·ев·óй (rail) fish bolt

~, раз·рыв·н·óй explosive bolt

~, стóпор·н·ый locking pin

~, у·пóр·н·ый stop, stay bolt

~, уш·кóв·ый eye bolt

~, фундамéнт·н·ый holding-down bolt, foundation bolt

~, чёрн·ый black bolt, rough-finished bolt

~, шарнúр·н·ый link bolt, swing bolt

~, шатýн·н·ый (M/T) connecting-rod bolt, big-end bolt

болт·áнк·а, -и, (*g pl*) **-н·ок·,** (*f*) (air) bump, rough air

болт·á·ть, -ют, (*imp*) stir; dangle; chatter; **-ся,** dangle, hang loosely; hang about, lounge about

болт·лúв·ый, -ая, -ое, (*a*) talkative, garrulous

болт·ов·óй, -áя, -óе, (*a*) of **болт·**

болто·кóв·очн·ая машúн·а (*f*) bolt-heading machine

болт·óвщик·, -а, (*m*) bolter-down (rolling mill operative)

болто·за·клёп·очн·ый, -ая, -ое, (*a*) bolt-and-rivet, bolting-and-rivetting

болто·рéз·н·ый, -ая, -ое,   (*a*) bolt threading

~ стан·óк·   (*m*) bolt-threading machine

болт·ýнь·я, -и, (*g pl*) **-ни·й,** (*f*) scrambled eggs; chatterbox (woman)

болт·ýшк·а, -и,   (*g pl*) **-шек·,** (*f*) mash

бóл·ь, -и, (*f*) pain, ache

бóльверк·, -а,   (*m*) (civ eng) camp sheeting, sheet piling, retaining wall

боль·н·áя, -óй, (*f decl as adj*) patient, female patient

боль·нíц·а, -ы, (*i*) **-ей,** (*f*) hospital

боль·нíчн·ый, -ая, -ое, (*a*) of **боль·-нíц·а**

~ лист·óк·   (*m*) medical certificate

бóль·н·о   (*predic adj*) it is painful; (*adv*) painfully; extremely, terribly

боль·н·óй, -áя, -óе, (*a*) sick, diseased, ill; (*m decl as adj*) patient, male patient

Бóльцман·а, по·сто·я́нн·ая (*f*) (phys) Boltzmann's constant

бóльш·е (*comp* of **больш·óй**) bigger, larger, greater; more (than)

больше·берц·óв·ый, -ая, -ое,   (*a*) (anat) tibial

больше·голóв·ый, -ая, -ое, (*a*) (zool) macrocephalic, large-headed

бóльш·ий, -ая, -ее, (*a*) larger, greater, bigger, higher

больш·инств·ó, -á, (*n*) majority, greater part, most

больш·óй, -áя, -óе, (*a*) large, great, big, high, major

~ бýкв·а (*f*) capital letter

~ дорóг·а (*f*) highway, highroad

Больш·óй Пёс·   (*m*) (astron) Canis Major

бóлюс·, -а, (*m*) (bot) bole

бóмб·а, -ы, (*f*) bomb

~, áтом·н·ая   atom bomb, A-bomb, nuclear fission weapon

~, броне·бóй·н·ая   armour-piercing bomb

~, водо·рóд·н·ая   hydrogen bomb, H-bomb

~, вулкан·и́ческ·ая   (geol) volcanic bomb, bomb

~, глуб·и́нн·ая   (naut) depth charge/bomb; (air) anti-submarine bomb

~, дел·éни·я   fission bomb

~, за·жиг·áтельн·ая   incendiary bomb

~, калори·метр·и́ческ·ая   (chem) bomb calorimeter

~, малер·а   bomb calorimeter

бóмб·а

~, номинáль·н·ая áтом·н·ая   nominal atomic bomb (equiv. 20,000 tons of TNT)

~, о·свет·и́тельн·ая   flash bomb

~, о·с·кóл·очн·ая   anti-personnel bomb, fragmentation bomb

~, противо·лóд·очн·ая   (naut) depth charge/bomb; (air) anti-submarine bomb

~, свет·я́щ·ая   aircraft flare

~, термо·я́дер·н·ая   thermonuclear bomb, nuclear fission bomb

бомбáж·, -а, (*m*) (food) blowing, bulging (of cans)

бомбард·и́р·, -а, (*m*) (air) bomb aimer; (obs) bombardier

бомбард·ировáни·е, -я, (*n*) *see* **бомбард·ировк·а**

бомбард·ир·овáть, -ýют, (*imp*) bomb; bombard

бомбард·ирóвк·а, -и, (*g pl*) **-в·ок·,** (*f*) (air) bombing; (gunn) bombardment

~, ио́н·н·ая   (nucl) ion bombardment, back bombardment

~, обрáт·н·ая   (nucl) back-bombardment

~, пере·крéст·н·ая   (nucl) cross bombardment

бомбард·ирóвочн·ый, -ая, -ое,   (*a*) bomber, bombing, bombardment

бомбард·ирóвщик·, -а,   (*m*) (a/c) bomber

~, áтом·н·ый   atom bomber; nuclear-powered bomber

~, блíж·н·ий   light bomber

~, дáль·н·ий   heavy bomber

бомбард·ирýем·ый, -ая, -ое, (*pres part pass*) of **бомбард·ир·овáть;** (nucl) bombarded, struck, target

~ част·и́ц·а (*f*) (nucl) struck particle

бомбард·ирýющ·ий, -ая, -ее, (*pres part act*) of **бомбард·ир·овáть;** (nucl) incident

~ част·и́ц·а (*f*) (nucl) incident particle

бомб·ёжк·а, -и, (*g pl*) **-жек·,** (*f*) (mil, air) bombing

бомб·ирóвк·а, -и, (*f*) (rubber) crown, crowning (of rolls)

бомб·и́ть, -я́т, (*imp*) (air) bomb

бомбо·вóз·, -а, (*m*) (obs) = **бомбард·ирóвщик·**

бóмб·ов·ый, -ая, -ое, (*a*) of **бóмб·а**

~ от·сéк·   (*m*) (a/c) bomb bay

~ плутóн·и·й   (*m*) (nucl) bomb-grade plutonium

бомбо·держ·а́тел·ь, -я, (*m*) (a/c) bomb rack/carrier

бомбо·дро́м, -а, (*m*) bombing range

бомбо·мёт·, -а, (*m*) (naut) depth-charge thrower

~, реакти́в·н·ый antisubmarine mortar

бомбо·мет·а́ни·е, -я, (*n*) bombing; depth-charge throwing

~, выс·о́тн·ое high-level bombing

~, групп·ов·о́е pattern bombing

бомбо·мет·а́тельн·ый, -ая, -ое, (*a*) bomb-throwing; depth-charge throwing

бомбо·не·про·бив·а́ем·ый, -ая, -ое, (*a*) bomb-proof

бомбо·от·се́к·, -а, (*m*) (a/c) bomb bay

бомбо·при·це́л·, -а, (*m*) (a/c) bomb sight

бомбо·с·бра́с·ывател·ь, -я, (*m*) (naut) depth-charge rails; (air) bomb-release gear

бомбо·у·бе́ж·ищ·е, -а, (*i*) **-ем,** (*n*) air-raid shelter

бомбо·у·кры́т·и·е, -я, (*n*) air-raid shelter

Боме́ (*m indecl*) Beaumé (surname)

~, гра́дус·ы (*m pl*) degrees Beaumé, °Bé

Бомэ́, *see* **Боме́**

бо́н·, -а, (*m*) (naut) boom, harbour boom

бо́н·а, -ы, (*f*) (fin) bond, bill; money order; (*pl*) paper money

бона́нц·а, -ы, (*f*) (min) bonanza

бонда́р·н·о-тока́р·н·ый стан·о́к· (*m*) cooper's lathe

бонда́р·н·ый, -ая, -ое, (*a*) cooper's, coopering, cooperage

~ про·из·во́д·ств·о (*n*) cooperage

бонда́р·ь, -я́, (*m*) cooper

бонда́р·н·я, -и, (*g pl*) **-р·ен·,** (*f*) cooperage

бондериз·а́ци·я, -и, (*f*) (met) bonderizing

бо́нд·ов·ый, -ая, -ое, (*a*) bonded (customs)

~ с·кла́д· (*m*) bonded warehouse

бонит·ёр·, -а, (*m*) (agr) grader

бонит·е́т·, -а, (*m*) estimated productivity (of a forest etc.)

~ место·про·из·раст·а́ни·я (agr) soil site index

бонит·иро́вк·а, -и, (*g pl*) **-вок·,** (*f*) (agr) judging, valuation, graduation

бонит·иро́вщик·, -а, (*m*) classifier

бонифик·а́ци·я, -и, (*f*) (com) bonification

~, обра́т·н·ая retrobonification

Бонне́ (*m indecl*) Bonnet (surname)

бо́н·ов·ый, -ая, -ое, (*a*) of **бон·**

~ за·гражд·е́ни·е (*n*) (naut) boom defence

~ су́д·н·о (*n*) boom-defence vessel

бо́р·, -а, (*m*) (chem) boron, B; pine-wood; (bot) milletgrass, *Milium*; **Бо́р·, -а,** (*m*) Bohr (surname) *see* **бор·а́**

~, азо́т·ист·ый boron nitride

~, трёх·фто́р·ист·ый boron trifluoride

~, фто́р·ист·ый fluoboric acid

бор·а́, -ы́, (*f*) (meteor) bora

Бо́р·а, а́том· (phys) Bohr's atom

~, магнето́н· Bohr magneton

бораго́ (*m indecl*) (bot) borage, *Borago*

боразо́н·, -а, (*m*) (chem) borazon (cubic form of boron nitride)

бор·а́л·, -а, (*m*) *see* **бор·а́л·ь**

бор·алки́л·, -а, (*m*) (chem) boron alkyl, borin

бор·а́л·ь, -я, (*m*) (nucl, met) boral

бор·а́н·, -а, (*m*) (chem) borane, boron hydride

бор·ари́л·, -а, (*m*) boron aryl

бор·а́т·, -а, (*m*) (chem) borate

~ на́тр·и·я sodium borate

бора́т·н·ый, -ая, -ое, (*a*) of **бор·а́т·**

~ крон· (*m*) borosilicate crown (glass)

борато·фтористо·водо·ро́д·н·ая кис·л·от·а́ (*f*) boratofluoric acid

бор·аци́т·, -а, (*m*) (min) boracite, boric spar

бо́ргес·, -а, (*m*) (print) bourgeois, 9-point type

бордо́ (*n indecl*) claret-type wine; claret (rare); wine-coloured, claret-coloured; Bordeaux (dye)

бордо́с·ск·ая жи́дк·ост·ь (*f*) (hortic) Bordeaux mixture

бо́рдс·ы, -ов, (*m pl*) (com) boards (timber)

бордю́р·, -а, (*m*) kerb (of road); edging, border; (astron) limb

боре·а́льн·ый, -ая, -ое, (*a*) (biol) boreal

бор·е́ц·, (*g*) **бор·ц·а́,** (*m*) fighter, champion (of causes); wrestler (sport)

бор·и́д·, -а, (*m*) (chem) boride

бор·и́л·, -а, (*m*) (chem) boryl

бор·и́ровани·е, -я, (*n*) (met) boronizing, borating

бор·и́рованн·ый, -ая, -ое, (*pres part pass*) borated, boronized

бо́р·ист·ый, -ая, -ое,   (*a*) (chem) boride of; boron

~ **алюмин·и·й** (*m*) aluminium boride

~ **ста́л·ь** (*f*) (met) boron steel

бор·ка́льк·, -а, (*m*) calcium boride

бор·маши́н·а, -ы, (*f*) dentist's drill

бормот·а́ни·е, -я, (*n*) (*v n*) *see* **бормот·а́ть;** (telecom) babble

бормот·а́ть, -о́ч·ут, (*imp*) mutter, mumble, (telecom) babble

бор·мет·и́л·, -а, (*m*) (chem) trimethyl-borine

бормо́ч·ущ·ий, -ая, -ее, (*pres part act*) of **бормот·а́ть**

борн·, -а, (*m*) (elec) battery terminal

борнео́л·, -а, (*m*) (chem) borneol

борни́л·, -а, (*m*) (chem) bornyl

~**, хло́р·ист·ый** bornyl chloride

борнил·хлори́д·, -а, (*m*) bornyl chloride

борни́т·, -а, (*m*) (min) bornite (a copper ore)

бо́рн·овск·ое при·ближ·е́ни·е (*n*) (math) Born approximation

борно·ки́сл·ый, -ая, -ое, (*a*) borate of

~ **ба́р·и·й** (*m*) barium borate

борно·эти́л·ов·ый эфи́р· (*m*) borate ester

бо́р·н·ый, -ая, -ое, (*a*) boric, boracic

~ **ан·гидр·и́д·** (*m*) boric anhydride, boron oxide, boron trioxide

~ **ка́мер·а** (*f*) (nucl) boron ionization chamber

~ **кисл·от·а́** (*f*) boric acid, boracic acid

бо́ров·, -а, (*nom pl*) **-а́,** (*g pl*) **-о́в,** (*m*) flue; **бо́ров·, -а,** (*nom pl*) **-ы,** (*g pl*) **-о́в,** (*m*) (agr) hog

бор·ови́к·, -а́, (*m*) (bot) cep, edible boletus, *Boletus edulis*

боров·ко́в·ый, -ая, -ое, (*a*) of **бо́-ров·о́к·**

боро·водо·ро́д·, -а, (*m*) hydroboron, boron hydride, borane

бор·ов·о́й, -а́я, -о́е, (*a*) of **бор·;** pinewood

бо·ров·о́к·, -в·к·а́, (*m*) bridge wall (of boiler); baffle plate

бо́р·овск·ий, -ая, -ое, (*a*) (nucl) Bohr

бор·о́вш·ийся, -аяся, -ееся, (*past part act*) of **бор·о́ться**

бо́ров·ый, -ая, -ое, (*a*) of **бо́ров·;** *see also* **бор·ов·о́й**

бород·а́, (*acc sing*) бо́род·у, (*g sing*) бород·ы́, (*nom pl*) бо́род·ы, (*g pl*) боро́д·, (*f*) beard, whiskers

борода́в·к·а, -и, (*g pl*) -в·ок·, (*f*) (med) wart, verruca

борода́в·чат·ый, -ая, -ое, (*a*) warty

бород·а́т·ый, -ая, -ое, (*a*) bearded; (bot) barbed, barbate

боро́д·к·а, -и, (*g pl*) -д·ок·, (*f*) (*dim*) of **бород·а́;** barb, tuft; (zool) beard; (ornith) barbule

бород·о́к·, -д·к·а́, (*m*) punch, broach, drift (tool)

бород·о́чн·ый, -ая, -ое, (*a*) tufty, bearded

~ **трос·** (*m*) second-grade hemp rope

**борозд·а́, -ы, (*nom pl*) бо́розд·ы, (*g pl*) боро́зд·, (*see also* боро́зд·к·а) furrow, groove; (anat) fissure, sulcus; frog (on a brick); crazy cracking (in gun bore)

~**, по·жа́бер·н·ая** (zool) endostyle

борозд·и́ть, -я́т, (*imp*) furrow, plough; make a groove, channel

борозд·к·а, -и, (*g pl*) -д·ок·, (*f*) (*dim*) of **борозд·а́** (q.v.); groove, slot, furrow, channel

~**, дожд·ев·а́я** rain rill

борозд·ов·о́й, -а́я, -о́е, (*a*) (anat) grooved, furrowed, sulcate

борозд·о·де́л·ател·ь, -я, (*m*) furrower

борозд·о·ме́р·, -а, (*m*) (agr) furrow measuring rod

боро́зд·чатост·ь, -и, (*f*) striation, sulcation

боро́зд·чат·ый, -ая, -ое, (*a*) grooved, furrowed, sulcate; striated

борозж·ённ·ый, -ая, -ое, (*past part pass*) of **борозд·и́ть**

борон·а́, -ы, (*acc sing*) бо́рон·у, (*nom pl*) бо́рон·ы, (*g pl*) боро́н·, (*f*) (agr) harrow

~**, ди́ск·ов·ая** (agr) disk harrow

~**, пруж·и́нн·ая** (agr) spring-tooth harrow

~**, рыча́ж·н·ая** (agr) lever harrow

боронатрокальци́т·, -а, (*m*) (min) boronatrocalcite, ulexite

борон·е́ни·е, -я, (*n*) harrowing

борон·и́ть, -я́т, (*imp*) harrow

борон·ова́ть, -у́ют, (*imp*) *see* **борон·и́ть**

бор·орган·и́ческ·ий, -ая, -ое, (*a*) (chem) organo-boron

боро·силика́т·, -а, (*m*) borosilicate

боро·ско́п·, -а, (*m*) (nucl) borescope

**бор·о́ться, (*pres 3rd sing*) бо́р·ется, (*pres 3rd pl*) бо́р·ются, (*imp*) struggle, strive, fight; wrestle (sport)

боро·фтористо·водо·ро́д·н·ая   **ки-сл·от·а́**   (*f*) hydrofluoboric acid

борт·, -а, (*nom pl*) **-а́,** (*g pl*) **-о́в,** (*m*) edge, rim, border; side (of ship etc.), board (e.g. "on board"); bead (of tire); bort (industrial diamond); lapel (of clothing)

   за́ борт· overboard (motion)

   за бо́рт·ом outboard, overboard (position)

~, за́д·н·ий (M/T) tail-board

~, ле́в·ый port side

   на́ борт· onboard (motion)

   на борт·у́ onboard (position)

~, над·во́д·н·ый (naut) freeboard

~ о́ борт· (naut) alongside

~, пра́в·ый (naut) starboard side

борт- (*root*) (air) flight, flying, airborne; (rocket) rocket (as opposed to ground)

борт·журна́л·, -а, (*m*) (air) log-book, aircraft log-book

бо́рт·ик·, -а, (*m*) (text) welt; (build) upturn

борт·инжене́р·, -а, (*m*) (air) flight engineer

борт·меха́ник·, -а, (*m*) (air) flight mechanic

борт·ова́льн·ый пресс· (*m*) flanging press

борт·ов·о́й, -а́я, -о́е, (*a*) of **борт·** (a/c) flight, flying, airborne

~ журна́л· (*m*) (air) flight/navigation log

~ за́д·н·ий шварто́в· (*m*) (naut) after head spring

~ за·ря́д·к·а (*f*) (air, elec) charging *in situ*

~ ка́ч·к·а (*f*) rolling (ship motion)

~ зна́к· (*m*) (naut) pennant number

~ ли́н·и·я корабл·я́ (*f*) (shipb) plan outline

~ об·ору́д·ovani·e (*n*) (air) airborne equipment

~ ого́н·ь (*m*) (air) navigation light; (naut) side light

~ пере·да́ч·а (*f*) final drive (of tractor etc.)

~ радио·ста́нци·я (*f*) (air) aircraft radio set

~ стри́нгер· (*m*) (shipb) side stringer

~ фрикцио́н· (*m*) steering clutch (tractor)

борто·ре́з·к·а, -и, (*g pl*) **-з·ок·,** (*f*) debeader; debeading (tyres)

борт·про·вод·ни́к·, -а́, (*m*) (air) steward

борт·про·вод·ни́ц·а, -ы, (*f*) (air) stewardess, air hostess

бор·триалки́л·, -а, (*m*) boron trialkyl

бор·триари́л·, -а, (*m*) boron triaryl

борули́н·, -а, (*m*) borulin (an asbo-bitumen damp-proofing material)

бор·ц·ы́, (*nom pl*) of **бор·е́ц·**

бор·шта́нг·а, -и, (*f*) boring bar

бо́рщ·, -а́, (*i*) **-о́м,** (*m*) borsch (beetroot and cabbage soup)

бор·ьб·а́, -ы́, (*f*) struggle, fight, fighting; prevention, countermeasures; control (of pests); wrestling (sport)

~ с по·жа́р·ами fire prevention, fire fighting

бор·ю́щ·ийся, -аяся, -ееся, (*pres part act*) see **бор·о́ться**

бос·ик·о́м (*adv*) barefoot

босо·но́г·ий, -ая, -ое, (*a*) barefooted

босо·но́ж·к·а, -и, (*g pl*) **-ж·ек·,** (*f*) sandal (shoe)

бостони́т·, -а, (*m*) (geol) bostonite

бо́т·, -а, (*m*) boat, cutter

~, ло́цман·ск·ий pilot boat

бота́н·ик·, -а, (*m*) botanist

бота́н·ик·а, -и, (*f*) botany

ботан·и́ческ·ий, -ая, -ое, (*a*) botanical

ботв·а́, -ы́, (*f*) (agr) haulm, top

ботво·ре́з·, -а, (*m*) (agr) beet-topping machine

бо́т·дек·, -а, (*m*) (naut) boat deck

бо́т·ик·, -а, (*m*) galosh, overshoe; (naut) jolly boat

боти́н·ок·, -н·к·а, (*m*) boot, bootee

бо́т·ник·, -а, (*m*) boat deck

ботрио·ли́т·, -а, (*m*) (min) botryolite

ботрио·мик·о́з·, -а, (*m*) (vet) botryo-mycosis

ботр·и́ческ·ий, -ая, -ое, (*a*) (bot) racemose, botrytic

ботро·де́ндр·ов·ые, -ых, (*pl decl as adj*) (bot) bothrodendrons, *Bothrodendraceae*

ботто́м·, втор·о́й (*m*) (met) nickel bottom

ботул·и́зм·, -а, (*m*) (med) botulism

Бо́форт·а, балл· (*m*) (meteor) degree on Beaufort scale, wind strength

~, шкал·а́ (*f*) (meteor) Beaufort scale

бо́цман·, -а, (*nom pl*) **-а́,** (*m*) (naut) boatswain; boatswain's mate

боч·а́р·, -а, (*m*) cooper

боч·ёнок·, -нк·а, (*m*) (*dim*) of **бо́ч·к·а**

бо́ч·к·а, -и, (*g pl*) **-ч·ек·,** (*f*) barrel, cask, keg; (air) roll; (naut) mooring buoy; leader pass (metal rolling)

~ вал·к·а́ (met) roll body

бо́ч·к·а
~, вос·ход·я́щ·ая   (air) upward roll
~, за·ме́дл·енн·ая   (air) slow roll
~, при·ча́ль·н·ая   mooring buoy
бо́ч·ков·ый, -ая, -ое, (*a*) of **бо́ч·к·а**
бо́ч·к·о́м (*adv*) sideways
бочко·обра́з·н·ый, -ая, -ое, (*a*) barrel-shaped, pan (of rivet heads etc.); (biol) doliiform, dolioform
боч·о́к·, -ч·к·а́, (*m*) flank (of carcase)
боч·о́нк·ом (*adv*) V-formation, in V-formation
боч·о́нок·, -нк·а, (*m*) (*dim*) of **бо́ч·-к·а**; keg, small barrel
Боэ́ци·й, -я, (*m*) Boetius, Boethius (surname)
бо·я́ться (*imp* + *genitive*) (*pres 1st, 3rd sing, 3rd pl*) **бо·ю́сь, бо·и́тся, бо·я́тся** fear, be afraid of, dread
бра (*n indecl*) bracket, wall bracket
Браве́, за·ко́н· (*m*) (cryst) Bravais law
~, реше́т·к·а   (*f*) (cryst) Bravais lattice
бра́·вш·ий, -ая, -ее, (*past part act*) of **бр·ать**
бра́г·а, -и, (*f*) small beer, home-brewed beer; (naut) towing strop
браги́т·, -а, (*m*) (min) bragite
браго·пере·го́н·н·ый аппара́т· (*m*) (chem) beer still
брадзо́т·, -а, (*m*) (vet) bradsot, braxy
бради·кард·и́·я, -и, (*f*) (med) bradycardia
бради·кине́з·и·я, -и, (*f*) (med) bradykinesia
бра́ж·к·а, -и, (*g pl*) -ж·ек·, (*f*) (chem) beer; must (winemaking)
брази́ль·ск·ий, -ая, -ое, (*a*) Brazilian, Brazil
бра́йл·евск·ий, -ая, -ое, (*a*) braille
Бра́йл·ь, -я, (*m*) Braille (surname)
Бра́йт·ов·а, бол·е́зн·ь (*f*) (med) Bright's disease
бра́йтсток·, -а, (*m*) (oil) bright stock
бра́к·, -а, (*m*) marriage; defect, defective part, reject, waste
бракебуши́т·, -а, (*m*) (min) brackebuschite
брак·ёр·, -а, (*m*) = **брак·о́вщик·** inspector, sorter and counter (of rejects)
брак·ера́ж·, -а, (*i*) -ем, (*m*) inspection, sorting (for rejects)
браке́т·а, -ы, (*f*) bracket
браке́т·н·ый, -ая, -ое, (*a*) of **браке́т·а**

брак·о́ванн·ый, -ая, -ое, (*past part pass*) of **брак·ова́ть**; condemned, rejected, unserviceable
брак·ова́ть, -у́ют, (*imp*) reject, condemn, scrap
брак·о́вк·а, -и, (*g pl*) -вок·, (*f*), (*v n*) of **брак·ова́ть**; rejection, condemnation
брак·о́вщик·, -а, (*m*) inspector, sorter and counter (of rejects)
брако·де́л·, -а, (*m*) bad workman, bungler
браконье́р·, -а, (*m*) poacher
бракте́·я, -и, (*f*) (bot) bract
бр·ал, (*past masc sing*) of **бр·ать**
бран·вахт·а, -ы, (*f*) see **бранд·ва́хт·а**
бранд·ва́хт·а, -ы, (*f*) fire float; guard ship; dredger depot ship; (mil) bridge protection group
бра́нд·ер·, -а, (*m*) blockship
бранд·майо́р·, -а, (*m*) chief fire officer (Brit), fire chief (U.S.)
бранд·ма́уер·, -а, (*m*) (build) fire wall, fire-resisting wall
бранд·ме́йстер·, -а, (*m*) = **бранд·-майо́р·**
бранд·спо́йт·, -а, (*m*) fire-hose nozzle; manual fire pump
браннери́т·, -а, (*m*) (min) brannerite
бр·а́нн·ый, -ая, -ое, (*past part pass*) of **бр·ать**
бра́с·, -а, (*m*) (naut) brace (cordage)
брасле́т·, -а, (*m*) bracelet
брат·, -а, (*nom pl*) **бра́т·ья,** (*g pl*) **бра́т·ьев,** (*m*) brother
бра́т·ск·ий, -ая, -ое, (*a*) fraternal, brotherly; Bratsk (town in Siberia)
бр·ать, (*pres 1st, 3rd sing, 3rd pl*) **бер·у́, бер·ёт, бер·у́т,** (*past masc, fem, neut*) **бр·а́л, бр·ала́, бр·а́ло** (*imp*) take; **-ся** undertake, set about, start (work etc.)
~ взайм·ы́   borrow
~ во в·ним·а́ни·е   draw attention to
~ ку́рс· set out for; (naut) shape course
~ на·ло́г·и   tax
~ под· аре́ст· arrest, put under arrest
бра́т·ья, (*nom pl*) of **брат·**
Бра́ун·, -а, (*m*) Brown (surname, English origin); Braun (surname, German origin)
Бра́унер·, -а, (*m*) Brauner (surname)
Бра́унинг·, -а, (*m*) Browning (surname)
брауни́т·, -а, (*m*) (min) brownite
бра́ун·овск·ий, -ая, -ое, (*a*) and see **бро́ун·овск·ий** (phys) Brownian, Braunian

Бра́уэр·, -а, *(m)* Brouwer (surname, Dutch origin)

брахи·антиклина́л·ь, -и, *(f)* (geol) brachy-anticline

брахи·би·пирами́д·а, -ы, *(f)* (cryst) brachydipyramid

брахи·диагона́ль·н·ый, -ая, -ое, *(a)* brachydiagonal

брахи·до́м·а, -ы, *(f)* (cryst) brachy-dome

брахи·дом·ати́ческ·ий, -ая, -ое, *(a)* (cryst) brachydomatic

брахи·о́л·а, -ы, *(f)* (biol) brachiole

брахио·ля́ри·я, -и, *(f)* (zool) brachiolaria

брахио·по́д·ы, -ов, *(m pl)* (zool) brachiopods, *Brachiopoda*

брахи·о́с·ь, -и, *(f)* brachy-axis

брахи·пирами́д·а, -ы, *(f)* (cryst) brachypyramid

брахи·цефа́ль·н·ый, -ая, -ое, *(a)* (zool) brachycephalic

брахисто·хро́н·а, -ы, *(f)* (math) brachistochrone

бра́ч·н·ый, -ая, -ое, *(a)* marriage, conjugal; (zool) breeding, nuptial, mating

~ **на·ря́д·** *(m)* (zool) breeding dress

~ **о·пер·е́ни·е** *(n)* (zool) nuptial plumage

~ **пери́од·** *(m)* (zool) mating season

бра́·шпил·ь, -я, *(m)* (naut) windlass

бревён·чат·ый, -ая, -ое, *(a)* timbered, log

брев·и·й, -я, *(m)* (chem) brevium, protoactinium isotope 234

бревн·о́, -а́, *(nom pl)* **брёвн·а,** *(g pl)* **брёвен·,** *(n)* log

бревно·ка́т·к·а, -и, *(g pl)* **-т·ок·,** *(f)* log elevator

бревно·с·бра́с·ыватэл·ь, -я, *(m)* log kicker

бревно·с·пу́ск·, -а, *(m)* log run (by-pass to hydroelectric works)

бревно·та́ск·а, -и, *(f)* log haul/slip/way

бревно·у·кла́д·чик·, -а, *(m)* log stacker

бреггери́т·, -а, *(m)* (min) broeggerite

бред·, -а, *(m)* delirium

бред·ен·ь, -д·н·я, *(m)* (fish) drag net

бре́д·ивш·ий, -ая, -ее, *(past part act)* of **бре́д·ить**

бре́д·ить, -ят, *(imp)* be delirious, rave

бред·ов·о́й, -а́я, -о́е, *(a)* of **бред·**

бред·у́щ·ий, -ая, -ое, *(pres part act)* of **брес·ти́**

бре́д·ш·ий, -ая, -ее, *(past part act)* of **брес·ти́**

бре́д·ящ·ий, -ая, -ее, *(pres part act)* of **бре́д·ить**

бр·е́ем·ый, -ая, -ое, *(pres part pass)* of **бр·и́ть**

брезе́нт·, -а, *(m)* tarpaulin

брезент·и́р·овать, -уют, *(imp and perf)* (naut) batten down (hatches)

брейд-вы́мпел·, -а, *(m)* (naut) broad pennant, senior officer's pennant

брейнери́т·, -а, *(m)* (min) breunnerite

бре́й·ск·ий, -ая, -ое, *(a)* (biol) brea

Бре́йс·а, компенс·а́тор· *(m)* (chem) Brace (half-shadow) compensator

Брейт·а-Ви́гнер·а, фо́рмул·а *(f)* (nucl) Breit–Wigner formula

бре́йт·овск·ий, -ая, -ое, *(a)* (math) Breit, Breit's

бреква́тер·, -а, *(m)* breakwater

бре́кер·, -а, *(m)* breaker, breaker strip (of tire)

брекч·и́рованн·ый, -ая, -ое, *(a)* (geol) brecciated

бре́кчи·я, -и, *(f)* (geol) breccia

~ **тр·е́ни·я** (geol) fault breccia

брело́к·, -а, *(m)* suspender

бре́мен·и, *(g, d, p, sing)* of **бре́м·я**

бре́мсберг·, -а, *(m)* (min) gravity plane/incline

бре́м·я, *(g, d, p, sing)* **бре́мен·и,** *(nom pl)* **бремен·а́,** *(n)* burden, load

Бре́нкен·а, при·бо́р· *(m)* (oil) Brenken flashpoint apparatus

брес·ти́, *(pres 3rd sing, 3rd pl)* **бред·ёт, бред·у́т,** *(past masc, fem sing)* **брё·л, бре·ла́** *(imp deter)* make one's way, go with difficulty

брестро́п·, -а, *(m)* (naut) breast rope

брете́ль·к·а, -и, *(g pl)* **-л·ек·,** *(f)* shoulder strap (of under garment)

бретон·ск·ий, -ая, -ое, *(a)* (geol) Bretonian

бре́штук·, -а, *(m)* (naut) breast hook

бре́ш·ь, -и, *(f)* breach, gap, break

бр·е́ющ·ий, -ая, -ее, *(pres part act)* of **бр·и́ть**

~ **по·лёт·** *(m)* (air) low-level flight

Брианшо́н·а, теоре́м·а *(f)* (math) Brianchon theorem

бр·и́вш·ий, -ая, -ее, *(past part act)* of **бр·и́ть**

бри́г·, -а, *(m)* (naut) brig

брига́д·а, -ы, *(f)* gang, squad, crew, team; (mil) brigade; (naut) squadron

~**, авар·и́йн·ая** emergency crew

брига́д·а
~, **го́рно·с·пас·а́тельн·ая** mine rescue team
~, **компле́кс·н·ая** team, group, co-ordinating group
~, **про·из·во́д·ственн·ая** production team
~, **сквоз·н·а́я** start-to-finish production team
бригад·и́р·, -а, (*m*) chargeman, chargehand; (rail) ganger
~ **пут·и́** (rail) ganger
бри́ггс·ов· логари́фм· (*m*) (math) Briggsian logarithm
бри́г·ов· логари́фм· (*m*) (math) Briggsian logarithm
бри́дел·ь, -я, (*m*) (naut) bridle
бри́дер·, -а, (*m*) (nucl) breeder
бри́з·, -а, (*m*) breeze
~, **бере́г·ов·о́й** land breeze
~, **го́р·н·ый** mountain breeze
~, **дн·евн·о́й** sea breeze
~, **дол·и́нн·ый** valley breeze
~, **лед·нико́в·ый** glacier breeze
~, **ноч·н·о́й** land breeze
бриза́нт·ност·ь, -и, (*f*) (expl) disruptive capacity, brisance
бриза́нт·н·ый, -ая, -ое, (*a*) high-explosive, HE
брике́т·, -а, (*m*) briquette, pellet
брикет·и́ровани·е, -я, (*n*) briquetting
Бри́кнер·а, цикл· (*m*) (meteor) Brückner cycle
Бри́кс·а, арео́·метр· (*m*) Brix hydrometer (sugar ind.)
бриллиа́нт·, -а, (*m*) brilliant, brilliant-cut gem; diamond proper, cut diamond, diamond; (print) brilliant, 4-point type
бриллиа́нт·ов·ый, -ая, -ое, (*a*) of **бриллиа́нт·**; brilliant (of dyes etc.)
~ **зе́лен·ь** (*f*) brilliant green (dye)
~ **фосфи́н·** (*m*) brilliant phosphine (dye)
Бриллюэ́н·а, зо́н·а (*f*) (nucl) Brillouin zone
брилья́нт·, -а, (*m*) *see* **бриллиа́нт·**
брилья́нт·ност·ь, -и, (*f*) brilliance, brilliancy, brightness
Бри́н·а, спо́соб· (*m*) (chem) Brin's process
Бринэ́лл·ь, -я, (*d*) **-ю,** (*m*) Brinell (surname, Swedish origin)
ме́тод· Бринэ́лл·я (*m*) (met) Brinell hardness testing
твёрд·ост·ь по Бринэ́лл·ю Brinell hardness

брио·зо́·й, -я, (*nom pl*) **-и,** (*m*) (zool) bryozoan, (*pl*) bryozoans, *Bryozoa*
брио·ло́г·и·я, -и, (*f*) (bot) bryology
бри́слинг·, -а, (*m*) (fish) brisling
бристо́л·ь, -я, (*m*) Bristol board
бристо́ль·ск·ий карто́н· (*m*) Bristol board
брита́н·к·а, -и, (*f*) British woman; (elec) Brittania splice, Brittania joint (method of jointing copper conductor)
брита́н·ск·ий, -ая, -ое, (*a*) British
~ **един·и́ц·а тепл·а́** (*f*) British thermal unit, BTU
бр·и́тв·а, -ы, (*f*) razor; blade
~, **под·по́ль·н·ая** (agr) scuffle knife
бр·и́т·ый, -ая, -ое, (*past part pass*) of **бр·и́ть;** shaven
бр·и́ть, (*pres 3rd sing, 3rd pl*) **бр·е́ет, бр·е́ют,** shave
бро́в·к·а, -и, (*g pl*) **-в·ок·,** (*f*) brow, lip; (rail) edge, edge of track
бро́в·ь, -и, (*g pl*) **-е́й,** (*f*) (anat) eyebrow
бро́д·, -а, (*m*) ford
брод·и́л·о, -а, (*n*) ferment
брод·и́льн·ый, -ая, -ое, (*a*) fermenting, fermentation
брод·и́ть, (*pres 3rd sing, 3rd pl*) **бро́д·ит, бро́д·ят,** (*imp indet*) wander; (*imp*) ferment
брод·я́ч·ий, -ая, -ее, (*a*) erratic, vagrant
брож·е́ни·е, -я, (*n*) fermentation, ferment
~, **спирт·ов·о́е** (bot) alcoholic fermentation
бро́йлер·, -а, (*m*) (agr) broiler
бро́кер·, -а, (*m*) (fin) broker
бро́кер·ск·ая коми́сс·и·я (*f*) brokerage
бром·, -а, (*m*) (chem) bromine, Br
~, **хло́р·ист·ый** bromine chloride
~, **циа́н·ист·ый** cyanogen bromide
бром·алки́л·, -а, (*m*) (chem) alkyl bromide
бром·ангидри́д·, -а, (*m*) (chem) acid bromide
~ **се́р·н·ой кисл·от·ы́** (chem) sulfuryl bromide
бром·а́т·, -а, (*m*) (chem) bromate
~ **ка́л·и·я** (chem) potassium bromate
бромато·ме́тр·и·я, -и, (*f*) (chem) bromatometry, bromate titration
бром·ацето́н·, -а, (*m*) (chem) bromacetone
бром·бензо́й·н·ая кисл·от·а́ (*f*) (chem) bromobenzoic acid

бром·бензо́л·, -а,   *(m)* (chem) brom-benzene

бром·гидра́т·, -а,   *(m)* (chem) hydro-bromide, hydrogen bromide

бромели́н·, -а,   *(m)* bromelin (enzyme found in pineapples)

бром·за·мещ·ённ·ый, -ая, -ое,   *(a)* (chem) bromo-

~ **кисл·от·а́**   *(f)* bromo-acid

бром·и́д·, -а,   *(m)* (chem) bromide

~ **ка́л·и·я**   potassium bromide

~ **ка́льц·и·я**   calcium bromide

~ **ра́д·и·я**   radium bromide

бром·и́рован·и·е, -я,   *(n)* (chem) bro-mination, bromating

бром·и́рованн·ый, -ая, -ое,   (*past part pass*) brominated

бромисто·водо·ро́д·н·ый, -ая, -ое, *(a)* (chem) hydrobromic

бро́м·ист·ый, -ая, -ое, *(a)* bromous; bromine, bromide; bromide of (-ous)

~ **водо·ро́д·**   *(m)* hydrogen bromide

~ **ка́л·и·й**   *(m)* potassium bromide

~ **кисл·от·а́**   *(f)* bromous acid

~ **ме́д·ь**   *(f)* cuprous bromide

~ **на́тр·и·й**   *(m)* sodium bromide

~ **от·печа́т·ок·**   *(m)* (phot) bromide print

~ **эти́л·**   *(m)* ethyl bromide

броми́т·, -а,   *(m)* (min) bromite

бром·крезо́л·, -а,   *(m)* (chem) brom-cresol

~**, пурпу́р·н·ый**   bromcresol purple

бром·мета́н·, -а,   *(m)* methyl bromide

бром·новати́ст·ый, -ая, -ое,   *(a)* hypobromous

бромноватисто·ки́сл·ый, -ая, -ое, *(a)* hypobromite of

бромновато·ки́сл·ый, -ая, -ое,   *(a)* bromate of

бром·новат·ый, -ая, -ое, *(a)* bromic, bromate of

~ **кисл·от·а́**   *(f)* bromic acid

бро́м·н·ый, -ая, -ое,   *(a)* bromine; bromide of (-ic)

~ **вод·а́**   *(f)* bromine water

~ **кисл·от·а́**   *(f)* perbromic acid

~ **ме́д·ь**   *(f)* cupric bromide

~ **числ·о́**   *(n)* bromine number/value

бромо·водо·ро́д·, -а,   *(m)* hydrogen bromide

бром·о́·кис·ь, -и,   *(f)* oxybromide

бромо·ма́сл·ян·ая печа́т·ь   *(f)* (phot) bromoil process

бромо́н·иев·ый, -ая, -ое,   *(a)* (chem) bromonium

бромо·у́ксус·н·ая кисл·от·а́   *(f)* bromo-acetic acid

бромо·фо́рм·, -а,   *(m)* (chem, min) bromoform

бромо·серебр·ян·ый, -ая, -ое,   *(a)* (phot, chem) silver bromide

бром·стиро́л·, -а,   *(m)* bromostyrene

бром·странди́т·, -а,   *(m)* (min) brom-strandite

бром·фено́л·, -а,   *(m)* bromphenol

бром·циа́н·, -а,   *(m)* cyanogen bromide

бром·эти́л·, -а,   *(m)* ethyl bromide

броне-   *(root)*   armoured/armored, armour/armor

броне·автомоби́л·ь, -я,   *(m)* armoured/armored car

броне·бо́й·н·ый, -ая, -ое,   *(a)* armour/armor-piercing

брон·еви́к·, -а,   *(m)* armoured/armored car

брон·еви́ст·, -а,   *(m)* crewman (of armoured car)

брон·ев·о́й, -а́я, -о́е,   *(a)* armoured/armored, armour/armor

броне·ка́тер·, -а,   *(nom pl)* **-а́,** *(g pl)* **-о́в,** *(m)* motor gunboat

броне·ле́нт·а, -ы,   *(f)* (elec) armouring/armoring tape

броне·маши́н·а, -ы,   *(f)* armoured/armored vehicle

броне·но́с·ец·, -с·ц·а,   *(i)* **с·ц·ем,** *(m)* (zool) armadillo, *(pl)* *Dasypodidae*; (obs naut) armour-clad battleship, pre-dreadnought battleship

броне·про́·волок·а, -и,   *(f)* (elec) armouring/armoring wire

броне·стекл·о́, -а́, *(nom pl)* **стёкл·а,** *(g pl)* **стёкол·,**   *(n)* bullet-resistant glass, bullet-proof glass

броне·транспорт·ёр·, -а, *(m)* armour-ed personnel carrier

бро́нз·а, -ы,   *(f)* bronze

~**, анти·фрикцио́н·н·ая**   anti-friction bronze, bearing bronze

~**, бери́лл·иев·ая**   beryllium bronze

~**, без·олов·я́нист·ая**   tinless bronze, high-tensile brass

~**, дефо́рм·и́руем·ая**   wrought bronze

~**, ка́дм·иев·ая**   cadmium copper

~**, колоко́ль·н·ая**   bell bronze/metal

~**, лит·е́йн·ая**   casting bronze

~**, марганц·о́в·ая**   manganese copper

~**, маши́н·н·ая**   gun metal

~**, моне́т·н·ая**   coinage bronze

~**, олов·я́нист·ая**   tin bronze

~**, под·ши́п·ников·ая**   bearing bronze

~**, пу́ш·ечн·ая**   gunmetal

бро́нз·а

~, фо́сфор·ист·ая   phosphor-bronze

бронз·ирова́льн·ая маши́н·а   (*f*) (print) bronzing machine

бронз·ирова́ни·е, -я,   (*n*) bronzing

бронзи́т·, -а,   (*m*) (min) bronzite

бро́нз·ов·ый, -ая, -ое,   (*a*) of **бро́нз·а** bronze; bronzed, tanned

~ бол·е́зн·ь   (*f*) (med) Addison's disease

~ век·   (*m*) Bronze Age

брон·и́рова́ни·е, -я,   (*n*) reservation, reserve; **брон·ирова́ни·е, -я,** (*n*) armo(u)ring, armo(u)r-plating, protection; (expl) inhibiting, restricting

брон·и́рова́нн·ый, -ая, -ое,   reserved, in reserve; **брон·ирова́нн·ый, -ая, -ое,** armo(u)red, armo(u)r plated; (expl) inhibited, restricted

брон·и́р·овать, -уют,   (*imp*) reserve, book; (fin) assure; **брон·ир·ова́ть, -у́ют,** (*imp*) armour, armor; (expl) inhibit, restrict

бронто·за́вр·, -а,   (*m*) (pal) brontosaur, *Brontosaurus*

бронто·ме́тр·, -а,   (*m*) (meter) brontometer

бро́нх·, -а,   (*m*) (anat) bronchus, (*pl*) bronchi

бронхи·а́льн·ый, -ая, -ое,   (*a*) bronchial

бронх·и́т·, -а,   (*m*) (med) bronchitis

бронхо·граф·и́·я, -и,   (*f*) (med) bronchography

бронхо·скоп·и́·я, -и,   (*f*) (med) bronchoscopy

бронхо·эктаз·и́·я, -и,   (*f*) (med) bronchiectasis, bronchiectasia

брон·я́, -и́,   (*f*) armour, armor; **бро́н·я, -и,** (*f*) reservation, booking, advance allocation

брос·а́ни·е, -я,   (*n*) (*v n*) see **брос·а́·ть;** projection; abandoning, abandonment

ли́ния брос·а́ни·я   (*f*) line of departure (ballistics)

у́гол· брос·а́ни·я   (*m*) (gunn) launching angle (ballistic missile)

брос·а́·ть, -ют,   (*imp*) throw, cast, hurl, fling; drop, leave/let go; give up, stop doing

бро́с·ить, -ят,   (*perf*) see **брос·а́·ть**

брос·ов·о́й, -а́я, -о́е,   see **бро́с·ов·ый**

бро́с·ов·ый, -ая, -ое,   (*a*) worthless

~ пла́в·к·а   (*f*) (met) off-heat

~ экспо́рт·   (*m*) (fin) dumping

брос·о́к·, -с·к·а́,   (*m*) throw, kick; rush, charge, surge; sprint (sport)

~ то́к·а   (elec) current surge

брото·криста́лл·, -а,   (*m*) brotocrystal

бро́ун·овск·ий, -ая, -ое,   (*a*) = **бра́ун·овск·ий** Brown (surname), Brownian

~ движ·е́ни·е   (*n*) (phys) Brownian movement

броханти́т·, -а,   (*m*) (min) brochantite

брохио·дро́м·н·ый, -ая, -ое,   (*a*) (bot) brochiodromous, brochiodromic

броше́   (*n indecl*) (text) bronché

бро́ш·енн·ый, -ая, -ое,   (*past part pass*) of **бро́с·ить,** see **брос·а́·ть** (*imp*); abandoned, deserted

броши́р·о́вк·а, -и,   (*f*) see **брошю·р·о́вк·а**

брошю́р·а, -ы,   (*f*) brochure, pamphlet, booklet

брошюр·ова́льн·ый, -ая, -ое,   (*a*) (print) stitching

брошюр·ова́ть, -у́ют,   (*imp*) (print) stitch

брошю́р·о́вк·а, -и,   (*g pl*) **-вок·,** (*f*) (print) stitching

бру́дер·, -а,   (*m*) brooder (for poultry)

бруки́т·, -а,   (*m*) (min) brookite

бру́с·, -а,   (*nom pl*) **бру́с·ья,** (*g pl*) **бру́с·ьев,** (*m*) (civ eng) beam, joist; (mech eng) bar

~, гельмпо́рт·н·ый   (shipb) helm-port transom

~, за·по́р·н·ый   locking bar

~, колесо·от·бо́й·н·ый   bridge curbing

~, крив·о́й   curved bar (strength tests)

~, на·сте́н·н·ый   (build) wall plate

~, па́ль·цев·ый   (agr) finger bar

~, пере·во́д·н·ые   (rail) crossing timbers

~, под·но́ж·н·ый   stretcher (of boat)

~, под·планши́р·н·ый   (shipb) gunwhale

~, при·ва́ль·н·ый   (shipb) rubbing strake, thwart rising

~, при·жи́м·н·ый   (shipb) paddle wing/walk

~, при·ча́ль·н·ый   (shipb) permanent fender

~, про·во́д·ков·ый   (mech eng) guide bar, rest bar

~, про·до́ль·н·ый   (a/c) secondary longitudinal

~, рас·по́р·н·ый   distance/gauge bar

~ -скли́з·, на·правл·я́ющий   guide/slide bar

~ с·це́п·к·и   drawbar

~, эквивале́нт·н·ый   (shipb) equivalent girder

бру́си́т·, -а, (*m*) (min) brucite

брус·ко́в·ый, -ая, -ое, (*a*) of бру-с·о́к·

брус·о́вк·а, -и, (*g pl*) -вок·, (*f*) carpenter's file

брус·о́к·, -с·к·а́, (*m*) bar; 3 in. square (wood); (nucl) slug

~, то́пл·ивн·ый (nucl) fuel slug

~, точ·и́льн·ый abrasive stick, grinding slip, oil stone, whetstone

~, шлиф·ова́льн·ый = брус·о́·к· точ·и́льн·ый

бру́ствер·, -а, (*m*) (build) parapet, breastwork

брус·ча́тк·а, -и, (*g pl*) -ток·, (*f*) (*dim*) of брус·; paving block

брус·ча́т·ый, -ая, -ое, (*a*) of брус·-ча́тк·а

~ структу́р·а (*f*) (geol) rodding/mullion structure

бру́с·ья (*nom pl*) of брус·

бру́тто (*indecl*) gross, gross weight

бру́тто-реги́стр·ов·ая в·мест·и́-мост·ь (*f*) (naut) gross tonnage

бруцелл·ёз·, -а, (*m*) (med) brucellosis, undulant fever, Malta fever

бруц·и́н·, -а, (*m*) (chem) brucine

брыже́·ечн·ый, -ая, -ое, (*a*) (anat) mesenteric

брыже́й·к·а, -и, (*g pl*) -е́·ек·, (*f*) (anat) mesentery, mesenterium

бры́зг·ал·о, -а, (*n*) coolings sprays

бры́зг·алк·а, -и, (*g pl*) -лок·, (*f*) spray nozzle, sprayer

брызг·а́льн·ый, -ая, -ое, (*a*) spray, splash, sprinkle

~ бассе́йн· (*m*) spray cooling pond

бры́зг·альц·е, -а, (*n*) (zool) spiracle

бры́зг·а·ть, -ют, or бры́зж·ут, (*imp*) spray, splash, spatter, sprinkle

бры́зг·и (*nom pl*), (*g pl*) брызг· spray, splash; sparks; (met) splash, slopping, spatter

брызг·ови́к·, -а́, (*m*) (M/T) mudguard

брызго·за·щищ·ённ·ый, -ая, -ое, (*a*) spray/hose-proof

брызго·не·про·ниц·а́ем·ый, -ая, -ое, (*a*) splash/drip-proof

брызго·у·лов·и́тел·ь, -я, (*m*) spray trap

бры́з·н·уть, -ут, (*perf*) spray, splash, spatter, sprinkle; spurt out, gush out

Брэ́гг·а, у́гол· (*m*) Bragg angle (X-ray diffraction)

Брэ́гг·а-Ву́льф·а, у·сло́в·и·е (*n*) Bragg law (X-rays)

Брэ́кет·а, сер·и·я (*f*) Brackett series (spectral analysis)

брю́кв·а, -ы, (*f*) swede, Swedish turnip, rutabaga, *Brassica napus esculenta*

брю́к·и (*nom pl*), (*g pl*) брюк·; trousers, plus-fours (Brit), pants, knickerbockers (US), breeches

~ -клёш·, -а, (*m*) (text) ducks

брюссе́ль·ск·ая капу́ст·а (*f*) (bot) Brussels sprouts, *Brassica oleracea gemmifera*

Брю́стер·а, пласт·и́н·ы (*pl*) (opt) Brewster's bands

брю́х·о, -а, (*n*) abdomen, belly, maw

брюхо·но́г·ие, -их, (*pl decl as adj*) (zool) gastropods, *Gastropoda*

брюш·и́н·а, -ы, (*f*) (anat) peritoneum

брюш·и́нн·ый, -ая, -ое, (*a*) (anat) peritoneal

брюш·к·о́, -а́, (*nom pl*) -и́, (*g pl*) -о́в, (*n*) abdomen, belly

~ ло́паст·и blade face (turbines etc.)

брюш·н·о́й, -а́я, -о́е, (*a*) abdominal, (zool) ventral, (med) alvine

~ ао́рт·а (*f*) (zool) ventral aorta

~ по́л·ост·ь (*f*) (anat) abdominal cavity

~ тиф· (*m*) (med) typhoid, typhoid/enteric fever

бубо́н·, -а, (*m*) (med) bubo

бубо́н·н·ый, -ая, -ое, (*a*) bubonic

~ чум·а́ (*f*) (med) bubonic plague

Буво́, синте́з·а (*f*) (chem) Bouveault aldehyde synthesis

Буво́-Блáн·а, вос·становл·éни·е (*n*) (chem) Bouveault and Blanc reduction

Буге, грави·метр·и́ческ·ая анома́л·и·я Bouguer effect

бу́гел·ь, -я, (*m*) hoop, bow, collar, yoke; mild steel hoop (on head of timber pile); stirrup; (elec) pantograph

~, эксентри́к·ов·ый eccentric strap

Буге́р·а-Лáмберт·а-Бэ́р·а, за·ко́н· (*m*) (phys) Lambert's law of absorption, Bouguer's law

буго́р·, -гр·а́, (*m*) hillock, hummock; protuberance; (med) tubercle

бугор·о́к·, -р·к·а́, (*m*) node, tubercle, pimple (on a surface)

буго́р·чат·ый, -ая, -ое, (*a*) knobby, knobbly; (med) tuberculous, tubercular; (bot) tuberous

бугр·и́стост·ь, -и, (*f*) pimpling (of surfaces)

бугр·и́ст·ый, -ая, -ое, (*a*) hillocky, hummocky; knobbly, pimply, wrinkled; torous, torose

бу́ден·ь, -дн·я   (*m*) *see* **бу́дн·и**

бу́д·ет,   (*fut 3rd sing*) of **бы́·ть** will be (+ *infin*) (*future imperfective*)

буд·и́льник·, -а,   (*m*) alarm clock

буд·и́ть, -ят,   (*imp*) wake, awaken, arouse

бу́д·к·а, -и,   (*g pl*) **-д·ок·,**   (*f*) cabin, box, kiosk, booth

~, англ·и́йск·ая   (meteor) Stevenson screen

~, карау́ль·н·ая   sentry box

~, про·ход·н·а́я   check point

~, психро·метр·и́ческ·ая   (meteor) Stevenson screen

~, раз·гово́р·н·ая   telephone/call box, telephone booth

~, телефо́н·н·ая   telephone/call box, telephone booth

бу́дн·и   (*nom pl*), (*g pl*) **-ей,**   (*m*) work-day(s), working day(s)

бу́дн·ичн·ый, -ая, -ое,   (*a*) of **бу́д-н·и;** dull, prosaic

бу́дто   (*conj*) as if, as though, apparently, seemingly

бу́д·ут, (*fut 3rd pl*) of **бы́·ть** will be (+ *infin*) (*future imperfective*)

бу́д·учи   (*pres gerund*) (+ *instr*) being, while, as

бу́д·ущ·ий, -ая, -ее,   (*a*) future, coming, next (of time); (*n decl as adj*) the future

бу́д·ьте,   (*pl imper*) of **бы́·ть** be

бу·ёк,   (*g sing*) **буй·к·а́,**   (*m*) (*dim*) of **бу·й** (q.v.)

~, свет·я́щ·ийся   (naut) calcium light, flashing lifebuoy

бу́·ен·,   (*masc s f*) of **бу́й·н·ый**

бу́ер·, -а, (*nom pl*) **-а́,** (*g pl*) **-о́в,**   (*m*) ice yacht

бу́ж·, -а́, (*i*) **-о́м,**   (*m*) (med) bougie; (a/c) split reinforcement tube

бужен·и́н·а, -ы,   (*f*) pork (usually pickled)

буж·у́, (*pres 1st sing*) of **буд·и́ть**

бузин·а́, -ы́,   (*f*) (bot) elder, *Sambucus*; pith

бузи́н·н·ый, -ая, -ое,   (*a*) (bot, horol) pith

~ ша́р·ик·   (*m*) (horol) pith ball

бузу́н·, -а́,   (*m*) rock salt

бу́·й, -я,   (*nom pl*) **-и́,**   (*g pl*) **-ёв,** (*m*) (naut) buoy

~ -обо·знач·а́тел·ь, -я,   (*m*) dan/ marker buoy

~, о·свещ·а́ем·ый   light buoy

~, радио·гидро·акуст·и́ческ·ий radiosonobuoy

бу́·й

~, свет·я́щ·ийся   light buoy

~, сигна́ль·н·ый   indicator buoy

~, с·пас·а́тельн·ый   lifebuoy (with light)

~, тупо·кон·е́чн·ый   can buoy

бу́й·н·ый, -ая, -ое,   (*a*) (*s f*) **бу́·ен·, бу́й·н·а, бу́й·н·о,** violent, furious, wild; lush (of vegetation)

~ сорок·ов·ы́е   (*pl*) (meteor) the roaring forties

бу́йреп·, -а,   (*m*) (naut) buoy-rope

бу́к·, -а,   (*m*) (bot) beech, *Fagus*

бу́кв·а, -ы,   (*f*) letter

~, курси́в·н·ая   (print) italic letter

~, над·стро́ч·н·ая   (print) superior letter, superscript

~, под·стро́ч·н·ая   (print) inferior letter, subscript

~, про·пис·н·а́я   (print) capital letter, upper case letter, majuscule

~, строч·н·а́я   small letter, lower case letter

букв·а́льн·о   (*adv*) literally, word for word, verbatim

букв·а́льн·ый, -ая, -ое,   (*a*), (*s f*) **-лен·, -льн·а,**   literal, verbatim

букв·а́рн·ый, -ая, -ое,   (*a*) alphabetic

бу́кв·енн·ый, -ая, -ое,   (*a*) of **бу́кв·а**

букво·е́д·, -а,   (*m*) pedant

букво·е́д·ств·о, -а,   (*n*) pedantry

букво·печа́т·ающ·ий, -ая, -ее,   (*a*) (telecom) printing

~ аппара́т· teleprinter

буке́т·, -а,   (*m*) bouquet, bunch of flowers; bouquet, aroma; clump (of plants)

буке́т·ик·, -а,   (*m*) (bot) cyathium

букет·иро́вк·а, -и,   (*f*) (agr) thinning

бу́ки   (*n indecl*) letter Б

букини́ст·, -а,   (*m*) second-hand bookseller

бу́ккер·, -а,   (*m*) drill plough/plow

Бу́кки, блéнд·а   (*f*) (opt) Potter–Bucky grid (diffraction grating)

букко·ка́мфор·а, -ы,   (*f*) (chem) buchu camphor

бу́к·ов·ый, -ая, -ое,   (*a*) of **бук·;** (as pl noun) (bot) *Fagaceae*

~ ма́сл·о   (*n*) beechnut oil

бу́кс·, -а,   (*m*) (bot) box, *Buxus sempervirens*

бу́кс·а, -ы,   (*f*) (rail) axle box

букси́р·, -а,   (*m*) tug, tugboat, towboat; towing hawser, tow-rope, towline

~, ре́йд·ов·ый   harbour tug

букси́р

∼ -толк·а́ч·, -а́,   (i) -о́м,   (m) pusher tug/towboat

букси́р·н·ый, -ая, -ое,   (a) tow, towing

букси́р·овани·е, -я,   (n) see **букси-р·о́вк·а**

букси́р·овать, -у́ют,   (imp) tow, tug, have in tow

буксир·о́вк·а, -и,   (g pl) -вок·,   (f) towing, tow

∼ толк·а́ни·ем   push-towing

буксир·о́вщик·, -а,   (m) tower, (air) towing aircraft

букс·ова́ни·е, -я,   (n) wheel slipping/spinning

букс·ова́ть, -у́ют,   (imp) slip, spin (of wheels)

бу́кс·ов·ый, -ая, -ое,   (a) of **бу́кс·а**; of **букс·**

булав·а́, -ы́,   (f) club, mace; (biol) clava

була́в·к·а, -и,   (g pl) -в·ок·,   (f) pin

∼, англ·и́йск·ая   safety pin

булаво·ви́д·н·ый, -ая, -ое,   (a) club-shaped, (biol) clavate

була́в·очн·ый, -ая, -ое,   (a) of **бу-ла́в·а**, **була́в·к·а**; (bot) clavulate

була́н·ый, -ая, -ое,   (a) cream colo(u)red, pale yellow; bay (of horses)

булга́р·а, -ы,   (f) box leather

бу́л·ев·а, а́лгебр·а   (f) (math) Boolean algebra

∼, фу́нкци·я   (f) (math) Boolean function

бу́л·ев·ый, -ая, -ое,   (a) (math) Boolean

були́н·ь, -я,   (m) (naut) bowline

бу́л·к·а, -и,   (g pl) -л·ок·,   (f) roll (of bread)

бу́л·очн·ая, -ой,   (f decl as adj) baker's shop

булы́ж·ник·, -а,   (m) cobbles, cobble-stones

бу́л·ь, -я,   (m) (horol, instr) boule; (shipb) bilge, (pl) blisters, side bulges; buhl, boule (furniture); **Бу́л·ь**, -я, (m) Boole (surname)

бу́льб·, -а,   (m) (met) bulb, bulb bar

бу́льб·а, -ы,   (f) = **бульб·**

бу́льб·ов·ый, -ая, -ое,   (a) (met) bulb, bulb bar

∼ ста́л·ь   (f) bulb plate

бульб·спри́нклер·, -а, (m) bulb-type sprinkler (fire extinguisher)

бульва́р·, -а,   (m) avenue, boulevard

бульдо́зер·, -а,   (m) bulldozer; (met) rolled section bending machine

бульдозер·и́ст·, -а,   (m) bulldozer driver

бульдо́зер·н·ый, -ая, -ое,   (a) bulldozer

бу́ль·к·а, -и,   (f) (instr) boule; pellet

бульо́н·, -а, (m) broth, bouillon

∼, мясо·пепто́н·н·ый   (bact) peptone bouillon/broth, nutrient broth

бу́м·, -а,   (m) (fin) boom, sensation

бума́г·а, -и,   (f) paper; document; cotton material/thread (rare); (pl) (fin) securities

∼, арм·и́рованн·ая   burlap-lined paper

∼, афи́ш·н·ая   poster paper

∼, бактери·ци́д·н·ая   bactericidal paper

∼, бромо·сере́бр·ян·ая   (phot) bromide paper

∼, бюва́р·н·ая   blotting paper

∼, влаго·про́ч·н·ая   wet-strength paper, weatherproof paper

∼, газе́т·н·ая   newsprint

∼, два́·жды логари́фм·и́ческ·ая   (math) log-log paper

∼, диагра́мм·н·ая   (instr) chart paper, recording paper

∼, ка́рт·очн·ая   Bristol board

∼, конденс·а́торн·ая   (elec) capacitor tissue

∼, ла́кмус·ов·ая   (chem) litmus paper

∼, лист·ов·а́я   (print) flat paper

∼, логари́фм·и́ческ·ая   (math) logarithmic coordinate paper, log paper

∼ маши́н·н·ой гла́д·кост·и   machine-finish paper

∼, милли·метр·о́в·ая   squared/plotting/graph paper

∼, нажда́ч·н·ая   emery paper

∼, но́т·н·ая   music printing

∼, обёрт·очн·ая   wrapping paper

∼, об·ло́ж·ечн·ая   book-cover paper

∼, обо́й·н·ая   wall paper

∼, по́люс·н·ая   (elec) pole-finding paper

∼, почт·о́в·ая   note paper, letter paper

∼, про·зр·а́чн·ая   tracing paper

∼, про·мок·а́тельн·ая   blotting paper

∼, реакти́в·н·ая   (chem) test paper

∼, само·вир·и́рующ·ая   (phot) self-toning paper

∼, свето·за·щи́т·н·ая   (phot) backing paper

∼, свето·чувств·и́тельн·ая   photographic paper, sensitized paper

∼, сенсибилиз·и́рованн·ая   (phot) colour-sensitized paper

∼, телефо́н·н·ая   telephone cable paper

∼, тряп·и́чн·ая   rag paper

бума́г·а
~, у·пак·о́вочн·ая wrapping paper
~, фильтр·ова́льн·ая filter paper
~, фо́рзац·н·ая board paper
~, хлоп·ча́т·ая cotton
~, хро́м·ов·ая chromo paper
~, це́н·н·ые (*pl*) (fin) securities
~, эмиссио́н·н·ая banknote paper
бумаго·де́л·ательн·ая маши́н·а (*f*)
   papermaking machine
бумаго·держ·а́тел·ь, -я, (*m*) paper clip;
   (fin) holder of securities, bondholder
бумаго·ре́з·альн·ый, -ая, -ое, (*a*)
   paper-cutting
~ **трёх·нож·ев·ая маши́н·а** (*f*)
   (print) three-edge trimmer
бумаго·ре́з·ательн·ый, -ая, -ое, (*a*)
   paper-cutting
~ **маши́н·а** (*f*) guillotine
бума́ж·к·а, -и, (*g pl*) **-ж·ек·,** (*f*) (*dim*)
   of **бума́г·а;** piece of paper; note
   (money)
бума́ж·ник·, -а, (*m*) paper-maker;
   wallet, pocket book, billfold
бума́ж·н·ый, -ая, -ое, (*a*) paper;
   paper-making; (elec) impregnated
   paper; cotton
~ **конденс·а́тор·** (*m*) (elec) impregnated-paper capacitor
~ **лит·ь·ё** (*n*) papier-maché
~ **ма́сс·а** (*f*) paper pulp
~ **маши́н·а** (*f*) paper-making machine
~ **пря́ж·а** (*f*) cotton yarn
~ **тк·ан·ь** (*f*) cotton cloth
~ **шпага́т·** (*m*) paper twine
бумазе́·я, -и, (*f*) (text) fustian
бумлитиз·, -а, (*m*) papier-maché insulating material
бумофеноли́т·, -а, (*m*) *see* **гетина́кс·**
бу́н·а, -ы, (*f*) buna (synth rubber);
   dike, jetty
Бу́нзен·, -а, (*m*) Bunsen (surname,
   German origin)
~ **-Роско, за·ко́н·** (*m*) (chem) Bunsen–Roscoe law
бу́нкер·, -а, (*nom pl*) **-а́,** (*g pl*) **-о́в,**
   (*m*) bunker; hopper, bin
~, **грунт·ов·о́й** hopper, bunker (of
   dredger)
~, **жи́дк·ий** bunker oil, bunker
~ **-пит·а́тел·ь, -я,** (*m*) feed hopper
~, **подъ·ём·н·ый** (build) portable
   ready-use mortar hopper
бу́нкер·н·ый, -ая, -ое, (*a*) of **бу́нкер·**
бункер·ова́ть, -у́ют, (*imp and perf*),
   bunker, oil; coal, fuel; **-ся** bunker,
   take on oil, oil; take on coal, coal

бун·одо́нт·н·ый, -ая, -ое, (*a*) (zool)
   bunodont
буно·лоф·одо́нт·н·ый, -ая, -ое, (*a*)
   (zool) bunolophodont
буно·селен·одо́нт·н·ый, -ая, -ое, (*a*)
   (zool) bunoselenodont
бу́нт·, -а́, (*m*) packet, bale, bundle;
   бу́нт·, -а, (*m*) riot, mutiny
бунт·ова́ть, -у́ют, (*imp*) make up into
   bundles/packs; riot, mutiny
буняковск·ая не·ра́вен·ств·о   (*n*)
   (math) Schwarz inequality
бу́р·, -а, (*m*) auger, hand borer; jumper,
   bit (of rock drill)
~ **Ги́ллер·а** Hiller auger
~, **земл·ян·о́й** earth auger
~ **Розанов·а** Rosanov auger
~, **руч·н·о́й** auger
~, **руч·н·о́й таре́ль·чат·ый** posthole
   auger
~ **Сытин·а** Sytin auger
бур·а́, -ы́, (*f*) (chem) borax, sodium
   borate
бур·а́в·, -а́, (*m*) gimlet, auger, drill;
   post-hole digger, hole digger
бур·а́вчик·, -а, (*m*) gimlet
   пра́в·ил·о бур·а́вчик·а (*n*) (elec)
   corkscrew/screwdriver rule
бура́к·, -а, (*m*) (bot, food) sugar beet
бур·а́н·, -а, (*m*) (meteor) snow storm
бура́т·, -а, (*m*) (min) revolving screen
бура́ч·ник·, -а, (*m*) (bot) borage,
   Borage
бура́ч·н·ый, -ая, -ое, (*a*) of **бура́к·**
бурд·а́, -ы́, (*f*) slops
буре·ло́м·, -а, (*m*) windbreak; wind
   slash (timber)
бур·е́ни·е, -я, (*n*) drilling, boring
~, **алма́з·н·ое** diamond drilling
~, **вращ·а́тельн·ое** rotary drilling
~, **дву·ство́ль·н·ое** duplex drilling
~, **дроб·ов·о́е** shot boring/drilling
~ **жело́н·к·ой** bailer drilling
~, **зонд·иро́вочн·ое** exploratory drilling
~, **кана́т·н·о-уда́р·н·ое** cable-tool
   drilling, percussion drilling
~, **карт·иро́вочн·ое** core drilling,
   coring
~, **ке́рн·ов·ое** core drilling, coring
~, **коло́н·ков·ое** core drilling, coring
~, **на·пра́вл·енн·ое** directional drilling
~, **о·по́р·н·ое** prospect boring
~, **перфора́тор·н·ое** rotary pneumatic drilling

бур·е́ни·е
~, **раз·ве́д·очн·ое** probe/exploratory drilling
~, **структу́р·н·ое** structural drilling
~, **уда́р·н·ое** percussion drilling
~, **уда́р·н·о-кана́т·н·ое** cable-tool drilling, churn drilling
~, **уда́р·н·о-шта́нг·ов·ое** rod-tool drilling
~, **шта́нг·ов·ое** rod drilling, rod-tool drilling
~, **эксплуатацио́н·н·ое** (oil) exploitation drilling
бур·е́·ть, -ют, (*imp*) brown, become/grow/turn/appear/show brown
буржуа́з·н·ый, -ая, -ое, (*a*), (*s f*) **-з·ен·, -з·н·а,** bourgeois, non-communist, western
бур·и́льн·ый, -ая, -ое, (*a*) boring, drilling
~ **молот·о́к·** (*m*) (min) hammer drill, jack hammer
~ **труб·а́** (*f*) drill pipe
бур·и́льщик·, -а, (*m*) (oil) drill runner, rough neck; (min) machine drill operator
бур·и́ть, -ят, (*imp*) bore, drill
бурл·е́ни·е, -я, (*n*) swirling, churning, seething
бурл·и́ть, -я́т, (*imp*) swirl, churn, seethe
бурнони́т·, -а, (*m*) (min) burnonite, wheel-ore
бу́р·н·ый, -ая, -ое, (*a*), (*s f*) **-р·ен·, -р·н·а, -р·н·о,** stormy, rough, violent, vigorous
бур·ова́т·ый, -ая, -ое, (*a*) brownish
буро·взр·ыв·н·а́я рабо́т·а (*f*) blast-hole drilling
бур·ов·о́й, -а́я, -о́е, (*a*) boring, drilling, bore; (*fem as noun*) borehole
~ **вы́ш·к·а** (*f*) drilling derrick
~ **на·кон·е́чник·** (*m*) drilling bit
~ **скваж·ин·а** (*f*) borehole
~ **стан·о́к** (*m*) drilling rig
~ **у·стано́в·к·а** (*f*) drilling rig (includes power supply etc.)
~ **шта́нг·а** (*f*) drill rod (surface boring); (min) jumper drill
~ **ша́ш·к·а** (*f*) (expl) cylindrical charge
буро·за·пра́в·очн·ый стан·о́к· (*m*) drill sharpener, drill sharpening machine
буро·зём·, -а, (*m*) (agr) brown soil, brunizem
бу́рс·а, -ы, (*f*) (anat) bursa

бурс·а́льн·ый, -ая, -ое, (*a*) bursal
бу́рт·, -а, (*m*) (mech eng) collar; clamp (for potatoes etc.)
~, **от·риц·а́тельн·ый** negative collar, collar hole
~, **по·лож·и́тельн·ый** positive collar
~, **ступ·е́нчат·ый** thrust collar, end collar
бу́рт·ик·, -а, (*m*) rim, collar, edge, bead; (naut) rubbing strake/strip
бурт·ова́ни·е, -я, (*n*) (agr) clamping, storing in clamps
буру́н·, -а́, (*m*) breakers, surf, wave
бурундук·, -а́, (*m*) (zool) ground squirrel, *Tamias sibiricus*; (naut) after-guy
бурунду́ч·н·ый, -ая, -ое, (*a*) of **бурунду́к·**
~ **руд·а́** (*f*) banded ore (zincblende in association with galenite)
бу́р·щик·, -а, (*m*) machine drill operator; (oil) drill runner, rough neck
бу́р·ый, -ая, -ое, (*a*), (*s f*) **бур·, бур·а́, бур·о́** brown
~ **желез·ня́к·** (*m*) (min) limonite
~ **у́гол·ь** (*m*) brown coal, lignite
~ **шпат·** (*m*) (min) aukenite
бурья́н·, -а, (*m*) (agr) weed
бу́р·я, -и, (*f*) storm
бус·, -а, (*m*) necklace
бу́с·ин·а, -ы, (*f*) bead
бусино·ви́д·н·ый, -ая, -ое, (*a*) bead-like
буссо́л·ь, -и, (*f*) surveying compass; (naut) bearing compass
~, **артиллер·и́йск·ая** (gunn) director
~, **го́р·н·ая** dipping compass, clinometer
~ **на штати́в·е** mining transit
бу́стер·, -а, (*m*) booster, positive booster
~ **сумм·а́рн·ой тя́г·и** (air) series actuator
~, **фи́дер·н·ый** (elec) tail-end booster
бу́т·, -а, (*m*) (min) rough stone, quarry-stone; (civ eng) rubble, hard core
бута·ди·е́н·, -а, (*m*) (chem) butadiene, erythrene
~ **-нитри́ль·н·ый, каучу́к·** (*m*) butadiene-acrylonitrile rubber
бутадие́н·ов·ый, -ая, -ое, (*a*) of **бута·ди·е́н·**
бутадио́н·, -а, (*m*) (pharm) phenyl butazone
бут·а́н·, -а, (*m*) (chem) butane
бутан·ди·кислот·а́, -ы, (*f*) butane diacid

бута́н·ов·ая кисл·от·а́   (*f*) butanoic acid

бутан·о́л·, -а,   (*m*) butanol, *n*-butyl alcohol

бутан·о́н·, -а,   (*m*) butanone

бута́р·а, -ы,   (*f*) (min) washing trommel/drum; gold washer

бутафо́р·и·я, -и,   (*f*) (theat) props, properties; window dressing

бутва́р·, -а,   (*m*) (plast) butvar, polyvinylbutyral

бут·е́н·, -а,   (*m*) (chem) butene, butylene

бутилен·глико́л·ь, -я,   (*m*) butanediol

бутен·ди·кислот·а́, -ы,   (*f*) butene diacid

буте́н·ов·ая кисл·от·а́   (*f*) butenoic acid

бутербро́д·, -а,   (*m*) (food) open sandwich

бут·и́л·, -а,   (*m*) (chem) butyl

~, бро́м·ист·ый   butyl bromide

бутил·ацета́т·, -а,   (*m*) butyl acetate

бутил·е́н·, -а,   (*m*) butylene, butene

бутил·каучу́к·, -а,   (*m*) butyl rubber

бути́л·ов·ый, -ая, -ое,   (*a*) (chem) butyl

~ спи́рт·   (*m*) butanol, *n*-butyl alcohol

~ эфи́р· у́ксус·н·ой кисл·от·ы́   (*m*) butyl acetate

бутил·целло·зо́льв·, -а,   (*m*) butyl cellosolve

бут·и́н·, -а,   (*m*) butyne, butine

бутин·дио́л·, -а,   (*m*) (chem) butynediol

бути́н·ов·ая кисл·от·а́   (*f*) butinoic acid

бутир·а́л·ь, -я,   (*m*) (plast) polyvinyl butyral

бутир·а́т·, -а,   (*m*) (chem) butyrate

~ на́тр·и·я   sodium butyrate

бутир·и́л·, -а,   (*m*) (chem) butyryl

бутиро·лакто́н·, -а,   (*m*) (chem) butyrolactone

бутиро́·метр·, -а,   (*m*) (food) butyrometer

бут·и́ть·, -я́т,   (*imp*) fill with rubble

буто·бето́н·, -а,   (*m*) rubble concrete

бу́т·ов·ый, -ая, -ое,   (*a*) of **бут·**

~ ка́мен·ь   (*m*) rubble, hard core

бут·окси-   (*int component*) (chem) butoxy-

буто́н·, -а,   (*m*) flower bud, bud

бут л·к·а, -и,   (*g pl*) **-л·ок·,** (*f*) bottle

бутылко·но́с·, -а,   (*m*) (zool) bottle-nosed whale, *Hyperoodon rostratus*

бутыло·мо́·ечн·ая маши́н·а   (*f*) bottle-washing machine

буты́л·очн·ый, -ая, -ое,   (*a*) bottle

~ ка́мен·ь   (*m*) (min) bottlestone, moldavite, water chrysolite

буты́л·ь, -и,   (*f*) large bottle

~, о·плет·ённ·ая   carboy, demijohn

~ пери·ско́п·а   top of periscope

буты́ль·н·ый, -ая, -ое,   (*a*) of **буты́л·ь**

бут·я́щ·ий, -ая, -ее,   (*pres part act*) of **бут·и́ть**

бу́фер·, -а,   (*nom pl*) **-а́,** (*g pl*) **-о́в,** (*m*) (rail) buffer; (M/T) bumper; (chem) buffer solution

бу́фер·ност·ь, -и,   (*f*) buffering power (of soil)

бу́фер·н·ый, -ая, -ое,   of **бу́фер·**

~ с·ме́с·ь   (*f*) (electrochem) buffer reagent, buffer

буфе́т·, -а,   (*m*) buffet; sideboard

буфе́т·чик·, -а,   barman; (naut) steward

бухга́лтер·, -а,   (*m*) accountant, bookkeeper

бухгалте́р·и·я, -и,   (*f*) accountancy, book-keeping; accounts department

~, двой·н·а́я   double-entry book-keeping

бухга́лтер·ск·ий, -ая, -ое,   (*a*) accounting, accounts, book-keeping

бу́хт·а, -ы,   (*f*) (geog) bay; coil (of rope etc.)

~, по·хо́д·н·ая   coil of rope

бу́хт·ов·ый, -ая, -ое,   (*a*) of **бу́хт·а;** embayed (of coastlines etc.)

бу́хт·очк·а, -и,   (*g pl*) **-чек·,** (*f*) (geog) cove

бучард·а, -ы,   (*f*) tooth axe

бу́ч·ени·е, -я,   (*n*) (*v n*) of **бу́ч·ить** (text) steeping, bucking, lyeing; deliming (leather); **буч·ёни·е, -я,** (*n*) (*v n*) of **бут·и́ть** filling with rubble

буч·и́льник·, -а,   (*m*) steeping vat; (text, paper) kier, bucking kier; clothes boiler (laundering)

буч·и́льн·ый, -ая, -ое,   (*a*) steeping, bucking, lyeing

~ котёл·   (*m*) (text, paper) kier, bucking kier

бу́ч·ить, -ят,   (*imp*) steep in lye, boil in lime/soda water

буч·у́,   (*pres 1st sing*) of **бут·и́ть; бу́ч·у,** (*pres 1st sing*) of **бу́ч·ить**

бу́шел·ь, -я,   (*m*) bushel (36·37 l. in Brit., 35·24 l. in U.S.)

Бушеро, мото́р·   (*m*) (elec) double squirrel-cage motor, boucherot motor

бушла́т·, -а,   (*m*) pea coat (U.S.), British warm (Brit)

~, капк·о́в·ый   life jacket (kapok filled)

бушпри́т·, -а, (*m*) (naut) bowsprit

бы (*particle*) (*with past tense*) should, would; **где бы** wherever; **как бы** as if; **чем бы** no matter how; **что бы ни** whatever

быв·а́·ло (*past neut sing*) of **быв·а́·ть;** used to

быв·а́лост·ь, -и, (*f*) experience

быв·а́л·ый, -ая, -ое, (*a*) experienced, veteran; worldly-wise

быв·а́·ть, -ют, (*imp*) be, be sometimes; visit, go to; happen, occur

бы́·вш·ий, -ая, -ее, (*past part act*) of **бы·ть;** former, one-time, ex-, late

бык·, -а́, (*m*) (zool) bull, ox; pier (of bridge); (met) cutout

∼ -го́рден·ь, -я, (*m*) (naut) buntline

бы·л (*past masc sing*) of **бы·ть** was

бы·ла́ (*past fem sing*) of **бы·ть** was

бы́·ли (*past pl*) of **бы·ть** were

были́н·к·а, -и, (*g pl*) **-н·ок·,** (*f*) (bot) blade

бы́·ло (*past neut sing*) of **бы·ть** was

быстр·ин·а́, -ы́, (*nom pl*) **-и́н·ы,** (*g pl*) **-и́н,** (*f*) (ocean) race; (river) rapids, swift course

быстро- (*root*) quick, quickly, fast

быстро·гор·я́щ·ий, -ая, -ее, (*a*) (expl) instantaneous, quick-burning

быстро·дви́ж·ущ·ийся, -аяся, -ееся, (*a*) fast-moving, rapid, fast

быстро·де́й·ствующ·ий, -ая, -ее, (*a*) high-speed, quick-acting, fast

быстро·лет·я́щ·ий, -ая, -ее, (*a*) fast-flying, fast-moving, fast

быстро·от·пуск·а́ющ·ий, -ая, -ее, (*a*) quick-release

быстро·разъ·ём·н·ый, -ая, -ое, (*a*) quick-release

быстро·рас·пад·а́ющ·ийся, -аяся, -ееся, (*a*) short-lived

быстро·ре́ж·ущ·ая стал·ь (*f*) high-speed steel

быстро·с·ме́н·н·ый, -ая, -ое, (*a*) quick-change

быстро·с·мык·а́ющ·ийся, -аяся, -ееся, (*a*) instantaneous (of coupling and jointing devices)

быстро·со́х·нущ·ий, -ая, -ее, (*a*) quick-drying

быстро·с·хва́т·ывающ·ийся, -аяся, -ееся, (*a*) quick-setting (of cement)

быстр·от·а́, -ы́, (*f*) speed, rapidity, velocity, quickness

быстро·тверд·е́ющ·ий, -ая, -ее, (*a*) quick-hardening, quick-setting

быстро·то́к·, -а, (*m*) (civ eng) channel-type spillway, bye-channel, bye-wash; chute (for timber etc.)

быстро·хо́д·ност·ь, -и, (*f*) rapidity **коэффицие́нт· быстро·хо́д·ност·и** (*m*) power-speed coefficient (of air screw); specific speed (of turbines etc.)

быстро·хо́д·н·ый, -ая, -ое, (*a*), (*s f*) **-д·ен·, -д·н·а,** high-speed, fast, fast-running, quick

бы́стр·ый, -ая, -ое, (*a*), (*s f*) **быстр·, быстр·а́, быстр·о́** quick, fast, rapid

быт·, -а, (*m*) life, mode of life

быт·ов·о́й, -а́я, -о́е, (*a*) of **быт·;** domestic, living, everyday

∼ вод·а́ (*f*) domestic water

∼ газ· (*m*) household/domestic gas

∼ об·слу́ж·ивани·е (*n*) public utilities and social services

∼ с·то́ч·н·ые во́д·ы (*pl*) domestic sewage

бытовни́т·, -а, (*m*) (min) bytownite

бы·ть, (*no present tense,* **есть** = there is/are) (*imp*); be; (*fut 3rd pl*) **бу́д·ут** will be; (*past 3rd pl*) **бы́·ли** were

быч·а́ч·ий, -ья, -ье, (*a*) bovine, bull, ox

бы́ч·ий, -ья, -ье, (*a*) bovine, bull, ox

быч·и́н·а, -ы, (*f*) oxhide

быч·о́к·, -ч·к·а́, (*m*) (civ eng) buttress; (agr) bullcalf

бье́ф·, -а, (*m*) reach (of river or canal)

∼, ве́рх·н·ий upper reach, upstream side, headwater; (hydroelec) head bay

∼, ни́ж·н·ий lower reach, downstream side; (hydroelec) tail bay **у́·ровен·ь ни́ж·н·его бье́фа** (*m*) lower water level

бь·ю́щ·ий, -ая, -ее, (*pres part act*) of **би·ть;** pulsating, beating; untrue, not true (of rotating parts)

бэв (*abbr*) = **биллио́н·электро́н·во́льт·** (nucl) Bev, billion electron volts; 10^9 electron volts

бэва·тро́н·, -а, (*m*) (nucl) bevatron

бэр (*abbr*) = **био·ло́г·и́ческ·ий экви·ва́лент· рентге́н·а** (nucl) REM, rem, roentgen equivalent man

бюва́р·, -а, (*m*) blotting pad

бюдже́т·, -а, (*m*) (fin) budget

бюк-го́рден·ь, -я, (*m*) (naut) buntline

бюкс·а, -ы, (*f*) sample bottle; (chem) weighing bottle

бюллете́н·ь, -я, (*m*) bulletin, report; medical certificate; ballot paper

∼, метеоро·ло́г·и́ческ·ий weather report

бю́л·ь, -я, (*m*) projecting neck

Бю́ргерс·а, ве́ктор· (*m*) (cryst) Burgers vector

бюре́т·к·а, -и, (*g pl*) -т·ок·, (*f*) (chem) burette, dropping glass

бюро́ (*n indecl*) bureau, office

~, констру́ктор·ск·ое design bureau/office

~ на·хо́д·ок lost property office

бюро́

~, о́·пыт·н·о-констру́ктор·ск·ое experimental design bureau

бюрократ·и́ческ·ий, -ая, -ое, (*a*) bureaucratic

бюстга́льтер·, -а, (*m*) brassiere, bust, bodice

Бю́хнер·а, воро́н·к·а (*f*) (chem) Buchner funnel

бя́з·ь, -и, (*f*) (text) coarse calico

В

в (*prep* +*prepositional case indicating location etc.* +*acc case indicating movement etc.*) in, into, to, on, onto; at; per, in a; as

~ **кон·ц·е́** at the end

~ **полтор·а́ ра́з·а бо́льш·е** half as much/many/big again, one and a half times as much/many/big

~ **слу́ч·а·е** in case

~ **суббо́т·у** on Saturday

~ **то́ч·к·е** at the point

~ **тре́ть·ем час·у́** between two and three o'clock

~ **тр·ёх киломе́тр·ах от** at a distance of 3 kilometres from, 3 kilometres from

~ **три ра́з·а бо́льш·е** three times as much/many/big

три ра́з·а в ден·ь three times a day

~ **три ра́з·а ме́ньш·е** a third as many/much/big

~ **три час·а́** at three o'clock

~ **фу́нкци·и** as a function of

в. (*abbr*) = **вольт·** (elec) V, volt; = **вос·то́к·** E, East

в- (*prefix* = **во-**, **въ** *particular significances*: (1) movement into; (2) movement up (*usually followed by preposition* **на**); (3) penetration (*words ending in* **-ся**)

ва (*abbr*) = **вольт·ампер·** Va, va, voltampere

вавеллит·, **-а**, (*m*) (min) wavellite

ва́г·а, **-и**, (*f*) weighing machine; crowbar, lever

Ва́гнер·а, реа́кци·я (*f*) (chem) Wagner reaction

вагнерит·, **-а**, (*m*) (min) wagnerite

ваго́н·, **-а**, (*m*) wagon, coach, van, car, carriage, truck

~, **бага́ж·н·ый** (rail) luggage and parcel van, baggage car (U.S.)

~, **вентиляц·ио́нн·ый** (rail) ventilated wagon/car

~ **-вес·ы́**, **-о́в**, (*m pl*) (rail) scale car

~, **груз·ов·о́й** (rail) freight wagon/car, goods wagon/truck

~, **жёстк·ий** (rail) coach with wooden seating

~, **за·гру́з·очн·ый** charging car (furnaces)

~, **изо·те́рм·и́ческ·ий** (rail) insulated wagon/car

ваго́н

~, **кры́·т·ый** (rail) covered wagon/van

~, **купе́й·н·ый** (rail) compartmented passenger coach/car

~ **-лёд·ник·**, **-а**, (*m*) (rail) ice-cooled wagon/car

~ **ме́ст·н·ого со·общ·е́ни·я** (rail) local coach/car

~, **мя́гк·ий** coach with upholstered seating

~, **мото́р·н·ый** (rail) motor coach/car

~, **от·кры́·т·ый** open wagon, gondola (U.S.)

~, **пассажи́р·ск·ий** (rail) passenger coach/car

~ **-платфо́рм·а** **-ы**, (*f*) (rail) flat, flat car/wagon

~, **при́·город·н·ый** (rail) suburban coach/car

~, **при·цеп·н·а́я** (rail) trailer coach

~ **прям·о́го со·общ·е́ни·я** (rail) through coach/car

~ **-рестора́н·**, **-а**, (*m*) (rail) dining car

~, **с·бо́р·н·ый** (rail) mixed wagon-load

~, **сп·а́льн·ый** (rail) sleeping car

~, **това́р·н·ый** (rail) goods wagon, freight car

~, **тормоз·н·о́й** (rail) brake van/car

~ **-транспорт·ёр·**, **-а**, (*m*) (rail) well wagon

~, **туш·и́льн·ый** quenching car (coke ovens)

~, **у·сло́в·н·ый** (rail) wagon unit

~ **-цисте́рн·а**, **-ы**, (*f*) (rail) tank wagon/car

вагоне́т·к·а, **-и**, (*g pl*) **-т·ок·**, (*f*) (rail) trolley car, narrow gauge wagon; trolley, carrier (ropeway); (min) tub, car, wagon

~, **о·про·кид·н·а́я** (rail) side-tipping wagon

~, **руд·ни́чн·ая** (min) tub, mine car

ваго́н·ник·, **-а**, (*m*) (rail) wagon examiner

ваго́н·н·ый, **-ая**, **-ое**, (*a*) of **ваго́н·**

~ **лист·** (*m*) waybill

вагоно·вож·а́т·ый, **-ого**, (*m decl as adj*) (rail) motorman; tramdriver

вагоно·за·медл·и́тел·ь, **-я**, (*m*) (rail) retarder, wagon retarder, car retarder (U.S.)

вагоно·оборо́т·, **-а**, (*m*) (rail) turnround, wagon traffic

вагоно·о·про·кид·ыватель·ь, -я, (*m*) (rail) wagon tipper/tippler

вагран·к·а, -и, (*g pl*) **-н·ок·,** (*f*) cupola, cupola furnace

вагра́н·очн·ый, -ая, -ое, (*a*) of **вагра́н·к·а**

вагра́н·щик·, -а, (*m*) cupola furnace-man

ва́д·, -а, (*m*) (min) wad, bog manganese

вадо́з·н·ая вод·а́ (*f*) (geol) vadose water

важ·не́йш·ий, -ая, -ее, (*a*) most important

ва́ж·н·о (*adv*) importantly; (*predic adj*) it is important

ва́ж·ност·ь, -и, (*f*) importance

ва́ж·н·ый, -ая, -ое, (*a*) (*sf*) **ва́ж·ен·, важ·н·а́, ва́ж·н·о** important

ва́з·а, -ы, (*f*) vase

вазели́н·, -а, (*m*) petrolatum, petroleum jelly; (pharm) Vaseline, vaseline-base ointment

~, бо́р·н·ый boric ointment

вазели́н·ов·ый, -ая, -ое, (*a*) of **вазели́н·**

~ ма́сл·о (*n*) liquid paraffin

вазо·констрикти́в·н·ый, -ая, -ое, (*a*) (med) vasoconstrictive

вазо·мото́р·н·ый, -ая, -ое, (*a*) (physiol) vasomotor

вазо·пресс—и́н·, -а, (*m*) (biochem) vasopressin

вазо·тон·и́ческ·ий, -ая, -ое, (*a*) (med) vasotonic

вайерба́р·, -а, (*m*) (met) wire-bar

ва́йм·а, -ы, (*f*) joiner's clamp

ва́й·я, -и, (*f*) (bot) frond

вака́нси·я, -и, (*f*) vacancy, vacant site/place

~, двой·н·а́я (cryst) bivacancy, double vacancy

~, па́р·н·ые (*pl*) (nucl) pairs of vacancies, vacancy pairs

~, с·ло́ж·н·ые (*pl*) complex vacancies

вака́нт·н·ый, -ая, -ое, (*a*) vacant

ва́кк·а, -и, (*f*) (geol) wacke

~, се́р·ая (geol) greywacke

Ва́ккенродер·а, рас·тво́р· (*m*) (chem) Wackenroder's solution

ва́кс·а, -ы, (*f*) shoe polish

ваку·о́л·ь, -и, (*f*) *see* **ваку·о́л·я**

ваку·о́л·я, -и, (*f*) (biol) vacuole

~, пище·вар·и́тельн·ая (zool) food vacuole

вакуолиз·а́ци·я, -и, (*f*) (biol) vacuolization

ва́куум·, -а, (*m*) vacuum

~, абсолю́т·н·ый perfect vacuum

в ва́куум·е *in va uo*; vacuum (*as adj*)

~, выс·о́к·ий high vacuum

~, глуб·о́к·ий high vacuum

~ -на·со́с·, -а, (*m*) air-pump; vacuum pump

~ -на·со́с·, водо·кольц·ев·о́й water-ring air-pump (Brit), liquid-piston air (U.S.)

~ -на·со́с·, Гэ́де Gaede molecular pump

~ -на·со́с·, диффузио́н·н·ый diffusion high-vacuum pump

~ -на·со́с·, мо́кр·ый wet air-pump

~ -на·со́с·, молекуля́р·н·ый molecular pump

~ -на·со́с·, объ·ём·н·ый backing pump

~ -на·со́с·, пласт·и́нчат·ый vane air-pump, sliding vane rotary air-pump

~ -на·со́с·, ртутно·диффузио́н·н·ый mercury diffusion pump

~ -на·со́с·, сух·о́й dry air-pump

~, ни́з·к·ий rough vacuum

~, парциа́ль·н·ый partial vacuum

~, пред·вар·и́тельн·ый preliminary vacuum

~, пре·де́ль·н·ый ultimate vacuum

~ -фи́льтр·, -а, (*m*) vacuum filter

~ -фи́льтр·, бараба́н·н·ый rotary filter, drum vacuum filter

~, част·и́чн·ый partial vacuum

~, чёрн·ый non-conducting vacuum

вакуум·и́ровани·е, -я, (*n*) (*v n*) *see* **вакуум·ир·овать** vacuum compaction (of concrete); (chem) evacuation; degassing (molten metal)

вакуум·и́р·овать, -уют, evacuate

вакуум·ме́тр·, -а, (*m*) vacuum-gauge

~, гру́б·ый vacuum indicator

вакуум·на·со́с·, -а, (*m*) *see* **ва́куум-на·со́с·**

ва́куум·н·ый, -ая, -ое, (*a*) of **ва́куум·**

~ про·во́д· (*m*) vacuum line/manifold

вакуум·пло́т·н·ый, -ая, -ое, (*a*) vacuum-tight

вакци́н·а, -ы, (*f*) vaccine

~ БЦЖ BCG vaccine

~, много·вале́нт·н·ая multipartial vaccine

~, сибире·я́зв·енн·ая anthrax vaccine

~, с·ло́ж·н·ая mixed vaccine

вакцин·а
~ **Ферми** Fermi's anti-rabies vaccine, Semple vaccine
~, **яйч·н·ая** chick-embryo vaccine
вакцин·а́ци·я, -и, (*f*) vaccination
вакцин·ейри́н·, -а, (*m*) vaccineurine
вакци́н·н·ый, -ая, -ое, (*a*) vaccine, vaccinal
вакцино·профила́ктик·а, -и, (*f*) vaccinoprophylaxis
вакцино·фо́б·и·я, -и, (*f*) vaccinophobia
вал·, -а, (*nom pl*) **-ы́,** (*g pl*) **-о́в,** (*m*) shaft, spindle; roller; bank embankment, rampart; (geol) swell, arch; (gunn) barrage
~ **бараба́н·а** (horol) barrel arbor
~, **бок·ов·о́й** (geol) marginal rampart
~, **вед·у́щ·ий** driving/input shaft
~, **вс·по·мог·а́тельн·ый** intermediate shaft
~, **втор·и́чн·ый** main shaft (vehicle gear box)
~, **ги́б·к·ий** flexible shaft
~, **гла́в·н·ый** (М/Т) layshaft; eight-frame shaft (cinema projector)
~, **греб·н·о́й** (shipb) propeller shaft, tail shaft
~, **греб·е́нчат·ый** thrust/collar shaft
~, **де́йдвуд·н·ый** (shipb) tail intermediate shaft
~, **карда́н·н·ый** (М/Т) propeller shaft; (shipb) universal shaft
~, **ка́р·ов·ый** (geol) rock-step
~, **коле́н·чат·ый** crankshaft
~, **корен·н·о́й** crankshaft, mainshaft; (gunn) elevating pinion gear
~, **ла́в·ов·ый** (geol) lava billow
ли́н·и·я вал·о́в (*f*) (mech eng) line shafting
~, **мно́го·о·по́р·н·ый трансмис·сио́н·н·ый** line shaft
~, **огн·ев·о́й** (gunn) barrage
~ **от·бо́р·а мо́щ·ност·и** power take-off shaft
~, **перв·и́чн·ый** (М/Т) primary shaft; input shaft (tractor)
~, **пере·да́т·очн·ый** transmission/primary shaft
~, **плющ·и́льн·ый** (text) calender roller
~, **по·движ·н·о́й** (gunn) moving barrage
~ **по́д·очн·ой пе́ч·и** rabble-driving shaft (of a multihearth roasting furnace)
~, **по́л·ый** hollow/quill shaft

вал
~, **при·вод·н·о́й** driving shaft
~, **про·меж·у́точн·ый** intermediate/lay/primary shaft
~, **прост·о́й** (gunn) HE barrage
~, **рас·пре·дел·и́тельн·ый** camshaft
~, **турби́н·ы** turbine spindle
~, **у·по́р·н·ый** thrust shaft
вал·е́жин·а, -ы, (*f*) fallen tree, deadwood
вал·е́жк·а, -и, (*f*) (air) wing-tip stalling
вал·е́жник·, -а, (*m*) dead standing wood
вал·ёк, (*g sing*) **валь к·а́,** (*m*) (*and see* **вал·о́к·**); loom (of oar); (print) roll
ва́л·енок·, -н·к·а, (*nom pl*) **ва́л·енк·и** felt boot
вале́нт·ност·ь, -и, (*f*) (chem) valence, valency
~, **максима́ль·н·ая** absolute valency
вале́нт·н·ый, -ая, -ое, (*a*) valence, valency; (*as component*) -valent
~ **зо́н·а** (*f*) (phys) valence band
~ **со·сто·я́ни·е** (*n*) (phys, chem) valent state
~ **электр·о́н·** (*m*) valency/valence electron
валенциани́т·, -а, (*m*) (min) valencianite
ва́л·ен·ый, -ая, -ое, (*a*) = **ва́л·я·н·ый** felt
валер·ами́д·, -а, (*m*) (chem) valeramide
валериа́н·ов·ый, -ая, -ое, (*a*) valerian, valerianic, valeric; (*pl as noun*) (bot) *Valerianaceae*
~ **кисл·от·а́** (*f*) valeric acid
валер·и́л·, -а, (*m*) (chem) valeryl
~, **хло́р·ист·ый** valeryl chloride
валерья́н·ов·ый, -ая, -ое, (*a*) *see* **валериа́н·ов·ый**
вал·е́ц, -ль·ц·а́, (*m*) roller, roll (*see pl* **валь·ц·ы́**)
ва́л·ик·, -а, (*m*) (*see also* **вал·**); spindle, shaft, bearing pin; roller; platen (of typewriter); bolster; (zool) torus, torulus
~, **горизонт·а́льн·ый** (gunn) central training gear shaft
~, **кра́с·очн·ый** (print) inker, ink roller
~, **кулач·ко́в·ый** camshaft
~, **моч·и́льн·ый** (print) dampening roller
~, **о·по́р·н·ый** carrying/supporting roller

ва́л·ик

~, **ос·ев·о́й** (diesel) rocker shaft

~, **от·жиг·а́ющ·ий** annealing weld

~, **пере·во́д·н·ый** (horol) set arbor

~, **плющ·и́льн·ые** (*pl*) (text) condensing calender

~, **при·ём·н·ый** (text) take-in roller

~, **со·един·и́тельн·ый** pin (roller chain)

~, **съ·ём·н·ый** (text) doffer/removing roller/roll

~, **ход·ов·о́й** feed shaft/rod (of machine tool)

вал·и́л·, -а, (*m*) (chem) valyl

вал·и́н·, -а, (*m*) (chem) valine

вал·и́ть, -я́т, (*imp*) bowl over, roll over, overthrow, throw down, throw off; fell (timber); bundle, pack, jumble; flock (of crowds) (**вал·ят**); belch, pour (of smoke); shift, shelve (responsibility)

ва́л·ичн·ый, -ая, -ое, (*a*) of **ва́л·ик**

ва́л·к·а, -и, (*f*) *cf* **вал·о́к·, ва́л·к·ий** (text) fulling, milling

ва́л·к·ий, -ая, -ое, (*a*) **ва́л·ок, вал··к·а́, ва́л·к·о** unsteady; (naut) crank

вал·кова́ни·е, -я, (*n*) (agr) swathing

валко·обра́ч·иватель·ь, -я, (*m*) (agr) side delivery rake and swath turner

ва́л·кост·ь, -и, (*f*) (air) rolling; (naut) tendency to roll, crankness

вал·ов·о́й, -ая, -ое, (*a*) gross

~ **анали́з·** (*m*) (chem) bulk analysis

~ **в·мест·и́мост·ь** (*f*) (naut) gross tonnage

~ **грузо·подъ·ём·ност·ь** (*f*) (naut) deadweight carrying capacity

~ **проду́кци·я** (*f*) (fin) gross output

~ **товаро·оборо́т·** (*m*) (fin) gross turnover

~ **фо́сфор·** (*m*) (chem) total phosphorus

вал·о́к·, -л·к·а́, (*nom pl*) **-л·к·и́,** (*m*) *cf* **ва́л·к·а, ва́л·к·ий** (mech) roll, roller; (agr) swathe

~, **вы·тя́г·ивающ·ие** (*pl*) withdrawal rolls (continuous casting)

~, **гла́д·к·ий** (met) plain/flat roll

~, **двух·сло́й·н·ый** (met) composite/duplex roll

~, **дроб·и́льн·ый** crushing roll (ore dressing)

~, **за·ги́б·очн·ый** bending roll

~, **калибр·о́ванн·ый** (met) grooved roll

~, **ко́в·очн·ый** gap roll

~, **корен·н·о́й** (met) main roll

вал·о́к

~, **лист·ов·о́й** (met) plain/flat roll

~, **мя́гк·ий** (met) grain roll

~, **на·пла́вл·енн·ый** built-up roll

~, **на·правл·я́юш·ий** guide roll (continuous casting)

~, **на·сеч·ённ·ый** (met) ragged roll

~, **об·жим·н·о́й** (met) cogging/bloom roll

~, **о·по́р·н·ый** (met) back-up/support roll

~, **пере·да́т·очн·ый** dummy roll

~, **под·готов·и́тельн·ый** (met) bloom roll

~, **полу·твёрд·ый** (met) indefinite chill roll

~, **рабо́ч·ий** (met) work roll

~, **ручь·ев·о́й** (met) grooved roll

~, **сорт·ов·о́й** (met) grooved roll

~, **сталь·н·о́й** steel-base roll

~, **твёрд·ый** (met) definite chill roll, clear chill roll

~, **тя́·нущ·ий** withdrawing roll (continuous casting)

~, **чугу́н·н·ый** iron-base roll

вало·про·во́д·, -а, (*m*) shafting

вал·у́н·, -а, (*m*) boulder

вал·у́нник·, -а, (*m*) cobble, roundstone

вал·у́нн·ый, -ая, -ое, (*a*) boulder; rubbly

~ **су·гли́н·ок·** boulder clay

вальвуло·то́м·, -а, (*m*) (med) valvulotome

Ва́льден·а, пра́в·ил·о (*n*) Walden's rule (viscosity)

ва́льден·овск·ое обращ·е́ни·е (*n*) (chem) Walden inversion

ва́льм·а, -ы, (*f*) hip, hipped end (of a roof)

ва́льм·ов·ый, -ая, -ое, (*a*) hipped (of roof)

вальпурѓит·, -а, (*m*) (min) walpurgite

вальц·ева́ни·е, -я, (*n*) *see* **валь·ц·о́вк·а**

вальц·ева́ть, -у́ют, (*imp*) roll; roll, forge; flatten (between rolls); crush (between rolls); masticate, mill (rubber)

вальц·ев·о́й, -а́я, -о́е, (*a*) of **вальц·ы́, вальц·о́вк·а**

~ **дроб·и́лк·а** (*f*) crushing rolls

вальце·де́к·ов·ый стан·о́к (*m*) shelling/hulling machine

вальце·тока́р·н·ый стан·о́к (*m*) roll-turning lathe, roll lathe

вальц·о́ванн·ый, -ая, -ое, (*past part pass*) rolled; roll-forged; (rubber) masticated, milled

вальц·о́вк·а, -и, (*f*) rolling; roll forging; flattening (between rolls); mastication, milling (of rubber); crushing (between rolls)

вальц·о́вщик·, -а, (*m*) (met roll) catcher; (rubber) mill operator

вальц·о́в·ый, -ая, -ое, (*a*) of **вальц·ы́, вальц·о́вк·а**

вальц·ы́, -о́в, (*m pl*) rolls, rollers; (rubber) mill, roll mill

~, вы·пуск·н·ы́е (rubber) final mill; (text) delivery rolls

~, ги́б·очн·ые (met) bending rolls

~, ко́в·очн·ые (met) forging rolls

~, листо·ги́б·очн·ые bending rolls (for metal sheet or plate)

ва́ль·чат·ый, -ая, ое, (*a*) ridged

валь·я́н, -а, (*m*) (text) take-up roller

валю́т·а, -ы, (*f*) (fin) currency

~, дефици́т·н·ая hard currency

валю́т·н·ый, -ая, -ое, (*a*) currency, monetary

~ ку́рс· (*m*) (fin) rate of exchange

~ о·гово́р·к·а (*f*) currency clause

вал·я́льн·о-во́йлоч·н·ое про·из·во́д··ств·о (*n*) felt making

вал·я́льн·ый, -ая, -ое, (*a*) fulling, milling

~ маши́н·а (*f*) fulling machine

вал·я́льн·я, -и, (*g pl*) **-лен·,** (*f*) (text) fulling mill

вал·я́ни·е, -я, (*n*) (text) milling, fulling

ва́л·янн·ый, -ая, -ое, (*past part pass*) of **вал·я́·ть**

ва́л·ян·ый, -ая, -ое, (*a*) (text) felt

вал·я́·ть, -ют, (*imp*) roll, roll across, roll along; (text) felt, full, mill

вам, (*pron*) (*dative*) of **вы** to you, for you etc.

ва́ми, (*pron*) (*inst*) of **вы**

ван for surnames of Dutch origin; trace entry as if written as one word

ванад·а́т·, -а, (*m*) (chem) vanadate

~ ура́н·а uranium vanadate

ванадато·ме́тр·и·я, -и, (*f*) (chem) vanadate titrimetry

ванадиево·ки́сл·ый, -ая, -ое, (*a*) (chem) vanadate of

~ на́тр·и·й (*m*) sodium vanadate

ванад·иев·ый, -ая, -ое, (*a*) vanadium; (chem) vanadic

~ кисл·от·а́ (*f*) vanadic acid

~ ста́л·ь (*f*) vanadium steel

вана́д·и·й, -я, (*m*) (chem) vanadium, V

~, сер·ни́ст·ый vanadium sulphide

ванад·и́л·, -а, (*m*) vanadyl

~, хло́р·ист·ый vanadyl chloride

ванад·ини́т·, -а, (*m*) (min) vanadinite

ванад·и́т·, -а, (*m*) (min) vanadite

вана́д·ист·ый, -ая, -ое, (*a*) (chem) vanadous

ванг·пу́тенс·, -а, (*m*) (shipb) channel plate

ван-де-Гра́аф·а, у·скор·и́тел·ь (phys) (*m*) Van de Graaf generator

ванденбранди́т·, -а, (*m*) (min) vandenbrandite

ван дер Ва́альс·, -а, (*m*) Van der Waals (surname, Dutch origin)

у·равн·е́ни·е вандерва́алс·а (*n*) (phys) Van der Waals equation

вандерва́альс·ов·ы си́л·ы (*f pl*) (phys) Van der Waals forces

вандерва́альс·ов·ый, -ая, -ое, (*a*) (phys) Van der Waals

ванил·и́н·, -а, (*m*) (chem) vanillin

вани́л·ь, -и, (*f*) vanilla

ва́нн·а, -ы, (*f*) bath, vat; (phot) tray, bath

~, дуб·я́щ·ий (phot) hardening bath

~, о·свет·я́ющ·ая (phot) clearing bath

~, с·ва́р·очн·ая weld pool

~, термо·стат·и́руем·ая thermostatically controlled bath, constant-temperature bath

~, электро·лит·и́ческ·ая electrolytic cell

ва́ннер·, -а, (*m*) (min) vanner

ва́нн·очк·а, -и, (*g pl*) **-чек·,** (*dim*) of **ва́нн·а** (q.v.)

ваноксѝт·, -а, (*m*) (min) vanoxite

ва́нт·а, -ы, (*f*) (naut) shroud

Ван·сля́йк·а, ме́тод· (*m*) Van Slyke method (rubber)

ва́нт·а, -ы, (*f*) (naut) shroud

вант-Го́фф·а, за·ко́н· (*m*) (chem) van't Hoff's law

вант·пу́тенс·, -а, (*m*) *see* **ванг·пу́тенс·**

ванту́з·, -а, (*m*) air relief cock

ва́пор·, -а, (*m*) steam cylinder oil

вапор·иза́тор·, -а, (*m*) vaporizer

вапор·иза́ци·я, -и, (*f*) vaporization

вапори́·метр·, -а, (*m*) (chem) vaporimeter

ва́р·, -а, (*m*) pitch, pine tar

~, сапо́ж·н·ый cobbler's wax

вар (*abbr*) = **вольт·а́мпер·реакти́в··н·ый** (elec) var, reactive volt-ampere

вар·éни·е, -я,   (*n*) = **вáр·к·а** (*v n*) *see* **вар·и́ть**

вар·ён·ый, -ая, -ое, (*a*) boiled; (text) scoured

вар·éнь·е, -я, (*n*) = **вар·éни·е** (food) jam

вари·áнт·, -а, (*m*) version, variant; modification, alternative method

вари·áнт·а, -ы, (*f*) (biol) variant; (math) sequence

~, в·раст·áющ·ая (math) increasing sequence

~, о·гран·и́ченн·ый (math) bounded sequence

вари́áнт·ност·ь, -и, (*f*) variability, variance

вари·áтор·, -а, (*m*) regulator; (elec) buncher, input resonator

~ коэффициéнт·ов coefficient setter (of computer)

вариациóн·н·ый, -ая, -ое, (*a*) variation, variational

~ ис·числ·éни·е (*n*) (math) calculus of variations

~ мéтод· (*m*) (math) variation method

~ стáнц·и·я (*f*) magnetic variation measuring station

вари·áци·я, -и, (*f*) variation

~, времен·н·áя time variation

~, индивидуáль·н·ые (*pl*) fluctuating variations

~, магни́т·н·ая magnetic variation

~, сóлн·ечн·ая solar-connected variation (cosmic rays)

вариетéт·, -а, (*m*) variety

варикóз·н·ый, -ая, -ое, (*a*) varicose

варимю́, лáмп·а (*f*) (rad) variable-mu valve/tube

вари·óл·а, -и, (*f*) (med) smallpox, variola

варио·ли́т·, -а, (*m*) (geol) variolite

вари·óл·ь, -и, (*f*) (ent) variole

варио́·метр·, -а, (*m*) (instr) variometer

~, авиациóн·н·ый (air) rate-of-climb indicator

~, гравитациóн·н·ый gravity variometer

~, крут·и́льн·ый torsion balance

~, одно·ни́т·н·ый unifilar variometer

варистор·, -а, (*m*) (elec) varistor

варисци́т·, -а, (*m*) (min) variscite

вари·трóн·, -а, (*m*) (nucl) varitron

вар·и́ть, -ят, (*imp*) boil; cook; make (steel, glass); brew (beer); (food, phot) digest; (text) scour; cure (rubber)

вáр·к·а, -и, (*g pl*) **-р·ок,** (*f*) (*v n*) *see* **вар·и́ть**

вáр·ниц·а, -ы, (*i*) **-ей,** (*f*) saltworks; salt pan

варо·обрáз·н·ый, -ая, -ое, (*a*) pitch-like

вáр·очн·ый, -ая, -ое, (*a*) of **вáр·к·а**

варь·и́ровани·е, -я, (*n*) (*v n*) of **варь·и́р·овать**

~ электрóн·ов (elec) bunching

варь·и́р·овать, -уют, (*imp*) vary, modify, diversify

вас, (*pron*) (*acc/gen pl*) of **вы**

вáт·а, -ы, (*f*) cotton wool/waste, wadding

~, гóр·н·ая rock wool

~, минерáль·н·ая mineral wool

~, стекл·я́нн·ая glass wool

~, шлáк·ов·ая slag wool

вáтер·, -а, (*m*) (text) ring spinning frame

~, крут·и́льн·ый (text) ring doubling frame

ватервéйс·, -а, (*m*) (naut) waterway

ватержакéт·, -а, (*m*) water-jacketed blast furnace

ватерклозéт·, -а, (*m*) water closet, lavatory

ватерли́н·и·я, -и, (*f*) (naut) waterline

~, груз·ов·áя load line

ватермаши́н·а, -ы, (*f*) = **вáтер·**

ватерпáс·, -а, (*m*) level, spirit-level

вати́н·, -а, (*m*) (text) stockinet

вáтман·, -а, (*m*) Whatman's paper

вáт·ник·, -а, (*m*) quilted jacket

вáт·н·ый, -ая, -ое, (*a*) cotton wool, wadding, quilted

~ о·де·я́л·о (*n*) quilt

вáт·очн·ая маши́н·а (*f*) (text) intermediate scribbler

вáтт·, -а, (*g pl*) **вáтт·ов,** (*m*) (elec) watt

~ -секу́нд·а, -ы, (*f*) watt-second

~ -чáс·, -а, (*m*) watt-hour

ватт·мéтр·, -а, (*m*) (elec) wattmeter

вáтт·н·ый, -ая, -ое, (*a*) (elec) active, wattful

вáфл·я, -и, (*g pl*) **-фел·ь** (*f*) (food) wafer, waffle

вáхт·а, -ы, (*f*) (naut) watch (duty)

~, при·ём·н·ая listening watch

~, по·хóд·н·ая watch under way

  с·да·вáть вáхт·у turn over the watch

  сто·я́ть на вáхт·е to be on watch

~, ход·ов·áя watch under way

~, я́кор·н·ая anchor watch

ва́хт·енн·ый, -ая, -ое, (*a*) of ва́хт·а
~ журна́л· (*m*) ship's log (book), deck log
~ команд·и́р· (*m*) officer of the watch
~ маши́н·н·ый журна́л· (*m*) engine log (book)
ваш·, -его, (*pron*) (*nom sing masc*) your; (*acc sing masc inanim*) your
ва́ш·а, -ей, (*pron*) (*nom sing fem*) of ваш·
ва́шгерд·, -а, (*m*) (min) buddle, concentrating pit
ва́ш·е, -его, (*pron*) (*nom/acc sing neut*) of ваш·
ва́шет·ы (*pl*) split hides (leather), splits
ва́ш·и, (*pron*) (*nom pl masc/fem/neut*) of ваш·; (*acc pl masc/fem/neut inanim*) of ваш·
ва·я́л·о, -а, (*n*) sculptor's chisel
ва·я́ни·е, -я, (*n*) sculpture
ва·я́тел·ь, -я, (*m*) sculptor
ва·я́·ть, -ют, (*imp*) sculpture
вб (*abbr*) = ве́бер· weber (unit of magnetic flux)
в·бег·а́·ть, -ют, (*imp*) run in/into
в·беж·а́ть (*pres 1st, 3rd sing, 3rd pl*) в·бег·у́, в·беж·и́т, в·бег·у́т (*perf*); *see* в·бег·а́·ть
в·бер·у́т, (*fut 3rd pl*) of во·бр·а́ть (*perf*) *see* в·бир·а́·ть (*imp*)
в·би́в·к·а, -и, (*v n*) *see* в·бив·а́·ть
в·бив·а́·ть, -ют, (*imp*) drive in, hammer in, knock in; sink (piles etc.)
в·бир·а́·ть, -ют, (*imp*) absorb, suck in, soak up; inhale
в·би́·т·ый, -ая, -ое, (*past part pass*) of в·би́·ть
в·би́·ть, (*fut 3rd sing pl*) во·бь·ёт, во·бь·ю́т (*perf*) *see* в·бив·а́·ть
в·близ·и́ (*adv*) near, not far off, off, nearby, in the neighbourhood of, very close to, closely
в·бо́к· (*adv*) sideways
в·бра́с·ыва·ть, -ют, (*imp*) throw in
в·бро́д· (*adv*) by wading/fording
в·брос, -а, (*m*) (*v n*) *see* в·бро́с·ить; remelt sugar
в·бро́с·ить, -ят, (*perf*) throw in
в·бро́ш·енн·ый, -ая, -ое, (*past part pass*) of в·бро́с·ить; thrown in
в·бры́зг·ивани·е, -я, (*n*) spraying/sprinkling in, injecting
в·бры́з·нут·ый, -ая, -ое, (*past part pass*) injected, sprayed in
ВВ (*abbr*) = вз·ры́в·чат·ое вещ·е·ств·о́ explosive

в·ва́л·ива·ть, -ют, (*imp*) throw in, pile in, heap into; -ся fall in, tumble in; become sunken, form a hollow
в·вал·и́ть, -ят, (*perf*) *see* в·ва́л·ива·ть
в·вед·е́ни·е, -я, (*n*) (*v n*) *see* в·вод·и́ть; introduction, insertion; preface, introduction; addition, incorporation; administration (of a drug etc.)
~ в эксплуата́ци·ю setting in operation, starting up; blowing in (a blast furnace)
в·вед·ённ·ый, -ая, -ое, (*past part pass*) of в·вес·ти́; inserted, incorporated, introduced, administered
~ изото́п· (*m*) incorporated isotope
в·вез·ти́, -у́т, (*past masc, fem*) в·вёз·, в·вез·ла́ (*perf*); *see* в·воз·и́ть
в·вё·л, (*past masc sing*) of в·вес·ти́
в·верг·а́·ть, -ют, (*imp*) fling in, plunge in
в·ве́рг·н·уть, -ут, (*past masc, fem*) в·ве́рг·, в·верг·л·а́, (*perf*); *see* в·верг·а́·ть
в·вер·енн·ый, -ая, -ое, (*past part pass*) of в·ве́р·ить; entrusted with
в·верж·е́ни·е, -я, (*n*) (*v n*) *see* в·верг·а́·ть
в·ве́р·ить, -ят, (*perf*) *see* в·вер·я́·ть
в·вер·н·у́ть, -у́т, (*perf*) *see* в·вёрт·ыва·ть
в·вёрт·ыва·ть, -ют, (*imp*) screw in/into
в·вёрт·ыш, -а, (*m*) cap screw, screw stopper
в·верх· (*adv*) up, upwards, upward
~ дн·ом upside down
~ по рек·е́ up river, up stream
в·верх·у́ (*adv*) above, overhead, at the top
в·вер·я́·ть, -ют, (*imp*) entrust; -ся trust in
в·вес·ти́, (*fut 3rd sing, pl*) в·вед·ёт, в·вед·у́т (*past masc, fem*) в·вёл, в·ве·ла́, (*perf*) *see* в·вод·и́ть (*imp*)
в·вид·у́ (*prep+genitive*) in view (of), as, whereas
в·винт·и́ть, -ят, (*perf*) *see* в·ви́н·ч·ива·ть (*imp*)
в·винч·енн·ый, -ая, -ое, (*past part pass*) of в·винт·и́ть; screwed in/up, engaged
в·ви́нч·ива·ть, -ют, (*imp*) screw in/into/up; -ся (*passive*) engage
в·во́д·, -а, (*m*), (*v n*) of в·вод·и́ть leading/bringing in; lead in, lead; input, intake, entrance, bushing; loading (of computer); (GW) gathering

в·во́д
~, конденса́тор·н·ый   (elec) capacitor bushing
   то́ч·к·а в·во́д·а   (f) (phys) insertion/injection point (of a beam)
   у·стро́й·ств·о в·во́д·а да́·нн·ых   (n) input device/unit (of computer)
в·вод·и́ть, -ят,   (imp) bring in, lead in, put in, introduce, insert, incorporate (an isotope); (med) administer
~ в де́й·стви·е   put in operation, start up, blow in (a blast furnace)
~ в ку́рс·   tell all about, put in the picture
~ в стро́·й   put into service/operation, commission (a ship)
в·во́д·н·ый, -ая, -ое,   (a) introductory, leading in; inlet, lead-in, input; (gram) parenthetic; (f decl as adj) (mil) tactical problem
~ у·стро́й·ств·о   (n) (elec) input unit
~ у·стро́й·ств·о на перфо·ка́рт·ах   (n) punched card reader (of computer)
в·вод·я́щ·ий, -ая, -ее,   (pres part act) of в·вод·и́ть; lead, feed
в·во́з·, -а,   (m) import, importation, bringing in
в·воз·и́ть, -ят,   (imp) bring in; import
в·во́з·н·ый, -ая, -ое,   (a) imported, import
в·вы́рт·н·ый, -ая, -ое,   (a) screw-in, screw
в·вы́с·ь,   (adv) upwards, up
в·вя́з·анн·ый, -ая, -ое,   (past part pass) of в·вяз·а́ть
в·вяз·а́ть   (pres 3rd sing pl) в·вя́ж·ет, в·вя́ж·ут,   (perf) see в·вя́з·ыва·ть
в·вя́з·ыва·ть, -ют,   (imp) knit in; involve, draw in, implicate; -ся (passive) meddle, interfere
в·ги́б·, -а,   (m) dent, concavity, inward bend, sagging, sag
в·гиб·а́·ть, -ют,   (imp) bend in/inwards
в·глад·ь,   (adv) flush
   об·ши́в·к·а в·гла́д·ь   (f) (shipb) planking with carvel work
   с·ва́р·к·а в·гла́д·ь   (f) flush welding
в·глу́б·ь,   (adv) deep into/down, in the depths of, inside
в·гляд·е́ться, -я́тся,   (perf) see в·гля́д·ыва·ться
в·гля́д·ыва·ться, -ются,   (imp) peer at, look closely
в·гон·я́·ть, -ют,   (imp) drive in, force in

в·гор·яч·у́ю,   (adv) hot, in the hot state
в·да·ва́ться, -ются,   (imp) penetrate, stick in; jut out, stick out; go into (details)
в·дав·и́ть, -ят,   (perf) see в·да́вл·ива·ть
в·давл·е́ни·е, -я,   (n), (v n) see в·да́вл·ива·ть; impression, depression
в·да́вл·енн·ый, -ая, -ое,   (past part pass) see в·да́вл·ива·ть
в·да́вл·ивани·е, -я,   (n) (v n) of в·да́вл·ива·ть   (q.v.);   indentation
в·да́вл·ива·ть, -ют,   (imp) press in, force inwards, indent, stave in; -ся (passive) cave in
в·да́лбл·ива·ть, -ют,   (imp) ram in, pack in
в·дал·ек·е́   (adv) see в·дал·и́
в·дал·и́   (adv) in the distance, far off, away from
в·да́л·ь,   (adv) into the distance
в·да́·ться,   (fut 1st, 3rd sing, 3rd pl), в·да́мся, в·да́стся, в·дад·у́тся, (perf); see в·да·ва́ться
в·да·ю́щ·ийся, -аяся, -ееся,   (pres part act) of в·да·ва́ться
~ теч·е́ни·е   (n) indraft, inward flow
в·двиг·а́·ть, -ют,   (imp) move in/into, slide on/in, slip on/in; -ся (pass) go in/into
в·дви́·нут·ый, -ая, -ое,   (past part pass) of в·дви́·н·уть; pushed in, slid in/on
в·дви́·н·уть, -ут,   (perf) see в·двиг·а́·ть.
в·дво́·е   (adv) twice, double; in two
~ ме́н·ьш·е   half as much/many/big
в·дво·ём,   (adv) two, two together, two alone
в·двой·н·е́   (adv) double, doubly
в·дев·а́·ть, -ют,   (imp) put in, insert; thread through
в·де́л·анн·ый, -ая, -ое,   (past part pass) see в·де́л·ыва·ть
в·де́л·а·ть, -ют,   (perf) see в·де́л·ыва·ть
в·де́л·ыва·ть, -ют,   (imp) fit/fix/build in, inlay, set (jewels); (elec) seal in
в·де́·н·ут,   (fut 3rd pl) of в·де́·ть
в·дёрг·ива·ть, -ют,   (imp) pass through, thread through, lace up
в·дёр·н·уть, -ут,   (perf) see в·дёрг·ива·ть
в·де́сят·еро   (adv) ten times, tenfold
в·де́·т·ый, -ая, -ое,   (past part pass) of в·де́·ть; inserted, threaded through

в·де́·ть (*fut 3rd sing pl*) в·де́·н·ет, в·де́·н·ут, (*perf*) *see* в·дев·а́·ть

в·до·ба́в·ок· (*adv*) in addition, besides, furthermore, moreover

в·до́·вол·ь (*adv*) in plenty

в·до·го́н·к·у (*adv*) in pursuit of, after

в·долб·и́ть, -я́т, (*perf*) *see* в·да́лб·л·ива·ть

в·долбл·ённ·ый, -ая, -ое, (*past part pass*) of в·долб·и́ть

в·до́л·ь (*prep + genitive*) along; (*adv*) lengthwise, longitudinally

~ бо́рт·а (naut) alongside

~ по (+ *dative*) along

~ су́д·н·а (naut) fore and aft

в·до́х·, -а, (*m*) inhalation, breath

в·дох·нове́ни·е, -я, (*n*) inspiration

в·дох·н·у́ть, -у́т, (*perf*) *see* в·дых·а́·ть

в·дре́безги (*adv*) to/into smithereens, to pieces

в·дру́г· (*adv*) suddenly

~, все all together, all at once

в·дув·а́ни·е, -я, (*n*) blowing in, injection; (med) insufflation

~ в ва́нн·у (met) lancing the bath

в·дув·а́·ть, -ют, (*imp*) blow in/into

в·дув·н·о́й, -а́я, -о́е, (*a*) intake, injection, induction

~ магистра́л·ь (*f*) ventilation induction main

в·ду́м·а·ться, -ются, (*perf*) *see* в·ду́·м·ыва·ться

в·ду́м·чив·ый, -ая, -ое, (*a*) thoughtful

в·ду́м·ыва·ться, -ются, (*imp*) consider, think over

в·ду́·н·уть, -ут, (*perf*) blow in/into (once)

в·ду́·т·ый, -ая, -ое, (*past part pass*) of в·ду́·ть

в·ду́·ть, -ют, (*perf*) blow in/into, inject (gas); (met) lance (with oxygen)

в·дых·а́ни·е, -я, (*n*) breathing in; inhalation, inspiration

в·дых·а́тельн·ый, -ая, -ое, (*a*) inspiratory

~ кла́пан· (*m*) respirator inlet valve

в·дых·а́·ть, -ют, (*imp*) inhale, inspire, breathe in

ве́бер·, -а, (*m*) weber (unit of magnetic flux)

Ве́бер·а-Фе́хнер·а, за·ко́н· (*m*) Weber–Fechner law (acoustics)

Ве́бер·а, фото́·метр· (*m*) Weber photometer

ве́бер·ов· аппара́т· (*m*) (zool) Weberian apparatus

вебстери́т·, -а, (*m*) (geol) websterite

вевли́нг·а, -и, (*f*) (naut) ratline

Вега́рд·а, пра́в·ил·о (*n*) (chem) Vegard's law

Ве́генер·а, гипо·те́з·а (*f*) (geol) Wegener's hypothesis, continental drift theory

вегет·ариа́нск·ий, -ая, -ое, (*a*) vegetarian

вегет·ати́вн·ый, -ая, -ое, (*a*) vegetative

вегет·ацио́нн·ый, -ая, -ое, (*a*) vegetative, vegetation

вегет·а́ци·я, -и, (*f*) vegetation; vegetative stage

вегет·и́рующ·ий, -ая, -ее, (*pres part act*); vegetative

-вед· (*root*) (specialist, subject named in first part of word)

ве́д·а·ть, -ют, (*imp*) know, know about; manage, deal with, be in charge of, run

вед·е́ни·е, -я, (*n*) conduct, conducting, carrying on (an activity); management, control, charge

~, бы·ть в be in charge (of)

-ве́д·ени·е, -я, (*n*) (*component*) (signifies branch of knowledge, specialization, science, -ology)

вед·ённ·ый, -ая, -ое, (*a*), (*s f*) ·д·ён·, ·д·ен·а́, -д·ен·о́, (*past part pass*) of вес·ти́

ведёр·н·ый, -ая, -ое, (*a*) of ведр·о́

ве́ди (*n indecl*) letter 'V'

вед·и́те, (*imper*) of вес·ти́

ве́д·ом·о (*adv*) knowing, knowingly, with knowledge

ве́д·омост·ь, -и, (*f*) list, catalogue; journal, gazette

~, дефе́кт·н·ая repair list

~, доро́ж·н·ая (rail) freight bill

~ координа́т· index of co-ordinators (cartography)

~, пере·да́т·очн·ая (rail) transfer freight bill

~, пост·ов·а́я (mil) guard orders

ве́д·омственн·ый, -ая, -ое, (*a*) departmental; official

ве́д·омств·о, -а, (*n*) department (esp. government department)

вед·о́м·ый, -ая, -ое, (*pres part pass*) of вес·ти́; driven; ве́д·ом·ый, -ая, -ое, (*a*) known

~ пило́т· (*m*) (air) second pilot

~ само·лёт· (*m*) supporting aircraft (in formation)

~ шестерн·я́ (*f*) driven pinion

вед·о́м·ый, -ая, -ое,
~ **элеме́нт·**   (*m*) follower (in transmission system)

ведр·о́, -а,   (*nom pl*) **вёдр·а,** (*g pl*) **вёдер·,** (*n*) bucket, pail; fair weather

вед·у́щ·ий, -ая, -ее,   (*pres part act*) of **вес·ти́**; leading, foremost; driving (transmitting motion); master; (*m decl as adj*) leader
~ **вал·**   (*m*) driving/input shaft
~ **кана́т·**   (*m*) traversing rope (of cableway)
~ **кон·е́ц·**   (*m*) driving side (of driving belt etc.)
~ **шестерн·я́**   (*f*) driving pinion

вёд·ш·ий, -ая, -ее,   (*past part act*) of **вес·ти́**

вед·ь   (*conj*) you see, you know, as a matter of fact, in fact (often omitted in English)

вед·я́   (*pres gerund*) of **вес·ти́**

ве́ер·, -а, (*nom pl*) **-а́,** (*g pl*) **-о́в,** (*m*) fan; (gunn) sheaf (U.S.), lines of fire (Brit)
~ **де́й·стви́тельн·ого по·раж·е́ни·я** distributed guns, open sheaf
~, **паралле́ль·н·ый**   guns parallel, parallel sheaf
~, **рас·ход·я́щ·ийся** distributed guns, open sheaf
~, **со·средо·то́ч·енн·ый**   guns concentrated, converging sheaf
~, **с·ход·я́щ·ийся**   guns concentrated, converging sheaf

вееро·обра́з·н·ый, -ая, -ое,   (*a*) fan-shaped, flabelliform, flabellate
~ **о́рган·**   (*m*) (zool) flabellum
~ **с·вод·**   (*m*) (arch) fan vaulting

ве́·ет   (*pres 3rd sing*) of **ве́·ять**

вёз·,   (*past masc sing*) of **вез·ти́**

везде́   (*adv*) everywhere

везде·при·су́т·ствующ·ий, -ая, -ее, (*a*) = **везде·су́щ·ий**

везде·су́щ·ий, -ая, -ее,   (*a*) ubiquitous, omnipresent

везде·хо́д·, -а,   (*m*) cross-country vehicle

везде·хо́д·н·ый, -ая, -ое,   (*a*) cross-country

вез·ённ·ый, -ая, -ое,   (*s f*) **-з·ён, -з·ен·а́, -з·ен·о́** (*past part pass*) of **вез·ти́** (*imp deter*), see **воз·и́ть** (*imp indet*)

везеро́·метр·, -а,   (*m*) weatherometer

вез·ёт,   (*pres 3rd sing*) of **вез·ти́**

везико·то́м·и·я, -и, (*f*) (med) vesicotomy

везико·фикс·а́ци·я, -и,   (*f*) (med) vesicofixation

вези́к·ул·а, -ы,   (*f*) (med) vesicle, vesicula

везик·ул·и́т·, -а, (*m*) (med) vesiculitis

везик·ул·я́рн·ый, -ая, -ое,   (*a*) (biol) vesicular

Вези́н·а, про·бир·а́тел·ь   (*m*) Vezin sampler (ore assaying)

вез·и́те   (*imper*) carry, take, convey, transport

вез·о́м·ый, -ая, -ое, (*pres part pass*) of **вез·ти́** (*imp deter*), *see* **воз·и́ть** (*imp indet*)

вез·ти́, -у́т, (*past, masc/fem sing*) **вёз·, вез·ла́,** (*imp deter*), *see* **воз·и́ть** (*imp indet*)

везувиа́н·, -а,   (*m*) (min) vesuvianite, idocrase

везуви́н·, -а, (*m*) Bismarck brown (dye)

вез·у́щ·ий, -ая, -ее, (*pres part act*) of **вез·ти́** (*imp deter*), *see* **воз·и́ть** (*imp indet*)

вёз·ш·ий, -ая, -ее,   (*past part act*) of **вез·ти́** (*imp deter*), *see* **воз·и́ть** (*imp indet*)

вез·я́   (*pres gerund*) carrying, taking, transporting, conveying

Ве́йл·я, боле́зн·ь   (*f*) (med) Weil's disease

Ве́йс·а, молекуля́р·н·ое по́л·е   (*n*) Weiss molecular field
~, **о́бласт·ь**   (*f*) domain (ferromagnetism)

ве́к·, -а, (*nom pl*) **-а́,** (*g pl*) **-о́в,**   (*m*) century; epoch, age
~, **вюрм·ск·ий**   (geol) Wurmian stage
~, **дре́вн·ий ка́мен·н·ый**   (geol) Palaeolithic period
~, **желе́з·н·ый**   (geol) Iron Age
~, **но́в·ый ка́мен·н·ый**   Neolithic period
~, **неолит·и́ческ·ий**   (geol) Neolithic period

ве́к·о, -а,   (*nom pl*) **-и,**   (*n*) eyelid

веко·ве́ч·н·ый, -ая, -ое,   (*a*) timeless, eternal

век·ов·о́й, -а́я, -о́е,   (*a*) secular, age-old
~ **ход·**   (*m*) secular change (magnetic variation)

вексел·е·да́·тел·ь, -я,   (*m*) (fin) drawer (of bill)

вексел·е·держ·а́тел·ь, -я,   (*m*) (fin) drawee

вексел·е·по·луч·а́тел·ь, -я,   (*m*) endorsee, indorsee

вексел·ь, -я, (*nom pl*) **-я́,** (*g pl*) **-éй,** (*m*) (fin) bill of exchange, bill

~, без·де́неж·н·ый accommodation bill

~, дру́ж·еск·ий accommodation bill

~ на инка́ссо bill for collection

~, пере·во́д·н·ый bill of exchange

ве́ксель·н·ый, -ая, -ое, (*a*) of **ве́ксел·ь**

вексельра́д·, -а, (*m*) (horol) minute wheel

ве́к ор·, -а, (*m*) (math) vector

~, аксиа́ль·н·ый (phys) pseudovector, axial vector

~, без·вихр·ев·о́й irrotational vector

~ воз·мущ·е́ни·я disturbance vector

~, волн·ов·о́й (phys) wave vector

~ вращ·е́ни·я spin vector

~, един·и́чн·ый (math) unit vector

~ крив·изн·ы́ (nucl) buckling vector

~ на·магни́ч·енност·и magnetic induction

~, ос·ев·о́й (phys) pseudovector

~ пут·ев·о́й ско́р·ост·и flight path vector

~, свобо́д·н·ый free vector

~, с·вя́з·анн·ый localized vector

~, со́б·ственн·ый (math) eigenvector

~ спи́н·а (nucl) spin vector

~ цвет·н·о́й полос·ы́ colour bar vector (television)

векториа́ль·ност·ь, -и, (*f*) vectorial feature

векториа́ль·н·ый, -ая, -ое, (*a*) vectorial

вектор·ме́р·н·ый, -ая, -ое, (*a*) vector-measuring

ве́ктор·н·ый, -ая, -ое, (*a*) vector, vectorial

~ ана́лиз· (*m*) (math) vector analysis

~ ис·числ·е́ни·е (*n*) (math) vector algebra and vector analysis

~ про·из·вед·е́ни·е (*n*) (math) vector product of two vectors

~ про·стра́н·ств·о (*n*) (math) vector space

~ фу́нкци·я (*f*) (math) scalar function of vectors

вектор·ско́п·, -а, (*m*) vectorscope (television)

вё·л, (*past masc sing*) of **вес·ти́**

ве·ла́, (*past fem sing*) of **вес·ти́**

веле́н·ев·ая бума́г·а (*f*) (paper) vellum, wove paper

велери́т·, -а, (*m*) (min) wöhlerite

вел·е́ть, -я́т, (*imp and perf*) (+ *dat*) order, command

вел·и́ (*past pl*) of **вес·ти́**

велик·а́н, -а, (*m*) giant

вели́к·ий, -ая, -ое, (*a*) (*s f*) **вели́к, велик·а́, вели́к·о** great, large

вели́к·о- (*root*) great, big

велико·ле́п·н·ый, -ая, -ое, (*a*) superb, magnificent, splendid

велихови́т·, -а, (*m*) velichovite (a bitumen)

велич·а́йш·ий, -ая, -ее, (*a*) greatest

велич·ин·а́, -ы́, (*nom pl*) **-ч·и́н·ы,** (*g pl*) **-ч·и́н·,** (*f*) magnitude, size, value, quantity, amount

~, абсолю́т·н·ая (math) absolute value

~, абсолю́т·н·ая боло·метр·и́ческ·ая (astron) absolute bolometric magnitude

~, абсолю́т·н·ая звёзд·н·ая (astron) absolute magnitude

~ атмосфе́р·н·ых по·ме́х· (rad) static level

~, без·раз·ме́р·н·ая non-dimensional value

~, ве́ктор·н·ая (math) vector, vectorial entity, vector quantity

~, вероя́т·не́йш·ая best possible value, most probable value

~, вз·ве́ш·енн·ая weighted value (statistics)

~, вс·по·мог·а́тельн·ая intermediate value

~, в·ход·н·а́я (autom) input

~, вы·ход·н·а́я (autom) output

~, гармон·и́чн·ая harmonic quantity

~, да́·нн·ая datum

~ девиа́ци·и amount of deviation

~, де́й·ствующ·ая virtual value; effective size

~, дискре́т·н·ая discrete quantity

~ за·ви́с·ящ·ая от ма́сс·ы mass-sensitive quantity

~ за·ря́д·к·и (phot) exposure loading

~, звёзд·н·ая (astron) stellar magnitude

~, зем·н·а́я ground value (of any parameter)

~ зерн·а́ grain size, (met) grain size number

~ из·мен·е́ни·я пе́ленг·а (nav etc.) bearing rate

~, иск·о́м·ая unknown value/quantity

~, компле́кс·н·о со·пряж·ённ·ая (math) complex conjugate value

~, кон·е́чн·ая finite quantity

~ максиму́м·а height of the maximum

велич·ин·а́

~ мёртв·ой зон·ы   (rad) skip distance

~, натура́ль·н·ая   full size, full scale

~, нату́р·н·ая   full size, full scale

~ ни́з·ш·их по·ряд·к·о́в   background value (of ore)

~, обра́т·н·ая   reciprocal value

~ о·ши́б·к·и   extent of error

~ перв·и́чн·ой   концентр·а́ци·и   background value (of ore)

~ по·ме́х·   (rad) noise level

~, по·сто·я́нн·ая   (*see also* по·сто·-я́нн·ая *as noun*); constant

~, при·вед·ённ·ая   corrected value

~ про·лёт·а   miss-distance

~, рас·чёт·н·ая   calculated value

~, регул·и́руем·ая   (autom) controlled quantity

~ сеч·е́ни·я, обра́т·н·ая   reciprocal cross section

~ си́л·ы тя́ж·ест·и   intensity of gravity

~, скаля́р·н·ая   scalar gravity

~, с·луч·а́йн·ая   (math) random variable, stochastic variable

~, со·пряж·ённ·ая   conjugate value

~ у·велич·е́ни·я   (phot) enlargement range

~, у·равн·ённ·ая   adjusted value

~, це́л·ая   integral

~, чи́сл·енн·ая   numerical value

~ эне́рг·и·й, сре́д·н·яя   median energy

вел·о́   (*past neut sing*) of **вес·ти́**

вело-   (*component*) bicycle, cycle

вело·апте́чк·а, -и,   (*f*) cycle repair outfit

вело·ка́мер·а, -ы,   (*f*) inner tube (of bicycle tire)

вело·по·кры́ш·к·а, -и,   (*g pl*) **-ш·ек·,**   (*f*)   outer cover (of bicycle tire)

вело·ра́м·а, -ы,   (*f*) bicycle frame

велосипе́д·, -а,   (*m*) bicycle; tricycle

велосипе́д·н·ое шасси́   (*n*) (a/c) tricycle undercarriage

вело·ши́н·а, -ы,   (*f*) bicycle tire

вельбо́т·, -а,   (*m*) (naut) whaler, whaleboat

вельве́т·, -а,   (*m*) (text) velveteen

Ве́льдон·а, спо́соб·   (*m*) Weldon's process (of chlorine production)

вельп·, -а,   (*m*) whelp (capstan)

Ве́льц-спо́соб·, -а,   (*m*) (met) Waelz process

велю́м·, -а,   (*m*) (biol) velum, veil

веля́р·н·ый, -ая, -ое,   (*a*) (ling) velar

ве́н·а, -ы,   (*f*) (anat) vein, vena

венге́р·ск·ий, -ая, -ое,   (*a*) Hungarian

Вене́р·а, -ы,   (*f*) (astron) Venus

венер·и́ческ·ий, -ая, -ое,   (*a*) (med) venereal

венеро·ло́г·и·я, -и,   (*f*) venereology

вен·е́ц, -н·ц·а́,   (*i*) **-н·ц·о́м,**   (*m*) corona; (bot) crown; (mech) ring, rim; (min) jacket set

~, зуб·ча́т·ый   toothed ring/rim, gearwheel rim; starter ring (of flywheel)

~, лопа́т·очн·ый   blading ring (of turbine)

~, луч·и́ст·ый   (zool) corona radiator

~, на·правл·я́ющ·ий   turbine stator blade (gas turbine)

~, прямо·зу́б·ый   spur rim

венециа́н·ск·ий, -ая, -ое,   (*a*) venetian

вен·е́чн·ый, -ая, -ое,   (*a*) coronate, crown; (med) coronary; coronal

ве́нзел·ь, -я,   (*nom pl*) **-я́,**   (*g pl*) **-е́й**   (*m*) initials, monogram

ве́н·ик·, -а,   (*m*) besom, broom

ве́н·ичек·, -чк·а,   (*m*) (*dim*) of **ве́н·ик·** (bot) panicle

веннер·, -а,   (*m*) (min) vanner

вен·о́зн·ый, -ая, -ое,   (*a*) venous

вен·о́к·, -н·к·а́,   (*m*) wreath, garland

вено·склер·о́з·, -а,   (*m*) (med) venosclerosis

Венса́н·а, пресс·   (*m*) mechanical press

ве́н·ск·ий, -ая, -ое,   (*a*) Viennese, Vienna

~ и́звест·ь   (*f*) French chalk

~ кле·й   (*m*) (chem) Vienna paste

вентил·и́ровани·е, -я,   (*n*) *see* **вентил·я́ци·я**

вентил·и́рованн·ый, -ая, -ое,   (*past part pass*) of **вентил·и́р·овать** ventilated, aired

вентил·и́р·овать, -уют,   (*imp*) ventilate, air

ве́нтил·ь, -я,   (*m*) (*see also* **кла́пан·**); valve

~, баро·регул·и́рующ·ий   barostat

~, в·пуск·н·о́й   inlet valve

~, вы·пуск·н·о́й   release valve

~, вы·ход·н·о́й   out gate (computers)

~, дро́ссель·н·ый   choke, restrictor (in pipe), throttle valve

~, за·по́р·н·ый   shut-off valve

~, иго́ль·чат·ый   needle valve

~, ио́н·н·ый   thermionic rectifier

~, кон·и́ческ·ий   plug valve

~, магни́т·н·ый   solenoid-operated valve

вéнтил·ь

~, мéдно·за·кúс·н·ый (elec) copper-oxide rectifier

~, пере·пýск·н·óй bypass valve; (hydr) unloading valve

~, про·дýв·очн·ый blow-off valve

~, раз·грýз·очн·ый relieving valve

~, с·лúв·н·óй drain valve

~, угл·ов·óй angle valve

~, электр·úческ·ий (elec) electrolytic valve/rectifier, electrochemical valve

вéнтиль·н·ый, -ая, -ое, (a) of **вéнтил·ь**

~ сло·й (m) (elec) barrier layer, rectifying layer

~ пре·образ·овáтел·ь (m) inverter, inverted rectifier

~ сéт·к·а (f) (elec) rectifier network

~ фóто·элемéнт· (m) photovoltaic cell, barrier-layer cell

вентил·я́тор·, -а, (m) fan; exhauster

~, вы́·тяж·н·óй exhauster, exhaust fan

~, котéль·н·ый boiler forced-draught fan

~, мéль·ничн·ый exhauster

~, мел·я́щ·ий air swept unit mill (coal burning)

~, на·гнет·áтельн·ый forced-draught fan

~, ос·ев·óй axial-flow fan

~, от·сáс·ывающ·ий sampling aspirator (of gas analyser)

~, туннéль·н·ый ducted fan

~, центро·бéж·н·ый centrifugal fan

вентил·я́торн·ый, -ая, -ое, (a) of **вентил·я́тор·**

вентил·яциóнн·ый, -ая, -ое, (a) ventilated, ventilation

~, труб·á (f) ventilation pipe; breathing pipe (oil fuel tank)

~ шáхт·а (f) ventilating trunk

вентил·я́ци·я, -и, (f) ventilation, airing; ventilation system

~, в·дув·н·áя ventilation supply system

~, вы́·тяж·н·áя exhaust/extraction ventilation

~, естéств·енн·ая natural ventilation

~, искýсств·енн·ая forced ventilation

~, при·тóч·н·ая plenum ventilation

~, при·тóч·н·о-вы́·тяж·н·áя balanced ventilation

вентр·áльн·ый, -ая, -ое, (a) ventral

вентро·тóм·и·я, -и, (f) (med) ventrotomy

вентро·фикс·áци·я, -и, (f) (med) ventrofixation, ventrifixation

вентýри, трýб·к·а (f) Venturi tube/meter

венце·обрáз·н·ый, -ая, -ое, (a) crown-shaped, coroniform

вен·цóв·ый, -ая, -ое, (a) coronated, corona

вéн·чик·, -а, (m) (dim) of **вен·éц·**; (bot) corolla

венчико·обрáз·н·ый, -ая, -ое, (a) corolliform

венчико·цвéт·н·ый, -ая, -ое, (a) (bot) corolliflorous

вéр·а, -ы, (f) belief, faith

вератр·úн·, -а, (m) (chem) veratrine

вератр·úл·, -а, (m) (chem) veratryl

вератр·óв·ая кисл·от·á (f) veratric acid

вéрб·а, -ы, (f) (see also úв·а) willow

вербáль·н·ый, -ая, -ое, (a) verbal

вербéн·а, -ы, (f) (bot) verbena

верблю́д·, -а, (m) camel

~, одно·гóрб·ый dromedary

верблю́ж·ий, -ья, -ье, (a) of **вер·блю́д·**

верб·овáть, -ýют, (imp) recruit, enlist

верб·óвк·а, -и, (g pl) -вок·, (f) recruitment

Верде, по·сто·я́нн·ая (f) Verdet's constant (light)

вердо·гемат·úн·, -а, (m) verdohematin

вердо·гемо·глоб·úн·, -а, (m) verdo-hemoglobin

вердо·пероксидáз·а, -ы, (f) verdoperoxidase

верёв·к·а, -и, (g pl) -вок·, (f) cord, rope, string, twine

верёв·очн·ый, -ая, -ое, (a) of ве-рёв·к·а

~ мнóго·угóль·ник (m) (math) circumscribed polygon

верен·úц·а, -ы, (i) -ей, (f) file, row, line, succession, train

~ úмпульс·ов (rad) pulse train

вéреск·, -а, (m) (bot) heath, heather, *Erica*

вéреск·ов·ый, -ая, -ое, (a) of вéреск·, (pl) (bot) *Ericaceae*

веретен·úц·а, -ы, (i) -ей, (f) (zool) slow worm, blind worm

веретено·обрáз·н·ый, -ая, -ое, (a) spindle-shaped, ellipsoidal

верет·ённ·ый, -ая, -ое, (a) of вере-т·ен·ó

~ мáсл·о (n) spindle oil

верет·ен·о́, -а, (*nom pl*) **-т·ён·а,** (*g pl*) **-т·ён·,** (*n*) spindle; shaft (of oar); shank (of anchor); control plunger (of recoil system)

веретено·обра́з·н·ый, -ая, -ое, (*a*) spindle-shaped, (biol) fusiform

вере·я́, -й, (*f*) gate post; (naut) wherry

верже́ (*n indecl*) (paper) laid lines
бума́г·а верже́ (*f*) laid paper

верико́н·, -а, (*m*) (elec) vericon, image vericon

вер·и́тельн·ая гра́мот·а (*f*) (dipl) letter of credence

вер·ить, -ят, (*imp*) believe

верифика́тор·, -а, (*m*) verifier

ве́ркбле·й, -я, (*m*) crude lead bullion

верли́т·, -а, (*m*) (geol) wehrlite

вермикули́т·, -а, (*m*) (min) vermiculite

вермильо́н·, -а, (*m*) (chem) vermilion

вермише́л·ь, -и, (*f*) vermicelli

ве́рмут·, -а, (*m*) vermouth

вер·н·е́е, (*comp*) of **ве́р·н·о, ве́р·н·ый** more correctly, or rather

вернери́т·, -а, (*m*) (min) wernerite

ве́р·н·о (*adv*) correctly, truly, right

ве́р·ност·ь, -и, (*f*) fidelity; correctness, accuracy

вер·н·у́ть, -у́т, (*perf*) give back, return; get back, recover; **-ся** return, come back

ве́р·н·ый, -ая, -ое, (*a*) correct, right, valid; faithful, loyal; reliable, sure

верньер·, -а, (*m*) vernier

верона́л·, -а, (*m*) (pharm) veronal

веро́н·ск·ая зе́лен·ь (*f*) Veronese green

вероя́т·и·е, -я, (*n*) probability, likelihood

вероя́т·н·о (*adv*) probably; (*predic adj*) it is probable

вероя́т·ностн·ый, -ая, -ое, (*a*) of **вероя́т·ност·ь**

~ зако́н· (*m*) probability law

~ проце́сс· (*m*) stochastic process

вероя́т·ност·ь, -и, (*f*) probability

~ вы́лет·а (nucl) escape probability

~ дел·е́ни·я (nucl) fission probability
коэффицие́нт вероя́т·ност·и (*m*) probability factor
табл·и́ц·а вероя́т·ност·и (*f*) (gunn) probability factor table

вероя́т·н·ый, -ая, -ое, (*a*), (*s f*) **-т·ен·, -т·н·а,** probable, likely

ве́рп·, -а, (*m*) (naut) kedge anchor

верп·ова́льн·ый, -ая, -ое, (*a*) (naut) warping
~ ло́д·к·а (*f*) boat fitted with a warping winch

верп·ова́ни·е, -я, (*n*) (naut) warping

верп·ова́ть, -у́ют, (*imp*) (naut) warp

версе́н·, -а, (*m*) = **версена́т·**

версена́т·, -а, (*m*) (chem) versene, ethylenediaminetetraacetic acid
~ на́тр·и·я versene, EDTA, sodium salt of ethylenediaminetetraacetic acid

ве́рси·я, -и, (*f*) version

ве́рсор·, -а, (*m*) (math) versor

верст·а́, -ы, (*nom pl*) **вёрст·ы,** (*f*) (obs) verst (3500 feet)

верста́к·, -а́, (*m*) work bench, carpenter's bench

верст·а́льщик·, -а, (*m*) (print) imposer, stone-hand, maker-up

верст·а́ни·е, -я, (*n*) (*v n*) *see* **верст·а́ть**

верст·а́тк·а, -и, (*g pl*) **-ток·,** (*f*) (print) composing stick

верст·а́·ть, -ю́т, (*imp*) (print) make up, impose

верста́ч·н·ый, -ая, -ое, (*a*) of **верста́к·**

вёрст·к·а, -и, (*g pl*) **-т·ок·,** (*f*) (print) imposition, make-up; proof in sheets

верст·ов·о́й, -а́я, -о́е, (*a*) of **верст·а́**

версьер·а, -ы, (*f*) (math) versiera, witch, witch of Agnesi

вёрт·ел·, -а, (*nom pl*) **-а́,** (*g pl*) **-о́в,** (*m*) spit (for roasting); (anat) trochanter

верт·е́льк·а, -и, (*f*) (text) warp loom

верт·е́ть, -ят, (*imp*) spin, rotate, revolve; **-ся** (*intrans*)

вертика́л·, -а, (*m*) (astron) vertical circle

~, пе́рв·ый prime vertical circle, prime vertical

вертика́л·ь, -и, (*f*) vertical line, vertical; vertical observation point

~, гравитацио́н·н·ая gravity vertical

~, ста́рт·ов·ая (rocket) origin vertical

вертика́ль·н·о (*adv*) vertically

~ -от·клон·я́ющ·ая пласт·и́н·а (*f*) Y-plate

~ -подъ·ём·н·ый мост· (*m*) lift/lifting bridge

~ -про·тя́ж·н·ый стан·о́к· (*m*) vertical broaching machine

~ -попере́чн·ый раз·ре́з· (*m*) cross-section

~ строг·а́льн·ый стан·о́к· (*m*) slotter, vertical shaper

вертика́ль·ност·ь, -и, (*f*) verticality, vertical position

вертика́ль·н·ый, -ая, -ое, (*a*) vertical

~ сто́й·к·и теодоли́т·а (*pl*) theodolite pillars

вертика́ль·н·ый, -ая, -ое
~ **съ·ём·к·а** (*f*) (surv) levelling
вертика́нт·, -а, (*m*) (rocket) roll stabilizer
вертлю́г·, -а, (*m*) swivel; (gunn)
saddle, central pivot; (ent) trochanter
вертлю́ж·н·о-ло́н·н·ый, -ая, -ое, (*a*)
(zool) cotylopubic
вертлю́ж·н·ый, -ая, -ое, (*a*) of
вертлю́г·
~ **гак·** (*m*) swivel hook
~ **об·уш·о́к·** (*m*) swivel eye
верто·лёт·, -а, (*m*) (a/c) helicopter
верт·у́шечн·ый, -ая, -ое, (*a*) of
верт·у́шк·а
~ **лаг·** (*m*) screw log
верт·у́шк·а, -и, (*g pl*) **-шек·,** (*f*)
rotator, spinner, impeller; current
meter (of liquids); revolving door
верт·я́щ·ийся, -аяся, -ееся, (*pres
part act*) of **верт·е́ться;** rotating,
revolving, spinning
~ **ого́н·ь** (*m*) revolving light
ве́рф·ь, -и, (*f*) shipyard
ве́рх·, -а, (*nom pl*) **-и́,** (*g pl*) **-о́в,**
(*m*) top, head, high, height, summit;
cover, outside; upper (of shoe) (*nom
pl*) **верх·а́;** upper storey; upper reaches (of river); (*pl*) bosses, brass,
management; high notes (music)
~, **с·клад·н·о́й** hood
верхне·брод·и́льн·ый, -ая, -ое, (*a*)
top-fermenting
верхне·дол·ев·о́й, -а́я, -о́е, (*a*) (anat)
supralobar
верхн·е·челюст·н·о́й, -а́я, -о́е, (*a*)
(anat) maxillary
ве́рх·н·ий, -яя, -ее, (*a*) top, upper,
upstream, overhead; outer, outdoor
(of clothes); (biol) apical
~ **с·ва́р·к·а** (*f*) overhead welding
~ **сло·й** (*m*) upper/top layer; top
lift (concrete)
верх·ня́к·, -а, (*m*) roof timber
верх·о́вн·ый, -ая, -ое, (*a*) supreme
~ **суд·** (*m*) (law) supreme court
верхо·вод·к·а, -и, (*g pl*) **-д·ок·,** (*f*)
vadose/kremastic water
верх·ов·о́й, -а́я, -о́е, (*a*) top, upper;
up-river, upstream; riding (on animal);
(*m decl as adj*) rider, horseman
~ **езд·а́** (*f*) riding (on horseback etc.)
верх·о́в·ый, -ая, -ое, (*a*) = **верх·о-
в·о́й**
верх·о́вь·е, -я, (*g pl*) **верх·о́вь·ев**
(*n*) head (of valley etc.), upper reaches (of river)

верхо·ла́з·, -а, (*m*) steeplejack, spider-
man
верх·о́м (*adv*) on horseback; astride,
on top of; **ве́рх·ом** (*adv*) along
the top, by way of the top; to the
top, right up (when filling)
верхо·пло́д·н·ый, -ая, -ое, (*a*) acro-
carpous
верхо·под·вес·н·о́й, -а́я, -о́е, (*a*)
(build) top-hung
верхо·цве́т·ник·, одно·луч·ев·о́й (*m*)
(bot) monochasium
верхушечно·цве́т·н·ый, -ая, -ое, (*a*)
(bot) terminiflorous
верх·у́шечн·ый, -ая, -ое, (*a*) of
верх·у́шк·а; apical, terminal
верх·у́шк·а, -и, (*g pl*) **-шек·,** (*f*)
top, apex, summit; leaders, management, bosses, brass
ве́рш·а, -и, (*i*) **-ей,** (*f*) (fish) bow net
верш·и́н·а, -ы, (*f*) summit, top,
peak, crest, apex
~ **анти·клин·а́л·и** (geol) apex/crest/
crown of anticline
~ **зуб·ц·а́** tooth point/crest
~ **криста́лл·а** apex of crystal
ли́ния верш·и́н· (*f*) face line
(of a saw)
~, **пирамид·а́льн·ая** (geol) horn
~ **рез·ьб·ы́** thread head
~ **угл·а́** (math) vertex of an angle
верш·и́нник·, -а, (*m*) top log (timber)
верш·и́нн·ый, -ая, -ое, (*a*) apical
~ **у́гол·** (*m*) (math) vertex/vertical
angle
верш·кова́ни·е, -я, (*n*) topping
верш·о́к·, -ш·к·а́, (*m*) (obs) vershok
($1\frac{3}{4}$ inches)
ве́с·, -а, (*m*) weight; (*pl*) weights;
scales (*see* **весы́**)
~, **а́том·н·ый** atomic weight
~, **бо·ев·о́й** weight in running order
(armoured car etc.)
~ **бру́тто** gross weight
~, **вал·ов·о́й** gross weight
~, **вз·лёт·н·ый** (a/c) take-off weight
~, **вы́·груж·енн·ый** (rocket) out-
turn weight
~ **за·пра́вл·енн·ой сту́п·ен·и** (rocket)
stage gross weight
~ **констру́кци·и** (a/c) structural
weight
~, **молекуля́р·н·ый** (chem) molecular weight
~ **на·гру́з·к·и** (air) disposable load
~, **на·сып·н·о́й** bulk density
~, **не·по́лн·ый** short weight

вес
~ **нéтто** net weight
~, **объ·ём·н·ый** volume weight, weight by volume, bulk density
~, **по·лёт·н·ый** (air) all-up weight
~, **пóлн·ый** gross weight
~, **пóлн·ый по·лёт·н·ый** (air) maximum all-up weight
~, **по·сáд·оч·н·ый** (air) landing weight
~, **пуст·óй** (com) tare weight; (air) weight empty; dry weight (of liquid-fuel rocket)
~, **суммáр·н·ый** total weight
~, **с·цеп·н·óй** adhesive weight
~ **тáр·ы** tare weight
~, **трóй·ск·ий** troy weight
~, **у·бóй·н·ый** (agr) dead weight, dressed weight
~, **у·дéль·н·ый** specific weight/gravity; density (in g/cm^3); importance
~, **чѝст·ый** net weight
весел·ѝть, -я́т, (*imp*) gladden, cheer; **-ся** enjoy oneself, have a good time
весёл·ый, -ая, -ое, (*a*) cheerful, jolly, merry
весéль·н·ый, -ая, -ое, (*a*) of **весл·ó**
весел·я́щ·ий, -ая, -ее, (*pres part act*) of **весел·ѝть**
~ **газ·** (*m*) laughing gas (nitrous oxide)
весéн·н·ий, -яя, -ее, (*a*) spring (season), vernal
~ **равно·дéн·стви·е** (*n*) vernal equinox
вéс·ить, -ят, (*imp*) (*intrans*) weigh
вéс·к·ий, -ая, -ое, (*a*), (*s f*) **-с·ок·, -с·к·а** high specific gravity
весл·ó, -а, (*nom pl*) **вёсл·а,** (*g pl*) **вёсел·,** (*n*) oar
~, **вáльков·ое** (naut) double banked oar
~, **пáр·н·ое** scull
весло·нóг·ий, -ая, -ое, (zool) steganopodous, (*pl as noun*) copepods, *Copepoda*
~ **рачк·ѝ** (*pl*) (zool) copepods, *Copepoda*
весн·á, -ы́, (*nom pl*) **вёсн·ы,** (*g pl*) **вёсен·,** (*f*) spring, spring time
весно·вс·пáш·к·а, -и, (*g pl*) **-ш·ек·,** (*f*) spring ploughing/plowing
весн·óй (*adv*) in spring
веснýш·чат·ый, -ая, -ое, (*a*) freckled
вес·ов·óй, -áя, -óе, (*a*) of **вес·**; by weight
~ **анáлиз·** (*m*) gravimetric analysis
~ **доз·áтор·** (*m*) weighing batcher
~ **дóл·я** (*f*) weight fraction

вес·ов·óй, -áя, -óе
~ **коэффициéнт·** (*m*) (elec) weight coefficient
~ **по·тóк·** (*m*) weight flow
~ **скóр·ост·ь** (*f*) weight flow rate
~ **фýнкци·я** (*f*) (math) weighting function
~ **чáст·ь** (*f*) part by weight, weight fraction
~ **элемéнт·** (*m*) balance (of aerodynamic scales)
вес·óмост·ь, -и, (*f*) weightiness, ponderability, positive *g*
вёс·óм·ый, -ая, -ое, (*a*) ponderable, weighable
вéст·, -а, (*m*) (naut) west; west wind
вес·тѝ, (*pres 3rd sing, pl*) **вед·ёт, вед·ут,** (*past sing, m f*) **вё·л, ве·лá,** (*imp deter*) lead, guide; conduct, direct, run, do; drive, steer, pilot; keep (paper records)
вестибуля́р·н·ый аппарáт· (*m*) (anat) vestibule
вестибю́л·ь, -я, (*m*) entrance-hall, lobby, vestibule; (zool) vestibule
вестѝм·о (*adv*) certainly
Вéстингауз·а, тóрмоз· (*m*) (rail) Westinghouse brake
вестѝнд·ск·ий аррорýт· (*m*) West Indian arrowroot
вéст·ник·, -а, (*m*) herald, messenger
вест·ов·óй, -áя, -óе, (*a*) signal, warning; (*m decl as adj*) messenger, orderly
вéст·ов·ый, -ая, -ое, (*a*) waste, overflow
~ **труб·á** (*f*) overflow, waste pipe
Вéстон·а, элемéнт· (*m*) (elec) Weston cell, Weston normal/standard cell
вéст·ь, -и, (*g pl*) **-éй,** (*f*) news, news item
вес·ы́ (*nom pl*), (*g pl*) **-óв,** (*m*) scales, balance, weighing machine
~, **автомобѝль·н·ые** weighbridge
~, **аналит·ѝческ·ие** analytical balance
~, **аэродинам·ѝческ·ие** aerodynamic balance
~, **вертикáль·н·ые магнѝт·н·ые** Z-variometer
~ **Вестфáл·я** Westphal balance
~, **горизонтáль·н·ые магнѝт·н·ые** H-variometer
~, **дву·рычáж·н·ые крут·ѝльн·ые** double-beam torsion balance
~ **дéмпфер·н·ые** damped balance
~, **десят·ѝчн·ые** decimal balance
~ **-доз·áтор·, -а,** (*m*) bagging scales, batching scales

вес·ы́

~, **кольц·ев·ы́е**   ring balance

~, **крут·и́льн·ые**   torsion balance

~, **магни́т·н·ые**   magnetometric balance, relative magnetometer

~, **металло·метр·и́ческ·ие**   metallometric balance

~, **много·компоне́нт·н·ые**   (dynam) multicomponent balance

~, **не·демпф·и́рованн·ые**   undamped balance

~, **одно·плеч·ие**   constant-load balance

~, **платфо́рм·енн·ые**   weighbridge; platform scales (ore dressing)

~, **про·би́р·н·ые**   assay balance

~, **пруж·и́нн·ые**   spring balance

~, **равно·плеч·ие**   equal-arm balance

~, **радиацио́н·н·ые**   radiation balance

~, **руч·н·ы́е**   hand scales

~, **рыча́ж·н·ые**   beam balance

~, **то́к·ов·ые**   (elec) ampere/current weigher/balance

~, **торзио́н·н·ые**   torsion balance

~, **то́ч·н·ые**   precision balance

вес·ь   (*pron*) (*nom sing masc, acc sing masc inanim*) all, the whole; (*noun*) all; (*neuter noun*) everything; (*pl noun*) everyone, everybody; (*adv*) *see* **вс·ё, вс·его́**

весьма́   (*adv*) highly, greatly, extremely

~ **чи́ст·ый**   superfine-finished

вет-   (*abbr of component*) veterinary

ветвисто·у́с·ые, -ых,   (*pl decl as adj*) (ent) *Cladocera*

ветв·и́ст·ый, -ая, -ое,   (*a*) branchy, branching, (bot) ramose, ramified, dendroid, cladose

ветв·и́ться, -я́тся,   (*imp*) branch, sprout

ветвл·е́ни·е, -я,   (*n*) branching, branch, sprouting, (bot) ramification

    коэффицие́нт· ветвл·е́ни·я   (*m*) branching ratio

вет·вра́ч·, -а́,   (*i*) **-о́м,**   (*m*) veterinary surgeon, veterinarian

ве́тв·ь, -и,   (*g pl*) **-е́й,**   (*f*) (*see also* **ве́т·к·а**) branch, bough (of tree); branch, leg

~, **вос·ход·я́щ·ая**   pre-vertex branch, ascending branch (of a curve, hyperbola); pre-vertex trajectory

~ **крив·о́й**   branch of a curve

~, **нис·ход·я́щ·ая**   (math) post-vertex branch, (rocket) post-vertex trajectory

~ **о́с·и**   sub-axis

ветв·я́щ·ийся, -аяся, -ееся,   (*pres part act*) of **ветв·и́ться** branching, diverging, dendritic

ве́тер·, -тр·а,   (*m*) wind

~, **баллист·и́ческ·ий**   ballistic wind; (gunn, meteor) equivalent constant wind

~, **бок·ов·о́й**   cross wind

~, **встре́ч·н·ый**   head wind

~, **госпо́д·ствующ·ий**   prevailing wind

~, **градие́нт·н·ый**   (meteor) gradient wind

~, **нис·ход·я́щ·ий**   catabatic wind

~, **по·пу́т·н·ый**   tail wind

~, **пре·облад·а́ющ·ий**   prevailing wind

    ско́р·ост·ь ве́тр·а   (*f*) wind velocity

    у́гол· ве́тр·а   (*m*) wind direction

~, **электр·и́ческ·ий**   (elec, med) static breeze

ветерина́р·, -а,   (*m*) veterinary surgeon, veterinarian

ветерина́р·и·я, -и,   (*f*) veterinary science

ве́т·к·а, -и,   (*g pl*) **-т·ок·,**   (*f*) (*dim*) of **ве́тв·ь** (q.v.); (rail) branch line, spur

вет·лазаре́т·, -а,   (*m*) veterinary hospital

ве́то   (*n indecl*) veto

ве́т·очк·а, -и,   (*g pl*) **-чек·,**   (*f*) twig, sprig, shoot

ветош·к·а, -и,   (*g pl*) **-шек·,**   (*f*) rags

ве́тош·ь, -и,   (*f*) *see* **вето́ш·к·а**

ве́тр·а,   (*g sing*) of **ве́тер·**

ве́тр·ениц·а, -ы,   (*i*) **-ей,**   (*f*) wind crack (in timber); (bot) anemone, *Anemone*

ветр·ов·о́й, -а́я, -о́е,   (*a*) of **ве́тер·**

~ **колпа́к·**   (*m*) wind gag (on microphone)

~ **крюч·о́к·**   (*m*) (build) casement catch

~ **на·гру́з·к·а**   (*f*) (civ eng) wind load

~ **стекл·о́**   (*m*) (M/T) windscreen, windshield

ветро·го́н·, -а,   (*m*) wind scoop

ветро·дви́г·ател·ь, -я,   (*m*) wind motor

ветро·ло́в·, -а,   (*m*) wind scoop

ветро·ло́м·, -а,   (*m*) windbreak

ветро·ме́р·, -а,   (*m*) anemometer

ветро·на·со́с·, -а,   (*m*) wind pump, wind-driven pump

ветро·о·пыл·я́ем·ый, -ая, -ое,   (*a*) anemophilous, wind-pollinated

ветро·у·каз·а́тел·ь, -я,   (*m*) (air) wind sock/cone

ветро·чёт·, -а,   (*m*) (air) drift computer

ветро·электро·ста́нци·я, -и,   (*f*) wind-motor electric generating set

ветр·я́к·, -а,   (*m*) windmill; wind motor

ветр·я́нк·а, -и,   (*f*) (horol) fly

ветр·ян·о́й, -а́я, -о́е, (*a*) (*see also* **ветр·ов·о́й**) wind

ве́тх·ий, -ая, -ое, (*a*), (*s f*) **ветх·, ветх·а́, ве́тх·о** shabby, dilapidated

ветчин·а́, -ы́, (*f*) ham

ветш·а́·ть, -ют, (*imp*) fall into decay, become dilapidated

ве́х·а, -и, (*f*) (surv) marker post, stake, beacon, landmark, point; (naut) spar buoy

~, тра́ль·н·ая (naut) dan buoy

ве́чер·, -а, (*nom pl*) **-а́,** (*g pl*) **-о́в,** (*m*) evening

вече́р·н·ий, -яя, -ее, (*a*) of **ве́чер·**

ве́чер·ом (*adv*) in the evening

ве́ч·н·о- (*root*) permanent, perma-, everlasting, ever-

вечно·зелён·ый, -ая, -ое . (*a*) evergreen

вечно·мёрз·л·ый, -ая, -ое, (*a*) permafrost, everfrozen

вечно·пла́в·ающ·ий·, -ая, -ее, (*a*) (zool) holoplanktonic

ве́ч·н·ый, -ая, -ое, (*a*), (*s f*) **-ч·ен·, -ч·н·а,** eternal, perpetual

~ дви́г·ател·ь (*m*) perpetuum mobile

~ мерзл·от·а́ (*f*) permafrost

ве́ш·алк·а, -и, (*f*) coat rack, hat stand

ве́ш·а·ть, -ют, (*imp*) hang, hang up; (*trans*) weigh

ве́ш·ани·е, -я, (*n*) (*v n*) of **ве́ш·а·ть**

веш·е́ни·е, -я, (*n*) (*v n*) *see* **веш·и́ть**

веш·и́ть, -я́т, (*imp*) (surv) trace, mark out, stake out

ве́ш·к·а, -и, (*g pl*) **-ш·ек·,** (*f*) (*dim*) of **ве́х·а;** (surv) field rod, skewer, stake

~, нивели́р·н·ая (*f*) (surv) levelling pole

ве́ш·у, (*pres 1st sing*) of **ве́с·ить**

вещ·а́ни·е, -я, (*n*) (telecom) broadcasting

вещ·а́тельн·ый, -ая, -ое, (*a*) (telecom) broadcasting

вещ·ев·о́й, -а́я, -о́е, (*a*) of **ве́щ·ь**

вещ·е́ственн·ый, -ая, -ое, (*a*), (*s f*) **-твен·, -твенн·а** material, substantial, real

~ до·каз·а́тельств·о (*n*) (law) material evidence

~ ча́ст·ь (*f*) real part

вещ·еств·о́, -а́, (*n*) (*see also associated adjectives*); substance, matter, material, stuff; agent

~, вяж·ущ·ий cement, binder

~, гор·ю́ч·ее combustible

~, дуб·и́льн·ое tannin

~, ис·тир·а́ющ·ее abrasive

вещ·еств·о́

~, кра́с·ящ·ее dyestuff

~, о·кра́ш·ивающ·ее pigment

~, о·сажд·а́ющ·ее precipitant

~, о·хлажд·а́ющ·ее coolant

~, пит·а́тельн·ое nutrient

~, раз·жиж·а́ющ·ее diluent

~, рас·твор·ённ·ое solute

~, рас·твор·и́м·ое soluble matter

~, рас·твор·я́ющ·ее solvent

~, реаги́р·ующ·ее reactant

~, с·вяз·у́ющ·ее binder

~, с·гущ·а́ющ·ее coagulant

~, с·ма́з·очн·ое lubricant

~, с·ма́ч·ивающ·ее wetting agent

ве́щ·н·ый, -ая, -ое, (*a*) physical (of objects); (law) estate

~ пра́в·о (*n*) (law) law of estate

ве́щ·ь, -и, (*nom pl*) **-и,** (*g pl*) **-е́й,** (*f*) thing, (*pl*) effects, things, estate

ве́·ющ·ий, -ая, -ее, (*pres part act*) of **ве́·ять**

ве́·ялк·а, -и, (*g pl*) **-лок·,** (*f*) winnowing machine

ве́·яльн·ый, -ая, -ое, (*a*) winnowing

ве́·ян·ый, -ая, -ое, (*a*) winnowed

ве́·ять, (*pres 3rd sing, pl*) **ве́·ет, ве́·ют,** (*imp*) blow, blow gently, waft; wave, flutter, stir (of flags etc.); (agr) winnow

в·жат·ь, (*pres 3rd sing, pl*) **во·жм·ёт, во·жм·у́т,** (*perf*) *see* **в·жим·а́·ть**

в·жиг·а́ни·е, -я, (*n*) burning in

в·жим·а́·ть, -ют, (*imp*) squeeze/press in

вз- (*prefix*) = **взо-, взъ-, вс-** *particular significances*: (1) movement upwards; (2) intensity, force, sudden onset; (3) completion of action, achievement of state or limit

в·за́д· (*adv*) back, backwards

~ и в·перёд· (*adv*) backwards and forwards, to and fro

в·зад·и́ (*adv*) behind

в·за·и́м·н·о (*adv*) mutually, reciprocally, to one another

в·за·и́м·н·о- (*component*) *see also* **в·за·-и́м·о-**

~ -до·полн·и́тельн·ый, -ая, -ое, (*a*) (math) mutually complementary

~ -одно·зна́ч·н·ый, -ая, -ое, (*a*) (math) one-to-one

~ -перпендикуля́р·н·ый, -ая, -ое, (*a*) perpendicular, perpendicular to one another

~ -поля́р·н·ый, -ая, -ое, (*a*) (math) polar reciprocal

в·за·и́м·н·о-
~ -про·ник·а́ющ·ий, -ая, -ее,   (*a*) interpenetrating
~ -прост·ы́е чи́сл·а (*pl*) (math) reciprocals
~ -с·вя́з·анн·ый, -ая, -ое,   (*a*) coupled, linked
в·за·и́м·ност·ь, -и,   (*f*) reciprocity, correlation (geometry), mutuality
при́нцип в·за·и́м·ност·и   (*m*) (elec) reciprocity principle
в·за·и́м·н·ый, -ая, -ое, (*a*), (*s f*) -м·ен·, -м·н·а mutual, reciprocal, with respect to one another, to one another
~ коррел·я́ци·я (*f*) cross-correlation
~ с·вяз·ь (*f*) interconnection, interdependence; coupling
~ страх·ова́ни·е (*n*) mutual insurance
~ у·ничтож·е́ни·е (*n*) cancellation
в·за·и́м·о- (*component*) *see also* в·за·-и́м·н·о-
взаимо·ви́д·имост·ь, -и, (*f*) intervisibility, mutual visibility
взаимо·де́й·стви·е, -я, (*n*) interaction, interplay, reciprocal action; (phys) coupling; cooperation; interference (of wells)
~ ква́нт·ов·ое (rad) quantized interaction
коэффицие́нт· взаимо·де́й·стви·я (*m*) (elec) interaction factor
~, об·ме́н·н·ое (nucl) exchange interaction
~, пере·крёст·н·ое cross coupling
~ спи́н·ов (nucl) spin-spin interaction
~, спин·орбит·а́льн·ое (nucl) spin-orbit interaction, j–j coupling
взаимо·де́й·ств·овать, -уют, (*imp*) interact
взаимо·за·ви́с·имост·ь, -и, (*f*) interdependence, interdependency
взаимо·за·ви́с·им·ый, -ая, -ое, (*a*) interdependent
взаимо·за·мен·я́емост·ь, -и, (*f*) interchangeability
взаимо·за·мен·я́ем·ый, -ая, -ое, (*a*) interchangeable
взаимо·за·мест·и́мост·ь, -и, (*f*) interchangeability
за·ко́н· взаимо·за·мест·и́мост·и (*m*) law of photographic reciprocity, Bunsen-Roscoe law
взаимо·за·мык·а́ни·е, -я, (*n*) interlocking (railway signals)

взаимо·инду́кци·я, -и, (*f*) (elec) mutual inductance
коэффицие́нт· взаимо·инду́кци·и (*m*) mutual induction coefficient
взаимо·ис·ключ·а́ющ·ий, -ая, -ее, (*a*) (math) mutually exclusive; incompatible
взаимо·ис·ключ·е́ни·е, -я, (*n*) mutual exclusion, incompatibility; antipathy (of minerals)
взаимо·от·нош·е́ни·е, -я, (*n*) interrelation, relationship
взаимо·по́·мощ·ь, -и, (*f*) mutual aid
взаимо·пре·вращ·а́ем·ый, -ая, -ое, (*a*) interconvertible
взаимо·пре·вращ·е́ни·е, -я, (*n*) interconversion
взаимо·про·ник·а́ющ·ий, -ая, -ее, (*a*) interpenetrating, mutually penetrating
взаимо·про·ник·нове́ни·е, -я, (*n*) mutual penetration
взаимо·с·вя́з·анн·ый, -ая, -ое, (*a*) interrelated, related, interconnected, connected, coupling
взаимо·с·вя́з·ь, -и, (*f*) interconnection, interrelation, intercoupling, coupling, interdependence; intercommunication
~ ма́сс·ы и эне́рг·и·и mass-energy interrelation
~, пере·крёст·н·ая cross-coupling
в·за·йм·ы́ (*adv*) as a loan
да́ть в·за·йм·ы́ lend
по·луч·и́ть в·за·йм·ы́ borrow
в·за·ме́н· (*prep + genitive*) instead, in exchange for, in return for
в·за·мо́к· (*adv*) as a lock, lockwise
со·един·е́ни·е в·за·мо́к (*n*) seaming (sheet metal)
в·за·пер·т·и́ (*adv*) locked-up
вз·ба́лт·ыва·ть, -ют, (*imp*) shake, shake up
вз·бег·а́·ть, -ют, (*imp*) run up
вз·беж·а́ть (*fut 3rd sing, pl*) вз·беж·и́т, вз·бег·у́т, (*perf*) *see* вз·бег·а́·ть
вз·беш·ённ·ый, -ая, -ое, (*a*) furious, infuriated
вз·бив·а́·ть, -ют, (*imp*) beat up, whip, whisk; shake up, fluff up
вз·бир·а́·ться, -ются (*imp*) climb, clamber
вз·би́·т·ый, -ая, -ое, (*past part pass*) of вз·би́·ть *see* вз·бив·а́·ть
вз·би́·ть (*fut 3rd sing, pl*) взо·бь·ёт, взо·бь·ю́т, (*perf*) *see* вз·бив·а́ть
вз·болт·а́·ть, -ют, (*perf*) shake, shake up

вз·бра́с·ыва·ть, -ют, (*imp*) throw up, toss up

вз·бро́с·, -а, (*m*) (geol) reverse fault, overfault, upthrow fault, upthrust; (naut) spray

вз·бро́с·ить, -ят, (*perf*) throw/toss up

вз·бро́ш·енн·ый, -ая, -ое, (*past part pass*) of вз·бро́с·ить; thrown up; (geol) upthrown, overfaulted

вз·бры́зг·ива·ть, -ют, (*imp*) sprinkle

вз·бры́зг·н·уть, -ут, (*perf*) *see* вз·бры́зг·ива·ть

вз·буд·ора́жива·ть, -ют, (*imp*) agitate, disturb

вз·бунт·ова́ть, -у́ют, (*perf*) *see* бунт·ова́ть

вз·бух·а́·ть, -ют, (*imp*) swell up/out

вз·бу́х·н·уть, -ут, (*past masc fem/sing*) вз·бу́х·, вз·бу́х·ла, (*perf*) *see* вз·бух·а́·ть

вз·ва́л·ива·ть, -ют, (*imp*) hoist onto/into, hoist and load

вз·вал·и́ть, -ят, (*perf*) *see* вз·ва́л·ива·ть

вз·вед·е́ни·е, -я, (*n*) (*v n*) *see* вз·во·д·и́ть

вз·вед·ённ·ый, -ая, -ое, (*past part pass*) *see* вз·вод·и́ть

вз·ве́с·ить, -ят, (*perf*) *see* вз·ве́ш·ива·ть

вз·вес·ти́ (*fut 3rd sing, pl*) вз·вед·ёт, вз·вед·у́т, (*past masc, fem*) вз·вёл, вз·вела́, (*perf*) *see* вз·вод·и́ть

вз·ве́с·ь, -и, (*f*) (chem) suspension, suspended matter

вз·ве́ш·енност·ь, -и, (*f*) suspended state, suspension

вз·ве́ш·енн·ый, -ая, -ое, (*past part pass*) of вз·ве́с·ить; weighed, weighted; (chem) suspended

~ велич·ин·а́ (*f*) weighted value (statistics)

вз·ве́ш·ивани·е, -я, (*n*) weighing, weighting

вз·ве́ш·ива·ть, -ют, (*imp*) (*trans*) weigh; weight; ponder, consider

вз·ве́ш·ивающ·ий, -ая, -ое, (*pres part act*) of вз·ве́ш·ива·ть; weighing, weighting

~ давл·е́ни·е (*n*) uplift

~ коэффицие́нт· (*m*) weighting coefficient

вз·вив·а́·ть, -ют, (*imp*) send spinning upwards; -ся (*pass*) soar (of birds); go up, be raised (of curtains etc.)

вз·ви·ть, (*fut 3rd sing, pl*) взо·вь·ёт, взо·вь·ю́т, (*perf*) *see* вз·вив·а́·ть

вз·во́д·, -а, (*m*) (mil) platoon; cocking (of gun)

~ уда́р·ник·а (gunn) inner cocking lever

вз·вод·и́ть, -ят, (*imp*) lead up; cock (levers etc.), re-set; impute (fault etc.); cock (guns), arm (weapons)

вз·во́з·, -а, (*m*) road up a mountain/hill

вз·во́й·к·а, -и, (*f*) (text) batching roll

вз·волн·о́ванн·ый, -ая, -ое, (*past part pass*) *see* волн·ова́ть agitated

вз·волн·ова́ть, -уют, (*perf*) *see* волн·ова́ть

вз·гля́д·, -а, (*m*) look, gaze, glance; view, opinion

~ в·перёд· (surv) foresight

~ на·за́д· (surv) backsight; retrospect, hindsight

вз·гля́д·ыва·ть, -ют, (*imp*) glance, look

вз·гля·н·у́ть, -ут, (*perf*) *see* вз·гля́д·ыва·ть

вз·го́р·ь·е, -я, (*g pl*) -р·ий, (*n*) hill

вз·громожд·а́·ть, -ют, (*imp*) heave on/onto, pile on/onto

вз·громозд·и́ть, -ят, (*perf*) *see* вз·громожд·а́·ть

вз·дёрг·ива·ть, -ют, (*imp*) hitch up, jerk up

вз·дёр·н·уть, -ут, (*perf*) *see* вз·дёр·г·ива·ть

вз·до́р·, -а, (*m*) nonsense

вз·дорож·а́·ть, -ют, (*perf*) become more expensive, rise in price

вз·дох·н·у́ть, -у́т, (*perf*) *see* вз·ды·х·а́·ть

вз·дра́г·ивани·е, -я, (*n*) tremor, shudder

вз·дув·а́·ть, -ют, (*imp*) inflate; -ся swell

вз·дув·а́ни·е, -я, (*n*) *see* вз·ду́т·и·е

вз·ду́т·и·е, -я, (*n*) inflation, swelling, bulge, blister

вз·ду́·т·ый, -ая, -ое, (*past part pass*) *see* вз·дув·а́·ть; swollen

вз·ду́·ть, -ют, (*perf*) *see* вз·дув·а́·ть

вз·дым·а́ни·е, -я, (*n*) (*v n*) *see* вз·ды·м·а́·ть; (geol) upthrust, upthrusting

вз·дым·а́·ть, -ют, (*imp*) raise; -ся rise, be raised, billow

вз·дых·а́·ть, -ют, (*imp*) breathe heavily/deeply, sigh

вз·им·а́ем·ый, -ая, -ое, (*pres part pass*) of вз·им·а́·ть; taxable

вз·им·а́ни·е, -я, (*n*) (*v n*) *see* **вз·им·а́·ть;** tax levy, collection

вз·им·а́·ть, -ют, (*imp*) (fin) levy, collect

вз·ла́м·ыва·ть, -ют, (*imp*) break up/open/through

вз·лез·а́·ть, -ют, (*imp*) climb up, scale

вз·ле́з·ть, -ут, (*perf*) *see* **вз·лез·а́·ть**

вз·лёт·, -а, (*m*) (air) take-off, (rocket) start

вз·лет·а́·ть, -ют, (*imp*) (air) take off

вз·лет·е́ть, -я́т, (*perf*) *see* **вз·лет·а́·ть**

вз·лёт·н·о-по·са́д·очн·ая полос·а́ (*f*) (air) runway

вз·ло́м·, -а, (*m*) (*v n*) *see* **вз·ла́м·ы·ва·ть**

вз·лом·а́·ть, -ют, (*perf*) *see* **вз·ла́·м·ыва·ть**

вз·ма́х·, -а, (*m*) stroke, sweep, movement, wave, (air) flapping

   у́гол· вз·ма́х·а (*m*) flapping angle (of helicopter rotor)

вз·ма́х·ива·ть, -ют, (*imp*) flourish, wave, flap

вз·мёт·, -а, (*m*) (agr) first ploughing, breaking

вз·мет·а́·ть, (*fut 3rd sing, pl*) **вз·ме́ч·ет, вз·ме́ч·ут,** (*perf*) throw up, fling up

вз·мет·н·у́ть, -ут, (*perf*) throw/fling up (once)

вз·мёт·ыва·ть, -ют, (*imp*) throw/fling up

вз·ме́ч·ут (*fut 3rd pl*) of **вз·мет·а́·ть**

вз·мо́р·ь·е, -я, (*n*) sea shore

вз·мут·и́ть, -я́т, (*perf*) *see* **вз·му·щ·а́·ть**

вз·му́ч·енн·ый, -ая, -ое, (*past part pass*) of **вз·мут·и́ть;** muddy, turbid, cloudy

вз·мущ·а́·ть, -ют, (*imp*) stir up, make turbid/muddy/cloudy

вз·му́ч·ивани·е, -я, (*n*) mixing (of liquids); (agr) roiling

вз·мыв·а́ни·е, -я, (*n*) (*v n*) of **вз·мыв·а́·ть;** ballooning (of aircraft)

вз·мыв·а́·ть, -ют, (*imp*) sweep up (in flight)

вз·мыл·е́ни·е, -я, (*n*) foaming, frothing

вз·мы́л·ить, -ят, (*perf*) cover with foam/froth/lather; **-ся** (*pass*) froth, foam, get in a lather

вз·мы·ть, (*fut 3rd sing, pl*) **вз·мо́·ет, вз·мо́·ют,** (*perf*) sweep up/upwards (in flight)

вз·но́с·, -а, (*m*) payment, fee, due

~, част·и́чн·ый (fin) instalment

вз·ну́зд·ыва·ть, -ют, (*imp*) bridle

взо- (*prefix*) = **вз-, взъ-, вс-** *particular significances*: (1) movement upwards; (2) intensity, force, sudden onset; (3) completion of action, achievement of state or limit

взо·бр·а́ться, (*fut 3rd sing, pl*) **вз·-бер·ётся, вз·бер·утся;** *see* **вз·-бир·а́·ться**

взо·бь·ю́т (*fut 3rd pl*) of **вз·би́·ть**

взо·вь·ю́т (*fut 3rd pl*) of **вз·ви·ть**

взо·гн·а́ть, (*fut 3rd sing, pl*) **вз·го́н·ит, вз·го́н·ят;** *see* **воз·гон·я́·ть**

взо·йд·у́т (*fut 3rd pl*) of **взо·йти́**

взо·йти́, (*fut 3rd sing, pl*) **взо·йд·ёт, взо·йд·у́т** (*past masc, fem sing*) **взо·-шёл, взо·шла́;** mount, ascend, climb; (astron) rise; (bot) sprout, germinate, arise, originate in, come from

в·зо́р·, -а, (*m*) glance

взо́·рв·анн·ый, -ая, -ое, (*past part pass*) of **взо·рв·а́ть;** blown up, exploded, burst

взо·рв·а́ть, -у́т, (*perf*) *see* **вз·рыв·а́·ть**

вз·рез·а́·ть, -ют, (*imp*) cut up; **вз·ре́·з·ать,** (*fut 3rd sing, pl*) **вз·ре́ж·ет, вз·ре́ж·ут,** (*perf*) cut up

вз·ре́ж·ут (*fut 3rd pl*) of **вз·ре́з·ать** (*perf*)

вз·ро́·ет (*fut 3rd sing*) of **вз·ры·ть**

вз·ро́сл·ост·ь, -и, (*f*) maturity, adulthood, adultness

вз·ро́с·л·ый, -ая, -ое, (*a*) adult, full-grown, ephebic, grown-up; (ent) imaginal

вз·ро́·ют (*fut 3rd pl*) of **вз·ры·ть**

вз·ры́в·, -а, (*m*) explosion, outburst, burst, blast

~, воз·ду́ш·н·ый air explosion/blast

~ на·пра́вл·енн·ый внутр· implosion

~, тепл·ов·о́й thermal shock

~, я́дер·н·ый nuclear explosion

~ ядр·а́ nuclear burst/fragmentation

вз·рыв·а́ни·е, -я, (*n*) (*v n*) *see* **вз·ры·в·а́·ть**

~ на вы́·брос· auger-hole blasting

~ на рыхл·е́ни·е demolition blasting

~ нару́ж·н·ого за·ря́д·а (min) air blasting

~, огн·ев·о́е (min) fuse blasting, cap-and-fuse blasting/fusing

~ шпу́р·а (min) hole firing

вз·рыв·а́тел·ь, -я, (*m*) (gunn) fuse/fuze; (expl) fuse, blasting fuse

~, бок·ов·о́й body fuse

~, голов·н·о́й (mil) nose fuse

~ дистанц·ио́нн·о-уда́р·н·ого дей·-стви·я (mil) time-and-percussion fuse вз·рыв·а́тел·ь

~, дистанц·ио́нн·ый   proximity fuse/fuze; (min) time fuse

~, до́н·н·ый   (mil) base fuse

~, инерц·ио́нн·ый   (mil) graze fuse/fuze

~, искр·ов·о́й   (expl) spark fuse

~ мг·нове́нн·ого де́й·стви·я   instantaneous fuse/fuze

~, не·конта́кт·н·ый   proximity fuse

~, ра́дио·локацио́н·н·ый   (mil) radar pulse fuse

~ уда́р·н·ого де́й·стви·я   percussion fuze/fuse

вз·рыв·а́·ть, -ют,   (imp) dig up; blow up (buildings etc.), blast (rocks etc.); explode, fire (explosive devices); shoot (holes and wells); -ся explode, be blown up; explode, detonate (of devices)

вз·рыв·ни́к, -а́,   (m) (min) powderman, blaster, shot firer, shooter

вз·рыв·н·о́й, -а́я, -о́е,   (a) explosive, explosion, blasting, blast

~ волн·а́   (f) shock wave, blast

~ воро́н·к·а   (f) crater

~ маши́н·к·а   (f) (min) firing/blasting machine

~ на·груж·е́ни·е   (n) (met) explosive shock loading

~ на·клеп·   (m) (met) shock hardening

~ пункт·   (m) shot-firing point (seismic geophysical prospecting)

~ рабо́т·ы   (pl) blasting

взрыво·без·о·па́с·н·ый, -ая, -ое,   (a) flame/explosion proof

взрыво·о·па́с·ност·ь, -и,   (f) explosion hazard/danger

взрыво·о·па́с·н·ый, -ая, -ое,   (a) dangerously explosive, highly inflammable

взрыво·сто́й·к·ий, -ая, -ое,   (a) blast-proof

вз·рыв·ча́тк·а, -и,   (g pl) -ток·,   (f) explosive (material); (GW) military load

~, я́дер·н·ая·   nuclear explosive

вз·ры́в·чатост·ь, -и,   (f) explosiveness

вз·ры́в·чат·ый, -ая, -ое,   (a) explosive

~ вещ·еств·о́   (n) explosive (material)

~ вещ·еств·о́, бриза́нт·н·ое   disruptive explosive

~ вещ·еств·о́, иници·и́рующ·ее   initiating/priming explosive

~ вещ·еств·о́, пласт·и́чн·ое   plastic explosive

~ вещ·еств·о́, пред·о·хран·и́тель·н·ое   (min) permissible/permitted explosive

вз·ры́в·чат·ый, -ая, -ое

~ вещ·еств·о́, эквивале́нт·н·ое   equal-strength explosive

вз·ры·ть,   (fut 3rd sing, pl) вз·ро́·ет, вз·ро́·ют,   (perf) dig up

вз·рыхл·и́ть, -я́т,   (perf) see вз·рыхл·я́·ть

вз·рыхл·я́·ть, -ют,   (imp) loosen (soil etc.)

взъ-   (prefix) = вз-, (q.v.)

в·зыв·а́·ть, -ют,   (imp) appeal

вз·ыск·а́ни·е, -я,   (n) penalty, punishment; (law) recovery, exaction

взя́·вш·ий, -ая, -ее,   (past part act) of взя́·ть   (perf) see бр·ать

взя́·ти·е, -я,   (n) taking

~ ке́рн·а   (oil) core sampling

~ проб·   sampling

взя́·тк·а, -и,   (g pl) -ток·,   (f) bribe; trick (at cards)

взя́·ток·, -тк·а,   (m) (agr) honey flow

взя́·т·ый, -ая, -ое,   (a), (s f) -я́·т·, -я́·т·а́, -я́·т·о (past part pass) of взя́·ть

взя́·ть, (fut 3rd sing, pl) возьм·ёт, возьм·у́т; see бр·ать (ся)

виаду́к·, -а,   (m) viaduct

вибр·а́тор·, -а,   (m) (see also генера́тор, резона́тор, анте́нна) vibrator; (elec) vibrator, oscillator, resonator, trembler; (rad) dipole

~, вертика́ль·н·ый   (rad) vertical dipole

~, гармон·и́ческ·ий   (phys) sine-wave generator

~, горизонта́ль·н·ый   (rad) horizontal dipole

~, кольц·ев·о́й   ring-type oscillator

~, кулач·ко́в·ый   cam-and-spring vibrator (ore dressing)

~ на·пра́вл·енн·ого де́й·стви·я   (civ eng) directional vibrator

~, магнито·стрикцио́н·н·ый   (phys) magnetostriction rod

~, нару́ж·н·ый   shutter/external vibrator (for concrete)

~, сантиметр·о́в·ый   (rad) microwave dipole

~, симметр·и́чн·ый   (rad) resonant antenna, Hertzian dipole, resonator

вибрацио́н·н·ый, -ая, -ое,   (a) vibration, vibratory, vibrator, vibrating, vibrating-reed, oscillating

~ вы·но́с·ливост·ь   (f) (met) endurance/fatigue limit

~ пре·образ·ова́тел·ь   (m)   (elec) vibrator, chopper

вибрацио́н·н·ый, -ая, -ое
~ **при·бо́р·** (*m*) (elec) vibrating-reed instrument
~ **регул·я́тор· на·пряж·е́ни·я** (*m*) vibrator-type voltage regulator
вибр·а́ци·я, -и, (*f*) vibration
вибрио́н·, -а, (*m*) (bact) vibrio
вибр·и́р·овать, -уют, (*imp*) vibrate, tremble
вибро- (*int component*) vibrating, oscillating, trembling
вибро·вы·прям·и́тел·ь, -я, (*m*) (elec) vibrating-reed rectifier
вибро́·граф·, -а, (*m*) vibrograph
вибро·гро́хот·, -а, (*m*) racking screen (concrete); (min) vibrating screen, vibrator
~, **бараба́н·н·ый стру́н·н·ый** drum-wire vibrating screen
вибро·да́т·чик·, -а, (*m*) (instr) vibration pick-up
вибро·дви́г·ател·ь, -я, (*m*) vibrator motor
вибро·за·хва́т·, -а, (*m*) vibratory grab (ground clearance)
вибро·игл·а́, -ы́, (*f*) needle vibrator, vibrating spear (concrete work)
вибро·изол·я́ци·я, -и, (*f*) vibration damping
вибро·кат·о́к·, -т·к·а́, (*m*) (print) vibrating roller
вибро·лопа́т·а, -ы, (*f*) spade vibrator (concrete work)
вибро́·метр·, -а, (*m*) vibrometer, vibration meter
вибро·площ·а́дк·а, -и, (*g pl*) **-док·,** (*f*) table/platform vibrator, vibroplate (concrete)
вибро·по·груж·е́ни·е, -я, (*n*) vibro pile-driving
вибро·пре·образ·ова́тел·ь, -я, (*m*) (elec) vibrator, chopper
вибро·про́ч·ност·ь, -и, (*f*) (met) endurance/fatigue limit
вибро·ре́й·к·а, -и, (*g pl*) **-ре́·ек·,** (*f*) rack vibrator (concrete)
вибро·сто́й·к·ий, -ая, -ое, (*a*) vibration-resistant
вибро·у·плот·не́ни·е, -я, (*n*) vibratory compaction (concrete)
вибро·у·сто́й·чивост·ь, -и, (*f*) resistance to vibration
вибро·шабло́н·, -а, (*m*) template vibrator (for concrete)
вива́р·и·й, -я, (*m*) (zool) vivarium
вивиани́т·, -а, (*m*) (min) vivianite
виви·се́кц·и·я, -и, (*f*) vivisection

ви́·вш·ий, -ая, -ее, (*past part act*) of **ви·ть**
Ви́гнер·а, эне́рги·я (*f*) (nucl) Wigner energy
~, **эффе́кт·** (*m*) (nucl) Wigner effect
виго́н·ь, -и, (*f*) (zool, text) vicuna
вид·, -а, (*m*) appearance, aspect; kind, sort, mode, (biol) species, (geol) facies; form, shape; view, prospect; sight
~, **бок·ов·о́й** side view, profile
 в вид·е in the form/shape of, as
 в вид·у́ in view of, because of, as
 им·е́·ть в вид·у́ have in mind, think of; intend, mean, mean to
~ **в пла́н·е** plan view
~, **вертика́ль·н·ый** elevation
~, **вне́ш·н·ий** exterior view; (bot) habit
~, **второ·степе́н·н·ый** minor species
~, **за·мещ·а́ющ·ий** (biol) vicarian species
~ **и́мпульс·ов** pulse mode
~, **канон·и́ческ·ий** (math) canonical form
~ **колеб·а́ни·я** mode
~, **не·со·верш·е́нн·ый** (gram) imperfective aspect
~, **по·все·ме́ст·ный** (biol) cosmopolitan species
~ **рас·па́д·а** (phys) mode of decay
~, **рели́кт·ов·ый** (biol) relict species
~ **с·бо́к·у** side view, side elevation
~ **с·за́д·и** rear view, back elevation
~ **симметр·и́·и** (cryst) class of symmetry
~ **с·пе́ред·и** front view, front elevation
~, **я́дер·н·ый** nuclear species
Вида́л·я, про́б·а (*f*) (med) Widal's test
ви́д·анн·ый, -ая, -ое, (*past part pass*) of **вид·а́·ть;** seen, often/commonly seen
вид·а́·ть, -ют, (*imp*) see often
ви́д·евш·ий, -ая, -ее, (*past part act*) *see* **ви́д·еть**
Ви́деман·а, эффе́кт· (*m*) Wiedemann effect (magnetostriction)
Ви́деман·а-Фра́нц·а, за·ко́н· (*m*) (phys) Wiedemann–Franz law
Ви́деман·а-Фра́нц·а, со·от·нош·е́·ни·е (*n*) (elec) Wiedemann–Franz ratio
ви́д·ени·е, -я, (*n*) (*v n*) of **ви́д·еть;** vision, sight; viewing
~, **прям·о́е** (phot) direct viewing
ви́д·енн·ый, -ая, -ое, (*past part pass*) *see* **ви́д·еть**

видео·каска́д·, под·рез·а́ющ·ий video clipper

видео·ми́кшер,· -а,   (*m*) (T/V) video mixer

видео·сигна́л·, -а,   (*m*) (rad) video-signal, video

~, не·строб·и́рованн·ый   raw video

видео·то́к·, -а,   (*m*) (T/V) video current

видео·у·сил·и́тел·ь, -я,   (*m*) video-amplifier

ви́дер·, -а,   (*m*) (agr) weeder

ви́д·еть, -ят,   (*imp*) see

вид·ико́н·, -а,   (*m*) vidicon, (T/V) photoconductor

ви́д·им·о   (*adv*) apparently; evidently

ви́д·имост·ь, -и,   (*f*) visibility

~ земл·и́   (air) visual contact with the ground

   зо́н·а ви́д·имост·и   (*f*) coverage pattern (radar)

~, ис·ключ·и́тельн·ая   (meteor) exceptional visibility (above 50,000 metres or above 27 miles)

~, о́чень плох·а́я (naut meteor) very bad visibility (0–3 cables)

~, о́чень сла́б·ая very poor visibility (1000–2000 metres)

~, о́чень хоро́ш·ая   (meteor) very good visibility (20,000–50,000 metres); (naut meteor) very good visibility (11–27 miles)

~, плох·а́я   bad visibility (200–1000 metres); (naut meteor) poor visibility (3 cables–1 mile)

~, по·сре́д·ственн·ая   moderate visibility (4000–7000 metres)

~, сла́б·ая   poor visibility (2000–4000 metres)

~, сре́д·н·яя   (meteor) moderate visibility (4000–7000 metres); (naut meteor) moderate visibility (1–5 miles)

~, хоро́ш·ая   (meteor) good visibility (13,000–20,000 metres); (naut meteor) good visibility (5–11 miles)

ви́д·им·ый, -ая, -ое, (*pres part pass*) see **ви́д·еть**; visible, apparent

~ реч·ь   (*f*) (elec) visible speech

видманште́тт·, -а, (*m*) (met) Widmann-stätten structure

видманште́ттен·ов·а, структу́р·а (*f*) (met) Widmannstätten structure

видманште́тт·ов·, -а, -о, (*possess adj*) = **видманште́ттен·ов·**

ви́д·н·о   (*predic adj*) it can be seen, it is evident; (*adv*) evidently

ви́д·ност·ь, -и,   (*f*) (phot) visibility/luminosity factor

ви́д·н·ый, -ая, -ое, (*a*) (*s f*) **ви́д·ен·, вид·н·а́, ви́д·н·о,** visible; prominent, distinguished, eminent; (*as component adj*) -shaped, -form, -oid

видо·образ·ова́ни·е, -я,   (*n*) (biol) speciation

вид·ов·о́й, -а́я, -о́е, (*a*) of **вид·** (q.v.); (biol) specific

~ объекти́в·   (*m*) (phot) landscape lens

~ раз·ли́ч·и·е (*n*) (biol) specific difference

видо·из·мен·е́ни·е, -я,   (*n*) modification; variant

видо·из·мен·и́ть, -я́т,   (*perf*) see **ви·до·из·мен·я́·ть**

видо·из·мен·я́емост·ь, -и,   (*f*) mutability, variability

видо·из·мен·я́·ть, -ют,   (*imp*) modify, alter, refashion

видо·иск·а́тел·ь, -я, (*m*) (*see also* **ви·зи́р·**) (phot) viewfinder

~, зеркал́ь·н·ый   (phot) reflex view-finder

~, иконо·метр·и́ческ·ий direct-vision viewfinder

~, я́щич·н·ый   ground-glass view-finder

видо·образ·ова́ни·е, -я,   (*n*) (biol) interspecific conversion, species formation

ви́д·ящ·ий, -ая, -ее,   (*pres part act*) of **ви́д·еть**

ви́ж·у   (*pres 1st sing*) of **ви́д·еть**

ви́з·а, -ы,   (*f*) visa

визави́   (*adv*) facing, opposite; (*m or f indecl*) vis-à-vis

визг·ли́в·ый, -ая, -ое,   (*a*) shrill, whining, screeching

визибилиметр·, -а,   (*m*) visibility meter

визи́г·а, -и,   (*f*) (food) dried spinal cord of sturgeon

визи́р·, -а,   (*m*) (*see also* **видо·иск·а́·тел·ь**); sight (of instrument); (phot) view-finder

~, маршру́т·н·ый   strip finder (aerial photography)

~, навигацио́н·н·ый   (air) drift meter

~, по·та́й·н·ой   (phot) spyglass view-finder

визи́р·к·а, -и, (*g pl*) **-р·ок·,**   (*f*) (instr, gunn) sight vane; (surv) ranging pole/rod

визи́р·н·ый, -ая, -ое,   (*a*) of **визи́р·**

~ тру́б·к·а   (*f*) (surv) telescopic view-finder; (gunn) collimator sighting

визи́р·овани·е, -я, (*n*) sight, sighting, aiming, alignment, pointing
   ли́н·и·я визи́р·овани·я (*f*) (surv etc.) line of sight, guide line, collimating line, transit line; (phot) viewing line
   ли́н·и·я визи́р·овани·я звезд·ы́ (*f*) star line
   у́гол· визи́р·овани·я (*m*) viewing/sighting angle
визи́т·, -а, (*m*) call, visit (formal)
визулиз·и́р·овать, -уют, (*imp and perf*) make visible, visualize, reveal (by some chemical etc. means)
визуа́ль·н·ый, -ая, -ое, (*a*) visual
виикит·, -а, (*m*) (min) wiikite
Винийр·а, явл·ёни·е (*n*) Villard effect (X-ray)
Вийс·у, йод·н·ое числ·о по (*n*) (chem) Wijs iodine number
ви́к·а, -и, (*f*) (bot) vetch, *Vicia*
   ~, мохна́т·ая hairy vetch, *V. villosa*
   ~, о·зи́м·ая winter vetch
   ~, паннон·ск·ая Hungarian vetch, *V. pannonica*
   ~, по·сев·н·а́я (bot) common vetch, *V. sativa*
   ~, яр·ов·а́я spring vetch
вика́р·н·ый, -ая, -ое, (*a*) (biol) vicarious
викалло́·й, -я, (*m*) (met) Vicalloy
викел·ёвк·а, -и, (*f*) (rubber) wrapping, rolling/winding up
ви́кел·ь, -я, (*m*) (rubber) rolled-up stock
Ви́ккерс·а, ме́тод· (*m*) (met) Vickers hardness test
   ~, спо́соб· (*m*) (met) Vickers hardness test
Виккерс·у, твёрд·ост·ь по Vickers hardness
ви́к·ов·ый, -ая, -ое, (*a*) (bot) vetch
   ~ с·мес·ь (*f*) (agr) vetch mix
ви·л (*past masc sing*) of **ви·ть**
вил·а (*f*) see **ви́лы·**; **ви·ла́** (*past fem sing*) of **ви·ть**
ви́·ли (*past pl*) of **ви·ть**
ви́л·к·а, -и, (*g pl*) **-л·ок·,** (*f*) fork; fork end, jaw (mechanical part); (elec) two-pin plug, plug; (gunn) bracket; (horol) crutch; (nucl) branching (of a decay)
   ~, а́нкер·н·ая (mech) lever fork
   ~, много·конта́кт·н·ая штепсель··н·ая (elec) multipin plug
   ~, на·гру́з·очн·ая (elec) battery-cell tester

ви́л·к·а
   ~, нул·ев·а́я (gunn) virtual straddle
   ~ от·во́д·к·и toggle
   ~ пере·ключ·ёни·я striker clutch
   ~, штёпсель·н·ая (elec) plug, two-pin plug
ви́л·ков·ый, -ая, -ое, (*a*) of **ви́л·к·а;** **вил·ко́в·ый, -ая, -ое,** (*a*) of **вил·о́к·**
вилко·обра́з·н·ый, -ая, -ое, (*a*) forked, bifurcate
Виллари, то́ч·к·а (*f*) Villari reversal point (magnetostriction)
   ~, эффе́кт· (*m*) Villari effect (magnetostriction)
виллеми́т·, -а, (*m*) (min) willemite
ви́ллис·, -а, (*m*) (M/T) jeep
вил·о́к·, -л·к·а́, (*m*) cabbage head
вило·обра́з·н·ый, -ая, -ое, (*a*) forked, furcate
ви́л·очк·а, -и, (*g pl*) **-чек·,** (*f*) wishbone
ви́л·очн·ый, -ая, -ое, (*a*) of **ви́л·к·а** forked, bifurcate
ви́лт·, -а, (*m*) (bot) wilt
ви́л·ы (*nom pl*), (*g pl*) **вил·,** (*f*) pitchfork, fork
виль·н·у́ть, -у́т, (*perf*) see **вил·я́·ть**
ви́ль·н·ый сено·на·гру́з·чик (*m*) fork hay-loader, rake-bar hay-loader
Ви́льсон·а, ка́мер·а (*f*) (nucl) Wilson cloud/expansion chamber
ви́ль·чат·ый, -ая, -ое, (*a*) forked, furcate
вил·я́·ть, -ют, (*imp*) yaw, weave (motion)
вин·а́, -ы, (*nom pl*) **-ы́,** (*f*) fault, guilt
Ви́н·а, за·ко́н· из·луч·е́ни·я (*m*) (phys) Wien radiation law
   ~, за·ко́н· с·мещ·е́ни·я (*m*) (phys) Wien displacement law
ви́ндзейл·ь, -я, (*m*) (naut) windsail
виндобо́н·ск·ий, -ая, -ое, (*a*) (geol) Vindobonian
виндро́уэр·, -а, (*m*) (agr) windrower
вин·и́л·, -а, (*m*) (chem) vinyl
   ~· хло́р·ист·ый vinyl chloride
винил·ацета́т·, -а, (*m*) vinyl acetate
винил·бензо́л·, -а, (*m*) vinyl benzene, styrene
винил·е́н·, -а, (*m*) vinylene
винил·иде́н·, -а, (*m*) vinylidene
   ~, хло́р·ист·ый vinylidene chloride
винил·иза́ци·я, -и, (*f*) vinylation
винил·и́т·, -а, (*m*) vinylite
винил·и́т·ов·ый, -ая, -ое, (*a*) vinylite
винил·карбазо́л·, -а, (*m*) vinyl carbazole

вини́л·ов·ый, -ая, -ое, (*m*) vinyl

~ **спирт·** (*m*) vinyl alcohol

винило́ги·я, -и, (*f*) vinylogy

винил·пирролидо́н·, -а, (*m*) vinylpyrrolidone

винил·хлори́д·, -а, (*m*) vinyl chloride

вини́ль·н·ый, -ая, -ое, (*a*) *see* **вини́л·ов·ый**

вини·пла́ст·, -а, (*m*) rigid PVC plastic

вин·и́тельн·ый пад·е́ж· (*m*) (gram) accusative case

вин·и́ть, -я́т, (*imp*) blame, accuse; **-ся** confess, admit guilt

ви́нкел·ь, -я, (*nom pl*) **-я́,** (*g pl*) **-ей,** (*m*) set square

ви́н·н·о- (*component*) (chem) -tartaric; -tartrate of

винно·желе́з·н·ая со́л·ь (*f*) ferric tartrate

винно·ка́л·иев·ая со́л·ь (*f*) potassium tartrate

винно·ка́льц·иев·ая со́л·ь (*f*) calcium tartrate

винно·ка́мен·н·ый, -ая, -ое, (*a*) (*see also* **ви́н·н·ый**) (chem) tartaric

~ **кисл·от·а́** (*f*) tartaric acid

винно·ки́сл·ый, -ая, -ое, (*a*) tartrate of

ви́н·н·ый, -ая, -ое, (*a*) wine, winy, vinous; tartaric

~ **ка́мен·ь** (*m*) cream of tartar; tartar, scale (on teeth)

~ **спирт·** (*m*) spirits of wine

~ **у́кскс·** (*m*) wine vinegar

~ **я́год·а** (*f*) fig

вин·о́, -а, (*nom pl*) **-а́,** (*n*) wine

~, **игр·и́ст·ое** sparkling wine

~, **крепл·ён·ое** fortified wine

~, **не·игр·и́ст·ое** still wine

~, **спирт·о́ванн·ое** brandy

~, **хлеб·н·ое** vodka

вин·ова́т·ый, -ая, -ое, (*a*) guilty, blameworthy; sorry (when accepting fault)

вин·о́вник·, -а, (*m*) culprit, guilty party

вин·о́вн·ый, -ая, -ое, (*a*) (*s f*) **-вен·, -вн·а** guilty

вино·го́н·н·ый, -ая, -ое, (*a*) distillatory, distilling

виногра́д·, -а, (*m*) grape, *Vitis* sp.; vine

~ **изабе́лл·а** *Vitis labrusca*

~, **культу́р·н·ый** *Vitis vinifera*

виногра́д·арств·о, -и, (*n*) viticulture, viniculture

виногра́д·ар·ь, -я, (*m*) viticulturalist, vigneron

виногра́д·ин·а, -ы, (*f*) grape

виногра́д·ник·, -а, (*m*) vineyard

виногра́д·н·ый, -ая, -ое, (*a*) grape, racemic; (*pl noun decl as adj*) *Vitaceae*

~ **кисл·от·а́** (*f*) racemic/tartaric acid

вино·де́л·, -а, (*m*) wine-maker

вино·де́л·и·е, -я, (*n*) wine-making

вино·ку́р·, -а, (*m*) distiller

вино·кур·е́ни·е, -я, (*n*) distillation

вино·ку́р·енн·ый, -ая, -ое, (*a*) distilling

~ **за·во́д·** (*m*) distillery

ви́нт·, -а́, (*m*) screw; propeller, screw, screw propeller; rotor (of helicopter)

~, **авиацио́н·н·ый** aircraft propeller, airscrew

~ **-автома́т·, -а,** (*m*) constant-speed propeller

~, **Архиме́д·ов·** Archimedean screw

~ **бара́ш·ков·ый** butterfly/thumb screw

~ **бес·кон·е́чн·ый** endless screw, screw/worm conveyor

~ **вед·у́щ·ий** driving/engaging screw

~ **ви́л·к·и мая́т·ник·а** (horol) beat screw

~, **воз·ду́ш·н·ый** airscrew, aircraft propeller

~, **ги́б·к·ий воз·ду́ш·н·ый** (a/c) non-rigid propeller/airscrew

~, **греб·н·о́й** propeller, screw propeller, ship's propeller

~, **двух·ло́паст·н·ый** twin-bladed propeller

~, **двух·оборо́т·н·ый** double-threaded screw

~, **дву·ход·ов·о́й** double-threaded screw

~, **дифференциа́ль·н·ый** hunter's screw

~, **за·жим·н·о́й** clamping/fastening screw

~, **за·креп·и́тельн·ый** fastening/ securing screw

~, **за·по́р·н·ый** plug/stopper screw

~, **зерка́ль·н·ый** (T/V) mirror screw

~ **из·мен·я́ем·ого ша́г·а** variable-pitch propeller

~, **каре́т·н·ый** coach/lag screw

~ **ко́рпус·а** (horol) dog screw

~ **ле́в·ого ша́г·а** left-handed propeller

~, **макро·метр·и́ческ·ий** coarse adjustment screw

~, **микро·метр·и́ческ·ий** fine adjustment screw, micrometer screw

ви́нт

~, на·жим·н·о́й   thrust/closing/forcing/pressing screw; (met roll) screw down

~, на·правл·я́ющ·ий   driving/engaging screw

~, на·тя́ж·н·о́й   turnbuckle screw

~ не·из·ме́н·н·ого ша́г·а   fixed-pitch propeller

~, нес·у́щ·ий   main rotor (of helicopter)

~ нивели́р·а   (surv) levelling screw

~, о·по́р·н·ый   foot screw

~, подъ·ём·н·ый   levelling/elevating/jack screw

~ по·сто·я́нн·ого ша́г·а   fixed-pitch propeller

~, по·та́й·н·о́й   countersunk screw

~ пра́в·ого ша́г·а   right-handed propeller

~, пра́в·ый   right-handed propeller

~, при·гн·а́нн·ый   reamed screw

~, про́б·н·ый   test screw

~, раз·жим·а́ющ·ий   expansion screw

~, реакти́в·н·ый нес·у́щ·ий   jet rotor (of helicopter)

~, реверси́в·н·ый   (a/c) reversible pitch propeller, braking propeller, negative thrust propeller

~, регул·иро́вочн·ый = винт·, ре·гул·и́рующ·ий

~, регул·и́рующ·ий   adjusting/setting/governing screw; (horol) quarter screw

~ регул·и́руем·ого ша́г·а   adjustable-pitch propeller

~, рул·ев·о́й   tail rotor (of helicopter)

~ руч·н·о́го у·правл·е́ни·я   manual control setting screw

~ с на·са́д·к·ой   shrouded propeller

~, само·на·рез·а́ющ·ий   self-tapping screw

~, сил·ов·о́й   forcing screw

~, с·крепл·я́ющ·ий   fixing screw

~, со·о́с·н·ый   contra-rotating propeller

~, сто́пор·н·ый   stop/set screw

~, с·тя́ж·н·о́й   coupling screw

~, суд·ов·о́й   propeller, ship's propeller

~, толк·а́ющ·ий   (a/c) pusher propeller

~, тормоз·н·о́й   braking propeller

~, тунне́ль·н·ый   ducted propeller

~, тя́·нущ·ий   (a/c) tractor propeller

~, у·по́р·н·ый   stop/limiting screw; (horol) banking screw

~, у·стано́в·очн·ый   adjusting/setting screw

ви́нт

~ -фикса́тор·, -а,   anchoring screw

~, флю́гер·н·ый   feathering propeller

~, хвост·ов·о́й   tail rotor (of helicopter)

~, ход·ов·о́й   feed/lead screw (of machine tool)

~, элевацио́н·н·ый   levelling screw

~, электр·и́ческ·ий   electric propeller

~, электро·механ·и́ческ·ий   electric propeller

винт·и́ть, -я́т,   (*imp*) screw; screw in; unscrew.

винт·ова́льн·ый, -ая, -ое,   (*a*) screw-cutting

~ доск·а́   (*f*) stock die, screw plate

винт·о́вк·а, -и,   (*g pl*) **-вок·,**   (*f*) (mil) rifle

винт·ов·о́й, -а́я, -о́е,   (*a*) screw, screwed, spiral, helical

~ движ·е́ни·е   (*n*) spiral motion

~ колёс·а   (*pl*) helical gears

~ ли́н·и·я   (*f*) (math) helix

~ на·ре́з·к·а   (*f*) screw thread

~ на·ре́з·ы   (*pl*) rifling

~ ос·ь   (*f*) screw axis

~ пере·да́·ч·а   (*f*) helical gear

~ рул·ев·о́й при·во́д·   (*m*) screw steering gear

~ с·мес·и́тел·ь   (*m*) mixing conveyor

~ со·един·е́ни·е   (*n*) union screw

винт·о́вочн·ый, -ая, -ое,   (*a*) of **винт·о́вк·а**

винто·кры́·л·, -а,   (*m*) rotary-wing aircraft, gyrodyne

винт·о́м   (*adv*) spirally

винто·мото́р·н·ая у·стано́в·к·а   (*f*) power plant (of propeller a/c)

винто·обра́з·н·ый, -ая, -ое,   (*a*) (*s f*) **-з·ен·, -з·н·а**   helical, spiral, coiled

винто·реакти́в·н·ый дви́г·ател·ь   (*m*) (a/c) turboprop

винто·ре́з·н·ый, -ая, -ое,   (*a*) screw-cutting

~ стан·о́к·   (*m*) screw-cutting lathe

винт·о-рул·ев·о́й компле́кс·   (*m*) stern arrangement (of ship's rudder and propeller)

винч·у́   (*pres 1st sing*) of **винт·и́ть**

виньет·и́ровани·е, -я,   (*n*) (phot) vignetting, masking

виньет·к·а, -и,   (*g pl*) **-т·ок·,**   (*f*) vignette, mask

ви́ньол·я, ре́льс   (*m*) (rail) Vignoles rail, flat-bottomed rail

виола́н·, -а,   (*m*) (min) violane

виол·ани́н·, -а,   (*m*) (chem) violanine

виофо́рм·, -а, (*m*) (chem) vioform
ви́р·а (*imperative*) (naut) hoist away!
вира́ж·, -а́, (*i*) -о́м, (*m*) turn, curve; banked curve (of road); **вира́ж·, -а,** (*m*) (phot) toning bath, toner
~, глуб·о́к·ий (air) steep turn
~, ме́д·н·ый copper toning bath
~, ме́ль·к·ий (air) shallow turn
~, про·трав·н·о́й (phot) dyeing/mordanting bath
ви́риал·а, теоре́м·а (*f*) (math) Virial theorem
ви́риаль·н·ый, -ая, -ое, (*a*) Virial
~ коэффицие́нт· (*m*) (phys) Virial coefficient
~ рае·лож·е́ни·е (*n*) (math) Virial series
виридин·, -а, (*m*) (min) viridine
вир·и́ровани·е, -я, (*n*) (phot) toning
~ про·травл·е́ни·ем (phot) mordanting
виртуа́ль·н·ый, -ая, -ое, (*a*) virtual
вируле́нт·ност·ь, -и, (*f*) virulence
вируле́нт·н·ый, -ая, -ое, (*a*), (*s f*) -т·ен·, -т·н·а virulent
ви́рус·, -а, (*m*) virus
ви́рус·н·ый, -ая, -ое, (*a*) virose, virous
вирусо́·лог·, -а, (*m*) virologist
вирусо·ло́г·и·я, -и, (*f*) virology
вис· (*past masc sing*) of **ви́с·нуть**
вис·е́ть, -я́т, (*imp*) hang, be suspended; hover (of helicopter)
ви́ски (*n indecl*) whisky; **вис·к·и́** (*nom pl*) of **вис·о́к·**
виско́з·а, -ы, (*f*) (chem) viscose
вискози·ме́тр·, -а, (*m*) viscometer, viscosimeter
виско́з·н·ый, -ая, -ое, (*a*) viscose
ви́с·ла (*past fem sing*) of **ви́с·нуть**
висло·пло́д·ник·, -а, (*m*) (bot) cremocarp
висло·у́х·ий, -ая, -ое, (*a*) lop-eared
ви́смут·, -а, (*m*) (chem) bismuth, Bi
~, азотно·ки́сл·ый bismuth nitrate
~, азотно·ки́сл·ый основ·н·о́й (chem) bismuth subnitrate
~, хлор·и́ст·ый bismuth trichloride
висмут·и́д·, -а, (*m*) bismuthide
висмут·и́л·, -а, (*m*) bismuthyl
~, бро́м·ист·ый bismuth oxybromide, bismuthyl bromide
висмут·и́н·, -а, (*m*) (min) bismuth glance, bismuthinite; (chem) bismuth hydride, bismuthine
ви́смут·ист·ый, -ая, -ое, (*a*) bismuthous

висмуто·водо·ро́д, -а, (*m*) bismuth hydride
ви́смут·ов·ый, -ая, -ое, (*a*) bismuthic, bismuth
~ бел·и́л·а (*pl*) bismuth litharge, pearl white
~ блеск· (*m*) (min) bismuth glance
~ об·ма́н·к·а (*f*) (min) bismuth silicate
~ о́хр·а (*f*) (min) bismuth ochre, bismite
~ спира́л·ь (*f*) (elec) bismuth spiral
ви́с·нуть, -ут, (*past masc, fem sing*) **ви́с·, ви́с·ла** hang, droop
вис·о́к·, -ск·а́, (*m*) (anat) temple
високо́сн·ый год· (*m*) leap year
вис·о́чн·ый, -ая, -ое, (*a*) (anat) temporal
висцер·а́льн·ый, -ая, -ое, (*a*) (anat) visceral
висцеро·мото́р·н·ый, -ая, -ое, (*a*) (anat) visceromotor
висци́н·, -а, (*m*) (chem) viscin
висци́н·ов·ое ма́сл·о (*n*) viscine oil
ви́с·ш·ий, -ая, -ее, (*pres part act*) of **ви́с·нуть**
вис·я́ч·ий, -ая, -ее, (*a*) hanging, suspended, pendent, pendant
~ бок· (*m*) (min) hanging wall
~ мост· (*m*) suspension bridge
~ плуг· (*m*) (agr) swing plough
~ систе́м·а (*f*) (build) roof truss
ВИТ (*abbr*) **врем·енно-и́мпульс·н·ая телегра́ф·и·я** PTM, pulse-time modulation telegraphy
вита́лл·и·й, -я, (*m*) (met) vitallium
витами́н·, -а, (*m*) vitamin
витамин·изи́рованн·ый, -ая, -ое, (*a*) (food) vitaminized, with vitamins added
витами́н·н·ый, -ая, -ое, (*a*) vitamin, vitaminous
витамин·о́зност·ь, -и, (*f*) vitamin content
витамин·о́зн·ый, -ая, -ое, (*a*) vitamin-rich
витами́т·, -а, (*m*) (min) withamite
вит·а́ни·е, -я, (*n*) (*v n*) of **ви·ть**
вителли́н·, -а, (*m*) (biol) vitelline; (chem) vitellin
витери́т·, -а, (*m*) (min) witherite
ви·т·о́й, -а́я, -о́е, (*a*) twisted, winding, spiral
~ пруж·и́н·а (*f*) helical spring
вит·о́к·, -т·к·а́, (*m*) turn, loop (of spiral); convolution, whorl
~, ампе́р- (elec) ampere-turn

вит·о́к
~, мёртв·ый (elec) dead turn
~, на·стро́·енн·ый (rad) resonance loop
~ с·вя́з·и (rad) pick-up loop, coupling loop; output coupling loop (of magnetron)
витр·е́н·, -а, (m) (min) vitrain
витр·и́н·а, -ы, (f) shop/display window; showcase
витро·фи́р·, -а, (m) (geol) vitrophyre, glass porphyry
ви́·т·ый, -ая, -ое, (a), (s f) **ви·т·, ви·т·а́, ви́·т·о** (past part pass) of **ви·ть**; twisted, wound, twined
ви·ть, (pres 3rd sing, pl) **вь·ёт, вь·ю́т,** (imp) twist; twist (strands into ropes); wind (wires into strands); **-ся** eddy, whirl, swirl; twine
вихр·ев·о́й, -а́я, -о́е, (a) vortical, vortex, eddy, turbulent
~ движ·е́ни·е (n) circulatory movement, eddy
~ доро́ж·к·а (f) (air) vortex street
~ ка́мер·а (f) (hydr) swirl/turbulence chamber
~ пелен·а́ (f) (dynam) vortex sheet
~ по́л·е (n) (elec) rotational field
~ про·стра́н·ств·о (n) (dynam) eddy zone
~ раз·мо́л· (m) vortex grinding
~ со·противл·е́ни·е (n) (dynam) eddy-making resistance
тео́ри·я вихр·ев·ы́х поло́с· (f) (dynam) vortex-lattice theory
~ ток· (m) (elec) eddy current
вихре·обра́з·ова́ни·е, -я, (n) turbulence, swirl, swirl formation
ви́хр·ь, -я, (m) whirlwind; whirl, eddy, vortex; (math) curl
~ ве́ктор·н·ого по́л·я (math) curl of a vector function
~ интенси́в·ност·и г (air) gamma-strength vortex
~, кон·цев·о́й (air) wing-tip vortex, tip eddy
~, меж·лопа́т·очн·ый (dynam) blade-to-blade vortex
~, нес·у́щ·ий (air) lift vortex
~, поля́р·н·ый (meteor) polar vortex; (aerodynam) jet airstream
~, при·со·един·ённ·ый bound vortex
~, раз·го́н·н·ый (air) starting vortex
~, свобо́д·н·ый free vortex
~, ско́р·ост·и (meteor) vorticity

ви́хрь
~, с·пу́т·н·ый (dynam) adjacent vortex
у·равн·е́ни·е ви́хр·а (n) vorticity equation
вице- (root) vice
вицина́л·ь, -я, (m) (cryst) vicinal face
вицина́ль·н·ый, -ая, -ое, (a) vicinal
~ гран·ь (f) (cryst) vicinal face
вишн·ёвк·а, -и, (g pl) **-вок·,** (f) cherry liqueur
~, душ·и́ст·ая Morello cherry
вишн·ёв·ый, -ая, -ое, (a) cherry
ви́шн·я, -и, (g pl) **-шен·,** (f) cherry, *Cerasus*
виш·у́ (pres 1st sing) of **вис·е́ть**
ви·я́ (pres gerund) of **ви·ть**
в·ка́л·ыва·ть, -ют, (imp) stick in/into, spike
в·ка́п·ыва·ть, -ют, (imp) dig in, entrench
в·кат·и́ть, -ят, (perf) see **в·ка́т·ыва·ть**
в·ка́т·ыва·ть, -ют, (imp) roll in, wheel in
ВКБ-ме́тод·, -а, (m) (nucl) Wenzel-Kramers-Brillouin method
вкл. (abbr) incl., inclusive, inclusively
в·кла́д·, -а, (m) deposit (in bank); investment; contribution
~, бес·сро́ч·н·ый (fin) demand deposit
~, основ·н·о́й main contribution
~, сро́ч·н·ый (fin) time deposit
~ эффе́кт·а за·ря́д·а (nucl) contribution of charge effect
в·кла́д·к·а, -и, (g pl) **-д·ок·,** (f) (mech eng) insert, brass; (print) supplementary sheet, inset
в·клад·н·о́й, -а́я, -о́е, (a) insert, insertion
~ лист· (m) (print) loose leaf, insert
в·кла́д·чик·, -а, (m) depositor
в·кла́д·ывани·е, -я, (n) (v n) see **в·кла́д·ыва·ть**
в·кла́д·ыва·ть, -ют, (imp) put in, insert; (fin) deposit; invest
в·кла́д·ыш·, -а, (i) **-ем,** (m) (mech eng) bush, bushing, brass, insert, shell bearing
в·кле́·ива·ть, -ют, (imp) glue/paste/stick in
в·кле́·ить, -ят, (perf) see **в·кле́·ива·ть**
в·кле́й·к·а, -и, (g pl) **-е́·ек·,** (f) (v n) see **в·кле́·ива·ть**; inset, insertion (glued in place); (cinema) timing tape
в·клёп·анн·ый, -ая, -ое, (past part pass) of **в·клеп·а́ть**; rivetted, rivetted in

в·клеп·а́·ть, -ют, (*perf*) *see* в·клёп·ы·ва·ть

в·клёп·ыва·ть, -ют, (*imp*) rivet, rivet in

в·кли́н·ива·ть, -ют, (*imp*) wedge in; form a wedge in

в·клин·и́ть, -я́т, (*perf*) *see* в·кли́н·и·ва·ть

в·ключ·а́тел·ь, -я, (*m*) switch

в·ключ·а́·ть, -ют, (*imp*) include; switch on, turn on; engage, start; throw in (clutch)

~ параллéль·н·о (elec) shunt

в·ключ·а́·я (*pres gerund*) of в·ключ·а́·ть; including, inclusive

в·ключ·éни·е, -я, (*n*) (*v n*) *see* в·клю·ч·а́ть; inclusion; (chem) occlusion; (elec) connection

~, капле·ви́д·н·ый elongated inclusion

~, каскáд·н·ое (elec) cascade connection

~, линзо·обрáз·н·ое (geol) lens

~, не·метáлл·и́ческ·ое (met) non-metallic inclusion, sonim

~ нéфт·и (geol) oil pocket

~, по·сторóн·н·ее (met) exogenous inclusion

сигнал· в·ключ·éни·я (*m*) (telecom, autom) connect signal

срéд·ств·о в·ключ·éни·я (*n*) engaging means/mechanism

~, ступ·éнчат·ое grading (telephony)

табл·и́ц·а в·ключ·éни·я (*f*) sequence table (switching algebra)

ýгол· в·ключ·éни·я (*m*) (elec) closing angle (of relay)

~, шлáк·ов·ое (met) slag inclusion, indigenous nonmetallic inclusion; (geol) entrapped slag

в·ключ·ённ·ый, -ая, -ое, (*past part pass*) *see* в·ключ·а́·ть; included, entrained, incorporated, occluded, embedded; (elec) switched on, connected on; (mech) engaged, in gear

в·ключ·и́тельн·о (*adv*) inclusive, inclusively

в·ключ·и́ть, -áт, (*perf*) *see* в·ключ·а́·ть

в·колáч·ива·ть, -ют, (*imp*) hammer in, drive in

в·колот·и́ть, -я́т, (*perf*) *see* в·колáч·и·ва·ть

в·кóл·от·ый, -ая, -ое, (*past part pass*) of в·кол·óть

в·кол·óть, -ют, (*perf*) stick in/into, spike

в·колóч·енн·ый, -ая, -ое, (*past part pass*) of в·колот·и́ть

в·кон·ец· (*adv*) completely, quite

в·коп·а́·ть, -ют, (*perf*) dig in, entrench

в·корен·и́ть, -я́т, (*perf*) *see* в·коре·н·я́·ть

в·корен·я́·ть, -ют, (*imp*) inculcate; -ся (bot) take root

в·корот·к·é (*adv*) in a short time, soon

в·кос·ь (*adv*) obliquely, on the slant, slanting

в·крáд·ыва·ться, -ются, (*imp*) steal/creep/slip in

в·крáп·ить, -я́т, (*perf*) *see* в·крáпл·и·ва·ть

в·крапл·éни·е, -я, (*n*) (*v n*) *see* в·крáп·л·ива·ть; impregnation, dissemination

в·крáпл·енник·, -а, (*m*) (geol) phenocryst, inset, impregnation

в·крáпл·ива·ть, -ют, (*imp*) speck, speckle, intersperse, pepper; (geol) impregnate, disseminate

в·крапл·я́·ть, -ют, (*imp*) = в·крáпл·ива·ть

в·крáс·ться, (*fut 3rd pl*) в·крад·ýтся, (*perf*) *see* в·крáд·ыва·ться

в·крáт·це (*adv*) in brief, briefly, in short

в·крест· (*adv*) across, crosswise

~ пад·éни·я (geol) across the pitch

~ про·стир·áни·я (geol) across the strike

в·крив·ь (*adv*) awry, crookedly

в·крут·и́ть, -я́т, (*perf*) screw/twist in

в·крут·ýю (*adv*) hard-boiled (of eggs)

в·крýч·енн·ый, -ая, -ое, (*past part pass*) of в·крут·и́ть

в·кýс·, -а, (*m*) taste, flavour, flavor; taste, fancy, liking (for)

в·кус·и́ть, -я́т, (*perf*) *see* в·куш·а́·ть

в·кýс·н·ый, -ая, -ое, (*a*), (*s f*) -с·ен, -с·н·á, -с·н·о tasty, appetizing, delicious, palatable

в·кус·ов·óй, -áя, -óе, (*a*) taste, gustatory; flavouring, flavoring

в·куш·а́·ть, -ют, (*imp*) partake of, taste, try (a food)

влáг·а, -и, (*f*) moisture, dampness

~, вал·ов·áя (bot) holard

в·лаг·áлищ·е, -а, (*i*) -ем, (*n*) (zool) vagina; (bot) sheath

в·лаг·áлищн·ый, -ая, -ое, (*a*) (bot) sheath, sheathed; (anat) vaginal

в·лаг·а́·ть, -ют, (*imp*) *see* в·клáд·ы·ва·ть

влáг·о= (*root*) moisture, hygro-

влаго·ём·кост·ь, -и, (*f*) moisture capacity/content

влаго·ём·кост·ь

~, вес·ов·ая   gravimetric moisture capacity

влаго·изол·яци·я, -и, (*f*) damp-proofing

влаго·люб·ив·ый, -ая, -ое, (*a*) (bot) hygrophile

влаго·мер·, -а, (*m*) moisture meter, hygrometer

влаго·не·про·ниц·аем·ый, -ая, -ое, (*a*) moisture-proof, damp-proof

влаго·оборо́т·, -а, (*m*) (meteor) hydrological cycle

влаго·от·дел·ител·ь, -я, (*m*) moisture separator

влаго·по·глот·ител·ь, -я, (*m*) desiccator, drying cell

влаго·со·держ·а́ни·е, -я, (*n*) moisture content

влаго·сто́й·к·ий, -ая, -ое, (*a*) damp/ moisture-proof/resistant

влаго·у·дал·ител·ь, -я, (*m*) moisture extractor, dehumidifier

влаго·у·пор·н·ый, -ая, -ое, (*a*) damp/ moisture-proof

влаго·чувств·и́тельн·ый, -ая, -ое, (*a*) moisture-sensitive

влад·е́лец·, -льц·а, (*i*) -льц·ем, (*m*) owner, proprietor ·

влад·е́ни·е, -я, (*n*) ownership, possession

влад·е́·ть, -ют, (*imp + instr*) own, possess; be able to use, know how to use

влаж·не́·ть, -ют, (*imp*) become humid

влажно·адиабат·и́ческ·ий, -ая, -ое, (*a*) (meteor) saturated adiabatic

вла́ж·ност·ь, -и, (*f*) humidity, moisture, dampness

~, абсолю́т·н·ая   (meteor) absolute humidity, humidity

~, го́р·н·ая   quarry water/sap, water of imbibition (in building stone)

~ на·сыщ·е́ни·я   saturated/saturation humidity

~, от·нос·и́тельн·ая   relative humidity

~, поле́з·н·ая   available moisture

~ угл·я́   coal humidity

~, у·де́ль·н·ая   specific humidity

вла́ж·н·ый, -ая, -ое, (*a*) humid, moist, damp

~ термо́·метр· (*m*) wet-bulb thermometer

в·ла́м·ыва·ться, -ются, (*imp*) break into

в·ла́п·у (*adv*) lap, overlap, with an overlap, scarf

в·ласк·у (*adv*) lap, lapped, overlap, with an overlap, scarf

вла́ст·ь, -и, (*nom pl*) -и, (*g pl*) -е́й, (*f*) power, authority

в·ле́в·о (*adv*) to the left; counterclockwise; (naut) to port

в·лез·а́ни·е, -я, (*n*) (*v n*) *see* в·лез·а́·ть; intrusion

в·лез·а́·ть, -ют, (*imp*) climb in/up, climb

в·ле́з·ть, -ут, (*past masc, fem*) в·ле́з·, в·ле́з·ла, (*perf*) *see* в·лез·а́·ть

влек·о́м·ый, -ая, -ое, (*pres part pass*) of влеч·ь (*imp*)

влек·у́щ·ий, -ая, -ое, (*pres part act*) *see* влеч·ь (*imp*)

влёк·ш·ий, -ая, -ее, (*past part act*) of влеч·ь (*imp*)

в·лет·а́·ть, -ют, (*imp*) fly in; (air) touch down

в·лет·а́ющ·ий, -ая, -ее, (*pres part act*) of в·лет·а́·ть; flying in, incoming, oncoming, entering

~ част·и́ц·а (*f*) (nucl) oncoming/incoming particle

в·лет·е́ть, -я́т, (*perf*) *see* в·лет·а́·ть

влеч·а́ (*pres gerund*) of влеч·ь

влеч·е́ни·е, -я, (*n*) tendency, bent, inclination; drag

сил·а влеч·е́ни·я (*f*) traction, tractive force

влеч·ённ·ый, -ая, -ое, (*past part pass*) of влеч·ь (*imp*)

влеч·ь, (*pres 3rd sing, pl*) влеч·ёт, влек·у́т, (*past masc, fem*) влёк·, влек·ла́, (*imp*) draw, attract, pull

~ за соб·о́й involve, entail

в·леч·у́ (*fut 1st sing*) of в·лет·е́ть

в·лив·а́ни·е, -я, (*n*) (*v n*) of в·лив·а́·ть

в·лив·а́·ть, -ют, (*imp*) pour in, infuse; -ся (*pass*); flow in, disembogue (of rivers)

влить, (*fut 3rd sing, pl*) во·ль·ёт, во·ль·ю́т; *see* в·лив·а́·ть

в·ли·я́ни·е, -я, (*n*) (*see also* эффе́кт· *and associated words*) influence, effect

~ за·па́зд·ывани·я delaying effect

~, индукти́в·н·ое induction effect

~ на бли́ж·н·ий кон·е́ц· near-end crosstalk (telephony)

~ на да́ль·н·ый кон·е́ц· far-end crosstalk (telephony)

о·ка́з·ыва·ть в·ли·я́ни·е affect, influence, exert an influence, have an effect

~ о·круж·а́ющ·ей сред·ы́ ambient effects

в·ли·я́ни·е
  под в·ли·я́ни·ем   by, by the agency of, on exposure to
~ **пя́тен·**   (eleç) patch effect
~ **с·пу́т·н·ой стру·и́**   (dynam) wake effect
~ **сте́н·ок·**   (dynam) wall constraint
~ **тр·е́ни·я**   friction effect
в·ли·я́·ть, -ют,   (*imp*) influence, affect
в·лож·е́ни·е, -я,   (*n*) enclosure (with letter etc.); (fin) investment
~**, долго·сро́ч·н·ое**   (fin) long-term investment
в·ло́ж·енн·ый, -ая, -ое, (*past part pass*) *see* **в·кла́д·ыва·ть;** enclosed, inserted; invested, deposited; (geol) inset, engrafted, nested
в·лож·и́ть, -ат,   (*perf*) *see* **в·кла́д·ыва·ть**
в·лом·и́ться, -ятся,   (*perf*) break into
в·ма́ж·ут   (*fut 3rd pl*) of **в·ма́з·ать**
в·ма́з·ать, (*fut 3rd sing, pl*) **в·ма́ж·ет, в·ма́ж·ут,** (*perf*); *see* **в·ма́з·ыва·ть**
в·ма́з·к·а, -и, (*g pl*) **-з·ок·,** (*f*) (*v n*) *see* **в·ма́з·ыва·ть;** puttying in; cementing in
в·ма́з·ыва·ть, -ют,   (*imp*) putty/ cement/mortar in; **-ся** get stuck
в·мен·и́ть, -ят, (*perf*) impute
в·мен·я́емост·ь, -и   (*f*) (law) responsibility
в·мен·я́ем·ый, -ая, -ое   (*pres part pass*) of **в·мен·я́·ть** (law) responsible, of sound mind
в·мен·я́·ть, -ют,   (*imp*) impute
в·мерз·а́ни·е, -я,   (*n*) freezing in
в·ме́ст·е   (*adv*) together
~ **с тем** at the same time; in addition, besides this, along with this
в·мест·и́лищ·е, -а, (*i*) **-ем,** (*n*) receptacle, reservoir, container, tank
в·мест·и́мост·ь, -и,   (*f*) capacity, capaciousness; tonnage (of ships)
~**, бру́тто-реги́стр·ов·ая**   gross tonnage
~**, вал·ов·а́я**   gross tonnage
~**, не́тто-реги́стр·ов·ая**   register tonnage
~**, чи́ст·ая**   net tonnage
~**, чи́ст·ая реги́стр·ов·ая**   net register tonnage
в·мес·и́ть, -ят,   (*perf*) knead/mix in
в·мест·и́ть, -ят,   (*perf*) *see* **в·мещ·а́·ть**
в·ме́ст·о (*adv*), (*prep+gen*) instead of, in place of
в·мет·а́·ть, -ют,   (*perf*) *see* **в·мёт·ыва·ть**

в·мёт·ыва·ть, -ют,   (*imp*) tack in/on (by sewing)
в·меш·а́тельств·о, -а,   (*n*) interference, intervention
в·меш·а́·ть, -ют,   (*perf*) *see* **в·ме́ш·и·ва·ть**
в·ме́ш·енн·ый, -ая, -ое,   (*past part pass*) of **в·мес·и́ть**
в·ме́ш·ива·ть, -ют,   (*imp*) mix in/up; implicate; **-ся** interfere, intervene, meddle
в·мещ·а́·ть, -ют,   (*imp*) contain, hold, enclose; have capacity/room for
в·мещ·а́ющ·ая по·ро́д·а   (*f*) (geol) enclosing rock
в·миг·   (*adv*) in an instant, in a moment
в·мокр·у́ю   (*adv*) wet
в·монт·и́рованн·ый, -ая, -ое, (*past part pass*) of **в·монт·и́р·овать** built in, fixed
в·монт·и́р·овать, -уют,   (*imp and perf*) build in
в·мур·о́ванн·ый, -ая, -ое, (*past part pass*) walled in, immured
в·мыв·а́ни·е, -я,   (*n*) inwash, washing in
в·мя́т·ин·а, -ы,   (*f*) dent, indentation, hollow; (geol) depression
в·мя́·т·ый, -ая, -ое, (*past part pass*) of **в·мя́·ть**
в·мя́·ть, (*fut 3rd sing, pl*) **во·мн·ёт, во·мн·у́т,** (*perf*) dent, indent, make a dent in
в·на·ва́л·к·у   (*adv*) in bulk
в·на·ём·   (*adv*) on hire, to let
~**, бр·ать**   hire, rent
~**, от·да·ва́ть**   hire out, let, rent
в·на·йм·ы́   (*adv*) *see* **в·на·ём·**
в·на·ки́д·к·у   (*adv*) thrown over the shoulders; with folds
в·на·кро́·й   (*adv*) lap, lapped
  со·един·е́ни·е в·на·кро́·й   (*n*) lapped joint
в·на·пу́ск·   (*adv*) = **в·на·кро́·й**
в·на·сып·н·у́ю   (*adv*) = **в·на·ва́л·к·у**
в·на·хлёст·к·у   *see* **в·на·кро́·й**
в·на·ча́·л·е   (*adv*) in the beginning, at first, first of all, to start with
вне   (*prep+gen*) outside, out of
вне- (*prefix*) *significance*: (1) located beyond the limits; (2) taking place outside or beyond the scope of
вне·габари́т·н·ый, -ая, -ое,   (*a*) off-gauge, oversize
~ **груз·** (*m*) (rail) off-gauge load
вне·галакт·и́ческ·ий, -ая, -ое,   (*a*) (astron) extra-galactic

вне·город·ск·о́й, -а́я, -о́е, (*a*) out-of-town, suburban, country

вне·до́мен·н·ый, -ая, -ое, (*a*) (met) outside the blast furnace

в·недр·е́ни·е, -я, (*n*) inculcation, introduction; intrusion, injection; (biol) intussusception

~, ва́куум·н·ое vacuum injection

~, ле́нт·очн·ое (geol) lenticular, intercalation

рас·тво́р· в·недр·е́ни·я (*m*) (met) interstitial solution

~ руд·ы́ (geol) ore injection

фа́з·а в·недр·е́ни·я (*f*) (phys, chem) interstitial phase

в·недр·ённ·ый, -ая, -ое, (*past part pass*) of **в·недр·и́ть** (cryst) interstitial

~ а́том· (*m*) interstitial atom

в·недр·и́ть, -я́т, (*perf*) *see* **в·недр·я́·ть**

в·недр·я́·ть, -ют, (*imp*) inculcate, instil; introduce, implant, intrude

внеза́п·н·ый, -ая, -ое, (*a*) sudden, unexpected

вне·за·ро́д·ышев·ый, -ая, -ое, (*a*) (zool) extra-embryonic

вне·зем·н·о́й, -а́я, -о́е, (*a*) extraterrestrial

вне·интергра́ль·н·ый, -ая, -ое, (*a*) (math) integrated

вне·капилля́р·н·ый, -ая, -ое, (*a*) extracapillary

вне·киш·е́чн·ый, -ая, -ое, (*a*) (anat) extra-intestinal, abenteric

вне·кле́т·очн·ый, -ая, -ое, (*a*) extracellular

вне·конкурре́нт·н·ый, -ая, -ое, (*a*) non-competitive

вне·лими́т·н·ый, -ая, -ое, (*a*) (fin) extraordinary, over and above the allotment/quota

~ капитало·в·лож·е́ни·е (*n*) extraordinary/unbudgeted capital investment

вне·о́с·ев·ый, -ая, -ое, (*a*) off-axis

вне·очерёд·ност·ь, -и, (*f*) priority

вне·о·черед·н·о́й, -а́я, -о́е, (*a*) priority, extraordinary, special

вне·па́зуш·н·ий, -яя, -ее, (*a*) (bot) extra-axillary

вне·пла́н·ов·ый, -ая, -ое, (*a*) not provided for by the plan, unenvisaged, unforeseen; surplus to the plan, above plan

вне·пло́д·ник·, -а, (*m*) exocarp, epicarp

вне·по́л·остн·ый, -ая, -ое, (*a*) extra-cavitous

в·нес·е́ни·е, -я, (*n*) carrying in, bringing in, introduction; entering, entry (on paper); (fin) paying in, deposit; application (of fertilizer)

~ ды́р·ок· hole injection (semiconductor)

вне·служ·е́бн·ый, -ая, -ое, (*a*) spare-time

~ вре́м·я (*n*) out-of-office hours

вне·с·ме́т·н·ый, -ая, -ое, (*a*) unbudgeted, outside the estimates

вне·со·су́д·ист·ый, -ая, -ое, (*a*) (zool) extravascular

вне·сте́н·очн·ый, -ая, -ое, (*a*) extramural; (biol) exothecal

в·нес·ти́, -у́т, (*past masc, fem*) **в·нёс·, в·нес·ла́** (*perf*) *see* **в·нос·и́ть**

вне·суд·е́бн·ый, -ая, -ое, (*a*) (law) extrajudicial

вне·цвет·ко́в·ый, -ая, -ое, (*a*) (bot) extra-floral

вне·це́нтр·енн·ый, -ая, -ое, (*a*) off-centre/center

вне·шка́ль·н·ый, -ая, -ое, (*a*) off the scale, inferred (of measurement)

вне·шко́ль·н·ый, -ая, -ое, (*a*) extra-scholastic, extra-mural (of studies)

~ образ·ова́ни·е (*n*) adult education

внешне·торг·о́в·ый, -ая, -ое, (*a*) foreign-trade

вне́ш·н·ий, -яя, -ее, (*a*) external, outside, exterior, extrinsic; outer, outside, outward; superficial, surface; foreign, extraneous

~ вид· (*m*) appearance

~ диа́метр· (*m*) outside/external diameter

~ об·луч·е́ни·е (*n*) external irradiation

~ торг·о́вл·я (*f*) foreign trade

вне́ш·ност·ь, -и, (*f*) exterior; appearance

вне·шта́т·н·ый, -ая, -ое, (*a*) supernumerary, unestablished (of personnel)

вне·я́дер·н·ый, -ая, -ое, (*a*) (phys) extranuclear; (chem) exocyclic

в·низ· (*adv*) down, downwards

~ по теч·е́ни·ю downstream

в·низ·у́ (*adv*) below; downstairs

ВНИИ (*abbr*) = **все·сою́з·н·ый на·-у́ч·н·о-ис·след·ова́тельск·ий ин·-ститу́т·** All Union Research Institute

в·ник·а́·ть, -ют, (*imp*) grasp the essentials, get to the heart of

в·ни́к·н·уть, -ут, (*perf*) *see* **в·ник·а́·ть**

в·ним·а́ни·е, -я, (*n*) attention, heed, regard, notice, note, consideration
  до·сто́й·н·ый в·ним·а́ни·я worth noting, worthy of attention
  обрат·и́ть в·ним·а́ни·е pay attention, notice; draw attention to; **-ся** attract attention, be noticeable
  при·ня́·ть во в·ним·а́ни·е allow for, remember (that), consider (that), bear in mind (that)
  сигна́л в·ним·а́ни·я (*m*) attention signal
  у·дел·и́ть в·ним·а́ни·е pay attention, put emphasis on, emphasize (a fact)

в·ним·а́тельн·ый, -ая, -ое, (*a*) attentive, intent; considerate

в·ни·чью́ (*adv*) drawn, in a draw (of contest)

в·нов·ь (*adv*) anew, again, once more; recently, just, lately

в·но́с·, -а, (*m*) (*v n*) *see* **в·нос·и́ть;** entry; introduction; (fin) deposit, payment; application (fertilizer)

в·нос·и́м·ый, -ая, -ое, (*pres part pass*) of **в·нос·и́ть;** inserted, for inserting/entering

в·нос·и́ть, -ят, (*imp*) carry in, bring in, introduce, enter (on paper); (fin) pay in; apply (fertilizers)

в·нош·у́ (*pres 1st sing*) of **в·нос·и́ть**

внутренне·се́т·чат·ый, -ая, -ое, (*a*) intrareticulate

вну́тр·енн·ий, -яя, -ее, (*a*) internal, inside, inner, interior, intrinsic; indoor, inboard, inland; home, domestic; (biol) endogenous
~ гор·е́ни·е (*n*) internal combustion
~ эне́рг·и·я (*f*) (phys) intrinsic/inner energy

вну́тр·енностн·ый, -ая, -ое, (*a*) interior; (anat) visceral

вну́тр·енност·ь, -и, (*f*) interior; (*pl*) (anat) intestines, viscera

внутр·и́ (*adv*) inside, on the inside; (*prep + gen*) inside, within

внутр·и- (*prefix*) *significance*: inside, within

внутри·а́том·н·ый, -ая, -ое, (*a*) intra-atomic, subatomic
~ част·и́ц·а (*f*) subatomic particle

внутри·брюш·н·о́й, -а́я, -о́е, (*a*) (anat) intraperitoneal

внутри·ве́н·н·ый, -ая, -ое, (*a*) intravenous, endovenous

внутри·ви́д·н·ый, -ая, -ое, (*a*) (biol) intraspecific

внутри·вид·ов·о́й, -а́я, -о́е, (*a*) intraspecific, interspecies

внутри·глаз·н·о́й, -а́я, -о́е, (*a*) (anat) intraocular

внутри·го́р·н·ый, -ая, -ое, (*a*) (geol) intermontane

внутри·желт·о́чн·ый, -ая, -ое, (*a*) (zool) intravitelline

внутри·за·во́д·ск·ий, -ая, -ое, (*a*) factory, works, factory internal
~ план·и́ровани·е (*n*) factory planning
~ тра́нспорт· (*m*) works/factory transport

внутри·зерн·ов·о́й, -а́я, -о́е, (*a*) intragranular, internal-grain, inside a grain, grain

внутри·кле́т·очн·ый, -ая, -ое, (*a*) (cyt) intracellular, endocellular

внутри·ко́ж·н·ый, -ая, -ое, (*a*) (anat) intradermal

внутри·кольц·ев·о́й, -а́я, -о́е, (*a*) endocyclic

внутри·компле́кс·н·ый, -ая, -ое, (*a*) (chem) chelate

внутри·ко́ст·н·ый, -ая, -ое, (*a*) (anat) endosteal

внутри·кристалли́т·н·ый = внутри·-кристалл·и́ческ·ий

внутри·кристалл·и́ческ·ий, -ая, -ое, (*a*) intracrystalline, crystal; transcrystalline
~ из·ло́м· (*m*) transcrystalline fracture

внутри·ма́сс·ов·ая о·са́д·к·а (*f*) (meteor) convectional precipitation

внутри·ма́т·очн·ый, -ая, -ое, (*a*) (anat) intra-uterine, endometrial

внутри·металл·и́ческ·ая с·вя́з·ь (*f*) metallic bond

внутри·молекуля́р·н·ый, -ая, -ое, (*a*) (phys) intramolecular, internal

внутри·орбита́ль·н·ый, -ая, -ое, (*a*) intra-orbital

внутри·о·хлажд·ённ·ый, -ая, -ое, (*a*) internally cooled

внутри·плевра́ль·н·ый, -ая, -ое, (*a*) (zool) intrapleural

внутри·пло́д·ник·, -а, (*m*) (bot) endocarp

внутри·пол·ост·н·о́й, -а́я, -о́е, (*a*) (med) intracavitary

внутри·пузы́р·н·ый, -ая, -ое, (*a*) (anat) intravesical

внутри·ра́м·очн·ая част·ь (*f*) body (of map)

внутри·реа́ктор·н·ый, -ая, -ое, (*a*) (nucl) in-pile

внутри·решёт·очн·ый, -ая, -ое,   (*a*) intralattice

внутри·род·ов·ой, -ая, -ое,   (*a*) intrageneric

внутри·систе́м·н·ый, -ая, -ое,   (*a*) intrasystem, internal

внутри·сорт·ов·ой, -а́я, -о́е,   (*a*) (biol) intravarietal

внутри·со·су́д·ист·ый, -ая, -ое,   (*a*) (anat) intravascular

внутри·тка́н·ев·ый, -ая, -ое,   (*a*) interstitial

внутри·формацио́н·н·ый, -ая, -ое,   (*a*) (geol) intraformational

внутри·череп·н·о́й, -а́я, -о́е,   (*a*) (zool) endocranial

внутри·шлиф·ова́льн·ый, -ая, -ое,   (*a*) internal-grinding

внутри·я́дер·н·ый, -ая, -ое,   (*a*) intranuclear; (chem) endocyclic

внутри·ячей·ков·ый, -ая, -ое,   (*a*) (phys) intracellular

внутр·ь   (*adv*) in, inside

внуш·а́·ть, -ют,   (*imp*) inspire, impress, convince

внуш·и́ть, -а́т,   (*perf*) *see* **внуш·а́·ть**

внуш·и́тельн·ый, -ая, -ое,   (*a*) (*s f*) **-лен·, -льн·а,** impressive

в·ня́т·н·ый, -ая, -ое,   (*a*) distinct

во   (*prep*) *see* **в**

во-   (*prefix*) = **в-, въ-** *particular significances*:   (1)   movement into; (2) movement up (*usually followed by preposition* **на**); (3) penetration (*woras ending in* **-ся**)

во́бл·а, -ы,   (*f*) (fish) Caspian roach, *Rutilus rutilus caspicus*

во·бр·а́ть,   (*fut 3rd sing, pl*) **в·бер·ёт, в·бер·у́т,**   (*perf*) *see* **в·бир·а́·ть**

во·влек·а́·ть, -ют,   (*imp*) draw in/into, involve, implicate; (chem, phys) entrain

во·влеч·е́ни·е, -я,   (*n*) (*v n*) *see* **во··влек·а́·ть**; entrainment

во·влеч·ь,   (*fut 3rd sing, pl*) **во·влеч·ёт, во·влек·у́т,**   (*perf*) *see* **во·влек·а́·ть**

во·вне́   (*adv*) outside

во·вну́тр·ь   (*adv*) *see* **внутр·ь**

во́·врем·я   (*adv*) in time, opportunely, timely, on time

во́·все   (*adv*) quite, completely

~ **не** not at all, by no means

во-втор·ы́х   (*conj*) secondly, in the second place

во·гн·а́ть,   (*fut 3rd sing, pl*) **в·го́н·ит, в·го́н·ят,**   (*perf*) drive/force in

во·гн·ут·о-   (*component*) concavo-

~ **-вы́·пук·л·ый, -ая, -ое,**   (*a*) concavo-convex

во́·гн·утост·ь, -и,   (*f*) concavity

во́·гн·ут·ый, -ая, -ое,   (*a*) concave; bent/pressed in

~ **бе́рег·**   (*m*) (geol) undercut slope

во·гн·у́ть, -ут,   (*perf*) bend in/inwards

вод. ст.   (*abbr*) = **во́д·н·ого столб·а** water gauge, water column (unit of pressure)

вод·а́, -ы́,   (*acc*) **во́д·у,** (*nom pl*) **во́д·ы,** (*g pl*) **вод·,** (*d pl*) **вод·а́м** water

~, **аммиа́ч·н·ая**   hydrous ammonia, ammonia liquor

~, **атмосфе́р·н·ая**   (civ eng) surface water (Brit), storm water (U.S.); (geol) meteoric water

~, **бро́м·н·ая**   aqua bromata, bromine water

~, **ве́рх·н·яя**   (geol) top/upper water

~, **верх·ов·а́я**   upstream water

~, **водо·про·во́д·н·ая**   tap water

~, **газ·иро́ванн·ая**   aerated water

~, **гидра́т·н·ая**   (chem) water of hydration/crystallization

~, **грунт·ов·а́я**   ground water, subsoil water

~, **до·ба́в·очн·ая**   (st eng) make-up water

~, **жёлт·ая**   (med) glaucoma

~, **жёстк·ая**   hard water

~, **за·бо́рт·н·ая**   (naut) sea water

~, **ис·коп·а́ем·ая**   (geol) fossil water

~, **конституцио́н·н·ая**   (chem) constitutional water, water of crystallization

~, **ко́нтур·н·ая**   (oil) edge water

~, **кра·ев·а́я**   (geol) edge water

~, **ма́л·ая**   (ocean) low water/tide

~, **ме́л·к·ая**   shallow/shoal water

~, **над·смо́ль·н·ая**   (chem) flushing liquor

~, **нефт·ян·а́я**   oil reservoir water

~ **нефт·ян·о́го место·рожд·е́ни·я** oil-field water

~, **ни́ж·н·яя**   (oil) bottom water

~, **ни́з·к·ая**   low water/tide

~, **низ·ов·а́я**   (hydro-elec) tail water

~, **о·пресн·ённ·ая**   distilled water

~ **о·рош·е́ни·я**   (chem) sparge water

~, **пит·а́тельн·ая**   (st eng) feed water

~, **пить·ев·а́я**   drinking/potable water

~, **пласт·ов·а́я**   (geol) stratal water

~, **плён·очн·ая**   (geol) adhesive/pellicular water

~, **по·греб·ённ·ая**   (geol) connate water

вод·а́

~, подо́·шв·енн·ая   (geol) bottom water

~, под·пи́т·очн·ая   (st eng) make-up water

~, под·смо́ль·н·ая   (chem) aqueous distillate

~, по́л·ая   (ocean) spring high tide

~, по́лн·ая   (ocean) high water/tide

~, по́р·ов·ая   (geol) connate water

~, пре́сн·ая   fresh/sweet water

~, про·меж·у́точн·ая   (ocean) height of the tide; (oil) intermediate water

~, про·мы́в·н·а́я   wash water; (chem) sparge water

~, род·нико́в·ая   spring water

~, са́м·ая ни́з·к·ая   (ocean) neap tide

~, свобо́д·н·ая   gravity/free water

~, с·вя́з·анн·ая   (chem) combined water

~, сел·ев·ы́е   (pl) (geog) runoff waters

~, серо·водо·ро́д·н·ая   sulphuretted water

~, с·по́р·н·ая   race (strong current of water)

~, с·то́ч·н·ые   (pl) sewage, effluent

~, сто·я́ч·ая   dead water; (naut) stand of the tide

~, сыр·а́я   raw/fresh water

~, терма́ль·н·ая   hot spring, thermal water

~, техно·лог·и́ческ·ая   process water

~, тяж·ёл·ая   heavy water, deuterium monoxide

~, чи́ст·ая   (ocean) ice-free water, open water

во·двор·е́ни·е, -я,   (n) settlement, establishment

во·двор·и́ть, -я́т,   (perf) see **во·двор·я́·ть**

во·двор·я́·ть, -ют,   (imp) install, settle, establish

вод·и́л·о, -а,   (n) lead, leash; sun wheel (planetary gear)

вод·и́тел·ь, -я,   (m) driver

вод·и́ть, -я́т,   (imp indet) lead, conduct, guide; drive, steer, pilot; keep (animals); **-ся** live, establish themselves (of wild animals); consort/associate with

вод·и́ч·, -а,   (m) (naut) pilot

во́д·к·а, -и,   (g pl) **-д·ок·,**   (f) vodka

~, кре́п·к·ая   (chem) aqua fortis, concentrated nitric acid

~, ца́р·ск·ая   (chem) aqua regia

во́д·ник·, -а,   (m) inland waterways worker

вод·н·о-зем·н·о́й, -а́я, -о́е,   (a) terraqueous

во́д·н·ый, -ая, -ое,   (a) water, aquatic; (chem) aqueous, hydrous; (geol) neptunian

~ арсена́т·   (m) hydrous arsenate

~ культу́р·а   (f) (bot) hydroponics

~ о́·кис·ь   (f) hydroxide

~ поду́ш·к·а   (f) water cushion; (oil) water bottom

~ пут·и́   (pl) inland waterways

~ рас·тво́р·   (m) (chem) aqueous solution

~ сол·ь   (f) hydrated salt

~ тра́нспорт·   (m) inland water transport

~ хозя́й·ств·о   (n) water conservation

во́д·о-   (root) water-, hydraulic-

водо·бо́·й, -я,   (m) (civ eng) water apron; (min) monitor, hydraulic monitor

водо·бо́й·н·ый коло́д·ец·   (m) (civ eng) stilling pit

водо·бо·я́зн·ь, -и,   (f) (med) hydrophobia, rabies

водо·во́д·, -а,   (m) aqueduct, water line/pipe/conduit/channel/flume

~, от·вод·я́щ·ий   (hydro-elec) conduit-type tail race

~, турби́н·н·ый   (hydro-elec) penstock

водо·вод·ян·о́й, -а́я, -о́е,   (a) water–water, water-to-water; (nucl) water-cooled, water-moderated

водо·во́з·, -а,   (m) water carrier

водо·воро́т·, -а,   (m) whirlpool, eddy; vortex, whirl

водо·вы·пуск·н·о́й, -а́я, -о́е,   (a) water-discharge, discharge

водо·гре́й·н·ая труб·а́   (f) generating tube (boiler)

водо·ём·, -а,   (m) reservoir, water tank, basin, pond

водо·ём·к·ий, -ая, -ое,   (a) watery, full of water, with high water content

водо·ём·кост·ь, -и,   (f) water/water-holding capacity

водо·за·бо́р·, -а,   (m) (hydr eng) intake

водо·за·бо́р·н·ое со·оруж·е́ни·е   (n) (civ eng) intake works

водо·за·пра́в·щик·, -а,   (m) (M/T) water bowser

водо·за·щищ·ённ·ый, -ая, -ое,   (a) (elec) hose-proof, splash-proof

водо·из·мещ·е́ни·е, -я,   (n) (shipb) displacement

коэффицие́нт· водо·из·мещ·е́ни·я   (m) (shipb) coefficient of displacement

водо·из·мещ·éни·е

~, **пóлн·ое**   (shipb) displacement at full load

~, **порожн·ём**   displacement at light draught

~, **стандáрт·н·ое**   standard displacement

водо·кáч·к·а, -и,   (*f*) pumping station, pump house

водо·кольц·ев·óй на·сóс·   (*m*) water-ring pump, liquid-piston pump

водо·крáс·, -а,   (*m*) (bot) frogbit, *Hydrocharis*

водо·крáс·ов·ые, -ых,   (*pl decl as adj*) (bot) *Hydrocharitaceae*

водо·лáз·, -а,   (*m*) diver

водо·лáз·н·ый, -ая, -ое,   (*a*) of **во·до·лáз·**

~ **бассéйн·**   (*m*) diving tank, test tank

~ **бот·**   (*m*) diving boat

водо·лé·й, -я,   (*m*) (astron) Aquarius; (naut) waterboat

водо·леч·éни·е, -я,   (*n*) (med) hydrotherapy, hydropathy

водо·люб·и́в·ый, -ая, -ое,   (*a*) hydrophilous

водо·масло·грéй·к·а, -и,   (*f*) (a/c) water-and-oil heater

водо·мéр·, -а,   (*m*) water meter, water gauge

~, **поршн·ев·óй**   positive water meter

~, **скор·остн·óй**   inferential water meter

водо·мéр·н·ый, -ая, -ое,   (*a*) of **во·до·мéр·**;   water-gauging/measuring/metering

~ **кран·**   (*m*) water gauge/cock

~ **пост·**   (*m*) water gauging post, depth post

~ **рéй·к·а**   (*f*) water board-gauge

~ **стекл·ó**   (*n*) water-gauge glass

водо·мёт·, -а,   (*m*) fountain

водо·на·грев·áтел·ь, -я,   (*m*) water heater

водо·на·лив·н·óе сýд·н·о   (*n*) waterboat

водо·на·пóр·н·ый, -ая, -ое,   (*a*) water-pressure, water

~ **бáшн·я**   (*f*) water tower

~ **режи́м·**   (*m*) (oil) water drive

водо·на·сы́щ·енн·ый, -ая, -ое,   (*a*) water-saturated, waterlogged

водо·не·про·ниц·áем·ый, -ая, -ое, (*a*) watertight; impermeable; (text) waterproof

водо·нóс·ность·, -и,   (*f*) rate of flow (of a river etc.)

водо·нóс·н·ый, -ая, -ое,   (*a*) water-bearing, aquiferous, water-vascular, hydrophoric

~ **форм·áци·я**   (*f*) (geol) water-bearing formation, aquifer

водо·оби́ль·н·ый, -ая, -ое,   (*a*) watery, full of water

водо·о·пресн·и́тел·ь, -я,   (*m*) water distilling plant

водо·от·вóд·, -а,   (*m*) drain, overflow

водо·от·вóд·н·ый, -ая, -ое,   (*a*) drain, drainage, draining

водо·от·вóд·чик·, -а,   (*m*) (st eng) steam trap

водо·от·дá·ч·а, -и,   (*f*) yield of water

водо·от·дел·и́тел·ь, -я,   (*m*) water trap, water separator; (chem) dehydrator

водо·от·ли́·в·, -а,   (*m*) (shipb) drainage

водо·от·лив·н·óй, -áя, -óе,   (*a*) water-pumping

~ **на·сóс·, авари́·йн·о-с·пас·áтель·н·ый** salvage pump

~ **систéм·а**   (*f*) (shipb) bilge system

водо·от·стóй·ник·, -а,   (*m*) water settling tank

водо·от·тáлк·ивающ·ий, -ая, -ее, (*a*) water-repelling, hydrophobe

водо·о·хлажд·áющ·ий, -ая, -ее,   (*a*) water-cooling

водо·о·хлажд·áем·ый, -ая, -ое,   (*a*) water-cooled

водо·о·хрáн·н·ые лес·á   (*pl*) river-protective plantation

водо·о·чи́ст·к·а, -и,   (*f*) water purifying/purification; (st eng) feed-water treatment

водо·пáд·, -а,   (*m*) waterfall

водо·пит·áющ·ий, -ая, -ее,   (*a*) water-supply, feed-water

водо·по·глощ·áющ·ий, -ая, -ее,   (*a*) hygroscopic, water-absorbing

водо·под·готóв·к·а, -и,   (*f*) water treatment

водо·подо·грев·áтел·ь, -я,   (*m*) water heater

водо·подъ·ём·н·ый, -ая, -ое,   (*a*) water-lifting

~ **труб·á**   (*f*) (hydr eng) rising main

водо·пó·й, -я,   (*m*) (agr) watering place

водо·при·ём·ник·, -а,   (*m*) (hydr eng) intake, water intake

водо·про·вóд·, -а,   (*m*) water pipe/piping/line; (zool) aqueduct

~, **глáв·н·ый**   water main

~, **магистрáль·н·ый**   water main

~, **сильви·ев·**   (zool) Sylvius aqueduct

водо·про·вод·н·ый, -ая, -ое, (*a*) water-main, mains, tap, mains water
~ вод·а́ (*f*) tap/mains water, town water
~ ста́нци·я (*f*) water works
водо·про·вод·чик·, -а, (*m*) plumber
водо·про·вод·я́щ·ий, -ая, -ее, (*a*) water-conducting
водо·про·ниц·а́ем·ый, -ая, -ое, (*a*) permeable, not watertight
~ пере·бо́р·к·а (*f*) (shipb) partition bulkhead
водо·про·то́к·, -а, (*m*) (shipb) water-course, waterway, limber hole
водо·про́ч·н·ый агрега́т· (*m*) (min) water-stable aggregate
водо·раз·бо́р·н·ый кран· (*m*) hydrant, water cock, water tap
водо·раз·де́л·, -а, (*m*) (geog) watershed, (geol) divide
ли́н·и·я водо·раз·де́л·ов (*f*) (math) divide/master line
водо·раз·де́ль·н·ый щит· (*m*) circulation augmentor (boiler)
водо·рас·твор·и́м·ый, -ая, -ое, (*a*) water-soluble
водо·ре́з·, -а, (*m*) (naut) cutwater; (civ eng) starling; (zool) skimmer, *Rhynchops*
водо·ро́д·, -а, (*m*) (chem) hydrogen, H
~, атом·а́рн·ый atomic hydrogen
~, бро́м·ист·ый hydrogen bromide
~, мышьяк·о́вист·ый arsine, hydride of arsenic
~, сверх·тяж·ёл·ый Tritium, T, $\mathrm{1H^3}$
~, сер·ни́ст·ый sulphuretted hydrogen, hydrogen sulphide
~, тяж·ёл·ый deuterium
~, фо́сфор·ист·ый hydrogen phosphide, phosphine
~, фто́р·ист·ый hydrogen fluoride
~, хло́р·ист·ый hydrogen chloride
~, циа́н·ист·ый hydrogen cyanide, prussic acid
водо·ро́д·ист·ый, -ая, -ое, (*a*) hydrogen; (chem) hydride of
~ ли́т·и·й (*m*) lithium hydride
водородно·во́д·н·ый об·ме́н· (*m*) hydrogen-water exchange
водородно·со·держ·а́щ·ий, -ая, -ее, (*a*) hydrogen-containing, hydrogenous
водо·ро́д·н·ый, -ая, -ое, (*a*) hydrogen, hydrogenous
~ по·каз·а́тел·ь (*m*) pH indicator, hydrogen ion indicator
водородо·по·до́б·н·ый, -ая, -ое, (*a*) hydrogen-like

во́до·рос·лев·ый, -ая, -ое, (*a*) of во́до·рос·л·ь
во́до·рос·л·ь, -и, (*f*) water plant; seaweed; alga
~, циа́н·ов·ые (*pl*) (bot) *Cyanophyceae*
водо·с·бо́р·, -а, (*m*) (geol) catchment area/basin
водо·с·бо́р·ник·, -а, (*m*) leakage drain; water header
водо·с·бо́р·н·ый, -ая, -ое, (*a*) water-collecting; (geol) catchment
~ кана́л· (*m*) leakage drain
~ коло́д·ец· (*m*) drainage sump
~ пло́щад·ь (*f*) (geol) catchment area
водо·с·бро́с·, -а, (*m*) (hydr eng) floodgates, discharge arrangements, waste weir
~, водо·с·ли́в·н·ый overfall spillway
~, сифо́н·н·ый syphon spillway
~, холост·о́й (hydr eng) wasteway
водо·с·ли́в·, -а, (*m*) spillway; weir
~, бок·ов·о́й (hydr eng) side-channel spillway
~, ша́хт·н·ый shaft spillway
водо·с·ли́в·н·ая плот·и́н·а (*f*) weir
водо·сло́·й, -я, (*m*) water stain (timber)
водо·снабж·е́ни·е, -я, (*n*) water supply
водо·со·держ·а́щ·ий, -ая, -ее, (*a*) water-bearing, aquiferous
водо·с·пу́ск·, -а, (*m*) water outlet; (hydro-elec) outlet, outlet conduit
водо·сто́й·к·ий, -ая, -ое, (*a*) water-proof/resistant
водо·с·то́к·, -а, (*m*) surface water sewerage (Brit), storm sewerage (U.S.); (geol) runoff, drain
~, канализацио́н·н·ый spoil sewerage
водо·столб·ов·о́й мото́р· (*m*) hydraulic motor
водо·с·то́ч·н·ый, -ая, -ое, (*a*) of водо·с·то́к·
~ жёлоб· (*m*) (build) gutter
~ се́т·ь (*f*) storm sewer system (U.S.), surface-water sewer system (Brit)
водо·стру́й·н·ый, -ая, -ое, (*a*) water-jet
~ на·со́с· (*m*) jet pump
~ эже́ктор· (*m*) water-jet ejector (jet pump)
водо·те́к·, -а, (*m*) leak
водо·те́ч·ност·ь, -и, (*f*) leakage, leaking
водо·то́к·, -а, (*m*) stream, water course, current of water
водо·то́ч·н·ый, -ая, -ое, (*a*) of во·до·то́к·

водо·тру́б·н·ый котёл·   (*m*) water-tube boiler

водо·у·ка·за́те·ль, -я,   (*m*) water-level indicator

водо·у·мягч·и́тел·ь, -я,   (*m*) water softener

водо·у·по́р·, а,   (*m*) (geol) water-confining stratum

водо·у·по́р·н·ый, -ая, -ое,   (*a*) water-resisting, water-repellent

водо·хран·и́лищ·е, -а,   (*n*) reservoir

водо·цеме́нт·н·ый фа́ктор·   (*m*) water–cement ratio

во́д·очн·ый, -ая, -ое,   (*a*) of **во́д·к·а**

водруж·а́·ть, -ют,   (*imp*) erect, set up, hoist (flag)

водруж·и́ть, -а́т,   (*perf*) *see* **водру·ж·а́·ть**

-вод·ств·о, -а,   (*n*) (*component*) -culture, -growing; -breeding

вод·яни́ст·ый, -ая, -ое,   (*a*) watery; (chem) aqueous, hydrous

вод·я́нк·а, -и,   (*f*) waterboat; (med) dropsy

вод·ян·о́й, -а́я, -о́е,   (*a*) water, aqueous, hydraulic; aquatic

~ **за·тво́р·** (*m*) hydraulic back-pressure valve; water seal

~ **за·щи́т·а** (*f*) (nucl) water shield

~ **магистра́л·ь** (*f*) (shipb) fire main

~ **меш·о́к·** (*m*) water pocket

~ **не́б·о** (*n*) (meteor) water sky

~ **про·стра́н·ств·о** (*n*) water space (boiler)

~ **руб·а́шк·а** (*f*) water jacket

~ **столб·** (*m*) water column

~ **числ·о́** (*n*) (phys) water equivalent

во·ева́ть, (*pres 3rd sing, pl*) **во·ю́ет, во·ю́ют,** (*imp*) wage war, make war; struggle with

во·еди́н·о (*adv*) together

воен·иза́ци·я, -и,   (*f*) putting on war basis; militarization

воен·кома́т·, -а,   (*m*) military service and mobilization office

воен·ко́р·, -а,   (*m*) war correspondent

воен·н·о- (*component*) = **воен·н·ый** war, military, armed forces

военно·воз·ду́ш·н·ый, -ая, -ое,   (*a*) air force, air

~ **си́л·ы** (*pl*) air force

военно·мор·ск·о́й, -а́я, -о́е,   (*a*) naval

~ **флот·** (*m*) navy

военно·обя́з·анн·ый, -ого, (*m decl as adj*) reservist

военно·пле́н·н·ый, -ого, (*m decl as adj*) prisoner of war

военно·слу́ж·ащ·ий, -его, (*m decl as adj*) serviceman

воен·н·ый, -ая, -ое, (*a*) military, war; (*m decl as adj*) serviceman, soldier

~ **за·во́д·** (*m*) munitions factory

~ **министе́р·ств·о** (*n*) War Ministry (Brit), War Department (U.S.)

~ **по·лож·е́ни·е** (*n*) martial law

~ **порт·** (*m*) naval port

~ **со·общ·е́ни·я** (*n*) lines of communication

~ **суд·** (*m*) court martial

во́·ет (*pres 3rd sing*) of **вы·ть**

вож·а́к·, -а́,   (*m*) leader; (fish) net-/herring-work warp

вож·а́тый, -ого, (*m decl as adj*) tram driver; youth leader

вожд·е́ни·е, -я,   (*n*) driving, steering, piloting

во́жд·ь, -я́,   (*m*) leader

вожж·а́, -и́, (*i*) **-о́й,** (*nom pl*) **во́ж·ж·и,** (*g pl*) **вожж·е́й,** (*f*) rein(s)

во·жм·у́ (*fut 1st sing*) of **в·жат·ь**

воз·, -а, (*nom pl*) **-ы́,** (*g pl*) **-о́в,** (*m*) cart, carriage, conveyance; cart-load, cartful

воз- (*prefix*) = **вос-** *particular significances*: (1) movement upwards; (2) renewed action; (3) action in response; (4) intensity, solemnity

воз·буд·и́мост·ь, -и,   (*f*) excitability; (med) irritability

воз·буд·и́м·ый, -ая, -ое,   (*a*) excitable; (med) irritable

воз·буд·и́тел·ь, -я, (*m*) exciter; (med) causative agent, stimulus; (anat) excitor

~ **бол·е́зн·и** (med) pathogen

~ **воз·буд·и́тел·я** (elec) pilot exciter

~, **ква́рц·ев·ый** (rad) crystal driver

воз·буд·и́тельн·ый, -ая, -ое, (*a*) exciting, exciter; (zool) excitatory

~ **агрега́т·** (*m*) (elec) exciter set

воз·буд·и́ть, -я́т, (*perf*) *see* **воз·буж·д·а́·ть**

воз·бужд·а́ем·ый, -ая, -ое, (*pres part pass*) of **воз·бужд·а́·ть;** excitable

воз·бужд·а́·ть, -ют, (*imp*) excite, stimulate, arouse, actuate, provoke; raise (a question), institute (proceedings)

~ **иск·** (law) bring/institute a suit

воз·бужд·а́ющ·ий, -ая, -ое, (*pres part act*) of **воз·бужд·а́·ть**

~ **об·мо́т·к·а** (*f*) (elec) exciter coil

~ **сре́д·ств·о** (*n*) (pharm) stimulant

воз·бужд· éни·е, -я, (*n*) excitation, stimulation; generation; excitement, agitation

~, кóнус·н·ое, (rad) gable illumination

~ пóл·я, (elec) field excitation

~ раз·ря́д·а (nucl) initiation of a discharge
реостáт· воз·бужд·éни·я (*m*) field rheostat, exciter field rheostat

~, удáр·н·ое (phys) impact collision, impulse excitation

~ фотóн·ами photon excitation
цеп·ь воз·бужд·éни·я (*f*) (elec) exciting circuit

воз·бужд·ённ·ый, -ая, -ое, (*past part pass*) *see* **воз·бужд·á·ть**

воз·вед·éни·е, -я, (*n*) erection, raising

~ во втор·ýю стéп·ен·ь (math) raising to the second power, squaring

~ в стéп·ен·ь (math) involution, raising to a power

воз·вед·ённ·ый, -ая, -ое, (*past part pass*) of **воз·вес·ти́**

воз·велич·ива·ть,-ют, (*imp*) extol, exalt

воз·велич·ить, -ат, (*perf*) *see* **воз·велич·ива·ть**

воз·вес·ти́, (*fut 3rd sing, pl*) **воз·вед·ёт, воз·вед·ýт,** (*past masc, fem sing*) **воз·вё·л, воз·ве·лá;** *see* **воз·вод·и́ть**

воз·вест·и́ть, -ят, (*perf*) *see* **воз·вещ·á·ть**

воз·вещ·á·ть, -ют, (*imp*) announce, proclaim, herald

воз·вод·и́ть, -ят, (*imp*) erect, build; elevate, promote; derive from, trace to; (math) raise

~ в квадрáт· (math) square, raise to the power of two

воз·врáт·, -а, (*m*) return; (met) recovery; (autom) reset; (math) regression; (fin) reimbursement, repayment, refund; (med) recurrence; re-entry (of rocket to atmosphere)
коэффициéнт· воз·врáт·а тепл·á (*m*) (a/c) reheat factor

~, руч·н·óй manual reset

воз·врат·и́ть, -ят, (*perf*) *see* **воз·вращ·á·ть**

воз·врáт·н·о-по·ступ·áтельн·ый, -ая, -ое, (*a*) reciprocating

воз·врáт·н·ый, -ая, -ое, (*a*) return; (gram) reflexive; (med) recurrent, relapsing; (math) recurrent

~ глагóл· (*m*) reflexive verb

~ коллéктор· (*m*) return header (pipes)

воз·врáт·н·ый, -ая, -ое

~ пóшлин·а (*f*) (fin) drawback

воз·вращ·á·ть, -ют, (*perf*) give/send/ bring back, return; (fin) repay, refund, reimburse; **-ся** (*pass*) go back, come back, return, revert

~ в по·втóр·н·ый цикл· recycle

воз·вращ·áющ·ийся, -аяся, -ееся, (*pres part act*) of **воз·вращ·á·ться;** returning, recurrent

воз·вращ·éни·е, -я, (*n*) return; recurrence; restitution (to owner)

~ пруж·и́н·ой (instr) spring control

воз·вы́с·ить, -ят, (*perf*) *see* **воз·выш·á·ть**

воз·выш·á·ть, -ют, (*imp*) raise; **-ся** rise, rise above

воз·выш·éни·е, -я, (*n*) (*v n*) *see* **воз·выш·á·ть;** rise; (geog) eminence, rising ground; increase; dais; elevation (of angles etc.); (hydrog) knoll

~ в стéп·ен·ь (math) raising to the power
ýгол· воз·выш·éни·я (*m*) angle of elevation, quadrant elevation, elevation

воз·выш·енност·ь, -и, (*f*) elevation, elevated position, (geog) hills, upland, heights; (ocean) rise, ridge

~, плóск·ая plateau

воз·выш·енн·ый, -ая, -ое, (*past part pass*) *see* **воз·выш·á·ть;** elevated, high, lofty

воз·глáв·ить, -ят, (*perf*) *see* **воз·главл·я́·ть**

воз·главл·я́·ть, -ют, (*imp*) head, lead, be at the head of

вóз·глас·, -а, (*m*) exclamation, cry

воз·глас·и́ть, -я́т, (*perf*) *see* **воз·глаш·á·ть**

воз·глаш·á·ть, -ют, (*imp*) proclaim

воз·гóн·, -а, (*m*) (chem) sublimate

воз·гóн·к·а, -и, (*f*) (chem) sublimation

воз·гóн·н·ый, -ая, -ое, (*a*) subliming

~ аппарáт· (*m*) sublimate

~ со·сýд· (*m*) subliming pot

воз·гон·я́·ть, -ют, (*imp*) (chem) sublimate

воз·гор·áемост·ь, -и, (*f*) inflammability, combustibility

воз·гор·áни·е, -я, (*n*) (*v n*) *see* **воз·гор·á·ться;** conflagration, deflagration

воз·гор·á·ться, -ются, (*imp*) catch fire, flare up

воз·гор·áющ·ийся, -аяся, -ееся, (*pres part act*) of **воз·гор·á·ться,** inflammable воз·гор·е́ться, -я́тся, (*perf*) *see* воз·-гор·а́·ться

воз·да·ва́ть, -ю́т, (*imp*) render

воз·да́·ть (*fut 3rd sing, pl*) возда́ст, воз·дад·у́т (*perf*); *see* воз·да·ва́ть

воз·двиг·а́·ть, -ю́т, (*imp*) erect

воз·дви́г·н·уть, -ут, (*perf*) *see* воз·-двиг·а́·ть

воз·де́й·стви·е, -я, (*n*) (*see also* де́й·-стви·е) influence, effect, action, attack

~, воз·мущ·а́ющ·ее disturbance; perturbing effect

~, в·ход·н·о́е (autom) input variable

~, динам·и́ческ·ое dynamic impact

~, коррозио́н·н·ое corrosive attack под воз·де́й·стви·ем (+ *gen*) exposed to

~, температу́р·н·ое thermal effect

~, цикл·и́ческ·ое температу́р·н·ое thermal cycling

воз·де́й·ств·овать, -уют, (*imp and perf*) influence; act on

воз·де́л·а́·ть, -ют, (*perf*) *see* воз·де́-л·ыва·ть

воз·де́л·ыва·ть, -ют, (*imp*) cultivate; till (soil)

воз·держ·а́ни·е, -я, (*n*) abstemiousness, temperateness; abstention, abstinence, continence

воз·де́рж·анност·ь, -и, (*f*) *see* воз·-держ·а́ни·е

воз·держ·а́ться, -а́тся·, (*perf*) *see* воз·де́рж·ива·ться

воз·де́рж·ива·ться, -ются, (*imp*) abstain, refrain, forbear

во́з·дух·, -а, (*m*) air

~, альвеоля́р·н·ый alveolar air

~ во́з·дух· (GW) air-to-air

~, воз·мущ·ённ·ый turbulent air

~, втор·и́чн·ый econdary air

~, за·тормож·ённ·ый brought-to-rest air

~, за·сто́й·н·ый stagnant air

~ земл·я́ (GW) air-to-surface

~, поля́р·н·ый (meteor) polar air; (aerodyn) jet airstream

~, не·с·поко́й·н·ый moving/disturbed air

~, норма́ль·н·ый standard air

~, перв·и́чн·ый primary air

~, раз·реж·ённ·ый (dynam) light/rarefied air

~, с·жа́·т·ый compressed air; (aerodyn) heavy air

~, троп·и́ческ·ий (meteor) tropical air

~, эжéкт·ирован·н·ый induced air

во́з·дух

~, экваториа́ль·н·ый (meteor) equatorial air

воздухо·во·влеч·éни·е, -я, (*n*) air entrainment

воздухо·во́д·, -а, (*m*) air flue

воздухо·во́з·, -а, (*m*) (min) air-powered locomotive, compressed-air engine locomotive

воздухо·воз·ду́ш·н·ый, -ая, -ое, (*a*) (mil) air-to-air

воздухо·ду́в·к·а, -и, (*g pl*) -в·ок·, (*f*) blower; scavenger (diesel)

воздухо·ду́в·н·ый, -ая, -ое, (*a*) of воздухо·ду́в·к·а; air-blast, blast

~ труб·а́ (*f*) (rail) blast pipe

воздухо·за·бо́р·ник·, -а, (*m*) (a/c) air intake/entry air scoop

~, с·мещ·ённ·ый (a/c) off-axis inlet

воздухо·за·бо́р·н·ый, -ая, -ое, (*a*) air-intake

воздухо·за·пра́в·щик·, -а, (*m*) air charging trolley

воздухо·мéр·, -а, (*m*) air meter, air-flow meter

воздухо·на·грев·а́тел·ь, -я, (*m*) (met) hot-blast stove

воздухо·не·про·ниц·а́ем·ый, -ая, -ое, (*a*) air-tight

воздухо·но́с·н·ый, -ая, -ое, (*a*) aeriferous

~ ткан·ь (*f*) (biol) respiratory tissue, aerenchyma, aerating tissue

воздухо·об·ма́н·, -а, (*m*) replacement of air (ventilation)

воздухо·от·раж·а́тел·ь, -я, (*m*) air deflector

воздухо·о·хлад·и́тел·ь, -я, (*m*) air cooler; (a/c) cold air unit

воздухо·о·хлад·и́тельн·ый, -ая, -ое, (*a*) air-cooling

воздухо·о·хлажд·а́ем·ый, -ая, -ое, (*a*) air-cooled

воздухо·о·чист·и́тел·ь, -я, (*m*) air cleaner/filter

воздухо·пла́в·ани·е, -я, (*n*) aerostatics, lighter-than-air aeronautics, aero-station

воздухо·подо·грев·а́тел·ь, -я, (*m*) air preheater

воздухо·при·ём·ник·, -а, (*m*) (a/c) air intake scoop

~, без·на·по́р·н·ый non-ramming air scoop

воздухо·про·во́д·, -а, (*m*) air line/lead/duct; air adaptor (gas turbine)

воздухо·про·вóд

~, кольц·ев·óй   bustle pipe (blast furnace)

~, под·фюзеля́ж·н·ый   (a/c) ventral duct

воздухо·про·тóк·, -а,   (*m*) (shipb) air course

воздухо·про·ниц·а́ем·ый, -ая, -ое, (*a*) not airtight; permeable to air

воздухо·пýск·, -а,   (*m*) (ICE) compressed-air starter

воздухо·рас·пре·дел·и́тел·ь, -я,   (*m*) air distributor; regulating feed valve (pneumatic brakes)

воздухо·с·бóр·ник·, -а, (*m*) air receiver

воздухо·эквивале́нт·н·ый, -ая, -ое, (*a*) air-equivalent

воз·дýш·ник·, -а,   (*m*) air vent

воз·дýш·н·о- (*component*) air, pneumatic

воздушно·гидравл·и́ческ·ий, -ая, -ое, (*a*) hydro-pneumatic

воздушно·деса́нт·н·ый, -ая, -ое, (*a*) (mil) airborne, airborne-landing

воздушно·звук·ов·óй, -а́я, -óе, (*a*) sound signalling

воздушно·реакти́в·н·ый, -ая, -ое, (*a*) thermal jet, jet (engine etc.)

воз·дýш·ност·ь, -и, (*f*) airiness, lightness

воздушно·сух·óй, -а́я, -óе, (*a*) air-dry

воздушно·тепл·ов·а́я противо·об·лед·ени́тельн·ая систе́м·а (*f*) (a/c) thermal de-icing system

воз·дýш·н·ый, -ая, -ое, (*a*) air, aerial; light, airy

~ аккумуля́тор· (*m*) air-loaded hydraulic accumulator, hydropneumatic accumulator

~ вы·ключ·а́тел·ь (*m*) (elec) air-blast breaker

~ за·слóн·к·а карбюра́тор·а (*f*) (ICE) choke

~ инфе́кци·я (*f*) airborne infection

~ желе́з·н·ая дорóг·а (*f*) elevated/overhead railway

~ ли́н·и·я (*f*) (elec) overhead line

~ ма́сс·а (*f*) (meter) air mass

~ маши́н·а (*f*) air engine

~ на·ка́т·ник· (*m*) (gunn) pneumatic recuperator

~ от·дел·е́ни·е (*n*) buoyancy chamber

~ по·тóк· (*m*) air stream

~ про·вóд·к·а (*f*) (elec) overhead wires

~ рас·пы́л·ивани·е (*n*) blast injection (diesel), air injection

~ само·пýск· (*m*) compressed-air starter

воз·дýш·н·ый, -ая, -ое

~ систе́м·а (*f*) (a/c) pneumatic system

~ трансформ·а́тор· (*m*) (elec) air-core transformer

~ тревóг·а (*f*) air raid warning

~ трýб·а́ (*f*) (shipb) ventilation trunk

~ хозя́й·ств·о (*n*) compressed air supply

~ шар· (*m*) balloon

~ я́щик· (*m*) air space (boiler); buoyancy tank (boat)

воз·дым·а́ни·е, -я, (*n*) bulging up

воз·зв·а́ть, (*fut 3rd sing, pl*) **воз·зов·ёт, воз·зов·ýт,** (*perf*) appeal

воз·зов·ýт (*fut 3rd pl*) of **воз·зв·а́ть**

воз·зр·е́ни·е, -я, (*n*) views, opinions, attitude

воз·и́ть, -́ят, (*imp indet*) carry (in vehicle), take, convey, transport cart; draw (vehicle); **-ся** busy oneself; mess/fuss about

вóз·к·а, -и, (*g pl*) **-з·ок·,** (*f*) carriage, transportation, cartage

воз·лаг·а́·ть, -ют, (*imp*) entrust, place (hopes etc.); lay, place (solemn occasion)

воз·лож·и́ть, -́ат, (*perf*) see **воз·лаг·а́·ть**

вóзле (*adv*) (*prep+gen*) by, beside

воз·леж·а́ть, -а́т, (*imp*) recline

воз·ле́чь, (*fut 3rd sing, pl*) **воз·ля́ж·ет, воз·ля́г·ут,** (*past sing*) **воз·лёг, воз·легла́,** (*perf*); see **воз·леж·а́ть**

воз·ме́зд·и·е, -я, (*n*) retribution, requital, punishment

воз·мест·и́ть, -ят, (*perf*) see **воз·мещ·а́·ть**

воз·мещ·а́·ть, -ют, (*imp*) compensate, refund, reimburse, replace

воз·мещ·е́ни·е, -я, (*n*) compensation, reimbursement, indemnification

воз·мóж·н·о (*adv*) possibly; it is possible

~ лýчш·е as well as possible

~ лýчш·ий the best possible

воз·мóж·ност·ь, -и, (*f*) possibility; opportunity, chance; means, resources

да·ть воз·мóж·н·ост·ь permit, enable, make possible, provide an opportunity of

воз·мóж·н·ый, -ая, -ое, (*a*) possible, feasible

~ пере·мещ·е́ни·я (*pl*) (mech) virtual displacements

при́нцип· воз·мóж·н·ых пере·мещ·е́ни·й (*m*) (mech) principle of virtual work

воз·мут·и́ть, -ят,   (*perf*) *see* **воз·му·щ·а́·ть**

воз·мущ·а́·ть, -ют,   (*imp*) perturb, disturb

воз·мущ·а́ющ·ий, -ая, -ее, (*pres part act*) of **воз·мущ·а́·ть**

~ **потенциа́л·**   (*m*) (elec) perturbing potential

~ **фа́ктор·**   (*m*) (phys) perturbation factor

воз·мущ·е́ни·е, -я,   (*n*) indignation; perturbation, disturbance

~ **гран·и́чн·ых у·сло́в·и·й**   (nucl) perturbation of boundary conditions

~, **из·ги́б·очн·ое**   (nucl) wriggling perturbation

~, **ме́ст·н·ое** (nucl) local perturbation

пло́ск·ост·ь воз·мущ·е́ни·я   (*f*) (air) Mach plane

тео́ри·я воз·мущ·е́ни·й (*f*) (phys) perturbation theory

у́гол· воз·мущ·е́ни·я   (*m*) (air) Mach angle

воз·на·гражд·е́ни·е, -я,   (*n*) reward, recompense, remuneration, fee

воз·ник·а́·ть, -ют, (*imp*) come into being, appear, arise; occur

воз·ник·нове́ни·е, -я,   (*n*) origin, beginning, genesis; emergence, occurrence, appearance, onset

~ **колеб·а́ни·й**   onset of oscillating **у·равн·е́ни·е воз·ник·нове́ни·я траекто́ри·и** (*n*) birth equation (radar)

воз·ни́к·н·уть, -ут, (*past masc sing*) **воз·ни́к·,** (*perf*); *see* **воз·ник·а́·ть**

воз·об·нов·и́ть, -ят, (*perf*) *see* **воз·об··новл·я́·ть**

воз·об·новл·е́ни·е, -я,   (*n*) renewing, recommencing, resuming

~ **проте́ктор·а**   retreading (of tires)

воз·об·новл·я́·ть, -ют,   (*imp*) renew, recommence, resume

воз·об·новл·я́ющ·ийся, -аяся, -ееся, (*pres part pass*); regenerative; recurrent

воз·раж·а́·ть, -ют, (*imp*) object

воз·раж·е́ни·е, -я,   (*n*) objection

воз·раз·и́ть, -я́т, (*perf*) *see* **воз·раж·а́·ть**

во́з·раст·, -а, (*m*) age

~, **абсолю́т·н·ый**   (geol) absolute age

~ **лун·ы́**   age of the moon

~ **нейтро́н·а**   neutron Fermi age

~, **пере·зре́·л·ый**   post-maturity

~ **по радио·акти́в·н·ому угле·ро́д·у** (nucl, geol) radiocarbon age

~ **по ра́ди·ю**   (geol, nucl) radium age

~ **по свин·ц·у́**   lead age

во́з·раст

~, **при·ли́в·а**   age of tide, retard of tide

~, **ста́р·ческ·ий**   senility, gerontic/senile age

воз·раст·а́ни·е, -я,   (*n*) growth, increase, gain augmentation; increment

~ **мо́щ·ност·и**   increase in power

~ **температу́р·ы**   temperature rise/increase

воз·раст·а́·ть, -ют,   (*imp*) grow, increase

воз·раст·а́ющ·ий, -ая, -ее, (*pres part act*) of **воз·раст·а́·ть**; increasing, growing, incremental; (math) ascending

воз·рас·ти́, (*fut 3rd sing, pl*) **воз·раст·ёт, воз·раст·у́т,** (*past masc sing*) **воз··ро́с·,** (*perf*); *see* **воз·раст·а́·ть**

воз·раст·н·о́й, -ая, -ое, (*a*) = **во́з··раст·н·ый**

во́з·раст·н·ый, -ая, -ое, (*a*) age

~ **у·равн·е́ни·е**   (*n*) (nucl) age-diffusion equation

воз·род·и́ть, -ят, (*perf*) *see* **воз·рожд··а́·ть**

воз·рожд·а́·ть, -ют,   (*imp*) revive, regenerate, reactivate, revitalize

воз·рожд·е́ни·е, -я,   (*n*) revival, rebirth, regeneration; renaissance; **воз··рожд·е́ни·е, -я,** (*n*) the Renaissance

воз·рожд·ённ·ый, -ая, -ое, (*past part pass*) *see* **воз·рожд·а́·ть**; resurgent

воз·ым·е́·ть, -ют,   (*perf*) have, conceive

возьм·и́те   (*perf imper*) of **взя́·ть,** *see* **бр·ать**

во́·ин·, -а,   (*m*) serviceman, soldier

во́ин·ск·ий, -ая, -ое,   (*a*) military, martial

~ **по·ви́н·ност·ь**   (*f*) conscription, compulsory military service

войн·ственн·ый, -ая, -ое,   (*a*) warlike, martial; bellicose

во́·й, -я,   (*m*) howl, wail; (rad) singing

во́йлок·, -а, (*m*) felt

войлоко·обра́з·н·ый, -ая, -ое,   (*a*) felt-like; (geol) diverse columnar

во́йлоч·н·ый, -ая, -ое,   (*a*) felt

войн·а́, -ы́, (*nom pl*) **во́й·н·ы, (*g pl*) вой·н·,** (*f*) war, warfare

~, **вели́к·ая оте́ч·ественн·ая** Second World War (as fought by U.S.S.R.)

вой·ск·а́ (*nom pl*), (*g pl*) **вой·ск·,** (*d pl*) **вой·ск·а́м,**   (*n*) forces, troops

войск·ов·о́й, -а́я, -о́е,   (*a*) military, forces, troops

во·йти́, (*fut 3rd sing, pl*) **во·йд·ёт, во·йд·у́т,** (*past masc sing*) **во·шёл,** (*perf*); *see* **в·ход·и́ть**

вока́бул·а, -ы, (*f*) (ling) vocable

вокализ·а́ци·я, -и, (*f*) (ling) vocalization

вокза́л·, -а, (*m*) (rail) station, main station, terminal station

вокодер·, -а, (*m*) (telecom) vocoder

во·кру́г· (*adv*) (*prep+gen*) round, around

во́л·, -а́, (*m*) bullock, ox

волды́р·еват·ый, -ая, -ое, (*a*) blistered, covered with blisters/bumps, pimply

волды́р·н·ый, -ая, -ое, (*a*) of **волды́р·ь**

волды́р·ь, -я́, (*m*) blister, bump, boil

вол·ев·о́й, -а́я, -о́е, (*a*) volitional; resolute, determined

во́л·ей-не·во́л·ей (*adv*) willy-nilly, against one's will

во́лк·, -а, (*nom pl*) **во́лк·и,** (*g pl*) **волк·о́в,** (*m*) (zool) wolf, *Canis lupus*; (astron) Wolf, Lupus

волк-маши́н·а, -ы, (*f*) (text) willow, willey

~, мор·ск·о́й (fish) bass, *Morone labrax*; (naut) old salt, experienced seaman

волластони́т·, -а, (*m*) (min) wollastonite

волн·а́, -ы, (*nom pl*) **во́лн·ы,** (*f*) wave, surge

~, ба́зис·н·ая (nucl) base surge

~, бег·у́щ·ая (elec) travelling wave; (dynam) wave of translation

~, блужд·а́ющ·ая (rad) stray wave

~, ветр·ов·а́я (hydrodyn) wind-generated wave

~, вз·рыв·н·а́я blast, shock wave

~, вну́тр·енн·яя internal wave; (geol) subsurface wave

~, гекто·метр·и́ческ·ая (rad) hectometric wave, medium wave

~, голов·н·а́я (dynam) bow wave, shell wave, leading wave

~, дву·го́рб·ая double-humped wave

~, деци·метр·о́в·ая (rad) decimetric wave; (*pl*) UHF

~, дека·метр·и́ческ·ая (rad) decametric wave, short wave

~, диффраг·и́рованн·ая diffracted wave

~, дли́н·н·ая (rad) long wave (above 999 metres)

~, ду́ль·н·ая muzzle wave (ballistics)

~, за·тух·а́ющ·ая damped/decaying wave

волн·а́

~, звук·ов·а́я sound wave

~, зем·н·а́я (rad) direct wave; (seismol) earthquake wave

~, знач·н·ая marking wave (telegraphy)

~, ионо·сфе́р·н·ая (rad) sky/ionospheric wave

~, коро́т·к·ая (rad) short wave, decametric wave

~ Ло́ве (seismol) Love wave

~, метр·о́в·ая (rad) metric wave

~, милли·метр·о́в·ая millimetric wave, (*pl*) EHF

~, модул·и́рованн·ая не·за·тух·а́ющ·ая (rad) modulated continuous wave

~, на·бег·а́ющ·ая incident wave

~, негати́в·н·ая (rad) compensation wave

~, не·за·тух·а́ющ·ая continuous/ undamped wave

~, не·от·раж·ённ·ая direct/non-reflected wave

~, не·с·вя́з·анн·ая (rad) unguided wave

~, нес·у́щ·ая (rad) carrier wave

~, нос·ов·а́я (ocean) bow wave

~, об·ме́н·н·ая alternating/transformed/composite wave

~, объ·ём·н·ая (seismol) bodily wave

~, основ·н·а́я (rad) fundamental wave; principal/dominant wave (in a waveguide)

~, от·раж·ённ·ая reflected/back wave, backward wave; (naut) back wash

~, па́д·ающ·ая incident wave

~, пад·е́ни·я с·на·ря́д·а (gunn) impact wave

~, парциа́ль·н·ая partial wave

~, пло́ск·ая plane wave

~, попере́ч·н·ая (seismol) transverse wave, secondary wave, S wave

~, попере́ч·н·ая электр·и́ческ·ая (rad) TE wave

~, попере́ч·н·ая магнит·н·ая (rad) TM wave

~, попере́ч·н·ая электро·магни́т·н·ая (rad) TEM wave

~, посыл·а́ем·ая emitted wave

~, пра́в·ильн·ая regular wave

~, пре·де́ль·н·ая edge wave, limiting wave

~, пре·ломл·ённ·ая refracted wave

~, пре·ры́в·ист·ая не·за·тух·а́ющ·ая interrupted continuous wave

~, пре·ход·я́щ·ая (elec) surge

волн·а́

~, про·бе́ль·н·ая   (rad) spacing wave

~, прогресси́в·н·ая   (hydrodyn) wave of translation

~, про·до́ль·н·ая   (seismol) longitudinal/primary/P wave

~, про·меж·у́точн·ая   intermediate wave (50–200 m)

~, про·стра́н·ственн·ая   (rad) ionospheric/indirect wave; (geol) space wave

~, прям·а́я   forward-travelling wave

~, прямо·уго́ль·н·ая   square wave

~, рабо́ч·ая   marking wave (telegraphy); (rad) transmitting wave

~ раз·реж·е́ни·я   rarefaction/rarification wave, low-pressure wave; (aerodynam) suction wave

~ раз·ры́в·а   distortional wave; (expl) burst wave

~, рас·се́·янн·ая   scattered/dissipated wave

~ рас·шир·е́ни·я   rarefaction wave

~, санти·метр·о́в·ая   (rad) centimetric wave; (pl) SHF

~, сверх·дли́н·н·ая   (rad) very long wave (above 3000 m)

~, с·вя́з·анн·ая   bound wave

~ с·дви́г·а   shear/displacement wave

~, сейсм·и́ческ·ая   seismic wave

~ с·жа́т·и·я   pressure/compression wave; plastic wave (blasting)

~, синусоида́ль·н·ая   sine/sinusoidal wave

~ с·мещ·е́ни·я   displacement wave

~, спи́н·ов·ая   spin wave

~, сре́д·н·яя   (rad) medium wave, hectometric wave

~, сто·я́ч·ая   (rad) standing wave; (ocean) stationary wave

~ теч·е́ни·я   translational wave

~, трохоида́ль·н·ая   (ocean) trochoid wave

~, уда́р·н·ая   shock wave

~, у·един·ённ·ая   solitary wave

~, ультра·коро́т·к·ая   (rad) ultrashort wave; (pl) VHF (also used for UHF upwards), wave band 10–0·001 m

~, у·площ·ённ·ая   flattened wave

~, у·стан·ови́вш·аяся   steady-state wave

~, фронта́ль·н·ая   (meteor) frontal wave

~, циркуля́р·н·о-поляр·изо́ванн·ая   circularly polarized wave

волн·е́ни·е, -я,   (n) agitation, disturbance; (meteor) sea disturbance, sea state, waves

~, жест·о́к·ое   very high sea

~, знач·и́тельн·ое   rough sea

~, ис·ключ·и́тельн·ое   phenomenal sea

~, кру́п·н·ое   very rough sea

~, лёг·к·ое   slight sea

~, не·с·поко́й·н·ое   rough sea

~, си́ль·н·ое   high sea

~, сла́б·ое   smooth sea

~, у·ме́р·енн·ое   moderate sea

шкал·а́ волн·е́ни·я   (f) (meteor) sea disturbance scale

волн·и́стост·ь, -и,   (f) waviness, undulation, ripple; wavy grain (in wood)

волн·и́ст·ый, -ая, -ое,   (a) wavy, undulating, rippling, curly; corrugated (roofing etc.) crimped (wire)

~ стал·ь   (f) corrugated steel sheets

волн·ова́ть, -у́ют,   (imp) agitate, stir, excite, disturb

волно·во́д·, -а,   (m) (rad) waveguide

~ -анте́нн·а, щел·ев·о́й   leaky waveguide

~, атмосфе́р·н·ый   atmospheric duct

~, пла́зм·енн·ый   plasma waveguide

~, с·жим·а́ем·ый   squeezable waveguide

~, со·глас·о́ванн·ый   matched waveguide

~, с·у́ж·иваюш·ий   tapered waveguide

волно·во́д·н·ый, -ая, -ое,   (a) of волно·во́д·

~ ли́н·и·я за·де́рж·к·и   (f) (elec) waveguide delay line

~ рас·про·стран·е́ни·е   (n) (rad) super-refraction

~ у·ча́ст·ок регул·и́руем·ого сеч·е́ни·я   (m) waveguide squeeze section

волн·ов·о́й, -а́я, -о́е,   (a) of волн·а́

~ движ·е́ни·е   (n) wave motion

~ паке́т·   (m) (nucl) wave packet/train

~ про·вод·и́мост·ь   (f) (elec) characteristic admittance

~ у·равн·е́ни·е   (n) wave equation

волново·корпускуля́р·н·ый дуали́зм·   (m) (nucl) wave-particle duality

волно·гра́ф·, -а,   (m) wave recorder

волно·иск·а́тел·ь, -я,   (m) (rad) wave detector, detector

волно·ло́м·, -а,   (m) breakwater

волно·ме́р·, -а,   (m) wavemeter, cymometer

~, гетероди́н·н·ый   heterodyne wavemeter

~, про·ход·н·о́й   two-entry wavemeter

волно·мér

~, **резонáнс·н·ый**   resonator wave-meter

волно·механ·úческ·ая теóр·и·я   (*f*) wave-mechanical theory

волно·об·наруж·úтел·ь, -я,   (*m*) (rad) coherer, wave detector, detector

волно·обрáз·н·ый, -ая, -ое,   (*a*) undulatory, undulating, wavy

волно·образ·úющ·ий, -ая, -ее,   (*a*) wave-generating (test tank)

волно·при·бóй·н·ый, -ая, -ое,   (*a*) (geol) wave-cut, wave-built

~ **террáс·а**   (*f*) wave-built terrace

~ **у·стýп·**   (*m*) (geol) wave-cut escarpment

волно·продýктор·, -а,   (*m*) wave generator (test tank)

волно·рéз·, -а,   (*m*) breakwater

волно·у·каз·áтел·ь, -я,   (*m*) (rad) wave detector, detector

волно·у·лов·úтел·ь, -я,   (*m*) (rad) wave detector, detector

вол·óв·ий, -ья, -ье,   (*a*) ox, oxen

~ **шкýр·а**   (*f*) ox-hide

вóлок·, -а,   (*m*) portage; die/draw plate; **волóк** (*past masc sing*) of **волóч·ь;** (*gen pl*) of **волóк·а**

волóк·а, -и,   (*f*) die/draw plate

волокúт·а, -ы,   (*f*) red tape

волóкн·а (*nom pl*) of **волокн·ó**

волокн·úн·а, -ы,   (*f*) (zool) fibrin

волокнисто·обрáз·н·ый, -ая, -ое, (*a*) fibroid

волокнисто·сéт·чат·ый, -ая, -ое, (*a*) reticulate-fibrous

волокн·úстост·ь, -и,   (*f*) fibrousness, stringiness

волокнисто·эласт·úческ·ий, -ая, -ое, (*a*) fibroelastic

волокн·úст·ый, -ая, -ое,   (*a*) fibrous, fibrillar, filaceous, filamentous, stringy

~ **структýр·а**   (*f*) (met) fibrous structure

волокн·ó, -а, (*nom pl*) **волóкн·а,** (*g pl*) **волóкон·,** (*n*) fibre, fiber, filament staple; (met) flow line, fibre, fiber

~, **древ·éсн·ое**   libriform fibre/fiber

~, **искýсств·енн·ое**   man-made fibre/fiber, rayon fibre/fiber

~, **казейн·ов·ое**   (text) milk wool

~, **клét·очн·ое**   cell filament

~, **коллагéн·ов·ое**   (biol) intercellular fibre/fiber, white fibre/fiber

~, **луб·ян·óе**   bast fibre/fiber

~, **медно·аммиáч·н·ый**   (text) cupr-ammonium fibre/fiber

волокн·ó

~, **моч·éнцев·ое**   (text) water-retted flax fibre/fiber

~, **перв·úчн·ое**   (biol) embryo fibre/fiber

~, **полу·древ·есúнн·ое**   (bot) meta-xylem

~, **стекл·я́нн·ое**   fiberglass

~, **стл·áнцев·ое**   (text) dew-retted flax fibre/fiber

~, **хим·úческ·ое**   man-made fibre/fiber

волокно·от·дел·úтел·ь, -я,   (*m*) *and see* **джин·** (text) gin

~, **пúль·н·ый**   gin saw, saw gin

вóлок·ом   (*adv*) by dragging/traction

волокóн·ц·е, -а,   (*n*) fibril

волокóн·чат·ый, -ая, -ое,   (*a*) filamentary

волок·ýш·а, -и,   (*f*) (agr) sweep/bunching/buck rake; (fish) tuck-net

~, **сен·н·áя**   hay sweep

волóк·ш·ий, -ая, -ее,   (*past part act*) of **волóч·ь**

волопáс·е, -а,   (*m*) (astron) Boötes

вóлос·, -а, (*nom pl*) **-ы,** (*g pl*) **волóс·,** (*d pl*) **волос·áм,** (*m*) hair

~, **óст·ев·ый**   guard hair (furs)

волос·úн·а, -ы,   (*f*) = **волос·овúн·а**

волос·áт·ый, -ая, -ое,   (*a*) hairy, hirsute; woolly, lanate; (biol) villose, villous

волос·úст·ый, -ая, -ое,   (*a*) hairy, pilose, capillate; (min) fibrous

волос·н·óй, -áя, -óе,   (*a*) capillary, pilar

волос·нóст·ь, -и,   (*f*) capillarity

волосо·вúд·н·ый, -ая, -ое,   (*a*) hair-like; (zool) trichoid; (biol) piliform

волос·овúн·а, -ы,   (*f*) (met) hair-line crack, flake

волосожáр·ы   (*pl*) (astron) Pleiades

волосо·нóс·н·ый, -ая, -ое,   (*a*) piliferous

волос·óк·, -с·к·á,   (*m*) hair, fine hair; filament; (horol) hair spring; (*pl*) (bot) trichome

волосо·обрáз·н·ый, -ая, -ое,   (*a*) hair-like; (zool) trichoid

волосо·с·гóн·н·ая машúн·а   (*f*) de-hairing machine (leather)

волос·ян·óй, -áя, -óе,   (*a*) hair; capillary

волоч·éни·е, -я,   (*n*) dragging, traction; (met) drawing, wire-drawing

волоч·úльн·ый, -ая, -ое,   (*a*) drawing, wire-drawing

волоч·и́льн·ый, -ая, -ое
~ **доск·а́** (*f*) (met) draw-plate
~ **стан·о́к·** (*m*) (met) draw bench
волоч·и́ть, -ат (*imp*) drag, draw, pull along, trail; (met) draw (wire)
волоч·ь, (*pres 3rd sing, pl*) **волоч·ёт, волок·у́т,** (*past masc sing*) **воло́к·,** (*imp*) drag, draw, pull along, trail
воло́ш·ск·ий оре́х· (*m*) walnut
волча́нк·а, -и, (*f*) (med) lupus
волч·е́ц·, -ч·ц·а́, (*i*) **-ч·ц·о́м,** (*m*) (bot) thistle
во́лч·ий, -ья, -ье, (*a*) wolf, lupine
во́лч·ников·ые, ых, (*pl decl as adj*) (bot) *Thymelaeaceae*
волч·о́к·, -ч·к·а́, (*m*) spinning top, top, gyroscope; (bot) sucker; (food) meat grinding machine, mincing machine; (met) crucible furnace
волч·о́нок·, -нк·а, (*nom pl*) **-ч·а́т·а,** (*g pl*) **-ч·а́т·,** (*m*) wolf cub
во́ль·нич·а·ть, -ют, (*imp*) take liberties
во́ль·н·о (*adv*) freely; (mil) at ease
вольно·на·ём·н·ый (*m decl as adj*) (mil) civilian; civilian worker (not under military or penal restraint)
~ **труд·** (*m*) hired labour
вольно·о·пре·дел·я́ющ·ийся, -егося, (*m decl as adj*) volunteer
вольно·слу́ш·ател·ь, -я, (*m*) un-attached/external student
во́ль·ност·ь, -и, (*f*) liberty, familiarity; licence (in conduct)
во́ль·н·ый, -ая, -ое, (*a*) (*s f*) **во́л·ен·, воль·н·а́, во́ль·н·о** free, unrestric-ted; impudent, familiar
~ **порт·** (*m*) free port
~ **райо́н·** (*m*) free zone
~ **с·клад·** (*m*) bonded warehouse
во́льт·, -а, (*g pl*) **во́льт·,** (*m*) (elec) volt, V
~ **секу́нд·а, -ы,** (*f*) volt-line (unit of magnetic flux = 10^3 maxwells)
вольт·а́ж·, -а, (*i*) **-ем,** (*m*) voltage
вольт·а́метр·, -а, (*m*) (elec) volta-meter
вольт·амметр·, -а, (*m*) (elec) volt-ammeter, voltmeter-ammeter
вольт·ампе́р·, -а, (*m*) volt-ampere
вольт·ампер·ме́тр·, -а, (*m*) volt-ammeter, voltmeter-ammeter
Вольт·а, потенциа́л· (*m*) (elec) Volt effect
вольт·ме́тр·, -а, (*m*) (elec) voltmeter
~, **амплиту́д·н·ый** peak/crest volt-meter

вольт·ме́тр
~, **генер·и́рующ·ий** generating volt-meter
~, **като́д·н·ый** electronic voltmeter, vacuum-tube voltmeter (U.S.), valve voltmeter
~, **компенсацио́н·н·ый** slide-back voltmeter
~, **коро́н·н·ый** corona voltmeter
~, **ла́мп·ов·ый** electronic voltmeter thermionic/valve voltmeter (Brit), va-cuum-tube voltmeter, VTVM (U.S.)
~, **много·ка́мер·н·ый** multi-cellular voltmeter
~, **пи́к·ов·ый** peak-load voltmeter
~, **ро́тор·н·ый** rotary voltmeter
~, **сере́бр·ян·ый** (elec) silver volta-meter
во́льт·ов·, -·ов·а, -·ов·о, (*possess adj*) voltaic
~ **столб·** (*m*) voltaic pile
~ **дуг·а́** (*f*) electric arc
вольто·до·ба́в·очн·ая маши́н·а (*f*) (elec) booster, positive booster
вольто́л·, -а, (*m*) voltol, voltalized oil
вольто·по·ниж·а́ющ·ая маши́н·а (*f*) (elec) negative booster
Во́льф·а, числ·о́ (*n*) Wolf sunspot number
Во́льф·ов·о те́л·о (*n*) (zool) Wolffian body
вольфра́м·, -а, (*m*) tungsten; wolfram (of ore)
~, **шести·хло́р·ист·ый** tungsten hexa-chloride
вольфрам·а́т·, -а, (*m*) (chem) tung-state
вольфрам·и́ровани·е, -я, (*n*) tungsten coating
вольфрам·и́т·, -а, (*m*) (min) wolf-ramite
вольфрамово·ки́сл·ый, -ая, -ое, (*a*) tungstate of
вольфра́м·ов·ый, -ая, -ое, (*a*) tung-sten; (chem) tungstic; wolfram (of ore)
~ **ка́мен·ь** (*m*) (min) scheelite
~ **кисл·от·а́** (*f*) tungstic acid
~ **ла́мп·очк·а** (*f*) (elec) tungsten lamp
~ **руд·а́** (*f*) wolfram, wolfram ore
~ **стал·ь** (*f*) tungsten steel
вольфсберги́т·, -а, (*m*) (min) wolfs-bergite, chalcostebite
волюме́тр·, -а, (*m*) volumeter
волюметр·и́ческ·ий, -ая, -ое, (*a*) volumetric
волюмино́·метр·, -а, (*m*) volumeter

волюмо·ме́тр·, -а,   (*m*) volumeter

волю́т·а, -ы,   (*f*) (arch) volute

волюти́н·, -а,   (*m*) (chem) volutine

во́л·я, -и,   (*f*) will; liberty

во·мн·у́т   (*fut 3rd pl*) of в·мя́·ть

вон   (*adv*) out, away; there, over there

вонз·а́·ть, -ют,   (*imp*) stick in, spike, pierce

вонз·и́ть, -ят,   (*perf*) *see* вонз·а́·ть

во́н·ь, -и,   (*f*) stink, stench

вон·ю́ч·ий, -ая, -ее,   (*a*) stinking, fetid

~ каме́д·ь   (*f*) (bot) asafetida, *Ferula asafetida*

вон·я́·ть, -ют,   (*imp*) stink, smell foul/fetid

во·ображ·а́ем·ый, -ая, -ое,   (*a*) imaginary, fictitious

во·ображ·е́ни·е, -я,   (*n*) imagination

во·ображ·а́·ть, -ют,   (*imp*) imagine, fancy; -ся (*pass*) seem

во·ображ·ённ·ый, -ая, -ое,   (*past part pass*) *see* во·ображ·а́·ть; imaginary

во·образ·и́м·ый, -ая, -ое,   (*a*) imaginable

во·образ·и́ть, -я́т,   (*perf*) *see* во·ображ·а́·ть

во·общ·е́   (*adv*) in general, generally, on the whole; always

~ не   not generally; not ... at all

во·о·душ·ев·и́ть, -я́т,   (*perf*) *see* во·о·душ·евля́·ть

во·о·душ·евля́·ть, -ют,   (*imp*) inspire, inspirit, encourage

во·оруж·а́·ть, -ют,   (*imp*) arm; equip, fit out, outfit

во·оруж·е́ни·е, -я,   (*n*) (*v n*) *see* во·оруж·а́·ть; armament, arms; equipment; rig (of sails)

~, авиацио́н·н·ое   aircraft armament

~, кос·о́е   (naut) fore-and-aft rig

во·оруж·ённ·ый, -ая, -ое,   (*past part pass*) *see* во·оруж·а́·ть armed; equipped, fitted out

~ глаз·   (*m*) aided eye

во·оруж·и́ть, -а́т,   (*perf*) *see* во·оруж·а́·ть

во·пе́рв·ых   (*conj*) firstly, in the first place

воп·и́ть, -я́т,   (*imp*) howl, wail

воп·ию́щ·ий, -ая, -ое, (*a*) scandalous, crying

во·плот·и́ть, -я́т,   (*perf*) *see* во·площ·а́·ть

во·площ·а́·ть, -ют,   (*imp*) embody, personify

во́пл·ь, -я,   (*m*) howl, wail, cry

вопреки́   (*adv*) in spite of, despite, notwithstanding

во·про́с·, -а,   (*m*) question

во·прос·и́тельн·ый, -ая, -ое,   (*a*) interrogative, enquiring, questioning

~ знак·   (*m*) question mark; (telecom) interrogative

во·про́с·ник·, -а,   (*m*) questionnaire; question paper

во́р·, -а,   (*g pl*) -о́в, (*m*) thief

во́рван·ь, -и,   (*f*) blubber-oil

воробьеви́т·, -а,   (*m*) (min) vorobyerite

вороб·е́·й, -б·ь·я́,   (*m*) (orn) sparrow, *Passer*

вороб·ьи́н·ый, -ая, -ое,   (*a*) (orn) sparrow, passerine

вор·ова́ть, -у́ют,   (*imp*) steal

вор·овск·о́й, -а́я, -о́е,   (*a*) thieves, thievish

во́рон·, -а,   (*m*) (orn) raven, *Corvus corax*

воро́н·а, -ы,   (*f*) (orn) crow

ворон·е́ни·е, -я,   (*n*) black shine; blueing, oxide coating (of steels)

ворон·ён·ый, -ая, -ое,   (*a*) burnished/polished/shining black; blued (of steels)

воро́н·ий, -ья, -ье,   (*a*) (orn) crow's, corvine

~ гнезд·о́   (*n*) (naut) crow's nest

воро́нк·а, -и,   (*g pl*) -нок·, (*f*) funnel (conical), cone; crater, shell hole; eddy; (geol) cone, conical depression, sink; (geog) sink hole

~, вихр·ев·а́я   vortex

~, за·гру́з·очн·ая   feed hopper

~, ка́бель·н·ая   (elec) cable pothead

~, конц·ев·а́я   (elec) cone-type terminal box

~, ли́т·ник·ов·ая   (met cast) runner basin

~, мёртв·ая   (rad) dead space

~, рас·пре·дел·и́тельн·ая   (met) tundish

~ с рас·тру́б·ом к·верх·у́   upright funnel

воронко·обра́з·н·ый, -ая, -ое,   (*a*) funnel/trumpet shaped

ворон·о́й, -а́я, -о́е,   (*a*) black

воро́н·ье   (*neut*) of воро́н·ий

воро́н·ья   (*fem*) of воро́н·ий

во́рот·, -а,   (*m*) collar; windlass, winch

воро́т·а   (*nom pl*), (*g pl*) воро́т·, (*n*) gate, gates; arch; goal (football)

~, време́н·н·ые   (elec) time gate

~, дву·ство́р·чат·ый   double gate(s); double-leaf gate

воро́т·а
~, **есте́ств·енн·ые** (geol) natural arch
~, **за·движ·н·ы́е** sliding/traversing caisson (dry dock)
~, **низ·ов·а́я** (civ eng) tail gate
~, **одно·ство́р·чат·ые** single-leaf gate(s), single gate(s)
~, **от·ка́т·н·ые** box gate (dry dock)
~, **ство́р·чат·ые** leaf-type gate (of docks and locks)
~, **шлюз·н·ые** lock gate
ворот·ни́к·, -а́, (m) collar, (biol) frill
ворот·ничо́к·, -чка́, (m) collar
воро́т·н·ый, -ая, -ое, (a) of **воро́т·а;** (anat) portal; **ворот·н·ый, -ая, -ое,** (a) of **во́рот·**
ворот·о́к·, -т·к·а́, (m) tap wrench, die stock; shoulder (leather)
во́рох·, -а, (nom pl) **-а́,** (g pl) **-о́в,** (m) pile, heap
воро́ч·а·ть, -ют, (imp) shift, manoeuvre (something heavy); control, direct; **-ся** rotate, turn over and over; move, stir
ворош·и́лк·а, -и, (g pl) **-лок·,** (f) agitator, stirrer; swath turner (hay-making); peat turner
ворош·и́ть, -а́т, (imp) stir, turn (hay, peat); **-ся** move about
во́рс·, -а, (m) (text) nap, pile
во́рс·а, -ы, (f) junk (cordage)
ворс·и́льн·ый, -ая, -ое, (a) see **ворс·ова́льн·ый**
ворс·и́льщик·, -а, (m) teaseler
ворс·и́нк·а, -и, (g pl) **-нок·,** (f) hair, nap, lint; fibre, fiber; (biol) villus
ворс·и́нчатост·ь, -и, (f) (biol) villosity
ворс·и́нчат·ый, -ая, -ое, (a) fleecy, lanate, villous, villose
~ **оболо́ч·к·а** (f) (zool) chorion
ворс·и́ст·ый, -ая, -ое, (a) fleecy, lanate, pubescent, villous, villose
ворс·и́т·, -а, (m) shoe quality velvet
ворс·и́ть, -ят, (imp) = **ворс·ова́ть**
ворс·ова́льн·ый, -ая, -ое, (a) teaseling, napping
~ **маши́н·а** (f) (text) napping and friezing machine
~ **ши́ш·к·а** (f) teasel
ворс·ова́ни·е, -я, (n) (text) napping, teaseling
ворс·ова́ть, -у́ют, (imp) (text) tease, raise a nap
ворс·ов·о́й, -а́я, -о́е, (a) of **ворс·**
~ **тк·ан·ь** (f) pile cloth

во́рс·ов·ый, -ая, -ое, (a) see **ворс·ов·о́й**
ворс·я́нк·а, -и, (g pl) **-нок·,** (f) (bot) teasel, *Dipsacus fullonum*
ворч·а́ть, -а́т, (imp) grumble, growl
вос- (prefix) = **воз-** *particular significances*: (1) movement upwards; (2) renewed action; (3) action in response; (4) intensity, solemnity
восем·на́·дцат·ь, -и, (f) eighteen
во́сем·ь, (g) **восьм·и́,** (i) **восьм·ью́** eight
во́семь·десят·, (g) **восьми́·десят·и,** (i) **восьмью́·десят·ью** eighty
восемь·со́т·, (g) **восьми·со́т·,** (i) **восьмью́·ст·а́ми,** (p) **восьми·ст·а́х** eight hundred
во́сем·ью (adv) eight times
во́ск·, -а, (m) wax; bees-wax
~, **го́р·н·ый** (min) ozokerite, mineral wax
~, **земл·ян·о́й** mineral wax
вос·кли́к·н·уть, -ут, (perf) see **вос·-клиц·а́·ть**
вос·клиц·а́ни·е, -я, (n) exclamation
вос·клиц·а́тельн·ый, -ая, -ое, (a) exclamatory, exclamation
~ **знак·** (m) exclamation mark
вос·клиц·а́·ть, -ют, (imp) exclaim, cry out, cry
воско·бо́·й, -я, (m) wax-refiner
воск·ови́н·а, -ы, (f) (zool) cere, ceroma
воск·ови́нн·ый, -ая, -ое, (a) cerous
воск·о́вк·а, -и, (g pl) **-вок·,** (f) waxed tracing paper
воск·ов·о́й, -а́я, -о́е, (a) wax, waxen, waxy, ceraceous
~ **желез·а́** (f) (ent) wax pocket
воско·но́с·н·ый, -ая, -ое, (a) (bot) ceriferous
воско·обра́з·н·ый, -ая, -ое, (a) wax-like, waxy; (bot) ceraceous
воско·то́п·к·а, -и, (g pl) **-п·ок·,** (f) wax-refinery
вос·крес·е́ни·е, -я, (n) resurrection; revival
вос·крес·е́нь·е, -я, (n) Sunday
вос·пал·е́ни·е, -я, (n) inflammation; (med) -itis
~ **киш·е́чник·а** enteritis
~ **лёг·к·их** pneumonia
~ **плёвр·ы** pleurisy
~ **со·су́д·ов** vasculitis
вос·пал·ённ·ый, -ая, -ое, (a) inflamed
вос·пал·и́тельн·ый, -ая, -ое, (a) inflammatory

вос·пит·а́ни·е, -я,   (*n*) education, training, upbringing, rearing

вос·пит·а́тельн·ый, -ая, -ое,   (*a*) educational, educative

вос·пит·а́·ть, -ют,   (*perf*) *see* **вос·пи́·т·ыва·ть**

вос·пи́т·ыва·ть, -ют,   (*imp*) educate, train, bring up, rear

вос·пламен·е́ни·е, -я,   (*n*) (*v n*) *see* **вос·пламен·я́·ть** ignition; setting alight; catching alight

~, за·медл·ённ·ое   delayed ignition

~ за·ря́д·а, за·тяж·н·о́е   (gunn) hang fire

~ от с·жа́т·и·я   compression ignition

~, перёд·н·ее   (rocket) head-end ignition

температу́р·а вос·пламен·е́ни·я (*f*) fire point (of oil)

вос·пламен·и́тел·ь, -я,   (*m*) (expl) igniter, igniting charge

вос·пламен·и́тельн·ый, -ая, -ое,   (*a*) igniting

~ голо́в·к·а   (*f*) match head (of detonator)

~ со·ста́в·   (*m*) (expl) flash composition

~ тру́б·к·а   (*f*) (gunn) vent tube

вос·пламен·и́ть(ся), -я́т(ся),   (*perf*) *see* **вос·пламен·я́·ть(ся)**

вос·пламен·я́емост·ь, -и,   (*f*) inflammability, ignitability

вос·пламен·я́ем·ый, -ая, -ое,   (*a*) inflammable

вос·пламен·я́·ть, -ют,   (*imp*) set on fire, ignite; **-ся** catch fire, ignite, flare/light up

вос·пламен·я́ющ·ий, -ая, -ее, (*pres part act*) of **вос·пламен·я́·ть**; (expl) initiating

~ капсю́ль·н·ый со·ста́в·   (expl) initiator

вос·пламен·я́ющ·ийся, -аяся, -ееся, (*pres part act*) of **вос·пламен·я́·ться** inflammable, combustible

вос·по́лн·ить, -ят, (*perf*) *see* **вос·полн·я́·ть**

вос·полн·я́·ть, -ют, (*imp*) make up/good, meet, fill (a deficiency)

вос·по́льз·оваться, -уются, (*perf*+*i*) use, make use of, take advantage of

вос·по·мин·а́ни·е, -я,   (*n*) recollection, memory, reminiscence

вос·препя́тств·овать, -уют, (*perf*+ *dat*) prevent, hinder

вос·прет·и́ть, -ят, (*perf*) *see* **вос·пре·щ·а́·ть**

вос·прещ·а́·ть, -ют,   (*imp* + *dat*) prohibit, forbid

вос·при·и́м·чивост·ь, -и,   (*f*) receptivity, receptance, susceptibility; (physiol) perceptivity

~ диа·магни́т·н·ая   diamagnetic susceptibility

~ объ·ём·н·ая   volume susceptibility

вос·при·ним·а́ем·ый, -ая, -ое,   (*a*) perceptible, apprehensible

вос·при·ним·а́·ть, -ют,   (*imp*) take in/up, receive; perceive, apprehend, learn

вос·при·ним·а́ющ·ий, -ая, -ее, (*pres part act*) of **вос·при·ним·а́·ть** perceptive; (*i*) sensitive, sensing

вос·при·ня́·ть   (*perf*) (*fut 3rd sing, pl*) **вос·при́·м·ет, вос·при·м·у́т;** *see* **вос·при·ним·а́·ть**

вос·при·я́т·и·е, -я,   (*n*) (physiol) perception, sensation, impression

~, зр·и́тельн·ое   visual perception

~ цвет·о́в, зр·и́тельн·ое   chromatic sensation

вос·про·из·вед·е́ни·е, -я,   (*n*) reproduction

~ звук·а   sound reproduction, playback

вос·про·из·вед·ённ·ый, -ая, -ое, (*past part pass*) reproduced; (nucl) bred

вос·про·из·вес·ти́, -·вед·у́т, *see* **вос··про·из·вод·и́ть**

вос·про·из·вод·и́мост·ь, -и,   (*f*) reproducibility

вос·про·из·вод·и́м·ый, -ая, -ое, (*pres part pass*) *see* **вос·про·из·вод·и́ть**; reproducible

вос·про·из·вод·и́тельн·ый, -ая, -ое, (*a*) reproductive

вос·про·из·вод·и́ть, -ят, (*imp*) reproduce; breed (nuclear fuel)

вос·про·из·во́д·ств·о, -а,   (*n*) reproduction; conversion, breeding (of nuclear fuel)

зо́на вос·про·из·во́д·ства (*f*) (nucl) blanket region, breeding blanket, blanket

коэффицие́нт· вос·про·из·во́д·ств·а (*m*) (nucl) regeneration/reproduction factor conversion/breeding ratio

~ на бы́стр·ых нейтро́н·ах   (nucl) fast conversion

~ на тепл·ов·ы́х нейтро́н·ах   (nucl) thermal conversion

~ нейтро́н·ов   (nucl) neutron breeding

~ рас·ши́р·енн·ое   (nucl) breeding

вос·про·из·во́д·ств·о
~ сумма́р·н·ое   (nucl) complete breeding
вос·про·из·вод·я́щ·ий, -ая, -ое, (*pres part act*) of вос·про·из·вод·и́ть; reproducing, breeding, fertile
~ перфора́тор·   (*m*) reproducing punch (for punched cards)
~ сред·а́   (*f*) (nucl) fertile medium
вос·проти́в·иться, -ятся,   (*perf*) oppose, object to, resist
вос·со·един·е́ни·е, -я,   (*n*) reunification (of which speaker/writer approves); recombing (of ions)
вос·со·зда·ва́ть, -ю́т, (*imp*) re-create, reconstruct
вос·со·зда́·ть   (*perf*) (*fut 3rd sing, pl*) вос·со·зда́ст, вос·со·здаду́т; *see* вос·со·зда·ва́ть
вос·ста·ва́ть, -ю́т,   (*imp*) rise, rise up; revolt, rebel
вос·ста́в·ить, -ят,   (*perf*) *see* вос·ставл·я́·ть
вос·ставл·я́·ть, -ю́т,   (*imp*) erect
вос·стана́вл·иваемост·ь, -и,   (*f*) restorability, reconstructibility; (chem) reducibility
вос·стана́вл·иваем·ый, -ая, -ое, (*pres part pass*) of вос·стана́вл·ива·ть
вос·стана́вл·ива·ть, -ют,   (*imp*) restore, re-establish, rehabilitate, reconstruct, regenerate; recall, recollect; (math) erect, draw (a perpendicular); (chem) reduce, deoxidize
вос·стана́вл·иваю́щ·ий, -ая, -ее, (*pres part act*) of вос·стана́вл·ива·ть
~ моме́нт·   (*m*) (mech) righting moment
~ си́л·а   (*f*) restoring force
~ спосо́б·ност·ь   (*f*) recovery characteristics; (chem) reducing power
вос·ста́·ни·е, -я,   (*n*) revolt, uprising, insurrection
~ пласт·а́   (geol) rise of seam
по в ос·ста́·ни·ю   (*adv*) (min) up, upgrade, upwards
вос·станов·и́м·ый, -ая, -ое,   (*a*) (chem) reducible
вос·станов·и́тел·ь, -я,   (*m*) restorer; (chem) reducing agent, reducer; (rad) regenerator
вос·станов·и́тельн·ый, -ая, -ое,   (*a*) reducing, reduction; restoring
~ спосо́б·ност·ь   (*f*) (chem) reducing power
вос·станов·и́ть, -ят,   (*perf*) *see* во·с·стана́вл·ива·ть

вос·становл·е́ни·е, -я,   (*n*) restoration, re-establishment, reconstruction, restitution; recovery, reclamation, regeneration; (chem) reduction; (math) erection
~ водо·ро́д·ом   hydrogen reduction
вре́м·я вос·становл·е́ни·я   (*f*) recovery time
~ давл·е́ни·я   pressure recovery
~, като́д·н·ое   cathodic reduction
коэффицие́нт· вос·становл·е́ни·я (*m*) (mech) coefficient of restitution
крива́я вос·становл·е́ни·я давл·е́ния   (*f*) (oil) pressure build-up curve
~ сигна́л·а   (rad) signal restoration
~ тепл·а́   heat recovery
~, у·пру́г·ое   elastic recovery (text) crease resistance
вос·стано́вл·енн·ый, -ая, -ое, (*past part pass*) *see* вос·стана́вл·ива·ть; restored, recovered; (chem) reduced
~ руд·а́   (*f*) reduced ore
~ ши́н·а   (*f*) (M/T) retreaded tire
вос·становл·я́·ть, -ют,   (*imp*) = вос·-стана́вл·ива·ть
вос·ста́·ть, -нут,   (*perf*) *see* вос·ста·-ва́ть
вос·ста·ю́щ·ий, -ая, -ее,   (*pres part act*) of вос·ста·ва́ть; (min) uphill, up; (*m as adj*) raise, rise
~ вы·рабо́т·к·а   (*f*) (min) upset
вос·то́к·, -а,   (*m*) last
востоко·ве́д·, -а,   (*m*) orientalist
востоко·вед·е́ни·е, -я,   (*n*) oriental studies
вос·то́рг·, -а,   (*m*) delight, enthusiasm
вос·то́рж·енн·ый, -ая, -ое,   (*a*) enthusiastic
вос·торж·еств·ова́ть, -у́ют,   (*perf*) triumph over
вос·то́ч·н·ый, -ая, -ое,   (*a*) east, eastern, oriental
вос·тре́б·овани·е, -я,   (*n*) (*v n*) *see* вос·тре́б·овать; claim, demand
до вос·тре́б·овани·я   poste restante (Brit), general delivery (U.S.)
вос·тре́б·овать, -уют,   (*perf*) claim
вос·хвал·е́ни·е, -я,   (*n*) eulogy, praise
вос·хит·и́тельн·ый, -ая, -ое,   (*a*) delightful, exquisite, delicious
вос·хищ·а́·ться, -ются,   (*imp + i*) admire, be enraptured about
вос·хищ·е́ни·е, -я,   (*n*) admiration, delight
вос·хо́д·, -а,   (*m*) rising, rise, ascent
~ со́лн·ц·а   sunrise

вос·ход·и́ть, -́ят, (*imp*) rise, ascend; originate in, come from, arise

вос·ход·я́щ·ий, -ая, -ее, (*pres part act*) of **восход·и́ть;** ascending, rising; upward; (*m decl as adj*) (min) upcast, uptake

~ **бо́ч·к·а** (*f*) (air) upward roll

~ **скольж·е́ни·е** (*n*) (mech) climbing slip; upslide motion

~ **ток во́з·дух·а** (*m*) (air) thermal

вос·хожд·е́ни·е, -я, (*n*) ascent

~, прям·о́е (navig) right ascension

восьм·ери́к·, -а́, (*m*) an eight (containing eight); eight-stranded rope; packet of eight

восьм·ери́чн·ый, -ая, -ое, (*a*) (math) octonary

~ **с·числ·е́ни·е** (*n*) (*in gen as adj*) octal (computers)

восьм·ёрк·а, -и, (*g pl*) **-рок,** (*f*) an eight, the eight, a number eight, figure of eight; S-hook; two-lobe rotary pump

восьм·ерни́к·, -а́, (*m*) eightling, eightfold twin

во́сьм·ер·о, -ы́х, eight, a set of eight

восьм·ёрочн·ый, -ая, -ое, (*a*) figure-of-eight

~ **на·со́с·** (*m*) two-lobe rotary pump

восьми- (*root*) eight, octo-, octa-

восьми·вале́нт·н·ый, -ая, -ое·, (*a*) octavalent

восьми·гра́н·ник·, -а, (*m*) octahedron

восьми·гра́н·н·ый, -ая, -ое, (*a*) octahedral, eight-sided

восьми·десяти- (*component*) eighty-

восьми·деся́т·ый, -ая, -ое, (*a*) eightieth

восьми·кра́т·н·ый, -ая, -ое, (*a*) eightfold, octuple

восьми·ле́т·н·ый, -ая, -ое, (*a*) octennial, eight-year-, eight-year-old

восьми·но́г·, -а, (*m*) (zool) octopus

восьми·со́т·ый, -ая, -ое, (*a*) eight-hundredth

восьми·уго́ль·ник·, -а, (*m*) octagon

восьми·уго́ль·н·ый, -ая, -ое, (*a*) octagonal

восьми·час·ов·о́й, -а́я, -о́е, (*a*) eight-hour; eight-o'clock

восьм·о́й, -а́я, -о́е, (*a*) eighth

восьм·у́шк·а, -и, (*g pl*) **-шек·,** (*f*) (paper) octavo

вот (*particle*) there, there is, here (indicating)

~ **и всё** that is all

вот·и́р·овать, -уют, (*imp and perf*) vote, pass a vote

во·тк·а́ть, -у́т, (*perf*) interweave

во·тк·н·у́ть, -у́т, (*perf*) *see* **в·тык·а́·ть**

во·тр·у́т (*fut 3rd pl*) of **в·тер·е́ть**

вофати́т·, -а, (*m*) wolfatit, wolfatite (water softeners)

вош·ь, (*g*) **вши,** (*i*) **во́шью,** (*nom pl*) **вши,** (*g pl*) **вшей,** (*f*) (ent) louse, *Anoplura*

~ **ка́рп·ов·ая** fish louse, *Branchima*

во·шь·ю́т (*fut 3rd pl*) of **в·ши·ть**

вощ·а́нк·а, -и, (*g pl*) **-нок,** (*f*) wax-paper

вощ·е́ни·е, -я, (*n*) waxing, putting on wax

вощ·ён·ый, -ая, -ое, (*a*) waxed

~ **бума́г·а** (*f*) waxed paper

вощ·и́ть, -а́т, (*imp*) wax; polish with wax

во́·ющ·ий, -ая, -ее, (*pres part act*) of **вы·ть;** howling

вою́·ющ·ий, -ая, -ее, (*pres part act*) of **во·ева́ть;** belligerent; (*m decl as adj*) belligerent

во́·я (*pres gerund*) howling

в·пад·а́·ть, -ют, (*imp*) fall/flow/discharge into

в·пад·е́ни·е, -я, (*n*) (*v n*) *see* **в·пад·а́·ть;** inflow; confluence, mouth, issue (river)

в·па́д·ин·а, -ы, (*f*) indentation, hollow, cavity; socket; (geog) depression basin; (ocean) deep, trough; valley (of a curve)

~ **зу́б·а** tooth space (gears)

~, ка́рст·ов·ая (geol) karst cavity

~, кра·ев·а́я (ocean) foredeep

~, перед·ов·а́я (ocean) foredeep

~, тектон·и́ческ·ая (geol) tectonic basin

в·па́·ивани·е, -я, (*n*) (*v n*) *see* **впа́·ива·ть**

в·па́·ива·ть, -ют, (*imp*) solder in; (elec) seal in

в·па·й, -я, (*m*) soldering in, part soldered in; (elec) seal in, press lead

в·па́·йк·а, -и, (*g pl*) **в·па́·ек·,** (*f*) sealing/soldering in; sealed/soldered in part

в·па́л·ост·ь -и, (*f*) hollowness, concavity

в·па́л·ый, -ая, -ое, (*a*) hollow, sunken, depressed, concave

в·паралле́л·ь (*adv*) in parallel, parallel

в·пас·ть (*fut 3rd sing, pl*) **в·пад·ёт, в·пад·у́т,** (*past masc sing*) **в·па́·л,** (*perf*); *see* **в·пад·а́·ть**

впа·я́·ть, -ют, (*perf*) solder in, seal in

в·перв·ы́е (*adv*) for the first time, the first

в·пере·вя́з·к·у (*adv*) broken, staggered

    с·ты́к·и в·пере·вя́з·к·у (*m pl*) break/broken joints

в·перёд· (*adv*) forward, ahead; fast (of clocks); in future, henceforth; (math) progressive, forward

~, ма́л·ый (naut) slow speed ahead

~, по́лн·ый (naut) full speed ahead

~, сре́д·н·ий (naut) half speed ahead

в·перед·и́ (*adv*) (*prep + gen*) in front of, ahead of, before; in future

в·перед·смотр·я́щ·ий, -его, (*m decl as adj*) (naut) lookout, masthead lookout

в·пере·кро́·й (*adv*) lap, lapped

в·пере·кры́ш·к·у (*adv*) lap, lapped

в·пере·ме́ж·к·у (*adv*) alternately, in alternating/staggered order, staggered

в·пере·ме́ш·к·у (*adv*) helter-skelter, in a state of disorder

в·печа́т·анн·ый, -ая, -ое, (*past part pass*) overprinted, printed over/in

в·печат·ле́ни·е, -я, (*n*) impression, imprint, cast

в·печат·ля́ющ·ий, -ая, -ее, (*pres part act*) impressive

в·печа́т·ывани·е, -я, (*n*) (phot) printing in; (print) overprinting

в·пив·а́·ть, -ют, (*imp*) absorb, imbibe; **-ся** pierce, stick into

в·пи́с·анн·ый, -ая, -ое, (*past part pass*) *see* **в·пи́с·ыва·ть;** entered, written in; (math) inscribed

~ круг· (*m*) (math) encircle, inscribed circle

~ угол (*m*) (math) inscribed angle

в·пис·а́ть (*fut 3rd sing, pl*) **в·пиш·ет, в·пи́ш·ут,** (*perf*); *see* **в·пи́с·ыва·ть**

в·пи́с·ыва·ть, -ют, (*imp*) write in, enter; (math) inscribe

в·пи́т·анн·ый, -ая, -ое, (*past part pass*) *see* **в·пи́т·ыва·ть;** absorbed, taken up/in

в·пит·а́·ть, -ют, (*perf*) *see* **в·пи́т·ыва·ть**

в·пи́т·ываемост·ь, -и, (*f*) (print) receptivity, ink receptivity penetration

в·пи́т·ывани·е, -я, (*n*) absorption, seepage, imbibition, soaking

в·пи́т·ыва·ть, -ют, drink in, imbibe, absorb, take up; **-ся** (*passive*) soak in

в·пи́·ться (*fut 3rd sing, pl*) **во·пь·ёт·ся, во·пь·ю́тся,** (*perf*); *see* **в·пи·ва́·ться**

в·пи́х·ива·ть, -ют, (*imp*) push/cram/stuff in

в·пих·н·у́ть, -у́т, (*perf*) *see* **в·пи́х·ива·ть**

в·пла́в·ить, -ят, (*perf*) *see* **в·плавл·я́·ть**

в·плавл·я́·ть, -ют, (*imp*) melt/fuse in/into

в·плав·ь (*adv*) swimming, by swimming

в·пласт·о́ванн·ый, -ая, -ое, (*past part pass*) interbedded, intercalated, intraformational

в·плес·ти́, -ет·у́т, (*perf*) *see* **в·плет·а́·ть**

в·плет·а́ни·е, -я, (*n*) interweaving; implicating, implication

в·плет·а́·ть, -ют, (*imp*) intertwine, interlace, plait; implicate, involve

в·плет·ённ·ый, -ая, -ое, (*past part pass*) *see* **в·плет·а́·ть**

~ структу́р·а (*f*) (geol) interwoven structure

в·плот·н·у́ю (*adv*) close, closely, against, up against

в·плот·ь до (*adv*) up to, right up to, down to

в·плыв·а́·ть, -ют, (*imp*) swim/float in

в·плы·ть, -ыв·у́т, (*perf*) *see* **в·плы·ва́·ть**

в·по·ва́л·к·у (*adv*) side by side, in a row

в·пол·го́лос·а (*adv*) in an undertone, quietly

в·пол·дре́в·а (*adv*)

    в·ру́б·к·а в·пол·дре́в·а (*f*) cross-halved joint (timber), lipped joint

    со·един·е́ни·е в·пол·дре́в·а (*n*) halving (carpentry joint)

в·полз·а́·ть, -ют, (*imp*) crawl in/into

в·полз·ти́, -ут, (*perf*) *see* **в·полз·а́·ть**

в·полн·е́ (*adv*) fully, entirely, completely, quite

в·пол·оборо́т·а (*adv*) half-turned, halfway, half-way round, through/by 180°

в·пол·овин·у (*adv*) half, by half

в·полу·на·хлёст·к·у (*adv*) half-lap, scarf (of joints)

в·полу·по·та́й (*adv*) half-countersunk (of screws etc.)

в·по·сле́д·н·ие (*adv*) for the last time

в·по·сле́д·стви·и (*adv*) later, later on, subsequently, afterwards

в·по·та́й (*adv*) flush, level

в·пра́в·е (*adv*) justified, right, in the right; on/to the right в·пра́в·ить, -ят, (*perf*) *see* в·правл·я́·ть

в·пра́в·к·а, -и, (*g pl*) -в·ок·, (*f*) setting, resetting

в·правл·я́·ть, -ют, (*imp*) (med) set, reduce; (tech) set, insert

в·пра́в·о (*adv*) to the right, to starboard; clockwise

~ вращ·а́ющ·ийся dextrorotatory

в·пред·ь (*adv*) henceforth, in the future, from now on

~ до pending, until

в·пресс·о́ванн·ый, -ая, -ое, (*past part pass*) pressed in

в·при·ты́к· (*adv*) flush, against, end-to-end; butt

в·прок· (*adv*) for preservation/store/keeping

в·про́ч·ем· (*conj*) though, all the same, incidentally

в·пры́г·ива·ть, -ют, (*imp*) jump in/into

в·пры́г·н·уть, -ут, (*perf*) *see* в·пры́г·ива·ть

в·прыск·, -а, (*m*) injection, shot

~, на·пра́вл·енн·ый vectored injection

~, не·по·сре́д·ственн·ый direct injection

~, противо·то́ч·н·ый contra-injection, contra-stream injection

~ с за·вихр·е́ни·ем swirl injection

в·пры́ск·ивани·е, -я, (*n*) injection

в·пры́ск·иватель·, -я, (*m*) injector

в·пры́ск·ива·ть, -ют, (*imp*) inject, squirt in, spray in

в·пры́с·н·уть, -ут, (*perf*) *see* в·пры́с·к·ива·ть

в·пуск·, -а, (*m*) admittance, admission; intake, inlet

~ га́з·а gas injection/intake

~, одно·сторо́н·н·ий single admission

в·пуск·а́·ть, -ют, (*imp*) let in, take in, admit

в·пуск·н·о́й, -а́я, -о́е, (*a*) entrance; intake, inlet

~ кла́пан· (*m*) intake valve, inlet valve

~ кон·е́ц· (*m*) lead-in, feed end

в·пуст·и́ть, -ят, (*perf*) *see* в·пуск·а́·ть

в·пуст·у́ю (*adv*) to no purpose, for nothing

в·пу́т·анн·ый, -ая, -ое, (*past part pass*) of в·пу́т·ить, involved, implicated

в·пу́т·а·ть, -ют, (*imp*) *see* в·пу́т·ыва·ть

в·пу́т·ыва·ть, -ют, (*imp*) entangle, enmesh; involve, implicate

в·пу́щ·енн·ый, -ая, -ое, (*past part pass*) *see* в·пуск·а́·ть; admitted; injected

в·пя́т·ер·о (*adv*) five times

в·пят·ер·о́м (*adv*) five together

в·пя́т·ых (*adv*) fifthly

враг·, -а́, (*m*) enemy, foe, opponent

вражд·е́бн·ый, -ая, -ое, (*a*) hostile, inimical, enemy

в·раз·бе́ж·к·у (*adv*) in broken order, staggered

с·ты́к·и в·раз·бе́ж·к·у (*m pl*) break/broken joints

в·раз·би́в·к·у (*adv*) at random, haphazardly

в·раз·бро́д· (*adv*) separately, disunitedly

в·раз·бро́с· (*adv*) random, randomly, haphazard

в·раз·ре́з· (*adv*) across

в·раз·ум·и́тельн·ый, -ая, -ое, (*a*) intelligible, comprehensible

в·разъ·ём· (*adv*) *see* разъ·ём·

шит·ь·ё в·разъ·ём· (*n*) (print) flexible sewing

в·рас·пло́х· (*adv*) unawares, by surprise

в·рас·се́ч·к·у (*adv*) (elec) in series

в·рас·сы́п·н·у́ю (*adv*) in confusion/disorder, in all directions

в·раст·а́ни·е, -я, (*n*) ingrowth, growing in; (geol) intergrowth

в·раст·а́·ть, -ют, (*imp*) grow in

в·рас·ти́, -ст·у́т, (*perf*) *see* в·раст·а́·ть

в·рас·тя́ж·к·у (*adv*) at full length; slowly, with emphasis (of speech)

в·рас·ще́п· (*adv*) split, forked

со·един·е́ни·е в·рас·ще́п· (*n*) split joint (hammer welding)

вр·ать, -ут, (*imp*) lie, fib; be wrong/inaccurate

врач·, -а́, (*i*) -о́м, (*m*) (med) physician, doctor

~, ветерина́р·н·ый veterinary surgeon, vet

~, зуб·н·о́й dentist

врач·е́бн·ый, -ая, -ое, (*a*) medical

вращ·а́тел·ь, -я, (*m*) rotator, spinner

вращ·а́тельн·ый, -ая, -ое, (*a*) rotary, rotatory, revolving, turning

~ бур·е́ни·е (*n*) rotary drilling

~ движ·е́ни·е (*n*) rotation, rotary motion

~ стол (*m*) revolving table

вращ·а́·ть, -ют, (*imp*) rotate, turn; -ся gyrate, revolve, rotate, turn

вращ·а́ющ·ий, -ая, -ее, (*pres part act*) of **вращ·а́·ть**

~ **момéнт·** (*m*) torque

вращ·а́ющ·ийся, -аяся, -ееся, (*pres part act*) of **вращаться;** rotary, rotatory

~ **в·лéво** levorotatory

~ **в·пра́во** dextrorotatory

~ **контáкт·н·ое у·стрóй·ств·о** (*n*) (elec) rotary junction

    мéтод· вращ·áющ·егося крис-та́лл·а (*m*) rotating crystal method (of X-ray diffraction analysis)

~ **печ·ь** (*f*) rotary kiln

~ **ядр·ó** (*n*) (nucl) spinning nucleus

вращ·éни·е, -я, (*n*) rotation; revolution; gyration

~ **áтом·ов** atomic rotation

~, **лéв·ое** counter-clockwise rotation, anti-clockwise rotation

    момéнт· вращ·éни·я (*m*) angular momentum

    плóск·ост·ь вращ·éни·я (*f*) plane of rotation

~ **по áзимут·у** azimuthal rotation

~, **прáв·ое** clockwise rotation

врд (*abbr*) = **воздушно·реакти́в·-н·ый дви́г·ател·ь** (*m*) jet engine

вред·, -á, (*m*) harm, injury, damage

вред·и́тел·ь, -я, (*m*) pest, vermin, blight; wrecker, saboteur

врéд·ност·ь, -и, (*f*) hazard, danger (to health etc.)

врéд·н·о (*adv*) see **врéд·н·ый;** it is harmful, it is bad

врéд·н·ый, -ая, -ое, (*s f*) **-д·ен·, -д·н·á, -д·н·о,** harmful, injurious, noxious

~ **дéй·стви·е** (*n*) damage

~ **про·стрáн·ств·о** (*n*) (mech eng) dead space; excess clearance (in cylinder)

в·рéз·анн·ый, -ая, -ое, (*past part pass*) of **в·рéз·ать;** notched, incised

в·рез·áни·е, -я, (*n*) cutting, cutting in, incision

в·рез·á·ть, -ют, (*imp*), **в·рéз·ать, -éж·ут,** (*perf*) cut in, mortise, fit (into cut)

в·рéз·к·а, -и, (*g pl*) **-з·ок·,** (*f*) see **в·рез·áни·е;** (surv) inset map

в·рез·н·óй, -áя, -óе, (*a*) mortise, inset; notched

~ **за·дви́ж·к·а** (*f*) (build) flush bolt

~ **за·мóк·** (*m*) mortise lock

времен·á (*nom pl*) of **врéм·я**

времен·áми (*adv*) at times, now and then, from time to time

врéмен·н·о (*adv*) temporarily, provisionally

~ **ис·полн·я́ющ·ий обя́з·анност·ь** acting (in post)

времен·н·óй, -áя, -óе, (*a*) (*see also* **врéмен·н·ый**) temporal, time, time-dependent

~ **дел·éни·е** (*n*) (telecom) time-division allocation

~ **диагрáмм·а** (*f*) timing chart (computers)

~ **и́мпульс·** (*m*) timing/time pulse

~ **код·** (*m*) pulse-length code

~ **раз·вёрт·к·а** (*f*) time base (cathode ray tubes)

~ **ход·** (*m*) time dependence (of a curve)

врéмен·н·ый, -ая, -ое, (*a*) (*see also* **времен·н·óй**) temporary, provisional, interim; (naut) jury; ephemeral

~ **за·кóн·** (*m*) provisional law

~ **по·стрóй·к·а** (*f*) temporary building/erection

~ **мéр·а** (*f*) interim/temporary measure, makeshift arrangement

врéм·я, (*g, d, l*) **врéмен·и,** (*i*) **врé-мен·ем,** (*nom pl*) **времен·á,** (*g pl*) **времён·,** (*d pl*) **времен·áм,** (*n*); time, period of time; (gram) tense

~ **без·дéй·стви·я** idle time

    в на·сто·я́щ·ее врéм·я currently, presently, (U.S.) at the present time, present-day

~, **в своё** formerly; at the proper time; in due course

    во врéм·я during, in the course of

~ **воз·ник·новéни·я** origin time

~ **воз·врáт·а** recovery time; reset time (of computer)

~, **все·ми́р·н·ое** universal time

~ **вы́·бег·а и о·станóв·к·а** start-stop time (magnetic tape)

~ **гóд·а** season

~, **гражд·áнск·ое** civil time

~, **гри́нвич·еск·ое** Greenwich time

~, **декрéт·н·ое** daylight-saving time, summer time (zone time plus one hour)

    до на·сто·я́щ·его врéмен·и to date, so far, until/till now

~ **до раз·руш·éни·я** (mech) time to rupture, time to failure

~ **жи́зн·и** lifetime

~, **зá·дан·н·ое** prescribed time; present time

~ **за·пáзд·ывани·я** lag, time lag

врéм·я

~ за·пáзд·ывани·я, пóлн·ое total lag

~ за·пáзд·ывани·я процéсс·а (autom) plant lag, process lag

~ за·пáзд·ывани·я регул·и́рова·ни·я (autom) control system lag

~ за·пýск·а starting-up period

~, звёзд·н·ое sidereal time

~, зонáль·н·ое zone time

~, инерциóн·н·ое (rocket) coasting time

~, и́стин·н·ое apparent time

~, и́стин·н·ое сóлн·ечн·ое apparent solar time

~, каменно·угóль·н·ое (geol) Carboniferous time

коэффициéнт· врéмен·и (m) time factor

ли́ния врéмен·и (f) time base (radar etc.)

~, мéст·н·ое local time

~, на·сто·я́щ·ее present time; (geol) recent time

~ о·кон·чáни·я при·ём·а (telecom) time of receipt

~, о·пóр·н·ое reference time

~ пере·дáч·и (telecom) time of dispatch

~ под·пис·áни·я (telecom) time of origin, date time group

~, пояс·н·óе zone time

~, пояс·н·óе декрéт·н·ое standard time

~ при·бы́т·и·я, рас·чёт·н·ое (nav) estimated time of arrival

~ при·бы́т·и·я, факт·и́ческ·ое (nav) actual time of arrival

~ при·ём·а (telecom) time of receipt; resting time (radar)

~ при·ём·а в кáсс·е (telecom) time of handing in

~, про·лёт·н·ое transit time (of an electron)

сигнáл· врéмен·и (m) time signal

~, сидерáль·ное sidereal time

~, срéд·н·ее mean time, median time

~, срéд·н·ее гри́нвич·ск·ое Greenwich Mean Time

~, сýд·ов·óе ship time

~, с·чёт·а count time

~ эксплуат·áци·и life, useful life

время·и́мпульс·н·ая модул·я́ци·я (elec) pulse-time modulation, PTM

время·ис·числ·éни·е, -я, (n) chronology

время·пре·про·вожд·éни·е, -я, (n) pastime

вр·ёт (pres 3rd sing) of **вр·ать**

в·рóвен·ь (adv) flush with, flush, level with, level

в·рóд·е (prep + gen) like, not unlike, such as

в·рó·ет (fut 3rd sing) of **в·ры́·ть**

в·рожд·ённ·ый, -ая, -ое, (a) inborn, congenital, innate, inherent

в·роз·ь (adv) separately, apart

в·рóс·ш·ий, -ая, -ее, (past part act) see **в·раст·á·ть** ingrown; (geol) intergrown

в·рó·ют (fut 3rd pl) of **в·ры́·ть**

в·рýб·, -а, (m) (min) cut

~, клин·ов·óй (min) wedge cut

~, пирамидáль·н·ый (min) centre/pyramid cut

~, прям·óй (expl) burn cut

~, щел·ев·óй line cut

в·руб·á·ть, -ют, (imp) cut in/into

в·руб·и́ть, -ят, (perf) see **в·руб·á·ть**

в·рýб·к·а, -и, (g pl) -б·ок·, (f) (v n) see **в·руб·á·ть;** (carpentry) incision, groove, cut

в·рýбл·енн·ый, -ая, -ое, (past part pass) see **в·руб·á·ть;** grooved, mortised

в·рýб·машин·и́ст·, -а, (m) (min) cutting machineman

в·рýб·ов·ый, -ая, -ое, (a) cutting

~ маши́н·а (f) (min) cutting machine

~ маши́н·а, ди́ск·ов·ая (f) (min) disc cutting machine

~ маши́н·а, узко·за·бóй·н·ая (min) shortwalling machine

~ маши́н·а, штáнг·ов·ая (min) bar cutting machine

вр·ут (pres 3rd pl) of **вр·ать**

вруч·éни·е, -я, (n) handing, handing in/over, presentation

в·руч·н·ýю (adv) by hand, manually

вр·ýщ·ий, -ая, -ее, (pres part act) of **вр·ать**

в·рыв·á·ть, -ют, (imp) dig in/into; -ся (passive) break in, enter by force

в·ры·ть, -рó·ют, (perf) see **в·рыв·á·ть**

вряд́ли (adv) scarcely, hardly, it is doubtful whether

вс- (prefix) = **вз-, взо-, взъ-** particular significances: (1) movement upwards; (2) intensity, force, sudden onset; (3) completion of action, achievement of state or limit

в·сад·и́ть, -я́т, (perf) see **в·сáж·и·ва·ть**

в·сáж·енн·ый, -ая, -ое, (past part pass) see **в·сáж·ива·ть**

в·са́ж·ива·ть, -ют, (*imp*) thrust/plunge/stick into

в·са́л·ивани·е, -я, (*n*) (geol) salting in

в·са́с·ываем·ый, -ая, -ое, (*pres part pass*) of в·са́с·ыва·ть; absorbable

в·са́с·ывани·е, -я, (*n*) suction, intake; absorption

    ход в·са́с·ывани·я (*m*) (ICE) admission/induction stroke

в·са́с·ыва·ть, -ют, (*imp*) suck, take in; absorb

в·са́с·ывающ·ий, -ая, -ее, (*pres part act*) of в·са́с·ыва·ть; suction, sucking, absorbing

~ кла́пан· (*m*) inlet/intake valve; suction valve

~ магистра́л·ь (*f*) induction main

~ рук·а́в· (*m*) suction hose

~ труб·а́ (*f*) intake pipe; (hydro-elec) draft tube

вс·е (*pron*) (*nom pl, acc pl inanim of* весь) all

вс·ё (*pron*) (*nom/acc sing neut of* весь) all, the whole; (*adv*) always, all the time, constantly, still

~ вре́м·я all the time

~ ещё still

все- (*root*) all, omni-

все·воз·мо́ж·н·ый, -ая, -ое, (*a*) all sorts/kinds of, every possible

все·во́лн·ов·ый, -ая, -ое, (*a*) (rad) all-wave

все·гда́ (*adv*) always

вс·его́ (*pron*) (*gen sing masc/neut, acc sing masc anim of* весь) all, the whole; (*adv*) in all, in total, all together

вс·ей (*pron*) (*gen/dat/instr/prep*) of весь

всек (*abbr*) = во́льт·секу́нд·а volt-second

вселе́нн·ая, -ой, (*f decl as adj*) universe, macrocosm

вс·ем (*pron*) (*instr sing masc/neut, dat plur masc/fem/neut*) of весь

вс·ём (*pron*) (*prep sing masc/neut*) of весь

всём во in all ways

все·ме́р·н·о (*adv*) in every way

все·ме́р·н·ый, -ая, -ое, (*a*) every kind of, all possible

в·се́м·ер·о (*adv*) seven times

в·сем·ер·о́м (*adv*) seven together

вс·е́ми (*pron*) (*instr plur*) of весь

все·ми́р·н·ый, -ая, -ое, (*a*) world, world-wide, universal

~ тяг·оте́нни·е (*n*) gravitation

вс·ему́ (*pron, dat sing masc/neut*) of весь

все·на·пра́вл·енн·ый, -ая, -ое, (*a*) omnidirectional

все·на·ро́д·н·ый, -ая, -ое, (*a*) national, public

все·о́бщ·ий, -ая, -ее, (*a*) general, universal

все·режи́м·н·ый, -ая, -ое, (*a*) variable, variable-speed

все·со·ю́з·н·ый, -ая, -ое, (*a*) All-Union

все·сторо́н·н·ий, -яя, -ее, (*a*) all-round, comprehensive, thorough; (phys) cubic, volumetric, cubical

~ с·жим·а́емост·ь (*f*) cubic compressibility

всё-таки (*conj*) all the same, however, still

вс·ех (*pron*) (*gen/prep/acc anim*) of весь

все·це́л·о (*adv*) wholly, entirely

все·ча́с·н·о (*adv*) hourly

все·я́д·н·ый, -ая, -ое, (*a*) omnivorous

в·скак·а́ть, -а́ч·ут, (*imp*) jump/spring into

в·ска́к·ива·ть, -ют, (*imp*) gallop/jump/leap into/onto/upon

в·ска́льз·ыва·ть, -ют, (*imp*) slide/slip into

вс·ка́п·ыва·ть, -ют, (*imp*) dig up, excavate

вс·кара́бк·а·ться, -ются, (*perf*) *see* вс·кара́бк·ива·ться

вс·кара́бк·ива·ться, -ются, (*imp*) scramble/clamber up/upon

вс·ка́рмл·ива·ть, -ют, (*imp*) bring up rear

вс·ки́д·ыва·ть, -ют, (*imp*) throw up, jerk up

вс·ки́·н·уть, -ут, (*perf*) *see* вс·ки́д·ыва·ть

вс·кип·а́ни·е, -я, (*n*) boiling up, bubbling up, effervescence; priming (in boiler)

вс·кип·а́·ть, -ют, (*imp*) boil up/over, come to the boil

вс·кип·е́ть, -я́т, (*perf*) *see* вс·кип·а́·ть

вс·кипят·и́ть, -я́т, (*perf*) boil, bring to the boil

вс·кипяч·ённ·ый, -ая, -ое, (*past part pass*) of вс·кипят·и́ть

вс·колы́х·ива·ть, -ют, (*imp*) rock, agitate

вс·колы́х·н·у́ть, -у́т, (*perf*) *see* вс·колы́х·ива·ть

вс·ко́п·анн·ый, -ая, -ое, (*past part pass*) of вс·коп·а́·ть excavated, dug up

вс·коп·а́·ть, -ют, (*perf*) dig up, excavate

в·ско́р·е (*adv*) soon/shortly after, soon

вс·корм·и́ть, ´ят, (*perf*) *see* вс·ка́рм·л·ива·ть

в·скоч·и́ть, ´ат, (*perf*) *see* в·ска́к·и·ва·ть

вс·крыв·а́·ть, -ют, (*imp*) open, open up; reveal, disclose; (geog) discover; (min) strip; (anat) dissect; (med) lance

вс·кры́т·и·е, -я, (*n*) (*v n*) *see* вс·кры·в·а́·ть; (bot) dehiscence; break up, debacle (of river ice); section, autopsy (on corpse)

~ жи́л·ы (min) resuing

~ на·но́с·ов (min) stripping overburden

вс·кры·ть, -ро́·ют (*perf*) *see* вс·кры·в·а́·ть

вс·кры́·т·ый, -ая, -ое, (*past part pass*) *see* вс·крыв·а́·ть

~ руд·а́ (*f*) developed ore

вс·кры́·ш·а, -и, (*f*) (min, civ eng) overburden; stripping, overburden stripping

    коэффицие́нт· вс·кры́·ш·и (*m*) (min) overburden ratio

~, ма́л·ая (min) thin overburden

вс·кры́ш·к·а, -и, (*f*) (min) stripping

вс·крыш·н·о́й, -ая, -ое, (*a*) (min) overburden

~ рабо́т·а (*f*) (min) removing the overburden, stripping

в·след· (*adv*) after, following

~ за behind, following

в·сле́д·стви·е (*prep* + *gen*) owing to, in consequence of, on account of, through, as a result of

    рабо́т·а в·сем·у́ю (*f*) work by feel/touch, blind work

в·слеп·у́ю (*adv*) blindly

в·слух· (*adv*) aloud

в·слу́ш·а·ться, -ются, (*perf*) *see* в·слу́ш·ива·ться

в·слу́ш·ива·ться, -ются, (*imp*) listen attentively

ВСМ (*abbr*) = высоко·молеку·ля́р·н·ое со·един·е́ни·е (*n*) macromolecular compound

все·объ·ём·л·ющ·ий, -ая, -ее, (*a*) all-embracing, comprehensive, universal

все·про·ник·а́ни·е, -я, (*n*) pervasion

все·росс·и́йск·ий, -ая, -ое, (*a*) All-Russian

в·серьёз· (*adv*) seriously, in earnest

Всес. (*abbr*) = Все·со·ю́з·н·ый All-Union

все·све́т·н·ый, -ая, -ое, (*a*) cosmopolitan

в·сма́тр·ива·ться, -ются, scrutinize, peer at

в·смотр·е́ться, ´ятся, (*perf*) *see* в·сма́тр·ива·ться

всмя́тку (*adv*) soft-boiled (of eggs)

в·сов·а́ть, -су·ю́т, (*perf*) *see* в·со́в·ы·ва·ть

в·со́в·ыва·ть, -ют, (*imp*) poke/thrust/shove in

в·со́с·анн·ый, -ая, -ое, (*past part pass*) *see* в·са́с·ыва·ть; sucked in, pulled in, drawn in

в·сос·а́ть, -у́т, (*perf*) *see* в·са́с·ыва·ть

вс·па́х·анн·ый, -ая, -ое, (*past part pass*) *see* вс·па́х·ива·ть; ploughed, plowed, tilled

вс·пах·а́ть, -па́ш·ут, (*perf*) *see* вс·па́х·ива·ть

вс·па́х·ива·ть, -ют, (*imp*) plough, plow, till, turn up

вс·па́ш·к·а, -и, (*g pl*) -ш·ек·, (*f*) ploughing, plowing

~, по·вто́р·н·ая backset

вс·пе́н·енн·ый, -ая, -ое, (*a*) (*past part pass*) *see* вс·пе́н·ива·ть frothed, foamed

вс·пе́н·иватель·, -я, (*m*) frother, foaming/frothing agent; (rubber) foaming/blowing agent

вс·пе́н·ива·ть, -ют, (*imp*) froth, make foam; -ся foam, froth, lather (of soap)

вс·пе́н·ить, -ят, (*perf*) *see* вс·пе́н·и·ва·ть

вс·плеск·, -а, (*m*) splash; bump (of elec. current); flash-up (of power)

вс·плеск·ивани·е, -я, (*n*) (*v n*) *see* вс·плёск·ива·ть

вс·плёск·ива·ть, -ют, (*imp*) splash

вс·плес·н·у́ть, -у́т, (*perf*) *see* вс·плё·ск·ива·ть

всплош·н·у́ю, (*adv*) one after the other in succession

вс·плыв·а́емост·ь, -и, (*f*) buoyancy, flotation

вс·плыв·а́ни·е, -я, (*n*) surfacing, floating to the surface

вс·плыв·а́·ть, -ют, (*imp*) come to the surface, surface

вс·плыв·а́ющ·ий, -ая, -ее, (*pres part act*) of вс·плыв·а́·ть; surfacing, buoyant; (chem) supernatant

вс·плы́т·и·е, -я, (*n*) surfacing

вс·плы·ть, ´плыв·у́т, (*perf*) *see* вс·плыв·а́·ть

вс·пола́ск·ива·ть, -ют, (*imp*) rinse out

вс·полз·а́ни·е, -я, (*n*) (*v n*) *see* вс·полз·а́·ть

вс·полз·а́·ть, -ют, (*imp*) climb/creep up

вс·полос·н·у́ть, -у́т, (*perf*) *see* вс·по·ласк·ива·ть

вс·по́л·ь·е, -я, (*n*) (agr) ridge

вс·по·мин·а́·ть, -ют, (*imp*) remember, recollect, recall; -ся come back to (the mind)

вс·по́·мн·ить, -ят, (*perf*) *see* вс·по··мин·а́·ть

вс·по·мог·а́тельн·ый, -ая, -ое, (*a*) auxiliary; secondary, subsidiary; booster, boosting

~ материа́л·ы (*pl*) indirect materials (production)

вс·пры́ск·ивани·е, -я, (*n*) (*v n*) *see* вс·пры́ск·ива·ть; injection

вс·пры́ск·ива·ть, -ют, (*imp*) sprinkle, moisten; inject

вс·пры́с·н·уть, -ут, (*perf*) *see* вс·пры́·ск·ива·ть

вс·пух·а́·ть, -ют, (*imp*) swell, become swollen

вс·пу́х·л·ый, -ая, -ое, (*a*) swollen, inflated, (med) tumid

вс·пу́х·н·уть, -ут, (*perf*) *see* вс·пу·х·а́·ть

вс·пу́х·ш·ий, -ая, -ее, (*past part act*) swollen, inflated, (med) tumid

вс·пу́ч·енн·ый, -ая, -ое, (*past part pass*) *see* вс·пу́ч·ива·ть; swollen, bulged, inflated, expanded; (geol) heaved

вс·пу́ч·ивани·е, -я, (*n*) swelling, bulging, bloating, intumescence; (geol) heave, heaving, heaves

вс·пу́ч·ива·ться, -ются, (*imp*) swell, bulge, distend

вс·пу́ч·ивающ·ийся, -аяся, -ееся, (*pres part act*) of вс·пу́ч·ива·ться; expansible

вс·пу́ч·ин·а, -ы, (*f*) bulge, swelling, blister, heave

вс·пу́ч·иться, -атся, (*perf*) *see* вс·пу́·ч·ива·ться

вс·пы́х·ивани·е, -я, (*n*) (*v n*) *see* вс·пы́х·ива·ть

вс·пы́х·ива·ть, -ют, (*imp*) flash, flare up, burst into flames; (chem) deflagrate

вс·пы́х·ивающ·ий, -ая, -ее, (*pres part act*) of вс·пы́х·ива·ть

~ о·кра́с·к·а (*f*) (zool) flash colours/colors

вс·пы́х·н·уть, -ут, (*perf*) *see* вс·пы́·х·ива·ть

вс·пы́ш·ечн·ый, -ая, -ое, (*a*) of вс·пы́ш·к·а; flash, flashing, flare

вс·пы́ш·к·а, -и, (*g pl*) -ш·ек·, (*f*) flash; (chem) deflagration; outbreak; (nucl) burst; spike (on a curve)

~ акти́в·ност·и (phys) burst of activity

~, нейтро́н·н·ая neutron burst

~, обра́т·н·ая (ICE) blowback

~ по Бренкен·у (oil) Brenken flash point

~ све́т·а flash of light, light flash

~, свет·ов·а́я (*f*) light pulse

~, скор·остн·а́я (phot) high-speed flash

~, со́лн·ечн·ая (astron) solar flare температу́р·а вс·пы́ш·к·и (oil) flash point

~, тепл·ов·а́я (phys) thermal spike

в·ста·ва́ни·е, -я, (*n*) (*v n*) of в·ста·-ва́ть

в·ста·ва́ть, -ю́т, (*imp*) get/stand up; rise, arise

в·ста́в·ить, -ят, (*perf*) *see* в·ставл·я́·ть

в·ста́в·к·а, -и, (*g pl*) -в·ок·, (*f*) (*v n*) *see* в·ставл·я́·ть; insert, insertion; insertion/packing piece; (zool) intercalation

~, за·мк·о́в·ая locking piece знак· в·ста́в·к·и (*m*) (print) caret (insert symbol)

~, ка́бель·н·ая (telecom) cable insert

~, пла́в·к·ая (elec) fuse element, fuse

~, разъ·един·и́тельн·ая (elec) isolating/disconnecting link

~, цилиндр·и́ческ·ая (shipb) parallel middle body

в·ставл·е́ни·е, -я, (*n*) (*v n*) *see* в·ставл·я́·ть; insertion, introduction; (zool) intercalation

в·ста́вл·енн·ый, -ая, -ое, (*part past pass*) *see* в·ставл·я́·ть; inserted, fitted in, set in

в·ставл·я́·ть, -ют, (*imp*) insert, introduce, fix in, put in, install, bed; (zool) intercalate

в·став·н·о́й, -а́я, -о́е, (*a*) insert, insertion; inserted, set in; plug in; (zool) intercalary

~ на·со́с· (*m*) (oil) insert pump

~ рабо́ч·ая втул́·к·а (*f*) cylinder liner

~ ра́м·а (*f*) double window frame

в·ста́в·очн·ый, -ая, -ое, (*a*) insertion; insert, inserted; (zool) intercalary

в·ста́·ть, -нут, (*perf*) *see* в·ста·ва́ть

в·стра́·ива·ть, -ют, (*imp*) build in/into; **-ся** be incorporated in

встрет·ить, -ят, (*perf*) *see* **встреч·а́·ть**

встреч·а, -и, (*i*) **-ей,** (*f*) meeting, encounter

    то́ч·к·а встреч·и (*f*) (gunn) point of impact; (rocket) collision point

    у́гол· встреч·и (*m*) angle of incidence, angle of impact (of missiles)

встреч·а́·ть, -ют, (*imp*) (*trans*) meet, encounter; **-ся** (*intrans*) meet, encounter, be met with; be found, be met, occur

встреч·н·о-дви́ж·ущ·иеся по́рш·н·и opposed pistons

~ -паралле́ль·н·ый anti-parallel

встреч·н·ый, -ая, -ое, (*a*) meeting, counter, coming from the opposite direction; (nav) reciprocal; (elec) opposing, opposite

~ ве́тер· (*m*) head wind

в·стро́·енн·ый, -ая, -ое, (*past part pass*) of **в·стро́·ить** built in, fixed, incorporated

в·стро́·ить, -ят, (*perf*) build in, incorporate

вс·тря́х·ивани·е, -я, (*n*) (*v n*) *see* **вс·тря́х·ива·ть**

вс·тря́х·ивател·ь, -я, (*m*) shaker (ore dressing); (telecom) scrambler

вс·тря́х·ива·ть, -ют, (*imp*) joggle, shake up; (telecom) scramble

вс·тря́х·ивающ·ий, -ая, -ее, (*pres part act*) of **вс·тря́х·ива·ть**; vibrating, jolting, jogging

~ форм·о́вочн·ая маши́н·а (*f*) (cast) jolting molding machine

вс·трях·н·у́ть, -у́т, (*perf*) joggle, shake up; scramble

в·ступ·а́·ть, -ют, (*imp*) enter; **-ся** intercede

в·ступ·а́ющ·ий, -ая, -ее, (*pres part act*) of **в·ступ·а́·ть**; entering, incoming

в·ступ·и́тельн·ый, -ая, -ое, (*a*) entrance; introductory, opening

в·ступ·и́ть, -ят, (*perf*) *see* **в·ступ·а́·ть**

~ в де́й·стви·е come into force (regulations etc.); start up (equipment, plan, organizations)

~ в реа́кци·ю (chem) enter into reaction

~ в си́л·у come into force

в·ступл·е́ни·е, -я, (*n*) entry; introduction, preamble; prelude (music); (geol) arrival

в·с·тык· (*adv*) butt; butt joint

    со·един·е́ни·е в·с·тык· (*n*) butt joint

~ с дв·умя́ на·кла́д·к·ами double strapped joint

~ с одн·о́й на·кла́д·к·ой single strapped joint

в·су́·н·уть, -ут, (*perf*) poke/thrust/shove in

в·сух·у́ю (*adv*) dry, in the dry state

вс·хо́д·, -а, (*m*) (bot) shoot, sprout

вс·ход·и́ть, -ят, (*imp*) mount, ascend, climb; (astron) rise; (bot) sprout, germinate

вс·хо́ж·ест·ь, -и, (*f*) germination, germinating power

в·сы́п·ать, -лют, (*perf*) **в·сып·а́·ть, -ют,** (*imp*) pour in (solids)

в·сып·а́ни·е, -я, (*n*) (*v n*) *see* **в·сы·п·а́·ть**

в·сыр·у́ю, (*adv*) damp, wet, in the wet state

вс·ю (*pron*) (*acc sing fem*) of **весь**

всю́ду (*adv*) everywhere

вс·я (*pron*) (*nom sing fem*) of **весь**

вся́·к·ий, -ая, -ое, (*a*) any, every; (*n decl as adj*) anything

вся́·ческ·и (*adv*) in every possible way

вся́·ческ·ий, -ая, -ое, (*a*) of every kind, every kind of, all kinds of

вт (*abbr*) = **ватт·** W, w, watt

в·тавр· (*adv*) T-, tee-

    со·един·е́ни·е в·тавр· (*n*) T-joint

в·та́й·н·е (*adv*) in secret, secretly

в·та́лк·ивани·е, -я, (*n*) (*v n*) *see* **в·та́лк·ива·ть**

в·та́лк·ивател·ь, -я, (*m*) pusher

в·та́лк·ива·ть, -ют, (*imp*) push/shove in

в·та́пл·ива·ть, -ют, (*imp*) melt in, embed (in wax etc.)

в·та́пт·ывани·е, -я, (*n*) (*v n*) of **в·та́пт·ыва·ть**

в·та́пт·ыва·ть, -ют, (*imp*) trample down/in

в·та́ск·ивани·е, -я, (*n*) (*v n*) of **в·та́ск·ива·ть**

в·та́ск·ива·ть, -ют, (*imp*) drag in/into

в·та́ч·анн·ый, -ая, -ое, (*past part pass*) *see* **в·та́ч·ива·ть**

в·тач·а́·ть, -ют, (*perf*) *see* **в·та́ч·ива·ть**

в·та́ч·ива·ть, -ют, (*imp*) stitch in/into

в·та́ч·к·а, -и, (*g pl*) **-ч·ек·,** (*f*) (*v n*) *see* **в·та́ч·ива·ть**; stitched-in/oversewn-in piece

в·та́ч·к·у (*adv*)

    шитьё в·та́ч·к·у oversewing

в·тащ·и́ть, -́ат,　(*perf*) drag in/into
в·тек·а́ни·е, -я,　(*n*) inflow, influx
в·тек·а́·ть, -ют,　(*imp*) flow in
в·тек·а́ющ·ий, -ая, -ее,　(*pres part act*) of в·тек·а́·ть; influent, inflowing
в·тер·е́ть,　(*fut 3rd sing, pl*) во·тр·ёт, во·тр·у́т,　(*past masc sing*) в·тёр·, (*perf*); rub in
в·теч·е́ни·е, -я,　(*n*) *see* в·тек·а́ни·е
в·теч·ь,　(*fut 3rd sing, pl*) в·теч·ёт, в·тек·у́т,　(*past masc sing*) в·тёк·, (*perf*); flow in
в·тир·а́·ть, -ют,　(*imp*) rub in; mix in (rubber); -ся (*pass*) push/worm one's way in
в·ти́ск·ива·ть, -ют,　(*imp*) squeeze/cram in
в·ти́с·н·уть, -ут,　(*perf*) *see* в·ти́ск·ива·ть
в·толк·н·у́ть, -у́т,　(*perf*) push/shove in
в·толк·ова́ть, -у́ют,　(*perf*) *see* в·толк·о́выва·ть
в·толк·о́выва·ть, -ют,　(*imp*) make understand, din/ram in, explain
в·топ·и́ть, -́ят,　(*perf*) *see* в·та́пл·ива·ть
в·топт·а́ть, -то́пч·ут,　(*perf*) trample down
в·торг·а́·ться, -ются,　(*imp*) invade, intrude, encroach, trespass, (geol) irrupt
в·то́рг·н·уться, -утся,　(*perf*) *see* в·торг·а́·ться
в·торе́ц·　(*adv*) (mech eng) face
~, об·то́ч·ка　facing
в·торж·е́ни·е, -я,　(*n*) (*v n*) *see* в·торг·а́·ться; encroachment, intrusion, trespass; incursion, invasion, irruption
втор·ить, -ят,　(*imp*) echo, repeat; take second part
втор·и́чн·о　(*adv*) a second time, for the second time, again
вторично-　(*component*) secondary, deutero-, sub-
вторично-из·ве́рж·енн·ый, -ая, -ое,　(*a*) (geol) secondary eruptive
вторично·раз·ветвл·я́ющ·ийся, -аяся, -ееся,　(*a*) subdichotomous
вторично·ро́т·ые, -ых,　(*pl decl as adj*) (zool) *Deuterostomato*
втор·и́чн·ый, -ая, -ое,　(*a*) secondary, second-order; second; (chem) dibasic
~ вал·　(*m*) main shaft (vehicle gear box)
~ об·мо́т·к·а　(*f*) (elec) secondary winding

втор·и́чн·ый, -ая, -ое
~ мета́лл·　(*m*) secondary metal
~ рот·　(*m*) (zool) deuterostoma
~ сол·ь　(*f*) (chem) caustic salt
~ спирт·　(*m*) secondary alcohol
вто́р·ник·, -а,　(*m*) Tuesday
втор·о-　(*root*) second, second-order secondary, deutero-
второ·бра́ч·н·ый, -ая, -ое,　(*a*) (biol) deuterogamous
второ·го́д·ник·, -а,　(*m*) second-year student
втор·о́й, -а́я, -о́е,　(*a*) second
~ д·но　(*n*) (shipb) inner bottom, tank top
~ о́·черед·ь　(*f*) (telecom) second degree of priority
~ член·　(*m*) consequent (of a ratio)
второ·ку́рс·ник·, -а,　(*m*) second-year student
второ·о·черед·н·о́й, -а́я, -о́е,　(*a*) secondary; second-stage
в·тороп·я́х　(*adv*) hastily, hurriedly
второ·раз·ря́д·н·ый, -ая, -ое,　(*a*) second-rate
второ·со́рт·н·ый, -ая, -ое,　(*a*) second quality/grade
второ·степе́н·н·ый, -ая, -ое,　(*a*) secondary, minor
~ вулка́н·　(*m*) subordinate volcano
в·трамб·о́выва·ть, -ют,　(*imp*) tamp, tamp in
в·тре́т·их　(*adv*) thirdly, in the third place
в·тро́·е　(*adv*) three times, threefold
в·тро·ём　(*adv*) three, three together
в·трой·не́　(*adv*) three times as much
втс　(*abbr*) = ватт-секу́нд·а W/sec, w/sec, watt/second
в·туг·у́ю　(*adv*) taut, tight
вы·бир·а́·ть в·туг·у́ю　haul taut
вту́з·, -а,　(*m*) = вы́с·ш·ее техн·и́ческ·ое уч·е́бн·ое за·вед·е́ни·е technical college, technical institute
вту́з·овск·ий, -ая, -ое,　(*a*) of вту́з·
вту́л·к·а, -и,　(*g pl*) -л·ок·, (*f*) bush, bushing; bung, plug; sleeve, liner, barrel; hose coupling tail
~, в·став·н·а́я　cylinder liner
~ дейдвуд·н·ой труб·ы́　(shipb) stern-tube bush, engineer's tube
~, до́н·н·ая　base plug (ammunition)
~, ка́псюль·н·ая　primer (cartridge)
~, конду́ктор·н·ая　jig bushing
~, кон·и́ческ·ая　cone bush
~, на·жи́м·н·ая　gland (stuffing box)
~, на·правл·я́ющ·ая　valve guide

вту́л·к·а

~, от·раж·а́тельн·ая spill deflector (diesel)

~, па́луб·н·ая (shipb) deck plate

~, пла́в·ающ·ая floating bush

~, по·воро́т·н·ая control sleeve (diesel)

~, раз·ре́з·н·а́я split bearing/sleeve

~, рабо́ч·ая (ICE) cylinder liner

~, рас·по́р·н·ая distance piece/push; gear retainer (diesel unit injector)

~, регул·и́рующ·ая throttling bush (recoil mechanism)

~, с·ме́н·н·ая slip/loose bush

~ толк·а́тел·я, на·правл·я́ющ·ая follower guide (diesel)

~, цил́индр·ов·ая liner, barrel, sleeve (of ICE cylinder)

вту́л·очк·а, -и, (g pl) **-чек·,** (f) (dim) of **вту́л·к·а**

вту́л·очн·о-ро́лик·ов·ая цеп·ь (f) bush roller chain

вту́л·очн·ый, -ая, -ое, (a) of **вту́л·к·а**

в·туп·и́к· see **туп·и́к·**

вт.ч (abbr) = **ватт.час** W. hr., w. hr., watt-hour

в·тык·а́·ть, -ют, (imp) stick/push/drive in (something pointed)

в·тя́г·ивани·е, -я, (n) (v n) of **в·тя́г·ива·ть;** retracting

сила в·тя́гивани·я (f) compressive force

в·тя́г·ива·ть, -ют, (imp) pull/draw/haul in; involve, implicate; **-ся** (pass) retract; become inured to

в·тя́г·ивающ·ийся, -аяся, -ееся, (pres part act) of **в·тяг·ива́·ться;** retractable

в·тяж·н·о́й, -а́я, -о́е, (a) suction; retractable

в·тя́·нут·ый, -ая, -ое, (past part pass) see **в·тя́г·ива·ть**

в·тя·н·у́ть, -у́т, (perf) see **в·тя́г·и·ва·ть**

вуал·и́ровани·е, -я, (n) (phot) fogging

вуа́л·ь, -и, (f) veil; (phot) fog

~, ди·хро·и́ча·ая (phot) dichroic fog

~, свет·ов·а́я (phot) light fog

Ву́д·а, с·плав· (m) Wood's alloy

Вудруф·а, шпо́н·к·а (mech eng) Woodruff Key

вуз·, -а, (m) = **вы́с·ш·ее уч·е́б·н·ое за·вед·е́ни·е** higher educational establishment, university, institute, college

ву́з·ов·ец, -в·ц·а, (m) student (at **вуз·** q.v.)

ву́з·овск·ий, -ая, -ое, (a) of **вуз·**

вулка́н·, -а, (m) volcano

~, двой·н·о́й cone-in-cone volcano

~, гряз·ев·о́й (geog) mud volcano

~, с·ло́ж·н·ый composite volcano

вулканиза́т·, -а, (m) (rubber) vulcanizate

вулканиза́тор·, -а, (m) vulcanizer

~, кольц·ев·о́й annular vulcanizer

вулканизацио́н·н·ый, -ая, -ое, (a) of **вулканиз·а́ци·я**

~, пресс, (m) vulcanizing press

~, пресс·, мно́го-эта́ж·н·ый (m) multi-daylight vulcanization press

~ сре́д·ств·о (n) curative

вулканиз·а́ци·я, -и, (f) vulcanization, curing, cure, cross-linking (polymers)

~ гор·я́ч·им во́з·дух·ом dry-heat cure/vulcanization

~, ка́мер·н·ая hot-air chamber vulcanization

~, коро́т·к·ая quick cure

~, полу·хло́р·ист·ый acid cure

~, прежде-вре́мен·н·ая scorching

~, сверх-бы́стр·ая (rubber) flash cure

вулканиз·и́рованн·ый, -ая, -ое, (past part pass) of **вулканиз·и́р·овать;** vulcanized, cured

вулканиз·и́р·овать, -уют, (imp and perf) vulcanize, cure

вулканиз·и́р·ующ·ийся, -аяся, -ееся, (pres part act) of **вулканиз·и́р·о·ваться;** curable

вулкан·и́зм·, -а, (m) (geol) volcanism

вулканиз·о́вщик·, -а, (m) vulcanizer (operator)

вулканиз·у́емост·ь, -и, (f) rate of cure

вулкани́т·, -а, (m) (min) vulcanite; (rubber) vulcanite

вулкан·и́ческ·ий, -ая, -ое, (a) volcanic, igneous, eruptive

вулкан·о́ид·, -а, (m) mud volcano

вулкано·ли́т·, -а, (m) volcanic ejectamenta

вулкано́·лог·, -а, (m) volcanologist

вулкано·ло́г·и·я, -и, (f) volcanology

вулкафо́р·, -а, (m) vulcafor (rubber accelerator)

вульзини́т·, -а, (m) (geol) vulsinite

вульпини́т·, -а, (m) (min) vulpinite

вульфени́т·, -а, (m) (min) wulfenite

в·хо́д·, -а, (m) entry, entrance; inlet, input

~, лабири́нт·н·ый maze entrance

на в·хо́д·е (adv) at entry, inlet

у́гол· в·хо́д·а (m) angle of entry, angle of incidence (seismic radiations)

в·ход·и́ть, -ят, *(imp)* enter, go/come in; (math) be contained in, be in, be included in, appear in

~ **в со·ста́в·** be a member of; form a part of, be part of

в числ·о́ ... в·ход·ят among ... are

в·ход·н·о́й, -а́я, -о́е, *(a)* entrance; input

~ **велич·ин·а́** *(f)* (autom) input

~ **информ·а́ци·я** *(f)* (autom) input information

~ **у·стро́й·ств·о** *(n)* (elec) input unit, input equipment

~ **у·стро́й·ств·о, кла́виш·н·ое** *(n)* input typewriter (of computer)

в·ход·я́щ·ий, -ая, -ее, *(pres part act)* of **в·ход·и́ть;** entering, incoming; (math) reentrant; male (elec fittings); *(f decl as adj)* incoming paper (office work)

~ **у́гол·** *(m)* reentrant angle

в·хожд·е́ни·е, -я, *(n)* *(v n) see* **в·хо·ди́ть**

в·холмл·ённост·ь, -и, *(f)* (geog) ruggedness

в·холод·н·у́ю *(adv)* cold, in the cold state

в·холост·у́ю *(adv)* idle, free (producing no useful work)

~ **рабо́т·а** *(f)* idling (of engines)

в·цеп·и́ться, -ятся, *(perf) see* **в·цепл·я́·ться**

в·цепл·я́·ться, -ются, *(imp)* grasp, seize; hold fast, cling on to

вч *(abbr)* = **выс·о́к·ая част·от·а́,** hf, HF, high-frequency; (telecom) rf, RF, radio-frequency

вчера́ *(adv)* yesterday

вчера́·шн·ий, -яя, -ее, *(a)* yesterday's

в·черн·е́ *(adv)* in the rough

в·че́твер·о *(adv)* four times

в·четвер·о́м *(adv)* four, four together

в-четвёр·т·ых *(adv)* fourthly

в·чист·у́ю *(adv)* in final form

в·чит·а́·ться, -ются, *(perf) see* **в·чи́т·ыва·ться**

в·чи́т·ыва·ться, -ются, *(imp)* read carefully

вш·а́ми *(instr pl)* of **вош·ь**

в·ше́ст·ер·о *(adv)* six times

в·шест·ер·о́м *(adv)* six, six together

вш·и *(nom pl)* of **вош·ь** lice

в·шив·а́ни·е, -я, *(n)* *(v n)* of **в·ши·ва́·ть**

в·шив·а́·ть, -ют, *(imp)* sew in

в·ши́в·е·ть, -ют, *(imp)* become infested with lice

в·шив·н·о́й, -а́я, -о́е, *(a)* sewn in

в·ши́в·ост·ь, -и, *(f)* lousiness; (med) pediculosis

в·ши́в·ый, -ая, -ое, *(a)* louse-infested, lousy, pediculous

~ **се́м·я** *(n)* (bot, pharm) sabadilla seed, *Sabadilla officinalis*

в·ши́р·ь *(adv)* in breadth

в·ши·ть, (fut 3rd sing, pl) во·шь·ёт, во·шь·ю́т, *(perf)* sew in

в·шпу́нт· *(adv)*

со·един·е́ни·е в·шпу́нт· *(n)* tonguing (carpentry)

въ- *(prefix)* = **в-, во-** *particular significances:* (1) movement into; (2) movement up *(usually followed by preposition* **на***);* (3) penetration *(words ending in* **-ся***)*

въ·ед·а́·ться, -ются, *(imp)* eat into, corrode

въ·е́д·ут *(fut 3rd pl)* of **въ·е́х·ать**

въ·е́д·чив·ый, -ая, -ое, *(a)* corrosive

въ·е́зд·, -а, *(m)* entrance, vehicle entrance, drive, entry, driving-in; (rail) approach ramp

у́гол· въ·е́зд·а *(m)* (rail) angle of approach

въ·езд·н·о́й, -а́я, -о́е, *(a)* entrance, drive-in; approach

въ·езж·а́·ть, -ют, *(imp)* enter, drive in/up

въ·е́ст·ься, (fut 3rd sing, pl) въ·е́стся, въ·ед·я́тся, (past masc sing) въ·е́лся, (perf); see **въ·ед·а́·ться**

въ·е́х·ать, -е́д·ут, *(perf)* enter, drive in/up

вы *(pron)* you

вы- *(prefix) signifies:* (1) motion out of; (2) undoing, taking out/off, removal; (3) achievement by means of the action; (4) absolute completion of action; (5) **-ся** absolute completion of action or action performed to necessary extent

вы·ба́н·ить, -ят, *(perf)* wash out

вы́·бег·, -а, *(m)* running out, running down, run down (of engines); (instr) overshoot, overrun, overswing

~ **част·от·ы́** frequency drift

вы·бег·а́·ть, -ют, *(imp)* run out

вы́·беж·ать, (fut 3rd sing, pl) вы́·беж·ит, вы́·бег·ут, (perf); see **вы·бег·а́·ть**

вы·бе́л·ивани·е, -я, *(n)* *(v n) see* **вы·бе́л·ивать**

вы·бе́л·ива·ть, -ют, *(imp)* whiten, bleach, blanch, whitewash

вы́·бел·ить, -ят, (*perf*) whiten, bleach, whitewash, blanch

вы·бив·а́льн·ый, -ая, -ое, (*a*) knocking-out

вы·бив·а́ни·е, -я, (*n*) (*v n*) *see* **вы·бив·а́·ть;** (nucl) knockout

вы·бив·а́·ть, -ют, (*imp*) knock/kick out, dislodge; stamp; hammer out

вы́·бив·к·а, -и, (*g pl*) **-в·ок·,** (*f*) (*v n*) *see* **вы·бив·а́·ть**

вы́·бив·щик·, -а, (*m*) (cast) knocker-out

вы·бир·а́тел·ь, -я, (*m*) (*v n*) *see* **вы·бир·а́·ть;** selector (telegraphy)

вы·бир·а́·ть, -ют, (*imp*) choose, select, pick out, elect; take out (patent etc.); dig up; pull/get out; haul/heave, shorten in; **-ся** move from/away, get out

~ **ру́ч·к·у на себя́** (air) bring the stick back

~ **слаб·ин·у́** take up the slack, haul in the slack

вы́·би·т·ый, -ая, -ое, (*past part pass*) *see* **вы·бив·а́·ть;** knocked out, dislodged; hammered out

~ **а́том** (*m*) knocked-on atom

вы́·би·ть, -бь·ют, (*perf*) *see* **вы·бив·а́·ть**

вы́·бленк·а, -и, (*f*) (naut) ratline

вы́·бо·ин·а, -ы, (*f*) dent; pothole (in road)

вы́·бо·й, -я, (*m*) = **вы́·бо·ин·а**

вы́·бой·к·а, -и, (*g pl*) **-бо·ек·,** (*f*) (text) single-colour printing; (min) knocking out timbers, removing timbers; siftings, boltings (flour)

вы́·бор·, -а, (*m*) (*v n*) *see* **вы·бир·а́·ть;** choice, selection, discrimination; option; (*pl*) election(s); (math) sample

~, **амплиту́д·н·ый** (autom) amplitude discriminator

~, **групп·ов·о́й** group discrimination (telemetering)

~ **ка́др·а** viewing (a ciné film)

~ **моме́нт·а вре́мен·и** (autom) timing, time-point selection

~, **поля́р·н·ый** polarity discriminator

вы́·бор·к·а, -и, (*g pl*) **-р·ок·,** (*f*) pulling/getting out; hauling/heaving in; extract, excerpt; (math) sample, sampling

~, **бес·по·втор́·н·ая** random sampling

вре́м·я вы́·бор·к·и (*f*) access time (computers)

вы́·бор·н·ый, -ая, -ое, (*a*) chosen, sample; elected

~ **тип·** (*m*) (zool) lectotype

вы́·бор·очн·ый, -ая, -ое, (*a*) selective; (math) sampling; access (computers)

~ **обра́з·ец·** (*m*) random sample

~ **про·ве́р·к·а** (*f*) spot check

~ **цили́ндр·** (*m*) access cylinder (computers)

вы·бра́ж·ивани·е, -я, (*n*) (*v n*) of **вы·бра́ж·ива·ть;** (chem) degree of fermentation

вы·бра́ж·ива·ть, -ют, (*imp*) wander over; ferment fully

вы́·брак·овать, -уют, (*perf*) discard, reject

вы·брак·о́вк·а, -и, (*g pl*) **-вок·,** (*f*) rejecting; (genet) roguing, culling

вы·бра́с·ывани·е, -я, (*n*) (*v n*) *see* **вы·бра́с·ыва·ть;** ejection, expulsion; discarding, throwing away; throwing out; twisting, bending (in railway lines)

~ **а́том·а** (cryst) discomposition

~ **за́ борт·** jettisoning

~ **ма́сл·а** oil whip

вы·бра́с·ыватель·, -я, (*m*) ejector, extractor

вы·бра́с·ыва·ть, -ют, (*imp*) throw out, eject, emit, jettison; eliminate, reject, suppress, abandon, discard; **-ся** (*pass*); (air) bale out

вы́·бр·ать, (*fut 3rd sing, pl*) **вы́·бер·ёт, вы́·бер·ут,** (*perf*) *see* **вы·бир·а́·ть**

вы́·бр·анн·ый, -ая, -ое, (*past part pass*) of **вы́·бр·ать**

вы́·бр·ит·ый, -ая, -ое, (*past part pass*) of **вы́·бр·ить** shaven

вы́·бр·ить, -еют, (*perf*) shave, shave off

вы́·брод·ивш·ий, -ая, -ее, (*past part act*) *see* **вы·бра́ж·ива·ть;** fully fermented, fermented-out

вы́·брод·ить, -ят, (*perf*) *see* **вы·бра́ж·ива·ть**

вы́·брос·, -а, (*m*) (*v n*) *see* **вы·бра́с·ыва·ть;** ejection, rejection; outburst, outbreak; (nucl) singe, excursion; (oil) blowout; overshoot, overshooting; (elec) overpeak, peaked trace, pip, blip (CRT); (*pl*) (naut) flotsam

~, **внеза́п·н·ый,** (min) outburst, burst, break

~ **га́з·а** gas blowout

~ **да́ль·ност·и** (rad) range pip

~, **кон·е́чн·ый** (rad) undershot

~ **мо́щ·ност·и** surge/burst of power; (nucl) excursion of power

вы́·брос

~, пуск·ов·о́й   (rad) trigger pip

~, со·глас·у́ющ·ий   (rad) timing pip

у́гол· вы́·брос·а (*m*) angle of emission

вы́·брос·ить, -ят,   (*perf*) *see* **вы·бра́с·ыва·ть**

вы́·брос·к·а, -и,   (*g pl*) -с·ок·, (*f*) (*v n*) *see* **вы·бра́с·ыва·ть;** drop (of paratroops etc.)

вы́·брош·енн·ый, -ая, -ое, (*past part pass*) *see* **вы·бра́с·ыва·ть;** discarded, ejected, eliminated, rejected

вы·бу́р·ива·ть, -ют,   (*imp*) bore/drill out

вы́·бур·ить, -ят,   (*perf*) *see* **вы·бу́р·и·ва·ть**

вы́·буч·ить, -ат,   (*perf*) boil in caustic soda

вы·быв·а́·ть, -ют, (*imp*) leave, go away

вы́·бы·ть (*fut 3rd sing, pl*) **вы́·буд·ет, вы·буд·ут,** (*past masc sing*) **вы́·бы·л,** (*perf*) *see* **вы·быв·а́·ть**

вы́·вал·, -а,   (*m*) (geol) inrush

вы·ва́л·ива·ть, -ют,   (*imp*) throw/ tumble/turn out; **-ся** collapse, fall in/out; (geol) rush in

вы́·вал·ить, -ят, (*imp*) *see* **вы·ва́л·и·ва·ть**

вы́·вар·енн·ый, -ая, -ое, (*past part pass*) *see* **вы·ва́р·ива·ть;** boiled, extracted; digested

вы·ва́р·ивани·е, -я, (*n*) (*v n*) *see* **вы·ва́р·ива·ть**

вы·ва́р·ива·ть, -ют, (*imp*) boil down, extract by boiling; (chem) digest; (text) scour

вы́·вар·ить, -ят, (*perf*) *see* **вы·ва́р·и·ва·ть**

вы́·вар·к·а, -и, (*g pl*) -р·ок·, (*f*) (*v n*) *see* **вы·ва́р·ива·ть;** (*pl*) residue, concentrate (after boiling)

вы́·вед·а·ть, -ют, (*perf*) investigate, find out

вы·вед·е́ни·е, -я, (*n*) (*v n*) *see* **вы·вод·и́ть;** withdrawal, removal, extraction; deduction; elimination

вы́·вед·енн·ый, -ая, -ое, (*past part pass*) *see* **вы·вод·и́ть;** drawn out, extracted; eliminated, withdrawn; (math) deduced, arrived, derived

вы·ве́д·ыва·ть, -ют, (*imp*) investigate, find out

вы́·вез·ти, -ут, (*perf*) *see* **вы·воз·и́ть**

вы́·вер·енн·ый, -ая, -ое, (*past part pass*) *see* **вы·вер·я́·ть;** checked, calibrated, adjusted

вы́·вер·ить, -ят, (*perf*) *see* **вы·вер·я́·ть**

вы́·вер·к·а ·, -и, (*g pl*) -р·ок·, (*f*) (*v n*) *see* **вы·вер·я́·ть;** check, adjustment, regulation, calibration, alignment

вы́·вер·нут·ый, -ая, -ое, (*past part pass*) *see* **вы·вёрт·ыва·ть;** unscrewed, undone (of screws)

вы́·вер·н·уть, -ут, (*perf*) *see* **вы·вёрт·ыва·ть**

вы·вёрт·ыва·ть, -ют, (*imp*) unscrew; twist, wrench; **-ся** come unscrewed

вы·вер·я́·ть, -ют, (*imp*) check, adjust, regulate

вы́·вес·ить, -ят, (*perf*) hang out; weigh

вы́·вес·к·а, -и, (*g pl*) -с·ок·, (*f*) weighing; sign, signboard

вы́·вес·ти, (*fut 3rd sing, pl*) **вы́·вед·ет, вы́·вед·ут,** (*past masc sing*) **вы́·ве·л,** (*perf*); *see* **вы·вод·и́ть**

~ на орби́т·у   put into orbit

вы́·ветр·ел·ый, -ая, -ое, (*a*) weathered

вы́·ветр·енн·ый, -ая, -ое, (*past part pass*) *see* **вы·ве́тр·ива·ть;** weathered, eroded; aired; aerated (wines)

вы·ве́тр·ивани·е, -я, (*n*) airing; weathering, erosion; (cryst) efflorescence; aeration (of wine)

~, ветр·ов·о́е   (*geol*) aeolation, deflation

~, снег·ов·о́е   nivation

~ у́гл·я   coal slacking

вы·ве́тр·ива·ть, -ют, (*imp*) weather, erode; air; aerate (wines etc.); **-ся** (*pass*) become weathered/eroded

вы·ве́тр·ивающ·ийся, -аяся, -ееся, (*pres part act*) of **вы·ве́тр·ива·ться;** efflorescent (of crystals)

вы́·ветр·ивш·ийся, -аяся, -ееся, (*past part act*) *see* **вы·ве́тр·ива·ться;** weathered

вы́·ветр·ить, -ят, (*perf*) *see* **вы·ве́т·р·ива·ть**

вы́·веш·анн·ый, -ая, -ое, (*past part pass*) *see* **вы·ве́ш·ива·ть**

вы·ве́ш·ивани·е, -я, (*n*) (*v n*) *see* **вы·ве́ш·ива·ть**

вы·ве́ш·ива·ть, -ют, (*imp*) hang out; weigh

вы́·винт·ить, -ят, (*perf*) *see* **вы·ви́нч·ива·ть**

вы·ви́нч·ива·ть, -ют, (*imp*) unscrew; **-ся** come unscrewed, work loose (of screws)

вы́·винч·енн·ый, -ая, -ое, (*past part pass*) *see* **вы·ви́нч·ива·ть;** unscrewed, loosened, loose (of screws)

вы́·вих·, -а, (*m*) (med) dislocation

вы́·ви́х·ива·ть, -ют, (*imp*) (med) dislocate, sprain

вы́·вих·н·уть, -ут,   (*perf*) *see* вы·ви́·х·ива·ть

вы́·вод·, -а,   (*m*) (*v n*) *see* вы·вод·и́ть; withdrawal, extraction; conclusion, deduction, derivation; (elec) output (lead), tap (of transformer), outlet bushing; (air) recovery

~ като́д·а   (elec) cathode lead

~ пуч·к·а́   (nucl) beam extraction/ rejection

~, с·де́л·а·ть   conclude, draw a conclusion

~ сте́ржн·я   (nucl) withdrawal of rod

у́гол·вы́·вод·а   (*m*) (air) approach angle

~ у·равн·е́ни·я   (math) derivation of an equation

~, электро́д·н·ый   electrode lead (transistor)

вы·вод·и́ть, -́ят,   (*imp*) lead out, take out, withdraw, remove, extract; conclude, deduce, infer, derive; destroy, exterminate, eliminate; hatch, breed; depict, portray; -ся (*passive*); fall into disuse, go out of use; (biol) become extinct

вы́·вод·ков·ый, -ая, -ое,   (*a*) proliferous

вы·вод·н·о́й, -а́я, -о́е,   (*a*) output, discharge; (anat) excretory, exhalent

~ брус·о́к·   (*m*) (elec) connector/ terminal bar

~ у·стро́й·ств·о   (*n*) (nucl) extractor

~ у·стро́й·ств·о на перфо·ка́рт·ах   (*n*) punched card output (of computer)

~ у·стро́й·ств·о, фото·печа́т·аю·щ·ее   (*n*) photographic (output) printer

вы́·вод·ок·, -д·к·а,   (*m*) hatch, brood

вы·вод·я́щ·ий, -ая, -ее,   (*pres part act*) of вы·вод·и́ть; outgoing; discharging, excreting, excretory

вы́·воз·, -а,   (*m*) taking, sending (by vehicle); exports

вы·воз·и́ть, -́ят,   (*imp*) send, take, bring (by vehicle); export

вы·воз·н·о́й, -а́я, -о́е,   (*a*) export

вы·вола́к·ивани·е, -я,   (*n*) (*v n*) *see* вы·вола́к·ива·ть

вы·вола́к·ива·ть, -ют,   (*imp*) drag out

вы́·волоч·ь,   (*fut 3rd sing, pl*) вы́·волоч·ет, вы́·волок·ут,   (*past masc sing*) вы́·волок·,   (*perf*); *see* вы·вола́к·ива·ть

вы·вора́ч·ива·ть, -ют,   (*imp*) (text) turn out, turn (e.g. a suit); evert

вы·вора́ч·ива·ть

~ наизна́нку   (text) turn, turn inside out

вы́·ворот·, -а,   (*m*) (*v n*) *see* вы·во·ра́ч·ива·ть; (med) eversion

вы́·ворот·н·ый, -ая, -ое,   (*a*) turnable (of garments); turning (in sewing); (med) eversion

вы́·вш·ий, -ая, -ее,   (*past part act*) of вы·ть

вы́·гар·, -а,   (*m*) slag, dross

вы́·гар·к·и   (*nom pl*), (*g pl*) -ов (*m*) slag, dross

вы́·гиб·, -а,   (*m*) curve, bend, buckling, camber

вы·гиб·а́ни·е, -я,   (*n*) bending, warping

вы·гиб·а́·ть, -ют,   (*imp*) bend; -ся (*pass*) buckle

вы́·глад·ить, -ят,   (*perf*) вы·гла́·ж·ива·ть

вы·гла́ж·ива·ть, -ют,   (*imp*) smooth, level; iron, press (clothes etc.) -ся become smooth/level

вы·глубл·е́ни·е, -я,   (*n*) raising, shallowing-up

вы́·гляд·еть, -ят,   (*imp*) look, have the appearance, appear; (*perf*) notice, discover

вы·гля́д·ыва·ть, -ют,   (*imp*) look out; become visible, appear

вы́·гля·н·уть, -ут,   (*perf*) *see* вы·гля́·д·ыва·ть

вы́·гн·ать,   (*fut 3rd sing, pl*) вы́·гон·ит, вы́·гон·ят,   (*perf*); *see* вы·гон·я́·ть

вы́·гн·ут·ый, -ая, -ое,   (*past part pass*) *see* вы·гиб·а́·ть; bent, curved, cambered, convex

вы́·гн·уть, -ут,   (*perf*) *see* вы·гиб·а́·ть

вы·гова́р·ивани·е, -я,   (*n*) (*v n*) *see* вы·гова́р·ива·ть

вы·гова́р·ива·ть, -ют,   (*imp*) articulate, pronounce, utter; reprimand, rebuke; stipulate

вы́·говор·, -а,   (*m*) pronunciation, articulation; reprimand, rebuke

вы́·говор·ить, -ят,   (*perf*) *see* вы·го·ва́р·ива·ть

вы́·год·а, -ы,   (*f*) benefit, advantage, interest

фу́нкци·я вы́·год·ы   (*f*) (math) utility function

вы́·год·ност·ь, -и,   (*f*) profitability, advantageousness; utility

вы́·год·н·ый, -ая, -ое,   (*a*) (*s f*) -д·ен·, -д·н·а advantageous, profitable, favourable вы́·гон·, -а,   (*m*) driving out; pasture

вы́·гон·к·а, -и,   (*g pl*) -н·ок·,   (*f*) (*v n*) *see* вы·гон·я́·ть; distillation; forcing (of plants)

вы·го́н·щик·, -а,   (*m*) beater (shooting)

вы·гон·я́·ть, -ют,   (*imp*) drive out, expel; distil; force (plants)

вы·гора́ж·ива·ть, -ют,   (*imp*) fence/ rail/hedge/divide off

вы·гор·а́ни·е, -я,   (*n*) burning, burning-up; burnup; depletion (of nuclear fuel)

~, глуб·о́к·ое   (nucl) high burnup

~, не·глуб·о́к·ое   (nucl) low burnup

~ се́р·ы   (met) desulphurization, desulfurization

вы·гор·а́·ть, -ют,   (*imp*) burn down/ out/up; fade, bleach (in sunlight)

вы́·гор·евш·ий, -ая, -ее,   (*past part act*) *see* вы·гор·а́·ть

вы́·гор·ел·ый, -ая, -ое,   (*a*) faded (by sun)

вы́·гор·еть, -ят,   (*perf*) *see* вы·гор·а́·ть

вы́·город·ить, -ят,   (*perf*) *see* вы·гора́ж·ива·ть

вы́·город·к·а, -и,   (*g pl*) -д·ок·,   (*f*) compartment

вы́·грав·ир·овать, -уют,   (*perf*) engrave; etch

вы́·гран·ить, -ят,   (*perf*) *see* вы·гра́н·ива·ть

вы·гра́н·ива·ть, -ют,   (*imp*) cut (gem stones)

вы́·греб·, -а,   (*m*) raking out, drawing (a fire); cesspool

вы·греб·а́·ть, -ют,   (*imp*) rake out, draw (a fire); clean out; row out/ against

вы́·греб·н·о́й, -а́я, -о́е,   (*a*) of вы́·-греб·

вы́·грес·ти,   (*fut 3rd sing, pl*) вы́·греб·ет, вы́·греб·ут,   (*past masc sing*) вы́·греб·,   (*perf*); *see* вы·греб·а́·ть

вы·груж·а́·ть, -ют,   (*imp*) unload, discharge, disembark, detrain

вы́·груз·ить, -ят,   (*perf*) *see* вы·груж·а́·ть

вы́·груз·к·а, -и,   (*g pl*) -з·ок·,   (*f*) (*v n*) *see* вы·груж·а́·ть

вы́·груз·н·о́й, -а́я, -о́е,   (*a*) = вы́·-груз·очн·ый

вы́·груз·очн·ый, -ая, -ое,   (*a*) of вы́·груз·к·а; discharge

вы·грыз·а́·ть, -ют,   (*imp*) gnaw/bite out

вы́·грыз·енн·ый, -ая, -ое,   (*past part pass*) *see* вы·грыз·а́·ть; gnawed, bitten

вы́·грыз·ть, -ут,   (*perf*) *see* вы·грыз·а́·ть

вы·да·ва́·ть, -ют, (*imp*) give out, issue, distribute, serve/hand out, present; give up, extradite; (fin) draw (a bill on); give away, betray, reveal; -ся (*pass*) be conspicuous/distinguished; protrude, jut out; occur, happen to be

вы́·дав·ить, -ят,   (*perf*) *see* вы·да́в·л·ива·ть

вы́·дав·к·а, -и,   (*f*) (*v n*) *see* вы·да́в·л·ива·ть; (met) spinning

вы·да́вл·иван·и·е, -я,   (*n*) extrusion; squirting

вы·да́вл·ива·ть, -ют,   (*imp*) press/ squeeze out, extrude; force, break, smash, burst

вы·да́вл·иваю·щ·ий, -ая, -ее,   (*pres part act*) of вы·да́вл·ива·ть

вы·да́·иван·и·е, -я,   (*n*) (agr) milking

вы·да́лбл·ива·ть, -ют,   (*imp*) hollow/ chisel/chip out; learn by heart

вы́·да·ть,   (*fut 3rd sing, pl*) вы́·даст·, вы́·дадут,   (*perf*); *see* вы·да·ва́·ть

вы́·да·ч·а, -и,   (*f*) (*v n*) *see* вы·да·ва́·ть устро́й·ств·о вы́·да·ч·и резулта́т·ов (*n*) output unit/device (computers)

вы·да·ю́щ·ийся, -аяся, -ееся,   (*pres part act*) of вы·да·ва́·ться; outstanding, eminent, prominent; projecting, protruding, salient

вы́·двиг·, -а,   (*m*) protrusion, outthrust

вы·двиг·а́·ть, -ют,   (*imp*) move/pull/ push out; advance, put forward, promote; nominate, propose; -ся (*passive voice*); (*imp only*) move/slide in and out; rise, come to the fore

вы·движ·е́ни·е, -я,   (*n*) advancement, promotion; extension

~ обьекти́в·а   lens extension

вы·движ·н·о́й, -а́я, -о́е,   (*a*) extensible, telescopic; sliding (e.g. drawer), pull-out

вы́·дви·н·уть, -ут,   (*perf*) *see* вы·двиг·а́·ть

вы́·дел·, -а,   (*m*) portion, share (of property)

вы́·дел·анн·ый, -ая, -ое,   (*past part pass*) *see* вы·де́л·ыва·ть

вы́·дел·а·ть, -ют, (*perf*) *see* вы·де́л·ыва·ть

вы·дел·éни·е, -я,   (*n*) (*v n*) *see* **вы·де·л·я́·ть;** (chem) isolation, separation, parting, precipitation, liberation, evolution, emission; (physiol) excretion, secretion; exudation; (print) display

~ **га́з·а**   (geol) gas emission/emanation; (chem) gas liberation/evolution, gassing

~ **кисло·ро́д·а**   deoxidation

~ **криста́лл·ов**   crystallization

~, **металло·но́с·н·ое**   (geol) metalliferous exhalation

~ **при́·мес·ей**   (met) precipitation of impurities

~ **тепл·а́**   liberation of heat

~ **фа́з·ы**   (met) precipitation of a phase

~ **част·от·ы́**   (rad) frequency discrimination

~ **шри́фт·ом**   (print) display work/composition

~, **электро·лит·и́ческ·ое**   (met) electrolytic parting

~ **эне́рг·и·и**   release of energy

вы́·дел·енн·ый, -ая, -ое,   (*past part pass*) *see* **вы·дел·я́·ть**

~ **сеч·éни·е** (*n*) detail section (drawing)

~ **с·чёт·чик·ом** (instr) counter-defined

вы́·дел·ител·ь,   -я,   (*m***)** separator, discriminator

вы·дел·и́тельн·ый, -ая, -ое,   (*a*) separating, discriminating;   (biol) secretory, excretory

~ **систéм·а** (*f*) (zool) excretory system

вы́·дел·ить, -ят,   (*perf*) *see* **вы·де·л·я́·ть**

вы́·дел·к·а, -и,   (*g pl*) **-л·ок·,**   (*f*) (*v n*) *see* **вы·де́л·ыва·ть**

вы·де́л·ыва·ть, -ют,   (*imp*) make; treat, dress (skins etc.)

вы·дел·я́·ть, -ют,   (*imp*) separate, single out, isolate; allot, apportion, share out; discriminate, mark out, distinguish; (print) display; **-ся** (*pass*) escape, emanate, precipitate

вы·дел·я́ем·ый, -ая, -ое,   (*pres part pass*) of **вы·дел·я́·ть;** separable; separating

вы·дёрг·ивани·е, -я,   (*n*) (*v n*) *see* **вы·дёрг·ива·ть;** extraction; surface lifting (of paper)

вы·дёрг·ива·ть, -ют,   (*imp*) extract, draw/pull out

вы́·держ·анн·ый, -ая, -ое, (*past part pass*) *see* **вы·де́рж·ива·ть;** delayed, postponed; laid aside, set aside; seasoned (timber), mature (paper); (met) held, soaked; (nucl) stored, cooled; sustained, maintained

вы́·держ·анн·ый, -ая, -ое

~ **у́·ровен·ь** (*m*) (geol) recurrence level

вы́·держ·ать, -ат,   (*perf*) *see* **вы·де́р·ж·ива·ть**

вы·де́рж·ивани·е, -я,   (*n*) (*v n*) *see* **вы·де́рж·ива·ть;** (air) holding down; holding, seasoning, setting aside; curing (concrete); storage (for nuclear decay), cooling (for nuclear decay)

~ **над земл·е́й**   (air) holding off

~ **от·хо́д·ов**   (nucl) holdup of waste

вы·де́рж·ива·ть, -ют,   (*imp*) withstand, endure, hold out; pass (tests etc.); persist, maintain, hold to; mature, ripen, season (by keeping), age; (nucl) store, hold, cool; (chem) leave standing

вы́·держ·к·а, -и,   (*g pl*) **-ж·ек·,** (*f*) *see* **вы·де́рж·ивани·е;** endurance, staying power, stamina, tenacity; extract, excerpt, quotation; seasoning, maturing; (met) holding, soaking; (phot) exposure, time element, shutter speed

~ **врéмен·и**   time element/delay

вы́·держ·к·и врéм·я   (elec) delay time; (met) soaking/holding time; (nucl) decay/cooling time; (rubber) curing/freeze time

на вы́·держ·к·у   at random

с вы́·держ·к·ой врéмен·и   delay, delayed action

вы́·дер·н·уть, -ут,   (*perf*) extract, pull out

вы·дир·а́·ть, -ют,   (*imp*) tear out

вы́·долб·ить, -ят,   (*perf*) *see* **вы·да́лб·л·ива·ть**

вы́·долбл·енн·ый, -ая, -ое,   (*past part pass*) *see* **вы·да́лбл·ива·ть;** hollowedout

вы́·дох·, -а,   (*m*) (physiol) expiration, exhalation

вы́·дох·н·уть, -ут,   (*perf*) *see* **вы·ды·х·а́·ть**

вы́др·а, -ы,   (*f*) (zool) otter, *Lutra lutra*; otter fur; holing waste (metal working)

~, **мор·ск·а́я**   sea otter, *Enhydra lutris*

вы́·др·ать, (*fut 3rd sing, pl*) **вы́·дер·ет, вы́·дер·ут,**   (*perf*) flog, thrash (*imp*), (*pres 3rd sing, pl as above*) tear out

вы́·дрессир·овать, -уют,   (*perf*) train, school

вы́·дуб·ить, -ят,   (*perf*) tan

вы́·дубл·енн·ый, -ая, -ое,   (*past part pass*) of **вы́·дуб·ить**

~ **ко́ж·а** (*f*) leather

вы·дув·а́льщик, -а,   (*m*) glass-blower

вы·дув·а́ни·е, -я,   (*n*) blow-off, blow-out, deflation

~, магни́т·н·ое   (elec) magnetic blow-out

вы·дув·а́·ть, -ют,   (*imp*) blow out; blow (glass etc.); deflate

вы́·дув·к·а, -и,   (*g pl*) -в·ок·,   (*f*) blowing out, blow-off, blowout; deflation

вы·дув·н·о́й, -а́я, -о́е,   (*a*) blown; blowing, blow

вы́·дум·анн·ый, -ая, -ое,   (*past part pass*); thought-up, invented, concocted

вы́·дум·к·а, -и,   (*g pl*) -м·ок·,   (*f*) fabrication, invention; gadget, device

вы́·ду·ть, -ют,   (*perf*) *see* вы·дув·а́·ть

вы·дых·а́ни·е, -я,   (*n*) exhalation, expiration

вы·дых·а́тельн·ый, -ая, -ое,   (*a*) expiratory

вы·дых·а́·ть, -ют,   (*imp*) breathe out, exhale; -ся (*passive*); become spent; lose force; become flat (of drinks); lose fragrance/smell

вы·дых·а́ющ·ийся, -аяся, -ееся, (*pres part act*) of вы·дых·а́·ться; fading

вы·ед·а́ни·е, -я,   (*n*) (*v n*) of вы·е·д·а́·ть; corrosion

вы·ед·а́·ть, -ют, (*imp*) corrode, eat away

вы́·езд·, -а, (*m*) (*v n*) *see* вы·езж·а́·ть; departure; "out" (vehicular exit); turnout (of horses)

вы́·езд·ить, -ят,   (*perf*) break in, break in and train (horses)

вы́·езд·к·а, -и,   (*v n*) *see* вы́·езд·ить

вы·езд·н·о́й, -ая, -ое, (*a*) of вы́·езд·; travelling, visiting (of law courts, theatre companies etc.)

~ ло́шад·ь   (*f*) saddle-horse

~ се́сси·я суд·а́   (*f*) (law) assizes, circuit court session

вы·езж·а́·ть, -ют,   (*imp*) leave, depart, drive out; break in, train (horses)

вы́·ем·к·а, -и,   (*g pl*) -м·ок·,   (*f*) (*v n*) *see* вы·ним·а́·ть; (min) cutting, winning, excavation, slicing cut; (civ eng) cutting, dugout, ditch; notch, groove, hollow; (law) seizure

~, глуб·о́к·ая   (min) deep cut

~ гру́нт·а   excavation

~, кра·ев·а́я   (biol) emargination

~ ла́в·ами   (min) longwall mining, longwall

~, о́рт·ов·ая   (min) side slicing

~, радиа́ль·н·ая   (min) radial top-slicing

вы́·ем·к·а

~, сло·ев·а́я   (min) top slicing

~, сплош·н·а́я   (min) longwall, longwall system

~ туннел·я, порта́ль·н·ая   (rail min) tunnel-approach cutting

~, у·сту́п·н·ая   (min) bench working/cut

вы·ем·н·о́й, -а́я, -о́е,   (*a*) for cutting out, cutting, notching

вы́·ем·очн·ый, -ая, -ое,   (*a*) of вы́·ем·к·а; (min) stoping

вы́·ем·чат·ый, -ая, -ое,   (*a*) grooved, notched, indented; (biol) sinuate, emarginate, retuse

вы́·ест·ь,   (*fut 3rd sing, pl*) вы́·ест, вы́·ед·ят, (*past masc sing*) вы́·ел, (*perf*); corrode, eat away

вы́·ехать,   (*fut 3rd sing, pl*) вы́·едет, вы́·едут, (*perf*); leave, depart, drive out

вы́·жат·ь, (*fut 3rd sing, pl*) вы́·жм·ет, вы́·жм·ут, (*perf*) *see* вы·жим·а́·ть; (*fut 3rd sing, pl*) выжнет, выжнут, (*perf*); *see* вы·жин·а́·ть

вы́·жат·ый, -ая, -ое, (*past part pass*) *see* вы·жим·а́·ть

~ сок·   (*m*) expressed juice

вы́·жд·ать, -ут,   (*perf*) *see* вы·жи·д·а́·ть

вы́·жеч·ь, (*fut 3rd sing, pl*) вы́·жж·ет, вы́·жг·ут, (*past masc, fem sing*) вы́·жег·, вы́·жг·ла, (*perf*); *see* вы··жиг·а́·ть

вы·жив·а́емост·ь, -и,   (*f*) survival rate, survival, hardiness

вы·жив·а́ни·е, -я, (*n*) survival

вы·жив·а́·ть, -ют, (*imp*) survive, live through; outlive

вы́·жиг·, -а, (*m*) (*v n*) *see* вы·жиг·а́·ть; yield from one firing (e.g. of coke, lime etc.)

вы·жиг·а́ни·е, -я, (*n*) (*v n*) of вы··жиг·а́·ть; (med) cauterization; (vet) firing

глуб·ин·а́ вы·жиг·а́ни·я (*f*) (nucl) burnup fraction, burning depth

вы·жиг·а́·ть, -ют,   (*imp*) burn out/down/through; brand, burn on; burn (bricks); (med) cauterize; (vet) fire

вы·жид·а́тельн·ый, -ая, -ое, (*a*) expectant; waiting, temporizing

вы·жид·а́·ть, -ют, (*imp*) wait for/until

вы·жим·а́ни·е, -я, (*n*) (*v n*) *see* вы··жим·а́·ть; expressing; lifting (by frost)

вы·жим·а́тел·ь, -я, (*m*) squeezer, expresser

вы·жим·а́·ть, -ют, (*imp*) squeeze/press/wring out, express; (geol) pinch, pinch out, lift, heave (e.g. from frost)

вы́·жим·к·а, -и, (*g pl*) **-м·ок·,** (*f*) *see* **вы·жим·а́ни·е;** (*pl m or f*) pressed residue, pomace, cake, husks, marc (of grapes); overflow, spew (from press moulding)

вы·жин·а́·ть, -ют, (*imp*) mow (in certain area)

вы́·жи·ть, (*fut 3rd sing, pl*) **вы́·жив·ет, вы́·жив·ут** (*perf*); *see* **вы·жи·в·а́·ть**

вы́·зв·анн·ый, -ая, -ое, (*past part pass*) *see* **вы·зыв·а́·ть;** induced, caused, due to; (telecom) called

вы́·зв·ать (*fut 3rd sing, pl*) **вы́·зов·ет, вы́·зов·ут,** (*perf*); *see* **вы·зыв·а́·ть**

вы́·звол·ить, -ят, (*perf*) *see* **вы·зво·л·я́·ть**

вы·звол·я́·ть (*imp*) help/get out, rescue

вы·здоро́вл·ивани·е, -я, (*n*) recovery, convalescence

вы·здоро́вл·ива·ть, -ют, (*imp*) get well/better, convalesce, recover

вы́·здоров·е·ть, -ют, (*perf*) *see* **вы·-здоро́вл·ива·ть**

вы́·зов·, -а, (*m*) (*v n*) *see* **вы·зыв·а́·ть;** (telecom) call; challenge; (law) summons, subpoena; (mil) call-up

~, авар·и́йн·ый, (telecom) distress call

~, инду́ктор·н·ый magneto calling system (telephony)

~, ло́ж·н·ый rub-out signal (telephony)

~, не·по́лн·ый incompletely dialled call (telephony)

~, прям·о́й ringdown, generator signalling (telephony)

у·каз·а́тел·ь вы́·зов· (*m*) (telecom) call indicator/display

~, фон·и́ческ·ий (telecom) buzzer calling/call

вы·зола́ч·ива·ть, -ют, (*imp*) gild

вы́·золот·ить, -ят, (*perf*) *see* **вы·зо·ла́ч·ива·ть**

вы́·золоч·енн·ый, -ая, -ое, (*past part pass*) *see* **вы·золо́ч·ива·ть**

вы·зрев·а́ни·е, -я, (*n*) (*v n*) of **вы·-зрев·а́·ть;** ripening, maturing, ageing

вы·зрев·а́·ть, -ют, (*imp*) grow ripe, ripen, mature

вы́·зре·ть, -ют, (*perf*) *see* **вы·зрев·а́·ть**

вы·зыв·а́ем·ый, -ая, -ое, (*pres part pass*) of **вы·зыв·а́·ть;** called (telephony); evoked; caused

~ сторон·а́ (*f*) called party (telephony)

вы·зыв·а́·ть, -ют, (*imp*) call, call out/up, challenge (mil), call by telephone, ring up; (law) summons, issue a summons against, subpoena; cause, provoke, induce, give rise to; **-ся** (*pass*) volunteer

вы·зыв·а́ющ·ий, ая, -ое, (*pres part act*) of **вы·зыв·а́·ть;** calling; causing, bringing about, provoking

~ сторон·а́ (*f*) calling party

~ су́дорог·и (*n*) (med) convulsant

вы·зыв·н·о́й, -а́я, -о́е, (*a*) call, calling

~ кно́п·к·а (*f*) call/calling button

вы́·игр·а·ть, -ют, (*perf*) *see* **вы·йг·р·ыва·ть**

вы́·игр·ыва·ть, -ют, (*imp*) win, gain; benefit from

вы́·игр·ыш·, -а, (*i*) **-ем,** (*m*) winnings, prize, gain, advantage, payoff

~ анте́нн·ы (rad) antenna gain

~, в·нос·и́м·ый (rad) insertion gain

коэффицие́нт· вы́·игр·ыш·а (*m*) (nucl) advantage factor

фу́нкци·я вы́·игр·ыш·а (*f*) (math) payoff function

вы́·игр·ышн·ый, -ая, -ое, (*a*) winning; lottery

вы́·иск·, -а, (*m*) search

вы́·иск·ать, (*fut 3rd sing, pl*) **вы́·ищ·ет, вы́·ищ·ут,** (*perf*); *see* **вы·-йск·ива·ть**

вы́·иск·ива·ть, -ют, (*imp*) search/seek/hunt/find out

вы́·йти, (*fut 3rd sing, pl*) **вы́·йд·ет, вы́·йд·ут,** (*past masc, fem sing*) **вы́·-шел, вы́·шла,** (*perf*); *see* **вы·хо·д·и́ть**

вы·каз·ать, (*perf*) (*fut 3rd sing, pl*) **вы́·каж·ет, вы́·каж·ут,** (*perf*); *see* **вы·ка́з·ыва·ть**

вы·каз·н·о́й, -а́я, -о́е, (*a*) exhibited, manifested

вы·ка́з·ыва·ть, -ют, (*imp*) display, manifest

вы·ка́л·ыва·ть, -ют, (*imp*) prick/spike out

вы·ка́п·ыва·ть, -ют, (*imp*) dig up/out, exhume

вы́·карабк·а·ться, -ются, (*perf*) *see* **вы·кара́бк·ива·ться**

вы·кара́бк·ива·ться, -ются, (*imp*) clamber/scramble out, extricate oneself

вы́·кат·а·ть, -ют, (*perf*) roll flat; press/iron (with rollers); mangle (laundry)

вы́·кат·ить, -ят, (*perf*) *see* **вы·ка́т·ы·ва·ть**

вы·ка́т·ыва·ть, -ют, (*imp*) roll/wheel out

вы·ка́ч·анн·ый, -ая, -ое, (*past part pass*) *see* **вы·ка́ч·ива·ть**

вы́·кач·а·ть -ют, (*perf*) *see* **вы·ка́·ч·ива·ть**

вы́·кач·енн·ый, -ая, -ое, (*past part pass*) *see* **вы·ка́т·ыва·ть**

вы·ка́ч·ива·ть, -ют, (*imp*) pump out, evacuate, extort

вы·ка́ш·ива·ть, -ют, (*imp*) mow

вы·ка́шл·ива·ть, -ют, (*imp*) expectorate, cough up

вы́·кашл·я·ть, -ют, (*perf*) *see* **вы··ка́шл·ива·ть**

вы́·кид·, -а, (*m*) one-throw sample

вы·ки́д·ыва·ть, -ют, (*imp*) throw out, discard, eliminate, reject; (med) have a miscarriage, (vet) abort

вы́·кид·ыш·, -а, (*i*) **-ем,** (*m*) miscarriage; abortion

вы́·ки·нут·ый, -ая, -ое, (*past part pass*) *see* **вы·ки́д·ыва·ть** с **вы́·ки·нут·ой** discarding, except for

вы́·ки·н·уть, -ут, (*perf*) *see* **вы·ки́·д·ыва·ть**

вы·кип·а́·ть, -ют, (*imp*) boil away

вы́·кип·еть, -ят, (*perf*) *see* **вы·ки·п·а́·ть**

вы́·кипят·ить, -ят, (*perf*) scald; clean by boiling, boil, sterilize

вы́·клад·к·а, -и, (*g pl*) **-д·ок·,** (*f*) (*v n*) *see* **вы·кла́д·ыва·ть;** laying, laying out; (mil) pack, kit; (*usually plur*) calculations, computations, mathematical operations

вы·кла́д·ыва·ть, -ют, (*imp*) lay, lay out, unpack; spread, surface with

вы́·кл·евать, -юют, (*perf*) *see* **вы··клёв·ыва·ть**

вы·клёв·ыва·ть, -ют, (*imp*) peck out/up; **-ся** hatch

вы·клик·а́·ть, -ют, (*imp*) call out

вы́·клик·н·уть, -ут, (*perf*) *see* **вы··клик·а́·ть**

вы·кли́н·ивани·е, -я, (*n*) crowding/squeezing out, tapering/petering out; (cryst) suppression

~, взаи́м·н·ое intertonguing

вы·кли́н·ивающ·ий, -ая, -ое, (*pres part act*) squeezing/crowding out; (cryst) suppressing

вы·кли́н·ивающ·ийся, -аяся, -ееся, (*pres part act*) tapering away, wedge-shaped, gash

~ тре́щ·ин·а (*f*) (geol) gash joint

вы·ключ·а́ем·ый, -ая, -ое, (*pres part pass*) *see* **вы·ключ·а́·ть**

вы·ключ·а́тел·ь, -я, (*m*) (elec) switch, breaker, circuit-breaker; releasing device

~, авто·га́з·ов·ый (elec) gas-blast breaker

~, ба́к·ов·ый high oil-content circuit-breaker, bulk-oil circuit-breaker

~, быстро·де́й·ствующ·ий (elec) quick-break switch

~, вод·ян·о́й water circuit-breaker

~, воз·ду́ш·н·ый air-blast breaker

~, вы·движ·н·о́й withdrawable circuit-breaker

~, газо·генер·и́рующ·ий (elec) gas-blast breaker

~, двер·н·о́й door switch

~, жи́дк·ост·н·ый liquid-immersed circuit breaker

~, кон·е́чн·ый limit switch

~, конц·ев·о́й limit switch

~, мало·ём·кост·н·ый anti-capacity switch

~, мало·ма́сл·ян·ый (elec) orthojector circuit breaker

~, ма́сл·ян·ый oil switch/breaker

~, между·ши́н·н·ый (elec) bus-coupler switch

~ на·ка́л·а filament switch

~, паке́т·н·ый packet rotary switch

~, пере·кид·н·о́й tumbler switch

~ пит·а́ни·я power switch

~, пред·о·хран·и́тельн·ый (st eng) emergency governor

~, пут·ев·о́й tappet switch; (rail) track switch

~, секцио́н·н·ый section switch

~, стациона́р·н·ый fixed breaker

~, флаж·ко́в·ый trip switch

~, час·ов·о́й time switch

~, шино·со·един·и́тельн·ый bus-coupler switch

~, электро·магни́т·н·ый electromagnetic switch

вы·ключ·а́·ть, -ют, (*imp*) turn off, switch off, cut out/off, disconnect; put out of gear, disengage; shut off (steam), shut down (nucl reactor)

вы·ключ·е́ни·е, -я, (*n*) (*v n*) *see* **вы··ключ·а́·ть** (elec) shut-off, trip; (rocket) cutoff; (nucl) shut-down, scram

вы́ключéние

~, автомати́ческое (elec) automatic cut-out

~, максима́льное (elec) overload trip

~, минима́льное no-load trip

~ с замедлéнием time-lag trip

сигна́л выключéния (*m*) cutoff/shutoff signal

цепь выключéния (*f*) shut-off circuit, trip circuit

вы́ключенный, -ая, -ое, (*past part pass*) *see* **выключáть**; out-of-gear, disengaged, turned-off, switched-off, disconnected

вы́ключить, -ат, (*perf*) *see* **вы́ключáть**

вы́ключка, -и, (*f*) (print) justifying, justification, spacing

~ набóра, плóтная (print) closed spacing

~, слáбая loose justification

~ строк (print) justification

вы́клюют (*fut 3rd pl*) of **вы́клевать**

вы́ковать (*fut 3rd sing, pl*) **вы́кует, вы́куют,** (*perf*); *see* **выкóвывать**

выкóвывать, -ют, (*imp*) forge, hammer out

выколáчиватель, -я, (*m*) (print) planer

выколáчивать, -ют, (*imp*) knock/beat out

вы́колотить, -ят, (*perf*) *see* **выколáчивать**

вы́колотка, -и, (*g pl*) **-ток,** (*f*) (mech eng) knockout, drift (tool); centre key; lap stone (leather)

вы́колотый, -ая, -ое, (*past part pass*) of **вы́колоть**

вы́колоть, -ют, (*perf*) prick/spike out

вы́колоченный, -ая, -ое, (*past part pass*) *see* **выколáчивать**

вы́копать, -ют, (*perf*) *see* **выка́пывать**

выкопи́рование, -я, (*n*) (phot) printing out

выкорчевать, -уют, (*perf*) *see* **выкорчёвывать**

выкорчёвывать, -ют, (*imp*) grub/stub up, uproot

вы́косить, -ят, (*perf*) mow

вы́кошенный, -ая, -ое, (*past part pass*) of **вы́косить**; mown, mowed

выкра́дывать, -ют, (*imp*) steal; (elec) tap

выкра́ивать, -ют, (*imp*) cut out (patterns etc.)

вы́красить, -ят, (*perf*) *see* **выкра́шивать**

выкра́сть (*fut 3rd sing, pl*) **вы́крадет, вы́крадут,** (*past masc sing*) **вы́крал,** (*perf*); steal; tap (electric current etc.)

выкра́шивание, -я, (*n*) (*v n*) of **выкра́шивать(ся);** (met) pitting

~, нача́льное corrective/incipient pitting

~, ограни́ченное corrective/incipient pitting

~, прогресси́вное destructive pitting

~, тóчечное pitting

выкра́шивать, -ют, (*imp*) paint, dye; **-ся** take/hold a colour, crumble, break up, break off

выкристаллизова́ние, -я, (*n*) crystallizing; efflorescence

вы́кристаллизовать, -уют, (*perf*) *see* **выкристаллизо́вывать**

выкристаллизо́вывание, -я, (*n*) efflorescence

выкристаллизо́вывать, -ют, (*imp*) crystallize, crystallize out; **-ся** (*pass*) effloresce

вы́кроить, -ят, (*perf*) *see* **выкра́ивать**

вы́кройка, -и, (*f*) pattern (guide to cutting)

вы́крошиться, -атся, (*perf*) *see* **выкра́шиваться;** crumble, crumble away, break up, break off (e.g. screws)

вы́кружка, -и, (*f*) fillet, recess, chamber; rounding off; (shipb) boss plate, boss

выкру́живать, -ют, (*imp*) round off, give a round shape

вы́крутить, -ят, (*perf*) *see* **выкру́чивать**

выкру́чивать, -ют, (*imp*) twist, lay, form by twisting; unscrew; **-ся** (*pass*) come unscrewed

вы́куп, -а, (*m*) (fin) redemption

выкупа́ть, -ют, (*imp*) (fin) redeem; **выкупа́ть, -ют,** (*perf*) bath, bathe

вы́купить, -ят, (*perf*) (fin) redeem

вы́купленный, -ая, -ое, (*past part pass*) of **вы́купить**

выкупнóй, -а́я, -óе, (*a*) of **вы́куп**

~ пра́во (*n*) right of redemption

вы́кует (*fut 3rd sing*) of **вы́ковать**

выку́ривать, -ют, (*imp*) smoke out, use up (tobacco)

вы́·кур·ить, -ят, (*perf*) *see* вы·ку́·р·ива·ть

вы́·куют (*fut 3rd pl*) of вы́·ков·ать

вы́·лавл·ива·ть, -ют, (*imp*) fish/get out, catch; recover, get back

вы́·лаз·к·а, -и, (*g pl*) -з·ок·, (*f*) sortie, raid

вы́·лам·ыва·ть, -ют, (*imp*) break in/out/down/open

вы́·легч·а·ть, -ют, (*perf*) loosen, relieve; (agr) geld

~ вальц·ы́ open up rolls

вы́·легч·ивани·е (*v n*) *see* вы́·легч·а·ть; lightening (paper)

вы́·леж·ать, -ат, (*perf*) *see* вы·ле́ж·ива·ть

вы·ле́ж·ивани·е, -я, (*n*) keeping, standing; ageing, seasoning

вы·ле́ж·ива·ть, -ют, (*imp*) spend a time in bed, be bedridden; -ся mature, ripen, season, age

вы́·лёж·к·а, -и, (*f*) *see* вы·ле́ж·ивани·е

вы·лез·а́·ть, -ют, (*imp*) climb/clamber out; get out, alight; wear, become worn (of fur); come/drop out (of hair)

вы́·лез·ти = вы́лезть

вы́·лез·ть (*perf*) *see* вы·лез·а́·ть

вы́·лет·, -а, (*m*) (*v n*) *see* вы·лет·а́·ть; departure; (nucl) emission, escape; overhang; throat clearance, sweep (of lathe); boom, outrigger; gap

~, бо·ев·о́й (air) operational mission, sortie

~ вы́·брос·а radius of ejection
ли́ни·я вы́·лет·а (*f*) line of departure (ballistics)

~ пре́сс·а ram overhang
точка вы́·лет·а (*f*) (gunn) origin
у́гол· вы́·лет·а (*m*) jump (ballistics); (phys) angle of emission

вы·лет·а́·ть, -ют, (*imp*) fly out/off/away, depart; leak out (gases), escape, emerge

вы·лет·а́ющ·ий, -ая, -ее, (*pres part act*) of вы·лет·а́·ть; outgoing, exit, emerging (of flying particles), escaping (of flying particles)

вы́·лет·еть, -ят, (*perf*) *see* вы·лет·а́·ть

вы·ле́ч·ивани·е, -я, (*n*) (*v n*) *see* вы·ле́ч·ива·ть; recovery

вы·ле́ч·ива·ть, -ют, (*imp*) cure, heal; -ся be healed, cured, recover

вы́·леч·ить, -ат, (*perf*) *see* вы·ле́ч·и·ва·ть

вы́·леч·у (*fut 1st sing*) of вы́·лет·еть; (*fut 1st sing*) of вы́·леч·ить

вы·лив·а́ни·е, -я, (*n*) (*v n*) *see* вы·лив·а́·ть

вы·лив·а́·ть, -ют, (*imp*) pour out, decant; (met) cast; -ся (*pass*) pour/run/flow out

вы́·лин·я·ть, -ют, (*perf*) fade; moult; shed hairs

вы́·ли·т·ый, -ая, -ое, (*past part pass*) *see* вы·лив·а́·ть; poured, decanted; (met) cast

вы́·ли·ть, (*fut 3rd sing, pl*) вы́·ль·ет, вы́·ль·ют, (*perf*); *see* вы·лив·а́·ть

вы́·лов·ить, -ят, (*perf*) *see* вы·ла́вл·ива·ть

вы́·ловл·енн·ый, -ая, -ое, (*past part pass*) *see* вы·ла́вл·ива·ть

вы́·лож·енн·ый, -ая, -ое, (*past part pass*) *see* вы·кла́д·ыва·ть; laid out; lined

вы́·лож·ить, -ат, (*perf*) *see* вы·кла́д·ыва·ть

вы́·лом·, -а, (*m*) (*v n*) *see* вы·ла́м·ыва·ть

вы́·лом·а·ть, -ют, (*perf*) *see* вы·ла́м·ыва·ть

вы́·лом·к·а, -и, (*g pl*) -м·ок·, (*f*) (*v n*) *see* вы·ла́м·ыва·ть

вы́·лощ·ить, -ат, (*perf*) polish

вы́·луд·ить, -ят, (*perf*) *see* вы·лу́ж·ива·ть

вы́·луж·енн·ый, -ая, -ое, (*past part pass*) *see* вы·лу́ж·ива·ть; tinned, tin-plated

вы·лу́ж·ива·ть, -ют, (*imp*) tin, coat with tin

вы́·луп·ить, -ят, (*perf*) *see* вы·луп·л·я́·ть

вы́·лупл·я́·ть, -ют, (*imp*) shell, remove shell, bark (a tree); -ся hatch

вы·лу́щ·ени·е, -я, (*n*) (*v n*) *see* вы́·лу·щ·ить (med) enucleation

вы·лу́щ·ива·ть, -ют, (*imp*) shell, remove shell, husk; (med) enucleate

вы́·лущ·ить, -ат, (*perf*) *see* вы·лу́·щ·ива·ть

вы́·маз·ать, (*fut 3rd sing, pl*) вы́·маж·ет, вы́·маж·ут, (*perf*); *see* вы·ма́з·ыва·ть

вы·ма́з·ыва·ть, -ют, (*imp*) smear, daub, cover with oil/grease/paint

вы·ма́н·ива·ть, -ют, (*imp*) entice, lure; swindle

вы́·ман·ить, -ят, (*perf*) *see* вы·ма́·н·ива·ть

вы́·мар·а·ть, -ют, (*perf*) *see* вы·ма́·р·ыва·ть

вы·ма́р·ивани·е, -я, (*n*) expurgation

вы·ма́р·ывани·е, -я, (*n*) (*v n*) of вы·ма́р·ыва·ть

вы·ма́р·ыва·ть, -ют, (*imp*) dirty, soil, smear; strike/cross out

вы·ма́сл·ива·ть, -ют, (*imp*) oil, butter

вы·ма́т·ыва·ть, -ют, (*imp*) unwind, unreel, pay out, wind out; dissipate, deplete

вы·ма́ч·ивани·е, -я, (*n*) (*v n*) of вы·ма́ч·ива·ть

~ льн·а́ flax retting

вы·ма́ч·ива·ть, -ют, (*imp*) soak, drench; steep; ret (flax); (zool) macerate

вы·ма́щ·ива·ть, -ют, (*imp*) pave

вымбо́вк·а, -и, (*g pl*) -вок·, (*f*) (naut) capstan bar

вы́·мен·, -а, (*m*) barter, exchange

вы·ме́н·ива·ть, -ют, (*imp*) barter, exchange

вы́·мен·я·ть, -ют, (*perf*) see вы·ме́н·ива·ть

вы́·мер· (*past masc sing*) of вы́·мер·еть

вы́·мер·еть, (*fut 3rd sing, pl*) вы́·м·рет, вы́·мр·ут, (*past masc sing*) вы́·мер·, (*perf*); see вы·мир·а́·ть

вы́·мер·енн·ый, -ая, -ое, (*past part pass*) of вы́·мер·ить

вы·мерз·а́ни·е, -я, (*n*) (*v n*) of вы··мерз·а́·ть; (bot) winterkilling

вы·мерз·а́·ть, -ют, (*imp*) be killed/destroyed by frost/cold; freeze through, turn into solid ice

вы́·мерз·н·уть, -ут, (*perf*) see вы··мерз·а́·ть

вы́·мерз·л·ый, -ая, -ое, (*a*) congealed, frozen; killed by frost

вы́·мер·ивш·ий, -ая, -ее, (*past part act*) of вы́·мер·ить

вы́·мер·ить, -ят, (*perf*) measure

вы́·мер·ш·ий, -ая, -ее, (*past part act*) see вы·мир·а́·ть; (biol) extinct

вы·мер·я́·ть, -ют, (*imp*) measure

вы́·мес·ить, -ят, (*perf*) knead, mash

вы́·мес·ти, (*fut 3rd sing, pl*) вы́·мет·ет, вы́·мет·ут, (*past masc sing*) вы́·мел, (*perf*); sweep out

вы·мет·а́·ть, -ют, (*imp*) sweep out; (*perf*) edge, buttonhole (with thread); throw/pay/let/ease out; (fish) spawn

вы́·мет·енн·ый, -ая, -ое, (*past part pass*) of вы́·мес·ти

вы́·мет·к·и (*nom pl*) small horse hides

вы́·мет·ш·ий, -ая, -ее, (*past part act*) of вы́·мес·ти

вы·мёт·ывани·е, -я, (*n*) (*v n*) of вы··мёт·ыва·ть; (bot) tasseling, heading

вы·мёт·ыва·ть, -ют, (*imp*) edge, buttonhole (with thread); throw/pay/let/ease out; spawn

вы́·меш·енн·ый, -ая, -ое, (*past part pass*) of вы́·мес·ить; see вы·ме́ш·ива·ть

вы·ме́ш·ива·ть, -ют, (*imp*) knead, mash

вы·мир·а́ни·е, -я, (*n*) dying out, extinction

вы·мир·а́·ть, -ют, (*imp*) die out, become extinct

вы·мог·а́тельск·ий, -ая, -ое, (*a*) extortionate

вы́·мо·ин·а, -ы, (*f*) gully

вы·мок·а́·ть, -ют, (*imp*) be soaked/drenched; be steeped; be retted (of flax)

вы́·мок·н·уть, -ут, (*perf*) see вы·мо·к·а́·ть

вы́·мол·, -а, (*m*) cleaning, final grinding (flour)

~ от·руб·е́й cleaning bran

вы·мола́ч·ива·ть, -ют, (*imp*) thresh

вы́·молот·ить, -ят, (*perf*) see вы·мо·ла́ч·ива·ть

вы·мора́ж·ивани·е, -я, (*n*) (*v n*) of вы·мора́ж·ива·ть

вы·мора́ж·ива·ть, -ют, (*imp*) freeze out, freeze, kill by freezing

вы́·мор·ить, -ят, (*perf*) exterminate

вы́·морож·енн·ый, -ая, -ое, (*past part pass*) see вы·мора́ж·ива·ть

вы́·мороз·ить, -ят, (*perf*) see вы·мо·ра́ж·ива·ть

вы́·мор·очн·ый, -ая, -ое, (*a*) (law) escheated

~ им·у́ществ·о (*n*) (law) escheat

вы́·мост·ить, -ят, (*perf*) see вы·ма́щ·ива·ть

вы́·мот·анн·ый, -ая, -ое, (*past part pass*) of вы́·мот·а·ть

вы́·мот·а·ть, -ют, (*perf*) unwind, unreel, pay out; deplete

вы́·моч·енн·ый, -ая, -ое, (*past part pass*) see вы·ма́ч·ива·ть; soaked, steeped, retted (flax)

вы́·моч·ить, -ат, (*perf*) see вы·ма́ч·ива·ть

вы́·моч·к·а, -и, (*f*) (*v n*) see вы·ма́ч·ива·ть

вы́·мощ·енн·ый, -ая, -ое, (*past part pass*) see вы·ма́щ·ива·ть

вы́мпел·, -а, (*m*) (naut) pennant; ship, unit; (air) message container

~, от·ве́т·н·ый answering pennant

вы́мпель·н·ый, -ая, -ое, (*a*) of **вы́м·пел·**
~ флаг· (*m*) pennant
вы·мыв·а́ем·ый, -ая, -ое, (*pres part pass*) of **вы·мыв·а́·ть;** (chem) extractable, strippable
вы·мыв·а́ни·е, -я, (*n*) washing, washing-out, leaching; (geol) outwash, erosion
вы·мыв·а́·ть, -ют, (*imp*) wash, wash out/away/off; hollow out
вы·мыв·а́ющ·ий, -ая, -ее, (*pres part act*) of **вы·мыв·а́·ть**
~ аге́нт· (*m*) washing agent, (chem) stripping agent
вы́·мысел·, -сл·а, (*m*) fiction, fabrication
вы́·мы·ть, (*fut 3rd sing, pl*) **вы́·мо·ет, вы́·мо·ют,** (*perf*); see **вы·мыв·а́·ть**
вы́·мышл·енн·ый, -ая, -ое, (*past part pass*) imaginary, fictitious
вы́м·я, (*g, d, p sing*) **вы́мен·и,** (*n*) udder
вы·на́ш·ива·ть, -ют, (*imp*) carry (unborn child); perfect (a plan etc.)
вы́·нес·енн·ый, -ая, -ое, (*past part pass*) see **вы·нос·и́ть;** outlying, remote, extended, extension; tolerated, endured
вы́·нес·ти, -ут, (*perf*) see **вы·нос·и́ть**
вы·ним·а́ни·е, -я, (*n*) (*v n*) see **вы·ним·а́·ть**
вы·ним·а́·ть, -ют, (*imp*) take out, extract
вы́ном·, -а, (*m*) exudate
вы́·нос·, -а, (*m*) (*v n*) see **вы·нос·и́ть;** removal; drift; remote/forward placing; (geol) effluse, evacuation, debris cone
~ ба́лк·и overhanging length (of beam)
~ пес·к·а́ (oil) sloughing of sand
~ рек·и́ river drift
   у́гол· вы́·нос·а (*m*) stagger angle (of a/c undercarriage)
   у́гол· бок·ов·о́го вы́·нос·а (*m*) (air) sideward stagger angle
~, хим·и́ческ·ий (geol) chemical subtraction
вы·нос·и́м·ый, -ая, -ое, (*pres part pass*) of **вы·нос·и́ть** (*imp*); endurable
вы·нос·и́тел·ь, -я, (*m*) (oil) entrainer
вы·нос·и́ть, ⁀ят, (*imp*) carry/take out/off/away; endure, stand, bear; announce, bring out (findings etc.); -ся (*pass*); rush out; **вы́·нос·ить, -ят** (*perf*) see **вы·на́ш·ива·ть**

вы́·нос·к·а, -и, (*g pl*) **-с·ок·,** (*f*) carrying out, removal; (print) note, footnote, marginal note
вы·но́с·ливост·ь, -и, (*f*) endurance, staying power, tolerance; durability (of materials)
~, вибрацио́н·н·ая (met) endurance limit, fatigue limit
   пре·де́л· вы·но́с·ливост·и (*m*) (met) fatigue/endurance limit
~ при много·кра́т·н·ом из·ги́б·е (rubber) flexing strength
~, радиацио́н·н·ая (nucl) radiation tolerance
вы·нос·н·о́й, -а́я, -о́е, (*a*) remote, outlying; extension
~ индика́тор· (*m*) remote indicator
~ элеме́нт·, ве́рх·н·ий (*m*) (print) ascender
~ элеме́нт·, ни́ж·н·ий (*m*) (print) descender, tail
вы́·нош·енн·ый, -ая, -ое, (*past part pass*) see **вы·на́ш·ива·ть;** mature; worn-out, threadbare
вы́·нуд·ить, -ят, (*perf*) see **вы·ну·жд·а́·ть**
вы·нужд·а́·ть, -ют, (*imp*) compel, force, oblige
вы́·нужд·енн·ый, -ая, -ое, (*past part pass*) see **вы·нужд·а́·ть**
~ колеб·а́ни·е (*n*) forced oscillation
~ структу́р·а (*f*) (geol) coercive structure
вы́·н·уть, -ут, (*perf*) take out, extract
вы́·ныр·н·уть, -ут, (*perf*) surface, come to the surface again
вы́·нюх·а·ть, -ют, (*perf*) see **вы·ню́х·ива·ть**
вы·ню́х·ива·ть, -ют, (*imp*) smell/nose out
вы́·па·вш·ий, -ая, -ее, (*past part act*) see **вы·пад·а́·ть;** precipitated
вы́·пад·, -а, (*m*) (*v n*) see **вы·пад·а́·ть;** lunge, thrust
~ раст·е́ни·й plant mortality
вы·пад·а́·ть, -ют, (*imp*) fall out, fall, (chem) precipitate; occur
вы·пад·е́ни·е, -я, (*n*) fallout, falling out; precipitation; (med) prolapse
~ о·са́д·к·а (chem) precipitation, deposition
~ о·са́д·к·ов (meteor) rainfall, precipitation
~ радио·акти́в·н·ой пы́л·и radioactive fallout
~ са́ж·и soot fall
~ хло́пь·ями (chem) flocculation

вы·па́л·ива·ть, -ют, (*imp*) fire, shoot

вы́·пал·ить, -ят, (*perf*) fire, shoot

вы́·па́л·ывани·е, -я, (*n*) weeding

вы́·па́л·ыва·ть, -ют, (*imp*) weed, weed out

вы́·пар·енн·ый, -ая, -ое, (*past part pass*) *see* вы·па́р·ива·ть; evaporated

вы·па́р·ивани·е, -я, (*n*) evaporation; steam cleaning; concentration (by evaporation)

вы·па́р·иватель·, -я, (*m*) evaporation

вы·па́р·ива·ть, -ют, (*imp*) evaporate; steam (for cleaning etc.)

вы́·пар·ить, -ят, (*perf*) *see* вы·па́р·и·ва·ть

вы́·пар·к·а, -и, (*g pl*) -р·ок·, (*f*) evaporation; evaporator station; steam cleaning

вы·па́р·ник·, -а, (*m*) evaporator

~, трех·ко́рпус·н·ый, triple-effect evaporator

вы·пар·н·о́й, -а́я, -о́е, (*a*) evaporating; steaming

~ аппара́т· (*m*) evaporator

~ аппара́т·, много·ко́рпус·н·ый multiple-effect evaporator

~ аппара́т·, плён·очн·ый film-type evaporator

~ аппара́т·, четырёх·ко́рпус·н·ый quadruple-effect evaporator

~ аппара́т·, шести·ко́рпус·н·ый sextuple-effect evaporator

вы·па́р·ыва·ть, -ют, (*imp*) rip out (along a seam); gut (an animal)

вы́·пас·, -а, (*m*) (agr) grazing, pasturage, keep

вы́·пас·ть, (*fut 3rd sing, pl*) вы́·пад·ет, вы́·пад·ут, (*past masc sing*) вы́·па·л, (*perf*); *see* вы·пад·а́·ть

вы́·пах·ивани·е, -я, (*n*) ploughing, scouring

вы́·пачк·а·ть, -ют, (*perf*) dirty, soil, make a mess

вы·пек·а́ни·е, -я, (*n*) (*v n*) *see* вы·пе·к·а́·ть

вы·пек·а́·ть, -ют, (*imp*) bake, bake well

вы·пе́н·ива·ть, -ют, (*imp*) froth out; clear of froth/scum; -ся (*pass*) froth up/over

вы́·пер·еть, (*fut 3rd sing, pl*) вы́·пр·ет, вы́·пр·ут, (*perf*); *see* вы·пир·а́·ть

вы·печа́т·ывани·е, -я, (*n*) (phot) printing out

вы́·печ·енн·ый, -ая, -ое, (*past part pass*) of вы́·печ·ь

вы́·печ·к·а, -и, (*g pl*) -ч·ек·, (*f*) (*v n*) *see* вы́·печ·ь

вы́·печ·ь, (*fut 3rd sing, pl*) вы́·печ·ет, вы́·пек·ут, (*past masc sing*) вы́·пек·, (*perf*); bake, bake well

вы·пив·а́·ть, -ют, (*imp*) drink, drink up

вы́·пил·енн·ый, -ая, -ое, (*past part pass*) *see* вы·пи́л·ива·ть

вы·пи́л·ива·ть, -ют, (*imp*) saw, saw out

вы́·пил·ить, -ят, (*perf*) *see* вы·пи́л·и·ва·ть

вы·пир·а́ни·е, -я, (*n*) pushing/squeezing out; bulging, protrusion; heaving (of seedlings); lifting (by frost)

   призм·а вы·пир·а́ни·я (*f*) overturning/tilting triangle (retaining wall calculations)

вы·пир·а́·ть, -ют, (*imp*) push/squeeze out; bulge, protrude

вы́·пис·ать, (*perf*) (*fut 3rd sing, pl*) вы́·пиш·ет, вы́·пиш·ут; *see* вы·пи́с·ыва·ть

вы́·пис·к·а, -и, (*g pl*) -с·ок·, (*f*) copy, extract

~ с·чёт·а (fin) statement of account

вы·пис·н·о́й, -а́я, -о́е, (*a*) copied, extracted; ordered by post/mail; mail-order

вы·пи́с·ыва·ть, -ют, (*imp*) write/trace out; copy out, extract; write for, order; write/strike off, discharge

вы́·пи·ть, (*fut 3rd sing, pl*) вы́·пь·ет, вы́·пь·ют, (*perf*); drink, have a drink

вы́·плав·ить, -ят, (*perf*) *see* вы·плав·л·я́·ть

вы́·плав·к·а, -и, (*g pl*) -в·ок·, (*f*) *see* вы·плавл·е́ни·е; melt, smelted metal

вы·плавл·е́ни·е, -я, (*n*) (*v n*) *see* вы·плавл·я́·ть; (geol) fusion

вы·плавл·я́·ть, -ют, (*imp*) smelt, melt, melt out

вы́·плат·а, -ы, (*f*) (*v n*) *see* вы́·пла·т·ить

вы́·плат·ить, -ят, (*perf*) *see* вы·пла́·ч·ива·ть

вы·пла́ч·ива·ть, -ют, (*imp*) pay, pay out/off

вы·плёв·ыва·ть, -ют, (*imp*) spit out

вы·плёск·ива·ть, -ют, (*imp*) splash out/over, spill

вы́·плес·н·уть, -ут, (*perf*) splash out/over, spill

вы́·плес·ти, (*fut 3rd sing, pl*) вы́·плет·ет, вы́·плет·ут (*past masc sing*) вы́·пле·л (*perf*); *see* вы·плет·а́·ть

вы·плет·а́·ть, -ют, (*imp*) weave, braid

вы́·плод·, -а, (*m*) insect production/breeding

вы́·плыв·, -а,   (*m*) (rubber) spew

вы·плыв·а́·ть, -ют,   (*imp*) swim out; surface

вы́·плы·ть,   (*fut 3rd sing, pl*) **вы́·плыв·ет, вы́·плыв·ут** (*perf*); *see* **вы·-плыв·а́·ть**

вы́·плю·н·уть, -ут,   (*perf*) spit out

вы·по·ла́ж·ива·ть, -ют,   (*imp*) flatten/smooth out

вы·пола́ск·ива·ть, -ют,   (*imp*) rinse, rinse out

вы·полз·а́·ть, -ют,   (*imp*) creep/crawl out

вы́·полз·ти, -ут,   (*perf*) *see* **вы·полз·а́·ть**

вы·полн·éни·е, -я,   (*n*) execution, fulfilment, implementation, carrying out; (geol) infilling, filling

вы́·полн·енност·ь, -и,   (*f*) plumpness, filling

вы́·полн·енн·ый, -ая, -ое, (*past part pass*) *see* **вы·полн·я́·ть**; filled-out, well filled, solid, plump, filled (of grain etc.)

вы·полн·и́мост·ь, -и,   (*f*) feasibility, practicability

вы·полн·и́м·ый, -ая, -ое, (*a*) feasible, practicable

вы́·полн·и·ть, -ят, (*perf*) *see* **вы·полн·я́·ть**

вы·полн·я́·ть, -ют, (*imp*) carry out, execute, fulfil, perform (functions, operations), implement (instructions); satisfy, meet (conditions etc.); fill up; **-ся** (*pass*)

~ **вы·числ·éни·е**   perform/carry out computations, make calculations

~ **из·мер·éни·е**   take measurements, measure

~ **при·каз·а́ни·е**   execute an order

~ **у·сло́в·и·е**   satisfy a condition

~ **фу́нкци·ю**   act as (of instruments, gadgets etc.)

вы́·по·ло́ж·енн·ый, -ая, -ое, (*past part pass*) *see* **вы·по·ла́ж·ива·ть,** flattened, flattened-out

вы́·полоск·ать, (*fut 3rd sing, pl*) **вы́·-полощ·ет, вы́·полощ·ут,** (*perf*); *see* **вы·пола́ск·ива·ть**

вы́·пол·оть, -ют, (*perf*) weed, weed out

вы́·пор·, -а, (*m*) vent; (met cast) air gate, riser

вы́·пор·оток·, -тк·а, (*m*) unborn (skins, fur)

вы́·пор·оть, -ют, (*perf*) *see* **вы·па́-р·ыва·ть**

вы́·пот·, -а,   (*m*) exudate, effusion; sweating, sweat; suppuration

вы·пот·ева́ни·е, -я,   (*n*) exudation, sweating, bleeding, oozing, suppuration

вы·пот·ева́·ть, -ют,   (*imp*) sweat, sweat out; ooze, suppurate

вы·прав·и́тельн·ый, -ая, -ое,   (*a*) correcting, straightening, rectifying

~ **рабо́т·ы**   (*pl*) contraction works (river engineering)

вы́·прав·ить, -ят,   (*perf*) *see* **вы·-правл·я́·ть**

вы·правл·éни·е, -я,   (*n*) (*v n*) *see* **вы·правл·я́·ть**

вы·правл·я́·ть, -ют, (*imp*) straighten; correct, put right, rectify; get, obtain (official document)

вы·пра́ст·ыва·ть, -ют, (*imp*) empty; work free

вы·пра́ш·ива·ть, -ют, (*imp*) get/obtain by asking, solicit

вы́·пресс·овать, -уют, (*perf*) press/squeeze out

вы·пресс·о́вк·а, -и, (*g pl*) **-вок·,** (*f*) (met) flash (from pressing); spew, overflow (plastics, rubber)

вы·пресс·о́выва·ть, -ют,   (*imp*) squeeze/press out

вы́·прос·ить, -ят, (*perf*) *see* **вы·пра́-ш·ива·ть**

вы́·прост·а·ть, -ют, (*perf*) *see* **вы·-пра́ст·ыва·ть**

вы́·прош·енн·ый, -ая, -ое, (*past part pass*) *see* **вы·пра́ш·ива·ть**

вы·пры́г·ива·ть, -ют, (*imp*) jump/spring out

вы́·прыг·н·уть, -ут, (*perf*) *see* **вы·-пры́г·ива·ть**

вы·пряг·а́·ть, -ют, (*imp*) unharness

вы́·пряж·енн·ый, -ая, -ое, (*past part pass*) *see* **вы·пряг·а́·ть**

вы·прям·и́тел·ь, -я,   (*m*) (elec) rectifier

~, **ва́куум·н·ый,** thermionic/vacuum-tube rectifier

~, **вибрацио́н·н·ый**   vibrating rectifier

~, **газо·тро́н·н·ый**   discharge-tube rectifier

~, **двух·полу·перио́д·н·ый**   full-wave rectifier

~, **жи́дк·ий** liquid electrolytic rectifier

~, **кено·тро́н·н·ый** kenotron

~, **конта́кт·н·ый** contact rectifier

~, **купр·о́кс·н·ый** copper-oxide rectifier

вы·прям·и́тел·ь

~, ла́мп·ов·ый   thermionic/vacuum-tube rectifier

~, магнетро́н·н·ый   rectifier/rectifying magnetron

~, медно·за́·кис·н·ый   copper-oxide rectifier

~, механ·и́ческ·ий   mechanical rectifier

~, мо́ст·иков·ый   bridge rectifier

~, одно·полу·пери́од·н·ый   half-wave rectifier

~, одно·фа́з·н·ый   single-pass rectifier

~, пут·ев·о́й   (rail) track rectifier

~, ртУ́т·н·ый   mercury rectifier, mercury-arc rectifier

~, сверх·мо́щ·н·ый   pumped ignitron

~, сверх·про·вод·я́щ·ий   superconducting rectifier

~, селе́н·ов·ый   selenium rectifier

~, сух·о́й   dry-disk rectifier

~, твёрд·ый   metallic/dry-disk/barrier-layer rectifier

~, электро́н·н·ый   thermionic/vacuum-tube rectifier

вы·прям·и́тельн·ый, -ая, -ое,   (a) rectifier, rectifying

~ ла́мп·а   (f) rectifier valve

вы́·прям·ить, -ят,   (perf) see вы··прямл·я́·ть

вы·прямл·е́ни·е, -я,   (n) straightening; (elec) rectification

~, двух·полу·пери́од·н·ое   (elec) full-wave rectification

коэффицие́нт· вы·прямл·е́ни·я (m) rectification factor

~, лин·е́йн·ое   (elec) linear rectification

~, одно·полу·пери́од·н·ое   half-wave rectification

вы·прямл·я́·ть, -ют,   (imp) straighten; (elec) rectify

вы́·пряч·ь,   (fut 3rd sing, pl) вы́··пряж·ет, вы́·пряг·ут (past masc sing) вы́·пряг·, (perf); unharness

вы́·пук·лин·а, -ы,   (f) (bot) umbo

вы́·пук·л·о   (adv) in relief; (as prefix) convexo-

вы́·пук·л·о-во́·гн·ут·ый, -ая, -ое, (a) convexo-concave

вы́·пук·л·о-изо́·гн·ут·ый, -ая, -ое, (a) unwarped

вы́·пук·л·о-пло́ск·ий, -ая, -ое,   (a) convexo-plane

вы́·пук·лост·ь, -и,   (f) convexity; bulge, protruberance; camber (of roll etc.); relief

вы́·пук·л·ый, -ая, -ое,   (a) convex, cambered, domed; bulged, bulging, protuberant; raised

~ бе́рег·   (m) (geog) slip-off slope

~ шов·   (m) convex weld

вы́·пуск·, -а,   (m) letting out, release; output, delivery; outlet, escape, exhaust, emission, discharge; (met) tapping; issue (loan etc.), number (periodical); graduates, qualifiers (completing at same time)

~, по·сле́д·н·ий   latest type

~ руд·ы́   (min) draw

~ шасси́   (air) lowering the undercarriage

вы·пуск·а́ни·е, -я,   (n) (v n) see вы··пуск·а́·ть

вы·пуск·а́·ть, -ют,   (imp) let out; let go, release; let down, lower; launch; turn out, produce qualified (personnel); issue, publish; put on the market, deliver (goods); leave out, omit; (met) tap

вы·пуск·а́ю·щий, -ая, -ее,   (pres part act) of вы·пуск·а́·ть

~ част·ь   (f) (print) delivery

вы·пуск·ни́к·, -а́,   (m) graduating student; top-class man, final-year student

вы·пуск·н·о́й, -а́я, -о́е,   (a) of вы́··пуск·

~ анте́нн·а   (f) (air) trailing aerial

~ кла́пан·   (m) exhaust valve

~ класс·   (m) graduating class

~ колле́ктор·   (m) exhaust manifold/collector

~ кран·   (m) discharge cock

~ экза́мен·   (m) qualifying exam, finals

вы́·пуст·ить, -ят,   (perf) see вы·пуск·а́·ть

вы́·пут·а·ть, -ют,   (perf) see вы·пу́·т·ыва·ть

вы́·пу́т·ыва·ть, -ют,   (imp) disentangle

вы́·пуч·енн·ый, -ая, -ое,   (past part pass) see вы·пу́ч·ива·ть; bulged, bulging, swollen out, puffed-up

вы·пу́ч·ивани·е, -я,   (n) (v n) of вы··пу́ч·ива·ть; (geol) unwarping

вы·пу́ч·ива·ть, -ют,   (imp) bulge, puff up

вы́·пуч·ин·а, -ы,   (f) bulge

вы́·пуч·ить, -ат,   (perf) see вы·пу́·ч·ива·ть

вы́·пуш·к·а, -и,   (f) (text) fur edging, fur collar etc.

вы́·пущ·енн·ый, -ая, -ое,   (past part pass) see вы·пуск·а́·ть

вы́·пущ·енн·ый
~ **акционе́р·н·ый капита́л·** (*m*) (fin) issued share capital
~ **шасси́** (*n*) (a/c) undercarriage down
вы́·пят·ить, -ят, (*perf*) *see* **вы·пя́·ч·ива·ть**
вы·пя́ч·ива·ть, -ют, (*imp*) stick/thrust out
вы·раба́т·ыва·ть, -ют, (*imp*) manufacture, produce, make; work out, draw up (plan etc.); (min) work/mine out, exhaust; earn
вы́·работ·анн·ый, -ая, -ое, (*past part pass*) *see* **вы·раба́т·ыва·ть;** (geol) eroded, erosion
вы́·работ·а·ть, -ют, (*perf*) *see* **вы··раба́т·ыва·ть**
вы́·работ·к·а, -и, (*g pl*) **-т·ок·,** (*f*) development, working out, drawing up; production, manufacture; (elec) generation; (min) working
~, **го́р·н·ые** (*pl*) mine workings
~, **де́й·ствующ·ие** (*pl*) (min) active workings
но́рм·а вы́·работ·к·и (*f*) output standard
~, **от·кры́т·ая го́р·н·ая** opencast mining/working
~, **под·готов·и́тельн·ая** (min) development working
~, **раз·ве́д·очн·ая** (min) prospecting/ proving hole
вы́·равн·енн·ый, -ая, -ое, (*past part pass*) *see* **вы·ра́вн·ива·ть**
вы·ра́вн·ивани·е, -я, (*n*) (*v n*) of **вы·ра́вн·ива·ть;** alignment, line-up; (air) flattening out, round-out; (nucl) flattening; (geol) planation
~ **да́н·н·ых** (rad) data smoothing
~ **ём·кост·ей** (elec) capacitance balancing
~, **зо́н·н·ое** (met) zone levelling
~, **лед·нико́в·ое** (geol) glacial planation
~ **по·то́к·а** (nucl) flux flattening, flattening
систе́м·а вы·ра́вн·ивани·я кре́н·а и дифере́нт·а (*f*) (mar eng) trimming system
вы·ра́вн·ивател·ь, -я, (*m*) leveller, equalizer
~ **за·тух·а́ни·я** attenuation equalizer
вы·ра́вн·ивательн·ый, -ая, -ое, (*a*) levelling; equalizing
~ **со·оруж·е́ни·я** (*pl*) training works (rivers etc.)

вы·ра́вн·ива·ть, -ют, (*imp*) level, even, smooth out; put in line, align, train (a river); equalize, trim, balance
вы·ра́вн·ивающ·ий, -ая, -ее, (*pres part act*) of **вы·ра́вн·ива·ть**
~ **конденса́тор·** (*m*) (elec) equalizing capacitor
~ **механ·и́зм·** (*m*) stabilizing gear; levelling gear
~ **спосо́б·ност·ь** (*f*) evenness (of dye)
~ **схе́м·а** (*f*) (elec) equalizing/equalizer network
вы·раж·а́·ть, -ют, (*imp*) express
вы·раж·е́ни·е, -я, (*n*) expression; (math) formula, form, term, expression
~, **не·о·пре·дел·ённ·ое** (math) indeterminate form
~, **под·интегра́ль·н·ое** (math) integrand
вы́·раж·енн·ый, -ая, -ое, (*past part pass*) *see* **вы·раж·а́·ть;** expressed; pronounced, marked
~ **через** expressed in, expressed in terms of
~, **чёт·к·о** distinct, clearly defined
вы·раз·и́тельн·ый, -ая, -ое, (*a*) (*s f*) **-лен·, -льн·а** expressive; significant
вы́·раз·ить, -ят, (*perf*) *see* **вы·раж·а́·ть**
вы·раст·а́ни·е, -я, (*n*) growing, growth, growing up
вы·раст·а́·ть, -ют, (*imp*) grow, increase, grow up/out of
вы́·раст·и, -ут, (*past masc sing*) **вы́·рос·,** (*perf*); *see* **вы·раст·а́·ть**
вы́·раст·ить, -ят, (*perf*) *see* **вы·ра́·щ·ива·ть**
вы·ра́щ·ивани·е, -я (*n*) (*v n*) of **вы··ра́щ·ива·ть;** cultivation
~ **криста́лл·ов** growing crystals, crystal growing
вы·ра́щ·ива·ть, -ют, (*imp*) cultivate, raise, grow; train, prepare
вы́·рв·ать, -ут, (*perf*) tear/pull out/ away
вы́·рез·, -а, (*m*) (*see also* **вы́·рез·к·а**); cut, notch, opening, groove, cut-out
~ **встре́ч·н·ым ко́нус·ом** two-sided chamfer cavity
~ **нару́ж·н·ым ко́нус·ом** outside chamfer cavity
вы́·рез·-зу́б·, -а, (*m*) ratchet tooth
вы́·рез·анн·ый, -ая, -ое, (*past part pass*) *see* **вы́·рез·ать**

вы·рез·а́·ть, -ют, (*imp*) **вы́·рез·ать,** (*fut 3rd pl*) **вы́·реж·ут,** (*perf*); cut out; engrave, carve; (med) excise; massacre

вы́·рез·к·а, -и, (*g pl*) **-з·ок·,** (*f*) (*see also* **вы́·рез·**); cutting out, blanking, notch, notching; carving, engraving; cutting, clipping, (newspaper); (anat) incisura; fillet, sirloin (meat)

~ **вну́тр·енн·им ко́нус·ом** inside chamfer cutting

~ **встре́ч·н·ым ко́нус·ом** two-sided chamfer cutting

~ **ступ·е́ньк·ами в ра́м·к·у** rectangular step-cutting

вы·рез·н·о́й, -а́я, -о́е, (*a*) of **вы́·рез,** **вы́·рез·к·а**

вы·ре́з·ывани·е, -я, (*n*) cutting out; engraving, carving; (med) excision

вы·ре́з·ыва·ть, -ют, (*imp*) *see* **вы́·ре·з·ать**

вы́·ровн·я·ть, -ют, (*perf*) *see* **вы·ра́вн·ива·ть**

вы́·ровн·енн·ый, -ая, -ое, (*past part pass*) *see* **вы·ра́вн·ива·ть**

вы́·род·ивш·ийся, -аяся, -ееся, (*past part act*) *see* **вы·рожд·а́·ться;** degenerate, degenerated, vestigial

вы́·род·иться, -ятся, (*perf*) *see* **вы··рожд·а́·ться**

вы́·род·ок·, -д·к·а, (*m*) (biol) degenerate

вы·рожд·а́·ться, -ются, (*imp*) degenerate

вы·рожд·а́ющ·ийся, -аяся, -ееся, (*pres part act*) of **вы·рожд·а́·ться** degenerating, degenerative

вы·рожд·е́ни·е, -я, (*n*) degeneration; deterioration, degradation

~ **кру́п·н·ой по·ро́д·ы** dwarfing

~ **у́·ровн·ей энерг·и·и** (phys) degeneracy of energy levels

вы·рожд·ённ·ый, -ая, -ое, (*a*) degenerate; (math) confluent

~ **электр·и́ческ·ая схе́м·а** (*f*) network diagram, relay switching circuit diagram (switching algebra)

вы́·рос· (*past masc sing*) of **вы́·раст·и**

вы́·рост·, -а, (*m*) excrescence, protuberance, growth, apophysis

~, **корн·ев·о́й** root tubercle

вы·рост·н·о́й, -а́л, -о́е, (*a*) breeding, nursery

вы́·рост·ок, -т·к·а, (*m*) kip, yearling hide (leather)

вы́·рос·ш·ий, -ая, -ее, (*past part act*) *see* **вы·раст·а́·ть**

вы́·руб·, -а, (*m*) *see* **вы́·руб·к·а**

вы·руб·а́·ть, -ют, (*imp*) chop/cut down, fell; chop/cut out; chip

~ **на пре́сс·е** punch (metal)

вы́·руб·ить, -ят, (*perf*) *see* **вы·руб·а́·ть**

вы́·руб·к·а, -и, (*g pl*) **-б·ок·,** (*f*) (*v n*) *see* **вы·руб·а́·ть;** (met forge) blanking, cutting out, punching, notching; hole, cut, notch; felling area (timber)

вы́·рубл·енн·ый, -ая, -ое, (*past part pass*) *see* **вы·руб·а́·ть**

вы·руб·н·о́й, -а́я, -о́е, (*a*) of **вы́··руб·к·а**

~ **штамп·** (*m*) blanking die, blanking tool

вы·ру́л·ива·ть, -ют, (*imp*) (air) taxi, taxi out

вы́·рул·ить, -ят, (*perf*) *see* **вы·ру́·л·ива·ть**

вы·руч·а́·ть, -ют, (*imp*) help out, rescue; make, gain, net (profit etc.)

вы́·руч·ить, -ат, (*perf*) *see* **вы·ру·ч·а́·ть**

вы́·руч·к·а, -и, (*g pl*) **-ч·ек·,** (*f*) (fin) receipts, proceeds; assistance, aid

~, **чи́ст·ая** (com) net avails/proceeds

вы́·рыв·, -а, (*m*) broken section, partial section (of drawing)

вы·рыв·а́ни·е, -я, (*n*) (*v n*) of **вы·рыв·а́·ть**

~, **потенциа́ль·н·ое** field emission

~ **электро́н·ов** electron emission

вы·рыв·а́·ть, -ют, (*imp*) tear/pull out/away; dig up/out, exhume

вы́·ры·ть, (*fut 3rd sing, pl*) **вы́·ро·ет, вы́·ро·ют,** (*perf*); dig up/out, exhume

вы́·сад·ить, -ят, (*perf*) *see* **вы·са́ж·и·ва·ть**

вы́·сад·к·а, -и, (*g pl*) **-д·ок·,** (*f*) (*v n*) *see* **вы·са́ж·ива·ть;** disembarkation; planting out; (met forge) upsetting, heading; bedding plant

~, **холо́д·н·ая** (met forge) cold heading

вы́·сад·очн·ый, -ая, -ое, (*a*) of **вы́··сад·к·а**

~ **маши́н·а** (*f*) (met forge) heading/upsetting machine

вы́·саж·енн·ый, -ая, -ое, (*past part pass*) *see* **вы·са́ж·ива·ть**

вы·са́ж·ива·ть, -ют, (*imp*) set down, put/let out (from vehicle), put ashore, land, disembark (from ship), detrain, debus (troops); (hortic) transplant; break/smash in; (met) upset, head; **-ся** (*pass*); get down/out, alight; precipitate, settle out

вы·са́л·ивани·е, -я,   (*n*) (chem) salting out

вы·са́л·иватель, -я, (*m*) salt-removing agent, salting-out agent

вы·са́с·ывани·е, -я, (*n*) (*v n*) of **вы·са́с·ыва·ть**

вы·са́с·ыва·ть, -ют, (*imp*) suck out, exhaust, evacuate

вы·са́ч·ивани·е, -я, (*n*) seepage, leaking out, oozing out; show (of oil); outcrop (of water)

вы·све́рл·ива·ть, -ют, (*imp*) drill, bore; ream

вы́·сверл·ить, -ят, (*perf*) see **вы́·све́рл·ива·ть**

вы·све́ч·ивани·е, -я, (*n*) afterglow, fluorescence; (nucl) de-excitation
вре́м·я вы·све́ч·ивани·я (*f*) fluorescent life, de-excitation time

вы́·свобод·ить, -ят, (*perf*) see **вы·свобожд·а́·ть**

вы·свобожд·а́·ть, -ют, (*imp*) free, let out, release, disengage

вы́·се·в·, -а, (*m*) (agr) sowing; seeding
~, ле́нт·очн·ый band seeding

вы·сев·а́·ть, -ют, (*imp*) (agr) sow

вы·сев·а́ющ·ий, -ая, -ее, (*pres part act*) of **вы·сев·а́·ть**; sowing
~ аппара́т·, кату́ш·ечн·ый fluted-wheel feed mechanism
~ аппара́т·, ло́ж·ечн·ый cup feed mechanism (of seed drill)
~ аппара́т, щёт·очн·ый brush feed mechanism

вы́·сев·к·и, (*nom pl*), (*g pl*) -в·ок·, (*f*) chaff, siftings, screenings; dross (of coal)

вы́·сев·н·о́й, -а́я, -о́е, (*a*) of **вы́·се·в·**

вы·сед·а́ни·е, -я, (*n*) (nucl) fall-out

вы·се́·ива·ть, -ют, (*imp*) see **вы·сев·а́·ть**

вы·сек·а́·ть, -ют, (*imp*) carve, hew; sculpture, cut

вы·сел·е́ни·е, -я, (*n*) eviction

вы́·сел·ить, -ят, (*perf*) see **вы·сел·я́·ть**

вы·сел·я́·ть, -ют, (*imp*) evict; move from/to

вы́·серебр·ить, -ят, (*perf*) silver-plate

вы́·сеч·к·а, -и, (*g pl*) -ч·ек·, (*f*) (*v n*) see **вы·сек·а́·ть**; (min) cutting, trench; (met forge) cutting out/off; (*pl*) punching(s), chip(s)

вы́·сечь, (*fut 3rd sing, pl*) вы́·сеч·ет, вы́·сек·ут, (*past masc sing*) вы́·сек·, (*perf*); see **вы·сек·а́·ть**

вы́·се·ять, -ют, (*perf*) sowing; seeding

вы·си́ж·ива·ть, -ют, (*imp*) stay, sit out; hatch, incubate, brood

вы́·сид·еть, -ят, (*perf*) see **вы·си́ж·ива·ть**

вы́с·иться, -ятся, (*imp*) rise/tower above (of stationary objects)

вы·ска́бл·ива·ть, -ют, (*imp*) scrape, rub down; erase, rub out

вы́·с·каз·анн·ый, -ая, -ое, (*past part pass*) see **вы·с·ка́з·ыва·ть**; expressed, formulated, stated

вы·с·ка́з·ывани·е, -я, (*n*) (*v n*) see **вы·с·ка́з·ыва·ть**

вы́·с·каз·ать, (*fut 3rd sing, pl*) вы́·с·каж·ет, вы́·с·каж·ут, (*perf*); see **вы·с·ка́з·ыва·ть**

вы·с·ка́з·ыва·ть, -ют, (*imp*) state, express, declare, formulate; **-ся** speak out, speak one's mind

вы·ска́к·ива·ть, -ют, (*imp*) jump/spring out; (naut) run onto, ground on; (math) deviate from

вы·ска́льз·ыва·ть, -ют, (*imp*) slip out

вы·скобл·енн·ый, -ая, -ое, (*past part pass*) see **вы·ска́бл·ива·ть**

вы́·скобл·ить, -ят, (*perf*) see **вы·ска́бл·ива·ть**

вы́·скольз·н·уть, -ут, (*perf*) see **вы·ска́льз·ыва·ть**

вы́·скоч·ить, -ат, (*perf*) see **вы·ска́к·ива·ть**

вы·скреб·а́·ть, -ют, (*imp*) scrape/scratch off/out; rake out

вы́·скрес·ти, (*fut 3rd sing, pl*) вы́·скреб·ет, вы́·скреб·ут, (*past masc sing*) вы́·скреб·, (*perf*); see **вы·скреб·а́·ть**

вы́·сл·ать, (*fut 3rd sing, pl*) вы́·шл·ет, вы́·шл·ют, (*past masc sing*) вы́·сл·ал, (*perf*); see **вы·сыл·а́·ть**

вы́·след·ить, -ят, (*perf*) see **вы·сле́ж·ива·ть**

вы·сле́ж·ивани·е, -я, (*n*) (*v n*) of **вы·сле́ж·ива·ть**

вы·сле́ж·ива·ть, -ют, (*imp*) trace, track, come/be on the track of

вы́·слуг·а, -и, (*f*) period/term of service/office, service

вы·слу́ж·ива·ть, -ют, (*imp*) serve a term; receive for service, qualify for

вы́·служ·ить, -ат, (*perf*) see **вы·слу́ж·ива·ть**

вы́·слуш·а·ть, -ют, (*perf*) see **вы·слу́ш·ива·ть**

вы·слу́ш·ивани·е, -я, (*n*) listening to, hearing out; (med) auscultation

вы·слу́ш·ива·ть, -ют, *(imp)* listen to, hear out; (med) sound

вы·сма́л·ива·ть, -ют, *(imp)* tar

вы·сма́тр·ива·ть, -ют, *(imp)* look out for, spy out

вы·сме́·ива·ть, -ют, *(imp)* ridicule, deride

вы́·сме·ять, -ют, *(perf) see* **вы·сме́·и·ва·ть**

вы́·смол·ить, -ят, *(perf) see* **вы·сма́·л·ива·ть**

вы́·смотр·еть, -ят, *(perf) see* **вы·сма́·тр·ива·ть**

вы·со́в·ыва·ть, -ют, *(imp)* poke/thrust/shove out; **-ся** protrude, stick out, project

выс·о́к·ий, -ая, -ое, *(s f)* **выс·о́к·, выс·ок·а́, выс·о́к·о;** high, tall, elevated; high-temperature (of heat treatments)

~ давл·е́ни·е *(n)* high pressure; *(gen as adj)* high-pressure, HP

выс·о́к·о- *(component)* high-, highly

высоко·автома́т·изи́рованн·ый, -ая, -ое, *(a)* highly automated

высоко·акти́в·н·ый, -ая, -ое, *(a)* highly active, high-activity; (nucl) hot

высоко·ва́куум·н·ый, -ая, -ое, *(a)* high-vacuum

высоко·во́льт·н·ый, -ая, -ое, *(a)* high-voltage, high-potential

высоко·вя́з·к·ий, -ая, -ое, *(a)* high-viscosity

высоко·глино·зём·ист·ый, -ая, -ое, *(a)* high-silica

высоко·го́р·н·ый, -ая, -ое, *(a)* high-mountain; (biol etc.) Alpine; (surv etc.) high-level

высоко·диспе́рс·н·ый, -ая, -ое, *(a)* finely dispersed, highly disperse, very fine

высоко·добр·о́тн·ый, -ая, -ое, *(a)* high-quality; (elec) high-Q

высоко·зо́ль·н·ый, -ая, -ое, *(a)* high ash-content, high-ash

высоко·интенси́тив·н·ый по·то́к *(m)* (phys) high flux

высоко·ка́честв·енн·ый, -ая, -ое, *(a)* high quality, high-grade, high-class, high-fidelity (sound reproduction)

высоко·квалифиц·и́рованн·ый, -ая, -ое, *(a)* skilled, highly skilled, with advanced qualifications

высоко·коэрцити́в·н·ый, -ая, -ое, *(a)* high-coercivity

~ с·плав· *(m)* magnet alloy

высоко·кремн·и́ст·ый, -ая, -ое, *(a)* high silicon-content, high-silicon; (met) silicon

~ стал·ь *(f)* silicon steel

высоко·куч·ев·ы́е облак·а́ *(pl)* alto-cumulus clouds

высоко·леги́р·ованн·ый, -ая, -ое, *(a)* high-alloy, high (of alloys)

высоко·магни́т·н·ый с·пла́в· *(m)* magnet alloy

высоко·марганц·о́в·ый, -ая, -ое, *(a)* high manganese-content, high-manganese; (met) manganese

высоко·моду́ль·н·ый, -ая, -ое, *(a)* high-modulus

высоко·молекуля́р·н·ый, -ая, -ое, *(a)* high-molecular, high-molecular-weight

~ со·един·е́ни·е *(n)* macromolecular compound, high polymer, high-molecular-weight compound

высоко·мо́щ·н·ый, -ая, -ое, *(a)* heavy-duty

высоко·на·по́лн·енн·ый, -ая, -ое, *(a)* heavily loaded (rubber)

высоко·на·по́р·н·ый, -ая, -ое, *(a)* (hydr eng) high-head

высоко·объ·ём·н·ый, -ая, -ое, *(a)* high-bulk, bulky

высоко·окта́н·ов·ый, -ая, -ое, *(a)* (oil) high-octane

высоко·ом·и́ческ·ий, -ая, -ое, *(a)* (elec) high-resistance/impedance

высоко·пла́в·к·ий, -ая, -ое, *(a)* high-melting

высоко·пла́н·, -а, *(m)* high-wing aircraft/monoplane

высоко·полиме́р·, -а, *(m)* (chem) high polymer, macromolecular compound

высоко·полиме́р·н·ый, -ая, -ое, *(a)* high-polymer, polymer

высоко·про·из·вод·и́тельн·ый, -ая, -ое, *(a)* highly productive, high-output, high-performance

высоко·про́ч·н·ый, -ая, -ое, *(a)* high-strength, high-duty, high-test, high-tenacity (fibres), high-tensile

высоко·са́хар·ист·ый, -ая, -ое, *(a)* high-sugar

высоко·свето·си́ль·н·ый, -ая, -ое, *(a)* (phot) ultra-rapid

высоко·сер·ни́ст·ый, -ая, -ое, *(a)* high-sulphur

высоко·ско́р·остн·ый, -ая, -ое, *(a)* high-speed

высоко·сло·и́ст·ые облак·а́ *(pl)* alto-stratus clouds

высоко·со́рт·н·ый, -ая, -ое, (*a*) high-grade, high-class, high-test

высоко·ство́ль·н·ый, -ая, -ое, (*a*) high-standing, high (of timber); (*gen as adj*) direct-seeding

~ **лес·н·о́е хозя́й·ств·о** (*n*) direct-seeding forest/plantation

высоко·температу́р·н·ый, -ая, -ое, (*a*) high-temperature

высоко·тепло·про·во́д·н·ый, -ая, -ое, (*a*) high-thermal conductivity

высоко·това́р·н·ый, -ая, -ое, (*a*) producing a high marketable surplus

высоко·то́ч·н·ый, -ая, -ое, (*a*) high-precision

высоко·уваж·а́ем·ый, -ая, -ое, (*a*) dear (in formal correspondence)

высоко·угле·ро́д·ист·ый, -ая, -ое, (*a*) high-carbon

высоко·ударно·про́ч·н·ый, -ая, -ое, (*a*) high-impact

высоко·у·рож·а́йн·ый, -ая, -ое, (*a*) (agr) fertile, high-yielding

высоко·хро́м·ист·ый, -ая, -ое, (*a*) high-chromium; (met) chromium

высоко·част·о́тн·ый, -ая, -ое, (*a*) high-frequency, hf, HF; (telecom) radio-frequency, rf, RF

высоко·чувств·и́тельн·ый, -ая, -ое, (*a*) highly sensitive, high-sensitivity

высоко·шир·о́тн·ый, -ая, -ое, (*a*) high-latitude, high-latitudinal

высоко·эласт·и́чн·ый, -ая, -ое, (*a*) high-elasticity, highly elastic

вы·сола́ж·ивани·е, -я, (*m*) sweeting off

вы́·сос·ать, -ут, (*perf*) *see* **вы·са́с·ыва·ть**

выс·от·а́, -ы́, (*nom pl*) **-о́т·ы,** (*g pl*) **-о́т·,** (*f*) height, altitude; pitch (of sound); thickness (of rolled piece); depth

~, **абсолю́т·н·ая** height above sea level, absolute height

~ **ба́л·к·и** (civ eng) depth of beam

~, **близ·меридиона́ль·н·ая** ex-meridian altitude

~, **больш·а́я** high altitude

~ **бо́рт·а** (shipb) moulded depth

~ **бо́рт·а, реги́стр·ов·ая** registered depth (of ship)

~ **в·са́с·ывани·я** (hydr eng) suction head

~ **в·са́с·ывани·я, геометр·и́ческ·ая** static suction lift

~ **вы́·ступ·а зу́б·а** addendum (gears)

~, **геометр·и́ческ·ая** true height; total static head (pumps)

выс·от·а́

~ **голо́в·к·и** addendum (of gear wheel)

~ **голо́в·к·и, хорда́ль·н·ая** chordal addendum (of gear wheel)

~ **давл·е́ни·я** piezometric height/level

~, **де́й·ствующ·ая** effective height

~, **за·кры́·т·ая** shut height (of power press)

~ **за·лёт·а** aircraft flying height

~ **зуб·ц·а́** depth of tooth (gears)

~ **кристаллиз·а́тор·а** length of mould (continuous casting)

~, **магни́т·н·ая** magnetometer height

~, **магнито·метр·и́ческ·ая** magneto-meter height

~, **максима́ль·н·ая воз·мо́ж·н·ая** (a/c) ceiling

~, **меридиона́ль·н·ая** meridian altitude

~, **метацентр·и́ческ·ая** metacentric height

~ **на́·сып·и** (civ eng) depth of fill

~, **на·ча́ль·н·ая по·пере́ч·н·ая ме·тацентр·и́ческ·ая** transverse metacentric height

~, **не·больш·а́я** low altitude

~ **но́ж·к·и** dedendum (gears)

~, **орто·метр·и́ческ·ая** orthometric height, height above sea level

~, **от·нос·и́тельн·ая** (air) relative height above airfield of take-off

~ **подъ·ём·а** (surv) elevation; altitude (of a balloon)

~ **подъ·ём·а, по́лн·ая** total head (of pump)

~ **под·этаж·а́** (min) sublevel interval

~ **по́лн·ых вод·, сре́д·н·яя** mean high water (tides)

~, **при·бо́р·н·ая ис·пра́вл·енн·ая** (air) calibrated altitude

~ **при́зм·ы** altitude of prism

~, **рабо́ч·ая** operational height

~, **рас·чёт·н·ая** (a/c) rated altitude

~ **с·бра́с·ывани·я** (air) height of release, dropping height

~ **с·бро́с·а** (geol) fault throw, heave

~ **сеч·е́ни·я** (surv) vertical interval; cross-sectional height

~ **то́ч·к·и сто́·яни·я** (surv) height of site

~ **траекто́р·и·и** maximum ordinate of trajectory

~ **у́·ровн·я** height of metal level, height of metal surface (continuous casting)

~ **у́ст·ь·я сква́ж·ин·ы** (oil) elevation of well

выс·от·а́
~, факт·и́ческ·ая   true height
~, хорда́ль·н·ая   chordal addendum (gear wheel)
~ шв·а, рас·чёт·н·ая   (weld) throat
выс·о́тност·ь, -и,   (f) height, altitude; design altitude; height power factor (of an a/c engine)
выс·о́тн·ый, -ая, -ое,   (a) of выс·о·т·а́; high-altitude, high-level; tall, multistorey (of house); pitch (of sound)
~ кольц·о́   (n) (rad) height circle
~ корре́ктор·   (m) (a/c) altitude control
~ от·ме́т·ка   (f) (surv) ground elevation mark, individual height/elevation
~ по·лёт·   (m) high-altitude flight
~ ход·   (m) height dependance (of a curve)
высото·гра́ф·, -а,   (m) (air) recording altimeter
высото·ме́р·, -а,   (m) altimeter; height finder, height-measuring instrument
~, и́мпульс·н·ый   pulsed altimeter
~, само·пи́ш·ущ·ий   recording altimeter, altitude recorder
высото·пи́с·ец·, -с·ц·а,   (m) (air) recording altimeter
вы́·сох·нуть, -нут,   (past masc sing) вы́·сох·; see вы·сых·а́·ть
вы́·сох·ш·ий, -ая, -ее,   (past part act) see вы·сых·а́·ть; shrivelled
выс·оча́йш·ий, -ая, -ее,   (superlative) of выс·о́к·ий
вы́·сп·аться, -ятся,   (perf) sleep out, have a good sleep
вы·с·пра́ш·ива·ть, -ют,   (imp) question, interrogate
вы́·с·прос·ить, -ят,   (perf) see вы·с·пра́ш·ива·ть
вы́·став·ить, -ят,   (perf) see вы·ста·вл·я́·ть
вы́·став·к·а, -и,   (g pl) -в·ок·, (f) exhibition, show, display; setting (instruments)
вы́·ставл·енн·ый, -ая, -ое,   (past part pass) see вы·ставл·я́·ть
вы·ставл·я́·ть, -ют,   (imp) put/place out/in front; put forward, nominate; set out, exhibit, show, expose; dismiss, sack, fire
вы·став·н·о́й, -а́я, -о́е,   (a) exhibiting, showing
вы́·став·очн·ый, -ая, -ое,   (a) of вы́·став·к·а
~ зал·   (m) exhibition hall, showroom

вы·ста́·ива·ть, -ют,   (imp) stand, remain standing; droop (photographic emulsions); -ся mature
вы·стил·а́·ть, -ют,   (imp) surface, cover, line, pave
вы·стил·а́ющ·ий, -ая, -ее,   (pres part act) of вы·стил·а́·ть
~ ткан·ь   (m) (biol) investing tissue, cameral mantle
вы́·стил·к·а, -и,   (g pl) -л·ок·, (f) (v n) see вы·стил·а́·ть
вы́·с·тир·а·ть, -ют,   (perf) wash, launder
вы́·стл·ать,   (fut 3rd sing, pl) вы́·стел·ет, вы́·стел·ют, (perf); see вы·стил·а́·ть
вы́·сто·ять, -ят,   (perf) see вы·ста́·ива·ть
вы·стра́·иван·и·е, -я,   (n) (v n) of вы·стра́·ива·ть; alignment
вы·стра́·ива·ть, -ют,   (imp) form/draw/line up, set out
вы́·стрел·, -а,   (m) shot, discharge; round (ammunition); boom (crane etc.)
~, за·тяж·н·о́й   hang fire
~, мино·мёт·н·ый   mortar bomb
~, прежде·вре́мен·н·ый   premature
~, прям·о́й   close-range shot
~, уч·е́бн·ый   drill round
~, холост·о́й   blank round
вы·стре́л·ива·ть, -ют,   (imp) see вы́·стрел·ить, (perf), вы́·стрел·я·ть (perf)
вы́·стрел·ить, -ят,   (perf) shoot, fire
вы́·стрел·я·ть, -ют,   (perf) expend (ammunition), fire off; destroy, wipe out (by shooting), shoot (all)
вы·стриг·а́·ть, -ют,   (imp) clip, shear
вы́·стрич·ь,   (fut 3rd sing, pl) вы́·стриж·ет, вы́·стриг·ут, (past masc sing) вы́·стриг·, (perf); see вы·стриг·а́·ть
вы́·строг·а·ть, -ют,   (perf) plane
вы́·стро·енн·ый, -ая, -ое,   (past part pass) of вы́·стро·ить
~ ядр·а   (pl) aligned nuclei
вы́·стро·ить, -ят,   (perf) form/draw/line up, align, aline, set out; build
вы́·строч·ить, -ат,   (perf) hemstitch
вы́·струг·а·ть, -ют,   (perf) plane
вы́·студ·ить, -ят,   (perf) see вы·сту́·ж·ива·ть
вы·сту́ж·ива·ть, -ют,   (imp) chill, cool
вы́·стук·а·ть, -ют,   (perf) see вы·сту́·к·ива·ть

вы·сту́к·ивани·е, -я, (*n*) tapping out; (med) percussion

вы·сту́к·ива·ть, -ют, (*imp*) tap out; (med) tap

вы́·ступ·, -а, (*m*) projection, protrusion, protuberance; lug, tab; salient; overhang, overshot; apron (of glacier)

~, банда́ж·н·ый tenon (turbine)
ли́н·и·я вы́ступ·ов (*f*) face line of teeth (saws)

~ фла́н·ц·а flange ridge

вы·ступ·а́·ть, -ют, (*imp*) come forward/out, advance; appear, perform, play (before public); make (a public act); project, jut/stick out

вы·ступ·а́юш·ий, -ая, -ее, (*pres part act*) of вы·ступ·а́·ть; protruding, projecting, jutting out, overhanging; salient; (bot) emersed; prominent, outstanding

~ пояс·о́к· (*m*) oversailing course (bricklaying)

~ част·ь (*f*) outshot, overhang

вы́·ступ·ить, -ят, (*perf*) *see* вы·ступ·а́·ть

вы·ступл·е́ни·е, -я, (*n*) public appearance; speech, address; performance, action; protest

вы́·су·н·уть, -ут, (*perf*) poke/thrust/shove out

вы́·суш·енн·ый, -ая, -ое, (*past part pass*) *see* вы·су́ш·ива·ть; dried, desiccated, seasoned (timber)

вы·суш·ива́ни·е, -я, (*n*) (*v n*) of вы·су́ш·ива·ть; desiccation

вы·су́ш·ива·ть, -ют, (*imp*) dry, dry up, desiccate, exsiccate; cure, season

вы́·суш·ить, -ат, (*perf*) *see* вы·су́ш·и·ва·ть

вы́·с·чит·а·ть, -ют, (*perf*) *see* вы·с·чи́т·ыва·ть

вы·с·чи́т·ыва·ть, -ют, (*imp*) calculate

вы́с·ш·ий, -ая, -ее, (*a*) higher, superior, upper; highest, supreme, uppermost

~ главно·кома́нд·овани·е (*n*) (mil) supreme command

~ пилот·а́ж· (*m*) acrobatics

~ то́ч·к·а (*f*) peak, climax, apex; acme

~ уч·ёбн·ое за·вед·е́ни·е (*n*) higher educational establishment, college

~ шко́л·а (*f*) high school (U.S.)

вы·сыл·а́·ть, -ют, (*imp*) send, send out; expel, deport

вы́·сыл·к·а, -и, (*g pl*) -л·ок·, (*f*) dispatch, sending; expulsion, deportation, banishment

вы·сып·а́ни·е, -я, (*n*) (*v n*) *see* вы·сып·а́·ть

вы́·сып·анн·ый, -ая, -ое, (*a*) (*past part pass*) *see* вы·сып·а́·ть

вы·сып·а́·ть, -ют, (*imp*); вы́·сып·ать, (*fut 3rd sing, pl*) вы́·сыпл·ет, вы́·сыпл·ют, (*perf*); pour, pour out, spill, tip out (of solids); break out (of rash); -ся (*pass*) (*imp only*) sleep out, have a good sleep

вы·сых·а́·ть, -ют, (*imp*) dry, dry up/out; wither

вы·сых·а́юш·ий, -ая, -ее, (*pres part act*) of вы·сых·а́·ть; drying, siccative

вы·та́лк·ивани·е, -я, (*n*) (*v n*) of вы·та́лк·ива·ть; ejection

~, ве́рх·н·ее (mech) top ejection

вы·та́лк·ивател·ь, -я, (*m*) ejector, discharger, pusher, push-rod, ejector pin

~, печ·н·о́й furnace discharger

~, фрикцио́н·н·ый friction discharger

вы·та́лк·ива·ть, -ют, (*imp*) push/shove/force/jog out, eject

вы·та́пл·ивани·е, -я, (*n*) (*v n*) of вы·та́пл·ива·ть; (met) liquation

вы·та́пл·ива·ть, -ют, (*imp*) turn on the heating, heat, light (a stove); separate, clarify (by heating); (met) liquate

вы·та́пт·ыва·ть, -ют, (*imp*) trample down

вы·та́ск·ивател·ь, -я, (*m*) extractor

вы·та́ск·ива·ть, -ют, (*imp*) drag/pull out, withdraw, extract; steal, pinch

вы·та́ч·ива·ть, -ют, (*imp*) turn (on lathe); bore, hollow out; neck

вы́·тач·к·а, -и, (*g pl*) -ч·ек·, (*f*) tuck

вы́·тащ·ить, -ат, (*perf*) *see* вы·та́ск·ива·ть

вы·тек·а́ни·е, -я, (*n*) discharge, flowing out, efflux, effluence, escape; (plast) flash groove (on mould)

вы·тек·а́·ть, -ют, (*imp*) flow/run out, drain away; rise, have its source (of river); follow, result, ensue

вы·тек·а́юш·ий, -ая, -ее, (*a*) (*pres part act*) of вы·тек·а́·ть; effluent; resulting, resultant

вы́·тер·еть, (*fut 3rd sing, pl*) вы́·тр·ет, вы́·тр·ут, (*past masc sing*) вы́·тер·, (*perf*); *see* вы·тир·а́·ть

вы́·тер·т·ый, -ая, -ое, (*past part pass*) *see* вы·тир·а́·ть; threadbare

~ нефт·и вод·о́й (oil) water drive

вы́·тес·ать, (*fut 3rd sing, pl*) **вы́·те-ш·ет, вы́·теш·ут,** (*perf*); *see* **вы́·тёс·ыва·ть**

вы·тесн·éни·е, -я, (*n*) (*v n*) *see* **вы·-тесн·я́·ть;** displacement, expulsion, supplanting; (chem) substitution, replacement; (cinema) wipe, wiping

~ **тóк·а** (elec) surface leakage

тóлщ·ин·á вы·тесн·éни·я (*f*) (dyn) displacement thickness

вы·тесн·и́тел·ь, -я, (*m*) rotating piston, plunger (rotating plunger pump)

вы́·тесн·ить, -ят, (*perf*) *see* **вы·тесн·я́·ть**

вы·тесн·я́·ть, -ют, (*imp*) force/crowd out, eject, displace, dislodge, expel; supplant, substitute

вы́·тёс·ыва·ть, -ют, (*imp*) cut/hew/ carve out, trim

вы́·теч·ь, (*fut 3rd sing, pl*) **вы́·теч·ет, вы́·тек·ут,** (*past masc sing*) **вы́·тек·,** (*perf*); *see* **вы·тек·á·ть**

вы́·теш·ут (*fut 3rd pl*) of **вы́·тес·ать**

вы·тир·á·ть, -ют, (*imp*) wipe off/out, wipe dry; wear, wear out; **-ся** (*pass*) become worn/threadbare

вы́·тисн·ить, -ят, (*perf*) *see* **вы·тисн·я́·ть**

вы·тисн·я́·ть, -ют, (*imp*) stamp, imprint, impress

вы́·тк·ать, -ут, (*perf*) weave

вы́·толк·а·ть, -ют, (*perf*) *see* **вы·тáлк·ива·ть** push/shove/force/jog out

вы́·толк·н·уть, -ут, (*perf*) give a push/ shove/jog out

вы́·топ·ить, -ят, (*perf*) *see* **вы·тáпл·ива·ть**

вы́·топ·к·а, -и, (*g pl*) **-п·ок·,** (*f*) (*v n*) *see* **вы·тáпл·ива·ть;** (*pl*) scum, dross

вы́·топл·енн·ый, -ая, -ое, (*past part pass*) *see* **вы·тáпл·ива·ть**

вы́·топт·ать, (*fut 3rd sing, pl*) **вы́·топ-ч·ет, вы́·топч·ут,** (*perf*); trample down

вы́·торг·овать, -уют, (*perf*) *see* **вы·торг·óвыва·ть**

вы·торг·óвыва·ть, -ют, (*imp*) net clear (by trading); bargain, beat down

вы́·точ·енн·ый, -ая, -ое, (*past part pass*) *see* **вы·тáч·ива·ть**

вы́·точ·ить, -ат, (*perf*) *see* **вы·тáч·ива·ть**

вы́·точ·к·а, -и, (*g pl*) **-ч·ек·,** (*f*) (*v n*) *see* **вы·тáч·ива·ть** groove, recess; (mech) neck, neck journal worm hole (timber)

вы́·точ·к·а

~ **канáв·ок·** necking (with lathe)

~, **кольц·ев·áя** neck

вы́·трав·ить, -ят, (*perf*) *see* **вы·трав-л·я́·ть**

вы́·трав·к·а, -и, (*g pl*) **-в·ок·,** (*f*); (text) discharge printing

вы·травл·éни·е, -я, (*n*) (*v n*) *see* **вы·травл·я́·ть;** extermination; etching; chemical removal; trampling down

вы·трáвл·ивани·е, -я, (*n*) *see* **вы·-травл·éни·е**

вы·травл·я́·ть, -ют, (*imp*) exterminate, poison; etch, reveal by etching; remove (stains); trample down; (naut) veer, pay out

вы·трав·н·óй, -áя, -óе, (*a*) discharge-printing; chemical-removal; etching

вы́·треб·овать, -уют, (*perf*) demand, obtain; summon, send for

вы́·трезв·ить, -ят, (*perf*) *see* **вы·-трезвл·я́·ть**

вы·трезвл·я́·ть, -ют, (*imp*) sober, make sober

вы·тряс·á·ть, -ют, (*imp*) shake out

вы́·тряс·ти, -ут, (*perf*) *see* **вы·тря-с·á·ть**

вы·трях·ива·ть, -ют, (*imp*) shake out

вы́·трях·н·уть, -ут, (*perf*) *see* **вы·-трях·ива·ть**

вы·ть, (*pres 3rd sing, pl*) **вó·ет, вó·ют,** (*imp*); howl, wail

вы·тя́г·ивани·е, -я, (*n*) (*see also* **вы́·-тяж·к·а**); drawing out, stretching, extending, extension, extent; pulling out, withdrawal; extraction, exhaustion; drawing (glass, textiles)

~ **из рас·плáв·а** crystal pulling, Czochralski method (of growing crystals)

мóдул·ь вы·тя́г·ивани·я (*m*) (mech) modulus of stretch

~ **парашю́т·а** parachute opening/ deployment

вы·тя́г·ива·ть, -ют, (*imp*) draw out, stretch, extend; pull out, withdraw, extract, exhaust; **-ся** (*pass*) extend

вы·тя́г·ивающ·ий, -ая, -ее, (*pres part act*) of **вы·тя́г·ива·ть**

~ **рóл·ик·и** (*pl*) (met) withdrawing rolls

вы·тяж·éни·е, -я, (*n*) (*see also* **вы·тя́-г·ивани·е**); (med) traction, traction therapy вы́·тяж·к·а, -и,   (*g pl*) -ж·ек·, (*f*) *see* вы·тя́г·ивани·е; (met) drawing, drawing down; deep drawing, clipping (in dies); draft (bar rolling); stretch ratio (of tire cord); (chem) extract, extraction; (text) draft; (rail) shunting neck

~ в шта́мп·е   (met) dishing, cupping

~, во́д·н·ая   (chem) aqueous extract

~, глуб·о́к·ая   (met, plast) deep drawing
испыта́ние· на вы́·тяж·к·у   (*n*) (met) cupping test

~, ки́сл·ая   (chem) acid extract
коэффицие́нт· вы́·тяж·к·и   (*m*) (met) elongation factor, cupping ductility ratio

~, кузн·е́чн·ая   (met) fullering, drawing down

~ не́фт·и   oil extraction

~, от·нос·и́тельн·ая   (met) percentage elongation

~ при волоч·е́ни·и   wire-drawing

~, раст·и́тельн·ая   vegetable extract
сорт·иро́вк·а на вы́·тяж·к·е   (*f*) (rail) flat shunting

~, спирт·ов·а́я   (chem) alcoholic extract

~, терм·и́ческ·ая   (rubber) heat stretching
штамп·о́вк·а с вы́·тяж·к·ой   (*f*) (plast) stretch fuming (pressing)

вы·тяж·н·о́й, -ая, -ое,   (*a*) of вы́·-тяж·к·а, вы·тя́г·ивани·е

~ аппара́т·   (*m*) (text) drawing head/rollers

~ вентил·я́ци·я   (*f*) exhaust ventilation

~ кольц·о́   (*n*) ripcord ring (parachute)

~ маши́н·а   (*f*) (text) drafter

~ пресс·   (*m*) (met) drawing press

~ трос·   (*m*) rip-cord (parachute)

~ тя́г·а   (*f*) induced draught

~ шкаф·   (*m*) (chem) fume cabinet

вы́·тя·нутост·ь, -и,   (*f*) elongation, extensiveness

вы́·тя·нут·ый, -ая, -ое,   (*past part pass*) *see* вы·тя́г·ива·ть; stretched, elongated, prolonged, extensive; (math) prolate; exhausted (air)

вы́·тя·н·уть, -ут,   (*perf*) *see* вы·тя́г·ива·ть

вы·у́ч·ива·ть, -ют,   (*imp*) learn; teach, train; -ся (*pass*) learn

вы́·уч·ить, -ат,   (*perf*) *see* вы·у́ч·ива·ть

вы́·уч·к·а, -и,   (*g pl*) -ч·ек·, (*f*) teaching, training

вы·ха́ж·ива·ть, -ют,   (*imp*) look after, care for, nurse; grow, rear, raise; bring up (person); (naut) shorten in

вы́·хват·ить, -ят,   (*perf*) *see* вы·хва́т·ыва·ть

вы·хва́т·ыва·ть, -ют,   (*imp*) snatch out/away

вы́·хвач·енн·ый, -ая, -ое,   (*past part pass*) *see* вы·хва́т·ыва·ть

вы́·хлоп·, -а,   (*m*) exhaust, discharge

вы·хлоп·н·о́й, -а́я, -о́е,   (*a*) exhaust, discharge

~ раз·ря́д·ник·   (*m*) (elec) expulsion gap

~ газ·   (*m*) exhaust gas; (*pl*) (ICE) exhaust

вы́·ход·, -а,   (*m*) (*v n*) of вы·ход·и́ть; departure, emergence; excursion, trip; exit, way out, outlet, delivery (end); (geol) outcrop; (print) appearance, issue, number; show (of oil); yield, output; outcome, result

~ актёр·а   (theat) entrance, entry; (cinema) step-in

~, вал·ов·о́й   gross yield, total yield

~ го́д·н·ого   output of sound goods etc.

~, дро́б·н·ый   fractional yield

~, за·пас·н·о́й   emergency exit

~ из бо́·я   (mil) disengagement, withdrawal

~ из стро́·я   breakdown, failure, going out of action

~, изо́·гн·ут·ый   (geol) arcuate outcrop

~, контро́ль·н·ый   (shipb) final trial trip
па вы́·ход·е   at the output/outlet

~ на по·са́д·к·у   (air) landing approach

~ на сыр·ь·ё   yield on charge

~ не́фт·и   oil show

~ о·бедн·ённ·ого проду́кт·а   (nucl) stripped output

~ о·с·ко́л·к·ов   (nucl) fission yield

~ па́р·а   escape of steam, steam outflow

~ пласт·а́   (geol) outcrop

~ пото́к·у   (elec, chem) current efficiency

~, пол·ев·о́й   (mil) field exercise

~ поли·ме́р·а   (chem) polymer yield

~ пуч·к·а́   beam escape
рабо́т·а вы́·ход·а   (*f*) work function (of electrons)

~ сцинтилл·я́ци·й   scintillation response

вы́·ход

~, у·бо́й·н·ый　dressing percentage (of meat)

ý·гол· вы́·ход·а　(*m*) angle of emergence/departure (of waves, beams)

~ шла́к·а　(nucl) poisoning yield

вы́·ход·ить, -ят,　(*perf*) *see* **вы·ха́·ж·ива·ть; вы·ход·и́ть, -ят,** (*imp*) go out, emerge, leave, put to sea; appear, be published, be issued; run out (of things or time), be up (time); be, prove to be, turn out (as a result); be (by origin); overlook, look out, face (of windows etc.)

~ из стро́·я　break down, go out of action/service, go off stream

вы·ход·н·о́й, -а́я, -о́е,　(*a*) outgoing, going-out, departure; outlet, exit; publication, issue; output, yield

~ ве́нтил·ь　(*m*) outgate (computers)

~ ден·ь　(*m*) day off, rest day, holiday

~ каска́д·　(*m*) (elec) output stage

~ лист·　(*m*) (print) title page

~ с·ве́д·ени·я　(*pl*) (print) publisher's imprint

~ сигна́л·　(*m*) (rail) departure signal

~ у·стро́й·ств·о　(*n*) (elec) output unit

вы·ход·я́щ·ий, -ая, -ее,　(*pres part act*) of **вы·ход·и́ть,** (*imp*); outgoing, coming out, emergent; (geol) outcropping

вы·хожд·ени·е, -я,　(*n*) (*v n*) *see* **вы·хо·д·и́ть,** (*imp*); emergence

вы́·хож·енн·ый, -ая, -ое,　(*past part pass*) *see* **вы·ха́ж·ива·ть**

вы·холаж·ивани·е, -я,　(*n*) cooling, chilling; heat loss

вы·холаж·ива·ть, -ют,　(*imp*) cool, chill

вы·хола́щ·ива·ть, -ют,　(*imp*) castrate, geld

вы́·холод·ить, -ят,　(*perf*) *see* **вы·холаж·ива·ть**

вы́·холост·ить, -ят,　(*perf*) *see* **вы·хола́щ·ива·ть**

вы́·холощ·енн·ый, -ая, -ое,　(*past part pass*) *see* **вы·хола́щ·ива·ть**

выхухол·ь, -я,　(*m*) (zool) Russian Desman, *Desmana moschata*

вы́·царап·а·ть, -ют,　(*perf*) *see* **вы·цара́п·ыва·ть**

вы·цара́п·ыва·ть, -ют,　(*imp*) scratch, scratch out; **-ся** (*pass*) scrape through, have a narrow escape

вы́·цвест·и,　(*fut 3rd sing, pl*) **вы́·цвет·ет, вы́·цвет·ут,** (*past masc sing*) **вы́·цве·л,** (*perf*); *see* **вы·цвет·а́·ть**

вы́·цвет·, -а,　(*m*) efflorescence

вы·цвет·а́емост·ь, -и,　(*f*) blooming (of rubber)

вы·цвет·а́ни·е, -я,　(*n*) (*v n*) of **вы·цвет·а́·ть**; efflorescence; blooming (of rubber)

вы·цвет·а́·ть, -ют,　(*imp*) fade, lose colour/color; (chem, geol) effloresce; (bot) finish blooming

вы́·цед·ить, -ят,　(*perf*) *see* **вы·це́ж·и·ва·ть**

вы·це́ж·ива·ть, -ют,　(*imp*) decant, filter, strain

вы·чёрк·ива·ть, -ют,　(*imp*) cross/strike/cut out, expunge, delete

вы́·черк·н·уть, -ут,　(*perf*) *see* **вы·чёрк·ива·ть**

вы́·черп·ать, -ют,　(*perf*) *see* **вы·чёр·п·ыва·ть**

вы·чёрп·ыва·ть, -ют,　(*imp*) bail out, scoop out; exhaust (supply)

вы́·черт·ить, -ят,　(*perf*) *see* **вы·чёр·ч·ива·ть**

вы·чёрч·ивани·е, -я,　(*n*) (*v n*) of **вы·чёрч·ива·ть** tracing, drawing, plotting

вы·чёрч·иватель·, -я,　(*m*) plotter, tracer

вы·чёрч·ива·ть, -ют,　(*imp*) draw, trace, plot

вы́·чес·ать,　(*fut 3rd sing, pl*) **вы́·чеш·ет, вы́·чеш·ут,** (*perf*); comb out

вы́·чес·к·а, -и,　(*g pl*) **-с·ок·,**　(*f*) combing out; (*pl*) combings

вы́·чест·ь,　(*fut 3rd sing, pl*) **вы́·чт·ет, вы́·чт·ут,** (*past masc sing*) **вы́·чел,** (*perf*); *see* **вы·чит·а́·ть**

вы·чёс·ыва·ть, -ют,　(*imp*) comb out

вы́·чет·, -а,　(*m*) (*v n*) *see* **вы·чит·а́·ть;** deduction; (math) residue

за вы́·чет·ом　less, minus, deducting

~, квадрат·и́чн·ый　(math) residue of the second order

вы́·чет·н·ый, -я, -ое,　(*a*) of **вы́·чет·;** (math) residual

вы́·чет·ш·ий, -ая, -ее,　(*past part act*) *see* **вы·чит·а́·ть**

вы́·чеш·ут　(*fut 3rd pl*) of **вы́·чес·ать**

вы·числ·е́ни·е, -я,　(*n*) calculation, calculating, computation, computing, reckoning

~, камера́ль·н·ое　(surv) office calculations

~, у·равн·и́тельн·ое　compensating calculation

вы́·числ·енн·ый, -ая, -ое,　(*past part pass*) *see* **вы·числ·я́·ть;** calculated, estimated, computed

вы́·числ·енн·ый, -ая, -ое
~ ме́сто·на·хожд·е́ни·я (*n*) observed position (from celestial observation)
вы·числ·и́м·ый, -ая, -ое, (*a*) comput-able
вы·числ·и́тел·ь, -я, (*m*) *see* вы·чи·сл·и́тельн·ая маши́н·а; calcula-tor, computer
вы·числ·и́тельн·ый, -ая, -ое, (*a*) calculating, computing
~ маши́н·а (*f*) computer, computing/calculating machine
~ маши́н·а дискре́т·н·ого де́й·ст·ви·я digital computer
~ маши́н·а, кла́виш·н·ая key-res-ponsive adding machine
~ маши́н·а, модел·и́рующ·ая analog/analogue computer
~ маши́н·а не·пре·ры́в·н·ого де́й·стви·я analog/analogue computer
~ маши́н·а, перфораци́он·н·ая punched-card machine, punched-type machine
~ маши́н·а, рыча́ж·н·ая lever-responsive adding machine
~ маши́н·а, у·правл·я́ющ·ая control computer
~ маши́н·а, цифр·ов·а́я digital computer
~ перфора́тор· (*m*) calculating punch
~ у·стро́й·ств·о = вы·числ·и́тельн·ая маши́н·а computer, comput-ing device
вы́·числ·ить, -ят, (*perf*) *see* вы·числ·я́·ть
вы·числ·я́·ть, -ют, (*imp*) calculate, compute
вы́·чист·ить, -ят, (*perf*) *see* вы·чищ·а́·ть
вы·чит·а́ем·ый, -ая, -ое, (*pres part pass*) of вы·чит·а́·ть; to be deducted/subtracted; deductible; (*n decl as adj*) (math) subtrahend
вы·чит·а́ни·е, -я, (*n*) subtraction; deduction
вы́·чит·а·ть, -ют, (*perf*) read out; read, check (proofs etc.); вы·чи·т·а́·ть, -ют, (*imp*) subtract, deduct
вы·чи́т·ыва·ть, -ют, (*imp*) *see* вы́·чит·а·ть
вы·чищ·а́·ть, -ют, (*imp*) clean out; purge, expel
вы·член·е́ни·е, -я, (*n*) exarticulation; (bio) singling out
вы́·чт·енн·ый, -ая, -ое, (*past part pass*) *see* вы·чит·а́·ть

вы́·чур·н·ый, -ая, -ое, (*a*) fanciful, pretentious, affected
вы́·шаг·а·ть, -ют, (*perf*) *see* вы·ша́·г·ива·ть
вы·ша́г·ива·ть, -ют, (*imp*) pace, pace out
вы·швы́р·ива·ть, -ют, (*imp*) fling/hurl/chuck out
вы́·швыр·н·уть, -ут, (*perf*) *see* вы·швы́р·ива·ть
вы́·швыр·я·ть, -ют, (*perf*) throw out, eject
вы́ш·е (*comp*) higher, taller; (*adv*) (*prep + gen*) above, over, beyond
~ по теч·е́ни·ю upstream
выше- (*root*) above
выше·из·ло́ж·енн·ый, -ая, -ое, (*a*) the above, above-stated, the foregoing
выше·леж·а́щ·ий, -ая, -ое, (*a*) overlying, superimposed, superposed, superincumbent, superjacent; upper
выше·на́·зв·анн·ый, -ая, -ое, (*a*) the above, above-named, aforesaid
выше·о·зна́ч·енн·ый, -ая, -ое, (*a*) aforesaid, above-mentioned
выше·при·вед·ённ·ый, -ая, -ое, (*a*) cited above, already cited
выше·с·ка́з·анн·ый, -ая, -ое, (*a*) aforesaid
выше·сто·я́щ·ий, -ая, -ее, (*a*) higher, superior (authority etc.)
выше·у·ка́з·анн·ый, -ая, -ое, (*a*) pointed out above, above-mentioned
выше·у·по·мя́·нут·ый, -ая, -ое, (*a*) above-mentioned
вы·шиб·а́ни·е, -я, (*n*) (*v n*) *see* вы·шиб·а́·ть
~ а́том·а (cryst) discomposition
вы·шиб·а́·ть, -ют, (*imp*) knock/kick/chuck out
вы́·шиб·ить, -ут, (*perf*) *see* вы·ши·б·а́·ть
вы́·шибл·енн·ый, -ая, -ое, (*past part pass*) *see* вы·шиб·а́·ть
вы́·шиб·н·ый за·ря́д· (*m*) (gunn) bursting charge
вы·шив·а́льн·ый, -ая, -ое, (*a*) em-broidering
вы·шив·а́ни·е, -я, (*n*) embroidering
вы·шив·а́·ть, -ют, (*imp*) embroider
вы́·шив·к·а, -и, (*g pl*) -в·ок·, (*f*) em-broidery
вы·шив·н·о́й, -а́я, -о́е, (*a*) embroidered
выш·ин·а́, -ы́, (*nom pl*) -ин·ы, (*g pl*) -и́н·, (*f*) height
вы́·ши·ть (*fut 3rd sing, pl*) вы́·шь·ет, вы́·шь·ют, (*perf*); embroider

вы́ш·к·а, -и, *(g pl)* **вы́ш·ек·,** *(f)* tower; (oil) derrick

~, автомоби́ль·н·ая tower wagon

~, бур·ов·а́я drilling derrick

~, диспе́тчер·ск·ая traffic control tower

~, кома́нд·н·ая control tower (at airport)

~, ста́рт·ов·ая (rocket) launching tower

вы́·школ·ить, -ят, *(perf)* school, discipline

вы́·шл·ет *(fut 3rd sing)* of **вы́·сл·ать**

вы́·шлиф·овать, -уют, *(perf) see* **вы·шлиф·о́выва·ть**

вы·шлиф·о́выва·ть, -ют, *(imp)* polish

вы́·шл·ют *(fut 3rd pl)* of **вы́·сл·ать**

вы·штукату́р·ива·ть, -ют, *(imp)* plaster, stucco

вы́·штукату́р·ить, -ят, *(perf) see* **вы·штукату́р·ива·ть**

вы́·шут·ить, -ят, *(perf) see* **вы·шу́·ч·ива·ть**

вы·шу́ч·ива·ть, -ют, *(imp)* ridicule, make fun of

вы·щела́ч·ивани·е, -я, *(n)* leaching, lixiviation

~, автокла́в·н·ое pressure leaching

вы·щела́ч·ива·ть, -ют, *(imp)* leach, lixiviate

вы́·щелоч·ить, -ат, *(perf) see* **вы··щела́ч·ива·ть**

вы́·щелоч·енн·ый, -ая, -ое, *(past part pass) see* **вы·щела́ч·ива·ть** leached, extracted

вы́·щип·ать *(fut 3rd sing, pl)* **вы́··щипл·ет, вы́·щипл·ют,** *(perf); see* **вы·щи́п·ыва·ть**

вы·щи́п·ыва·ть, -ют, *(imp)* pull out, pluck

вы́·щуп·а·ть, -ют, *(perf)* feel for, probe

вы·щу́п·ывани·е, -я, *(n) (v n)* of **вы··щу́п·ыва·ть**

вы·щу́п·ыва·ть, -ют, *(imp)* feel for, probe

вы́·яв·и́тел·ь, -я, *(m)* detector

вы́·яв·ить, -ят, *(perf) see* **вы·яв·л·я́·ть**

вы·явл·е́ни·е, -я, *(n) (v n) see* **вы··явл·я́·ть** appearance; development; (instr) visualization

вы·явл·я́емост·ь, -и, *(f)* detectability

вы·явл·я́·ть, -ют, *(imp)* make clear, reveal; expose, show up; visualize, reveal by chemical treatment (etc.); **-ся** *(pass)* become clear, be revealed

вы·ясн·е́ни·е, -я, *(n)* explanation; elucidation, making clear, clarification; finding out

вы́·ясн·ить, -ят, *(perf) see* **вы·ясн·я́·ть**

вы·ясн·я́·ть, -ют, *(imp)* elucidate, make clear, clarify; find out, ascertain; **-ся** *(pass)* turn out, prove to be

вь·ет *(3rd sing pres)* of **ви·ть**

вь·ю́г·а, -и, *(f)* blizzard, snow storm

вь·ю́жн·ый, -ая, -ое, *(a)* of **вь·ю́г·а**

вь·ю́к, -а, *(m)* pack, pack saddle

вь·ю́н·, -а́, *(m)* (fish) groundling

вь·юнко́в·ые *(pl decl as adj)* (bot) *Convolvulaceae*

вь·юно́к·, -нк·а́, *(m)* (bot) bindweed *Convolvulus*

вью́ч·ить, -ат, *(imp)* load up a pack animal

вью́ч·н·ый, -ая, -ое, *(a)* of **вь·ю́к·**

вь·ю́шк·а, -и, *(g pl)* **-шек·,** *(f)* reel; damper

вь·ю́щ·ий, -ая, -ее, *(pres part act)* of **ви·ть; -ся** coiling, curling, winding, vinaceous; (bot) climbing

вюсти́т·, -а, *(m)* (min) wüstite

вя́ж·ет *(pres 3rd sing)* of **вяз·а́ть**

вя́ж·ущ·ее *(n decl as adj)* binder, binding agent, cement

~, извест·ко́в·ое calcaneous cement

вя́ж·ущ·ий, -ая, -ее, *(pres part act)* of **вяз·а́ть;** *(see also* **вя́ж·ущ·ее)** *(as noun)* binding, tying, knitting; astringent, tart

~ в·ку́с· *(m)* astringent taste

~ сре́д·ств·о *(n)* astringent; binding agent, cement; *(see also* **вя́ж·ущ·ее)** *(as noun)*

вяз·, -а, *(m)* (bot) elm, *Ulmus*

вяз·а́льн·ый, -ая, -ое, *(a)* knitting, crocheting; binding, banding

~ маши́н·а *(f)* knitting machine; binding machine, wiring machine; banding machine

вяз·а́льщик·, -а, *(m)* tier, binder; knitter

вяз·а́ни·е, -я, *(n) (v n) see* **вяз·а́ть;** knitting (work), crochet-work

вяз·а́нк·а, -и, *(g pl)* **-нок·,** *(f)* bundle, bunch, sheaf

вя́з·анн·ый, -ая, -ое, *(past part pass)* of **вяз·а́ть**

~ силика́т· *(m)* (geol) inosilicate

вя́з·ан·ый, -ая, -ое, *(a)* knitted; crocheted

вяз·а́ть, (*pres 3rd sing, pl*) **вя́ж·ет, вя́ж·ут,** (*imp*); bind, tie up, tie; knit crochet; be astringent; **-ся** (*pass*) be in accord, agree/tally with

вязи́г·а, -и, (*f*) *see* **визи́г·а**

вя́з·к·а, -и, (*g pl*) **-з·ок·,** (*f*) binding, tying; knitting, crocheting

вя́з·к·ий, -ая, -ое, (*a*) viscous, viscid, glutinous; (met) tough, ductile

~ **из·ло́м·** (*m*) (met) ductile fracture

~ **теч·éни·е** (*n*) viscous flow

вя́з·кост·ь, -и, (*f*) viscosity; (met) toughness

~, **вну́тр·енн·яя** intrinsic/internal viscosity, (plastics) limiting viscosity index

~, **втор·и́чн·ая** bulk viscosity

~, **жёст·к·ая** firmo-viscosity

~, **кинемат·и́ческ·ая** (oil) kinematic viscosity; viscosity/density ratio (rubber)

    коэффициéнт· вя́з·кост·и (*m*) viscosity coefficient, coefficient of viscosity

    коэффициéнт· вя́з·кост·и, втор·о́й (*m*) (dynam) cross-velocity coefficient

    логари́фм· характерист·и́ческ·ой вя́з·кост·и (*m*) log viscosity index

~, **магни́т·н·ая** magnetic viscosity

~, **объ·ём·н·ая** bulk viscosity

~, **пласт·и́ческ·ая** plastic viscosity

~, **по Энглер·у** Engler viscosity

~, **со́б·ственн·ая** intrinsic viscosity

~, **структу́р·н·ая** structural viscosity

~, **турбулéнт·н·ая** eddy viscosity

~ **у́гл·я** coal rheology

вя́з·кост·ь

~, **уда́р·н·ая** (met) impact toughness/strength, notch toughness

~, **у·дéль·н·ая** specific viscosity

~, **у·дéль·н·ая уда́р·н·ая** (met) impact toughness number, notched bar value; impact strength (of rubber)

~, **у·пру́г·ая** elastic viscosity

~, **характерист·и́ческ·ая** characteristic viscosity

~, **цикл·и́ческ·ая** (phys) damping capacity, internal friction, mechanical hysteresis loss

    числ·о́ вяз·кост·и, предéл·ьн·ое (*n*) limiting viscosity index

вязко·тек·у́ч·ий, -ая, -ее, (*a*) viscous-flow

визко·у·пру́г·ий, -ая, -ое, (*a*) visco-elastic

вя́з·н·уть, -ут, (*imp*) get stuck, get bogged down

вя́з·ов·ый, -ая, -ое, (*a*) of **вяз·**

вя́л·ени·е, -я, (*n*) (*v n*) *see* **вя́л·ить**

вя́л·енн·ый, -ая, -ое, (*past part pass*) of **вя́л·ить;** (food) sun-cured/dried, jerked

вя́л·ить, -ят, (*imp*) (food) sun/dry-cure, dry, jerk

вя́л·ост·ь, -и, (*f*) apathy, flabbiness, limpness; (phot) flatness, lack of contrast

вя́л·ый, -ая, -ое, (*a*) sluggish, flaccid, limp, dull; withered, drooping (of plants); (phot) flat

~ **по·гран·и́чн·ый сло́·й** (*m*) (dynam) slow boundary layer

вян·уть, -ут, (*imp*) wilt, droop, wither

Г

г. (*abbr*) (*see also* **грамм·**) g, gm, gram, gramme (unit of mass); **Г.** g, gram (unit of force)

габа́р·, -а, (*m*) (naut) barge frame

габарди́н·, -а, (*m*) (text) gabardine

габари́т·, -а, (*m*) clearance; (rail) clearance diagram; (*pl*) overall dimensions, size

~ ли́н·и·и ground clearance (of overhead transmission line)

~, под·мо́ст·ов·ый bridge clearance

габари́т·н·ый, -ая, -ое, (*a*) of **габари́т·;** overall

~ воро́т·а (*pl*) (rail) loading gauge

~ выс·от·а́ (*f*) (build) headroom

~ длин·а́ (*f*) (shipb) length overall

~ шир·ин·а́ (*f*) (shipb) extreme breadth

~ раз·ме́р· (*m*) overall dimension

~ черт·ёж· (*m*) outline drawing

га́ббро (*indecl*) (geol) gabbro

габбро·ви́д·н·ый, -ая, -ое, (*a*) gabbroid

га́ббр·ов·ый, -ая, -ое, (*a*) gabbro, gabbroic

Га́бер·а, проце́сс· (*m*) (chem) Haber process (nitrogen fixation)

габио́н·, -а, (*m*) (civ eng) gabion

га́битет·, -а, (*m*) (biol) habitat

га́битус·, -а, (*m*) habitus, habit

~, волокн·и́ст·ый (cryst) fibrous habit

га́ван·ск·ий, -ая, -ое, (*a*) of **га́ван·ь**

га́ван·ь, -и, (*f*) harbour, port

~, вое́н·н·ая naval anchorage

~, во́ль·н·ая free port

~, свобо́д·н·ая free port

гавер·си́нус·, -а, (*m*) (math) haversine

га́г·а, -и, (*f*) (orn) eider, *Somateria mollissima*

гагари́нит·, -а, (*m*) (min) gagarinite

гага́т·, -а, (*m*) (min) jet

гага́т·ов·ый, -ая, -ое, (*a*) of **гага́т·**

гаг·а́чч·ий, -ья, -ье, (*a*) of **га́г·а**

~ пух· (*m*) eiderdown

гад·, -а, (*m*) (zool) reptile

гад·а́ни·е·, -я, (*n*) (*v n*) *see* **гад·а́·ть;** guesswork

гад·а́тельн·ый, -ая, -ое, (*a*) hypothetical, conjectural, guessed

гад·а́·ть, -ют, (*imp*) guess, conjecture

га́д·к·ий, -ая, -ое, (*a*) loathsome, foul, nasty

гадоли́н·и·й, -я, (*m*) (chem) gadolinium, Gd

гадолин·и́т·, -а, (*m*) (min) gadolinite

гадоли́н·ов·ый, -ая, -ое, (*a*) gadolinium

~ земл·я́ (*f*) (min) gadolinia

гадр·о́м·, -а, (*m*) = **хадр·о́м·,** (bot) hadrome, hadrom

Га́дфильд·а, проце́сс· (*m*) (met) Hadfield process

гад·ю́к·а, -и, (*f*) (zool) viper, adder

гад·ю́ч·ий, -ья, -ье, (*a*) of **гад·ю́к·а**

га́·ечн·ый, -ая, -ое, (*a*) of **га́·йк·а**

~ ключ· (*m*) spanner, wrench

~ ключ·, от·кры́·т·ый (*m*) double-end spanner, open-end wrench

~ ключ·, раз·движ·н·о́й (*m*) adjustable spanner

га́ж·е, (*comp*) of **га́д·к·ий**

гаж·е́ни·е, -я, (*n*) degassing, outgassing

~ сте́н·ок· (nucl) wall outgassing

газ·, -а, (*m*) gas; (med) wind; (text) gauze, gossamer

~, акти́в·н·ый reactive gas

~, бе́д·н·ый (oil) dry gas

~, благо·ро́д·н·ый rare/inert/noble gas

~, бога́т·ый (oil) wet gas

~, боло́т·н·ый marsh gas, methane

~, бу́фер·н·ый (oil) cushion gas

~, в·ключ·ённ·ый included gas

~, вла́ж·н·ый (oil) wet gas

~, вод·ян·о́й water blue gas

~, воз·ду́ш·н·ый producer gas; air gas

~, высоко·сер·ни́ст·ый (oil) sour gas

~, вы·хлоп·н·о́й exhaust gas

~, вя́з·к·ий imperfect gas

~, генера́тор·н·ый producer gas

~, голо́в·очн·ый (oil) casing-head gas

~, гор·ю́ч·ий flammable/fuel gas

~, грем·у́ч·ий (min) fire-damp, detonating gas

~, дым·ов·о́й flue/stack gas

~, есте́ств·енн·ый natural gas, rock gas

~, жи́дк·ий liquified/bottled/compressed gas

~, жи́р·н·ый wet/rich gas

~, за·во́д·ск·ий refinery gas

~, за·ключ·ённ·ый occluded gas

~, идеа́ль·н·ый perfect gas

~, индифере́нт·н·ый inert/chemically inactive gas

~, ине́рт·н·ый inert/rare/noble gas

газ

~, **ко́кс·ов·ый** coke-oven gas
~, **колош·ников·ый** blast-furnace gas
~ **кре́кинг·а** cracked gas
~, **ма́л·ый** (ICE) idling
~, **масло·ро́д·н·ый** olefiant gas
~, **на·пряж·ённ·ый** tensioned gas
~, **не·де́·ятельн·ый** rare/inert/noble gas
~, **не·идеа́ль·н·ый** imperfect gas
~, **нейтра́ль·н·ый** inert gas
~, **не·конденс·и́рующ·ийся** fixed gas
~, **нефт·ян·о́й** natural gas
~, **о·свобожд·ённ·ый** resurgent gas
~, **от·бензи́н·енн·ый** (oil) dry gas
~, **от·рабо́т·анн·ый** waste gas, exhaust
~ **Пи́нч·а** (oil) Pintsch gas
~ **пиро·ли́з·а** cracked gas
~, **пласт·ов·о́й** (oil) formation gas
~, **пло́т·н·ый** solid gas
~, **по·глощ·ённ·ый** occluded gas
~, **по́лн·ый** (ICE) full-throttle, (M/T) accelerator hard down
~, **полу·вод·ян·о́й** semi-water gas
~, **полу·ко́кс·ов·ый** rich gas, carbonization gas (from low temperature distillation)
~, **по·пу́т·н·ый** (oil) casing-head gas
~, **при·ро́д·н·ый** natural gas
~, **раз·бав·и́тел·ь, -я,** (*m*) carrier gas (metal heat treatment etc.)
~, **раз·реж·ённ·ый** dilute/rarefied gas
~, **рас·твор·ённ·ый** dissolved/solution gas
~, **реа́ль·н·ый** imperfect gas
~, **ре́д·к·ий** rare/inert/noble gas
~, **руд·ни́чн·ый** fire-damp, detonating gas
~, **свет·и́льн·ый** town gas
~, **сер·ни́ст·ый** sulphurous gas, sulphur oxide/dioxide
~, **с·жа́т·ый** pressure/compressed gas; liquified gas
~, **с·жи́ж·енн·ый** liquified/bottled/compressed gas
~, **с·жим·а́ем·ый** coercible gas
~, **си́н·ий** blue water gas
~, **с·коп·и́вш·ийся** occluded gas
~, **слезо·точ·и́в·ый** tear gas
~, **со·пу́т·ствующ·ий** (oil) casing-head gas
~, **сорб·и́рованн·ый** retained gas
~, **спонта́н·н·ый** spontaneous gas
~, **сух·о́й** dry gas
~, **то́п·очн·ый** flue gas
~, **то́щ·ий** lean gas

газ

~, **уга́р·н·ый** carbon monoxide/oxide, carbonyl
~, **угле·водо·ро́д·н·ый** hydrocarbon gas, carburetted hydrogen gas
~, **угле·ки́сл·ый** carbon dioxide, carbonic acid gas
~, **у·ду́ш·лив·ый** choke-gas, choke-damp
~, **у·плот·ня́ющ·ий** blanketing gas
~, **фреат·и́ческ·ий** phreatic gas
~, **фумаро́ль·н·ый** fumarolic/fumarole gas
~, **электро́н·н·ый** (phys) electron gas
газ·а́тор·, -а, (*m*) wickless pressure lamp, primus stove
газ·го́льдер·, -а, (*m*) gas-holder
газе́т·а, -ы, (*f*) newspaper
~, **свет·ов·а́я** news reel
газе́т·н·ый, -ая, -ое, (*a*) of **газе́т·а**
~ **бума́г·а** (*f*) (paper) newsprint
газ·и́р·овать, -уют, (*imp*) aerate; emit gas, bubble up
газ·и́рованн·ый, -ая, -ое, (*past part pass*) of **газ·и́р·овать**; aerated; carbonated (of drinks)
гази·фик·а́тор·, -а, (*m*) vaporizer; gas fuel specialist; gasification engineer
гази·фик·а́ци·я, -и, (*f*) gasification, gas manufacture; conversion to gas (of engines, plant); gas supply (to towns etc.); vaporization
~, **под·зе́м·н·ая** (min) underground gasification
гази·фиц·и́р·овать, -уют, (*imp and perf*) lay on gas, supply with gas, supply gas
газ·ли́фт·, -а, (*m*) gas lift (pumping)
га́з·н·ый, -ая, -ое, (*a*) gas, gassy
газо·анализ·а́тор·, -а, (*m*) gas analyser
~, **денси·метр·и́ческ·ий** densimetric gas analyser
~, **магни́т·н·ый** magnetic oxygen gas analyser
~, **опт·и́ческ·ий** optical gas analyser
~, **термо·магни́т·н·ый** magnetic-wind oxygen analyser
газо·балло́н·, -а, (*m*) gas cylinder/bottle
газо·без·о·па́с·н·ый, -ая, -ое, (*a*) gas-proof
газо·бензи́н·ов·ый за·во́д·, (*m*) natural gasoline extraction plant
газо·бето́н·, -а, (*m*) gas concrete
газ·ова́ни·е, -я, (*n*) gassing, ebullition, bubbling up

газ·ова́ть, -у́ют, (*imp*) emit gas, bubble up

газ·ов·и́к·, -а́, (*m*) gas producer/generator; gasworks engineer

газ·овщи́к·, -а́, (*m*) gas/gasification engineer; gas fitter

газо·вы́·дел·е́ни·е, -я, (*n*) gas evolution, ebullition, gassing

га́з·ов·ый, -ая, -ое, (*a*) gas, gaseous, gassy; gas-filled; gauze

~ за·во́д· (*m*) gas works

~ ла́мп·а (*f*) (elec) gas-filled filament lamp

ме́тод· га́з·ов·ого на·сыщ·е́ни·я (*m*) transpiration method (of measuring vapour pressure)

~ по·сто·я́нн·ая (*f*) (phys) gas constant

~ раз·ря́д·ник· (*m*) (elec) spark-gap tube, gas-discharge tube

~ рул·ь (*m*) (rocket) jet vane

~ со·сто·я́ни·е (*n*) gaseous state

~ то́пл·ив·о (*n*) gas/gaseous fuel

газо·воз·ду́ш·н·ый, -ая, -ое, (*a*) air-gas

газо·га́з·ов·ый, -ая, -ое, (*a*) gas-to-gas (of heat exchanges etc.)

газо·ге́н·, -а, (*m*) gasogen

газо·генера́тор·, -а, (*m*) gas producer/generator; (rocket) gas generator

газо·генера́тор·н·ый, -ая, -ое, (*a*) of газо·генера́тор·

газо·генери́р·ующ·ий вы·ключ·а́·тел·ь (*m*) (elec) gas-blast breaker

газо·гра́ф·и·я, -и, (*f*) (phot) vapography

газо·ди́зел·ь, -я, (*m*) (ICE) dual-fuel engine

газо·дина́м·ик·а, -и, (*f*) gas dynamics

газо·ду́в·к·а, -и, (*g pl*) -в·ок·, (*f*) gas blower, blower

~, ротацио́н·н·ая positive-displacement blower

газо·золо·бето́н·, -а, (*m*) gas-ash concrete

газ·о́йл·ь, -я, (*m*) gas-oil, diesel fuel

газо·кал·и́льн·ый, -ая, -ое, (*a*) (gas) incandescent

~ се́т·к·а (*f*) incandescent/gas mantle

газ·ол·и́н, -а, (*m*) gasoline; natural gasoline

газо·ме́р·, -а, (*m*) gas meter

газо́·метр·, -а, (*m*) gasometer (lab equipment) gas meter

газо·метр·и́·я, -и, (*f*) gasometry

газо·мото·во́з·, -а, (*m*) gas-driven locomotive

газо·мото́р·, -а, (*m*) gas engine

газо́н·, -а, (*m*) lawn; turf

газо·на·по́лн·енн·ый, -ая, -ое, (*a*) gas-filled (of devices); gas-expanded (of materials)

газо·на·угле·ро́ж·ивани·е, -я, (*n*) (met) gas carbonizing

газо·не·про·ниц·а́ем·ый, -ая, -ое, (*a*) gas-tight, gas-proof

газоно·кос·и́лк·а, -и, (*f*) lawn mower

газо·но́с·ност·ь, -и, (*f*) (geol) presence of gas; gas content

газо·но́с·н·ый, -ая, -ое, (*a*) gas-bearing/containing

газо·обра́з·н·ый, -ая, -ое, (*a*) gaseous, vaporized, vapor/vapour-type

газо·об·ме́н·, -а, (*m*) gas exchange

газо·образ·н·ый, -ая, -ое, (*a*) gaseous

~ ге́л·и·й (*m*) helium gas

газо·образ·ова́ни·е, -я, (*n*) gasification; gas generation; gas evolution; vaporization

газо·образ·ова́тел·ь, -я, (*m*) gas developing agent, blowing/foaming agent

газо·образ·у́ющ·ий, -ая, -ее, (*a*) gas-developing/forming; foaming

газо·от·во́д·, -а, (*m*) gas outlet/bleeder; (min) gas drain

газо·от·дел·е́ни·е, -я, (*n*) gas evolution

газо·от·дел·и́тел·ь, -я, (*m*) (chem) gas separator

газо·о·чист·и́тел·ь, -я, (*m*) (chem) scrubber, gas scrubber

газо·о·чи́ст·к·а, -и, (*g pl*) -т·ок·, (*f*) gas purifying

газо·пло́т·ност·ь, -и, (*f*) gas-tightness

газо·пло́т·н·ый, -ая, -ое, (*a*) gas-proof/tight

газо·по·глот·и́тел·ь, -я, (*m*) (elec) getter

газо·по·глощ·е́ни·е, -я, (*n*) gas occlusion

газо·по́лн·ый, -ая, -ое, (*a*) gas-filled

газо·при·ём·н·ый, -ая, -ое, (*a*) gas-collecting

газо·про·во́д·, -а, (*m*) gas pipeline, gas line/conduit

газо·про·во́д·н·ый, -ая, -ое, (*a*) of газо·про·во́д·; gas (of pipes etc.)

газо·про·во́д·чик·, -а, (*m*) gas fitter

газо·про·мыв·а́тел·ь, -я, (*m*) gas washer

газо·про·мыв·н·о́й, -а́я, -о́е, (*a*) gas-washing/scrubbing

газо·про·ниц·а́емост·ь, -и, (*f*) gas permeability/penetrability

газо·про·ниц·а́ем·ый, -ая, -ое,   (*a*) permeable to gas, gas-permeable

газо·раз·ря́д·н·ый, -ая, -ое,   (*a*) (elec) gas-discharge

~ тру́б·к·а   (*f*) (elec) gas-discharge tube/valve

газо·рас·пре·дел·е́ни·е, -я,   (*n*) valve gear (furnaces, ICE etc.)

~, золот·нико́в·ое   (ICE) sleeve-valve gear

газо·ре́з·к·а, -и,   (*g pl*) **-з·ок·,**   (*f*) (weld) gas cutting; (met) flame cutter, cutoff station (continuous casting)

газо·с·бо́р·ник·, -а,   (*m*) gas collector; hydraulic main (of gas producer)

газо·све́т·н·ый, -ая, -ое,   (*a*) (elec) glow-discharge

~ ла́мп·а   (*f*) glow-discharge lamp

газо·с·жиг·а́тельн·ый, -ая, -ое,   (*a*) gas-burning, gas-fired

газо·со́с·, -а,   (*m*) gas exhauster

газо·стекл·о́, -а,   (*n*) foam glass

газо·те́хн·ик·а, -и,   (*f*) gas engineering

газо·тро́н·, -а,   (*m*) (elec) discharge-tube rectifier

газо·турби́н·а, -ы,   (*f*) gas turbine

газо·у·бе́ж·ищ·е, -а,   (*n*) gas-proof air-raid shelter

газо·у·де́рж·ивающ·ая спосо́б·н·ость   (*f*) gas-retaining power

газо·у·лов·и́тел·ь, -я,   (*m*) gas trap

газо·у·по́р·н·ый, -ая, -ое,   (*a*) gas-proof

газо·хо́д·, -а,   (*m*) gas line/duct/flue; gas-fuelled motorship

газо·хран·и́лищ·е, -а,   (*n*) gas-holder

газ·я́щ·ий, -ая, -ее,   (*a*) gassy, gassing

Га́игер·, -а,   (*m*) *see* **Ге́йгер·** Geiger (surname)

гайдингери́т·, -а,   (*m*) (min) haidingerite

гайдро́п·, -а,   (*m*) (naut) guide rope

га́·йк·а, -и,   (*g pl*) **га́·ек;**   (*f*) nut, female screw; coupling, hose coupling

~ -бара́ш·ек·, -ш·к·а,   (*m*) wing/fly/butterfly nut

~, быстро·с·мык·а́ющ·аяся   instantaneous coupling

~, винт·ов·а́я   screw coupling

~, выс·о́к·ая   high nut

~, глух·а́я   box nut, blind nut

~, ду́ль·н·ая   sealing collar (gun barrel)

~, за·жи́м·н·ая   clamp/locking nut

~, кле́мм·ов·ая   split nut

~, колпач·ко́в·ая   dome nut, nut cap

~, кольц·ев·а́я   ring nut

~, коро́н·чат·ая   castle/castellated nut

га́·йк·а

~ ко́рпус·а форсу́н·к·и   spray valve nut

~, ма́т·очн·ая   clasp nut

~, на·кид·н·а́я   adapter nut; nozzle cap nut (diesel injecter)

~, полу-   half-coupling

~, про·рез·н·а́я   slotted nut

~ рук·ав·а́   hose coupling

~, с·го́н·н·ая   nut cap

~, со·един·и́тельн·ая   coupling

~, сто́пор·н·ая   check/lock nut

~, с·тяж·н·а́я   (ICE) spray-valve nut; turnbuckle

~, у·по́р·н·ая   stop nut

гайко·вы́·сад·очн·ый автома́т·   (*m*) automatic nut-upsetting machine

гайко·на·рез·н·о́й, -а́я, -о́е,   (*a*) nut-tapping

гайко·рез·н·а́я маши́н·а   (*f*) nut-cutting machine

гаймори́т·, -а,   (*m*) (med) highmoritis

гайперко,   (*indecl*) Hyperco (magnetic alloy)

гайперник·, -а,   (*m*) (met) Hypernik

гайтоно·га́м·и·я, -и,   (*f*) (bot) geitonogamy

га́к·, -а,   (*m*) (*see also* **крюк·**); (naut) hook

~ бо́ч·ечн·ого стро́п·а   can hook

~, вертлю́ж·н·ый   swivel hook

~, глаго́ль·н·ый   (naut) Senhouse slip

~, караби́н·н·ый   spring hook

~, от·кид·н·о́й   slip hook

~, по·вёр·нут·ый   plain hook

~, под·рыв·н·о́й   case hook

~, прост·о́й   tackle hook

~, с·клад·н·о́й   calliper hook, clasp hook

~, тормоз·н·о́й   (air) arrester hook

~, хвост·ов·о́й   arrester hook (carrier-borne aircraft)

гакабо́рт·, -а,   (*m*) (naut) taffrail

галакт·а́з·а, -ы,   (*f*) (biochem) galactase

галакт·а́н·, -а,   (*m*) (chem) galactan

гала́кт·ик·а, -и,   (*f*) (astron) galaxy

галакт·и́т·, -а,   (*m*) (min) galactite

галакт·и́ческ·ий, -ая, -ое,   (*a*) (astron) galactic

галакт·о́з·, -а,   (*m*) (zool) galactosis, lactation

галакт·о́з·а, -ы,   (*f*) (chem) galactose

галакт·ози́д·, -а,   (*m*) (chem) galactoside

галакто́·метр·, -а,   (*m*) (food) galactometer

галактон·ов·ая кисл·от·а́ (*f*) galactomic acid, pentahydroxyhexoic acid

галакто·ре́·я, -и, (*f*) (med) galactorrhea

гала·ли́т·, -а, (*m*) (plast) galalith

гала́н·ить, -ят, (*imp*) scull (with one oar)

галантере́·йн·ый, -ая, -ое, (*a*) of **галантере́·я**

галантере́·я, -и, (*f*) haberdashery

га́лбвинд·, -а, (*m*) (naut) beam wind

галга́н·, -а, (*m*) *see* **калга́н·**

галев·о, -а, (*n*) (text) heald

галени́т·, -а, (*m*) (min) galena, lead glance

гале́н·ов·ы препара́т·ы (*pl*) (pharm) galenics, galenicals

гале́н·ов·ые препара́т·ы (*pl*) (pharm) galenics, galenicals

галере́·я, -и, (*f*) gallery; culvert (of dry docks, locks etc.)

~, при·ём·н·ая filling culvert (of a dry dock)

~, шлюз·ов·а́я gravity culvert

гале́т·а, -ы, (*f*) hard tack, ship's biscuit

гале́т·н·ый, -ая, -ое, (*a*) flat, biscuit

~ батаре́·я (*f*) (elec) flat dry battery

га́л·ечник·, -а, (*m*) shingle, coarse gravel; pebble-bed

га́л·ечн·ый, -ая, -ое, (*a*) pebble, shingle

~ ме́ль·ниц·а (*f*) pebble mill (ore dressing)

Галиле́·я, зр·и́тельн·ая труб·а́ (*f*) Galilean telescope

гали·планкто́н·, -а, (*m*) (biol) haloplankton, haliplankton

галипо́т·, -а, (*m*) (chem) galipot

гал·и́т·, -а, (*m*) halite, common salt, rock salt

га́лл·, -а, (*m*) (biol) gall

галлере́·я, -и, (*f*) *see* **галере́·я**

га́лл·иев·ый, -ая, -ое, (*a*) gallium, (chem) gallic

галл·и·й, -я, (*m*) (chem) gallium, Ga

~, дву·хлор·ист·ый gallium dichloride

галлово·ки́сл·ый, -ая, -ое, (*a*) gallate of

~ желе́з·о (*n*) iron gallate

галлово·мети́л·ов·ый эфи́р· (*m*) methyl gallate

га́лл·ов·ый, -ая, -ое, (*a*) gallium, (chem) gallic

~ кисл·от·а́ (*f*) (chem) gallic acid

галло·дуб·и́льн·ая кисл·от·а́ (*f*) gallotannic acid

галлоизи́т·, -а, (*m*) (min) halloysite

галло·ксант·и́н·, -а, (*m*) (biochem) galloxanthin

галло́н·, -а, (*m*) gallon

галло·циан·и́н·, -а, (*m*) (chem) gallocyanine

галлуази́т·, -а, (*m*) (min) halloysite

Га́лл·я, цеп·и (*pl*) (mech) driving chain

галлюцин·и́р·овать, -уют, (*imp*) suffer from hallucinations

галме́·й, -я, (*m*) (min) calamine, hemimorphite, smithsonite, galmei

галме́·йн·ый, -ая, -ое, (*a*) calamine, hemimorphite, smithsonite, galmei

гало́ (*n indecl*) (astron) halo

гало·ва́кс·, -а, (*m*) (elec) halowax

гало·ге́н·, -а, (*m*) (chem) halogen

галоген·ангидри́д·, -а, (*m*) acid halide

галоген·водо·ро́д·, -а, (*m*) halogen hydride

гало·ген·е́з·, -а, (*m*) (geol) halogenesis

гало·ген·и́д·, -а, (*m*) (chem) halide, halogenide

гало·ген·и́ровани·е, -я, (*n*) (chem) halogenation

гало·ге́н·н·ый, -ая, -ое, (*a*) halogen, haloid, halide

~ кисл·от·а́ (*f*) halogen/haloid acid

гал·о́ид·, -а, (*m*) (chem) haloid, halide, halogen

галоид·алки́л·, -а, (*m*) alkyl halide

галоид·ангидри́д·, -а, (*m*) acid halide

галоид·за·мещ·ённ·ый, -ая, -ое, (*a*) halogenated, halogen

~ кисл·от·а́ (*f*) halogen acid

~ угле·водо·ро́д· (*m*) halocarbon

галоид·и́ровани·е, -я, (*n*) halogenation

гал·о́ид·н·ый, -ая, -ое, (*a*) (chem) halogen, haloid, halide

галоидо·водоро́д·, -а, (*m*) halogen hydride

галоидо·водоро́д·н·ая кисл·от·а́ (*f*) hydrohalic/halohydric acid

галоидо·кисл·от·а́, -ы, (*f*) halogen acid

галоидо·орган·и́ческ·ий, -ая, -ое, (*a*) halo-organic

галоидо·углеро́д·, -а, (*m*) halocarbon

гало·ли́т·, -а, (*m*) (geol) halogen rock

гало́·метр·, -а, (*m*) (chem) halometer

гало·метр·и́·я, -и, (*f*) halometry

галоп·и́р·овать, -уют, (*imp*) gallop

гало́с·, -а, (*m*) (astron) halo

гало·трихи́т·, -а, (*m*) (min) halotrichite, iron alum

гало·фи́т·, -а, (*m*) (bot) halophyte

гало·фо́рм·енн·ая реа́кци·я   (*f*) (chem) haloform reaction

гало·хром·и́зм·, -а,   (*m*) (chem) halochromism

гало·хром·и́·я, -и,   (*f*) halochromism

гало́ш·а, -и,   (*f*) overshoe, overboot, galosh (footwear)

~, кле·ён·ая   hand-built galosh

~, штамп·о́ванн·ая   stamped galosh

га́лс·, -а,   (*m*) (naut) tack, lap, run; line (of soundings)

га́лс·ов·ый, -ая, -ое,   (*a*) of **галс·**

га́лстук·, -а,   (*m*) tie, necktie

га́лтел·ь, -я,   (*m*) fillet

га́лтель·н·ый рез·е́ц·   (*m*) corner tool

галт·ова́ни·е, -я,   (*n*) *see* **галт·о́вк·а**

галт·о́вк·а, -и,   (*f*) barrel finishing, tumbling, tumbler polishing

галу́н·, -á,   (*m*) lace (metallic), galloon, braid

га́лфвинд·, -а,   (*m*) (naut) beam wind

гальваниз·а́ци·я, -и,   (*f*) (met) galvanization

гальваниз·и́р·овать, -уют,   (*imp and perf*) (met) galvanize; electroplate

гальван·и́зм·, -а,   (*m*) (met) galvanism

гальван·и́ческ·ий, -ая, -ое,   (*a*) galvanic

~ элеме́нт·   (*m*) primary cell (elec), galvanic cell (obs)

~ элеме́нт·, коррозио́н·н·ый   (*m*) (met) local/localized galvanic cell

гальвано·ка́уст·ик·а, -и,   (*f*) (met) galvanocautery

гальвано·магни́т·н·ый, -ая, -ое, (*a*) galvanomagnetic

гальвано́·метр·, -а, (*m*) (elec) galvanometer

~, а·стат·и́ческ·ий   astatic galvanometer

~, баллист·и́ческ·ий   ballistic galvanometer

~, би·фил·я́рн·ый   bifilar galvanometer

~, вибрацио́нн·ый   vibration galvanometer, vibrating-reed galvanometer

~, зерка́ль·н·ый   mirror/reflecting galvanometer

~, крут·и́льн·ый   torsion galvanometer

~, нул·ев·о́й   zero galvanometer

~, си́нус-   sine galvanometer

~, стре́л·очн·ый   moving-magnet needle galvanometer

~, стру́н·н·ый   string galvanometer

~, та́нгенс-   tangent galvanometer

гальвано·пла́ст·ик·а, -и,   (*f*) (met) electroforming; (print) electrotyping; galvanoplasty

гальвано·пласт·и́ческ·ий, -ая, -ое, (*a*) of **гальвано·пла́ст·ик·а**

гальвано·ско́п·, -а,   (*m*) (elec) galvanoscope

гальвано·сте́г·, -а,   (*m*) electroplater (man)

гальвано·сте́г·и·я, -и,   (*f*) electroplating; (chem) galvanostegy

гальвано·стерео·ти́п·, -а,   (*m*) (print) electrotype

гальвано·стерео·ти́п·и·я, -и,   (*f*) (print) electrotyping

гальвано·та́ксис·, -а,   (*m*) (biol) galvanotaxis, galvanotropism

гальвано·те́хн·ик·а, -и,   (*f*) (met) electrodeposition technology

гальвано·фо́рм·, -а,   (*m*) (plastic) electroformed mould/mold

га́ль·к·а, -и,   (*f*) pebble, pebbles, shingle

галько·ви́д·н·ый, -ая, -ое, (*a*) calculiform, pebble-shaped

гальпини́т·, -а,   (*m*) (min) johannite

Га́льтон·а, за·ко́н· (*m*) (gen) Galton's law

галью́н·, -а,   (*m*) (naut) heads, ship's lavatory

гама́к·, -á,   (*m*) hammock

гама́ш·и (*nom pl*), (*g pl*) **-а́ш·,**   (*f*) leggings, gaiters

га́мбир·, -а,   (*m*) (chem) gambir, pale catechu

гамбу́зи·я, -и,   (*f*) (fish) *Gambusia offinis*

гаме́т·а, -ы,   (*f*) (biol) gamete

гамет·а́нги·й, -я,   (*m*) (biol) gametangium

Га́мильтон·а, опера́тор·   (*m*) (math, phys) Hamiltonian operator, Hamiltonian

гамильтониа́н·, -а, (*m*) Hamiltonian, Hamiltonian function/operator

~, по́лн·ый   complete hamiltonian

га́мм·а, -ы, (*f*) gamma (Greek letter)

~ -акти́в·ност·ь, -и,   (*f*) (nucl) gamma activity

~ бес·кон·е́чност·и   (phot) infinity gamma, gamma infinity

~ -дефекто·скоп·и́·я, -и, (*f*) gamma-ray flaw detection

~ -из·луч·а́тел·ь, -я,   (*m*) gamma-emitter/radiator

~ -из·луч·е́ни·е, -я,   (*n*) (nucl) gamma radiation

га́мм·а

~ -из·луч·е́ни·е дел·е́ни·я   fission gamma radiation

~ -каротта́ж·, -а,   (*m*) gamma-logging, (oil) gamma-ray well-logging

~ -ква́нт·, -а,   (*m*) (nucl) gamma quantum

~ -ли́н·и·я, -и,   (*f*) gamma peak

~ -луч·и́   (*nom pl*), (*g pl*) -е́й,   (*m*) gamma rays

~ об·луч·е́ни·е, -я,   (*n*) gamma irradiation/bombardment

~, результ·и́рующ·ая   (phot, TV) overall gamma

~ -реле́   (*n indecl*) gamma relay

~ свет·л·ых тон·о́в   (phot) high key

~ -с·ни́м·ок·, -м·к·а,   (*m*) (nucl) gammagraph

~ сре́д·н·их тон·о́в   (phot) normal key, plain key

~ -съ·ём·к·а, -и,   (*g pl*) -м·ок·, (*f*) (nucl) gamma-radiation survey

~ тёмн·ых тон·о́в   (phot) low key

~ тон·о́в   (phot) key, gamma

~ -у·ста́нов·к·а, -и,   (*g pl*) -в·ок·, (*f*) gamma-ray source

~ -фа́з·а, -ы,   (*f*) (met) gamma phase

гамма·гра́мм·а, -ы,   (*f*) (nucl) gamma-graph

гамма·гра́ф·и·я, -и,   (*f*) gamma radiography

гаммо́·граф·, -а,   (*m*) *see* **гамма· гра́мм·а**

Гаммет·а, у·равн·е́ни·е   (*n*) (chem) Hammett equation

гамо·ген·е́з·, -а,   (*m*) (biol) gamo-genesis

гамо·сте́л·и·я, -и,   (*f*) (bot) gamostely

гамо·троп·и́зм·, -а,   (*m*) (biol) gamo-tropism

га́нгли·й, -я,   (*m*) (anat) ganglion

гангре́н·а, -ы,   (*f*) (med) gangrene

га́ндшпуг·, -а,   (*m*) (naut) handspike

га́нистер·, -а,   (*m*) (min) ganister, gannister

гани́т·, -а,   (*m*) (min) gahnite, zinc-spinel

ган·о́ид·н·ый, -ая, -ое,   (*a*) (zool) ganoid

~ чешу·я́   (*f*) (fish) ganoid scale

гано·и́н·, -а,   (*m*) (biol) ganoine, ganoin

гантеле·обра́з·н·ый, -ая, -ое,   (*a*) dumb-bell-shaped/like

ганте́л·ь, -и,   (*f*) dumb-bell

га́ншпуг·, -а,   (*m*) (naut) handspike

гап·, -а,   (*m*) gap

гапл·о́ид·, -а,   (*m*) (gen) haploid

гапл·о́нт·, -а,   (*m*) (gen) haplont

гапло·спор·и́ди·и   (*nom pl*), (*g pl*) -й, (*f*) (zool) *Haplosporidia*

гапло·сте́л·а, -ы,   (*f*) (gen) haplostele

гапло·ти́п·, -а,   (*m*) (gen) haplotype

гапто·троп·и́зм·, -а,   (*m*) (bot) hapto-tropism

гара́ж·, -а́,   (*i*) -о́м, (*m*) garage

гаранси́н·, -а,   (*m*) garancine (dye)

гара́нт·, -а,   (*m*) guarantor, guarantee

гарант·и́йн·ый, -ая, -ое,   (*a*) guaran-tee

~ пись·м·о́   (*n*) letter of indemñity

гарант·и́р·овать, -уют,   (*imp and perf*) guarantee

гара́нт·и·я, -и,   (*f*) guarantee

~, авари́·йн·ая,   (mar ins) average bond

Га́рда, по́мп·а   (*f*) hand-operated double-acting piston pump

гардама́н·, -а,   (*m*) (naut) sailmaker's palm

гардеро́б·, -а,   (*m*) cloak-room; ward-robe, coat hooks

гарди́н·а, -ы,   (*f*) curtain

гарди́н·н·ая маши́н·а   (*f*) lace-cur-tain machine

гардино́л·ь, -я,   (*m*) (chem) gardinol, sodium lamyl sulphate

гар·ев·о́й, -а́я, -о́е,   (*a*) ashes; cinder

га́рк·а·ть, -ют,   (*imp*) shout/bark at

га́рк·н·уть, -у·,   (*perf*) *see* **га́рк·а·ть**

гарми́н·, -а,   (*m*) (chem) harmin

гармодо·тро́н·, -а,   (*m*) (elec) harmo-dotron

гармониз·и́р·овать, -уют,   (*imp and perf*) harmonize

гармо́н·ик·а, -и,   (*f*) harmonic; ac-cordion, concertina

   коэффицие́нт· гармо́н·ик·   (*m*) (elec) distortion factor

~, объ·ём·н·ая   solid harmonic

гармон·и́ческий, -ая, -ое,   (*a*) har-monic, harmonious; (*n decl as adj*) (math) harmonic

~ фу́нкци·я   (*f*) (math) harmonic

~ фу́нкци·я, то́ч·ечн·ая   (*f*) point harmonic

гармото́м·, -а,   (*m*) (min) harmotome

гарниери́т·, -а,   (*m*) *see* **гарньери́т·**

гарнизо́н·, -а,   (*m*) garrison

гарни́р·, -а,   (*m*) (food) garnish

гарниту́р·, -а,   (*m*) *see* **гарниту́р·а**

гарниту́р·а, -ы,   (*f*) set, suite, outfit; (build) furniture; mountings (on large equipment), fittings; (print) type face, fount

гарниту́р·а
~, **иго́ль·чат·ая**   (text) card clothing
~ **котл·а́**   boiler mountings
гарниту́р·н·ый, -ая, -ое,   (*a*) of **гарниту́р·а**
~ **пере·плет·е́ни·е**   (*n*) (text) tabby weave, plain/calico weave
гарньери́т·, -а,   (*m*) (min) garnierite
га́рпиус·, -а,   (*m*) rosin, colophony
га́рпиус·н·ый, -ая, -ое,   (*a*) rosin, colophony
гарпу́н·, -а́,   (*m*) harpoon
Га́ррис·а, спо́соб·   (*m*) Harris process (of softening lead)
га́рт·, -а,   (*m*) (print) type-metal
гартбле́·й, -я,   (*m*) hard/antimonial lead
Га́ртле·я, трёх·то́ч·ечн·ый генера́тор·   (*m*) (rad) Hartley oscillator
Га́ртман·а, фо́рмул·а   (*f*) (opt) Hartmann dispersion formula
га́рт·ов·ый, -ая, -ое,   (*a*) of **гарт·**
га́рус·, -а,   (*m*) (text) worsted yarn; worsted-finish cotton cloth
га́р·ь, -и,   (*f*) ashes, cinders; burnt clearing (in forest)
гаря́нск·ое су́д·н·о   (*n*) wooden cargo barge/raft
гас· (*past masc sing*) of **га́с·нуть**
гас·и́тел·ь, -я,   (*m*) damper, extinguisher, quencher; (hydr eng) apron
~, **динам·и́ческ·ий**   dynamic damper
~ **колеб·а́ни·я**   vibration/oscillation damper; (a/c) shimmy damper
~, **ма́ятник·ов·ый**   pendulum damper
~ **фосфоресц·е́нци·и**   scotophor
гас·и́ть, -я́т,   (*imp*) extinguish, put out; damp down, reduce; clear (computer); slake (lime); (fin) cancel, liquidate
га́с·нуть, -нут,   (*imp*) go/die out/down
гаспиц·ы   (*pl*) (naut) hawse pieces/ timbers
гастеро·мицет·ы (*nom pl*), (*g pl*) **-ов,** (*m*) (bot) *Gasteromycetes*
гастингси́т·, -а,   (*m*) (min) hastingsite
гастр·алг·и́·я, -и,   (*f*) (med) gastralgia
гастр·и́н·, -а,   (*m*) (chem) gastrin, histamine
гастр·и́т·, -а,   (*m*) (med) gastritis
гастр·и́ческ·ий, -ая, -ое,   (*a*) gastric
гастро·васкуля́р·н·ый, -ая, -ое,   (*a*) (zool) gastrovascular
гастрол·и́р·овать, -уют,   (*imp*) tour, play on tour (theatre)
гастро·ном·и́ческ·ий, -ая, -ое,   (*a*) gastronomic, gastronomical

гастро·трих·и (*nom pl*), (*g pl*) **-ов,** (*m*) (zool) *Gastrotricha*
гастро·энтер·и́т·, -а,   (*m*) (med) gastro-enteritis
га́стр·ул·а, -ы,   (*f*) (zool) gastrula
гастр·эктом·и́·я, -и,   (*f*) (med) gastrectomy
га́с·ш·ий, -ая, -ее,   (*past part act*) of **га́с·нуть**
гас·я́щ·ий, -ая, -ее,   (*pres part act*) of **гас·и́ть**
~ **схе́м·а**   (*f*) (elec) quenching circuit
г-ат   (*abbr*) = **гра́мм·а́том** (phys) g.at., gram-atom
гат·и́ть, -я́т,   (*imp*) (civ eng) lay a corduroy road
гатгетти́т·, -а,   (*m*) (min) hatchettite, hatchettine
гату́р·а, -ы,   (*f*) (print) hachure
гатчетоли́т·, -а,   (*m*) (min) hatchetolite
гат·ь, -и,   (*f*) (civ/mil eng) corduroy road
га́убиц·а, -ы,   (*i*) **-ей,** (*f*) (gunn) howitzer
гаультер·иев·ое ма́сл·о   (*n*) (chem) gaultheria/wintergreen oil
гауптва́хт·а, -ы,   (*f*) (mil) guardhouse
гаусманни́т·, -а,   (*m*) (min) hausmannite
га́усс·, -а,   (*m*) gauss (unit of magnetic induction)
Га́усс·а, рас·пре·дел·е́ни·е   (*n*) (phys) Gaussian/normal distribution
га́усс·ов·, -а, -о,   (*possess adj*) Gaussian
га́усс·овск·ий, -ая, -ое,   (*a*) Gaussian
гаусто́р·и·й, -я,   (*m*) (bot) haustorium
га́уч·, -а,   (*m*) (paper) couch
~ **-ва́л·, -а,**   (*m*) (paper) couch-roll
~ **-ва́л·, от·са́с·ывающ·ий**   (paper) suction couch, suction couch-roll
~ **-пре́сс·, -а,**   (*m*) (paper) couch-press, press-couch
гауэри́т·, -а,   (*m*) (min) haverite
га́фель-гарде́л·ь, -я,   (*m*) (naut) throat halyard
га́фель·н·ый, -ая, -ое,   (*a*) (naut) gaff
га́фн·и·й, -я,   (*m*) (chem) hafnium, Hf
гафн·и́л·, -а,   (*m*) (chem) hafnyl
га́ч·, -а,   (*m*) (chem) slack wax
~, **парафи́н·ов·ый**   slack wax
гаш·е́ни·е, -я,   (*n*) (*v n*) *see* **гас·и́ть;** cancellation, suppression
~ **интенси́в·ност·и**   cancellation of intensities (of oscillations)
~ **пучк·а́**   (phys) beam suppression

гаш·ён·ый, -ая, -ое, (*a*) slaked

~ и́звест·ь (*f*) slaked lime

гашéт·к·а, -и, (*g pl*) **-т·ок·,** (*f*) trigger (small arms); button (of firing mechanism)

гаши́ш·, -а, (*i*) **-ем,** (*m*) hashish

гáшпил·ь, -я, (*m*) reel (leather making)

гашю́р·а, -ы, (*f*) (print) hachure, dash

Гаюой, за·кóн· (*m*) (cryst) Hauy's law

гаюини́т·, -а, (*m*) (min) hauynite

гб (*abbr*) = **ги́лберт·** (phys) Gb, gilbert

гвайю́л·а, -ы, (*f*) *see* **гваю́л·а**

гвáрд·éйск·ий, -ая, -ое, (*a*) (mil) guards; rocket-launcher

~ мино·мёт· (*m*) (mil) rocket-launcher

гваю́л·а, -ы, (*f*) (bot) guayule, *Parthenium argentatum*; guayule rubber

гвая́к·ов·ый, -ая, -ое, (*a*) (chem) guaiac

~ дéрев·о (*m*) (bot) guaiacum, guajacum, lignum vitae

гваяк·óл·, -а, (*m*) (chem) guaiacol

гваяцети́н·, -а, (*m*) (pharm) guaiacetin

гвеяри́т·, -а, (*m*) (min) guejarite

гвозд·ев·óй, -áя, -óе, (*a*) of **гвóзд·ь**

гвозде·дёр·, -а, (*m*) tack/nail extractor

гвóзд·ик·, -а, (*m*) (*see also* **гвозд·ь**) tack

~, луж·ён·ый tin tack

гвозди́к·а, -и, (*f*) (food) clove; (bot) carnation, pink, *Dianthus*

гвозд·и́льн·ый, -ая, -ое, (*a*) nail, nail-making

~ за·вóд· (*m*) nailery

гвозд·и́льщик·, -а, (*m*) nail-machine operator

гвозд·и́м·ый, -ая, -ое, (*pres part pass*) of **гвозд·и́ть;** nailable

гвозди́ч·н·ый, -ая, -ое, (*a*) of **гвозди́к·а;** (*pl decl as adj*) (bot) pink family, *Caryophyllaceae*

~ мáсл·о (*n*) oil of cloves

~ пéрец· (*m*) (bot) allspice, *Pimento officinalis*

гвоздо·дёр, -а, (*m*) = **гвозде·дёр·**

гвóзд·ь, -я, (*nom pl*) **гвóзд·и,** (*g pl*) **гвозд·éй,** (*m*) nail, stud, pin

~, паркéт·н·ый brad

гвозд·ян·óй, -áя, -óе, (*a*) of **гвóзд·ь**

гвт (*abbr*) = **гектовáтт·** Hw, hectowatt

гг. (*abbr*) = **гóд·ы** years; = **горóд·á** towns, cities

где (*adv*) where

где́ бы ни wherever

где бы то ни́ было no matter where

где-ли́бо somewhere, anywhere (choice implied)

где́-нибудь somewhere, anywhere (vague)

где́-то somewhere (precise)

геветти́т·, -а, (*m*) (min) hewettite

гéве·я, -и, (*f*) (bot) rubber tree, *Hevea brasiliensis*

гёдел·евск·ий, -ая, -ое, (*a*) (math) Godel

геденберги́т·, -а, (*m*) (min) hedenbergite

геди·фáн·, -а, (*m*) (min) hedyphane

гедонáл·, -а, (*m*) (pharm) hedonal

гедри́т·, -а, (*m*) (min) gedrite

гé·ев·ое, мáсл·о (*n*) ghee

гезéнк·, -а, (*m*) (min) winze

Гéйгер·а, с·чёт·чик· (*m*) (phys) Geiger-Müller counter, Geiger counter

Гéйгер·а-Мюллер·а, с·чёт·чик· (*m*) Geiger-Müller counter, Geiger counter

Гéйгер-Нéттол·а, за·кóн· (*m*) Geiger-Nutall law

гéйгер·овск·ий, -ая, -ое, (*a*) Geiger, Geiger-Müller

Гéйзенберг·а, пред·ставл·éни·е (*n*) (phys) Heisenberg representation

гéйзенберг·овск·ий при́нцип· (*m*) (phys) Heisenberg principle, indeterminancy principle

гéйзер·, -а, (*m*) (geol) geyser

гейзери́т·, -а, (*m*) (min) geyserite, siliceous sinter

гейланди́т·, -а, (*m*) (min) heulandite

Гей-Луссáк·а, бáшн·я (*f*) (chem) Gay-Lussac tower

~, за·кóн· (*m*) (phys) Gay-Lussac's law, Charles's law; Gay-Lussac's law of volumes

гейлюси́т·, -а, (*m*) (min) gaylussite

Гéйслер·а, трýб·к·а (*f*) (elec) Geissler tube

гéйслер·ов·ы с·плáв·ы (*pl*) (met) Heusler alloys

гéкл·я, -и, (*f*) (text) flax comb

геклинг·маши́н·а, -ы, (*f*) (text) hackling machine (flax)

гекса·гидро·бензóл·, -а, (*m*) (chem) hexahydrobenzene, cyclohexane

гекса·гидро·фенóл·, -а,   (*m*) (chem) hexahydrophenol, cyclohexanol

гекса·гѝр·, -а,   (*m*) *see* **гекса·гѝр·а**

гекса·гѝр·а, -ы,   (*f*) hexagyre, (cryst) sixth-order symmetry axis

гекса·гѝр·н·ый, -ая, -ое,   (*a*) hexagyroid; (cryst) hexagonal

гекса·гир·óид·, -а,   (*m*) (cryst) hexagyroid, sixth-order axis of rotary inversion

гекса·гонáль·н·ый, -ая, -ое,   (*a*) hexagonal

~ систéм·а   (*f*) (cryst) hexagonal system

гекса·дек·áн·, -а,   (*m*) hexadecane

гекса·дек·éн·, -а,   (*m*) hexadecene

гекса·декен·ди·кисл·от·á, -ы,   (*f*) hexadecenedioic acid

гекса·дец·éн·, -а,   (*m*) hexadecene

гекса·дец·ил·éн·, -а,   (*m*) hexadecylene

гексакис-   (*component*) (cryst) hexakis, six times

гексакис·октá·эдр·, -а,   (*m*) (math, cryst) hexakisoctahedron, hexoctahedron

гексакоз·áн·, -а,   (*m*) (chem) *n*-hexacosane

гекс·áл·, -а,   (*m*) (chem) hexal

гекса·мéр·, -а,   (*m*) (chem) hexamer

гекса·мета·фосфáт·, -а,   (*m*) (chem) hexametaphosphate

гекса·метилéн·, -а,   (*m*) (chem) hexamethylene, cyclohexane

гекса·метилен·бенз·амѝд·, -а,   (*m*) (chem) hexamethylenebenzamide, hexamide

гекса·метилен·тетрамѝн·, -а,   (*m*) (chem) hexamethylenetetramine, hexamine

гекс·амидѝн·, -а,   (*m*) (pharm) hexamidine, mysoline

гекс·áн·, -а,   (*m*) (chem) hexane

гекса·нафтéн·, -а,   (*m*) cyclohexene

гексан·ди·кислот·á, -ы́,   (*f*) hexanedioic acid, hexane diacid

гексáн·ов·ый, -ая, -ое,   (*a*) hexane, hexanoic

~ кисл·от·á   (*f*) hexanoic acid, caproic acid

гекса·окси·бензóл·, -а,   (*m*) (chem) hexaphenol

гекса·плóид·, -а,   (*m*) (gen) hexaploid

гекс·áстр·, -а,   (*m*) (bot) hexaster

гекса·фтор·ѝд·, -а,   (*m*) (chem) hexafluoride

гекса·хлор·áн·, -а,   (*m*) (chem) hexachloran, hexachlorcyclohexane

гекса·хлор·бензóл·, -а,   (*m*) hexachlorobenzene

гекса·хлор·цикло·гексáн·, -а,   (*m*) (chem) hexachlorocyclohexane, hexachloran

гекса·хлор·этáн·, -а,   (*m*) (chem) carbon trichloride, (pharm) hexachlorethane

гексá·эдр·, -а,   (*m*) (math) hexahedron

гекса·эдр·ѝт·, -а,   (*m*) (min) hexahedrite

гекса·эдр·ѝческ·ий, -ая, -ое,   (*a*) hexahedral

гекса·этил·ди·óлов·о, -а,   (*n*) distannic ethide

гекса·этил·ди·свин·éц·, -н·ц·á,   (*m*) diplumbic ethide

гекс·éн·, -а,   (*m*) (chem) hexene

гексенáл·, -а,   (*m*) (pharm) hexobarbitone sodium, *Evipan*

гекса·стѝл·, -а,   (*m*) (arch) hexastyle

гекс·ѝл·, -а,   (*m*) (chem) hexyl

гексил·резорцѝн·, -а,   (*m*) hexyl resorcinol

гекс·ѝн·, -а,   (*m*) (chem) hexyne, hexine

гексѝт·, -а,   (*m*) (chem) hexitol

гексо-   (*int component*) hexo-, six-

гексо·гéн·, -а,   (*m*) (expl) hexogen, cyclonite

гекс·óд·, -а,   (*m*) (elec) hexode

гекс·óз·а, -ы,   (*f*) (chem) hexose

гексо·кинáз·а, -ы,   (*f*) (biochem) hexokinase

гекс·окта·эдр·ѝческ·ий, -ая, -ое, (*a*) hexoctahedral

гексóни·й, -я,   (*m*) (chem) hexonium, hexamethonium

гексóн·ов·ая кисл·от·á   (*f*) hexonic acid

гекс·урóн·ов·ая кисл·от·á   (*f*) (chem) hexuronic acid

гект·áр·, -а,   (*m*) hectare (10,000 m²)

гект·ѝческ·ий, -ая, -ое,   (*a*) hectic

гекто-   (*int component*) hecto-

гекто·вáтт·, -а,   (*g pl*) -вáтт·,   (*m*) (elec) hectowatt

гектó·граф·, -а,   (*m*) hectograph

гекто·лѝтр·, -а,   (*m*) hectolitre, hectoliter

гекто·мéтр·, -а,   (*m*) hectometre, hectometer

гекто·метр·ѝческ·ий, -ая, -ое,   (*a*) hectometric

~ волн·á   (*f*) (rad) hectometric wave, medium wave

гекто·пьéз·а, -ы,   (*f*) (mech) hectopiezo

гелени́т·, -а, *(m)* (min) gehlenite; helenite (coal)

гелен·о́к·, -н·к·а, *(m)* instep (of shoe)

гелен·о́чн·ый, -ая, -ое, *(a)* instep (of shoe)

геле·обра́з·н·ый, -ая, -ое, *(a)* jelly-like, gelatinous

геле·образ·ова́ни·е, -я, *(n)* gel-forming/formation, gelation, gelatination

гел·и, *(nom pl)* of **гел·ь**

гелиант·и́н·, -а, *(m)* (chem) helianthin, helianthine

ге́ли·ев·ый, -ая, -ое, *(a)* (chem) helium

гелие·подо́б·н·ый, -ая, -ое, *(a)* helium-like

ге́ли·й, -я, *(m)* (chem) helium, He

гелик·о́ид·, -а, *(m)* helicoid

гелик·оид·а́льн·ый, -ая, -ое, *(a)* helicoidal, spiral

гелико·пте́р·, -а, *(m) see* **верто·лёт,** helicopter

гели́кс·, -а, *(m)* helix

гелио- *(int component)* helio-, sun, solar

гелио·гравю́р·а, -ы, *(f)* (print) heliogravure

гелио́·граф·, -а, *(m)* heliograph (signalling instrument); (meteor) solar radiation recorder; (astron) solar telescope

гелио·гра́ф·и·я, -и, *(f)* solar science, study of the sun

гелиодо́р·, -а, *(m)* (min) heliodor

гелио́·метр·, -а, *(m)* (astron) heliometer

гелио·ско́п·, -а, *(m)* (astron) helioscope

гелио·ста́т·, -а, *(m)* (astron) heliostat

гелио·сфе́р·а, -ы, *(f)* (astron) heliosphere

гелио·терап·и́·я, -и, *(f)* (med) heliotherapy

гелио·те́хн·ик·а, -и, *(f)* solar radiation engineering, solar radiation utilization

гелио·тро́п·, -а, *(m)* (bot, min) heliotrope; (min) bloodstone

гелио·троп·и́зм·, -а, *(m)* (bot) heliotropism

гелио·троп·и́н·, -а, *(m)* (chem) heliotropin, piperonal

гелио·у·стано́в·к·а, ·и, *(g pl)* **-вок·,** *(f)* solar energy equipment/installation/plant

гелио·фи́з·ик·а, -и, *(f)* solar physics

гелио·фи́т·, -а, *(m)* (bot) heliophyte

гелио·центр·и́ческ·ий, -ая, -ое, *(a)* (astron) heliocentric

гели·фик·а́ци·я, -и, *(f)* gelification

гели·фиц·и́р·ованн·ый, -ая, -ое, *(a)* gelified

гелихри́зум·, -а, *(m)* (bot) *Helichrisum*

геллебо́р·, -а, *(m)* (bot) hellebor, *Helleborus*

гело·фи́т·, -а, *(m)* (bot) helophyte

ге́л·ь, -я, *(m)* (chem) gel

~, силика́т·н·ый silica gel

Гёльдер·а, не·ра́вен·ств·о *(n)* (math) Holder's inequality

гельземи́н·, -а, *(m)* (chem) gelsemine

Ге́льмгольц·а, кольц·а́ *(pl)* (elec) Helmholtz coil

гель·ме́тр·, -а, *(m)* (phys) gel-meter, gelometer

гельминто́·граф·, -а, *(m)* (med) helminthograph

гельминт·о́з·, -а, *(m)* (med) helminthosis

гельминто·ло́г·и·я, -и, *(f)* helminthology

гельминто·ци́д·, -а, *(m)* (pharm) helminthocide

гельмпо́рт·, -а, *(m)* (naut) helm-port, rudder trunk

гельмпо́рт·н·ый бру́с· *(m)* (shipb) helm-port transom

гельмпо́рт·ов·ая труб·а́ *(f)* (shipb) rudder tube, rudder trunk

гем- *(int component)* (biochem) haem-, hem-

гем·агглютин·а́ци·я, -и, *(f)* (med) haemagglutination, hemagglutination

гем·ангио́м·а, -ы, *(f)* (med) haemangioma, hemangioma

гемат·и́н·, -а, *(m)* (biochem) haematin, hematin

гемат·и́т·, -а, *(m)* (min) haematite, hematite, specular iron

гемато- *(int component)* (med, zool) haemato-, hemato-

гемато·бла́ст·, -а, *(m)* (zool) haematoblast, hematoblast

гемато·ге́н·, -а, *(m)* (pharm) haematogen, hematogen

гемато·ге́н·н·ый, -ая, -ое, *(a)* (zool) haematogenous, hematogenous

гемат·оид·и́н·, -а, *(m)* haematoidine, hematoidin, bilirubin

гемат·оксил·и́н·, -а, *(m)* (chem) haematoxylin, hematoxylin

гемато·ло́г·и·я, -и, *(f)* (med) haematology, hematology

гемат·о́м·а, -ы, (*f*) (*med*) haematoma, hematoma

гемато·миел·и́·я, -и, (*f*) (med) haematomyelia, hematomyelia

гемато·порфири́н·, -а, (*m*) (chem) haematoporphyrin, hematoporphyrin

гемато·хро́м·, -а, (*m*) (biol) haematochrome, hematochrome

гемат·ур·и́·я, -и, (*f*) (med) haematuria, hematuria

гемеллито́л·, -а, (*m*) (chem) hemellitene

гемерал·оп·и́·я, -и, (*f*) (med) hemeralopia

геми- (*int component*) hemi-, half

геми·класт·и́ческ·ий, -ая, -ое, (*a*) (geol) hemiclastic

геми·кристалл·и́ческ·ий, -ая, -ое, (*a*) hemicrystalline

геми·мелли́т·овая кисл·от·а́ (*f*) (chem) hemimellitic/hemellitic acid

геми·морф·и́зм,· -а, (*m*) (cryst) hemimorphism

геми·морф·и́т·, -а, (*m*) (min) hemimorphite, calamine

геми·морф·и́·я, -и, (*f*) hemimorphy

геми·мо́рф·н·ый, -ая, -ое, (*a*) hemimorphic

гем·и́н·, -а, (*m*) (chem) haemin, hemin, hematin chloride

геми·орто·дро́м·а, -ы, (*f*) (cryst) rhombic prism

геми·пар·е́з·, -а, (*m*) (med) hemiparesis

геми·пинак·о́ид·, -а, (*m*) (cryst) hemipinacoid

геми·пе́н·овая кисл·от·а́ (*f*) (chem) hemipinic acid

геми·пирами́д·а, -ы, (*f*) (cryst) hemipyramid

геми·плег·и́·я, -и, (*f*) (med) hemiplegia

геми·при́зм·а, -ы, (*f*) (cryst) hemiprism

геми·син·дро́м·, -а, (*m*) (med) haemosyndrom, hemosyndrom

геми·сфер·и́ческ·ий, -ая, -ое, (*a*) hemispherical

геми·целлюло́з·а, -ы, (*f*) hemicellulose

геми·цикл·и́ческ·ий, -ая, -ое, (*a*) hemicyclic

геми́·эдр·, -а, (*m*) (cryst) hemihedron

геми·эдр·и́ческ·ий, -ая, -ое, (*a*) (cryst) hemihedral

ге́мм·а, -ы, (*f*) (biol) gemma; (print) intaglio, cameo

гемма·по·до́б·н·ый, -ая, -ое, (*a*) (zool) gemmiform

гемма́·ци·я, -и, (*f*) (biol) gemmation

ге́мм·ул·а, -ы, (*f*) (zool) gemmule

гемо·глоб·и́н·, -а, (*m*) (zool) haemoglobin, hemoglobin

гемо·глобино́·метр·, -а, (*m*) (med) haemoglobinometer, hemoglobinometer

гемо·глобин·ур·и́йн·ая лихора́д·к·а (*f*) (med) black water fever

гемо·глобин·ур·и́·я, -и, (*f*) (med) haemoglobinuria, hemoglobinuria

гемо·гра́мм·а, -ы, (*f*) (med) haemogram, hemogram

гемо·дина́м·и·ка, -и, (*f*) haemodynamics, hemodynamics

гемо́·лиз·, -а, (*m*) (med) haemolysis, hemolysis, haemocytolysis

гемо·лиз·и́н·, -а, (*m*) (biochem) haemolysin, hemolysin

гемо·лит·и́ческ·ий, -ая, -ое, (*a*) (med) haemolytic, hemolytic

гемо·по́эз·, -а, (*m*) (zool) haemapoiesis, hemapoiesis

гемо·рраг·и́ческ·ий, -ая, -ое, (*a*) (med) haemorrhagic, hemorrhagic

гемо·рраг·и́·я, -и, (*f*) (med) haemorrhage, hemorrhage

гемо·рроида́льн·ый, -ая, -ое, (*a*) haemorrhoidal, hemorrhoidal

гемо·рро́·й, -я, (*m*) (med) haemorrhoid, hemorrhoid

гемо·спор·и́ди·и (*nom pl*) (*g pl*) **-й,** (*f*) (zool) *Haemosporidia, Hemosporidia*

гемо·спор·ид·и́н·, -а, (*m*) (vet, pharm) haemosporidin, hemosporidin

гемо·спор·иди·о́з·, -а, (*m*) (vet) haemosporidiosis

гемо·фил·и́·я, -и, (*f*) (med) haemophilia, hemophilia

гемо·це́л·ь, -и, (*f*) (zool) haemocoel, hemocoel

ге́н·, -а, (*m*) (gen) gene

генеа·ло́г·и·я, -и, (*f*) genealogy, pedigree

ге́незис·, -а, (*m*) genesis, origin

генейкоза́н·, -а, (*m*) (chem) heneicosane

генейкоза́н·ов·ая кисл·от·а́ (*f*) heneicosoic acid

генекен·, -а, (*m*) (bot) henequen, Yucatan seisal

генера́л·, -а, (*m*) (mil) general

генера́л-лейтена́нт·, -а, (*m*) lieutenant-general

генера́л-майо́р·, -а, (*m*) major-general

генера́л-полк-о́вник-, -а, *(m)* colonel-general

генерал-иза́ци-я, -и, *(f)* generalization; distribution (cartographic)

генерал-ите́т-, -а, *(m)* (mil) generals, brass-hats

генера́ль-н-ый, -ая, -ое, *(a)* general

генера́ль-ск-ий, -ая, -ое, *(a)* general's

генера́ль-ств-о, -а, *(n)* general's rank

генер-ати́вн-ый, -ая, -ое, *(a)* generative

генер-а́тор-, -а, *(m)* (elec) generator, oscillator; (gas) generator, producer

~, **авто-модул-и́руем-ый** self-pulsed oscillator

~, **а-син-хро́н-н-ый** (elec) alternator

~, **ацетиле́н-ов-ый** (weld) acetylene generator

~, **блок-и́рующ-ий** (elec) blocking oscillator

~, **бала́нс-ов-ый** balanced oscillator

~ **времен-н-о́й раз-вёрт-к-и** (elec) time-base generator

~ **выс-о́к-ой част-от-ы** (elec) high-frequency alternator

~ **га́з-а** gas generator/producer

~ **га́з-а, свобо́д-н-о-поршн-ев-о́й** (mar eng) free-piston gasifier

~, **гидро-турби́н-н-ый** hydroelectric generator

~, **двух-по́люс-н-ый** (elec) two-terminal oscillator

~, **двух-та́кт-н-ый** push-pull oscillator

~, **динатро́н-н-ый** (elec) dynatron oscillator

~, **дуг-ов-о́й** (elec) arc converter; (rad) arc generator

~, **за-да-ю́щ-ий** (rad) self-oscillator; (autom) timing pulse generator, clock, master oscillator

~, **звук-ов-о́й** (elec) audio-frequency oscillator

~ **и́мпульс-ов** pulse generator/oscillator; impulse generator

~ **квант-у́ющ-их и́мпульс-ов** quantization pulse generator (computers)

~, **ква́рц-ев-ый** crystal/quartz oscillator

~, **клистро́н-н-ый** klystron oscillator

~ **колеб-а́ни-й** oscillator

~, **компа́унд-н-ый** (elec) compound-wound generator

~ **кратко-време́н-н-ых и́мпульс-ов** impulse generator

~, **ла́мп-ов-ый** (elec) valve oscillator

~, **магнетро́н-н-ый** magnetron oscillator

генер-а́тор

~, **магнито-электр-и́ческ-ий** magneto-alternator, magneto-electric generator

~ **ме́т-ок- време́н-и** time-mark generator (cathode-ray tubes)

~, **много-каска́д-н-ый** multivibrator

~ **модул-и́рующ-его де́й-стви-я** (rad) correction signal generator, modulating oscillator

~, **молекуля́р-н-ый** (phys) maser oscillator

~ **мост-ов-о́й схе́м-ы** bridge oscillator

~ **мо́щ-н-ых и́мпульс-ов** surge generator

~ **на би-е́ни-ях** beat-frequency oscillator

~ **о-по́р-н-ых част-о́т** reference-frequency oscillator

~, **паро-турби́н-н-ый** turbogenerator

~ **пере-ме́н-н-ого то́к-а** (elec) alternator, A.C. generator

~ **пило-обра́з-н-ого на-пряж-е́ни-я** saw-tooth generator

~ **по-вы́ш-енн-ой част-от-ы́** HF power generator

~ **под давл-е́ни-ем** pressurized generator

~, **под-ваго́н-н-ый** (rad) axle generator

~ **по-сто-я́нн-ого то́к-а** (elec) direct-current generator

~ **по-сто-я́нн-ого-пере-ме́нн-ого то-к-а** double-current generator

~ **раз-вёрт-к-и** (elec) sweep generator

~ **раз-вёрт-к-и а́зимут-а** azimuth sweep generator (radar)

~, **раз-вёрт-очн-ый** time-base generator, sweep generator

~ **раз-ры́в-н-ых колеб-а́ни-й, ре-лаксацио́н-н-ый** relaxation oscillator

~, **релаксацио́н-н-ый** (elec) relaxation oscillator

~ **с от-се́ч-к-ой** squegging generator

~ **с фа́з-ов-ым с-дви́г-ом** phase-shift oscillator

~ **синхрониз-и́р-ующ-ий** (autom) synchrogenerator

~ **син-хро́н-н-ый** (elec) synchronous generator

~ **станда́рт-н-ых сигна́л-ов** standard-signal generator

~ **строб-и́мпульс-ов** (elec) gate generator

~, **такт-и́рующ-ий** (elec) timing-pulse generator, clock

генер·а́тор
~, та́кт·ов·ых и́мпульс·ов   clock (computer)
~ то́к·а   (elec) dynamo
~, тона́ль·н·ый   (rad) audio oscillator
~, трёх·то́ч·ечн·ый   Colpitts oscillator, Hartley oscillator, Meissner oscillator
~, фа́з·о-и́мпульс·н·ый   pulse-position generator
~, шум·ов·о́й   flatter generator
~, электро́н·н·ый   (elec) valve oscillator
~, этало́н·н·ый ква́рц·ев·ый   standard crystal generator
генера́тор·н·ый, -ая, -ое,   (a) of **генер·а́тор·**
~ газ·   (m) producer gas
~ ла́мп·а   (f) (elec) oscillator valve
генератри́с·а, -ы,   (f) (math) generatrix
генер·а́ци·я, -и,   (f) generation, production, formation; (rad) oscillation
~, парази́т·н·ая   spurious oscillation
~, пре·ры́в·ист·ая   squitter
генер·и́ровани·е, -я,   (n) see **генер·а́ци·я**
~ и́мпульс·ов   pulsing
генер·и́рованн·ый, -ая, -ое,   (past part pass) of **генер·и́р·овать;** generated
генер·и́р·овать, -уют,   (imp) generate
генер·и́рующ·ий, -ая, -ое,   (pres part act) of **генер·и́р·овать;** (rad) self-oscillating
генет·ик·, -а,   (m) geneticist
генет·ик·а, -и,   (f) genetics
генет·и́ческ·ий, -ая, -ое,   (a) genetic
гени·а́льн·ый, -ая, -ое,   (a) of **ге́ни·й;** ingenious, brilliant
ге́ни·й, -я,   (m) genius
ге́нн·а, -ы,   (f) henna
ге́н·н·ый, -ая, -ое,   (a) of **ген·;** gene
-ген·н·ый, -ая, -ое,   (component of adj) -genic, -genous
ген·о́м·, -а,   (m) (gen) genome, genom
ген·о́м·и·я, -и,   (f) (biol) genomy
гено·ме́р·, -а,   (m) (gen) genomere
гено·син·ти́п·, -а,   (m) (biol) genosyntype
гено·со́м·а, -ы,   (f) (genet) genosome
гено·ти́п·, -а,   (m) (biol) genotype
гено·ци́д·, -а,   (m) genocide
ген·пла́н·, -а,   (m) general plan, overall plan/view
ге́нри   (m indecl) (elec) henry (unit of inductance)
Ге́нри, за·ко́н·   (m) (chem) Henry's law

генри·ме́тр·, -а,   (m) (elec) henrymeter
гентиан·и́н·, -а,   (m) (chem) gentianin
гентриаконта́н·, -а,   (m) (chem) hentriacontane
генциан·виоле́т·, -а,   (m) gentian violet (dye)
ген·шта́б·, -а,   (m) (mil) general staff
ген·штаб·и́ст·, -а,   (m) officer on the general staff
гео·биот·и́ческ·ий, -ая, -ое,   (a) (biol) geobiotic
гео·бота́н·ик·а, -и,   (f) geobotany, phytogeography
гео·гидро·ло́г·и·я, -и,   (f) geohydrology
гео·гно́з·и·я, -и,   (f) geognosy
гео·граф·, -а,   (m) geographer
гео·граф·и́ческ·ий, -ая, -ое,   (a) geographical
гео·гра́ф·и·я, -и,   (f) geography
гео·дез·и́ст·, -а,   (m) geodesist, geodetic surveyor
гео·дез·и́ческ·ий, -ая, -ое,   (a) geodetic, geodesic
~, констру́кци·я   (f) (a/c) geodetic construction
~ о·кру́ж·ност·ь   (f) (math) geodesic circle
гео·де́з·и·я, -и,   (f) geodesy, geodetic surveying
геоди·ме́тр·, -а,   (m) geodimeter
ге·о́ид·, -а,   (m) (surv) geoid
гео·изо·те́рм·а, -ы,   (f) (geol) geoisotherm
гео·карп·и́ческ·ий, -ая, -ое,   (a) (bot) geocarpic
гео́·лог·, -а,   (m) geologist
гео·лог·и́ческ·ий, -ая, -ое,   (a) geological, geologist's
~ ко́мпас·   (m) geologist's compass
гео·ло́г·и·я, -и,   (f) geology
~, инжене́р·н·ая   geological engineering
геолого·раз·ве́д·очн·ый, -ая, -ое,   (a) geological prospecting
гео·магни́т·н·ый, -ая, -ое,   (a) geomagnetic
гео́·метр·, -а,   (m) geometrician
гео·метр·а́льн·ый, -ая, -ое,   (a) geometric
гео·метр·и́ческ·ий, -ая, -ое,   (a) geometrical
~ о́пт·ик·а   (f) geometrical optics
гео·ме́тр·и·я, -и,   (f) geometry
~, аффи́н·н·ая   affine geometry
~, лин·е́йн·ая   line geometry
~, на·черт·а́тельн·ая   projective geometry

гео·ме́тр·и·я
~, **не·евкли́д·ов·ая**   non-Euclidean geometry
~ **по·лож·е́ни·я**   topology
~, **прое́ктив·н·ая**   projective geometry
~, **Ри́ман·ов·а**   Riemannian geometry
~ **тр·ёх из·мер·е́ни·й**   three-dimensional geometry
~, **эвкли́д·ов·ая**   Euclidean geometry
гео·морфо·ло́г·и·я, -и,   (f) geomorphology
гео·наст·и́ческ·ий, -ая, -ое,   (a) (bot) geonastic
гео·на́ст·и·я, -и,   (f) (bot) geonasty
гео·потенциа́л·, -а,   (m) geopotential
гео·син·клин·а́л·ь, -и,   (f) (geol) geosyncline
гео·строф·и́ческ·ий, -ая, -ое,   (a) (meteor) geostrophic
гео·сфе́р·а, -ы,   (f) geosphere
гео·те́рм·ик·а, -и,   (f) (geol) study of geothermic conditions
гео·терм·и́ческ·ий, -ая, -ое,   (a) geothermal
~ **градие́нт·**   (m) geothermal gradient
гео·фи́з·ик·, -а, (m) geophysicist
гео·фи́з·ик·а, -и,   (f) geophysics
гео·физ·и́ческ·ий, -ая, -оё,   (a) geophysical
гео·фи́т·, -а,   (m) (bot) geophyte
гео·фо́н·, -а,   (m) (geol) geophone
гео·хи́м·ик·, -а,   (m) geochemist
гео·хим·и́ческ·ий, -ая, -ое,   (a) geochemical
гео·хи́м·и·я, -и,   (f) geochemistry
гео·хроно·лог·и́ческ·ий, -ая, -ое, (a) geologic-time, geochronological
гео·центр·и́ческ·ий, -ая, -ое,   (a) (astron) geocentric
гепари́н·, -а,   (m) (chem) heparin
гепат·и́т·, -а,   (m) (med) hepatitis; (min) hepatite
гепато·ло́г·и·я, -и,   (f) (med) hepatology
гепат·о́м·а, -ы,   (f)   (med) hepatoma
гепато·мега́л·и·я, -и,   (f) (med) hepatomegaly
гепта·гон·а́льн·ый, -ая, -ое,   (a) heptagonal
гепта·дек·а́н·, -а,   (m) (chem) heptadecane
гепта·дека́н·ов·ая кисл·от·а́ (f) heptadecanoic acid
гепта·ме́р·, -а, (m) (chem) heptamer
гепта·ме́р·н·ый, -ая, -ое,   (a) heptamer, heptameric; (bot) heptamerous
гепт·а́н·, -а,   (m) (chem) heptane

гептана́л·ь, -я,   (m) heptanal, heptaldehyde, heptyl aldehyde
гента́н·ов·ая кисл·от·а́   (f) heptanoic acid
гепта·хло́р·, -а,   (m) heptachlor
гепт·е́н·, -а,   (m) heptene, heptylene
гепт·и́л·, -а,   (m) heptyl
гепти́л·ов·ый спи́рт·   (m) heptyl alcohol
гепт·о́д·, -а,   (m) (elec) heptode, pentagrid
гепт·о́з·а, -ы,   (f) (chem) heptose
герания́л·, -а,   (m) (chem) geranial, citral
герании́ол·, -а,   (m) (chem) geraniol
гера́н·ь, -и,   (f) (bot) geranium
ге́рб·, -а́,   (m) crest, armorial bearings; (math) arms, heads
гербариз·и́р·овать, -уют,   (imp) herbarize, botanize
герба́р·и·й, -я,   (m) herbarium
герби·ци́д·, -а,   (m) herbicide, weed-killer
~ **из·бир·а́тельн·ого   де́й·стви·я** selective weed-killer
ге́рб·ов·ый, -ая, -ое,   (a) of **герб·**; heraldic; (com, law) stamp
~ **ма́рк·а**   (f) stamp
~ **с·бор·**   (m) stamp duty
гергардти́т·, -а,   (m) (min) gerhardtite
герко·га́м·и·я, -и,   (f) (bot) hercogamy
геркуле́с·, -а,   (m) (astron) Hercules; rolled oats
герман·а́т·, -а,   (m) (chem) germanate
германие·водо·ро́д·, -а,   (m) germanium hydride
герма́н·иев·ый, -ая, -ое,   (a) germanium; (chem) germanic
~ **ди·о́д·**   (m) (elec) germanium diode
~ **кисл·от·а́**   (f) germanic acid
германие·фторо·водо·ро́д·ная   кисл·от·а́   (f) (chem) hydrofluogermanic acid
герма́н·и·й, -я,   (m) (chem) germanium, Ge
~, **сер·ни́ст·ый**   germanium sulphide
герман·и́т·, -а,   (m) (min) germanite
герма́н·ск·ий, -ая, -ое,   (a) (ling) Teutonic, Germanic; German
гермафроди́т·, -а,   (m) (biol) hermaphrodite
герметиз·а́ци·я, -и,   (f) sealing, pressure-sealing, making air tight; (elec) canning, potting; (a/c) pressurization
герметиз·и́рованн·ый, -ая -ое, (past part pass); sealed, hermetically sealed, canned; (elec) potted

гермéт·ик·, -а,   (*m*) sealing compound

гермет·и́ческ·ий, -ая, -ое,   (*a*) hermetic, hermetically sealed, air-tight, water-tight, canned, encapsulated; (elec) submersible, water-tight; (a/c) pressurized

гермет·и́чност·ь, -и,   (*f*) tightness, air-tightness, sealing, seal

гермет·и́чн·ый, -ая, -ое,   (*a*) = **гермет·и́ческ·ий**

гермо·кабин·а, -ы,   (*f*) (a/c) pressurized cabin

гермо·кáмер·а, -ы,   (*f*) sealed/pressurized chamber

героúн·, -а,   (*m*) (chem) heroin

геро́·й, -я,   (*m*) hero

геро́тор·н·ый на·со́с·   (*m*) helical rotor pump

герпето·ло́г·и·я, -и,   (*f*) (zool) herpetology

герпето·фа́ун·а, -ы,   (*f*) (zool) herpetofauna

герсдорфи́т·, -а,   (*m*) (min) gersdorffite

Геру́, печ·ь систéм·ы   (*m*) (met) Heroult furnace

~, спо́соб·   (*m*) Heroult process (aluminium production)

герунд·и́в·, -а,   (*m*) (gram) gerundive

геру́нд·и·й, -я,   (*m*) (gram) gerund

ге́рц·, -а,   (*i*) **-ем,**   (*m*) Hertz, hertz; (phys) cycle per second, c/s

    Ге́рц·а дипо́л·ь   (*m*) (rad) Hertzian dipole/radiator

герценбергúт·, -а,   (*m*) (min) herzenbergite

герцинúт·, -а,   (*m*) (min) hereynite, green spinel

Ге́рцшпрунг·а Ре́ссел·а диагра́мм·а   (*f*) (astron) Hertzsprung Russell diagram

Ге́ршел·а, эффéкт·   (*m*) (phot) Herschel effect

гесперидúн·, -а,   (*m*) (chem) hesperidin

Ге́сс·а, за·ко́н·   (*m*) (chem) Hess's law

гéссен·ск·ая му́х·а   (*f*) (ent) Hessian fly, *Mayetiola destructor*

Гессиáн·, -а,   (*m*) (math) Hessian

гессúт·, -а,   (*m*) (min) hessite

гетеро·азео·тро́п·н·ый, -ая, -ое,   (*a*) (chem) heteroazeotropic

гетеро·ауксúн·, -а,   (*m*) (bot) heterauxin

гетеро·валéнт·н·ый, -ая, -ое,   (*a*) heterovalent

гетеро·гамéт·а, -ы,   (*f*) (zool) heterogamete

гетеро·гáм·и·я, -и,   (*f*) (gen) heterogamy

гетеро·гáм·н·ый, -ая, -ое,   (*a*) (biol) heterogamous

гетеро·ген·изáци·я -и,   (*f*) heterogenization

гетеро·ген·éз·, -а,   (*m*) (zool) heterogenesis

гетеро·ген·и́·я, -и,   (*f*) (zool) heterogeny

гетеро·ге́н·ност·ь, -и,   (*f*) heterogeneity

гетеро·ге́н·н·ый, -ая, -ое,   (*a*) heterogeneous

гетеро·го́н·и·я, -и,   (*f*) (zool) heterogony

гетеро·десм·и́чност·ь, -и,   (*f*) (phys) heterodesmic nature

гетеро·десм·и́чн·ый, -ая, -ое,   (*a*) heterodesmic

гетеро·ди́н·, -а,   (*m*) (rad) heterodyne

~, интерполяцио́н·н·ый   heterodyne interpolation oscillator

~, клистро́н·н·ый   klystron oscillator

гетеро·зиго́т·ност·ь, -и,   (*f*) (gen) heterozygosis, heterosis

гетерꞓ·зиго́т·н·ый, -ая, -ое,   (*a*) (gen) heterozygous

гетеро́зис·, -а,   (*m*) (gen) heterosis, heterozygosis

гетеро·ли́т·, -а,   (*m*) (min) heterolite, haeterolite

гетеро·лит·и́ческ·ий, -ая, -ое,   (*a*) (chem) heterolytic

гетеро·мо́рф·н·ый, -ая, -ое,   (*a*) (bot) heteromorphous, (zool) heteromorphic

гетероп·и́ческ·ий, -ая, -ое,   (*a*) (geol) heteropic

гетеро·поли·кислот·а́, -ы́,   (*f*) (chem) heteropoly acid

гетеро·поля́р·н·ый, -ая, -ое,   (*a*) heteropolar

гетеро·стил·и́я, -и,   (*f*) (gen) heterostyly

гетеро·талл·и́зм·, -а,   (*m*) (bot) heterothallism

гетеро·тро́ф·н·ый, -ая, -ое,   (*a*) (bot) heterotrophic

гетеро·хро́м·ый, -ая, -ое,   (*a*) (phot) heterochromatic

гетеро·хромат·и́ческ·ий, -ая, -ое,   (*a*) (phot) heterochromatic

гетеро·хро́н·ност·ь, -и,   (*f*) (zool) heterochronism

гетеро·це́ль·н·ый, -ая, -ое,   (*a*) (zool) heterocoelous

гетеро·цеп·н·о́й, -а́я, -о́е,   (*a*) (chem) heterochain

гетеро·церк·а́льн·ый, -ая, -ое,   (*a*) (zool) heterocercal

гетеро·цикл·и́ческ·ий, -ая, -ое,   (*a*) (chem) heterocyclic

гетина́кс·, -а,   (*m*) getinax (paper/bakelite plastic)

гети́т·, -а,   (*m*) (min) goethite

ге́ттер·, -а,   (*m*) (elec) getter

∼ -ио́н·н·ый на·со́с·   (*m*) (phys) ion-getter pump

∼ -на·со́с·, -а,   (*m*) (phys) pump-getter

Ге́фнер·а, свеч·а́   (*f*) (opt) Hefner candle

гиал·и́нов·ый, -ая, -ое,   (*a*) hyaline, hyaloid; glassy, vitreous; transparent, clear, free from inclusions

гиал·и́т·, -а,   (*m*) (min) hyalite, water opal

гиало·кристалл·и́ческ·ий, -ая, -ое, (*a*) hyalocrystalline

гиало·гра́ф·и·я, -и,   (*f*) (phot) hyalography

гиало·ме́р·, -а,   (*m*) (biol) hyalomere

гиало·пили́тов·ая структу́р·а   (*f*) (geol) hyalopilitic texture

гиало·пла́зм·а, -ы,   (*f*) (biol) hyaloplasm

гиало·со́м·а, -ы,   (*f*) (genet) hyalosome

гиало·фа́н·, -а,   (*m*) (min) hyalophane

гиал·у́р·ов·ая кисл·от·а́   (*f*) (chem) hyalmic acid

гиал·урон·ида́з·а, -ы,   (*f*) (chem) hyaluronidase

гиал·уро́н·ов·ая кисл·от·а́   (*f*) (chem) hyaluronic acid

гиаци́нт·, -а,   (*m*) (bot) hyacinth; zircon, hyacinth (jewel)

ги́б·, -а,   (*m*) bend, bending; (*past masc sing*) of **ги́б·нуть**

гиббере́лл·а, -ы,   (*f*) (bot) gibberela

гиббере́лл·и́н·, -а,   (*m*) (chem) gibberellin

Ги́ббс·а, пра́в·ил·о фаз·   (*n*) (chem) Gibbs' phase rule

∼, явл·е́ни·е   (*n*) (math) Gibbs phenomenon

гиббси́т·, -а,   (*m*) (min) gibbsite

ги́ббс·ов·ый анса́мбл·ь   (*m*) (math) Gibbs ensemble

ги́б·ел·ь, -и,   (*f*) wreck, ruin, destruction, loss; (biol) death, deletion

∼, по́лн·ая   total loss

ги́б·ельн·ый, -ая, -ое,   (*a*) disastrous, ruinous

гиберелли́н, -а,   (*m*) = **гибберелл·и́н**

ги́б·к·а, -и,   (*f*) bending

ги́б·к·ий, -ая, -ое,   (*a*) flexible, pliable, supple

ги́б·кост·ь, -и,   (*f*) flexibility, pliability, suppleness; (met) flexure

ги́б·нуть, -нут,   (*past masc sing*) **ги́б·,** (*imp*); perish

ги́б·очн·ый, -ая, -ое,   (*a*) bending

∼ маши́н·а   (*f*) (met) rolled-section bending machine

∼ шта́мп·   (*m*) bending tool

гибри́д·, -а,   (*m*) hybrid

∼, при·ви́в·очн·ый   graft hybrid, plant chimaera

гибрид·иза́ци·я, -и,   (*f*) hybridization, cross-fertilization, cross-breeding

гибрид·изи́р·овать, -уют,   (*imp*) hybridize, cross, interbreed

гибри́д·н·ый, -ая, -ое,   (*a*) hybrid

ги́б·че   (*comp*) of **ги́б·к·ий**

ги́б·ш·ий, -ая, -ее,   (*past part act*) of **ги́б·нуть**

ги́б·щик·, -а,   (*m*) bender (metal worker)

гиг·, -а,   (*m*) (naut) gig

гига-   (*int component*) (math) Giga, g

гига́нт·, -а,   (*m*) giant

гигант·и́зм·, -а,   (*m*) (biol) gigantism, (med) giantism

гиганто·ли́т·, -а,   (*m*) (min) gigantolite

гиганто·ци́т·, -а,   (*m*) (bot) gigantocyte

гига́нт·ск·ий, -ая, -ое,   (*a*) giant, gigantic, huge

∼ резона́нс·   (*m*) (phys) giant resonance

гигие́н·а, -ы,   (*f*) hygiene, sanitation; (med) environmental physiology

гигиен·и́ческ·ий, -ая, -ое,   (*a*) of **гигие́н·а**; hygienic, sanitary

гигр·и́н·, -а,   (*m*) (chem) hygrine

гигри́н·ов·ая кисл·от·а́   (*f*) (chem) hygric acid

гигро́·граф·, -а,   (*m*) (meteor) hygrograph

гигро́·метр·, -а,   (*m*) (meteor) hygrometer

∼, волос·н·о́й   hair hygrometer

∼, конденсацио́н·н·ый   dew-point hygrometer

гигро·ме́тр·и·я, -и,   (*f*) (meteor) hygrometry

гигро·ско́п·, -а,   (*m*) hygroscope

гигро·скоп·и́ческ·ий, -ая, -ое,   (*a*) hygroscopic, moisture/water-absorbing/absorbent

гигро·скоп·и́чност·ь, -и, (*f*) hygroscopicity, hygroscopic nature, moisture/water-absorbing capacity

гигро·скоп·и́чн·ый, -ая, -ое, (*a*) hygroscopic, moisture/water-absorbing/absorbent

гигро·терм·и́ческ·ий, -ая, -ое, (*a*) hygrothermal

гигро·фи́ль·н·ый, -ая, -ое, (*a*) (bot) hygrophile, hygrophilous

гигро·фи́т·, -а, (*m*) (bot) hygrophyte

гигро·ста́т·, -а, (*m*) hygrostat

гид·, -а, (*m*) guide (tourism etc.)

гидантойн·, -а, (*m*) (chem) hydantoin

гидат·о́д·, -а, (*m*) (bot) hydathode, water stoma

гидато·морф·и́зм·, -а, (*m*) (bot) hydatomorphism

гидно·ка́рп·ов·ая кисл·от·а́ (*f*) (chem) hydnocarpic acid

гидра́вл·ик·, -а, (*m*) hydraulic engineer

гидра́вл·ик·а, -и, (*f*) hydraulics

гидравл·и́ческ·ий, -ая, -ое, (*a*) hydraulic, fluid, water, water-operated

~ ис·пыт·а́ни·е (*n*) hydraulic test

~ кру́п·ност·ь (*f*) (geol) sinking/fall velocity

~ пере·да́·ч·а (*f*) hydraulic transmission gear

~ ра́диус· (*m*) hydraulic mean depth

~ уда́р· (*m*) water hammer

гидравл·и́чност·ь, -и, (*f*) hydraulicity (of cements etc.)

гидраз·и́д·, -а, (*m*) (chem) hydrazide

~ кисл·от·ы́ acid hydrazide

гидр·аз·и́н·, -а, (*m*) (chem) hydrazine

гидразо·бензо́л·, -а, (*m*) hydrazobenzene

гидразо·кислот·а́, -ы́, (*f*) hydrazoic acid

гидраз·о́н·, -а, (*m*) (chem) hydrazone

гидразо·со·един·ени·е, -я, (*n*) (chem) hydrazocompound

гидр·акри́л·ов·ая кисл·от·а́ (*f*) hydracrylic acid

гидр·а́нт·, -а, (*m*) hydrant

гидр·аргилли́т·, -а, (*m*) (min) hydragillite, gibbsite

гидр·асти́н·, -а, (*m*) (pharm) hydrastine

гидра́стис·, -а, (*m*) (bot) golden seal, *Hydrastis canadensis*

гидр·а́т·, -а, (*m*) (chem) hydrate

~ за́·кис·и (lower or -ous) hydroxide, hydroprotoxide

гидр·а́т

~ за́·кис·и желе́з·а ferrous hydroxide/hydrate

~ за́·кис·и ме́д·и cuprous hydroxide, copper hydroprotoxide

~ о́·кис·и (higher or -ic) hydroxide

~ о́·кис·и ка́ли·я potassium hydroxide

~ о́·кис·и ме́д·и (chem) cupric hydroxide

~ о́·кис·и о́лов·а stannic hydroxide

гидрат·а́ци·я, -и, (*f*) hydration

гидрат·и́рованн·ый, -ая, -ое, (*past part pass*) of **гидрат·и́р·овать;** hydrated

гидрат·и́р·овать, -уют, (*imp and perf*) hydrate

гидрат·и́руемост·ь, -и, (*f*) (chem) hydratability

гидрат·и́рующ·ийся, -аяся, -ееся, (*a*) (chem) hydratable

гидра́т·н·ый, -ая, -ое, (*a*) hydrated, hydrate, hydration

~ вод·а́ (*f*) (chem) water of crystallization

гидр·атро́п·н·ая кисл·от·а́ (*f*) hydratropic acid

гидр·атро́п·ов·ый альдеги́д· (*m*) hydratropaldehyde

гидр·ациони́т·, -а, (*m*) (min) evaporite

гидр·еми́·я, -и, (*f*) (med) hydraemia, hydremia

гидр·и́д·, -а, (*m*) (chem) hydride

~ ли́ти·я lithium hydride

гидр·инда́н·, -а, (*m*) (chem) hydrindane

гидр·инде́н·, -а, (*m*) (chem) hydrindene

гидр·и́ровани·е, -я, (*n*) *see* **гидро·-ген·иза́ци·я**

гидр·и́рованн·ый, -ая, -ое, (*past part pass*); hydrogenated, hydrogenized

гидр·и́рующ·ий, -ая, -ее, (*pres part act*); hydrogenating, hydrogenant; **-ся** hydrogenated, hydrogenable

гидро- (*prefix*) hydro-, hydraulic, water, underwater

гидро·авин·но́с·ец·, -с·ц·а, (*m*) seaplane carrier

гидро·авиа·тра́нспорт·, -а, (*m*) seaplane tender

гидро·авиа́ци·я, -и, (*f*) seaplanes, flying boats; seaplanes and flying boats

гидро·автома́т·, -а, (*m*) airlift (pump)

ги́дро·аккумуляцио́н·н·ый, -ая, -ое, (*a*) water-storage

ги́дро·аккумуля́тор·, -а, (*m*) hydraulic accumulator

ги́дро·аку́ст·ик·, -а, (*m*) (naut) sonar operator

ги́дро·аку́ст·ик·а, -и, (*f*) underwater acoustics, hydroacoustics

ги́дро·акуст·и́ческ·ий, -ая, -ое, (*a*) hydroacoustic, underwater sound

~ лока́тор· (*m*) sonar equipment

ги́дро·аромат·и́ческ·ий, -ая, -ое, (*a*) (chem) hydroaromatic

ги́дро·аэропо́рт·, -а, (*m*) seaplane port

ги́дро·би·о́с·, -а, (*m*) (biol) hydrobiosis

ги́дро·борци́т·, -а, (*m*) (min) hydroboracite

ги́дро·бу́р·, -а, (*m*) (min) hydro-drill

ги́дро·гало́ид·, -а, (*m*) (chem) hydrohaloid

ги́дро·ге́л·ь, -я, (*m*) (min) hydrogel

ги́дро·гемат·и́т·, -а, (*m*) (min) turgite, hydrohaematite

ги́дро·генера́тор·, -а, (*m*) hydroelectric generator

ги́дро·гениз·а́т·, -а, (*m*) hydrogenate, hydrogenated product

ги́дро·гениз·а́ци·я, -и, (*f*) hydrogenation

~, деструкти́в·н·ая destructive hydrogenation, hydrogen cracking

ги́дро·гениз·и́рованн·ый, -ая, -ое, (*past part pass*); hydrogenated, hydrogenized

~ жир· (*m*) hydrogenated fat

ги́дро·гениз·и́рующ·ий, -ая, -ее, (*pres part act*); hydrogenating, hydrogenant; **-ся** hydrogenable

ги́дро·гениз·о́ванн·ый, -ая, -ое, (*past part pass*); hydrogenated

ги́дро·ге́н·н·ый, -ая, -ое, (*a*) hydrogenous, hydrogenic, hydratogenous

ги́дро·гено́·лиз·, -а, (*m*) hydrogenolysis, destructive hydrogenation, hydrogen cracking

ги́дро·гео·ло́г·и·я, -и, (*f*) hydrogeology

ги́дро·гёти́т·, -а, (*m*) (min) hydrogoethite

ги́дро·гра́ф·, -а, (*m*) hydrographer; (instr) hydrograph

~, едини́чн·ый unit hydrograph

ги́дро·граф·и́ческ·ий, -ая, -ое, (*a*) hydrographic, hydrographical

~ сет·ь (*f*) (geol) drainage

ги́дро·граф·и́ческ·ий, -ая, -ое

~ су́д·н·о (*n*) (naut) survey vessel

ги́дро·гра́ф·и·я, -и, (*f*) hydrography

ги́дро·дина́м·ик·а, -и, (*f*) hydrodynamics, fluid dynamics

~, магни́т·н·ая magnetohydrodynamics

ги́дро·динам·и́ческ·ий, -ая, -ое, (*a*) hydrodynamic

~ пере·да́·ч·а (*f*) hydrodynamic transmission gear

ги́дро·до·бы́·ч·а, -и, (*f*) (min) hydraulicking, hydraulic winning/mining

ги́дро·домкра́т·, -а, (*m*) hydraulic jack

ги́дро·за·тво́р·, -а, (*m*) hydraulic/water seal

ги́дро·земле·со́с·, -а, (*m*) hydraulic suction dredge(r)

ги́дро·золо·у·дал·е́ни·е, -я, (*n*) hydraulic ash removal

ги́дро·зо́л·ь, -и, (*f*) (chem) hydrosol

гидр·о́ид·н·ые, -ых, (*pl decl as adj*) (zool) hydrozoans, *Hydrozoa*

гидр·о́ид, -а, (*m*) (zool) hydroid; (*pl*) hydroids, *Hydroidea*

ги́дро·изо·ба́т·а, -ы, (*f*) (geol) hydroisobath, isobath of water table

ги́дро·изо·ги́пс·а, -ы, (*f*) (zool) contour of water table

ги́дро·изоля́ци·я, -и, (*f*) water/damp proofing

ги́дро·изо·пье́з·а, -ы, (*f*) (surv) line joining artesian wells of equal pressure

ги́дро·иоди́д·, -а, (*m*) (chem) hydroiodide

ги́дро·ион·иза́тор·, -а, (*m*) hydroionizer

ги́дро·кана́л·, -а, (*m*) water channel

ги́дро·карбона́т·, -а, (*m*) (chem) bicarbonate, hydrocarbonate

ги́дро·каули́с·, -а, (*m*) (zool) hydrocaulis

ги́дро·каучу́к·, -а, (*m*) hydrorubber

ги́дро·классифика́тор·, -а, (*m*) (min) hydraulic classifier

ги́дро·класт·и́ческ·ий, -ая, -ое, (*a*) (geol) hydroclastic

ги́дро·кодиме́р·, -а, (*m*) (chem) hydro-codimer

ги́дро·кор·и́чн·ая кисл·от·а́ (*f*) (chem) hydrocinnamic acid

ги́дро·кры·л·о́, -а́, (*nom pl*) **-кры́·л·ья,** (*g pl*) **-ьев,** (*n*) (shipb) hydrofoil

ги́дро·кры́л·ышк·о, -а, (*n*) = **гид·ро·кры·л·о́** (shipb) hydrofoil

гидро́кс·, -а, (*m*) (expl) hydrox

гидрокс·а́м·ов·ая кисл·от·а́   (*f*) (chem) hydroxamic acid

гидроксл·апати́т·, -а,   (*m*) (min) hydroxyapatite

гидр·окси́д·, -а,   (*m*) (chem) hydroxide

гидр·окс·и́л·, -а,   (*m*) (chem) hydroxyl

гидроксил·ами́н·, -а,   (*m*) hydroxylamine

гидроксил·и́ровани·е, -я,   (*n*) (chem) hydroxilation

гидр·окс·и́ль·н·ый, -ая, -ое,   (*a*) hydroxyl

гидрокс·о́ни·й, -я,   (*m*) (chem) hydroxonium ion

гидро·ла́з·а, -ы,   (*f*) (biochem) hydrolase

гидро·лакко·ли́т·, -а,   (*m*) (geol) hydrolaccolite

гидро́·лиз·, -а,   (*m*) (chem) hydrolysis

гидро·лиз·а́т·, -а,   (*m*) hydrolysate

гидро́·лиз·н·ый, -ая, -ое,   (*a*) hydrolysis

~ **аппара́т·**   (*m*) hydrolyser

~ **спирт·**   (*m*) (chem) ethyl alcohol, ethanol

гидро·ли́т·, -а,   (*m*) (chem) hydrolith, calcium hydride; hydrolyte

гидро·лит·и́ческ·ий, -ая, -ое,   (*a*) hydrolytic

гидро́·лог·, -а,   (*m*) hydrologer, hydrologist

гидро·лог·и́ческ·ий, -ая, -ое,   (*a*) hydrologic(al)

~ **режи́м·**   (*m*) hydrologic regime

~ **сет·ь**   (*f*) stream-gauging network

~ **цикл·**   (*m*) hydrologic cycle

гидро·ло́г·и·я, -и,   (*f*) hydrology

гидро·лока́тор·, -а,   (*m*) (naut) sonar, asdic, underwater sound locator

гидро·локацио́н·н·ый, -ая, -ое,   (*a*) underwater sound-location, sonar, asdic

гидро·лока́ци·я, -и,   (*f*) underwater sound location, sonar, asdic

гидр·о́л·ь, -я,   (*m*) (chem) hydrol

гидро·магнези́т·, -а,   (*m*) (geol) hydromagnesite

гидро·ма́сс·а, -ы,   (*f*) pulp, sludge, slush

гидро·меду́з·а, -ы,   (*f*) (zool) hydromedusa, (*pl*) *Hydromedusae*

гидро·металл·у́рги·я, -и,   (*f*) hydrometallurgy

гидро·метео́р·, -а,   (*m*) (meteor) hydrometeor

гидро·метеоро·ло́г·и́ческ·ий, -ая, -ое,   (*a*) hydrometeorological

гидро·метеоро·ло́г·и·я, -и,   (*f*) hydrometeorology, hydrology and meteorology

гидро́·метр·, -а,   (*m*) hydrometer

гидро·метр·и́ческ·ий, -ая, -ое,   (*a*) stream-gauging, water-gauging, water-measuring

~ **верт·у́шк·а**   (*f*) (hydr) current meter

~ **ста́нци·я**   (*f*) (surv) measuring station, stream-gauging station

~ **с·твор·**   (*m*) stream-gauging section

гидро·ме́тр·и·я, -и,   (*f*) (surv) stream gauging, water flow gauging

гидро·механ·иза́тор·, -а,   (*m*) hydraulic excavation engineer

гидро·механ·иза́ци·я, -и,   (*f*) (civ eng) hydraulicking, hydraulic mining, hydraulic excavation, sluicing

гидро·меха́н·ик·а, -и,   (*f*) fluid mechanics

гидро·монито́р·, -а,   (*m*) (civ eng) hydraulic giant/excavator

гидро·мото́р·, -а,   (*m*) hydraulic motor, hydraulic actuator

гидро·му́фт·а, -ы,   (*f*) (mech eng) hydraulic coupling, variable speed gear

гидр·о́н·, -а,   (*m*) (chem) hydrone

гидро·настура́н·, -а,   (*m*) (min) hydromanite

гидро·нефр·о́з·, -а,   (*m*) (med) hydronephrosis

гидро·обес·пы́л·ивани·е, -я,   (*n*) water-spray dust prevention

гидро·о·богащ·е́ни·е, -я,   (*n*) sink-float ore dressing

гидро·о́·кис·ь, -и,   (*f*) hydrated oxide, hydroxide

~ **желе́з·а**   iron hydroxide, ferric hydrate

~ **ка́ли·я**   caustic potash

гидро·о·чи́ст·к·а, -и,   (*f*) (oil) hydro-refining, hydrodesulphurization

гидро·па́т·, -а,   (*m*) (med) hydropathist

гидро·пере·да́·ч·а, -и,   (*f*) hydraulic transmission gear

~, **объ·ём·н·ая**   hydrostatic transmission

гидро·пе́ре·кис·ь, -и,   (*f*) hydrogen peroxide, hydroperoxide

~ **на́три·я**   sodium hydrogen peroxide

гидро·пла́н·, -а,   (*m*) *see* **гидро·са·мо·лёт·**

гидро·пневмат·и́ческ·ий, -ая, -ое,   (*a*) hydropneumatic

гидро·подъ·ём·, -а,   (*m*) hydraulic lift

гидро·по́чт·а, -ы,   (*f*) (nucl) hydraulic rabbit

гидро·при·во́д·, -а,   (*m*) hydraulic drive

гидро·пульс·а́тор·, -а,   (*m*) hydraulic pulsator (materials testing)

гидро·пу́льт·, -а,   (*m*) spray-gun

гидро·раз·бив·а́тел·ь, -я,   (*m*) (paper) hydrapulper

гидро·сальпи́нкс·, -а,   (*m*) (med) hydrosalpinx

гидро·само·лёт·, -а,   (*m*) seaplane

~, лод·оч·ный   flying boat

~, по·плав·ко́в·ый   seaplane, float plane

гидро·сепар·а́тор·, -а,   (*m*) (min) sink-float separator

гидро·сер·ни́ст·ый, -ая, -ое,   (*a*) hydrosulphurous, hydrosulphite of

~ кисл·от·а́   (*f*) hyposulphurous acid

гидро·се́ть, -и,   (*f*) (geol) drainage system

гидро·силика́т·н·ый, -ая, -ое,   (*a*) hydrosilicate

гидро·систе́м·а, -ы,   (*f*) hydraulic system

гидро·ско́п·, -а,   (*m*) hydroscope

гидро·слюд·а́, -ы́,   (*f*) (min) hydromica

гидро·с·ме́с·ь, -и,   (*f*) hydraulic liquid/fluid/mixture; water-material mixture, sludge, pulp

гидро·со́л·ь, -и,   (*f*) (chem) acid salt

гидро·со́м·а, -ы,   (*f*) (zool) hydrosome

гидро·со·оруж·е́ни·е, -я,   (*n*) hydraulic engineering installation/works

гидро·ста́нци·я, -и,   (*f*) hydro-electric power station

гидро·ста́т·ик·а, -и,   (*f*) hydrostatics

гидро·стат·и́ческ·ий, -ая, -ое,   (*a*) hydrostatic

гидро·сульф·а́т, -а,   (*m*) (chem) bisulphate

гидро·сульф·и́д·, -а,   (*m*) hydrosulphide

гидро·сульф·и́т·, -а,   (*m*) hyposulphite $(-S_2O_4)$; bisulphite $(-SO_3)$

~ на́три·я   (*f*) sodium hyposulphite

гидро·сфе́р·а, -ы,   (*f*) (geog) hydrosphere

гидро·с·цепл·е́ни·е, -я,   (*n*) hydraulic coupling, turbo-coupling

гидро·та́ксис·, -а,   (*m*) (biol) hydrotaxis

гидро·терм·а́льн·ый, -ая, -ое,   (*a*) (geol) hydrothermal

гидро·терм·и́ческ·ий, -ая, -ое,   (*a*) hydrothermal

гидро·термо·об·рабо́т·к·а древ·е·си́н·ы   (*f*) wood steaming, steam treatment of wood

гидро·те́хн·ик·, -а,   (*m*) water engineer

гидро·те́хн·ик·а, -и,   (*f*) water engineering

гидро·техн·и́ческ·ий, -ая, -ое,   (*a*) water engineering

~ со·оруж·е́ни·е·   (*n*) water-engineering works/construction

гидро·ти́п·н·ая печа́т·ь   (*f*) (phot) hydrotype printing

гидро·тори́т·, -а,   (*m*) (min) hydrothorite

гидро·то́рмоз·, -а,   (*m*) hydraulic brake; water brake; absorption dynamometer

гидро·то́рф·, -а,   (*m*) hydraulic peat winning

гидро·транспорт·ёр·, -а,   (*m*) hydraulic conveyor

гидро·трансформ·а́тор·, -а,   (*m*) (mech eng) torque converter

гидро·троп·и́·я, -и,   (*f*) (biol) hydrotropism

гидро·тро́п·н·ый, -ая, -ое,   (*a*) (biol) hydrotropic

гидро·турби́н·а, -ы,   (*f*) (*see* **турби́н·а**); water/hydraulic turbine

гидро·у́зел·, -зл·а́,   (*m*) (civ eng) water engineering system/scheme

~, водо·за·бо́р·н·ый   impounding scheme

~, высоко·на·по́р·н·ый   high-head scheme

~, гидро·энергет·и́ческ·ий   hydro-electric works

~, регул·и́рующ·ий   river-contraction system

гидро·у·сил·и́тел·ь, -я,   (*m*) hydraulic intensifier/amplifier

гидро·фа́н·, -а,   (*m*) (min) hydrophane

гидро·фи́з·ик·а, -и,   (*f*) physics of the hydrosphere, physical hydrology

гидро·фи́ль·н·ый, -ая, -ое,   (*a*) hydrophilous, hydrophilic

гидро·фи́т·, -а,   (*m*) (bot) hydrophyte

гидро·фиц·и́рованн·ый, -ая, -ое,   (*past part pass*); hydraulic, hydraulically operated

~ тока́р·н·о-дав·и́льн·ый стан·о́к·   hydrospinning machine

гидро·фо́б·, -а,   (*m*) (bot) hydrophobe

гидро·фоб·иза́тор·, -а,   (*m*) hydrophobizing agent

гидро·фо́б·и·я, -и,   (*f*) (med) hydrophobia

гидро·фо́б·ност·ь, -и, (*f*) hydrophobic nature

гидро·фо́б·н·ый, -ая, -ое, (*a*) hydrophobic

гидро·фо́н·, -а, (*m*) (naut) hydrophone

гидро·форма́т·, -а, (*m*) (oil) hydroformate

гидро·фо́рминг·, -а, (*m*) (oil) hydroforming

гидро·хи́м·и·я, -и, (*f*) geochemistry of the hydrosphere, chemical hydrology

гидро·хин·и́н·, -а, (*m*) (chem) hydroquinine

гидро·хин·о́н·, -а, (*m*) (chem) hydroquinone

гидро·целлюло́з·а, -ы, (*f*) hydrocellulose

гидро·це́л·ь, -и, (*f*) (med) hydrocoele

гидро·церусси́т·, -а, (*m*) (min) hydrocerussite

гидро·циклó́н·, -а, (*m*) hydraulic cyclone (flotation)

гидро·цинки́т·, -а, (*m*) (min) hydrozincate, zinc bloom

гидро·экстра́ктор·, -а, (*m*) (chem eng) hydroextractor, whizzer

гидро·элева́тор·, -а, (*m*) waterjet pump

гидро·электр·и́ческ·ий, -ая, -ое, (*a*) hydroelectric

гидро·электро·ста́нци·я, -и, (*f*) hydroelectric power station

~, на·со́с·н·о-аккумуля́тив·н·ая pumped storage hydroelectric station

~, при·плот·и́нн·ая hydroelectric power station downstream from dam

гидро·энерге́т·ик·а, -и, (*f*) hydraulic power engineering

гидрю́р·, -а, (*m*) (chem) hydrogenation product

гиле́·я, -и, (*f*) tropical forest

Гиллемен·а, ли́ни·я (*f*) (telecom) Guillemine line

гило·фи́т·, -а, (*m*) (bot) hylophyte

ги́льберт·, -а, (*m*) (elec) gilbert (unit of magneto-electric force); **Ги́льберт·** (math) Hilbert

Ги́льберт·ов·о про·стра́н·ств·о (*n*) (math) Hilbert space

гильз·а, -ы, (*f*) (ICE) linear, barrel sleeve; check sleeve (for testing springs); (met) bottle, hollow, shell; (gun) cartridge, case; cigarette wrapper

~, мо́кр·ая (ICE) wet liner

~, про·ши́т·ая (met) hollow

~, сух·а́я (ICE) dry liner

ги́льз·а

~, экстракцио́н·н·ая (gunn) extraction thimble

ги́льз·ов·ый, -ая, -ое, (*a*) of **ги́льз·а**

гильзо·у·ла́вл·иватель·ь, -я, (*m*) (gunn) extractor

гильоти́н·а, -ы, (*f*) guillotine

гильоти́н·н·ые но́ж·ниц·ы (*pl*) guillotine shears

гильош·и́ровани·е, -я, (*n*) guilloche work (decorative)

гильпини́т·, -а, (*m*) (min) gilpinite

гильсони́т·, -а, (*m*) (min) gilsonite, mintaite

гиме́н·и·й, -я, (*m*) (bot) hymenium

гимено·мице́т·, -а, (*m*) (bot) hymenomycete

гимено·фо́р·, -а, (*m*) (bot) hymenophone

ги́мн·, -а, (*m*) anthem

гимназ·и́ст·, -а, (*m*) grammar-school boy

гимна́ст·, -а, (*m*) gymnast

гимни́т·, -а, (*m*) (min) gymnite

гимно·спо́р·а, -ы, (*f*) (biol) gymnospore

гимно·ци́т·, -а, (*m*) (cyt) gymnocyte

гин·а́з·а, -ы, (*f*) (gen) gynase

гин·андро·фо́р·, -а, (*m*) (biol) gynandrophore

гингив·и́т·, -а, (*m*) (med) gingivitis

гинеко·ло́г·и·я, -и, (*f*) (med) gynaecology

гинеко·фо́б·и·я, -и, (*f*) (med) gynaecophobia

гин·еце́·й, -я, (*m*) (bot) gynaeceum

ги́н·и (*nom pl*), (*g pl*) **-ей** (naut) winding tackle, purchase

ги́нкг·о, -а, (*n*) (bot) ginks, kew tree, *Ginko biloba*

гино·ген·е́з·, -а, (*m*) (biol) gynogenesis

гино·сте́г·и·й, -я, (*m*) (bot) gynostegium

гино·сте́м·и·й, -я, (*m*) (bot) gynostemium

ги́нц·ы (*pl*) (naut) gun tackle, small tackle

гинь, *see* **ги́н·и**

Ги́нье, зе́лен·ь (*f*) (chem) Guignet's green

Ги́нье–Пре́стон·а, зон·а (*f*) (met) Guinier-Preston zone

гио́ид·, -а, (*m*) (zool) hyoid

гио́ид·н·ый, -ая, -ое, (*a*) (zool) hyoid

гио·дезокси·хо́л·евая кисл·от·а́ (*f*) hyodesoxycholic acid гио·стил·и́·я, -и, (*f*) (zool) hyostyly

гиосциам·и́н·, -а, (*m*) (chem) hyoscyamine

гио·хо́л·овая кисл·от·а́ (*f*) hyocholic acid

гип·абисса́ль·н·ый, -ая, -ое, (*a*) (geol) hypabyssal

гип·аллело·мо́рф·, -а, (*m*) (biol) hypallelomorph

гип·а́нт·и·й, -я, (*m*) (bot) hypanthium

гипе́р·бол·а, -ы, (*f*) (math) hyperbola

~, равно·бо́ч·н·ая equilateral/rectangular hyperbola

гипер·бол·и́ческ·ий, -ая, -ое, (*a*) (math) hyperbolic

гипер·бол·и́чн·ый, -ая, -ое, (*a*) *see* **гипер·бол·и́ческ·ий**

гипер·бол·о́ид·, -а, (*m*) (math) hyperboloid

~, дву·пол·остн·о́й (math) hyperboloid of two sheets

~, одно·пол·остн·о́й (math) hyperboloid of one sheet

гипер·ге́н·н·ый, -ая, -ое, (*a*) (geol) supergene

гипер·гео·метр·и́ческ·ий, -ая, -ое, (*a*) (math) hypergeometric

гипер·глик·ем·и́·я, -и, (*f*) (med) hyperglycaemia

гипер·ди́н·, -а, (*m*) (rad) hyperdyne

гипер·ем·и́·я, -и, (*f*) (med) hyperemia

гипер·и́т·, -а, (*m*) (min) hypersthene-gabbro; norite; (obs) hyperite

гипер·квант·ова́ни·е, -я, (*n*) (nucl) hyperquantization

гипер·кин·е́з·, -а, (*m*) (med) hyperkinesia

гипер·конъюг·а́ци·я, -и, (*f*) hyperconjugation

гипер·о́л·ь, -я, (*m*) (pharm) hyperole, ortizone

гипер·о́н·, -а, (*m*) (nucl) hyperon, super proton

гипер·пла́в·к·ий, -ая, -ое, (*a*) hyperfusible

гипер·плаз·и́ческ·ий, -ая, -ое, (*a*) hyperplastic

гипер·пла́з·и·я, -и, (*f*) (biol) hyperplasia, overgrowth

гипер·пло́ск·ост·ь, -и, (*f*) (math) hyperplane

гипер·по·ве́рх·ност·ь, -и, (*f*) (phys) hypersurface

гипер·сенсибилиз·а́ци·я, -и, (*f*) (phot) hypersensitization

гипер·со́л·, -а, (*m*) (pharm) hypersole

гипер·сте́н·, -а, (*m*) (min) hypersthene

гипер·те́л·и·я, -и, (*f*) (zool) hypertely

гипер·стен·и́т·, -а, (*m*) (min) hypersthenite

гипер·тон·и́ческ·ий, -ая, -ое, (*a*) hypertonic

гипер·тон·и́·я, -и, (*f*) (med) hypertension

гипер·троф·и́ческ·ий, -ая, -ое, (*a*) (zool) hypertrophic

гипер·ядр·о́, -а, (*nom pl*) -я́др·а, (*g pl*) -я́дер·, (*n*) (nucl) hyperfragment, hypernucleus

гипно́·з·, -а, (*m*) hypnosis

гипнот·изи́р·овать, -уют, (*imp*) hypnotize

гипнот·и́ческ·ий, -ая, -ое, (*a*) hypnotic

гипо·бром·и́т·, -а, (*m*) (chem) hypobromite

гипо·витамин·о́з·, -а, (*m*) (med) hypovitaminosis

гипо·ге́н·н·ый, -ая, -ое, (*a*) (geol) hypogenetic; (bot) hypogenous

гипо·де́рм·а, -ы, (*f*) (bot) hypodermis, hypoderm

гипо·дермат·о́з·, -а, (*m*) (med) hypodermatosis

гипо́ид·н·ый, -ая, -ое, (*a*) hypoid

~ пере·да́ч·а (*f*) hypoid gear

гипоиод·и́т·, -а, (*m*) (chem) hypoiodite

гипо·ко́тил·ь, -я, (*m*) (bot) hypocotyl

гипо·кристалл·и́ческ·ий, -ая, -ое, (*a*) hypocrystalline

гипо·ксант·и́н·, -а, (*m*) (biochem) hypoxanthine

гипо·на́ст·и·я, -и, (*f*) (bot) hyponasty

гипо·нитри́т·, -а, (*m*) (chem) hyponitrite

гипо·пио́н·, -а, (*m*) (med) hypopyon

гипо·пла́з·и·я, -и, (*f*) (biol) hypoplasia

гипо·пло́ид·, -а, (*m*) (cyt) hypoploid

гипо·ста́з·, -а, (*m*) (med) hypostasis

гипо·сти́л·ь, -и, (*f*) (arch) hypostyle

гипо·сульфи́т·, -а, (*m*) (phot, chem) hyposulphite, hypo

гипо́·тез·а, -ы, (*f*) hypothesis

гипо·тену́з·а, -ы, (*f*) (math) hypotenuse

гипо·терм·а́льн·ый, -ая, -ое, (*a*) (geol) hypothermal

гипо·терм·и́·я, -и, (*f*) hypothermy

гипо·тет·и́ческ·ий, -ая, -ое, (*a*) hypothetic(al)

гипо·тет·и́чн·ый, -ая, -ое, (*a*) hypothetic(al)

гипо·тон·и́ческ·ий, -ая, -ое, (*a*) (chem) hypotonic

гипо·тон·и́·я, -и, (*f*) (med) hypotension

гипо·трохо́ид·а, -ы, (*f*) (math) hypotrochoid

гипо·фи́з·, -а, (*m*) (biol) hypophysis

гипо·физ·и́н·, -а, (*m*) (biochem) hypophysin, pituitary extract

гипо·фосф·а́т·, -а, (*m*) (chem) subphosphate

гипо·фосф·и́т·, -а, (*m*) (chem) hypophosphite

гипо·хлор·и́т·, -а, (*m*) (chem) hypochlorite

гипо·це́нтр·, -а, (*m*) (seismol) hypocentre, hypocenter, focus; (zool) hypocentrum

гипо·цикл·о́ид·а, -ы, (*f*) (math) hypocycloid

гиппо·пота́м·, -а, (*m*) (zool) hippopotamus

гипп·у́р·ов·ая кисл·от·а́ (*f*) (chem) hippuric acid

гипп·ур·и́т·, -а, (*m*) (pal) hippurite

гипп·ур·и́тов·ый, -ая, -ое, (*a*) (zool) hippuritic

ги́пс·, -а, (*m*) (min) gypsum; plaster, plaster cast

~, без·во́д·н·ый anhydrite

~, натура́ль·н·ый land plaster

~, полу·во́д·н·ый plaster of Paris

~, штукату́р·н·ый plaster of Paris, alabaster

гипсо·бето́н·, -а, (*m*) gypsum concrete

гипс·ова́ть, -у́ют, (*imp*) (agr) spread gypsum; (build) plaster; (med) put on plaster, put into plaster

ги́пс·ов·ый, -ая, -ое, (*a*) of ги́пс·

~ вяж·ущ·ий, материа́л·; (*m*) (build) gypsum binding material

~ по·вя́з·к·а (*f*) (med) plaster bandage

~ с·ле́п·к·а (*f*) plaster copy (of sculptures), plaster cast

гипсо·граф·и́ческ·ий, -ая, -ое, (*a*) hypsographic

~ крив·а́я (*f*) (geog) hypsographic curve

гипс·одо́нт·, -а, (*m*) (zool) hypsodont

гипсо·метр·и́ческ·ий, -ая, -ое, (*a*) hypsometer, hypsometric

~ шкал·а́ (*f*) (surv) layer system/scale

гипсо·ме́тр·и·я, -и, (*f*) (phys) hypsometry

гипсо·но́с·н·ый, -ая, -ое, (*a*) (geol) gypsiferous

гипсо·термо́метр·, -а, (*m*) (phys) hypsometer

гипсо·фи́л·, -а, (*m*) (bot) gypsophil

гипсо·фи́лл·, -а, (*m*) (bot) hypsophyll

гипсо·хро́м·, -а, (*m*) (chem) hypsochrome

ги́р·а, -ы, (*f*) (cryst) symmetry axis

гирак·одо́нт·, -а, (*m*) (zool) hyracodont, (*pl*) *Hyracodontidae*

гирако·те́р·и·й, -я, (*m*) (pal) hyracotherium

гира́кс·, -а, (*m*) (pal) hyrax

гир·ацио́н·н·ый, -ая, -ое, (*a*) gyratory

ги́рл·о, -а, (*n*) (geog) delta

гирля́нд·а, -ы, (*f*) garland, festoon, chain

~ изоля́гор·ов (elec) suspension insulators, insulator chain

ги́р·н·ый, -ая, -ое, (*a*) of ги́р·я

гиро́·бус·, -а, (*m*) gyro-driven bus

гиро·вертика́л·ь, -и, (*f*) vertical gyro; (air) gyrohorizon, artificial horizon

гиро·вертика́нт·, -а, (*m*) (rocket) roll stabilizer

гиро·гони́т·, -а, (*m*) (pal) gyrogonite

гиро·горизо́нт·, -а, (*m*) artificial horizon, gyrohorizon; (rocket) pitch control gyro

гиро·да́т·чик·, -а, (*m*) (a/c) gyroscope

гиро·ди́н·, -а, (*m*) (air) gyrodyne

гиро́ид·а, -ы, (*f*) (cryst) gyroid, rotation-inversion axis

гироидно·плана́ль·н·ая сим·ме́тр·и·я (*f*) (cryst) gyroid symmetry

гиро·ка́мер·а, -ы, (*f*) gyro-casing (compass)

гиро·ко́мпас·, -а, (*m*) gyrocompass

гиро·магни́т·н·ый, -ая, -ое, (*a*) gyromagnetic

~ от·нош·е́ни·е (*n*) gyromagnetic ratio; (nucl) g-factor

гиро·ма́ят·ник·, -а, (*m*) gyropendulum, three-frame gyroscope; (air) artificial horizon

гиро·полу·ко́мпас·, -а, (*m*) (air) directional gyro

гиро·рул·ев·о́й, -о́го, (*m decl as adj*) (naut) gyro-pilot, automatic steerer, iron quartermaster

гиро·ско́п·, -а, (*m*) gyroscope

~ двух·степе́н·н·ый two-frame gyroscope

~ на·правл·е́ни·я directional gyro

ги·ро·ско́п
~, от·ста·ю́щ·ий displacement gyroscope
~, трех·степе́н·н·ый three-frame gyroscope
~, у·равно·ве́ш·енн·ый free gyroscope
ги·ро·скоп·и́ческ·ий, -ая, -ое, (*a*) (*see also* **гиро-**) gyroscopic
~ систе́м·а (*f*) sensitive element (of gyro)
ги·ро·стабилиз·а́тор·, -а, (*m*) gyroscopic stabilizer, gyrostabilizer
ги·ро·сфе́р·а, -ы, (*f*) gyrosphere (compass)
ги·ро·тахо́·метр·, -а, (*m*) (a/c) rate gyro
ги·ро·тро́п·н·ый, -ая, -ое, (*a*) gyrotropic
ги·ро·у·с·поко́·ител·ь, -я, (*m*) (naut) gyrostabilizer
гируди́н·, -а, (*m*) (biochem) hirudin
ги́р·я, -и, (*g pl*) **гир·ь,** (*f*) weight (for scales etc.)
~, образ·цо́в·ая standard weight
ги́ря-ре́йтер·, -а, (*m*) rider (of chemical balance)
гист·ами́н·, -а, (*m*) (biochem) histamine
гистер·е́зис·, -а, (*m*) (phys) hysteresis
~ вращ·е́ни·я rotational hysteresis
~, ди·электр·и́ческ·ий dielectric hysteresis
~, магни́т·н·ый magnetic hysteresis
~, полз·у́ч·ий viscous hysteresis, magnetic creep
~, тепл·ов·о́й thermal hysteresis
~, у·пру́г·ий mechanical hysteresis
гистер·е́зисн·ый, -ая, -ое, (*a*) hysteresis
~ дви́г·ател·ь (*m*) (elec) hysteresis motor
гистеро·ге́н·н·ый, -ая, -ое, (*a*) (zool) hysterogenic
гистеро·магмат·и́ческ·ий, -ая, -ое, (*a*) (geol) hysteromagmatic
гистеро·па́т·ия, -и, (*f*) (med) hysteropathy
гистеро·то́м·и·я, -и, (*f*) (med) hysterectomy
гистер·эк·то́м·и·я, -и, (*f*) (med) hysterectomy
гист·иди́н·, -а, (*m*) (biochem) histidine
гистио·ген·е́з·, -а, (*m*) (zool) histogenesis
гистио·ци́т·, -а, (*m*) (biol) histiocyte, histocyte

гисто·бла́ст·, -а, (*m*) (biol) histoblast
гисто·гемат·и́н·, -а, (*m*) (biochem) histohaematin
гисто·гра́мм·а, -ы, (*f*) (math) histogram
гисто·ло́г·и·я, -и, (*f*) (zool) histology
гист·о́н·, -а, (*m*) (chem) histone
гита́р·а, -ы, (*f*) quadrant, quadrant plate, wheel plate (on machine tools)
ги́тов·, -а, (*m*) (naut) brail
ги́тч·, -а, (*m*) hitch, hitching device
гиф·а, -ы, (*f*) (bot) hypha, filament
ги́ч·к·а, -и, (*g pl*) **-ч·ек·,** (*f*) (naut) gig
г-кисл·от·а́, -ы́, (*f*) (chem) G-acid
гл. (*abbr*) = **глав·а́** chapter
глабе́л·ь, -и, (*f*) (anat) glabella
глав- (*prefix*) head, chief, main, principal
глав·а́, -ы́, (*nom pl*) **гла́в·ы,** (*f*) head, chief; (print) chapter; (arch) cupola
~ госуда́рств·а (polit) head of state
~ прав·и́тельств·а head of government (in U.K. — Prime Minister, in U.S.— President, in U.S.S.R.—Chairman of Council of Ministers)
глав·бух·, -а, (*m*) (*abbr*) of **гла́в·н·ый бухга́лтер·;** chief accountant
глав·вра́ч·, -а́, (*i*) **-о́м,** (*m*) head physician
гла́в·к·, -а, (*m*) (*abbr*) of **гла́в·н·ый комите́т·;** chief directorate (in ministries etc.), main/central board/administration
главк- (*int component*) *see* **глаук-**
главно·кома́нд·ующ·ий, -его, (*m decl as adj*) Commander-in-Chief
верх·о́вн·ый главно·кома́н·дую·щ·ий (mil) supreme commander
гла́в·н·ый, -ая, -ое, (*a*) main, chief, head, principal
~ ба́з·а (*f*) (mil) main base, depot
~ инжене́р·, (*m*) chief engineer
~ кни́г·а (*f*) (com) ledger
~ норма́л·ь (*f*) (math) principal normal
~ гла́в·н·ым о́браз·ом (*adv*) primarily, chiefly, mainly, principally, for the most part
~ ос·ь (*f*) principal axis
~ пло́ск·ост·ь (*f*) (opt) principal plane
~ реда́ктор· (*m*) editor-in-chief
~ то́ч·к·а (*f*) (opt) principal point
~ у·правл·е́ни·е (*n*) chief directorate (in ministeries etc.), main/central board/administration
~ член· (*m*) (math) dominant term

глав·пище·пром·, -а, (*m*) (*abbr*) of **глáв·н·ое у·правл·éни·е пищ·е·в·óй про·мы́шл·енност·и;** chief/main directorate/administration of the food industry

-глав·ый, -ая, -ое, (*component of adj*) -headed

глагери́т·, -а, (*m*) (min) halloysite

глагóл·, -а, (*m*) (gram) verb

глагóл·ь, -я, (*m*) L-shaped object (obs)

~ -гак·, -а, (*m*) (naut) Senhouse slip

глад·и́лк·а, -и, (*g pl*) **-лок·,** (*f*) burnisher (engraving tool); flatter (forging tool); float (plasterer's tool); (naut) rubber (sailmaking)

глад·и́льн·ый, -ая, -ое, (*a*) ironing, smoothing

~ пресс· (*m*) ironing press (laundry)

глáд·ить, -ят, (*imp*) iron, press; stroke, smooth

глáд·к·ий, -ая, -ое, (*a*) smooth, even, plane, psilate; (text) unfigured, plain; crushed (leather); facile, fluent; (anat) unstriated

~ сéктор· (*m*) plain portion, interruption (in a screw)

гладко- (*prefix*) smooth, flush, psilo-

гладко·пáлуб·н·ый, -ая, -ое, (*a*) (shipb) flush-decked

гладко·по·крóв·н·ый, -ая, -ое, (*a*) psilotegillate

гладко·про·свéт·н·ый, -ая, -ое, (*a*) psiloluminate

гладко·стéн·н·ый, -ая, -ое, (*a*) (gunn) smooth-bore

глáд·кост·ь, -и, (*f*) smoothness; fluency

глáд·ь, -и, (*f*) glassy surface; (text) satin-stitch, plain stitch (knitting)

глáж·е, (*comp*) of **глáд·к·ий**

глáж·ени·е, -я, (*n*) = **глáж·ень·е**

глáж·енн·ый, -ая, -ое, (*past part pass*) of **глáд·ить**

глáж·ен·ый, -ая, -ое, (*a*) ironed, smoothed

глáж·ень·е, -я, (*n*) ironing, pressing, smoothing

глаз·, -а, (*nom pl*) **-á,** (*g pl*) **глаз·,** (*d pl*) **-áм,** (*m*) eye

~ бýр·и (meteor) vortex, eye of storm

~, во·оруж·ённ·ый aided eye

~, не·во·оруж·ённ·ый naked/unaided eye

~, прост·óй (anat) ocellus

~, с·лóж·н·ый (ent) compound eye

~ фýрм·ы tuyere nose (of blast furnace)

глазéт·, -а, (*m*) silk brocade

глаз·ировáни·е, -я, (*n*) (*v n*) *see* **глаз·ир·овáть**

глаз·ирóванн·ый, -ая, -ое, (*past part pass*) of **глаз·ир·овáть**

глаз·ир·овáть, -ýют, (*imp and perf*) glaze; ice (with sugar), candy (fruit)

глаз·ирóвк·а, -и, (*g pl*) **-вок,** (*f*) (*v n*) of **глаз·ир·овáть**

глаз·ни́ц·а, -ы, (*i*) **-ей,** (*f*) (anat) orbit, eye socket

глаз·н·óй, -áя, -óе, (*a*) eye, ocular, ophthalmic, optic

~ плáст·ик·а (*f*) (zool) ocular plate

глазо·дви́г·ательн·ый, -ая, -ое, (*a*) (anat) oculomotor

глаз·óк·, -з·к·á, (*m*) (*dim*) of **глаз·;** eyelet; peephole; (zool) ocellus

~, волоч·и́льн·ый nib (of wiredrawing die), die insert

~, смотр·ов·óй sight glass

глазо·мéр·, -а, (*m*) estimation by sight, visual estimation

глазо·мéр·н·ый, -ая, -ое, (*a*) by eye, visual (of measurement)

глазо·обрáз·н·ый, -ая, -ое, (*a*) oculiform

глаз·ýнь·я, -и, (*g pl*) **-ни·й,** (*f*) fried eggs

глазур·овáть·, -ýют, (*imp and perf*) glaze

глазур·óванн·ый, -ая, -ое, (*past part pass*) of **глазур·овáть;** glazed (surface)

глазýр·ь, -и, (*f*) (ceram) glaze; icing (of sugar); (food) glazing syrup

-глáз·ый, -ая, -ое, (*adj component*) -eyed

глáнд·а, -ы, (*f*) (anat) gland, (*see also* **желез·á**); (pop) tonsil, *see* **мин·дáл·ин·а**

глáсис·, -а, (*m*) (arch) glacis

глас·и́ть, -я́т, (*imp*) state, read, run, say (of documents)

глáс·н·ый, -ая, -ое, (*a*) public, open; vowel; (*m decl as adj*) vowel

глаубери́т·, -а, (*m*) (min) glauberite

глáубер·ов·а сол·ь (*f*) (min) Glauber salt, mirabilite

глаук·óм·а, -ы, (*f*) (med) glaucoma

глаукодóт·, -а, (*m*) (min) glaucodot(e)

глаукон·и́т·, -а, (*m*) (min) glauconite

глауко·фáн·, -а, (*m*) (min) glaucophane

глéб·а, -ы, (*f*) (bot) gleba

глé·ев·ый, -ая, -ое, (*a*) gley

глé·й, -я, (*m*) gley (soil)

глёт·, -а,   (*m*) litharge

~, свин·цо́в·ый   litharge

гле́тчер·, -а,   (*m*) mountain/Alpine glacier

глиади́н·, -а,   (*m*) (biochem) gliadin

глик·ем·и́·я, -и,   (*f*) (med) glycaemia

глико·ге́н·, -а,   (*m*) (chem) glycogen, animal starch

гликоз·ами́н·, -а,   (*m*) (biochem) glycosamine

гликоз·и́д·, -а,   (*m*) glycoside

гликоз·ур·и́·я, -и,   (*f*) (med) glucosuria, glycosuria

глико·ко́лл·, -а,   (*m*) (chem) glycocoll

глико·ко́л·, -а,   (*m*) glycocoll, glycine

глик·о́л·, -а,   (*m*) glycol

глико́л·ев·ый, -ая, -ое,   (*a*) glycol, glycolic

~ альдеги́д·,   (*m*) glycol aldehyde

~ кисл·от·а́   (*f*) glycolic acid

гликол·и́д·, -а,   (*m*) glycolide

глико́·лиз·, -а,   (*m*) (chem) glycolysis

глико·лит·и́ческ·ий, -ая, -ое,   (*a*) (biol) glycolitic

глик·о́л·ь, -я,   (*m*) (chem) glycol

глико·хо́л·ев·ая кисл·от·а́   (*f*) glycocholic acid

глин·а, -ы,   (*f*) clay

~, бе́л·ая   kaolin, china clay

~, кирпи́ч·н·ая   brick clay/earth

~, ко́м·ов·ая   ball clay

~, купоро́с·н·ая   alum earth

~, легко·пла́в·к·ий   plastic clay

~, ле́нт·очн·ая   varved clay

~, огне·у·по́р·н·ая   fire-clay, refractory clay

~, пласт·и́чн·ая   ball clay

~, под·стил·а́ющ·ая   (min) underclay

~, сукно·ва́ль·н·ая   fuller's earth

~, туго·пла́в·к·ая   semi-refractory plastic clay

~, фарфо́р·ов·ая   kaolin, china clay, porcelain clay

~, форм·о́вочн·ая   (met cast) clay binder

глин·иза́ци·я, -и,   (*f*) (geol) argillization; (oil) mudding off

гли́н·ист·о-желе́з·ист·ый, -ая, -ое,   (*a*) (min) argillo-ferruginous

~ -пес·ча́н·ый, -ая, -ое,   (*a*) (min) argillo-arenaceous

гли́н·истост·ь, -и,   (*f*) clayiness

гли́н·ист·ый, -ая, -ое,   (*a*) clay, clayey, argillaceous

~ рас·тво́р·,   (*m*) (oil) drilling fluid, mud

глин·и́т·, -а,   (*m*) (build) burnt clay, burnt ballast

глино·бето́н·, -а,   (*m*) (build) clay puddle

глин·ова́ни·е, -я,   (*n*) claying

глино·за·во́д·, -а,   (*m*) (oil) mud plant

глино·зём·, -а,   (*m*) (min) alumina, aluminium oxide

глино·зём·ист·ый, -ая, -ое,   (*a*) aluminous; (geol) aluminiferous

~ цеме́нт·,   (*m*) aluminous cement

глино·зём·н·ый, -ая, -ое,   (*a*) aluminous

глино·меш·а́лк·а, -и,   (*g pl*) -лок·,   (*f*) (agr) clay mixer; (oil) mud mixer

глино·мя́·лк·а, -и,   (*g pl*) -лок·,   (*f*) pug mill

гли́нт·, -а,   (*m*) (geol) clint, grike

глинтве́йн·, -а,   (*m*) mulled wine

гли́н·ян·ый, -ая, -ое,   (*a*) clay; earthenware

~ ма́сс·а   (*f*) (ceram) clay slip

~ рас·тво́р·,   (*m*) (oil) drilling fluid, mud

гли·окса́л·ь, -я,   (*m*) (chem) glyoxal

гли·окса́л·ев·ая кисл·от·а́   (*f*) glyoxalic/glyoxylic acid

гли·оксал·и́н·, -а,   (*m*) glyoxaline

глипта́л·ь, -я,   (*m*) *see* **глифта́л·ь**

глипт·одо́нт·, -а,   (*m*) (pal) glyptodont

глипто·ли́т·, -а,   (*m*) (min) glyptolith

глипто·морф·о́з·, -а,   (*m*) (geol) crystal mould

глисса́д·а, -ы,   (*f*) (air) glide path; (nav) glide slope

глисса́д·н·ый, -ая, -ое,   (*a*) of **глисса́д·а**

~ радио·при·ём·ник·   (*m*) (air nav) glide-slope receiver

~ у·стро́й·ств·о   (*n*) glide-slope facility

глисса́ж·н·ый, -ая, -ое,   (*a*) glide, slide

гли́ссер·, -а,   (*m*) (shipb) hydroplane, skimming vessel

глисс·и́ровани·е, -я,   (*n*) hydroplaning

глисс·и́рующ·ий, -ая, -ее,   (*pres part act*); hydroplaning

~ по·ве́рх·ност·ь   (*f*) (shipb) hydroplane (surface)

гли́ст·, -а́,   (*m*) (zool) intestinal worm, helminth

~ ов·е́ц·,   (zool) fluke

глист·а́, -ы́,   (*f*) *see* **гли́ст·**

глисто·го́н·н·ый, -ая, -ое,   (*a*) (med) vermifuge, vermifugal; (*n decl as adj*) (pharm) vermifuge, anthelminthic

глисто·го́н·н·ый, -ая, -ое
~ сре́д·ств·о   (*n*) (pharm) vermifuge, anthelmintic, anthelminthic
глисто·у·бив·а́ющ·ий, -ая, -ее,   (*a*) vermicidal
глифта́л·ев·ый, -ая, -ое,   (*a*) glyptal
~ смол·а́   (*f*) glyptal resin (plastics)
глифта́л·ь, -я,   (*m*) glyptal
глицер·и́д·, -а,   (*m*) (chem) glyceride
глицер·и́н·, -а,   (*m*) (chem) glycerin(e), glycerol
глицери́н·ов·ый, -ая, -ое,   (*a*) glycerine, glycerol; (chem) glyceric
~ альдеги́д·,   (*m*) glyceraldehyde
~ вод·а́   (*f*) glycerine sweet water
~ кисл·от·а́   (*f*) glyceric acid, $\alpha\beta$-dioxypropionic acid
глицерино·фосфа́т·, -а,   (*m*) *see* глицеро·фосфа́т·
глицери́н·тринитра́т·, -а,   (*m*) nitroglycerine
глицер·о́з·а, -ы,   (*f*) glycerose
глицер·о́л·ь, -я,   (*m*) (pharm) glycerol
глицеро·фосфа́т·, -а,   (*m*) (pharm) glycerophosphate
глицеро·фосфа́т·н·ая кисл·от·а́   (*f*) glycerophosphoric acid
глицеро·фосфе́н·, -а,   (*m*) (pharm) glycerophosphene
глиц·и́д·, -а,   (*m*) glycide
глиц·и́д·н·ый, -ая, -ое,   (*a*) (chem) glycidic
глиц·и́н·, -а,   (*m*) (chem) glycine, aminacetic acid, glycocole, paraoxyaminocetic acid
глици́н·и·й, -я,   (*m*) (chem) glucinium, Gl, beryllium, Be
глици́н·и·я, -и,   (*f*) (bot) wistaria, *Wistaria*
гли́·я, -и,   (*f*) (zool) glia
глоб·а́льн·ый, -ая, -ое,   (*a*) global
глоба́р·а, штифт·   (*m*) globar (infrared radiation source)
глобигери́н·а, -ы,   (*f*) (ocean) globigerina
глобигери́н·ов·ый, -ая, -ое,   (*a*) globigerinal
глоб·и́н·, -а,   (*m*) (biol) globin
глоб·о́ид·, -а,   (*m*) (bot) globoid
глоб·оид·а́льн·ый, -ая, -ое,   (*a*) globoid, globoidal
~ за·цепл·е́ни·е   (*n*) double-throated worm mating, cone drive worm mating
~ черв·я́к·   (*m*) Hindley screw
глоб·о́ид·н·ый, -ая, -ое,   (*a*) *see* глоб·оид·а́льн·ый

глоб·ул·а, -ы,   (*f*) globule
глоб·ули́н·, -а,   (*m*) (chem) globulin
глоб·ули́т·, -а,   (*m*) (geol) globulite
глоб·уля́рн·ый, -ая, -ое,   (*a*) globular
гло́б·ус·, -а,   (*m*) globe
~ звёзд·н·ого не́б·а   celestial globe
Гло́вер·а, ба́шн·я   (*f*) (chem) Glover tower
~, кисл·от·а́   (*f*) (chem) Glover acid, tower acid
глод·а́ть   (*pres 3rd sing, pl*) гло́ж·ет, гло́ж·ут, (*imp*); gnaw, nibble
гло́ж·ущ·ий, -ая, -ее, (*pres part act*) of глод·а́ть
гломеро·гранули́т·ов·ый, -ая, -ое, (*a*) (geol) glomerogranular
гломеро·кристалл·и́ческ·ий,   -ая, -ое,   (*a*) (geol) glomerocrystalline
гломеруло·нефри́т·, -а,   (*m*) (med) glomerulonephritis
гло́мус·, -а,   (*m*) (anat) glomus
глори·я, -и,   (*f*) (meteor) glory
глосс·а́ри·й, -я,   (*m*) glossary; (zool) glossarium
глосс·и́т·, -а,   (*m*) (med) glossitis
глоссо·па́т·и·я, -и,   (*f*) (med) glossopathy
глоссо·рраф·и·я, -и,   (*f*) (med) glossorrhaphy
глоссо·те́к·а, -и,   (*f*) (zool) glossotheca
глот·а́ни·е, -я,   (*n*) (*v n*) *see* глот·а́·ть; deglutition, ingestion
глот·а́тельн·ый, -ая, -ое,   (*a*) swallowing, deglutitory, deglutitive
глот·а́·ть, -ют,   (*imp*) swallow, gulp
гло́т·к·а, -и,   (*g pl*) -т·ок·,   (*f*) (anat) pharynx; (pop) gullet, throat
глот·о́к·, -т·к·а́,   (*m*) swallow; mouthfull, sip, gulp
гло́т·очн·ый, -ая, -ое,   (*a*) (zool) pharyngeal
глохи́д·и·й, -я,   (*m*) (biol) glochidium
гло́х·нуть, -нут,   (*past masc sing*) глох,   (*imp*) become/grow deaf; subside, grow faint, die away (of sound); grow wild, run to seed (of plants)
гло́х·ш·ий, -ая, -ее,   (*past part act*) of гло́х·нуть
глу́б·же   (*comp*)   of   глуб·о́к·ий   deeper profounder
глуб·ин·а́, -ы́,   (*nom pl*) -и́н·ы, (*g pl*) -и́н·,   (*f*) depth; profundity; degree, extent
~ в·ру́б·а   depth of cut, cutting depth
~ вс·па́ш·к·и   draught (ploughing)

глуб·ин·а́
~ вы·гор·а́ния·   (nucl) burn-up fraction, burning depth
~, высоко·температу́р·н·ая   (geol) hot interior
~ горизо́нт·а   dip of the horizon
~ конве́рс·и·и   degree of conversion
~ модул·я́ци·й   (rad) modulation percentage/depth
~, от·ве́с·н·ая   vertical depth
~ полимериз·а́ци·и   degree of polymerization
~ скин-сло́·я   (elec) skin depth
~ фо́кус·а   (phot, opt) depth of focus
глуб·и́нн·ый, -ая, -ое,   (a) depth; deep-seated, (geol) hypogene, bathygenic, plutonic, (ocean) abyssal, abysmal
~ бо́мб·а   (f) (mil) depth charge
~ до́з·а   (f) (med) depth dose
~ на·со́с·   (m) deep-well pump
~ по·ро́д·а   (f) (geol) abyssal rock, deep-seated rock, plutonic rock
~, по·сто́·янн·о   (zool) holobenthic
глубино·ме́р·, -а,   (m) depth gauge
глуб·о́к·ий, -ая, -ое,   (a) deep, profound
~ о·хлажд·е́ни·е   (n) deep freezing
~ печа́т·ь   (f) (print) intaglio
глубоко·во́д·н·ый, -ая, -ое,   (a) deep-water/sea; (zool) bathophilous, bathybic; (geol) bathypelagic, (ocean) abyssal
глубоко·за·лег·а́ющ·ий, -ая, -ее,   (a) deep-seated
глубоко·корн·ев·о́й, -а́я, -о́е,   (a) (bot) deep-rooting
глубоко·мы́сл·енн·ый, -ая, -ое,   (a) profound, penetrating, deep-thinking
глубо́ко·уваж·а́ем·ый, -ая, -ое,   (a) dear (formal correspondence)
глубо·ме́р·, -а,   (m) (ocean) barhometer, bathymeter
глук·уро́н·ов·ая кисл·от·а́   (f) (chem) glucuronic acid
глуп·е́·ть, -ют,   (imp) become/grow stupid/foolish
глу́п·ост·ь, -и,   (f) foolishness, stupidity; nonsense
глу́п·ый, -ая, -ое,   (a) foolish, stupid, silly
глуп·ы́ш·, -а́,   (i) -о́м,   (m) (zool) fulmar, *Fulmarus glacialis*
глут-   (*component*) = **глют-** (q.v.)
глут·ами́н·, -а,   (m) see **глют·ами́н·**
глут·ами́н·ов·ая кисл·от·а́   (f) glutamic acid

глут·а́ров·ый, -ая, -ое,   (a) (chem) glutaric
~ кисл·от·а́   (f) glutaric acid
глута·ти·о́н·, -а,   (m) (chem) glutathione
глуха́р·ь, -я́,   (m) coach screw; (zool) glass snake, *Ophisaurus apodus*; capercailye, capercaillie
~, обык·нове́нн·ый   (zool) capercaillie, *Tetrao urogallus*
глухо·до́н·н·ый, -ая, -ое,   (a) closed-bottom
глухо·за·зёмл·енн·ый, -ая, -ое,   (a) (elec) directly earthed/grounded
глух·о́й, -а́я, -о́е,   (a) deaf; toneless; fully enclosed/shut in, blind, blank; remote, solitary, lonely; overgrown, wild, dense; slack, dead (of activity)
~ арка́д·а   (f) (build) blind arcade
~ кон·е́ц·   (m) dead end
~ пере·сеч·е́ни·е   (n) (rail) fixed crossing
~ рето́рт·а   (f) stop-end retort
~ со·гла́с·н·ый   (m) (ling) voiceless consonant
глухо·нем·о́й, -а́я, -о́е,   (a) deaf-mute, surdo-mute
глух·от·а́, -ы́,   (f) deafness, surdity
глуш·е́ни·е, -я,   (n) (v n) see **глу·ш·и́ть**
глуш·и́тел·ь, -я,   (m) silencer, muffler, damper, suppressor; (rad) jammer; opacifier (glass)
~ вы́·хлоп·а   exhaust silencer
~ по́·иск·а   (rad) search jammer
~ шу́м·а   sound suppressor
глуш·и́ть, -а́т,   (imp) make deaf, prevent hearing; silence, damp, drown, jam (sounds etc.); switch/turn off, extinguish; choke, stifle (plants etc.)
глу́ш·ь, -и́,   (f) thicket
глы́б·а, -ы,   (f) block, lump, chunk
~, земл·ян·а́я   clod
~, лед·ян·а́я   ice block; small iceberg (at sea)
глыб·и́ст·ый, -ая, -ое,   (a) lumpy, chunky, clody
глы́б·к·а, -и,   (f) (dim) of **глы́б·а**
глы́б·ов·ый, -ая, -ое,   (a) block, in blocks, en bloc
глыбо·дро́б·, -а,   (m) (agr) clod breaker
глюко·гепто́н·ов·ая кисл·от·а́   (f) glucoheptonic acid
глюко́з·а, -ы,   (f) (chem) glucose, grape sugar, dextrose
глюкоз·ами́н·, -а,   (m) (biochem) glucosamine
глюкоз·а́н·, -а,   (m) glucosan

глюкоз·е́н·, -а,   (*m*) glucosene
глюкоз·и́д·, -а,   (*m*) glucoside
глюкоз·ида́з·а, -ы,   (*f*) glucosidase
глюкон·а́т·, -а,   (*m*) gluconate
глюкон·ов·ый, -ая, -ое,   (*a*) gluconic, glucate of
~ кисл·от·а́   (*f*) gluconic acid
глюко·проте·и́д·, -а,   (*m*) gluco-protein, glycoprotein
глюко·проте·и́н·, -а,   (*m*) (chem) glucoprotein, glycoprotein
глюко·фо́р·, -а,   (*m*) glucophore
глюк·уро́н·ов·ая   кисл·от·а́   (*f*) glucuronic acid
глют-   (*component*) = глут-   (q.v.)
глют·ами́н·, -а,   (*m*) (chem) glutamine
глют·амина́з·а, -ы,   (*f*) (biochem) glutaminase
глют·ами́н·ов·ая   кисл·от·а́   (*f*) glutamic acid
глюта·ти·о́н·, -а,   (*m*) (chem) glutathione
глют·ели́н·, -а,   (*m*) (chem) glutelin
глют·е́н·, -а,   (*m*) (food) gluten
~, кукуру́з·н·ый   corn gluten
глют·ени́н·, -а,   (*m*) glutenin
гляд·е́лк·а, -и,   (*g pl*) -лок·,   (*f*) sight hole, inspection hole
гляд·е́ть, -я́т,   (*imp*) look at; show from behind/under, peep out; face, look out/onto/towards; look like, resemble; look after, see to
гляж·у́   (*pres 1st sing*) of гляд·е́ть
глян·ец·, -н·ц·а,   (*m*) gloss, lustre, polish
гля́·н·уть, -ут,   (*perf*) throw/cast a glance/look at
глянц·еви́т·ый, -ая, -ое,   (*a*) glossy, lustrous
гля́нц·ев·ый, -ая, -ое,   (*a*) glossy, lustrous, shiny; polishing
глянц·о́ван·ый, -ая, -ое,   (*a*) polished, glossed
гляци·а́льн·ый, -ая, -ое,   (*a*) glacial
гляци·ге́н·н·ый, -ая, -ое,   (*a*) (geol) glacigenous
гляцио·аллюви·а́льн·ый, -ая, -ое, (*a*) (geol) glacio-alluvial
гляцио·ло́г·и·я, -и,   (*f*) glaciology
гмелин·и́т·, -а,   (*m*) (min) gmelinite
ГМК   (*abbr*) of гидрази́д· малеи́н·ов·ой кисл·от·ы́;   (chem) maleic hydrazide
г-моль   (*abbr*) of грамм-моль; gram-molecule, mole
ГМТ   (*abbr*) = гекса·метилин·тетра·ми́н·   (chem) hexamethylenetetramine

гн   (*abbr*) of ге́нри;   (elec) henry
гн·а́вш·ий, -ая, -ее,   (*past part act*) of гн·ать
гн·а́нн·ый   (*past part pass*) of гн·ать
гнат·и́т·, -а,   (*m*) (med) gnathitis
гнато·со́м·а, -ы,   (*f*) (zool) gnathosome
гнато·те́к·а, -и,   (*f*) (zool) gnathotheca
гн·ать   (*pres 3rd sing, pl*) го́н·ит, го́н·ят,   (*imp deter*); drive, drive on/out, urge on; chase, pursue, drive off; distil; race
гнев·, -а,   (*m*) anger
гне́в·н·ый, -ая, -ое,   (*a*) angry, irate
гнед·о́й, -а́я, -о́е,   (*a*) bay (coloured)
гнёзд·ност·ь, -и,   (*f*) number of cavities (of a plastics mould)
гнезд·о́, -а́,   (*nom pl*) гнёзд·а   (*g pl*) гнёзд,   (*n*) nest, pocket, socket, recess, housing (e.g. of beam); (bact) nidus; jack (of telephone switchboard); cavity (in moulding/molding etc.); box (of type case); (ling) family
~, за·мо́ч·н·ое   (gunn) breech bush
~ кла́пан·а   valve seat
~, клин·ов·о́е   (gunn) breech mortice
~, паралле́ль·н·ое   branching jack (telephones)
~, по·след·ова́тельн·ое   break jack (telephones)
~, ру́д·н·ое   (geol) ore pocket
~, семен·н·о́е   (bot) medulla, pith
~ син·хрон·иза́ци·и   lock-out jack (telephones)
~ шип·а   mortise (carpentry)
гнезд·ова́ни·е, -я,   (*n*) (zool) nesting; (bact etc.) nidification
гнезд·ов·о́й, -а́я, -о́е,   (*a*) of гнезд·о́
~ по·се́в·   (*m*) (agr) hole planting/sewing
гнезио·га́м·и·я, -и,   (*f*) (gen) gnesiogamy
гне́йс·, -а,   (*m*) (geol) gneiss
~, с·ло́ж·н·ый   migmatite gneiss
~, сре́д·н·ий   amphoteric gneiss
~, шест·ова́т·ый   dendricit gneiss
гне́йс·ов·ый, -ая, -ое,   (*a*) (geol) gneissic, gneissose
гнес·ти́   (*pres 3rd sing, pl*) гнет·ёт, гнет·у́т, (*imp*); oppress, weigh upon
гнёт·, -а,   (*m*) press; pressure, weight; oppression
гнет·у́щ·ий, -ая, -ее,   (*a*) (*pres part act*) of гнес·ти́
гни́·вш·ий, -ая, -ее,   (*a*) (*past part act*) of гни·ть; rotten, decayed, putrified, decomposed
гни́д·а, -ы,   (*f*) (ent) nit

гни·éни·е, -я, (*n*) (*v n*) of **гни·ть;** rotting, decay, decomposition, putrefaction, putrescence

гни·л·óй, -áя, -óе, (*a*) rotten, putrid, putrescent, decayed, decomposed, carious

гнило·крóв·и·е, -я, (*n*) septicaemia, blood poisoning

гнил·остност·ь, -и, (*f*) putrefactiveness

гнил·остн·ый, -ая, -ое, (*a*) putrefactive, saprogenic

гнил·ост·ь, -и, (*f*) decay, putrefaction, rottenness

гнил·ýшк·а, -и, (*g pl*) **-шек·,** (*f*) punk, piece of rotten wood

гнил·ь, -и, (*f*) rot, mould, mold; rotten stuff, rot

~, верш·и́нн·ая (hort) blossom-end rot

~, кóр·ков·ая rind-rot (of cheese)

гни·ть, -ют, (*imp*) rot, decay, putrefy, decompose

гни·ющ·ий, -ая, -ее, (*a*) (*pres part act*) of **гни·ть**

гное·ви́д·н·ый, -ая, -ое, (*a*) puriform, puruloid

гное·крóв·и·е, -я, (*n*) (med) pyaemia

гное·рóд·н·ый, -ая, -ое, (*a*) (med) pyogenic

гное·теч·éни·е, -я, (*n*) (med) pyorrhea, suppuration

гное·точ·и́в·ый, -ая, -ое, (*a*) suppurative

гно·и́ть, -я́т, (*imp*) rot, let rot; **-ся** suppurate, fester

гнó·й, -я, (*m*) pus, ichor

гной·ни́к·, -á, (*m*) abscess

гнóй·н·ый, -ая, -ое, (*a*) suppurative, purulent

гномон·и́ческ·ий, -ая, -ое, (*a*) gnomonic

~ проéкци·я (*f*) (surv, cryst) gnomonic projection

гно·я́щ·ий, -ая, -ее, (*pres part act*) of **гно·и́ть; -ся** festering, purulent

гнýс·, -а, (*m*) gnat

гнус·áв·ый, -ая, -ое, (*a*) nasal

гнýс·н·ый, -ая, -ое, (*a*) infamous, vile

гн·ýт·ый, -ая, -ое, (*past part pass*) of **гн·уть;** curved

гн·уть, -ут, (*imp*) bend, bow, make curved

гн·утл·ё, -я, (*n*) bending, forming (e.g. wood)

гоаци́н·, -а, (*m*) (zool) hoatzin

г-обрáз·н·ый, -ая, -ое, (*a*) L-shaped

~ антéнн·а (*f*) (rad) inverted-L antenna

говарди́т·, -а, (*m*) (min) howardite

гóвор·, -а, (*m*) sound of talking/voices; accent, pronunciation; (ling) dialect

говор·и́ть, -я́т, (*imp*) say, speak, talk, tell

не говор·я́ not to mention, to say nothing of

это говор·и́т о том this means/shows that

говор·ли́в·ый, -ая, -ое, (*a*) talkative, garrulous

говóр·н·ый, -ая, -ое, (*a*) talking, for talking/speaking

~ рас·трýб· (*m*) mouthpiece (telephones)

говор·я́щ·ий, -ая, -ее, (*pres part act*) of **говор·и́ть**

~ автомáт· (*m*) automatic telephone operator

говя́д·ин·а, -ы, (*f*) beef

говя́ж·ий, -ья, -ье, (*a*) beef

год·, -а, (*nom pl*) **гóд·ы** or **год·á,** (*g pl*) **-óв,** (*m*) year

~, бéссел·ев· Besselian year

~, високóс·н·ый leap year

~, гидро·лог·и́ческ·ий water year

~, грегориáн·ск·ий Gregorian year

~, звёзд·н·ый sidereal year

~, лýн·н·ый lunar year

Между·на·рóд·н·ый Гео·физ·и́·ческ·ий Год· (МГГ) International Geophysical Year (I.G.Y.)

~, не·пóлн·ый deficient year

~, от·чёт·н·ый fiscal/financial year

~, прост·óй common year (365 days)

~, свет·ов·óй (astron) light year

~, сидер·и́ческ·ий sidereal year

~, сóлн·ечн·ый solar year

~, троп·и́ческ·ий tropical year

~, уч·éбн·ый academic year (autumn–summer)

~, фикти́в·н·ый fictitious year

~, юлиáн·ск·ий Julian year

год·и́ться, -я́тся, (*imp*) be suitable for, suit, serve, fit, do

год·и́чн·ый, -ая, -ое, (*a*) year's; annual, yearly

~ кольц·ó (*n*) annual ring (of tree)

~ слó·й (*m*) annual ring (of tree)

гóд·ност·ь, -и, (*f*) fitness, suitability, serviceability; validity (of document)

~ к плáв·ани·ю seaworthiness

~ к по·лёт·ам airworthiness

гóд·н·ый, -ая, -ое, (*a*) fit, suitable, serviceable, sound; valid

год·ови́к·, -á, (*m*) yearling

год·ов·óй, -áя, -óе, (*a*) annual, yearly

год·овщи́н·а, -ы, (*f*) anniversary

годо́·граф·, -а, (*m*) hodograph, travel-time curve

~, вертика́ль·н·ый time-depth curve

~, на·гон·я́ющ·ий overtaking travel-time curve

~, обра́т·н·ый inverse travel-time curve

~, с·во́д·н·ый resulting travel-time curve

годо·ско́п·, -а, (*m*) (nucl) hodoscope

го́йм·а, -ы, (*f*) (min) prop, pit prop

гола́вл·ь, -я, (*m*) (fish) chub, *Leuciscus cephalus*

гол·аркт·и́ческ·ая о́бласт·ь (*f*) (zool) holarctic region

голен·а́ст·ый, -ая, -ое, (*a*) long-legged; (*pl noun decl as adj*) (zool) *Ciconiiformes*

голен·и́щ·е, -а, (*n*) upper (of boot)

го́лен·ь, -и, (*f*) shin, shank, (anat) crus

гол·е́ц·, -ль·ц·а́, (*i*) **-ль·ц·о́м,** (*m*) (fish) char, *Salvelinus*; (geol) bald mountain/peak

Голиа́ф·а, патро́н· (*m*) (elec) goliath Edison screw cap

голи́к·, -а́, (*m*) broom, besom; (naut) brush

голла́ндер·, -а, (*m*) (paper) hollander

голла́нд·ск·ий, -ая, -ое, (*a*) Dutch, Netherlands'

~ череп·и́ц·а (*f*) (build) pantile

го́лм·и·й, -я, (*m*) *see* **го́льм·и·й**

голо·бе́нтос·, -а, (*m*) (ocean) holobenthos

голо·бласт·и́ческ·ий, -ая, -ое, (*a*) (zool) holoblastic

голов·а́, -ы́, (*acc*) **го́лов·у,** (*nom pl*) **го́лов·ы,** (*see also* **го́лов·к·а**) (*f*); head; (anat) caput; (zool) cephalon

~ са́хар·у sugar loaf

~ сы́р·у a cheese

голов·а́стик·, -а, (*m*) (zool) tadpole

голов·а́ст·ый, -ая, -ое, (*a*) large-headed

голо́в·к·а·, -и, (*g pl*) **-в·ок·,** (*f*) (*dim*) of **голов·а́;** head; (anat) capitellum, caput; (bot) capitulum; (mech eng) head, cap, end, top; port (of open hearth furnace); (chem) fraction, cut; vamp (of shoe); (horol) winding square

~, абрази́в·н·ая abrasive tip

~, бо·ев·а́я (mil) warhead

~, больш·а́я (M/T) big end

~, ве́рх·н·яя (M/T) small end

голо́в·к·а

~, ви́ль·чат·ая double-top end (of connecting rod)

~, вос·про·из·вод·я́щ·ая playback/reproducing head (sound recorder)

~, грануля́тор·н·ая (plast) pelletizing head

~, двух·кана́ль·н·ая double-gap head (of tape recorder)

~, дел·и́тельн·ая (mech) dividing head

~ дистилла́т·а (chem) top fraction/cut

~, за́д·н·яя back header (boiler)

~, за·жи́м·н·ая clamping head; bending head (of bending press)

~, за·кла́д·н·а́я rivet head

~, за·мык·а́ющ·ая river point, closing head

~ за́·пис·и-чт·е́ни·я read-write head (of tape recorder)

~, за·пи́с·ывающ·е-вос·про·из·вод·-я́щ·ая read-write head (of tape recorder)

~, за·пи́с·ывающ·ая magnetic cutter (disc recording); record head (of tape recorder)

~, из·мер·и́тельн·ая (instr) measuring head

~, калориза́тор·н·ая combustion torch, hot bulb (of semi-diesel)

~, ко́д·ов·ая (electr) information head (of magnetic drum)

~, кон·и́ческ·ая pan-head (of rivet)

~, криво·ши́п·н·ая (mech eng) big end, connecting-rod end, crank-pin end

~ лу́к·а onion

~, марке́р·н·ая index head (of magnetic drum)

~, ни́ж·н·яя (mech eng) big end

~, пере́д·н·яя front header (boiler)

~, полу·кру́гл·ая half round, button head (screw); snap head (rivet)

~, полу·по·та́й·н·а́я raised ˙ counter-sunk head (rivet), fillister head (screw)

~, по·та́й·н·а́я flat countersunk head (rivet)

~, раз·ры́хл·и́тельн·ая suction-dredger cutter head

~, разъ·ё́м·н·ая split-type bearing

~, рас·пыл·и́тельн·ая (rocket) injection assembly

~, револьве́р·н·ая turret head (lathe etc.)

~, резьбо·на·ка́т·н·ая thread-rolling attachment (to lathe)

голо́в·к·а

~, с·ва́р·очн·ая welding blowpipe

~, семен·н·а́я seed pod, boll

~ с·тир·а́ни·я erase head (of recorder)

~, с·тир·а́ющ·ая erase head (of tape recorder)

~, с·чи́т·ывающ·ая reproduce head (of tape recorder)

~ сы́р·а a cheese

 у́гол· голо́в·к·и (*m*) addendum angle (gear teeth)

~, цемент·и́рующ·ая cementing head (for boreholes)

~, цилиндр·и́ческ·ая cheese head (screw)

~ цили́ндр·а (ICE) cylinder head

~ чеснок·а́ garlic clove

голов·н·о́й, -а́я, -о́е, (*a*) head, cephalic; leading, advanced, prototype; head, end, nose

~ вз·рыв·а́тел·ь (*m*) (amm) nose fuse

~ за·во́д· (*m*) pilot production plant

~ образ·е́ц· (*m*) prototype

~ о·хран·е́ни·е (*n*) (mil) advance guard

~ под·ши́п·ник· (*m*) (mech eng) small-end bearing

~ при́зм·а (*f*) (opt) head prism

~ сери́·йн·ый само·лёт· (*m*) first-production aircraft (after prototypes etc.)

~ телефо́н· (*m*) head phone

~ у́зел· со·оруж·е́ни·й (*m*) (hydr eng) headworks

~ част·ь (*f*) top, top end/part; (mil) warhead

голов·н·я́, -й, (*g pl*) **-е́й,** (*f*) (agr) smut, blight; charred log

голово·гру́д·ь, -и, (*f*) (zool) cephalo-thorax

голово·круж·е́ни·е, -я, (*n*) giddiness, (med) vertigo

голово·но́г·ие, -их, (*pl decl as adj*) (zool) cephalopods

голово·хо́рд·ов·ые, -ых, (*pl decl as adj*) (zool) cephalochords

голо́в·чат·ый, -ая, -ое, (*a*) bulbous; (bot) capitate; (zool) cephalated

голо·гаме́т·а, -ы, (*f*) (gen) hologamete

голо·га́м·и·я, -и, (*f*) (gen) hologamy

го́лод·, -а, (*m*) hunger, (med) fames; famine

голод·а́ни·е, -я, (*n*) starvation, fasting, malnutrition; deficiency

~, кисло·ро́д·н·ое oxygen deficiency

голод·а́·ть, -ют, (*imp*) starve, fast

голо́д·н·ый, -ая, -ое, (*a*) hungry

голо·жа́бер·н·ый, -ая, -ое, (*a*) (zool) nudibranchiate

голо·зёр·н·ый, -ая, -ое, (*a*) hulled (of grains, corn etc.)

голо·зима́з·а, -ы, (*f*) (biochem) holysymase

голо·кри́н·ов·ый, -ая, -ое, (*a*) (med) holocrine

голо·кристалл·и́ческ·ий, -ая, -ое, (*a*) holycrystalline

голо·лёд·, -а, (*m*) *see* **голо·лёд·иц·а**

голо·лёд·иц·а, -ы, (*i*) **-ей,** (*f*) ice crust, glazed frost

голо·мо́рф·н·ый, -ая, -ое, (*a*) (math) holomorphic

голо·но́г·ий, -ая, -ое, (*a*) bare-legged; (*pl decl as adj*) (zool) *Nudipeda*

голо·но́м·н·ый, -ая, -ое, (*a*) (mech) holonomic

~ систе́м·а (*f*) (mech) holonomic system

голо·пло́д·н·ый, -ая, -ое, (*a*) (bot) gymnocarpous

го́лос·, -а, (*nom pl*) **-а́,** (*g pl*) **-о́в,** (*m*) voice; vote, ballot; voice, part (music)

голо·семен·н·о́й, -а́я, -о́е, (*pl*) **-ые,** *see* **голо·сем·я́нн·ый**

голо·сем·я́нн·ый, -ая, -ое, (*a*) (bot) gymnospermous; (*pl decl as a*) *Gymnospermae*

голо·сло́в·н·ый, -ая, -ое, (*a*) unsubstantiated, unproven, unfounded

голос·ни́к·, -а́, (*m*) (arch, acous) resonator

голос·ова́ни·е, -я, (*n*) (*v n*) of **голо·с·ова́ть**

голос·ова́ть, -у́ют, (*imp*) vote; put to the vote, vote on

голос·ов·о́й, -а́я, -о́е, (*a*) vocal

~ щёл·ь (*f*) (anat) glottis

голо·сти́л·и·я, -и, (*f*) (fish) holostyly

голо·сто́м·н·ый, -ая, -ое, (*a*) (zool) holostomate, holostamatous

голо·ти́п·, -а, (*m*) (zool) holotype

голоту́р·и·я, -и, (*f*) (zool) sea cucumber, (*pl*) *Holothurioidea*

голо·ферме́нт·, -а, (*m*) (chem) holoenzyme

голо·фи́т·н·ый, -ая, -ое, (*a*) (biol) holophytic

голо·це́н·, -а, (*m*) (geol) Holocene

голо́·эдр·, -а, (*m*) (cryst) holohedron

голо·эдр·и́ческ·ий, -ая, -ое, (*a*) holohedral

голо·эдр·и·я, -и, *(f)* (cryst) holohedral forms, holohedry

го́лтел·ь, -я, *(m) see* **га́лтел·ь**

голт·о́вк·а, -и, *(f)* tumbling (cleaning with abrasive)

голуб·е́·ть, -ют, *(imp)* look/appear/show azure/blue; turn azure/blue

голуб·изн·а́, -ы́, *(f)* azure, sky/pale/light blue

голуб·и́к·а, -и, *(f)* (bot) blueberry, *Vaccinium*

голуб·и́н·ый, -ая, -ое, *(a)* of **го́луб·ь**; pigeon, dove

голуб·ни́ц·а, -ы, *(f)* = **голуб·и́к·а;** (naut) water course, limber hole

голуб·ова́т·ый, -ая, -ое, *(a)* pale bluish

голуб·о́й, -а́я, -о́е, *(a)* light/pale/sky blue, blue, azure

∼ кит· *(m)* (zool) blue whale, *Balaenoptera musculus*

∼, метиле́н·ов·ый *(chem)* methylene blue

го́луб·ь, -я, *(g pl)* **-е́й,** *(m)* pigeon, dove

го́л·ый, -ая, -ое, *(a)* *(s f)* **гол·,** **гол·а́, го́л·о** naked, bare, uncovered, exposed; (elec) uninsulated, bare

гол·ы́ш·, -а́, *(i)* **-о́м,** *(m)* pebble, scree, gravel

го́л·ь, -и, *(f)* nakedness, poverty

гол·ь·ё, -я́, *(n)* raw hide (leather)

Го́льджи, аппара́т· *(m)* (cyt) Golgi apparatus

го́льм·и·й, -я, *(m)* (chem) holmium, Ho

гольян·, -а, *(m)* (fish) minnow, *Phoxinus*

гомео·морф·и́зм·, -а, *(m)* (math) homeomorphism

гомео·мо́рф·н·ый, -ая, -ое, *(a)* (genet) homeomorphic

гомео·па́т·, -а, *(m)* homeopath

гомео·пат·и́ческ·ий, -ая, -ое, *(a)* homeopathic

гомео·па́т·и·я, -и, *(f)* (med) homeopathy

гомео·поля́р·ност·ь, -и, *(f)* (chem) homopolarity, covalence

гомео·поля́р·н·ый, -ая, -ое, *(a)* homeopolar, homopolar, covalent

гомео·ста́з·, -а, *(m)* (biol) homeostosis

гомео·тип·и́ческ·ий, -ая, -ое, *(a)* (biol) homeotypic

гомео·траусмат·и́ческ·ий, -ая, -ое, *(a)* (geol) homeothrausmatic

гоми *(n indecl)* (bot) Italian millet, *Setaria italica*

гоммо́з·, -а, *(m)* (bot) gummosis

гомо- *(component)* homo-

гомо·а́том·н·ый, -ая, -ое, *(a)* (chem) homoatomic

гомо·ген·е́з·, -а, *(m)* (biol) homogenesis

гомо·ген·иза́тор·, -а, *(m)* homogenizer, emulsifier (of liquids)

гомо·ген·иза́ци·я, -и, *(f)* homogenization, homogenizing

гомо·ген·изи́рованн·ый, -ая, -ое, *(past part pass)* homogenized

гомо·ге́н·ност·ь, -и, *(f)* homogeneity

гомо·ге́н·н·ый, -ая, -ое, *(a)* homogeneous

∼ реа́ктор· *(m)* (nucl) homogeneous reactor

гомо·зиго́т·а, -ы, *(f)* (gen) homozygote

гомо·зиго́т·н·ый, -ая, -ое, *(a)* (biol) homozygous

гомойо·те́рм·н·ый, -ая, -ое, *(a)* (zool) homothermous, homoiothermal, warm-blooded

гомо·клина́л·ь, -и, *(f)* (geol) homocline

гомо́·лог·, -а, *(m)* homologue; (chem) homologous compound

гомо·лог·и́ческ·ий, -ая, -ое, *(a)* homologous

∼ вари·а́ци·я *(f)* (bot) homologous variation

∼ ряд· *(m)* (chem) homologous series

гомо·лог·и́чн·ый, -ая, -ое, *(a)* homologous

гомо·ло́г·и·я, -и, *(f)* (biol) homology

гомо·морф·и́зм·, -а, *(m)* (biol) homomorphism

гомо·мо́рф·н·ый, -ая, -ое, *(a)* (cyt) homomorphic

гом·о́ним·, -а, *(m)* homonym

гомо·пикна́ль·н·ый, -ая, -ое, *(a)* (geol) homopycnal

гомо·пла́зи·я, -и, *(f)* (zool) homoplasy, homoplasia

гомо·полиме́р·, -а, *(m)* (chem) homopolymer

гомо·поля́р·н·ый, -ая, -ое, *(a)* (chem) homopolar

гомо·сина́псис·, -а, *(m)* (cyt) homosynapsis

гомо·та́ксис·, -а, *(m)* homotaxis

гомо·талл·и́ческ·ий, -ая, -ое, *(a)* (bot) homothallic

гомо·тет·и́чн·ый, -ая, -ое, *(a)* (math) homothetic

гомо·топ·и́ческ·ий, -ая, -ое, *(a)* (math) homotopic

гомо·ци́кл·, -а,   (*m*) (chem) homoatomic ring
гомо·цикл·и́чн·ый, -ая, -ое,   (*a*)
(chem) homocyclic
гомо·энергет·и́ческ·ий, -ая, -ое,   (*a*)
(dynam) homoenergetic
го́н·, -а,   (*m*) chase, pursuit; (agr) run,
pass; (zool) rut
гон·а́д·а, -ы,   (*f*) (gen) gonad
гон·адо·троп·и́н·, -ая, -ое,   (*a*) (gen)
gonadotropin
гон·адо·тро́п·н·ый, -ая, -ое,   (*a*)
gonadotropic
го́нг·, -а,   (*m*) gong
гондо́л·а, -ы,   (*f*) (rail) gondola; (a/c)
nacelle, housing; car (of airship)
гон·е́ц·, -н·ц·а́, (*i*) -н·ц·о́м, (*m*) express
messenger
гониати́т·, -а,   (*m*) (min) goniatite
гониди·а́льн·ый, -ая, -ое,   (*a*) (biol)
gonidial
гонидио·фи́лл·, -а,   (*m*) (bot) gonidio-
phyll
гонидио·фо́р·, -а,   (*m*) (bot) gonidio-
phore, conidiophore
гон·и́д·и·я, -и,   (*f*) (bot) gonidium
гонимо·бла́ст·, -а,   (*m*) (bot) gonimo-
blast
гон·и́м·ый, -ая, -ое,   (*a*) (*pres part
pass*) of гн·ать
гонио́·метр·, -а,   (*m*) goniometer
~, дву·кру́ж·н·ый   (cryst) two-circle
goniometer
~, одно·на·пра́вл·енн·ая   (rad)
azimuth-indicating goniometer
~, при·клад·н·о́й   (cryst) contact
goniometer
гонио·скоп·и́·я, -и,   (*f*) gonioscopy
го́н·ит   (*3rd sing pres*) of гн·ать
-го́н·и·я, -и,   (*f component*) -gony (reproduction, generation)
го́н·к·а, -и, (*g pl*) -н·ок·, (*f*) (*v n*) *see*
гон·я́·ть; hurry, haste; race, chase;
flotation, raft (timber)
~, греб·н·а́я   boat race
гоно·зо́ид·, -а,   (*m*) (zool) gonozooid
гон·о́к·, -н·к·а́,   (*m*) (text) picker
гоно·ко́кк·, -а,   (*m*) (bact) gonococcus
гонора́р·, -а,   (*m*) fee
~, а́втор·ск·ий   (print) royalties
гоно·ре́·я, -и,   (*f*) (med) gonorrhea
го́н·очн·ый, -ая, -ое,   (*a*) racing
го́нт·, -а,   (*m*) (build) shingles
гонт·ов·о́й, -а́я, -о́е,   (*a*) of гонт·;
(build) shingled
гонча́р·, -а́,   (*m*) (ceram) potter

гонча́р·н·ый, -ая, -ое,   (*a*) potter's;
earthenware
~ круг·,   (*m*) potter's wheel
~ из·де́л·и·я   (*pl*) pottery, earthenware
~ про·из·во́д·ств·о   (*n*) pottery,
ceramics
гонча́р·ств·о, -а,   (*m*) pottery
го́н·щик·, -а,   (*m*) racer
гон·я́·ть, -ют,   (*imp indet*) drive,
drive on/out, urge on; (chem) distil;
race (of engine revolutions); -ся chase
after, pursue
гон·я́щ·ий, -ая, -ее,   (*pres part act*)
of гн·ать
гон·я́щ·ий, -ая, -ее,   (*pres part act*)
of гон·я́·ть
гопе́йт·, -а,   (*m*) (min) hopeite
гопкали́т·, -а,   (*m*) (chem) hopcalite
гор·а́, -ы́, (*acc*) го́р·у, (*nom pl*) го́р·ы,
(*g pl*) гор·, (*d pl*) -а́м, (*f*) mountain,
hill
~, лед·ян·а́я   iceberg
~ -свид·е́тел·ь   (*f*) (geol) island
mountain, inselberg
гора́здо·   (*adv*) much, far, by far
го́рб·, -а́,   (*m*) hump, bulge, protuberance
горб·а́тик·, -а,   (*m*) curved plane
(carpenter's tool)
горб·а́т·ый, -ая, -ое,   (*a*) humped,
humpbacked, gibbous
горб·а́ч·, -а́, (*i*) -о́м,   (*m*) (zool)
humpback (whale), *Megaptera nodosa*;
curved plane (carpenter's tool)
го́рб·ить, -ят,   (*imp*) hump, hunch,
arch
горбо·но́с·ый, -ая, -ое,   (*a*) hook-
nosed
горб·у́ш·а, -и, (*i*) -ей, (*f*) (fish) pink
salmon, humpback salmon, *Oncorhynchus gorbuscha*; (agr) sickle
горб·у́шк·а, -и,   (*g pl*) -шек·, (*f*)
crust (of bread)
горб·ылёк·, -лк·а́,   (*m*) (build) mullion
горб·ы́л·ь, -я́,   (*m*) offcut (timber)
горбул·я, -и,   (*f*) (arch) gargoyle
гордейн·, -а,   (*m*) (chem) hordein
горд·ели́в·ый, -ая, -ое,   (*a*) proud,
haughty
го́рден·ь, -я,   (*m*) (naut) gantline,
single whip
~, подъ·ём·н·ый   (naut) recovery
pendant
горд·и́ться, -я́тся,   (*imp*) (+ *instr*)
be proud of, take pride in
гордови́н·а, -ы,   (*f*) (bot)   arrowroot,
Viburnum lantana

горд·ый, -ая, -ое, (*a*) proud
гор·е, -я, (*n*) grief, sorrow
гор·евш·ий, -ая, -ее, (*past part act*) of **гор·еть**
гор·елк·а, -и, (*g pl*) **-лок·,** (*f*) burner; (weld) welding blowpipe, torch, cutting blowpipe, torch; (elec) quartz mercury-vapour lamp
~, за·вихр·ивающ·ая turbulent-type burner (of boiler)
~, лаборатор·н·ая Bunsen burner
~, муфель·н·ая muffle burner (pulverized fuel)
~, прямо·точ·н·ая long-flame burner
~, турбулент·н·ая turbulent-type burner (for pulverized fuel)
~, эконом·ическ·ая (gas) pilot burner
гор·елочн·ый, -ая, -ое, (*a*) of **гор·елк·а**
гор·ел·ый, -ая, -ое, (*a*) burnt, scorched
горельеф·, -а, (*m*) high relief, (arch) alto-rilievo
гор·ени·е, -я, (*n*) (*v n*) of **гор·еть;** burning, combustion
~, у·станов·ивш·ееся sustained combustion
гор·естн·ый, -ая, -ое, (*a*) sorrowful, sad
гор·еть, -ят, (*imp*) burn, be on fire; shine, be lit/on (of lights)
гор·ец·, -р·ц·а, (*i*) **-р·ц·ем,** (*m*) mountain dweller, highlander; (bot) *Polygonum*
горечавк·а, -и, (*f*) (bot) gentian, *Gentiana*
гореч·ь, -и, (*f*) bitter taste; acrid smell; bitterness
горж·а, -и, (*f*) (geog) gorge, defile
горж·усь (*pres 1st sing*) of **горд·иться**
горизонт·, -а, (*m*) horizon, level, layer
~, абсолют·н·ый (surv etc.) datum plane, (min) datum level
~ верх·н·его бьеф·а (civ eng) upstream water line, upper pool elevation
~, вид·им·ый visible/apparent/sensible/terrestrial horizon
~ визир·овани·я (*f*) (surv) elevation head
~ вод·ы (*geol*) water table/level
~, глав·н·ый (min) main level
~, ист·инн·ый true/rational/celestial horizon
    лини·я горизонт·а (*f*) (surv) map/ground parallel

горизонт
~, марк·ирующ·ий (min) indicator/key horizon
~, нефте·нос·н·ый (geol) oil horizon
~ ниж·н·его бьеф·а (civ eng) downstream water line, lower pool elevation
~ около·ствол·н·ого двор·а (min) bottom/shaft level
~, от·кат·очн·ая (min) haulage level
~, ртут·н·ый (surv) artificial horizon
~ штольн·и (min) adit/gallery level
горизонтал·ь, -и, (*f*) horizontal, horizontal line; (surv) contour, contour line
~, вс·по·мог·ательн·ый (surv) intermediate contour
~, геоидал·н·ая (surv) geoid contour
~, глав·н·ая isometric parallel (air survey photog)
~, нул·ев·ая initial level
~, об·общ·ённ·ая (surv) generalized contour
~, стро·ительн·ая (civ eng) base line
горизонтально·ков·очн·ая машин·а (*f*) (met) upsetter, horizontal forging machine
горизонтал·н·о-про·тяж·н·ый стан·ок· (*m*) horizontal broaching machine)
горизонтал·ност·ь, -и, (*f*) horizontal position
горизонтал·н·ый, -ая, -ое, (*a*), (*s f*) **-л·ен·, -ль·н·а,** horizontal, flat-lying, aclinal
~ дал·ност·ь (*f*) (rad) horizontal range, (gunn) plan range
~ на·вод·к·а (*f*) (gunn) training
~ по·лёт· (*m*) (air) horizontal/translational flight
горизонт·ировани·е, -я, (*n*) (surv) levelling, setting true to perpendicular
горизонт·ир·овать, -уют, (*imp*) level, level off, flatten
горизонт·ирующ·ий, -ая, -ее, (*pres part act*) of **горизонт·ир·овать;** (surv) cross-levelling
гор·ист·ый, -ая, -ое, (*a*) mountainous, hilly, upland
гор·к·а, -и, (*g pl*) **-р·ок·,** (*f*) hill, hillock; (air) zoom, zooming; (rail) hump (shunting); (agr) gravity cleaner
~, потенциал·н·ая (elec) potential hill
~ ролл·а (paper) back-fall, weir, descent plate
гор·к·н·уть, -ут, (*imp*) turn/become bitter/rancid

го́рл·о, -а, (*n*) throat; neck (narrow opening); (geog) sound, strait
~, дых·а́тельн·ое (anat) windpipe
горл·ови́н·а, -ы, (*f*) neck, mouth (narrow opening), orifice; manhole; vent (of volcano); (gunn) breech opening
горл·ов·о́й, -а́я, -о́е, (*a*) of **го́рл·о;** (ling) guttural
горло·пёр·ый, -ая, -ое, (*a*) (anat) jugular
го́рл·ышк·о, -а, (*n*) (*dim*) of **го́рл·о; горл·ови́н·а**
горля́нк·а, -и, (*f*) (bot) bottle gourd, *Lagenaria vulgaris*
гормо·го́н·и·й, -я, (*m*) (bot) hormogonium
горм·о́н·, -а, (*m*) (biol) hormone
гормон·а́льн·ый, -ая, -ое, (*a*) hormone, hormonic
гормо́н·н·ый, -ая, -ое, (*a*) hormone, hormonic
гормо·проте́йн·, -а, (*m*) hormoprotein
гормо·ци́ст·, -а, (*m*) (bot) hormocyst
го́рн·, -а, (*m*) (met) hearth, forge; hearth furnace; well (of blast furnace); bugle, horn
~, гонча́р·н·ый potter's kiln
~, за·клёп·очн·ый brazier
~, кузн·е́чн·ый forge
~, плав·и́льн·ый (met) crucible furnace
~, ти́гель·н·ый (met) crucible furnace
~, тума́н·н·ый (naut) foghorn
горнбленди́т·, -а, (*m*) (min) hornblendite
горн·и́ст·, -а, (*m*) bugler
горн·ов·о́й, -а́я, -о́е, (*a*) of **горн·;** (*m decl as adj*) furnaceman
~ пла́в·к·а (*f*) Scotch-hearth smelting (of lead)
горно·за·во́д·ск·ий, -ая, -ое, (*a*) mining, mining industry
горно·о·богат·и́тельн·ый, -ая, -ое, (*a*) ore-dressing
горно·про·мы́шл·енн·ый, -ая, -ое, (*a*) mining, mining-industry
горно·рабо́ч·ий, -его, (*m decl as adj*) miner
горно·ру́д·н·ый, -ая, -ое, (*a*) ore-mining, metalliferous-mining
горно·с·пас·а́тельн·ый, -ая, -ое, (*a*) mine-rescue
горноста́·й, -я, (*m*) (zool) stoat (summer), ermine (winter), *Mustela erminea*

горно·техн·и́ческ·ий, -ая, -ое, (*a*) mine, mining, mining engineering
горнтимберс·, -а, (*m*) (shipb) horn timbers
го́р·н·ый, -ая, -ое, (*a*) mountain, mountainous; mine, mining, mineral; rock; (bot) montane
~ ва́т·а (*f*) (min) rock wool
~ воск· (*m*) ozocerite, mineral wax
~ вы́·работ·к·а (*f*) mine working
~ давл·е́ни·е (*n*) (min) underground/ ground pressure
~ де́л·о (*n*) mining
~ инжене́р· (*m*) mining engineer
~ ко́ж·а (*f*) mountain leather (a variety of asbestos)
~ комба́йн· (*m*) (min) cutter-loader
~ ко́мпас· (*m*) (min) geologist's compass
~ лён· (*m*) mineral flax, amianthus (variety of asbestos)
~ мук·а́ (*f*) (min) diatomite
~ по·ро́д·а (*f*) (min) rock
~ про́б·к·а (*f*) mountain cork (a variety of asbestos)
~ про·хо́д· (*m*) mountain pass
~ систе́м·а (*f*) (geog) range of mountains
~ хруста́л·ь (*m*) (min) rock/mountain crystal
~ эффе́кт· (*m*) (rad) high terrain effect
гор·ня́к·, -а, (*m*) miner
го́род·, -а, (*nom pl*) **-а́,** (*g pl*) **-о́в,** (*m*) town, city
город·о́к·, -д·к·а́, (*m*) (*dim*) of **го́род·**
~, вое́н·н·ый military station/camp, cantonment
~, гимнаст·и́ческ·ий athletics/gymnastics ground
~, университе́т·ск·ий university buildings
город·ск·о́й, -а́я, -о́е, (*a*) town, city, urban, municipal
~ со·ве́т· (*m*) municipal/town/city council
~ тра́нспорт· (*m*) municipal/city transport
~ хозя́·йств·о (*n*) public utilities
городчато·цвет·ко́в·ый, -ая, -ое, (*a*) (bot) crenatiflorous, crenate-flowered
горо́д·чат·ый, -ал, -ое, (*a*) crenate, notched; crenulate
горо·образ·ова́ни·е, -я, (*n*) orogenesis, orogeny, mountain building/ making

горо·образ·у́ющ·ий, -ая, -ее,   (*a*) orogenic

горо·фи́ль·н·ый, -ая, -ое, (*a*) (met) horophile, horophole

горо́х·, -а, (*m*) (bot) pea, *Pisum*

~, бара́н·ий   (bot) chick pea, *Cicer arietinum*

горохо·ви́д·н·ый, -ая, -ое, (*a*) pisiform; (geol) pisolitic

горо́х·овик·, -а, (*m*) (bot) *Caragana*

горо́х·ов·ый, -ая, -ое, (*a*) pea

го́р·очн·ый, -ая, -ое, (*a*) of **го́р·к·а**

~ сорт·иро́вочн·ая ста́нци·я   (*f*) (rail) hump shunting yard

горо́ш·ек·, -ш·к·а, (*m*) (*dim*) of **горо́х·**; (bot) vetch, *Vicia*

го́рст·, -а, (*m*) (geol) horst

го́рст·очк·а, -и, (*g pl*) **-чек·,** (*f*) (*dim*) of **го́рст·ь**

го́рст·ь, -и, (*nom pl*) **-и,** (*g pl*) **-е́й** (*f*) hollow, cup (of hand); handful

горта́н·н·ый, -ая, -ое, (*a*) (ling) gutteral; (anat) laryngeal

горта́н·ь, -и, (*f*) (anat) larynx

горч·и́ть, -а́т, (*imp*) taste bitter/rancid

горч·и́ц·а, -ы, (*i*) **-ей,** (*f*) (bot) mustard, *Sinapis*

горч·и́чник·, -а, (*m*) (med) sinapism, mustard plaster

горч·и́чн·ый, -ая, -ое, (*a*) mustard

~ ма́сл·о (*n*) (chem) mustard oil

го́р·ше, (*comp*) of **го́рьк·ий**

горш·е́чник·, -а, (*m*) potter

горш·е́чн·ый, -ая, -ое, (*a*) pottery, potter's

~ ка́мен·ь (*m*) (min) soapstone, steatite

горш·о́к·, -ш·к·а́ (*m*) pot, vessel

~, вы́·гон·очн·ый   (hortic) seedling pot

~, вы·хлоп·н·о́й   (M/T) exhaust silencer

~, конденсацио́н·н·ый (st eng) steam trap

го́рьк·ий, -ая, -ое, (*a*) bitter

~ сол·ь (*f*) (min) epsomite, Epsom salts

горьк·ова́т·ый, -ая, -ое, (*a*) rather bitter, bitterish

горько·зём·, -а, (*m*) (min) bitter spar, dolomite

горько·минда́ль·н·ый, -ая, -ое, (*a*) bitter-almond

горько·пло́д·н·ый, -ая, -ое, (*a*) (bot) picrocarpous, bitter-fruited

гор·ю́ч·ее   (*n decl as adj*) (*see also* **то́пл·ив·о**) fuel

~, по·дви́ж·н·ое (nucl) fluidized fuel

~, я́дер·н·ое nuclear fuel

гор·ю́чест·ь, -и,   (*f*) combustibility, inflammability

гор·ю́ч·ий, -ая, -ее, (*a*) (*see also* **гор·ю́ч·ее**) (*as noun*); combustible, inflammable, fuel

~ газ· (*m*) fuel gas

~ с·мес·ь (*f*) gas mixture, fuel mixture

горяче·ка́т·ан·ый, -ая, -ое,   (*a*) (met) hot-rolled

гор·я́ч·ий, -ая, -ее, (*a*), (*s f*) **-я́ч·, -яч·а́;** hot

гор·я́чк·а, -и, (*g pl*) **-чек·,** (*f*) fever

~, бе́л·ая (med) delirium tremens

гор·я́щ·ий, -ая, -ее, (*pres part act*) of **гор·е́ть;** burning

гос- (*abbr component*) = **госуда́р·ст·венн·ый** state

гос·ба́нк·, -а, (*m*) (fin) State Bank

гослари́т·, -а, (*m*) (min) goslarite

госпитал·иза́ци·я, -и,   (*f*) (*v n*) *see* **госпитал·изи́р·овать;** hospitalization, hospital treatment

госпитал·изи́р·овать, -уют,   (*imp and perf*) hospitalize, treat in hospital

го́спитал·ь, -я, (*nom pl*) **-и,** (*g pl*) **-е́й,** (*m*) hospital, military hospital

~, плов·у́ч·ий hospital ship

~, пол·ев·о́й (mil) field ambulance

гос·пла́н·, -а, (*m*) State Planning Committee

господ·а́ (*nom pl*) of **господ·и́н, госпож·а́;** gentlemen, ladies, ladies and gentlemen; Sirs (in correspondence); Messrs (with surnames)

господ·и́н·, -а, (*nom pl*) **-од·а́,** (*g pl*) **-о́д,** (*d pl*) **-од·а́м,** gentleman; Mr. (with surname); Sir (address without name)

госпо́д·ств·о, -а,   (*n*) supremacy, domination; prevalence, predominance

госпо́д·ств·овать, -уют, (*imp*) rule, command, dominate; prevail, predominate

госпо́д·ствующ·ий, -ая, -ее, (*pres part act*) of **госпо́д·ств·овать;** dominant, prevailing, prevalent

госпож·а́, -и́, (*i*) **-о́й,** (*f*) lady; Mrs (with surname); Madam (address without name)

госсип·и́н·, -а, (*m*) (chem) gossypine

госсип·о́з·а, -ы, (*f*) (chem) gossypose

госси́п·о́л·, **-а**, (*m*) (chem) gossypol

гост·, **-а**, (*m*) (*abbr*) of **госуда́р·ст-венн·ый общесою́з·н·ый стан-да́рт·** State Standard Specification, GOST

го́сте·при·и́м·ый, **-ая**, **-ое**, (*a*) hospitable

го́сте·при·и́м·ств·о, **-а**, (*n*) hospitality

гост·и́н·ая, **-ой**, (*f decl as adj*) drawing/sitting room; reception room; suite of furniture

гост·и́ниц·а, **-ы**, (*i*) **-ей**, (*f*) hotel

гост·и́р·овать, **-ую́т**, (*imp*) specify in a GOST standard, issue a GOST on (cf **гост**)

гост·и́ть, **-я́т**, (*imp*) stay with, be on a visit

го́ст·ь, **-я**, (*nom pl*) **-и**, (*g pl*) **-е́й**, (*m*) guest, visitor

госуда́р·ственн·ый, **-ая**, **-ое**, (*a*) state, government; national, public; official

~ аппара́т· (*m*) the apparatus/machinery of government

~ долг· (*m*) (fin) national debt

~ пра́в·о (*n*) public law

~ стро́·й (*m*) political organization, political system

~ та́й·н·а (*f*) official secret

~ флаг· (*m*) national flag; (mil) colours

~ язы́к· (*m*) official language

госуда́р·ств·о, **-а**, (*n*) state, sovereign state/land

гот·и́ческ·ий, **-ая**, **-ое**, (*a*) (arch) Gothic

готов·а́льн·я, **-и**, (*g pl*) **-лен·**, (*f*) case of drawing instruments

гото́в·ить, **-ят**, (*imp*) prepare, make/get ready; train; prepare/get (a meal)

гото́в·ност·ь, **-и**, (*f*) readiness, preparedness; maturity (of rubber)

~, бо·ев·а́я (mil) readiness (for action)

~, по·хо́д·н·ая (naut) notice of readiness to get under way

гото́в·ый, **-ая**, **-ое**, (*a*) ready, prepared; finished (of manufactures)

го́т·ск·ий, **-ая**, **-ое**, (*a*) Gothic

гофр·и́рованн·ый, **-ая**, **-ое**, (*past part pass*) of **гофр·ир·ова́ть**; corrugated, goffered, crimped, fluted

гофр·и́р·ова́ть, **-у́ют**, (*imp*) crimp, (text) goffer, (met) corrugate

гош-фо́рм·а, **-ы**, (*f*) gauche-form (rubber)

граб·, **-а**, (*m*) (bot) hornbeam, *Carpinus*

граб·ёж·, **-еж·а́**, (*i*) **-еж·о́м**, (*m*) robbery

гра́бель·н·ый, **-ая**, **-ое**, (*a*) rake, rabble

гра́бен·, **-а**, (*m*) (geol) graben, fault trough

граб·ить, **-ят**, (*imp*) rob, plunder, loot

грабл·енн·ый, **-ая**, **-ое**, (*past part pass*) of **гра́б·ить**

грабл·и (*nom pl*), (*g pl*) **-бел·ь** or **-бл·ей** (agr) rake

~, бок·ов·ы́е side-delivery rake

~, ко́н·н·ые horse rake

~, по·пере́ч·н·ые dump rake

грабл·и́н·а, **-ы**, (*f*) rake

грабшти́хел·ь, **-я**, (*m*) engraving tool, graver, burin

граве́р·, **-а**, (*m*) engraver; etcher

гра́вие·мо́·ечн·о-сорт·иро́вочн·ый бараба́н· (*m*) wet gravel-screening trommel

гравие·мо́й·к·а, **-и**, (*g pl*) **-о·ек·**, (*f*) gravel washer

гра́в·и·й, **-я**, (*m*) gravel, grit

~, угл·ова́т·о-зерн·и́ст·ый grit

грав·и́йн·ый, **-ая**, **-ое**, (*a*) gravel, grit; gravelly, tophaceous

гравила́т·, **-а**, (*m*) (bot) avens, geum, *Geum*

грави·ме́тр·, **-а**, (*m*) (geophys) gravimeter

~, а·стаз·и́рованн·ый astatic balance gravimeter, unstable gravimeter

~, не·а·стаз·и́рованн·ый spring balance gravimeter, stable gravimeter

грави·ме́тр·и́ческ·ий, **-ая**, **-ое**, (*a*) gravimetric, gravimeter

грави·ме́тр·и·я, **-и**, (*f*) gravimetry

грави·раз·ве́д·к·а, **-и**, (*f*) (geophys) gravitation prospecting

грав·ирова́льн·ый, **-ая**, **-ое**, (*a*) engraver's, engraving; etcher's, etching

~ доск·а́ (*f*) plate

~ рез·е́ц· (*m*) (print) graver, burin

грав·ирова́ни·е, **-я**, (*m*) engraving; etching

грав·ир·ова́ть, **-у́ют**, (*imp*) engrave; etch

грави·сфе́р·а, **-ы**, (*f*) (astron) gravisphere

грави·тахео́·метр·, **-а**, (*m*) gravi-tacheometer

гравитацио́н·н·ый, **-ая**, **-ое**, (*a*) (phys) gravitational; (civ eng) gravity

гравитацио́н·н·ый, -ая, -ое
~ **ме́тод· о·богащ·е́ни·я**   (*m*) gravity method of ore dressing
~ **плот·и́н·а**   (*f*) gravity dam
гравита́ци·я, -и,   (*f*) (phys) gravitation
гравю́р·а, -ы,   (*f*) engraving, print; etching
град·, -а,   (*m*) degree (of measurement); (meteor) hail
град·ацио́н·н·ый, -ая, -ое,   (*a*) gradation, graduation
град·а́ци·я, -и,   (*f*) graduation, grading
градие́нт·, -а,   (*m*) gradient
~**, адиабат·и́ческ·ий**   (meteor) adiabatic lapse rate, adiabatic gradient
~**, вертика́ль·н·ый**   lapse rate (of temperature etc.)
~ **вла́ж·ност·и**   moisture gradient
~ **давл·е́ни·я**   pressure gradient
~**, на·по́р·н·ый**   pressure gradient
~ **ско́р·ост·и**   lapse rate
градие́нт·н·ый ве́тер·   (*m*) (meteor) gradient wind
градиенто́·метр·, -а,   (*m*) (surv) gradienter
гра́дин·а, -ы,   (*f*) (meteor) hailstone; (med) chalazion
градио́·метр·, -а,   (*m*) (surv) gradiometer
гради́рн·я, -и,   (*g pl*) **-рен·,**   (*f*) water-cooling tower
~**, ба́шен·н·ая**   natural-draught cooling tower
~**, вентиля́тор·н·ая**   mechanical/ induced cooling tower
~**, от·кры́·т·ая**   open/atmospheric cooling tower
градо·би́т·и·е   hail damage
град·ов·о́й, -а́я, -о́е,   (*a*) hail
~ **систе́м·а**   (*f*) (surv) centesimal system
градо·стро·и́тел·ь, -я,   (*m*) town-planning architect
градо·стро·и́тель·ств·о, -а,   (*n*) town planning and building
граду·и́ровани·е, -я,   (*n*) *see* **граду·и·ро́вк·а**
граду·и́р·овать, -уют,   (*imp and perf*) graduate, mark out a scale
граду·иро́вк·а, -и,   (*g pl*) **-вок·,**   (*f* (*v n*) *see* **граду·и́р·овать;** graduation; (elec) calibration
гра́дус·, -а,   (*m*) degree (of measurement); proof (alcohol)
~ **кисл·о́тност·и**   acidity number
~**, электр·и́ческ·ий**   electrical degree

гра́дус·ник·, -а,   (*m*) (horol) index (pop) thermometer
гра́дус·н·ый, -ая, -ое,   (*a*) of **гра́дус·;** grade
~ **се́т·к·а**   (*f*) (surv) grade grid/frame
гражд·ани́н·, -а,   (*nom pl*) **гра́жд·а·н·е,**   (*g pl*) **гра́жд·ан·,**   (*m*) citizen
гражд·а́нк·а, -и,   (*g pl*) **-нок·,**   (*f*) citizen (female)
гражд·а́нск·ий, -ая, -ое,   (*a*) civil
гражд·а́нств·о, -а,   (*n*) citizenship
грале́кс·, -а,   (*m*) Gralex (leather substitute)
грами·цид·и́н·, -а,   (*m*) (pharm) gramicidine
Грамм, -а,   (*m*) Gramme (surname, Belgian origin)
Гра́мм·а, кольц·ев·о́й я́кор·ь   (elec) Gramme ring armature
гра́мм·, -а,   (*g pl*) **гра́мм·ов,**   (*m*) gram, gramme
~ **-а́том·, -а,**   (*m*) gram-atom
~ **-ио́н·, -а,**   (*m*) gram-ion
~ **-кало́р·и·я, -и,**   (*f*) gram-calorie, small calorie
~ **-моле́кул·а, -ы,**   (*f*) gram-molecule, mole
~· **-молекуля́р·н·ый, -ая, -ое,**   (*a*) gram-molecular, molar
~ **-эквивале́нт·, -а,**   (*m*) gram-equivalent
-гра́мм·а, -ы,   (*f component*) -gram, -gramme
грамма́т·ик·, -а,   (*m*) grammarian
грамма́т·ик·а, -и,   (*f*) grammar
граммат·и́ст·, -а,   (*m*) grammarian
грамм·о́в·ый, -ая, -ое,   (*a*) of **грамм·**
~ **коли́честв·о**   (*n*) (nucl) gram scale
-гра́мм·ов·ый, -ая, -ое,   (*adj component*) -gram, -gramme
граммо́л·ь, -и,   (*f*) gram-molecule, mole
граммо·фо́н·, -а,   (*m*) gramophone, phonograph
граммо·фо́н·н· ый, -ая, -ое,   (*a*) gramophone, phonograph
~ **пласт·и́нк·а**   (*f*) gramophone record, disc
гра́мот·а, -ы,   (*f*) reading and writing; deed, certificate, instrument
~**, вер·и́тельн·ая**   (dipl) letter of credence, diplomatic credentials
гра́мот·ност·ь, -и,   (*f*) literacy; grammatical correctness; competence
~**, техн·и́ческ·ая**   technical competence, acquired technical skill
гра́мот·н·ый, -ая, -ое,   (*a*) literate; competent

грам·от·риц·а́тельн·ый, -ая. -ое, (*a*) (bact) Gram-negative

грам·пласт·и́нк·а, -и, (*g pl*) **-нок·,** (*f*) gramophone record, disc

грам·по·лож·и́тельн·ый, -ая, -ое, (*a*) (bact) Gram-positive

гра́н·, -а, (*g pl*) **гран·,** (*m*) grain (avoirdupois or apothecaries)

гранат·, -а, (*m*) (min) garnet; (bot) pomegranate, *Punica granatum*

~, изве́ст·ков·ый carbuncle, red garnet

грана́т·а, -ы, (*f*) (mil) grenade, shell

грана́т·н·ый, -ая, -ое, (*a*) (mil) grenade, shell

грана́т·ов·ый, -ая, -ое, (*a*) of **гра-на́т·;** cherry-coloured, cerise

гранато·мёт·, -а, (*m*) (mil) grenade discharger

~, руж·е́йн·ый (mil) grenade discharger

грандио́з·н·ый, -ая, -ое, (*a*) huge, immense, powerful, majestic

гран·е́ни·е, -я, (*n*) (*v n*) *see* **гран·и́ть**

гран·ён·ый, -ая, -ое, (*a*) faceted, cut

~ стекл·о́ (*n*) cut glass

гране·центр·и́рованн·ый, -ая, -ое, (*a*) (cryst) face-centred/centered

гран·и́льник·, -а, (*m*) (print) mezzotint graver

гран·и́льн·ый, -ая, -ое, (*a*) of **гра-н·е́ни·е;** lapidary (of gems)

~ де́л·о (*n*) gem cutting, lapidary cutting

гран·и́т·, -а, (*m*) (min) granite

гранито·ви́д·н·ый, -ая, -ое, (*a*) (geol) granitoid

гранито·гне́йс·, -а, (*m*) granite gneiss

гранито́л·ь, -я, (*m*) granitol (leather substitute)

гран·и́ть, -я́т, (*imp*) facet, cut (gems etc.)

гран·и́ц·а, -ы, (*i*) **-ей,** (*f*) boundary, border, frontier, limit

~, абсолю́т·н·ая absolute boundary, (math) absolute limit

~, ве́рх·н·яя upper limit; (math) upper bound (of a set)

~, во·образ·а́ем·ая fictitious boundary

~ зёрн·а (met) grain boundary

~ кип·е́ни·я фра́кци·и cut point (distillation)

~ нефт·и и вод·ы́ (geol) oil-water surface

~, обою́д·н·ая mutual boundary

~ мно́ж·еств·а (math) bound of a set

гран·и́ц·а

~, пло́ск·ая plane boundary

~ по·ве́рх·н·ост·и (met) interface

~ по·глощ·е́ни·я в ка́дм·и·и (nucl) cadmium cutoff

~ раз·де́л·а separation boundary, interface (of two media etc.)

~ раз·де́л·а фаз· (chem) interface

~, снег·ов·а́я (meteor) snow-line

~ стру·и́ (dynam) jet edge

~ у·сто́й·чивост·и (dynam) stability boundary

~, фи́рн·ов·ая (geol) firn/névé line

гран·и́ч·ить, -ат, (*imp*) border on, be contiguous to

гран·и́чн·ый, -ая, -ое, (*a*) of **гра-н·и́ц·а;** boundary, border

~ сло́·й (*m*) boundary layer

~ то́ч·к·а (*f*) (math) end point; interior point (of a set)

~ у·сло́в·и·е (*n*) boundary/limiting condition

гра́н·к·а, -и, (*g pl*) **-н·ок·,** (*f*) (print) galley; (print) proof in galley, galley proof, slip proof

-гра́н·ник·, -а, (*m component*) -hedron

гра́н·н·ый, -ая, -ое, (*a*) of **гра́н·ь**

грано·бласт·и́ческ·ий, -ая, -ое, (*a*) (geol) granoblastic

гран·ови́т·ый, -ая, -ое, (*a*) faceted

грано·диори́т·, -а, (*m*) (min) granodiorite

гранозан·, -а, (*m*) granosan (mercurial fungicide)

грано·фи́р·, -а, (*m*) (geol) granophyre

гра́н·ул·а, -ы, (*f*) granule, (plast) pellet

~, хромат·и́нов·ая (cyt) chromomere

гранул·и́рованн·ый, -ая, -ое, (*past part pass*) of **гранул·и́р·овать;** granulated, (plast) pelletized

гранул·и́р·овать, -уют, (*imp and perf*) granulate, (plast) pelletize

гранул·ём·а, -ы, (*f*) (med) granuloma

гранул·и́ровани·е, -я, (*n*) granulation, pelletizing, nodulizing

гранул·и́рующ·ий, -ая, -ее, (*pres part act*) of **гранул·и́р·овать**

~ голо́в·к·а (*f*) pelletizing head

гранул·и́т·, -а, (*m*) (min) granulite

грануло·метр·и́ческ·ий, -ая, -ое, (*a*) granulometric, grain sizing

~ со·ста́в· (*m*) grain/particle-size distribution

грануло·ме́тр·и·я, -и, (*f*) granulometry, grain-size classification

грануло·ци́т·, -а, (*m*) (cyt) granulocyte

гранул·я́тор·, -а, (*m*) granulator, (plast) pelletizer

гранул·я́ционн·ый, -ая, -ое, (*a*) granulation, (plast) pelletization

~ **ткан·ь** (*f*) (med) granulation tissue

гранул·я́ци·я, -и, (*f*) granulation; (plast) pelletizing; (astron) granules, willow leaves

гра́н·ь, -и, (*f*) border, boundary, verge, edge; side, facet, face

~, **бо·ев·а́я** (gunn) driving edge

~, **бок·ов·а́я** lateral face (of prism)

~, **ве́рх·н·яя** (math) upper bound

~, **верш·и́нн·ая** (cryst) cap/top face

~, **за́д·н·яя** flank (of cutting tool)

~ **криста́лл·а** crystal face

~ **много·гра́н·ник·а** (math) face of a polyhedron

~ **мно́ж·еств·а** (math) bound of a set

~, **пере́д·н·яя** face (of cutting tool)

~, **то́ч·н·ая ве́рх·н·яя** (math) least upper bound

~, **то́ч·н·ая ни́ж·н·яя** (math) greatest lower bound

~, **холост·а́я** (gunn) loading edge

граптоли́т·, -а, (*m*) (pal) graptolite

гра́т·, -а, (*m*) burr, whisker (on machine parts), flash (on mouldings/ moldings)

граува́кк·а, -и, (*f*) (geol) greywacke

граф·а́, -ы́, (*f*) (print) column

графеко́н·, -а, (*m*) (elec) graphecon, picture storage tube

гра́ф·ик·, -а, (*m*) graph; graphical chart, diagram; plot; time-table, schedule, roster; graphic artist

~ **движ·е́ни·я по·езд·о́в** (rail) train movement graph

~ **дежу́р·ств** duty table/roster

~ **хо́д·а рабо́т·** progress chart

~, **эксплуатаци·о́н·н·ый** operating schedule

гра́ф·ик·а, -и, (*f*) graphics, the graphic art

граф·и́лк·а, -и, (*g pl*) **-лок·,** (*f*) (eng) marking tool

граф·и́т·, -а, (*m*) (min) graphite, plumbago, black lead

~, **лит·е́йн·ый** foundry graphite

~, **пласт·и́нчат·ый** lamellar graphite

~, **реа́ктор·н·ый** (nucl) reactor-grade graphite

~, **шаро·ви́д·н·ый** (met) spheroidal graphite

графи́т·иза́ци·я, -и, (*f*) graphitization

графи́т·изи́рованн·ый, -ая, -ое, (*past part pass*) graphitized

графи́т·ов·ый, -ая, -ое, (*a*) graphite, graphitic; (nucl) graphite-moderated

~ **кисл·от·а́** (*f*) (chem) graphitic acid

граф·и́ть, -я́т, (*imp*) rule, square, draw lines, graph

граф·и́ческ·ий, -ая, -ое, (*a*) graphic, graphical

-гра́ф·и·я, -и, (*f component*) -graphy

графл·ённ·ый, -ая, -ое, (*past part pass*) of **граф·и́ть**

графл·ён·ый, -ая, -ое, (*a*) ruled, squared (lined paper)

графо·ма́н·и·я, -и, (*f*) (med) graphomania

графо·стати́ст·ик·а, -и, (*f*) graphical statistics

гра́ф·ств·о, -а, (*m*) (geog) county

гра́ч·, -а́, (*i*) **-о́м,** (*m*) (zool) rook, *Corvus frugilegus*

грёб· (*verb, 1st 2nd 3rd masc past*) of **грес·ти́**

греб·ёнк·а, -и, (*g pl*) **-нок·,** (*f*) comb; rack; thread chaser, chasing tool

~, **зубо·ре́з·н·ая** rack-type gear-planing cutter

~, **рас·пре·дел·и́тельн·ая** glass manifold (laboratory ware)

~, **снег·ов·а́я** (air) ski grating

греб·енн·о́й, -а́я, -о́е, (*a*) (text) combing, hackling

~ **маши́н·а** (*f*) combing machine (for wool); hackling machine (for flax, hemp etc.)

греб·ённ·ый, -ая, -ое, (*past part pass*) of **грес·ти́**; raked; rowed, paddled

гребенча́то·жа́бер·н·ые, -ых, (*pl decl as adj*) (pal) ctenobranchs, *Ctenobranchia*

гребенча́то·пло́д·н·ый, -ая, -ое, (*a*) (bot) lophocarpus

греб·е́нчат·ый, -ая, -ое, (*a*) comb, comb-like, pectinate, pectineal, crested

~ **вал·** (*m*) thrust shaft, collar shaft

гре́б·ен·ь, -б·н·я, (*m*) comb; (ocean, geog) ridge, crest; (mech eng) collar, feather, toe (of cam); (meteor) ridge, wedge; (text) card; hackle (for flax); comb crest (of birds); (zool) pecten, loph

~ **банда́ж·а** flange (of railway wheel)

~, **берег·ов·а́я** (ocean) shore ice ridge

~, **раз·ре́з·ывающ·ий** (met roll) tongue, dividing tongue

~, **топо·гра́ф·и́ческ·ий** (surv) crest-line, ridge-line

греб·е́ц·, -б·ц·а́, (*i*) **-б·ц·о́м,** (*m*) oarsman, rower

греб·ешо́к·, -шк·а́,   (*m*) (*dim*) of **гре́б·ен·ь** (q.v.);(zool) scallop, *Pecten ponticus*

греб·ешко́в·ый, -ая, -ое,   (*a*) of **греб·ешо́к·** ; pectineal

греб·ли́   (*past pl*) of **грес·ти́**

греб·л·о́, -а́,   (*n*) spatula

греб·ко́в·ый, -ая, -ое,   (*a*) of **греб·о́к·**

гре́бл·я, -и,   (*f*) rowing

~, байда́р·очн·ая   canoeing, paddling

~, па́р·н·ая   sculling (with two oars)

~, рас·па́ш·н·ая   sweep rowing

гребн·ева́ни·е, -я,   (*n*) (hortic) ridging

гребне·ви́д·н·ый, -ая, -ое,   (*a*) comb/ridge-shaped, pectiniform

гребн·еви́к·, -а́,   (*m*) (bot) dogstail, *Cynosurus*; (*pl*) (zool) ctenophorans, *Ctenophora*

гребн·ев·о́й, -а́я, -о́е,   (*a*) (agr) furrow ridge; pectineal

гребне·чес·а́льн·ая маши́н·а   (*f*) (text) combing machine/engine

греб·н·о́й, -а́я, -о́е,   (*a*) rowing, paddling, paddle

~ вал·   (*m*) propeller shaft

~ винт·   (*m*) screw propeller

~ го́н·к·а   (*f*) boat race

~ колес·о́   (*n*) paddle-wheel

греб·о́к·, -б·к·а́,   (*m*) paddle; stroke (of oar); rabble (furnace rake); (agr) rake; strickle (measure leveller)

греб·н·у́ть, -у́т,   (*perf*) have a row/paddle/pull

греб·у́щ·ий, -ая, -ее,   (*pres part act*) of **грес·ти́**

греб·я́   (*pres gerund*) of **грес·ти́**

гре́в·ш·ий, -ая, -ее,   (*past part act*) see **гре·ть**

грегами́т·, -а,   (*m*) (min) grahamite

грегари́н·ы   (*pl*) (zool) Gregarinida

гре́·ем·ый, -ая, -ое,   (*pres part pass*) see **гре·ть**

гре́йдер·, -а,   (*m*) (civ eng) grader

~ -элева́тор·, -а,   (*m*) (civ eng) elevating grader

гре́йзен·, -а,   (*m*) (min) greisen

грейзен·иза́ци·я, -и,   (*f*) (geol) greisenization

гре́йпфрут·, -а,   (*m*) (bot) grapefruit, *Citrus paradisi*

гре́йфер·, -а,   (*m*) (civ eng) grab, grab bucket

~, двух·челю́ст·н·о́й   clamshell, two-leafed grab bucket

~, мно́го·челю́ст·н·о́й   orange peel, multi-leafed grab bucket

гре́йфер

~, само·за·хва́т·ываю́щ·ий   automatic grab

гре́йфер·н·ый, -ая, -ое,   (*a*) of **гре́йфер·**

~ кран·   (*m*) grabbing/grab crane

~ механ·и́зм·   (*m*) (cinema) pull-down mechanism

~ экскава́тор·   (*m*) grab excavator

гре́·лк·а, -и,   (*g pl*) **-лок·,**   (*m*) heater, warming pan; hot-water bottle

грем·е́ть, -я́т,   (*imp*) thunder, resound, fulminate, rattle, rumble

грем·у́ч·ий, -ая, -ее,   (*a*) thundering, roaring, fulminating; (chem) fulminate of

~ газ·   (*m*) (min) firedamp, methane; oxyhydrogen gas

~ зо́лот·о   (*n*) (chem) fulminating gold

~ кисл·от·а́   (*f*) (chem) fulminic acid

~ рту́т·ь   (*f*) mercury fulminate

~ сту́д·ен·ь   (*m*) nitro-gelatine

грем·у́чник·, -а,   (*m*) (zool) rattlesnake

грён·а, -ы,   (*f*) silkworm eggs

гренади́лл·, -а,   (*m*) (bot) granadilla, passion fruit

грен·а́ж·, -а,   (*i*) **-ем,** (*m*) silkworm breeding

грен·а́жн·ый за·во́д·   (*m*) silkworm nursery

Грене́, элеме́нт·   (*m*) (elec) Grenet cell

греп·, -а,   (*m*) (shipb) gripe

грес·ти́,   (*pres 3rd sing, pl*) **греб·ёт, греб·у́т,** (*past masc sing*) **грёб·** (*imp deter*); rake; row, paddle, pull

гре́·т·ый, -ая, -ое,   (*past part pass*) of **гре·ть**

гре́·ть, -ют,   (*imp*) warm, heat, warm/heat up

гре́цк·ий оре́х·   (*m*) walnut, Persian walnut

гре́ч·а, -и,   (*i*) **-ей,** (*f*) buckwheat

гре́ч·еск·ий, -ая, -ое,   (*a*) Greek, (arch) Grecian

грече·ру́ш·а́льн·ый стан·о́к·   (*m*) shelling/hulling machine

греч·и́х·а, -и,   (*f*) (bot) buckwheat, *Polygonum fagopyrum*

греч·и́шн·ый, -ая, -ое,   (*a*) of **греч·и́х·а**

греч·нев·ый, -ая, -ое,   (*a*) buckwheat

~ круп·а́   (*f*) buckwheat groats

греш·и́ть, -а́т,   (*imp*) sin, transgress, violate (laws, truth etc.)

гре́·ющ·ий, -ая, -ее,   (*pres part act*) see **гре·ть**

гриб·, -á,   (*m*)   (*see also* **гриб·óк·**) fungus, mould, mushroom, toadstool

~, базиди·áльн·ый basidiomycete

~ -зóнт·ик·, -а,   (*m*) (bot) parasol mushroom, *Lepiota*

~, не·со·верш·éнн·ые   (*pl*) *Fungi Imperfecti*

~, нúз·ш·ий eumycete

~, сýм·чат·ый ascomycete

~, трýб·чат·ый cep, *Boletus*

~, цáр·ск·ий fly agaric, *Amanita muscaria*

гриб·к·а, -и,   (*g pl*) **-б·ок·,**   (*f*) (plast) button-shaped test piece

гриб·кóв·ый, -ая -ое,   (*a*) of **гриб·óк·**

гриб·н·óй, -áя, -óе,   (*a*) of **гриб·**

грибо·вúд·н·ый, -ая, -ое,   (*a*) mushroom, mushroom-shaped, fungoid, fungiform

~ клáпан·   (*m*) (mech) mushroom valve

~ тéл·о   (*n*) (zool) mushroom body

грибо·вóд·н·я, -и,   (*f*) (bot) mycotheca

гриб·óк·, -б·к·а,   (*m*) (*dim*) of **гриб·**

~, брод·úльн·ый (bot) yeast plant

~, луч·úст·ый (bot) actinomycete

~, плесн·ев·óй mould

грибо·кóрен·ь, -рн·я,   (*m*) (bot) mycorrhiza, mycorhiza

грибо·у·стóй·чив·ый, -ая, -ое,   (*a*) fungus-resistant

грúв·а, -ы,   (*f*) (zool) mane, crest

грúвенник·, -а,   (*m*) 10-kopeck piece

грúв·ист·ый, -ая, -ое,   (*a*) (zool) long-maned/haired

грúдлик·, -а,   (*m*) (elec) grid leak

гризáйл·ь, -я,   (*m*) grisaille (art)

грикvalendúт·, -а,   (*m*) (min) griqualandite

~, о·кремн·ённ·ый (min) tiger's eye

грим·, -а,   (*m*) make-up, grease paint

Грúн·а, фóрмул·ы (*pl*) (math) Green's formulae

грúнд·а, -ы,   (*f*) (zool) pilot whale, *Globicephalus malus*

гриниáн·, -а,   (*m*) (math) Green's function

грúн·овск·ая фýнкци·я (*f*) (math) Green's function

гринокúт·, -а,   (*m*) (min) greenockite

Гриньяр·а, реáкци·я (*f*) (chem) Grignard reaction

Гриньяр·а, со·единéни·е (*n*) (chem) Grignard reagent

грúпп·, -а,   (*m*) (med) influenza

грипп·óзн·ый, -ая, -ое,   (*a*) influenzal

грúф·, -а,   (*m*) signature stamp; (zool) (*pl*) *Falconiformes*

~ секрéт·ност·и (print) security classification stamp

грúф·а·, -ы,   (*f*) lithium acetyl salicylate

грúфел·ь, -я,   (*m*) slate-pencil

грúфель·н·ый, -ая, -ое,   (*a*) of **грúфел·ь**

~ слáн·ец·   (*m*) writing-slate, grapholite

гроб·, -а,   (*nom pl*) **-ы́,** (*g pl*) **-óв,**   (*m*) coffin

Грóве, элемéнт·   (*m*) (elec) Grove cell

Грóвер·а, шáйб·а   (*f*) (mech eng) Grover washer

грóг·, -а,   (*m*) grog (drink)

гроденáпл·ев·ое пере·плет·éни·е (*n*) (text) plain weave, calico weave

грож·ý (*pres 1st sing*) of **гроз·úть**

гроз·á, -ы,   (*nom pl*) **грóз·ы,** (*g pl*) **гроз·,** (*f*) thunderstorm; terror, threat, menace

грозде·вúд·н·ый, -ая, -ое,   (*a*) botryoidal, botryose, clustered, racemose, grape-like

грóзд·ь, -и,   (*nom pl*) **грóзд·и** or **грóзд·ья,** (*g pl*) **грозд·éй** or **грóзд·ьев,** (*f m*) cluster, raceme; bunch (of grapes)

гроз·úть, -я́т,   (*imp*) threaten

грóз·н·ый, -ая, -ое,   (*a*) menacing, threatening; terrible, formidable

гроз·ов·óй, -áя, -óе,   (*a*) thunder, storm, (elec) lightning

~ раз·ря́д·ник· (*m*) lightning arrester

~ электр·úчеств·о (*n*) atmospheric electricity

грозо·от·мéт·чик·, -а,   (*m*) (rad) storm indicator

гром·, -а,   (*nom pl*) **грóм·ы,** (*g pl*) **-óв,** (*m*) thunder

громáд·н·ый, -ая, -ое,   (*a*) huge, immense, enormous

гром·úть, -я́т,   (*imp*) destroy, annihilate, defeat, rout

грóм·к·ий, -ая, -ое, (*a*) loud; famous, celebrated; bombastic

громко·говор·úтел·ь, -я,   (*m*) loudspeaker

~, динам·úческ·ий dynamic/moving coil loudspeaker

~, кóмнат·н·ый subscriber's loudspeaker (broadcast relay system)

~, на·прáвл·енн·ый directional loudspeaker

громко·говор·и́тел·ь

~, **пневмат·и́ческ·ий** pneumatic/compressed air loudspeaker

~ **прям·о́го из·луч·е́ни·я** direct-radiator loudspeaker

~, **пьезо·электр·и́ческ·ий** piezoelectric/crystal loudspeaker

~, **ру́пор·н·ый** horn loudspeaker

~, **электро·динам·и́ческ·ий** dynamic/moving coil loudspéaker

~, **электро·магни́т·н·ый** magnetic armature loudspeaker, electromagnetic loudspeaker

гро́м·кост·ь, -и, (*f*) loudness, sound intensity, (rad etc.) volume

   регул·и́ровани·е гро́м·кост·и (*n*) (rad) volume control

   регуля́тор· гро́м·кост·и (*m*) (rad) volume/gain control

   у́·ровен·ь гро́м·кост·и (*m*) loudness/noise level

гром·ов·о́й, -а́я, -о́е, (*a*) thunder, thunderous

громозд·и́ть, -я́т, (*imp*) pile/stack/heap up

громо́зд·к·ий, -ая, -ое, (*a*) bulky, awkward, cumbersome

~ **товáр·ы** (*pl*) (rail) measurement goods

громо·от·во́д·, -а, (*m*) lightning conductor/rod

гро́м·че (*comp*) of **гро́м·к·ий;** louder

громых·а́·ть, -ют, (*imp*) rumble

громых·н·у́ть, -у́т, (*perf*) give/make a rumble

гро́сс·, -а, (*m*) (com) gross (144)

гроссбу́х·, -а, (*m*) ledger

гроссме́йстер·, -а, (*m*) grand master (at chess etc.)

гроссуляри́т·, -а, (*m*) (min) grossularite

грот·, -а, (*m*) grotto, cave; (naut) main course, mainsail

гроте́ск·, -а, (*m*) (print) grotesque, printer's gothic

гро́т·ов·ый, -ая, -ое, (*a*) of **грот·**

~ **па́рус·** (*m*) (naut) mainsail

гро́хот·, -а, (*m*) crash, din, rattle, roar; screen, sieve, sifter, riddle

~, **бараба́н·н·ый** trommel screen

~, **вибрацио́н·н·ый** vibrating screen

~, **инерцио́н·н·ый** unbalanced-pulley vibrating screen

~, **кач·а́ющ·ийся** shaking screen

~, **колосни́к·ов·ый** bar screen, grizzly

~, **кольц·ев·о́й** ring grizzly

гро́хот

~, **круг·ов·о́й** revolving screen

~, **мно́го·я́рус·н·ый** multi-deck screen

~, **скреб·ко́в·ый** drag screen

~, **электро·вибрацио́н·н·ый** electromagnetic vibrating screen

грохот·а́нь·е, -я, (*n*) (*v n*) of **гро·хот·а́ть**

грохот·а́ть (*pres 3rd sing, pl*) **гро·хо́ч·ет, грохо́ч·ут,** (*imp*); crash, rumble, roar

грохот·и́ть, -я́т, (*imp*) screen, sift, riddle

грохоч·е́ни·е, -я, (*n*) (*v n*) of **гро·хот·и́ть**

гру́ббер·, -а, (*m*) (agr) grubber, scuffler

груб·е́·ть, -ют, (*imp*) coarsen

груб·и́ть, -я́т, (*imp*) be rude/coarse

грубо·волокн·и́ст·ый, -ая, -ое, (*a*) coarse-fibred; coarse-grained

грубо·дро́бл·ен·ый, -ая, -ое, (*a*) coarse-ground

грубо·зерн·и́ст·ый, -ая, -ое, (*a*) coarse-grained

грубо·моза́ич·н·ый, -ая, -ое, (*a*) granoblastic

грубо·об·ло́м·очн·ый, -ая, -ое, (*a*) coarsely fragmented/broken

грубо·пор·и́ст·ый, -ая, -ое, (*a*) coarse-pored, macroporous

гру́б·ый, -ая, -ое, (*a*) rough, coarse, crude; rough-finished

~ **от·с·чёт·** (*m*) (instr) coarse reading

~ **при·бли́ж·е́ни·е** (*n*) crude/rough approximation

~ **регул·иро́вк·а** (*f*) coarse adjustment

гру́д·а, -ы, (*f*) heap, pile

груд·и́н·а, -ы, (*f*) (anat) breast-bone, sternum

груд·и́нк·а, -и, (*g pl*) **-нок·,** (*f*) breast (of meat)

груд·и́нн·ый, -ая, -ое, (*a*) (anat) sternal

груд·ни́ц·а, -ы, (*i*) **-ей,** (*f*) (med) mastitis; (text) breast beam

груд·н·о́й, -а́я, -о́е, (*a*) breast, chest, thoracic, pectoral

~ **кле́т·к·а** (*f*) (anat) chest, thorax

~ **про·то́к·** (*m*) (anat) thoracic duct

грудо·брюш·н·ый, -ая, -ое, (*a*) (anat) thoracoabdominal

~ **пре·гра́д·а** (*f*) (anat) diaphragm

гру́д·ь, -и, -й, (*nom pl*) **-и,** (*g pl*) **-е́й,** (*f*) (anat) breast, chest, pectus; (entomol) thorax

гру́ж·енн·ый, -ая, -ое, (*past part pass*) of **груз·и́ть**

груж·ён·ый, -ая, -ое, (*a*) loaded, laden

гру́з·, -а, (*m*) weight, flyweight, bob, plummet; load, cargo, freight; burden; (text) let-off

~, водо·ла́з·н·ые (*pl*) (naut) diver's weighted belt

~, генера́ль·н·ый general cargo

~, ки́п·ов·ый bale cargo/goods

~, компенсацио́н·н·ый compensator weight, compensating weight

~, кубато́р·н·ый cubiform cargo

~, кус·ков·о́й cubiform cargo

~, масс·о́в·ый bulk cargo/freight/goods

~, меш·ко́в·ый bag cargo

~, на·ва́л·очн·ый bulk cargo/freight (non-free-flowing)

~, на·лив·н·о́й liquid cargo/freight

~, на·сып·н·о́й free-flowing bulk cargo/freight

~, объ·ём·н·ый measurement cargo/freight

~, огне·о·па́с·н·ый hazardous cargo/freight

~, парце́ль·н·ый parcel cargo

~, поле́з·н·ый (air) disposable load

~, рабо́ч·ий useful load

~, с·пас·ённ·ый (naut) salvage

~, та́р·н·ый general cargo/freight (packed)

~, у·пак·о́вочн·ый general cargo/freight (packed)

~, шту́ч·н·ый piece freight

~, экспорт·и́руем·ый outward cargo/freight

гру́зд·ь, -я́, (*nom pl*) **гру́зд·и,** (*g pl*) **-е́й,** (*m*) (bot) milk/pepper mushroom, *Lactarius*

~, на·сто·я́щ·ий (bot) *Lactarius resimus*

~, пе́реч·н·ый (bot) *Lactarius piperatus*

гру́з·ик·, -а, (*m*) (*dim*) of **груз·**

~ регуля́тор·а governor weight, flyweight

груз·и́л·о, -а, (*n*) sinker

груз·и́нск·ий, -ая, -ое, (*a*) Georgian (of U.S.S.R. republic)

груз·и́ть, -я́т, (*imp*) load (a vehicle, train, ship); ship (a cargo); **-ся** take on (cargo); get on (train etc.); go aboard (ship)

грузо- (*root*) cargo-, freight-, load-

груз·ов·и́к·, -а́, (*m*) lorry, truck

грузо·в·мест·и́мост·ь, -и, (*f*) load/carrying capacity, load-carrying capacity

~ для шту́ч·н·ых гру́з·ов (naut) bale capacity/cubic/space

~ на́·сып·ью grain capacity/cubic/space

~ су́д·н·а (naut) stowage factor

груз·ов·о́й, -а́я, -о́е, (*a*) goods, cargo-carrying, freight, cargo; load, loading, gravity; weight-loaded

~, аккумуля́тор· (*m*) (hydr) weight-loaded accumulator

~ ватерли́н·и·я (*f*) (shipb) load waterline

~ движ·е́ни·е (*n*) (rail) goods traffic

~ ма́рк·а (*f*) (shipb) load line

~ план· (*m*) (naut) cargo plan

~ площа́д·к·а (*f*) (naut) tea-board

~ при·во́д· (*m*) gravity-operating gear

~ раз·ме́р· (*m*) (naut) displacement curve

~ с·бо́р· (*m*) wharfage

~ ста́нци·я (*f*) (rail) goods station/yard

~ су́д·н·о (*n*) cargo vessel, freighter

~ тра́верс· (*m*) load bar (conveying gear)

~ шкал·а́ (*f*) (naut) vertical displacement of deadweight scales

грузо·за·хва́т·н·ый, -ая, -ое, (*a*) load-lifting/hoisting

~ при·способл·е́ни·е (*n*) lifting/suspending attachments (to cranes, hoists etc.)

~ у·стро́й·ств·о (*n*) hoisting gear

грузо·на·пряж·ённост·ь, -и, (*f*) (rail) density of goods/freight traffic

грузо·оборо́т·, -а, (*m*) (rail) goods/freight turnround/traffic

грузо·от·прав·и́тел·ь, -я, (*m*) shipper, consigner

грузо-пассажи́р·ск·ое су́д·н·о (*n*) (naut) passenger-carrying freighter, cargo liner

грузо·подъ·ём·ност·ь, -и, (*f*) load/lifting/carrying capacity, capacity (of a hoist)

~ вал·ов·а́я (shipb) deadweight carrying capacity

~ -де́двейт· (*m*) (shipb) deadweight carrying capacity

~, поле́з·н·ая (shipb) net capacity

~, по́лн·ая (shipb) dead-weight capacity/tonnage

~, чи́ст·ая (shipb) useful dead-weight, cargo-carrying capacity

грузо·по·луч·а́тел·ь, -я, (*m*) consignee
грузо·по·то́к·, -а, (*m*) freight traffic (over a period of time)
гру́з·чик·, -а, (*m*) stevedore, docker, loader
грундбу́кс·а, -ы, (*f*) (mech eng) collar/neck bushing
грунт·, -а, (*m*) soil, earth, ground; subsoil; primer, ground paint; sea bottom
~ -асфа́льт·, -а, (*m*) (build) soil-asphalt
~, жи́дк·ий (ocean) thermal layer
~, полу·ска́ль·н·ый (civ eng) cohesive soil
~, ры́хл·ый friable soil, loose ground; (civ eng) cohesionless/granular soil
~, ска́ль·н·ый (civ eng) rock
грунт·ла́к·, -а, (*m*) primer (painting)
грунто·бло́к·, -а, (*m*) (build) soil block (bound with cement etc.)
грунто́в·, -а, (*m*) (naut) gripe
грунт·ова́льн·ый, -ая, -ое, (*a*) priming, coating
~ маши́н·а (*f*) (paper) coating machine
грунт·ова́ни·е, -я, (*n*) priming (painting)
грунт·ова́ть, -у́ют, (*imp*) apply prime/priming coat (for paint)
грунто·ве́д·ени·е, -я, (*n*) soil science
грунт·о́вк·а, -и, (*g pl*) **-вок·,** (*f*) primer, priming coat, prime coating, ground (for paint)
грунт·ов·о́й, -а́я, -о́е, (*a*) of **грунт·**
~ вод·а́ (*f*) subsurface/underground water
~ доро́г·а (*f*) dirt/earth/unsurfaced road
~ кра́с·к·а (*f*) primer, priming paint, undercoat
~ трюм· (*m*) dredger hopper
грунто·за·це́п·, -а, (*m*) (civ eng) grouser; cleat (of tyre)
грунто·но́с·, -а, (*m*) (min etc.) sampler, core lifter
~, бок·ов·о́й side-wall sample taker
~, стрел·я́ющ·ий coring gun
грунто·но́с·к·а, -и, (*g pl*) **-с·ок·,** (*f*) = **грунто·но́с·**
грунто·про·во́д·, -а, (*m*) delivery pipeline (dredger)
грунт·ро́п·ы (*pl*) (naut) ground rope
гру́пп·а, -ы, (*f*) (*see also under associated adjectives*); group, party, team, cluster, series, train; (plast) system, agents

гру́пп·а
~, а́бел·ев·ая (math) Abelian group
~, авар·и́йн·ая breakdown gang; (naut) damage-control party
~ а́том·ов (phys) cluster of atoms
~ вака́нс·и·й (phys) cluster of vacancies
~, коммутати́в·н·ая (math) commutative group
~, внутри·комплéкс·н·ая (chem) chelate group
~ за·медл·éни·я (nucl) slowing-down group
~ о·са́д·очн·ых пласт·о́в (geol) sedimentary complex
~, пере·хо́д·н·ая transition series/group
~, по·жа́р·н·ая fire-fighting party
~ стержн·éй (nucl) rod group
~ пре·образ·ова́ни·я (math) permutation group
~, черн·ов·а́я (met roll) roughing train
~, чист·ов·а́я (met roll) finishing train
~, таксо·ном·и́ческ·ая (biol) phylum
групп·ирова́ни·е, -я, (*n*) grouping, bunching, clustering
~ пуч·к·а́ (nucl) beam stacking
~ фаз· (phys) phase bunching
групп·ирова́тел·ь, -я, (*m*) (elec) buncher
групп·ир·ова́ть, -у́ют, (*imp*) group; classify; bunch
групп·иро́вк·а, -и, (*g pl*) **-вок·,** (*f*) group, grouping, classification, bunching
групп·ов·о́й, -а́я, -о́е, (*a*) group; series
~ вз·лёт· (*m*) (air) formation take-off
~ при·во́д· (*m*) group drive
~ щит· (*m*) (elec) distribution panel, distribution board
гру́ст·н·ый, -ая, -ое, (*a*) melancholy, sad
гру́ш·а, -и, (*i*) **-ей,** (*f*) pear; (bot) pear tree, *Pyrus*; yoke screw
~, земл·ян·а́я Jerusalem artichoke, *Helianthus tuberosus*
~, рези́н·ов·ая rubber bulb
груше·ви́д·н·ый, -ая, -ое, (*a*) pear-shaped, pyriform
гры́ж·а, -и, (*i*) **-ей,** (*f*) (med) hernia
грыз·ть, -у́т, (*imp*) gnaw, nibble
грыз·у́н·, -а́, (*m*) (zool) rodent
грыз·у́щ·ий, -ая, -ее, (*a*) (*pres part act*) of **грыз·ть**
грюнери́т·, -а, (*m*) (min) grunerite
грюнштéйн·, -а, (*m*) (min) greenstone

Грюнейзен·а, за·кóн·　　(*m*) (phys) Grüneisen law, Grüneisen second rule

гряд·á, -ы́,　(*nom pl*) **гря́ды,** (*g pl*) **гряд,** (*d pl*) **-áм,** (*f*) (geog) ridge, range; bank (of clouds); (hortic) bed, seedbed, bed ridge

гря́д·ил·ь, -я,　(*m*) beam (of plough)

грядо·дéл·ател·ь, -я,　(*m*) (agr) ridger

гряд·у́щ·ий, -ая, -ее, (*a*) impending, approaching, succeeding

гряз·еви́к·, -á,　(*m*) mud trap/drum, sludge pan

гряз·ев·óй, -áя, -óе,　(*a*) mud, slime

∼ корóб·к·а　(*f*) mud box (of pump)

∼ на·сóс·　(*m*) mud pump

грязе·от·стóй·ник·, -а,　(*m*) mud sump/drum, sediment tank

грязе·у·лов·и́тел·ь, -я,　(*m*) (met cast) slag trap

гряз·н·и́ть, -я́т,　(*imp*) soil, dirty

гря́з·н·ый, -ая, -ое,　(*a*) dirty, filthy, muddy

гря́з·ь, -и,　(*f*) mud, slime, sludge; dirt, filth

∼, густ·áя　sludge

∼, и́л·ист·ая　ooze

∼, тóнк·ая　slurry

∼, сатурацио́н·н·ая　slime

гря́·н·уть, -ут,　(*perf*) break/burst out

гс.　(*abbr*) of **гáусс·** gauss (unit of magnetic flux density)

гуанáк·о, -а,　(*n*) (zool) guanaco, huanaco, *Lama guanicoe*

гуан·иди́н·, -а, (*m*) (biochem) guanidine

гуан·и́л·ов·ая кисл·от·á　(*f*) guanylic acid

гуан·и́н·, -а, (*m*) (chem) guanine

гуáно　(*n indecl*) (agr) guano

гуано·фóр·, -а,　(*m*) (cyt) guanophore

гуари́　(*indecl*) (naut) gunter rig

гуáш·ь, -и,　(*f*) gouache; gouache paint

гуая́к·, -а,　(*m*) (bot) lignum vitae, *Guaiacum officinale*

губ·á, -ы́,　(*nom pl*) **гу́б·ы,** (*g pl*) **губ·,** (*d pl*) **-áм,** (*f*) lip; (geog) gulf, bay, inlet; (biol) labium; labellum; tiller (of boat); (chem) jaw (of manipulator)

∼, вéрх·н·яя　upper lip; (entomol) labrum (and of gastropods)

∼, нару́ж·н·ая　labrum (of gastropods etc.)

∼, ни́ж·н·яя lower lip; (entomol) labium (and of gastropods)

губернáтор·, -а,　(*m*) governor

губ·и́тельн·ый, -ая, -ое, (*a*) ruinous, pernicious, destructive

губ·и́ть, ⸗ят,　(*imp*) ruin, spoil

гу́б·к·а, -и,　(*g pl*) **-б·ок·,** (*f*) (zool) sponge; sponge (cellular rubber); foam (of latex); jaw (of vice)

∼, в·клад·н·áя　jaw plate (of vice)

∼, на·клад·н·áя　vice clamp

∼, по·движ·н·áя　sliding jaw, moving jaw (of vice)

∼ руб·и́льник·а　jaw of a knife switch

губ·н·óй, -áя, -óе,　(*a*) lip, labial

∼ гармóн·ик·а　(*f*) mouth-organ

губо·нóг·ие, -их,　(*pl decl as adj*) (zool) chilopods, *Chilopoda*

губо·цвéт·н·ые, -ых,　(*pl decl as adj*) (bot) mint family, *Labiatae*

гу́б·чатост·ь, -и,　(*f*) sponginess

гу́б·чат·ый, -ая, -ое,　(*a*) sponge, spongy; foam (of latex)

∼ желéз·о　(*n*) (met) sponge iron

∼ плáтин·а　(*f*) spongy platinum

гугели́т·, -а,　(*m*) (min) hugelite

гуд·éни·е, -я,　(*n*) humming, buzzing, drone; hooting

гуд·éть, -я́т,　(*imp*) buzz, hum, drone; hoot

гуд·óк·, -д·к·á,　(*m*) hooter, whistle, horn (motor car), blast (on whistle, horn etc.)

гудрóн·, -а,　(*m*) tar, tar oil, asphaltum oil

∼, асфáльт·ов·ый　asphaltic bitumen (Brit), asphalt (U.S.)

∼, вáпор·н·ый　residual superheated steam cylinder oil

∼, ки́сл·ый　(oil) acid sludge, acid tar

∼, мáсл·ян·ый　(oil) naphthenic residue

∼, парафи́н·ов·ый　paraffinous residue

гудрон·áтор·, -а,　(*m*) (civ eng) bitumen/tar/asphalt sprayer

гудрон·и́р·овать, -уют,　(*imp and perf*) tar, asphalt

гудрóн·н·ый, -ая, -ое, (*a*) of **гудрóн·**

∼ вискози·мéтр·　(*m*) tar viscosimeter

гуж·ев·óй, -áя, -óе, (*a*) animal-drawn/hauled; animal transport

∼ дорóг·а　(*f*) cart road

∼ трáнспорт·　(*m*) cart transport, animal-hauled transport

∼ трел·ёвк·а　(*f*) animal logging

гужóн·, -а,　(*m*) tap, rivet

гу́з·а, -ы,　(*f*) cotton boll

гузо·лóм·к·а, -и,　(*g pl*) **-м·ок·,** (*f*) (agr) boll breaker

Гу́к·а, за·кóн·　(*m*) (mech) Hooke's law

гул·, -а,　(*m*) rumble, rumbling, roaring, roar; hum, buzz

гу́л·к·ий, -ая, -ое,   (*a*) resounding, booming; resonant

~ **ка́мер·а**   (*f*) (acous) reverberation chamber

гул·я́·ть, -ют,   (*imp*)   go for/take a walk/stroll

гуманита́р·н·ый, -ая, -ое, (*a*) humanitarian

гума́н·н·ый, -ая, -ое,   (*a*) humane

гума́т·, -а,   (*m*) (chem) humate

гу́мбо   (*m indecl*) (min) gumbo

гумённ·ый, -ого,   (*m decl as adj*) threshing floor

гумери́т·, -а,   (*m*) (min) humerite

гуми́д·н·ый, -ая, -ое,   (*a*) humid

гумидо·ста́т·, -а,   (*m*) humidifier, humidity controller

гум·и́нов·ый, -ая, -ое,   (*a*) (bot) humus, humic

~ **кисл·от·а́**   (*f*) (chem) humic acid

гуми́т·, -а,   (*m*) (min) humite

гуми·фик·а́ци·я, -и,   (*f*) (bot) humification

гу́мм·а, -ы,   (*f*) (med) gumma

Гу́ммер·а, гро́хот·   (*m*) Hummer screen

гу́мми   (*n indecl*) gum

гуммиара́бик·, -а,   (*m*) gum arabic

гуммигу́т·, -а,   (*m*) gamboge gum

гуммила́к·, -а,   (*m*) shellac

гу́мми-печа́т·ь, -и,   (*f*) (phot) gum printing

гумм·и́ровани·е, -я,   (*n*) rubberizing

гумм·и́рованн·ый, -ая, -ое,   (*past part pass*) rubberized, rubber-lined/covered/coated

гумми́т·, -а,   (*m*) (min) gummite

гумм·о́з·, -а,   (*m*) (bot) gummosis

гумн·о́, -а́,   (*nom pl*) **гу́мн·а,** (*g pl*) **гу́мен,** (*d pl*) **гу́мн·ам,** (agr) threshing floor

гумо·аммофо́с·, -а,   (*m*) humoammophos (fertilizer)

гумо·ге́н·, -а, (*m*) humogen (fertilizer)

гумо·ли́т·ов·ый, -ая, -ое,   (*a*) (geol) humolith

гу́мус·, -а,   (*m*) (agr) humus

~, **ки́сл·ый**   raw humus

гу́рт·, -а́,   (*m*) drove, herd, flock

гус·а́к·, -а́,   (*m*) (orn) gander

гуса́р·ик·, -а,   (*m*) rider (on a weighing machine)

гус·ёк·, -сь·к·а́,   (*m*) gosling; gooseneck (piping); (arch) ogee, flute

гу́сен·иц·а, -ы,   (*i*) -ей,   (*f*) (zool) caterpillar, grub; track, caterpillar track, crawler belt

гу́сен·ичн·ый, -ая, -ое,   (*a*) of **гу́сен·иц·а**

~ **полотн·о́**   (*n*) track (of tracked vehicles)

~ **тра́ктор·**   (*m*) crawler/caterpillar tractor

гус·ён·ок·, -нк·а,   (*nom pl*) **-с·ят·а,** (*g pl*) **-с·ят·,** (*m*) gosling

гус·и́н·ый, -ая, -ое,   (*a*) goose

~ **стро́·й**   (*m*) (air) arrowhead formation

густ·е́·ть, -ют,   (*imp*) become/grow thicker/thick

густ·о́й, -а́я, -о́е,   (*a*) dense, thick

~ **с·ма́з·к·а**   (*f*) grease lubrication

густо·зерн·и́ст·ый, -ая, -ое,   (*a*) close-grained, densely grained

густо·ли́ст·ый, -ая, -ое,   (*a*) (bot) dasyphyllous, densifolius, densely leaved

густо·пло́д·н·ый, -ая, -ое,   (*a*) (bot) pycnocarpous, dasycarpous, closely fruited

густ·от·а́, -ы́,   (*f*) thickness, density; richness, deepness

гус·ы́н·я, -и,   (*f*) goose (female)

гу́сь, -я,   (*nom pl*) **-и,** (*g pl*) **-е́й,** (*m*) (orn) goose, gander, *Anser*

гусь·к·о́м   (*adv*) in single file; in line ahead

~, **клас·ть я́кор·ь**   (naut) back anchor

гутали́н·, -а,   (*m*) shoe polish/cream

гуттапе́рч·а, -и,   (*i*) -ей, (*f*) gutta-percha

гутт·а́ци·я, -и,   (*f*) (bot) guttation

гу́щ·а, -и,   (*i*) -ей (*f*) grounds, lees, sediment; thicket

гу́щ·е   (*comp*) of **густо́й** thicker, denser

гц.   (*abbr*) of **герц·;** c/s, cycles per second

гюбнери́т·, -а,   (*m*) (min) huebnerite

гю́зинг·, -а,   (*m*) (naut) houseline, spun yarn

Гюйгенс·а, окуля́р·   (*m*) (opt) Huygen's eyepiece

гюйс·, -а,   (*m*) (naut) jack

гюйsubstaншто́к·, -а,   (*m*) (naut) jack staff

гэв.   (*abbr*) of **гига·электорн·во́льт** Gev, gigaelectronvolt

гэс.   (*abbr*) of **гидро·электро·ста́н·ци·я** (q.v.) hydroelectric power station

Д

Д (*abbr*) of **деба́·й** debye (unit of dipole moment)

да (*adv*) yes; no (*in confirmation of negation*); (*as particle indicates stress*); (*conj*) and, but

~ **здра́в·ству·ет**   long live!

дав·а́вш·ий, -ая, -ее, (*past part act*) of **да·ва́ть**

дав·а́ем·ый, -ая, -ое, (*pres part pass*) of **да·ва́ть**

дав·а́йте, (*imper*) of **да·ва́ть;** let us, let's

да·ва́ть, -ю́т, (*imp*) give, afford, provide, offer; produce, yield; let, allow; send, transmit

~ **воз·мо́ж·ност·ь**   permit, make possible, provide an opportunity of

~ **звон·о́к·**   ring a bell, give a ring

~ **машин·е**   start the engine, starting

~ **ног·у** (air) give rudder

~ **по́чк·и** (bot) bud

~ **ру́ч·к·у** (air) push forward; pull back (control column)

~ **теч·ь**   spring a leak

~ **ход·**   get (something) going; (naut) get under way; (law) prosecute

~ **у·са́д·к·у**   shrink

давиди́т·, -а, (*m*) (min) davidite

дав·и́лк·а, -и, (*g pl*) **-лок·,** (*f*) press, mill (for extracting juice)

~, **тростник·о́в·ая**   cane-mill (sugar production)

дав·и́л·о, -а, (*n*) juice-press; pressing weight

дав·и́льник·, -а, (*m*) forming bar (in metal spinning process)

дав·и́льн·ый, -ая, -ое, (*a*) press, pressing; (met) spinning

~ **стан·о́к·** (*m*) (met) spinning machine

~ **про·из·во́д·ств·о** (*n*) juice pressing; (met) spinning

дав·и́льн·я, -и, (*g pl*) **-лен·,** (*f*) juice/wine press; juice/wine pressing building/shed

дав·и́ть, -я́т, (*imp*) press, crush, squeeze, weight on, lie heavy on; **-ся** (*pass*); choke, suffocate, choke down; (met) spin

давл·е́ни·е, -я, (*n*) pressure

~, **атмо·сфе́р·н·ое**   atmospheric pressure

~, **баро·метр·и́ческ·ое**   barometric pressure (in mm Hg)

~, **ва́куум-метр ческ ((**   vacuum pressure, subatmospheric pressure

давл·е́ни·е

~, **ве́рх·н·ее**   upper pressure, pressure from above; (*g as adj*) downstroke (e.g. of presses); (met roll) over-draught

~, **вну́тр·енн·ее**   internal pressure, intrinsic pressure; inflation pressure

~ **внутр·ь**   inward pressure

~ **во́з·дух·а**   air pressure, atmospheric pressure

~ **в·пры́ск·а** (ICE) induction pressure

~ **в·пу́ск·а**   admission pressure

~, **все·сторо́н·н·ее**   (min) confined pressure; volumetric compression

~, **выс·о́к·ое**   high pressure (*frequently used in gen as adj*)

~, **гидро·стат·и́ческ·ое**   fluid/hydro-static pressure

~, **го́р·н·ое**   underground, ground pressure

~, **динам·и́ческ·ое**   dynamic pressure; (mech) impact pressure

~, **за·бо́й·н·ое**   (oil) bottom-hole pressure

~ **зву́к·а**   acoustic/sound/excess pressure

~, **из·бы́т·очн·ое**   gauge pressure, psig; (acous) excess pressure; over-pressure, extra pressure

~ **из·луч·е́ни·я** (phys) radiation pressure

~, **инерцио́н·н·ое**   mass pressure

~ **ис·теч·е́ни·я**   pressure of fluidity (of solids)

ли́н·и·я давл·е́ни·я (*f*) (mech) line of action

~, **мано·метр·и́ческ·ое**   gauge pressure

~ **мета́лл·а на ва́л·к·и**   (met roll) rolling load, roll load, roll-separating force

~ **на в·хо́д·е**   inlet pressure

~ **на вы́·ход·е**   outlet pressure

~ **на·сы́щ·енн·ого па́р·а**   vapour pressure

~, **на·ча́ль·н·ое**   (geol) virgin pressure (in stratum)

~, **ни́ж·н·ее** (met roll) lower pressure, pressure from below, (*g as adj*) upstroke (of presses); (met roll) under-draught

~, **ни́з·к·ое**   low pressure (*frequently used in gen as adj*)

давл·éни·е

~, **нормáль·н·ое** (meteor) barometric/ standard pressure; (civ eng) pressure at right angles

~, **обрáт·н·ое** back pressure

~ **о·круж·áющ·ей сред·ы́** ambient pressure

~, **о·пóр·н·ое** bearing pressure, abutment pressure (soil mechanics)

~, **осмот·и́ческ·ое** (chem) osmotic pressure

~, **о·стáт·очн·ое** vacuum pressure

~, **от·с·жи́м·н·ое** squeeze roll pressure (rubber)

~, **пласт·ов·óе** (geol) formation pressure

   под давл·éни·ем under pressure, pressurized, pressure-

   под давл·éни·ем, за·ли́·т·ый (civ eng) pressure-grouted

   под давл·éни·ем, лит·ь·ё (met) pressure casting

~, **пóр·ов·ое** (geol) interstitial pressure, pore-water pressure

~, **пояс·н·óе** circumferential pressure

~, **пре·дéль·н·ое** refusal pressure (of pump); (geol) ultimate stress

~ **пресс·овáни·я** (met) compacting pressure

~ **при за·кры́т·и·и** sealing-off pressure; (plast) moulding pressure

~, **рабóч·ее** effective pressure

~, **раз·вив·áем·ое** developed pressure

~, **раз·ры́в·н·ое** bursting pressure
регулáтор· давл·éни·я (m) pressure regulator

~ **с·гор·áни·я** (ICE) pressure of combustion

~ **с·жáт·и·я** (ICE) pressure of compression

~, **с·жим·áющ·ее** compressional stress

~, **скóр·ости·ое** velocity pressure

~, **срéд·н·ее индикáтор·н·ое** (eng) mean effective pressure

~, **срéд·н·ее эффекти́в·н·ое** (eng) mean effective pressure

~, **стационáр·н·ое** steady-state pressure

~ **с·цепл·éни·я** cohesive pressure

~, **с·плю́щ·ивающ·ее** collapsing pressure (of tubes)

~, **у·дéль·н·ое** unit pressure

~, **у·скор·я́ющ·ее** acceleration pressure (of a jet)

~, **физ·и́ческ·ое** barometric pressure (measured in mm Hg)

дáвл·енн·ый, -ая, -ое, (past part pass) of **дав·и́ть**

дáвл·ен·ый, -ая, -ое, (a) squashed

давн·éнько (adv) for a long time

дáвн·ий, -яя, -ее, (a) ancient, old; long-standing, well established

давн·ó (adv) long ago, a long time ago; for a long time

давно·про·шéд·ш·ий, -ая, -ее, (a) remote (in time); (gram) pluperfect

дáвн·ост·ь, -и, (f) remoteness (in time), antiquity; (law) prescription

давн·ы́м-давн·ó (adv) very long ago, ages ago

дá·вш·ий, -ая, -ее, (past part act) of **да·ть** (perf) see **да·вáть** (imp)

дагерроти́п·, -а, (m) (phot) daguerrotype

дад·и́м (fut 1st pl) of **да·ть** (perf) see **да·вáть** (imp)

дад·у́т (fut 3rd pl) of **да·ть** (perf) see **да·вáть** (imp)

да·ёт (pres 3rd sing) of **да·вáть**

дá·же (adv) even

дáйк·а, -и, (f) (geol) dike

~, **об·лóм·очн·ая** clastic dike

~, **сек·у́щ·ая** transverse dike

дá·йте (imper) of **да·ть**

дакéит, -а, (m) (min) dakeite

дáкен·овск·ий рас·твóр· (m) (chem) Dakin's solution

дакрио·аден·и́т·, -а, (m) (med) dacryoadenitis

дакрио·цист·и́т·, -а, (m) (med) dacryocystitis

дактил·и́т·, -а, (m) (med) dactylitis

дактило·зóид·, -а, (m) (zool) dactylozoid

дактило·пóр·а, -ы, (f) (zool) dactylopore

дактило·скоп·и́р·овать, -уют, (imp and perf) take the fingerprints of

дактило·скоп·и́·я, -и, (f) (med) dactyloscopy, (law) finger-printing

да·л (past masc sing) of **да·ть** (perf) see **да·вáть** (imp)

да·лá (past fem sing) of **да·ть** (perf) see **да·вáть** (imp)

Даламбéр·а, при́нцип· (m) (mech) D'Alembert principle

дáл·ее (comp and adv) further, farther

~, **и так** and so on, and so forth, etc.

дал·ёк·ий far, far away, distant, remote; alien to, different from, far from

дал·ек·ó (adv) far, far off; by far

~ **не** far from, by no means, not at all

далеко·лет·а́ющ·ий, -ая, -ее, (*a*) long-range (of rockets etc.)

да́·ли (*past pl*) of да·ть (*perf*) *see* да·ва́ть (*imp*); (*gen*) of даль

далмат·и́нск·ий, -ая, -ое, (*a*) (geol) Dalmatian

далма́т·ск·ий, -ая, -ое, (*a*) (geog) Dalmatian

~ ко́рен·ь (*m*) (pharm) pyrethrum root

да́·ло (*past neut sing*) of да·ть (*perf*) *see* да·ва́ть (*imp*)

да́л·ь, -и, (*f*) distance

дальне·вос·то́ч·н·ый, -ая, -ое, (*a*) Far-Eastern

дальне·де́й·ствующ·ий, -ая, -ее, (*a*) remote-acting

даль·не́йш·ий, -ая, -ее, (*a*) further, subsequent
  в даль·не́йш·ем in future, henceforth, below (i.e. further in a book etc.); subsequently

да́ль·н·ий, -яя, -ее, (*a*) distant, remote, far; long-range, long-distance; (telecom) trunk

~ де́й·стви·е (*n*) long range; remote operation; (*g as adj*) remote, remote-operating; long-range

~ об·наруж·е́ни·я early warning (radar)

~ при·во́д·н·ая ста́нци·я (*f*) (air) outer marker beacon

дально- (*component*) distant, far, remote, tele-, long, long-distant; distance

дально·бо́й·ност·ь, -и, (*f*) range (of a weapon)

дально·бо́й·н·ый, -ая, -ое, (*a*) (gunn, rocket) long-range

дально·ви́д·ный, -ая, -ое, (*a*) far-sighted, far-seeing, prescient

дально·де́й·ствующ·ий, -ая, -ее, (*a*) long-range; remote-operating

дально·зо́р·к·ий, -ая, -ое, (*a*) long-sighted, hyperopic

дально·зо́р·кост·ь, -и, (*f*) long sight, (med) hyperopia, hypermetropia

дально·ме́р·, -а, (*m*) rangefinder

~, внутри·ба́з·н·ый coincidence rangefinder

~, звук·ов·о́й sonar set

дально·ме́р·н·ый, -ая, -ое, (*a*) of дально·ме́р·

дально·ме́р·щик·, -а, (*m*) range-taker

да́ль·ност·ь, -и, (*f*) distance, range; remoteness, farness

~ ви́д·имост·и visibility range

да́ль·ност·ь
~, горизонта́ль·н·ая (rad) horizontal range; (gunn) plan range

~ де́й·стви·я range

~ де́й·стви·я радио·лока́тор·а radar range

лин·и·я да́ль·ност·и (*f*) (rad) range line

~, на·зе́м·н·ая ground range

~, на·кло́н·н·ая (mil) slant range

~ на·сти́ль·ност·и point-blank range

~ об·наруж·е́ни·я detection range

~ пла́в·ани·я range, endurance (of a ship)

~ по·раж·а́ющ·его де́й·стви·я (nucl) damaging range

~ по·лёт·а (a/c, rocket) range, flying range

~ по·лёт·а с·на·ря́д·а missile range

~, по́лн·ая горизонта́ль·н·ая (nucl) range to point of graze

~, по·сто·я́нн·ая constant/fixed range

~, при·це́ль·н·ая (gunn) sighting range

~ рас·сто·я́ни·я distance, range

сигна́л· да́ль·ност·и (*m*) (rad, navig) range-correction signal

~, сре́д·н·яя intermediate range; (mil, rad) medium range

~, тек·у́щ·ая present range

дально·у·правл·я́ем·ый, -ая, -ое, (*a*) remote-controlled

Дальто́н·а, за·ко́н· (*m*) (chem) Dalton's law

дальтон·и́зм·, -а, (*m*) colour-blindness, (med) daltonism

дальто́н·ик·, -а, (*m*) colour-blind person, (med) daltonian

да́ль·ше (*comp and adv*) farther, further

дам (*fut 1st sing*) of да·ть (*perf*)

да́м·а, -ы, (*f*) lady; queen (at cards)

дама́ск·, -а, (*m*) Damascus steel

дама́сск·ий, -ая, -ое, (*a*) Damascus; Damascus-steel

да́мб·а, -ы, (*f*) (civ eng) barrage, levee, dyke

~, о·град·и́тельн·ая dyke

~, по·пере́ч·н·ая groyne, jetty

~, про·до́ль·н·ая levee, longitudinal dyke

~, струе·на·правл·я́ющ·ая training dyke/wall

дамижа́н·, -а, (*m*) demijohn (large flask)

да́м·к·а, -и, (*g pl*) -м·ок·, (*f*) king (at chess)

дамма́р·а, -ы, (*f*) (bot, chem) dammar

данбури́т·, -а, (*m*) (min) danburite

Да́ниел·я, элеме́нт· (*m*) (elec) Daniell cell

даннемори́т·, -а, (*m*) (min) dannemorite

да́н·н·ые, -ых, (*pl decl as adj*) data, findings, figures, information; grounds (in reasoning)

~, вс·по·мог·а́тельн·ые contributary evidence

~ в·ход·н·ые incoming data (computers), (rad) data in

~, дискре́т·н·ые sampled data (computers)

~, ис·ход·я́щ·ие (rad) data out

~, рабо́ч·ие operational/operating data

~ характер·и́стик·и performance figures

~, цифр·ов·ы́е figures, data figures, numerical data

да́·нный, -ая, -ое, (*past part pass*) of да·ть (*perf*) see да·ва́ть; (*pl decl as adj*) see да́н·н·ые

~ велич·ин·а́ (*f*) (math) datum, given value/quantity

~ с·луч·а́й (*m*) present/this case, case on hand

~ с·равн·е́ни·я datum line

~ температу́р·а (*f*) given temperature, temperature in question

да́нсинг·, -а, (*m*) (arch) dance hall/floor

дар·, -а, (*nom pl*) -ы́, (*g pl*) -о́в, (*m*) gift

дарвин·и́зм·, -а, (*m*) (biol) Darwinism

дар·и́тел·ь, -я, (*m*) donor, giver

дар·и́ть, -я́т, (*imp*) make a present, give, grant, present to

дар·ова́ни·е, -я, (*n*) talent, gift

дар·ов·о́й, -а́я, -о́е, (*a*) gratuitous, free

да́р·ом (*adv*) free, free of charge, gratis; to no purpose, in vain

не да́р·ом not in vain, not for nothing

дарси́ (*m indecl*) darcy (unit of permeability of a porous material)

Дарси́, за·ко́н· (*m*) (hydr) D'Arcy's law

дарсонвал·иза́ци·я, -и, (*f*) (med, elec) high frequency treatment, D'Arsonvalism

да́р·ственн·ый акт· (*m*) (law) grant, deed of settlement

даст (*fut 3rd sing*) of дат·ь (*perf*)

да́т·а, -ы, (*f*) date

дата·тро́н·, -а, (*m*) (autom) datatron

да́·тельн·ый, -ая, -ое, (*a*) (gram) dative

дат·и́ровани·е, -я, (*n*) dating, age determination

дат·и́р·овать, -уют, (*imp and perf*) date; determine the date/age of; -ся (*pass*) belong to (a certain date/era)

дат·и́руем·ый, -ая, -ое, (*pres part pass*) of дат·и́р·овать; datable

дато·ли́т·, -а, (*m*) (min) datolite

да́т·ск·ий, -ая, -ое, (*a*) Danish

да́т·очн·ый, -ая, -ое, (*a*) of да́т·а

датури́н·, -а, (*m*) (chem) daturine, hyoscyamine

да́т·чик·, -а, (*m*) (instr) primary/sensing element/unit; gauge; transmitter, transducer

~ воз·ду́ш·н·ой ско́р·ост·и air mileage unit, a.m.u.

~ выс·от·ы́ altitude data unit (of rocket)

~ давл·е́ни·я pressure cell, pressure-sensitive element, mechanical pressure gauge

~, дифференциа́ль·н·ый ртут·н·ый differential-pressure mercury transmitter

~ и́мпульс·ов impulser

~, индукти́в·н·ый inductance pickup, variable inductance pickup

~, магнито·модуляцио́н·н·ый magnetic modulation pickup

~ на·ча́л·а от·с·чёт·а starting-pulse generator

~, нейтро́н·н·ый neutron-sensitive element

~ пере·гру́з·ок· accelerometer

~ по·то́к·а flow-sensing element

~, програ́мм·н·ый (electr) programme control unit (of computer)

~, рео·ста́т·н·ый potentiometer transmitter

~, сельси́н·· selsyn/synchro-transmitter

~, сумм·и́рующ·ий summating transmitter

~ температу́р·ы temperature pickup/sensing element

~, тензо·метр·и́ческ·ий strain gauge

~ угл·а́ скольж·е́ни·я (a/c) yaw detector

~ угл·ов·о́го по·лож·е́ни·я angular position pickup

~ Хо́лл·а Hall pickup

да·ть, (*fut 1st, 3rd sing, 1st 2nd 3rd pl*) дам, даст, дад·и́м, дад·и́те, дад·у́т (*perf*) see да·ва́ть

даусони́т·, -а, *(m)* (min) dawsonite

даутéрм·, -а, *(m)* (chem) Dowtherm

дáфни·я, -и, *(f)* (ent) water flea, *Daphnia*

дáхпризм·а, -ы, *(f)* (opt) Amici/roof prism

даци́т·, -а, *(m)* (min) dacite

дá·ч·а, -и, *(f)* giving; (agr) rate; plot, lot; country house, country cottage, weekend cottage, dacha

~, сýт·очн·ая (agr) daily rate

дá·ющ·ий, -ая, -ее, *(pres part act)* of **да·вáть;** *(n decl as adj)* see **дáт·чик·**

~ при·бóр· *and see* **дáт·чик·** (autom) transmitter, data unit

дб *(abbr)* of **деци·бéл·** (elec) decibel (unit of power-level difference)

дв·а two

двадцати·грáн·ник·, -а, *(m)* (math, cryst) icosahedron

двадцати·дн·éвн·ый, -ая, -ое, *(a)* twenty-day

двадцати·лéт·и·е, -я, *(n)* twentieth anniversary/birthday; twenty years, twenty-year period

двадцати·мéст·н·ый, -ая, -ое, *(a)* twenty-seat/berth/place

двадцати·пяти·лéт·н·ий, -яя, -ее, *(a)* twenty-five-year

двадцати·угóль·ник·, -а, *(m)* (math, cryst) icosagon

двадцати·четырёх·грáн·ник·, -а, *(m)* (cryst) trisoctahedron

два·дцáт·ый, -ая, -ое, *(a)* twentieth

двá·дцат·ь, -й, *(i)* **-ью,** *(f)* twenty

дважды- *(component)* double

двá·жды *(adv)* twice

~ маг·и́ческ·ий изо·тóп· (nucl) double magic isotope

дв·е *(fem)* of **дв·а** two

двенадцати·вáл·ков·ый стан· *(m)* (met roll) Rohn mill, twelve-high mill

двенадцати·грáн·ник·, -а, *(m)* dodecahedron

двенадцати·пёрст·н·ый, -ая, -ое, *(a)* (anat) duodenal

двенадцати·пéстик·ов·ый, -ая, -ое, *(a)* (bot) dodecagynous

двенадцати·угóль·ник·, -а, *(m)* dodecagon

двенадцати·час·ов·óй, -áя, -óе, *(a)* twelve-hour; twelve o'clock

две·нá·дцат·ый, -ая, -ое, *(a)* twelfth

две·нá·дцат·ь, -и, *(i)* **-ью,** *(f)* twelve

двере·съ·ём·н·ое у·стрóй·ств·о *(n)* door extractor (coke oven)

двере·экстрáктор·, -а, *(m)* door extractor (coke oven)

двéр·к·а, -и, *(g pl)* **-р·ок·,** *(f)* *(dim)* of **двер·ь**

двер·н·óй, -áя, -óе, *(a)* door

~ вы·ключ·áтел·ь *(m)* (elec) door switch

~ на·клáд·к·а *(f)* hasp and staple

~ про·ём· *(m)* doorway

двéр·ц·а, -ы, *(i)* **-ей;** *(g pl)* **-р·ец·** *(f)* *(dim)* of **двер·ь;** door, hatch

~, тóп·очн·ая fire door (of boiler)

двер·ь, -и, *(nom pl)* **-и,** *(g pl)* **-éй,** *(d pl)* **-я́м,** *(f)* door

~, раз·движ·н·áя sliding door

~, штóр·н·ая roller-blind door

двé·сти, *(g, d, i, p)* **двух·сóт, двум·стáм, двумя·стáми, двух·стáх;** two hundred

дви́г·ател·ь, -я, *(m)* *(see also under associated adjectives)* engine, motor, prime mover; (autom) actuator

~, а·син·хрóн·н·ый (elec) asynchronous/induction motor

~ без над·дý·в·а unsupercharged engine

~, бензи́н·ов·ый petrol/gasoline engine

~, бес·клáпан·н·ый sleeve-valve engine

~, бес·компрéссор·н·ый solid/air-less-injection engine

~, бес·кривоши́п·н·ый axial engine

~, быстро·хóд·н·ый high-speed engine

~ внýтр·енн·его с·гор·áни·я internal combustion engine

~, водо·столб·ов·óй hydraulic motor

~, воз·дýш·н·о-реакти́в·н·ый jet engine, thermal jet engine

~ воз·дýш·н·ого о·хлажд·éни·я air-cooled engine

~, газо·турби́н·ный gas-turbine engine

~ -генерáтор·, -а, *(m)* (elec) motor-generator

~ двой·н·óго дéй·стви·я double-acting engine

~ двой·н·óго раз·реш·éни·я reciprocating compound engine

~ двой·н·óго рас·шир·éни·я double-expansion engine

~, двух·контýр·н·ый трубо·реакти́в·н·ый by-pass turbo-jet engine, ducted-fan turbo-jet engine

~, двух·ря́д·н·ый double-bank engine

~, двух·тáкт·н·ый two-stroke engine

двиг·ател·ь

~, **жи́дк·остн·ый ракет·н·ый** liquid-propellant rocket engine

~, **звездо·обра́з·н·ый** (air) radial engine

~, **золотни́к·ов·ый** sleeve-valve engine

~, **калор·иза́торн·ый** semi-diesel

~, **кант·ова́льн·ый** (elec) barring motor

~, **коллектор·н·ый** (elec) commutator motor

~, **компа́унд·н·ый** (elec) compound-wound motor

~, **комплекс·н·ый** compound engine

~, **конденса́тор·н·ый** (elec) capacitor motor

~, **коромы́сл·ов·ый** (st eng) beam engine

~, **коротко·за·мк·ну́т·ый** (elec) squirrel-cage motor

~, **крейцко́пф·н·ый** crosshead engine

~, **ма́рш·ев·ый** (rocket) sustainer, sustainer motor, cruise propulsion unit

~, **много·бло́ч·н·ый** multibank engine

~ **много·кра́т·н·ого рас·шир·ени·я** multiple-expansion engine

~, **много·оборо́т·н·ый** high-speed engine

~, **момент·н·ый** torque motor

~, **мо́щ·н·ый** (elec) large power motor; high power/duty engine

~, **на·вес·н·о́й** outboard engine

~ **на·жи́м·н·ого у·стро́й·ств·а** (met roll) screw-down motor

~, **нефт·ян·о́й** oil engine

~, **одно·кра́т·н·ого рас·шир·ени·я** single-expansion engine, simple steam engine, simple reciprocating engine

~, **ориент·и́руем·ый** (rocket) gimbaled motor

~, **ориент·и́рующ·ийся** (rocket) swivelling motor

~, **основ·н·о́й** (rocket) sustainer motor

~, **перв·и́чн·ый** prime mover

~, **пневмат·и́ческ·ий** air motor, compressed-air motor

~, **под·вес·н·о́й** (rocket) pivoted motor

~, **под·синхрон·н·ый реакти́в·н·ый** (elec) subsynchronous reluctance motor

~, **порох·ов·о́й ракет·н·ый** solid-propellant rocket engine

~, **поршн·ев·о́й** piston/reciprocating engine

двиг·ател·ь

~, **поршн·ев·о́й пневмат·и́ческ·ий** piston-type compressed-air motor, reciprocating-type compressed-air motor

~ **прост·о́го де́й·стви·я** single-acting engine

~ **прост·о́го рас·шир·ени·я** single-expansion engine, simple steam engine, simple reciprocating engine

~, **прямо·то́ч·н·ый воздушно реакти́в·н·ый** ram jet engine

~, **пульс·и́рующ·ий воздушно·ре·акти́в·н·ый** pulse-jet engine

~, **ракет·н·ый** rocket engine/motor

~, **реакти́в·н·ый** (*see also* **реакти́в·н·ый двиг·а́тел·ь**) jet engine; reaction-propulsion engine (as generalization); (elec) reluctance motor

~, **реверси́в·н·ый** directly reversible engine (diesel)

~, **регул·и́руем·ый** (elec) variable-speed motor

~, **реду́ктор·н·ый** geared engine/motor

~, **репульсио́н·н·ый** (elec) repulsion motor

~, **ротацио́н·н·ый** vane-type air motor, turbine-type air motor

~, **рул·ев·о́й** (naut) steering engine

~, **ря́д·н·ый** (ICE) in-line engine

~ **с вну́тр·енн·ым смесе·образ·о·ва́ни·ем** compression-ignition engine, diesel

~ **с вос·пламен·ени·ем от с·жа́т·и·я** compression-ignition engine

~ **с над·ду́·в·ом** supercharged engine

~ **-с·вя́з·к·а, -и,** (*f*) (rocket) multiple barrel/rocket motors

~, **серие́с·н·ый** (elec) series-characteristic motor, series motor

~, **синхрон·и́рующ·ий** (elec) synchro

~, **синхро́н·н·ый** (elec) synchronous motor

~ **с·ме́ш·анн·ого с·гор·а́ни·я** engine working on dual cycle

~, **соленои́д·н·ый** solenoid-operated actuator (servo-systems)

~, **ста́рт·ов·ый** (elec) starter motor

~, **ста́рт·ов·ый реакти́в·н·ый** (a/c) rocket-assisted take-off unit

~, **стацпонар·н·ый** stationary engine

~, **суд·ов·о́й** marine engine

~ **тверд·ого то́плив·а, ракет·н·ый** solid-propellant rocket engine

~ **тёмн·ого жи́дк·ого то́плив·а** heavy-oil engine

дви́г·ател·ь
~, **тепл·ов·о́й** heat engine
~, **терм·и́ческ·ий** heat engine
~, **тихо·хо́д·н·ый** slow-speed engine
~ **трёх·крат·н·ого рас·шир·е́ни·я** triple expansion engine
~, **тро́нк·ов·ый** trunk-piston engine
~, **турбо·вентиля́тор·н·ый реакти́·в·н·ый** turbo-fan jet engine
~, **турбо·винт·ов·о́й** (air) turbo-prop engine
~, **турбо·компре́ссор·н·ый воз·ду́·ш·н·о-реакти́в·н·ый** turbo-compressor jet engine
~, **турбо·реакти́в·н·ый** and see **турбо·реакти́в·н·ый дви́г·ател·ь** (air) turbo-jet engine
~ **тяжёл·ого жи́дк·ого то́плив·а** heavy-oil engine
~, **фла́нц·ев·ый** (elec) flanged motor
~ **четвёр·н·ого рас·шир·е́ни·я** quadruple-expansion engine
~ **универса́ль·н·ый** universal motor
~, **четырёх·та́кт·н·ый** four-stroke engine
~, **шунт·ов·о́й** (elec) shunt-wound motor
~, **электро·магни́т·н·ый** solenoid-operated actuator (servos)
дви́г·ательн·ый, -ая, -ое, (a) of **дви́г·ател·ь**; (anat) motor
~ **нерв·** (m) (anat) motor nerve
дви́г·а·ть, -ют, ог дви́ж·ут, (imp) (trans) move, work, operate (machinery etc.), set in motion; **-ся** (pass); (intrans) move, move off, work, operate
движ·е́ни·е, -я, (n) movement, motion, travel, flow; propulsion; traffic
~, **без·вихр·ев·о́е** (hydr) viscous/laminar/streamline/steady flow
~ **бок·ов·о́е** lateral movement/travel **у·равн·е́ни·е бок·ов·о́го движ·е́·ни·я** (n) (aerodynam) lateral equation
~, **бро́ун·овск·ое** (phys) Brownian movement
~, **вихр·ев·о́е** (hydr) vortex/eddy motion/flow
~, **волн·ов·о́е** wave motion, oscillation
~, **во́ль·н·ые** (pl) Swedish drill
~ **га́з·а** gas flow
~ **гру́з·ов** goods/cargo/freight traffic
~, **двух·пу́т·н·ое** two-lane traffic
~, **до·ба́в·очн·ое** increment motion
~, **железно·доро́ж·н·ое** railway traffic
~ **зем·н·о́й о́с·и** (astron) nutation

движ·е́ни·е
~, **и́ст·инн·ое** (astron) proper motion
~, **капилля́р·н·ое** capillary migration
~ **материк·о́в** (geol) continental drift
~, **мах·ов·о́е** flapping
~, **не·у·станов·и́вш·ееся** unsteady flow
~, **обра́т·н·ое** counter motion
~, **пере·но́с·н·ое** translatory movement
~, **пло́ск·ое** translation, translational motion
~ **по о́с·и X** X-motion
~ **по фа́з·е** phase motion
~ **по́люс·ов земл·и́** (astron) nutation
~, **профессиона́ль·н·ое** trades-union movement
~, **раз·ры́в·н·ое** discontinuous motion
~ **рас·твор·и́тел·я** solvent development
~, **с·ло́ж·н·ое** absolute motion
~ **со·противл·е́ни·я** resistance movement
~, **угл·ов·о́е** angular motion
~, **у́лич·н·ое,** traffic, vehicular traffic
дви́ж·имост·ь, -и, (f) (law) movables, movable property
дви́ж·им·ый, -ая, -ое, (pres part pass) of **дви́г·а·ть** movable
дви́ж·ител·ь, -я, (m) (mech) propeller, propelling agent, propulsive agent
~, **кры́ль·чат·ый** (shipb) Voith–Schneider propeller, cycloidal propeller (U.S.)
~, **реакти́в·н·ый** (a/c) reaction nozzle
движ·о́к·, -ж·к·а́, (m) cursor, slide; (elec) contact blade
дви́ж·ущ·ий, -ая, -ее, (pres part act) of **дви́г·а·ть**; motive, moving
~ **си́л·а** (f) motive force
дви́ж·ущ·ийся, -аяся, -ееся, (pres part act) of **дви́г·аться** moving, working, running
дви́н·уть, -ут, (perf) see **дви́г·а·ть**
дво́·е, -и́х, (d) -и́м, (i) -и́ми or **двумя́ (p) -и́х,** two, a couple, a pair
двое·бра́ч·и·е, -я, (n) (law) bigamy
двое·ду́ш·и·е, -я, (n) duplicity, double dealing
дво·е́ни·е, -я, (n) (v n) of **дво·и́ть**
двое·то́ч·и·е, -я, (n) (gram) colon
дво·и́лн·о-мер·и́тельн·ая маши́н·а (f) (text) folding and measuring machine
дво·и́льн·ый, -ая, -ое, doubling, folding into two; halving, dividing/splitting into two

дво·и́льн·ый, -ая, -ое
~ маши́н·а   (*f*) splitting machine (leather)
дво·и́м   (*dat*) of дво́·е
дво·и́ми   (*instr*) of дво́·е
дво·и́ть, -я́т,   (*imp*) divide into two; double
дво·и́х   (*gen*) of дво́·е
дво·и́чн·о-десят·и́чн·ая систе́м·а   (*f*) binary-coded decimal system (computers)
дво·и́чн·о-код·и́рованн·ый, -ая, -ое, (*a*) binary-coded, binary-coded notation (computers)
дво·и́чн·о-пяти·ри́чн·ый код·   (*m*) biquinary code (computers)
дво·и́чн·ый, -ая, -ое,   (*a*) binary
~ знак·   (*m*) bit, binary digit
~ систе́м·а с·числ·е́ни·я   (*f*) binary number system
~ с·чёт·н·ое у·стро́й·ств·о   (*n*) binary counter
~ числ·о́   (*n*) bit, binary digit
двойк·а, -и,   (*g pl*) дво́·ек·, (*f*) two, a two, a number two; two-piece suit, two-oared boat; poor (class marking); (agr) second ploughing; (sport) doubles match
двой·ни́к·, -а́,   (*m*) double, twin
~, четвер·н·о́й   fourling
двойник·нове́ни·е, -я,   (*n*) (cryst) twinning
двойник·ова́ни·е, -я,   (*n*) (cryst) twinning
двойник·ованн·ый, -ая, -ое,   (*a*) twinned
двой·ни́ков·ый, -ая, -ое,   (*a*) twin' twinning
~ про·сло́й·к·а   (*f*) (cryst) twin band
двой·ников·ая про·сло́й·к·а за·кли́·н·ив·ш·аяся   (cryst) twin wedge
~ ~ клино·обра́з·н·ая   (cryst) twin wedge
двой·н·о́й, -а́я, -о́е,   (*a*) double, twofold, dual, binary, duplex; (bot) geminate, binate, paired
~ бухгалте́ри·я   (*f*) double-entry book-keeping
~ де́й·стви·е   (*n*) double action; (gen) double-acting (*used as adj*)
~ дио́д·пента́д·   (*m*) (elec) twin/double diode pentode
~ пере·воро́т·   (*m*) (air) barrel roll
~ с·вя́з·ь   (*f*) (chem) double bond
~ систе́м·а   (*f*) binary system
~ с·мес·ь   (*f*) binary mixture

двой·н·о́й, -а́я, -о́е
~ со·един·е́ни·е   (*n*) (chem) binary compound
~ с·раст·а́ни·е   (*n*) (cryst) twinning
~ ту́мблер·   (*m*) (elec) double-toggle actuator/switch
~ тя́г·а   (*f*) four-wheel drive
~ у·правл·е́ни·е   (*n*) dual control
~ цикл·   (*m*) binary cycle
~ элеме́нт·   (*m*) (math) fixed point
двой·н·я, -и,   (*g pl*) дво́·ен·, (*f*) twins
двой·ственност·ь, -и,   (*f*) duality; duplicity
за·ко́н· двой·ствен·лост·и   (*m*) duality principle
двой·ственн·ый, -ая, -ое,   (*a*) dual; double-faced
двой·чат·ый, -ая, -ое,   (*a*) didymous; (bot) geminate
двор·, -а́,   (*m*) court, courtyard, yard
~, груз·ов·о́й   (rail) freight/goods yard
~, огн·ев·о́й   (rocket) firing bay
~, около·ство́ль·н·ый   (min) shaft station/bottom
~, руд·ни́чн·ый   stock yard (of blast furnace); (min) shaft bottom/station
~, ското·при·го́н·н·ый   (agr) stockyard
~, ши́хт·ов·ый   (met) stockyard
двор·е́ц·, -р·ц·а́,   (*i*) -р·ц·о́м, (*m*) palace
дво́р·ик·, -а,   (*m*) (*dim*) of двор·
~, плеохро·и́чн·ый   (min) pleochroic halo
~ у́сть·иц·а, за́д·н·ий   (bot) stomatal back cavity
дво́р·ник·, -а,   (*m*) concierge, door-keeper, porter
дво·я́к·ий, -ая, -ое,   (*a*) double, ambiguous, in two ways, of two kinds
дво·я́к·о   (*adv*) in two ways, of two kinds, doubly
дво·як·о-   (*component*) (*see also* дв·у·, дв·ух-) double-, bi-, di-
двояко·во́·гн·ут·ый, -ая, -ое,   (*a*) double concave, biconcave, concavo-concave
двояко·вы́·пук·л·ый, -ая, -ое,   (*a*) double convex, lenbicular, biconvex, convexo-convex
двояко·горо́д·чат·ый, -ая, -ое,   (*a*) (bot) bicrenate
двояко·ды́ш·ащ·ие, -их,   (*m decl as adj*) (zool) dipnoans, lung-fishes, *Dipnoi, Dipneusti*
двояко·за·зу́бр·енн·ый   (bot) bicrenate
двояко·из·о́·гн·ут·ый, -ая, -ое,   (*a*) birecurvate

двояко·пе́р·ист·ый, -ая, -ое, (*a*) bipinnate

двояко·пре·ломл·я́ющ·ий, -ая, -ое, (*a*) (opt) birefringent, double-refracting

двояко·пи́ль·чат·ый, -ая, -ое, (*a*) (bot) biserrate, biseriate, double serrate

дв·у- (*component*) (*see also* дво·я́к·о-, дв·у́х·-) two-, bi-, di-, double, diplo-

дву·аперту́р·н·ый, -ая, -ое, (*a*) diorate

дву·а́том·н·ый, -ая, -ое, (*a*) diatomic, biatomic

~ основ·а́ни·е (*n*) (chem) diacid base

дву·бо·ев·а́я молот·к·а́ (*f*) double-faced hammer

дву·боро́зд·чат·ый, -ая, -ое, (*a*) (bot) bisulcate

дву·бром·за·мещ·ённ·ое со·един·е́-ни·е (*n*) dibromo compound

дву·бро́м·ист·ый, -ая, -ое, (*a*) di-bromide, bromide (lower or -ous)

дву·брюш·н·о́й, -а́я, -о́е, (*a*) (zool) digastric

дву·вале́нт·н·ост·ь, -и, (*f*) (chem) bivalence

дву·вале́нт·н·ый, -ая, -ое, (*a*) bi-valent, divalent

дву·вал·ко́в·ый, -ая, -ое, (*a*) two-roll

дву·вариа́нт·н·ый, -ая, -ое, (*a*) (chem) bivariant

дву·ветв·и́ст·ый, -ая, -ое, (*a*) (biol) biramose, biramous

дву·ви́д·н·ый, -ая, -ое, (*a*) (biol) dimorphous, dimorphic

дву·винно·ки́сл·ый, -ая, -ое, (*a*) (chem) bitartrate of

дву·втор·и́чн·ый, -ая, -ое, (*a*) di-secondary

дву·гало́ид·н·ый, -ая, -ое, (*a*) (chem) dihalide

дву·гаме́т·н·ый, -ая, -ое, (*a*) (gen) digametic

дву·гла́в·ый, -ая, -ое, (*a*) (zool) bi-cipital, two-/double-headed

~ мы́шц·а (*f*) (anat) biceps

дву·гла́с·н·ый, -ая, -ое, (*a*) (ling) diphthong

дву·гнёзд·н·ый, -ая, -ое, (*a*) (bot) bilocular

дву·го́рб·ый, -ая, -ое, (*a*) two-humped; (rad) double-humped; (math) double-peak (of curves)

дву·гра́н·н·ый, -ая, -ое, (*a*) dihedral, two-sided

дву·гу́б·ый, -ая, -ое, (*a*) (bot) bi-labiate

дву·до́ль·н·ый, -ая, -ое, (*a*) (bot) dicotyledonous; (*m decl as adj*) di-cotyledon

дву·до́м·н·ый, -ая, -ое, (*a*) (bot) dioecious, diclinous

дву·дуг·ов·о́й, -а́я, -о́е, (*a*) double-arc

~ с·ва́р·к·а (*f*) double-arc welding

дву·ду́ж·н·ый, -ая, -ое, (*a*) two-arched; (zool) diapsid

дву·ды́ш·ащ·ие, -их, (*pl decl as adj*) = двояко·ды́ш·ащ·ие (q.v.)

дву·жа́бер·н·ый, -ая, -ое, (*a*) (zool) dibranchiate; (*pl decl as adj*) dibran-chiates, *Dibranchiata*

дву·жгу́тик·ов·ый, -ая, -ое, (*a*) (biol) biflagellate, biciliate

дву·же́н·н·ый, -ая, -ое, (*a*) (bot) digynous

дву·за·мещ·ённ·ый, -ая, -ое, (*a*) (chem) disubstituted, secondary

~ сол·ь фо́сфор·н·ой кисл·от·ы́ (*f*) secondary phosphate

~ фосфа́т· ка́льци·я (*m*) (chem) dicalcium phosphate

дву·за·ря́д·н·ый, -ая, -ое, (*a*) double-charged, doubly charged

дву·зерн·я́нк·а, -и, (*g pl*) -нок·, (*f*) (bot) emmer, *Triticum dicoccum*

дву·зна́ч·ност·ь, -и, (*f*) ambiguity

дву·зна́ч·н·ый, -ая, -ое, (*a*) two-digit; ambiguous

~ сигна́л· (*m*) (rail) two-aspect signal

дву·зу́б·ый, -ая, -ое, (*a*) (zool) bi-dentate, two-toothed

дву·карбокси́ль·н·ый, -ая, -ое, (*a*) (chem) dicarboxylic

дву·кил·ев·о́й, -а́я, -о́е, (*a*) (bot) bicarinate; (shipb) twin-keel

дву·ки́ль·н·ый, -ая, -ое, (*a*) (bot) bicarinate; (shipb) twin-keel

дву·кисл·о́тн·ый, -ая, -ое, (*a*) (chem) diacid

дву·ки́сл·ый, -ая, -ое, (*a*) (chem) diacid

дву·колёс·н·ый, -ая, -ое, (*a*) two-wheeled

дву·ко́л·к·а, -и, (*g pl*) -л·ок·, (*f*) dog-cart

дву·кон·и́ческ·ий, -ая, -ое, (*a*) biconic

дву·короб·очн·ый, -ая, -ое, (*a*) bicapsular

дву·кра́т·н·ый, -ая, -ое, (*a*) double, twofold; (telecom) two-channel

дву·кру́ж·н·ый, -ая, -ое, (*a*) two-circle

дву·кры́л·ый, -ая, -ое, (*a*) dipterous, two-winged; (*pl decl as adj*) (ent) dipterons, *Diptera*

~ семафо́р· (*m*) (rail) two-blade semaphore

дву·лепе́ст·н·ый, -ая, -ое, (*a*) (bot) dipetalous

дву·ле́т·ник·, -а, (*m*) (bot) biennial

дву·ле́т·н·ий, -яя, -ее, (*a*) biennial; bi-annual .

дву·ли́ст·н·ый, -ая, -ое, (*a*) (bot) bifoliate, diphyllous

дву·лист·о́чков·ый, -ая, -ое, (*a*) (bot) bifoliolate

дву·ло́паст·н·ый, -ая, -ое, (*a*) bilobate, two-lobed; (mech eng) twin/two-bladed, two-vaned

дву·луче·пре·ломл·е́ни·е, -я, (*n*) (opt) birefringence

дв·ум (*dat*) of **дв·а** two

дву·ме́р·н·ый, -ая, -ое, (*a*) two-dimensional

дву·мешк·о́в·ый, -ая, -ое, (*a*) (bot) bisaccate

дву·молекул·я́рн·ый, -ая, -ое, (*a*) bimolecular, dimolecular

двум·ста́м (*dat*) of **две́·сти**

дву·му́скуль·н·ый, -ая, -ое, (*a*) (zool) dimyaric, dimyarian

дв·умя (*instr*) of **дв·а** two

двумя·ста́ми (*instr*) of **две́·сти**

дву·ни́т·н·ый, -ая, -ое, (*a*) bifilar

дву·но́г·а, -и, (*f*) bipod

дву·но́г·ий, -ая, -ое, (*a*) two-legged, (zool) biped, bipedal

дву·о́·кис·ь, -и, (*f*) (chem) dioxide, (-ous) oxide

~ то́ри·я thorium dioxide, thoria

~ угле·ро́д·а carbon dioxide

дву·осно́в·н·ый, -ая, -ое, (*a*) (chem) dibasic; diatomic

~ кисл·от·а́ (*f*) diacid

дву·о́с·н·ый, -ая, -ое, (*a*) biaxial

дву·от·раж·е́ни·е, -я, (*n*) double reflection

дву·па́л·ый, -ая, -ое, (*a*) (zool) didactylous

двупарно·но́г·ие, -их, (*pl decl as adj*) (zool) *Diplopoda*

двупарно·рез·цо́в·ые, -ых, (*pl decl as adj*) (zool) *Duplicidentata*

дву·па́р·н·ый, -ая, -ое, (*a*) (bot) bijugate, bijugous, bigeminate

дву·перв·и́чн·ый, -ая, -ое, (*a*) diprimary

дву·пе́р·ист·ый, -ая, -ое, (*a*) (bot) bipinnate, (zool) bipenniform

дву·пле́ч·ий, -ая, -ее, (*a*) (mech) double/two-arm

дву·пол·ости·о́й, -а́я, -о́е, (*a*) two/double-cavity

~ гиперболо́ид· (*m*) (math) hyperboloid of two sheets

дву·по́л·ый, -ая, -ое, (*a*) bisexual

дву·по́л·ь·е, -я, (*n*) (agr) two-field crop rotation

дву·по́люс·н·ый, -ая, -ое, (*a*) *see* **двух·по́люс·н·ый**

дву·пре·ломл·е́ни·е, -я, (*n*) (opt) birefringence

дву·пре·ломл·я́ющ·ий, -ая, -ее, (*a*) birefringent, double-refracting

дву·раз·де́ль·н·ый, -ая, -ое, (*a*) (bot) bifid, bipartite

дву·рез·цо́в·ые, -ых, (*pl decl as adj*) (zool) *Diprodontia*

дву·ро́г·ий, -ая, -ое, (*a*) two-horned, (zool) bicornate, bicornute

дву·ро́т·ый, -ая, -ое, (*a*) two-mouthed, (zool) distomous

дву·ру́ч·н·ый, -ая, -ое, (*a*) two-handed, bimanous; (med) bimanual

дву·ря́д·н·ый, -ая, -ое, (*a*) two/double-row/bank; biserial, two-series; (bot) distichous, bifarious, biseriate

~ звездо·обра́з·н·ый дви́г·ател·ь (*m*) (a/c) double-row radial engine

~ си́т·о (*n*) double-deck screen

дву·све́т·н·ый, -ая, -ое, (*a*) having two tiers of windows; having windows on opposite sides

дву·с·вя́з·н·ый, -ая, -ое, (*a*) (math) double connected

дву·семено·до́ль·н·ый, -ая, -ое, (*a*) (bot) dicotyledonous

дву·сём·я́нк·а, -и, (*g pl*) **-нок,** (*f*) (bot) diachaenium

дву·сернисто·ки́сл·ый, -ая, -ое, (*a*) (chem) bisulphate of

дву·сер·ни́ст·ый, -ая, -ое, (*a*) (chem) disulphide of, (-ous) sulphide

~ желе́з·о (*n*) iron disulphide, ferrous sulphide

~ ура́н· (*m*) uranous sulphide

дву·серно·ки́сл·ый, -ая, -ое, (*a*) bisulphate of

дву·се́р·н·ый, -ая, -ое, (*a*) (chem) disulphuric, disulfuric

дву·силици́д·, -а, (*m*) disilicide

дву·с·ка́т·н·ый, -ая, -ое, (*a*) having two sloping surfaces, ridged

~ колес·о́ (*n*) twin-tyred wheel

~ кры́·ш·а (*f*) ridge roof

дву·с·кос·н·о́й, -а́я, -о́е, (*a*) double-bevel

дву·с·лог·ов·о́й, -а́я, -о́е, (*a*) disyllabic

дву·с·ло́ж·н·ый, -ая, -ое, (*a*) disyllabic

дву·сло́й·н·ый, -ая, -ое, (*a*) two-layer, double-layer, two-ply; (biol) diploblastic; (met) duplex, clad

~ **атмосфе́р·а** (*f*) (meteor) two-level atmosphere

~ **о·плёт·к·а** (*f*) double braiding

дву·с·ме́н·н·ый, -ая, -ое, (*a*) two-shift

дву·с·мы́сл·енн·ый, -ая, -ое, (*a*) ambiguous, equivocal

дву·ство́ль·н·ый, -ая, -ое, (*a*) double-barrelled

дву·с·тво́р·чат·ый, -ая, -ое, (*a*) two-leaf, double-leaf (of doors), double-sash (of windows); double (of doors, gates); (biol) bivalve, bivalvular; (anat) bicuspid

~ **моллю́ск·и** (zool) bivalves, *Bivalvia, Lamellibranchia*

дву·сторо́н·н·ий, -яя, -ее, (*a*) bilateral, two-sided, two-way, two-faced, bifaced, bifacial, double-end

~ **в·ход·** (*m*) double entry (of compressors etc.)

~ **де́й·стви·е** (*n*) (gen) two-action (*used as adj*)

~ **кали́бр·** (*m*) go-and-not-go gauge

~ **ключ·** (*m*) (elec) two-position switch

~ **котёл·** (*m*) double-ended boiler **ли́н·и·я дву·сторо́н·н·его де́й··стви·я** (*f*) (telecom) two-action line

~ **ми́шен·ь** (*f*) (TV) two-sided target

~ **при·бо́р·** (*m*) centre-zero measuring instrument

~ **с·вя́з·ь** (*f*) (telecom) two-way communications, duplex channel

~ **шкал·а́** (*f*) (instr) centre-zero scale

дву·ступ·е́нчат·ый, -ая, -ое, (*a*) two-stage; two-step

дву·та́вр·о́, -а́, (*n*) (met) joist, RSJ beam, I-beam, H-beam

дву·та́вр·о́в·ый, -ая, -ое, (*a*) of **дву·та́вр·о́**

~ **ба́лк·а** (*f*) = **дву·та́вр·о́**

дву·тре́т·и́чн·ый, -ая, -ое, (*a*) (chem) di-tertiary

дву·тычи́н·ков·ый, -ая, -ое, (*a*) (bot) diandrous

дву·угле·ки́сл·ый, -ая, -ое, (*a*) (chem) bicarbonate of

дву·уго́ль·ник·, -а, (*m*) (math) lune

дву·узл·ов·о́й, -а́я, -о́е, (*a*) binodal

дву·уксусно·ки́сл·ый, -ая, -ое, (*a*) (chem) diacetate of

дву·у́ст·к·а, -и, (*g pl*) **-т·ок·,** (*f*) (zool) fluke

дву·утро́б·к·а, -и, (*g pl*) **-б·ок·,** (*f*) (zool) didelphis, marsupial

дву·утро́б·н·ый, -ая, -ое, (*a*) (anat) didelphic, amphidelphic

дв·ух (*gen/prep*) of **дв·а**

дв·ух- (*component*) (*see also* **дв·у-, дво··як·о-**) two-, bi-, di-, double, twin

двух·а́том·н·ый, -ая, -ое, (*a*) diatomic, biatomic

двух·ба́лль·н·ый, -ая, -ое, (*a*) 2-degree, No.2

двух·вале́нт·н·ый, -ая, -ое, (*a*) (chem) bivalent, divalent

двух·вал·ко́в·ый, -ая, -ое, (*a*) two-high (of rolling mill); two-roll; two-bowl

дву·вальц·о́в·ый, -ая, -ое, (*a*) twin-drum

двух·вид·ов·о́й, -а́я, -о́е, (*a*) two-way; (elec) two-mode

двух·винт·ов·о́й, -а́я, -о́е, (*a*) twin-screw; twin-propeller

дву·хвост·к·и, -т·ок·, (*f pl*) (ent) *Diplura*

двух·год·и́чн·ый, -ая, -ое, (*a*) two-year, biennial

двух·год·ова́льн·ый, -ая, -ое, (*a*) two-year-old

двух·групп·ов·о́й, -а́я, -о́е, (*a*) two-group

двух·дека́д·н·ый, -ая, -ое, (*a*) twenty-day

двух·жи́ль·н·ый, -ая, -ое, (*a*) (elec) two-core, two-conductor, twin

двух·за·мещ·ённ·ый, -ая, -ое, (*a*) *see* **дву·за·мещ·ённ·ый**

двух·за·хо́д·н·ый, -ая, -ое, (*a*) double (of screw threads)

двух·зве́н·ник·, -а, (*m*) (a/c) torsion link (of undercarriage)

двух·зве́н·н·ый, -ая, -ое, (*a*) two-link, two-section, two-unit

~ **фильтр·** (*m*) (elec) two-section/mesh filter

двух·и́мпульс·н·ый, -ая, -ое, (*a*) two-pulse; (autom) two-factor

двух·ка́льц·иев·ый ферри́т· (*m*) dicalcium ferrite

двух·ка́мер·н·ый, -ая, -ое, (*a*) two-chamber/compartment/cell; (bot) bilocular

двух·кана́ль·н·ый, -ая, -ое, (*a*) two-channel, two-groove

~ **голо́в·к·а** (*f*) double-gap head (of tape recorder)

двух·каска́д·н·ый, -ая, -ое, (*a*) (elec) two-stage

двух·километр·о́вк·а, -и, (*g pl*) **-вок·,** (*f*) scale 1/200 000 map

двух·ко́вш·ев·ый, -ая, -ое, (*a*) twin-bucket

двух·кол·е́йн·ый, -ая, -ое, (*a*) (rail) double-track

двух·коле́н·чат·ый, -ая, -ое, (*a*) (mech eng) double-throw

двух·компоне́нт·н·ый, -ая, -ое, (*a*) bi-/two-component, binary

~ **систе́м·а** (*f*) binary system

~ **то́пл·ив·о** (rocket) bi-propellant

двух·консо́ль·н·ый, -ая, -ое, (*a*) double-cantilever

двух·конта́кт·н·ая ви́л·к·а (*f*) (elec) two-pin plug

двух·конту́р·н·ый, -ая, -ое, (*a*) (elec) two-circuit; double-loop; (ICE) dual cycle; limbate

~ **клистро́н·** (*m*) (rad) double-resonator klystron

~ **турбо·реакти́в·н·ый дви́г·ател·ь** (*m*) (a/c) turbo-fan, ducted fan, turbine engine

двух·копы́т·н·ый, -ая, -ое, (*a*) (zool) cloven-footed

двух·ле́т·и·е, -я, (*n*) two years, two-year period

двух·ле́т·н·ий, -яя, -ее, (*a*) two-year, bi-annual; two-year-old

двух·лин·е́йн·ая раз·вёрт·к·а (*f*) (elec) two-trace sweep

двух·ло́паст·н·ый, -ая, -ое, (*a*) *see* **дву·ло́паст·н·ый**

~ **винт·** (*m*) twin-bladed propeller

~ **долот·о́** (*n*) (oil) two-way bit

дву·хло́р·ист·ый, -ая, -ое, (*a*) (chem) dichloride of, chloride (lower, -ous)

~ **цирко́ни·й** (*m*) zirconium dichloride/bichloride

~ **ма́рган·ец·** (*m*) manganese dichloride, manganous chloride

дву·хлор·за·мещ·ённ·ое со·един·е́·ни·е (*n*) dichloro compound

дву·хлор·о́·кис·ь, -и, (*f*) oxydichloride

двух·луч·ев·о́й, -а́я, -о́е, (*a*) double-beam, dual-beam; (astron) two-pronged

~ **спектро·скоп·и́·я** (*f*) double-beam spectroscopy

двух·ма́чт·ов·ый, -ая, -ое, (*a*) two-masted

двух·маши́н·н·ый, -ая, -ое, (*a*) twin-engined

~ **аппара́т·** (*m*) (text) double card machine

двух·ме́р·н·ый, -ая, -ое, (*a*) two-dimensional

~ **схе́м·а** (*f*) (elec) network analyzer

двух·ме́ст·н·ый, -ая, -ое, (*a*) two-seater/berth/place

двух·ме́сяч·н·ый, -ая, -ое, (*a*) two-month, bimonthly; two-month-old

двух·минера́ль·н·ый, -ая, -ое, (*a*) (geol) binary, bimineral

двух·мото́р·н·ый, -ая, -ое, (*a*) twin-engined, two-engined

двух·неде́ль·н·ый, -ая, -ое, (*a*) fortnightly, two-week; two-week-old

двух·об·мо́т·очн·ый, -ая, -ое, (*a*) (elec) double-wound

двух·оборо́т·н·ый, -ая, -ое, (*a*) double (of screw threads)

двух·о·по́р·н·ый кла́пан· (*m*) double-beat/seat valve

двух·о́с·н·ый, -ая, -ое, (*a*) biaxial

двух·от·ва́ль·н·ый снего·о·чист·и́·тел·ь (*m*) double-cheek snow plough

двух·отве́рст·н·ый, -ая, -ое, (*a*) two-hole, two-entry

двух·па́луб·н·ый, -ая, -ое, (*a*) (naut) double-decked, two-decked

двух·пли́т·н·ый, -ая, -ое, (*a*) two-platen (of presses)

двух·позицио́н·н·ый, -ая, -ое, (*a*) (autom) two-position/step

~ **элеме́нт·** (*m*) (electr) bi-stable unit, binary cell

двух·по́лос·н·ый, -ая, -ое, (*a*) two-strip, two-band

двух·полу·перио́д·н·ый, -ая, -ое, (*a*) (rad) full-wave

~ **вы·прям·и́тел·ь** (*m*) (elec) full-wave rectifier

двух·по́люс·ник·, -а, (*m*) (elec) branch, two-terminal element

двух·по́люс·н·ый, -ая, -ое, (*a*) bipolar; (elec) two/double-pole, two-terminal; (telecom) double-current

~ **вы·ключ·а́тел·ь** (*m*) double-pole switch

~ **телеграф·и́ровани·е** (*n*) double-current telegraphy

двух·по·то́ч·н·ый, -ая, -ое, (*a*) two-flow, double-flow

двух·пре·де́ль·н·ый, -ая, -ое, (*a*) (instr) double-range

двух·про·вод·н·о́й, -а́я, -о́е, (*a*) (telecom) two-wire, open-wire, twin (of cables etc.)

~ **ли́н·и·я** (*f*) two-wire circuit

двух·про·лёт·н·ый, -ая, -ое, (*a*) (elec) double-transit; (civ eng) double-bay/span

двух·про·чёс·н·ый, аппара́т· (*m*) (text) double card machine

двух·пу́т·н·ый, -ая, -ое, (*a*) two-lane; (rail) two-track, double-line

двух·пя́т·очн·ый, -ая, -ое, (*a*) (text) two-butt (of latch needle)

двух·раз·ры́в·н·ый, -ая, -ое, (*a*) (elec) double-break

двух·рёбер·н·ый, -ая, -ое, (*a*) bicostate

двух·ребо́рд·н·ый, -ая, -ое, (*a*) double-rim/flange, U-shaped (of edges)

~ **колес·о́** (*n*) double-flange wheel

двух·ре́·ечн·ый, -ая, -ое, (*a*) two-rake

~ **классифика́тор·** (*m*) duplex classifier (ore dressing)

двух·режи́м·н·ый, -ая, -ое, (*a*) two-mode

двух·хромово·ки́сл·ый, -ая, -ое, (*a*) (chem) bichromate of, dichromate of

~ **ка́ли·й** (*m*) potassium bichromate/dichromate

дву·хро́м·ов·ый, -ая, -ое, (*a*) (chem) dichromic, bichromic

~ **кисл·от·а́** (*f*) (chem) dichromic/bichromic acid

двух·ро́тор·н·ый, -ая, -ое, (*a*) two-rotor, two-lobe (pumps)

двух·ру́чь·ев·ый, -ая, -ое, (*a*) two strand (continuous casting)

двух·ря́д·н·ый, -ая, -ое, (*a*) see **дву··ря́д·н·ый**

двух·секцио́н·н·ый, -ая, -ое, (*a*) two-section, two-unit, two-compartment

двух·с·ка́т·н·ый, -ая, -ое, (*a*) see **дву·с·ка́т·н·ый**

двух·сло́й·н·ый, -ая, -ое, (*a*) see **дву·сло́й·н·ый**

двух·со́т· (*gen*) of **две́·сти**

двух·сот·ле́т·и·е, -я, (*n*) bicentenary

двух·со́т·ый, -ая, -ое, (*a*) two-hundredth

двух·с·ре́з·н·ый, -ая, -ое, (*a*) double-shear

двух·ста́х (*prep*) of **две́·сти**

двух·с·тво́р·чат·ый, -ая, -ое, (*a*) see **дву·с·тво́р·чат·ый**

двух·степе́н·н·ый, -ая, -ое, (*a*) two-stage

~ **гиро·ско́п·** (*m*) two-frame gyroscope

двух·сто́·ечн·ый, -ая, -ое, (*a*) two-pole, double-pole (masts etc.); (mech eng) double-sided, double-standard

двух·сторо́н·н·ий, -яя, -ее, (*a*) see **дву·сторо́н·н·ий**

двух·стре́л·очн·ый, -ая, -ое, (*a*) two-pointer

двух·ступ·е́ньчат·ый, -ая, -ое, (*a*) two-stage; two-step

двух·су́т·очн·ый, -ая, -ое, (*a*) 48-hour, two-day

двух·тавр·о́в·ый, -ая, -ое, (*a*) of **дву·тавр·о́**

двух·та́кт·н·ый, -ая, -ое, (*a*) (ICE) two-stroke, two-cycle; (electr) push-pull

~ **магни́т·н·ый у·сил·и́тел·ь** (*m*) push-pull magnetic amplifier

двух·тари́ф·н·ый с·чёт·чик· (*m*) (elec) double-tariff/rate meter

двух·то́м·ник·, -а, (*m*) two-volume edition/work

двух·тона́ль·н·ый, -ая, -ое, (*a*) two-tone (sound)

двух·тон·ов·о́й, -а́я, -о́е, (*a*) (print) two-tone (colour)

двух·то́ч·ечн·ый, -ая, -ое, (*a*) double-dot, double-point

двух·фа́з·н·ый, -ая, -ое, (*a*) two-phase, biphase

двух·фонту́р·н·ый, -ая, -ое, (*a*) two-bed (circular knitting machine)

двух·ход·ов·о́й, -а́я, -о́е, (*a*) two-way, two-pass; double-throw (elec. switches); double (of screw threads)

~ **кран·** (*m*) two-way cock

двух·цве́т·н·ый, -ая, -ое, (*a*) two-colour, dichromatic

двух·це́л·ев·ый, -ая, -ое, (*a*) dual-purpose

двух·центр·ов·о́й, -а́я, -о́е, (*a*) two-point

двух·час·ов·о́й, -а́я, -о́е, (*a*) two-hour; two-o'clock

двух·част·и́чн·ые си́л·ы (*pl*) (phys) two-body forces

двух·ша́г·ов·ый винт· (*m*) two-pitch propeller

двух·электро́д·н·ый, -ая, -ое, (*a*) two-electrode, two-element

двух·эта́ж·н·ый, -ая, -ое, (*a*) (min) two-level, two-stage, two-deck, two-storey двухъ·я́дер·н·ый, -ая, -ое, (*a*) binucleate

двухъ·я́рус·н·ый, -ая, -ое, (*a*) two-layer, two-storey, double-deck, two-level

дву·цикл·и́ческ·ий, -ая, -ое, (*a*) dicyclic

дву·чле́н·, -а, (*m*) (math) binomial

дву·чле́н·ист·ый, -ая, -ое, (*a*) (zool) biarticulate

дву·чле́н·н·ый, -ая, -ое, (*a*) (math) binomial

дву·шка́ль·н·ый, -ая, -ое, (*a*) double-scale/dial; double-range

дву·шки́в·н·ый влок· (*m*) double block, two-sheave block

дву·шле́йф·ов·ый, -ая, -ое, (*a*) (rad) two-stub

дву·язы́ч·н·ый, -ая, -ое, (*a*) bilingual

ДДТ DDT (insecticide)

де- (*component*) de-, dis-

де·автома́т·иза́ци·я, -и, (*a*) breaking a habit

де·азо́т·иза́тор·, -а, (*m*) (chem) denitrator

де·алкил·ирова́ни·е, -я, (*n*) (chem) dealkylation

де·аэра́тор·, -а, (*m*) (st eng) deaerator, deaerating heater

~, атмосфе́р·н·ый atmospheric-pressure deaerating heater

~, терм·и́ческ·ий feed-water deaerator

деба́·евск·ий, -ая, -ое, (*a*) of Деба́·й Debye

~ спектр· (*m*) Debye spectrum

~ температу́р·а (*f*) (phys) Debye temperature

дебае·гра́мм·а, -ы, (*f*) (spectr) powder photograph/pattern

Деба́·й, -я, (*m*) Debye; debye (unit of dipole moment)

мéтод Деба́·я–Ше́ррер·а (*m*) (spectr) Debye–Scherrer method

де·бала́нс·, -а, (*m*) disbalance

дебарка́дер·, -а, (*m*) landing stage, floating station (rivers)

де·бензине́ (*indecl*) wash oil (for gas washing)

де́бет·, -а, (*m*) (fin) debit

дебет·ова́ть, -у́ют, (*imp and perf*) (fin) debit

дебет·о́в·ый, -ая, -ое, (*a*) debit

~ ави́зо (*n*) debit note

~ са́льдо (*n*) debit balance

дебит·, -а, (*m*) discharge, flow, yield, output

~ сква́ж·ин·ы (oil) well yield/output

дебито·ме́р·, -а, (*m*) (oil) well-head flow meter

дебито́р·, -а, (*m*) debtor; (*pl*) debit side

де·блок·иро́вк·а, -а, (*f*) clearing, unblocking; (mil) raising a blockade; (elec) release

Де́бнер·а–Ми́ллер·а, реа́кци·я (*f*) (chem) Doebner–Miller synthesis

дебю́т·, -а, (*m*) debut; opening (at chess)

де Бро́йл·я, во́лн·ы (*f pl*) (nucl) de Broglie waves

де·вальва́ци·я, -и, (*f*) (fin) devaluation

~, от·кры́·т·ая currency reform

~, с·кры́·т·ая devaluation, reduction of official rate of exchange

Дева́л·я, ис·пыт·а́ни·е (*n*) (civ eng) Deval attrition test

дев·а́·ть, -ю́т, (*imp*, *indet*) put, put away; put up, accommodate; -ся disappear; find accommodation

Девере́, агита́тор· (*m*) Devereux agitator

девиа́тор·, -а, (*m*) (mech) deviator; compass adjuster

девиацио́н·н·ый, -ая, -ое, (*a*) deviation

~ зна́к·и (*pl*) deviation marks (for compass)

~ пеленг·а́тор· (*m*) (air) bearing compass

~ при·бо́р· (*m*) (a/c) deviation microadjuster

~ рабо́т·ы (*pl*) (navig) compass swinging

деви·а́ци·я, -и, (*f*) deviation, error

~, ко́мпас·н·ая compass deviation

~, кре́н·ов·ая heeling error (compass)

~, четвёрт·н·ая quadrantal deviation

деви́з·, -а, (*m*) motto, slogan

деви́з·а, -ы, (*f*) (fin) currency

девинди́т·, -а, (*m*) (min) dewindite

девио́·метр·, -а, (*m*) (ocean) magnetic deviation meter

де·витри·фик·а́ци·я, -и, (*f*) (geol) devitrification

дев·и́ч·ий, -ья, -ье, (*a*) girlish; (med) virginal

дево́н·, -а, (*m*) (geol) Devonian

дево́н·ск·ий, -ая, -ое, (*a*) (geol) Devonian

дев·о́чк·а, -и, (*g pl*) -чек·, (*f*) girl

де́в·ственн·ый, -ая, -ое, (*a*) virgin, virginal

де·вулканиз·а́т·, -а, (*m*) devulcanizate

де·вулканиз·а́ци·я, -и, (*f*) devulcanization, reclaiming (rubber)

де́·вш·ий, -ая, -ее, (*a*) (*past part act*) of **де·ть** (*perf*) *see* **дев·а́·ть** (*imp*)

девяно́ст·а (*gen, dat, instr, prep*) of **девяно́ст·о**

девяно́ст·о, -а, (*n*) ninety

девяносто·дн·е́вн·ый, -ая, -ое, (*a*) ninety-day

девяносто·ле́т·н·ий, -яя, -ее, (*a*) ninety-year; ninety year-old

девяно́ст·ый, -ая, -ое, (*a*) ninetieth

девят·ери́к·, -а́, (*m*) nineling

де́вят·ер·о, -ы́х, nine, nine together

девяти·зве́н·н·ый поли·ме́р· (*m*) (chem) nonamer

девяти·кра́т·н·ый, -ая, -ое, (*a*) ninefold, nine times, nonuple

девяти·со́т·ый, -ая, -ое, (*a*) nine-hundredth

девяти·уго́ль·ник·, -а, (*m*) nonagon, (math) enneagon

девяти·час·ов·о́й, -а́я, -о́е, (*a*) nine-hour; nine o'clock

девяти·чле́н·н·ое кольц·о́ (*n*) (chem) nonatomic ring

девя́т·к·а, -и, (*g pl*) **-т·ок·,** (*m*) a nine, a number nine

девятнадцати·дн·е́вн·ый, -ая, -ое, (*a*) nineteen-day

девятнадцати·ме́ст·н·ый, -ая, -ое, (*a*) nineteen-seat/berth/place

девят·на́·дцат·ый, -ая, -ое, (*a*) nineteenth

девят·на́·дцат·ь, -и, (*i*) **-ью,** nineteen

девя́т·ый, -ая, -ое, (*a*) ninth

де́вят·ь, -и́, (*i*) **-ью́** nine

девять·со́т·, (*gen, dat, instr, prep*) **девяти·со́т, девяти·ста́м, девятью·ста́ми, девяти·ста́х** nine hundred

де·газ·а́тор·, -а, (*m*) (chem) degasifier, stripper; decontaminator (civil defence)

де·газ·а́ци·я, -и, (*f*) (*v n*) *see* **де·газ·и́р·овать;** decontamination

де·газ·ацио́нн·ый, -ая, -ое, (*a*) degassing; decontamination, gas cleansing; gas freeing (of tankers etc.)

~ пункт· (*m*) CW decontamination centre

де·газ·и́р·овать, -уют, (*imp and perf*) degas, strip; decontaminate

де·газ·и́рующ·ий, -ая, -ее, (*pres part act*) of **де·газ·и́р·овать**

~ вещ·еств·о́ (*n*) anti-gas compound

де·гази·фик·а́ци·я, -и, (*f*) degasification

де·генера́т·, -а, (*m*) degenerate

де·генерат·и́вн·ый, -ая, -ое, (*a*) degenerative

де·генер·а́ци·я, -и, (*f*) degeneration

де·генер·и́р·овать (*imp and perf*) degenerate

де·гидр·а́з·а, -ы, (*f*) (chem) dehydrogenase

де·гидрат·абиети́н·ов·ая кисл·от·а́ (*f*) dehydroabietic acid

де·гидрат·а́ци·я, -и, (*f*) dehydration, removal/loss of water

де·гидр·а́тор·, -а, (*m*) dehydrator

де·гидр·аце́т·ов·ая кисл·от·а́ (*f*) dehydracetic acid

де·гидр·и́ровани·е, -я, (*n*) dehydrogenation, removal of hydrogen

де·гидр·и́рованн·ый, -ая, -ое, (*past part pass*) of **де·гидр·и́р·овать** dehydrogenated

де·гидр·и́р·овать, -уют, (*imp*) dehydrogenate

де·гидрогениз·а́ци·я, -и, (*f*) (chem) dehydrogenation, removal of hydrogen

де·гидр·окси- (*component*) (chem) dehydroxy-

дегорж·а́ж·, -а, (*m*) disgorging (of sparkling wine)

дегорж·и́ровани·е, -я, (*n*) *see* **дегорж·а́ж·**

дёгот·ь, -гт·я, (*m*) tar, pitch

~, каменно·уго́ль·н·ый coal tar, tar pitch

~, ко́кс·ов·ый coal tar

~, перв·и́чн·ый primary tar

дегра́ (*indecl*) degras, moellen (fat from sheepskins); sod oil (similar, manufactured)

де·град·а́ци·я, -и, (*f*) degradation

де·град·и́р·овать, -уют, (*imp and perf*) degrade, degradate

дёгте·бето́н·, -а, (*m*) tarmac, tar concrete

дегт·я́рн·ый, -ая, -ое, (*a*) tar, tarry, pitch

дегт·я́рн·я, -и, (*g pl*) **-рен·,** (*f*) tar works

дегуст·а́тор·, -а, (*m*) taster

дегуст·а́ци·я, -и, (*f*) tasting

дегуст·и́р·овать, -уют, (*imp and perf*) taste

де́двейт·, -а, (*m*) (naut) deadweight

~, грузо·подъ·ём·ност·ь deadweight capacity

де́декинд·ов·о сеч·е́ни·е (*n*) (math) Dedekind cut

де·доломит·иза́ци·я, -и, (*f*) (geol) dedolomitisation

дедукти·в·н·ый, -ая, -ое, (*a*) deductive

дедукци·я, -и, (*f*) deduction

дедуц·и́р·овать, -уют, (*imp and perf*) deduce

дее·при·ча́ст·и·е, -я, (*n*) (gram) gerund

дее·спо́соб·н·ый, -ая, -ое, (*a*) able to function; (law) capable

деж·а́, -и, (*i*) **-о́й,** (*nom pl*) **дёж·и,** (*g pl*) **деж·е́й** bowl, pan through

дежёкци·я, -и, (*f*) dejection

дежу́р·ить, -ят, (*imp*) be on duty

дежу́р·н·ый, -ая, -ое, (*a*) duty, on duty; (*m decl as adj*) person on duty, person in charge, attendant (of a machine), (mil) duty officer/NCO

~ **блю́д·о** (*n*) plat du jour, today's dish

~, **операти́в·н·ый** (*m*) duty operations officer

~ **по ста́нци·и** (*m*) (rail) assistant station master

~ **топли́в·н·ый на·со́с·** (*m*) (naut) fuel oil service pump

дежу́р·ств·о, -а, (*n*) duty, being on duty, attendance

дез- (*component*) dis-, de-, des-

дез·аву·и́р·овать, -уют, (*imp and perf*) repudiate, disavow

дез·акси́аль·н·ый, -ая, -ое, (*a*) (mech eng) offset

дез·актив·а́тор, -а, (*m*) (nucl) decontamination apparatus

дез·актив·а́ци·я, -и, (*f*) (chem) deactivation; radioactive decontamination

~ **по·ве́рх·ност·и** surface decontamination

~ **уда́р·н·ая** collision decontamination

дез·амин·а́з·а, -ы, (*f*) (biochem) desaminase

дез·амин·и́ровани·е, -я, (*n*) (chem) deamination

дезерт·и́р·овать, -уют, (*imp and perf*) desert

дез·и́л·, -а, (*m*) (chem) desyl

дез·инсе́кци·я, -и, (*f*) disinfestation

дез·инсекцио́н·н·ый, -ая, -ое, (*a*) disinfestation

дез·интегр·а́тор·, -а, (*m*) disintegrator, disintegrating mill; (chem) mechanical gas washer

дез·интегр·ацио́н·н·ый, -ая, -ое, (*a*) disintegration, disintegrating

дез·интегр·а́ци·я, -и, disintegration

дез·инфекцио́н·н·ый, -ая, -ое, (*a*) disinfection, disinfecting

~ **сре́д·ств·о** (*n*) disinfectant

дез·инфе́кци·я, -и, (*f*) disinfection

дез·инфиц·и́р·овать, -уют, (*imp and perf*) disinfect

дез·инфиц·и́р·ующ·ий, -ая, -ее, (*pres part act*) of **дез·инфиц·и́р·овать**

~ **сре́д·ств·о** disinfectant

дез·информ·ацио́н·н·ый, -ая, -ое, (*a*) false

дез·информ·а́ци·я, -и, (*f*) misinformation

дез·информ·и́р·овать, -уют, (*imp and perf*) misinform

дез·одор·а́нт, -а, (*m*) = **дез·одор·а́тор·**

дез·одор·а́тор·, -а, (*m*) deodorant, deodorizer, destinker

дез·одор·иза́ци·я, -и, (*f*) deodorization

дез·одор·и́р·овать, -уют, (*imp and perf*) deodorize

дез·окси·риб·о́з·а, -ы, (*f*) (chem) desoxyribose

дез·окси·со·един·е́ни·е, -я, (*n*) (chem) desoxy compound

дез·организ·а́ци·я, -и, (*f*) disorganization

дез·организ·ова́ть, -уют, (*imp and perf*) disorganize

дез·ориент·а́ци·я, -и, (*f*) disorientation, misorientation

дез·ориент·и́рованн·ый, -ая, -ое, (*past part pass*) of **дез·ориент·и́р·овать**

дез·ориент·и́р·овать, -у́ют, (*imp and perf*) disorientate, confuse

де́йдву·д·, -а, (*m*) (shipb) deadwood

де́йдвуд·н·ый, -ая, -ое, (*a*) (shipb) deadwood

~ **вал·** (*m*) (shipb) tail intermediate shaft

~ **труб·а́** (*f*) stern/shipbuilder's tube

де́йк·а, -и, (*f*) (geol) dike

де́й·ственн·ый, -ая, -ое, (*a*) effective, operative, active

де́й·стви·е, -я, (*n*) (*see also associated words*) action, activity, operation, working, running; action, effect, influence; (math) operation; (theat) act; (*and frequently used adjectivally, noun and adj being in gen*)

~, **бриза́нт·н·ое** (*expl*) brisance

в·вес·ти́ в де́й·стви·е implement, carry into effect (an order etc.), commission, start up (works, plant etc.)

~, **вое́н·н·ые** (*pl*) hostilities, military operations

дей· стви·е
~, **дво·ичн·ое** (autom) binary operation
~, **дискрéт·н·ое** digital action; (in gen) digital (used as adj)
~, **за·мéдл·енн·ое** delayed action
~, **за·медл·я́ющ·ее** moderating effect
~, **за·пóр·н·ое** barrier/blocking action
~ **из·луч·éни·я** (phys) effect of radiation, exposure to radiation
~, **не·прáв·ильн·ое** malfunctioning
  под дéй·стви·ем under; by, induced by
  под дéй·стви·ем из·луч·éни·я under radiation, by exposure to radiation, by radiation
~, **по·лéз·н·ое** and see **по·лéз·н·ое дéй·стви·е** efficiency (of a machine)
  при·вес·ти́ в дéй·стви·е start up (a machine), actuate
  у́гол· дéй·стви·я (m) angle of action, working approach angle (machine tools)
  центр· дéй·стви·я (m) (meteor) centre/center of action
дей·стви́тельн·о (adv) really, indeed, actually, in fact (confirming); and indeed
дей·стви́тельност·ь, -и, (f) reality, fact; effectiveness, efficacy
  в дей·стви́тельност·и in fact/reality/practice, actually (contrasting)
дей·стви́тельн·ый, -ая, -ое, (a), (sf) **-лен·, -льн·а,** actual, real, true; effective, efficacious; valid, in force; active
~ **мáсс·а** (f) (chem) active mass
~ **мéст·о** (n) (air) ground position
~ **на·пряж·éни·е** (n) (mech) effective/true stress
~ **слу́ж·б·а** (f) (mil) active service
~ **числ·ó** (n) (math) real number
~ **член·** (m) member, full member (of organization)
дéй·ств·овать, -уют, (imp) operate, work, act, do, function, run; take effect; **-ся** (passive); be occupied
~ **на** act upon, work on, have an effect/influence on influence
дéй·ств·ующ·ий, -ая, -ее, (pres part act) of **дéй·ств·овать** active, effective; in force, valid; current, operating, in operation, on-stream
~ **áрм·и·я** (f) (mil) field forces
  за·кóн· дéй·ствующ·их масс· (m) (chem) law of mass action

дéй·ств·ующ·ий, -ая, -ее
~ **ли́ц·а** (pl) (theat) dramatis personae, cast
~ **си́л·а** (f) (mech) action/external force
~ **цен·á** (f) current price
дейтер·и́д·, -а, (m) (chem) deuteride
~ **ли́т·и·я** lithium deuteride
дейтéр·иев·ый, -ая, -ое, (a) (chem) deuterium
дейтер·изáци·я, -и, (f) (nucl) deuterating, deuteration
дейтер·и́з·овать, -уют, (imp and perf) (chem) deuterate
дейтéр·и·й, -я, (m) (chem) deuterium, heavy hydrogen, D
дейтер·и́рованн·ый, -ая, -ое, (a) deuterated
дейтер·и́ческ·ий, -ая, -ое, (a) deuteric, secondary
дейтеро·мóрф·н·ый, -ая, -ое, (a) deuteromorphic
дейтер·óн·, -а, (m) (phys) deuteron, deutron
дейтеро·ó·кис·ь, -и, (f) (chem) deuteroxide, heavy water
дейтеро·плáзм·а, -ы, (f) (gen) deuteroplasm, deutoplasm
дейт·óн·, -а, (m) (phys) deuteron, deuton
дейтóн·н·ый, -ая, -ое, (a) deuteron, heavy water
дейто·плáзм·а, -ы, (f) (gen) deuteroplasma, deutoplasma
дейтр·óн·, -а, (m) (phys) deuteron, deuton
дéк·а, -и, (f) (agr) concave (of thresher) sounding board (music)
дека- (component) deca-, ten-
декáбр·ь, -я́, (m) December
дека·грáмм·, -а, (m) decagram
декáд·а, -ы, (f) ten-day period, ten days; (math) decade; (instr) decade device
~, **пере·с·чёт·н·ая** decade scaler
~ **со·противл·éни·я** (elec) decade resistance box
~, **холост·áя** idle/vacant level (vacuum tubes)
декáд·н·ый, -ая, -ое, (a) of **декáд·а**
~ **магази́н·** (m) (elec) decade box
~ **схéм·а** (f) (elec) decade scaler, decade scaling circuit
декали́н·, -а, (m) (chem) decalin, decahydronaphthalene
дека·ли́тр·, -а, (m) decalitre, **decaliter**

де·кальк·и́р·овать, -уют, (*imp and perf*) (print) take/make a transfer

де·калько·ма́н·и·я, -и, (*f*) (print) transfer, decalcomania

де·кальци·фик·а́ци·я, -и, (*f*) decalcification

декаментониу́м·, -а, (*m*) (pharm) decamenthonium

дека·ме́тр·, -а, (*m*) decametre, decameter

дека·метр·и́ческ·ий, -ая, -ое, (*a*) decametric

~ волн·а́ (*f*) (rad) decametric wave, short wave (10–100 m)

дека́н·, -а, (*m*) dean, president of faculty (of university etc.); (chem) decane

декан·а́т·, -а, (*m*) dean's office, faculty administration

декант·а́тор·, -а, (*m*) (chem) settler, settling tank, decanter

декант·а́ци·я, -и, (*f*) (chem) decantation

декапи́р·, -а, (*m*) pickled iron/steel

декапир·ова́ни·е, -я, (*n*) (met) pickling

декапи́р·ованн·ый, -ая, -ое, (*past part pass*) (met) pickled

декапир·о́вк·а, -и, (*g pl*) -вок·, (*f*) (met) pickling

дека·пло́ид·, -а, (*m*) (gen) decaploid

дека·по́д·, -а, (*m*) (zool) decapod, (*pl*) decapods, *Decapoda*

де·карбоксил·и́ровани·е, -я, (*n*) (biochem) decarboxylizing, decarboxylation

де·карбон·иза́ци·я, -и, (*f*) (ICE) decarbonization; (met) decarburization

Дека́рт·а, лист· (*m*) (math) folium of Descartes

дека́рт·ов·, -а, -о, (*possess adj*) (math) Cartesian, Descartes'

дека́рт·ов·ый, -ая, -ое, (*a*) (math) Cartesian

декарт·ир·ова́ть, -у́ют, (*imp and perf*) (text) decatize

декат·иро́вк·а, -и, (*f*) (text) decatizing

дека·тро́н·, -а, (*m*) (nucl) decatron

дека́·эдр·, -ы, (*f*) (math) decahedron

де·квали·фик·а́ци·я, -и, (*f*) loss of qualification/skill/status

де·квали·фиц·и́р·оваться, -уются, (*imp and perf*) lose a qualification/skill, be no longer qualified

де́кел·ь, -я, (*m*) (print) tympan

~, жёстк·ий (print) hard packing

деклар·а́ци·я, -и, (*f*) declaration; declaration, manifest, bill (customs)

~, мор·ск·а́я maritime declaration

~ по·дав·а́ем·ая тамо́ж·н·е bill of entry

~ по от·хо́д·у entry outwards

~ по при·хо́д·у entry inwards

~, тамо́ж·енн·ая bill of entry

деклар·и́р·овать, -уют, (*imp and perf*) declare, proclaim

деклин·а́тор·, -а, (*m*) (instr) declinometer

деклин·а́ци·я, -и, (*f*) (astron surv) declination

деклино́·метр·, -а, (*m*) declinometer

деклуази́т·, -а, (*m*) (min) descloizite

де·коге́рер·, -а, (*m*) (rad) decoherer, anticoherer

деко́кт·, -а, (*m*) (pharm) decoction

декольт·и́рованн·ый, -ая, -ое, (*a*) low-necked, décolleté

де·компрессио́н·н·ый, -ая, -ое, (*a*) (physiol) decompression

~ за·бол·ева́ни·е (*n*) (med) decompression sickness, caisson disease, the bends

де·компре́сс·и·я, -и, (*f*) decompression

де·компре́сс·ор·, -а, (*m*) (ICE) decompression mechanism

де·конъ·юг·а́ци·я, -и, (*f*) (cyt) deconjugation

декорати́в·н·ый, -ая, -ое, (*a*) decorative, ornamental; (mil) disguise (camouflage)

декор·а́ци·я, -и, (*f*) (theat) scenery, décor

декор·и́ровани·е, -я, (*n*) decoration

~ дис·лок·а́ци·й (met) decoration of dislocations

де·кортик·а́ци·я, -и, (*f*) (med) decortication; stripping

декреме́нт·, -а, (*m*) (math) decrement

~ за·тух·а́ни·я (phys) decrement of decay, attenuation

~, логарифм·и́ческ·ий logarithmic decrement

декреме́тр·, -а, (*m*) (phys) decremeter

декре́т·, -а, (*m*) decree

декрет·и́р·овать, -уют, (*imp and perf*) decree

декре́т·н·ый, -ая, -ое, (*a*) of декре́т·

~ вре́м·я (*n*) daylight-saving time, summer time (zone time plus one hour)

~ от·пу́ск· (*m*) maternity leave

де́ксел·ь, -я, (*m*) adze

дексио·тро́п·н·ый, -ая, -ое, (*a*) (zool) dexiotropic

декстр·а́н, -а, (*m*) (chem) dextran

декстр·и́н·, -а, (*m*) (chem) dextrin, starch gum

декстрин·а́з·а, -ы, (*f*) (biochem) dextrinase

декстрин·иза́тор·, -а, (*m*) (chem) dextrinizing tank

декстрин·а́ци·я, -и, (*f*) (chem) dextrinizing

декстр·о́з·а, -ы, (*f*) (chem) dextrose

декстро·ка́рд·и·я, -и, (*f*) (anat) dextrocardia

декстро·пима́р·ов·ая кисл·от·а́ (*f*) (chem) dextropimaric acid

де·л (*past masc sing*) of **де·ть** (*perf*), *see* **дев·а́·ть** (*imp*); (*g pl*) of **де́л·о** (q.v.)

де·ламин·а́ци·я, -и, (*f*) (biol) delamination

де́л·ани·е, -я, (*n*) (*v n*) of **де́л·а·ть**

де́л·анн·ый, -ая, -ое, (*past part pass*) of **де́л·а·ть;** simulated, feigned, artificial

де́л·а·ть, -ют, (*imp*) (*see also* **де́л·а·ть·ся**) make, do

~ паралле́ль·н·ый make parallel

де́л·а·ться, -ются, (*imp*) (+ *instr*) become, be getting; happen, be going on

делега́т·, -а, (*m*) delegate

~ с·вя́з·и (mil) liaison officer

делег·а́ци·я, -и, (*f*) delegation

делег·и́р·овать, -уют, (*imp and perf*) send/elect as delegate

дел·ёж·, -а́, (*i*) **-о́м,** (*m*) division, partition, sharing, distribution

дел·е́ни·е, -я, (*n*) (*v n*) of **дел·и́ть;** division, dividing; (mech eng) indexing; fission, fissioning; division, unit, point (on scale)

~ без о·ста́т·к·а (math) exact division

~, вз·ры́в·н·о́е (nucl) explosive fission

~, вы́·нужд·енн·ое (nucl) induced fission

~ кле́т·к·и (cyt) cell-division, fission

~ на бло́к·и (phys) lumping

~ на ка́мер·ы concameration

~ на три ча́ст·и (nucl) ternary fission; (math) tripartition

~ на у·ча́ст·к·и (math) sectionalization

~ о·кру́ж·ност·и cyclotomy

у·равн·е́ни·е дел·е́ни·я о·кру́ж··ност·и (*n*) cyclotomic equation

дел·е́ни·е

~, по·сле́д·овательн·ое (mech eng) consecutive indexing

~, страти·гра́ф·и́ческ·ое (geol) stratigraphic division

~ фото́н·ами (nucl) photofission

цеп·ь дел·е́ни·я (*f*) (autom) counting-down circuit

~ част·от·ы́ и́мпульс·ов (elec) scaling down

~ шкал·ы́ (instr) scale division/graduation

-де́л·и·е, -я, (*n component*) -making, manufacture

деликат·н·ый, -ая, -ое, (*a*) delicate, tactful

дел·и́м·ое, -ого, (*n decl as adj*) (math) dividend

дел·и́мост·ь, -и, (*f*) divisibility; (nucl) fissionability

дел·и́м·ый, -ая, -ое, (*pres part pass*) of **дел·и́ть;** divisible; fissionable; (*see also* **дел·и́м·ое**) (*as noun*)

делинтер·о́вк·а, -и, (*f*) (text) delinting

дел·и́тел·ь, -я, (*m*) divider; (math) divisor

~, желоб·чат·ый (min) riffle

~, наи·бо́льш·ий о́бщ·ий (math) greatest common divisor/measure, GCD

~ на·пряж·е́ни·я (elec) voltage divider, volt box, potential divider

~, норма́ль·н·ый (math) normal divisor, invariant subgroup

~ част·от·ы́ (rad) frequency divider

дел·и́тельн·ый, -ая, -ое, (*a*) dividing; indexing; fission

~ голо́в·к·а (*f*) dividing/indexing head (machine tool attachment)

~ ка́мер·а (*f*) (nucl) fission chamber

~ маши́н·а (*f*) (eng) dividing engine

~ скля́н·к·а (*f*) (chem) separatory flask

~ стан·о́к· (*m*) deep resawing machine

~ у·стро́й·ств·о (*n*) dividing/indexing device (machine tools); divider (of computer)

дел·и́ть, -ят, (*imp*) divide; **-ся** (*pass*); (*intrans*) divide; share; tell, impart, confide in

~ на (math) divide by

~ по·пол·а́м halve

~ по́·ровн·у divide into equal parts

де́л·о, -а, (*nom pl*) **-а́,** (*d pl*) **-а́м,** (*n*) affair, business, matter, cause; thing; file (of papers), dossier; (law) case, action; mil) engagement

дел·о
~, **ба́нк·ов·ое** banking
в са́м·ом де́л·е actually, really
~ **в том что** the point is that
~, **го́р·н·ое** mining
~, **из·да́т·ельск·ое** publishing
~, **им·е́·ть** have to do with, deal with
~, **констру́ктор·ск·ое** designing
~, **маркше́йдер·ск·ое** mine/underground surveying
~, **ру́д·н·ое** mining, metalliferous mining
~, **холод·и́льн·ое** refrigeration
дел·ови́т·ый, -ая, -ое, (*a*) efficient, business-like
дел·ов·о́й, -а́я, -о́е, (*a*) business; business-like
~ **древ·еси́н·а** (*f*) commercial timbers
дело·про·из·вод·и́тел·ь, -я, (*m*) clerk, filing/correspondence clerk
дело·про·из·во́д·ств·о, -а, (*n*) clerical work
делоренци́т·, -а, (*m*) (min) delorenzite
де́ль·н·ый, -ая, -ое, (*a*) business-like, efficient, capable; commercial
~ **ве́щ·и** (*pl*) (naut) deck fittings
~ **лес·** (*m*) commercial timber
де́льт·а, -ы, (*f*) delta, δ, Δ (Greek); (geog) delta
~ **-древ·еси́н·а, -ы,** (*f*) hot-pressed resin plywood
~ **-желе́з·о, -а,** (*n*) (met) delta iron
~ **-лу́ч·, -а,** (*m*) (phys) delta ray
~ **-мета́лл·, -а,** (*m*) (obs) delta metal, high-strength brass
~ **-фу́нкци·я, -и,** (*f*) (phys) delta function of Dirac
~ **-част·и́ц·а, -ы,** (*f*) (phys) delta particle
дельти́диум·, -а, (*m*) (zool) deltidium
дель·ти́р·и·й, -я, (*m*) (zool) delthyrium
де́льт·ов·ый, -ая, -ое, (*a*) deltaic, deltoid
~ **от·лож·е́ни·е** (*n*) (geol) deltaic deposit
дельто·ви́д·н·ый, -ая,ʼ -ое, (*a*) deltoid
дельт·о́ид·, -а, (*m*) deltoid
дельто́·эдр·, -а, (*m*) (cryst) deltoid dodecahedron
дельфи́н·, -а, (*m*) (zool) dolphin
~, **черно·мо́р·ск·ий** (zool) common dolphin, *Delphinus delphis*
дельфин·а́т·, -а, (*m*) (chem) delphinate

дельфин·и́н·, -а, (*m*) (chem) delphinine
делюв·и·й, -я, (*m*) (geol) rock waste, sliderock; talus, scree
дел·я́нк·а, -и, (*g pl*) **-нок·,** (*f*) (agr) plot (of land), allotment
дел·я́щ·ийся, -аяся, -ееся, (*pres part act*) of **дел·и́ться;** (nucl) fissionable, fissile
демант·о́ид·, -а, (*m*) (min) demantoid
демарк·ацио́нн·ый, -ая, -ое, (*a*) demarcation
демарк·а́ци·я, -и, (*f*) demarcation
дема́рш·, -а, (*i*) **-ем,** (*m*) (dipl) démarche
де·маск·и́р·овать, -уют, (*imp and perf*) remove camouflage
де·материал·иза́ци·я, -и, (*f*) (phys) dematerialization
деме́нци·я, -и, (*f*) (med) dementia
демерса́ль·н·ый, -ая, -ое, (*a*) (zool) demersal
деметил·и́ровани·е, -я, (*n*) demethylation
демидови́т·, -а, (*m*) (min) demidovite
де·миэлин·иза́ци·я, -и, (*f*) (biol) demyelination
де·мобилиз·а́ци·я, -и, (*f*) demobilization; (biochem) fixation
де·мобилиз·ова́ть, -у́ют, (*imp and perf*) demobilize
демо·гра́ф·и·я, -и, (*f*) demography
де·милитариз·ова́ть, -у́ют, (*imp and perf*) demilitarize
де·минерализ·а́тор·, -а, (*m*) demineralizer
~, **об·во́д·н·ый** by-pass demineralizer
де·минерализ·а́ци·я, -и, (*f*) demineralization
деми·сезо́н·н·ый, -ая, -ое, (*a*) between-seasons, demi-saison (of clothes)
де·модуля́тор·, -а, (*m*) (rad) demodulator, frequency changer
~ **сигна́л·а цве́т·ност·и** (TV) chrominance demodulator
де·модул·я́ци·я, -и, (*f*) demodulation
демо·крат·и́ческ·ий, -ая, -ое, (*a*) democratic
демонстр·ати́вн·ый, -ая, -ое, (*a*) demonstrative; demonstrated, illustrated
демонстр·ацио́нн·ый, -ая, -ое, (*a*) demonstration
~ **зал·** (*m*) showroom

демонстр·а́ци·я, -и, (*f*) demonstration; (mil) feint attack, demonstration

демонстр·и́р·овать, -уют, (*imp and perf*) demonstrate; (cinema) show

демонт·а́ж, -а, (*m*) dismantling, stripping

демонт·и́р·овать, -уют, (*imp and perf*) dismantle, strip

де·морф·и́зм, -а, (*m*) demorphism

дё́мпинг·, -а, (*m*) (econ) dumping

дё́мпинг·ов·ый, -ая, -ое, (*a*) (econ) dumping

дё́мпфер·, -а, (*m*) damper, vibration damper, torsional vibration damper

~, вяз·костн·ый viscously damped absorber

~, динам·и́ческ·ий dynamic damper

~, фрикцио́н·н·ый friction drag damper, friction damper

дё́мпфер·н·ый, -ая, -ое, (*a*) damper, damping

~ ла́мп·а (*f*) (TV) damper tube

~ об·мо́т·к·а (*f*) (elec) damper, damper/damping winding

демпф·ирова́ни·е, -я, (*n*) damping, vibration damping; shock absorption

демпф·и́рующ·ий, -ая, -ее, (*pres part act*) damping

~ с·плав· (*m*) (met) high damping-capacity alloy

де·мультиплё́ксер·, -а, (*m*) (rad) demultiplexer

де·мультиплик·а́тор·, -а, (*m*) (M/T) auxiliary gearbox (Brit), auxiliary transmission (U.S.)

де·мультиплик·а́ци·я, -и, (*f*) (mech) gearing down

де·натур·а́т·, -а, (*m*) (chem) denatured spirit/alcohol; methylated spirit

де·натур·а́тор·, -а, (*m*) (chem) denaturant

де·натур·а́ци·я, -и, (*f*) denaturing (alcohol); denaturation (of proteins)

де·натур·и́рованн·ый, -ая, -ое, (*past part pass*) of **де·натур·и́р·овать;** denatured

де·натур·и́р·овать, -уют, (*imp and perf*) (chem) denature

де·натур·и́рующ·ий, -ая, -ее, (*pres part act*) of **де·натур·и́р·овать;** denaturant

~ сре́д·ств·о (*n*) denaturant

дендиро́л·ь, -я, (*m*) (paper) dandy roll

дендр·и́т·, -а, (*m*) dendrite; (cryst) dendrite, arborescent crystal

дендр·и́т·н·ый, -ая, -ое, (*a*) dendritic, tree-like

дендр·и́т·ов·ый, -ая, -ое, (*a*) dendrite, dendritic

дендро́·граф·, -а, (*m*) (bot) dendrograph

дендр·о́ид·н·ый, -ая, -ое, (*a*) (bot) dendroid

дендро·ло́г·и·я, -и, (*f*) (bot) dendrology

дендро́·метр·, -а, (*m*) dendrograph (for measuring width of trees)

де́неж·н·ый, -ая, -ое, (*a*) money, monetary

~ обращ·е́ни·е (*n*) monetary circulation, currency

де·нерв·а́ци·я, -и, (*f*) (biol) denervation

де·нитр·ацио́нн·ый, -ая, -ое, (*a*) denitrating, denitration

де·нитр·а́ци·я, -и, (*f*) denitrating, denitration

де·нитр·и́ровани·е, -я, (*n*) denitrating, denitration

де·нитр·и́р·овать, -уют, (*imp and perf*) denitrate

де·нитри·фик·а́ци·я, -и, (*f*) (bot) denitrification

денитр·ова́ть, -у́ют, (*imp and perf*) denitrate

денонс·и́р·овать, -уют, (*imp and perf*) (dipl) denounce

денси·ме́тр·, -а, (*m*) densimeter (dielectrics); (phys) hydrometer

денсито́·метр·, -а, (*m*) (phot) densitometer, opacity meter

дент·а́льн·ый, -ая, -ое, (*a*) dental

дент·и́кул·а, -ы, (*f*) (arch) dentil

дент·и́н·, -а, (*m*) (zool) dentine

де·нуд·а́ци·я, -и, (*f*) (geol) denudation

ден·ь, дн·я, (*nom pl*) **дн·и,** (*g pl*) **дн·ей,** (*m*) day

~, вы·ход·н·о́й rest-day, day off

~, рас·чё́т·н·ый pay-day
тре́ть·его дн·я (*gen*) the day before yesterday

де́ньг·и (*nom pl*), (*g pl*) **де́нег·,** (*d pl*) **деньг·а́м,** (*f pl*) money

~, на·ли́ч·н·ые ready money, ready cash

~, про·до·во́ль·ствен·н·о-пут·ев·ы́е travel allowance, subsistence allowance

де́нь·е (*indecl*) (text) denier

де́н·ьте (*imper*) of **де·ть,** (*perf*) *see* **дев·а́·ть** (*imp*) де·парафин·иза́ци·я, -и, (*f*) deparaffination

департа́мент·, -а, (*m*) department

депе́ш·а, -и, (*i*) -ей, (*f*) dispatch, message

де·пил·я́тори·й, -я, (*m*) depilatory (compound)

де·пил·я́ци·я, -и, (*f*) (bot) depilation

де·плазмо·ли́з·, -а, (*m*) (bot) deplasmolysis

де·план·а́ци·я, -и, (*f*) (cryst) plane shift/displacement

депо́ (*n indecl*) depot, station

~, кру́г·л·ое (rail) roundhouse

~, локомоти́в·н·ое (rail) locomotive depot, roundhouse, locomotive shed

~, пожа́р·н·ое fire station

~, прямо·уго́ль·н·ое (mil) oblong engine shed

депози́т·, -а, (*m*) (fin) deposit

депози́тор·, -а, (*m*) (fin) depositor

де·полимер·а́з·а, -ы, (*f*) (chem) depolymerase

де·полимер·иза́ци·я, -и, (*f*) (chem) depolymerization

де·поляриз·а́тор·, -а, (*m*) (elec) depolarizer

де·поляриз·а́ци·я, -и, (*f*) depolarization

депон·е́нт·, -а, (*m*) (fin) depositor

депон·и́р·овать, -уют, (*imp and perf*) (fin) deposit

де·пресс·а́нт·, -а, (*m*) depressant, depressing agent (ore dressing)

де·пресс·и́вн·ый, -ая, -ое, (*a*) depressed, depressing; (econ) depression

де·пре́сс·и·я, -и, (*f*) depression

~ горизо́нт·а (surv) dip of the horizon

де·пре́сс·ор·, -а, (*m*) (anat) depressor; depressant, depressing agent; (oil) pour-point depressant

де·пре́ссор·н·ый, рефле́кс· (*m*) (anat) depressor reflex

депси́д·, -а, (*m*) (chem) depside

депут·а́т·, -а, (*m*) deputy, elected representative

депут·а́ци·я, -и, (*f*) deputation

дератиз·а́ци·я, -и, (*f*) deratization, rat-catching, rodent extermination

дёрг·ани·е, -я, (*n*) (*v n*) see дёрг·а·ть

дёрг·а·ть, -ют, (*imp*) tug, jerk, pull; twitch; worry, harass

дереве́·й, -я, (*m*) (bot) yarrow, milfoil, *Achillea millefolium*

дерев·е́ни·е, -я, (*n*) lignification

дерев·ен·е́·ть, -ют, (*imp*) lignify, become wood

дерев·е́нск·ий, -ая, -ое, (*a*) rural, country; village

дерев·н·я, -и, (*g pl*) -в·е́н·ь, (*f*) village; countryside, country

де́рев·о, -а, (*nom pl*) дере́в·ья, (*g pl*) дере́в·ьев, (*n*) (*and see under associated adjectives*) tree; wood, timber

~, кни́ц·ев·ое (naut) compass timber

~, кра́с·н·ое mahogany

~, кро́н·ист·ое standard tree

~, лес·н·о́е forest tree

~, ли́ст·венн·ое deciduous tree

~, ранго́ут·н·ое (naut) spar

~, стро·ев·о́е timber tree

~, хво́й·н·ое conifer, coniferous tree

~, чёрн·ое ebony

~, чешу́й·чат·ое (geol) lepidodendron

дерево·бето́н·, -а, (*m*) cement wood

дерево·об·де́л·очник·, -а, (*m*) woodworker

дерево·об·де́л·очн·ый, -ая, -ое, (*a*) woodworking

~ мастер·ск·а́я (*f*) woodworking shop; (shipb) shipwright's shop

~ стан·о́к· (*m*) woodworking machine

дерево·об·раба́т·ывающ·ий, -ая, -ее, (*a*) woodworking

~ стан·о́к· (*m*) woodworking machine

дерево·ре́ж·ущ·ий стан·о́к· (*m*) sawmilling machine, breaking-down saw

де́рев·ц·е, -а, (*n*) see дерев·ц·о́

дерев·ц·о́, -а́, (*n*) (bot) sapling

дере́в·ья (*nom pl*) of де́рев·о

дерев·яни́ст·ый, -ая, -ое, (*a*) (bot) ligneous, woody

дерев·я́нн·ый, -ая, -ое, (*a*) wood, wooden, timber

дерев·я́шк·а, -и, (*g pl*) -шек·, (*f*) stump, piece of wood

дёрен·, -а, (*m*) (bot) dogwood, *Cornus*

дер·ёт (*pres 3rd sing*) of др·ать

держ·а́в·а, -ы, (*f*) state, power

держ·а́вк·а, -и, (*g pl*) -вок·, (*f*) holder, stand, support; tool post (of a machine tool); sting (aerodynamic model tests)

держ·а́вн·ый, -ая, -ое, (*a*) reigning, ruling

держ·а́тел·ь, -я, (*m*) holder, carrier, support

держ·а́ть, -а́т, (*imp*) hold, keep, retain; be supported by, -ся (+ *gen*) be supported by, be held up by, adhere, stick to, hold to, keep to; behave, conduct oneself; last, continue for

~, экза́мен· take an examination

дёрз·к·ий, -ая, -ое,   (*a*) impertinent, insolent; daring, bold

Дéри, двиг·ател·ь   (*m*) (elec) Deri motor

дерив·áт·, -а,   (*m*) derivative

дерив·ациóнн·ый, -ая, -ое,   (*a*) (civ eng) diversion

~ канáл·   (*m*) (civ eng) diversion cut, bye-channel; aqueduct, conduit (hydroelectric power stations)

дерив·áци·я, -и,   (*f*) diversion (of water from rivers etc.); (air, mil) drift; (ling) derivation

деривó·метр·, -а,   (*m*) drift gauge

дéрм·а, -ы,   (*f*) (zool) derma, dermis, derm

дермат·ин·, -а,   (*m*) American cloth, American leather (Brit), enamelled cloth, leatherette

дермат·ит·, -а,   (*m*) (med) dermatitis

дермато·гéн·, -а,   (*m*) (bot) dermatogen

дермато·зóм·а, -ы,   (*f*) (bot) dermatosome

дермат·óл·, -а,   (*m*) (chem) dermatol, bismuth subgallate

дермато·лóг·и·я, -и,   (*f*) (med) dermatology

дерма·тóм·, -а,   (*m*) (biol) dermatome, dermatomere

дермо·блáст·, -а,   (*m*) (zool) dermoblast

дерм·óид·, -а,   (*m*) (med) dermoid

дёрн·, -а,   (*m*) sod, turf

дерн·é·ть, -ют,   (*imp*) turf

дерн·ин·а, -ы,   (*f*) turf, root mat, sod

дерно·вид·н·ый, -ая, -ое,   (*a*) (bot) caespitose, cespitose

дернóво-под·зóл·ист·ый, -ая, -ое, (*a*) sod-podzolic (of soil)

дерн·óв·ый, -ая, -ое,   (*a*) sod, turf, cespititious

дерно·рéз·, -а,   (*m*) turf-cutter

дерно·с·ним·, -а,   (*m*) (agr) skim-coulter

дёр·н·уть, -ут,   (*perf*) *see* **дёрг·а·ть**

дéррик·, -а,   (*m*) derrick

~ -крáн·, -а,   (*m*) derricking jib-crane

деррис·, -а,   (*m*) (chem) derris

дер·ущ·ий, -ая, -ее,   (*pres part act*) of **др·ать**

дерюг·а, -и,   (*f*) hessian, sacking, sackcloth

десаксáж·, -а,   (*m*) (eng) offsetting

десáнт·, -а,   (*m*) (mil) landing (offensive); landing party/force

~, воз·дýш·н·ый   airborne landing

~, корабéль·н·ый   (naut) landing party

десáнт

~, мор·ск·óй   sea-borne landing

~, тáнк·ов·ый   tank-borne infantry

десáнт·ник·и, -ов,   (*m pl*) paratroopers, airborne troops; troops/forces taking part in landing (sea-borne)

десáнт·н·ый, -ая, -ое,   (*a*) (mil) landing

~ срéд·ств·а   (*pl*) (mil) landing craft

~ трáнспорт·   (*m*) landing ship

де·сатур·áци·я, -и,   (*f*) desaturation

дёсен·   (*gen pl*) of **десн·á**

дёсен·н·ый, -ая, -ое,   (*a*) (anat) gingival

де·сенсибилиз·áци·я, -и,   (*f*) (phot, med) desensitization

де·сенсибилиз·áтор·, -а,   (*m*) desensitizer

десéрт·н·ый, -ая, -ое,   (*a*) dessert

~ вин·ó   (*n*) sweet wine; cordial

де·сик·áнт·, -а,   (*m*) (chem) desiccant

де·силик·áци·я, -и,   (*f*) desilication

де·силиц·ир·овать, -уют,   (*imp and perf*) desiliconize

де·сквам·áци·я, -и,   (*f*) desquamation, peeling

десклоизит·, -а,   (*m*) (min) descloizite

дескрáйбер·, -а,   (*m*) (rail) describer

десмин·, -а,   (*m*) (min) desmine, stilbite

десмо·гéн·, -а,   (*m*) (biol) desmogen

десмо·гнат·ическ·ий, -ая, -ое,   (*a*) (zool) desmognathous

десмо·лáз·а, -ы,   (*f*) (biol) desmolase

десмо·лиз·, -а,   (*m*) (chem) desmolysis

десмо·троп·и·я, -и,   (*f*) (chem) desmotropism, desmotropy

десмо·трóп·н·ый, -ая, -ое,   (*a*) (chem) desmotropic

десм·ург·ическ·ий, -ая, -ое,   (*a*) (med) bandaging

десм·ург·и·я, -и,   (*f*) (med) bandaging

десн·á, -ы,   (*nom pl*) **дёсн·ы,** (*g pl*) **дёсен·,** (*d pl*) **дёсн·ам** (anat) gum, gingiva

де·сóрб·ер·, -а,   (*m*) (chem) desorber, stripper, stripping column

де·сóрб·ци·я, -и,   (*f*) (chem) desorption, stripping

дес·сик·áнт·   *see* **де·сик·áнт·**

дест·ев·óй, -áя, -óе,   (*a*) of **дéст·ь**

дестилл·áт·, -а,   (*m*) distillate

~ давл·éни·я   (oil) pressure distillate

~, мáсл·ян·ый   lubricating-oil distillate

дестилл·ирован·н·ый, -ая, -ое,   (*past part pass*) distilled

дестилл·яци·я, -и,   (*f*) *and see* **пере·-гóн·к·а** distillation

дестилл·я́ци·я
~, молекуля́р·н·ая   molecular distillation

деструкти́в·н·ый, -ая, -ое,   (*a*) destructive; degradative (polymers)

деструкт·и́рованн·ый, -ая, -ое,   (*a*) broken-down (polymers)

дестру́кц·и·я, -и,   (*f*) destruction; degradation, breakdown (polymers)

де́ст·ь, -и,   (*nom pl*) **-и,** (*g pl*) **-е́й,** quire (of paper)

де·сульфур·а́тор·, -а,   (*m*) desulphurizer

де·сульфур·иза́ци·я, -и,   (*f*) desulphurization

де·сульфур·а́ци·я, -и,   (*f*) desulphurization

десят·ери́чн·ый, -ая, -ое,   (*a*) tenfold, ten times as big

десят·ерн·о́й, -а́я, -о́е,   (*a*) tenfold, ten times as big

де́сят·ер·о, -ых,   (*numeral*) ten, ten together

десяти·во́д·н·ый, -ая, -ое,   (*a*) (chem) decahydrate

десяти·гра́н·ник·, -а,   (*m*) decahedron

десяти·дн·е́вк·а, -и,   (*g pl*) **-вок·,** (*f*) ten days, ten-day period

десяти·километр·о́вк·а, -и,   (*g pl*) **-вок·,** (*f*) (surv) 10 km-to-1 cm scale map

десяти·клави́ш·н·ый, -ая, -ое,   (*a*) ten-key; ten-column (of adding machines)

десяти·кра́т·н·ый, -ая, -ое,   (*a*) tenfold, ten times

десяти·ле́т·и·е, -я,   (*n*) decade, ten-year period; tenth anniversary/birthday

десяти·ле́т·н·ий, -яя, -ее,   (*a*) ten-year, decennial, ten-year-old

десят·и́н·а, -ы,   (*f*) tenth part, tithe

десят·и́нн·ый, -ая, -ое,   (*a*) tenth part

десяти·но́г·ие, -их,   (*pl decl as adj*) (zool) *Decapoda*

десяти·уго́ль·ник·, -а,   (*m*) (math) decagon

десят·и́чн·ый, -ая, -ое,   (*a*) (math) decimal

~ дроб·ь   (*f*) decimal fraction

~ знак·   (*m*) decimal point

~ масшта́б·   (*m*) (surv) natural scale

деся́т·к·а, -и,   (*g pl*) **-т·ок·,** (*f*) a ten, a tenner, a number ten; ten-rouble note

деся́т·ник·, -а,   (*m*) chargehand

деся́т·ок·, -т·к·а,   (*m*) ten (*collective noun*)

деся́т·ый, -ая, -ое,   (*a*) tenth

де́сят·ь, -и,   (*i*) **-ью́,** (*numeral*) ten

дет-   (*component*) = **де́т·ск·ий** (q.v.)

детал·иза́ци·я, -и,   (*f*) elaboration, itemization

детал·изи́р·овать, -уют,   (*imp and perf*) detail, elaborate, itemize

детал·из·ова́ть, -у́ют,   (*imp and perf*) see **детал·изи́р·овать**

детáл·ь, -и,   (*f*) detail; part, component

~, креп·ёжн·ые   (*pl*) fastenings

детáль·н·ый, -ая, -ое,   (*a*) of **детáл·ь;** detailed

~ черт·ёж·   (*m*) (mech eng) detail drawing

детáндер·, -а,   (*m*) expansion engine; reducing valve

детв·á, -ы́,   (*f*) grub, larva (of bees)

дет·до́м·, -а,   (*m*) children's home, orphanage

дет·е́й   (*gen*) of **де́т·и** (*pl*)

детект·и́ровани·е, -я,   (*n*) (rad) detection, demodulation, rectification

~, ано́д·н·ое   (rad) plate-circuit detection

~, се́т·очн·ое   (elec) grid detection

детéктор·, -а,   (*m*) (rad) rectifier, demodulator, detector

~, кристалл·и́ческ·ий   crystal detector/rectifier

~, ла́мп·ов·ый   electron-tube detector

~, лин·е́йн·ый   linear-response detector

~, пе́рв·ый   first/heterodyne detector

~, черенко́в·ый   Čerenkov detector

детéктор·н·ый, -ая, -ое,   (*a*) of **детéктор·**

~ коэффициéнт·   (*m*) (rad) detector coefficient

~ при·ём·ник·   (*m*) (rad) crystal receiver

дет·ёныш·, -а,   (*i*) **-ем,** (*m*) (zool) young one, cub, calf

детергéнт·, -а,   (*m*) detergent

детерминáнт·, -а,   (*m*) (math) determinant; (gen) determinant, heredity factor

де́т·и (*nom pl*), (*g pl*) **-е́й,** (*d pl*) **де́т·ям,** (*i pl*) **де́т·ьми́,** (*p pl*) **де́т·ях** children

де́т·ищ·е, -а,   (*n*) creation, work, brain child

де·токсик·а́ци·я, -и,   (*f*) detoxication

детон·а́тор·, -а,   (*m*) *and see* **электро·-детона́тор·** detonator, detonating charge

детон·ацио́нн·ый, -ая, -ое, *(a)* detonation, detonating; (ICE) knock, detonation

~ с·гор·а́ни·е *(n)* (ICE) detonation

~ сто́й·кост·ь *(f)* anti-knock value (liquid fuels)

детон·а́ци·я, -и, *(f)* detonation; (ICE) detonation, knock

детон·и́р·овать, -уют, *(imp)* detonate; (ICE) detonate, knock; be out of tune (music)

детон·и́рующ·ий, -ая, -ее, *(pres part act) see* **детон·и́р·овать**

~ шнур· *(m)* (expl) instantaneous detonating fuse

детоно́·метр·, -а, *(m)* (oil) detonation meter

дето·ро́д·н·ый, -ая, -ое, *(a)* (anat) genital, child-bearing

дето·рожд·е́ни·е, -я, *(n)* parturition, child-bearing

дето·у·би́й·ств·о, -а, *(n)* infanticide

детри́т·, -а, *(m)* detritus

~, о́сп·енн·ый (pharm) calf-lymph

детри́т·ов·ый, -ая, -ое, *(a)* detrital

де́тритус·, -а, *(m)* (geol) detritus

дет·са́д·, -а, *(m)* kindergarten, nursery school

де́т·ск·ий, -ая, -ое, *(a)* (*see also* **дет-**) child's, children's; infantile, puerile; (med) paediatric, pediatric

~ ме́ст·о *(n)* (anat) placenta, afterbirth

~ с·ме́рт·ност·ь *(f)* infant mortality

де́т·ств·о, -а, *(n)* childhood, infancy

де́·т·ый, -ая, -ое, *(a)* (*past part pass*) of **де·ть**

де·ть (*fut 3rd sing, pl*) **де́н·ет, де́н·ут,** (*past masc sing*) **де·л,** (*perf*) *see* **де·в·а́·ть**

деутеро- (*component*) *see also* **дейтеро-**

деутеро·ген·е́з·, -а, *(m)* (biol) deuterogenesis

деутеро·пла́зм·а, -ы, *(f)* (gen) deuteroplasm, deutoplasm

де-фа́кто (*adv*) *de facto*

дефек·а́т·, -а, *(m)* (agr) defecate, defecation mud

дефек·а́тор·, -а, *(m)* defection vessel/pan (sugar refining)

дефек·ацио́нн·ый, -ая, -ое, *(a)* defecation; defection (sugar)

дефек·а́ци·я, -и, *(f)* (zool) defecation, defaecation; defection (sugar refining)

дефе́кт·, -а, *(m)* defect, imperfection, fault

дефе́кт

~, конструкти́в·н·ый design/structural defect

~ ма́сс·ы (phys) mass defect

~, по·ве́рх·ностн·ый surface defect

~ решёт·к·и (cryst) lattice imperfection

~, то́ч·ечн·ый (cryst) point defect

~ у·пак·о́вк·и (cryst) stacking fault

дефект·и́вн·ый, -ая, -ое, *(a)* (med) abnormal, deficient, subnormal, handicapped

дефе́кт·ност·ь, -и, *(f)* imperfection, defectiveness

дефе́кт·н·ый, -ая, -ое, *(a)*, *(s f)* **-т·ен·, -т·н·а** defect; defective, faulty, imperfect; bad (copy, print, etc.)

~ ве́д·омост·ь *(f)* repair note

~ числ·о́ *(n)* (math) defective/deficiency number

дефекто·ло́г·и·я, -и, *(f)* (med) study of handicapped children

дефекто·ско́п·, -а, *(m)* flaw/fault detector

~, абсорбцио́н·н·ый absorption flaw detector

~, магни́т·н·ый magnetic flaw detector

дефекто·скоп·и́·я, -и, *(f)* flaw/fault/crack detection

~, индукцио́н·н·ая eddy-current testing/inspection

~, луч·ев·а́я radio examination/flaw detection

~, люминисце́нт·н·ая luminescence flaw detection

~, магни́т·н·ая magnetic crack detection

~ ме́тод·ом магни́т·н·ого порош·к·а́ magnetic-particle testing/inspecting

~, ультра·звук·ов·а́я ultrasonic flaw detection

дефекто·ско́п·н·ый, -ая, -ое, *(a)* of **дефекто·ско́п·**

~ теле́ж·к·а *(f)* (rail) mobile flaw detector

дефере́нт·, -а, *(m)* (astron) deferent, epicycle

деферриза́ци·я, -и, *(f)* deferrization

дефибре́р·, -а, *(m)* (paper) grinder

дефибре́р·н·ый ка́мен·ь *(m)* (paper) grinder

дефибрин·и́ровани·е, -я, *(n)* (med) defibrination

дефибр·и́ровани·е, -я, *(n)* (paper) grinding

дефиле́ (*n indecl*) (mil) defile

дефил·и́р·овать, -уют, (*imp*) (mil etc.) march past

дефини́ци·я, -и, (*f*) definition

дефи́с·, -а, (*m*) hyphen

дефици́т·, -а, (*m*) deficit; deficiency

~ **вла́ж·ност·и** (meteor) saturation deficit

дефици́т·н·ый, -ая, -ое, (*a*) (fin) losing, running at a loss; scarce, in short supply

де·флагр·а́тор·, -а, (*m*) (chem) deflagrator, deflagrating spoon

де·флагр·а́ци·я, -и, (*f*) (chem) deflagration; (ICE) detonation

де·флегм·а́тор·, -а, (*m*) (chem) dephlegmator, fractionating column, reflux condenser

де·флегм·атиза́тор·, -а, (*m*) (expl) desensitizer

де·флегм·а́ци·я, -и, (*f*) dephlegmation; (chem) reflux, refluxing

    коэффицие́нт· де·флегм·а́ци·й (*m*) reflux ratio

де·флегм·и́р·овать, -уют, (*imp and perf*) (chem) reflux

дефле́ктор·, -а, (*m*) deflector, baffle; compass deviation indicator; cowl (on chimney, ventilating shaft etc.)

~ **гидро·монито́р·а** (min) monitor deflector

де·флоккул·я́ци·я, -и, (*f*) deflocculation

де·флокул·я́ци·я, -и, (*f*) *see* **де·флок·кул·я́ци·я**

де·флор·а́нт·, -а, (*m*) (agr chem) defoliant

дефля́ци·я, -и, (*f*) deflation

Дефо, жёстк·ост·ь по (*f*) (plast) Defo value

де·фокус·иро́вк·а, -и, (*f*) defocusing

де·фоли·а́нт·, -а, (*m*) (agr chem) defoliant

лефо́·метр·, -а, (*m*) (plast) Defo meter

де·форм·ацио́нн·ый, -ая, -ое, (*a*) deformation, strain

де·форм·а́ци·я, -и, (*f*) deformation, strain

~, **абсолю́т·н·ая** (mech) extension

~, **близко·одно·вре́мен·н·ая** (geol) contemporary strain

~, **време́н·н·ая** (rubber) subpermanent set

~, **вз·ры́в·а** (met) high-velocity deformation

~, **выс·о́тн·ая** (met roll) vertical deformation

де·форм·а́ци·я

~ **двойник·ова́ни·я** (met) deformation by twinning

~, **знако·пере·ме́н·н·ая** flexing deformation/strain

~ **из·ги́б·а** bending/flexural strain/deformation

    коэффицие́нт· по·пере́ч·н·ой де·форм·а́ци·и (*m*) Poisson's ratio

~, **крит·и́ческ·ая** critical deformation/strain; (geol) ultimate strain

~ **круч·е́ни·я** shear/torsional strain/deformation

    ли́н·и·и де·форм·а́ци·и (*pl*) (met) stretcher strains, strain figures

~, **механ·и́ческ·ая** strain

    мо́дуль де·форм·а́ци·и (*m*) (mech) modulus of resilience

~ **на площа́д·ке теку́ч·ест·и** yield point elongation

~, **не·одно·ро́д·н·ая** inhomogeneous deformation/strain

~, **не·обрат·и́м·ая** irreversible deformation, permanent set

~, **не·равно·ме́р·н·ая** inhomogeneous/non-uniform deformation/strain

~, **объ·ём·н·ая** volume deformation

~, **одно·ро́д·н·ая** homogeneous strain/deformation

~, **о·ста́т·очн·ая** (met) plastic deformation, permanent set; (geol) residual deformation

~, **от·нос·и́тельн·ая** (met) unit strain; (geol) relative strain; (plast) strain ratio, strain

~, **пласт·и́ческ·ая** plastic deformation

~, **пло́ск·ая** uniaxial/plane strain/deformation

~, **по·вто́р·н·ая** a repeated/second deformation

~ **полз·у́чест·и** (met) total creep

~, **по·пере́ч·н·ая** lateral/transverse strain/contraction

~, **по·сто·я́нн·ая** residual/permanent deformation

~ **при ...** *see* **де·форм·а́ци·я** (+*gen*)

~ **при рас·тяж·е́ни·и** tensile strain, extension

~, **при·гран·и́чн·ая** boundary deformation

~, **про·до́ль·н·ая** stretch

~, **равно·ме́р·н·ая** homogeneous/uniform deformation/strain

~, **раз·ры́в·н·ая** rupture, failure by rupture, rupturing deformation

~ **с·дви́г·а** shear, shear strain; (geol) deformation shift

де·форм·а́ци·я

~ с·жа́т·и·я   compressive strain/deformation

~, скачко·обра́з·н·ая   heterogeneous deformation, non-uniform deformation

~ скольж·е́ни·я   deformation by slip

~ с·ре́з·а   shearing strain/deformation

~, сто́й·к·ая   persistent deformation

~, тек·у́ч·ая   flow, viscous flow

температу́р·а де·форм·а́ци·и   (*f*) softening point (of refractories)

~ теч·е́ни·я, вя́з·к·ая   (plast) viscous flow deformation

~, угл·ов·а́я   angular strain

~, у·проч·нённ·ая   rigid deformation

~, у·пру́г·ая   (met) elastic strain; (plast) resilient/elastic deformation

~, у·са́д·очн·ая   shrinkage strain

~, экзо·ге́н·н·ая   exogenic deformation

~, эндо·ге́н·н·ая   endogenic deformation

де·форм·и́р·овать, -уют,   (*imp and perf*) deform

де·форм·и́руемост·ь, -и,   (*f*) deformability

де·форм·и́руем·ый, -ая, -ое,   (*pres part pass*) of **де·форм·и́р·овать;** deformable

~, цикл·и́ческ·и   subject to deformation under cyclic load

де·форм·и́рующ·ий, -ая, -ее,   (*pres part act*) of **де·форм·и́р·овать;** dazzle (of camouflage)

~ на·пряж·е́н·ие   (*n*) flow stress

де·фосфирил·и́ровани·е, -я,   (*n*) (chem) dephosphyrilation

де·фосфор·а́ци·я, -и,   (*f*) (chem) dephosphoration

де·фрост·а́ци·я, -и,   (*f*) defrosting (refrigeration)

де·фро́ст·ер·, -а,   (*m*) defroster

дехени́т·, -а,   (*m*) (min) dechenite

де·хлор·а́ци·я, -и,   (*f*) dechlorination

де·хлор·и́ровани·е, -я,   (*n*) dechlorination

дехоли́н·, -а,   (*m*) (pharm) sodium dehydrochlorate

де·целеро́·метр·, -а,   (*m*) decelerometer

де·централиз·а́ци·я, -и,   (*f*) decentralization

де·центр·и́рованн·ый, -ая, -ое,   (*past part pass*) decentralized

деци-   (*component*) deci- (10^{-1})

деци·бе́л·, -а,   (*m*) (phys) decibel

деци·бе́л·ь, -я,   (*m*) see **деци·бе́л·**

деци·гра́мм·, -а,   (*g pl*) **-ов,** (*m*) decigramme, decigram

дец·и́л·, -а,   (*m*) (chem) decyl, decatyl

децел·азелаина́т·, -а,   (*m*) decyl azelate

децил·е́н·, -а,   (*m*) decylene

деци·ли́тр·, -а,   (*m*) decilitre, deciliter

деци́л·ов·ый, -ая, -ое,   (*a*) decyl

~ спирт·   (*m*) decyl alcohol, decanol

децима́ль·н·ый, -ая, -ое,   (*a*) decimal

деци·ме́тр·, -а,   (*m*) decimetre, decimeter

деци·метр·о́в·ый, -ая, -ое,   (*a*) decimetric, decimetre, decimeter

деци·не́пер·, -а,   (*m*) (telecom) decineper

деци́н·, -а,   (*m*) (chem) decine, decyne

деци́н·ов·ая кисл·от·а́   (*f*) decynic acid

деци·норма́ль·н·ый, -ая, -ое,   (*a*) (chem) decinormal

дешев·е́·ть, -ют,   (*imp*) become cheaper, fall in price

дешев·и́зн·а, -ы,   (*f*) cheapness, low prices

дешёв·к·а, -и,   (*f*) low price; bargain

дешёв·ый, -ая, -ое,   (*a*), (*s f*) **дёшев·, дешев·а́, дёшев·о,** cheap, low (of price)

дешени́т·, -а,   (*m*) (min) dechenite

де·шифр·а́тор·, -а,   (*m*) decoder; decoder, translator (telegraphy); (rad) decoder, converter unit; reader (punched cards etc.); (autom) decoder, tree

~, пере·крёст·н·ый   (autom) cross-over tree

~, реле́·йн·о-конта́кт·н·ый   (autom) relay and contact tree

~, электр·и́ческ·ий   mechanical reader

де·шифр·и́ровани·е, -я,   (*n*) decoding, deciphering; (phot) interpreting, interpretation

де·шифр·и́р·овать, -уют,   (*imp and perf*) decode, decipher; (phot) interpret

де·шифр·ова́ни·е, -я,   (*n*) = **де·шифр·и́ровани·е**

де·шифр·о́ванн·ый, -ая, -ое,   (*past part pass*) decoded, deciphered; (phot) interpreted

де·шифр·о́вк·а, -и,   (*f*) = **де·шифр·и́ровани·е**

де·шифр·о́вщик·, -а,   (*m*) decoder; interpreter (of survey photos etc.)

де·шлам·а́ци·я, -и,   (*f*) desliming

де·эмульг·а́тор·, -а,   (*m*) (oil) demulsifying reagent

де·эмульг·а́ци·я, -и,   (*f*) demulsification

числ·о́ де·эмульг·а́ци·и   (*n*) (oil) demulsification number

де·этан·иза́ци·я, -и,   (*f*) (chem) deethanation

де-ю́ре   (*adv*) *de jure*

де·я́ни·е, -я,   (*n*) deed, act

де́·ятел·ь, -я,   (*m*) prominent person, distinguished figure

~, госуда́р·ственн·ый   politician, government figure

~ на·у́к·и, за·слу́ж·енн·ый   Distinguished Scientist (title)

де́·ятельност·ь, -и,   (*f*) work, activity, activities, occupation, profession, career

де́·ятельн·ый, -ая, -ое,   (*a*) active, energetic

дж   (*abbr*) of **джо́ул·ь**   joule (unit of work)

джейра́н·, -а,   (*m*) (zool) *Gasella subgutturosa*

джак·, -а,   (*m*) (*see also* **джек·**) (text) jack frame

джалма́йт·, -а,   (*m*) (min) djalmaite

джек·, -а,   (*m*) (min) hammer drill, jackhammer; (elec) jack

джем·, -а,   (*m*) (food) jam

дже́мпер·, -а,   (*m*) jumper, pullover (clothing)

джемсони́т·, -а,   (*m*) (min) jamesonite

джеспили́т·, -а,   (*m*) (min) jaspilite

джет·, -а,   (*m*) (min) jet

джиг·, -а,   (*m*) (min) tail-rope incline

~ -пульс·а́тор·, -а,   (*m*) pulsator jig (ore dressing)

джи́ггер·, -а,   (*m*) jigger (fishing boat)

джи́гер·, -а,   (*m*) jigger (dyeing machine)

джи́д·а, -ы,   (*f*) (bot) elaeagnus

джилпини́т·, -а,   (*m*) (min) gilpinite

джин·   (*m*) (text) gin

~, ва́л·ичн·ый   roller gin

~, воздухо·ду́в·н·ый   air-blast gin

~, пи́ль·н·ый   saw gin

джин·и́р·овать, -уют,   (*imp and perf*) (text) gin

Джо́лли, вес·ы́   Jolly balance

джо́ул·ь, -я,   (*m*) (phys) joule

Джо́ул·я, за·ко́н·   (*m*) Joule's law of thermodynamics

~ -Ле́нц·а, за·ко́н·   (*m*) (elec) Joule's law

~ -То́мсон·а, эффе́кт·   (*m*) Joule-Thomson/Kelvin effect

джугар·а́, -ы́,   (*f*) (bot) *Sorghum cernuum*

джу́т·, -а,   (*m*) (bot) jute, *Corchorus*

джуто·плющ·и́лк·а, -и,   (*g pl*) **-лок·,** (*f*) jute softener

дзе́т·а, -ы,   (*f*) zeta (Greek letter) ζ, Z

~ -потенциа́л·, -а,   (*m*) electrokinetic potential

дзере́н   (zool) *Gasella gutturosa*

ди-   (*component*) di-, bi-, two-, twin-, double

диа·ба́з·, -а,   (*m*) (min) diabase

диа·ба́з·ов·ый, -ая, -ое,   (*a*) diabase; diabasic

~ ка́мен·ь   (*m*) (geol) greenstone

диабе́т·, -а,   (*m*) (med) diabetes

диа·ген·е́з·, -а,   (*m*) (geol) diagenesis, lithification

диа́·гноз·, -а,   (*m*) diagnosis

диа·гно́ст·, -а,   (*m*) (med) diagnostician

диа·гно́ст·ик·, -а,   (*m*) diagnostician

диа·гно́ст·ик·а, -и,   (*f*) diagnostics; diagnosis

диа·гност·ику́м·, -а,   (*m*) (bact) diagnosticum

диа·гност·и́ческ·ий, -ая, -ое,   (*a*) diagnostic

диа·гон·а́л·ь, -и,   (*f*) (math) diagonal; (text) diagonal canton

~, про·стра́н·ственн·ая   (math) body diagonal

диа·гон·а́льн·о-ре́з·ательн·ая ма·ши́н·а   (*f*) bias-cutting machine (rubber)

диа·гон·а́льн·ый, -ая, -ое,   (*a*) diagonal

~ систе́м·а   (*f*) (min) rill stoping

диа·гра́мм·а, -ы,   (*f*) diagram, graph, scheme, chart, pattern

~, времен·н·а́я   timing chart (of computer)

~ де·фо́рм·а́ци·и   (mech) stress–strain diagram

~ из·луч·е́ни·я   (rad) radiation pattern, cardioid diagram

~, индика́тор·н·ый   (eng) indicator diagram

~ ис·теч·е́ни·я   (phys) effluogram

~, круг·ов·а́я   (elec) circle diagram

~ Кюри́   (nucl) Curie plot

~ на·пра́вл·енност·и   (rad) polar diagram

~ пла́в·кост·и   (met) constitution equilibrium diagram

~, поля́р·н·ая   (rad) polar diagram

~ пу́льс·а   (med) sphygmogram

диа·грáмм·а
~ **равно·вéс·и·я** (met) equilibrium/ constitution diagram
~ **рас·пре·дел·éни·я** porting diagram (of valveless engine)
~ **рас·пре·дел·éни·я по·тóк·а** (elec) flux pattern
~ **рас·тяж·éни·я** (met) load-elongation curve
~ **Рúд·а** (shipb) curve of stability
~ **со·сто·я́ни·я** (met) constitution diagram
~ **стат·и́ческ·ой о·стóй·чивост·и** (shipb) curve of stability
~, **стóлб·чат·ая** block diagram
~, **структýр·н·ая** (elec) flow diagram
~ **фаз·** (phys) phase diagram
~ **фаз· газо·рас·пре·дел·éни·я** (ICE) valve-opening diagram
~, **характерист·и́ческ·ая** performance diagram
диа·грáмм·н·ый, -ая, -ое, (a) of **диа·грáмм·а**
~ **бумáг·а** (f) (instr) chart paper
диа·грáф·, -а, (m) (instr) diagraph
диад·а, -ы, (f) (gen) dyad, bivalent chromosome
ди·аз·о- (component) (chem) diazo-, disazo-
диазо·амин·о- (component) (chem) diazoamino-
диазоамино·бензол·, -а, (m) diazoamino benzene
диазоамино·со·един·éни·е, -я, (n) diazoamino compound
диазо·грýпп·а, -ы, (f) (chem) diazo group
диаз·ол·и́н·, -а, (m) diazoline
диаз·óл·ь, -я, (m) (chem) diazole
диазо·метáн·, -а, (m) (chem) diazomethane
диаз·óниев·ый, -ая, -ое, (a) diazonium
диаз·óни·й, -я, (m) diazonium
диазо·со·един·éни·е, -я, (n) (chem) diazo compound
диазо·сóл·ь, -и, (f) diazonium salt
ди·азот·áт·, -а, (m) diazotate, diazoate
диазо·тúп·и·я, -и, (f) (phot) diazotype
ди·азот·и́ровани·е, -я, (n) (chem) diazotization, diazotizing
ди·азот·и́р·овать, -уют, (imp and perf) diazotize
диа·кáрб·, -а, (m) (pharm) diacarb, diamox, acetazolamide
диакис·додекá·эдр·, -а, (m) (cryst) diakisdodecahedron, diploid

диа·клáз·, -а, (m) (geol) diaclase
диа·крит·и́ческ·ий, -ая, -ое, (a) (ling) diacritic, diacritical
ди·актин·и́ческ·ий, -ая, -ое, (a) (phys) diactonic
диалéкт·, -а, (m) (ling) dialect
диалект·áльн·ый, -ая, -ое, (a) (ling) dialect, dialectal
диалéкт·ик·, -а, (m) dialectical theorist
диалéкт·ик·а, -и, (f) dialectics
диалект·и́ческ·ий, -ая, -ое, (a) dialectical
диалéкт·н·ый, -ая, -ое, (a) (ling) dialect, dialectal
диá·лиз·, -а, (m) (chem) dialysis
диа·лиз·áт·, -а, (m) (chem) dialysate
диа·лиз·áтор·, -а, (m) (chem) dialyser
диали·стел·и́·я, -и, (f) (bot) dialystely
диа·лит·и́ческ·ий, -ая, -ое, (a) dialytic
ди·алкúл·, -а, (m) (chem) diakyl
диаллáг·, -а, (m) (min) diallage
диаллаг·и́т·, -а, (m) (min) diallagite
ди·аллúл·, -а, (m) (chem) diallyl
ди·аллúл·ов·ый, -ая, -ое, (a) diallyl
диаллил·тио·моч·евúн·а, -ы, (f) (chem) diallyl thiourea
диалогúт·, -а, (m) (min) dialogite
ди·ал·ýр·ов·ая кисл·от·á (f) (chem) dialuric acid
ди·альдегúд·, -а, (m) (chem) dialdehyde
диа·лóг·, -а, (m) dialogue
диа·магнет·и́зм·, -а, (m) (phys) diamagnetism
диа·магнúт·н·ый, -ая, -ое, (a) diamagnetic
диамáнт·, -а (m) diamond
диá·метр·, -а, (m) diameter
~ **в мúдел·е** (a/c) maximum diameter of fuselage
~ **в свет·ý** bore, internal diameter
~, **внýтр·енн·ий** internal diameter, bore (of pipe); minor/core diameter (of thread)
~ **вы·ход·н·óго сеч·éни·я** exit diameter (of nozzle)
~, **крит·и́ческ·ий** critical size/diameter
~, **наи·мéньш·ий** throat diameter (of worm gear)
~, **нарýж·н·ый** outside diameter: nominal/major diameter (of thread)
~, **рабóч·ий** effective/working diameter
~ **релé·йн·ого на·бóр·а** (elec) table of components (switching algebra)

диа́·метр
~, **сре́д·н·ий**   mean diameter; pitch/effective/angle diameter (of thread)
диа·метр·а́льн·о·противо·по·ло́ж·-н·ый, -ая, -ое,   (*a*) diametrically opposite, antipodal
диа·метр·а́льн·ый, -ая, -ое, (*a*) diameter, diametric, diametral, centre-line
~ **пло́ск·ост·ь**   (*f*) (shipb) centreline, middle line
ди·ами́д·, -а,   (*m*) (chem) diamide, hydrazine hydrate
ди·амид·о- (*component*) (chem) diamido-
диамидо·фено́л·, -а,   (*m*) diamido-phenol, phenol diamide
ди·амил·- (*component*) (chem) diamyl-
ди·ами́н·, -а, (*m*) (chem) diamine
ди·амин·о- (*component*) diamine, di-amino-
диамино·фено́л·, -а,   (*m*) diamino-phenol
диамино·эта́н·, -а,   (*m*) ethylene diamine
ди·аммо·фо́с·, -а,   (*m*) (agr chem) diammophos, diammonium phosphate
диа́н·, -а, (*m*) (chem) diane, dihydroxy diphenyl propane
диа·пазо́н·, -а, (*m*) range, band
~ **волн·** (rad) waveband
~, **динам·и́ческ·ий**   dynamic range (tape recorder)
~, **дроссел·и́ровани·я**   throttling range
~, **об·зо́р·н·ый**   (rad) search band
~, **пуск·ов·о́й**   star-up range
~ **центр·о́вок·**   (air) CG limits, limits of centre of gravity
диа·пазо́н·н·ый, -ая, -ое, (*a*) of **диа·пазо́н·**
~ **анте́нн·а** (*f*) wide-band antenna
диа·па́уз·а, -ы, (*f*) (biol) diapause
диа·пед·е́з·, -а, (*m*) (zool) diapedesis
диа·пир·и́зм·, -а, (*m*) (geol) dia-pirism
диа·пи́р·ов·ый, -ая, -ое, (*a*) (geol) diapiric
диа·позити́в·, -а, (*m*) (phot) diaposi-tive, lantern slide, colour transparency
диа·позити́в·н·ый, -ая, -ое, (*a*) of **диа·позити́в·**
ди·апофи́з·, -а, (*m*) (biol) diapophysis
диа·прое́ктор·, -а, (*m*) (phot) slide projector
ди·апси́д·н·ый, -ая, -ое, (*a*) (zool) diapsid
ди·ари́л·, -а, (*m*) (chem) diaryl

диаспо́р·, -а, (*m*) (min) diaspore
диа·спо́р·а, -ы, (*f*) (bot) diaspore
диа·ста́з·а, -ы, (*f*) (biochem) diastase
диастази́с·, -а, (*m*) (physiol) diastasis, (*pl*) diastases
диастат·и́ческ·ий, -ая, -ое, (*a*) (biochem) diastatic
диасте́м·а, -ы, (*f*) (anat) diastem
диа·сти́л·ь, -я, (*m*) (arch) diastyle
диа́·стол·а, -ы, (*f*) (physiol) diastole
диа·стол·и́ческ·ий шум· (*m*) (med) diastolic murmur
диа·строф·и́зм·, -а, (*m*) (geol) dia-strophism
диа·схи́ст·ов·ый, -ая, -ое, (*a*) (cyt) diaschystic
диа·те́з·, -а, (*m*) (med) diathesis, (*pl*) diatheses
диа·терм·и́чн·ый, -ая, -ое, (*a*) (phys) diathermanous; (med) diathermic
диа·терм·и́чност·ь, -и, (*f*) diather-mancy
диа·терм·и́·я, -и, (*f*) (med) diathermy
диа·термо·коагул·я́ци·я, -и, (*f*) (med) electrodesiccation, diathermic coagulation
диа·том·е́·я, -и, (*f*) (biol) diatom, (*pl*) *Diatomeae*
диа·том·и́т·, -а, (*m*) (min) diatomite; diatomaceous earth
диа·то́м·н·ый, -ая, -ое, (*a*) diatom-aceous, diatomic
диа·то́м·ов·ый, -ая, -ое, (*a*) (biol) diatom, diatomaceous; (*pl*) diatoms, *Diatomaceae*
~ **ил·** (*m*) (ocean) diatom ooze
диа·тре́м·а, -ы, (*f*) (geol) diatreme, volcanic vent
диа·троп·и́зм·, -а, (*m*) (bot) diatrop-ism
диа·фа́н·, -а, (*m*) transparent print on glass; unglazed pottery
диа·фано́·метр·, -а, (*m*) (phys) dia-phanometer
диа·фано·ско́п·, -а, (*m*) diaphano-scope; visibility meter
диа·фи́з·, -а, (*m*) (anat) diaphysis
диа·фи́льм·, -а, (*m*) (phot) diapositive film, diapositive film strip
диа·фо́н·, -а, (*m*) diaphone (acoustics)
диа·фра́гм·а, -ы, (*f*) diaphragm; (opt) diaphragm, stop; (rad) iris, waveguide diaphragm
~, **аперту́р·н·ая** (phot) aperture stop; objective aperture (of microscope)
~, **гофр·и́рованн·ая** (acous) pleated diaphragm

диа·фра́гм·а
~, **де́й·ствующ·ая** aperture stop
~, **ири́с·ов·ая** (phot) iris diaphragm, iris shutter frame
   коэффицие́нт· диафра́гм·ы (*m*) (dynam) orifice coefficient
~, **о́стр·ая** thin-plate orifice
~ **по́л·я зр·е́ни·я** (opt) field stop
~ **с·вя́з·и** (rad) coupling iris
~, **спин·н·а́я** (anat) dorsal diaphragm
диа·фра́гм·енн·ый, -ая, -ое, (*a*) of **диа·фра́гм·а**
~ **коро́б·к·а** (*f*) (electrochem) cathode box
диа·фрагм·и́ровани·е, -я, (*n*) (phot) stopping down, reducing the aperture
диа·фра́гм·ов·ый, -ая, -ое, (*a*) of **диа·фра́гм·а**
диа·фтор·е́з·, -а, (*m*) (geol) diaphthoresis
диа·хил·о́н·, -а, (*m*) (pharm) diachylon
ди·ацет·ами́д·, -а, (*m*) (chem) diacetamide
ди·ацет·а́т·, -а, (*m*) (chem) diacetate
ди·ацет·и́л·, -а, (*m*) (chem) diacetyl
ди·ацети́ль·н·ый, -ая, -ое, (*a*) diacetyl; diacetylated
ди·ацет·о́н·, -а, (*m*) diacetone
ди·ацето́н·ов·ый спирт·, (*m*) diacetone alcohol
диа·ши́ст·ов·ый, -ая, -ое, (*a*) diashistic
ди·бенз·и́л·, -а, (*m*) (chem) dibenzyl
ди·бензо·и́л·, -а, (*m*) (chem) dibenzoyl
ди·бор·а́н·, -а, (*m*) (chem) diborane
ди·бром·- (*component*) (chem) dibromo-
дибром·бензо́л·, -а, (*m*) dibromobenzene
дибром·этиле́н·, -а, (*m*) (chem) ethylene dibromide
дибром·за·мещ·ённ·ый, -ая, -ое, (*a*) dibromo-
ди·бути́л·, -а, (*m*) (chem) dibutyl
ди·бути́л·ов·ый, -ая, -ое, (*a*) (chem) dibutyl
ди·бутил·фтала́т·, -а, (*m*) dibutyl-phthalate
ди·вакци́н·а, -ы, (*f*) (med) bivalent vaccine
дива́н·, -а, (*m*) sofa, settee
ди·ванад·и́л·, -а, (*m*) (chem) divanadyl
ди·вариа́нт·н·ый, -ая, -ое, (*a*) bivariant
ди·варик·а́тор·, -а, (*m*) (zool) divaricator

ди·верг·е́нт·н·ый, -ая, -ое, (*a*) divergent
ди·верг·е́нци·я, -и, (*f*) divergence
диверс·а́нт·, -а, (*m*) saboteur, wrecker
диве́рс·и·я, -и, (*f*) sabotage; (mil) diversion (tactical)
диве́ртер·, -а, (*m*) (elec) diverter
дивиде́нд·, -а, (*m*) (fin) dividend
дивизио́н·, -а, (*m*) (mil) division, squadron, flotilla; battery (artillery), battalion (U.S.)
дивизио́н·н·ый, -ая, -ое, (*a*) of **ди·визио́н·**
диви́зи·я, -и, (*f*) division (army); large tactical force (naval)
ди·вини́л·, -а, (*m*) (chem) divinyl, (plast) butadiene
ди·винил·ацетиле́н·, -а, (*m*) (chem) divinylacetylene
ди·вини́л·ов·ый, -ая, -ое, (*a*) divinyl, butadiene
~ **каучу́к·** (*m*) butadiene rubber
дивинил-стиро́ль·н·ый, -ая, -ое, (*a*) (plast) butadiene-styrene
ди·ви́н·н·ая кисл·от·а́ (*f*) ditartaric acid
ди́в·н·ый, -ая, -ое, (*a*) marvellous, wonderful, amazing
ди·га́лл·ов·ая кисл·от·а́ (*f*) (chem) digallic acid
ди·гало́ид·, -а, (*m*) (chem) dihalide
ди·галоидо·про·из·во́д·н·ый, -ая, -ое, (*a*) dihalide derivative
ди·гедра́ль·н·ый, -ая, -ое, (*a*) dihedral
ди·гекса·гон·а́льн·ый, -ая, -ое, (*a*) (cryst) dihexagonal
ди·ген·е́з·, -а, (*m*) (zool) digenesis
ди·гидрази́н·, -а, (*m*) (chem) diamidogen
ди·гидра́т·, -а, (*m*) dihydrate
ди·гидри́т·, -а, (*m*) (geol) dihydrite
ди·гидро- (*component*) (chem) dihydro-
ди·гидр·окси- (*component*) (chem) dihydroxy-
дигидрокси·фенилалани́н·, -а, (*m*) (chem) dihydroxyphenyl-alanine
ди·ги́р·а, -ы, (*f*) (cryst) digyre, second-order symmetry axis
ди·ги́р·н·ый, -ая, -ое, (*a*) (cryst) digyric, orthorhombic
дигитал·и́н·, -а, (*m*) (chem) digitalin
дигит·а́лис·, -а, (*m*) (bot, pharm) digitalis
дигит·они́н·, -а, (*m*) (chem) digitonin
ди·гра́ф·, -а, (*m*) (math) digraph, directed graph

ди·деси́л·, -а, (*m*) (chem) didecyl
дидим·и·й, -я, (*m*) (obs chem) didymium, Di
дидимо·ли́т·, -а, (*m*) (min) didymolite
ди·додека́·эдр·, -а, (*m*) didodecahedron
дие́з·, -а, (*m*) sharp (music)
дие́н·, -а, (*m*) (chem) diene, diolefine
~, цикл·и́чн·ый cycloalkadiene
диен·ов·ый, -ая, -ое, (*a*) (chem) diene
~ син·тéз· (*m*) diene synthesis
дие́т·а, -ы, (*f*) diet
диет·е́тик·а, -и, (*f*) dietetics
диет·ети́ческ·ий, -ая, -ое, (*a*) dietetic, dietary
диза́жио (*n indecl*) discount (of stocks and shares)
ди·за·мещ·ённ·ый, -ая, -ое, (*a*) (chem) di-substituted
дизанали́т·, -а, (*m*) (min) dysanalyte
дизеле·стро·е́ни·е, -я, (*n*) diesel engineering/construction
ди́зел·ь, -я, (*m*) diesel, diesel engine
~, бес·компре́ссор·н·ый solid-injection diesel
~, компре́ссор·н·ый air-injection diesel, air-blast diesel
~, суд·ов·о́й marine diesel
~ -электро·хо́д·, -а, (*m*) diesel-electric motor ship
ди́зель·н·ый, -ая, -ое, (*a*) diesel
~ то́пл·ив·о (*n*) diesel fuel (for stationary diesels)
диз·ентер·и́·я, -и, (*f*) (med) dysentery
диз·ъ·ю́нкт·ност·ь, -и, (*f*) (math) disjunction, disjointness
диз·ъ·ю́нкт·н·ый, -ая, -ое, (*a*) disjoint, disjunct
диз·ъ·ю́нкц·ио́нн·ый, -ая, -ое, (*a*) disjunctive
диз·ъ·ю́нкц·и·я, -и, (*f*) disjunction
~, раз·дел·и́тельн·ая (math) exclusive disjunction
ди·изо- (*component*) (chem) di-iso-, diiso-
диизо·бути́л·, -а, (*m*) di-isobutyl
диизо·никотиноил·гидрази́д·, -а, (*m*) (pharm) di-isonicotinoyl hydrazide
диизо·окти́л·, -а, (*m*) di-isooctyl, diisoctyl
диими́д·, -а, (*m*) di-imide, diimide
дими́н·, -а, (*m*) di-imine, diimine
ди·имино- (*component*) di-imino-, diimino-
ди·иодо- (*component*) di-iodo-, diiodo-
дииодо·анили́н·, -а, (*m*) di-iodoaniline, diiodoaniline

Ди́к·а, реа́кци·я (*f*) Dick test (for scarlet fever)
ди·ка́йн·, -а, (*m*) (pharm) dicaine
ди·карио́н·, -а, (*m*) (cyt) dicaryon, dikaryon
ди·кете́н·, -а, (*m*) (chem) diketene
ди·кето- (*component*) (chem) diketo- (containing two ketone groups)
ди·кето́н·, -а, (*m*) (chem) diketone
дикето·пиперази́н·, -а, (*m*) (chem) diketopiperazine
ди́к·ий, -ая, -ое, (*a*) wild, savage, feral; shy, unsociable
дико·раст·у́щ·ий, -ая, -ее, (*a*) (bot) wild
дикт·а́нт·, -а, (*m*) dictation
диктат·у́р·а, -ы, (*f*) dictatorship
диктио·ге́н·, -а, (*m*) (bot) dictyogen
диктио·каул·ёз·, -а, (*m*) (vet) dictyocaulosis
диктио·кин·ёз·, -а, (*m*) (cyt) dictyokinesis
диктио·стéл·а, -ы, (*f*) (bot) dictyostele
дикт·ова́ть, -у́ют, (*imp*) dictate
дикт·о́вк·а, -и, (*g pl*) -вок·, (*f*) dictation
ди́ктор·, -а, (*m*) announcer (broadcasting)
ди́ктор·ск·ий, -ая, -ое, (*a*) announcer's
дикто·фо́н·, -а, (*m*) dictaphone
ди́кц·и·я, -и, (*f*) diction, (ling) enunciation, articulation
дилат·а́ци·я, -и, (*f*) dilatation
дилато·гра́мм·а, -ы, (*f*) (phys) dilation curve
дилато́·метр·, -а, (*m*) (phys) dilatometer
дилато·ме́тр·и·я, -и, (*f*) (phys) dilatometry
диле́мм·а, -ы, (*f*) dilemma
дилен·ы (*pl*) deal ends (timber)
дилета́нт·ск·ий, -ая, -ое, (*a*) amateurish
ди́льс·, -а, (*m*) deal (timber)
ди·люви·а́льн·ый, -ая, -ое, (*a*) (geol) diluvial
ди·лю́ви·й, -я, (*m*) (obs geol) drift, diluvium
диля́тор·, -а, (*m*) (zool) dilator
диме·д·о́н·, -а, (*m*) (chem) dimedon, dimethyl-dihydro-resorcinol
димедро́л·, -а, (*m*) (pharm) dimedrol, diphenhydramine hydrochloride
ди·ме́р·, -а, (*m*) (chem) dimer

ди·ме́р·н·ый, -ая, -ое,   (*a*) (chem) dimeric

ди·мер·иза́ци·я, -и,   (*f*) (chem) dimerization

ди·метил- (*component*) dimethyl-

диметил·ами́н·, -а,   (*m*) dimethylamine

диметил·бутади́н·, -а,   (*m*) dimethylbutadiene

диметил·глиокси́м·, -а,   (*m*) dimethylglyoxime

диметил·кето́н·, -а,   (*m*) dimethyl ketone, acetone

диметил·кето́н·иза́ци·я, -и,   (*f*) (chem) acetonation

диметил·сульфа́т·, -а,   (*m*) (chem) dimethyl sulphate

димети́л·ов·ый, -ая, -ое,   (*a*) dimethyl

ди·мето́кси- (*component*) (chem) dimethoxy-

ди·метр·одо́н·, -а,   (*m*) (pal) dimetrodon

ди·моно́·эдр·, -а,   (*m*) (cryst) dimonohedron

ди·мо́рф·изм·, -а,   (*m*) dimorphism

ди·мо́рф·н·ый, -ая, -ое,   (*a*) dimorphous, dimorphic

дин (*abbr*) of **ди́н·а**

ди́н·а, -ы,   (*f*) dyne (unit of force)

дина́м·ик·, -а,   (*m*) dynamic/moving-coil loudspeaker

дина́м·ик·а, -ы,   (*f*) dynamics

~ **взаимо·де́й·стви·я** cross-coupling dynamics

~, **га́з·ов·ая** gas dynamics, compressible-flow aerodynamics

~ **не·свобо́д·н·ой систе́м·ы** rigid-body dynamics

~, **о́бщ·ая** gross dynamics

~ **твёрд·ого те́л·а** rigid dynamics, rigid-body dynamics

~, **элемента́р·н·ая** simple dynamics

динам·и́т·, -а,   (*m*) (expl) dynamite

динами́т·н·ый, -ая, -ое,   (*a*) (expl) dynamite

динам·и́ческ·ий, -ая, -ое,   (*a*) dynamic

динам·и́чн·ый, -ая, -ое,   (*a*) dynamic

дина́м·н·ый, -ая, -ое,   (*a*) dynamo

дина́мо (*n or f indecl*) dynamo

динамо́·граф·, -а,   (*m*) dynamograph, recording dynamometer

динамо·маши́н·а, -ы,   (*f*) dynamo, D.C. generator

динамо́·метр·, -а,   (*m*) dynamometer

~, **инструмента́ль·н·ый** electric torsion meter

динамо́·метр

~, **колц·ев·о́й** proof-ring dynamometer

~, **крут·и́льн·ый** torque dynamometer

~, **по·глощ·а́ющ·ий** absorption dynamometer

~, **трансмиссио́н·н·ый** transmission dynamometer

~, **тя́г·ов·ый** traction dynamometer

динамо·метр·и́ческ·ий, -ая, -ое,   (*a*) of **динамо́·метр·**

~ **вту́л·к·а**   (*f*) torque meter

~ **теле́ж·к·а**   (*f*) (rail) rail dynamometer car

динамо́н·, -а,   (*m*) (expl) dynammon

динамо·пласт·и́ческ·ий, -ая, -ое,   (*a*) dynamoplastic

дина·мото́р·, -а,   (*m*) dynamotor, rotary transformer

дина́нт·ск·ий от·де́л·   (*m*) (geol) Dinantian series

ди́нас·, -а,   (*m*) dinas brick (refractory)

ди́нас·ов·ый, -ая, -ое,   (*a*) dinas, dinas-brick

ди·на́тр·и·й, -я,   (*m*) (chem) dibasic sodium

дина·тро́н·, -а,   (*m*) (elec) dynatron

ди·нафт- (*component*) (chem) dinaphtho-

ди·нейтро́н·, -а,   (*m*) (phys) dineutron

ди·нитро- (*component*) (chem) dinitro-

динитро·бензо́л·, -а,   (*m*) (expl) dinitrobenzene

динитро·глицери́н·, -а,   (*m*) (expl) dinitroglycerine

динитрозил·гало́ид·, -а,   (*m*) dinitrosyl halide

динитрозил·хлори́д, -а,   (*m*) dinitrosyl chloride

динитро·толуо́л·, -а,   (*m*) (chem) dinitrotoluene

дин·о́д·, -а,   (*m*) (elec) dynode

дино·за́вр·, -а,   (*m*) (pal) dinosaur

~, **птице·та́з·ов·ый** ornithischian dinosaur

~, **ящеро·та́з·ов·ый** saurichian dinosaur

дино·те́р·и·й, -я,   (*m*) (pal) dinothere

дино·цера́т·ы, -ов,   (*m pl*) (pal) *Dinocerata*

ди·нукло́н·, -а,   (*m*) (phys) dinucleon

ди·о́д·, -а, (*m*)   (elec) diode

~, **двой·н·о́й** double diode, binode

~, **де́мпфер·н·ый** damping diode

~, **кристалл·и́ческ·ий** crystal diode, semiconductor rectifier

~, **от·ключ·а́ющ·ий** isolating/isolation diode

ди·о́д

~, о·гран·и́чивающ·ий   bootstrap diode

~, от·сек·а́ющ·ий   pick-off diode

~ -пент·о́д·, -а, (*m*) diode-pentode

~, плоск·остн·о́й   junction diode

~ -три·о́д·, -а, (*m*) diode-triode

~, фикс·и́рующ·ий   clamping diode

~, шум·ов·о́й   noise diode

~, эквивале́нт·н·ый   dummy diode

дио́д·н·ый, -ая, -ое, (*a*) (elec) diode

~ ключ· (*m*) gate trigger diode

ди·окс·а́н·, -а, (*m*) (chem) dioxane

ди·окси- (*component*) dioxy-, dihydroxy-

диокси·азобензо́л·, -а, (*m*) (chem) azophenol

диокси·бензо́л·, -а, (*m*) dihydroxybenzene

диокси·ви́н·н·ая кисл·от·а́ (*f*) dioxytartaric acid

ди·окси́м·, -а, (*m*) dioxime

диокси·толуо́л·, -а, (*m*) (chem) dihydroxytoluene

ди·окти́л·, -а, (*m*) (chem) dioctyl

диокто·фи́м·, -а, (*m*) (zool) dioctophyme

ди·о́л·, -а, (*m*) (chem) diene, diolefin

ди·олефи́н·, -а, (*m*) (chem) diolefine, diene

**дион
и́н·**, -а, (*m*) (pharm) dionine

ди·опси́д·, -а, (*m*) (min) diopside

ди·опт·а́з·, -а, (*m*) (min) dioptase, emerald copper

ди·о́птр·, -а, (*m*) peep sight, sightingvane

ди·оптр·и́·я, -и, (*f*) dioptre, diopter, dioptrie, D (unit of power of a lens)

ди·оптр·и́ческ·ий, -ая, -ое, (*a*) dioptric

диори́т·, -а, (*m*) (min) diorite, (build) black granite

ди·оти́ческ·ий, -ая, -ое, (*a*) binaural, diotic

Диофа́нт·, -а, (*m*) Diophantus (Greek mathematician)

диофа́нт·ов·ый, -ая, -ое, (*a*) (math) diophantine

ди·пенте́н·, -а, (*m*) (chem) dipentene, terpene

дипептид·а́з·а, -ы, (*f*) (biochem) dipeptidase

ди·пи́р·, -а, (*m*) (min) dipyre, dipyrite

ди·пирами́д·а, -ы, (*f*) (cryst) dipyramid

ди·пириди́л·, -а, (*m*) (chem) dipyridyl

ди·пле́кс·н·ый, -ая, -ое, (*a*) (telecom) diplex

ди·пле́кс·н·ый, -ая, -ое

~ при·ём· (*m*) diplex/double reception

ди·плеу́р·ул·а, -ы, (*f*) (zool) dipleurula

дипл·о́ид·н·ый, -ая, -ое, (*a*) (cyt) diploid

дипло·ко́кк·, -а, (*m*) (bact) *Diplococcus*

дипло́м·, -а, (*m*) diploma, certificate, degree (university)

диплом·а́нт·, -а, student preparing for final exams, final-year student

диплома́т·, -а, (*m*) diplomat

дипломат·и́ческ·ий, -ая, -ое, (*a*) diplomatic

~ по́чт·а (*f*) diplomatic bag

диплома́т·и·я, -и, (*f*) diplomacy

диплом·и́рованн·ый, -ая, -ое, (*past part pass*) of **диплом·и́р·овать** graduate, possessing a diploma/certificate

диплом·и́р·овать, -уют, (*imp and perf*) present a diploma/degree/ certificate

дипло́м·н·ый, -ая, -ое, (*a*) of **дипло́м·**

~ рабо́т·а (*f*) work presented for a diploma etc., thesis

дипл·опи́·я, -и, (*f*) (med) diplopia

дипло́т·, -а, (*m*) (naut) deep-sea lead

диплотли́н·ь, -я, (*m*) long-hand leadline (50–155 m)

ди·по́л·ь, -и, (*f*) *and see* **анте́нна** dipole; (rad) dipole antenna/radiator

~, магни́т·н·ый   magnetic dipole

мом·е́нт· ди·по́л·я (*m*) (phys) dipole moment

~, при·ём·н·ый   pick-up dipole

дипо́ль·н·ый, -ая, -ое, (*a*) dipole

~ моме́нт· (*m*) (phys) dipole moment

ди·поля́р·н·ый, -ая, -ое, (*a*) dipolar, bipolar

дипрокси́д·, -а, (*m*) (chem) diproxide, diisopropylxanthogen disulphide

ди·прот·одо́нт·н·ый, -ая, -ое, (*a*) (zool) diprotodont

ди·пропи́л·, -а, (*m*) (chem) dipropyl

ди·прото́н·, -а, (*m*) (phys) diproton

ди·та́нк·, -а, (*m*) (shipb) deep tank

Дира́к·а, у·равн·е́ни·е (*n*) Dirac equation (quantum mechanics)

дирак·о́вск·ий, -ая, -ое, (*a*) Dirac

дирам·а́ци·я, -и, (*f*) (math) diramation

директи́в·а, -ы, (*f*) directive, instructions

директи́в·н·ый, -ая, -ое, (*a*) directive

дире́ктор·, -а, (*nom pl*) -а́, (*g pl*) -о́в, (*m*) director, manager; head-master/mistress; director (teleph); (rad) director; (autom) transmitter
~ за·во́д·а works manager, factory manager

директора́т·, -а, (*m*) directorate, board

директориа́ль·н·ый, -ая, -ое, (*a*) (math) directorial

дире́ктор·н·ый, -ая, -ое, (*a*) (rad) director
~ анте́нн·а (*f*) (rad) Yagi aerial/antenna

дире́ктор·ск·ий, -ая, -ое, (*a*) manager's, directorial

директр·и́с·а, -ы, (*f*) directress, head-mistress; (math) directrix
~ стрел·ьб·ы́ (gunn) directrix, mean line of fire

дирекцио́н·н·ый, -ая, -ое, (*a*) of дире́кци·я; directional
~ у́гол· (*m*) (surv) bearing/position/grid angle

дире́кци·я, -и, (*f*) directorate, board, management

дирижа́бл·ь, -я, (*m*) airship
~ жёстк·ой систе́м·ы rigid airship, dirigible
~ полу·жёстк·ой систе́м·ы semi-rigid airship
~ мя́гк·ой систе́м·ы non-rigid air-ship, blimp

дириж·ёр·, -а, (*m*) conductor (of music); (cinema) musical director

дириж·и́р·овать, -уют, (*imp*) conduct (music)

Дирихле́, интегра́л· (*m*) (math) Dirichlet integral

дирода́н·, -а, (*m*) (chem) dithiocyanogen

дис·азо- (*component*) (chem) disazo-

ди·са́хар·и́д·, -а, (*m*) (chem) disacchar-ide

дис·бала́нс·, -а, (*m*) disbalance, un-balance

дис·гармон·и́р·овать, -уют, (*imp*) be out of tune, be inharmonious, conflict/clash with

дис·ген·и́ческ·ий, -ая, -ое, (*a*) (zool) dysgenic

дис·грег·а́ци·я, -и, (*f*) disintegration

ди·сила́н·, -а, (*m*) (chem) disilane

ди·силици́д·, -а, (*m*) (chem) disilicide
~ молибде́н·а molybdenum disilicide

ди́ск·, -а, (*m*) disc, disk, plate; (air) bat (landing); discus (sport)
~, аб·актин·а́льн·ый (zool) abactinal shield

ди́ск
~, имагина́ль·н·ый (zool) imaginal bud/disc
~ колес·а́ wheel disc/web
~, контро́ль·н·ый (nucl) control disc
~ ма́ятник·а (horol) pendulum bob
~, раз·гру́з·очн·ый hydraulic-balanc-ing disc
~, ступ·е́нчат·ый (horol) snail
~ -у·каз·а́тел·ь, -я, (*m*) indicator disc

ди́скант·, -а, (*m*) treble (music)

дискант·о́в·ый, -ая, -ое, (*a*) treble (music)

дис·квали·фик·а́ци·я, -и, (*f*) disquali-fication

дис·квали·фиц·и́р·овать, -уют, (*imp and perf*) disqualify; strike off the rolls

дис·кине́з·и·я, -и, (*f*) (med) dyskinesia

диск·ова́ни·е, -я, (*n*) (agr) discing; dialling (telephones)

диско·ви́д·н·ый, -ая, -ое, (*a*) disc-shaped, (zool) discal

ди́ск·ов·ый, -ая, -ое, (*a*) disc, disk
~ но́ж·ниц·ы (*pl*) rotary shear(s)

диск·оид·а́льн·ый, -ая, -ое, (*a*) discoid; patelloid

диско·меду́з·ы (*nom pl*), (*f*) (zool) *Discomedusae*

диско·мице́т·ы (*nom pl*), (*g pl*) -ов, (*m*) (bot) *Discomycetes*

диско́нт·, -а, (*m*) (fin) discount

дисконт·и́р·овать, -уют, (*imp and perf*) (fin) discount

диско·обра́з·н·ый, -ая, -ое, (*a*) disc-shaped, plate-like, (zool) discal

диско·пла́зм·а, -ы, (*f*) (biol) discoplasm

диско·по·до́б·н·ый, -ая, -ое, (*a*) discoid

дис·кредит·и́р·овать, -уют, (*imp and perf*) discredit

дискрепа́нт·ност·ь, -и, (*f*) discrepancy

дискре́т·н·о-не·пре·ры́в·н·ый пре-·образ·ова́тел·ь (*m*) digital-to-analogue computer

дискре́т·ност·ь, -и, (*f*) discreteness, discontinuity, discontinuousness, in-dividual nature, individuality, dis-tinctiveness

дискре́т·н·ый, -ая, -ое, (*a*) discrete, separate, individual; digital (compu-ters)
~ де́й·стви·е (*n*) (*in gen*) digital (*as adj*)
~ с·чёт· (*m*) digital computation
~ у·правл·е́ни·е (*n*) digital process control

дискримин·а́нт·, -а,   (*m*) (math) discriminant

дискримин·а́тор·, -а,   (*m*) (elec) discriminator

~, амплиту́д·н·ый   pulse-height discriminator/selector

~ и́мпульс·ов   pulse discriminator

~, фа́з·ов·ый   phase discriminator

~, част·о́тн·ый   (rad) frequency discriminator

дискримин·и́р·овать, -уют,   (*imp and perf*) discriminate

дискуссио́н·н·ый, -ая, -ое, (*a*) debate, debating; debatable, controversial, under discussion

дискусс·и́р·овать, -уют,   (*imp and perf*) debate, discuss

диску́сс·и·я, -и,   (*f*) debate, public discussion

дискут·и́р·овать, -уют,   (*imp and perf*) debate, discuss

дис·лок·а́ци·я, -и,   (*f*) dislocation; (geol) displacement, dislocation; (mil) stationing, deployment, strategic disposition (of troops), basing (of ships)

~, винт·ов·а́я   (cryst) screw dislocation

~, двой·ни́ков·ая   (cryst) twinning dislocation

~, кра·ев·а́я   (cryst) edge dislocation

~, лин·е́йн·ая   (cryst) edge dislocation

~, пликати́в·н·ая   (geol) folding

~, с·кла́д·чат·ая   (geol) plicative dislocation

дис·лоц·и́р·овать, -уют,   (*imp and perf*) dislocate

дис·ме́тр·и·я, -и,   (*f*) (med) dysmetria

диспансе́р·, -а,   (*m*) dispensary, public health clinic

диспа́ш·а, -и,   (*f*) (mar ins) average statement

диспаш·е́р·, -а,   (*m*) (mar ins) average adjuster

дис·пепс·и́·я, -и,   (*f*) (med) dyspepsia

дисперг·а́тор·, -а,   (*m*) (chem) disperser, dispersing agent

дисперг·е́нт·, -а,   (*m*) dispersion medium

дисперг·из·а́ци·я, -и,   (*f*) dispersion

дисперг·и́рования·е, -я,   (*n*) (chem) dispersion

дисперг·и́рованн·ый, -ая, -ое, (*past part pass*) dispersed

дисперг·и́рующ·ий, -ая, -ее, (*pres part act*) dispersive, dispersing

дисперсионно·тверд·е́ющ·ий с·пла́в· (*m*) (met) dispersion-hardening alloy

дисперсио́н·н·о-у·проч·ня́ем·ая стал·ь   (*f*) dispersion-hardening steel, artificial-ageing steel

дисперсио́н·н·ый, -ая, -ое,   (*a*) dispersion, dispersing

~ вз·ве́с·ь   (*f*) dispersion slurry

~ при́зм·а   (*f*) (opt) dispersing prism

~ спосо́б·ност·ь   (*f*) (opt) dispersive power

~ тверд·е́ни·е   (*n*) (met) dispersion hardening, artificial ageing, precipitation hardening

диспе́рс·и·я, -и,   (*f*) dispersion; variance, variability, standard deviation (statistics)

~, анома́ль·н·ая   anomalous dispersion (of light)

~, вращ·а́тельн·ая   rotatory dispersion (of light)

~, генет·и́ческ·ая   genetic variance

диспе́рс·ност·ь, -и,   (*f*) dispersion, dispersity, degree of fineness, particle size

диспе́рс·н·ый, -ая, -ое,   (*a*) disperse, dispersed, fine

~ фа́з·а   (*f*) (chem) dispersed phase

диспе́тчер·, -а,   (*m*) progress chaser, dispatcher (U.S.); controller (of transport movement), dispatcher; load dispatcher (elec supply)

~, по·езд·н·о́й   (rail) traffic controller

диспетчер·иза́ци·я, -и,   (*f*) progress system (Brit), dispatching (U.S.); supervizory control (elec supply); centralized traffic control

диспе́тчер·ск·ая, -ой,   (*f decl as adj*) control room

диспе́тчер·ск·ий, -ая, -ое,   (*a*) of **диспе́тчер·;** *see* **диспе́тчер·ск·ая** (*as noun*)

~ бюро́   (*n*) progress department (Brit), dispatching office (U.S.)

~ гра́ф·ик·   (*m*) (rail) traffic control graph; progress chart

~ пункт·   (*m*) control room/tower

~ пункт·, по·дви́ж·н·ый,   (*m*) (air) mobile traffic control point

~ радио·лок·а́тор·   (*m*) airfield control radar

~ с·вя́з·ь   (*f*) works communications

~ сигнал·иза́ци·я   (*f*) special-subscriber telephone signalling

~ централ·иза́ци·я   (*f*) (rail) centralized train control

дис·пози́ц·и·я, -и, (*f*) (mil) disposition, plan of disposition, station

дис·прози·й, -я,   (*m*) (chem) dysprosium, Dy

дис·пропорц·и·я, -и,   (*f*) disproportion, unevenness

дис·пропорц·ионӥровани·е, -я,   (*n*) (chem) disproportionation

дӥспут·, -а,   (*m*) public debate, disputation

дис·сект·ор·, -а,   (*m*) (TV) image dissector, dissector tube

дис·сепи·мент·, -а,   (*m*) (zool) dissepiment

диссерт·аци·я, -и,   (*f*) thesis (for degree etc.); dissertation

дис·симил·яци·я, -и,   (*f*) dissimilation

диссипатӥв·н·ый, -ая, -ое,   (*a*) dissipative

диссипа́ци·я, -и,   (*f*) dissipation

диссона́нс·, -а,   (*m*) discord, dissonance

дис·соци·а́тор·, -а,   (*m*) (chem) cracking plant

~, аммиа́ч·н·ый   ammonia cracking plant

диссоци·а́ци·я, -и,   (*f*) dissociation

~, электро·лит·и́ческ·ая   electrolytic dissociation

дис·соци·и́р·овать, -уют,   (*imp and perf*) dissociate, (chem) break down

дист·а́льн·ый, -ая, -ое,   (*a*) (biol) distil

диста́нт·н·ый, -ая, -ое,   (*a*) distant

~ реце́птор·   (*m*) (zool) distant receptor

дистанцион·и́ровани·е, -я,   (*n*) spacing; (acous) pitching

дистанцио́н·но   (*adv*) by remote control

дистанцио́н·н·ый, -ая, -ое,   (*a*) of **дист́анци·я**; distant, distance, remote; remote-operated/controlled; distant-reading, transmitting (of instruments); (expl) time

~ автома́т·ик·а   (*f*) remote-control feedback automatics

~ вз·рыв·а́тел·ь   (*m*) (mil) time fuse

~ контро́л·ь   (*m*) remote checking

~ термо́·метр·   (*m*) transmitting/distant-reading thermometer

~ у·правл·е́ни·е   (*n*) remote control

диста́нци·я, -и,   (*f*) distance; interval, space; (mil) range; (rail) track length, division; sector (air, shipping routes etc.)

тысяч·н·ая диста́нци·и   (*f*) (gunn) mil

шкал·а́ диста́нци·и   (*f*) (mil) range scale

ди·стеари́н·, -а,   (*m*) (chem) distearin

дис·тект·и́ческ·ий, -ая, -ое,   (*a*) (chem) dystectic

дисте́н·, -а,   (*m*) (min) disthene, kyanite

дистилл·а́т·, -а,   (*m*) (*see also* **дестилл·а́т·**) distillate

дистилл·ёр·, -а,   (*m*) ammonia still

дистилл·и́ровани·е, -я,   (*n*) *and see* **дестилл·я́ци·я** *and* **пере·го́н·к·а** distillation

дистилл·и́р·овать, -уют,   (*imp and perf*) distil

дистилл·я́т·, -а,   (*m*) = **дистилл·а́т·**

дистилл·я́тор·, -а́,   (*m*) (chem) still

дистилл·я́ци·я   (*see also* **дестилл·я́ци·я** *and* **пере·го́н·к·а**) distillation

ди·стильбе́н·, -а,   (*m*) (chem) distilbene

дисто́рс·и·я, -и,   (*f*) (opt) distortion, lens distortion

дистрибути́в·ност·ь, -и,   (*f*) (math) distributivity

дистрибути́в·н·ый, -ая, -ое,   (*a*) (math) distributive

дистрибю́тор·, -а,   (*m*) distributor (cotton breaking)

дис·троф·и́·я, -и,   (*f*) (biol) dystrophy

дис·тро́ф·н·ый, -ая, -ое,   (*a*) dystrophic

ди·сульф·а́н·, -а,   (*m*) (pharm) disulphan, disulfan, 4-sulphanil-sulphanilamide

ди·сульф·а́т·, -а,   (*m*) disulphate, disulfate

ди·сульф·и́д·, -а,   (*m*) (chem) disulphide, disulfide

ди·сульфо·кисл·от·а́   (*f*) (chem) disulphonic acid, disulfonic acid

дисципли́н·а, -ы,   (*f*) discipline; discipline, branch (of learning)

дисциплин·а́рн·ый, -ая, -ое,   (*a*) disciplinary

~ у·ста́в· войн·ск·ий   articles of war

ди·терпе́н·, -а,   (*m*) (chem) diterpene

ди·тетра·го́н·, -а,   (*m*) (cryst) ditetragon

ди·тетра·гон·а́льн·ый, -ая, -ое,   (*a*) (cryst) ditetragonal

ди·тетра́·эдр·, -а,   (*m*) (cryst) ditetrahedron

дити·азани́н·, -а,   (*m*) (chem) dithiazanine

дити·ени́л·, -а,   (*m*) (chem) dithienyl

ди·ти·зо́н·, -а,   (*m*) (chem) dithizone, diphenyl-thiocarbazone

дитио·карбам·а́т·, -а,   (*m*) (chem) dithiocarbamate

дитио·кисл·от·а́, -ы,   (*f*) (chem) dithionic/thionthiolic acid ди·тион·а́т·, -а, (*m*) (chem) dithionate, hyposulphate

ди·тион·и́т·, -а, (*m*) dithionite

ди·тио́н·ов·ый, -ая, -ое, (*a*) dithionic; hyposulphate of, dithionate of

~ кисл·от·а́ (*f*) dithionic acid, hyposulphuric acid

дитио·у́голь·н·ая кисл·от·а́ (*f*) dithiocarbonic acid

дитио·фо́с·, -а, (*m*) dithiophos (insecticide)

дитио·фосфа́т·, -а, (*m*) dithiophosphate

дитрази́н·, -а, (*m*) (pharm) ditrazine, diethylcarbamazine citrate

ди·три·гон·а́льн·ый, -ая, -ое, (*a*) (cryst) ditrigonal

дитро́йт·, -а, (*m*) (geol) ditroite

дитце́йт·, -а, (*m*) (min) dietzeite

ди́тчер·, -а, (*m*) (civ eng) ditcher

дитя́ (*n*) see де́т·и child

ди·ур·е́з·, -а, (*m*) (med) diuresis

ди·урет·и́н·, -а, (*m*) (pharm) diuretin

ди·урет·и́ческ·ий, -ая, -ое, (*a*) (pharm) diuretic

ди·фен·а́т·, -а, (*m*) (chem) diphenate

ди·фен·и́л·, -а, (*m*) (chem) diphenyl, phenylbenzene

дифенил·ами́н·, -а, (*m*) diphenylamine

дифенил·бензо́л·, -а, (*m*) diphenyl benzene

дифенил·гуаниди́н·, -а, (*m*) diphenylguanidine

дифенил·е́н·, -а, (*m*) diphenylene

дифенилен·ов·ая пере·групп·иро́в·к·а (*f*) (chem) benzidine rearrangement

дифенил·и́н·, -а, (*m*) diphenyline

дифенил·кето́н·, -а, (*m*) diphenyl ketone, benzophenone

дифенил·мета́н·, -а, (*m*) diphenylmethane

дифени́л·ов·ый, -ая, -ое, (*a*) diphenyl

~ эфи́р· (*m*) diphenyl ether; diphenyl ester, diphenylate

дифенил·окси́д·, -а, (*m*) diphenyl oxide

дифенил·тио·моч·еви́н·а, -ы, (*f*) diphenylthiourea

дифенил·хлор·арси́н·, -а, (*m*) diphenylchlorarsine

дифенил·циан·арси́н·, -а, (*m*) diphenylcyanoarsine

ди·фен·и́н·, -а, (*m*) (pharm) diphenine

ди·фе́н·ов·ый, -ая, -ое, (*a*) (chem) diphenic; diphenate of

дифере́нт·, -а, (*m*) see дифферент·

диференциа́л· (*m*) see дифференциа́л·

ди·фи́ль·ност·ь, -и, (*f*) bifilarity, bifilar design/construction

дифло́н·, -а, (*m*) (plast) Diflon

диф·мано́метр·, -а, (*m*) manometer, differential manometer

~, колоко́ль·н·ый inverted-bell manometer

~, по·пла́в·ков·ый float-chamber manometer

~, сильфо́н·н·ый bellows-type manometer

~ -рас·ходо·ме́р·, -а, (*m*) differential manometer flow meter

ди·фосге́н·, -а, (*m*) (chem) diphosgene

дифраг·и́р·овать, -уют, (*imp and perf*) diffract

дифракто́·метр·, -а, (*m*) (phys, instr) diffractometer

дифра́кц·и·я, -и, (*f*) diffraction

дифракц·ио́нн·ый, -ая, -ое, (*a*) diffraction

~ карти́н·а (*f*) (spectr) diffraction pattern

~ полос·а́ (*f*) diffraction fringe

~ решёт·к·а (*f*) (opt) diffraction grating

~ решёт·к·а, регуля́р·н·ая (*f*) blazed grating

дифтер·и́йн·ый, -ая, -ое, (*a*) (med) diphtherial, diptheroid, diphtheroid

дифтер·и́т·, -а, (*m*) (med) diphtheria

дифтери́т·н·ый, -ая, -ое, (*a*) diphtheria

дифтер·и́·я, -и, (*f*) diphtheria

дифто́нг·, -а, (*m*) (ling) diphthong

ди·фторо·фо́сфор·н·ая кисл·от·а́ (*f*) (chem) difluorophosphoric acid

диффама́ци·я, -и, (*f*) (law) libel, defamation

диффере́нт·, -а, (*m*) trim (of ship etc.)

диффере́нт·н·ый, -ая, -ое, (*a*) (naut) trimming, trim

дифферент·о́вочн·ый, -ая, -ое, (*a*) (naut) trimming

дифференто́·метр·, -а, (*m*) trim indicator

дифференциа́л·, -а, (*m*) (math, eng) differential; differential gear

~, кон·и́ческ·ий bevel-gear differential

дифференциа́ль·н·ый, -ая, -ое, (*a*) differential

~ **ис·числ·е́ни·е** (*n*) (math) differential calculus

~ **механ·и́зм·** (*m*) (eng) differential motion

~ **тари́ф·** (*m*) differential rate (of charges)

~ **у·равн·е́ни·е с ча́ст·н·ыми про·-из·во́д·н·ыми** (*n*) (math) partial differential equation

дифференци·а́т·, -а, (*m*) differentiate

дифференци·а́тор·, -а, (*m*) (autom) differentiator, differentiating circuits

дифференци·а́ци·я, -и, (*f*) differentiation

~, **ликвацио́н·н·ая** (chem) liquation

~ **по пло́т·ност·и** (geol) gravitational differentiation

~, **фракцио́н·н·ая** fractional differential, fractionation

дифференц·и́ровани·е, -я, (*n*) (math) differentiation

дифференц·иро́вк·а, -и, (*g pl*) **-вок·,** (*f*) (biol) differentiation

дифференц·и́р·овать, -уют, (*imp and perf*) differentiate

дифференц·и́рующ·ий, -ая, -ее, (*pres part act*) of **дифференц·и́р·о·вать**; differentiating

~ **цеп·ь** (*f*) = **дифференци·а́тор·**

диффракто́·метр·, -а, (*m*) = **ди-фракто́·метр·**

диффракц·ио́нн·ый, -ая, -ое, (*a*) see **дифракц·ио́нн·ый**

диффра́кц·и·я, -и, (*f*) see **дифра́к-ц·и·я**

диффуз·а́т·, -а, (*m*) diffusate

диффузи́в·ност·ь, -и, (*f*) diffusivity

диффузи́в·н·ый, -ая, -ое, (*a*) diffusive

диффузио́н·н·ый, -ая, -ое, (*a*) diffusion

~ **аппара́т·** (*m*) (chem) diffusion vessel

~ **ка́мер·а** (*f*) (nucl) cloud chamber

~ **металл·иза́ци·я** (*f*) (met) diffusion coating

~ **по·дви́ж·ност·ь** (*f*) diffusion movement

~ **по·кры́т·и·е** (*n*) (met) diffusion coating

~ **рас·сто·я́ни·е** (*n*) diffusion length (semiconductors)

~ **сло́·й** (*m*) (electrochem) diffusion layer

~ **спосо́б·ност·ь** (*f*) diffusivity

диффуз·и́р·овать, -уют, (*imp and perf*) diffuse

диффу́з·и·я, -и, (*f*) (phys, chem) diffusion

~ **в твёрд·ом фа́з·е** (chem, phys) solid diffusion

~, **взаи́м·н·ая** interdiffusion

~ **внутр·ь** inwards diffusion

~, **внутри·кристалл·и́тн·ая** diffusion in crystals

~, **вос·ход·я́щ·ая** uphill diffusion **за·ко́н· диффу́з·и·й** (*m*) (phys) law of diffusion, Fick's law

~ **к конц·а́м** end diffusion **коэффицие́нт· диффу́з·и·и** (*m*) diffusion coefficient

~, **обра́т·н·ая** back diffusion

~, **объ·ём·н·ая** bulk diffusion

~ **под давл·е́ни·ем** pressure diffusion

~ **по трубк·е** (met) pipe diffusion

~ **че́рез пере·горо́д·к·у** barrier diffusion

диффузно·рас·се́·янн·ый, -ая, -ое, (*a*) diffusely scattered

диффу́з·ност·ь, -и, (*f*) diffusivity, diffuseness

диффу́з·н·ый, -ая, -ое, (*a*) diffuse, scattered

~ **рас·се́·яни·е** (*n*) diffuse scattering

~ **тума́н·ност·ь** (*f*) (astron) diffuse/ irregular nebula

~ **фон·** (*m*) diffusion background

~ **ядр·о́** (*n*) (cyt) diffuse nucleus

~ **се́ри·я** (*f*) (opt) diffuse series

диффу́з·ор·, -а, (*m*) diffusor, diffuser; venturi tube, expanding duct, choke tube; (chem) diffusion vessel; (rad) cone

~, **в·ход·н·о́й** inlet diffuser

~, **двух·скач·ко́в·ый** (dynam) double-shock diffuser

~, **лопа́т·очн·ый** (dynam) vaned diffuser

~, **нару́ж·н·ый** (dynam) free-air diffuser

~, **рас·шир·я́ющ·ийся** (dynam) divergent diffuser

диффунд·и́р·овать, -уют, (*imp and perf*) diffuse (of gases and liquids)

диффунд·и́руем·ый, -ая, -ое, (*pres part pass*) of **диффунд·и́р·овать**; diffusible

ди·хлор- (*component*) (chem) dichlor-, dichloro-, dichloride

дихлора́ль·моч·еви́н·а, -ы, (*f*) (chem) dichloral urea

дихлор·ами́н·, -а, (*m*) dichloramine
дихлор·ацето́н·, -а, (*m*) acetone dichloride
дихлор·бензо́л·, -а, (*m*) dichlorobenzene
дихлор·дифенил·трихлорэта́н·, -а, (*m*) (chem) dichlorodiphenyltrichloroethane, DDT
дихлор·эта́н·, -а, (*m*) (chem) dichloroethane
дихо·га́м·и·я, -и, (*f*) (bot) dichogamy
дихо·га́м·н·ый, -ая, -ое, (*a*) (bot) dichogamous
дихо·том·и́ческ·ий, -ая, -ое, (*a*) (bot) dichotomous
дихо·том·и́·я, -и, (*f*) (bot) dichotomy
ди·хро·и́зм·, -а, (*m*) (opt) dichroism
ди·хро·и́чн·ый, -ая, -ое, (*a*) dichroic
~ вуа́л·ь (*f*) (phot) dichroic fog
ди·хрома́т·, -а, (*m*) (chem) dichromate
ди·хромат·и́зм·, -а, (*m*) (opt) dichromatism
ди·хромат·и́ческ·ий, -ая, -ое, (*a*) dichromatic
ди·центр·и́н·, -а, (*m*) (chem) dicentrine
ди·циа́н·, -а, (*m*) (chem) cyanogen
ди·циан·диами́д·, -а, (*m*) (chem) dicyandiamide
ди·цикл·и́ческ·ий, -ая, -ое, (*a*) dicyclic
дич·ь, -и, (*f*) game (animals and birds); wilderness; (pop) nonsense, rubbish
ди·цин·одо́нт·, -а, (*m*) (pal) dicynodont
ди·эдр·, -а, (*m*) (cryst) dihedron
ди·эдр·и́ческ·ий, -ая, -ое, (*a*) dihedral; sphenoid
~ без·о́с·н·ый (cryst) monoclinic antihemihedral, monoclinic dodomatic
~ ос·ев·о́й (cryst) monoclinic sphenoidal, hemimorphic
ди·эле́ктр·ик·, -а, (*m*) (elec) dielectric
ди·электр·и́ческ·ий, -ая, -ое, (*a*) dielectric
коэффицие́нт· ди·электр·и́ческ·их по·те́р·ь (*m*) dielectric loss factor
~ по·сто·я́нн·ая (*f*) (elec) dielectric constant, relative permittivity, specific inductive capacity
~ по·те́р·я (*f*) dielectric loss
~ про·ниц·а́емост·ь (*f*) permittivity; absolute dielectric constant

ди·электр·и́ческ·ий, -ая, -ое
~ про·ниц·а́емост·ь, компле́кс·н·ая (*f*) complex permittivity, relative complex dielectric constant, complex capacitivity
~ про·ниц·а́емост·ь, от·нос·и́тель·н·ая (*f*) relative permittivity, specific inductive capacitance, relative capacitivity, dielectric constant
диэ́та, -ы, (*f*) *see* дие́т·а
ди·этанолами́н·, -а, (*m*) (chem) diethanolamine
ди·эти́л·, -а, (*m*) (chem) diethyl
диэтил·ами́н·, -а, (*m*) diethylamine
ди·этиле́н·, -а, (*m*) diethylene
диэтил·карбона́т·, -а, (*m*) diethyl carbonate
ди·эти́л·ов·ый, -ая, -ое, (*a*) diethyl
~ эфи́р· (*m*) diethyl ether, ether
~ эфир· янта́р·н·ой кисл·от·ы́ diethyl succinate
диэтил·стильбэстро́л·, -а, (*m*) (pharm) diethylstilbesterol
диэтил·сульфа́т·, -а, (*m*) diethyl sulphate
дк (*abbr*) of дека- (*component*) deca-
дкг (*abbr*) of дека·гра́мм· decagram
дкл (*abbr*) of дека·ли́тр· decalitre, decaliter
дл (*abbr*) of деци·ли́тр· decilitre, deciliter
дл (*abbr*) of длин·а́ length
длане·ви́д·н·ый, -ая, -ое, (*a*) (bot) palmate
длане·ли́ст·н·ый, -ая, -ое, (*a*) (bot) cheirophyllous
длане·раз·де́ль·н·ый, -ая, -ое, (*a*) (bot) palmatipartite
дл·и́вш·ий, -ая, -ее, (*past part act*) of дл·ить
длин·а́, -ы́, (*f*) length
в длин·у́ lengthwise
~ волн·ы́ (phys) wavelength
~, габари́т·н·ая (shipb) length overall
~ диффу́з·и·и (phys) diffusion length
~ кон·е́чн·ая finite length
~, наи·бо́льш·ая length overall, overall length
~, обра́т·н·ая inverse/reciprocal length
~ от·ре́з·к·а прям·о́й (math) length of a straight line segment
по длин·е́ lengthwise
~, поле́з·н·ая nett length; effective length; length over flanges (pipes)
~, по́лн·ая overall length
~, рас·па́д·н·ая (nucl) cascade/radiation length

длин·а́

~, рас·чёт·н·ая   gage/gauge length (of metal test piece)

~, регистр·ов·ая   registered length (of ship)

~, свобо́д·н·ая   (mech) overhanging/unsupported length

~ х·о́д·а   travel, stroke (of piston)

длин·е́·ть, -ют,   (*imp*) lengthen, increase in length, become longer

дли́н·н·о   (*adv*) long, at length

длин·нова́т·ый, -ая, -ое,   (*a*) rather long; lengthy

длинно·во́лн·ов·ый, -ая, -ое,   (*a*) long-wave

длинно·волокн·и́ст·ый, -ая, -ое,   (*a*) (text) long-staple, long-fibred

длинно·голо́в·ый, -ая, -ое,   (*a*) (zool) dolichocephalic

длинно·ме́р·н·ый, -ая, -ое,   (*a*) lengthy

длинно·пла́мен·н·ый у́гол·ь   (*m*) long-flame coal

длинно·пло́д·н·ый, -ая, -ое,   (*a*) (bot) dolichocarpous

длинно·про·бе́ж·н·ый, -ая, -ое,   (*a*) long-range (of atomic particles)

дли́н·н·ый, -ая, -ое,   (*a*) long, lengthy

~ ме́р·к·а   (*f*) (text) long stick

длино·ме́р·, -а,   (*m*) measuring machine, length measuring machine

дл·и́тельност·ь, -и,   (*f*) duration, time, period, length (of time)

~ в·ключ·е́ни·я   (elec) on-time

~ гор·я́ч·их про·сто́·ев   downtime for hot repairs (furnaces etc.)

~ до раз·руш·е́ни·е   (met) time-to-rupture

~ за́д·н·его фро́нт·а и́мпульс·а   (phys) pulse-delay time

~ и́мпульс·а   pulse width

~ пере·хо́д·н·ого режи́м·а   transient time

дл·и́тельн·ый, -ая, -ое,   (*a*) long, lengthy, protracted, sustained, long-time/term, prolonged, slow

~ про́ч·ност·ь   (*f*) (met) long-time strength, creep limit

ис·пыт·а́ни·е на дл·и́тельн·ую про́ч·ност·ь   (*n*) stress-rupture test

крив·а́я дл·и́тельн·ой про́ч·ност·и   (*f*) stress-rupture curve

пре·де́л· дл·и́тельн·ой про́ч·ност·и   (*m*) rupture stress (in creep)

~ раз·руш·е́ни·е   (*n*) (met) delayed fracture

дл·и́ть, -я́т,   (*imp*) prolong, protract, delay; **-ся** last/continue a long time

для   (*prep + genitive*) for, to

~ того́ чтобы   in order to, so as to, to

дл·я́щ·ийся, -аяся, -ееся,   (*pres part act*) of **дл·и́ться**; lasting

дм   (*abbr*) of **деци·ме́тр·** decimetre; (*abbr*) of **дюйм·** inch

дн   (*abbr*) of **ди́н·а** dyne

дне   (*prepos*) of **ден·ь, дн·о**

дн·ева́льн·ый, -ого   (*m decl as adj*) (mil) orderly; man on duty, dutyman, -keeper

дн·ёвк·а, -и,   (*g pl*) **-вок·;** day's rest

дн·евни́к·, -а́,   (*m*) diary, journal

дн·евн·о́й, -а́я, -о́е,   (*a*) day, daytime, daily

    коэффицие́нт· дн·евн·о́го све́т·а   (*m*) daylight factor

~ по·ве́рх·ност·ь   (*f*) (build) day

~ свет·   (*m*) daylight

~ спекта́кл·ь   (*m*) (theat) matinée

-дн·е́вн·ый, -ая, -ое,   (*adj component*) -day

дн·ей   (*gen pl*) of **ден·ь**

дн·ём   (*adv*) by day, in the daytime; (*instr*) of **ден·ь**

дн·и   (*nom pl*) of **ден·ь**

дн·и́щ·е, -а,   (*i*) **-ем,** (*n*) (*see also* **дн·о**) bottom, base; end-plate (of boilers); crown, head (of piston); face (of CRT); tabula (of corals)

дн·ищев·о́й, -а́я, -о́е,   (*a*) of **дн·и́щ·е**

~, втор·о́е   double bottom; (shipb) inner-bottom plating

~ дол·и́н·ы   (geol) valley floor

~, и́л·ист·ое   (geol) siltground, clay ground; (ocean) muddy/silty bottom

~, нару́ж·н·ое   (shipb) outer-bottom plating

~ океа́н·а   ocean bed

~ по́ршн·я   piston crown/head

~, ступ·е́нчат·ое   (ocean) shelving bottom

дн·о, -а,   (*nom pl*) **до́н·ья,** (*g pl*) **до́ньев,** (*n*) (*see also* **дн·и́щ·е**); bottom, base, bed, floor; face (CRT); (med) fundus

дно·у·глуб·и́тел·ь, -я   (*m*) dredger, dredge

дно·у·глуб·и́тель·н·ый, -ая, -ое,   (*a*) dredging

~ рабо́т·ы   (*pl*) dredging

~ с·на·ря́д·   (*m*) dredger

дно·черп·а́тел·ь, -я,   (*m*) (ocean) bottom sampler

дн·ю   (*dat*) of **ден·ь**

дн·я́   (*gen*) of ден·ь

до   (*prep+gen*) (*relating to movement, location etc.*) to, up to, down to, as far as, till, until; (*relating to numbers, quantities etc.*) to, up to, not more than, not over, not exceeding; (*relating to time*) to, up to, before, until

∼ сих пор·   hitherto, until now

∼ тех пор· как   until

до   (*n indecl*) C (music)

∼ -диез   C sharp

до-   (*as prefix*) (*maintains the same senses as the preposition. When it is implied that an action is carried out to the end, it is often rendered "to finish" cooking, reading etc.*) pre-

до·ба́в·ить, -ят, (*perf*) *see* до·бавл·я́·ть

до·ба́в·к·а, -и,   (*g pl*) -вок·, (*f*) (*v n*) *see* до·бавл·я́·ть; addition

до·бавл·е́ни·е, -я,   (*n*) (*v n*) *see* до·бавл·я́·ть; addition, supplementation; addendum, appendix, supplement

до·ба́вл·енн·ый, -ая, -ое,   (*past part pass*) of до·бавл·я́·ть

до·бавл·я́·ть, -ют,   (*imp*) add, add to, admix, supplement, boost

до·ба́в·ок·, -в·к·а,   (*m*) addition, additive, addition agent

до·ба́в·очн·ый, -ая, -ое,   (*a*) additional, supplementary, extra, extension (telephones), (elec) booster

∼ аге́нт·   (*m*) addition agent

∼ батар·е́·я   (*f*) (elec) booster battery

∼ материа́л·ы   (*pl*) (met) addition agents

∼ но́мер·   (*m*) extension number (telephone)

∼ по́люс·   (*m*) (elec) interpole

∼ сопл·ов·о́й кла́пан·   (*m*) load nozzle valve (turbine)

∼ то́п·ов·ый ого́н·ь   (*m*) (naut) after steaming light

до·бег·а́ни·е, -я,   (*n*) (*v n*) *see* до·бег·а́·ть; (ICE) lag

до·бег·а́·ть, -ют,   (*imp*) run to/as far as/till

до·беж·а́ть   (*fut 3rd sing, pl*) до·бе·ж·и́т, до·бег·у́т (*perf*); *see* до·бег·а́·ть

до·бел·а́   (*adv*) till white; till white hot, to white heat; till spotlessly clean

∼, рас·кал·ённ·ый   (*a*) white hot

до·бив·а́·ть, -ют, (*imp*) kill off, finish off, deal the final blow; -ся (+ *gen*) achieve, obtain, secure; (*imp only*) try to achieve/obtain/secure, strive to/for, endeavour to

до·бир·а́·ть, -ют,   (*imp*) finish collecting/taking/gathering; collect/take (a certain amount); (print) finish setting up/typesetting; -ся reach, get to/as far as

до·би́·ть   (*fut 3rd sing, pl*) до·бь·ёт, до·бь·ют (*perf*); *see* до·бив·а́·ть

до́·блест·н·ый, -ая, -ое,   (*a*) valiant

до·браж·е́ни·е, -я,   (*n*) secondary fermentation

до·бра́ж·ивани·е, -я,   (*n*) *see* до·браж·е́ни·е

до·браж·а́·ть, -ют,   (*perf*) finish fermenting, complete a fermentation

до́·бр·анн·ый, -ая, -ое,   (*past part pass*) of до·бир·а́·ть

до·бра́с·ыва·ть, -ют,   (*imp*) throw to/as far as

до·бр·а́ть   (*fut 3rd sing, pl*) до·бер·ёт, до·бер·у́т, (*perf*); *see* до·бир·а́·ть

добр·о́, -а́,   (*n*) good; property; D (phonetic alphabet)

добро·во́л·ец·, -ль·ц·а, (*m*) volunteer; (mil) regular, continuous service man

добро·во́ль·н·ый, -ая, -ое, (*a*) voluntary

добро·во́ль·ческ·ий, -ая, -ое,   (*a*) volunteer, voluntary

добро·ду́ш·н·ый, -ая, -ое, (*a*) good-natured

добро·жел·а́тельн·ый, -ая, -ое, (*a*) benevolent, well-wishing

добро·ка́честв·енност·ь, -и, (*f*) good quality, quality; (elec) figure/factor of merit

добро·ка́честв·енн·ый, -ая, -ое, (*a*) good-quality; (med) benign

до·бро́с·ить, -ят, (*perf*) *see* до·бра́с·ыва·ть

добро·со́вест·н·ый, -ая, -ое,   (*a*) honest, conscientious

добр·от·а́, -ы,   (*f*) kindness, goodness

добр·о́тност·ь, -и,   (*f*) quality, high-quality/durability; (phys) energy factor, Q factor, Q, factor of merit

добр·о́тн·ый, -ая, -ое,   (*a*) high-Q, good-quality

до·бро́ш·енн·ый, -ая, -ое,   (*past part pass*) *see* до·бра́с·ыва·ть

до́бр·ый, -ая, -ое,   (*a*) (*s f*) добр·, добр·а́, до́бр·о, до́бр·ы kind, good

до·бу́д·ут   (*fut 3rd pl*) of до·бы́·ть

до·б*ыв·а́ни·е, -я,   (*n*) (*v n*) *see* до·быв·а́·ть; recovery

до·быв·а́·ть, -ют,   (*imp*) get, obtain, procure; (min) win, mine, extract

до·быв·а́ющ·ий, -ая, -ее, (*pres part act*) of **до·быв·а́·ть;** extractive

~ **про·мы́шл·енность** (*f*) primary production industries; extractive industry

до·бы́·ть (*fut 3rd sing, pl*) **до·бу́д·ет, до·бу́д·ут,** (*perf*); *see* **до·быв·а́·ть**

до·бы́·ч·а, -и, (*i*) **-ей,** (*f*) (*v n*) *see* **до·быв·а́·ть;** recovery, output, yield, production; loot, plunder, bag (hunting), catch (fishing), (naut) prize

~, **втор·и́чн·ая** (oil) secondary recovery

~ **нефт·и** crude/oil recovery/production

~, **от·кры́·т·ая** (min) opencast/open-cut/pit/open-pit mining

~, **су́т·очн·ая** (min) daily output/yield

~ **у́гл·я** coal recovery; coal mining/winning

до·ва́р·енн·ый, -ая, -ое, (*past part pass*) of **до·ва́р·ива·ть;** boiled, cooked, digested

до·ва́р·ива·ть, -ют, (*imp*) finish (cooking/boiling), cook/boil (through/completely)

до·вар·и́ть, -я́т, (*perf*) *see* **до·ва́р·и·ва·ть**

Дове, при́зм·а (*f*) (opt) Dove prism

до·вед·е́ни·е, -я, (*n*) (*v n*) *see* **до·вод·и́ть;** (math) reduction

до·вед·ённ·ый, -ая, -ое, (*past part pass*) *see* **до·вод·и́ть** brought to, finished

до·вез·ти́, -у́т, (*perf*) *see* **до·вод·и́ть**

до·ве́р·енность, -и, (*f*) warrant, power of attorney

до·вер·и́тельн·ый, -ая, -ое, (*a*) confidential, fiducial; (math) confidence

~ **вероя́т·ность** (*f*) (math) confidence level

~ **гран·и́ц·а** (*f*) (math) confidence limit; (*pl*) confidence interval

подо·ве́р·енност·и per pro, for, on behalf of (when signing)

до·ве́р·енн·ый, -ая, -ое, (*past part pass*) of **до·ве́р·ить;** entrusted; (*m decl as adj*) agent, proxy

до·ве́р·и·е, -я, (*n*) faith, confidence

до·вер·и́тел·ь, -я, (*m*) principal (in business commitments etc.)

~ **рас·пре·дел·е́ни·е** (*n*) fiducial distribution

до·ве́р·ить, -ят, (*perf*) entrust

Дове́р·ов· порош·о́к· (*m*) (pharm) Dover's powder

до́·верх·у (*adv*) up to the top, to the brim

до·ве́р·чивост·ь, -и, (*f*) credulity, trust

до·ве́р·чив·ый, -ая, -ое, (*a*) credulous, gullible; inclined to trust, trusting

до·верш·а́·ть, -ют, (*imp*) complete, accomplish, finish, consummate

до·верш·е́ни·е, -я, (*n*) completion, accomplishment, consummation

до·верш·и́ть, -а́т, (*perf*) *see* **до·вер·ш·а́·ть**

до·вер·я́·ть, -ют, (*imp*) entrust; trust, believe in, rely on

до·ве́с·ить, -ят, (*perf*) *see* **до·ве́ш·и·ва·ть**

до·ве́с·ок·, -с·к·а, (*m*) makeweight

до·вес·ти́ (*fut 3rd sing, pl*) **до·вед·ёт, до·вед·у́т,** (*past masc sing*) **до·вёл,** (*perf*); *see* **до·вод·и́ть**

до·ве́ш·ива·ть, -ют, (*imp*) make up the weight of

до·винт·и́ть, -я́т, (*perf*) *see* **до·ви́нч·ива·ть**

до·ви́нч·ива·ть, -ют, (*imp*) screw up/home/in

до́·вод·, -а, (*m*) reason, argument (in support of proof)

до·вод·и́ть, -я́т, (*imp*) lead/bring to/up to/as far as; get ready (new machinery); finish (by machine processes); refine (OH furnace); (math) reduce to; **-ся** (*pass*) have occasion to

~ **до ми́нимум·а** reduce to the minimum, minimize

до·во́д·к·а, -и, (*g pl*) **-д·ок·,** (*f*) (*v n*) *see* **до·вод·и́ть;** polishing, fine finishing (machining); refining (OH furnace); final concentration (ore dressing)

до·вод·очн·ый, -ая, -ое, (*a*) of **до··во́д·к·а**

~ **па́ст·а** (*f*) polishing/buffing paste

~ **стан·о́к·** (*m*) lapping machine

до·воен·н·ый, -ая, -ое, (*a*) prewar

до·вож·у́ (*pres 1st sing*) of **до·вод·и́ть**

до·во́льн·о (*adv*) enough; rather, fairly, quite a

до·во́льн·ый, -ая, -ое, (*a*), (*s f*) **-л·ен·, -ль·н·а** satisfied, content

до·во́ль·стви·е, -я, (*n*) (mil) allowance(s), ration

до·воль·ств·о, -а, (*n*) satisfaction, contentment; prosperity

до·во́ль·ств·овать, -уют, (*imp*) satisfy, supply; **-ся** be satisfied/content

до·вулканиз·а́ци·я, -и, (*f*) after-vulcanization/cure

до·вы́·бор·ы (*nom pl*), (*g pl*) **-ов**, (*m*) by-election, supplementary election

до·гад·а́·ться, -ются (*perf*) guess

до·га́д·к·а, -и, (*g pl*) **-д·ок·**, (*f*) guess, conjecture, surmise

до·га́д·лив·ый, -ая, -ое, (*a*) quick/sharp-witted, shrewd

до·га́д·ыва·ться, -ются, (*imp*) guess; surmise, conjecture, suspect

доггéр·, -а, (*m*) (geol) Dogger

до·гла́д·ить, -ят, (*perf*) *see* **до·гла́·ж·ива·ть**

до·гла́ж·ива·ть, -ют, (*imp*) finish (ironing), iron (completely)

до·гляд·е́ть, -я́т, (*perf*) *see* **до·гля́·д·ыва·ть**

до·гля́д·ыва·ть, -ют, (*imp*) watch/see to the end, finish watching

до·гляж·у́ (*fut 1st sing*) of **до·гляд·е́ть**

дóгм·а, -ы, (*f*) dogma

дóгмат·, -а, (*m*) dogma

догмат·и́ческ·ий, -ая, -ое, (*a*) dogmatic

до·гн·а́ть (*fut 3rd sing, pl*) **до·гóн·ит, до·гóн·ят,** (*perf*); *see* **до·гон·я́·ть**

до·гнив·а́·ть, -ют, (*imp*) go completely rotten, rot through/away, finish rotting

до·гни́·ть, -ют, (*perf*) *see* **до·гнив·а́·ть**

до·гова́р·ива·ть, -ют, (*imp*) finish speaking, conclude (speaking); **-ся** negotiate, arrange for, come to an understanding/agreement; fly to extremes, lose sense of proportion

до·говóр·, -а, (*m*) (com) agreement, contract; pact, treaty

~, арéнд·н·ый lease

~, бодмер·éйн·ый (mar ins) bottomry bond

до·говор·ённост·ь, -и, (*f*) understanding, arrangement

до·говор·и́ть, -я́т, (*perf*) *see* **до·гова́р·ива·ть**

до·говóр·ник·, -а, (*m*) contract worker

до·говóр·н·ый, -ая, -ое, (*a*) of **до·говóр·**

до·гол·а́ (*adv*) naked, till naked, to the skin

до·гóн·к·а, -и, (*g pl*) **-н·ок·**, (*f*) overtaking, catching up, driving (animals)

до·гон·я́·ть, -ют, (*imp*) catch up with; drive to (of animals etc., on foot)

до·гор·а́ни·е, -я, (*n*) (*v n*) *see* **до·го·р·а́·ть;** (ICE) after-burning

до·гор·а́·ть, -ют, (*imp*) burn out/down

до·гор·е́ть, -я́т, (*perf*) *see* **до·гор·а́·ть**

до·греб·а́·ть, -ют, (*imp*) finish (raking), rake completely; row to/as far as

до·грес·ти́ (*fut 3rd sing, pl*) **до·греб·ёт, до·греб·у́т,** (*past masc sing*) **до·грёб;** (*perf*); *see* **до·греб·а́·ть**

до·груж·а́·ть, -ют, (*imp*) finish loading; bring up to full load, load till full, fill up; add to the load

до·груз·и́ть, -я́т, (*perf*) *see* **до·груж·а́·ть**

до·гру́з·к·а, -и, (*g pl*) **-з·ок·**, (*f*) (*v n*) *see* **до·груж·а́·ть;** additional charge (furnaces etc.)

до·да·ва́·ть, -ю́т, (*imp*) give the remaining, give the rest, make up (to a larger quantity etc.)

до·да́·ивани·е, -я, (*n*) stripping (in milking)

дó·да·нн·ый, -ая, -ое, (*past part pass*) *see* **до·да·ва́·ть;** made up, added

до·да́·ть (*fut 1st, 3rd sing, 1st, 2nd, 3rd pl*) **до·да́м, до·да́ст, до·дади́м, додади́·те, до·даду́т,** (*perf*); *see* **до·да·ва́·ть**

додек·а́н·, -а, (*m*) (chem) dodecane

додекан·а́л·, -а, (*m*) dodecanal, lauraldehyde

додека́н·ов·ая кисл·от·а́ (*f*) dodecanic/dodecoic acid

додекатилéн·, -а, (*m*) (chem) dodecatylene

додека́·эдр·, -а, (*m*) (math) dodecahedron

додека·эдр·и́ческ·ий, -ая, -ое, (*a*) (cryst) dodecahedral, rhombic

додек·éн·, -а, (*m*) (chem) dodecene

додекéн·ов·ая кисл·от·а́ (*f*) lauroleic/dodecenoic acid

додеки́н·ов·ая кисл·от·а́ (*f*) dodecynoic acid

до·де́л·анн·ый, -ая, -ое, (*past part pass*) *see* **до·де́л·ыва·ть;** finished, completed, ready

до·де́л·а·ть, -ют, (*perf*) *see* **до·де́л·ыва·ть**

до·де́л·к·а, -и, (*g pl*) **-л·ок·**, (*f*) (*v n*) *see* **до·де́л·ыва·ть**

до·де́л·ыва·ть, -ют, (*imp*) finish/complete (work)

додец·и́л·, -а, (*m*) (chem) dodecyl

додеци́л·ов·ый спирт· (*m*) dodecyl alcohol, dodecanol

до·ду́бл·ивани·е, -я, (*n*) final tanning

до·ду́м·а·ть, -ют, (*perf*) *see* **до·ду́м·ыва·ть**

до·ду́м·ыва·ть, -ют, (*imp*) finish considering/thinking about; **-ся** come to the conclusion; think up, evolve (a plan etc.)

до·е́·вш·ий, -ая, -ее, (*past part act*) of до·е́ст·ь

до·ед·а́·ть, -ют, (*imp*) finish eating; eat up

до·е́д·ут (*fut 3rd pl*) of до·е́х·ать

до·езж·а́·ть, -ют, (*imp*) reach, arrive at, drive up to

до·е́ни·е, -я, (*n*) milking

до·е́ст·ь (*fut 1st, 3rd sing, 1st, 2nd, 3rd pl*) до·е́м, до·е́ст·, до·ед·и́м, до·ед·и́те, до·ед·я́т, (*past masc sing*) до·е́л, (*perf*); see до·ед·а́·ть

до·е́х·ать (*fut 3rd sing, pl*) до·е́д·ет, до·е́д·ут, (*perf*); see до·езж·а́·ть

до·жа́р·ива·ть, -ют, (*imp*) finish (frying/grilling/roasting), roast/grill/fry (completely)

до·жа́т·ь (*fut 3rd sing, pl*) до·жн·ёт, до·жн·у́т, (*perf*); finish reaping; (*fut 3rd sing, pl*) до·жм·ёт, до·-жм·у́т, (*perf*); press home, press (completely)

до·жг·у́т (*fut 3rd pl*) of до·же́ч·ь

до·жд·а́ться, -у́тся (*perf*) (+ *gen*) wait until, wait for

дожд·ева́льн·ый, -ая, -ое, (*a*) (agr) sprinkler, sprinkling, overhead irrigation

дожд·ева́ни·е, -я, (*n*) (agr) sprinkling, overhead irrigation, sprinkler irrigation; dropwise separation (centrifugal casting)

дожд·ева́тел·ь, -я, (*m*) sprinkle r overhead irrigator

дожд·еви́к·, -а, (*m*) (bot) puffball *Lycoperdon*; raincoat

дожд·ев·о́й, -а́я, -о́е, (*a*) rain, pluvial, hyetal; spray; (*pl decl as adj*) nimbus (clouds)

~ мороз·и́лк·а (*f*) spray freezer

~ эква́тор· (*m*) hyetal equator

дожде·ме́р·, -а, (*m*) rain-gauge, pluviometer, ombrometer, hyetometer

~, само·пи́ш·ущ·ий (meteor) hyetograph, pluviograph

дожде·ме́р·н·ый, -ая, -ое, (*a*) rainfall, rain-measuring

~ ведр·о́ (*n*) rain-gauge receiver/collector

дожде·при·ём·ник·, -а, (*m*) street inlet (in gutter); (agr) rain gully

дожде·пис·е́ц, -с·ц·а́, (*m*) (instr) hyetograph, pluviograph

дожд·ли́в·ый, -ая, -ое, (*a*) rainy, showery

до́жд·ь, -я́, (*m*) rain; rainfall

~, ка́мен·н·ый (geol) rock rain/fall; meteoric rain

~, лед·ян·о́й sleet

~, ливн·ев·о́й rain shower(s)

~, ме́лк·ий drizzle, fine rain

~, морос·я́щ·ий drizzle, light rain

~, об·ло́ж·н·о́й steady/continuous rain

~, пе́пель·н·ый (seismol) ash shower

~, пере·меж·а́ющ·ийся intermittent rain

~, про·ли́в·н·о́й torrential rain

до·же́ч·ь (*fut 3rd sing, pl*) до·жж·ёт, до·жг·у́т, (*past masc sing*) до·жёг·, (*perf*); see до·жиг·а́·ть

до·жж·ённ·ый, -ая, -ое, (*past part pass*) see до·жиг·а́·ть; burnt up

до·жив·а́ни·е, -я, (*n*) (*v n*) see до·-жив·а́ть; survival

до·жив·а́·ть, -ют, (*imp*) survive, live till; stay till

до·жиг·а́ни·е, -я, (*n*) (*v n*) of до·-жиг·а́·ть; after-burning (jet engines); complete combustion

до·жиг·а́тел·ь, -я, (*m*) (a/c) after-burner

до·жиг·а́·ть, -ют, (*imp*) burn up, finish (burning), burn (completely)

до·жид·а́·ться, -ются, (*imp*) (+ *gen*) wait until/for, await

до·жим·а́·ть, -ют, (*imp*) press home, press (completely)

до·жин·а́·ть, -ют, (*imp*) finish (reaping), reap (completely)

до·жи́т·ь (*fut 3rd sing, pl*) до·жив·ёт, до·жив·у́т, (*past masc sing*) до́·жи·л, (*perf*); see до·жив·а́·ть

до·жм·у́т (*fut 3rd pl*) of до·жа́т·ь

до·жн·у́т (*fut 3rd pl*) of до·жа́т·ь

до́з·а, -ы, (*f*) dose, draught, potion; portion, measured amount, dosage

~ вс·пры́ск·а shot (injection)

~, воз·раст·а́ющ·ая (nucl) incremental dose

~ на вы́·ход·е (nucl) exit dose

~ пол·ови́нн·ой вы·жив·а́емост·и (nucl) median lethal dose

~, сумм·а́рн·ая cumulative dose

~, тка́н·ев·ая (nucl) tissue dose

дозад·о́ (*indecl*) (print) dos-à-dos

до·за·пра́в·к·а, -и, (*f*) refuelling, topping up

до·за́р·ивани·е, -я, (*n*) (hortic) after-ripening, artificial ripening

до·за·ря́д·, -а, (*m*) (elec) milking

до·за·ря́д·к·а, -и, (*g pl*) -д·ок·, (*f*) charging, topping up the charge

до·за·ряж·а́ющ·ий генер·а́тор· (*m*) (elec) milking generator, milker

доз·а́тор·, -а, (*m*) metering device/hopper/feeder, dispenser; batcher, batch meter, gauge box; (min) scale pocket, measuring pocket

~, вес·ов·о́й weighing batcher

~ вре́мен·и timing unit

~ изо·то́п·ов (nucl) isotope dispenser

~, ле́нт·очн·ый continuous batcher

~ скип·ов·о́го подъ·ём·а (min) skip measuring pocket/bin

до·звёзд·н·ый, -ая, -ое, (*a*) (phys) prestellar

до·звол·е́ни·е, -я, (*n*) permission

до·зво́л·енн·ый, -ая, -ое, (*past part pass*); permissable, allowed, legal

~ орби́т·а (*f*) (phys) allowed orbit

до·зво́н·иться, -ятся, (*perf*) get through (on telephone), get an answer

до·звук·ов·о́й, -а́я, -о́е, (*a*) subsonic

доз·ёр·, -а, (*m*) *see* **доз·а́тор·**

дози·ме́тр·, -а, (*m*) (phys) dosimeter, radiation dosimeter

~, групп·ов·о́й stationary dosimeter

~, контро́ль·н·ый monitor

~ ме́ст·ност·и area monitor

~, пло́ск·ий плён·очн·ый film badge

~, по́·иск·ов·ый survey dosimeter

~, полу·про·вод·ников·ый transistorized dosimeter

~ -свето·фо́р·, -а, (*m*) go-not-go radiation detector

дози·метр·и́ст·, -а, (*m*) health physicist, radiation supervisor

дози·метр·и́ческ·ий, -ая, -ое, (*a*) dosimetric, health/radiation monitoring, radiac

~ при·бо́р· (*m*) (nucl) radiac instrument

дози·ме́тр·и·я, -и, (*f*) dosimetry, radiation monitoring

доз·и́рование·, -я, (*n*) = **доз·иро́вк·а**

доз·и́рованн·ый, -ая, -ое, (*past part pass*) of **доз·и́р·овать**

доз·и́р·овать, -уют, (*imp and perf*) dose out, apportion, proportion; batch, meter

доз·иро́вк·а, -и, (*g pl*) **-во́к·,** (*f*) dosage, proportion, amount; dosing, proportioning, batching, metering; feed; (nucl) monitoring

~, вес·ов·а́я weight feed

доз·иро́вочн·ый, -ая, -ое, (*a*) of **доз·иро́вк·а**

доз·иро́вщик·, -а, (*m*) (civ eng) batcher, gauge box

до·зна·ва́ться, -ю́тся, (*imp*) ascertain

до·зна́·ни·е, -я, (*n*) (law) examination (pre-trial)

до·зна́·ться, -ю́тся (*perf*) ascertain

до·зо́р·, -а, (*m*) patrol

до·зо́р·н·ый, -ая, -ое, (*a*) patrol, scout

до·зрев·а́ни·е, -я, (*n*) ripening in store, maturing, artificial ripening

до·зрев·а́тел·ь, -я, (*m*) maturing tank (for rubber)

до·зрев·а́·ть, -ют, (*imp*) ripen, mature

до·зре́·л·ый, -ая, -ое, (*a*) ripe, mature, fully ripened, fully matured

до·зре́·ть, -ют, (*perf*) *see* **до·зрев·а́ть**

до·и́льн·ый, -ая, -ое, (*a*) (agr) milking

до·иск·а́ться (*fut 3rd sing, pl*) **до·и́щ·ется до·и́щ·утся,** (*perf*); *see* **до·и́ск·ива·ться**

до·и́ск·ива·ться, -ются, (*imp*) seek out, discover (something known): ascertain

до·ис·пар·е́ни·е, -я, (*n*) residual evaporation

до·истор·и́ческ·ий, -ая, -ое, (*a*) prehistoric

до·и́ть, -я́т, (*imp*) milk; **-ся** (*pass*) give milk

до·и́щ·утся (*fut 3rd pl*) of **до·иск·а́ться**

до́й·к·а, -и, (*g pl*) **до́·ек·,** (*f*) milking

дой·ни́к·, -а́, (*m*) milk pail

до·йти́ (*fut 3rd sing, pl*) **до·йдет, до·йду́т,** (*past masc sing*) **до·шёл,** (*perf*); *see* **до·ход·и́ть**

ДОК (*abbr*) of **дез·окси·кортико·стеро́н·** (pharm) desoxycorticosterone

док·, -а, (*m*) dock

~, мо́кр·ый wet dock

~, на·лив·н·о́й building basin, dry dock with artificial flooding

~, не·симметр·и́чн·ый offshore floating dock

~, плов·у́ч·ий floating dock

~, при·лив·н·о́й tidal dock

~, секциона́ль·н·ый sectional dock

~, само·подъ·ём·н·ый self-docking floating dock

~, стро·и́тельн·ый building dock

~, судо·ремо́нт·н·ый graving dock

~, сух·о́й dry dock

до·ка́дм·иев·ый, -ая, -ое, (*a*) (chem) sub-cadmium

до·ка́з·анн·ый, -ая, -ое, (*past part pass*) *see* до·ка́з·ыва·ть; proven, proved

до·каз·а́тельн·ый, -ая, -ое, (*a*) (*s f*) -лен·, -льн·а, demonstrative, conclusive, convincing

до·каз·а́тельственн·ый, -ая, -ое, (*a*) of до·каз·а́тельств·о

~ пра́в·о (*n*) law of evidence

до·каз·а́тельств·о, -а, (*n*) proof; evidence; (math) demonstration

до·каз·а́ть (*fut 3rd sing, pl*) до·ка́ж·ет, до·ка́ж·ут, (*perf*); *see* до··ка́з·ыва·ть

что и тре́б·овалось до·каз·а́ть (math) Q.E.D.

до·ка́з·ыва·ть, -ют, (*imp*) prove, demonstrate

до·ка́л·ива·ть, -ют, (*imp*) heat right through; (met) quench fully

до·кал·и́ть, -я́т, (*perf*) *see* до·ка́л·ыва·ть

до·ка́л·ыва·ть, -ют, (*imp*) finish chopping; chop up

до·ка́нч·ивани·е, -я, (*n*) completion, completing, finish, finishing

до·ка́нч·ива·ть, -ют, (*imp*) complete, finish

до·ка́п·ыва·ть, -ют, (*imp*) finish (digging), dig (completely); dig to/as far as; -ся dig up, uncover, discover, unearth

до·ка́рмл·ива·ть, -ют, (*imp*) finish feeding, feed (completely), feed until

до·кат·и́ться, -я́тся, (*perf*) *see* до··ка́т·ыва·ться

до·ка́т·ыва·ться, -ются, (*imp*) roll to/up to/as far as; be heard

до·ка́ш·ива·ть, -ют, (*imp*) finish mowing, mow completely

до·кембр·и́йск·ий, -ая, -ое, (*a*) (geol) pre-cambrian

до́кер·, -а, (*m*) docker, longshoreman, dockyard workman

до·ки́д·ыва·ть, -ют, (*imp*) throw to/as far as

до·ки́н·уть, -ут, (*perf*) *see* до·ки́д·ыва·ть

до·кла́д·, -а, (*m*) report; lecture, address; paper (on academic subject etc.)

до·кла́д·н·о́й, -а́я, -о́е, (*a*) of до··кла́д·

~ за·пи́с·к·а (*f*) memorandum

до·кла́д·чик·, -а, (*m*) lecturer, speaker; reporting officer

до·кла́д·ыва·ть, -ют, (*imp*) report, make a report; present a paper; announce; (math) add to

до·кле́·енн·ый, -ая, -ое, (*past part pass*) *see* до·кле́·ива·ть

до·кле́·ива·ть, -ют, (*imp*) glue up

до·кле́·ить, -ят, (*perf*) *see* до·кле́·и·ва·ть

до·кле́т·очн·ый, -ая, -ое, (*a*) (cyt) precellular

до·ков·а́ть (*fut 3rd sing, pl*) до·куёт, до·кую́т, (*perf*) finish forging, finish hammering out; до́к·овать, -уют, (*imp and perf*) (naut) dock

до·ко́в·ыва·ть, -ют, (*imp*) finish forging, finish hammering out

до́к·ов·ый, -ая, -ое, (*a*) (naut) dock, docking

~ с·бо́р·ы (*pl*) dock dues

докоза́н·, -а, (*m*) (chem) docosane

докоза́н·ов·ая кисл·от·а́ (*f*) docosanoic/behenic acid

до·кол·о́ть, -ют, (*perf*) *see* до·ка́·л·ыва·ть

до·ко́нч·ить, -а́т, (*perf*) *see* до·ка́нч·ива·ть

до·коп·а́ть, -ют, (*perf*) *see* до·ка́·п·ыва·ть

до·корм·и́ть, -я́т, (*perf*) *see* до·ка́рм·л·ива·ть

до·кос·и́ть, -я́т, (*perf*) *see* до·ка́·ш·ива·ть

до·ко́ш·енн·ый, -ая, -ое, (*past part pass*) *see* до·ка́ш·ива·ть

до·кра́с·ить, -я́т, (*perf*) *see* до·кра́·ш·ива·ть

до·крас·н·а́ (*adv*) till red; red-hot, to red heat

до·кра́ш·ива·ть, -ют, (*imp*) finish (painting), paint (completely); paint to/up to/as far as; (chem) counter-stain

до·крит·и́ческ·ий, -ая, -ое, (*a*) sub-critical; (air) before stalling

до·кру́т·к·а, -и, (*g pl*) -т·ок·, (*f*) after-twist, twisting (mule spinning)

до́ктор·, -а, (*nom pl*) -а́, (*g pl*) -о́в, (*m*) docto

доктор·а́нт·, -а, (*m*) graduate studying for doctor's degree, predoctorial graduate

доктор·а́т·, -а, (*m*) doctorate, doctor's degree

до́ктор·ск·ий, -ая, -ое, (*a*) doctor's, doctoral

~ диссерт·а́ци·я (*f*) doctoral thesis

~ про́б·а (*f*) doctor test (gasoline)

до́ктор·ск·ий, -ая, -ое
~ **рас·тво́р·** (*m*) (oil) doctor solution
~ **сте́п·ен·ь** (*f*) doctorate, doctor's degree
доктри́н·а, -ы, (*f*) doctrine
докуме́нт·, -а, (*m*) document, deed, paper; identity card/paper, pass; (law) instrument
~, **за·да·ю́щ·ий** (autom) input information document
~, **по·гру́з·оч·ный** shipping document
~, **товаро·рас·по·ряд·и́тельн·ый** document of title (to goods)
докуме́нт·а́льн·ый, -ая, -ое, (*a*) document, documentary
~ **фильм·** (*m*) (cinema) documentary film
докуме́нт·а́ци·я, -и, (*f*) documentation; documents, papers
докуме́нт·и́р·овать, -уют, (*imp and perf*) document
до·куп·а́·ть, -ют, (*imp*) buy in addition, make an additional purchase; (*perf*) finish bathing, bath completely; bath/bathe till
до·куп·и́ть, -я́т, (*perf*) buy in addition, make an additional purchase
до·ку́п·к·а, -и, (*g pl*) **-п·ок·,** (*f*) additional purchase
до·ку́пл·енн·ый, -ая, -ое, (*past part pass*) of **до·куп·и́ть**
доку́ч·лив·ый, -ая, -ое, (*a*) tiresome, annoying
дол· (*abbr*) of **долг·от·а́** longitude
до·ла́м·ыва·ть, -ют, (*imp*) break (completely); break up to; break off (at one go)
дола́ц·, -а, (*m*) (geol) doline
долб·ёжн·ый, -ая, -ое, (*a*) mortising (wood), slotting (metal)
~ **стан·о́к·** (*m*) vertical shaper, slotter, slotting machine; mortising machine (for wood)
долб·и́ть, -я́т, (*imp*) gouge, chisel; mortise; learn by heart
долбл·е́ни·е, -я, (*n*) slotting; mortising
долб·я́к·, -а́, (*m*) gear shaper-cutter; mortising cutter
до́лг·, -а, (*nom pl*) **-и́,** (*g pl*) **-о́в,** (*m*) debt; duty, obligation
~, **без·на·дёж·н·ый** bad debt
   в долг· on credit, on trust
~, **госуда́р·ственн·ый** national/public debt
~, **не·у·пла́ч·енн·ый** outstanding debt

до́лг
~, **при·сужд·ённ·ый** judgement debt
~, **фунд·и́рованн·ый** funded debt
до́лг·ий, -ая, -ое, (*a*) long
~ **звук·** (*m*) dash, long (sound signalling)
до́лг·о (*adv*) long, for a long time
долго- (*component*) long-
долг·ова́т·ый, -ая, -ое, (*a*) lengthy, rather long
долго·ве́ч·ност·ь, -и, (*f*) longevity; life, durability (of a material); (met) time-to-rupture (creep test), life, endurance (fatigue test)
~, **цикли́ч·еск·ая** (*f*) endurance (fatigue test)
долго·ве́ч·н·ый, -ая, -ое, (*a*) durable, lasting; long-lived, long-life
долг·ов·о́й, -а́я, -о́е, (*a*) of **долг·**
~ **обяз·а́тельств·о** (*n*) (fin) promissory note
долго·вре́мен·н·ый, -ая, -ое, (*a*) lasting, long-term, long-time; permanent
~ **за·по·мин·а́ющ·ее у·стро́й·ств·о** (*n*) external store (computer)
~ **со·противл·е́ни·е** (*n*) (met) creep limit
долго·жив·у́щ·ий, -ая, -ее, (*a*) long-life, long-lived
долго·звуч·а́щ·ий, -ая, -ее (*a*) long-playing (of sound records etc.)
долго·игр·а́ющ·ий, -ая, -ее, (*a*) long-playing
долго·ле́т·и·е, -я, (*n*) longevity
долго·ле́т·н·ий, -яя, -ее, (*a*) long, extending over many years, long-standing
долго·но́с·ик·, -а, (*m*) (ent) weevil
~, **хло́п·ков·ый** boll weevil, *Anthonomus grandis*
долго·по́л·ый, -ая, -ое, (*a*) long-skirted
долго·сро́ч·н·ый, -ая, -ое, (*a*) long, long-term, continuous; (fin) long-dated
~ **прогно́з·** (*m*) (meteor) long range forecast
долг·от·а́, -ы́, (*nom pl*) **-о́т·ы,** (*g pl*) **-о́т·,** (*f*) length; longitude
~, **галакт·и́ческ·ая** (astron) galactic longitude
~, **гео·граф·и́ческ·ая** terrestrial/geographical longitude
~, **гео·центр·и́ческ·ая** (astron) geocentric longitude

долг·от·á
~, изо·метр·и́ческ·ая (surv) conformal longitude
~, неб·éсн·ая celestial longitude
~, при·вед·ённ·ая (surv) reduced longitude
~, эклипт·и́ческ·ая (astron) celestial longitude
долг·óтн·ый, -ая, -ое, (a) longitudinal
дол·ев·óй, -áя, -óе, (a) longitudinal, lengthwise; fractional, part
~ плáт·а (f) royalties
до·лед·никóв·ый, -ая, -ое, (a) (geol) preglacial
дóл·ее (comp and adv) longer
до·лез·á·ть, -ют, (imp) climb to/up to/as far as
до·лéз·ть, -ут, (perf) see **до·лез·á·ть**
долери́т·, -а, (m) (min) dolerite
до·лет·á·ть, -ют, (imp) fly to/as far as, reach
до·лет·éть, -ят, (perf) see **до·лет·á·ть**
до·лéч·ива·ть, -ют, (imp) heal; complete the cure of
до·леч·и́ть, -ат, (perf) see **до·лéч·и·ва·ть**
дóл·ечн·ый, -ая, -ое, (a) (bot) lobular, lobate
до·леч·ý (fut 1st sing) of **до·лет·éть**; (fut 1st sing) of **до·леч·и́ть**
должж·á·ть, -ют, (imp) borrow from
дóлжж·ен·, -ж·н·á, -ж·н·ó, (predic adj) owe; (+ infinitive) must, have to, ought to
 дóлж·ен· бы·л бы·ть should have been, ought/was to have been
 должж·н·ó бы́·ло (+ infinitive) was bound to
 должж·н·ó бы·ть is supposed to be, ought to be, should be, must be
должж·ни́к·, -á, (m) debtor
~, не·со·сто·я́тельн·ый (fin) insolvent
должж·ностн·óй, -áя, -óе, (a) official
~ лиц·ó (n) official
~ пре·ступл·éни·е (n) (law) malfeasance
дóлжж·ност·ь, -и, (nom pl) -и, (g pl) -éй, (f) post, office, function, appointment, employment, duty
~, ис·полн·я́ющ·ий acting
дóлжж·н·ый, -ая, -ое, (a) due, right, proper
~, штáт·н·ая permanent appointment, established appointment

до·лив·á·ть, -ют, (imp) pour till fill to, add, replenish
дол·и́н·а, -ы, (f) valley
~ волн·ы́ wave trough
~, гóр·н·ая (geol) intermontane valley
~, за·кры́·т·ая blind valley
~ за·топл·éни·я flood plain
~, кáрст·ов·ая (geol) sinkhole
~, потенциáль·н·ая (phys) potential valley/trough
~ раз·ли́·в·а alluvial/flood plain
~ раз·мы́·в·а wash-out plain
~, сквоз·н·áя (geol) watergap
дол·и́нн·ый, -ая, -ое, (a) of **дол·и́н·а**; valley
до·ли́·т·ый, -ая, -ое, (past part pass) see **до·лив·á·ть**; poured up, added to, poured up to
до·ли́·ть (fut 3rd sing, pl) **до·ль·ёт, до·ль·ю́т,** (perf); see **до·лив·á·ть**
долихо·зáвр·, -а, (m) (pal) dolichosaur
долихо·цефáл·, -а, (m) (zool) dolichocephal
долихо·цефáль·н·ый, -ая, -ое, (a) (zool) dolichocephalic, long-headed
дóллар·, -а, (m) (fin) dollar
до·лож·éни·е, -я, (n) deposition
до·лóж·енн·ый, -ая, -ое, (past part pass) see **до·клáд·ыва·ть**
до·лож·и́ть, -áт, (perf) see **до·клáд·ыва·ть**
долóй (adv) away/down/off with! (cry of opposition)
 шáпк·и долóй hats off!
до·лом·á·ть, -ют, (perf) break (completely, beyond repair); break up to
доломи́т·, -а, (m) (min) dolomite, pearl spar
доломит·изáци·я, -и, (f) (geol) dolomitization
доломи́т·н·ый, -ая, -ое, (a) dolomite, dolomitic
до·лом·и́ть, -я́т, (perf) break off (at one go)
доломóл·, -а, (m) (chem) dolomol, magnesium stearate
долореси́т·, -а, (m) (min) doloresite
долот·н·óй, -áя, -óе, (a) of **долот·ó**
долот·ó, -á, (nom pl) -óт·а, (g pl) -óт·, (n) chisel, gouge; bit (drilling, boring)
~, глад·и́льн·ое broad chisel
~, двух·лóпаст·н·ое two-way bit
~, двух·шарóш·ечн·ое two-core bit
~, долб·ёжн·ое ripping chisel
~, колóн·ков·ое (oil) core bit

долот·о́

~, кресто·обра́з·н·ое four-wing/point bit

~, трёх·гра́н·н·ое cornel chisel

~, трёх·шаро́ш·ечн·ое three-core bit

~, шаро́ш·ечн·ое Hawthorn roller bit, rolling-cutter bit

долото·ви́д·н·ый, -ая, -ое, (a) dolabriform, axe/hatchet shaped

долото·лепе́ст·н·ый, -ая, -ое, (a) (bot) dolabripetalous

долото·обра́з·н·ый, -ая, -ое, (a) dolabriform, axe/hatchet shaped

доло́т·чат·ый, -ая, -ое, (a) chisel

~ бур· (m) chisel/bull bit

до·ль·ёт (fut 3rd sing) of **до·ли́·ть**

до́ль·к·а, -и, (g pl) **-л·ек·,** (f) lobule, (bot) lacinule; slice, section

~ плод·а́ (bot) mericarp

до́ль·ков·ый, -ая, -ое, (a) (biol) lobular

-доль·н·ый, -ая, -ое, (adj component) -lobe (e.g. двух·до́ль·н·ый two-lobe)

до́ль·чатост·ь, -и, (f) lobulation

до́ль·чат·ый, -ая, -ое, (a) lobed, lobate, lobulated

до́ль·ш·е (comp) longer

до́л·я, -и, (nom pl) **-и,** (g pl) **-е́й,** (f) part, portion, share, fraction; lobe; fate, lot

~, а́том·н·ая atomic fraction

~, вес·ов·а́я weight fraction

~, за·ты́л·очн·ая (anat) occipital lobe

~, миллио́н·н·ая part per million

до́м·, -а, (nom pl) **-а́,** (g pl) **-о́в,** (m) house; home; club, social centre; domatic crystal

~, с·бо́р·н·ый prefabricated house

~, станда́рт·н·ый standardized house, standard-pattern house

~, щитово·с·бо́р·н·ый prefabricated sectional house, frame-and-panel house

~, щи́т·ов·ый frame-and-panel house

до́м·а (gen, sing) of **дом·**; (adv) at home; **дом·а́** (nom pl) of **дом·**

до·ма́л·ыва·ть, -ют, (imp) finish grinding, grind up

домат·и́ческ·ий, -ая, -ое, (a) domatic

дома́ш·н·ий, -яя, -ее, (a) domestic, house, home

домейки́т·, -а, (m) (min) domeykite

до·мел·ов·о́й, -а́я, -о́е, (a) (geol) Pre-Cretaceous

доме́н·, -а, (m) domain (ferromagnetism); **до́мен·** (g pl) of **до́мн·а** (q.v.)

~, за·мык·а́ющ·ий flux-closure domain

до́мен·н·ый, -ая, -ое, (a) (met) blast-furnace, ironmaking; **доме́н·н·ый, -ая, -ое,** (a) domain (ferromagnetism)

~ гран·и́ц·а (f) domain wall

~ мета́лл· (m) hot metal (in blast furnace)

~ печ·ь (f) blast furnace

~ про·из·во́д·ств·о (n) ironmaking

~ структу́р·а (f) domain structure, magnetic structure

~ цех· (m) blast furnace

до́мен·щик·, -а, (m) blast-furnace worker/operative/engineer

до·ме́р·ива·ть, -ют, (imp) finish measuring/surveying, measure/survey (completely)

до·ме́р·ить, -ят, (perf) see **до·ме́р·ива·ть**

до·мес·и́ть, -ят, (perf) see **до·ме́ш·ива·ть**

до·мес·ти́ (fut 3rd sing, pl) **до·мет·ёт, до·мет·у́т,** (past masc sing) **до·мёл,** (perf); see **до·мет·а́·ть**

до·мет·а́·ть, -ют, (imp) finish sweeping, sweep (completely); sweep to/up to/as far as; finish overstitching/overcasting, (perf) of **до·мёт·ыва·ть**; finish throwing; throw to/as far as

до·меш·а́·ть, -ют, (perf) see **до·ме́ш·ива·ть**

до·ме́ш·ива·ть, -ют, (imp) finish mixing/kneading, mix, knead (completely), finish puddling, puddle (of clay)

до́м·ик·, -а, (m) (dim) of **дом·**; (nucl) castle; lorica (of protozoans)

~, вегетацио́н·н·ый (hortic) cloche; greenhouse, glasshouse

домина́нт·а, -ы, (f) dominant; (biol) dominance, dominant character

домина́нт·н·ый, -ая, -ое, (a) dominant, prevalent

домини́р·овать, -уют, (imp) dominate, predominate

домкра́т·, -а, (m) jack, lifting jack

~, буты́л·очн·ый bottle/screw jack

~, винт·ов·о́й screw jack

~, в·стро́·ечн·ый built-in jack

~, гидравл·и́ческ·ий hydraulic jack

~, ка́рлик·ов·ый planer jack

~, ре́·ечн·ый rack-and-pinion jack

~, рыча́ж·н·о-ре́·ечн·ый rack-and-lever jack

до́мн·а, -ы, (g pl) **до́мен·,** (f) blast furnace

домо·влад·е́лец·, -льц·а, (m) house-owner

домо·вод·ств·о, -а,   (*n*) domestic science; household management
дом·о́в·ый, -ая, -ое, (*a*) house, housing
~ гриб· (*m*) dry rot
дом·о́й (*adv*) home, homewards
до·мола́ч·ива·ть, -ют, (*imp*) finish (threshing), thresh (thoroughly/completely)
до·молот·и́ть, -я́т, (*perf*) *see* **до·мола́ч·ива·ть**
до·мо́л·от·ый, -ая, -ое, (*past part pass*) *see* **до·ма́л·ыва·ть**
до·мол·о́ть (*fut 3rd sing, pl*) **до·ме́л·ет, до·ме́л·ют,** (*perf*); *see* **до·ма́л·ыва·ть**
до·моло́ч·енн·ый, -ая, -ое, (*past part pass*) *see* **до·мола́ч·ива·ть**
домо·ро́щ·енн·ый, -ая, -ое, (*a*) home-grown/produced
домо·стро·е́ни·е, -я, (*n*) house-building
домо·стро·и́тел·ь, -я, (*m*) builder
домо·стро·и́тельн·ый, -ая, -ое, (*a*) house-building
домо·у·правл·е́ни·е, -я, (*n*) house manager's office; estate office
до·мыв·а́·ть, -ют, (*imp*) finish (washing), wash (completely); wash to/as far as
до́·мысел·, -сл·а, (*m*) conjecture
до·мы́·ть (*fut 3rd sing, pl*) **до·мо́·ет, до·мо́·ют,** (*perf*); *see* **до·мыв·а́·ть**
до·наг·а́ (*adv*) naked, till naked, to the skin
до·на·груж·ённ·ое со·сто·я́ни·е (*n*) preload state
до·на́ш·ива·ть, -ют, (*imp*) wear out
донба́сс·к·ий, -ая, -ое, (*a*) Donets, Donets-coalfield
до·не́льзя (*adv*) to the uttermost limit
до·нес·е́ни·е, -я, (*n*) report, dispatch
до·нес·ти́, -у́т, (*perf*) *see* **до·нос·и́ть**
доне́цк·ий, -ая, -ое, (*a*) (geog) Donetz
до́·низ·у (*adv*) to the bottom
до́нк·а, -и, (*g pl*) **-нок·,** (*f*) steam-driven piston pump, donkey pump
до́нник·, -а, (*m*) (bot) sweet clover, *Melilotus*
до́н·н·ый, -ая, -ое, (*a*) of **дн·о**; bottom, base; (biol) benthic, bottom-dwelling; ground
~ вз·ры́в·а́тел·ь (*m*) (gunn) base percussion fuse
~ доск·а́ (*f*) bottom board (of boat)
~ кла́пан· (*m*) foot valve (of a pump)
~ о·ста́т·ок· (*m*) (chem) bottoms, residue, sludge
до́нор·, -а, (*m*) donor

до́нор
~, электро́н·н·ый (phys, chem) electron donor
до́нор·н·о-акце́птор·н·ый, -ая, -ое, (*a*) (phys chem) donor–acceptor
~ с·вя́з·ь (*f*) donor–acceptor bond
до́нор·н·ый, -ая, -ое, (*a*) donor
~ при́·мес·ь (*f*) donor impurity, *n*-type impurity (semiconductors)
до́нор·ск·ий, -ая, -ое, (*a*) (med) blood-donor
до·но́с·, -а, (*m*) denunciation
до·нос·и́ть, -я́т, (*imp*) report, inform of; carry up to; denounce; **-ся** (*pass*) be heard, reach (of sound); (*perf*) of **до·на́ш·ива·ть** wear out
до·но́с·чик·, -а, (*m*) informer
до·но́ш·енн·ый, -ая, -ое, (*past part pass*) *see* **до·на́ш·ива·ть**
дон·ск·о́й, -а́я, -о́е, (*a*) (geog) Don, river Don
до́н·ышк·о, -а, (*g pl*) **-шек·,** (*n*) (*dim*) of **дн·о**; bottom, base, end plate
донье-вы·рез·н·о́й стан·о́к· (*m*) barrel-head circling machine
до́н·ья (*nom pl*) of **дн·о**
до·об·ору́д·овать, -уют, (*imp and perf*) fit out, equip completely
до·о·кисл·е́ни·е, -я, (*n*) further oxidation
до·о·ко́р·к·а, -и, (*g pl*) **-р·ок·,** (*f*) rebarking
до·о·пре·дел·е́ни·е, -я, (*n*) supplement to a definition
до·от·ка́з·а (*adv*) home, right in, as far as possible
до·о·хлад·и́тел·ь, -я, (*m*) after-cooler
до·о·чи́ст·к·а, -и, (*g pl*) **-т·ок·,** (*f*) final cleaning/purification
доп. (*abbr*) of **до·полн·е́ни·е** supplement
до·палео·зо́й·ск·ий, -ая, -ое, (*a*) (geol) Pre-Paleozoic
до·пек·а́·ть, -ют, (*imp*) finish (baking), bake (completely); bake till/enough; weary, plague
до·пе́ч·ь (*fut 3rd sing, pl*) **до·печ·ёт, до·пек·у́т,** (*past masc sing*) **до·пёк·,** (*perf*); *see* **до·пек·а́·ть;**
до·пив·а́·ть, -ют, (*imp*) drink up, finish (drinking)
до·пис·а́ть (*fut 3rd sing, pl*) **до·пи́ш·ет, до·пи́ш·ут,** (*perf*); *see* **до·пи́с·ыва·ть**
до·пи́с·ыва·ть, -ют, (*imp*) finish (writing); write to/as far as

до·пи́·ть (*fut 3rd sing, pl*) **до·пь·ёт, до·пь·ю́т,** (*past masc sing*) **до́·пи·л** (*perf*); *see* **до·пив·а́·ть**

до·пла́т·а, -ы, (*f*) additional payment, supplement; (rail) excess fare

до·плат·и́ть, -ят, (*perf*) *see* **до·пла́·ч·ива·ть**

до·пла́ч·ива·ть, -ют, (*imp*) pay in addition, pay the outstanding/remaining

до·плё·л (*past masc sing*) of **до·плес·ти́**

До́плер·а, эффе́кт· (*m*) (phys) Doppler effect

∼, явл·е́ни·е (*n*) (phys) Doppler effect

до́плер·овск·ий, -ая, -ое, (*a*) Doppler

∼ у·шир·е́ни·е (*n*) Doppler broadening

∼ эффе́кт· (*m*) Doppler effect

до·плес·ти́ (*fut 3rd sing, pl*) **до·плет·ёт, до·плет·у́т,** (*past masc sing*) **до·плё·л,** (*perf*); *see* **до·плет·а́·ть**

до·плет·а́·ть, -ют, (*imp*) finish (weaving/plaiting); weave/plait to/up to/as far as; **-ся** drag oneself to

до·плыв·а́·ть, -ют, (*imp*) swim to, reach (by swimming), float to

до·плы́·ть (*fut 3rd sing, pl*) **до·плы·в·ёт, до·плыв·у́т,** (*perf*); *see* **до·плыв·а́·ть**

до·по́длин·н·ый, -ая, -ое, (*a*) authentic

до·полз·а́·ть, -ют, (*imp*) crawl to/as far as

до·полз·ти́, -у́т, (*perf*) *see* **до·полз·а́·ть**

до·полн·е́ни·е, -я, (*n*) addition, supplement, complement; (gram) object

∼, алгебра·и́ческ·ое (math) cofactor, signed minor

∼ для на·клон·е́ни·я codeclination

∼ до (math) complement to, complement with respect to, completing

∼ до 9 (math) 9 complement
и́мпульс· до·полн·е́ни·я (*m*) complement pulse (computer)

∼, ко́св·енн·ое (gram) indirect object

до·по́лн·енн·ый, -ая, -ое, (*past part pass*) *see* **до·полн·я́·ть;** complemented

до·полн·и́тельн·о (*adv*) in addition

до·полн·и́тельн·ый, -ая, -ое, (*a*) supplementary, complementary, additional, additive, extra, booster; re-

∼ велич·ин·а́ (*f*) (math) complement

∼ иониз·а́ци·я (*f*) reignition (in a radiation counter tube)

∼ множ·и́тел·ь (*m*) (math) cofactor

∼ нейтро́н· (*m*) extra neutron

∼ у́гол· (*m*) complementary angle

∼ у·сло́в·и·е (*n*) subsidiary/supplementary condition

до·полн·и́тельн·ый

∼ шир·от·а́ (*f*) (math) colatitude

до·по́лн·ить, -ят, (*perf*) *see* **до·полн·я́·ть**

∼ квадра́т· (math) complete the square

до·полн·я́·ть, -ют, (*imp*) complete, supplement, add, amplify

до·по·луч·а́·ть, -ют, (*imp*) receive in addition, receive the remaining/outstanding

до·по·луч·и́ть, -а́т, (*perf*) *see* **до·по·луч·а́·ть**

до́ппель·дуо́ (*indecl*) (met roll) double-duo

До́пплер·, -а, (*m*) *see* **До́пплер·**

допплеpи́т·, -а, (*m*) dopplerite, pitch peat

до·пра́ш·ива·ть, -ют, (*imp*) question, interrogate

до·пре·де́ль·н·ый, -ая, -ое, (*a*) limiting

до·про·бо́й·н·ый, -ая, -ое, (*a*) (elec) preconduction

до·про́с·, -а, (*m*) interrogation, questioning

до·прос·и́ть, -я́т, (*perf*) *see* **до·пра́ш·ива·ть**

до·про·явл·е́ни·е, -я, (*n*) (phot) developing, development

до·пры́г·ива·ть, -ют, (*imp*) jump to/as far as/as high as

до·пры́г·н·уть, -ут, (*perf*) *see* **до·пры́г·ива·ть**

до́·пуск·, -а, (*m*) admittance, admission; (mech eng) tolerance, allowance, permissible error/variation; clearance, authorization, security clearance

∼, двух·сторо́н·н·ий tolerance plus or minus

∼ на с·мещ·е́ни·е location tolerance

∼, стро́г·ий close tolerance

∼, то́ч·н·ый close tolerance

до·пуск·а́ем·ый, -ая, -ое, (*pres part pass*) of **до·пуск·а́·ть;** permissible, allowable

∼ за·зо́р· (*m*) permissible clearance

∼ на·пряж·е́ни·е (*n*) allowable unit stress, working stress

до·пуск·а́·ть, -ют, (*imp*) admit, allow, permit, tolerate; postulate, admit, accept, assume

до·пуст·и́мост·ь, -и, (*f*) admissibility, validity

до·пуст·и́м·ый, -ая, -ое, (*a*) permissible, admissible, allowable, safe, maximum permissible, acceptable

до·пуст·и́м·ый, -ая, -ое
~ коли́честв·о обращ·е́ни·й *(n)* read-around ratio (computer)
~ на·гру́з·к·а *(f)* permissible load, safe load
~ на·гру́з·к·а то́к·а *(f)* (elec) current-carrying capacity
до·пуст·и́ть, -я́т, *(perf) see* **до·пу·ск·а́·ть**
до·пущ·е́ни·е, -я, *(n)* assumption, admission, hypothesis; permission
до·пу́щ·енн·ый, -ая, -ое, *(past part pass) see* **до·пуск·а́·ть;** admitted, supposed
до·пыт·а́·ться, -ются, *(perf) see* **до·пы́т·ыва·ться**
до·пы́т·ыва·ться, -ются, *(imp)* find out, elicit; *(imp only)* try to find out/elicit
до·пь·ю́т *(fut 3rd pl)* of **до·пи́·ть**
до·раба́т·ыва·ть, -ют, *(imp)* finish, finish off; develop, work up; work on, work through
до·рабо́т·а·ть, -ют, *(perf) see* **до·раба́т·ыва·ть**
до·рабо́т·к·а, -и, *(g pl)* **-т·ок·,** *(f)* *(v n) see* **до·раба́т·ыва·ть;** development; modification
~, механ·и́ческ·ая additional machining
до·раз·лож·е́ни·е, -я, *(n)* (chem) final decomposition
до·рв·а́ть, -у́т, *(perf)* pick/pluck all; tear away all, tear away to the end
до·револю́цио́н·н·ый, -ая, -ое, *(a)* pre-revolutionary (especially pre Nov. 1917)
до·ре́з·ать *(fut 3rd sing, pl)* **до·ре́ж·ет, до·ре́ж·ут,** *(perf); see* **до·ре́з·ыва·ть**
до·ре́з·ыва·ть, -ют, *(imp)* finish (cutting), cut (to the end); cut to/as far as
до·релятиви́ст·ск·ий, -ая, -ое, *(a)* prerelativistic
до·ремо́нт·н·ый, -ая, -ое, *(a)* prior to repair/repairing
дорз- *(component) (see also* **дорс-)** dors- (back, dorsal)
дорз·а́льн·ый, -ая, -ое, *(a)* (zool) dorsal, tergal
дорзи·вентр·а́льн·ый, -ая, -ое, *(a)* (bot) dorsiventral
дори́ч·еск·ий, -ая, -ое, *(a)* (arch) Doric
~ о́рдер· *(m)* (arch) Doric order
дорн·, -а, *(m)* (met) piercing mandrel (tube rolling); core die (casting); core (tyre-making)

дорн
~, вулканизацио́н·н·ый vulcanizing/curing core/arm
До́рн·а, эффе́кт· *(m)* (electrochem) Dorn effect, sedimentation potential
дорн·ов·о́й, -а́я, -о́е, *(a)* of **дорн·**
~ по·кры́ш·к·а core-built tyre casing
дорно·держ·а́тел·ь, -я, *(m)* core holder (tyre-building); (met) spider (of extruding machine); mandrel support bar (tube piercing)
до́р·н·ый, -ая, -ое, *(a)* split, cracked
доро́г·а, -и, *(f)* road, way; journey
в доро́г·е en route, on the way
да·ва́ть доро́г·у make way for
~, желе́з·н·ая railway, railroad
~, кана́т·н·ая aerial ropeway/cableway
~, на·сы́п·н·а́я causeway
~, шоссе́·йн·ая highway, main road
до́рог·о *(adv)* dear, expensive
дорог·ови́зн·а, -ы, *(f)* expensiveness, high prices
дорог·о́й, -а́я, -о́е, *(a)* *(s f)* **до́рог·, дорог·а́, до́рог·о** expensive, costly, dear; dear (affectionate and formal address); *(adv)* on the way
дорого·сто·я́щ·ий, -ая, -ее, *(a)* expensive
до·ро́·ет *(fut 3rd sing)* of **до·ры́·ть**
дорож·а́ть, -ют, *(imp)* become more expensive, rise in price
доро́ж·е *(comp)* dearer, more expensive
дорож·и́ть, -а́т, *(imp)* (+ *instr*) value; take care of; **-ся** overcharge, ask too high a price
доро́ж·к·а, -и, *(g pl)* **-ж·ек·,** *(f)* *(dim)* of **доро́г·а;** path, way, walk, track, strip, runway; stair carpet
~, бег·ов·а́я race (of bearing); wheel path (of tracked vehicle); tread (of tyre)
~, вз·лёт·но-по·са́д·очн·ая (air) landing strip
~, вихр·ев·а́я (dynam) vortex street
~, звук·ов·а́я (cinema) sound track
~, кил·ев·а́я (shipb) keel blocks
~, ко́д·ов·ая information track (magnetic drum)
~, ма́ркер·н·ая index/marker track (magnetic drum)
~, рул·ёжн·ая (air) taxi-way
~, синхрониз·и́рующ·ая timing track (of magnetic drum)
~, с·пуск·ов·а́я (shipb) groundway
доро́ж·ник·, -а, *(m)* road-builder; highways engineer; box plane (carpenter's tool)

дорожно·стро́·и́тельн·ый, -ая, -ое,
(*a*) roadmaking, roadbuilding

доро́ж·н·ый, -ая, -ое, (*a*) of **доро́-
г·а,** and **доро́ж·к·а;** travelling (for
travel)

до·ро́·ют (*fut 3rd pl*) of **до·ры́·ть**

До́рр·а, классифика́тор· (*m*) (min)
Dorr classifier

дорс- (*component*) (*see also*) **дорз-** dors-
(back, dorsal)

дорс·а́льн·ый, -ая, -ое, (*a*) (zool)
dorsal, tergal

дорсо·вентр·а́льн·ый, -ая, -ое, (*a*)
(bot) dorsoventral

дорсо·спин·а́льн·ый, -ая, -ое, (*a*)
dorsospinal

до·руб·а́·ть, -ют, (*imp*) finish (cutting), cut (to the end); cut to/as far as

до·руб·и́ть, -я́т, (*perf*) *see* **до·руб·а́·ть**

до·ру́б·к·а, -и, (*g pl*) **-б·ок·,** (*f*) (*v n*),
see **до·руб·а́·ть**

до·ру́д·н·ый, -ая, -ое, (*a*) (geol) pre-
mineral, pre-ore

~ с·бро́с· (*m*) (geol) pre-mineral fault

до·рыв·а́·ть, -ют, (*imp*) finish (digging), dig (completely); dig to, dig
as far as; pick/pluck all; tear away all,
tear away to the end

до·ры́·ть (*fut 3rd sing, pl*) **до·ро́·ет,
до·ро́·ют,** (*perf*); dig (completely),
finish (digging); dig to, dig as far as

доса́д·а, -ы, (*f*) annoyance, vexation;
disappointment

досад·и́ть, -я́т, (*perf*) *see* **досажд·а́·ть;
до·сад·и́ть , -я́т,** (*perf*) *see* **до·са́-
ж·ива·ть**

доса́д·н·о (*predic, adj*) it is annoying,
it is a pity

доса́д·н·ый, -ая, -ое, (*a*) annoying,
vexatious, displeasing

досажд·а́·ть, -ют, (*imp*) annoy, irritate, vex

до·са́ж·ива·ть, -ют, (*imp*) finish
(planting), plant (completely); plant
more

до·са́л·ива·ть, -ют, (*imp*) add more
salt; finish salting down

до·свет·ов·о́й, -а́я, -о́е, (*a*) below
the speed of light, slower than light

до·синхро́н·н·ый, -ая, -ое, (*a*)
hyposynchronous

доск·а́, -и́, (*acc*) до́ск·у, (*nom pl*)
до́ск·и, (*g pl*) досо́к·, (*d pl*) дос-
к·а́м, (*f*) board, plank; plate; cover
(of book)

~, волоч·и́льн·ая (met) multi-hole
wire-drawing die

доск·а́

~, вы·ве́р·очн·ая straight-edge, pattern

~, ёл·ов·ая deal, fir wood

~, за·спи́н·н·ая (print) backboard

~, кла́сс·н·ая blackboard

~, ма́л·очн·ая bevel board

~, на·сти́л·очн·ая chess (of bridge)

~, планше́т·н·ая plotting table

~, под·моде́ль·н·ая (met cast) odd-
side, turnover board

~, пол·ев·а́я slade (of plough); grass-
board, landside

~, при·бо́р·н·ая instrument panel

~ слан·и (*pl*) (naut) bottom boards
(of a boat)

~, сосн·о́в·ая deal, pine wood

~, с·тряс·н·ая tossing board

~, тра́нц·ев·ая (shipb) transom

~, треп·а́льн·ая (text) scutching board

~, тру́б·н·ая tube sheet (of boiler)

~, у·кло́н·н·ая (shipb) declivity board

~, фунда́мент·н·ая foundation/seat-
ing plate

~, чёрн·ая black list

~, шпу́нт·ов·ая match-board

~, штéпсель·н·ая (elec) plug board,
switchboard; problem board (of computer)

до·с·каз·а́ть (*fut 3rd sing, pl*) **до·с·ка́-
ж·ет, до·с·ка́ж·ут,** (*perf*); *see* **до·-
с·ка́з·ыва·ть**

до·с·ка́з·ыва·ть, -ют, (*imp*) finish
telling/relating; tell/relate as far as/
to/up to

доск·о́в·ый, -ая, -ое, (*a*) of **доск·а́**

~ фут· (*m*) board-foot (timber)

доскона́ль·н·ый, -ая, -ое, (*a*)
thorough, detailed

доско·по·гру́з·очн·ая маши́н·а (*f*)
plant stacker

доско·у·кла́д·очн·ая маши́н·а (*f*)
timber stacker

до́·сл·анн·ый, -ая, -ое, (*past part
pass*) *see* **до·сыл·а́·ть**

до·сл·а́ть (*fut 3rd sing, pl*) **до·шл·ёт,
до·шл·ю́т,** (*perf*); *see* **до·сыл·а́·ть**

до·сле́д·овать, -ют, (*imp and perf*)
re-examine; complete the examination of

до·сло́в·н·о (*adv*) word-for-word, literally, verbatim

до·сло́в·н·ый, -ая, -ое, (*a*) literal,
verbatim

до·слу́ж·ива·ть, -ют, (*imp*) work/
serve till/for; **-ся** rise to (in rank);
qualify for (pension etc.)

до·служ·и́ть, -а́т, (*perf*) *see* до·слу́·ж·ива·ть

до·слу́ш·а·ть, -ют, (*perf*) *see* до··слу́ш·ива·ть

до·слу́ш·ива·ть, -ют, (*imp*) listen to the end, hear out; listen till/as far as

до·сма́тр·ива·ть, -ют, (*imp*) look/ watch till/as far as/to the end; inspect, examine

до·смо́тр·, -а, (*m*) inspection, examination

до·смо́тр·енн·ый, -ая, -ое, (*past part pass*) *see* до·сма́тр·ива·ть

до·смотр·е́ть, -я́т, (*perf*) *see* до··сма́тр·ива·ть

до·смо́тр·ов·ый, -ая, -ое, (*a*) examination, survey

~ офице́р· (*m*) (naut) boarding officer (customs)

~ ро́с·пис·ь (*f*) customs examination list

~ с·вид·е́тельств·о (*n*) customs survey report

~ су́д·н·о (*n*) examination vessel (customs)

до·смо́тр·щик·, -а, (*m*) customs officer/ surveyor, preventive officer

до·сол·и́ть, -я́т, (*perf*) finish salting down; add more salt

до·со́х·н·уть, -ут, (*perf*) dry up

до·спе·ва́ни·е, -я, (*n*) ripening, maturing

до·с·пе́н·ивани·е, -я, (*n*) final foaming

до·средне·ордови́к·ск·ий, -ая, -ое, (*a*) (geol) premedial ordivician

до·сро́ч·н·ый, -ая, -ое, (*a*) premature, ahead of schedule, early, before due

до ста·ва́ть, -ю́т, (*imp*) get, obtain; reach, touch; be enough, suffice; -ся fall to one's lot

~ дн·о́ (naut) take soundings

до·ста́·в·ить, -ят, (*perf*) *see* до·став·л·я́·ть

до·ста́в·к·а, -и, (*g pl*) -в·ок·, (*f*) delivery, supply, conveyance

до·ставл·я́·ть, -ют, (*imp*) supply, deliver, convey; give, cause, afford

до·ста́т·ок·, -т·к·а, (*m*) sufficiency; prosperity, comfort (of living standard)

до·ста́т·очн·о (*adv*) sufficiently, quite

до·ста́т·очн·ый, -ая, -ое, (*a*) sufficient, enough, adequate; prosperous, well-to-do

до·ста́·ть (*fut 3rd sing, pl*) до·ста́·н·ет, до·ста́·н·ут; *see* до·ста·ва́ть

до·стиг·а́емост·ь, -и, (*f*) accessibility, practicability

до·стиг·а́ем·ый, -ая, -ое, (*pres part pass*) of до·стиг·а́·ть; accessible, within reach, practicable

до·стиг·а́·ть, -ют, (*imp*) (+ *gen*) achieve, attain, reach

до·сти́г·нут·ый, -ая, -ое, (*past part pass*) *see* до·стиг·а́·ть

до·сти́г·нуть, -нут, (*past masc sing*) до·сти́г· or до·сти́г·нул (*perf* + *gen*); *see* до·стиг·а́·ть

до·стиж·е́ни·е, -я, (*n*) achievement, attainment

до·стиж·и́мост·ь, -и, (*f*) accessibility, practicability

до·стиж·и́м·ый, -ая, -ое, (*a*) attainable, accessible, practicable, within reach

до·сти́ч·ь (*fut 3rd sing, pl*) до·сти́г·-нет, до·сти́г·нут, (*past masc sing*) до·сти́г·, (*perf* + *gen*); *see* до·стиг·а́·ть

досто·ве́рно (*adv*) for certain/sure

досто·ве́р·ност·ь, -и, (*f*) reliability, authenticity

    гран·и́ц·а досто·ве́р·ност·и (*f*) confidence limit

    коэффицие́нт· досто·ве́р·ност·и (*m*) (math) confidence coefficient

досто·ве́р·н·ый, -ая, -ое, (*a*) reliable, authentic, certain, positive

~ за·па́с·ы (*pl*) (min) proven/positive resources

досто́ин·ств·о, -а, (*n*) dignity; merit, virtue

~, ма́л·ое (fin) small denomination

досто́йн·о (*adv*) suitably, properly, befittingly

досто́йн·ый, -ая, -ое, (*a*) deserving, worthy, worth; merited, deserved

досто·при·меч·а́тельност·ь, -и, (*f*) notability, remarkability; tourist attraction, sight(s)

досто·при·меч·а́тельн·ый, -ая, -ое, (*a*) remarkable, notable

до·сто·я́ни·е, -я, (*n*) property

до·стра́·ива·ть, -ют, (*imp*) finish (building); (shipb) fit out

до·стро́·енн·ый, -ая, -ое, (*a*) (*past part pass*) *see* до·стра́·ива·ть; finished (of buildings); fitted out (of ships); filled (of energy levels)

до·стро́·ечн·ый, -ая, -ое, (*a*) of до··стро́й·к·а

~ цех· (*m*) (shipb) fitting-out shop

до·стро́·ить, -ят, (*perf*) *see* **до·стра́·-ива·ть**

до·стро́й·к·а, -и, (*g pl*) **-о́·ек·,** (*f*) final building work, completion; (phys) filling (a level etc.); (shipb) fitting-out

до́·сту́п·, -а, (*m*) access, entrance, approach, inlet

до·сту́п·н·о (*adv*) easily, readily, simply

до·сту́п·ност·ь, -и, (*f*) accessibility, approachability, availability; intelligibility, comprehensibility

~, хим·и́ческ·ая (agrochem) chemical availability

до·сту́п·н·ый, -ая, -ое, (*a*) accessible, approachable, usable, available; popular; understandable, simple, comprehensible

досу́г·, -а, (*m*) leisure

до́·сух·а (*adv*) dry, till dry, to dryness

до·су́ш·ива·ть, -ют, (*imp*) dry out; air (laundry); **-ся** dry up

до·суш·и́ть, -а́т, (*perf*) *see* **до·су́-ш·ива·ть**

до·с·чит·а́·ть, -ют, (*perf*) *see* **до·-с·чи́т·ыва·ть**

до·с·чи́т·ыва·ть, -ют, (*imp*) finish counting, count to

до·сыл·а́тел·ь, -я, (*m*) rammer; ejector (machine tool)

до·сыл·а́·ть, -ют, (*imp*) send the rest, send on; ram home

до·сы́п·ать, -пл·ют, (*perf*), **до·сы-п·а́·ть, -ют,** (*imp*), fill (with grains); pour to/up to, add; (*imp*) of **до·-сп·а́ть** sleep till/through

до·сы́п·к·а, -и, (*f*) (*v n*), *see* **до·сы́-п·ать**; addition (of grains)

до́·сыт·а, (*adv*) till satisfied/satiated/replete

до·сых·а́·ть, -ют, (*imp*) dry up

до·сю́да (*adv*) as far as this, up to here

до·сяг·а́емост·ь, -и, (*f*) reach, range; attainability

до·сяг·а́ем·ый, -ая, -ое, (*pres part pass*) of **до·сяг·а́·ть**; attainable, achievable, accessible, within reach

до·сяг·а́·ть, -ют, (*imp*) attain, achieve, reach

до·сяг·н·у́ть, -у́т, (*perf*) *see* **до·ся-г·а́·ть**

до·та́ск·ива·ть, -ют, (*imp*) drag/pull to/as far as; carry to/as far as

дота́ци·я, -и, (*f*) grant, subsidy

до·та́щ·енн·ый, -ая, -ое, (*past part pass*) *see* **до·та́ск·ива·ть**

до·тащ·и́ть, -а́т, (*perf*) *see* **до·та́с-к·ива·ть**

до·тек·а́·ть, -ют, (*imp*) flow up to, flow as far as

до·тектон·и́ческ·ий, -ая, -ое, (*a*) (geol) pretectonic

до·те́ч·ь (*fut 3rd sing, pl*) **до·теч·ёт, до·тек·у́т,** (*past masc sing*) **до·тёк·** (*perf*); *see* **до·тек·а́·ть**

до·тл·а́ (*adv*) utterly, completely, out

дото́шн·ый, -ая, -ое, (*a*) meticulous

до·тра́г·ива·ться, -ются (*imp*) touch

до·тро́·н·уться, -утся, (*perf*) *see* **до·-тра́г·ива·ться**

до·ту́да (*adv*) up to there, that far

до·тя́г·ива·ть, -ют, (*imp*) draw/drag/haul to/up to/as far as; hold/last out; **-ся** reach

до·тя·н·у́ть, -у́т, (*perf*) *see* **до·тя-г·ива·ть**

до·у·по·ря́д·очени·е, -я, (*n*) (phys) partial ordering

До́усон·а, газ· (*m*) Dowson gas

до·у́ч·ива·ть, -ют, (*imp*) complete a study, finish learning/studying; learn/study up to/till; finish teaching; teach until/up to; **-ся** learn/study up to/until; perfect one's knowledge, complete one's education

~ до серед·и́н·ы learn by heart

до·уч·и́ть, -а́т, (*perf*) *see* **до·у́ч·и-ва·ть**

дох· (*past masc sing*) of **до́х·нуть**

дох·а́, -и́, (*nom pl*) **до́х·и,** (*f*) fur coat

до́х·л·ый, -ая, -ое, (*a*) dead (of animal)

до́х·нуть, -нут, (*past masc sing*) **дох·,** (*perf*) breathe (in and out); sigh; die (of animals)

до·хо́д·, -а, (*m*) income, revenue, receipts

~, вал·ов·о́й gross income

~, не·про·из·во́д·ственн·ый unearned income

~, не·труд·ов·о́й unearned income

~, ре́нт·н·ый investment income

~, чи́ст·ый clear profit(s); net income

до·ход·и́ть, -я́т, (*imp*) reach, come to; amount to

до·хо́д·ност·ь, -и, (*f*) profitability, earning capacity

до·хо́д·н·ый, -ая, -ое, (*a*) (*s f*) **-д·ен·, -д·н·а,** income-producing, revenue-producing; profitable, paying

до·хо́д·чив·ый, -ая, -ое, (*a*) intelligible, simple

до·хож·у́ (*pres 1st sing*) of **до·ход·и́ть**

до·цвес·ти́ (*fut 3rd sing, pl*) **до·цвет·ёт, до·цвет·у́т,** (*past masc sing*) **до·цвёл,** (*perf*); *see* **до·цвет·а́·ть**

до·цвет·а́·ть, -ют, (*imp*) cease flowering; fade, wither

доцент·, -а, (*m*) reader (in university)

доцент·у́р·а, -ы, (*f*) readership

до́чер·и (*g, d, p sing, nom pl*) of **доч·ь**

дочёр·н·ий, -яя, -ее, (*a*) daughter; (nucl) daughter, product

~ акти́в·ность (*f*) (nucl) daughter activity

~ вещ·еств·о́ (*n*) (nucl) daughter, decay product

~ пред·при·я́т·и·е (*n*) subsidiary, branch (of business)

до·черт·и́ть, -я́т, (*perf*) *see* **до·чёр·ч·ива·ть**

до·чёрч·ива·ть, -ют, (*imp*) finish (drawing); draw to/as far as

до́·чист·а (*adv*) till perfectly clean; till highly finished; completely

до·чи́ст·ить, -ят, (*perf*) *see* **до·чи·щ·а́·ть**

до·чит·а́·ть, -ют, (*perf*) *see* **до·чи́т·ыва·ть**

до·чи́т·ыва·ть, -ют, (*imp*) finish (reading); read till/up to/as far as

до·чищ·а́·ть, -ют, (*imp*) finish cleaning; clean till/up to/as far as

до́ч·к·а, -и, (*g pl*) **-ч·ек·,** (*f*) (*dim*) of **доч·ь**

доч·ь (*g, d, p*) **до́чер·и,** (*i*) **до́чер·ью,** (*nom pl*) **до́чер·и,** (*g pl*) **дочер·е́й·,** (*f*) daughter

до·шёд·ш·ий, -ая, -ее, (*past part act*) of **до·йти́**

до·шёл (*past masc sing*) of **до·йти́**

до·шив·а́·ть, -ют, (*imp*) finish (sewing); sew to/as far as

до·ши́·т·ый, -ая, -ое, (*past part pass*) **до·шив·а́·ть;** finished (of sewn articles), sewn

до·ши́·ть (*fut 3rd sing, pl*) **до·шь·ёт, до·шь·ю́т,** (*perf*); *see* **до·шив·а́·ть**

до·шко́ль·н·ый, -ая, -ое, (*a*) pre-school, nursery

до·шл·ёт (*fut 3rd sing*) of **до·сл·а́ть**

до́·шл·ый, -ая, -ое, (*a*) experienced, skilled

до·шл·ю́т (*fut 3rd pl*) of **до·сл·а́ть**

дошни́к·, -а́, (*m*) (food) sauerkraut vat

до·шь·ю́т (*fut 3rd pl*) of **до·ши́·ть**

дощ·а́ник·, -а, (*m*) flat-bottom boat

дощ·а́т·ый, -ая, -ое, (*a*) plank, made of planks, board, wooden

~ кры·л·о́ (*n*) (naut) leeboard

дощ·а́т·ый, -ая,-ое

~ пере·горо́д·к·а (*f*) wooden partition

дощ·е́чк·а, -и, (*g pl*) **-чек·,** (*f*) (*dim*) of **доск·а́;** (met) plate; tally-plate, name-plate

~, парке́т·н·ая (*f*) parquet block

до·юсти́р·овать, -уют, (*perf*) (instr) reset

до·эвтект·и́ческ·ий, -ая, -ое, (*a*) (met) hypoeutectic

до·эвтекто́ид·н·ый, -ая, -ое, (*a*) (met) hypoeutectoid

до·я́р·, -а, (*m*) milker

до·я́рк·а, -и, (*g pl*) **-рок·,** (*f*) milker, milkmaid

др. (*abbr*) of **друг·и́е** (*pl*) others, the others/rest

драви́т·, -а, (*m*) (min) dravite

дра́г·а, -и, (*f*) (*see also* **зем·с·на·ря́д·**) drag, dredge, dredger (for mining gold etc.)

~, цеп·н·ы́е (*pl*) drag chains

~, щип·цов·а́я clamshell dredger/dredge

дра́г·ер, -а, (*m*) *see* **дра́г·а**

драг·и́ровани·е, -я, (*n*) (min) dredging, dragging

драгла́йн·, -а, (*m*) (civ eng) dragline, dragline excavator

драго·це́н·н·ый, -ая, -ое, (*a*) precious; oriental (of minerals)

~ ка́мен·ь (*m*) (*see also* **ка́мен·ь**) gem, gem stone

дра·ёк·, -а, (*m*) dowel, fid

драже́ (*n indecl*) (pharm) dragée; drop (small round sweet)

драж·и́ровани·е, -я, (*n*) coating

дра́ж·н·ый, -ая, -ое, (*a*) dredging, dragging

~ полиго́н· (*m*) (min) placer area (for working by dredger)

~ рабо́т·ы (*pl*) (min) dredging

дразн·е́ни·е, -я, (*n*) (*v n*) (*see* **дразн·и́ть**); (met) poling; stirring

дразн·и́ть, -я́т, (*imp*) tease; excite

драймо·фи́т·, -а, (*m*) (bot) drymophyte

драко́н·, -а, (*m*) dragon

драко́н·ов·а кров·ь (*f*) (chem) dragon's blood

драмат·и́ческ·ий, -ая, -ое, (*a*) dramatic

драмат·у́рг·, -а, (*m*) playwright, dramatist

дра́н·иц·а, -ы, (*f*) *see* **дра́н·к·а**

дра́н·к·а, -и, (*g pl*) **-н·ок·,** (*f*) (build) shingle, shide, lath, lathine

~, кро́вель·н·ая roofing shingle/shide

дра́н·к·а
~, **штукату́р·н·ая** lath
дра́н·очн·ый, -ая, -ое, (*a*) of **дра́н··к·а**
дра́·н·ый, -ая, -ое, (*a*) torn, ragged; break (flour milling)
дра́н·ь, -и, (*f*) *see* **дра́н·к·а**
драп·, -а, (*m*) thick woollen cloth
драп·ир·ова́ть, -у́ют, (*imp*) drape
дра́тв·а, -ы, (*f*) wax-end (for leather sewing)
др·ать, (*pres 3rd s ng, pl*) **дер·ёт, де-р·у́т,** (*imp*) tear, strip off; flog, thrash; irritate; **-ся** fight
дра́хм·а, -ы, (*f*) drachm, dram
~, **аптека́р·ск·ая** apothecaries, drachm
дра́ч·ев·ый, -ая, -ое, (*a*) bastard (of tools etc.)
~ **на·пи́ль·ник·** (*m*) bastard file
~ **на·се́ч·к·а** (*f*) bastard cut (of tools)
дребезж·а́ни·е, -я, (*n*) (*v n*) of **дребез-ж·а́ть**
дребезж·а́ть, -а́т, (*imp*) rattle, clink, jingle, jar
древес·и́н·а, -ы, (*f*) wood, timber, wood pulp; (bot) xylem
~, **бала́нс·ов·ая** pulp wood
~, **весе́н·н·яя** early wood, spring wood
~, **втор·и́чн·ая** (bot) secondary xylem/wood
~, **дел·ов·а́я** commercial timbers
~, **де́льта-** laminated wood
~, **дров·ян·а́я** fuelwood
~, **кольце·по́р·ист·ая** ring-porous wood
~, **ле́т·н·яя** summer wood
~, **матёр·ая** close-grained wood
~, **осе́н·н·яя** late wood
~, **перв·и́чн·ая** (bot) primary xylem/wood
~, **рассеяно·по́р·ов·ая** diffuse porous wood
~, **ядр·о́в·ая** heart-wood
древес·и́нник·, -а, (*m*) (ent) pinhole borer, *Xyloterus*
древес·и́нн·ый, -ая, -ое, (*a*) *see* **древе́с·н·ый**
древесино·вед·е́ни·е, -я, (*n*) wood technology
древе́с·н·о- (*component*) wood, xylo-
древесно·волокн·и́ст·ый, -ая, -ое, (*a*) wood-fibre, xylem
древесно·кольц·ев·о́й, -ая, -ое, (*a*) tree-ring
древесно·ли́ст·н·ый, -ая, -ое, (*a*) (bot) woody-leaved, xylophyllous
древесно·ма́сс·н·ый, -ая, -ое, (*a*) wood-pulp, pulp

древесно·пло́д·н·ый, -ая, -ое, (*a*) (bot) xylocarpous
древесно·сло·и́ст·ый, -ая, -ое, (*a*) laminated, laminated-wood
~ **пла́стик·** (*m*) laminated wood
древесно·смол·ян·о́й, -а́я, -о́е, (*a*) wood-resin
древесно·стру́ж·ев·ая плит·а́ (*f*) wood-particle board
древесно·у́голь·н·ый, -ая, -ое, (*a*) charcoal
древесно·у́ксус·н·ый, -ая, -ое, (*a*) pyroligneous
~ **кисл·от·а́** (*f*) (chem) pyroligneous acid
~ **порош·о́к·** (*m*) pyrolignite of lime
древесно·уксусно·ки́сл·ый, -ая, -ое, (*a*) pyrolignite of
древе́с·н·ый, -ая, -ое, (*a*) wood, woody, ligneous; arboreal
~ **воск·** (*m*) (hortic) grafting wax
~ **ва́т·а** (*f*) wood wool, wood shavings
~ **карто́н·** (*m*) fibreboard
~ **ма́сл·о** (*n*) tung oil, China wood oil
~ **ма́сс·а** (*f*) wood pulp, pulpwood
~ **мук·а́** (*f*) wood flour
~ **на·сажд·е́ни·е** (*n*) plantation, arbo-retum
~ **смол·а́** (*f*) wood tar/resin
~ **спирт·** (*m*) wood alcohol, methyl alcohol
~ **стру́ж·к·а** (*f*) wood wool/shavings
~ **у́ксус·** (*m*) wood vinegar, pyro-ligneous acid
дре́вк·о, -а, (*nom pl*) **-и,** (*g pl*) **-ов,** (*n*) pole, staff, shaft
древн·е- (*component*) palaeo-, paleo-, old, ancient
древне·вулкан·и́ческ·ий, -ая, -ое, (*a*) (seismol) paleovolcanic
древне·гре́ч·еск·ий, -ая, -ое, (*a*) ancient Greek
древне·кры́·л·ые (*pl decl as adj*) (pal) palaeopters, *Palaeoptera*
древне·ру́с·ск·ий, -ая, -ое, (*a*) old Russian
дре́вн·ий, лл, -ее, (*a*) ancient
~ **голоце́н·** (*m*) (geol) Palaeoholocene
древо·ва́л·, -а, (*m*) tree grubbing tool
древо·ви́д·н·ый, -ая, -ое, (*a*) arbor-escent, dendroid, dendritic
древо·во́д·ств·о, -а, (*n*) arboriculture
древо·гры́з·, -а, (*m*) false powder-post beetle, (*pl*) *Lyctidae*
древо·грыз·у́щ·ий, -ая, -ее, (*a*) (zool) woodcutting

древо·на·сажд·éни·е, -я,   (*n*) tree planting, afforestation

древо·стó·й, -я, (*m*) standing timber, stand

древо·тóч·ец·, -ч·ц·а,   (*m*) carpenter moth, (*pl*) *Cossidae*

~, мор·ск·óй   teredo, shipworm

дрезúн·а, -ы, (*f*) (rail) trolley, hand car

дрейкáнтер·, -а, (*m*) (geol) dreikanter

дрéйф·, -а,   (*m*) drift; (autom) drift, offset, droop; (naut, air) drift, leeway

~ нул·я́ (instr) zero drift

~ температýр·ы (phys) temperature drift

дрейф·овáть, -ýют, (*imp*) drift; drag (of vessel at anchor)

дрéйф·ов·ый, -ая, -ое, (*a*) drift

дрейф·ýющ·ий, -ая, -ее, (*pres part act*) of **дрейф·овáть**; drifting, drift

дрéк·, -а, (*m*) boat's anchor

дректóв·, -а, (*m*) (naut) boat's cable

дрéл·ь, -и, (*f*) drill

дрем·áть, -л·ют, (*imp*) doze, slumber

дрéмл·ющ·ий, -ая, -ее, (*pres part act*) of **дрем·áть**; somnolent, slumbering; inactive

дрем·óт·а, -ы, (*f*) somnolence, drowsiness

дрем·óтн·ый, -ая, -ое, (*a*) somnolent, drowsy

дрем·ýч·ий, -ая, -ее, (*a*) dense, thick (of forests etc.)

дрéн·, -а, (*m*) *see* **дрéн·а**

дрéн·а, -ы, (*f*) drain, land drain, field drain

~, гончáр·н·ая   tile drain

~, крот·óв·ая   mole drain

~, фащúн·н·ая   faggot drain, fascine ditch

дрен·áж·, -а, (*i*) **-ем,** (*m*) drainage, land drainage, draining; (med) drain

~, за·кры́·т·ый   under-drainage

~, из·бы́т·очн·ый   overdrainage

~, крот·óв·ый   mole draining

~, от·кры́·т·ая   open-cut drain

~, щел·ев·óй   fissure/crevice drainage

~, электро·магнúт·н·ый (elec) electrical drainage, cathodic protection

дренаж·úр·овать, -уют, (*imp and perf*) drain

дренáж·н·ый, -ая, -ое, (*a*) drain, drainage

~ труб·á (*f*) (civ eng) drain tile

дрен·úрованн·ый, -ая, -ое, (*past part pass*) of **дрен·úр·овать**; drained

дрен·úр·овать, -уют, (*imp and perf*) drain

дрéнчер·, -а, (*m*) drencher (fire fighting)

дресв·á, -ы,   (*f*) (geol) rotten stone, druss, gruss

дресв·я́н·ый, -ая, -ое,   (*a*) rotten stone, druss, gruss

дресс·ир·овáть, -ýют,   (*imp*) train, school (animals)

дресс·ирóвк·а, -и, (*g pl*) **-вок·,** (*f*) training (of animals); (met) skin-pass rolling, temper rolling

дресс·ирóвочн·ый, -ая, -ое,   (*a*) of **дресс·ирóвк·а**

дрéшер·, -а, (*m*) thrasher (papermaking)

дримо·фúт·, -а, (*m*) (bot) drymophyte

дрио·питéк·, -а, (*m*) (pal) dryopithecus

дрипп·, -а, (*m*) steam trap

дрúфтер·, -а, (*m*) (fish) drifter; (min) drifter (rock-drill)

дробе·мёт·, -а, (*m*) (met) shot peener/blaster

дробе·мёт·н·ый, -ая, -óе,   (*a*) shot-blasting, peening

~ об·рабóт·к·а (*f*) (met) peening, shot peening, shot blasting

дробе·стрý·йн·ый, -ая, -ое,   (*a*) shot-blasting, peening

~ об·рабóт·к·а   (*f*) peening, shot-peening/blasting

~ у·станóв·к·а (*f*) shot-peening equipment

дрóб·и   (*g, d, p sing, nom, acc pl*) of **дроб·ь**

дроб·úлк·а, -и, (*g pl*) **-лок·,** (*f*) crusher, breaker, grinder, crushing mill

~, вал·кóв·ая   roll breaker, crushing roll

~, вальц·óв·ая   roll breaker, crushing roll

~, дúск·ов·ая   disc crusher

~, зерн·ов·áя (agr) grain grinder

~, зуб·чáт·ая   (min) toothed-roll crusher

~, кóнус·н·ая (min) gyratory crusher, gyratory cone crusher

~, коротко·кóнус·н·ая   (min) short-head cone crusher

~ крýп·н·ого дробл·éни·я   boulder crusher

~, кулач·кóв·ая   (agr) hammer mill

~, молот·кóв·ая   (min) hammer crusher/mill

~, редукциóн·н·ая   reduction crusher

~, щек·óв·ая (min) jaw crusher, jaw breaker

дроб·úльн·ый, -ая, -ое, (*a*) crushing, grinding

дроб·и́н·а, -ы, (*f*) ground material (product of grinding), grains; pellet; (agr) mash

~, пив·н·а́я brewing waste, spent grains

дроб·и́нк·а, -и, (*g pl*) **-нок·,** (*f*) (*dim*) of **дроб·и́н·а**

дроб·и́ть, -я́т, (*imp*) grind, crush, crush/grind/break up; **-ся** (*passive*) crumble

дробл·е́ни·е, -я, (*n*) crushing, breaking, grinding, crushing/breaking/dividing up; fragmentation; (zool) segmentation

~ до́з·ы (nucl) dose fragmentation

~ част·от·ы́ (rad) frequency splitting

дробл·ён·ый, -ая, -ое, (*a*) crushed, broken, ground, crumbled, fragmented

дроб·н·о- (*component*) (math) fractional; (phys) sub-; (meteor) fracto-

дробно·а́том·н·ый, -ая, -ое, (*a*) (nucl) subatomic

дробно·дожд·ев·о́й, -а́я, -о́е, (*a*) (meteor) fractonimbus

дробно·квадра́т·н·ый, -ая, -ое, (*a*) (math) quadratic fractional

дробно·лин·е́йн·ый, -ая, -ое, (*a*) (math) linear-fractional

дроб·н·ый, -ая, -ое, (*a*) fractional

~ вы́·ход· (*m*) (chem) fractional yield

~ пере·го́н·к·а (*f*) (chem) fractional distillation

~ числ·о́ (*n*) (math) fraction

дроб·ови́к·, -а́, (*m*) (gunn) fowling piece

дроб·ов·о́й, -а́я, -о́е, (*a*) shot

~ шум· (*m*) (elec) shot noise, Schottky/Schrot noise/effect

~ эффе́кт· (*m*) (elec) noise, Schottky/Schrot noise/effect

дробс·, -а, (*m*) indicator drop (telephone switchboard)

дроб·ь, -и, (*nom pl*) **дроб·и,** (*g pl*) **-е́й,** (*f*) (math) fraction; shot, pellets; (naut) "stand-to"

~, бес·кон·е́чн·ая не·пре·ры́в·н·ая (math) non-terminating continued fraction

~, десят·и́чн·ый decimal fraction

~, ка́мен·н·ая (geol) detritus

~, кон·е́чн·ая terminating fraction

~, кру́п·н·ая buck-shot, grape-shot

~, ме́лк·ая duck-shot

~, не·пра́в·ильн·ая improper fraction

~, не·со·крат·и́м·ая unit fraction, non-reducible fraction

~, период·и́ческ·ая recurring continued fraction

дроб·ь

~, под·ход·я́щ·ая convergent of a continuous fraction

~, пра́в·ильн·ая proper fraction

~, прост·а́я vulgar fraction

~, прост·е́йш·ие (*pl*) partial fractions

~, цеп·н·ы́е (*pl*) continued fractions

черт·а́ в дроб·ь (*f*) fraction stroke

~, элемента́р·н·ые (*pl*) partial fractions

дроб·я́нк·а, -и, (*g pl*) **-нок·,** (*f*) (bot) *Schizophyta*

дроб·я́щ·ий, -ая, -ее, (*pres part act*) of **дроб·и́ть**

~ эффе́кт· (*m*) shattering effect

дров·а́ (*nom pl*), (*g pl*) **дров·,** (*d pl*) **-а́м,** firewood, wood

дрово·ко́ль·н·ый, стан·о́к· (*m*) wood chopping/cleaving machine

дров·ян·о́й, -а́я, -о́е, (*a*) firewood, wood

дро́г·и (*nom pl*), (*g pl*) **дрог·** dray, cart

дро́г·н·уть, -ут, (*imp*) shiver; (perf) quiver, shudder, tremble

дрож·а́ни·е, -я, (*n*) shivering, shaking, vibration, tremor; (air) buffeting

~ и́мпульс·ов (rad) pulse jitter

дрож·а́ть, -а́т, (*imp*) shake, tremble, shiver, quiver

дрож·а́щ·ий, -ая, -ее, (*pres part act*) of **дрож·а́ть**

дрожж·ев·о́й, -а́я, -о́е, (*a*) yeast

~ гу́щ·а (*f*) wine lees

~ нуклеон·ов·ая кисл·от·а́ (*f*) (biochem) ribonucleonic acid

дро́жж·и (*nom pl*), (*g pl*) **-е́й** yeast(s), *Saccharomycetaceae*

~, ви́н·н·ые wine yeast, *Saccharomyces ellipsoideus*

~, корм·ов·ы́е food/fodder yeast

~, пив·н·ы́е brewer's yeast, *Saccharomyces cerevisiae*

~, хлебо·пек·а́рн·ые baker's yeast, *Saccharomyces cereviciae*

~, хмел·ев·ы́е brewer's yeast

дро́жж·ь, -и, (*f*) trembling, shivering; tremor, quaver

дрозо·ме́тр·, -а, (*m*) (meteor) drosometer

дрозо·фи́л·а, -ы, (*f*) (ent) fruit fly, *Drosophila*

дрок·, -а, (*m*) (bot) broom, *Genista*

дромаде́р·, -а, (*m*) (zool) dromedery

дромге́д·, -а, (*m*) (naut) capstan drumhead

дрос·, -а, (*m*) (met) skimmings, dross

дро́сс·, -а, (*m*) (met) skimmings, dross

дроссел·и́ровани·е, -я, *(n)* (ICE) throttling; (elec) choking; (dynam) orificing

дроссел·и́р·овать, -у́ют, *(imp and perf)* (elec) choke; (ICE) throttle down

дроссел·и́рующ·ий, -ая, -ее, *(pres part act)* of **дроссел·и́р·овать;** throttling, choking, orificing

~ **игл·а́** *(f)* valve needle

дро́ссел·ь, -я, *(m)* (ICE etc.) throttle, choke; restrictor; (elec) choke, choke coil

~ **на·сыщ·е́ни·я** (autom) transductor, saturable-core choke, saturable-core reactor

~, с·ты́к·ов·о́й (elec rail) impedance bond

~, уда́р·н·ый (elec) ringing choke

~ **фи́льтр·а** (elec) smoothing choke

дро́ссель·н·ый, -ая, -ое, *(a)* throttle, choke

~ **за·сло́н·к·а** *(f)* (ICE) choke, strangler; baffle plate

~ **кла́пан·** *(m)* (ICE) throttle

~ **с·ты́к·** *(m)* (rail, elec) impedance bond

дро́сс·ов·ый, -ая, -ое, *(a)* (met) dross

дро́тик·, -а, *(m)* dart, javelin

~, лов·и́льн·ый (oil) rope spear

дру́г·, -а, *(nom pl)* **друз·ья́,** *(g pl)* **-зе́й,** *(m)* friend; *(s f)* of **друг·о́й**

~ **дру́г·а** each other, one another

~ **за дру́г·ом** one after another

~ **от дру́г·а** one from another, from one another

~ **про́тив дру́г·а** opposite each other, opposite one another

друг·о́й, -а́я, -о́е, *(a)* other, another, different; *(as noun)* another, *(pl)* others

в друг·о́м ме́ст·е *(adv)* elsewhere, somewhere else

ни тот ни друг·о́й neither, neither of them, neither the one nor the other

ни·кто́ друг·о́й no other, none other

тот и друг·о́й both, both of them, both the one and the other

Дру́де, у·равн·е́ни·е, -я, *(n)* (chem) Drude equation

дру́ж·а́щ·ий, -ая, -ее, *(pres part act)* of **дру́ж·и́ть**

дру́ж·б·а, -ы, *(f)* friendship

дру́ж·еск·ий, -ая, -ое, *(a)* friend's, friendly

дру́ж·ественн·ый, -ая, -ое, *(a)* *(s f)* **-вен·, -венн·а** friendly, amicable

дру́ж·и́ть, -а́т, *(imp)* be friends, be on friendly terms

дру́ж·н·о *(adv)* amicably, in a friendly manner; simultaneously, together

дру́ж·н·ый, -ая, -ое, *(a)* friendly, amicable; simultaneous, harmonious

~ **чи́сл·а** *(pl)* (math) amicable numbers

дру́з·а, -ы, *(f)* (cryst) druse; duster

друзо·ви́д·н·ый, -ая, -ое, *(a)* drusoid

дру́з·ов·ый, -ая, -ое, *(a)* (min) drusy

друз·ья́ *(nom pl)* of **друг·**

друмли́н·, -а, *(m)* (geol) drumlin

дры́г·а́ни·е, -я, *(n)* *(v n)* of **дры́г·а·ть**

дры́г·а·ть, -ют, *(imp)* jerk

дры́г·н·уть, -ут, *(perf)* jerk, give a jerk

дря́б·л·ый, -ая, -ое, *(a)* flabby, flaccid

дря́х·л·ый, -ая, -ое, *(a)* decrepit, senile

ДСП *(abbr)* of **древесно·сло·и́ст·ый пла́стик·** laminated wood

дуал·и́зм·, -а, *(m)* duality

~, волн·ов·о·част·и́чн·ый (nucl) wave-particle duality

дуал·исти́ческ·ий, -ая, -ое, *(a)* dualistic

дуа́ль·н·ый, -ая, -ое, *(a)* dual

дуа́нт·, -а, *(m)* (nucl) duant, dee, D-shaped electrode

ду́б·, -а, *(nom pl)* **-ы́,** *(g pl)* **-о́в,** *(m)* (bot) oak, oak-tree, *Quercus*

~, австр·и́йск·ий turkey oak, *Quercus cerris*

~, бархат·и́ст·ый black oak, *Quercus velutina*

~, бе́л·ый white oak, *Quercus alba*

~, боло́т·н·ый pin oak, *Quercus palustris*

~, вирги́н·ск·ий live oak, *Quercus virginiana*

~, груз·и́нск·ий Iberian oak, *Quercus iberica*

~, дву·цве́т·н·ый swamp white oak, *Quercus bicolor*

~, зи́м·н·ий durmast oak, *Quercus petraea*

~, иво·ли́ст·н·ый willow oak, *Quercus phellos*; red oak lumber

~, кашта́н·ов·ый chestnut oak, *Quercus prinus*

~, крупно·пло́д·н·ый bur oak, *Quercus macrocarpa*

ду́б

~, ла́вро·ли́ст·н·ый   laurel oak, *Quercus laurifolia*; red oak lumber

~, ле́т·н·ий   English oak, *Quercus robur, Quercus pedunculata*

~, пуш·и́ст·ый   pubescent oak, *Quercus pubescens*

~, серпо·ви́д·н·ый   southern red oak, *Quercus falcata*

~, сидяче·цве́т·н·ый   durmast oak, *Quercus sessiliflora*

~, ска́ль·н·ый   durmast oak, *Quercus petraea*

~, череш·ча́т·ый   English oak, *Quercus robur*

~, чёрн·ый   water oak, *Quercus nigra*; red oak lumber

~, шарла́х·ов·ый   scarlet oak, *Quercus coccinea*

дубильно·ки́сл·ый, -ая, -ое,   (*a*) (chem) tannate of

~ ка́льц·и·й   (*m*) calcium tannate

дуб·и́льн·ый, -ая, ое,   (*a*) tanning, tannic

~ вещ·еств·о́   (*n*) tannin

~ кисл·от·а́   (*f*) tannic acid

~ кор·а́   (*f*) tanbark

дуб·и́льн·я, -и,   (*g pl*) **-лен·,** (*f*) tannery

дуб·и́льщик·, -а,   (*m*) tanner

дуб·и́н·а, -ы,   (*f*) club, cudgel

дубино·обра́з·н·ый, -ая, -ое,   (*a*) clavate, calviform

дуб·и́тел·ь, -я,   (*m*) tanning agent; (phot) hardener

дуб·и́ть, -я́т,   (*imp*) tan

дубл·е́ни·е, -я,   (*n*) tanning; (phot) hardening, intensifying

~, квас·цо́в·ое   tawing (tanning process)

~, ла́й·ков·ое   tawing, alum tanning

~, хро́м·ов·ое   chrome tanning

дубл·ён·ый, -ая, -ое,   (*a*) tanned

дублёр·, -а,   (*m*) (met) doubler, doubling machine; (theat) understudy, dubbing actor

дубле́т·, -а,   (*m*) duplicate; doublet

дубле́т·н·ый, -ая, -ое,   (*a*) doublet

~ рас·сто·я́ни·е   (*n*) (spectr) doublet separation

дублик·а́т·, -а,   (*m*) duplicate, copy, replica

дублик·а́тор·, -а,   (*m*) copying attachment (of lathe); automatic spacing table, spacer (punching machine)

дубл·и́рованн·ый, -ая, -ое,   (*past part pass*) of **дубл·и́р·овать**

дубл·и́р·овать, -уют,   (*imp*) duplicate, make a copy, fold; (theat) understudy, double; dub (a film); repeat (readings from an instrument); double, ply up (rubber); (naut) pass a signal

дубл·и́р·о́вщик·, -а,   (*m*) (met) doubler

дубль·негати́в·, -а,   (*m*) (cinema) dupe negative

дубля́ж·, -а,   (*m*) (cinema) dubbing

дуб·о́в·ый, -ая, -ое,   (*a*) oak, quercoid

дуботзи́н·, -а,   (*m*) (chem) duboisine

дуб·я́щ·ий, -ая, -ее,   (*pres part act*) of **дуб·и́ть**; tanning

ду́·вш·ий, -ая, -ее,   (*pres part act*) of **ду·ть**

дуг·а́, -и́,   (*nom pl*) **ду́г·и,** (*g pl*) **дуг·** (*f*) arc, arch, arcus, bow; (mil) bulge, salient

~ больш·о́го кру́г·а   (nav) great circle

~ выс·о́к·ой интенси́в·ност·и   (elec) high-current density arc, high-intensity arc

~, высоко·температу́р·н·ая   с·жа́·т·ая   (weld) plasma jet

~ горизо́нт·а   arc of visibility

~, жа́бер·н·ая   (zool) branchial/gill arch

~ за·жиг·а́·ть дуг·у́   strike an arc

~ за·хва́т·а   (met roll) contact arc

~ за·цепл·е́ни·я   arc of action/contact (of gear wheels)

~, подо́·шв·енн·ая   (anat) plantar arch

~, по·ю́щ·ая   (elec) singing/duddle arc

~ с·жа́т·и·я   arc of compression

~, су́мер·ечн·ая   twilight circle

~, угло·ме́р·н·ая   (surv) bearing arc

~, у·пряж·н·а́я   shaft-bow (of cart)

~ э́ллипс·а   elliptic arc

дуг·ов·о́й, -а́я, -о́е,   (*a*) arc, arched

~ с·ва́р·к·а   (*f*) arc welding

дуго·обра́з·н·ый, -ая, -ое,   (*a*) arched, arching, arcuate

дуго·гас·и́тельн·ый, -ая, -ое,   (*a*) (elec) arc-suppression/suppressing

дуго·гас·я́щ·ая кату́ш·к·а   (*f*) arc suppression coil

дуго·сто́й·к·ий, -ая, -ое,   (*a*) (elec) non-arcing

ду́дк·а, -и,   (*g pl*) **-д·ок·,** (*f*) pipe, fife; (min) open pit; (naut) bo'sun's pipe

ду́·ет   (*pres 3rd sing*) of **ду·ть**

ду́ж·к·а, -и,   (*g pl*) **-ж·ек·,** (*f*) (*dim*) of **дуг·а́**; handle; (zool) arculus; wishbone дуктило·метр·, -а, (*m*) ductility testing apparatus (for bituminous etc., materials)

дуктиль·н·ый, -ая, -ое, (*a*) ductile

дул·о, -а, (*n*) muzzle (of gun); (*past neut sing*) of ду·ть

дуль·н·ый, -ая, -ое, (*a*) muzzle

~ скор·ост·ь (*f*) initial velocity (ballistics)

дуль·ц·е, -а, (*n*) mouth of cartridge case

дульцин·, -а, (*m*) (chem) dulcin, sucrol

дум·а·ть, -ют, (*imp*) think

думмис·, -а, (*m*) dummy/balance piston

думпкар·, -а, (*m*) (rail) dump car, tipper wagon

дунит·, -а, (*m*) (geol) dunite

ду·новёни·е, -я, (*n*) puff, whiff

ду·н·уть, -ут, (*perf*) puff, give a blow, blow (once)

дуо (*n indecl*) (met roll) two-high

~, двой·н·ое (met roll) double-duo, Dowlais mill

~ -стан·, -а, (*m*) two-high mill

дуотал·, -а, (*m*) (chem) quiacol carbonate

дуплекс·, -а, (*m*) duplex

~ -авто·тип·и·я, -и, (*f*) (phot) double-transfer autotype/carbon process

~ -процёсс·, -а, (*m*) (met) duplexing

дуплекс·н·ый, -ая, -ое, (*a*) duplex

дупл·ист·ый, -ая, -ое, (*a*) hollow

дупл·о, -а, (*nom pl*) дупл·а, (*g pl*) дупел·, (*n*) hollow

дур·ак·, -а, (*m*) fool

дурал·иев·ый, -ая, -ое, (*a*) dur-alumin

дур·ал·ь, -я, (*m*) (met) duralumin

дур·алюмин·и·й, -я, (*m*) duralumin

дуран·металл·, -а, (*m*) durana metal

дуранд·а, -ы, (*f*) (agr) oil cake

дур·ацк·ий, -ая, -ое, (*a*) foolish, stupid

дурит·, -а, (*m*) (min) durain

дурман·, -а, (*m*) (bot) thorn apple, *Datura*

дурн·ё·ть, -ют, (*imp*) become/get/grow bad/evil/ugly

дурн·ик·, -а, (*m*) (bot) whortleberry, *Vaccimium uliginosum*

дурн·иц·а, -ы, (*f*) = дурн·ик·

дурн·ой, -ая, -ое, (*a*) (*s f*) дурен·, дурн·а, дурн·о bad, evil, ugly

~ стал·ь (*f*) wild steel

дурн·от·а, -ы, (*f*) giddiness, vertigo

дурол·, -а, (*m*) (chem) durene

дуро·метр·, -а, (*m*) rubber hardness meter

дуршлаг·, -а, (*m*) colander

дуст·, -а, (*m*) insecticide powder

дутик·, -а, (*m*) loess doll

ду·т·ый, -ая, -ое, (*past part pass*) of ду·ть; blown, blown up, hollow; inflated; exaggerated

ду·ть, -ют, (*imp*) blow

дут·ь·ё, -я, (*n*) draught, draft, blast, blowing

~, магнит·н·ое (elec) magnetic blow-out

~, обогащ·ённ·ое (met) oxygen-enriched blast

~, холод·н·ое cold blast клапан· холод·н·ого дут·ь·я (*m*) cold-blast valve (of a blast furnace)

дут·ьев·ой, -ая, -ое, (*a*) draught, draft; blast

~ у·строй·ств·о (*n*) forced draught system

дух·, -а, (*m*) spirit; morale, courage; (*pl*) perfume, scent; (spectr) satellite line

~ от решёт·к·и (spectr) grating satellite

дух·овк·а, -и, (*g pl*) -вок·, (*f*) oven

дух·овн·ый, -ая, -ое, (*a*) spiritual, moral

дух·ов·ой, -ая, -ое, (*a*) air, air-operated, wind

~ оркёстр· (*m*) brass band (music)

~ печ·ь (*f*) oven

~ пистолёт· (*m*) air pistol

дух·от·а, -ы, (*f*) stuffy air; closeness, oppressive damp heat

душ·, -а, (*i*) -ем, shower, shower-bath; (med) douche

душ·а, -и, (*acc*) душ·у, (*i*) душ·ой, (*nom pl*) душ·и, (*g pl*) душ·, (*d pl*) душ·ам, (*f*) soul, mind

~ на·сел·ёни·я head of population

душ·ев·ая, -ой, (*f decl as adj*) shower-baths

душевно·боль·н·ой, -ая, -ое, (*a*) mentally sick, mentally deranged, insane; (*m or f decl as adj*) mental case

душ·евн·ый, -ая, -ое, (*a*) of душ·а; mental; sincere, cordial, heartfelt

душ·ир·овать, -уют, (*imp and perf*) douche, sprinkle, shower

душ·ирующ·ий, -ая, -ее, (*pres part act*) of душ·ир·овать

~ трубо·провод· (*m*) sprinkler pipe

душ·и́ст·ый, -ая, -ое, (*a*) fragrant, aromatic, sweet, sweet-smelling

~ вещ·еств·о́ (*n*) perfume, scented substance

душ·и́ть, -а́т, (*imp*) suffocate, stifle; scent, perfume

душ·ни́к·, -а́, (*m*) (met cast) venting rod/wire; control cock (of a central heating system)

ду́ш·н·ый, -ая, -ое, (*a*) (meteor) sultry, close, oppressive

ду́·ющ·ий, -ая, -ее, (*pres part act*) of ду·ть blowing

ду́·ющ·ийся, -аяся, -ееся, (geol) heaving, swelling

ду́·я (*pres gerund*) of ду·ть blowing

дьюар·, -а, (*m*) Dewar/vacuum/thermos flask

Дьюар·а, со·су́д (*m*) Dewar/vacuum/thermos flask

дым·, -а, (*nom pl*) -ы́, (*g pl*) -о́в, (*m*) smoke, fume

дым·а́р·ь, -я́, (*m*) smoke generator

дым·за·ве́с·а, -ы, (*f*) smoke screen

дым·и́ть, -я́т, (*imp*) smoke, make smoke, fume

ды́м·к·а, -и, (*g pl*) -м·ок·, (*f*) (meteor) dry haze; mist, haze (in processes)

дымно·ме́р·, -а, (*m*) smoke gauge/meter

~, электр·и́ческ·ий optical density smoke meter

ды́м·н·ый, -ая, -ое, (*a*) smoke, smoky, smoking, fumy, fuming

~ по́рох· (*m*) (expl) black powder, gunpowder

дым·ов·о́й, -а́я, -о́е, (*a*) smoke

~ газ· (*m*) flue gas

~ за·ве́с·а (*f*) smoke screen

~ кла́пан· (*m*) stack valve (of furnace)

~ труб·а́ (*f*) chimney, smoke stack, funnel

~ я́щик· (*m*) (rail) smoke box

дым·о·га́р·н·ый, -ая, -ое, (*a*) fire-tube (of boilers)

~ котёл· (*m*) fire-tube boiler

~ тру́б·к·а (*f*) fire tube

дым·о́к·, -м·к·а́ (*m*) puff of smoke

дымо·не·про·пц·а́ем·ый, -ая, -ое, (*a*) smoke-tight

дымо·об·наруж·и́тел·ь, -я, (*m*) smoke detector

дымо·образ·ова́ни·е, -я, (*n*) smoking, fuming

дымо·образ·у́ющ·ее вещ·еств·о́ (*n*) smoke agent

дымо·со́с·, -а, (*m*) exhaust fan, flue gas extractor

дымо·хо́д·, -а, (*m*) flue, uptake

ды́м·чатост·ь, -и, (*f*) smokiness, haze

ды́м·чат·ый, -ая, -ое, (*a*) smoke-coloured, smoked (glass), ash-grey

~ кварц· (*m*) (min) smoky quartz, cairngorm

дым·я́щ·ий, -ая, -ее, (*pres part act*) of дым·и́ть; smoking, fuming

~ кисл·от·а́ (*f*) (chem) fuming acid

ды́н·н·ый, -ая, -ое, (*a*) melon, musk-melon, cantaloupe

~ де́рев·о (*n*) (bot) papaya, *Carica papaya*

ды́н·я, -и, (*g pl*) дын·ь, (*f*) (bot) muskmelon, cantaloupe, *Cucumis melo*

дыр·а́, -ы́, (*nom pl*) ды́р·ы, (*f*) hole, perforation, cavity, (zool) foramen

ды́р·к·а, -и, (*g pl*) -р·ок·, (*f*) (*dim*) of дыр·а́; (phys) hole

в·вед·е́ни·е ды́р·ок· (*n*) hole injection (semiconductors)

~ электро́н·н·ая electron hole

дыро·ко́л·, -а, (*m*) puncher

дыро·ме́р·, -а, (*m*) hole gauge/gage

дыро·про·бив·н·о́й, -а́я, -о́е, (*a*) punching, hole-punching

~ стан·о́к· (*m*) punching machine

дыро·про·би́в·очн·ый пресс· (*m*) punch, hole-puncher

ды́р·очк·а, -и, (*g pl*) -чек·, (*f*) (*dim*) of дыр·а́

ды́р·очник·, -а, (*m*) (zool) foraminifer

ды́р·очн·ый, -ая, -ое, (*a*) of ды́р·к·а hole, hole-type

~ пере·мещ·е́ни·е (*n*) hole migration (semiconductors)

~ по·дви́ж·ност·ь (*f*) hole mobility

~ про·вод·и́мост·ь (*f*) *p*-type/hole-type conductivity, hole conduction (of semiconductors)

дырчато·апертур·н·ый, -ая, -ое, (*a*) foraminorate

ды́р·чат·ый, -ая, -ое, (*a*) perforated, holey, foraminated

дыр·я́в·ить, -ят, (*imp*) make a hole in, perforate

дыр·я́в·ый, -ая, -ое, (*a*) holed, perforated

ды́х·ал·о, -а, (*n*) blowhole; (zool) windpipe

ды́х·альц·е, -а, (*nom pl*) -льц·а, (*g pl*) -лец·, (*n*) (zool) spiracle

дых·а́ни·е, -я, (*n*) respiration, breathing

дых·а́тельн·ый, -ая, -ое, (*a*) respiratory; (eng) breather

дых·а́тельн·ый, -ая,-ое
~ **аппара́т·**   (*m*) respirator, breathing apparatus
~ **кла́пан·**   (*m*) breather valve (oil storage tanks); throttle valve (turbo-pumps)
~ **меш·о́к·**   (*m*) (oil) tank balloon
~ **пресс·**   (*m*) veneer-drying press
дыш·а́ть, -ат,   (*imp*) breathe, respire
ды́шл·о, -а,   (*n*) pole, shaft; draft bar (of cart etc.)
~, **вед·у́щ·ее**   connecting rod
~, **поршн·ев·о́е**   connecting rod
~ **при·це́п·а**   trailer tongue
~, **с·цеп·н·о́е**   coupling rod
Дэкин·а-Вест·а, реа́кци·я   (*f*) (chem) Dakin–West reaction
дю́бел·ь, -я,   (*nom pl*) **-я́,** (*g pl*) **-е́й,** (*m*) (build) rawlplug
дю́жин·а, -ы,   (*f*) dozen
дю́з·а, -ы,   (*f*) spray nozzle
дюйм·, -а,   (*m*) inch (2·54 cm)
дюйм·о́вк·а, -и,   (*g pl*) **-вок·,** (*f*) one-inch plank; one-inch nail
дюйм·о́в·ый, -ая, -ое,   (*a*) one-inch
-дюйм·о́в·ый, -ая, -ое,   (*adj component*) -inch
дю́кер·, -а, (*m*) (hydr) sag pipe, inverted siphon, culvert siphon
Дюло́нг·а и Пти, за·ко́н·   (*m*) (chem) Dulong and Petit's law

Дюм·а́, ме́тод·   (*m*) Dumas method for nitrogen
дюмонти́т·, -а,   (*m*) (min) dumontite
дю́н·а, -ы,   (*f*) dune, sand drift
~, **блужд·а́ющ·ая**   travelling/wandering dune
Дюпе́н·а, индикатри́с·а   (*f*) (math) Dupin's indicatrix
дюр·а́л·ь, -я,   (*m*) (met) duralumin
дюралюми́н·, -а, (*m*) (met) duralumin
дюралюми́ни·й, -я,   (*m*) (met) duralumin
дюрано́л·, -а,   (*m*) duranol (dye)
дюре́н·, -а,   (*m*) (min) durain
дюре́н·ов·ый, -ая, -ое,   (*a*) (min) durain
Дю́ринг·а, пра́в·ил·о (*n*) (chem) Dühring's rule
дюри́т·, -а,   (*m*) hose joint (piping); (min) durite
дюри́т·ов·ый, -ая, -ое,   (*a*) hose, hose-type; durite
~ **му́фт·а**   (*f*) hose joint
дюро́·метр·, -а,   (*m*) durometer, hardness tester
Дюфо́р·а, эффе́кт·   (*m*) (phys) Dufour effect
дя́гил·ев·ый, -ая, -ое,   (*a*) (bot) angelica
дя́гил·ь, -я,   (*m*) (bot) angelica, *Angelica archangelica*
дя́д·я, -и, (*nom pl*) **-и** *or* **-ья́,** (*g pl*) **-ей** *or* **-ьёв,** (*m*) uncle

Е

ев·ге́н·ик·а, -и, (*f*) (gen) eugenics

ев·ген·и́ческ·ий, -ая, -ое, (*a*) eugenic

евге́н·ов·ый блеск· (*m*) (min) polybasite

евдо́шк·а, -и, (*g pl*) **-шек·,** (*f*) (fish) umbra, *Umbra krameri*

евдо́шк·ов·ые, -ых, (*pl decl as adj*) (fish) *Umbridae*

евкли́д·ов·а, гео·ме́тр·и·я (*f*) Euclidean geometry

евр·а́зи·я, -и, (*f*) (geog) Eurasia

евре́й·ск·ий, -ая, -ое, (*a*) Jewish

евро·азиа́т·ск·ий, -ая, -ое, (*a*) (geog) Euroasiatic

европ·е́ец·, -е́йц·а, (*i*) **-е́йц·ем,** (*m*) European

европ·е́йск·ий, -ая, -ое, (*a*) European

евро́п·и·й, -я, (*m*) (chem) europium, Eu

евста́хи·ев·а труб·а́ (*f*) (anat) Eustachian tube

е́вш·ий, -ая, -ее, (*past part act*) of **ест·ь** (*imp*)

е́гер·ь, -я, (*nom pl*) **-я́,** (*g pl*) **-е́й,** (*m*) huntsman; (mil) Jäger, soldier (of mountain troops)

еги́пет·ск·ий, -ая, -ое, (*a*) Egyptian

его́ (*pron*) (*acc/gen sing*) of **он/оно́** him, it

ед. (*abbr*) of **един·и́ц·а** unit

ед·а́, -ы́, (*f*) food

едва́ (*adv*) hardly, scarcely, only just, no sooner than

~ ... как hardly ... than

~ ли hardly, scarcely, not likely, doubtful

~ не nearly, almost

е́д·енн·ый, -ая, -ое, (*a*) (*past part pass*) of **ест·ь** (*imp*)

е́д·ет (*pres 3rd sing*) of **е́х·ать** (*imp deter*), *see* **е́зд·ить** (*imp indet*)

ед·и́м (*pres 1st pl*) of **ест·ь** (*imp*)

един·е́ни·е, -я, (*n*) unity

един·и́ц·а, -ы, (*i*) **-е́й,** (*f*) unit; (math) unity, one; individual, person; very bad (school marking)

~, абсолю́т·н·ая (phys) absolute unit

~ а́том·н·ого ве́с·а (phys) atomic weight unit, awu

~ а́том·н·ой ма́сс·ы (phys) atomic mass unit

~, брита́н·ск·ая тепл·ов·а́я British thermal unit

в един·и́ц·ах in units of, in terms of

един·и́ц·а

~, вес·ов·а́я weight unit; (g) per unit weight

~ де́й·стви·я, коша́ч·ья (pharm) cat unit

~ де́й·стви·я пеницилли́н·а international unit (of penicillin)

за един·и́ц·у per unit

~, имму́н·н·ая (bact) Felton's unit

~ кольц·а́ (math) unit element of a ring

~ ма́сс·ы mass unit

на един·и́ц·у per unit

~, оксфо́рд·ск·ая Oxford unit (of penicillin)

~ пере·гово́р·а message unit (telephones)

~ пере·но́с·а (phys) transfer unit

~ по́л·я (math) unit element of a field

~, при·вед·ённ·ая (math) normalized unit

~ с·во́д·а unit roof (of furnace)

~ част·от·ы́ пере·крёст·а (gen) cross-over unit

~, энергет·и́ческ·ая energy unit

един·и́чност·ь, -и, (*f*) singleness

един·и́чн·ый, -ая, -ое, (*a*) unit; single, individual, unique; solitary

~ объ·ём· (*m*) unit volume

~ пло́щад·ь (*f*) unit area

едино- (*component*) mono-, uni-

едино·бра́ч·и·е, -я, (*n*) (biol) monogamy

едино·бра́ч·н·ый, -ая, -ое, (*a*) (zool) monogamous

едино·вре́мен·н·о (*adv*) once only; simultaneously, at the same time

едино·вре́мен·н·ый, -ая, -ое, (*a*) simultaneous, synchronous, isosynchronous; once only, one-time

едино·гла́с·н·ый, -ая, -ое, (*a*) unanimous

едино·ду́ш·н·ый, -ая, -ое, (*a*) unanimous

едино·кро́в·н·ый, -ая, -ое, (*a*) (gen) consanguineous

едино·ди́ч·н·ый, -ая, -ое, (*a*) personal, individual

едино·мы́шл·енник·, -а, (*m*) person of the same opinion (as); adherent, supporter

едино·мы́шл·енн·ый, -ая, -ое, (*a*) like-minded, having the same opinion

едино·на·сле́д·и·е, -я, (*n*) (law) male entail

едино·на·ча́л·и·е, -я,   (*n*) personal/single authority

едино·обра́з·и·е, -я,   (*n*) uniformity

едино·обра́з·н·ый, -ая, -ое,   (*a*) uniform

едино·ро́г·, -а, (*m*) (astron) Unicorn, Monoceros; (zool) narwhal, *Monodon monoceros*

едино·утро́б·н·ый, -ая, -ое,   (*a*) uterine

еди́н·ственност·ь, -и,   (*f*) uniqueness

еди́н·ственн·ый, -ая, -ое, (*a*) only, sole, unique

~ **знач·е́ни·е** (*n*) (math) unique value

~ **числ·о́** (*n*) (gram) singular

еди́н·ств·о, -а, (*n*) unity, unanimity

еди́н·ый, -ая, -ое, (*a*) single, only, sole; united, unified; common, standard

~ **высоко·во́льт·н·ая се́т·ь** (*f*) grid (elec, power supply)

~ **цен·а́** (*f*) standard/unified price

е́д·к·ий, -ая, -ое, (*a*) caustic, corrosive; acrid, pungent

~ **газ·** (*m*) corrosive gas

~ **ка́ли** (*n*) (chem) caustic potash, potassium hydroxide

~ **натр·** (*m*) caustic soda, sodium hydroxide

ед·кова́т·ый, -ая, -ое, (*a*) (chem) subacrid

е́д·кост·ь, -и, (*f*) causticity

едо́к·, -а́, (*m*) consumer

ед·о́м·ый, -ая, -ое, (*pres part pass*) of **ест·ь** (*imp*); edible

е́д·ут (*pres 3rd pl*) of **е́х·ать** (*imp deter*) *see* **е́зд·ить** (*imp indet*)

е́д·учи (*pres gerund*) of **е́х·ать** (*imp deter*) *see* **е́зд·ить** (*imp indet*)

е́д·ущ·ий, -ая, -ее, (*pres part act*) of **е́х·ать** (*imp deter*) *see* **е́зд·ить** (*imp indet*)

ед·я́ (*pres gerund*) of **ест·ь** (*imp*)

ед·я́т (*pres 3rd pl*) of **ест·ь** (*imp*)

ед·я́щ·ий, -ая, -ее, (*pres part act*) of **ест·ь** (*imp*)

её (*pron acc/gen sing*) of **она́** her, it

ёж·, -а́, (*i*) **-о́м,** (*m*) (zool) hedgehog, *Erinaceus*; (mil) barbed-wire entanglement, anti-tank obstacle

~, **мор·ск·о́й** sea-urchin, echinus

еж·а́, -ы́, (*f*) (bot) orchard grass, *Dactylis*

еже- (*component*) every

еж·еви́к·а, -и, (*f*) (bot) blackberry, *Rubus*

еж·еви́чн·ый, -ая, -ое, (*a*) blackberry, *Rubus*

~ **кор·а́** (*f*) (pharm) rubus

еже·го́д·ник·, -а, (*m*) almanac, yearbook, annual

~, **авиацио́н·н·ый астроном·и́ческ·ий** (nav) air almanac

~, **мор·ск·о́й астроном·и́ческ·ий** (nav) nautical almanac

еже·го́д·н·о (*adv*) yearly, annually

еже·го́д·н·ый, -ая, -ое, (*a*) yearly, annual

еже·голо́в·к·а, -и, (*f*) (bot) bur reed, *Sparganium*

еже·голо́в·ник·, -а, (*m*) *see* **еже·голо́в·к·а**

еже·дн·е́вн·ый, -ая, -ое, (*a*) daily; everyday, ordinary

еже·кварта́ль·н·о (*adv*) quarterly, every 3 months

еже·кварта́ль·н·ый, -ая, -ое, (*a*) quarterly, every 3 months

ежеки́т·, -а, (*m*) (min) jezekite

е́жели (*conj*) if, in case

еже·ме́сяч·н·ый, -ая, -ое, (*a*) monthly

еже·мину́т·н·о (*adv*) every minute; incessantly

еже·мину́т·н·ый, -ая, -ое, (*a*) every minute, incessant

еже·неде́ль·ник·, -а, (*n*) (print) weekly

еже·неде́ль·н·ый, -ая, -ое, (*a*) weekly

еже·но́щ·н·ый, -ая, -ое, (*a*) nightly, every night

еже·су́т·очн·о (*adv*) every day, every 24 hours, diurnal

еже·су́т·очн·ый, -ая, -ое, (*a*) diurnal

еже·ча́с·н·о (*adv*) hourly, every hour

еже·ча́с·н·ый, -ая, -ое, (*a*) hourly, every hour

еж·о́в·ый, -ая, -ое, (*a*) of **ёж·** (bot) echinate; (zool) echinoid

~ **трансформа́тор·** (*m*) (rad) hedgehog transformer

езд·а́, -ы́, (*f*) ride, riding, drive, driving, travel (by vehicle); (rail) service, run, running

~, **верх·ов·а́я** horseback riding

~, **кольц·ев·а́я** (rail) turn-around service, loop run

~, **петл·ев·а́я** (rail) looping

е́зд·ить, -ят, (*imp indet*) go (in vehicle), ride, drive, travel

е́зд·к·а, -и, (*g pl*) **-д·ок·,** (*f*) delivery trip, round

езд·о́к·, -а́, (*m*) rider

е́зж·у (*pres 1st sing*) of **е́зд·ить**

ей (*pron, dat/instr sing*) of **она́** (*dat*) to her/it; (*instr*) by/with etc. her/it

ел· (*past masc sing*) of **ест·ь** (*imp*)

ёл·а (*past fem sing*) of **ест·ь** (*imp*)

ёл·а, -ы, (*f*) (naut) yole (fishing boat)

ела́н·ь, -и, (*f*) glade

ёле (*adv*) hardly, scarcely

ёл·ев·ый, -ая, -ое, (*a*) of **ел·ь**; (*pl decl as adj*) *Abietineae*

ел·е́ц·, ел·ьц·а́, (*i*) **ел·ьц·о́м** (*m*) (fish) dace, *Leuciscus leuciscus*

ёл·и (*past pl*) of **ест·ь** (*imp*)

ёл·к·а, -и, (*g pl*) **ёлок·,** (*dim*) of **ел·ь** (q.v.); (oil) Christmas tree; herringbone (brickwork); chevron tread pattern (tyres)

ёл·о (*past neut sing*) of **ест·ь**

ел·о́в·ый, -ая, -ое, (*a*) of **ел·ь**

ёл·очн·ый, -ая, -ое, (*a*) of **ёл·к·а, ел·ь;** (cryst) dendritic

ел·ь, -и, (*f*) (bot) spruce, *Picea*

~, голуб·а́я blue spruce, *Picea pungens*

~, кол·ю́ч·ая Colorado blue spruce, *Picea pungens*

~, кра́с·н·ая red spruce, *Picea rubens*

~, обык·нове́нн·ая (bot) Norway spruce, *Picea abies*

~, серебр·и́ст·ая Sitka spruce, *Picea sitchensis*

~, ситх·и́нск·ая Sitka spruce, *Picea sitchensis*

~, чёрн·ая black spruce, *Picea mariana*

ёль·ник·, -а, (*m*) spruce wood/grove

ем (*pres 1st sing*) of **ест·ь** (*imp*)

ём·к·ий, -ая, -ое, (*a*) capacious, roomy, large-capacity, large

ём·костн·о-резисти́в·н·ый, -ая, -ое, (*a*) (elec) capacitance-resistance

ём·костн·о-с·вя́з·анн·ый, -ая, -ое, (*a*) (elec) capacitance-coupled

ём·костн·ый, -ая, -ое, (*a*) (elec) capacitance, capacitive, capacity

~ за·па́зд·ывани·е (*n*) (autom) capacity lag

~ с·вяз·ь (*f*) capacity/capacitance/capacitive coupling

ём·кост·ь, -и, (*f*) capacity, cubic content; (elec) capacitance; tank, container

~, вре́д·н·ая stray capacitance

~, вы·ход·н·а́я (elec) output capacitance

~ конденса́тор·а (elec) capacitance

коэффицие́нт· ём·кост·и (*m*) (autom) capacity

~, меж·электро́д·н·ая (elec) interelectrode capacitance

~, обра́т·н·ая (elec) elastance

ём·кост·ь

~, парази́т·н·ая spurious/stray capacitance

~ пере·хо́д·а junction capacitance

~ рас·се́·яни·я (elec) stray capacitance

~, со·средо·то́ч·енн·ая lumped capacitance

~, част·и́чн·ая direct capacitance

ему́ (*pron, dat sing*) of **он/оно́** to him/it

енами́н·, -а, (*m*) (chem) enamine

ендов·а́, -ы́, (*f*) (build) valley

ено́л·, -а, (*m*) (chem) enol

епанч·а́, -и́, (*i*) **-о́й,** (*f*) (biol) mantle

ёренс-та́л·и (*nom pl*), (*g pl*) **-ей,** (naut) guy tackle

ерет·и́ческ·ий, -ая, -ое, (*a*) heretical

ерунд·о́в·ый, -ая, -ое, (*a*) absurd, nonsensical

ерун·о́к·, -н·к·а́, (*m*) bevel (carpenter's tool)

ёрш·, ерш·а́, (*i*) **ерш·о́м,** (*m*) brush, wire brush; (fish) ruff, *Acerina cernua*; jag/rag-bolt; beer with vodka

~, одно·ро́г·ий (oil) rope spear

ершова́тк·а, -и, (*f*) (fish) dab, *Limanda limanda*

еры́ (*n indecl*) letter "ы"

ёр·ь, -я, (*m*) soft sign "ь" (Russian letter)

е́сли (*conj*) if

~ бы if (+ *subjunctive*)

~ бы не but for, were it not for

~ не unless

~ то́лько provided, provided that, if only

~ только не unless

~ уже́ if anything

ест (*pres 3rd sing*) of **ест·ь** (*imp*)

есте́ств·енник·, -а, (*m*) naturalist; (obs) scientist

есте́ств·енн·о (*adv*) naturally, of course

есте́ств·енн·ый, -ая, -ое, (*a*) natural, native, spontaneous; inherent

~ магнет·и́зм· (*m*) spontaneous magnetism

~ сред·а́ (*f*) (biol) habitat

естеств·о́, -а́, (*n*) nature, substance

естество·ве́д·, -а, (*m*) naturalist, scientist

естество·ве́д·ени·е, -я, (*n*) natural sciences

естество·зн·а́ни·е, -я, (*n*) natural sciences

естество·ис·пыт·а́тел·ь, -я, (*m*) naturalist

ест·ь, (*pres 1st, 3rd sing, 1st, 2nd, 3rd pl*) **ем, ест, ед·и́м, ед·и́те, ед·я́т,** (*past masc sing*) **ел,** (*imp*) eat; corrode, eat away; (*indecl*) there is/are; (as response) very good, aye aye, yes, all right, O.K.

ефре́йтор·, -а, (*m*) (mil) lance corporal (Brit), Private 1st class (U.S.); leading aircraftman

е́х·авш·ий, -ая, -ее, (*past part act*) of **е́х·ать**

е́х·ать, (*pres 3rd sing, pl*) **е́д·ет, е́д·ут,** (*imp deter*) *see* **е́зд·ить** (*imp indet*)

ехи́д·н·ый, -ая, -ое, (*a*) malicious, spiteful

е́шьте (*imper*) of **ест·ь** (*imp*)

ещё (*adv*) still, yet; as long ago as, only

~ **бо́льш·е** further, still further/more

~ **в** even in; as long ago as (of time)

~ **не** not yet

~ **не·да́вн·о** only recently

~ **раз·** once more, once again

е́ю (*pron, instr sing*) of **она́** by/with etc. her/it

Ж

ж. (*abbr*) of **журна́л·** (print) journal

жа́б·а, -ы, (*f*) (med) quinsy, tonsilitis, angina; (zool) toad, *Bufo vulgaris*

~, груд·н·а́я (med) angina pectoris

жаберно·ды́ш·ащ·ий, -ая, -ее, (*a*) (zool) gill-breathing

жа́бер·н·ый, -ая, -ое, (*a*) (zool) branchiate, branchial, gill, (*pl as noun*) branchiates, *Branchiata*

~ кры́ш·к·а (*f*) (zool) operculum, opercular apparatus

~ мешк·и́ (*pl*) (zool) visceral clefts, gill clefts/slits

~ щел·ь (*f*) (zool) gill-slit/cleft, branchial slit/cleft

жа́б·ий, -ья, -ье, (*a*) of **жа́б·а**

~ глаз· (*m*) (min) toads-eye tin

жа́бник·, -а, (*m*) (bot) *Ficaria*

жа́бр·а, -ы, (*f*) (*see also* **жа́бр·ы**); (a/c) sponson

жабре́·й, -я, (*m*) (bot) hemp-nettle, *Galeopsis*

жабро- (*component*) branchi-, branchio-

жабро·но́г·ие, -их, (*pl decl as adj*) (zool) *Branchiopoda*

~, рако·обра́з·н·ые *Anostraca*, fairy shrimps, brine shrimps

жа́бр·ы (*nom pl*), (*g pl*) **жабр·,** (*f*) gill, gills, branchia, branchiae; (a/c) sponsons

жавел·ев·а вод·а́ (*f*) (chem) Javelle/ Javel water

жавел·ь, -я, (*m*) *see* **жавел·ев·а вод·а́**

жа́воронок·, -нк·а, (*m*) (orn) lark

жа́·вш·ий, -ая, -ее, (*pres part act*) of **жат·ь** (all meanings)

жад·, -а, (*m*) (min) jade

жадеит·, -а, (*m*) (min) jadeite

жа́д·н·ый, -ая, -ое, (*a*) greedy, avid

жа́жд·а, -ы, (*f*) thirst, craving

жа́жд·ать, -ают, *or* **-ут,** (*imp + gen*) crave, thirst

жакет·, -а, (*m*) jacket; (met cast) mould jacket

жакка́рд·а, маши́н·а (*f*) (text) Jacquard machine

жакка́рд·ов·ый, -ая, -ое, (*a*) (text) Jacquard

жал (*past masc sing*) of **жат·ь** (all meanings)

жал·е́·ть, -ют, (*imp*) be sorry that, feel sorry for, pity

   не жал·е́·ть spare no, grudge no (effort etc.)

жа́л·и (*past pl*) of **жат·ь** (all meanings)

жа́л·ить, -ят, (*imp*) sting, bite

жа́л·к·ий, -ая, -ое, (*a*) pitiful, miserable

жа́л·о, -а, (*n*) (ent) sting, stinger; dart; finger, pin, needle; (amm) needle (fuse), striker; (*past neut 3rd sing*) of **жат·ь** (1) and (2)

жа́л·об·а, -ы, (*f*) complaint

   кни́г·а жа́лоб· и пред·лож·е́ни·й (*f*) complaints and suggestions book

жа́л·обн·ый, -ая, -ое, (*a*) complaints, complaint, complaining

жа́л·ованн·ый, -ая, -ое, (*past part pass*) of **жа́л·овать**; granted, bestowed, conferred on

жа́л·овань·е, -я, (*n*) salary, pay

жа́л·овать, -уют, (*imp*) grant, give, bestow; like, favour, favor; **-ся** complain, put in a complaint

жа́л·остн·ый, -ая, -ое, (*a*) regrettable, deplorable

жаль (*impersonal*) **ему́ жаль** he is sorry

жа́ль·н·ый, -ая, -ое, (*a*) of **жа́л·о**

жалюзи́ (*n indecl*) venetian shutters/ blinds, jalousies, louvres, louvers

жалюз·и́йн·ый, -ая, -ое, (*a*) of **жалюзи́**

~ золо·у·лов·и́тел·ь (*m*) baffle separator (of dust catcher)

жалюз·н·ый, -ая, -ое, (*a*), *see* **жалюз·и́йн·ый**

Жа́мен·а, интерферо́·метр· (*m*) (opt) Jamin interferometer

жа́нр·, -а, (*m*) genre

жа́нр·ов·ый, -ая, -ое, (*a*) of **жанр·**

жа́р·, -а, (*m*) heat; (med) fever, temperature

жар·а́, -ы́, (*f*) (meteor) heat, hot weather

жарго́н·, -а, (*m*) jargon, slang; (min) jargon, jargoon

жарго́н·н·ый, -ая, -ое, (*a*) jargon, slang; (min) jargon, jargoon

жа́р·ен·ый, -ая, -ое, (*a*) fried; grilled; roasted

жа́р·ень·е, -я, (*n*) (*v n*) *see* **жа́р·ить**

жа́р·ить, -ят, (*imp*) fry; grill; roast

жа́р·к·ий, -ая, -ое, (*a*) hot; (*see also* **жар·к·о́е**) (*as noun*)

~ по́яс· (*m*) (geog) Torrid Zone

жар·к·о́е, -о́го, (*m decl as adj*) (food) a fry/roast/grill

жар·о́вн·я, -и, (*g pl*) **-вен·,** (*f*) brazier
жар·ов·о́й, -а́я, -о́е, (*a*) heat, fire
~ труб·а́ (*f*) fire/furnace tube (of boiler); flame tube (of jet engine)
жаро·вы·но́с·ливост·ь, -и, (*f*) (biol) heat-resistance
жаро·по·ниж·а́ющ·ий, -ая, -ее, (*a*) (med) antipyretic, febrifugal
жаро·про́ч·ност·ь, -и, (*f*) heat resistance; (met) creep resistance, heat resistance, resistance to creep and scaling
жаро·про́ч·н·ый, -ая, -ое, (*a*) fireproof/resistant; (met) creep-and-scaling resistant, heat-resistant, high-temperature
~ с·плав· (*m*) heat-resistant/resisting alloy
~ стал·ь (*f*) heat-resistant steel, high temperature steel
жаро·сто́й·к·ий, -ая, -ое, (*a*) heat-resistant, heat-proof; (met) scaling-resistant
жаро·сто́й·кост·ь, -и, (*f*) heat resistance; (met) resistance to scaling
жаро·тру́б·н·ый, котёл· (*m*) firetube boiler
жаро·у·по́р·ност·ь, -и, (*f*) heat resistance, refractoriness; (met) resistance to scaling
жаро·у·по́р·н·ый, -ая, -ое, (*a*) heat-resistant, refractory; (met) scaling resistant
жасми́н·, -а, (*m*) (bot) jasmine, *Jasminium*
жасми́н·н·ый, -ая, -ое, (*a*) of **жас-ми́н·**
жасмо́н·, -а, (*m*) (chem) jasmone
жа́т·в·а, -ы, (*f*) reaping; crop, harvest (gathering)
жа́т·венн·ый, -ая, -ое, (*a*) of **жа́-т·в·а**
жа́т·к·а, -и, (*g pl*) **-т·ок·,** (*f*) (*see also* **жн·е·я́**) (agr) harvester; reaper; sandal strap
~, ряд·ов·а́я (agr) windrower
~ -само·с·бро́с·к·а, -и, (*f*) (agr) self-delivery reaper
~ -снопо·вяз·а́лк·а, -и, (*f*) (agr) reaper-and-binder
жа́т·ый, -ая, -ое, (*past part pass*) of **жат·ь** (all meanings)
жат·ь (1) (*pres 3rd sing, pl*) **жм·ёт, жм·ут,** (*imp*) squeeze, compress, press; (2) (*pres 3rd sing, pl*) **жн·ёт, жн·ут,** (*imp*) reap, harvest, crop, cut (crops)

жба́н·, -а, (*m*) can, tub
жва́ка-га́лс·, -а, (*m*) (naut) cable slip
жв·а́л·о, -а, (*n*) (zool) mandible
жв·а́чк·а, -и, (*g pl*) **-чек·,** (*f*) (zool) rumination, cud; (naut) painting rag
жв·а́чн·ый, -ая, -ое, (*a*) (zool) ruminant, chewing, (*pl as noun*) ruminants, *Ruminantia*
жг·ли (*past pl*) of **жеч·ь**
жгу́т, -а́, (*m*) plait, braid, braided strap; (med) tourniquet; oakum; roving (rubber); (ent) pedipalp; rolled-up vortex sheet; (*pres 3rd pl*) of **жеч·ь**
~, вихр·ев·о́й vortex core (*fluid flow*)
~, карам́ель·н·ый soft caramel
~ про·вод·о́в (elec) bunched conductors
жгу́т·ик·, -а, (*m*) (*dim*) of **жгут·;** (bot) cilium; (zool) flagellum
жгутико·ви́д·н·ый, -ая, -ое, (*a*) flagelliform
жгу́тик·ов·ый, -ая, -ое, (*a*) (zool) flagellate, (*pl as noun*) flagellates, *Flagellata*
жгутико·но́с·ц·ы, -ев, (*m pl*) (zool) *Flagellata, Mastigophora*
жгуто·но́г·ие, -их, (*pl decl as adj*) (ent) *Pedipalpi*
жгуто·ре́з·к·а, -и, (*g pl*) **-з·ок·,** (*f*) (text) staple cutter
жг·у́чи (*pres gerund*) of **жеч·ь**
жг·у́ч·ий, -ая, -ее, (*a*) burning, smarting; caustic, corrosive; hot (of tastes); poignant
жг·у́щ·ий, -ая, -ее, (*pres part act*) of **жеч·ь**
ж.-д. (*abbr*) of **железно·доро́ж·н·ый** railway, railroad
жд·а́нн·ый, -ая, -ое, (*past part pass*) of **жд·ать**
жд·ать (*pres 3rd sing, pl*) **жд·ёт, жд·ут,** (*imp*) (+ *gen*) wait for, await, expect
жд·ёт (*pres 3rd sing*) of **жд·ать**
жд·и́те (*imper*) of **жд·ать** (telecom) wait, stand by
жд·у́щ·ий, -ая, -ее, (*a*) (*pres part act*) of **жд·ать**; slave, driven
~ раз·вёрт·к·а (*f*) slave sweep (oscillographs etc.)
же (*conj*) but, and, however; (*particle giving emphasis to preceding word*); (*particle signifying sameness*)
  оди́н и тот же one and the same, the same
  он же the very same, the self same
  так же also

же
> **так·ой же**   the same
> **тогда же**   at the same time
> **тот же**   the same
> **точ·н·о так же**   in exactly the same way

жев·ани·е, -я,   (*n*) (*v n*) *see* **жев·ать;** mastication

жев·ательн·ый, -ая, -ое,   (*a*) masticatory, chewing

жев·ать   (*prés 3rd sing, pl*) **жу·ёт, жу·ют,** (*imp*) chew, masticate

жёг· (*past masc sing*) of **жеч·ь**

жёг·ш·ий, -ая, -ее,   (*a*) (*past part act*) of **жеч·ь**

жезл·, -а,   (*m*) baton, rod; (rail) staff; (surv) base-line rule
> ∼ **электро·жезл·ов·ой систем·ы** (rail) token

жезл·ов·ой, -ая, -ое,   (*a*) of **жезл·**
> ∼ **систем·а**   (*f*) (rail) token system, staff system

жезло·об·мен·иватель·, -я,   (*m*) (rail) token exchanger, staff exchanger

жел·аем·ый, -ая, -ое,   (*pres part pass*) of **жел·а·ть;** desired; desirable

жел·ани·е, -я,   (*n*) (*v n*) of **жел·а·ть;** wish, desire
> **по жел·ани·ю**   as desired, if desired, optional

жел·анн·ый, -ая, -ое,   (*past part pass*) of **жел·а·ть;** desired, wanted, wished for

жел·ательн·ый, -ая, -ое,   (*a*) desirable

желатин·, -а,   (*m*) *see* **желатин·а**

желатин·а, -ы,   (*f*) gelatin
> ∼, **мясо·пептон·н·ая**   (bact) nutrient gelatin
> ∼, **фотограф·ическ·ая**   (phot) bichromated gelatin

желатин·изатор·, -а,   (*m*) gelling agent

желатин·изаци·я, -и,   (*f*) gelatination, gelatinizing, gelation

желатин·ировани·е, -я,   (*n*) gelatinizing, gelatination, gelling

желатин·ированн·ый, -ая, -ое, (*past part pass*) of **желатин·ир·овать;** gelled, gelated

желатин·ир·овать, -ую·т,   (*imp and perf*) gelatinize, gelate, gel

желатин·н·ый, -ая, -ое,   (*a*) gelatin

желатин·ов·ый, -ая, -ое,   (*a*) gelatin, gelatinous

желатиноз·н·ый, -ая, -ое,   (*a*) gelatinous

желатино·образ·н·ый, -ая, -ое,   (*a*) jelly-like, gelatinoid

желатино·по·доб·н·ый, -ая, -ое,   (*a*) gelatinous, jelly-like, gelatinoid

жел·а·ть, -ют, (*imp*) (+ *gen*) wish, desire

желв·ак·, -а,   (*m*) nodule, concretion; (plast) bulge; (med) scirrhus

желв·ачн·ый, -ая, -ое,   (*a*) nodular

желе   (*n indecl*) jelly

желез·а, -ы,   (*nom pl*) **желез·ы,** (*g pl*) **желёз·,** (*f*) (*and see under associated words*) (zool) gland
> ∼, **миндале·вид·н·ая**   tonsil
> ∼, **молоч·н·ая**   mammary gland
> ∼, **над·поч·ечн·ая**   suprarenal gland
> ∼, **щито·вид·н·ая**   thyroid gland

желез·ин·а, -ы,   (*f*) piece of iron

желез·ист·о- (*component*) (chem) ferro-, ferrous, ferrite

железисто·желез·н·ый, -ая, -ое,   (*a*) (chem) ferroso-ferric

железисто·кисл·ый, -ая, -ое,   (*a*) ferrite of
> ∼ **натр·и·й**   (*m*) sodium ferrite

железисто·синеродисто·водород·-н·ая кисл·от·а   (*f*) ferrocyanic acid

железисто·синерод·ист·ый, -ая -ое,   (*a*) ferrocyanide of
> ∼ **кал·и·й**   (*m*) potassium ferrocyanide, potassium hexacyanoferrate

желез·ист·ый, -ая, -ое,   (*a*) iron; ferrous, ferruginous, ferriphous, chalybeate (of water); (biol) glandular
> ∼ **ис·точ·ник·**   (*m*) chalybeate spring
> ∼ **кисл·от·а**   (*f*) ferrous acid
> ∼ **слан·ец·**   (*m*) (geol) ferruginous schist
> ∼ **ткан·ь**   (*f*) (zool) glandular tissue/ epithelium

желез·к·а, -и,   (*g pl*) **-з·ок·,** (*f*) (*dim*) of **желез·о;** iron/metal piece/plate

желез·к·а, -и,   (*g pl*) **-з·ок·,** (*f*) (*dim*) of **желез·а;** (anat) glandule

желез·к·о, -а,   (*g pl*) **-з·ок·,** (*n*) cutter, cutting iron (of planes etc.)

желез·нени·е, -я,   (*n*) (met) iron coating/plating

желез·о- (*component*) (*see also* **желез·о-**) iron

железно·дорож·н·ый, -ая, -ое,   (*a*) railway, railroad

железно·кисл·ый, -ая, -ое,   (*a*) (chem) ferrate of

желез·н·ый, -ая, -ое,   (*a*) iron, ferric
> ∼ **блеск·**   (*m*) (min) iron glance, specular iron
> ∼ **дерев·о**   (*n*) ironwood
> ∼ **дорог·а**   (*f*) railway, railroad

желе́з·н·ый, -ая, -ое

~ квасц·ы́ (*pl*) (min) iron alum, halotrichite

~ кисл·от·а́ (*f*) ferric acid

~ колчеда́н· (*m*) (min) pyrites, iron pyrites

~ купоро́с· (*m*) iron vitriol

~ руд·а́ (*f*) (min) iron ore, iron stone

~ слюд·к·а (*f*) (min) iron glance, haematite, hematite

~ шля́п·а (*f*) (min) gossan

~ шпат· (*m*) (min) siderite

желез·ня́к·, -а́, (*m*) iron ore, iron stone; (bot) ironwood; (bot) Jerusalem sage, *Phlomis*; hard stock (brick)

~, бу́р·ый (min) limonite, brown iron ore

~, кра́с·н·ый specular iron ore

~, магни́т·н·ый magnetic iron ore

~, шпа́т·ов·ый spathic iron ore

желе́з·о, -а, (*n*) (chem) iron, ferrum, Fe; (com) steel, mild steel

~, азотно·ки́сл·ое (chem) iron nitrate

~, болт·ов·о́е round iron bars

~, вана́д·иев·ое ferrovanadium

~, вольфрам·и́ст·ое ferrotungsten

~, двух·вале́нт·н·ое iron (II), ferrous iron, (*in gen as adj*) ferrous

~, декап·и́рованн·ое (met) pickled iron

за́·кис·ь желе́з·а (*f*) (chem) ferrous oxide, iron monoxide/protoxide

за́·кис·ь-о́кис·ь желе́з·а (*f*) ferrosoferric oxide, magnetic oxide of iron

~, карбони́ль·н·ое carbonyl iron

~, ко́в·к·ое wrought iron

~, кремн·и́ст·ый ferro-silicon

~, кро́вель·н·ое sheet iron; (civ eng) roofing iron

~, лист·ов·о́е plate iron

~, ли́т·ое ingot iron

о́·кис·ь желе́з·а (*f*) (chem) ferric oxide, iron sesquioxide

~, о·цинк·о́ванн·ое galvanized iron

~, само·ро́д·н·ое native iron

~, с·ва́р·очн·ое wrought iron

~, сер·ни́ст·ое ferrous sulphide

~, серно·ки́сл·ое iron sulphate; ferrous sulphate, FeS; ferric sulphate Fe_2S_3

со·един·е́ни·е трёх·вале́нт·н·ого желе́з·а (*n*) ferric compound

~, техн·и́ческ·ое mild steel

~, трёх·вале́нт·н·ое iron (III), (*in gen used as adj*) ferric

~, хло́р·н·ое ferric chloride

желе́з·о- (*component*) (*see also* желе́з·н·о-) iron, ferro-, (chem) ferri-, ferric

железо·бакте́р·и·я, -и, (*f*) (bact) iron bacteria

железо·бето́н·, -а, (*m*) reinforced/ferro-concrete

~, моно·ли́т·н·ый cast in-situ ferro-concrete

~, с·бо́р·н·ый precast ferroconcrete

железо·бето́н·н·ый, -ая, -ое, (*a*) ferroconcrete

~ из·де́л·и·я (*pl*) precast ferroconcrete products

~ констру́кци·и (*pl*) ferroconcrete structural members

железо·водо·ро́д·н·ое со·противл·е́ни·е (*n*) (elec) barretter

железо·ру́д·н·ый, -ая, -ое, (*a*) iron-ore

железо·синеродисто·водоро́д·н·ая кисл·от·а́ (*f*) ferricyanic acid

железо·синеро́д·и́ст·ый, -ая, -ое, (*a*) ferricyanide of

~ желе́з·о (*n*) ferrous ferricyanide

~ ка́ли·й (*m*) potassium ferricyanide

~ мед·ь (*f*) copper/cupric ferrocyanide

железо·со·держ·а́щ·ий, -ая, -ее, (*a*) ferriferous, ferruginous, chalybeate (water)

железо·углерод·и́ст·ый с·плав· (*m*) (met) ferro-alloy

железо·хро́м·ов·ый, -ая, -ое, (*a*) ferrochrome

желе́·йн·ый, -ая, -ое, (*a*) of желе́; jelly

жёлоб·, -а, (*nom pl*) желоб·а́, (*g pl*) желоб·о́в, (*m*) channel, groove, riffle, trough, gutter, chute

~, под·со́ч·ечн·ый talang (rubber tapping)

~, при·ём·н·ый feed trough, delivery trough

~, раз·ли́в·очн·ый (met) runner

~, шар·ов·о́й race track, ball/roller path (of bearing)

желоб·и́ть, -ят, (*imp*) flute, groove

желоб·кова́т·ый, -ая, -ое, (*a*) grooved, channelled; (bot) valleculate

желобо·брю́х·ие моллю́ск·и (*pl*) (zool) *Solenogastres*

желоб·о́к·, -б·к·а́, (*m*) (*dim*) of жёлоб·; slot, groove, flute, furrow, riffle; groove (gun barrel rifling); (bot) vallecula

желобо·шлиф·ова́льн·ый стан·о́к· (*m*) groove-grinding machine

желоб·ча́т·ый, -ая, -ое, (*a*) grooved, fluted, channelled; groove/trough-shaped, U-shaped

~ дел·и́тел·ь (*m*) (min) riffle

жело́н·к·а, -и, (*g pl*) -н·ок·, (*f*) sand pump, shell pump, (civ eng) sludger

жёлт·ая, -ой (*f decl as adj*) yellow (pigment)

~, ка́дм·иев·ая cadmium yellow

желт·е́·ть, -ют, (*imp*) go/turn yellow; look/show/appear yellow

желт·ова́т·ый, -ая, -ое, (*a*) yellowish, sallow

жёлто·зелён·ый, -ая, -ое, (*a*) yellowy-green

желто·зём·, -а, (*m*) yellow-podzolic soil

желт·о́к·, -т·к·а́, (*m*) yolk (of egg), vitellus

желто·кал·и́льн·ый, -ая, -ое, (*a*) yellow (of hot things)

жёлто·кори́чнев·ый, -ая, -ое, (*a*) yellowy-brown

желто·пло́д·н·ый, -ая, -ое, (*a*) xanthocarpous

желт·о́чник·, -а, (*m*) (zool) yoke gland, vitellarium

желт·о́чн·ый, -ая, -ое, (*a*) (biol) vitelline, egg-yellow

~ про́б·к·а (*f*) (zool) yoke plug

~ про·то́к· (*m*) (zool) yolk stalk

желт·у́х·а, -и, (*f*) (med) jaundice, icterus; (bot) yellows

желт·у́шн·ый, -ая, -ое, (*a*) of желт·у́х·а

жёлт·ый, -ая, -ое, (*a*) yellow; (*as f noun*) yellow (pigment)

~ вод·а́ (*f*) (med) glaucoma

~ ко́рен·ь (*f*) (pharm bot) hydrastis

~ лихора́д·к·а (*f*) (med) yellow fever/jack

~ пятн·о́ (*n*) (zool) yellow spot, macula lutea

желт·я́нк·а, -и, (*g pl*) -нок·, (*f*) (bot) dyer's weed, *Reseda luteola*

желуд·ёв·ый, -ая, -ое, (*a*) acorn

желу́д·ок·, -д·к·а, (*m*) (anat) stomach, venter; crop, craw (of birds)

желу́д·очек·, -чк·а, (*m*) (anat) ventricle

желудочно·киш·е́чн·ый, -ая, -ое, (*a*) (anat) gastro-intestinal

желу́д·очн·ый, -ая, -ое, (*a*) (anat) gastric

жёлуд·ь, -я, (*nom pl*) -и, (*g pl*) -е́й, (*m*) acorn; (elec) acorn valve/tube

~, мор·ск·о́й (zool) sessile barnacle, *Balanus balanoides*

желче- (*component*) chol-, chole-, cholo-

желче·го́н·н·ый, -ая, -ое, (*a*) (pharm) cholagogue, bile-expelling

желч·е́ни·е, -я, (*n*) yellowing; yellow fermentation (of tobacco)

желче·образ·ова́ни·е, -я, (*n*) (med) biligenesis

желчно·ка́мен·н·ый, -ая, -ое, (*a*) (med) cholelithic

~ бол·е́зн·ь (*f*) (med) cholelithiasis

жёлч·н·ый, -ая, -ое, (*a*) bile, biliary, gall

~ кисл·от·а́ (*f*) (biochem) bile acid

~ про·то́к· (*m*) (anat) bile duct

~ пузы́р·ь (*m*) (zool) gall bladder

жёлч·ь, -и, (*f*) (anat) bile, gall

жёмчуг·, -а, (*m*) pearl

жемчу́ж·ин·а, -ы, (*f*) pearl; (met) bead

жемчу́ж·ниц·а, -ы, (*f*) pearl oyster

жемчу́ж·н·ый, -ая, -ое, (*a*) pearl, pearly

жен·а́, -ы́, (*nom pl*) жён·ы, (*g pl*) жён·, (*f*) wife

женѐв·ск·ий, -ая, -ое, (*a*) Geneva

жен·и́ться, -ятся, (*imp and perf*) marry

жено·по·до́б·н·ый, -ая, -ое, (*a*) effeminate

жён·ск·ий, -ая, -ое, (*a*) female, feminine

жён·щин·а, -ы, (*f*) woman

женьшён·ь, -и, (*f*) (bot) ginseng, *Panax ginseng*

жеод·а, -ы, (*f*) (geol) geode, vug

Жера́р·а, реа́кци·я, (*f*) (chem) Gerhardt reaction/test

жерд·ев·о́й, -а́я, -о́е, (*a*) of жерд·ь

жёрд·ь, -и, (*nom pl*) -и, (*g pl*) -е́й, (*f*) pole, rod

жерд·ня́к·, -а́, (*m*) plantation of young trees for poles

жеребе́йк·а, -и, (*g pl*) -е́ек·, (*f*) (met cast) chaplet

жереб·ёнок·, -нк·а, (*nom pl*) -б·я́т·а, (*g pl*) -б·я́т, (*m*) (zool) foal

жереб·е́ц·, -б·ц·а́, (*i*) -б·ц·о́м, (*m*) (zool) stallion

жёрех·, -а, (*m*) (zool) *Aspius aspius* (a carp like fish)

жерл·о́, -а́, (*nom pl*) жёрл·а, (*g pl*) жерл·, (*n*) orifice; (gunn) muzzle pipe, vent (of volcano)

жёрнов·, -а, (*nom pl*) -а́, (*g pl*) -о́в (*m*) millstone

жернов·о́й, -а́я, -о́е, (*a*) millstone

~ по·ста́в (*m*) set of millstones

же́ртв·а, -ы, (*f*) sacrifice; victim

же́ртв·овать, -уют, (*imp*)(+ *instr*) sacrifice

жерýх·а, -и,   (*f*) (bot) water cress

жест·, -а,   (*m*) gesture

жёст·к·ий, -ая, -ое,   (*a*) hard, rigid, stiff; tough, strong; fixed; rigorous, severe, stringent, drastic

~ **вагóн·**   (*m*) (rail) coach/car with wooden seating

~ **модифик·áци·я**   (*f*) (phys) stable modification

~ **о·пóр·а**   (*f*) (elec) rigid support; (mech eng) fixed foot

~ **плáстик·**   (*m*) rigid plastic

~ **трос·**   (*m*) steel wire rope

~ **трéб·овани·я**   (*pl*) stringent requirements

~ **флáн·ец·**   (*m*) fixed flange

~ **с·вяз·ь**   (*f*) (eng) rigid connection

жестк·овáт·ый, -ая, -ое,   (*a*) stiff, rather rigid

жестко·крý·л·ый, -ая, -ое,   (*a*) (zool) coleopterous, (*pl as noun*) *Coleoptera*

жёстк·ост·ь, -и,   (*f*) rigidity, stiffness; hardness (of water)

~ **вод·ы́, óбщ·ая**   total hardness of water

~ **из·гúб·а**   (mech) flexural rigidity

~ **из·луч·éни·я**   (phys) hardness of radiation

~ **круч·éни·я**   (mech) torsional rigidity

~, **магнúт·н·ая**   (phys) magnetic rigidity

~ **нá с·двиг·**   (mech) shear stiffness/modulus

    центр· жёстк·ост·и   (*m*) (mech) shear/flexural centre/center

~, **электр·úческ·ая**   (elec) elastance

жестóк·ий, -ая, -ое,   (*a*) cruel, brutal; severe

жестоко·фокус·úрующ·ий, -ая, -ое,   (*a*) strong-focussing

жёст·че   (*comp*) of **жёст·к·ий**

жестч·éни·е, -я,   (*n*) stiffening; (elec) ageing; hardening (of nuclear particles)

жест·ь, -и,   (*f*) thin sheetmetal; tin plate

~, **бéл·ая**   tin plate, terne plate

~, **луж·ён·ая**   tin plate

~, **чёрн·ая**   black plate, iron plate

жест·я́ник·, -а,   (*m*) tinsmith; metal-container fabricator

жест·я́ницк·ий, -ая, -ое,   (*a*) of **жест·ь**

~ **про·из·вóд·ств·о**   (*n*) manufacture of metal containers

жест·я́ничн·ый, -ая, -ое,   (*a*) of **жест·я́ник·**

жест·я́нк·а, -и,   (*g pl*) **-нок·,** (*f*) tin, can, metal/tin can/box, metal container

жестяно·бáн·очн·ое про·из·вóд·ств·о   (*n*) manufacture of food preserving cans

жест·ян·óй, -áя, -óе,   (*a*) of **жест·ь**

жест·я́ночн·ый, -ая, -ое,   (*a*) of **жест·я́нк·а**

жест·я́нщик·, -а,   (*m*) metal container fabricator

жетóн·, -а,   (*m*) badge, button; token, counter

жеч·ь   (*pres 3rd sing, pl*) **жж·ёт, жг·ут,** (*past masc/fem sing*) **жёг·, жг·ла,** (*imp*) burn; burn up/down; bake (bricks etc.), roast (ore etc.)

жжéние, -я,   (*n*) (*v n*) of **жеч·ь**

жж·ённ·ый, -ая, -ое,   (*past part pass*) of **жеч·ь**

жж·ён·ый, -ая, -ое,   (*a*) burnt; roasted

жж·ёт   (*pres 3rd sing*) of **жеч·ь**

жив·   (*predic adj masc*) of **жив·óй**

жив·ёт   (*pres 3rd sing*) of **жи·ть**

жив·ёте   letter "**ж**"; (*pres 2nd pl*) of **жи·ть**

живéт·ск·ий, я́рус·   (*m*) (geol) Givetian stage

жив·éц·, -в·ц·а,   (*i*) **-в·ц·óм,** (*m*) (fish) bait

жив·úтельн·ый, -ая, -ое,   (*a*) invigorating, revivifying

жив·úть, -я́т,   (*imp*) vivify, revive, bring to life

жив·úц·а, -ы,   (*i*) **-ц·ей,** (*f*) turpentine

жив·úчн·ый, -ая, -ое,   (*a*) of **жив·úц·а**

жúв·ност·ь, -и,   (*f*) small animals/creatures; domestic poultry

живо·дёр·, -а,   (*m*) flayer

жив·óй, -áя, -óе,   (*a*) alive, living, live, vital; lively, animated, keen; vivid (of colours etc.)

~ **сеч·éни·е**   (*n*) (hydr) cross section, cross sectional area

    плóщ·ад·ь жив·óго сеч·éни·я   (*f*) discharge area (of rivers)

~ **сеч·éни·е судо·хóд·н·ое**   (*n*) navigation clearance (of a waterway)

~ **сúл·а**   (*f*) manpower; kinetic energy

~ **сúл·а удáр·а**   (*f*) impact pressure

жúвокост·ь, -и,   (*f*) (bot pharm) delphinium, staphisagria

жúво·пис·ь, -и,   (*f*) painting (art)

живо·род·я́щ·ий, -ая, -ее, (*a*) (biol) viviparous, (*pl as noun*) viviparous animals

живо·рожд·е́ни·е, -я, (*n*) (zool) viviparity; (bot) vivipary

живо·сеч·е́ни·е, -я, (*n*) vivisection

жи́в·ост·ь, -и, (*f*) liveliness, animation, vivacity

живо́т·, -а, (*m*) (anat) abdomen, belly

живо·твор·и́ть, -я́т, (*imp*) vivify, bring to life

живо·тво́р·н·ый, -ая, -ое, (*a*) (biol) vivifying

живо́т·ик·, -а, (*m*) (*dim*) of **живо́т·**

~ игл·ы́ needle spring (of Jacquard weaving machine)

животно·во́д·, -а, (*m*) stock/animal breeder

животно·во́д·ств·о, -а, (*n*) animal husbandry; stock breeding

животно·во́д·ческ·ий, -ая, -ое, (*a*) of **животно·во́д·ств·о**; stock

живо́т·н·ое, -ого (*n decl as adj*) animal, beast

~, бес·позвон·о́чн·ое invertebrate

~, вью́ч·н·ое pack animal

~, млеко·пит·а́ющ·ее mammal

~, тя́г·ов·ое draught/draft animal

животно·раст·е́ни·е, -я, (*n*) (zool) zoophyte

живо́т·н·ый, -ая, -ое, (*a*) animal; (*see also* **живо́т·н·ое**) (*as noun*)

~ электр·и́честв·о (*n*) animal electricity

жив·у́чест·ь, -и, (*f*) vitality; ability to keep alive, viability, tenacity; (naut) ability to keep afloat; (mech eng) life

систе́м·а жив·у́чест·и (*f*) (naut) damage control system

жив·у́ч·ий, -ая, -ее, (*a*) viable, tenacious of life

жив·у́щ·ий, -ая, -ее, (*pres part act*) of **жи·ть** living

жи́в·чик·, -а, (*m*) (gen) spermatozoon; (bot) spermatozoid

жи́в·ш·ий, -ая, -ее, (*past part act*) of **жи·ть**

жив·я́ (*pres gerund*) of **жи·ть**; living, while living, alive

жиг·а́лк·а, -и, (*g pl*) **-лок·,** (*f*) (ent) stable fly, *Stomoxys calcitrans*

жи́дк·ий, -ая, -ое, (*a*) fluid, liquid

~ бу́нкер· (*m*) bunker oil

~ во́з·дух· (*m*) liquified air

~ грунт· (*m*) (ocean) thermal layer

жи́дк·ий, -ая, -ое

~ мета́лл· (*m*) liquid/moulten/molten metal

~ плав· (*m*) liquid melt, melt

~ систе́м·а (*f*) (phys, chem) fluid system

~ со·сто·я́ни·е (*n*) liquid state

~ стекл·о́ (*n*) (chem) waterglass

~ те́л·о (*n*) liquid, fluid

~ то́плив·о (*n*) liquid fuel

~ цикл· (*m*) fluid cycle

жидко·металл·и́ческ·ий, -ая, -ое, (*a*) liquid-metal

жидко·пла́в·к·ий, -ая, -ое, (*a*) liquid, fluid

жидко·пла́в·кост·ь, -и, (*f*) liquidity, fluidity

жидко·стекл·ова́т·ый, -ая, -ое, (*a*) semivitreous, liquevitreous

жидкосте·ме́р·, -а, (*m*) fluidimeter

жидкостно·реакти́в·н·ый дзи́г·а·тел·ь (*m*) (a/c) liquid-fuel rocket motor

жи́дк·остн·ый, -ая, -ое, (*a*) liquid, fluid; liquid-immersed; (ICE) liquid-fuel, running on liquid fuel; (instr) liquid column/filled/expansion; fluid-flow

~ за·де́рж·к·а (*f*) liquid hold-up

~ за·тво́р· (*m*) liquid/hydraulic seal

~ ма́но·метр· (*m*) liquid-column pressure gauge/gage

~ плён·к·а (*f*) fluid film

~ со·един·е́ни·е (*n*) liquid junction

жи́дк·ост·ь, -и, (*f*) liquid, fluid, liquor; fluidity

~, баро·кли́н·н·ая baroclinic fluid

~, вордо́с·ск·ая (agr chem) Bordeaux mixture

~, вя́з·к·ая viscous fluid, Newtonian fluid

~, вяз·к·о-у·пру́г·ая viscous-elastic fluid/liquid

~, идеа́ль·н·ая ideal/perfect fluid, inviscid fluid

меха́н·ик·а жи́дк·ост·и (*f*) fluid mechanics

~, не·по·дви́ж·н·ая (hydrodyn) fluid at rest

~, не·с·жиж·а́ем·ая incompressible fluid

~, одно·компоне́нт·н·ая single fluid

~, от·рабо́т·анн·ая spent/discharge liquor/liquid

~, о·хлажд·а́ющ·ая coolant; cutting fluid (machine tools)

жи́дк·ост·ь
~, **по·крыв·а́ющ·ая** supernatant liquor
~, **про·мы́в·очн·ая** wash liquid; (oil) drilling fluid
~, **рабо́ч·ая** (hydr) power/working/hydraulic fluid/liquid; braking fluid (of brake); pressure fluid (of press) **экстра́кци·я жи́дк·ост·и жи́дк·ост·ью** (f) liquid-liquid extraction
~, **я́дер·н·ая** nuclear fluid
жидко·тек·у́чест·ь, -и, (f) (phys) fluidity; (met) castability
жидко·тек·у́щ·ий, -ая, -ее, (a) fluid, fluid-flow
жи́ж·а, -и, (i) **-ей,** (f) liquor, liquid
~, **на·во́з·н·ая** liquid manure
жиже·при·ём·ник·, -а, (m) liquid manure sump/tank
жиже·с·бо́р·ник·, -а, (m) see **жиже·при·ём·ник·**
жизне·де́·ятельност·ь, -и, (f) (biol) vital activity; occupation
жизне·де́·ятельн·ый, -ая, -ое, (a) active, vital
жи́зн·енност·ь, -и, (f) (biol) life, vitality
жи́зн·енн·ый, -ая, -ое, (a) life's; vital
жизне·о·пис·а́ни·е, -я, (n) biography
жизне·спосо́б·ност·ь, -и, (f) (biol) viability; (plast) life, storage/shelf life
жизне·спосо́б·н·ый, -ая, -ое, (a) viable, vital
жизне·сто́й·к·ий, -ая, -ое, (a) resilient, enduring
жизн·ь, -и, (f) life, vita
жиклёр·, -а, (m) jet
~ **карбюра́тор·а** carburettor jet
~ **ма́л·ых оборо́т·ов** (ICE) slow-speed jet, idle jet
~, **пуск·ов·о́й** (ICE) slow-speed jet, idle jet
~ **холост·о́го хо́д·а** idle jet
жи·л (past masc sing) of **жи·ть**
жи́л·а, -ы, (f) (pop) vein, sinew, tendon; (geol) vein, lode, seam; (elec) core, strand (of cable); (med) catgut
~, **бога́т·ая** (min) pay streak
~, **изол·и́рованн·ая** core (of cable)
~, **ква́рц·ев·ая** (geol) quartz reef
~, **конта́кт·н·ая** (geol) contact lode
~, **не·про·мы́шл·енн·ая** (min) dead lode, unkindly lode
~, **сух·а́я** (anat) sinew
~, **токо·про·вод·я́щ·ая** (elec) conductor (of cable)

жи́л·а
~, **трещ·и́нн·ая** (min) true lode
~, **у·плот·нён·н·ая** (elec) segmental-strand conductor
жиле́т·, -а, (m) waistcoat (Brit), vest (U.S.)
~, **с·пас·а́тельн·ый** life jacket
жи́л·ист·ый, -ая, -ое, (a) sinewy; heavily veined
жил·и́щ·е, -а, (i) **-ем,** (n) dwelling, residence, abode, living quarters
жил·и́щн·о-бы́т·ов·о́й, -а́я, -о́е, (a) housing, domestic
жил·и́щн·о-коммуна́ль·н·ый от·де́л· (m) accommodation and housing department
жил·и́щн·ый, -ая, -ое, (a) housing
жи́л·к·а, -и, (g pl) **-л·ок·,** (f) (dim) of **жи́л·а;** (biol) venule, nervure, nerve; fibre, vein, fiber; (min) vein, stringer
~, **рези́н·ов·ая** elastic thread
жил·кова́ни·е, -я, (n) venation, veining, neuration
жил·кова́т·ый, -ая, -ое, (a) veined; fibrous, stringy
жи·л·о́й, -а́я, -о́е, (a) dwelling, living, residential
~ **дом·** (m) house, dwelling house
~ **по·мещ·е́ни·е** (n) accommodation, living space/quarters
жи́л·очк·а, -и, (g pl) **-чек·,** (f) (dim) of **жи́л·к·а;** fibril, string (of a fibre)
жил·стро·и́тельств·о, -а, (n) house building
жил·фо́нд·, -а, (m) housing allocation
жил·ь·ё, -я́, (n) domicile, residence; habitation
жи́ль·н·ый, -ая, -ое, (a) (min) vein
~ **по·ро́д·а** (f) vein material; lode, rock, matrix
жир·, -а, (nom pl) **-ы́,** (g pl) **-о́в,** (m) fat
~, **гидр·и́рованн·ый** hydrogenated fat
~, **жи́дк·ий** (food) soft oil, oleo oil
~, **под·ко́ж·н·ый** (zool) blubber
~, **полу·твёрд·ый** lard, butter
~, **ры́б·ий** fish oil
~, **свин·о́й** (food) lard
~, **твёрд·ый** tallow; (food) hard fat, oleostearine
жира́нт·, -а, (m) (fin) endorser, indorser
жира́т·, -а, (m) (fin) endorsee, indorsee
жира́тор·н·ый, -ая, -ое, (a) gyratory
жира́ф·, -а, (m) (zool) giraffe
жир·е́·ть, -ют, (imp) fatten, grow fat

жир·н·о- (*component*) fatty, aliphatic

жирно·аромат·и́ческ·ий, -ая, -ое, (*a*) (chem) fatty-aromatic

жирно·моло́ч·ност·ь, -и, (*f*) fat content, butter fat content (of milk)

жирно·моло́ч·н·ый, -ая, -ое, (*a*) high-grade dairy (of cattle)

жи́р·ност·ь, -и, (*f*) fat content

жи́р·н·ый, -ая, -ое, (*a*) fat, greasy, oily; rich (in content), fertile (of soil); (chem) fatty, aliphatic; thick, (print) bold, black, heavy

~ **газ·** (*m*) rich gas

~ **кисл·от·а́** (*f*) (chem) fatty acid

~ **со·един·е́ни·е** (*n*) (chem) aliphatic compound

жи́ро (*n indecl*) (fin) endorsement, indorsement

жиро- (*component*) fat, fatty, aliphatic, lipo-; gyro- (*and see* **гиро-**); (fin) (implies the use of crossed cheques and other non-cash methods of payment)

жир·ова́ни·е, -я, (*n*) greasing, grease impregnation, fat-liquoring (leather)

жир·ови́к·, -а́, (*m*) (min) soapstone, steatite; (med) lipoma

жир·о́вк·а, -и, (*g pl*) **-вок·,** (*f*) fat liquor (leather processing)

жир·ов·о́й, -а́я, -о́е, (*a*) fat, fatty, adipose, lipoid, aliphatic

~ **вещ·еств·о́** (*n*) fatty matter, fat

~ **ткан·ь** (*f*) (zool) adipose tissue

жиро·во́ск·, -а, (*m*) (anat) adipocere

жиро·клино́·метр·, -а, (*m*) (air) artificial horizon

жиро·ко́мпас·, -а, (*m*) *see* **гиро·-ко́мпас·**

жиро·ло́в·к·а, -и, (*g pl*) **-в·ок·,** (*f*) fat collector, grease trap (sewerage systems)

жиро·моло́ч·ност·ь, -и, (*f*) *see* **жирно·моло́ч·ност·ь**

жиро·не·про·ниц·а́ем·ый, -ая, -ое, (*a*) greaseproof

жиро·оборо́т·, -а, (*m*) (fin) paper transaction

жиро·об·раба́т·ывающ·ий аппара́т· (*m*) animal fat processing plant

жиро·пла́н·, -а, (*m*) gyroplane

жиро·по·до́б·н·ый, -ая, -ое, (*a*) lipoid

жиро·по́т·, -а, (*m*) suint, (text) wool sweat

жиро·при·ка́з·, -а, (*n*) (fin) banker's order

жиро·рас·твор·и́м·ый, -ая, -ое, (*a*) fat soluble, (biol) liposoluble

жиро·рас·твор·я́ющ·ий, -ая, -ее, (*a*) fat-dissolving

жиро·рас·чёт·, -а, (*m*) (fin) non-cash settlement, paper transaction

жиро·рас·цепл·е́ни·е, -я, (*n*) (bio-chem) adipolysis, lipolysis

жиро·ре́ктор·, -а, (*m*) (a/c) artificial horizon

жиро·ско́п·, -а, *see* **гиро·ско́п·**

жиро·со·держ·а́щ·ий, -ая, -ее, (*a*) fat-containing, fatty, adipose

жиро·ста́т·ик·а, -и, (*f*) gyrostatics

жиро·стат·и́ческ·ий, -ая, -ое, (*a*) gyrostatic

жиро·топл·е́ни·е, -я, (*n*) rendering (of fats)

жиро·че́к·, -а, (*m*) (fin) crossed cheque/check

жи́т·ел·ь, -я, (*m*) inhabitant, dweller, resident

жит·ня́к·, -а, (*m*) (bot) wheatgrass, *Agropyron*

жи́т·о, -а, (*n*) (agr) corn

жи́·ть (*pres 3rd sing, pl*) **жив·ёт, жив·у́т,** (*imp*) live

жмак·, -а, (*m*) *see* **жмых·** (agr) oilcake

жм·ёт (*pres 3rd sing*) of **жат·ь** (*see sub-entries* **жм-**)

жм·у́чи (*pres gerund*) of **жат·ь** (*see sub-entries* **жм-**)

жм·у́щ·ий, -ая, -ее, (*pres part act*) of **жат·ь** (*see sub-entries* **жм-**)

жмых·, -а́, (*m*) (agr) oilcake

~**, клещеви́н·н·ый** castor cake, castor pomace

~**, льн·ян·а́я** linseed oil cake

жмых·о́в·ый, -ая, -ое, (*a*) of **жмых·**; extracted

~ **мук·а́** extracted meal

жмыхо·дроб·и́лк·а, -и, (*g pl*) **-лок·,** (*f*) oilcake breaker

жн·е́йк·а, -и, (*g pl*) **-еек·,** (*f*) *see* **жа́т·-к·а, жн·е·я́** reaper

жн·ёт (*pres 3rd sing*) of **жат·ь** (*see sub-entries* **жн-**)

жн·е·я́, -и́, (*see also* **жа́т·к·а**) (agr) reaper

жн·е·я́-лобо·гре́·йк·а, -и, (*f*) (agr) manual delivery reaper

жн·е·я́-снопо·вяз·а́лк·а, -и, (*f*) (agr) reaper-binder

жн·и́в·о, -а, (*n*) (agr) stubble

жн·ивь·ё, -я́, (*n*) (agr) stubble field, harvested cornfield жн·у́щ·ий, -ая, -ее, (*pres part act*) of жат·ь (*see sub-entries under* жн-)

жн·я́ (*pres gerund*) of жат·ь (*see sub-entries under* жн-)

жоке́·й, -я, (*m*) jockey

жо́лоб·, -а, (*m*) *see* жёлоб·

жо́луд·ь -я, (*m*) *see* жёлуд·ь

жо́м·, -а, (*m*) squeezer; pulp, press pulp; bagasse (sugar cane/beet/pulp)

жо́м·ов·ый, -ая, -ое, (*a*) of жом·

жорда́н·ов·ый, -ая, -ое, (*a*) (math) Jordan

жр·ать, -ут, (*imp*) guzzle, gorge

ЖРД (*abbr*) of жи́дк·остн·о-реакти́в·н·ый дви́г·ател·ь liquid-fuel rocket motor

жреб·и·й, -я, (*m*) lot (for chance selection)

жр·у́щ·ий, -ая, -ее, (*pres part act*) of жр·ать

жу·ёт (*pres 3rd sing*) of жев·а́ть

жу́ж·елиц·а, -ы, (*f*) (ent) ground beetle, (*pl*) *Carabidae*

жужж·а́л·а, -ы, (*f*) (ent) beefly, (*pl*) *Bombyliidae*

жужж·а́льц·а, -ы, (*f*) (ent) halteres

жужж·а́ни·е, -я, (*n*) (*v n*) *see* жужж·а́ть

жужж·а́ть, -а́т, (*imp*) hum, buzz, drone

жу́к·, -а́, (*m*) (ent) beetle

~, карто́фель·н·ый Colorado beetle, *Doryphora decemlineata*

~, хле́б·н·ый corn weevil, *Calandra granaria*

жу́ль·нича·ть, -ют, (*imp*) swindle, cheat

жу́ль·ническ·ий, -ая, -ое, (*a*) fraudulent

жура́вл·ь, -я́, (*m*) (zool) crane, *Grus*; microphone boom; (mech eng) sweep; (astron) Crane, Grus

~, кран- mast-jib crane

жура́вчик·, -а, (*m*) (geol) loess doll, lime nodule

Журде́н·а, при́нцип· (*m*) (mech) Jourdain principle

жур·и́ть, -я́т, (*imp*) reprove, rebuke

журна́л·, -а, (*m*) periodical, magazine, journal; journal, register, diary; log, log-book

~, борт·ов·о́й (air) flight log

~, ва́хт·енн·ый (naut) log book, deck log

~, в·ход·я́щ·ий "In" register

~ за·сед·а́ни·й minute-book

~, ис·ход·я́щ·ий "Out" register

~, суд·ов·о́й ship's log book

журнал·и́ст·, -а, (*m*) journalist

журнал·и́стик·а, -и, (*f*) journalism; periodical publications

журч·а́лк·а, -и, (*g pl*) -лок·, (*f*) hover fly, (*pl*) *Syrphidae*; bulb fly, *Eumerus*

журч·а́ть, -а́т, (*imp*) gurgle, splash, tinkle (of water)

жу́т·к·ий, -ая, -ое, (*a*) awful, awe-inspiring, sinister

жуч·о́к·, -ч·к·а́, (*m*) (*dim*) of жук·

жу·ю́щ·ий, -ая, -ее, (*pres part act*) of жев·а́ть

жюри́ (*n indecl*) judges, jury (of competition etc.); law jury (not in U.S.S.R.)

З

з (*abbr*) of **за́·пад·н·ый** western, West

за (*prep+accusative or instrumental*) behind, beyond, the other side of, outside, at, after (*acc with movement, instr with location or relative position*); for; by (use of part of an object); because of, owing to (*instr*); for, at (of distance) (*acc*); during, for (of time), within, in (if time limit imposed)

~ **год** a year ago

~ **город·ом** out of town, outside the town

~ **и про́тив** for and against

~ **ис·ключ·éни·ем** with the exception of, except for

ни за что by no means; under no circumstances

~ **от·су́т·стви·ем** for the lack of, owing/due to the lack/absence of

о́·черед·ь за ва́ми your turn

~ **пят·ь рубл·éй** for five roubles

сест·ь за стол· sit down at/beside/behind the table

слóв·о за ва́ми speak now, will you now speak

ста·ть за пульт· у·правл·éни·я go to the control desk, man the control desk

~ **с·чёт·** at the expense of; due to, because of

что за what sort of, what

~ **что чтоб** so that, in order that

шаг за ша́г·ом step by step

за- (*prefix*) *signifies*: (1) beginning of action; (2) movement to beyond, behind; (3) movement into, particularly as part of other activity; (4) formation of perfective aspect and derived imperfectives, hence often sense of action completed or to be completed; (5) action to extreme degree, excess (with and without **-ся**); (6) (*as adj prefix*) location beyond, on the far side of

за·аплод·и́р·овать, -уют, (*perf*) start applauding/clapping, break into applause

за·аренд·ова́ть, -у́ют, (*perf*) *see* **за·-аренд·óвыва·ть**

за·аренд·óвыва·ть, -ют, (*imp*) rent, lease

за·аррет·и́р·овать, -уют, (*perf*) cage (a gyro)

за·асфальт·и́р·овать, -уют, (*perf*) asphalt, cover with asphalt

за·атлáнт·и́ческ·ий, -ая, -ое, (*a*) transatlantic

за·бáвл·я́·ть, -ют, (*imp*) amuse

за·бáв·н·ый, -ая, -ое, (*a*) amusing, funny

за·баллот·и́р·овать, -уют, (*perf*) *see* **за·баллот·ир·óвыва·ть**

за·баллот·ир·óвыва·ть, -ют, (*imp*) not elect, vote against, blackball

забаст·ова́ть, -у́ют, (*perf*) go on strike, come out on strike, strike

забастóвк·а, -и, (*g pl*) **-вок·,** (*f*) strike

~, **италь·я́нск·ая** sit-down strike

за·бв·éни·е, -я, (*n*) oblivion

за·бéг·, -а, (*m*) (*v n*) *see* **за·бег·á·ть;** run, sprint (over set distance); overshoot, overswing (of pointer)

за·бéг·, -а, (*m*) race, run, sprint (if short)

за·бег·á·ть, -ют, (*imp*) run in/off/away; hurry in and out; outrun, run past; (*perf*) begin to run to and for

~ **в·перёд·** get ahead; get ahead of schedule

за·беж·á·ть (*fut 3rd sing, pl*) **за·беж·и́т, за·бег·у́т,** (*perf*); *see* **за·бег·á·ть**

за·беж·к·а, -и, (*f*) (naut) kedging boat

за·бéл·ива·ть, -ют, (*imp*) whiten

за·бел·и́ть, -я́т, (*perf*) *see* **за·бéл·ива·ть**

зá·берег·, -а, (*m*) (ocean) submerged off-shore ice, submerged land-fast ice

~, **вод·ян·óй** off-shore water (in ice)

~, **сквоз·н·óй вод·ян·óй** clear off-shore water (caused by spring thaw)

зá·береж·ь·е, -я, (*n*) *see* **зá·берег·**

за·берéмен·е·ть, -ют, (*perf*) become pregnant

за·бер·у́т (*fut 3rd pl*) of **за·бр·á·ть**

за·бетон·и́рованн·ый, -ая, -ое, (*past part pass*); concreted, concreted in, concrete-enveloped

за·бив·á·ть, -ют, (*imp*) knock/drive/hammer in; stuff, fill in, pack (cracks etc.); board up; choke up, choke, jam; obscure, confuse; slaughter, kill

за·би́в·к·а, -и, (*g pl*) **-в·ок·,** (*f*) (*v n*) of **за·бив·á·ть**

~ **кост·ы́л·ей** spiking

~ **свá·й** pile driving

сигнáл за·би́в·к·и (*m*) (telecom) erase/erasure signal

за·бив·н·óй, -áя, -óе, (*a*) packing, stuffing; knocking, driving

~ креп·ь (*f*) (min) spilling, spills, spiles, forepoling

за·бинт·овáть, -ýют, (*perf*) *see* **за·бинт·óвыва·ть**

за·бинт·óвыва·ть, -ют, (*imp*) bandage; bind, lag

за·бир·á·ть, -ют, (*imp*) take in/up; (naut) heave/shorten in; collect, gather; **-ся** (*pass*); get into/onto, penetrate

за·бúр·к·а, -и, (*f*) fencing, safety fencing (round pits etc.)

за·бú·т·ый, -ая, -ое, (*past part pass*) *see* **за·бив·á·ть**

за·бú·ть (*fut 3rd sing, pl*) **за·бь·ёт, за·бь·ют,** (*perf*); *see* **за·бив·á·ть;** begin to beat; (oil) gush

за·благо·врéмен·н·о (*adv*) in good time

за·благо·врéмен·ност·ь прогнóз·а (*f*) term of forecast

за·блест·éть (*fut 3rd pl*) **за·блéщ·ут** *or* **за·блест·я́т** (*perf*) shine, become shiny

за·блéщ·ут (*fut 3rd pl*) of **за·блест·éть**

за·блуд·úться, -я́тся, (*perf*) get/ become lost, lose one's way

за·блужд·á·ться, -ются, (*imp*) be mistaken, be in error

за·блужд·éни·е, -я, (*n*) error, delusion, fallacy

за·бó·ечн·ый, -ая, -ое, (*a*) of **за·бóй·-к·а**

~ материáл· (*m*) (min expl) stemming

за·бó·й, -я, (*m*) (min) face, stope; bottom (of bore hole); slaughter

~, вéрх·н·ий (min) high face

~, дéй·ствующ·ий (min) working face

~ кáмер·ы (min) room face

~, корóт·к·ий (min) shortwall
лúн·и·я за·бó·ев· целик·óв (*f*) (min) advance line

~, о·чúст·н·óй (min) stope

~, под·готóв·úтельн·ый shortwall

~, сплош·н·óй (min) continuous stope

~, ýз·к·ий (min) shortwall

за·бóй·к·а, -и, (*f*) tamping (for blasting); stemming (material)

за·бóй·н·ый, -ая, -ое, (*a*) of **за·бó·й**

~ давл·éни·е (*n*) bottom-hole pressure (oil well)

~ рабóч·ий (*m*) (min) faceman, faceworker

за·бóй·щик·, -а, (*m*) (min) miner, cutter; slaughterman

за·болáч·ивани·е, -я, (*n*) swamping, flooding, bogging

за·болáч·ива·ться, -ются, (*imp*) become boggy

за·бол·евáемост·ь, -и, (*f*) morbidity

за·бол·евáни·е, -я, (*n*) disease, illness, sickness

~, за·рáз·н·ое infectious disease

~, луч·ев·óе radiation sickness

~, профессионáль·н·ое occupational disease

за·бол·евá·ть, -ют, (*imp*) fall/become ill; begin to ache, ache

за·бол·é·ть, -ют, (*perf*) (+*instr*) fall/ become ill; **за·бол·éть, -я́т,** (*perf*) begin to ache, ache

зá·болон·ник·, -а, (*m*) (ent) bark beetle, *Scolytus*

зá·болон·ь, -и, (*f*) sapwood, alburnum

за·болóч·енн·ый, -ая, -ое, (*a*) swampy, boggy

за·болóт·иться, -я́тся, (*perf*) *see* **за·-болáч·ива·ться**

за·бóр·, -а, (*m*) fence, hedge, partition, enclosure; intake, collection; purchases, order (in shop)

~ вóз·дух·а air intake

за·бóр·к·а, -и, (*f*) partition

за·бóр·н·ый, -ая, -ое, (*a*) of **за·бóр·;** gathering, collecting

~ кнúж·к·а (*f*) ration book; (obs) credit account book

за·борон·úть, -я́т, (*perf*) (agr) harrow, finish harrowing

за·бóрт·н·ый, -ая, -ое, (*a*) outboard, overboard, outside (a ship)

~ арматýр·а (*f*) sea valves

~ вод·á (*f*) sea water

~ клáпан· (*m*) (shipb) sea connection, Kingston (valve)

~ трап· (*m*) accommodation ladder

за·борт·óвк·а, -и, (*f*) beading

забóт·а, -ы, (*f*) worry, care, trouble, anxiety

забóт·ить, -я́т, (*imp*) cause worry/ anxiety, trouble; **-ся** (*pass*); care for, look after

забóт·ливост·ь, -и, (*f*) solicitude, care, thoughtfulness

забóч·у (*pres 1st sing*) of **забóт·ить**

за·брак·óванн·ый, -ая, -ое, (*past part pass*) of **за·брак·овáть;** reject, scrap

за·брак·овáть, -ýют, (*perf*) reject, condemn, scrap

за·брак·о́вк·а, -и, (*f*) (*v n*) *see* **за·бра-к·ова́ть;** rejection

за́·бр·анн·ый, -ая, -ое, (*past part pass*) *see* **за·бир·а́·ть**

за·бра́с·ывани·е, -я, (*n*) (*v n*) *see* **за·бра́с·ыва·ть**

за·бра́с·ыва·ть, -ют, (*imp*) throw, throw away, abandon; deliver, supply; throw in, fill, cover up with; bespatter, shower with

за·бр·а́ть (*fut 3rd sing, pl*) **за·бер·ёт, за·бер·у́т,** (*perf*); *see* **за·бир·а́·ть**

за·бред·а́·ть, -ют, (*imp*) stray, wander; wander in

за·бре́д·ить, -ят, (*perf*) become delirious

за·бред·у́т (*fut 3rd pl*) of **за·брес·ти́**

за·бре́зж·ить, -ат, (*perf*) begin to gleam

за·брес·ти́ (*fut 3rd sing, pl*) **за·бре-д·ёт, за·бред·у́т,** (*past masc sing*) **за·брёл,** (*perf*) stray, wander; wander in

за·брон·и́рованн·ый, -ая, -ое, (*past part pass*) of **за·брон·и́р·овать;** book-ed, reserved; **за·брон·иро́ван·ный, -ая, -ое,** (*past part pass*) of **за·брон·ир·ова́ть;** armoured, armored, armour/armor-plated

за·брон·и́р·овать, -уют, (*perf*) reserve, book; **за·брон·ир·ова́ть, -у́ют,** (*perf*) armour, armor

за·бро́с·, -а, (*v n*) *see* **за·бра́с·ыва·ть;** neglect; jump, sudden move, (*instr*) overswing

~ **оборо́т·ов** (mech) overspeed

за·брос·а́·ть, -ют, (*perf*) throw in, fill, cover up with; bespatter, shower with

за·бро́с·ить, -ят, (*perf*) throw; throw away, abandon, give up; neglect; deliver, supply

за·бро́ш·енн·ый, -ая, -ое, (*past part pass*) of **за·бро́с·ить;** derelict, neglected

за·бры́зг·а·ть, -ют, or **за·бры́зж·ут,** (*perf*) *see* **за·бры́зг·ива·ть**

за·бры́зг·ива·ть, -ют, (*imp*) splash with, bespatter

за·бу́д·ем (*fut 1st pl*) of **за·бы́·ть**

за·бу́р·ива·ть, -ют, (*imp*) start boring/drilling; drill/bore into; plough into

~ **шпур·** (min) collar a hole

за·бур·и́ть, -я́т, (*perf*) *see* **за·бу́р·и-ва·ть**

за·бу́р·к·а, -и, (*f*) (oil) spudding in

забуру́н·ь, -и, (*f*) (geog) sandbank, bar

за·бурья́н·енн·ый, -ая, -ое, (*a*) weedy, weed-infested

за·бут·и́ть, -я́т, (*perf*) (civ eng) fill with rubble

за·бут·о́вк·а, -и, (*g pl*) **-вок·,** (*f*) (civ eng) filling with rubble; rubble packing, hard core

за·бут·о́вщик·, -а, (*m*) (min) packer

за·быв·а́·ть, -ют, (*imp*) forget; **-ся** doze off; become oblivious to one's surroundings; fall into deep concentration; lose self control; forget, be forgotten

за·бы́в·чивост·ь, -и, (*f*) forgetfulness, absent-mindedness; (med) amnesia

за·бы́·ть (*fut 3rd sing, pl*) **за·бу́д·ет, за·бу́д·ут,** (*perf*); *see* **за·быв·а́·ть**

за·бы́т·ь·ё, -я́, (*n*) drowsiness; loss of memory; trance, deep concentration; loss of self control

за·бь·ю́т (*fut 3rd pl*) of **за·би́·ть**

зав (*abbr*) of **за·ве́д·ующ·ий;** (*abbr*) of **за·во́д·ск·ий**

за·ва́л·, -а, (*m*) heap, pile, obstruction, landslide, avalanche, fall; (med) obstruction, constipation; (air) flapping; steep slope/drop (of a curve)

~ **борт·о́в** (naut) tumble home

~ **-та́л·и, -ей,** (*pl*) (naut) guy (of derrick)

за·ва́л·енн·ый, -ая, -ое, (*past part pass*) of **аа·вал·и́ть;** heaped/piled/filled/clogged up; fallen in, fallen

за·ва́л·ива·ть, -ют, (*imp*) heap/fill/block up with, clog, choke; tumble, knock/roll over; **-ся** (*pass*); fall, fall down/in/behind, slip down/behind

~ **я́кор·ь по по·хо́д·н·ому** (naut) secure the anchor for sea

за·вал·и́ть, -я́т, (*perf*) *see* **за·ва́л·и-ва·ть**

за·ва́л·к·а, -и, -л·ок·, (*f*) charging, filling (with non-liquid material); (met) cold charge, (furnace) cold charging

за·ва́л·очн·ый, -ая, -ое, (*a*) of **за-·ва́л·к·а**

~ **маши́н·а** (*f*) (met) charging machine

~ **маши́н·а выс·о́к·ого ти́п·а, на-·по́ль·н·ая** (*f*) high ground-type charging machine

~ **маши́н·а кра́н·ов·ая** (*f*) overhead-type charging machine

~ **маши́н·а мост·ов·а́я шаржи́р·-н·ая** (*f*) overhead-type charging machine

за·ва́л·очн·ый, -ая, -ое
~ маши́н·а, на·по́ль·н·ая (*f*) ground-type charging machine
~ маши́н·а, под·вес·н·а́я (*f*) overhead-type charging machine
~ окн·о́ (*n*) charging door
за·ва́р·енн·ый, -ая -ое, (*past part pass*) *see* за·ва́р·ива·ть; scalded; (met) welded/closed up; sealed in
за·ва́р·ива·ть, -ют, (*imp*) pour boiling water on to, scald, brew (tea etc.); weld up, close up
за·вар·и́ть, -ят, (*perf*) *see* за·ва́р·ива·ть
за·ва́р·к·а, -и, (*g pl*) -р·ок·, (*f*) (*v n*) *see* за·ва́р·ива·ть; welding up, repair welding; brazing; weld, seal
~, холо́д·н·ая brazing
за·вед·е́ни·е, -я, (*n*) establishment, institution
~, вы́с·ш·ее уч·е́бн·ое higher educational establishment, university
~, уч·е́бн·ое educational/training establishment
за·ве́д·овать, -уют, (*imp*) (+*instr*) manage, superintend
за·ве́д·ом·о (*adv*) knowingly, deliberately; for certain, *a priori*; as is well known
за·ве́д·ующ·ий, -ая, -ее, (*pres part act*) of за·ве́д·овать; managing, superintending; (*m decl as adj*) manager, head, superintendent, officer in charge
~ институ́т·ом director of the/an institute
~ про·до·во́ль·стви·ем (naut) purser
~ уч·е́бн·ой ча́ст·ью director of studies
за·вез·ти́, -у́т, (*perf*) *see* за·воз·и́ть
за·верб·ова́ть, -у́ют, (*perf*) recruit, enlist
за·вер·е́ни·е, -я, (*n*) (*v n*) *see* за·-вер·я́·ть; assurance, certification, confirmation
за·ве́р·ить, -ят, (*perf*) *see* за·вер·я́·ть
за·вер·н·у́ть, -у́т, (*perf*) *see* за·вёр·т·ыва·ть
за·верт·е́·ть, -ют, (*perf*) rotate, revolve, spin; begin to rotate/revolve/spin
за·вёрт·к·а, -и, (*g pl*) -т·ок·, (*f*) (*v n*) *see* за·вёрт·ыва·ть; wrapping, wrapper; door bolt
за·вёрт·очн·ый, -ая, -ое, (*a*) of за·вёрт·к·а
~ маши́н·а (*f*) wrapping machine

за·вёрт·ывани·е, -я, (*n*) (*v n*) *see* за·вёрт·ыва·ть; involution
~ на ли́н·и·ю (math) involution on a line
за·вёрт·ыва·ть, -ют, (*imp*) wrap/parcel up; drop in at, call at; screw up; turn off (tap); turn/tuck/roll up; (naut) make fast, belay; (*intrans*) turn
за·ве́рт·ь, -и, (*f*) eddy, whirlpool
за·верш·а́·ть, -ют, (*imp*) complete, conclude, finish, consummate
за·верш·е́ни·е, -я, (*n*) completion, consummation, conclusion
за·верш·ённ·ый, -ая, -ое, (*past part pass*) *see* за·верш·а́·ть; complete, completed, final; perfect
за·верш·и́·ть, -ат, (*perf*) *see* за·вер-ш·а́·ть
за·вер·я́·ть, -ют, (*imp*) assure; confirm (subordinate's action); witness, attest, certify
за·ве́с·а, -ы, (*f*) curtain, screen, veil
~, дым·ов·а́я smoke screen
за·ве́с·ить, -ят, (*perf*) *see* за·ве́ш·и·ва·ть
за·вес·ти́ (*fut 3rd sing, pl*) за·вед·ёт, за·вед·у́т, (*past masc sing*) за·вёл, (*perf*); *see* за·вод·и́ть
за·ве́т·, -а, (*m*) testament, legacy
за·ве́ш·ивани·е, -я, (*n*) curtaining, screening
за·ве́ш·ива·ть, -ют, (*imp*) curtain, screen
за·вещ·а́ни·е, -я, (*n*) will, testament
за·вещ·а́тел·ь, -я, (*m*) testator
за·вещ·а́тельн·ый, -ая, -ое, (*a*) testamentary
за·вещ·а́·ть, -ют, (*imp and perf*) bequeath
за·вив·а́ни·е, -я, (*n*) (*v n*) of за·вив·а́·ть
за·вив·а́·ть, -ют, (*imp*) curl, curl up, twist, wave
за·ви́в·к·а, -и, (*f*) curling, waving, twirling; curl (of hair)
за·ви́в·очн·ый, -ая, -ое, (*a*) of за·-ви́в·к·а
~ маши́н·а (*f*) (food) can-sealing machine, edge-curling machine
за·ви́д·еть, -ят, (*perf*) catch sight of
за·ви́д·н·ый, -ая, -ое, (*a*) enviable
за·ви́д·овать, -уют, (*imp*) (+ *dat*) envy
за·виз·и́р·овать, -уют, (*perf*) grant a visa; enter a visa (in passport etc.)
за·вил·я́·ть, -ют, (*perf*) yaw, weave (motion); begin to yaw/weave

за·винт·и́ть, -я́т, (*perf*) *see* **за·ви́нч·ива·ть**

за·ви́нч·ива·ть, -ют, (*imp*) screw up

за·вис·а́ни·е, -я, (*n*) hovering (in the air); sticking, hanging (e.g. of charges in a high furnace)

за·вис·а́·ть, -ют, (*imp*) hover, hang in mid air, be suspended in mid air; stick, hang fire

за·ви́с·еть, -ят, (*imp*) depend (on), be dependent upon/on

за·ви́с·имост·ь, -и, (*f*) dependence, relation

в за·ви́с·имост·и as a function of, versus, against (of graphs)

~, времен·н·а́я time dependence

~, не·пере·хо́д·н·ая intransitive dependence

~ от вре́мен·и time dependence

~ температу́р·ы от вре́мен·и time–temperature relation/curve, heating curve, time versus temperature

за·ви́с·им·ый, -ая, -ое, (*a*) dependent

за́вист·ь, -и, (*f*) envy

за·ви́с·ящ·ий, -ая, -ее, (*pres part act*) of **за·ви́с·еть;** depending, dependent

за·ви·т·о́й, -а́я, -о́е, (*a*) waved, curled, spiraled, twisted; (bot) involute

за·вит·о́к·, -т·к·а́, (*m*) curl, convolution, whorl, loop (of spiral); (bot) bostryx; (anat) helix; (arch) volute, scroll

за·ви́·т·ый, -ая, -ое, (*past part pass*) of **за·ви́·ть;** waved, curled, spiraled

за·ви́·ть (*fut 3rd sing, pl*) **за·вь·ёт, за·вь·ю́т,** (*past masc sing*) **за·ви́·л** (*perf*) curl, wave

за·вихр·éни·е, -я, (*n*) eddy, eddying, churning; turbulence, swirl, whirl

за·вихр·ённост·ь, -и, (*f*) vorticity

~, равно·ме́р·н·ая uniform vorticity

за·вихр·и́тел·ь, -я, (*m*) swirler

~, лопа́т·очн·ый swirl vane

зав·ко́м·, -а, (*m*) **за·во́д·ск·ий ко·мите́т·** works committee

завко́м·овск·ий, -ая, -ое, (*a*) of **зав·ко́м·**

зав·ко́н·, -а, (*m*) **за·ве́д·ующ·ий конто́р·ой** office manager

за·влад·ева́·ть, -ют, (*imp*) take possession, seize; (mil) capture

за·влад·е́·ть, -ют, (*perf*) *see* **за·влад·ева́·ть**

за·влек·а́·ть, -ют, (*imp*) entice, lure; (phys) capture

за·влеч·ённ·ый, -ая, -ое, (*past part pass*) of **за·вле́ч·ь;** enticed, allured, lured; (phys) captured

за·вле́ч·ь (*fut 3rd sing, pl*) **за·вле·ч·ёт, за·влек·у́т** (*past masc sing*) **за·влёк·** (*perf*); *see* **за·влек·а́·ть**

за·во́д·, -а, (*m*) factory, works, plant, mill; winding, winding up; winding mechanism; wind (period of operation of wound mechanism); (print) impression; (fish) net

~, вино·ку́р·енн·ый distillery, brandy distillery

~ -из·гото́в·и́тел·ь, -я, (*m*) manufacturer, factory of origin

~, ко́н·н·ый stud (for horses)

~, консе́рв·н·ый cannery

~, кре́кинг- (oil) cracking plant, cracking refinery

~, ледо·де́л·ательн·ый ice-making plant

~, лесо·пи́ль·н·ый saw-mill

~, лит·е́йн·ый foundry

~, льно·во́д·ческ·ий flax-mill

~, машино·стро·и́тельн·ый engineering works/plant

~, нефте·пере·го́н·н·ый oil refinery

~, пере·го́н·н·ый distillery

~, пере·движ·н·о́й асфальто·бето́н·н·ый (civ eng) portable asphalt plant

~, пиво·ва́р·енн·ый brewery

~ -по·ста́в·щик·, -а, (*m*) supplier, supplying factory

~, рыбо·во́д·н·ый fish farm

~, са́хар·н·ый sugar mill

~, с·бо́р·н·ый package factory

~ -с·ме́ж·ник·, -а, (*m*) subcontractor, subcontracting factory

~, судо·стро·и́тельн·ый shipyard and engineering works

~, тексти́ль·н·ый (text) mill

~, хлопко·о·чист·и́тельн·ый (text) ginnery

за·вод·и́ть, -ят, (*imp*) bring, lead, take to; establish, set up, found; (horol) wind up; (ICE) crank

за·во́д·к·а, -и, (*f*) winding up; (ICE) cranking

за·вод·не́ни·е, -я, (*n*) flooding, (oil) water-flooding

~, внутри·ко́нтур·н·ое (oil) boundary/contour flooding

~, за·ко́нтур·н·ое (oil) circular water flooding, circular flood

~, площад·н·о́е (oil) line flooding/flood

за·вод·нённ·ый, -ая, -ое, (*past part pass*) *see* **за·вод·ня́·ть;** flooded

за·вод·н·и́ть, -я́т, (*perf*) *see* **за·вод·-ня́·ть**

за·вод·н·о́й, -а́я, -ое, (*a*) winding, cranking; clockwork

~ голо́в·к·а (*f*) winder

за·вод·ня́·ть, -ют, (*imp*) flood

за·во́д·ск·ий, -ая, -ое, (*a*) of **за·во́д·**

~ ма́рк·а (*f*) trademark, nameplate

~ лаборато́ри·я (*f*) works laboratory

за·во́д·чик·, -а, (*m*) manufacturer, factory owner

за́·вод·ь, -и, (*f*) backwater

за·во·ева́ни·е, -я, (*n*) conquest; achievement

за·во·ева́ть, -юют, (*perf*) conquer

за·во́·ет (*fut 3rd sing*) of **за·вы́·ть**

за·во́з·, -а, (*m*) delivery; (naut) kedge and warp; laying out (of kedge anchor)

за·воз·и́ть, -·ят, (*imp*) deliver, supply, put/set down, leave; lay out (anchor)

за·вола́к·ивани·е, -я, (*n*) (*v n*) *see* **за·вола́к·ива·ть**

за·вола́к·ива·ть, -ют, (*imp*) cloud over; (med) heal up, close

за·волн·ова́ться, -у́ются, (*perf*) become agitated, begin to fret

за·воло́ч·ь (*fut 3rd sing, pl*) **за·воло́-ч·ёт, за·волок·у́т** (*past masc sing*) **за·воло́к·** (*perf*); *see* **за·вола́к·и-ва·ть**

за·вора́ч·ива·ть, -ют, (*imp*) **за·вёр-т·ыва·ть**

за́·ворот·, -а, (*m*) (*v n*) *see* **за·воро-т·и́ть;** bend, fold; (zool) doublure; **за·воро́т·, -а,** (*m*) sharp turn

~ ко́рк·и (*pl*) (met cast) rippled surface

~ кро́м·ок· в·низ· turn-down (of tire)

~, холо́д·н·ый (met cast) cold shut/lap, teeming arrest

за·ворот·и́ть, -я́т, (*perf*) turn; turn/tuck/roll up/back, double

за·вороч·енн·ый, -ая, -ое, (*past part pass*) of **за·ворот·и́ть**

за·во́·ют (*fut 3rd pl*) of **за·вы́·ть**

за·во·ю́ют (*fut 3rd pl*) of **за·во·-ева́ть**

за́втра (*adv*) tomorrow

за́втрак·, -а, (*m*) breakfast; lunch

за́втра·шн·ий, -яя, -ее, (*a*) of **за́втра**

за·вуал·и́р·ова·ть, -уют, (*imp*) veil

за·вулканиз·и́р·ова·ть, -уют, (*perf*) cure in (rubber)

за·вш·и́ве·ть, -ют, (*perf*) become lousy

за·выв·а́ни·е, -я, (*n*) howling

за·выв·а́·ть, -ют, (*imp*) howl

за·вы́с·ить, -·ят, (*perf*) *see* **за·вы-ш·а́·ть**

за·вы́·ть (*fut 3rd sing, pl*) **за·во́·ет, за·во́·ют,** (*perf*) begin to howl, start howling

за·выш·а́·ть, -ют, (*imp*) overstate, exaggerate, overestimate

за·выш·енн·ый, -ая, -ое, (*past part pass*) *see* **за·выш·а́·ть;** overstated, exaggerated; too high

за·вь·ю́т (*fut 3rd pl*) of **за·ви́·ть**

за·вяд·а́·ть, -ют, (*imp*) fade, wither, wilt

за·вя́д·ш·ий, -ая, -ее, (*past part act*) of **за·вя́·нуть**

за·вя́ж·ут (*fut 3rd pl*) of **за·вяз·а́ть** (*perf*)

за·вя́з·анн·ый, -ая, -ое, (*past part pass*) of **за·вяз·а́ть** (*perf*)

за·вяз·а́·ть, -ют, (*imp*) get stuck, stick, get bogged down; **за·вяз·а́ть,** (*fut 3rd sing, pl*) **за·вя́ж·ет, за·вя́-ж·ут** (*perf*) tie up, tie; start, set up/in; **-ся** (*pass*); set (of fruit)

за·вя́з·к·а, -и, (*g pl*) **-з·ок·,** (*f*) string, binding, lacing; plot (of story)

за·вяз·н·о́й, -а́я, -о́е, (*a*) tying, binding

за·вя́з·нуть, -нут, (*past masc sing*) **за·вя́з·,** (*perf*) get stuck, stick, get bogged down

за·вя́з·ыва·ть, -ют, (*imp*) tie up, tie; start, set up/in; **-ся** (*pass*); set (of fruit)

за́·вяз·ь, -и, (*f*) (bot) ovary; setting (of fruit)

за·вя́л·ить, -ят, (*perf*) dry, dry-cure

за·вя́·нуть, -нут, (*past masc sing*) **за·-вя́·л,** (*perf*) fade, wither, wilt

за·гад·а́·ть, -ют, (*perf*) *see* **за·га́д·ы-ва·ть**

за·га́д·ить, -ят, (*perf*) *see* **за·га́ж·и-ва·ть**

за·га́д·к·а, -и, (*g pl*) **-д·ок·,** (*f*) riddle, enigma

за·га́д·очн·ый, -ая, -ое, (*a*) puzzle, riddle; puzzling

за·га́д·ыва·ть, -ют, (*imp*) make/think up a riddle; guess, conjecture

за·га́ж·ива·ть, -ют, (*imp*) make dirty/foul, pollute

за·газ·и́рованн·ый, -ая, -ое, (*past part pass*) (min) gassed, polluted/con-taminated with gas

за·га́р·, -а, (*m*) sun-tan, sunburn; tarnish, (geol) varnish
~ пусты́н·и (geol) desert varnish
за·га́с·и́ть, ⁼ят, (*perf*) extinguish, put out
за·га́ш·енн·ый, -ая, -ое, (*past part pass*) of за·гас·и́ть
за·ги́б·, -а, (*m*) bend, crease, fold; bending, creasing, folding; crown (of hook); dog-ear (of page); deviation (improper conduct)
~, ис·пыт·а́ни·е на bend/bending test/testing
эффе́кт· за·ги́б·а (*m*) (phys) hook effect
за·гиб·а́ни·е, -я, (*n*) (*v n*) of за·гиб·а́·ть
~ кро́м·к·и bending
за·гиб·а́·ть, -ют, (*imp*) bend, crease, fold
за·ги́б·очн·ый, -ая, -ое, (*a*) bending
~ стан·о́к· (*m*) (met) bending machine
за·гипнотиз·и́р·овать, -уют, (*perf*) hypnotize
за·гипс·о́ванн·ый, -ая, -ое, (*past part pass*) of за·гипс·ова́ть; plastered
за·гипс·ова́ть, -у́ют, (*perf*) plaster
за·гла́в·и·е, -я, (*n*) title, heading
за·гла́в·н·ый, -ая, -ое, (*a*) title, capital (letters)
за·гла́д·ить, -ят, (*perf*) see за·гла́ж·ива·ть
за·гла́ж·ива·ть, -ют, (*imp*) smooth; smooth over, put right
за·глаз·ни́чн·ый, -ая, -ое, (*a*) (zool) postorbital
за·гла́з·н·о (*adv*) without seeing, without having seen
за·гла́т·ыва·ть, -ют, (*imp*) swallow, ingest
за·глин·изи́рованн·ый, -ая, -ое, (*past part pass*) of за·глин·изи́р·овать; (oil) mudded off, mudded
за·глин·изи́р·овать, -уют, (*imp and perf*) (oil) mud-off
за·глот·а́·ть, -ют, (*perf*) see за·гла́т·ыва·ть
за·гло́х·л·ый, -ая, -ое, (*a*) overgrown
за·гло́х·нуть, -нут, (*past masc sing*) за·гло́х· or за·гло́х·нул (*perf*) die away, peter out (of sounds); (hort) be grown over, get smothered in weeds
за·глуб·и́ть, -ят, (*perf*) see за·глуб·л·я́·ть

за·глубл·е́ни·е, -я, (*n*) (*v n*) see за·глубл·я́·ть; (agr) draught (of implements)
за·глубл·я́·ть, -ют, (*imp*) put deeper/deep at/down to
за·глуш·а́·ть, -ют, (*imp*) muffle, suppress, deaden, damp, smother, stifle; alleviate, soothe; overgrow (of plants)
за·глуш·е́ни·е, -я, (*n*) (*v n*) see за·глуш·а́·ть
за·глуш·ённ·ый, -ая, -ое, (*past part pass*) see за·глуш·а́·ть
за·глуш·и́ть, -а́т, (*perf*) see за·глу·ш·а́·ть
за·глу́ш·к·а, -и, (*g pl*) -ш·ек·, (*f*) plug, stopper, blank flange (pipe fitting), end-piece, end-cap
за·гля́д·ыва·ть, -ют, (*imp*) glance, look in/into
за·гля·н·у́ть, ⁼ут, (*perf*) see за·гля́д·ыва·ть
за·гна́·ива·ться, -ются, (*imp*) fester
за·гн·а́ть (*fut 3rd sing, pl*) за·го́·н·ит, за·го́н·ят, (*perf*); see за·гон·я́·ть
за·гнив·а́ни·е, -я, (*n*) rotting, decay; (med) suppuration, festering
за·гнив·а́·ть, -ют, (*imp*) rot, decay; suppurate, fester
за·гни́·ть, -ю́т, (*perf*) see за·гни·в·а́·ть
за·гно·и́ться, -я́тся, (*perf*) fester
за́·гн·ут·ый, -ая, -ое, (*past part pass*) of за·гн·у́ть; bent, creased, folded; recurved, recurvate
за·гн·у́ть, -у́т, (*perf*) see за·гиб·а́·ть
за·гова́р·ива·ть, -ют, (*imp*) address, harangue, talk incessantly; enter into conversation with, start/begin a conversation
за́·говор·, -а, (*m*) conspiracy, plot
за·говор·и́ть, -я́т, (*perf*) begin to speak/talk; address, harangue, talk incessantly
за·гово́р·щик·, -а, (*m*) conspirator, plotter
за·гол·и́ть, -я́т, (*perf*) bare
за·голо́в·ник·, -а, (*m*) head rest
за·голо́в·ок·, -в·к·а, (*m*) headline, heading, title
за·гол·я́·ть, -ют, (*imp*) bare
за·го́н·, -а, (*m*) (*v n*) see за·гон·я́·ть; pen, pound, enclosure, paddock
за·го́н·к·а, -и, (*g pl*) -н·ок·, (*f*) (*v n*) see за·гон·я́·ть; (text) scribbler
за·го́н·н·ый, -ая, -ое, (*a*) of за·го́н·
~ маши́н·а) (text) scribbler

за·го́н·н·ый, -ая, -ое
~ паст·ьб·а́ (*f*) strip pasture
за·гон·я́·ть, -ют, (*imp*) drive/force in; herd in; (sport) score a goal; tire out, exhaust
за·гора́ж·ивани·е, -я, (*n*) (*v n*) of за·гора́ж·ива·ть
за·гора́ж·ива·ть, -ют, (*imp*) shut/fence in, enclose; bar, obstruct
за·гор·а́ни·е, -я, (*n*) (*v n*) of за·гор·а́·ть; ignition, firing
за·гор·а́·ть, -ют, (*imp*) acquire a suntan; -ся catch fire; be on fire
за·гор·е́л·ый, -ая, -ое, (*a*) sunburnt, tanned
за·гор·е́ть, -я́т, (*perf*) see за·гор·а́·ть
за·город·и́ть, -я́т, (*perf*) see за·гора́ж·ива·ть
за·горо́д·к·а, -и, (*g pl*) -д·ок·, (*f*) fence, partition, enclosure
за́·город·н·ый, -ая, -ое, (*a*) out-of-town, country
за·гота́вл·ива·ть, -ют, (*imp*) store, stock, lay in; prepare
за·готов·и́тел·ь, -я, (*m*) government purchasing agent, stock-piling agent
за·готов·и́тельн·ый, -ая, -ое, (*a*) storing, stock piling; preparing
~ цех· (*m*) (shipb) metallurgical shop
за·готов·ить, -ят, (*perf*) see за·гота́вл·ива·ть
за·гото́в·к·а, -и, (*g pl*) -в·ок·, (*f*) stocking, stock-piling, procurement; semi-finished product, blank, preform, piece, stock; pre-arrangement (polymers)
~, авто·ка́мер·н·ая semi-finished tire
~, куз·не́чн·ая forging stock
~, на·ка́т·анн·ая roll (of rubber)
~, об·раба́т·ываем·ая work, work-piece
~, ос·ев·а́я axle billet
за·готовл·е́ни·е, -я, (*n*) (*v n*) see за·гота́вл·ива·ть
за·гото́вл·енн·ый, -ая, -ое, (*a*) (*past part pass*) see за·гота́вл·ива·ть
за·готовл·я́·ть, -ют, (*imp*) see за·гота́вл·ива·ть
за·гото́в·очн·ый, -ая, -ое, (*a*) of за·гото́в·к·а
~ стан· (*m*) (met roll) roughing mill, breaking-down mill
загот·пу́нкт·, -а, (*m*) (*abbr*) of за·готов·и́тельн·ый пункт· storage place
за·град·и́тел·ь, -я, (*m*) (naut) minelayer; (rad) wavetrap, rejector; (telecom) trap

за·град·и́тел·ь
~, ми́н·н·ый minelayer
~, сет·ев·о́й (naut) netlayer
за·град·и́тельн·ый, -ая, -ое, (*a*) of за·град·и́тел·ь; (naut) minelaying; boundary; (gunn) barrage; (rad) rejecting, trapping
~ огон·ь (*m*) boundary light; (gunn) barrage
~ фильтр· (*m*) (telecom) absorption trap
заград·и́ть, -я́т, (*perf*) see за·гражд·а́·ть
за·гражд·а́·ть, -ют, (*imp*) block, bar, obstruct; fence in, shut in, enclose
за·гражд·а́ющ·ий, -ая, -ее, (*pres part act*) of за·гражд·а́·ть
~ ко́нтур· (*m*) (telecom) band-elimination filter, wave trap, rejector
~ сло·й (*m*) (elec) barrier/blocking layer
~ фильтр· (*m*) (telecom) band-elimination filter, wave trap, rejector
за·гражд·е́ни·е, -я, (*n*) (*v n*) see за·гражд·а́·ть; obstruction, barrier, barrage; (telecom) rejection, blocking
~, бо́н·ов·ое harbour boom; (naut) boom defence
~, воз·ду́ш·н·ое (mil) balloon barrage
~, инженéр·н·ое (mil) obstacle
~, ми́н·н·ое (naut) mine field
~, поле·за·щи́т·н·ое (agr) windbreak, shelter belt
за·гражд·ённ·ый, -ая, -ое, (*past part pass*) see за·гражд·а́·ть
за·гран·и́ц·а, -ы, (*f*) foreign countries
за·гран·и́ц·ей (*adv*) abroad
за·гран·и́чн·ый, -ая, -ое, (*a*) foreign
~ па́спорт· (*m*) passport
за·греб·, -а, (*m*) sandbank, bar
за·греб·а́·ть, -ют, (*imp*) rake in/up/together; bank (of fires); accumulate, gather in
за·греб·н·о́й, -ая, -ое, (*a*) stroke (rowing); (*as noun*) stroke (oarsman)
за·грем·е́ть, -я́т, (*perf*) begin to thunder; fall with a crash
за·грес·ти́ (*fut 3rd sing, pl*) за·греб·ёт, за·греб·у́т, (*past masc sing*) за·грёб·, (*perf*); see за·греб·а́·ть
за·грим·ир·ова́ть, -у́ют, (*perf*) make up, apply make-up
за·громожд·а́·ть, -ют, (*imp*) encumber, block up, jam
за·громожд·е́ни·е, -я, (*n*) blockage, jam
за·громозд·и́ть, -я́т, (*perf*) see за·громожд·а́·ть

за·грохот·а́ть   (*fut 3rd sing, pl*) **за·гро-хо́ч·ет, за·грохо́ч·ут,**   (*perf*) begin to rumble/rattle, start rumbling/rattling

за·груб·е́ни·е, -я,   (*v n*) of **за·гру-б·е́·ть**; roughening (of the skin)

за·груб·е́·ть, -ют,   (*perf*) coarsen, become coarse/callous

за·груд·и́нн·ый, -ая, -ое,   (*a*) (anat) substernal

за·груж·а́тел·ь, -я,   (*m*) loader, charger (of furnaces etc.)

~, пруж·и́нн·ый   (a/c) spring-loaded booster

за·груж·а́·ть, -ют,   (*imp*) load, charge (furnaces etc.), stoke (fires)

за·гру́ж·енност·ь, -и,   (*f*) load, charge; employment, utilization; (statistical) utilization factor

за·гру́ж·енн·ый, -ая, -ое,   (*past part pass*) *see* **за·груж·а́·ть**; loaded, charged; employed, working

за·груз·и́ть, -я́т,   (*perf*) *see* **за·гру-ж·а́·ть**

за·гру́з·к·а, -и,   (*g pl*) **-з·ок·,** (*f*) load, charge; loading, charging, stoking; utilization, employment (of machine)

~, по́лн·ая   full order-book (of factory)

за·гру́з·очн·ый, -ая, -ое,   (*a*) loading, charging

~ ваго́н·   (*m*) charging car

~ ка́мер·а   (*f*) hopper (of acetylene generator)

~ маши́н·а   (*f*) charging machine

~ отве́рст·и·е   (*n*) fire door (of furnace)

за·грунт·ова́ть, -у́ют,   (*perf*) *see* **за·-грунт·о́выва·ть**

за·грунт·о́выва·ть, -ют,   (*imp*) apply primer/prime coating

за·грунт·о́ванн·ый, -ая, -ое,   (*past part pass*) of **за·грунт·ова́ть**; primed (painting)

за·гряз·не́ни·е, -я,   (*n*) soiling, dirtying, pollution, contamination; (chem) impurity

~, атмосфе́р·н·ое   atmospheric pollution

~ като́д·а   (electrochem) cathode poisoning

~, минера́ль·н·ое   mineral impurities

~, радио·акти́в·н·ое   radioactive contamination

за·гряз·нённ·ый, -ая, -ое,   (*past part pass*) *see* **за·гряз·ня́·ть**

за·гряз·н·и́ть, -я́т, (*perf*) *see* **за·гряз·-ня́·ть**

за·гряз·ня́·ть, -ют,   (*imp*) dirty, soil, pollute, contaminate

за́гс·, -а, (*m*) (*abbr*) = **за́·пис·ь а́кт·ов гражд·а́нск·ого со·сто·я́ни·я** registrar's office

загс·и́р·оваться, -уются,   (*imp and perf*) get married at registrar's office

за·губ·и́ть, -я́т, (*perf*) ruin; waste

за·гу́б·к·а, -и, (*g pl*) **-б·ок·,** (*f*) mouthpiece

за·гу́бл·енн·ый, -ая, -ое,   (*past part pass*) of **за·губ·и́ть**

за·гуд·е́ть, -я́т,   (*perf*) drone, hoot; begin to drone/hoot

за·густ·ева́ни·е, -я,   (*n*) (*v n*) *see* **за·-густ·е́·ть**; thickening, concentration; livering (of paint)

за·густ·е́·ть, -ют,   (*perf*) get/become thick, solidify

за·густ·и́тел·ь, -я,   (*m*) (chem) thickening agent, stiffener

за́д·, -а, (*nom pl*) **-ы́,** (*g pl*) **-о́в,** (*m*) back, rear; (anat) buttock, clunis

за·да·ва́ть, -ю́т, (*imp*) set, give, assign, pose (a question), define (a problem)

~ програ́мм·у (autom) programme

за·дав·и́ть, -я́т, (*perf*) crush; run over, knock down

за·да́вл·енн·ый, -ая, -ое,   (*past part pass*) of **за·дав·и́ть**

за·да́н·и·е, -я, (*n*) (*see also* **за·да́·ч·а**); task, assignment, mission, requirement, setting; (autom) control point

~, пла́н·ов·ое   target (planned)

~ програ́мм·ы (autom) programming

~, такт·и́ческ·ое (mil) tactical assignment

~ фу́нкци·и   (math) definition of a function

~ фу́нкци·и, я́в·н·ое (math) explicit definition of a function

за́·да·нн·ый, -ая, -ое, (*past part pass*) *see* **за·да·ва́ть**; prescribed, given, assigned, set; preset, predetermined

~ вре́м·я о·переж·е́ни·я   (*n*) (GW) procurement lead time

лин·и·я за́·да·нн·ого пу́т·и   (*f*) (air navig) required track

~ знач·е́ни·е (*n*) (autom) set point

~ то́ч·к·а (*f*) preset/set/prescribed point

за·да́т·ок, -т·к·а,   (*m*) (fin) deposit; (*pl*) instincts, inclination, inherited properties

за·да́т·чик·, -а,   (*m*) (autom) setting mechanism, control-point setting mechanism, reference input element; master (of a manipulator)

за·да́т·чик
  механи́зм за·да́т·чик·а (*m*) (autom) control-point setting mechanism
  сигна́л· за·да́т·чик·а (*m*) demand signal; (autom) reference input
~ **угл·о́в с·клон·ени·й** variation setting corrector (of gyro)

за·да́ть(*fut*) **за·да́м, за·да́шь, за·да́ст, за·дади́м, за·дади́ть, за·даду́т,** (*perf*); *see* **за·да·ва́ть**

за·да́·ч·а, -и, (*i*) **-ей,** (*f*) (*see also* **за·да́·н·и·е**); task, problem, question, job, mission; feed, feeding, entry
~, **гран·и́чн·ая** (math) boundary value problem
~ **дв·ух тел·** (mech) two-body problem
~, **кра·ев·а́я** (phys) boundary value problem
~ **о четыр·ёх кра́с·к·ах** (math) four-colour problem
~, **пло́ск·ая** (phys) plane problem
~ **по** a problem in/of

за·да́ч·ник·, -а, (*m*) (math) book of problems

за·да·ю́щ·ий, -ая, -ее, (*pres part act*) of **за·да·ва́ть;** (autom) input, master
~ **механ·и́зм·** (*m*) (autom) control point setting mechanism

за·двиг·а́·ть, -ют, (*imp*) move to/into, push to/home, bolt, bar, slide/draw to, shut, close, cover up; **за·дви́г·а·ть, -ют,** begin to move, start moving, shifting

за·дви́ж·к·а, -и, (*g pl*) **-ж·ек·,** (*f*) bolt (for doors etc.); damper, shutter, cut-off plate; gate/sluice valve
~, **клинкет·н·ая** gate/sluice valve
~, **кольц·ев·а́я** needle valve

за·движ·н·о́й, -а́я, -о́е, (*a*) sliding, slidable
~ **двер·ь** (*f*) sliding door
~ **штык·** (*m*) (naut) rolling hitch

за·дви́·нут·ый, -ая, -ое, (*past part pass*) *see* **за·двиг·а́·ть**

за·дви́·н·уть, -ут, (*perf*) *see* **за·двиг·а́·ть**

за·дев·а́·ть, -ют, (*imp*) touch, graze, brush/knock against; (naut) foul; (*perf*) mislay

за·де́л·, -а, (*m*) process stock; **за·де́·л** (*past masc sing*) of **за·де́·ть**
~, **оборо́т·н·ый** turnover process stock

за·де́л·анн·ый, -ая, -ое, (*past part pass*) *see* **за·де́л·ыва·ть**
~ **в ка́псул·у** encapsulated

за·де́л·а·ть, -ют, (*perf*) *see* **за·де́л·ыва·ть**

за·де́л·к·а, -и, (*f*) (*v n*) *see* **за·де́л·ыва·ть**
  глуб·ин·а́ за·де́л·к·и (*f*) (agr) planting depth; setting depth (mining timbers)
  моме́нт· за·де́л·к·и (*m*) moment at fixed end (of a beam), end-restraint moment

за·де́л·ыва·ть, -ют, (*imp*) close/stuff/ stop up, seal, patch; do/pack/fasten up; embed, cover in/up, build in/up

за·де́·н·ут (*fut 3rd pl*) of **за·де́·ть**

за·дёрг·а·ть, -ют, (*perf*) begin to pull; pester, worry

за·дёрг·ива·ть, -ют, (*imp*) pull/draw to/across

за·дер·ёт (*fut 3rd sing*) of **за·др·а́ть**

за·держ·а́ни·е, -я, (*n*) (*v n*) *see* **за·де́рж·ива·ть;** detention, arrest; delay; (med) retention, suppression

за·де́рж·анн·ый, -ая, -ое, (*past part pass*) *see* **за·де́рж·ива·ть**

за·держ·а́тел·ь, -я, (*m*) retaining/ delaying device
~ **ве́рх·н·его реги́стр·а** shift lock (of typewriter)

за·держ·а́ть, -ат, (*perf*) *see* **за·де́рж·ива·ть**

за·держ·ива́ни·е, -я, (*n*) (*v n*) of **за·де́рж·ива·ть;** (chem) inhibition

за·де́рж·ива·ть, -ют, (*imp*) detain, delay, keep, retard, arrest, inhibit, withhold (payment), be behind with, hold back; **-ся** (*pass*); linger, lag

за·де́рж·иваюш·ий, -ая, -ее, (*pres part act*) of **за·де́рж·ива·ть**
~ **де́й·стви·е** (*n*) delaying effect; (chem) inhibiting effect
~ **потенциа́л·** (*m*) (elec) counter potential
~ **се́т·к·а** (*f*) pick-up plate (vacuum tube)
~ **си́т·о** (*n*) retaining screen
~ **сло́·й** (*m*) (phys) trapping layer
~ **спосо́б·ност·ь** (*f*) (elec) retentivity, remanence

за·де́рж·к·а, -и, (*g pl*) **-ж·ек·,** (*f*) (*v n*) *see* **за·де́рж·ива·ть;** delay, lag, detention, holding back; interruption, hold-up
~ **вос·пламен·е́ни·я** ignition delay/ lag
~ **вре́мен·и** lag, time lag
~ **за·жиг·а́ни·я** ignition delay/lag
  ли́н·и·я за·де́рж·ки (*f*) (elec) delay line

за·де́рж·к·а

~ **о·гиб·а́ющ·ей** (elec) envelope delay
    схе́м·а за·де́рж·к·и (*f*) (autom)
    delay circuit
    цеп·ь за·де́рж·к·и (*f*) (autom)
    inhibit circuit

за·де́рж·ник·, -а, (*m*) arrestor

за·дёр·н·уть, -ут, (*perf*) *see* **за·дёр-**
    г·ива·ть

задер·у́т (*fut 3rd pl*) of **за·др·а́ть**

за·де́·т·ый, -ая, -ое, (*past part pass*)
    see **за·дев·а́·ть** (*imp*)

за·де́·ть (*fut 3rd sing, pl*) **за·де́·н·ет,**
    за·де́·н·ут, (*past masc sing*) **за·де́·л,**
    (*perf*); *see* **за·дев·а́·ть** (*imp*)

за·ди́р·, -а, (*m*) *see* **за·дир·а́ни·е**

за·дир·а́ни·е, -я, (*n*) (*v n*) of **за·ди-**
    р·а́·ть

за·дир·а́·ть, -ют, (*imp*) score, scratch,
    tear up, seize (of engines); raise,
    lift; (air) stall; pick a quarrel, start
    quarrelling; **-ся** (*pass*) ride up

за́д·н·е- (*component*) rear, postero-,
    post-, meta-, opisth-, opistho-

задне·вис·о́чн·ый, -ая, -ое, (*a*)
    (anat) post-temporal

задне·вну́тр·енн·ый, -ая, -ое, (*a*)
    posterointernal

задне·гру́д·ь, -и, (*f*) (ent) metathorax

задне·жа́бер·н·ые, -ых, (*pl decl as*
    adj) opisthobranchiates, (fish) *Opistho-*
    branchia

задне·колёс·ник·, -а, (*m*) stern wheeler
    (ship)

задне·ло́б·н·ый, -ая, -ое, (*a*) (zool)
    postfrontal

задне·нару́ж·н·ый, -ая, -ое, (*a*)
    posteroexternal

задне·про·хо́д·н·ый, -ая, -ое, (*a*)
    (anat) anal

задне·темен·н·о́й, -а́я, -о́е, (*a*)
    (zool) post parietal

за́д·н·ий, -яя, -ее, (*a*) back, rear, hind
    tail, posterior, (naut etc.) stern, after
    aft; trailing (of wave shapes)

~ **брюш·н·о́й** (*masc adj*) (zool) postero-
    ventral

~ **кро́м·к·а** (*f*) (a/c) trailing edge

~ **план·** (*m*) background

~ **про·хо́д·** (*m*) (anat) anus

~ **с·рез·** (*m*) trailing edge (of a pulse)

~ **сторон·а́** (*f*) reverse side, back

~ **фронт·** (*m*) trailing edge (of pulse);
    back casing (of boiler)

~ **ход·** (*m*) reversing, backing, (naut)
    sternway

за́д·ник·, -а, (*m*) counter (of a shoe);
    (theat) back drop; (elec) heel, back,
    leaving/trailing edge

за·до́лг·о (*adv*) long before, long in
    advance

за·до́лж·а́·ть, -ют, (*perf*) borrow from;
    be in debt

за·до́лж·енност·ь, -и, (*f*) indebtedness,
    debts, liabilities

за́д·ом (*adv*) backwards

за·до́р·ин·а, -ы, (*f*) scratch, score

за·до́р·инк·а, -и, (*g pl*) **-нок·,** (*f*) =
    за·до́р·ин·а

за·дох·н·у́ться, -у́тся, (*perf*) *see* **за·ды-**
    х·а́·ться

за·дра́·ек· (*g pl*) of **за·дра́·йк·а**

за·дра́·ива·ть, -ют, (*imp*) (naut)
    batten down, clip up, close

за·дра́·ить, -ят, (*perf*) *see* **за·дра́·и-**
    ва·ть

за·дра́·йк·а, -и, (*g pl*) **-а́·ек·,** (*f*) catch,
    clip, bolt

за́·др·анн·ый, -ая, -ое, (*past part*
    pass) *see* **за·дир·а́·ть**

за·др·а́ть (*fut 3rd sing, pl*) **задерёт,**
    задеру́т, (*perf*); *see* **за·дир·а́·ть**

за·дребезж·а́·ть, -а́т, (*perf*) rattle,
    jingle; begin to rattle/jingle

за·дрем·а́·ть, -л·ют, (*perf*) doze off

за·дрож·а́·ть, -а́т, (*perf*) shiver, shake,
    tremble, quiver; begin to shiver/
    shake etc.

за·дроссел·и́р·овать, -уют, (*perf*)
    throttle back

за·дры́г·а·ть, -ют, (*perf*) jerk, twitch;
    begin to jerk/twitch

за·дубл·ён·ый, -ая, -ое, (*a*) tanned

за·дув·а́ни·е, -я, (*n*) (*v n*) *see* **за·ду-**
    в·а́·ть

за·дув·а́·ть, -ют, (*imp*) blow, blow
    into/onto; blow in (blast furnace);
    blow out, extinguish

за·ду́в·к·а, -и, (*g pl*) **-в·ок·,** (*f*) blowing
    in, starting up (a blast furnace)

за·ду́в·очн·ый, -ая, -ое, (*a*) of **за·-**
    ду́в·к·а

за·ду́м·анн·ый, -ая, -ое, (*past part*
    pass) *see* **за·ду́м·ыва·ть;** conceived,
    planned; well-conceived/planned

за·ду́м·а·ть, -ют, (*perf*) *see* **за·ду́-**
    м·ыва·ть

за·ду́м·чив·ый, -ая, -ое, (*a*) thought-
    ful, pensive

за·ду́м·ыва·ть, -ют, (*imp*) conceive, plan

за·ду́·т·ый, -ая, -ое, (*past part pass*)
    of **за·ду́·ть;** blown, blown out;
    blown in started up (blast furnace)

за·ду́·ть, -ют, (*perf*) *see* **за·дув·а́·ть;** begin to blow

за·душ·е́вн·ый, -ая, -ое, (*a*) sincere

за·душ·е́ни·е, -я, (*n*) choking, suffocation

за·душ·и́ть, ⸗ат, (*perf*) *see* **душ·и́ть**

за·дым·и́ть, -я́т, (*perf*) smoke, emit smoke, fume; begin to smoke/fume; screen with smoke

за·дымл·ённост·ь, -и, (*f*) smoke content

за·дымл·е́ни·е, -я, (*n*) (*v n*) *see* **за·дым·и́ть**

за·дымл·ённ·ый, -ая, -ое, (*past part pass*) of **за·дым·и́ть;** smoke-filled

за·дымл·я́·ть, -ют, (*imp*) screen with smoke

за·дых·а́·ться, -ются, (*imp*) choke, suffocate

за·дыш·а́ть, ⸗ат, (*perf*) begin to breathe

за·ед·а́ни·е, -я, (*n*) (*v n*) of **за·ед·а́·ть;** jamming

за·ед·а́·ть, -ют, (*imp*) eat, eat in; catch, jam, stick, (mech eng) seize; (naut) foul (a rope)

за·е́зд·, -а, (*m*) (*v n*) *see* **за·езж·а́·ть;** event

за·езж·а́·ть, -ют, (*imp*) drop in, call on/for; intercept, drive ahead of; turn

за·ём·, (*g*) **за́·йм·а,** (*m*) (fin) loan

~, вы́·игр·ышн·ый lottery loan
с·де́л·а·ть за·ём· raise a loan

за·ём·н·ый, -ая, -ое, (*a*) of **за·ём·**
~ пись·м·о́ (*n*) IOU

за·ём·щик·, -а, (*m*) borrower

заесть (*fut*) **за·е́м, за·е́шь, за·е́ст, за·ед·и́м, за·ед·и́те, за·ед·я́т,** (*past masc sing*) **за·е́л,** (*perf*); *see*, **за·ед·а́·ть**

за·е́х·ать (*fut 3rd sing, pl*) **за·е́д·ет, за·е́д·ут,** (*perf*); *see* **за·езж·а́·ть**

за·жа́р·ива·ть, -ют, (*imp*) fry; grill; roast; **-ся** be ready fried/grilled/roasted

за·жа́р·ить, -ят, (*perf*) *see* **за·жа́р·ива·ть**

за·жа́т·и·е, -я, (*n*) pressing, squeezing

за·жа́т·ь (*fut 3rd sing, pl*) **за·жм·ёт, за·жм·у́т,** (*past masc sing*) **за·жа́·л,** (*perf*); *see* **за·жим·а́·ть**

за·жечь·ь, (*fut 3rd sing, pl*) **за··жж·ёт, за·жг·у́т,** (*past masc sing*) **за·жёг·;** *see* **за·жиг·а́·ть**

за·жж·ённ·ый, -ая, -ое, (*past part pass*) of **за·жечь·ь,** *see* **за·жиг·а́·ть**

за·жив·а́ни·е, -я, (*n*) healing

за·жив·а́·ть, -ют, (*imp*) heal

за·жив·и́ть, -я́т, (*perf*) heal

за·живл·е́ни·е, -я, (*n*) healing

за·живл·я́·ть, -ют, (*imp*) heal

за́·жи·в·о (*adv*) alive

за·жиг·а́лк·а, -и, (*g pl*) **-лок·,** (*f*) cigarette lighter; incendiary bomb

за·жиг·а́ни·е, -я, (*n*) ignition; setting fire to; striking (an arc)

~, батар·е́йн·ое (ICE) battery ignition

~, втор·и́чн·ое (elec) reignition (of counter tube)

~, двой·н·ое duel ignition

~, за·ме́дл·енн·ое delayed ignition

~, за·тяж·н·о́е (a/c) hang fire

~, катал·и́ческ·ое catalytic ignition

~, кату́ш·ечн·ое coil ignition

~, много·и́скр·ов·ое multipoint ignition

~ на от·ры́в· make-and-break ignition

~, обра́т·н·ое backfiring; (elec) flashback, arcing back

~ от магне́то (ICE) magneto ignition

~ от магнето выс·о́к·ого на·пряж·е́ни·я (ICE) high-tension ignition

~, ступ·е́нчат·ое (a/c) multistage ignition

у́гол· за·жиг·а́ни·я ignition/firing angle

~ электро·за·па́л·ом (a/c) hot-wire ignition

за·жиг·а́тел·ь, -я, (*m*) igniter; inciter

за·жиг·а́тельн·ый, -ая, -ое, (*a*) igniting, ignition, incendiary

~ вещ·еств·о́ (*n*) incendiary

за·жиг·а́·ть, -ют, (*imp*) set fire to, ignite, light, kindle; switch on (a light); incite; **-ся** (*pass*); catch fire, light up

~ дуг·у́ (weld) strike an arc

за·жи́м·, -а, (*m*) (*v n*) *see* **за·жим·а́·ть;** clip, clamp, clamping screw; chuck (machine tools); (elec) terminal, screw terminal; pinchcock (rubber tubing); contact jaw (of welder)

~, винт·ов·о́й screw clamp/chuck

~, вы·пуск·а́ющ·ий (elec) free-centre-type clamp

~ за·земл·е́ни·я (elec) earth/ground clamp/terminal

~, мину́с·ов·ый negative terminal

~, на·тяж·н·о́й strain clamp; (elec) dead-end clamp

за·жи́м

~, под·де́рж·ивающ·ий (elec) suspension clamp

~, по·воро́т·н·ый pivoted clamp

~, при·со·един·и́тельн·ый circuit terminal/clamp

~, пруж·и́нн·ый spring chuck; pinchcock clamp (for rubber tubing); clip

~, ца́нг·ов·ый collet chuck

~, шарни́р·н·ый toggle clamp

~ элеме́нт·а (elec) cell terminal

за·жим·а́·ть, -ют, (*imp*) clutch, squeeze, clamp; block, stop up (by squeezing); suppress

за·жим·н·о́й, -а́я, -о́е, (*a*) *see* **за·-жи́м·н·ый**

за·жи́м·н·ый, -ая, -ое, (*a*) clamping, gripping

~ кле́щ·и (*pl*) clip

~ патро́н· (*m*) chuck

~ скоб·а́ (*f*) binding grip/clip

за·жи́т·очн·ый, -ая, -ое, (*a*) prosperous, well-to-do

за·жи́·ть (*fut 3rd sing, pl*) **за·жив·ёт, за·жив·у́т,** (*perf*) recover, get well, heal (of wounds); begin to live

за·жм·у́т (*fut 3rd pl*) of **за·жа́т·ь**

за·жо́р·, -а, (*m*) ice-dam/jam

за·жо́р·а, -ы, (*f*) *see* **за́·жо́р·**

за·жужж·а́ть, -а́т, (*perf*) hum, buzz, drone; begin to hum/buzz/drone

за·звен·е́ть, -я́т, (*perf*) ring, jingle; begin to ring/jingle; start ringing/jingling

за·звон·и́ть, -я́т, (*perf*) ring; begin to ring, start ringing

за·звуч·а́ть, -а́т, (*perf*) sound, resound; begin to sound/resound

за·зелен·е́·ть, -ют, (*perf*) turn green

за·зелен·и́ть, -я́т, (*perf*) colour/color green

за·земл·е́ни·е, -я, (*n*) (elec) earth, earthing, ground, grounding

~, глух·о́е dead earth/ground

~, жёстк·ое dead earth/ground

за·земл·ённ·ый, -ая, -ое, (*past part pass*) *see* **за·земл·я́·ть;** (elec) earthed, grounded

~ цеп·ь (*f*) earthed/grounded circuit

за·земл·и́тел·ь, -я, (*m*) (elec) earthing/grounding connection, earth/ground plate

за·земл·и́тельн·ый, -ая, -ое, (*a*) earthing, grounding

за·земл·и́ть, -я́т, (*perf*) *see* **за·земл·я́·ть**

за·земл·я́·ть, -ют, (*imp*) (elec) earth, ground

за·зим·ова́ть, -у́ют, (*perf*) winter, spend the winter

за·зо́р·, -а, (*m*) clearance, gap, allowance; play, backlash; nip (of rolls)

~, воз·ду́ш·н·ый air gap

~, га́з·ов·ый (*met*) vapour/vapor/gas/air gap

~, до·пуст·и́м·ый allowance, permissible clearance

~, кольц·ев·о́й annular/radial clearance

~, между·венц·о́в·ый rim clearance

~, пред·о·хран·и́тельн·ый (horol) expansion gap

~, радиа́ль·н·ый diameter/radial clearance

~ по голо́в·к·ам tip clearance (gears)

за·зу́бр·енн·ый, -ая, -ое, (*past part pass*) *see* **за·зу́бр·ива·ть;** notched, indented, jagged; (bot) serrate, serrated

за·зу́бр·ива·ть, -ют, (*imp*) notch, indent

за·зу́бр·ин·а, -ы, (*f*) notch, indentation; grat (of glacier)

за·зубр·и́ть, ⸗ят, (*perf*) *see* **за·зу́б·р·ива·ть**

за·игр·а́·ть, -ют, (*perf*) begin to play, start playing; begin to sparkle, start sparkling

за·ик·а́ни·е, -я, (*n*) stuttering, stammering

за·ил·е́ни·е, -я, (*n*) silting up

за·ил·ива́ни·е, -я, (*n*) (*v n*) of **за·-и́л·ива·ть;** (min) mud injection (to extinguish fires)

за·и́л·ива·ть, -ют, (*imp*) fill with mud, inject mud; **-ся** silt up

за·и́л·ить, -ят, (*perf*) *see* **за·и́л·и·ва·ть**

заимо·да́в·ец·, -в·ц·а, (*i*) **-в·ц·ем,** (*m*) (fin) creditor

заимо·обра́з·н·ый, -ая, -ое, (*a*) on loan, on credit, borrowed

за·и́м·ствовани·е, -я, (*n*) adoption; borrowing, copying

за·и́м·ствованн·ый, -ая, -ое, (*past part pass*) of **за·и́м·ств·овать**

~ сло́в·о (*n*) (ling) loan-word

за·и́м·ств·овать, -уют, (*imp and perf*) adopt, take on/over/from, borrow

за·и́нд·евел·ый, -ая, -ое, (*a*) covered with hoar frost/rime, frosted, rimy

за·и́нд·ева́ни·е, -я, (*n*) frosting

за·интерес·ова́ть, -у́ют, (*perf*) interest; **-ся** (+ *instr*) become/be interested in

за·и́скр·иться, -ятся, (*perf*) sparkle; begin to sparkle

за́·йм·а (*g*) of **за·ём·**

за́·йм·ёт (*fut 3rd sing*) of **за·ня́·ть**

за́·йм·ищ·е, -а, (*i*) **-ем,** (*n*) flood plain

займо·держ·а́тел·ь, -я, (*m*) (fin) loan-holder

за·йм·у́т (*fut 3rd pl*) of **за·ня́·ть**

за́·йм·ы (*nom pl*) of **за·ём·**

за·йти́ (*fut 3rd sing, pl*) **за·йд·ёт, за·ид·у́т,** (*past masc sing*) **за·шёл,** (*perf*); *see* **за·ход·и́ть**

за́йц·а (*g*) of **за́яц·**

за́йч·ик·, -а, (*m*) (*dim*) of **за́яц·**; light spot, spot; white horses, white crests (on waves)

за·кабо́л·ивани·е, -я, (*n*) (naut) mousing (of hook)

за·ка́дм·иев·ый, -ая, -ое, (*a*) (nucl) epicadmium

за·кадм·и́рованн·ый, -ая, -ое, (*a*) cadmium coated/plated

за·ка́ж·ут (*fut 3rd pl*) of **за·каз·а́ть**

за·ка́з·, -а, (*m*) (com) order

за·ка́з·анн·ый, -ая, -ое, (*past part pass*) of **за·каз·а́ть**; made-to-order, bespoke

за·каз·а́ть (*fut 3rd sing, pl*) **за·ка́ж·ет, за·ка́ж·ут** (*perf*); *see* **за·ка́з·ыва·ть**

за·каз·н·о́й, -а́я, -о́е, (*a*) made to order, ordered, bespoke; registered (of mail)

за·ка́з·чик·, -а, (*m*) customer

за·ка́з·ыва·ть, -ют, (*imp*) (com) order, have made

за·ка́л·, -а, (*m*) (*v n*) *see* **за·ка́л·и·ва·ть**

за·кал·ённост·ь, -и, (*f*) toughness, fitness

за·кал·ённ·ый, -ая, -ое, (*past part pass*), *see* **за·ка́л·ива·ть**; (met) hardened, quench-hardened, (met plast) quenched; tough, fully trained, hardy (e.g. of athletes), seasoned (soldier); hardened-off (of plants)

за·ка́л·иваемост·ь, -и, (*f*) quench hardness (of steel)

за·ка́л·ива·ть, -ют, (*imp*) harden, toughen; (met) harden, quench-harden, quench; harden off (plants)

за·кал·и́ть, -ят, (*perf*) *see* **за·ка́л·и·ва·ть**

за·ка́л·к·а, -и, (*g pl*) **-д·ок·,** (*f*) (*v n*) *see* **за·ка́л·ива·ть**; temper, calibre (of people); (met) quench-hardening, hardening, quenching; (plast) quenching

~, воз·ду́ш·н·ая (met) air hardening

~, втор·и́чн·ая (met) secondary hardening

~, выс·о́к·ая (met) high-temperature quench-hardening

~, высоко·част·о́тн·ая (met) induction hardening

~, газо·пла́мен·н·ая (met) flame hardening

~, двой·н·а́я (met) double grain-refining, regenerative quenching

~, дифференциа́ль·н·ая (met) differential quenching

~, изо·терм·и́ческ·ая (met) isothermal quenching, isothermal heat treating, austempering

~, изо·терм·и́ческ·ая на мартенси́т· (met) martempering

~, изо·терм·и́ческ·ая сорби́т (met) isothermal quench-hardening to sorbite

~, индукцио́н·н·ая (met) induction hardening

~, и́стин·н·ая (met) true quench hardening

~ ма́гм·ы (geol) magma quenching

~, ме́ст·н·ая (met) differential quenching, differential hardening, local quenching

~, не·по́лн·ая (met) partial quench-hardening

~, пла́мен·н·ая (met) flame hardening

~, по·ве́рх·ностн·ая (met) surface hardening

~ под давл·е́ни·ем (met) pressure hardening

~, по́лн·ая (met) full quench-hardening

~, пре·ры́в·ист·ая interrupted quenching, time quenching

~ с под·сту́ж·ивани·ем hardening with retarded cooling

~ с само·о́т·пуск·ом (met) time quenching

~, све́т·л·ая (met) bright quench-hardening

~, сквоз·н·а́я (met) through hardening

~, стру́й·чат·ая (met) spray quenching/hardening

~, ступ·е́нчат·ая (met) martempering

~, физ·и́ческ·ая toughening up, training (of men); fitness, temper, calibre (of trained men)

за·ка́л·к·а

~, **чи́ст·ая** (met) quench hardening in a protective atmosphere

за·ка́л·очн·ый, -ая, -ое, (*a*) of **за·ка́л·к·а;** quenching, quench-hardening

~ **на·пряж·ёни·е** (*n*) (nucl) quenching stress

~ **сред·а́** (*f*) (met) quenching medium

за·ка́л·ыва·ть, -ют, (*imp*) stab, pin; slaughter, butcher

за·кал·я́·ть, -ют, (*imp*) *see* **за·ка́л·ива·ть**

за·ка́нч·ива·ть, -ют, (*imp*) finish, end, complete

за·ка́п·а·ть, -ют, (*perf*) begin to drip; pour drops; spot, become spotted

за·ка́пч·ива·ть, -ют, (*imp*) smoke, cure; blacken with smoke

за·ка́п·ывани·е, -я, (*n*) (*v n*) of **за·ка́п·ыва·ть;** burial, interment; ground/land disposal (of radioactive waste)

за·ка́п·ыва·ть, -ют, (*imp*) bury, dig in; fill in

за·ка́рмл·ивани·е, -я, (*n*) (*v n*) of **за·ка́рмл·ива·ть;** (agr) fattening

за·ка́рмл·ива·ть, -ют, (*imp*) fatten, feed up, overfeed

за·ка́рст·ова́ни·е, -я, (*n*) (geol) Karsting

за·ка́рст·ова́ть, -у́ют, (*perf*) *see* **за·карст·о́выва·ть**

за·карст·о́выва·ть, -ют, (*imp*) (geol) karst

за·ка́т·, -а, (*m*) (astron) set; lap, overlap (defect in rolled metal)

~ **со́лн·ц·а** sunset

за·кат·а́·ть, -ют, (*perf*) begin to roll, start rolling; roll in/up

за·кат·и́ть, -ят, (*perf*) *see* **за·ка́т·ыва·ть**

за·ка́т·к·а, -и, (*g pl*) **-т·ок·,** (*f*) (*v n*) *see* **за·кат·а́·ть;** seaming, sealing (preserving cans); beading, collaring, wrapping (rubber)

за·ка́т·очн·ый, -ая, -ое, (*a*) of **за·ка́т·к·а**

~ **маши́н·а** (*f*) double seamer (canning), wrapping machine (rubber)

~ **ткан·ь** (*f*) wrapping (rubber)

за·ка́т·ыва·ть, -ют, (*imp*) (astron) set; roll under/behind; roll in/up; wind up

за·кач·а́·ть, -ют, (*perf*) rock, swing; begin to rock/swing, start rocking/ swinging; pump to/into

за·ка́ч·енн·ый, -ая, -ое, (*past part pass*) *see* **за·ка́т·ыва·ть**

за·ква́нт·ова́ть, -у́ют, (*perf*) (phys) quantize

за·ква́с·ить, -ят, (*perf*) *see* **за·ква́ш·ива·ть**

за·ква́с·к·а, -и, (*g pl*) **-с·ок·,** (*f*) ferment; leaven, barm, top yeast

за·ква́ш·енн·ый, -ая, -ое, (*past part pass*) of **за·ква́с·ить;** leavened, fermented

за·ква́ш·ива·ть, -ют, (*imp*) ferment, leaven (dough)

за·кид·а́·ть, -ют, (*perf*) bespatter, shower; fill, fill in/up with, cover

за·ки́д·ыва·ть, -ют, (*imp*) bespatter, shower; fill, fill in/up with, cover; throw back/up

~ **но́к·и** (naut) brace sharp

за·ки́·н·уть, -ут, (*perf*) throw back/up

за·кип·а́ни·е, -я, (*n*) (*v n*) of **за·кип·а́·ть;** commencement/start of boil

за·кип·а́·ть, -ют, (*imp*) start boiling, begin to boil, come to the boil; be on the boil, simmer

за·кип·е́ть, -я́т, (*perf*) *see* **за·кип·а́·ть**

за·кир·ова́ни·е, -я, (*n*) (geol) brea impregnation

за·кир·о́ванн·ый песо́к· (*m*) (geol) brea

за·кис·а́·ть, -ют, (*imp*) turn sour, acquire acidity

за·кисл·е́ни·е, -я, (*n*) overacidity

за·ки́сл·ый, -ая, -ое, (*a*) sour; overacid

за·ки́с·нуть, -нут, (*past masc sing*) **за·ки́с·,** (*perf*); *see* **за·кис·а́·ть**

за́·кис·н·ый, -ая, -ое, (*a*) of **за́·кис·ь**

~ **сол·ь** (*f*) (chem) lower (-ous) salt, monoxide, protoxide

~ **желе́з·о** (*n*) ferrous oxide, iron protoxide/monoxide

за·ки́с·ш·ий, -ая, -ее, (*past part act*) of **за·ки́с·нуть** sour, soured

за́·кис·ь, -и, (*f*) oxide (lower *or* -ous), protoxide

~ **азо́т·а** nitrous oxide

~ **желе́з·а** ferrous oxide, iron protoxide/monoxide

~ **ни́кел·я** nickel oxide/protoxide

~ **-о́·кис·ь, -и,** (*f*) (mixed -ous/-ic) oxide

~ **-о́·кис·ь желе́з·а** ferrosoferric/ ferriferous oxide

~ **-о́·кис·ь ко́бальт·а** cobalto-cobaltic/cobaltosic oxide

~ **-о́·кис·ь ура́н·а** uranous-uranic oxide

за·кла́д·, -а, (*m*) pawning, mortgaging; pledge, pawn

за·кла́д·к·а, -и, (*g pl*) -д·ок·, (*f*) laying, placing (concrete); (min) filling, fill, gobbing, gob; (shipb) laying ·down; (chem) batch, charge; charging, stocking; bookmark; latch (of door)

~ ва́р·очн·ой ка́мер·ы   bagging (tire)

~, гидравл·и́ческ·ая (min) hydraulic stowing/gobbing/filling

за·клад·н·о́й, -а́я, -о́е, (*a*) mortgage, pawn; laying, insation

~ голо́в·к·а (*f*) rivet head

за·кла́д·очн·ый, -ая, -ое,   (*a*) of за·кла́д·к·а

~ маши́н·а (*f*) (min) filling/gob-stowing machine

за·кла́д·ыва·ть, -ют, (*imp*) put, lay; pile up, heap up; pawn, mortgage; harness (a horse)

за·кл·ева́ть (*fut 3rd sing, pl*) за·кл·юёт, за·кл·юю́т (*perf*); begin to bite, rise (of fish); peck to death

за·кл·ёвыва·ть, -ют, (*imp*) peck to death

за·кле́·енн·ый, -ая, -ое, (*past part pass*) see за·кле́·ива·ть

за·кле́·ива·ть, -ют, (*imp*) glue/stick/seal up

за·кле́·ить, -ят, (*perf*) see за·кле́·ива·ть

за·клеим·ённ·ый, -ая, -ое, (*past part pass*) of за·клейм·и́ть; branded, stamped, trade-marked

за·клейм·и́ть, -я́т, (*perf*) brand, stamp

за·клёп·анн·ый, -ая, -ое, (*past part pass*) of за·клеп·а́·ть; riveted

за·клеп·а́·ть, -ют, (*perf*) rivet

за·клёп·к·а, -и, (*g pl*) -п·ок·, (*f*) rivet; riveting

~, бонда́р·н·ая   flat-head rivet

~, бочко·обра́з·н·ая   pan-head rivet

~, вз·рыв·н·а́я   explosive rivet

~, грибо·ви́д·н·ая   mushroom-head rivet

~, кон·и́ческ·ая   cone-head rivet

~, одно·сторо́н·н·яя   blind rivet

~, пло́ск·ая   flat-head rivet

~, пло́ск·о-вы́·пук·л·ая   mushroom-head rivet, cup-head rivet

~, по·та́й·н·ая   countersunk-head rivet

~, тру́б·чат·ая   snap-head hollow rivet

за·клёп·очн·ый, -ая, -ое, (*a*) rivet, riveting

~ шов· (*m*) riveted/rivet seam/connection

за·клёп·очн·ый, -ая, -ое

~ шов·, цеп·н·о́й   chain-rivet seam

за·клёп·ыва·ть, -ют, (*imp*) rivet

за·клин·е́ни·е, -я, (*n*) (*v n*) of за·кли·н·и́ть

   у́гол· за·клин·е́ни·я (*m*) angle of incidence, rigging angle (of a/c wing)

за·кли́н·ива·ть, -ют, (*imp*) wedge; key, key on; squeeze out, displace; -ся become wedged, jam, stick; be displaced; become wedge-shaped

за·клин·и́ть, -я́т, (*perf*) see за·кли́·н·ива·ть

за·клокот·а́ть (*fut 3rd sing, pl*) за·кло·ко́ч·ет, за·клоко́ч·ут, (*perf*); bubble; begin to bubble

за·ключ·а́·ть, -ют, (*imp*) confine, shut in, enclose; include, contain; conclude, complete, finish; conclude/make (contract, treaty); conclude, draw conclusions; -ся (*pass*); be contained in; consist in, amount to

за·ключ·е́ни·е, -я, (*n*) conclusion(s), findings, resolution; inclusion, enclosure; imprisonment, custody

~ в оболо́ч·к·у   canning, sheathing, encasing

~, вы́·вес·ти   come to the conclusion, conclude, draw the conclusion, deduce

за·ключ·ённ·ый, -ая, -ое, (*past part pass*) see за·ключ·а́·ть; (*m decl as adj*) prisoner

~ пла́зм·а (*f*) (phys) confined plasma

за·ключ·и́тельн·ый, -ая, -ое, (*a*) concluding, final

за·ключ·и́ть, -а́т, (*perf*) see за·клю·ч·а́·ть

за·клю́ч·к·а, -и, (*f*) (print) quoin

за·ков·а́ть, (*fut 3rd sing, pl*) за·ку·ёт, за·ку·ю́т (*perf*); begin to forge; put in chains

за·кокс·о́ванн·ый, -ая, -ое, (*past part pass*) coked, coked up

за·ко́л·, -а, (*m*) fishweir; slaughtering

за·кола́ч·ива·ть, -ют, (*imp*) nail up/down, board up

за·колеб·а́ться, ⁻лятся, (*perf*) hesitate, waver; begin to hesitate/waver

за·колот·и́ть, ⁻ят, (*perf*) see за·кола́·ч·ива·ть

за·кол·о́т·ый, -ая, -ое, (*past part pass*) of за·кол·о́ть

за·кол·о́ть, ⁻ют, (*perf*) see за·ка́л·ы·ва·ть

за·колоч·енн·ый, -ая, -ое, (*past part pass*) of за·колот·и́ть

за·кольц·ева́ть, -у́ют, (*perf*) link up/together; ring (trees, birds etc.)

за·комел·и́стост·ь, -и, (*f*) bulging butt (in trees)

за·ко́н·, -а, (*m*) (*see also associated words*); law, principle; law, act, statute

~ **бол́ьш·и́х чи́сел·**   law of large numbers (statistics)

~ **ве́тр·а, бар·и́ческ·ий**   (meteor) Buys–Ballot's law

~ **взаимо·за·мест·и́мост·и**   reciprocity law

~ **воз·раст·а́ни·я энтропи́·и**   (phys) law of degradation of energy

~ **де́й·ствующ·их масс·**   (chem) law of mass action

~, **ква́нт·ов·ый**   quantum law

~ **кра́т·н·ых от·нош·е́ни·й**   (chem) law of multiple proportions

~ **норма́ль·н·ого рас·пре·дел·е́ни·я о·ши́б·ок·**   normal law of error, law of error

~ **обра́т·н·ых квадра́т·ов**   law of inverse squares, inverse square law

~ **обра́т·н·ых куб·о́в**   law of inverse cubes

~ **о·ши́б·ок·**   law of error, normal law of error

~ **парциа́ль·н·ых давл·е́ни·й**   law of partial pressures

~ **площад·е́й**   (astron) law of areas, law of equal areas

~ **по·до́б·и·я**   similarity law

~ **по·сто·я́нн·ых от·нош·е́ни·й**   (chem) law of constant proportion, law of definite proportions

~ **по·сто·я́нств·а, со·ста́в·а**   (chem) law of constant proportion, law of definite proportions

~ **пре·ломл·е́ни·я**   law of refraction

~ **с·мещ·е́ни·я**   displacement law

~, **степ·енн·о́й**   (math) power law

~ **со·хран·е́ни·я мате́р·и·и**   law of conservation of matter

~ **со·хран·е́ни·я эне́рг·и·и**   law of conservation of energy

~ **термо·дина́м·ик·и, втор·о́й**   second law of thermodynamics

~ **у·мень·ш·а́ющ·ихся воз·мещ·е́ни·й**   (econ) law of diminishing returns

~ **хим·и́ческ·их па·ёв**   (chem) law of multiple proportions

за·ко́н·ник·, -а, (*m*) legislator; lawyer

за·ко́н·ност·ь, -и, (*f*) legality, lawfulness; law; validity, legitimacy

за·ко́н·н·ый, -ая, -ое, (*a*) legal, lawful, legitimate, valid

законо·ве́д·, -а, (*m*) jurist

законо·да́·тел·ь, -я, (*m*) legislator

законо·да́·тельн·ый, -ая, -ое, (*a*) legislative

законо·да́·тельств·о, -а, (*n*) legislation

законо·ме́р·ност·ь, -и, (*f*) regularity, conformity (with law), laws governing (a phenomenon); mechanism (of a phenomenon)

законо·ме́р·н·ый, -ая, -ое, (*a*) regular, natural, conforming to law

~ **явл·е́ни·е** (*n*) natural phenomenon

за·конопа́т·ить, -ят, (*perf*) caulk

за·конопа́ч·енн·ый, -ая, -ое, (*past part pass*) of **за·конопа́т·ить;** caulked

за·конопа́ч·ива·ть, -ют, (*imp*) caulk

законо·по·лож·е́ни·е, -я, (*n*) (law) statute

законо·прое́кт·, -а, (*m*) bill, draft law

за·консерв·и́р·овать, -у́ют, (*perf*) (food) preserve, can, bottle, pot; lay up, put into reserve

за·контракт·ова́ть, -у́ют, (*perf*) *see* **за·контракт·о́выва·ть**

за·контракт·о́выва·ть, -ют, (*imp*) enter into/make a contract

за·ко́нтур·н·ый, -ая, -ое, (*a*) peripheral

~ **вод·а́** (*f*) (oil geol) edge water

~ **за·вод·не́ни·е** (*n*) (oil) circular water flooding, circular flood

за·кон·цо́вк·а, -и, (*g pl*) **-вок·,** (*f*) end, tip

за·ко́нч·ить, -ат, (*perf*) finish, end, complete

за·коп·а́·ть, -ют, (*perf*) *see* **за·ка́п·ыва·ть**

за·копт·е́л·ый, -ая, -ое, (*a*) sooty

за·копт·и́ть, -я́т, (*perf*) *see* **за·ка́пч·ива·ть**

за·коп·у́шк·а, -и, (*g pl*) **-шек·,** (*f*) (min) test pit

за·копч·ённ·ый, -ая, -ое, (*past part pass*) *see* **за·ка́пч·ива·ть**

за·корач·ивател·ь, -я, (*m*) (elec) contact bar

за·коро́ч·енн·ый, -ая, -ое, (*past part pass*) (elec) shorted-out

за·корен·е́л·ый, -ая, -ое, (*a*) inveterate

за·корм·и́ть, -я́т, (*perf*) *see* **за·ка́рм·л·ива·ть**

за·косн·е́·ть, -ют, (*perf*) stagnate; stiffen

за·кост·ене́·ть, -ют, (*perf*) become numbed/stiff

за·коч·ене́л·ый, -ая, -ое, (*a*) numb/ stiff with cold

за·кра́·ин·а, -ы, (*f*) edge, rim, border; tip; (plast) flange; (print) ridge shoulder, flange

за·кра́·йк·а, -и, (*f*) selvage

за·кра́с·ить, -ят, (*perf*) paint

за·кра́с·к·а, -и, (*g pl*) -с·ок·, (*f*) painting; filling, filling in (cartography); tint

за·крас·не́·ть, -ют, (*perf*) show red; blush, flush, go/turn red

за·крахма́л·ивани·е, -я, (*n*) starching

за·кра́ш·ива·ть, -ют, (*imp*) paint

за·креп·и́тел·ь, -я, (*m*) fastener; fixative, fixing agent

за·креп·и́ть, -я́т, (*perf*) see за·крепл·я́·ть

за·крепл·е́ни·е, -я, (*m*) fastening, fixing, securing; fastenings; fixing (photographs, dyes)

∼ грунт·ов (geol) soil fastening

∼ дис·лок·аци·и (cryst) dislocation locking

    то́ч·к·а за·крепл·е́ни·я (*f*) (cryst) anchored point

за·крепл·я́·ть, -ют, (*imp*) fasten, secure, attach, consolidate; allot, assign; (phot) fix; (med) be constipated

за·крепл·я́ющ·ий, -ая, -ее, (*pres part act*) of за·крепл·я́·ть

∼ маши́н·а (*f*) (text) pocket/buttonhole facing machine

за·кристаллиз·ова́ть, -уют, (*perf*) crystallize

за·крит·и́ческ·ий, -ая, -ое, (*a*) supercritical; (air) beyond stalling

за·кро́·ет (*fut 3rd sing*) of за·кры́·ть

за·кро·е́чн·ый, -ая, -ое, (*a*) cutting-out

за·кро́·й, -я, (*m*) cut, cutting, cutting out

за·кро́й·н·ый, -ая, -ое, (*a*) cutting, cutting-out

∼ маши́н·а (*f*) cutter, cutting-out machine (clothing)

за·кро́й·щик·, -а, (*m*) cutter (e.g. of clothes)

за́кром·, -а, (*nom pl*) -á, (*g pl*) -óв, (*m*) bin, corn bin

за·кро́·ют, (*fut 3rd pl*) of за·кры́·ть

за·кругл·е́ни·е, -я, (*n*) rounding, curving; curvature

∼, скул·ов·ое (shipb) turn of the bilge

за·кругл·ённ·ый, -ая, -ое, (*past part pass*) see за·кругл·я́·ть; rounded, curved, round; polished (phrases etc.)

за·кругл·и́ть, -я́т, (*perf*) see за·круг-л·я́·ть

за·кругл·я́·ть, -ют, (*imp*) round off, make round/curved

за·круж·и́ть, -а́т, (*perf*) spin, rotate; circle, wheel round; begin to spin/ rotate; begin to circle/wheel round; -ся (*pass*); get dizzy from turning

за·крут·и́ть, -я́т, (*perf*) screw on/in/ up, tighten (nuts); twist, twirl; begin to twist/twirl/whirl/spin

за·кру́т·к·а, -и, -т·ок·, (*f*) (*v n*) see за·кру́ч·ива·ть; twist; (med) tourniquet

∼, аэро·динам·и́ческ·ая aerodynamic twist

∼, от·риц·а́тельн·ая (dynam) wash-out

∼, по·лож·и́тельн·ая (dynam) wash-in

за·круч·ённ·ый, -ая, -ое, (*past part pass*) see за·кру́ч·ива·ть; twisted, turned, twirled, rolled-up/over

∼ кро́м·к·а (*f*) rolled edge (of paper)

за·кру́ч·ивание, -я, (*n*) twisting, twirling, twist; (chem) spiralization

    у́гол· за·кру́ч·ивани·я (*m*) angle of twist

за·кру́ч·ива·ть, -ют, (*imp*) twist, twirl

за·крыв·а́·ть, -ют, (*imp*) close, shut, cover; close/shut down/up/in

за·кры́л·ок·, -л·к·а, (*m*) (a/c) flap

∼, прост·о́й plain flap

∼, раз·рез·н·о́й (a/c) split flap

∼, реакти́в·н·ый (a/c) jet flap

    вре́м·я за·кры́т·и·я (*n*) change time (of servomechanism)

∼, щел·ев·о́й slotted flap

за·кры́т·и·е, -я, (*n*) closing, shutting; shutting up, closing down; shelter, cover

∼ ка́др·а (cinema) fade-out

    сигна́л· за·кры́т·и·я связ·и (*m*) (telecom) end-of-work/copy signal

за·кры́·т·ый, -ая, -ое, (*past part pass*) see за·крыв·а́·ть; shut, closed; enclosed

∼ дре́н·а (*f*) blind drain

∼ при́·бы·л·ь (*f*) (met cast) blind riser

∼ я́щик· (*m*) (nucl) black box

за·кры́·ть (*fut 3rd sing, pl*) за·кро́·ет, за·кро́·ют, (*perf*); see за·крыв·а́·ть

заксо·грамм·а (*f*) (spectr) Sachs pattern

за·ку́кл·ива·ться, -ются, (*imp*) (ent) pupate

за·кули́с·н·ый, -ая, -ое, (*a*) behind-the-scenes

за·куп·а́·ть, -ют, *(imp)* buy in/up, buy a stock of

за·куп·и́ть, ⸗ят, *(perf) see* **за·куп·а́·ть**

за·ку́п·к·а, -и, *(g pl)* **-п·ок·,** *(f)* buying in/up; purchase

~, ма́сс·ов·ая bulk buying/purchase

~, централиз·о́ванн·ая bulk buying/ purchase

за·ку́пл·енн·ый, -ая, -ое, *(past part pass) see* **за·куп·а́·ть**

за·ку́пор·енн·ый, -ая, -ое, *(past part pass) see* **за·ку́пор·ива·ть;** pent-up, sealed, capped

за·ку́пор·ивани·е, -я, *(n) (v n) see* **за·ку́пор·ива·ть**

за·ку́пор·ива·ть, -ют, *(imp)* stuff/ stop/bury/plug/clog/cork up

за·ку́пор·ить, -ят, *(perf) see* **за·ку́по·р·ива·ть**

за·ку́пор·к·а, -и, *(g pl)* **-р·ок·,** *(f)* *(v n) see* **за·ку́пор·ива·ть**

~ со·су́д·ов (med) embolism

за·ку́п·очн·ый, -ая, -ое, *(a)* of **за·ку́п·к·а**

за·ку́р·ива·ть, -ют, *(imp)* light (tobacco), begin to smoke

за·кур·и́ть, ⸗ят, *(perf) see* **за·ку́р·ива·ть**

за·кус·и́ть, ⸗ят, *(perf)* eat/take a snack

за·ку́с·к·а, -и, *(g pl)* **-с·ок·,** *(f)* hors-d'oeuvre, cocktail savoury, snack

за·ку́с·очн·ая, -ой, *(f decl as adj)* snack bar

за·ку́т·а·ть, -ют, *(perf) see* **за·ку́т·ы·ва·ть**

за·ку́т·ыва·ть, -ют, *(imp)* muffle, wrap/tuck up

за·ку́ш·енн·ый, -ая, -ое, *(past part pass)* of **за·кус·и́ть**

зал·, -а, *(m)* room, hall

~, мото́р·н·ый motor room

~ о·жид·а́ни·я waiting room

~, турби́н·н·ый turbine room

за·ла́м·ыва·ть, -ют, *(imp)* start breaking up/crushing; overcharge

за·лат·а́·ть, -ют, *(perf)* patch up

за·лёг· *(past masc sing)* of **за·ле́ч·ь**

за·лег·а́ни·е, -я, *(n) (v n) see* **за·лег·а́·ть;** (geol) occurrence, bedding

глуб·ин·а́ за·лег·а́ни·я *(f)* (min) depth of occurrence, cover thickness

~, изо·клина́ль·н·ое isoclinal folding

~, линзо·ви́д·н·ое (geol) lensing

~, не·со·глас·н·ое (geol) unconformity

~, паралле́ль·н·ое parallel bedding

~ пласт·а́ (geol) bedding, position of a bed

за·лег·а́·ть, -ют, *(imp)* lie, lie up, hide; (geol) occur

~ над (geol) overlay

~ ни́ж·е (geol) underlay

за·лег·а́ющ·ий, -ая, -ое, *(pres part act)* of **за·лег·а́·ть**

~, глуб·о́к·о (geol) deep-seated

за·лёг·ш·ий, -ая, -ее, *(past part act)* of **за·ле́ч·ь**

за·лед·ене́·ть, -ют, *(perf) (intrans)* freeze, be frozen, ice-up, get iced up

за·лед·ен·и́ть, -я́т, *(trans)* freeze, ice up

за·леж·а́·ться, -а́тся, *(perf) see* **за·лёж·ива·ться**

за·лёж·ива·ться, -ются, *(imp)* lie a long time; be unsaleable, have no market; become stale

за́·леж·н·ый, -ая, -ое, *(a)* of **за́·леж·ь**

за·леж·а́л·ый, -ая, -ое, *(a)* stale, old, shop-soiled

~ систе́м·а земле· де́л·и·я *(f)* ley farming

за́·леж·ь, -и, *(f)* (geol) deposit, bed, field; (com) stale goods, old goods; (agr) long-fallow land, lea, ley

~, нефт·ян·а́я oil pool

~, ру́д·н·ая ore bed/run/shoot

~ у́гл·я coal deposit

за·лез·а́·ть, -ют, *(imp)* climb, climb up; creep into

за·ле́з·ть, -ут, *(past masc sing)* **за·ле́з·,** *(perf); see* **за·лез·а́·ть**

за·леп·и́ть, -ят, *(perf) see* **за·леп·л·я́·ть**

за·лепл·я́·ть, -ют, *(imp)* paste/glue up, pack (with grease etc.)

за·лес·е́ни·е, -я, *(n)* afforestation

за·лес·и́ть, -я́т, *(perf)* afforest, plant forests

за·лёт·, -а, *(m)* flight

за·лет·а́·ть, -ют, *(imp)* fly into/behind/ beyond; (air) land en route, make an intermediate stop

за·лет·е́ть, -я́т, *(perf) see* **за·лет·а́·ть**

за·лёт·н·ый, -ая, -ое, *(a)* flown in

за·ле́ч·енн·ый, -ал, -ое, *(past part pass)* of **за·леч·и́ть**

за·ле́ч·ивани·е, -я, *(n)* healing, curing; closing up, healing

за·ле́ч·ива·ть, -ют, *(imp)* heal, cure; **-ся** *(pass);* heal, close up (wounds, cracks etc.)

за·леч·и́ть, ⸗ат, *(perf) see* **за·ле́ч·и·ва·ть**

за·леч·у́   (*fut 1st sing*) of **за·лет·е́ть;** (*fut 1st sing*) of **за·леч·и́ть**

за·ле́ч·ь   (*fut 3rd sing, pl*) **за·ля́ж·ет, за·ля́г·ут,** (*past masc sing*) **за·лёг·,** (*perf*); lie down; hide, lie up

за·ли́в·, -а,   (*m*) (geog) bay, cove, inlet, gulf; filling (rubber)

за·лив·а́ни·е, -я,   (*n*) (*v n*) of **за·лив·а́·ть**

за·лив·а́·ть, -ют,   (*imp*) flood, pour over, inundate; pour, pour in, prime; cast (rubber); lay, spread (asphalt etc.); extinguish, put out with water (fires)

за·лив·и́н·а, -ы,   (*f*) protrusion; (met) bleeder (defect in an ingot)

за·ли́в·к·а, -и,   (*g pl*) **-в·ок·,** (*f*) pouring over, flooding; pouring in, filling; (met) hot metal charging; (elec) potting; babbitting (a bearing); laying, spreading (asphalt etc.)

~ **в·сух·у́ю**   (met) dry-sand casting

~ **в·сыр·у́ю**   (met) green-sand casting

~ **гало́ш·**   (rubber) mending galoshes

~ **форм·**   (met) casting
форм·ова́ни·е за·ли́в·к·ой   (*n*) slush moulding/molding (rubber)

~ **чугун·а́**   (met) addition of hot metal (to OH furnace etc.)

за·лив·н·о́й, -а́я, -о́е,   (*a*) flood, flooding; pouring, priming; (food) jellied, in aspic

~ **луг·**   (*m*) (agr) water meadow

за·ли́в·очн·ый, -ая, -ое,   (*a*) of **за·ли́в·к·а**

~ **кран·**   (*m*) (met) hot-metal charging crane

~ **проб·к·а**   (*f*) priming plug (pumps etc.)

~ **со·ста́в·**   (*m*) (elec) filling compound

за·ли́в·чик·, -а,   (*m*) (geog) creek

за·ли́в·щик·, -а,   (*m*) pourer (in foundry etc.)

за·ли́з·, -а,   (*m*) fillet, fairing

за·ли́·т·ый, -ая, -ое,   (*past part pass*) see **за·лив·а́·ть**

~ **компа́унд·ом**   (elec) potted

~ **препара́т·,**   (*m*) embedded preparation

за·ли́·ть   (*fut 3rd sing, pl*) **за·ль·ёт, за·ль·ю́т,** (*perf*); see **за·лив·а́·ть**

за·лич·а́·ть, -ют,   (*perf*) set grain (leather)

за·ло́г·, -а,   (*m*) (fin) deposit, pledge, security, mortgage; (gram) voice

~, **дей·стви́тельн·ый**   active voice

~, **страд·а́тельн·ый**   passive voice

за·ло́г·ов·ый, -ая, -ое,   (*a*) of **за·ло́г·**

~ **това́р·**   (*m*) bonded goods

за·лож·е́ни·е, -я,   (*n*) putting, laying; location (of oil wells); (surv) horizontal equivalent

за·ло́ж·енн·ый, -ая, -ое,   (*past part pass*) of **за·лож·и́ть**

~ **дета́л·ь**   (*f*) (plast) embedded part, insert

за·лож·и́ть, -ат,   (*perf*) put, lay; arrange; deposit

за·ло́м·, -а,   (*m*) jam (lumbering); (fish) Caspian herring; (text) crease

за·лом·и́ть, -ят,   (*perf*) see **за·ла́м·ы·ва·ть**

за́лп·, -а,   (*m*) salvo, volley

за́лп·ов·ый, -ая, -ое,   (*a*) of **за́лп·**

за·луж·е́ни·е, -я,   (*n*) (*v n*) of **за·луж·и́ть**

за·луж·и́ть, -а́т,   (*perf*) (agr) sow to grass, turn into meadow/pasture

за́льбанд·, -а,   (*m*) (min) gouge, flucan

за·ль·ю́т   (*fut 3rd pl*) of **за·ли́·ть**

зам-   (*component*) **за·мест·и́тел·ь** deputy

за·ма́ж·ут   (*fut 3rd pl*) of **за·ма́з·ать**

за·ма́з·ать   (*fut 3rd sing, pl*) **за·ма́·ж·ет, за·ма́з·ут** (*perf*); see **за·ма́з·ыва·ть**

за·ма́з·к·а, -и,   (*f*) putty, paste, cement; puttying, smearing; (ceram) luting

~, **желе́з·н·ая**   iron cement

~, **су́рик·ов·ая**   red lead cement

за·ма́з·ыва·ть, -ют,   (*imp*) putty, daub, besmear, spread over

за·ма́лч·ива·ть, -ют,   (*imp*) keep quiet about, refrain from acknowledgement

за·ма́н·ива·ть, -ют,   (*imp*) entice, lure, decoy

за·ман·и́ть, -ят,   (*perf*) see **за·ма́·н·ива·ть**

за·марин·ова́ть, -у́ют,   (*perf*) pickle, marinade

за·маск·ир·ова́ть, -у́ют,   (*perf*) camouflage, disguise

за·ма́сл·енн·ый, -ая, -ое,   (*past part pass*) see **за·ма́сл·ива·ть**

за·ма́сл·иватель·, -я,   (*m*) (plast) lubricant, oiling agent

за·ма́сл·ива·ть, -ют,   (*imp*) oil, lubricate, make oily/greasy

за·ма́сл·ить, -ят,   (*perf*) see **за·ма́·сл·ива·ть**

за·ма́т·ыва·ть, -ют,   (*imp*) coil, reel, roll/wind up

за·ма́х·, -а, *(m)* *(v n)* see **за·мах·а́ть**

за·мах·а́ть *(fut 3rd sing, pl)* **за·ма́-ш·ет, за·ма́ш·ут,** *(perf)*; begin to wave/wag/flap

за·ма́ч·ива·ть, -ют, *(imp)* wet, soak, damp, moisten; ret (hemp etc.); liquor (malt)

за·ма́ш·ут *(fut 3rd pl)* of **за·мах·а́ть**

за·ма́щ·ива·ть, -ют, *(imp)* pave

за·ма́·яться, -ются, *(perf)* be exhausted/tired out

за·медл·е́ни·е, -я, *(n)* slowing down, moderation, reduction of speed, deceleration; delay, lag

   длин·а́ за·медл·е́ни·я *(f)* (phys) moderation length

~ **до тепл·ов·о́й сќор·ост·и** (nucl) thermalization

   коэффицие́нт· за·медл·е́ни·я *(m)* moderating ratio

   тео́р·и·я за·медл·е́ни·я *(f)* (nucl) slowing-down theory

за·ме́дл·енн·ый, -ая, -ое, *(past part pass)* see **за·медл·я́·ть**; slow-speed

~ **де́й·стви·е** *(n)* delayed action, *(in gen as a)* delayed-action

за·медл·и́тел·ь, -я, *(m)* retarder, delay mechanism/element; (chem) inhibitor; (nucl) moderator; (phot) restrainer, retarder, inhibitor

~, **ваго́н·н·ый** (rail) retarder, retarding mechanism

~ **вулканиз·а́ци·и** vulcanization retarder

~ **корро́з·и·и** corrosion inhibitor

за·ме́дл·ить, -ят, *(perf)* see **за·медл·я́·ть**

за·медл·я́·ть, -ют *(imp)* slow down, reduce speed, retard, decelerate; restrain, moderate, inhibit; delay

за·мельк·а́·ть, -ют, *(perf)* flash, gleam

за·ме́н·а, -ы, *(f)* substitution, replacing, replacement, (law) commutation

~ **строк· столб·ц·а́ми** (math) interchanging rows and columns

~ **то́пл·ив·а** refuelling

за·мен·и́м·ый, -ая, -ое, *(a)* replaceable

за·мен·и́тел·ь, -я, *(m)* substitute, substitute material, alternative material, alternative

~, **кож·е́венн·ый** leather substitute, artificial leather

за·мен·и́ть, ²ят, *(perf)* see **за·мен·я́·ть**

за·мен·я́емост·ь, -и, *(f)* replaceability, exchangeability

за·мен·я́ем·ый, -ая, -ое, *(pres part pass)* of **за·мен·я́·ть**; exchangeable, replaceable

за·мен·я́·ть, -ют, *(imp)* substitute, replace

за·ме́р·, -а, *(m)* measuring survey, surveying; reading (of instrument); **за́·мер·** *(past masc sing)* of **за·ме·р·е́ть**

~, **маркше́йдер·ск·ий** (min) underground survey

за·мер·е́ть *(fut 3rd sing, pl)* **за·мр·ёт, за·мр·у́т,** *(past masc sing)* **за́·мер·,** *(perf)*; see **за·мир·а́·ть**

за·мерз·а́ни·е, -я, *(n)* freezing, congealing

   температу́р·а за·мерз·а́ни·я *(f)* congealing/freezing point

за·мерз·а́·ть, -ют, *(imp)* *(trans and intrans)* freeze, congeal

за·мёрз·нуть, -нут, *(past masc sing)* **за·мёрз·,** *(perf)*; see **за·мерз·а́·ть**

за·ме́р·ить, -ят, *(perf)* see **за·ме·р·я́·ть**

за·ме́р·н·ый, -ая, -ое, *(a)* gauge, gage, measuring

~ **ре́·йк·а** (f) gauge rod, dip stick

за·ме́р·ш·ий, -ая, -ее, *(past part act)* of **за·мер·е́ть** *(perf)*; see **за·мир·а́·ть** *(imp)*

за·мер·я́·ть, -ют, *(imp)* gauge, measure

за·ме́с·, -а, *(v n)* see **за·мес·и́ть**; mix, batch

за·мес·и́ть, ²ят, *(perf)* mix, knead

за·ме́с·очн·ый, -ая, -ое, *(a)* mixing, batching

~ **маши́н·а** *(f)* mixer, (ceram) blunger

за·мест·и́тел·ь, -я, *(m)* deputy, substitute; proxy; (chem) substituent

~ **дире́ктор·а** deputy director/manager/head

~ **пред·сед·а́тел·я** vice-chairman

за·мест·и́тельн·ый, -ая, -ое, *(a)* deputizing; compensating; deputy

~ **систе́м·а** *(f)* (naut) compensating system

за·мест·и́ть, -я́т, *(perf)* see **за·ме-щ·а́·ть**

за·мёт·, -а, *(m)* cast, casting, shooting (fishing nets)

за·мёт·анн·ый, -ая, -ое, *(past part pass)* of **за·мет·а́·ть** *(perf)*

за·мет·а́·ть, -ют, *(imp)* sweep up; cover/sprinkle over; *(perf)* sew/tack up

за·ме́т·ить, -ят, *(perf)* see **за·ме-ч·а́·ть**

за·мéт·к·а, -и, (*g pl*) -т·ок·, (*f*) note, memorandum; notice; mark; за·мёт·к·а, -и, (*g pl*) -т·ок·, (*f*) sewing up, tacking

~, пут·ев·ы́е (*pl*) itinerary

за·мéт·н·ый, -ая, -ое, (*a*) noticeable, marked, appreciable, pronounced; notable, outstanding

за·мéт·ыва·ть, -ют, (*imp*) sew/tack up

за·меч·áни·е, -я, (*n*) remark, remarks, note

за·меч·áтельн·ый, -ая, -ое, (*a*) remarkable, striking

за·меч·á·ть, -ют, (*imp*) note, notice, observe; (*intrans*) remark

за·мéш·анн·ый, -ая, -ое, (*past part pass*) of за·меш·á·ть

за·меш·áтельств·о, -а, (*n*) confusion, disorder

за·меш·á·ть, -ют, (*perf*) involve, entangle, mix up

за·мéш·енн·ый, -ая, -ое, (*past part pass*) of за·мес·и́ть

за·мéш·ива·ть, -ют, (*imp*) involve, entangle, mix up; mix, knead

за·мещ·áем·ый, -ая, -ое, (*pres part pass*) of за·мещ·á·ть; replaceable, displaceable

за·мещ·á·ть, -ют, (*imp*) deputize, act for, substitute; replace, substitute, insert (math)

за·мещ·áющ·ий, -ая, -ое, (*pres part act*) of за·мещ·á·ть; substitution, substitutional, substituent

за·мещ·áющ·ийся, -аяся, -ееся, (*pres part act*) of за·мéщ·а·ться; replaceable

за·мещ·éни·е, -я, (*n*) substitution, replacement

мéтод· за·мещ·éни·я (*m*) substitution method (of measurement)

рас·твóр· за·мещ·éни·я (*m*) (chem phys) substitutional solution

реáкци·я за·мещ·éни·я (*f*) (chem) substitutional reaction

систéм·а за·мещ·éни·я (*f*) (mar eng) compensating system

цеп·ь за·мещ·éни·я (*f*) (elec) equivalent circuit

за·мещ·ённ·ый, -ая, -ое, (*past part pass*) see за·мещ·á·ть; (*m decl as adj*) (chem) substituted form

~ в ядр·é (chem) nucleus-substituted

за·миг·á·ть, -ют, (*perf*) begin to blink/wink

за·ми́н·к·а, -и, (*g pl*) -н·ок·, (*f*) hitch, difficulty

за·мир·áни·е, -я, (*n*) (*v n*) of за·мир·á·ть

за·мир·á·ть, -ют, (*imp*) die down/away, fade; stop beating (of heart)

зá·мк·нутост·ь, -и, (*f*) reticence; state/property of being closed, closure

зá·мк·нут·ый, -ая, -ое, (*past part pass*) of за·мк·н·у́ть; closed, locked; (geog) land-locked

~ на·корóт·к·о (elec) short circuited

~ пруж·и́н·а (*f*) flat spring

~ систéм·а (*f*) closed system; (autom) closed-loop system

систéм·а зá·мк·нут·ого цикл·а (*f*) (autom) monitored control system

~ стрó·й (*m*) (mil) close order/formation

~ цéп·ь (*f*) (elec) closed circuit

~ цикл· (*m*) closed cycle; (autom) closed loop; (chem) closed circuit

за·мк·н·у́ть, -ут, (*perf*) close, lock; (elec) make/close circuit

зá·мк·ов·ый, -ая, -ое, (*a*) of зá·мок·; за·мк·óв·ый, -ая, -ое, (*a*) of за·мóк·

за·мó·ет (*fut 3rd sing*) of за·мы́·ть

за·мóк·, -·мк·á, (*m*) lock, locking device; scarf, scarf joint; (arch) keystone; (gunn) breech mechanism; (zool) hinge; за·мóк· (*past masc sing*) of за·мóк·нуть; зá·мок·, -·мк·а, (*m*) castle

~, америкáн·ск·ий Yale lock

~, бур·и́льн·ый (oil) tool joint

~, вис·я́ч·ий padlock

~, в·рез·н·óй mortise lock

~, гá·ечн·ый lock nut

~, канáт·н·ый (oil) rope socket; cable joint

~ клáпан·а valve lock ring

~, на·клад·н·óй rim lock

~, цили́ндр·ов·ый Yale lock

~ шасси́ (a/c) undercarriage safety lock

~, штык·ов·óй bayonet joint

за·мок·á·ть, -ют, (*imp*) get soaked/wet through

за·мóк·нуть, -нут, (*past masc sing*) за·мóк·; see за·мок·á·ть

за·мóк·нуть, -нут, (*past masc sing*) змок; see за·мок·á·ть

за·молк·á·ть, -ют, (*imp*) fall silent, stop speaking

за·мóлк·нуть, -нут, (*past masc sing*) за·мóлк· (*perf*); see за·молк·á·ть

за·молч·á·ть, -áт, (*perf*) see за·мол·к·á·ть; see за·мáлч·ива·ть за·мора́ж·ивани·е, -я, (*n*) freezing, chilling, refrigerating, congealing; (nucl) quenching

~ гру́нт·ов freezing method (shaft sinking)

за·мора́ж·ива·ть, -ют, (*imp*) freeze, refrigerate, ice, chill

за·морг·а́·ть, -ют, (*perf*) begin to blink

за·мор·ённ·ый, -ая, -ое, (*past part pass*) of за·мор·и́ть; emaciated, starved

за·мор·и́ть, -я́т, (*perf*) underfeed, starve

за·мо́р·к·а, -и, (*f*) stoving, stifling (silk production)

за·моро́ж·енн·ый, -ая, -ое, (*a*) (*past part pass*) see за·мора́ж·ива·ть; frozen, iced; refrigerated; (plast) frozen-in, (phys) quenched

~ орбита́ль·н·ый моме́нт· (*m*) (nucl) quenched orbital moment

за·моро́з·ить, -ят, (*perf*) see за·мора́ж·ива·ть

за́·мороз·к·и, -ов, (*m pl*) night frosts

за·морос·и́ть (*fut 3rd sing*) -и́т, (*perf*) drizzle; begin to drizzle

за·мост·и́ть, -я́т, (*perf*) see за·ма́щ·ива·ть

за·мот·а́·ть, -ют, (*perf*) see за·ма́т·ыва·ть

за·моч·и́ть, -ат, (*perf*) see за·ма́ч·ива·ть

за·мо́ч·к·а, -и, (*g pl*) -ч·ек·, (*f*) wetting, soaking, dampening, moistening; retting (linen, hemp etc.)

за·мо́ч·н·ый, -ая, -ое, (*a*) of за·мо́к·, (*m decl as a*) (gunn) breechworker

~ ша́йб·а (*f*) lock washer/nut

~ гнезд·о́ (*n*) (gunn) breech bush

~ сква́ж·ин·а (*f*) keyhole

за·мощ·ённ·ый, -ая, -ое, (*past part pass*) of за·мост·и́ть (*perf*) see за·· ма́щ·ива·ть (*imp*)

за·мо́·ют (*fut 3rd pl*) of за·мы́·ть

за·мр·у́т (*fut 3rd pl*) of за·мер·е́ть

за·му́ж·н·ий, -яя, -ее, (*a*) married (of women)

за·мур·ова́·ть, -у́ют, (*perf*) see за·му·р·о́выва·ть

за·мур·о́выва·ть, -ют, (*imp*) brick up, immure, wall-in

за·му́сор·енност·ь, -и, (*f*) degree of contamination, impurity content

за·мут·и́ть, -я́т, (*perf*) make cloudy/muddy/turbid

за·мут·нённ·ый, -ая, -ое, (*past part pass*) of за·мут·не́·ть; cloudy, clouded, muddy, turbid

за·мут·не́·ть, -ют, (*perf*) become cloudy/muddy/turbid

за́мш·а, -и, (*i*) -ей, (*f*) chamois leather, shammy, suede

замш·ева́ни·е, -я, (*n*) chamoising, suede finishing

за́мш·ев·ый, -ая, -ое, (*a*) of за́мш·а

за·мыв·а́ни·е, -я, (*n*) (*v n*) of за·мы·в·а́·ть

за·мыв·а́·ть, -ют, (*imp*) wash out/off (marks); wash smooth (of streams, waves etc.)

за·мык·а́ни·е, -я, (*n*) (*v n*) see за·мы·к·а́·ть; (math) closure; (elec) short circuit, shorting, fault

~ в кольц·о́ (chem) cyclizing

~, коро́т·к·ое (elec) short circuit

~ на земл·ю́ (elec) fault to ground, earth fault

~ про·ги́б·а (geol) trough closure

за·мык·а́тел·ь, -я, (*m*) (elec) contactor, contact maker; (expl) switch, firing unit; abutment (rotary pumps)

за·мык·а́·ть, -ют, (*imp*) close, lock; (elec) make/close circuit

за·мык·а́ющ·ий, -ая, -ее, (*pres part act*) of за·мык·а́·ть; closing, closure; contacting; flux-closure (ferromagnetism)

~ голо́в·к·а за·клёп·к·и rivet point

~ доме́н· (*m*) (phys) domain of closure

за́·мысел·, -сл·а, (*m*) plan, intention, design

за·мы́сл·ить, -ят, (*perf*) see за·мышл·я́·ть

за·мысл·ова́т·ый, -ая, -ое, (*a*) intricate, complicated

за·мышл·я́·ть, -ют, (*imp*) plan, design, intend

за·мы́·ть (*fut 3rd sing, pl*) за·мо́·ет, за·мо́·ют, (*perf*); see за·мыв·а́·ть

за́·на·вес·, -а, (*m*) curtain, hanging screen

за·на·ве́с·ить, -ят, (*perf*) see за·на·ве́ш·ива·ть

за·на·ве́ш·ива·ть, -ют, (*imp*) curtain, curtain off

за·на·ря́д·к·а, -и, (*g pl*) -д·ок·, (*f*) request indent

За́ндмейер·а реа́кци·я (*f*) (chem) Sandmeyer's reaction

за́ндр·, -а, (*m*) frontal apron (of glacier)

за·не·во́л·ивани·е, -я, (*n*) compression (of spring)

за·нес·éни·е, -я, *(n)* *(v n)* *see* **за·нос·и́ть**

за·нес·ти́, -ут, *(past masc sing)* **за·нёс·,** *(perf)*; *see* **за·нос·и́ть**

за·ниж·á·ть, -ют, *(imp)* understate, put too low, underestimate

за·ниж·éни·е, -я, *(n)* understatement

за·ни́ж·енн·ый, -ая, -ое, *(past part pass)* of **за·ни́з·ить**; too low, too small

за·ни́з·ить, -ят, *(perf)* *see* **за·ниж·á·ть**

за·ним·áтельн·ый, -ая, -ое, *(a)* interesting; entertaining

за·ним·á·ть, -ют, *(imp)* occupy; engage, interest; **-ся** (+ *instr*) be occupied with, be engaged/interested in, be doing

зá·нов·о *(adv)* anew, again

заноз·а, -ы, *(f)* splinter

за·нóр·ыш·, -а, *(m)* (geol) cavity

за·нóс·, -а, *(m)* drift, accumulation; (М/Т) skidding, skid

~, снéж·н·ый snow drift

за·нос·и́ть, ´ят, *(imp)* bring/deliver to; record, enter, register, put down; carry (an infection); drift, accumulate (of sand, snow etc.); (М/Т) skid

за·нул·éни·е, -я, *(n)* (elec) neutral grounding/earthing

за·ноч·евá·ть, -у́ют, *(perf)* stay/spend the night

за·нумер·овá·ть, -у́ют, *(perf)* *see* **за··нумер·óвыва·ть**

за·нумер·óвыва·ть, -ют, *(imp)* number

за·ня́т·и·е, -я, *(n)* occupation, business, employment, pursuit; studies, lessons, instruction

за·ня́т·ост·ь, -и, *(f)* state of being busy/engaged

   релé за·ня́т·ост·и *(n)* busy relay (telephones)

   сигнáл за·ня́т·ост·и *(m)* engaged/ busy tone/signal (telephones)

за·ня·т·óй, -áя, -óе, *(a)* busy, having a lot of work

зá·ня·т·ый, -ая, -ое, *(past part pass)* of **за·ня́·ть**; occupied, engaged, employed; (nucl) filled

~ обол·óчк·а *(f)* (nucl) filled shell

за·ня́·ть *(fut 3rd sing, pl)* **за·йм·ёт, за·йм·у́т,** *(perf)*; *see* **за·ним·á·ть**

за·одн·ó *(adv)* in agreement/concert, together; as well as, together with

за·остр·éни·е, -я, *(n)* *(v n)* *see* **за··остр·я́·ть**; taper

   у́гол· за·остр·éни·я *(m)* throat angle, end lip angle, lip angle (of cutting tool)

за·остр·ённ·ый, -ая, -ое, *(past part pass)* *see* **за·остр·я́·ть**; pointed, sharp, (bot) acuminate; tapered

за·остр·и́ть, -ят, *(perf)* *see* **за·остр·я́·ть**

за·остр·я́·ть, -ют, *(imp)* sharpen, point, make pointed, taper down; **-ся** become pointed/tapered/sharpened, taper

за·остр·я́ющ·ий, -ая, -ее, *(pres part act)* of **за·остр·я́·ть**; (elec) peaking

~ схéм·а *(f)* peaking circuit

за·óч·ник· -а, *(m)* correspondence student

за·óч·н·о *(adv)* without seeing, in the absence of; by correspondence; (law) by default

за·óч·н·ый, -ая, -ое, *(a)* *in absentio,* by default; correspondence (of studies)

~ институ́т· *(m)* correspondence school

~ при·говóр· *(m)* (law) judgement by default

зап *(abbr)* of **зá·пад·н·ый** western

за·пá·вш·ий, -ая, -ее, *(past part act)* of **за·пáс·ть**

зá·пад·, -а, *(m)* West, Occident

за·пад·áни·е, -я, *(n)* *(v n)* of **за·па·д·á·ть**; attenuation

за·пад·á·ть, -ют, *(imp)* fall/drop behind/back, (astron) set, sink; *(perf)* begin to fall/drop/sink

за·пад·ёт *(fut 3rd sing)* of **за·пáс·ть**

за·пá́д·ин·а, -ы, *(f)* (geol) hole, sink hole; (bot) patella

западно·европ·éйск·ий, -ая, -ое, *(a)* West-European

зá·пад·н·ый, -ая, -ое, *(a)* west, western, occidental

за·пад·н·я́, -й, *(f)* drop trap (e.g. for animals)

за·пáзд·ывани·е, -я, *(n)* delay, retardation; time lag, lag; interval (between successive operations)

~ вы́·стрел·а (gunn) firing interval

~, дистанциóн·н·ое (autom) distance velocity lag, transportation lag, dead-time lag

~, ём·костн·ое (autom) capacity lag

~ из·мер·и́тельн·ой систéм·ы (autom) measurement lag

~ ору́ж·и·я (gunn) mechanical interval

~, пере·дáт·очн·ое = з. дистанциóн··н·ое

~, пере·хóд·н·ое (autom) transfer lag

~ реаг·и́ровани·я response lag

за·па́зд·ывани·е

~ стрел·к·а́ (gunn) personal·interval у́гол· за·па́зд·ывани·я (*m*) (eng) delay/lag angle

~, чи́ст·ое = з. дистанцио́н·н·ое явл·е́ни·е за·па́зд·ывани·я (*n*) lag effect

за·па́зд·ыва·ть, -ют, (*imp*) be late, lag, be retarded

за·па́зд·ывающ·ий, -ая, -ое, (*pres part act*) of за·па́зд·ыва·ть; retarding, retarded; delayed

~ нейтро́н· (*m*) delayed neutron

~ обра́т·н·ая с·вяз·ь (*f*) (autom) retarding feedback

~ э́хо (*n*) delayed echo

за·па́·ива·ть, -ют, (*imp*) solder; (elec) seal, seal off

за·па́·йк·а, -и, (*g pl*) -а́·ек·, (*f*) soldering; sealing, seal, sealing off/in (of glass vessels etc.)

за·пак·ова́ть, -у́ют, (*perf*) *see* за·пак·о́выва·ть

за·пак·о́выва·ть, -ют, (*imp*) pack/wrap up

за·па́л·, -а, (*m*) (ICE) ignition; (expl) detonator; (gunn) vent, touch hole, tubeflash channel; (hort) heat scorching; (nucl) seed, seeding; broken wind (of horses); за·па́·л (*past masc sing*) of за·па́с·ть

зо́н·а за·па́л·а (*f*) (nucl) seed region

за·па́л·ива·ть, -ют, (*imp*) ignite, light

за·пал·и́ть, -я́т, (*perf*) *see* за·па́л·ива·ть

за·па́ль·ник·, -а, (*m*) igniter, (a/c) torch igniter, cartridge starter

за·па́ль·н·ый, -ая, -ое, (*a*) ignition, igniting; detonating

~ патро́н· (*m*) (expl) detonator; (a/c) cartridge starter, starting cartridge; (ICE) torch igniter, combustion torch

~ с·бо́р·к·а (*f*) (nucl) seed assembly/bundle

~ свеч·а́ (*f*) (ICE) spark plug

~ стака́н· (*m*) (expl) primer

~ шар· (*m*) hot bulb (of semi-diesel)

за·па́ль·щик·, -а, (*m*) (min) shot firer, powderman

за́пан·ь, -и, (*f*) boom (on logging river)

~, ледо·за·щи́т·н·ая ice boom

за·паралле́ль·н·ый, -ая, -ое, (*a*) parallel-connected

за·па́р·ива·ть, -ют, (*imp*) steam

за·па́р·ить, -ят, (*perf*) *see* за·па́р·ива·ть

за·па́р·к·а, -и, (*g pl*) -р·ок·, (*f*) steaming; (text) steam printing; (text) steamer, steam-printing plant

за·па́р·ник·, -а, (*m*) (agr) steaming plant, steamer

за·па́р·н·о́й, -а́я, -о́е, (*a*) steam, steaming

за·па́с·, -а, (*m*) stock, supply, reserve, bank; за·па́с· (*past masc sing*) of за·па́с·ти́

~ водо·из·мещ·е́ни·я (shipb) moulded displacement allowance

коэффицие́нт· за·па́с·а (*m*) (elec) assurance factor

коэффицие́нт· за·па́с·а про́ч·ност·и (*m*) safety factor

~ на·де́ж·ност·и safety margin

~, не·при·кос·нове́нн·ый emergency supplies/ration, reserve stores

~, не·раз·ве́д·анн·ые (*pl*) (geol) possible reserves

~ плов·у́чест·и (naut) reserve of buoyancy, buoyancy

~ подъ·ём·н·ой си́л·ы lift margin

~ про́ч·ност·и safety factor/margin/coefficient

~, това́р·н·ые (*pl*) (com) stock-in-trade

~ то́пл·ив·а fuel supplies/stock; (naut) fuel capacity

~ тя́г·и reserve thrust (jet engines)

~ у·сто́й·чивост·и margin of stability

~ элеме́нт·ов (math) domain

за·пас·а́ни·е, -я, (*n*) (*v n*) of за·пас·а́·ть; storage

за·пас·а́·ть, -ют, (*imp*) store/stock up, lay in a stock/store

за·пас·ённ·ый, -ая, -ое, (*past part pass*) of за·пас·ти́ (*perf*); *see* за·пас·а́·ть (*imp*); stored, saved up

за·пас·н·о́й, -а́я, -о́е, (*a*) *see* за·па́с·н·ый

за·па́с·н·ый, -ая, -ое, (*a*) spare, reserve, emergency; (mil) replacement

~ аэродро́м· (*m*) alternative aerodrome

~ вы́·ход· (*m*) emergency exit

~ полк· (*m*) replacement regiment

~ част·ь (*f*) spare part

~ я́кор·ь (*m*) (naut) sheet anchor

за·пас·ти́, -у́т, (*past masc sing*) за·па́с·, (*perf*); *see* за·пас·а́·ть

за·па́с·ть (*fut 3rd sing, pl*) за·пад·ёт, за·пад·у́т, (*past masc sing*) за·па́·л, (*perf*) drop/fall behind/back, (astron) sink, set

за·пах·, -а, (*m*) smell, scent, odour, odor; **за·пах·, -а,** (*m*) overlap, wrap (of clothes); **за·пах·** (*past masc sing*) of **за·пах·нуть**

за·пах·ать (*fut 3rd sing, pl*) **за·паш·ет, за·паш·ут,** (*perf*) plough, plough in/under, plow, plow in/under

за·пах·ива·ть, -ют, (*imp*) plough, plough in/under, plow, plow in/under; wrap/draw closer/over

за·пах·нуть, -нут, (*past masc sing*) **за·пах·** (*perf*) begin to smell, give off a smell; **за·пах·н·уть, -ут,** (*perf*) wrap over, wrap/draw closer

за·паш·ет (*fut 3rd sing*) of **за·пах·ать**

за·паш·к·а, -и, (*g pl*) **-ш·ек·,** (*f*) ploughing in/under, plowing in/under; ploughing land, plowing land, tillage

за·паш·ник·, -а, (*m*) (agr) ridger, shallow plough/plow

за·па·янн·ый, -ая, -ое, (*past part pass*) of **за·па·я·ть**

за·па·я·ть, -ют, (*perf*) solder; seal, seal off (glass, elec wires); sweat on

за·пек·а·ть, -ют, (*imp*) bake, bake hard, bake on/in; **-ся** (*pass*); clot, coagulate (of blood)

за·пёк·ш·ий, -ая, -ее, (*past part act*) of **за·печ·ь** (*perf*); *see* **за·пек·а·ть** (*imp*)

за·пеленг·ова́ть, -у́ют, (*perf*) take a bearing, take bearings

за·пен·ить, -ят, (*perf*) cover with foam; **-ся** begin to foam/froth

за·пер·е́ть (*fut 3rd sing, pl*) **за·пр·ёт, за·пр·у́т,** (*past masc sing*) **за́·пер·,** (*perf*); *see* **за·пир·а́·ть**

за́·пер·т·ый, -ая, -ое, (*past part pass*) of **за·пер·е́ть**; (elec) blocked, barred, cut off, ungated

~ **режи́м·** (*m*) (elec) cutoff condition

за·пёр·ш·ий, -ая, -ее, (*past part act*) of **за·пер·е́ть**

за·печа́т·а·ть, -ют, (*perf*) seal, put a seal on; seal, seal up, close with a seal; begin to seal

за·печатл·ева́·ть, -ют, (*imp*) imprint, impress

за·печатл·е́·ть, -ют, (*perf*) *see* **за·печатл·ева́·ть**

за·печа́т·ыва·ть, -ют, (*imp*) seal, put a seal on; seal, seal up

за·печ·ённ·ый, -ая, -ое, (*past part pass*) *see* **за·пек·а́·ть**

за·пе́ч·к·а, -и, (*f*) baking (varnish)

за·пе́ч·н·ый, -ая, -ое, (*a*) behind/in the vicinity of a furnace/stove/fireplace, chimney-back

~ **котёл·** (*m*) waste-heat boiler

за·пе́ч·ь (*fut 3rd sing, pl*) **за·печ·ёт, за·пек·у́т,** (*past masc sing*) **за·пёк·,** (*perf*); *see* **за·пек·а́·ть**

за·пи́л·, -а, (*m*) gash, notch (made by saw etc.)

за·пил·ённ·ый, -ая, -ое, (*past part pass*) notched (with saw etc.)

за·пин·а́·ться, -ются, (*imp*) hesitate, falter

за·пир·а́ни·е, -я, (*n*) closing, locking; (elec) blocking (valves/vacuum tubes etc.) (TV) blanking, blanketing; choking, choke-out (of wind tunnel)

~ **пуч·к·а́** (elec) beam suppression/ blanking

~ **по·то́к·а** (dynam) flow choking

за·пир·а́тельн·ый, -ая, -ое, (*a*) *and see* **за·пир·а́ющ·ий**; locking; blocking

за·пир·а́·ть, -ют, (*imp*) lock, lock in, lock up, seal up; shut off, cut off; (elec) block, blank, suppress

за·пир·а́ющ·ий, -ая, -ее, (*pres part act*) of **за·пир·а́·ть**; (elec) blocking, blanking, cutoff, barrier

~ **за·мы́ч·к·а** (*f*) (rail) control lock

~ **ко́нтур·** (*m*) (elec) band-stop filter

~ **с·ло́·й** (*m*) (elec) blocking layer

за·пи́с·анн·ый, -ая, -ое, (*past part pass*) *see* **за·пи́с·ыва·ть**; recorded; (math) given, written

за·пис·а́ть (*fut 3rd sing, pl*) **запи́шет, запи́шут,** (*perf*); *see* **за·пи́с·ыва·ть**

за·пи́с·к·а, -и, (*g pl*) **-с·ок·,** (*f*) note, memorandum, (*pl*) transactions (of learned society)

~, лю́к·ов·ая (naut) inspections certificate

за·пис·н·о́й, -а́я, -о́е, (*a*) note, writing; regular, inveterate

~ **кни́ж·к·а** (*f*) notebook

за·пи́с·ывани·е, -я, (*n*) (*v n*) *see* **за·пи́с·ыва·ть**

за·пи́с·ыва·ть, -ют, (*imp*) write down, take notes; register, enter; record (sound); **-ся** (*pass*); sign up/on, join; make an appointment

за·пи́с·ывающ·ий, -ая, -ее, (*pres part act*) of **за·пи́с·ыва·ть**; recording (of instruments)

~ **блок·** (*m*) (instr) recording element

~ **у·стро́й·ств·о** (*n*) (instr) recorder

за́·пис·ь, -и, (*f*) entry, record, registration, recording; (math) notation

~ **а́кт·ов гражд·а́нск·ого со·сто·я́·ни·я (ЗАГС)** registrar's office

~, **дво·и́чн·ая** (autom) binary notation

~, **двой·н·а́я** (com) double entry

~, **десят·и́чн·ая** (autom) decimal notation

~ **зву́к·а** *see* **звуко·за́·пис·ь** sound recording

~ **интенси́в·ност·и** intensity trace (X-ray)

~, **норма́ль·н·ая** (autom) normal notation

~, **опера́тор·н·ый** (math) differential operator form

~, **черн·и́льн·ая** ink/pen recording

~, **четвер·и́чн·ая** (autom) quaternary notation, tetral notation

~, **шестнадцат·ери́чн·ая** (autom) sexadecimal notation

за·пих·а́·ть, -ют, (*perf*) begin to stuff/cram in

за·пи́х·ива·ть, -ют, (*imp*) cram/stuff in

за·пих·н·у́ть, -ут, (*perf*) *see* **за·пи́х·ива·ть**

за·пи́ш·ут (*fut 3rd pl*) of **за·пис·а́ть**

за·план·и́р·овать, -уют, (*perf*) plan, include in plan, envisage in plan

за·пла́т·а, -ы, (*f*) (text) patch

за·пла́т·анн·ый, -ая, -ое, (*past part pass*) of **за·плат·а́·ть**; patched

за·плат·а́·ть, -ют, (*perf*) (text) patch

за·плат·и́ть, -ят, (*perf*) pay

за·пла́ч·енн·ый, -ая, -ое, (*past part pass*) of **за·плат·и́ть**; paid

за·пл·ева́·ть, -юют, (*perf*) *see* **за·пл·ё·выва·ть**

за·пл·ёвыва·ть, -ют, (*imp*) spit at

за·пле́с·к·а, -и, (*g pl*) **-с·ок·,** (*f*) (met) double skin (continuous casting)

за·плеск·а́·ть, -ют, (*perf*) splash/spill on to

за·пле́сн·евел·ый, -ая, -ое, *a*) mouldy, moldy, mildewed

за·пле́сн·евени·е, -я, (*n*) (*v n*) *see* **за·пле́сн·еве·ть**; formation of mould/mold

за·пле́сн·еве·ть, -ют, (*perf*) go/grow mouldy/moldy/musty

за·плес·ти́ (*fut 3rd sing, pl*) **за·плет·т·ёт, за·плет·у́т,** (*past masc sing*) **за·плё·л** (*perf*); *see* **за·плет·а́·ть**

за·плет·а́·ть, -ют, (*imp*) plait, braid

за·пле́ч·ик·, -а, (*m*) shoulder; fixing lug; (*pl*) bosh (of a blast furnace); (print) shoulders

за·пле́ч·н·ый, -ая, -ое, (*a*) shoulder, shoulder-borne

~ **мешо́к·** (*m*) rucksack, knapsack

за·пле́ч·ь·е, -я, (*g pl*) **-ч·и·й,** (*n*) (anat) shoulder-blade, scapula

за·пломб·ир·ова́·ть, -у́ют, (*perf*) *see* **за·пломб·иро́выва·ть**

за·пломб·иро́выва·ть, -ют, (*imp*) seal, attach metal seal; stop (tooth)

за·плы́·в·, -а, (*m*) boat race

за·плыв·а́ни·е, -я, (*n*) (*v n*) of **за·плыв·а́·ть**

за·плыв·а́·ть, -ют, (*imp*) swim/float in/to/beyond; become filled up (with), fill up (with); get covered with, become bloated

за·плы́·ть (*fut 3rd pl*) **за·плыв·у́т,** (*perf*); *see,* **за·плыв·а́·ть**

за·пн·у́ться, -у́тся, (*perf*) hesitate, falter

за·по·ве́д·ник·, -а, (*m*) reservation, sanctuary, national park

за·под·лиц·о́ (*adv*) flush with

со·един·е́ни·е за·под·лиц·о́ (*n*) flush-filled joint

за·позд·а́л·ый, -ая, -ое, (*a*) belated, delayed

за·позд·а́·ть, -ют, (*perf*) *see* **за·па́зд·ыва·ть**

за·по́лз·а·ть, -ют, (*perf*) (*imp*) crawl/creep into/under; (*perf*) start/begin to crawl

за·полз·ти́, -у́т, (*past masc sing*) **за·по́лз·,** (*perf*) crawl/creep into/under

за·полимери́з·овать, -уют, (*perf*) (chem) polymerize

за·пол·и́ст·ый, -ая, -ое, (*a*) spready (leather)

за·полн·е́ни·е, -я, (*n*) filling, filling in/up; (phys) occupation

коэффицие́нт· за·полн·е́ни·я (*m*) duty cycle factor, duty factor (of pulses); space factor (of coil); (shipb) block coefficient

коэффицие́нт· за·полн·е́ни·я ме́д·ью (*m*) (elec) copper factor

~ **оболо́ч·к·и** (nucl) shell filling

за·по́лн·енш·ый, -ая, -ое, (*past part pass*) *see* **за·полн·я́·ть**

~ **со·сто·я́ни·е** (*n*) (phys) occupied state

~ **зо́н·а** (*f*) (phys) filled band

за·полн·и́тел·ь, -я, (*m*) filler; aggregate (concrete etc.)

~, **сил·ов·о́й** load-bearing filler (rubber)

за·по́лн·ить, -ят, (*perf*) *see* **за·пол·н·я́·ть**

за·полн·я́·ть, -ют, (*imp*) fill, fill in/up; (phys) occupy

за·поля́р·н·ый, -ая, -ое, (*a*) (geog) arctic

за·поля́р·ь·е, -я, (*n*) (geog) the Arctic

за·по·мин·а́ни·е, -я, (*n*) memorizing, remembering; storage (computers)

за·по·мин·а́·ть, -ют, (*imp*) memorize, remember

за·по·мин·а́ющ·ий, -ая, -ее, (*pres part act*) of **за·по·мин·а́·ть**; memory, storage (of computers)

~ **блок·,** (*m*) storage unit (computers)

~ **систе́м·а** (*f*) memory/storage system (computers)

~ **у·стро́й·ств·о** (*n*) store (of computer)

~ **у·стро́й·ств·о, долго·вре́мен·н·ое** (*n*) external store

~ **у·стро́й·ств·о на барабáн·е** (*n*) magnetic-drum store

~ **у·стро́й·ств·о на кондесáтор·ах** (*n*) capacitor store (computer)

~ **у·стро́й·ств·о на ли́ниях задéрж·ки** (*n*) delay-line store

~ **у·стро́й·ств·о на фе́ррит·ах** (*n*) ferrite-core store

~ **у·стро́й·ств·о, операти́в·н·ое** (*n*) internal store, working store

~ **у·стро́й·ств·о паралле́ль·н·ого дéй·стви·я** parallel store

~ **у·стро́й·ств·о, про·меж·у́точн·ое** (*n*) buffer store

~ **у·стро́й·ств·о, раз·руш·а́ющ·ееся** (*n*) volatile store

~ **у·стро́й·ств·о, рециркуляцио́н·н·ое** (*n*) dynamic store

~ **у·стро́й·ств·о с не·по·срéд·ствен·н·ой вы́·бор·к·и** (*n*) zero-access store

~ **у·стро́й·ств·о, со·хран·я́ющ·ееся** (*n*) permanent store, non-volatile store

~ **у·стро́йство, тви́стор·но·е** (*n*) twistor memory element

за·по́·мн·ить, -ят, (*perf*) *see* **за·по··мин·а́·ть**

зáпон·к·а, -и, (*g pl*) **-н·ок·,** (*f*) stud, collar-button, cuff-link; button

за·по́р·, -а, (*m*) bolt, locking bolt; locking, bolting; (med) constipation

за·по́р·н·ый, -ая, -ое, (*a*) locking, shutting, stopping, stop; (elec) barrier

~ **клáпан·** (*m*) (eng) stop valve

~ **кольц·о́** (*n*) locking ring

~ **сло́·й** (*m*) (elec) barrier blocking layer

за·порош·и́ть, -а́т, (*perf*) powder/ dust with

за·пот·ева́·ть, -ют, (*imp*) mist up, become misted, sweat

за·пот·éл·ый, -ая, -ое, (*a*) misted, sweating, sweaty, covered with condensation

за·пот·é·ть, -ют, (*perf*) *see* **за·пот·ева́·ть**

за·прав·и́л·а, -ы, (*f*) boss

за·прáв·ить, -ят, (*perf*) *see* **за·правл·я́·ть**

за·прáв·к·а, -и, (*g pl*) **-в·ок·,** (*f*) fuelling, refuelling, filling; servicing, setting up (machinery); dressing, grinding, sharpening (tools); feeding, starting (a rolling mill); fettling (an open-hearth furnace); dressing, manuring (soil); dressing, flavouring (food); charge, batch (rubber)

за·прáвл·енн·ый, -ая, -ое, (*past part pass*) *see* **за·правл·я́·ть**

за·правл·я́·ть, -ют, (*imp*) fuel, refuel, fill up; get ready, prepare, set up (machinery); grind, sharpen, dress (tools); (food) flavour with, season, dress; add; tuck in; thread in; thread up

за·прáв·очн·ый, -ая, -ое, (*a*) of **за··прáв·к·а**

~ **маши́н·а** (*f*) (met) fettling machine

~ **стан·о́к·** (*m*) (oil) tool-dressing machine

~ **стáнци·я** (*f*) (M/T) filling station

за·прáв·ск·ий, -ая, -ое, (*a*) true' real; regular

за·прáв·щик·, -а, (*m*) (air) bowser, refueller, aircraft tender

за·прáш·ива·ть, -ют, (*imp*) enquire; (com) overcharge, ask too much

за·пресс·о́ванн·ый, -ая, -ое, (*past part pass*) of **за·пресс·овáть;** pressed; (elec) packaged, moulded, molded

за·пресс·овáть, -у́ют, (*perf*) press; (elec) package, mould, mold

за·пресс·о́вк·а, -и, (*g pl*) **-вок·,** (*f*) pressed composition; pressurizing; pressing in, press-fitting; pressing, moulding, molding

за·прéт·, -а, (*m*) *and see* **за·прещ·é·ни·е;** ban, prohibition; (phys) exclusion; **за·пр·ёт** (*fut 3rd sing*) of **за··пер·éть**

за·прет·и́тельн·ый, -ая, -ое, (*a*) prohibitive

за·прет·и́ть, -я́т, (*perf*) *see* **за·пре·щ·á·ть**

за·прѐт·н·ый, -ая, -ое, (*a*) forbidden, prohibited; (phys) exclusion

~ зо́н·а (*f*) forbidden area; (phys) exclusion area

за·прещ·а́·ть, -ют, (*imp*) forbid, ban, prohibit

за·прещ·ѐни·е, -я, (*n*) (*v n*) *see* **за·прещ·а́·ть;** (phys) forbidden transition

~, суд·ѐбн·ое (law) injunction

за·прещ·ённост·ь, -и, (*f*) (phys) forbiddenness

за·прещ·ённ·ый, -ая, -ое, (*past part pass*) of **за·прет·и́ть;** forbidden, prohibited, banned

~ зо́н·а (*f*) (phys) forbidden band (of energy levels)

~, сла́б·о (phys) unfavoured, unfavored

~ у·ча́ст·ок· (*m*) (phys) forbidden band (of energy levels)

за·при·мѐт·ить, -ят, (*perf*) spot, notice

за·при·мѐч·енн·ый, -ая, -ое, (*past part pass*) of **за·при·мѐт·ить**

за·при·хо́д·овать, -уют, (*perf*) (fin) debit

за·программ·и́р·овать, -уют, (*perf*) programme, program

за·про·да·ва́ть, -ют, (*imp*) (fin) sell forward, sell for future delivery, contract to supply

за·про·да́·ж·а, -и, (*i*) **-ей,** (*f*) sale for future delivery, contract to supply

за·про·да́ть (*fut*) **за·про·да́м, за·про·да́шь, за·про·да́ст, за·про·дади́м, за·про·дади́те, запрода·ду́т;** *see* **за·про·да·ва́ть,**

за·проект·и́р·овать, -уют (*perf*) plan, form/launch project

за·про·ки́д·ыва·ть, -ют, (*imp*) throw back, tilt; overturn

за·про·ки́·н·уть, -ут, (*perf*) *see* **за·про·ки́д·ыва·ть**

за·про́с·, -а, (*m*) inquiry, interrogation; overcharging; (*pl*) requirements, needs

~ и́мпульс·ами (rad) pulse interrogation

за·прос·и́ть, -я́т, (*perf*) *see* **за·пра́·ш·ива·ть**

за·про́с·н·ый, -ая, -ое, (*a*) of **за·про́с·**

~ бланк· (*m*) request form

за·про́с·чик·, -а, (*m*) interrogator (radar)

~ -от·вѐт·чик·, -а, (*m*) (a/c rad) interrogator-responder

за·протест·ова́ть, -у́ют, (*perf*) protest, voice a protest, raise an objection

за·протокол·и́р·овать, -уют, (*imp*) enter in minutes/records

за·про́ш·енн·ый, -ая, -ое, (*past part pass*) of **за·прос·и́ть** (*perf*); *see* **за·пра́ш·ива·ть** (*imp*)

за·про·я́вл·енн·ый, -ая, -ое, (*past part pass*) (phot) over-developed

за·пру́д·а, -ы, (*f*) (*v n*) *see* **за·пру́·ж·ива·ть;** (hydr eng) dam, bar; artificial pond/lake, reservoir

за·пруд·и́ть, -я́т, (*perf*) *see* **за·пру́·ж·ива·ть**

за·пру́д·н·ый, -ая, -ое, (*a*) impounded (of waters)

за·пру́ж·ива·ть, -ют, (*imp*) dam, dam up

за·пр·у́т (*fut 3rd pl*) of **за·пер·ѐть**

за·пры́г·а·ть, -ют, (*perf*) jump; begin to jump

за·пряг·а́·ть, -ют, (*imp*) harness, yoke

за·пря́г·ш·ий, -ая, -ее, (*past part act*) of **за·пря́ч·ь**

за·пряж·ённ·ый, -ая, -ое, (*past part pass*) of **за·пря́ч·ь**

за·пря́ж·к·а, -и, (*g pl*) **-ж·ек·,** (*f*) harnessing; team (of draught animals)

за·пря́т·ать (*fut 3rd sing, pl*) **за·пря́ч·ет, за·пря́ч·ут,** (*perf*); *see* **за·пря́т·ыва·ть**

за·пря́т·ыва·ть, -ют, (*imp*) hide, conceal

за·пря́ч·ут (*fut 3rd pl*) of **за·пря́т·ать**

за·пря́ч·ь (*fut 3rd sing, pl*) **за·пряж·ёт, за·пряг·у́т,** (*past masc sing*) **за·пря́г·,** (*perf*); harness, yoke

за·пуг·а́·ть, -ют, (*perf*) *see* **за·пу́г·ива·ть**

за·пу́г·ива·ть, -ют, (*imp*) intimidate

за́·пуск·, -а, (*m*) (*v n*) *see* **за·пус·к·а́·ть;** start-up

~ втор·о́й сту́п·ен·и (rocket) second stage firing

~, жёстк·ий (rocket) hard start

~, залп·ов·ый battery launching (of rockets)

~ три́ггер·а (elec) triggering

за·пуск·а́·ть, -ют, (*imp*) put/shove/thrust in; fling; (ICE) start; launch (rocket), fire (rocket engine); (elec) start, trigger; neglect

за·пуст·ѐни·е, -я, (*n*) wilderness; state of neglect

за·пуст·é·ть, -ют, (*perf*) go to waste, turn into a wilderness

за·пуст·и́ть, -́ят, (*perf*) *see* **за·пуск·á·ть**

за·пу́т·анност·ь, -и, (*f*) complexity, complication; confusion, muddle

за·пу́т·анн·ый, -ая, -ое, (*past part pass*) *see* **за·пу́т·ыва·ть;** complicated, muddled, confused, tangled; (naut) foul

за·пу́т·а·ть, -ют, (*perf*) *see* **за·пу́т·ыва·ть**

за·пу́т·ыва·ть, -ют, (*imp*) tangle, muddle, confuse; (naut) foul

за·пух·á·ть, -ют, (*imp*) be swollen

за·пу́х·нуть, -нут, (*past masc sing*) **за·пу́х·,** (*perf*); *see* **за·пух·á·ть**

за·пу́щ·енност·ь, -и, (*f*) neglect

за·пу́щ·енн·ый, -ая, -ое, (*past part pass*) *see* **за·пуск·á·ть**

зап·част·и, -éй, (*f pl*) (*abbr*) of **за·пáс·н·ые част·и;** spare parts

за·пыл·á·ть, -ют, (*perf*) blaze/flare up

за·пыл·ённост·ь, -и, (*f*) dust content

за·пы́л·ива·ть, -ют, (*imp*) cover with dust, dust, make dusty

за·пыл·и́ть, -ят, (*perf*) *see* **за·пы́л·ива·ть**

за·пя́ст·н·ый, -ая, -ое, (*a*) (anat) carpal, wrist

за·пя́ст·ь·е, -я, (*n*) (anat) carpus, wrist

запят·áя, -óй, (*f decl as adj*) comma; decimal point

~, дво·и́чн·ая binary point (computers)

~, плáв·ающ·ая floating decimal point

~, фикс·и́рованн·ая fixed point

за·пятн·á·ть, -ют, (*perf*) spot, stain, soil

за·рабáт·ыва·ть, -ют, (*imp*) earn; **-ся** (*pass*); overwork

за·рабóт·а·ть, -ют, (*perf*) begin to work, start work; earn; **-ся** (*pass*); overwork

зá·работ·н·ый, -ая, -ое, (*a*) earned, earning

~ плáт·а (*f*) wages, pay, salary

зá·работ·ок·, -т·к·а, (*m*) earnings

за·рáвн·ива·ть, -ют, (*imp*) level, even up, flatten, smooth

за·раж·á·ть, -ют, (*imp*) infect, contaminate

за·раж·éни·е, -я, (*n*) infection, contamination, contagion

за·раж·ённ·ый, -ая, -ое, (*past part pass*) *see* **за·раж·á·ть;** infected, contaminated

за·рáз· (*adv*) at a go/stroke/sitting

за·рáз·а, -ы, (*f*) infection, contagion

за·раз·и́тельн·ый, -ая, -ое, (*a*) infectious, contagious, contaminating

за·раз·и́ть, -ят, (*perf*) *see* **за·раж·á·ть**

заразих·а, -и, (*f*) (bot) broom rape, *Orobanche*

за·рáз·н·ый, -ая, -ое, (*a*) infectious, contagious

за·рáм·очн·ый, -ая, -ое, (*a*) edge, border (maps, drawings etc.)

~ дá·нн·ые (*pl*) legendary data (on maps etc.)

~ част·ь (*f*) (surv) edge, border (of map sheet)

~ част·ь, восто́ч·н·ая right-hand edge (of map sheet)

~ част·ь, зá·пад·н·ая left-hand edge (of map sheet)

~ част·ь, сéвер·н·ая top border/margin (of map sheet)

за·рáн·ее (*adv*) beforehand

за·раст·á·ть, -ют, (*imp*) get/be overgrown; heal, close (of wounds)

за·раст·и́, -у́т, (*past masc sing*) **за·ро́с·,** (*perf*); *see* **за·раст·á·ть**

за·раст·и́ть, -ят, (*perf*) *see* **за·рáщ·ива·ть**

за·ращ·ённ·ый, -ая, -ое, (*past part pass*) *see* **за·рáщ·ива·ть;** grown over, overgrown

за·рáщ·ива·ть, -ют, (*imp*) grow over, plant all over, cultivate over

за·рев·é·ть, -у́т, (*perf*) roar; begin to roar

зáрев·о, -а, (*n*) glow

за·регистр·и́ровани·е, -я, (*n*) registration; record

за·регистр·и́р·овать, -уют, (*perf*) register, record

~ за·я́в·к·у file a patent claim

за·ре́ж·ут (*fut 3rd pl*) of **за·ре́з·ать** (*perf*)

за·ре́з·, -а, (*m*) slaughter

за·рез·á·ть, -ют, (*imp*), **за·ре́з·ать,** (*fut 3rd sing, pl*) **за·ре́ж·ет, за·ре́ж·ут,** (*perf*) slaughter, butcher

за·ре́з·ыва·ть, -ют, (*imp*), slaughter, butcher

за·ржáв·е·ть, -ют, (*perf*) become/get rusty

за·ржáвл·енн·ый, -ая, -ое, (*a*) rusty, corroded

за·рис·овá·ть, -у́ют, (*perf*) *see* **за·рис·о́выва·ть**

за·рис·óвыва·ть, -ют, (*imp*) sketch

зар·нíц·а, -ы, (*i*) **-ей,** (*f*) summer/sheet lightning

зар·нíч·ный, -ая, -ое, (*a*) of **зар·-нíц·а**

за·ровн·я́·ть, -ют, (*perf*) level, even up

за·род·и́ть, -ят, (*perf*) *see* **за·рожд·á·ть**

за·рóд·ыш·, -а, (*i*) **-ем,** (*m*) (biol) embryo, foetus, fetus; (bot) germ; (cryst) nucleus, embryo

образ·овáни·е за·рóд·ыш·ем (*n*) (cryst) nucleation

~ **рекристаллиз·áци·и** (cryst) recrystallization nucleus

~ **стéбл·я** (bot) plumule

за·рóд·ышев·ый, -ая, -ое, (*a*) embryonic, embryo; (cryst) nucleus, nucleating, nucleation, embryo; (bot) germ; incipient

~ **мук·á** (*f*) germ meal

~ **трéщ·ин·а** (*f*) (cryst) embryo crack

~ **центр·** (*m*) (cryst) nucleation centre

за·рó·ет (*fut 3rd sing*) of **за·ры́·ть**

за·рожд·á·ть, -ют, (*imp*) engender, conceive

за·рожд·éни·е, -я, (*n*) conception, origin, generation, genesis; onset, commencement; (cryst) nucleation; initiation (of chain reaction)

~, **само·про·из·вóль·н·ое** (biol) autogenesis, autogeny, spontaneous generation

за·рожд·ённ·ый, -ая, -ое, (*past part pass*) *see* **за·рожд·á·ть**

за·рокот·á·ть (*fut 3rd sing, pl*) **за·ро·кóч·ет, за·рокóч·ут,** (*perf*) rumble; begin to rumble

зá·росл·ь, -и, (*f*) thicket, brushwood

зá·рост·ок·, -т·к·а, (*m*) (bot) prothallus

за·рóс·ш·ий, -ая, -ее, (*past part act*) of **за·раст·и́**

за·рó·ют (*fut 3rd pl*) of **за·ры́·ть**

зар·плáт·а, -ы, (*f*) (*abbr*) of **зá·работ·н·ая плáт·а;** wages, pay, salary

за·рýб·, -а, (*m*) notch, indentation, chop

за·руб·á·ть, -ют, (*imp*) notch (with axe etc.); (min) make a cut, cut; cut down, kill

за·руб·áшечн·ое про·стрáн·ств·о (*n*) water jacket (of furnace etc.)

за·руб·éжн·ый, -ая, -ое, (*a*) foreign; beyond the border

за·рýб·ин·а, -ы, (*f*) *see* **за·рýб·**

за·руб·и́ть, -я́т, (*perf*) *see* **за·руб·á·ть**

за·рýб·к·а, -и, (*g pl*) **-б·ок·,** (*f*) notch, nick, incision, shop; (met forge) notching; (autom) digit (in unit code); (min) cutting, cut

~, **вéрх·н·яя** (min) top cutting/cut

~, **нíж·н·яя** (min) bottom cutting/cut

за·руб·цóванн·ый, -ая, -ое, (*a*) (med) cicatrized

за·руг·á·ть, -ют, (*perf*) abuse

за·ры́б·ить, -ят, (*perf*) *see* **за·рыб·л·я́·ть**

за·рыбл·я́·ть, -ют, (*imp*) stock with fish

за·рыв·á·ть, -ют, (*imp*) bury, dig in; **-ся** bury oneself, burrow; (mil) dig in; (naut) pitch heavily

за·ры́·ть (*fut 3rd sing, pl*) **за·рó·ет, за·рó·ют,** (*past masc sing*) **за·ры́·л,** (*perf*); *see* **за·рыв·á·ть**

зар·я́, -и́, (*acc*) **-ю́,** *or* **зóр·ю,** (*nom pl*) **зóр·и,** (*g pl*) **зор·ь** (*f*) glow

~, **вечéр·н·яя** sunset colours, afterglow, sunset

~, **ýтр·енн·яя** sunrise colours, sunrise

за·ря́д·, -а, (*m*) charge

~, **бо·ев·óй** (gunn) full charge, propellant charge; (rocket) pay load

~, **брон·ирóванн·ый порох·ов·óй** (rocket) restricted burning charge

~, **внýтр·енн·ий** (expl) confined charge

~, **двой·н·óй порох·ов·óй** (rocket) duplex load

~, **за·трáв·очный** (nucl) bare charge

~, **из·бы́т·очн·ый** (phys) electrical charge

~, **кáмер·н·ый** tunnel-blasting charge

~, **колóн·ков·ый** (expl) column charge

~, **котл·óв·ый** (expl) concentrated charge

~, **кумулятíв·н·ый** (expl) shaped charge

~, **ли·т·óй** (rocket) cast charge

~, **много·шáш·ечн·ый** (rocket) multigrain charge

~, **много·канáль·н·ый порох·ов·óй** (rocket) multiperforated grain

~, **на·вед·ённ·ый** (elec) induced charge

~, **плóск·ий** (expl) slab charge

~, **порох·ов·óй** (rocket) propellant/solid charge/filling

~, **при·сáс·ывающ·ий** (mil) limpet charge

~, **про·стрáн·ственн·ый** (elec) space charge

~ **радикáл·а** (phys chem) charge on the radical

~, **раз·рыв·н·óй** (mil) bursting charge

за·ря́д
~, **результ·и́рующ·ий**   net charge
~, **с·вя́з·анн·ый** (elec) bound charge
~, **то́ч·ечн·ый** (elec) point charge
~ **у а́том·а** (phys chem) charge on the atom
~, **у·равн·и́тельн·ый** (elec) battery-equalizing charge
~, **фуга́с·н·ый** (min) booster charge
~, **холост·о́й** (expl) blank charge
~, **ядр·а́** (phys) nuclear charge
за·ря́д·и́ть, -я́т, (perf) charge, load; do something incessantly/repeatedly
за·ря́д·к·а, -и, (g pl) **-д·ок·,** (f) charging, loading; (sport) training, drill
~, **борт·ов·а́я** (a/c) charging in situ
за·ря́д·ност·ь, -и, (f) size of charge; (phys chem) electrovalence
за·ря́д·н·ый, -ая, -ое, (a) of **за·ря́д·, за·ря́д·к·а**
~ **ка́мер·а** (f) (elec) charging station; (expl) charge chamber
~ **от·дел·е́ни·е** (n) (mil) warhead
~ **ста́нци·я** (f) (elec) battery-charging set, battery charger
~ **ток·** (m) (elec) charging current
~ **у·стро́й·ств·о** (n) (elec) battery charger
~ **шту́цер·** (m) charging point (gas cylinders etc.)
за·ря́д·ов·ый, -ая, -ое, (a) of **за·ря́д·**
~ **сим·ме́тр·и·я** (f) (nucl) charge symmetry
за·ря́д·о-не·за·ви́с·им·ый, -ая, -ое, (a) (phys) charge-independent/invariant
за·ряж·а́ни·е, -я, (n) (v n) see **за·ря·ж·а́·ть,** (see also **за·ряж·е́ни·е**); ammunition
~, **карту́з·н·ое** BL ammunition
~, **патро́н·н·ое** QF fixed ammunition
за·ряж·а́·ть, -ют, (imp) charge; load (a gun); stoke (a fire)
за·ря́ж·еннос·т·ь, -и, (f) charged state
за·ряж·е́ни·е, -я, (n) loading, charging; (expl) charging, charging up; (mil) ammunition
за·ря́ж·енн·ый, -ая, -ое, (past part pass) of **за·ряд·и́ть** charged, loaded
за·са́д·а, -ы, (f) ambush
за·сад·и́ть, -я́т, (perf) see **за·са́ж·и·ва·ть**
за·са́ж·ивани·е, -я, (n) (v n) of **за·са́·ж·ива·ть;** sooting-up
за·са́ж·ива·ть, -ют, (imp) plant, sow, seed; drive/plunge/stick into; put in, shut/lock up; put/set to

за·са́л·ива·ть, -ют, (imp) grease, make greasy/dirty; salt, corn, pickle (in brine)
за·са́л·ить, -ят, (perf) grease, make greasy/dirty
за·са́р·ива·ть, -ют, (imp) see **за·со·р·я́·ть**
за·са́с·ыва·ть, -ют, (imp) suck in, engulf, swallow up
за·са́хар·енн·ый, -ая, -ое, (past part pass) see **за·са́хар·ива·ть;** (food) candied, sugared, crystallized
за·са́хар·ива·ть, -ют, (imp) (food) candy, sugar, crystallize
за·са́хар·ить, -ят, (perf) see **за·саха·р·ива·ть**
за·сверк·а́·ть, -ют, (perf) sparkle; begin to sparkle
за·свет·и́ть, -я́т, (perf) light, light up; expose (film)
за·све́т·к·а, -и, (g pl) **-т·ок·,** (f) (phot) exposure, lightmark/fog; (rad) gating
    и́мпульс· за·све́т·к·и (rad) gate
за·све́ч·енн·ый, -ая, -ое, (past part pass) of **за·свет·и́ть;** (phot) light-struck, fogged
за·с·вид·е́тельствовани·е, -я, (n) (v n) of **за·с·вид·е́тельств·овать;** authentication, certification
за·с·вид·е́тельств·овать, -уют, (perf) testify, witness
за·свист·а́ть (fut 3rd sing, pl) **за·сви́щ·ет, за·сви́щ·ут,** (perf) begin to whistle, start whistling
за·свист·е́ть, -я́т, (perf) begin to whistle, start whistling
за·сви́щ·ет (fut 3rd sing) of **за·свис·т·а́ть**
за·се́·в·, -а, (m) (v n) see **за·сев·а́·ть;** (agr) sown area; spill (computers)
за·сев·а́·ть, -ют, (imp) sow, seed, plant; inoculate (with bacteria)
за·се́·вш·ий, -ая, -ее, (past part act) of **за·се́ст·ь**
за·сед·а́ни·е, -я, (n) sitting, meeting, session
за·сед·а́·ть, -ют, (imp) sit, meet, hold a session; stay embedded/stuck in
за·се́·ет (fut 3rd sing) of **за·се́·ять**
за·се́к·а, -и, (f) (mil) abatis
за·сек·а́·ть, -ют, (imp) notch; intersect; locate (by intersections); (nav) fix
за·секре́т·ить, -ят, (perf) see **за·се·кре́ч·ива·ть**
за·секре́ч·енн·ый, -ая, -ое, (past part pass) see **за·секре́ч·ива·ть;** classified (of material); cleared (of people)

за·секре́ч·ивани·е, -я,   (*n*) (*v n*) of за·секре́ч·ива·ть; security classification

за·секре́ч·ивател·ь, -я,   (*m*) (telecom) secrecy device/system, scrambler

за·секре́ч·ива·ть, -ют,   (*imp*) classify, give security grading to, make secret; clear, give a security pass to, give access to secret material

за·се́к·ш·ий, -ая, -ее,   (*past part act*) of за·се́ч·ь

за·се́·л (*past masc sing*) of за·се́ст·ь

за·сел·е́ни·е, -я,   (*n*) (*v n*) *see* за·се·л·я́·ть

за·сел·ённост·ь, -и,   (*f*) population

за·сел·и́ть, -я́т,   (*perf*) *see* за·сел·я́·ть

за·сел·я́·ть, -ют,   (*imp*) populate, settle, tenant, quarter, occupy (houses)

за·серебр·и́ть, -я́т, (*perf*) make silvery; -ся look silvery; begin to look silvery

за·се́ст·ь (*fut 3rd sing, pl*) за·ся́д·ет, за·ся́д·ут, (*past masc sing*) за·се́·л, (*perf*) sit down to; settle, establish oneself; become lodged, get stuck in; lie (in ambush etc.)

за·се́ч·к·а, -и, (*g pl*) -ч·ек·, (*f*) notch, indentation, making an indentation; (surv) intersection; (nav) fix; (print) serif, hair stroke

~ вре́мен·и (nav) timing

~, звук·ов·а́я   sound ranging

~, обра́т·н·ая   (surv) resection

за·се́ч·ь (*fut 3rd sing, pl*) за·сеч·ёт, за·сек·у́т, (*past masc sing*) за·се́к·, (*perf*); *see* за·сек·а́·ть

за·се́·янн·ый, -ая, -ое,   (*past part pass*) of за·се́·ять; (hortic) sown

за·се́·ять, -ют, (*perf*) (hortic) sow

за·син·е́·ть, -ют,   (*perf*) appear/look/ show blue/dark blue

за·син·и́ть, -я́т,   (*perf*) colour/color/ make dark blue

за·си·я́·ть, -ют,   (*perf*) shine; begin to shine

за·си́л·ь·е, -я,   (*n*) preponderance, predominance

за·сл·а́ть (*fut 3rd sing, pl*) за·шл·ёт, за·шл·ю́т, (*perf*); *see* за·сыл·а́·ть

за·сло́н·, -а, (*m*) barrier; (mil) screen; (agr) shelter belt; furnace/oven door

за·слон·е́ни·е, -я, (*n*) (*v n*) *see* за·слон·я́·ть

у́гол· за·слон·е́ни·я   (*m*) cutoff angle (illumination)

за·слон·и́ть, -я́т,   (*perf*) *see* за·слон·я́·ть

за·сло́н·к·а, -и,   (*g pl*) -н·ок·, (*f*) slide, baffle plate, damper (for draught regulation); furnace/oven door

~, дро́ссель·н·ая   (ICE) throttle valve, choke; butterfly valve; baffle plate

~, за·темн·я́ющ·ая   (phot) dark screening slide

~ карбюра́тор·а, воз·ду́ш·н·ая (ICE) choke

~, ма́ятник·ов·ая   pendulous vane (artificial horizon)

~, по·воро́т·н·ая   butterfly valve

~, регул·и́рующ·ая   baffle, flopper (pneumatics)

~, скольз·я́щ·ий   damper

за·слон·я́·ть, -ют,   (*imp*) screen, shield, cover

за·слу́г·а, -и, (*f*) service, distinguished service

за·слу́ж·енн·ый, -ая, -ое, (*past part pass*) *see* за·слу́ж·ива·ть; distinguished, meritorious

~ де́·ятел·ь   (*m*) distinguished worker (title)

~ де́·ятел·ь на·у́к·и   (*m*) distinguished scientist

за·слу́ж·ива·ть, -ют,   (*imp*) deserve, merit, be worthy of

за·служ·и́ть, -ат,   (*perf*) *see* за·слу́ж·ива·ть

за·слу́ш·а·ть, -ют,   (*perf*) *see* за·слу́ш·ива·ть

за·слу́ш·ива·ть, -ют,   (*imp*) hear (officially)

за·слу́ш·ивани·е, -я,   (*n*) hearing

за·слы́ш·ать, -ат,   (*perf*) hear, catch the sound of

за·сме·я́ть, -ю́т,   (*perf*) ridicule; -ся begin to laugh; laugh

за·смол·е́ни·е, -я, (*n*) (*v n*) of за·смол·и́ть

за·смол·и́ть, -я́т,   (*perf*) tar, pitch

за·снеж·ённ·ый, -ая, -ое, (*a*) snowedup, snow-covered

за·с·ни́м·ут (*fut 3rd pl*) of за·с·ня́·ть

за·сн·у́ть, -у́т,   (*perf*) fall asleep, go to sleep

за·с·ня́·ть (*fut 3rd sing, pl*) за·с·ни́м·ет, за·с·ни́м·ут, (*perf*) photograph, take a photograph

за·со́в·, -а, (*m*) locking bar/bolt

за·со́в·ыва·ть, -ют,   (*imp*) shove/push/ thrust in

за·со́л·, -а, (*m*) (*v n*) *see* за·сол·и́ть; salted preserves

за·сол·е́ни·е, -я, (*n*) (*v n*) *see* за·сол·и́ть; salinization (of soil) за·со́л·енност·ь, -и, (*f*) salinity

за·со́л·енн·ый, -ая, -ое, (*past part pass*) of за·сол·и́ть; saline

за·сол·и́ть, -я́т, (*perf*) salt, corn, pickle (in brine)

за·со́л·к·а, -и, (*g pl*) -л·ок·, (*f*) salting, pickling; brine, pickle

за·со́р·, -а, (*m*) (met) (*pl*) non-metallic inclusions

за·сор·е́ни·е, -я, (*n*) (*v n*) *see* за·сор·я́·ть; clutter, grass (on radar screen); pollution, contamination; constipation

за·сор·ённост·ь, -и, (*f*) weediness (of soil)

за·сор·и́ть, -я́т, (*perf*) *see* за·сор·я́·ть

за·сор·я́·ть, -ю́т, (*imp*) choke, obstruct, stop up; (med) make constipated

за·со́с·, -а, (*m*) (*v n*) *see* за·сос·а́ть; intake

за·сос·а́ть, -у́т, (*perf*) suck in, engulf, swallow up

за·со́х·нуть, -нут, (*past masc sing*) за·со́х·, (*perf*) dry up, wither

за·соч·и́ть, -а́тся, (*perf*) begin to ooze

за·спеш·и́ть, -а́т, (*perf*) begin to hurry

за·спи́н·н·ый, -ая, -ое, (*a*) back

за·спирт·ова́ть, -у́ют, (*perf*) *see* за·спирт·о́выва·ть

за·спирт·о́выва·ть, -ют, (*imp*) preserve in alcohol, alcoholize

за·ста́·в·а, -ы, (*f*) gate, gates, barrier, toll-gates

~, сторож·ев·а́я (mil) outpost

за·ста·ва́ть, -ю́т, (*imp*) succeed in seeing, find, catch

за·ста́·в·ить, -ят, (*perf*) *see* за·ставл·я́·ть

за·ста́в·к·а, -и, (*g pl*) -в·ок·, (*f*) (print) head-piece

за·ставл·я́·ть, -ю́т, (*imp*) compel, force, make, induce; cram, fill, block up

за·ста́·ива·ться, -ются, (*imp*) stagnate, stand too long; get stale

за·ста́·ть, -нут, (*perf*) *see* за·ста·ва́ть

за·стёг·ивани·е, -я, (*n*) (text) fastening

за·стёг·ива·ть, -ют, (*imp*) do up, fasten

за·стег·н·у́ть, -у́т, (*perf*) *see* за·стё·г·ива·ть

за·стёж·к·а, -и, (*g pl*) -ж·ек·, (*f*) (text) fastener

~ мо́лни·я zip-fastener, slide-fastener, zipper

за·стекл·ённ·ый, -ая, -ое, (*past part pass*) *see* за·стекл·я́·ть; glazed, fitted with glass

за·стекл·и́ть, -я́т, (*perf*) *see* за·стекл·я́·ть

за·стекл·ова́ни·е, -я, (*n*) glazing; (plast) transition to glass, second-order transition, glass transition

температу́р·а за·стекл·ова́ни·я (*f*) (plast) glass point/temperature, second-order transition point

за·стекл·о́ванн·ый, -ая, -ое, (*past part pass*) glazed, fitted with glass; (plast) glass-like, glassy

за·стекл·я́·ть, -ю́т, (*imp*) glaze, fit with glass

за·сте́л·ют (*fut 3rd pl*) of за·стл·а́ть

за·стиг·а́·ть, -ю́т, (*imp*) catch, come upon

за·сти́г·нуть, -нут, (*past masc sing*) за·сти́г·, (*perf*); *see* за·стиг·а́·ть

за·стил·а́·ть, -ю́т, (*imp*) spread over, cover

за·сти́ч·ь (*fut 3rd sing, pl*) за·сти́г·нет, за·сти́г·нут, (*past masc sing*) за··сти́г·, (*perf*); *see* за·стиг·а́·ть

за·стл·а́ть (*perf*) (*fut 3rd sing, pl*) за··сте́л·ет, за·сте́л·ют, (*perf*); *see* за·стил·а́·ть

за·сто́·й, -я, (*m*) standstill, stagnation, (med) stasis

~ кро́в·и (med) haemostasis

за·сто́й·н·ый, -ая, -ое, (*a*) of за·сто́·й; immobile; stagnant (of water); sluggish

~ зо́н·а (*f*) dig-in zone (metal machining)

за·столб·и́ть, -я́т, (*perf*) peg, peg/stake out

за·стопо́р·ивани·е, -я, (*n*) (*v n*) *see* за·сто́пор·ива·ть

за·сто́пор·ива·ть, -ют, (*imp*) stop up, stop, cut off, lock; cage (gyros)

за·сто́пор·ить, -ят, (*perf*) *see* за·сто́пор·ива·ть

за·сто·я́ться, -я́тся, (*perf*) stagnate, stand too long; get stale

за·стра́·ива·ть, -ют, (*imp*) build over, cover with buildings, erect buildings on, develop (land)

за·страх·ова́ть, -у́ют, (*perf*) *see* за··страх·о́выва·ть

за·страх·о́выва·ть, -ют, (*imp*) insure

за·стр·ева́·ть, -ю́т, (*imp*) stick, get stuck, jam

~ во льд·у́ (naut) be fast in ice, be set by ice

за·стрел·и́ть, -ят, (*perf*) shoot, kill; (a/c) shoot down

за·стро́·енн·ый, -ая, -ое, (*past part pass*) *see* за·стра́·ива·ть; built-up, covered with buildings, developed
~ у·ча́ст·ок· (*m*) built-up area

за·стро́·ить, -ят, (*perf*) *see* за·стра́·-ива·ть

за·стро́й·к·а, -и, (*g pl*) -о́·ек·, (*f*) (civ eng) development
ли́н·и·я за·стро́й·к·и (*f*) building line
~ ме́ст·ност·и development of a site; site coverage
пло́щ·ад·ь за·стро́й·к·и (*f*) (phys) site coverage

за·стро́п·к·а, -и, (*f*) slinging, putting a sling round

за·стр·я́·ть, -нут, (*perf*) *see* за·-стр·ева́·ть

за·студ·и́ть, -́ят, (*perf*) *see* за·сту́-ж·ива·ть

за·студ·нева́ни·е, -я, (*n*) jellying, jellification; (chem) gelatination

за·студ·нева́·ть, -ют, (*imp*) (chem) gel; (food) make into a jelly

за·сту́ж·ивани·е, -я, (*n*) (*v n*) of за·-сту́ж·ива·ть

за·сту́ж·ива·ть, -ют, (*imp*) chill; expose to cold; -ся (*pass*); catch cold, get cold, get a chill

за́·ступ·, -а, (*m*) spade

за·стуч·а́ть, -а́т, (*perf*) begin to knock; knock

за·стыв·а́ни·е, -я, (*n*) (*v n*) of за·-стыв·а́·ть
температу́р·а за·стыв·а́ни·я (*f*) pour point (oils, fuels etc.)

за·стыв·а́·ть, -ют, (*imp*) congeal, thicken, harden, set (on cooling)

за·стыв·а́ющ·ий, -ая, -ее, (*pres part act*) of за·стыв·а́·ть; (plast) highly viscous

за·сты́·вш·ий, -ая, -ее, (*past part act*) of за·сты́·ть (*perf*), *see* за·сты-в·а́·ть (*imp*); congealed, solidified

за·сты́·н·уть, -ут, (*perf*) *see* за·сты-в·а́·ть

за·сты́·ть, -нут, (*perf*) *see* за·сты-в·а́·ть

за·су́·н·уть, -ут, (*perf*) *see* за·со́в·ы-ва·ть

за́·сух·а, -и, (*f*) drought

засухо·у·сто́й·чив·ый, -ая, -ое, (*a*) (bot) drought-resistant, xerophile

за·су́ш·ива·ть, -ют, (*imp*) dry up, shrivel

за·суш·и́ть, -́ат, (*perf*) *see* за·су́-ш·ива·ть

за·су́ш·лив·ый, -ая, -ое, (*a*) arid, drought, droughty, xeric

за·с·чит·а́·ть, -ют, (*perf*) *see* за·-с·чи́т·ыва·ть

за·с·чи́т·ыва·ть, -ют, (*imp*) take into consideration, allow for, take into account

за·сыл·а́·ть, -ют, (*imp*) send, send astray; infiltrate (agents etc.)

за·сы́п·ать, -лют, (*perf*) fill in/up (with grains etc.); strew, sprinkle;
за·сып·а́·ть, -ют, (*imp*) (as above); fall asleep

за·сы́п·к·а, -и, (*g pl*) -п·ок·, (*f*) (*v n*) *see* за·сы́п·ать (*perf*); (civ eng) filling; backfilling (of trench etc.) charging (a furnace)

за·сып·н·о́й, -а́я, -о́е, (*a*) of за·-сы́п·к·а
~ аппара́т· (*m*) bell-and-hopper, bell-and-cone, charging apparatus (blast furnace)

за·сых·а́·ть, -ют, (*imp*) dry up, wither

за·ся́д·ут (*fut 3rd pl*) of за·се́ст·ь

за·та́пл·ива·ть, -ют, (*imp*) light/make a fire, light (stoves etc.), kindle; start heating (houses etc.)

за·та́пт·ыва·ть, -ют, (*imp*) trample down/under foot

за·та́ск·ива·ть, -ют, (*imp*) drag in/ to/away

за·та́ч·ива·ть, -ют, (*imp*) sharpen, grind

за·тащ·и́ть, -́ат, (*perf*) *see* за·та́с-к·ива·ть

за·тверд·ева́ни·е, -я, (*n*) solidification, solidifying, setting, induration; (plast) transition to glass-like state; (met) freezing, solidifying
температу́р·а за·тверд·ева́ни·я (*f*) (met) freezing/solidification point; (plast) glass point/temperature

за·тверд·ева́·ть, -ют, (*imp*) become hard, solidify, set, indurate; (met) freeze, solidify; (plast) become glass-like

за·тверд·е́вш·ий, -ая, -ее, (*past part act*) *see* за·тверд·ева́·ть; solidified, set, congealed; (plast) glassy, glass-like

за·тверд·е́лост·ь, -и, (*f*) firmness, hardness

за·тверд·е́л·ый, -ая, -ое, (*a*) *see* за·-тверд·е́вш·ий; firm

за·тверд·е́ни·е, -я, (*n*) *see* за·твер-д·ева́ни·е; (med) sclerosis за·тверд·е́·ть, -ют, (*perf*) *see* за·-тверд·ева́·ть

за·тверд·и́ть, -я́т, (*perf*) *see* за·-тве́рж·ива·ть

за·тве́рж·ива·ть, -ют, (*imp*) learn by heart, commit to memory

за·тво́р·, -а, (*m*) bolt, bar; (gunn) breech mechanism, bolt (small arms); seal; (hydr) gate; door (of hoppers etc.); (phot) shutter

~, бу́нкер·н·ый   bin/hopper door

~, ва́куум·н·ый   vacuum seal

~, вальц·о́в·ый   (hydr eng) drum gate

~, вод·ян·о́й   water/hydraulic seal, (st eng) seal pot, hydraulic backpressure valve

~, вращ·а́ющ·ийся   (phot) rotating shutter

~, гермет·и́ческ·ий   hermetic seal

~, гидро·техн·и́ческ·ий   spillway gate, water/flood gate

~, клин·ов·о́й   (gunn) QF breech

~, нару́ж·н·ый што́р·н·ый   (phot) roller-blind shutter

~, пло́ск·ий   (hydr eng) slide gate

~, поршн·ев·о́й   (gunn) BL breech

~, пред·о·хран·и́тельн·ый   safety lock

~, рту́т·н·ый   (chem) mercury seal/cutoff

~, сегме́нт·н·ый   (hydr eng) radial gate

~, се́ктор·н·ый   (civ eng) sector regulator

~, скольз·я́щ·ий   (hydr eng) slide gate, vertical-lift gate

~, с·тво́р·чат·ый   (phot) leaf/flicker shutter; flap gate

~ фо́рм·ы   (plast) mould/mold closure

~, центра́ль·н·ый   (phot) diaphragm shutter

~, шар·ов·о́й   (mech) ball check; (phot) spherical shutter

~, што́р·н·ый   (phot) focal-plane shutter

~, щел·ев·о́й   (phot) slit/slot/split shutter

за·тво́р·енный, -ая, -ое, (*past part pass*) of за·твор·и́ть; shut, closed

за·твор·и́ть, -́ят, (*perf*) *see* за·тво-р·я́·ть

за·твор·я́·ть, -ют, (*imp*) shut, close

за·тво́р·н·ый, -ая, -ое, (*a*) of за·-тво́р·

за·тев·а́·ть, -ют, (*imp*) venture, undertake

за·тёк·, -а, (*m*) flow

за·тек·а́ни·е, -я, (*n*) (*v n*) of за·те-к·а́·ть; inflowing

глуб·ин·а́ за·тек·а́ни·я (*f*) (plast) penetration

за·тек·а́·ть, -ют, (*imp*) flow/leak in/into

за·те́м (*adv and conj*) then, and then, thereupon, and after this

за·темн·е́ни·е, -я, (*n*) (*v n*) *see* за·-темн·я́·ть; blacking out, black-out; (cinema) fade-out

за·темн·ённ·ый, -ая, -ое, (*past part pass*) *see* за·темн·я́·ть; darkened, dark, obscured; blacked-out

~ по́л·е (*n*) (opt) dark field

за·темн·е́·ть, -ют, (*perf*) grow/get/become dark; look/appear/show dark

за·темн·и́тел·ь, -я, (*m*) dimmer

за·темн·и́ть, -я́т, (*perf*) *see* за·темн·я́·ть

за·темн·я́·ть, -ют, (*imp*) darken, obscure; black out; -ся become dark/obscure

за·темн·я́ющ·ий, -ая, -ее, (*pres part act*) of за·темн·я́·ть; (TV) blanking

за·тен·е́ни·е, -я, (*n*) (aerodynam) shading, shadowing; blanket, blanketing

~, аэродинам·и́ческ·ое   (air) blanketing effect

за·тен·и́тел·ь, -я, (*m*) shade; (cinema) nigger

за·тен·и́ть, -я́т, (*perf*) *see* за·тен·я́·ть

за·тен·я́·ть, -ют, (*imp*) shade, overshade

за·тер·е́ть (*fut 3rd sing, pl*) за·тр·ёт, за·тр·у́т, (*past masc sing*) за·тёр·, (*perf*); *see* за·тир·а́·ть

за·тер·я́·ть, -ют, (*perf*) mislay, lose

за·тёс·, -а, (*m*) blaze (mark chipped out on a tree)

за·тес·а́ть (*fut 3rd sing, pl*) за·те́ш·ет, за·те́ш·ут, (*perf*); *see* за·тёс·ы·ва·ть

за·тёс·к·а, -и, (*f*) *see* за·тёс·

за·тёс·ыва·ть, -ют, (*imp*) rough-hew; blaze (make marks on trees)

за·те́чь (*fut 3rd sing, pl*) за·теч·ёт, за·тек·у́т, (*past masc sing*) за·тёк·, (*perf*); *see* за·тек·а́·ть

за·те́ш·ут (*fut 3rd pl*) of за·тес·а́ть

за·те́·ять, -ют, (*perf*) *see* за·тев·а́·ть

за·тир·а́ни·е, -я, (*n*) (*v n*) of за·ти-р·а́·ть; rubbing, grinding; mashing (in brewing)

за·тир·а́·ть, -ют, (*imp*) rub down/over/off; jam, stick; (naut) become fast/bound; soil, dirty

за·ти́р·к·а, -и, (*g pl*) **-р·ок·,** (*f*) (build) trowelling

за·ти́ск·ива·ть, -ют, (*imp*) squeeze in

за·ти́с·н·уть, -ут, (*perf*) *see* **за·ти́ск·ива·ть**

за·тих·а́·ть, -ют, (*imp*) quieten, become quiet, abate; calm down, become calmer/calm, die down/away

за·ти́х·нуть, -нут, (*past masc sing*) **за·ти́х·,** (*perf*) *see* **за·тих·а́·ть**

за·ти́ш·ь·е, -я, (*n*) calm, lull, slack зо́н·а за·ти́ш·ь·я (*f*) (meteor) calm

за·тк·а́ть, -у́т, (*perf*) weave in, interweave

за·тк·н·у́ть, -у́т, (*perf*) *see* **за·ты·к·а́·ть**

за·тк·у́т (*fut 3rd pl*) of **за·тк·а́ть**

за·тм·ева́·ть, -ют, (*imp*) darken, obscure, eclipse, cover; **-ся** (*pass*)

за·тм·ева́ющ·ийся, -аяся, -ееся, (*pres part act*) of **за·тм·ева́·ться;** occulting

~ огó́н·ь (*m*) (nav) occulting light

за·тм·éни·е, -я, (*n*) (astron) eclipse

~, гла́в·н·ое primary eclipse

~, кольце·обра́з·н·ое annular/ring-shaped eclipse

~, лу́н·н·ое lunar eclipse

~, пóлн·ое total eclipse

~, сóлн·ечн·ое solar eclipse

~, ча́ст·н·ое partial eclipse

за·тм·и́ть, -я́т, (*perf*) *see* **за·тм·ева́·ть**

за·тó (*conj*) whereas, on the other hand, to make up for

за·това́р·ивани·е, -я, (*n*) overstocking; glut

за·толк·а́·ть, -ют, (*perf*) shove/push in/into

за·тóн·, -а, (*m*) backwater, creek

за·тóн·ин·а, -ы, (*f*) (*dim*) of **за·тóн·**

за·тон·у́·ть, -ут, (*perf*) sink, become submerged

за·топ·и́ть, -я́т, (*perf*) make a fire, kindle, light a fire, start heating (houses etc.); flood, submerge, sink, scuttle (a ship)

за·топл·éни·е, -я, (*n*) (*v n*) *see* **за·топ·л·я́·ть** кла́пан· за·топл·éни·я (*m*) flooding valve систéм·а за·топл·éни·я (*f*) (mar eng) flooding system

за·топл·я́·ть, -ют, (*imp*) flood, inundate, submerge; sink, scuttle (a ship)

за·топт·а́ть (*fut 3rd sing, pl*) **за·тóп·ч·ет, за·тóпч·ут,** (*perf*) trample down/under foot

за·тóр·, -а, (*m*) jam, obstruction, jamming, blocking; (elec) blocking; mash (for beer etc.)

~, лед·ян·óй ice jam

~, хмел·ев·óй mash (for beer)

за·тормож·ённ·ый, -ая, -ое, (*past part pass*) of **за·тормоз·и́ть**

~ пере·хóд· (*m*) (phys) hindered transition

за·тормоз·и́ть, -я́т, (*perf*) brake, apply the brake; slow down, retard; inhibit

за·тóр·н·ый, -ая, -ое, (*a*) of **за·тóр·**

~ котёл· (*m*) masher, mashing tub/kettle/vat (brewing)

~ чан· (*m*) mash tun (brewing)

за·тороп·и́ться, -я́тся, (*perf*) begin to hurry

за·точ·и́ть, -а́т, (*perf*) sharpen, grind

за·тóч·к·а, -и, (*g pl*) **-ч·ек·,** (*f*) sharpening, tool grinding

~, анодно·механ·и́ческ·ая electric-contact tool grinding

за·точ·нóй, -áя, -óе, (*a*) *see* **за·тóч·н·ый**

за·тóч·н·ый, -ая, -ое, (*a*) sharpening, tool-grinding

~ стан·óк· (*m*) tool grinder, tool grinding machine, tool sharpener

за·трав·и́ть, -я́т, (*perf*) catch, kill (with dogs); bait, harass

за·тра́в·к·а, -и, (*g pl*) **-в·ок·,** (*f*) igniter (obsolete), initiator; arousing, interest; (cryst) seeding, inoculation; (met) dummy (continuous casting), dummy billet/bar/slab; (chem) seeding polymer

за·тра́вл·енн·ый, -ая, -ое, (*past part pass*) of **за·трав·и́ть;** seeded, inoculated

за·тра́в·очн·ый, -ая, -ое, (*a*) of **за·тра́в·к·а**

~ ис·тóч·ник· (*m*) (nucl) source of neutrons

~ криста́лл· (*m*) (cryst) inoculator seed

~ ма́сс·а (*f*) (nucl) bare mass

за·тра́г·ива·ть, -ют, (*imp*) touch, affect; touch upon, mention

за·тра́т·а, -ы, (*f*) expenditure, expense, expenses; (phys) dissipation

~, кóс·венн·ые (*pl*) (fin) overhead expenses

~, прям·ы́е (*pl*) (fin) prime cost

~ энéрг·и·и dissipation of energy

за·тра́т·ить, -я́т, (*perf*) *see* **за·тра́ч·ива·ть**

за·тра́ч·енн·ый, -ая, -ое, (*past part pass*) *see* **за·тра́ч·ива·ть**

за·тра́ч·ива·ть, -ют, (*imp*) spend, expend, consume, dissipate

за·тре́б·овать, -уют, (*perf*) request, require, ask for

за·трепет·а́ть (*perf*) (*fut 3rd sing, pl*) **за·трепе́щ·ет, за·трепе́щ·ут** palpitate, begin to palpitate

за·тр·ёт (*fut 3rd sing*) of **за·тер·е́ть**

за·трещ·а́ть, -а́т, (*perf*) crack, begin to crack, start cracking

за·тро́·н·уть, -ут, (*perf*) *see* **за·тра́·г·ива·ть**

за·труд·не́ни·е, -я, (*n*) difficulty, embarrassment

за·труд·нённ·ый, -ая, -ое, (*past part pass*) *see* **за·труд·ня́·ть;** difficult; (phys) unfavoured, hindered

~ **пере·хо́д·** (*m*) (phys) unfavoured transition

за·труд·ни́тельност·ь, -и, (*f*) difficulty

за·труд·ни́тельн·ый, -ая, -ое, (*a*) difficult

за·труд·н·и́ть, -я́т, (*perf*) *see* **за·труд·-ня́·ть**

за·труд·ня́·ть, -ют, (*imp*) impede, hamper, make difficult

за·тр·у́т (*fut 3rd pl*) of **за·тер·е́ть**

за·тума́н·ива·ть, -ют, (*imp*) fog, cloud, dim

за·тума́н·ить, -ят, (*perf*) *see* **за·тума́-н·ива·ть**

за·туп·и́ть, ⸗ят, (*perf*) *see* **за·тупл·я́·ть**

за·ту́пл·енн·ый, -ая, -ое, (*past part pass*) *see* **за·тупл·я́·ть;** blunted; dulled

за·тупл·я́·ть, -ют, (*imp*) blunt, dull

за·тух·а́ни·е, -я, (*n*) (*v n*) of **за·ту-х·а́·ть;** attenuation (of radiations); damping, decay (of vibrations)

~ **антённ·ы** antenna damping

~ **бок·ов·ы́х полос·** (rad) sideband attenuation

~ **в про·стра́н·ств·е** (rad) free-space attenuation

~ **в с·ме́ж·н·ом кана́л·е** (rad) adjacent-channel attenuation

вре́м·я за·тух·а́ни·я (*n*) damping time

~ **вязк·ост·н·ое** (phys) viscous decay

~ **гармо́н·ик·** (phys) decay of harmonics

~ **из·луч·е́ни·я** attenuation of radiation; radiation damping

~ **колеб·а́ни·й** vibration/oscillation damping/decay

за·тух·а́ни·е

~ **ко́нтур·а** (elec) Q damping factor **коэффициéнт· за·тух·а́ни·я** (*m*) (elec) attenuation coefficient/factor; (mech) damping coefficient/factor

~, **лин·е́йн·ое** line damping, line attenuation

~ **люмин·есцéнци·и** (nucl) scintillation decay

мо́дул·ь за·тух·а́ни·я (*m*) (phys) modulus of decay

~ **нейтро́н·н·ого по·то́к·а** neutron attenuation

~, **о·ста́т·очн·ое** (telecom) overall line attenuation

~, **пере·хо́д·н·ое** crosstalk attenuation (telephones)

~, **радиа́ль·н·ое** radial decay

~, **по·вто́р·н·ое** (telecom) iterative attenuation

~ **свеч·е́ни·я** decay of luminescence

~, **со́б·ственн·ое** (elec) attenuation equivalent, attenuation constant (of a network)

~ **цéп·и** circuit damping

~, **част·о́тн·ое** frequency attenuation

~, **экспоненциáль·н·ое** exponential decay/damping

за·тух·а́·ть, -ют, (*imp*) die down, fade, decay, become attenuated, be damped, go out

за·тух·а́ющ·ий, -ая, -ее, (*pres part act*) of **за·тух·а́·ть;** fading, damped; attenuated (of radiations)

~ **колеб·а́ни·я** (*pl*) damped oscillations

~ **регул·и́ровани·е** (*n*) (autom) damped oscillation control

за·ту́х·нуть, -нут, (*past masc sing*) **за·ту́х·,** (*perf*); *see* **за·тух·а́·ть**

за·туш·и́ть, ⸗ат, (*perf*) extinguish, put out

за́тхл·ый, -ая, -ое, (*a*) musty

за·тык·а́·ть, -ют, (*imp*) plug, stop up; thrust/stuff in

за·ты́л·ова́ни·е, -я, (*n*) (mech eng) backing-off, relieving

за·ты́л·о́вочн·ый стан·о́к· (*m*) relieving lathe, backing-off lathe

за·ты́л·ок·, -л·к·а, (*m*) (anat) occiput, back of the head; back (of a cutting tool)

~, **с·ним·а́·ть** (mech eng) back off, relieve

за·ты́л·очн·ый, -ая, -ое, (*a*) of **за·ты́-л·ок·** (anat) occipital, nuchal

~ **тесем·к·а** (*f*) head band (respirator's etc.)

за·ты́ль·ник·, -а, (*m*) back plate

за·ты́ч·к·а, -и, (*g pl*) **-ч·ек·**, (*f*) plug, bung, stopper, spigot

за·тя́г·ива́ни·е, -я, (*n*) (*v n*) of **за·тя́·г·ива·ть;** (aerodyn) pitch-up; (rad) pulling, frequency pulling; lasting (shoemaking)

∼ пере·хо́д·а transition delay (fluid flow)

за·тя́г·ива·ть, -ют, (*imp*) tighten, draw in/together; heal (of wounds etc.); stretch over, cover; delay, draw out, prolong

за·тя́ж·к·а, -и, (*g pl*) **-ж·ек·**, (*f*) (*v n*) *see* **за·тя́г·ива·ть;** (build) tie-beam; (min) lag. lagging; lasting (shoemaking)

∼ кро́вл·и (min) head lagging

момéнт за·тя́ж·к·и (*m*) tightening torque (on screws)

∼, не·сплош·н·а́я (min) open lagging

∼, сплош·н·а́я (min) tight lagging

за·тя́ж·н·о́й, -а́я, -о́е, (*a*) protracted, drawn-out, slow, lingering; tightening

∼ вы́·стрел· (*m*) (gunn) hang fire

∼ прыж·о́к· (*m*) free-fall jump (by parachute)

за·тя·н·у́ть, ⌐ут, (*perf*) *see* **за·тя́г·и·ва·ть**

за·тя́·нут·ый, -ая, -ое, (*past part pass*) *see* **за·тя́г·ива·ть**

заусéн·ец·, -н·ц·а, (*i*) **-н·ц·ем** (*m*) hang-nail; projection; (met) fin (rolling); burr (machining), seam (casting), flash (forging/welding defect); (plast) spew, flash

заусéни·ц·а, -ы, (*i*) **-ей,** (*f*) *see* **заусéн·ец·**

за·у́ч·ива·ть, -ют, (*imp*) learn by heart

за·уч·и́ть, ⌐ат, (*perf*) *see* **за·у́ч·ива·ть**

за·у́ш·ниц·а, -ы, (*i*) **-ей,** (*f*) (med) mumps

за·у́ш·н·ый, -ая, -ое, (*a*) (anat) parotid

за·фикс·и́р·овать, -уют, (*perf*) *see* **фикс·и́р·овать**

за·фрахт·ова́ть, -у́ют, (*perf*) *see* **за··фрахт·о́выва·ть**

за·фрахт·о́выва·ть, -ют, (*imp*) charter (a ship)

за·ха́ж·ива·ть, -ют, (*imp*) be accustomed to call on, be a frequent visitor

за·хва́т·, -а, (*m*) seizure, capture; grip, clamp, claw, clasp, gripping device; spread, coverage; bite, nip (of rolls); (gunn) gun guides, training clips; (chem) entrapping

за·хва́т

∼, бес·полéз·н·ый (nucl) non-productive capture

∼ бы́стр·ого нейтрóн·а (nucl) fast-neutron capture

∼, врéд·н·ый (nucl) non-reproductive capture

∼, за·жим·н·о́й vice-type grip

∼, за·мк·о́в·ый locking grip

∼, инфильтрациóн·н·ый seepage area

∼ манипуля́тор·а tongs, manipulator fingers

∼ мезóн·а (nucl) meson capture

момéнт· за·хва́т·а (*m*) (met roll) moment of contact

∼, рабóч·ий operating width (of tools)

∼ с дел·éни·ем (nucl) fission capture

у́гол· за·хва́т·а (*m*) (met roll) contact angle; angle of nip (crushing rolls)

у́гол· за·хва́т·а, максима́ль·н·ый (*m*) angle of bite (max)

∼ цéл·и (mil rad) lock-on

центр· за·хва́т·а (*m*) (nucl) trapping centre/center

за·хват·и́ть, ⌐я́т, (*perf*) *see* **за·хва́·т·ыва·ть**

за·хва́т·к·а, -и, (*f*) (mech eng) dog, clutching attachment; catch, checking/catching device

∼, па́д·ающ·ая disappearing dog

за·хва́т·ническ·ий, -ая, -ое, (*a*) predatory

за·хва́т·н·ый, -ая, -ое, (*a*) of **за·хва́т·;** (nucl) capture

за·хва́т·чик·, -а, (*m*) invader; (chem) acceptor; (geog) pirate river, diverter

за·хва́т·ывани·е, -я, (*n*) (*v n*) of **за·хва́т·ыва·ть;** (rad) capture, capture effect

за·хва́т·ыва·ть, -ют, (*imp*) seize, capture, take, catch, entrap, entrain; (nucl) capture, trap

за·хва́т·ывающ·ий, -ая, -ее, (*pres part act*) of **за·хва́т·ыва·ть**

∼ рек·а́ (*f*) (geog) pirate river

за·хва́ч·енн·ый, -ая, -ое, (*past part pass*) *see* **за·хва́т·ыва·ть**

за·хир·é·ть, -ют, (*perf*) become sickly/weak

за·хлеб·н·у́ть, -у́т, (*perf*) *see* **за·хлё·б·ыва·ть**

за·хлéб·ыва·ть, -ют, (*imp*) gulp, swallow; flood (of distillation column)

за·хлёст·ывани·е, -я, (*n*) (*v n*) of **за·хлёст·ыва·ть;** entanglement (of wires)

за·хлест·áть   (*fut 3rd sing, pl*) **за·- хлéщ·ет, за·хлéщ·ут,**   (*perf*) *see* **за·хлёст·ыва·ть**

за·хлёст·ыва·ть, -ют   (*imp*) whip, lash, begin to whip/lash; wrap

за·хлóп·а·ть , -ют,   (*perf*) begin to clap/flap; clap, flap

за·хлóп·к·а, -и,   (*g pl*) **-п·ок,** (*f*) flap, flap valve

за·хлóп·н·уть, -ут,   (*perf*) *see* **за·- хлóп·ыва·ть**

за·хлóп·ывани·е, -я,   (*n*) (*v n*) of **за·- хлóп·ыва·ть;** (cryst) welding

за·хлóп·ыва·ть, -ют,   (*imp*) slam

за·хóд·, -а,   (*m*) (*v n*) *see* **за·ход·úть;** (naut) call (at a port), entry (into port); (air) approach, approach turn; entry (of screw threads)

∼, гелиакт·úческ·ий   (astron) helical setting

глуб·ин·á за·хóд·а   (*f*) working depth (of gear teeth)

∼ за   overshoot, overswing (of a pointer)

зóн·а за·хóд·а   (*f*) approach zone (to air fields)

∼ сóлн·ц·а   sunset

за·ход·úть, -·ят,   (*imp*) go/get to; call at/on, call in passing; go round/ behind; (air) approach; (astron) set

за·хóд·к·а, -и, (*g pl*) **-д·ок·,** (*f*) (min) progress cycle

за·хóд·ност·ь, -и,   (*f*) number of entries (of a screw thread)

за·хóд·н·ый, -ая, -ое,   (*a*) of **за·хóд·**

∼ чáст·ь   (*f*) entry, entry part (e.g. of screw conveyer)

за·хожд·éни·е, -я,   (*n*) wheeling (manœuvre); piping the side, sounding off (naval)

за·хож·ý   (*fut 1st sing*) of **за·ход·úть**

за·хорон·éни·е, -я, (*n*) burial, burying

за·хорон·úть, -·ят, (*perf*) bury, dig in

за·хот·éть   (*fut tense*) **за·хоч·ý, за·- хóч·ешь, за·хóч·ет, за·хот·úм, за·хот·úте, за·хот·ят,** (*perf + gen*) want, like to

за·хóч·ет   (*fut 3rd sing*) of **за·хот·éть**

за·хрип·éть, -·ят,   (*perf*) begin to wheeze

за·хром·á·ть, -ют,   (*perf*) begin to limp

за·цвест·ú   (*perf*) (*fut 3rd sing, pl*) **за·цвет·ёт, за·цвет·ýт,** (*past masc sing*) **за·цвёл;** *see* **за·цвет·á·ть**

за·цвет·á·ть, -ют,   (*imp*) be in bloom/ flower; come into bloom

за·цéп·, -а,   (*m*) hook, latch, stop, detainer, detent; hooking, catching, engaging

за·цéп·а, -ы,   (*f*) hook, latch, stop, detainer, detent

за·цéп·ить, -·ят,   (*perf*) *see* **за·цеп- л·я́·ть**

за·цéп·к·а, -и, (*g pl*) **-п·ок·,** (*f*) (*v n*) *see* **за·цепл·я́·ть;** catch, hook, stop, checking device; (horol) detent, detainer

за·цепл·éни·е, -я,   (*n*) (*v n*) *see* **за·- цепл·я́·ть;** meshing, engagement (gears); gear, gearing, gear teeth

∼, внýтр·енн·ее   internal/inside gearing/toothing

длин·á за·цепл·éни·я   (*f*) length of contact (gears)

лúн·и·я за·цепл·éни·я   (*f*) line of action (of gears)

пóлюс· за·цепл·éни·я   (*m*) pitch point (of a gear)

ýгол· за·цепл·éни·я   (*m*) generating angle (gear cutting); angle of obliquity (gears)

∼, цéв·очн·ое   lantern-wheel gear, mangle gear

∼, черв·я́чн·ое   worm gear/gearing

∼, эвольвéнт·н·ое   involute gearing/ gear teeth

за·цепл·я́·ть, -ют,   (*imp*) catch/hook on/up, engage, mesh, put into gear

за·част·ýю   (*adv*) often, frequently

за·чáт·и·е, -я,   (*n*) conception

за·чáт·ок·, -т·к·а,   (*m*) embryo, rudiment; (cryst) nucleus

за·чáт·очн·ый, -ая, -ое,   (*a*) of **за·- чáт·ок·**

за·чекáн·енн·ый, -ая, -ое,   (*past part pass*) caulked-in

за·чёл   (*past masc sing*) of **за·чéст·ь**

за·чéм   (*adv*) what for, why

за·чёрк·ивани·е, -я,   (*n*) (*v n*) of **за·чёрк·ива·ть;** deletion

за·чёрк·ива·ть, -ют,   (*imp*) cross/ strike out, delete

за·черк·н·ýть, -ýт,   (*perf*) *see* **за·- чёрк·ива·ть**

за·черн·é·ть, -ют,   (*perf*) look/appear/ show black; turn/go black

за·черн·úть, -·я́т,   (*perf*) blacken, colour/color black

за·черп·н·ýть, -ýт,   (*perf*) *see* **за·- чéрп·ыва·ть**

за·чéрп·ыва·ть, -ют,   (*imp*) scoop

за·черств·é·л·ый, -ая, -ое,   (*a*) stale

за·чес·а́ть   (*fut 3rd sing, pl*) за·чё́·ш·ет, за·че́ш·ут (*perf*) comb, begin combing

за·че́ст·ь   (*fut 3rd sing, pl*) за·чт·ёт, за·чт·у́т, (*past masc sing*) за·чёл, (*perf*); see за·чи́т·ыва·ть

за·чёс·ыва·ть, -ют,   (*imp*) comb, begin combing

за·чёт·, -а, (*m*) examination, test; part payment, instalment, payment on account

за·чёт·н·ый, -ая, -ое, (*a*) of за·чёт·

~ кни́ж·к·а   (*f*) student's academic record book

за·че́ш·ут (*fut 3rd pl*) of за·чес·а́ть

за·чин·а́тел·ь, -я,   (*m*) founder, initiator

за·чи́н·ива·ть, -ют, (*imp*) mend, patch

за·чин·и́ть, -́ят, (*perf*) see за·чи́н·ива·ть

за·чи́сл·ить, -ят, (*perf*) see за·чис·л·я́·ть

за·числ·я́·ть, -ют,   (*imp*) enter (on list), include; enlist, enrol, take on

за·чи́ст·к·а, -и,   (*g pl*) -т·ок·, (*f*) (met) surface dressing; (met forge) trimming; (plast) trimming

~ наждак·а́ми   grinding

~, огн·ев·а́я   (met) flame scarfing, hot deseaming

за·чит·а́·ть, -ют,   (*perf*) read out; begin reading; wear out by reading

за·чи́т·ыва·ть, -ют,   (*imp*) read out; (fin) take on account; take into account; pass, accept

за·чла́   (*past fem sing*) of за·че́ст·ь (*perf*); see за·чи́т·ыва·ть (*imp*)

за·чт·ённ·ый, -ая, -ое,   (*past part pass*) of за·че́ст·ь (*perf*); see за·чи́т·ыва·ть (*imp*)

за·шаг·а́·ть, -ют,   (*perf*) begin to pace, pace

за·шевел·и́ть, -́ят,   (*perf*) begin to stir/move; stir, move

за·ше́·ек·, -е́·йк·а,   (*m*) nape of neck

за·шиб·а́·ть, -ют,   (*imp*) hurt, bruise

за·шиб·и́ть, -у́т,   (*perf*) see за·шиб·а́·ть

за·шив·а́ни·е, -я,   (*n*) (*v n*) of за·шив·а́·ть; (med) suture

~ по́ч·к·и   (med) nephrorrhaphy

за·шив·а́·ть, -ют,   (*imp*) sew up; line

за·шип·е́ть, -я́т, (*perf*) begin to hiss, hiss

за·ши́·ть   (*fut 3rd sing, pl*) за·шь·ёт, за·шь·ю́т, (*perf*); see за·шив·а́·ть

за·шифр·о́ванн·ый, -ая, -ое,   (*past part pass*) of за·шифр·ова́ть; encoded; encyphered

за·шифр·ова́ть, -у́ют,   (*perf*) see за·шифр·о́выва·ть

за·шифр·о́выва·ть, -ют,   (*imp*) encypher; encode

за·шка́л·ивани·е, -я,   (*n*) (instr) off-scale reading

за·шкал·и́ть, -я́т,   (*perf*) (instr) go off scale

за·шлак·о́вывани·е, -я,   (*n*) slagging

за·шл·ю́т (*fut 3rd pl*) of за·сл·а́ть

за·шнур·ова́ть, -у́ют,   (*perf*) see за·шнур·о́выва·ть

за·шнур·о́выва·ть, -ют,   (*imp*) lace up

за·шпакл·ева́ть, -ю́ют,   (*perf*) see за·шпакл·ёвыва·ть

за·шпакл·ёвыва·ть, -ют,   (*imp*) putty in, apply putty, fix in with putty

за·шпи́л·ива·ть, -ют,   (*imp*) pin up

за·шпи́л·ить, -ят,   (*perf*) see за·шпи́л·ива·ть

за·шплинт·ова́ть, -у́ют,   (*perf*) see за·шплинт·о́выва·ть

за·шплинт·о́выва·ть, -ют,   (*imp*) cotter, secure with pins/cotters; tighten with cotters

за·штамп·ова́ть, -у́ют,   (*perf*) stamp, punch, press, drop forge

за·штемпел·ева́ть, -ю́ют, (*perf*) stamp

за·штил·е́вш·ий, -ая, -ее, (*past part act*) of за·штил·е́·ть; (naut) becalmed

за·штил·е́·ть, -ют,   (*perf*) (meteor) become/be calm

за·што́п·а·ть, -ют, (*perf*) darn

за·штрих·о́ванн·ый, -ая, -ое, (*past part pass*) of за·штрих·ова́ть; hatched (in drawing)

~ на́·ис·кос·ь   cross-hatched

за·штрих·ова́ть, -у́ют,   (*perf*) shade, hatch (in drawing)

за·штукату́р·ить, -ят,   (*perf*) plaster up

за·шум·е́ть, -я́т,   (*perf*) begin to make a noise

за·шунт·и́р·ова·ть, -уют, (*perf*) (elec) shunt, switch

за·шь·ю́т (*fut 3rd pl*) of за·ши́·ть

за·щёлк·а, -и,   (*g pl*) -лок·, (*f*) latch, satch, pawl

~, па́д·ающ·ая   latch

за·щёлк·ива·ть, -ют,   (*imp*) latch, fasten with a latch

за·щёлк·н·уть, -ут,   (*perf*) see за·щёлк·ива·ть

за·щем·и́ть, -я́т, (*perf*) *see* **за·щем-
л·я́·ть**

за·щемл·е́ни·е, -я, (*n*) sticking, jamming, pinching
  момéнт за·щемл·е́ни·я (*m*) (mech)
  restraining moment, moment at fixed
  end (of a beam)

за·щемл·ённ·ый, -ая, -ое, (*past part
pass*) *see* **за·щемл·я́·ть;** pinched,
jammed, stuck; restrained, fixed

~ кон·е́ц· (*m*) (build) fixed end

за·щемл·я́·ть, -ют, (*imp*) jam, stick,
pinch

за·щи́т·а, -ы, (*f*) protection, defence,
defense; (nucl) shield, shielding

~, ано́д·н·ая (met) anodic protection

~, био·ло́г·и́ческ·ая (nucl) biological
shield; biological methods of protection

~, дистанцио́н·н·ая (elec) remote-
control relay protection

~, земл·ян·а́я (elec) leakage protection

~, максима́ль·н·ая то́к·ов·ая (elec)
overcurrent protection

~ от пере·на·пряж·е́ни·й (elec) overvoltage protection

~ по́чв·ы soil conservation

~ проте́ктор·ами (met) cathodic
protection, zinc block protection

~, противо·хим·и́ческ·ая (mil) anti-
gas protection

~ от с·верх·то́к·ов (elec) overcurrent protection

~, реле́·йн·ая (elec) relay protection

~, селекти́в·н·ая (elec) discriminating protection
  при·бо́р· за·щи́т·ы хвост·а́ (*m*)
  (a/c rad) backward-looking warning
  device

за·щит·и́тельн·ый, -ая, -ое, (*a*)
protecting, protective; (law) defending

~ реч·ь (*f*) (law) defending counsel's
address

за·щит·и́ть, -я́т, (*perf*) *see* **за·щи-
щ·а́·ть**

за·щи́т·ник·, -а, (*m*) defender, protector, protective device; (law) defence counsel

за·щи́т·н·ый, -ая, -ое, (*a*) protective,
shielding; defending

~ кож·у́х· (*m*) sheathing; blade shield
(of turbine); (nucl) can

~ лес·н·ы́е по́лос·ы (*pl*) (agr) protective forest plantations

~ оч·к·и́ (*pl*) goggles

за·щи́т·н·ый, -ая, -ое

~ тро́с· (*m*) (elec) overhead ground
wire

~ цвет· (*m*) protective colouring/
coloring, khaki

~ электро́д·ы (*pl*) (electro chem) protective devices

за·щищ·а́·ть, -ют, (*imp*) protect,
defend; speak in support of; (law)
plead

за·щищ·ённ·ый, -ая, -ое, (*past part
pass*) *see* **за·щищ·а́·ть;** protected,
shielded

~ маши́н·а (*f*) (elec) protected machine

за·эвтект·и́ческ·ий, -ая, -ое, (*a*)
hypereutectic

за·эвтекто́ид·н·ый, -ая, -ое, (*a*)
hypereutectoid

за·яв·и́тел·ь, -я, (*m*) applicant

за·яв·и́ть, ⸗ят, (*perf*) *see* **за·явл·я́·ть**

за·я́в·к·а, -и, (*g pl*) **-в·ок·,** (*f*) claim,
demand, application; indent, requisi-
tion, application form

за·явл·е́ни·е, -я, (*n*) declaration, statement; application

за·явл·я́·ть, -ют, (*imp*) declare, announce; claim

~ жа́л·об·у make a complaint

за·я́кор·енн·ый, -ая, -ое, (*a*) an-
chored

за́яц·, (*g*) **за́йц·а,** (*i*) **за́йц·ем,** (*m*)
(zool) hare, *Lepus*; stowaway

~, мор·ск·о́й bearded seal, *Erignathus barbatus*; *Aplysia* (shellfish)

за́яч·ий, -ья, -ье, (*a*) (zool) hare,
hare's

зв·а́вш·ий, -ая, -ее, (*past part act*)
of **зв·ать**

зв·а́ни·е, -я, (*n*) rank, title

зв·а́тельн·ый, -ая, -ое, (gram) vo-
cative

зв·ать (*pres 3rd sing, pl*) **зов·ёт, зо-
в·у́т,** (*past masc sing*) **звал,** (*imp*)
call, call upon; invite, ask

звезд·а́, -ы́, (*nom pl*) **звёзд·ы,** (*f*)
star; (cyt) aster

~, двой·н·а́я (astron) binary star

~, двух·луч·ев·а́я two-pronged star

~, за·тм·ённ·ая пере·мéн·н·ая
eclipsing variable star

~, мор·ск·а́я (zool) starfish, (*pl*)
Asteroidea

~, но́в·ая (astron) nova

~, пере·мéн·н·ая variable star

~, по·вто́р·н·ая repeating star
  со·един·éни·е в звезд·у́ (*n*) (elec)
  star connection

звезд·á
~, у·велич·ивающ·аяся accreting star
звёзд·н·ый, -ая, -ое, (*a*) star, stellar, starry, starlit, sidereal
~ велич·ин·á (*f*) star magnitude
~ врéм·я (*n*) sidereal time
~ глóбус· (*m*) star globe
звездо·обрáз·н·ый, -ая, -ое, (*a*) star-shaped, astroid, radial, (biol) stellate; (*f decl as adj*) (math) astroid
~ двиг·áтел·ь (*m*) radial engine
звездо·образ·овáни·е, -я, (*n*) (nucl) star production
звезд·плáв·ани·е, -я, (*n*) astronautics, cosmonautics
звездо·чёт·ы (*nom pl*), (*g pl*) **-ов,** (*m*) (fish) *Uranoscopidae*
звёзд·очк·а, -и, (*g pl*) **-чек·,** (*f*) (*dim*) of **звезд·á;** sprocket, chain/star wheel; (horol) star wheel; (print) asterisk; (naut) snug (capstan), gypsy (windlass)
~ гýсен·иц·ы crawler wheel
звезд·чáтк·а, -и, (*g pl*) **-ток·,** (*f*) **звёзд·очк·а;** (bot) starwort, *Stellaria*
~, срéд·н·яя (bot) chickweed, *Stellaria media*
звёзд·чатост·ь, -и, (*f*) (med) asterism
звёзд·чат·ый, -ая, -ое, (*a*) star-shaped, astroid, stellar, stellate, star-like
звен·éть, -я́т, (*imp intrans*) ring, jingle, clang, clink
звен·ó, -á, (*nom pl*) **звéн·ья,** (*g pl*) **звéн·ьев,** (*n*) link; team, group, section, flight (3 aircraft); (M/T) track shoe; (telecom) network; (autom) component element; (chem) unit, link, linkage, bond
~, бок·ов·óе (chem) side unit (of polymer chain)
~, Г-óбраз·н·ое (telecom) L-network
~, не·раз·бóр·н·ое (M/T) one-piece track shoe
~, О-óбраз·н·ое (telecom) o-network
~, пере·крёст·н·ое (telecom) lattice network
~, П-óбраз·н·ое (telecom) pi-network
~, по·втор·яющ·ее·ся repeating unit (polymers)
~, Х·óбраз·н·ое (telecom) lattice network
звен·ов·óй, -áя, -óе, (*a*) of **звен·ó**
звéн·чат·ый, -ая, -ое, (*a*) link, made of links
~ кардáн· (*m*) ring joint, flexible universal joint

звен·ьев·óй, -áя, óе, (*a*) of **звен·ó;** (*noun decl as adj*) team leader
звéн·ья (*nom pl*) of **звен·ó**
звен·я́щ·ий, -ая, -ее, (*pres part act*) of **звен·éть;** *see also* **звон·к·ий**
звер·и́нец·, -нц·а, (*i*) **-нц·ем,** (*m*) menagerie
звер·и́н·ый, -ая, -ое, (*a*) of **звер·ь**
зверо·бó·й, -я, (*m*) trapper, hunter; (naut) sealer; (bot) St. John's wort
зверо·бóй·н·ый, -ая, -ое, (*a*) hunting, trapping; sealing; (*pl as noun*) (bot) *Hypericaceae*
~ сýд·н·о (*n*) sealer
зверо·вóд·ств·о, -а, (*n*) fur farming
зверо·зýб·ые, -ых, (*pl decl as adj*) (pal) theriodonts, *Theriodontia*
зверо·лóв·, -а, (*m*) trapper
зверо·нóг·ие, -их, (*pl decl as adj*) (pal) theropods, *Theropoda*
звер·ь, -я, (*nom pl*) **-и,** (*g pl*) **-éй,** (*m*) wild animal/beast
~, пуш·н·óй fur-bearing animal
звон·, -а, (*m*) peal, chime, ringing
звон·и́ть, -я́т, (*imp trans*) ring, clang
звóн·к·ий, -ая, -ое, (*a*) ringing, clear, sonorous; (ling) voiced
~ монéт·а (*f*) (fin) specie
звон·óк·, -н·к·á, (*m*) bell
~, коротко·за·мык·áющ·ийся (elec) vibrating bell
~ -пре·рыв·áтел·ь, -я, (*m*) vibrating bell
~ -удáр·ник·, -а, (*m*) single-stroke bell
звýк·, -а, (*m*) sound
~, глáс·н·ый (gram) vowel
~, со·глáс·н·ый (gram) consonant
звук·ов·óй, -áя, -óе, (*a*) sound; (rad) audio; acoustical, acoustic
~ блóк· (*m*) (cinema) sound head
~ генерáтор· (*m*) (elec) audio-frequency oscillator, a-f oscillator
~ дорóж·к·а (*f*) sound track
~ част·от·á (*f*) audio frequency
звуко·вос·про·из·вед·éни·е, -я, (*n*) audio/sound reproduction
звуко·глуш·и́тел·ь, -л, (*m*) silencer
звуко·грáмм·а, -ы, (*f*) sound-signalled message
звуко·за·пи́с·ывател·ь, -я, (*m*) sound recorder
звуко·зá·пис·ь, -и, (*f*) sound recording
~, механ·и́ческ·ая acoustical sound recording
~, опт·и́ческ·ая optical sound recording

звуко·за́·пис·ь
~, **фото·граф·и́ческ·ая** optical sound recording
~, **электро·механ·и́ческ·ая** electrical sound recording
звуко·зо́нд·, -а, (*m*) (geol) sonoprobe
звуко·изол·я́ци·я, -а, (*f*) sound proofing/insulation
   коэффициéнт· звуко·изол·я́ци·и (*m*) acoustical reduction factor, acoustic noise reduction coefficient
звуко·из·луч·а́тел·ь, -я, (*m*) acoustic radiator
звуко·лéнт·а, -ы, (*f*) sound recording tape
звуко·мéр·н·ый, -ая, -ое, (*a*) noise/ sound measuring
~ **ка́мер·а** (*f*) noise measurement cabin
звуко·метр·и́ческ·ий, -ая, -ое, (*a*) sound-ranging
~ **ста́нци·я** (*f*) sound-ranging set/ equipment, sound locator
звуко·мéтр·и·я, -и, (*f*) sound ranging/ location
звуко·не·про·ниц·а́ем·ый, -ая, -ое, (*a*) soundproof
звуко·нос·и́тел·ь, -я, (*m*) record (disc, tape or wire)
звуко·опера́тор·, -а, (*m*) (cinema) sound producer
звуко·от·раж·а́тел·ь, -я, (*m*) (build) sounding board
звуко·от·раж·éни·е, -я, (*n*) sound/ acoustic reflection
   коэффициéнт· звуко·от·раж·é·ни·я (*m*) acoustic reflection factor
зву́ко·пис·ь, -и, (*f*) *see* **звуко·за́·пис·ь**
звуко·по·глот·и́тел·ь, -я, (*m*) muffler, sound absorber
звуко·по·глощ·а́ющ·ий, -ая, -ее, (*a*) sound-absorbant, muffling; (build) acoustic
звуко·по·глощ·éни·е, -я, (*n*) sound/ acoustic absorption, muffling
   коэфициéнт· звуко·по·глощ·é·ни·я (*m*) acoustical/absorption factor
звуко·под·раж·а́тельн·ый, -ая, -ое, (*a*) onomatopoeic
звуко·пре·ломл·éни·е, -я, (*n*) acoustic refraction
звуко·при·ём·ник·, -а, (*m*) sound pickup
звуко·про·во́д·, -а, (*n*) delay line (ultrasonics)
звуко·прово́д·н·ый, -ая, -ое, (*a*) sound-conducting

звуко·про·из·вед·éни·е, -я, (*n*) acoustic reproduction
звуко·про·ниц·а́емост·ь, -и, (*f*) acoustic/sound transmission
   коэффициéнт· звуко·про·ниц·а́емост·и (*m*) acoustic/sound transmission factor
звуко·с·ним·а́тел·ь, -я, (*m*) acoustic/ sound pick-up/box
звуко·съ·ём·очн·ый, -ая, -ое, (*a*) sound, sound-pickup
~ **ка́мер·а** (*f*) (cinema) sound camera
звуко·у·ла́вл·иватеl·ь, -я, (*m*) sound locator
звуко·част·о́тн·ый, -ая, -ое, (*a*) (rad) audio, audio-frequency
звуко·у·сил·и́тел·ь, -я, (*m*) sound amplifier
звуч·а́ни·е, -я, (*n*) (*v n*) of **звуч·а́ть;** vibration (of sound); sound; (ling) phonation
звуч·а́ть, -а́т, (*imp*) sound, resound, ring, be heard
звуч·а́щ·ий, -ая, -ее, (*pres part act*) of **звуч·а́ть**
зву́ч·н·ый, -ая, -ое, (*a*) sonorous
звя́к·ань·е, -я, (*n*) tinkling
з-д (*abbr*) of **за·во́д·**
з.д. (*abbr*) of **за́·пад·н·ая долг·от·а́** western longitude, longitude west
зда́·ни·е, -я, (*n*) building
~, **жил·о́е** residential building, living accommodation, block of flats, house
~, **карка́с·н·ое** framed building
~, **маши́н·н·ое** (hydroelec) powerhouse
здесь (*adv*) here; local (on letter)
здéш·н·ий, -яя, -ее, (*a*) local, of this locality
здоро́в·аться, -ются, (*imp*) greet, say "how do you do"
здоро́в·о well done! fine!; **здоро́в·о** healthily
здоро́в·ый, -ая, -ое, (*a*) healthy; good (for the health); strong, sound
~ **древес·и́н·а** (*f*) sound wood
здоро́в·ь·е, -я, (*n*) health
здра́в·иц·а, -ы, (*i*) **-ей,** (*f*) toast (to drink)
здра́в·ниц·а, -ы, (*i*) **-ей,** (*f*) sanatorium
здра́в·о (*adv*) reasonably, sensibly
здраво·мы́сл·ящ·ий, -ая, -ее, (*a*) sensible, sound, sane
здраво·о·хран·éни·е, -я, (*n*) public health, public health service
здрав·от·дéл·, -а, (*m*) health department, public-health department

здрав·пу́нкт·, -а, (*m*) clinic, health clinic

здра́в·ств·уйте how do you do!, good day! (etc.)

здра́в·ый, -ая, -ое, (*a*) sensible, sound

~ с·мысл· (*m*) common sense

зеа·ксанти́н·, -а, (*m*) (biochem) zeaxanthin

зе́бр·а, -ы, (*f*) (zool) zebra, *Equus zebra*

зёв·, -а, (*m*) (anat) pharynx, throat; (mech eng) jaw, jaw opening, mouth; opening, gap; (text) shed; span, mouth (of a wrench)

~ валк·о́в (met roll) roll gap; roll-gap zone (cont casting)

~, за́·мк·нут·ый box end/mouth (of spanner)

~, от·кры́·т·ый (text) plain shed

~, полу·от·кры́·т·ый (text) compound shed

~, чи́ст·ый (text) clear shed

зев·а́·ть, -ют, (*imp*) yawn

зев·н·у́ть, -у́т, (*perf*) *see* **зев·а́·ть**

зев·о́к·, -в·к·а́, (*m*) yawn

зево·образ·ова́ни·е, -я, (*n*) (text) shedding, shed formation

~ каре́т·к·ой dobby shedding

зев·о́т·а, -ы, (*f*) yawning

Зе́гер·а, ко́нус· (*m*) Seger/fusion cone (for temperature measurement)

Зе́ебек·а, эффект (*m*) Seebeck effect (thermoelectricity)

~, явл·е́ни·е (*n*) Seebeck effect (thermo-electricity)

Зе́еман·а, явл·е́ни·е (*n*) (spectr) Zeeman effect

зеер·, -а, (*m*) cage/barrel (of an oil press)

зе́ер·н·ый пресс (*m*) oil extraction press

зе́йгер·н·ый, -ая, -ое, (*a*) (met) liquation

зе́йгер·ова́ни·е, -я, (*n*) (met) liquation

зелен·е́·ть, -ют, (*imp*) turn green; appear/look/show green

зелен·н·о́й, -а́я, -о́е, (*a*) of **зе́лен·ь**

~ культу́р·ы (*pl*) green vegetables

зелен·ова́т·ый, -ая, -ое, (*a*) greenish

зелено·гла́з·к·а, -и, (*g pl*) **-з·ок·,** (*f*) (ent) green-eyed fly, *Chlorops taeniopus*

зелён·о-жёлт·ый, -ая, -ое, (*a*) greenish yellow, olive

зелёно·ка́мен·н·ый, -ая, -ое, (*a*) (geol) greenstone

зелен·у́шк·а, -и, (*g pl*) **-шек·,** (*f*) (fish) goldsinny, *Crenilabrus*; (orn) greenfinch, *Chloris chloris*; (*pl*) (ent) *Dolichopodidae*

зелен·щи́к·, -а́, (*m*) greengrocer

зелён·ый, -ая, -ое, (*a*) green

~ зо́н·а (*f*) (arch) green belt

~ стро·и́тельств·о (*n*) landscape gardening

зе́лен·ь, -и, (*f*) green (pigment); greens, green vegetables; verdure

~, брилли́ант·ов·ая (med) brilliant green

~, веро́н·ск·ая Veronese green

~ ги́нье Guignet's green, chrome green

~, малахи́т·ов·ая malachite green

~, ме́д·н·ая (min) green carbonate of copper, malachite

~, пари́ж·ск·ая Paris green

~, швейнфурт·ск·ая Schweinfurt green

~, щелк·овск·ая emerald green

Зельмейер·а, фо́рмул·а (*f*) (opt) Sellmeier's dispersion formula

зе́льц·, -а, (*i*) **-ем,** (*m*) Sulze (a sausage)

земе́ль·н·ый, -ая, -ое, (*a*) land, agrarian

~ со́б·ственност·ь (*f*) real estate

зем·карава́н·, -а, (*m*) dredger string, string of dredgers and auxiliaries

земле- (*component*) earth, land; (met) sand

земле·вед·е́ни·е, -я, (*n*) physical geography

земле·влад·е́лец·, -льц·а, (*m*) landowner, landlord

земле·де́л·, -а, (*m*) (met cast) sand mixer

земле·де́л·ец·, -ль·ц·а, (*m*) farmer, cultivator, agriculturalist

земле·де́л·и·е, -я, (*n*) agriculture, farming, cropping

земле·де́л·к·а, -и, (*g pl*) **-л·ок·,** (*f*) (met cast) sand-mixing bay

земле·де́ль·ческ·ий, -ая, -ое, (*a*) agricultural

земле·ко́п·, -а, (*m*) navvy

земле·ко́п·н·ый, -ая, -ое, (*a*) digging

земле·ме́р·, -а, (*m*) land surveyor

земле·ме́р·н·ый, -ая, -ое, (*a*) surveying, surveyor's

земле·по́льз·овани·е, -я, (*n*) land tenure

земле·при·готов·и́тельн·ая маши́н·а (*f*) (met cast) sand-mixing machine

земле·ро́й·к·а, -и, (*g pl*) **-о́·ек·,** (*f*) (zool) shrew, *Sorex*, (*pl*) *Soricidae*

земле·ро́й·н·ый, -ая, -ое, (*a*) digging, excavating, earthmoving

земле·ро́й·н·ый, -ая, -ое
~ **маши́н·а** (*f*) earthmover
земле·со́с·, -а, (*m*) sand pump; suction/hydraulic dredger/dredge
~, **рефулёр·н·ый** shore-delivery suction dredger
~, **само·от·во́з·н·ый** self-discharging suction dredger; hopper suction dredger
земле·со́с·н·ый, -ая, -ое, (*a*) suction/hydraulic dredging
~ **с·на·ря́д·** (*m*) suction/hydraulic dredger
земле·тряс·е́ни·е, -я, (*n*) earthquake
земле·у·плот·ни́тел·ь, -я, (*m*) (agr) roller
земле·у·стро́й·ств·о, -а, (*n*) land management
земле·черп·а́лк·а, -и, (*g pl*) **-лок·,** (*f*) mechanical dredger/dredge, digging dredger/dredge
земле·черп·а́ни·е, -я, (*n*) dredging
земле·черп·а́тельн·ый, -ая, -ое, (*a*) (*see also* **зем·снаря́д·**); dredge, dredger, dredging
~ **карава́н·** (*m*) dredger string
~ **с·на·ря́д·** (*n*) mechanical dredge/dredger
земл·и́ст·ый, -ая, -ое, (*a*) earthy
земл·я́, -и́, (*acc*) **зе́мл·ю,** (*nom pl*) **зе́мл·и,** (*g pl*) **земе́л·ь,** (*d pl*) **зе́мл·ям** earth, land, ground, soil; (astron etc.) the Earth; (met cast) sand; (telecom) letter "Z"
~ **-во́з·дух·** (GW) ground-to-air, surface-to-air
~, **инфузо́р·н·ая** infusorial earth, diatomaceous earth
~, **ре́д·к·ая** (chem) rare earth
~, **форм·о́вочн·ая** (met cast) moulding/molding sand
~, **цел·и́нн·ая** virgin land
~, **щелоч·н·а́я** alkali/alkaline earth
земл·я́к·, -а́, (*m*) fellow-countryman/villager/townsman
земл·яни́к·а, -и, (*f*) (bot) strawberry, *Fragaria*
земл·яни́чник·, -а, (*m*) (bot) Arbutus
земл·я́нк·а, -и, (*g pl*) **-нок·,** (*f*) dug-out
земл·ян·о́й, -а́я, -о́е, (*a*) earth, earthen; ground; sand
~ **гру́ш·а** (*f*) (bot) Jerusalem artichoke, *Helianthus tuberosus*
~ **оре́х·** (*m*) ground nut, earth nut, pea-nut, *Arachis hypogea L.*
~ **полотн·о́** (*n*) road bed

земл·ян·о́й, -а́я, -о́е
~ **рабо́т·ы** (*pl*) earth-moving, excavation
~ **ра́ковин·а** (*f*) (met) sand inclusion
~ **ток·** (*m*) (elec) earth/ground current
зе́м·ник·, -а, (*m*) (min) ground coal
земно·во́д·н·ый, -ая, -ое, (*a*) amphibious, amphibian; (*n decl as adj*) (zool) (*pl*) amphibians, *Amphibia*
зем·н·о́й, -а́я, -о́е, (*a*) earth, earthly, terrestrial, telluric; ground
~ **кор·а́** (*f*) (geol) lithosphere
~ **магнет·и́зм·** (*m*) terrestrial magnetism, geomagnetism
~ **обеспеч·е́ни·е самолёто·вожд·е́ни·я** (air) ground control service
~ **о́с·ь** (*f*) the Earth's axis
~ **радио·пеленг·а́тор·** (*m*) ground D/F station
~ **ток·** (*m*) (elec) telluric/ground current
зем·с·на·ря́д·, -а, (*m*) **земле·черп·а́тельн·ый/земле·со́с·н·ый с·на·ряд·** *see* **с·на·ряд** (*and associated adj*); dredge, dredger
~, **гре́йфер·н·ый** grab/grapple dredger/dredge
~, **много·ко́вш·ев·ый** ladder/elevator dredger/dredge
~, **одно·черп·а́ков·ый** dipper dredger/dredge
~, **шта́нг·ов·ый** dipper dredger/dredge
зензу́бел·ь, -я, (*m*) rabbet plane, rabbetting plane (carpenter's tool)
зени́т·, -а, (*m*) zenith
~ **-телеско́п·, -а,** (*m*) (astron) zenith telescope
зенит·а́льн·ый, -ая, -ое, (*a*) zenithal
зени́т·н·ый, -ая, -ое, (*a*) (astron) zenith; (mil) anti-aircraft
~ **от·раж·а́тельн·ая труб·а́** (*f*) (astron) photographic zenith tube/telescope
~ **пу́шк·а** (*f*) anti-aircraft gun
~ **рас·сто·я́ни·е** (*n*) (astron) zenith distance
зе́нкер·, -а, (*m*) countersink, counterbore
~, **комбин·и́рованн·ый** combination counterbore
~, **кон·и́ческ·ий** countersink
~, **ступ·е́нчат·ый** step counterbore
зенкер·ова́ни·е, -я, (*n*) countersinking, counterboring
зенкер·ова́ть, -у́ют, (*imp*) countersink, counterbore
зенк·ова́ни·е, -я, (*n*) reaming
зенк·ова́ть, -у́ют, (*imp*) ream

зенк·о́вк·а, -и, (g pl) **-вок·,** (f) reamer (tool)

зе́ркал·о, -а, (n) mirror, looking glass; (med) speculum, (met) spiegel; sliding face, seat (of slide valve); (rad) reflector; surface, surface area, area

~, больш·о́е index glass (sextant)

~, вибр·и́рующ·ее rocking mirror (of telescope)

~, во́д·н·ое (geog) water table/plane

~, глаз·н·о́е ophthalmoscope

~ гор·е́ни·я firebox surface

~ грунт·ов·ы́х вод· (geog) ground water table

~, ма́л·ое horizon glass (sextant)

~, много·ря́д·н·ое (rad) mattress reflector

~, парабол·и́ческ·ое parabolic reflector

~, полу·про·зр·а́чн·ое half-silvered mirror

~, свето·дел·и́тельн·ое dichroic mirror

~ скольж·е́ни·я (geol) slickensides

зеркально·от·раж·ённ·ый, -ая, -ое, (a) mirror-image

зерка́ль·н·ый, -ая, -ое, (a) mirror, mirror-like, specular, reflecting, highly polished; (rad) image

~ из·ображ·е́ни·е (n) mirror image

~ ис·то́ч·ник· (m) (nucl) image source

~ ка́мер·а (f) (phot) reflex camera

~ кана́л· (m) (rad) image

~ по·ме́х·а (f) (rad) image interference

~ част·от·а́ (f) (rad) image frequency

~ чугу́н· (m) (met) spiegeleisen, spiegel iron

зерн·е́ни·е, -я, (n) granulation

зерн·ён·ый, -ая, -ое, (a) grained, granulated; beaded

~ катализа́тор· (m) bead catalyst

зерн·и́льн·ый, -ая, -ое, (a) graining

зерн·и́стост·ь, -и, (f) grain size/fineness, granularity, graininess

~, на·сле́д·ственн·ая (met) inherent grain size, McQuaid-Ehn grain size

зерн·и́ст·ый, -ая, -ое, (a) granular, granulated, grained, grainy

зерн·и́ть, -и́т, (imp) granulate

зерн·о́, -а́, (nom pl) **зёрн·а,** (g pl) **зёрен·,** (n) grain, granule; (bot) kernel; (agr) seed; (met) grain, crystallite

~, вальц·о́в·ое grain effect (on plastics, leather etc.)

велич·ин·а́ зерн·а́ (f) (met) grain size number

~, дей·стви́тельн·ое actual grain size

зерн·о́

~, кру́п·н·ое coarse grain

~, на·ли́т·ое (agr) laden kernel

~, на·сле́д·ственн·ое (met) McQuaid-Ehn grain size, inherent grain size

~, не·реб·р·ов·ое (met) chaotically oriented grain

~, про·до·во́ль·ственн·ое bread grain

~, стекло·ви́д·н·ое vitreous grain

у·круп·не́ни·е зерн·а́ (n) (met) grain growth

зерно·аспира́тор·, -а, (m) (agr) grain cleaner

зерно·боб·о́в·ые культу́р·ы (pl) pulse crops

зерно·ви́д·н·ый, -ая, -ое, (a) granular

зерн·о́вк·а, -и, (g pl) **-вок·,** (f) (bot) caryopsis; (ent) weevil, pea/bean weevil

зерно·во́з·, -а, (m) grain carrier (ship)

зерно·во́з·к·а, -и, (f) grain wagon

зерн·ов·о́й, -а́я, -о́е, (a) of **зерн·о́**

~ зла́к·и (pl) cereals

зерно·дроб·и́лк·а, -и, (g pl) **-лок·,** (f) cornmill

зерно·от·дел·и́тел·ь, -я, (m) kernel scourer, grain separator

зерно·о·чист·и́тельн·ый, -ая, -ое, (a) grain-cleaning, winnowing

~ маши́н·а (f) winnower, winnowing machine

зерно·пу́льт·, -а, (m) grain projector, grain throwing plant

зерно·суш·и́лк·а, -и, (g pl) **-лок·,** (f) grain dryer

зерно·хран·и́лищ·е, -а, (n) granary, barn

зерно·я́д·н·ый, -ая, -ое, (a) (zool) granivorous

зёрн·ышк·о, -а, (g pl) **-шек·,** (n) (dim) of **зерн·о́,** granule

зеро́ (n indecl) zero

зерото́р·, -а, (m) slab of dry ice (used in refrigeration)

зет·, -а, (m) Zed, Zee, Z

зе́та (f indecl) zeta (Greek)

зет·ме́тр·, -а, (m) (elec) impedance bridge

зе́т·ов·ый, -а, -ое, (a) Z-, Z-shaped; Zed/Zee-shaped; (f decl as adj) (met) Z-bar, Zed/Zee bar

зиг·апофи́з·, -а, (m) (zool) zygapophysis

Зи́гбан·а, спектро́·граф· (m) Siegbahn spectrograph

зигза́г·, -а, (m) zigzag

зигза́г·маши́н·а, -ы, (*f*) zigzag/swing-needle sewing machine

зиг·маши́н·а, -ы, (*f*) two-roll bending machine

зиго·мо́рф·н·ый, -ая, -ое, (*a*) zygomorphic, sygomorphous

зиго·спо́р·а, -ы, (*f*) (gen) zygospore

зиго́т·а, -ы, (*f*) (gen) zygote

зиго·фа́з·а, -ы, (*f*) (bot) zygophase

зиго·фи́лл·ов·ые, -ых, (*pl decl as adj*) (bot) bean-caper family, *Zygophyllaceae*

зиго·фи́т·, -а, (*m*) (bot) zygophyte

зик·маши́н·а, -ы, (*f*) *see* **зиг·маши́н·а**

зим·а́, -ы́, (*nom pl*) **-зи́м·ы,** (*f*) winter

зим·а́з·а, -ы, (*f*) (biochem) zymase

зи́м·н·ый, -яя, -ее, (*a*) winter, hibernal

зимо- (*component*) winter; (biochem) zymo-

зим·ова́ни·е, -я, (*n*) wintering; hibernation

зим·ова́ть, -у́ют, (*imp*) winter, spend the winter; hibernate

зим·о́вк·а, -и, (*f*) wintering; winter quarters; polar station; hibernation

зим·о́вник·, -а, (*m*) winter hive; (bot) hellebore

зим·о́вочн·ый, -ая, -ое, (*a*) hibernating

зим о́вь·е, -я, (*n*) winter quarters

зимо·ге́н·, -а, (*m*) (biochem) zymogen, proferment

зимози́·метр·, -а, (*m*) zymosimeter

зим·о́й (*adv*) in winter

зимо·ли́з·, -а, (*m*) (biochem) zymolysis

зимо·ло́г·и·я, -и, (*f*) (biochem) zymology

зимо·сто́й·к·ий, -ая, -ое, (*a*) (bot) winter-hardy, hardy

зи·я́ни·е, -я, (*n*) hiatus, gap, gaping

зи·я́·ть, -ют, (*imp*) gape

зла́к·, -а, (*m*) (agr) grass, cereal; (*pl*) (bot) grasses, *Gramineae*

~, корм·ов·ы́е (*pl*) fodder grasses

~, пищ·ев·ы́е (*pl*) cereals

зла́к·ов·ый, -ая, -ое, (*a*) grassy, herbaceous; cereal, grain

зла́т·к·и (*nom pl*), (*g pl*) **-т·ок·,** (*f*) (ent) metallic wood-boring beetles, *Buprestidue*

злато- (*root*) (*see also* **золото-**) gold-, chryso-

злато·гла́з·к·а, -и, (*f*) (ent) chrysop, *Chrysopa vulgaris*

злато·гу́з·к·а, -и, (*g pl*) **-з·ок·,** (*f*) (ent) brown-tail moth, *Euproctis chrysorrhoea, Nygmia phaeorrhoea*

злато·и́скр·, -а, (*m*) (min) goldstone, aventurine

злато·кро́т·, -а, (*m*) (zool) golden mole; (*pl*) *Chrysochloridae*

злато·цве́т·, -а, (*m*) *see* **золото·-цве́т·**

злато·цве́т·н·ый, -ая, -ое, (*a*) golden, gold-coloured/colored

зл·о, -а, (*n*) evil, harm, wrong

злобо·дн·е́вн·ый, -ая, -ое, (*a*) topical

зло·ве́щ·ий, -ая, -ее, (*a*) ominous, sinister

зло·во́н·и·е, -я, (*n*) stink, stench, bad smell/odour/odor

зло·во́н·н·ый, -ая, -ое, (*a*) stinking, malodorous, fetid, smelly

зло·вре́д·н·ый, -ая, -ое, (*a*) harmful, noxious

зло·ка́честв·енн·ый, -ая, -ое, (*a*) malignant

зло·ключ·е́ни·е, -я, (*n*) mishap; misfortune

зло·по·лу́ч·н·ый, -ая, -ое, (*a*) ill-fated

зл·о́стн·ый, -ая, -ое, (*a*) malicious

~ банкро́т· (*m*) fraudulent bankrupt

зло·у·по·треб·и́ть, -я́т, (*perf*) (+ *instr*) *see* **зло·у·по·требл·я́·ть**

зло·у·по·требл·я́·ть, -ют, (*imp*) (+ *instr*) misuse, abuse

змее·ви́д·н·ый, -ая, -ое, (*a*) serpentine, sinuous

зме·еви́к·, -а́, (*m*) coil, spiral tube; (bot) snake weed, *Polygonum bistorta*; (min) serpentine

~, пар·ов·о́й steam coil, steam heater coil

~, ребр·и́ст·ый gilled tubing, finned coil (refrigeration etc.)

зме·евико́в·ый, -ая, -ое, (*a*) of **зме·еви́к·**

змее·голо́в·, -а, (*m*) snake-head fish, *Ophicephalus*

змее·голо́в·ник·, -а, (*m*) (bot) dragon's head, *Dracocephalum*

змее·хво́ст·к·а, -и, (*g pl*) **-т·ок·,** (*f*) (zool) ophiuroid

зме·и́н·ый, -ая, -ое, (*a*) snake, snake's

зме·и́ст·ый, -ая, -ое, (*a*) snaky, sinuous, tortuous

зме·и́ться, -я́тся, (*imp*) snake, wriggle, writhe, twist, coil, wind; (geol) serpentinize

зме·й, -я, (*m*) (*see also* **зме·я́**); kite; dragon, serpent

зме·йк·а, -и, (*g pl*) **зме·ек·,** (*f*) (*dim*) of **зме·я́;** (air) "snake", weaving; (agr) spiral separator; (mil) broken file

змей·ков·ый, -ая, -ое, (*a*) kite

~ аэроста́т· (*m*) kite balloon

зме·я́, -й, (*nom pl*) **зме́·и,** (*g pl*) **зме·й,** (*f*) (zool) snake, *Serpentes*, *Ophidia*; (astron) Serpent, Serpens

зна́к·, -а, (*m*) sign, signal, token, symbol, mark, badge; (gram) stop

~, ба́нк·ов·ый (fin) banknote

~, берег·ов·о́й landmark

~, вод·ян·о́й (paper) watermark

~ во·про́с·а question mark, (telecom) interrogative

~, вос·клиц·а́тельн·ый exclamation mark

~ вы́·зов·а (telecom) call-up sign

~, дво·и́чн·ый (math) bit, binary digit

~, девиацио́н·н·ые (*pl*) (naut) deviation marks, leading marks for compass correction

~, десят·и́чн·ый (math) decimal place

~ зв·а́ни·я badge/mark of rank

~ интегра́л·а (math) integral sign

~ ио́н·а (phys) ionic charge

~ ко́рн·я (math) radical sign

~, ли́ч·н·ый identification card/disc

~, математ·и́ческ·ий mathematical symbol

~, меж·ев·о́й boundary/land mark

~, навигацио́н·н·ый navigation mark

~ не·ра́вен·ств·а (math) inequality sign

~, но́мер·н·ый (M/T) number plate

~, о·по·зна·ва́тельн·ый recognition signal; (navig) landmark; (*pl*) aircraft identification marks

~, о·по́р·н·ый (met cast) core seat/marker

~ о·пре·дел·и́тел·я (math) determinant notation

~, ос·ев·о́й river navigation buoy

~ от·ли́ч·и·я decoration, insignia

~, от·мен·и́тельн·ый (telecom) annual sign

~, плов·у́ч·ий seamark

~, под·стро́ч·н·ый (print) subscript sign, subscript

пра́в·ил·о зна́к·ов (*n*) (math) rule of signs

~, пред·о·стерег·а́тельн·ый (naut) warning light

зна́к

~ пре·пин·а́ни·я punctuation mark, punctuation sign

~, раз·дел·и́тельн·ый (telecom) long break sign

~, служ·е́бн·ый (telecom) procedure sign

~ со·един·е́ни·я junction mark (inland navig)

~ уда́р·а percussion/impact mark

~, у·сло́в·н·ый symbol, conventional sign

~, фабри́ч·н·ый trade-mark

~, хим·и́ческ·ий chemical symbol

зна́к·ов·ый, -ая, -ое, (*a*) of **знак·**

~ ли́н·и·я (*f*) (math) directed line

~ фу́нкци·я (*f*) (math) signum function

знако·инве́ртор·, -а, (*m*) sign inverter (computers)

знако́м·ить, -ят, (*imp*) acquaint, introduce; familiarize

знако́м·ств·о, -а, (*n*) acquaintance

знако́м·ый, -ая, -ое, (*a*) familiar; acquainted; (*noun decl as adj*) acquaintance

знако·о·пре·дел·ённ·ый, -ая, -ое, (*a*) of fixed sign, fixed

знако·пере·ме́н·н·ый, -ая, -ое, (*a*) sign-changing/variable, alternating

~ гру́пп·а (*f*) (math) alternating group

знако·по·сто·я́нн·ый, -ая, -ое, (*a*) of constant sign, constant

знамён·а (*nom, acc pl*) of **знам·я**

знамен·а́тел·ь, -я, (*m*) (math) denominator

~ масшта́б·а scale number (on a map)

~, о́бщ·ий common denominator

~ прогре́сс·и·и (math) common ratio of a progression

знамен·а́тельн·ый, -ая, -ое, (*a*) significant, portentous; denominative

знамен·и (*g, d, p sing*) of **знам·я**

знамен·и́т·ый, -ая, -ое, (*a*) famous, celebrated, eminent

знам·я (*g, d, p sing*) **зна́мен·и,** (*i sing*) **зна́мен·ем,** (*nom, acc pl*) **знамён·а,** (*g pl*) **знамён·,** (*n*) banner, standard; (mil) colours

зна́·ни·е, -я, (*n*) knowledge

зна́т·н·ый, -ая, -ое, (*a*) distinguished, notable

знат·о́к·, -а́, (*m*) expert, connoisseur

зна́·ть, -ют, (*imp*) know

знач·ащ·ий, -ая, -ее, (*pres part act*) of **знач·ить**; significant

~ цифр· (*m*) (math) significant digit

знач·ени·е, -я, (*n*) (*see also* **вели-ч·ин·а**); significance, importance; (math) value, figure; meaning (of a word)

~, абсолют·н·ое absolute value

~, амплитуд·н·ое peak value

~, без·раз·мер·н·ое non-dimensional value

~ в язык·е semantics

~, глав·н·ое (math) principal value

~, год·ов·ое annual value

~, гран·ичн·ое boundary value

~ дел·ени·я scale value

~, край·н·ее limiting value

~, максималь·н·ое maximum value, (elec) peak value

~ от·клон·ени·я significance of a deviation

~, порог·ов·ое threshold value

~, соб·ственн·ое (math) eigenvalue, proper value

~, стационар·н·ое steady state value

знач·имост·ь, -и, (*f*) significance

знач·ительн·ый, -ая, -ое, (*a*) considerable, significant

в знач·ительн·ой степ·ен·и to a considerable degree/extent, largely, much

знач·ить, -ат, (*imp*) mean, signify

знач·кист·, -а, (*m*) badgeman, badge holder

знач·ност·ь, -и, (*f*) (chem) atomicity

знач·н·ый, -ая, -ое, (*a*) of **знак·**; (telecom) marking

~ волн·а (*f*) (telecom) marking wave

~ цифр·а (*f*) (math) significant figure

знач·ок·, -ч·к·а, (*m*) (*dim*) of **знак·**; mark, sign, index; badge

~, под·строч·н·ый (print) subscript

зна·ющ·ий, -ая, -ее, (*pres part act*) of **зна·ть**; expert, skilled, informed

зноб·ить (*pres 3rd sing*) **-ит,** (*imp*) (med) shiver, have a shivering fit

зно·й, -я, (*m*) intense heat, sultriness

зной·н·ый, -ая, -ое, (*a*) torrid, sultry

зо·ант·ари·и (*nom pl*), (*g pl*) **-й,** (*m*) (zool) zoantharians, *Zoantharia*

зоб·, -а, (*nom pl*) **-ы,** (*g pl*) **-ов,** (*m*) crop, craw (of birds etc.); (med) goitre, goiter, struma

зоб·н·ый, -ая, -ое, (*a*) of **зоб·**

~ желез·а (*f*) (biol) thymus

зоб·, -а, (*m*) call

зов·ом·ый, -ая, -ое, (*pres part pass*) of **зв·ать**

зов·ущ·ий, -ая, -ее, (*pres part act*) of **зв·ать**

зодиак·, -а, (*m*) (astron) Zodiac

зодиак·альн·ый свет· (*m*) (astron) zodiacal light

зод·ческ·ий, -ая, -ое, (*a*) architecture, architectural

зод·честв·о, -а, (*n*) architecture

зод·ч·ий, -его, (*m decl as adj*) architect

зое·а, -ы, (*f*) (zool) zoea

зол·а, -ы, (*f*) ash, ashes, cinder

~, лет·уч·ая light ash

зол·ени·е, -я, (*n*) liming (leather)

зол·ист·ый, -ая, -ое, (*a*) ash, ashen, ashy

зол·ов·ой, -ая, -ое, (*a*) of **зол·а**

~ под·вал· (*m*) ash basement (boiler plant etc.)

золот·инк·а, -и, (*g pl*) **-нок·,** (*f*) speck of gold

золот·ист·о- (*component*) gold, (chem) auro-, aurous

золотисто·жёлт·ый, -ая, -ое, (*a*) golden yellow

золот·ист·ый, -ая, -ое, (*a*) golden, aureate; (chem) aurous

золот·ить, -ят, (*imp*) gild

золот·ник·, -а, (*m*) slide valve; (obs) zolotnik (approx 4·26 gr)

~, короб·чат·ый steam chest

~, масл·ян·ый lubrication slide valve

~ у·правл·ени·я (hydr) control valve

золот·ников·ый, -ая, -ое, (*a*) of **золот·ник·**

~ клапан· (*m*) (ICE) sleeve valve

~ короб·к·а (*f*) steam chest

~ рас·пре·дел·ительн·ый механ·изм· (*m*) (ICE) sleeve-valve gear

~ регулятор· скор·ост·и (*m*) slide-valve governor (of turbine)

~ тяг·а (*f*) slide rod

~ у·правл·ени·е (*n*) valve operational control

золот·о, -а, (*n*) (chem) gold, Au

~, грем·уч·ее fulminating gold

~, лигатур·н·ое base bullion

~, одно·хлор·ист·ое chloride gold

~, сер·нист·ое aurous sulphide

~, серно·кисл·ый auric sulphate

~, сусаль·н·ое gold leaf; artificial gold leaf, gold-coloured leaf

~, теллур·ист·ое (min) calaverite, gold telluride

~, трёх·хлор·ист·ое (chem) chlor-auride

зо́лот·о

~, **хло́р·н·ое** auric chloride

~, **чи́ст·ое** fine bullion/gold

золото·до·бы́·ч·а, -и, (*f*) goldmining

золот·о́й, -а́я, -о́е, (*a*) gold, golden, (chem) auric

~ **кисл·от·а́** (*f*) auric acid/hydroxide

~ **сеч·е́ни·е** (*n*) (math) golden/harmonic section of a line

~ **числ·о́** (*n*) (chem) gold number

золото·ки́сл·ый, -ая, -ое, (*a*) (chem) aurate of

~ **на́тр·и·й** (*m*) sodium aurate

золото·но́с·н·ый, -ая, -ое, (*a*) auriferous, gold-bearing

золото·хлори́сто·водоро́д·н·ая кисл·от·а́ (*f*) (chem) chloroauric acid

золото·цве́т·, -а, (*m*) (bot) pyrethrum

золот·у́х·а, -и, (*f*) (med) scrofula

золот·у́шн·ый, -ая, -ое, (*a*) (med) scrofulous

золо·у·дал·е́ни·е, -я, (*n*) ash handling/disposal

золо·у·ла́вл·ивани·е, -я, (*n*) dust collection

золо·у·лов·и́тел·ь, -я, (*m*) dust catcher/collector, flue dust extraction plant

~, **жалюз·и́йн·ый** baffle separator (dust catcher)

золоч·е́ни·е, -я, (*n*) gilding; gold plating

золоч·ён·ый, -ая, -ое, (*a*) gilded, gilt, gold plated

золоч·ённ·ый, -ая, -ое, (*past part pass*) of **золот·и́ть**; gilded

зо́л·ь, -я, (*m*) (chem) sol

зо́ль·ник·, -а, (*m*) ashpit; lime pit/reel (leather); (bot) cineraria, *Senecio cineraria*

зо́ль·ност·ь, -и, (*f*) ash content

зо́ль·н·ый, -ая, -ое, (*a*) ash, ashen, cinder; limed (leather)

зо́н·а, -ы, (*f*) (*and see under associated words*), zone, region, band, belt

~, **акти́в·н·ая** active zone/section; (nucl) core, fuel core, nuclear core

~, **вале́нт·н·ая** (phys) valence band

~ **во·ры́в·н·ого на·клеп·а** (met) shock-hardened zone

~ **вы́·держ·к·и** (met) soaking zone (furnace)

~ **вы·ра́вн·ивани·я температу́р·ы** (met) soaking zone (furnace)

~ **выс·о́к·их температу́р·** (met) heating zone (furnace)

зо́н·а

~, **за·по́лн·енн·ая** (phys) filled band

~, **за·пре́т·н·ая** prohibited area; exclusion area (of nuclear reactor)

~, **за·прещ·ённ·ая** forbidden zone; (phys) forbidden band

~, **лёт·н·ая** (air) movement area

~, **ма́л·еньк·ая** zonule, zonula

~, **мёртв·ая** dead/blind zone/belt; (rad) zone of silence

~, **метод·и́ческ·ая** (met) preheating zone (furnace)

~ **о·конч·а́тельн·ого на·гре́·в·а** (met) heating zone (furnace)

~ **под·хо́д·ов** (air) approach funnel

~, **по·раж·а́ем·ая** (gunn) danger zone

~, **при·ко́рн·ев·а́я** (bot) rhizosphere

~, **при·ли́в·н·ая** intertidal zone

~ **про·вод·и́мост·и** (phys) conduction band

~, **равно·сигна́ль·н·ая** (rad) equisignal zone (of beam)

~, **рыбо·ло́в·н·ая** fishery limits

~, **сейсм·и́ческ·ая** seismic belt

~, **со·мн·и́тельн·ая** (naut) danger zone, dangerous area

~, **фронта́ль·н·ая** (meteor) frontal surface

зона́ль·ност·ь, -и, (*f*) zonality, zonation, zoning

~, **обра́т·н·ая** oscillatory zoning

зона́ль·н·ый, -ая, -ое, (*a*) zonal, zonary

зо́нд·, -а, (*m*) probe, sonde, sound; (meteor) sounding balloon

~, **магни́т·н·ый** magnetic probe

~, **потенциа́ль·н·ый** (elec) potential probe

~, **при·ём·н·ый** pickup probe

зонд·а́ж·, -а, (*m*) *see* **зонд·и́ровани·е**

зонд·и́ровани·е, -я, (*n*) probing, sounding

зонд·и́р·овать, -уют, (*imp*) sound, probe

зонд·иро́вочн·ый, -ая, -ое, (*a*) *see* **зонд·и́ровани·е**

зонд·и́рующ·ий и́мпульс· (*m*) transmitted pulse, main pulse/bang (radar)

зо́н·н·ый, -ая, -ое, (*a*) of **зо́н·а** zonal, zone

~ **пла́в·к·а** (*f*) (melt) zone refining

~ **структу́р·а** (*f*) (cryst) band structure

~ **тео́р·и·я** (*f*) (phys) band theory

зон·ов·о́й, -а́я, -о́е, (*a*) of **зо́н·а**

зоно·о·кра́ш·енн·ый, -ая, -ое, (*a*) colour/color-zoned

зо́нт·, -а́, *(m)* umbrella, sunshade, awning, canopy; hood (furnace); spray hood (motor boat); (phot) screen, hood

зо́нт·ик·, -а, *(m)* umbrella, sunshade; (bot) umbel, umbella

зонтико·обра́з·н·ый, -ая, -ое, *(a)* umbrella-shaped, umbellate

зо́нт·ичн·ый, -эя, -ое, *(a)* umbrella, umbrella-shaped; (bot) umbellate, umbelliferous, *(pl as noun)* *Umbelliferae*

зо́н·ул·а, -ы, *(f)* zonule, zonula

зоо·бе́нтос·, -а, *(m)* (zool) zoobenthos

зоо·гео·гра́ф·и·я, -и, *(f)* zoogeography

зоо·гле́·я, -и, *(f)* (bact) zoogloea

зоо·е́ци·я, -и, *(f)* (zool) zooecium

зо·о́ид·, -а, *(m)* (zool) zooid

зоо·ксант·е́лл·а, -ы, *(f)* (biol) zooxanthella

зоо·ли́т·, -а, *(m)* (pal) zoolith

зоо́·лог·, -а, *(m)* zoologist

зоо·лог·и́ческ·ий, -ая, -ое, *(a)* zoological

зоо·ло́г·и·я, -и, *(f)* zoology

зоо·но́з·, -а, *(m)* (med) zoonosis

зоо·планкто́н·, -а, *(m)* (zool) zooplankton

зоо·спо́р·а, -ы, *(f)* (biol) zoospore

зоо·стери́н·, -а, *(m)* (biochem) zoosterol

зоо·те́хн·ик·а, -и, *(f)* (zool) zootechnics, zootechny, livestock/animal rearing/husbandry

зоо·те́хн·и·я, -и, *(f)* zootechny, livestock/animal rearing/husbandry

зоо·фи́т·, -а, *(m)* (zool) zoophyte

зоо·фо́б·и·я, -и, *(f)* (med) zoophobia

зоо·хо́р·и·я, -и, *(f)* (bot) zoochory

зоо·эритр·и́н·, -а, *(m)* (biochem) zoonerythrin, zooerythrin

зо́р·и *(nom pl)* of **зар·я́**

зо́р·к·ий, -ая, -ое, *(a)* sharp-sighted; vigilant

зо́р·кост·ь, -и, *(f)* visual acuity

зо́р·я, -и, *(f)* (bot) lovage, *Levisticum officinale*

зостер·а, -и, *(f)* (bot) eel grass, *Zostera marina*

зр·ачко́в·ый, -ая, -ое, *(a)* pupil, pupillary

зр·а́чок·, -чк·а́, *(m)* (anat) pupil

~, вы·ход·н·о́й (rad) exit pupil

зр·е́лищ·е, -а, *(n)* spectacle, sight

зре́·лост·ь, -и, *(f)* ripeness, maturity

аттеста́т· зре́·лост·и *(m)* certificate of general secondary education

~, пол·ов·а́я sexual maturity, puberty

зре́·льник·, -а, *(m)* (text) steam ager, steamer, ager

~, вос·станов·и́тельн·ый vat ager

~, о·кисл·и́тельн·ый acid ager

~, скор·остн·о́й quick steamer

зре́·льн·я, -и, *(g pl)*, **-лен·,** *(f)* (text) steaming apparatus

зре́·л·ый, -ая, -ое, *(a)* ripe, mature

зр·е́ни·е, -я, *(n)* vision, sight, eyesight; view

~, бок·ов·о́е peripheral vision

по́л·е зр·е́ни·я *(n)* field of view

то́ч·к·а зр·е́ни·я *(f)* point of view

у́гол· зр·е́ни·я *(m)* (opt) viewing angle, (phot) lens/camera angle

зре́·ть, -ют, *(imp)* ripen, mature, age;

зр·еть, -ят, *(imp)* see, look at, view

зр·и́тел·ь, -я, *(m)* spectator, viewer

зр·и́тельн·ый, -ая, -ое, *(a)* visual, optic

~ зал· *(m)* (theatre) auditorium

~ нерв· *(m)* (anat) optic nerve

~ сигнал·иза́ци·я *(f)* visual signalling

~ труб·а́ *(f)* telescope

зря *(adv)* to no purpose, for nothing, in vain

зуб, -а, *(nom pl anatomical sense* **зу́б·ы,** *mechanical sense* **зу́б·ья)** *(g pl anatomical sense* **-о́в,** *mechanical sense* **зу́б·ьев)** tooth; (anat) tooth, dens; point, jag, tine; land (of drill body); crack, flaw (in ice)

~, в·став·н·ы́е *(pl)* false teeth, denture

~ контргре́йфер·а (phot) pilot/fixed pin

~, корен·н·о́й (anat) molar tooth

~, кос·о́й skew tooth

~, о·чищ·а́ющ·ий raker, raker tooth (on a saw)

~, пере́д·н·ий (anat) incisor tooth

толщ·ин·а́ зу́б·а *(f)* chordal thickness (gears)

~ триб·а (horol) leaf

~, угло·винт·ов·о́й double-helical tooth

~, угл·ов·о́й (mech) herringbone tooth

~, у·по́р·н·ый (mech) stopping tooth

~, швейца́р·ск·ий (mech) club tooth

~, эвольве́нт·н·ый (mech) involute tooth

зуб·а́р·ь, -и, *(m)* toothed plane (carpenter's tool)

зуб·а́ст·ый, -ая, -ое, *(a)* large-toothed; *(pl)* (zool) *Cyprinodontidae*

~ ка́рп·ы *(pl)* killi fishes, *Cyprinodontiformes*

зуб·а́тк·а, -и, (*g pl*) **-ток·,** (*f*) bush hammer; (zool) wolf-fish, (*pl*) *Anarhichadidae*

зуб·е́ц·, -б·ц·а́, (*i*) **-б·ц·о́м,** (*m*) (mech eng) tooth, cog, spur, prong, jag, (biol) tooth, tine, point; (arch) merlon

зуб·и́л·о, -а, (*n*) chisel, cold chisel, sett (blacksmiths)

~, куз·не́чн·ое anvil chisel

~, под·руб·н·о́е bottom chisel

~, прям·о́е straight-edged sett, chisel sett

руб·и́ть зуб·и́л·ом chisel

~, слеса́р·н·ое cold chisel

~, фасо́н·н·ое gouge sett

зуб·и́льн·ый, -ая, -ое, (*a*) of **зуб·и́л·о**

зу́б·не·, -а, (*m*) (*dim*) of **зуб·;** (zool) denticle

зуб·н·о́й, -а́я, -о́е, (*a*) tooth, dental

~ бараба́н· (*m*) sprocket

~ врач· (*m*) dentist

~ проте́з· (*m*) false teeth, denture

~ те́хн·ик· (*m*) dental mechanic

зубо·ви́д·н·ый, -ая, -ое, (*a*) toothed, tooth-like, odontoid

зуб·ов·о́й, -а́я, -о́е, (*a*) tooth, toothed

~ по́яс· (*m*) cutting plank (carpentry)

зубо·врач·е́бн·ый, -ая, -ое, (*a*) dental

зубо·долб·ёжн·ый, стан·о́к· (*m*) gear-shaping machine

зубо·за·кругл·я́ющ·ий стан·о́к· (*m*) gear tooth chamfering machine

зубо·из·мер·и́тельн·ый при·бо́р· (*m*) gear-measuring instrument

зуб·о́к·, -б·к·а́, (*nom pl*) **зу́б·к·и** (anat), **зуб·к·и́** (mech), (*g pl*) **зу́б·ок·** (anat), **зуб·о́в** (mech), (*m*); (*dim*) of **зуб;** (min) cutter pick; clove (of garlic)

~, ре́ж·ущ·ий (min) chisel (of coal cutter)

зубо·на·рез·н·о́й стан·о́к· (*m*) saw-cutting machine

зубо·от·де́л·очн·ый станок gear-finishing machine

зубо·проте́з·н·ый, -ая, -ое, (*a*) denture, false teeth

~ с·плав· (*m*) dental alloy

зубо·про·тяж·н·о́й стан·о́к· (*m*) gear-broaching machine

зубо·ре́з·н·ый стан·о́к· (*m*) gear-cutting machine

зубо·строг·а́льн·ый стан·о́к· (*m*) gear shaper

зубо·фре́зер·н·ый стан·о́к· (*m*) gear-milling machine

зубо·шлиф·ов·а́льн·ый стан·о́к· (*m*) gear-grinding machine

зу́бр·, -а, (*m*) (zool) aurochs, *Bison bonasus*; (pop) diehard

зубр·и́л·о, -а, (*n*) **зуб·и́л·о**

зубр·и́ть, -я́т, (*imp*) make/cut notches/ teeth; learn mechanically, cram

зубр·о́вк·а, -и, (*g pl*) **-вок·,** (*f*) (bot) sweet-grass, *Hierochloe*; zubrovka (vodka flavoured with above)

зуб·ча́тк·а, -и, (*g pl*) **-ток·,** (*f*) sprocket; gear wheel, gear; indented chisel; (bot) bartsia, *Odontites*

~ с винт·о́м rack and pinion

зуб·ча́т·ый, -ая, -ое, (*a*) (mech eng) toothed, cogged; gear, gear-wheel; serrate, serrated; (biol) tooth, toothed, dentate

~ бараба́н· (*m*) (cinema) sprocket

~ вен·е́ц· (*m*) crown wheel, toothed ring

~ за·цепл·е́ни·е (*n*) gear, gearing

~ колес·о́ (*n*) gear wheel, toothed wheel

~ колес·о́, винт·ов·о́е (*n*) spiral gear

~ колес·о́, кон·и́ческ·ое (*n*) bevel gear

~ колес·о́, косо·зу́б·ое (*n*) single helical gear

~ колес·о́, прямо·зу́б·ое (*n*) spin gear

~ колес·о́, цилиндр·и́ческ·ое (*n*) spin gear

~ колес·о́, шеврон·н·ое (*n*) double-helical gearwheel

~ на·со́с· (*m*) spur-gear pump

~ па́р·а (*f*) gearwheel and pinion

~ пере·да́ч·а (*f*) gearing, gear drive, spur-gearing

~ ре́·йк·а (*f*) rake

зу́б·чик·, -а, (*m*) *see* **зуб·е́ц·;** (biol) denticle; (arch) dentil

зу́б·ья (*nom pl*) of **зуб**

зуд·, -а, (*m*) itching, (met) pruritus

зу́д·ен·ь, -д·н·я, (*m*) (ent) itch mite

зуд·е́ть, -я́т, (*imp*) itch

зу́ммер·, -а, (*m*) buzzer

~ за́·нят·ост·и engaged buzzer/signal (telephones), busy tone

~ пре·рыв·а́тел·ь, -я, (*m*) (telecom) buzzer

зумм·и́ровани·е, -я, (*n*) buzzing

зу́мпф·, -а, (*m*) *see* **зумф·**

зу́мф·, -а, (*m*) sump, pit, (min) draining pit

зунд·, -а, (*m*) (geog) sound, strait

зы́б·к·ий, -ая, -ое, (*a*) unsteady, unstable; vacillating

зы́б·кост·ь, -и, (*f*) unsteadiness, fluctuations, vacillation
зы́б·у́ч·ий, -ая, -ее, (*a*) *see* зы́б·к·ий
~ песк·и́ (*pl*) quicksands, shifting sand
зы́б·ь, -и, (*f*) (ocean) swell, surge
~, мёртв·ая (ocean) ground swell
зю́йд·, -а, (*m*) (naut) south
зю́йд·вест·, -а, (*m*) south west

зю́йд·ов·ый, -ая, -ое, (*a*) south
зя́б·к·ий, -ая, -ое, (*a*) sensitive to cold, chilly
зя́б·нуть, -нут, (*past masc sing*) зяб·, (*imp*) feel the cold, be sensitive to cold; suffer from the cold
зя́б·ь, -и, (*f*) fall/winter plowing/ploughing; fall/winter-plowed/ploughed field

И

и (*conj*) and; also, even; (*particle*) (implies emphasis)

~ **др.** (*abbr*) of **и друг·и́е**; *et al*, and collaborators

~ **... и** both ... and, ... as well as ...

~ **проч.** *see* **и про́ч·ее**

~ **про́ч·ее** etc.

~ **т.д.** (*abbr*) of **и так да́л·ее**; and so on/forth etc.

~ **т.п.** (*abbr*) of **и тому́ по·до́б·н·ое**; and the like, and so on/forth etc.

и-кисл·от·а́, -ы, (*f*) (chem) J-acid

иатро́н·, -а, (*m*) (rad) iatron

иатро·хи́м·и·я, -и, (*f*) iatrochemistry

Ибн-си́н·а, -ы, (*f*) Avicenna

и́бо (*conj*) for, because, since

ИВ (*abbr*) of **и́ндекс· вя́з·кост·и**; viscosity index

и́в·а, -ы, (*f*) (bot) willow, *Salix*

~, **пурпу́р·н·ая** purple osier willow, *Salix purpurea*

ива́н-да-ма́рь·я, -и, (*f*) (bot) cow wheat, *Melampyrum nemorosum*

ива́н·ов· хлеб· (*m*) (bot) carob, carob tree, *Ceratonia siliqua*

~ **черв·я́к·** (*m*) (ent) firefly

ив·ня́к·, -а, (*m*) (agr) osier bed

и́в·ов·ый, -ая, -ое, (*a*) of **и́в·а**; (*pl as noun*) (bot) willow family, *Salicaceae*

и́волг·а, -и, (*f*) (zool) oriole, *Oriolus oriolus*

игл·а́, -ы́, (*nom pl*) **и́гл·ы,** (*f*) needle, spicule, spike, thorn, quill; indenter (of hardness tester); small fishing vessel

~, **амортиз·и́рующ·ая** (rocket) landing spike

~, **вяз·а́льн·ая** knitting-needle

~, **за·по́р·н·ая** needle valve

~, **за·ря́д·н·ая** (nucl) spray point

~, **кол·ечн·ая** (naut) bolt-rope needle

~, **крюч·ко́в·ая** beard needle (knitting machine)

~, **на·жи́м·н·ая** (ICE) nozzle valve rod

~ **реакти́в·н·ого сопл·а́** throat bullet (of jet engine)

игл·и́ст·ый, -ая, -ое, (*a*) prickly, bristly, echinate, thorny, spiculate; containing acicular crystals

игл·ова́т·ый, -ая, -ое, (*a*) prickly, spiky, thorny; needle-shaped, acicular

игло·ви́д·н·ый, -ая, -ое, (*a*) needle-shaped, acicular, (bot) acerose, acerous

игло·держ·а́тел·ь, -я, (*m*) needle holder

игло·ко́ж·ие, -их, (*pl decl as adj*) (zool) echinoderms, *Echinodermata*

игло·обра́з·н·ый, -ая, -ое, (*a*) needle-shaped, acicular

игло·фи́льтр·, -а, (*m*) (civ eng) well-point

игни·тро́н·, -а, (*m*) (rad) ignitron

~, **раз·бо́р·н·ый** pumped ignitron

игни·тро́н·н·ый, -ая, -ое, (*a*) ignitron

~ **пре·рыв·а́тел·ь** (*m*) (weld) ignitron control

игнор·и́р·овать, -уют, (*imp and perf*) ignore, disregard, neglect

игнор·и́руем·ый, -ая, -ое, (*pres part pass*) of **игнор·и́р·овать**; negligible

и́г·о, -а, (*n*) yoke

иго́л·к·а, -и, (*g pl*) **-л·ок·,** (*f*) see **игл·а́**

иго́л·очк·а, -а, (*g pl*) **-чек·,** (*f*) (*dim*) of **иго́л·к·а, игл·а́**

иго́ль·ник·, -а, (*m*) needle-case

иго́ль·ниц·а, -ы, (*i*) **-ей,** (*f*) needle cylinder (circular knitting machine)

иго́ль·н·ый, -ая, -ое, (*a*) of **игл·а́**

иго́ль·чат·ый, -ая, -ое, (*a*) needle-shaped, acicular, spicular, (bot) acerose, acerous; (text) card, carding

~ **криста́лл·** (*m*) acicular crystal

~ **ле́нт·а** (*f*) (text) card clothing

~ **под·ши́п·ник·** (*m*) needle roller bearing

~ **треп·а́л·о** (*n*) (text) card beater

игр·а́, -ы́, (*nom pl*) **и́гры,** (*f*) play, game; acting, performance; (mech eng) play, give, backlash; bubbling, sparkling (of wine); sparkling (of diamonds etc.)

~, **матри́ч·н·ая** (math) matrix game

игр·а́·ть, -ют, (*imp*) play; act, perform; bubble, sparkle (of wine)

~ **рол·ь** play a part, be of importance, contribute to

игр·и́ст·ый, -ая, -ое, (*a*) sparkling (of wine etc.), frothy, foaming, fizzy

идеа́л·, -а, (*m*) ideal

идеал·исти́ческ·ий, -ая, -ое, (*a*) idealistic

идеа́ль·н·ый, -ая, -ое, (*a*) ideal, perfect

~ **газ·** (*m*) (phys) perfect/ideal gas

в идеа́ль·н·ом слу́ч·а·е ideally

иде́·йн·ый, -ая, -ое, (*a*) of **иде́·я**; ideological

ид·ём (*pres 2nd pl*) of **ид·ти́**

идем·потéнт·н·ый, -ая, -ое, (*a*) (math) idempotent

идем·фáктор·, -а, (*m*) (math) idem factor

иденти·фик·áтор·, -а, (*m*) identifier (computers)

иденти·фик·ациóн·н·ый, -ая, -ое, (*a*) identification

иденти·фик·áци·я, -и, (*f*) identification

иденти·фиц·и́р·овать, -уют, (*imp and perf*) identify

идент·и́чност·ь, -и, (*f*) identity
~ **мн·éни·й** identity of views

идент·и́чн·ый, -ая, -ое, (*a*) identical

идео·лог·и́ческ·ий, -ая, -ое, (*a*) ideological

идео·мотóр·н·ый, -ая, -ое, (*a*) (med) ideomotor

идео·френ·и́·я, -и, (*f*) (med) ideophrenia

ид·ёт (*pres 3rd sing*) of **ид·ти́**

идé·я, -и, (*f*) idea, concept

идио·блáст·, -а, (*m*) idioblast

идио·зóм·а, -ы, (*f*) (gen) idiozome

идиóм·а, -ы, (*f*) (ling) idiom

идиомат·и́ческ·ий, -ая, -ое, (*a*) idiomatic

идио·мóрф·н·ый, -ая, -ое, (*a*) (min) idiomorphic, euhedral

идио·плáзм·а, -ы, (*f*) (cyt) idioplasm

идио·синкраз·и́·я, -и, (*f*) (biol) idiosyncrasy

идио·стат·и́ческ·ий, -ая, -ое, (*a*) idiostatic

идио·ти́п·, -а, (*m*) (gen) idiotype

идиот·и́ческ·ий, -ая, -ое, (*a*) idiotic

идиот·и́·я, -и, (*f*) idiocy

идио·фан·и́зм·, -а, (*m*) idiophanism

идио·фáн·н·ый, -ая, -ое, (*a*) (cryst) idiophanous

идио·хромат·и́ческ·ий, -ая, -ое, (*a*) idiochromatic

ид·и́те (*plur imper*) of **ид·ти́**

идитóл·, -а, (*m*) (plast) Iditol

ид·óз·а, -ы, (*f*) (biochem) idose

идо·крáз·, -а, (*m*) (min) idocrase, vesuvianite

ид·ти́, -у́т, (*imp deter*) go, come; walk; take place, proceed, go on; (mech etc.) run, work, operate, move, travel; fall (of rain, snow)

ид·у́щ·ий, -ая, -ее, (*pres part act*) of **ид·ти́**

иезаконити́н·, -а, (*m*) (pharm) jesoconotine

иени́т·, -а, (*m*) (min) ilvaite, yenite

иéн·ск·ое стекл·ó (*n*) Jena glass

иерó·глиф·, -а, (*m*) hieroglyph

иеро·глиф·и́ческ·ий, -ая, -ое, (*a*) hieroglyphic

иждивéн·ец·, -н·ц·а, (*i*) **-н·ц·ем,** (*m*) dependant

из- (*as verbal prefix*) *signifies*: (1) movement from inside/within; (2) spread of action to whole object, in all directions; (3) action leading to extreme limit, completeness; (4) complete expenditure of object; (5) **-ся** acquisition/loss of quality/attribute as a result of repeated action

из (*prep + gen*) out of, of, from

изадри́н·, -а, (*m*) (chem) isadrine, isopronaline

из·алло·бáр·а, -ы, (*f*) (meteor) isallobar

из·алло·бар·и́ческ·ий, -ая, -ое, (*a*) isallobaric

из·алло·ги́пс·, -а, (*m*) (meteor) isallohypse

из·алло·тéрм·а, -ы, (*f*) (meteor) isallotherm

из·аномáл·а, -ы, (*f*) (meteor) isanomal, isanomal line

изат·и́н·, -а, (*m*) (chem) isatin

изафени́н·, -а, (*m*) (pharm) isaphenin

изб·á, -ы́, (*nom pl*) **и́зб·ы,** (*f*) peasant house

из·бáв·им·ый, -ая, -ое, (*a*) (*pres part pass*) *see* **из·бавл·я́·ть**; (math) removable

из·бáв·ить, -ят, (*perf*) *see* **из·бавл·я́·ть**

из·бавл·я́·ть, -ют, (*imp*) save from, rid, release; **-ся** (*pass*) get rid of

из·бег·áни·е, -я, (*n*) (*v n*) of **из·бег·á·ть**; avoidance, evasion
зóн·а из·бег·áни·я (astron) zone of avoidance

из·бег·á·ть, -ют, (*imp*)(*+ gen*) avoid, evade; **из·бéг·а·ть, -ют,** (*perf*) run about

из·бéг·нуть, -нут, (*past masc sing*) **из·бéг·** (*perf*); *see* **из·бег·á·ть**

из·беж·áни·е, -я, (*n*) avoidance; evasion, escape
~ **резонáнс·н·ого за·хвáт·а** (nucl) resonance escape

из·беж·áть (*fut 3rd sing, pl*) **из·-беж·и́т, из·бег·у́т,** (*perf*)(*+ gen*); *see* **из·бег·á·ть**

из·бер·у́т (*fut 3rd pl*) of **из·бр·áть**

из·би·éни·е, -я, (*n*) beat, beat up, assault, (law) assault and battery

из·бир·а́ни·е, -я, (*n*) choosing, electing, selecting; (autom) discrimination

~, групп·ов·о́е (autom) group discrimination

~, ко́д·ов·ое pulse-code discrimination

~, комбинацио́н·н·ое multi-channel discrimination

~, мно́го·ступ·е́нчат·ое group discrimination

~, одно·ступ·е́нчат·ое single-stage discrimination

из·бир·а́тел·ь, -я, (*m*) elector, voter; selector

из·бир·а́тельност·ь, -и, (*f*) selectivity, discrimination

из·бир·а́тельн·ый, -ая, -ое, (*a*) election, electoral; selective, discriminating, discrimination

~ с·вяз·ь (*f*) (telecom) selective calling telephone system

~ систе́м·а (*f*) electoral system; (telecom) selector/selection system

~ схе́м·а (*f*) (elec) selective circuit; tree, decoder (computers)

из·бир·а́·ть, -ют, (*imp*) choose, elect, select

из·бо́·ин·а, -ы, (*f*) (agr) oilcake

из·бо́р·ник·, -а, (*m*) (print) selected works

из·борожд·е́ни·е, -я, (*n*) groove, furrow; (geol) striation, streak

из·борозд·и́ть, -я́т, (*perf*) furrow, plough; make a groove/channel

из·бр·а́ни·е, -я, (*n*) (*v n*) *see* **из·бир·а́·ть**; selection; election

и́з·бр·анн·ый, -ая, -ое, (*past part pass*) of **из·бр·а́ть**; chosen, elected, selected

~ про·из·вед·е́ни·я (*pl*) selected works

из·бр·а́ть (*fut 3rd sing, pl*) **из·бер·ёт, из·бер·у́т,** (*perf*); *see* **из·бир·а́·ть**

из·бы́т·ок·, -т·к·а, (*m*) surplus, excess, margin; abundance, plenty

коэффицие́нт· из·бы́т·к·а во́з·дух·а (*m*) (ICE) air–fuel ratio

~ мо́щ·ност·и (ICE) margin of power

~ нейтро́н·ов (phys) neutron excess

~ фа́з·ы phase margin

из·бы́т·очн·ый, -ая, -ое, (*a*) surplus, excess, excessive; redundant

из·ва·я́ни·е, -я, (*n*) sculpture

из·ве́д·а·ть, -ют, (*perf*) *see* **из·ве́д·ыва·ть**

из·вед·ённ·ый, -ая, -ое, (*past part pass*) of **из·вес·ти́** (*perf*), *see* **из··вод·и́ть** (*imp*)

из·ве́д·ыва·ть, -ют, (*imp*) experience, come to know

из·вёл (*past masc sing*) of **из·вес·ти́**

из·верг·а́·ть, -ют, (*imp*) eject, throw out, disgorge, discharge; **-ся** (*pass*) erupt, be in eruption (of volcano)

из·ве́рг·нуть, -нут, (*past masc sing*) **из·ве́рг·** (*perf*); *see* **из·верг·а́·ть**

из·верж·е́ни·е, -я, (*n*) eruption

из·ве́рж·енн·ый, -ая, -ое, (*past part pass*) *see* **из·верг·а́·ть**; eruptive, volcanic; igneous

~ по·ро́д·а (*f*) (geol) igneous rock

из·вер·н·у́ться, -у́тся, (*perf*) dodge

из·вёрт·ыва·ться, -ются, (*imp*) dodge

известе·гас·и́лк·а, -и, (*g pl*) **-лок·,** (*f*) lime slaker

из·вес·ти́ (*fut 3rd sing, pl*) **из·вед·ёт, из·вед·у́т,** (*past masc sing*) **из·вёл,** (*perf*); *see* **из·вод·и́ть**

из·ве́ст·и·е, -я, (*n*) news, information; (*pl*) proceedings, transactions (of learned society)

из·вест·и́тел·ь, -я, (*m*) (rail) annunciator

~, по·жа́р·н·ый fire-alarm call box

из·вест·и́ть, -я́т, (*perf*) inform, notify

извёст·к·а, -и, (*g pl*) **-т·ок·,** (*f*) *see* **и́звест·ь**

извест·кова́ни·е, -я, (*n*) (agr) liming, spreading lime; (geol) liming, calcification

извест·ко́в·о-желе́з·н·ая руд·а́ (*f*) calcareous iron ore

~ -пес·ча́н·ый кирпи́ч· (*m*) silica brick

извест·ко́вост·ь, -и, (*f*) calcareousness, lime content

извест·ко́в·ый, -ая, -ое, (*a*) lime, calcareous, calciferous

~ молок·о́ (*n*) (chem) milk of lime

~ пятн·а (*pl*) lime blast (leather)

~ сели́тр·а (*f*) nitrate of lime, calcium nitrate

~ туф· (*m*) (min) calcareous tufa

~ флюс· (*m*) limestone flux

известко·лю́б·, -а, (*m*) (bot) gypsophyte, calciphyte, calcicole

известко·об·жиг·а́тельн·ая печ·ь (*f*) lime/limestone kiln

из·ве́ст·н·о (*predic*) of **из·ве́ст·н·ый**

~ из·ве́ст·н·о as is well known, it is common knowledge, as we know

из·ве́ст·ност·ь, -и, (*f*) knowledge (of facts); reputation, repute, fame

при·вес·ти́ в из·ве́ст·ност·ь make known/public

из·ве́ст·ност·ь
   ста́·в·ить в из·ве́ст·ност·ь inform, bring to the knowledge of
из·ве́ст·н·ый, -ая, -ое, (*a*) (*s f*) -т·ен·, -т·н·а, well-known, noted; notorious; certain, particular, limited
извест·ня́к·, -а́, (*m*) limestone
~, доломит·изи́рованн·ый dolomite, dolomite rock
~, кремн·и́ст·ый cherty limestone
~, об·ло́м·очн·ый (min) calcarenite
извест·няко́в·ый, -ая, -ое, (*a*) calcareous
и́звест·ь, -и, (*f*) lime
~, без·во́д·н·ая серно·ки́сл·ая (min) anhydrite
~, бел·и́льн·ая chloride of lime, bleaching lime
~, ве́н·ск·ая french chalk
~, воз·ду́ш·н·ая air-hardening lime
~, гаш·ён·ая slaked lime
~, гидравл·и́ческ·ая hydraulic lime
~, е́д·к·ая caustic lime
~, жж·ён·ая quicklime, burnt lime
~, ком·ов·а́я quicklime, caustic lime
~, натро́н·н·ая soda lime
~, не·гаш·ён·ая quicklime, caustic lime
~, серно·ки́сл·ая (min) gypsum, calcium sulphate
~, хло́р·н·ая chloride of lime
из·вещ·а́·ть, -ют, (*imp*) inform, notify
из·вещ·е́ни·е, -я, (*n*) (*v n*) see из·вещ·а́·ть; notice, notification
~ море·плав·а́тел·ям (naut) Notices to Mariners
из·ви́·в·, -а, (*m*) see из·ви́л·ин·а
из·вив·а́·ть, -ют, (*imp*) coil, wind, curve
из·ви́л·ин·а, -ы, (*f*) bend, crook, convolution, (geog) meander
из·ви́л·ист·ый, -ая, -ое, (*a*) winding, meandering, sinuous
из·вин·и́ть, -я́т, (*perf*) see из·вин·я́·ть
из·вин·я́·ть, -ют, (*imp*) excuse; -ся apologize
из·ви́·ть (*fut 3rd sing, pl*) изо·вь·ёт, изо·вь·ю́т, (*perf*); see из·вив·а́·ть
из·влек·а́·ть (*imp*) extract, derive, elicit, recover, remove; withdraw, draw
~ ко́рен·ь (math) extract/take/find a root
из·влеч·е́ни·е, -я, (*n*) extraction, recovery; withdrawal, removal; abstract, excerpt; (math) evolution

из·влеч·е́ни·е
~ кисл·от·о́й (chem) aid recovery
~ ко́рн·я (math) evolution, extraction of a root
~ кре́п·и (min) drawing timbers/props, timber/prop extraction/removal
~, про·мы́шл·енн·се (min, met) commercial recovery
~ рас·твор·и́тел·ем (chem) solvent extraction
из·влеч·ённ·ый, -ая, -ое, (*past part pass*) see из·влек·а́·ть
~ ци́фр·а (*f*) extracted digit (computers)
из·вле́ч·ь (*fut 3rd sing, pl*) из·влеч·ёт, из·влек·у́т, (*past masc sing*) из·влёк·, (*perf*); see из·влек·а́·ть
из·вне́ (*adv*) from outside
из·вод·и́ть, -я́т, (*imp*) use up, expend, exhaust
из·вора́ч·ива·ться, -ются, (*imp*) dodge
из·врат·и́ть, -я́т, (*perf*) see из·враща́·ть
из·вращ·а́·ть, -ют, (*imp*) distort, misinterpret; pervert, spoil
из·вращ·е́ни·е, -я, (*n*) distortion, misinterpretation
из·га́д·ить, -ят, (*perf*) see из·га́ж·и·ва·ть
из·га́ж·ива·ть, -ют, (*imp*) foul, spoil
и́з·гар·ь, -и, (*f*) (met) hammer/forge scale
из·ги́б·, -а, (*m*) bend, bending, flexure, flexion; (math) knee, bend, bend point (on a curve); elbow (of a pipe)
~, знако·пере·ме́н·н·ый reverse bending, flexing
~, ис·пыт·а́ни·е на (mech) transverse bend test
~, много·кра́т·н·ый flexing
момѐнт из·ги́б·а (*m*) bending moment
~, по·ло́г·ий slight/shallow bend; (geol) broad warp
~, про·до́ль·н·ый buckling
~ с·кла́д·к·и (geol) knee of fold
центр· из·ги́б·а (*m*) (mech) flexural/shear centre/center
из·гиб·а́·ть, -ют, (*imp*) bend
из·гиб·а́ющ·ий момѐнт· (*m*) bending moment
изгибо·у·сто́й·чивост·ь, -и, (*f*) flexural stability, flexing resistance
из·гла́д·ить, -ят, (*perf*) see из·гла́·ж·ива·ть

из·гла́ж·ивани·е, -я,   *(n) (v n)* of **из·гла́ж·ива·ть**

из·гла́ж·ива·ть, -ют,   *(imp)* erase, efface, rub/smooth/iron out; **-ся** *(pass)*; fade/die out

из·гн·а́ть *(fut 3rd sing, pl)* **из·го́н·ит, из·го́н·ят,** *(perf)*; *see* **из·гон·я́·ть**

из·гон·я́·ть, -ют,   *(imp)* expel, drive out, exile

и́з·город·ь, -и,   *(f)* fence

~, жив·а́я   green fence, hedge

из·гото́вл·ива·ть, -ют,   *(imp)* make, manufacture, prepare, make/get ready

из·готов·и́тел·ь, -я,   *(m)* manufacturer, producer

из·гото́в·ить, -ят, *(perf)* make, manufacture, prepare, make/get ready; **-ся** *(pass)*; get ready

из·готовл·е́ни·е, -я,   *(n)* manufacture, making, preparation, production

из·гото́вл·енн·ый, -ая, -ое,   *(past part pass)* of **из·гото́в·ить**

из·готовл·я́·ть, -ют,   *(imp)* make, manufacture, produce

из·грыз·а́·ть, -ют,   *(imp)* gnaw/nibble to pieces

из·гры́з·ть, -у́т, *(past masc sing)* **из··гры́з·** *(perf)*; *see* **из·грыз·а́·ть**

из·да·ва́·ть, -ют,   *(imp)* publish; promulgate, issue; make, emit (sounds), make, give off (smells)

и́з·давн·а   *(adv)* long since/ago, a long time ago

из·дал·ек·а́   *(adv)* from far away

и́з·дал·и   *(adv)* from far away

из·да́н·и·е, -я, *(n)* publication, promulgation; edition

~, вре́мен·н·ое   interim/advanced edition

~, на́·бр·анн·ое   (print) edition in boards

~ не·под·леж·а́щ·ее о·глаш·е́ни·ю   classified/restricted/secret publication

из·да́·тел·ь, -я, *(m)* publisher

из·да́·тельств·о, -а,   *(n)* publishing house, publisher's

из·да́·ть *(fut tense)* **из·да́м, из·да́шь, из·да́ст, из·дади́м, из·дади́те, из··даду́т,** *(perf)*; *see* **из·да·ва́·ть**

из·дев·а́·тельств·о, -а,   *(n)* mockery

из·дев·а́·ться, -ются,   *(imp)* jeer at, taunt, mock, ridicule

из·де́л·и·е, -я,   *(n)* product, manufactured product/article, *(pl)* goods, wares, manufactures; make, manufacture

~, с·ва́р·н·ое   weldment

из·дер·ёт *(fut 3rd sing)* of **изо·др·а́ть** *(perf)*, *see* **из·дир·а́·ть** *(imp)*

из·держ·а́ть, -ат, *(perf)* *see* **из·де́р·ж·ива·ть**

из·де́рж·ива·ть, -ют,   *(imp)* expend, spend

из·де́рж·к·и *(nom pl)*, *(g pl)* **-ж·ек·,** *(f)* expenses, costs

из·дер·у́т *(fut 3rd pl)* of **изо·др·а́ть** *(perf)*, *see* **из·дир·а́·ть** *(imp)*

из·дир·а́·ть, -ют,   *(imp)* tear, rend

из·до́х·нуть, -нут,   *(past masc sing)* **из·до́х·** *(perf)*; *see* **из·дых·а́·ть**

из·дых·а́·ть, -ют,   *(imp)* die (of animals)

из·ентро́п·а, -ы,   *(f)* (meteor) isentrope

из·ентроп·и́ческ·ий, -ая, -ое, isentropic

из·ентро́п·н·ый, -ая, -ое,   *(a)* isentropic

из·жа́р·ить, -ят, *(perf)* fry; grill; roast

из·жев·а́ть *(fut 3rd sing, pl)* **из·жуёт, из·жую́т,** *(perf)*; *see* **из·жёв·ыва·ть**

из·жёв·ыва·ть, -ют,   *(imp)* chew to pieces

и́з·желт·а-бу́р·ый, -ая, -ое,   *(a)* brown with a yellowish tint, brown tinged with yellow

~ -кра́с·н·ый, -ая, -ое,   *(a)* red with a yellowish tint, red tinged with yellow

из·жив·а́·ть, -ют,   *(imp)* get rid of, eradicate, overcome (faults etc.); live out one's life; **-ся** exhaust oneself, use up one's resources, give one's whole life to

из·жи́·ть *(fut 3rd sing, pl)* **из·жив·ёт, из·жив·у́т,** *(perf)*; *see* **из·жив·а́·ть**

из·жо́г·а, -и, *(f)* (med) pyrosis, heartburn

из·жую́т *(fut 3rd pl)* of **из·жев·а́ть**

из-за   *(prep+gen)* from behind/beyond; because of, on account of, through

изи́ди·й, -я,   *(m)* (bot) isidium

из·ла́вл·ива·ть, -ют,   *(imp)* catch

из·лаг·а́тельн·ый, -ая, -ое,   *(a)* explanatory

из·лаг·а́·ть, -ют,   *(imp)* expound, lay/set out, state, present (results, data etc.)

из·ла́м·ыва·ть, -ют,   *(imp)* break to pieces, break up

из·лёт·, -а,   *(m)* final moment of flight

пу́л·я на из·лёт·е *(f)* spent bullet

из·леч·е́ни·е, -я, (*n*) (med) recovery, cure, treatment

из·ле́ч·ива·ть, -ют, (*imp*) (med) cure

из·леч·и́м·ый, -ая, -ое, (*pres part pass*) of из·леч·и́ть; curable

из·леч·и́ть, -͘ат, (*perf*) cure

из·лив·а́·ться, -ются, (*imp*) pour/flow out, effuse, gush/well out

из·ли·вш·ийся, -аяся, -ееся, (*a*) (*past part act*) *see* из·лив·а́·ться; issuing, effusive; (geol) extrusive

из·ли́·ться (*fut 3rd sing, pl*) изо·ль·ёт·ся, изо·ль·ю́тся, (*perf*); *see* из·лив·а́·ться

из·ли́ш·ек·, -ш·к·а, (*m*) surplus, excess

~ ве́с·а overweight

из·ли́ш·н·ий, -яя, -ее, (*a*) excessive, superfluous, unnecessary

из·ли·я́ни·е, -я, (*n*) outpouring, outflow, effusion, discharge

из·лов·и́ть, -͘ят, (*perf*) catch

из·лов·ча́·ться, -ются, (*imp*) contrive, manage

из·лов·ч·и́ться, -а́тся, (*perf*) *see* из·лов·ча́·ться

из·лож·е́ни·е, -я, (*n*) exposition, presentation, essay, account, statement

из·ло́ж·енн·ый, -ая, -ое, (*past part pass*) *see* из·лаг·а́·ть

из·лож·и́ть, -ат, (*perf*) *see* из·лаг·а́·ть

из·ло́ж·ниц·а, -ы, (*i*) -ей, (*f*) (met) ingot mould/mold

~, глухо·до́н·н·ая solid/closed-bottom mould/mold

~, лист·ов·а́я slabbing ingot mould/mold

~, сквоз·н·а́я split/two-piece mould/mold

~, у·ши́р·енн·ая к·ве́рх·у big-end-up mould/mold

из·ло́ж·н·я, -и, (*f*) *see* из·ло́ж·ниц·а

из·ло́м·, -а, (*m*) fracture, break; (math) salient point (of a curve)

~, внутри·кристалл·и́ческ·ий transcrystalline fracture

~, волокн·и́ст·ый fibrous/woody fracture

~, вя́з·к·ий tough fracture

~, камне·ви́д·н·ый conchoidal fracture

~, крупно·зерн·и́ст·ый granular fracture

~, меж·кристалл·и́ческ·ий inter-crystalline fracture

~, нафтали́н·н·ый silky fracture

~, пласт·и́нчат·ый lamellar/foliated fracture

из·ло́м

~, ра́к·овист·ый conchoidal fracture

~, стру́й·чат·ый wavy fracture

~, хру́п·к·ий brittle fracture; (plast) brittle break

~, ши́фер·н·ый flaky fracture

из·ло́м·анн·ый, -ая, -ое, (*past part pass*) of из·лом·а́·ть; broken, fractured

из·лом·а́·ть, -ют, (*perf*) break up, break to pieces

из·лом·и́ть, -͘ят, (*perf*) break, fracture

из·луч·а́ем·ый, -ая, -ое, (*pres part pass*) of из·луч·а́·ть; radiated, emitted

из·луч·а́тел·ь, -я, (*m*) emitter, radiator; oscillator (hydrophone), transmitter (echo-sounder); (rad) radiating element, waveguide, mouth (aerial)

~, звук·ов·о́й acoustic radiator

~, ма́сс·ов·ый (rad) mass radiator

~, по́лн·ый (phys) black body

~, ру́пор·н·ый (rad) horn radiator

~, то́ч·ечн·ый (nucl) point-source radiator

из·луч·а́тельн·ый, -ая, -ое, (*a*) radiating, radiative, radiant

~ спосо́б·ност·ь (*f*) emissivity, emittance

из·луч·а́·ть, -ют, (*imp*) radiate, emit

из·луч·е́ни·е, -я, (*n*) radiation, emission, emanation

~, жёстк·ое hard/high-energy radiation

~, косм·и́ческ·ое cosmic radiation

коэффицие́нт· из·луч·е́ни·я (*m*) radiation factor

~, мульти·по́ль·н·ое multipole radiation

~, мя́гк·ое soft/low-energy radiation

~, обра́т·н·ое reradiation

~, от·раж·ённ·ое back-scattered radiation

~, по·бо́ч·н·ое spurious radiation

~, рас·се́·янн·ое stray/scattered radiation

~, свет·ов·о́е luminous radiation

~, селекти́в·н·ое selective emission

~, слабо·про·ник·а́ющ·ее (nucl) short-range radiation

~, тормоз·н·о́е (nucl) bremsstrahlung, braking radiation

~, характер·исти́ческ·ое characteristic radiation

из·луч·ённ·ый, -ая, -ое, (*past part pass*) of из·луч·и́ть; emitted, radiated

из·лу́ч·ин·а, -ы, (*f*) bend, curve, meander

из·лу́ч·ист·ый, -ая, -ое, (*a*) winding, full of curves, tortuous

из·луч·и́ть, -ат, (*perf*) *see* **из·луч·а́·ть**

из·ма́ж·ут (*fut 3rd pl*) of **из·ма́з·ать**

из·ма́з·ать (*fut 3rd sing, pl*) **из·ма́ж·ет, из·ма́ж·ут,** (*perf*); *see* **из·ма́з·ы·ва·ть**

из·ма́з·ыва·ть, -ют, (*imp*) smear, dirty; use up (paint, grease etc.)

из·ма́л·ыва·ть, -ют, (*imp*) grind/crush up

из·ма́т·ыва·ть, -ют, (*imp*) exhaust, wear out

из·ме́л·ет (*fut 3rd sing*) of **из·мол·о́ть**

из·мельч·а́ни·е, -я, (*n*) (*v n*) of **из··мельч·а́·ть**; nanism, dwarfing

из·мельч·а́·ть, -ют, (*imp*) grind, crumble, break up, reduce to fragments, mince, disintegrate; (*perf*) become small/smaller; become shallow/shallower

из·мельч·е́ни·е, -я, (*n*) (*v n*) *see* **из··мельч·а́·ть** (*imp*); reduction, comminution, fragmentation; refinement (of a crystalline structure, grain etc.) **коэффицие́нт· из·мельч·е́ни·я** (*m*) reduction ratio

из·мельч·ённ·ый, -ая, -ое, (*past part pass*) *see* **из·мельч·а́·ть**; fragmented, finely divided, reduced

из·мельч·и́тел·ь, -я, (*m*) grinder, pulverizer, shredder, mincer

из·мельч·и́ть, -а́т, (*perf*) grind, crumble, break up, reduce to fragments, mince, disintegrate

из·ме́л·ют (*fut 3rd pl*) of **из·мол·о́ть**

из·ме́н·а, -ы, (*f*) treachery, betrayal

∼, госуда́р·ственн·ая (law) treason

из·мен·е́ни·е, -я, (*n*) alteration, modification, variation, change

∼ а́дрес·а (autom) address modification

∼, век·ов·о́е (astron) secular variation, (geol) secular change

∼ во вре́мен·и time variation

∼ знак·а (phys) reversal, (math) sign modification

∼ масшта́б·а rescaling

∼, морфо·лог·и́ческ·ое morphological variation

∼ ориент·а́ци·я спи́н·а (nucl) spin flip

∼ пад·е́ни·я (geol) reversal of dip

∼, пере·но́с·н·ое (math) convective change

∼, пере·хо́д·н·ые (*pl*) (phys) transients

∼, ре́зк·ое abrupt change, jump

∼ фа́з·ы на 180° (phys) phase reversal

из·мен·е́ни·е

∼ чёт·ност·и (math) parity transition

∼ чувств·и́тельност·и (instr) sensitivity shift/drift

из·мен·ённ·ый, -ая, -ое, (*past part pass*) *see* **из·мен·я́·ть**

из·мен·и́тел·ь знак·а (*m*) inverter (computers)

из·мен·и́ть, -ят, (*perf*) *see* **из·мен·я́·ть**

из·ме́н·чивост·ь, -и, (*f*) variability, variation, mutability, mutation **коэффицие́нт· из·ме́н·чивост·и** (*m*) (math) coefficient of variation

∼, пре·ры́в·ист·ая discontinuous variation

∼, скачко·обра́з·н·ая (gen) saltatory variation, mutation, saltation

из·ме́н·чив·ый, -ая, -ое, (*a*) variable, changeable, mutable

из·мен·я́емост·ь, -и, (*f*) variability, changeability

из·мен·я́ем·ый, -ая, -ое, (*pres part pass*) of **из·мен·я́·ть**; variable, changeable, mutable; subject to change

из·мен·я́·ть, -ют, (*imp*) vary, alter, modify, change; betray; **-ся** (*pass*) vary, fluctuate

из·мен·я́ющ·ийся, -аяся, -ееся, (*pres part act*) of **из·мен·я́·ться**; varying, variable

из·мер·е́ни·е, -я, measuring, measurement, mensuration, metering, gauging, gaging; surveying; (math) dimension; (*as last component*) -metry

∼, бел·ко́в·ое (chem) albuminometry

∼ выс·о́т· hypsometry

∼ глуб·ин·ы́ depth-taking, sounding (in water)

∼ да́ль·ност·и ranging

∼ му́т·ност·и turbidimetry

∼ от·нос·и́тельн·ым ме́тод·ом comparison-method of measurement

∼, про·из·вод·и́ть make/take measurements

∼ рас·хо́д·а вод·ы́ stream gauging

∼, сква́ж·инн·ое (oil) well shooting

∼ угл·о́в (surv) angular surveying

∼ у́ровн·я level gauging

∼ щёлоч·ност·и alkalimetry

из·ме́р·енн·ый, -ая, -ое, (*past part pass*) *see* **из·мер·я́·ть**

∼ а́зимут· (*m*) (nav) observed azimuth

из·мер·и́м·ый, -ая, -ое, (*a*) measurable

из·мер·и́тел·ь, -я, (*m*) (*and see under associated words*) meter, gauge; (math) index, code number, characteristic

из·мер·и́тел·ь
~ **брож·éни·я** (chem) zymometer
~ **ви́д·имост·и** visibility meter
~ **вращ·áющ·его момéнт·а** torque meter
~ **врéмен·и** timer
~ **ды́м·а** smoke gauge/gage
~ **объ·ём·а** volume meter
~ **о·свещ·ённост·и** (phys) lux meter
~ **по·то́к·а** flux meter
~ **при·ли́·в·а** tide-gauge
из·мер·и́тельн·ый, -ая, -ое, (a) measuring; standard, reference/measuring standard
~ **ём·кост·ь** (f) (elec) standard capacitance
~ **индукти́в·ност·ь** (f) (elec) standard inductance
~ **кольц·о́** (n) (rad) range marker circle
~ **при·бо́р·** (m) measuring instrument
~ **со·противл·éни·е** (n) (elec) standard resistance
~ **трансформ·áтор·** (m) (elec) instrument transformer
~ **труб·á** (f) sounding pipe
из·мéр·ить, -ят, (perf) see **из·мер·я́·ть**
из·мер·я́·ть, -ют, (imp) measure, gauge
из·мер·я́ем·ый, -ая, -ое, (pres part pass) of **из·мер·я́·ть;** measurable
из·мождд·éни·е, -я, (n) exhaustion, enfeeblement
из·мок·á·ть, -ют, (imp) become soaked/drenched
из·мо́к·нуть, -нут, (past masc sing) **из·мо́к·** (perf); see **из·мок·á·ть**
из·мол·о́ть (fut 3rd sing, pl) **из·мéл·ет, из·мéл·ют,** (perf); see **из·мáл·ы·ва·ть**
и́з·мороз·ь, -и, (f) (meteor) rime, hoar frost
и́з·морос·ь, -и, (f) (meteor) drizzle
из·мот·á·ть, -ют, (perf) see **из·мáт·ы·ва·ть**
из·му́ч·ить, -ат, (perf) exhaust, wear out
из·мы́л·ива·ть, -ют, (imp) use up (soap)
из·мы́л·ить, -ят, (perf) see **из·мы́л·и·ва·ть**
из·мышл·éни·е, -я, (n) figment, fabrication (of imagination)
из·мя́·ть (fut 3rd sing, pl) **изо·мн·ёт, изо·мн·у́т,** (perf); rumple, crumple
изна́нк·а, -и, (f) underside, reverse, wrong side, back
из·на·си́л·овать, -уют, (perf) rape

из·на·ча́ль·н·ый, -ая, -ое, (a) primordial
из·на́ш·иваемост·ь, -и, (f) wearability, wear resistance
из·на́ш·ивани·е, -я, (n) (v n) of **из·на́ш·ива·ть;** wear, wearing, wearing out, deterioration, depreciation
    ис·пыт·áни·е на из·на́ш·ивани·е (n) wear testing, abrasion test
из·на́ш·ива·ть, -ют, (imp) wear, wear out, erode
из·не·мож·éни·е, -я, (n) (med) exhaustion, breakdown
из·но́с·, -а, (m) wear, abrasion, wear-and-tear
~, **абрази́в·н·ый** abrasive wear
~, **авар·и́йн·ый** very quick wear (of machinery)
~, **эксплуатаци́он·н·ый** service wear
из·нос·и́ть, ⸗ят, (perf) see **из·на́ш·ива·ть**
из·носо·сто́й·к·ий, -ая, -ое, (a) wear-resistant, durable (of a material); sturdy (of a machine)
из·носо·сто́й·кост·ь, -и, (f) wear resistance
из·носо·у·сто́й·чив·ый, -ая, -ое, (a) wear-resistant
из·но́ш·енност·ь, -и, (f) wear, state/degree of wear
из·но́ш·енн·ый, -ая, -ое, (past part pass) see **из·на́ш·ива·ть;** worn, worn out
из·нур·éние, -а, (n) exhaustion, emaciation
из·нутр·и́ (adv) from inside/within
изо (prep + gen) **из** out of, of, from; used instead of **из** (1) before words beginning with two or more consonants, the first of which is **р** or **л**; (2) before the words **весь, всякий;** (3) in certain expressions; e.g. **изо дня в день** from day to day
изо- (prefix) **из-,** used instead of **из-;** (1) before **й;** (2) before two or more consonants; (3) before consonant and **ь** (soft sign); iso- (Greek equal, same)
изо·англютини́н·, -а, (m) (biochem) isoagglutinin
изо·алка́н·, -а, (m) (chem) isoalkane
изо·алке́н·, -а, (m) (chem) isoalkene
изо·ами́л·, -а, (m) (chem) isoamyl
изо·ами́л·ов·ый, -ая, -ое, isoamyl
изо·áнт·, -а, (m) vegetation limit line (on a map)
изо·ба́з·а, -ы, (f) (geol) isobase

изо·бáлл·, -а, (*m*) (ocean) line of equal ice density

изо·бáр·, -а, (*m*) (nucl) isobar, (*pl*) isobars, isobares

~, зеркáль·н·ый (nucl) mirror/Wigner isobar

изо·бáр·а, -ы, (*f*) (meteor, chem) isobar

изо·бар·íческ·ий, -ая, -ое, (*a*) isobaric, isobarometric; constant-pressure

изо·бáр·н·ый, -ая, -ое, (*a*) (nucl) isobar, isobaric

изобаро·раз·лúв·очн·ая машúн·а (*f*) champagne bottling machine

изо·бáт·а, -ы, (*f*) (ocean) isobath, depth/sounding line, bathymetric curve

изо·бати·тéрм·а, -ы, (*f*) isobathytherm

изо·би·латер·áльн·ый, -ая, -ое, (*a*) (bot) isobilateral

из·обúль·н·ый, -ая, -ое, (*a*) abundant, plentiful

из·об·лич·á·ть, -ют, (*imp*) expose, unmask, show up; (*imp only*) reveal, betray

из·об·лич·úть, -áт, (*perf*) *see* **из·об·ли·ч·á·ть**

изо·борнеóл·, -а, (*m*) (chem) isoborneol

из·об·раж·á·ть, -ют, (*imp*) depict, portray, represent; **-ся** (*pass*); show, appear

из·об·раж·éни·е, -я, (*n*) image, picture; graph, graphing, graphical representation; presenting, presentation, representation, description (math); map

~ в· натýр·е life-size/same-size ratio (cartography etc.)

~, граф·úческ·ое graphical representation, graph

~, диффракциóн·н·ое (phys) diffraction image

~, кóнтур·н·ое (surv) planimetric plotting

~, масштáб·н·ое scale representation **метóд· из·об·раж·éни·я** (*m*) (phys) method of images

~, мн·úм·ое (opt) false image

~, о·стáт·очн·ое (phys) afterimage

~, пере·вёр·нут·ое (opt) inverted image

~, полу·тóн·ов·ое (print) half-tone picture **сигнáл· из·об·раж·éни·я** (*m*) picture signal (facsimile, TV)

~, с·кры́·т·ое (opt) latent image

из·об·раж·éни·е

~, тен·ев·óе roentgenogram, skiagraph, skiagram

~ фýнкци·и по Лаплáс·у (*f*) (math) Laplace transform

из·об·раз·úтельн·ый, -ая, -ое, (*a*) representational, figurative, graphic

~ искýсств·а (*pl*) fine arts

из·об·раз·úть, -я́т, (*perf*) *see* **из·об·раж·á·ть**

из·обрес·тú (*fut 3rd sing, pl*) **из·об·рет·ёт, из·обрет·ýт,** (*past masc sing*) **из·обрёл,** (*perf*); *see* **из·обре·т·á·ть**

из·обрет·áтел·ь, -я, (*m*) inventor

из·обрет·áтельн·ый, -ая, -ое, (*a*) inventive, ingenious, resourceful

из·обрет·áтельск·ий, -ая, -ое, (*a*) inventor's

из·обрет·áтельств·о, -а, (*n*) invention, inventiveness, inventions

из·обрет·á·ть, -ют, (*imp*) invent, devise

из·обрет·éни·е, -я, (*n*) invention

изо·брóнт·а, -ы, (*f*) (meteor) isobront, isoceraunic line

изо·бутáн·, -а, (*m*) (chem) isobutane

изо·бутéн·, -а, (*m*) (chem) isobutylene

изо·бутилéн·, -а, (*m*) (chem) isobutylene

изо·бутилéн·ов·ый, -ая, -ое, (*a*) isobutylene

изо·бутúл·ов·ый, -ая, -ое, (*a*) isobutyl

изо·валерья́н·ов·ая кисл·от·á (*f*) (chem) isovaleric acid, isovalerianic acid

изо·вéл·а, -ы, (*f*) (meteor) isovel, isotach

изо·вóл·а, -ы, (*f*) *see* **изо·вóл·ь**

изо·вóл·ь, -и, (*f*) (geol) isovol, isoanthracite lime

изо·вь·ю́т (*fut 3rd pl*) of **из·вúть**

изо·галúн·а, -ы, (*f*) (ocean) isohaline

изо·гамéт·а, -ы, (*f*) (biol) isogamete

изо·гáм·и·я, -и, (*f*) (gen) isogamy

изо·гáмм·а, -ы, (*f*) (geophys) isogam

изо·гáм·н·ый, -ая, -ое, (*a*) isogamous

изо·гексúл·, -а, (*m*) (chem) isohexyl

изо·гéл·а, -ы, (*f*) (meteor) isohel

изо·гéн·н·ый, -ая, -ое, (*a*) (bot) isogenic

изо·гео·тéрм·а, -ы, (*f*) (meteor) isogeotherm

изо·гидр·úческ·ий, -ая, -ое, (*a*) (chem) isohydric

изо·гиет·а, -ы,　(*f*) (meteor) isohyet, isohyetal line

изо·гипс·а, -ы,　(*f*) isohypse, height contour (on maps)

изо·гир·а, -ы,　(*f*) (opt) isogyre

изо·глосс·а, -ы,　(*f*) (ling) isogloss

изо́·гн·утост·ь, -и,　(*f*) curvature, camber, flexion; curviness, sinuosity

изо́·гн·ут·ый, -ая, -ое,　(*a*) (*past part pass*) *see* из·гиб·а́·ть; curved, bent; sinuous, winding; joggled (of metal plates)

изо·гн·у́ть, -у́т,　(*perf*) *see* из·ги-б·а́·ть

изо·го́н·а, -ы,　(*f*) (geophys) isogon, isogonic line

изо·гон·а́л·ь, -и,　(*f*) (math) isogonal, isogonal line

изо·гон·а́льн·ый, -ая, -ое,　(*a*) (math) isogonal

изо·гра́д·а, -ы,　(*f*) isograd

изо·гра́д·н·ый, -ая, -ое,　(*a*) (min) isograde, isofacial

изо·диа·фе́р·, -а,　(*m*) (nucl) isodiaphere

изо·ди·мо́рф·н·ый, -ая, -ое,　(*a*) (chem) isodimorphous

изо·ди́н·а, -ы,　(*f*) *see* изо·дина́м·а

изо·дина́м·а, -ы,　(*f*) (geophys) isodynamic line, isogam

изо·до́з·а, -ы,　(*f*) isodose

изо·дозо́·граф·, -а,　(*m*) isodose chart, isodose-rate chart

изо·др·а́ть　(*fut 3rd sing, pl*) из·де-р·ёт, из·дер·у́т, (*perf*); tear, rend

изо·дро́м·, -а,　(*m*) (autom) floating component

вре́м·я　изо·дро́м·а　(*n*) (autom) floating time

механи́зм·　изо·дро́м·а　(*m*) (autom) floating control mechanism

изо·дро́м·н·ый, -ая, -ое,　(*a*) of изо·дро́м·

~　механ·и́зм·　(*m*) (autom) floating control mechanism; (hydr) restoring mechanism

~　регул·и́ровани·е　(*n*) (autom) proportional speed floating control, proportional control

~　со·ставл·я́ющ·ая　(*f*) (autom) floating component

изо·йти́　(*perf*) become exhausted from, become exhausted from loss of, (*fut 3rd sing, pl*) изо·йд·ёт, изо·йд·у́т, (*past masc sing*) изо·шёл

изо·каучу́к·, -а,　(*m*) isorubber

изо·кли́н·а, -ы,　(*f*) (geophys) isoclinic line

изо·клин·а́л·ь, -и,　(*f*) (geol) isocline, isoclinal fold

изо·кор·и́чн·ая　кисл·от·а́　(*f*) isocinnaminic acid

изо·ла́к·, -а,　(*m*) (elec) insulating varnish

изо·лейци́н·, -а,　(*m*) (biochem) isoleucine

изо·лецит·а́льн·ый, -ая, -ое,　(*a*) (zool) isolecithal

изо·ли́н·и·я, -и,　(*f*) isometric line, isoline; (surv) contour line; (chem) iso-curve

~　по·то́к·а　(nucl) isoflux

изол·и́ровани·е, -я,　(*n*) *see* изол·я́-ци·я

изол·и́рованн·ый, -ая, -ое,　(*a*) (*past part pass*) of изол·и́р·овать; insulated; isolated

изол·и́р·овать, -уют,　(*imp and perf*) isolate; insulate

изол·иро́вочн·ый, -ая, -ое,　(*a*) insulating

изол·и́рующ·ий, -ая, -ее,　(*pres part act*) of изол·и́р·овать; isolating; insulating; self-contained

~　материа́л·　(*m*) insulating material

~　само·с·пас·а́тел·ь　(*m*) (min) self-contained breathing apparatus

изо·ло́г·, -а,　(*m*) (chem) isologue, isolog

изо·лог·и́ческ·ий, -ая, -ое,　(*a*) (chem) isologous

изо·ль·ю́тся　(*fut 3rd pl*) of из·ли́·ться

изо·лю́кс·, -а,　(*m*) (phys) isolux, isolux curve/line

изол·я́тор·, -а,　(*m*) (elec) insulator; (med) isolation ward

~　бра́к·а　rejects bin

~, в·во́д·н·ый　lead-in insulator

~, двух·ю́б·очн·ый　double-shed insulator, double-cone insulator

~, о·по́р·н·ый　support-type insulator, insulating support, stand-off insulator

~, оре́ш·ков·ый　shell-type insulator

~, под·вес·н·о́й　suspension insulator

~, под·кла́д·очн·ый　accumulator insulator; stand insulator

~, про·ход·н·о́й　bushing, lead-in insulator

~　тре́т·ь·его　ре́льс·а　(rail) conductor-rail insulator

~, штыр·ев·о́й　pin-type insulator

изол·яцио́нн·о-про·пуск·н·о́й пункт·　(*m*) (med) anti-epidemic check point

изол·яцио́нн·ый, -ая, -ое, (*a*) insulating, insulation; isolating, isolation

изол·я́ци·я, -и,   (*f*) insulation; isolation

∼, гидро- water-proofing

∼, звук·ов·а́я sound proofing/insulation

∼, про·стра́н·ственн·ая (genet) distance isolation

∼, тепл·ов·а́я heat insulation

изо·ма́сл·ян·ый, -ая, -ое, (*a*) (chem) isobutyric

∼ альдеги́д· (*m*) isobutyraldehyde

∼ кисл·от·а́ (*f*) (chem) isobutyric acid

изо·ме́р·, -а, (*m*) (chem) isomer

∼ -тра́нс·, -а, (*m*) (chem) trans-isomer

∼ -ци́с·, -а, (*m*) cis-isomer

изо·мер·иза́т·, -а, (*m*) (oil) isomerizate

изо·мер·иза́ци·я, -и, (*f*) (chem) isomerization

изо·мер·изо́ванн·ый, -ая, -ое, (*past part pass*); (chem) isomerized

изо·мер·и́·я, -и, (*f*) (chem) isomerism, isomery

∼, динам·и́ческ·ая (chem) dynamic isomerism

∼ по·лож·е́ни·я position isomerism

∼, про·стра́н·ственн·ая space isomerism

∼ радика́л·ов metamerism

∼, син-анти syn-anti isomerism

∼, скеле́т·н·ая geometrical isomerism

изо·ме́р·н·ый, -ая, -ое, (*a*) (chem) isomeric, (bot) isomerous

изо·метр·и́ческ·ий, -ая, -ое, (*a*) isometric

изо·ме́тр·и·я, -и, (*f*) isometry

изо·мн·у́т (*fut 3rd pl*) of **из·мя́·ть**

изо·морф·и́зм·, -а, (*m*) isomorphism

изо·мо́рф·н·ый, -ая, -ое, (*a*) isomorphous, isomorphic

∼ с·мес·ь (*f*) (chem) isomorphous mixture, mixed crystal

изо·моч·еви́н·а, -ы, (*f*) (chem) isourea

изо·не́ф·а, -ы, (*f*) (meteor) isoneph

изо·нитри́л·, -а, (*m*) (chem) isonitrile, isocyanide

изо·нитро·со·един·е́ни·е, -я, (*n*) (chem) isonitrocompound

изо·окта́н·, -а, (*m*) (chem) isooctane

изо·осмот·и́ческ·ий, -ая, -ое, (*a*) (chem) isotonic

изо·па́г·а, -ы, (*f*) (geog) equiglacial line

изо·пента́н·, -а, (*m*) (chem) isopentane

изо·пери·метр·и́ческ·ий, -ая, -ое, (*a*) (math) isoperimetric

изо·пе́рм·, -а, (*m*) Isoperm (magnetically soft alloy)

изо·пиест·и́ческ·ий, -ая, -ое, (*a*) (chem) isopiestic

изо·пи́кн·а, -ы, (*f*) isopycnic, isopyknic

изо·пле́р·а, -ы, (*f*) (chem) isochore

изо·пле́т·а, -ы, (*f*) (geol math) isopleth

изо·пло́т·ност·ь, -и, (*f*) isodensity

изо·поли·кислот·а́, -ы́, (*f*) isopolyacid

изо·по́р·а, -ы, (*f*) (geophys) isoporic line

изо·пре́н·, -а, (*m*) (chem) isoprene, 2-methyl-1,3-butadiene

изо·пропи́л·, -а, (*m*) (chem) isopropyl

изо·пропил·бензо́л·, -а, (*m*) isopropyl-benzene, (chem) cumene

изо·пропил·иде́н·, -а, (*m*) (chem) isopropylidene

изо·пропил·нитра́т·, -а, (*m*) isopropylnitrate

изо·пропи́л·ов·ый, -ая, -ое, (*a*) isopropyl

∼ спирт· (*m*) (chem) isopropyl alcohol

изо·пье́ст·а, -ы, (*f*) (chem) isopiestic line

изо́·рв·анн·ый, -ая, -ое, (*past part pass*) of **изо·рв·а́ть**; torn, ragged

изо·рв·а́ть, -у́т, (*perf*) tear up; split (a sail)

изо·рода́н·ов·ый, -ая, -ое, (*a*) isothiocyanic, isothiocynate of

∼ кисл·от·а́ (*f*) isothiocyanic acid

∼ эфи́р· (*m*) isothiocyanate

изо·сафро́л·, -а, (*m*) (chem) isosafrole

изо·се́йсм·а, -ы, (*f*) (geophys) isoseismal line

изо·се́йст·а, -ы, (*f*) (geophys) isoseismal line

изо·скле́р·а, -ы, (*f*) isosclere

изо·ско́п·, -а, (*m*) (instr) isoscope

изо·стаз·и́·я, -и, (*f*) (geol) isostasy

  тео́р·и·я изо·стаз·и́·и (*f*) isostatic theory

изо·стат·и́ческ·ий, -ая, -ое, (*a*) isostatic

изо·сте́р·а, -ы, (*f*) (chem) isostere

изо·стер·и́ческ·ий, -ая, -ое, (*a*) isosteric

изо·такт·и́ческ·ий, -ая, -ое, (*a*) (chem) isotactic

изо·та́х·а, -ы, (*f*) (meteor) isotach, isotachic line, (hydr) curve of equal velocity

изо·тени·скóп·, -а,   (*m*) isoteniscope, vapour pressure measuring device

изо·тéр·а, -ы, (*f*) (geophys) isothere

изо·тéрм·а, -ы, (*f*) isotherm, isothermic/isothermal line/curve

∼ адсóрбци·и   adsorption isotherm

изо·терм·и́ческ·ий, -ая, -ое, (*a*) isothermal, isothermic; insulated

∼ вагóн· (*m*) (rail) insulated wagon/car

изо·тóн·, -а, (*m*) (nucl) isotone

изо·тон·и́ческ·ий, -ая, -ое, (*a*) (chem) isotonic

изо·тóн·н·ый, -ая, -ое, (*a*) isotonic

изо·тóп·, -а, (*m*) isotope

∼, радио·акти́в·н·ый   radioactive isotope, radioisotope

∼, стаби́ль·н·ый   stable isotope

изо·топ·и́ческ·ий, -ая, -ое, (*a*) isotope, isotopic

∼ числ·ó (*n*) isotopic number, neutron excess

изо·тóп·н·ый, -ая, -ое, (*a*) isotope, isotopic

∼ с·двиг· (*m*) isotope shift

изо·тр·ёт (*fut 3rd sing*) of **ис·тер·éть**

изо·трóн·, -а, (*m*) (elec) isotron

изо·троп·и́ческ·ий, -ая, -ое, (*a*) isotropic

изо·троп·и́·я, -и, (*f*) (phys) isotropy, isotropism

изо·трóп·ност·ь, -и, (*f*) isotropy, isotropism

изо·трóп·н·ый, -ая, -ое, (*a*) isotropic

изо·тр·ýт (*fut 3rd pl*) of **ис·тер·éть**

изо·фациáль·н·ый, -ая, -ое, (*a*) (phys) isofacial

изо·фéн·, -а, (*m*) (meteor) isophene

изо·фóрминг·, -а, (*m*) (oil) isoforming

изо·фóт·а, -ы, (*f*) (phys) isolux/isophot line/curve

изо·хáзм·а, -ы, (*f*) (geophys) isochasm

изо·хéл·а, -ы, (*f*) (zool) isochela

изо·химéн·а, -ы, (*f*) (meteor) isocheim, line of equal winter temperatures

изо·хиноли́н·, -а, (*m*) (chem) isoquinoline

изо·хиóн·а, -ы, (*f*) (meteor) isochion, equal-show line

изо·хóр·а, -ы, (*f*) (phys, chem) isochore

∼ реáкци·и (chem) reaction isochore

изо·хор·и́ческ·ий, -ая, -ое, (*a*) isochore, isochoric

изо·хóр·н·ый, -ая, -ое, (*a*) isochore, isochoric

изо·хромáт·а, -ы, (*f*) isochromatic line

изо·хромат·и́ческ·ий, -ая, -ое, (*a*) isochromatic

изо·хрóн·а, -ы, (*f*) (astron etc.) isochrone, isochron, isochronic curve

изо·хрон·и́зм·, -а, (*m*) (phys) isochronism

изо·хрóн·ност·ь, -и, (*f*) (phys) isochronism

изо·хрóн·н·ый, -ая, -ое, (*a*) isochronous

изо·циáн·ов·ый, -ая, -ое, (*a*) (chem) isocyanic

∼ кисл·от·á (*f*) isocyanic acid

∼ эфи́р· (*m*) isocyanate

изо·цикл·и́ческ·ий, -ая, -ое, (*a*) (chem) isocyclic

изо·шéд·ш·ий, -ая, -ее, (*past part act*) of **изо·йти́**

из·ощр·и́ть, -я́т, (*perf*) see **из·ощр·я́ть**

из·ощр·я́·ть, -ют, (*imp*) develop, cultivate, sharpen (mind, abilities etc.)

изо·электр·и́ческ·ий, -ая, -ое, (*a*) isoelectric

изо·электрóн·н·ый, -ая, -ое, (*a*) isoelectronic

изо·янтáр·н·ая кисл·от·á (*f*) isosuccinic acid

из-под (*prep + gen*) from under; from (place of origin); of (e.g. **ме·шóк· из-под мук·и́** a sack of flour)

из-подо (*prep*) see **из-под**

израз·éц·, -з·ц·á, (*i*) **-з·ц·óм,** (*m*) wall tile, glazed tile

израз·цóв·ый, -ая, -ое, (*a*) of **израз·éц·**

из·рас·хóд·овать, -уют, (*perf*) expend, use up, consume

и́з·ред·к·а (*adv*) now and then, occasionally

из·реж·ённ·ый, -ая, -ое, (*a*) (bot) spare, thin

из·рéж·ет (*fut 3rd sing*) of **из·рéз·ать**

из·рéз·анн·ый, -ая, -ое, (*a*) (*past part pass*) see **из·рéз·ыва·ть**; (geog) rugged

из·рéз·ать (*fut 3rd sing, pl*) **из·рéж·ет, из·рéж·ут,** (*perf*); see **из·рéз·ыва·ть**

из·рéз·ыва·ть, -ют, (*imp*) cut to pieces; cut up, divided; scratch/incise heavily

из·рек·á·ть, -ют, (*imp*) declaim, pronounce

из·реч· éни·е, -я,   (*n*) dictum, apophthegm, saying

из·réч·ь (*fut 3rd sing, pl*) из·реч·ёт, из·рек·у́т, (*past masc sing*) из·рёк·, (*perf*); *see* из·рек·á·ть

из·решет·и́ть, -я́т,   (*perf*) *see* из·решéч·ива·ть

из·решéч·ива·ть, -ют, (*imp*) perforate, riddle with holes

из·рис·овáть, -у́ют, (*perf*) *see* из·рис·óвыва·ть

из·рис·óвыва·ть, -ют,   (*imp*) draw/ scribble all over; use up (drawing materials)

из·рó·ют (*fut 3rd pl*) of из·ры́·ть

из·руб·á·ть, -ют,   (*imp*) chop up; (mil) cut to pieces

из·руб·и́ть, ⸗я́т, (*perf*) chop up; (mil) cut to pieces

из·ру́бл·енн·ый, -ая, -ое, (*past part pass*) of из·руб·и́ть

из·рыв·á·ть, -ют,   (*imp*) dig up, dig all over

из·ры́·ть (*fut 3rd sing, pl*) из·рó·ет, из·рó·ют, (*perf*); *see* из·рыв·á·ть

из·ря́д·н·о (*adv*) rather, fairly well

из·ум·и́тельн·ый, -ая, -ое, (*a*) amazing, astounding, wonderful

из·ум·и́ть, -я́т, (*perf*) *see* из·умл·я́·ть

из·умл·я́·ть, -ют, (*imp*) amaze, astound

изумру́д·, -а,   (*m*) emerald

из·уч·á·ть, -ют, (*imp*) study; learn

из·уч·éни·е, -я, (*n*) study, studying

из·у́ч·енност·ь, -и,   (*f*) coverage (of studies)

из·у́ч·енн·ый, -ая, -ое,   (*past part pass*) of из·уч·и́ть; studied; learnt

из·уч·и́ть, ⸗ат, (*perf*) study; learn

изъ- (*prefix*) *see* из-

изъ·ед·á·ть, -ют, (*imp*) eat, eat into, corrode

изъ·éд·енн·ый, -ая, -ое, (*past part pass*) *see* изъ·ед·á·ть; pitted

изъ·éзд·ить, -ят,   (*perf*) travel all over

изъ·éзж·енн·ый, -ая, -ое, (*past part pass*) of изъ·éзд·ить; rutted, worn (of roads etc.)

изъ·éст·ь (*fut 3rd sing, pl*) изъ·éст, изъ·ед·я́т, (*past masc sing*) изъ·éл, (*perf*); *see* изъ·ед·á·ть

изъ·яв·и́тельн·ый, -ая, -ое, (*a*) (gram) indicative

изъ·яв·и́ть, ⸗я́т, (*perf*) *see* изъ·явл·я́·ть

изъ·явл·я́·ть, -ют, (*imp*) express

изъ·язв·и́ть, -я́т, (*perf*) *see* изъ·язвл·я́·ть

изъ·язвл·я́·ть, -ют, (*imp*) ulcerate

изъ·я́н·, -а, (*m*) flaw, defect

изъ·я́т·и·е, -я, (*n*) withdrawal, removal; exception

изъ·я́т·ь (*fut 3rd sing, pl*) из·ы́м·ет, из·ы́м·ут, (*perf*); *see* из·ым·á·ть

из·ым·á·ть, -ют,   (*imp*) withdraw, remove

из·ыск·áни·е, -я, (*n*) finding, procuring; project investigation/survey; (geol) prospecting; (*pl*) research

~, бур·ов·óе   borehole surveying

~, эконом·и́ческ·ое economic survey

из·ыск·áтел·ь, -я,   (*m*) project surveyor

из·ыск·áть (*fut 3rd sing, pl*) из·ы́щ·ет, из·ы́щ·ут, (*perf*); *see* из·ы́ск·ива·ть

из·ы́ск·ива·ть, -ют, (*imp*) search out, procure, find; (geol) prospect for; (*imp only*) carry out research on

из·ы́щ·ут (*fut 3rd pl*) of из·ыск·áть

из·энтроп·и́ческ·ий, -ая, -ое,   (*a*) (phys) isentropic

из·энтрóп·н·ый, -ая, -ое, (*a*) isentropic

изюм·, -а, (*m*) raisins, sultanas, currants (dried grapes)

изюм·ин·а, -ы,   (*f*) raisin, sultana, currant (a dried grape)

изя́щ·н·ый, -ая, -ое,   (*a*) refined, elegant, graceful

ийоли́т·, -а, (*m*) (geol) ijolite

ИК (*abbr*) of инфра·крáс·н·ый infrared, IR

ик·á·ть, -ют, (*imp*) hiccup

и-кисл·от·á, -ы́,   (*f*) (chem) j-acid

ик·н·у́ть, -у́т, (*perf*) *see* ик·á·ть

икóн·а, -ы, (*f*) icon

иконо·скóп·, -а,   (*m*) (TV) iconoscope

~, без·инерциóн·н·ый   (TV) non-storage camera tube

икóр·н·ый, -ая, -ое, (*a*) of икр·á

икосá·эдр·, -а, (*m*) (cryst) icosahedron

икоса·эдр·и́ческ·ий, -ая, -ое,   (*a*) icosahedral

икоси·тетрá·эдр·, -а, (*m*) (cryst etc.) icositetrahedron

ик·óт·а, -ы, (*f*) hiccup

ико·ти́п·, -а, (*m*) (biol) icotype

икр·á, -ы́, (*nom pl*) и́кр·ы, (*f*) (fish) roe, spawn, ova; caviar; calf (of leg)

~, зерн·и́ст·ая fresh caviar

~, пáюс·н·ая pressed/preserved caviar

икр·и́нк·а, -и, (*g pl*) -нок·, (*f*) (fish) roe corn, ovum икр·и́ст·ый, -ая, -ое, (*a*) roed, heavy with roe

икр·и́ться, -я́тся, (*imp and perf*) spawn

икро·мёт·, -а, (*n*) see икро·мет·а́ни·е

икро·мет·а́ни·е, -я, (*n*) spawning

икр·ян·о́й, -а́я, -о́е, (*a*) roe; containing roe

~ ка́мен·ь (*m*) (min) oolite, egg/roe stone

йкс·, -а, (*m*) x (latin letter)

иксио·ли́т·, -а, (*m*) (min) ixiolite

иксо·ли́т·, -а, (*m*) (min) ixolyte

иктидо·за́вр·, -а, (*m*) (pal) ictidosaur

йл·, -а, (*m*) mud, silt, ooze; sludge

~, акти́в·н·ый (san eng) activated sludge

~, гру́б·ый (geol) silt

~, зоо·ге́н·н·ый (geol) ooze

коэффицие́нт· и́л·а (*m*) (san eng) sludge index

~, птеро·по́д·ов·ый (geol) pteropod ooze

йлем·, -а, (*m*) (nucl) ylem

и́ли (*conj*) or

~ ... и́ли either ... or

и́л·ист·ый, -ая, -ое, (*a*) muddy, silty, oozy

иллини·й, -я, (*m*) (chem) illinium, Il, promethium

иллит·, -а, (*m*) (min) illite

иллю́ви·й, -я, (*m*) (geol) illuvium, illuvial deposits

иллюви·а́льн·ый, -ая, -ое, (*a*) illuvial

иллюзион·и́ст·, -а, (*m*) conjurer, conjuror

иллю́зи·я, -и, (*f*) illusion

иллюмин·а́тор·, -а, (*m*) illumination engineer; (naut) porthole, scuttle, light

~, глух·о́й fixed light

~ па́лубный deck light

иллюмин·ацио́нн·ый, -ая, -ое, (*a*) of иллюмин·а́ци·я

иллюмин·а́ци·я, -и, (*f*) illuminating, floodlighting

иллюмин·и́р·овать, -уют, (*imp and perf*) floodlight, illuminate (buildings etc.)

иллюмин·ова́ть, -у́ют, (*imp and perf*) see иллюмин·и́р·овать

иллюстр·ати́вн·ый, -ая, -ое, (*a*) illustrative

иллюстр·а́тор·, -а, (*m*) illustrator

иллюстр·а́ци·я, -и, (*f*) illustration

иллюстр·и́р·овать, -уют, (*imp and perf*) illustrate

ил·ова́т·ый, -ая, -ое, (*a*) rather/somewhat muddy, containing mud

и́л·овк·а, -и, (*f*) (geol) loam, siltstone

и́л·ов·ый, -ая, -ое, (*a*) of ил·

~ площа́д·к·а (*f*) sludge bed (sewage systems)

ило·со́с·, -а, (*m*) (san eng) sludge disposal lorry

ильва́ит·, -а, (*m*) (min) ilvaite

и́льм·, -а, (*m*) (bot) elm, *Ulmus*

~, америка́н·ск·ий white elm, American elm

~, шотла́нд·ск·ий wych elm, Scotch elm, *Ulmus glabra*

ильмени́т·, -а, (*m*) (min) ilmenite

ильмено·рути́л·, -а, (*m*) (min) ilmenorutile

и́льм·ов·ый, -ая, -ое, (*a*) of и́льм·

им (*pron, instr sing m/n*) of он/оно, by/with etc. him/it; (*pron, dat plur m/f/n*) of он/она, оно to them

имагина́ль·н·ый, -ая, -ое, (*a*) (zool) imaginal

има́г·о, -а, (*n*) (ent) imago

имбиб·и́ци·я, -и, (*f*) imbibition

имби́р·ь, -я, (*m*) (bot) ginger, *Zingiber*

имбрикацио́н·н·ый, -ая, -ое, (*a*) (biol) imbricated

и́мен·и (*g, d, p sing*) of и́м·я

за·во́д· и́мен·и Ле́нин·а (*m*) the Lenin works/factory

им·е́ни·е, -я, (*n*) landed property, country estate

имен·и́н·ы (*nom pl*), (*g pl*) -н·и́н·, (*f*) name-day

имен·и́тельн·ый, -ая, -ое, (*a*) (gram) nominative

и́мен·н·о (*adv*) namely, just, exactly, precisely

имен·н·о́й, -а́я, -о́е, (*a*) named, nominal, bearing the owner's name, inscribed

~ це́н·н·ые бума́г·и (*pl*) (fin) inscribed stock

имен·ова́ни·е, -я, (*n*) naming; denomination

имен·о́ванн·ый, -ая, -ое, (*past part pass*) of имен·ова́ть; denominate

~ числ·о́ (*n*) (math) denominate number

имен·ова́ть, -у́ют, (*imp*) name, call, denominate

им·е́·ть, -ют, (*imp*) have; -ся (*impersonal*) be, exist

~ в вид·у́ have in mind

~ вид· be in the form/shape (of)

~ ме́ст·о take place, occur

им·е́·ется (*pres 3rd sing*) of им·е́·ться; there is/are

им·е́юш·ий, -ая, -ее, (*pres part act*) of им·е́·ть having

им·е́юш·ийся, -аяся, -ееся, (*pres part act*) of им·е́·ться; available

и́ми (*pron, instr plur*) of он/она́/оно́ by/with etc. them

ими́д·, -а, (*m*) (chem) imide

имидазо́л·, -а, (*m*) (chem) iminazole, imidazole, glyoxaline

имидо- (*component*) imido-, imino-

имидо-эфи́р·, -а, (*m*) iminoester

имидо·азоло́н·, -а, (*m*) iminazolone

имино- (*component*) *see* имидо-

имино·со·елин·е́ни·е, -я, (*n*) (chem) iminocompound

имит·а́тор·, -а, (*m*) imitator; simulator; trainer

~ по·лёт·а (a/c) flight simulator

имит·а́ци·я, -и, (*f*) imitation; simulation

~ траекто́р·и·и по·лёт·а (rocket) trajectory simulation

имит·и́р·овать, -уют, (*imp*) imitate; simulate

имит·и́рующ·ий, -ая, -ое, imitating, simulating; imitative

иммельма́н·, -а, (*m*) (air) Immelmann turn

иммерсио́н·н·ый, -ая, -ое, (*a*) immersion

~ спо́соб· (*m*) (phys) immersion method of determining refractive index

имме́рси·я, -и, (*f*) immersion

иммигр·а́ци·я, -и, (*f*) immigration

иммигр·и́р·овать, -уют, (*imp and perf*) immigrate

иммитанс·, -а, (*m*) (elec) immittance

иммобилиз·а́ци·я, -и, (*f*) immobilization

иммуниз·а́ци·я, -и, (*f*) immunization

иммуниз·и́р·овать, -уют, (*imp and perf*) immunize

иммун·ите́т·, -а, (*m*) immunity

имму́н·ност·ь, -и, (*f*) immunity

имму́н·н·ый, -ая, -ое, (*a*) immune

~ един·и́ц·а (*f*) (pharm) Felton's unit

иммуно·ге́н·н·ый, -ая, -ое, (*a*) immunogenous

иммуно·ло́г·и·я, -и, (*f*) immunology

иммуно·терап·и́·я, -и, (*f*) (med) immunotherapy

иммуно·хи́м·и·я, -и, (*f*) immunochemistry

иммун·те́л·о, -а, (*n*) (zool) immune body

импеда́нс·, -а, (*m*) (elec) impedance

~, акуст·и́ческ·ий acoustic impedance

импеда́нс

~, механ·и́ческ·ий mechanical impedance

~ на·магни́ч·ивани·я reluctance

~, по·вто́р·н·ый iterative impedance

~ у то́ч·к·и пит·а́ни·я driving-point impedance

импеда́нс·н·ый, -ая, -ое, (*a*) (elec) impedance

~, коэффицие́нт· (*m*) impedance factor

импе́ллер·, -а, (*m*) (*see also* кры́ль·-ча́тк·а) impeller

императи́в·, -а, (*m*) (gram) imperative

империал·и́зм·, -а, (*m*) imperialism

импе́ри·я, -и, (*f*) empire

импи́дор·, -а, (*m*) (elec) impedor

имплант·а́ци·я, -и, (*f*) (med) implantation

импло́з·и·я, -и, (*f*) implosion

импон·и́р·овать, -уют, (*imp*) command respect, impress

имп. (*abbr*) of и́мпульс·, impulse, pulse, momentum

и́мпорт·, -а, (*m*) (com) import; (econ) imports

импорт·ёр·, -а, (*m*) importer

импорт·и́р·овать, -уют, (*imp and perf*) (com) import

и́мпорт·н·ый, -ая, -ое, (*a*) import, imported

импоте́нт·ност·ь, -и, (*f*) impotence

импоте́нц·и·я, -и, (*f*) impotence

имрегн·а́ци·я, -и, (*f*) impregnation

импрессионист·и́ческ·ий, -ая, -ое, (*a*) impressionistic

импровиз·и́р·овать, -уют, (*imp and perf*) improvise, extemporise

и́мпульс·, -а, (*m*) impulse, pulse momentum (of a particle)

вид· и́мпульс·ов (*m*) pulse mode

~, в·ключ·а́ющ·ий turn-on pulse

~, в·ход·н·о́й input/incoming pulse

~, вы·ключ·а́ющ·ий turn-off pulse

~, гас·я́щ·ий (rad) quenching/blanking pulse

~, дву·на·пра́вл·енн·ый bidirectional pulse

~, за·да·ю́щ·ий master pulse

за·ко́н· со·хран·е́ни·я и́мпульс·а (*m*) law of conservation of momentum

~ за·про́с·а (rad) interrogation pulse

~, за·темн·я́ющ·ий (TV, rad) blanking pulse/signal

~, зна́ч·н·ый marking pulse (telegraphy)

~, зонд·и́рующ·ий (rad) transmitter pulse

и́мпульс
~, зуб·ча́т·ый   serrated pulse
~, ка́др·ов·ый   (TV) broad pulse
~, -кома́нд·а, -ы,   (f) (elec) command pulse
~, кратко·вре́мен·н·ый   narrow pulse
~, крут·о́й   steep-sided pulse
~ на·ча́л·а от·с·чёт·а   zero time pulse (of an instrument)
~, на·ча́ль·н·ый   (rad) transmitter pulse; (telecom) transmitted pulse
~, нейтро́н·н·ый   (nucl) neutron burst
~, не́рв·н·ый   (med) neural impulse
~, одно·по́люс·н·ый   (rad) single-polarity pulse
~, от·пир·а́ющ·ий   (elec) enabling pulse
~, от·раж·ённ·ый   (rad) echo
~, пило·обра́з·н·ый   sawtooth pulse
~, по́лн·ый   (phys) total momentum; (elec) overall pulse
~, поро́г·ов·ый   momentum cutoff
~, про·пуск·а́ющ·ий   gate/gating pulse
про·стра́н·ств·о и́мпульс·ов (n) (phys) momentum space
~ с·бро́с·а   reset pulse (computers)
~, селе́ктор·н·ый   selector pulse, gate (electronic)
се́р·и·я и́мпульс·ов (f) pulse train
~ си́л·ы   (mech) impulse
~, синхрониз·и́рующ·ий   (autom) master/timing/clock pulse
~, со·пряж·ённ·ый   conjugate momentum
~ с·чёт·а   count pulse
~, такт·и́рующ·ий   timing pulse, clock pulse
~, та́кт·ов·ый   clock pulse
~, то́к·ов·ый   (autom) make pulse
тольщ·ин·а́ потер·и и́мпульс·а (f) momentum thickness
~, уда́р·н·ый   (mech) impulse
~, у́з·к·ий   spike
~, у·коро́ч·енн·ый   chopped pulse
~, у·правл·я́ющ·ий   (autom) driving pulse
~, у·ра́вн·ивающ·ий   (TV) equalizing pulse
~ -э́х·о, -а,   (n) (rad) echo
импульс·а́тор·, -а,   (m) (elec) pulsing device, pulser
импульси́·метр·, -а,   (m) dynamometer
и́мпульс·н·о-ко́д·ов·ый, -ая, -ое, (a) pulse-code
~ -модул·и́рованн·ый, -ая, -ое, (a) pulsed, pulse-modulated

и́мпульс·н·ый, -ая, -ое, (a) impulse, pulse, pulsing, pulsed, pulse-driven; momentum
~ дви́г·ател·ь   (m) (elec) step-by-step motor
~ деформ·а́ци·я   (f) (met) high-velocity deformation
~ комбин·а́ци·я   (f) pulse mode
~ коро́н·а   (f) corona discharge
~ магни́т·н·ое по́л·е   (n) pulsed magnetic field
~ модул·я́ци·я   (f) pulse modulation
~ на·груж·е́ни·е   (n) (met) sudden/impulsive loading
~ нес·у́щ·ая   (f) pulse carrier
~ по·сы́л·к·а   (f) pulse train
~ пред·ставл·е́ни·е   (n) (phys) momentum representation
~ про́ч·н·ост·ь   (f) (met) impact strength
~ радио·с·вя́з·ь   (f) pulse radio
~ раз·ря́д·   (m) pulsed discharge
~ режи́м·   (m) pulse operation
с и́мпульс·н·ым режи́м·ом ра·бо́т·ы   pulse-operated
~ систе́м·а   (f) (autom) sampled-data system
~ систе́м·а, ко́д·ов·ая   (f) (autom) pulse-code/count system
~ те́хн·ик·а   (f) pulse technology
~ цеп·ь   (f) impulse circuit (telephones)
~ част·о́тн·ая систе́м·а   (f) (autom) pulse-code system, pulse-count system
импульсо·ви́д·н·ый, -ая, -ое,   (a) pulselike, pulsed
импульсо·за·остр·я́ющ·ий, -ая, -ое (a) pulse-sharpening
импф·и́ровани·е, -я,   (n) sulphuric/hydrochloric acid water treatment
им·у́щественн·ый, -ая, -ое, (a) of им·у́ществ·о
им·у́ществ·о, -а, (n) property, belongings
~, дви́ж·им·ое   movables, movable goods
~, не·дви́ж·им·ое   real estate, realty
им·у́щ·ий, -ая, -ее, (a) propertied
и́м·я (g, d, p sing) и́мен·и, (nom pl) имен·а́, (g pl) име́н·, (d pl) имен·а́м, (n) name
~, при·лаг·а́тельн·ое   (gram) adjective
~, сущ·естви́тельн·ос   (gram) noun, substantive
ин·актив·а́ци·я, -и, (f) inactivation

ин·а́че   (*adv*) differently, otherwise;
(*conj*) or, otherwise, else
  так· и́ли ина́че one way or another,
somehow
инбри́динг·, -а,   (*m*) (agr) inbreeding
ин·вагин·а́ци·я, -и,   (*f*) (zool) inva-
gination
инва́зи·я, -и,   (*f*) invasion, infestation
инвали́д·, -а,   (*m*) invalid, disabled
person
~ **труд·а́**   disabled worker
инвали́д·ност·ь, -и,   (*f*) (med) dis-
ability
инвали́д·н·ый, -ая, -ое,   (*a*) of **ин-
вали́д·, инвали́д·ност·ь**
инва́р·, -а,   (*m*) (met) Invar
инвариа́нт·, -а,   (*m*) (math) invariant
factor, invariant
инвариа́нт·ност·ь, -и,   (*f*) invariance
инва́р·ов·ый, -ая, -ое,   (*a*) of **инва́р·**
инвентариз·а́ци·я, -и,   (*f*) inventory
compilation, entering in an inventory; stocktaking, checking an inventory
инвентариз·и́р·овать, -уют,   (*imp
and perf*) *see* **инвентариз·ова́ть**
инвентариз·ова́ть, -у́ют,   (*imp and
perf*) make/compile an inventory,
enter in an inventory; check an inventory, take stock
инвента́р·н·ый, -ая, -ое,   (*a*) inventory, stock; accession (to libraries)
~ **но́мер·**   (*m*) accession number
(libraries)
инвента́р·ь, -я́,   (*m*) inventory, stock
~, **сельско·хозя́й·ственн·ый** agricultural implements
~, **торг·о́в·ый** stock-in-trade
инверсио́н·н·ый, -ая, -ое,   (*a*) inversion
инве́рс·и·я, -и,   (*f*) inversion
  опера́тор· инве́рс·и·и   (*m*) (phys)
space inversion/parity operator
инве́рс·ор·, -а,   (*m*) inversor (math
instrument)
инверт·а́з·а, -ы,   (*f*) (biochem) inver-
tase, saccharase, sucrase
инверт·и́ровани·е, -я,   (*n*) inversion;
reversal (binary coding); (elec) conversion from dc to ac
инверт·и́рованн·ый, -ая, -ое,   (*past
part pass*) of **инверт·и́р·овать;** inverted, invert
~ **са́хар·**   (*m*) (chem) invert sugar
инверт·и́р·овать, -уют,   (*imp and
perf*) invert

инве́рт·ор·, -а,   (*m*) (elec) inverter,
inverted rectifier, inverted convertor
инвест·и́р·овать, -уют,   (*imp and
perf*) (fin) invest
инволю́т·а, -ы,   (*f*) (math) involute,
involute function
инволю́т·н·ый, -ая, -ое,   (*a*) involute
инволюто́р·н·ый, -ая, -ое,   (*a*) in-
volutionary
инволю́ци·я, -и,   (*f*) involution
ингал·я́тор·, -а,   (*m*) inhaler
ингал·я́ци·я, -и,   (*f*) inhalation
ингиб·и́р·овать, -уют,   (*imp and perf*)
inhibit
ингиби́тор·, -а,   (*m*) inhibitor
ингредие́нт·, -а,   (*m*) ingredient, component
ингресси́в·н·ый, -ая, -ое,   (*a*) in-
gressive, entering, incoming
ингре́сс·и·я, -и,   (*f*) ingression
индами́н·, -а,   (*m*) (chem) indamine
инд·а́н·, -а,   (*m*) (chem) indan
инда**ни́л·, -а,**   (*m*) indanyl
инд·антре́н·, -а,   (*m*) (chem) indanthrene
инд·антре́н·ов·ый, -ая, -ое,   (*a*) of
инд·антре́н·
и́ндев·е·ть, -ют,   (*imp*) become covered
with rime/hoar frost
инде́·йк·а, -и,   (*g pl*) **-е́·ек·,**   (*f*) turkey
индейко·во́д·ств·о, -а,   (*n*)   turkey
farming/rearing
инде́й·ск·ий, -ая, -ое,   (*a*) American-
Indian
и́ндекс·, -а,   (*m*) index; (math) index,
index number; subscript, superscript
~, **ве́рх·н·ий**   (print) superscript
~ **вя́з·кост·и**   viscosity index
~, **не·по·дви́ж·н·ый**   index mark
~, **ни́ж·н·ий**   (print) subscript
~ **пре·ломл·е́ни·я**   (opt) refractive
index, index of refraction
~ **себе·сто́·имост·и, индивидуа́ль-
н·ый**   (econ) unit cost index-number
~ **себе·сто́·имост·и, о́бщ·ий**   (econ)
production costs index-number
и́ндекс·н·ый, -ая, -ое,   (*a*) index
инде́н·, -а,   (*m*) (chem) indene
инде́нтор·, -а,   (*m*) indenter
индиви́д·, -а,   (*m*) (biol) individual;
(chem) compound; element
индивидуализ·и́р·овать, -уют,   (*imp
and perf*) individualize, classify; suit to
индивидуа́ль·н·ый, -ая, -ое,   (*a*)
individual, personal
~ **ли́н·и·я**   (*f*) direct line (telephones)
индиви́дуум·, -а,   (*m*) individual

инди́го (*n indecl*) (chem) indigo
инди́г·ов·ый, -ая, -ое, (*a*) indigo
индигозо́л·, -а, (*m*) (chem) indigosol
индиго́ид·н·ый, крас·и́тел·ь (*m*) indigoid dye
индиго́ид·, -а, (*m*) indigoid (dye)
индиго·карми́н·, -а, (*m*) (chem) indigo carmine
индиго·но́с·к·а, -и, (*g pl*) **-с·ок·,** (*f*) (bot) indigo, indigo plant, *Indigofera*
индиго·ли́т·, -а, (*m*) (min) indigolite
индиготи́н·, -а, (*m*) (chem) indigotin
и́нд·иев·ый, -ая, -ое, (*a*) (chem) indium
и́нд·и·й, -я, (*m*) (chem) indium, In
инди́й·ск·ий, -ая, -ое, (*a*) Indian
индика́н·, -а, (*m*) (chem) indican
~, жив·о́тн·ый (chem) indican; (biochem) indoxyl-sulphuric acid
~, раст·и́тельн·ый (bot) indican, indoxyl glucoside
индика́тор·, -а, (*m*) (*see also* **у·ка·з·а́тел·ь**) indicator; (rad) display, indicator; (a/c) indicator, engine gauge unit; guide card (card index); (nucl) tracer
~ вéс·а (oil) drillometer
~ да́ль·ност·и range indicator; (rad) "A" display
~, динамо·метр·и́ческ·ий (ICE) indicator
~ круг·ов·о́го об·зо́р·а (rad) plan position indicator, PPI
~ маршру́т·а (rail) route-indicating signal
~ много·компоне́нт·н·ый (nucl) multiple tracer
~ на·стро́й·к·и, электро́н·н·ый (rad) magic eye, tuning indicator, electron ray tube
~, об·зо́р·н·ый (rad) plan position indicator
~ по·са́д·к·и, зерка́ль·н·ый (a/c) deck-landing mirror sight
~, трёх·стре́л·очн·ый (a/c) three-sectional engine gauge unit
~, час·ов·о́й dial gauge
индика́тор·н·ый, -ая, -ое, (*a*) of **индика́тор·**; indicated, indicating
~ диагра́мм·а (*f*) (ICE) indicator diagram
~ до́з·а (*f*) (nucl) tracer dose
~ мо́щ·ност·ь (*f*) (ICE) indicated horse power
индикатри́с·а, -ы, (*f*) (math) indicatrix; function

индикатри́с·а
~, вы́·тя·нут·ая (cryst) prolate indicatrix
~, опт·и́ческ·ая (cryst) optical indicatrix
~, с·плюс·нут·ая (cryst) oblate indicatrix
индик·а́ци·я, -и, (*f*) display, presentation, indication
~, на·гля́д·н·ая display presentation
~ о·ши́б·ок· error indication
~, радиолокацио́н·н·ая radar display
~ рас·со·глас·ова́ни·я error indication, matching-error indication
индитро́н·, -а, (*m*) (elec) inditron
индифферéнт·ност·ь, -и, (*f*) indifference
индифферéнт·н·ый, -ая, -ое, (*a*) indifferent
индиц·и́рова́ни·е, -я, (*n*) indication; identification (X-ray diffraction technique)
индо·евро́п·éйск·ий, -ая, -ое, (*a*) Indo-European
инд·окси́л·, -а, (*m*) (chem) indoxyl
инд·о́л·, -а, (*m*) (chem) indole
индолил·у́ксус·н·ая кисл·от·а́ (*f*) indoleacetic acid
индосс·амéнт·, -а, (*m*) (fin) endorsement
индосс·а́нт·, -а, (*m*) (fin) endorser
индосс·а́тор·, -а, (*m*) (fin) endorsee
индосс·и́р·овать, -уют, (*imp and perf*) (fin) endorse
индо·фено́л·, -а, (*m*) (chem) indophenol
инду́зи·й, -я, (*m*) (bot) indusium
индукта́нс·, -а, (*m*) (elec) inductance, coefficient of inductance, inductance factor
индукти́в·ност·ь, -и, (*f*) (elec) inductance; inductance factor; inductivity
индукти́в·н·ый, -ая, -ое, (*a*) inductive, inductance
~ на·гру́з·к·а (*f*) inductance load
~ с·вяз·ь (*f*) (rad) inductance coupling
~ со·противл·éни·е (*n*) (elec) inductance; (air) inducted drag
~ со·противл·éни·е рас·сé·яни·я (*n*) leakage reactance
~ тормож·éни·е (*n*) induction drag (magnetism)
индукт·и́рованн·ый, -ая, -ое, (*past part pass*) induced
инду́ктор·, -а, (*m*) (chem) inductor; (elec) inductor, field magnet; magneto generator (telephony); inductance coil

индукто́р·и·й, -я, (*m*) (elec) induction coil

инду́ктор·н·ый, -ая, -ое, (*a*) of **инду́ктор·**

~ **маши́н·а** (*f*) (elec) inductor generator

индукцио́н·н·ый, -ая, -ое, (*a*) induction

~ **дефекто·скоп·и́·я** (*f*) eddy-current flow detection

~ **кат·у́шк·а** (*f*) (elec) induction coil

~ **маши́н·а** (*f*) induction motor

~ **на·гре́·в·** (*m*) induction heating

~ **регуля́тор·** (*m*) induction regulator, induction-voltage regulator

инду́кци·я, -и, (*f*) induction; (elec) induction, inducing (electric charges); field density; (chem) induction period; induction

~, взаи́м·н·ая mutual induction

~, магни́т·н·ая magnetic induction (flux density)

~, о·ста́т·очн·ая магни́т·н·ая (phys) remanence, remanant induction

~, электро·магни́т·н·ая electromagnetic induction

индули́н·, -а, (*m*) (chem) induline

инду́с·ск·ий, -ая, -ое, (*a*) Hindu

индуста́р·, -а, (*m*) (phot) industar lens (similar to Tessar)

индустриализ·и́р·овать, -уют, (*imp and perf*) industrialize

индустриализ·ова́ть, -у́ют, (*imp and perf*) *see* **индустриализ·и́р·овать**

индустриа́ль·н·ый, -ая, -ое, (*a*) industrial

инду́стри·я, -и, (*f*) (econ) industry

индуц·и́ров·ать (*imp and perf*) induce

индю́к·, -а́, (*m*) (zool) turkey-cock, *Meleagris gallopavo*

и́не·й, -я, (*m*) rime, hoar frost

инее·ви́д·н·ый, -ая, -ое, (*a*) frosted (finish)

ине́рт·ност·ь, -и, (*f*) inertness, inertia

ине́рт·н·ый, -ая, -ое, (*a*) inert

~ **газ·** (*m*) rare/inert gas

инерциа́ль·н·ый, -ая, -ое, (*a*) inertial

~ **систе́м·а** (*f*) inertial system, (math) inertial reference frame

~ **систе́м·а координа́т·** (math) inertial coordinate system

инерцио́н·ност·ь, -и, (*f*) time lag/constant, persistence

~ **из·об·раж·е́ни·я** (TV) image persistence

~ **экра́н·а** (TV) screen speed

инерцио́н·н·ый, -ая, -ое, (*a*) inertia, inertion, inertial

~ **вре́м·я** (*n*) (rocket) coast/coasting time

~ **гро́хот·** (*m*) unbalanced-pulley vibrating screen

~ **навига́ци·я** (*f*) inertial navigation

~ **пред·о·хран·и́тел·ь** (*m*) inertia switch

ине́рци·я, -и, (*f*) inertia

~, виртуа́ль·н·ая virtual inertia

~, вращ·е́ни·я rotary inertia

моме́нт инерци·и (*m*) moment of inertia

моме́нт· инерци·и сече́ни·я (*m*) moment of inertia of area

моме́нт инерци·и те́л·а (*m*) moment of inertia of mass

си́л·а инерци·и (*f*) inertial force

це́нтр инерци·и (*m*) (mech) centre/center of mass

центр·о·бе́ж·н·ый моме́нт инерци·и (*m*) (aerodyn) deviation moment

инж. (*abbr*) of **инжене́р·, инжене́р·н·ый**

инже́ктор·, -а, (*m*) injector; inspirator (a type of jet pump)

инжене́р·, -а, (*m*) engineer

~, го́р·н·ый mining engineer

~ **-констру́ктор·, -а,** (*m*) design engineer

~ **-меха́н·ик·, -а,** (*m*) mechanical engineer; shipyard trial's engineer

~ **-стро·и́тел·ь, -я,** (*m*) civil engineer

инжене́р·н·о-техн·и́ческ·ие рабо́т·-ник·и (*pl*) engineering and technical staff

инжене́р·н·ый, -ая, -ое, (*a*) *of* **инжене́р**

~ **се́т·и** (*pl*) mains services (of factory etc.)

инжене́р·ств·о, -а, (*n*) engineering profession

инжи́р·, -а, (*m*) (bot) fig, *Ficus carica*

инициа́л·, -а, (*m*) initial, initial letter

~, от·те́н·очн·ый (print) shadow initial

~, прост·о́й (print) plant initial

инициати́в·а, -ы, (*f*) initiative

инициати́в·н·ый, -ая, -ое, (*a*) having/demanding initiative

инициа́тор·, -а, (*m*) initiator, pioneer, author (of a new idea/technique); (chem) initiator

иници·и́р·овани·е (*n*) initiation

иници·и́р·овать, -уют, (*imp and perf*) initiate, trigger

иници·и́рующ·ий, -ая, -ее, (*pres part act*) of **иници·и́р·овать;** initiating

~ **вз·ры́в·чат·ое веществ·о́** (*n*) (expl) initiator

ин·капсул·я́ци·я, -и, (*f*) encapsulation

инкасс·а́тор·, -а, (*m*) (fin) collector

инкасс·и́ровани·е, -я, (*n*) fin) collection

инка́ссо (*n indecl*) (fin) collection, collection of payments

инклин·а́тор·, -а, (*m*) (geophys) dip needle/circle, inclinometer

~**, индукцио́н·н·ый** dip/earth inductor, induction inclinometer, earth inductor compass

~**, стрел·о́чн·ый** dip circle

инклино́·метр·, -а, (*m*) (oil) deviation recorder

инклю́зи·я, -и, (*f*) inclusion

инконгруэ́нт·н·ый, -ая, -ое, (*a*) incongruent

инконел·ь, -я, (*m*) (met) Inconel

инкорпор·и́р·овать, -уют, (*imp and perf*) incorporate

инкорпор·а́ци·я, -и, (*f*) incorporation

инкреме́нт·, -а, (*m*) increment

инкре́т·, -а, (*m*) (anat) hormone

инкрето́рн·ый, -ая, -ое, (*a*) incretory

~ **желез·а́** (*f*) (anat) endocrine gland

инкре́ци·я, -и, (*f*) incretion

инкруст·а́ци·я, -и, (*f*) incrustation; inlay (in furniture), marquetry

инкруст·и́р·овать, -уют, (*imp and perf*) incrust; inlay

инкуба́тор·, -а, (*m*) incubator

инкубато́р·н·ый, -я, (*m*) incubator house

инкубацио́н·н·ый, -ая, -ое, (*a*) incubation

~ **пери́од·** (*m*) incubation period, incubation; (met) cooling period

инкуба́ци·я, -и, (*f*) incubation

ин·нерв·а́ци·я, -и, (*f*) (biol) innervation

ино·ви́д·н·ый, -ая, -ое, (*a*) different, different kind of

ино·гда́ (*adv*) sometimes

ино·горо́д·н·ий, -яя, -ее, (*a*) in/of/ from another town/city, from elsewhere; (telecom) not local; inter-city

иноз·и́н·, -а, (*m*) (chem) inosine, hypoxanthine

иноз·и́н·ов·ая кисл·от·а́ (*f*) inosinic acid

иноз·и́т·, -а, (*m*) (chem) inositol, inosite

ин·о́й, -а́я, -о́е, (*a*) other, another, different; some, certain

инокул·и́р·овать, -уют, (*imp and perf*) inoculate

инокул·я́тор·, -а, (*m*) (met) inoculant; (bact) propagator

инокул·я́ци·я, -и, (*f*) inoculation

ино·ро́д·н·ый, -ая, -ое, (*a*) foreign, alien

~ **те́л·о** (*n*) (med) foreign body

ино·с·каз·а́тельн·ый, -ая, -ое, (*a*) allegorical

ино·стра́н·ец·, -н·ц·а, (*m*) foreigner

ино·стра́н·н·ый, -ая, -ое, (*a*) foreign

ино·цера́м·, -а, (*m*) (pal) Inoceramus

ино·язы́ч·н·ый, -ая, -ое, (*a*) foreign-language; speaking a foreign language

инсеквэ́нт·н·ый, -ая, -ое, (*a*) insequent

инсекти·си́д·, -а, (*m*) *see* **инсекти·- ци́д·**

инсекти·ци́д·, -а, (*m*) insecticide

~**, внутри·раст·и́тельн·ый** systematic insecticide

инсекти·ци́д·н·ый, -ая, -ое, (*a*) insecticidal

инсол·я́ци·я, -и, (*f*) insolation, exposure to the sun

инспект·и́р·овать, -уют, (*imp*) inspect

инспе́ктор·, -а, (*m*) inspector; marine superintendent

инспе́кци·я, -и, (*f*) inspection; inspectorate

~**, мор·ск·а́я** marine inspectorate

инспир·а́ци·я, -и, (*f*) inspiration, incitement

инспир·и́р·овать, -уют, (*imp and perf*) inspire, incite

инста́нци·я, -и, (*f*) level (of authority)

~**, выше·сто·я́щ·ая** higher/superior authority

суд· втор·о́й инста́нци·и (*m*) (law) court of appellate jurisdiction

суд· пе́рв·ой инста́нци·и (*m*) (law) court of original jurisdiction, court of first instance

инсти́нкт·, -а, (*m*) instinct

институ́т·, -а, (*m*) institute; institution

~**, на·у́ч·н·о-ис·сле́д·овательск·ий** scientific research institute

инструкт·а́ж·, -а, (*m*) instructing; instruction

инструкт·и́р·овать, -уют, (*imp and perf*) instruct

инстру́кт·ор·, -а, (*m*) instructor

инстру́кц·и·я, -и, (*f*) instructions

инструме́нт·, -а,   (*m*) tool(s); instrument; (met rolling) rolls

~, лов·и́льн·ый (oil) fishing tools

~, с·ва́р·очн·ый electrode holder

~, универса́ль·н·ый (surv astron) altazimuth

инструмента́ль·н·ый, -ая, -ое,   (*a*) tool, instrument; instrumental (music)

~ от·де́л· (*m*) tool department

~ хозя́й·ств·о (*n*) tool organization

~ цех· (*m*) tool shop

инструмента́ль·щик·, -а,   (*m*) tool-maker

инструме́нт·а́ри·й, -я,   (*m*) set of tools/instruments

инструме́нт·и́р·овать, -уют,   (*imp and perf*) orchestrate (music)

инсули́н·, -а, (*m*) (pharm) insulin

инсу́льт·, -а, (*m*) (med) insult, insultus

инсуфл·я́ци·я, -и, (*f*) insufflation

инсцен·и́р·овать, -уют,   (*imp and perf*) dramatize, stage

инта́рси·я, -и, (*f*) inlay, marquetry (in furniture etc.)

интегра́л·, -а, (*m*) (math) integral

~ вероя́т·ност·и о·ши́б·ок· error function

~, кра́т·н·ый iterated/multiple integral

~, криво·лин·е́йн·ый integral over a curve

~, не·о·пре·дел·ённ·ый indefinite integral

~, не·со́б·ственн·ый improper integral

~, объ·ём·н·ый space integral

~, о·пре·дел·ённ·ый definite integral

~ о·ши́б·ок· (math) error function

~, ча́ст·н·ый particular integral

интегра́ль·н·ый, -ая, -ое, (*a*) (math) integral, integrated; (instr) integrating

~ ис·числ·е́ни·е (*n*) integral calculus

~ о·ста́т·ок· (*m*) (math) integral residue

~ сеч·е́ни·е (*n*) (phys) integrated cross section

интегра́тор·, -а, (*m*) (math) integrator; (elec) integrator, integrating circuit

~, ди́ск·ов·ый фрикцио́н·н·ый wheel-and-disc integrator

~, лоб·ов·о́й фрикцио́н·н·ый wheel-and-disc integrator

интегра́ф·, -а, (*m*) (math) integraph

интегр·а́ци·я, -и, (*f*) integration

интегр·и́ровани·е, -я, (*n*) integration

~ по integration over/by

~ по част·я́м integration by parts

интегр·и́р·овать, -уют,   (*imp and perf*) (math) integrate

интегр·и́руем·ый, -ая, -ое,   (*pres part pass*) of **интегр·и́р·овать**; integrable, integrand

интегр·и́рующ·ий, -ая, -ее, (*pres part act*) of **интегр·и́р·овать**; integrating

~ у·стро́й·ств·о (*n*) integrator

интегро·дифференциа́ль·н·ое у·равн·е́ни·е (*n*) integrodifferential equation

интегро·степ·е́нн·ый, -ая, -ое,   (*a*) (math) integral-power

интегуме́нт·, -а, (*m*) (biol) integument

интелле́кт·, -а, (*m*) intellect

интеллектуа́ль·н·ый, -ая, -ое,   (*a*) intellectual

интеллиге́нт·, -а,   (*m*) intellectual, member of the intelligentsia

интенда́нт·, -а, (*m*) (mil) commissary, supply officer

интенда́нт·ск·ий, -ая, -ое,   (*a*) (mil) supply, commissary

~ слу́ж·б·а (*f*) (mil) supply service

интенси́в·ност·ь, -и,   (*f*) intensity, strength; rate (of a motion, operation)

~ движ·е́ни·я traffic density

~ дел·е́ни·я (nucl) fission rate

~ за·хва́т·а (nucl) capture rate

из·мер·и́тел·ь интенси́в·ност·и (*m*) = **интенси́·метр·**

~ из·но́с·а rate of wear

~ ис·то́ч·ник·а (nucl) source strength

интенси́в·н·ый, -ая, -ое, (*a*) intensive, vigorous, rapid

интенси́·метр·, -а, (*m*) (nucl) ratemeter

интенси·фик·а́тор·, -а, (*m*) intensifier

интенси·фиц·и́р·овать, -уют,   (*imp and perf*) intensify

интер- (*component*) *and see* **меж-, между-,** inter-

интер·амбуля́кр·, -а, (*m*) (zool) inter-ambulacrum

интерва́л·, -а,   (*m*) interval, space, distance; range

~ и́мпульс·ов (phys) momentum range; pulse time

~ мо́щ·ност·ей range of power levels

~ про·пуск·а́ни·я сигна́л·а (elec) gate width

~ пре·вращ·е́ни·я (met) transformation range

~, с·ме́ж·н·ый (math) open interval

интервью·и́р·овать, -уют, (*imp and perf*) interview

интер·гляциа́л·, -а, (*m*) (geol) Interglacial period

интер·грануля́р·н·ый, -ая, -ое,   (*a*) intergranular

интере́с·, -а, (*m*) interest
интере́с·н·ый, -ая, -ое, (*a*) interesting
интерес·ова́ть, -у́ют, (*imp*) interest; -ся (+*instr*) be interested in
интер·каля́р·н·ый, -ая, -ое, (*a*) (biol) intercalary
интер·кварт·и́льн·ый, -ая, -ое, (*a*) interquartile
интер·кин·е́з·, -а, (*m*) (cyt) interkinesis
интер·костёль·н·ый, -ая, -ое, (*a*) (shipb) intercostal
интер·кристалли́т·н·ый, -ая, -ое, (*a*) (cryst) intercrystalline, intergranular
интер·линья́ж, -а, (*m*) (print) interlinear blank/space
интер·металл·и́ческ·ий, -ая, -ое, (*a*) intermetallic
инте́рн·, -а, (*m*) intern, student doctor (in a hospital)
интерна́т·, -а, (*m*) boarding school, residential college; (med) internship
~, шко́ль·н·ый boarding school
интернациона́ль·н·ый, -ая, -ое, (*a*) international
интеро·це́птор·, -а, (*m*) (zool) interoceptor
интерпелл·и́р·овать, -уют, (*imp and perf*) interpellate
интерпол·и́р·овать, -уют, (*imp and perf*) interpolate
интерпол·и́рующ·ий, -ая, -ее, (*pres part act*) of интерпол·и́р·овать; interpolating, interpolation
~ у·стро́й·ств·о (*n*) (autom) interpolator
интерпол·яцио́нн·ый, -ая, -ое, (*a*) interpolation
интерпол·я́ци·я, -и, (*f*) interpolation
~ на·за́д· regressive interpolation
интерпрет·а́ци·я, -и, (*f*) interpretation
интерпрет·е́р·, -а, (*m*) interpreter (punched cards etc.)
интерпрет·и́р·овать, -уют, (*imp and perf*) interpret, explain
интерсе́птор·, -а, (*m*) (a/c) interceptor, spoiler
интерсерта́ль·н·ый, -ая, -ое, (*a*) (geol) intersertal
интерстициа́ль·н·ый, -ая, -ое, (*a*) interstitial, intersticial
интер·ти́п·, -а, (*m*) (print) intertype
интерференц·ио́нн·ый, -ая, -ое, (*a*) interference
~ фигу́р·а (*f*) (crys opt) interference figure
интерфере́нц·и·я, -и, (*f*) interference

интерферо́·метр·, -а, (*m*) (opt) interferometer
~, акуст·и́ческ·ий acoustic interferometer
интерферо·ме́тр·и·я, -и, (*f*) interferometry
интер·це́птор·, -а, (*m*) (a/c) spoiler, interceptor
интерье́р·, -а, (*m*) (arch) interior
инти́м·а, -ы, (*f*) (zool) intima
инти́м·н·ый, -ая, -ое, (*a*) intimate
инти́н·а, -ы, (*f*) (bot) intine
интоксик·а́ци·я, -и, (*f*) intoxication
интра- (*component*) *and see* внутри- intra-
интра·грануля́р·н·ый, -ая, -ое, (*a*) intragranular
интра·целлюля́р·н·ый, -ая, -ое, (*a*) intracellular
интриг·ова́ть, -у́ют, (*imp*) carry on an intrigue; intrigue, rouse the interest of
интроду́кци·я, -и, (*f*) introduction
интро́рз·н·ый, -ая, -ое, (*a*) (bot) introrse
интруд·и́р·овать, -уют, (*imp and perf*) intrude
интрузи́в·н·ый, -ая, -ое, (*a*) intrusive
интру́зи·я, -и, (*f*) (geol) intrusion
интуб·а́ци·я, -и, (*f*) (med) intubation
интуи́ци·я, -и, (*f*) intuition
интуссусце́щи·я, -и, (*f*) (bot) intussusception
инул·а́з·а, -ы, (*f*) (biochem) inulase
инул·и́н·, -а, (*m*) (biochem) inulin
инфанти́ль·н·ый, -ая, -ое, (*a*) infantile
инфа́ркт·, -а, (*m*) (med) infarct
инфекцио́н·н·ый, -ая, -ое, (*a*) (med) infectious
инфе́кци·я, -и, (*f*) infection
~, конта́кт·н·ая contagion, immediate contagion
~, пыл·ев·а́я dust-borne infection
и́нфикс·, -а, (*m*) (gram) infix
ин·фильтр·а́т·, -а, (*m*) infiltrate
ин·фильтр·а́ци·я, -и, (*f*) infiltration, seepage
инфинити́в·, -а, (*m*) (gram) infinitive
инфиц·и́руемост·ь, -и, (*f*) infection risk
инфлюэ́нц·а, -ы, (*f*) (*see also* грипп·) influenza
~ лошад·е́й (vet) equine influenza
~ свин·е́й (vet) swine influenza
инфлюэ́нт·а, -ы, (*f*) influence line
инфля́ци·я, -и, (*f*) (econ) inflation
информ·а́тор·, -а, (*m*) informant

информ·а́ци·я, -и, (*f*) information, data

~, в·ход·н·а́я input information (computers)

информ·бюро́ (*n indecl*) information bureau

информ·и́р·овать, -уют, (*imp and perf*) inform

инфра·зву́к·, -а, (*m*) infrasonic sound, sound wave of subsonic frequency

инфра·звук·ов·о́й, -а́я, -о́е, (*a*) infrasonic, subsonic, subaudio

~ част·от·а́ (*f*) infrasonic/subsonic frequency

инфра·кра́с·н·ый, -ая, -ое, (*a*) infrared

инфра·нейстон·, -а, (*m*) (zool) infra-neuston

инфу́з·, -а, (*m*) infusion

инфу́з·и·я, -и, (*f*) infusion

инфузо́р·и·я, -и, (*f*) (zool) infusorian, (*pl*) *Infusoria*

инфузо́р·н·ый, -ая, -ое, (*a*) infusorial

инциде́нт·, -а, (*m*) incident

инциде́нт·ност·ь, -и, (*f*) incidence

    коэффицие́нт· инциде́нт·ност·и (*m*) (math) incidence number

инциде́нт·н·ый, -ая, -ое, (*a*) incident, (math) coincident

инци́з·и·я, -и, (*f*) (med) incision

инцу́хт·, -а, (*m*) (gen) inbreeding

инъекцио́н·н·ый, -ая, -ое, (*a*) injection

инъе́кци·я, -и, (*f*) injection

иоганни́т·, -а, (*m*) (min) johannite

иогимби́н·, -а, (*m*) (chem) yohimbine

ио́д·, -а, (*m*) (chem) iodine, I

~, пяти·фто́р·ист·ый iodine pentafluoride

~, трёх·хло́р·ист·ый iodine trichloride

~, хло́р·ист·ый iodine monochloride/chloride

иод·алки́л·, -а, (*m*) alkyl iodide

иод·ангидри́д·, -а, (*m*) iodic anhydride, acid iodide

иод·а́т·, -а, (*m*) iodate

иод·и́д·, -а, (*m*) iodide

~ бери́лли·я beryllium iodide

иод·и́р·овать, -уют, (*imp and perf*) iodinate, iodize

иодисто·водо·ро́д·н·ый, -ая, -ое, (*a*) hydriodide of

~ кисл·от·а́ (*f*) hydriodic acid

иодисто·ки́сл·ый, -ая, -ое, (*a*) iodous; iodite of

ио́д·ист·ый, -ая, -ое, (*a*) iodide of (lower or -ous)

~ водо·ро́д· (*m*) hydrogen iodide

~ ка́л·и·й (*m*) potassium iodide

ио́д·ист·ый, -ая, -ое

~ мета́лл· (*m*) metallic iodide

~ мет·и́л· (*m*) methyl iodide

иод·и́т·, -а, (*m*) iodite

иод·новати́ст·ая кисл·от·а́ (*f*) hypoiodous acid

иодновато·ки́сл·ый, -ая, -ое, (*a*) iodate of

иод·нова́т·ый, -ая, -ое, (*a*) iodic

~ ангидри́д· (*m*) iodic anhydride

~ кисл·от·а́ (*f*) iodic acid

иодно·ки́сл·ый, -ая, -ое, (*a*) periodate (of)

ио́д·н·ый, -ая, -ое, (*a*) iodine; iodide of (higher or -ic)

~ кисл·от·а́ (*f*) periodic acid

~ ртут·ь (*f*) mercuric iodide

~ числ·о́ (*n*) iodine value

иодо·водоро́д·, -а, (*m*) hydrogen iodide

иодозо- (*component*) (chem) iodoso-

иодо·крахма́л·, -а, (*m*) iodized starch

иод·окси- (*component*) (chem) iodoxy-

иодо·метр·и́ческ·ий, -ая, -ое, (*a*) iodometric

иодо·ме́тр·и·я, -и, (*f*) iodometry

иод·опс·и́н·, -а, (*m*) (biochem) iodopsin

иодно·сере́бр·ян·ый, -ая, -ое, (*a*) (phot) silver iodide

иодо·фо́рм·, -а, (*m*) (chem) iodoform

иод·стирол·, -я, (*m*) iodostyrene

иод·циа́н·, -а, (*m*) iodocyanogen

ио́л·, -а, (*m*) (naut) yawl

ио́л·а, -ы, (*f*) (naut) yole

ио́н·, -а, (*m*) (phys) ion

~, амфоте́р·н·ый zwitterion

~ водо·ро́д·а hydrogen ion, hydrion

~, за·хва́ч·енн·ый trapped ion

~, ис·хо́д·н·ый parent ion

~ индика́тор·а tracer ion

~, молекуля́р·н·ый ionized molecule, ion

~ от·да́·ч·и recoil ion

~, парази́т·н·ый spurious ion

~, сре́д·н·ий hybrid ion, zwitterion

ион·дипо́ль·н·ая с·вяз·ь (*f*) ion-dipole interaction

ио́н·иев·ый, -ая, -ое, (*a*) (chem) ionium, Io

иониз·а́тор·, -а, (*m*) ionizer, ionizing agent

ионизацио́н·н·ый, -ая, -ое, (*a*) ionization, ionizing

~ ка́мер·а (*f*) (nucl) ionization/ion chamber

~ ка́мер·а, норма́ль·н·ая free-air ionization chamber

~ толч·о́к· (*m*) ionization burst

иониз·а́ци·я, -и, (*f*) (phys) ionization
~, по·ве́рх·ностн·ая surface ionization
~, уда́р·н·ая impact ionization
иониз·и́р·овать, -уют, (*imp*) ionize
иониз·ова́ть, -у́ют, (*imp*) ionize
иониз·и́рующ·ий, -ая, -ее, (*pres part act*) of **иониз·и́р·овать** ionizing
~ с·толк·нове́ни·е (*n*) ionizing collision
ио́н·и·й, -я, (*m*) (chem) ionium, Io
ион·и́йск·ий, -ая, -ое, (*a*) (arch) Ionic
иони́т·, -а, (*m*) (chem) ion exchanger, ion-exchange resin, ionite
иони́т·ов·ый, -ая, -ое, (*a*) of **иони́т·**
ион·и́ческ·ий, -ая, -ое, (*a*) (arch) Ionic
~ о́рдер· (*m*) (arch) Ionic order
ионо·об·ме́н·ник·, -а, (*m*) ion-exchanger
ио́н·н·ый, -ая, -ое, (*a*) (phys) ion, ionic; thermionic
~ атмосфе́р·а (*f*) ion cloud
~ венти́л·ь (*m*) ionic/thermionic rectifier
~ да́т·чик· (*m*) thermionic transmitter
~ ис·то́ч·ник· (*m*) ion source
~ ла́мп·а (*f*) ionic/electronic valve, rectifier tube/valve
~ пуч·о́к· (*m*) ionic/ion beam
~ с·вяз·ь (*f*) ionic bond
~ си́л·а (*f*) ionic strength
~ тео́р·и·я (*f*) ionics
ионо·ге́н·, -а, (*m*) ionogen
ионо́·метр·, -а, (*m*) ionometer, roentgenometer
ионо́н·, -а, (*m*) (chem) ionone
ионо·об·ме́н·ник·, -а, (*m*) ion exchanger, ion-exchange resin
ионо·об·ме́н·н·ый, -ая, -ое, (*a*) (chem) ion-exchange
ионо·па́уз·а, -ы, (*f*) ionopause
ионо·сфе́р·а, -ы, (*f*) ionosphere
~ Юпи́тер·а (astron) Jovian ionosphere
ионо·терап·и́·я, -и, (*f*) medical ionization
ионто·форе́з·, -а, (*m*) (med) iontophoresis
ио́т·а, -ы, (*f*) iota (Greek)
иохимби́н·, -а, (*m*) (pharm) yohimbine
ипоме·и́н·, -а, (*m*) (pharm) ipomein
ипоте́к·а, -и, (*f*) (fin) mortgage
ипоте́ч·н·ый, -ая, -ое, (*a*) of **ипоте́к·а**
~ банк· (*m*) (fin) building society (Brit), savings and loan association (U.S.)

иппо·дро́м·, -а, (*m*) race track/course, horse-racing track
и пр. (*abbr*) of **и про́ч·ее** etc., and so on, and the rest
ипри́т·, -а, (*m*) mustard gas, yperite
и́псилон·, -а, (*m*) upsilon (Greek); (ent) black cutworm, *Agrotis ypsilon*
ира́зер·, -а, (*m*) (phys) iraser
ирид·есценци·я, -и, (*f*) iridescence
ири́д·иев·ый, -ая, -ое, (*a*) (chem) iridium, Ir
ири́д·и·й, -я, (*m*) (chem) iridium, Ir
ирид·и́ст·ый, -ая, -ое, (*a*) iridous
ирид·осми́н·, -а, (*m*) (min) iridosmine
иридо·цикли́т·, -а, (*m*) (med) iridocyclites
иридо·ци́т·, -а, (*m*) (zool) iridocyte
ириза́ци·я, -и, (*f*) (meteor) irisation, iridescence
и́рис·, -а, (*m*) (bot) iris; (opt) iris; (pharm) orris; **ири́с·, -а,** (*m*) (food) toffee
ири́с·ов·ый, -ая, -ое, (*a*) of **и́рис·**
~ диафра́гм·а (*f*) (phot) iris shutter frame
ири́т·, -а, (*m*) (med) iritis
ирла́нд·ск·ий, -ая, -ое, (*a*) Irish
ирон·и́ческ·ий, -ая, -ое, (*a*) ironical
ирради·а́ци·я, -и, (*f*) (*see also* **об·луч·е́ни·е**); irradiation
иррациона́ль·н·ый, -ая, -ое, (*a*) irrational
~ числ·о́ (*n*) (math) irrational number
ирригацио́н·н·ый, -ая, -ое, (*a*) irrigation
ИСЗ (*abbr*) of **иску́с·ственн·ый с·пу́т·ник· земл·и́** artificial earth satellite
и́ск·, -а, (*m*) (law) suit, action
предъ·яв·и́ть, иск· bring an action against, sue
иск·а́вш·ий, -ая, -ее, (*past part act*) of **иск·а́ть**
ис·каж·а́·ть, -ют, (*imp*) distort
ис·каж·а́юш·ий, -ая, -ее, (*pres part act*) of **ис·каж·а́·ть**
~ о·кра́ш·ивани·е (*n*) dazzle painting (camouflage)
ис·каж·е́ни·е, -я, (*m*) distortion
~, амплиту́д·н·ое (rad) attenuation distortion
коэффицие́нт· ис·каж·е́ни·я (*m*) distortion factor
~, не·лин·е́йн·ое non-linear distortion
~, о·гиб·а́юще·е (rad) envelope delay distortion
~, пере·крёст·н·ое intermodulation distortion

ис·каж·éни·е

~, **подушко·обрáз·н·ое** (rad) pin-cushion distortion

~, **трапецоидáль·н·ое** (rad) trapezoidal keystone distortion

~, **фáз·ов·ое** phase distortion; (telecom) phase-delay distortion

ис·каж·ённ·ый, -ая, -ое, (*past part pass*) of **ис·каз·úть;** distorted

ис·каз·úть, -ят, (*perf*) distort

ис·кáл·ыва·ть, -ют, (*imp*) prick/stab all over

иск·áни·е, -я, (*n*) search, searching; (telecom) hunting, selection

~ **в одн·óй декáд·е** selection on one level, rotary hunting (telephony)

~ **вы́·зов·а** finding (telephony)

~, **свобóд·н·ое** hunting, trunk hunting (telephony)

~, **шáг·ов·ое** step-by-step selection (telephony)

úск·анн·ый, -ая, -ое, (*past part pass*) of **иск·áть;** (telecom) sought, looked for, dialled

ис·кáп·ыва·ть, -ют, (*imp*) dig up

иск·áтел·ь, -я, (*m*) seeker, finder; (telecom) selector, finder; searcher (radar); scanner (scintillation); (phot) view finder

~, **вращ·áтельн·ый** uniselector (telephony)

~ **вы́·зов·ов** linefinder (telephony)

~, **координáт·н·ая** crossbar selector/switch (telephones)

~, **лин·éйн·ый** first selector (telephony)

~, **машúн·н·ый** power-driven selector (telephony)

~ **между·горóд·н·ых лин·и·й** trunk-offering selector (telephones)

~ **пéрв·ой бýкв·ы** A-digit selector (telephones)

~, **пéрв·ый групп·ов·óй** district selector (telephones)

~, **по·глощ·áющ·ий** digit absorber (telephones)

~, **подъ·ём·о·вращ·áтельн·ый** two-motion selector (telephony)

~, **регúст·ов·ый** A-digit selector

~, **транзúт·н·ый** tandem selector (telephones)

~, **шáг·ов·ый** step-by-step selector

иск·áть (*pres 3rd sing, pl*) **úщ·ет, úщ·ут,** (*imp*)(+ *gen*) look/search for, seek; (law) claim

~ **трáл·ом за** drag for

иск·áтельн·ый, -ая, -ое, (*a*) searching, finding, finder

ис·ключ·á·ть, -ют, (*imp*) exclude, except; eliminate, rule out, preclude

ис·ключ·éни·е, -я, (*n*) exclusion; exception; (math) elimination

за ис·ключ·éни·ем except for, with exception of, other than

~ **под·станóв·к·ой** (math) elimination by substitution

прúнцип ис·ключ·éни·я (*m*) (phys) exclusion principle

~ **фóн·а** (phys) background cancellation

ис·ключ·úтельн·ый, -ая, -ое, (*a*) exceptional; sole, exclusive

ис·ключ·úть, -áт, (*perf*) *see* **ис·-ключ·á·ть**

ис·ковéрк·а·ть, -ют, (*perf*) distort, corrupt

иск·ов·óй, -áя, -óе, (*a*) of **úск·**

ис·кол·óть, -ют, (*perf*) prick/stab all over

ис·кóм·ка·ть, -ют, (*perf*) reduce to a lump/clot, crumple up

иск·óм·ый, -ая, -ое, (*pres part pass*) of **иск·áть;** sought for, desired, to be determined; (*n decl as adj*) (math) unknown quantity

~ **велич·ин·á** (*f*) (math) value sought, value to be determined, unknown quantity

ис·кóн·н·ый, -ая, -ое, (*a*) primordial, primeval; indigenous (of people); ancient (of towns etc.)

ис·коп·áем·ое, -ого, (*n decl as adj*) fossil; mineral

~, **жив·óе** (zool) living fossil

~, **мéлк·ое** (pal) microfossil

~, **не·рýд·н·ое** gauge/non-metallic mineral

~, **по·лéз·н·ое** mineral, useful/commercial mineral

~, **рýд·н·ое** ore mineral

~, **руко·вод·я́щ·ее** index fossil

ис·коп·áем·ый, -ая, -ое, (*pres part pass*) of **ис·коп·á·ть;** mined, dug up; mineral; fossilized (*see also* **ис·коп·áем·ое** *as noun*)

~ **ýгол·ь** coal

ис·коп·á·ть, -ют, (*perf*) dig up

ис·корен·éни·е, -я, (*n*) eradication, extirpation, extermination, uprooting

ис·корен·úть, -я́т, (*perf*) *see* **ис·ко-рен·я́·ть**

ис·корен·я́·ть, -ют, (*imp*) eradicate, extirpate

и́с·кос·а (*adv*) aslant

и́скр·а, -ы, (*f*) spark

~, гас·и́м·ая (rad) quenched spark

~, за·па́ль·н·ая ignition spark

~, не по́лн·ост·ью за·тух·а́ющ·ая arcing spark

~, тон·а́льн·ая (elec) singing spark

искр·е́ни·е, -я, (*n*) sparking

и́скрен·н·о пре́·да·нн·ый вам yours faithfully/sincerely

и́скрен·н·ий, -яя, -ее, (*a*) sincere, candid, frank

ис·крив·и́ть, -я́т, (*perf*) *see* **ис·кри·вл·я́·ть**

ис·кривл·е́ни·е, -я, (*n*) (*v n*) *see* **ис·-кривл·я́·ть**; bend; distortion, warping

~ позвон·о́чник·а (med) curvature of the spine

~ сква́ж·ин·ы (oil) well deviation

ис·кривл·ённ·ый, -ая, -ое, (*past part pass*) *see* **ис·кривл·я́·ть**

ис·кривл·я́·ть, -ют, (*imp*) bend, crook, curve; distort, twist, warp

искр·и́ст·ый, -ая, -ое, (*a*) sparking

искр·и́ть, -я́т, (*imp*) spark; **-ся** sparkle, scintillate

искр·ов·о́й, -а́я, -о́е, (*a*) spark, sparking

~ потенциа́л· (*m*) (elec) sparking potential

~ про́б·а (*f*) (met) spark test

~ пере·кры́т·и·е (*n*) (elec) sparkover, flashover

~ про·меж·у́ток· (*m*) (elec) spark gap arrester; spark gap

~ раз·ря́д· (*m*) (elec) spark, spark discharge

~ раз·ря́д·ник· (*m*) (rad) spark gap; (elec) spark gap arrester

искро·гас·и́тел·ь, -я, (*m*) spark extinguisher/arrester/killer

искро·гас·и́тельн·ый, -ая, -ое, (*a*) (elec) spark-extinguishing/quenching, blowout, quenched

~ ко́нтур· (*m*) (elec) quenching circuit

искро·гаш·е́ни·е, -я, (*n*) spark extinguishing, (elec) blow-out, spark blow-out

искро·ме́р·, -а, (*m*) (phys) scintillometer

искро·образ·ова́ни·е, -я, (*n*) sparking

искро·сто́й·к·ий, -ая, -ое, (*a*) non-sparking, non-arcing

искро·у·держ·а́тел·ь, -я, (*m*) spark arrester

искро·у·каз·а́тел·ь, -я, (*m*) spark detector

искро·у·лов·и́тел·ь, -я, (*m*) spark arrester

ис·крош·и́ть, -а́т, (*perf*) crumble, break to pieces

ис·кус·а́·ть -ют, (*perf*) bite/sting all over

ис·кус·и́ть, -я́т, (*perf*) *see* **ис·куш·а́·ть**

искус·н·о (*adv*) skilfully

искус·н·ый, -ая, -ое, (*a*) skilful, expert

искус·ственн·ый, -ая, -ое, (*a*) artificial, man-made

~ волокн·о́ (*n*) man-made fibre

~ дых·а́ни·е (*n*) (med) artificial respiration

~ зу́б·ы (*pl*) false teeth

~ ко́ж·а (*f*) artificial leather, leather substitute

~ са́л·о (*n*) hydrogenated fat

~ со·оруж·е́ни·я (*pl*) (rail) track auxiliary works/structures

~ с·пу́т·ник· (*m*) (rocket) artificial satellite (*see* **с·пу́т·ник·**)

~ тя́г·а (*f*) mechanical draught

~ шёлк· (*m*) rayon, artificial silk

искус·ств·о, -а, (*n*) art; skill, artifice

~, воен·н·ое military science

про·из·вед·е́ни·е искус·ств·а (*n*) work of art

ис·куш·а́·ть, -ют, (*imp*) put to the test, prove

ИСЛ (*abbr*) of **искус·ственн·ый с·пу́т·ник· лун·ы́** artificial moon satellite

исла́нд·ск·ий, -ая, -ое, (*a*) Iceland, Icelandic

~ мох· (*m*) (bot) Iceland moss, *Cetraria islandica*

~ шпат· (*m*) (min) Iceland spar

испа́н·ск·ий, -ая, -ое, (*a*) Spanish

ис·пар·е́ни·е, -я, (*n*) (*v n*) *see* **ис·па·р·я́·ть**; evaporation, vaporization, volatilization; (bot) transpiration

ис·пар·ённ·ый, -ая, -ое, (*past part pass*) *see* **ис·пар·я́·ть**; evaporated, vaporized, volatilized

ис·па́р·ин·а, -ы, (*f*) perspiration, sweat; condensation

ис·пар·и́тел·ь, -я, (*m*) evaporator, vaporizer; (meteor) evaporimeter, evaporometer, evaporation gauge

~, вертика́ль·н·ый (chem) vertical shell evaporator

~, горизонта́ль·н·ый (chem) horizontal shell evaporator

~, куб·ов·о́й pot-type evaporator

ис·пар·и́тел·ь

~, **мно́го·ступ·е́нчат·ый** multi-effect evaporator

~, **паро·тру́б·н·ый** horizontal shell evaporator

ис·пар·и́тел·ьност·ь, -и, (*f*) evaporation rate (of boilers)

ис·пар·и́тел·ьн·ый, -ая, -ое, (*a*) evaporator, evaporative, evaporating

~ **спосо́б·ност·ь** (*f*) calorific value (fuels, refrigeration media); evaporation rate (boilers)

~ **у·стано́в·к·а** (*f*) evaporator

ис·пар·и́ть, -я́т, (*perf*) *see* **ис·пар·я́·ть**

ис·пар·я́емост·ь, -и, (*f*) volatility, evaporability

ис·пар·я́ем·ый, -ая, -ое, (*pres part pass*) of **ис·пар·я́·ть;** volatile

ис·пар·я́·ть, -ют, (*imp*) evaporate, volatilize; **-ся** be evaporated, vaporize

ис·па́чк·а·ть, -ют, (*perf*) soil, dirty, stain

ис·пёк·ш·ий, -ая, -ее, (*past part act*) of **ис·пе́ч·ь**

ис·пепел·е́ни·е, -я, (*m*) incineration

ис·пепел·и́ть, -я́т, (*perf*) *see* **ис·пепел·я́·ть**

ис·пепел·я́·ть, -ют, (*imp*) incinerate, reduce to ashes

ис·пестр·и́ть, -я́т, (*perf*) *see* **ис·пест·я́·ть**

ис·пест·я́·ть, -ют, (*imp*) variegate, mottle, speckle

ис·пе́ч·ь (*fut 3rd sing, pl*) **ис·печ·ёт, ис·пек·у́т,** (*pas· masc sing*) **ис·пёк·,** (*perf*) bake

ис·пещр·и́ть, -я́т, (*perf*) *see* **ис·пещ·р·я́·ть**

ис·пещр·я́·ть, -ют, (*imp*) variegate, mottle, speckle

ис·пис·а́ть (*fut 3rd sing, pl*) **ис·пи́ш·ет, ис·пи́ш·ут,** (*perf*) *see* **ис·пи́с·ыва·ть**

ис·пи́с·ыва·ть, -ют, (*imp*) cover/fill with writing; use up (by writing)

ис·пи́ш·ут (*fut 3rd pl*) of **ис·пис·а́ть**

ис·под·ни́з·у (*adv*) from below

ис·по́д·ник·, -а, (*m*) (met) bed die

исполи́н·, -а, (*m*) giant

исполи́н·ов·ый, -ая, -ое, (*a*) of **исполи́н·**

~ **котёл·** (*m*) (geol) pot hole

исполи́н·ск·ий, -ая, -ое, (*a*) gigantic

исполко́м·, -а, (*m*) (*abbr*) of **ис·полн·и́тельн·ый комите́т·** executive committee

ис·полн·е́ни·е, -я, (*n*) (*v n*) *see* **ис·полн·я́·ть;** execution, fulfilment (of orders, instructions), discharge (of duties)

ис·полн·енн·ый, -ая, -ое, (*past part pass*) *see* **ис·полн·я́·ть**

ис·полн·и́мост·ь, -и, (*f*) feasibility, practicability

ис·полн·и́м·ый, -ая, -ое, (*pres part pass*) of **ис·по́лн·ить;** feasible, practicable

ис·полн·и́тел·ь, -я, (*m*) executive, executor; performer, artist

~, **суд·е́бн·ый** bailiff

ис·полн·и́тельн·ый, -ая, -ое, (*a*) executive; (autom) final, slave

~ **лист·** (*m*) (law) court order, writ of execution

~ **о́рган·** (*m*) executive body; (autom) final control element

~ **о́рган·, пере·двиг·а́ющ·ий** (*m*) transfer final-control element

~ **рыча́г·** (*m*) slave arm (of a manipulator)

~ **у·стро́й·ств·о** (*n*) (autom) final control element; (instr) recording/indicating unit

~ **у·стро́й·ств·о, а·стат·иче́ск·ое** (*n*) floating final-control element

~ **цили́ндр·** (*m*) receiver cylinder (of telemotor)

~ **черт·ёж·** (*m*) working drawing

ис·по́лн·ить, -ят, (*perf*) *see* **ис·полн·я́·ть**

ис·полн·я́·ть, -ют, (*imp*) carry out, execute, discharge, fulfil; perform, play

ис·полн·я́ющ·ий, -ая, -ее, (*pres part act*) of **ис·полн·я́·ть**

~ **обя́з·анност·и** acting (rank)

ис·полос·ова́ть, -у́ют, (*perf*) cross out; cut to pieces

ис·по́льз·овани·е, -я, (*n*) use, utilization, employment

   коэффицие́нт· ис·по́льз·овани·я (*m*) duty/utilization factor (of plant, machinery etc.)

   коэффицие́нт· ис·по́льз·овани·я электро́д·а (*m*) (weld) deposition efficiency

   проце́нт· ис·по́льз·овани·я (*m*) efficiency; utilization

ис·по́льз·овать, -у́ют, (*imp and perf*) utilize, make use of, use, employ

~ **максима́ль·н·о** make the most of, get the best out of (machines)

~ **по·вто́р·н·о** use again; recycle

ис·по́рт·ить, -ят, (*perf*) spoil, damage, harm, put out of order; **-ся** (*passive*) get damaged, be out of order; go bad, rot, deteriorate

ис·по́рч·енн·ый, -ая, -ое, (*past part pass*) of **ис·по́рт·ить;** spoiled, spoilt, rotten, tainted

ис·прав·и́м·ый, -ая, -ое, (*pres part pass*) of **ис·пра́в·ить;** rectifiable, remediable

ис·прав·и́тельн·ый, -ая, -ое, (*a*) correctional, corrective

ис·правл·е́ни·е, -я, (*n*) (*v n*) see **ис·-правл·я́·ть;** rectification; (print) emendation, amendment, correction

ис·пра́вл·енн·ый, -ая, -ое, (*past part pass*) see **ис·правл·я́·ть**

~ из·да́н·и·е (*n*) revised edition

~ шир·от·а́ (*f*) observed latitude

ис·пра́в·ить, -ят, (*perf*) see **ис·правл·я́·ть**

ис·правл·я́·ть, -ют, (*imp*) put right, repair, rectify, remedy; (print) revise (an edition), amend, emendate (a text)

ис·пра́в·ност·ь, -и, (*f*) good condition/ repair/working order; conscientious-ness

ис·пра́в·н·ый, -ая, -ое, (*a*) in good repair/working order, sound; con-scientious

~ со·сто·я́ни·е (*n*) good working order

ис·праж·не́ни·е, -я (*n*) (physiol) defecation, motion; (*pl*) faeces, excre-ment

ис·пра́ш·ива·ть, -ют, (*imp*) obtain (by soliciting); solicit

ис·про́б·овать, -уют, (*perf*) test, try out; experience

ис·прос·и́ть, ⸓ят, (*perf*) obtain (by soliciting)

ис·про́ш·енн·ый, -ая, -ое, (*past part pass*) of **ис·прос·и́ть**

ис·пуг·а́·ть, -ют, (*perf*) frighten; **-ся** (+ *gen*) be frightened of

ис·пуск·а́ем·ый, -ая, -ое, (*pres part pass*) of **ис·пуск·а́·ть**

ис·пуск·а́ни·е, -я, (*n*) emission, ema-nation

ис·пуск·а́тел·ь, -я, (*m*) emitter

ис·пуск·а́тельн·ый, -ая, -ое, (*a*) emitting, emissive

~ спосо́б·ност·ь (*f*) (phys) emissive power

ис·пуск·а́·ть, -ют, (*imp*) emit; **-ся** (*pass*); emanate

ис·пуст·и́ть, ⸓ят, (*perf*) see **ис·пуск·а́·ть**

ис·пу́щ·енн·ый, -ая, -ое, (*past part pass*) of **ис·пуст·и́ть**

ис·пыт·а́вш·ий, -ая, -ое, (*past part act*) see **ис·пы́т·ыва·ть**

~ на·гре́·в· heated

~ цикл· cycled

ис·пыт·а́ни·е, -я, (*n*) (*and see under associated words*) test, testing; assay; examination; trial(s) (of machinery)

~, био·лог·и́ческ·ое (pharm) bioassay

~ в нату́р·е (met) testing on an un-notched piece

~, внутри·реа́ктор·н·ое (nucl) in-pile test

~, в·ступ·и́тельн·ое entrance exami-nation (academic)

~, госуда́р·ственн·ое official test/ trials, contract acceptance trials

~, дл·и́тельн·ое long-term/time test/ testing

~ до· раз·руш·е́ни·я test to destruc-tion

~, за·во́д·ск·ое works/factory/shop test; (shipb) contractor's/builder's trials

~ магни́т·н·ым порош·к·о́м mag-netic particle testing/inspection

~, моде́ль·н·ое scale-model test, mock-up test; simulation test

~ на ... (+ *acc*) (*see also associated words and* **и.** *plus adj form*), test for ..., ... test; (+ *dat*) testing on/in, testing in a state of

~ на га́з·у (ICE) test under own power

~ на гермет·и́чност·ь leak test

~ на из·ги́б· deflection test, (met) transverse-bend test/testing; flex-test-ing (of rubber)

~ на кисл·о́тност·ь acid test

~ на круч·е́ни·е shear/torsion test/ testing

~ на ме́ст·е field/site test/testing

~ на моде́л·и model/mock-up test

~ на рас·тяж·е́ни·е tensile testing

~ на с·дви́г· shear/torsion test/testing

~ на с·жа́т·и·е compression testing

~ на· ско́р·ост·ь speed test/trial(s)

~ на ход·у́ running test

~ на уда́р·н·ый из·ги́б· impact test, notched-bar test

~ на уда́р·н·ое рас·тяж·е́ни·е ten-sile impact test

, ~ на у·ста́л·ост·ь fatigue test

~, нату́р·н·ое full-scale test/trials; (met) testing on unnotched specimen

~, по·ве́р·очн·ое check test

~, порцио́н·н·ое batch testing

ис·пыт·а́ни·е
~, при·ём·очн·ое   acceptance test/trial(s)
~, прогресси́в·н·ое   (shipb) progressive speed/standardization trial(s)
~, рабо́ч·ее   service test
~, стат·и́ческ·ое   static test/testing
~, сте́нд·ов·ое   bench test/testing
~, сте́нд·ов·ое   огн·ев·о́е   (rocket) static firing
~, техно·лог·и́ческ·ое   (met) forge/workshop test
~, ти́п·ов·ое   type-approval test
~ травл·е́ни·ем   etching test
~, уда́р·н·ое   shock/impact/dynamic test/testing
~, ход·ов·о́е   running trial(s); (shipb) sea trial(s)
~, холост·о́е   no-load test; check testing
~, шва́рт·ов·ое   (shipb) basin/dock trial(s)
~, эксплуатацио́н·н·ое   performance test/trials; service test
ис·пы́т·анн·ый, -ая, -ое,   (past part pass) see ис·пы́т·ыва·ть; tried, proven
ис·пыт·а́тел·ь, -я,   (m) tester
~ изол·я́ци·й   (elec) insulation tester
лёт·чик·-ис·пыт·а́тел·ь, -я,   (m) test pilot
те́хн·ик·-ис·пыт·а́тел·ь -я,   (m) testing officer
~ у́·ровн·ей   (surv) level trier
ис·пыт·а́тельн·ый, -ая, -ое,   (a) test, testing, trial; experimental
~ програ́мм·а   (f) test routine
~ ста́нци·я   (f) experimental station
~ стенд·   (m) test bed/bench
ис·пыт·а́·ть, -ют,   (perf) see ис·пы́т·ыва·ть
ис·пыт·у́ем·ый, -ая, -ое,   (a) test, under test, undergoing, being subjected to; (m decl as adj) examinee
~ вещ·еств·о́   (n) test material
ис·пы́т·ыва·ть, -ют,   (imp) test, try out, assay; experience, undergo, be subjected to
~ на   test for
ис·сек·а́·ть, -ют,   (imp) (med) resect
и́с·сер·а-   (component) tinged/shot with grey/gray, with a greyish/grayish tint/sheen
ис·сеч·е́ни·е, -я,   (n) (med) resection, excision
ис·се́ч·ь   (fut 3rd sing, pl) ис·сеч·ёт, ис·сек·у́т, (past masc sing) ис·се́к·, (perf); see ис·сек·а́·ть

и́с·сиз·а-   (component) tinged/shot with bluish-grey/gray
и́с·син·я-   (component) shot/tinged with blue, with a bluish tint/sheen
ис·сле́д·овани·е, -я,   (n) investigation; research, study; exploration; examination, analysis; research paper
~, вед·у́щ·ее   pilot study/research
~, на·у́ч·н·ое   scientific research
~, пробле́м·н·ое   basic research
~, рентгено·граф·и́ческ·ое   X-ray diffraction analysis
~, хим·и́ческ·ое   chemical analysis
ис·сле́д·ованн·ый, -ая, -ое,   (past part pass) of ис·сле́д·овать
ис·сле́д·овател·ь, -я,   (m) investigator, research worker; explorer
ис·сле́д·овательск·ий, -ая, -ое,   (a) investigation; research; exploratory
~ рабо́т·а   research work
ис·сле́д·овать, -уют,   (imp and perf) investigate, study; explore; examine, analyse
ис·сле́д·уем·ый, -ая, -ое,   (pres part pass) of ис·сле́д·овать; under investigation/test, experimental, studied, covered, test, in question
~ об·раз·е́ц·   (m) test specimen
~ с·плав·   (m) the alloy in question, the alloy under study
иссо́п·, -а,   (m) (bot) hyssop, *Hyssop officinalis*
иссо́п·н·ый, -ая, -ое,   (a) (bot) hyssop
ис·со́х·нуть, -нут,   (past masc sing) ис·со́х·, (perf) see ис·сых·а́·ть
ис·суш·а́·ть, -ют,   (imp) dehydrate, dry, desiccate; -ся (pass); wither, dry up
ис·су́ш·ива·ть, -ют,   (imp) see ис·-суш·а́·ть
ис·суш·и́ть, -ат,   (perf) see ис·су-ш·а́·ть
ис·сых·а́·ть, -ют,   (imp) dry, dry up, wither, shrivel
иссяк·а́·ть, -ют,   (imp) dry up; run out, come to an end
исся́к·нуть, -нут,   (past masc sing) иссяк·, (perf) see иссяк·а́·ть
ист   (abbr) of ис·то́ч·ник· source
ис·та́пл·ива·ть, -ют,   (imp) heat, have the heating on, light (a stove etc.); use in heating; melt
ис·та́пт·ыва·ть, -ют,   (imp) trample, trample on
ис·таск·а́·ть, -ют,   (perf) see ис·-та́ск·ива·ть
ис·та́ск·ива·ть, -ют,   (imp) wear out

ис·та́ч·ива·ть, -ют, (*imp*) wear down by sharpening

ис·тек·а́·ть, -ют, (*imp*) run out, elapse, expire; run out, discharge, flow out, emanate, effuse

ис·тек·ш·ий, -ая, -ее, (*past part act*) of **ис·те́ч·ь** (*perf*), see **ис·тек·а́·ть** (*imp*); past, last

ис·тер·е́ть (*fut 3rd sing, pl*) **изо·-тр·ёт, изо·тр·у́т,** (*past masc sing*) **ис·тёр·,** (*perf*); see **ис·тир·а́·ть**

истер·и́ческ·ий, -ая, -ое, (*a*) hysterical

истер·и́чн·ый, -ая, -ое, (*a*) hysterical

ис·тёр·т·ый, -ая, -ое, (*past part pass*) of **ис·тер·е́ть;** worn

ист·е́ц·, ист·ц·а́, (*i*) **ист·ц·о́м,** (*m*) (law) plaintiff; petitioner

ис·теч·е́ни·е, -я, (*n*) efflux, discharge, emission, outflow, escape; expiration, expiry, lapse (of time)

    коэффицие́нт· ис·те́ч·е́ни·я (*m*) (hydr) coefficient of discharge

~, не·свобо́д·н·ое (hydr) submerged discharge

ис·те́чь (*fut 3rd sing, pl*) **ис·теч·ёт, ис·тек·у́т,** (*past masc sing*) **ис·-тёк·,** (*perf*); see **ис·тек·а́·ть**

йст·ин·а, -ы, (*f*) truth

йст·инн·ый, -ая, -ое, (*a*) true

~ горизо́нт· на·блюд·а́тел·я (*m*) celestial/rational horizon

~ меридиéн· (*m*) geographical meridian

~ со́лн·ечн·ое вре́м·я true/apparent solar time

~ су́тк·и (*pl*) solar day

ис·тир·а́емост·ь, -и, (*f*) wearability, wear, abrasion

ис·тир·а́ни·е, -я, (*n*) (*v n*) of **ис·-тир·а́·ть;** attrition; abrasion, wear

~ вод·о́й water abrasion

    ис·пыт·а́ни·е на ис·тир·а́ни·е (*n*) (build) attrition test

~ с·ги́б·ом (text) flex abrasion

    со·противл·е́ни·е ис·тир·а́ни·ю (*n*) abrasive resistance

ис·тир·а́·ть, -ют, (*imp*) grind up; wear down/away, abrade

ис·тл·ева́·ть, -ют, (*imp*) moulder, decay to dust, rot away; be reduced to ash

ис·тл·е́вш·ий, -ая, -ее, (*past part act*) of **ис·тл·е́·ть;** mouldy, moldy, rotten, decayed

ис·тл·е́·ть, -ют, (*perf*) see **ис·тл·е-ва́·ть**

исто́д·, -а, (*m*) (bot) milkwort, *Polygala*

ис·то́к·, -а, (*m*) (*v n*) see **ис·тек·а́·ть;** source (of river)

ис·токо·обра́з·н·ый, -ая, -ое, (*a*) (phys) source-like

ис·толк·ова́ни·е, -я, (*n*) interpretation, construction

ис·толк·ова́тел·ь, -я, (*m*) exponent

ис·толк·ова́·ть, -у́ют, (*perf*) see **ис·-толк·о́выва·ть**

ис·толк·о́выва·ть, -ют, (*imp*) interpret, construe, expound

ис·толк·у́т (*fut 3rd pl*) of **ис·толо́ч·ь**

ис·толо́ч·ь (*fut 3rd sing, pl*) **ис·тол-ч·ёт, ис·толк·у́т,** (*past masc sing*) **ис·толо́к·,** (*perf*); pound, stamp

ис·толч·ённ·ый, -ая, -ое, (*past part pass*) of **ис·толо́ч·ь;** crushed, broken up

ис·то́м·а, -ы, (*f*) fatigue

ис·том·и́ть, -я́т, (*perf*) see **ис·том-л·я́·ть**

ис·томл·е́ни·е, -я, (*n*) exhaustion

ис·томл·я́·ть, -ют, (*imp*) weary, fatigue, exhaust, tire out

ис·топ·и́ть, -я́т, (*perf*) see **ис·та́п-л·ива·ть**

ис·топ·ни́к·, -а́, (*m*) stoker, boiler/furnace-man

ис·топт·а́ть (*fut 3rd sing, pl*) **ис·-то́пч·ет, ис·то́пч·ут,** (*perf*); see **ис·-та́пт·ыва·ть**

ис·торг·а́·ть, -ют, (*imp*) eject, expel, erupt

ис·то́рг·нуть, -нут, (*past masc sing*) **ис·то́рг·,** (*perf*); see **ис·торг·а́·ть**

исто́р·ик·, -а, (*m*) historian

исто́р·ик·о- (*component*) **истор·и́-ческ·ий** historical

истор·и́ческ·ий, -ая, -ое, (*a*) historical; historic

истор·и́чн·ый, -ая, -ое, (*a*) historically accurate/correct/well-founded

истор·и́чност·ь, -и, (*f*) historicity

исто́р·и·я, -и, (*f*) history; story, event, happening; unpleasantness, trouble, business

ис·точ·а́·ть, -ют, (*imp*) emit, give off

ис·точ·и́ть, -а́т, (*perf*) sharpen away, wear down by sharpening; eat through, cover with holes

ис·то́ч·ник·, -а, (*m*) source, origin; spring, well

~, ко́бальт·ов·ый (nucl) cobalt bomb

~, лин·е́йн·ый (phys) line source

ис·то́ч·ник

~ **нейтро́н·ов** (phys) neutron source

~, **не·капт·и́рованн·ый** wild spring

~ **пит·а́ни·я** (elec) power supply, power supply unit

источнико·ве́д·ени·е, -я, (n) historiography

ис·тощ·а́·ть, -ют, (imp) exhaust, wear out, deplete

ис·тощ·ени·е, -я, (n) exhaustion, depletion, impoverishment

ис·тощ·и́ть, -ат, (perf) see **ис·то·щ·а́·ть**

ис·тра́т·ить, -ят, (perf) see **ис·тра́·ч·ива·ть**

ис·тра́ч·ива·ть, -ют, (imp) expend, spend

ис·треб·и́тел·ь, -я, (m) destroyer, exterminator; (air) fighter

~ **-пере·хва́т·чик·, -а,** (m) (air) interceptor-fighter

ис·треб·и́тельн·ый, -ая, -ое, (a) destructive; (air) fighter

ис·треб·и́ть, -я́т, (perf) see **ис·треб·л·я́·ть**

ис·требл·ени·е, -я, (n) (v n) see **ис··требл·я́·ть**; destruction, extermination

ис·требл·я́·ть, -ют, (imp) destroy, exterminate, extirpate, annihilate

ис·тре́б·овать, -уют, (perf) demand, demand and obtain

ис·трёп·анн·ый, -ая, -ое, (past part pass) see **ис·трёп·ыва·ть**; ragged, frayed, torn

ис·треп·а́ть (fut 3rd sing, pl) **ис··тре́пл·ет, ис·тре́пл·ют,** (perf); see **ис·трёп·ыва·ть**

ис·трёп·ыва·ть, -ют, (imp) fray, tear, wear to rags

и́ст·ый, -ая, -ое, (a) real, true

ис·ты́к·а·ть, -ют, (perf) pierce all over, perforate, pin/stud all over

ис·хлопот·а́ть (fut 3rd sing, pl) **ис··хлопо́ч·ет, ис·хлопо́ч·ут,** (perf), obtain by fussing

ис·хо́д·, -а, (m) outcome, issue, result

ис·ход́ай·ств·овать, -уют, (perf) apply for and obtain

ис·ход·и́ть, -ят, (imp) go/come out (from), emanate/emerge from; proceed from, proceed on the basis of; run out, expire (time); go/walk all over

ис·ход·я́ из (pres gerund and prep) proceeding from, on the basis of, from

~ **из э́т·ого** hence

ис·хо́д·н·ый, -ая, -ое, (a) initial, original, starting; parent, base, reference

~ **материа́л·,** (m) starting material, stock (material for processing)

~ **пункт·** (m) point of departure, basis (of an argument); departure point, starting point; initial position

~ **по·лож·е́ни·е** (n) initial position, starting point

~ **со·сто·я́ни·е** (n) initial/original state

~ **ядр·о́** (n) (nucl) parent nucleûs

ис·ход·я́щ·ий, -ая, -ее, (pres part act) of **ис·ход·и́ть**; outgoing; (f decl as adj) outgoing/out papers (office work)

~ **но́мер·** (m) reference number

ис·худ·а́л·ый, -ая, -ое, emaciated

ис·цара́п·а·ть, -ют, (perf) see **ис··цара́п·ыва·ть**

ис·цара́п·ыва·ть, -ют, (imp) scratch all over, cover with scratches, striate

ис·цел·и́ть, -я́т, (perf) see **ис· цел·я́·ть**

ис·цел·я́·ть, -ют, (imp) heal, cure; **-ся** (pass); recover

ис·ча́х·нуть, -нут, (past masc sing) **ис·ча́х·,** (perf) waste away

исчез·а́ни·е, -я, (n) disappearing, disappearance

исчез·а́·ть, -ют, (imp) disappear, vanish; (biol) become extinct

исчез·нове́ни·е, -я, (n) disappearance; (biol) extinction

исче́з·нуть, -нут, (past masc sing) **ис·че́з·,** (perf); see **исчез·а́·ть**

ис·че́рп·а·ть, -ют, (perf) see **ис·че́р·п·ыва·ть**

ис·че́рп·ыва·ть, -ют, (imp) exhaust, use up; complete, bring to an end; settle (a question)

ис·че́рп·ывающ·ий, -ая, -ее, (pres part act) of **ис·че́рп·ыва·ть**; exhaustive

ис·черт·и́ть, -я́т, (perf) see **ис·че́р·ч·ива·ть**

ис·че́рч·ива·ть, -ют, (imp) cover with drawings/lines/symbols, striate; use up (in drawing)

ис·числ·е́ни·е, -я, (n) calculation; (math) calculus

~, **бари·центр·и́ческ·ое** (math) barycentric calculus

~ **бес·кон·е́чн·ома́л·ых** infinitesimal calculus

~, **вариацио́н·н·ое** calculus of variations

ис·числ·éни·е
~, вéктор·н·ое (math) vector algebra and vector analysis
~ вероя́т·ност·и calculus of probability
~, дифференциáль·н·ое differential calculus
~ кон·éчн·ых рáзн·ост·ей calculus of variables
~, опера́тор·н·ое operator theory
~, операцио́н·н·ое operational calculus
ис·чи́сл·енн·ый, -ая, -ое, *(a) (past part pass)* of **ис·чи́сл·ить;** calculated; predicted
ис·чи́сл·ить, -ят, *(perf) see* **ис·чи·сл·я́·ть**
ис·числ·я́·ть, -ют, *(imp)* calculate, estimate; **-ся** *(pass);* amount/come to
ис·штрих·о́ванн·ый, -ая, -ое, *(past part pass)* striated, scratched
и-т *(abbr)* of **институ́т·** institute
и·та́к *(conj)* thus, consequently, so
италья́н·ск·ий, -ая, -ое, *(a)* Italian
~ за·баст·о́вк·а *(f)* stay-in strike; go-slow strike
~ бухгалтéр·и·я *(f)* double-entry bookkeeping
и т. д. = и так да́л·ее and so on/forth, etc.
итерацио́н·н·ый, -ая, -ое, *(a)* iteration
итера́ци·я, -и, *(f)* iteration
ито́г·, -а, *(m)* total, sum; result
в ито́г·е summing up; finally
~, о́бщ·ий grand total
~ по смéт·е estimated total
~, ча́ст·н·ый sub-total
ито́г·ов·ый, -ая, -ое, *(a)* total, sum, summing-up, concluding
~ перфо·ка́рт·а *(f)* summary punched card
и т. п. = и тому́ по·до́б·н·ое etc., et cetera, and the like
итсегéк·, -а, *(m)* (bot) *Anabasis aphylla*

иттéрб·и·й, -я, *(m)* (chem) ytterbium, Yb
итти́ *(imp deter)* **идти́**
иттриали́т·, -а, *(m)* (min) yttrialite
и́ттр·иев·ый, -ая, -ое, *(a)* of **и́т-тр·и·й**
и́ттр·и·й, -я, *(m)* (chem) yttrium, Y
иттрогумми́т·, -а, *(m)* (min) yttrogummite
иттро·крази́т·, -а, *(m)* (min) yttro-crasite
иттро·флюори́т·, -а, *(m)* (min) yttro-fluorite
иттро·цери́т·, -а, *(m)* (min) yttro-cerite
иттро·тантали́т·, -а, *(m)* (min) yttro-tantalite
их *(pron, acc/gen plur)* of **он, она́; оно́,** them, of them, their
ихно·ло́г·и·я, -и, *(f)* (geol) ichnology
ихо́р·, -а, *(m)* (geol) ichor
ихти·о́з·, -а, *(m)* (med) ichthyosis, fishskin disease
ихтио·за́вр·, -а, *(m)* (pal) ichthyosaur
ихтио·ло́г·и·я, -и, *(f)* (fish) ichthyology
~, техн·и́ческ·ая industrial ichthyology
ихти·о́рнис·, -а, *(m)* (pal) Ichthyornis
ихтио·стега́ли·и *(nom pl),* *(g pl)* **-й,** *(f)* (pal) Ichthyostegalia
и́шиас·, -а, *(m)* (med) sciatica
ишикава́йт·, -а, *(m)* (min) ishikawaite
иш·хан·, -а, *(m)* (fish) *Salmo ischchan* (Lake Sevan trout)
ищ·а́ *(pres gerund)* of **иск·а́ть**
ищ·éйк·а, -и, *(g pl)* **-éек·,** *(f)* bloodhound
и́щ·ущ·ий, -ая, -ее, *(pres part act)* of **иск·а́ть**
ию́л·ь, -я, *(m)* July
ию́ль·ск·ий, -ая, -ое, *(a)* July
ию́н·ь, -я, *(m)* June
ию́нь·ск·ий, -ая, -ое, *(a)* June

Й

йе́мен·ск·ий, -ая, -ое, (*a*) (geog) Yemen, Yemeni, Yemenite

йогу́рт·, -а, (*m*) (food) yogurt

йод·, -а, (*m*) (*and cognate words*) *see* ио́д· (*and cognate words*)

йо́н·, -а, (*m*) *see* ио́н·

йо́т·, -а, (*m*) letter J, letter й

йо́т·а, -ы, (*f*) iota (Greek letter)

йота·структу́р·а, -ы, (*f*) (geol) iota-structure

К

к (*prep* + *dative*) towards, to; by (time, date); for (a future event); in commemoration of (anniversaries etc.)

~ **во·про́с·у** concerning the question, (**к** often omitted in translation); the question/problem of

~ **обе́д·у** for dinner

~ **со·жал·е́ни·ю** unfortunately

~ **сча́ст·ью** luckily, happily

~ **тому́же** moreover, besides, in addition

~ **чему́?** what for? why?

к (*abbr*) of **куло́н·** coulomb

К-за·хва́т·, -а, (*m*) (phys) K-capture

К-оболо́ч·к·а, -и, (*f*) K-shell

К-со·сто·я́ни·е, -я, (*n*) K-state

ка (*n indecl*) letter K; (*abbr*) of **кило·ам-пе́р·** Ка, КА, kiloampere

кабаля́р·, -а, (*m*) (naut) messenger

каба́н·, -а́, (*m*) (zool) wild boar/pig, *Sus scrofa*; block, lump; (a/c) cabane

каба́н·ий, -ья, -ье, (*a*) of **каба́н·**

каба́н·чик·, -а, (*m*) (a/c) horn

кабарг·а́, -и́, (*g pl*) **-ро́г·,** (*f*) (zool) musk deer, *Moschus moschiferus*

кабарг·о́в·ый, -ая, -ое, (*a*) of **кабар-г·а́**

кабач·о́к·, -ч·к·а́, (*m*) (bot) marrow-type pumpkin, *Cucurbita pepo*

кабел·ёк·, -л·ьк·а́, (*m*) (*dim*) of **ка́-бел·ь**; (elec) pigtail

кабеле·про·во́д·, -а, (*m*) cable duct

кабеле·у·кла́д·чик·, -а, (*m*) cable-layer

ка́бел·ь, -я, (*m*) cable; electric cable

~, **асфа́льт·и́рованн·ый** (elec) served lead-covered cable

~, **брон·иро́ванн·ый** (elec) armoured cable

~, **в·во́д·н·ый** lead-in cable, lead-in

~, **вед·у́щ·ий** (elec) leader/piloting/guide cable

~, **воз·ду́ш·н·ый** (elec) overhead/aerial cable

~, **газо·на·по́лн·енн·ый** (elec) gas-filled cable

~, **газо·по́лн·ый** (elec) gas-filled cable

~, **го́л·ый** (elec) plain cable

~, **двух·жи́ль·н·ый** (elec) two-core cable, twin cable

~, **желе́з·н·ый** cable-laid rope

~ **за·де́рж·к·и** (elec) delay cable

~, **ко·аксиа́ль·н·ый** (elec) coaxial cable

ка́бел·ь

~, **ковр·о́в·ый** (elec) braided cable

~, **комбин·и́рованн·ый** (telecom) composite cable

~, **контро́ль·н·ый** (elec) control cable

~, **кон·центр·и́ческ·ий** (elec) concentric cable

~, **ко́рдель·н·ый** (telecom) paper-cored cable, dry-core cable

~ **-кра́н·, -а,** (*m*) cableway, aerial ropeway

~ **-кра́н·, пере·движ·н·о́й** travelling cableway

~ **-кра́н·, стациона́р·н·ый** fixed cableway

~, **краруп·изо́ванн·ый** (telecom) continuously loaded cable, Krarup loading cable

~, **масло·на·полн·енн·ый** (elec) oil-filled cable

~, **масло·по́лн·ый** (elec) oil-filled cable

~, **много·жи́ль·н·ый** (elec) multicore cable

~, **мор·ск·о́й** (telecom) submarine cable; cable for ship use

~, **одно·жи́ль·н·ый** (elec) single-core cable

~, **о·свет·и́тельн·ый** (elec) light/lighting cable

~, **о·свинц·о́ванн·ый** (elec) lead-covered cable

~, **пал·и́льн·ый** (min) shot-firing cable

~ **па́р·н·ой с·кру́т·к·и** (telecom) paired cable

~, **педа́ль·н·ый** (rail) track cable

~ **-пла́н·, -а,** (*m*) (elec) cable layout

~, **под·во́д·н·ый** (elec) submarine cable, underwater cable

~, **пупин·изи́рованн·ый** (telecom) coil-loaded cable, Pupin cable

~ **с·вя́з·и** telecommunications cable

~, **сил·ов·о́й** power cable

~ **с пояс·н·о́й изол·я́ци·ей** (elec) belted cable

~, **симметр·и́чн·ый** (telecom) balanced cable

~, **со·о́с·н·ый** (elec) coaxial cable

~, **сух·о́й** (elec) SL cable

~, **трёх·жи́ль·н·ый** (elec) three-core cable

~, **ша́хт·н·ый** (elec) mine cable

~, **экран·и́рованн·ый** (elec) screened cable, H-type cable

кабель·н·ый, -ая, -ое, (*a*) of **ка́бел·ь**
~ **в·ста́в·к·а** (*f*) (telecom) cable insert
~ **коло́д·ец·** (*m*) (elec) draw-in box
~ **кран·** (*m*) *see* **ка́бел·ь** *subentry* **кра́н·** cableway
~ **ла́п·к·а** (*f*) (elec) cable grip
~ **ма́сс·а** (*f*) (elec) cable-filling compound
~ **му́фт·а** (*f*) (elec) cable-jointing device
~ **на·кон·е́чник·** (*m*) (naut) sweating thimble
~ **рабо́т·а** (*f*) cable lay (of rope)
~ **су́д·н·о** (*n*) cable ship
~ **шабло́н·** (*m*) (telecom) cable form
ка́бельтов·, -а, (*nom pl*) **-ы,** (*g pl*) **-ых,** (*m*) (naut) rope hawser; (naut) cable (608 feet)
ка́бель·щик·, -а, (*m*) cable-layer; cable-marker
кабеста́н·, -а, (*m*) capstan
каби́н·а, -ы, (*f*) cabin, cockpit, cab, car, compartment, booth, capsule
~, **гермет·и́чн·ая** pressure cabin/capsule
~, **пассажи́р·ск·ая** (a/c) passenger compartment
~ **пило́т·а** (a/c) pilot's cockpit
кабине́т·, -а, (*m*) study, office; laboratory, classroom; surgery, consulting room; (polit) cabinet
~, **зубо·врач·е́бн·ый** dental surgery
~, **радио·граф·и́ческ·ий** X-ray room
каби́н·к·а, -и, (*g pl*) **-н·ок·,** (*f*) (*dim*) of **каби́н·а**
кабл·и́ровани·е, -я, (*n*) (elec) cabling
кабло·гра́мм·а, -ы, (*f*) (telecom) cablegram, cable (message)
каблу́к·, -а́, (*m*) heel (of shoe)
каблу́ч·н·ый, -ая, -ое, (*a*) of **каблу́к·**
~ **стан·о́к·** (*m*) heel-making machine
каблуч·о́к·, -ч·к·а́, (*m*) (*dim*) of **каблу́к·**; (arch) heel, heel strap
ка́болк·а, -и, (*g pl*) **-лок·,** (*f*) yarn (of rope)
кабота́ж·, -а, (*i*) **-ем,** (*m*) coasting trade; coastal navigation, inshore navigation
~, **больш·о́й** (naut) inter-sea domestic traffic
~, **ма́л·ый** coasting, coastal traffic
кабота́ж·ник·, -а, (*m*) (naut) coaster
кабота́ж·н·ый, -ая, -ое, (*a*) of **кабота́ж·**
~ **су́д·н·о** (*n*) coasting vessel, coaster
кабошо́н·, -а, (*m*) cabochon (of gem)

кабриоле́т·, -а, (*m*) (M/T) cabriolet, sunshine roof, convertible coupé
кабр·и́ровани·е, -я, (*n*) (air) tail heaviness, pitch-up, nose-up
ка́в·а, -ы, (*f*) (bot) kava, *Piper methysticum*
кава́·йн·ый, -ая, -ое, (*a*) of **ка́в·а**
~ **кисл·от·а́** (*f*) (chem) kavaic acid
кавале́р·, -а, (*m*) holder (of decoration); partner (of lady at dance)
кавале́р·и·я, -и, (*f*) cavalry
кавалье́р·, -а, (*m*) (civ eng) earth bank
кавард·а́к·, -а́, (*m*) mess, muddle
кавас́аки (*indecl*) Kavasaki (fishing vessel used in Far East)
каве́рн·а, -ы, (*f*) cavity, pocket, vesicle; cavern
каверн·о́зн·ый, -ая, -ое, (*a*) (geol) cavernous; interstitial (of water); honey combed, porous, spongy
каверно·ме́р·, -а, (*m*) well calliper
кавитацио́н·н·ый, -ая, -ое, (*a*) of **кавита́ци·я**
~ **пузы́р·ь** (*m*) cavitation bubble/void
~ **труб·а́** (*f*) cavitation test tunnel
~ **числ·о́** (*n*) cavitation coefficient
кавита́ци·я, -и, (*f*) (hydrodynam) cavitation
коэффициен́т· кавита́ци·и (*m*) void coefficient
кавка́з·ск·ий, -ая, -ое, (*a*) Caucasian
~ **черн·и́к·а** (*f*) (bot) Caucasian bilberry, *Vaccinium arctostaphylos* (a tea substitute)
кавы́ч·к·а, -и, (*g pl*) **-ч·ек·,** (*f*) inverted comma, quotation mark
кага́т·, -а, (*m*) (agr) clamp, storage pit (for potatoes etc.)
кагат·и́ровани·е, -я, (*n*) (agr) clamping, storing in a clamp
кадаве́рин·, -а, (*m*) (chem) cadaverine, pentamethylenediamine
када́стр·, -а, (*m*) (surv) cadastre, cadastral survey
када́стр·ов·ый, -ая, -ое, (*a*) cadastral
каде́нци·я, -и, (*f*) cadence
каде́т·, -а, (*m*) cadet
кадино́л·, -а, (*m*) (chem) cadinal
кадио́н·, -а, (*m*) (chem) radion, para-nitrobenzene-diazoaminoazobenzene
ка́д·к·а, -и, (*g pl*) **-д·ок·,** (*f*) tub, vat
ка́дм·иев·ый, -ая, -ое, (*a*) (chem) cadmium, cadmic, cadmous
~ **жёлт·ая** (*f*) cadmium yellow (pigment)
~ **от·нош·е́ни·е** (*n*) (nucl) cadmium ratio

кадмие·со·держ·а́щ·ий, -ая, -ее, (*a*) cadmiferous, cadmium-containing

ка́дм·и·й, -я, (*m*) (chem) cadmium, Cd

~, бро́м·ист·ый cadmium bromide

~, жёлт·ый cadmium yellow

~, ио́д·ист·ый cadmium iodide

~, сер·ни́ст·ый cadmium sulphide

~, уксусно·ки́сл·ый cadmium acetate

кадм·и́ровани·е, -я, (*n*) cadmium plating

кадм·и́рованн·ый, -ая, -ое, (*past part pass*), cadmium-plated

ка́др·, -а, (*m*) (cinema) frame, still, set, setting; (elec) plane (of ferrite core store); (*pl only*) **ка́др·ы** cadres, key personnel; (mil) regulars

~, в·став·н·о́й (cinema) cut-in

~, звук·ов·о́й (cinema) proportional frame

~, монта́ж·н·ый (cinema) cut

~ раз·вёрт·к·и (TV) scanning frame

~, сцена́р·н·ый (cinema) set

кадр·и́ровани·е, -я, (*n*) (cinema) framing, cutting

ка́др·ов·ый, -ая, -ое, (*a*) of **ка́др·** and **ка́др·ы**

~ офице́р· (*m*) regular officer

~ раз·вёрт·к·а (*f*) (TV) frame scan **регул·иро́вк·а ка́др·ов·ой син·хрониз·а́ци·и** (*f*) (TV) vertical hold control

кадро·ме́р·, -а, (*m*) (cinema) frame counter

кад·у́шк·а, -и, (*g pl*) **-шек·,** (*f*) wooden bucket, tub

кады́к·, -а́, (*m*) (anat) Adam's apple, prominentia laryngea

каём·к·а, -и, (*g pl*) **-м·ок·,** (*f*) (*dim*) of **кайм·а́;** narrow border/edging, rim margin, fringe

каём·чат·ый, -ая, -ое, (*a*) bordered, edged, limbate

каепу́т·, -а, (*m*) *see* **каепу́т·ов·ое де́·рев·о**

каепу́т·ов·ый, -ая, -ое, (*a*) (bot) cajeput, cajuput

~ де́рев·о (*n*) (bot) cajuput, *Melaleuca leucadendron*

~ ма́сл·о (*n*) (pharm) cajeput oil

каждо·дн·е́вн·ый, -ая, -ое, (*a*) daily, everyday, diurnal

ка́жд·ый, -ая, -ое, (*a*) each, every; (as pronoun) everyone

ка́ж·ется (*pres 3rd sing*) of **каз·а́ться;** it seems/appears

ка́ж·ущ·ийся, -аяся, -ееся, (*pres part act*) of **каз·а́ться;** apparent

ка́ж·ущ·ийся, -аяся, -ееся

~ мо́щ·ност·ь (*f*) (elec) apparent power

ка́з·а, -ы, (*f*) bin (wine cellar)

каз·а́вш·ийся, -аяся, -ееся, (*past part act*) of **каз·а́ться**

каза́к·, -а́, (*nom pl*) **-и́,** (*m*) Cossack

каза́рк·а, -и, (*g pl*) **-рок·,** (*f*) (ent) snout beetle, *Rhynchites bacchus*; (zool) brant, (*pl*) *Branta*

каза́рм·а, -ы, (*f*) barracks

каза́рм·енн·ый, -ая, -се, (*a*) of **каза́рм·а**

каз·а́ться (*pres 3rd sing, pl*) **ка́ж·ется, ка́ж·утся,** (*imp*)(+*instr*), seem, appear

каза́х·ск·ий, -ая, -ое, (*a*) Kazakh

каза́ц·к·ий, -ая, -ое, (*a*) Cossack

каза́ч·ий, -ья, -ье, (*a*) Cossack

казеи́н·, -а, (*m*) (chem) casein

~, раст·и́тельн·ый vegetable casein, legumen

казеин·а́з·а, -ы, (*f*) (biochem) caseinase

казеин·а́т·, -а, (*m*) (chem) caseinate

казеи́н·ов·ый, -ая, -ое, (*a*) casein, caseinic

~ кисл·от·а́ (*f*) caseinic acid

~ кле́·й (*m*) casein glue

казеино́·ге́н·, -а, (*m*) (biochem) caseinogen

каземат·, -а, (*m*) (mil) casemate

казён·ник·, -а, (*m*) (gunn) breech ring

казён·н·ый, -ая, -ое, (*a*) public, fiscal, state; bureaucratic, formal; (gunn) breech

~ де́ньг·и (*pl*) public money, public funds

~ с·рез· (*m*) (gunn) breech end

~ част·ь (*f*) (gunn) breech

казнач·е́·й, -я, (*m*) treasurer; (mil) paymaster

казнач·е́йств·о, -а, (*n*) treasury

казн·и́ть, -я́т, (*imp and perf*) execute, put to death; (*imp only*) brand/lash with (scorn etc.)

ка́зн·ь, -и, (*f*) capital punishment, execution

~, с·ме́рт·н·ая capital punishment; death (as sentence)

казоли́т·, -а, (*m*) (min) kasolite

казуар·и́н·а, -ы, (*f*) (bot) beefwood, *Casuarina*

казуар·и́нов·ые, -ых, (*pl decl as a*) (bot) *Casuarinaceae*

казуист·и́ческ·ий, -ая, -ое, (*a*) casuistic

ка́зус·, -а, (*m*) (law) special case, *casus*

~ бе́лли (law) *casus belli*

ка́зус

~ фе́дерис (law) *casus foederis*

ка́зус·н·ый, -ая, -ое, (*a*) exceptional, involved, unusual; (law) *casus*

кайк·, -а́, (*m*) (naut) caique

каинит·, -а, (*m*) (min) kainite

кайек·, -а, (*m*) Kayak

кайе́н·ск·ий пе́рец· (*m*) cayenne pepper

кайл·а́, -ы́, (*f*) (min) pick, hack

кайл·о́, -а́, (*n*) *see* **кайл·а́**

кайл·о́в·ый, -ая, -ое, (*a*) of **кайл·а́, кайл·о́**

кайм·а́, -ы́, (*f*) border, fringe, edging

~, капилля́р·н·ая capillary fringe

~, коррозио́н·н·ая corrosion rim

~, цвет·н·а́я (phot) fringing, bleeding, colouring

кайма́к, -а́, (*m*) clotted cream

кайма́н·, -а, (*m*) (zool) caiman, cayman

кайма́н·ов·ый, -ая, -ое, (*a*) of **кайма́н·**

~ ры́б·а (*f*) (fish) bill-fish, *Lepidosteus*

~ череп·а́х·а (*f*) (zool) turtle, (*pl*) *Chelydridae*

кайно·зо́·й, -я, (*m*) (geol) Cenozoic, Cainozoic, Caenosoic

кайно·зо́й·ск·ий, -ая, -ое, (*a*) of **кайно·зо́·й**

кайно·ти́п·н·ый, -ая, -ое, (*a*) (geol) cenotypal, cainotypic

ка́йр·а, -ы, (*f*) (orn) guillemot, *Uria*

кайс·а́, -ы́, (*f*) dried apricot

как (*adv*) how; what; (*conj*) as, like; when, since

как бу́дто as though/if

ка́к бы as if, as though

ка́к бы не lest, in case

ка́к бы ни however

как в·дру́г when suddenly

ка́к же certainly, undoubtedly

как и as with, as also

ка́к-либо somehow

ка́к не but

ка́к ни however

ка́к-нибудь (*adv*) somehow; anyhow, carelessly; sometime

как ра́з exactly, just as, as soon as

как ... так и both ... and, just as ... so also

ка́к-то somehow; how; once

как то́лько as soon as

кака́о (*n indecl*) (bot) cocoa, cacao, *Theobroma cacao*

~ -бо́бб·а, -ы, (*f*) cacao pod, cacao husk

~ -бо́б·, -а́, (*m*) cacao/cocoa bean

~ -ма́сл·о, -а, (*n*) cocoa butter

како (*n indecl*) letter K

какод·и́л·, -а, (*m*) (chem) cacodyl, trimethyl arsine

о́·кис·ь какод·и́л·а cacodylic oxide, dimethyl-arsenic oxide

какодил·а́т·, -а, (*m*) cacodylate

какодилово·ки́сл·ый, -ая, -ое, (*a*) cacodylate

~ на́тр·и·й (*m*) sodium salt of cacodyl, sodium cacodylate

какоди́л·ов·ый, -ая, -ое, (*a*) cacodyl, cacodylate

~ кисл·от·а́ (*f*) cacodylic acid

како·кла́з·, -а, (*m*) (min) cacoclase, cacoclasite

как·о́й, -а́я, -о́е, (*a*) (*pronoun*) what; such ... as

~ -либо *see* **как·о́й-нибудь**

~ -нибудь some, some kind of

~ .-то some, someone; a kind of

~ ... ни whatever

ка́комицл·, -а, (*m*) (zool) cacomistle, ring-tail cat, *Bassariscus astutus*

како·тел·и́н·, -а, (*m*) (chem) cacotheline

ка́ктус·, -а, (*m*) (bot) cactus

ка́ктус·ов·ый, -ая, -ое, (*a*) (bot) cactus; (*pl*) cactus family, *Cactaceae*

ка́л·, -а, (*m*) faeces, feces, excrement

кал. (*abbr*) of **кало́р·и·я** calorie, cal

ка́лабар·ск·ий боб· (*m*) (bot) calabar bean, *Physostigma venenosum*

калавери́т·, -а, (*m*) calaverite

калами́н·, -а, (*m*) (min) calamine, smithsonite, hemimorphite

калами́т·, -а, (*m*) calamite (fossil)

калами́ямнк·а, -и, (*g pl*) -нок·, (*f*) (text) calamance

кала́н·, -а, (*m*) (zool) sea otter, *Enhydra lutris*

кала́ндр·, -а, (*m*) calender, calendering machine, calender roll

~, гофр·ирова́льн·ый (paper) embossing calender

~, лист·ов·о́й (paper) sheet calender

~, лощ·и́льн·ый (paper) friction glazing calender

~, серебр·и́ст·ый (text) schreiner calender

каландр·ирова́ни·е, -я, (*n*) calendering

каландр·ова́ни·е, -я, (*n*) calendering

каландр·ова́ть, -у́ют, (*imp and perf*) calender

каландр·о́вщик·, -а, (*m*) calender man/operator

каланч·а́, -и́, (*i*) -о́й, (*f*) watch-tower

калга́н·, -а, (*m*) (bot) galanga, China root, *Alpinia officinalis*; tormentil, *Potentilla*

кал·ева́ть, -ю́т, (*imp*) bead (carpentry)

кал·ёвк·а, -и, (*g pl*) **-вок·,** (*f*) bead, beading (carpentry); bead-plane

~, двой·н·а́я double bead/beading

кал·ёвочн·ый, -ая, -ое, (*a*) of **кал·ёвк·а**

~ стан·о́к· (*m*) moulder/molder (woodworking machine)

кал·ейдо·ско́п·, -а, (*m*) kaleidoscope

калёк·а, -и, (*f*) cripple

календа́р·н·ый, -ая, -ое, of **кален·да́р·ь**

~ числ·о́ (*n*) date

календа́р·ь, -я, (*m*) calendar

кал· е́ни·е, -я, (*n*) (*v n*) of **кал·и́ть;** incandescence, heat

~, бе́л·ое white heat

~, кра́с·н·ое red heat

кален·и́ц·а, -ы, (*f*) lehr (glassmaking)

кал·ён·ый, -ая, -ое, (*a*) red-hot; roasted

кале́ч·ить, -ат, (*imp*) cripple, maim

ка́ли (*n indecl*) (chem, min) potash

~, азотно·ки́сл·ое nitre, niter

~, дву·угле·ки́сл·ое potassium bicarbonate

~, е́д·к·ое caustic potash, potassium hydroxide

кали·аппара́т·, -а, (*m*) (chem) potash bulb

кали́бр·, -а, (*m*) gauge, gage; (gunn) bore, calibre, caliber; (met roll) pass

~, ба́л·очн·ый beam pass

~ брак·о́вщик·а inspection gauge/gage

~ валк·а́ (met roll) roll pass/hole

~, вы·тяж·н·о́й (met roll) breakdown pass

~, двух·сторо́н·н·ий go-and-not-go gauge/gage

~, за·кры́·т·ый (met roll) closed pass

~, квадра́т·н·ый (met roll) square pass

~, кольц·ев·о́й ring gauge/gage

~, кон·и́ческ·ий plug gauge/gage

~, контро́ль·н·ый master gauge/gage; (met roll) controlling pass

~, кру́п·н·ый (gunn) heavy calibre/caliber

ли́н·и·я кали́бр·ов (*f*) (met roll) pass line

~, ма́л·ый (gunn) small calibre/caliber

кали́бр

~, на·кло́н·н·ый (met roll) twisting pass; diagonal pass

~, не·про·хо́д·н·о́й not/no-go gauge/gage

~, об·жи́м·н·ый (met roll) breakdown pass

~, от·де́л·очн·ый (met roll) finishing pass

~, от·кры́·т·ый (met roll) open pass

~, пли́т·очн·ый block gauge/gage

~, па́р·н·ый mating gauge/gage

~, под·гото́в·и́тельн·ый (met roll) intermediate pass, roughing pass

~, пре·де́ль·н·ый limit gauge/gage

~, при·ём·н·ый inspection gauge/gage (buyer's)

~ -про́б·к·а, -и, (*f*) plug gauge/gage

~, про·хо́д·н·о́й go-gauge/gage

~, прямо·уго́ль·н·ый (met roll) box pass

~, рабо́ч·ий shop gauge/gage, working gauge/gage

~, раз·дви́ж·н·о́й calliper, calliper gauge/gage

~, раз·ре́з·н·о́й angular wire gauge/gage

~, ребр·ов·о́й (met roll) edging pass

~, резьб·ов·о́й screw gauge/gage

~, ромб·и́ческ·ий (met roll) diamond pass

систе́м·а кали́бр·ов (*f*) (roll) pass sequence

~, сре́д·н·ий (gunn) medium calibre

~, черн·ов·о́й (met roll) roughing pass

~, чист·ов·о́й (met roll) finishing pass

~, я́щич·н·ый (met roll) box pass

калибр·а́тор·, -а, (*m*) calibrator

~, ква́рц·ев·ый (rad) crystal controlled oscillator

калибрацио́н·н·ый, -ая, -ое, (*a*) calibration

~ кольц·о́ (*n*) calibration marker circle (radar)

калибр·и́рова́ни·е, -я, (*n*) *see* **калибр·ова́ни·е**

калибр·ова́ни·е, -я, (*n*) (*and see* **калибр·о́вк·а**) calibration; sorting, checking, gauging, gaging, sizing

калибр·о́ванн·ый, -ая, -ое, (*past part pass*) of **калибр·ова́ть;** calibrated; graduated; gauged, gaged; screened, sized, (met) precision-forged, sized; bright-drawn; cut (of rolls); thickness-gauged (rubber sheet)

калибр·о́ванн·ый, -ая, -ое
~ **стал·ь** (*f*) bright-drawn steel, cold-drawn steel
калибр·ова́ть, -у́ют, (*imp*) calibrate, graduate (a scale-reading instrument); gauge, gage, measure, check (sizes); design (rolling mill rolls)
калибр·о́вк·а, -и, (*g pl*) **-вок·,** (*f*) (*and see* **калибр·ова́ни·е**) calibration, graduation, gauging, gaging; (met) roll designing; (met) sizing, precision forging; (agr) grading (by size)
~**, взайм·н·ая** intercalibration
~ **по тепл·у́** thermal calibration
~**, раз·вёр·нут·ая** (met roll) butterfly passes
~ **-чека́н·к·а, -и,** (*f*) precision forging, sizing
калибр·о́вочн·ый, -ая, -ое, (*a*) of **калибр·о́вк·а**
~ **штамп·** (*m*) (met forge) sizing die
калибр·ов·ый, -ая, -ое, (*a*) of **ка·ли́бр·**
~ **хозя́й·ств·о** (*n*) metrology, metrology room, metrology department
калибро·ме́р·, -а, (*m*) thickness gauge/gage; (gunn) calibre/caliber gauge/gage
калибро́·метр·, -а, (*m*) wire-and-sheet gauge/gage
ка́л·иев·ый, -ая, -ое, (*a*) *see* **ка·ли́йн·ый**
ка́л·и·й, -я, (*m*) (chem) potassium, K
~**, азотно·ки́сл·ый** potassium nitrate
~**, марганцово·ки́сл·ый** potassium permanganate
~**, о́·кис·н·ый** potash
~**, сернисто·ки́сл·ый** potassium sulphite
~**, серно·ки́сл·ый** potassium sulphate
~**, угле·ки́сл·ый** potassium carbonate
~**, фто́р·ист·ый** potassium fluoride
~**, хло́р·ист·ый** potassium chloride
~**, хлорновато·ки́сл·ый** potassium chlorate
~**, хлорно·ки́сл·ый** potassium perchlorate
~**, циа́н·ист·ый** potassium cyanide
~**, шавело·ки́сл·ый** potassium oxalate
~ **-на́тр·и·й, винно·ки́сл·ый** sodium potassium tartrate, Rochelle salt
кал·и́йн·ый, -ая, -ое, (*a*) potassic, potassium; potash
~ **сели́тр·а** (*f*) potassium nitrate, saltpetre, saltpeter, nitre, niter
~ **сол·ь** (*f*) potash salt
~ **щёлок·** (*m*) caustic potash

калик·антем·и́·я, -и, (*f*) (bot) calycanthemy
ка́ликс·, -а, (*m*) (bot) calyx
кал·и́льн·ый, -ая, -ое, (*a*) incandescent, hot
~ **голо́в·к·а** (*f*) hot bulb (surface ignition)
~ **ла́мп·а** (*f*) incandescent light/mantle (e.g. gas)
~ **о·свещ·е́ни·е** (*n*) incandescent lighting
~ **свеч·е́ни·е** (*n*) (nucl) candoluminescence
кали·магне́з·и·я, -и, (*f*) (agr chem) sulphomag
калимм·а, -ы, (*f*) (biol) kalymma
кали́н·а, -ы, (*f*) (bot) viburnum
калини́т·, -а, (*m*) (min) kalinite
калиптро·ге́н·, -а, (*m*) (bot) calyptrogen
кали·сте́г·и·я, -и, (*f*) (bot) bindweed, *Calystegia*
кали́т·к·а, -и, (*g pl*) **-т·ок·,** (*f*) wicket gate
кал·и́ть, -я́т, (*imp*) heat, incandesce, bring to red heat; roast
калифо́рн·и·й, -я, (*m*) (chem) californium, Cf
ка́лк·а, -и, (*f*) fat-liquoring (leather)
калка́н·, -а, (*m*) (fish) *Rhombus maeoticus*
каллаи́т·, -а, (*m*) (min) callaite
каллёз·, -а, (*m*) (anat) callus, callosity
каллёз·а, -ы, (*f*) (biochem) callose
калли·гра́ф·и·я, -и, (*f*) calligraphy
калли·ка́рп·а, -ы, (*f*) (bot) French mulberry, *Callicarpa*
каллиро·тро́н·, -а, (*m*) (elec) kallitron, negative-resistance tube
калли·ти́п·и·я, -и, (*f*) (phot) kallitype
Ка́лло, элеме́нт· (*m*) (elec) Callaud cell
ка́ллус·, -а, (*m*) (bot, med) callus, callosity
Калльé, коэффицие́нт· (*m*) (phys) Callier quotient, Q factor
ка́ллюс·, -а, (*m*) *see* **ка́ллус·**
ка́л·ов·ый, -ая, -ое, (*a*) of **кал·**
каломел·ь, -и, (*f*) (chem) calomel, mercurous chloride
ка́ломельн·ый, -ая, -ое, (*a*) of **ка́ломел·ь**
~ **электро́д·** (*m*) calomel electrode, calomel half-cell
калор·есце́нци·я, -и, (*f*) (phys) caloreșcence

калор·изáтор·, -а, (*m*) (ICE) hot bulb; (chem eng) heater

калор·изáторн·ый, -ая, -ое, (*a*) of калор·изáтор·

~ двúг·ател·ь (*m*) semi-diesel engine, hot-bulb engine

калор·изáци·я, -и, (*f*) (met) calorizing, alitizing, aluminizing

калор·úйност·ь, -и, (*f*) calorific value, caloricity

калорú·метр·, -а, (*m*) (phys) calorimeter

~, лед·ян·óй   Bunsen ice calorimeter

калори·метр·úческ·ий, -ая, -ое, (*a*) calorimeter, calorimetric

~ бóмб·а (*f*) bomb calorimeter

~ рас·ходо·мéр·   (*m*) calorimetric flowmeter

калори·мéтр·и·я, -и, (*f*) (phys) calorimetry

~, про·тóч·н·ая   flow calorimetry

калорú·фер·, -а, (*m*) calorifier

калор·úческ·ий, -ая, -ое, (*a*) of калóр·и·я

калóр·и·я, -и, (*f*) (phys) calorie

~, больш·áя   great calorie, kilo-calorie

~, британ·ск·ая   British Thermal Unit

~, мáл·ая   small calorie, gram-calorie

~, срéд·н·яя   mean calorie

кало·тúп·и·я, -и, (*f*) (phot) calotype

калóтт·а, -ы, (*f*) (build) calotte

калóш·а, -и, (*f*) *see* галóш·а

калýг·а, -и, (*f*) (fish) hausen, *Acipenser huso* (a type of sturgeon)

калýж·ниц·а, -ы, (*f*) (bot) buttercup, *Caltha*

калýфер·, -а, (*m*) (bot) *Pyrethrum balsamita* (a culinary herb)

кальгóн·, -а, (*m*) (chem) calgon, hexametasulphate

кальдéр·а, -ы, (*f*) (geol) caldera

кáлье (*indecl*) (food) curds

кáльк·а, -и, (*g pl*) кáлек·, (*f*) tracing paper, tracing cloth; tracing; (ling) loan-translation

кальк·úр·овать, -уют, (*imp and perf*) trace, copy; (ling) make/use a loan-translation

калькул·úр·овать, -уют, (*imp*) calculate costs, estimate, give an estimate

калькул·я́тор·, -а, (*m*) (phot) calculator; cost estimator/accountant

калькул·я́ци·я, -и, (*f*) cost accounting, costing, estimating

калькуляциóн·н·ый, -ая, -ое, (*a*) of калькул·я́ци·я

кальмáр·, -а, (*m*) (fish) squid, *Teuthoidea decapoda*

кáльцекс·, -а, (*m*) (pharm) calcex

кальцеóл·ов·ый, -ая, -ое, (*a*) calceolate, slipper-shaped

кальце·фúль·н·ый, -ая, -ое, (*a*) (bot) calciphilous, calcicolous

кальце·фóб·н·ый, -ая, -ое, (*a*) (bot) calciphobous

кáльц·иев·ый, -ая, -ое, (*a*) calcium

~ вóз·раст· (*m*) (geol) calcium age

~ селúтр·а (*f*) calcium nitrate

кальцие·терм·úческ·ий, -ая, -ое, (*a*) (met) calcium-reduced

~ вос·становл·éни·е (*n*) (met) calcium thermal reduction

кáльц·и·й, -я, (*m*) (chem) calcium, Ca

~, азотно·кúсл·ый   calcium nitrate

~, водо·рóд·ист·ый   calcium hydride

~, угле·кúсл·ый   calcium carbonate

~, фтóр·ист·ый   calcium fluoride

~, хлор·úст·ый   calcium chloride

кальц·úн·, -а, (*m*) calcine

кальц·инáци·я, -и, (*f*) (chem met) calcination, calcining

кальцин·úровани·е, -я, (*n*) *see* каль·ц·инáци·я

кальцин·úрованн·ый, -ая, -ое, (*past part pass*) of кальцин·úр·овать; calcined, roasted

~ сóд·а (*f*) soda ash, carbonate of soda

кальцин·úр·овать, -уют, (*imp*) calcine, roast

кальцио·вольбортúт·, -а, (*m*) (min) calciovolborthite

кальцио·торúт, -а, (*m*) (nucl) calciothorite

кальц·úт·, -а, (*m*) (min) calcite, calcspar

кальци·фер·óл·, -а, (*m*) (chem) calciferol, vitamin D_2

кальци·фик·áци·я, -и, (*f*) (bot geol) calcification

кальци·фóб·, -а, (*m*) (bot) calciphobe

кальц·сóд·а, -ы, (*f*) *see* кальцин·úрованн·ая сóд·а

калы́шк·а, -и, (*g pl*) -шек·, (*f*) kink (rope)

калю·трóн·, -а, (*m*) (phys) calutron

кал·ю́ют (*pres 3rd pl*) of кал·евáть

кáм·, -а, (*m*) (geol) kame

кáм·а, лед·ян·áя (*f*) (meteor) brash, mush (ice)

камацúт·, -а, (*m*) (min) kamacite

кáмбал·а, -ы, (*f*) (fish) flounder

камбало·обрáз·н·ые, -ых, (*pl decl as adj*) (fish) *Pleuronectidae* камбиа́ль·н·ый, -ая, -ое, (*a*) (bot) cambial

камби·й, -я, (*m*) (bot) cambium

~, про́б·ков·ый cork-producing layer, cork cambium

камби·обра́з·н·ый, -ая, -ое, (*a*) cambiform

камбо́дж·а, -и, (*f*) (chem) cambogia, gamboge gum

ка́мбуз·, -а, (*m*) (naut) galley, galley range

камбо́ль·н·ый, -ая, -ое, (*a*) (text) worsted

камеде·но́с·н·ый, -ая, -ое, (*a*) gum-miferous

камеде·обра́з·н·ый, -ая, -ое, (*a*) gummy, gum-like

камеде·теч·е́ни·е, -я, (*n*) (bot) gum-mosis, exudation of sap

каме́д·ист·ый, -ая, -ое, (*a*) gummy

каме́д·н·ый, -ая, -ое, (*a*) gum

~ де́рев·о (*n*) (bot) yellow pine, *Pinus ponderosa*; gum tree, *Liquidambar styraciflua*

каме́д·ь, -и, (*f*) gum, resin

~, арав·и́йск·ая gum arabic

~, древе́с·н·ая wood gum, xylan

~, копа́л·ов·ая copal

ка́мел·ь, -и, (*f*) (hyd eng) camel

камен·е́·ть, -ют, (*imp*) petrify, become petrified

камен·и́ст·ый, -ая, -ое, (*a*) stony, lithic

~ кле́т·к·а (*f*) (bot) stone-cell, sclereide

каменно·уго́ль·н·ый, -ая, -ое, (*a*) coal; (geol) carboniferous

~ смол·а́ (*f*) coal tar

ка́мен·н·ый, -ая, -ое, (*a*) stone, stony, hard, rock, lithoidal; jewel (in measuring instruments)

~ де́рев·о (*n*) (bot) hackberry, *Celtis*

~ кла́д·к·а (*f*) bricklaying; brickwork

~ кун·и́ц·а (*f*) (zool) beech marten, *Martes foina*

~ мук·а́ (*f*) rock flour

~ сол·ь (*f*) rock salt, halite

~ у́гол·ь (*m*) coal

камено- (*component*) (*and see also* камне-) stone, rock

камено·ло́м·н·я, -и, (*g pl*) -м·ен·, (*f*) quarry

камено·на·бро́с·н·ый, -ая, -ое, (*a*) rock-fill (of dams etc.)

камено·тёс·, -а, (*m*) stone-mason

камено·тёс·н·ый, -ая, -ое, (*a*) stone-mason's, stone-cutting

ка́мен·щик·, -а, (*m*) bricklayer, mason

ка́мен·ь, -мн·я, (*nom pl*) ка́мн·и, (*g pl*) камн·е́й, (*m*) stone, rock; (mech) block; (med) calculus, stone; (instr) jewel

~, бето́н·н·ый concrete building brick, concrete block

~, бу́т·ов·ый quarrystone, (civ eng) rubble, hard core

~, ви́н·н·ый (chem) tartar, cream of tartar

~ втор·о́го по·ря́д·к·а, драго·це́н·-н·ый semiprecious stone

~, за·мо́ч·н·ый (arch) keystone, arch-stone

~, коте́ль·н·ый boiler scale

~, кули́с·н·ый (st eng) slide block, slide link; guide shoe

~, лиз·а́льн·ый (agr) salt/mineral lick

~ мотыл·я́ (eng) crank sliding block

~, мы́ль·н·ый (min) soapstone, steatite

~, на·клад·н·о́й (horol) cap/end jewel

~, олов·я́нн·ый (min) tin-stone, cassiterite

~ пе́рв·ого по·ря́д·к·а, драго·це́н·-н·ый precious stone

~ -плитня́к·, -а́, (*m*) flagstone, stone slab

~, под·во́д·н·ый reef

~, по·де́л·очн·ый цвет·н·о́й gem mineral

~ поко́·я (horol) locking jewel

~, пале́т·н·ый (horol) pallet jewel

~, рв·а́н·ый (civ eng) rubble, hard core

~, рво́т·н·ый (chem) tatar emetic, potassium antimonyl tartrate

~, тёс·ан·ый (build) ashlar

~, точ·и́льн·ый whetstone, hone

~ тре́т·ьего по·ря́д·к·а, драго·це́н·-н·ый semiprecious stone

~, шла́к·ов·ый slag brick

~, шту́ч·н·ый (build) ashlar

каме́н·ья (*nom pl*) of ка́мен·ь

ка́мер·а, -ы, (*f*) chamber, compartment, cell, office, room; (min) chamber, room, stall; header (boiler); (phot X ray) camera; inner tube (of tyre)

~, автомоби́ль·н·ая (M/T) inner tube

~, био·терм·и́ческ·ая artificial composting building (sewerage)

~, ва́р·очн·ая (rubber) curing tube/bag

ка́мер·а

~, Ви́льсон·а (*f*) (nucl) cloud/expansion chamber

~, вихр·ев·а́я (ICE) turbulence chamber, swirl chamber

~ вла́ж·ност·и (nucl) wet chamber, sweat box

~, вод·ян·а́я water box (of condenser)

~, воз·ду́ш·н·о-вс·по·мог·а́тельн·ая (ICE) air cell

~ вы́·держ·к·и (nucl) decay chamber

~, выс·о́тн·ая pressure chamber

~, вы·тяж·н·а́я (rubber) stretching chamber

~, гас·и́тельн·ая (elec) oil-blast chamber

~, гиро·скоп·и́ческ·ая gyro-compass casing

~, гор·я́ч·ая (nucl) hot cave/cell

~, груз·ов·а́я lorry/truck inner tube

~ давл·е́ни·я pressure chamber

~ дел·е́ни·я (nucl) fission chamber

~ до·жиг·а́ни·я (a/c) afterburner

~, доро́ж·н·ая (phot) stand camera

~, за·гру́з·очн·ая hopper

~, за́д·н·яя back header (boiler)

~, за·па́ль·н·ая (a/c) igniter chamber

~, за·ря́д·н·ая (elec) battery charging station; (min) charge chamber

~, зерка́ль·н·ая (phot) reflex camera

~, ионизацио́н·н·ая (phys) ionization chamber

~ кокс·ова́ни·я coking chamber/cell

~, конденсацио́н·н·ая (nucl phys) cloud chamber

~, конта́кт·н·ая (chem) catalyst chamber

~, копт·и́льн·ая smokehouse, smoke-drying room

~, мало·формати́в·н·ый (phot) 35 mm camera

~, ма́рш·ев·ая (rocket) cruising thrust chamber

~ ма́рш·ев·ого двиг·а́тел·я (rocket) sustainer chamber

~, микро·фото·на·са́д·очн·ая (phot) photomicrography attachment

~, на·са́д·н·ая photographic eyepiece (microscope attachment)

~ не·по·сре́д·ственн·ого в·пры́ск·а (ICE) direct-injection chamber

~ -обску́р·а, -ы, (*f*) (phot) camera obscura

~, огн·ев·а́я combustion chamber (boiler)

~ от·бо́р·а steam extraction casing (of turbine)

ка́мер·а

~, от·ли́в·н·а́я outlet chamber

~, от·вод·я́щ·ая (met) outlet chamber (continuous casting)

~, о·чи́ст·н·ая (min) room, bord, stall; (nucl) decontamination cell

~, пере́д·н·яя front header (boiler)

~, под·вод·я́щ·ая (met) supply chamber (continuous casting)

~, по·пла́в·ков·ая (ICE) float chamber, fuel chamber, fuel bowl (of carburettor)

~, при·ём·н·ая inlet chamber

~ раз·лож·е́ни·я (chem) decomposition chamber, catalyst chamber

~, рас·пре·дел·и́тельн·ая distribution header (boiler)

~ с·гор·а́ни·я (ICE) combustion chamber; combustor (of ram jet); explosion chamber (of pulse jet)

~ с·гор·а́ни·я, не·раз·дел·ённ·ая open combustion chamber, direct-injection combustion chamber

~ с·гор·а́ни·я с пред·ка́мер·ой pre-combustion chamber

~, семен·н·а́я (bot) seed vessel

~ с·жа́т·и·я (ICE) combustion chamber

~, с·мес·и́тельн·ая mixing chamber

~, с·пас·а́тельн·ая escape chamber (submarines)

~, спира́ль·н·ая volute chamber (centrifugal pump)

~ тума́н·ов (nucl) cloud chamber

~, у·равн·и́тельн·ая surge chamber

~ у·скор·и́тел·я (nucl) acceleration tube, accelerating chamber

~, форса́ж·н·ая (a/c) afterburner

~, шлюз·ов·а́я air lock (underwater working)

~, шумо·глуш·и́тельн·ая silencing chamber

камер·а́льн·ый, -ая, -ое, (*a*) office/laboratory (of work associated with scientific expeditions, surveys etc.)

ка́мер·к·а, -и, (*g pl*) **-р·ок·,** (*f*) ⟨*dim*⟩ of **ка́мер·а;** chamberlet

камерно·столб·ов·о́й, -а́я, -о́е, (*a*) (min) room-and-pillar, bord-and-pillar, pillar-and-stall

ка́мер·н·ый, -ая, -ое, (*a*) of **ка́мер·а**

~ за·ря́д· (*m*) (min) tunnel-blasting charge

~ спо́соб· (*m*) chamber process (of sulphuric acid production)

~ систе́м·а (*f*) (min) room-and-pillar, bord-and-pillar, pillar-and-stall камеро́н·, -а, (*m*) (naut) Cameron pump

камеро·сто́м·а, -ы, (*m*) (bot) camerostome

камерто́н·, -а, (*m*) tuning fork

камерто́н·н·ый, -ая, -ое, (*a*) of камерто́н·

~ генера́тор· (*m*) (rad) tuning fork generator/oscillator

ками́н·, -а, (*m*) fire place

~, электр·и́ческ·ий electric radiator/fire

камло́т·, -а, (*m*) (text) coarse serge

камло́т·ов·ый, -ая, -ое, (*a*) of камло́т·

камне- (*root*) stone, litho

камне·ви́д·н·ый, -ая, -ое, (*a*) lithoidal, stone-like

камне·дроб·и́лк·а, -и, (*g pl*) -лок·, (*f*) stone/rock crusher, disintegrating mill

камне·дробл·е́ни·е, -я, (*n*) stone crushing; (med) lithotripsy

камне·ло́м·к·а, -и, (*g pl*) -м·ок·, (*f*) (bot) saxifrage, *Saxifraga*

камне·об·рабо́т·к·а, -и, (*f*) stone-masonry

камне·ре́з·, -а, (*m*) stone-cutter

камне·сеч·е́ни·е, -я, (*n*) (med) lithotomy

камне·то́ч·ец·, -ч·ц·а, (*m*) (zool) stone borer, *Pholas dactylus*

камне·то́ч·ечн·ый, -ая, -ое, (*a*) stone-boring; (zool) lithotomous

камне·у·бо́р·очн·ая маши́н·а (*f*) (agr) stone picker

ка́мн·и (*nom pl*) of ка́мен·ь

ка́мор·а, -ы, (*f*) cavity (of shell); chamber (of gun); камо́р·а, -ы, (*f*) (print) superscript

кампен·е́йск·ий, -ая, -ое, (*a*) of кампа́н·и·я

кампа́н·и·я, -и, (*f*) campaign; campaign, run, operating period (of furnaces, reactors etc.); (naut) commission, voyage

кампа́н·ск·ий я́рус· (*m*) (geol) Campanian

кампеш·, -а, (*m*) *see* кампе́ш·ев·ое де́рев·о

кампе́ш·ев·ый, -ая, -ое, (*a*) logwood

~ де́рев·о (*n*) logwood, *Haematoxylon campechianum*

кампили́т·, -а, (*m*) (min) kampylite, campylite

кампило·тро́п·н·ый, -ая, -ое, (*a*) (bot) campylotropous

камполо́н·, -а, (*m*) (med) campolon

камс·а́, -ы́, (*f*) (fish) anchovy, *Engraulis encrasicholus*

кампто·дро́м·н·ый, -ая, -ое, (*a*) (bot) camptodromic

камуфле́т·, -а, (*m*) (mil) camouflet; unpleasant surprise

камуфл·и́ровани·е, -я, (*n*) *see* камуфл·я́ж·

камуфл·я́ж·, -а, (*m*) (mil) dazzle painting

ка́м·ушек·, -шк·а, (*m*) (*dim*) of ка́мен·ь

камф·а́н·, -а, (*m*) camphane

камфа́н·ов·ая кисл·от·а́ (*f*) camphanic acid

камфар·а́, -ы́, (*f*) camphor

камфа́р·н·ый, -ая, -ое, (*a*) camphor, camphoric

~ де́рев·о (*n*) (bot) camphor laurel, *Cinnamomum camphora*

~ кисл·от·а́ (*f*) (chem) camphoric acid

~ лавр· (*m*) (bot) camphor laurel, *Cinnamomum camphora*

~ ма́сл·о (*n*) camphor oil

камфаро́см·а, -ы, (*f*) (bot) camphorosma

камф·е́н·, -а, (*m*) (chem) camphene

камф·и́л·, -а, (*m*) camphyl

камф·о́н·, -а, (*m*) camphone

камфо́н·ов·ый, -ая, -ое, (*a*) camphone, camphonic

ка́мфор·а, -ы, (*f*) *see* камфар·а́

камфора́н·ов·ый, -ая, -ое, (*a*) camphoranic

камфо́р·н·ый, -ая, -ое, (*a*) *see* камфа́р·н·ый

камфоро́н·ов·ая кисл·от·а́ (*f*) camphoronic acid

камфоро́см·а, -ы, (*f*) *see* камфаро́см·а

камча́т·к·а, -и, (*g pl*) -т·ок· (*f*) (text) damask

камча́т·н·ая ткан·ь (*f*) (text) damask

камча́т·ск·ий краб· (*m*) (zool) Kamchatka crab, *Paralithodes camtschatika*

камы́ш·, -а́, (*i*) -о́м, (*m*) (bot) rush, reed; cane

камыше·кос·и́лк·а, -и, (*g pl*) -лок·, (*f*) reed mower, weed cutting launch

камыш·о́в·ый, -ая, -ое, (*a*) rush, rushy, rattan, juncaceous

камыш·и́т·, -а, (*m*) (build) rush pressboard

ка́мь·я, -и, (*f*) single-seat boat, dinghy

кана́в·а, -ы, (*f*) ditch, trench; inspection pit (in garages etc.)

~, водо·с·то́ч·н·ая gutter

кана́в·а
~, при·доро́ж·н·ая   cuvette, ditch
~, с·то́ч·н·ая   gutter, drain, sewer
кана́в·к·а, -и, (*g pl*) **-в·ок·,** (*f*) (*dim*) of **кана́в·а;** groove, slot; (agr) drill; flute (in drill etc.)
~, кольц·ев·а́я   (M/T) piston-ring groove
~, по·сев·н·а́я   (agr) sowdrill
   у́гол· кана́в·к·и   (*m*) groove angle (sound recording)
~, шпо́н·очн·ая   taper key slot, saddle key groove
канаво·коп·а́тел·ь, -я, (*m*) trencher, ditcher, ditch digger
канаво·о·чист·и́тел·ь, -я, (*m*) ditcher, ditch cleaner
кана́в·очник·, -а, (*m*) fluting/chasing chisel (woodwork)
кана́д·ск·ий, -ая, -ое, (*a*) Canadian
~ бальза́м·   (*m*) Canadian balsam
кана́л·, -а, (*m*) canal; channel, duct, conduit; lane (e.g. in ice and astron); (anat) canal, canalis; (nucl) channel, hole
~, био·лог·и́ческ·ий   biological hole (nucl reactor)
~, вертика́ль·н·ый   downtake, uptake
~, водо·с·бо́р·н·ый   leakage drain
~ вы·ключ·е́ни·я   (nucl) shut-down channel
~, глух·о́й   (nucl) thimble chamber
~ да́ль·ност·и   (rocket) range channel
~, деривацио́н·н·ый   (hydroelec) aqueduct, conduit, diversion canal
~, дрена́ж·н·ый   (agr) drain, gutter
~, дым·ов·о́й   chimney flue
~, за·ли́в·очн·ый   (plast) sprue channel (of mould)
~, зол·ов·ый   water trough (boiler plant)
~, ка́бель·н·ый   (elec) cable duct
~, ли́мбер-борт·ов·о́й   (shipb) limber
~, ма́сл·ян·ый   (bot) vitta
~, мор·ск·о́й   seaway
~ мундштук·а́   die channel (of extrusion press)
~, обра́т·н·ый   return duct
~, о·рос·и́тельн·ый   (agr) irrigation channel/canal
~, от·вод·я́щ·ий   (hydr eng) tail race
~, пар·ов·о́й   steam port (of reciprocating engine)
~ пере·да́·ч·и   (telecom) channel
~, под·вод·я́щ·ий   (hydroelec) diversion canal, aqueduct, conduit
~ про·дув·а́ни·я   (geol) blow-hole

кана́л
~ про·меж·у́точн·ой част·от·ы́   (rad) intermediate frequency amplifier, i-f amplifier
~, раз·вод·я́щ·ий   (plast) runner
~ с·вя́з·и   (telecom) channel
~ ствол·а́   bore (of gun)
~ сте́ржн·я   (nucl) rod passage
~, судо·хо́д·н·ый   navigable canal
~, эксперимента́ль·н·ый   (nucl) beam hole
кана́л·ец·, -лца, (*m*) (*dim*) of **кана́л·;** (anat) canalicule, canaliculus
канализ·ацио́нн·ый, -ая, -ое, (*a*) sewerage, sewer, sewage
канализ·а́ци·я, -и, (*f*) sewerage, sewerage system, drainage; (elec) underground cable system
~, ли́в·нев·ая   surface water sewerage, storm water sewerage
канализ·и́р·овать, -уют, (*imp and perf*) lay a sewer/sewerage system
кана́л·ов·ый, -ая, -ое, (*a*) of **кана́л·**
~ луч·и́   (*pl*) (phys) canal rays, positive rays
~ у·сил·и́тел·ь   (*m*) window amplifier
~ эффе́кт·   (*m*) (nucl) channeling
кана́ль·н·ый, -ая, -ое, (*a*) of **кана́л·**
кана́ль·чат·ый, -ая, -ое, (*a*) (zool) canalicular
канар·е́ечник·, -а, (*m*) (bot) Canary grass, *Phalaris*
канар·е́ечн·ый, -ая, -ое, (*a*) canary, canary-coloured
канар·е́йк·а, -и, (*g pl*) **-е́ек·,** (*f*) (orn) canary, canary bird, *Serinus canarius*
кана́т·, -а, (*m*) (*and see* **трос·**) cable, hawser
~, бе́л·ный   white rope
~, вед·у́щ·ий   traversing rope (cableway)
~, гла́в·н·ый   fixed cable (of cableway)
~, жёстк·ий   steel-wire rope
~, за·кры́·т·ый   locked-coil rope
~ ка́бель·н·ой с·ви́в·к·и   cable-laid rope
~, не·рас·кру́ч·ивающ·ийся   non-rotating construction rope
~, нес·у́щ·ий   fixed cable (of cableway)
~, овало·пря́д·н·ый   flattened-strand rope, oval-strand rope
~ ордина́р·н·ой с·ви́в·к·и   ordinary-lay rope
~ от·бо́й·н·ый   (min) balance rope
~, пло́ск·ий   flat rope
~, при·ча́ль·н·ый   (naut) mooring line/hawser

кана́т
~, про́·волоч·н·ый   wire rope, cable
~ прям·о́й с·ви́в·к·и   regular-laid rope
~, раст·и́тельн·ый   fibre/fiber rope
~, руд·ни́чн·ый   mining rope
~, смо́ль·н·ый   tarred rope
~, сталь·н·о́й   steel wire rope
~ тро́й·н·ой с·ви́в·к·и   cable-laid rope
~ тро́с·ов·ой рабо́т·ы   hawser-laid rope
~, тя́г·ов·ый   traction rope; traversing rope (of cableway)
кана́т·ик·, -а,   (*m*) (*dim*) of **кана́т·**; (anat) funiculus
~, анте́нн·ый   (rad) aerial cable
канатно·скреб·ко́в·ый экскава́тор·   (*m*) dragline excavator
канатно·скре́пер·н·ая у·стано́в·к·а   (*f*) scraper-chain conveyor
кана́т·н·ый, -ая, -ое,   (*a*) rope, cable
~ бур·е́ни·е   (*n*) cable-tool drilling
~ доро́г·а   (*f*) cableway, ropeway, rope-haulage way
канато·у·кла́д·чик·, -а,   (*m*) layering idler (on winding drum)
канв·а́, -ы́,   (*f*) embroidery/tapestry canvas
кандида́т·, -а,   (*m*) candidate
~ на·у́к·   degree (approx. equivalent to B.Sc. or B.Sc./M.A.)
кандида́т·ск·ий, -ая, -ое,   (*a*) candidate's
~ диссерта́ци·я   (*f*) B.A./B.Sc. dissertation
~ минимум·   (*m*) B.A./B.Sc. course
кандидат·у́р·а, -ы,   (*f*) candidature, candidacy
канегр·а, -ы,   (*f*) (bot) tanner's dock, *Rumex hymenosepalus*
канефо́р·а, -ы,   (*f*) (arch) caryatid
кани́кул·ы (*nom pl*), (*g pl*) -ул,   (*f*) vacation, holidays
кандицид·и́н·, -а,   (*m*) (pharm) candicidin
кани́стр·а, -ы,   (*f*) can, canister
канител·ь, -я,   (*m*) (text) bullion, bullion fringe; (M/T) wire-mesh element (of air filter); laborious/protracted proceeding
канифа́с·, -а,   (*m*) (text) dimity
канифа́с-бло́к·, -а,   (*m*) (naut) snatch block
канифо́л·ь, -и,   (*f*) colophony, rosin
канкрини́т·, -а,   (*m*) (min) cancrinite
ка́нн·а, -ы,   (*f*) (zool) eland, *Taurotragus oryx*, (bot) canna

ка́нн·а
~, съ·ед·о́бн·ая   arrowroot, *Canna edulis*
каннклю́р·а, -ы,   (*f*) (build) flute, (*pl*) fluting
Канница́ро, реа́кци·я   (*f*) (chem) Cannizzaro reaction
кано́н·, -а,   (*m*) (print) canon
канонёр·к·а, -и,   (*g pl*) -р·ок·,   (*f*) gunboat
канонёр·ск·ий, -ая, -ое,   (*a*) of **канонёр·к·а**
~ ло́д·к·а   (*f*) gunboat
канон·и́ческ·ий, -ая, -ое,   (*a*) (math) canonical
кано́пус·, -а,   (*m*) (astron) Canopus
ка́нт·, -а,   (*m*) edging, piping (on clothes); square (of book); welt (of shoe)
кантарид·и́н·, -а,   (*m*) (biochem) cantharidin
кантарид·и́н·ов·ая кисл·от·а́   (*f*) cantharidinic acid
кантилле́вер·н·ый, -ая, -ое,   (*a*) cantilever
кант·ова́льн·ый, -ая, -ое,   (*a*) canting, tilting, turning, manipulating
~ голо́в·к·а   (*f*) manipulator (forging)
~ при·способл·е́ни·е   (*n*) (weld) rotating jig
кант·ова́ни·е, -я,   (*n*) canting, tilting, turning, manipulating
кант·ова́тел·ь, -я,   (*m*) (met roll) tilter, tilting cradle, manipulator, turn-over device; (weld) positioner, manipulator
~, крюк·ов·о́й   lifting-finger manipulator
~ лист·о́в   turnover arms
~, ро́л·иков·ый   bobbin-type manipulator
~, ступ·е́нчат·ый   stationary-type manipulator
кант·ова́ть, -уют,   (*imp*) edge, pipe (by sewing); tilt, cant, manipulate; bevel
кант·о́вк·а, -и,   (*g pl*) -вок·,   (*f*) edging, piping; tilting, manipulating; switching over (coke ovens)
кант·о́вочн·ый, -ая, -ое,   (*a*) of **кант·о́вк·а**
~ башма́к·   (*m*) (weld) rocking positioner
~ кольц·о́   (*n*) (weld) frame rotator
ка́нтор·ск·ое мно́ж·еств·о   (*n*) (math) Cantor set, Cantor discontinuum
кану́фер·, -а,   (*m*) *see* **калу́фер·**

ка́нцеллинг·, -а, (*m*) (com) cancelling date (of charter party)

канцелля́т·н·ый, -ая, -ое, (*a*) cancellous, cancellated

канцеля́р·и́ст·, -а, (*m*) bureaucrat; clerk

канцеля́р·и·я, -и, (*f*) office

канцеля́р·ск·ий, -ая, -ое, (*a*) office, clerical

~ **при·над·ле́ж·ност·и** (*pl*) office equipment

канцеро·ге́н·, -а, (*m*) (med) carcinogen

канцеро·ге́н·н·ый, -ая, -ое, (*a*) (med) carcinogenic

канцеро·ло́г·и·я, -и, (*f*) (med) cancerology

каньо́н·, -а, (*m*) canyon

каню́л·я, -и, (*f*) (med) cannula

каоли́н·, -а, (*m*) (min) kaolin, china clay

~, мя́гк·ий lithomarge

каолин·иза́ци·я, -и, (*f*) (geol) kaolinization

каолин·и́т·, -а, (*m*) (min) kaolinite

ка́п·, -а, (*m*) (bot) excrescence, burr, burl, nodule; (naut) sliding hatch cover; (geog) cape

ка́п·а, -ы, (*f*) (elec) packaging seal, sealed end (of cable)

ка́п·авш·ий, ·-ая, -ее, (*past part act*) of **ка́п·а·ть, ка́п·ать**

кападáйн·, -а, (*m*) (elec) capadyne

ка́п·ани·е, -я, (*n*) (*v n*) of **ка́п·а·ть, ка́п·ать**

ка́п·ань·е, -я, (*n*) (*v n*) of **ка́п·а·ть, ка́п·ать**

капаситáнс·, -а, (*m*) (elec) capacitance

ка́п·а·ть, -ют, (*imp*) drip, dribble; **ка́п·ать, -лют,** (*imp*); drop (a medicine), place drops

кап·ёж·, -а́, (*i*) **-óм,** (*m*) (geol) water drip

капел·и́ровани·е, -я, (*n*) *see* **капелл·и́ровани·е**

капелл·и́ровани·е, -я, (*n*) (met) cupellation

капелл·я́ци·я, -и, (*f*) (met) cupellation

кап·éл·ь, -и, (*f*) (met) cupel; dripping snow, dripping (of melting snow); **ка́п·ел·ь** (*g pl*) of **ка́пл·я**

~, леж·áщ·ая (phys) stationary drop

~, не·по·дви́ж·н·ая (phys) stationary drop

ка́п·ельк·а, (*g pl*) **-лек·,** (*f*) (*dim*) of **ка́пл·я;** droplet

ка́п·ельник·, -и, (*m*) *see* **ка́п·ельниц·а**

ка́п·ельниц·а, -ы, (*i*) **-ц·ей,** (*f*) drip pan; (chem) dropper, dropping bottle; dripcock

ка́п·ельн·о·жи́дк·ий, -ая, -ое, (*a*) liquid-drop; liquid enough to drip; (meteor) water

ка́п·ельн·ый, -ая, -ое, (*a*) drip, dripping, drop, dropping; (nucl) liquid-drop; tiny

~ **анали́з·** (*m*) spot test analysis

~ **реáкци·я** (*f*) spot reaction

~ **под·за·ря́д·к·а** (*f*) (elec) trickle charging

~ **теóр·и·я** (*f*) (nucl) liquid-drop theory

ка́пер·ец·, -рца, (*m*) *see* **ка́перс·**

ка́перс·, -а, (*m*) (bot) caper, *Capparis*

ка́перс·ник·, -а, (*n*) *see* **ка́перс·**

капилл·и́ци·й, -я, (*m*) (bot) capillitium

капилл·я́р·, -а, (*m*) capillary tube; (anat) capillary

капилляри·мéтр·, -а, (*m*) capillary-tube surface-tension meter

капилля́р·ност·ь, -и, (*f*) (phys) capillarity, capillary attraction

капилля́р·н·ый, -ая, -ое, (*a*) capillary

~ **про·сáч·ивани·е** (*n*) capillary penetration

~ **электрó·метр·** (*m*) (chem) capillary electrometer

капитáл·, -а, (*m*) (econ) capital

~, оборóт·н·ый, working/floating capital

~, основ·н·óй fixed capital, stock

~, пере·мéн·н·ый variable capital

капитализ·áци·я, -и, (*f*) capitalization

капитализ·и́р·овать, -уют, (*imp and perf*) (fin) capitalize

капиталист·и́ческ·ий, -ая, -ое, (*a*) capitalist

капитало·в·лож·éни·е, -я, (*n*) capital investment, investment

капитáль·н·ый, -ая, -ое, (*a*) capital, major, principal, fundamental

~ **ремóнт·** (*m*) capital repairs, major overhaul

~ **стен·á** (*f*) (build) main wall

~ **стро·и́тельств·о** (*n*) capital construction

капитáн·, -а, (*m*) (mil) captain; (naut) master, skipper

~, с·дáт·очн·ый (shipb) acceptance trials captain

капита́н

~ втор·о́го ра́нг·а captain 2nd rank (naval)

~ -лейтена́нт·, -а, (*m*) captain-lieutenant (naval)

~ пе́рв·ого ра́нг·а captain 1st rank (naval)

~ тре́т·ьего ра́н·га captain 3rd rank (naval)

капита́н·ск·ий, -ая, -ое, (*a*) of **капита́н·**

~ мо́ст·ик· (*m*) (shipb) captain's bridge, navigating bridge

~ по·каз·а́ни·е (*n*) manifest (of cargo)

капите́л·ь, -и, (*f*) (arch) capital; (print) small capitals

капите́ль·н·ый, -ая, -ое, (*a*) of **капите́л·ь**

~ бу́кв·а (*f*) (print) small capital

капитул·и́р·овать, -уют, (*imp and perf*) capitulate

Капи́ц·ы, скач·о́к· (*m*) (phys) Kapitza temperature jump

капка́н·, -а, (*m*) trap (for wild animals)

кап·ко́в·ый, -ая, -ое, (*a*) kapok

Капла́н·а, турби́н·а (*f*) Kaplan water turbine

ка́пл·е- (*root*) drip-, drop-

капле·за·щищ·ённ·ый, -ая, -ое, (*a*) (elec eng) drip-proof

капле·от·дел·и́тел·ь, -я, (*m*) steam trap

капле·пад·е́ни·е, -я, (*n*) drip, dripping, drop, dropping (of drips)

температу́р·а капле·пад·е́ни·я (*f*) drop point (of grease etc.)

капле·у·каз·а́тел·ь, -я, (*m*) drip-feed/sight-feed lubricator

каплу́н·, -а́, (*m*) (agr) capon

ка́пл·ющ·ий, -ая, -ее, (*pres part act*) of **ка́п·ать**

ка́пл·я, -и, (*g pl*) **ка́п·ел·ь,** (*f*) drop, drip; (pharm) drops

~ ла́в·ы (geol) lava-tear

капно·ско́п·, -а, (*m*) smoke gauge/gage

ка́п·н·уть, -ут, (*perf*) drop, let fall a drop, place a drop

кап·о́к·, -п·к·а́, (*m*) (text) kapok; (bot) kapok/silk-cotton tree, *Ceiba pentandra*

капони́р·, -а, (*m*) aircraft pen; (mil) double caponier

ка́пор·, -а, (*m*) hood (clothing)

капо́рк·а, -и, (*f*) *see* **ива́н-ча́й**

капо́т·, -а, (*m*) cowling, cowl, hood; (М/Т) bonnet

~ дви́г·ател·я (М/Т) bonnet, (a/c) engine cowling

~, от·кид·н·о́й hinged cowling

капот·а́ж·, -а, (*i*) **-ем,** (*m*) *see* **капот·и́ровани·е**

капот·и́рова́ни·е, -я, (*n*) (air) nosing over; fixing a cowl/hood

ка́пп·, -а (*m*) kapp (a serge fabric)

ка́пп·а, -ы, (*f*) kappa, К (Greek)

крив·а́я Ка́пп·а (*f*) (math) Kappa curve

капри́з·нича·ть, -ют, (*imp*) be capricious

капри́з·н·ый, -ая, -ое, (*a*) capricious

капри́л·, -а, (*m*) (chem) capryl

капри́л·ов·ый, -ая, -ое, (*a*) capryl, caprylic; caprylate (of)

~ спирт· (*m*) capryl alcohol

капри́н·ов·ый, -ая, -ое, (*a*) capric; caprate (of)

~ альдеги́д· (*m*) capraldehyde

~ кисл·от·а́ (*f*) capric acid

капри·фик·а́ци·я, -и, (*f*) (bot) caprification

капроа́т·, -а, (*m*) (chem) caproate

капро·лакта́м·, -а, (*m*) (chem) caprolactam

капро́н·, -а, (*m*) capron (synthetic fibre), polycaprolactam

капро́н·ов·ый, -ая, -ое, (*a*) capron

~ кисл·от·а́ (chem) caproic acid

капроново·ки́сл·ый, -ая, -ое, (*a*) caproate (of)

капсаици́н·, -а, (*m*) (pharm) capsaicin

ка́псел·ь, -я, (*m*) (ceram) saggar, sagger

ка́псул·а, -ы, (*f*) (*see also* **ка́псюл·ь**) capsule

ка́псуль·н·ый, -ая, -ое, (*a*) canned, sealed in, encapsulated

ка́псюл·ь, -я, (*m*) (*see also* **ка́псул·ь**) (expl) detonator, percussion cap; inset (of microphone)

~ -вос·пламен·и́тел·ь, -я, (*m*) percussion cap

~ -детона́тор·, -а, (*m*) detonator, disruptive detonator, blasting cap

ка́псюль·н·ый, -ая, -ое, (*a*) of **ка́псюл·ь**

~ втул·к·а (*f*) (expl) primer

капт·а́ж·, -а, (*i*) **-ем,** (*m*) capping, captation (of wells); (geol) catchment

капта́кс·, -а, (*m*) (chem) mercapto-benzothiazole, captax

каптáл·, -а, (*m*) headband (bookbinding)

~, нúж·н·ий tail cap (of book)

каптенáрмус·, -а, (*m*) (mil) quartermaster sergeant

каптáн·, -а, (*m*) (chem) captan

капýст·а, -ы, (*f*) (bot) cabbage, *Brassica oleracea*

~, брюссель·ск·ая Brussels sprout, *Brassica oleracea gemmifera*

~, кúсл·ая sauerkraut

~, кочáн·н·ая head cabbage, cabbage

~, мор·ск·áя sea kale, *Crambe maritima*

~, спáрж·ев·ая broccoli

~, цвет·н·áя cauliflower

капýст·ник·, -а, (*m*) cabbage field

капýст·ниц·а, -ы, (*i*) **-ей,** (*f*) (ent) cabbage butterfly, *Pieris brassicae*

капýст·н·ый, -ая, -ое, (*a*) of **капýст·а**

~ мол·ь (*f*) (ent) diamondback moth, *Plutella maculipennis*

~ мýх·а (*f*) cabbage maggot, *Hylemya brassicae*

~ сóв·к·а (*f*) cabbage moth, *Barathra brassicae*

~ тл·я (*f*) cabbage aphis, *Brevicoryne brassicae*

капуцúн·, -а, (*m*) (bot) nasturtium

капюшóн·, -а, (*m*) hood, cowl

кáр·, -а, (*m*) (geog) corrie, cirque

карабúн·, -а, (*m*) (mil) carbine; spring/snap/latch hook

каребúн·н·ый крюч·óк· (*m*) safety/ swivel hook

карабúн·чик·, -а, (*m*) = **карабúн·**

карáбк·а·ться, -ются, (*imp*) clamber

караваe·обрáз·н·ый, -ая, -ое, (*a*) bun-shaped

каравá·й, -я, (*m*) round loaf

каравáн·, -а, (*m*) caravan (of pack animals); (naut) tow string; convoy, convoyed vessels; (min) pile of peat, peat rick

~, земле·черп·áтельн·ый (naut) dredger string

карагáн·, -а, (*m*) (bot) elm, *Ulmus densa*

карадрúн·а, -ы, (*f*) (ent) beet army worm, *Laphygma exempta*

кар·áем·ый, -ая, -ое, (*pres part pass*) of **кар·á·ть;** punishable

каракáтиц·а, -ы, (*i*) **-ей,** (*f*) (zool) cuttlefish

карáк·ов·ый, -ая, -ое, (*a*) dark bay, dun

карáкул·ь, -я, (*m*) Persian lamb, karakul (fur)

каракýль·ск·ая по·рóд·а (*f*) karakul sheep

каракульч·á, -й, (*i*) **-óй,** (*f*) astrakhan (fur)

карамеле·вáр·очн·ая стáнци·я (*f*) (food) caramel boiler

карамéл·ь, -и, (*f*) caramel, burnt sugar

карáн·, -а, (*m*) (chem) carane

карандáш·, -á, (*i*) **-óм,** (*m*) pencil

~, винт·ов·óй propelling pencil

~, хим·úческ·ий (*m*) indelible pencil

карандáш·ев·ый, -ая, -ое, (*a*) of **карандáш·**

~ дéрев·о (*n*) (bot) red cedar, *Juniperus oxycedrus, J. virginiana*

карандáш·н·ый, -ая, -ое, (*a*) pencil, pencil-type

карантúн·, -а, (*m*) quarantine

~, держ·áть be in quarantine

~, с·ним·á·ть declare the end of quarantine; (naut) grant pratique

карантúн·н·ый, -ая, -ое, (*a*) of **карантúн·**

~ патéнт· (*m*) (naut) bill of health

~ прó·пуск· (*m*) (naut) pratique

~ с·вид·éтельств·о (*n*) bill of health

карапáкс·, -а, (*m*) (zool) carapace

карапáс·н·ый, -ая, -ое, (*a*) of **карапáкс·**

~ пáлуб·а (*f*) (shipb) turtle-back deck, whale back

карапýз·ик·и (*nom pl*), (*g pl*) **-ов,** (*m*) (ent) black beetles, *Histeridae*

карасик·, -а, (*m*) crosscut file

карáс·ь, -я, (*m*) (fish) crusian carp, *Carassius carassius*

~, мор·ск·óй (fish) sea bream

карáт·, -а, (*m*) carat

~, метр·úческ·ий metric carat (200 mg)

кар·áтельн·ый, -ая, -ое, (*a*) punitive

кар·á·ть, -ют, (*imp*) punish

караýл·, -а, (*m*) guard

~, бр·áть на present arms

~, пол·ев·óй (mil) outguard

~, по·чёт·н·ый guard of honour

караýл·ить, -ят, (*imp*) guard

карáчки, ста·ть на get down on all fours

карб- (*component*) (chem) carb-, carbo-

карб·азóл·, -а, (*m*) (chem) carbazole

карб·амáт·, -а, (*m*) (chem) carbamate

карб·амúд·, -а, (chem) carbamide, urea

карб·ами́д·н·ый, -ая, -ое, (*a*) carbamide
~ **смол·а́** (*f*) urea resin
карб·ами́л·, -а, (*m*) (chem) carbamyl
карб·ами́л·ов·ый, -ая, -ое, (*a*) carbamyl
карб·ами́н·ов·ая кисл·от·а́ (*f*) (chem) carbamic acid
карбамино́во·ки́сл·ый, -ая, -ое, (*a*) carbamate (of)
карб·ани́л·, -а, (*m*) carbamil, phenyl cyanate
карб·анили́д·, -а, (*m*) carbanilide, diphenyl urea
карба́т·, -а, (*m*) karbate (a plastic)
карб·е́н·, -а, (*m*) (chem) carbene
карб·и́д·, -а, (*m*) (chem) carbide
~ **бо́р·а** boron carbide
~ **вольфра́м·а** tungsten carbide
~ **желе́з·а** (met) cementite
~ **ка́льц·и·я** calcium carbide
~ **кре́мн·и·я** silicon carbide, carborundum
~ **цирко́н·и·я** zirconium carbide
карбидиз·а́ци·я, -и, (*f*) (met) carbidizing
карбидиз·о́ванн·ый, -ая, -ое, (*past part pass*); carbidized
карби́д·н·ый, -ая, -ое, (*a*) carbide
карбидо·вольфра́м·ов·ый, -ая, -ое, (*a*) tungsten carbide
карбидо·образ·у́ющ·ий, -ая, -ее, (*a*) carbide-forming
~ **элеме́нт·** (*m*) (met) carbide former
карбил·ами́н·, -а, (*m*) carbylamine, isocyanide
карби́л·ов·ая кисл·от·а́ (*f*) carbylic acid
карбин·о́л·, -а, (*m*) (chem) carbinol, methyl alcohol
карбито́л·, -а, (*m*) carbitol, diethylene glycol
карбо- (*component*) carbo-
карбо·ангидра́з·а, -ы, (*f*) (chem) carbonic anhydrase
карбо·гемоглоби́н·, -а, (*m*) carbohaemoglobin
карбо·ге́н·, -а, (*m*) (med) carbogen
карбо·гидра́з·а, -ы, (*f*) (biochem) carbohydrase
карбо́·граф·, -а, (*m*) (phot) carbograph
карб·о́ид·, -а, (*m*) carboid
карб·окси·гемоглоби́н·, -а, (*m*) (biochem) carboxyhaemoglobin
карб·окси́л·, -а, (*m*) (chem) carboxyl
карб·оксил·а́з·а, -ы, (*f*) (biochem) carboxylase

карб·оксила́т·н·ый, -ая, -ое, (*a*) carboxylated, carboxyl-containing
карб·оксил·и́рованн·ый, -ая, -ое, (*past part pass*); carboxylated
карбоксил·со·держ·а́щ·ий, -ая, -ее, (*a*) carboxyl-containing
карб·окси́ль·н·ый, -ая, -ое, (*a*) carboxy, carboxylic
карбокси·метил·целлюло́з·а, -ы, (*f*) carboxymethyl cellulose
карбол·е́н·, -а, (*m*) carbolene (activated charcoal)
карболи́неум·, -а, (*m*) carbolineum (wood preservative)
карбо·ли́т·, -а, (*m*) (plast) carbolite
карбо́л·к·а, -и, (*f*) carbolic acid, phenol
карбо́л·овк·а, -и, (*f*) carbolic acid, phenol
карбо́л·ов·ый, -ая, -ое, (*a*) carbolic, phenolic, phenic
~ **кисл·от·а́** (*f*) carbolic acid, phenol
карболо́·й, -я, (*m*) (met) carboloy
карбо́·метр·, -а, (*m*) carbometer
карбо́н·, -а, (*m*) (geol) Carboniferous period
карбона́до (*m indecl*) carbonado (industrial diamond)
карбон·а́т·, -а, (*m*) (chem) carbonate
~ **аммо́н·и·я** ammonium carbonate
~ **-апати́т·, -а,** (*m*) carbonate-apatite
~ **ра́д·и·я** radium carbonate
карбонат·иза́ци·я, -и, (*f*) (geol) formation of natural carbonates, carbonation
карбона́т·ност·ь, -и, (*f*) carbonate content
карбона́т·н·ый, -ая, -ое, (*a*) carbonate; carbonaceous
карбониз·а́ци·я, -и, (*f*) carbonation (of liquids), carbon saturation; carbonization (of solids)
карбониз·и́рованн·ый, -ая, -ое, (*past part pass*) of **карбониз·и́р·овать;** carbonated (of liquids), carbonized (of solids)
карбониз·и́р·овать, -уют, (*imp and perf*) carbonify (of liquids); carbonize (of solids)
карбон·и́л·, -а, (*m*) (chem) carbonyl
~ **ни́кел·я** nickel carbonyl
~, **хло́р·ист·ый** carbonyl chloride, phosgene
карбон·и́ль·н·ый, -ая, -ое, (*a*) carbonyl, carbonylic
~ **желе́з·о** (*n*) carbonyl iron
~ **спо́соб·** (*m*) (met) Mond process (of producing pure nickel)

карбо́н·ов·ый, -ая, -ое, (*a*) carbon, carbonaceous; (geol) Carboniferous

~ **кисл·от·а́** (*f*) organic acid (loosely); fatty acid

карбору́нд·, -а, (*m*) carborundum

карбо·терм·и́ческ·ий спо́соб· (*m*) (met) carbothermic process, reduction with carbon

карбо·тита́н·, -а, (*m*) (met) titanium pig

карбо·холи́н·, -а, (*m*) (pharm) carbocholin

карбо·цеп·н·о́й, -а́я, -о́е, (*a*) (chem) carbon-chain

карбо·цикл·и́ческ·ий, -ая, -ое, (*a*) (chem) carbocyclic, isocyclic

Ка́рбро-проце́сс·, -а, (*m*) (phot) Carbro process

карб·у́нкул·, -а, (*m*) (min) carbuncle; (vet) anthrax

~, **шум·я́щ·ий** (vet) symptomatic anthrax, Rauschbrand

карбюра́тор·, -а, (*m*) carburettor, carburetor

~, **в·пры́ск·ивающ·ий** spraying carburettor

~, **двух·диффу́зор·н·ый** double-barrel carburettor

~, **двух·жиклёр·н·ый** double-jet carburettor

карбюра́тор·н·ый, -ая, -ое, (*a*) (M/T) carburettor; carburizing

карбюр·а́ци·я, -и, (*f*) (M/T) carburation, carburetion

карбюриз·а́тор·, -а, (*m*) (met) carburizing agent

карбюр·и́рованн·ый, -ая, -ое, (*past part pass*) of **карбюр·и́р·овать;** carburetted; carburized

карбюр·и́р·овать, -уют, (*imp and perf*) carburate, carburet; carburize

карв·акр·о́л·, -а, (*m*) (chem) carvacrol

карв·е́н·, -а, (*m*) (chem) carvene

карв·о́н·, -а, (*m*) (chem) carvone

ка́рго (*n indecl*) cargo

карго·пла́н·, -а, (*m*) cargo plan

ка́рд·а, -ы, (*f*) (text) card clothing

кардамо́н·, -а, (*m*) (bot) cardamon, cardamom, *Elettaria cardamomum*

карда́н·, -а, (*m*) Cardan/universal joint; (horol) gimbals

~, **жёстк·ий** Hardy–Spicer joint

~, **звен·чат·ый** flexible universal joint, ring joint

~, **иго́ль·чат·ый** mechanical universal joint, needle-bearing universal joint

~, **мя́гк·ий** flexible/resilient universal joint

карда́н

~, **син·хро́н·н·ый** constant-velocity universal joint

карда́н·н·ый, -ая, -ое, (*a*) of **карда́н·**

~ **вал·** (*m*) universal shaft, propeller shaft

~ **механ·и́зм·** *see* **карда́н·**

~ **пере·да́·ч·а** (*f*) (M/T) propeller shaft and universal joints

~ **шарни́р·** (*m*) (*and see* **карда́н·**) universal joint

карда́н·ов·о кольц·о́ (*n*) gimbal ring (of compass)

карда́н·ов·ый, -ая, -ое, (*a*) of **карда́н·;** *see also* **карда́н·н·ый**

~ **кольц·о́** (*n*) (horol) gimbal ring

~ **под·ве́с·** (*m*) (horol) gimbals

кардина́ль·н·ый, -ая, -ое, (*a*) cardinal

~ **пло́ск·ост·ь** (*f*) (opt) cardinal plane

~ **то́ч·к·а** (*f*) (opt) cardinal point

~ **числ·о́** (*n*) (math) cardinal, cardinal number

кардио- (*component*) cardio-, heart-

кардио·гра́мм·а , -ы, (*f*) (med) electrocardiogramme, cardiogram

кардио́·граф·, -а, (*m*) (med) electrocardiograph

карди·о́ид·а, -ы, (*f*) (math) cardioid; (rad) cardioid diagram

карди·о́ид·н·ый, -ая, -ое, (*a*) (med) cardioid; (mech) heart, heart-shaped

~ **кулач·о́к·** (*m*) heart cam

кардиотра́ст·, -а, (*m*) cardiotrast (X-ray contrast medium)

ка́рд·н·ый, -ая, -ое, (*a*) of **ка́рд·а** (text) card, carding

кардо·ле́нт·а, -ы, (*f*) (text) card clothing

кардо·чес·а́льн·ый, -ая, -ое, (*a*) (text) carding

~ **маши́н·а** (*f*) (text) carding engine

кардо·чес·а́ни·е, -я, (*n*) (text) carding

каре́ (*n indecl*) square, square-shaped object

каре́т·а, -ы, (*f*) carriage

каре́т·к·а, -и, (*g pl*) **-т·ок·,** (*f*) (*dim*) of **каре́т·а;** (mech) carriage, pallet, saddle; (text) dobby; (M/T) gear cluster, crank assembly (bicycle)

~, **ремизо·подъ·ём·н·ая** (text) dobby

кари- (*component*) cary-, kary-

кариати́д·а, -ы, (*f*) (arch) caryatid

ка́р·ий, -яя, -ее, (*a*) brown, hazel

кари́н·а, -ы, (*f*) (biol) carina

кари·о́з·, -а, (*m*) (med) caries

кари·о́з·н·ый, -ая, -ое, (*a*) carious, cariose

карио·кин·ез·, -а,   (*m*) (cyt) karyokinesis

карио·лимф·а, -ы,   (*f*) (bot) karyolymph

карио·лог·и·я, -и,   (*f*) (cyt) karyology, nuclear cytology

кари·опс·, -а,   (*m*) (bot) caryopsis

карио·филл·ен·, -а,   (*m*) (biochem) caryophyllene

карио·церит·, -а,   (*m*) (min) caryocerite

кари·я, -и,   (*f*) (bot) hickory, *Carya*

Кариус·а, метод·   (*m*) (chem) Carius' method

каркас·, -а,   (*m*) (build) carcass, framework, frame, skeleton frame; (elec) chassis; carcass (of tyre); (bot) hackberry, *Celtis*

~ кат·ушк·и   (elec) coil form/bobbin

~ само·лёт·а   (a/c) airframe

каркас·н·о-панель·н·ая констру́к·ци·я   (*f*) (build) frame-and-panel construction

каркас·ност·ь, -и,   (*f*) body stiffness (of tyre)

каркас·н·ый, -ая, -ое,   (*a*) of **каркас·**

~ констру́кци·я   (*f*) (build) framework, carcassing; frame construction

карленгс·, -а,   (*m*) *see* **карлингс·**

карлик·, -а,   (*m*) dwarf

~, бел·ый   (astron) white dwarf

карлик·ов·ый, -ая, -ое,   (*a*) dwarf, dwarfish, midget, stunted

карлинг·, -а,   (*m*) (geog) horn, horn peak

карлингс·, -а,   (*m*) (shipb) deck girder, carling

карман·, -а,   (*m*) pocket

карман·н·ый, -ая, -ое,   (*a*) pocket

кармаш·ек·, -ш·к·а,   (*m*) (bot) pocket

кармин·, -а,   (*m*) carmine (pigment)

кармин·н·ый, -ая, -ое,   (*a*) carmine

карналлит·, -а,   (*m*) (min) carnallite

карнауб·а, -ы,   (*f*) (bot) Brazilian wax palm, *Copernicia cerifera*

карнауб·ск·ий, -ая, -ое,   (*a*) of **карнауб·а**

~ воск·   (*m*) carnauba wax, Brazil wax

карниз·, -а,   (*m*) (arch) cornice; (geol) bench

Карно, цикл·   (*m*) (eng) Carnot cycle (heat engines)

карно·завр·, -а,   (*m*) (pal) *Carnosaurus*

карн·озин·, -а,   (*m*) (chem) carnosine, β-alanil-*l*-histidine

карнотит·, -а,   (*m*) (min) carnotite

Каро, кисл·от·а   (*f*) (chem) Caro's acid, permonosulphuric acid

кар·ов·ый, -ая, -ое,   (*a*) of **кар·**

кар·ов·ый, -ая, -ое

~ озеро   (*n*) corrie/cirque lake, tarn

карот·аж·, -а,   (*i*) -ем,   (*m*) (geol) well logging

~, газ·ов·ый   (oil) mud logging

~, гамм·а-   gamma-ray logging

~, нейтрон·н·ый   neutron logging

~, рад·иев·ый   radioactive logging

~, радио·актив·н·ый   radioactive logging

~, терм·ическ·ий   geothermal logging

~, электр·ическ·ий   electrical logging

карот·аж·н·ый, -ая, -ое,   (*a*) logging, well-logging

~ станци·я   (*f*) well-logging unit

каротел·ь, -и,   stump-rooted carrot

каротид·н·ый, -ая, -ое,   (*a*) (anat) carotid

карот·ин·, -а,   (*m*) (chem) carotine

карот·ин·оид·, -а,   (*m*) (biochem) carotenoid

карот·ир·овать, -уют,   (*imp or perf*) log (boreholes)

каротт·аж·, -а,   (*m*) *see* **карот·аж·**

карп·, -а,   (*m*) (fish) carp, *Cyprinus carpio*

карп·ов·ый, -ая, -ое,   (*a*) carp, cyprinid; (*pl as noun*) Cyprinids, *Cyprinidae*

карпо·ген·, -а,   (*m*) (bot) carpogen

карпо·ед·, -а,   (*m*) fish louse, *Branchiura*

карпо·образ·н·ые, -ых,   (*pl decl as adj*) (fish) *Cypriniformes*

карпо·спор·а, -ы,   (*f*) (bot) carpospore

карпо·спор·анги·й, -я,   (*m*) (bot) carposporangium

карр·а, -ы,   (*f*) resin blaze (timber)

карролит·, -а,   (*m*) (min) carrollite

каррот·аж·, -а,   (*m*) *see* **карот·аж·**

карсино·трон·, -а,   (*m*) (elec) carcinotron

карст·, -а,   (*m*) (geog) karst

карст·ов·ый, -ая, -ое,   (*a*) of **карст·**

~ ворон·к·а   (*f*) (geog) sink-hole

~ явл·ени·е   (*n*) (geog) karst region, karstland

карт·а, -ы,   (*f*) map, chart; card (playing); peat drainage channel

~ бар·ическ·ой топо·граф·и·и   (meteor) isobaric chart

~, бланк·ов·ая   outline map

~, гео·граф·ическ·ая   map

~, глазо·мер·н·ая   pace-and-compass traverse map

~ квадрат·ов   gridded chart

~, климат·ическ·ая   climatic map

~, кольц·ев·ая   (meteor) regional chart

~, контур·н·ая   blank map

ка́рт·а
~, **магни́т·н·ая**   magnetic map/chart
~ **микро·движ·ёни·й**   simo chart, simultaneous-motion-cycle chart
~, **мор·ск·а́я**   (naut) chart
~ **не́б·а**   star chart
~, **нем·а́я**   outline map
~, **об·зо́р·н·ая**   sketch map
~, **по·лёт·н·ая**   aeronautical chart
~, **по·ли́ст·н·ая**   sheet/quadrangle map
~ **пра́в·ой и ле́в·ой ру́к·и** two-handed chart (time and motion study)
~, **пут·ев·а́я**   road map; navigational chart
~, **синопт·и́ческ·ая**   (meteor) synoptic chart
~, **с·пра́в·очн·ая**   non-navigational chart
у́гол· ка́рт·ы   (m) (nav) grivation
~, **ча́ст·н·ая**   (naut) coastal chart
картамейн·, -а,   (m) (chem) carthamein
картами́н·, -а,   (n) (chem) carthamin
картезиа́н·ск·ий, -ая, -ое,   (a) Cartesian; Carthusian
картел·и́р·овать, -уют,   (imp and perf) (fin) form a cartel
карте́л·ь, -я,   (m) (fin) cartel
ка́ртер·, -а,   (m) gear casing; crankcase
~, **мо́кр·ый**   (M/T) wet sump
~ **рул·ев·о́го у·правл·е́ни·я**   (M/T) steering box
~, **сух·о́й**   (M/T) dry sump
~ **с·цепл·е́ни·я**   (M/T) clutch case
ка́ртер·н·ый, -ая, -ое,   (a) of **ка́ртер·**
~ **ма́сл·о**   (n) crankcase oil
карте́чч·ин·а, -ы,   (m) shot (for game guns)
карте́ч·ь, -и,   (f) shrapnel; buck-shot
карт·и́н·а, -ы,   (f) (see also **диагра́м·м·а**); picture, pattern; (theat) scene
~, **волн·ов·а́я**   (dynam) wave pattern
~ **движ·е́ни·я**   (dynam) motion pattern
~, **дифракцио́н·н·ый**   diffraction pattern (X-ray)
~, **ка́честв·енн·ая**   qualitative picture
~ **с·ры́в·а по·то́к·а**   (dynam) stall pattern
~ **теч·е́ни·я**   (dynam) flow pattern
карт·и́нк·а, -и,   (g pl) **-нок·,** (f) (dim) of **карт·и́н·а**
~, **пере·во́д·н·ая**   transfer (picture)
карти́н·н·ый, -ая, -ое,   (a) of **карт·и́н·а;** pictorial; picturesque
карт·и́р·овать, -уют,   (imp and perf) map
карто- (component) card, carto-
карто·гра́мм·а, -ы,   (f) collation map

карто·граф·и́р·овать, -уют,   (imp) map, chart
карто·граф·и́ческ·ий, -ая, -ое,   (a) cartographic(al)
карто·гра́ф·и·я, -и,   (f) cartography
карто·держ·а́тел·ь, -я,   (m) map-holder
карто́·метр·, -а,   (m) (surv) cartometer, curvometer
карто́н·, -а,   (m) paperboard, cardboard, pasteboard, millboard, board; (text) card (in a jacquard machine)
~, **асбе́ст·ов·ый**   asbestos panel, sheet asbestos
~, **каре́т·очн·ый**   (text) dobby card
~, **коро́б·очн·ый**   boxboard
~, **лак·иро́ванн·ый**   glazed cardboard
~, **обув·н·о́й**   boot/shoe board
~, **пере·плёт·н·ый**   triplex board
~, **электро·техн·и́ческ·ий**   electric-insulating cardboard
картон·а́ж·, -а,   cardboard articles/products
картон·а́ж·н·ый, -ая, -ое,   (a) of **картон·а́ж·**
карто́н·к·а, -и,   (g pl) **-н·ок·,** (f) carton, cardboard box, box; piece of cardboard
карто́н·н·ый, -ая, -ое,   (a) of **карто́н·**
картоно·де́л·ательн·ая маши́н·а   (f) paperboard machine
картоно·рез·а́льн·ая маши́н·а   (f) card cutter, cardboard cutter
карто·о·бес·пе́ч·ени·е, -я,   (n) map coverage
карто·те́к·а, -ы,   (f) card index
картофеле·коп·а́тел·ь, -я,   (m) potato digger/lifter/picker
~, **элева́тор·н·ый**   elevator potato digger
картофеле·мо́й·к·а, -и,   (g pl) **-о́·ек·,** (f) potato washer
картофеле·саж·а́лк·а, -и,   (g pl) **-лок·,** (f) potato planter
картофеле·сорт·иро́вк·а, -и,   (g pl) **-вок·,** (f) potato grader
картофеле·тёр·к·а, -и,   (g pl) **-р·ок·,** (f) potato-grinding mill
картофеле·чи́ст·к·а, -и,   (g pl) **-т·ок·,** (f) potato-peeling machine
карто́фел·ин·а, -ы,   (f) potato tuber
карто́фел·ь, -я,   (m) potato(es), *Solanum tuberosum*
карто́фель·н·ый, -ая, -ос,   (a) potato
~ **баци́лл·а**   (f) *Bacillus mesentericus*
~ **гриб·о́к·**   (m) potato blight, *Phytophthora infestans*

картофель·н·ый, -ая, -ое

~ **жук·** (*m*) (zool) colorado beetle, *Leptinotarsa decemlineata*

~ **мук·á** (*f*) potato flour

~ **немат·óд·а** (*f*) golden nematode (potato pest), *Heterodera rostoch·ensis*

~ **пáл·очк·а** (*f*) *Bacillus mesentericus*

кáрт·очк·а, -и, (*g pl*) **-чек·,** (*f*) card, index card, record card

~, **перфорациóн·н·ая** punched card

~, **раз·дел·йтельн·ая** guide card

~, **стрел·кóв·ая** (mil) range card

~, **техн·йческ·ая** (M/T, a/c) servicing schedule

~, **хроно·метр·áж·н·ая** time card

кáрт·очн·ый, -ая, -ое, (*a*) of **кáрт·очк·а**

картýз·, -á, (*m*) cap; (mil) cartridge bag, powder bag

картýш·, -а, (*i*) **-ем,** (*m*) (arch) cartouche

карт·ýшк·а, -и, (*f*) (naut) compass card

карýб·а, -ы, (*f*) (bot) carob, *Ceratonia siliqua*

кар·ýнкул·, -а, (*m*) (biol) caruncle

карусéл·ь, -и, (*f*) roundabout, turntable; (met) casting wheel

карусéль·н·о-фрéзер·н·ый стан·óк· (*m*) vertical revolving milling machine, rotary-table milling machine

карусéль·н·ый, -ая, -ое, (*a*) revolving, rotary, rotating, rotating-feed

~ **печ·ь** (*f*) rotary-hearth furnace

~ **стан·óк·** (*m*) turning and boring mill, turning mill, vertical turning and boring mill

карцино·гéн·, -а, (*m*) (med) carcinogen

карцино·лóг·и·я, -и, (*f*) (zool) carcinology

карцин·óм·а, -ы, (*f*) (med) carcinoma

карцино·трóн·, -а, (*m*) (rad) carcinotron

карьéр·, -а, (*m*) quarry, open-cast mine, surface mine, open working/cut; stretch gallop (horse riding)

карьéр·а, -ы, (*f*) career

карьéр·н·ый, -ая, -ое, (*a*) of **карьéр·, карьéр·а**

кас·áни·е, -я, (*n*) (*v n*) see **кас·á·ться;** contact, tangency

~ **земл·й** (air) touch-down

лйн·и·я кас·áни·я (*f*) tangent, line of contact

по·вéрх·ност·ь кас·áни·я (*f*) tangential surface

тóч·к·а кас·áни·я (*f*) (math) point of contact/tangency, (surv) tangential point

кас·áтельн·ая, -ой, (*f decl as adj*) tangent

~ **в тóч·к·е пере·гйб·а** inflectional tangent (of a curve)

кас·áтельн·о (*prep*) (+*gen*) about, concerning, touching on

кас·áтельн·ый, -ая, -ое, (*a*) see also **кас·áтельн·ая** (*as noun*); (math) tangent, tangential; (mech) shear, shearing; concerning, touching

~ **кóнус·** (*m*) (math) tangent cone

~ **круг·** (*m*) (math) escribed circle, excircle

~ **на·пряж·éни·е** (*n*) (mech) shear stress

~ **плóск·ост·ь** (*f*) (math) tangent plane

касáтик·, -а, (*m*) (bot) iris, *Iris*

касатйк·ов·ые, -ых, (*pl decl as adj*) (bot) iris family, *Iridaceae*

касáт·к·а, -и, (*g pl*) **-т·ок·,** (*f*) (zool) killer whale, *Orca orca, Grampus orca*; (orn) swallow, *Hirundinidae*

кас·á·ться, -ются, (*imp*) (+*gen*) touch; touch upon; concern, affect, apply to

что кас·á·ется as for, regarding, as concerns, with regard to

кас·áющ·ийся, -аяся, -ееся, (*pres part act*) of **кас·á·ться;** pertinent; tangent

кáск·а, -и, (*g pl*) **кáсок·,** (*f*) helmet

каскáд·, -а, (*m*) cascade; (rad) stage; series (of hydroelectric stations); (chem) stage; (elec) cascade connection, cascade

~, **а·син·хрóн·н·ый** cascade-connected asynchronous motors

~, **бýфер·н·ый** (rad) buffer amplifier/stage

~, **в·ход·н·óй** input stage

~, **вы·ход·н·óй** output stage

~, **за·да·ющ·ий** (rad) driver/driving stage/amplifier

~, **квадратýр·н·ый** (nucl) squared-off cascade

~, **модулятор·н·ый** modulated amplifier/stage

~, **о·кон·éчн·ый** output stage

~, **про·пуск·áющ·ий** gate/gating stage

~, **рас·кáч·ивающ·ий** (elec) driver

~, **регенератйв·н·ый** (chem) stripping stage/section

~, **скор·остн·óй** (elec) cascade-controlled motors

~ **у·сил·éни·я** amplifier, amplifier stage

~ **электро·машйн·н·ый** (elec) cascade, cascade connection

каска́д·н·о (*adv*) (elec) in series
каска́д·н·ый, -ая, -ое, (*a*) of **каска́д·** (elec) cascade-connected, cascade, series-connected
~ **в·ключ·е́ни·е** (*n*) (elec) cascading; cascade circuit, cascade connection
~ **пре·образ·ова́тел·ь** (*m*) (elec) cascade converter
~ **со·един·е́ни·е** (*n*) (elec) cascade connection
~ **ли́в·ен·ь** (*m*) (nucl) cascade shower
каскари́лл·а, -ы, (*f*) (bot) cascarilla, cascarilla bark, bark of *Croton eluteria*
ка́ско (*n indecl*) (mar) hull insurance
каско́д·, -а, (*m*) (elec) cascode, cascode amplifier
касп·и́йск·ий, -ая, -ое, (*a*) (geog) Caspian
ка́сс·а, -ы, (*f*) cash box, till; cash desk, ticket office; cash; (print) type case
~, **биле́т·н·ая** (rail) booking office; (theat) box office
~ **взаимо·по́·мощ·и** (fin) friendly society
~, **на·бо́р·н·ая** (print) type case
~, **с·берег·а́тельн·ая** savings bank
касса́в·а, -ы, (*f*) cassava, manioca starch
кассацио́н·н·ый, -ая, -ое, (*a*) (law) cassation
~ **суд·** (*m*) (law) court of appeal/cassation
касса́ци·я, -и, (*f*) (law) cassation
кассе́т·а, -ы, (*f*) cassette, holder; (phot) cassette, matazine, dark slide; caisson
~, **тепло·вы·дел·я́ющ·ая** fuel assembly (nucl reactor)
Ка́сс·иев· пу́рпур· (*m*) (chem) purple of Cassius
ка́сс·иев·ый, -ая, -ое, (*a*) cassia, senna
Касси́ни, овал· (math) oval of Cassini
кассиопе́·й, -я, (*m*) (chem) cassiopeium, Cp
кассиопе́·я, -и, (*f*) (astron) *Cassiopeia*
касси́р·, -а, (*m*) cashier
касс·и́р·овать, -уют, (*imp and perf*) (law) annul
касситери́т·, -а, (*m*) (min) cassiterite, tin stone
ка́сси·я, -и, (*f*) (bot) senna, *Cassia*
ка́сс·ов·ый, -ая, -ое, (*a*) of **ка́сс·а**
~ **аппара́т·** (*m*) cash register
~ **кни́г·а** (*f*) cash-book
~ **на·ли́ч·ност·ь** (*f*) cash in hand
ка́ст·а, -ы, (*f*) (ent) caste

касто́р·, -а, (*m*) (text) castor
касто́р·к·а, -и, (*f*) castor oil
касто́р·ник·, -а, (*m*) (bot) castor oil plant, *Ricinus communis*
касто́р·ов·ый, -ая, -ое, (*a*) castor
~ **ма́сл·о** (*n*) castor oil
кастр·а́ци·я, -и, (*f*) castration, gelding
кастр·и́р·овать, -уют, (*imp and perf*) castrate, geld
кастрю́л·я, -и, (*g pl*) **-юл·ь** (*f*) pan, saucepan
ката- (*component*) kata-, cata-
ката·бат·и́ческ·ий, -ая, -ое, (*a*) (meteor) katabatic
ката·бол·и́зм·, -а, (*m*) (biol) katabolism, catabolism
катаво́тр·, -а, (*m*) (geol) sink/swallow hole
ката·ген·е́з·, -а, (*m*) (zool) katagenesis
ката·ди·оптр·и́ческ·ий, -ая, -ое, (*a*) (opt) catadioptic
~ **систе́м·а** (*f*) (opt) catadioptric system
ката·дро́м·н·ый, -ая, -ое, (*a*) (biol) catadromous, katadromous
~ **мигр·а́ци·я** (*f*) (fish) catadromous migration
ката·кла́з·, -а, (*m*) (geol) cataclasm
ката·класт·и́ческ·ий, -ая, -ое, (*a*) (geol) cataclastic, kataclastic
~ **структу́р·а** (*f*) (geol) cataclastic structure
ката·кли́зм·, -а, (*m*) cataclysm
катал·а́з·а, -ы, (*f*) (chem) catalase
ката·ле́пс·и·я, -и, (*f*) (med) catalepsy
ката́·лиз·, -а, (*m*) (chem) catalysis
~, **ге́теро·ге́н·н·ый** heterogeneous catalysis
~, **гомо·ге́н·н·ый** homogeneous catalysis
~, **от·риц·а́тельн·ый** negative catalysis
ката·лиз·а́тор·, -а, (*m*) catalyst
~, **от·риц·а́тельн·ый** (chem) anticatalyst
~, **пла́в·ающ·ий** suspended catalyst
~, **по·движ·н·о́й** moving-bed catalyst
~, **стациона́р·н·ый** fixed-bed catalyst
~, **тек·у́ч·ий** fluidized catalyst
катали́н·, -а, (*m*) (plast) catalin
ката·лит·и́ческ·ий, -ая, -ое, (*a*) catalytic, catalyst, catalysing
~ **яд·** (*m*) (chem) catalytic/catalyst poison
ката·ло́г·, -а, (*m*) catalogue, catalog

каталогиз·а́тор·, -а, (*m*) cataloguer, cataloger

каталогиз·и́р·овать, -уют, (*imp and perf*) catalogue, catalog

катало́ж·н·ая, -ой, (*f decl as adj*) catalogue/catalog room

катало́ж·н·ый, -ая, -ое, (*a*) see also **катало́ж·н·ая** (*as noun*); catalogue, catalog, cataloguing, cataloging

~ **я́щик·** (*m*) catalogue/catalog index cabinet

ката́льп·а, -ы, (*f*) catalpa (tree)

ката·морф·и́зм·, -а, (*m*) (geol) katamorphism

ката·мо́рф·н·ый, -ая, -ое, (*a*) katamorphic

кат·а́ни·е, -я, (*n*) (*v n*) of **кат·а́·ть**

ка́т·анк·а, -и, (*g pl*) **-нок·,** (*f*) (met) wire rod; (*pl*) felt boots

ка́т·анн·ый, -ая, -ое, (*past part pass*) of **кат·а́·ть;** *see also* **ка́т·ан·ый**

ка́т·ан·ый, -ая, -ое, (*a*) (met) rolled; mangled (of laundry)

ката·пла́зм·а, -ы, (*f*) (med) cataplasm

катапу́льт·а, -ы, (*f*) catapult

катапульт·и́руем·ый, -ая, -ое, (*pres part pass*); catapult, catapult-ejected

~ **сид·е́ни·е** (*n*) (a/c) ejection seat

ката́·р, -а, (*m*) *see* **ката́·рр·**

ката·ра́кт·, -а, (*m*) (geog) cataract; (hydr) dashpot

~, **у·пру́г·о-при·со·един·ённ·ый** dashpot and spring

ката·ра́кт·а, -ы, (*f*) (med) cataract

ката·р·а́льн·ый, -ая, -ое, (*a*) *see* **ката·рр·а́льн·ый**

катаро́·метр·, -а, (*m*) (chem, phys) katharometer

ката́·рр·, -а, (*m*) (med) catarrh

ката·рр·а́льн·ый, -ая, -ое, (*a*) (med) catarrhal

катарт·и́н·, -а, (*m*) (pharm) cathartin

ката·стро́ф·а, -ы, (*f*) disaster, catastrophe; crash

ката·строф·и́ческ·ий, -ая, -ое, (*a*) of **ката·стро́ф·а;** catastrophic

ката·термо́·метр·, -а, (*m*) (phys) kata thermometer

кат·а́·ть, -ют, (*imp indet*) roll/wheel (along), roll (out), roll (into a ball/roll), mangle (laundry); take for a drive/outing, take out; **-ся** (*pass*); roll (*intrans*), move; go for a drive/outing

ката́ться на конька́х skate

ката́ться на рул·е́ (naut) yaw

ката·форе́з·, -а, (*m*) (chem) cataphoresis

ката·форе́з·н·ый, -ая, -ое, (*a*) cataphoretic

ката·фро́нт·, -а, (*m*) (meteor) katafront, surface of subsidence

ката·хто́н·н·ый, -ая, -ое, (*m*) katachthonous

категор·и́ческ·ий, -ая, -ое, (*a*) categorical

категор·и·я, -и, (*f*) category, class

катен·а́рн·ый, -ая, -ое, (*a*) (math) catenary

~ **крив·а́я** (*f*) (math) catenary curve

катен·о́ид·, -а, (*m*) (math) catenoid

ка́тер·, -а, (*nom pl*) **-а́,** (*g pl*) **-о́в,** (*m*) boat, motorboat, launch

~, **букси́р·н·ый** towing launch, small tug

~, **мото́р·н·ый** motor boat

~, **пар·ов·о́й** steam boat

~, **по·сы́ль·н·ый** despatch boat, tender

~, **про·ме́р·н·ый** sounding boat

~, **разъ·езд·н·о́й** harbour launch

ка́тер·н·ый, -ая, -ое, (*a*) boat's; small, light (of equipment)

ка́тет·, -а, (*m*) (math) cathetus, side/leg of a right-angled triangle

катет·ер·, -а, (*m*) (med) catheter

катето́·метр·, -а, (*m*) (phys) cathetometer, reading microscope/telescope

катетро́н·, -а, (*m*) (elec) cathetron

катехи́н·, -а, (*m*) (chem) catechol

ка́теху (*n indecl*) (biochem) catechu, cutch

кати́н·, -а, (*m*) (pharm) catine

кат·ио́н·, -а, (*m*) (elec, chem) cation

кат·ион·и́т·, -а, (*m*) cationite, cation exchanger, cation exchange resin

кат·ион·и́т·ов·ый, -ая, -ое, (*a*) of **кат·ио́н·**

кат·ион·и́ровани·е вод·ы́ (*n*) cation-exchange water softening

катионо·об·ме́н·ник·, -а, (*m*) cation exchanger

кат·и́ть, -ят, (*imp deter*) *see* **кат·а́·ть**

кат·к·и́ (*nom pl*) of **кат·о́к·**

кат·кова́ни·е, -я, (*n*) rolling

кат·ко́в·ый, -ая, -ое, (*a*) of **кат·о́к·** roller, rolling

кат·о́д·, -а, (*m*) (elec) cathode

~, **бори́д·н·ый** boride cathode

~, **вс·по·мог·а́тельн·ый** (electrochem) starting-sheet blank

~, **гор·я́ч·ий** hot/thermionic cathode

~, **де́й·ствующ·ий** virtual cathode

кат·о́д

~, дуг·ов·о́й   arc cathode

~, жидко·металл·и́ческ·ий   pool cathode

~ ко́св·енн·ого на·ка́л·а   indirectly heated cathode

~, мета́лл·о-плён·очн·ый   metal-film cathode

~, металло·по́р·ист·ый   dispenser cathode

~, на·ка́л·иваем·ый   filamentary/hot cathode, directly heated cathode

~, окси́д·н·ый   oxide-coated cathode

~, подо·гре́в·н·ый   heater-type cathode

~, ртут·н·ый   mercury-pool cathode

~, се́т·чат·ый   reticular cathode

~, тор·и́рованн·ый   thoriated filament

катодно·луч·ев·о́й, -а́я, -о́е,   (*a*) (*and see* электронно·луч·ев·о́й) cathode-ray

като́д·н·ый, -ая, -ое,   (*a*) cathode, cathodic; cathode-ray

~ вольт·ме́тр·   (*m*) cathode-ray voltmeter

~ за·щи́т·а   (*f*) cathodic protection (against corrosion)

~ лист·   (*m*) (electrochem) starting sheet

~ луч·   (*m*) cathode ray

~ осно́в·а   (*f*) (electrochem) starting sheet blank

~ осцилло́·граф·   (*m*) cathode-ray oscillograph

~ пад·е́ни·е   (*n*) cathode drop

~ пятн·о́   (*n*) cathode spot

~ рас·пыл·е́ни·е   (*n*) cathode sputtering

~ с·мещ·е́ни·е   (*n*) cathode bias

~ тём·н·ое про·стра́н·ств·о, втор·о́е   (*n*) cathode dark space, cathode sheath, Crookes dark space

~ тём·н·ое про·стра́н·ств·о, пе́рв·ое   (*n*) Aston dark space

катодо·люмин·есце́нци·я, -и,   cathodoluminescence, cathode luminescence

катодо·про·вод·и́мост·ь, -и,   (*f*) cathode conductivity

катодо·фосфор·есце́нци·я, -и,   (*f*) cathodephosphorescence

кат·о́к·, -т·к·а́,   (*m*) roller, roll; skating rink; mangle (laundry); rolling jack (leather)

~ гу́сен·иц·ы   track roller/wheel (of tracked vehicle)

~, доро́ж·н·ый   (civ eng) road roller

кат·о́к

~, ко́ль·чат·ый   (agr) ring roll, Cambridge roll, corrugated roll

~, кулач·ко́в·ый   sheepsfoot roller

~, о·по́р·н·ый   bogie wheel, track roller, truck wheel (tracked tractor)

~, под·де́рж·иваю́щ·ий   jockey roller

~, подо́·шв·енн·ый   sole-leather roller

~, под·по·ве́рх·ностн·ый   (civ eng) subsurface packer

~ -у·плот·ни́тел·ь, -я,   (*m*) (civ eng) packer

католи́т·, -а,   (*m*) (electrochem) catholyte

кат·о́птр·ик·а, -и,   (*f*) (opt) catoptrics

кат·оптр·и́ческ·ий, -ая, -ое,   (*a*) (opt) catoptric

ка́торг·а, -и,   (*f*) (law) penal servitude, hard labour

ка́торж·н·ый, -ая, -ое,   (*a*) of ка́торг·а

кат·о́чек·, -ч·к·а,   (*m*) (*dim*) of кат·о́к·; roller, roll, wheel

катра́н·, -а,   (*m*) (bot) sea kale, *Crambe maritima*; (fish) catran, *Acanthias acanthias* fam. *Squalidae*

ка́тт·а, -ы,   (*f*) (zool) ring-tailed lemur, *Lemur catta*

кат·у́н·, -а,   (*m*) roll, ball

кат·у́ч·ий, -ая, -ее,   (*a*) rolling

~ ба́л·к·а   (*f*) overhead traveller, overhead travelling crane

кат·у́шечн·ый, -ая, -ое,   (*a*) of кат·у́шк·а

кат·у́шк·а, -и, (*g pl*) -шек·,   (*f*) reel, bobbin, spool; (elec) coil; small capstan, (oil) cathead

~ воз·бужд·е́ни·я   field coil

~, высоко·част·о́тн·ая   high-frequency choke

~, пупи́н·овск·ая   (telecom) Pupin/loading coil

~ -да́т·чик·, -а,   (*m*) pick-up coil

~ за·жиг·а́ни·я   (M/T) ignition coil

~, звук·ов·а́я   voice coil (loud speaker)

~ индукти́в·ност·и   inductance coil, inductor

~, инду́ктор·н·ая   inductor/field coil

~, индукцио́н·н·ая,   induction/field coil; (M/T) ignition coil

~, искро·гас·и́тельн·ая   (elec) spark killer

~, корзи́н·очн·ая   basket coil

~ обра́т·н·ой с·вя́з·и   (rad) reaction coil

~, по·движ·н·а́я   moving coil; voice coil (loudspeaker)

кат·у́шк·а

~, при·ём·н·ая (rad) detector coil

~, противо·де́й·ствующ·ая bucking coil

~, реакти́в·н·ая reactance coil, choking coil, choke

~ с корзи́н·очн·ой об·мо́т·к·ой basket coil

~ с·бра́с·ывани·я drop coil

~ с·вя́з·и coupling coil

~, секцион·и́рованн·ая (rad) sectional coil; (nucl) tapped coil

~ с·жа́т·и·я (nucl) collapse coil

~ со·противл·е́ни·я (elec) standard resistance, resistance standard

~, терм·и́ческ·ая (telecom) heat coil

~, торо́ид·аль·н·ая toroid/doughnut coil

~ электро·магни́т·н·ого дут·ь·я́ гиро·сфе́р·ы internal-repulsion coil, centralizing coil (gyro compass)

кат·я́щ·ийся, -аяся, -ееся, (*pres part act*) of **кат·и́ться**

кауд·а́льн·ый, -ая, -ое, (*a*) (zool) caudal

каули·фло́р·и·я, -и, (*f*) (bot) cauliflory

каул·о́ид·, -а, (*m*) (bot) cauloid

ка́упер·, -а, (*m*) (met) Cowper hot-blast stove

кау́ри (*indecl*) kauri gum

кау́ри-бутано́л·ов·ый по·каз·а́тел·ь (*m*) (oil chem) kauri-butanol number

кау́р·ый, -ая, -ое, (*a*) light brown/chestnut

ка́устик·, -а, (*m*) (chem) caustic, caustic soda; caustic, caustic potash

~, кра́с·н·ый caustic bottom

ка́уст·ик·а, -и, (*f*) (opt) caustic, caustic curve

каусти·фика́ци·я, -и, (*f*) (chem) lime-caustic processing/process (manufacture of caustic soda)

кауст·и́ческ·ий, -ая, -ое, (*a*) of **ка́устик·** and **ка́уст·ик·а;** caustic

~ по·ве́рх·ност·ь (*f*) (opt) caustic curve, caustic

~ хру́п·кост·ь (*f*) (met) caustic embrittlement

каусто·био·ли́т·, -а, (*m*) (geol) caustobiolith, mineral fuel

каусто·ли́т·, -а, (*m*) (geol) caustolith

каучу́к·, -а, (*m*) rubber, raw rubber, caoutchouc, rubber hydrocarbon

~, бутадие́н·ов·ый (rubber) buna, butadiene rubber

~, бутадие́н-стиро́ль·н·ый butadiene-styrene rubber, SBR

каучу́к

~, высоко·температу́р·н·ый "hot" rubber

~, гваю́л·ов·ый guayule rubber

~, дивини́л·ов·ый butadiene rubber

~, жёстк·ий stiff rubber

~, за·моро́ж·енн·ый chilled rubber

~, изо·цианат·н·ый urethane rubber

~, кре́мн·ий-орган·и́ческ·ий silicone rubber

~, масло·на·по́лн·енн·ый oil-extended rubber

~, минера́ль·н·ый (min) mineral caoutchouc, elaterite

~ на́тр·и·й-бутадие́н·ов·ый (rubber) sodium-catalysed butadiene rubber

~ натура́ль·н·ый natural rubber

~, низко·температу́р·н·ый "cold" rubber

~, нитрил·акри́ль·н·ый nitrile rubber

~, плантацио́н·н·ый estate/plantation rubber

~, рог·ов·о́й hard rubber, ebonite

каучуко·во́д·ств·о, -а, (*n*) rubber cultivation/farming

каучу́к·ов·ый, -ая, -ое, (*a*) rubber

~ де́рев·о (*n*) (bot) india rubber plant, *Ficus elastica*

каучуко·но́с·, -а, (*m*) rubber plant

каучуко·но́с·н·ый, -ая, -ое, (*a*) rubber-bearing

каучуко·обра́з·н·ый, -ая, -ое, (*a*) rubbery, rubber-like

кафе́ (*n indecl*) café

ка́федр·а, -ы, (*f*) department, chair (universities etc.); lecturer's desk

ка́фел·ь, -я, glazed tile

кач·а́лк·а, -и, (*g pl*) **-лок·,** (*f*) rocker, rocking device; swinging link (diesel); (oil) pumping jack

кач·а́ни·е, -я, (*n*) rocking, swinging; oscillation, swaying, fluctuation; wobbling, fluttering; swing (pendulum); pumping

~ нес·у́щ·ей (rad) carrier flutter

част·от·а́ кач·а́ни·я (*f*) (a/c) wobble frequency

~ част·от·ы́ (telecom) frequency swing

кач·а́тельн·о-со·член·ённ·ый, -ая, -ое, (*a*) hinged

кач·а́тельн·ый, -ая, -ое, (*a*) (*and see* **кач·а́ющ·ийся**) rocking, swinging

~ движ·е́ни·е (*n*) rocking/swinging motion

кач·а́·ть, -ют, (*imp*) rock, shake, vibrate, wobble; oscillate, swing; pump

кач·а́ющ·ийся, -аяся, -ееся, (*pres part act*) *see* **кач·а́·ть**

~ **антéнн·а** (*f*) rockinghorse antenna (radar)

~ **грóхот·** (*m*) (min) shaker screen

~ **диагрáмм·а** (*f*) scanning diagram (radar)

~ **мéль·ниц·а** (*f*) vibratory mill

~ **площáд·к·а** (*f*) tilting platform

~ **печ·ь** (*f*) tilting furnace

~ **разъ·един·и́тел·ь** (*m*) (elec) rocker-type isolating switch

~ **труб·á** (*f*) swinging link (diesel)

кач·éни·е, -я, (*n*) rolling

под·ши́п·ник· кач·éни·я (*m*) (mech) rolling-contact bearing

кáчеств·енн·ый, -ая, -ое, (*a*) quality, high-grade; qualitative

~ **анáлиз·** (*m*) (chem) qualitative analysis

~ **стал·ь** (*f*) quality steel

кáчеств·о, -а, (*n*) quality; (*pl*) character, nature, properties

~, **аэро·динам·и́ческ·ое** lift–drag ratio

в кáч·еств·е as, in the capacity of

~, **вес·ов·óе** (rocket) structural efficiency factor

~, **в·кус·ов·óе** palatability

~, **вы́с·ш·ее** top quality

~ **рабóт·ы** workmanship

~, **эксплуатациóн·н·ое** performance

качи́м·, -а, (*m*) (bot) gypsophila

кáч·к·а, -и, (*g pl*) **-ч·ек·,** (*f*) tossing, rolling, pitching

~, **бок·ов·áя** (naut) rolling

~, **борт·ов·áя** rolling

~, **вертикáль·н·ая** (naut) heaving

~, **кил·ев·áя** (naut) pitching

~, **про·дóль·н·ая** pitching

кач·н·у́ть, -у́т, (*perf single action*) *see* **кач·á·ть**

кач·у́ (*pres 1st sing*) of **кат·и́ть**

кáш·а, -и, (*i*) **-ей,** (*f*) porridge; semi-liquid mass, mush, mash; muddle, hash; (rad) mush

кашалóт·, -а, (*m*) (zool) sperm whale, cachalot, *Physeter catodon*

кáшел·ь, -шл·я, (*m*) (med) cough, tussis

кашеми́р·, -а, (*m*) (text) cashmere

каш·ировáльн·ый, -ая, -ое, (*a*) of **каш·ирóвк·а**

каш·ировáни·е, -я, (*n*) = **каш·и·рóвк·а**

каш·ирóвк·а, -и, (*f*) (print) backing; (phot) vignetting

кáш·иц·а, -ы, (*i*) **-ц·ей,** (*f*) (dim) of **кáш·а;** thin porridge; paste

~, **бумáж·н·ая** paper pulp

~, **пищ·ев·áя** (zool) chyme

кáш·к·а, -и, (*g pl*) **-ш·ек·,** (*f*) (pop) clover

кáшл·яни·е, -я, (*n*) coughing, cough

кáшл·ян·уть, -ут, (*perf*) give a cough

кáшл·я·ть, -ют, (*imp*) cough; have a cough

каштáн·, -а, (*m*) (bot) chestnut, *Castanea*

~, **америкáн·ск·ий** sweet chestnut, *Castanea dentata*

~, **на·сто·я́щ·ий** chestnut, *Castanea sativa*

каштáн·ов·ый, -ая, -ое, (*a*) chestnut; chestnut-coloured/colored

кашу́ (*n indecl*) (biochem) catechu, cutch

кашэ́ (*n indecl*) (phot) mask, vignette

каю́т·а, -ы, (*f*) cabin, room, state-room

каю́т-компáни·я, -и, (*f*) mess, mess-room; (naval) wardroom

кбм (*abbr*) of **куб·и́ческ·ий метр·** cbm, m^3, cubic metre/meter

кв (*abbr*) of **кило·вóльт·** kV, kv, kilovolt

ква (*abbr*) of **кило·вольт·ампéр·** kVA, kva, kilovoltampere

квадрáнт·, -а, (*m*) quadrant; (gunn) clinometer

механи́зм· квадрáнт·а (*m*) (text) strapping motion, governor motion

квадранти́д·ы (*nom pl*), (*g pl*) **-ов,** (*m*) (astron) quadrantids

квадрáнт·н·ый, -ая, -ое, (*a*) quadrant (of devices); quadrantae (of values etc.)

квадрáт·, -а, (*m*) square; (cartog) grid square; (print) 48-point (unit of measurement); (print) quad, quadrat; (oil) kelly, kelly bar

воз·вод·и́ть в квадрáт· (math) square

закóн· обрáт·н·ых квадрáт·ов (*m*) (math) inverse-square law

~, **лати́н·ск·ий** (math) Latin square

мéтод наи·мéнь·ш·их квадрáт·ов (*m*) (math) method of least squares

~, **обрáт·н·ый** inverse square

~ **пегáс·а** (astron) Pegasus, The Square

~, **пóлн·ый** perfect square

сумм·и́руем·ый в квадрáт·е (math) quadratically integrable

квадрат·и́ческ·ий, -ая, -ое, (*a*) *see* **квадрат·и́чн·ый**

квадрат·и́чн·ый, -ая, -ое, (*a*) (math) quadratic, quadric, of the second degree; square-law, root-mean-square

~ **зако́н·** (*m*) square law

~ **о·ши́б·к·а** (*f*) (math) root-mean-square deviation, standard deviation

~ **сре́д·н·ее** (*n*) (math) quadratic mean, mean square

~ **фо́рм·а** (*f*) (math) quadratic form

~ **фо́рм·а, по·лож·и́тельн·ая о·пре-·дел·ённ·ая** (math) positive definite quadratic form

квадра́т·н·о-гне́зд·ов·а́я по·са́д·к·а (*f*) (agr) square-pocket sowing

квадра́т·н·ый, -ая, -ое, (*a*) square, quadratic; (cryst) tetragonal

~ **ко́рен·ь** (*m*) (math) square root

~ **метр·** (*m*) square metre

~ **о·кла́д·** (*m*) (min) square set

~ **систе́м·а** (*f*) (cryst) quadratic/tetragonal system

   систе́м·а квадра́т·н·ых о·кла́-д·ов (*f*) (min) square-set stoping

~ **у·равн·е́ни·е** (*n*) (math) quadratic equation

квадр·а́тор·, -а, (*m*) (autom) quadrator, square-law-function generator

квадр·атри́с·а, -ы, (*f*) (math) quadratrix

квадратро́н·, -а, (*m*) (elec) quadratron

квадрат·у́р·а, -ы, (*f*) (math) quadrature, squaring, finding the square

~ **кру́г·а** (math) squaring the circle, quadrature of a circle

квадрат·у́рн·ый, -ая, -ое, (*a*) quadrature, squaring, integration, integrating

~ **при·ли́·в·** (*m*) (ocean) neap tide

~ **фо́рмул·а** (*f*) (math) formula of integration

квадриллио́н·, -а, (*m*) (math) quadrillion, 10^{15} (U.S.S.R, U.S.A), 10^{24} (Britain, Germany)

квадри·ме́р·, -а, (*m*) (chem) quadripolymer

квадр·и́ровани·е, -я, (*n*) quadrature, squaring, integrating

квадр·и́руем·ый, -ая, -ое, (*pres part pass*); (math) integrable, squarable

квадру́·плекс·, -а, (*m*) quadruplex, form-fold

квадру·по́л·ь, -я, (*m*) quadrupole, quadripole

квадру·по́ль·н·ый, -ая, -ое, (*a*) quadrupole, quadripole

квадру·по́ль·н·ый, -ая, -ое

~ **взаимо·де́й·стви·е** (*n*) (phys) quadrupole interaction

квази·ана·лит·и́ческ·ий, -ая, -ое, (*a*) (math) quasi-analytic

квази·линеа́р·н·ый, -ая, -ое, (*a*) quasilinear

квази·лин·е́йн·ый, -ая, -ое, (*a*) quasilinear

квази·моно·хромат·и́ческ·ий, -ая, -ое, (*a*) quasi-monochromatic

квази·резона́нс·, -а, (*m*) quasi-resonance

квази·решёт·к·а, -и, (*g pl*) **-т·ок·,** (*f*) (rad) quasi-array

квази·с·вя́з·анн·ый, -ая, -ое, (*a*) (chem phys) quasibound

квази·стат·и́ческ·ий, -ая, -ое, (*a*) quasi-static

квази·стациона́р·н·ый, -ая, -ое, (*a*) quasi-stationary

квази·твёрд·ый, -ая, -ое, (*a*) quasi-solid

квази·у·пру́г·ий, -ая, -ое, (*a*) quasi-elastic

ква́кш·а, -и, (*i*) **-ш·ей,** (*f*) tree frog, (*pl*) *Hylidae*

квалите́кс·, -а, (*m*) qualitex (natural latex concentrate)

квалифик·ацио́нн·ый, -ая, -ое, (*a*) qualified, skilled

квалифик·а́ци·я, -и, (*f*) qualifying; qualification(s); skill, specialization

квалифиц·и́рованн·ый, -ая, -ое, (*past part pass*); qualified; skilled, trained, specialized

ква́нт·, -а, (*m*) (phys) quantum

~ **де́й·стви·я** quantum of action

~ **по́л·я** field quantum

~ **све́т·а** photon, light quantum

~ **эне́рг·и·и** quantum of energy, energy quantum

ква́нт·а, -ы, (*f*) *see* **ква́нт·**

квант·иза́ци·я, -и, (*f*) *see* **квант·о·ва́ни·е**

квант·ова́ни·е, -я, (*n*) quantization

~ **с вы·дел·е́ни·ем дискре́т·н·ып то́ч·ек·** quantization by discrete digital units

   част·от·а́ квант·ова́ни·я (*f*) sampling rate

ква́нт·ованн·ый, -ая, -ое, (*past part pass*) of **ква́нт·овать**; (phys) quantized

ква́нт·овать, -уют, (*imp and perf*) (math) quantize

кванто·механ·и́ческ·ий, -ая, -ое, (*a*) quantum-mechanical

квант·ов·ый, -ая, -ое, (*a*) quantum

~ **вы·ход·** (*m*) quantum yield

~ **механ·ик·а** (*f*) quantum mechanics

~ **числ·ó** (*n*) quantum number

~ **числ·ó, внутр·енн·ее** total angular-momentum quantum number

квантó·метр·, -а, (*m*) (nucl) quanto-meter

квáнтор·, -а, (*m*) (math) quantifier

квáнт·ующ·ее у·стрóй·ств·о (*n*) quantizer

квар (*abbr*) of **реактúв·н·ый кило·-вольт·ампéр·** (phys) kvar, reactive kilovoltampere

квáрт·а, -ы, (*f*) quart; fourth (music)

квартáл·, -а, (*m*) quarter (of year); block, city block, block of houses; compartment (forestry)

квáртер·, -а, (*m*) (geol) Quaternary period/system

квартúл·ь, -я, (*m*) (math) quartile

квартúль·н·ый, -ая, -ое, (*a*) quartile

квартúр·а, -ы, (*f*) apartment, flat; (mil) billet, quarters

~, **глáв·н·ая** (mil) general headquarters

квартúр·н·ый, -ая, -ое, (*a*) of **квар-тúр·а**

квáрто (*n indecl*) (met roll) four-high; (print) quarto

~ **-стáн·, -а,** (*m*) four-high mill

кварт·овáни·е, -я, (*n*) quartering; inquartation (gold and silver assay)

кварт·овáть, -ýют, (*imp*) quarter

кварт·плáт·а, -ы, (*f*) rent

кварц·, -а, (*i*) **-ем,** (*m*) quartz; (elec) piezoelectric crystal, quartz crystal

~, **а·мóрф·н·ый** fusible quartz

~, **дáвл·енн·ый** stressed quartz

~, **плáвл·ен·ый** vitreosil, fused quartz

квáрц·ев·ый, -ая, -ое, (*a*) quartz; (elec, rad) crystal, quartz-crystal, piezoelectric

~ **диорúт·** (*m*) (min) quartz-diorite

~ **стабил·изáци·я** (*f*) (rad) crystal control

~ **трахúт·** (*m*) (min) rhyolite

~ **фúльтр·** (*m*) sand filter (for water treatment); (rad) crystal filter

~ **час·ы́** (*pl*) quartz clock, quartz-crystal clock

кварце·нóс·н·ый, -ая, -ое, (*a*) quartzi-ferous, quartzose

кварц·úт·, -а, (*m*) (geol) quartzite

квáс·, -а, (*nom pl*) **-ы́,** (*g pl*) **-óв,** kvass (beverage)

квáс·ить, -ят, (*imp*) make sour; pickle; leaven

квасц·евáть, -ýют, (*imp*) taw (leather)

квасц·óванн·ый, -ая, -ое, (*a*) alum-tanned, tawed (leather)

квасц·óвщик·, -а, (*m*) alum tanner, tawer

квасц·óв·ый, -ая, -ое, (*a*) alum

~ **земл·я́** (*f*) (min) alumina

~ **кáмен·ь** (*m*) (min) alunite, alum-stone

~ **руд·á, блёк·л·ая** (*f*) (min) alunite

~ **слáн·ец·** (*m*) (geol) alum shale

квасц·ы́ (*nom pl*), (*g pl*) **-óв,** (*m*) (chem) alum

~, **железо·аммóн·иев·ые** ferriam-monium sulphate/sulfate

~, **жж·ён·ые** burnt alum

кватерниóн·, -а, (*m*) (math) quaternion

квáш·а, -и, (*f*) (food) leaven

квáш·ени·е, -я, (*n*) fermentation, fer-menting

квáш·енн·ый, -ая, -ое, (*past part pass*) of **квáс·ить**

квáш·ен·ый, -ая, -ое, (*a*) sour, fer-mented, leavened

~ **капýст·а** (*f*) sauerkraut

кваш·н·я́, -й, (*g pl*) **-éй,** (*f*) kneading/ dough trough

квебрахúт·, -а, (*m*) (chem) quebrachi-tol

квебрáхо (*n indecl*) (chem) quebracho; (bot) quebracho, *Aspidosperma*

квебрáх·ов·ая стрýж·к·а (*f*) que-bracho chip (leather)

к·вéрх·у (*adv*) upwards, up

кверцúт·, -а, (*m*) (chem) quercitol

кверцитрóн·, -а, (*m*) (bot) quercitron, dyer's oak, *Quercus tinctoria*

квершлáг·, -а, (*m*) (min) cross-cut, cross-entry, cross-drift

квéст·а, -ы, (*f*) (geol) cuesta

квúлл·, -а, (*m*) (mech) quill

квúнсленд·ск·ий арророýт· (*m*) Queensland arrowroot

квинтáл·, -а, (*n*) quintal (100 kg)

~, **метр·úческ·ий** quintal (100 kg)

~, **англ·úйск·ий** cental, short hun-dredweight (45·36 kg, 100 lb)

квинтéт·, -а, (*m*) quintet

квинтиллиóн·, -а, (*m*) (math) quintil-lion, 10^{18} (U.S.S.R, U.S.A.), 10^{30} (Britain, Germany)

квинтильóн·, -а, (*m*) *see* **квинтил-лиóн·**

квит· *or* **квúты** (*indecl*) (bot) quince, *Cydonia*

квитанциóн·н·ый, -ая, -ое, (*a*) of **квитáнци·я**

квита́нци·я, -и, (*f*) receipt, receipt form
квит·о́в·ый, -ая, -ое, (*a*) of **квит**·
кво́рум·, -а, (*m*) quorum
кво́т·а, -ы, (*f*) quota
квт (*abbr*) of **кило·ва́тт**· kW, kw, kilowatt
квт-ч (*abbr*) of **кило·ва́тт·ча́с** kWh,
kwh, kilowatt-hour
кг (*abbr*) of **кило·гра́мм·а** kg, kilogram, kilogramme
кГ (*abbr*) of **килогра́мм-си́л·а** kg,
kilogram (unit of force)
к/г (*abbr*) of **кулон· на грамм**· C/g
coulomb per gram
кГм (*abbr*) of **кило·граммо·ме́тр**·
kgm, kilogram-meter (unit of work)
кг.м² (*abbr*) of **килогра́мм-квадра́т·-
н·ый метр**· kgm², kilogram-meter
squared
кг/м³ (*abbr*) of **килогра́мм на куб·и́-
ческ·ий метр**· kg/m³, kilogram per
cubic meter
кГм/сек (*abbr*) of **килогра́мм-си́ла-
ме́тр· в секу́нд·у** kilogram meter
per second
кг.сек/кг (*abbr*) of **килогра́мм-се-
ку́нд·а на килогра́мм** kgsec/kg,
kilogram-second per kilogram
кгс.м (*abbr*) of **килогра́мм-си́ла-
метр**· kilogram weight meter
кгс/м² (*abbr*) of **килогра́мм-си́ла на
квадра́т·н·ый ме́тр**· kilogram per
square meter
кгц (*abbr*) of **кило·ге́рц**· kc/s, kcs,
kilocycles per second
кдж (*abbr*) of **кило·джо́ул·ь** kilojoule
ке́гл·ь, -я, (*m*) (print) point size, body
size
ке́гл·я, -и, (*f*) ninepin, skittle
ке́др·, -а, (*m*) (bot) cedar, *Cedrus*
~, атла́с·ск·ий Atlas cedar, *Cedrus
atlantica*
~, гимала́й·ск·ий deodar cedar, *Cedrus
deodara*
~, лива́н·ск·ий cedar of Lebanon,
Cedrus Libani
~, сиби́р·ск·ий Swiss stone pine,
Pinus cembra
кедр·о́вник·, -а, (*m*) (bot) dwarf pine,
Pinus pumila
кедр·о́в·ый, -ая, -ое, (*a*) of **кедр**·
~ стл·а́нник· (*m*) (bot) dwarf pine,
Pinus pumila
ке́д·ы (*nom pl*), (*g pl*) **кед**·, (*f*) rubber-
soled sports boots
ке́йпер·, -а, (*m*) (geol) keuper marl
ке́к·, -а, (*m*) filter cake

ке́кс·, -а, (*m*) plum cake
кеку́р·, -а, (*m*) (geol) pinnacle rock
кела́т·н·ый, -ая, -ое, (*a*) *see* **хела́т·-
н·ый**
келе́ри·я, -и, (*f*) (bot) koeleria
кело́ид·, -а, (*m*) (biol) keloid
Ке́львин·а, шкал·а́ (*f*) (phys) Kelvin
scale
ке́льм·а, -ы, (*f*) bricklayers trowel
ке́льт·и·й, -я, (*m*) (chem) celtum, Ct
кем (*pron instr*) of **кто** by/with whom
ке́мбридж·ск·ий, -ая, -ое, (*a*) Cambridge
ке́мбр·и·й, -я, (*m*) (geol) Cambrian,
Cambrian period/system
кембр·и́йск·ий, ая, -ое, (*a*) (geol)
Cambrian
ке́мбрик·, -а, (*m*) (elec) cambric,
varnished cambric
кемига́м·, -а, (*m*) (plast) chemigum
кена́ф·, -а, (*m*) (bot) Bombay hemp,
Deccan hemp, *Hibiscus cannabinus*
кенгуру́ (*m indecl*) (zool) kangaroo,
(*pl*) *Macropodidae*
кенды́р·ь, -я, (*m*) apocynum, kendyr,
(bot) *Apocynum venetum*, *Apocynum
sibirica* (a type of dog-bane)
кенне́л·ев·ый, -ая, -ое, (*a*) cannel
(coal)
кенне́ль-богхе́д·, -а, (*m*) cannel-bog-
head (coal)
кено- (*component*) *see also* **кайно**-
кено·зо́й·ск·ий, -ая, -ое, (*a*) (geol)
Caenozoic, Cenozoic
кено·зо́ид·, -а, (*m*) (pal) cenozooid
кено·тро́н·, -а, (*m*) (elec) kenotron,
hot-cathode diode
кенотро́н·н·ый вы·прям·и́тел·ь (*m*)
(elec) kenotron rectifier
кентер·, -а, (*m*) (naut) lugged joining
shackle
Ке́плер·а, зако́н·ы (*pl*) (astron) Kep-
ler's laws
~, зр·и́тельн·ая труб·а́ (*f*) Keplerian
telescope, astronomical telescope
ке́плер·овск·ий, -ая, ·ое, (*a*) Keplerian
кер·аги́рит·, -а, (*m*) (min) ceragyrite,
horn silver
керази́н·, -а, (*m*) (chem) kerasin
кераме́т·, -а, (*m*) ceramet, metal
ceramic
керамзи́т·, -а, (*m*) "keramzit" (a light-
weight concrete aggregate)
керамзито·бето́н·, -а, (*m*) keramzit
concrete
кера́м·ик·, -а, (*m*) ceramist

кера́м·ик·а, -и, (*f*) ceramics

~, электро·техн·и́ческ·ая electro-ceramics, electrical ceramics

кера́м·иков·ый, -ая, -ое, (*a*) ceramic

керам·и́ческ·ий, -ая, -ое, (*a*) ceramic

керат·и́н·, -а, (*m*) (zool) keratin

керат·иниза́ци·я, -и, (*f*) (zool) keratinization, horn formation

керат·и́нов·ый, -ая, -ое, (*a*) keratin, horny

керат·и́т·, -а, (*m*) (med) keratitis

керат·о́з·, -а, (*m*) (med) keratosis

керато·пла́ст·ик·а, -и, (*f*) (med) keratoplasty, cornea grafting

керато·фи́р·, -а, (*m*) (geol) keratophyre

ке́рвел·ь, -я, (*m*) (bot) chervil, *Anthriscus cereifolium*; chervil, *Chaerophyllum bulbosum*

кери́т·, -а, (*m*) kerite

кермези́т·, -а, (*m*) (min) kermezite, pyrostibnite

кермёк·, -а, (*m*) (bot) thrift, statice

кермёс·, -а, (*m*) (bot) pokeberry, pokeweed, *Phytolacca*

кермёт·, -а, (*m*) metal ceramic, cermet, ceramet

кёрн·, -а, (*m*) (min) core, drill core; (elec) cathode base

кёрнер·, -а, (*m*) centre/center punch; centre dot (mark made by centre punch)

керни́т·, -а, (*m*) (min) kernite

кёрн·овать, -уют, (*imp*) prick punch, centre/center punch

кёрн·ов·ый, -ая, -ое, (*a*) (min) core

керно·ко́л·, -а, (*m*) (min) core splitter

керно·подъ·ём·ник·, -а, (*m*) (min) core-lifter

керно·рв·а́тел·ь, -я, (*m*) (min) core-lifter

керога́з·, -а, (*m*) oil pressure stove

керо·ге́н·, -а, (*m*) (geol) kerogen

керона́фт·, -а, (*m*) keronapht (kerosene-naphthalene compound used to preserve hides)

кероси́н·, -а, (*m*) kerosene, paraffin (Brit. pop)

~, авиацио́н·н·ый aircraft kerosene, jet fuel, aviation turbine fuel, AVTUR

~, ла́к·ов·ый white spirit

~, о·свет·и́тельн·ый illumination kerosene, lamp kerosene/fuel/oil

~, тра́ктор·н·ый tractor kerosene, vaporizing oil

кероси́н·к·а, -и, (*g pl*) **-н·ок·,** (*f*) oil stove

кероси́н·ов·ый, -ая, -ое, (*a*) kerosene, oil, paraffin

~ ла́мп·а (*f*) oil lamp

керосино·ре́з·, -а, (*m*) (weld) kerosene cutter

Ке́рр·а, явл·е́ни·е (*n*) (elec) Kerr effect

~, яче́·йк·а (*f*) (opt) Kerr cell

керсанти́т·, -а, (*m*) (min) kersantite

ке́сар·ев·о сеч·е́ни·е (*n*) (med) Caesarean section

ке́сар·ск·ое сеч·е́ни·е (*n*) (med) Caesarean section

кессо́н·, -а, (*m*) caisson, cofferdam; cooling water jacket (metallurgical furnaces); doghouse (OH furnace)

~ кры́·л·а (a/c) wing section

кессо́н·н·ый, -ая, -ое, (*a*) of **кессо́н·**

~ бол·е́зн·ь (*f*) caisson disease, the bends, diver's paralysis

~ констру́кци·я (*f*) (a/c) monocoque structure/construction

~ про·хо́д·к·а шахт· (*f*) (min) caisson shaft sinking

ке́т·а, -ы, (*f*) (fish) chum, keta, dog salmon, fall, *Oncorhynchus keta*

кетгу́т·, -а, (*m*) cat gut

кет·е́н·, -а, (*m*) (chem) ketene

кет·ими́н·, -а, (*m*) (chem) ketimin

кето·альдеги́д·, -а, (*m*) aldehyde ketone

ке́т·ов·ый, -ая, -ое, (*a*) of **ке́т·а**

~ икр·а́ red caviar

кето·гексо́з·а, -ы, (*f*) (chem) ketohexose

кето·ге́н·н·ый, -ая, -ое, (*a*) (med) ketogenic

кето·глюта́р·ов·ая кисл·от·а́ (*f*) ketoglutaric acid

кето·ено́ль·н·ая тауто·ме́р·и·я (*f*) (chem) keto-enolic tautomerism

кет·о́з·а, -ы, (*f*) (chem) ketose

кето·кисл·от·а́, -ы́, (*f*) (chem) keto-acid

кет·окси́м·, -а, (*m*) (chem) ketoxime

кето́н·, -а, (*m*) (chem) ketone

кетон·ацета́л·, -а, (*m*) acetal ketone

кетоно·кисл·от·а́, -ы́, (*f*) *see* **кето·-кисл·от·а́**

кетоно·спи́рт·, -а, (*m*) (chem) benzoin

ке́ттель·н·ая маши́н·а (*f*) (text) linking/looping machine, binding-off machine

кеттл·ёвк·а, -и, (*f*) (text) linking, binding-off, looping

ке́тчер·, -а, (*m*) (oil) tubing catcher

кефал- (*component*) *see also* **цефал-**

кефа́л·ий, -ья, -ье,   (*a*) of **кефа́л·ь**

кефа́л·ев·ый, -ая, -ое,   (*a*) of **кефа́л·ь**

кефал·иза́ци·я, -и,   (*f*) (zool) cephalization

кефал·и́н·, -а,   (*m*) (biochem) cephalin

кефало·ме́тр·и·я, -и,   (*f*) cephalometry

кефа́л·ь, -и,   (*f*) (fish) grey mullet, *Mugilidae*

кефаэл·и́н·, -а,   (*m*) (biochem) cephaeline

кефи́р·, -а,   (*m*) (food, chem) kefir

ке́ч·, -а,   (*m*) (naut) ketch

кианиз·а́ци·я, -и,   (*f*) (build) kyanizing

киан·и́т·, -а,   (*m*) (min) cyanite, kyanite, disthene

киано·трих·и́т·, -а,   (*m*) (min) cyanotrichite

киберне́т·ик·, -а,   (*m*) cyberneticist, cybernetician

киберне́т·ик·а, -и,   (*f*) cybernetics

кибернет·и́ческ·ий, -ая, -ое,   (*a*) cybernetic

кив·а́·ть, -ют,   (*imp*) nod

кив·н·у́ть, -у́т,   (*perf*) give a nod, nod

кид·а́ни·е, -я,   (*n*) (*v n*) of **кид·а́·ть**

кид·а́·ть, -ют,   (*imp*) fling, throw, cast

ки́жуч·, -а,   (*m*) (fish) coho, silver salmon, *Oncorhynchus kisutch*

кизельгу́р·, -а,   (*m*) (min) kieselguhr, diatomite

кизери́т·, -а,   (*m*) (min) kieserite

кизи́л·, -а,   (*m*) (bot) dogwood, *Cornus*

~, на·сто·я́щ·ий (bot) Cornelian cherry, *Cornus mas*

кизи́ль·ник·, -а,   (*m*) (bot) cotoneaster

кил·а́, -ы́,   (*f*) (agr) clubroot

кил·ева́ни·е, -я,   (*n*) (naut) careening

кил·ева́т·ый, -ая, -ое,   (*a*) keeled

кил·ева́тост·ь, -и,   (*f*) keel shape/formation

килевидно·пере·горо́д·чат·ый, -ая, -ое,   (*a*) (geol) carinimurate

киле·ви́д·н·ый, -ая, -ое,   (*a*) carinate, keeled

кил·ев·о́й, -а́я, -о́е,   (*a*) keel, (bot) carinal

~ по́яс·   (*m*) (shipb) garboard strake

киле·гру́д·н·ый, -ая, -ое,   (*a*) cristate, cristiform, carinate

киленба́нк·а, -и,   (*g pl*) **-нок·,** (*f*) (naut) careening wharf

кило́ (*n indecl*) kilogram

кило- (*component*) kilo-

кило·ба́рн·, -а,   (*m*) (nucl) kilobarn, kb

кило·ва́р·, -а,   (*m*) (phys) kilovar, kVAR, reactive kilovolt-ampere

кило·вар·ча́с·, -а,   (*m*) kilovar hour, reactive kilovolt-ampere-hour

кило·ва́тт·, -а,   (*m*) (elec) kilowatt, kW, kw

кило·ва́тт-ча́с·, -а,   (*m*) kilowatt-hour, kWh, kwh, kwhr

кило·во́льт·, -а,   (*m*) (elec) kilovolt, kV, kv

кило·во́льт·ампе́р·, -а,   (*m*) kilovolt-ampere, kVA, kva

кило·во́льт·ме́тр·, -а,   (*m*) (instr) kilovoltmeter

~, пи́к·ов·ый   peak-reading kilovoltmeter

кило·ге́рц·, -а,   (*i*) **-ем,** (*m*) kilocycles per second, kc/s, kcs

кило·гра́мм·, -а,   (*m*) kilogramme, kilogram, kg

кило·гра́мм-кало́р·и·я, -и,   (*f*) kilocalorie, large calorie, kcal

кило·гра́мм-моле́кул·а, -ы,   (*f*) kilogram molecule, kmol

кило·гра́мм-мо́л·ь, -и,   (*f*) kilogram molecule, kmol

кило·граммо·ме́тр·, -а,   (*m*) kilogram-meter, kgm

кило·джо́ул·ь, -я,   (*m*) kilojoule (unit of work)

кило·кало́р·и·я, -и,   (*f*) kilocalorie, large calorie, kcal

кило·кюри́ (*n indecl*) (nucl) kilocurie, kc

кило·ли́тр·, -а,   (*m*) kilolitre, kiloliter, kl

кило·ме́тр·, -а,   (*m*) kilometre, kilometer, km

кило·метр·а́ж·, -а,   (*m*) number of kilometres; kilometerage, life (of tyres etc.)

кило·метр·о́вк·а, -и,   (*g pl*) **-вок·,** (*f*) 1/100,000 scale map

кило·но́мер·, -а,   (*m*) (text) kilogramme, number

кило·резерфо́рд·, -а,   (*m*) (nucl) kilorutherford

кило·рентге́н·, -а,   (*g pl*) **-ген·,** (*m*) kiloroentgen

килоте́кс·, -а,   (*m*) (text) kilotex

кило·электрон·во́льт·, -а,   (*m*) kilo-electron-volt, kev

ки́л·ь, -я,   (*m*) (shipb) keel; (a/c) fin, rudder fin; (biol) carina, keel

~, бок·ов·о́й   bilge keel

~, брус·ко́в·ый   bar keel

ки́л·ь
~, вы·движ·н·о́й   sliding/drop keel, centre board, sliding keel
~, горизонта́ль·н·ый   flat keel, keel plate
~, о·пуск·н·о́й   sliding/drop keel, centre board
~, пло́ск·ий   keel plate, flat keel
~, ро́вн·ый   even keel
~ синклина́л·и   (geol) syncline trough
~, скул·ов·о́й   bilge rail
~, сло́й·чат·ый   side-bar keel,
киль·бло́к·, -а,   (m) (shipb) keelblock
киль·ва́тер·, -а,   (m) (naut) wake
киль·ва́тер·н·ый, -ая, -ое,   (a) stern, wake
~ ого́н·ь   (m) stern light
~ стру·я́   (f) wake
ки́ль·к·а, -и,   (g pl) -л·ек·,   (f) (fish) sprat, *Sprattus*
кильсо́н·, -а,   (m) (shipb, a/c) keelson
~, бок·ов·о́й   side keelson
~, интеркосте́ль·н·ый   (shipb) intercostal, longitudinal
КИМ   (abbr) of ко́д·ов·о·и́мпульс·-н·ая модул·я́ци·я   (f) PCM, pulse-code modulation
кимберли́т·, -а,   (m) (geol) kimberlite, blue earth
киммер·и́йск·ий, -ая, -ое,   (a) (geol) kimmeridgean, kimmeridge
кимо́·граф·, -а,   (m) (med) kymograph, cymograph
кин·а́з·а, -ы,   (f) (biochem) kinase
кингсто́н·, -а,   (m) (shipb) Kingston valve
кинема́т·ик·а, -и,   (f) kinematics
~ ви́хр·ей   vorticity kinematics
~ пло́ск·ого движ·е́ни·я   plane kinematics
кинемат·и́ческ·ий, -ая, -ое,   (a) kinematic
~ схе́м·а   (f) skeleton diagram, functional diagram; diagram of gears
кинемато́·граф·, -а,   (m) ciné camera, cinematograph camera
кинема́то·гра́ф·и·я, -и,   (f) cinematography
кине·ско́п·, -а,   (m) (T/V) kinescope
кин·естези́·я, -и,   (f) (med) kinaesthesia, kinesthesia
кин·естет·и́ческ·ий, -ая, -ое,   (a) kinaesthetic, kinesthetic
~ ощущ·е́ни·я   (pl) kinaesthetic sensations
кине́т·ик·а, -и,   (f) kinetics
~ жи́дк·ост·и   fluid kinetics

кине́т·ик·а
~ тума́н·ност·и   (astron) nebular kinetics
кинет·и́ческ·ий, -ая, -ое,   (a) kinetic
~ у·равн·е́ни·е   (n) (phys) equation of motion
кинето·ге́незис·, -а,   (m) (geol) kinetogenesis
кинето·мета·морф·и́зм·, -а,   (m) (geol) kinetometamorphism
кинжа́л·, -а,   (m) dagger
кинжа́ль·н·ый, -ая, -ое,   (a) of кинжа́л·
~ ого́н·ь   (m) (mil) short-range concentrated fire
кинка́н·, -а,   (m) (bot) kumquat, *Fortunella*
кино́   (n indecl) cinema; cinematography
кино-   (component) cinema, cine-, kine-, kino, motion picture, movie; cyno-
кино́а   (indecl) (bot) guinea, *Chenopodium quinoa*
кино·аппара́т·н·ая, -ой,   (f decl as adj) (cinema) projector room
кино·ателье́   (n indecl) film studio, motion picture studio (U.S.)
ки́новар·ь, -и,   (f) (min) cinnabar
кино·журна́л·, -а,   (m) news-reel
кино·журнал·и́ст·, -а,   (m) (cinema) script writer
кино·ка́др·, -а,   (m) (cinema) frame, individual picture
кино·карт·и́н·а, -ы,   (f) (cinema) film, motion picture (U.S.)
кино·коп·ирова́льн·ый аппара́т·   (m) (cinema) copying machine, printing machine
кин·о́л·, -а,   (m) (phot chem) quinol
кино·ло́г·и·я, -и,   (f) (zool) cynology
кино·механ·и́зм·, ленто·про·тя́ж·-н·ый   (m) pull-down mechanism (of camera)
кино·меха́н·ик·, -а,   (m) (cinema) maintenance engineer
кино·монт·а́ж·, -а,   (m) (cinema) mounting, montage (U.S.)
кино·опера́тор·, -а,   (m) (cinema) chief cameraman
кино·пере·дви́ж·к·а, -и,   (f) (cinema) portable projection set, portable cinema equipment
кино·плён·к·а, -и,   (g pl) -н·ок·,   (f) see кино·фото·плён·к·а
кино·прое́ктор·, -а,   (m) (cinema) projector
~, узко·плён·очн·ый   16-mm projector

кино·прое́ктор

~, широко·плён·оч·н·ый 35-mm projector

кино·про·ка́т·, -а, (*m*) (cinema) release and circulation

кино·режисс·ёр·, -а, (*m*) (cinema) director

кино·репорт·а́ж·, -а, (*i*) -ем, (*m*) (cinema) making news films/reels

кино·сеа́нс·, -а, (*m*) (cinema) performance, showing

кино·сту́ди·я, -и, (*f*) film studio, motion picture studio (U.S.)

кино·сцена́ри·й, -я, (*m*) (cinema) scenario, script, screen play

кино·съ·ём·к·а, комбин·и́рованн·ая (*f*) (cinema) multiple photography

кино·съ·ём·оч·н·ый аппара́т· (*m*) film camera, motion-picture, camera, cine camera

кино·у·станóв·к·а, -и, (*g pl*) -вок·, (*f*) cinema/movie projection equipment

кино·фик·а́ци·я, -и, (*f*) development of the film/motion-picture industry, development and organization of film facilities

кино·фи́льм·, -а, (*m*) (cinema) film, movie, motion picture (U.S.)

кино·фото·плён·к·а, -и, (*g pl*) -н·ок·, (*f*) (cinema) cinematograph film, film

~, гор·ю́ч·ая (cinema) inflammable (cellulose) film

~, не·гор·ю́ч·ая (cinema) non-inflammable film

~, у́з·к·ая 16-mm film

~, шир·óк·ая 35-mm film

кин·урен·и́н·, -а, (*m*) (chem) kynurenin

кú·н·уть, -ут, (*perf*) fling, throw, cast

кио́ск·, -а, (*m*) kiosk, booth, stall

киоск·ёр·, -а, (*m*) kiosk attendant

ки́п·, -а, (*m*) gorge, groove (of a sheave); score (of a block); (naut) fairlead

КИП (*abbr*) of **контро́ль·н·о-из·ме·р·и́тель·н·ый при·бо́р·** (*m*) control and measuring instrument, instrumentation

ки́п·а, -ы, (*f*) (text) bale; pile, stack

кипари́с·, -а, (*m*) (bot) cypress, *Cupressus*

кипари́с·ов·ый, -ая, -ое, (*a*) of **кипари́с·** (*pl as noun*) (bot) *Cupressaceae*

кипе́л·к·а, -и, (*g pl*) -л·ок·, (*f*) quicklime, caustic lime

кип·е́ни·е, -я, (*n*) boiling; bubbling

зóн·а кип·е́ни·я (*f*) steam void

~, интенси́в·н·ое violent boiling

~, объ·ём·н·ое bulk boiling

~, пузы́р·чат·ое nucleate boiling

температу́р·а кип·е́ни·я (*f*) boiling point

тóч·к·а кип·е́ни·я (*f*) boiling point

~, чи́ст·ое quiescent boil

кипер·, -а, (*m*) (text) twill

кипер·н·ый, -ая, -ое, (*a*) of **кипер·**

~ лéнт·а (*f*) twill tape (for insulating)

кип·е́ть, -я́т, (*imp*) boil; seethe, bubble

ки́п·н·ый, -ая, -ое, (*a*) of **ки́п·а**

~ пресс· (*m*) (text) baler

ки́п·овк·а, -и, (*f*) baling

ки́п·ов·ый, -ая, -ое, (*a*) of **ки́п·а, кип·**

~ груз· (*m*) bale cargo

~ пла́н·к·а (*f*) chock; (naut) chock fairlead

кип·ор·, -а, (*m*) packaging repairer

кипо·раз·рыв·а́тел·ь, -я, (*m*) (text) bale breaker

Ки́пп·а, при·бо́р· (*m*) (chem) Kipp's apparatus

кипре́гел·ь, -я, (*m*) (surv) alidade, telescopic alidade

кипре́·й, -я, (*m*) (bot) willow-herb, *Epilobium*

кипре́й·ник·, -а, (*m*) *see* **кипре́·й**

кипре́й·н·ые, -ых, (*pl decl as adj*) (bot) *Onagraceae, Oenotheraceae*

кипрелé (*n indecl*) (elec) trigger relay

ки́пр·ск·ий, -ая, -ое, (*a*) (geog) Cyprus, Cyprian, Cypriot

ки́пс·, -а, (*m*) kip (leather)

кип·у́ч·ий, -ая, -ее, (*a*) boiling, seething, bubbling

кипят·и́льник·, -а, (*m*) boiler, water boiler

кипят·и́льн·ый, -ая, -ое, (*a*) boiler, boiling

~ бак· (*m*) boiler

кипят·и́ть, -я́т, (*imp*) boil, bring to the boil

кипят·óк·, -тк·а, (*m*) boiling water

кипяч·е́ни·е, -я, (*n*) boiling

кипяч·ённ·ый, -ая, -ое, (*past part pass*) of **кипят·и́ть**

кипяч·ён·ый, -ая, -ое, (*a*) boiled

кип·я́щ·ий, -ая, -ее, (*pres part act*) of **кип·е́ть**

~ слó·й (*m*) fluidized bed (ore roasting etc.)

~ ста́л·ь (*f*) rimming steel

ки́р·, -а, (*m*) (geol) brea, kir

кирз·а, -ы, (*f*) (text) kersey

кири́лл·иц·а, -ы, (*i*) **-ей,** (*f*) Cyrillic alphabet

кирк·а́, -и́, (*g pl*) **ки́рок·,** (*f*) pickaxe, (min) pick; scraper, hoe

кирказо́н·, -а, (*m*) (bot) Dutchman's pipe, birthwort, *Aristolochia*

кирказо́н·ов·ые, -ых, (*pl noun decl as adj*) (bot) birthwort family, *Aristolochiaceae*

кирк·о́вщик·, -а, (*m*) (mech) scarifier

кирк·о́в·ый, -ая, -ое, (*a*) of **кирк·а́;** scarifying

кирко·моты́г·а, -и, (*f*) pickaxe, pick

кирпи́ч·, -а́, (*i*) **-о́м,** (*m*) brick

~, вагра́н·очн·ый circle

~, воз·ду́ш·н·ый adobe, sun-dried brick

~, гильбе́рт·ов·ый (math) Hilbert parallelotope

~, дина́с·ов·ый Dinas brick

~, до·пресс·о́ванн·ый repressed brick

~, ды́р·чат·ый perforated brick

~, за·мык·а́ющ·ий closer

~, карни́з·н·ый capping brick

~, кли́н·чат·ый arch/key/compass brick

~, крупно·по́р·ист·ый highly porous brick

~, лёгк·ий light-weight brick

~, лиц·ев·о́й facing brick

~, мя́гк·ий soft/fixing brick

~, не·до·жж·ённ·ый light brick

~, не·обо·жж·ённ·ый unburned/air brick; green brick

~, об·легч·ённ·ый light-weight brick

~, обо·жж·ённ·ый burnt

~, обык·нове́нн·ый common brick

~, огне·у·по́р·н·ый firebrick

~, пусто·те́л·ый cored brick, hollow brick, hollow block

~, силика́т·н·ый sand-lime brick, silica brick

~, сплош·н·о́й solid/full brick

~, стро·и́тельн·ый common brick, building brick

~ -сыр·е́ц·, -р·ц·а́, (*m*) raw brick, green brick, cob

~, твёрд·ый engineering brick

~, тыч·ко́в·ый header/inbond brick

~, хроми́т·ов·ый chromite brick

~, хромо·магнези́т·ов·ый chrome-magnesite brick

~, шамо́т·н·ый fireclay brick

~, шамо́т·н·о-андези́т·ов·ый fireclay-andalusite brick

~, шпу́нт·ов·ый lug brick

кирпи́ч·н·ый, -ая, -ое, (*a*) of **кирпи́ч·;** brick-red, brownish red

~ кла́д·к·а (*f*) brickwork

~ за·во́д· (*m*) brickyard, brickworks

кирро·ли́т·, -а, (*m*) (min) cirrolite

киртоли́т·, -а, (*m*) (min) cyrtolite

Ки́рхгоф·а, зако́н·ы (*pl*) (phys) Kirchhoff's laws

кис· (*past masc sing*) of **ки́с·нуть**

кисе́л·евани·е, -я, (*n*) drenching, acidification (leather treatment)

кисе́л·ь, -я́, (*m*) kissel (Russian sweet dish); semi-liquid/sticky mass; drench (for leather)

кисе·я́, -и́, (*f*) (text) muslin

кисл·и́ц·а, -ы, (*f*) (bot) oxalis

кисл·о́ванн·ый, -ая, -ое, (*a*) (*past part pass*) of **кисл·ова́ть;** sour, acidulated; treated with acid

кисл·ова́т·ый, -ая, -ое, (*a*) sourish, acidulous

кисл·ова́ть, -у́ют, (*imp and perf*) treat with acid; acidulate, sour

кисл·о́вк·а, -и, (*g pl*) **-вок·,** (*f*) acidification; (text) souring

~ по́сле буч·е́ни·я (text) grey/lime souring

кисло·моло́ч·н·ый, -ая, -ое, (*a*) sour/fermented milk

~ проду́кт· (*m*) fermented-milk product

кисло·ро́д·, -а, (*m*) (chem) oxygen, O

~ во́з·дух·а atmospheric oxygen

кисло·ро́д·н·о-ацетиле́н·ов·ый, -ая, -ое, (*a*) oxy-acetylene

кисло·ро́д·н·о-дых·а́тельн·ый при·бо́р· (*m*) oxygen breathing apparatus

кисло·ро́д·н·о-за·ря́д·н·ая ста́нци·я (*f*) (a/c) oxygen cylinder trolley

кисло·ро́д·н·ый, -ая, -ое, (*a*) oxygen, oxygenous, oxy-

~ кисл·от·а́ (*f*) oxy-acid

~ ре́з·к·а (*f*) (weld) oxyacetylene cutting

кислородо·от·щепл·я́ющ·ий, -ая, -ее, (*a*) oxygen removing, deoxidizing

кисло·сла́д·к·ий, -ая, -ое, (*a*) sweet-sour

кисло·с·ли́в·очн·ое ма́сл·о (*n*) whey butter

кисл·от·а́, -ы́, (*nom pl*) **-о́т·ы,** (*g pl*) **-о́т·,** (*f*) (*and see under qualifying adj*) (chem) acid; sourness, acidity

~, азо́т·н·ая (*f*) nitric acid

~, ба́шен·н·ая (chem) tower/Glover acid

~, без·во́д·н·ая acid anhydride

~, дым·я́щ·ая fuming acid

кисл·от·а́

~, ка́мер·н·ая chamber acid

~, моно·над·се́р·н·ая Caro's acid

~, одно·осно́в·н·ая monobasic acid

кислотно·рас·твор·и́м·ый, -ая, -ое, (*a*) soluble in acids

кисл·о́тност·ь, -и, (*f*) acidity

кисл·о́тн·ый, -ая, -ое, (*a*) (chem) acid, acidic

~ аккумуля́тор· (*m*) (elec) lead-acid accumulator

~ о·ста́т·ок· (*m*) acid radical

~ числ·о́ (*n*) (chem) pH number

кислото·ме́р·, -а, (*m*) (chem) acidimeter

кислото·но́с·н·ый, -ая, -ое, (*a*) acidiferous

кислото·образ·ова́тел·ь, -я, (*m*) acidifier

кислото·по·до́б·н·ый, -ая, -ое, (*a*) acidoid, acid-like

кислото·сто́й·к·ий, -ая, -ое, (*a*) acid-resistant

кислото·у·по́р·н·ый, -ая, -ое, (*a*) acid-resistant

ки́сл·ый, -ая, -ое, (*a*) sour; (chem) acid, acidic, bi-, hydro-

~ серно·ки́сл·ый на́тр·и·й (*m*) sodium bisulphate/bisulfate, sodium hydrosulphate/hydrosulfate

~ сол·ь (*f*) acid salt, bi-salt

~ щавелово·ки́сл·ый ка́л·и·й (*m*) potassium bioxalase

~ щавелово·ки́сл·ая сол·ь (*f*) bioxalate

~ эфи́р· (*m*) acid ester

ки́с·нуть, -нут, (*past masc sing*) **кис·,** (*imp*) turn sour

кист·а́, -ы́, (*f*) (*see also* **цист·а́**) (med) cyst

кист·ева́ни·е, -я, (*n*) putting on with a brush, brushing

кисте·ви́д·н·ый, -ая, -ое, (*a*) (bot) racemose, clustering

кист·ев·о́й, -а́я, -о́е, (*a*) of **кист·ь**

~ раз·ря́д· (*m*) (elec) brush discharge

кисте·но́с·н·ый, -ая, -ое, (*a*) racemiferous

кист·и́ст·ый, -ая, -ое, (*a*) bunchy, (bot) racemose

ки́ст·очк·а, -и, (*g pl*) **-чек·,** (*f*) (*dim*) of **кист·ь**

кисточко·ви́д·н·ый, -ая, -ое, (*a*) (bot) brush-like

ки́ст·ь, -и, (*nom pl*) **-и,** (*g pl*) **-е́й,** (*f*) brush (painter's); bunch, cluster; tassel; (anat) wrist; (bot) raceme

ки́ст·ь

~, мах·ов·а́я (*f*) ground brush, pound brush

~, пер·ов·а́я fitch, liner (paint brush)

~, по·бе́л·очн·ая distemper brush

~, фил·ёночн·ая fitch (paint brush)

ки́с·ш·ий, -ая, -ее, (*past part act*) of **ки́с·нуть**

кит·, -а́, (*m*) (zool) whale, *Cetus*

~, гренла́нд·ск·ий Greenland whale

~, се́р·ый (zool) Pacific grey whale, *Rhachianectes glaucus*

~, си́н·ий blue whale, *Balaenoptera musculus*; *Sibbaldus musculus*

~, ю́ж·н·ый Southern right whale, *Balaena australis, Eubalaena glacialis*

китае·ве́д·ени·е, -я, (*n*) sinology

кита·и́ст·, -а, (*m*) sinologist

кита·и́стик·а, -и, (*f*) sinology

кита́й·к·а, -и, (*g pl*) **-а́ек·,** (*f*) (text) nankeen; (bot) Chinese apple, *Malus prunifolia*

кита́й·ск·ий, -ая, -ое, (*a*) Chinese

~ акти́в·н·ая то́ч·к·а (*f*) (med) acupuncture point

ки́тел·ь, -я, (*m*) tunic, jacket

кито·ба́з·а, -ы, (*f*) whale factory ship

кито·бо́·ец·, -о́й·ц·а, (*i*) **-о́й·ц·ем,** (*m*) whale catcher

кито·бо́·й, -я, (*m*) whaler

кито·бо́й·н·ый, -ая, '-ое, (*a*) whaling

~ плов·у́ч·ая ба́з·а (*f*) whale factory-ship

кито·ви́д·н·ый, -ая, -ое, (*a*) cetacean, cetaceous, whale-like

ки́т·ов·ый, -ая, -ое, (*a*) of **кит·**

~ жир· (*m*) blubber

~ ус· (*m*) baleen, whalebone

кито·ло́в·н·ый, -ая, -ое, (*a*) = **кито·бо́й·н·ый**

кито·обра́з·н·ый, -ая, -ое, (*a*) (zool) cetacean, cetaceous, (*pl as noun*) *Cetacea*

киш·е́ть, -а́т, (*imp*) swarm/teem with

киш·е́чник·, -а, (*m*) (anat) intestine

кишечно·ды́ш·ащи·е, -их, (*pl decl as adj*) (zool) acorn/tongue worms, *Enteropneusta*

кишечно·пол·ост·н·ы́е, -ы́х, (*pl decl as adj*) (zool) *Coelenterata*

киш·е́чн·ый, -ая, -ое, (*a*) intestinal, enteric

киш·к·а́, -и́, (*g pl*) **-ш·о́к·,** intestine, gut

~, двенадцати·пе́рст·н·ая (*f*) (anat) duodenum

~, обод·о́чн·ая (anat) colon

~, прям·а́я rectum

киш·к·а́
~, то́лст·ая   large intestine
~, то́нк·ая   small intestine
~, то́щ·ая   jejunum
кишми́ш·, -а́, (*i*) **-о́м,** (*m*) seedless grape; currant
киш·нец·, -нц·а, (*m*) (bot) coriander
киян·к·а, -и, (*g pl*) **-нок·,** mallet
ккал (*abbr*) of **кило·калор·и·я** (*f*) kcal, large calorie
ккал/г.атом (*abbr*) of **кило·калор·и·я на грамм-а́том** kcal/g atom, kilocalorie per gram atomic weight
ккал/град (*abbr*) of **кило·калор·и·я на гра́дус·** kcal/deg, kilocalorie per degree
ккал/кг (*abbr*) of **кило·калор·и·я на кило·гра́мм·** kcal/kg, kilocalorie per kilogram
ккал/ч (*abbr*) of **кило·калор·и·я в час·** kcal/hr, kilocalorie per hour
ккюри (*abbr*) of **кило·кюри́** kc, kcurie, kilocurie
кл (*abbr*) of **кило·литр·** kl, kilolitre
клаваци́н·, -а, (*m*) (pharm) clavacin, patulin
клавиату́р·а, -ы, (*f*) keyboard
кла́виш·, -а, (*m*) = **кла́виш·а**
кла́виш·а, -и, (*i*) **-ей,** (*f*) key (typewriters, pianos etc.)
кла́виш·н·ый, -ая, -ое, (*a*) key, keyboard
~ маши́н·а (*f*) key-responsive adding machine
~ у·стро́й·ств·о (*n*) keyboard (typewriter, computers etc.)
кла́·вш·ий, -ая, -ее, (*past part act*) of **кла́с·ть**
кла́д·бищ·е, -а, (*n*) cemetery, burial ground
кла́д·енн·ый, -ая, -ое, (*past part pass*) of **кла́с·ть**
кла́д·ен·ый, -ая, -ое, (*a*) castrated
кла́д·ёт (*pres 3rd sing*) of **кла́с·ть**
кла́д·к·а, -и, (*g pl*) **-д·ок·,** (*f*) laying (object produced by laying); (build) masonry, walling; clutch (of eggs)
~, бу́т·ов·ая   rough walling
~, ка́мен·н·ая   masonry; bricklaying
~, кирпи́ч·н·ая   brickwork
~, об·лиц·о́вочн·ая   (build) bastard ashlar
~, цеп·н·а́я   (build) English bond
клад·ов·а́я, -о́й, (*f decl as adj*) storeroom, store; larder, pantry
клад·о́вк·а, -и, (*g pl*) **-вок·,** (*f*) see **клад·ов·а́я**

клад·овщи́к·, -а, (*m*) storekeeper, warehouseman
кладо·генёз·, -а, (*m*) (biol) cladogenesis
клад·о́ди·й, -я, (*m*) (bot) cladode, phylloclade
клад·о́м·ый, -ая, -ое, (*pres part pass*) of **кла́с·ть**
кладо́ни·я, -и, (*f*) (bot) cladonia
кладо·села́х·и·и (*nom pl*), (*g pl*) **-й,** (*f*) (pal) *Cladoselachii*
кладо·спор·ио́з·, -а, (*m*) (hortic) scab
кладо́·фор·а, -ы, (*f*) (bot) cladophora
клад·у́щ·ий, -ая, -ее, (*pres part act*) of **кла́с·ть**
кла́д·ь, -и, (*f*) load; luggage
кла́д·я (*pres gerund*) of **кла́с·ть**
кла́йден·а, явл·е́ни·е (*n*) (phot) Claydon effect
Кла́йзен·а, конденс·а́ци·я (*f*) (chem) Claisen condensation
кла·л (*past masc sing*) of **кла́с·ть**
кла́н·я·ться, -ются (*imp*) bow
кла́пан·, -а, (*m*) valve (*see also* **вéн·тил·ь, за·сло́н·к·а**); (anat) valve, valvula; flap; palm (sailmaking); thumb index (book); (elec) gate; drop, drop indicator/annunciator (telephones)
~, в·дых·а́тельн·ый inlet valve (respirator)
~, вéнтиль·н·ый   lift valve
~ вентил·я́ци·и   vent valve
~, воз·вра́т·н·ый   back-pressure valve
~, воз·ду́ш·н·ый   раз·гру́з·очн·ый snort valve (submarine)
~, в·пуск·н·о́й   (ICE) inlet/intake valve
~, в·са́с·ывающ·ий   inlet/intake/suction valve
~, вы·брос·н·о́й   blow-down valve
~, вы·зыв·н·о́й   calling drop (telephones)
~, вы·пуск·н·о́й   exhaust/discharge/release valve
~, вы·хлоп·н·о́й   exhaust valve
~, декомпрессио́н·н·ый   decompressor (diesel)
~, га́з·ов·ый   дро́ссель·н·ый   burner valve
~, до·ба́в·очн·ый сопл·ов·о́й overload nozzle valve (turbo prop)
~, дро́ссель·н·ый   throttle valve
~, дым·ов·о́й   stack valve (OH furnace)
~, дых·а́тельн·ый   throttle valve
~, за·пир·а́ющ·ий   stop valve
~, за·по́р·н·ый   stop valve
~, за·кло́п·очн·ый   flap valve
~, иго́ль·чат·ый   needle valve

кла́пан

~, **изол·и́рующ·ий** isolating valve

~, **и́мпульс·н·ый** (elec) pulse gate

~, **контро́ль·н·ый** control valve, pilot valve

~, **кон·цев·о́й** end-seating valve

~, **маневр·о́в·ый** (st eng) manoeuvring valve

~, **много·ход·ов·о́й** multiple valve

~, **на·гнет·а́тельн·ый** delivery valve

~, **не·воз·врат·н·о-у·правл·я́ем·ый** screw-down non-return valve

~, **не·воз·врат·н·ый** non-return valve

~ **ни́ж·н·его про·дув·а́ни·я** blow-off valve

~, **обра́т·н·ый** flat check valve; reflux valve

~, **от·кид·н·о́й** (rail) clack valve

~, **от·бо́й·н·ый** ring-off indicator (telephones)

~, **от·пуск·н·о́й** (rail) release valve

~ **-от·сек·а́тел·ь, -я,** (m) cut-off valve

~, **от·се́ч·н·ый** check valve

~, **пар·ов·о́й** pressure relief valve (of radiator)

~, **пере·во́д·н·ый** reversing valve

~, **пере·пуск·н·о́й** unloading valve, by-pass valve

~, **пит·а́тельн·ый** feed water valve

~, **пласт·и́нчат·ый** flat check valve; ring-plate valve

~, **по·воро́т·н·ый** reversing valve

~, **под·вес·н·о́й** overhead valve

~, **по·пла́в·ков·ый** float valve

~, **пред·о·хран·и́тельн·ый** (st eng) relief/safety valve; (weld) back-flow check valve

~, **при·ём·н·ый** foot valve (pump)

~, **про́б·н·ый** (st eng) test cock, gauge cock

~, **противо·пере·гру́з·очн·ый** (a/c) anti-g valve

~, **про·ход·н·о́й** globe valve

~, **прям·о́й** globe valve

~, **пуск·ов·о́й** starting valve

~, **раз·гру́ж·енн·ый** relieved valve

~, **раз·гру́з·очн·ый** relief/unloading valve

~, **раз·общ·и́тельн·ый** stop valve, isolating valve

~ **ра́нц·а** satchel/haversack flap

~, **регул·я́торн·ый** regulating valve; (st eng) throttle valve

~, **регул·и́рующ·ий** regulator, regulating valve, pressure-regulating/reducing valve, throttling valve (pressure systems)

кла́пан

~, **редукцио́н·н·ый** reducing valve, regulator, throttling/expansion valve

~, **с·лив·н·о́й** overflow valve

~, **с·мес·и́тельн·ый** mixing valve

~ **-снорт·, -а,** (m) snort valve (of a blast furnace)

~, **со·глас·у́ющ·ий** (a/c) brake-relay valve

~, **соленои́д·н·ый** solenoid valve

~, **сопл·ов·о́й** nozzle valve

~, **сто́пор·н·ый** stop valve

~, **таре́ль·чат·ый** disc valve

~, **угл·ов·о́й** angle valve

~, **у·по́р·н·ый** check valve

~, **шар·ов·о́й** ball valve

~ **эконо́м·и·и** economic valve

~, **экран·и́рованн·ый** masked inlet valve (diesel)

кла́пан·н·ый, -ая, -ое, (a) of **кла́пан·**; valve, valvular; flap; drop

~ **коро́б·к·а** (f) valve box/chest

~ **пруж·и́н·а** (f) valve spring

~ **рас·пре·дел·е́ни·е** (n) valve control/gear/regulation

Клапейро́н·а, у·равн·е́ни·е (n) (chem) Clapeyron equation, Clausius–Clapeyron equation

кларе́н·, -а, (m) (min) clarain

кларкеи́т·, -и, (m) (min) clarkeite

кла́сс·, -а, (m) class; grade, form (school etc.); type; (pal) genus

~ **чи́сел·** (math) rest class, residue

кла́ссик·, -а, (m) authority (person); classical scholar

класс·иро́вщик·, -а, (m) classer, grader

классифик·а́тор·, -а, (m) (min chem) classifier

~, **воз·ду́ш·н·ый** air classifier/elutriator

~, **двух·ре́·ечн·ый** duplex classifier

~, **ко́нус·н·ый** cone classifier

~, **коры́т·н·ый** launder/tank classifier

~, **много·де́к·ов·ый** multi-deck classifier

~, **одно·ре́·ечн·ый** simplex rake classifier

~, **ре́·ечн·ый** Dorr/rake classifier

~, **спира́ль·н·ый** Akins classifier, spiral classifier

~, **ча́ш·ечн·ый** bowl classifier

~, **четырёх·ре́·ечн·ый** quadruplex classifier

классифик·ацио́нн·ый, -ая, -ое, (a) of **классифик·а́ци·я**

~ **о́бщ·еств·о** (n) registry of shipping

классифик·а́ци·я, -и, (*f*) classification, grading

~, системат·и́ческ·ая (biol) taxonomy

классифиц·и́р·овать, -уют, (*imp and perf*) classify, class, grade; rate

класс·и́ческ·ий, -ая, -ое, (*a*) classic(al)

кла́сс·н·ый, -ая, -ое, (*a*) (*see also* **кла́сс·ов·ый**) class; high-class, high-grade, of a class/grade, graded

~ ваго́н· (*m*) (rail) passenger coach/car

~ доск·а́ (*f*) blackboard

~ каю́та (*f*) private cabin

~ ко́мнат·а (*f*) classroom

~ руко·вод·и́тел·ь (*m*) form master (schools)

~ специал·и́ст· (*m*) qualified man, tradesman, specialist

кла́сс·ов·ый, -ая, -ое, (*a*) class

~ бор·ьб·а́ (*f*) the class struggle, class warfare (political, social)

~ раз·би·е́ни·е (*n*) (math) partition into classes

кла́стер·, -а, (*m*) (cryst) cluster

класт·и́ческ·ий, -ая, -ое, (*a*) clastic

~ по·ро́д·а (*f*) (geol) clastic rock

класто·кристалл·и́ческ·ий, -ая, -ое, (*a*) clastocrystalline

класто·мо́рф·н·ый, -ая, -ое, (*a*) clastomorphic

клас·ть, (*pres 3rd sing, pl*) **клад·ёт, клад·у́т,** (*past masc sing*) **кла·л,** (*imp*) lay, put, set, place

клатра́т·ы (*nom pl*), (*g pl*) **-ов,** (*m*) (chem) clathrate compounds

Кла́удиус·а, клет·к·и (zool) (*pl*) Claudius' cells

Кла́узиус·а-Клапе́йрон·а, у·равн·е́ни·е (*n*) (chem) Clausius–Clapeyron equation

Кла́узиус·а-Мосо́тти, фо́рмул·а (*f*) (elec) Clausius–Mosotti relationship

кла́узул·а, -ы, (*f*) (law) clause, stipulation, proviso

клаузоне́ (*indecl*) *cloisonné* (interior decoration)

клебе·ма́сс·а, -ы, (*f*) (build) mastic

клёв·, -а, (*m*) (fish) bite, biting

клев·а́ни·е, -я, (*n*) biting, pecking; (autom) pumping, hunting, cycling

клёвант·, -а, (*m*) (naut) toggle

~, металл·и́ческ·ий Inglefield clip

кл·ева́ть, -юю́т, (*imp*) peck (of birds), bite (of fish)

Кле́ве, кисл·от·а́ (*f*) (chem) Cleve's acid

клеве́ит·, -а, (*m*) (min) cleveite

клевеланди́т·, -а, (*m*) (min) cleavelandite

кле́вер·, -а, (*nom pl*) **-а́,** (*g pl*) **-о́в,** (*m*) (bot) clover, *Trifolium*

~, под·зе́м·н·ый sub-clover, *Trifolium subterraneum*

~, ро́з·ов·ый alsike clover, *Trifolium hybridum*

клеверо·тёр·к·а, -и, (*g pl*) **-р·ок·,** (*f*) (agr) clover huller

клевет·а́ть (*pres 3rd sing, pl*) **клеве́щ·ет, клеве́щ·ут,** (*imp*) slander

клев·ец·, -в·ц·а, (*m*) (mech) tooth

клев·о́к·, -в·к·а́, (*m*) peck; burst (of projectile); notcher, incisor

клее·ва́р·, -а, (*m*) glue maker

клее·вар·е́ни·е, -я, (*n*) glue boiling

клее·ва́р·к·а, -и, (*f*) glue boiler

клее·ва́р·ный, -ая, -ое, (*a*) glue-boiling, glue

кле·ев·о́й, -а́я, -о́е, (*a*) glue, adhesive

~ вальц·ы́ (*pl*) glue spreader

~ кра́с·к·а (*f*) size, glue size

~ рези́н·а (*f*) rubber adhesive

клее·меш·а́лк·а, -и, (*g pl*) **-лок·,** (*f*) paste/dough mixer (rubber)

клее·на·нос·я́щ·ий, -ая, -ее, (*a*) glue spreading

кле·ёнк·а, -и, (*g pl*) **-нок·,** (*f*) oilskin (garment); oil cloth

клё·енн·ый, -ая, -ое, (*past part pass*) of **клё·ить;** glued, stuck

~ фане́р·а (*f*) plywood

~ холст· (*m*) buckram

кле·ёнчат·ый, -ая, -ое, (*a*) oilskin, oil-cloth

~ костю́м· (*m*) oilskin suit

кле·ён·ый, -ая, -ое, (*a*) adhesive, sticky; glued, sized, bonded; glue/size impregnated

кле·и́лн·ый, -ая, -ое, (*a*) gluing, sizing, pasting

клё·ить, -ят, (*imp*) glue, stick, gum; **-ся** (*pass*); become sticky; get on, make progress, catch on (colloq)

клё·й, -я, (*m*) glue, adhesive, gum, size

~, бел·у́ж·ий isinglass

~, казеи́н·ов·ый casein glue

~, кост·ян·о́й bone glue

~, мал·я́рн·ый painter's putty

~, мездр·о́в·ый hide glue

~, мор·ск·о́й marine glue

~, муч·н·о́й paste

~, мяс·н·о́й sarcocolla

~, плён·очн·ый film glue

кле́·й

~, пти́ч·ий   bird lime

~, раст·и́тельн·ый   gum, mucilage

~, рези́н·ов·ый   rubber cement

~ шу́б·н·ый   hide glue

клей·да·ю́щ·ие волóкн·а   (*pl*) (biol) intercellular fibres/fibers, white fibres/fibers

клей́·к·а, -и,   (*g pl*) **кле́·ек·,** (*f*) sticking, bonding, gluing; building up (tyres etc.)

клей́·к·ий, -ая, -ое,   (*a*) sticky, adhesive, tacky

клей·кови́н·а, -ы,   (*f*) gluten

клей·кови́нн·ый, -ая, -ое,   (*a*) gluten

клейко·ме́р·, -а,   (*m*) (instr) tackmeter

клей́·кост·ь, -и,   (*f*) stickiness, adhesiveness, tackiness

клейм·éни·е, -я,   (*n*) branding, marking, stamping

клейм·и́ть, -я́т,   (*imp*) brand, mark, stamp

клейм·ó, -á,   (*nom pl*) **кле́йм·а,** (*n*) brand, mark, stamp, seal

~, про·би́р·н·ое   hall mark, assay mark

~, фабри́ч·н·ое   trade mark

Кле́йн·а, по·ве́рх·ность·ь   (*f*) (math) Klein bottle

кле́йстер·, -а,   (*m*) paste, adhesive paste

клейсто·га́м·н·ый, -ая, -ое,   (*a*) (bot) cleistogamic

клейсто·ка́рп·и·й, -я,   (*m*) (bot) cleistocarp

клема́тис·, -а,   (*m*) (bot) clematis

кле́мм·а, -ы,   (*f*) clamp, clip; (elec) terminal

~, вы·ход·н·а́я   output terminal

кле́мм·ов·ый, -ая, -ое,   (*a*) of **кле́мм·а**

~ га́·йк·а   (*f*) split nut

кле́мшел·, -а,   (*m*) (civ eng) clamshell, grab

клён·, -а,   (*m*) (bot) maple, *Acer*

~, кра́с·н·ый   red maple, *Acer rubrum*

~, ложно·плата́н·ов·ый   (bot) sycamore, *Acer pseudoplatanus*

~, платано·ви́д·н·ый   Norway maple, *Acer platanoides*

~, са́хар·н·ый,   maple (wood); (bot) sugar/rock maple, *Acer saccharum*

~, пол·ев·о́й   small maple, *Acer campestre*

~, серебр·и́ст·ый   silver maple, *Acer saccharinum*

клен·о́в·ый, -ая, -ое,   (*a*) of **клён·**; (chem) aceric; (*pl decl as noun*) maple family, *Aceraceae*

клеп·а́льн·ый, -ая, -ое,   (*a*) rivet, riveting

~ молот·о́к·   (*m*) riveting hammer

~ об·жи́м·к·а   (*f*) rivet header

~ пресс·   (*m*) squeeze riveter

~ скоб·а́   (*f*) yoke riveter

клеп·а́льщик·, -а,   (*m*) riveter

клёп·анн·ый, -ая, -ое,   (*past part pass*) of **клеп·а́·ть;** riveted

клеп·а́·ть, -ют,   (*imp*) rivet; **клеп·а́ть, ‑лют,** (*imp*) malign

клёп·к·а, -и,   (*g pl*) **-п·ок·,** (*f*) riveting; stave, dowel

~, вз·рыв·н·а́я   explosion riveting

~, одно·сторо́н·н·яя   blind riveting

~ -строг·а́льн·ый стан·о́к·   (*m*) stave-cutting machine

клер·, -а,   (*m*) (telecom) plain language

клере́т·, бе́л·ый   (*m*) white clairette (a wine grape)

Клерó, у·равн·éни·е   (*n*)   (math) Clairaut's (differential) equation

клер·ова́ть, -у́ют, (*imp and perf*) clarify (sugar), refine

клер·о́вочн·ый котёл·   (*m*) blow-up pan (sugar)

клёст·, -а,   (*m*) (zool) crossbill, *Loxia*

клёт·к·а, -и,   (*g pl*) **-т·ок·,** (*f*) cage, coop, hutch; crate, crib; (biol) cell; (shipb) double bottom; gridiron; square (of squared paper)

~, бе́л·ич·ья   (elec) squirrel cage

~, верх·у́шечн·ая   apical cell

~, груд·н·а́я   (anat) chest, thorax

~, за·ро́д·ышев·ая   (biol) germ cell

~, камен·и́ст·ая   (bot) stone-cell, scleroid

~, мерц·а́тельн·ая   flame cell

~, палиса́д·н·ая   (bot) palisade cell

~, пыль·цев·а́я   (bot) pollen grain

~, с·ло́ж·н·ая   compound cell

~, со·про·вожд·а́ющ·ая   companion cell

~, яйц·ев·а́я   (zool) ovule

клетко·обра́з·н·ый, -ая, -ое,   (*a*) cytoid, cell-like

клетн·ева́ни·е, -я,   (*n*) lagging (pipe); parcelling, serving (rope)

клетн·ева́ть, -ю́ют,   (*imp*) lag (pipes); serve, parcel (ropes)

клетн·еви́н·а, -ы,   (*f*) lagging (of pipes); serving, parcelling (of rope)

клет·о́чк·а, -и,   (*g pl*) **-чек·,** (*f*) (*dim*) of **кле́т·к·а;** (biol) cellule

клеточко·обра́з·н·ый, -ая, -ое, (*a*) cellulated, cellule-like

клёт·очн·ый, -ая, -ое, (*a*) cage; cell, cellular

~ **ко́ж·иц·а** (*f*) (biol) cell wall

клет·ча́тк·а, -и, (*g pl*) **-ток·,** (*f*) (*see also* **целлюло́з·а**) cellulose, cellular tissue

клет·ча́тков·ый, -ая, -ое, (*a*) (*see also* **целлюло́з·н·ый**) cellulose

клёт·чат·ый, -ая, -ое, (*a*) cellular; squared (of paper etc.)

~ **бума́г·а** (*f*) squared paper

клёт·ь, -и, (*nom pl*) **-и,** (*g pl*) **-ей,** (*f*) store room; (min) cage; (met roll) stand

~, **двадцати·вал·ко́в·ая** (met roll) Sendzimir mill

~, **Ла́ут·а** (met roll) Lauth mill

~, **мно́го·вал·ко́в·ая** (met roll) multi-roll stand

~, **об·жи́м·н·ая** breaking-down stand

~, **о·про·ки́д·н·а́я** (min) tilting-deck cage

~, **под·гото́в·и́тельн·ая** (met) intermediate stand

~, **подъ·ём·н·ая** (min) cage, hoisting cage

~, **рабо́ч·ая** roll/mill stand; saddle (of Rockrite machine)

ли́н·и·я рабо́ч·их клёт·ей (*f*) (met roll) open train

~ **-сто́й·л·о, -а,** (*n*) horse box (for loading horses)

~, **тя́·нущ·ая** (met) withdrawing rolls (continuous casting)

~, **шестерён·н·ая** (met roll) pinion stand

клешне·ви́д·н·ый, -ая, -ое, (*a*) chelate, cheliform

клешне·обра́з·н·ый, -ая, -ое, (*a*) chelate, cheliform

клешн·я́, -и́, (*g pl*) **-ей,** (*f*) claw, nipper, (zool) chela

клещ·, -а́, (*i*) **-о́м,** (*m*) (*see also* **клёщ·и, клёщ·ик·**); (ent) mite, tick, *Acarina*

~, **крово·со́с·у́щ·ий пти́ч·ий** chicken mite, *Dermanyssus gallinae*

~, **чес·о́точн·ый** itch mite, *Sarcoptes scabiei*

клещ·е·ви́д·н·ый, -ая, -ое, (*a*) (chem) chelate

клещ·еви́н·а, -ы, (*f*) (bot) castor oil plant, caster bean plant, *Ricinus communis*; (met) gripping end of a forging piece

клещ·еви́нн·ый, -ая, -ое, (*a*) of **клещ·еви́н·а**

клещ·ев·о́й, -а́я, -о́е, (*a*) of **клещ·, клёщ·и**

клёщ·и (*nom pl*) of **клещ·,** (*g pl*) **-ей,** (*m*) pincers, tongs, nippers; (ent) mites, ticks, *Acarina*

~, **гре́йфер·е·ые** lever tongs

~ **Ди́тца·,** (elec) clip-on ammeter

~, **куз·не́чн·ые** blacksmith's tongs

~, **тру́б·н·ые** pipe tongs

клёщ·ик·, -а, (*m*) (*see also* **клещ·**) (ent) mite

~, **паути́н·н·ый** spider mite, *Epitetranychus*

~, **пуз·а́т·ый** louse mite, *Pediculoides ventricosus*

клё·ящ·ий, -ая, -ее, (*pres part act*) of **клё·ить**; sticking, bonding

кле·я́нк·а, -и, (*g pl*) **-нок·,** (*f*) glue jar/jug

кливáж·, -а, (*i*) **-ем,** (*m*) (geol) cleavage

кли́вер·, -а, (*m*) (naut) jib

клидоно́·граф·, -а, (*m*) (elec) klydonograph, Lichtenburg figure camera, surge voltage recorder

клие́нт·, -а, (*m*) (com) client, customer

клиенте́л·а, -ы, (*f*) clientele, clients, customers

клиент·у́р·а, -ы, (*f*) clientele, clients, customers

кли́зм·а, -ы, (*f*) (med) enema, clysma

кдизм·и́ческ·ий, -ая, -ое, (*a*) (med) clysmic

кли́к·а, -и, (*f*) clique

кли́к·ать, (*pres 3rd sing, pl*) **кли́ч·ет, кли́ч·ут,** (*imp*) call

кли́к·н·уть, -ут, (*perf*) give a call, call out

клима·гра́мм·а, -ы, (*f*) (meteor) climatic diagram

кли́макс·, -а, (*m*) climax

климакте́р·и·й, -я, (*m*) (med) climacteric

кли́мат·, -а, (*m*) climate

климат·иза́тор·, -а, (*m*) climatizer

клима·ти́п·, -а, (*m*) (bot) climatype

климат·и́ческ·ий, -ая, -ое, (*a*) climatic

климато·ло́г·и·я, -и, (*f*) climatology

климато·терап·и́·я, -и, (*f*) (med) climatotherapy

кли́н·, -а, (*nom pl*) **кли́н·ья,** (*g pl*) **-ьев·,** (*m*) wedge, wedge chock, cotter; (mech) key; (meteor) ridge, tongue wedge; V, vee; (gunn) breech block; panel (of parachute)

~, **бар·и́ческ·ий** (meteor) high pressure ridge

клин

~, **в·клад·н·ой** (mech eng) gib

~, **вращ·а́ющ·ийся** (opt) rotating wedge

~, **горизонта́ль·н·ый** (mech eng) side wedge

~, **двой·н·о́й** (mech) gib-and-cotter

~, **кач·а́ющ·ийся** (opt) swinging wedge

~, **ма́сл·ян·ый** (rail) oil wedge, oil/lubricating clearance

~, **опт·и́ческ·ий** optical wedge

~ **по·вы́ш·енн·ого давл·е́ни·я** (meteor) high-pressure ridge

~, **раз·бо́р·н·ый** (mech) loosening wedge

~, **ступ·е́ньчат·ый** (phot) stepped/photometric wedge

~, **у·по́р·н·ый** (mech eng) grip wedge

~, **фо́то·метр·и́ческ·ий** (opt) photometric/stepped wedge

~, **шпац·ио́н·ный** (print) set-spacing wedge

~, **шпунт·ов·о́й** (mech) sunk key

~, **юстир·о́вочн·ый** (opt) correction wedge

кли́нгер·, -а, (*m*) Klinger glass (on a boiler)

клингери́т·, -а, (*m*) (plast) klingerite

кли́н·ец·, -н·ц·а, (*m*) (civ eng) key

кли́н·ик·а, -и, (*f*) clinic

клин·и́ческ·ий, -ая, -ое, (*a*) clinical

кли́нкер·, -а, (*nom pl*) **-а́,** (*g pl*) **-о́в,** (*m*) slag, clinker; clinker brick; (shipb) clinker

кли́нкер·н·ый, -ая, -ое, (*a*) of **кли́нкер·**

~ **за·во́д·** (*m*) clinker brick works

клинке́т·, -а, (*m*) (shipb) sluice valve, gate valve

~, **с·пуск·н·о́й** (shipb) bulkhead sluice

клин·ко́в·ый, -ая, -ое, (*a*) of **клин·о́к·;** blade, knife

клинко·обра́з·н·ый, -ая, -ое, (*a*) bladed

клино·ви́д·ност·ь, -и, (*f*) wedge angle

клино·ви́д·н·ый, -ая, -ое, (*a*) wedge-shaped, cuneiform, sphenoid, cuneate, V-shaped, V-, vee-

клин·ов·о́й, -а́я, -о́е, (*a*) wedge, key, tapered, V-, vee-

~ **за·клю́ч·к·а** (*f*) (print) wedge, V

~ **реме́н·ь** (*m*) V-belt

~ **со·един·е́ни·е** (*n*) key coupling

клино·гуми́т·, -а, (*m*) (min) clinohumite

клино·до́м·а, -ы, (*f*) (cryst) rhombic prism

клин·о́к·, -н·к·а́, (*m*) blade, cutting edge, knife section

клино·кине́з·, -а, (*m*) (biol) klinokinesis

клино·ли́ст·н·ые, -ых, (*pl decl as adj*) (pal, bot) *Sphenophyllales*

клино́·метр·, -а, (*m*) clinometer

клино·обра́з·н·ый, -ая, -ое, (*a*) V/wedge-shaped, sphenoid, cuneiform

клино·педио́н·, -а, (*m*) (cryst) klinopedion

клино·ремё́н·н·ый, -ая, -ое, (*a*) V-belt

~ **пере·да́·ч·а** (*f*) V-belt transmission

клино·ро́мб·и́ческ·ий, -ая, -ое, (*a*) (cryst) monoclinic

клино·ромбо·эдр·и́ческ·ий, -ая, -ое, (*a*) (cryst) triclinic

клино·спор·а́нг·и·й, -я, (*m*) (bot) clinosporangium

клино·ста́т·, -а, (*m*) (bot) klinostat

клино·хло́р·, -а, (*m*) (min) clinochlore

клино·эдр·и́ческ·ий, -ая, -ое, (*a*) clinohedral

клиноэнстати́т·, -а, (*m*) (min) clinoenstatite

клинтони́т·, -а, (*m*) (min) clintonite

кли́пер·, -а, (*m*) (naut) clipper

кли́пп·, -а, (*m*) (geol) klippe

кли́ренс·, -а, (*m*) (M/T) ground/road clearance

кли́ринг·, -а, (*m*) (fin) clearing

клирфа́ктор·, -а, (*m*) (elec) klirrfactor, non-linear distortion factor

клис·тро́н·, -а, (*m*) (elec) klystron

~, **двух·ко́нтур·н·ый** double cavity/resonator klystron

~, **мно́го·резона́тор·н·ый** multi-cavity klystron

~, **от·раж·а́тельн·ый** reflex klystron

~, **прямо·лёт·н·ый** multi-cavity klystron

клистро́н·н·ый, -ая, -ое, (*a*) of **клистро́н·**

кли́фстон·, -а, (*m*) (geol) true chalk

клиц·а, -ы, (*i*) **-ей,** (*f*) cleat, clamp

~, **крест·о́в·ая** crossing cleat

кли́ч·к·а, -и, (*g pl*) **-ч·ек·,** (*f*) nickname; cover-name; name (as personal name of animal)

клич·ущ·ий, -ая, -ее, (*pres part act*) of **кли́к·ать**

клише́ (*n indecl*) (print) cliché, stereotype block, block, plate, engraving plate

~, **полу·то́н·ов·ое** half-tone block

клише́

~, рас·тво́р·ов·ое   half-tone block

клм (*abbr*) of **кило·лю́мен·** kilolumen

клоа́к·а, -и,   (*f*) cloaca, cesspool; (zool) cloaca

клоа́ч·н·ые, -ых,   (*pl decl as adj*) (zool) monotremes, *Monotremata*

Кло́д·а, спо́соб·   (*m*) Claude method (of liquefying air)

клок·, -а,   (*nom pl*) **кло́ч·ья** *or* **клок·и́,** (*g pl*) **кло́ч·ьев** *or* **клок·о́в,** (*m*) tuft, flock; shred, bit, piece

клокот·а́ть   (*pres 3rd sing, pl*) **клоко́ч·ет, клоко́ч·ут,** (*imp*) bubble, gurgle

клон·, -а,   (*m*) (gen) clone

клон·и́ть, -ят,   (*imp*) incline; **-ся** bow; near, approach

кло́н·ов·ый, -ая, -ое,   (*a*) (gen) clone

клоп·, -а́,   (*m*) (ent) bug, (*pl*) *Hemiptera, Heteroptera*

клоп·о́вник·, -а,   (*m*) (bot) peppergrass, *Lepidium ruderale*

кло́пфер·, -а,   (*m*) sounder, acoustic dial (telephony)

кло́т·, -а,   (*m*) (naut) truck (of mast)

кло́т·ик·, -а,   (*m*) (naut) truck (of mast)

клото́ид·а, -ы,   (*f*) (opt) Cornu spiral

клоч·кова́ни·е, -я,   (*n*) flocculation

клоч·кова́т·ый, -ая, -ое,   (*a*) flocculent

клоч·о́к·, -ч·к·а́,   (*m*) scrap, wisp, shred; flake, flock

кло́ч·ья   (*nom pl*) of **клок·**

клуб·, -а,   (*nom pl*) **-ы,** (*g pl*) **-ов,** (*m*) club (social); **клуб·, -а,** (*nom pl*) **-ы́,** (*g pl*) **-о́в,** (*m*) cloud (of steam etc.)

клуб·енёк·, -ньк·а́,   (*m*) (*dim*) of **клу́б·ен·ь;** tubercle, nodule, knob

клу́б·ен·ь, -б·н·я,   (*g pl*) **-б·н·ей,** (*m*) (bot) tuber, bulb

клуб·еньков·ый, -ая, -ое,   (*a*) of **клу́б·ен·ь**

~ бактер·и·и   (*pl*) (bot) symbiotic bacteria

клуб·и́ть, -ят,   (*imp*) send up in clouds; **-ся** rise in clouds

клубко·обра́з·н·ый, -ая, -ое,   (*a*) (biol) glomerate

клубне·ви́д·н·ый, -ая, -ое,   (*a*) tuberoid, tuberous, bulbous

клуб·нев·о́й, -а́я, -о́е,   (*a*) tuberous, tuber, bulbous, bulb

клубне·но́с·н·ый, -ая, -ое,   (*a*) tuberiferous, bulbiferous

клубне·лу́к·овиц·а, -ы,   (*i*) **-ей,** (*f*) (bot) corm, bulbo-tuber

клубне·пло́д·, -а,   (*m*) (agr) root, root-crop, (bot) tuber

клуб·ни́к·а, -и,   (*f*) (bot) garden strawberry, *Fragaria rosaceae*

клуб·ни́чн·ый, -ая, -ое,   (*a*) strawberry

клуб·о́к·, -б·к·а,   (*m*) clew, ball (of thread); (med) glomus; tangle, muddle; (chem) coil

на·мо́т·к·а в клуб·к·и́   (*f*) (text) balling

с·но́в·к·а с клуб·к·о́в   (*f*) (text) ball warping

клуб·о́чек·, -чк·а,   (*m*) (*dim*) of **клу·б·о́к·;** (zool) glomerule, glomerulus

~, семен·н·о́й   (bot) seed ball

клуб·о́чн·ый, -ая, -ое,   (*a*) of **клуб·о́к·**

~ маши́н·а   (*f*) (text) balling frame

Клу́зиус·а, коло́нн·а   (*f*) (nucl) Clusius column

клу́мб·а, -ы,   (*f*) flower bed

клу́н·я, -и,   (*f*) (agr) threshing shed/barn

клу́п·ик·, а,   (*m*) (*see*) **клу́пп·**

клу́пп·, -а,   (*m*) tap wrench, die stock, screw stock

клык·, -а́,   (*m*) canine tooth; tusk; (horol) detent

клюв·, -а,   (*m*) (zool) beak, bill, rostrum

клюво·голо́в·ые, -ых,   (*pl decl as adj*) (zool) *Rhynchocephalia*

клюво·обра́з·н·ый, -ая, -ое,   (*a*) rostrate, rostral, beak-shaped

клюз·, -а,   (*m*) (naut) hawse pipe

~, па́луб·н·ый   (naut) navel/deck pipe

~, шварт·о́в·ый   mooring pipe

~, я́кор·н·ый   hawse pipe

клюз·а, -ы,   (*f*) (geol) defile

клюк·а́, -и́,   (*f*) crutch

клю́кв·а, -ы,   (*f*) (bot) cranberry, *Vaccinium*

~, крупно·пло́д·н·ая   cultivated cranberry, *Vaccinium macrocarpum*

клю́·н·уть, -ут,   (*perf*) take a peck/bite

ключ·, -а́,   (*i*) **-о́м,** (*m*) key; wrench, spanner; (oil) tongs; (geog) spring, source; (elec) switch

~, ви́л·очн·ый   fork/straddle wrench

~, воз·вращ·а́ющ·ийся   (telecom) non-locking key

~, вы́·зыв·н·о́й   (telecom) calling/ringing key

~, га́·ечн·ый   spanner, wrench

~, гор·я́ч·ий   (geol) thermal, hot spring

~, двенадцати·гра́н·н·ый   bi-hexagonal spanner

ключ

~, **дву·сторо́н·н·ий** (telecom) bug key

~, **за·мык·а́ющ·ий** (telecom) locking key

~ **за·пу́ск·а** (rocket) launching switch

~, **зуб·ча́т·ый** lug

~ **ко́д·а** code key

~, **на·ки́д·н·о́й** ring spanner, box wrench (U.S.)

~, **на·кла́д·н·о́й** ring spanner, box wrench

~, **одно·сторо́н·н·ий** single-head wrench

~, **о·про́с·н·ый** (telecom) listening key

~, **от·бо́й·н·ый** (telecom) clearing key

~, **от·кры́·т·ый га́·ечн·ый** double-end spanner, open-end wrench

~, **пере·гово́р·н·о-вы·зыв·н·о́й** (telecom) speak-buzz key

~, **пере·да́т·очн·ый** transmitting/ morse key

~, **раз·во́д·н·о́й** adjustable/monkey wrench

~, **раз·дви́ж·н·о́й га́·ечн·ый** adjustable wrench

~, **селе́ктор·н·ый** (telecom) selector

~, **тар·и́рованн·ый** torque wrench

~, **телегра́ф·н·ый** morse/telegraph key

~, **тор·цев·о́й** box spanner; (horol) winder

~, **тру́б·н·ый** pipe wrench

~, **универса́ль·н·ый** adjustable wrench

~, **францу́з·ск·ий** monkey wrench

~, **цеп·н·о́й** chain tongs, chain pipe wrench

~, **шве́д·ск·ий** adjustable wrench/ spanner

~, **шарни́р·н·ый** (oil) hinged-jaw tongs

~, **штифт·ов·ый** peg spanner

ключ·ев·о́й, -а́я, -о́е, (*a*) of **ключ·**

~ **ка́мен·ь** (*m*) (build) keystone

~ **коммута́тор·** (*m*) (telecom) keyboard/box

ключ·и́ц·а, -ы, (*i*) **-е́й,** (*f*) (anat) clavicle, collarbone

ключ·и́чн·ый, -ая, -ое, (*a*) clavicular

кля́·вш·ий, -ая, -ее, (*past part act*) of **кля́·сть**

кля́кс·а, -ы, (*f*) blot, ink blot

кля́·л (*past masc sing*) of **кля́·сть**

кля́мер·а, -ы, (*f*) cramp iron

кля́ммер·а, -ы, (*f*) (build) roll joint, roll

клян·у́щ·ий, -ая, -ее, (*pres part act*) of **кля́·сть**

кля́п·, -а, (*m*) gag

кля́ск·а, -и, (*f*) (rad) blur

кля́·сть (*pres 3rd sing, pl*) **клян·ёт, клян·у́т,** (*imp*) curse; **-ся** swear

кля́т·в·а, -ы, (*f*) vow, oath

клятво·пре·ступл·е́ни·е, -я, (*n*) perjury

кля́ч·к·а, -и, (*g pl*) **-ч·ок·,** (*f*) chewing gum

км (*abbr*) of **кило·ме́тр·** kilometre, kilometer

кне́хт·, -а, (*m*) (naut) bollard

~, **крест·о́в·ый** staghorn

кни́г·а, -и, (*f*) book

~, **а́дрес·н·ая** directory

~, **вес·ов·а́я** (shipb) book of weights

~, **гла́в·н·ая** (fin) ledger

~, **пла́з·ов·ая** (shipb) table of mould offsets

~, **с·пра́в·очн·ая** reference book

книго·держ·а́тел·ь, -я, (*m*) book-end

книго·е́д·, -а, (*m*) (zool) book louse, *Troctes divinatorius*

книго·хран·и́лищ·е, -а, (*n*) storage (in library), shelving; major library

книдо·спор·и́ди·и (*nom pl*), (*g pl*) **-й,** (*f*) (zool) *Cnidosporidia*

книдо·фо́р·, -а, (*m*) (zool) cnidophore

книдо·ци́л·ь, -я, (*m*) (zool) cnidocil

кни́ж·ечк·а, -и, (*g pl*) **-чок·,** (*f*) booklet

кни́ж·к·а, -и, (*g pl*) **-ж·ек·,** (*f*) (*dim*) of **кни́г·а;** (zool) psalterium, omasum, manyplies

~, **за·пис·н·а́я** note-book

~, **рас·чёт·н·ая** pay book

~, **с·берег·а́тельн·ая** savings-bank book

кни́ж·н·ый, -ая, -ое, (*a*) of **кни́г·а**

~ **вош·ь** (*f*) (zool) book louse

~ **знак·** (*m*) bookmark, ex libris

к·ни́з·у (*adv*) downwards

кни́ц·а, -ы, (*i*) **-е́й,** (*f*) (shipb etc.) knee, knee plate

~, **би́мс·ов·ая** beam knee

~, **горизонта́ль·н·ая** stringer plate

~, **кно́п·ов·ая** deadwood

~, **скул·ов·а́я** tank margin bracket

кно́п·, -а, (*m*) (naut) knot; knee (boats)

~, **корм·ов·о́й** stern knee

~, **нос·ов·о́й** fore knee, fore deadwood

кно́пер·, -а, (*m*) oak/tan gall

кноп·к·а, -и, (*g pl*) **-п·ок·,** (*f*) button, push button, knob; drawing pin, tack; snap fastener, stud fastener

~, бо·ев·ая (mil) firing button

~, на·жат·ая depressed button

~ реш·ени·я compute button (computers)

кноп·очн·ый, -ая, -ое, (*a*) of **кноп·-к·а**

~ у·правл·ени·е (*n*) pushbutton control

кнопперс·, -а, (*m*) gall nuts, gall

Кнудсен·а, мано·метр· (*m*) Knudsen pressure gauge

кнут·, -а, (*m*) whip

княвдигед·, -а, (*m*) (shipb) knee of the head

княжик·, -а, (*m*) (bot) clematis

ко (*prep + dat*) *see* **к**

ко- (*component*) co-

коагул·аз·а, -ы, (*f*) (biochem) coagulase

коагул·ировани·е, -я, (*n*) coagulation, coalescence

коагул·ированн·ый ·ая, -ое, (*past part pass*) of **коагул·ир·овать**; coagulated, coalesced

коагул·ир·овать, -уют, (*imp and perf*) coagulate, coalesce

коагулюм·, -а, (*m*) coagulum

коагул·янт·, -а, (*m*) coagulant

коагул·ят·, -а, (*m*) coagulate, coagulation

коагул·яци·я, -и, (*f*) coagulation, coalescence; agglomeration (carbon blade)

коаксиаль·ност·ь, -и, (*f*) coaxial arrangement

коаксиаль·н·ый, -ая, -ое, (*a*) coaxial

~ кабел·ь (*m*) (elec) coaxial cable

коалесц·енци·я, -и, (*f*) coalescence

коалици·я, -и, (*f*) coalition

коацерв·ат·, -а, (*m*) (chem) coacervate

коацерв·аци·я, -и, (*f*) (chem) coacervation

кобаламин·, -а, (*m*) (biochem) cobalamin, vitamin B_{12}

кобальт·, -а, (*m*) (chem) cobalt, Co

~, дву·сер·нист·ый cobalt disulphide

~, мышьяк·овист·ый white cobalt

~, син·ий cobalt blue

кобальт·амин·, -а, (*m*) (chem) cobaltammine

кобальт·ин·, -а, (*m*) (min) cobaltine, cobaltite, cobalt glance

кобальт·ировани·е, -я, (*n*) cobalt plating

кобальт·ированн·ый, -ая, -ое, (*past part pass*) cobalt-plated

кобальт·ист·ый, -ая, -ое, (*a*) cobaltous

кобальт·ит·, -а, (*m*) (min) cobaltite, cobalt glance

кобальт·ов·ый, -ая, -ое, (*a*) cobalt, cobaltic, cobalt-containing, cobaltiferous

~ блеск· (*m*) (min) cobalt glance, cobaltite

~ зелен·ь (*f*) (chem) Rinman's green

~ цвет·ы (*pl*) (min) cobalt bloom, erythrite

кобур·а, -ы, (*f*) holster, leather case

кобыл·а, -ы, (*f*) mare

~, жереб·ая brood mare

кобыл·ий, -ья, -ье, (*a*) of **кобыл·а**

кобыл·иц·а, -ы, (*i*) **-ей,** (*f*) filly

кобыл·к·а, -и, (*g pl*) **-л·ок·,** (*f*) (ent) cricket, (*pl*) *Cicadina*

~, дерев·янн·ая (build) wood nog/brick

ковалент·ност·ь, -и, (*f*) (phys) covalence

ковалент·н·ый, -ая, -ое, (*a*) (phys) covalence, covalent

ков·ал·о, -а, (*n*) blacksmith's hammer

ков·альн·я, -и, (*f*) smithy, forge

ков·анн·ый, -ая, -ое, (*past part pass*) of **ков·ать**

ков·ан·ый, -ая, -ое, (*a*) forged, hammered, wrought, beaten

ковар·, -а, (*m*) (met) Kovar, Fernico

коваригант·, -а, (*m*) (math) covariant

коваригант·ност·ь, -и, (*f*) (math) covariance

коваригант·н·ый, -ая, -ое, (*a*) (math) covariant

коварнацион·н·ый, -ая, -ое, (*a*) (math) covariance

ков·ать (*pres 3rd sing, pl*) **ку·ёт, ку·ют,** (*imp*) forge, hammer; shoe (a horse)

ковеллин·, -а, (*m*) *see* **ковеллит·**

ковеллит·, -а, (*m*) (min) covellite, indigo copper

ковёр·, -вр·а, (*m*) carpet, mat

~ Серпинского (math) Sierpinski's set

~, сфагн·ов·ый (bot) sphagnum mat

коверк·а·ть, -ют, (*imp*) mangle, twist, distort

коверкот·, -а, (*m*) (text) covert coating

коверси́нус·, -а,   (*m*) (math) coversed sine

ко́в·к·а, -и,   (*g pl*) **-в·ок·,** (*f*) forging; (agr) shoeing

~, руч·н·а́я   hand forging

~ в шта́мп·ах   drop forging

~, свобо́д·н·ая   hammer/smith forging

ко́в·к·ий, -ая, -ое,   (*a*) malleable

~ чугу́н·   (*m*) malleable iron

ко́в·кост·ь, -и,   (*f*) malleability

ко́в·оч·н·о-про·ка́т·н·ый ста́н·   (*m*) forging rolls

ковочно·штамп·о́вочн·ый, -ая, -ое, (*a*) drop-forging

~ маши́н·а   (*f*) drop-forging press

~ пресс·   (*m*) drop-forging press

ко́в·очн·ый, -ая, -ое,   (*a*) forging

~ вальц·ы́   (*pl*) forging rolls, preparing rolls

~ маши́н·а   (*f*) forging machine

~ маши́н·а, горизонта́ль·н·ая   horizontal forging machine, upsetting machine

~ маши́н·а, кулач·ко́в·ая   crank-type upsetting machine

~ мо́лот·   (*m*) forging hammer

коври́г·а, -и,   (*f*) large round loaf

ко́вр·ик·, -а,   (*m*) (*dim*) of **ковёр·**; mat

~ -доро́ж·к·а, -и,   (*f*) strip/narrow carpet; (a/c) protection mat

ковр·о́вщик·, -а,   (*m*) carpet-maker

ковр·о́в·ый, -ая, -ое,   (*a*) carpet; carpeted

ковро·тк·а́цк·ий, -ая, -ое,   (*a*) carpet-weaving

~ стан·о́к·   (*m*) carpet loom

ковш·, -а́,   (*i*) **-о́м,**   (*m*) ladle, scoop, dipper; bucket, pail; (geog) basin (natural or artificial)

~, до·стро́·ечн·ый   (shipb) fitting-out basin

~, за·гру́з·очн·ый   (met) loading tray/trough, feed hopper

~, лит·е́йн·ый   (met) casting ladle

~ -лопа́т·а, -ы,   (*f*) (build) mortar ladle

~, про·меж·у́точн·ый   (*m*) (met) tundish, pouring basket

~ с нос·к·о́м   (met) lip-pour ladle

~, стале·раз·ли́в·очн·ый   steel casting ladle, steel/metal ladle

~, шла́к·ов·ый   cinder ladle, slag ladle

~ экскава́тор·а   excavator bucket

ковш·ев·о́й, -а́я, -о́е,   (*a*) *see* **ковш·о́в·ый**

ковш·о́в·ый, -ая, -ое,   (*a*) of **ковш·**

~ турби́н·а   (*f*) Pelton bucket with turbine

~ элева́тор·   (*m*) bucket conveyor, bucket elevator

ковы́л·ь, -я,   (*m*) (bot) feather grass, *Stipa*

ковыр·н·у́ть, -у́т,   (*perf*) *see* **ковыр·я́·ть**

ковыр·я́·ть, -ют,   (*imp*) pick at, tinker with

ко·гда́   (*adv*) when; (*conj*) when, while, as

     ко·гда́-либо   sometime

     ко·гда́-нибудь   sometime

     ко·гда́·так   if so, in that case

     ко·гда́-то   one time, once

когези·я, -и,   (*f*) (phys) cohesion

когере́нт·, -а,   (*m*) coherent, adherent

когере́нт·ност·ь, -и,   (*f*) coherence, cohesion

когере́нт·н·ый, -ая, -ое,   (*a*) coherent, coherence, cohesive

~ рас·се́·яни·е   (*n*) coherent scattering (light, X-rays etc.)

коге́рер·, -а,   (*m*) (rad) coherer

кого́   (*pron, acc/gen*) of **кто**; who, of who/whom

ко·гомо·топ·и́ческ·ий, -ая, -ое,   (*a*) (math) cohomotopy

ко·гомо·то́п·и·я, -и,   (*f*) (math) cohomotopy

когóрт·а, -ы,   (*m*) (bot) cohort

когот·н·о́й, -а́я, -о́е,   (*a*) (biol) unguiculate, clawed; (*pl as noun*) (zool); unguiculates, *Unguiculata*]

когот·о́к·, -т·к·а́,   (*m*) (*dim*) of **ко́гот·ь**

ко́гот·ь, -гт·я,   (*nom pl*) **-и,**   (*g pl*) **-е́й,**   (*m*) claw, talon, nail

когредие́нт·н·ый, -ая, -ое,   (*a*) (math) covariant

когте·о́бра́з·н·ый, -ая, -ое,   (*a*) claw-shaped, unguiform

когте·ход·я́щ·ий, -ая, -ее,   (*a*) (zool) unguligrade; (*as noun*) unguligrade

ко́гт·и   (*nom pl*) of **ко́гот·ь**; pole climbers, climbing irons, grapplers

ко́д·, -а,   (*m*) code, code book; (elec) code (binary systems); word (computer stores)

~ воз·ра́т·а каре́т·к·и   (autom) carriage-return code

~ информа́ц·и·и   character code (computers)

~, времен·н·о́й   (elec) pulse-length code

~, дво·и́чн·о-пяти·ри́чн·ый   biquinary code (computers)

код
~, дво·и́чн·ый binary code
~ кома́нд· instruction code (computers)
~, мно́го·а́дрес·н·ый multiple-address code (computers)
~, не·равно·ме́р·н·ый (telecom) unequal-length code
~, одно·а́дрес·н·ый single-address code (computers)
~ опера́ц·и·и operation code (computers)
~, пяти·зна́ч·н·ый (telecom) five-unit code
~, равно·ме́р·н·ый (telecom) equal length code
~ с из·ли́ш·к·ом 3 excess-three code (computers)
~, со·кращ·ённ·ый (telecom) brevity code
~, схе́м·н·ый (elec) circuit-state code
~, телегра́ф·н·ый (telecom) cable code
~, част·о́тн·ый (telecom) pulse-frequency code
~, числ·ов·о́й (telecom) pulse-count code
кодегидра́з·а, -ы, (f) (chem) codehydrogenase
кодейн·, -а, (m) (pharm) codeine
ко́декс·, -а, (m) (law) code
~, у·голо́в·н·ый criminal code
ко́дер·, -а, (m) coder
код·и́ев·ый, -ая, -ое, (a) (microbiol) codiacean, codiaceous
код·и́ровани·е, -я, (n) encoding, coding
~, матр·и́чн·ое step-matrix coding
код·и́рованн·ый, -ая, -ое, (past part pass) of **код·и́р·овать;** coded, encoded
код·и́р·овать, -уют, (imp and perf) encode, code (computers)
кодифик·а́ци·я, -и, (f) (law) codification
кодифиц·и́р·овать, -уют, (imp) codify
ко́д·ов·о-и́мпульс·н·ая модуля́ци·я (f) pulse-code modulation, pulse-count modulation
ко́д·ов·ый, -ая, -ое, (a) of **код·**
~ голо́в·к·а (f) (electr) information head (magnetic drum)
~ из·бир·а́ни·е (n) pulse-code discrimination
~ форм·ирова́тел·ь (m) word-pulse shaper (computer)
кодо·обра́з·ова́ни·е, -я, (n) coding, encoding
ко́е-где́ (adv) here and there
ко́е-ка́к (adv) anyhow

ко́е-како́й, -а́я, -о́е, (a) some
ко́е-кто́, ко́е-кого́, (pron) somebody
ко́е-куда́ (adv) somewhere
ко́е-что́, ко́е-чего́, (pron) something
ко́·ечн·ый, -ая, -ое, (a) hammock; (med) stretcher/cot case
ко́ж·а, -и, (i) **-ей,** (m) (anat) skin, cutis, derma; hide, leather; peel, rind
~, бара́н·ья sheepskin
~, бу́йвол·ов·ая buff hide
~, вы·де́л·анн·ая leather
~, жёстк·ая rough skin
~, иску́с·ственн·ая leather substitute, artificial leather
~, ла́к·ов·ая patent leather
~, не·дубл·ён·ая untanned hide, raw hide
~, сыро·мя́т·н·ая raw hide
~, теля́ч·ья kip, kipskin
ко́ж·ан·ый, -ая, -ое, (a) leather, leathern
кож·е́венн·ый, -ая, -ое, (a) tanning, leather-dressing, leather
~ за·во́д· (m) tannery
~ това́р· (m) leather article/goods
~ сыр·ь·ё (n) hides and skins
кож·е́вник·, -а, (m) tanner; (bot) sumas, *Rhus*
коже·е́д·, -а, (m) (zool) skin beetle, *Dermestida*
кож·за·мен·и́тел·ь, -я, (m) artificial leather, leather substitute
кожими́т·, -а, (m) "kozhimit" (a leather substitute)
кожисто·кры́·л·ые, -ых, (pl decl as adj) (ent) *Dermaptera*
ко́ж·ист·ый, -ая, -ое, (a) leathery
ко́ж·иц·а, -ы, (i) **-ей,** (f) thin skin, film; (biol) pellicle, cuticle
~ колбас·ы́ sausage skin
~, ресни́т·чат·ая ciliated epidermis
ко́ж·ник·, -а, (m) dermatologist
ко́ж·н·ый, -ая, -ое, (a) skin, cutaneous, dermal
~ по·кро́в· (m) (biol) integument
кож·сыр·ь·ё, -я, (n) hides and skins, raw leather
кож·ур·а́, -ы́, (f) rind, skin, pod, shell, husk
кож·у́х·, -а́, (m) casing, housing, jacket, cowl, sheath, mantle; (rail) outer shell of firebox; hearth jacket (furnaces)
~, гермет·и́ческ·ий (nucl) can, jacket
~, за·щи́т·н·ый protective cover; blade shield (turbine)
~, спира́ль·н·ый volute casing

коз·а́, -ы́, *(nom pl)* **ко́з·ы,** *(g pl)* **коз·,** *(f)* (agr) she-goat, nanny-goat, (zool) milk goat, *Capra hircus*; (min) timber skip

~, анго́р·ск·ая *(f)* Angora goat

~, ди́к·ая *(f)* (zool) roe, *Capreolus capreolus*

козёл·, -зл·а́, *(m)* goat, billy-goat; (met) salamander, bear; *(pl)* trestle, sheerlegs, saw horse, vaulting horse; (air) bounces; (mil) stack, pile (of rifles etc.)

~, ка́мен·н·ый *see* **козе·ро́г·** (zool)

козе·ро́г·, -а, *(m)* (zool) ibex, wild goat, *Capra ibex*; (astron) Capricorn

ко́з·ий, -ья, -ье, *(a)* goat, caprine

козима́з·а, -ы, *(f)* (biochem) cozymase

козл·ёнок·, -нк·а, *(nom pl)* **-я́т·а,** *(g pl)* **-ят·,** *(m)* (zool) kid

ко́зл·ик·, -а, *(m)* kidskin

козло·боро́д·ник, -а, *(m)* (bot) salsify, *Tragopogon*

козл·ов·о́й, -а́я, -о́е, *(a)* of **козёл·;** *see also* **козл·о́в·ый**

~ кран· *(m)* gantry crane

козл·о́в·ый, -ая, -ое, *(a)* *(see also* **козл·ов·о́й**) goatskin

козл·ы́ *(nom pl)* of **козёл·, ко́злы** *(nom pl)*, *(g pl)* **ко́зел,** box (of coach)

козл·я́т·а *(nom pl)* of **козл·ёнок·**

козу́л·я, -и, *(f)* (zool) roe deer, *Capreolus capreolus*

козыр·ёк·, -рь·к·а́, *(m)* peak (of cap), eyeshade; shroud, deflector, baffle plate; (M/T) sun vizor; a/c windscreen, windshield

ко́зыр·ь, -я, *(m)* trump, trump card

коипу́ `** *(indecl)* (zool) coypu, swamp beaver, *Myocastor coypus*

ко́ир·, -а, *(m)* coir (coconut fibre)

ко́й- *see* **ко́е-**

ко́·йк·а, -и, *(g pl)* **ко́·ек·,** *(f)* hammock; bunk, sleeping berth

~, под·ве́с·н·ая hammock

ко́йлер·, -а, *(m)* (text) coiler

койо́т·, -а, (zool) coyote, *Canis latrans*

ко́йр·, -а, *(m)* *see* **ко́ир·**

ко́к·, -а, *(m)* (naut) ships cook; (a/c) spinner, nosecap

~, в·ход·н·о́й nose cowling (gas turbine)

~, за́д·н·ий (a/c) turtleback

ко́к·а, -и, *(f)* (bot) coca, *Erythroxylon coca*

ко́к·ев·ый, -ая, -ое, *(a)* of **ко́к·а**

~ кисл·от·а́ *(f)* cocacic acid

кокаи́н·, -а, *(m)* (pharm) cocaine

кокаи́н·ов·ый, -ая, -ое, *(a)* of **кокаи́н·**

~ куст· *(m)* (bot) coca, *Erythroxylon coca*

ко·карбоксила́з·а, -ы, *(f)* (chem) cocarboxylase

коке́т·к·а, -и, *(g pl)* **-т·ок·,** *(f)* yoke (of dress)

коки́л·ь, -я, *(m)* (cast) permanent/metal/shell mould/mold

коки́ль·н·ый, -ая, -ое, *(a)* of **коки́л·ь**

~ кра́с·к·а *(f)* (met cast) blacking

~ лит·ь·ё *(n)* metal/shell-mould casting

ко́кк·, -а, *(m)* (bact) coccus, *(pl)* cocci, *Coccaceae*

кокко·ли́т·, -а, *(m)* (pal) coccolith

кокко·лито·фор·и́д·а, -ы, *(f)* (zool) coccolithophorid, *(pl)* *Coccolithophorida, Coccolithidae*

кокко·сфе́р·а, -ы, *(f)* (pal) coccosphere

коклю́ш·, -а, *(i)* **-ем,** *(m)* (med) whooping cough, cough pertussis

коклю́ш·к·а, -и, *(g pl)* **-ш·ек·,** *(f)* bobbin

ко́кон·, -а, *(m)* chrysalis

~, за·со́х·ш·ий calcined cocoon

~, шелк·ови́чн·ый cocoon (silk)

коконо·мот·а́ни·е, -я, *(n)* silk reeling

коконо·пря́д·, -а, *(m)* (zool) tent caterpillar, Lapit moth, *(pl)* *Lasiocampidae*

коконо·суш·и́лк·а, -и, *(g pl)* **-лок·,** *(f)* silk cocoon storing centre

кокор·, -а, *(m)* (mil) cartridge container

коко́р·а, -ы, *(f)* tree stump; (shipb) knee

коко́с·, -а, *(m)* (bot) coconut, *Cocos nucifera*

коко́с·ов·ый, -ая, -ое, *(a)* coconut; coir

~ па́льм·а *(f)* (bot) coconut palm, *Cocus nucifera*

кок-сагы́з·, -а, *(m)* (bot) kok-saghyz, *Taraxacum kok-saghyz*

ко́кс·, -а, *(m)* coke

~, до́мен·н·ый metallurgical coke

~, есте́ств·енн·ый native coke

~, лит·е́йн·ый foundry-grade coke

~, нефт·ян·о́й (chem) petroleum coke

~, у́голь·н·ый coal coke

кокса́ль·н·ый, -ая, -ое, *(a)* (biol) coxal

кокс·и́к·, -а́, *(m)* coke fines

кокси́т·, -а, *(m)* (biol) coxite; (med) coxitis

кокс·ова́льн·ый, -ая, -ое, *(a)* *see* **ко́кс·ов·ый;** coking

кокс·ова́ни·е, -я, *(n)* coking

кокс·ова́ть, -у́ют, *(imp)* coke; **-ся,** *(pass)* coke

кокс·о́в·ый, -ая, -ое, (*a*) coke, coking, coke-oven

~ газ· (*m*) coke-oven gas

~ ме́лоч·и (*pl*) breeze coke

~ печ·ь (*f*) coke oven

~ у́гол·ь (*m*) coking coal

~ числ·о́ (*n*) coking value

коксо·вы·та́лк·ивател·ь, -я, (*m*) coke pusher

коксо·образ·ова́ни·е, -я, (*n*) (chem) coke formation

коксо·хим·и́ческ·ий, -ая, -ое, (*a*) coking by-product

~ про·мы́шл·енност·ь (*f*) coking by-product industry

кокс·у́емост·ь, -и, (*f*) cokeability, coking power (of coals)

кокцидио́з·, -а, (*m*) (vet) coccidiosis

кокци́н·, -а, (*m*) (phot chem) coccin

кол·, -а́, (*nom pl*) **-ы́,** (*g pl*) **-о́в,** (*m*) or (*nom pl*) **кол·ья,** (*g pl*) **-ев,** (*m*) stake, picket; (hortic) dibble, dibber

~, а́нкер·н·ый anchoring picket

~, нивели́р·н·ый (surv) levelling peg

~, саж·а́льн·ый (hortic) dibber

ко́л·а, -ы, (*f*) (bot) cola, *Cola acuminata*

колами́н·, -а, (*m*) colamine, (chem) monoethanolamine

ко́лб·а, -ы, (*f*) (chem) flask, retort; (elec) envelope, bulb (lamp, valve)

~, ме́р·н·ая graduated/measuring flask

~, от·со́с·н·ая filter flask

колбас·а́, -ы́, (*nom pl*) **-а́с·ы,** (*g pl*) **-а́с·,** (*f*) sausage

колба́с·н·ый, -ая, -ое, (*a*) of **колбас·а́**

~ из·де́л·и·я (*pl*) sausages

~ шприц· (*m*) sausage machine

колбо·ви́д·н·ый, -ая, -ое, (*a*) (bot) phialine

колеб·а́ни·е, -я, (*n*) oscillation; (mech) vibration; fluctuation, variation (of values)

~, акуст·и́ческ·ое acoustic vibration

~, вале́нт·н·ое (phys, chem) stretching vibration, bond-stretching vibration

~, век·ов·о́е (astron) secular variation

~, вы́·нужд·енн·ое forced oscillation, forced vibration

~, год·ов·о́е (astron) annual variation

~, за·тух·а́ющ·ее damped oscillation, damped vibration, decaying vibration

~, из·ги́б·очн·ое transverse vibration, flexural vibration

~ кли́мат·а climate fluctuation

~, кристалл·и́ческ·ое (phys) lattice vibration

колеб·а́ни·е

~, крут·и́льн·ое torsional vibration/oscillation

~ ма́ятник·а pendulum swing/oscillation

~ механ·и́ческ·ое mechanical vibration

~ на·гру́з·к·и (mech) load variation

~, не·за·тух·а́ющ·ее stable/sustained/undamped vibration/oscillation

~, не·пло́ск·ое out-of-plane vibration

~, норма́ль·н·ое free oscillation/vibration

~, парази́т·н·ое (rad) parasitic/spurious oscillation, squeggers

~, параметр·и́ческ·ое (math) parametric variation

~ пит·а́ни·я (*pl*) (elec) power-line fluctuation

~ пла́зм·ы plasma oscillation

~ по с·ре́з·у (cryst) shear oscillation

~, по·пере́ч·н·ое transverse vibration/oscillation

~, по·ступ·а́тельн·ое translation vibration

~ по́чв·ы earth tremor

~, пре·ры́в·ист·ое (rad) squegging, squagging, squitter

~, прецесси́он·е·ое gyroscopic wobbling

~, про·до́ль·н·ое longitudinal vibration/oscillation; (a/c) pitching oscillation

~, раз·ры́в·н·ое (elec) relaxation oscillation

~, релаксацио́н·н·ое (elec) relaxation oscillation

~, свобо́д·н·ое free oscillation/vibration, natural oscillation/vibration

~, сейсм·и́ческ·ое seismic oscillation, seism

~, со́б·ственн·ое free oscillation/vibration, natural oscillation/vibration

~, су́т·очн·ое (astron) diurnal variation

~ у·пру́г·их систе́м· (mech) vibration of elastic bodies

~, фуго́ид·н·ое (autom) hunting, phugoid/hunting oscillation, long-period oscillation

~ фу́нкци·и (math) variation/oscillation of a function

центр· колеб·а́ни·й (*m*) (phys) centre/center of oscillation

част·от·а́ со́б·ственн·ых колеб·а́ни·й (*f*) free/natural frequency

~, эвстат·и́ческ·ое (geol) eustatic change

колеб·а́ни·е
~, электр·и́ческ·ое electric oscillation
колеб·а́тельн·ый, -ая, -ое, (a) (elec) oscillating, oscillatory; (mech) vibrating; (math) fluctuating, varying
~ ко́нтур· (m) (elec) oscillatory circuit
~ пере·хо́д· (m) (phys) vibrational transition
~ по́л·е (n) (elec) oscillating field
колеб·а́ть (pres 3rd sing, pl) коле́бл·ет, коле́бл·ют, (imp) oscillate, vibrate, sway; fluctuate, vary; hesitate, waver; -ся (pass or reflex)
колеб·а́ться в пре·де́л·ах fluctuate/vary between, lie in the range
коле́бл·емост·ь, -и, (f) (math) scatterration, dispersion
коле́бл·ем·ый, -ая, -ое, (pres part pass) of колеб·а́ть
коле́бл·енн·ый, -ая, -ое, (past part pass) of колеб·а́ть
коле́бл·ющ·ий, -ая, -ее, (pres part act) of колеб·а́ть
коле́бл·ющ·ийся, -аяся, -ееся, (pres part act) of колеб·а́ться; variable
колеб·н·у́ть, -у́т, (perf single action) see колеб·а́ть
коле́й·ност·ь, -и, (f) traffic width (of a road carriageway)
-коле́й·н·ый, -ая, -ое, (adj component) -gauge, -gage, -track, -way
коле́й·н·ый, -ая, -ое, (a) wheel-track
~ мост· (m) (mil) wheel-track bridge
колеманит·, -а, (m) (min) colemanite
коленко́р·, -а, (m) bookbinder's calico, bookcloth
~, пере·плёт·н·ый bookbinder's cloth
коле́н·н·о-рыча́ж·н·ый механ·и́зм· (m) toggle
коле́н·н·ый, -ая, -ое, (a) of коле́н·о
~ суста́в· (m) (anat) knee joint
~ ча́ш·ечк·а (f) (anat) patella
коле́н·о, -а, (nom g pl) of knees; -и, -ей, (nom g pl) of joints, links; коле́н·ья, коле́н·ьев, (n) (anat) knee, genu; bend; knee, elbow (pipes); (mech) crank, crank web and pin; (bot) node, knob, joint; (gen) branch, line, generation; (geol) limb
~ волно·про во́д·а (rad) waveguide bend
~, над·семено·до́ль·н·ое (bot) epicotyl
по коле́н·и в вод·е́ up to the knees in water, knee-deep in water
~, пог·семено·до́ль·н·ое (bot) hypocotyl

коле́н·о
~ рек·и́ river bend
~ с·бро́с·а (geol) fault node
ста·ть на коле́н·и kneel, kneel down
сто·я́ть на коле́н·ях kneel, be on ones knees
колено·обра́з·н·ый, -ая, -ое, (a) geniculate
коле́н·чат·ый, -ая, -ое, (a) (med) crank, cranked; bent, elbow; (biol) geniculate
~ вал· (m) crankshaft
~ рыча́г· (m) crank
~ сквоз·ник· (m) double-entry passageway
~ телеско́п· (m) elbow telescope
коле́н·ья (nom pl) of коле́н·о
~ труб·ы́ knees/elbows/bends of pipe
колео́·птер·, -а, (m) (a/c) coleopter; (ent) coleopteron, (pl) Coleoptera
колео́·птил·е, -я, (n) (bot) coleoptile
колео·ри́з·а, -ы, (f) (bot) coleorhiza
ко́лер·, -а, (m) (build) colour, color, paint (ready for use); colouring, coloring, caramel, sugar dye (winemaking); (vet) staggers
коле́с·ик·о, -а, (n) (dim) of колес·о́; caster, castor (on furniture)
коле́с·ник·, -а, (m) wheelwright; wheel unit
~, конта́кт·н·ый trolley (gyro compass)
коле́с·н·о- (component) see колес·о́
коле́с·н·ый, -ая, -ое, (a) wheel, wheeled
~ па́р·а (f) wheel pair/set
~ маз·ь (f) (rail) wheel grease
~ су́д·н·о (n) paddle-wheel boat
~ фо́рмул·а (f) (rail) wheel arrangement
~ центр· (m) boss, wheel centre
колес·о́, -а, (nom pl) -ёс·а, (g pl) -ёс, (n) wheel
~, а́нкер·н·ое (horol) escape wheel
~, бараба́н·н·ое (horol) main wheel
~, бег·у́нков·ое (rail) front carrying wheel
~, без·банда́ж·н·ое (rail) solid wheel
~, бло́ч·н·ое (horol) great wheel
~, борозд·н·о́е furrow wheel (of plough)
~ бо́·я (horol) cam wheel
~, вед·о́м·ое driven wheel
~, вед·у́щ·ее driving wheel; driving sprocket (of tracked vehicle)
~, ве́ксель·н·ое (horol) minute wheel

колес·о́

~, винт·ов·о́е helical gear wheel

~, гимнаст·и́ческ·ое gym wheel

~, греб·н·о́е (shipb) paddle wheel

~, зеркáль·н·ое (TV) mirror wheel/drum

~, зуб·чáт·ое (*see under* **зуб·чáт·ый**) gear wheel, toothed wheel

~, косты́ль·н·ое (a/c) tailwheel

~, кры́ль·чат·ое (mar eng) Voith-Schneider propeller, cycloidal propeller (U.S.)

~, ло́паст·н·ое impeller vane wheel (turbines etc.)

~, лопáт·очн·ое impeller, vane wheel

~, мах·ов·о́е flywheel, handwheel; governor; balance wheel (sewing machine)

~, мину́т·н·ое (horol) minute wheel

~, на·ли́в·н·о́е (hydr) overshot wheel

~, на·прáвл·я́ющ·ее diffuser (turbines etc.); guiding wheel, idler wheel

~, на·тя́ж·н·о́е jockey wheel

~, нос·ов·о́е nose wheel

~, от·ки́д·н·о́е с·по́р·н·ое jockey wheel (of semitrailer)

~, парази́т·н·ое idler wheel

~, под·дéрж·ивающ·ее (rail) carrying wheel

~, пол·ев·о́е (agr) land wheel (of plough)

~, рабóч·ее impeller (centrif pump)

~, рул·ев·о́е steering wheel

~, само·центр·и́рующ·ееся swivelling/castor/caster wheel

~, слéд·ящ·ее phantom (gyro compass)

~, с·пáр·енн·ые (*pl*) coupled wheels

~, с·цеп·н·ы́е (*pl*) coupled wheels

~, с·чёт·н·ое (horol) locking plate

~, трещ·о́точн·ое (horol) escape wheel

~, фон·и́ческ·ое (telecom) phonic wheel

~, хвост·ов·о́е (a/c) tail wheel

~, ход·ов·о́е (mech) travelling wheel; (M/T) road wheel; (horol) escapement/escape wheel, balance/verge wheel (of wall clocks)

~, храп·ов·о́е ratchet, ratchet wheel

~, цéв·очн·ое lantern pinion

~, цéльно·кáт·ан·ое (rail) solid wheel

~, цеп·н·о́е sprocket, sprocket wheel

~, черв·я́чн·ое worm wheel

~, штифт·ов·о́е (horol) cam wheel

колесо·обрáз·н·ый, -ая, -ое, (*a*) wheel-shaped, trochoid; (biol) trochate, trochal

колесо·от·бо́·й, -я, (*m*) curbing, wheel guides (on a bridge etc.)

колесо·про·кáт·н·ый, стан· (*m*) (met) wheel-rolling mill, wheel mill

колесо·токáр·н·ый стан·о́к· (*m*) (met) wheel lathe

ко́л·ет (*pres 3rd sing*) of **кол·о́ть**

кол·éчк·о, -а, (*g pl*) **-чек·,** (*n*) (*dim*) of **кольц·о́**; ringlet

кол·е·я́, -й, (*f*) rut (in road etc.); wheel track, gauge, gage, track

~, у́з·к·ая (rail) narrow gauge/gage

~, шир·о́к·ая (rail) broad gauge/gage

ко́ли (*indecl*) (*adv*) when; (*conj*) if, since, as; (biol) coli

коли·баци́лл·а, -ы, (*f*) (bact) coli-bacillus

коли·и́ндекс·, -а, (*m*) colibacillus density

ко́лик·и (*nom pl*), (*g pl*) **-ик·,** (*f*) (med) colic

ко·лин·éйн·ый, -ая, -ое, (*a*) colinear, collinear

колистати́н·, -а, (*m*) (pharm) coli-statin

кол·и́т·, -а, (*m*) (med) colitis

коли·тéст·, -а, (*m*) (bact) colibacillus density

коли·ти́тр·, -а, (*m*) (bact) colibacillus index

коли·фáг·, -а, (*m*) (bact) coliphage

коли́честв·енн·ый, -ая, -ое, (*a*) quantitative

~ анáлиз· (*m*) (chem) quantitative analysis; (met) assay

~ регул·и́ровани·е (*n*) quantitative control; nozzle-control governing (steam turbine)

~ числ·о́ (*n*) (math) cardinal number

коли́честв·о, -а, (*n*) quantity, amount

~, вес·о́м·ое ponderable amount

~ движ·éни·я (mech) momentum

 закóн· коли́честв·а движ·éни·я (*m*) law of linear momentum

~ движ·éни·я, пере·нос·и́м·ое (phys) drift momentum

~, индикáтор·н·ое (nucl) tracer amount

~ информ·áци·и information content

~, на·ли́ч·н·ое inventory, amount to hand, stock on hand

~ облак·о́в (meteor) cloud amount

~ обращ·éни·я, до·пуст·и́м·ое (autom) read-around ratio (of storage tube)

~ про·пуск·áем·ого материáл·а throughput

коли́честв·о

~, **у́льтра·ма́л·ое**   trace quantity/amount

ко́л·к·а, -и, (*g pl*) **-л·ок·,** (*f*) chopping, chipping

ко́л·к·ий, -ая, -ое, (*a*) splitting, cleaving; easily split

колла·ге́н·, -а, (*m*) (chem, zool) collagen, collogen, ossein

колла́пс·, -а, (*m*) collapse

колларго́л·, -а, (*m*) (pharm) argentum colloidale

коллатера́ль·н·ый, -ая, -ое, (*a*) collateral

коллактиви́т·, -а, (*m*) collactivit (used in sugar making)

колле́г·а, -и, (*f*) colleague, associate

коллегиа́ль·н·ый, -ая, -ое, (*a*) collegiate, joint

колле́г·и·я, -и, (*f*) collegium, college, board

коллекти́в·, -а, (*m*) collective, association, organization, members (of an organization); (phot) field lens

коллективиз·и́рованн·ый, -ая, -ое, (*past part pass*) of **коллективиз·и́р·овать;** collectivized

~ **электро́н·** (*m*) (phys) covalence electron, shared electron

коллективиз·и́р·овать, -уют, (*imp and perf*) collectivize

коллекти́в·н·ый, -ая, -ое, (*a*) collective, joint, cooperative

~ **эффе́кт·** (*m*) (phys) cooperative effect

колле́ктор·, -а, (*m*) collector, receiver (air), manifold; (elec) commutator; drum, header (boilers); air receiver (compr air); nozzle box (jet eng); sewer

~, **ве́рх·н·ий** (M/T) top tank

~, **вод·ян·о́й** water drum (boiler)

~, **в·ход·н·о́й** (st eng) inlet header

~, **вы·пуск·н·о́й** (ICE) exhaust manifold/collector

~, **вы·хлоп·н·о́й** (ICE) exhaust manifold

~, **вы·ход·н·о́й** (st eng) outlet header

~ **за·жиг·а́ни·я** (э/о) ignition wiring manifold

~, **кольц·ев·о́й вы·хлоп·н·о́й** (ICE) exhaust ring

~, **ли́в·нев·ый** surface water sewer, storm sewer (U.S.)

~, **ли́т·ников·ый** (met cast) dirt trap

~ **на·сы́щ·енн·ого па́р·а** boiler steam header, saturated steam header

колле́ктор

~ **нефт·и** (geol) petroleum reservoir; (oil) gathering main

~, **ни́ж·н·ий** bottom tank

~, **о·богат·и́тельн·ый** flotation collector, collecting agent (ore dressing)

~, **пар·ов·о́й** steam drum (boiler)

~ **паро·пере·грев·а́тел·я** (st eng) superheater header

~ **подо·гре́в·а во́з·дух·а** (ICE) heating muff

~, **с·бо́р·н·ый** (air) collector tank

~, **с·у́ж·ивающ·ийся** (dynam) contracting effuser

~, **холо́д·н·ый** cold collector (refrigeration)

~ **экра́н·а** water-wall header (boiler)

~ **электро́н·ов** (nucl) electron collector

колле́ктор·н·ый, -ая, -ое, (*a*) of **колле́ктор·**

~ **вту́л·к·а** (*f*) commutator hub/sleeve/shell/bush

~ **генера́тор·** (*m*) (elec) D.C. generator

~ **дви́г·ател·ь** (*m*) (elec) commutator motor

~ **звезд·а́** (*f*) (elec) commutator spider

~ **пере·хо́д·** (*m*) collector junction (semiconductors)

~ **пульс·а́ци·я** (elec) commutator ripple

коллекцион·е́р·, -а, (*m*) collector

коллекцион·и́р·овать, -уют, (*imp*) collect, form a collection

колле́кци·я, -и, (*f*) collection, set

коллен·и·я, -и, (*f*) (pal) collenia

колл·енхи́м·а, -ы, (*f*) (bot) collenchyma

коллима́тор·, -а, (*m*) (opt) collimator

~ **пуч·к·а́** beam collimator

коллимацио́н·н·ый, -ая, -ое, (*a*) (opt) collimation, collimator

~ **пло́ск·ост·ь** (*f*) (surv) plane of collimation

коллим·а́ци·я, -и, (*f*) collimation

коллим·и́р·овать, -уют, (*imp and perf*) collimate

коллинеа́р·ност·ь, -и, (*f*) collinearity

коллинеа́р·н·ый, -ая, -ое, (*a*) collinear

коллинеацио́н·н·ый, -ая, -ое, (*a*) collineatory

коллинеа́ци·я, -и, (*f*) collineation

колл·о́ди·й, -я, (*m*) (chem) collodion, collodium

колл·одио́н·н·ый, -ая, -ое, (*a*) of **колл·о́ди·й**

колл·о́ид·, -а, (*m*) (chem) colloid

~, **жи́дк·ий** colloidal solution

колл·о́ид·н·ый, -ая, -ое, (*a*) colloid, colloidal

колло́квиум·, -а, (*m*) oral examination/test

колло·ксили́н·, -а, (*m*) (chem) colloxylin

колло·мо́рф·н·ый, -ая, -ое, (*a*) (geol) colloform

колло·ти́п·и·я, -и, (*f*) (print) collotype, photogelatin printing

колло·фа́н·, -а, (*m*) (geol) collophane

кол·лю́в·и·й, -я, (*m*) (geol) colluvium, colluvial deposits

ко́лоб·, -а, (*nom pl*) -а́, (*g pl*) -о́в, (*m*) (agr) oil cake; (ocean) consolidated ice

коло·воро́т·, -а, (*m*) brace, ratchet brace; whirlpool

~, груд·н·о́й breast brace

коло·вра́т·к·а, -и, (*g pl*) -ток·, (*f*) (zool) rotifer, (*pl*) rotifers, *Rotifera*

коло·вра́т·н·ый, -ая, -ое, (*a*) rotary, rotative, gyratory

~ движ·е́ни·е (*n*) gyration

коло·вращ·е́ни·е, -я, (*n*) circular motion, gyration

кол·о́бш·ий, -ая, -ее, (*past part act*) of кол·о́ть

ко·логари́фм·, -а, (*m*) (math) cologarithm

коло́д·а, -и, (*f*) log; hollowed log, dugout; bole (of tree); pack of cards

коло́дез·н·ый, -ая, -ое, (*a*) of коло́д·ец·

~ вод·а́ (*f*) well water

~ су́дн·о (*n*) well-deck vessel

коло́д·ец·, -д·ц·а, (*i*) -д·ц·ем, (*m*) well, pit; sump; (naut) well deck; (build) shaft foundation

~, бур·ов·о́й driven well

~, водо·бо́й·н·ый (hydroelec) stilling pool

~, водо·с·бо́р·н·ый drainage sump

~, дози·метр·и́ческ·ий (nucl) monitoring well

~, за·крепл·ённ·ый cased well

~, ка́бель·н·ый (elec) draw-in box/pit

~, ка́рст·ов·ый (geol) sinkhole

~, ла́мп·ов·ый (build) lamphole

~, на·грев·а́тельн·ый (met) soaking pit, holding furnace

~, от·сто́й·н·ый drain, sink

~, по·глот·и́тельн·ый (build) soakaway

~, рас·шир·и́тельн·ый expansion trunk (of tanker)

коло́д·ец

~, регенерати́в·н·ый (met) regenerative pit

~, рекуперати́в·н·ый (met) recuperative pit

~, с·бо́р·н·ый (naut) bilge well

~, смотр·ов·о́й manhole, drop manhole; (M/T etc.) inspection pit

~, тру́б·чат·ый driven/drilled well

~, ша́хт·н·ый dug well

~, змшерск·ий Imhoff tank (sewage treatment)

коло́д·к·а, -и, (*g pl*) -д·ок·, (*f*) block, chock; shoe tree, last

~, за·жи́м·н·ая grip cheek (of hoisting block)

~, при·це́ль·н·ая backsight bed (of rifle)

~, рас·пре·дел·и́тельн·ая distributor segment (of magnets)

~, рези́н·ов·ая rubber pedal (bicycle)

~, сто́пор·н·ая stop block, (a/c) chock

~, тормоз·н·а́я brake block; (a/c) chock

~, у́голь·н·ая (elec) carbon block

~, у·по́р·н·ая thrust block, backing block, stop block

~, штéпсель·н·ая (elec) plug, socket

коло́д·очк·а, -и, (*g pl*) -чек·, (*f*) (*dim*) of коло́д·к·а

коло́д·очн·ый, -ая, -ое, (*a*) of коло́д·к·а

~ тормоз· (*m*) brake block

коло́д·цев·ый, -ая, -ое, (*a*) of коло́д·ец·

~ печ·ь (*f*) (met) soaking pit, holding furnace

колодце·коп·а́тел·ь, -я, (*m*) well borer

кол·о́к·, -л·к·а́, (*m*) peg, pin

колока́зи·я, -и, (*f*) (bot) dasheen, taro, *Colocasia esculenta*

колокви́нт·, -а, (*m*) (bot) bitter apple, coloquinth, colocinth, *Citrullus colocynthis*

ко́локол·, -а, (*nom pl*) -а́, (*g pl*) -о́в, (*m*) bell

~, водо·ла́з·н·ый diving bell

~, лов·и́льн·ый (oil) slip socket; (civ eng) beche

~, с·пас·а́тельн·ый escape bell (submarine)

~, суд·ов·о́й (naut) ship's bell

колоколо·обра́з·н·ый, -ая, -ое, (*a*) *see* колоколько·обра́з·н·ий

колокольно·обра́з·н·ый, -ая, -ое, (*a*) bell-shaped, belled-out, (bot) campaniform, campanulate

колоко́ль·н·ый, -ая, -ое, (*a*) of **ко́локол·**

колоко́ль·н·я, -и, (*g pl*) **-л·ен·,** (*f*) bell tower

колоко́ль·чик·, -а, (*m*) (*dim*) of **ко́локол·**; (bot) campanula; (*pl*) xylophone, glockenspiel

колоко́ль·чиков·ые, -ых, (*pl decl as adj*) (bot) Campanulales, *Campanulaceae*

ко́ло·маз·ь, -и, (*f*) wheel grease

коломя́нк·о, -а, (*n*) (text) calamanco

колониа́ль·н·ый, -ая, -ое, (*a*) colonial

коло́ни·я, -и, (*f*) colony

~, куст·и́ст·ая (biol) branching colony

коло́н·к·а, -и, (*g pl*) **-н·ок·,** (*f*) (*dim*) of **коло́нн·а;** column; (min) core

~, бензи́н·ов·ая petrol/gasoline pump (for refuelling M/T)

~, водо·за·бо́р·н·ая public water fountain

~ волос·к·а́ (horol) stud

~, га́з·ов·ая instantaneous gas water heater, geyser

~ груз·ов·о́й стрел·ы́ samson post

~ и́мпульс·а (horol) impulse pin

~ ка́п·ел·ь band of drops

~, масло·раз·да́т·очн·ая oil dispenser (for supplying M/T)

~ нес·у́щ·его ви́нт·а rotor mast (of helicopter)

~, о·гран·ичи́тельн·ая banking piece (of a balance bridge)

~, по·глот·и́тельн·ая (chem) absorption bottle

~, ру́д·н·ая (geol) ore chute

~, рул·ев·а́я (M/T) steering column

~, с·лив·н·а́я (hydr) overflow pipe

~ у·правл·е́ни·я (a/c) control column

~, штурва́ль·н·ая (a/c) control column

коло́н·ков·ый, -ая, -ое, (*a*) of **коло́н·к·а;** column, column-mounted, core

колон·ко́в·ый, -ая, -ое, (*a*) of **колон·о́к·**

~ бур·е́ни·е (*n*) (geol) core drilling

~ за·ря́д· (*m*) (expl) column charge

~ перфора́тор· (*m*) (min) drifter, drifter drill

колонко·обра́з·н·ый, -ая, -ое, (*a*) columner

коло́нн·а, -ы, (*f*) column; (chem) tower, column; string (of drill pipe)

~, барбота́ж·н·ая (chem) bubble-cap tower, plate tower

~, вы·нос·н·а́я от·па́р·н·ая (oil) side stream stripper, side-stripping column

коло́нн·а

~, дистилляцио́н·н·ая (chem) distillation column

~ звен·ьев (air) flights line astern

~, ион·и́ческ·ая (arch) column of the Ionic order

~, ионо·об·ме́н·н·ая ion-exchange column

~, колпач·ко́в·ая (chem) bubble-cap column/tower

~, кори́нф·ск·ая (arch) column of the Corinthian order

~, на·са́д·очн·ая (chem) packed tower/column

~ пар· (air) section line astern

~, по·глот·и́тельн·ая (chem) scrubber, absorption tower/column

~, про·меж·у́точн·ая (oil) water string

~, раз·дел·и́тельн·ая (chem) fractionating tower

~, рас·пы́л·ивающ·ая (chem) spray tower

~, эксплуатацио́н·н·ая (oil) oil string

~, эпюрацио́н·н·ая (chem) fractionating column/tower

колонна́д·а, -ы, (*f*) (arch) colonnade

коло́нн·ый, -ая, -ое, (*a*) of **коло́нн·а**

ко́л·онн·ый, -ая, -ое, (*past part pass*) of **кол·о́ть;** *see* **ко́л·от·ый**

колон·о́к·, -н·к·а́, (*m*) (zool) kolinsky, Siberian ferret, *Mustela sibirica*

коло́н·ок· (*g pl*) of **коло́н·к·а**

колон·ти́тул·, -а, (*m*) (print) running heading

колон·ци́фр·а, -ы, (*m*) (print) page number, folio

~, чёт·н·ая (print) even number

колон·што́йн·, -а, (*m*) (horol) roller jewel

колорадои́т·, -а, (*m*) (min) coloradoite

колора́д·ск·ий жук· (*m*) (zool) Colorado beetle, *Leptinotarsa decemlineata*

колори́·метр·, -а, (*m*) colorimeter

~, визуа́ль·н·ый visual colorimeter

~, трёх·цве́т·н·ый tricolorimeter

~, хим·и́ческ·ий chemical colorimeter

колори·метр·и́ческ·ий, -ая, -ое, (*a*) colorimetric

колори·ме́тр·и·я, -и, (*f*) colorimetry, colorimetric analysis

колор·и́ст·, -а, (*m*) (text etc.) colourist, colorist

колори́т·, -а, (*m*) colouring, coloring, selection of colours/colors

колорсти́л·, -а, (*m*) (phot) Colorstil

ко́лос·, -а, (*nom pl*) **колос·ья,** (*g pl*)
колос·ьев (*m*) (bot) ear, spike
(astron) Spica
~, по·ни́к·л·ый nodding ear
~, ры́хл·ый lax ear
колос·и́ст·ый, -ая, -ое, (*a*) heavy
eared (corn, wheat etc.)
колос·и́ться, -я́тся, (*imp*) (bot) form
ear(s)
колосни́к·, -а́, (*m*) firebar, (*pl*) grating,
firebars; (theat) flies, fly gallery;
trap (of solid fuel rocket)
колосни́к·ов·ый, -ая, -ое, (*a*) of
колосни́к·
~ гро́хот· (*m*) (min) grizzly
~ решёт·к·а (*f*) grate, firegrate
(boiler)
колос·ня́к·, -а, (*m*) (bot) wild-rye,
Elymus
колосо·ви́д·н·ый, -ая, -ое, (*a*) spiked,
(bot) spicate; herringbone
колос·ов·о́й, -а́я, -о́е, (*a*) eared;
(*pl as noun*) cereals
колос·о́к·, -с·к·а́, (*m*) (*dim*) of **ко́лос·;**
spikelet
колосса́ль·н·ый, -ая, -ое, (*a*) colossal
колос·ья (*nom pl*) of **ко́лос·**
колот·и́л·о, -а, (*n*) beater
колот·и́ть, -я́т, (*imp*) beat/knock on;
break; scutch (flax)
колот·у́шк·а, -и, (*g pl*) **-шек·,** (*f*)
mallet, mall, maul, beetle; (naut)
heaving-line weight
ко́л·от·ый, -ая, -ое, (*past part pass*)
of **кол·о́ть**
~ лесо·материа́л· (*m*) chopped wood
кол·о́ть, -ют, (*imp*) stab, prick, pierce
(with spike etc.); slaughter; split,
chop, cleave; crack open
колофо́н·, -а, (*m*) (print) colophon
колофон·и́т·, -а, (*m*) (min) colophonite
колоци́нт·, -а, (*m*) *see* **колоквинт·**
колоцинт·и́н·, -а, (*m*) (chem) colo-
cynthin
коло́ч·ен·ый, -ая, -ое, (*a*) broken
коло́ш·а, -и, (*i*) **-ей,** (*f*) (met) charge
~, рабо́ч·ая charge
~, холост·а́я bed charge (cupola)
колош·е́ни·е, -я, (*n*) (bot) ear formation, heading
колош·ни́к·, -а́, (*m*) throat, top (of
blast furnace)
колош·ни́ков·ый, -ая, -ое, (*a*) of
колош·ни́к·
~ газ· (*m*) top gases, blast-furnace gas
~ окн·о́ (*n*) charging door
~ площа́д·к·а (*f*) charging platform

колпа́к·, -а, (*m*) hood, cover, cap,
cowl; protection cap, safety cap; shade,
globe (light fitting); horn (sea mine)
~ анте́нн·ы radome, radar dome
~ астро·лю́к·а (a/c) astrodome
~, воз·ду́ш·н·ый air chamber (of
pump)
~, в·са́с·ывающ·ий suction air chamber (of pump)
~, грибо·ви́д·н·ый mushroom top
(ventilator head)
~, ко́ль·чат·ый *Rozites caperata*
(edible fungus)
~, на·гнет·а́тельн·ый discharge air
chamber (of pump)
~, раз·ры́в·н·о́й tear-off cap
~ ступ·и́ц·ы (М/Т) hub cap
колпач·ко́в·ый, -ая, -ое, (*a*) of
колпач·о́к·
~ га́·йк·а (*f*) dome nut
колпач·о́к·, -ч·к·а́, (*m*) (*dim*) of
колпа́к·; (gas) mantle; (chem) bubble cap
~, воз·ду́ш·н·ый air shroud (jet
engine burner)
колумб·и·й, -я, (*m*) (chem) columbium,
Cb, niobium, Nb
колумб·и́т·, -а, (*m*) (min) columbite
колуме́лл·а, -ы, (*f*) (biol) columella
кол·у́н·, -а́, (*m*) wood chopper/cleaver,
wedge axe
колхамеи́н·, -а, (*m*) (chem) colcha-
meine
колхами́н·, -а, (*m*) colchamin
колхик·у́м·, -а, (*m*) (bot) colchicum,
Colchicum
колхиц·и́н·, -а, (*m*) (chem) colchicine
колхо́з·, -а, (*m*) kolkhoz, collective
farm
колхо́з·ник·, -а, (*m*) collective-farm
worker
колчеда́н·, -а, (*m*) (min) pyrites
~, волос·и́ст·ый capillary pyrite, millerite
~, желе́з·н·ый iron pyrite, pyrite
~, луч·и́ст·ый (min) white iron pyrite, marcasite
~, ме́д·н·ый copper pyrites, chalcopyrite
~, мышьяк·о́вист·ый lollingite, arsenide of iron
~, мышяк·о́в·ый (min) arsenopyrite,
mispickel
~, се́р·н·ый iron pyrites
колыб·е́л·ь, -и, (*f*) cradle
колыб·е́льк·а, -и, (*f*) (*dim*) of **ко-
лыб·е́л·ь**

колых·а́·ть, -ют, (*pres 3rd sing, pl*) колы́ш·ет, колы́ш·ут, (*imp*) wobble, rock, sway

колых·н·у́ть, -у́т, (*perf, single action*) see колых·а́·ть

ко́л·ышек·, -шк·а, (*m*) (*dim*) of кол·; peg; (naut) thole pin; (hortic) dibber

колы́ш·к·а, -и, (*f*) (naut) sheepshank

колы́ш·ущ·ий, -ая, -ее, (*pres part act*) of колых·а́·ть

коль (*indecl*) (*conj*) though, if, when; (*adv*) how much/many; ко́л·ь, -я, (*m*) (naut) line coiled by hand

~ -ско́р·о as soon as

кольдкре́м·, -а, (*m*) (pharm) cold cream

ко́льз·ев·ое ма́сл·о (*n*) colza oil

колькотар·, -а, (*m*) colcotar, Venetian red

ко́льм·, -а, (*m*) (geol) kolm

кольмата́ж·, -а, (*m*) silt deposition, colmatage

коль·н·у́ть, -у́т, (*perf*) give a stab/prick

кольпо·ско́п·, -а, (*m*) (med) colposcope

кольра́би (*f indecl*) (bot) kohlrabi, *Brassica oleracea caulorapa*

Ко́льрауш·а, зако́н· (*m*) (chem) Kohlrausch's law

кольтер·, -а, (*m*) (agr) coulter

кольц·ева́ни·е, -я, (*n*) ringing, putting/forming a ring round

~ бензо·по·да́·ч·и (ICE) fuel cross-feed

кольц·ева́ть, -у́ют, (*imp*) ring, band, girdle

кольце·ви́д·н·ый, -ая, -ое, (*a*) ring-shaped, annular, annulate

кольц·ев·о́й, -а́я, -о́е, (*a*) ring, ring-shaped, annular, circular

~ воздухо·про·во́д· (*m*) bustle pipe (blast furnace)

~ десят·и́чн·ый с·чёт·чик· (*m*) (instr) ring-of-ten counter

~ пряд·и́льн·ая маши́н·а (*f*) (text) ring-spinning frame

~ структу́р·а (*f*) (chem) ring/cyclic structure, ring

кольцекрут·и́льн·ая маши́н·а (*f*) (text) ring-doubling frame

кольце·обра́з·н·ый, -ая, -ое, (*a*) ring-shaped, annular, annulate

кольц·о́, -а́, (*i*) -о́м, (*n*) (*nom pl*) ко́льц·а, (*g pl*) коле́ц·, (*d pl*) коль·ц·ам ring, hoop

кольц·о́

~, ассоциати́в·н·ое (math) associative ring

~, бензо́ль·н·ое (chem) benzene ring

~, борт·ов·о́е bead ring (of tyre)

~, газо·у·плот·ни́тельн·ое (mech) gas ring, gasket

~, год·и́чн·ое annual ring (timber)

~, дистанцио́н·н·ое distance piece; time ring (of fuse)

~ жёстк·ост·и stiffening/reinforcing ring

~, за·мкн·ут·ое (chem) closed ring за·мык·а́ни·е кольц·а́ (*n*) (chem) cycling, ring closure

~, за·по́р·н·ое locking ring

~, калибрацио́н·н·ое range ring (radar)

~, коммутати́в·н·ое (math) commutative ring

~, ко́нус·н·ое V-ring

~, кулач·ко́в·ое cam ring

~, на·би́в·очн·ое gasket, packing ring

~, на·жи́м·н·ое clamp, clamping ring

~ на·ча́л·а от·чёт·а (rad) zero ring

~, о·по́р·н·ое carrier ring; mantle (of blast furnace)

~, плеохро·и́ческ·ое (min) pleochroic halo

~, под·клад·н·о́е bottom ring (of piston ring)

~, про·кла́д·очн·ое gasket ring

~, пруж·и́нящ·ее сто́пор·н·ое circlip (of wrist pin)

~, раз·вод·н·о́е split ring

~, раз·жи́м·н·ое expanding ring раз·ры́·в· кольц·а́ (*m*) (chem) ring breakage

~, рас·по́р·н·ое distance collar/ring рас·кры́т·и·е кольц·а́ (*n*) (chem) ring opening

~, регул·иро́вочн·ое adjusting ring с·вяз·ь в кольц·е́ (*f*) (chem) cyclic bond

~, сепара́тор·н·ое roller ring (of bearings)

~, со·бир·а́тельн·ое (elec) slip ring

~, сопл·ов·о́е (rocket) thrust ring

~, с·тяж·н·о́е shrink ring, strengthening ring

~, у́голь·н·ое carbon ring

~, у·плот·ни́тельн·ое gasket, packing ring; compression/piston ring

~, у·по́р·н·ое thrust collar

~, у·стано́в·очн·ое adjusting ring, distance ring, check ring; collar (shaft)

~, фаз·и́рующ·ее (elec) phasing ring

кольчат·ый, -ая, -ое, (*a*) ring, ringed, annulate; (biol) hispid; (zool) annelid; (*pl*) (zool) *Annelida*

кольцец·ы (*nom pl*), (*g pl*) **-ов,** (*m*) (zool) annelids, *Annelida*

коль·я (*nom pl*) of **кол·**

колюр·, -а, (*m*) (astron) colure

колюри·я, -и, (*f*) (bot) clove-root plant, *Coluria*

кол·ют (*pres 3rd sing*) of **кол·оть**

колюче·голов·ые, -ых, (*pl decl as adj*) (zool) *Acanthocephala*

колюче·лист·н·ый, -ая, -ое, (*a*) prickly-leaved

колюче·пёр·ые, -ых, (*pl decl as adj*) (fish) *Acanthopterygii*

кол·юч·ий, -ая, -ее, (*a*) spiked, barbed, prickly, thorny; (bot, zool) acanaceous

~ акул·а (*f*) (fish) catran, *Acanthias acanthias Squalida*)

~ про·волок·а (*f*) barbed wire

кол·ючк·а, -и, (*g pl*) **-чек·,** (*f*) prickle, needle, spine, thorn, acantha; barb, burr

кол·юшк·а, -и, (*g pl*) **-шек·,** (*f*) (fish) stickle-back, (*pl*) *Gasterosteidae*

кол·ющ·ий, -ая, -ее, (*pres part act*) of **кол·оть**

кол·я (*pres gerund*) of **кол·оть**

коляск·а, -и, (*g pl*) **-сок·,** (*f*) carriage; pram, perambulator; (M/T) sidecar

ком·, -а, (*nom pl*) **ком·ья,** (*g pl*) **ком·ьев,** (*m*) clot, clod, lump

ком (*prepositional*) of **кто**; (*abbr*) of **кило·ом·** kohm

ком- (*component*) (*abbr*) of **коммунист·ическ·ий** communist; (*abbr*) of **команд·н·ый**; (*abbr*) of **команд·ир·**

ком·а, -ы, (*f*) (opt) coma, oblique spherical abberation; (med) coma

ко·магмат·ическ·ий, -ая, -ое, (*a*) (geol) comagmatic

~ област·ь (*f*) (geol) comagmatic assemblage

команд·а, -ы, (*f*) command, order; command (in charge); crew, team, squad, party, detachment, ship's company; (autom) instruction, order, command

~, авар·ийн·ая (air) crash crew, fire tender crew

~ за·пуск·а (rocket) launching crew

~ из·мен·ени·я (autom) address-modifier instruction

~, на·зём·н·ая (air) ground crew

команд·а
~, по·жар·н·ая fire brigade

~ с·равн·ени·я (autom) comparison instruction

~, старт·ов·ая (rocket) launching crew

~, трёх·адрес·н·ая (autom) three-address instruction

~ у·слов·н·ого пере·ход·а (autom) conditional instruction

команд·ир·, -а, (*m*) (mil) commanding officer, commander, officer in charge, officer commanding; captain (of warship); leader, person in charge, i/c

~, вахт·енн·ый (naut) officer of the watch

~ звен·а (air) flight commander

команд·ирова́ни·е, -я, (*n*) (*v n*) see **команд·ир·ова́ть**

команд·иро́ванн·ый, -ая, -ое, (*past part pass*) of **команд·ир·ова́ть**

команд·ир·ова́ть, -у́ют, (*imp and perf*) send (a person on a mission, business trip, detached duty); **-ся** (*pass*); be/go on detached duty, a mission, a business trip etc.

команд·иро́вк·а, -и, (*g pl*) **-вок·,** (*f*) business trip, mission, detached duty

командиро́вочн·ый, -ая, -ое, (*a*) of **команд·иро́вк·а**; (*pl as noun*) travel allowance, detached duty allowance, expenses of business trip

команд·ир·ск·ий, -ая, -ое, (*a*) of **команд·ир·**

кома́нд·н·о-даль·номе́р·н·ый пост· (*m*) (gunn) director control tower

кома́нд·н·о-диспе́тчер·ск·ий пункт· (*m*) (air) air traffic control point/tower, control tower

кома́нд·н·ый, -ая, -ое, (*a*) of **кома́нд·а**; control; (instr) main, master

~ агрега́т· (*m*) (a/c) self-contained automatic control system

~ вы́ш·к·а (*f*) (air) control tower

~ при·бо́р· (*m*) (autom) control instrument

~ пункт· (*m*) (mil) control position, command point, advanced HQ

~ пункт· на·вед·ени·я (*m*) (mil) GCI station (radar)

~ сигнал·иза́ци·я (*f*) progress control/despatching signalling system (factories); (rail) traffic control signalling system; load despatch signalling system (power stations)

кома́нд·н·ый, -ая, -ое

~ **со·ста́·в·** (*m*) (mil) officers of fighting units, executive officers, executive body

кома́ндо-аппара́т·, -а, (*m*) (elec) control switch

кома́ндо-контроллёр·, -а, (*m*) (elec) master controller

кома́нд·овани·е, -я, (*n*) command, commanding; command, headquarters

~, **верх·о́вн·ое** (mil) supreme command, general headquarters

кома́нд·овать, -уют, (*imp*) (+ *instr*) give orders, command; be in command, command; be higher than, command

командо́р·, -а, (*m*) commodore (foreign navies); commander (of an order); manager, steward (of sporting event)

кома́нд·ующ·ий, -его, (*m decl as adj*) (+ *instr*) general officer commanding (army); senior officer, commander (navy); commandant, commander-in-chief

~ **флот·ом** Commander-in-Chief, C-in-C (of fleet)

коман·ов·ая кисл·от·а́ (*f*) (chem) comanic acid

кома́р·, -а́, (*m*) (ent) mosquito, (*pl*) *Culicidae*

~, **мадяр·и́йн·ый** *Anopheles mosquito*

~, **обык·нове́нн·ый** house mosquito, *Culex pipiens*

кома́р·ик·, -а, (*m*) (ent) midge, *Chironomus*

комба́йн·, -а, (*m*) combine (a machine which performs more than one type of operation); (agr) combine harvester/picker/lifter; (min) cutter-loader, continuous miner; (fish) trawler-drifter

~, **вы́·ем·очн·ый** (min) continuous miner

~, **картофеле·у·бо́р·очн·ый** (agr) potato harvester

~, **кукурузо·у·бо́р·очн·ый** (agr) corn picker-husker

~, **свекло·у·бо́р·очн·ый** (agr) beet lifter

~, **силосо·у·бо́р·очн·ый** (agr) silage harvester

~, **торфо·у·бо́р·очн·ый** (min) peat harvester

~, **узко·за·хва́т·н·ый** (agr) small-cut combine

комба́йн·ер·, -а, (*m*) combine driver/operator

комбайн·и́ровани·е, -я, (*n*) combining, working a combine; (agr) combining, combine harvesting

комби·жи́р·, -а, (*m*) (food) lard compound

комби·ко́рм·, -а, (*nom pl*) **-а́,** (*g pl*) **-о́в,** (*m*) (agr) balanced rations, balanced fodder

комби·ко́рм·ов·ый, -ая, -ое, (*a*) of **комби·ко́рм·**

комбина́т·, -а, (*m*) (econ) combine, industrial group, group

комбина́тор·, -а, (*m*) schemer; governor (water turbine); (telecom) combiner

комбинато́р·ик·а, -и, (*f*) (math) combinatorial analysis

комбинато́р·н·ый, -ая, -ое, (*a*) (math) combinatorial

~ **ана́лиз·** (*m*) (math) combinatorial analysis

комбинацио́н·н·ый, -ая, -ое, (*a*) combination, combinative

~ **из·бир·а́ни·е** (*n*) (elec) multichannel discrimination

~ **при́нцип·** (*m*) (opt) combination principle

~ **тон·** (*m*) (acous) combination tone

комбин·а́ци·я, -и, (*f*) combination; combinations (garment)

комбинезо́н·, -а, (*m*) overalls, one-piece working suit

~, **лёт·н·ый** flying suit

комбин·и́ровани·е, -я, (*n*) (*v n*) see **комбин·и́р·овать**

комбин·и́рованн·ый, -ая, -ое, (*a*) (*past part pass*) of **комбин·и́р·овать;** combined, multiple, compound, composite

~ **жир·ы́** (*pl*) compound fats

~ **ла́мп·а** (*f*) multiple-function lamp

~ **плоско·губ·ц·ы** (*pl*) cutting pliers

~ **регул·я́тор·** (*m*) (autom) rate-action controller, derivative controller

~ **трос·** (*m*) fibre-and-wire rope

комбин·и́р·овать, -уют, (*imp*) combine, compound, co-ordinate, put/group together, arrange; form a combine/group

ко́мел·ь, -мл·я, (*m*) butt end, butt log, butt (timber); (a/c) blade root

~ **-бло́к·, -а,** (*m*) (naut) sister block, fiddle block

~ **ло́паст·и** blade shank

коменда́нт·, -а, (*m*) (mil) commandant; person in charge; (mil rail) railway transport officer; water/sea transport officer

комендо́р·, -а, (*m*) seaman gunner, gunnery rating

комендо́р·ск·ий, -ая, -ое, (*a*) of **комендо́р·**

комéт·а, -ы, (*f*) (astron) comet

~, пред·вы́·числ·енн·ая predicted comet

кометáр·н·ый, -ая, -ое, (*a*) cometary, comet-shaped

~ тумáн·ност·ь (*f*) (astron) comet-shaped nebula

комéт·н·ый, -ая, -ое, (*a*) comet, cometary

~ газ· (*m*) (astron) cometary gas

кóмингс·, -а, (*m*) (shipb) coaming

комиссáр·, -а, (*m*) commissar

~, на·рóд·н·ый People's Commissar (Soviet Government Minister from 1917 to 1946)

комиссариáт·, -а, (*m*) commissariat

~, на·рóд·н·ый People's Commissariat (Soviet Government Ministry from 1917 to 1946)

комиссионéр·, -а, (*m*) broker, agent, commissioner

комиссиóн·н·ый, -ая, -ое, (*a*) of **комúсси·я**

комúсси·я, -и, (*f*) commission, committee, board

~, врач·éбн·о-экспертúз·н·ая medical board

~, при·ём·н·ая selection board; acceptance commission

~, слéд·ствени·ая board/committee of enquiry

комисс·у́р·а, -ы, (*f*) (biol) commissure

комитéнт·, -а, (*m*) (law) principal, client

комитéт·, -а, (*m*) committee

~, вед·у́щ·ий steering committee

~ со·дéй·стви·я board of assistance

кóм·ка·ть, -ют, (*imp*) crumple, bunch up, form into a lump/clot; make a mess/hash of

ком·ковáни·е, -я, (*n*) clotting, caking; nodulizing

ком·ковáт·ый, -ая, -ое, (*a*) clotted, lumpy

кóмл·евост·ь, -и, (*f*) bulging butt (in trees)

кóмл·ев·ый, -ая, -ое, (*a*) of **кóмел·ь**

~ част·ь (*f*) root (of turbine blade)

коммандúт·н·ый, -ая, -ое, (*a*) (com) joint-stock

комменсалúзм·, -а, (*m*) (biol) commensalism

коммент·áри·й, -я, (*m*) commentary

коммент·áтор·, -а, (*m*) commentator

коммент·úр·овать, -уют, (*imp and perf*) make/compile a commentary; comment

коммерсáнт·, -а, (*m*) merchant, businessman

коммéрц·и·я, -и, (*f*) commerce

коммéрч·еск·ий, -ая, -ое, (*a*) commercial, economic, pay

~ акт· (*m*) carrier's statement

~ груз· (*m*) merchandise

~ за·гру́з·к·а (*f*) (air) pay load

~ су́д·н·о (*n*) merchant vessel

коммивояжёр·, -а, (*m*) commercial traveller, travelling salesman

коммýн·а, ы, (*f*) commune

коммунáль·н·ый, -ая, -ое, (*a*) municipal, public, communal

~ автомобúл·ь (*m*) municipal vehicle, highways vehicle

~ хозя́й·ств·о (*n*) public services, public utility services

коммуник·áци·я, -и, (*f*) communications (*see also* **связь**); (mil) lines of communication; (build) underground mains, services (water, gas etc.)

коммунист·úческ·ий, -ая, -ое, (*a*) communist

коммутáнт·, -а, (*m*) (math) commutant

коммутатúв·ност·ь, -и, (*f*) (math) commutativity

коммутатúв·н·ый, -ая, -ое, (*a*) (math) commutative, satisfying the commutative law

коммутáтор·, -а, (*m*) (elec) commutation switch; switchboard (telephones); commutator (telegraph); (math) commutator

~, антéнн·ый (rad) T-R switch

~, комáнд·н·ый (telecom) grouped-exchange switchboard

~, ламéль·н·ый cordless switchboard

~, лин·éйн·о-батар·éйн·ый (telecom) concentrator switchboard

~, телегрáф·н·ый telegraph commutator

~, центрáль·н·ый (telecom) main exchange switchboard

~, шаг·ов·óй step-by-step switch

~, электрóн·н·ый electronic commutation switch

~, электронно·луч·ев·óй beamdeflection selector tube

~, элемéнт·н·ый battery regulating switchboard (telephones)

коммута́тор·н·ый, -ая, -ое, (*a*) of **коммута́тор·**

~ зал· (*m*) (telecom) switchboard room

коммутацио́н·н·ый, -ая, -ое, (*a*) of **коммут·а́ци·я**

~ аппарат·у́р·а (*f*) (elec) switching apparatus, switches, switchgear

~ доск·а́ (*f*) switchboard

коммут·а́ци·я, -и, (*f*) (elec) commutation (D.C. generators); power distribution system; (telecom) switching system; switching

~, автомат·и́ческ·ая machine switching

~, реле́·йн·ая relay switching

коммут·и́ровани·е, -я, (*n*) (telecom) switching

коммюнике́ (*n indecl*) communiqué

ко́мнат·а, -ы, (*f*) room

ко́мнат·н·ый, -ая, -ое, (*a*) room, indoor

~ раст·е́ни·е (*n*) indoor plant

~ температу́р·а (*f*) room temperature

ком·ов·о́й, -а́я, -о́е, (*a*) lump, block, clot

комо́д·, -а, (*m*) chest-of-drawers

ком·о́к·, -м·к·а́, (*m*) (*dim*) of **ком·;** lump, clot

комо́л·, -а, (*m*) (met) Comol

комо́л·ый, -ая, -ое, (*a*) hornless

~ живо́т·н·ое (*n*) (agr) pollard

ком·о́чек·, -чк·а, (*m*) (*dim*) of **ком·о́к·;** pellet

компа́кт·, -а, (*m*) (math) compact set

~, с·вяз·н·о́й (math) compact connected set, continuum

компа́кт·ност·ь, -и, (*f*) compactness

компа́кт·н·ый, -ая, -ое, (*a*) compact

компа́ндор·, -а, (*m*) compandor (telephones)

компа́н·и·я, -и, (*f*) company

компан·ова́ться, -у́ются, (*imp*) be composed (of), consist (of/in)

компаньо́н·, -а, (*m*) companion; (com) partner

компара́тор·, -а, (*m*) (instr) comparator; comparison instrument/meter

~, амплиту́д·н·ый (elec) amplitude comparator

~, фа́з·ов·ый (elec) phase meter; (navig) raydist

компар·и́ровани·е, -я, (*n*) comparison

ко́мпас·, -а, (*m*) compass

~, вис·я́ч·ий hanging compass

~, гео·ло́г·и́ческ·ий geologist's compass

~, гиро·магни́т·н·ый gyromagnetic compass

ко́мпас

~, гиро·скоп·и́ческ·ий gyro-compass

~, гла́в·н·ый standard compass

~, го́р·н·ый (min) surveyor's/surveying compass

~, дистанцио́н·н·ый remote-indicating compass

~, индукцио́н·н·ый flux-gate compass

~, магни́т·н·ый magnetic compass

~, осно́в·н·о́й master compass

~ -по·втор·и́тел·ь, -я, (*m*) gyro-repeater

~, под·вес·н·о́й hanging compass

~, пут·ев·о́й steering compass

~, со́лн·ечн·ый solar compass

~ -у·каз·а́тел·ь, -я, (*m*) gyro-repeater

~, универса́ль·н·ый гео·ло́г·и́ческ·ий (surv) pocket transit

~, этало́н·н·ый (air) landing compass

ко́мпас·н·ый, -ая, -ое, (*a*) of **ко́мпас·**

~ раст·е́ни·е (*n*) (bot) compass plant

компа́унд·, -а, (*m*) compound; compound cooking fats

~ -маши́н·а, -ы, (*f*) (st eng) compound (-expansion) engine; (elec) compound (-wound) machine

~, про·пи́т·очн·ый impregnating compound

компа́унд·н·ый, -ая, -ое, (*a*) compound, compounded; (st eng) compound (-expansion); (elec) compound (-wound)

~ генера́тор· (*m*) (elec) compound (-wound) generator

~ об·мо́т·к·а (*f*) (elec) compound winding

компаунд·и́ровани·е, -я, (*n*) compounding

компе́ндиум·, -а, (*m*) compendium

компенс·а́тор·, -а, (*m*) compensator; (elec) phase modifier; (eng) expansion joint/bend/piece

~, а·син·хро́н·н·ый (elec) asynchronous phase modifier

~ Ба́бин·е (opt) Babinet's compensator

~, гн·у́т·ый expansion bend (of pipe)

~, ди́ск·ов·ый globe-type expansion joint (pipe)

~, ли́нз·ов·ый corrugated expansion joint (pipe)

~, лиро·обра́з·н·ый lyre-type expansion bend (pipe)

~, пруж·и́нн·ый (a/c) spring tab

~, са́ль·ников·ый slip joint (pipes)

~, син·хро́н·н·ый (elec) synchronous phase modifier

компенс·а́тор
~, с·кла́д·чат·ый corrugated- expansion bend (pipe)
компенс·ацио́нн·ый, -ая, -ое, (*a*) compensating, balancing
~ жи́клер· (*m*) compensating jet
~ об·мо́т·к·а (*f*) (elec) compensating winding
компенс·а́ци·я, -и, (*f*) compensation, balancing
~, вес·ов·а́я (air) mass balance
~, рог·ов·а́я horn balance
~ с·дви́г·а фаз· (telecom) phase compensation
компенс·и́р·овать, -уют, (*imp and perf*) compensate, balance; make up for, offset
компенс·и́рующ·ий, -ая, -ее, (*pres part act*) of **компенс·и́р·овать;** compensating, compensator
компете́нт·ност·ь, -и, (*f*) competence
компете́нт·н·ый, -ая, -ое, (*a*) competent
компил·я́тор·, -а, (*m*) compiler
компил·я́ци·я, -и, (*f*) compilation
компланар·ност·ь, -и, (*f*) coplanarity
компланар·н·ый, -ая, -ое, (*a*) coplanar
ко́мплекс·, -а, (*m*) complex, combination, arrangement; (math) complex number, set, group; group aggregate (of atoms)
ко́мплекс·ност·ь, -и, (*f*) complexity; heterogeneity (of soil)
ко́мплекс·н·ый, -ая, -ое, (*a*) complex, multiple, compound, composite, combined; comprehensive, co-ordinated
~ брига́д·а (*f*) team (industrial)
~ дви́г·ател·ь (*m*) compound engine (diesel)
~ об·уч·е́ни·е (*n*) comprehensive education
~ со·един·е́ни·е (*n*) (chem) complex compound
~ сол·ь (*f*) (chem) complex salt
~ числ·о́ (*n*) (math) complex number
комплексо·образ·ова́ни·е, -я, (*n*) (chem, phys) complexing; (geol) complexion formation
комплексо·образ·ова́тел·ь, -я, (*m*) complexing agent
компле́кт·, -а, (*m*) (complete) set, outfit, kit; complement, staff, establishment
~, год·ов·о́й yearly file

комплект·а́тор·, -а, (*m*) assembly-unit collector (industrial); acquisitions librarian
компле́кт·ност·ь, -и, (*f*) completeness; synchronization (of manufacturing process)
компле́кт·н·ый, -ая, -ое, (*a*) complete, fully equipped; enclosed (elec machine)
~ по·ста́в·к·а (*f*) supply in full sets
~ рас·пре·дел·и́тельн·ое у·стро́й·ств·о (*n*) (elec) cubicle switchgear
комплект·ова́ни·е, -я, (*n*) (*v n*) of **комплект·ова́ть;** recruitment; acquisition (library)
комплект·ова́ть, -у́ют, (*imp*) make up a set, complete; supply with equipment; staff, recruit, man; (print) gather
комплеме́нт·, -а, (*m*) (bact) complement
комплиме́нт·, -а, (*m*) compliment
компози́т·н·ый, -ая, -ое, (*a*) composite
композ·и́тор·, -а, (*m*) composer (of music)
композ·и́ци·я, -и, (*f*) composition; furnish (papermaking)
компон·е́нт·, -а, (*m*) component, constituent
~ гор·ю́ч·его (rocket) propellant component
~, гор·ю́ч·ий combustible constituent
~, модифиц·и́рующ·ий modifier
~, о·кисл·я́ющ·ий (rocket) oxidant
компон·ова́ть, -у́ют, (*imp*) put together, make up, arrange
компон·о́вк·а, -и, (*g pl*) **-вок·,** (*f*) putting together, making up, arranging; layout, arrangement
~ маши́н· arrangement of engines
компон·о́вочн·ый, -ая, -ое, (*a*) of **компон·о́вк·а**
~ черт·ёж· (*m*) layout drawing
компо́ст·, -а, (*m*) (agr) compost
~, лист·ов·о́й leaf-mould
компо́стер·, -а, (*m*) ticket punch
компост·и́р·овать, -уют, (*imp*) punch (tickets etc.); (agr) compost, make into compost; spread compost
компо́ст·н·ый, -ая, -ое, (*a*) compost
компо́т·, -а, (*m*) stewed fruit
компре́сс·, -а, (*m*) (med) compress
компресс·ио́нн·ый, -ая, -ое, (*a*) compression
компре́сс·и·я, -и, (*f*) compression
~ реч·и (telecom) speech compression

компре́сс·н·ый, -ая, -ое, (*a*) of ком-
прéсс·

компрессо́·метр·, -а, (*m*) compresso-
meter

компрéссор·, -а, (*m*) compressor;
(gunn) recoil absorber

~, водо·кольц·ев·óй water-ring compressor, liquid-piston compressor

~ двух·сторóн·н·его в·хóд·а double-
entry compressor

~, двух·ступ·éньчат·ый double-acting reciprocative compressor; two-
stage rotary compressor

~ одно·сторóн·н·его в·хóд·а single-
entry compressor

~, ос·ев·óй axial compressor

~, пласт·и́нчат·ый sliding-vane compressor

~, поршн·ев·óй reciprocating compressor

~ прост·óго дéй·стви·я single-acting compressor

~, ротациóн·н·ый rotary compressor

компрéссор·н·ый, -ая, -ое, (*a*) compressor, compression, air-injection
(diesel); (*f decl as adj*) compressor
house/room

~ у·станóв·к·а (*f*) (civ eng) compressor unit, compressor set

комприм·и́р·овать, -уют, (*imp*) compress

комприм·и́рованн·ый, -ая, -ое, (*past
part pass*) of комприм·и́р·овать;
compressed

компрометáнт·н·ый, -ая, -ое, (*a*)
compromising, damaging, unfavourable

компромет·и́р·овать, -уют, (*imp*)
compromise; show in an unfavourable
light; -ся (*pass*); act foolishly

компромет·áци·я, -и, (*f*) (*v n*) *see*
компромет·и́р·овать

компроми́сс·, -а, (*m*) compromise,
agreement

компроми́сс·н·ый, -ая, -ое, (*a*) of
компроми́сс·

Кóмптон·а, эффéкт· (*m*) (nucl) Compton effect

~, явл·éни·е (*n*) (nucl) Compton effect

кóмптон·овск·ое рас·сé·яни·е (*n*)
(nucl) Compton effect

комсомóл·, -а, (*m*) Young Communist
League

комсомóл·ец·, -ль·ц·а, (*m*) member
of Young Communist League, Young
Communist

ком·со·стá·в·, -а, (*m*) = комáнд·н·ый
со·стá·в·

комý (*pron dat*) of кто to who/whom

комфóрт·, -а, (*m*) comfort
зóна комфóрт·а (*f*) comfort zone

комфортáбель·н·ый, -ая, -ое, (*a*)
comfortable

комфортáбель·ност·ь, -и, (*f*) comfort; smoothness (of motion)

конваллери́н·, -а, (*m*) (chem) con-
vallerin

конвéйер·, -а, (*m*) (*and see* транс-
порт·ёр·) conveyor, conveyer

~, вибрациóн·н·ый shaker/vibrating
conveyor

~, винт·ов·óй screw conveyor

~, инерциóн·н·ый vibrating/shaker
conveyor

~, кач·áющ·ийся shaker conveyor

~, ковш·ев·óй bucket elevator

~, лéнт·очн·ый belt conveyor

~, лот·кóв·ый troughed-belt conveyor,
pan conveyor

~, пере·движ·н·óй mobile conveyor

~, пласт·и́нчат·ый slat/apron conveyor

~, по·да·ю́щ·ий feed conveyor

~, под·вес·н·óй overhead trolley conveyor; overhead chain conveyor

~, портáль·н·ый bridge conveyor

~, с·бóр·очн·ый assembly line

~, скреб·кóв·ый scraper/flight conveyor, scraper-chain conveyor, drag-
chain, drag

~, толк·áющ·ий thrust conveyor

~, фáртуч·н·ый apron/slat conveyor

~, телéж·ечн·ый foundry railway,
flat-truck conveyor

~, цéп·очн·о·плáн·чат·ый chain and
slat conveyor

~, черв·я́чн·ый screw conveyor

~, штáнг·ов·ый push-bar conveyor

конвéйер·н·ый, -ая, -ое, (*a*) conveyor,
conveying, travelling; assembly-line

~ печ·ь (*f*) continuous furnace

конвекти́в·н·ый, -ая, -ое, (*a*) convective

конвекциóн·н·ый, -ая, -ое, (*a*) con-
vection(al), convective

конвéкци·я, -и, (*f*) convection

конвергéнт·ност·ь, -и, (*f*) convergence

конвергéнт·н·ый, -ая, -ое, (*a*) convergent

конвергéнци·я, -и, (*f*) convergence

конверсиóн·н·ый, -ая, -ое, (*a*) conversion

конве́рс·и·я, -и, (*f*) conversion

~, взаи́м·н·ая interconversion

  коэффицие́нт· вне́шн·ей конве́рс·и·и (*m*) external conversion ratio

  коэффицие́нт· конве́рс·и·и (*m*) conversion ratio

~ на К-оболо́ч·к·е (phys) K-conversion

конве́рт·, -а, (*n*) envelope

конве́ртер·, -а, (*m*) (met) converter, Bessemer converter; (rad) converter (rare)

конверт·и́ровани·е, -я, (*n*) (met) conversion; Bessemerizing (copper)

конве́рт·н·ый, -ая, -ое, (*a*) of **конве́рт·**

конверто·пла́н·, -а, (*m*) (a/c) converto-plane

конво·и́р·овать, -у́ют, (*imp and perf*) convoy, escort

конво́·й, -я, (*m*) escort, convoy

конво́й·н·ый, -ая, -ое, (*a*) of **конво́·й**

конволами́н·, -а, (*m*) (chem) convolamine

конволвули́н·, -а, (*m*) (chem) convolvulin

конволю́т·н·ый, -ая, -ое, (*a*) convolute

конвульси́в·н·ый, -ая, -ое, (*a*) convulsive

конву́льс·и·я, -и, (*f*) convulsion

конгломер·а́т·, -а, (*m*) (geol) conglomerate; conglomeration; (cryst) aggregate

конгломер·а́ци·я, -и, (*f*) conglomeration, glomeration

конгре́сс·, -а, (*m*) congress

конгруэ́нт·ност·ь, -и, (*f*) (math) congruence, Euclidean congruence

  аксио́м·а конгруэ́нт·ност·и (*f*) (math) Euclid's 4th axiom

конгруэ́нт·н·ый, -ая, -ое, (*a*) (math) congruent

конгруэ́нц·и·я, -и, (*f*) (math) linear congruence

конденса́т·, -а, (*m*) condensate, water of condensation

конденса́т·н·ый, -ая, -ое, (*a*) condensate

~ на·со́с· (*m*) condensate pump

конденсато·про·во́д·, -а, (*m*) condensate return piping/main (steam heating)

конденс·а́тор·, -а, (*m*) (*and see* **холод·и́льник·**) (elec) capacitor; (chem eng) condenser

конденс·а́тор

~, баро·метр·и́ческ·ий barometric-type jet condenser

~, блок·иро́вочн·ый (elec) blocking capacitor

~, бума́ж·н·ый (elec) impregnated-paper capacitor

~, вибрацио́н·н·ый (elec) vibrating capacitor

~, воз·ду́ш·н·ый (elec) air capacitor

~, вы́·ра́вн·ивающ·ий (elec) padder

~ гор·я́ч·его о·рош·е́ни·я (oil) reflux condenser

~, дифференциа́ль·н·ый (elec) differential capacitor

~, до·ба́в·очн·ый (nucl) after condenser

~, за·град·и́тельн·ый (elec) counter-current capacitor

~, из·мер·и́тельн·ый (elec) capacitance meter

~, искро·гас·я́щ·ий (elec) spark capacitor

~, квадрат·и́чн·ый (rad) straight-line wavelength capacitor, square-law capacitor

~, керам·и́ческ·ий (elec) ceramic capacitor

~ Ке́рр·а (elec, opt) Kerr cell

~, кожухо·тру́б·н·ый (chem eng) shell-and-tube condenser

~, мано·метр·и́ческ·ий (elec) pressure capacitor

~, мо́кр·ый wet condenser

~, на·копл·я́ющ·ий memory capacitor

~, нейтро·ди́н·н·ый (elec) neutrodyne

~ обра́т·н·ой с·вя́з·и (rad) reaction capacitor

~, обрат·н·ый reflux condenser

~, о·рос·и́тель·н·ый atmospheric evaporative condenser

~ пере·ме́н·н·ой ём·кост·и (elec) variable capacitor

~, ι ере·ме́н·н·ый variable capacitor, continuously adjustable capacitor

~, по·ве́рх·ности·ый surface condenser

~, под·стро́·ечн·ый (elec) trimmer capacitor

~, полу·пере·ме́н·н·ый (elec) semi-variable capacitor, padder, trimmer

~, пред·вар·и́тельн·ый (nucl) pre-condenser

~, противо·то́ч·п·ый reflex condenser

~, прямо·волн·ов·о́й straight-line wavelength capacitor, square-law capacitor

конденс·а́тор
~, **прямо·ём·костн·ый** straight-line capacitor
~ **рас·стро́й·к·и** (rad) detuning capacitor
~, **рулóн·н·ый** (elec) Mansbridge capacitor
~, **само·ис·правл·я́ющ·ийся** self-healing capacitor
~ **с·вя́з·и** (elec) coupling capacitor
~, **симметр·и́рующ·ий** balancing capacitor
~, **сим·метр·и́чн·ый** balanced capacitor
~, **син·хрóн·н·ый** synchronous capacitor
~ **с·меш·éни·я** contact/jet condenser
~, **стру́й·н·ый** contact/jet condenser
~, **сух·óй** (chem eng) dry condenser; (elec) dry-electrolytic capacitor
~ **тóч·н·ой на·стро́й·к·и** (rad) band-spreading capacitor, bandspreader
~, **шунт·и́рующ·ий** by-pass capacitor
~, **электро·лит·и́ческ·ий** electrolytic capacitor
~, **электро·машѝн·н·ый** (elec) phase modifier
~, **элемéнт·н·ый** double-pipe condenser
конденса́тор·н·о-диóд·н·ый, -ая, -ое, (a) (elec) diode-capacitor
конденса́тор·н·ый, -ая, -ое, (a) of **конденс·а́тор·**
~ **ма́сл·о** (n) (elec) capacitor-grade insulating oil
~ **фильтр·** (m) anti-interference capacitor
конденсациóн·н·ый, -ая, -ое, (a) of **конденс·а́ци·я**
~ **горш·óк·** (m) steam trap
конденс·а́ци·я, -и, (f) condensation
~, **ароматизациóн·н·ая** (chem) aromatic condensation
~, **ка́пель·н·ая** dropwise condensation
~ **на иóн·ах** ion condensation
~, **сложно·эфѝр·н·ая** (chem) aceto-acetic ester condensation
кондéнс·ер·, -а, (m) (text) condenser
кондéнс·и́р·овать, -уют, (imp and perf) condense (steam etc.); store up
кондéнс·ор·, -а, (m) (opt) condenser
~ **тём·н·ого пóл·я** dark-field condenser
кондéнсор·н·ый, -ая, -ое, (a) (opt) condenser, condensing
~ **лѝнз·а** (f) (opt) condensing lens, condenser lens

конди́тер·, -а, (m) confectioner
конди́тер·ск·ая, -ой, (f decl as adj) confectioner's
конди́тер·ск·ий, -ая, -ое, (a) of **конди́тер·**; (see also **конди́тер·ск·ая**) (as noun)
~ **из·дéл·и·я** (pl) confectionery
кондициóн·éр·, -а, (m) air conditioning plant; conditioner (grain)
кондициóн·и́ровани·е, -я, (n) conditioning
~ **вóз·дух·а** air conditioning
кондициóн·и́рованн·ый, -ая, ˋ-ое, (past part pass) conditioned
~ **пря́ж·а** (f) (text) conditioned yarn
кондициóн·н·ый, -ая, -ое, (a) of **конди́ци·я;** up to standard, in (good) condition, according to specification; prepared
~ **концентра́т·** (m) (min) clean concentrate
~ **пар·** (m) standard/specified-parameter steam
~ **по при́·мес·ям** (met) impurities specified
конди́ци·я, -и, (f) condition, standard, specification, technical condition
~ **зерн·а́** grain condition
конду́ктанс·, -а, (m) (elec) conductance
кондукти́в·ност·ь, -и, (f) (elec) conductivity
кондукти́в·н·ый, -ая, -ое, (a) (elec) conductive
кондуктó·метр·, -а, (m) (elec) conductivity bridge
кондукто·метр·и́ческ·ий, -ая, -ое, (a) (chem, elec) conductimetric
~ **титр·ова́ни·е** (n) (chem) conductimetric titration
кондукто·мéтр·и·я, -и, (f) (chem) conductimetric analysis
конду́ктор·, -а, (m) (rail) conductor (U.S.), guard (Britain); (mech) jig; (oil) conductor pipe
~, **дѝск·ов·ый** plate jig
~, **с·двó·енн·ый** (shipb) double half-section jig
~, **тормоз·н·óй** (rail) brakeman
конду́ктор·н·ый, -ая, -ое, (a) of **конду́ктор·**
кондукциóн·н·ый, -ая, -ое, (a) conductive, conduction, conducting
коне·вóд·, -а, (m) horse-breeder
коне·за·вóд·, -а, (m) stud farm, stables
кон·ёк, -нь·к·а́, (m) (dim) of **кон·ь;** hobby; (build) ridge board/piece; skate; (zool) pipit, *Anthus*

кон·е́ц·, **-н·ц·а́**, (*i*) **-н·ц·о́м**, (*m*) end; end point; terminus (of glacier); (naut) rope's end, line, pendant

~, брос·а́тельн·ый (naut) heaving line

~, букси́р·н·ый (naut) towing pendant
в кон·ц·е́ кон·ц·о́в in the end, finally
в о́ба кон·ц·а́ there and back
в оди́н· кон·е́ц one-way (journey)

~, в·ход·н·о́й input end

~, высоко·энергет·и́ческ·ий high-energy tail (of a spectrum)

~, вы·ход·н·о́й outlet end
до кон·ц·а́ fully, to the very end

~ интерва́л·а (math) end point of an interval

~ кип·е́ни·я (chem) end point, final/end boiling point

~, корен·н·о́й standing part (of rope)

~ кры·л·а́ (a/c) wing tip

~ ло́паст·и blade tip (turbine rotors etc.)
под·кон·е́ц· finally, towards the end

~, рабо́ч·ий cutting tip (of cutting tool)

~ раз·гру́з·к·и delivery end

~, сигна́ль·н·ый breastrope (diving)

~ титр·а́ци·и (chem) end point (of a titration)

~, ход·ов·о́й (naut) running part

кон·е́чн·о (*adv*) of course, naturally

конечно·ра́з·ностн·ый, **-ая**, **-ое**, (*a*) (math) finite-difference

кон·е́чност·ь, **-и**, (*f*) finiteness; extremity, tip; (zool) limb

~, лист·ов·а́я (bot) leaf tip

кон·е́чн·ый, **-ая**, **-ое**, (*a*) final, last, end, limit; eventual, ultimate; (math) finite, limited
в кон·е́чн·ом с·чёт·е in the end, eventually, in the final analysis; considering everything, all things considered

~ велич·ин·а́ (*f*) finite quantity/value

~ вероя́т·ност·ь (*f*) (math) finite probability

~ вы·ключ·а́тел·ь (*m*) limit switch

~ знач·е́ни·е (*n*) (math) final value; finite value

~ при·ращ·е́ни·е (*n*) (math)
фо́рмул·а кон·е́чн·ых при·ращ·е́ни·я (*f*) (math) Lagrange's form of the remainder for Taylor's theorem

~ ра́з·ност·ь (*f*) (math) finite difference

кон·е́чн·ый, **-ая**, **-ое**

~ ско́р·ост·ь (*f*) final/terminal velocity

~ ста́нци·я (*f*) terminus, terminal

кон·и́ди·й, **-я**, (*m*) *see* **кон·и́ди·я**

конидио·фо́р·, **-а**, (*m*) (bot) conidiophore

кон·и́ди·я, **-и**, (*f*) (bot) conidium

конии́н·, **-а**, (*m*) (biochem) coniine

коник·о́ид·, **-а**, (*m*) (math) conicoid

кони́·метр·, **-а**, (*m*) conimeter, konometer (for measuring dust content of air)

кон·и́н·а, **-ы**, (*f*) horse-flesh

кони·фери́л·, **-а**, (*m*) (chem) coniferyl

кони·фери́н·, **-а**, (*m*) (chem) coniferin

кон·и́ческ·ий, **-ая**, **-ое**, (*a*) conic(al), tapered; bevel-gear, bevel

~ болт· (*m*) taper bolt

~ голо́в·к·а (*f*) pan head (rivet)

~ пере·да́·ч·а (*f*) bevel drive

~ по·ве́рх·ност·ь (*f*) (math) conical surface

~ ро́л·ик· (*m*) tapered roller

~ сеч·е́ни·е (*n*) (math) conic section

~ шестерн·я́ (*f*) bevel gear, bevel wheel

кон·и́чност·ь, **-и**, (*f*) conicity, conicalness; angle of taper

конкретиз·а́ци·я, **-и**, (*f*) precise/clear definition

конкретиз·и́р·овать, **-уют**, (*imp and perf*) define more precisely/accurately

конкре́т·н·ый, **-ая**, **-ое**, (*a*) concrete, specific; real, actual

конкрецио́н·н·ый, **-ая**, **-ое**, (*a*) concretionary

конкре́ц·и·я, **-и**, (*f*) concretion

конкурент·, **-а**, (*m*) competitor, rival

конкуре́нт·н·ый, **-ая**, **-ое**, (*a*) competitive

конкуре́нц·и·я, **-и**, (*f*) rivalry, competition

~ га́мма-пере·хо́д·ов (phys) gamma-gamma competition

~, меж·вид·ов·а́я (gen) inter-varietic competition

конкур·и́р·овать, **-уют**, (*imp*) compete

конкур·и́рующ·ий, **-ая**, **-ее**, (*pres part act*) of **конкур·и́р·овать**; competitive

ко́нкурс·, **-а**, (*m*) competition; (fin) procedure in bankruptcy

ко́нкурс·н·ый, **-ая**, **-ое**, (*a*) of **ко́нкурс·**

~ экза́мен· (*m*) competitive examination

ко́н·ниц·а, -ы, (*f*) cavalry

конно·за·во́д·ств·о, -а, (*n*) horse-breeding

конно·спорти́в·н·ый, -ая, -ое, (*a*) horse-racing; show (of jumping etc.)

ко́н·н·ый, -ая, -ое, (*a*) horse; horse-drawn

Конова́лов·а, пе́рв·ый зако́н· (*m*) (phys) Konovalov's/Konovaloff's rule

ко́но·вяз·ь, -и, (*f*) picket line (for horses)

коно·кра́д·, -а, (*m*) horse-thief

коно́д·, -а, (*m*) (math) conoid

коно·до́нт·, -а, (*m*) (pal) conodont

кон·о́ид·, -а, (*m*) (math) conoid

коноп·а́т·ить, -ят, (*imp*) caulk, calk

коноп·а́тк·а, -и, (*g pl*) **-т·ок·,** (*f*) caulking tool; caulking material; caulking (completed work)

коноп·а́чени·е, -я, (*n*) (*v n*) of **коноп·а́т·ить** caulking

конопле·во́д·ств·о, -а, (*m*) hemp growing

конопл·ёв·ый, -ых, (*pl decl as adj*) (bot) hemp family, *Cannabinaceae*

конопле·мя́л·к·а, -и, (*g pl*) **-л·ок·,** (*f*) hemp-crushing mill

конопл·я́, -и́, (*f*) (bot) hemp, *Cannabis sativa*

~, инд·и́йск·ая (bot) jute, *Corchorus*

конопл·я́ник·, -а, (*m*) hemp field

конопл·я́н·ый, -ая, -ое, (*a*) hemp, hempen

~ ма́сл·о (*n*) hempseed oil

коносаме́нт·, -а, (*m*) (com) bill of lading

~, борт·ов·о́й on-board bill of lading

~, имен·н·о́й straight bill of lading

~, мор·ск·о́й ocean bill of lading

~, не·чи́ст·ый claused/foul bill of lading

~, сквоз·н·о́й through/transhipment bill of lading

~, чи́ст·ый clean bill of lading

коно·ско́п·, -а, (*m*) (cryst) conoscope

коно·скоп·и́ческ·ий, -ая, -ое, (*a*) conoscopic

~ Фигу́р·а (*f*) (cryst) interference figure

коно·скоп·и́·я, -и, (*f*) (cryst) conoscopy

консекве́нт·н·ый, -ая, -ое, (*a*) consequent

~ дол·и́н·а (*f*) (geol) consequent valley

консеку́ти́в·н·ый, -ая, -ое, (*a*) consecutive

консе́рв·, *see* **консе́рв·ы**

консерва́нт·, -а, (*m*) (food) preservative

консервати́в·н·ый, -ая, -ое, (*a*) conservative

~ колеб·а́тель·н·ое звен·о́ (*n*) (autom) conservative oscillatory stage

~ систе́м·а (*f*) (mech) conservative system

консерва́тор·, -а, (*m*) conservator; conservative (person)

~ трансформ·а́тор·а oil conservator (of transformer)

консервато́ри·я, -и, (*f*) conservatoire/ academy of music

консерв·а́ци·я, -и, (*f*) preservation; state of preservation; temporary closing, putting into reserve; mothballing

консерв·и́рованн·ый, -ая, -ое, (*past part pass*) of **консерв·и́р·овать;** preserved; tinned, canned, bottled etc.

консерв·и́р·овать, -уют, (*imp and perf*) (food) conserve, can, tin, bottle, pot, preserve; preserve, protect against rot, decay etc., put with reserve; mothball

консе́рв·н·ый, -ая, -ое, (*a*) (food) preserving, canning, tinning etc.

~ за·во́д· (*m*) canning factory; food preserving factory

консе́рв·ы (*nom pl*), (*g pl*) **-ов,** (*m*) preserves, canned/tinned/potted/bottled/pickled food

~, моло́ч·н·ые preserved milk products

консерта́ль·н·ый, -ая, -ое, (*a*) consertal

консигн·а́нт·, -а, (*m*) (comm) consignor, consigner

консигн·а́тор·, -а, (*m*) consignee

консигн·ацио́нн·ый, -ая, -ое, (*a*) of **консигн·а́ци·я;** consignment

консигн·а́ци·я, -и, (*f*) (comm) contract of consignment

конси́лиум·, -а, (*m*) (med) consilium, consultation

консисте́нт·н·ый, -ая, -ое, (*a*) thick (of liquids), firm, solid, compact

~ с·ма́з·к·а (*f*) lubricating grease

консисте́нци·я, -и, (*f*) consistence, consistency

консисто́·метр·, -а, (*m*) (chem) consistometer

ко́н·ск·ий, -ая, -ое, (*a*) horse, equine

~ боб· (*m*) (bot) broad bean

консолид·а́ци·я, -и, (*f*) consolidation

консолид·и́рованн·ый, -ая, -ое, (*past part pass*) of **консолид·и́р·овать;** consolidated, permanent

~ за·ём· (*m*) (fin) permanent loan

консолид·и́р·овать, -уют, (*imp and perf*) consolidate

консо́л·ь, -и, (*f*) console (control panel); cantilever, bracket; overhang (of beam etc.)

длин·а́ консо́л·и (*f*) length of overhang, unsupported length

~ кры·л·а́ (a/c) main plane

консо́ль·н·о (*adv*) cantilever-wise, cantilever

консо́льно·фре́зер·н·ый стан·о́к· (*m*) column-and-knee milling machine

консо́ль·н·ый, -ая, -ое, (*a*) of **консо́л·ь**

~ мост· (*m*) cantilever bridge

консоме́ (*n indecl*) (food) consommé

консона́нс·, -а, (*m*) consonance

консона́нт·н·ый, -ая, -ое, (*a*) consonant

консо́рциум·, -а, (*m*) (econ) consortium

конспе́кт·, -а, (*m*) conspectus, prospectus, synopsis, summary

конспект·и́вн·ый, -ая, -ое, (*a*) concise

конспект·и́р·овать, -уют, (*imp*) summarize

конста́нт·а, -ы, (*f*) (*and see* **по·сто··я́нн·ая** *and associated words*) (math) constant

константа́н·, -а, (*m*) (met) Constantan

константа́н·ов·ый, -ая, -ое, (*a*) Constantan

конста́нт·ност·ь, -и, (*f*) constancy

конста́нт·н·ый, -ая, -ое, (*a*) of **конста́нт·а;** constant

констат·и́р·овать, -уют, (*imp and perf*) ascertain, establish; certify

констелл·я́ци·я, -и, (*f*) constellation

конститу·и́р·овать, -уют, (*imp and perf*) constitute

конститути́в·н·ый, -ая, -ое, (*a*) constitutive

конституцио́н·н·ый, -ая, -ое, (*a*) constitutional

конститу́ци·я, -и, (*f*) constitution

констру·и́ровани·е, -я, (*n*) (*v n*) of **констру·и́р·овать**

при́нцип· констру·и́ровани·я реа́ктор·а (*m*) (nucl) reactor concept

констру·и́р·овать, -уют, (*imp*) design; form (an organization)

конструкти́в·ност·ь, -и, (*f*) constructiveness; constructability, efficiency/satisfactoriness of design

конструкти́в·н·ый, -ая -ое, (*a*) constructive; constructional; design

~ черт·ёж· (*m*) design drawing

констру́ктор·, -а, (*m*) designer, designing engineer

~, авиацио́н·н·ый aircraft designer; aero-engine designer

~ корабл·е́й naval architect/constructor

~ -черт·ёжник·, -а, (*m*) draughtsman

констру́ктор·ск·ий, -ая, -ое, (*a*) of **констру́ктор·**

~ бюро́ (*n*) drawing office; design bureau

конструкцио́н·н·ый, -ая, -ое, (*a*) of **констру́кц·и·я**

констру́кц·и·я, -и, (*f*) construction; design; (civ eng) structure, structural member

~, бло́ч·н·ая modular design

~, карка́с·н·ое (build) framework, frame construction

ко́нсул·, -а, (*m*) (dipl) consul

~, генера́ль·н·ый consul-general

ко́нсуль·ск·ий, -ая, -ое, (*a*) consular

ко́нсуль·ств·о, -а, (*n*) (dipl) consulate

консульт·а́нт·, -а, (*m*) consultant; tutor (in university etc.)

консультати́в·н·ый, -ая, -ое, (*a*) consultative, advisory

консульт·а́ци·я, -и, (*f*) consultation; tutorial (at university etc.); consultation office/room

консульт·и́р·овать, -уют, (*imp*) advise, consult; **-ся** consult

конта́г·и·й, -я, (*m*) contagion

конта́кт·, -а, (*m*) contact; contact area/surface; (chem) contact action, catalysis, catalyst; (elec) contact, contact stud

~, вод·о-нефт·ян·о́й (geol) oil-water surface

~, вс·по·мог·а́тельн·ый (elec) auxiliary contact

~, гла́в·н·ый (elec) main contact

~, двер·н·о́й (elec) door trip/switch

~, за́д·н·ий (elec) back/break contact, (normally) closed contact

~, за·мык·а́ющ·ий (elec) "make" contact

~ извест·няк·а́ и сла́нц·а (geol) limeschist contact

конта́кт

~, **искро·гас·и́тельн·ый** (elec) arcing contact

~, **и́свр·ящ·ий** (elec) sparking contact

~, **крыл·ов·о́й** (rail elec) semaphore contact

~, **мост·ов·о́й** (elec) bridging contact

~, **нож·н·о́й** (elec) foot/floor contact

~, **норма́ль·н·о за́·мк·нут·ый** (telecom) (normally) closed contact, back contact

~, **норма́ль·н·о разо́·мк·нут·ый** (telecom) (normally) open contact, make/front contact

~ **об·гор·а́ни·я** (elec) arcing contact

~, **пере·кид·н·о́й** (elec) break-before-make contact

~, **пло́ск·остн·ый** flat contact

~, **по·дви́ж·н·ый** (elec) moving contact

~, **рабо́ч·ий** (elec) "make" contact

~, **разо́·мк·нут·ый** (elec) dead contact

~, **рту́т·н·ый** (elec) mercury contact

~, **скольз·я́щ·ий** (elec) sliding contact, wiping contact

~, **со·гла́с·н·ый** (geol) conformable contact

~, **с·поко́й·н·о-рабо́ч·ий** (elec) make-and-break contact

~, **тл·е́ющ·ий** (elec) glow switch

~, **то́ч·ечн·ый** point contact

~, **фронт·ов·о́й** (telecom) front contact, relay front contact, (normally) open contact

~, **щёт·очн·ый** (elec) brush contact

конта́кт·н·ый, -ая, -ое, (a) contact; (chem) catalytic, contact

~ **вещ·еств·о́** (n) (chem) contact/catalytic agent, catalyst

~ **гру́пп·а** (f) spring assembly (in an electric relay)

~ **кисл·от·а́** (f) (chem) contact acid, contact tower sulphuric acid

~ **место·рожд·е́ни·е** (n) (geol) contact vein

~ **нож·** (m) (elec) contact/member blade (of a switch)

~ **о·сажд·е́ни·е** (n) (met) electroplating

~ **печа́т·ь** (f) (phot) contact printing

~ **по́л·е** (n) (telecom) bank of contacts

~ **пруж·и́н·а** (f) (elec) contact clip

~ **с·ва́р·к·а** (f) resistance welding

~ **сет·ь** (f) traction power-supply system (railways, trams or trolley-buses)

~ **след·я́щ·ая систе́м·а** (f) (autom) step-by-step follow-up system

конта́кт·н·ый, -ая, -ое

~ **фильт·р·а́ци·я** (f) contact filtration (sewage)

~ **чан·** (m) (min) contact tank, conditioning tank

конта́кт·ов·ый, -ая, -ое, (a) see **конта́кт·н·ый**

конта́ктор·, -а, (m) (elec) contactor

~, **групп·ов·о́й** (elec) contactor group

~, **лин·е́йн·ый** (elec) mains contactor

контамин·а́ци·я, -и, (f) contamination

конте́йнер·, -а, (m) (rail etc.) container; rabbit (in nuclear reactor); receptacle, vessel, container

континге́нт·, -а, (m) contingent; quota (goods)

контингент·и́ровани·е, -я, (n) (comm) quantitative regulation (of imports), putting on a quota

континге́нт·н·ый, -ая, -ое, (a) of **континге́нт·**

контине́нт·, -а, (m) (geog) continent

континента́ль·н·ый, -ая, -ое, (a) continental

~ **аркт·и́ческ·ий во́з·дух·** (m) (meteor) continental arctic air

континуа́ль·н·ый, -ая, -ое, (a) continual

конти́нуум, -а, (m) (math) continuum

конти́нуум·н·ый, -ая, -ое, (a) (dynam) continuum, continuous

ко́нто (n indecl) (fin) account

контокорре́нт·, -а, (m) (fin) current account

конто́р·а, -ы, (f) office, bureau

~, **почт·о́в·ая** post office

конто́р·к·а, -и, (g pl) **-р·ок·,** (f) high desk

конто́р·ск·ий, -ая, -ое, (a) of **конто́р·а**

конто́р·щик·, -а, (m) clerk

конта- (component) counter, contra

контр- (component) counter-, contra-

контраба́нд·а, -ы, (f) contraband, smuggling

контра·вале́нт·ност·ь, -и, (f) (chem) contravalence

контра·вариа́нт·ност·ь, -и, (f) (math) contravariance

контра·вариа́нт·н·ый, -ая, -ое, (a) contravariant

контра·винт·, -а, (m) countershaft

контраге́нт·, -а, (m) contracting party; sub-contractor

контра·гредие́нт·ност·ь, -и, (f) = **контра·вариа́нт·ност·ь**

контр-адмира́л·, -а, (m) Rear Admiral

контражу́р·, -а, *(m)* (phot) silhouetting

контра́кт·, -а, *(m)* contract

контра́кт·ацио́нн·ый, -ая, -ое, *(a)* contracting, contract-negotiating

контра́кт·а́ци·я, -и, *(f)* contracting, concluding a contract, entering into a contract, negotiating a contract

контра́кт·н·ый, -ая, -ое, *(a)* contract, as contracted, according to contract

контра́кт·ова́ть, -у́ют, *(imp)* contract for/to; sign up (workpeople)

контра́кт·ов·ый, -ая, -ое, *(a)* contract

контракту́р·а, -ы, *(f)* (med) contractura

контракцио́н·н·ый, -ая, -ое, *(a)* contractional

контра·пу́нкт·, -а, *(m)* counterpoint (music)

контра́ст·, -а, *(m)* contrast

~, выс·о́к·ий (phot) vigorous/high contrast, hard grade

~, ма́л·ый (phot) flat/soft grade

контра́ст·и́ровани·е, -я, *(n)* contrasting

контра́ст·и́р·овать, -у́ют, *(imp)* contrast

контра́ст·ност·ь, -и, *(f)* (phot) hardness, contrast, contrastiness, contrast grade; (TV) contrast

~ из·об·раж·е́ни·я (TV) picture contrast

коэффицие́нт· контра́ст·ност·и *(m)* (TV) gamma

~, поро́г·ов·ая (TV) threshold contrast

регул·и́ровани·е контра́ст·ност·и *(n)* (TV) contrast control

контра́ст·н·ый, -ая, -ое, *(a)* contrasty, contrasting, high-contrast

контр·ата́к·а, -и, *(f)* (mil) counter attack

контра·ти́п·, -а, *(m)* (cinema) dupe, duplicate negative

контра·ти́п·и́ровани·е, -я, *(n)* (cinema) duplicating

контра·фа́кц·и·я, -и, *(f)* copyright infringement

контра·фо́рс·, -а, *(m) see* **контр·фо́рс·**

контр·бу́кс·а, -ы, *(f)* (rail) keep

контр·га́·йк·а, -и, *(f)* lock nut, check nut

контр·гру́з·, -а, *(m)* counterweight, counterbalance

контр·жи́·л·а, -ы, *(f)* (geol) counterlode

контрибу́ци·я, -и, *(f)* contribution

ко́нтр·ить, -ят, *(imp)* (mech) secure, fasten, lock (nuts etc.)

контр·кали́бр·, -а, *(m)* master gauge/gage

контр·ку́рс·, -а, *(m)* (nav) opposite course

контр·на·ступл·е́ни·е, -я, *(n)* (mil) counter-offensive

контрове́рз·а, -ы, *(f)* controversy

контрол·ёр·, -а, *(m)* (*and see* **контро́ллер·**) controller; inspector (quality control); ticket collector

контрол·и́рованн·ый, -ая, -ое, *(past part pass)* of **контрол·и́р·овать**

контрол·и́р·овать, -у́ют, *(imp)* inspect, check, supervise, exercise control over

контрол·и́руем·ый, -ая, -ое, *(pres part pass)* of **контрол·и́р·овать;** controllable, controlled

~ атмосфе́р·а *(f)* (met) controlled atmosphere, protective atmosphere

контрол·и́рующ·ий, -ая, -ее, *(pres part act)* of **контрол·и́р·овать;** monitoring, supervising

контро́ллер·, -а, *(m)* (elec) controller

~, бараба́н·н·ый drum controller

~, конта́ктор·н·ый contactor controller

~, кулач·ко́в·ый camshaft controller

~ машин·и́ст·а (rail) master controller

~, пло́ск·ий faceplate controller

~, реоста́т·н·ый rheostatic controller

~, схе́м·н·ый circuit controller

контро́л·ь, -я, *(m)* supervision; check, control, checking, monitoring; inspection

~, автома́т·и́ческ·ий automatic control, automatic inspection/checking/ quality control

~, лет·у́ч·ий floor inspection (quality control)

~, медици́н·ск·ий medical examination

~, о·бег·а́ющ·ий (autom) scan checking, multipoint scan checking

~, о·конч·а́тельн·ый screening inspection (quality control)

~, операцио́н·н·ый working inspection

~ пут·и́ (air) track check

~, радиацио́н·н·ый (nucl) radiation monitoring

~ с·вя́з·и (telecom) monitoring

~, тепл·ов·о́й thermal process control

~, техн·и́ческ·ий quality control

~ у́·ровн·я level gauging/gaging

контро́ль·ник·, -а, *(m)* verifier, verifying key punch (punched card systems)

контро́ль·н·о-из·мер·и́тельн·ые при·бо́р·ы (*pl*) control and measuring instruments, instrumentation

контро́ль·н·о-ка́сс·ов·ый аппара́т· (*m*) cash register

контро́ль·н·о-по·ве́р·очн·ый пункт· (*m*) gauge checking point (quality control)

контро́ль·н·о-про·пуск·н·о́й пункт· (*m*) check point; traffic control point

контро́ль·н·о-рас·пре·дел·и́тельн·ый пункт· (*m*) check and distribution point

контро́ль·н·о-ревизио́н·н·ая коми́с·с·и·я (*f*) control and auditing commission

контро́ль·н·о-сорт·иро́вочн·ый ав·тома́т· (*m*) automatic inspecting and sorting machine

контро́ль·н·о-счи́т·ивающ·ее у·с·тро́й·ств·о (*n*) (autom) checking device

контро́ль·н·ый, -ая, -ое, (*a*) of **кон·тро́л·ь**; reference

∼ в·ход·н·ые велич·и́н·ы (*pl*) reference input (values) (computers)

∼ выс·от·а́ (*f*) (TV) reference height (of image)

∼ ма́стер· (*m*) foreman inspector

∼ лист· (*m*) check sheet; servicing form (M/T, a/c etc.)

∼ ме́т·к·и (*pl*) (print) collation marks, registers, back marks

∼ о́рган· (*m*) (autom) inspection element

∼ при·бо́р· (*m*) reference/check meter/gauge/gage; (nucl) monitoring device

∼ при·способл·е́ни·я (*pl*) inspection instruments/gauges/gages

∼ сигна́л· (*m*) pilot signal

∼ то́ч·к·а (*f*) reference/check point

∼ у́·ровен·ь (*m*) reference level

∼ у·стро́й·ств·о (*n*) checking device

контр·па́р·, -а, (*m*) counter-steam, back-steam

контр·под·гото́в·к·а, -и, (*g pl*) **-в·ок·,** (*f*) (mil) counterpreparation

контр·пред·лож·е́ни·е, -я, (*n*) counter-proposal

контр·прете́нзи·я, -и, (*f*) counter-claim

контр·при·во́д·, -а, (*m*) (mech) countershaft

контр·при·ка́з·, -а, (*m*) countermand

контр·пруж·и́н·а, -ы, (*f*) recoil spring

контр·раз·ве́д·к·а, -и, (*g pl*) **-д·ок·,** (*f*) counter intelligence, security service

контр·ре́льс·, -а, (*m*) (rail) wheelguide guard rail

контр·те́л·о, -а, (*n*) abradant

контр·ти́мберс·, -а, (*m*) (shipb) counter timber

контр·фо́рс·, -а, (*m*) (civ eng) buttress, abutment, abamurus, counterfort; bar, stud (in chain cable)

контр·што́к·, -а, (*m*) (gunn) control plunger

контр·эска́рп·, -а, (*m*) (mil) vertical drop

контр·этало́н·, -а, (*m*) master check gauge/gage

ко́нтр·ящ·ий, -ая, -ее, (*pres part act*) of **ко́нтр·ить**; locking (of nuts etc.)

конту́ж·енн·ый, -ая, -ое, (*past part pass*) of **конту́з·ить**; (med) contused

конту́ж·ен·ый, -ая, -ое, (*a*) bruised, contused

конту́з·ить, -ят, (*perf*) bruise, (med) contuse

конту́з·и·я, -и, (*f*) (med) contusion

ко́нтур·, -а, (*m*) (*see also* **цеп·ь**); contour, periphery, outline; (elec) circuit; (autom) mesh, loop, circuit; (rad) filter circuit, filter, circuit

∼, ано́д·н·ый anode circuit, plate circuit

∼, а·период·и́ческ·ий aperiodic circuit

∼, бала́нс·н·ый balancing network

∼, берег·ов·о́й (geog) margin of coast

∼, втор·и́чн·ый secondary circuit

∼, двух·та́кт·н·ый push-pull circuit

∼, демпф·и́рующ·ий buffer circuit

∼, дро́ссель·н·ый (rad) choke filter

∼, за·гражд·а́ющ·ий (rad) band-stop filter

∼ за·да́т·чик·а мо́щ·ност·и power demand circuit

∼, за·да·ю́щ·ий (autom) driver circuit

∼, за·пир·а́ющ·ий (rad) band-stop filter

∼, колеб·а́тельн·ый oscillatory circuit

∼, корре́кт·и́рующ·ий (autom) closed correction circuit

∼, мо́ст·н·ый (autom) minor loop

∼, меш·а́ющ·ий parasitic circuit

∼, на·стро́·енн·ый tuned circuit

∼, не·колеб·а́тельн·ый aperiodic circuit

∼ обра́т·н·ой с·вя́з·и (autom) feedback loop

∼, о·по́р·н·ый (surv) index contour

∼, от·пир·а́ющ·ий (rad) gate opener

∼, перв·и́чн·ый primary circuit

ко́нтур

~ **-про́б·к·а, -и,** (*g pl*) **-в·ок·,** (*f*) band-stop filter

~, **про·пуск·а́ющ·ий** acceptor circuit

~, **резона́нс·н·ый** resonant circuit

~, **с·вя́з·анн·ый** coupled circuit

~, **с·гла́ж·ивающ·ий** smoothing circuit

~, **уда́р·н·ый** impulsing circuit

ко́нтур·н·ый, -ая, -ое, (*a*) of **ко́нтур·**; (surv) planimetric

~ **черт·ёж·** (*m*) outline drawing

~ **интегра́л·** (*m*) contour integral

конул·я́ри·и (*nom pl*), (*g pl*) **-й,** (*f*) (pal) *Conularia*

ко́нус·, -а, (*m*) cone; (zool) conus; (air) drogue, sleeve target; hub (water turbine)

~, **больш·о́й** big bell (of blast furnace)

~ **воз·му́щ·ени·й** (dynam) Mach cone

~, **вод·ян·о́й** (st eng) combining tube

~, **в·хо́д·н·о́й** taper

у́гол· в·хо́д·н·о́го ко́нус·а (*m*) angle of taper

~ **вы́·нос·а** (geol) alluvial fan

~ **за·вихр·и́тел·я** swirl cone

~ **Зе́гер·а** Seger/fusion cone

~, **зе́йгер·овск·ий** Seger/fusion cone (for temperature measurement in furnaces)

~, **кас·а́тельн·ый** (math) tangential cone

~, **круг·ов·о́й** (math) circular cone

~, **ма́л·ый** little bell (blast furnace)

~ **молч·а́ни·я** (acous) cone of silence

~, **на·гнет·а́тельн·ый** (st eng) delivery tube (injector)

~ **на·раст·а́ни·я** (bot) apical/vegetative cone

~, **обра́т·н·ый** inverse taper

~ **-об·тек·а́тел·ь, -я,** (*m*) bullet (jet engine)

~, **о·сад·и́тельн·ый** (min) cone classifier

~, **от·де́л·очн·ый** (mech eng) adjustable/taper parallel

~, **пар·ов·о́й** steam tube (of injector)

~, **по·дви́ж·н·о́й** (a/c) variable-area nozzle

~ **раз·лёт·а** (gunn) cone of dispersion

~, **рас·шир·я́ющ·ийся** expansion cone (of wind tunnel)

~ **реакти́в·н·ого сопл·а́** exhaust cone (of jet engine)

~, **с·мес·и́тельн·ый** (st eng) combining tube (of injector)

~, **то́нк·ий** slender cone

ко́нус

~ **тр·е́ни·я** angle of friction

~, **у·сеч·ённ·ый** (math) frustrum of a cone, conical frustrum, frustrum, truncated cone

~, **у·стано́в·очн·ый** locating cone/cup (machine tool attachment)

~, **черн·ов·о́й** (met roll) die reduction angle, disapproach angle

ко́нус·ност·ь, -и, (*f*) taper, coning, conicity

у́гол· ко́нус·ност·и (*m*) coning angle (helicopter rotor); draft angle (of foundrymold)

ко́нус·н·ый, -ая, -ое, (*a*) cone, conic(al), tapered

конусо·ви́д·н·ый, -ая, -ое, (*a*) *see* **конусо·обра́з·н·ый**

конусо·обра́з·н·ый, -ая, -ое, (*a*) cone-shaped, coniform, conoid, conical, tapered

конфекцио́н·н·ый, -ая, -ое, (*a*) made-up, readymade

конфе́кци·я, -и, (*f*) (text) putting together, making up; building (assembling rubber articles)

конфере́нци·я, -и, (*f*) conference

конфе́т·а, -ы, (*f*) sweet, candy

конфе́т·н·ый, -ая, -ое, (*a*) of **конфе́т·а**

конфигурацио́н·н·ый, -ая, -ое, (*a*) configuration, configurational

конфигур·а́ци·я, -и, (*f*) configuration

конфина́ль·ност·ь, -и, (*f*) confinality

конфирм·ова́ть, -у́ют, (*imp and perf*) confirm, ratify

конфиск·ова́ть, -у́ют, (*imp and perf*) confiscate, seize

конфли́кт·, -а, (*m*) conflict

конфли́кт·н·ый, -ая, -ое, (*a*) of **конфли́кт·**

~ **коми́сс·и·я** dispute committee

конфока́ль·н·ый, -ая, -ое, (*a*) (*and see* **со·фо́кус·н·ый**) confocal

конфо́р·к·а, -и, (*g pl*) **-р·ок·,** (*f*) ring (on cooking stove)

конфо́рм·ност·ь, -и, (*f*) conformity

конфо́рм·н·ый, -ая, -ое, (*a*) (math) conformal

~ **от·об·раж·е́ни·е** (*n*) conformal transformation/map

~ **пре·об·раз·ова́ни·е** (*n*) conformal transformation/map

конфу́з·, -а, (*m*) embarrassment, awkwardness

конфу́зор·, -а, (*m*) converging/convergent duct/nozzle, converging tube reducer

ко́нх·, -а, (*m*) (zool) conch

ко́нх·а, -и, (*f*) (arch) concha

конхи·оли́н·, -а, (*m*) (zool) conchiolin

конх·и́т·, -а, (*m*) (min) conchite

конх·о́ид·а, -ы, (*f*) (math) conchoid, conchoid of Nikodemus

конх·о́ид·а́льн·ый, -ая, -ое, (*a*) conchoidal

конхо·метр·и́ческ·ий, -ая, -ое, (*a*) conchometrical

кон·цев·о́й, -а́я, -о́е, (*a*) of **кон·е́ц·**; terminal, limit

~ **вы·ключ·а́тел·ь** (*m*) (elec) limit switch

~ **му́фт·а** (*f*) (elec) cable pothead

~ **реакти́в·н·ое сопл·о́** (*n*) blade-tip reaction nozzle (helicopter)

концентр·а́т·, -а, (*m*) concentrate

~**, коллекти́в·н·ый** (min) bulk concentrate

~**, кондицио́н·н·ый** (min) clean concentrate

~**, кон·е́чн·ый** (min) final concentrate

концентр·а́тор·, -а, (*m*) concentrator

~**, вращ·а́ющ·ийся** (min, chem) rotary concentrator

~ **на·пряж·е́ни·я** (mech) stress raiser/concentrator

~**, телегра́ф·н·ый** (telecom) (line) concentrator

концентрацио́н·н·ый, -ая, -ое, (*a*) concentration, concentrating

концентр·а́ци·я, -и, (*f*) concentration

~**, а́том·н·ая** atomic percentage

~ **по ма́сс·у** mass concentration

~ **на·пряж·е́ни·я** stress concentration

~ **по у·де́ль·н·ому ве́с·у** (min) gravity concentration

~**, объ·ём·н·ая** volume concentration, concentration by volume

~ **при́·мес·и** impurity concentration/percentage

~ **узл·о́в** cross-link density (polymers)

концентр·и́рованн·ый, -ая, -ое, (*past part pass*) of **концентр·и́р·овать**; concentrated

концентр·и́р·овать, -уют, (*imp*) concentrate

концентр·и́ческ·ий, -ая, -ое, (*a*) concentric

концентр·и́чност·ь, -и, (*f*) concentricity

конце́пци·я, -и, (*f*) concept; (gen) conception

конце·равн·и́тельн·ая маши́н·а (*f*) equalizer (cooperage)

конце́рн·, -а, (*m*) (econ) concern, major group of companies

конце́рт·, -а, (*m*) concert (performance); concerto (music); concerted action, concert

концерт·а́нт·, -а, (*m*) performing musician

концерти́но (*n indecl*) concertino (music); concertina

концерт·и́р·овать, -уют, (*imp*) perform at concerts (music), give concerts

концерт·ме́йстер·, -а, (*m*) first violin; accompanist

концессионе́р·, -а, (*m*) concessionaire

конце́сс·и·я, -и, (*f*) concession

кон·цо́вк·а, -и, (*g pl*) **-вок·,** (*f*) (print) tail piece

конч·а́·ть, -ют, (*imp*) finish, end

ко́нч·ик, -а, (*m*) tip, apex, end

конч·и́н·а, -ы, (*f*) death, decease

ко́нч·ить, -ат, (*perf*) see **конч·а́·ть**

конъекту́р·а, -ы, (*f*) (print) conjecture

конъюг·а́т·, -а, (*m*) (biol) conjugate

конъюг·а́ци·я, -и, (*f*) (biol) conjugation

конъюг·и́р·ованн·ый, -ая, -ое, (*see also* **со·пряж·ённ·ый**) conjugate, conjugated, paired, coupled

конъюнкти́в·, -а, (*m*) (gram) conjunctive

конъюнкти́в·а, -ы, (*f*) (zool) conjunctiva

конъюнктиви́т·, -а, (*m*) (med) conjunctivitis

конъюнкти́в·н·ый, -ая, -ое, (*a*) conjunctive

конъюнкту́р·а, -ы, (*f*) situation; (econ) state of the market

кон·ь, -я́, (*nom pl*) **-и,** (*g pl*) **-е́й,** (*m*) horse, stallion; knight (at chess); (fish) *Hemibarbus labeo*

конь·к·и́ (*nom pl*), (*g pl*) **-о́в,** (*m*) skates

конько·бе́ж·н·ый, -ая, -ое, (*a*) skating

коньяк·, -а́, (*m*) brandy

коньяч·н·ый, -ая, -ое, (*a*) of **коньяк·**

~ **ма́сл·о** (*n*) cognac oil, ethyl oenanthate

коню́ш·н·я, -и, (*g pl*) **-ш·ен·,** (*f*) stable

ко·объ·ём·, -а, (*m*) co-content

кооперати́в·, -а, (*m*) co-operative, co-operative society; co-operative store/shop

кооперати́в·н·ый, -ая, -ое, *(a)* co-operative

коопер·а́тор·, -а, *(m)* co-operator

коопер·а́ци·я, -и, *(f)* co-operation; co-operative society/system; co-operative store/shop

коопер·и́р·овать, -уют, *(imp and perf)* form a co-operative society; bring in/admit to a co-operative society

коопт·и́р·овать, -уют, *(imp and perf)* co-opt

координа́т·а, -ы, *(f)* (math) coordinate

~, в·ход·н·а́я (autom) input

~, Дека́рт·ов·ы *(pl)* Cartesian coordinates

~ звен·а́, в·ход·н·а́я *(f)* (autom) component input

~ када́стр·ов·ые cadastral coordinates

~, косо·уго́ль·н·ые *(pl)* oblique coordinates

на·ча́·л·о координа́т· *(n)* coordinate root/origin

~, на·ча́ль·н·ые *(pl)* origin coordinates

~, не·по́лн·ые *(pl)* short coordinates

~, пла́н·ов·ые *(pl)* plane coordinates

~, плю́с- *(pl)* positive coordinates

~, поля́р·н·ые *(pl)* polar coordinates

~, прямо·уго́ль·н·ые *(pl)* orthogonal coordinates

систе́м·а координа́т· *(f)* (math) frame of reference/axes, coordinate system

~, угл·ов·а́я angular coordinate

~, у·сло́в·н·ые *(pl)* (surv) false coordinates

~, цилиндр·и́ческ·ие *(pl)* cylindrical coordinates

координа́т·ник·, -а, *(m)* (air) traversing equipment/gear

координа́т·н·о-рас·то́ч·н·ый ста·н·о́к· *(m)* drill-layout machine, jig-boring machine

координа́т·н·ый, -ая, -ое, *(a)* co-ordinate

~ ме́р·к·а *(f)* romer (map grid)

~ навиг·а́ци·я *(f)* (air) grid navigation

~ систе́м·а *(f)* crossbar system (telephones)

координато́·граф·, -а, *(m)* (surv) coordinatograph

координато·ме́р·, -а, *(m)* (surv) romer

координа́тор·, -а, *(m)* coordinator; (air nav) vector resolving apparatus

координацио́н·н·ый, -ая, -ое, *(m)* coordination

~ гру́пп·а *(f)* (chem) coordination group, ligand

~ сфе́р·а *(f)* (phys) coordination sphere

~ числ·о́ *(n)* (phys) coordination number

коордни·а́ци·я, -и, *(f)* coordination

~, ближ·а́ющ·ая (phys) nearest-neighbour coordination

координ·и́ровани·е, -я, *(n)* coordinating

ру́ч·к·а координ·и́ровани·я раз··воро́т·а *(f)* (a/c) main turn control

координ·и́р·овать, -уют, *(imp and perf)* coordinate

коп. *(abbr)* of **копе́·йк·а** copeck(s)

копа́й·ск·ий, -ая, -ое, *(a)* copaiba; copaivic

~ бальза́м· *(m)* (chem) copaiba balsam

~ кисл·от·а́ *(f)* copaivic acid

копа́л·, -а, *(m)* copal (resin); (min) copaline, copalite

копал·и́н·, -а, *(m)* (chem) copalin

копа́л·ов·ый, -ая, -ое, *(a)* copal, copalic

копа́ль·н·ый, -ая, -ое, *(a)* copal

коп·а́ни·е, -я, *(n)* digging, excavating

коп·а́тел·ь *(m)* digger

коп·а́тельн·ый, -ая, -ое, *(a)* digging; (biol) fossorial

коп·а́·ть, -ют, *(imp)* dig, excavate; dig up; **-ся** *(pass)*; rummage, scratch about

копе́·йк·а, -и, *(g pl)* **-е́·ек·,** *(f)* (fin) copeck (1/100th of rouble)

ко́пел·ь, -я, *(m)* (met) copel

копе·по́д·ы *(nom pl)*, *(g pl)* **-ов,** *(m)* (zool) copepods, *Copepoda*

копёр·, -пр·а́, *(m)* pile driver; drop hammer, impact breaker; impact testing machine; (min) headgear, headframe

~, копр·о́в·ый piling frame, pile-driving frame; (met) impact testing machine, impact tester

~, крут·и́льн·ый torsion impact tester

~, маятни́к·ов·ый pendulum impact tester

~, про·хо́д·ческ·ий (min) sinking frame

~ Ша́рпи Charpy impact tester

ко́п·и *(nom pl)*, *(g pl)* **-ей,** *(f)* (obs) mines, pits коп·и́йн·ый, -ая, -ое, (a) of ко́п·и·я

коп·и́лк·а, -и, (f) collection box

коп·и́льник·, -а, (m) (met) mixer/receiver (of cupola)

коп·и́р·, -а, (m) (mech) master, former plate, guide block; (agr) contour follower (ploughing); (plast) former

коп·и́рк·а, -и, (g pl) -р·ок·, (f) carbon paper, copying-paper

коп·и́рн·ый, -ая, -ое, (a) of коп·и́р·; copy

коп·ирова́льн·о-множ·и́тельн·ая маши́н·а (f) (print) photo-composing machine

коп·ирова́льн·о-фре́зер·н·ый стан·о́к· (m) tracer/copy(ing) milling machine

коп·ирова́льн·о-шлиф·ова́льн·ый стан·о́к· (m) profile grinding machine

коп·ирова́льн·ый, -ая, -ое, (a) copy, copying, profiling; (phot) printing

~ аппара́т· (m) (phot) printer

~ маши́н·а (f) (print) copying machine

~ стан·о́к· (m) copying lathe, profiling machine; (phot) contact printing machine

~ у·стро́й·ств·о (n) copy turning attachment

коп·ирова́ни·е, -я, (n) (v n) of ко·пи́р·овать

коп·и́р·овать, -уют, (imp) copy; duplicate, reproduce; imitate

коп·иро́вщик·, -а, (m) copyist

коп·и́рующ·ий, -ая, -ее, (pres part act) of коп·и́р·овать; slave (of manipulator)

коп·и́тел·ь, -я, (m) storer, storage device

коп·и́ть, -я́т, (imp) collect up, store, accumulate

ко́п·и·я, -и, (f) copy, duplicate, replica; (phot) print

~, си́н·яя blueprint

ко́п·к·а, -и, (f) digging, excavating

ко·плана́р·ност·ь, -и, (f) (math) coplanarity

ко·плана́р·н·ый, -ая, -ое, (a) (math) coplanar

ко́пл·енн·ый, -ая, -ое, (past part pass) of коп·и́ть

копн·а́, -ы́, (nom pl) ко́пн·ы, (g pl) -пён·, (f) shock, stook (of corn sheaves)

копн·и́тел·ь, -я, (m) (agr) stacker

копн·и́ть, -я́т, (imp) stack, heap up

копно·во́з·, -а, (m) (agr) stacking combine

коп·н·у́ть, -у́т, (perf) dig, dig up

ко·полимериз·а́ци·я, -и, (f) see со·полимериз·а́ци·я

ко́пот·н·ый, -ая, -ое, (a) sooty, smoky

ко́пот·ь, -и, (f) soot, lamp-black

ко́пр·а, -ы, (f) (food) copra

копр·о́в·ый, -ая, -ое, (a) of копёр·

~ цех· (m) scrap breaking shop

~ шкив· (m) (min) pulley wheel, hoisting sheave

копро·ли́т·, -а, (m) (min) coprolite; (med) coprolith

копро·ста́з·, -а, (m) (med) coprostasis

копро·стери́н·, -а, (m) (chem) coprosterol

копро·фа́г·, -а, (m) (zool) coprophage

копт·е́ть, -я́т, (imp) become sooty, become covered with soot

копт·и́льн·ый, -ая, -ое, (a) curing, smoking

копт·и́льн·я, -и, (g pl) -лен·, (f) smoke-house/room/shed

копт·и́ть, -я́т, (imp) (intrans) smoke, emit smoke; soot up; (food) smoke, cure

копул·и́р·овать, -уют, (imp and perf) (hortic) make a tongue graft, graft; copulate

копул·иро́вк·а, -и, (g pl) -вок·, (f) (v n) of копул·и́р·овать; (hortic) tongue graft

копул·я́ци·я, -и, (f) (gen) copulation; (chem) coupling reaction

копункта́ль·н·ый, -ая, -ое, (a) (math) copunctal

копч·е́ни·е, -я, (n) (v n) of копт·и́ть; (pl) smoked foods

копч·ённ·ый, -ая, -ое, (past part pass) of копт·и́ть

копч·ёност·и (nom pl), (g pl) -ей, (f) smoked foods

копч·ён·ый, -ая, -ое, (a) smoked, cured

ко́пчик·, -а, (m) (anat) ообсуд, tail-bone

ко́пчик·ов·ый, -ая, -ое, (a) coccygeal

копч·у́шк·а, -и, (g pl) -шек·, (f) smoked fish

копы́л·ок·, -л·к·а, (m) (shipb) riband key

копы́л·ь·я (pl) (shipb) launching poppets

копы́т·ен·ь, -т·н·я, (m) (bot) asarum

копы́т·н·ый, -ая, -ое,   (*a*) hoof, hoofed, ungulate; (*pl*) (zool) ungulates, *Ungulata*

~ ма́сл·о   (*n*) (chem) neat's foot oil

копы́т·о, -а,   (*n*) hoof

ко́п·ь, -и,   (*f*) (*obs pl*) mines, pits

коп·ь·ё, -я́,   (*nom pl*) **ко́п·ь·я,** (*g pl*) **ко́п·и·й,** (*d pl*) **ко́п·ь·ям** (*n*) spear, lance

~, кисло·ро́д·н·ое   oxygen lance

копье·ви́д·н·ый, -ая, -ое,   (*a*) spear-shaped, hastate, lanceolate, spicular

коп·я́щ·ий, -ая, -ее,   (*past part act*) of **коп·и́ть**

кор·а́, -ы́,   (*f*) crust, shell, (biol) cortex, (bot) bark, rind, peel

~, дуб·и́льн·ая   tan bark

~, мозг·ов·а́я   (anat) cerebral cortex

~, площад·н·а́я,   (geol) areal crust

~, хи́н·н·ая   chinona bark

корабе́ль·н·ый, -ая, -ое,   (*a*) ship, ship's, marine; shipborne, seaborne

~ авиа́ци·я   (*f*) shipborne aircraft

~ архитект·у́р·а   (*f*) naval architecture

~ инженер·   (*m*) constructor (naval)

~ лес·   (*m*) shipbuilding timber

~ ма́клер·   (*m*) shipbroker

~ че́рв·ь   (*m*) (zool) toredo, marine borer

корабле·вожд·е́ни·е, -я,   (*n*) navigation (of ship), seamanship

корабле·круш·е́ни·е, -я,   (*n*) shipwreck

корабле·стро·е́ни·е, -я,   (*n*) shipbuilding

корабле·стро·и́тел·ь, -я,   (*m*) shipbuilder

корабле·стро·и́тельн·ый, -ая, -ое,   (*a*) shipbuilding

кора́бл·ь, -я́,   (*m*) ship; (arch) nave

~, вое́н·н·ый   warship

~ -во́з·дух· (GW) ship-to-air

~, воз·ду́ш·н·ый   airship; large aircraft

~, голов·н·о́й   (shipb) prototype; leading ship (of formation)

~, косм·и́ческ·ий   (rocket) space vehicle

~, лин·е́йн·ый   battleship

~, -ми́шен·ь, -и,   (*f*) target ship

~, один·о́чн·ый   vessel proceeding independently

~, одно·ти́п·н·ый   sister ship

~, па́рус·н·ый   full rigged ship

~ -стенд·, о́·пыт·ов·ый   trials ship

~, сторож·ев·о́й   escort vessel

~, уч·е́бн·ый   training ship

~, флагма́н·ск·ий   flag ship

корак·о́ид·, -а,   (*m*) (zool) coracoid

кора́лл·, -а,   (*m*) coral

коралл·и́н·а, -ы,   (*f*) coralline alga, nullipore

коралл·и́нов·ый, -ая, -ое,   (*a*) coralline

кора́лл·ов·ый, -ая, -ое,   (*a*) coral

корац·и́т·, -а,   (*m*) (min) coracite

корвуз·и́т·, -а,   (*m*) (min) corvusite

ко́рд·, -а,   (*m*) cord (for automobile tyres); (text) corduroy

~, об·рез·и́н·енн·ый   rubberized cord

~, про·пи́т·анн·ый   dipped cord

~, ре́д·к·ий   open-weave cord

~, суро́в·ый   untreated cord

~, хло́п·ков·ый   cotton cord

кордаи́т·ы (*nom pl*), (*g pl*) **-ов,**   (*m*) (pal bot) cordaites, *Cordaitales*

кордиами́н·, -а,   (*m*) (pharm) cordiamine

кордиери́т·, -а,   (*m*) (min) cordierite, iolite

кордили́н·а, -ы,   (*f*) (bot) cordyline

корди́т·, -а,   (*m*) (expl) cordite

ко́рд·н·ый, -ая, -ое,   (*a*) of **ко́рд·**

корд·ов·о́й, -а́я, -о́е,   (*a*) of **ко́рд·**

кордо́н·, -а,   (*m*) cordon

~, веер·обра́з·н·ый   (hortic) fan-trained cordon

корёж·ить, -ат,   (*imp*) bend, warp

корен·а́ст·ый, -ая, -ое,   (*a*) thickset, stumpy

корен·и́ться, -я́тся,   (*imp*) be rooted in

корен·н·о́й, -а́я, -о́е, (*a*) radical, trunk, main; native, indigenous

~ вал· (*m*) (mech) crankshaft, mainshaft

~ рас·со́л· (*m*) deep-well brine

~ теч·е́ни·е (*n*) main stream (river)

~ место·рожд·е́ни·е   (*n*) (geol) lode

~ по·ро́д·а (*f*) bedrock

ко́рен·ь, -рн·я,   (*nom pl*) **ко́рн·и,** (*g pl*) **корн·е́й,**   (*m*) root; zero (of a polynomial); heel (of railway points/switch)

~, гла́в·н·ый   (bot) tap-root, main root

~, квадра́т·н·ый   (math) square root

~, кра́тн·ост·и *n*   (math) nth root

~, кра́тн·ый   (math) multiple root

~, кры́·л·а   (a/c) wing root

~, куб·и́ческ·ий   (math) cube root

~, ли́ш·н·ий   (math) extraneous root

~, перв·и́чн·ый   primordial root

~, при·да́т·очн·ый   (bot) adventitious root

~, со·в·пад·а́ющ·ие   (*pl*) (math) equal roots

~, со·пряж·ённ·ые   (*pl*) (math) conjugate roots

~, стержн·ев·о́й   (bot) tap-root

корен·ья (*nom pl*), *see* **корен·ь**; (food) roots

кореш·óк·, -ш·к·á, (*m*) (bot) rootlet, radicle; back, backbone, spine (of book); counterfoil

~ гриб·á mushroom stalk

~, за·чáт·очн·ый (bot) radicle

корж·, -á, (*m*) (min) jutting-out rock in roof of a coal seam

корзи́н·а, -ы, (*f*) basket; cage

корзи́н·к·а, -и, (*g pl*) **-н·ок·,** (*f*) basket; (bot) calathide, anthodium

корзи́н·очн·ый, -ая, -ое, (*a*) of **корзи́н·к·а**

корзи́н·чат·ый, -ая, -ое, (*a*) basket, cage

кориáндр·, -а, (*m*) (bot) coriander, *Coriandrum sativum*

кориáндр·ов·ый, -ая, -ое, (*a*) coriander

~ мáсл·о (*n*) coriandrol

коридóр·, -а, (*m*) corridor, passage, gangway

~ греб·н·óго вáл·а (shipb) shaft passage

~ по·ни́ж·енн·ого давл·éни·я (meteor) col

~ трубо·про·вóд·ов (shipb) pipe space

кори́нф·ск·ий, -ая, -ое, (*a*) Corinthian

~ óрдер· (*m*) (arch) Corinthian order

Кориоли́с·а, си́л·а (*f*) (mech) Coriolis force

Кориоли́с·а, у·скор·éни·е (*n*) (dynam) Coriolis acceleration

корифé·й, -я, (*m*) coryphaeus, leading figure

кориф·одóнт·, -а, (*m*) (pal) coryphodont

кор·и́ц·а, -ы, (*i*) **-ей,** (*f*) (bot) cinnamon, *Cinnamomum zeylanicum*

коричнев·áт·ый, -ая, -ое, (*a*) brownish

кори́чнев·ый, -ая, -ое, (*a*) brown

~ оснóв·н·óй (*m*) basic brown

коричне·зём·, -а, (*m*) korichnezem (soil)

кор·и́чник·, -а, (*m*) (bot) cinnamon, *Cinnamomum*

корично·бути́л·ов·ый эфи́р (*m*) butyl cinnamate

корично·ки́сл·ый, -ая, -ое, (*a*) cinnamate of

кор·и́чн·ый, -ая, -ое, (*a*) cinnamon, cinnamic

~ альдеги́д· (*m*) cinnamic aldehyde, cinnamaldehyde

кор·и́чн·ый, -ая, -ое

~ дéрев·о (*n*) *see* **кор·и́ц·а**

~ кисл·от·á (*f*) (chem) cinnamic acid

кóр·к·а, -и, (*g pl*) **кóр·ок·,** (*f*) (*dim*) of **кор·á**; crust, scab; rind, peel; (met) shell, skin, chill zone

~, лед·ян·áя ice rind

кор·ковáт·ый, -ая, -ое, (*a*) subereous

корко·ви́д·н·ый, -ая, -ое, (*a*) crustate

кóр·ков·ый, -ая, -ое, (*a*) of **кóр·к·а**; crustal, crusted, cortical

корколи́т·, -а, (*m*) (build) corkboard

корм·, -а, (*nom pl*) **-á,** (*g pl*) **-óв,** (*m*) (agr) forage, fodder, feed, food (for animals)

~, жи́дк·ий feeding slop

корм·á, -ы́, (*f*) (naut) stern

корм·ёжк·а, -и, (*g pl*) **-жек·,** (*f*) feed, feeding

корм·и́ть, -ят, (*imp*) feed

кормл·éни·е, -я, (*n*) feeding

корм·ов·óй, -áя, -óе, (*a*) of **корм·** and **корм·á**

~ под·зóр· (*m*) (shipb) counter

~ с·вес· (*m*) (shipb) counter

~ трав·á (*f*) fodder grass

~ част·ь (*f*) (a/c) afterbody; after part, stern (of ship)

~ эквивалéнт· (*m*) fodder equivalent

кормо·за·пáр·ник·, -а, (*m*) (agr) food/ fodder steamer

кормо·меш·áлк·а, -и, (*g pl*) **-лок·,** (*f*) (agr) feed mixer

кормо·при·готов·и́тельн·ая маши́н·а (*f*) fodder processing machine

кормо·фило·гéн·и·я, -и, (*f*) (gen) cormophylogeny

кормо·фи́т·, -а, (*m*) (bot) cormophyte

корм·у́шк·а, -и, (*g pl*) **-шек·,** (*f*) trough, feed trough/rack

корн·еви́щ·е, -а, (*n*) (bot) rhyzome, rootstock

корн·ев·óй, -áя, -óе, (*a*) root, radical

корне·голóв·ые, -ых, (*pl decl as adj*) (zool) rhizocephalans, *Rhizocephala*

корне·éд·, -а, (*m*) (zool) root-borer, (*pl*) *Hepialidae*

корне·иск·áтел·ь, -я, (*m*) root finder (computer)

корне·клубне·мóй·к·а, -и, (*g pl*) **-ó·ек·,** (*f*) vegetable washing machine

корне·нóж·к·а, -и, (*g pl*) **-ж·ек·,** (*f*) (zool) rhizopodium, rhizopod

корне·плóд·, -а, (*m*) root crop, root vegetable

корне·рéз·к·а, -и, (*g pl*) **-з·ок·,** (*f*) (agr) root cutter

корниш ·о́н·, -а, (*m*) gherkin

корню́ри (*pl*) hot gas con
coke oven) duits (of

Корню́, спира́л·ь (*f*) (opt) Cornu spiral

Корню́-Га́ртман·а фо́рмул·а (*f*)
(opt) Cornu–Hartman formula

ко́роб·, -а, (*nom pl*) **-а́,** (*g pl*) **-о́в,** (*m*)
box, chest; basket, hamper; duct

~, от·вод·я́щ·ий exhaust duct

коро́б·ить, -ят, (*imp*) warp, buckle

коро́б·к·а, -и, (*g pl*) **-б·ок·,** (*f*) box;
(elec) distribution box; (build) shell

~ агрега́т·ов (a/c) accessory gear-box

~, анеро́ид·н·ая aneroid capsule

~, воз·ду́ш·н·ая wind box, air belt
(cupola furnace)

~, гряз·ев·а́я mud box (bilge pump)

~, двер·н·а́я (build) door-frame

~, диафра́гм·енн·ая (electrochem)
cathode box

~, дым·ов·а́я smoke box (boiler)

~, золот·нико́в·ая steam chest

~, кла́пан·н·ая (st eng) valve box,
valve chest

~, крест·о́в·ая ceiling rose (light
fitting)

~, ла́мп·ов·ая (elec) lights distribution
box

~, магази́н·н·ая magazine (of rifle)

~, магистра́ль·н·ая (elec) mains distribution box

~, на·би́в·очн·ая (st eng) stuffing box

~, ни́ж·н·яя (rail) keep

~, огн·ев·а́я combustion chamber;
(rail) inner firebox

~ от·бо́р·а мо́щ·ност·и (М/Т) transfer case

~, от·ветв·и́тельн·ая (elec) distribution box

~, паро·рас·пре·дел·и́тельн·ая steam
box

~, пере·да́т·очн·ая (М/Т) transfer
case

~ пере·да́·ч· gear-box

~ пере·да́·ч·, до·полн·и́тельн·ая
(М/Т) auxiliary gear-box (Brit), auxiliary transmission (U.S.)

~ пере·да́·ч·, много·ступ·е́нчат·ая
multiple-speed transmission

~ по·да́·ч· gear-box (machine tools)

~, пред·о·хран·и́тельн·ая (elec) fuse
box

~ при·во́д·ов gear-box

~, рас·пре·дел·и́тельн·ая distribution box; oil box lubricator (diesel)

~, регул·я́торн·ая control box, control panel (elec generator)

коро́б·к·а

~ са́ль·ник·а (st eng) stuffing box

~ скор·ост·е́й gear-box

~, сопл·ов·а́я nozzle box (turbine)

~, стволь·н·ая breech (of rifle)

коробл·е́ни·е, -я, (*n*) buckling; warping
(wood; acoustics)

коро́б·очк·а, -и, (*g pl*) **-чек·,** (*f*) (*dim*)
of **коро́б·к·а**; (bot) capsule, boll, pod

~, больш·а́я (a/c) bad-weather-timed
circuit, big box (for landing)

коро́б·очн·ый, -ая, -ое, (*a*) of **ко-
ро́б·к·а**

коро́б·чат·ый, -ая, -ое, (*a*) box-
shaped, box-like

~ ста́л·ь (*f*) channel bars (steel)

коро́в·а, -ы, (*f*) cow

~, у·до́й·н·ая good milker (of cow)

коро́в·ий, -ья, -ье, (*a*) cow, cow's,
bovine

~ горо́х· (*m*) (bot) common cow pea,
Vigna sinensis

кор·ов·о́й, -а́я, -о́е, (*a*) crustal, cortical

коро·е́д·, -а, (*m*) (ent) bark beetle, (*pl*)
Ipidae

корол·е́в·а, -ы, (*f*) queen

корол·е́вск·ий, -ая, -ое, (*a*) royal;
king's (at chess)

корол·е́вств·о, -а, (*n*) kingdom

корол·ёк·, -ль·к·а́, (*m*) (*dim*) of **ко-
ро́л·ь**; (met) regulus, biscuit (of
reduced metal); bead, bullion bead
(assay); (*pl*) splash (casting defect);
metal inclusion in slag; (zool) gold-
crest, kinglet, *Regulus*; (bot) red-
juice orange, *Citrus sinensis sanguinea*

коро́л·ь, -я́, (*m*) king

король·к·и́ (*nom pl*) of **корол·ёк·**

коромы́сл·о, -а, (*g pl*) **-сел·,** (*n*) yoke;
beam (of scales); rocker, rocker arm,
balance arm, balancing lever; (ent)
dragonfly

коро́н·а, -ы, (*f*) crown, corona

по·те́р·и от коро́н·ы (*pl*) (elec)
corona power loss

корон·а́льн·ый, -ая, -ое, (*a*) coronal;
(zool) coronal, frontal

~ ли́н·и·и (*pl*) (astron) coronal promi-
nences

корон·а́рн·ый, -ая, -ое, (*a*) (zool)
coronary

коро́н·и·й, -я, (*m*) (astron) coronium

корон·и́ровани·е, -я, (*n*) corona discharge

корон·и́рующ·ий, -ая, -ее, (*a*) (elec)
corona, corona-discharge

~ электро́д· (*m*) corona electrode

коро́н·к·а, -и, (*g pl*) **-н·ок·,** (*f*) crown; (min) crown bit

~, кресто·обра́з·н·ая cross bit

коронко·с·ним·а́тел·ь, -я, (*m*) crown remover (dentistry)

коро́н·н·ый, -ая, -ое, (*a*) of **коро́н·а** and **коро́н·к·а**

~ раз·ря́д· (*m*) (elec) corona discharge

~ част·ь (*f*) crown (of tyre)

короно́·граф·, -а, (*m*) (astron) corona-graph

короно·обра́з·н·ый, -ая, -ое, (*a*) coronal

коро́н·чат·ый, -ая, -ое, (*a*) crown-shaped, castellated

~ га́·йк·а (*f*) castellated nut

коро·об·ди́р·к·а, -и, (*g pl*) **-р·ок·,** (*f*) barker (forestry)

коро·об·ди́р·н·ый, -ая, -ое, (*a*) bark-stripping

коро́ст·а, -ы, (*f*) scab

коросте́л·ь, -я, (*f*) (orn) corn crake, *Crex crex*

корот·а́·ть, -ют, (*imp*) while away (time), spend (time)

коро́т·к·ий, -ая, -ое, (*a*) short

~ вз·лёт· (*m*) **и по·са́д·к·а** (*f*) (air) short take-off and landing, STOL

~ волн·а́ (*f*) short wave

~ за·мык·а́ни·е (*n*) (elec) short circuit

~ за·мык·а́ни·е, вит·ко́в·ое (*n*) turn-to-turn short circuit

~ за·мык·а́ни·е у·станов·и́вш·ееся (*n*) sustained short circuit

~ звук· (*m*) dot, short (sound signal)

~ ме́р·к·а (*f*) (text) short stick

~ про́·блеск· (*m*) dot, short (light signal)

коро́т·к·о- (*component*) short-, brachy-, brevi-

коротко·боро́зд·н·ый, -ая, -ое, (*a*) (zool) brevicolpate

коротк·ова́т·ый, -ая, -ое, (*a*) rather short

коротко·волн·ови́к·, -а́, (*m*) radio ham/amateur

коротко·во́лн·ов·ый, -ая, -ое, (*a*) short-wave

коротко·волокн·и́ст·ый, -ая, -ое, (*a*) short-fibred, short-staple

коротко·голо́в·ост·ь, -и, (*f*) (zool) brachycephaly

коротко·вы́·держ·анн·ый, -ая, -ое, (*a*) (nucl) short-decayed

коротко·де́й·стви·е, -я, (*n*) short-range action

коротко·за·ме́дл·енн·ое де́й·стви·е (*n*) short-period delay (of detonators etc.)

коротко·жив·у́щ·ий, -ая, -ее, (*a*) short-lived

коротко·за́·мк·нут·ый, -ая, -ое, (*a*) (elec) short circuited, shorted, squirrel-cage

~ а·син·хро́н·н·ая маши́н·а (*f*) repulsion-induction motor

~ об·мо́т·к·а (*f*) (elec) squirrel-cage winding

коротко·за·мык·а́тел·ь, -я, (*m*) short circuiting device

коротко·за·мык·а́ющ·ийся, -аяся, -ееся, (*a*) short-circuiting

коротко·ко́нус·н·ый, -ая, -ое, (*a*) brevicone

~ дроб·и́лк·а (*f*) short-head crusher

коротко·кры́·лый, -ая, -ое, (*a*) short-winged, (ent) brachypterous

коротко·ли́ст·н·ый, -ая, -ое, (*a*) short-leaved, brevifoliate

коротко·метра́ж·н·ый, -ая, -ое, (*a*) short (of films)

коротко·над·кры́·л·ый, -ая, -ое, (*a*) brachylytrous

коротко·но́г·ий, -ая, -ое, (*a*) brachypedous

коротко·пла́мен·н·ый, -ая, -ое, (*a*) short-flame (coal etc.)

коротко·пле́ч·ий, -ая, -ее, (*a*) short-armed, (mech) short-arm

коротко·про·бе́ж·н·ый, -ая, -ое, (*a*) short-run, (nucl) short-range

коротко·у́с·ый, -ая, -ое, (*a*) short-whiskered, (biol) brachycerous

коротко·хво́ст·ый, -ая, -ое, (*a*) bob-tailed, short-tailed, brevicaudate

коро́т·ыш·, -а́, (*i*) **-о́м,** (*m*) bolster (of bridge)

коро́ч·е (*comp*) of **коро́т·к·ий** shorter

ко́р·очк·а, -и, (*g pl*) **-чек·,** (*f*) (*dim*) of **ко́р·к·а**; shell, skin; incrustation

ко́рпи·я, -и, (*f*) lint

корпора́ц·и·я, -и, (*f*) corporation

ко́рпус·, -а, (anat) (*nom pl*) **-ы,** (*g pl*) **-ов,** (other) (*nom pl*) **-а́,** (*g pl*) **-о́в,** (*m*) body; housing, case, casing, shell; hull (ship, tank etc.); (shipb) body plan; (build) building, building block; (print) long primer; (mil) corps

~, герметиз·и́рованн·ый pressurized body

~ двиг·а́тел·я motor body

~, дипломат·и́ческ·ий diplomatic corps

ко́рпус
~ конденс·а́тор·а condenser shell
~ котл·а́ boiler shell
~, лёгк·ий outer hull (submarine)
~ под·ши́п·ник·а bearing shell
~, про́ч·н·ый (submarine) pressure hull
~ с·на·ря́д·а missile body
~ турби́н·ы stator, turbine casing
корпусо·образ·у́ющ·ая машин·а (f) can-forming machine
корпу́скул·а, -ы, (f) corpuscle
корпускуля́р·н·ый, -ая, -ое, (a) corpuscular
~ по·то́к· (m) (phys) corpuscular stream
корпус·н·о́й, -а́я, -о́е, (a) (mil) corps
ко́рпус·н·ый, -ая, -ое, (a) of ко́рпус·
корпусо·до·стро́·ечн·ый, -ая, -ое, (a) (shipb) fitting-out
корпусо·с·бо́р·очн·ый цех· (m) (shipb) hull assembly shop
коррег·и́ровани·е, -я, (n) correction
коррег·и́р·овать, -уют, (imp) correct
корректи́в·, -а, (m) amendment, correction
коррект·и́р·овать, -уют, (imp) correct; (gunn) adjust (fire), spot
коррект·иро́вк·а, -и, (g pl) -вок·, (f) correction, correcting; (mil) spotting
коррект·иро́вочн·ый, -ая, -ое, (a) corrective, correcting; (mil) spotting
коррект·иро́вщик·, -а, (m) (mil) spotter, spotter aircraft
коррект·и́рующ·ий, -ая, -ее, (pres part act) of коррект·и́р·овать; correcting, corrective
~ воз·де́й·стви·е (n) corrective action
~ звен·о́ (n) (autom) correction feedback device
~ у·стро́й·ств·о, по·след·ова́тель·н·ое (n) series-connected correction feedback device
корре́кт·н·ый, -ая, -ое, (a) correct, proper (of behaviour)
корре́ктор·, -а, (nom pl) -а́, (g pl) -о́в, (m) adjuster (instruments); (surv nav) corrector; (print) corrector of the press, proof-reader
~, выс·о́тн·ый (a/c) altitude mixture control
~ гиро·ко́мпас·а gyrocompass speed corrector
~ горизонта́ль·н·ого паралла́кс·а (gunn) convergence corrector
~ дерива́ци·и (gunn) drift corrector

корре́ктор
~ нул·я́ (instr) zero adjuster
~, пут·ев·о́й (air nav) track corrector
~ цвето·дел·е́ни·я (phot, TV) colour/color corrector
коррект·у́р·а, -ы, (f) correction; (print) proof reading; proof sheet, proof
~ ло́ци·и (naut) hydrographic note
корректу́р·н·ый, -ая, -ое, (a) of коррект·у́р·а
~ знак· (m) (print) proof-reader's mark
~ стан·о́к· (m) proof press
коррекцио́н·н·ый, -ая, -ое, (a) correction
корре́кци·я, -и, (f) correction
~ га́з·а throttle (helicopter)
~ квадра́нт·н·ой о·ши́б·к·и (nav) quadrantal correction
коэффицие́нт· корре́кци·и (m) extension coefficient (gearing)
~ нул·я́ (autom) zero offset correction
~, част·о́тн·ая (elec) frequency correction
корреляти́в·н·ый, -ая, -ое, (a) correlative
~ пре·об·раз·ова́ни·е (n) (math) correlation
корреля́тор·, -а, (m) correlation computer
корреляцио́н·н·ый, -ая, -ое, (a) of корреля́ци·я
~ от·нош·е́ни·е (n) (math) correlation ratio
корреля́ци·я, -и, (f) correlation
коэффицие́нт· корреля́ци·й (m) (math) correlation coefficient
~, криво·лин·е́йн·ая (math) curvilinear correlation
~, мно́ж·ественн·ая (math) multiple correlation
~, прямо·лин·е́йн·ая (math) linear correlation
~ по вре́мен·и time correlation
~, у·дал·ённ·ая long-range correlation
~, ча́ст·н·ая (math) partial correlation
корреспонде́нт·, -а, (m) correspondent
корреспонде́нт·ск·ий, -ая, -ое, (a) of корреспонде́нт· and корреспонде́нци·я
корреспонд·е́нци·я, -и, (f) correspondence, mail; report, dispatch
~, за·ка́з·н·а́я registered mail
~, ис·ход·я́щ·ая outgoing mail
~, прост·а́я non-registered mail

корридо́р·, -а, (*m*) *see* **коридо́р·**
корро́з·ийност·ь, -и, (*f*) corrosivity
корро́з·ийн·ый, -ая, -ое, (*a*) corrosive, corrosion
коррозио́нно·сто́й·к·ий, -ая, -ое, (*a*) corrosion-resistant
коррозио́н·н·ый, -ая, -ое, (*a*) corrosion
~ **рас·тре́ск·ивани·е** (*n*) (met) season cracking, corrosion cracking
~ **тре́щ·ин·а** (*f*) corrosion crack
~ **у·ста́л·ост·ь** (*f*) (met) corrosion fatigue
~ **я́зв·а** (*f*) (met) pitting
корро́з·и·я, -и, (*f*) corrosion
~, **атмосфе́р·н·ая** atmospheric corrosion
~ **блужд·а́ющ·ими то́к·ами** stray-current corrosion
~, **бы́стр·о воз·раст·а́ющ·ая** run-away corrosion
~, **межкристалли́т·н·ая** intergranular/intercrystalline corrosion
~, **ме́ст·н·ая** local corrosion
~, **механ·и́ческ·ая** stress corrosion
~, **нож·ев·а́я** knife-line corrosion
~, **оспо·ви́д·н·ая** pitting
~ **пя́тн·ами** local corrosion
~, **рас·тре́ск·ивающ·ая** season/corrosion cracking
~, **сплош·н·а́я** general surface corrosion
~, **электро·хим·и́ческ·ий** cathodic corrosion
корса́к·, -а́, (*m*) (zool) corsac, *Vulpes corsac*
корсе́т·, -а, (*m*) corset (garment)
ко́рт·, -а, (*m*) tennis court
Ко́рт·а, на·са́д·к·а (*f*) (mar eng) Kort nozzle
кортизо́н·, -а, (*m*) (pharm) cortisone
ко́ртик·, -а, (*m*) dirk
кортик·а́льн·ый, -ая, -ое, (*a*) cortical
кортико·стеро́ид·, -а, (*m*) (biochem) corticosteroid
кортико·стеро́н·, -а, (*m*) (biochem) corticosterone
корти́н·, -а, (*m*) (pharm) cortin
ко́рточки (*pl*) knees drawn up
сесть на ко́рточки squat down
сид·е́ть на ко́рточках squat
кору́нд·, -а, (*m*) (min) corundum
~, **жёлт·ый** oriental topaz
~, **зелён·ый** oriental emerald
~, **кра́с·н·ый** ruby
~, **си́н·ий** (min) sapphire

корч·а, -и, (*i*) **-ей,** (*nom pl*) **-и,** (*g pl*) **-ей,** (*f*) tree stump and root; (*pl*) (med) cramp
корч·ева́льн·ая маши́н·а (*f*) stump/root puller
корч·ева́ни·е, -я, (*n*) rooting out, uprooting; clearing ground
корч·ева́тел·ь, -я, (*m*) (civ eng) grubber, rootdozer, stump/bush-puller
корч·ева́ть, -у́ют, (*imp*) stub, root/grub up/out
корч·ёвк·а, -и, (*g pl*) **-вок·,** (*f*) = **корч·ева́ни·е**
корч·у́ющ·ий, -ая, -ее, (*pres part act*) of **корч·ева́ть**
ко́ршун·, -а, (*m*) (orn) kite, *Milvus*
коры́ст·н·ый, -ая, -ое, (*a*) mercenary
корысто·люб·и́в·ый, -ая, -ое, (*a*) self-interested
коры́т·н·ый, -ая, -ое, (*a*) trough-shaped, trough
~ **стал·ь** (*f*) channel bars (steel)
коры́т·о, -а, (*n*) trough, pan; face (of spade)
ко́р·ь, -и, (*f*) (med) measles
кор·ь·ё, -я́, (*n*) bark, tanbark
ко́р·юшк·а, -и, (*g pl*) **-шек·,** (*f*) (fish) smelt, (*pl*) *Osmeridae*
кор·я́в·ый, -ая, -ое, (*a*) uneven, rough; pock-marked; clumsy
кор·я́г·а, -и, (*f*) snag, stump and root
кос·а́, -ы́, (*nom pl*) **ко́с·ы,** (*f*) scythe; (geog) spit; plait, braid, tress
~' **мор·ск·а́я** (geog) off-shore bar
кос·а́р·ь, -я́, (*m*) mower, scytheman; chopper (tool)
ко́с·венн·ый, -ая, -ое, (*a*) indirect, oblique
~ **на·клон·éни·е** (*n*) indirect declension
~ **пад·éж·** (*m*) (gram) oblique case
косéканс·, -а, (*m*) (math) cosecant, cosec, $\cos^{-1}$
кос·и́лк·а, -и, (*g pl*) **-лок·,** (*f*) mower, mowing machine
~, **трёх·слéд·н·ая** triple-gang mower
кос·и́н·а, -ы, (*f*) obliquity
ко́синус·, -а, (*m*) (math) cosine, cos
~, **на·правл·я́ющ·ий** direction cosine
~ **фи** (elec) cos φ, power factor (of a circuit)
косинус·о́ид·, -а, (*m*) (math) cosine curve
кос·и́ть, -я́т, (*imp*) mow, mow down; **-я́т,** put on a slant, tilt; squint, look sideways; **-ся** (*pass*); slant, slope

кос·и́ц·а, -и, (*i*) -ей, (*f*) braid, plait; string, rope (of articles strung together)

косма́т·ый, -ая, -ое, (*a*) shaggy

космет·ик·а, -и, (*f*) cosmetics

космет·и́ческ·ий, -ая, -ое, (*a*) cosmetic

косм·и́ческ·ий, -ая, -ое, (*a*) cosmic, space

~ био·ло́г·и·я (*f*) space biology

~ по·сто·я́нн·ая (*f*) cosmological constant

~ луч·и́ (*pl*) cosmic rays

космо·го́н·и·я, -и, (*f*) (astron) cosmogony

~, планёт·н·ая planetary cosmogony

космо·гра́ф·и·я, -и, (*f*) cosmography

космо·дро́м·, -а, (*m*) spaceport

косм·о́ид·н·ый, -ая, -ое, (*a*) cosmoid

~ чешу·я́ (*f*) (fish) cosmoid scale

космо·лог·и́ческ·ий, -ая, -ое, (*a*) cosmological

космо·ло́г·и·я, -и, (*f*) cosmology

~, не·релятиви́ст·ск·ая non-relativistic cosmology

космо·на́вт·, -а, (*m*) cosmonaut, spaceman

космо·на́вт·ик·а, -и, (*f*) cosmonautics, astronautics, space travel

космо·поли́т·, -а, (*m*) (biol) cosmopolitan, cosmopolite

ко́смос·, -а, (*m*) (astron, bot) cosmos

космо·тро́н·, -а, (*m*) (nucl) cosmotron

косн·е́·ть, -ют, (*imp*) stagnate; stiffen

ко́сн·ост·ь, -и, (*f*) stagnation; backwardness

кос·н·у́ться, -у́тся, (*perf*) (+ *gen*) *see* кас·а́·ться

ко́сн·ый, -ая, -ое, (*a*) unprogressive, conservative-minded, reactionary, backward, stubborn; stagnant, inert

кос·о (*adv*) slantwise, aslant, obliquely

косо·бо́к·, -а, (*m*) (fish) turbot, *Rhombus maximus*

косо·бо́к·ий, -ая, -ое, (*a*) sloping-sided; bent

кос·ова́т·ый, -ая, -ое, (*a*) slightly askew; slightly sloping

кос·ови́ц·а, -ы, (*i*) -ей, (*f*) haymaking

кос·ови́чник·, -а, (*m*) (min) cross cut

косо·гла́з·и·е, -я, (*n*) (med) squint, strabismus

косо·гла́з·ый, -ая, -ое, (*a*) squint-eyed, strabismal

косо·го́р·, -а, (*m*) hillside, slope, declivity

косо·зу́б·ый, -ая, -ое, (*a*) helical, twisted, chevron

~ колес·о́ (*n*) helical gear

кос·о́й, -а́я, -о́е, (*a*) slanting, sloping, oblique, skew, diagonal; squinting; indirect

~ выс·от·а́ (*f*) slant height

~ лин·е́йчат·ая по·ве́рх·ност·ь (*f*) (math) non-developable surface

~ ли́н·и·и (*pl*) skew lines

~ с·ты́к· (*m*) (build) skew joint

~ у́гол· (*m*) oblique angle

~ шов· (*m*) (weld) diagonal fillet weld

косо·ла́п·ост·ь, -и, (*f*) (med) club-foot, talipes

косо·ла́п·ый, -ая, -ое, (*a*) (med) club-footed, talipedic

косо·ли́ст·н·ый, -ая, -ое, (*a*) (bot) plagiophyllous

косо·на·ре́з·анн·ый, -ая, -ое, (*a*) (text) cut on the cross, bias-cut

косо·пло́д·н·ый, -ая, -ое, (*a*) (bot) loxocarpous

косо·при·це́ль·н·ый, -ая, -ое, (*a*) oblique (artillery fire)

косо·све́т·, -а, (*m*) angle-reflector lighting unit

косо·симметр·и́чн·ый, -ая, -ое, (*a*) skew-symmetric

косо·сло·и́ст·ый, -ая, -ое, (*a*) (geol) cross-bedded

косо·сло́·й, -я, (*m*) cross-grain (in wood); (geol) cross-bed

косо·сло́й·н·ый, -ая, -ое, (*a*) cross-grained

косо·уго́ль·н·ый, -ая, -ое, (*a*) (math) oblique-angled, oblique; bevel, scalene

косоу́р·, -а, (*m*) (build) bridge/notch board

косо·у·сеч·ённ·ый, -ая, -ое, (*a*) truncated

ко́ст·е- (*component*) bone, oste, osteo-

косте·дроб·и́лк·а, -и, (*f*) bone-grinder

кост·ене́·ть, -ют, (*imp*) ossify

косте·образ·у́ющ·ий, -ая, -ее, (*a*) (zool) osteogenic, ossifying

костёр, -тр·а́, (*m*) bonfire; (min) chock, cog; (bot) brome grass, *Bromus*

косте·язы́ч·н·ые, -ых, (*pl decl as adj*) (fish) osteoglossids

Кости́нск·ого, явл·е́ни·е (*n*) (phot) Kostinsky effect

кост·и́ст·ый, -ая, -ое, (*a*) osseous, bony; (*pl as noun*) (pal) teleosts, *Teleostei*

кост·ля́в·ый, -ая, -ое, (*a*) gaunt, bony

костно·пузы́р·н·ые, -ых, (*pl decl as adj*) (fish) ostariophysans, *Ostariophysi*

костно·щит·ко́в·ые, -ых, (*pl decl as adj*) (pal) osteostracans, *Osteostraci*

ко́ст·н·ый, -ая, -ое, (*a*) (*and see* ко́ст·ян·о́й) bone, bony, osseous

~ ма́сл·о (*n*) bone oil, Dippel's oil

~ че́реп· (*m*) (zool) osteocranium

косто·е́д·а, -ы, (*f*) (med) caries

косто·пра́в·, -а, (*m*) osteopath, bone-setter

ко́ст·очк·а, -и, (*g pl*) -чек·, (*f*) (anat) ossicle; (bot) kernel, stone, pip

косточко·вы·бив·а́тел·ь, -я, (*n*) (food) stoner, seeder

костр·а́, -ы́, (*f*) (text) shive, shove, tow, chaff

костр·и́к·а, -и, (*f*) *see* костр·а́

костр·и́нк·а, -и, (*g pl*) -нок·, (*f*) *see* костр·а́

костро·бето́н·, -а, (*m*) flax-tow concrete

костр·о́в·ый, -ая, -ое, (*a*) of кос·тёр· and костр·а́

~ креп·ь (*f*) (min) cribbing

костыле·дёр·, -а, (*m*) (rail) spike drawer

костыле·за·би́в·щик·, -а, (*m*) (rail) spike driver

костыл·ёк·, -ль·к·а, (*m*) (*dim*) of кост·ы́л·ь

кост·ы́л·ь, -я, (*m*) crutch; spike, drive, cramp; (a/c) tail skid

~, пут·ев·о́й (rail) dog spike/nail

ко́ст·ь, -и, (*nom pl*) -и, (*g pl*) -е́й, (*i pl*) -я́ми, (*f*) (anat) bone, os

~, жж·ён·ая bone black (pigment), bone ash

~, из·мельч·ённ·ая bone meal (fertilizer)

~, ло́кт·ев·а́я (anat) ulna

~, осно́в·н·а́я basal/basilar bone

~, слон·о́в·ая ivory

костю́м, -а, (*m*) suit, costume, dress

~, выс·о́тн·ый (air) pressure suit

~, гермет·и́чн·ый pressure suit

~, компенс·и́рующ·ий anti-*g* suit

~, проти́во·пере·гру́з·очн·ый anti-*g* suit

кост·я́к, -а, (*m*) (anat) skeleton

кост·яни́к·а, -и, (*f*) (bot) stone berry, *Rubus saxatilis*; (ent) wireworm, click beetle larva

кост·я́нк·а, -и, (*g pl*) -нок·, (*f*) (bot) drupe

кост·ян·о́й, -а́я, -о́е, (*a*) of ко́ст·ь; *and see* ко́ст·н·ый

~ мук·а́ (*f*) bone meal

~ фарфо́р· (*m*) bone china

косу́л·я, -и, (*g pl*) -у́л·ь, (*f*) (zool) roe deer, *Capreolus capreolus*

кос·ы́нк·а, -и, (*g pl*) -нок·, (*f*) (mech) corner/gusset plate, bracket joint

кос·ьб·а́, -ы́, (*f*) mowing, scything

кос·я́к·, -а́, (*m*) jamb, cant, cheek; door post; (zool) shoal, school, herd, flock

кот·, -а́, (*m*) (zool) tom-cat, *Felis domestica*

кота́нгенс·, -а, (*m*) (math) cotangent

котёл·, -тл·а́, (*m*) boiler, kettle; (nucl) reactor; (hydr) accumulator; (min) shot chamber (blasting hole); (geol) pot-hole, kettle

~, батар·е́йн·ые (*pl*) elephant boiler

~, биту́м·н·ый (civ eng) tar boiler

~, ва́р·очн·ый (chem, food etc.) digester; curing pan (rubber)

~, водо·тру́б·н·ый water-tube boiler

~, воз·ду́ш·н·ый (hydr) air-loaded accumulator, hydro-pneumatic accumulator

~, вулканизацио́н·н·ый (rubber) vulcanizing kettle/pan

~, газо·тру́б·н·ый fire-tube boiler, centre-flue boiler

~, гипсо·ва́р·очн·ый (chem) gypsum calciner

~, го́л·ый (nucl) bare pile without reflector

~, гомо·ге́н·н·ый (nucl) homogeneous reactor

~, двух·сторо́н·н·ий цилиндр·и́ческ·ий double-ended Scotch boiler

~, дымо·га́р·н·ый economic boiler, multitubular boiler

~, жаро·тру́б·н·ый Lancashire boiler, Cornish boiler

~, за·пе́ч·н·ый waste-heat boiler

~ Ко́хран·а (st eng) Cochran boiler

~ Ла-Мо́нт·а (st eng) La Mont boiler

~, лед·нико́в·ый (geol) glacial pot-hole

~ Лёфлер·а (st eng) Loeffler boiler

~, масло·воз·ду́ш·н·ый (hydr) air-loaded accumulator, hydropneumatic accumulator

~, мыло·ва́р·енн·ый soap kettle/pan

~, на·по́р·н·ый (hydr) pressure tank, air-loaded accumulator

котёл

~, оборо́т·н·ый Scotch boiler, marine return-tube cylindrical boiler

~, огне·тру́б·н·ый fire-tube boiler

~, одно·бараба́н·н·ый single-drum boiler

~, пар·ов·о́й boiler, steam generator

~, прямо·то́ч·н·ый once-through boiler

~, сатурацио́н·н·ый saturation boiler (sugar-making)

~, секцио́н·н·ый sectional (water-tube) boiler

~, сёр·н·ый (geol) sulphur cauldron

~, сто·я́ч·ий vertical boiler

~, суд·ов·о́й секцио́н·н·ый водо--тру́б·н·ый Babcock and Wilcox boiler

~, суд·ов·о́й цилиндр·и́ческ·ий marine/Scotch boiler

~, транспорта́бель·н·ый transportable boiler

~, тре·уго́ль·н·ый three-drum boiler

~, трёх·бараба́н·н·ый three-drum boiler

~ -утилиз·а́тор·, -а, (*m*) waste-heat boiler

~, шатр·о́в·ый three-drum boiler

~, шотла́нд·ск·ий Scotch/marine boiler

~, я́дер·н·ый nuclear reactor

~ Я́рроу Yarrow boiler

котел·о́к·, -л·к·а́, (*m*) (*dim*) of **котёл·**; compass bowl; bowler hat, derby (U.S.); (mil) mess tin

коте́ль·н·ая, -ой, (*f decl as adj*) boiler room, boiler house

коте́ль·н·ый, -ая, -ое, (*a*) of **котёл·**; *see also* **коте́ль·н·ая** (*as noun*)

~ ка́мен·ь (*m*) boiler scale

~ кла́д·к·а (*f*) boiler setting

~ машин·и́ст· (*m*) boilerman

~ цех· (*m*) boiler shop

коте́ль·щик·, -а, (*m*) boilermaker

котида́ль·н·ый, -ая, -ое, (*a*) (ocean) cotidal

~ ка́рт·а (*f*) cotidal chart

~ ли́н·н·я cotidal line

кот·ик·, -а, (*m*) (*dim*) of **кот·**; (zool) fur-seal; sealskin; (bot) clover, *Trifolium arvense*

котило·за́вр·, -а, (*m*) (pal) cotylosaur, (*pl*) *Cotylosauria*

коти́п·, -а, (*m*) (bot) cotype

кот·и́р·овать, -уют, (*imp and perf*) (fin) quote

кот·иро́вк·а, -и, (*g pl*) **-вок·,** (*f*) (fin) quotation

котле́т·а, -ы, (*f*) (food) rissole

~, от·би́в·н·а́я cutlet, chop

котлето·де́л·ательн·ая маши́н·а (*f*) rissole machine

котл·ова́н·, -а, (*m*) (civ eng) foundation pit, excavation

котл·ова́нн·ый, -ая, -ое, (*a*) of **котл·ова́н·**

котл·ови́н·а, -ы, (*f*) (geog) basin, hollow, doline; (ocean) trough

котл·ови́нн·ый, -ая, -ое, (*a*) of **котл·ови́н·а**

котл·о́в·ый, -ая, -ое, (*a*) of **котёл·**

~ за·ря́д· (*m*) (min) concentrated charge

котло·над·зо́р·, -а, (*m*) boiler inspection; boiler inspectorate

котло·стро·е́ни·е, -я, (*n*) boilermaking

котло·стро·и́тельн·ый, -ая, -ое, (*a*) boilermaking

котон·иза́ци·я, -и, (*f*) (text) cottonizing

котон·и́н·, -а, (*m*) (text) cottonin, cottonized fibre/fiber

котто́н-маши́н·а, -ы, (*f*) (text) Cotton machine, flat-bed knitting-machine

котто́н-муто́н·а, явл·е́ни·е (*n*) (phys) Cotton–Mouton effect

котто́н·н·ый, -ая, -ое, (*a*) (text) Cotton, flat-bed knitting

~ маши́н·а (*f*) *see* **котто́н-маши́н·а**

кото́р·ый, -ая, -ое, (*pron*) which, who

Ко́ттрелл·а, атмосфе́р·а (*f*) (phys) Cottrell atmosphere

~ облак·а́ (*pl*) (phys) Cottrell atmosphere

ко́уш·, -а, (*m*) (naut) thimble eye

ко́фе (*m indecl*) coffee (drink); coffee bean; (bot) coffee, *Coffea*

~, со́·ев·ый soybean coffee

кофеи́н·, -а, (*m*) (pharm) caffeine

кофе·и́нк·а, -и, (*g pl*) **-нок·,** (*f*) coffee bean

кофе·и́нов·ый, -ая, -ое, (*a*) caffeine

кофе́·йник·, -а, (*m*) coffee pot

кофе́·йн·ый, -ая, -ое, (*a*) coffee

ко·ферме́нт·, -а, (*m*) (chem) coferment

ко·фу́нкци·я, -и, (*f*) (math) cofunction

кофферда́м·, -а, (*m*) coffer-dam

коффини́т·, -а, (*m*) (min) coffinite

кохе́рер·, -а, (*m*) (rad) coherer

коча́н·, -а́, *or* **-чн·а́,** (*m*) head (of cabbage)

коч·ева́ть, -у́ют, (*imp*) roam, rove; lead a nomadic life; (zool) migrate

коч·ёвк·а, -и, (*g pl*) **-вок·,** (*f*) roaming; migration

коч·ев·ой, -áя, -óе, (*a*) nomad, nomadic; (zool) migratory

кочегáр·, -а, (*m*) stoker, fireman

кочегáр·к·а, -и, (*g pl*) **-рок·,** (*f*) stokehold, stokehole

кочегáр·н·ый, -ая, -ое, (*a*) of **кочегáр·**

кочедыж·ников·ые, -ых, (*pl decl as adj*) (bot) *Polypodiaceae*

кочен·é·ть, -ют, (*imp*) grow stiff/numb

кóчен·ь, -чня́, (*m*) *see* **кочáн·** and **кочерыж·к·а**

кочерг·á, -и́, (*g pl*) **-рёг·,** (*f*) poker

кочерыж·к·а, -и, (*g pl*) **-ж·ек·,** (*f*) cabbage heart

кóч·к·а, -и, (*g pl*) **-ч·ек·,** (*f*) hummock, hillock, tussock

коч·ковáт·ый, -ая, -ое, (*a*) hummocky

коч·у́ющ·ий, -ая, -ое, (*pres part act*) of **коч·евáть;** nomadic

~ орýд·и·е (*n*) (mil) roving gun

кош·áч·ий, -ья, -ье, (*a*) cat, catlike, feline; (*pl used as noun*) *Felidae*

~ лемýр· (*m*) (zool) ring-tailed lemur, *Lemur catta*

кошёл·ь, -я́, (*m*) purse, wallet; (naut) raft; ammunition net

кошенúл·ев·ый, -ая, -ое, (*a*) cochineal, cochenillic

кошенúл·ь, -и, (*f*) (ent) cochineal, *Dactylopius coccus*

Кошú, не·рáвен·ств·о (*n*) (math) Cauchy's inequality

Кошú-Рúман·а, у·равн·éни·я (*pl*) (math) Cauchy–Riemann equations

кóш·к·а, -и, (*g pl*) **-ш·ек·,** (*f*) she-cat, cat; grapnel; pulley block; carriage, crab, trolley (of crane); (*pl*) climbing irons, grapplers

кошмáр·, -а, (*m*) nightmare

коэнзúм·, -а, (*m*) coenzyme

коэрцитúв·ност·ь, -и, (*f*) (phys) coercivity

коэрцитúв·н·ый, -ая, -ое, (*a*) coercive

~ сúл·а (*f*) coercive force

коэрцитú·метр·, -а, (*m*) coercimeter

коэффициéнт·, -а, (*m*) (*and see associated words*); coefficient, factor, ratio

~, без·раз·мéр·н·ый non-dimensional coefficient

~ вос·становл·éни·я restoring coefficient, recovery factor; coefficient of restitution (elasticity)

~ вос·становл·éни·я температý·р·ы temperature recovery factor

~ дилюци·и dilution coefficient

коэффициéнт

~, динам· úческ·ий dynamic factor; coefficient of impact

~ до·вéр·и·я confidence coefficient

~ за·грýз·к·и load factor

~ за·полн·éни·я coefficient of charge

~ за·полн·éни·я нес·ýщ·его вúнт·á rotor solidity ratio (helicopter)

~ из·мéн·чивост·и coefficient of variation

~ ис·пóльз·овани·я duty factor, output coefficient

~ ис·пóльз·овани·я ус·танóв·к·и plant factor

~ конвéрси·и conversion factor

~ мóщ·ност·и power factor

~ на·един·úц·у мáсс·ы mass coefficient

~ на·прáвл·енн·ого дéй·стви·я (rad) antenna gain

~ пере·хóд·а conversion factor

~ подъ·ём·н·ой сúл·ы lift coefficient

~ полéз·н·ого дéй·стви·я efficiency, performance

~ полéз·н·ого дéй·стви·я, индикá·тор·н·ый actual efficiency, indicated efficiency

~ полéз·н·ого дéй·стви·я котл·á boiler efficiency

~ полéз·н·ого дéй·стви·я, механ·úческ·ий mechanical efficiency

~ полéз·н·ого дéй·стви·я, óбщ·ий overall efficiency

~ полéз·н·ого дéй·стви·я, от·нос·ú·тельн·ый relative efficiency

~ полéз·н·ого дéй·стви·я, резуль·т·úрующ·ий net efficiency

~ полéз·н·ого дéй·стви·я, терм·úческ·ий ideal thermal efficiency, air standard efficiency

~ полéз·н·ого дéй·стви·я, эконом·úческ·ий overall efficiency

~ полéз·н·ого дéй·стви·я, эффектúв·н·ый overall efficiency

~, по·прáв·очн·ый correction factor

~, порóг·ов·ый threshold coefficient

~ по·ряд·ков·ой коррелЯци·и rank correlation coefficient

~ раз·брóс·а scatter coefficient

~ рас·шир·éни·я expansion ratio

~ с·жим·áемост·и compressibility factor

~ скольж·éни·я slip factor

~, стáр·ш·ий leading coefficient

~ тЯг·и thrust force coefficient

кпд (*abbr*) of **коэффициéнт· полéз·н·ого дéй·стви·я** efficiency

краб·, -а, (*m*) (zool) crab, *Brachyura*

~, голуб·ой blue crab, *Callinectes sapidus*

~, зелён·ый common shore crab, *Carcinus maenas*

~, камчат·ск·ий Kamchatka crab, *Paralithodes camtschatica*

~, сухо·пут·н·ый land crab, *Gecarcinus lateralis*

крабо·лов·, -а, (*m*) crabber (boat, man)

кра·вш·ий, -ая, -ее, (*pres part act*) of **крас·ть**

краг·, -а, (*m*) (geol) crag

краг·а, -и, (*f*) collar; (*pl*) leggings

крад·енн·ый, -ая, -ое, (*past part pass*) of **крас·ть**; stolen

крад·ен·ый, -ая, -ое, (*a*) stolen

крад·ёт (*pres 3rd sing*) of **крас·ть**

крад·ущ·ий, -ая, -ее, (*pres part act*) of **крас·ть**; stealing

крае·вед·ени·е, -я, (*n*) regional study

кра·ев·ой, -ая, -ое, (*a*) of **кра́·й**; (bot) marginal

~ за·да́·ч·а (*f*) (math) boundary-value problem

~ знач·ени·е (*n*) (math) boundary value

~ по·ток· (*m*) (TV) edging, fringing

~ угол· (*m*) (phys) angle of contact, contact angle

~ у·слов·и·е (*n*) boundary condition

~ эффект· (*m*) (elec) fringe/edge effect

крае·уголь·н·ый ка́мен·ь (*m*) corner stone

кра́ж·а, -и, (*i*) **-ей,** (*f*) theft, (law) larceny

кра́·й, -я, (*nom pl*) **кра·я́,** (*g pl*) **-ёв,** (*m*) edge, rim, margin, brim, border, boundary; land, territory, province, region

~, вы·ём·чат·ый (bot) sinuate margin

~, за·моч·н·ый hinge line

~, из·вил·ист·ый (bot) sinuate margin

~ ко́нус·а (geol) skirt of a cone

~ кра́тер·а crater lip/rim

~, от·жим·н·ый land (of plastics mould)

~ по·глощ·ени·я (phys) absorption edge/limit/ discontinuity

~, с·мыч·н·ой hinge line

крайзель·компас·, -а, (*m*) gyro compass

рай·испол·ком, -а, (*m*) Regional Executive Committee

рай·н·е (*adv*) extremely

кра́и·н·ий, -яя, -ее, (*a*) extreme, last, ultimate, outer

~ знач·ени·е (*n*) limit

~ мер·ы (*pl*) extreme/severe measures
по край·н·ей мер·е at least, at any rate

~ с·луч·а·й (*m*) emergency, last resort

край·ност·ь, -и, (*f*) extremity, extreme

кра·л (*past masc sing*) of **крас·ть**

кра́мбал·, -а, (*m*) (naut) cathead

кра́мбол·, -а, (*m*) (naut) cathead

Кра́мер·а·, теорем·а (*f*) (math) Cramer's rule

Кра́мерс·а, теорем·а (*f*) (phys) Kramers theorem

кра́н·, -а, (*m*) cock, tap, faucet; crane

~, аккумуля́тор·н·ый electric run-about crane

~ -бал·к·а, -и, (*f*) single-rail overhead travelling crane; (naut) cathead

~, ба́шен·н·ый tower crane

~, берег·ов·ой harbour crane, quayside crane

~, водо·мер·н·ый water gauge-cock

~, водо·раз·бор·н·ый hydrant

~, воз·ду́ш·н·ый air cock; air tap, pet cock

~, грейфер·н·ый grab crane

~, двух·ход·ов·ой two-way cock

~ гермет·иза́ци·и (a/c) pressurization control/switch

~ -де́ррик·, -а, (*m*) derrick-jib crane

~, дро́ссель·н·ый throttle valve

~ -жура́вл·ь, -я́, (*m*) mast-jib crane

~, за·пор·н·ый stopcock

~, ка́бель·н·ый cableway

~, кантилле́вер·н·ый cantilever crane

~, кат·у́ч·ий non-revolving crane

~, козл·о́в·ый gantry crane

~, колод·цев·ый (met) ingot pit crane, soaking pit crane

~, консо́ль·н·ый overhanging-beam jacking crane, cantilever crane, arm crane (garage)

~, кон·цев·ой (rail) coupling cock

~, магнит·н·ый magnet crane

~ машин·ист·а (rail) driver's brake valve

~, монта́ж·н·ый (civ eng) erecting crane

~, мост·ов·ой overhead travelling crane

~, на·стен·н·ый по·ворот·н·ый wall-mounted rotary jib crane

~, на·тяж·н·ой plug cock

~, одно·бал·очн·ый single-rail overhead travelling crane

кра́н
~, плов·у́ч·ий floating crane
~, по·воро́т·н·ый revolving crane
~, полз·у́ч·ий climbing tower crane
~, полно·по·воро́т·н·ый swing crane
~, полу·порта́ль·н·ый semi-portal crane
~, по·плав·ко́в·ый ball cock
~, порта́ль·н·ый portal crane
~, про́б·н·ый test cock, gauge/gage cock
~, про·дув·а́тельн·ый (st eng) drain cock
~, про·хо́д·н·о́й globe cock, straight-through cock
~, са́ль·ников·ый screwed-gland cock, packed-gland cock
~, с·лив·н·о́й drain cock, draw-off cock
~, с·пуск·н·о́й drain cock, draw-off cock
~, стрел·ов·о́й derrick crane, derrick-ing jib crane
~, стрел·ов·о́й по·воро́т·н·ый rota-ry-jib crane
~, тормоз·н·о́й (М/Т) brake control valve
~, угл·ов·о́й angle cock
~ у·правл·е́ни·я шасси́ (а/с) under-carriage selector lever
~, э́ллинг·ов·ый shipbuilding crane
кра́н·ец·, -н·ц·а, (i) -н·ц·ем, (m) (naut) fender; ready-use ammunition locker
~, мя́гк·ий coir fender
~, твёрд·ый spar fender
крани·а́льн·ый, -ая, -ое, (a) cranial
кра́н·ик·, -а, (m) (dim) of кран·
кранио·ло́г·и·я, -и, (f) craniology
кранио·ме́тр·и·я, -и, (f) craniometry
кра́н·ов·ый, -ая, -ое, (a) of кран·
~ ба́л·к·а (f) crane jib/boom
~ фе́рм·а (f) crane structure
кра́п·, -а, (m) specks, spots
кра́п·ать, -пл·ют, (imp) spot, speckle
крапи́в·а, -ы, (f) (bot) nettle, *Urtica*
~, жг·у́ч·ая stinging nettle, *Urtica urens*
крапи́в·ниц·а, -ы, (i) -ей, (f) (med) urticaria, nettle rash; (ent) thistle butterfly, *Vanessa urticae*
крапи́в·н·ый, -ая, -ое, (a) nettle; urticant, urticating; (pl as noun) (bot) nettle family, *Urticaceae*
кра́п·ин·а, -ы, (f) speck, speckle, spot
кра́п·инк·а, -и, (g pl) -нок, (f) speck, speckle, spot
~, цвет·н·а́я tracer, coloured thread (elec cable)

крапла́к·, -а, (m) madder lake (pigment)
кра́пл·енн·ый, -ая, -ое, (past part pass) of кра́п·ать; speckled, spotted
кра́пп·, -а, (m) (bot) madder, *Rubia tinctorum*
кра́п·чат·ый, -ая, -ое, (a) speckled, mottled, flecked, spotted
крарупиз·а́ци·я, -и, (f) (telecom) Krarup loading, distributed induc-tance, continuous loading
крарупиз·и́рованн·ый, -ая, -ое, (a) continuously loaded, Krarup
крас·а́вк·а, -и, (f) (bot) belladonna, *Atropa belladonna*
кра́сен· (masc) (s f) of кра́сн·ый
крас·и́в·ее (comp adj) more beautiful/handsome, finer
крас·и́в·ый, -ая, -ое, (a) beautiful, handsome
крас·и́льн·ый, -ая, -ое, (a) dye, dyeing
крас·и́льн·я, -и, (g pl) -лен·, (f) dye-works
крас·и́тел·ь, -я, (m) dye, dyestuff
~, ази́н·ов·ый azine dye
~, азо·кисл·о́тн·ый acid azo dyestuff
~, акриди́н·ов·ый acridine dye
~, бензантро́н·ов·ый benzanthrone dye
~, вы·трав·н·о́й dischargeable dye
~, диспе́рс·н·ый disperse dye
~, жиро·рас·твор·и́м·ый fat dye
~, кисл·о́тн·ый acid dye
~, ку́б·ов·ый vat dyestuff
~, лед·ян·о́й (chem) ice colour, ingrain dye
~, основ·н·о́й basic dye
~, печа́т·н·ый (text) printing colour
~, по·кры́в·н·ый grain-side (leather) dye
~, про·трав·н·о́й mordant dye
~, прям·о́й direct dye
~, реакционно·спосо́б·н·ый fibre reactive dye
~, сер·ни́ст·ый (chem) sulphide dye stuff
~, три·фенил·мета́н·ов·ый triphe-nylmethane dyestuff
~, холо́д·н·ый (chem) ice colour/colo
кра́с·ить, -ят, (imp) colour, colo paint; dye, stain
кра́с·к·а, -и, (g pl) -с·ок·, (f) pain colour, color, pigment; (print) ink
~, звуко·по·глощ·а́ющ·ая acousti antinoise paint
~, казеи́н·ов·ая casein paint
~, кле·ев·а́я distemper

крáс·к·а
~, крó·ющ·ая covering paint
~, лáк·ов·ая enamel
~, лит·éйн·ая (met cast) mould wash, facing
~, не·об·раст·áющ·ая (shipb) anti-fouling paint/compound
~, печáт·н·ая printer's ink
~, термо·чувств·и́тельн·ая therm-index paint
~, тёрт·ая paste paint
~, трéск·ающ·аяся crackle paint
~, фасáд·н·ая brick and concrete paint
~, форм·óвочн·ая (met cast) mould wash, facing
~, эмáл·ев·ая enamel paint
~, эмульсиóн·н·ая emulsion paint
краско·меш·áлк·а, -и, (g pl) -лок·, (f) paint mixer
краско·на·гнет·áтельн·ый бак· (m) paint container (for spray gun)
краско·пýльт·, -а, (m) paint spray, paint spraying machine, paint spray gun
краско·рас·пыл·и́тел·ь, -я, (m) paint spray
краско·тёр·к·а, -и, (g pl) -р·ок·, (f) paint mill
красн·é·ть, -ют, (imp) redden, flush, grow/turn red; show red
красно·арм·éец·, -éйц·а, (i) -éйц·ем, (m) Red Army soldier, Russian soldier
красно·бýр·ый, -ая, -ое, (a) russet
красн·овáт·ый, -ая, -ое, (a) reddish
красно·дерéв·ец·, -в·ц·а, (i) -в·ц·ем, (m) cabinet-maker
красно·дерéв·щик·, -а, (m) cabinet-maker
красно·дýб·н·ый, -ая, -ое, (a) vegetable-tanned (leather)
красно·зём·, -а, (m) (agr) krasnozem, red soil; (geol) terra rossa
~, карбонáт·н·ый calcareous red soil, krasnozem
красно·знамён·н·ый, -ая, -ое, (a) Red-Banner (title)
красно·кáл·, -а, (m) red heat
красно·лéс·ь·е, -я, (n) pine forest
красно·лóм·, -а, (m) (met) hot/red shortness
красно·лóм·к·ий, -ая, -ое, (a) (met) hot/red short
красно·лóм·кост·ь, -и, (f) (met) hot/red shortness
красно·плóд·н·ый, -ая, -ое, (a) (bot) erythrocarpous

красно·реч·и́в·ый, -ая, -ое, (a) eloquent
красно·стóй·к·ий, -ая, -ое, (a) (met) red-hard
красно·стóй·кост·ь, -и, (f) (met) red hardness
красн·от·á, -ы́, (f) redness
красн·ýх·а, (f) (med) German measles; (bot) red rot
крáсн·ый, -ая, -ое, (a), (s f) -сен·, -сн·а, -сн·о, red
~ дéрев·о (n) mahogany
~ желéз·няк· (m) (min) hematite
~ кров·ян·áя сол·ь (f) (chem) potassium ferricyanide
~ пятн·ó (n) hot spot
крас·от·á, -ы́, (nom pl) -óт·ы, (g pl) -óт·, (f) beauty
крáс·очн·ый, -ая, -ое, (a) paint; colour, color; colourful, coloured, colorful, colored; (print) ink
крáс·ть (pres 3rd sing, pl) крад·ёт, крад·ýт, (past masc sing) кра·л (imp) steal
крáс·ящ·ий, -ая, -ее, (pres part act) of крáс·ить
крáтер·, -а, (m) (geol) crater; (elec) arc crater, crater
~, по·бóч·н·ый (geol) subordinate vent
крáтер·н·ый, -ая, -ое, (a) of крáтер·
~ лáмп·а (f) carbon-arc lamp
кратеро·обрáз·н·ый, -ая, -ое, (a) crater-shaped, crateriform
крáт·к·ий, -ая, -ое, (a) short, brief, concise
кратко·врéмен·ност·ь, -и, (f) brevity, short duration
кратко·врéмен·н·ый, -ая, -ое, (a) short, brief, momentary; short-time, short-term, short-lived; transitory
кратко·срóч·ност·ь, -и, (f) shortness, brevity; short term
кратко·срóч·н·ый, -ая, -ое, (a) short-term; (fin) short-dated
крáт·кост·ь, -и, (f) brevity, conciseness
крáтн·ое, -ого, (n decl as adj) (math) multiple
~, наи·мéн·ш·ее óбщ·ее (math) least common multiple
~, цéл·ое integral multiple
крáтн·ост·ь, -и, (f) multiplicity factor, multiplying factor; frequency; expansion ratio (in cellular rubbers etc.)
~ об·мéн·а air changes per hour (ventilation)

крáтн·ост·ь

~ **о·хлажд·éни·я** condenser capacity (expressed as kg cooling water/kg condensed steam)

~ **свето·фи́льтр·а** (phot) filter factor

~ **тóк·а вы·плавл·éни·я** (elec) fusing factor

-крáт·н·ый, -ая, -ое, (*adj component*) -times, -fold

крáтн·ый, -ая, -ое, (*a*) (*see also* **крáтн·ое**) (*as noun*); (math) multiple; divisible

~ **интегрáл·** (*m*) (math) iterated/multiple integral

~ **об·раз·овáни·е** (*n*) multiple production

за·кóн· крáтн·ых от·нош·éни·й (*m*) law of multiple proportions

~ **тóч·к·а** (*f*) multiple point

кратóн·, -а, (*m*) (geol) craton

крат·чáйш·ий, -ая, -ее, (*a*) shortest, briefest

крафт-бумáг·а, -и, (*f*) Kraft paper

крáх·, -а, (*m*) failure

крахмáл·, -а, (*m*) starch

~**, живóт·н·ый** animal starch

~**, кукурýз·н·ый** corn starch

~**, мáйс·ов·ый** corn starch

~**, марáнт·ов·ый** arrowroot starch

крахмáл·ени·е, -я, (*n*) starching

~**, одно·стóрóн·н·ее** back starching

~ **по·груж·éни·ем** slop starching

крахмáл·истост·ь, -и, (*f*) starchiness

крахмáл·ист·ый, -ая, -ое, (*a*) starchy, amyloid

крахмáл·к·а, -и, (*f*) starching

крахмáль·н·ый, -ая, -ое, (*a*) starch, starchy, amylaceous

~ **пáток·а** (*f*) starch syrup, glucose

крац·евáльн·ый стан·óк· (*m*) (met) scratch-brushing machine

крац·евáни·е, -я, (*n*) (met) scratch brushing

крац·óвк·а, -и, (*f*) (met) scratch brushing

крáш·ени·е, -я, (*n*) dyeing

крáш·енн·ый, -ая, -ое, (*past part pass*) of **крáс·ить**

крáш·ен·ый, -ая, -ое, (*a*) painted, coloured, stained, dyed

креат·и́н·, -а, (*m*) (biochem) creatine

креат·ини́н·, -а, (*m*) (biochem) creatinine

креат·и́нов·ый, -ая, -ое, (*a*) creatinic

креатин·ур·и́·я, -и, (*f*) (med) creatinurea

креатин·фосфáт·, -а, (*m*) creatine phosphate

кревéт·к·а, -и, (*g pl*) **-т·ок·,** (*f*) (zool) shrimp, (*pl*) *Natantia*

крéдит·, -а, (*m*) credit, credit side (in accounting)

креди́т·, -а, (*m*) (fin) credit

~**, под·товáр·н·ый** commercial credit

про·дáж·а в креди́т· (*f*) credit sale

креди́т·н·ый, -ая, -ое, (*a*) of **креди́т·**

кредит·овáть, -ýют, (*imp and perf*) (fin) credit, give/grant a credit

крéдит·ов·ый, -ая, -ое, (*a*) of **крéдит·**

~ **сáльдо** (*n*) credit balance

кредитóр·, -а, (*m*) creditor; (*pl*) credit side (of account)

кредито·спосóб·н·ый, -ая, -ое, (*a*) (fin) solvent

крез·и́л·, -а, (*m*) (chem) cresyl

крези́л·ов·ый, -ая, -ое, (*a*) cresylic

~ **кисл·от·á** (*f*) cresylic acid

~ **спирт·** (*m*) cresol

крез·óл·, -а, (*m*) (chem) cresol

~**, техн·и́ческ·ий** cresylic acid

крезолáт·, -а, (*m*) cresylate

крезóль·н·ый, -ая, -ое, (*a*) cresol

~ **смол·á** (*f*) cresol resin

крезоти́н·ов·ый, -ая, -ое, (*a*) cresotinic, cresotic

~ **кисл·от·á** (*f*) cresotinic/cresotic acid

крéйсер·, -а, (*nom pl*) **-á,** *or* **-ы,** (*g pl*) **-óв** *or* **-ов,** (*m*) cruiser; cruising yacht

~**.-нос·и́тел·ь, -я,** (*m*) guided missile cruiser

крéйсер·ск·ий, -ая, -ое, (*a*) cruiser, cruising

~ **по·лож·éни·е** (*n*) (submarine) diving trim

~ **скóр·ост·ь** (*f*) (a/c) best speed for range; (naut) economical cruising speed

крéйцкопф·, -а, (*m*) (mech) cross-head, slide block

~**, за·кры́·т·ый** closed/box cross-head

~**, одно·стóрóн·н·ий** single-bar cross-head

крéйцкопф·н·ый, -ая, -ое, (*a*) cross-head, slide-block

~ **дви́г·атель·** (*n*) cross-head engine

крейцмéйсел·ь, -я, (*m*) cross-cut chisel

крéкер·, -а, (*m*) (chem) cracking plant; (*pl*) (food) crispbread, crackers

~**, аммиáч·н·ый** (weld) ammonia cracking plant

кре́кинг·, -а, (*m*) (chem) cracking; (*as prefix*) cracking, cracked

~ -бензи́н·, -а, (*m*) cracked gasoline

~ -биту́м·, -а, (*m*) cracked asphalt, pitch-type asphaltic bitumen

~ -газ·, -а, (*m*) cracking/cracked gas

~ -дестилла́т·, -а, (*m*) cracking distillate

~, катали́т·и́ческ·ий catalytic cracking

~ -мазу́т·, -а, (*m*) cracked residue

~ -о·ста́т·ок·, -т·к·а, (*m*) (oil) cracking/cracked residuum

~, терм·и́ческ·ий thermal cracking

~ -у·стано́в·к·а, -и, (*g pl*) **-в·ок·,** (*f*) cracking plant/unit

кре́м·, -а, (*m*) cream; (food) cream-filling

~, ланоли́н·ов·ый, (pharm) lanoline cream

~, сапо́ж·н·ый boot polish, shoe cream

кремалье́р·а, -ы, (*f*) rack and pinion

кремато́р·и·й, -я, (*m*) crematorium

кремён·ь, -мн·я́, (*m*) flint

кремл·ёвск·ий, -ая, -ое, (*a*) of **кремл·ь**

кремл·ь, -я́, (*m*) Kremlin

кре́мн·е- (*component*) silico-

кремне·водо·ро́д·, -а, (*m*) hydrosilicon, silicohydride, silicon hydride, silane

кремне·вольфра́м·ов·ая кисл·от·а́ (*f*) (chem) silicotungstic acid

кремн·ёв·ый, -ая, -ое, (*a*) flint, flinty; (chem) silicon, silicic; (geol) siliceous

~ ангидри́д· (*m*) silica, silicon dioxide

~ га́ль·к·а (*f*) flint pebble

~ кисл·от·а́ (*f*) silicic acid

~ на́·кип·ь (*f*) (geol) siliceous sinter

кремне·зём·, -а, (*m*) (min) silica, silicon dioxide

кремне·зём·ист·ый, -ая, -ое, (*a*) siliceous, silicic

~ по·ро́д·а (*f*) silicic rock

кремне·каучу́к·, -а, (*m*) silicone rubber

кремне·кисл·от·а́, -ы́, (*f*) silicic acid

кремне·ки́сл·ый, -ая, -ое, (*a*) (chem) silicic, silicate of

~ бери́лл·и·й (*m*) beryllium silicate

~ ка́ли·й (*m*) potassium silicate

~ на́тр·и·й (*m*) sodium silicate

кремне·орган·и́ческ·ий, -ая, -ое, (*a*) (*and see* **кре́мний-орга́н·и́ческ·ий**) silicone, organosilicon

кремне·угле·водоро́д·, -а, (*a*) silicane

кремне·фтори́д·, -а, (*m*) silicon fluoride

кремне·фтори́сто·водоро́дно·кисл·ый, -ая, -ое, (*a*) fluorosilicate (of), fluosilicate (of)

кремне·фтори́сто·водоро́д·н·ый, -ая, -ое, (*a*) hydrofluorosilicic, fluorosilicic, fluosilicic

~ кисл·от·а́ (*f*) (chem) hydro/fluorosilicic acid, fluosilicic acid

кремне·фто́р·ист·ый, -ая, -ое, (*a*) fluosilicate of

~ ба́р·и·й (*m*) barium fluosilicate

кре́мн·иев·ый, -ая, -ое, (*a*) silicon, siliceous

~ бро́нз·а (*f*) (met) silicon bronze

кре́мн·и·й, -я, (*m*) (chem) silicon, Si

~, азо́т·ист·ый silicon nitride

~, четырёх·фто́р·ист·ый silicon tetrafluoride

~, четырёх·хло́р·ист·ый silicon tetrachloride

кре́мний-орга́н·и́ческ·ий, -ая, -ое, (*a*) silicone

~ поли·ме́р·ы (*pl*) (chem) silicones

~ со·един·е́ни·е (*n*) organosilicon compound

кремн·и́ст·ый, -ая, -ое, (*a*) flinty; siliceous, silicon; silicide

~ ма́рган·ец· (*m*) manganese silicide

~ мета́лл· (*m*) silicide

~ по·ро́д·а (*f*) (geol) siliceous rock

~ ста́л·ь (*f*) silicon steel

кре́м·ов·ый, -ая, -ое, (*a*) of **крем·**; cream-coloured

кре́н·, -а, (*m*) (naut) list, heel; (a/c) bank, banking, rolling

моме́нт· кре́н·а (*m*) (aerodynam) rolling moment; (shipb) heeling moment

у́гол· кре́н·а (*m*) (naut) angle of heel; (air) angle of bank/roll, rolling angle; (gunn) level angle

кренг·ова́ни·е, -я, (*n*) (naut) careening

кренг·ова́ть, -у́ют, (*imp*) (naut) careen

кре́н·ев·ый, -ая, -ое, (*a*) of **крен·ь**

~ древес·и́н·а (*f*) glassy wood

крен·е́ни·е, -я, (*n*) (*v n*) of **крен·и́ть**

крен·и́ть, -я́т, (*imp*) roll/heel over; **-ся** list, heel over (ship); bank (a/c)

крен·ова́ни·е, -я, (*n*) *see* **кренг·о·ва́ни·е**

кре́ндел·ь, -я, (*nom pl*) **-и,** (*g pl*) **-е́й,** (*m*) (food) pretzel

крен·ов·о́й, -а́я, -о́е, (*a*) of **крен·**

~ магни́т· (*m*) heeling magnet (compass)

~ девиа́ци·я (*f*) heeling error

крен·ов·ый, -ая, -ое, (*a*) (*see also* крен·ов·ой) crenic
~ кисл·от·а (*f*) (chem) crenic acid
крено·мер·, -а, (*m*) (naut) inclinometer, clinometer
крено·метр· -а, (*m*) (naut) inclinometer, clinometer
крен·ь, -и, (*f*) = крен·ев·ая древес·ин·а glassy wood
кре·одонт, -а, (*m*) (pal) creodont
креоз·ол·, -а, (*m*) (chem) creosol
креозот·, -а, (*m*) (chem) creosote, creosote oil
креолин·, -а, (*m*) (pharm) creolin
креп·, -а, (*m*) (text) crepe; креп· (*past masc sing*) of креп·нуть
~, свет·л·ый pale crepe
креп·ёж·, -еж·а, (*i*) -еж·ом, (*m*) fastenings, fasteners
креп·ёжн·ый, -ая, -ое, (*m*) securing, fastening; (min) support
~ детал·ь (*f*) (eng) fastening
~ лес· (*m*) (min) pit props, mining timber, pitwood, timber props, timber supports
~ материал· (*m*) (min) supports, props
~ рам·а (*f*) (min) set, frame set
крепе·у·клад·чик·, -а, (*m*) (min) support placer
креп·ильщик·, -а, (*m*) (min) timberman, prop erector
креп·ированн·ый, -ая, -ое, (*a*) crepe, crimped
~ бумаг·а (*f*) crepe-paper
крепит·аци·я, -и, (*f*) (med) crepitation, crepitus
креп·ител·ь, -я, (*m*) binder, binding agent
~, лит·ейн·ый (met) core binder
креп·ительн·ый, -ая, -ое, (*a*) strengthening; binding
креп·ить, -ят, (*imp*) fasten, secure; strengthen; (min) support, timber; (naut) lash, make fast; furl, strike (sails)
креп·к·ий, -ая, -ое, (*a*) strong, firm, sturdy, robust
~ вод·к·а (*f*) (chem) aqua fortis, concentrated nitric acid
~ щёл·ок (*m*) strong lye
крепл·ени·е, -я, (*n*) fastening, fixing, securing; fastenings; (min) supporting, supports, timber; (oil) casing (action and tubing); bonding (rubber); fortifying, strengthening (wines)
~, тюбинг·ов·ое (min) tubbing, circular steel support

крепл·ени·е
~ лопат·ок· blade fixing (turbine)
крепл·ённ·ый, -ая, -ое, (*past part pass*) of креп·ить
креп·нуть, -ут, (*past masc sing*) креп· (*imp*) get stronger/harder
креп·ов·ый, -ая, -ое, (*a*) of креп·
креп·ок·, (*m*) (*s f*) of креп·к·ий
креп·остн·ой, -ая, -ое, (*a*) of кре·п·ост·ь; serf (historical)
креп·ост·ь, -и, (*f*) strength; fortress, stronghold; deed of possession, title deed
коэффициент· креп·ост·и (*m*) tensile strength
~, рабоч·ая working load (rope)
~, раз·рыв·н·ая breaking stress (rope)
креп·ча·ть, -ют, (*imp*) get stronger/harder
креп·ч·е (*comp*) of креп·к·ий
креп·ш·ий, -ая, -ее, (*past part act*) of креп·нуть
креп·ь, -и, (*f*) (min) supports, set cribbing, props
~, за·бив·н·ая spills, spilling, forepoling
~, за·бой·щицик·ая face supports
~, костр·ов·ая cribbing
~, о·пуск·н·ая drop shaft
~, по·дат·лив·ая yielding props
~, при·за·бой·н·ая face/stope supports
~, с·вод·чат·ая arch support, steel arches
креп·ящ·ий, -ая, -ее, (*pres part act*) of креп·ить; strengthening, fastening
кресл·о, -а, (*nom pl*) -а, (*g pl*) -сел·, (*n*) armchair, chair, seat
~, катапульт·н·ое (a/c) ejection seat
кресл·овин·а, -ы, (*f*) (geog) cirque, corrie
кресс·, -а, (*m*) (bot) cress
~, вод·ян·ой water cress, *Nasturtium officinale*
~, о·город·н·ый garden cress, *Lepidium sativum*
крест·, -а, (*m*) cross
~ нит·ей (surv) crosshair; cross wires (of gunsight)
крест·ец·, -т·ц·а, (*i*) -т·ц·ом, (*m*) (anat) sacrum; shock, stook (of corn etc.)
крест·ик·, -а, (*m*) (*dim*) of крест·; (print) dagger
крест·н·ый, -ая, -ое, (*a*) of крест·
кресто·вид·н·ый, -ая, -ое, (*a*) cruciform

крест·овик·, -а,   (*m*) (bot) groundsel, *Senecio*

крест·овин·а, -ы,   (*f*) cross, cross-piece, cross-plate; turnstile; (rail) frog; spider (of universal joint), fourway union (pipes)

~ **кардан·а** trunnion cross

~ **с зуб·чат·ым обод·ом** azimuth gearwheel and frame (of gyrocompass)

крест·ов·ый, -ая, -ое, (*a*) of **крест·**

~ **с·вод·** (*m*) groin arch

крестоморит·, -а, (*m*) (min) crestomoreite

кресто·образ·н·ый, -ая, -ое, (*a*) cruciform, cross-shaped, (biol) cruciate

кресто·цвет·н·ые, -ых, (*pl decl as adj*) (bot) crucifers, *Cruciferae*

крест·цов·ый, -ая, -ое, (*a*) of **крест·ец·**

кресть·янск·ий, -ая, -ое, (*a*) peasant

кретин·, -а, (*m*) (med) cretin

кретон·, -а, (*m*) (text) cretonne

креш·ер·, -а, (*m*) (gunn) crusher gauge/gage

крибблер·, -а, (*m*) cribble, screen, sieve

крив·ая, -ой, (*f decl as adj*) (*and see associated words*) curve

~ **блеск·а** (*f*) (astron) light curve

~ **вероят·ност·и** probability curve

~, **во·об·раж·аем·ая** eye-fitted curve

~ **втор·ого по·ряд·к·а** quadratic curve

~, **граду·ировочн·ая** calibration curve

~, **за·мк·нут·ая** closed curve

~, **катенар·н·ая** catenary curve

~ **крат·чайш·его с·пуск·а** brachistochrone

~, **крут·ая** abrupt/steep curve

~, **кумулятив·н·ая** cumulative frequency curve

~, **лево·руч·н·ая** left-hand curve

~, **на·правл·яющ·ая** directrix

~ **на·раст·ани·я** build-up curve, growth/rise curve

~, **не·прав·ильн·ая** irregular curve

~, **о·гиб·ающ·ая** envelope curve

~, **остров·н·ая** peaky curve

~ **о·шиб·ок·** error function

~ **Пеано** Peano's curve

~ **пере·вод·н·ая** (rail) closure rail

~ **пере·гиб·ов** (math) curve through points of inflection

~, **пере·ход·н·ая** transition curve, adjustment curve, connection curve; (eng) fillet

крив·ая

~, **плав·н·ая** smooth curve

~, **плоск·ая** (math) curve in a plane, plane curve

~, **по·каз·ательн·ая** exponential curve

~, **по·лог·ая** flat curve

~, **про·стран·ственн·ая** (math) space curve, twisted curve, (air) skewed curve

~, **рабоч·ая** performance curve

~ **раз·гон·а** transient curve

~, **раз·ностн·ая** difference curve

~, **рас·пад·ающ·аяся** bipartite curve

~ **рас·пре·дел·ени·я** distribution curve

~ **синус·а** sine curve

~, **S-образ·н·ая** sigmoid curve

~, **степен·н·ая** exponential curve

~, **сумм·арн·ая** cumulative curve

~, **тауто·хрон·н·ая** tautochrone

~, **чёт·к·ая** well-defined curve

~, **экстремаль·н·ая** curve with a maximum

крив·изн·а, -ы, (*f*) curvature, camber; crookedness; curve

~, **верх·н·яя** (a/c) upper camber

~, **втор·ая** torsion

~ **кры·л·а** (a/c) wing camber

лин·и·я крив·изн·ы (*f*) line of curvature

~ **плоск·ой крив·ой** curvature of a plane curve

~, **полн·ая** total/gaussian curvature

~, **по·лож·ительн·ая** (dynam) positive camber

~, **про·стран·ственн·ая** compound curvature

~ **сеч·ени·я лопаст·и** blade-section camber

центр· крив·изн·ы (*m*) centre/center of curvature

крив·ить, -ят, (*imp*) bend, distort, curve

криво·бок·ий, -ая, -ое, (*a*) lop-sided

крив·ой, -ая, -ое, (*a*) (*see also* **крив·ая**) (*as noun*) curved; crooked

криво·лин·ейн·ый, -ая, -ое, (*a*) curvilinear

криво·лопаст·н·ый, -ая, -ое, (*a*) (bot) cyrtolobous

криво·нож·к·а, -и, (*g pl*) **-ж·ек·,** (*f*) contour pen

криво·нос·ый, -ая, -ое, -(*a*) crook-nosed

криво·рот·ый, -ая, -ое, (*a*) wry-mouthed

криво·шип·, -а, (*m*) crank, crank throw

~, **кон·цев·ой** overhung crank

криво·шип·н·о-коромы́сл·ов·ый меха́н·изм· (*m*) crank and rocker mechanism

криво·шип·н·о-кули́с·н·ый меха́н·изм· (*m*) crank and slider mechanism

криво·шип·н·ый, -ая, -ое, (*a*) crank, cranked, crank-operated/driven

кри́з·, -а, (*m*) *see* **кри́зис·**

кри́зис·, -а, (*m*) crisis; (econ) depression

~, волн·ов·о́й (aerodynam) shock stall

кри́к·, -а, (*m*) cry, shout

кри́к·н·уть, -ут, (*perf*) shout, cry, give a shout/cry

криминал·и́стик·а, -и, (*f*) criminal law

кримино·ло́г·и·я, -и, (*f*) criminology

крин·оид·е́·я, -и, (*f*) (zool) crinoid, (*pl*) *Crinoidea*

крино·ли́н·, -а, (*m*) (naut) crinoline, rudder shield

кри́н·ум·, -а, (*m*) (bot) crinum

крио·ге́н·н·ый, -ая, -ое, (*a*) cryogenic

~ те́хн·ик·а (*f*) cryogenic engineering, low-temperature engineering

крио·гидр·а́т·, -а, (*m*) cryohydrate

крио·скоп·, -а, (*m*) cryoscope

крио·скоп·и́ческ·ий, -ая, -ое, (*a*) cryoscope, cryoscopic

крио·скоп·и́·я, -и, (*f*) (chem) cryoscopy, cryoscopic method

крио·ста́т·, -а, (*m*) cryostat

крио·тро́н·, -а, (*m*) (elec) cryotron

крио·фи́ль·н·ый, -ая, -ое, (*a*) (biol) cryophil, cryophilic

кри́п·, -а, (*m*) (*and see* **полз·у́чест·ь**) (mech) creep

преде́л· кри́п·а (*m*) limiting creep stress

крипо·у·сто́й·чивост·ь, -и, (*f*) creep resistance

кри́пт·а, -ы, (*f*) crypt

~, ли́беркюн·ов·а (anat) Lieberkuhn's crypt

крипт·и́ческ·ий, -ая, -ое, (*a*) cryptic

~ о·кра́с·к·а (*f*) (zool) cryptic coloration

крипто·га́м·, -а, (*m*) (bot) cryptogam

крипто·ге́н·н·ый, -ая, -ое, (*a*) cryptogenic

крипто·гра́мм·а, -ы, (*f*) cryptogram

крипто·гра́ф·и·я, -и, (*f*) cryptography

крипто·кристалл·и́ческ·ий, -ая, -ое, (*a*) cryptocrystalline

крипто́л·, -а, (*m*) (chem) Kryptol

крипто·ме́р·, -а, (*m*) (gen) cryptomere

крипто·ме́р·и·я, -и, (*f*) (bot) Japanese cedar, *Cryptomeria japonica*

крипт·о́н·, -а, (*m*) (chem) krypton, Kr

крипто·перти́т·, -а, (*m*) (min) cryptoperthite

крипт·орх·и́зм·, -а, (*m*) (biol) cryptorchism

крипто·фи́т·, -а, (*m*) (bot) cryptophyte

крипсель-маши́н·а, -ы, (*f*) graining machine (leather)

кристади́н·, -а, (*m*) (rad) crystal detector

криста́лл·, -а, (*m*) (*see also under associated words*) crystal

~, вале́нт·н·ый, valence crystal

~, вращ·а́ющ·егося rotating crystal **ме́тод· вращ·а́ющ·егося криста́лл·а** (*m*) rotation method (of structural analysis)

~, генер·и́рующ·ий (rad) oscillating crystal

~, гране·центр·и́рованн·ый куб·и́ческ·ий face-centred cubic crystal

~, двой·нико́в·ый twin (crystal)

~, двояко·пре·ломл·я́ющ·ий birefringent crystal

~, дву·о́с·н·ый biaxial crystal

~, дендри́т·н·ый dendrite, arborescent crystal

~, ёл·очн·ый dendrite, arborescent crystal

~, жи́дк·ий (chem) liquid crystal, crystalline liquid, anisotropic liquid

~, за·рожд·а́ющ·ийся incipient crystal

~, за·тра́в·очн·ый seed, inoculating crystal

~, иго́ль·чат·ый acicular/needle crystal

~, изо́·гн·ут·ый bent crystal

~, ка́мер·н·ый chamber crystal

~, ле́в·ый left-handed crystal

~, недо·раз·ви́·вш·ийся immature crystal

~, не·со·верш·е́нн·ый imperfect crystal

~, ните·ви́д·н·ый whisker crystal, crystal whisker

~, объёмно·центр·и́рованн·ый куб·и́ческ·ий body-centered cubic crystal

~, одно·о́с·н·ый uniaxial crystal

~ с кана́л·ом well-type crystal

кристаллиз·áтор·, -a, *(m)* (chem) crystallizer, crystallizing tank/dish; (met) mould, mold (continuous casting)

~, **выс·óк·ий** (met) long mould/mold
~, **ни́з·к·ий** (met) short mould/mold

кристаллиз·ациóнн·ый, -ая, -ое, *(a)* crystallization, crystallizing

~ **вод·á** *(f)* (chem) water of crystallization

кристаллиз·áци·я, -и, *(m)* crystallization; (met) freezing, solidification

~, **втор·и́чн·ая** (met) recrystallization
~, **второ·степéн·н·ая** minor crystallization
~, **дрóб·н·ая** fractional crystallization
~, **перв·и́чн·ая** primary crystallization; (met) freezing, solidification
~, **с·лóж·н·ая** confused crystallization

температу́р·а кристаллиз·áци·и *(f)* cloud point (of oils)

кристаллиз·и́р·овать, -уют, *(imp and perf)* crystallize

кристаллиз·óванн·ый, -ая, -ое, *(past part pass)* of **кристаллиз·овáть;** crystallized

кристаллиз·овáть, -уют, *(imp and perf)* crystallize

кристáлл·ик·, -a, *(m)* *(dim)* of **кри-стáлл·;** crystalline/crystal particle

кристалл·и́т·, -a, *(m)* crystallite, crystal, grain

~, **перв·и́чн·ый** (met) primary crystal

кристаллит·н·ый, -ая, -ое, *(a)* crystallite, crystal, grain

кристалл·и́ческ·ий, -ая, -ое, *(a)* crystalline, crystal

~ **вещ·еств·ó** *(n)* crystalline solid
~ **детéктор·** *(m)* (rad) crystal detector
~ **решёт·к·а** *(f)* crystal lattice
~ **слáн·ец·** *(m)* (geol) crystalline schist
~ **триóд·** *(m)* (elec) transistor
~ **фиолéт·ов·ый** (chem) crystal violet

кристалл·и́чност·ь, -и, *(f)* extent of crystallization, crystallinity

кристалло·блáст·и́ческ·ий, -ая, -ое, *(a)* (geol) crystalloblastic

кристалло·гидрáт·, -a, *(m)* crystalline hydrate

кристаллóг·и·я, -и, *(f)* crystallogy

кристалло·грáмм·а, -ы, *(f)* crystallogram, X-ray diffraction pattern

кристаллó·граф·, -a, *(m)* crystallographer

кристалло·граф·и́ческ·ий, -ая, -ое, *(a)* crystallographic

кристалло·грáф·и·я, -и, *(f)* crystallography

кристалл·óз·а, -ы, *(f)* (chem) crystallose, soluble gluside

кристалл·óид·, -a, *(m)* crystalloid

кристалло·нóс·н·ый, -ая, -ое, *(a)* crystalliferous

кристалло·образ·овáни·е, -я, *(n)* crystal formation

кристалло·óпт·ик·а, -и, *(f)* crystal optics

кристалло·плáст·и́ческ·ий, -ая, -ое, *(a)* crystalloplastic

кристалло·фóсфор·, -a, *(m)* crystal phosphor

кристалло·хи́м·и·я, -и, *(f)* crystal chemistry

кристáль·н·ый, -ая, -ое, *(a)* crystal; crystal clear

кристобали́т·, -a, *(m)* (min) cristobalite

Кристóффел·я, си́мвол· *(m)* (math) Christoffel symbol

кри́т·, -a, *(m)* crith (unit of mass)

критéри·й, -я, *(m)* criterion

~ **знáч·имост·и** (math) significance test

кри́тик·, -a, *(m)* critic

кри́т·ик·а, -и, *(f)* criticism, review (of a book)

критик·овáть, -у́ют· *(imp)* criticize, review

критици́зм·, -a, *(m)* critical attitude

крит·и́ческ·ий, -ая, -ое, *(a)* critical

~ **температу́р·а** *(f)* critical temperature

~ **тóч·к·а** *(f)* critical point; (met) critical point/temperature; transformation point

~ **числ·ó оборóт·ов** *(n)* critical speed (of engine)

крит·и́чност·ь, -и, *(f)* criticality

коэффициéнт· крит·и́чност·и *(m)* criticality factor

крит·мáсс·а, -ы, *(f)* (nucl) crit, critical mass

кри́ц·а, -ы, *(i)* **-ей,** *(f)* (met) ball, puddle, puddle iron, pellet, wrought iron

крич·áть, -áт, *(imp)* shout, cry

крично·ру́д·н·ый процéсс· *(m)* (met) Krupp–Renn process

кри́ч·но·ый, -ая, -ое, *(a)* of **кри́ц·а**

~ **горн·** *(m)* (met) finery, Lancashire hearth

~ **желéз·о** *(n)* wrought iron

~ **пере·дéл·** *(m)* (met) Lancashire hearth process, Walloon process

крó·в·, -a, *(m)* shelter

кров·а́вик·, -а,   (*m*) (min) haematite, hematite

кров·а́в·ый, -ая, -ое,   (*a*) bloody; blood-red, red

крова́т·н·ый, -ая, -ое,   (*a*) bed, bedstead

крова́т·ь, -и,   (*f*) bed, bedstead

кро́в·е-   (*component*) (*see also* **кро́в·о-**) blood, sangui-, haemo-, haema-, haemato-, hemo,- hema-, hemato-

крове·за·мен·и́тел·ь, -я,   (*m*) (med) blood substitute

кро́вель·к·а, -и, (*g pl*) **-лек·,** (*f*) (bot) aril

кро́вель·н·ый, -ая, -ое,   (*a*) roofing, roof-covering

~ желе́з·о   (*n*) sheet iron

~ щеп·а́   (*f*) (build) shingles

~ рабо́т·ы   (*pl*) roofing, roof covering, covering roofs

крове·но́с·н·ый, -ая, -ое,   (*a*) (med) sanguiferous, blood

~ со·су́д·   (*m*) blood vessel

крове·тво́р·н·ый, -ая, -ое,   (*a*) (zool) haemopoietic, hemopoietic

кро́вл·я, -и, (*g pl*) **-вел·ь,** (*f*) (build) roofing, roof covering; (min) roof, back, hanging wall

кро́в·ност·ь, -и,   (*f*) (agr) pedigree

кро́в·н·ый, -ая, -ое,   (*a*) blood, haematic, hematic; thorough-bred

кро́в·о-   (*component*) (*see also* **кро́в·е-**) blood, sangui-, haemo-, hemo-,

крово·из·ли·я́ни·е, -я, (*n*) (med) extravasation; internal haemorrhage

крово·обращ·е́ни·е, -я, (*n*) circulation of the blood, (physiol) circulation

крово·о·стана́вл·ивающ·ий, -ая, -ое, (*a*) (med) styptic, anti-haemorrhagic

крово·под·тёк·, -а, (*m*) (med) bruise, ecchymosis

крово·пуск·а́ни·е, -я, (*n*) (med) bleeding

крово·со́с·к·а, -и, (*g pl*) **-с·ок·,** (*f*) (ent) tick, (*pl*) *Hippoboscidae*

крово·твор·е́ни·е, -я, (*n*) (zool) haemapoiesis, hemapoiesis, haematogenesis, hematogenesis

крово·теч·е́ни·е, -я, (*n*) (med) haemorrhage, hemorrhage, bleeding

крово·точ·и́ть, -а́т, (*imp*) bleed

крово·то́ч·ност·ь, -и,   (*f*) (med) haemophilia, hemophilia

кро́в·ь, -и, (*nom pl*) **-и,** (*g pl*) **-е́й,** (*f*) blood

~, драко́н·ов·а (chem) dragon's blood

~, за·пёк·ш·аяся cruor

~, кра́сн·ая erythrocytes

кров·яни́ст·ый, -ая, -ое,   (*a*) bloody, full of blood

кров·ян·о́й, -а́я, -о́е,   (*a*) blood; (zool) haemal, hemal, haematinic

~ сол·ь, жёлт·ая (*f*) potassium ferrocyanide

~ сол·ь, кра́сн·ая (*f*) potassium ferricyanide

~ у́гол·ь (*m*) blood charcoal

кро́·ет (*pres 3rd sing*) of **кры·ть** (*imp*)

кро́·енн·ый, -ая, -ое, (*past part pass*) of **кро·и́ть**

кро·и́ть, -я́т, (*imp*) cut out, cut (to a pattern)

кро́й·к·а, -и, (*g pl*) **кро́·ек·,** (*f*) cutting out

кро·и́те (*pl imper*) of **кро·и́ть**

кро́·йте (*pl imper*) of **кры·ть**

кроки́ (*n indecl*) sketch, sketch-map

крокидоли́т·, -а,   (*m*) (min) crocidolite

крок·иро́вк·а, -и, (*f*) (surv) field sketching

крокоди́л·, -а,   (*m*) (zool) crocodile, *Crocodylus*

крокоизи́т·, -а,   (*m*) *see* **крокойт·**

крокойт·, -а,   (*m*) (min) crocoit, crocoisite

кро́кус·, -а,   (*m*) rouge, crocus; (bot) crocus

кро́лик·, -а,   (*m*) (zool) rabbit, *Lepus*

кро́лич·ий, -ья, -ье,   (*a*) rabbit

кроманьо́н·ец·, -н·ц·а, (*i*) **-н·ц·ем,** (*m*) Cro-Magnon man

кро́ме (*prep + gen*) except; besides

~ того́ moreover, furthermore

кро́м·к·а, -и, (*g pl*) **-м·ок·,** (*f*) edge, border, rim, hem, shoulder; (text) selvage, list; (met) toe (of roll, section etc.)

~, вс·по·мог·а́тельн·ая end-cutting edge (of tool)

~, гла́в·н·ая side-cutting edge (of tool)

~, за́д·н·яя trailing edge; back edge (of tool)

~, звук·ов·а́я (aerodyn) sonic sweep

~ льд·а́ (ocean) ice edge, sea bar

~, от·рез·н·а́я (text) false selvage

~, пере́д·н·яя leading edge

~, по·пере́ч·н·ая chisel edge (of a drill)

~, ре́ж·ущ·ая cutting edge (of tool); lip (of a drill)

~, с·кош·ённ·ая chamfered edge

кромко·за·ви́в·очн·ая маши́н·а (*f*) can-sealing machine, flange/edge curling machine

кромко·за·ги́б·очн·ый, -ая, -ое, (*a*) beading (metal working)

~ **при·спосо̆бл·е́ни·е** (*n*) beader

кромко·крош·и́тел·ь, -я, (*m*) (met) side scrap cutter

кромко·строг·а́льн·ый стан·о́к· (*m*) (met) edge-planing machine, edger

кромко·фуг·а́льн·ый стан·о́к· (*m*) stave-jointing machine

кро́млех·, -а, (*m*) cromlech

кро́м·очн·ый, -ая, -ое, (*a*) of **кро́м·-к·а**

кромс·а́·ть, -ют, (*imp*) shred

крон·, -а, (*m*) yellow chromate (pigment); crown glass

~, **жёлт·ый** chrome yellow

~, **зелён·ый** chrome green, Guignets green

~, **свин·цо́в·ый** chrome yellow, lead chromate

~, **цинк·о́в·ый** zinc yellow (pigment), zinc chrome

кро́н·а, -ы, (*f*) top (of tree); krone (coin)

кронблок·, -а, (*m*) crownblock

кронгла́с·, -а, (*m*) crown glass

кронгла́сс·, -а, (*m*) *see* **кронгла́с·**

Кро́некер·а, си́мвол· (*m*) Kronecker's delta symbol

кронен·про́б·к·а, -и, (*g pl*) -б·ок·, (*f*) crown cork/cap (of bottle)

кронци́ркул·ь, -я, (*m*) callipers

~ **-за·клёп·очник·,** -а, (*m*) compasses (drawing instruments)

кронште́йн·, -а, (*m*) bracket, cantilever, angle bracket, (arch) corbel

~ **греб·н·о́го ва́л·а** (shipb) A bracket

~ **под·мото́р·н·ой ра́м·ы** (a/c) motor mount

~, **угл·ов·о́й** cornet bracket

~, **ца́пф·енн·ый** trunnion bracket

кроп·и́ть, -я́т, (*imp*) sprinkle, spray, spot (of rain)

кропот·ли́в·ый, -ая, -ое, (*a*) painstaking, laborious, meticulous

крос·пе́лент·, -а, (*m*) (surv, nav) cross bearing

кро́сс·, -а, (*m*) main distribution frame (telephones); cross-country race; (gen) cross

кроссбри́динг·, -а, (*m*) (gen) cross-breeding

кро́ссинг·, -а, (*m*) cross-over (piping etc.)

кросс·иро́вк·а, -и, (*g pl*) -вок·, (*f*) cross-connection field (telephones)

кросс-модул·я́ци·я, -и, (*f*) (rad) cross modulation

кроссо́вер·, -а, (*m*) (elec) crossover; (gen) crossover

кро́сс·ов·ый, -ая, -ое, (*a*) of **кросс·**

~ **про́·вод·** (*m*) jumper (telephones)

крот·, -а́, (*m*) (zool) mole, (*pl*) *Talpidae*; moleskin

кротал·я́ри·я, -и, (*f*) (bot) crotalaria

кротил хлори́д, -а, (*m*) (chem) crotyl chloride

крот·ова́ни·е, -я, (*n*) (agr) mole draining

крот·о́вин·а, -ы, (*f*) mole-hill

крот·о́в·ый, -ая, -ое, (*a*) of **крот·**

~ **дрен·а́ж·** (*m*) (agr) mole draining/drainage

крото·дрена́ж·н·ый, -ая, -ое, (*a*) mole-drainage, mole

~ **плуг·** (*m*) (agr) mole plough/plow

кротон·, -а, (*m*) (bot) croton

кротон·и́л·, -а, (*m*) (chem) crotonyl

кротон·ов·ый, -ая, -ое, (*a*) croton, crotonic

~ **альдеги́д·** (*m*) (chem) crotonaldehyde

~ **кисл·от·а́** (*f*) crotonic acid

~ **ма́сл·о** (*n*) croton oil

крох·а́, -и́, (*nom pl*) -о́х·и, (*d pl*) -а́м, (*g pl*) **крох·,** (*f*) crumb, scrap; (*pl*) scrapings, remains

крохал·ь, -я́, (*m*) (orn) merganser, *Mergus*

кро́х·отн·ый, -ая, -ое, (*a*) diminutive

кро́ш·енн·ый, -ая, -ое, (*past part pass*) of **крош·и́ть**

кро́ш·ен·ый, -ая, -ое, (*a*) crumbled, shredded, minced

крош·и́льн·ый, -ая, -ое, (*a*) crumbling, mincing, shredding

~ **стан·о́к·** (*m*) tobacco shredding machine

крош·и́мост·ь, -и, (*f*) friability

крош·и́ть, -я́т, (*imp*) crumble, shred, mince; **-ся** (*pass*); crumble, disintegrate

кро́ш·к·а, -и, (*g pl*) -ш·ек·, (*f*) (*v n*) *see* **крош·и́ть;** crumb; powder, chips

~, **алма́з·н·ая** diamond chips

~, **мра́мор·н·ая** marble chips

пресс·кро́ш·к·а, -и, (*f*) (plast) coarse moulding powder

кро́·ющ·ий, -ая, -ее, (*pres part act*) of **кры·ть**

~ **спосо́б·ност·ь** (*f*) hiding/covering power, coverage (of paints, photographic emulsion etc.)

кру́г·, -а, (*nom pl*) **-и́,** (*g pl*) **-о́в,** (*m*) circle, ring, coil, round, disc; sphere, range; (air) circuit; (met) round bar, round; lap (sport)

~, абрази́в·н·ый abrasive disc/wheel

~, азимут·а́льн·ый (surv, nav) azimuth circle, verge ring (compass)

~, артиллер·и́йск·ий (mil) director, aiming circle

~, больш·о́й (nav) great circle

~, вертика́ль·н·ый (surv, astron etc.) verticle circle

~, горизонта́ль·н·ый (surv) horizontal circle

~ ине́рц·и·и (mech) momental ellipsoid

~, кас·а́тельн·ый (math) tangential/inscribed circle (of a triangle)

~ крив·изн·ы́ (math) circle of curvature

~ крово·обращ·е́ни·я (physiol) circulation system

~, ма́л·ый (nav) small circle

~, на on average

~ на·пряж·е́ни·я (mech) Mohr's circle, Mohr's stress circle

~ о·жид·а́ни·я (air) holding circuit

~, пло́ск·ий disc wheel, flat wheel (grinding)

~, по·воро́т·н·ый turntable, turnplate

~, поля́р·н·ый polar/Arctic/Antarctic circle

~, пояс·н·о́й zonal circle

~ ра́вн·ых выс·о́т· (astron) circle of equal altitude, circle of position

~ с·клон·е́ни·я (astron) small circle of declination

~, со·при·кас·а́ющ·ийся (math) osculating circle

~, с·пас·а́тельн·ый (naut) lifebuoy

~ с·ход·и́мост·и (math) circle of convergence

~ -таре́л·к·а, -и, (*g pl*) **-л·ок·,** (*f*) dished wheel (grinding tool)

~, час·ов·о́й (surv) circle of declination

~ -ча́ш·к·а, -и, (*g pl*) **-ш·ек·,** (*f*) cup wheel (grinding tool)

~, шлиф·ова́льн·ый grinding wheel

кругл·е́ни·е, -я, (*n*) (*v n*) *see* **круг·л·и́ть;** backing (books)

кругл·е́·ть, -ют, (*imp*) grow/become round

кругл·и́льн·ый, -ая, -ое, (*a*) rounding

~ стан·о́к· (*m*) (print) backing machine

кругл·и́ть, -я́т, (*imp*) make round

кру́гл·о- (*component*) round, circular

кругло·бо́рт·н·ый, -ая, -ое, (*a*) (shipb) round-bilge

кругло·вяз·а́льн·ый автома́т· (*m*) automatic circular knitting machine

кругл·ова́т·ый, -ая, -ое, (*a*) roundish, rather round, slightly rounded

кругло·год·и́чн·ый, -ая, -ое, (*a*) year-round, all-year-round

кругло·год·ов·о́й, -а́я, -о́е, (*a*) year-round, all-year-round

кругло·гу́б·ц·ы (*nom pl*), (*g pl*) **-ев,** (*m*) round pliers

кругло·зуб·ча́т·ый, -ая, -ое, (*a*) (bot) crenate

кругло·ли́ст·н·ый, -ая, -ое, (*a*) (bot) rotundifolious

кругло·па́л·очн·ый стан·о́к· (*m*) wood dowelling machine

кругло·пи́ль·н·ый, -ая, -ое, (*a*) circular-sawing

~ стан·о́к· (*m*) circular saw

~ стан·о́к·, ребр·о́в·ый roller-feed circular saw

кругло·ресни́ч·н·ые, -ых, (*pl decl as adj*), (zool) *Peritrichida*

кругло·ро́т·ый, -ая, -ое, (*a*) (zool) cyclostomous, (*as noun*) cyclostome, (*pl*) *Cyclostomata*

кругло·су́т·очн·ый, -ая, -ое, (*a*) twenty-four-hour, round-the-clock

кругло·чул·о́чн·ый автома́т· (*m*) automatic tubular knitting machine

кругло·шлиф·ова́льн·ый стан·о́к· (*m*) circular grinding machine, plain cylindrical grinding machine

кру́гл·ый, -ая, -ое, (*a*) round, circular; whole (time)

~ депо́ (*n*) (rail) roundhouse

кру́г·ым с·чёт·ом in round figures

~ профи́л·ь (*m*) (met) round

~ руч·ни́к· (*m*) sash tool (paint brush)

кругл·я́к·, -а́, (*m*) round timber

круг·ов·о́й, -а́я, -о́е, (*a*) circular, all-round, cyclic

~ движ·е́ни·е (*n*) circular/rotary motion, rotation, circling, gyration

~ дву·уго́ль·ник· (*m*) crescent

~ де́й·стви·е (*n*) (rad) non-directive action, all-round action; (*used in gen as adj*) non-directive, all-round

~ диагра́мм·а (*f*) (elec) circle diagram; (st eng) valve diagram

~ много·уго́ль·ник· (*m*) arc polygon

~ пере·но́с· (*m*) end-around carry (computers)

~ проце́сс· (*m*) (phys, chem) cycle

~ ско́р·ост·ь (*f*) angular velocity

круг·ов·о́й, -а́я, -о́е

~ **фу́нкци·я** (*f*) (math) inverse trigonometric function, antitrigonometric function

круго·воро́т·, -а, (*m*) rotation, cycle, turnover

~ **азо́т·а** (biol) nitrogen cycle

~ **вод·ы́** (geog) hydrological cycle
вре́мя круго·воро́т·а (*n*) (com) turnover time

круго·вращ·а́тельн·ый, -ая, -ое, (*a*) circulatory, gyrating

круго·вращ·е́ни·е, -я, (*n*) circling

круго·дел·ённ·ый, -ая, -ое, (*a*) (math) cyclotomic

круго·зо́р·, -а, (*m*) outlook, view(s); range of vision, lookout sector

круг·о́м (*adv and prep+gen*) round, around; about turn/face (order)

круго·оборо́т·, -а, (*m*) circulation

~ **по·то́к·ов** convection

круго·обра́з·н·ый, -ая, -ое, (*a*) circular, round

круго·поля́р·н·ый, -ая, -ое, (*a*) circumpolar

круго·све́т·н·ый, -ая, -ое, (*a*) round-the-world

~ **пла́в·ани·е** (*n*) round-the-world voyage, circumnavigation

круж·а́л·о, -а, (*n*) riddle; (build) arch staging, centring

кру́ж·ев·о, -а, (*nom pl*) **круж·ев·а́, (*g pl*) -ев,** (*n*) lace

круж·е́ни·е, -я, (*n*) (*v n*) of **круж·и́ть;** gyration

круж·и́ть, -а́т, (*imp*) spin, whirl, circle, twist, go round, rotate in a circle, gyrate

кру́ж·к·а, -и, (*g pl*) **-ж·ек·,** (*f*) mug; can; (med) douche, irrigator

кру́ж·н·ый, -ая, -ое, (*a*) circuitous, roundabout

круж·о́к·, -ж·к·а́, (*m*) (*dim*) of **круг·** circle, society; washer; disc

~ **рас·се́·яни·я, наи·ме́нь·ш·ий** (phot) least circle of confusion

крукеси́т·, -а, (*m*) (min) crookesite

кру́кс·ов·о тём·н·ое про·стра́н·ств·о (*n*) (phys) Crook's dark space

круно́д·а, -ы, (*f*) (math) crunode

круп·, -а, (*m*) (med, vet) croup

круп·а́, -ы́, (*f*) hulled grain, groats

~, **гре́ч·нев·ая** buckwheat groats

~, **лед·ян·а́я** small hail

~, **ма́н·н·ая** wheat groats

~, **перл·о́в·ая** pearl barley

~, **сне́ж·н·ая** (meteor) soft hail

круп·и́нк·а, -и, (*g pl*) **-нок·,** (*f*) grain, granule

крупи́·т·чат·ый, -ая, -ое, (*a*) grainy, granular, granulated

~ **мук·а́** (*f*) 10% extraction flour, granular flour

круп·и́ц·а, -ы, (*i*) **-ей,** (*f*) crumb, grain

кру́п·к·а, -и, (*g pl*) **-п·ок·,** (*f*) grain, grit; (food) semolina; (bot) whitlow grass, *Draba*

крупн·е́·ть, -ют, (*imp*) grow, increase, become larger/heavier/coarser

кру́пн·о- (*component*) large, coarse, macro-, megalo-

крупно·би́·т·ый, -ая, -ое, (*a*) broken into large pieces, lumpy

~ **лёд·** (*m*) light floe (ice)

крупно·габари́т·н·ый, -ая, -ое, (*a*) large, large-size

крупно·зём·, -а, (*m*) coarse earth

крупно·зерн·и́ст·ый, -ая, -ое, (*a*) coarse-grained, macrogranular

крупно·кали́ер·н·ый, -ая, -ое, (*a*) (gunn) heavy-calibre

крупно·кристалл·и́ческ·ий, -ая, -ое, (*a*) macrocrystalline, (min) phanerocrystalline

крупно·масшта́б·н·ый, -ая, -ое, (*a*) large-scale

крупно·ме́р·н·ый, -ая, -ое, (*a*) large-scale, large

крупно·мо́лот·ый, -ая, -ое, (*a*) coarse-ground, coarse

крупно·пане́ль·н·ый, -ая, -ое, (*a*) large-panel

~ **стро·и́тельств·о** (*n*) frame-and-panel construction

крупно·пло́д·н·ый, -ая, -ое, (*a*) (bot) large-fruited, megalocarpous

крупно·по́р·ист·ый, -ая, -ое, (*a*) macroporous, large-pored

крупно·раз·ме́р·н·ый, -ая, -ое, (*a*) large, big

крупно·сер·и́йн·ый, -ая, -ое, (*a*) large-scale, multiple (production manufacture)

крупно·со́рт·н·ый, -ая, -ое, (*a*) large-variety; (met) heavy-section

~ **стан·** (*m*) (met) heavy-section mill

~ **стал·ь** (*f*) heavy steel sections

крупно·тонна́ж·н·ый, -ая, -ое, (*a*) large (of ships), large-tonnage

кру́пн·ост·ь, -и, (*f*) size, coarseness; fineness (of sand, cement, particles etc.)

~ **гра́в·и·я** size of gravel

крупно·ячé·йн·ый, -ая, -ое, (*a*) wide-meshed, large-meshed

крýпн·ый, -ая, -ое, (*a*) large, heavy, coarse; great, major, important, significant

~ **калúбр·** (*m*) (gunn) heavy calibre (larger than 155 mm or 100 mm in the case of anti-aircraft guns)

~ **план·** (*m*) (cinema) close-up

~ **рог·áт·ый скот·** (*m*) cattle

крупо·вé·ечн·ый, -ая, -ое, (*a*) groat-winnowing

крупо·вé·йк·а, -и, (*g pl*) **-вé·ек·,** (*f*) (agr) winnower

крупо·дёр·к·а, -и, (*g pl*) **-р·ок·,** (*f*) (agr) hulling machine

круп·óзн·ый, -ая, -ое, (*a*) (med, vet) croup

крупóн·, -а, (*m*) butt, crop (leather)

крупо·рýш·к·а, -и, (*g pl*) **-ш·ек·,** (*f*) (agr) hulling mill

круп·чáтк·а, -и, (*g pl*) **-ток·,** (*f*) 10% extraction flour, granular flour, sharps

круп·чáт·ый, -ая, -ое, (*a*) grainy, large-grained, granular

круп·ян· óй, -áя, -óе, (*a*) of **круп·á**

крусти·фик·áци·я, -и, (*f*) (geol) crustification

крут·изн·á, -ы́, (*f*) steepness; steep slope; (elec) transconductance

~, **меж·электрóд·н·ая актúв·н·ая** interelectrode transconductance

~, **меж·электрóд·н·ая комплéкс·-н·ая** interelectrode transadmittance

~, **пере·мéн·н·ая** (elec) variable mu

~, **по·сто·я́нн·ая** uniform twist (rifling)

~, **прогрессúв·н·ая** gaining/increasing twist (of rifling)

~ **с·кáт·а** angle of slope, gradient

крут·úльн·ый, -ая, -ое, (*a*) twisting, torsional, torsion

~ **аппарáт·** (*m*) (text) twisting mechanism

~ **вáтер·** (*m*) (text) doubling frame, twisting frame

~ **вáтер·, кольц·ев·óй** ring-doubling frame

~ **вес·ы́** (*pl*) torsion balance

~ **колеб·áни·е** (*n*) torsional vibration/oscillation

~ **машúн·а** (*f*) (text) doubling frame

~ **фáбрик·а** (*f*) (text) doubling mill

крут·úть, -я́т, (*imp*) twist, twirl, spin, turn; whirl, gyrate, spin

крýт·к·а, -и, (*f*) twist, twisting; (text) twist

~, **выс·óк·ая** (text) hard twist

круткó·метр·, -а, (*m*) (text) twist counter

крут·óй, -áя, -óе, (*a*) steep; abrupt, sudden, sharp; stern, drastic, rigorous

~ **úмпульс·** (*m*) steep pulse

~ **кипят·óк·** (*m*) water boiling strongly, strong boil

~ **тéст·о** (*n*) thick dough

~ **яйц·ó** (*n*) hard-boiled egg

круто·пáд·ающ·ий, -ая, -ее, (*a*) steeply falling/dipping, steep

круто·с·клáд·чат·ый, -ая, -ое, (*a*) closely folded

крут·я́щ·ий, -ая, -ее, (*pres part act*) of **крут·úть**

~ **момéнт·** (*m*) turning moment, torque

крýч·а, -и, (*i*) **-ей,** (*f*) steep slope

круч·éни·е, -я, (*n*) torsion, twisting; (text) doubling

ис·пыт·áни·е на круч·éни·е (*n*) torsion test

момéнт· круч·éни·я (*m*) turning moment, torque

на·пряж·éни·е при круч·éни·и (*n*) torsion stress

круч·ённ·ый, -ая, -ое, (*past part pass*) of **крут·úть**; twisted; spun (e.g. of balls in games)

круш·éни·е, -я, (*n*) crash, wreck; ruin

круш·úлк·а, -и, (*g pl*) **-лок·,** (*f*) crusher, grinder, disintegrator

круш·úн·а, -ы, (*f*) (bot) buckthorn, *Rhamnus*

круш·úнн·ые, -ых, (*pl decl as adj*) (bot) *Rhamnaceae*

круш·úть, -áт, (*imp*) smash, shatter

круш·úнов·ые, -ых, (*pl decl as adj*) *see* **круш·úнн·ые**

кры́·вш·ий, -ая, -ее, (*past part act*) of **кры·ть**

крыж·, -á, (*i*) **-óм·,** (*m*) cross; (naut) racking, seizing

крыжóвник·, -а, (*m*) (bot) gooseberry, *Grossularia*

крыл (*past masc sing*) of **кры·ть**

крыл·áтк·а, -и, (*g pl*) **-ток·,** (*f*) (*see also* **крыль·чáтк·а**); impeller (of pump etc.); (bot) samara; (zool) banded seal, *Histriophoca fasciata*

крылáто·о·каймл·ённ·ый, -ая, -ое, (*a*) (bot) peripterate

крыл·а́т·ый, -ая, -ое, (*a*) of **кры·л·о́;** winged, alate

крыле́чк·о, -а, (*nom pl*) **-и,** (*g pl*) **-чек·,** (*n*) (*dim*) of **крыль·ц·о́**

крыл·о́, -а́, (*nom pl*) **кры́·л·ья, -ьев,** (*n*) wing; vane, impeller; arm, sail (of windmill); (M/T) mudguard, wing, fender; bead (of tire); (geol) limb, leg, side, wall; **кры́·ло** (*verb 3rd sing neut past*) of **кры·ть**

~, **гидро·план·и́рующ·ее** hydrofoil

~, **дощ·а́т·ое** (naut) leeboard

~, **кольц·ев·о́е** (a/c) annular wing

~ **кон·е́чн·ого раз·ма́х·а** (a/c) finite-span wing

~ **лемех·а́** (agr) share wing

~ **ма́л·ого у·длин·е́ни·я** (a/c) low-aspect-ratio configuration

~, **не·стрело·ви́д·н·ое** (a/c) unswept/straight wing

~, **низко·рас·по·ло́ж·енн·ое** (a/c) low wing

~ **обра́т·н·ой стрело·ви́д·ност·и** swept-forward wing

~, **ова́ль·н·ое** (a/c) elliptical wing

~, **о·пущ·ённ·ое** (geol) downthrown side

~, **по·воро́т·н·ое** (a/c) pivoting wing

~, **под·во́д·н·ое** (shipb) hydrofoil

~, **по́д·нят·ое** (geol) upthrown side

~, **по́лн·ое** (a/c) gross wing

~, **при·по́д·нят·ое** (geol) upthrown side (of a fault)

~, **прям·о́е** straight wing

~ **прям·о́й стрело·ви́д·ност·и** swept-back wing

~, **раз·ре́з·н·о́е** (a/c) slotted aerofoil

~ **резона́нс·а** (*pl*) (phys) resonance tails

~ **с·бро́с·а** (geol) side/limb of fault, fault wall

~ **с·во́д·а** (geol) arch limb

~, **серпо·ви́д·н·ое** crescent wing

~, **скольз·я́щ·ее** oblique airfoil

~, **стрело·ви́д·н·ое** (a/c) swept wing

~ **структу́р·ы** (geol) flank
теори·я крыл·а́ (*f*) airfoil theory, wing section theory

~, **трапеце·ви́д·н·ое** (a/c) tapered wing

~, **тре·уго́ль·н·ое** (a/c) delta wing

~, **эллипт·и́ческ·ое** (a/c) elliptical wing

крыл·ова́тост·ь, -и, (*f*) twist (timber defect)

крыло·ви́д·н·ый, -ая, -ое, (*a*) wing-shaped, aliform, pteroid; (anat) pterygoid

крыло·не́б·н·ый, -ая, -ое, (*a*) (anat) pterygopalatine

крыло·но́г·ие, -их, (*pl decl as adj*) (zool) pteropods, *Pteropoda*

крыло·сем·я́нн·ый, -ая, -ое, (*a*) (bot) pterospermous

крыло·у́х·ие, -их, (*pl decl as adj*) (ent) *Pterygota*

кры́л·ышк·о, -и, (*g pl*) **-шек·,** (*n*) (*dim*) of **кры·л·о́·**

крыль·ц·о́, -а́, (*i*) **-ом,** (*nom pl*) **кры́ль·ц·а,** (*g pl*) **-л·е́ц·,** (*n*) (build) porch

крыль·ча́тк·а, -и, (*g pl*) **-ток·,** (*f*) impeller; vane (pumps, compressors); (horol) fly

~, **дву·сторо́н·н·яя** double-entry impeller, double-sided impeller

~ **на·гнет·а́тел·я** supercharger impeller

~, **одно·сторо́н·н·яя** single-entry impeller, single-sided impeller

~ **с·мес·и́тел·я** mixing impeller

крыль·ча́т·ый, -ая, -ое, (*a*) wing, winged, vaned

~ **колес·о́** (*n*); (shipb) Voith–Schneider propeller; impeller; cycloidal propeller

~ **на·со́с·** (*m*) wing pump

~ **рас·ходо·ме́р·** (*m*) turbine flow-meter

кры́л·ья (*nom pl*) of **крыл·о́**

кры́м·ск·ий, -ая, -ое, (*a*) Crimean

кры́с·а, -ы, (*f*) (zool) rat, *Rattus*

~, **бобр·о́в·ая** (zool) coypu, swamp beaver, *Myocaster coypus*

~, **канцеля́р·ск·ая** (pop) pen-pusher, office rat

~, **кенгур·о́в·ая** (zool) rat kangaroo, potoroo, (*pl*) *Potoroinae*

~, **му́скус·н·ая** (zool) muskrat, *Ondatra zibethica, Fiber zibethicus*

~, **се́р·ая** (zool) brown rat, Norway rat, *Rattus norvegicus*

~, **чёрн·ая** (zool) house rat, black rat, *Rattus rattus*

крыс·ёнок·, -нк·а, (*nom pl*) **-с·я́т·а·,** (*g pl*) **-с·я́т;** (*m*) young rat

крыс·и́д·, -а, (*m*) rat poison

кры́с·ий, -ья, -ье, (*a*) rat, rat's

крыс·и́н·ый, -ая, -ое, (*a*) rat, rat's

крысо·ло́в·, -а, (*m*) rat-catcher

крысо·ло́в·к·а, -и, (*g pl*) **-в·ок·,** (*f*) rat-trap

кры́·т·ый, -ая, -ое, (*past part pass*) of **кры·ть**

кры·ть, (*pres 3rd sing, pl*) **кро́·ет, кро́·ют,** (*past masc sing*) **кры·л,** (*imp*), cover; roof

кры́·ш·а, -и, (*i*) **-ей,** (*f*) roof

~, **гот·и́ческ·ая** hammer-beam roof

~, **дву·с·ка́т·н·ая** (build) ridge roof

~, **жа́бер·н·ая** (zool) operculum

~, **одно·с·ка́т·н·ая** lean-to roof, pent roof

~, **манса́рд·н·ая** (build) mansard roof, curb roof, gambrel roof

~, **шатр·о́в·ая** (build) hipped roof

кры́ш·ечк·а, -и, (*g pl*) **-чек·,** (*f*) (*dim*) of **кры́ш·к·а;** (biol) operculum, opercle

крышечко·ви́д·н·ый, -ая, -ое, (*a*) (biol) operculiform

крыше·ви́д·н·ый, -ая, -ое, (*a*) roof-like, tectiform

кры́ш·ечн·ый, -ая, -ое, (*a*) covering, roofing

кры́ш·к·а, -и, (*g pl*) **-ш·ек·,** (*f*) lid, cover, hood, cap, door; (zool) tegmen; (print) case

~, **брон·ев·а́я** (shipb) deadlight

~, **га́з·ов·ая** gas flap (type of valve)

~, **глух·а́я** (shipb) deadlight

~ **лю́к·а** (shipb) hatch cover

~, **пере·плёт·н·ая** bookbinding case

~ **под·ши́п·ник·а** bearing cap, keep

~, **рас·пре·дел·и́тельн·ая** oil filter element retaining plate

~ **стан·и́н·ы** (met roll) housing cap

~, **цили́ндр·ов·ая** (ICE) cylinder head

~, **што́рм·ов·а́я** (shipb) deadlight

крышко·де́л·ательн·ая маши́н·а (*f*) (print) case-making machine

крю́йс·, -а, (*m*) (naut) mizzen

крю́йс-пе́ленг·, -а, (*m*) (nav) cross-bearing

крюк·, -а́, (*m*) hook, crook, finger (of manipulation); detour

~ **и серьг·а́** hook and eye

~, **от·по́р·н·ый** boathook

~, **дву·ро́г·ий** double hook

~, **одно·ро́г·ий** single hook

~, **о·стана́вл·ивающ·ий** arrester hook

~, **от·кид·н·о́й** slip hook

~, **с·клад·н·о́й** clasp hook

~, **у·пряж·н·о́й** coupling hook, draw hook

крюк·ов·о́й, -а́я, -о́е, (*a*) of **крюк·**

крю́ч·ечник·, -а, (*m*) hooker-in (rolling mill operative)

крюч·кова́т·ый, -ая, -ое, (*a*) hooked, crooked; (bot) hamate hamose

крючко·ви́д·н·ый, -ая, -ое, (*a*) hook-like; (bot) uncate, uncinate

крюч·ко́в·ый, -ая, -ое, (*a*) hook, hooked

~ **ре́йк·а** (*f*) (hydr) hook gauge

крючко·обра́з·н·ый, -ая, -ое, (*a*) hook-shaped, unciform

крюч·о́к·, -ч·к·а́, (*m*) (*dim*) of **крюк·;** hook, catch, claw; fishhook; cramp iron; (zool) hamulus; (zool) uncus

~ **бараба́н·а** (horol) mainspring hook

~, **ветр·ов·о́й** casement catch (window)

~, **око́н·н·ый** casement catch (window)

~, **с·пуск·ов·о́й** trigger (gun)

крюч·о́чн·ый, -ая, -ое, (*a*) of **крюч·о́к·;** (bot) glochidiate

крю́ч·ья (*nom pl*) of **крюк·**

крюшо́н·, -а, (*m*) white-wine cup (drink)

кряж·, -а, (*i*) **-ем,** (*m*) (geog) ridge; block (of wood)

~, **под·во́д·н·ый** (ocean) ridge

кря́к·ань·е, -я, (*n*) quack, quacking; grunt

кря́к·в·а, -ы, (*f*) (orn) mallard, *Anas platyrhynchos*

ксант·а́т·, -а, (*m*) (chem) xanthate

ксант·гидро́л·, -а, (*m*) xanthydrol

ксант·е́ин·, -а, (*m*) (bot) xanthein

ксант·е́н·, -а, (*m*) (chem) xanthene

ксанте́н·ов·ый, -ая, -ое, (*a*) xanthene

ксант·и́н·, -а, (*m*) (biochem) xanthine

ксантин·дегидра́з·а, -ы, (*f*) xanthine dehydrogenase

ксантин·оксида́з·а, -ы, (*f*) xanthine oxidase

ксант·и́ровани·е, -я, (*n*) (text) xanthation

ксанто·гена́т·, -а, (*m*) (chem) xanthate

ксанто·ге́н·ов·ый, -ая, -ое, (*a*) xanthic, xanthate

~ **кисл·от·а́** (*f*) xanthic acid

~ **эфи́р·,** (*m*) ethyl xanthate

ксант·о́м·а, -ы, (*f*) (med) xanthoma

ксант·о́н·, -а, (*m*) (chem) xanthone

ксанто·проте́ин·ов·ая реа́кц·и·я (*f*) (chem) xantho-protein reaction

ксант·опс·и́·я, -и, (*f*) (med) xanthopsia

ксанто·сидери́т·, -а, (*m*) (min) xanthosiderite

ксанто·фи́лл·, -а, (*m*) (biochem) xanthophyll

ксе́н·и·я, -и, (*f*) (bot) xenia

ксено·га́м·и·я, -и, (*f*) (biol) xenogamy

ксено·ли́т·, -а, (*m*) (geol) xenolith

ксен·о́н·, -а, (*m*) (chem) xenon, Xe

ксено·кри́ст·, -а, (*m*) (geol) xenocryst

ксено·ти́м·, -а, (*m*) (min) xenotime

ксеро·гра́ф·и·я, -и, (*f*) (print) xerography

ксеро·де́рм·а, -ы, (*f*) (med) xeroderma

ксер·о́з·, -а, (*m*) (med) xerosis, xerophthalmia

ксеро·мо́рф·н·ый, -ая, -ое, (*a*) xeromorphic

ксеро·радио·гра́ф·и·я, -и, (*f*) (med) xeroradiography

ксеро·фи́ль·н·ый, -ая, -ое, (*a*) (biol) xerophilous

ксеро·фи́т·, -а, (*m*) (bot) xerophyte

ксеро·фо́рм·, -а, (*m*) (pharm) xeroform, bismuth tribromphenate

ксер·офтальм·и́·я, -и, (*f*) (med) xerophthalmia

кси (*indecl*) xi, ξ (Greek)

ксил·а́н·, -а, (*m*) (biochem) xylan

ксил·е́м·а, -ы, (*f*) (bot) xylem

ксил·ено́л·, -а, (*m*) (chem) xylenol

ксил·иди́н·, -а, (*m*) aminoxylene

ксил·и́л·, -а, (*m*) xylyl; (expl) trinitroxylene, TNX

ксило·гра́ф·и·я, -и, (*f*) xylography; xylograph

ксил·о́з·а, -ы, (*f*) (chem) xylose

ксил·о́л·, -а, (*m*) (chem) xylene

ксило·ли́т·, -а, (*m*) (build) magnesite-bonded wood-shaving flooring

ксило́·метр·, -а, (*m*) xylometer, forestry volumenometer

кста́ти (*adv*) opportunely, at the right moment; at the same time, in passing, incidentally; (as opening remark) by the way, incidentally

к-та (*abbr*) of **кисл·от·а́** acid

ктен·и́ди·й, -я, (*m*) (zool) ctenidium

ктено·пла́н·а, -ы, (*f*) (zool) *Ctenoplana*

ктено·со́м·а, -ы, (*f*) (gen) ctenosome

ктено·фо́р·ы (*nom pl*), (*g pl*) **-ов,** (*m*) (zool) ctenophores, *Ctenophora*

ктето·ло́г·и·я, -и, (*f*) (biol) ctetology

кто (*pron*) who

    кто́ бы не whoever ...!

    кто́ бы ни whoever

    кто́ ... кто́ some ... some

    кто́-либо *see* **кто́-нибудь**

    кто́-нибудь somebody, someone, anybody, anyone

    кто́-то somebody, someone

    кто́-то друго́й somebody/someone else

    тот кто ... he who the man who

куб·, -а, (*nom pl*) **-ы́,** (*g pl*) **-о́в,** (*m*) cube, hexahedron; (chem) still; vat (dyeing); (*as abbr*) **куб·и́ческ·ий** cubic(al)

~, воз·вод·и́ть в (math) cube, raise to the power of 3

~, инди́г·ов·ый indigo vat

~, пар·ов·о́й steam-heated still

~, пере·го́н·н·ый (chem) still

~, прост·о́й (cryst) simple cube, primitive cube

~, тру́б·чат·ый (chem) pipe still

кубаре·ви́д·н·ый, -ая, -ое, (*a*) top-shaped, turbinate

куб·а́р·ь, -я́, (*m*) top (toy)

куб·ату́р·а, -ы, (*f*) cubic content/capacity, cubage

ку́б·ик·, -а, (*m*) (*dim*) of **куб·;** (civ eng) brick, block, cube

~, шта́мп·ов·ый stamping die

куб·и́ческ·ий, -ая, -ое, (*a*) cubic(al); (math) cube

~ ко́рен·ь (*m*) (math) cube root

~ пара́бол·а (*f*) (math) cubical parabola

~ систе́м·а (*f*) (cryst) cubic system

~ у·равн·е́ни·е (*n*) (math) cubic equation

ку́б·ков·ый, -ая, -ое, (*a*) of **ку́б·ок·**

кубо·ви́д·н·ый, -ая, -ое, (*a*) cubiform, cuboid

кубо·восьми·гра́н·ник·, -а, (*m*) (cryst) cubooctahedron

куб·ов·о́й, -а́я, -о́е, (*a*) (*and see* **ку́-б·ов·ый**) still; (*f decl as adj*) still-room

~ батар·е́·я (*f*) (oil) battery of stills

~ у·стано́в·к·а (*f*) still, distillation plant

~ у·стано́в·к·а период·и́ческ·ая (*f*) batch still

ку́б·ов·ый, -ая, -ое, (*a*) (*and see* **ку-б·ов·о́й**) vat; indigo-blue

~ краш·ёни·е (*n*) vat dyeing

~ си́н·ий vat-blue

ку́б·ок·, -б·к·а, (*m*) cup (as prize)

кубо·ме́тр·, -а, (*m*) cubic metre

~, норма́ль·н·ый normal cubic meter

ку́брик·, -а, (*m*) (naut) crew space, mess deck

куб·ы́шк·а, -и, (*g pl*) **-шек·,** (*f*) pitcher, jug; (bot) water-lily, *Nuphar*

кува́лд·а, -ы, (*f*) sledge hammer

кувел·я́ци·я, -и, (*f*) (min) tubbing

кувши́н·, -а, (*m*) jug, pitcher

кувши́н·к·а, -и, (*g pl*) **-н·ок·,** (*f*) (bot) water-lily, *Nymphaea*

кувши́н·чик·, -а, (*m*) (bot) ascidium

куда́ where, where to, whither
   куда́ бы ни wherever
   куда́-либо *see* **куда́-нибудь**
   куда́-нибудь somewhere, anywhere
   куда́-то somewhere

куде́л·ь, -и, (*f*) (text) tow, codilla

кудря́в·ый, -ая, -ое, (*a*) curly (of hair)

кудря́ш·, -а́, (*i*) **-о́м,** (*m*) (bot) crown flax, *Linium usitatissimum*

ку·ёт (*pres 3rd sing*) of **ков·а́ть**

кузба́сс·, -а, (*m*) **кузне́цк·ий бас-се́йн·** (geog) the Kuznetsk Basin

кузба́сс·ла́к·, -а, (*m*) black varnish, bituminous varnish

куз·не́ц·, -а́, (*i*) **-о́м,** (*m*) blacksmith

куз·не́цк·ий, -ая, -ое, (*a*) Kuznetsk (coal fields etc.)

куз·не́чик·, -а, (*m*) (ent) grasshopper, (*pl*) *Orthoptera*

куз·не́чн·о-с·ва́р·очн·ый, -ая, -ое, (*a*) hammer-welded

куз·не́чн·о-штамп·о́вочн·ый, -ая, -ое, (*a*) forging

~ автома́т· (*m*) horizontal forging machine, upsetter

куз·не́чн·ый, -ая, -ое, (*a*) forge, forging, smith's, blacksmith's

~ с·ва́р·к·а (*f*) hammer welding

~ цех· (*m*) forge, smithy

ку́з·ниц·а, -ы, (*i*), **-ей,** (*f*) forge, forging shop, smithy

ку́з·ничн·ый, -ая, -ое, (*a*) forge

ку́зов·, -а, (*nom pl*) **-а́,** (*g pl*) **-о́в,** (*m*) (M/T) body; hood; basket

~, само·с·ва́ль·н·ый tipping body (of lorry)

ку́зьк·а, -и, (*g pl*) **-зек·,** (*f*) (ent) grain beetle, *Anisoplia austriaca*

кукерси́т·, -а, (*m*) (min) kuckersite (a combustible shale)

ку́кл·а, -ы, (*g pl*) **-кол·,** (*f*) doll, puppet; slug (of rubber)

куколе·от·бо́р·ник·, -а, (*m*) (agr) cockle separator

ку́кол·к·а, -и, (*g pl*) **-л·ок·,** (*f*) (*dim*) of **ку́кл·а;** (ent) chrysalis, pupa, nympha

ку́кол·ь, -я, (*m*) (bot) cockle, corncockle, *Agrostemma*

ку́коль·н·ый, -ая, -ое, (*a*) of **ку́кл·а**

~ теа́тр· (*m*) puppet theatre

кукуру́з·а, -ы, (*f*) (bot) maize, Indian corn (Brit), corn (U.S.), *Zea Mays*

~, зубо·ви́д·н·ая dent corn, *Z. Mays* var. *indentata*

кукуру́з·а

~, крахма́л·ист·ая starchy corn, *Z. Mays* var. *amylacea*

~, кремн·и́ст·ая flint corn, *Z. Mays* var. *indurata*

~, ло́п·ающ·аяся pop corn, *Zea Mays* var. *everta*

~, са́хар·н·ая sweet corn, *Z. Mays* var. *saccharata*

кукуру́з·н·ый, -ая, -ое, (*a*) of **кукуру́з·а**

~ ма́сл·о (*n*) corn oil

кукурузо·лущ·и́лк·а, -и, (*g pl*) **-лок·,** (*f*) (agr) corn sheller

кукурузо·сило·у·бо́р·очн·ый комба́йн· (*m*) maize and silage combine harvester

кукурузо·у·бо́р·очн·ая маши́н·а (*f*) corn picker, corn picker-sheller

кукурузо·хран·и́лищ·е, -а, (*n*) corn crib

куку́ш·к·а, -и, (*g pl*) **-ш·ек·,** (*f*) (orn) cuckoo, (*pl*) *Cuculidae*; (min) dinky

кукушко·обра́з·н·ые, -ых, (*pl decl as adj*) (orn) *Cuculiformes*

кул. (*abbr*) of **куло́н·** coulomb

кула́к·, -а, (*m*) (*and see* **кулач·о́к·**) (anat) fist; cam, pawl; crusher hammer; rich peasant, kulak; (*pl*) (min) keps, keeps

кулач·ко́в·ый, -ая, -ое, (*a*) cam, cam-operated, camshaft

~ вал· (*m*) camshaft

~ контролле́р· (*m*) (elec) camshaft controller

~ му́фт·а (*f*) dog-tooth clutch

~ ша́йб·а (*f*) cam ring

кулач·о́к·, -ч·к·а́, (*m*) (*dim*) of **кула́к·;** lobe (of cam shaft); jaw (of lathe chuck); wiper, cam, tappet (in a weaving machine)

~, бараба́н·н·ый (horol) cam wheel

~, двух·ди́ск·ов·ый double-profile cam

~, двух·ступ·е́нчат·ый double-lift cam

~, карди·о́ил·н·ый heart cam

~, па́з·ов·ый covered cam

~, толк·а́ющ·ий push cam

~, цилиндр·и́ческ·ий covered-drum cam

кул·ев·о́й, -а́я, -о́е, (*a*) of **кул·ь**

кулина́р·, -а, (*m*) (food) chef

кулина́р·и·я, -и, (*f*) cookery, culinary art

кулина́р·н·ый, -ая, -ое, (*a*) culinary

кули́р·н·ый, -ая, -ое, (*a*) (text) sinker, sinking

кули́р·овать, -уют,   (*imp*) sink (knitting machine operation)

кули́р·ующ·ая пла́т·ин·а (*f*) (text) sinker

кули́с·а, -ы,   (*f*) link, slotted guide plate; coulisse (carpentry); (a/c) windbreak strip; (*pl*) wings (theatre)

за кули́с·ами behind the scenes

~, за·кры́·т·ая closed link

кули́с·н·ый, -ая, -ое, (*a*) of **кули́с·а;** imbricate

~ ка́мен·ь (*m*) link/slide block

~ механ·и́зм· (*m*) link motion/gear, linkage

~ пар· (*m*) (agr) coulisse/slot fallow

куло́н·, -а (*m*) (elec) coulomb; pendent (jewelry)

Куло́н·а, вес·ы́ (*pl*) torsion balance

~, за·ко́н· (*m*) (elec) Coulomb's law

кулоно́·метр·, -а, (*m*) (elec) coulometer

куло́н·овск·ий, -ая, -ое, (*a*) coulomb

~ взаимо·де́й·стви·е (*n*) (phys) coulomb interaction

кулуа́р·, -а, (*m*) shoot (of dredger); (*pl*) lobby

кул·ь, -я́, (*m*) bag, sack

кульмин·а́ци·я, -и, (*f*) culmination

~, ни́ж·н·яя (astron) lower culmination

ку́льт·, -а, (*m*) cult, sect

культ- (*component*) (*abbr*) of **культ·у́р·н·ый** educational; cultural

культив·а́тор·, -а, (*m*) (agr) cultivator, harrow

~, ди́ск·ов·ый disk harrow

культив·а́ци·я, -а, (*f*) cultivation; (agr) harrowing; (biol) culture

культив·и́ровани·е, -я, (*n*) *see* **культив·а́ци·я**

~ в вис·я́ч·ей ка́пл·е (biol) liquid suspension culture

культив·и́р·овать, -уют, (*imp*) cultivate; harrow

ку́льт·ов·ый, -ая, -ое, (*a*) religious, church

культ·по·хо́д·, -а, (*m*) cultural outing

культ·у́р·а, -ы, (*f*) culture; (agr) crop

~, си́лос·н·ая silage crop

~, техн·и́ческ·ая (agr) industrial crop

~ тк·а́н·и (biol) tissue culture

~, чи́ст·ая pure culture (of bacteria etc.)

культ·у́рность·, -и, (*f*) culture, standard of culture/education/development

культ·у́рн·ый, -ая, -ое, (*a*) cultural; cultured, cultivated; educated

культ·я́, -й, (*f*) (med) stump

кумари́н·, -а, (*m*) (chem) coumarin

кумари́н·ов·ая кисл·от·а́ (*f*) coumaric/coumarinic acid

кумаро́н·, -а, (*m*) (chem) coumarone

кума́ч·, -а́, (*i*) **-о́м·,** (*m*) red calico

ку·ме́тр·, -а, (*m*) (elec) Q meter

ку́мж·а, -ы,   (*f*) (fish) brown trout, *Salmo trutta*

куми́н·ов·ый, -ая, -ое, (*a*) (chem) cumic, cuminic

~ альдеги́д· (*m*) cumic aldehyde, cuminol

кумква́т·, -а, (*m*) (bot) Kumquat, *Fortunella*

куммингтони́т·, -а, (*m*) (min) cummingtonite

кумо́л·, -а, (*m*) (chem) cumene, isopropyl-benzene

кумул·и́рованн·ый, -ая, -ое, (*a*) accumulated

кумуло·фи́р·ов·ый, -ая, -ое, (*a*) (geol) cumulophyric

кумул·яти́вн·ый, -ая, -ое, (*a*) cumulative; (expl) hollow-charge; (geol) cumulose

~ за·ря́д· (*m*) (expl) hollow charge

кумы́с·, -а, (*m*) (food) koumiss

куна́к·, -а́, (*m*) (bot) Italian millet, *Setaria italica*

ку́ндж·а, -ы, (*f*) (fish) char, *Salveliners*

Ку́ндт·а, пыл·ев·ы́е фигу́р·ы (acous) Kundt's tube

кунжу́т , -а, (*m*) (bot) sesame, *Sesamum*

~, вос·то́ч·ный sesame, *Sesamum indicum*

кунжу́т·н·ый, -ая, -ое, (*a*) sesame; (*pl*) Pedaliaceae

~, ма́сл·о (*n*) sesame oil

кунниа́л·ь, -я, (*m*) (met) Cunial (Cu–Ni–Al alloy)

кунико (*indecl*) Cu–Ni–Co alloy

куни́фе (*indecl*) Cu–Ni–Fe alloy

кун·и́ц·а, -ы, (*i*) **-цей,** (*f*) (zool) marten, *Martes*

~, америка́н·ск·ая Hudson Bay marten, *Martes americana*

~, благо·ро́д·н·ая pine marten, *Martes martes*

~, ка́мен·н·ая (zool) beech marten, *Martes foina*

~, лес·н·а́я (zool) pine marten, *Martes martes*

кунци́т·, -а, (*m*) (min) Kunzite (a spodumene)

купа́ж·, -а, (*m*) blending (of wines etc.)

купаж·и́ровани·е, -я, (*n*) *see* **купа́ж·**

купаж·и́р·овать, -уют, (*imp*) blend (teas, wines etc.)

купа́л·ь, -я, (*m*) (met) Cupal, Cu–Al duplex metal

куп·а́льн·ый, -ая, -ое, (*a*) bathing

куп·а́льн·я, -и, (*g pl*) **-лен·,** (*f*) bath house

куп·а́·ть, -ют, (*imp*) bath, bathe

купе́ (*n indecl*) (rail) compartment; (М/Т) coupé

купел·и́рован·и·е, -я, (*n*) (met) cupellation

купе́л·ь, -и, (*f*) (met) cupel

купел·я́ци·я, -и, (*f*) (met) cupellation

купе́н·а, -ы, (*f*) (bot) Solomons seal, *Polygonatum*

Ку́пер·а, эффе́кт· (*m*) (phys) Cooper effect

Ку́пер·а-Ю́итт·а, ла́мп·а (*f*) Cooper–Hewitt lamp

куп·е́ц·, -п·ц·а́, (*m*) merchant

куп·е́ческ·ий, -ая, -ое, (*a*) of **куп·е́ц·**

куп·и́ть, -ят, (*perf*) buy

ку́пл·енн·ый, -ая, -ое, (*past part pass*) of **куп·и́ть**

ку́пл·я, -и, (*f*) purchase

ку́пол·, -а, (*nom pl*) **-а́, -о́в,** (*m*) cupola, dome; canopy (parachute)

∼, сол·ян·о́й (geol) salt dome/plug

купол·о·обра́з·н·ый, -ая, -ое, (*a*) dome-like, dome-shaped, cupola-shaped, quaquaversal

∼ манже́т·а (*f*) chevron packing

∼ с·кла́д·к·а (*f*) (geol) quaquaversal fold

купо́н·, -а, (*m*) coupon

купоро́с·, -а, (*m*) (min) vitriol, hydrous sulphate

∼, желе́з·н·ый (min) copperas, melanterite, green vitriol; (chem) ferrous sulphate

∼, ме́д·н·ый (min) blue vitriol, hydrous sulphate of copper; (chem) copper sulphate

∼, ни́кел·ев·ый hydrous nickel sulphate

∼, ци́нк·ов·ый (min) white vitriol, goslasite; hydrous zinc sulphate

купоро́с·н·ый, -ая, -ое, (*a*) of **купоро́с·;** vitriolic

∼ ма́сл·о (*n*) oil of vitriol, sulphuric acid

купр·а́т·, -а, (*m*) (chem) cuprate

купре·и́н·, -а, (*m*) (biochem) cupreine

купр·и́т·, -а, (*m*) (min) cuprite

купр·о́кс·н·ый, -ая, -ое, (*a*) copper-oxide

купр·о́н·, -а, (*m*) (chem) cupron, alpha benzoinoxyme

∼ -элеме́нт·, -а, (*m*) caustic soda cell

купро·отуни́т·, -а, (*m*) (min) cupro-autunite

купро·склодовски́т·, -а, (*m*) (min) cuprosklodowskite

купро·урани́т·, -а, (*m*) (min) cupro-manite, torbenite

купфермерито́л·ь, -я, (*m*) (agr) Bordeaux mixture

купферо́н·, -а, (*m*) (chem) cupferron

купферона́т·, -а, (*m*) (chem) cupferronate

ку́п·ч·ая, -ей, (*f decl as adj*) title, deed

купы́р·ь, -я, (*m*) (bot) chervil, *Anthriscus*

купю́р·а, -ы, (*f*) cut, omission (in a play etc.); (fin) denomination; (math) line segment, interval

кур (*gen pl*) of **ку́р·иц·а,** *see* **ку́р·ы**

кураг·а́, -и́, (*f*) dried stoneless apricots

кура́нт·, -а, (*m*) paint grinding stick

кура́ре (*n indecl*) (pharm) curare

кураре·по·до́б·н·ый, -ая, -ое, (*a*) curare-like

ку́рбел·ь, -я, (*m*) knob

курви·ме́тр·, -а, (*m*) (surv) opisometer, map measurer

курга́н·, -а, (*m*) barrow, tumulus, burial mound

∼, меж·ев·о́й landmark

кур·е́ни·е, -я, (*n*) smoking

кур·и́н·ый, -ая, -ое, (*a*) of **ку́р·иц·а;** (*pl as noun*) (zool) *Galliformes*; (bot) *Gallinaceae*

∼ прос·о (*n*) (bot) Japanese millet, *Echinochloa crusgalli*

кур·и́тельн·ый, -ая, -ое, (*a*) smoking

кур·и́ть, -ят, (*imp*) smoke; burn incense; distil

ку́р·иц·а, -ы, (*i*) **-ей,** (*nom pl*) **-ку́р·ы,** (*g pl*) **кур·,** (*f*) hen, chicken

кур·ко́в·ый, -ая, -ое, (*a*) of **кур·о́к·**

куркум·а, -ы, (*f*) (bot) tumeric, *Curcuma longa*

куркуми́н·, -а, (*m*) curcumine (dye)

курку́м·ов·ый, -ая, -ое, (*a*) turmeric, curcuma

∼ ма́сл·о (*n*) curcuma oil, turmeric

кур·н·о́й, -а́я, -о́е, (*a*) smoky; chimneyless

куро·во́д·ств·о, -а, (*n*) poultry breeding/farming

кур·о́к·, -р·к·а́, (*m*) cock, cocking piece, trigger

∼, гидравл·и́ческ·ий (shipb) hydraulic-release launch trigger

куро́·метр·, -а, *(m)* curometer (rubber)

куропа́т·к·а, -и, *(g pl)* **-т·ок·,** *(f)* (orn) partridge

~, се́р·ая Hungarian partridge, *Perdix perdix*

~, тундр·ян·áя ptarmigan, *Lagopus mutus*

~, шотла́нд·ск·ая red grouse, *Lagopus scoticus*

куро́рт·, -а, *(m)* health resort

ку́рс·, -а, *(m)* (nav) course, heading; course (instructional); (fin) rate of exchange

в ку́рс·е informed, well informed, in the picture

~, валю́т·н·ый (fin) rate of exchange

~, ве́ксель·н·ый rate of exchange

~, встречно·пере·сек·а́ющ·ий (nav) collision course

~, встре́ч·н·ый (nav) head-on course

~, генера́ль·н·ый (nav) course made good

~, держ·а́ть head for

~, за́·да·нн·ый predetermined course

ли́н·и·я ку́рс·а *(f)* (naut) course steered; (air) course line

~, магни́т·н·ый magnetic course/ heading

~ обра́т·н·о-по·са́д·очн·ый (air) runway reciprocal heading

~, орто·дро́м·н·ый great circle heading

~ сле́д·овани·я (nav) heading

~ у·со·верш·е́нствовани·я course of further instruction, advanced-level course

~ це́л·и (mil) target course

курс·а́нт·, -а, *(m)* student; (mil) cadet

курси́в·, -а, *(m)* (print) italics, italic type

курси́в·н·ый, -ая, -ое, *(a)* italic

курси́в·ом *(adv)* in italics

курс·и́р·овать, -уют, *(imp)* (naut) ply (between)

курс·ов·о́й, -а́я, -о́е, *(a)* of **ку́рс·**

~ кольц·о́ *(n)* (naut) lubber ring

~ ра́дио·мая́к· *(m)* (air) landing beam

ку́рс·о-глисса́д·н·ая систе́м·а *(f)* (air) glide-path/localizer landing system

курсо́·граф·, -а, *(m)* (nav) course recorder

курсо·за·да́т·чик·, -а, *(m)* (air) heading setter

курсо·про·кла́д·чик·, -а, *(m)* (nav) course recorder

курсо·у·каз·а́тел·ь, -я, *(m)* (a/c) direction indicator

курта́ж·, -а, *(i)* **-ем,** *(m)* brokerage

ку́рт·к·а, -и, *(g pl)* **-т·ок·,** *(f)* jacket

~, с·пас·а́тельн·ая life jacket

курча́в·ост·ь, -и, *(f)* curl (plant disease)

курча́в·ый, -ая, -ое, *(a)* curly, crimped; (geol) ice-dressed, sheep-back

ку́р·ы *(nom pl)* of **ку́р·иц·а;** poultry, chickens, hens

курьёз·, -а, *(m)* curiosity, strange/ curious occurrence

курьёз·н·ый, -ая, -ое, *(a)* strange, curious, odd, funny

кус·ков·о́й, -а́я, -о́е, *(a)* lump

~ са́хар· *(m)* lump sugar

курье́р·, -а, *(m)* messenger, courier

курье́р·ск·ий, -ая, -ое, *(a)* of **курье́р·**

~ по́·езд· *(m)* (rail) fast train

кур·я́щ·ий, -ая, -ее, *(pres part act)* of **кур·и́ть;** *(m decl as adj)* smoker

кус·а́·ть, -ют, *(imp)* bite; sting

кус·а́чк·и *(nom pl),* *(g pl)* **-чек·,** *(f)* cutting pliers, wire cutters, nippers

кус·о́к·, -с·к·а́, *(m)* piece, bit, lump

кус·о́чек·, -чк·а, *(m)* *(dim)* of **кус·о́к·**

кус·о́чн·о *(adv)* piecewise

кусочно·гла́д·к·ий, -ая, -ое, *(a)* piecewise smooth/continuous

кус·о́чн·о-по·сто·я́нн·ая фу́нкци·я *(f)* step function

куст·, -а́, *(m)* bush, shrub; root (of potatoes); group (of small organizations); cluster (of points)

куст·а́рник·, -а, *(m)* bush, shrub; undergrowth, brushwood

куст·а́рников·ый, -ая, -ое, *(a)* of **куст·а́рник·**

куста́р·н·ый, -ая, -ое, *(a)* handicraft, hand-made, home-made

~ про·мы́шл·енност·ь *(f)* domestic/ cottage industry

куста́р·ь, -я́, *(m)* handicraftsman working at home

куст·и́ст·ый, -ая, -ое, *(a)* bushy

куст·и́ться, -я́тся, *(imp)* (bot) branch out, tiller

куст·ова́ни·е, -я, *(n)* (elec) interconnection

куст·ова́т·ый, -ая, -ое, *(a)* (bot) shrub-like, fruticose, fruticulose

куст·ов·о́й, -а́я, -о́е, *(a)* of **куст·**

кусто́д·а, -ы, *(f)* (print) carry-over word, catch word

кусто·ре́з·, -а,   (*m*) (agr) bush cutter, brush cutter

ку́т·ан·ый, -ая, -ое,   (*a*) muffled/wrapped up

ку́т·а·ть, -ют,   (*imp*) muffle/wrap up

кути́·кул·а, -ы,   (*f*) (biol) cuticle, cuticula

кут·и́н·, -а,   (*m*) (bot) cutin

кут·иниза́ци·я, -и,   (*f*) (biol) cutinization

кут·ини́т·, -а,   (*m*) (min) cutinite

кут·о́к·, -т·к·а́,   (*m*) (min) back of ore block

ку́ттер·, -а,   (*m*) (naut) cutter

кух·а́рк·а, -и,   (*g pl*) **-рок·,** (*f*) cook

ку́х·н·я, -и,   (*g pl*) **ку́х·он·ь,** (*f*) kitchen, (naut) galley; cooking, cuisine; boiler

∼, оча́ж·н·ая по·хо́д·н·ая   soyer boiler

ку́х·онн·ый, -ая, -ое,   (*a*) of **ку́х·н·я**

ку́ч·а, -и,   (*i*) **-ей,** (*f*) heap, pile

ку́ч·ев·о-   (meteor) cumulo-, cumuli-

ку́ч·ев·о-дожд·ев·ы́е, -ых,   (*pl decl as a*) (meteor) cumulo-nimbus

куч·ев·о́й, -а́я, -о́е,   (*a*) (meteor) cumulus

∼ облак·а́   (*pl*) (meteor) cumulus

Ку́черов·а, реа́кци·я   (*f*) (chem) Kucherov reaction

ку́ч·ност·ь, -и,   (*f*) close grouping, accuracy (of aim); small spread (of missiles); cluster, clustering

ку́ч·н·ый, -ая, -ое,   (*a*) of **ку́ч·а;** (geol) glomeroblastic

кушáк·, -á,   (*m*) sash, girdle

ку́ш·анье, -я,   (*n*) dish (food)

ку́ш·а·ть, -ют,   (*imp*) eat, dine, take a meal (polite)

кушéт·к·а, -и,   (*g pl*) **-т·ок·,** (*f*) couch

кущ·éни·е, -я,   (*n*) (bot) bushing out, tillering

ку·ю́щ·ий, -ая, -ее,   (*pres part act*) of **ков·а́ть**

кХ   (phys) kX, kilo-X (= 1·00202 Å)

Кье́льдал·я, колб·а́   (*f*) (chem) Kjeldahl flask

∼, мéтод·   (*m*) (chem) Kjeldahl's method

кэб-сигнáл·, -а,   (*m*) (rail) cab signal

кэв   (*abbr*) of **кило-электрон·во́льт·** (nucl) kev, kilo-electron-volt

кэватро́н·, -а,   (*m*) (nucl) kevatron

Кэ́ли, чи́сл·а   (*pl*) Cayley numbers

кэннели́т·, -а,   (*m*) (min) cannel coal

кэ́прок·, -а,   (*m*) (geol) cap rock

кэ́ч·, -а,   (*m*) (naut) ketch

кю́бел·ь, -я,   (*m*) bucket

кювел·я́ци·я, -и,   (*f*) *see* **кувел·я́ци·я**

кювéт·, -а,   (*m*) side drain (roads, rail)

∼, водо·с·бо́р·н·ый   drain

кювéт·а, -ы,   (*f*) (phot) tray, dish; (nucl) cell, trough

кювéт·к·а, -и,   (*g pl*) **-т·ок·,** (*f*) *see* **кювéт·а**

кюль·вальц·ы́   (*nom pl*), (*g pl*) **-о́в,** (*m*) (chem) drum flaker

кюрасо́   (*n indecl*) curaçao (liqueur)

кюри́   (*indecl*) (nucl) curie (unit of measurement)

Кюри́, то́ч·к·а   (*f*) (met) Curie point, ferromagnetic transformation point

Кюри́-Ве́йс·а, за·ко́н·   (*m*) Curie-Weiss law (of ferromagnetism)

Кюри́-Ву́льф·а, пра́в·ил·о   (*n*) (phys) Curie law

Кюри́-Шевено́, вес·ы́   (*pl*) (phys) Curie–Cheveneau balance

кю́р·иев·ый, -ая, -ое,   (*a*) (chem) curium

кю́р·и·й, -я,   (*m*) (chem) curium, Cm

кюри́т·, -а,   (*m*) (min) curite

кюри·терап·и́·я, -и,   (*f*) (med) Curie therapy, radium treatment

кяри́з·, -а,   (*m*) irrigation tunnel

Л

л. (*abbr*) **литр·** l, litre, liter
лаба́з·, -а, (*m*) (agr) granary, barn
лаба́з·ник·, -а, (*m*) corn merchant; (bot) dropwort, *Spiraea filipendula*
лабарда́н·, -а, (*m*) salted cod fillet
лабарра́к·ов·а вод·а́ (*f*) (phot, chem) Labarraque's solution, Javel water
лабиализ·ова́ть, -у́ют, (*imp and perf*) (ling) labialize
лабиа́ль·н·ый, -ая, -ое, (*a*) labial
лаби́ль·ност·ь, -и, (*f*) lability
лаби́ль·н·ый, -ая, -ое, (*a*) labile, unstable
лабио·дента́ль·н·ый, -ая, -ое, (*a*) (ling) labiodental
лабири́нт·, -а, (*m*) labyrinth, labyrinth packing/washer
∼, радиа́ль·н·ый (mech) radial-clearance labyrinth
лабири́нт·ов·ый, -ая, -ое, (*a*) labyrinth
∼ аппара́т· (*m*) (zool) labyrinth
∼ у·плот·не́ни·е (*n*) (mech) labyrinth packing
лабиринт·одо́нт·, -а, (*m*) (pal) labyrinthodont
лабор·а́нт·, -а, (*m*) laboratory assistant/technician
лабор·ато́ри·я, -и, (*f*) laboratory; (photog) dark room
∼, высоко·го́р·н·ая mountain-top laboratory
∼, гор·я́ч·ая (nucl) hot laboratory
∼, ста́рт·ов·ая mobile aircraft-fuel testing laboratory
лаборато́р·н·ый, -ая, -ое, (*a*) laboratory, laboratorial
лабрадо́р·, -а, (*m*) (min) labradorite
лабрадори́т·, -а, (*m*) (min) labradorite
лабро·ци́т·, -а, (*m*) (biol) labrocyte
ла́в·а, -ы, (*f*) (geol) lava; (min) longwall, long face
    вы·ём·к·а ла́в·ами (*f*) longwall, longwalling
    вы·ём·к·а ла́в·ами прям·ы́м хо́д·ом (*f*) advancing longwall
    вы·ём·к·а ла́в·ами обра́т·н·ым хо́д·ом (*f*) retreating longwall
Лава́л·я, сопл·о́ (*n*) Laval nozzle (turbines)
лава́нд·а, -ы, (*f*) (bot) lavender, *Lavandula*
лава́нд·ов·ый, -ая, -ое, (*a*) of **лава́нд·а;** lavender-blue
∼ ко́пи·я (*f*) (phot) lavender print

лави́н·а, -ы, (*f*) avalanche
∼, электро́н·н·ая (phys) avalanche, Townsend ionization
лави́н·н·ый, -ая, -ое, (*a*) avalanche, cascade
∼ длин·а́ (*f*) (nucl) cascade/radiation length
∼ един·и́ш·а (*f*) (nucl) cascade unit
∼ про·бо́·й (*m*) avalanche breakdown (of transistors)
лав·и́р·овать, -уют, (*imp*) (naut) tack, ply; avoid, find a way round
ла́в·к·а, -и, (*g pl*) **-вок·,** (*f*) bench, wall seat; shop, kiosk; (naut) bank, sandbank
ла́в·ов·ый, -ая, -ое, (*a*) of **ла́в·а**
ла́в·очк·а, -и, (*g pl*) **-чек·,** (*f*) (*dim*) of **ла́в·к·а**
ла́вр·, -а, (*m*) (bot) laurel, *Laurus*; (food) bay
∼, благо·ро́д·н·ый bay, sweet laurel, *Laurus nobilis*
∼, камфа́р·н·ый camphor, camphor tree, *Cinnamomum camphora*
∼, кор·и́чн·ый cinnamon, *Cinnamomum zeylanicum*
лавра́к·, -а, (*m*) (fish) bass, *Morone labrax*
лавро·вишн·ёв·ый, -ая, -ое, (*a*) of **лавро·ви́шн·я**
лавро·ви́шн·я, -и, (*g pl*) **-шен·,** (*f*) (bot) laurel cherry, cherry laurel, laurocerasus
лавре́нть·ев·ская с·кла́д·чатост·ь (*f*) (geol) Laurentian granites
ла́вр·ов·ый, -ая, -ое, (*a*) of **ла́вр·;** (*pl*) (bot) *Lauraceae*
лавсо́ни·я, -и, (*f*) (bot) henna, *Lawsonia inermis*
лаг·, -а, (*m*) (naut) log; broadside, beam (of ship)
∼, верт·у́шечн·ый screw log
∼, воз·ду́ш·н·ый (a/c) distance recorder
∼, гидравл·и́ческ·ий pressure log
∼, голла́нд·ск·ий Dutchman's log
∼, до́н·н·ый ground log
∼, за·бо́рт·н·ый screw log
∼, руч·н·о́й common log
    ла́г·ом к (naut) beam on to
∼, прост·о́й common log
лаг·а, -и, (*f*) (build) floor joist
ла́гер·н·ый, -ая, -ое, (*a*) of **ла́гер·ь**

Лагéрр·а, мнóго·член· (*m*) (math) Laguerre polynomial

лáгер·ь, -я, (mil) (*nom pl*) **ꞌя,** (*g pl*) **-éй,** faction; (*nom pl*) **-и,** (*g pl*) **-ей,** (*m*) camp

~, барáч·н·ый hutted camp

~, по·хóд·н·ый bivouac

лаглúн·ь, -я, (*m*) (naut) log line

лагó·метр·, -а, (*m*) (a/c) distance counter

лаг·офтáльм·, -а, (*m*) (med) lagophthalmus, lagophthalmos

Лагрáнж·а, интерполяциóн·н·ая фóрмул·а (*f*) (math) Lagrange's formula of interpolation

~, мéтод· мнóж·ител·ей (*m*) (math) Lagrange's method of multipliers

~, мнóж·ител·ь (*m*) (math) Lagrange's multiplier

~, оперáтор· (*m*) (math) Lagrangian operator, Lagrangian

~,,у·равн·éни·е (*n*) Lagrange's theorem; (mech) Lagrange's equation

~, фóрмул·а (*f*) Lagrange's form of the remainder for Taylor's theorem

~, фýнкци·я (*f*) (mech) Lagrangian function, Lagrange's density function, kinetic potential

лагрáнж·ев·ый, -ая, -ое, (*a*) Lagrangian, Lagrange

лагранжиáн·, -а, (*m*) (math) Lagrangian, Lagrangian operator

лагýн·, -а, (*m*) drinking water container, water urn

лагýн·а, -ы, (*f*) (geog) lagoon

лагýн·н·ый, -ая, -ое, (*a*) of **лагýн·а**

~ от·лож·éни·е (*n*) lagoonal deposition

лáд·, -а, (*nom pl*) **-ы́,** (*g pl*) **-óв,** (*m*) harmony, concord; way, manner; stop (musical instrument); (cast) mould joint face

лáдан·, -а, (*m*) (biochem) benzoin gum

лáдан·ник·, -а, (*m*) (bot) rockrose, ladanum cistus, *Cistus ladaniferus*; (*pl*) Cistaceae

ладúн·ск·ий ярус· (*m*) (geol) Ladinian stage

лáд·ить, -ят, (*imp*) be on good terms with, get on well with; put in order/right

лáд·н·о (*adv*) very well, all right, O.K.

лáд·ов·ый, -ая, -ое, (*a*) of **лад·**

лáдож·ск·ая свúт·а (*f*) (geol) Ladogan deposit

ладóн·н·ый, -ая, -ое, (*a*) (anat) palm, palmar

ладóн·чат·ый, -ая, -ое, (*a*) (anat) palmate

ладóн·ь, -и, (*f*) (anat) palm

ладыг·а, -и, (*f*) trunnion bracket

ладье·вúд·н·ый, -ая, -ое, (*a*) scaphoid, carinate, boat-shaped

ладье·нóг·ие, -их, (*pl decl as adj*) (pal) scaphopods, *Scaphopoda*

ладь·я́, -й, (*g pl*) **ладé·й** (*f*) (obs) boat; (anat) scapha; rook, castle (at chess)

лá·ек· (*g pl*) of **ла·йк·а**

лá·ечник·, -а, (*m*) alum tanner, tower (leather)

лáж·, -а, (*i*) **-ем,** (*m*) (fin) agio

лáж·ива·ть, -ют, (*imp*) *see* **лáд·ить**

лáж·у (*pres 1st sing*) of **лáд·ить**

лáз·, -а, (*m*) manhole

лáз·ань·е, -я, (*n*) climbing

лазарéт·, -а, (*m*) (mil) sick bay, medical reception station, dispensary

лáз·а·ть, -ют, (*imp*) of **лáз·ить**

лаз·éйк·а, -и, (*g pl*) **-éек·,** (*f*) hole, way through; loop-hole

лáзер·, -а, (*m*) (phys) laser

лáз·ить, -ят, (*imp indet*) climb, clamber

лазóрник·, -а, (*m*) (bot) day flower, *Commelina communis*

лазулúт·, -а, (*m*) (min) lazulite

лазур·úт·, -а, (*m*) (min) lazurite

лазýр·ник·, -а, (*m*) (bot) laser wort, *Laserpitium*

лазýр·н·ый, -ая, -ое, (*a*) of **лазýр·ь**

лазýр·ь, -и, (*f*) azure, sky-blue

~, берлúн·ск·ая Prussian/Berlin blue

~, мéд·н·ая (min) azurite

лáз·ящ·ий, -ая, -ее, (*pres part act*) of **лáз·ить**; (biol) scansorial, scandent, climbing

лá·й, -я, (*m*) (zool) bark, barking

лáйб·а, -ы, (*f*) (naut) laiba, laiva, two-masted vessel

лáйд·а, -ы, (*f*) (geog) laida

лá·йк·а, -и, (*g pl*) **лá·ек·,** (*d pl*) **лá·й·к·ам,** (*f*) alum/tawed leather; laika, eskimo dog

лá·йков·ый, -ая, -ое, (*a*) of **ла·йк·а**

~ дубл·éни·е (*n*) tawing, alum tanning

~ об·жóр· (*m*) tawing paste (leather)

лáйм·, -а, (*m*) (bot) lime, *Citrus aurantifolium*

лáйм·а, -ы, (*f*) (bot) Lima bean, *Phaseolus lunatus*

лáйман·а, сéри·я (*f*) (phys) Lyman series

лаймква́т·, -а, *(m)* (bot) limequat (lime/kumquat hybrid)
ла́йнер·, -а, *(m)* (naut) liner, passenger ship; liner (oil wells)
~, воз·ду́ш·н·ый airliner
ла́к·, -а, *(m)* varnish, lacquer, lac; lake (colour); (ICE) gummy deposit
~, бакели́т·ов·ый bakelite varnish
~ воз·ду́ш·н·ой су́ш·к·и air-drying varnish/lacquer
~, глифта́л·ев·ый glyptal/alkyd varnish
~ гор·я́ч·ей су́ш·к·и baking varnish/lacquer
~, гор·я́ч·ий (photog) hot varnish
~, жи́р·н·ый long oil varnish
~, каменно·уго́ль·н·ый bituminous varnish, black varnish
~, кита́й·ск·ий lac, lacquer
~, кле́·ящ·ий sticking varnish
~, крас·и́льн·ый lithographic ink
~, ма́сл·ян·ый oil varnish
~, ма́т·ов·ый (photog) matte varnish, retouching varnish
~ пе́ч·н·ой су́ш·к·и stoving/baking varnish, thermosetting varnish
~, пол·ов·о́й floor varnish, varnish stain
~, про·пи́т·очн·ый impregnating varnish
~, спирт·ов·о́й spirit varnish
~, у·зо́р·чат·ый effect varnish/lacquer
~, холо́д·н·ый (phot) cold varnish
~ холо́д·н·ой су́ш·к·и air-drying varnish/lacquer
~, шелла́ч·н·ый shellac varnish, French polish
~, шлиф·ова́льн·ый flatting varnish
лак·иро́ванн·ый, -ая, -ое, *(a)* varnished, lacquered; patent (leather)
лак·ир·ова́ть, -у́ют, *(imp)* varnish, lacquer
лак·иро́вочн·ый, -ая, -ое, *(a)* varnishing, lacquering
лакко·ли́т·, -а, *(m)* (geol) laccolith
лакмо́ид·, -а, *(m)* (chem) resorcin blue
ла́кмус·, -а, *(m)* (chem) litmus
ла́кмус·ов·ый, -ая, -ое, *(a)* litmus
~ бума́г·а *(f)* litmus paper
~ на·сто́йк·а *(f)* litmus tincture
Лакова́на, ·ва́·я *(f)* (civ eng) Lackawanna sheet pile
ла́к·ов·ый, -ая, -ое, *(a)* varnish, lacquer; (ICE) gum, gummy
~ де́рев·о *(n)* (bot) varnish tree, lacquer tree, *Rhus vernicifera*

ла́к·ов·ый, -ая, -ое
~ плён·к·а *(f)* varnish film; (ICE) gummy matter film
лако·кра́с·очн·ый, -ая, -ое, *(a)* paint, paint and varnish, varnish
лакон·и́ческ·ий, -ая, -ое, *(a)* laconic, brief
лакон·и́чн·ый, -ая, -ое, *(a)* laconic, brief
лако·но́с·, -а, *(m)* (bot) poke weed, *Phytolacca*
лако·образ·ова́ни·е, -я, *(n)* (ICE) gummy matter deposition
лако·стекло·ткан·ь, -и, *(f)* varnished glass cloth
лако·ткан·ь, -и, *(f)* varnished cloth; (elec) varnished insulating fabric
лако·шёлк·, -а, *(m)* (elec) varnished insulating silk
лакрим·а́тор·, -а, *(m)* (chem) lachrymator
лакр·и́ц·а, -ы, *(f)* (bot) liquorice
лакр·и́чник·, -а, *(m)* (bot) liquorice, *Glycyrrhiza glabra*
лакт·а́з·а, -ы, *(f)* (biochem) lactase
лактаза́м·, -а, *(m)* (chem) lactazam
лакт·альбуми́н·, -а, *(m)* (biochem) lactalbumin
лакта́м·, -а, *(m)* (chem) lactam
лакта́м·н·ый, -ая, -ое, *(a)* lactam
~ кольц·о́ *(n)* (chem) lactam ring
лакт·а́т·, -а, *(m)* (chem) lactate
лакт·ацио́нн·ый, -ая, -ое, *(a)* lactation
лакт·а́ци·я, -и, *(f)* lactation
лакт·и́д·, -а, *(m)* (chem) lactide
лакт·и́л·, -а, *(m)* lactyl
лактил·моло́ч·н·ая кисл·от·а́ *(f)* lacto-lactic acid
лакти́м·, -а, *(m)* lactim
лакти́м·н·ый, -ая, -ое, *(a)* lactim
~ фо́рм·а *(f)* lactim formation
лакто·баци́лл·а, -ы, *(f)* (bact) *Lactobacillus*
лакто·бацилл·и́н·, -а, *(m)* yoghurt
лакто·бути́ро́·метр·, -а, *(m)* lactobutyrometer
лакто·ге́н·н·ый, -ая, -ое, *(a)* lactogenic
лакто·денси́·метр·, -а, *(m)* lactometer
лакт·о́з·а, -ы, *(f)* (biochem) lactose
лакто́·метр·, -а, *(m)* lactometer
лакт·о́н·, -а, *(m)* (chem) lactone
лакт·ониза́ци·я, -и, *(f)* lactonation, lactonization
лакт·о́нов·ый, -ая, -ое, *(a)* lactone, lactonic
лакто·ско́п·, -а, *(m)* lactoscope

лакто·флав·и́н·, -а, (*m*) (biochem) lactoflavine, vitamin B₂

лакт·укáри·й, -я, (*m*) (bot) lactucarium, lactuca opium

лакýн·а, -ы, (*f*) (biol) lacuna; gap
  теорéм·а о лакýн·ах (*f*) (math) gap theorem

лакун·áрн·ый, -ая, -ое, (*a*) (biol) lacunary, lacunar, lacunose; gap

лáм·а, -ы, (*f*) (zool) llama, *Lama glama*

ламантúн·, -а, (*m*) (zool) manatee, *Manati*

ламарк·úзм·, -а, (*m*) (biol) Lamarckism

лáмбд·а, -ы, (*f*) lambda, λ (Greek)

лáмберт·, -а, (*m*) lambert (unit of brightness)

Лáмберт·а, за·кóн· (*m*) (opt) Lambert's law

ламé (*n indecl*) (text) lamé

Ламé, по·сто·я́нн·ые (*pl*) (math) Lamé's constants

~, фýнкци·и (*pl*) (math) Lamé's functions

ламéл·ь, -я, (*m*) (biol) lamella; (telecom) commutator bar/segment, bank contact; (text) lam

ламéль·н·ый, -ая, -ое, (*a*) of **ламéл·ь**

~ при·бóр· (*m*) (text) warp stop motion

ламин·áриев·ый, -ая, -ое, (*a*) laminarian

~ вóдо·росл·и (*pl*) (mar bot) kelps, brown seaweed, *Laminaria*

ламин·áри·я, -и, (*f*) (bot) laminaria

ламин·áрн·ый, -ая, -ое, (*a*) laminar, lamellar

~ движ·éни·е (*n*) (hydr) laminar flow, viscous flow

ламин·áци·я, -и, (*f*) lamination

Ламóнт·а, за·кóн· (*m*) (elec) Lamont's law

~, котёл· (*m*) La Mont boiler

лáмп·а, -ы, (*f*) lamp; (elec) electric lamp, bulb; valve, tube, vacuum tube, electron tube

~, áвто·электрóн·н·ая cold-cathode tube

~, бактери·цúд·н·ая (med) bactericidal lamp

~ бег·ýщ·их волн· travelling-wave tube

~, бес·цóколь·н·ая baseless valve/tube

~ бес·шýм·н·ой регул·úровани·я грóм·кост·и squelch tube

~, би·спирáль·н·ая coiled-coil lamp

~, вéнтиль·н·ая gated-beam valve/tube

лáмп·а

~, вс·пы́ш·ечн·ая (phot) flash lamp/bulb

~, вы·зы́в·н·áя line lamp (teleph)

~, вы·прям·и́тельн·ая rectifier valve/tube

~, вы·хóд·н·áя output valve/tube

~, гáзо·пóлн·ая gas-filled valve/tube; gas-filled lamp

~, гáзо·раз·ря́д·н·ая gas-discharge tube; gas-discharge lamp

~, гáзо·свéт·н·ая (elec) glow-discharge lamp

~, генерáтор·н·ая oscillator valve/tube

~, голов·н·áя (min) cap/head lamp

~, дел·и́тельн·ая scaling tube/valve

~, дуг·ов·áя arc lamp

~, жёстк·ая hard tube/valve

~ -жёлуд·ь, -я, (*m*) acorn valve/tube

~, за·да·ю́щ·ая driver tube

~, за·по·мин·áющ·ая storage/memory tube

~, за·свéч·ивающ·ая (cinema) exposure lamp

~, индикáтор·н·ая indicator valve/tube

~, карбúд·н·ая acetylene lamp

~, квáрц·ев·ая quartz mercury-vapour lamp

~, комбин·и́рованн·ая multiple-function valve/tube

~, корон·и́рующ·ая corona tube

~, крáтер·н·ая carbon-arc lamp

~, луч·ев·áя beam-pentode/tube

~, люмин·есцéнтн·ая fluorescent lamp, daylight lamp

~, масштáб·н·ая scaling tube/valve

~, мат·и́рованн·ая pearl/frosted lamp/bulb

~, мая́ч·кóв·ая lighthouse valve/tube, disc-seal valve/tube

~, металл·и́ческ·ая metallized valve/tube, spray-shielded valve/tube

~, метáлло·керáм·и́ческ·ая stacked electron tube

~, мнóго·сéт·очн·ая multi-electrode tube/valve

~ на·кáл·ивани·я filament/incandescent lamp, electric light bulb

~, нáтр·иев·ая sodium-discharge lamp, sodium-vapour lamp

~, нúзко·вóльт·н·ая battery lamp/bulb

~ обрáт·н·ой волн·ы́ carcinotron

~, о·грáн·ичи́тельн·ая (TV) clipper tube

ла́мп·а

~, **орбита́ль·н·о-луч·ев·а́я**   orbital-beam tube

~, **от·бо́й·н·ая**   (telecom) clearing lamp

~, **па́ль·чиков·ая**   miniature/bantam tube/valve; miniature bulb

~, **па·я́льн·ая**   blowlamp

~, **пере·гру́ж·енн·ая**   overdriver tube/valve

~, **подо·гре́в·н·ая**   heater-type cathode, indirectly heated cathode

~, **подо·грев·а́тельн·ая**   heating lamp/stove (for engines)

~, **под·све́ч·ивающ·ая**   (TV, cinema) exciter lamp

~, **пред·о·кон·е́чн·ая**   driver tube

~, **пред·о·хран·и́тельн·ая**   (min) safety lamp

~, **пре·образ·ова́тельн·ая**   converter tube/valve

~, **про·пуск·а́ющ·ая**   gate valve/tube

~, **про·све́ч·ивающ·ая**   (cinema) exciting lamp

~ **прям·о́го на·ка́л·а**   directly heated valve/tube

~, **пугов·и́чн·ая**   door knob tube

~, **пяти·се́т·очн·ая**   pentode

~, **пяти-электро́д·н·ая**   pentode

~, **раз·ря́д·н·ая**   discharge lamp; discharge valve/tube, thyratron

~, **реоста́т·н·ая**   resistance lamp

~, **реакти́в·н·ая**   valve reactor, reactance tube

~, **рту́т·н·ая**   mercury-vapour lamp, mercury-arc lamp

~, **ртутно·ква́рц·ев·ая**   (med) mercury-vapour lamp

~ **с** (and see under associated words)

~ **с больш·и́м коэффицие́нт·ом у·сил·е́ни·я**   high-mu tube/valve

~ **с дре́йф·ов·ым про·стра́н·ств·ом**   drift tube

~ **с пере·ме́н·н·ой крут·изн·о́й**   variable-mu valve/tube

~ **с по·пере́ч·н·ым по́л·ем**   beam-deflection tube

~ **с по·пере́ч·н·ым у·правл·е́ни·ем**   beam-deflection tube

~, **сигна́ль·н·ая**   indicating lamp, annunciator lamp, pilot lamp

~, **сигна́ль·н·о-про·бле́ск·ов·ая**   signalling lantern

~, **с·мес·и́тельн·ая**   (elec) mixer tube/valve

~, **смотр·ов·а́я**   inspection lamp

~ **со·в·мещ·е́ни·я**   coincidence tube

~, **софи́т·н·ая**   strip/horizon lamp

ла́мп·а

~ **с·равн·е́ни·я**   comparison lamp (photo-metering)

~, **стабилиз·у́ющ·ая**   voltage-regulator valve/tube

~, **с·чёт·н·ая**   counter tube/valve

~, **тл·е́ющ·ая**   glow-discharge tube/valve; glow-discharge lamp

~, **то́ч·ечн·ая**   point-source lamp, spot lamp/light

~, **трех·электро́д·н·ая**   triode

~, **у́голь·н·ая**   carbon-filament lamp

~, **у·правл·я́ющ·ая**   control tube/valve

~, **у·сил·и́тельн·ая**   amplifier tube/valve

~, **фикс·и́рующ·ая**   (TV) clamper tube

~, **четырёх·электро́д·н·ая**   tetrode

~, **чит·а́ющ·ая**   (cinema) exciter lamp

~, **шести·электро́д·н·ая**   hexode

~, **экран·и́рованн·ая**   screen-grid valve/tube, tetrode

~, **электро́н·н·ая**   thermonic valve/tube, electron/vacuum tube

~, **электро·метр·и́ческ·ая**   electrometer tube

~, **электронно·луч·ев·а́я**   electron-beam tube/valve, electron-indicator tube/valve, electron-ray tube/valve

лампа́с·, -а, (m) (mil) trouser stripe

ла́мп·ов·ый, -ая, -ое, (a) of **ла́мп·а;** thermionic, electronic, vacuum; (f as noun) (min) relighting room

~ **генера́тор·** (m) thermionic/valve oscillator, vacuum-tube oscillator; thermionic generator

~ **колпач·о́к·** (m) lamp cap

~ **у·сил·и́тел·ь** (m) thermionic amplifier/magnifier, valve amplifier, vacuum-tube amplifier

~ **цо́кол·ь** (m) valve/tube base

лампо·держ·а́тел·ь, -я, (m) valve/tube holder; lamp holder

ла́мп·очк·а, -и, (g pl) **-чок·,** (f) (dim) of **ла́мп·а**

~, **авар·и́йн·ая сигна́ль·н·ая**   warning light

~, **контро́ль·н·ая сигна́ль·н·ая**   indicator light

лампридо·обра́з·н·ые, -ых, (pl decl as a) (zool) Lampridiformes

лампро·филл·и́т·, -а, (m) (geol) lamprophyllite

лампро·фи́р·, -а, (m) (geol) lamprophyre

ланарки́т·, -а, (m) (min) lanarkite

лангбейни́т·, -а, (m) (min) langbeinite

лангу́ст·, -а, (m) (fish) spring lobster, langouste. Palinurus

лангу́ст·а, -ы,   (*f*) of **лангу́ст·**

Ланда́у, диа·магнет·и́зм   (*m*) Landau diamagnetism

Ланда́у-Ли́фшиц·а, у·равн·е́ни·е   (*n*) (phys) Landaú–Lifshits equation

Ланде́, мно́ж·ител·ь   (*m*) (phys) Lande factor, Lande *g* factor

ландша́фт·, -а,   (*m*) landscape; (geog) topography, terrain, country

ла́ндыш·, -а,   (*i*) (*m*) (bot) lily of the valley, *Convallaria*

Ланжеве́н·а, ио́н·   (*m*) (phys) Langevin ion

ла́н·ий, -ья, -ье,   (*a*) of **ла́н·ь**

лан·оли́н·, -а,   (*m*) (pharm) lanolin

ланта́н·, -а,   (*m*) (chem) lanthanum, La

лантан·и́д·, -а,   (*m*) lanthanide

лантан·о́ид·, -а,   (*m*) lanthanide

лану́г·о, -а,   (*n*) (zool) lanugo, pre-natal hair

ланце́т·, -а,   (*m*) (med) lancet; stylus

ланце́т·ник·, -а,   (*m*) (zool) lancelet, *Branchiostoma*

ланцето·ви́д·н·ый, -ая, -ое,   (*a*) (bot) lanceolate

ла́н·ь, -и,   (*f*) (zool) fallow deer, *Dama dama*

ла́п·а, -ы,   (*f*) paw, pad, foot; lug, foot, arm; dovetail, tenon; tine, cultivator tooth, sweep (of plough); fluke (of anchor)

~, гус·и́н·ая   (agr) goose foot

~, пере·греб·а́ющ·ая   rabble arm (ore roasting furnace)

~, пере́д·н·яя   (zool) paw, front leg

~, перепо́н·чат·ая   (zool) webbed foot

~, плоско·ре́з·н·ая   (agr) rigid tine, scuffle knife

~, подо́ль·н·ая плоско·ре́з·н·ая   (agr) flat sweep

~, почво·у·глуб·и́тельн·ая   subsoiling tine

~, рыхл·и́тельн·ая   (agr) cultivator tooth

~, стре́ль·чат·ая   sweet, central shovel (of cultivator)

лапаро·скоп·и́·я, -и,   (*f*) (med) laparoscopy

лапаро·то́м·и·я, -и,   (*f*) (med) laparotomy

лапид·а́рн·ый, -ая, -ое,   (*a*) lapidary

лапи́лл·и   (*nom pl*) (geol) lapilla

лапи́н·а, -ы,   (*f*) (bot) wingnut, *Pterocarya*

лапинг-проце́сс·, -а,   (*m*) (mech) lapping

лапинг·ова́ни·е, -я,   (*n*) lapping

ла́п·к·а, -и,   (*g pl*) **-п·ок·,**   (*f*) (*dim*) of **ла́п·а;** lug; grip, draw vice, pawl; tang (of tool); leg (ropes); (zool) tarsus

~, при·жи́м·н·ая   (agr) sickle/knife clip; (plast) presser/suction foot

Лапла́с·а, а́зимут·   (*m*) (surv) Laplace azimuth

~, опера́тор·   (*m*) (math, phys) Laplacian operator, Laplacian

~, пре·о́браз·ова́ни·е   (*n*) (math) Laplace's transform

~, у·равн·е́ни·е   (*n*) (math) Laplace's differential equation, Laplace equation

лаплас·иа́н·, -а,   (*m*) (math, phys) Laplacian, Laplacian operator; Laplacian, negative buckling (reactor theory)

лапла́с·ов· пункт·   (*m*) (surv) Laplace point

Лапо́рт·а, пра́в·ил·о чёт·ност·и   (*n*) (phys) Laporte parity rule

лап·ча́тк·а, -и,   (*g pl*) **-ток·,**   (*f*) (bot) cinquefoil, *Potentilla*

лапчато·но́г·, -а,   (*m*) (orn) sun grebe, *Heliornis*

лапчато·но́г·ий, -ая, -ое,   (*a*) webfooted, (orn) palmiped

ла́п·чат·ый, -ая, -ое,   (*a*) of **ла́п·а** and **ла́п·к·а;** (zool) web footed, palmate; (bot) digitate, pedate

лапш·а́, -и́,   (*i*) **-о́й,**   (*f*) (food) noodles; noodle soup; (*pl*) finely-cut rubber

лапш·а́-ры́б·а, -ы,   (*f*) glass fish, *Salangichtys microdon*

лапше·ре́з·н·ая маши́н·а   (*f*) (food) noodle machine

ла́рг·а, -и,   (*f*) larga seal, *Phoca largha*

лар·ев·о́й, -а́я, -о́е,   (*a*) of **лар·ь**

лар·ёк·, -рь·к·а́,   (*m*) (*dim*) of **лар·ь**

ларинг·и́т·, -а,   (*m*) (med) laryngitis

ларинго·ско́п·, -а,   (*m*) (med) laryngoscope

ларинго·фиссу́р·а, -ы,   (*f*) (med) laryngofissure, thyrotomy

ларинго·фо́н·, -а,   (*m*) (rad) throat microphone, laryngophone

Ла́рмор·а, прецесс·и·я   (*f*) (phys) Larmor precession

ларни́т·, -а,   (*m*) (min) larnite

лар·ь, -я́,   (*m*) bin, bunker, stowage, pocket

~, ру́д·н·ый   (min) ore pocket

ла́ск·а, -и,   (*g pl*) **ла́сок·,**   (*f*) (zool) weasel, *Mustela nivalis*; chamfer, lap; (*g pl*) **ласк·,**   (*f*) caress, endearment

с·ва́р·ива·ть в ла́ск·у   lap-weld

ласк·а́тельн·ый, -ая, -ое, (*a*) endearing, caressing
~ **и́м·я** (*n*) (gram) term of endearment
ласкир·ь, -я, (*m*) (fish) sea bream
ла́ст·, -а, (*m*) (zool) flipper, flopper, fin
ла́стик·, -а, (*m*) (text) lasting; eraser, rubber
ла́стов·ен·ь,-в·н·я, (*m*) (bot) swallow wort, *Cynanchum*
ла́стов·нев·ые, -ых, (*pl decl as adj*) (bot) milk weeds, *Asclepiadaceae*
ла́стов·ник·, -а, (*m*) *see* **ла́стов·ен·ь**
ласто·но́г·ие, -их, (*pl decl as adj*) (zool) pinnipeds, *Pinnipedia*
ла́сточ·к·а, -и, (*g pl*) **-ч·ек·,** (*f*) (orn) swallow, (*pl*) *Hirundinidae*
ла́сточ·кин·, -а, -о, (*a*) swallow, swallow's
~ **хвост·** (*m*) dovetail (joint, woodw); (ent) *Papilio machaon*
ла́сточ·ник·, -а, (*m*) *see* **ла́стов·ен·ь**
ла́т·а, -ы, (*g pl*) **лат·,** (*f*) (*pl*) (build) lathing, laths
латв·и́йск·ий, -ая, -ое, (*a*) Latvian
ла́текс·, -а, (*m*) (rubber) latex
~, **де·газ·и́рованн·ый** stripped latex
ла́текс·н·ый, -ая, -ое, (*a*) latex
лате́нт·н·ый, -ая, -ое, (*a*) latent
латенси·фик·а́ци·я, -и, (*f*) (phot) latensification
латер·а́льн·ый, -ая, -ое, (*a*) lateral
латер·иза́ци·я, -и, (*f*) (geol) laterization
латер·и́т·, -а, (*m*) (geol) laterite
латерит·иза́ци·я, -и, (*f*) (geol) laterization
латери́т·н·ый, -ая, -ое, (*a*) lateritic (of soils)
лати́н·ск·ий, -ая, -ое, (*a*) Latin, Roman; (naut) lateen (sail)
~ **шрифт·** (*m*) (print) Roman characters
~ **язы́к·** (*m*) Latin
лату́к·, -а, (*m*) (bot) lettuce, *Lactuca*
~ **-роме́н·, -а,** (*m*) cos/romaine lettuce, *Lactuca sativa* var. *longifolia*
~ **-сала́т·, -а,** (*m*) lettuce, cultivated lettuce, *Lactuca sativa*
~ **-сала́т·, коча́н·н·ый** head lettuce, *Lactuca sativa* var. *capitata*
~ **-сала́т·, спа́рж·ев·ый** stem/asparagus lettuce, *Lactuca sativa* var. *asparaginu*
лату́к·ов·ый, -ая, -ое, (*a*) lettuce; (chem) lactucic
латун·и́ровани·е, -я, (*n*) brass plating
латун·и́рованн·ый, -ая, -ое, (*past part pass*) brass-plated

лату́н·н·ый, -ая, -ое, (*a*) brass
лату́н·ь, -и, (*f*) brass
~, **автома́т·н·ая** free-cutting brass
~, **алюми́н·иев·ая** aluminium brass
~, **высоко·ме́д·ист·ая** rich low brass
~, **деформ·и́руем·ая** high/forging/wrought brass
~, **жёлт·ая** yellow brass
~, **лит·е́йн·ая** casting brass
~, **марган·цеви́ст·ая** manganese brass/bronze, high-tensile brass, complex brass
~, **марган·цо́в·ая** manganese bronze/brass, complex brass, high-tensile brass
~, **мор·ск·а́я** naval brass
~, **ни́кел·ев·ая** nickel brass
~, **обык·нове́нн·ая** 60–40 brass
~, **олов·я́нн·ая** naval brass/bronze
~ **от·ве́т·ственн·ого на·знач·е́ни·я** 70–30 brass
~, **патро́н·н·ая** cartridge brass
~, **свинц·о́вист·ая** leaded brass
~, **торг·о́в·ая** market brass
~, **час·ов·а́я** clock brass
ла́т·ы (*nom pl*) of **ла́т·а**
латы́ш·ск·ий, -ая, -ое, (*a*) Lettish
лаубани́т·, -а, (*m*) (min) laubanite
лаудани́н·, -а, (*m*) (pharm) laudanine
лаурдали́т·, -а, (*m*) (min) laurdalite
лауреа́т·, -а, (*m*) laureate (title)
лаудану́м·, -а, (*m*) (pharm) laudanum
лаури́л·, -а, (*m*) (chem) lauryl
лаури́л·ов·ый, -ая, -ое, (*a*) lauryl
~ **спирт·** (*m*) lauryl/dodecyl alcohol
лаури́н·, -а, (*m*) (chem) laurin
лаури́н·ов·ый, -ая, -ое, (*a*) (chem) lauric
~ **альдеги́д·** (*m*) lauraldehyde, lauric aldehyde
~ **кисл·от·а́** (*f*) (chem) lauric acid
лауриони́т·, -а, (*m*) (min) laurionite
Ла́ут·а, клет·ь (*f*) (met roll) Lauth mill
лаутари́т·, -а, (*m*) (min) lautarite
лауэ·гра́мм·а, -ы, (*f*) (spectr) Laue photograph/pattern
лафе́т·, -а, (*m*) gun carriage; (timber) plank; (metal) plate; (oil) spider
~ **стре́л·очн·ого пере·во́д·а** (rail) switch side-plate
лафе́т·н·ый, -ая, -ое, (*a*) of **лафе́т·**
~ **ствол·** (*m*) gun; monitor (on a fireboat)
лаха́р·, -а, (*m*) (geol) lahar
ла́хт·а, -ы, (*f*) small cove, bay

лахта́к·, -и, (*m*) (zool) bearded seal, *Erignathus barbatus*

ла́цборт·, -а, (*m*) *see* **ла́цпорт·**

ла́цкан·, -а, (*nom pl*) **-ы,** (*g pl*) **-ов,** (*m*) (text) lapel

ла́цкан·н·ый, -ая, -ое, (*a*) lapel

ла́цпорт·, -а, (*m*) (naut) side port, cargo port

ла́·яни·е, -я, (*n*) barking (of dogs etc.)

ла́·ят·ь, -ют, (*imp*) bark (of dogs)

лб·а (*g*) of **лоб·**

лг·а́вш·ий, -ая, -ее, (*past part act*) of **лг·ать**

лг·ать, (*pres 3rd sing, pl*) **лж·ёт, лг·ут,** (*past masc sing*) **лг·ал,** (*imp*) lie, tell lies

лг·у́щ·ий, -ая, -ее, (*pres part act*) of **лг·ать**

Лебе́г·а, интегра́л· (*m*) (math) Lebesgue integral

~, ме́р·а (*f*) (math) Lebesgue measure

лебе́г·ов·ый, -ая, -ое, (*a*) (math) Lebesgue

лебед·а́, -ы́, (*f*) (bot) saltbush, *Atriplex*

лебёд·к·а, -и, (*g pl*) **-д·ок·,** (*f*) winch, windlass, (min) winding engine, winder, hoist; (zool) female swan

~, бараба́н·н·ая drum winder

~, двух·бараба́н·н·ая two-drum winch/hoist, double-drum winch/hoist

~, коло́н·ков·ый column-mounted winch/hoist

~, му́сор·н·ая (naut) ash hoist

~, кант·о́вочн·ая valve tilting gear (gas producer)

~, много·бараба́н·н·ая multi-drum winch

~, на·сте́н·н·ая wall/bracket winch

~, пар·ов·а́я steam winch

~, подъ·ём·н·ая (oil) draw works

~, стрел·ов·а́я boom hoist/winch, jib hoist/winch

~, тра́л·ов·ая trawl winch

~, трел·ёвочн·ая cable skidder

~, фрикцио́н·н·ая friction-drive winch/winder, friction hoist

~, шпил·ев·а́я capstan

лебед·о́в·ые, -ых, (*pl decl as adj*) (bot) goosefoot family, *Chenopodiaceae*

лебёд·очн·ый, -ая, -ое, (*a*) of **лебёд·к·а**

лебёд·чик·, -а, (*m*) winding engine man, winch man

ле́бед·ь, -я, (*nom pl*) **-и,** (*g pl*) **-е́й,** (*m*) (orn) swan, *Cygnus*; (astron) Cygnus

лебез·а́, -ы́, (*f*) caulking iron

лебердин·а, -ы, (*f*) (shipb) bottom frame template

Лебла́н·а, спо́соб· (*m*) (chem) Leblanc process

~, схе́м·а (*f*) (elec) Leblanc connection

лебя́ж·ий, -ья, -ье, (*a*) of **ле́бед·ь**

лев (*g sing*) **льв·а,** (*m*) (zool) lion, *Felis leo*; (astron) Leo, the lion

~, мор·ск·о́й (zool) sea lion, *Zalophus*

~, тлёв·ый (ent) aphis lion, chrysop, *Chrysopa vulgaris*

лев·а́н·, -а, (*m*) (chem) levan

левант·и́йск·ий, -ая, -ое, (*a*) levant, levantine

леве́йт·, -а, (*m*) (min) lowite

левиафа́н·, -а, (*m*) (text) leviathan washer

левини́т·, -а, (*m*) (min) levynite

левкани́да, одно·то́ч·ечн·ая (*f*) (ent) army worm, *Leucania unipuncta*

левко́·й, -я, (*m*) (bot) stock, *Mathiola*

лево- (*root*) left, left-hand, laevo-, levo-

лево·борт·ов·о́й, -а́я, -о́е, (*a*) left-side; (naut, air) port, port-side

лево·вращ·а́ющ·ий, -ая, -ее, (*a*) (phys, chem) laevorotatory, levorotatory; (mech) counter-clockwise, left-handed

~ изо·ме́р· (*m*) (chem) laevoisomer, levoisomer

~ со·един·е́ни·е (*n*) (chem) laevo-compound, levo-compound

лево·идеа́ль·н·ый, -ая, -ое, (*a*) (math) left-ideal

лево·пима́р·ов·ая кисл·от·а́ (*f*) (chem) levopimaric acid

лево·сторо́н·н·ий, -ая, -ее, (*a*) left-side, sinistral; (naut, air) port, port-side

лев·ул·ёз·а, -ы, (*f*) (chem) laevulose, levulose

левули́н·ов·ый, -ая, -ое, (*a*) levulin, levulinic

~ кисл·от·а́ (*f*) levulinic acid

лев·ш·а́, -й, (*i*) **-о́й,** (*nom pl*) **-и,** (*m or f*) left-hander

ле́в·ый, -ая, -ое, (*a*) left, left-hand; anticlockwise; (naut) port

пра́в·ил·о ле́в·ой ру́к·и (*n*) (elec) left-hand rule, Fleming's left-hand rule

лёг· (*past masc sing*) of **леч·ь** (*perf*) *see* **лож·и́ться** (*imp*)

легал·иза́ци·я, -и, (*f*) legalization; (comm) certification

лега́ль·н·ый, -ая, -ое, (*a*) legal

леге́нд·а, -ы, (*f*) legend; (print) legend, caption

леги́р·овани·е, -я, (*n*) (met) alloying

леги́р·ованн·ый, -ая, -ое, (*past part pass*) alloyed, alloy

~ **стал·ь** (*f*) alloy steel

леги́р·ующ·ий, -ая, -ее, (*pres part act*) alloying

~ **материа́л·** (*m*) alloying element, alloy-forming element

лёгк·ий, -ая, -ое, (*a*) (*see also* **лёг·к·о́е**) (*as noun*) light, light-weight; easy, light; slight

~ **ма́сл·о** (*n*) light oil

~ **мета́лл·** (*m*) light metal, low-density metal

~ **на·магни́ч·ивани·е** (*n*) (elec) easy magnetization

~ **то́пл·ив·о** (*n*) light fractions (petroleum)

ле́гко- (*root*) light, lightweight; easy, easily, readily; slightly

легко·ве́р·н·ый, -ая, -ое, (*a*) gullible

легко·водо·ла́з·, -а, (*m*) shallow water diver, skin/independent diver

легк·ов·о́й, -а́я, -о́е, (*a*) passenger (of vehicles)

~ **автомоби́л·ь** (*m*) motor car, automobile

легко·вос·пламен·я́ющ·ийся, -аяся -еэся, (*a*) highly inflammable

легко·до·сту́п·н·ый, -ая, -ое, (*a*) readily available, easy to get/find

лёгк·ое, -ого, (*n decl as adj*) (anat) lung, pulmo

легко·зо́ль·н·ый, -ая, -ое, (*a*) light-ash

легко·иониз·и́рующ·ий, -ая, -ее, (*a*) easily ionized

легко·кип·я́щ·ий, -ая, -ее, (*a*) low-boiling

легко·лет·у́ч·ий, -ая, -ее, (*a*) highly volatile

легко·мы́сл·енн·ый, -ая, -ое, (*a*) frivolous

легко·па́луб·н·ый, -ая, -ое, (*a*) (shipb) awning deck

легко·пла́в·к·ий, -ая, -ое, (*a*) low-melting point, fusible

легко·по·дви́ж·н·ый, -ая, -ое, (*a*) highly mobile

легко·полимери́з·ующ·ий, -ая, -ое, (*a*) readily polymerized

легко·раз·лич·и́м·ый, -ая, -ое, (*a*) easily distinguishable

легко·рас·твор·и́м·ый, -ая, -ое, (*a*) readily soluble

лёгк·ост·ь, -и, (*f*) lightness; easiness, ease; (naut) heaving-line weight

легко·сып·у́ч·ий, -ая, -ее, (*a*) free-flowing

легко·сь·ём·н·ый, -ая, -ое, (*a*) easily removable

легко·хо́д·ов·ый, -ая, -ое, (*a*) free-moving, free-running

~ **по·са́д·к·а** (*f*) (mech) free fit

лег·ли́ (*past pl*) of **леч·ь** (*perf*) *see* **лож·и́ться** (*imp*)

лёг·очн·ый, -ая, -ое, (*a*) (anat) pulmonary, lung; (zool) pulmonate, (*pl*) *Pulmonata*

~ **автома́т·** (*m*) oxygen breathing apparatus

~ **моллю́ск·и** (*pl*) (zool) *Pulmonata*

лёг·очниц·а, -ы, (*f*) (bot) lungwort, *Pulmonaria officinalis*

легуми́н·, -а, (*m*) (biochem) legumine

лёгч·е (*comp*) of **лёгк·ий**

лёг·ш·ий, -ая, -ее, (*past part act*) of **леч·ь** (*perf*), *see* **лож·и́ться** (*imp*)

лёд, (*g sing*) **льд·а,** (*m*) ice

~, **берег·ов·о́й** ice foot/belt/wall

~, **бли́н·чат·ый** pancake ice

~, **внутри·во́д·н·ый** frazil ice

~, **гле́тчер·н·ый** glacier ice

~, **гнил·о́й** rotten ice

~, **до́н·н·ый** bottom ice

~, **дре́вн·ий поля́р·н·ый** paleocrystic ice

~, **дрейф·ов·ый** drift/floe ice

~, **за·лив·н·о́й** bay ice

~, **зерн·и́ст·ый** névé, firn

~, **ис·коп·а́ем·ый** fossil ice

~, **крупно·би́·т·ый** light floe ice

~, **материк·о́в·ый** land ice, terrestrial ice

~, **мелко·би́·т·ый** very light floe ice

~, **много·ле́т·н·ый** old ice

~, **молод·о́й** fresh/young ice

~, **мор·ск·о́й** sea ice

~, **на·бив·н·о́й** (ocean) ridge, layered ice due to rafting

~, **не·по·дви́ж·н·ый** stagnant ice

~, **одно·ле́т·н·ий** winter ice

~, **от·ступ·а́ющ·ий** waning ice

~, **па́к·ов·ый** pack ice

~, **про·зр·а́чн·ый** black ice

~, **по́р·ов·ый** interstitial ice

~, **раз·реж·ённ·ый** open pack, open pack ice

~, **ре́д·к·ий** drift ice

~, **ро́вн·ый** level ice

лёд
~, с·плоч·ённ·ый close pack ice
~, сплош·н·ой consolidated pack ice
~, ста́р·ый paleocrystic ice
~, сто·я́ч·ий ground ice
~, сух·о́й (chem) dry ice, solid carbon dioxide
~, фи́рн·ов·ый firn, firn ice, névé
ледебури́т·, -а, (m) (met) ledeburite
ледгили́т·, -а, (m) (min) ledhillite
лед·ене́·ть, -ют, (imp intrans) freeze, congeal
лед·ене́ц·, -нц·а́, (i) -нц·о́м, (m) (food) fruit-drop, acid-drop, sugar candy
лед·ен·и́ть, -я́т, (imp trans) freeze, chill
ледери́н·, -а, (m) leatheroid, leather cloth, artificial leather
лед·ни́к·, -а́, (m) (geog) glacier, ice form; **лёд·ник·**, -а, (m) cold room/store, ice box, refrigerator
~ втор·о́го раз·ря́д·а corrie glacier
~, ка́р·ов·ый corrie glacier
~ перв·о́го раз·ря́д·а alpine glacier
лед·ников·ый, -ая, -ое, (a) glacial, englacial; ice box/room/store, refrigerator
~ абра́зи·я (f) glacial abrasion
~ ве́тер· (m) glacier breeze
~ пери́од· (m) (geol) Ice Age
~ стол· (m) glacier table
~ язы́к· (m) glacier snout
лед·никовь·е, -я, (n) (geol) glaciation, glacial period
ледо·ве́д·ени·е, -я, (n) cryology
лед·ови́т·ый, -ая, -ое, (a) icy
~ океа́н· (m) Arctic Ocean
лед·о́в·ый, -ая, -ое, (a) ice
~ об·стано́в·к·а (f) (naut) ice conditions
~ об·ши́в·к·а (f) (shipb) ice belt
~ по́яс· (m) (shipb) ice belt
~ раз·ве́д·к·а (f) ice reconnaissance
ледо·де́л·ательн·ый, ая, -ое, (a) ice-making
ледо·ко́л·, -а, (m) (naut) ice-breaker
~, лин·е́йн·ый large ice breaker
ледо·ко́ль·н·ый, -ая, -ое, (a) ice-breaker
~ паро·хо́д· (m) ice-breaking steamship
ледо·ло́м·, -а, (m) (civ eng) ice apron/breaker
ледо·образ·ова́ни·е, -я, (n) formation of ice

ледо·па́д·, -а, (m) (geog) steeply sloping glacier, ice fall
ледо·раз·де́л·, -а, (m) ice divide
ледо·ре́з·, -а, (m) (civ eng) ice apron, ice-breaker (river bridges); (naut) ice-breaker (horizontal action only)
ледо·ру́б·, -а, (m) ice axe, ice pick
ледо·с·бро́с·, -а, (m) (civ eng) ice spillway
ледо·пу́ск·, -а, (m) (civ eng) ice chute
ледо·ста́·в·, -а, (m) icing up, freezing over (of waters)
ледо·фо́рм·а, -ы, (f) ice mould/mold
ледо·хо́д·, -а, (m) débâcle (on rivers), ice-gang
лед·ы́шк·а, -и, (g pl) -шек·, (f) piece of ice
лед·я́нк·а, -и, (g pl) -нок·, (f) ice boat
Ледю́к·а, ток· (m) (elec) Leduc current
лед·ян·о́й, -а́я, -о́е, (a) ice, icy, glacial
~ тор·а́ (f) iceberg
~ до́жд·ь (m) sleet
~ игл·а́ (f) ice crystal/needle
~ ко́р·к·а (f) ice rind
~ крас·и́тел·ь (m) (chem) ice dye/colour/color
~ по·кро́в· (m) (geog) ice cover/sheet/cap
~ по́л·е (n) (ocean) field ice, ice field
~ стён·к·а (f) (min) frost wall
~ у́ксус·н·ая кисл·от·а́ (f) (chem) glacial acetic acid
ле́ер·, -а, (nom pl) -а́, (g pl) -о́в, (m) (naut) stay, manrope
~, што́рм·ов·о́й lifeline
ле́ер·н·ый, -ая -ое, (a) of **ле́ер·**
~ о·гражд·е́ни·е (n) guard rails/chain
~ со·общ·е́ни·е (n) breech buoy transfer
~ у·стро́й·ств·о (n) guard rail/chain
леж·а́л·ый, -ая, -ое, (a) stale, old
Лежа́ндр·а, много·чле́н·ы (pl) (math) Legendre's polynomials, Legendre's coefficients
~, си́мвол·, (m) (math) Legendre's symbol
леж·а́ть, (pres 3rd sing, pl) **леж·и́т**, **леж·а́т**, (imp) lie, be lying down, be recumbent; be (situated)
~ в дре́йф·е (naut) lie to
леж·а́ч·ий, -ая, -ее, (a) lying, recumbent
~ бок· (m) (min) footwall, underwall

леж·а́щ·ий, -ая, -ее, (*pres part act*) of **леж·а́ть**
~ **капе́л·ь** (*f*) (phys) stationary drop
~ **между** (geol) interjacent
ле́ж·бищ·е, -а, (*i*) **-ем,** (*n*) lay place, store
ле́ж·ен·ь, -ж·н·я, (*m*) (build, min) ledger, stay block; (civ eng) foundation beam, ground sill, sill, sleeper
~, **бе́рег·ов·о́й** bank seat (of bridge)
~, **про·до́ль·н·ый** (geol) longitudinal sill
лёж·к·а, -и, (*f*) keeping; maturing (by keeping); (zool) lay place, lair
лёж·кост·ь, -и, (*f*) (food) keeping capacity
ле́зв·и·е, -я, (*n*) blade, edge, cutting edge; razor blade
лез·ти́ (*imp deter*) of **лез·ть**
лез·ть, (*pres 3rd sing, pl*) **ле́з·ет, ле́з·ут,** (*imp deter*) climb, clamber, get; creep; reach/feel in (with the hand); fit; press on, thrust forward; fall/come out (of hair)
ле́з·ущ·ий, -ая, -ее, (*pres part act*) of **лез·ть**
ле́з·ш·ий, -ая, -ее, (*past part act*) of **лез·ть**
лейа́с·, -а, (*m*) (geol) Liassic, Lias
Ле́йбниц·а, фо́рмул·а (*f*) (math) Leibnitz formula/theorem
лейдейс·, -а, (*m*) (com) lay-days
ле́йден·ск·ая ба́н·к·а (*f*) (elec) Leyden jar
ле́·йк·а, -и, (*g pl*) **ле́·ек·,** (*f*) pourer, funnel; watering-can; (naut) boiler; (phot) Leica
~, **за·лив·н·а́я** priming funnel (pumps)
лейк·еми́·я, -и, (*f*) (med) leukemia, leucaemia
лейк·о́з· = лейк·еми́·я
лейко·кра́т·ов·ый, -ая, -ое, (*a*) (geol) leucocratic
лейко·ксе́н·, -а, (*m*) (min) leucoxene
лейко·ку́б·ов·ый, -ая, -ое, (*a*) (chem) leuco-vat
лейко·пени́·я, -и, (*f*) (med) leuco-penia, leukopenia
лейко·основ·а́ни·е, -я, (*n*) (chem) leuco base
лейко·пла́ст·, -а, (*m*) (cyt) leukoplast
лейко·пла́стыр·ь, -я, (*m*) (med) white plaster
лейко·со·един·е́ни·е, -я, (*n*) (chem) leuco compound
лейко·ци́т·, -а, (*m*) (biol) leukocyte, leucocyte

лейко·цит·о́з·, -а, (*m*) (med) leuco-cytosis, leukocytosis
ле́йна-сели́тр·а, -ы, (*f*) (chem) Leuna saltpetre
ле́йнер·, -а, (*m*) liner; (gunn) inner A tube, liner
~, **свобо́д·н·ый** (gunn) loose liner
~, **цилиндр·и́ческ·ий** parallel liner
лейн·и́ровани·е, -я, (*n*) (gunn) relining
ле́йст·а, -ы, (*f*) (geol) lath
ле́й·те (*plur imper*) of **ли·ть**
лейтена́нт·, -а, (*m*) lieutenant
~, **мла́д·ш·ий** junior lieutenant
~, **ста́р·ш·ий** senior lieutenant
лейц·и́н·, -а, (*m*) (chem) leucine
лейц·и́т·, -а, (*m*) (min) leucite
лейц·ити́т·, -а, (*m*) (min) leucitite
лейци́т·ов·ый, -ая, -ое, (*a*) leucitic
лейцито·фи́р·, -а, (*m*) (geol) leucito-phyre
лейшман·ио́з·, -а, (*m*) (med) leishman-iosis
лейшма́н·и·я, -и, (*f*) (zool) *Leishmania*
лека́л·о, -а, (*n*) curve, french curve (for drawing); mould, template; (a/c) control surface locking band
лека́ль·н·ый, -ая, -ое, (*a*) of **лека́л·о**
~ **рабо́т·ы** (*pl*) (mech) gauge/gage work
лека́ль·щик·, -а, (*m*) gauger, gager
ле́кар·ск·ий, -ая, -ое, (*a*) of **ле́кар·ь**
лека́р·ственн·ый, -ая, -ое, (*a*) medi-cinal
~ **ве́щ·еств·о** (*n*) pharmaceutical
~ **раст·е́ни·е** (*n*) medicinal herb
~ **со·един·е́ни·е** (*n*) drug
лека́р·ств·о, -а, (*n*) medicine; **ле́кар·ство, -а,** (*n*) medicine, medical profession
ле́кар·ь, -я, (*m*) doctor
Лекланше́, элеме́нт· (*m*) (elec) Le-clanché cell
ле́кс·ик·а, -и, (*f*) vocabulary
лексико́·граф·, -а, (*m*) lexicographer, dictionary maker
лексико·гра́ф·и·я, -и, (*f*) lexicography
лексик·о́н·, -а, (*m*) vocabulary; (obs) lexicon, dictionary
лекс·и́ческ·ий, -ая, -ое, (*a*) (ling) lexical
ле́ктор·, -а, (*m*) lecturer
лекто́р·и·й, -я, (*m*) lecture-organizing office; lecture hall
лекцио́н·н·ый, -ая, -ое, (*a*) of **ле́кци·я**
ле́кци·я, -и, (*f*) lecture
лёллинги́т·, -а, (*m*) (min) lollingite
ле́мех·, -а, (*m*) (agr) share, plough share, plow share

ле́мм·а, -ы, (*f*) (math) lemma

ле́мминг·, -а, (*m*) (zool) lemming, *Lemmus*

лемна́ч·, -а, (*m*) hand-formed brick

лемниска́т·а, -ы, (*f*) (math) lemniscate, Bernoulli's lemniscate

лемонгра́сс·ов·ое ма́сл·о (*n*) (chem) lemon-grass oil

лемпа́ч·, -а, (*m*) adobe, sun-dried brick

лему́р·, -а, (*m*) (zool) lemur, (*pl*) *Lemuroidea*

лён, (*g sing*) **льн·а,** (*m*) (bot) flax, *Linum*; (text) linen

~, го́р·н·ый asbestos

~ -долг·унец́·, -унц·а́, (*m*) fibre/fiber flax

~ -кудря́ш·, -а́, (*i*) **-ш·о́м,** (*m*) seed flax

~ -мо́ч·енец·, -нц·а, (*m*) water-retted flax fibre/fiber

~, мя́·т·ый broken flax fibre/fiber

~, обык·нове́нн·ый flax, *Linum usitatissimum*

~, рос·ов·ая dew-retted flax fibre/fiber

~, семен·н·о́й flaxseed, flax for seed

~ -стл·а́нец·, -нц·а, (*i*) **-нц·ем,** (*m*) dew-retted flax fibre/fiber

~ -сыр·е́ц·, -р·ц·а́, (*i*) **-р·ц·о́м,** (*m*) broken flax fibre/fiber

~ тепл·ов·о́й мо́ч·к·и heat-retted flax fibre/fiber

~, трёп·ан·ый scutched flax

~, чёс·ан·ый hackled flax

Ле́нгмюр·а, фо́рмул·а (*f*) (elec) Langmuir–Child law

лен·и́вец·, -вц·а, (*i*) **-вц·ем,** (*m*) idler, idle/lazy person; (zool) sloth; (mech) idler, idler wheel/sprocket

лен·и́в·ый, -ая, -ое, (*a*) lazy

лен·и́ться, -ятся, (*imp*) be lazy/idle

ле́нкер·, -а, (*m*) guide (drawing instrument)

Ле́ннард-Джо́нс·а, потенциа́л· (*m*) (phys) Lennard–Jones potential

лён·ов·ые, -ых, (*pl decl as adj*) (bot) flax family, *Linaceae*

ле́нт·а, -ы, (*f*) ribbon, tape, band, belt, strip; (met) narrow strip; (text) sliver, chart (of a recording instrument)

~, авиацио́н·н·ая (a/c) fabric tape, wing tape

~, банда́ж·н·ая rim band/strip (of wheel); shrouding strip (of turbine)

~, борт·ов·а́я chafer strip (rubber)

~, гре́б·енн·ая ше́рст·ян·ая (text) wool tops

ле́нт·а

~, двой·н·а́я duplex belt, two-ply belt (conveyor)

~, гу́сен·ичн·ая (M/T) crawler belt, caterpillar chain

~, за·мк·ну́т·ая (autom) tape loop

~, за·ря́д·н·ая (nucl) charging belt

~, иго́ль·чат·ая (text) card clothing

~, изоляцио́н·н·ая (elec) insulating tape

~, ки́пер·н·ая twill/herringbone tape

~, конве́йер·н·ая conveyor belt

~ кома́нд· order tape (computers)

~ коэффицие́нт·ов coefficient tape (computers)

~, крыль·ев·а́я bead strip (rubber)

~, ли́п·к·ая adhesive tape

~, лот·ков·а́я troughed conveyor belt

~, магни́т·н·ая magnetic tape

~, ме́р·н·ая tape measure, measuring/surveying tape

~, обо́д·н·а́я rim band/strip (of wheel)

~, об·мо́т·очн·ая binding tape

~, о́бруч·н·ая barrel band

~, паспо́рт·н·ая (cinema) light-change band

~, перфор·о́ванн·ая perforated tape, punched tape

~, породо·от·бо́р·н·ая (min) hand-picking beet

~, програ́мм·н·ая (autom) programme tape

~, пружи́н·н·ая spring steel strip

~, пуле·мёт·н·ая (gunn) cartridge belt

~ рас·ча́л·к·и (a/c) bracing strip

~, рези́н·ов·ая rubber belting strip

~, рессо́р·н·ая spring-steel strip

~, с·кле́·ечн·ая (cinema) splicing tape

~, скреб·ко́в·ая scraper conveyor belt, flight belt

~, смол·ян·а́я (elec) bituminized insulating tape

~, сталь·н·а́я narrow steel strip

~, с·тёр·т·ая erased tape

~ транспо́ртер·а conveyor belt

~, у·сил·и́тельн·ая reinforcing tape

~ ферро́до (M/T) Ferrodo brake lining

~, чес·а́льн·ая (text) card sliver

~, числ·ов·а́я number tape (computers)

~, чи́ст·ая blank/raw tape (of tape recorder)

лент·е́ц·, -т·ц·а́, (*m*) (zool) tapeworm

лентикуля́р·н·ый, -ая, -ое, (*a*) lenticular, lens-shaped

ленто·мот·а́льн·ая маши́н·а (*f*) (text) braid bobbin winder

ленто·об·мо́т·очн·ый, -ая, -ое, (*a*) taping

ленто·от·ли́в·очн·ый, -ая, -ое, (*a*) sheet-forming (rubber)

ленто·про·ка́т·н·ый, стан· (*m*) (met) narrow strip mill

ленто·про·тя́ж·н·ый, -ая, -ое, (*a*) tape-drawing/winding

~ механ·и́зм· (*m*) tape transport (of tape recorder etc.), feeding mechanism (ciné projector), chart/paper roller (recording instr)

ленто·с·ва́р·очн·ый, -ая, -ое, (*a*) seam-welding

ленто·со·един·и́тельн·ая· маши́н·а (*f*) (text) sliver lapping machine

ленто·тк·а́цк·ий стан·о́к· (*m*) (text) ribbon loom

ленто·фре́зер·н·ый стан·о́к· (*m*) (mech) band milling machine

ле́нт·очк·а, -и, (*g pl*) **-чек·,** (*f*) (*dim*) of **ле́нт·а;** margin (of drill)

~, креп·и́тельн·ая bead flipper (rubber)

ленточно·пи́ль·н·ый стан·о́к· (*m*) band saw

ле́нт·очн·ый, -ая, -ое, (*a*) of **ле́нт·а** and **ле́нт·очк·а**

~ гре́бен·н·ая маши́н·а (*f*) (text) sliver combing machine

~ конве́йер· (*m*) belt conveyor

~ маши́н·а (*f*) (text) drawing frame

~ пил·а́ (*f*) band saw

~ само·пи́с·ец· (*m*) strip-chart recorder

~ то́рмоз· (*m*) band brake

~ черв·ь (*f*) (zool) tapeworm, *Cestoid*

Ле́нц·а, зако́н· (*m*) (elec) Lentz's law

~, пра́в·ил·о (*n*) (elec) Lentz's law

~ -Джо́ул·я, зако́н· (*m*) (elec) Joule's law

лен·ь, -и, (*f*) laziness, indolence

леони́т·, -а, (*m*) (min) leonite

леопа́рд·, -а, (*m*) (zool) leopard, *Felis pardus*

~, мор·ск·о́й (zool) sea leopard, *Hydrurga leptonyx*

лео·тро́п·н·ый, -ая, -ое, (*a*) (zool) laeotropic, leiotropic, anti-clockwise, sinistral

лепестко·ви́д·н·ый, -ая, -ое, (*a*) petal-shaped, petaloid

лепест·ко́в·ый, -ая, -ое, (*a*) petaline, (bot) petalous, petalled

лепест·о́к·, -т·к·а́, (*m*) (bot) petal; (rad) lobe; arris (on a brick)

~, бок·ов·о́й (rad) side lobe

~, жа́бер·н·ый (fish) gill fringe

~, основ·н·о́й (rad) major/main lobe **пере·ключ·е́ни·е лепест·к·о́в** (*n*) (rad) lobing

~ по·бо́ч·н·ого из·луч·е́ни·я (rad) side lobe

леп·ёшк·а, -и, (*g pl*) **-шек·,** (*f*) (pharm) tablet; cake (of coal), lump (of slag)

лепешко·обра́з·н·ый, -ая, -ое, (*a*) oblate, muffin-shaped

лепид·и́н·, -а, (*m*) (chem) lepidine

лепидо·де́ндрон·, -а, (*m*) (pal, bot) *Lepidodendron*

лепидо·крокит·, -а, (*m*) (min) lepidocrocite

лепидо·ли́т·, -а, (*m*) (min) lepidolite, lithia-mica

лепидо·мела́н·, -а, (*m*) (min) lepidomelane

леп·и́ть, ⁀ят, (*imp*) model, shape, fashion, mould; stick on

леп·н·о́й, -а́я, -о́е, (*a*) modelling, modeling, shaping, shaped, plastic

~ искус·ств·о (*n*) plastic art

лепо·ци́т·, -а, (*m*) (cyt) lepocyte

ле́пр·а, -ы, (*m*) (med) leprosy

лепр·о́зн·ый, -ая, -ое, (*a*) leprous

лепроз·о́ри·й, -я, (*m*) leprosy hospital

лепт·и́т·, -а, (*m*) (geol) leptite

лепт·о́м·, -а, (*m*) (bot) leptome, leptom

лепт·о́м·а *see* **лепт·о́м·**

лепто·меду́з·ы (*pl*) (zool) *Leptomedusae*

лепт·о́н·, -а, (*m*) (phys) lepton

лепто́н·н·ый, -ая, -ое, (*a*) (phys) lepton·

лепто·спир·о́з·, -а, (*m*) (med) leptospirosis

~, желт·у́шн·ый spirochetal jaundice

лепто·флоэ́м·а, -ы, (*f*) (bot) leptophloem

лепто·хлор·и́т·, -а, (*m*) (min) leptochlorite

ле́птус·, -а, (*m*) (ent) leptus

лерци́т·, -а, (*m*) (min) lherzite

ле́рк·, -и, (*f*) screw-thread die

лернео·по́д·, -а, (*m*) (fish) lernaeopod

лернео·це́р·а, -ы, (*f*) (zool) lernaeocera

лес·, -а, (*nom pl*) **лес·а́,** (*g pl*) **-о́в,** (*m*) wood, woods, forest, silva, sylva; (*sing only*) timber(s), wood; (*pl only*) scaffolding

~, корабе́ль·н·ый shipbuilding timber

~, креп·ёжн·ый mining timber, pit props

~, пил·ён·ый sawn/converted timber

лес

~, **стро·ев·о́й** ṣtanding timber

~, **сыр·о́й** green wood

лёс·а, -ы, (*nom pl*) **лёс·ы,** (*g pl*) **лёс·,** (*f*) fishing line; *see also* **лес·**

лес·и́ст·ый, -ая, -ое, (*a*) wooded

лес·ни́к·, -а́, (*m*) forester, woodman

лес·н·о́й, -а́я, -о́е, (*a*) of **лес·**; wood, forest, sylvan, silvan

~ **масси́в·** (*m*) forest

~ **про·мы́шл·енност·ь** (*f*) timber/lumber industry

~ **ре́йд·** (*m*) timber/lumber trap (paper mill)

~ **хозя́й·ств·о** (*n*) forestry

лесо·ве́дени·е, -я, (*n*) sylvics

лесо·во́д·ств·о, -а, (*n*) sylviculture, arboriculture

лесо·во́д·ческ·ий, -ая, -ое, (*a*) of **лесо·во́д·ств·о**

лесо·во́з·, -а, (*m*) timber carrier (ship); log transporter, logging truck

лесо·во́з·н·ый, -ая, -ое, (*a*) logging, timber-carrying

~ **авто·маши́н·а** (*f*) timber carrier, logging truck

лесо·воз·об·новл·е́ни·е, -я, (*n*) forest regeneration, reforestation

лес·ов·о́й, -а́я, -о́е, (*a*) wood, forest

лесо·за·во́д·, -а, (*m*) timber mill

лесо·за·гото́в·к·а, -и, (*f*) lumbering, logging

лесо·за·готов·и́тельн·ый, -ая, -ое, (*a*) lumber, lumbering

лесо·за·щи́т·а, -ы, (*f*) forest-protection

лесо·материа́л·ы, -ов, (*m pl*) lumber, timber, wood

~, **креп·ёжн·ые** mining timbers, pit props

~, **боч·а́рн·ый** barrel wood

лесо·мелиор·а́ци·я, -и, (*f*) forest reclamation

лесо·на·сажд·е́ни·е, -я, (*n*) afforestation

лесо·о·хран·е́ни·е, -я, (*n*) forest protection

лесо·па́рк·, -а, (*m*) national park

лесо·пи́л·к·а, -и, (*g pl*) **-лок·,** (*f*) sawmill; saw frame

лесо·пи́ль·н·ый, -ая, -ое, (*a*) saw-milling

~ **ра́м·а** (*f*) frame saw, sash gang saw

~ **стан·о́к· ра́м·н·ый** (*m*) frame saw, sash gang saw

лесо·пит·о́мник·, -а, (*m*) forest tree nursery

лесо·по·гру́з·очн·ый, -ая, -ое, (*a*) timber/lumber-loading

~ **маши́н·а** (*f*) timber/log loader

лесо·по·са́д·очн·ый, -ая, -ое, (*a*) forest/tree planting

~ **маши́н·а** (*f*) tree planting machine

лесо·про·мы́шл·енност·ь, -и, (*f*) timber industry

лесо·пу́нкт·, -а, (*m*) forestry station, lumber camp

лесо·раз·вед·е́ни·е, -я, (*n*) afforestation

~, **поле·за·щи́т·н·ое** protective afforestation

лесо·ру́б·, -а, (*m*) woodcutter, lumberman, timber filler

лесо·се́к·а, -и, (*f*) clearing, cutting/felling area (in forest)

лесо·се́ч·н·ый, -ая, -ое, (*a*) of **лесо·се́к·а**

лесо·семен·н·о́е де́л·о (*n*) forestry seed collection and storage

лесо·спи́рт·, -а, (*m*) wood spirit, methyl alcohol

лесо·с·пла́в·, -а, (*m*) timber flotation/rafting, logging

лесо·с·пу́ск·, -а, (*m*) timber slide

лесо·степ·н·о́й, -а́я, -о́е, (*a*) forest-steppe (transitional zone)

лесо·сте́п·ь, -и, (*f*) forest-steppe

лесо·суш·и́лк·а, -и, (*g pl*) **-лок·,** (*f*) timber drying plant, timber kiln

лесо·та́ск·а, -и, (*g pl*) **-сок·,** (*f*) log hauler and skidder

лесо·техн·и́ческ·ий, -ая, -ое, (*a*) wood-technology

лесо·ту́ндр·а, -ы, (*f*) forest-tundra (transitional zone)

лесо·у·кла́д·чик·, -а, (*m*) timber stacker

лесо·у·стро́й·ств·о, -а, (*n*) forest management

лесо·хи́м·и·я, -и, (*f*) wood chemistry

лесо·хозя́й·ств·о, -а, (*n*) forestry

леспеде́ц·а, -ы, (*f*) (bot) lespedeza

лёсс·, -а, (*m*) (geol) loess

лесс·иро́вк·а, -и, (*g pl*) **-вок·,** (*f*) glazing, scumbling (painting)

лёсс·ов·ый, -ая, -ое, (*a*) of **лёсс·**

ле́стниц·а, -ы, (*i*) **-ц·ей,** (*f*) stairs, staircase, ladder

~, **верёв·очн·ая** rope ladder

~, **колыш·ко́в·ая** Jacob's ladder

~, **пред·го́р·н·ая** (geol) piedmonttreppen

~, **при·ста́в·н·а́я** lean-to ladder

~, **с·клад·н·а́я** step ladder

лéстнич·н·ый, -ая, -ое, (*a*) of лéст-
ниц·а; scalariform
~ клéт·к·а (*f*) (build) staircase cage/
well
~ марш· (*m*) flight of stairs
лéст·н·ый, -ая, -ое, (*a*) flattering;
complimentary
лёт·, -а, (*m*) flight, flying
на лет·ý on the wing, in flight
лет·á, (*g pl*) лет·, (*d pl*) лет·áм, (*pl*)
of год·; years, age
летáл·ь, -и, (*f*) lethal gene
летáль·ност·ь, -и, (*f*) lethality, death
rate
летáль·н·ый, -ая, -ое, (*a*) lethal
лет·áни·е, -я, (*n*) flying
летарг·úческ·ий, -ая, -ое, (*a*) (med)
lethargic
летарг·ú·я, -и, (*f*) lethargy; (phys)
mean logarithmic energy decrement/
loss
лет·áтельн·ый, -ая, -ое, (*a*) flying
лет·á·ть, -ют, (*imp indet*) fly
лет·éть, -ят, (*imp deter*) fly
лёт·к·а, -и, (*g pl*) -т·ок·, (*f*) notch
(of blast furnace); bee-gate (of hive)
~, козл·óв·ая salamander notch (blast
furnace)
~, чугýн·н·ая iron notch (blast furnace), taphole
~, шлáк·ов·ая slag, cinder notch,
slag hole
-лет·и·е, -я, (*n word ending*) anniversary
лёт·н·е- (*component*) aestival, summer
лёт·н·е-осéн·н·ий, -яя, -ее, (*a*) aes-
tivo-autumnal
лет·н·ий, -яя, -ее, (*a*) summer, aestival, estival; (*as suffix*) -year, -year old
лёт·ник·, -а, (*m*) (bot) annual
лёт·н·о- (*component*) air-, flying-
~ -подъ·ём·н·ый со·стá·в· (*m*) air
crew
лёт·н·ый, -ая, -ое, (*a*) flying, flight
~ зóн·а (*f*) movement area (aerodrome)
~ ис·пыт·áни·е (*n*) flight testing
~ пóл·е (*n*) landing area (aerodrome)
~ со·стáв· (*m*) aircrew
лéт·о, -а, (*n*) summer
лет·овáть, -ýют, (*imp*) spend the
summer, summer, (zool) aestivate
лéт·ом (*adv*) in summer
лéто·пис·ь, -и, (*f*) chronicle, annals;
(geol) records
лéто·росл·ь, -и, (*f*) (bot) shoot, sprout,
sucker

лето·с·числ·éни·е, -я, (*n*) chronology,
time-table
~, гео·лог·úческ·ое geochronology,
geologic time-table
лет·óк·, -т·к·á, (*m*) bee-gate (of hive
лет·ýчест·ь, -и, (*f*) volatility
лет·ýч·ий, -ая, -ее, (*a*) (chem) volatile; flying
~ вещ·еств·о (*n*) volatile matter
~ мыш·ь (*f*) (zool) bat, (*pl*) *Micro-
chiroptera*; hurricane lamp
~ нóж·ниц·ы (*pl*) (met) flying shears
~ рыб·а (*pl*) (fish) flying fish, *Exo-
coetus volitans*
лет·ýчк·а, -и, (*g pl*) -чек·, (*f*) leaflet;
(bot) egret; (rail) shuttle train; mobile team, flying squad; (oil) liner
лёт·чик·, -а, (*m*) airman, pilot, avia-
tor, flier
~ -ис·пыт·áтел·ь, -я, (*m*) test pilot
~ -наблюд·áтел·ь, -я, (*m*) air ob-
server/navigator
лет·яг·а, -ы, (*f*) (zool) flying squirrel,
Pteromys volans
лет·ящ·ий, -ая, -ее, (*pres part act*)
of лет·éть
Лéхер·а, систéм·а (*f*) (elec) Lecher
system
лéхер·овск·ий, -ая, -ое, (*a*) of Лé-
хер·
лецит·úн·, -а, (*m*) (biochem) lecithin
лецитин·áз·а, -ы, (*f*) (biochem) leci-
thinase
лецито·протéин·, -а, (*m*) (biochem)
lecithoprotein
леч·éбниц·а, -ы, (*f*) hospital, specialist
hospital
~, глаз·н·áя ophthalmic hospital
леч·éбн·ый, -ая, -ое, (*a*) healing,
therapeutic
~ пит·áни·е (*n*) diet
~ физ·культ·ýр·а (*f*) physiotherapy
леч·éни·е, -я, (*n*) (med) treatment,
therapy, cure
~, луч·ев·óе radiotherapy
лéч·енн·ый, -ая, -ое, (*past part pass*)
of леч·úть; treated
леч·úть, -ат, (*imp*) (med) treat, cure
леч·ý (*pres 1st sing*) of лет·éть; (*pres
1st sing*) of леч·úть
леч·ь, (*fut 3rd sing, pl*) ляж·ет, ля-
г·ут, (*past masc sing*) лёг· (*perf
intrans*); see лож·úться
Ле Шательé, прúнцип· (*m*) (phys)
Le Chatelier-Braun principle
лещ·, -á, (*i*) -щ·óм, (*m*) (fish) bream,
Abramis brama

ле́щад·ь, -и,   (*f*) (met) hearth, well, hearth bottom, pad

лещи́н·а, -ы,   (*f*) (bot) filbert, hazel, hazelnut, *Corylus*

лже- (*component*) false, pseudo-

лже·ака́ци·я, -и,   (*f*) (bot) false acacia, *Robinia pseudoacacia*

лже·гриб·ни́ц·а, -ы,   (*f*) (bot) pseudo-mycelium

лже·на·у́к·а, -и,   (*f*) pseudo-science

лже·прися́г·а, -и,   (*f*) (law) perjury

лж·ёт (*pres 3rd sing*) of **лг·а́ть** (*imp*)

лже·пу́г·а, -и,   (*f*) (bot) Douglas fir, *Pseudotsuga*

лж·и (*g sing*) of **лож·ь**

лж·и́в·ый, -ая, -ое,   (*a*) lying, mendacious; false, deceitful

ли (*interrogative particle, not translated*); (*conj*) whether, if

Ли гру́пп·а   (*f*) (math) Lie group

лиа́н·а, -ы,   (*f*) (bot) liana, liane

ли́беркюн·ов·ы кри́пт·ы   (*pl*) (zool) Lieberkuhn's crypts

Ли́берман·а, реа́кция   (*f*) (chem) Liebermann–Burchard test

либетени́т·, -а,   (*m*) (min) libethenite

либиги́т·, -а,   (*m*) (min) liebigite

либи́до (*n*) (biol) libido

ли́бо (*conj*) or; (*as particle implies indefiniteness* or *option*, e.g. **кто-либо** someone, anyone, anyone you like)

~ ... ли́бо either ... or

либр·а́ци·я, -и,   (*f*) (astron) libration

либри·фо́рм·, -а,   (*m*) (bot) libriform fibre

лива́н·ск·ий, -ая, -ое,   (*a*) Levantine

~ кедр·   (*m*) (bot) cedar of Lebanon, *Cedrus Libani*

ли́в·ен·ь, -в·н·я,   (*m*) (meteor) heavy shower, downpour; (nucl) shower

~, у́з·к·ий   (nucl) narrow shower, jet

~, шир·о́к·ий   (nucl) extended shower

ли́вер·, -а,   (*m*) (food) offal; siphoning tube

лив·и́йск·ий, -ая, -ое,   (*a*) Libyan

ливисто́н·а, -ы,   (*f*) (bot) fan-palm, *Livistona*

ли́в·нев·ый, -ая, -ое, (*a*) of **ли́в·ен·ь**

~ дли́н·а (*f*) (nucl) shower length

~ един·и́ц·а (*f*) (nucl) shower unit

~ о·са́д·к·и (*pl*) (meteor) heavy shower

~ сет·ь (*f*) storm/surface-water sewer system

ли́·вш·ий, -ая, -ее, (*past part act*) of **ли·ть**

ли́г·а, -и, (*f*) league

лигату́р·а, -ы,   (*f*) (med) ligature; (met) hardener, master alloy; (print) logotype, bound letter

лигн·и́н·, -а,   (*m*) (bot) lignin

лигн·и́т·, -а,   (*m*) (geol) lignite, brown coal

лигно·волокн·и́ст·ый, -ая, -ое,   (*a*) lignin-and-fibre

лигно·сто́н·, -а,   (*m*) (build) lignostone

лигно·фо́л·ь, -я,   (*m*) laminated wood

лигно·целлюло́з·а, -ы,   (*f*) (biochem) lignocellulose

лигно·цери́н·ов·ая кисл·от·а́   (*f*) (chem) lignoceric acid

лигро́ин, -а,   (*m*) (oil) ligroin, heavy benzine

лигу́л·а, -ы,   (*f*) (bot) ligule; (zool) ligula

лигустру́м·, -а,   (*m*) (bot) *Ligustrum*

ли́дер·, -а,   (*m*) leader; (naut) flotilla leader

~ мо́лни·и step leader (lightning)

~, с·пас·а́тельн·ый fast salvage vessel

лидо́л·, -а,   (*m*) (pharm) lidol, demerol

ли́ж·ущ·ий, -ая, -ее, (*pres part act*) of **лиз·а́ть**

лиза́т·, -а,   (*m*) (biol) histolysate

лиз·а́ть, (*pres 3rd pl*) **ли́ж·ут,** (*imp*) lick

Лизега́нг·а, кольц·а́   (*pl*) (chem) Liesegang rings

лизе́н·а, -ы,   (*f*) (arch) lisena

лизи·ге́н·н·ый, -ая, -ое,   (*a*) (biol) lysigenic, lysigenous, lysigenetic

лизи́·метр·, -а,   (*m*) (agr) lysimeter

лиз·и́н·, -а,   (*m*) (chem) lysine; (zool) lysin

ли́зис·, -а,   (*m*) (med) lysis

лиз·н·у́ть, -у́т, (*perf*) *see* **лиз·а́ть**

лизо́л·, -а,   (*m*) (chem) lysol

лизо·ци́м·, -а,   (*m*) (biochem) lysozyme

лизо·фо́рм·, -а,   (*m*) (pharm) lysoform

лиз·у́нец·, -нц·а,   (*m*) (agr) salt brick, lick salt

лик·, -а, (*m*) (naut) bolt-rope; face

ликв·а́т·, -а,   (*m*) (met) segregate

ликв·а́ци·я, -и,   (*f*) (chem, geog) liquation; liquation (non-ferrous metallurgy); segregation (steelmaking)

~, внутри·кристалл·и́ческ·ая (met) coring

, га́зовая (met) segregation in blow-holes

~, дендри́т·н·ая (met) dendritic segregation

ликв·а́ци·я
~, **зона́ль·н·ая** (met) major segregation
~, **макро·скоп·и́ческ·ая** (met) macrosegregation, ingot segregation
~, **меж·кристалл·и́ческ·ая** (met) minor segregation
~, **микро·скоп·и́ческ·ая** (met) microsegregation
~, **норма́ль·н·ая** (met) normal segregation
~, **обра́т·н·ая** (met) negative/inverse segregation
~ **по с·ли́т·к·у** (met) ingot segregation
~ **по у·де́ль·н·ому ве́с·у** (met) layer segregation
ликвид·а́мбар·, -а, (m) (bot) sweet gum, *Liquidambar*
ликвид·а́тор·, -а, (m) liquidator; (rocket) destructor mechanism
ликвид·ацио́нн·ый, -ая, -ое, (a) of **ликвид·а́ци·я**
ликвид·а́ци·я, -и, (f) liquidation, elimination, winding up (a business)
ликвид·и́р·овать, -уют, (*imp and perf*) liquidate, eliminate, abolish; (com) put into liquidation, wind up; **-ся** (*pass*); (com) go into liquidation
ликвид·ком·, -а, (m) (com) winding-up commission
ликви́д·ност·ь, -и, (f) (econ) convertibility
ликви́д·н·ый, -ая, -ое, (a) (fin) fluid, convertible, accessible
ликви́д·ус·, -а, (m) (met) liquidus
ли́квор·, -а, (m) (physiol) liquor amnii
ликв·и́р·овать, -уют, (*imp*) liquate; **-ся** segregate out
ликёр·, -а, (m) liqueur
ли́к·ов·ый, -ая, -ое, (a) of **лик·**
лико·по́д·и·й, -я, (m) (bot) club moss, lycopod, *Lycopodium*
лико·пи́н·, -а, (m) (biochem) lycopene, lycopin, lycopersicin
лико·под·и́н·, -а, (m) (chem) lycopodine
лико·поди́н·ов·ая кисл·от·а́ (f) lycopodic acid
лик·тро́с·, -а, (m) (naut) bolt-rope
ли·л (*past masc sing*) of **ли·ть**
лил·е́йн·ые, -ых, (*pl decl as a*) (bot) lily family, *Liliaceae*
ли́л·и·я, -и, (f) (bot) lily, *Lilium*; (zool, pal) crinoid
~, **мор·ск·а́я** (zool) sea lily, (*pl*) *Crinoidea*

ли́·ло (*past neut sing*) of **ли·ть**
лил·ова́т·ый, -ая, -ое, (a) lilac-coloured, light-mauve
лил·о́в·ый, -ая, -ое, (a) lilac-coloured, light-mauve
лим·а, -ы, (f) (bot) Lima bean, *Phaseolus limensis*
лима́н·, -а, (m) (geog) liman, drowned estuary
лима́н·н·ый, -ая, -ое, (a) of **лима́н·**
лимаци́д·, -а, (m) slug poison
лимб·, -а, (m) limb; (instr) graduated circle, dial
~, **азимут·а́льн·ый** (surv) azimuth circle/dial
~ **буссо́л·и** declination arc, main azimuth plate (of magnetic compass)
~, **голов·н·о́й** (zool) cephalic limb
~, **экваториа́ль·н·ый** (astron) limb
лимбербо́рд·ов·ый, -ая, -ое, (a) (naut) limber
лимбурги́т·, -а, (m) (geol) limburgite
лиме́тт·а, -ы, (f) (bot) limetta (lemon/orange hybrid)
лими́т·, -а, (m) quota, limit, allotment; establishment (personnel)
лимит·а́нт·, -а, (m) (math) limitant
лимит·а́ци·я, -и, (f) (v n) see **лимит·и́р·овать**
лимит·и́р·овать, -уют, (*imp and perf*) fix/establish a limit/quota
лими́т·н·ый, -ая, -ое, (a) of **лими́т·**
~ **цен·а́** (f) price limit
лимни́·граф·, -а, (m) limnigraph, (hydr) water-level recording gauge/gage
лимно·биот·и́ческ·ий, -ая, -ое, (a) (zool) limnobiotic, fresh-water
лимно·ло́г·и·я, -и, (f) limnology
лимно·планкто́н·, -а, (m) (biol) limnoplankton
лимно·фи́ль·н·ый, -ая, -ое, (a) (biol) limnophilous
лимо́н·, -а, (m) (bot) lemon, *Citrus limonia*
~, **сла́д·к·ий** limetta
лимона́д·, -а, (m) (food) lemonade, lemon squash; fruit squash
лимон·е́н·, -а, (m) (biochem) limonene
лимон·и́т·, -а, (m) (min) limonite
лимо́н·ник·, -а, (m) (bot) magnolia vine, *Schizandra*
лимо́н·н·о- (*component*) (chem) citrate of, lemon-
лимонно·бути́л·ов·ый эфи́р· (m) butyl citrate

лимо́нно·ки́сл·ый, -ая, -ое, (*a*) (chem) citric, citric acid, citrate (of)

~ **ка́ли·й** (*m*) potassium citrate

~ **на́три·й** (*m*) sodium citrate

лимо́нно·рас·твор·и́м·ый, -ая, -ое, (*a*) (chem) citric-soluble

лимо́н·н·ый, -ая, -ое, (*a*) lemon, citrous; (chem) citric

~ **кисл·от·а́** (*f*) (chem) citric acid

~ **ма́сл·о** (*n*) lemon oil

лимоно·ви́д·н·ый, -ая, -ое, (*a*) lemon-shaped, citriform, limoniform

лимузи́н·, -а, (*m*) (M/T) limousine

ли́мф·а, -ы, (*f*) (biol) lymph

лимф·адени́т·, -а, (*m*) (med) lymphadenitis

лимф·анги́т·, -а, (*m*) (med) lymphangitis

лимф·ати́ческ·ий, -ая, -ое, (*a*) lymphatic, lymph

лимфо·ци́т·, -а, (*m*) (zool) lymphocyte

лимфо·цито́з·, -а, (*m*) (med) lymphocytosis

линали́л·, -а, (*m*) (chem) linalyl

линалил·ацета́т·, -а, (*m*) (chem) linalyl acetate

линалоо́л·, -а, (*m*) (biochem) linalool

линари́н·, -а, (*m*) (pharm) linarin

линари́т·, -а, (*m*) (min) linarite

лина́ри·я, -и, (*f*) (bot) toadflax, *Linaria*

лингв·и́стик·а, -и, (*f*) linguistics

~, **дескрипти́в·н·ая** descriptive linguistics

~, **матема́т·и́ческ·ая** mathematical linguistics

лингв·исти́ческ·ий, -ая, -ое, (*a*) linguistic

Ли́нде, спо́соб· (*m*) (chem) Linde process

линеар·иза́ци·я, -и, (*f*) linearization

линеа́р·ность·, -и, (*f*) linearity

линеа́р·н·ый, -ая, -ое, (*a*) (*and see* **лин·е́йн·ый**) linear

коэффицие́нт· линеа́р·н·ого рас·-шир·е́ни·я (*m*) (phys) coefficient of linear expansion

лин·ева́ни·е, -я, (*n*) (*v n*) of **лин·ева́ть**

лин·ева́ть, -ю́т, (*imp*) rule, make lines

лин·е́чн·ый, -ая, -ое, (*a*) of **лин·е́йк·а**

лин·е́йк·а, -и, (*g pl*) **-е́ек·,** (*f*) ruler, rule, straight-edge; ruled line; (tech) guide bar, guide, bar

~, **бок·ов·а́я** (met roll) side guide

~, **визи́р·н·ая** (surv) aiming rule, azimuth rule

лин·е́йк·а

~ **выс·о́т·** height arm/slide (airstereophotog)

~, **кат·я́щ·аяся** roller ruler

~, **комбина́тор·н·ая** (telecom) selector bar

~, **лека́ль·н·ая** toolmaker's straight-edge

~, **логари́фм·и́ческ·ая** slide rule

~ **манипуля́тор·а** (met roll) manipulator head, power side guard, ram

~, **масшта́б·н·ая** ruler, rule

~, **на·бо́р·н·ая** (print) setting rule

~, **навигацио́н·н·ая** navigator's protractor

~, **на·правл·я́ющ·ая** guide bar

~, **о·гран·ичи́тельн·ая** spacing strip

~, **по·ве́р·очн·ая** straight-edge

~, **при·во́д·очн·ая** (surv) register rule

~, **си́нус·н·ая** (eng) sine bar

~, **с·пуск·ов·а́я** (telecom) universal bar

~, **с·чёт·н·ая** slide rule

~, **штрих·о́ванн·ая** graduated rule

лин·е́йн·о-аппара́т·н·ый зал· (*m*) tandem office (telephones)

лин·е́йн·о-ро́лик·ов·ая маши́н·а (*f*) seam welder

лин·е́йн·о-поляриз·о́ванн·ый, -ая, -ое, (*a*) (phys) linearly polarized

лин·е́йность·, -и, (*f*) linearity

~ **строк·** line linearity

~ **характер·и́стик·и** (elec) linearity of response

лин·е́йн·о-у·по·ря́д·оченн·ый, -ая, -ое, (*a*) (phys) linearly ordered

лин·е́йн·ый, -ая, -ое, (*a*) linear, line; (bot) bandform

~ **ис·то́ч·ник·** (*m*) (phys) line source

~ **конта́ктор·** (*m*) (elec) mains contactor

~ **кора́бл·ь** (*m*) battleship

~ **раз·ме́р·** (*m*) linear dimension

~ **ста́нци·я** (*f*) (rail) way/wayside/field station

~ **суд·** (*m*) (law) transport-line court

~ **у·скор·и́тел·ь** (*m*) (nucl) linear accelerator

~ **фу́нкци·я** (*f*) (math) linear function

~ **част·ь** (*f*) linear group (computers)

~ **шум·** (*m*) (telecom) line noise

лин·е́йчатость·, -и, (*f*) lineation

лин·е́йчат·ый, -ая, -ое, (*a*) ruled, lined, line

~ **гео·ме́тр·и·я** (*f*) line geometry

~ **по·ве́рх·ность·ь** (*f*) (math) ruled surface

лин·е́йчат·ый, -ая, -ое
~ по·ве́рх·ност·ь, кос·а́я (math) skew ruled surface
~ по·ве́рх·ност·ь, раз·ве́рт·ываю·щ·аяся (math) developable ruled surface
лине·раке́т·а, -ы, (f) (naut) life-saving rocket
ли́нз·а, -ы, (f) (opt) lens; (phot) lens cap; (М/Т) lamp cap; (geol) lens, lenticle
~, анаморфо́т·н·ая (opt) anamorphote/distorting lens
~, вогнуто·вы́·пукл·ая (opt) positive meniscus lens
~, выпукло·во́·гн·ут·ая (opt) negative meniscus lens
~, вы́·тя·нут·ая (geol) elongated lens
~, двояко·во́·гн·ут·ая (opt) double concave lens
~, двояко·вы́·пукл·ая (opt) double convex lens
~, дист·а́рн·ая (phot) telephoto attachment lens
~ колчеда́н·а (min) sulphur kidney
~, конде́нсор·н·ая (opt) condenser lens
~, магни́т·н·ая (phys) magnetic lens
~, металл·и́ческ·ая (rad) antenna lens, metal lens
~, моло́ч·н·ая opal lamp camp
~, на·са́д·очн·ая (photog) attachment/portrait lens
~, от·риц·а́тельн·ая concave/negative lens
~ пири́т·а (geol) pyrite lens, lenticular pyrite
~, плоско·во́·гн·ут·ая (opt) plano-concave lens
~, плоско·вы́·пукл·ая (opt) plano-convex lens
~, по·ло́ж·и́тельн·ая (opt) convex/positive lens
~, прямо·уго́ль·н·ая от·раж·а́ю·щ·ая (opt) Perre lens
~, рас·се́·ивающ·ая (opt) dispersive/divergent lens
~, секцион·и́рованн·ая (opt) facetted lens
~, со·бир·а́тельн·ая (opt) collective/convex/convergent lens
~, ступ·е́нчат·ая (opt) Fresnel lens
~, электро·магни́т·н·ая (phys) electromagnetic lens
~, электро́н·н·ая electronic lens (cathode ray tubes)
~, электро·стат·и́ческ·ая (phys) electrostatic lens

линзо·ви́д·н·ый, -ая, -ое, (a) lenticular, lens-shaped, (geol) phacoidal
ли́нз·ов·ый, -ая, -ое, (a) of ли́нз·а
линзо·обра́з·н·ый, see линзо·ви́д·н·ый
ли́нз·очк·а, -и, (g pl) -чек·, (f) (geol) lenticule, lenticle
линиату́р·а ра́стр·а (f) (print, surv) screen size
линиме́нт·, -а, (m) (pharm) liniment
ли́н·и·я, -и, (f) (see also under associated words) line, (rad) link
~, абоне́нт·ск·ая subscriber's line (teleph), central office line, exchange line
~, а́виа- (air) air line
~, автомат·и́ческ·ий automation line
~, агон·и́ческ·ий (geog) agonic line
~ апси́д· (astron) line of apsides
~, бази́с·н·ая (surv) base line
~, ба́з·ов·ая base line, datum line
~, бала́нс·н·ая (telecom) compensation circuit
~, борт·ов·а́я (shipb) plan outline
~, визи́р·н·ая line of sight
~, винт·ов·а́я (math) helix; (mech) helical/screw line; thread, helix (of extruder)
~, вихр·ев·а́я (dynam) vortex line
~, водо·раз·дель·н·ая (math) master/watershed/divide line
~, воз·ду́ш·н·ая (elec) overhead line, air line, open-wire line, aerial transmission line; (air) air route
~, вращ·а́тельн·ая (spectr) rotational line
~, вы·нос·н·а́я extension line (on drawing)
~, гран·и́чн·ая boundary line, line of demarcation
~, груз·ов·а́я (shipb) load line
~, групп·ов·а́я party line (teleph)
~, двух·ни́т·очн·ая duplex production line, parallel-flow line
~, двух·про·вод·н·а́я (telecom) balanced line, twin line
~, дел·и́тельн·ая pitch line (of a curve)
~, до·ба́в·очн·ая extension (teleph)
~ за·де́рж·к·и (elec) delay line
~ за·де́рж·к·и, акуст·и́ческ·ая sonic delay line
~ за·де́рж·к·и, рту́т·н·ая mercury delay line
~, за·мык·а́ющ·ая (math) abutment/closing line
~, за́·ня·т·ая engaged/busy line (teleph)

ли́н·и·я

~, из·мер·я́ем·ая survey line

~, индивидуа́ль·н·ая direct line, individual line (teleph)

~, инфлюе́нт·н·ая influence line

~, иску́с·ственн·ое (telecom) artificial line

~, ка́бель·н·ая (elec) cable line, underground line

~, кома́нд·н·ая (rad) command link

~, кон·е́чн·ая terminated/finite line

~, контро́ль·н·ая (mech eng) proof/guide line, (surv) checking line; (telecom) monitoring circuit

~, крив·а́я (math) curved line, curve

~, ло́м·ан·ая (math) broken line

~, маршру́т·н·ая (air) flight line

~, мир·ов·а́я (math) material line

~, мн·и́м·ая image line

~ на шле́йф·ах (rad) stub-supported line

~, на·гру́ж·енн·ая terminated line (teleph)

~, на·пра́вл·енн·ая directed line

~, на·правл·я́ющ·ая directional line; (math) directrix; ruling (of a ruled surface)

~, не·одно·ро́д·н·ая (telecom) non-uniform line

~, нес·у́щ·ая (air) lifting line

~, нул·ев·а́я zero line, reference/datum line

~, объ·един·я́ющ·ая (elec) tie line

~, одно·ни́т·очн·ая single-flow production line

~, одно·ро́д·н·ая (elec) uniform line

~, ос·ев·а́я centre/center line

~, основ·н·а́я datum/base line

~, от·ве́с·н·ая plumb line

~, от·ка́ч·н·ая vacuum-pump line

~, пере·горо́д·очн·ая (biol) suture

~ пере·да́ч·и transmission line

~ пере·да́ч·и да́·нн·ых (autom) data line

~ пере·да́ч·и кома́нд· (telecom, autom) command line

~ пере·ме́н·ы да́т·ы internal date line

~, пит·а́ющ·ая supply line; (rad) feeder

~, по·то́ч·н·ая production/flow line

~, прям·а́я straight line

~, пункти́р·н·ая dotted line; (print) leader

~, раз·мы́·т·ая blurred line; (spectr) smeared/broadened line

~, разо·мк·ну́т·ая (elec) open-circuit line

ли́н·и·я

~, ра́стр·ов·ая (print) screen line

~, само·о·пыл·ённ·ая (agric) inbred line

~, сек·у́щ·ая (math) secant, secant line

~, сил·ов·а́я line of force

~, сла́б·ая (agric) mild strain

~, с·ливн·а́я drainage/discharge channel/pipe

~, сло·ев·а́я (cryst) layer line

~, со·един·и́тельн·ая connecting line; junction route (telephones)

~, сплош·н·а́я solid/continuous line

~, сре́д·н·яя middle/mean/centre line; (met roll) pitch line; halving line (range finder)

~, ста́н·очн·ая transfer line, automation line

~ то́к·а (dynam) flow line

~ то́к·а ве́ктор·н·ого по́л·я (math) vector line in a vector field

~, трансляцио́н·н·ая pick-up line (radio relay)

~, трансмиссио́н·н·ая (mech) transmission line, continuous line of shafting

~, у·каз·а́тельн·ая index line

~, у·пру́г·ая (mech) elastic line

~, у·сло́в·н·ая broken line

~, ход·ов·а́я (surv) leg

~, цеп·н·а́я (math) catenary curve, catenary

~, центр·ов·а́я centre/center line; line of centres/centers (gearing)

~, че́тверть·во́лн·ов·ая (rad) quarter wavelength line

~, шов·н·ая (biol) suture line

~, штри́х·ов·ая broken/dashed line

~, электр·и́ческ·ая си́л·ов·ая line of electric force

лин·ко́р·, **-а**, (m) (abbr) of **лин·е́йн·ый кора́бл·ь** battleship

линкру́ст·, **-а**, (m) lincrusta (a heavy type of wallpaper)

линне́йт·, **-а**, (m) (min) linnaeite

лин·о́ван·ый, **-ая**, **-ое**, (a) lined, ruled

лин·ова́ть, **-у́ют**, (imp) rule, rule lines

лино·гравю́р·а, **-ы**, (f) (print) linoleum print

лин·окси́н·, **-а**, (m) (chem) linoxyn

лин·о́л·, **-а**, (m) (chem) linol

линол·ев·ая кисл·от·а́ (f) (chem) linoleic acid

лино́леум·, **-а**, (m) linoleum

линоти́п·, **-а**, (m) (print) linotype

ли́нтер·, **-а**, (m) (text) linter, cotton linter; linter cleaning machine

лин·ь, -я́, (*m*) (fish) tench, *Tinca tinca;* (naut) line (rope), ratline

~, сигна́ль·н·ый breast-rope (diving)

~, с·пас·а́тельн·ый life-line

ли́ньк·а, -и, (*f*) (zool) moulting, molting

лин·ю́ч·ий, -ая, -ее, (*a*) non-fade-resistant, fading, fugitive (dyes); moulting

лин·я́·ть, -ют, (*imp*) (text) fade, lose colour, run; (zool) moult

лио·ли́з·, -а, (*m*) (chem) lyolysis

лио·со́рпци·я, -и, (*f*) (chem) lyosorption

лио·тро́п·н·ый, -ая, -ое, (*a*) (chem) lyotropic

лио·филиз·и́р·овать, -уют, (*imp*) (chem) lyophilize

лио·фи́ль·н·ый, -ая, -ое, (*a*) (chem) lyophilic

лио·фо́б·н·ый, -ая, -ое, (*a*) (chem) lyophobic

лио·хро́м·ы (*pl*) (chem) lyochromes

лип (*past masc sing*) of **ли́п·нуть**

ли́п·а, -ы, (*f*) (bot) lime, linden, basswood, *Tilia;* (build) basswood, whitewood

~, америка́н·ск·ая basswood, American linden, *Tilia americana*

лип·а́з·а, -ы, (*f*) (biochem) lipaze

липари́т·, -а, (*m*) (geol) liparite

лип·еми́·я, -и, (*f*) (med) lipaemia

лип·и́д·, -а, (*m*) (biochem) lipid, lipoid

ли́п·к·ий, -ая, -ое, (*a*) of **ли́п·кост·ь**

ли́п·кост·ь, -и, (*f*) stickiness, tackiness, gumminess, adhesiveness; oiliness (of lubricating oils)

Ли́пман·а, проце́сс· (*m*) (phot) Lippmann's process

ли́п·нуть, -нут, (*past masc sing*) **лип·** or **ли́п·нул,** (*imp*) stick, adhere

лип·ня́к·, -а́, (*m*) linden wood/forest

ли́п·ов·ый, -ая, -ое, (*a*) of **ли́п·а;** lime blossom; (*pl*) *Tiliaceae*

~ ма́сл·о (*n*) linden oil

липо·де·гидр·а́з·а, -ы, (*f*) (chem) lipohydrogenase

лип·о́ид·, -а, (*m*) (biochem) lipid, lipoid

лип·оксида́з·а, -ы, (*f*) (biochem) lipoxidase

липо·ли́з·, -а, (*m*) (chem) lipolysis

липо·лит·и́ческ·ий, -ая, -ое, (*a*) liplytic

лип·о́м·а, -ы, (*f*) (med) lipoma

липомат·о́з·, -а, (*m*) (med) lipomatosis

липо·проте́ид·, -а, (*m*) (biochem) lipoproteid

липо·хро́м·, -а, (*m*) (chem) lipochrome

лип·у́ч·ий, -ая, -ее, (*a*) sticky

лип·у́чк·а, -и, (*g pl*) **-чек·,** (*f*) (bot) stickseed, *Lappula*

Ли́пшиц·а, у·сло́в·и·е (*n*) (math) Lipschitz condition

ли́р·а, -ы, (*f*) lyre; (zool) lyre, lyra; (astron) Lyra, the Lyre; lira (Italian currency); U-support/joint; apron (carpentry); swivel plate (on planing machine)

лири́д·ы (*pl*) (astron) Lynds

лирио·де́ндрон·, -а, (*m*) (bot) tulip tree, yellow-poplar, *Liriodendron tulipifera*

лироко́ни́т·, -а, (*m*) (min) liroconite

лиро·обра́з·н·ый, -ая, -ое, (*a*) U-shaped, lyre-type, lyriform, lyrate

лис·а́, -ы́, (*f*) *see* **лис·и́ц·а**

ли́с·ий, -ья, -ье, (*a*) of **лис·и́ц·а**

лис·и́ц·а, -ы, (*i*) **-ц·ей,** (*f*) (zool) fox, *Vulpes*

лис·и́чк·а, -и, (*g pl*) **-чек·,** (*f*) (*dim*) of **лис·и́ц·а;** (bot) chanterelle, *Cantharellus cibarius*

~, мор·ск·а́я (fish) sea poacher, (*pl*) *Agonidae*

лисо·во́д·ств·о, -а, (*n*) fox fur farming

лисо·хво́ст·, -а, (*m*) (bot) foxtail, *Alopecurus*

Лисса́жу́, фигу́р·а (phys) Lissajous figure

лист·, -а́, (*nom pl*) **-ы́,** (*g pl*) **-о́в,** (*m*) leaf, sheet (of manufactures); (met) plate, sheet, strip; **лист·, -а,** (*nom pl*) **-ья,** (*g pl*) **-ьев,** (*m*) (bot) leaf

~, а́втор·ск·ий (print) 40,000 ens

~, влаг·а́лищн·ый (bot) coleoptile

~ дека́рт·а (math) folium of Descartes

~, диало́г·ов·ый (cinema) cue/dialogue sheet

~, за·гла́в·н·ый (print) title page

~ зла́к·а (bot) blade

~ ка́рт·ы (surv) chart sheet, map sheet

~, като́д·н·ый (electromet) starting sheet

~, контро́ль·н·ый inspection/check list/sheet; (a/c) servicing form

~, коте́ль·н·ый boiler plate

~, кра́·йн·ый между·до́н·н·ый (shipb) margin plate

~, маршру́т·н·ый route sheet (transport etc.)

~, монта́ж·н·ый (cinema) cue/dialogue sheet

~, на·крыв·а́ющ·ий (shipb) outer strake

~ па́льм·ы (bot) frond

~ пол·ев·о́й съ·ём·к·и survey sheet

лист

~, **при·лег·а́ющ·ий** (shipb) inner strake

~, **пут·ев·о́й** work ticket (goods vehicles)

~, **с·бо́р·н·ый** (print) table of contents; map index

~, **с·ме́ж·н·ый** adjacent sheet (maps)

~, **сталь·н·о́й** (met) steel plate

~, **судо·стро·и́тельн·ый** (met) ship plate

~, **ти́туль·н·ый** (print) title page; (build) authorized schedule of works

~, **то́лст·ый** (met) plate

~, **то́нк·ий** (met) sheet, strip

~, **трансформа́тор·н·ый** (met elec) transformer sheet/steel

~, **у·хва́т·н·ый** (rail) saddle plate

~, **шине́ль·н·ый** wrapper plate (boiler)

лист·а́ж·, -а́, (i) **-ж·о́м·,** (m) (print) number of pages

лист·а́·ть, -ю́т, (imp) turn over (pages), thumb through

лист·в·а́, -ы́, (f) (bot) foliage

листвени́т·, -а, (m) (min) listvenite

ли́ст·венниц·а, -ы, (i) **-ц·ей·,** (f) (bot) larch, *Larix*

~, **америка́н·ск·ая** tamarack, hackmatack, *Larix laricina*

~, **европ·е́йск·ая** European larch, *Larix decidua*

~, **за́·пад·н·ая** western larch, *Larix occidentalis*

~, **о·пад·а́ющ·ая** European larch, *Larix decidua*

лист·венни́чн·ый, -ая, -ое, (a) larch, tamarack

~ **гу́б·к·а** (f) (pharm) agaric

ли́ст·венн·ый, -ая, -ое, (a) leafed, leaved, deciduous

~ **де́рев·о** (n) deciduous tree

~ **о́вощ·и** (pl) (food) greens, green vegetables

~ **по·ро́д·а** (f) hardwood

лист·вя́к·, -а́, (m) (bot) thallus

листел·ь, -я, (m) (arch) listel, facette

ли́стер·овани·е, -я, (n) (agr) lister sowing, listering

листо- (root) leaf, sheet, plate, (bot) phyllo-, foli-

листо·бло́ш·к·а, -и, (g pl) **-ш·ек·,** (f) (ent) leaf hopper, psylla, (pl) *Psyllidae*

лист·ова́льн·ый, -ая, -ое, (a) sheet-forming

лист·ова́ни·е, -я, (n) forming into sheets, sheeting, flatting

лист·о́ванн·ый, -ая, -ое, (past part pass) of **лист·ова́ть;** sheeted, in sheets

лист·ова́т·ый, -ая, -ое, (a) foliaceous, foliated

лист·ова́ть, -у́ют, (imp) turn the pages of, thumb through

листо·вёрт·к·а, -и, (g pl) **-т·ок·,** (f) (ent) leaf roller, tortrix moth, (pl) *Tortricidae*

листо·ви́д·н·ый, -ая, -ое, (a) leaf-shaped, foliate, phylloid

~ **жа́бр·а** (f) (zool) phyllobranchia

лист·о́вк·а, -и, (g pl) **-вок·,** (f) (print) leaflet; (bot) follicle; parmelia (foliose lichen)

лист·ов·о́й, -а́я, -о́е, (a) of **лист·;** foliate, foliar, (bot) foliose

~ **бума́г·а** (f) (print) flat paper

~ **мета́лл·** (m) sheet metal, plate metal

~ **след·** (m) (bot) leaf trace

~ **щел·ь** (f) (bot) foliar/leaf gap

листо·ги́б·очн·ый, -ая, -ое, (a) sheet-metal-bending

~ **маши́н·а** (f) (met) bending machine

~ **маши́н·а трёх·вал·ко́в·ая** (met) roll bending machine

~ **пресс·** (m) die-bending press

листо·гры́з·, -а, (m) see **листо·е́д·**

листо·е́д·, -а, (m) (ent) leaf beetle, (pl) *Chrysomelidae*

лист·о́к·, -т·к·а́, (m) (dim) of **лист·;** leaflet, sheet, form, certificate, chit; (print) leaf

~ **не·трудо·спосо́б·ност·и** medical certificate, doctor's certificate

листо·коло́с·ник·, -а, (m) (bot) bamboo, *Phyllostachis*

листо·но́г·ие, -их, (pl decl as adj) (ent) phyllopods, *Phyllopoda*

листо·но́с·, -а, (m) (zool) spear-nosed bat, *Phyllostomus*

листо·обра́з·н·ый, -ая, -ое, (a) leaf-shaped, foliate, phylloid, (bot) foliose

листо·па́д·, -а, (m) leaf fall, exfoliation, defoliation

листо·под·бо́роч·ная маши́н·а (f) (print) gathering machine, gatherer

листо·пра́в·ильн·ая маши́н·а (f) (met) straightening machine, levelling machine

листо·про·ка́т·н·ый стан· (m) (met) plate/sheet mill

листо·рас·по·ло́ж·е́ни·е, -я, (n) (bot) phyllotaxis

листо·рéз·н·ый стан·óк· (*m*) (met) shearing mill

листо·с·лож·éни·е, -я, (*n*) (bot) vernation

листо·стéбель·н·ый, -ая, -ое, (*a*) (bot) cormophytic

лист·óчек·, -чк·а, (*m*) (*dim*) of **лист·óк·;** (bot) leaflet, pinna; flake, scale

~, втор·úчн·ый (bot) pinnule

лист·óчн·ый, -ая, -ое, (*a*) of **лист·óк·**

листо·штамп·óвочн·ый, -ая, -ое, (*a*) stamping, sheet-metal stamping

листо·яд·н·ый, -ая, -ое, (*a*) (zool) leaf-eating, browsing

лúст·ья (*nom pl*) of **лист·** leaf (bot)

лит·éйн·ая, -ой, (*f decl as adj*) foundry

лит·éйн·ый, -ая, -ое, (*a*) casting, founding, foundry (*see also* **лит·éйн·ая**) (*as noun*)

~ двор· (*m*) casting pit

~ кокс· (*m*) foundry coke

~ машúн·а компрéссор·н·ая (*f*) pressure diecasting machine, cold-chamber diecasting machine

~ машúн·а пóрш·нев·ая piston diecasting machine, submerged hot-chamber machine

~ про·из·вóд·ств·о (*n*) foundry work

~ цех· (*m*) foundry

~ чугýн· (*m*) foundry pig iron, foundry iron

лит·éйщик·, -а, (*m*) (met) founder, foundryman

лúтер·, -а, (*m*) travelling warrant

лúтер·а, -ы, (*f*) letter (of the alphabet); (print) type

~, с·лог·ов·áя (print) logotype

литератýр·а, -ы, (*f*) literature

литератýр·н·ый, -ая, -ое, (*a*) literary; literature; published

~ дá·нн·ые (*pl*) published data

лúтер·н·ый, -ая, -ое, (*a*) lettered, enumerated by letters of alphabet

лит·иáз·, -а, (*m*) (med) lithiasis

лúт·иев·ый, -ая, -ое, (*a*) of **лúт·и·й**

~ слюд·á (*f*) (min) lithia mica, lepidolite

лúт·и·й, -я, (*m*) (chem) lithium, Li

~, азóт·ист·ый lithium nitride

~, водо·рóд·ист·ый lithium hydride **ó·кис·ь лúт·и·я** (*f*) (chem) lithia, lithium monoxide

литионúт·, -а, (*m*) (min) lithionite

литиофилúт·, -а, (*m*) (min) lithiophilite

литификáци·я, -и, (*f*) (geol) lithification

лúт·ник·, -а, (*m*) (met cast) gating, gate; pouring down-tube (continuous casting)

~, вихр·ев·óй swirl gate

~, в·пуск·н·óй (plast) gate, inlet

~, дúск·ов·ый (plast) diaphragm gate

~, нúж·н·ый bottom gate

~, под·вод·я́щ·ий in-gate, gate

~, сифóн·н·ый bottom gating

~, ход·ов·óй dirt trap

~, центрáль·н·ый sprue, sprue gating

лúт·ников·ый, -ая, -ое, (*a*) of **лúт·ник·**

~ ворóн·к·а (*f*) pouring basin, runner basin, runner cup

~ пит·áтел·ь (*m*) gate, in-gate

~ систéм·а (*f*) gating, gating system

~ сто·я́к· (*m*) downgate, sprue

~ ход· (*m*) runner

~ чáш·а (*f*) pouring basin, runner basin, runner cup

лито- (*root*) litho-

литóв·ск·ий, -ая, -ое, (*a*) Lithuanian

лито·ген·éз·, -а, (*m*) (geol) lithogenesis

лито·генет·úческ·ий, -ая, -ое, (*a*) (geol) lithogenous

литó·граф·, -а, (*m*) lithographer

лито·грáф·и·я, -и, (*f*) (print) lithography, lithographic printing process; lithograph

~, цвет·н·áя lithotint

лито·грáф·ск·ий, -ая, -ое, (*a*) of **лито·грáф·и·я**

~ кáмен·ь (*m*) (geol) lithographic stone

~ óт·тиск· (*m*) lithograph

лит·óид·н·ый, -ая, -ое, (*a*) lithoidal

ли·т·óй, -áя, -óе, (*f*) (met) cast

лито·лóг·и·я, -и, (*f*) (geol) lithology

лито·пóн·, -а, (*m*) lithopone (pigment)

лито·птéрн·ы (*pl*) (pal) litopterns, *Litopterna*

литорáл·ь, -и, (*f*) (ocean) eulittoral zone; littoral system

~, нúж·н·яя sublittoral zone

литорáль·н·ый, -ая, -ое, (*a*) littoral

~ от·лож·éни·я (*pl*) (geol) littoral deposits

лито·сфéр·а, -ы, (*f*) (geol) lithosphere

лито·тáмн·и·я (*f*) (pal) liththamnion

лито·фáци·я, -и, (*f*) (geol) lithofacies

лито·фúз·а, -ы, (*f*) (geol) lithophysa, stone bubble, (*pl*) *lithophysae*

лито·фúль·н·ый, -ая, -ое, (*a*) lithophile, lithophilous

лито·цúст·, -а, (*m*) (biol) lithocyst

лúтр·, -а, (*m*) litre, liter, l

литр·а́ж·, **-а**, (*i*) **-ж·о́м·**, (*m*) capacity, size (in litres), (ICE) cubic capacity

литр·о́в·ый, **-ая**, **-ое**, (*a*) litre, per litre, liter, per liter

~ **буты́л·к·а** (*f*) litre/liter bottle, one-litre/liter bottle

~ **вес·** (*m*) weight per litre/liter

~ **мо́щ·ност·ь** (*f*) (M/T) brake horse power per litre/liter

литцендра́т·, **-а**, (*m*) *see* **лицендра́т·**

ли́·т·ый, **-ая**, **-ое**, (*past part pass*) of **ли·ть**

ли·ть, (*pres 3rd sing, pl*) **ль·ёт**, **ль·ют**, (*imp*) pour; (met) cast, found

лит·ь·ё, **-я́**, (*n*) casting, founding; casting, cast object, cast

~, **без·ли́т·ник·ов·ое** (plast) hot-gate moulding

~, **бума́ж·н·ое** papier-maché

~ **в зе́мл·ю** loam/sand casting

~ **в коки́л·ь** permanent-mould casting, gravity die casting

~ **в ко́р·ков·ые фо́рм·ы** shell moulding/molding

~ **в оболо́ч·ков·ую фо́рм·у** shell moulding/molding

~ **в по·сто·я́нн·ую фо́рм·у** permanent-mould casting

~ **в сыр·ую фо́рм·у** undried mould casting

~, **горизонта́ль·н·ое** (plast) horizontal moulding

~, **ка́мен·н·ое** artificial/cast stone

~, **коки́ль·н·ое** gravity die casting

~, **лег·и́рованн·ое** (met) alloy casting

~, **не·пре·ры́в·н·ое** (met) continuous casting

~ **по вы·плавл·я́ем·ым моде́л·ям** investment casting, lost-wax process

~ **под давл·е́ни·ем** pressure die casting, die casting (U.S.); (plast) injection moulding/molding

~, **про́б·н·ое** (met) pouring test

~, **то́ч·ечн·ое** (plast) hot-gate/runner moulding

~, **то́ч·н·ое** precision casting

~, **фасо́н·н·ое** mould/mold casting

~, **центро·бе́ж·н·ое** centrifugal casting

~, **чугу́н·н·ое** iron casting, cast iron

лить·ев·о́й, **-а́я**, **-о́е**, (*a*) (met) casting; (plast) moulding, molding, moulded, molded

~ **материа́л·**, (*m*) (plast) injection-moulding material

~ **маши́н·а** (*f*) (plast) injection moulding machine

Лиуви́лл·я, **теоре́м·а** (*f*) (math) Liouville's theorem

лифт·, **-а**, (*m*) lift, hoist, elevator

~, **плу́нжер·н·ый** (oil) plunger lift

лифт·ов·о́й, **-а́я**, **-о́е**, of **лифт·**

~ **труб·а́** (*f*) (oil) lifting pipe

лиф·чик·, **-а**, (*m*) (text) brassiere

лих·а́ч·, **-а́**, (*i*) **-ч·о́м**, (*m*) keen/intrepid person

лихен·а́з·а, **-ы**, (*f*) (biochem) lichenase

лихен·и́н·, **-а**, (*m*) (biochem) lichenin

лихено·ло́г·и·я, **-и**, (*f*) [(bot) lichenology

лихено·ме́тр·и·я, **-и**, (*f*) (geol) lichenometry

лихора́д·ить, **-ят**, (*imp*) (med) have a fever, be feverish

лихора́д·к·а, **-и**, (*g pl*) **-д·ок·**, (*f*) (med) fever, febris

~, **гемо·глобин·ур·и́йн·ая** blackwater fever

~, **жёлт·ая** yellow fever

~ **ку** Q fever

лихора́д·очн·ый, **-ая**, **-ое**, (*a*) feverish, febrile, pyretic

Лихтенбе́рг·а, **фигу́р·а** (*f*) (elec) Lichtenberg figure

ли́хтер·, **-а**, (*m*) (naut) lighter, dumb lighter

ли́хтер·н·ый, **-ая**, **-ое**, (*a*) lighter

~ **с·бор·** (*m*) lighterage

лихтло́х·, **-а**, (*m*) (min) auxiliary shaft

лиц·ева́льн·ый, **-ая**, **-ое**, (*a*) facing, (text) turning

лиц·ева́ть, **-у́ют**, (*imp*) turn (cloth in remaking garment)

лиц·ев·о́й, **-а́я**, **-о́е**, (*a*) face, facial, facing, front; personal

~ **кирпи́ч·** (*m*) facing brick

~ **сторон·а́** (*f*) right side (of a material); (arch) façade, front; obverse (of coin); grain side (of leather)

~ **с·чёт·** (*m*) (fin) personal account

лице·ме́р·н·ый, **-ая**, **-ое**, (*a*) hypocritical

лицендра́т·, **-а**, (*m*) (rad) litzendraht, litz wire, litz

лиценц·ио́нн·ый, **-ая**, **-ое**, (*a*) of **лице́нц·и·я**

лице́нц·и·я, **-и**, (*f*) (econ) licence

ли́ци·й, **-я**, (*m*) (bot) *Lycium*

лиц·о́, **-а́**, (*nom pl*) **-а́**, (*g pl*) **лиц·** (*n*) (anat) face, facies; person; front, (text etc.) right side; character, image

~, **до́лж·ности·ое** official, officer

лиц·о́вк·а, -и, (*g pl*) **-вок·,** (*f*) smooth-cut file (tool)

лиц·у́ющ·ий, -ая, -ее, (*pres part act*) of **лиц·ева́ть**

лич·и́н·а, -ы, (*f*) mask, guise; keyhole escutcheon

лич·и́нк·а, -и, (*g pl*) **-нок·,** (*f*) (*dim*) of **лич·и́н·а;** (zool) larva, grub

~, бо·ев·а́я bolt head (of a rifle)

лич·и́нник·, -а, (*m*) (ent) caterpillar, *Scorpiurus*

лич·и́ночн·ый, -ая, -ое, (*a*) (zool) larval

ли́ч·ник·, -а, (*m*) striking plate (of a lock)

ли́ч·н·о (*adv*) personally, in person

лич·н·о́й, -а́я, -о́е, (*a*) face, facial; smooth, fine (of serrated surfaces)

ли́ч·ность·, -и, (*f*) personality, person, individual

ли́ч·н·ый, -ая, -ое, (*a*) personal

~ знак· (*m*) identity card/disk

~ о·ши́б·к·а (*f*) (instr) human error, personal equation

~ салю́т· (*m*) salute of guns (to royalty, VIP's etc.)

~ со·ста́в· (*m*) personnel, staff

~ у·равн·е́ни·е (*n*) (instr) personal equation

лишае·ви́д·н·ый, -ая, -ое, (*a*) (bot) lichenoid; (med) herpetic

лиша́·й, -я́, (*m*) (bot) lichen, *Lichenes*; (med) herpes

лиша́й·ник·, -а, (*m*) (bot) lichen, lichenes

лиша́й·ников·ый, -ая, -ое, (*a*) of **лиша́й·ник·**

лиша́й·н·ый, -ая, -ое, (*a*) (bot) lichen; (med) herpetic

лиш·а́·ть, -ют, (*imp*) (+ *gen*) deprive of; **-ся** (*pass*); lose

лиш·е́ни·е, -я, (*n*) deprivation, forfeiture; privation, hardship

лиш·ённ·ый, -ая, -ое, (*past part pass*) of **лиш·и́ть** (*perf*); *see* **лиш·а́·ть** (*imp*); devoid, void, deprived; without, lacking, -free, apo-

~ дисто́рси·и (opt) distortion-free

~ хромат·и́ческ·ой аберра́ци·и (opt) apochromatic

лиш·и́ть, -а́т, (*perf*) (+ *gen*) *see* **лиш·а́·ть**

ли́ш·н·ий, -яя, -ее, (*a*) superfluous, excessive, supernumerary

с ли́ш·н·им and a bit, more than

лишь (*adv*) only, but; (*conj*) as soon as

~ бы if only, provided that

лишь

~ то́лько no sooner than

ли·я́ (*pres gerund*) of **ли·ть**

лм (*abbr*) of **лю́мен·** lumen (unit of luminous flux)

лоб, лб·а, (*m*) (anat) forehead, sinciput

~ на·дви́г·а (geol) front of thrust

~ с·кла́д·к·и (geol) crown

лоба́н·, -а, (*m*) (fish) gray mullet, *Mugil cephalus*

лобел·и́н·, -а, (*m*) (chem) lobeline

лобе́л·и·я, -и, (*f*) (bot) lobelia

ло́бзик·, -а, (*m*) compass/fret/scroll saw

ло́бзик·ов·ый, -ая, -ое, (*a*) of **ло́бзик·**

~ стан·о́к· (*m*) fret saw machine, jig saw machine

лоб·ко́в·ый, -ая, -ое, (*a*) (anat) pubic

~ кост·ь (*f*) (anat) pubis

ло́б·н·ый, -ая, -ое, (*a*) (anat) frontal

лоб·ов·о́й, -а́я, -о́е, (*a*) frontal, front, face; head-on

~ со·противл·е́ни·е (*n*) (dynam) head resistance, drag

коэффицие́нт· лоб·ов·о́го со·про·тивл·е́ни·я (*m*) (dynam) drag coefficient

лобо·гре́·йк·а, -и, (*g pl*) **-е́·ек·,** (*f*) (agr) manual-delivery reaper

лоб·о́к·, -б·к·а́, (*m*) (anat) pubis

лобо·по́д·и·й, -я, (*m*) (zool) lobopodium

лобо·ток·а́рн·ый стан·о́к· (*m*) face lathe

лобо·фре́зер·н·ый стан·о́к· (*m*) face-milling machine

ло́в·, -а, (*m*) catching, trapping, snaring, fishing

~ за·па́д·н·ей (fish) trapping

~, контро́ль·н·ый test fishing

~, коше́ль·ко́в·ый (fish) purse seining

~ на блесн·у́ (fish) trolling

~, тра́л·ов·ый (fish) trawling

Ло́ве, волн·а́ (*f*) (seismol) Love wave

лов·е́ц·, -в·ц·а́, (*i*) **-в·ц·о́м,** (*m*) fisherman; trapper

лов·е́цк·ий, -ая, -ое, (*a*) of **лов·е́ц·**

Ловибо́нд·а, колори́·метр· (*m*) Loribond tintometer

лов·и́льн·ый, -ая, -ое, (*a*) fishing, catching, trapping

~ дро́т·ик· (*m*) (oil) rope spear

~ инструме́нт· (*m*) (oil) fishing tools

~ колоко́л· (*m*) (oil) bell socket

~ ме́т·чик· (*m*) (oil) fishing tap

~ рабо́т·ы (*pl*) (oil) fishing operations, fishing

лов·и́тел·ь, -я, (*m*) catch, catcher; (min) kep, keps

лов·и́ть, -ят, (*imp*) catch; pick up, recover (floating object etc.)

ло́в·к·ий, -ая, -ое, (*a*) dexterous, deft, adroit

ло́в·кост·ь, -и, (*f*) dexterity, skill

ло́вл·енн·ый, -ая, -ое, (*past part pass*) of **лов·и́ть**

ло́вл·я, -и, (*i*) **ло́вел·ь**, (*f*) catching, trapping, snaring, fishing, recovering (from water)

лов·у́ч·ий, -ая, -ее, (*a*) grasping, prehensile

лов·у́шк·а, -и, (*g pl*) **-шек·**, (*f*) trap, snare; (mil) decoy

~, гидравл·и́ческ·ая (min) hydraulic trap, hydraulic gold trap

~, ио́н·н·ая (phys) ion trap

~, не·за·мк·ну́т·ая (geol) misclosure

~, нефт·ян·а́я (geol) petroleum trap

~, резона́нс·н·ая (elec) resonance trap

~, тепл·ов·а́я (phys) thermal trap

ло́в·ч·ий, -ая, -ее, (*a*) trapping, trap, trappers

ловчорри́т·, -а, (*m*) (min) lovchorrit

лог·, -а, (*nom pl*) **-а́**, (*g pl*) **-о́в**, (*m*) (geog) ravine; (math) log, logarithm; (ending) -logist

логари́фм·, -а, (*m*) (math) logarithm

~, десят·и́чн·ый common logarithm, Briggs' logarithm

~, натура́ль·н·ый natural/Napierian logarithm

~ от·нош·е́ни·я, сре́д·н·ий logarithmic mean difference

логари́фм·ик·а, -и, (*f*) (math) graph of a logarithmic function

логарифм·и́ровани·е, -я, (*n*) taking the logarithm, logarithmic operation

логарифм·и́рованн·ый, -ая, -ое, (*past part pass*) of **логарифм·и́р·овать**; in logarithmic form

логарифм·и́р·овать, -уют, (*imp*) take the logarithm (of)

логарифм·и́ческ·ий, -ая, -ое, (*a*) logarithmic

~ интенси́·метр· (*m*) (nucl) log-counting ratemeter

~ про·из·во́д·н·ая (*f*) logarithmic derivative

~ спира́л·ь (*f*) logistic/logarithmic spiral

ло́г·ик·, -а, (*m*) logician

ло́г·ик·а, -и, (*f*) logic

логи́ст·ик·а, -и, (*f*) logistics

логист·и́ческ·ий, -ая, -ое, (*a*) logistic, logistical

логици́зм·, -а, (*m*) logistics

лог·и́ческ·ий, -ая, -ое, (*a*) logical, logic

~ схе́м·а (*f*) logical circuit (computers)

~ у·стро́й·ств·о (*n*) (autom) logic unit

~ част·ь (*f*) logic (of computers)

лог·и́чност·ь, -и, (*f*) logicality, logic

лог·и́чн·ый, -ая, -ое, (*a*) logical, logic

ло́г·овищ·е, -а, (*i*) **-щем**, (*n*) lair, den

ло́г·ов·о, -а, (*n*) lair, den

лого́·метр·, -а, (*m*) (instr) ratiometer

~, магнито·электр·и́ческ·ий permanent-magnet moving coil ratiometer

лого·метр·и́рующ·ий, -ая, -ее, (*a*) ratiometric

лого·пе́д·, -а, (*m*) (med) speech therapist

лого·пе́д·и·я, -и, (*f*) (med) logopedia, logopedics

лого·ти́п·, -а, (*m*) (print) logotype

лод·, -а, (*m*) load (of timber; 50 cu ft)

ло́джи·я, -и, (*f*) (arch) loggia

лоди́к·ул·а, -ы, (*f*) (bot) lodicule

ло́д·к·а, -и, (*g pl*) **-д·ок·**, (*f*) boat

~, кругло·бо́рт·н·ая round-bilge boat

~, лет·а́ющ·ая flying boat

~, на·ду́в·н·а́я inflatable boat

~, под·во́д·н·ая (*see* **под·ло́д·к·а**) submarine

~, при·бо́й·н·ая surf boat

~, про·ре́з·н·а́я fish-hold boat

~ с о́стр·ой скул·о́й hard-chine boat

ло́д·очк·а, -и, (*g pl*) **-чек·**, (*f*) (*dim*) of **ло́д·к·а**; dish, tray; drawing boat, debitense (glassmaking)

~, плати́н·ов·ая (geol) platinum boat

ло́д·очник·, -а, (*m*) boatman

Ло́дочников·а, эффе́кт· (*m*) dispersion effect

ло́д·очн·ый, -ая, -ое, (*a*) of **ло́д·к·а**

лоды́ж·к·а, -и, (*g pl*) **-жек·**, (*f*) (anat) ankle bone, malleolus; (mech) catch, cam

ло́ж·а, -и, (*i*) **-ж·ей**, (*f*) rifle stock; (theat) box

лож·а́щ·ийся, -аяся, -ееся, (*pres part act*) of **лож·и́ться**

лож·би́н·а, -ы, (*f*) hollow, cavity, trough; (meteor) trough of low pressure

лож·би́нк·а, -и, (*g pl*) **-нок·**, (*f*) (*dim*) of **лож·би́н·а**; (geol) stria, rill, furrow

ло́ж·е, -а, (*i*) **-ж·ем**, (*n*) bed; (geog) bed, floor

~, реч·н·о́е (geog) river bed

~ с·бро́с·а (geol) foot of fault

~, семен·н·о́е (hortic) seed bed

лож·ечниц·а, -ы, (*f*) (bot) spoonwort, *Cochlearia*

лож·ечн·ый, -ая, -ое, (*a*) spoon, spoon-shaped, cup-shaped; bed, bed-shaped

лож·и́ться, -а́тся, (*imp intrans*) lie down/flat/horizontal, lie (in a certain manner); fall (of shadow, snow etc.); (navig) turn to/onto

~ в дрейф· (naut) heave to

лож·к·а, -и, (*g pl*) **-ж·ек·,** (*f*) spoon, ladle; (met cast) slick, slicker, sleeker; (min) auger; (*g sing of* **лож·óк·**)

~, бур·ов·а́я (min) earth auger

лóж·н·о- (*component*) (*see also* **лже-**), pseudo-, mock, false

ложно·жа́бр·а, -ы, (*f*) (zool) pseudo-branch

ложно·лопато·нóс·, -а, (*m*) mock sturgeon, (fish) *Pseudoscaphirhynchus*

ложно·лук·ови́ц·а, -ы, (*f*) (bot) pseudobulb

ложно·нóж·к·а, -и, (*g pl*) **-ж·ек·,** (*f*) (zool) pseudopodium

ложно·равно·лóпаст·н·ый, -ая, -ое, (*a*) homocercal

лóж·ност·ь, -и, (*f*) falsity

лóж·н·ый, -ая, -ое, (*a*) false, spurious, feint, dummy, pseudo-, mock

~ муч·ни́ст·ая рос·а́ (*f*) downy mildew (grape disease)

~ сóл·н·ц·е (*n*) (meteor) parhelion, mock sun

лож·óк·, -ж·к·а́, (*m*) stretcher (brick)

лож·ь, (*g sing*) **лжи,** (*f*) falsehood, untruth

лоз·а́, -ы́, (*f*) (bot) almond-leaf willow, *Salix triandra*

~, виногра́д·н·ая (bot) vine, grape vine, *Vitis vinifera*

лоз·ня́к·, -а́, (*m*) willow bush

лóзунг·, -а, (*m*) slogan

локализ·а́тор·, -а, (*m*) localizer

локализ·а́ция·, -и, (*f*) localization

локализ·и́р·овать, -уют, (*imp and perf*) *see* **локализ·ова́ть**

локализ·óванн·ый, -ая, -ое, (*past part pass*) of **локализ·ова́ть**; localized

локализ·ова́ть, -у́ют, (*imp*) localize

лока́ль·ност·ь, -и, (*f*) localization, locality, localness

коэффициéнт· лока́ль·ност·и (*m*) coefficient of localization

лока́ль·н·ый, -ая, -ое, (*a*) local

лок·а́тор·, -а, (*m*) locator; radar set; sonar set

~, гидро·акуст·и́ческ·ий sonar set

локаут·и́р·овать, -уют, (*imp anp perf*) lock out (industrial dispute)

лок·а́ци·я, -и, (*f*) location, locating, finding

локо·моби́л·ь, -я, (*m*) locomobile, portable steam engine

~, стациона́р·н·ый semi-portable steam engine

локо·моти́в·, -а, (*m*) (rail) locomotive

~, голов·н·óй hauling locomotive

~ -толк·а́ч·, -а́, (*m*) (min) bank locomotive

локо·моти́в·н·ый, -ая, -ое, (*a*) of **локо·моти́в·**

локо·мотóр·н·ый, -ая, -ое, (*a*) locomotor

лóкон·, -а, (*m*) curl, curve

локон Аньéзи (*m*) (math) witch of Agnesi, versiera

локон·óид·, -а, (*m*) (math) witch-shape

локот·ни́к·, -а́, (*m*) arm rest

лóкот·ь, -кт·я, (*nom pl*) **лóкт·и,** (*g pl*) **-éй,** (*m*) (anat) elbow, forearm

локсо·дрóм·а, -ы, (*f*) **локсо·дрóм·и·я**

локсо·дром·и́ческ·ий, -ая, -ое, (*a*) loxodromic

локсо·дрóм·и·я, -и, (*f*) (surv) loxodromic curve, loxodrome, Mercator track; (navig) rhumb line

локт·ев·óй, -а́я, -óe, (*a*) (anat) elbow, ulnar

~ движ·éни·е (*n*) elbow motion (of a manipulator)

лок·у́л·а, -ы, (*f*) (biol) loculus, locule

лóм·, -а, (*nom pl*) **-ы,** (*g pl*) **-óв,** (*m*) crowbar; scrap

~, желéз·н·ый scrap iron

лом·а́лк·а, -и, (*g pl*) **-лок·,** (*f*) (text) fibre-breaking machine

ломано·лин·éйн·ая траектóр·и·я (*f*) (air) dog-leg trajectory/path

лóм·ан·ая, -ой, (*f decl as a*) (math) broken line

лóм·ан·ый, -ая, -ое, (*a*) broken; (*see also* **лóм·ан·ая**) (*as noun*)

~ профи́л·ь (*m*) broken contour/outline; (*in g as adj*) broken

лом·а́·ть, -ют, (*imp*) break in pieces/two, break up; murder (language, piece of poetry etc.); **-ся** (*passive and intrans*) fracture

~ гóлов·у rack one's brains

лóм·ик·, -а, (*f*) ratchet bar

лом·и́ть, ´ят, (*imp*) break, break up, break in pieces/two; have rheumatism in, suffer from rheumatic pains; rush forward/ahead/at

лом·к·а, -и, (*g pl*) **-м·ок·,** (*f*) breaking

лом·к·ий, -ая, -ое, (*a*) breakable, fragile, brittle, friable, short

лом·кост·ь, -и, (*f*) fragility, brittleness, friability, (met) shortness

лом·ов·ой, -ая, -ое, (*a*) draft, dray (of animals)

~ лошад·ь (*f*) dray (horse)

ломо·нос·, -а,· (*m*) (bot) clematis

Ломоносов·а, закон· (*m*) (phys) law of conservation of matter

ломоносовит·, -а, (*m*) lomonosovite

ломонтит·, -а, (*m*) (min) laumontite

лом·от·а, -ы, (*f*) rheumatic pain

лом·отн·ый, -ая, -ое, (*a*) rheumatic

лом·от·ь, -мт·я, (*nom pl*) **лом·т·й,** (*g pl*) **-ей,** (*m*) slice

лом·тик·, -а, (*m*) (*dim*) of **лом·от·ь**

лонгкулуар·, -а, (*m*) long shoot (dredgers)

лонгосалинг·, -а, (*m*) (naut) trestle-table, trestle-tree

лонгулит·, -а, (*m*) (min) longulite

Лондон·ов·о у·равн·ени·е (*n*) (phys) London equation

лонжерон·, -а, (*m*) (a/c) spar (of wing); (М/Т) side member

~, не·раз·рез·н·ой continuous spar

~, перед·н·ый front spar

лопар·ь, -я, (*m*) (naut) fall (rope); **лопарь, -я,** (*m*) Lapp, Laplander

лопаст·ност·ь, -и, (*f*) lobation

лопаст·н·ый, -ая, -ое, (*a*) of **лопаст·ь;** (bot etc.) lobed, lobate, laciniate; (mech) rotodynamic

~ водо·мер· (*m*) vane-type water gauge

~ колес·о (*n*) paddle wheel; impeller (pumps etc.)

~ меш·алк·а (*f*) paddle stirrer

~ на·сос· (*m*) rotodynamic pump

лопаст·ь, -и, (*f*) blade, vane, fan, paddle, bucket, palm; (biol, geol) lobe

~ винт·а propeller blade

~ нес·ущ·его винт·а rotor blade (of helicopter)

лопат·а, -ы, (*f*) spade, shovel

~, кругл·ая round-point shovel

~, механ·ическ·ая power shovel

~, обрат·н·ая back-acter, back-acting shovel/digger

~, пневмат·ическ·ая pneumatic shovel/chisel

~, прям·ая face shovel, face power shovel

лопат·к·а, -и, (*g pl*) **-т·ок·,** (*f*) (*dim*) of **лопат·а;** shovel, trowel; blade, vane (pump, turbine etc.); bucket (water turbine); (anat) shoulder blade, scapula; agitator (of mixing machine); (naut) serving board (cordage); (*pl*) blading

~, америкáн·ск·ая plasterer's trowel

~, двух·сторон·н·ая dumb-bell test piece (rubber)

~, монтаж·н·ая tire mounting tool

~, на·правл·яющ·ая guide blade/vane (turbines); (*pl*) cascade (of wind tunnel)

~ на·правл·яющ·его аппарат·а stator blade/vane (turbines)

~, рабоч·ая rotor/moving blade (turbine)

~ сопл·ов·ого аппарат·а stator blade, nozzle guide blade (turbine)

лопатко·образ·н·ый, -ая, -ое, (*a*) spatulate

лопато·ног·и·е, -их, (*pl decl as adj*) (zool) tusk shells, scaphopods, *Scaphopoda*

лопато·нос·, -а, (*m*) (fish) shovel-headed sturgeon, *Scaphirhynchus*

лопат·очн·ый, -ая, -ое, (*a*) of **лопат·к·а;** rotodynamic; scapular

~ вен·ец· (*m*) blading ring (turbine)

~ машин·а (*f*) rotodynamic machine

лоп·а·ться, -ются, (*imp*) burst, split, break

лоп·ающ·ийся, -аяся, -ееся, (*pres part act*) *see* **лоп·а·ться;** (bot) dissilient

лоп·н·уть, -ут, (*perf*) *see* **лоп·а·ться**

лополит·, -а, (*m*) (geol) lopolith

лопух·, -á, (*m*) (bot) burdock, *Arctium*

лоран·, -а, (*m*) Loran (radionavigation system)

Лоран·а, ряд· (*m*) (math) Laurent expansion

Лоренц·а, пре·образ·овáни·е (*n*) (math) Lorentz transformation

Лоренц·а, сил·а (*f*) (phys) Lorentz force

Лоренц-Лоренц·а, формул·а (*f*) (phys) Lorentz–Lorenz formula/law/equation, Lorenz–Lorenz formula/law/equation

Лоренц-Фицджéральд·а со·кращ·éни·е (*n*) (phys) Lorentz–Fitzgerald contraction, Lorentz contraction

лос·ий, -ья, -ье, (*a*) of **лос·ь**

лос·ин·а, -ы, (*f*) elk-skin

лос·ин·ый, -ая, -ое, (*a*) elk, moose

лоск·, -а, (*m*) lustre, gloss, (paper etc.) glaze

лоскут·, -а́, (*nom pl*) **-ы́,** *or* **-ья,** (*g pl*) **-о́в,** *or* **-ьев,** (*m*) (text) rag, cloth rag; remnants; shred (rubber)

~, трикота́ж·н·ый stockinette remnants

лосн·и́ст·ый, -ая, -ое, (*a*) glossy, shiny, sleek

лосос·ёв·ый, -ая, -ое, (*a*) salmon, (*pl decl as a*) salmon family, *Salmonidae*

лосос·и́н·а, -ы, (*f*) (food) salmon

лосос·и́нн·ый, -ая, -ое, (*a*) (food) salmon

ло́сос·ь, -я, (*m*) (fish) salmon, (*pl*) *Salmonidae*

~, благо·ро́д·н·ый (fish) trout, *Salmo*

~, далне·вос·то́ч·н·ый Pacific salmon, *Oncorhynchus*

~, озёр·н·ый lake salmon

~, со́б·ственн·о Atlantic salmon, *Salmo salar*

лос·ь, -я, (*nom pl*) **-и, -е́й,** (*m*) (zool) moose, elk, *Alces alces*

лось·штаг·, -а, (*m*) (naut) preventer stay

лот·, -а, (*m*) (naut) lead, sounding lead

~, механ·и́ческ·ий sounding machine

~ То́мсон·а (naut) Kelvin's sounding machine

лот·аппара́т·, -а, (*m*) (surv) vertical collimator

лотере́·я, -и, (*f*) lottery

лот·ко́в·ый, -ая, -ое, (*a*) of **лот·о́к·;** pan-type, pan

~ транспортёр· (*m*) pan conveyor, troughed belt conveyor

ло́т·лин·ь, -я, (*m*) (naut) lead line, sounding line

лот·овать, -ую́т, (*imp*) (naut) take soundings

лот·ов·о́й, -о́го, (*m decl as a*) (naut) leadsman

лот·о́к·, -т·к·а́, (*m*) tray, trough, chute, shoot, flume; (build) gutter; (min) launder; (naut) loading tray (cargo)

~, лесо·о·пла́в·н·ый (hydro elec) log run

~, ме́ль·нич·н·ый mill-race, mill-course

ло́тос·, -а, (*m*) (bot) lotus, *Nelumbium*

лофо·три́х·н·ый, -ая, -ое, (*a*) (bact) lophotrichous

лофо·фо́р·, -а, (*m*) (zool) lophophore

лох·ов·о́й, -а́я, -о́е, (*a*) of **лох·**

лох·, -а, (*m*) (bot) oleaster, *Elaeagnus*; (fish) salmon grils

лох·а́нк·а, -и, (*g pl*) **-нок·,** (*f*) tub, basin; (anat) pelvis

лох·а́нн·ый, -ая, -ое, (*a*) tub, basin; (anat) pelvis

лохм·а́т·ый, -ая, -ое, (*a*) shaggy

ло́х·ов·ый, -ая, -ое, (*a*) of **лох·**

лохштейн·, -а, (*m*) (horol) jewel

ло́ц·и·я, -и, (*f*) (naut) sailing directions, pilot (book)

ло́ц·ман·, -а, (*m*) (naut) pilot; (zool) pilot fish, *Naucrates ductor*

ло́ц·ман·ск·ий, -ая, -ое, (*a*) of **ло́ц·ман·**

~ бот· (*m*) pilot boat

~ с·бор· (*m*) pilotage, pilotage fee

~ про·во́д·к·а (*f*) piloting, pilotage

ло́ц·ман·ств·о, -а, (*n*) piloting, pilotage

лоц·ме́йстер·, -а, (*m*) chief pilot

лошад·и́н·ый, -ая, -ое, (*a*) horse, equine; (*pl as noun*) *Equidae*, equids

~ си́л·а (*f*) horse-power, HP

ло́шад·ь, -и, (*nom pl*) **-и,** (*g pl*) **-е́й,** (*f*) (zool) horse, *Equus*

~, верх·ов·а́я saddle/riding horse, hack

~, дом·а́шн·яя domestic horse, *Equus caballus*

~, за·во́д·ск·ая stud horse

~, за·пал·ён·ая broken winded horse

~ Пржева́льск·ого (zool) Przhevalski's wild horse, *Equus przewalskii*

~, рабо́ч·ая (*f*) dray

~, спорти́в·н·ая race horse

лош·а́к·, -а́, (*m*) (zool) hinny, *Equus hinnus*

Ло́шмидт·а, числ·о́ (*n*) (chem) Loschmidt number

лощ·е́ни·е, -я, (*n*) glazing, polishing (leather, paper)

лощ·ён·ый, -ая, -ое, (*a*) polished, glazed, glossy

~ бума́г·а (*f*) glazed paper

лощ·и́льн·ый, -ая, -ое, (*a*) polishing, glazing

~ маши́н·а (*f*) leather rolling/polishing machine

лощ·и́н·а, -ы, (*f*) ravine

лощ·и́ть, -а́т, (*imp*) polish, gloss, glaze

л.с. (*abbr*) HP, horse power

луб·, -а, (*nom pl*) **лу́б·ья,** (*g pl*) **-ьев,** (*m*) (bot) bast, liber, phloem

луб·ов·о́й, -а́я, -о́е, (*a*) of **луб·**

лубо·е́д·, -а, (*m*) (ent) bark beetle, (*pl*) *Ipidae*

луб·о́к·, -б·к·а́, (*m*) (bot) bast, liber, phloem; splint

луб·о́чн·ый, -ая, -ое, (*a*) of **луб·о́к·**

лубрика́тор·, -а, (*m*) lubricator, oiler, lubricating machine

луб·ян·о́й, -а́я, -о́е, (*a*) of **луб·**

~ волокн·о́ (*n*) bast fibre

лу́г·, -а, (*nom pl*) **луг·а́,** (*g pl*) **-о́в,** (*m*) (agr) meadow, grassland, pasture

~, за·ли́в·н·о́й water meadow

луг·ов·и́к·, -а́, (*m*) (bot) hairgrass, *Deschampsia*

луго·во́д·ств·о, -а, (*n*) grassland/pasture cultivation

луг·ов·о́й, -а́я, -о́е, (*a*) of **луг·**

~ волк· (*m*) (zool) prairie wolf, coyote

~ руд·а́ (*f*) (min) meadow ore

~ соба́ч·к·а (*f*) (zool) prairie dog

луго·мелиор·ати́вн·ая ста́нц·и·я (*f*) park/pool for meadow/prairie cultivation machinery

луго·мелиор·а́ци·я, -и, (*f*) prairie reclamation, meadow/grassland improvement/cultivation

лу́д·а, -ы, (*f*) tinning alloy; (geog) rocky shoal/islet

луд·и́льн·ый, -ая, -ое, (*a*) tin, tin plating

луд·и́льщик·, -а, (*m*) tinsmith, tinman

луд·и́ть, -я́т, (*imp*) tin, plate with tin

лудло́в·ск·ий я́рус· (*m*) (geol) Ludlow Beds, Ludlorian series

лу́ж·а, -и, (*i*) **-ж·ей,** (*f*) puddle, pool

луж·а́йк·а, -и, (*g pl*) **-а́ек·,** (*f*) grass plot

луж·а́нк·а, -и, (*g pl*) **-нок·,** (*f*) (zool) river snail, *Viviparus*

луж·е́ни·е, -я, (*n*) (met) tinning, plating/coating with tin

~, гальван·и́ческ·ое electrolytic tinning

~, гор·я́ч·ее hot tinning

луж·ён·ый, -ая, -ое, (*a*) tinned, tinplated, coated with tin

луж·о́к·, -жка́, (*m*) (*dim*) of **луг·**

лузг·а́, -и́, (*f*) (bot) husk, shell, hull

лузго·ве́·йк·а, -и, (*g pl*) **-е́·ек·,** (*f*) (agr) hull/husk aspirator

лу́к·, -а, (*m*) (bot) onion, *Allium*; bow (weapon)

~, ду́д·чат·ый Welsh onion, *Allium fistulosum*

~, медве́ж·ий bear's onion/garlic, *Allium ursinum*

лу́к

~, мор·ск·о́й (bot) squill, sea onion, *Urginea maritima*

~ -пор·е́·й, -я, (*m*) leek, *Allium porrum*

~ -ре́з·анец·, -нц·а, (*m*) chives, *Allium schoenoprasum*

~, ре́п·чат·ый common onion, *Allium cepa*

~ -сев·о́к·, -в·к·а́, (*m*) onion set

~ -чесно́к·, -а́, (*m*) garlic, *Allium sativum*

~ -шало́т·, -а, (*m*) shallot, *Allium ascalonicum*

лук·а́, -и́, (*nom pl*) **лу́к·и,** (*g pl*) **лук·** (*f*) bend, curve; saddle bow

~, реч·н·а́я (geog) meander, river bend

лук·а́в·ый, -ая, -ое, (*a*) cunning

лу́к·овиц·а, -ы, (*f*) (bot) bulb, bulbs; (food) onion; (arch) onion-shaped dome

~, обон·я́тельн·ая (anat) olfactory bulb

лу́к·овичн·ый, -ая, -ое, (*a*) (bot) bulb, bulbous, bulbaceous, onion

лу́к·овник·, -а, (*m*) (bot) spikesedge, spikerush, *Eleocharis*

лу́к·ов·ый, -ая, -ое, (*a*) of **лук·**

~ му́х·а (*f*) (ent) onion maggot, onion fly, *Hylemya/Chortophila antiqua*

луко·обра́з·н·ый, -ая, -ое, (*a*) onion-shaped, bow-arched, bulb-like, bulbiform

Лун·а́, -ы́, (*nom pl*) **лу́н·ы,** (*g pl*) **лун·,** (*f*) moon; (math) lune

~, ло́ж·н·ая (meteor) paraselene, mock moon

~, молод·а́я waxing moon

~ на ущёрб·е waning moon

~, по́лн·ая full moon

лунати́зм·, -а, (*m*) (med) somnambulism, sleep-walking

Лу́нг·е, нитро́·метр· (*m*) (chem) Lunge nitrometer

лу́н·к·а, -и, (*g pl*) **-нок·,** (*f*) hole, pit, hollow, dimple; (bot) alveolus, alveola; (met) crater, liquid pool (continuous casting); (min) edge/foot hole; (zool) lunule, lunula

глуб·ин·а́ лу́н·к·и (*f*) crater depth, depth of liquid-phase (continuous casting)

~, шар·ов·а́я (math) line of sphere

лункери́т·, -а, (*m*) Lunkerite (met) (an antipiping compound)

лунко·коп· átel·ь, -я, (*m*) (hort) dibber, dibbler; hole-digger

лун·н·ый, -ая, -ое, (*a*) moon, lunar

~ про·меж·у́ток· (*m*) lunitidal interval

~ ка́мен·ь (*m*) (min) moonstone

~ су́т·к·и (*pl*) lunar day

луно·обра́з·н·ый, -ая, -ое, (*a*) crescent-shaped

лун·очк·а, -и, (*g pl*) **-чек·,** (*f*) (*dim*) of **лу́н·к·а;** (anat) alveolus

лун·очн·ый, -ая, -ое, (*a*) of **лу́н·к·а;** alveolar, alveolate

~ по·са́д·к·а (*f*) (hortic) hole planting

лу́нчато·зуб·ые, -ых, (*pl decl as a*) (zool) selenodonts

лу́п·а, -ы, (*f*) magnifying glass/lens

~ вре́мен·и (cinema) high-speed slow-motion camera

~, фокус·иро́вочн·ая (phot) focusing glass

лупи́н·, -а, (*m*) *see* **люпи́н·**

лупини́н·, -а, (*m*) (chem) lupinine

луп·и́ть, -ят, (*imp*) peel, skin; thrash

лупули́н·, -а, (*m*) (biochem) lupulin

луска́ч·, -а, (*m*) (agr) corn shelelr

лут·, -а, (*m*) (zool) leatherback turtle, *Dermochelys coriacea*

лутеи́н·, -а, (*m*) (biochem) lutein

лутеоли́н·, -а, (*m*) (chem) luteolin

луте́ц·и·й, -я, (*m*) (chem) lutetium, lutecium, Lu

лутиди́н·, -а, (*m*) (chem) lutidin, dimethylpyridine

лути ди́н·ов·ая кисл·от·а́ (*f*) lutidinic acid

лу́ток·, -а, (*m*) (zool) white merganser, *Mergus albellus*

луф·а́р·ь, -я, (*m*) bluefish, *Pomatomus saltatrix*

Лу́фф·а, рас·тво́р· (*m*) (chem) Luff reagent

лу́ци- (*prefix*) *see* **лю́ци-**

луч·, -а́, (*i*) **-ч·о́м,** (*nom pl*) **-и́,** (*g pl*) **-е́й,** (*m*) (phys) ray, beam; (biol) arm; (math) ray, half-line

луч, актин·и́чн·ый (*m*) (phot) actinic ray

~, бе́г·ающ·ий (TV) scanning beam

~, вед·у́щ·ий guide beam

~, ве́ер·н·ый fan beam

~, визи́р·н·ый (opt) collimating ray

~ визи́р·овани·я line of sight

~, вращ·а́ющ·ийся rotating/rotary beam

~, жёстк·ий (nucl) hard ray

~, зем·н·о́й (rad) ground ray

луч

~ зр·е́ни·я (surv) visual ray, line of sight

~, кана́л·ов·ый (nucl) positive ray

~, косм·и́ческ·ий cosmic ray

~, на·вод·я́щ·ий guide beam

~, на·кло́н·н·ый (rad) slant aerial

~, не·обык·нове́нн·ый (opt) extraordinary ray

~, нож·ев·о́й (rad) beavertail beam

~, обык·нове́нн·ый (opt) ordinary ray

~, одно·имён·н·ый (math) conjugate ray

~, остро·на·пра́вл·енн·ый pencil beam

~, о́стр·ый narrow beam

~, па́д·ающ·ий incident ray

~, пла́зм·енн·ый plasma beam

~, пре·ры́в·ист·ый (rad) chopped beam

~, проект·и́рующ·ий image ray

~, раз·вёрт·ывающ·ий (TV) scanning beam

~, рас·се́·янн·ые (*pl*) scattered rays

~, рентге́н·овск·ий X-ray

~ све́т·а light ray

~, с·мещ·ённ·ый offset beam

~, со·про·вожд·а́ющ·ий locked-on beam

~, с·тир·а́ющ·ий scan-off beam, play-off beam

~ у·правл·е́ни·я с·на·ря́д·а (rocket) gathering beam

~, у·правл·я́ющ·ий (air nav) lead beam

~, электро́н·н·ый electron beam, electron ray, cathode beam

~, элемента́р·н·ый (opt) central ray

луче- (*root*) ray, radiating, actino-; beam

луч·еви́к·и (*pl*) (zool) radiolarians, *Radiolaria*

луч·ев·о́й, -а́я, -о́е, (*a*) of **луч·;** radial, radiating, radiation

~ бол·е́зн·ь (*f*) radiation sickness

~ кост·ь (*f*) (zool) radius

~ ско́р·ост·ь (*f*) (astron) radial velocity, line-of-sight velocity

~ тетрод· (*m*) (elec) beam tetrode

луче·за·пя́ст·н·ый, -ая, -ое, (*a*) (anat) radiocarpal

луче·за́р·ност·ь, -и, (*f*) radiance

луче·за́р·н·ый, -ая, -ое, (*a*) radiant

луче·ис·пуск·а́емост·ь, -и, (*f*) (phys) emittance, emissivity

луче·ис·пуск·а́ни·е, -я,   (*n*) (*and see* **из·луч·е́ни·е**) radiation

луче·ис·пуск·а́тельн·ый, -ая, -ое, (*a*) radiant, radiating, emitting

луче·ис·пуск·а́ющ·ая спосо́б·ност·ь (*f*) radiating power, radiant emissivity/ emittance

луче·локт·ев·о́й, -а́я, -о́е, (*a*) (anat) radio-ulnar

луч·е́ни·е, -я,   (*n*) (fish) torching

луче·образ·у́ющ·ий, -ая, -ее, (*a*) beam-forming

луче·о·гран·и́чивающ·ий, -ая, -ее, (*a*) beam-confining

луче·пёр·ые, -ых, (*pl decl as adj*) ray-fin fishes, rayfins, *Actinopterygii*

луче·плав·нико́в·ые, -ых, (*pl decl as adj*) *see* **луче·пёр·ые**

луче·пре·ломл·е́ни·е, -я, (*n*) (opt) refraction, refringence

~, двой·н·о́е (opt) birefringence

луч·и́н·а, -ы, (*f*) wood chip/shaving

луч·и́нк·и (*nom pl*), (*g pl*) **-нок·,** (*f*) (print) flyer

лучисто·гри́б·ков·ый, -ая, -ое, (*a*) (bot) actinomycetic

луч·и́стот·ь, -и, (*f*) radiance

луч·и́ст·ый, -ая, -ое, (*a*) radiant, radial, actinic, (bot) actinodromous

~ гриб·к·и́ (*pl*) (bact) *Actinomycetales*

~ ка́мен·ь (*m*) (min) actinolite

~ о·топл·е́ни·е (*n*) radiant heating

~ эне́рг·и·я (*f*) (phys) radiant energy

луч·и́ть, -ат, (*imp*) (fish) torch; **-ся** beam, shine

луч·ко́в·ый, -ая, -ое, (*a*) bow, bow-shaped

~ пил·а́ (*f*) frame saw

лу́чш·е (*comp*) better

лу́чш·ий, -ая, -ее, (*a*) better **в лу́чш·ем слу́ч·а·е** at best

лущ·е́ни·е, -я, (*n*) (*v n*) *see* **лущ·и́ть**

лущ·и́льник·, -а, (*m*) (agr) stubble breaker, surface plough/plow

лущ·и́льн·ый, -ая, -ое, (*a*) shelling, hulling, husking, scaling, peeling; scuffling

~ стан·о́к· (*m*) rotary veneer-cutting machine (woodworking)

лущ·и́ть, -а́т, (*imp*) (food) shell, husk, hull, scale, pod; (agr) scuffle

лы́ж·а, -и, (*i*) **-ж·ей,** (*f*) ski; snow shoe; (elec) collecting shoe

~, скольз·я́щ·ая ski

~, ступ·а́ющ·ая snow shoe

лыж·и́н·а, -ы, (*f*) ski

лы́ж·н·ый, -ая, -ое, (*a*) of **лы́ж·а**

лыж·н·я́, -й, (*f*) ski-track

лы́к·о, -а, (*n*) (bot) raffia, bast, phloem

лы́к·ов·ый, -ая, -ое, (*a*) of **лы́к·о**

лыс·е́·ть, -ют, (*imp*) go/grow bald

лы́с·ин·а, -ы, (*f*) bald spot, (med) area celsi

лы́с·к·а, -и, (*g pl*) **-сок·,** (*f*) (mech) flat keyway

лыс·у́х·а, -и, (*f*) (zool) coot, bald coot, *Fulica atra*

лы́с·ый, -ая, -ое, (*a*) bald

лыч·а́, -й, (*f*) cherry plum, *Prunus divaricata*

льв·ёнок·, -нк·а, (*nom pl*) **льв·я́т·а,** (*g pl*) **-ят·,** (*m*) lion cub

льв·и́нк·и (*nom pl*), (*g pl*) **-нок·,** (*f*) (zool) soldier flies, *Stratiomyidae*

льв·и́н·ый, -ая, -ое, (*a*) lion's, leonine

~ зуб· (*m*) (bot) dandelion, *Taraxacum dens-leonis*

льв·и́ц·а, -ы, (*f*) lioness

льв·я́т·а (*nom pl*) of **льв·ёнок·**

льг·о́т·а, -ы, (*f*) privilege, exception, preference

льг·о́тн·ый, -ая, -ое, (*a*) privilege, preferential

~ биле́т· (*m*) free/complementary ticket

~ дн·и (*pl*) (com) days of grace

~ цен·а́ (*f*) preferential price

льд·а (*gen*) of **лёд**

льд·и́н·а, -ы, (*f*) (ocean) ice floe; block of ice

льд·и́нк·а, -и, (*g pl*) **-нок·,** (*f*) piece of ice

льд·и́ст·ый, -ая, -ое, (*a*) icy, iced over/up

льд·о- (*root*) ice-

льдо·генера́тор·, -а, (*m*) ice-making plant

льдо·с·кал·ыва́тел·ь, -я, (*m*) ice leveller

льдо·сол·ян·о́е о·хлажд·е́ни·е (*n*) brine refrigeration

льдо·те́хн·ик·а, -и, (*f*) ice technology

льдо·фо́рм·а, -ы, (*f*) ice mould/mold (refrigeration)

льдо·хран·и́лищ·е, -а, (*i*) **-щ·ем,** (*n*) ice house/store

льд·ы́ (*pl*) *see* **лёд**

лье́ж·ск·ая номенклату́р·а (*f*) (org chem) Liege nomenclature

ль·ёт (*pres 3rd sing*) of **ли·ть** (*imp*)

льн·а (*gen*) of **лён**

льно- (*root*) flax, linen, linseed

льно·во́д·, -а, (*m*) flax grower/farmer

льно·во́д·ств·о, -а, (*n*) flax cultivation/ growing

льно·за·во́д·, -а, (*m*) flax-fibre/fiber processing plant

льно·клеверо·тёр·к·а, -и, (*g pl*) **-р·ок·,** (*f*) flax-hulling machine

льно·комба́йн·, -а, (*m*) flax-harvesting machine

льно·куде́л·ь, -и, (*f*) flax tow

льно·молот·и́лк·а, -и, (*g pl*) **-лок·,** (*f*) flax thresher

льно·мя́·лк·а, -и, (*g pl*) **-лок·,** (*f*) flax breaker

льно·пряд·е́ни·е, -я, (*n*) flax/linen fibre spinning

льно·соло́м·к·а, -и, (*g pl*) **-м·ок·,** (*f*) flax straw/stems

льно·тереб·и́лк·а, -и, (*g pl*) **-лок·,** (*f*) flax puller, flax-pulling machine

льно·треп·а́лк·а, -и, (*g pl*) **-лок·,** (*f*) flax scutcher

льно·треп·а́льн·ый, -ая, -ое, (*a*) flax-scutching

льно·трест·а́ -ы́, (*f*) flax stock

льно·трещ·о́тк·а, -и, (*g pl*) **-ток·,** (*f*) linseed sorter/separator

льно·чес·а́льн·ый, -ая, -ое, (*a*) flax-hackling

ль·н·уть, -ут, (*imp*) stick/cling to

льн·я́нк·а, -и, (*g pl*) **-нок·,** (*f*) (bot) toadflax, *Linaria*

льн·ян·о́й, -а́я, -о́е, (*a*) flax, flaxen, linen, linseed

~ **ма́сл·о** (*n*) linseed oil

~ **мук·а́** (*f*) linseed meal

~ **се́м·я** (*n*) linseed, flax seed

~ **се́·ялк·а** (*f*) (agr) flax-sowing machine

~ **ткан·ь** (*f*) (text) linen

льняно·ки́сл·ый, -ая, -ое, (*a*) (chem) linoleate (of)

льст·и́в·ый, -ая, -ое, (*a*) flattering

Льюис·а, тео́ри·я (*f*) (phys) Lewis's theory

ль·ю́щ·ий, -ая, -ее, (*pres part act*) of **ли·ть**

льял·о, -а, (*n*) (naut) bilge

лья́ль·н·ый, -ая, -ое, (*a*) of **лья́л·о**

~ **лючи́н·ы** (*pl*) (naut) limber boards

Лэмб·а-Ризерфорд·а, о́·пыт· (*m*) (phys) Lamb-Retherford experiment

лэмб·о́вск·ий, с·двиг· (*m*) (phys) Lamb shift, Lamb-Retherford shift

Лэнгмю́р·а, за·ко́н· (*m*) (phys) Langmuir law

люб·е́знича·ть, -ют, (*imp*) be pleasant, talk pleasantly, be nice (to)

люб·е́зн·ый, -ая, -ое, (*a*) amiable, pleasant, nice

люб·и́м·ый, -ая, -ое, (*a*) favourite; beloved

люб·и́стик·, -а, (*m*) (bot) lovage, *Levisticum officinale*

люб·и́тел·ь, -я, (*m*) amateur; enthusiast, fan

люб·и́тельск·ий, -ая, -ое, (*a*) of **люб·и́тел·ь**

люб·и́ть, -ят, (*imp*) love, like

люб·ова́ться, -у́ются, (*imp*) (+*instr*) admire

люб·о́вн·ый, -ая, -ое, (*a*) amorous, loving

люб·о́в·ь, -б·в·и́, (*i*) **люб·о́в·ью,** (*f*) love, amour

любо·зн·а́тельн·ый, -ая, -ое, (*a*) inquisitive, curious

люб·о́й, -а́я, -о́е, (*a*) any, either (of two); (*as noun*) anyone

любо·пы́т·н·ый, -ая, -ое, (*a*) curious

любрике́тинг·, -а, (*m*) "Lubricating" (a high-grade medium machine oil)

люверс·, -а, (*m*) louvre; (naut) grommet, eyelet

лю́ггер·, -а, (*m*) (naut) lugger

Лю́дерс·а, ли́н·и·и (*pl*) (met) Lüders lines, slip bands, flow lines

лю́д·и (*nom pl*), (*g pl*) **-е́й,** people; (mil) men

лю́д·н·ый, -ая, -ое, (*a*) populous, crowded (with people)

людо·е́д·ств·о, -а, (*n*) anthropophagy, cannibalism

люд·ск·о́й, -а́я, -о́е, (*a*) of **лю́д·и;** human

~ **материа́л·** (*m*) manpower

~ **со·ста́·в·** (*m*) personnel, staff

люизи́т·, -а, (*m*) lewisite (poison gas)

люк·, -а, (*m*) hatch, trap, manhole, access hole, (min) chute; trapdoor

~, **бо́мб·ов·ые** (*pl*) (a/c) bomb opening

~, **в·ход·н·о́й** access hatch (submarines)

~, **вы·пуск·н·о́й** (min) drawing/pull chute

~, **груз·ов·о́й** (naut) cargo hatch

~, **за·гру́з·очн·ый** charging/filling hatch

~, **решёт·чат·ый** hatch grating

~, **ру́б·очн·ый** conning tower hatch (submarine)

~, **свет·ов·о́й** skylight

~, **смотр·ов·о́й** inspection hatch/hole, peephole

~, **с·пас·а́тельн·ый** escape hatch (submarine)

люк·ов·о́й, -о́го, (*m decl as adj*) (min) chuteman

лю́к·ов·ый, -ая, -ое, (*a*) of **лю́к·**
~ **за·пи́с·к·а** (*f*) (naut) certificate of inspection
~ **за·тво́р·** (*m*) trap door, (min) chute door
лю́кс·, -а, (*m*) lux (unit of illumination); de luxe
Лю́ксембург-Го́рковск·ий эффе́кт· (*m*) (phys) Luxembourg effect
лю́кс·ме́тр·, -а, (*m*) (opt) luxmeter, luxometer
лю́л·ечн·ый, -ая, -ое, (*a*) of **лю́ль·к·а;** swinging-tray
лю́ль·к·а, -и, (*g pl*) **-л·ек·,** (*f*) cradle; pipe (for smoking)
~ **-кре́сл·о, -а,** (*n*) (naut) bosun's chair
люмбо·коло·стом·и́·я, -и, (*f*) (med) lumbocolostomy
лю́мен·, -а, (*m*) lumen, lu (unit of luminous flux)
~ **-ча́с·, -а,** (*m*) lumen-hour, lu-hr
люмено́·метр·, -а, (*m*) (opt) lumen-meter
люме́рг·, -а, (*m*) lumerg (unit of light energy)
люмина́л·, -а, (*m*) (pharm) luminal
люмин·есце́нтн·ый, -ая, -ое, (*a*) (phys) luminescent, luminescence
~ **актива́тор·** (*m*) luminescent activator
~ **дефекто·скоп·и́·я** (*f*) luminescence flaw detection
~ **с·чёт·чик·** (*m*) (phys) scintillation counter
люмин·есце́нци·я, -и, (*f*) luminescence
люмин·есци́рующ·ий, -ая, -ее, (*a*) luminous, luminescent
люмино·ско́п·, -а, (*m*) (instr) lumino-scope
люмино·флави́н·, -а, (*m*) (chem) luminoflavin
люмино·фо́р·, -а, (*m*) (chem) lumino-phore, phosphor
~, **равно·энергет·и́ческ·ий** equal-energy luminophore
~, **с·ло́ж·н·ый** multiple/complex luminophore
Лю́ммер·а-Бро́дхун·а, куб·и́к· (*m*) Lummer–Brodhun photometer
Лю́ммер·а-Ге́рке, пласт·и́нк·а (*f*) (opt) Lummer–Gehrcke interfero-meter
люмни́т·, -а, (*m*) (nucl) lumnite
люмпе́нус·, -а, (*m*) (zool) lump fish, *Lumpenus*
люне́т·, -а, (*m*) (arch) lunette; (mech) steady, back stay/rest (lathe attach-ment)

люне́т
~, **по·дви́ж·н·ый** travelling steady (of lathe)
люнкери́т·, -а, (*m*) *see* **лункери́т·**
люпи́н·, -а, (*m*) (bot) lupin, *Lupinus*
люпоз·о́ри·й, -я, (*m*) (med) lupus sanatorium
лю́пус·, -а, (*m*) (med) lupus
лю́стр·а, -ы, (*f*) chandelier, (min) cluster lamp/light
~, **про·хо́д·ческ·ая** (min) shaft light
люстр·и́н·, -а, (*m*) (text) lustrine
люте·и́н·, -а, (*m*) (biochem) lutein
люте·инизи́рующ·ий, -ая, -ее, (*a*) (genet) luteinizing, luteinic
люте·и́нов·ый, -ая, -ое, (*a*) lutein
~ **кле́т·к·и** (*pl*) (zool) lutein cells
лютео·стеро́н·, -а, (*m*) (biochem) luteosterone, progesterone
люте́ц·и·й, -я, (*m*) (chem) lutetium, lutecium, Lu
лю́т·ик·, -а, (*m*) (bot) buttercup, *Ranunculus*
лю́тик·ов·ые, -ых, (*pl decl as a*) (bot) crawfoot family, *Ranunculaceae*
люф·а́, -ы́, (*f*) (bot) loofah, luffa, *Luffa*
лю́фт·, -а, (*m*) air gap
люфф·а́, -ы́, (*f*) *see* **люф·а́**
люце́рн·а, -ы, (*f*) (bot) lucerne (Brit), alfalfa (U.S.), *Medicago*
~, **по·се́в·н·ая** alfalfa, *Medicago sativa*
люц·и́т·, -а, (*m*) lucite (transparent plastic)
люци·фера́з·а, -ы, (*f*) (biochem) luciferase
люци·фери́н·, -а, (*m*) (biochem) luciferin
лю́ч·ек·, -ч·к·а, (*m*) (dim) of **люк·**
~, **смотр·ов·о́й** vizor
люч·и́н·а, -ы, (*f*) (naut) hatch cover
~, **лья́ль·н·ые** (*pl*) limber boards
ляг·а́·ть, -ют, (*imp*) kick (of animals); **-ся** kick (habitually)
ляг·н·у́ть, -ут, (*perf*) *see* **ляг·а́·ть**
ля́г·ут· (*fut 3rd pl*) of **леч·ь** (*perf*), *see* **лож·и́ться** (*imp*)
лягуш·а́ч·ий, -ья, -ье, (*a*) of **ля-гу́ш·к·а**
~ **един·и́ц·а** (*f*) (pharm) frog unit
лягу́ш·ечник·, -а, (*m*) (bot) frogbit, *Hydrocharis morsus-ranae*
лягу́ш·к·а, -и, (*g pl*) **-ш·ек·,** (*f*) (zool) frog, *Rana*; diaphragm pump; draw tongs/vice/vise, eccentric clamp, grip; helical door spring; (naut) ferro-concrete mooring anchor
~, **ка́бель·н·ая** (elec) cable grip

ляд·а, -ы, (*f*) closing trap, trap door
ля́двен·ец·, -н·ц·а, (*m*) (bot) Dakota
  vetch, lotus
∼, рог·а́т·ый bird's foot trefoil, *Lotus
  corniculatus*
ля́ж·ет· (*fut 3rd sing*) of **леч·ь** (*perf*),
  see **лож·и́ться**
ля́ж·к·а, -и, (*g pl*) **-ж·ек·,** (*f*) (zool)
  haunch
ля́зг·а·ть, -ют, (*imp*) clang, clank
ля́зг·н·уть, -ут, (*perf*) *see* **ля́зг·а·ть**
ля́йм·, -а, (*m*) (bot) lime, *Citrus amanti-
  folium*
ляллема́нц·иев·ый, -ая, -ое, (*a*) of
  ляллема́нц·и·я
∼ ма́сл·о (*n*) lallemantia oil
ляллема́нц·и·я, -и, (*f*) (bot) *Lallemantia iberica*
ля́мбд·а, -ы, (*f*) λ, lambda (Greek
  letter)

ля́мбд·а
∼ -то́ч·к·а, -и, (*f*) (phys) lambda
  point, λ-point
лямин- *see* **ламин-**
ля́мин·а, -ы, (*f*) lamina
ля́минг·ов·а ма́сс·а (*f*) Laming mass
  (by-product of coke-oven gas purification; used as raw material for
  sulphur production)
ля́м·к·а, -и, (*g pl*) **-м·ок·,** (*f*) strap,
  lifting band; drag rope (barges)
ля́п·алк·а, -и, (*g pl*) **-лок·,** (*f*) (hortic)
  tamper
ля́пис·, -а, (*m*) (chem) silver nitrate
∼ -лазу́р·ь, -и, (*f*) (min) lapis lazuli
ля́рд·, -а, (*m*) (food) lard
ля́рд·ов·ый, -ая, -ое, (*a*) lard
ляринги́т·, -а, (*m*) *see* **ларинги́т·**
лятиме́р·и·я, -и, (*f*) (fish) coelacanth,
  Latimeria chalumnae
лятиля́мин·а, -ы, (*f*) (zool) latilamina

М

М (*abbr*) = **Ма́ха**   (air) Mach (number)

М (*abbr*) = **ме́га**   Mega-, m

М (*abbr*) = **Москва́**   Moscow

м (*abbr*) = **метр**   metre, meter, m

М-ме́тр·, -а,   (*m*) (air) Mach-number meter/gauge/gage

М-числ·о́, -а,   (*n*) (air) Mach number

ма (*abbr*) = **милли·ампе́р·** mA, milliampere, milliamp

мавзол·е́·й, -я,   (*m*) (arch) mausoleum

магази́н·, -а,   (*m*) (com) shop, store; magazine, warehouse; magazine

~ ём·кост·ей   (elec) capacitor box

~ за·тух·а́ни·я   (elec) attenuator pad

~ индукти́в·ност·ей   (elec) inductor

~, магни́т·н·ый   built-up magnet

~ со·противл·е́ни·я   (elec) resistance box

~ со·противл·е́ни·я, рыча́ж·н·ый (elec) dial-switch resistance box

~, универса́ль·н·ый   (com) department store

магазин·и́ровани·е, -я,   (*n*) (min) shrinkage stoping

магази́н·н·ый, -ая, -ое, (*a*) of **мага-зи́н·**

~ коро́б·к·а   (*f*) magazine (rifle)

~ систе́м·а   (*f*) (min) shrinkage stoping

магди́но   (*n indecl*) (elec) magdyno

маги́стр·, -а,   (*m*) master (non-Soviet academic degree)

магистра́л·ь, -и,   (*f*) main, main line, line, supply line; (telecom) trunk line; highway; base line (river velocity metrology)

~, ваго́н·н·ая   (elec rail) bus-line

~, вы·тяж·н·а́я   (shipb) exhaust main

~, гла́в·н·ая   main gallery (pressure lubrication)

~, диферент·о́вочн·ая   (shipb) trimming line

~, за́·мк·нут·ая   (elec) ring main

~, кольц·ев·а́я   (elec) ring main

~, на·гнет·а́тельн·ая   force/pressure piping/main; (shipb) oil-fuel service piping

~, на·по́р·н·ая   pumping/pressure/force main

~, по·да·ю́щ·ая   supply main

~, раз·ветвл·ённ·ая   (rad) herringbone system, Christmas-tree system

~, с·лив·н·а́я   main return pipe

~, тормоз·н·а́я   (rail) train pipe

магистра́л·ь

~, туп·ик·о́в·ая   dead-end main; (rad) herringbone system, Christmas-tree system

магистра́л·ь·н·ый, -ая, -ое,   (*a*) of **магистра́л·ь;** main, arterial

~ аппара́т координи́рованн·ого регули́ровани·я   (*m*) synchronized traffic lights

~ коро́б·к·а   (*f*) (elec) mains distribution box

~ с·вяз·ь   (*f*) (telecom) trunkline service

маги́ческ·ий, -ая, -ое,   (*a*) magic

~ глаз·   (*m*) (rad) magic eye

ма́гм·а, -ы,   (*f*) (geol) magma

магмат·и́ческ·ий, -ая, -ое,   (*a*) (geol) magmatic

магнаво́льт·, -а,   (*m*) (elec) magnavolt-type cascade exciter

магна́л·иев·ый, -ая, -ое,   (*a*) (met) magnalium

магна́л·и·й, -я,   (*m*) (met) magnalium

магнал·и́т·, -а,   (*m*) (met) magnalite

магнез·иа́льн·ый, -ая, -ое,   (*a*) magnesia, magnesic, magnesian

~ вяз·ущ·ий материа́л·   (*m*) Sorel's cement, magnesia cement

~ с·мес·ь   (*f*) (chem) magnesia mixture

магнез·и́т·, -а,   (*m*) (min) magnesite

магнези́т·ов·ый, -ая, -ое,   (*a*) of **магнез·и́т·**

~ кирпи́ч·   (*m*) magnesite brick

магнезит·хроми́т·, -а,   (*m*) chrome-magnesite

магне́з·и·я, -и,   (*f*) magnesia, magnesium oxide

~, англ·и́йск·ая   sulphate of magnesium

~, бе́л·ая   magnesia alba

~, жж·ённ·ая   burnt magnesia, magnesium oxide

~, угле·ки́сл·ая   basic magnesium carbonate

магнеси́н·, -а,   (*m*) (elec) magnesyn, delta-wound single-phase synchro

магнеста́т·, -а,   (*m*) (elec) magnestat, magnetic amplifier

магнет·и́зм·, -а,   (*m*) magnetism

~, есте́ств·енн·ый   natural/spontaneous magnetism

~, зем·н·о́й   terrestrial magnetism, geomagnetism

магнет·и́к·, -а,   (*m*) magnet, magnetic material

магнет·и́т·, -а,   (*m*) (min) magnetite, magnetic iron ore

магнети́т·ов·ый, -ая, -ое,   (*a*) of **магнет·и́т·**

магнет·и́ческ·ий, -ая, -ое, (*a*) magnetic

магне́то (*n indecl*) magneto, magneto-electric generator; (*as prefix*) *see* **магнито-**

магнето́·метр·, -а, *see* **магнито́·-метр·**

магнетон·, -а, (*m*) magnetor (unit of magnetic moment)

∼ Бо́р·а Bohr magneton

магнето·плюмби́т·, -а, (*m*) (min) magnetoplumbite

магнетор·, -а, (*m*) magnettor

магнето·стат·и́ческ·ий, -ая, -ое, (*a*) magnetostatic

магнето·хи́м·и·я, -и, (*f*) magneto-chemistry

магнетро́н·, -а, (*m*) (elec) magnetron

∼, двух·раз·рез·н·о́й two-segment magnetron

∼, лопа́т·очн·ый vane-anode magnetron

∼, много·ка́мер·н·ой multiresonator/multicavity magnetron

∼, много·раз·рез·н·о́й multisegment magnetron

∼, много·резона́тор·н·ый multicavity/multiresonator magnetron

∼, разно·резона́тор·н·ый rising-sun magnetron

∼, резона́тор·н·ый cavity magnetron

∼ с раз·рез·н·ы́м ано́д·ом split-anode magnetron

∼ со с·вя́з·к·ами strapped anode

∼, стерж·нев·о́й interdigital magnetron

∼, щел·ев·о́й slot-anode magnetron

магнетро́н·н·ый, -ая, -ое, (*a*) of **магнетро́н·**

ма́гн·иев·ый, -ая, -ое, (*a*) of **ма́г-н·и·й**

ма́гн·и·й, -я, (*m*) (chem) magnesium, Mg

∼, азо́т·ист·ый magnesium nitride

∼, серноватисто·ки́сл·ый magnesium thiosulphate

∼, серно·ки́сл·ый magnesium sulphate

∼, угле·ки́сл·ый magnesium carbonate

∼, хло́р·ист·ый magnesium chloride

∼, хлорно·ки́сл·ый magnesium perchlorate

магний·алки́ль·н·ый, -ая, -ое, (*a*) (chem) magnesium alkyl

магний·орган·и́ческ·ий, -ая, -ое, (*a*) (chem) organomagnesium

магнико (*n indecl*) (met) Magnico (alloy)

магнико́н·, -а, (*m*) (elec) magnicon

магнисто́р·, -а, (*m*) (elec) magnistor

магни́т·, -а, (*m*) magnet

∼, без·желе́з·н·ый air-core magnet

∼, воз·бужд·а́ющ·ий field magnet

∼, есте́ств·енн·ый lodestone

∼, за·ду́в·очн·ый blow-out/blowing magnet

∼, за·медл·я́ющ·ий timing magnet

∼, из·мер·и́тельн·ый permanent magnet (of compass)

∼, крен·ов·о́й heeling magnet (of ship's compass)

∼, подково·обра́з·н·ый horseshoe magnet

∼, по·сто·я́нн·ый permanent magnet

∼, пут·ев·о́й (autom) application magnet

∼, стержн·ев·о́й bar/axial magnet

∼, тормоз·н·о́й drag magnet

∼, электр·и́ческ·ий electromagnet

магни́т·ик·, -а, (*m*) small magnet

∼, элемента́р·н·ый elementary magnet

магни́т·н·о-жёстк·ий, -ая, -ое, (*a*) magnetically hard

∼ -мя́гк·ий, -ая, -ое, (*a*) magnetically soft

магни́т·ност·ь, -и, (*f*) magnetization; magnetizability

магни́т·н·ый, -ая, -ое, (*a*) magnetic, magnet

∼ анома́ли·я, -и, (*f*) magnetic anomaly

∼ вес·ы́ (*pl*) magnetometric balance

∼ воз·мущ·е́ни·е (*n*) magnetic perturbation

∼ вос·при·и́м·чивост·ь (*f*) magnetic susceptibility, coefficient of magnetization

∼ вя́з·кост·ь (*f*) magnetic viscosity/after-effect

∼ гидро·дина́мик·а (*f*) magneto hydrodynamics

∼ гистере́зис· (*m*) magnetic hysteresis

∼ голо́в·к·а (*f*) magnetic head (of recording machine)

∼ двой·н·о́й сло́·й (*m*) magnetic shell

∼ дубле́т· (*m*) magnetic dipole/doublet

магни́т·н·ый, -ая, -ое
~ **дут·ъ·ё** (*n*) magnetic blowout
~ **желез·няк·** (*m*) (min) magnetite, magnetic iron ore
~ **жёстк·ост·ь** (*f*) magnetic rigidity
~ **за·щи́т·а** (*f*) magnetic screening/shielding
~ **инду́кц·и·я** (*f*) magnetic induction, flux density
~ **колчеда́н·** (*m*) (min) pyrrhotite, magnetic pyrite
~ **лист·о́к** (*m*) magnetic shell
~ **моме́нт·** (*m*) magnetic moment
~ **на·клон·е́ни·е** (*n*) dip, magnetic inclination
~ **на·пряж·ённост·ь** (*f*) magnetic field strength, magnetizing force
~ **на·сыщ·е́ни·е** (*n*) saturation induction, magnetic saturation
~ **от·клон·е́ни·е** (*n*) magnetic deflection
~ **по́л·е** (*n*) magnetic field
~ **по́л·е, вращ·а́ющ·ееся** (*n*) rotating field
~ **по́л·е рас·се́·яни·я** (*n*) magnetic stray field
~ **по́люс·** (*m*) magnetic pole
~ **по́люс· се́вер·н·ый** (*m*) (geog) magnetic north
~ **поляриз·а́емост·ь** (*f*) magnetic polarizability
~ **после·де́й·стви·е** (*n*) magnetic after-effect
~ **по·то́к·** (*n*) magnetic flux
~ **про·вод·и́мост·ь** (*f*) permeance, magnetic conductance
~ **про·ниц·а́емост·ь** (*f*) permeability, magnetic permeability
~ **рас·се́·яни·е** (*n*) (elec) magnetic leakage, leakage flux; (nucl) magnetic scattering
  коэффицие́нт· магни́т·н·ого рас·-се́·яни·я (*m*) (elec eng) leakage factor/coefficient
~ **си́л·а** (*f*) magnetic force
~ **с·клон·е́ни·е** (*n*) magnetic declination/variation
  ка́рт·а магни́т·н·ых с·клон·е́ни·й (*f*) (surv) magnetic map
~ **со·противл·е́ни·е** (*n*) magnetic resistivity, reluctivity, reluctance
~ **с·полз·а́ни·е** (*n*) magnetic creeping
~ **стал·ь** (*f*) magnet steel
~ **ста́нц·и·я у·правл·е́ни·я** (*f*) electric-motor control set
~ **у·сил·е́ни·е** (*n*) magnetic intensification

магни́т·н·ый, -ая, -о е
~ **у·с·поко·е́ни·е** (*n*) magnetic damping
~ **у·с·поко·и́тел·ь** (*m*) eddy-current brake
~ **цеп·ь** (*f*) magnetic circuit
магнито- (*component*) magneto, magnetic
магнито·акти́в·н·ый, -ая, -ое, (*a*) magnetoactive
магнито·газо·дина́мик·а, -и, (*f*) magneto-gas dynamics
магнито·гидро·динам·и́ческ·ий, -ая, -ое, (*a*) (phys) magnetohydrodynamic
магнито·гра́мм·а, -ы, (*f*) magnetograph recording; (met) temperature/magnetization curve
магнито́·граф·, -а, (*f*) magnetograph
магнито·дви́ж·ущ·ая си́л·а (*f*) magnetomotive force
магнито·дина́мик·а, -и, (*f*) (phys) magnetohydrodynamics
магнито·ди·эле́ктр·ик·, -а, (*m*) (phys) ferrite
магнито·калор·и́ческ·ий, -ая, -ое, (*a*) magnetocaloric
~ **эффе́кт·** (*m*) magnetocaloric effect
магнито́·метр·, -а, (*m*) magnetometer
~, **вибрацио́н·н·ый** vibrating-coil magnetometer
~, **дефле́ктор·н·ый** deflection magnetometer
~, **крут·и́льн·ый** torque-coil magnetometer
~ **с на·сыщ·ённ·ым серд·е́чник·ом** flux-gate magnetometer
магнито·метр·и́ческ·ий, -ая, -ое, (*a*) magnetometer, magnetometric
магнито·ме́тр·и·я, -и, (*f*) magnetometry; (min) magnetic prospecting
магнито·механ·и́ческ·ий, -ая, -ое, (*a*) magnetomechanical
~ **от·нош·е́ни·е** (*n*) magnetomechanical/gyromagnetic ratio
магнито·модуляцио́н·н·ый, -ая, -ое, (*a*) magnetic modulation
магнито·нос·и́тел·ь, -я, (*m*) magnetic recording medium
магнито·о́пт·ик·а, -и, (*f*) magneto-optics
магнито·опт·и́ческ·ий, -ая, -ое, (*a*) magneto-optical
магнито·про́·вод·, -а, (*m*) (elec) magnetic core
магнито·раз·ве́д·к·а, -и, (*f*) (geol) magnetic prospecting

магни́то·стат·и́ческ·ий, -ая, -ое, (*a*) magnetostatic

магни́то·стрикцио́н·н·ый, -ая, -ое, (*a*) magnetostriction, magnetostrictive

магни́то·стри́кц·и·я, -и, (*f*) magnetostriction

~, объ·ём·н·ая Barrett effect, volume magnetostriction

магни́то·у·пру́г·и·й, -ая, -ое, (*a*) magnetoelastic

магни́то·тепл·ов·о́й, -а́я, -о́е, (*a*) magnetothermal

магни́то·фо́н·, -а, (*m*) magnetic sound-recorder, tape-recorder, wire-recorder, magnetophone

магни́то·хи́м·и·я, -и, (*f*) magneto-chemistry

магни́то·электр·и́ческ·ий, -ая, -ое, (*a*) magneto-electric (of machines); dynamic, moving-coil (of instruments)

~ маши́н·а (*f*) magneto electric generator, magneto-alternator

магни́то·эмульсио́н·н·ая му́фт·а (*f*) (mech eng) magnetic fluid clutch

магно́л·и́ев·ые, -ых, (*pl decl as adj*) (bot) magnolia family

магноли́т·, -а, (*m*) magnesite flooring cement

магно́л·и·я, -и, (*f*) (bot) magnolia

магно́н·, -а, (*m*) (nucl) magnon

Ма́гнус·а, эффе́кт· (*m*) (dynam) Magnus effect

мадапола́м·, -а, (*m*) (text) madapolam

маде́р·а, -ы, (*f*) Madeira, Madeira-type wine

ма́ж·а (*pres gerund*) of **ма́з·ать**

ма́ж·ет (*pres 3rd sing*) of **ма́з·ать**

мажеф·, -а, (*m*) Mazhef (manganese-iron phosphate antirust compound)

мажо́р·, -а, (*m*) major key (music)

мажора́нт·а, -ы, (*f*) (math) majorante

мажора́нт·н·ый, -ая, -ое, (*a*) (math) majorant

мажор·и́ровани·е, -я, (*n*) majorization

мажор·и́р·овать, -уют, (*imp and perf*) (math) to be/to find a majorant

мажор·и́рованн·ый, -ая, -ое, (*past part pass*) of **мажор·и́р·овать;** having as a majorant

ма́ж·ущ·ий, -ая, -ее, (*pres part act*) of **ма́з·ать**

ма́з·анн·ый, -ая, -ое, (*past part pass*) of **ма́з·ать**

ма́з·ать, (*pres 3rd sing, pl*) **ма́ж·ет, ма́ж·ут,** (*imp*) grease, smear, coat (with oil etc.), daub

мазе·обра́з·н·ый, -ая, -ое, (*a*) greasy, salve-like, unctious

ма́зер·, -а, (*m*) (phys instr) maser

маз·н·у́ть, -у́т, (*perf, single action*) *see* **ма́з·ать**

маз·о́к·, -з·к·а́, (*m*) smear, dab, daub

мазони́т·, -а, (*m*) Masonite (a fibreboard)

мазу́т·, -а, (*m*) (oil) residual oil; fuel oil

~, кре́кинг- cracked residue

~, ма́сл·ян·ый lubricating oil stock

~ пря́м·ой го́н·к·и topped crude

~, с·ма́з·очн·ый unrefined residual lubricating oil

~ -то́пл·ив·о fuel oil

~, то́п·очн·ый furnace fuel oil; boiler fuel

~, фло́т·ск·ий marine oil fuel, bunker fuel

мазу́т·н·ый, -ая, -ое, (*a*) of **мазу́т·**

~ о·топл·е́ние (*n*) oil firing

мазу́то·про·во́д·, -а, (*m*) residual oil transportation equipment

ма́з·ь, -и, (*f*) grease; ointment, liniment, salve, unquent

~, графи́т·н·ая graphited grease

~, консисте́нт·н·ая grease

~, копы́т·н·ая (pharm) neat's foot oil

~, рту́т·н·ая ointment of mercury

ма́йс·, -а, (*m*) *see* **кукуру́з·а**

ма́йс·ов·ый, -ая, -ое, (*a*) of **маис·**

ма́·й, -я, (*m*) May

майда́н·, -а, (*m*) tar pit

Ма́йер·а, у·равн·е́ни·е (*n*) (phys) Mayer equation

Ма́йкельсон·а, эшело́н· (*m*) (opt) Michelson interferometer

майо́лик·а, -и, (*f*) china-metal

майо́н·, -а, (*m*) intermediate link (bucket dredger)

майоне́з·, -а, (*m*) (food) mayonnaise, salad cream

майо́р·, -а, (*m*) (mil) major

майора́н·, -а, (*m*) (bot) marjoram, *Origanum majorana*

Майоран·а, си́ла (*f*) (phys) Majorana force

ма́й·ск·ий, -ая, -ое, (*a*) of **ма·й**

майтланди́т·, -а, (*m*) (min) mainlandite

ма́к·, -а, (*m*) (bot) poppy, *Papaver*

~, сно·тво́р·н·ый opium poppy, *Papaver somniferum*

макáк·а, -и, (*f*) (zool) macaque, *Macacus*

макалýб·а, -ы, (*f*) (geog) mud volcano

мак·áни·е, -я, (*n*) dipping, dip

мáк·ан·ый, -ая, -ое, (*a*) dipped

~ **из·дéл·и·е** (*n*) dipped article (rubber)

макарóн·н·ый, -ая, -ое, (*a*) of **макарóн·ы;** (food) noodles

~ **из·дéл·и·я** (*pl*) noodles (includes macaroni, spaghetti etc.)

макарóн·ы, -он, (*f pl*) (food) macaroni

мак·áтельн·ый, -ая, -ое, (*a*) dipping

мак·á·ть, -ют, (*imp*) dip in/into, dip

макéт·, -а, (*m*) model, mock-up, dummy, simulator

~ **кáрт·ы** (surv) map montage

макéт·н·ый, -ая, -ое, (*a*) of **макéт·;** simulated

макинтóш·, -а, (*m*) mackintosh, waterproof coat

макинтошúт·, -а, (*m*) (min) mackintoshite

Мáк-лебд·а, манó·метр· (*m*) McLeod (pressure) gauge/gage

мáклер·, -а, (*m*) (fin) broker

мáклер·ств·о, -а, (*n*) (fin) broking, brokerage (business)

макловúц·а, -ы, (*f*) thick-distemper brush

Маклорен·а, ряд· (*m*) (math) Maclaurin series

маклю́р·а, -ы, (*f*) (bot) osage-orange, *Maclura*

мáк·овк·а, -и, (*g pl*) **-вок·,** (*f*) poppy head; (arch) dome, cupola; crown, top, summit

мáк·ов·ый, -ая, -ое, (*a*) of **мак·;** (bot) poppy family, (*pl*) *Papaveraceae*

макрéл·ев·ые, -ых, (*pl decl as adj*) (fish) mackerels, *Scombridae*

макреле·щýк·и, -ук·, (*pl*) Hound fishes, (fish) *Belonidae*

макрéл·ь, -и, (*f*) (fish) mackerel, *Scomber*

макро·би·пирамúд·а, -ы, (*f*) (cryst) macrobipyramid

макро·дóм·а, -ы, (*f*) (cryst) macrodome, rhombic prism

макро·климáт·, -а, (*m*) (meteor) climate on a global scale, macroclimate

макро·коррóз·и·я, -и, (*f*) (met) general surface corrosion

макро·кóсм·, -а, (*m*) macrocosm

макро·мéр·, -а, (*m*) (biol) macromere

макрó·метр·, -а, (*m*) (surv) macrometer

макро·молéкул·а, -ы, (*f*) (chem) macromolecule

макро·нýклеус·, -а, (*m*) (zool) macronucleus

макро·на·пряж·éни·е, -я, (*n*) macrostress

макро·о·стáт·к·и, -т·ок·, (*pl*) macroremains

макро·óс·ь, -и, (*f*) (cryst) macro-axis

макро·педиóн·, -а, (*m*) (cryst) macropedion

макро·пирамúд·а, -ы, (*f*) (cryst) macropyramid, rhombic pyramid

макро·пóр·а, -ы, (*f*) macropore

макро·сегрег·áци·я, -и, (*f*) (met) macrosegregation

макро·скоп·úческ·ий, -ая, -ое, (*a*) macroscopic, megascopic

~ **винт·** (*m*) (instr) coarse-adjustment screw

макро·с·ним·óк·, -м·к·á, (*m*) (phot) macrograph, macrophotograph

макро·структ·ýр·а, -ы, (*f*) (met) macrostructure

макро·структ·ýрн·ое ис·слéд·овани·е (*n*) (met) macrography

макро·фáг·, -а, (*m*) (biol) macrophage

макро·фото·съ·ём·к·а, -и, (*g pl*) **-м·ок·,** (*f*) macrophotography

макро·шлúф·, -а, (*m*) (met) macrosection

мáксвелл·, -а, (*m*) maxwell (unit of magnetic flux)

~ **-вит·óк·, -т·к·á,** (*m*) (elec) Maxwell turn

Мáксвелл·а, за·кóн· рас·пре·дел·é·ни·я (*m*) (phys) Maxwellian distribution, Maxwell-Boltzmann distribution

~, **теорéм·а** (*f*) reciprocal theorem (structural mechanics)

~, **теóр·и·я** (*f*) (elec) Maxwell's rule

~, **у·равн·éни·я** (*pl*) (elec) Maxwell's equations

максвелл·мéтр·, -а, (*m*) (phys) maxwellmeter

мáксвелл·овск·ий, -ая, -ое, (*a*) (elec) Maxwellian

мáксим·а, -ы, (*f*) maxim

максимáль·н·о (*adv*) to the maximum extent, at a maximum, to/at the highest possible

максимáль·н·ый, -ая, -ое, (*a*) maximum; (elec) peak, overload

максима́ль·н·ый, -ая, -ое
~ **знач·е́ни·е** (*n*) maximum value, (elec) peak value
~ **реле́** (*n*) (elec) overcurrent/overload relay
~ **тари́ф** (*m*) (elec) maximum-demand tariff
~ **то́к·ов·ая за·щи́т·а** (*f*) (elec) overcurrent protection
макси́·метр·, -а, (*m*) injector nozzle tester (diesels)
ма́ксимум·, -а, (*m*) maximum, peak; (meteor) anticyclone
~**, азо́р·ск·ий** (meteor) Azores anticyclone
~**, бар·и́ческ·ий** (meteor) anticyclone
~**, горбо·обра́з·н·ый** hump (on a curve)
~**, диффракцио́н·н·ый** (phys) diffraction peak
~ **рас·се́·яни·я вперёд** (phys) forward peak
~**, ре́з·к·ий** spike (on a curve)
~**, сла́б·ый** flat peak (on a curve)
~ **слыш·и́мост·и** (rad) position of maximum signal strength
ма́кул·а, -ы, (*f*) (anat) macula
макулату́р·а, -ы, (*f*) (print) waste paper; trash
макулату́р·н·ый карто́н· (*m*) chip board
маку́х·а, -и, (*f*) oil/cattle cake
маку́ш·к·а, -и, (*g pl*) **-ш·ек·,** (*f*) top, summit, (anat) crown, vertex; apex (of shell); umbo (of mollusc etc.)
мала́г·а, -и, (*f*) Malaga, Malaga-type wine
малако·ло́г·и·я, -и, (*f*) (zool) malacology
малако·ме́тр·и·я, -и, (*f*) malacometry, measurement of softness and consistency
малако́н·, -а, (*m*) (min) malacon
малак·остра́к·и (*pl*) (zool) malacostracs, *Malacostraca*
малако·фил·и́·я, -и, (*f*) (bot) malacophily
малатио́н·, -а, (*m*) (chem) malathion
малахи́т·, -а, (*m*) (min) malachite
малахи́т·ов·ый, -ая, -ое, (*a*) malachite
малеа́т·, -а, (*m*) (chem) maleate
мал·ева́ть, -ю́ют, (*imp*) rough out (in paints), paint a cartoon
малеи́н·ов·ый, -ая, -ое, (*a*) (chem) malein, maleic
~ **кисл·от·а́** (*f*) (chem) maleic acid, *cis*-butanedoic acid

мал·е́йш·ий, -ая, -ее, (*a*) (*superl*) of **ма́л·ый**; least, slightest
мал·ёк·, -ль·к·а́, (*m*) *see* **маль·к·и́**
ма́л·еньк·ий, -ая, -ое, (*a*) small, little
мал·и́н·а, -ы, (*f*) (bot) raspberry, raspberry bush, *Rubus*
мали́н·н·ый, -ая, -ое, (*a*) of **мал·и́н·а**; raspberry
~ **жук·** (*m*) (ent) raspberry fly, *Trixagidae*
мали́н·овк·а, -и, (*g pl*) **-вок·,** (*f*) (orn) robin, *Erithacus rubecula*
мали́н·ов·ый, -ая, -ое, (*a*) of **мал·и́н·а**; raspberry; crimson, raspberry-coloured
ма́л·к·а, -и, (*g pl*) **-л·ок·,** (*f*) bevel (tool)
~**, раз·вод·н·а́я** standing bevel
~**, с·вод·н·ая** close bevel
малк·ова́ть, -у́ют, (*imp*) bevel
маллеи́н·, -а, (*m*) (pharm) mallein
маллеин·иза́ци·я, -и, (*f*) (vet) mallein inoculation
малли́л·, -а, (*m*) (chem) diallyl barbituric acid
ма́л·о (*adv*) few, little, not enough, insufficient, too little, not very much
мало- (*root*) low-; small-; little, rarely, seldom, hardly, insufficiently; under-, sparsely, poorly, -deficient, -poor; not very, only slightly, un-
мало·авар·и́йн·ое о·свещ·е́ни·е (*n*) (naut) emergency lighting
мало·акти́в·н·ый, -ая, -ое, (*a*) low-activity; (nucl) low-level, cold
мало·алкого́ль·н·ый, -ая, -ое, (*a*) low-alcohol
мало·ампе́р·н·ый, -ая, -ое, (*a*) (elec) low-amperage/current
мало·бел·ко́в·ый, -ая, -ое, (*a*) protein-deficient
мало·берцо́в·ый, -ая, -ое, (*a*) (anat) peroneal, fibular
~ **кост·ь** (*f*) (anat) fibula
мало·ва́жн·ый, -ая, -ое, (*a*) unimportant, of little importance, not very important
мало·ва́тт·н·ый, -ая, -ое, (*a*) (elec) low-watt, low-wattage
мало·вероя́т·н·ый, -ая, -ое, (*a*) improbable, unlikely
мало·ве́с·н·ый, -ая, -ое, (*a*) light, low-weight
мало·ве́тр·ен·ый, -ая, -ое, (*a*) (meteor) light
мало·ве́тр·и·е, -я, (*n*) (meteor) light air

мало·во́д·н·ый, -ая, -ое, (*a*) shallow, low (of water level); having little water; (geog) dry

~ пе́ре·кис·ь водо·ро́д·а (*f*) high-test hydrogen peroxide

мало·во́д·ь·е, -я, (*n*) water shortage, lack of water

мало·вы́·год·н·ый, -ая, -ое, (*a*) unprofitable, disadvantageous, hardly profitable

мало·вя́з·к·ий, -ая, -ое, (*a*) low-viscosity, non-viscous

мало·габари́т·н·ый, -ая, -ое, (*a*) small

мало·гра́мот·н·ый, -ая, -ое, (*a*) uneducated, semi-literate

мало·групп·ов·а́я тео́р·и·я диф-фу́з·и·я (*f*) (phys) few-group diffusion theory

мало·дей·стви́тельн·ый, -ая, -ое, (*a*) not very effective

мало·до·каз·а́тельн·ый, -ая, -ое, (*a*) not very convincing

мало·до·сту́п·н·ый, -ая, -ое, (*a*) hard to reach/get at, not very accessible, almost inaccessible

мало·до·хо́д·н·ый, -ая, -ое, (*a*) (com) low-yield, unprofitable, un-rewarding

мало·ём·кост·н·ый, -ая, -ое, (*a*) (elec) anti-capacitance

мало·за·ме́т·н·ый, -ая, -ое, (*a*) unobtrusive, barely noticeable, almost imperceptable

мало·за·сел·ённ·ый, -ая, -ое, (*a*) sparsely populated

мало·земе́л·ь·е, -я, (*n*) (agr) land shortage/hunger

мало·земе́ль·н·ый, -ая, -ое, (*a*) of мало·земе́л·ь·е

~ ту́ндр·а (*f*) (geog) desert tundra

мало·знако́м·ый, -ая, -ое, (*a*) little-known, not well-known

мало·знач·и́тельн·ый, -ая, -ое, (*a*) not important, unimportant, of no great significance

мало·зо́ль·н·ый, -ая, -ое, (*a*) low-ash

мало·из·ве́ст·н·ый, -ая, -ое, (*a*) little-known

мало·инерцио́н·н·ый, -ая, -ое, (*a*) (instr) low-inertia, quick-response

мало·интенси́в·н·ый, -ая, -ое, (*a*) low-intensity, weak

мало·ис·сле́д·ованн·ый, -а , -ое, (*a*) little-studied, insufficien ly investigated

мало·кали́бер·н·ый, -ая, -ое, (*a*) (gunn) small calibre/bore

мало·квалифиц·и́рованн·ый, -ая, -ое, (*a*) poorly/barely qualified; semi-skilled

мало·кро́в·и·е, -я, (*n*) (med) anaemia, anemia, oligemia

мало·кро́в·н·ый, -ая, -ое, (*a*) anaemic, anemic

мало·леги́р·ованн·ый, -ая, -ое, (*a*) (met) low-alloy

мало·ле́с·н·ый, -ая, -ое, (*a*) (geog) sparsely wooded

мало·ле́т·н·ий, -яя, -ее, (*a*) juvenile, young; (law) minor

мало·ле́т·ник·и (*m pl*) (bot) biennials and annuals

мало·литр·а́жн·ый, -ая, -ое, (*a*) (ICE) small (of small cubic capacity), low-powered

мало·лю́д·н·ый, -ая, -ое, (*a*) sparsely inhabited; almost deserted; poorly attended (of meetings)

мало·ма́сл·ян·ый, -ая, -ое, (*a*) oil-poor, low-oil-content

~ вы·ключ·а́тел·ь (*m*) (elec) low-oil-content circuit-breaker

мало·ме́р·н·ый, -ая, -ое, (*a*) small; undersized, too small

мало·мо́щ·н·ый, -ая, -ое, (*a*) weak, not strong enough; poor, lacking sufficient funds; low-power/duty (of machines); poor, shallow (of soils)

мало·на·сел·ённ·ый, -ая, -ое, (*a*) sparsely populated

малон·а́т·, -а, (*m*) (chem) malonate

мало́н·ов·ый, -ая, -ое, (*a*) (chem) malonic

~ кисл·от·а́ (*f*) (chem) malonic acid

~ эфи́р· (*m*) malonic esther

мало·на·де́ж·н·ый, -ая, -ое, (*a*) unreliable

мало·об·луч·ённ·ый, -ая, -ое, (*a*) slightly irradiated

мало·о·богащ·ённ·ый, -ая, -ое, (*a*) slightly enriched

мало·о·па́с·н·ый, -ая, -ое, (*a*) low-hazard

мало·о́·пыт·н·ый, -ая, -ое, (*a*) inexperienced

мало·основ·а́тельн·ый, -ая, -ое, (*a*) barely substantiated; unreliable

мало·от·ве́т·свенн·ый, -ая, -ое, (*a*) low-duty (of machines)

мало·поля́р·н·ый, -ая, -ое, (*a*) low-polarity

мало·по·ня́т·н·ый, -ая, -ое,   (*a*) obscure, difficult to understand

мало·продукти́в·н·ый, -ая, -ое,   (*a*) inefficient

мало·проце́нт·н·ый, -ая, -ое,   (*a*) low-yield; low-grade

мало·про́ч·н·ый, -ая, -ое,   (*a*) low-duty (of materials)

мало·ра́з·ви·т·ый, -ая, -ое, (*a*) under-developed, backward

мало·рас·пре·дел·ённ·ый, -ая, -ое, (*a*) (nucl) low-abundance

мало·рас·твор·и́м·ый, -ая, -ое, (*a*) barely soluble, only slightly soluble

мало·ресни́ч·н·ые, -ых, (*pl decl as a*) (zool) oligotrichida

мало·ро́сл·ый, -ая, -ое, (*a*) under-sized, stunted, dwarfish, nanoid

мало·све́т·и·е, -я,   (*n*) underlighting, insufficient lighting

мало·сер·ни́ст·ый, -ая, -ое, (*a*) low-sulphur/sulfur

мало·си́ль·н·ый, -ая, -ое,   (*a*) weak, not strong, low-power

мало·со·держ·а́тельн·ый, -ая, -ое, (*a*) superficial, lacking in content

мало·со́ль·н·ый, -ая, -ое, (*a*) (food) fresh-salted

мало·сто́й·к·ий, -ая, -ое, (*a*) (chem) not stable, non-persistant

мало·стро́ч·н·ый, -ая, -ое, (*a*) (TV) low-definition

ма́л·ост·ь, -и, (*f*) triviality

мало·тира́ж·н·ый, -ая, -ое,   (*a*) (print) small circulation, small (of edition)

мало·тонна́ж·н·ый, -ая, -ое,   (*a*) small (of small tonnage)

мало·угле·ро́д·ист·ый, -ая, -ое, (*a*) low-carbon; mild (of steel)

мало·у·бед·и́тельн·ый, -ая, -ое, (*a*) not very convincing

мало·у·по·треб·и́тельн·ый, -ая, -ое, (*a*) little-used, rarely used

мало·формати́в·н·ый, -ая, -ое, (*a*) (phot) 35-mm

мало·фо́сфор·ист·ый, -ая, -ое, (*a*) low-phosphorus

~ чугу́н· (*m*) hematite iron

мало·чи́сл·енн·ый, -ая, -ое, (*a*) few, scanty; small (of organizational units)

ма́л·очник·, -а, (*m*) (*dim*) of **ма́л·к·а**

ма́л·очн·ый, -ая, -ое, (*a*) of **ма́л·к·а**

~ доск·а́ (*f*) bevel board

мало·чувств·и́тельный, -ая, -ое, (*a*) low-sensitivity

мало·шу́м·н·ый, -ая, -ое, (*a*) low-noise, noiseless

ма́л·ый, -ая, -ое, (*a*) small, low, minor; slow (speed)

~, бес·кон·е́чн·о infinitesimal

в ма́л·ом locally

~ вод·а́ (*f*) (ocean) low tide

~ да́ль·ност·ь (*f*) short range

~ медве́д·иц·а (*f*) (astron) Ursa Minor, the little bear

~ ос·ь (*f*) minor axis

~ ос·ь рас·се́·ивани·я (*f*) (gunn) lateral axis of zone of dispersion

~ пёс· (*m*) (astron) Canis Minor

~ ход· (*m*) slow speed

ма́льв·а, -ы, (*f*) (bot) mallow, *Malva*

мальвиди́н·, -а, (*m*) (pharm) malvidin

маль·к·и́, (*nom pl*) of **мал·ёк·;** (fish) fry

маль·ко́в·ый, -ая, -ое, (*a*) of **маль·к·и́**

мальм·, -а, (*m*) (build) malm, washed clay

ма́льт·а, -ы, (*f*) (min) maltha, viscous bitumin; (geog) Malta

мальт·а́з·а, -ы, (*f*) (chem) maltase

мальт·и́йск·ий, -ая, -ое, (*a*) Maltese

~ крест· (*m*) Maltese cross (mechanism)

~ механи́зм·, (*m*) Maltese cross (mechanism)

мальт·о́з·а, -ы, (*f*) (chem) maltose

мальтузиа́н·ск·ий, -ая, -ое,   (*a*) (econ) Malthusian

мальхи́т·, -а, (*m*) (geol) malchite

ма́льц-экстра́кт·, -а, (*m*) (food) malt extract

ма́ль·чик·, -а, (*m*) boy

Малю́с·а, за·ко́н·, (*m*) (opt) Malus' law

мал·ю́тк·а, -и, (*g pl*) **-ток·,** (*f*) baby

мал·ю́ющ·ий, -ая, -ее, (*pres part act*) of **мал·ева́ть**

мал·я́р·, -а́, (*m*) (build) painter; paper-hanger

маляр·и́йн·ый, -ая, -ое, (*a*) malaria, malarial, anti-malaria

~ кома́р· (*m*) (ent) anopheline mosquito, *Anopheles*

маляр·и́·я, -и, (*f*) (med) malaria

мал·я́рн·ая (*f decl as a*) paint store

мал·я́рн·ый, -ая, -ое, (*m*) of **мал·я́р·**

~ рабо́т·ы (*pl*) (build) painting, interior decorating

мамерин·ец·, -инц·а, (*m*) packing strip/ring, seating strip

маммалио·ло́г·и·я, -и, (*f*) (zool) mammology

маммиля́рм·я, -и, (*f*) (bot) mamillaria

ма́монт·, -а, (*m*) (pal) mammoth, *Elephas primigenius*

манган·а́т·, -а, (*m*) (chem) manganate
~ **ка́ли** potassium manganate
манган·и́н·, -а, (*m*) (met) manganin
манган·и́т·, -а, (*m*) (min chem) manganite
манган·ози́т·, -а, (*m*) (min) manganosite
мангано·филли́т·, -а, (*m*) (min) manganophyllite
ма́нго (*n indecl*) (bot) mango, *Mangifera indica*
манго́льд·, -а, (*m*) (bot) spinach beet, *Beta vulgaris* var. *cicla*
ма́нгров·а, -ы, (*f*) (bot) mangrove
мангро́в·ый лес· (*m*) (bot) mangrove swamp
мангу́ст·а, -ы, (*m*) (zool) mongoose, *Herpestes*
мандари́н·, -а, (*m*) (bot) mandarin, *Citrus reticulata*
мандари́н·ов·ый, -ая, -ое, (*a*) mandarin
~ **ма́сл·о** (*n*) mandarin oil
манда́т·, -а, (*m*) mandate, warrant
мандельште́йн·, -а, (*m*) (geol) amygdale, amygdule
манди́бул·а, -ы, (*f*) (zool) mandible, mandibulum
мандраго́р·а, -ы, (*f*) (bot) mandrake
~, **лека́р·ственн·ая** (bot) mandrake, *Mandragora officinalis*
манеба́х·ск·ий за·ко́н· (*m*) (cryst) Manebach law
манёвр·, -а, (*m*) manoeuvre, maneuver; (rail) shunting, switching
~, **встре́ч·н·ый** counter-manoeuvre
манёвр·енность·, -и, (*f*) manoeuvrability, maneuverability
манёвр·енн·ый, -ая, -ое, (*a*) manoeuvre, maneuver, manoeuvring, maneuvering, manoeuvrable, maneuverable; tactical; (rail) shunting, switching
~ **парк·** (*m*) (rail) shunting/switching yard
~ **площа́д·к·а** (*f*) parking apron (airport)
~ **со·един·е́ни·е** (*n*) (mil) tactical force, task force
маневр·и́рова́ни·е, -я, (*n*) manoeuvring, maneuvering; (rail) shunting, switching
маневр·и́ст·, -а, (*m*) (naut) throttle watchkeeper
маневр·о́в·ый, -ая, -ое, (*a*) (rail) shunting, switching
~ **локомоти́в·,** (*m*) (rail) shunter, switcher

мане́ж·, -а, (*i*) **-ем,** (*m*) riding school
манеке́н·, -а, (*m*) dummy, tailor's dummy; lay figure; male mannequin
маннеке́н·щиц·а, -ы, (*i*) **-ей,** (*f*) mannequin
мане́р·а, -ы, (*f*) manner
манже́т *see* **манже́т·а**
манже́т·а, -ы, (*f*) cuff; (hydr) cup leather, seal, packing ring, gaiter (tire repairs)
~, **куполо·обра́з·н·ая** chevron packing
~, **п-обра́з·н·ая** U-ring packing
~, **уголь·ко́в·ая** cup packing
~, **ча́ш·ечн·ая** cup packing
~, **шевро́н·н·ая** chevron packing
манже́т·н·ый, -ая, -ое, (*a*) of **манже́т·а**
~ **за·ли́в·к·а** (*f*) (oil) cement basket
маниа́к·, -а, (*m*) maniac
мани́ль·ск·ий, -ая, -ое, (*a*) manila
манио́к·, съ·ед·о́бн·ый (*m*) (bot) cassava, manioc, *Manihot esculenta*
манио́к·а, -и, (*f*) of **манио́к·**
манипул·и́рова́ни·е, -я, (*n*) *see* **манипул·я́ци·я**
манипул·я́тор·, -а, (*m*) manipulator; (telecom) key, keyboard, keyer; (rad) sub-modulator; cock, cross-connecting valve (piping)
~ **бли́ж·н·его де́й·стви·я** (nucl) handler (laboratory work)
~, **коп·и́рующ·ий** master-slave manipulator
~, **координа́т·н·ый** rectilinear manipulator
~, **кран·ов·о́й** cock
~, **ро́лик·ов·ый** (met roll) roller manipulator
~ **с кант·ова́тел·ем** (met roll) universal manipulator
~, **с·ва́р·очн·ый** welding positioner
~, **част·о́тн·ый** (rad) frequency shift key
~, **шлиц·ев·о́й** (nucl) castle-type manipulator
манипуля́тор·н·ый, -ая, -ое, (*a*) of **манипул·я́тор·;** manipulating, positioning; (telecom) keying
манипуля́тор·щик·, -а, (*m*) (met roll) manipulator driver, manipulator
манипуляцио́н·н·ый, -ая, -ое, (*a*) manipulating, positioning; (telecom) keying
манипул·я́ци·я, -и, (*f*) manipulation, handling, positioning; (telecom) keying
~ **с·мещ·е́ни·ем** blocked-grid keying
~, **част·о́тн·ая** frequency-shift keying

ман·и́ть, -я́т, (*imp*) beckon; entice, lure

манифе́ст·, -а, (*m*) manifesto; (com) manifest

манифест·а́нт·, -а, (*m*) demonstrator (political)

манифест·а́ци·я, -и, (*f*) demonstration (political)

манифест·и́р·овать, -у́ют, (*imp*) take part in a political demonstration

мани́х·а, -ы, (*f*) slack water, slack tide

манихо́т·, -а, (*m*) (bot) manioc, manihot

ма́ни·я, -и, (*f*) (med) mania

~ велич·и·я megalomania

ма́нн·а, -ы, (*f*) (biochem) manna

манн·а́з·а, -ы, (*f*) (biochem) mannase

манн·а́н·, -а, (*m*) (chem) mannan

ма́нн·ая круп·а́ (*f*) (food) semolina

Ма́ннесман·а, спо́соб· (*m*) Mannesmann process (pipe rolling)

манни́т·, -а, (*m*) (chem) mannitol

манни́т·н·ый, -ая, -ое, (*a*) (chem) mannitol

манн·о́з·а, -ы, (*f*) (chem) mannose

манно́н·ов·ая кисл·от·а́ (*f*) (chem) mannonic acid

мано·вакуум·ме́тр·, -а, (*m*) (instr) compound gauge/gage, compound pressure and vacuum gauge/gage

мано·гра́ф·, -а, (*m*) recording pressure gauge/gage

мано·дета́ндер·, -а, (*m*) reducing valve

мано́·метр·, -а, (*m*) (tech) pressure gauge/gage; (phys) manometer; (elec) pressure transducer

~, абсолю́т·н·ый порш·нев·о́й absolute piston manometer

~, а́вто·ши́н·н·ый (M/T) tire gauge/gage

~, ва́куум·н·ый vacuum gauge/gage

~, гидравл·и́ческ·ий manometer, liquid-column gauge/gage

~, груз·ов·о́й dead-weight test apparatus

~, дифференциа́ль·н·ый *see* **дифманометр** differential manometer, manometer

~, жи́дк·ости·ый manometer, liquid-column gauge/gage

~, ионизацио́н·н·ый (phys) ionization gauge/gage

~, колоко́ль·н·ый bell-type gauge/gage

~, магнито·стрикцио́н·н·ый inductive-type pressure transducer

~, мембра́н·н·ый diaphragm pressure gauge/gage

мано́·метр

~, механ·и́ческ·ий expansible metallic-element gauge/gage

~, одно·тру́б·н·ый single-column manometer

~, порш·нев·о́й piston gauge/gage

~, пруж·и́нн·ый expansible metallic-element gauge/gage, hot-wire manometer

~, пьезо·электр·и́ческ·ий crystal pressure transducer

~, радио·акти́в·н·ый radioactive manometer

~, радио·метр·и́ческ·ий radiometric gauge/gage

~, резисти́в·н·ый resistance-wire strain gauge/gage; (elec) resistive pressure transducer

~, рту́т·н·о-жи́дк·ости·н·ый (phys) mercury gauge/gage

~ с на·копл·е́ни·ем (nucl) backing-space ionization gauge/gage

~ с при·ём·ник·ом transmitting pressure gauge/gage

~ с тру́б·чат·ой пруж·и́н·ой Bourdon-tube pressure gauge/gage

~, электр·и́ческ·ий electrical pressure transducer; strain gauge/gage

мано·метр·и́ческ·ий, -ая, -ое, (*a*) manometric, pressure

~, термо́·метр· (*m*) expansion thermometer

мано·ста́т·, -а, (*m*) manostat, pressure stabilizer

маноцити́н·, -а, (*m*) "Manotsitin" (antirust compound)

манса́рд·а, -ы, (*f*) (build) mansard, garrett; mansard roof

мант·и́йн·ый, -ая, -ое, (*a*) mantle; (zool) pallial

манти́сс·а, -ы, (*f*) (math) mantissa

ма́нт·и·я, -и, (*f*) cloak, mantle; (zool) mantle, pallium

~ земл·и́ (geol) mantle of Earth

~, но́в·ая (geol) neopallium

ману·скри́пт·, -а, (*m*) manuscript

ману·факту́р·а, -ы, (*f*) (obs) manufactory; dress-materials, drapery

мара́тор·н·ый, -ая, -ое, (*a*) of **мара́тр·а**

~ кольц·о́ (*n*) *see* **мара́тр·а**

мара́тр·а, -ы, (*f*) mantle (of blast furnace)

мара́тт·иев·ые, -ых, (*pl decl as adj*) (pal bot) *Marattiales*

мар·а́·ть, -ют, (*imp*) soil, stain, dirty; scribble

марблѝт·, -а, (*m*) marbled glass

ма́рган·ец·, -н·ц·а, (*i*) **-н·ц·ем,** (*m*) manganese, Mn

~, азо́т·ист·ый (*m*) (chem) manganese nitride

дву·о́·кис·ь ма́рган·ц·а (*f*) manganese dioxide

за́·кис·ь ма́рган·ц·а (*f*) manganous oxide

за́·кис·ь-о́·кис·ь ма́рган·ц·а (*f*) mangano-manganic oxide

о́·кис·ь ма́рган·ц·а (*f*) manganese sesquioxide

трёх·о·кис·ь ма́рган·ц·а (*f*) manganese sesquioxide

ма́рган·цев·ый, -ая, -ое, (*a*) *see* **марган·цо́в·ый**

марган·цовати́ст·ая кисл·от·а́ (*f*) manganous acid

марганцовисто·ки́сл·ый, -ая, -ое, (*a*) manganate of

~ ка́л·и·й (*m*) potassium manganate

марган·цо́вист·ый, -ая, -ое, (*a*) (chem) manganic; (met) manganese (low content), low-manganese, (min) manganiferous

~ кисл·от·а́ (*f*) manganic acid

~ стал·ь (*f*) low-manganese steel

марганцово·ки́сл·ый, -ая, -ое, (*a*) permanganate of

~ ка́л·и·й potassium permanganate

марган·цо́в·ый, -ая, -ое, (*a*) manganese, manganese-containing; (chem) permanganic

~ ангидри́д· (*m*) manganese heptoxide, permanganate anhydride

~ кисл·от·а́ (*f*) permanganic acid

~ стал·ь (*f*) manganese steel

~ шпат· (*m*) manganese spar, rhodochrosite

маргари́н·, -а, (*m*) (food) margarine

маргари́н·ов·ый, -ая, -ое, (*a*) margarine, margaric

~ кисл·от·а́ (*f*) (chem) margaric acid

маргари́т·, -а, (*m*) (min) margarite

маргари́т·к·а, -и, (*g pl*) **-т·ок·,** (*f*) (bot) daisy, *Bellis perennis*

марг·ина́л·и·я, -и, (*f*) (print) marginal note

Маргу́лес·а, у·равн·е́ни·е (*n*) (meteor) Margules equation

ма́р·ев·о, -а, (*n*) mirage; heat haze

ма́р·ев·ый, -ая, -ое, (*a*) (bot, pharm) chenopodium

маре́н·а, крас·и́льн·ая (*f*) (bot) madder, *Rubia tinctorum*

маре́н·ов·ые, -ых, (*pl decl as adj*) (bot) madder family, *Rubiaceae*

марео́·граф·, -а, (*m*) tide gauge/gage

марза́н·, -а, (*m*) (print) bearer, furniture

мариали́т·, -а, (*m*) (min) marialite

марина́д·, -а, (*m*) (food) marinade

мар·и́нк·а, -и, (*g pl*) **-нок·,** (*f*) (fish) marinka, *Schizothorax*

марин·ова́ни·е, -я, (*n*) (food) marinating, sousing

марин·о́ванн·ый, -ая, -ое, (*past part pass*) of **марин·ова́ть**; (food) marinated, soused

марин·ова́ть, -у́ют, (*imp*) (food) marinate, souse

марио·ти́п·и·я, -и, (*f*) (print) mariotype

Марио́тт·а, за·ко́н·, (*m*) (phys) Mariotte's/Boyle's law

ма́рк·а, -и, (*g pl*) **ма́рок·,** (*f*) marking, mark, tally, stamp; (com) brand, make; grade, sort, type; counter, chip; whipping (cordage)

~, груз·ов·а́я (naut) load line

~, почт·о́в·ая postage stamp

~ у·глубл·е́ни·я (naut) draught mark

~, фабри́ч·н·ая trademark, factory mark

~, фи́рм·енн·ая trademark, firm's mark

марказѝт·, -а, (*m*) (min) marcasite

маркёр·, -а, (*m*) marker

~, бли́ж·н·ий (air) inner marker

~, ве́ер·н·ый (air) fan marker

~, вне́ш·н·ий (air) outer marker

~, гран·и́чн·ый (air) boundary marker

~, да́ль·н·ий (air) outer marker

~, по·са́д·очн·ый (air) inner marker

маркёр·н·ый, -ая, -ое, (*a*) of **маркёр·**

~ голо́в·к·а (*f*) index head (of magnetic drum)

~ мая́к· (*m*) (air) marker beacon

марк·ирова́льн·ый, -ая, -ое, (*a*) marking, stamping, franking

~ маши́н·а (*f*) franking machine (for letters)

марк·иро́ванн·ый, -ая, -ое, (*past part pass*) marked, stamped, tagged

марк·иро́вк·а, -и, (*g pl*) **-вок·,** (*f*) marking, mark; brand, branding, stamping; (nucl) labelling, tagging

марк·иро́вочн·ый, -ая, -ое, (*a*) marking, stamping, tagging

маркита́нт·ск·ая ло́д·к·а (*f*) (naut) bumboat

ма́рк·ов·о не·ра́вен·ств·о   (*n*) (math) Markoff inequality

ма́рк·овск·ий проце́сс·   (*m*) (math) Markoff process

Марко́н·и, анте́нн·а   (*f*) (rad) Marconi beam antenna

маркше́йдер·, -а,   (*m*) mining surveyor, mine/underground surveyor

маркше́йдер·и́·я, -и,   (*f*) of **маркше́й-дер·ск·ое де́л·о**

маркше́йдер·ск·ий, -ая, -ое,   (*a*) of **маркше́йдер·**

∼ де́л·о   (*n*) mine/underground surveying

ма́рл·ев·ый, -ая, -ое, (*a*) (text) gauze

∼ салфе́т·к·а   (*f*) (med) gauze dressing

ма́рлин·ь, -я,   (*m*) (naut) marline

ма́рл·я, -и,   (*f*) (text) gauze

мармати́т·, -а,   (*m*) (min) marmatite

мармела́д·, -а,   (*m*) fruit jelly/candy

мармит·н·ая теле́ж·к·а   (*f*) hot food trolley

мармор·и́р·овать, -уют,   (*imp and perf*) marble, face with marble

мароке́н·, -а,   (*m*) (text) marocain

ма́р·очн·ый, -ая, -ое, (*a*) of **ма́рк·а;** branded

∼ вин·о́   (*n*) vintage wine

∼ стал·ь   (*f*) branded steel

маррубии́н·, -а,   (*m*) (pharm) marrubin

ма́рс·, -а,   (*m*) (naut) top, masthead

Марс, -а,   (*m*) (astron) Mars

марсал·а́, -ы́, (*f*) Marsala, Marsala-type wine

ма́рсел·ь, -я,   (*m*) (naut) topsail, square topsail

марсе́ль·ск·ий, -ая, -ое,   (*a*) (geog) Marseilles

марси́л·и·я, -и,   (*f*) (bot) pepperwort, *Marsilia*

ма́рс·ов·ый   (*m decl as adj*) (naut) rigger; topman

ма́рт·, -а,   (*m*) March (month)

марте́н·овск·ий, -ая, -ое,   (*a*) (met) open-hearth, Siemens–Martin, OH

∼ печ·ь   (*f*) open-hearth furnace

∼ печ·ь, кач·а́ющ·аяся   (*f*) tilting open-hearth furnace

∼ проце́сс· (*m*) Siemens–Martin process, open-hearth process

∼ проце́сс·, ки́сл·ый (*m*) acid open-hearth process

∼ проце́сс·, основ·н·о́й   (*m*) basic open-hearth process

∼ цех· (*m*) open-hearth furnace/shop

Ма́ртенс·у, тепло·сто́й·кост·ь по (*f*) Martens heat stability

мартенси́т·, -а,   (*m*) (met) martensite

мартенси́т·н·ый, -ая, -ое,   (*a*) (met) martensite, martensitic

∼ то́ч·к·а   (*f*) (met) martensitic transformation point

марти́т·, -а,   (*m*) (min) martite

ма́рт·овск·ий, -ая, -ое,   (*a*) of **март·**

март·ы́шк·а, -и,   (*g pl*) **-шек·,** (*f*) (zool) guenon, *Cercopithecus*

марципа́н·, -а,   (*m*) (food) marzipan

ма́рш·, -а,   (*i*) **-ем,** (*m*) march; flight (of stairs)

∼ -под·хо́д·, -а,   (*m*) (mil) approach march

∼, торж·е́ственн·ый   goosestep

Ма́рш·а спо́соб· (*m*) (chem) Marsh's test, Marsh–Berzelins test

ма́ршал·, -а, (*m*) Marshal·

ма́ршаль·ск·ий, -ая, -ое,   (*a*) of **ма́ршал·**

ма́рш·ев·ый, -ая, -ое, (*a*) of **ма́рш·;** (GW) cruise, cruising, sustaining; (mil) holding replacement (of units)

∼ дви́г·ател·ь   (*m*) (GW) cruising propulsion unit

∼ по·лёт· (*m*) (GW) cruise, cruising flight

∼ у·ча́ст·ок· траекто́р·и·й (*m*) (GW) midcourse

ма́рш·и, -ей, (*pl*) (geog) salt marshes

марш·и́р·ова́ть, -уют, (*imp*) march

маршру́т·, -а,   (*m*) route, itinerary; (air) course; (rail) road, route; through freight train, direct goods train

∼, за́·мк·нут·ый   (rail) shuttle freight train; locked route

∼, кольц·ев·о́й   shuttle freight train

∼ от·правл·е́ни·я   (rail) exit/departure road/route

∼ при·ём·а   (rail) entrance/reception road

∼, техн·и́ческ·ий   (rail) long-distance through freight train

маршрут·иза́ци·я, -и,   (*f*) (rail) through-freight organization, train-load system, freight liner system

маршрут·и́р·овать, -уют, (*imp and perf*) plan/lay down/organize/make out a route/itinerary; (rail) run a liner system

маршру́т·н·ый, -ая, -ое,   (*a*) of **маршру́т·**

∼ индика́тор· (*m*) (rail) route-indicating signal, route indicator

маршру́т·н·ый, -ая, -ое
~ **лист·** (*m*) follow-through sheet, route sheet (industrial production)
~ **от·пра́в·к·а** (*f*) liner consignment
~ **по́·езд·** (*m*) (rail) through freight train, freight liner
~ **по·лёт·** (*m*) (air) cross-country flight
~ **съ·ём·к·а** (*f*) (surv) route survey/ traverse; (air) strip survey
ма́р·ь, -и, (*f*) (bot) goosefoot, *Chenopodium*
~, **душ·и́ст·ая** Jerusalem oak, *Chenopodium botrys*
марья́ж·, -а, (*m*) (rocket) marriage
марьяли́т·, -а, (*m*) (min) marialite
марья́нник·, -а, (*m*) (bot) cow wheat, *Melampyrum*
ма́сер·, -а, (*m*) (*see* **ма́зер·**) (rad) maser
ма́ск·а, -и, (*g pl*) **-со́к·,** (*f*) mask, face guard, helmet, hood, mantlet; (ent) mask
~, **дел·и́тельн·ая** (phot) partial mask
~, **кисло·ро́д·н·ая** oxygen mask
~, **противо·га́з·ов·ая** gas mask
маск·иров́ани·е, -я, (*n*) *see* **маск·и·ро́вк·а**
маск·иро́ванн·ый, -ая, -ое, (*past part pass*) of **маск·ир·ова́ть**; masked; disguised, camouflaged
маск·ир·ова́ть, -у́ют, (*imp*) mask, disguise, conceal, camouflage
маск·иро́вк·а, -и, (*g pl*) **-во́к·,** (*f*) masking, disguise, camouflage
~, **звук·ов·а́я** (acous) masking, aural masking
~, **радиолокацио́н·н·ая** radar camouflage
~, **свет·ов·а́я** black-out
маск·иро́вочн·ый, -ая, -ое, (*a*) of **маск·иро́вк·а**; deceptive, concealing
~ **на·са́д·к·а фа́р·ы** (*f*) (М/Т) headlamp mask
~ **о·кра́ш·ивани·е** (*n*) camouflage painting
маск·иру́ющ·ий, -ая, -ее, (*pres part act*) *see* **маск·ир·ова́ть**
~ **эффе́кт·** (*m*) (acous) masking
маск·хала́т·, -а, (*m*) (mil) denison smock
масл·ёнк·а, -и, (*g pl*) **-нок·,** (*f*) oil can; lubricator; oiled paper; butter dish
~, **иго́ль·чат·ая** needle lubricator
~, **ка́пель·н·ая** drip lubricator
~, **колпа́ч·ков·ая** grease cup

масл·ёнк·а
~, **фити́ль·н·ая** wick-feed lubricator
~ **шта́уфер·а** grease cup
масл·ёнок·, -нк·а, (*nom pl*) **-ля́та,** (*g pl*) **-ля́т,** (*m*) (bot) *Boletus*
~, **зерн·и́ст·ый** *Boletus granulatus*
~ **лист·веннич·н·ый** *Boletus elegans*
~, **по́зд·н·ий** butter mushroom, *Boletus lutens*
ма́сл·ен·ый, -ая, -ое, (*a*) buttery, buttered; oily, greasy; oily, smooth, smarmy (figurative)
масл·и́н·а, -ы, (*f*) (bot) olive, *Olea*
~, **европ·е́йск·ая** (bot) evergreen olive, *Olea europeae*
масл·и́нн·ый, -ая, -ое, (*a*) of **масл·и́н·а**; (*pl*) olive family, *Oleaceae*
масл·и́нов·ый, -ая, -ое, (*a*) (bot) olive
ма́сл·ить, -ят, (*imp*) butter, grease
масл·и́чн·ый, -ая, -ое, (*a*) (bot etc.) oil-bearing, oil-producing, oleaginous
ма́сл·о, -а, (*nom pl*) **-л·а́,** (*g pl*) **-сел·,** (*n*) (*see also under qualifying adj*) oil; lubricating oil; butter; oils (art)
~, **авиацио́н·н·ое** aviation lubricating oil
~, **авто·тра́ктор·н·ое** automotive engine oil, motor and tractor engine oil
~, **бе́л·ое** white oil
~ **бензине́** benzolized wash oil
~, **боб·о́в·ое** soybean oil
~, **вазели́н·ов·ое** vaseline oil, fine lubricating oil
~, **ва́куум·н·ое** vacuum oil
~, **верет·ённ·ое** spindle oil
~, **вы·сых·а́юш·ее** drying oil
~, **вы́·щелоч·енн·ое** neutral oil
~ **дебензине́** debenzolized oil
~ **дестилла́т·н·ое** distillate oil
~, **ди́зель·н·ое** light diesel lubricating oil, high-speed diesel engine oil
~, **доро́ж·н·ое** road oil
~, **желуд·ёв·ое** acorn oil
~, **жи́р·н·ое** fatty/fixed oil
~, **изоляцио́н·н·ое** (elec) insulating oil
~, **касто́р·ов·ое** castor oil
~, **компаунд·и́рованн·ое** compounded oil
~, **компре́ссор·н·ое** compressor oil
~, **коро́в·ье** butter
~, **кост·ян·о́е** bone oil
~, **купоро́с·н·ое** (chem) oil of vitriol, sulphuric acid
~, **ла́к·ов·ое** drying oil
~, **лёг·к·ое** thin oil
~, **маши́н·н·ое** machine lubricating oil

ма́сл·о
~, **мото́р·н·ое**   heavy diesel lubricating oil; motor engine oil (rare)
~, **не·кондицио́н·н·ое**   refuse oil
~, **ос·ев·о́е**   axle oil
~, **о·ста́т·очн·ое**   residual oil
~, **от·сто́й·н·ое**   (ICE) sump oil
~, **парафи́н·ов·ое**   press oil
~, **парфюме́р·н·ое**   technical white oil
~, **по·глот·и́тельн·ое**   absorption oil, wash oil
~, **под·сы́р·н·ое**   sour cream butter
~, **раст·и́тельн·ое**   vegetable oil
~ **селекти́в·н·ой о·чи́ст·к·и**   select-ive-refined oil, solvent oil
~, **сиву́ш·н·ое**   fusel oil
~, **сла́н·цев·ое**   shale oil
~, **с·ли́в·очн·ое**   dairy/cream butter
~, **с·ма́з·очн·ое**   lubricating oil
~, **соля́р·ов·ое**   solar oil, diesel fuel
~, **суд·ов·о́е**   marine engine oil
~, **сульф·и́рованн·ое**   sulphonated oil
~, **топл·ён·ое**   renovated oil
~, **трансформа́тор·н·ое**   transformer oil
~, **ту́нг·ов·ое**   tung oil
~, **турби́н·н·ое**   turbine oil
~, **цили́ндр·ов·ое**   cylinder oil
~, **эфи́р·н·ое**   ethereal/essential oil
масло·бо́й·к·а, -и,   (g pl) **-о́·ек·,** (f) butter churn; oil mill, oil extraction plant
масло·бо́й·н·ый, -ая, -ое,   (a) vege-table oil extraction
масло·бо́й·н·я, -и,   (g pl) **-о́·ен·,** (f) oil mill, oil extraction plant
масл·о́вк·а, -и,   (f) (text) oiling
масло·водо·с·бо́р·ник·, -а,   (m) oil-and-water trap
масло·воз·ду́ш·н·ый котёл·   (m) (hydr) air-loaded accumulator, hydro-pneumatic accumulator
масло·де́л·и·е, -я,   (n) buttermaking, dairy work
масло·де́ль·н·я, -и,   (g pl) **-л·ен·,** (f) creamery, dairy
масло·до·бы́в·а́ющ·ий, -ая, -ее,   (a) oil-extraction
масло·ём·кост·ь, -и,   (f) oil absorption (of pigments)
масло·жир·ов·о́й, -а́я, -о́е,   (a) fat and oil, vegetable fats and oils
масло·за·во́д·, -а,   (m) creamery; oil mill
масло·за·пра́в·щик·, -а, (m) oil bowser
масло·из·готов·и́тел·ь, -я, (m) butter churn and worker

масло·из·мер·и́тельн·ый, -ая, -ое, (a) oil-level measuring, (instr) oil
~ **сте́рж·ен·ь**   (m) oil dipstick
масло·ме́р·, -а,   (m) oil gauge/gage
масло·ме́р·н·ый, -ая, -ое,   (a) oil-measuring/metering
масло·на·гнет·а́тел·ь, -я,   (m) oil gun
масло·на·по́лн·енн·ый, -ая, -ое,   (a) oil-filled; oil-extended (rubber)
масло·на·по́р·н·ый, -ая, -ое,   (a) (hydr) oil-pressure, pressurized oil
~ **у·стано́в·к·а** (f) (hydr) oil-pressure system; pressurized oil system
масло·на·правл·я́ющ·ий лот·о́к·   (m) oil baffle
масло·не·про·ниц·а́ем·ый, -ая, -ое, (a) oil-proof, oil-tight
масло·обра́з·н·ый, -ая, -ое,   (a) oily, greasy
масло·образ·у́ющ·ий, -ая, -ое,   (a) olefiant, oil-forming
масло·от·дел·и́тел·ь, -я,   (m) (mech eng) oil eliminator/expeller
масло·от·раж·а́тел·ь, -я,   (m) (mech eng) lubricant retainer, oil seal (for shaft), oil baffle
масло·от·сто́й·ник·, -а,   (m) oil sump
масло·о·хлад·и́тел·ь, -я,   (m) lubricat-ing oil cooler
масло·о·чист·и́тел·ь, -я,   (m) oil purifier/filter
масло·о·чи́ст·к·а, -а,   (f) oil cleaning/clarifying; oil refining
масло·при·ём·ник·, -а,   (m) (ICE) oil header
масло·про·во́д·, -а,   (m) oil pipe/line; oil lead
масло·ради́атор·, -а,   (m) oil cooler
масло·раз·да́т·очн·ый, -ая, -ое,   (a) oil-dispensing
~ **коло́н·к·а**   (f) oil dispenser
масло·рас·твор·и́м·ый, -ая, -ое,   (a) oil-soluble
масло·с·бо́р·ник·, -а,   (m) sump, oil sump, drip pan
масло·с·бо́р·ый, -ая, -ое,   (a) of **ма-сло·с·бо́р·ник·**
~ **бак·**   (m) oil sump tank
масло·с·бра́с·ыватель·, -я,   (m) oil thrower/slinger, oil flinger (electric motor)
масло·с·бра́с·ываю·щее кольц·о́ (n) (mech) scraper ring, oil regulating ring, oil ring
масло·с·ли́в·, -а,   (m) oil drain

масло·со·бир·а́ющ·ее кольц·о́ (*n*) (mech eng) oil ring, scraper ring, oil regulating ring

масло·сто́й·к·ий, -ая, -ое, (*a*) oil-resistant

масло·с·чёт·чик·, -а, (*m*) oil meter

масло·съ·ём·н·ое кольц·о́ (*n*) (mech) scraper ring, oil regulating ring, oil ring

масло·термо·сто́й·к·ий, -ая, -ое, (*a*) oil-and-heat resistant

маслотт·а, -ы, (*f*) ring pot

масло·у·каз·а́тел·ь, -я, (*m*) oil gauge/gage

масло·у·лов·и́тел·ь, -я, (*m*) oil interceptor/catcher; (st eng) oil separator/eliminator

масло·у·плот·ни́тельн·ый, -ая, -ое, (*a*) oil-seal

~ **кольц·о́** (*n*) oil-seal ring

масло·фас·о́вочн·ый автома́т· (*m*) (food) butter forming and packaging machine

масл·я́к·, -а́, (*m*) *see* **масл·ёнок·**

ма́сл·яник·, -а, (*m*) (elec) oil switch/breaker; (bot) *see* **масл·ёнок·**

масл·яни́стост·ь, -и, (*f*) oiliness; (bot) oil content (of oil-bearing plants)

масл·яни́ст·ый, -ая, -ое, (*a*) oily, oleaginous; buttery, (chem) butyrous

~ **смол·а́** (*f*) oleoresin

масляно·ки́сл·ый, -ая, -ое, (*a*) (chem) butyrate of

~ **брож·е́ни·е** (*n*) butyric fermentation

масляно·смол·ян·о́й, -а́я, -о́е, (*a*) oleoresinous

ма́сл·ян·ый, -ая, -ое, (*a*) of **ма́сл·о;** oil, lubricating-oil, lubricating; buttery, (chem) butyric; oil-extended (rubber); (elec) oil-immersed, oil-cooled, oil-break, oil-quenched

~ **альдеги́д·,** (*m*) (chem) butyric aldehyde, butyraldehyde

~ **бак·** (*m*) oil tank

~ **вы·ключ·а́тел·ь** (*m*) (elec) oil-break switch, oil switch, oil circuit breaker

~ **кисл·от·а́** (*f*) butyric acid

~ **кольц·о́** (*n*) scraper/oil ring

~ **лак·** (*m*) oil varnish

~ **на·со́с·** (*m*) lubricating-oil pump

~ **па́льм·а** (*f*) (bot) oil palm, *Elaeis guineensis*

~ **пред·о·хран·и́тел·ь** (*m*) (elec) oil-quenched fuse

~ **пятн·а́** (*pl*) oil slick

ма́сл·ян·ый

~ **сервомото́р·** (*m*) hydraulic servo-motor

ма́сс·а, -ы, (*f*) mass; body (of rock); (*pl*) the masses/people/public

~ **а́том·а** atomic mass

~, **бума́ж·н·ая** (paper) pulp

~, **воз·ду́ш·н·ая** (meteor) air mass

~, **де́й·ствующ·ая** *see* **за·ко́н·** *below*
дефе́кт· ма́сс·ы (*m*) (phys) mass defect

~, **древе́с·н·ая** wood pulp
за·ко́н· де́й·ствующ·их масс· (*m*) (chem) law of mass action

~, **зерн·ов·а́я** (agr) bulk grain

~, **из·бы́т·очн·ая** (phys) extra mass

~, **ине́рт·н·ая** (phys) mass, inertia mass

~, **ка́бель·н·ая** (elec) cable-filling compound

~, **мор·ск·а́я воз·ду́ш·н·ая** (meteor) maritime air mass

~, **осно́в·н·а́я** matrix, groundmass

~, **пасси́в·н·ая** (naut) salvage, salvaged cargo

~, **пласт·и́ческ·ая** plastic

~ **поко́·я** (phys) rest mass

~, **при·вед·ённ·ая** (phys) reduced mass

~, **при·со·един·ённ·ая** (phys) associated/virtual mass

~, **секу́нд·н·ая** (dynam) flux

~ **стру·и́** (dynam) flow mass

~, **тяжёл·ая** (mech) gravitational mass
центр· масс· (*m*) (mech) centre/center of mass

~, **эффекти́в·н·ая** (phys) effective mass

масса́ж·, -а, (*i*) -ем, (*m*) (med) massage

масса́ж·н·ый, -ая, -ое, (*a*) of **масса́ж·**

масси́в·, -а, (*m*) (geog) massif; concrete block, monolith

~ -**гига́нт·, -а,** (*m*) (civ eng) reinforced-concrete monolith

~, **лес·н·о́й** forest area/stand

~ **су́ш·и** (geol) land mass

масси́в·н·ый, -ая, -ое, (*a*) massive, bulky; bulk; solid; monolithic (of concrete structures)

~ **образ·е́ц·** (*m*) (met) massive/solid specimen (as opposed to a powder one)

массико́т·, -а, (*m*) (chem) massicot, litharge

масс·и́р·овать, -уют, (*imp and perf*) mass, concentrate; massage

масс·н·ый, -ая, -ое, (*a*) pulp, stock, stuff

~ ролл· (*m*) (paper) pulp-beater

масс·ов·ый, -ая, -ое, (*a*) mass, bulk; popular, for mass consumption

~ един·иц·а (*f*) (phys) mass unit

~ о·слабл·ени·е (*n*) (phys) mass absorption

~ пере·да·ч·а (*f*) (phys) mass transfer

~ про·бег· (*m*) (phys) mass range

~ сил·а (*f*) (mech) force

~ про·из·вод·ств·о (*n*) mass production

~ рас·ходо·мер· (*m*) mass flowmeter

~ числ·о (*n*) (phys) mass number

массо·пере·да·ч·а, -и, (*f*) (phys) mass transfer

массо·про·вод·, -а, (*m*) pulp pipeline/line

массо·про·вод·ност·ь, -и, (*f*) (phys) mass transfer

масс-спектр·, -а, (*m*) (phys) mass spectrum

масс-спектро·граф·, -а, (*m*) mass spectrograph/analyser

масс-спектро·метр·, -а, (*m*) mass spectrometer/analyser

масс-спектро·скоп·и·я, -и, (*f*) mass-spectroscopy

масс-эквивалент·, -а, (*m*) mass-equivalent

мастер·, -а, (*nom pl*) **мастер·а,** (*g pl*) **-ов,** (*m*) chargehand, foreman; master (qualified artisan), skilled, qualified; master (title awarded for high proficiency)

~, контрол·ь·н·ый foreman inspector (quality control)

~, про·бир·н·ый assay-master, assayer

~, сапож·н·ый master shoemaker, shoemaker

~, с·мен·н·ый shift foreman

~ час·ов·ых дел· master watchmaker, watchmaker

мастер-штамп·, -а, (*m*) (met forge) master die

мастер·ить, -ят, (*imp*) make, run up, knock together

мастер·ов·ой, -ая, -ое, (*a*) artisan's, tradesman's; (*m decl as a*) artisan, journeyman

мастер·ок·, -р·к·а, (*m*) brick trowel

мастер·ск·ая, -ой, (*f decl as a*) (*and see under associated words*) workshop, shop; repair shop; studio

~, за·точ·н·ая tool-grinding workshop

мастер·ск·ая

~, механ·ическ·ая engineering/engineer's shop

~, плов·уч·ая repair ship

мастер·ск·и (*adv*) skilfully, well

мастер·ск·ий, -ая, -ое, (*a*) of **мастер·;** (*see also* **мастер·ск·ая**) (*as noun*); *see also* **мастер·ск·ой**

мастер·ск·ой, -ая, -ое, (*a*) skilled, masterly

мастер·ств·о, -а, (*n*) mastery, craftsmanship, skill, high proficiency; trade, occupation

мастик·а, -и, (*f*) (build) mastic

~, битум·н·ая bitumastic

~ гор·яч·ая кровель·н·ая (build) mastic asphalt

~, цемент·н·ая mastic cement

мастик·ов·ый, -ая, -ое, (*a*) mastic

~ дерев·о (*n*) (bot) mastic tree, lentisk, *Pistacia lentiscus*

~ смол·а (*f*) mastic gum, mastich

мастикс·, -а, (*m*) mastic, mastic gum, mastich (a resin)

маст·ит·, -а, (*m*) (med) mastitis

мастич·н·ый *see* **мастик·ов·ый**

маст·один·и·я, -и, (*f*) (med) mastodynia

маст·одонт·, -а, (*m*) (pal) mastodon, mastodont

маст·оид·ит·, -а, (*m*) (med) mastoiditis

маст·ь, -и, (*nom pl*) **-и,** (*g pl*) **-ей,** (*f*) colour, color (of animal hair); suit (at cards)

масштаб·, -а, (*m*) scale
 в масштаб·е to scale
 в у·велич·енн·ом масштаб·е scaled up

~ времен·и time scale (computers); (aerodynam) air second

~ времен·и, ист·инн·ый real time (computers)

~ времен·и, натурал·ь·н·ый real time (computers)

~ времен·и, реал·ь·н·ый real time (computers)

~, лин·ейн·ый (surv) bar/drawn scale

~, натурал·ь·н·ый full scale, life size

~, у·велич·енн·ый scale of magnification; (*see* в у·велич·енн·ом мас·штаб·е *above*)

~, числ·енн·ый (surv) representative fraction

масштаб·ировани·е, -а, (*n*) (math) scaling

масштаб·н·ый, -ая, -ое, (*a*) scale, scaling

~ кольц·о (*n*) range scale (radar)

масштáб·н·ый, -ая, -ое

~ **коэффициéнт·**   (*m*) (phys) scale factor

~ **лáмп·а** (*f*) (elec) scaling tube

~ **лин·éйк·а** (*f*) (instr) rule, ruler

~ **модéл·ь** (*f*) scale model

~ **от·мéт·к·а** (*f*) calibration pip (radar)

~ **фáктор·** (*m*) (math) scale factor

~ **эффéкт·** (*m*) (phys) scale effect

мáт·, -а, (*m*) mat, matting; matte/matt surface/finish; checkmate

~, **битýм·н·ый** (build) bitumen damp-proof course

~, **тк·áнн·ый** (naut) sword mat

~, **шпиг·óванн·ый** (naut) thrum mat

мателóт·, -а, (*m*) (naut) adjacent ship (in formation)

математ·ик·, -а, (*m*) mathematician

математ·ик·а, -и, (*f*) mathematics

математ·ическ·ий, -ая, -ое, (*a*) mathematical

~ **машúн·а** (*f*) mathematical machine; computer, computing machine (broad classification)

~ **машúн·а, аналóг·ов·ая** (*f*) analogue/analog computer

~ **машúн·а дискрéт·н·ого дéй·ств·и·я** (*f*) digital computer

~ **машúн·а, модел·úрующ·ая** (*f*) analogue computer

~ **машúн·а не·пре·рýв·н·ого дéй·-ств·и·я** (*f*) analogue/analog computer

~ **машúн·а, электро·модел·úрую-щ·ая** (*f*) electronic analogue/analog computer

~ **модéл·ь** (*f*) simulator

материáл·, -а, (*m*) material; (*pl*) information, details, data, relevant papers

~, **ис·хóд·н·ый** material, raw material, stock (for given process), starting material

~, **мáт·очн·ый** foundation stock

~, **от·вáл·очн·ый** (geol) debris, talus

~, **перв·úчн·ый** raw/process/starting material

материализ·овáть, -ýют, (*imp and perf*) materialize

материало·вéд·ени·е, -я, (*n*) materiology, study of material

материало·ём·к·ий, -ая, -ое, (*a*) using/requiring/consuming a lot of materials, expensive/costly in materials (of processes)

материало·ём·кост·ь, -и, (*f*) consumption of materials (of processes)

материáль·н·о (*adv*) materially; (*as component*) material

~ **-техн·úческ·ий, -ая, -ое,** (*a*) supplies and equipment, equipment and materials, (mil) matériel

материáль·н·ый, -ая, -ое, (*a*) material, (phys) matter

~ **волн·á** (*f*) (phys) matter wave

~ **о·бес·пéч·ение** (*n*) benefit, pay (social insurance); supplies

~ **тóч·к·а** (*f*) (phys) mass point, particle

~ **част·ь** (*f*) materiel, equipment

матер·úк·, -á, (*m*) (*see also* **континéнт·**); (geog) mainland, continent; subsoil

матер·икóв·ый, -ая, -ое, (*a*) (geog) continental

~ **óт·мел·ь** (*f*) (ocean) continental shelf

~ **с·клон·** (*m*) (ocean) continental slope

матер·úнск·ий, -ая, -ое, (*a*) mother, maternal, parent

~ **ядр·ó** (*n*) parent nucleus

матéр·и·я, -и, (*f*) matter; material

мат·ировáни·е, -я, (*a*) (paper) matte/mat finishing; (text) dull finishing/finish; flatting, dead/flat finish (paint)

мат·ирóванн·ый, -ая, -ое, (*a*) mat, matte, matte-finished, (text) dull-finished, dull, dull-spun

мáт·иц·а, ы, (*i*) **-ей,** (*f*) (build) tie-beam, main beam

мáт·к·а, -и, (*g pl*) **-т·ок·,** (*f*) (zool) female parent; (anat) uterus; master batch (rubber); (min) timber/prop cap; (naut) depot ship; master compass

~, **сáж·ев·ая** black master batch (rubber)

~, **пчел·úн·ая** queen bee

матлóт·, -а, (*m*) *see* **мателóт·**

мáт·ов·ый, -ая, -ое, (*a*) matt, frosted (glass), dead, flat (paint), dull, dead (metals)

~ **от·дéл·к·а** (*f*) (met) dull finish; (paper) matte/mat finish

мáт·очник·, -а, (*m*) (chem) mother liquor; queen cell (bees); (bot) balm mint

мáт·очн·ый, -ая, -ое, (*a*) of **мáт·к·а, мáтр·иц·а;** (anat) uterine

~ **рас·твóр·** (*m*) (chem) mother liquor; (met) parent solution, matrix phase

~ **с·мес·ь** (*f*) (rubber) master batch

матрáс·, -а, (*f*) mattress

матрáц·, -а, (*i*) **-ц·ем** (*m*) mattress

ма́тр·иц·а, -ы, *(i)* **-ей,** *(f)* matrix; (met) die, female die; starting sheet (electrodeposition); (paper) mould, mold

~, венти́ль·н·ая gating matrix (computers)

~, волоч·и́льн·ая (met) wire-drawing die

~, вы·тяж·н·а́я (met) drawing die, deep-drawing die; reducing die

~, двух·ря́д·н·ая квадра́т·н·ая (math) two-by-two matrix

~, един·и́чн·ая (math) identity/unit matrix

~, като́д·н·ая starting sheet (electrodeposition)

~, квадра́т·н·ая (math) square matrix

~ коэффициéнт·ов коррел·я́ци·и (math) correlation matrix

~, лино·ти́п·н·ая (print) linotype matrix

~, не·о·со́б·енн·ая (math) non-singular matrix

~ -образ·éц·, -з·ц·а́, *(m)* (met) master die

~, обра́т·н·ая (math) inverse of a matrix

~, о·со́б·ая (math) singular matrix

~ пло́т·ност·и (phys) density matrix

~ рас·сé·яни·я (phys) scattering matrix

~, свин·цо́в·ая (print) lead mould/mold

~, транспон·и́рованн·ая (math) transpose of a matrix

~, унита́р·н·ая (math) unitary matrix

~, шифр·у́ющ·ая scrambling matrix (computers)

матрице·держ·а́тел·ь, -я, *(m)* matrix holder (of a press)

матриц·и́ровани·е, -я, *(n)* (math) matrixing; (print) matrix punching

ма́тр·ичн·ый, -ая, -ое, *(a)* matrix, step-matrix (computers); die

~ код·и́ровани·е *(n)* step-matrix coding (computer)

~ фа́з·а *(f)* (metal) matrix, matrix phase

матро́с·, -а, *(m)* seaman, sailor

~, ста́р·ш·ий able seaman

матро́с·ск·ий, -ая, -ое, *(a)* seaman's, sailor's

~ чемода́н· *(m)* kitbag

Матро́сов·а, то́рмоз· *(m)* (rail) Matrosov air brake

Маттисен·а, пра́в·ил·о *(n)* (elec) Matthieson standard

матуа́р·, -а, *(m)* matoir (engraving tool)

ма́тч·, -а, *(i)* **-ем,** *(m)* match (sport)

мат·ь, *(gen sing)* **ма́тер·и,** *(nom pl)* **ма́тер·и,** *(g pl)* **-éй,** *(f)* mother

мать-и-ма́чех·а, -и, *(f)* (bot) coltsfoot, *Tussilago farfara*

Матьё, фу́нкци·я *(f)* (math) Mathieu function

мауерла́т·, -а, *(m)* (build) wall plate

маф·и́ческ·ий, -ая, -ое, *(a)* (geol) mafic

мау́н·, -а, *(m)* (bot) valerian, *Valeriana*

ма́х·, -а, *(m)* stroke, single revolution

да·ть ма́х·у make a mistake

одн·и́м ма́х·ом at one stroke

Ма́х·а, числ·о́ *(n)* (air) Mach number

махаго́ни *(m indecl)* (bot) mahogany, *Swietenia mahagoni*

махаго́н·иев·ое де́рев·о *(n)* (bot) West Indies mahogany tree, *Swietenia mahagoni*

мах·а́льн·ый, -ая, -ое, *(a)* waving, flapping, swinging

мах·а́ни·е, -я, *(n)* waving, flapping, signalling, brandishing

мах·а́ть *(pres 3rd sing, pl)* **ма́ш·ет, ма́ш·ут,** *(imp)* wave, flap, wag, swing; *(intrans)* beat

мах·метр·, -а, *(m)* (a/c) Machmeter

мах·н·у́ть, -у́т, *(perf)* *see* **мах·а́ть**

мах·ови́к·, -а́, *(m)* flywheel; hand wheel; governor

~, разъ·ём·н·ый split flywheel

мах·ови́чн·ый, -ая, -ое, *(a)* of **мах·ови́к·**

~ при·во́д· *(m)* flywheel operating gear

мах·овичо́к·, -чк·а́, *(m)* hand wheel

мах·ов·о́й, -а́я, -о́е, *(a)* *(only used in certain combinations)*

~ колес·о́ *(n)* flywheel; hand wheel; governor

~ пér·ь·я *(pl)* (zool) wing feathers

~ эффéкт· *(m)* (mech) flywheel effect

Ма́х·а-Цендер·а, интерферо́·метр· *(m)* Mach-Zehnder interferometer

махо́р·к·а, -и, *(f)* (bot) common tobacco, *Nicotiana rustica*; makhorka, shag (coarse tobacco)

махо́р·очн·ый, -ая, -ое, *(a)* of **махо́р·к·а**

махр·о́в·ый, -ая, -ое, *(a)* (bot) double, double-petalled

~ ткан·ь *(f)* (text) Turkish towelling

мацера́л·, -а, *(m)* (min) maceral

мацер·ацио́нн·ый, -ая, -ое, (*a*) maceration

мацер·а́ци·я, -и, (*f*) maceration

мацер·и́р·овать, -уют, (*imp or perf*) macerate

ма́чт·а, -ы, (*f*) mast, pole, tower; spar tree

~ **-анте́нн·а, -ы,** (*f*) (rad) antenna/aerial mast

~, **бо·ев·а́я** conning tower (on a warship)

~, **монта́ж·н·ая** (build) gin mast/pole

~, **одно·дерев·ая** pole, polemast

~, **решёт·чат·ая** lattice mast

~, **с·клад·н·а́я** collapsible mast

~, **тре·но́г·ая** tripod mast

~, **фальш·и́в·ая** (naut) jury mast

ма́чт·ов·ый, -ая, -ое, (*a*) of **ма́чт·а**

ма́ш·, -а, (*m*) (bot) green gram, *Phaseolus mungo*

ма́ш·а (*pres gerund*) of **мах·а́ть**

ма́ш·ет (*pres 3rd sing*) of **мах·а́ть**

маши́н·а, -ы, (*f*) (*see associated adjectives*), (*see also* **дви́г·ател·ь**), machine, engine; (*pl*) machinery

~ **-анало́г·, -а,** (*m*) analogue computer

~ **-компа́унд·, -а,** (*m*) (st eng) compound expansion engine

~, **рабо́ч·ая** power tool

~ **-та́ндем·, -а,** (*m*) tandem engine

маши́н·иза́ци·я, -и, (*f*) mechanization

маши́н·изи́р·овать, -уют, (*imp or perf*) mechanize

маши́н·и́ст·, -а, (*m*) machine operator/driven; (rail) engine driver, engineer (U.S.); (naut) engine-room man/rating

~, **коте́ль·н·ый** boilerman

маши́н·и́стк·а, -и, (*g pl*) **-ток·,** (*f*) typist (woman)

маши́н·к·а, -и, (*g pl*) **-н·ок·,** (*f*) (*dim*) of **маши́н·а**

~, **пи́ш·ущ·ая** typewriter

~, **швей·н·ая** sewing machine

~, **шифр·ова́льн·ая** cypher machine

маши́н·н·о- (*component*) machine, machinery

маши́н·н·о-доро́ж·н·ый, -ая, -ое, (*a*) road-machine/machinery

~ **ста́нци·я** (*f*) road-machinery base/park

маши́н·н·о-мелиорати́в·н·ая ста́нци·я (*f*) land reclamation-machine base/park

маши́н·н·о-тракто́р·н·ая ста́нци·я (*f*) agricultural machine base/park

маши́н·н·ый, -ая, -ое, (*a*) machine, engine; mechanical; power-driven, mechanized

~ **бюро́** (*n*) typing office

~ **вы́·зов·** (*m*) power ringing (telephones)

~ **зал·** (*m*) engine room; (elec) turbine house; (print) machine room

~ **нул·ь** (*m*) round-off (computers)

~ **от·дел·е́ни·е** (*n*) engine room/compartment

машино·ве́д·, -а, (*m*) theoretical engineer

машино·ве́д·ени·е, -я, (*n*) engineering science, theoretical engineering

машино·пи́с·н·ый, -ая, -ое, (*a*) typewritten, typed; typing

~ **бюро́** (*n*) typing office/agency

маши́но·пис·ь, -и, (*f*) typescript; typing, typewriting

машино·по·де́л·очн·ый, -ая, -ое, (*a*) (met) engineering

машино·стро·е́ни·е, -я, (*n*) machine building/construction, engineering, industrial engineering; engineering industry

машино·стро·и́тел·ь, -я, (*m*) machine builder, industrial engineer

машино·стро·и́тельн·ый, -ая, -ое, (*a*) engineering, machine-building, industrial engineering

~ **за·во́д·,** (*m*) engineering works, machine-building plant

маши́н·съ·ём·, -а, (*m*) output per machine

маши́но-ча́с·, -а, (*m*) machine-hour

ма́ш·ущ·ий, -ая, -ее, (*pres part act*) *see* **мах·а́ть**; (dynam) flapping

мая́к·, -а́, (*m*) (*see also* **марке́р, радио·мая́к·**); standard, veteran (forestry)

~, **двух·луч·ев·о́й** (rad) double-modulation beacon

~, **ко́д·ов·ый** (rad) code beacon

~, **марке́р·н·ый** (air) boundary marker

~, **о·по·знав·а́тельн·ый** (rad) IFF beacon

~ **-от·ве́т·чик·, -а,** (*m*) (rad) racon, transponder beacon

~, **от·лич·и́тельн·ый свет·ов·о́й** landmark beacon

~, **плов·у́ч·ий** lightship, light vessel

~, **по·са́д·очн·ый** (air) landing beacon

маяк

~, пут·ев·óй аэро·навигацио́н·н·ый airway beacon

~, при·вод·н·óй (air) homing beacon

~, равно·сигна́ль·н·ый (rad) equi-signal beacon

~, радио·локацио́н·н·ый radar beacon, racon

~ -радио·пере·да́т·чик·, -а, (*m*) (rad) beacon transmitter

~, свет·ов·óй beacon light, light beacon, lighthouse

~, с·клад·н·óй folding-leg light buoy (reservoirs)

ма́ят·ник·, -а, (*m*) pendulum, bob; flyball, governor ball

~, баллист·и́чески·й ballistic pendulum

~, крут·и́льн·ый torsion pendulum; torsional oscillation machine

~, матема́т·и́ческ·ий simple pendulum

мето́д· крут·и́льн·ого ма́ят·ни-к·а (*m*) torsional oscillation method

~, с·вя́з·анн·ый connected pendulum

ма́ятник·ов·ый, -ая, -ое, (*a*) of **ма́ят·ник·**

~ копёр· (*m*) (met) pendulum impact tester

~ пил·а́ (*f*) pendulum cross-cut saw

~ при·бо́р· (*m*) pendulum apparatus (for gravity measurements); (met) pendulum impact tester

~ стан·о́к· (*m*) pendulum cross-cut saw

маяч·ко́в·ый, -ая, -ое, (*a*) of **маяк·**

~ ла́мп·а (*f*) (elec) lighthouse valve/tube; disk-seal valve

маяч·н·ый, -ая, -ое, (*a*) lighthouse, beacon

~ с·бор· (*m*) (naut) light dues

мб (*abbr*) = милли·ба́р· (meteor) mb, millibar; (nucl) милли·ба́рн· mb, millibarn

м. б. (*abbr*) = мо́ж·ет бы·ть maybe, perhaps

м-б (*abbr*) = масшта́б· scale

мбар (*abbr*) = милли·ба́р· (meteor) mb, millibar

мбарн (*abbr*) = милли·ба́рн· (nucl) mb, mbn, millibarn

МБР (*abbr*) = меж·континента́ль·-н·ая баллист·и́ческ·ая раке́т·а ICBM, intercontinental ballistic missile

мв (*abbr*) = милли·во́льт (elec) mV, mv, millivolt

Мв (*abbr*) = мега·во́льт· (elec) MV, Mv, megavolt

Мвт (*abbr*) = мега·ва́тт· (elec) MW, Mw, megawatt

мвт (*abbr*) = милли·ва́тт· (elec) mW, mw, milliwatt

Мвтч (*abbr*) = мега·ва́тт-час· (elec) MW/h, Mwhr, megawatt-hour

мг (*abbr*) = милли·гра́мм· mg, milligram; **мега,** mega-, M

мгвт *see* **Мвт**

мгвт-ч *see* **Мвтч**

МГГ (*abbr*) = между·на·ро́д·н·ый гео·физ·и́ческ·ий год· International Geophysical Year

мггц (*abbr*) = мега·ге́рц· Mc/s, mc/s, mcps, megacycles per second

мгл (*abbr*) = милли·гал· milligal (unit of acceleration)

мгл·а́, -ы́, (*f*) (meteor) dry haze

мгл·и́ст·ый, -ая, -ое, (*a*) misty, hazy

мгн (*abbr*) = милли·ге́нри (elec) mh, millihenry

мг·нове́ни·е, -я, (*n*) moment, instant

мг·нове́нн·о (*adv*) instantly, instantaneously; momentarily

~, крит·и́ческ·ий, -ая, -ое, (*a*) (nucl) prompt-critical

мг·нове́нн·ый, -ая, -ое, (*a*) instantaneous, momentary; (nucl) prompt

~ де́й·стви·е (*n*) instantaneous action; (*in gen as adj*) instantaneous

~ знач·е́ни·е (*n*) instantaneous value

~ ис·пар·е́ни·е (*n*) (chem) flash evaporation, (oil) flashing

~, почти́ (nucl) near-prompt

мгом (*abbr*) = мего́м· (elec) megohm, Mohm, meg

Мгц (*abbr*) = мега·ге́рц· Mc/s, mc/s, mcps, megacycles per second

Мдж (*abbr*) = мега·джо́ул·ь mega-joule (units of energy)

м.д.с. (*abbr*) = магнито·дви́ж·у-щ·ая си́л·а (phys) mmf, magneto-motive force

меа́ндр·, -а, (*m*) (geog) meander

ме́бел·ь, -и, (*f*) furniture

ме́бель·н·ый, -ая, -ое, (*a*) of **ме́-бел·ь**

Мёбиус·а лист· (*m*) (math) Mobius band/strip

мега- (*root*) mega-, meg-

мега·ва́рметр·, -а, (*m*) (elec) megavarmeter

мега·ва́тт·, -а, (*m*) megawatt, MW, Mw

мега·ва́тт-ча́с·, -а, *(m)* megawatt-hr, MW/h, Mwhr

мега·во́льт·, -а, *(m)* megavolt, MV, Mv

мега·ге́рц·, -а, *(i)* **-ем,** *(m)* megacycle-per-second, Mc/s

мега·джо́ул·ь, -я, *(m)* megajoule

мега·ди́н·а, -ы, *(f)* megadyne

мега·кариоци́т·, -а, *(m)* (zool) megakaryocyte

мега·кюри́ *(indecl)* (nucl) megacurie, Mc

мегало·бла́ст·, -а, *(m)* (biol) megaloblast

мега́·метр·, -а, *(m)* (astron) megameter

меганод·, -а, *(m)* (rad) meganode

мега·перм·, -а, *(m)* Megaperm (alloy)

мега·пи́р·, -а, *(m)* Megapyr (alloy)

мега·ра́д·, -а, *(m)* (nucl) megarad

мега·скоп·и́ческ·ий, -ая, -ое, *(a)* megascopic, visible to the naked eye

мега·спор·а́нги·й, -я, *(m)* (bot) megasporangium

мега·споро·фи́лл·, -а, *(m)* (bot) megasporophyll

мега·фанеро·фи́т·, -а, *(m)* megaphanerophyte (a tree over 30 m high)

мега·то́нн·а, -ы, *(f)* megaton

мега·то́н·н·ый, -ая, -ое, *(a)* megaton

мега·тро́н·, -а, *(m)* (elec) megatron, lighthouse valve/tube

мега·фо́н·, -а, *(m)* megaphone

мега·электрон·во́льт·, -а, *(m)* (nucl) mega-electron-volt, MeV, mev

мега·э́рг·, -а, *(m)* megaerg (unit of work)

ме́ггер·, -а, *(m)* megger, megohm meter

мег·о́м·, -а, *(m)* megohm, Mohm

мег·о́м·н·ый, -ая, -ое, *(a)* megohm

мег·ом·ме́тр·, -а, *(m)* megohmmeter, megger meter, megger

мег·э́рг·, -а, *(m)* megaerg (unit of work)

мед- *(abbr)* of **медици́н·ск·ий** medical

мёд·, -а, *(nom pl)* **-ы́,** *(g pl)* **-о́в,** *(m)* honey; syrup

~, арбу́з·н·ый melon syrup

меда́л·ь, -и, *(f)* medal

медальо́н·, -а, *(m)* medallion

медве́д·иц·а, -ы, *(i)* **-ей,** *(f)* she-bear

~, больш·а́я (astron) Great Bear, Ursa Major

~, ма́л·ая (astron) Ursa Minor, the Little Bear

медве́д·к·а, -и, *(g pl)* **-д·ок·,** *(f)* (zool) cricket, *Gryllotalpa;* punch (tool); (rail) hand truck/trolley, luggage trolley

медве́д·ь, -я, *(m)* (zool) bear, *Ursus*

~, бе́л·ый polar bear, *Ursus maritimus*

~, се́р·ый grizzly bear, *Ursus horribilis*

медве́ж·ий, -ья, -ье, *(a)* bear, bear's, ursine

~ у́х·о *(n)* (bot) mullein, *Verbascum thapsus*

~ я́год·а *(f)* (bot) bearberry, *Arctostaphylos uva ursi*

медв·я́н·ый, -ая, -ое, *(a)* smelling of honey, honey; made of honey, honey

меде- *(root)* *(see also* **ме́д·н·о-***)* copper, cupri-

меде·но́с·н·ый, -ая, -ое, *(a)* cupriferous, copper-bearing

меде·плав·и́льн·ый, -ая, -ое, *(a)* copper-smelting

меде·со·держ·а́щ·ий, -ая, -ее, *(a)* (geol) copper-bearing, cupriferous

меди·а́льн·ый, -ая, -ое, *(a)* median, medial, middle

меди·а́н·а, -ы, *(f)* (math) median

медиа́н·н·ый, -ая, -ое, *(a)* median

меди·астини́т·, -а, *(m)* (med) mediastinitis

ме́дик·, -а, *(m)* medical man, medico, doctor

медикаме́нт·ы, -ов, *(m pl)* drugs, medicines, medicaments

медикамент·о́зн·ый, -ая, -ое, *(a)* medicinal, medicine

~ сре́д·ств·а *(pl)* medicaments

ме́дико-санита́р·н·ый, -ая, -ое, *(a)* medical, health

ме́дико-суд·е́бн·ый, -ая, -ое, *(a)* (law) forensic medicine

ме́д·ист·о- *(component)* cupro-, copper

ме́д·ист·ый, -ая, -ое, *(a)* coppery; copper-bearing, cupriferous

~ ста́л·ь *(f)* copper-bearing steel

медици́н·а, -ы, *(f)* medicine (the science)

медици́н·ск·ий, -ая, -ое, *(a)* medical

ме́дл·енн·ый, -ая, -ое, *(a)* slow

медл·и́тельн·ый, -ая, -ое, *(a)* unhurried, slow

ме́дл·ить, -ят, *(imp)* delay, put off; linger

медмонти́т·, -а, *(m)* (min) medmontite, cupromontmorillonite

мед·не́ни·е, -я, *(n)* (met) coppering, copper cladding/deposition

мéд·ник·, -а,   (*m*) coppersmith
мéд·ницк·ий, -ая, -ое,   (*a*) of **мéд·-ник·**; (*f decl as a*) coppersmithy
~ **рабóт·ы**   (*pl*) coppersmithing, tinsmithing (broad classification)
мéд·н·о-   (*component*) cupr-, cupri-, cupro-, copper
медно·аммиáчн·ый, -ая, -ое,   (*a*) (chem) cuprammonium
~ **волокн·ó**   (*n*) (text) cuprammonium fibre/fiber
~ **шёлк·**   (*m*) (text) cupra/cuprammonium rayon
медно·за·кúс·н·ый, -ая, -ое,   (*a*) copper-oxide
медно·котéль·н·ый цех·   (*m*) coppersmith's and boilermaker's shop
мéд·н·о-крáс·н·ый, -ая, -ое,   (*a*) copper coloured/colored
медно·лит·éйн·ая, -ой,   (*f decl as a*) copper foundry
медно·лит·éйн·ый, -ая, -ое,   (*a*) copper-founding; (*see also* **медно·лит·éйн·ая**) (*as noun*)
~ **за·вóд·**   (*m*) copper foundry
медно·плав·úльн·ый, -ая, -ое,   (*a*) copper-smelting
~ **за·вóд·**   (*m*) copper works
медно·рýд·н·ый, -ая, -ое, (*a*) copper-ore, (min) copper
медно·цúнк·ов·ый, -ая, -ое,   (*a*) copper-zinc
~ **с·плав·**   (*m*) (met) cap copper
мéд·н·ый, -ая -ое, (*a*) copper
~ **блеск·**   (*m*) (min) copper glance, chalcosite
~ **зéлен·ь** (*f*) (min) green carbonate of copper, malachite
~ **индúго**   (*n indecl*) (min) indigo, copper, covellite
~ **купорóс·**   (*m*) (min) blue vitriol, copperas; (chem) copper sulphate
~ **лазýр·ь** (*f*) (min) azurite
~ **сéт·к·а** (*f*) (met) copper gauze/mesh
~ **уранúт·**   (*m*) (min) copper manite, torternite
~ **черн·ь** (*f*) (min) black copper, melaconite
~ **чáш·к·а** (*f*) (oil) copper dish
~ **числ·ó** (*n*) (chem) copper number
~ **штéйн·** (*m*) (met) copper matte
медо-   (*root*) honey, melli-
медо·вар·éни·е, -я, (*n*) honey extraction
мед·овúк·, -á, (*m*) (bot) nectary; hemp nettle, *Galeopsis tetrahit*
медо·вóд·, -а, (*m*) (bot) honey guide
мед·óв·ый, -ая, -ое, (*a*) honey

медо·гóн·к·а, -и,   (*g pl*) **-н·ок·,** (*f*) honey extractor/centrifuge
медо·éд·, -а, (*m*) (zool) mellivor
медо·нóс·н·ый, -ая, -ое,   (*a*) melliferous
~ **пчел·á**   (*f*) (ent) honey bee, *Apis mellifera*
медо·с·бóр·, -а,   (*m*) honey gathering; yield of honey
медо·смóтр·, -а, (*m*) medical examination
медо·сос·ýщ·ий, -ая, -ее,   (*a*) (zool) mellisugent
мед·пó·мощ·ь, -и,   (*f*) (*abbr*) of **меди·цúн·ск·ая пó·мощ·ь** medical aid/assistance
мед·сестр·á, -ы,   (*f*) (*abbr*) of **меди·цúн·ск·ая сестр·á** nurse
медýз·а, -ы, (*f*) (zool) medusa, jelly-fish
медулло·блáст·, -а, (*m*) (zool) medulliblast
медулл·я́рн·ый, -ая, -ое,   (*a*) (biol) medullary
медунúц·а, -ы, (*i*) **-ей,** (*f*) (bot) lungwort, *Pulmonaria*
~, **аптéч·н·ая**   (bot) lungwort, *Pulmonaria officinalis*
мед·ь, -и, (*f*) copper, Cu
~, **водо·рóд·ист·ая**   (chem) copper hydride
~, **гýб·чат·ая**   (met) copper sponge
~, **дву·хлóр·ист·ая**   cupric chloride
~, **жёлт·ая**   brass
   зá·кис·ь мед·и   cuprous oxide
~, **коммéр·ческ·ая**   commercial copper
~, **крáс·н·ая**   furnace/fire-refined copper
~, **кремн·úст·ая**   copper silicide
~, **одно·хлóр·ист·ая**   cuprous chloride
   ó·кис·ь мéд·и   (chem) cupric oxide
~, **под·óв·ая**   (met) copper bottoms
~, **серно·кúсл·ая**   copper sulphtae
~, **уксусно·кúсл·ая**   copper acetate
~, **хлóристая**   cuprous chloride
~, **хлóр·н·ая**   cupric chloride
~, **черн·ов·áя**   (met) blister copper (in Bessemerizing process), black copper (in blast furnace process)
медь·со·держ·áщ·ий, -ая, -ее,   (*a*) *see* **меде·со·держ·áщ·ий**
мед·янúц·а, -ы,   (*i*) **-ей,** (*f*) (ent) psylla, leaf hopper, (*pl*) *Psyllodea*
мед·я́нк·а, -и,   (*g pl*) **-нок,** (*f*) (chem) verdigris; (zool) grass-snake, *Coronella austriaca*
мед·я́шк·а, -и,   (*g pl*) **-шек·,** (*f*) copper coin; brasswork, brightwork

меж- (*root*) (*see also* **между-**) inter-, between

меж·а́, -и́, (*i*) **-о́й,** (*nom pl*) **ме́ж·и,** (*g pl*) **меж·,** (*f*) boundary; boundary strip

меж·а́том·н·ый, -ая, -ое, (*a*) interatomic, between atoms

~ **взаимо·де́й·стви·е** (*n*) (phys) atomic interaction

~ **рас·сто·я́ни·е** (*n*) (phys) interatomic distance, atomic spacing

меж·вид·ов·о́й, -а́я, -о́е, (*a*) (biol) interspecific

меж·го́р·н·ый, -ая, -ое, (*a*) (geog) intermontane, intermount

меж·дендри́т·н·ый, -ая, -ое, (*a*) (cryst) interdendritic, between dendrites

меж·до́ль·чат·ый, -ая, -ое, (*a*) interlobular

меж·доме́н·н·ый, -ая, -ое, (*a*) between domains, inter-domain (ferromagnetism)

междо·ме́т·и·е, -я, (*n*) (gram) interjection

междо·у́зл·и·е, -я, (*n*) (bot) internode; (cryst) interstitial site, interstice

междо·узл·ов·о́й, -а́я, -о́е, (*a*) internodal; interstitial

ме́жду (*prep* + *instr*) between, among

~ **про́ч·им** among other things, incidentally; by the way, in passing

~ **тём** meanwhile

~ **тём как** while

между- (*prefix*) (*see also* **меж-**) inter-, between

меж·дуа́нт·н·ый, -ая, -ое, (*a*) (instr) interdee

между·ве́д·омственн·ый, -ая, -ое, (*a*) interdepartmental

между·ве́н·цов·ый, -ая, -ое, (*a*) (mech eng) between rims

~ **за·зо́р·** (*m*) (mech eng) rim clearance

между·горо́д·н·ый, -ая, -ое, (*a*) intercity, interurban; (telecom) long-distance; trunk

между·го́р·ь·е, -я, (*n*) (geog) intermontane region, intermount

между·до́н·н·ый, -ая, -ое, (*a*) between bottoms

~ **про·стра́н·ств·о** (*n*) (shipb) double bottom

между·желе́з·н·ое про·стра́н·ств·о (*n*) (mech eng) air gap, clearance

между·звёзд·н·ый, -ая, -ое, (*a*) (astron) interstellar, intergalactic

между·зерн·и́ст·ый, -ая, -ое, (*a*) intergranular, between grains

между·леж·а́щ·ий, -ая, -ое, (*a*) lying between, in between, between

между·молекуля́р·н·ый, -ая, -ое, (*a*) (chem, phys) intermolecular

~ **конденс·а́ци·я** (*f*) (plast) intermolecular/external condensation

между·на·ро́д·н·ый, -ая, -ое, (*a*) international

~ **пра́в·о** (*n*) international law

между·па́р·н·ый, -ая, -ое, (*a*) between pairs, (elec) pair-to-pair

~ **ём·кост·ь** (*f*) pair-to-pair capacitance (telephones)

между·про·вод·н·о́й, -а́я, -о́е, (*a*) between wires, (elec) wire-to-wire

~ **рас·сто·я́ни·е** (*n*) wire spacing

между·пу́т·ь·е, -я, (*n*) (rail) intertrack

между·рёбер·н·ый, -ая, -ое, (*a*) intercostal

между·ре́ч·ь·е, -я, (*n*) (geog) interfluve

между·ря́д·ь·ё, -я, (*n*) (agr) width between rows, row spacing

между·слов·н·ый, -ая, -ое, (*a*) between words

~ **про·бе́л·** (*m*) (print) word spacing

между·сло́й·н·ый, -ая, -ое, (*a*) interlaminar, interleaving, interply

между·сто́лп·и·е, -я, (*n*) (arch) intercolumniation

между·стро́ч·н·ый, -ая, -ое, (*a*) interlinear, between lines

~ **рас·сто·я́ни·е** (*n*) line spacing

между·фа́з·н·ый, -ая, -ое, (*a*) interphase

между·ши́н·н·ый, -ая, -ое, (*a*) (elec) bus-to-bus, between busbars

~ **вы·ключ·а́тел·ь,** (*m*) (elec) bus tie

меж·ева́ни·е, -я, (*n*) boundary demarcation; land surveying, chorometry

меж·ев·о́й, -а́я, -о́е, (*a*) boundary; surveying

~ **знак·** (*m*) (surv) boundary mark

меж·е́нн·ый, -ая, -ое, (*a*) of **меже́н·ь**

меж·е́н·ь, -и, (*f*) low-water level (of river)

меж·за·во́д·ск·ий, -ая, -ое, (*a*) interfactory/works

меж·звёзд·н·ый, -ая, -ое, (*a*) (astron) interstellar

~ **сред·а́** (*f*) interstellar matter

меж·зерн·и́ст·ый, -ая, -ое, (*a*) intergranular, transgranular (of fractures etc.)

меж·каскáд·н·ый, -ая, -ое, (*a*) (elec) interstage

меж·клéт·ник·, -а, (*m*) (bot) intercellular space

∼, схизо·гéн·н·ый (bot) schizogenetic space

меж·клéт·н·ый, -ая, -ое, (*a*) (bot) intercellular

меж·клéт·очн·ый, -ая, -ое, (*a*) intercellular

меж·ключ·и́ц·а, -ы, (*f*) (zool) interclavicle

меж·континент·áльн·ый, -ая, -ое, (*a*) intercontinental

меж·кóст·н·ый, -ая, -ое, (*a*) intercostal

меж·кристалли́т·н·ый, -ая, -ое, (*a*) intercrystalline, intergranular; (met) boundary

меж·лед·никóв·ь·е (*n*) (geol) Interglacial, Interglacial period

меж·леж·áщ·ий, -ая, -ее, (*a*) (geol) intermediate; lying between, between

меж·лóпаст·н·ый, -ая, -ое, (*a*) (biol) interlobate, interlobar; between blades/vanes, vane-to-vane, blade-to-blade (turbines etc.)

меж·лопáт·очн·ый, -ая, -ое, (*a*) (anat) interscapular; blade-to-blade, between blades (turbines etc.)

∼ теч·éни·е (*n*) (air) blade-to-blade flow

меж·луч·ев·óй, -áя, -óе, (*a*) between rays/beams

меж·óст·ист·ый, -ая, -ое, (*a*) (anat) interspinal

меж·планéт·н·ый, -ая, -ое, (*a*) interplanetary

меж·плáч·ечн·ый, -ая, -ое, (*a*) interbundle

меж·плóд·ник·, -а, (*m*) (bot) mesocarp

меж·плоск·остн·óй, -áя, -óе, (*a*) interplanar

∼ рас·сто·я́ни·е (*n*) (phys) interplanar spacing

меж·по·вéрх·ностн·ый, -ая, -ое, (*a*) interfacial

меж·полóс·н·ый, -ая, -ое, (*a*) between bands/strips

меж·пóлюс·н·ый, -ая, -ое, (*a*) interpolar

меж·по·рóд·н·ый, -ая, -ое, (*a*) (biol) cross (breeding etc.)

меж·райóн·н·ый, -ая, -ое, (*a*) interregional

меж·рёбер·н·ый, -ая, -ое, (*a*) intercostal

меж·ремóнт·н·ый, -ая, -ое, (*a*) between repairs

меж·рóд·ов·óй, -áя, -óе, (*a*) (biol) intergeneric

меж·сóрт·ов·óй, -áя, -óе, (*a*) (biol) intervarietal

меж·станциóн·н·ый, -ая, -ое, (*a*) inter-office/exchange (telephones)

меж·у́т·очн·ый об·мéн· (*m*) (biol) metabolism

меж·цéнтр·ов·óй, -áя, -óе, (*a*) between centres/centers, centre-to-centre

∼ рас·сто·я́ни·е (*n*) centre distance (of gears etc.)

меж·электрóд·н·ый, -ая, -ое, (*a*) (elec) interelectrode, between electrodes

∼ изоля́ц·и·я (*f*) (elec) electrode insulation

∼ рас·сто·я́ни·е (*n*) (elec) electrode separation

меж·элемéнт·н·ое со·един·éни·е (*n*) (elec) inter-cell connector

мез- (*int component*) mes-, meso

мез·акóн·ов·ая кисл·от·á (*f*) (chem) mesaconic acid

мезг·á, -и́, (*f*) (bot) sapwood, alburnam; (food) pulp, *purée*

мездр·á, -ы́, (*f*) flesh, flesh side (leather)

мездр·éни·е, -я, (*n*) fleshing, scraping (of hides)

мездр·и́льн·ый, -ая, -ое, (*a*) fleshing

∼ маши́н·а (*f*) fleshing machine (leather)

мездр·óв·ый, -ая, -ое, (*a*) of **мездр·á;** hide

∼ клé·й (*m*) hide glue

∼ сторон·á (*f*) flesh side (of leather)

мездр·я́к·, -á, (*m*) fleshing knife

мездр·ян·óй, -áя, -óе, (*a*) *see* **мездр·óв·ый**

мез·ентéр·и·й, -я, (*m*) (anat) mesentery, mesenterium

мез·енхи́м·а, -ы, (*f*) (biol) mesenchyme, mesenchyma

мезити́л·, -а, (*m*) (chem) mesityl

мезитилéн·, -а, (*m*) (chem) mesitylene

мезити́н·, -а, (*m*) (min) mesitite

мезо- (*int component*) meso-, mes-, mesic

мезо·áтом·, -а, (*m*) (phys) mesic atom

мезо·блáст·, -а, (*m*) (biol) mesoblast

мезо·ви́н·н·ый, -ая, -ое, (*a*) (chem) mesotartaric

мезо·глé·я, -и, (*f*) (zool) mesogloea

мезо·дéрм·а, -ы, (*f*) (biol) mesoderm

мезо·динám·ик·а, -и, (*f*) meson dynamics

мезо·зóй·ск·ая э́р·а (*f*) (geol) Mesozoic

мезо·кáрп·и·й, -я, (*m*) (bot) mesocarp

мезо·кристалл·изáци·я, -и, (*f*) meso-crystallization

мез·оксáл·ев·ая кисл·от·á (*f*) (chem) mesoxalic acid

мезо·клúмат·, -а, (*m*) local climate

мезо·лúт·, -а, (*m*) (min) mesolite; (geol) Mesolithic

мезо·мéр·, -а, (*m*) (chem) mesomer

мезо·мéр·и·я, -и, (*f*) (chem) mesomerism

мезо·мéр·н·ый, -ая, -ое, (*a*) mesomeric

мезо·метеоро·лóг·и·я, -и, (*f*) mesometeorology

мезо·молéкул·а, -ы, (*f*) (phys) mesomolecule

мезо·мóрф·н·ый, -ая, -ое, (*a*) (chem) mesomorphous

мез·óн·, -а, (*m*) (phys) meson

мезо·нефрóс·, -а, (*m*) (zool) mesonephros

мезóн·и·й, -я, (*m*) (phys) mesonium

мезонúн·, -а, (*m*) (build) mezzanine, mezzanine floor

мезóн·н·ый, -ая, -ое, (*a*) meson, mesonic

мезо·пáуз·а, -ы, (*f*) (astron) mesopause

мезо·прóб·а, -ы, (*f*) (zool) mesoprobe

мезо·про·из·вóд·н·ое, -ого, (*n decl as a*) (chem) meso-derivative

мезо·сапрóб·а, -ы, (*f*) (bot) mesosaprobe

мезо·сфéр·а, -ы, (*f*) (astron) mesosphere

мезо·тéл·и·й, -я, (*m*) (zool) mesothelium

мезо·термáль·н·ый, -ая, -ое, (*a*) (geol) mesothermal

мезо·тóр·и·й, -я, (*m*) (chem nucl) mesothorium

мезо·трóн·, -а, (*m*) (nucl) meson, mesotron

мезо·трóф·н·ый, -ая, -ое, (*a*) (biol) mesotrophic

мезо·фанеро·фúт·, -а, (*m*) (bot) mesophanerophyte (tree 8–30 metres high)

мезо·фúлл·, -а, (*m*) (bot) mesophyll

мезо·фúт·, -а, (*m*) (bot) mesophyte

мейбóм·иев·ы жéлез·ы (*pl*) (zool) Meibomian glands

мейергóффер·úт·, -а, (*m*) (min) meyerhofferite

мейó·з·, -а, (*m*) (genet) meiosis

мейонúт·, -а, (*m*) (min) meionite

мейо·филл·ú·я, -и, (*f*) (bot) meiophylly

Мéйснер·а, эффéкт· (*m*) (phys) Meissner effect

мекамúн·, -а, (*m*) (pharm) mecamine

мéл·, -а, (*m*) (min) chalk; (geol) Cretaceous

мёл (*past masc sing*) of **мес·тú**

мелá (*past fem sing*) of **мес·тú**

мелáм·, -а, (*m*) (chem) melam

меламúн·, -а, (*m*) (chem) melamine

меламúн·ов·ая смол·á (*f*) melamine-formaldehyde resin

меламино·формальдетúд·н·ый, -ая, -ое, (*a*) melamine-formaldehyde

мелáнж·, -а, (*i*) **-ем,** (*m*) (food) bulk egg; (chem) mixed acid ($H_2SO_4 + HNO_3$); (text) melange

меланж·ёр·, -а, (*m*) mixer, melangeur (chocolate making)

мелан·úн·, -а, (*m*) (biochem) melanin

мелан·úт·, -а, (*m*) (min) melanite

мелан·óз·, -а, (*m*) (med) melanosis

мелано·крáт·ов·ый, -ая, -ое, (*a*) (geol) melanocratic

мелан·óм·а, -ы, (*f*) (med) melanoma

мелано·тúп·и·я, -и, (*f*) (phot) melanotype

мелано·фóр·, -а, (*m*) (biochem) melanophore

мелантерúт·, -а, (*m*) (min) melanterite

мелан·хол·ú·я, -и, (*f*) (med) melancholy, melancholia

мелáсс·а -ы, (*f*) molasses

мелафúр·, -а, (*m*) (geol) melaphyre

мел·éни·е, -я, (*n*) (*v n*) *see* **мел·úть;** chalking; (*v n*) *see* **мел·é·ть**

мел·ён·ый, -ая, -ое, (*a*) chalked

мéл·ет (*pres 3rd sing*) of **мол·óть**

мел·é·ть, -ют, (*imp*) become shallow, shoal, silt up

мелецитóз·а, -а, (*f*) (chem) melezitose

мели (*past plur*) of **мес·тú**

мели·би·óз·а, -ы, (*f*) (biochem) mellobiose

мели·лúт·, -а, (*m*) (min) melilite

мелин·úт·, -а, (*m*) (expl) melinite

мелино·фáн·, -а, (*m*) (min) melinophane

мелиор·áци·я, -и, (*f*) improvement (of natural states); land reclamation; weather modification

~, **вод·н·áя** water management

~ **клúмат·а** weather modification

~, **рыбо·вóд·н·ая** fisheries conservation

мелиор·и́р·овать, -уют, (*imp and perf*) improve, reclaim (land)

мелис·, -а, (*m*) sand sugar

мелисси́л·ов·ый, -ая, -ое, (*a*) melissyl, myricyl

~ спирт· (*m*) (chem) melissyl alcohol

мелисси́н·ов·ый, -ая, -ое, (*a*) (chem) melissic

~ кисл·от·а́ (*f*) (chem) melissic acid

мели́сс·ов·ый, -ая, -ое, (*a*) melissa, melissic

~ спирт· (*m*) melissyl/myricyl alcohol

мел·и́ст·ый, -ая, -ое, (*a*) (ocean) shelving, shoaling

мел·и́ть, -я́т, (*imp*) chalk, cover with chalk; pulverize, reduce to fragments; make small

мел·к·ий, -ая, -ое, (*a*) small, petty, fine; shallow

мел·к·о- (*component*) small, fine, finely

мелко·би́·т·ый, лёд· (*m*) very light floes (ice)

мелко·во́д·н·ый, -ая, -ое, (*a*) shallow, shallow-water

мелко·во́д·ь·е, -я, (*n*) shallows, shallow water

мелко·волокн·и́ст·ый, -ая, -ое, (*a*) close-grained (of wood), fine-fibrous

мелко·диспе́рс·н·ый, -ая, -ое, (*a*) finely dispersed/divided

мелко·дробл·ён·ый, -ая, -ое, (*a*) fine-ground, finely fragmented, crushed fine

мелко·за·зу́бр·енн·ый, -ая, -ое, (*a*) crenulate

мелко·зём·, -а, (*m*) (agr) fine earth, melkozem, (geol) aleurite, silt

мелко·зерн·и́ст·ый, -ая, -ое, (*a*) fine/close-grained

мелко·зу́б·к·а, -и, (*g pl*) **-б·ок·,** (*f*) smooth-cut file

мелко·зу́б·чат·ый, -ая, -ое, (*a*) denticulate, (biol) crenulate

мелко·зу́б·н·ый, -ая, -ое, (*a*) fine/close-toothed

мелко·корн·ев·о́й, -ая, -ое, (*a*) shallow-rooted

мелко·кали́бер·н·ый, -ая, -ое, (*a*) (gunn) small-bore

мелко·лепе́ст·ник·, -а, (*m*) (bot) flea-bane, *Erigeron*

мелко·мо́дуль·н·ый, -ая, -ое, (*a*) small-module

мелко·мо́р·ь·е, -я, (*n*) (ocean) shelf

мелко·пи́ль·чат·ый, -ая, -ое, (*a*) serrulate

мелко·по́р·ист·ый, -ая, -ое, (*a*) fine-pore, finely porous, fine-perforated

мелко·рас·пыл·ённ·ый, -ая, -ое, (*a*) finely dispersed/pulverized

мелко·сер·и́йн·ое про·из·во́д·ств·о (*n*) small-scale manufacture/production

мелко·сид·я́щ·ий, -ая, -ее, (*a*) (shipb) shallow-draught

мелко·сло́й·н·ый, -ая, -ое, (*a*) fine-grain (of wood); finely stratified

мелко·со́п·очник·, -а, (*m*) hilly/hillocky topography

мелко·со́рт·н·ое желе́з·о (*n*) (met) small bars and sections

мелко·со́рт·н·ый, стан· (*m*) (met) light section mill

мелко·това́р·н·ый, -ая, -ое, (*a*) (com) small-scale, small·

мелко·узл·ова́т·ый, -ая, -ое, (*a*) micronodular

мелко·яче́·ист·ый, -ая, -ое, (*a*) fine/close-mesh

мелко·яче́·йн·ый, -ая, -ое, (*a*) fine/close-mesh

Меллин·а, пре·образ·ова́ни·е (*n*) (math) Mellin inversion formula

мелл·и́т·, -а, (*m*) (min) mellite, honeystone

мелли́т·ов·ый, -ая, -ое, (*a*) (min) mellite; (chem) mellitic

~ кисл·от·а́ (*f*) (chem) mellitic acid

меллито́л·, -а, (*m*) (chem) mellitene, hexamethylbenzene

мело (*past neut sing*) of **мес·ти́**

мел·ова́ни·е, -я, (*n*) (paper) coating; compounding (rubber)

мел·о́ванн·ый, -ая, -ое, (*a*) chalked; (paper) coated; compounded (rubber)

~ бума́г·а (*f*) coated paper, chalk-overlay paper

мел·ов·а́я, -о́й, (*f decl as a*) compounding room (rubber)

мел·ов·о́й, -а́я, -о́е, (*a*) (*see also* **мел·о·в·а́я**) (*as noun*); chalk, chalky; (geol) Cretaceous

~ сво́й·ств·о (*n*) chalkiness

~ систе́м·а (*f*) (geol) Cretaceous system

мелод·и́чн·ый, -ая, -ое, (*a*) melodious

мел·о́к·, -л·к·а́, (*m*) piece of chalk, a chalk

мело·хо́рд·, -а, (*m*) (acous) melochord

мел·о́чн·ой, -а́я, -о́е, (*a*) *see* **ме́л·оч·н·ый**

ме́л·очн·ый, -ая, -ое, (*a*) petty, small/mean-minded

ме́л·оч·ь, -и, (*nom pl*) **-и,** (*g pl*) **-е́й,** (*f*) trifle, trifles, trivialities; small change (money); (tech) fines, smalls, shorts

~, ко́кс·ов·ая coke fines, breeze

~, ру́д·н·ая fine ore

~, у́голь·н·ая fine coal, slack, fines

ме́л·ь, -и, (*f*) shoal, shallow
　сад·и́ть на мел·ь (naut) beach, put aground; **-ся** run aground

мельк·а́ни·е, -я, (*n*) (*v n*) *see* **мельк·а́·ть;** flickering

мельк·а́·ть, -ю́т, (*imp*) appear for a moment; gleam, flash

мельк·н·у́ть, -у́т, (*perf*) *see* **мельк·а́·ть**

ме́льк·ом (*adv*) in passing, for a moment

ме́ль·ник·, -а, (*m*) miller

мельниковит·, -а, (*m*) (min) melnikovite

ме́ль·ниц·а, -ы, (*i*) **-ц·ей,** (*f*) mill, grinder; (paper) refiner; pulverizer (coal)

~, аэро·би́·льн·ая air-dispersion mill

~, бараба́н·н·ая tumbling mill

~, бег·унко́в·ая edge-runner mill

~, би́·льн·ая hammer mill

~, вал·ко́в·ая roller mill

~, вальц·о́в·ая roller mill

~, ветр·ян·а́я windmill

~, вод·ян·а́я watermill

~, га́л·ечн·ая pebble mill (ore dressing)

~, ди́ск·ов·ая attrition mill

~, жерн·ов·а́я attrition mill

~, каска́д·н·ая cascade mill

~, кач·а́ющ·аяся vibratory mill

~, колло́йд·н·ая colloid mill

~, кон·и́ческ·ая conical mill; (paper) conical refiner

~, корзи́н·чат·ая multiple-cage mill

~, кулач·ко́в·ая hammer mill

~, ма́сс·н·ая (paper) refiner

~, молот·ко́в·ая hammer mill

~, муко·мо́ль·н·ая flour mill

~, муко·се́й·н·ая bolting mill

~, об·ди́р·очн·ая (food) scouring mill

~, стерж·нев·а́я rod mill

~, стру́й·н·ая air-swept grinding mill

~, тру́б·н·ая tube mill

~, уда́р·н·о-ди́ск·ов·ая disintegrator, multiple-cage grinding mill

~, чил·и́йск·ая Chilian mill

~, шар·ов·а́я globe/ball mill, (paper) ball mill refiner

~, ша́хт·н·ая air-swept direct-fired pulverizer (for stoking furnaces)

ме́ль·ничн·ый, -ая, -ое, (*a*) mill, milling, grinding

мельхио́р·, -а, (*m*) 20 per cent cupronickel, melchior (precisely); cupronickel, German silver (in general)

мельхиор·иза́ци·я, -и, (*f*) nickeling (of gun barrels)

мель·ча́йш·ий, -ая, -ее, (*a*) smallest, minutest, finest

мельч·а́·ть, -ю́т, (*imp*) become smaller; lose significance, become trivial

ме́ль·ч·е (*comp*) of **ме́л·к·ий**

мельч·и́ть, -а́т, (*imp*) pulverize, grind up, mill, make smaller

ме́л·ющ·ий, -ая, -ее, (*pres part act*) of **мол·о́ть,** grinding, milling

~ вентиля́тор· (*m*) air-swept unit pulverizer (furnaces)

мел·я́ (*pres gerund*) of **мол·о́ть**

меля́сс·а, -ы, (*f*) *see* **мела́сс·а**

мембра́н·а, -ы, (*f*) membrane; (tech) diaphragm

~ вз·рыв·а́тел·я (gunn) needle disk (fuse)

~, у́голь·н·ая carbon diaphragm (telephone)

мембран·е́лл·а, -ы, (*f*) (zool) membranella

мембра́н·н·ый, -ая, -ое, (*a*) membrane, diaphragm; membranous

~ му́фт·а (*f*) flexible steel-member coupling

мемно·ско́п·, -а, (*m*) memnoscope

мемора́ндум·, -а, (*m*) memorandum

мемуа́р·ы, -ов, (*m pl*) memoirs

ме́н·а, -ы, (*f*) barter, exchange

менакани́т·, -а, (*m*) (min) menaccanite

менделев·и́й, -я, (*m*) mendelevium, Md (transuranium element)

менделев·и́т·, -а, (*m*) (min) mendelevite

Менделе́ев·а, систе́м·а элеме́нт·ов (*f*) (chem phys) periodic system/table, Mendeleev's table

менделе́ев·ск·ий, -ая, -ое, (*a*) Mendeleev, periodic

~ числ·о́ (*n*) (phys) atomic number

мендел·и́зм·, -а, (*m*) (genet) Mendelism

ме́н·ее (*comp*) of **ма́л·о;** less

~ вс·его́ least of all
　тем не ме́н·ее nevertheless

Менела·я, теоре́м·а (*f*) (math) Menelaus' theorem

ме́нзул·а, -ы, (*f*) (surv) plane table

ме́нзуль·н·ый, -ая, -ое, (*a*) plane-table

мензу́р·к·а, -и, (*f*) measuring glass, graduated cylinder

менили́т·, -а, (*m*) (min) menilite
менинги́т·, -а, (*m*) (med) meningitis
мени́ск·, -а, (*m*) meniscus
∼, от·риц·а́тельн·ый negative/diverging meniscus
∼, по·лож·и́тельн·ый positive/converging meniscus
∼, прост·о́й single meniscus
мени́ск·ов·ый, -ая, -ое, (*a*) of **мени́ск·**
мен·ов·о́й, -а́я, -о́е, (*a*) barter, exchange
∼ сто́·имост·ь (*f*) (com) exchange value
менструа́ль·н·ый, -ая, -ое, (*a*) (physiol) menstrual
ментадиен·, -а, (*m*) (chem) terpinene
мент·а́н·, -ы, (*m*) (chem) menthane
мент·ано́л·, -а, (*m*) (chem) menthanol
мент·е́н·, -а, (*m*) (chem) menthene
менте́нер·, -а, (*m*) (civ eng) maintainer, road maintenance vehicle
ментено́л·, -а, (*m*) (chem) terpineol
мент·и́л·, -а, (*m*) (chem) menthyl
мент·о́л·, -а, (*m*) (pharm) menthol
мент·о́н·, -а, (*m*) (chem) menthone
Менье́, теоре́м·а (*f*) (math) Meusnier's theorem
Менье́р·а, бол·е́зн·ь (*f*) (med) Ménière's disease
ме́нь·ш·е (*comp*) of **ма́л·ый, ма́л·еньк·ий;** smaller; less
ме́нь·ш·ий, -ая, -ее, (*a*) less, lesser, smaller, minor, inferior
мень·шинств·о́, -а́, (*n*) minority
меню́ (*n indecl*) (food) menu
меня́ (*pron, acc/gen*) of **я;** me
мен·я́льн·ый, -ая, -ое, (*a*) exchange (currency)
мен·я́·ть, -ют, (*imp*) change, alter; exchange
мен·я́ющий·ся, -аяся, -ееся, (*pres part act*) of **мен·я́·ться;** changing, varying, variable, fluctuating, alternating, live (load)
ме́р·а, -ы, (*f*) measure; unit of measurement
　　в знач·и́тельн·ой ме́р·е to a considerable extent, largely, highly
　　в из·ве́ст·н·ой ме́р·е to a known degree, to a certain extent
　　в ме́р·у reasonable, sufficiently, within bounds
∼ диспе́рс·и·и measure of dispersion
∼ за·щи́т·ы protective measure
∼, кон·цев·а́я end gauge/gage

ме́р·а
∼ мно́ж·еств·а (math) measure of a set
　　не в ме́р·у excessively, immoderately
　　по бо́ль·ш·ей ме́р·е at the most, at most
　　по кра́й·н·ей ме́р·е at least
　　по ме́р·е proportionately, in proportion to
　　по ме́р·е не·об·ход·и́мост·и as required
　　по ме́р·е того как as
　　по ме́р·е у·вели·че́ни·я … as … increases
∼, по·го́н·н·ая linear measure
∼ пред·о·сторо́ж·ност·и precaution, precautionary measure
　　при·ним·а́·ть ме́р·ы take/adopt measures
∼ социа́ль·н·ой за·щи́т·ы social security measure
∼ то́ч·ност·и (math) index of precision
мергел·ева́ни·е, -я, (*n*) (agr) marling
ме́ргел·ист·ый, -ая, -ое, (*a*) (geol) marly, marlaceous
ме́ргел·ь, -я, (*m*) (geol) marl
ме́ргель·н·ый, -ая, -ое, (*a*) (geol) marlaceous, marly
мерёж·а, -и, (*f*) (fish) seine, drag net
мерёж·к·а, -и, (*g pl*) **-ж·ек·,** (*f*) (text) hem-stitch
мере́·йн·ый, -ая, -ое, (*a*) of **мере·я́**
∼ маши́н·а (*f*) graining machine (leather)
∼ пресс· (*m*) boarding machine (leather)
мере·я́, -й, (*f*) natural grain (leather)
мёрз· (*past masc sing*) of **мёрз·нуть**
мёрз·к·ий, -ая, -ое, (*a*) vile
мёрзл·ост·ь, -и, (*f*) frozen state; congelation, congealment
мерзл·от·а́, -ы́, (*f*) frozen ground
∼, ве́ч·н·ая permafrost
мерзл·о́тн·ый, -ая, -ое, (*a*) of **мерз·л·от·а́**
мерзлото·ве́д·ени·е, -я, (*n*) geocryology
мерзлото·ме́р·, -а, (*m*) cryopedometer, frozen-ground depth gauge/gage
мёрзл·ый, -ая, -ое, (*a*) frozen, congealed
мерзл·я́к·, -а́, (*m*) frozen turf; person sensitive to cold
мерзл·я́тин·а, -ы, (*f*) frozen thing/object; person sensitive to cold
мёрз·нуть, -нут, (*past masc sing*) **мёрз,** (*imp intrans*) freeze

мерзоля́т·, -а, (*m*) (chem) sulphonate, sulponate

мёрз·ш·ий, -ая, -ее, (*past part act*) of **мёрз·нуть**

меридиа́н·, -а, (*m*) meridian

~, гри́нвич·ск·ий Greenwich meridian

~, и́ст·инн·ый geographical meridian

~, на·ча́ль·н·ый first meridian

~, неб·е́сн·ый celestial meridian

~, нул·ев·о́й zero meridian

~, у·сло́в·н·ый false meridian

меридиона́ль·н·ый, -ая, -ое, (*a*) meridional, meridian

~ выс·от·а́ (*f*) meridian altitude

~ част·ь (*f*) meridional part, mer-part

мери·ка́рп·, -а, (*m*) (bot) mericarp

мер·и́л·о, -а, (*n*) gauge, gage, standard, criterion

мер·и́льн·ый, -ая, -ое, (*a*) measuring, gauging, gaging

~ маши́н·а (*f*) (text) length-and-breadth measuring machine

мерино́с·, -а, (*m*) merino (sheep)

мери·сте́м·а, -ы, (*f*) (bot) meristem

мер·и́тельн·ый, -ая, -ое, (*a*) measuring

~ с·вид·е́тельств·о (*n*) (naut) tonnage certificate, certificate of measurement

мер·ить, -ят, (*imp*) measure, gauge; try on, fit (clothes for size)

мер·к·а, -и, (*g pl*) **-р·ок·,** (*f*) measure; measuring rod; yardstick

~, координа́т·н·ая (surv) romer

мерказоли́л·, -а, (*m*) (pharm) mercazolyl, mercazole

меркапта́л·, -а, (*m*) (chem) mercaptal

меркапта́н·, -а, (*m*) (chem) mercaptan

меркапти́д·, -а, (*m*) (chem) mercaptide

меркапто·бензо·тиазо́л·, -а, (*m*) (chem) mercaptobenzothiazole

меркапто́л·, -а, (*m*) (chem) mercaptol

меркапто·фо́с·, -а, (*m*) mercaptophos (insecticide)

мерка́тор·ск·ий, -ая, -ое, (*a*) Mercator's

~ ми́л·я (*f*) geographical mile

~ прое́кци·я (*f*) (surv) Mercator's projection

ме́рк·нуть, -нут, (*past masc sing*) **мерк·** *or* **мерк·нул,** (*imp*) fade, grow dark/dim/faint

меркуриали́зм·, -а, (*m*) (med) chronic mercurial poisoning

Меркури́·й, -я, (*m*) (astron) Mercury

меркури·ме́тр·и·я, -и, (*f*) mercurimetry, mercuric titration

меркур·и́ровани·е, -я, (*n*) (chem) mercuration, mercurization

меркуро·ме́тр·и·я, -и, (*f*) (chem) volumetric titration

мерл·у́шк·а, -и, (*f*) lambskin

мерл·у́шков·ый, -ая, -ое, (*a*) lambskin

ме́р·ник·, -а, (*m*) measurer, measuring hopper/vessel

ме́р·ност·ь, -и, (*f*) regularity, rhythm

ме́р·н·ый, -ая, -ое, (*a*) measuring, measured; (*as component*) -dimensional

~ ле́нт·а (*f*) measuring tape

~ ми́л·я (*f*) (shipb etc.) measured mile

~ по·су́д·а (*f*) (chem) measuring glass

меро·бла́ст·и́ческ·ий, -ая, -ое, (*a*) (zool) meroblastic

меро·га́м·и·я, -и, (*f*) (biol) merogamy

меро·ксе́н·, -а, (*m*) (min) meroxene

меро·мо́рф·н·ый, -ая, -ое, (*a*) meromorphic

~ фу́нкци·я (*f*) (math) meromorphic function

меро·планкто́н·, -а, (*ṁ*) (biol) meroplankton, seasonal plankton

меро·при·я́т·и·е, -я, (*n*) measure (administrative)

~, противо·локацио́н·н·ые (*pl*) radar countermeasures

меро·стома́т·ы, -ов, (*pl*) (zool) merostomes, *Merostomata*

меро·том·и́·я, -и, (*f*) (biol) merotomy

мерсериз·а́ци·я, -и, (*f*) (text) mercerization

мертв·е́·ть, -ют, (*imp*) become lifeless; become stiff/rigid/numb

мертв·е́ц·, -а, (*i*) **-·ц·о́м,** (*m*) dead person, corpse

мертво·е́д·, -а, (*m*) (ent) carrion beetle, (*pl*) *Silphidae*

мертво·рожд·ённ·ый, -ая, -ое, (*a*) (med) stillborn

мёртв·ый, -ая, -ое, (*a*) dead; stagnant (water)

~ вит·о́к· (*m*) (elec) dead turn, armature dead turn

~ воро́н·к·а (*f*) (rad) cone/zone of silence; (gunn) dead space

~ зо́н·а (*f*) (rad) cone/zone of silence

~ зыб·ь (*f*) (ocean) ground swell

~ про·стра́н·ств·о (*n*) (rad, gunn) dead ground/space

~ то́ч·к·а (*f*) dead centre/point

~ то́ч·к·а, ве́рх·н·яя (*f*) (mech eng) top dead centre

~ то́ч·к·а, ни́ж·н·яя (*f*) bottom dead centre

~ ход· (*m*) backlash, play

ме́ртел·ь, -я, (*m*) (build) lime mortar

мерц·а́ни·е, -я, (*n*) flickering, glimmering, twinkling; (astron) scintillation

~, **рас·се́·янн·ое** (astron) scattering scintillation

мерц·а́·ть, -ют, (*imp*) flicker, glimmer, twinkle

мерц·а́ющ·ий, -ая, -ее, (*pres part act*) of **мерц·а́·ть**; flickering, glimmering, twinkling

~ **метео́р·** (*m*) flickering meteor

ме́р·я·ть, -ют, (*imp*) **ме́р·ить**

месдо́з·а, -ы, (*f*) *see* **мессдо́з·а**

ме́с·ив·о, -а, (*n*) pulp, mash, mess

мес·и́лк·а, -и, (*g pl*) **-лок·,** (*f*) dough mixer, kneader, masticator

~ **-вз·би́в·а́лк·а, -и,** (*f*) (food) mixer and beater

мес·и́льн·ый, -ая, -ое, (*a*) kneading, puddling; mixing

~ **маши́н·а** (*f*) kneading machine, masticator, dough mixer

мес·и́ть, -́ят, (*imp*) knead, puddle; mix

Мёссбауер·а, еффе́кт· (*m*) (nucl) Mössbauer effect

мессдо́з·а, -ы, (*f*) (instr) hydraulic capsule

мес·ти́, (*pres 3rd sing, pl*) **мет·ёт, мет·у́т,** (*imp*) sweep

ме́ст·н·о (*adv*) locally

ме́ст·ност·ь, -и, (*f*) locality, place, area, country, terrain

~, **без·ориенти́р·н·ая** (nav) non-rapid-fixing area

~, **за·кры́·т·ая** (geog) close country

~, **от·кры́·т·ая** open country

~, **пере·се́ч·енн·ая** rolling/broken country

у́гол· ме́ст·ност·и (*m*) (gunn) angle of sight

ме́ст·н·ый, -ая, -ое, (*a*) local, localized; (*as component*) -berth, -seater, -place

~ **а́бсцесс·** (*m*) (med) localized abscess

~ **пад·е́ж·** (*m*) (gram) locative

~ **со·обш·е́ни·е** (*n*) (rail) local service

ме́ст·о, -а, (*n*) (*see also* **то́чка**) place, position, locus, site; seat, berth; piece (of luggage); passage (in book etc.)

~, **бо·ев·о́е** (*mil*) action station

~, **вака́нт·н·ое** (phys) vacant position, vacancy

~, **гео·ме́тр·и́ческ·ое** (math) locus

ме́ст·о

~, **дей·стви́тельн·ое** (air) ground position

~ **де́й·стви·я** scene of action

~, **де́т·ск·ое** (anat) placenta

~, **и́м·е·ть** take place, occur

~ **корн·е́й, гео·метр·и́ческ·ое** (math) root locus

~ **на·знач·е́ни·я** destination

~, **обсерв·о́ванн·ое** (nav) fix, observed position

~ **о·жид·а́ни·я** (air) holding point

~ **от·правл·е́ни·я** point of departure

~, **рас·чёт·н·ое** (nav) dead-reckoning position

~, **сп·а́льн·ое** sleeping berth

~ **сто·я́нк·и** (М/Т) parking place; (air) hard standing; (naut) moorings, anchorage

~, **с·числ·и́м·ое** (nav) dead-reckoning position

у́гол· ме́ст·а (*m*) elevation, tilt (radar)

у́гол· ме́ст·а це́л·и (*m*) (gunn) angle of sight

ме́сто·жи́т·ельств·о, -а, (*n*) residence, place of residence

ме́сто·им·е́ни·е, -я, (*n*) (gram) pronoun

ме́сто·на·хожд·е́ни·е, -я, (*n*) location, site, place; (nav) position; (geol) occurrence, locality

~, **вы́·числ·енн·ое** (nav) observed position

ме́сто·обит·а́ни·е, -я, (*n*) (biol) habitat

ме́сто·о·пре·дел·е́ни·е, -я, (*n*) (nav) fixing, position finding

ме́сто·по·лож·е́ни·е, -я, (*n*) situation, position, whereabouts, site; (nav) position

ме́сто·пре·быв·а́ни·е, -я, (*n*) place of residence

ме́сто·рас·по·лож·е́ни·е, -я, (*n*) location, situation, place

ме́сто·рожд·е́ни·е, -я, (*n*) (geol) deposit, bed; birthplace

~, **корен·н·о́е** (geol) lode

~, **рос·сып·н·о́е** placer deposit

~, **у́голь·н·ое** coal field

ме́сяц·, -а, (*i*) **-ем,** (*m*) month; moon

~, **аномалист·и́ческ·ий** anomalistic month

~, **висо·ко́с·н·ый** leap month

~, **дракон·и́ческ·ий** draconitic month

~, **календа́р·н·ый** calendar month

~, **сидер·и́ческ·ий** sidereal month

ме́сяц

~, сино́д·и́ческ·ий synodic/synodical/ lunar month

~, троп·и́ческ·ий tropical month

ме́сяч·ник·, -а, (*m*) month (nominal period for exhibition etc.); (bot) meadow grass, *Poa*

ме́сяч·н·ый, -ая, -ое, (*a*) monthly, month's; lunar, moon's

мет- (*int component*) (chem) meth-, methyl

мета- (*int component*) (chem, biol) meta

мета·антимона́т·, -а, (*m*) (chem) meta-antimonate

мета·арсена́т·, -а, (*m*) meta-arsenate

мета·бази́т·, -а, (*m*) (geol) metabasite

мета·био́з·, -а, (*m*) (biol) metabiosis

метаболи́зм·, -а, (*m*) (physiol) metabolism

мета·боли́т·, -а, (*m*) (biol, geol) metabolite

метабол·и́ческ·ий, -ая, -ое, (*a*) metabolic

метаболо́н·, -а, (*m*) (nucl) metabolon

мета·бора́т·, -а, (*m*) metaborate

мета·бо́р·н·ый, -ая, -ое, (*a*) (chem) metaboric

~ кисл·от·а́ (*f*) metaboric acid

мет·а́вш·ий, -ая, -ее, (*past part act*) of **мет·а́·ть**

мета·гала́ктик·а, -и, (*f*) (astron) metagalaxy

метагеветит·, -а, (*m*) (min) metahewetite

мета·гене́з·, -а, (*m*) (biol) metagenesis, alternation of generations

мет·а́ем·ый, -ая, -ое, (*pres part pass*) of **мет·а́·ть**

мета·ди́н·, -а, (*m*) (elec) metadyne, amplidyne generator, amplidyne

мета·кисл·от·а́, -ы́, (*f*) (chem) metaacid

мет·акрила́т·, -а, (*m*) (chem) methacrylate

мет·акри́л·ов·ая кисл·от·а́ (*f*) methacrylic acid

мет·акри́л·ов·ая смол·а́ (*f*) (plast) methyl metacrylate resin

мета·ксе́н·и·я, -и, (*f*) (bot) metaxenia

мета·ксиле́м·а, -ы, (*f*) (bot) metaxylem

мета́лл·, -а, (*m*) metal

~, благо·ро́д·н·ый noble metal

~, втор·и́чн·ый secondary metal

~, до́мен·н·ый liquid blast-furnace metal (steel making)

мета́лл

~, драго·це́н·н·ый precious metal

~, жи́дк·ий liquid/molten metal

~, лёгк·ий light metal

~, легко·пла́в·к·ий fusible metal

~, лист·ов·о́й flat-rolled metal, flat products

~, много·сер·ни́ст·ый (met) polysulphide

~, на·пла́вл·енн·ый weld/deposited metal

~, основ·н·о́й matrix, matrix metal; (weld) parent metal; base metal (of an alloy); backing metal (duplexing)

~, перв·и́чн·ый virgin metal

~, под·ши́п·ников·ый bearing metal

~, при·са́д·очн·ая (weld) added/filler metal

~, ре́д·к·ий rare metal, less-common metal

~, редко·земе́ль·н·ый rare-earth metals

~, сорт·ов·о́й sections, section metal, metal in rolled sections

~, техн·и́ческ·ий industrial metal, industrial-grade metal

~, туго·пла́в·к·ий infusible/refractory metal

~, тяжёл·ый base metal

~, цве́т·н·ый non-ferrous metal

~, цельно·решёт·чат·ый (build) expanded metal

~, чёрн·ый ferrous metal

~, щёлочно·земе́ль·н·ый alkali-earth metal

~, щелоч·н·о́й alkali metal

металл·и́д·, -а, (*m*) (chem) intermetallic compound

металлиз·а́тор·, -а, (*m*) metal spraygun

металлиз·а́ци·я, -и, (*f*) metallizing, metal coating; (elec) bonding (metal parts)

~, диффузио́н·н·ая (met) diffusion coating

~ рас·пыл·е́ни·ем metal spraying, metallization

металлиз·и́рованн·ый, -ая, -ое, (*a*) metallized; metal-coated/plated; (elec) bonded

~ ле́нт·ы (*pl*) window, chaff (radar obstacles)

~ фане́р·а (*f*) metal-faced plywood

металлиз·о́ванн·ый, -ая, -ое, (*a*) see **металлиз·и́рованн·ый**

металл·и́ческ·ий, -ая, -ое, (*a*) metal; metallic

металл·и́ческ·ий, -ая, -ое
~ **блеск·**　(*m*) metallic lustre/luster/glint
~ **порош·о́к·** (*m*) powder metal, metal powder
~ **ура́н·**　(*m*) uranium metal
металл·и́чност·ь, -и,　(*f*) metallicity, metallic properties
металл·микроско́п·, -а,　(*m*) metallographic microscope
металло·а́сбест·, -а,　(*m*) metallized asbestos
металло·ве́д·, -а, (*m*) metal physicist
металло·ве́д·ени·е, -я,　(*n*) physical metallurgy, metallography, science of metals
металло·ве́д·ческ·ий, -ая, -ое,　(*a*) of **металло·ве́д·ени·е**
металло·ви́д·н·ый, -ая, -ое,　(*a*) metalline
металло·ген·и́ческ·ий, -ая, -ое, (*a*) (geol) metallogenetic
~ **прови́нци·я** (*f*) (geol) metallogenetic province
металло·ге́н·и·я, -и, (*f*) (geol) metallogeny
металло́·граф·, -а, (*m*) metallographer
металло·граф·и́ческ·ий, -ая, -ое, (*a*) metallographic
металло·гра́ф·и·я, -и,　(*f*) metallography
металло·ём·кост·ь, -и, (*f*) metal content (of machinery etc.)
металл·о́ид·, -а, (*m*) (chem) metalloid, non-metal
металло·из·де́л·и·я, -й, (*n pl*) hardware, metal ware
металло·изо́л·, -а,　(*m*) (build) bituminized damp proofing foil
металло·иск·а́тел·ь, -я,　(*m*) metal detector/locator
металло·кера́м·ик·а, -и, (*f*) powder metallurgy; (nucl) metal-ceramics
металло·керам·и́ческ·ий, -ая, -ое, (*a*) powder metal; (nucl) metal ceramic, cermet, ceramet
~ **дета́л·ь** (*f*) powder metallurgy part; (nucl) cermet part, metal ceramic part
~ **ла́мп·а** (*f*) (elec) stacked-electron tube
~ **те́л·о** (*n*) (met) powder-metal compact
металло·констру́кци·я, -и, (*f*) metal structure
металло·ло́м·, -а, (*m*) scrap metal
металло·метр·и́ческ·ий, -ая, -ое, (*a*) metallometric

металло·метр·и́ческ·ий, -ая, -ое
~ **съ·ём·к·а,**　(*f*) (min) direct prospecting
металло·микроско́п·, -а,　(*m*) metallographical/microscope
металло·но́с·н·ый, -ая, -ое,　(*a*) metalliferous
металло·об·раба́т·ывающ·ий, -ая, -ее, (*a*) metal-working/machining
металло·о·плёт·к·а, -и, (*g pl*) **-т·ок·,** (*f*) metal braiding
мета́лло·-орган·и́ческ·ий, -ая, -ое, (*a*) (chem) metalloorganic, organometallic
металло·плён·очн·ый, -ая, -ое,　(*a*) metal-film
металло·по·кры́т·и·е, -я,　(*n*) metal coating
металло·при·ём·ник·, -а,　(*m*) well, crucible (blast furnace)
металло·ре́ж·ущ·ий, -ая, -ее,　(*a*) metal-working/machining
~ **инструме́нт·**　(*m*) metal-working tools
~ **стан·о́к·** (*m*) machine tool
металло·со·держ·а́щ·ий, -ая, -ее, (*a*) metalliferous, containing metal
металло·те́рм·и·я, -и,　(*m*) (met) metallothermic process
металло·фи́з·ик·а, -и,　(*f*) metal physics, physics of metals
металл·у́рг·, -а, (*m*) metallurgist
металл·ург·и́ческ·ий, -ая, -ое,　(*a*) metallurgical
металл·у́рги·я, -и,　(*f*) metallurgy, process/extraction metallurgy
~, **порош·ко́в·ая**　powder metallurgy
~, **цве́т·н·ая**　non-ferrous metallurgy; non-ferrous metals industry
~, **чёрн·ая**　ferrous metallurgy; iron and steel industry
мета·ло́ф·, -а, (*m*) (anat) metaloph
метальдеги́д·, -а, (*m*) (chem) metaldehyde, meta-aldehyde
мета·ме́р·, -а, (*m*) (biol) metamere
мета·мер·и́·я, -и, (*f*) (chem, biol) metamerism
мета·ме́р·н·ый, -ая, -ое, (*a*) metameric
метами́кт·н·ый, -ая, -ое, (*a*) (min) metamict
мета·морф·и́зм·, -а, (*m*) (geol) metamorphism
~, **вне́ш·н·ий** exomorphism
~, **вну́тр·енн·ий** endomorphism
~, **со·зид·а́тельн·ый** anamorphism

мета·морф·и́ческ·ий, -ая, -ое, (*a*) metamorphic

мета·морф·о́з·, -а, (*m*) metamorphosis

мета·морф·о́з·а, -ы, (*f*) metamorphosis

мет·а́н·, -а, (*m*) (chem) methane

мет·а́ни·е, -я, (*n*) (*v n*) of **мет·а́ть**

метани́л·ов·ый, -ая, -ое, (*a*) metanilic, metaniline, *m*-aniline, metanil

~ кисл·от·а́ (*f*) metanilic acid, meta-aniline-sulphonic acid, aniline-*m*-sulphonic acid

мёт·анн·ый, -ая, -ое, (*past part pass*) of **мет·а́ть**

мета́н·ов·ый, -ая, -ое, (*a*) (chem) methane, formic

~ кисл·от·а́ (*f*) formic acid, methanoic acid

метано́л·, -а, (*m*) methanol, methyl alcohol

метанол·и́з·, -а, (*m*) (chem) methanolysis

~ -сыр·е́ц·, -р·ц·а́, (*m*) unrectified methanol

метано́·метр·, -а, (*m*) methanometer, acoustic methane gas analyser

метано·от·го́н·н·ая коло́нн·а (*f*) (oil) demethanizer, demethanizing tower

метан·та́нк·, -а, (*m*) humus tank (sewage treatment)

мета·олов·я́нист·ый, -ая, -ое, (*a*) (chem) metastannic

мета·плаз·и́·я, -и, (*f*) (zool) metaplasia

мета·пла́зм·а, -ы, (*f*) (biol) metaplasma

мета·плевра́ль·н·ый, -ая, -ое, (*a*) (zool) metaplural

мета-по·лож·е́ни·е, -я, (*n*) (chem) meta-position

метароси́т·, -а, (*m*) (min) metarossite

мета·секво́й·я, -и, (*f*) (bot) dawn redwood, *Metasequoia*

мета·сику́л·а, -ы, (*f*) (bot) metasicule, metasicula

мета·со·един·е́ни·е, -я, (*n*) (chem) meta-compound

мета·сомати́зм·, -а, (*m*) (geol) metasomatism

мета·стаби́ль·ност·ь, -и, (*f*) metastability

мета·стаби́ль·н·ый, -ая, -ое, (*a*) metastable

мета·ста́з·, -а, (*m*) (med) metastasis

мета·стат·и́ческ·ий, -ая, -ое, (*a*) metastatic

мета·сто́м·а, -ы, (*f*) (zool) metastome, metastoma

мета·тарзаль·н·ый, -ая, -ое, (*a*) (zool) metatarsal

мет·а́тельн·ый, -ая, -ое, (*a*) throwing, flinging

~ вз·ры́в·чат·ое вещ·еств·о́ (*n*) (expl) propellant

~ с·на·ря́д· (*m*) missile, projectile

мета·те́ри·и, -ев, (*m pl*) (zool) metatherians, *Metatheria*

мета·ти́п·, -а, (*m*) (biol) metatype

метаторберни́т·, -а, (*m*) (min) metatorbernite

мет·а́ть, (*pres 3rd sing, pl*) **ме́ч·ет, ме́ч·ут,** (*imp*) throw, fling; **-ся,** rush about; **мет·а́·ть, -ют,** (*imp*), (text) baste

~ икр·у́ (fish) spawn

мета·у·сто́й·чивост·ь, -и, (*f*) metastability

мета·фа́з·а, -ы, (*f*) (cyt) metaphase

мета·фе́н·, -а, (*m*) (pharm) metaphen

мета·фор·и́ческ·ий, -ая, -ое, (*a*) metaphorical

метафо́с·, -а, (*m*) Metaphos (insecticide)

метафосф·а́т·, -а, (*m*) (chem) metaphosphate

метафосфо́р·н·ый, -ая, -ое, (*a*) (chem) metaphosphoric

~ кисл·от·а́ (*f*) (chem) metaphosphoric acid

метафосфорно·ки́сл·ый, -ая, -ое, (*a*) (chem) metaphosphate of

~ на́тр·и·й (*m*) sodium metaphosphate

мета·хромаз·и́·я, -и, (*f*) (biol) metachromasy

мета·хромати́н·, -а, (*m*) (biol) metachromatin

мета·хромат·и́ческ·ий, -ая, -ое, (*a*) metachromatic

метацейнери́т·, -а, (*m*) (min) metazeunerite

мета·це́нтр·, -а, (*m*) metacentre, metacenter (hydrostatics)

~, по·пере́ч·н·ый transverse metacentre/metacenter

~, про·до́ль·н·ый longitudinal metacentre/metacenter

мета·центр·и́ческ·ий, -ая, -ое, (*a*) metacentric

~ выс·от·а́ (*f*) metacentric height

~ выс·от·а́, на·ча́ль·н·ая по·пере́ч·н·ая (*f*) transverse metacentric height

мет·ацети́н·, **-а,** (*m*) (pharm) metha-
cetin

мета·цимо́л·, **-а,** (*m*) (chem) meta-
cymene

мет·гемо·глоби́н·, **-а,** (*m*) (biochem)
methaemoglobin, methemoglobin

мёт·ел· (*g pl*) of **мет·л·а́**

мет·ёлк·а, -и, (*g pl*) **-лок·,** (*f*) broom;
(bot) pannicle

мете́л·ь, -и, (*f*) blizzard, snowstorm

~, низ·ов·а́я (meteor) blowing snow

мете́ль·чат·ый, -ая, -ое, (*a*) panicu-
late

мете́ль·н·ый, -ая, -ое, (*a*) blizzard,
snowstorm; broom

мете́н·, -а, (*m*) (chem) methene

мет·ённ·ый, -ая, -ое, (*past part pass*)
of **мес·ти́;** swept

метео- (*abbr*) of **метеоро·лог·и́чес-
к·ий** meteorological

метео·зо́нд·, -а, (*m*) meteorological
sounding balloon

метео·об·стано́в·к·а, -и, (*f*) weather
conditions

метео́р·, -а, (*m*) (astron) meteor

~, кра́т·н·ый multiple meteor

~, па́д·ающ·ий incident meteor

~, петл·я́ющ·ий looping meteor

~, эклиптика́ль·н·ый ecliptical
meteor

метеори́зм, -а, (*m*) (med) meteorism

метеори́т·, -а, (*m*) (astron etc.) meteorite

метеори́т·ик·а, -и, (*f*) (astron etc.)
meteoritics, meteoritology

метеори́т·н·ый, -ая, -ое, (*a*) mete-
oritic, meteorite

метео́р·н·ый, -ая, -ое, (*a*) meteoric,
meteor

~ патру́л·ь (*m*) meteor camera

~ по·то́к· (*m*) (astron) meteor/mete-
oric shower/stream

~ след· (*m*) meteor train/trail

~ те́л·о (*n*) meteoroid

метеоро·гра́мм·а, -ы, (*f*) (meteor)
meteorogram, meteorograph recording

метеоро́·граф·, -а, (*m*) (meteor) meteorograph

метеоро́·лог·, -а, (*m*) meteorologist

метеоро·лог·и́ческ·ий, -ая, -ое, (*a*)
meteorological, weather

метеоро·ло́г·и·я, -и, (*f*) meteorology

~, высоко·го́р·н·ая alpine meteorology

~, синопт·и́ческ·ая synoptic meteorology, weather forecasting

метео·с·во́д·к·а, -и, (*g pl*) **-д·ок·,**
(*f*) weather report

метео·ста́нц·и·я, -и, (*m*) weather
station

метео·у·сло́в·и·я, -й, (*n pl*) weather
conditions

~, прост·о́е (air) VFR weather
conditions

метео·ша́р·, -а, (*m*) meteorological
balloon

мет·ёт (*pres 3rd sing*) of **мес·ти́**

мети́з·ы, -ов, (*m pl*) (*abbr*) of **метал-
л·и́ческ·ие из·де́л·и·я;** ironmon-
gery, metal ware

метиза́ци·я, -и, (*f*) (agr) cross-
breeding

мети́з·н·ый, -ая, -ое, (*a*) of **мети́з·ы**

мети́л·, -а, (*m*) (chem) methyl

~, хло́р·ист·ый methyl chloride, chlo-
romethane (refrigerant)

метил·акрила́т·, -а, (*m*) (chem)
methacrylate

метил·акри́л·ов·ый, -ая, -ое, (*a*)
(chem) methacrylic

~ кисл·от·а́ (*f*) methacrylic acid

метил·а́л·, -а, (*m*) (chem) methylal,
dimethylformal

метилалли́л·, -а, (*m*) (chem) methallyl

метил·а́л·ь, -я, (*m*) methylal, dimethylformal

метил·ами́н·, -а, (*m*) (chem) methyl-
amine

метил·ано́н·, -а, (*m*) methylanone,
methyl cyclohexanone

метил·арсе́н·ов·ая кисл·от·а́ (*f*)
arrhenic acid

метил·арсеново·ки́сл·ый, -ая, -ое,
(*a*) methylarsenate (of)

метил·а́т·, -а, (*m*) methylate

метил·бутадие́н·, -а, (*m*) isoprene

метил·винил·пириди́н·, -а, (*m*)
methylvinylpyridine

метил·виоле́т·, -а, (*m*) methyl violet
(dyestuff)

метил·гексали́н·, -а, (*m*) methyl-
hexalin, methyl cyclohexane

метил·глиокса́л·ь, -я, (*m*) methyl
glyoxal

метил·дихлорарси́н·, -а, (*m*) methyl-
dichlorarsine

метил·е́н·, -а, (*m*) (chem) methylene
methene

метиле́н·ов·ый, -ая, -ое, (*a*) methylene

~ гру́пп·а, мо́ст·иков·ая (*f*) methylene bridge

~ си́н·ий (*m*) (med) alkaline methylene blue

метилéн·ов·ый, -ая, -ое
~ **голуб·óй** (*m*) methylene blue (dye-stuff)
метил·ировáни·е, -я, (*n*) (chem) methylation
метил·каучýк·, -а, (*m*) methyl rubber
метил·метакрилáт·, -а, (*m*) methyl methacrylate
метил·ов·ый, -ая, -ое, (*a*) of **метил·**
~ **грýпп·а** (*f*) methyl group
~ **спирт·** (*m*) methyl alcohol, methanol, wood alcohol
~ **эфир·** (*m*) methyl ether/ester
~ **эфир· метакрил·ов·ой** (*m*) methyl methacrylate
~ **эфир· ýголь·н·ой кисл·от·ы** (*m*) methyl carbonate
~ **эфир· фенóл·а** (*m*) methyl phenate/phenoxide, anisole
метилолполиамид·, -а, (*m*) methylolpolyamide
метил·орáнж·, -а, (*m*) (chem) methyl orange
метил·рóт·, -а, (*m*) methyl red
метил·стирóл·, -а, (*m*) methyl styrene
метил·тестостерóн·, -а, (*m*) (pharm) methyltestosterone
метил·тиоурацил·, -а, (*m*) (pharm) methylthiouracil
метил·фенилдихлорсилáн·, -а, (*m*) methylphenyldichlorosilane
метил·цикло·пентáн·, -а, (*m*) methyl-cyclopentane
метил·этил·кетóн·, -а, (*m*) methyl ethyl ketone
метин·, -а, (*m*) methine, methylidyne
метинóн·, -а, (*m*) (pharm) methinone, menadione
метионил·, -а, (*m*) methionyl
метионин·, -а, (*m*) methionine
метиóн·ов·ая кисл·от·á (*f*) methionic acid, methyl disulphonic acid, methylene sulphonic acid
метис·, -а, (*m*) (zool) mongrel; half-caste; (bot) hybrid
метис·áци·я, -и, (*f*) intermarriage, interbreeding, cross-breeding; (bot) hybridizing
мéт·ить, -ят, (*imp*) aim; mark, label, tag
мéт·к·а, -и, (*g pl*) **-т·ок·,** (*f*) mark, marking, notch; (nucl) label, tag; blaze (on a tree)
~ **врéмен·и** (TV) time marking/trace
~, **инéртн·ая** (met) marker

мéт·к·а
~, **по·перéч·н·ая** cross line (bombsight)
~ **част·от·ы** (TV) frequency pip/blip/marker
мéт·к·ий, -ая, -ое, (*a*) well-aimed, (gunn) accurate; apt (figurative)
мéт·кост·ь, -и, (*f*) accuracy
мет·л·á, -ы, (*nom pl*) **мёт·л·ы,** (*g pl*) **мёт·ел·,** (*f*) broom
метлах·ск·ая плит·á (*f*) (build) vitreous floor tile
мет·н·ýть, -ýт, (*perf single action*) *see* **мет·áть**
мéтод·, -а, (*m*) (*and see under associated names and adj*) method, technique; (*as prefix*) methods
~, **кáпель·н·ый** drop analysis
~ **о·шиб·ок·** error method (computers)
~ **под·бóр·а** trial-and-error
метóд·ик·а, -и, (*f*) procedure, method; technique
~ **рабóт·ы** procedure
метод·ист·, -а, (*m*) educationalist, methodologist; methods specialist, methods engineer
метод·ическ·ий, -ая, -ое, (*a*) methodical, systematic, steady; methods
~ **зóн·а** (*f*) preheating zone (furnace)
~ **печ·ь** (*f*) continuous furnace
методо·лóг·и·я, -и, (*f*) methodology
метокси- (*int component*) (chem) methoxy-
метоксил·, -а, (*m*) (chem) methoxyl
метóл·, -а, (*m*) (chem phot) metol, *p*-methylaminol sulphate
метóл·ов·ый, -ая, -ое, (*a*) of **метóл·**
метохинóн·, -а, (*m*) (chem phot) metoquinone
мéтр·, -а, (*m*) metre, meter; metre/meter rule; (*as component*) -meter
метр·áж·, -а, (*m*) length in metres/meters; metrix area; (cinema) reelage (of film)
метр·ампéр·, -а, (*m*) (rad) metre-ampere, (*pl*) radiation constant (of antenna)
метранпáж·, -а, (*i*) **-ем,** (*m*) (print) clicker, maker-up
метрехóн·, -а, (*m*) (elec) metrechon, half-picture storage tube
мéтр·ик·а, -и, (*f*) birth certificate; (math) metrics, metric
~, **индефинит·н·ая** indefinite metric (space-time physics)

метр·и́ческ·ий, -ая, -ое, (*a*) metric

~ **кни́г·а** (*f*) register of births, marriages and deaths

~ **про·стра́н·ств·о** (*n*) (math) metric space

метро́ (*n indecl*) underground railway, the underground (Brit), the subway (U.S.)

метро·ампе́р·, -а, (*m*) (rad) metre-ampere

метр·о́вк·а, -и, (*g pl*) **-вок·,** (*f*) metre/meter rule

метр·ов·о́й, -а́я, -о́е, (*a*) *see* **метр·о́в·ый**

метр·о́в·ый, -ая, -ое, (*a*) metre, metric

~ **волн·а́** (*f*) (rad) metric wave

метро́·лог·, -а, (*m*) metrologist

метро·ло́г·и·я, -и, (*f*) metrology

метро·ме́р·, -а, (*m*) measuring machine, length measuring machine

метро́н·, -а, (*m*) (cyb) metron

метро·но́м·, -а, (*m*) metronome

метрополите́н·, -а, (*m*) underground railway, the underground (Brit), the subway (U.S.)

мет·у́щ·ий, -ая, -ее, (*pres part act*) of **мес·ти́;** sweeping

ме́т·чик·, -а, (*m*) tap, screw tap; marker (person)

~, **лов·и́льн·ый** (oil) tap, fishing tap

~, **ма́т·очн·ый** master/die tap

мёт·ш·ий, -ая, -ее, (*past part act*) of **мес·ти́**

мет·я́ (*pres gerund*) of **мес·ти́;** sweeping

мех·, -а, (*nom pl*) **-а́,** (*g pl*) **-о́в,** (*m*) fur, skin; **мех·, -а,** (*nom pl*) **-и́,** (*g pl*) **-о́в,** (*m*) skin (container for wine etc.); (*pl*) bellows

механиз·а́тор·, -а, (*m*) mechanizer, process mechanizer; (agr) machinery maintenance engineer

механиз·а́ци·я, -и, (*f*) mechanization; mechanism, device

~ **крыл·а́** (a/c) lift augmentation devices

механиз·и́рованн·ый, -ая, -ое, (*past part pass*), mechanized, mechanical; power, powered, motorized

~ **инструме́нт·** (*m*) power tools

механ·и́зм·, -а, (*m*) (*see under qualifying adjectives*) mechanism, mechanisms, gear, device; (*pl*) machinery, engines

~, **вс·по·мог·а́тельн·ые** (*pl*) (shipb) auxiliary machinery

~ **вы́·бор·к·и** access mechanism (computers)

механ·и́зм

~, **гла́в·н·ые** (*pl*) (shipb) main machinery, main engines

~, **час·ов·о́й** clockwork, clockwork mechanism

меха́н·ик·, -а, (*m*) maintenance engineer; mechanic, mechanician

~ ~ **при·бор·и́ст·, -а,** (*m*) instrument repairer

меха́н·ик·а, -и, (*f*) mechanics

~, **волн·ов·а́я** quantum mechanics

~ **грунт·о́в** soil mechanics

~ **жи́дк·ост·и** fluid mechanics

~, **ква́нт·ов·ая** quantum mechanics

~, **неб·е́сн·ая** celestial mechanics

~, **ньюто́н·овск·ая** Newtonian mechanics

~, **релятиви́ст·ск·ая** relativistic mechanics

~ **спош·н·ы́х сред·** continuum mechanics, classical field theory

механ·и́ческ·ий, -ая, -ое, (*a*) mechanical, machine, power-driven, engineering

~ **коэффициент· по·ле́з·н·ого де́й·стви·я** (*m*) mechanical efficiency

~ **мастер·ск·а́я** (*f*) engineering shop, maintenance shop

~ **об·рабо́т·к·а** (*f*) (mech eng) machining

~ **регуля́тор·** (*m*) (autom) mechanical control gear

~ **сво́й·ств·о** (*n*) mechanical property

~ **со·ста́в·** (*m*) (phys) mechanical composition, texture

~ **то́рмоз·** (*m*) power brake

~ **цех·** (*m*) engineering shop; maintenance shop

~ **цех·, больш·о́й** (shipb) main machinery shop

~ **цех·, ма́л·ый** (shipb) auxiliary machinery shop

~ **эквивале́нт·** (*m*) (phys) mechanical equivalent

механоли́т·, -а, (*m*) kitchen machine, "mix-master"

механо·про́ч·ност·ь, -и, (*f*) crushing strength (of soil)

механо·реце́птор·, -а, (*m*) (biol, elec) mechanoreceptor

механо·с·бо́р·очн·ый, -ая, -ое, (*a*) machine-assembly/assembling

механо·стри́кц·и·я, -и, (*f*) (phys) mechanostriction

мех·и́ (*pl*) of **мех·**

мех·ов·о́й, -а́я, -о́е, (*a*) of **мех·**

мех·овщи́к·, -а́, (*m*) furrier

мех·о́м·, -а, (*m*) = **механ·и́ческ·ий ом·** (phys) mechanical ohm

мех·соста́в·, -а, (*m*) *see* **механ·и́ческ·ий со·ста́·в·**

меч·, -а́, (*i*) **-о́м,** (*m*) sword

~ **-ры́б·а, -ы,** (*f*) sword fish, *Xiphias gladius*

меч·а́ (*pres gerund*) of **мет·а́ть**

мече·ви́д·н·ый, -ая, -ое, (*a*) sword-shaped, gladiate, ensiform, xiphoid

меч·е́ни·е, -я, (*n*) (*v n*) *see* **ме́т·ить;** marking; (nucl) tagging, labelling, labeling

ме́ч·енн·ый, -ая, -ое, (*past part pass*) *see* **ме́т·ить;** *see also* **ме́ч·ен·ый**

ме́ч·ен·ый, -ая, -ое, (*a*) marked, labelled, labeled, tagged

~ **а́том·** (*m*) (nucl) tagged/labelled atom

~ **химика́т·, -а,** (*m*) chemical tracer

ме́ч·ет (*pres 3rd sing*) of **мет·а́ть**

мече·хво́ст·ые, -ых, (*pl decl as adj*) (zool, pal) xiposurans, *Xiphosura*

мечт·а́, -ы́, (*f*) dream, vision, aspiration, hope

меч·у́ (*pres 1st sing*) of **мет·а́ть, ме́ч·у,** (*pres 1st sing*) of **ме́т·ить**

ме́ч·ущ·ий, -ая, -ее, (*pres part act*) of **мет·а́ть**

меш·, -а, (*m*) mesh (measuring unit; no. of holes to 1 inch)

меш·а́лк·а, -и, (*g pl*) **-лок·,** (*f*) mixer, stirrer, agitator

~, **стекл·я́нн·ая** stirring rod

меш·а́льн·ый, -ая, -ое, (*a*) stirring, agitating, mixing

меш·а́ни·е, -я, (*n*) (*v n*) *see* **меш·а́·ть**

ме́ш·анк·а, -и, (*g pl*) **-нок·,** (*f*) seed mixture; mash (for feeding livestock)

меш·а́·ть, -ют, (*imp*) (+ *dat*), hinder, interfere, disturb, prevent; (rad) jam; (+ *acc*) stir, agitate, mix, blend

меш·а́ющ·ий, -ая, -ее, (*pres part act*) of **меш·а́·ть;** interfering, interference; preventing

~ **и́мпульс·** (*m*) interference pulse

~ **кра·й** (*m*) interference fringe

~ **ста́нц·и·я,** (*f*) (rad) jamming station

ме́ш·енн·ый, -ая, -ое, (*past part pass*) of **мес·и́ть**

ме́шк·а·ть, -ют, (*imp*) delay, be slow in; linger, loiter

мешко·ви́д·н·ый, -ая, -ое, (*a*) bag/sack shaped, sacciform, sacculate

мешк·ови́н·а, -ы, (*f*) (text) burlap, sackcloth

мешко·вы́·бив·а́тел·ь, -я, (*m*) bag beater

меш·ко́в·ый, -ая, -ое, (*a*) of **меш·о́к·**

~ **груз·** (*m*) bag cargo/freight

мешко·вы́·колач·ива́тел·ь, -я, (*m*) bag beater

мешко·на·сы́п·а́тел·ь, -я, (*m*) (mech) bag filler

мешко·обра́з·н·ый, -ая, -ое, (*a*) *see* **мешко·ви́д·н·ый**

мешко·по·гру́з·чик·, -а, (*m*) bag-cargo loader, bag loader

мешко·ро́т·ые, -ых, (*pl decl as adj*) (fish) gulpers, *Saccopharyngiformes*

ме́шкот·н·ый, -ая, -ое, (*a*) sluggish

меш·о́к·, -ш·к·а́, (*m*) bag, sack; (biol) sac

~, **воз·ду́ш·н·ый** air lock, air pocket; air bag (tire vulcanization); (biol) air sac

~, **като́д·н·ый** (electrochem) cathode bag

~, **парашют·н·о·деса́нт·н·ый** (mil) supply-dropping container

~ **с песк·о́м** sandbag

~, **с·лёз·н·ый** (anat) lachrymal sac

меш·о́тчат·ый, -ая, -ое, (*a*) of **ме·ш·о́чек·**

меш·о́чек·, -чк·а, (*m*) (*dim*) of **ме·ш·о́к·;** (bot) utricle, (biol) follicle, saccule, sacculus

в меш·о́чек· medium (of boiled eggs)

меш·о́чн·ый, -ая, -ое, (*a*) of **меш·о́к·**

ми (*n indecl*) mu, μ (Greek); E, mi (music)

Ми, тео́р·и·я (*f*) (phys) Mie scattering theory

миа́зм·а, -ы, (*f*) miasma

миазмат·и́ческ·ий, -ая, -ое, (*a*) miasmatic, miasmal

миарсено́л·, -а, (*m*) (pharm) myarsenol, sulpharsphenamine

миароли́т·ов·ый, -ая, -ое, (*a*) (geol) miarolitic

миаски́т·, -а, (*m*) (geol) miaskite

ми́г·, -а, (*m*) moment, instant

миг·а́лк·а, -и, (*g pl*) **-лок·,** (*f*) (naut) flashing light; shut-off valve (of dust collecting apparatus)

миг·а́ни·е, -я, (*n*) flickering, winking, blinking

миг·а́·ть, -ют, (*imp*) blink, flicker, twinkle, wink

миг·а́ющ·ий, -ая, -ее, (*pres part act*) *see* **миг·а́·ть;** flashing (of lights); (phys) pulsed

~ **пуч·о́к·** (*m*) (phys) pulsed beam

миг·а́ющ·ий, -ая, -ее
~ **фона́р·ь** (*m*) blinker, blinking/flashing light/lamp
мигматиз·а́ци·я, -и, (*f*) (geol) migmatization
мигмат·и́т·, -а, (*m*) (geol) migmatite
мигр·ацио́нн·ый, -ая, -ое, (*a*) migration
мигр·а́ци·я, -и, (*f*) migration
~, **бок·ов·а́я** (geol) lateral migration
мигре́н·ь, -и, (*f*) (med) migraine
мигр·и́р·овать, -уют, (*imp and perf*) migrate
ми́дел·ев·ый, -ая, -ое, (*a*) of **ми́дел·ь**
~ **сеч·е́ни·е** (*n*) (shipb) midship frame; (phys) midsection, centre/center section
ми́дел·ь, -я, (*m*) (shipb) ́midship section; (a/c) mid section
~ **-шпанго́ут·, -а,** (*m*) (shipb) midship frame; (a/c) middle frame
миел- (*int component*) (*see also* **миэл-**) myel-, myelo-
миел·и́т·, -а, (*m*) (med) myelitis
миел·о́м·а, -ы, (*f*) (med) myeloma
миз·и́д·а, -ы, (*f*) (zool) opossum shrimp, mysid, (*pl*) *Mysida*
мизи́н·ец·, -н·ц·а, (*i*) **-ц·ем,** (*m*) little finger/toe
мизо·га́м·и·я, -и, (*f*) (biol) misogamy
мизо·стом·и́д·ы (*pl*) (zool) myzostomids, *Myzostomida*
микале́кс·, -а, (*m*) mycalex (a dielectric)
мика·ле́нт·а, -ы, (*f*) mica tape
микани́т·, -а, (*m*) micanite (a dielectric)
~, **жаро·у·по́р·н·ый** heater-plate micanite
~, **колле́ктор·н·ый** (elec) commutator micanite
~, **форм·о́вочн·ый** moulding/molding plate micanite
микафо́л·и·й, -я, (*m*) micafolium
мико- (*int component*) myco-
мико·де́рм·а, -ы, (*f*) (food, bact) mycoderm
мик·о́з·, -а, (*m*) (med) mycosis
мико·ло́г·и·я, -и, (*f*) (bot) mycology
мико·ри́з·а, -ы, (*f*) (bot) mycorrhiza, mycorhiza
микро- (*int component*) micro-; small; (elec) fractional-horsepower
микро·ампе́р·, -а, (*m*) (elec) microampere, μa, μA
микро·анали́з·, -а, (*m*) (chem) microanalysis

микро·анали́т·и́ческ·ий, -ая, -ое, (*a*) micro-analytical
микро́·б·, -а, (*m*) microbe
микро·ба́р·, -а, (*m*) microbar (unit of pressure)
микро·баро́·граф·, -а, (*m*) (meteor) microbarograph
микро·био·ло́г·и́ческ·ий, -ая, -ое, (*a*) microbiological
микро·био·ло́г·и·я, -и, (*f*) microbiology
микро́·б·н·ый, -ая, -ое, (*a*) microbe, microbial
микро·бюре́т·к·а, -и, (*g pl*) **-т·ок·,** (*f*) microburette
микро́·ва́льц·ы, -о́в, (*m pl*) micro-mill
микро·вес·ы́, -о́в, (*m pl*) microbalance
микро·ви́нт·, -а, (*m*) (instr) fine adjustment screw, micrometer screw
микро·волн·а́, -ы́, (*f*) (rad) microwave
микро·во́лн·ов·ый, -ая, -ое, (*a*) (rad) microwave
микро·во́льт·, -а, (*m*) (elec) microvolt, μv, μV
микро·вы·ключ·а́тел·ь, -я, (*m*) microswitch
микро·гальван·и́ческ·ий элеме́нт· (*m*) (met) local/localized galvanic cell (corrosion)
микро·ге́нри (*n*) (elec) microhenry
микро·гео·ме́тр·и·я, -и, (*f*) surface roughness, micro-roughness; study of surface roughness
микро·гли́·я, -и, (*f*) (zool) microglia
микро·гра́мм·а, -ы, (*f*) microgram (unit of mass)
микро·грани́т·н·ый, -ая, -ое, (*a*) (geol) microgranitic
микро·граф·и́ческ·ий, -ая, -ое, (*a*) micrographic
микро·гэ́с·, -а, (*m*) (elec) hydro-electric generating set
микро·дви́г·ател·ь, -я, (*m*) (elec) fractional horsepower motor
микро·доз·иро́вк·а, -и, (*g pl*) **-вок·,** (*f*) micrometering
микро·за́вр·, -а, (*m*) (pal) microsaur
микро·зерн·о́, -а́, (*n*) micro-grain
микро·зерн·и́стост·ь, -и, (*f*) microgranular structure
микро·из·мер·е́ни·е, -я, (*n*) micrometering
микро·интерферо́·метр·, -а, (*m*) (opt) microinterferometer
микро·ис·коп·а́ем·ое, -ого, (*n decl as adj*) (pal) microfossil

микро·ис·след·овани·е, -я, (*n*) microanalysis

микро·канон·и́ческ·ий, -ая, -ое, (*a*)
microcanonical

микро·кало́р·и·я, -и, (*f*) (phys)
small calorie

микро·ка́тор·, -а, (*m*) (eng) spring-
actuated contact comparator

микро·кино·съ·ём·к·а, -и, (*g pl*)
-м·ок·, (*f*) cinemicrography

микро·кли́мат·, -а, (*m*) microclimate

микро·кли́н·, -а, (*m*) (min) microcline

микро·ко́кк·, -а, (*m*) (bact) micro-
coccus

микро·концентр·а́ци·я, -и, (*f*) (chem)
trace concentration

микро·ко́п·и·я, -и, (*f*) (phot) positive
microfilm, microfilm print

микро·ко́см·, -а, (*m*) microcosm

микро·кристалл·и́ческ·ий, -ая, -ое,
(*a*) microcrystalline

микро·кудо́н·, -а, (*m*) (phys) microcoulomb

микро·кюри́ (*n indecl*) (nucl) microcurie, μc

микро·ли́т·, -а, (*m*) sintered aluminium oxide (for ceramic tool tips);
(geol) microlite

микро·луч·ев·о́й, -ая, -ое, (*a*) (phys)
microray

микро·лю́кс·, -а, (*m*) (phot) microlux

микр·о́м·, -а, (*m*) (elec) microhm

микро·манипуля́тор·, -а, (*m*) micro-
manipulator

микро·мано́·метр·, -а, (*m*) (instr)
micromanometer

микро·маши́н·а, -ы, (*f*) (elec) fractional horse-power machine

микро·ме́р·, -а, (*m*) (eng) comparator, contact comparator; (zool)
micromere

~, индукти́в·н·ый electrical contact
comparator

~, пневмат·и́ческ·ий pneumatic comparator

~, рыча́ж·н·о-зу́б·чат·ый mechanical comparator

микро́·метр·, -а, (*m*) (phys instr)
micrometer; (eng) micrometer gauge/
gage, micrometer caliper

~, нит·ян·о́й cross-wire micrometer,
filar micrometer

~, окуля́р·н·ый eyepiece micrometer

~, позицио́н·н·ый (astron) position-
angle micrometer

~, резь·бов·о́й (eng) screw-thread
micrometer gauge

микро·метр·и́ческ·ий винт· (*m*) fine-
adjustment screw, micrometer screw

микромикро- (*int component*) micro-
micro-

микромикро·фара́д·а, -ы, (*f*) (elec)
micromicrofarad (10^{-12} farad), μμf

микро·милли·ме́тр·, -а, (*m*) micro-
millimetre

микро·ми́р·, -а, (*m*) microcosm

микро·мо́ (*indecl*) (elec) micromho

микро́н·, -а, (*m*) micron (10^{-6} m), μ

микро·на·пряж·е́ни·е, -я, (*n*) micro-
stress

микро·на·са́д·ок·, -д·к·а, (*m*) (dynam) stanton tube

микро·не́кс·, -а, (*m*) micronex (activated gas black)

микро·не·ро́вн·ост·ь, -и, (*f*) micro-
roughness, surface roughness

микро·ну́клеус·, -а, (*m*) (zool) micro-
nucleus

микро·объекти́в·, -а, (*m*) microscope objective (lens)

микро·органи́зм·, -а, (*m*) (biol)
microorganism

микро·палеонто·ло́г·и·я, -и, (*f*)
micropaleontology

микро·па́р·а, -ы, (*f*) (elec) micro-
couple, microcell

микро·пегмати́т·ов·ый, -ая, -ое, (*a*)
(geol) micropegmate, micropegma-
titic

микро·пере·ключ·а́тел·ь, -я, (*m*)
micro change-over switch

микро·перти́т·, -а, (*m*) (geol) micro-
perthite

микро·пи́ле (*n*) (biol) micropyle

микро·пиро́·метр·, -а, (*m*) (instr)
micropyrometer

микро·по́р·а, -ы, (*f*) micropore;
microscopic capillary

микро·по́р·истост·ь, -и, (*f*) micro-
porosity

микро·по́р·ист·ый, -ая, -ое, (*a*)
microporous

микро·препара́т·, -а, (*m*) slide, mount
(for microscope)

микро·при́·мес·ь, -и, (*f*) (met) trace
impurity

микро·прое́ктор·, -а, (*m*) (opt)
microprojector

микро·прое́кц·и·я, -и, (*f*) microscope projection, projection microscopy

микро·пуч·о́к·, -ч·к·а́, (*m*) microbeam

микро·радио·спектро·ско́п·и·я, -и,
(*f*) (rad) microwave spectroscopy

микро·рельеф·, -а,   (*m*) (met) microrelief, microroughness; (geog) microtopography, microrelief

микро·рентгено·граф·и·я, -и,   (*f*) (phys) X-ray diffraction microscopy

микро·с·двиг·, -а,   (*m*) (geol) microshift, minute movement

микро·сейсм·, -а,   (*m*) (geophys) microseism

микро·секунд·а, -ы,   (*f*) microsecond

микро·син·, -а,   (*m*) (elec) microsyn

микро·скоп·, -а,   (*m*) microscope

~, авто·электрон·н·ый auto-electron microscope; field-emission microscope

~, бинокуля́р·н·ый binocular microscope

~, из·мер·и́тель·н·ый tool microscope

~, интерференцио́н·н·ый interference microscope

~, ис·след·ова́тель·ск·ий laboratory microscope

~, миг·а́ющ·ий (astron) blink microscope

~, поляризацио́н·н·ый polarizing microscope

~, проекцио́н·н·ый projection microscope

~, прост·о́й simple microscope, magnifying glass

~, ра́стр·ов·ый электро́н·н·ый scanning electron microscope

~, рис·ова́льн·ый projection microscope

~, све́тл·ый light microscope

~, с·ло́ж·н·ый compound microscope

~ с·равн·е́ни·я comparison microscope

~, фазово·контра́ст·н·ый phase-contrast microscope

~, электро́н·н·ый (*see also* **электро́нмикро·ско́п·**) electron microscope

микро·скоп·и́ческ·ий, -ая, -ое,   (*a*) microscope; microscopic

микро·с·ни́м·ок·, -м·к·а,   (*m*) photomicrograph

микро·со́м·а, -ы,   (*f*) (cyt) microsome

микро·спо́р·а, -ы,   (*f*) (bot) microspore

микро·спора́нг·и·й, -я,   (*m*) (bot) microsporangium

микро·споро·фи́лл·, -а,   (*m*) (bot) microsporophyll

микро·стро·е́ни·е, -я,   (*n*) microstructure

микро·структу́р·а, -ы,   (*f*) microstructure

микро·съ·ём·к·а, -и,   (*g pl*) **-м·ок·,** (*f*) photomicrography

микро·та́ст·, -а,   (*m*) (instr) contact comparator

микро·твердо·ме́р·, -а,   (*m*) microhardness tester

микро·твёрд·ост·ь, -и,   (*f*) microhardness

микро·телефо́н·, -а,   (*m*) telephone handset

микро·то́м·, -а,   (*m*) (instr) microtome

микро·тро́н·, -а,   (*m*) (elec) microtron

микро·уда́р·н·ый, -ая, -ое,   (*a*) micropulsating

~ де́й·стви·е (*n*) (met) micropulsating load, micropulsation

микро·у·добр·е́ни·е, -я,   (*n*) microfertilizer, trace fertilizer

микро·фа́г·, -а, (*m*) (zool) microphage

микро·фанеро·фи́т·, -а,   (*m*) (bot) microphanerophyte (tree 2–8 metres high)

микро·фара́д·а, -ы,   (*f*) (elec) microfarad

микро·фильм·и́ровани·е, -я,   (*n*) microfilming, microphotography

микро·фо́н·, -а, (*m*) microphone

~ градие́нт·а давл·е́ни·я pressure-gradient microphone

~ давл·е́ни·я pressure microphone

~, дву·на·пра́вл·енн·ый bidirectional microphone

~, двух·та́кт·н·ый push-pull microphone

~, из·мер·и́тельн·ый standard microphone

~, ка́псюль·н·ый telephone transmitter, button microphone

~, конденса́тор·н·ый condenser/electrostatic microphone

~, конта́кт·н·ый contact microphone

~, ле́нт·очн·ый ribbon microphone

~, миниато́р·н·ый lapel microphone

~, на·пра́вл·енн·ый directional microphone

~, пьезо·кристалл·и́ческ·ий piezo microphone, crystal microphone

~, скор·остн·о́й velocity microphone

~, тепл·ов·о́й thermal/hot-wire microphone

~, терм·и́ческ·ий thermal/hot-wire microphone

~, тл·е́ющ·ий glow-discharge microphone

~, у́голь·н·ый carbon microphone

~, шаро·обра́з·н·ый eight-ball microphone

микро·фо́н

~, электро·динам·и́ческ·ий moving-coil/electrodynamic microphone

~, электо·магни́т·н·ый electromagnetic microphone

микро·фо́н·н·ый, -ая, -ое, (*a*) microphone, microphonic

микро·фото·гра́мм·а, -ы, (*f*) micro-photometer recording

микро·фото·гра́ф·и·я, -и, (*f*) photomicrography; photomicrograph

микро·фото·коп·и́ровани·е, -я, (*n*) microfilming, microphotography

микро·фото·ко́п·и·я, -и, (*f*) microfilm print, positive microfilm

микро·фото́·метр·, -а, (*m*) micro-photometer

микро·фото·на·са́д·к·а, -и, (*g pl*) -д·ок·, (*f*) microscope photographic-attachment

микро·фото·плён·к·а, -и, (*g pl*) -н·ок·, (*f*) negative microfilm, microfilm

микро·фото·с·ни́м·ок·, -м·к·а, (*m*) photomicrograph

микро·фото·съ·ём·к·а, -и, (*g pl*) -м·ок·, (*f*) photomicrography, photomicrographing

микро·фото·у·стано́в·к·а, -и, (*g pl*) -вок·, (*f*) photomicrography apparatus

микр·офталм·и́·я, -и, (*f*) (zool) microphthalmia

микро·хи́м·и·я, -и, (*f*) microchemistry

микро·хирург·и́·я, -и, (*f*) micro-surgery

микро·цефа́л·и·я, -и, (*f*) (zool) microcephaly

микро·част·и́ц·а, -ы, (*f*) (nucl) elementary particle

микро·чит·а́тельн·ый аппара́т· (*m*) microfilm viewer

микро·шли́ф·, -а, (*m*) (met) micro-section

микро·элеме́нт·, -а, (*m*) (biol) trace element, microelement, rare element

микр·ург·и́·я, -и, (*f*) microdissection

микс·оде́м·а, -ы, (*m*) (med) myxoedema

ми́ксер·, -а, (*m*) (met) mixer, holding furnace

~ -пе́ч·ь, -и, (*f*) (met) holding furnace, mixer

миксо- (*int component*) mixo-, мухо-

миксо·бакте́р·и·я, -и, (*f*) myxo-bacterium

миксо·вариаци·я, -и, (*f*) (gen) mixo-variation

микс·оматоз·, -ы, (*f*) (zool) myxo-matosis

миксо·мице́т·, -а, (*m*) (bot) myxo-mycete, (*pl*) *Myxomycetes*

миксо·пло́йд·и·я, -и, (*f*) (genet) mixoploidy

миксо·по́д·и·й, -я, (*m*) (zool) myxo-podium

миксо·тро́ф·н·ый, -ая, -ое, (*a*) (bot) mixotrophic

миксту́р·а, -ы, (*f*) (pharm) mixture

ми́кшер·, -а, (*m*) (TV) mixer

микш·и́р·овать, -уют, (*imp and perf*) (cinema) mix (sound)

милиа́р·н·ый, -ая, -ое, (*a*) (med) miliary

милитарист·и́ческ·ий, -ая, -ое, (*a*) militaristic

милиционе́р·, -а, (*m*) policeman (in U.S.S.R.)

милици́·я, -и, (*f*) police (in U.S.S.R.); territorial conscript army

миллери́т·, -а, (*m*) (min) millerite, capillary pyrite

Ми́ллер·а, и́ндекс· (*m*) (cryst) Miller index

ми́ллер·овск·ий и́ндекс· (*m*) (cryst) Miller index

милли- (*int component*) milli-

милли·ампе́р·, -а, (*m*) milliampere, mA, ma

милли·ампер·ме́тр·, -а, (*m*) (elec) milliammeter

милли·а́рд·, -а, (*m*) milliard (10^9) (Brit); billion (10^9) (U.S.)

милли·ба́р·, -а, (*m*) millibar (unit of pressure)

милли·ба́рн·, -а, (*m*) (nucl) millibarn, mb

милли·ва́тт·, -а, (*m*) (elec) milliwatt, mW, mw

милли·во́льт·, -а, (*m*) (elec) millivolt, mV, mv

милли·га́л·, -а, (*m*) milligal (unit of acceleration = 0·001 cm/sec/sec)

милли·ге́нри (*m indecl*) (elec) milli-henry

милли·гра́мм·, -а, (*m*) milligram, mg; mG (unit of force)

милли·да́рси (*n indecl*) (oil) millidarcy

милли·кюри́ (*n indecl*) (nucl) millicurie, mC, mc

Ми́лликен·а, о́·пыт· (*m*) (phys) Millikan test

милли·ли́тр·, -а, (*m*) millilitre/milli-liter, ml

милли·ме́тр·, -а, (*m*) millimetre, millimeter, mm

~ **ртут·н·ого столб·а́** millimeter Hg (unit of pressure)

милли·метр·о́вк·а, -и, (*g pl*) **-вок·,** (*f*) (surv) drawing grid

милли·метр·о́в·ый, -ая, -ое, (*a*) of **милли·ме́тр·**

~ **бума́г·а** (*f*) graph paper, millimeter graph paper

милли·микро́н·, -а, (*m*) millimicron, $m\mu$, 10^{-6} mm

милли·мо́л·ь, -и, (*f*) (chem) millimol, millimole

милли·о́м·, -а, (*m*) (elec) milliohm, mO

миллио́н·, -а, (*m*) million

миллио́н·к·а, -и, (*g pl*) **-н·ок·,** (*f*) millionth-scale map

миллио́н·н·ый, -ая, -ое, (*a*) millionth; million(s) strong; worth million(s)

~ **до́л·я** (*f*) part per million

милли·ра́д·, -а, (*m*) (nucl) millirad, mrad

милли·резерфо́рд·, -а, (*m*) (nucl) millirutherford

милли·рентге́н·, -а, (*m*) (nucl) milliroentgen, mr

милли·секу́нд·а, -ы, (*f*) millisecond, ms, msec

милли·эквивале́нт·, -а, (*m*) (phys) milligram-equivalent, meq

Милло́н·а, реа́кци·я, (*f*) (chem) Millon's reaction

милло́н·ов·а реа́кци·я (*f*) (chem) Millon's reaction

милони́т·, -а, (*m*) (geol) mylonite

ми́лори (*indecl*) (print) milori

мил·ь, -я, (*m*) mil (10^{-3} inch)

мильба́рс·, -а, (*m*) (met) puddled bar

ми́льдью (*f indecl*) downy mildew (disease of grapes)

ми́ль·н·ый, -ая, -ое, (*a*) of **ми́л·я**

ми́л·ый, -ая, -ое, (*a*) nice, deal, kind

ми́л·я, -и, (*f*) mile

~, **англ·и́йск·ая** statute mile (1·609 km)

~, **мерка́тор·ск·ая** geographical mile

~, **ме́р·н·ая** measured mile

~, **метр·и́ческ·ая** kilometre, kilometer

~, **мор·ск·а́я** nautical mile, sea mile (in U.S.S.R. 1,852 m)

~, **станда́рт·н·ая** standard mile

~, **стату́т·н·ая** statute mile

~, **сухо·пу́т·н·ая** statute mile

~, **у·ста́в·н·ая** statute mile (1·609 km)

мимео·гра́ф·, -а, (*m*) mimeograph

миметези́т·, -а, (*m*) (min) mimetesite, mimetite

мимет·и́зм·, -а, (*m*) (zool) mimicry

мимет·и́т·, -а, (*m*) (min) mimetite

мимет·и́ческ·ий, -ая, -ое, (*a*) mimetic

ми́мик·а, -и, (*f*) mimicry

мимикр·и́·я, -и, (*f*) (biol) protective coloration

мим·и́ческ·ий, -ая, -ое, (*a*) (zool) mimic, mimetic

ми́мо (*adv or prep + gen*) past, by **по·пад·а́·ть ми́мо** to miss (target etc.)

мимо·е́зж·ий, -ая, -ее, (*a*) driving/passing by; by-pass, detour

мимо·лёт·н·ый, -ая, -ое, (*a*) transient, fleeting

мимо·хо́д·ом (*adv*) in passing, on/by the way

мин (*abbr*) = **мину́т·а; ми́нимум·;** minute, min; minimum, min

ми́н·а, -ы, (*f*) mine (weapon); (mil) mortar bomb/shell

~, **авиацио́н·н·ая** (naut) aircraft-laid mine

~, **гидро·динам·и́ческ·ий** (naut) oyster/pressure mine

~, **комбин·и́рованн·ая** (naut) combined-influence mine

~, **корабе́ль·н·ая** (naut) surface-laid mine

~, **о·свет·и́тельн·ая** illuminating mortar bomb

~, **сухо·пу́т·н·ая** land mine

минасрагри́т·, -а, (*m*) (min) minasragrite

миндале·ви́д·н·ый, -ая, -ое, (*a*) almond-shaped, amygdaloid

~ **желез·а́** (*f*) (zool) tonsil

минда́л·ев·ый, -ая, -ое, (*a*) of **минда́л·ь**

минда́л·ин·а, -ы, (*f*) (food) almond, almond nut; (anat) tonsil, amygdala; (geol) amygdule, amygdale

минда́л·ь, -я́, (*m*) (bot) almond, *Prunus amygdalus*

~, **го́рь·к·ий** bitter almond

минда́ль·н·ый, -ая, -ое, (*a*) almond, mandelic, amygdalic, amygdaloid; (*pl*) *Prunus* family

~ **кисл·от·а́** (*f*) (chem) mandelic/amygdalic acid

~ **молок·о́** (*n*) almond oil; (pharm) almond emulsion

ми́ндел·ь, -я, (*m*) (geol) Mindel

~ **-рисс·, -а,** (*m*) (geol) Mindel-Riss

минёр·, -а,   (*m*) (mil) miner; (naut) underwater weapons rating; (ent) miner, *Mineola*

минера́л·, -а,   (*m*) mineral

~, акцессо́р·н·ый accessory mineral

~, второ·сте́пен·н·ый accessory mineral

~, компле́кс·н·ый   complex mineral

~, осно́в·н·о́й essential mineral

~, по·ка́з·а́тельн·ый illustrative mineral

~, про·мы́шл·енн·ый   commercial mineral

~, путе·во́д·н·ый   tracer mineral

~, со·пу́т·ствующ·ий   associated/ancillary mineral

минерализ·а́тор·, -а, (*m*) (geol) mineralizer

минерализ·а́ци·я, -и,   (*f*) mineralization

минерализ·о́ванн·ый, -ая, -ое, (*past part pass*) of **минерализ·ова́ть;** mineralized; containing salts

минерализ·ова́ть, -у́ют,   (*imp and perf*) mineralize

минерало́г·, -а, (*m*) mineralogist

минерало́г·и́ческ·ий, -ая, -ое, (*a*) mineralogical

минерало́г·и·я, -и, (*f*) mineralogy

минерало·кера́м·ик·а, -и,   (*f*) (geol) crystallites; cermet, ceramet, oxide/metal ceramic

минера́ль·н·ый, -ая, -ое, (*a*) of **минера́л·**

~ ва́т·а (*f*) mineral wool

~ ма́сл·о (*n*) petroleum

минётт·а, -ы, (*f*) (geol) minette

миниатю́р·а, -ы, (*f*) miniature

миниатю́р·н·ый, -ая, -ое, (*a*) of **миниатю́р·а;** minute, diminutive, midget

минима́кс·, -а, (*m*) minimax

минима́ль·ност·ь, -и,   (*f*) (math) minimality

минима́ль·н·ый, -ая, -ое, (*a*) minimum, least

~ реле́ (*n*) (elec) undercurrent relay

мини́метр·, -а,   (*m*) (instr) vertical comparator, contact comparator

мини́метр·и·я, -и,   (*f*) (instr) comparison measurement

минмиз·а́тор·, -а, (*m*) minimum-seeking computer

минимиз·а́ци·я, -и,   (*f*) minimizing; (autom) minimum-seeking

минимиз·и́р·овать, -уют, (*imp*) minimize; (math) reduce to a minimum

ми́нимум·, -а,   (*m*) minimum

~, бар·и́ческ·ий (meteor) trough of low pressure

~, втор·и́чн·ый (meteor) secondary low

~ слыш·и́мост·и   (rad) minimum signal strength

~, ча́ст·н·ый (meteor) secondary low

мин·и́р·овать, -уют,   (*imp and perf*) mine, lay mines

министе́р·ск·ий, -ая, -ое,   (*a*) ministerial

министе́р·ств·о, -а,   (*n*) ministry (Soviet institutions), ministry, department, board, office, treasury, (foreign institutions at ministerial level)

~ торго́вл·и Ministry of Trade; Board of Trade (in Britain)

мини́стр·, -а, (*m*) Minister (in U.S.S.R); Minister, Secretary, Chancellor (etc. according to office outside U.S.S.R.)

миниу́м·, -а,   (*m*) (chem) minium, red lead

Минко́вск·ого, про·стра́н·ств·о   (*n*) (math) Minkowski distance function

ми́н·н·ый, -ая, -ое,   (*a*) (mil) mine, mining

~ за·гра́д·и́тел·ь   (*m*) (naut) mine-layer

~ за·гражд·е́ни·е,   (*n*) minefield

~ тра́ль·щик·   (*m*) (naut) mine-sweeper

мин·ова́ть, -у́ют, (*imp and perf*) pass, pass by, miss, avoid

мино́г·а, -и,   (*f*) (zool) lamprey, *Petromyzon*

мино·иск·а́тел·ь, -я,   (*m*) (mil) mine detector

мино·мёт·, -а, (*m*) mortar (weapon)

~, реакти́в·н·ый   mortar

мино·но́с·ец·, -с·ц·а, (*i*) **-с·ц·ем,** (*m*) (obs) torpedo-boat destroyer; (a/c) minelaying aircraft

~, эскадр·е́нн·ый (naut) destroyer

мино́р·, -а, (*m*) (math) minor; minor key (music)

минта́·й, -я, (*m*) (fish) mintai, *Theragra chalcogramma*

мин·у́вш·ий, -ая, -ее, (*past part act*) of **мин·ова́ть;** past, last; (*n decl as adj*) the past

ми́нус·, -а, (*m*) minus, negative; shortcoming

~ -координа́т·ы (*pl*) negative co-ordinates

ми́нус·ов·ый, -ая, -ое, (*a*) of **ми́нус·**

~ про́·вод· (*m*) (elec) negative wire

мину́т·а, -ы, (*f*) minute

мину́т·ник·, -а, (*m*) flour grain (abrasives)

мину́т·н·ый, -ая, -ое, (*a*) of **мину́т·а;** momentary

мин·у́ть, -ут, (*perf*) pass, pass by, miss, avoid

мио- (*int component*) mio-, myo-

мио·бла́ст·, -а, (*m*) (zool) myoblast

мио·ге́н·, -а, (*m*) (biochem) myogen

мио·гео·синклин·а́л·ь, -и, (*f*) (geol) miogeosyncline

мио·гипси́н·а, -ы, (*f*) (pal) miogypsine, (*pl*) *Miogypsina*

мио·гипсини́д·а, -ы, (*f*) (pal) miogypsinid, (*pl*) *Miogypsinidae*

мио·глоби́н·, -а, (*m*) (biochem) myoglobin

ми·о́з·, -а, (*m*) (med) myosis, miosis

мио·зи́н·, -а, (*m*) (biochem) myosin

ми·ози́т·, -а, (*m*) (med) myositis

мио·ка́рд·, -а, (*m*) (anat) myocardium

мио·кард·и́т·, -а, (*m*) (med) myocarditis

мио·клон·и́·я, -и, (*f*) (med) myoclonus

мио·ко́мм·а, -ы, (*f*) (zool) myocomma

мио·ло́г·и·я, -и, (*f*) myology

мио́л·ь, -я, (*m*) (pharm) myolum

ми·о́м·а, -ы, (*f*) (med) myoma

мио·ме́р·, -а, (*m*) (anat) myomere

мио·не́м·а, -ы, (*f*) zool myoneme

мио·пат·и́·я, -и, (*f*) (med) myopathy

ми·опи́·я, -и, (*f*) (med) myopia, short sight

мио·се́пт·а, -ы, (*f*) (anat) myoseptum

мио·то́м·, -а, (*m*) (anat) myotome

мио·тон·и́·я, -и, (*f*) (med) myotonia atrophica

мио·фо́р·, -а, (*m*) (zool) myophore

мио·це́н·, -а, (*m*) (geol) Miocene

мио·ци́т·, -а, (*m*) (zool) myocyte

миnола́м·, -а, (*m*) (plastic) mipolam

мипо́р·а, -ы, (*f*) Mipora (microporous rubber)

ми́р·, -а, (*nom pl*) **-ы́,** (*g pl*) **-о́в,** (*m*) world, universe; peace; (biol etc.) kingdom

ми́р·а, -ы, (*f*) (opt, TV) optical focusing/tuning chart/pattern

мирабе́л·ь, -и, (*f*) (bot) mirabelle plum, *Prunus cerasifera*

мирабили́т·, -а, (*m*) (min) mirabilite, Glauber's salt

мира́ж·, -а, (*i*) **-ем,** (*m*) mirage

~, ве́рх·н·ий (rad) looming, superior mirage

~, ни́ж·н·ий (rad) inferior mirage

мирба́н·ов·ое ма́сл·о (*n*) mirbane oil, nitrobenzene

мириа- (*int component*) myria- (10^4)

мириа·ва́тт·, -а, (*m*) myriawatt

мириа́метр·, -а, (*m*) (instr) myriameter

миринг·и́т·, -а, (*m*) (med) myringitis

мири́сти́н·, -а, (*m*) (chem) myristin

мирристи́н·ов·ый, -ая, -ое, (*a*) myristin, myristic

~ кисл·от·а́ (*f*) (chem) myristic acid

мир·и́ть, -я́т, (*imp*) reconcile

мирици́л·, -а, (*m*) (chem) myricyl

мирици́л·ов·ый спирт· (*m*) (chem) myricyl alcohol

ми́ри·я, -и, (*f*) (math) myria, 10^4

мирмек·и́т·, -а, (*m*) (min) myrmekite

мирмек·и́тов·ый, -ая, -ое, (*a*) (geol) myrmekitic

мирмеко·фи́ль·н·ый, -ая, -ое, (*a*) (bot) myrmecophilous

мирмеко·хо́р·и·я, -и, (*f*) (bot) myrmecochory

ми́р·н·ый, -ая, -ое, (*a*) peace; peaceful, placid

миро·бала́н·, -а, (*m*) (bot) myrobalan (fruit of the tree *Terminalia chebula*); myrobalan extract (leather taning)

миро·воз·зр·е́ни·е, -я, (*n*) outlook, views, attitude, standpoint

мир·ов·о́й, -а́я, -о́е, (*a*) world; mundane; (math) material

~ про·стра́н·ств·о (*n*) (phys) outer space

~ ли́н·и·я (*f*) world line

~ то́ч·к·а (*f*) (math) material point; (phys) world point

миро·зи́н·, -а, (*m*) (biochem) myrosin

ми́рт·, -а, (*m*) (bot) myrtle, *Myrtus*

миртена́л·, -а, (*m*) (chem) myrtenal

миртéн·ов·ый, -ая, -ое, (*a*) myrtenic

миртен·о́л·, -а, (*m*) (chem) myrtenol

ми́рт·ов·ый, -ая, -ое, (*a*) (bot) myrtle; (*pl*) myrtle/eucalyptus family, *Myrtaceae*

мирце́н·, -а, (*m*) (chem) myrcene

мир·я́щ·ий, -ая, -ее, (*pres part act*) of **мир·и́ть**

Мисим·а, с·плав· (*m*) (met) Mishima alloy

ми́с·к·а, -и, (*g pl*) **-с·ок·,** (*f*) bowl, basin

миспи́кел·ь, -я, (*m*) (min) mispickel, arsenopyrite

ми́сси·я, -и, (*f*) mission

мистифик·а́ци·я, -и, (*f*) deception, hoax, mystification

мистра́л·ь, -я, (*m*) (meteor) mistral

ми́тинг·, -а, (*m*) meeting, rally

миткал·ёв·ый, -ая, -ое, (*a*) of мит-ка́л·ь

митка́л·ь, -я, (*m*) (text) calico

мито·генет·и́ческ·ий, -ая, -ое, (*a*) (cyt) mitogenetic

~ луч· (*m*) (bot) mitogenetic ray

мито́·з·, -а, (*m*) (cyt) mitosis, karyokinesis

мит·оти́ческ·ий, -ая, -ое, (*a*) (cyt) mitotic

мито·хо́ндр·и·я, -и, (*f*) (cyt) mitochondria

ми́ттел·ь, -я, (*m*) (print) English

Митчерлих·а, за·ко́н·, (*m*) Mitscherlich's law (of isomorphism)

миф·и́ческ·ий, -ая, -ое, (*a*) mythical

Михлер·а, кето́н· (*m*) (chem) Michler's ketone

мицеле·ви́д·н·ый, -ая, -ое, (*a*) (bot) mycelium-shaped

мице́л·и·й, -я, (*m*) (bot) mycelium

мице́лл·а, -ы, (*f*) (chem) micelle

Мичелл·а, под·ши́п·ник· (*m*)' (mech eng) Michell bearing

Мичи, мето́д· (*m*) (oil) Michie sludge test

ми́чман·, -а, (*m*) (naut) warrant officer

мичу́рин·ск·ий, -ая, -ое, (*a*) (biol) Michurin

мишён·н·ый, -ая, -ое, (*a*) of мишён·ь

мишён·ь, -и, (*f*) target, practice target; (TV) mosaic

~, глаз·н·а́я eye vane (compass)

~, под·вес·н·а́я (air) drogue

~, пояс·н·а́я waist-figure target

~, пред·ме́т·н·ая object vane (compass)

~, рост·ов·а́я full-height figure target

~, фигу́р·н·ая figure target

мишур·а́, -ы́, (*f*) tinsel

мишу́р·н·ый, -ая, -ое, (*a*) tinsel

миэл- (*int component*) (*see also* миел-) myel-

миэл·и́н·, -а, (*m*) (anat) myelin

миэл·и́нов·ая оболо́ч·к·а (*f*) (anat) medullary sheath, myelin sheath

миэло·ци́т·, -а, (*m*) (zool) myelocyte

мк (*abbr*) of микро-, micro-

МК (*abbr*) of между·на·ро́д·н·ая коми́сс·и·я International Commission

мка (*abbr*) of микро·ампе́р· μa (U.S.), μA (Brit), microampere

мкбар (*abbr*) of микро·ба́р·, μb, microbar

мкв (*abbr*) of микро·ва́тт· μw (U.S.), μW (Brit), microwatt

мкг (*abbr*) of микро·гра́мм· μg, microgram

мкгн (*abbr*) of микро·генри (elec) microhenry

МКГС систе́м·а (*f*) (mech) MKS/mks system, metre-kilogram-second system (of units of measurement)

МКГСС = метр-килогра́мм (сила)-секунда (mech) MKS, mks (system of units of measurement)

мкк (*abbr*) of микро·куло́н· (elec) microcoulomb

мкмкф (*abbr*) of микро·микро·фара́д·а (elec) pf (U.S.), pF (Brit), micro-microfarad

мккюри (*abbr*) of микро·кюри́, μc, μC, (phys) microcurie

мкл (*abbr*) of микро·ли́тр· microlitre, microliter

мком (*abbr*) of микр·о́м· (elec) ohm, microhm

мкс (*abbr*) of ма́ксвелл· (elec) mx, maxwell

МКС (*abbr*) of метр-килогра́мм (массы)-секу́нда (mech) MKS, mks (system of units of measurement)

МКСА (elec) MKSA, mksa (metre-kilogram-second-ampere system)

МКСГ = метр-килогра́мм-секу́нда-гра́дус Ке́львин·а (*m*) (heat) MKS, mks (systems of units of measurement related to °K)

мкф (*abbr*) of микро·фара́д·а (elec) μf (U.S.), μF, mF (Brit), microfarad

мкюри (*abbr*) of милли·кюри́ (nucl) millicurie, mC, mc

Мкюри (*abbr*) of мега·кюри́ MC, mc, megacurie

мл (*abbr*) of милли·ли́тр· ml, millilitre, milliliter

млад·е́нец·, -нц·а, (*i*) -нц·ем, (*m*) infant, baby

мла́д·ш·ий, -ая, -ее, (*a*) younger, minor; youngest; junior

млеко·но́с·н·ый, -ая, -ое, (*a*) (biol) lactiferous

млеко·пит·а́ющ·ее, -ого, (*n decl as adj*) (zool) mammal; (*pl*) mammals, *Mammalia*

~, траво·я́д·н·ое herbivore

~, хи́щ·н·ое carnivore

млеко·пит·а́ющ·ий, -ая, -ее, (*a*) (*see also* млеко·пит·а́ющ·ее) (*as noun*) (zool) mammalian

мле́ч·ник·, -а, (*m*) (bot) lactiferous cell

млéч·н·ый, -ая, -ое, (*a*) lactic, lacteal

~ **пут·ь** (*m*) (astron) the Milky Way; galaxy

~ **сок·** (*m*) (bot) latex; (physiol) chyle

~ **со·сýд·** (*m*) (bot) lactiferous cell/element

млн (*abbr*) of **миллиóн·** million

млынóк·, -а, (*m*) (agr) winnower

мм (*abbr*) of **милли·мéтр·** mm, millimetre, millimeter

ММВ (*abbr*) of **милли·метр·óв·ая волн·á** (rad) millimetre wave

мм. вод. ст. (*abbr*) of **милли·мéтр·ы вод·н·óго столб·á** (*pl*) mm of water column (unit of pressure)

ммк (*abbr*) of millimicron, millimicro-, mu, mμ

мм. рт. ст. (*abbr*) of **милли·мéтр· ртýт·н·ого столб·á** (*m*) mm Hg, millimetres of mercury (unit of pressure)

мн (*abbr*) of **магнúт·н·ое на·сыщ·é·ни·е** (*n*) (phys) magnetic saturation

мне (*pron, dat*) of **я,** to me; (*pron, prep*) of **я,** (about etc.) me

мнемóн·ик·а, -и, (*f*) mnemonics

мнемон·úческ·ий, -ая, -ое, (*a*) mnemonic

~ **щит·** (*m*) graphic control panel/chart

мнемо·схéм·а, -ы, (*f*) graphic control panel/chart

мнемо·тéхн·ик·а, -и, (*f*) mnemonics

мн·éни·е, -я, (*n*) opinion

мн·ёт (*pres 3rd sing*) of **мя·ть**

мн·úм·ый, -ая, -ое, (*a*) imaginary, supposed; pretended false

~ **част·ь** (*f*) (math) imaginary part

~ **част·ь комплéкс·н·ого числ·á** (*f*) (math) imaginary part of a complex number

~ **числ·ó** (*n*) (math) imaginary number

~ **из·ображ·éни·е** (*n*) (opt) virtual image

мн·úте (*plur imper*) of **мя·ть**

мн·ить, -ят, (*imp*) think, be of the opinion that

мнóг·ий, -ая, -ое, (*a*) many, a lot of; (*n as noun*) much, a great deal; (*pl as noun*) many people

во мнóг·ом in many ways

мнóго (+ *gen*) much, many, plenty of, a lot of; (*adv*) much, a great deal, a lot

много- (*root*) multi-, multiple-, poly-

мнóго·áкт·н·ый, -ая, -ое, (*a*) (nucl) multi-event, plural

мнóго·ампéр·н·ый, -ая, -ое, (*a*) (elec) heavy-current

мнóго·áтом·н·ый, -ая, -ое, (*a*) polyatomic; polyacid, polyacidic (base); polybasic, polyhydric (alcohol), polyvalent

~ **олефúн·** (*m*) (chem) long-chain olefine

~ **основ·áни·е** (*n*) (chem) polyacid base

~ **спирт·** (*m*) (chem) polyhydric alcohol

мнóго·блóч·н·ый, -ая, -ое, (*a*) multibank (engines)

мнóго·борóзд·чат·ый, -ая, -ое, (*a*) multisulcate, much furrowed

мнóго·брáт·н·ый, -ая, -ое, (*a*) (bot) polyadelphous

мнóго·брáч·н·ый, -ая, -ое, (*a*) (biol) polygamous

мнóго·бугóр·чат·ые, -ых, (*pl decl as adj*) (pal) multituberculates, *Multituberculata*

мнóго·валéнт·ност·ь, -и, (*f*) multivalence, polyvalence

мнóго·валéнт·н·ый, -ая, -ое, (*a*) multivalent, polyvalent

мнóго·вариáнт·н·ый, -ая, -ое, (*a*) multivariant

мног·овáт·о (*adv*) rather too much

мнóго·вáтт·н·ый, -ая, -ое, (*a*) (elec) high-watt/duty

мнóго·верш·úнн·ый, -ая, -ое, (*a*) polyconic

мнóго·вúд·н·ый, -ая, -ое, (*a*) (rad) multimode; (bot) multiform

мнóго·винт·ов·óй, -áя, -óе, (*a*) multirotor (of helicopters)

мног о·вал·кóв·ый, -ая, -ое, (*a*) multiroll

мнóго·вит·кóв·ый, -ая, -ое, (*a*) multiturn (coil)

мнóго·вóд·н·ый, -ая, -ое, (*a*) (chem) polyhydrate; (geog) abounding in water; (med) hydramnious

мнóго·галогéн·н·ый, -ая, -ое, (*a*) (chem) polyhalide, polyhalogenated

мнóго·гнёзд·н·ый, -ая, -ое, (*a*) (biol) multilocular, multiloculate; multijack (telephones); multicavity; multislide (vacuum tubes)

~ **фóрм·а** (*f*) (plast, met) multicavity mould/mold

мнóго·грáн·ник·, -а, (*m*) (math) polyhedron; multifacet prism

мнóго·грáн·н·ый, -ая, -ое, (*a*) polyhedral; many-sided, versatile; (mech eng) multiple-cornered (lathe work)

мно́го·голо́в·очн·ый, -ая, -ое, (*a*)
multiple-head

мно́го·диапазо́н·н·ый, -ая, -ое, (*a*)
(rad) multirange, multiband

мно́го·ди́ск·ов·ый, -ая, -ое, (*a*)
(mech) multiplate, multidisc

мно́го·до́ль·чат·ый, -ая, -ое, (*a*)
(biol) multilobate

мно́го·доме́н·н·ый, -ая, -ое, (*a*)
(phys) polydomain

мно́го·до́м·н·ый, -ая, -ое, (*a*) (bot)
polygamous

мно́го·доро́ж·н·ый, -ая, -ое, (*a*)
multitrack (recording systems)

мно́го·ды́р·чат·ый, -ая, -ое, (*a*)
perforated

мно́г·ое *see* **мно́г·ий**

мно́го·ём·костн·ый, -ая, -ое, (*a*)
multicapacity

мно́го·жгу́тик·ов·ый, -ая, -ое, (*a*)
multiflagellate, polymastigote; (*pl decl
as adj*) (zool) *Polymastigida, Poly-
mastigina*

мно́го·жён·ств·о, -а, (*n*) (biol) poly-
gyny

мно́го·жи́ль·н·ый, -ая, -ое, (*a*)
multistrand; (elec) multicore

~ **ка́бел·ь** (*m*) (elec) multicore cable

~ **шнур·** (*n*) (elec) multiple cord

мно́го·за·зо́р·н·ый, -ая, -ое, (*a*)
multigap, multiple-gap

мно́го·за·мещ·ённ·ый, -ая, -ое, (*a*)
(chem) polysubstituted

мно́го·за·ря́д·н·ый, -ая, -ое, (*a*)
(nucl) multiply charged

мно́го·за·хо́д·н·ый, -ая, -ое, (*a*)
multistart, multi-entry, multiple (of
screw threads, spirals, helixes etc.);
multiple-thread (of screws)

мно́го·зве́н·н·ый, -ая, -ое, (*a*) multi-
link; (rad) iterated

мно́го·зву́ч·ий, -ая, -ее, (*a*) poly-
phonic

мно́го·знач·и́тельн·ый, -ая, -ое, (*a*)
significant, very important

мно́го·зна́ч·ност·ь, -и, (*f*) multiplicity; ambiguity

мно́го·опа́ч·н·ый, -ая, -ое, (*a*) multi-
digit; multiple-valued; ambiguous;
(gram) polysemantic

мно́го·зо́ль·н·ый, -ая, -ое, (*a*)
high-ash

мно́го·инерцио́н·н·ый, -ая, -ое, (*a*)
multilag

мно́го·ио́д·ист·ый, -ая, -ое, (*a*)
(chem) polyiodo-

мно́го·ка́мер·н·ый, -ая, -ое, (*a*)
multichamber, multicellular, multicell, multicavity; (zool) polythala-
mous; (biol) multiloculate, multilocular

~ **магнетро́н·** (*m*) (elec) multicavity
magnetron

~ **дви́г·ател·ь** (*m*) multichamber
rocket engine

~ **печ·ь** (*f*) (met) multiple-hearth
furnace; (chem) multicell heater

мно́го·камн·ев·ый, -ая, -ое, (*a*)
multiwheel (grinding)

мно́го·кана́ль·н·ый, -ая, -ое, (*a*)
(elec) multichannel; multiple-duct;
polysleeve; (expl) multiperforated

~ **канал·иза́ци·я** (*f*) multiple-duct
conduit (underground wiring systems)

~ **пере·да́т·чик·** (*m*) (rad) multichannel transmitter

мно́го·каска́д·н·ый, -ая, -ое, (*a*)
(elec) multistage

мно́го·кисл·о́тн·ый, -ая, -ое, (*a*)
(chem) polyacid, polyacidic

мно́го·ки́сл·ый, -ая, -ое, (*a*) (chem)
polyhydric

мно́го·кле́т·очн·ые, -ых, (*pl decl as
adj*) (zool) *Metazoa*

мно́го·кле́т·очн·ый, -ая, -ое, (*a*)
multicell, (bot) multicellular

мно́го·ко́вш·ев·ый, -ая, -ое, (*a*)
multibucket, bucket-ladder (excava-
tors, dredges); multicut (of trench-
ers etc.)

мно́го·коле́й·н·ый, -ая, -ое, (*a*)
multiwheel (of vehicles); multitrack/
lane/line (of roads, railroads)

мно́го·коле́н·чат·ые, -ых, (*pl as a*)
(zool) sea spiders, *Pycnogonida*

мно́го·колёс·н·ый, -ая, -ое, (*a*)
multiple-impeller (pumps)

мно́го·конта́кт·н·ый, -ая, -ое, (*a*)
multipoint, multicontact

мно́го·ко́нтур·н·ый, -ая, -ое, (*a*)
multistream (jet engines); (autom)
multiloop, multiple-loop

мно́го·ко́рпус·н·ый, -ая, -ое, (*a*)
multiple, multiple-unit

мно́го·кра́с·очн·ый, -ая, -ое, (*a*)
polychromatic, multicoloured, multi-
colored

мно́го·кра́т·н·о (*adv*) repeatedly

мно́го·кра́т·ност·ь, -и, (*f*) multiplicity, recurrence; (rad) multiplexing,
multiplex

~, **видо·и́мпульс·н·ый** (rad) pulse-
mode multiplex

мно́го·кра́т·н·ый, -ая, -ое, (*a*) repeated, reiterated; multiple, plural; multiple-effect; (telecom) multiplex; multichannel (radar)

~ **ис·пар·е́ни·е** (*n*) multiple-effect evaporation

~ **кана́л·** (*m*) (telecom) multichannel

~ **ра́дио·пере·да́·ч·а** (*f*) multiplex radio transmission

~ **разры́в** (*m*) (elec) multiple break; (*in gen as adj*) multibreak

~ **со·един·е́ни·е** (*n*) multiple connection

~ **телегра́ф·и́ровани·е** (*n*) multiplex telegraphy

мно́го·лезв·и́йн·ый, -ая, -ое, (*a*) multipoint (of metal-cutting tools)

мно́го·лепест·ко́в·ый, -ая, -ое, (*a*) (bot) polypetalous

мно́го·ле́т·н·ий, -яя, -ее, (*a*) several years old, of many years, long-term, long-standing; (bot) perennial

~ **лёд·** (*m*) (ocean) polar ice

мно́го·ле́т·ник·, -а, (*m*) (bot) perennial

мно́го·лин·е́йн·ый, -ая, -ое, (*a*) multiple-line, (spectr) band, multilinear

мно́го·ли́ст·н·ый, -ая, -ое, (*a*) (bot) multifoliate, polyphillous; multisheet, multisheeted, multifoil

~ **по·ве́рх·ност·ь** (*f*) (math) multi-sheeted surface

мно́го·ло́паст·н·ый, -ая, -ое, (*a*) multiblade (pumps etc.); (dynam) multilobe; palmate, multilobate; multifid

~ **на·со́с·** (*m*) sliding-vane rotary pump

мно́го·луч·ев·о́й, -а́я, -о́е, (*a*) multibeam, multiple-beam; multiwire (antennas); multitrace (CRT technique)

~ **ис·то́ч·ник·** (*m*) (nucl) multiple source

мно́го·лю́д·н·ый, -ая, -ое, (*a*) populous, densely populated; crowded

мно́го·ме́р·н·ый, -ая, -ое, (*a*) multidimensional, multivariate, multiple

мно́го·ме́ст·н·ый, -ая, -ое, (*a*) multiseat/berth/place

мно́го·молекуля́р·н·ый, -ая, -ое, (*a*) multimolecular, polymolecular

~ **сло́·й** (*m*) (chem) multilayer

мно́го·мото́р·н·ый, -ая, -ое, (*a*) multi-engined

мно́го·му́ж·еств·о, -а, (*n*) (biol) polyandry

мно́го·му́ж·н·ый, -ая, -ое, (*a*) polyandrous

мно́го·ни́т·очн·ый, -ая, -ое, (*a*) multiple-thread, multiple-strand, multiple

мно́го·ни́т·н·ый, -ая, -ое, (*a*) multiple-thread; (elec) multiple-wire

мно́го·но́ж·к·а, -и, (*g pl*) **-ж·ек·,** (*f*) (zool) myriapod, (*pl*) *Myriapoda*

мно́го·обещ·а́ющ·ий, -ая, -ее, (*a*) promising, hopeful

мно́го·оборо́т·н·ая суш·и́лк·а (*f*) multipass dryer

мно́го·обра́з·и·е, -я, (*n*) variety, diversity; (math) manifold; (biol) polymorphism

мно́го·обра́з·н·ый, -ая, -ое, (*a*) diverse, varied; (math) manifold; (biol) polymorphic, polymorphous

мно́го·осно́в·н·ый, -ая, -ое, (*a*) (chem) polybasic (acids)

мно́го·па́л·ост·ь, -и, (*f*) (zool) polydactyly, polydactylism

мно́го·петл·ев·о́й, -а́я, -о́е, (*a*) multiple-loop; compound (cycle)

мно́го·пе́ч·н·ый, -ая, -ое, (*a*) (oil) multiple-coil

мно́го·пли́т·н·ый, -ая, -ое, (*a*) multiplate, multiplaten, multiple-stage

мно́го·пло́д·и·е, -я, (*n*) (med) multiple pregnancy, polytoky; (bot) polycarpy

мно́го·пло́д·н·ый, -ая, -ое, (*a*) (zool) multiparous, polytokous; (bot) polycarpic

мно́го·позицио́н·н·ый, -ая, -ое, (*a*) multiposition, multistep

мно́го·полос·н·ый, -ая, -ое, (*a*) (rad) multiband

мно́го·по́ль·н·ый, -ая, -ое, (*a*) (agr) multiplefield, multiplecrop

~ **сево·оборо́т·** (*m*) (agr) multiple-crop rotation

мно́го·по́лост·н·ый, -ая, -ое, (*a*) multicavity, multiple-cavity

мно́го·по́люс·ник·, -а, (*m*) (telecom) multiterminal network, network

~, **акти́в·н·ый** (telecom) active network

мно́го·по́люс·н·ый, -ая, -ое, (*a*) multipole, multipolar; (telecom) multiterminal

мно́го·по́ст·ов·о́й, -а́я, -о́е, (*a*) (weld) multi-operator

мно́го·пре·де́ль·н·ый, -ая, -ое, (*a*) multirange

мно́го·проже́ктор·н·ый, -ая, -ое, (*a*) multigun (vacuum tube)

мно́го·про·лёт·н·ый, -ая, -ое, (*a*) (civ eng) multispan, multiple-arch, multiple-bay

мно́го·пря́д·н·ый, -ая, -ое, (*a*) multistrand

мно́го·про·хо́д·н·о́й, -а́я, -о́е, (*a*) multipass

мно́го·пу́т·н·ый, -ая, -ое, (*a*) (rad) multipath; multilane (roads); (rail) multitrack, multiple-line

мно́го·ра́з·ов·ый, -ая, -ое, (*a*) repeated, frequent

мно́го·раз·де́ль·н·ый, -ая, -ое, (*a*) (bot) multifid, decompound

мно́го·раз·ре́з·н·ый, -ая, -ое, (*a*) multisection; multislot, multisegment (magnetrons)

мно́го·режи́м·н·ый, -ая, -ое, (*a*) multichangeover

мно́го·резона́тор·н·ый, -ая, -ое, (*a*) multiresonator

мно́го·рез·цо́в·ый, -ая, -ое, (*a*) multiple-tool (of machine tools); (zool) polyprotodont

~ **держ·а́вк·а,** (*f*) (mech eng) gang tool

~ **стан·о́к·** (*m*) multiple-tool lathe

мно́го·ручь·ев·о́й, -а́я, -о́е, (*a*) multigroove

~ **штамп·** (*m*) multipass forging die

мно́го·ря́д·ник·, -а, (*m*) (bot) holly fern

мно́го·ря́д·н·ый, -ая, -ое, (*a*) multiple-row; (ICE) multibank; multilane (of roads); (bot) polystichous, multiseriate; (rad) mattress

мно́го·с·вя́з·н·ый, -ая, -ое, (*a*) (math) multiple-connected

мно́го·секцио́н·н·ый, -ая, -ое, (*a*) multisection; (rail) multiple-unit

мно́го·семя·до́ль·н·ый, -ая, -ое, (*a*) (bot) polycotyledonous

мно́го·сем·я́нн·ый, -ая, -ое, (*a*) (bot) polyspermous

мно́го·сер·ни́ст·ый, -ая, -ое, (*a*) (chem) polysulphide of

мно́го·се́т·очн·ый, -ая, -ое, (*a*) (elec) multi-electrode, multi-element, multigrid

мно́го·с·кач·ко́в·ый, -ая, -ое, (*a*) (dynam) multishock

мно́го·скор·остн·о́й, -а́я, -о́е, (*a*) multiple-speed, multispeed, multivelocity

мно́го·с·ло́ж·н·ый, -ая, -ое, (*a*) (ling) polysyllabic; (bot) decompound; (med) multipartial

мно́го·сло́й·н·ый, -ая, -ое, (*a*) multilayer, multiply, multistratal, laminated; multiple lift (of roads); multiwall (of vertical objects); multiple (of layered or stacked objects)

~ **меш·о́к·** (*m*) (paper) multiwall sack

мно́го·сопл·о́в·о́й, -ая, -ое, (*a*) multinozzle; multiple-hole (of injectors etc.)

мно́го·со́рт·н·ый, -ая, -ое, (*a*) multigrade

мно́го·сторо́н·н·ий, -ая, -ое, (*a*) (com etc.) multilateral; versatile, many-sided; (math) polygonal

мно́го·сту́п·е́нчат·ый, -ая, -ое, (*a*) multistage, multiple-stage, multistep; (math) group

~ **из·бир·а́ни·е** (*n*) (autom) group discrimination

~ **раз·дел·е́ни·е** (*a*) multiple separation

мно́го·тари́ф·н·ый, -ая, -ое, (*a*) (com) multirate

~ **с·чёт·чик·** (*m*) (instr) multirate meter

мно́го·то́м·н·ый, -ая, -ое, (*a*) (print) multivolume; voluminous

мно́го·то́ч·ечн·ый, -ая, -ое, (*a*) multipoint; (weld) multiple-point/projection

~ **га́з·ов·ая коло́н·к·а** (*f*) multipoint gas water heater

мно́го·то́ч·и·е, -я, (*n*) dots (punctuation etc.)

мно́го·у·важ·а́ем·ый, -ая, -ое, (*a*) Dear (in formal correspondence)

мно́го·уго́ль·ник·, -а, (*m*) (math) polygon

~, **пра́в·ильн·ый** regular polygon

~ **сил·** (mech) polygon of forces

мно́го·уго́ль·н·ый, -ая, -ое, (*a*) (math) polygonal; (elec) mesh (connections, circuits)

мно́го·уда́р·н·ый, -ая, -ое, (*a*) fast-hitting

мно́го·фа́з·н·ый, -ая, -ое, (*a*) polyphase, multiphase

мно́го·фо́рм·ност·ь, -и, (*f*) polymorphism

мно́го·функциона́ль·н·ый, -ая, ое, (*a*) multipurpose

мно́го·хо́д·ов·о́й, -а́я, -о́е, (*a*) multipass; multiway, multipath; multiple-entry, multistart (screw threads)

~ **конденса́тор·** (*m*) multipass condenser

мно́го·ход·ов·о́й, -а́я, -о́е

~ **пере·да́·ч·а,** *(f)* (rad) multipath transmission

мно́го·цвет·ко́в·ый, -ая, -ое, *(a)* (bot) multiflorous, floribunda, polyanthous

мно́го·цве́т·н·ый, -ая, -ое, *(a)* polychromatic, many-coloured

мно́го·цел·ев·о́й, -ая, -ое, *(a)* multipurpose

мно́го·част·и́чн·ый, -ая, -ое, *(a)* multiparticle, many-particle; many-body

~ **си́л·ы** *(pl)* (phys) many-body forces

~ **систе́м·а** *(f)* (phys) many-particle system

мно́го·челн·о́чн·ый, -ая, -ое, *(a)* (text) multishuttle

мно́го·чи́сл·енн·ый, -ая, -ое, *(a)* numerous, multitudinous, enumerable

мно́го·чле́н·, -а, *(m)* (math) polynomial

~, **сфер·и́ческ·ие** *(pl)* Legendre polynomials

мно́го·чле́н·н·ый, -ая, -ое, *(a)* polynomial; many-termed, many-place (logarithms)

мно́го·шамо́т·, -а, *(m)* high-fireclay refractory; (ceram) high-grog

мно́го·шка́ль·н·ый, -ая, -ое, *(a)* (instr) multirange, multiple

мно́го·шпи́ндель·н·ый, -ая, -ое, *(a)* multispindle

мно́го·щет·и́нков·ые, -ых, *(pl decl as a)* (zool) *Polychaeta*

мно́го·электро́д·н·ый, -ая, -ое, *(a)* multi-electrode

мно́го·элеме́нт·н·ый, -ая, -ое, *(a)* multi-unit, multicomponent

мно́го·эта́ж·н·ый, -ая, -ое, *(a)* multistorey, multideck; multiplaten (presses etc.)

~ **вулканиз·ацио́нн·ый пресс·** *(m)* multidaylight vulcanization press

мно́го·я́дер·н·ый, -ая, -ое, *(a)* polynuclear, polynucleate; (chem) polycyclic

~ **цикл·и́чн·ый** (chem) polycyclic

мно́го·я́д·ност·ь, -и, *(f)* (biol) polyphagism

мно́го·я́рус·н·ый, -ая, -ое, *(a)* (geol) multistage

~ **анте́нн·а** *(f)* (rad) stacked-dipole antenna array

мно́го·яче́·ист·ый, -ая, -ое, *(a)* multicellular

мно́го·яче́й·ков·ый, -ая, -ое, *(a)* cellular, multicell, multicellular

мно́ж·ественност·ь, -и, *(f)* plurality, multiplicity

мно́ж·ественн·ый, -ая, -ое, *(a)* plural, multiple

мно́ж·еств·о, -а, *(n)* great number; (math) set; (phys) aggregate, aggregation

~ **а́том·ов** aggregate of atoms

~, **за́·мк·нут·ое** (math) closed set

~, **из·мер·и́м·ое** (math) measurable set

~, **ка́нтор·ск·ое** (math) Cantor set/discontinuum

~, **от·кры́·т·ое** (math) open set

~, **пло́т·н·ое** (math) dense set

~, **с·вя́з·н·ое** (math) connected set

~, **со·верш·ённ·ый** (math) perfect set

~, **то́ч·ечн·ое** set of points

~, **част·и́чн·о у·по·ря́д·оченн·ое** (math) partially ordered set

множ·и́м·ое, -ого, *(n decl as adj)* (math) multiplicand

множ·и́тел·ь, -я, *(m)* (math) multiplier, factor

~, **вес·ов·о́й** weighting factor

~, **еди́н·ый масшта́б·н·ый** single scale factor

~ **за·тух·а́ни·я** attenuation factor

~, **масшта́б·н·ый** scale factor

~, **пере·во́д·н·ый** (math) conversion factor

~, **по·пра́в·очн·ый** correction factor

множ·и́тельн·ый, -ая, -ое, *(a)* multiplying, multiplication; (print) duplicating

~ **аппара́т·** *(m)* (print) duplicator

~ **механи́зм·** *(m)* mechanical multiplier

~ **перфора́тор·** *(m)* punched-card multiplier

~ **у·стро́й·ств·о** *(n)* multiplier (computers)

~ **у·стро́й·ств·о механ·и́ческ·ого ти́п·а,** *(n)* linkage multiplier

мно́ж·ить, -ат, *(imp)* multiply; increase

мно́й *(pron, instr)* of **я** by/with etc. me

мн·у́щ·ий, -ая, -ее, *(pres part act)* of **мя·ть**

мн·я *(pres gerund)* of **мя·ть**

мо (elec) mho (unit of conductance)

мо́а *(f indecl)* (pal) moa

мобилизацио́н·н·ый, -ая, -ое, *(a)* mobilization

мобилиз·а́ци·я, -и, *(f)* mobilization

моби́ль·ност·ь, -и, (*f*) mobility

моби́ль·н·ый, -ая, -ое, (*a*) mobile

мог· (*past masc sing*) of **моч·ь** was able, could

мога́р·, -а, (*m*) (bot) foxtail millet, *Setaria italica*

моги́л·а, -ы, (*m*) grave

моги́ль·ник·, -а, (*m*) cemetery, graveyard; (nucl) burial ground

мог·ла́ (*past fem sing*) of **моч·ь** was able, could

мог·ли́ (*past plur*) of **моч·ь**

мог·ло́ (*past neut sing*) of **моч·ь**

мог·у́ч·ий, -ая, -ее, (*a*) powerful

мог·у́щественн·ый, -ая, -ое, (*a*) powerful, potent

мог·у́ществ·о, -а, (*n*) power, might, potency

мог·у́щ·ий, -ая, -ее, (*pres part act*) of **моч·ь**; able, being able, who/which is able, who/which can; powerful, potent

мо́г·ш·ий, -ая, -ее, (*past part act*) of **моч·ь**; able, having been able, who/which was able, who/which had been able, who/which could

мо́д·а, -ы, (*f*) fashion (clothes etc.); (math) mode

мода́ль·ност·ь, -и, (*f*) modality

мода́ль·н·ый, -ая, -ое, (*a*) modal

модел·и́ровани·е, -я, (*n*) modelling, model testing/analysis, model test/study,. simulation; analogue/analog computation

~ **траекто́р·и·и** trajectory simulation

~ **у·сло́в·и·й по·лёт·а** flight simulation

~ **це́л·и** target simulation

модел·и́р·овать, -уют, (*imp and perf*) model, simulate

модел·иро́вк·а, -и, (*g pl*) **-вок·,** (*f*) modelling (art)

модел·и́рующ·ий, -ая, -ее, (*pres part act*) *see* **модел·и́р·овать**

~ **вы·числ·и́тельн·ая маши́н·а** (*f*) (*see also* **вы·числ·и́тельн·ая маши́н·а**) analogue computer

~ **маши́н·а** (*f*) simulator

~ **систе́м·а** (*f*) analogue, analog

~ **у·стро́й·ств·о** (*n*) simulator, analyser

моде́л·ь, -и, (*f*) model, pattern, simulator, analyser; mould/mold (for a relief map)

~ **в раз·ре́з·е** cutaway model

~**, динам·и́ческ·и по·до́бн·ая** dynamic model

моде́л·ь

~ **жидк·ой ка́пл·и** (phys) liquid-drop model

~**, масшта́б·н·ая** scale model

~**, металл·и́ческ·ая** (met cast) metal pattern

~**, много·ме́р·н·ая** multi-dimensional simulator

~**, нату́р·н·ая** full-scale model

~**, обра́т·н·ая** negative mould/mold (of a relief map)

~**, равно·ве́с·н·ая** (astron) steady-state model

~**, се́т·очн·ая** (elec) network analyser

~**, тру́б·н·ая** wind tunnel model

~ **ули́т·к·и** cochlear model

~**, уч·е́бн·ая** dummy, dummy model

~ **цеп·е́й** (elec) network analyser

~**, электро́н·н·ая** (phys) electronic simulator

моде́ль·н·ый, -ая, -ое, (*a*) of **моде́л·ь**; fashionable, model (of clothes)

~ **цех·** (*m*) (met cast) pattern-making shop

моде́ль·щик·, -а, (*m*) (met cast) patternmaker

моде́ль·щиц·а, -ы, (*f*) (met) pattern-maker (female); model milliner

модера́тор·, -а, (*m*) moderator; soft pedal (of a piano)

модерниз·а́ци·я, -и, (*f*) modernization

модерниз·и́р·овать, -уют, (*imp and perf*) modernize

модильо́н·, -а, (*m*) (build) modillion (type of ornamental bracket)

модифик·а́тор·, -а, (*m*) (met) modifying addition (non-ferrous); inoculant (of cast iron)

модифик·а́ци·я, -и, (*f*) modification, modifying; inoculation (of cast iron), inoculating

модифиц·и́ровани·е, -я, (*n*) *see* **модифик·а́ци·я**

модифиц·и́рованн·ый, -ая, -ое, (*past part pass*), modified; inoculated (of) cast iron)

мо́д·н·ый, -ая, -ое, (*a*) fashionable, stylish

модул·и́ровани·е, -я, (*n*) *see* **моду·ля́ци·я**

модул·и́рованн·ый, -ая, -ое, (*past part pass*) of **модул·и́р·овать**; modulated

~ **по част·от·е́** frequency-modulated

модул·и́р·овать, -уют, (*imp*) modulate

модуло́·метр·, -а, (*m*) (elec) modulation meter

мо́дул·ь, -я, (*m*) (*see also under associated words*) modulus; module

~ ва́нн·ы vat ratio (dyeing) (weight of material : weight of dye)

~ кас·а́тельн·ой   у·пру́г·ост·и (mech) modulus of elasticity in shear, modulus of rigidity

~ круч·е́ни·я torsional modulus

~ норма́ль·н·ой у·пру́г·ост·и (mech) modulus of elasticity, Young's modulus

~, объ·ём·н·ый (mech) bulk modulus
по мо́дул·ю (math) module, taken absolutely

~ раз·ме́р·ност·и modulus of dimensionality

~ с·дви́г·а (mech) modulus of rigidity, modulus of elasticity in shear, shear modulus
со·от·нош·е́ни·е мо́дул·ей у·пру́г·ост·и (*n*) modular ratio (of concrete and steel)

~ у·пру́г·ост·и modulus of elasticity, Young's modulus

~ у·пру́г·ост·и втор·о́го ро́д·а (mech) shear modulus

~ у·пру́г·ост·и пе́рв·ого ро́д·а (mech) Young's modulus

мо́дуль·н·ый, -ая, -ое, (*a*) modulus, modula

~ систе́м·а (*f*) (eng) modular system (of construction), building-block system (of construction)

модул·я́рн·ый, -ая, -ое, (*a*) modular

модул·я́тор·, -а, (*m*) (elec) modulator; buncher (klystron)

~, бала́нс·н·ый balanced modulator

~, и́мпульс·н·ый pulse modulator

~, искр·ов·о́й spark-gap modulator

~, ла́мп·ов·ый tube modulator, modulator valve

~, лин·е́йн·ый linear modulator

~, реакти́в·н·ый reactance modulator

~ с реакти́в·н·ой ла́мп·ой reactance-tube modulator

~, свет·ов·о́й light modulator/chopper

~ сигна́л·ов цве́т·ност·и (TV) chrominance modulator

~, тона́ль·н·ый voice-frequency telegraph modulator

~ экспози́ц·и·и (phot) sensitizer

модул·я́торн·ый, -ая, -ое, (*a*) modulator, modulated

модул·я́ци·я, -и, (*f*) (rad, mus) modulation

~, амплиту́д·н·ая amplitude modulation, AM

модул·я́ци·я

~, амплиту́д·н·о-и́мпульс·н·ая pulse-amplitude modulation, PAM

~, амплиту́д·н·о-и́мпульс·н·ая-ча́ст·отн·ая pulse-amplitude modulation-frequency modulation, PAM-FM

~, взаи́м·н·ая intermodulation

~ ви́нта *see* **частот·а модул·я́ции винт·а**

~ в·низ downward modulation

~ врем·я-и́мпульс·н·ая pulse-time modulation, PTM
глуб·ин·а модул·я́ци·и (*f*) modulation percentage/depth

~, и́мпульс·н·ая pulse modulation

~, ко́д·ов·о-и́мпульс·н·ая pulse-code modulation, pulse-count modulation, PCM
коэффицие́нт· модул·я́ци·й (*m*) modulation factor/depth

~, много·кра́т·н·ая multiple modulation

~ на тре́т·ью се́т·к·у (TV) screen-grid/suppressor-grid modulation

~, одно·по́люс·н·ая single-sideband modulation

~, одно·сторо́н·н·яя врем·я-и́м·пульс·н·ая single-edge pulse-time modulation

~, о·по́р·н·ая reference modulation

~, пере·кре́ст·н·ая cross modulation

~ по кон·и́ческ·ом скан·и́рова·ни·ем (rad) lobing

~ по по·сле́д·овательн·ой схе́м·ой, ано́д·н·ая series modulation

~ прямо·уго́ль·н·ыми и́мпульс·а·ми (TV) square-wave modulation

~ пуч·к·а́ beam modulation

~ све́т·а light modulation

~, се́т·очн·ая grid modulation

~, скор·ости·а́я velocity modulation

~ с·мещ·е́ни·ем (rad) grid bias modulation

~, тона́ль·н·ая audio modulation

~, трой·н·а́я triple modulation

~, фа́з·н·ая phase modulation

~ фо́н·ом hum modulation

~, фа́зо-и́мпульс·н·ая pulse-position modulation, PPM

~, фото·телегра́ф·н·ая facsimile modulation
част·от·а́ модул·я́ци·и ви́нт·а (*f*) (a/c) blade frequency

~, част·о́тн·ая frequency modulation, FM

модул·яци́·я

~, част·о́тн·о-и́мпульс·н·ая pulse-frequency modulation, PFM

~, шир·о́тн·о-и́мпульс·н·ая pulse-width/length modulation, pulse-duration modulation, PDM

~, шум·ов·а́я hum modulation

~ я́р·кост·и (TV) intensity modulation

мо́дус·, -а, (*m*) modus

моё (*pron, nom/acc n*) of **мой;** my

мо́·евк·а, -и, (*g pl*) **-вок·,** (*f*) (zool) kittiwake, *Rissa tridactyla*

мо́·ек· (*g pl*) of **мо́й·к·а**

мо́·ет (*pres 3rd sing*) of **мы́·ть**

мо́·ечн·ый, -ая, -ое, (*a*) washing, wash

~ желёбк·а (*f*) (min) trough washer

~ маши́н·а (*f*) washer, washer/cleaning machine (industrial, not domestic)

~ цех· (*m*) (min) washery

мож·а́ (*pres gerund*) of **моч·ь** being able

мо́ж·ет (*pres 3rd sing*) of **моч·ь** is able, can

~ бы́·ть (*adv*) perhaps, maybe, it may be

можжеве́л·ов·ый, -ая, -ое, (*a*) juniper

~ во́д·к·а (*f*) Holland gin

~ спирт· (*m*) spirit of juniper

можжеве́ль·ник·, -а, (*m*) (bot) juniper, *Juniperus*

~, кра́сн·ый (bot) red cedar, *Juniperus oxycedrus*

мо́ж·н·о (*predic*) it is possible, it/one can/may

 как мо́ж·н·о скор·е́е as quickly as possible

моза·за́вр·, -а, (*m*) (pal) mosasaur, mosasaurus

моза́ик·а, -и, (*f*) mosaic

~, лист·ов·а́я (bot) leaf/foliate mosaic

мозаи́ч·ност·ь, -и, (*f*) (cryst) mosaic development/structure

 у́гол·мозаи́ч·ност·и (*m*) (cryst) mosaic angle

мозаи́ч·н·ый, -ая, -ое, (*a*) mosaic

~ блок· (*m*) (cryst) mosaic block

~ фото·като́п· (*m*) (TV) mosaic

мо́зг·, -а, (*nom pl*) **-и́,** (*g pl*) **-о́в,** (*m*) (anat) brain, cerebrum, encephalon; (anat) marrow, medulla

~, голов·н·о́й brain, cerebrum

~, ко́ст·н·ый bone marrow, medulla ossium

~, спин·н·о́й spinal cord

мо́згл·ый, -ая, -ое, (*a*) damp, unwholesome

мозг·ови́к·, ов·е́ч·ий (*m*) (zool) gid-parasite, *Multiceps multiceps* (parasitic worm)

мозг·ов·о́й, -а́я, -о́е, (*a*) (anat) brain, cerebral, encephalic; (anat) marrow, medullary

~ кост·ь (*f*) marrow bone

~ оболо́ч·к·а, мя́гк·ая (*f*) cerebral pia mater

~ оболо́ч·к·а, паути́н·н·ая arachnoid mater of the brain

~ оболо́ч·к·а, твёрд·ая cerebral dura mater

мозж·ечо́к·, -чк·а́, (*m*) (anat) cerebellum

Мо́зли, за·ко́н· (*m*) (phys) Moseley's law

мозоле·но́г·ие, -их, (*pl decl as adj*) (zool) tylopods, *Tylopoda*

мозо́л·истост·ь, -и, (*f*) callosity

мозо́л·ист·ый, -ая, -ое, (*a*) callous, callose

мозо́л·ь, -и, (*f*) callosity, callose, corn, (med) tyloma

~, ко́ст·н·ая (med) callus

мозо́ль·н·ый, -ая, -ое, (*a*) callous, callose, corn

~ пласты́р·ь (*m*) (pharm) corn-plaster

мои́ (*pron, nom plur, acc plur inanim*) of **мой**

мой (*pron*) my, mine; (*sing imper*) of **мы́·ть**

мо́й·к·а, -и, (*g pl*) **мо́·ек·,** (*f*) (*v n*) of **мы́·ть**; washing; washer; washery

мо́·йте (*plur imper*) of **мы́·ть**

мо́кко (*n indecl*) Mocha, Mocha coffee

мо́к·нуть, -нут, (*past masc sing*) **мок·,** (*imp*) become wet/damp; soak, steep

мокр·е́ц·, -а́, (*i*) **-о́м,** (*m*) (vet) malanders; biting midge, (*pl*) *Ceratopogonidae*

мокр·и́ц·а, -ы, (*i*) **-ей,** (*f*) (bot) chickweed, *Stellaria media*; woodlouse, (*pl*) *Oniscoidea*

мокр·и́чник·, -а, (*m*) *see* **мокр·и́ц·а** (bot)

мо́кр·о (*adv*) wet, damp, moist

мокр·ова́т·ый, -ая, -ое, (*a*) moist, damp

мокро·воз·ду́ш·н·ый, -ая, -ое, (*a*) wet-air, steam-jet

мокр·от·а́, -ы́, (*f*) wetness, dampness, moisture; **мокро́та, -ы,** (*f*) (anat) mucus, sputum

мокр·о́тн·ый, -ая, -ое, (*a*) (zool) mucous

мо́кр·ый, -ая, -ое, (*a*) wet, moist, damp; (food) fresh
~ **прови́з·и·я** (*f*) (food) fresh provisions
~ **пут·ь** (*m*) wet process (analysis etc.)
~ **спо́соб·** (*m*) plastic process (of brickmaking)
~ **элеме́нт·** (*m*) (elec) wet cell
мо́л·, -а, (*m*) mole (marine works)
мол (*abbr*) of **моле́кула, молеку-ля́рный** molecule, molecular, mol
мол. % (*abbr*) of molar %
мол. вес (*abbr*) of **молекуля́р·н·ый вес·** molecular weight
мола́ль·ност·ь, -и, (*f*) (chem) molality
мола́ль·н·ый, -ая, -ое, (*f*) (chem) molal
~ **рас·тво́р·** (*m*) molal solution
мола́сс·а, -ы, (*f*) (geol) molasse
молв·а́, -ы́, (*f*) rumour
молдави́т·, -а, (*m*) (min) moldavite
мол·ев·о́й, -а́я, -о́е, (*a*) stream-driving
~ **с·плав·** (*m*) stream driving (logging)
моле́·кул·а, -ы, (*m*) molecule
~ **-акце́птор·, -а,** (*m*) (phys) acceptor-molecule
~, **ните·ви́д·н·ая** chain molecule
~, **перв·и́чн·ая** unit/parent molecule
~, **поля́р·н·ая** polar molecule
молекуля́р·ност·ь, -и, (*f*) molecularity
молекуля́р·н·ый, -ая, -ое, (*a*) molecular
~ **вес·** (*m*) molecular weight
~ **пуч·о́к·** (*m*) molecular beam
молески́н·, -а, (*m*) (text) moleskin
молибд·а́т·, -а, (*m*) (chem) molybdate
~ **ка́льц·и·я** (chem) calcium molybdate
молибд·е́н·, -а, (*m*) (chem) molybdenum, Mo
~, **сер·ни́ст·ый** molybdenum disulphide
молибден·и́л·, -а, (*m*) (chem) molybdenyl
молибден·и́ст·ый, -ая, -ое, (*a*) (chem) molybdenous; (met) low-molybdenum
молибден·и́т·, -а, (*a*) (min) molybdenite
молибденово·ки́сл·ый, -ая, -ое, (*a*) molybdate of
~ **аммо́н·и·й** (*m*) ammonium molybdate
молибде́н·ов·ый, -ая, -ое, (*a*) molybdenum; (chem) molybdic
~ **ангидри́д·** (*m*) molybdic oxide
~ **блеск·** (*m*) (min) molybdenite
~ **кисл·от·а́** (*f*) molybdic acid
~ **стал·а** (*f*) molybdenum steel
молибд·и́т·, -а, (*m*) (min) molybdite, ferrimolybdite, molybdic ochre

мол·иза́ци·я, -и, (*f*) (chem) solvation
моллеру́п·, -а, (*m*) duplex-plunger lubricator
моллю́ск·, -а, (*m*) (*see associated adjectives*), (zool) mollusc, mollusk, (*pl*) *Mollusca*
моллюск·оиде́·и, -й, (*f pl*) (zool) molluscoids, (*pl*) *Molluscoidea*
моллюско·обра́з·н·ые, -ых, (*pl decl as a*) (zool) molluscoids, *Molluscoidea*
моллюско·я́д·н·ый, -ая, -ое, (*a*) (zool) molluscivorous
молние·от·во́д·, -а, (*m*) (elec) lightning conductor/rod/mast
~, **стерж·нев·о́й** lightning rod
~, **тро́с·ов·ый** overhead ground wire
молние·но́с·н·ый, -ая, -ое, (*a*) lightning, lightning-speed, quick as lightning
молн·и́р·овать, -уют, (*imp and perf*) (telecom) wire, inform by express telegram
мо́лн·и·я, -и, (*f*) lightning; express (telegram); zipper (fastener)
~, **зигзаго·обра́з·н·ая** forked lightning
~, **ле́нт·очн·ая** ribbon lightning
~, **лин·е́йн·ая** streak lightning
~, **полос·а́т·ая** sheet lightning
~, **чёт·очн·ая** beaded lightning
~, **шар·ов·а́я** ball/globular lightning
мол·о́вш·ий, -ая, -ее, (*past part act*) of **мол·о́ть**
молод·ёжн·ый, -ая, -ое, (*a*) youth
молод·ёж·ь, -и, (*f*) youth, young people
молод·и́к·, -а, (*m*) young ice
моло́д·к·а, -и, (*g pl*) **-д·ок·,** (*f*) (agr) pullet
молод·ня́к·, -а́, (*m*) (agr) young stock; saplings, young wood
молод·о́й, -а́я, -о́е, (*a*) young
мо́лод·ост·ь, -и, (*f*) youth
мо́лод·ь, -и, (*f*) (zool) the young; (bot) new shoots; (fish) fry
моло́зив·о, -а, (*n*) (zool) first milk, colostrum
молока́н·, -а, (*m*) (bot) lettuce, *Lactuca*
моло́к·и, (*nom pl*) (*g pl*) **моло́к·** (*f*) (fish) melt, soft roe
молок·о́, -а́, (*n*) milk
~, **ацидо·фи́ль·н·ое** acidophilous milk
~, **го́р·н·ое** (min) rock milk, agaric mineral
~, **извест·ко́в·ое** (chem) milk of lime
~, **па́р·н·ое** fresh milk
~, **с·гущ·ённ·ое** condensed milk
~, **с·ня·т·о́е** skimmed milk

молок·о́
~, **сух·о́е** dried milk
~, **то́щ·ее** skim milk
молоко·го́н·н·ый, -ая, -ое, (*a*) (zool) lactiferous, lactific
молоко·ме́р·, -а, (*m*) milk measurer/weigher
молоко·о·хлад·и́тел·ь, -я, (*m*) milk cooler
молоко·со·держ·а́щ·ий, -ая, -ее, (*a*) lactiferous
мо́лот·, -а, (*m*) hammer; forging machine/hammer
~, **а́р·очн·ый** arch-form hammer
~, **бес·ша́бот·н·ый** counterblow hammer
~, **воз·ду́ш·н·ый** compressed-air hammer
~ **двойн·о́го де́й·стви·я** double-acting drop-hammer
~, **ко́в·очн·ый** free-forging hammer
~, **криво·ши́п·н·о-ремён·н·ый** kick stamp
~, **криво·ши́п·н·о-рыча́ж·н·ый** helve hammer
~, **мост·ов·о́й** independent/bridge-type drop-hammer
~, **одно·сто́·ечн·ый** overhanging hammer
~, **па́ро-воз·ду́ш·н·ый** steam hammer
~, **при·во́д·н·о·пневмат·и́ческ·ий** pneumatic power hammer
~ **прост·о́го де́й·стви·я** single-acting drop-hammer
~, **пруж·и́нн·ый** spring hammer
~ **с доск·о́й** board drop hammer
~, **уда́р·н·ый** jumper bit
~, **фрикцио́н·н·о-винт·ов·о́й** friction screw press
~, **фрикцио́н·н·ый** board/friction-drop hammer; friction forging press
~, **штамп·о́вочн·ый** drop hammer, drop-forging hammer, stamping hammer
молот·и́лк·а, -и, (*g pl*) **-лок·,** (*f*) (agr) thrasher, thresher
молот·и́льн·ый, -ая, -ое, (*a*) (agr) thrashing, threshing
молот·и́ть, -я́т, (*imp*) (agr) thrash, thresh
молот·ко́в·ый, -ая, -ое, (*a*) of **молот·о́к·**
~ **дроб·и́лк·а** (*f*) hammer crusher
молото·бо́·ец, -о́й·ц·а, (*i*) **-о́й·ц·ем,** (*m*) (met) hammerman, striker (smith's mate)

молото·бо́й·н·а, -ы, (*f*) hammer/forge scale
молот·ови́щ·е, -а, (*i*) **-щ·ем,** (*n*) hammer shaft/handle
молот·ов·о́й, -а́я, -о́е, (*a*) hammer, percussive
молот·о́к·, -т·к·а́, (*m*) hammer
~, **бо·ев·о́й** sledge hammer
~, **бур·и́льн·ый** (min) hammer drill, plugger
~, **дерев·я́нн·ый** mallet
~·**-кир·очк·а, -и,** (*g pl*) **-чек·,** (*f*) bricklayer's hammer
~, **клеп·а́льн·ый** riveting hammer
~, **колон·ков·ый бур·и́льн·ый** (min) drifter
~, **много·уда́р·н·ый клеп·а́льн·ый** fast-hitting riveting hammer
~, **одно·уда́р·н·ый клеп·а́льн·ый** one-shot riveting hammer
~, **от·бо́й·н·ый** mechanical pick
~, **пневмат·и́ческ·ий** pneumatic hammer
~, **редко·уда́р·н·ый клеп·а́льн·ый** slow-hitting riveting hammer
~, **руб·и́льн·ый** chipping and caulking hammer
~, **сва́й·н·ый** piling hammer
~ **с кру́гл·ым бой·к·о́м** ball-pane hammer
~, **слеса́р·н·ый** bench/fitter's hammer
~, **штукату́р·н·ый** (build) lath hammer
молот·о́чек·, -к·а, (*m*) (*dim*) of **молот·о́к·**; (anat) malleus
~, **регистр·и́рующ·ий** printing hammer
мо́л·от·ый, -ая, -ое, (*past part pass*) of **мол·о́ть**; ground, milled
мол·о́ть, (*pres 3rd sing, pl*) **ме́л·ет, ме́л·ют,** (*imp*) grind, mill
молот·ьб·а́, -ы́, (*f*) thrashing, threshing
молоча́·й, -я, (*m*) (bot) milk-wort, *Euphorbia*
молоча́й·н·ые, -ых, (*pl decl as adj*) (bot) spurges, *Euphorbiaceae*
моло́ч·н·ая, -ой, (*f decl as a*) dairy
моло́ч·н·о- (*component*) milk, lactate, lactic
молочно·бе́л·ый, -ая, -ое, (*a*) milk-white
молочно·ки́сл·ый, -ая, -ое, (*a*) (chem) lactic, lactate of, lactic acid
~ **аммо́н·и·й** (*m*) ammonium lactate
~ **брож· éни·е** (*n*) lactic fermentation

молочно·кисл·ый, -ая, -ое
~ **бактер·и·и** (*pl*) lactobacillaceae, sugar fermenters
~ **кальц·и·й** (*m*) calcium lactate
~ **ряд·** (*m*) lactic acid series
молоч·ност·ь, -и, (*f*) (agr) milk-producing capacity; lactescence, milkiness
молоч·н·ый, -ая, -ое, (*a*) (*see also* **молоч·н·ая**) (*as noun*); milk, milky, dairy; opal (of glass); (chem) lactic; (bot) lacteous
~ **дерев·о** (*n*) (bot) *Galactodendron utile*
~ **желез·а** (*f*) (zool) mammary gland
~ **кисл·от·а** (*f*) (chem) lactic acid
~ **кисл·от·а брож·ени·я** (*f*) (chem) fermentation lactic acid
~ **кисл·от·а, лево·вращ·ающ·ая** L-lactic acid, sarcolactic/paralactic acid
~ **кисл·от·а, право·вращ·ающ·ая** D-lactic acid
~ **машин·а** (*f*) dairy machine
~ **скот·** (*m*) dairy cows
~ **сыворот·к·а** (*f*) whey
молч·а (*adv*) silently, in silence; **мол-ч·а** (*pres gerund*) of **молч·ать**
молч·алив·ый, -ая, -ое, (*a*) taciturn, silent; tacit
молч·ани·е, -я, (*n*) silence
   зон·а молч·ани·я (*f*) (rad) silent zone
молч·ать, (*pres 3rd sing, pl*) **молч·ит, молч·ат,** (*imp*) be/keep/remain silent
молч·ащ·ий, -ая, -ее, (*pres part act*) of **молч·ать**; (rad) quiescent, idle
мол·ь, -и, (*m*) (chem) mole, mol, gram-molecule, g-mole, gram-molecular weight, gram-mole; kilogram mole, kilogram molecule; (ent) moth, *Tinea*; stream-driven timber
мольберт·, -а, (*m*) easel
Молье, диаграмм·а (*f*) (eng) Mollier steam diagram
моль·н·о-объ·ём·н·ая концентр·а-ци·я (*f*) (chem) molarity
моль·ност·ь, -и, (*f*) molarity
моль·н·ый, -ая, -ое, (*a*) (*and see* **моляр·н·ый**) (chem) molar, mole
~ **дол·я** (*f*) molar fraction
моляль·ност·ь, -и, (*f*) = **молаль·-ност·ь**
моляль·н·ый *see* **молаль·н·ый**
моляр·, -а, (*m*) (anat) molar, molar tooth
моляр·изаци·я, -и, (*f*) (chem) solvation

моляр·ност·ь, -и, (*f*) (chem) molarity
моляр·н·ый, -ая, -ое, (*a*) (*and see* **моль·н·ый**) molar, mole, gram-molecular
~ **концентр·аци·я** (*f*) (chem) molar concentration, molarity
~ **концентр·аци·я, от·нос·ительн·ая** (*f*) mole ratio
~ **поляриз·аци·я** (*f*) (phys) molecular/molar polarization
~ **рефракц·и·я** (*f*) (phys) molecular/molar refraction
~ **электро·про·вод·ност·ь** (*f*) (phys) molecular conductivity
момене, числ·о (*n*) (chem) Maumene number
момент·, -а, (*m*) (*see also under associated words*); (mech) moment, momentum
~, **баланс·ирующ·ий** trimming moment
~, **вос·станавл·ивающ·ий** restoring/righting moment
~, **вращ·ающ·ий** (mech) torque
~ **вращ·ени·я** angular momentum
~ **двиг·ател·я** motor torque
~, **до·полн·ительн·ый** secondary moment
~, **из·гиб·ающ·ий** bending moment, moment of deflection
~, **кинет·ическ·ий** moment of momentum
~ **количеств·а движ·ени·я** moment of momentum
~, **крут·ящ·ий** torque, turning moment
~, **мах·ов·ой** flywheel polar inertia; moment of gyration, flywheel moment
~, **механ·ическ·ий** moment
~, **о·пор·н·ый** moment at support
~, **орбиталь·н·ый** (phys) angular/orbital momentum
~, **прецессион·н·ый** precession
~, **синхрониз·ирующ·ий** (elec) synchronizing torque
~ **старт·а** (autom) initial time
~, **стат·ическ·ий** moment of force
~ **финиш·а** (autom) terminal time
момент·альн·ый, -ая, -ое, (*a*) instant, instantaneous; momentary
момент·н·ый, -ая, -ое, (*a*) moment, torque; instantaneous, momentary
~ **двиг·ател·ь** (*m*) (elec) torque motor
~ **нож·** (*m*) (elec) flicker blade (knife switch)
~ **характер·истик·а** (*f*) flywheel effect

моментó·метр·, -а, (*m*) (math) planimeter

монáрд·а, -ы, (*f*) (bot) horsemint, *Monarda*

монард·ѝн·, -а, (*m*) (chem) monardin

монацѝт·, -а, (*m*) (min) monazite

Мóнд·а, газ· (*m*) Mond gas

монел·ь, -я, (*m*) Monel metal

монéт·а, -ы, (*f*) coin

~, звóн·к·ая (fin) specie

монет·нѝк·, -а, (*m*) coin duct, coin-in-slot mechanism

монет·нѝц·а, -ы, (*f*) coin box

монéт·н·ый, -ая, -ое, (*f*) coin, coinage, coin-operated

~ автомáт· (*m*) coin-operated machine, automatic vending machine

~ двор· (*m*) (fin) mint

~ паритéт· (*m*) mint parity

монжýс· (*m*) (chem) montejus, automatic displacement elevator

мониезиóз·, -а, (*m*) (vet) Moniezia infection

монилиóз·, -а, (*m*) (med) moniliasis

монитóр·, -а, (*m*) monitor

~, гидравл·ѝческ·ий (min) monitor, hydraulic/monitor giant

~, фóльг·ов·ый (nucl) foil monitor

моно- (*int component*) mono-, one-, single-

моно·áзо·крас·ѝтел·ь, -я, (*m*) (chem) mono-azo dye

моно·арѝль·н·ый, -ая, -ое, (*a*) (chem) monoarylated

моно·ацѝль·н·ый, -ая, -ое, (*a*) (chem) monoacylated

моно·блóк·, -а, (*m*) monobloc, monoblock; monoblock container

моно·вакцѝн·а, -ы, (*f*) (med) univalent/monovalent vaccine

моно·вариáнт·н·ый, -ая, -ое, (*a*) monovariant

моно·волокн·ó, -á, (*n*) (plast) monofilament

моно·генет·ѝческ·ий, -ая, -ое, (*a*) monogenetic

моно·гéн·н·ый, -ая, -ое, (*a*) (math) monogenic; (geol) monomineralic

моно·гидрáт·, -а, (*m*) (chem) monohydric salt

моно·гѝр·а, -ы, (*f*) (cryst) monogyre, axis of first-order symmetry

моно·грáф·и·я, -и, (*m*) (print) monograph

моно дисперс·н·ый, -ая, -ое, (*a*) monodisperse

моно·дрóм·н·ый, -ая, -ое, (*a*) (math) monodromic

мон·óз·а, -ы, (*f*) (chem) monose, monosaccharose

моно·за·мещ·ённ·ый, -ая, -ое, (*a*) (chem) mono-substituted

моно·иод·ýксус·н·ая кисл·от·á (*f*) (chem) monoiodo-acetic acid

моно·карбóн·ов·ая кисл·от·á (*f*) (chem) monocarboxylic acid

моно·карп·ѝческ·ий, -ая, -ое, (*a*) (bot) monocarpic

мон·ó·кис·ь, -и, (*f*) (chem) monoxide

моно·клин·áл·ь, -и, (*f*) (geol) monocline

моно·клин·ѝческ·ий, -ая, -ое, (*a*) (bot) monoclinous

моно·клѝн·н·ый, -ая, -ое, (*a*) monoclinic

~ систéм·а (*f*) (cryst) monoclinic system

мон·óкл·ь, -я, (*m*) (opt) monode, simple meniscus lens, converging meniscus lens

моно·кóк·, -а, (*m*) (a/c) monocoque

моно·крепѝд·н·ый, -ая, -ое, (*a*) (zool) monocrepid

моно·кристáлл·, -а, (*m*) monocrystal, single crystal

моно·кристалл·ѝческ·ий, -ая, -ое, (*a*) of **моно·кристáлл·**

моно·кристáль·н·ый, -ая, -ое, (*a*) of **моно·кристáлл·**

моно·культýр·а, -ы, (*f*) (agr) one-crop system

мон·окуляр·н·ый, -ая, -ое, (*a*) (opt) monocular

моно·лѝт·, -а, (*m*) monolith

~, пóчв·енн·ый soil monolith, core sample

моно·лѝт·к·а, -и, (*g pl*) **-т·ок·,** (*f*) monolith sole, one-piece rubber shoe sole

моно·лѝт·н·ый, -ая, -ое, (*a*) monolithic; solid (not porous); made in one piece

монóм·, -а, (*m*) (math) monomial

моно·мéр·, -а, (*m*) (chem) monomer

~, за·полимериз·óванн·ый converted monomer

моно·мéр·н·ый, -ая, -ое, (*a*) (chem) monomeric

моно·молекуляр·н·ый, -ая, -ое, (*a*) monomolecular

~ слó·й (*m*) (chem) monomolecular layer

моно·мóрф·н·ый, -ая, -ое, (*a*) (zool) monomorphic; (cryst) monomorphous

моно·над·се́р·н·ая кисл·от·а́ (*f*) (chem) monopersulphuric acid

моно·нукле·о́з·, -а, (*m*) (med) mononucleosis

моно·о́·кис·ь, -и, (*f*) (chem) monoxide

моно·па́к·, -а, (*m*) (phot) monopack

моно·плако·фо́р·а, -ы, (*f*) (zool) (*pl*) monoplacophorans, *Monoplacophora*

моно·пла́н·, -а, (*m*) (a/c) monoplane

моно·плег·и́·я, -и, (*m*) (med) monoplegia

моно·по́д·и·й, -я, (*m*) (bot) monopodium

моно·по́л·и·я, -и, (*f*) monopoly

моно·по́л·ь, -и, (*f*) (phys) monopole, single pole

моно·по́ль·н·ый, -ая, -ое, (*a*) monopolistic, monopoly

моно·поля́р·н·ый, -ая, -ое, (*a*) monopolar

моно·ре́льс·, -а, (*m*) monorail, monorail transporter

моно·ре́льс·ов·ый пут·ь (*m*) monorail

моно·сахари́д·, -а, (*m*) (chem) monosaccharide

моно·сила́н·, -ы, (*m*) silicomethane

моно·симметр·и́ческ·ий, -ая, -ое, (*a*) (bot) monosymmetrical; (cryst) monoclinic

моно·ско́п·, -а, (*m*) (TV) monoscope, monotron, phasmajector

моно·сло́·й, -я, (*m*) (chem) monolayer, monomolecular layer

моно·ти́п·, -а, (*m*) (print etc.) monotype

моно·тип·и́чн·ый, -ая, -ое, (*a*) (biol) monotypic

моно·ти́п·н·ый, -ая, -ое, (*a*) (print) monotype; one-design

моно·то́н·н·о (*adv*) monotonically, smoothly, steadily

моно·то́н·ност·ь, -и, (*f*) monotony

моно·то́н·н·ый, -ая, -ое, (*a*) (math) monotonic, monotone, steady

~ у·меньш·е́ни·е (*n*) steady decrease

моно·тро́н·, -а, (*m*) (TV) monotron, monoscope, phasmajector; monotron hardness tester

моно·троп·и́ческ·ий, -ая, -ое, (*a*) (phys, chem) monotropic

моно·троп·и́·я, -и, (*f*) (phys, chem) monotropism

моно·тро́п·н·ый, -ая, -ое, (*a*) monotropic

моно·фил·ети́ческ·ий, -ая, -ое, (*a*) (genet) monophyletic

моно·ха́з·и·й, -я, (*m*) (bot) monochasium

моно·хлор- (*int component*) (chem) monochlor-, monochloro-

моно·хо́рд·, -а, (*m*) (phys) monochord

моно·хромат·иза́ци·я, -и, (*f*) monochromatization

моно·хромат·и́ческ·ий, -ая, -ое, (*a*) monochromatic

моно·хрома́тор·, -а, (*m*) (spectr) monochromator

~, без·щел·ев·о́й micropolarimeter

моно·хро́м·н·ый, -ая, -ое, (*a*) (phot) monochrome, monochromatic

моно·цикл·и́ческ·ий, -ая, -ое, (*a*) monocyclic

моно·ци́т·, -а, (*m*) (zool) monocyte

моно́·эдр·, -а, (*m*) (math) monohedron, hemihedron

моно·эдр·и́ческ·ий, -ая, -ое, (*a*) monohedric, monohedral; (cryst) triclinic hemihedral

моно·эфи́р·, -а, (*m*) (chem) monoester; mono-ether

монта́ж·, -а́, (*i*) **-ж·о́м,** (*m*) (mech) mounting, installing, fitting, installation; (civ eng) erecting, erection; (elec) wiring; (shipb) fitting out, outfitting; laying (pipes); (phot) montage; (cinema) montage, editing, cutting; mosaic (of air photographs)

монтаж·ёр·, -а, (*m*) (cinema) editor, cutter; mosaicker (of air survey photographs)

монта́ж·ник·, -а, (*m*) (civ eng) erector; installer

монта́ж·н·ый, -ая, -ое, (*a*) of **монта́ж·**

~ стан·о́к· (*m*) engine erecting stand

~ стол· (*m*) (cinema) editing and cutting stand

~ схе́м·а (*f*) (elec) wiring diagram; laying diagram (pipe systems)

~ табл·и́ц·а (*f*) (elec) sag table (overhead lines)

~ цех· (*m*) erecting shop; (shipb) fitting-out shop

~ черт·ёж· (*m*) installation drawing

монта́н-воск·, -а, (*m*) (min) montan wax

монтани́л·ов·ый, -ая, -ое, (*a*) (chem) montanyl

монтани́т·, -а, (*m*) montanite

монта́н·ов·ая кисл·от·а́ (*f*) montanic acid

монтебрази́т·, -а, (*m*) (min) montebrasite

монтежю́ *see* **монжу́с**

монт·ёр·, -а,   (*m*) fitter, maintenance man, mechanic

монт·и́ровани·е, -я,   (*n*) mounting, installing, erecting

монт·и́р·овать, -уют,   (*imp*) (mech) mount, install, fit; (civ eng) erect; (cinema) edit, cut; mount (pictures); mosaic (air survey pictures)

монт·иро́вочн·ый, -ая, -ое,   (*a*) decor (theat)

~ част·ь   (*f*) (theat) decor department

монтицелли́т·, -а,   (*m*) (min) monticellite

монтмориллони́т·, -а,   (*m*) (min) montmorillonite

монтросейт·, -а,   (*m*) (min) montroseite

монумента́ль·н·ый, -ая, -ое,   (*a*) monumental

монцони́т·, -а,   (*m*) (geol) monzonite

Мо́ос·а, шкал·а́   (*f*) Mohs' scale (of hardness)

Мо́р·а, сол·ь   (*f*) (chem) Mohr's salt

мора́ль·н·ый, -ая, -ое, (*a*) moral

мора́тор·и·й, -я, (*m*) (fin) moratorium

мо́рг·, -а, (*m*) mortuary, morgue

моргани́т·, -а,   (*m*) (min) morganite

морг·а́·ть, -ют,   (*imp*) blink; wink; flicker (of lights)

морг·н·у́ть, -у́т, (*perf*) *see* **морг·а́·ть**

мо́рд·а, -ы,   (*f*) muzzle, snout

мо́р·е, -я,   (*nom pl*) **-я́,** (*g pl*) **-е́й,** (*n*) sea; (astron) mare (*pl*) maria (on the moon)

~, ка́мен·н·ое   (geol) stone placer

~, не·с·поко́й·н·ое   rough sea

~, о·кра́и·н·ое   land-locked sea

~, от·кры́·т·ое   open sea, high seas

~, со·верш·ённ·о с·поко́й·н·ое   calm sea, glassy sea

~, с·поко́й·н·ое   calm sea, rippled sea

мо́р·ем   (*adv*) by sea

море́н·а, -ы,   (*f*) (geol) moraine

~, основ·н·а́я   ground moraine

мор·е́ни·е, -я,   (*n*) (*v n*) *see* **мор·и́ть**

море́н·н·ый, -ая, -ое,   (*a*) (geol) moraine, morainic

мор·ённ·ый, -ая, -ое,   (*past part pass*) of **мор·и́ть**

мор·ён·ый, -ая, -ое,   (*a*) fumed, smoked; stained (wood); fired unglazed (earthenware)

море·пла́в·ани·е, -я,   (*n*) seafaring, sailing

море·пла́в·ател·ь, -я,   (*m*) mariner, seafarer, sailor

море·пла́в·ательн·ый, -ая, -ое,   (*a*) seafaring

море·ста́·в·, -а,   (*m*) freezing-up season (of sea)

море·тряс·е́ни·е, -я,   (*n*) seaquake

море·хо́д·, -а,   (*m*) mariner, seafaring

море·хо́д·ност·ь, -и,   (*f*) (naut) seaworthiness; navigability

море·хо́д·н·ый, -ая, -ое,   (*a*) nautical, navigating, navigational, seagoing; seaworthy

~ инструме́нт·   (*m*) navigational instrument

~ табл·и́ц·ы   (*pl*) nautical tables, nautical almanac

~ уч·и́лищ·е   (*n*) nautical school

морж·, -а́, (*i*) **-о́м,** (*m*) (zool) walrus, morse, *Odobenus rosmarus*

Мо́рзе, а́збук·а   (*f*) Morse code

мор·и́лк·а, -и,   (*g pl*) **-лок·,** (*f*) wood stain; fume/smoke/stifling jar (e.g. for insect collection)

мори́н·, -а,   (*m*) (chem) morine (a yellow dye)

морио́н·, -а,   (*m*) (min) morion

мор·ист·ée,   (*adv*) further out to sea

мор·и́ть, -я́т,   (*imp*) exterminate, kill off; fume, smoke; stain (wood)

морк·о́в·ь, -и,   (*f*) carrot

мор·ов·о́й, -а́я, -о́е,   (*a*) pestilential, plague

моро́ж·ениц·а, -ы,   (*f*) (food) ice/ice-cream machine

моро́ж·ен·ое, -ого,   (*n decl as adj*) (food) ice-cream, ice

моро́ж·енн·ый, -ая, -ое,   (*a*) of **моро́ж·ен·ое**

моро́ж·ен·ый, -ая, -ое,   (*a*) frozen, chilled

моро́з·, -а,   (*m*) frost; freezing weather; frosted finish

мороз·и́лк·а, -и,   (*g pl*) **-лок·,** (*f*) freezing chamber, freezer

мороз·и́льн·ый, -ая, -ое,   (*a*) freezing

мороз·и́ть, -я́т,   (*imp*) freeze, ice, chill

моро́з·н·ый, -ая, -ое,   (*a*) frosty, frost

морозо·бо́·ин·а, -ы,   (*f*) (geol) frost crack/cleft/fissure; frost shake (timber)

морозо·бо́·й, -я,   (*m*) (hortic etc.) winterkilling

морозо·сто́й·к·ий, -ая, -ое,   (*a*) cold/frost-proof/resistant

морозо·сто́й·кост·ь, -и,   (*f*) frost resistance

морозо·у·сто́й·чив·ый, -ая, -ое,   (*a*) frost-resistant

морос·и́ть, (*pres 3rd sing*) **-и́т,** (*imp*) (meteor) drizzle
мо́рос·ь, -и, (*f*) (meteor) drizzle
мо́рс·, -а, (*m*) fruit juice (as drink)
мор·ск·о́й, -а́я, -о́е, (*a*) sea, marine, nautical; maritime; merchant naval, merchant marine; naval (*abbr* of **военно-морской)**
~ **авиа́ци·я** (*f*) naval air arm
~ **а́гент·ств·о** (*n*) (com) shipping office
~ **а́нгел·** (*m*) (zool) angel-fish, *Squatina*
~ **кара́с·ь** (*m*) (fish) sea bream
~ **кли́мат·** (*m*) maritime climate
~ **ми́л·я** (*f*) nautical mile
~ **модел·и́зм·** (*m*) model boat building/sailing
~ **о́кун·ь** (*m*) (fish) rose fish, *Sebastes marinus*
~ **планкто́н·** (*m*) (biol) haliplankton
~ **пра́в·о** (*n*) maritime law
~ **пра́кт·ик·а** (*f*) seamanship
~ **приз·** (*m*) (law) prize of war
~ **свин·к·а** (*f*) (zool) guinea pig
~ **свин·ь·я́** (*f*) (zool) porpoise, *Phocaena phocaena*
~ **соба́к·а** (*f*) (zool) dogfish, *Scylliorhinus*
~ **соба́ч·к·а** (*f*) (zool) blenny, blennoid fish
~ **флот·** (*m*) merchant marine/navy
~ **язы́к·** (*m*) (fish) Dover sole, (*pl*) *Soleidae*
морти́р·а, -ы, (*f*) (shipb) propeller shaft boss, stern gland; (mil) mortar
морти́р·к·а, -и, (*g pl*) **-р·ок·,** (*f*) grenade discharger (rifle)
морти́р·н·ый, -ая, -ое, (*a*) of **морти́р·а;** (gunn) upper-register (firing)
морф·аллакси́с·, -а, (*m*) (zool) morphallaxis
мо́рф·и·й, -я, (*m*) *see* **морф·и́н·**
морф·и́н·, -а, (*m*) (chem) morphine
морфо́·лог·, -а, (*m*) morphologist
морфо·ло́г·и·я, -и, (*f*) morphology
морфо·ме́тр·ия·, -и, (*f*) morphometry
морфо·троп·и́·я, -и, (*f*) (cryst) morphotropy
морщ·и́н·а, -ы, (*f*) wrinkle, crease, (zool) ruga, (geol) corrugation, wrinkle
морщ·и́нист·ый, -ая, -ое, (*a*) wrinkled, creased, puckered, rugate, (biol) rugous, rugose
мор·я́к·, -а, (*m*) sailor, seaman
мор·я́щ·ий, -ая, -ее, (*pres part act*) of **мор·и́ть**
мос- (*abbr*) of **моск·о́вск·ий** = Moscow

москате́л·ь, -и, (*f*) chemical product (as goods including paints, oils etc.)
москате́ль·н·ый, -ая, -ое, (*a*) of **москате́л·ь**
москате́ль·щик·, -а, (*m*) druggist (U.S.), chemist (Brit); paint merchant
моски́т·, -а, (*m*) (ent) sand fly, *Phlebotomus*
моски́т·н·ый, -ая, -ое, (*a*) of **моски́т.**
~ **лихора́д·к·а** (*f*) phlebotomous/ sandfly fever
моск·о́вск·ий, -ая, -ое, (*a*) Moscow
мосла́к·, -а́, (*m*) (anat) fetlock
мост·, -а́, (*nom pl*) **-ы́,** (*g pl*) **-о́в,** (*m*) bridge; (elec) bridge circuit, bridge; (M/T) axle; (biol) pons
~, **а́р·очн·ый** arched bridge
~, **вед·у́щ·ий** (M/T) driving axle
~, **вис·я́ч·ий** suspension bridge
~, **воз·ду́ш·н·ый** cross-over (for pipes etc.); (air) air lift
~, **двой·н·о́й** (elec) double bridge, Kelvin bridge
~, **за́д·н·ий** (M/T) back axle
~, **консо́ль·н·ый** cantilever bridge
~, **на·вод·н·о́й** temporary bridge
~, **на·кло́н·н·ый** inclined skip hoist (for furnaces etc.)
~, **на·плав·н·о́й** floating bridge
~, **не·раз·ре́з·н·ый** continuous-span bridge
~, **не·уравно·ве́ш·енн·ый** (elec) unbalanced bridge, out-of-balance bridge
~, **одно·под·ко́с·н·ый** single-lock bridge
~, **пере́д·н·ий** (M/T) front axle
~, **пере·ки́д·н·о́й** crossover (piping etc.)
~, **плашко́ут·н·ый** pontoon bridge
~, **по·воро́т·н·ый** swing/turning bridge
~, **по·дви́ж·н·о́й** movable bridge
~, **под·де́рж·ивающ·ий** (M/T) dead/ stationary axle
~, **под·ко́с·н·ый** frame bridge
~, **подъ·ём·н·ый** vertical-lift bridge, lifting bridge
~, **понто́н·н·ый** pontoon bridge
~, **раз·вод·н·о́й** swing/draw/bascule bridge
~, **раз·движ·н·о́й** opening/movable bridge
~, **раз·ре́з·н·о́й** independent-span bridge; (M/T) articulated rear axle
~, **ри́гель·н·о-под·ко́с·н·ый** double lock bridge

мост

~, **тепл·ов·о́й** (elec) resistance thermo-meter bridge

~, **уравно·ве́ш·енн·ый** (elec) balanced bridge

~, **частото·ме́р·н·ый** frequency bridge

~, **четырёх·пле́ч·н·ый** (elec) four-arm bridge

~, **электро́н·н·ый** (elec) valve/vacuum-tube bridge

мо́ст·ик, -а, (m) (dim) of **мост·**; (naut) bridge; (elec) bridge, bridge circuit; bridge wire; bow-shaped straight edge (tool)

~ **Ви́н·а** (elec) Wien bridge

~, **ве́рх·н·ий** (shipb) compass platform

~, **из·мер·и́тельн·ый** (elec) measuring bridge, bridge

~, **капита́н·ск·ий** (shipb) captain's bridge

~, **мего́м·н·ый** (elec) Megohm bridge

~, **навигацио́н·н·ый** (shipb) navigating bridge

~ **на·ка́л·ивани·я** (expl) bridge wire (of detonator)

~ **от·сто·я́ни·я** distance piece (in aerial photography)

~, **плазмат·и́ческ·ий** (bot) plasmodesm

~, **про·до́ль·н·ый** (shipb) catwalk, fore-and-aft bridge

~ **То́мсон·а** (elec) Kelvin bridge

~ **Уи́тстон·а** (elec) Wheatstone bridge

~, **ход·ов·о́й** (shipb) fore bridge

~, **шторм·ов·о́й** (shipb) hurricane deck

мо́ст·иков·ый, -ая, -ое, (a) bridge, bridge-type

мост·и́ть, -я́т, (imp) pave, surface

мост·ов·а́я, -о́й, (f decl as adj), (civ eng) pavement, block pavement

~, **брус·ча́т·ая** sett pavement

~, **булы́ж·н·ая** cobble pavement

~, **из·борожд·ённ·ая** (geol) striated pavement

~, **моза́йк·ов·ая** mosaic paving, radial sett pavement

мост·ов·о́й, -а́я, -о́е, (a) of **мост·**; see also **мост·ов·а́л**

~ **кран·** (m) overhead travelling crane

~ **пере·хо́д·** (m) (civ eng) bridge works; (elec) bridge transition

~ **фе́рм·а** (f) (civ eng) bridge truss/beam/girder

мосто·по́·езд·, -а, (m) (rail) bridge-building train

мосто·стро·е́ни·е, -я, (n) bridge building/engineering

мосто·у·кла́д·чик·, -а, (m) bridgelayer (machine)

мост·овь·ё, -я́, (n) upper leather hide

мот·а́лк·а, -и, (g pl) **-лок·,** (f) (met roll) coiler, reeler, reel; (text) winder

~, **про́·волоч·н·ая** (met roll) rod reel

~, **ро́л·иков·ая** (met roll) roll-type coiler

~, **сорт·ов·а́я** (met roll) small-section reel

~ **с ос·ев·о́й по·да́·ч·и** (met roll) laying reel

~ **с тангенциа́ль·н·ой по·да́·ч·и** (met roll) pouring reel

мот·а́льн·ый, -ая, -ое, (a) winding, reeling

~ **маши́н·а** (f) (text) reeler, winder

мот·а́·ть, -ют, (imp) wind, reel; shake, wag

моти́в·, -а, (m) motive; motif (art); tune (music)

мотив·иро́вк·а, -и, (g pl) **-вок·,** (f) explanation, justification; motivation

мо́т·к·а, -и, (g pl) **-т·ок·,** (f) winding, reeling

мот·н·у́ть, -у́т, (perf) give a shake/wag

мотн·я́, -и́, (f) (fish) bag (of seine net)

мото- (int component) moto- (with lat, gr roots); motor-, engine-; motorcycle-; motorized

мото·ваго́н·, -а, (m) (rail) motor coach/car

мото·вело·ра́м·а, -ы, (f) autocycle frame

мото·велосипе́д·, -а, (m) autocycle, motor-mounted bicycle

мото·ви́л·о, -а, (n) coiler, reel

~, **трост·и́льн·ое** (text) doubling reel

мото·во́з·, -а, (m) (rail) low-power I.C. locomotive, diesel shunter

мото·дрези́н·а, -ы, (f) (rail) motor-trolley

мото·дро́м·, -а, (m) motor-cycle race track

мот·о́к·, -т·к·а́, (m) hank, skein

мото·компре́ссор·, -а, (m) motor-driven compressor

мото·кро́сс·, -а, (m) motocross, cross-country motor car race

мото·лопа́т·а, -ы, (f) (civ eng) power shovel

мото·невро́н·, -а, (m) (zool) moto-neuron

мото·пехо́т·а, -ы, (f) (mil) motorized infantry

мото·планёр·, -а, (m) (air) motor-assisted glider

мото́р·, -а, (*m*) (*see also* дви́г·ател·ъ, электро·дви́г·ател·ь) motor; engine

~ -генера́тор·, -а, (*m*) (elec) motor-generator

~, двух·ря́д·н·ый звездо·обра́з·н·ый (a/c) double-row radial engine, two-row radial engine

~, одно·ря́д·н·ый звездо·обра́з·н·ый (a/c) single-row radial engine

~, пневмат·и́ческ·ий pneumatic/air motor

~, под·вес·н·о́й (shipb) outboard motor/engine

~, раз·го́н·н·ый starting motor

~, реакти́в·н·ый синхро́н·н·ый (elec) repulsion motor

~, ря́д·н·ый (a/c) in-line engine

мото·ра́м·а, -ы, (*f*) (a/c) engine mounting

моториз·о́ванн·ый, -ая, -ое, (*a*) motorized

мотор·и́ст·, -а, (*m*) motor mechanic

мото́р·к·а, -ы, (*g pl*) -р·ок·, (*f*) motor boat

мото́р·н·ый, -ая, -ое, (*a*) (mech) motor, engine; (biol) motor

~ су́д·н·о (*n*) motorship

~ то́пл·ив·о (*n*) heavy diesel fuel

~ шум· (*m*) (rad) motorboating

мото·ро́ллер·, -а, (*m*) (M/T) motor scooter

мото·у·станов·к·а, -и, (*g pl*) -вок·, (*f*) (a/c) engine installation

мото·ци́кл·, -а, (*m*) motor cycle

мото́р·чик·, -а, (*m*) (*dim*) of мото́р·; (elec) fractional-horsepower motor

Мо́тт·а, фо́рмул·а (*f*) (phys) Mott scattering formula

моттрами́т·, -а, (*m*) (min) mottramite

моты́г·а, -и, (*f*) (agr) hoe

~, вращ·а́ющ·аяся rotary hoe/cultivator

моты́ж·ени·е, -я, (*n*) hoeing

моты́л·ев·ый, -ая, -ое, (*a*) of мо·ты́л·ь

~ под·ши́п·ник· (*m*) (M/T) large-end bearing

~ ше́й·к·а (*f*) crankpin

мотыл·ёк·, -ль·к·а́, (*m*) moth, butterfly

моты́л·ь, -я́, (*m*) crank; (ent) mosquito larva

мотыль·ко́в·ый, -ая, -ое, (*a*) butterfly-like/shaped; (bot) papilionaceous, (*pl*) Papilionaceae

мох·, (*g sing*) мх·а *or* мо́х·а, (*nom pl*) мх·и, (*g pl*) мх·ов, (*m*) (bot) moss, *Muscus*, (*pl*) mosses, *Bryophyta*

мох

~, печён·оч·н·ый (bot) liverwort

мохн·а́т·ый, -ая, -ое, (*a*) hairy, wooly, shaggy, (bot) pilose, tomentose

мохо (*n indecl*) (geophys) moko

мох·ови́к·, -а, (*m*) (min) moss agate

мох·ов·о́й, -а́я, -о́е, (*a*) moss, mossy

мохо́л·, -а, (*m*) (geophys) mohole

мохо·обра́з·н·ый, -ая, -ое, (*a*) (bot) mossy, bryophitic; (*pl decl as adj*) (bot) bryophytes, *Bryophyta*

Мохорови́чин·а, по·ве́рх·ност·ь (*f*) (geol) Mohorovicic discontinuity, moho

~ сква́ж·ин·а (*f*) Mohorovicic hole, mohole

моцио́н·, -а, (*m*) exercise walk

моч·а́, -и́, (*f*) urine; (*pres gerund*) of моч·и́ть

вы·дел·е́ни·е моч·и́ urinary secretion

моча́л·ин·а, -ы, (*f*) filament

моча́л·ить, -ят, (*imp*) separate into fibres/fibers

моча́л·к·а, -и, (*g pl*) -л·ок·, (*f*) wisp, clump

моча́л·о, -а, (*n*) bast

моч·еви́н·а, -ы, (*f*) (chem) urea, carbamide

моч·еви́н·о-формальдеги́д·н·ая смол·а́ (*f*) urea formaldehyde resin

моч·ев·о́й, -а́я, -о́е, (*a*) (med) urinary, urine; (chem) uric

~ ка́мн·и (*pl*) (med) gravel

~ кисл·от·а́ (*f*) (chem) uric acid

~ пес·о́к· (*m*) (med) gravel

~ пузы́р·ь (*f*) (med) bladder

моче·го́н·н·ый, -ая, -ое, (*a*) (med) diuretic

моче·из·нур·е́ни·е, -я, (*n*) (med) diabetes

моче·ис·пуск·а́ни·е, -я, (*n*) (physiol) diuresis, urination

моче·ки́сл·ый, -ая, -ое, (*a*) (chem) uric, urate (of)

~ сол·ь (*f*) (chem) urate

моч·е́нец·, -нц·а, (*m*) retted flax fibre/fiber

~, тепл·ов·о́й heat-retted flax fibre/fiber

моч·е́ни·е, -я, (*n*) (*v n*) *see* моч·и́ть

моч·е́нцев·ый, -ая, -ое, (*a*) of моч·е́нец·

~ волокн·о́ (*n*) water-retted flax fibre/fiber

моч·ён·ый, -ая, -ое, (*a*) soaked, steeped, wetted, retted

мо́ч·енн·ый, -ая, -ое,   (*past part pass*) of **моч·и́ть**

моче·пол·ов·о́й, -а́я, -о́е,   (*a*) (physiol) urogenital

моче·то́ч·ник·, -а,   (*m*) (anat) ureter

моч·и́ть, ⸗ат,   (*imp*) wet, damp, moisten; soak, ret (of flax); **-ся** (*pass*); urinate

мо́ч·к·а, -и,   (*g pl*) **моч·ек·,** (*f*) soaking; retting, rotting (flax, straw); lobe (of ear); (bot) filament

~, рос·я́н·ая   dew retting

~, тепл·ов·а́я   heat retting

~, холодно·во́д·н·ая   water retting

моч·ков́ат·ый, -ая, -ое,   (*a*) (bot) filamentous

моч·ь, (*pres 3rd sing, pl*) **мо́ж·ет, мо́г·ут,** (*past masc sing*) **мог·,** (*imp*) be able, can; **моч·ь, -и,** (*f*) strength, might, power

мо́ш·к·а, -и,   (*g pl*) **-ш·ек·,** (*f*) (ent) black fly, (*pl*) *Simuliidae*

мош·о́нк·а, -и,   (*g pl*) **-нок·,** (*f*) (anat) scrotum

мощ·е́ни·е, -я,   (*n*) paving, surfacing

мощ·ён·ый, -ая, -ое,   (*a*) paved, surfaced

мощ·ённ·ый, -ая, -ое, (*past part pass*) of **мост·и́ть**

мо́щ·ност·ь, -и, (*f*) power, horse-power; capacity; (geol) thickness (of stratum etc.); (phys) rate, power; (a/c) output

~, акти́в·н·ая   (elec) active power

~ бок·ов·о́й полос·ы́   (rad, TV) sideband power

~, больш·а́я high power; (*g as adj*) high-power/duty, powerful, heavy-duty

~ в ва́тт·ах (elec) wattage, power

~ в и́мпульс·е (rad) peak pulse power

~ воз·бужд·е́ни·я се́т·к·и grid driving power (vacuum tubes)

~, гибри́д·н·ая (genet) hybrid vigour, heterosis

~, дей·стви́тельн·ая available power, effective power

~, дл·и́тельн·ая continuous power

~ до́з·ы (nucl) dose rate

~, до·пуст·и́м·ая power-carrying capacity

~ звук·а (phys) sound power

~ из·луч·е́ни·я (phys) radiation intensity

~ и́мпульс·а (phys) pulse power

~, индика́тор·н·ая indicated horse-power

ис·пыт·а́ни·е на мощ·ност·ь (*n*) efficiency test, output test

мо́щ·ност·ь

~, ка́ж·ущ·аяся (elec) apparent power

~ котл·а́ boiler capacity

коэффицие́нт· мо́щ·ност·и (*m*) (elec) power factor

~, кре́йсер·ск·ая (a/c) rocket, cruising output

крив·а́я мо́щ·ност·и (*f*) efficiency/power curve

~, литр·о́в·ая (ICE) brake horse-power per litre/liter

~, ма́л·ая low power; (*g as adj*) low-power/duty, light-duty

~, мг·нове́нн·ая (elec) instantaneous power

~ мно́ж·еств·а (math) power of a set, cardinal number

~ на вал·у́ shaft horse-power

~ на вы́·ход·е (elec) power output, output power

~ на то́рмоз·е (ICE) brake horse-power

~, номина́ль·н·ая rated power/output, rating

~ об·луч·е́ни·я (phys) irradiation/exposure rate

~, о́бщ·ая (geol) aggregate thickness

~, о·ста́т·очн·ая shut-down power (of nuclear reactor)

~, от·дав·а́ем·ая available power

~, пи́к·ов·ая (elec) peak power

~, под·вод·и́м·ая (elec) power input

~, поле́з·н·ая useful/net power

~, по·сто·я́нн·ая (geol) uniform thickness

~, по·требл·а́ем·ая (elec) intake power, consumed power

~, по·тре́б·н·ая (elec) power demand, power required, required power

~, про·из·во́д·ственн·ая productive capacity, theoretical capacity

~, про·мы́шл·енн·ая (geol) minable thickness (of a seam)

~, рас·по·лаг·а́ем·ая available power

~ рас·се́·яни·я dissipated power

~, реакти́в·н·ая (elec) wattless power

~, сре́д·н·яя (elec) active power

~, тормоз·н·а́я brake horse-power

~, тя́г·ов·ая thrust horse-power

~, у·де́ль·н·ая (nucl) power density

~, эффекти́в·н·ая (ICE) effective horse-power, shaft horse-power; (a/c) propeller output

мо́щ·н·ый, -ая, -ое,   (*a*) powerful, high-power, heavy/high-duty, heavy (machine); (geol, min) thick (of stratum)

мо́щ·ь, -и, (f) power
мою (pron, acc sing f) of мой; my; (pres 1st sing) of мы·ть (imp); wash
мо́·ющ·ий, -ая, -ее, (pres part act) of мы·ть; washing, (chem) detergent
~ сре́д·ств·о (n) (chem) detergent
мо·я́ (pres gerund) of мы·ть; washing; (pron, nom sing f) of мой, my
мпз (abbr) of милли·пуа́з· millipoise (unity of viscosity)
мр (abbr) of милли·рентге́н· (nucl) mr, milliroentgen
мра́з·ь, -и, (f) dirt, filth
мра́к·, -а, (m) gloom
мра́мор·, -а, (m) marble
мра́мор·н·ый, -ая, -ое, (a) of мра́мор·
~ кро́ш·к·а (f) marble chips
мра́ч·н·ый, -ая, -ое, (a) gloomy, sombre
м·резерфо́рд· (abbr) of милли·резер·фо́рд· (nucl) millirutherford
мс (abbr) of милли·секу́нд·а ms, msec, millisecond
мС (abbr) of милли·кюри́ (nucl) mC, millicurie
мсб (abbr) of милли·сти́льб· (phys) millistilb
Мст (abbr) of марте́н·овск·ая стал·ь (met) OH steel
мст·ить, -ят, (imp) avenge, take revenge
МТС (abbr) of маши́н·н·о-тра́ктор·н·ая ста́нци·я agricultural machine base/park; (phys) = метр-то́нн·а секу́нд·а metre/meter-ton-second (system of units)
муа́р·, -а, (m) (text) moiré
Муа́вр·а, фо́рмул·а (f) (math) de Moivre relation
мудр·ён·ый, -ая, -ое, (a) abstruse, difficult, tricky; ingenious
му́др·ост·ь, -и, (f) wisdom
му́др·ый, -ая, -ое, (a) wise
му́ж·, -а, (nom pl) муж·ья́, (g pl) муж·е́й, (m) husband
му́ж·еск·ий, -ая, -ое, (a) (gram) masculine
му́ж·ественн·ый, -ая, -ое, (a) courageous, manly
му́ж·еств·о, -а, (n) courage
муж·ск·о́й, -а́я, -о́е, (a) masculine, male, men's
муж·чи́н·а, -ы, (f) man, male
муж·ья́ (pl) of муж·
музе́·й, -я, (m) museum
музе́й·н·ый, -ая, -ое, (a) of музе́·й

му́зык·а, -и, (f) music
музык·а́нт·, -а, (m) musician, bandsman
мука́, -и́, (f) flour, meal; му́ка, -и, (f) torment, suffering
~, древ·е́сн·ая wood flour
~, жмых·о́в·ая cake/extracted meal
~, инфузо́р·н·ая (geol) diatomaceous/infusoriol earth
~, корм·ов·а́я offals (flour by-product)
~, кост·ян·а́я bone meal
~, круп·и́тчат·ая 10% extraction flour, granular flour
~, об·ди́р·н·ая sifted flour
~, о·бо́й·н·ая scoured flour
~, от·се́в·н·ая sifted flour/meal
~, па́р·ен·ая wholemeal
~, се́·ян·ая dressed flour
~, про·бо·и́ст·ая speckly flour
~, прост·а́я plain flour
~, рези́н·ов·ая rubber crumb
~, со́·ев·ая soya flour
~, сорт·ов·а́я graded flour
~, це́ль·н·ая wholemeal flour
~, экстракцио́н·н·ая extracted flour
муко·е́д·, -а, (m) (ent) saw-tooth grain beetle, Oryzaephilus surinamensis
мук·о́ид·, -а, (m) (biochem) mucoid
муко·меш·а́тел·ь, -я, (m) flour blender
муко·мо́л·, -а, (m) miller
муко·мо́ль·н·ый, -ая, -ое, (a) milling, flour milling
муко·мо́ль·н·я, -и, (f) flour mill
мукон·ов·ый, -ая, -ое, (a) (chem) muconic
~ кисл·от·а́ (f) muconic acid
муко·про·се́·иватель·ь, (m) flour sieving machine
муко·протеи́н·, -а, (m) (biochem) mucoprotein
мукори́н·, -а, (m) (biochem) mucorin
муко·целлюло́з·а, -ы, (f) (biochem) mucocellulose
муксу́н·, -а, (m) (zool) muksun, whitefish, Coregonus muksun
му́л·, -а, (m) (zool) mule
мул·ев·о́й, -а́я, -о́е, (a) see мол·ев·о́й
мулине́т·к·а, -и, (g pl) -т·ок·, (f) (air) windmill, windmill propeller; fan brake
му́лл·, -а, (m) (text) mull
муллит·, -а, (m) (min) mullite
му́льд·а, -ы, (f) (rubber) mould, mold; charging box (metallurgical furnace); (geol) trough, syncline

мульти-   (*int component*) multi

мульти·валéнт·н·ый, -ая, -ое,   (*a*) (chem, phys) multivalent, polyvalent

мульти·вариáнт·н·ый, -ая, -ое,   (*a*) multivariant, polyvariant

мульти·вибрáтор·, -а,   (*a*) (elec) multivibrator, flip-flop (computers)

~, вед·у́щ·ий   drive flip-flop

~ выс·от·ы́   height flip-flop

~, жд·у́щ·ий   flip-flop (computers); delay multivibrator

~, не·стабúль·н·ый   astable multivibrator

~, одно·периóд·н·ый   single-shop multivibrator, gate/monostable multivibrator

~, одно·рáз·ов·ый = **м. одно·периóд·н·ый**

~, одно·тáкт·н·ый = **м. одно·периóд·н·ый**

~ от·мéт·к·и   mark flip-flop

мульти·плáн·, -а,   (*m*) (a/c) multiplane

мульти·плéкс·, -а,   (*m*) (telecom) multiplex printing equipment; (phot) multiplex projector

мульти·плéт·, -а,   (*m*) (spectr) multiplet

~ по мáсс·е   (phys) mass multiplet

мультиплéт·ност·ь, -и,   (*f*) multiplicity

мульти·пликáтор·, -а,   (*m*) (elec) multiplier; (hydr) hydraulic intensifier; (cinema) animator, animated cartoon artist; (phot) multiplying camera; (phys) pressure intensifier/multiplier

~, паро·гидравл·úческ·ий   steam-hydraulic intensifier

мультипликациóн·н·ый, -ая, -ое, (*a*) of **мультиплик·áци·я**

~ фильм·   (*m*) (cinema) animated cartoon

мультиплик·áци·я, -и,   (*f*) (cinema) cartoon, animated cartoon

мульти·пóл·ь, -я,   (*m*) (elec) multipole

мульти·пóль·ност·ь, -и,   (*f*) (phys) multipolarity, multiple order

мульти·пóль·н·ый, -ая, -ое,   (*a*) of **мульти·пóл·ь**

мульти·рот·áци·я, -и,   (*f*) (chem) mutarotation, multirotation

мульти·скóп·, -а,   (*m*) (elec) multiscope

мульти·у·стóй·чив·ый, -ая, -ое,   (*a*) (autom) multi-stable

мульти·циклóн·, -а,   (*m*) multicyclone (gas cleaning apparatus)

му́льч·а, -и,   (*f*) (agr) mulch

мульч·úровани·е, -я,   (*n*) (agr) mulching

мулю́р·а, -ы,   (*f*) (arch) molding

муля́ж·, -á,   (*i*) **-ж·óм,**   (*m*) cast, mould, plaster cast/mould

му·метáлл·, -а,   (*m*) (met) Mu metal

муми·фик·áци·я, -и,   (*f*) mummification

му́ми·я, -и,   (*f*) (pal) mummy; (min) mummy, red ocher (as pigment)

мундúр·, -а,   (*m*) uniform, uniform dress

мундштýк·, -á,   (*m*) mouthpiece, nozzle; outer die (extrusion press); cigarette/cigar holder; curb, curb bit (harness)

~, рéж·ущ·ий   jet (of gas cutter head)

~, пряд·úльн·ый   spinneret

мунц·метáлл·, -а,   (*m*) (met) Muntz metal, yellow brass

мур·á, -ы́,   (*f*) rubbish

мурав·á, -ы́,   (*f*) (ceram) glaze

мурав·é·й, -в·ь·я́,   (*m*) (ent) ant, (*pl*) *Formicidae*

~, бéл·ый   termite, (*pl*) *Isoptera*

мурав·éйник·, -а,   (*m*) ant-hill

мурáв·ить, -ят,   (*imp*) (ceram) glaze

мурáвл·ен·ый, -ая, -ое,   (*a*) glazed

муравь·éд·, -а,   (*m*) (zool) anteater, *Myrmecophagus*

муравье·жýк·, -а,   (*m*) (ent) ant beetle, chequer beetle, *Cleroid*

муравьино·аммóн·иев·ая сол·ь   (*f*) ammonium formate

муравьино·бутúл·ов·ый эфúр·   (*m*) (chem) butyl formate

муравьино·кúс·л·ый, -ая, -ое,   (*a*) formate of

~ нáтр·и·й   (*m*) sodium formate

муравь·úн·ый, -ая, -ое,   (*a*) ant, (chem) formic

~ альдегúд·   (*m*) formic aldehyde, formaldehyde, methanal

~ кисл·от·á   (*f*) formic acid

~ лев·   (*m*) (zool) ant lion, *Myrmeleon formicarius*

мураше·éд·, -а,   (*m*) (zool) numbat, *Myrmecobius fasciatus*

мурáш·ник·, -а,   (*m*) (geol) myrmekite

мурéн·а, -ы,   (*f*) (fish) moray, *Muraena*

му́синг·, -а,   (*m*) (naut) mousing, standing Turk's head on a light

мусúнг·ов·ый элевáтор·   (*m*) endless-whip hoist

мускардúн·а, -ы,   (*f*) (vet) muscardine

мускари́н·, -а,   (*m*) (chem) muscarine

муска́т·, -а,   (*m*) muscated (wine); muscadine grape; nutmeg

муска́т·ник·, -а,   (*m*) (bot) nutmeg, *Myristica fragrans*

муска́т·н·ый, -ая, -ое,   (*a*) of **муска́т·, муска́т·ник·**

~ **оре́х·**   (*m*) nutmeg

~ **шалф·е́·й**   (*m*) (bot) clary sage, *Salvia sclarea*

мусковит·, -а,   (*m*) (min) muscovite, white mica

муско́н·, -а,   (*m*) (chem) muskon, methylcyclopentadecanone

му́скул·, -а,   (*m*) (anat) muscle

мускул·ату́р·а, -ы,   (*f*) (zool) musculature

му́скул·ист·ый, -ая, -ое,   (*a*) sinewy

му́скуль·н·ый, -ая, -ое,   (*a*) muscular

му́скус·, -а,   (*m*) (bot) musk

~, **а́мбр·ов·ый**   (chem) musk ambrette

му́скус·н·ый, -ая, -ое,   (*a*) of **му́скус·**

~ **же́лез·ы**   (*pl*) (zool) musk glands

му́слин·, -а,   (*m*) (text) muslin

му́сор·, -а,   (*m*) rubbish, refuse, garbage; debris

~, **у́голь·н·ый**   cinders

му́сор·н·ый, -ая, -ое,   (*a*) of **му́сор·**

~ **эже́ктор·**   (*m*) ash ejector (of furnace)

~ **я́щик·**   (*m*) dust bin, ash can

мусоро·во́з·, -а,   (*m*) garbage wagon, dust cart

мусоро·с·жиг·а́тел·ь, -я,   (*m*) incinerator

мусоро·с·жиг·а́тельн·ая ста́нц·и·я   (*f*) destructor station (electric power station burning refuse)

му́сс·, -а,   (*m*) (food) mousse

мусс·и́р·овать, -уют,   (*imp*) (food) whip; sparkle, froth (of beverages); exaggerate

муссо́н·, -а,   (*m*) (meteor) monsoon

мута·ге́н·, -а,   (*m*) (biochem) mutagen

мут·а́з·а, -ы,   (*f*) (biochem) mutase

мут·а́нт·, -а,   (*m*) (biol) mutant

мута·рот·а́ци·я, и,   (*f*) (chem) mutarotation

мут·ацио́нн·ый, -ая, -ое,   (*a*) mutation

мут·а́ци·я, -и,   (*f*) (biol) mutation

мут·и́рующ·ий, -ая, -ее,   (*a*) mutant

мут·и́ть, -я́т,   (*imp*) make cloudy/muddy/dull

Мутман·а, жи́дк·ост·ь   (*f*) (chem) Muthmann reagent (for separating minerals), acetylene tetrabromide

мут·не́·ть, -ют,   (*imp*) become muddy/cloudy/dull

мутно·ме́р·, -а,   (*m*) turbidimeter

му́т·ност·ь, -и,   (*f*) turbidity, cloudiness

му́т·н·ый, -ая, -ое,   (*a*) turbid, cloudy, muddy, dull

мут·о́вк·а, -и,   (*g pl*) **-вок·,**   (*f*) (bot) verticil; beater, whipper, churn beater

мут·о́вчат·ый, -ая, -ое,   (*a*) (bot) verticillate

муту·али́зм·, -а,   (*m*) (biol) mutualism

мут·ь, -и,   (*f*) lees, sediment; cloud, haze, mist; slime, sludge, suspended matter

му́фел·ь, -я,   (*m*) (met) muffle furnace; chamber (of muffle furnace)

му́фель·н·ый, -ая, -ое,   (*a*) muffle

~ **гор·е́лк·а**   (*f*) pulverized-fuel muffle burner

му́фт·а, -ы,   (*f*) (mech) coupling, clutch; jointing device, sleeve (for pipes, cables etc.); (elec) junction box

~ **Би́бби**   Bibby coupling

~, **ги́б·к·ая**   flexible coupling

~, **гидравл·и́ческ·ая**   hydraulic coupling

~, **глух·а́я**   solid-sleeve coupling, rigid/muff coupling

~ **Дже́нни**   Williams–Janney gear, variable-speed gear

~, **ди́ск·ов·ая**   flexible-disk coupling

~, **дюри́т·ов·ая**   hose-type joint, hose joint (for pipes)

~, **жёстк·ая**   rigid/solid coupling

~, **зуб·ча́т·ая**   gear coupling

~, **ка́бель·н·ая**   (elec) cable jointing device

~, **кана́т·н·ая**   taper coupling (for wire rope)

~, **компенс·и́рующ·ая**   flexible coupling

~, **ко́нус·н·ая фрикцио́н·н·ая**   cone friction clutch

~ **кон·цев·а́я**   (elec) terminal box, cable pothead

~, **корен·н·а́я**   (met roll) main coupling

~, **кулач·ко́в·ая**   dog-tooth clutch, dog clutch, claw coupling/clutch

~, **ма́чт·ов·ая**   (elec) pole-mounted terminal box

~, **мембра́н·н·ая**   flexible steel-member coupling

~, **об·го́н·н·ая**   overrunning clutch

му́фт·а

~ **оборо́т·ов**   magnetic clutch; stabilizing unit (of stabilized mechanisms)

~, **о·гран·ичи́тельн·ая**   muff coupling

~ **Ольдге́йм·а**   Oldham's coupling, double-slider coupling

~ **о·переж·е́ни·я за·жиг·а́ни·я**   injection timing coupling (diesels)

~, **от·ветв·и́тельн·ая**   (elec) dividing box

~, **пере·хо́д·н·ая**   reducer (pipe joint)

~, **по·вод·ко́в·ая**   flanged/faceplate coupling

~, **по·дви́ж·н·ая**   flexible coupling

~, **по·пере́ч·н·о-с·вёрт·н·ая**   flanged/faceplate coupling

~, **порош·ко́в·ая**   magnetic powder clutch

~, **по·сто·я́нн·ая**   permanent coupling

~, **про·до́ль·н·о-с·вёрт·н·ая**   split-muff/box coupling

~, **прям·а́я**   coupling

~, **раз·общ·и́тельн·ая**   clutch

~, **реве́рс·и́вн·ая**   (mar eng) reverse gear

~, **само·за·пир·а́ющ·аяся**   self-locking clutch

~ **с·цепл·е́ни·я**   clutch

~, **с·цеп·н·а́я**   clutch

~ **тр·е́ни·я**   friction clutch

~, **тро́й·ников·ая**   (elec) dividing box

~, **у·пру́г·ая**   flexible coupling

~, **фла́н·цев·ая**   flanged/faceplate coupling

~, **фрикцио́н·н·ая**   friction clutch

~, **фрикцио́н·н·ая ди́ск·ов·ая**   disk clutch, plate friction clutch

~, **цеп·н·а́я**   chain coupling

~, **шарни́р·н·ая**   universal-joint coupling

~, **шпи́н·н·о-пневмат·и́ческ·ая**   pneumatic clutch

~, **эласт·и́чн·ая**   flexible coupling

~, **электо·магни́т·н·ая**   electric coupling, electro-magnetic coupling; magnetic clutch

му́фт·ов·ый, -ая, -ое,   (*a*) of **му́фт·а**

~ **армату́р·а**   (*f*) coupled pipe fitting

~ **со·един·е́ни·е**   (*n*) pipe coupling

му́х·а, -и,   (*f*) (ent) fly; nailing template

~, **дом·ов·а́я**   domestic fly, *Muscina stabulans*

~, **ко́мнат·н·ая**   (zool) domestic fly, *Musca domestica*

му́х·а

~, **морк·о́вн·ая**   carrot fly, *Psylla rosae*

мухо·ло́в·к·а, -и,   (*g pl*) **-в·ок·,** (*f*) (orn) flycatcher, (*pl*) *Muscicapidae*; (ent) centipede

мухо·мо́р·, -а,   (*m*) (bot) fly agaric, *Amanita muscaria*

муц·и́н·, -а,   (*m*) (biochem) mucin

муч·е́ни·е, -я,   (*n*) pain, suffering; torment, agony

му́ч·ить, -ат,   (*imp*) torture, torment, worry, cause/inflict pain; **-ся** (*pass*); be in pain, suffer

му́ч·к·а, -и,   (*f*) middlings (flour milling)

муч·ни́ст·ый, -ая, -ое,   (*a*) mealy, floury, farinaceous

~ **рос·а́**   (*f*) (bot) mildew

~ **черв·е́ц·**   (*m*) (ent) mealy bug, (*pl*) *Pseudococcidae*

муч·н·о́й, -а́я, -о́е,   (*a*) flour, meal, farinaceous

~ **клещ·**   (*m*) (ent) flour mite, *Tyroglyphus farinae*

му́ш·к·а, -и,   (*g pl*) **-ш·ек·,** (*f*) (*dim*) of **му́х·а**; (surv) front sight, foresight, sighting point

му́шкар·ь, -я,   (*m*) *see* **му́шкел·ь**

му́шкел·ь, -я,   (*m*) mallet

мушмул·а́, обык·нове́нн·ая   (bot) medlar, *Mespilus germanica*

мф   (*abbr*) of **милли·фо́т·** (phot) milliphot; (*abbr*)=**миллифара́д** (phys) mf, millifarad

мхи   (*pl*) of **мох·**

м-ц   (*abbr*) of **ме́сяц·** month

м/ч   (*abbr*) of **ме́тр·ов в час** m/hr, metres per hour

мч·ать,   (*pres 3rd sing, pl*) **мч·ит, мч·ат,** rush, speed along

м-числ·о́, -а́,   (*n*) (air) Mach number, M number

мш·а́нк·а, -и,   (*g pl*) **-нок·,** (*f*) (bot) pearlweed, pearlwort, *Sagina*; (zool) moss animal, bryozoan, (*pl*) *Bryozoa*

мш·и́ст·ый, -ая, -ое,   (*a*) mossy

мщ·е́ни·е, -я,   (*n*) revenge

мы   (*pron*) we

мы́·вш·ий, -ая, -ее,   (*past part act*) of **мы·ть**

мыкани́ц·а, -ы,   (*f*) (text) ripple, hand ripple

мы́к·а·ть, -ют,   (*imp*) (text) ripple

мы·л   (*past masc sing*) of **мы·ть**

мы́л·ени·е, -я,   (*n*) soaping, lathering

мы́л·енн·ый, -ая, -ое,   (*past part pass*) of **мы́л·ить**; soaped, lathered

мы́л·ец·, -л·ц·а, (*m*) (med) suppository
мы́·ли (*past plur*) of мы·ть
мы́л·ить, -ят, (*imp*) soap, lather
мы́л·к·ий, -ая, -ое, (*a*) freely lathering
мы́л·о, -а, (*n*) soap; lather; (*verb 3rd sing past*) of мы·ть
~, алюми́н·иев·ое   aluminium soap
~, го́р·н·ое   (min) saponite, bowlingite
~, зелён·ое   soft/green soap
~, кали́й·н·ое   soft soap, potassium soap
~, канифо́ль·н·ое   rosin soap
~, кле·ев·о́е   middle/gum soap
~, медици́н·ск·ое   medicated soap
~, металл·и́ческ·ое   metallic soap
~, ядр·о́в·ое   neat soap
мыло·ва́р·, -а, (*m*) soap maker/boiler/manufacturer
мыло·вар·е́ни·е, -я, (*n*) soapmaking
мыло·ва́р·енн·ый, -ая, -ое, (*a*) soap-making, soap
~ за·во́д· (*m*) soap works
мы́л·овк·а, -и, (*f*) soaping, washing, lathering
мыло·на́фт·, -а, (*m*) naphthene soap
мыло·холод·и́льн·ый пресс· (*m*) soap frame
мы́ль·ниц·а, -ы, (*i*) -ц·ей, (*f*) soap-holder
мы́ль·н·ый, -ая, -ое, (*a*) soap, soapy, soponaceous
~ де́рев·о (*n*) (bot) soap berry, *Sapindus*
~ ка́м·ен·ь (*m*) (min) soapstone
~ кор·а́ (*f*) (bot) soap bark, *Quillaja Saponaria*
~ ко́р·ен·ь (*m*) soap root
~ крем· (*m*) shaving soap
мы́ль·нянк·а, -и, (*g pl*) -нок·, (*f*) (bot) soapwort, *Saponaria*
~, лека́р·ственн·ая   (bot) soapwort, *Saponacia officinalis*
мыс·, -а, (*m*) (geog) cape, point, headland, foreland, promontory; (anat) promontory, promontorium
мы́сл·енн·ый, -ая, -ое, (*a*) mental, imaginary
мысле́те (*n indecl*) letter М
мы́сл·им·ый, -ая, -ое, (*a*) conceivable, thinkable
мысл·и́тел·ь, -я, (*m*) thinker
мысл·и́тельн·ый, -ая, -ое, (*a*) thought, thinking
мы́сл·ить, -ят, (*imp*) think, conceive
мы́сл·ь, -и, (*f*) thought, idea
    им·е́·ть в мы́сл·ях   have in mind, be thinking of
мыт·, -а, (*m*) (vet) strangles; moult, molt (of poultry)

мы́·т·ый, -ая, -ое, (*past part pass*) of мы·ть
мы·ть, (*pres 3rd sing, pl*) мо́·ет, мо́·ют, (*imp*) wash
мыт·ь·ё, -я́, (*n*) wash, washing
мыша́тник·, -а, (*m*) (bot) thermopsis
мыше·ви́д·к·а, -и, (*g pl*) -д·ок·, (*f*) (zool) marsupial mouse, *Phascogal*
мыше·ло́в·к·а, -и, (*g pl*) -в·ок·, (*f*) mouse-trap
мыше·обра́з·н·ые, -ых, (*pl decl as adj*) *Muridae*
мыше·хвост·, -а, (*m*) (bot) *Hordeum*
мы́ш·ечн·ый, -ая, -ое, (*a*) muscular, muscle
мы́ш·ий, -ья, -ье, (*a*) *see* мыш·и́н·ый
мыш·и́н·ый, -ая, -ое, (*a*) of мыш·ь (zool) (*pl*) *Myoidea*
~ горо́ш·ек· (*m*) (bot) vetch, *Vicia cracca*
мы́ш·к·а, -и, (*g pl*) -ш·ек·, (*f*) (*dim*) of мыш·ь; = мы́ш·ц·а
    под·мы́ш·к·ой   under the arm
мышл·е́ни·е, -я, (*n*) (*v n*) of мы́сл·ить; thinking, thought
мы́ш·ц·а, -ы, (*i*) -ей, (*f*) (anat) muscle
~, гла́д·к·ая   unstriped muscle
~, скеле́т·н·ая   (anat) striped muscle
мыш·ь, -и, (*nom pl*) -и, (*g pl*) -е́й, (*f*) (zool) mouse, *Mus*
~, лет·у́ч·ая   bat, (*pl*) *Chiroptera*
~, пес·ча́н·ая   (zool) shrew, *Sorex*
мышь·я́к·, -а, (*m*) (chem) arsenic, As
~, бе́л·ый   white arsenic
~, кра́сн·ый   (min) realgar
~, се́р·ый   grey arsenic
мышьяковисто·ки́сл·ый, -ая, -ое, (*a*) (chem) arsenite of
~ ка́льц·и·й (*m*) calcium arsenite
мышьяк·ови́ст·ый, -ая, -ое, (*a*) arsenic, arsenous, arsenious, arsenide of
~ ангидри́д· (*m*) white arsenic, arsenic trioxide
~ водо·ро́д· (*m*) arsine, hydride of arsenic
~ кисл·от·а́ (*f*) (chem) arsenious/arsenous acid
~ колчеда́н· (*m*) (min) iron arsenide, löllingite
~ мета́лл· (*m*) (chem) arsenide
мышьяково·ки́сл·ый, -ая, -ое, (*a*) arsenate (of)
~ ка́льц·и·й (*m*) calcium arsenate
мышьяк·о́в·ый, -ая, -ое, (*a*) arsenic, arsenical
~ ангидри́д· (*m*) arsenic pentoxide

мышьяк·о́в·ый, -ая, -ое
~ кисл·от·а́   (*f*) arsenic acid
~ колчеда́н·   (*m*) (min) arsenical pyrite, mispickel
~ руд·а́, блёкл·ая   (*f*) tennantite
мышь·я́к-орган·и́ческ·ое со·един·е́·ни·е   (*n*) organo-arsenic compound
мы́щел·ок·, -л·к·а,   (*m*) (zool) condyle
Мэв   (*abbr*) of мега·электро́н·-во́льт· (phys) MeV, Mev, million electron volts
мекв   (*abbr*) of милли·эквивале́нт· (phys) milliequivalent
мю   mu, μ (Greek)
мюзе́ль·н·ая маши́н·а   (*f*) museler, champagne bottle wiring machine
мюзле   (*n indecl*) cork wiring (for champagne bottles)
мю́ллер·ов кана́л·   (*m*) (zool) Müllerian duct
мю́ллер·ов·ы волокн·а́   (*pl*) (anat) Müller's fibres/fibers
мюль-маши́н·а, -ы,   (*f*) (text) mule
мюльтипль·н·ая ра́м·а   (*f*) bank (autom telephones)
мю-мезо́н·, -а,   (*m*) (nucl) mu-meson, muon
мюо́н·, -а,   (*m*) muon, mu-meson
мюо́н·и·й, -я,   (*m*) muonium
мюрг·, -а,   (*m*) murgue (unit of resistance of mine workings)
мя́·вш·ий, -ая, -ее,   (*past part act*) of мя·ть
мя́гк·ий, -ая, -ое,   (*a*) soft; mild, gentle
~ ваго́н·   (*m*) (rail) first-class carriage (with upholstered seats etc.)
~ древес·и́н·а   (*f*) softwood
~ карда́н·   (*m*) flexible universal joint
мя́гк·ост·ь, -и,   (*f*) softness; gentleness
мягко·но́г·ие, -их,   (*pl decl as adj*) (pal) malacopods, *Malacopoda*
мягко·те́л·ый, -ая, -ое,   (*a*) flabby; (zool) malacoid; (*pl as adj*) molluscs, *Mollusca*
мя́гч·е   (*comp*) of мя́гк·ий, мя́гк·о; softer, more softly
мягч·е́ни·е, -я,   (*n*) softening, mollification; bating (leather)
мягч·и́тель·, -я,   (*m*) softener, plasticizer, mollifier

мягч·и́тельн·ый, -ая, -ое,   (*a*) softening; (med) emollient
мягч·и́ть, -а́т,   (*imp*) soften
мязг·а́, -и,   (*f*) (agr) pulp cake
мяк·и́н·а, -ы,   (*f*) (agr) chaff
мя́к·иш·, -а,   (*i*) -ш·ем, (*m*) crumb (soft bread)
мя́к·отн·ый, -ая, -ое,   (*a*) pulpy
~ оболо́ч·к·а   (*f*) (anat) medullary sheath
мя́к·от·ь, -и,   (*f*) pulp; flesh; furl (of sail)
~, порох·ов·а́я   (expl) black blasting powder
мя́·лк·а, -и,   (*g pl*) -лок·, (*f*) pulper, crusher, breaker; milling machine (for leather); flax breaker
мя́·льн·о-треп·а́льн·ая маши́н·а   (*f*) breaker (flax etc.)
мяс·и́ст·ый, -ая, -ое,   (*a*) fleshy, meaty; pulpy; succulent
мяс·ни́к·, -а́,   (*m*) butcher (shopkeeper)
мяс·н·о́й, -а́я, -о́е,   (*a*) meat, flesh, carneous
мя́с·о, -а,   (*n*) meat flesh
~, сол·ён·ое   cured/corned meat
мясо·консе́рв·н·ый, -ая, -ое,   (*a*) meat-packing/preserving
мясо·пепто́н·н·ый, -ая, -ое,   (*a*) (med) nutrient
~ а́гар·   (*m*) nutrient agar
мясо·ру́б·к·а, -и,   (*g pl*) -б·ок·, (*f*) (food) mincing machine
мя́т·а, -ы,   (*f*) (bot) mint, *Mentha*
мяте́ж·н·ый, -ая, -ое,   (*a*) rebellious, mutinous
мя́тлик·, -а,   (*m*) (bot) meadow grass, *Poa*
~, луг·ов·о́й   blue grass, *Poa pratensis*
мя́т·н·ый, -ая, -ое,   (*a*) mint, peppermint
мя́·т·ый, -ая, -ое,   (*past part pass*) of мя·ть
~ лён·   (*m*) broken flax
~ пар·   (*m*) exhaust steam
мя·ть   (*pres 3rd sing, pl*) мн·ёт, мн·ут, (*imp*) crumple, crease; knead (dough etc.); break (flax etc.); mill (leather)
мят·ь·ё, -я́,   (*n*) (*v n*) see мя·ть
мяч·, -а́,   (*i*) -ч·о́м, (*m*) ball

Н

н   (*abbr*) of **ньютóн·** N, newton (unit of force = 10^5 dynes)

н-   (*abbr*) of **нáно-** nano-, millimicro, 10^{-9}

н-обрáз·н·ый, -ая, -ое,   (*a*) H-shaped, H-

на   (*prep*) (+*loc*) on, upon; in, at (*with certain nouns*); with (including, based on); on, in, during (of time); (+ *acc*) on, onto, to; for (of time); (math) by; in, into (results)

~ **буксúр·е**   in tow

~ **вéс·**   by weight

~ **вод·é**   on water; water-type (*as adj*)

~ **мéст·е**   *in situ*

~ **мéтр· корóч·е**   a meter shorter

~ **транзúстор·ах**   transistorized

~ **ход·ý** under way, while in motion

~ **хóлод·ý**   in the cold, cold

~ **четы́р·е, у·мнóж·ить** multiply by four

на·бáв·ить, -ят, (*perf*) *see* **на·бавл·я́·ть**

на·бáв·к·а, -и, (*g pl*) **-в·ок·,** (*f*) addition, increase; surcharge; bonus

на·бавл·я́·ть, -ют, (*imp*) add, add to, increase

на·бáв·очн·ый, -ая, -ое,   (*a*) extra, additional

набáт·, -а, (*m*) alarm, alarm bell

на·бéг·, -а, (*m*) raid; inroad (of pests etc.); incursion

на·бег·áни·е, -я, (*n*) (*v n*) of **на·бег·á·ть;** climbing (of a drive chain or belt); running against/on to

на·бéг·а·ть, (*perf*), **на·бег·á·ть** (*imp*), run at/into/onto, dash up against, dash over, climb; get up, begin to blow (of wind)

на·бег·áющ·ий, -ая, -ее, (*pres part act*) of **на·бег·á·ть;** leading, climbing; inflowing

~ **крá·й** (*m*) leading edge (of a pulse)

на·бéг·ов·ый, -ая, -ое, (*a*) (tech) running

на·беж·áть, (*fut 3rd sing, pl*) **на·беж·úт, на·бег·ýт,** (*perf*); *see* **на·бéг·а·ть;** raid

на·бел·ó (*adv*) clean, fair

нá·береж·н·ая, -ой, (*f decl as adj*), embankment; quay; sea front

~ **-стéн·к·а, -и,** (*f*) quay wall

~ **-эстакáд·а, -ы,** (*f*) piled wharf

на·бер·ёт (*fut 3rd sing*) of **на·бр·áть**

на·бив·á·ть, -ют, (*imp*) fill, stuff, pack; tamp, ram in; (text) print; bind (cooperage)

на·бúв·к·а, -и, (*g pl*) **-в·ок·,** (*f*) packing, stuffing, filling; packing material; (text) block printing

на·бив·н·óй, -áя, -óе, (*a*) packed, stuffed, rammed; packing, stuffing, ramming; (text) printed, fast printed

~ **лёд·** (*m*) layered ice (formed by rafting)

~ **сúт·ец·** (*m*) printed calico

на·бúв·очн·ый, -ая, -ое, (*a*) packing, stuffing, ramming

~ **кольц·ó** (*n*) gasket, packing ring

~ **корóб·к·а** (*f*) stuffing box

на·бир·á·ть, -ют, (*imp*) gather, collect; recruit; (print) set up, compose; (autom) compile; dial (a telephone number)

~ **в·раз·ря́д·к·у** (print) space

~ **выс·от·ý** (a/c) gain height

~ **скóр·ост·ь** gain/gather speed, pick up speed

на·бú·т·ый, -ая, -ое, (*past part pass*) of **на·бú·ть;** packed, stuffed, rammed

на·бú·ть, (*fut 3rd sing, pl*) **на·бь·ёт, на·бь·ют,** (*perf*); *see* **на·бив·á·ть**

набла (*f indecl*) (math) nabla, del, Hamiltonian operator

~ **-оперáтор·, -а,** (*m*) Hamiltonian operator

на·блюд·áем·ый, -ая, -ое, (*a*) observable, observed, seen, experimental; under observation; being studied

на·блюд·áтел·ь, -я, (*m*) observer; look out, spotter; supervisor

на·блюд·áтельност·ь, -и, (*f*) observance, keenness of observation

на·блюд·áтельн·ый, -ая, -ое, (*a*) observation, observational; observant

~ **пост·** (*m*) (mil) observation post, look out position

на·блюд·á·ть, -ют, (*imp*) observe, watch; supervise; look after, see to

на·блюд·éни·е, -я, (*n*) observation, surveyance; look out, spotting; supervision, superintendence, control; (rad) watch

~ **в периóд· молч·áни·я** (naut, rad) distress frequency watch

лúн·и·я на·блюд·éни·я (*f*) (gunn) spotting line

на·блюд·е́ни·е

~, равно·то́ч·н·ые (*pl*) (surv) equal observations

~, со·пряж·ённ·ое (mil) cross observation

на·бо́й·к·а, -и, (*g pl*) **-о́·ек·,** (*f*) (met) furnace lining; (met cast) hand rammer; (text) block printed cloth; heel tap (of shoe)

~, ки́сл·ая (met) acid lining

~, основ·н·а́я (met) basic lining

на·бо́й·н·ый, -ая, -ое, (*a*) of **на·бо́й··к·а**

на·бо́й·щик·, -а, (*m*) packer, filler, rammer; (text) block printer

на́·бок· (*adv*) sideways; on one side

на·бо́р·, -а, (*m*) (*v n*) (*see* **на·бир·а́·ть**) set, collection, kit; collecting, gathering; engaging, recruiting, taking on; (print) type setting, composition, composing; (shipb) framing

~ а́том·ов (phys) collection of atoms

~ а́том·н·ых электро́н·ов (phys) complement of atomic electrons

~ выс·от·ы́ (air) climb

~ за·да́·ч· set up (computers)

~ пере·ме́н·н·ых (math) set of variables

~, по·пере́ч·н·ый (shipb) transverse framing; (a/c) lateral assembly

~ програ́мм·ы (autom) programming

~, про·до́ль·н·ый (shipb) longitudinal framing; (a/c) longitudinal assembly

~ си́мвол·ов language (of a computer)

~ скор·ост·е́й (phot) exposure-time marking

~, с·ме́ш·анн·ый (shipb) combination framing

~ у·равн·е́ни·й equation setup (computers)

~ це́р·и (tech) loop of chain

~ хромосо́м· (cyt) set of chromosomes

~ хромосо́м·, по́л·н·ый (cyt) genome, genom

~ шестер·ён· gear train

на·бо́р·н·ая, -ой, (*f decl as adj*) (print) type-setting room

на·бо́р·н·ый, -ая, -ое, (*a*) of **на·бо́р·;** (*see also* **на·бо́р·н·ая**) (*as noun*)

~ ка́сс·а (*f*) (print) job/type case

~ лин·е́йк·а (*f*) (telecom) combination/selector bar; (print) typesetting rule

~ маши́н·а (*f*) (print) composing machine, type-setting machine

~ рабо́т·а (*f*) (print) composition; inlay work (cabinet making etc.)

на·бо́р·щик·, -а, (*m*) (print) compositor, type-setter

на́·бр·ан·а, -ы, (*f*) (print) typeset

на́·бр·анн·ый, -ая, -ое, (*a*) (*past part pass*) of **на·бр·а́ть,** (*perf*); *see* **на·бир·а́·ть** (*imp*)

на·бра́с·ывани·е, -я, (*n*) (*v n*) *see* **на·бра́с·ыва·ть**

на·бра́с·ыва·ть, -ют, (*imp*) toss/throw on/onto/over; fill, heap, pile (by throwing); sketch, outline, draft, jot down; **-ся** fall/pounce on; be filled/heaped/piled (by throwing); be sketched/outlined/drafted/jotted down

на·бр·а́ть, (*fut 3rd sing, pl*) **на·бер·ёт, на·бер·у́т,** (*perf*); *see* **на·бир·а́·ть**

на·бре́д·ш·ий, -ая, -ее, (*past part act*) of **на·брес·ти́**

на·брес·ти́, (*fut 3rd sing, pl*) **на·бред·ёт, на·бред·у́т,** (*perf*); come across, hit/stumble/happen upon

на·бро́с·анн·ый, -ая, -ое, (*past part pass*) of **на·брос·а́ть**

на·брос·а́ть, -ют, (*perf*) fill, heap, pile (by throwing); sketch, outline, draft, jot down; **-ся** (*pass*)

на·бро́с·ить, -ят, (*perf*) toss/throw on/onto/over; **-ся** fall/pounce on

на·бро́с·к·а, -и, (*f*) (*v n*) *see* **на·брос·а́·ть, на·бро́с·ить;** (geol) talus

~, ка́мен·н·ая rubble

на·брос·н·о́й, -а́я, -о́е, (*a*) (civ eng) rubble, rubble-filled

на·бро́с·ок·, -с·к·а́, (*m*) sketch, outline, draft

на·бро́ш·енн·ый, -ая, -ое, (*past part pass*) of **на·бро́с·ить**

на·бры́зг· (*adv*) sputtering, by sputtering

на·бры́зг·а·ть, -ют, (*perf*) *see* **на·бры́зг·ива·ть**

на·бры́зг·ива·ть, -ют, (*imp*) sprinkle

на·брю́ш·ник·, -а, (*m*) abdominal band

на·бух·а́емост·ь, -и, (*f*) swelling, swelling capacity

на·бух·а́ни·е, -я, (*n*) swelling; bulking, moisture expansion (of concrete)

~, вес·ов·о́е swelling by weight

коэфицие́нт· на·бух·а́ни·я (*m*) swelling/expansion factor

на·бух·а́·ть, -ют, (*imp*) swell

на·бу́х·нуть, -нут, (*past masc sing*) **на·бу́х·,** (*perf*) *see* **на·бух·а́·ть**

на́·был·ь, -и, (*f*) (geol) thrust

на·бь·ю́т, (*fut 3rd pl*) of **на·би́·ть**

нава́г·а, -и, (*f*) (fish) navaga, *Eleginas navaga*

на·вáкс·ить, -ят, (*perf*) polish, black

на·вáкш·енн·ый, -ая, -ое, (*past part pass*) of **на·вáкс·ить**

на·вáл·, -а, (*m*) (*v n*) *see* **на·вáл·и·ва·ть;** build-up (in metal machining operations); bump (rowing)

на·вáл·ива·ть, -ют, (*imp*) load/pile/heave onto/on/in; **-ся** (*pass*); fall upon, attack; lean on

на·вал·и́ть, -ят, (*perf*) *see* **на·вáл·и·ва·ть**

на·вáл·к·а, -и, (*g pl*) **-л·ок·,** (*f*) loading, heaping

на·вáл·ом (*adv*) in bulk

навало·от·бóй·щик·, -а, (*m*) (min) cutter-shoveller

на·вáл·очн·ый, -ая, -ое, (*a*) heaving, piling, loading; bulk, bulky

~ груз· (*m*) bulk/bulky cargo/freight

~ машúн·а (*f*) (min) mechanical loader

на·вальц·óванн·ый, -ая, -ое, (*a*) rolled-on

на·вáль·щик·, -а, (*m*) (min) shoveller, loader

на·вáр·, -а, (*m*) (weld) built-up metal, hard facing; fat, fatty scum

на·вáр·енн·ый, -ая, -ое, (*past part pass*) *see* **на·вáр·ива·ть;** hard-faced, tipped (of tools); built-up (repairs by welding); fritted/burnt on (furnace linings)

на·вáр·ива·ть, -ют, (*imp*) (weld) hard face, apply a hard facing, build up; face, burn on, frit on

на·вáр·ист·ый, -ая, -ое, (*a*) (food) fatty, rich

на·вар·и́ть, -ят, (*perf*) *see* **на·вáр·и·ва·ть**

на·вáр·к·а, -и, (*f*) (met) fritting in, burning in (furnace); hard facing (by welding), building up, tipping (tools)

по на·вáр·к·е at the fritting level (in a furnace)

на·вар·н·óй, -áя, -óе, (*a*) (*and see* **на·вáр·н·ый**) welding, building-up, tipping

на·вáр·н·ый, -ая, -ое, (*a*) (food) fatty, rich

навахойт·, -а, (*m*) (min) navahoite

на·вед·éни·е, -я, (*n*) (*see also* **на·вóд·к·а**) guiding, guidance, leading, directing, direction; bringing on, inducing, induction; vectoring (airborne radar)

на·вед·éни·е

~, автонóм·н·ое (GW) pre-set guidance

~, инерциáль·н·ое (GW) inertial guidance

~ ис·треб·и́тел·ей (air, mil) fighter direction

~, комáнд·н·ое (GW) command guidance

~, луч·ев·óе (GW) beam guidance

~ по луч·ý (rad) beam rider guidance

~ по про·вод·áм (GW) wire guidance

~ по·мéх· (rad) noise induction

~, прогрáмм·н·ое (GW) pre-set guidance

~ с·прáв·ок· making enquiries; looking up

стáнц·и·я на·вед·éни·я ис·треб·и́тел·я (*f*) (mil, rad) ground-controlled interception station, G.C.I.

на·вед·ённ·ый, -ая, -ое, (*past part pass*) of **на·вес·ти́**

на·вез·ти́, -ýт, (*perf*) bring (a quantity)

на·вéк· (*adv*) for ever

на·вёл (*past masc sing*) of **на·вес·ти́** (*perf*) *see* **на·вод·и́ть** (*imp*)

на·верб·овáть, -ýют, (*perf*) enroll, recruit

на·вéр·н·о (*adv*) probably, most likely

на·вёр·нут·ый, -ая, -ое, (*past part pass*) *see* **на·вер·н·ýть**

на·вер·н·ýть, -ýт, (*perf*) wind round; screw on/up

на·вер·няк·á (*adv*) undoubtedly, for sure/certain

на·верст·á·ть, -ют, (*perf*) *see* **на··вёрст·ыва·ть**

на·вёрст·ыва·ть, -ют, (*imp*) make good/up for, compensate

на·верт·éть, -я́т, (*perf*) wind/reel on; drill; twist into shape

на·вёрт·к·а, -и, (*g pl*) **-т·ок·,** (*f*) screwing on/up, tightening (of nuts)

на·вёрт·ыва·ть, -ют, (*imp*) wind/reel on/round; screw on/up; drill; twist into shape

на·вéрх· (*adv*) up, upwards, upstairs; (naut) on deck

на·верх·ý (*adv*) above, overhead, upstairs

на·вéрч·енн·ый, -ая, -ое, (*past part pass*) of **на·верт·éть**

на·вéс·, -а, (*m*) canopy, tent, (arch) awning; (geol) overhand, shelter

~, акуст·и́ческ·ий acoustic canopy

на·ве́с
~ сен·н·о́й (*m*) (agr) dutch barn
на·ве́с·ить, -ят, (*perf trans*) hang, hang on/up, suspend
на·ве́с·к·а, -и, (*g pl*) **-с·ок·,** (*f*) (*v n*) (*see* **на·ве́с·ить**) (build) hinge, hinge plate; (chem) weighed quantity
на·вес·н·о́й, -а́я, -о́е, (*a*) (*note different meanings of* **на·ве́с·ный**); suspension, suspendable, suspended; (agr) tractor-mounted, mounted, toolbar
~ дви́г·ател·ь (*m*) (naut) outboard motor
~ пе́тл·я (*f*) (build) hinge
~ плуг· (*m*) toolbar/mounted plough/plow
на·ве́с·н·ый, -ая, -ое, (*a*) (*note different meanings of* **на·вес·н·о́й**); sheltering, overhanging, (arch) awning; (gunn) high/steep-trajectory
~ ого́н·ь (*m*) (gunn) steep/high-trajectory fire
~ па́луб·а (*f*) (shipb) shade deck
на·вес·ти́, (*fut 3rd sing*) **на·вед·ёт,** (*3rd pl*) **на·вед·у́т,** (*past masc sing*) **на·вёл,** (*perf*); *see* **на·вод·и́ть**
на·вест·и́ть, -ят, (*perf*) visit
на·вес·у́ (*adv*) in suspension; hanging
на·ве́тр·енн·ый, -ая, -ое, (*a*) windwood, exposed to the wind; (naut) weather
на·ве́ч·н·о (*adv*) for ever, for good
на·ве́ш·а·ть, -ют, (*perf*) hang up, suspend; weigh out
на·ве́ш·анн·ый, -ая, -ое, (*past part pass*) of **на·ве́ш·а·ть**
на·ве́ш·енн·ый, -ая, -ое, (*past part pass*) of **на·ве́с·ить**
на·ве́ш·ива·ть, -ют, (*imp*) hang up, hang, suspend; weight out; (agr) mount (on a tractor)
на·вещ·а́·ть, -ют, (*imp*) visit
на́взничь (*adv*) on the back, on ones back, face upwards, supine
на·вив·а́льн·ый, -ая, -ое, (*a*) (*see also* **на·ви́в·очн·ый**) winding, coiling
на·вив·а́·ть, -ют, (*imp*) wind on, coil; lay up (rope); twist (thread etc.)
на·ви́в·к·а, -и, (*f*) (*v n*) *see* **на·вив·а́·ть**; coiling (springs)
на·ви́в·очн·ый, -ая, -ое, (*a*) of **на·ви́в·к·а**
навиг·а́тор·, -а, (*m*) navigator
~, инерцио́н·н·ый inertial guidance system, inertial navigation system

навиг·ацио́нн·ый, -ая, -ое, (*a*) navigation, navigational
~ перио́д· (*m*) (naut) navigational period, ice-free period
~ ремо́нт· (*m*) maintenance at sea (by ship's crew)
~ со·оруж·е́ни·е (*n*) (civ eng) canalization works
навиг·а́ци·я, -и, (*f*) navigation; navigational period
~, инерцио́н·н·ая inertial navigation/guidance
~, координа́т·н·ая grid navigation
~ по при·бо́р·ам blind navigation
~, поля́р·н·ая omnibearing-distance navigation
на·ви́ль·ник·, -а, (*m*) (agr) forkfull
на·винт·и́ть, -я́т, (*perf*) *see* **на·ви́нч·ива·ть**
на·ви́нч·енн·ый, -ая, -ое, (*past part pass*) *see* **на·ви́нч·ива·ть**
на·ви́нч·ива·ть, -ют, (*imp*) screw on/up
на·вис·а́ни·е, -я, (*n*) (*v n*) of **на·вис·а́·ть**
на·вис·а́·ть, -ют, (*imp*) hang over, overhang; impend, threaten
на·вис·а́ющ·ий, -ая, -ее, (*pres part act*) *see* **на·вис·а́·ть**; overhanging, pendant
на·ви́с·нуть, -нут, (*past masc sing*) **на·ви́с·** (*perf*) *see* **на·вис·а́·ть**
на·ви́с·ш·ий, -ая, -ее, (*past part act*) of **на·ви́с·нуть**
на·ви́·т·ый, -ая, -ое, (*past part pass*) of **на·ви́·ть**; wound on, coiled, twisted, layed (rope)
на·ви́·ть, (*fut 3rd pl*) **на·вь·ю́т,** (*perf*) *see* **на·вив·а́·ть**
на·влек·а́·ть, -ют, (*imp*) incur, cause
на·вле́ч·ь, (*fut 3rd sing*) **на·влеч·ёт,** (*3rd pl*) **на·влек·у́т,** (*past masc sing*) **на·влек·,** (*perf*); *see* **на·влек·а́·ть**
на·вод·и́ть, -ят, (*imp*) direct, bring to; point/aim at; point out, indicate; cover, coat, put on, apply; throw across, build (bridges etc.); make (clean, beautiful etc.); induce, inspire, cause (emotion)
~ мат· make mat, produce a mat finish, put on a mat coating
~ на бес·кон·е́чност·ь (phot) focus at infinity
~ на мысл·ь suggest, give (someone) an idea
~ на ре́з·кост·ь (phot, opt) focus

на·вод·и́ть

~ **на с·твор· зна́к·ов** (surv) take in transit

~ **по·ря́д·ок·** put in order

~ **с·пра́в·к·у** make inquiries, look up

на·во́д·к·а, -и, (g pl) **-д·ок·,** (f) (v n) (see **на·вод·и́ть** and **на·вед·е́ни·е**); (gunn) aiming, laying, training; (phot) focusing; (telecom) pickup, induction, stray current

~, бок·ов·а́я (mil, air) drift setting

~, вертика́ль·н·ая (gunn) laying, laying for elevation

~, горизонта́ль·н·ая (gunn) laying for line/deflection, training

~ **на фо́кус·** (phot) focusing

~, не·прям·а́я (gunn) datum training, indirect laying

~, не·раз·де́ль·н·ая (mil) simultaneous laying

~, оруд·и́йн·ая gun aiming, (rad) fire control

~, пере·крёст·н·ая (telecom) cross-talk, crossfire

~ **под· цел·ь** aiming down

~ **по·ме́х·** (rad) noise induction, stray pickup

~, прям·а́я (rad) direct/stray pickup; (phot) direct focusing; (mil) direct/visual laying

ста́нц·и·я оруд·и́йн·ой на·во́д·к·и (f) fire control radar

то́ч·к·а на·во́д·к·и (f) aiming point

на·вод·не́ни·е, -я, (n) flooding; flood, inundation

на·вод·н·и́ть, -я́т, (perf) see **на·вод·-ня́·ть**

на·вод·ня́·ть, -ют, (imp) flood, inundate

на·вод·и́м·ый, -ая, -ое, (pres part pass) of **на·вод·и́ть**; (GW) guided

на·вод·н·о́й, -а́я, -о́е, (a) made-up, temporary

на·водора́ж·ивани·е, -я, (n) hydrogen absorption, hydrogenation

на·во́д·чик·, -а, (m) gunlayer; controller

~ **-на·блюд·а́тел·ь, -я,** (m) (mil) projector controller

~, пе́рв·ый (gunn) trainer

на·вод·я́щ·ий, -ая, -ее, (pres part act) of **на·вод·и́ть**; hinting, leading (of questions); (GW) guide, guiding

~ **луч·** (m) (GW) guide beam

на·вож·у́ (pres 1st sing) of **на·вод·и́ть**; (pres 1st sing) of **на·воз·и́ть** (imp); (fut 1st sing) of **на·воз·и́ть** (perf)

на·во́з·, -а, (m) manure, dung, muck

на·воз·и́ть, -́ят, (imp) bring; manure, spread manure; (perf) bring in, gather

на·во́з·ник·, -а, (m) (ent) dung beetle, tumblebug (pl) Coprinae

на·во́з·н·ый, -ая, -ое, (a) manure, dung

~ **жи́ж·а** (f) liquid manure

навозо·раз·бра́с·ывател·ь, -я, (m) manure/muck spreader

навозо·хран·и́лищ·е, -а, (i) **-щ·ем,** (n) manure pit/yard

наво́·й, -я, (m) (text) warp beam

пере·го́н·к·а на наво́·й (f) (text) beaming, turning-on

навой·н·ый, -ая, -ое, (a) of **наво́·й**

на·вола́к·ивани·е, -я, (n) (v n) see **на·вола́к·ива·ть**; (met) galling

на·вола́к·ива·ть, -ют, (imp) collect up, pile up, put together, keep bringing; cloud over

на́·волок·, -а, (m) (geol)· overthrust folding

на́·волок·а, -и, (f) pillow case/slip

на́·волоч·к·а, -и, (g pl) **-ч·ек·,** (f) (dim) of **на́·волок·а**

на·вора́ч·ива·ть, -ют, (imp) heap/pile up, make heaps/piles of

на·ворот·и́ть, -́ят, (perf) see **на·вора́-ч·ива·ть**

на·вощ·ённ·ый, -ая, -ое, (past part pass) of **на·вощ·и́ть**; waxed, cerated

на·вощ·и́ть, -а́т, (perf) wax

на·всегда́ (adv) for ever, for good

на·встре́ч·у (adv) (prep + dat) to meet, towards, from opposite directions

на·вы́·ворот· (adv) inside out

на́·вык·, -а, (m) habit, practice, (pl) skill

на·вы́·лет· (adv) right through

на·вы́·тяж·к·у (adv) (mil) at attention

на·вы́·нос· (adv) for taking away (e.g. of foodstuffs for sale)

Навье́-Сто́кс·а, у·равн·е́ни·е (dynam) Navier–Stokes equation

на·вь·ю́т (fut 3rd pl) of **на·ви́·ть**

на·вь·ю́чива·ть, -ют, (imp) load, saddle (a pack animal)

на·вь·ю́ч·ить, -ат, (perf) see **на·вь·ю́-чива·ть**

на·вяз·а́ть, -ют, (imp) stick, become lodged; (perf) **на·вяз·а́ть**; (fut 3rd pl) **на·вя́ж·ут**; tie on, fasten to; foist/impose on

на·вя́з·нуть, -нут, (past masc sing) **на·вя́з·** (perf) stick, become lodged

на·вя́з·ыва·ть, -ют, (imp) tie on, fasten to; foist on, impose on; knit (a number of)

на·га́д·ить, -ят, (*perf*) foul, dirty

на·га́ж·енн·ый, -ая, -ое, (*past part pass*) of **на·га́д·ить**

на·га́р·, -а, (*m*) scale, deposit; (ICE) gummy/carbon deposit, carbon

~, у·дал·я́·ть (ICE) decarburize

нагаро·образ·ова́ни·е, -я, (*n*) (ICE) carbon formation

на·гарт·о́ванн·ый, -ая, -ое, (*a*) (met) cold-worked

на·гарт·о́вк·а, -и, (*f*) (met) cold/work hardening, cold work

на́гел·ь, -я, (*m*) pin, dowel, peg; shoeing nail (horses)

на·гиб·а́·ть, -ют, (*imp*) bend, bend over, incline, bow

на·гла́з·ник·, -а, (*m*) (opt) eyecap, eyecup; eye flap, blinker (horses)

на·гла́з·н·ый, -ая, -ое, (*a*) eye

~ щит·о́к· (*m*) eye shield

на́·глух·о (*adv*) rightly, hermetically

на́гл·ый, -ая, -ое, (*a*) impudent, insolent

на·гля́д·н·о (*adv*) by demonstration, graphically

на·гля́д·ност·ь, -и, (*f*) clarity in illustration; visualization; clarity of representation

на·гля́д·н·ый, -ая, -ое, (*a*) visual, graphic, illustrated, demonstrated; obvious; easy to visualize

~ по·со́б·и·е (*n*) visual aid

на·гн·а́ть, (*fut 3rd pl*) **на·го́н·ят,** (*perf*) *see* **на·гон·я́·ть**

на·гнё·л, (*past masc sing*) of **на·гнес·ти́,** (*perf*); *see* **на·гнет·а́·ть** (*imp*)

на·гнес·ти́, (*fut 3rd pl*) **на·гнет·у́т,** (*perf*) *see* **на·гнет·а́·ть**

на·гнёт·, -а, (*m*) (vet) gall

на·гнет·а́ни·е, -я, (*n*) (*v n*) *see* **на·гнет·а́·ть**

    ход· на·гнет·а́ни·я (*m*) pressure stroke

на·гнет·а́тел·ь, -я, (*m*) (mech) blower, booster, force pump; (ICE) supercharger, blower

~ волно·во́д·а (rad) waveguide pump

~, объ·ём·н·ый positive-displacement supercharger

~, ос·ев·о́й axial-flow supercharger

на·гнет·а́тельн·ый, -ая, -ое, (*a*) force, forcing, pressure, injecting; discharge, delivery (pumps etc.)

~ зо́н·а (*f*) thrust zone (of extruder)

~ кла́пан· (*m*) delivery valve (of pump)

~ ко́нус· (*m*) delivery tube (of an injector)

на·гнет·а́тельн·ый, -ая, -ое

~ на·со́с· (*m*) force pump

~ па́·труб·ок· (*m*) discharge nozzle (of pump)

на·гнет·а́·ть, -ют, (*imp*) drive, force, inject, deliver, discharge (gas, liquid etc.); supercharge (an engine)

на·гнет·а́ющ·ий, -ая, -ее, (*pres part act*) of **на·гнет·а́·ть;** pressure

~ по·ве́рх·ност·ь (*f*) driving face (propeller)

на·гно·е́ни·е, -я, (*n*) (med) suppuration, festering

на·гн·у́ть, -у́т, (*perf*) *see* **на·гиб·а́·ть**

на·гова́р·ива·ть, -ют, (*imp*) say much/ a lot; record (speech), dictate (onto magnetic tape etc.); slander

на·говор·и́ть, -ят, (*perf*) *see* **на·гова́р·ива·ть**

наг·о́й, -а́я, -о́е, (*a*) naked, nude, bare

на́·гол·о́ (*adv*) naked, bare

на·голо́в·ник·, -а, (*m*) head, cap; driving cap (of a pile)

~ пи́ллерс·а (shipb) pillar-head runner

на·го́н·, -а, (*m*) (*v n*) *see* **на·гон·я́·ть;** surge, piling up (of water)

~, шторм·ов·о́й storm surge

на·гон·я́·ть, -ют, (*imp*) overtake, gain on, come up with, catch up with; drive, drive together, fit; distil (a quantity of)

на·гор·а́ (*adv*) (min) up to the surface

на·гора́ж·ива·ть, -ют, (*imp*) fence, partition, build a fence/partition; pile up

на·гор·е́ть, -я́т, (*perf*) form a carbon scale/deposit; burn/use up

на·го́р·н·ый, -ая, -ое, (*a*) mountainous, highland, upland

~ бе́рег· (*m*) high bank (of river)

на·горо́д·ить, -ят, (*perf*) *see* **на·гора́·ж·ива·ть**

на·го́р·ь·е, -я, (*n*) highland, upland

наг·от·а́, -ы́, (*f*) nakedness, nudity, bareness

на·гота́вл·ива·ть, -ют, (*imp*) lay in a supply of; (food) cook, prepare

на·гото́в·е (*adv*) at the ready, in readiness

на·гото́в·ить, -ят, (*perf*) *see* **на·гота́вл·ива·ть**

на·гра́д·а, -ы, (*f*) reward, award

на·град·и́ть, -я́т, (*perf*) *see* **на·гражд·а́·ть**

на·гра́д·к·а, -и, (*g pl*) **-д·ок·,** (*f*) back saw

на·град·н·ой, -ая, -ое, (*a*) of **на·гра́-д·а;** (*pl decl as adj*) award/reward (of money)

на·гражд·а́·ть, -ют, (*imp*) reward, award, decorate

на·гражд·е́ни·е, -я, (*n*) reward, award; presentation of an award

на·греб·а́·ть, -ют, (*imp*) rake together/up

на·гре́в·, -а, (*m*) heat, heating, warming up; reheating (metals for working)

~ **об·ши́в·к·и** skin heating

ис·пыт·а́ни·е на на·гре́в· (*n*) (elec) heat run

~ **под·ши́п·ник·а** hot bearing

на·грев·а́ни·е, -я, (*n*) see **на·гре́в·**

на·грев·а́тел·ь, -я, (*m*) heater; (met) reheater

~, **графи́т·ов·ый** graphite bar (electric induction furnace)

на·грев·а́тельн·ый, -ая, -ое, (*a*) heating, reheating

~ **печ·ь** (*f*) heating furnace; (met) reheating furnace, reheater

~ **плит·а́** (*f*) (elec) hot plate

~ **спира́л·ь** (*f*) spiral-coil heating element

на·грев·а́·ть (*imp*) heat, warm; reheat (metals for working); **-ся** (*pass*); get hot/warm, heat/warm up

на·грес·ти́, (*fut 3rd pl*) **на·греб·у́т,** (*past masc sing*) **на·грёб·,** (*perf*) rake together/up

на·гре́·т·ый, -ая, -ое, (*past part pass*) of **на·гре́·ть;** hot

на·гре́·ть, -ют, (*perf*) see **на·грев·а́·ть**

на·громожд·е́ни·е, -я, (*n*) accumulation, piling, pile-up, heaping, heap

на·громожд·а́·ть, -ют, (*imp*) pile/heap up, accumulate

на·громозд·и́ть, -я́т, (*perf*) see **на·громожд·а́·ть**

на·гру́б·о (*adv*) roughly

на·гру́д·ник·, -а, (*m*) breastplate; breast-collar (harness)

~, **де́т·ск·ий** bib

~, **с·пас·а́тельн·ый** life jacket

на·гру́д·н·ый, -ая, -ое, (*a*) breast

на·груж·а́емост·ь, -и, (*f*) load-carrying capacity

на·груж·а́·ть, -ют, (*imp*) load

на·груж·е́ни·е, -я, (*n*) = **на·гру́з·к·а** **коэффицие́нт· на·груж·е́ни·я** (*m*) load factor

на·гру́ж·енн·ый, -ая, -ое, (*past part pass*) of **на·груз·и́ть;** loaded, (mech) stressed

на·груз·и́ть, -я́т, (*perf*) see **на·груж·а́ть**

на·гру́з·к·а, -и, (*g pl*) **-з·ок·,** (*f*) load; loading; amount of data/details (on a map)

~, **акти́в·н·ая** (elec) resistive load

~, **бо·ев·а́я** (mil) operational load

~, **больш·а́я** excessive details (on a map)

~, **бы́т·ов·ая** (elec) domestic appliance load

~, **ветр·ов·а́я** (civ eng) wind force

~, **вне́ш·н·яя** (mech) external load, applied stress

~, **вре́мен·н·ая** (mech) live load

~, **вы·нос·н·а́я** remote load

~, **дл·и́тельн·ая** continuous/constant load, constantly acting load

~, **до·полн·и́тельн·ая** additional load, (mech) increment load; (elec) superimposed load

~, **един·и́чн·ая** unit load

~, **ём·костн·ая** (elec) capacitive/leading load

~, **знако·пере·ме́н·н·ая** alternating load

~, **индукти́в·н·ая** (elec) lagging load

~, **инерцио́н·н·ая** (mech) mass load

~, **иску́с·ственн·ая** (elec) artificial load

~, **комме́рч·еск·ая** pay load

~ **котл·а́** boiler capacity **коэффицие́нт· на·гру́з·к·и** (*m*) (elec) load factor

моме́нт· вре́мен·н·ой на·гру́з·к·и (*m*) (civ eng) live-load moment

моме́нт· на·гру́з·к·и (*m*) (mech) moment of load

~ **моме́нт·н·ая** (mech) torque load

~, **не·по́лн·ая** fractional load; underload

~, **номина́ль·н·ая** rated load

~, **о·кон·е́чн·ая** (elec) terminal load

~, **ом·и́ческ·ая** (elec) resistive load

~, **основ·н·а́я** (elec) base load

~, **пере·дав·а́ем·ая** transmitted load

~, **пере·ме́н·н·ая** (mech) live/variable load/stress

~, **период·и́ческ·ая** (mech) cyclic loading/stress

~, **пи́к·ов·ая** (elec) peak load

~, **пла́н·ов·ая** planimetric details (on a map)

~, **пла́т·н·ая** (air) pay load

~, **по·движ·н·а́я** live load

~, **под·ключ·ённ·ая** (elec) connected load

на·гру́з·к·а

~, полéз·н·ая service/useful load; (air) disposable load

~, по·сто·я́нн·ая (mech) steady/dead load, constant stress

~, пре·дéль·н·ая load capacity (of a/c etc.); (mech) limit load, ultimate load

~, пре·ры́в·и́ст·ая intermittent load

~, при·со·един·ённ·ая (elec) connected load

~, проéкт·н·ая load rating

~, пуч·к·óм (elec) beam load

~, равно·мéр·н·ая uniform load; (elec) balanced load

~, раз·руш·а́ющ·ая breaking load

~, рас·ход·у́ем·ая consumable load

~, со·глас·óванн·ая (elec) matched load

~, стат·и́ческ·ая static load

~ су́д·н·а (shipb) summary of weight of hull

~, тепл·ов·а́я thermal/heat load; heat input (furnaces)

~, толч·кóв·ая shock load

~, тя́г·ов·ая traction load

~, уда́р·н·ая (mech) shock, shock load

~, у·дéль·н·ая unit load

~, част·и́чн·ая fractional load

~, чи́ст·ая net load

~, эксплуатациóн·н·ая service load

на·гру́з·очн·ый, -ая, -ое, (a) loading, load

на·гру́з·чик·, -а, (m) loader

на·гу́л·, -а, (m) (agr) fattening (of cattle etc. by grazing)

на·гу́л·ива·ть, -ют, (imp) fatten (cattle etc. by grazing)

на·гул·я́·ть, -ют, (perf) see на·гу́·л·ива·ть

над (prep + instr) over, above; on top of

над- (prefix) above, over, super-; (biol) peri-, epi-; (chem) per-

над- (verbal prefix) signifies: (1) increase, addition; (2) incomplete action, action not extended to the whole object but only to its top part

на·да́в·ить, -ят, (perf) see на·да́вл·и·ва·ть

на·да́вл·ива·ть, -ют, (imp) press; squeeze out; squash, mash

над·ба́в·ить, -ят, (perf) see над·бав·л·я́·ть

над·ба́в·к·а, -и, (g pl) -в·ок·, (f) (v n) see над·бавл·я́·ть; surcharge, increase, bonus

над·бавл·я́·ть·, -ют, (imp) add, add to, increase

над·бензó·йн·ая кисл·от·а́ (chem) perbenzoic acid

над·борно·ки́сл·ый, -ая, -ое, (a) (chem) perborate (of)

над·бóр·н·ый, -ая, -ое, (a) (chem) perboric

~ кисл·от·а́ (f) (chem) perboric acid

над·брéкер·, -а, (m) top cushion (of tire)

над·брóв·н·ый, -ая, -ое, (a) (anat) superciliary

на·дви́г·, -а, (m) (geol) overthrust fault, overthrust, thrust fault, thrust

на·двиг·а́·ть, -ют, (imp) move/push/pull to/upto/over; (geol) thrust over, thrust; -ся (pass); approach, come up to, encroach

на·двиг·а́ющ·ийся, -аяся, -ееся, approaching, imminent

на·движ·н·óй, -а́я, -óе, (a) (for pulling/thrusting onto/to/over)

~ кон·éц· (m) bell end (of pipe)

на·дви́н·уть, -ут, (perf) see на·дви·г·а́·ть

над·вóд·н·ый, -ая, -ое, (a) above water, surface

~ борт· (m) (shipb) freeboard

~ част·ь (f) (shipb) upper works, dead works

на́·дво·е (adv) in two; with two meanings

на·двóр·н·ый, -ая, -ое, (a) outdoor

~ со·оруж·éни·е (n) outhouse

над·вяз·а́·ть, (fut 3rd pl) над·вя́ж·ут, (perf); see над·вя́з·ыва·ть

над·вя́з·ыва·ть, -ют, (imp) knit on; tie on, (naut) bond on

над·глаз·ни́чн·ый, -ая, -ое, (a) (anat) supra-orbital

над·глазу́р·н·ый, -ая, -ое, (a) (ceram) overglaze

над·гортáн·ник·, -а, (m) (anat) epiglottis

над·гортáн·н·ый, -ая, -ое, (a) glottal

над·грыз·а́·ть, -ют, (imp) nibble off/at, nibble a bit

над·гры́з·ть, -ут, (past masc sing) над·грыз·, (perf); see над·грыз·а́·ть

над·да·ва́·ть, -ют, (imp) increase, add, give some more

над·да́·ть (fut 3rd sing, pl) над·да́ст, над·даду́т, (perf); see над·да·ва́·ть

над·ду́·в·, -а, (m) (ICE) supercharging, boosting; pressurization (of tanks etc.)

над·ду́·в
    регуля́тор· над·ду́в·а   (*m*) (a/c) boost control

над·ду́в·очн·ый, -ая, -ое, (*a*) supercharging, boosting

~ **на·со́с·** (*m*) supercharger

на·дев·а́ни·е, -я, (*n*) putting/pulling on

на·дёв·анн·ый, -ая, -ое, (*past part pass*) of **на·дев·а́·ть;** worn, used, old, second-hand (of clothing)

на·дев·а́·ть, -ют, (*imp*) put/pull on; mount (e.g. a shoe on a last, a tire on a mould)

на·де́·ется (*pres 3rd sing*) of **на·де́·яться**

на·де́жд·а, -ы, (*f*) hope, expectation

на·дёж·ност·ь, -и, (*f*) reliability

на·дёж·н·ый, -ая, -ое, (*a*) reliable, dependable, trustworthy

на·де́л·, -а, (*m*) (*v n*) of **на·дел·я́·ть;** (agr) allotment, plot; **на·де́·л** (*past masc sing*) of **на·де́·ть**

на·де́л·анн·ый, -ая, -ое, (*past part pass*) of **на·де́л·а·ть**

на·де́л·а·ть, -ют, (*perf*) make a number of; commit/perpetrate (errors etc.); extend, add to

на·дел·е́ни·е, -я, (*n*) allotment, allocation

на·дел·ённ·ый, -ая, -ое, (*past part pass*) *see* **на·дел·я́·ть**

на·дел·и́ть, -я́т, (*perf*) *see* **на·дел·я́·ть**

на·де́л·к·а, -и, (*g pl*) **-л·ок·,** (*f*) extension, addition

~, **коро́б·чат·ая** cleat

~, **па́луб·н·ая** (shipb) superimposed structure

~, **рул·ев·а́я** (shipb) rudder enlargement plate

на·дел·я́·ть, -ют, (*imp*) allot; give, endow

на·де́н·ут (*fut 3rd pl*) of **на·де́·ть**

на·дёрг·а·ть, -ют, (*perf*) pluck, pull out (a quantity of)

на·дёрг·ива·ть, -ют, (*imp*) pluck, pull out (a quantity of); throw on (clothes etc.)

на·дер·ёт (*fut 3rd sing*) of **на·др·а́ть**

на·дёр·н·уть, -ут, (*perf*) throw on (clothes etc.)

на·дер·у́т (*fut 3rd pl*) of **на·др·а́ть**

на·де́·ть, (*fut 3rd pl*) **на·де́н·ут,** (*perf*) put/pull on

на·де́·яться, -ются, (*imp*) hope for/to; rely on, have confidence in

над·звук·ов·о́й, -а́я, -о́е, (*a*) supersonic

над·зе́м·н·ый, -ая, -ое, (*a*) above-ground, overhead, elevated; surface, ground-level

~ **желе́з·н·ая доро́г·а** (*f*) overhead/elevated railway/railroad

над·зир·а́тел·ь, -я, (*m*) supervisor, overseer

над·зир·а́·ть, -ют, (*imp*) supervise, oversee, superintend, control

над·зо́р·, -а, (*m*) supervision, inspection; inspectorate

~, **горно·техн·и́ческ·ий** mining inspectorate; mining supervision

над·ио́д·н·ый, -ая, -ое, (*a*) (chem) periodic, periodide (of a metal)

нади́р·, -а, (*m*) (astron) nadir

~, **зем·н·о́й** ground nadir

на·дир·а́·ть, -ют, (*imp*) tear off/up, strip, peel

над·кали́бер·н·ый с·на·ря́д·   (*m*) (mil) stick bomb

над·ка́л·ыва·ть, -ют, (*imp*) prick/pierce lightly; split slightly

над·кисл·от·а́, -ы́, (*f*) (chem) peracid, per acid

над·кла́сс·, -а, (*m*) superclass

над·ко́ж·иц·а, -ы, (*i*) **-ц·ей,** (*f*) (biol) cuticle

над·коле́н·ник·, -а, (*m*) (anat) patella

над·коле́н·н·ый, -ая, -ое, (*a*) (anat) patellar

над·кол·о́ть, -ют, (*perf*) *see* **над·ка́л·ыва·ть**

над·ко́ст·ниц·а, -ы, (*i*) **-ц·ей,** (*f*) (anat) periosteum

над·крит·и́ческ·ий, -ая, -ое, (*a*) above-critical, supercritical

над·крит·и́чност·ь, -и, (*f*) supercritical condition/state

над·кры́л·ь·е, -я, (*n*) (ent) elytrum

над·ла́м·ыва·ть, -ют, (*imp*) partly break, crack

над·леж·а́ть, (*pres 3rd sing*) **-и́т,** (*imp*) ought to, be necessary to, be due, (*as auxil verb*) should, must
    над·леж·и́т с·де́л·а·ть has/is/ought to be done, must be done

над·леж·а́щ·ий, -ая, -ее, (*pres part act*) *see* **над·леж·а́ть;** appropriate, proper, suitable, due

над·ло́м·, -а, (*m*) partial break, effraction

над·лом·а́·ть, -ют, (*perf*) *see* **над·ла́м·ыва·ть**

над·лом·и́ть, ⸗ят, (*perf*) *see* **над·ла́·м·ыва·ть**

над·мно́ж·еств·о, -а, (*n*) (math) including/comprehending set

над·молекуля́р·н·ый, -ая, -ое, (*a*) (phys, chem) supermolecular

над·молибде́н·ов·ый, -ая, -ое, (*a*) (chem) permolybdic

над·мы́щел·ок·, -л·к·а, (*m*) (zool) ipicondyle

над·ни́т·очник·, -а, (*m*) (text) fallen wire

на́до (*predic*) it is necessary; (*prefix*) = **над** (q.v.)

мне на́до I must/should/ought, it is necessary for me to, I have to

так на́до it must be so/thus

на́·доб·ност·ь, -и, (*f*) necessity, need

на́·доб·н·ый, -ая, -ое, (*a*) necessary

на·до·ед·а́·ть, -ют, (*imp*) bore, become boring/tiresome

на·до·е́д·лив·ый, -ая, -ое, (*a*) boring, tiresome

на·до·е́сть, (*fut 3rd sing, pl*) **на·до·е́ст, на·до·ед·я́т,** (*past masc sing*) **на·до·е́д,** (*perf*); *see* **на·до·ед·а́·ть**

на·до́·й, -я, (*m*) (agr) milk yield

на́·долб·ы, -лб·, (*f pl*) (mil) anti-tank obstructions, dragons' teeth

на·до́лг·о (*adv*) for a long time

на·до́м·н·ый, -ая, -ое, (*a*) at home, home, domestic, cottage (industry)

надо·рв·а́ть, -у́т, (*perf*) *see* **над·ры·ва́·ть**

над·от·ря́д·, -а, (*m*) (biol) superorder

надо·шь·ю́т (*fut 3rd pl*) of **над·шй·ть**

над·панел·ь, -и, (*f*) (elec instr) built-up panel, vertical instrument panel

над·па́р·ыва·ть, -ют, (*imp*) partly, unstitch/unpick/rip open/undone

над·пе́ре·кис·ь, -и, (*f*) (chem) hyperoxide

над·пест·и́чн·ый, -ая, -ое, (*a*) (bot) ipigynous

над·пи́л·, -а, (*m*) curf, notch (made with saw or file)

над·пи́л·ива·ть, -ют, (*imp*) saw/file into

над·пил·и́ть, ⸗ят, (*perf*) *see* **над·пи́л·ива·ть**

над·пис·а́ть, (*fut 3rd pl*) **над·пи́ш·ут,** (*perf*) *see* **над·пи́с·ыва·ть**

над·пи́с·ыва·ть, -ют, (*imp*) inscribe, superscribe, write above; write a caption/note/legend

на́д·пис·ь, -и, (*f*) inscription, superscription, caption, legend, notice, note; (cinema) title, caption

⸗, основ·н·а́я title block (on an engineering drawing)

⸗, пере·да́т·очн·ая (fin) endorsement

⸗, пред·о·стерег·а́ющ·ая warning/danger notice

⸗, трафаре́т·н·ая (tech) stencilling

над·пи́ш·ут (*fut 3rd pl*) of **над·пис·а́ть**

над·пло́д·ник·, -а, (*m*) (bot) ipicarp

над·по́л·е, -я, (*n*) (math) extension field

над·пор·о́ть, ⸗ют, (*perf*) *see* **над·па́р·ыва·ть**

над·по́ч·ечник·, -а, (*m*) (anat) adrenal gland

над·по́ч·ечн·ый, -ая, -ое, (*a*) (anat) suprarenal, adrenal

на·дра́·ива·ть, -ют, (*imp*) (naut) set up (rigging); clean, polish

на·дра́·ить, -ят, (*perf*) *see* **на·дра́·ива·ть**

над·ра́м·ник·, -а, (*m*) (M/T) underframe

на·др·а́ть, (*fut 3rd pl*) **на·дер·у́т,** (*perf*) tear off/up, strip, peel

над·рёбер·н·ый, -ая, -ое, (*a*) (anat) supracostal

над·ре́з·, -а, (*m*) cut, incision, notch, groove, gash, (math) cut

⸗, круг·ов·о́й (met) keyhole notch

⸗ Шарпи (met) keyhole notch, Charpy type-B notch

над·ре́з·анн·ый, -ая, -ое, (*past part pass*) of **над·ре́з·ать** (*perf*)

над·рез·а́·ть, -ют, (*imp*) *see* **над·ре́з·ыва·ть**

над·ре́з·ать, (*fut 3rd pl*) **над·ре́ж·ут,** (*perf*); *see* **над·ре́з·ыва·ть**

над·ре́з·к·а, -и, (*g pl*) **-з·ок·,** (*f*) (*v n*) of **над·ре́з·ыва·ть;** slitting, notching (sheet metal)

над·резона́нс·н·ый, -ая, -ое, (*a*) (phys) ipiresonance

над·ре́з·ыва·ть, -ют, (*imp*) incise, notch, groove, cut into/slightly

над·решёт·н·ый, -ая, -ое, (*a*) over-size (of sieved materials)

над·ру́б·, -а, (*m*) chip, chop, notch (with axe etc.), slash (with knife etc.); kerf (with saw)

над·руб·а́·ть, -ют, (*imp*) chop/hack/slash partly through, chip, notch

над·руб·и́ть, ⸗ят, (*perf*) *see* **над·ру·ба́·ть**

над·ру́д·н·ый, -ая, -ое, (*a*) (geol) supra-ore

над·ры́в·, -а, (*m*) (*v n*) *see* **над·ры-в·а́·ть;** (met) flake, fish eye, shatter, crack; (text) tear

над·рыв·а́·ть, -ют, (*imp*) tear slightly; strain, overtax

над·сек·а́·ть, -ют, (*imp*) notch, incise, mark

над·сем·е́йств·о, -а, (*n*) (biol) super-family

над·семя·до́ль·н·ое коле́н·о (*n*) (bot) ipicotyl

над·се́р·н·ая кисл·от·а́ (*f*) persulphuric/persulfuric acid

над·се́ч·ь, (*fut 3rd sing, pl*) **над·се-ч·ёт, над·сек·у́т,** (*past masc sing*) **над·се́к·,** (*perf*); *see* **над·сек·а́·ть**

над·синхро́н·н·ый, -ая, -ое, (*a*) (elec) supersynchronous, hypersynchronous

над·сма́тр·ива·ть, -ют, (*imp*) control, supervise, inspect

над·смо́тр·, -а, (*m*) control, supervision, inspection

над·со́л·ь, -и, (*f*) (chem) persalt

над·ста́в·ить, -ят, (*perf*) *see* **над·ставл·я́·ть**

над·ста́в·к·а, -и, (*g pl*) **-в·ок·,** (*f*) (*v n*) *see* **над·ставл·я́·ть;** extension, extension piece; adapter, built-up part; (met cast) feeder/feeding head, hot top, sinkhead

~, тёпл·ая при́·быль·н·ая (met cast) hot top, sinkhead, shrinkhead, refractory/feeder head

над·ста́вл·енн·ый, -ая, -ое, (*past part pass*) *see* **над·ставл·я́·ть**

над·ставл·я́·ть, ют, (*imp*) extend, lengthen, put on

над·стра́·ива·ть, -ют, (*imp*) build on top of, build over, build a superstructure on; build higher/upwards

над·стро́·ить, -ят, (*perf*) *see* **над·-стра́·ива·ть**

над·стро́й·к·а, -и, (*g pl*) **-о́·ек·,** (*f*) (*v n*) *see* **над·стра́·ива·ть;** superstructure; additional structure, extension, additional storey; (build) topping-out; building a superstructure; casing (of submarine)

~, сре́д·н·яя (shipb) midcastle

над·стро́ч·н·ый, -ая, -о е, (*a*) (print) above the line, superior

~ знак· (*m*) (print) superior sign, superscript

~ и́ндекс· (*m*) (print) superior figure, superscript

над·тепл·ов·о́й, -а́я, -о́е, (*a*) ipithermal

над·тона́ль·н·ый, -ая, -ое, (*a*) (telecom) super-audio

над·тре́с·нут·ый, -ая, -ое, (*past part pass*) scratched, cracked on the surface

на·ду́в·, -а, (*m*) (*v n*) *see* **на·дув·а́·ть;** drift

надув·а́ни·е, -я, (*n*) (*v n*) *see* **на·ду-в·а́·ть**

на·дув·а́·ть, -ют, (*imp*) inflate; blow (of wind etc.); swindle

на·дув·н·о́й, -а́я, -о́е, (*a*) inflatable

~ поду́ш·к·а (*f*) air cushion

над·у́голь·н·ая кисл·от·а́ (*f*) (chem) percarbonic acid

над·у́гольно·ки́сл·ый, -ая, -ое, (*a*) percarbonate (of)

над·у́ксус·н·ая кисл·от·а́ (*f*) (chem) peracetic acid

на·ду́ль·ник·, -а, (*m*) (gunn) muzzle cover

на·ду́ль·н·ый, -ая, -ое, (*a*) (gunn) muzzle

на·ду́м·анн·ый, -ая, -ое, (*past part pass*) of **на·ду́м·а·ть;** unreal, far-fetched

на·ду́м·а·ть, -ют, (*perf*) *see* **на·ду́-м·ыва·ть**

на·ду́м·ыва·ть, -ют, (*imp*) make up one's mind, decide; think up, invent

на·ду́·т·ый, -ая, -ое, (*past part pass*) of **на·ду́·ть;** inflated, blown up, exaggerated

на·ду́·ть, -ют, (*perf*) *see* **на·дув·а́·ть**

на́д·фил·ь, -и, (*f*) needle file

над·фосфо́р·н·ая кисл·от·а́ (*f*) (chem) perphosphoric acid

над·фюзеля́ж·н·ый, -ая, -ое, (*a*) (a/c) dorsal

над·хро́м·ов·ая кисл·от·а́ (*f*) (chem) perchromic acid

над·хрящ·ни́ц·а, -ы, (*f*) (anat) perichondrium

над·цирко́н·иев·ая кисл·от·а́ (*f*) (chem) perzirconic acid

над·череп·н·а́я оболо́ч·к·а (*f*) (anat) pericranium

над·чрев·н·а́я о́бласт·ь (*f*) (anat) epigastrium

над·ша́хт·н·ый, -ая, -ое, (*a*) (min) pithead, deckhead, head

над·шив·а́·ть, -ют, (*imp*) sew on to, lengthen

над·ши́·ть, (*fut 3rd pl*) **над·шь·ю́т,** (*perf*) *see* **над·шив·а́·ть**

на·дым·и́ть, -я́т, (*perf*) emit/make a lot of smoke, smoke (of stoves etc.)

на·един·е́   (*adv*) together in private, alone, privately

на·е́зд·, -а,   (*m*) (*v n*) (*see* на·езж·а́·ть) flying visit; (cinema) dollying in, tracking in

на·е́зд·ить, -ят,   (*perf*) *see* на·е́зж·и·ва·ть

на·е́зд·ник·, -а,   (*m*) rider; trainer (of horses); trick rider (in a circus); (ent) ichneumon fly, (*pl*) *Ichneumonoideae*

на·езж·а́ни·е, -я,   (*n*) running into, striking; interference, sticking, jamming (of moving parts)

на·езж·а́·ть, -ют,   (*imp*) drive/run/ bump into/over; arrive, drive up (of a number)

на·е́зж·ива·ть, -ют,   (*imp*) drive, cover (a distance); spend time driving, drive for; compact (surface by driving)

на·ём·, -а,   (*m*) hire, hiring, renting

на·ём·н·ый, -ая, -ое,   (*a*) hired, rented

на·е́х·ать, (*fut 3rd pl*) на·е́д·ут, (*perf*) *see* на·езж·а́·ть

на·жа́л·оваться, -уются, (*perf*) complain of/about

на·жа́р·ива·ть, -ют,   (*imp*) fry/roast/ grill (a quantity of)

на·жа́р·ить, -ят, (*perf*) *see* на·жа́р·и·ва·ть

на·жа́т·и·е, -я,   (*n*) (*v n*) *see* на·жи·м·а́·ть
сигна́л· на·жа́т·и·я   (*m*) (telecom) marking signal

на·жа́т·ь,   (*fut 3rd pl*) на·жм·у́т, (*perf*) *see* на·жим·а́·ть

на·жа́т·ь, (*fut 3rd pl*) на·жн·у́т, (*perf*) reap, harvest (a quantity of)

на·жг·у́т (*fut 3rd pl*) of на·же́ч·ь

нажда́к·, -а́,   (*m*) (min) emery

нажда́ч·н·ый, -ая, -ое,   (*a*) emery
~ бума́г·а   (*f*) emery paper
~ круг·   (*m*) grinding/emery wheel

на·жёг·ш·ий, -ая, -ее, (*past part act*) of на·же́ч·ь

на·же́ч·ь, (*fut 3rd sing pl*) на·жж·ёт, на·жг·у́т, (*past masc sing*) на·жёг·, (*perf*); *see* на·жиг·а́·ть

на·жж·ённ·ый, -ая, -ое, (*past part pass*) of на·же́ч·ь

на·жи́в·а, -ы, (*f*) gain, profit; bait

на·жив·а́·ть, -ют, (*imp*) save up; make (money, profit etc.); acquire, get contract; -ся (*pass*); get/become rich

на·жив·и́ть, -я́т, (*perf*) bait

на·жи́в·к·а, -и, (*g pl*) -в·ок·, (*f*) bait

на·живл·я́·ть, -ют, (*imp*) bait

на·жив·н·о́й, -а́я, -о́е,   (*a*) bait; accessible
~ снаст·ь   (*f*) fishing rod

на·жиг·а́·ть, -ют, (*imp*) burn, burn up (a quantity of); burn on, brand, mark

на·жи́м·, -а,   (*m*) pressing, pressure; pressure device
~ Про́ни   (eng) Prony brake

на·жим·а́·ть, -ют, (*imp*) press, press on/down, put pressure on, depress; squeeze out (juice etc.)
~ руко·я́т·к·у бд·и́тельност·и acknowledge (on a signal system)

на·жим·н·о́й, -а́я, -о́е,   (*a*) pressing/ holding down
~ винт·   (*m*) clamping/adjusting screw
~ га́·йк·а   (*f*) lock nut
~ де́й·стви·е   (*n*) (*in gen as adj*) contact-pressure
~ кно́п·к·а   (*f*) push button
~ механи́зм·   (*m*) (met roll) screwdown gear
~ у·стро́й·ств·о   (*n*) (met roll) screwdown

на·жи́м·н·ый, -ая, -ое,   (*a*) *see* на·жим·н·о́й

нажи́ть, (*fut 3rd pl*) на·жив·у́т, (*perf*) *see* на·жив·а́·ть

на·жм·у́т, (*fut 3rd pl*) of на·жа́т·ь, (*perf*); *see* на·жим·а́·ть (*imp*)

на·жн·у́т, (*fut 3rd pl*) of на·жа́т·ь reap, harvest

на·хо́р·, -а,   (*m*) (leather) bating

наз. (*abbr*) of на·зыв·а́ем·ый called

на·за́д·   (*adv*) back, backwards, in reverse; ago; (naut) astern
~, вз·я́т·ь   take back, retract
~, вз·гля́д·   retrospect, hindsight
~, два го́д·а тому́   two years ago
~, ма́л·ый   (naut) slow speed astern
~, по́лн·ый   (naut) full speed astern
~, сре́д·н·ий   (naut) half speed astern

на·зад·и́   (*adv*) behind

назализ·и́р·овать, -уют, (*imp and perf*) (ling) nasalize

назализ·ова́ть, -у́ют, (*imp or perf*) (ling) nasalize

наза́ль·н·ый, -ая, -ое,   (*a*) nasal

на·зв·а́ни·е, -я,   (*n*) name, title, appellation, designation
~, вид·ов·о́е   (biol) specific name
~, род·ов·о́е   (biol) generic name

на·зв·а́т·ь, (*fut 3rd pl*) на·зов·у́т, (*perf*) call, name; -ся (*+ instr*) be called/ named

на·зём·, -а, (*m*) dung, manure

назём·н·ый, -ая, -ое, (*a*) ground, surface, land, terrestrial

~ желéз·н·ая дорóг·а (*f*) surface railway/railroad

~ úне·й (*m*) (meteor) surface hoar, ground frost

ná·земь (*adv*) on the ground

на·знач·á·ть, -ют, (*imp*) name, nominate, designate, appoint; name, fix, set (of time etc.); name, quote (a price); (med) prescribe

на·знач·éни·е, -я, (*n*) appointment, nomination, designation; fixing, establishing; purpose, duty (of machines); (rail) destination; (med) prescription

~, двой·н·óе double purpose/duty; (*in gen as adj*) dual-purpose

~, óбщ·ее general purpose; (*in gen as adj*) general-purpose

~, о·сóб·ое special purpose; (*in gen as adj*) special-purpose

порт· на·знач·éни·я (*m*) port of destination

на·знáч·енн·ый, -ая, -ое, (*past part pass*) *see* **на·знач·á·ть**

на·знáч·ить, -ат, (*perf*) *see* **на·знач·á·ть**

на·зов·ýт (*fut 3rd pl*) of **на·зв·áть**

назóй·лив·ый, -ая, -ое, (*a*) importunate, repeatedly troublesome

на·зрев·áни·е, -я, (*n*) (*v n*) *see* **на·зрев·á·ть**

на·зрев·á·ть, -ют, (*imp*) ripen, become ripe, mature; become imminent, come to a head

на·зрé·ть, -ют, (*perf*) *see* **на·зрев·á·ть**

на·зубр·éнн·ый, -ая, -ое, (*a*) notched, serrated

на·зубр·ивá·ть, -ют, (*imp*) make notches

на·зубр·и́ть, -ят, (*perf*) *see* **на·зубр·ивá·ть**

на·зыв·áем·ый, -ая, -ое, (*pres part pass*) of **на·зыв·á·ть**; named, called **так на·зыв·áем·ый** so-called, as it is known, what is known as

на·зыв·á·ть, -ют, (*imp*) call, name; **-ся,** (+ *instr*) be called/named

наи- (*prefix*) (*used in forming the superlative of adjectives and adverbs*)

наи·бóл·ее (*adv*) the most

~ выс·óк·ий (*a*) highest

~ глуб·óк·ий (*a*) deepest, innermost

~ длúн·н·ый (*a*) longest

наи·бóльш·ий, -ая, -ее, (*a*) the greatest/largest/biggest, maximum, peak, extreme, overall

~ за·зóр· (*m*) maximum clearance

~ длин·á (*f*) overall length, length overall

~ знач·éни·е фýнкц·и·и (*n*) (math) maximum of a function

~ обрáт·н·ое на·пряж·éни·е (*n*) (elec) peak inverse voltage

~ óбщ·ий дел·úтел·ь (*m*) (math) greatest common divisor

наúв·н·ый, -ая, -ое, (*a*) naive

наи·вы·год·нéйш·ий, -ая, -ее, (*a*) optimum, most advantageous, best

~ скóр·ост·ь (*f*) (air) best flying speed

~ скóр·ост·ь на·бóр·а выс·от·ы́ (air) best climbing speed

наи·вы́с·ш·ий, -ая, -ее, (*a*) highest, utmost; very high

на·игр·á·ть, -ют, (*perf*) *see* **на·úгр·ыва·ть**

на·úгр·ыва·ть, -ют, (*imp*) record (music); play for some time (music)

на·изнáнку (*adv*) inside out

на·из·ýсть (*adv*) by heart, by rote

на·úл·ивани·е, -я, (*n*) colmatage, silt deposition

на·úл·ок·, -л·к·а, (*m*) silt deposition

наи·лýчш·ий, -ая, -ее, (*a*) best

наи·мéн·ее (*adv*) the least; the most un-

на·имен·овáни·е, -я, (*n*) name, title, designation, appellation; (math) denomination

~, кóд·ов·ое code name

~, у·слóв·н·ое code name

на·имен·овáть, -ýют, (*perf*) name, call, designate, denominate

наи·мéнь·ш·ий, -ая, -ее, (*a*) least; minimum

~ дéй·стви·е (*n*) least action

~ квадрáт· (*m*) (math) least square

~ при·нужд·éни·е (*n*) (mech) least constraint/curvature

прúнцип· наи·мéнь·ш·его дéй· стви·я (*m*) (phys) principle of least action

прúнцип· наи·мéнь·ш·его при·нужд·éни·я (*m*) (mech) principle of least constraint/curvature

~ рас·сто·я́ни·е (*n*) (phys) collision diameter

~ скóр·ост·ь хóд·а (*f*) (naut) steerage way

спóсоб· наи·мéнь·ш·их квадрáт·ов (*m*) (math) method of least squares

наи·скор·е́йш·ий, -ая, -ее, (*a*) fastest, quickest

на́·ис·кось (*adv*) across, crosswise, aslant, obliquely

наи·ху́д·ш·ий, -ая, -ее, (*a*) worst, the worst

на́·йд·енн·ый, -ая, -ое, (*past part pass*) of **на·йти́;** found; (*pl decl as adj*) findings

Найквист·а, фо́рмул·а (*f*) (elec) Nyquist formula

найло́н·, -а, (*m*) (text, chem) nylon

на·йм·ёт, (*fut 3rd sing*) of **на·ня́·ть**

на·йти́, (*fut 3rd sing, pl*) **на·йд·ёт, на·й·д·у́т,** (*past masc sing*) **на·шёл** (*perf*); *see* **на·ход·и́ть**

найто́в·, -а, (*m*) (naut) lashing, seizing

на·ка́з·, -а, (*m*) mandate, instructions

на·каз·а́ни·е, -я, (*n*) punishment; (law) sentence

на·каз·у́ем·ый, -ая, -ое, (*a*) (law) punishable, indictable

на·ка́л·, -а, (*m*) incandescence, intense heat

 батаре́·я на·ка́л·а (*f*) (elec) filament battery

~, бе́л·ый white heat

 на·пряж·е́ни·е на·ка́л·а (*n*) (elec) filament voltage

 нит·ь на·ка́л·а (*f*) (elec) filament

~, прям·о́й (elec) direct heating; (*in gen as adj*) directly heated

 те́л·о на·ка́л·а (*n*) filament (of electric lamp)

 цеп·ь на·ка́л·а (*f*) (elec) filament circuit

на·кал·ённ·ый, -ая, -ое, (*past part pass*) of **на·кал·и́ть;** incandescent, glowing

~ до·бел·а́ white-hot

~ до́·красн·а́ red hot

на·ка́л·иваем·ый, -ая, -ое, (*pres part pass*) of **на·ка́л·ива·ть**

~ катод· (*m*) (elec) directly heated cathode, filament

на·ка́л·ивани·е, -я, (*n*) incandescing, glowing

 ла́мп·а на·ка́л·ивани·я (*f*) (elec) incandescent/filament lamp

на·ка́л·ива·ть, -ют, (*imp*) make incandescent, subject to great heat, heat up; **-ся** (*pass*); glow, incandesce

на·кал·и́ть, -я́т, (*perf*) incandesce

на·кал·ыва́ни·е, -я, (*n*) (*v n*) *see* **на·ка́·л·ыва·ть;** indentation (hardness test)

на·ка́л·ыва·ть, -ют, (*imp*) prick, pierce; pin on/up; chop, split (wood etc.); slaughter (animals)

~ то́ч·к·у pin-point, locate/mark a point (on a map)

на·ка́ль·н·ый, -ая, -ое, (*a*) (elec) filament

~ ла́мп·а (*f*) (elec) filament lamp

на·кал·я́·ть, -ют, (*imp*) = **на·ка́л·и·ва·ть**

накану́не (*adv*) the day before

на·ка́п·а·ть, -ют, (*perf*) pour/measure out (drops); spot, make spots on

на·ка́пл·ива·ть, -ют, (*imp*) = **на·коп·л·я́·ть**

на·ка́пл·иваю·щий, -ая, -ее, (*pres part act*) of **на·ка́пл·ива·ть;** accumulating, amassing, piling up

~ сумм·а́тор· (*m*) accumulator (computers)

~ сумм·а́тор· дво·и́чн·ый (*m*) binary accumulator

~ сумм·а́тор· про·из·вед·е́ни·я (*m*) product accumulator

~ с·чёт·чик· (*m*) (*and see* **н. сумм·а́тор·** above) accumulator, accumulator register computers

на·ка́п·ыва·ть, -ют, (*imp*) pour/measure out (drops); spot, make spots on; dig, excavate (in several places); dig up (a quantity of)

на·ка́т·, -а, (*m*) (*v n*) *see* **на·ка́т·ыва·ть;** (build) counter floor; dead floor; run in/up (of wave on shore etc.); (gunn) run-out, recuperation; (met roll) scab

~, пруж·и́нн·ый spring return

на·ка́т·анн·ый, -ая, -ое, (*past part pass*) of **на·кат·а́·ть**

на·кат·а́·ть, -ют, (*perf*) (*see also* **на·ка́т·ывани·е**) roll up/along (a number of); make (by rolling); roll flat/smooth; imprint, knurl

на·кат·и́ть, -ят, (*perf*) roll onto; (gunn) recuperate; crop up, turn up unexpectedly

на·ка́т·к·а, -и, (*g pl*) **-т·ок·,** (*f*) knurling; rolling (screw threads); knurling tool; knurled surface; graining (leather); rolling on (rubber)

на·ка́т·ник·, -а, (*m*) (gunn) recuperator

на·ка́т·н·ый, -ая, -ое, (*a*) knurling; rolling, rolling-on

~ маши́н·а (*f*) boarding machine (leather)

на·ка́т·ывани·е, -я, (*n*) (*v n*) *see* **на··ка́т·ыва·ть;** rolling (screw thread); (phot) glazing

на·ка́т·ыва·ть, -ют, (*imp*) *see* (1) **на··кат·а́·ть;** (2) **на·кат·и́ть**

на·ка́ч·анн·ый, -ая, -ое, (*past part pass*) of **на·кач·а́·ть;** inflated, pumped

на·кач·а́·ть, -ют, (*perf*) *see* **на·ка́··ч·ива·ть**

на·ка́ч·енн·ый, -ая, -ое, (*past part pass*) of **на·кат·и́ть**

на·ка́ч·ива·ть, -ют, (*imp*) pump (a quantity of), pump up/full, inflate; deliver (by pump), pump

на·ка́ч·к·а, -и, (*g pl*) **-ч·ек·,** (*f*) (*v n*) *see* **на·ка́ч·ива·ть**

~, магни́т·н·ая (phys) magnetic pumping

на·ке́рн·ива·ть, -ют, (*imp*) (mech eng) punch, hole, punch a hole

на·керн·ова́ть, -у́ют, (*imp*) (mech eng) centre/center dot, punch a centre dot

на·кид·а́·ть, -ют, (*perf*) fill (by throwing); throw (a number of); throw on

на·ки́д·к·а, -и, (*g pl*) **-д·ок·,** (*f*) (*v n*) *see* **на·ки́д·ыва·ть**) surcharge, increase; cape (clothing); counterpane, bedspread

на·кид·н·о́й, -а́я, -о́е, (*a*) throwing, throw-on; surcharge, additional

~ га́·йк·а (*f*) adapter nut

на·ки́д·ыва·ть, -ют, (*imp*) throw on/over; add to, raise; **-ся,** fall upon, attack

на·ки́н·уть, -ут, (*perf*) *see* **на·ки́·д·ыва·ть**

на·кип·а́·ть, -ют, (*imp*) form a scum/scale (by boiling)

на·кип·е́ть, -я́т, (*perf*) *see* **на·кип·а́·ть**

на́·кип·ь, -и, (*f*) scum; scale, boiler scale; incrustation

~, жемчу́ж·н·ая (min) pearl sinter, geyserite

~, фумаро́л·ов·ая (geol) fumarole incrustation

на·кла́д·к·а, -и, (*g pl*) **-д·ок·,** (*f*) (*v n*) (*see* **на·кла́д·ыва·ть**) cover plate/strap/piece; butt strap (welded/riveted joint); fish plate/bar/piece (rails or beams); lining (clutch, brake); hasp-and-staple (gate fastening); (geol) lap, overlap

~, скул·ов·а́я (shipb) gusset plate

~, с·ты́к·ов·а́я fish plate/bar, shin, cover plate

на·кла́д·к·а

~ с·цепл·е́ни·я clutch lining

~ це́п·и link plate

на·клад·н·а́я, -о́й, (*f decl as adj*) consignment note; bill of lading

~ на груз· (air) bill of lading

~, тра́нспорт·н·ая bill of lading

на·клад·н·о́й, -а́я, -о́е, (*a*) (*see also* **на·клад·н·а́я**) (*as noun*); laid-on, superimposed, superposed; false, imitation

~ рас·хо́д·ы (*pl*) overhead expenses

~ серебр·о́ (*n*) silver plate

на·кла́д·ыва·ть, -ют, (*imp*) put/lay on/over/in, superimpose, superpose; fill (by putting/laying); put, lay (a number of); impose, inflict, make subject to, apply

~ протéктор· (*m*) retread (a tire)

на·кле́·ива·ть, -ют, (*imp*) glue/stick/paste on; glue, stick, paste (a number of)

на·кле́·ить, -ят, (*perf*) *see* **на·кле́·и·ва·ть**

на·кле́й·к·а, -и, (*g pl*) **-ле·ек·,** (*f*) (*v n*) *see* **на·кле́·ива·ть;** label, stick-on label; false beard/moustache/eyebrows (theatre)

на·кле́п·, -а, (*m*) (met) work hardening, cold work; peening

~, вз·ры́в·н·о́й (met) shock hardening

~, дробе·стру́й·н·ый (met) peening, shot peening

~, фа́з·ов·ый (met) precipitation hardening

на·клеп·а́·ть, -ют, (*perf*) *see* **на··клёп·ыва·ть**

на·клёп·ыва·ть, -ют, (*imp*) (met) work-harden, peen; rivet on

на·кло́н·, -а, (*m*) slope, incline, inclination, tilt, slant, rake; (geol) dip pitch; (civ eng) batter

~ и́мпульс·ов (elec) pulse droop/tilt

~ лопа́т·к·и blade inclination (of pumps, turbines etc.)

у́гол· на·кло́н·а (*m*) tilt/slope/dip angle, angle of inclination; (rail) gradient, (geol) pitch angle

у́гол· на·кло́н·а спира́л·я (*m*) helix angle (helical gear)

~ шкво́р·н·я в·бок· (М/Т) king-pin inclination

~ шкво́р·н·я на·за́д· (М/Т) castor/caster

~ шкво́р·н·я, про·до́ль·н·ый (М/Т) castor, caster

на·клон·е́ни·е, -я,   (*n*) inclination, batter, tilt, slant, dip; (gram) mood

~ горизо́нт·а (astron) dip of the horizon, apparent depression of the horizon

~, магни́т·н·ое   dip, magnetic inclination

на·клон·и́ть, –ят, (*perf*) *see* **на·клон·я́·ть**

на·кло́н·н·ая, -ой, (*f decl as adj*) (mech) oblique line

на·кло́н·ност·ь, -и, (*f*) tendency, inclination

на·кло́н·н·ый, -ая, -ое, (*a*) sloping, slanting, inclined, oblique, tilted; (rad) slant; (civ eng) batter, battered

~ да́ль·ност·ь (*f*) (rad) slant range

~ бур·е́ни·е (*n*) (oil) inclined boring

~ ис·то́ч·ник· (*m*) (nucl) inclined source

~ маши́н·а (*f*) inclined engine

~ сто́й·к·а (*f*) (civ eng) batter post

наклоно·ме́р·, -а, (*m*) tilt indicator/meter, batter level

на·клон·я́·ть, -ют, (*imp*) incline, tilt, tip, cant; decline; **-ся** (*pass*); bend/lean over

на·ков·а́ленк·а, -и, (*g pl*) **-н·ок·,** (*f*) (gunn) anvil

на·ков·а́льн·я, -и, (*g pl*) **-л·ен·,** (*f*) anvil; (anat) incus, anvil; (meteor) anvil cloud

~, дву·ро́г·ая double-bick anvil

на·ко́в·анн·ый, -ая, -ое, (*past part pass*) of **на·ков·а́ть**; forged-on, forged

на·ков·а́ть, (*fut 3rd pl*) **на·ку·ю́т,** (*perf*) forge on

на·ко́ж·н·ый, -ая, -ое, (*a*) (anat) epidermic, epidermal, epicutaneous; cutaneous

на·кола́ч·ива·ть, -ют, (*imp*) knock on/in, bind (cooperage)

на·коле́н·ник·, -а, (*m*) (anat) patella, knee-cap; knee piece (protection of knee)

на·коло́нн·ый, -ая, -ое, (*a*) (mech) post, post-mounted

на·колот·и́ть, –ят, (*perf*) *see* **на·кола́ч·ива·ть**

на·ко́л·от·ый, -ая, -ое, (*past part pass*) of **на·кол·о́ть**

на·кол·о́ть, –ют, (*perf*) *see* **на·ка́л·ыва·ть**

на·коло́ч·енн·ый, -ая, -ое, (*past part pass*) of **колот·и́ть**

на·кон·е́ц· (*adv*) at last finally

на·коне́ч·ник·, -а, (*m*) tip, point, end; head, cap; nozzle, spout; (elec) terminal, clip; tag; ferrule; penetrator, indenter (e.g. of hardness tester); (instr) contact member

~ баллист·и́ческ·ий (mil) ballistic cap

~, бур·ов·о́й drilling bit

~, ка́бель·н·ый sweating thimble (elec cable)

~ по·жа́р·н·ого ствол·а́ fire hose nozzle

~, по́люс·н·ый (elec) pole piece/tip

~, раз·да́т·очн·ый delivery nozzle (hose, pump)

на·коп·а́·ть, -ют, (*perf*) dig, excavate (in several places); dig up/out (a quantity of)

на·коп·и́тел·ь, -я, (*m*) storage device/element; store, storage, accumulator (computers); (rad) tank

~ да́·нн·ых (autom) information storage/store

~, ём·кост·н·ый (rad) capacitor tank

~, маршру́т·н·ый (rail) shunting programmer

~, по·сто·я́нн·ый (autom) permanent storage/store

~, реле́й·н·ый relay-operated accumulator (computers)

~ су́мм·ы sum storage/store (computers)

~, цифр·ов·о́й digital storage/store (computers)

~ эне́рг·и·и (rad) tank, tank circuit

на·коп·и́тельн·ый, -ая, -ое, (*a*) storage, reservoir; cumulative

на·коп·и́ть, –ят, (*perf*) *see* **на·ко·пл·я́·ть**

на·копл·е́ни·е, -я, (*n*) (*v n*) *see* **на·копл·я́·ть**; build-up, accumulation, pile-up; storage, accumulation

~ из·луч·е́ни·я (*n*) (nucl) radiation build-up

коэффицие́нт· на·копл·е́ни·я (*m*) build-up factor

ме́тод· на·копл·е́ни·я (*m*) (nucl) backing-space method

то́ч·к·а на·копл·е́ни·я (*f*) (math) accumulation/cluster/limit point

на·ко́пл·енн·ый,-ая,-ое, (*past part pass*) of **на·коп·и́ть**; stored, built-up, pent-up, piled-up, accumulated, cumulative

на·копл·я́·ть, -ют, (*imp*) accumulate, amass, save/pile/build up

на·копт·и́ть, -ят, (*perf*) smoke/cure (a quantity of)

на·ко́пч·енн·ый, -ая, -ое, (*past part pass*) of на·копт·и́ть

на·корм·и́ть, ⸗ят, (*perf*) feed

на́·корот·к·о (*adv*) short

~, за́·мк·нут·ый (elec) short-circuited

на·кос·и́ть, ⸗ят, (*perf*) mow, scythe, cut

на·кра́п·ыва·ть, -ют, (*imp*) spot, spit (with rain), fall (of occasional drops)

на·кра́с·ить, ⸗ят, (*perf*) *see* на·кра́·ш·ива·ть

на·крахма́л·енн·ый, -ая, -ое, (*past part pass*) of на·крахма́л·ить; starched; starchy

на·крахма́л·ить, -ят, (*perf*) starch

на·кра́ш·ива́емост·ь, -и, (*f*) (text) dye receptivity, dyeability

~, вну́тр·енн·яя built-in dyeability

на·кра́ш·ива·ть, -ют, (*imp*) make up (with cosmetics); paint

на·крен·ённ·ый, -ая, -ое, (*past part pass*) *see* на·крен·я́·ть; lopsided

накрени́ть, -я́т, (*perf*) *see* на·крен·я́·ть

на·крен·я́·ть, -ют, (*imp*) incline to one side, give a list to; -ся list, heel over; (air) bank

на́·креп·к·о (*adv*) fast, very firmly/tightly

на́·крест· (*adv*) crosswise

~, угл·ы́ леж·а́щ·ие (*pl*) (math) alternate angles

накри́т·, -а, (*m*) (min) nacrite

на·крич·а́·ть, -ат, (*perf*) shout, shout at

на·кро́·ет (*fut 3rd sing*) of на·кры́·ть

на·кро·и́т (*fut 3rd sing*) of на·кро·и́ть

на·кро·и́ть, -я́т, (*perf*) cut out (a number of)

на·кро́·й, -я, (*m*) *see* в·на·кро́·й; lap, overlap

на·кромс·а́·ть, -ют, (*perf*) shred (a quantity of)

на·крош·и́ть, ⸗ат, (*perf*) crumble

на·кро́·ют (*fut 3rd pl*) of на·кры́·ть

на·кро·я́т (*fut 3rd pl*) of на·кро·и́ть

на·крут·и́ть, ⸗ят, (*perf*) *see* на·кру́·ч·ива·ть

на·кру́ч·ива·ть, -ют, (*imp*) wind, reel; twist, lay up (rope etc.)

на·крыв·а́·ть, -ют, (*imp*) cover; (gunn) straddle; catch in the act/red handed

на·кры́в·к·а, -и, (*f*) (*v n*) *see* на·крыв·а́·ть; (build) plastering, rendering

на·кры́т·и·е, -я, (*n*) (math) covering; (gunn) straddle

на·кры́·т·ый, -ая, -ое, (*past part pass*) of на·кры́·ть; covered; (gunn) straddled

на·кры́·ть, (*fut 3rd pl*) на·кро́·ют, (*perf*) *see* на·крыв·а́·ть

накто́уз·, -а, (*m*) (naut) binnacle

на·ку·ёт (*fut 3rd sing*) of на·ков·а́ть

на·куп·а́·ть, -ют, (*imp*) buy, buy up/in (a quantity of)

на·куп·и́ть, ⸗ят, (*perf*) *see* на·куп·а́·ть

на·кур·и́ть, ⸗ят, (*perf*) distil (tar etc.); make smoky

на·ку·ю́т (*fut 3rd pl*) of на·ков·а́ть

на·ла́вл·ива·ть, -ют, (*imp*) catch (a quantity of)

на·лаг·а́·ть, -ют, (*imp*) lay on, impose, inflict, make subject to

на·ла́д·ить, -ят, (*perf*) *see* на·ла́ж·и·ва·ть

на·ла́д·к·а, -и, (*f*) (*v n*) *see* на·ла́ж·ива·ть; setting up (a machine tool)

на·ла́д·чик·, -а, (*m*) (mech eng) setter-up

на·ла́ж·ива·ть, -ют, (*imp*) put right, adjust, set up, tune up

на·лак·ир·ова́ть, -у́ют, (*perf*) varnish (a number of)

на·ле́в·о (*adv*) to the left, (naut) to port

на·лег·а́ни·е, -я, (*n*) (*v n*) *see* на·лег·а́·ть; superposition, super-imposition, coincidence (of lines etc.); cover

на·лег·а́·ть, -ют, (*imp*) lean on/against, overlie; set to, apply oneself to; urge, make (do something)

на·легк·е́ (*adv*) light, lightly (without much luggage); lightly clad/dressed

на·лёг·ш·ий, -ая, -ее, (*past part act*) of на·ле́ч·ь

на́·лед·ь, -и, (*f*) water above ice; ice crust, icing

на·лез·а́·ть, -ют, (*imp*) climb/clamber on/onto; move up to/onto; fit (of clothes)

на·ле́з·ть, -ут, (*perf*) *see* на·лез·а́·ть

на·леп·и́ть, ⸗ят, (*perf*) *see* на·леп·л·я́·ть

на·лепл·я́·ть, -ют, (*imp*) stick on

на·лёт·, -а, (*m*) thin coating, film, bloom, tarnish, incrustation, deposit; raid, robbery, attack, onslaught, inroad; (air) flying hours

~, гля́нц·ев·ый (geol) desert varnish

~, жи́дк·ий condensation film, condensate, sweat

на·лёт

~, общ·ий   (air) total flying hours

~, радио·актив·н·ый   radioactive deposit

на·лет·а·ть, -ют,   (*imp*) fly to/upto/ towards/into; attack, raid, fall upon; blow up, begin to blow (of wind); blow onto/to/against; (*imp only*) fly (a certain time or distance), acquire flying hours

на·лет·а́ющ·ий, -ая, -ее, (*pres part act*) of **на·лет·а́·ть;** (nucl) bombarding, impinging, incident, projectile

на·лет·е́ть, -я́т,   (*perf*) *see* **на·лет·а́·ть**

на·леч·у́   (*fut 1st sing*) of **на·лет·е́ть**

на·ле́чь   (*fut 3rd sing, pl*) **на·ля́ж·ет, на·ля́г·ут,** (*past masc sing*) **на·лёг·,** (*perf*); *see* **на·лег·а́·ть**

на·ли́·в, -а, (*m*) (*v n*) *see* **на·лив·а́·ть**

на·лив·а́·ть, -ют,   (*imp*) fill (with liquid), pour out/in; **-ся** (*pass*) take on, receive (fuel etc.); ripen, become juicy

на·ли́в·к·а, -и, (*g pl*) **-в·ок·,** (*f*) (*v n*) *see* **на·лив·а́·ть;** nalivka (alcoholic cordial), fruit liqueur

на·лив·н·о́й, -а́я, -о́е, (*a*) filling, pouring-in; liquid, liquid-transporting; water, water-operated; juicy; (shipb) floating-out (of building berths etc.)

~ ба́рж·а (*f*) tank barge

~ груз· (*m*) liquid cargo

~ ме́ль·ниц·а (*f*) water mill

~ отве́рст·и·е (*n*) filling hole

~ су́д·н·о (*n*) tanker, tank vessel

налим·, -а, (*m*) (fish) burbot, eel-point, *Lota lota*

на·лин·о́ванн·ый, -ая, -ое, (*past part pass*) of **на·лин·ова́ть**

на·лин·ова́ть, -у́ют, (*perf*) rule lines

на·лип·а́ни·е, -я, (*n*) (*v n*) of **на·лип·а́·ть**

на·лип·а́·ть, -ют, (*imp*) stick/adhere to

на·ли́п·нуть, -нут, (*past masc sing*) **на·ли́п·,** (*perf*) *see* **на·лип·а́·ть**

на·ли·т·о́й, -а́я, -о́е, (*a*) plump, full, juicy (of fruit); plump, well-filled, solid

на·ли́·т·ый, -ая, -ое, (*pass part pass*) *see* **на·лив·а́·ть**

на·ли́·ть, (*fut 3rd pl*) **на·ль·ю́т,** (*perf*) *see* **наливáть**

на·лиц·о́ (*adv*) present, available, to hand

на·ли́чч·и·е, -я, (*n*) presence, existence; availability

на·ли́чч·ник·, -а, (*m*) guard plate, escutcheon (door furniture); door/ window frame; truck-side bearing; (ent) clypeus

на·ли́чч·ност·ь, -и, (*f*) availability, presence; available resources

~, де́неж·н·ая cash, cash assets

~, ка́сс·ов·ая cash balance

на·ли́чч·н·ый, -ая, -ое, (*a*) available, present, to/on hand; (*pl as noun*) cash, ready money

~ де́ньг·и (*pl*) cash, cash on hand, ready money

~ коли́честв·о (*n*) available quantity, inventory, stock

~ рас·чёт· (*m*) cash payment

~ со·ста́в· (*m*) available/effective personnel, strength

на·лоб·ник·, -а, (*m*) facepiece (optical instruments); head band/strap

на·лов·и́ть, -я́т, (*perf*) catch (a quantity of)

на·ловч·и́ться, -а́тся, (*perf*) get the hang of, get skilled at

на·ло́г·, -а, (*m*) tax

~, натура́ль·н·ый tax in kind

~, по·до·хо́д·н·ый income tax

~ с оборо́т·а turnover tax

на·ло́г·ов·ый, -ая, -ое, (*a*) tax, taxation

~ об·лож·е́ни·е (*n*) taxation

~ сертифика́т· (*m*) tax reserve certificate (Brit), tax anticipation certificate (U.S.), tax note (U.S.)

налого·плат·е́льщик·, -а, (*m*) taxpayer

на·лож·е́ни·е, -я, (*n*) (*v n*) *see* **на·кла́д·ыва·ть;** application (of a force, current etc.); superposition; overlap; imposition

~ аре́ст·а (law) seizure

~ протéктор·а retreading (a tire)

~ шв·ов (med) suture, stitching

на·ло́ж·енн·ый, -ая, -ое, (*past part pass*) of **на·лож·и́ть;** applied, imposed; superimposed; cumulative (of multiples)

~ сеч·éни·е (*n*) section revolved in place (eng drawings)

~ ток· (*m*) (elec) superimposed current

на·лож·и́м·ый, -ая, -ое, (*a*) developable (surface)

~ по·гре́ш·ност·и (*pl*) cumulative error

на·лож·и́ть, -ат, *(perf) see* **на·кла́-д·ыва·ть**

на·локо́т·ник·, -а, *(m)* elbow piece (miner's jacket)

на·лощ·и́ть, -а́т, *(perf)* polish, glaze

на·лу́щ·ива·ть, -ют, *(imp)* shell, husk, hull, scale, pod (a quantity of)

на·лущ·и́ть, -а́т, *(perf) see* **на·лу́щ·и-ва·ть**

на·ль·ю́т *(fut 3rd pl)* of **на·ли́·ть**

на·ля́ж·ет *(fut 3rd sing)* of **на·ле́ч·ь**

на·ля́г·ут *(fut 3rd pl)* of **на·ле́ч·ь**

нам *(pron, dat)* of **мы** us, to us

на·магни́т·ить, -ят, *(perf) see* **на·маг-ни́ч·ива·ть**

на·магни́ч·енност·ь, -и, *(f)* (phys) magnetization, intensity of magnet-ization

~, на·сыщ·ённ·ая saturation magnet-ization/magnetism

~, о·ста́т·очн·ая residual magnetism/ magnetization, remanence

~, само·про·из·во́ль·н·ая sponta-neous magnetization/magnetism

на·магни́ч·енн·ый, -ая, -ое, *(past part pass)* of **на·магни́т·ить;** magnet-ized

на·магни́ч·ивани·е, -я, *(n)* (phys) magnetization, magnetizing

крив·а́я на·магни́ч·ивани·я *(f)* magnetization curve, B-H curve

~, лёгк·ое easy magnetization

ли́н·и·я на·магни́ч·ивани·я в во́з·дух·е *(f)* (elec) air line

на·магни́ч·ива·ть, -ют, *(imp)* magnet-ize

на·магни́ч·иваю́щ·ий, -ая, -ее, *(pres part act)* of **на·магни́ч·ива·ть;** magnetizing

~ по́л·е *(n)* magnetizing field

~ си́л·а *(f)* magnetomotive force

на·ма́з·анн·ый, -ая, -ое, *(past part pass) see* **на·ма́з·ыва·ть**

на·ма́з·ать, *(fut 3rd pl)* **на·ма́ж·ут,** *(perf) see* **на·ма́з·ыва·ть**

на·ма́з·к·а, -и, *(g pl)* **-з·ок·,** *(f)* coat-ing, pasting

на·маз·н·о́й, -а́я, -о́е, *(a)* coating, pasting

~ маши́н·а *(f)* lime pasting machine (leather)

на·ма́з·ыва·ть, -ют, *(imp)* smear, spread, daub, coat, cover

на́·маз·ь, -я, *(m)* lime paste (leather)

на·марин·ова́ть, -у́ют, *(perf)* (food) marinate (a quantity of)

на·ма́сл·ить, -ят, *(perf)* (food) add oil/butter

на·ма́т·ыва·ть, -ют, *(imp)* wind, coil, reel

на·ма́ч·ивани·е, -я, *(n) see* **моч·е́ни·е**

на·ма́ч·ива·ть, -ют, *(imp)* moisten, damp, soak

намёк·, -а, *(m)* hint, allusion

на·мёл *(past masc sing)* of **на·мес·ти́**

на·ме́л·ют *(fut 3rd pl)* of **на·мол·о́ть**

на·ме́н·ива·ть, -ют, *(imp)* receive/get in exchange; get change (for a sum of money)

на·мен·я́·ть, -ют, *(perf) see* **на·ме́н·и-ва·ть**

на·мер·ева́·ться, -ются, *(imp)* intend

на·ме́р·ени·е, -я, *(n)* intention, pur-pose

на·ме́р·енн·ый, -ая, -ое, *(past part pass) see* **на·ме́р·ива·ть;** intentional, deliberate, premeditated

на·мерз·а́·ть, -ют, *(imp)* freeze over/up

на·мёрз·нуть, -нут, *(past masc sing)* **на·мёрз·,** *(perf) see* **на·мерз·а́·ть**

на·ме́р·ива·ть, -ют, *(imp)* measure off/out an amount/a quantity of

на·ме́р·ить, -ят, *(perf) see* **на·ме́р·и-ва·ть**

на́·мертв·о *(adv)* dead, tightly, rigidly

на·мес·и́ть, -ят, *(perf)* knead/puddle (a quantity of)

на·мес·ти́, *(fut 3rd pl)* **наметут,** *(past masc sing)* **намёл,** *(perf)* sweep up/ together; form, drift (by action of wind, water etc.)

на·мёт·, -а, *(m)* *(v n) see* **на·мет·а́·ть;** net, casting net (for catching fish, birds, butterflies etc.); coat (of plaster)

на·мет·а́·ть, -ют, *(imp)* sweep up/ together; form, drift (by action of wind, water etc.); *(perf)* baste (sewing); throw on/onto (a quantity of); (zool) throw (give birth), spawn (of fishes); sketch, characterize, outline

на·мет·ённ·ый, -ая, -ое, *(past part pass)* of **на·мес·ти́**

на·мёт·ить, -ят, *(perf) see* **на·меч·а́·ть**

на·мёт·к·а, -и, *(g pl)* **-т·ок·,** *(f)* draft-ing, outlining; draft, outline (plan); basting (sewing); hole outline (on a drop stamping); (fish) net; (naut) mast clamp; (hydrog) sounding pole

на·мёт·ш·ий, -ая, -ее, *(past part act)* of **на·мес·ти́**

на·мёт·ыва·ть, -ют, *(imp)* (text) baste

на·меч·а́·ть, -ют, *(imp)* mark, sign, indicate; draft, outline, plan; aim

на·ме́ч·енн·ый, -ая, -ое, (*past part pass*) of **на·ме́т·ить;** (*perf*) *see* **на·ме·ч·а́·ть** (*imp*); marked, marked out; planned, projected, outlined; (text) basted

на·меш·а́·ть, -ют, (*perf*) mix in, admix

на·ме́ш·енн·ый, -ая, -ое, (*past part pass*) of **на·мес·и́ть**

на·ме́ш·ива·ть, -ют, (*imp*) knead/puddle (a quantity of); mix in, admix

нами (*pron, instr*) of **мы,** by/with etc. us

на·мно́г·о (*adv*) much, considerably

~ **вы́ш·ее** considerably higher, well above

на·мо́·ет (*fut 3rd sing*) of **на·мы́·ть**

на·мок·а́·ть, -ют, (*imp*) get wet/soaked

на·мо́к·нуть, -нут, (*past masc sing*) **на·мо́к·,** (*perf*) *see* **на·мок·а́·ть**

на·мо́к·ш·ий, -ая, -ее, (*past part act*) of **на·мо́к·нуть;** soaked, wet

на·мола́ч·ива·ть, -ют, (*imp*) thresh (a quantity of)

на·моло́т·, -а, (*n*) (agr) threshing yield

на·молот·и́ть, -ят, (*perf*) *see* **на·мо·ла́ч·ива·ть**

на·мо́л·от·ый, -ая, -ое, (*past part pass*) of **на·мол·о́ть**

на·мол·о́ть, (*fut 3rd pl*) **на·ме́л·ют,** (*perf*) grind, mill (a quantity of)

на·моло́ч·енн·ый, -ая, -ое, (*past part pass*) of **на·молот·и́ть**

на·мо́рд·ник·, -а, (*m*) (agr) muzzle

на·мора́ж·ива·ть, -ют, (*imp*) freeze (a quantity of); form an ice crust/coating

на·мороз·и́ть, -ят, (*perf*) *see* **на·мо·ра́ж·ива·ть**

на·мо́т·анн·ый, -ая, -ое, (*past part pass*) of **на·мот·а́·ть;** wound, wound-on, coiled, reeled

на·мот·а́·ть, -ют, (*perf*) wind, coil, reel

на·мо́т·к·а, -и, (*g pl*) **-т·ок·,** (*f*) (*v n*) (*see* **на·ма́т·ыва·ть**); (text) winding-on; (met) coiling

на·мо́т·очн·ый, -ая, -ое, (*a*) (met) coiling; (text) winding, winding-on

на·мо́ч·енн·ый, -ая, -ое, (*past part pass*) of **на·моч·и́ть;** wet, wetted; soaked

на·моч·и́ть, -ат, (*perf*) *see* **на·ма́ч·ива·ть**

на·мо́·ют (*fut 3rd pl*) of **на·мы́·ть**

на·му́сор·ить, -ят, (*perf*) leave/drop litter

на·му́ш·ник·, -а, (*m*) foresight protector (of rifle)

на·мы́в·, -а, (*m*) (geol) aggradation, alluviation, silting up; (civ eng) flood flanking, building up (a bank etc.)

на·мыв·а́·ть, -ют, (*imp*) wash (a quantity of); (geol) alluviate, aggrade, wash in

на·мыв·н·о́й, -а́я, -о́е, (*a*) (geol) alluvial, washed-in/up; (civ eng) hydraulic fill, built-up, flood-flanked

~ **плот·и́н·а** (*f*) hydraulic-fill dam

на·мы́л·ива·ть, -ют, (*imp*) soap, make soapy, rub with soap

на·мы́л·ить, -ят, (*perf*) *see* **на·мы́л·ива·ть**

на·мы́·ть, (*fut 3rd pl*) **на·мо́·ют,** (*perf*) *see* **на·мыв·а́·ть**

на́нбук·, -а, (*m*) (text) nankeen

на·нес·е́ни·е, -я, (*n*) (*v n*) *see* **на·но·с·и́ть;** applying, application; plotting, drawing, marking in; heaping up, drifting, forming drifts

~ **кра́с·к·и** painting; (print) inking

~ **раз·ме́р·ов** dimensioning (eng drawing)

~, **электо·лит·и́ческ·ое** electrodepositing

на·нес·ти́, -у́т, (*past masc sing*) **на·нёс·,** (*perf*) *see* **на·нос·и́ть**

на·низ·а́ть, (*fut 3rd pl*) **на·ни́ж·ут,** (*perf*) *see* **на·ни́з·ыва·ть**

нан·и́зм·, -а, (*m*) (med) nanism, dwarfism

на·ни́з·ыва·ть, -ют, (*imp*) string, thread

на·ним·а́тел·ь, -а, (*m*) hirer, employer; tenant

на·ним·а́·ть, -ют, (*imp*) hire, engage, take on (labour); hire, rent; **-ся** (*pass*); take up employment, contract to work for

на́н·к·а, -и, (*g pl*) **-н·ок·,** (*f*) (text) nankeen

нано- (*int component*) (math) nano-, millimicro-, 10^{-9}; (biol) nano-, dwarf

нано·планкто́н·, -а, (*m*) (biol) nanoplankton

нано·ампе́р·, -а, (*m*) (elec) nanoampere, 10^{-9} amp

на·но́с·, -а, (*m*) (*v n*) *see* **на·нос·и́ть;** aggradation; (*pl*) detritus, detrital deposit, load

~, **вз·ве́ш·енн·ый** suspended load, fine silt

~, **до́н·н·ые** (*pl*) (hydr) bottom load

~, **реч·н·о́й** river drift/load

~, **снёж·н·ый** snowdrift

нано·секу́нд·а, -ы,   (*f*) nanosecond (10⁻⁹ sec)

на·нос·и́ть, ⸗ят,   (*perf*) bring (a quantity of); drift, be carried; pile/heap up; bump into, strike; plot, draw, mark on/in; apply, cover/coat with; inflict; pay (formal visit); lay (some eggs)

~ в за·ви́с·имост·и от ... (math) plot against ...

~ пред·по·лож·и́тельн·о make a tentative plot

на·но́с·н·ый, -ая, -ое,   (*a*) alluvial, aggradational; uncharacteristic, untypical

нано·фанерофи́т·, -а,   (*m*) (bot) nanophanerophyte

на·но́ш·ённ·ый, -ая, -ое,   (*past part pass*) of **на·нос·и́ть**

на́нсук·, -а,   (*m*) (text) nainsook

на·ня́·ть, (*fut 3rd pl*) **на·йм·у́т,** (*perf*) *see* **на·ним·а́·ть**

на·обещ·а́·ть, -ют,   (*perf*) promise (much)

на·оборо́т· (*adv*) on the contrary; back to front, the wrong/other way round; inversely; in contrast

   и на·оборо́т· and vice versa

на·от·ре́з· (*adv*) flatly, point-blank

на·от·ко́с (*adv*) obliquely, aslant

на·о·щу́п· (*adv*) to the touch

на·па́·вш·ий, -ая, -ее, (*past part act*) of **на·па́с·ть**

на·па́д·, -а, (*m*) *see* **на·пад·е́ни·е;** infestation, onset (of insects)

на·па́д·а·ть, -ют, (*perf*) fall (in large amounts); **на·пад·а́·ть, -ют,** (*imp*) attack, fall upon; come upon, meet by chance

на·пад·е́ни·е, -я, (*n*) attack, assault, onset; forwards (football)

на·па́·ива·ть, -ют, (*imp*) solder, braze, solder on, braze on; give/provide drink

на·па́й·к·а, -и, (*g pl*) **-á·ек·,** (*f*) brazing/soldering on; soldered-on surface, soldered piece; sealing on (of glassware)

на·па́л (*past masc sing*) of **на·па́с·ть**

напа́лм·, -а, (*m*) napalm, jellied petrol

на·па́л·ок·, -л·к·а, (*m*) fingerstall, fingerguard

на·па́р·ник·, -а, (*m*) one of a pair; opposite number, paired worker

на·па́с·ть, (*fut 3rd pl*) **на·пад·у́т,** (*past masc sing*) **на·па́л,** (*perf*) *see* **на·пад·а́·ть,** (*imp*) **на·па́ст·ь, -и,** (*f*) misfortune, calamity

на·па́чк·а·ть, -ют, (*perf*) soil, dirty

на·па́·янн·ый, -ая, -ое, (*past part pass*) of **на·па·я́·ть;** soldered (on), brazed (on)

на·па·я́·ть, -ют, (*perf*) solder, solder on, braze, braze on

на·пе́в·, -а, (*m*) tune, melody

на·пев·а́·ть, -ют, (*imp*) sing, hum; record (singing)

на·пек·а́·ть, -ют, (*imp*) bake (a quantity of); get sunburnt

на·пёр· (*past masc sing*) of **на·пер·е́ть**

на·пере·бо́·й (*adv*) vying, in competition (with one another), interrupting (each other)

на·пере·ве́с· (*adv*) in a slanting position, aslant; at the ready (of firearms)

на·пере·го́н·к·и (*adv*) racing each other, trying to get ahead, in a race

на·перёд· (*adv*) in advance, beforehand

на·пере·ко́р· (*adv*) in defiance/disregard of

на·пере·ре́з· (*adv*) across, at a tangent

на·пере·ры́в· (*adv*) interrupting (one another), breaking in (on one another)

на·пер·е́ть, (*fut 3rd sing, pl*) **на·пр·ёт, на·пр·у́т,** (*past masc sing*) **на·пер·,** (*perf*); *see* **на·пир·а́·ть**

на·пере·чёт· (*adv*) without exception

на·пёр·ла (*past fem sing*) of **на·пер·е́ть**

на·пёрст·ок·, -т·к·а, (*m*) thimble

на·пёрст·ков·ый, -ая, -ое, (*a*) of **на·пёрст·ок·**

~ ка́мер·а (*f*) (nucl) thimble chamber

на·перст·я́нк·а, -и, (*g pl*) **-нок·,** (*f*) (bot) foxglove, *Digitalis purpurea*

на·пе́·ть, (*fut 3rd pl*) **на·по·ю́т,** (*perf*) *see* **на·пев·а́·ть**

на·печа́т·анн·ый, -ая, -ое, (*past part pass*) of **на·печа́т·а·ть**

на·печа́т·а·ть, -ют, (*perf*) print, type; be/have printed/published

на·пе́ч·ь, (*fut 3rd sing, pl*) **на·печ·ёт, на·пек·у́т,** (*past masc sing*) **на·пёк·** (*perf*); *see* **на·пек·а́·ть**

на·пи́л·енн·ый, -ая, -ое, (*past part pass*) of **на·пил·и́ть;** filed; sawn, sawn up

на·пи́л·ива·ть, -ют, (*imp*) saw, file (with several cuts), saw up

на·пил·и́ть, ⸗ят, (*perf*) *see* **на·пи́л·и·ва·ть**

на·пи́л·ок·, -л·к·а, (*m*) *see* **на·пи́ль·ник·**

на·пи́ль·ник·, -а, (*m*) file (tool)

~, ба́рхат·н·ый dead-smooth file

~, драчёв·ый bastard file

на·пи́ль·ник
~, лич·н·о́й   second-cut file
на·пир·а́·ть, -ют,   (*imp*) press, put pressure on, push, depress
на·пис·а́ть,   (*fut 3rd pl*) **на·пи́ш·ут,** (*perf*) write; paint (pictures)
на·пи́т·анн·ый, -ая, -ое,   (*past part pass*) of **на·пит·а́·ть;** impregnated, soaked
на·пит·а́·ть, -ют,   · (*perf*) saturate, impregnate; feed
на·пи́т·ок·, -т·к·а,   (*m*) drink, beverage
на·пи́т·ыва·ть, -ют,   (*imp*) saturate, impregnate
на·пих·а́·ть, -ют,   (*perf*) *see* **на·пи́х·и·ва·ть**
на·пи́х·ива·ть, -ют,   (*imp*) cram, stuff
на·пи́ш·ут   (*fut 3rd pl*) of **на·пис·а́ть**
на·пла́в·, -а,   (*m*) (fish) float
на·пла́в·а·ть, -ют,   (*perf*) swim, drift, float, sail (a certain distance)
на·пла́в·ить, -ят,   (*perf*) *see* **на·плавл·я́·ть**
на·пла́в·к·а, -и,   (*g pl*) **-в·ок·,** (*f*) (*v n*) *see* **на·плавл·я́·ть;** (weld) facing, build-up, hard facing (tools etc.), soft facing (bearing surfaces)
   пло́щ·ад·ь на·пла́в·к·и   (*f*) (weld) deposited metal zone
   про·из·вод·и́тельност·ь на·пла́в·к·и   (*f*) (weld) deposition rate
на·пла́вл·енн·ый, -ая, -ое,   (*past part pass*) of **на·пла́в·ить;** (weld) built-up, faced (parts); deposited, weld (material)
~ вал·о́к·   (*m*) built-up roll
~ мета́лл·   (*m*) (weld) deposited metal, weld metal
на·плавл·я́·ть, -ют,   (*imp*) melt (a quantity of); (weld) face, build up (a worn surface), hard face (tools), soft face (bearing surfaces)
на·плав·н·о́й, -а́я, -о́е,   (*a*) (civ eng) floating
~ мост·   (*m*) floating bridge, pontoon bridge
на·пласт·ова́ни·е, -я,   (*n*) (geol) bedding, stratification; superposition
на·пласт·о́ванн·ый, -ая, -ое,   (*past part pass*) of **на·пласт·о́выва·ть;** bedded, stratified
на·пласт·о́выва·ть, -ют,   (*imp*) bed, stratify, make into layers; **-ся** (*pass*); stratify, superimpose
на·плеск·а́·ть,   (*fut 3rd pl*) **на·плёщ·ут,** (*perf*) *see* **на·плёск·ива·ть**

на·плёск·ива·ть, -ют,   (*imp*) sprinkle, splash
на·плес·ти́,   (*fut 3rd pl*) **на·плет·у́т,** (*past masc sing*) **на·плё·л,** (*perf*); *see* **на·плет·а́·ть**
на·плет·а́·ть, -ют,   (*imp*) weave, plait (a quantity of)
на·плеч·ник·, -а,   (*m*) epaulette; shoulder guard
на·плеч·н·ый, -ая, -ое,   (*a*) shoulder (of clothing etc.)
на·плёщ·ут   (*fut 3rd pl*) of **на·плеск·а́ть**
на·плы́в·, -а,   (*m*) (*v n*) *see* **на·плы́в·а́·ть;** influx; (met) splash (on casting), collar (on weld); (plast) spew, flash; (bot) burr, excrescence, curly grain (timber); (cinema) fade-over, mix, dissolve
на·плыв·а́·ть, -ют,   (*imp*) swim, drift, float, sail upto/onto/up against, strike (of floating objects)
на·плыв·н·о́й, -а́я, -о́е,   (*a*) (geol) alluvial, washed in/up; (civ eng) hydraulic fill, flood-flanked
на·плы́·ть,   (*fut 3rd pl*) **на·плыв·у́т,** (*perf*) *see* **на·плыв·а́·ть**
на·плю́щ·ива·ть, -ют,   (*imp*) flatten (a quantity of)
на·плю́щ·ить, -ат,   (*perf*) *see* **на·плю́щ·ива·ть**
на·по·до́б·и·е   (*prep + gen*) like, not unlike, in a similar/like manner
на·по·ёт   (*fut 3rd sing*) of **на·пе́·ть**
на·по·и́ть, -ят,   (*perf*) give/provide drink
на·по·ка́з·   (*adv*) for show
на·полз·а́·ть, -ют,   (*imp*) crawl upto/up against/into; creep over, obscure from view
на·полз·ти́, -у́т,   (*perf*) *see* **на·полз·а́·ть**
на·полн·е́ни·е, -я,   (*n*) filling; (ICE) admission; loading (paper, rubber), filler content (rubber)
~ в ро́лл·е   (paper) beater loading
   коэффицие́нт· на·полн·е́ни·я   (*m*) (ICE) coefficient of admission, co-efficient of change, air/fuel ratio
~, по·втор·н·ое   refilling
на·по́лн·енн·ый, -ая, -ое,   (*past part pass*) of **на·по́лн·ить;** filled; inflated
~ вод·о́й   (geol) waterlogged
~ ла́текс·   (*m*) latex mix (rubber)
~ рези́н·а   (*f*) loaded rubber
на·полн·и́тель·, -я,   (*m*) filler; filling machine (canning factories); expander (of accumulator plates); builder (soap)

на·полн·и́тел·ь
~, акти́в·н·ый   active/reinforcing filler (rubber)
~, мало·акти́в·н·ый   semi-active filler
~, не·акти́в·н·ый   inert filler
на·по́лн·ить, -ят,   (perf) fill; inflate
на·полн·я́·ть, -ют,   (imp) fill; inflate
на·пол·ови́н·у   (adv) half
на·по́ль·н·ый, -ая, -ое,   (a) standing, standard, on the ground, ground-type
~ ла́мп·а   (f) standard lamp
на·по·мин·а́ни·е, -я,   (n) reminder; reminiscence
на·по·мин·а́·ть, -ют,   (imp) remind; resemble, be reminiscent of
на·по·мин·а́ющ·ий, -ая, -ее,   (pres part act) of на·по·мин·а́·ть; resembling, reminiscent of, like
на·по́·мн·ить, -ят,   (perf) of на·по·мин·а́·ть
на·по́р·, -а,   (m) (v n) see на·пир·а́·ть; pressure, thrust, (hydr) pressure head, head
~ бру́тто   total head
~, гео·метр·и́ческ·ий   total static head
~, гидравл·и́ческ·ий   hydraulic head
~, динам·и́ческ·ий   dynamic/kinetic head
~, ма́л·ый   low head
~, поле́з·н·ый   effective head
~, по́лн·ый   total head (pump)
~, по́лн·ый рабо́ч·ий   total head
~, рабо́ч·ий   working head
~, рас·чёт·н·ый   design head
~, скор·остн·о́й   (dynam) dynamic pressure/head; (air) form drag; (rocket) ram effect/compression
~, со·зд·а́вать   generate a head (of pressure)
~, стат·и́ческ·ий   static head/pressure
~, температу́р·н·ый   temperature drive/head
~, тепл·ов·о́й   thermal head
на·по́р·н·ый, -ая, -ое,   (a) of на·по́р·; pressure, pressurized; discharge, delivery (pumping)
~ реа́ктор·   (m) (nucl) pressurized reactor
~ котёл·   (m) (hydr) pressure tank, air-loaded accumulator
~ па́·труб·ок·   (m) discharge nozzle (pump)
~ рук·а́в·   (m) pressure hose
~ цисте́рн·а   (f) gravity tank

на·по́р·н·ый, -ая, -ое
~ труб·а́   (f) force pipe; (hydroelec) penstock
~ тру́б·к·а   (f) velocity head meter
~ трубо·про·во́д·   (m) penstock
на·по́рт·ить, -ят,   (perf) spoil
на·по́рч·енн·ый, -ая, -ое,   (past part pass) of на·по́рт·ить
на·по·ю́т   (fut 3rd pl) of на·пе́·ть
на·по·я́т   (fut 3rd pl) of на·по·и́ть
на·прав·и́тел·ь, -я,   (m) (mech) guide; diffuser (centrifugal pumps etc.); regulating ring, guide vanes (turbines)
на·пра́в·ить, -ят,   (perf) see на·прав·л·я́·ть
на·правл·е́ни·е, -я,   (n) (v n) see на·правл·я́·ть; direction; course, route, (ocean) set (of current); trend, tendency, line, school (of thought); (math) sense (of a vector); specialization; certificate of qualification
~ дифференциа́ци·и   differentiation trend
~, кристалло·граф·и́ческ·ое   crystallographic line
~, основ·н·о́е   zero line
~ о́с·и x x-direction
~ пад·е́ни·я   incident direction (of radiations)
~ по ко́мпас·у   (navig) compass heading
~ по·са́д·к·и   see ста́нц·и·я на·правл·е́ни·я по·са́д·к·и below
~, про·ве́ш·енн·ое   (surv) orienting line
~ рас·про·стран·е́ни·я   line of propagation
~, само·со·пряж·ённ·ое   (math) asymptote
~ с·вя́з·и   (telecom) routing; (mil) line of communications
~ спи́н·а   (phys) spin orientation/alignment
ста́нц·и·я на·правл·е́ни·я по·са́д·к·и   (f) precision approach radar
~ штрих·о́вк·и   (geol) orientation of stria
на·пра́вл·енност·ь -и,   (f) directivity, directional effect
диагра́мм·а на·пра́вл·енност·и   (f) (rad) polar diagram
коэфици́ент· на·пра́вл·енност·и   (m) (rad) directivity factor/efficiency
характер·и́стик·а на·пра́вл·енност·и   (f) (rad) directional/beam pattern, directional response pattern

на·пра́вл·енн·ый, -ая, -ое, (*past part pass*) of **на·пра́в·ить**

~ **антéнн·а** (*f*) directional aerial/antenna

~ **внутрь** inward, inwards

~ **дéй·стви·е** (*n*) (rad) directional action/effect; (*in gen as adj*) directional, directive

коэффициéнт· на·пра́вл·енн·ого дéй·стви·я (*m*) (rad) antenna gain

~ **с·вя́з·ь** (*f*) beam communications; (chem, phys) directional bond

на·правл·я́ем·ый, -ая, -ое, (*pres part pass*) of **на·правл·я́·ть**; (GW) guided, (pop) homing

на·правл·я́·ть, -ют, (*imp*) guide, direct, aim; send; **-ся** head for, make for, be bound for

на·правл·я́ющ·ая, -ей, (*f decl as adj*) guide, guide bar; (rocket) projector (launcher); (math) directrix; leader (piling frame)

~ **по·пла́в·к·а́** (*f*) float guide bar

~, **сот·ов·а́я** (rocket) open-frame projector

~, **цилиндр·и́ческ·ая** (rocket) barrel-type projector

на·правл·я́ющ·ий, -ая, -ее, (*pres part act*) of **на·правл·я́·ть**; guide; (*see also* **на·правл·я́ющ·ая**) (*as noun*)

~ **аппара́т·** (*m*) diffuser (centrifugal pump); regulating ring (water turbine), guide vanes

~ **вен·éц·** (*m*) stator blade (gas turbine)

~ **вту́л·к·а** (*f*) valve guide

~ **ко́синус·** (*m*) (math) direction cosine

~ **ли́н·и·я** (*f*) (math) directrix

~ **про́б·к·а** (*f*) positioning stop

~ **ра́дио-мая́к·** (*m*) landing-approach radio beacon

~ **с·твор·** (*m*) (navig) clearing line

~ **у·стро́й·ств·о** (*n*) guide; coke guide (coke oven)

~ **част·ь по́рш·н·я** (*f*) (ICE) piston skirt

~ **шарни́р·н·ый механи́зм·** (*m*) parallel-motion mechanism, straight-line mechanism

на·пра́в·о (*adv*) to the right, to the right of, (naut) to starboard, to starboard of; right turn! (as command)

бра·ть на·пра́в·о turn right, (naut) turn to starboard

напра́сн·о (*adv*) in vain, for nothing

напра́сн·ый, -ая, -ое, (*a*) useless, vain; unfounded, unnecessary

на·пра́ш·ива·ться, -ются, (*imp*) solicit; suggest itself (of an idea etc.)

на·пр·ёт (*fut 3rd sing*) of **на·пер·éть**

на·при·мéр· (*conj*) for example/instance, e.g.

на·про·ка́т· (*adv*) for hire, on hire

на·про·лёт· (*adv*) without intermission, all/right through

на·прос·и́ться, -я́тся, (*perf*) solicit

на·про́тив· (*adv and prep*) opposite; (*conj*) on the contrary; on the other hand, unlike (another event), in contrast to

на·пр·у́т (*fut 3rd pl*) of **на·пер·éть**

на·пряг·а́·ть, -ют, (*imp*) stretch, strain; span

на·пряж·éни·е, -я, (*n*) (*v n*) see **на·пряг·а́·ть**; effort; (mech) stress; (elec) voltage, tension

~ **ви́хр·я** (phys) vortex strength

~, **внéш·н·ее** (mech) external/applied stress; (elec) external voltage

~, **вну́тр·енн·ее** (mech) residual/internal stress; (elec) internal voltage

вну́тр·енн·ее на·пряж·éни·е в от·ли́в·к·е (met) casting stress

~, **втор·и́чн·ое** (mech) secondary/subsidiary stress; (elec) secondary winding voltage (transformers)

~, **в·ход·н·о́е** input voltage

~, **выс·о́к·ое** high voltage/tension

~, **дéй·ствующ·ее** (mech) active/actual stress; (elec) effective voltage

~, **деформи́р·ующ·ее** flow stress

~, **дл·и́тельн·ое** steady stress

~, **до·ба́в·очн·ое** (mech) additional stress/load; (elec) boosting voltage

~, **до·пуск·а́ем·ое** (met) allowable unit stress, (civ eng) permissible stress, safe working stress

~ **за·жиг·а́ни·я** starter breakdown voltage, striker voltage

~, **за·пир·а́ющ·ее** cutoff voltage

~, **за·ря́д·н·ое** flashover voltage

~, **знако·пере·мéнн·ое** alternating/reversing stress

~ **из·ги́б·а** bending stress

~, **и́мпульс·н·ое** (mech) impulsive stress/load; (elec) surge/shock voltage

~ **искр·éни·я** arcing voltage

~, **и́ст·инн·ое** (mech) true stress; (elec) true voltage

~, **кас·а́тельн·ое** (mech) shearing stress; viscous stress (fluid mechanics)

на·пряж·éни·е
~, квадратýр·н·ое   quadrature-axis voltage
~, контáкт·л·ое   (mech) bearing stress; (elec) contact potential
крив·áя на·пряж·éни·й   (f) (phys) line of intensities
~ круч·éни·я   shear/shearing stress
~, максимáль·н·ое обрáт·н·ое   peak inverse voltage
~, между·фáз·н·ое   (mech) interfacial stress, surface tension; (elec) voltage between phases
~, мокро·раз·ря́д·н·ое   wet flashover voltage
~ на вáнн·е   (electrochem) bath/tank voltage
~ на за·жи́м·ах   terminal voltage
~ на об·хóд·е   loop voltage
~ на по·вéрх·ност·и   (phys) surface tension
~ на стéн·к·е   wall stress
~, наи·бóльш·ее обрáт·н·ое   peak inverse voltage
~ на·кáл·а   filament voltage
~, на·чáль·н·ое   (mech) initial stress/load; (elec) initial voltage
~, не·пре·ры́в·н·о мен·я́ющ·ееся   analogue voltage
~, ни́з·к·ое   low voltage/tension
~, номинáль·н·ое   (mech) nominal stress; (elec) rated voltage
~, нормáль·н·ое   normal stress, direct stress, stress in tension or compression
~, обрáт·н·ое рабóч·ее   inverse voltage
~, объ·ём·н·ое   (mech) volumetric/cubical/hydrostatic stress; heating capacity (of furnace, in kcal/cub m/hr)
~, одно·рóд·н·ое   uniform stress
~, о·пóр·н·ое   reference voltage; Zener voltage (in a semiconductor)
~, о·стáт·очн·ое   (mech) residual/locked-up stress; (elec) residual voltage
~ от круч·éни·я   shear stress
~ отсéч·к·и   cutoff voltage
~, пере·мéн·н·ое   (mech) variable/alternating stress; (elec) alternating voltage
~, пи́к·ов·ое   peak/crest voltage
~, пило·обрáз·н·ое   saw-tooth voltage
~ по·вéрх·ност·и на·грé·в·а   rate of evaporation (of a boiler)
~, по·вéрх·ностн·ое   (mech) surface tension

на·пряж·éни·е
под на·пряж·éни·ем   (adv) (met) stressed, under stress, in the stressed state; (elec) alive, energized, live
~, по·сто·я́нн·ое   (mech) constant stress/load; (elec) direct voltage
~, пред·вар·и́тельн·ое   (civ eng) pre-stressing, prestress (concrete etc.)
~, пре·дéль·н·ое   breaking load, maximum stress
~ при с·дви́г·е   shear/shearing stress
~, при·вед·ённ·ое анóд·н·ое   composite controlling voltage
~, при·вед·ённ·ое критíч·еск·ое с·кáл·ывающ·ее   resolved critical shear stress
~, про·би́в·н·óе   breakdown voltage
~ про·бó·я   spark-over voltage
~, рабóч·ее   (mech) working stress, (elec) operating/running voltage, closed-circuit voltage (of battery or element)
~ раз·вёрт·к·и   sweep voltage (CRT)
~ раз·лож·éни·я   (electrochem) decomposition potential
~, раз·руш·áющ·ее   breaking stress/load
~, раз·ры́в·н·óе   tensile stress; (geol) tearing stress
~, рас·тя́г·ивáющ·ее   tensile stress
~, регул·ирóвочн·ое   calibration voltage
ряд· на·пряж·éни·й   (m) (chem) electromotive series
~, сéт·очн·ое   grid voltage
~, с·жим·áющ·ее   compressive stress
~, с·кáл·ывающ·ее   shearing stress
~ с·кольж·éни·я   sliding stress
~ с·мещ·éни·я   (mech) dislocation/shifting stress; (elec) bias voltage
~, срéд·н·ее   mean stress
~ с·рéз·а   shearing stress
~, структýр·н·ое   (met) transformation stress
~, тангенциáль·н·ое   shearing stress
~ тек·ýчест·и   yield stress
~, тепл·ов·óе   (met) thermal stress; heat release rate (of furnace, in kcal/cub m/hr)
~, терм·и́ческ·ое   thermal stress
~ теч·éни·я   flow stress
~ тóк·а   voltage
~, удáр·н·ое   shock/impact stress
~, у·правл·я́ющ·ее   control-circuit voltage
~, у·прýг·ое   elastic stress
~, у·сáд·очн·ое   shrinkage stress

на·пряж·éни·е

~, у·слóв·н·ое   unit stress

~ форм·óвк·и   (electrochem) formation voltage

~ холост·óго хóд·а   (elec) no-load voltage

~, факт· и́ческ·ое   (civ eng) working stress

~, цикл·и́ческ·ое   cyclic/alternating stress

~, эффекти́в·н·ое   true stress

на·пряж·ённост·ь, -и,   (*f*) intensity, tenseness, strength; tension

~ ви́хр·я   vorticity

~ за·вихр·éни·я   vorticity

ли́н·и·я на·пряж·ённост·и   (*f*) magnetic line of force

~ магни́т·н·ого пол·я   (phys) magnetic field intensity/strength, magnetizing force, magnetic intensity

~ огн·я́   (mil) rate of fire

~ пóл·я   (phys) field strength/intensity

~, про·би́в·н·áя   breakdown strength (of a dielectric)

~ сре́д·н·его пóл·я   electric field intensity

~ электр·и́ческ·ого пóл·я   electric field intensity

на·пряж·ённ·ый, -ая, -ое,   (*past part pass*) of **на·пря́ч·ь;** (mech) stressed

~ со·сто·я́ни·е   (*n*) stressed state, state of stress

на·пря́ч·ь (*fut 3rd sing, pl*) **напряжёт, напрягу́т,** (*past masc sing*) **напря́г,** (*perf*) stretch, tense; strain

на·пуг·á·ть, -ют, (*perf*) frighten, scare

на·пу́льс·ник·, -а, (*m*) wristlet, wristband

нá·пуск·, -а, (*m*) (*v n*) *see* **на·пуск·á·ть;** (agr) flooding, flood irrigation; (met) surplus/extra metal (to allow for subsequent machining of a forging or casting)

~ по полос·áм   border-strip irrigation

на·пуск·á·ть, -ют, (*imp*) admit, let/put in; let down/onto; affect, assume an air of

на·пуст·и́ть, -я́т, (*perf*) *see* **на·пуск·á·ть**

на·пу́т·а·ть, -ют, (*perf*) tangle, muddle, confuse

на·пу́т·ыва·ть, -ют, (*imp*) tangle, muddle, confuse

на·пух·á·ть, -ют, (*imp*) swell

на·пу́х·нуть, -нут, (*past masc sing*) **на·пу́х·,** (*perf*) *see* **на·пух·á·ть**

на·пу́щ·енн·ый, -ая, -ое, (*past part pass*) of **на·пуст·и́ть**

на·пыл·éни·е, -я, (*n*) (*v n*) *see* **на··пыл·и́ть;** (met) metal spray coating, vaporized metal coating

на·пыл·ённ·ый, -ая, -ое, (*past part pass*) of **на·пыл·и́ть;** (met) spray-coated, vaporized on; (elec) sputtered

на·пыл·и́ть, -я́т, (*perf*) raise the dust; (met) spray-coat; (elec) sputter on

напы́щенн·ый, -ая, -ое, (*a*) pompous

на·рабáт·ыва·ть, -ют, (*imp*) make (a quantity of); earn, make (money)

на·рабóт·а·ть, -ют, (*perf*) *see* **на·ра·бáт·ыва·ть**

на·равн·é (*adv*) on a level with, flush with; equally, on a par with

нарáльник·, -а, (*m*) furrow opener (of seed drill)

на·раст·áни·е, -я, (*n*) (*v n*) *see* **на·ра·ст·á·ть;** growth, increase, increment, rise, accretion, build-up

врéмя на·раст·áни·я (*f*) build-up time/period; rise time (of pulses)

коэффициéнт· на·раст·áни·я (*m*) growth factor

скóр·ост·ь на·раст·áни·я (*f*) rate of rise, build-up rate

~ тóк·а   (elec) current rise

на·раст·á·ть, -ют, (*imp*) grow, grow on the surface of, grow out of; increase, build up, accumulate

на·раст·и́, -у́т, (*past masc sing*) **на·рóс·,** (*perf*) *see* **на·раст·á·ть**

на·раст·и́ть, -я́т, (*perf*) *see* **на·рáщ·ива·ть**

на·ращ·éни·е, -я, (*n*) (*v n*) *see* **на·рá·щ·ива·ть;** increment, accretion

~, от·риц·áтельн·ое   (math) negative increment

на·ращ·ённ·ый, -ая, -ое, (*past part pass*) *see* **на·рáщ·ива·ть**

на·рáщ·ива·ть, -ют, (*imp*) (*trans*) grow, cultivate; add on, extend; (*imp only*) increase

на·рв·á·ть, -у́т, (*perf*) pick, pluck, gather; tear up (a quantity of); (min) blast out

нáрд·, -а, (*m*) (bot) citronella, *Cymbopogon nardus*

на·рéз·, -а, (*m*) groove, incision, cut, notch

на·рез·áни·е, -я, (*n*) (*v n*) *see* **на·ре·з·á·ть**

~ зуб·чáт·ых колёс·   gear cutting

на·рез·а́ни·е
~ **ре́·ек·** rack cutting
~ **рез·ьб·ы́** thread cutting
~ **рез·ьб·ы́ греб·ёнк·ой** screw thread·
chasing, thread chasing
на·рез·а́·ть, -ют, (*imp*) **на·ре́з·ать,**
(*fut 3rd pl*) **на·ре́ж·ут,** (*perf*); cut,
cut up/out, slice; make incisions
на·ре́з·к·а, -и, (*g pl*) **-зок,** (*f*) (*v n*)
see **на·рез·а́·ть;** screw thread; rifling (gun barrel); embossing
~, **винт·ов·а́я** screw thread
на·рез·н·о́й, -а́я, -о́е, (*a*) notched;
threaded, rifled, chased
~ **се́ктор·** (*m*) segment of screw thread
~ **част·ь** (*f*) (gunn) rifling
на·ре́з·ыва·ть, -ют, (*imp*) *see* **на·ре-
з·а́·ть**
на·ре́ч·и·е, -я, (*n*) (gram) adverb;
(ling) related dialects
нарза́н·, -а, (*m*) Narzan mineral water
на·рис·о́ванн·ый, -ая, -ое, (*past part
pass*) of **на·рис·ова́ть**
на·рис·ова́ть, -у́ют, (*perf*) draw,
depict, sketch
на·риц·а́тельн·ый, -ая, -ое, (*a*)
nominal
~ **сто́·имост·ь** (*f*) (fin) nominal/par/
face value
~ **цен·а́** (*f*) (fin) nominal/par/face
value
нарк·о́з·, -а, (*m*) narcosis; anaesthetics
нарко·ла́н·, -а, (*m*) (pharm) narkolan,
avertin, tribromethyl alcohol
нарко·ле́пси·я, -и, (*m*) (med) narco-
lepsy
нар·ко́м·, -а, (*m*) (obs) Peoples' Commissariat, (obs) Peoples' Commissar
нарко·ма́н·, -а, (*m*) drug addict
наркоти́з·овать, -ую т, (*imp and perf*)
(med) narcotize
нарко́тик·, -а, (*m*) (pharm) narcotic;
drug addict
наркоти́н·, -а, (*m*) (chem) narcotine
наркот·и́ческ·ий, -ая, -ое, (*a*) narcotic
~ **сре́д·ств·а** (*pl*) narcotics
на·ро́д·, -а, (*m*) people, nation
на·род·и́ть, -я́т, (*perf*) bear, give birth
to
на·ро́д·ност·ь, -и, (*f*) nationality,
national character
народно·хозя́й·ственн·ый, -ая, -ое,
(*a*) national-economic
на·ро́д·н·ый, -ая, -ое, (*a*) people's,
popular, folk; national
~ **фронт·** (*m*) (polit) popular front
~ **хозя́й·ств·о** (*n*) national economy

народо·ве́д·ени·е, -я, (*n*) (obs) ethnology
народо·на·сел·е́ни·е, -я, (*n*) national
population, population
на·ро́·ет (*fut 3rd sing*) of **на·ры́·ть**
на·рожд·а́·ть, -ют, (*perf*) bear, give
birth to
наро́жник·, -а, (*m*) (build) jack timber
на·ро́с· (*past masc sing*) of **на·раст·и́**
на·ро́ст·, -а, (*m*) (bot) growth, excrescence, nodule, node, wart; build-up
(of metal in machining)
на·ро́ст·ок·, -т·к·а, (*m*) (*dim*) of **на·-
ро́ст·;** (bot) ipiblast
на·ро́с·ш·ий, -ая, -ее, (*past part act*)
of **на·раст·и́**
наро́чн·о (*adv*) on purpose, purposely
на́рочн·ый, -ая, -ое, (*a*) intentional;
(*m decl as adj*) special messenger, telegram boy
на·ро́·ют (*fut 3rd pl*) of **на·ры́·ть**
на·руб·а́·ть, -ют, (*imp*) chop, chop
down, fell (a quantity of); chop out;
(min) hew, cut
на·руб·и́ть, -я́т, (*perf*) *see* **на·руб·а́·ть**
на·ру́б·к·а, -и, (*g pl*) **-б·ок·,** (*f*) (*v n*)
see **на·руб·а́·ть;** notch
нару́ж·н·ое, -ого, (*n decl as adj*) medicine for external application
нару́ж·ност·ь, -и, (*f*) exterior, outside; appearance
нару́ж·н·ый, -ая, -ое, (*a*) (*see also*
на·ру́ж·н·ое) (*as noun*); external,
outside, outer; (mech eng) female (of
mating parts); superficial
~ **с·хо́д·ств·о** (*n*) superficial likeness
нару́ж·у (*adv*) out, outwards, outside
на·рук·а́вник·, -а, (*m*) oversleeve,
sleeve protector
на·ру́ч·ник·и, -ов, (*m pl*) handcuffs
на·ру́ч·н·ый, -ая, -ое, (*a*) wrist
~ **час·ы́** (*pl*) wrist watch(es)
на·руш·а́·ть, -ют, (*imp*) disturb,
violate, infringe, break
не на·руш·а́·я о́бщ·ност·и without
loss of generality
на·руш·е́ни·е, -я, (*n*) (*v n*) *see* **на·ру-
ш·а́·ть;** infringement, breach, transgression; (phys) abnormality, disturbance, failure, damage; (cryst) imperfection
~ **ва́куум·а** (phys) deterioration in
vacuum
~ **де́·ятельност·и** (med) disorder
~ **до·гово́р·а** (law) breach of contract
~ **не·пре·ры́в·ност·и** discontinuity
~ **пра́в·ил·а** violation of a rule

на·руш· éни·е

~, радиаци́он·н·ое (nucl) radiation defect/damage

~ решёт·к·и (cryst) lattice disturbance/imperfection

~ структу́р·ы (phys) disarrangement of structure, imperfection

~ у·кла́д·к·и (cryst) stacking fault

на·ру́ш·енн·ый, -ая, -ое, (*past part pass*) *see* **на·руш·а́·ть**; (phys) disturbed, disrupted, upset; affected

на·руш·и́тел·ь, -я, (*m*) transgressor, violator

на·руш·и́ть, -ат, (*perf*) *see* **на·ру·ш·а́·ть**

нарцеи́н·, -а, (*m*) (pharm) narceine

на·ры́в·, -а, (*m*) (med) abscess, boil

на·рыв·а́·ть, -ют, (*imp*) pick, pluck, gather; tear up (a quantity of); (min) blast out; dig, dig up

на·ры́в·ник·, -а, (*m*) (ent) blister beetle, (*pl*) *Meloidae*

на·рыв·н·о́й, -а́я, -о́е, (*a*) vesicant, blistering; (pharm) boil, abscess

~ пласты́р·ь (*m*) (pharm) boil plaster/dressing

~ срéд·ств·о (*n*) (pharm) vesicant

на·ры́в·н·ый, -ая, -ое, (*a*) of **на·ры́в·**

на·ры́·ть, (*fut 3rd pl*) **на·ро́·ют,** (*perf*) dig, dig up

на·ря́д·, -а, (*m*) indent, warrant, order; dress; (mil) detail; patrol

~, бра́ч·н·ый (fish) spawning livery/dress

~ -за·ка́з·, -а, (*m*) production order

на·ряд·и́ть, -я́т, (*perf*) *see* **на·ря·ж·а́·ть**

на·ряд·у́ (*adv*) alongside, beside

~ с (*conj*) beside, besides

на·ря́д·чик·, -а, (*m*) (mil) detail assignment officer, man in charge

на·ряж·а́·ть, -ют, (*imp*) dress up; appoint, detail

нас (*pron, acc/gen/prep*) of **мы;** us

на·сад·и́ть, -я́т, (*perf*) *see* **на·са́ж·ива·ть** *and* **на·сажд·а́·ть**

на·са́д·к·а, -и, (*g pl*) **д·ок·,** (*f*) (*see also* **на·са́д·ок·**) putting/fitting on; filling, packing; (chem) packing, tower packing; (met) chequerwork, checkerwork (furnace); (mech eng) mask, hood, shroud; (phot) attachment; (fish) bait; capsill (of bridge)

~, грé·ющ·иеся (*pl*) (met) checkers on gas

~, двух·сторóн·н·ая (met) double-pass checkerwork

на·са́д·к·а

~, одно·оборóт·н·ая (met) single-pass checkerwork

~, поворóт·н·ая (mar eng) swivelling shroud

~, по·глощ·а́ющ·ая (rad) dry/sand load, power termination, absorption cell

~, хóрд·ов·ая (chem) tile packing

~, шести·гра́н·н·ая спира́ль·н·ая (chem) hexahelix packing

на·са́д·ок·, -д·к·а, (*m*) (*see also* **сопл·ó** *and* **на·са́д·к·а**) nozzle (letting out); headpiece (letting in); die (of an extruder)

~ Венту́ри (hydr) Venturi tube

~, дефлéктор·н·ый (rocket) deflector cap

~, коноида́ль·н·ый straight convergent nozzle

~, раз·бры́зг·ивающ·ий spray nozzle

~, рас·ход·я́щ·ийся divergent nozzle

~, су́ж·ающ·ийся convergent/contracting nozzle

~, с·ход·я́щ·ийся convergent nozzle **теóри·я на·са́д·к·ов пóлн·ого давл·éни·я** (*f*) (aerodyn) impact tube theory

на·са́д·очн·ый, -ая, -ое, (*a*) of **на·са́д·к·а** and **на·са́д·ок·**

~ ка́мер·а (*f*) photomicrography camera, camera attachment (to microscope)

~ колóнн·а (*f*) (chem) packed tower

~ ли́нз·а (*f*) (phot) lens attachment

на·саж·а́·ть, -ют, (*perf*) plant (in several operations), plant out; fill, pack in

на·сажд·а́·ть, -ют, (*imp*) introduce, inculcate, spread, implant

на·сажд·éни·е, -я, (*n*) introduction, inculcation; planting, planting out; plantation, stand (of timber)

~, вегетати́в·н·ое (hort) striking cuttings/shoots

~, семен·н·óе seeding

~, чи́ст·ое pure forest

на·сажд·ённ·ый, -ая, -ое, (*past part pass*) of **на·сажд·а́·ть**

на·са́ж·енн·ый, -ая, -ое, (*past part pass*) *see* **на·са́ж·ива·ть**

на·са́ж·ива·ть, -ют, (*imp*) plant, plant out; fill up, pack in, pack tight; fit tight, fit on tight

на·са́л·ива·ть, -ют, (*imp*) salt, pickle; grease, rub with fat

на·са́л·ить, -ят, (*perf*) grease, rub with fat

на·са́л·к·а, -и, (f) (v n) of на·са́л·и·ва·ть

на·са́с·ыва·ть, -ют, (imp) suck, draw by suction; pump full

на·са́хар·ить, -ят, (perf) add sugar, sugar

на·сверл·ённ·ый, -ая, -ое, (past part pass) of на·сверл·и́ть; perforated

на·свёрл·ива·ть, -ют, (imp) drill holes, drill (in a number of places)

на·сверл·и́ть, -ят, (perf) see на·свёр·л·ива·ть

на·све́ч·ивани·е, -я, (n) see вы·све́·ч·ивани·е

на·се́·вш·ий, -ая, -ее, (past part act) of на·се́ст·ь

на·сед·а́·ть, -ют, (imp) settle (of dust etc.); sit on, put one's weight on

на·се́д·к·а, -и, (g pl) -д·ок·, (f) broody hen

на·сек·а́льн·ый, -ая, -ое, (a) indenting, cutting, notching, scratching, hatching

~ маши́н·а (f) (text) card perforating machine; leather embossing machine

на·сек·а́·ть, -ют, (imp) incise, scratch, notch, carve; roughen, rag (surfaces); cut up fine/small

на·сек·о́м·ое, -ого, (n decl as adj) insect, (pl) Insecta

~, вз·ро́сл·ое imago, perfect insect

~, го́л·ое unarmoured insect

насекомо·о·пыл·я́ем·ый, -ая, -ое, (a) (bot) entomophilous

насекомо·я́д·н·ый, -ая, -ое, (a) insectivorous; (pl as noun) (zool) insectivores, Insectivora

на·се́·л (past masc sing) of на·се́ст·ь

на·сел·е́ни·е, -я, (n) (v n) see на·сел·я́·ть; population, inhabitants

на·сел·ённост·ь, -и, (f) population, population density

на·сел·и́ть, -я́т, (perf) see на·сел·я́·ть

на·сел·я́·ть, -ют, (imp) populate, inhabit; settle in

на·се́ст·, -а, (m) roost, perch (for chickens etc.)

на·се́ст·ь, (fut 3rd pl) на·ся́д·ут, (past masc sing) на·се́·л, (perf) find a seat, sit down (of large numbers); settle (of dust etc.); sit on, put one's weight on; на·се́ст·ь, -и, (f) see на·се́ст·

на·сеч·ённ·ый, -ая, -ое, (past part pass) see на·сек·а́·ть; incised, serrated, notched, hatched, ragged

~ вал·о́к· (m) (met roll) ragged roll

на·се́ч·к·а, -и, (g pl) -ч·ек·, (f) (v n) see на·сек·а́·ть; cut, teeth, set of teeth (on a file); inlay; ragging (rolling mill rolls etc.)

~, бархат·н·ая superfine cut (of file)

~, лич·н·ая fine/smooth cut (of file)

на·се́ч·ь, (fut 3rd sing, pl) на·сеч·ёт, на·сек·у́т, (past masc sing) на·се́к·, (perf); see на·сек·а́·ть

на·сид·е́ть, -я́т, (perf) see на·си́ж·и·ва·ть

на·си́ж·ива·ть, -ют, (imp) brood, hatch (eggs)

на·си́л·овать, -уют, (imp) coerce, force; rape

на·си́л·у (adv) with great difficulty; hardly

на·си́ль·н·о (adv) by force, under compulsion

на·си́ль·ственн·ый, -ая, -ое, (a) forcible

на·ска́бл·ива·ть, -ют, (imp) grate; scrape, shave, plane

на·ска́к·ива·ть, -ют, (imp) run into/onto, collide with, strike, smash into

на·сквозь (adv) through, right through, throughout

на·скобл·и́ть, -я́т, (perf) see на·ска́б·л·ива·ть

на·ско́лько (adv) as far as, so far as; how much, how

~ нам из·ве́ст·н·о as far as we know, to the best of our knowledge

на́·скор·о (adv) hurriedly, hastily

на·скоч·и́ть, -ат, (perf) see на·ска́к·и·ва·ть

на·скреб·а́·ть, -ют, (imp) scrape up/together

на·скрес·ти́, (fut 3rd pl) на·скреб·у́т, (past masc sing) на·скрёб·, (perf); see на·скреб·а́·ть

на·ску́ч·ить, -ат, (perf) bore, become tiresome

на·слажд·а́·ться, -ются, (imp + instr) be delighted with

на·слажд·е́ни·е, -я, (n) delight

на·сла́·иван·ие, -я, (n) (v n) see на·сла́·ива·ть; lamination, stratification

на·сла́·ива·ть, -ют, (imp) stratify, layer, make/arrange in layers, laminate; divide into layers, split

на·сл·а́ть, (fut 3rd pl) на·шл·ю́т, (perf) send (in several stages)

на·сле́д·и·е, -я, (n) heritage; inheritance, legacy

на·след·и́ть, -я́т, (perf) leave a trail, leave dirty marks behind на·сле́д·ник·, -а, (*m*) inheritor, successor, heir, legatee

на·сле́д·овани·е, -я, (*n*) inheritance, inheriting; succession

на·сле́д·овать, -уют, (*imp or perf*) inherit, succeed to

на·сле́д·ственност·ь, -и, (*f*) (genet) heredity, inheritance

~, матер·и́нск·ая (genet) maternal inheritance/heredity

на·сле́д·ственн·ый, -ая, -ое, (*a*) hereditary; inherited, inherent

на·сле́д·ств·о, -а, (*n*) (law etc.) inheritance, inherited property, legacy

на·след·у́емост·ь, -и, (*f*) heritability

на·сло·е́ни·е, -я, (*n*) (*v n*) *see* на·сла́·ива·ть; (geol) bedding, lamination, stratification

на·сло·и́ть, -я́т, (*perf*) *see* на·сла́·ива·ть

наслу́д·, -а, (*m*) (ocean) naslud, young pool ice

на·сма́л·ива·ть, -ют, (*imp*) tar

на́·с·мерт·ь (*adv*) to death, mortally

на·сме́ш·лив·ый, -ая, -ое, (*a*) derisive, mocking

на·смол·и́ть, -я́т, (*perf*) tar

на́сморк·, -а, (*m*) common cold, rhinitis, coryza

на·сола́ж·ивани·е, -я, (*n*) (geol) sweeting on

на·сол·и́ть, -я́т, (*perf*) salt; pickle in salt

на·со́с·, -а, (*m*) pump

~, аксиа́ль·н·ый axial-flow pump

~ Альве́йер·а wing pump

~, баланси́р·н·ый beam-operated pump

~, балла́ст·н·ый ballast pump

~ без на·правл·я́ющ·его колес·а́, центро·бе́ж·н·ый volute-type centrifugal pump

~, без·поршн·ев·о́й pulsating pump

~, ва́куум·н·ый vacuum pump; air pump

~, винт·ов·о́й screw pump

~, винт·ов·о́й пропе́ллор·н·ый screw-type propeller pump

~, винт·ов·о́й черв·я́чн·ый screw-type rotary pump

~ Ви́р·а Weir standard feed pump

~, вихр·ев·о́й turbine pump

~, водо·кольц·ев·о́й water-ring pump, liquid-piston pump

~, водо·от·ли́в·н·ый (mar eng) salvage pump; (min) drainage/dewatering pump

на·со́с

~, водо·стру́й·н·ый jet pump

~, воз·ду́ш·н·ый air pump; (ICE) priming pump

~, восьм·ёрочн·ый two-lobe rotary pump

~, в·са́с·ывающ·ий (min) suction pump, aspirator

~, вс·по·мог·а́тельн·ый service pump

~, в·став·н·о́й (oil) insert pump

~, высоко·на·по́р·н·ый high-head pump

~ вы·тесн·е́ни·я positive displacement pump

~, герота́р·н·ый helix pump

~, глуб·и́нн·ый deep-well borehole pump

~, глуб·о́к·ий deep-well/borehole pump

~, груз·ов·о́й cargo pump

~, гряз·ев·о́й (oil) slush/mud pump

~ двой·н·о́го де́й·стви·я double-acting pump

~, двух·ро́тор·н·ый lobar pump

~, дежу́р·н·ый то́пл·ивн·ый (mar eng) fuel-oil service pump

~, диагона́ль·н·ый screw pump

~, диафра́гм·ов·ый diaphragm pump

~, диффузио́н·н·ый diffusion pump

~, диффузио́н·н·о-конденсацио́н·-н·ый Langmuir condensation pump (high vacuum)

~, доз·а́торн·ый proportioning/metering pump

~ -доз·иро́вщик·, -а, (*m*) metering pump

~, до·полн·и́тельн·ый auxiliary pump

~, за·лив·н·о́й priming pump

~, золот·ни́ков·ый slide-valve pump; port-type pump (diesels)

~, зуб·ча́т·ый gear pump

~, зумпф·о́в·ый (min) gathering pump

~, ио́н·н·ый (phys) ion pump

~, коло·вра́т·н·ый lobar pump

~, конденса́т·н·ый condensate pump

~, кры́ль·чат·ый wing pump

~, ло́паст·н·ый rotodynamic pump

~, ло́паст·н·ый ротацио́н·н·ый sliding-vane pump

~, лопа́т·очн·ый rotodynamic pump

~, манжет·н·ый cup-packed plunger pump

~, ма́сл·ян·ый lubricating-oil pump

~, ма́сс·н·ый paper pump, stock pump

~, мембра́н·н·ый diaphragm pump

~, много·ло́паст·н·ый sliding-vane rotary pump

на·со́с

~, мно́го·ступе́н·чат·ый multistage centrifugal pump

~, мокро·воз·ду́ш·н·ый steam-jet/wet-air pump (of condenser)

~, молекуля́р·н·ый molecular pump

~, на·гнет·а́ющ·ий force/delivery pump; pressure pump, pressure lubrication pump; (min) filling pump

~, над·ду́в·очн·ый supercharger

~ не·по·сре́д·ственн·ого в·пры́ск·а direct injection pump

~, нефт·ян·о́й gasoline/petrol/oil pump; (mar eng) oil fuel pump

~ об·тек·а́ни·я rotodynamic pump

~, объ·ём·н·ый positive-displacement pump

~, один·а́рн·ого де́й·стви·я single-acting reciprocating pump

~, одно·колёс·н·ый single-stage centrifugal pump

~, одно·ро́тор·н·ый vane pump

~, одно·ступе́н·чат·ый single-stage pump

~, ос·ев·о́й axial-flow pump, propeller pump

~, от·ка́ч·ивающ·ий (a/c) scavenge pump, dump pump

~, от·са́с·ывающ·ий exhauster

~, о·хлажд·а́ющ·ий cooling-water pump

~, паро·воз·ду́ш·н·ый steam-driven compressed air pump

~, пар·ов·о́й steam-driven pump, steam pump

~, паро·ма́сл·ян·ый oil-vapour pump

~, паро·стру́й·н·ый diffusion pump

~, пере·ка́ч·ивающ·ий transfer pump

~ пере·ме́н·н·ой про·из·вод·и́тель·ност·и variable-delivery pump

~, пит·а́тельн·ый feed pump

~, пласт·и́нчат·ый vane pump, sliding-vane pump

~, плу́нжер·н·ый plunger pump

~, пневмат·и́ческ·ий pump with pneumatic drive; air-lift pump

~, пневмат·и́ческ·ий ка́мер·н·ый pneumatic conveyor

~, по·да·ю́щ·ий delivery pump

~, под·ка́ч·ивающ·ий (a/c) auxiliary fuel pump, booster pump; (M/T) fuel feed/transfer pump

~, поршень·ко́в·ый rotary piston pump

~, поршн·ев·о́й piston/reciprocating pump

на·со́с

~, поршн·ев·о́й ротацио́н·н·ый rotary piston pump

~, при·вод·н·о́й power pump

~ при·ем·и́стост·и (ICE) accelerator pump

~, про·ду́в·очн·ый scavenging pump

~, пропе́ллер·н·ый propeller pump, axial-flow pump

~ прост·о́го де́й·стви·я single-acting piston pump

~ прям·о́го де́й·стви·я direct-acting pump

~, пульсацио́н·н·ый jerk pump

~, радиа́льно·поршн·ев·о́й rotary piston pump

~, раз·реж·а́ющ·ий vacuum pump

~ регул·и́ру́ем·ой про·из·вод·и́тельност·и variable-delivery pump

~, ротати́в·н·ый rotary pump

~, ротацио́н·н·ый rotary pump

~, рту́т·н·ый mercury-vapour pump

~, ру́ч·н·о́й hand pump

~ с за·дви́ж·к·ами sliding-vane pump

~ с на·правл·я́ющ·им колес·о́м, центро·бе́ж·н·ый turbine pump, diffuser-type pump

~, само·в·са́с·ывающ·ий self-priming pump

~, с·дво́·енн·ый duplex pump

~, секцио́н·н·ый ring-type centrifugal pump

~, ска́ль·чат·ый plunger pump

~, с·мес·и́тельн·ый mixing pump

~, спира́ль·н·ый volute pump

~, стро́·енн·ый three-throw pump

~, стру́й·н·ый jet pump

~, то́пл·ивн·ый (st eng) oil-fuel pump; fuel injection pump (diesel)

~, то́пл·ив·о·под·ка́ч·ивающ·ий fuel supply/feed/transfer pump (diesels)

~, трёх·компоне́нт·н·ый triple pump, triple-component pump

~, трёх·кулач·ко́в·ый коло·вра́т·н·ый three-lobe rotary pump

~, турби́н·н·ый turbine pump, diffuser-type pump

~, турбо·конденса́т·н·ый turbine-driven condensate pump

~, турбо·ма́сл·ян·ый turbine-driven lubricating pump

~, турбо·циркуляцио́н·н·ый turbine-driven circulating pump

~, у·скор·и́тельн·ый accelerating pump

~, флю́гер·н·ый (a/c) feathering pump

на·со́с

~, фор·ва́куум·н·ый fore pump, fore-vacuum pump

~ -форсу́н·к·а, -и, (*g pl*) -н·ок·, (*f*) unit injector (diesel)

~, центро·бе́ж·н·о-вихр·ев·о́й combination pump

~, центро·бе́ж·н·ый centrifugal pump

~, цеп·н·о́й chain pump

~, циркуляцио́н·н·ый circulating pump

~, черв·я́ч·н·ый screw pump

~, ша́йб·ов·ый rotary piston pump

~, шест·ерённ·ый spur-gear pump, gear pump

~, ши́бер·н·ый vane pump, sliding-vane pump

~, шта́нг·ов·ый (oil) insert pump

на·сос·а́ть, -у́т, (*perf*) *see* на·са́с·ы·ва·ть

на·со́с·н·ая, -ой, (*f decl as adj*) pump house, pumping station

на·со́с·н·ый, -ая, -ое, (*a*) pump, pumping; (*see also* на·со́с·н·ая) (*as noun*)

~ по·те́р·и (*pl*) pump losses

на·со́х·нуть, (*fut 3rd sing*) -нет, (*past masc sing*) на·со́х·, (*perf*) *see* на·сых·а́·ть

на́·спех· (*adv*) in a hurry, hurriedly

на́ст·, -а, (*m*) snow crust

на·ста·ва́ть, -ю́т, (*imp*) come, begin, ensue, set in (of time, events etc.)

на·став·и́тельн·ый, -ая, -ое, (*a*) edifying, instructive

на·ста́в·ить, -ят, (*perf*) *see* на·ставл·я́·ть

на·ста́в·к·а, -и, (*g pl*) -в·ок·, (*f*) extension piece; adapter

на·ставл·е́ни·е, -я, (*n*) (*v n*) *see* на··ставл·я́·ть; precept, exhortation; instruction(s), manual (book)

~ по про·из·во́д·ств·у по·лёт·ов Air Traffic Regulations

~ по эксплоата́ци·и operating manual

на·ставл·я́·ть, -ю́т, (*imp*) put, put on; extend, lengthen; aim, point; admonish, exhort

на·ста́в·н·о́й, -а́я, -о́е, (*a*) extended, lengthened (by adding a piece); extension, adapter

на·ста·ёт (*pres 3rd sing*) of на·ста·ва́ть

на·ста́·ивани·е, -я, (*n*) (*v n*) *see* на··ста́·ива·ть; persistence; infusion

на·ста́·ива·ть, -ю́т, (*imp*) insist, persist; infuse, draw

на·ста́л·ивани·е, -я, (*n*) (met) steeling

на·ста́·ть, -нут, (*perf*) *see* на·ста··ва́ть

на·ста·ю́т (*pres 3rd pl*) of на·ста··ва́ть

на́·стежь (*adv*) wide, wide-open, open

на·сте́л·ет (*fut 3rd sing*) of на·стл·а́ть

на·сте́н·н·ый, -ая, -ое, (*a*) wall, wall-mounted

на·стиг·а́·ть, -ю́т, (*imp*) overtake

на·сти́г·нуть, -нут, (*past masc sing*) на·сти́г·, (*perf*) *see* на·стиг·а́·ть

на·сти́л·, -а, (*m*) (*v n*) *see* на··сти·л·а́·ть; (*see also* на·сти́л·к·а) flooring, planking, decking; trestles; (text) overlay, float

на·стил·а́ни·е, -я, (*n*) (*v n*) of на··стил·а́·ть; coating (with rubber)

на·стил·а́·ть, -ю́т, (*imp*) spread, spread out, lay; floor, plank, pave, deck

~ в·на·хлёст·к·у imbricate

на·сти́л·к·а, -и, (*g pl*) -л·ок·, (*f*) (*v n*) *see* на·стил·а́·ть; flooring, decking, planking, plating; (min) lagging

~, па́луб·н·ая deck plating

на·сти́л·очн·ый, -ая, -ое, (*a*) of на·сти́л·к·а

~ доск·а́ (*f*) (build) floorboard; chess (bridge)

на·сти́ль·ност·ь, -и, (*f*) horizontality, levelness; good bath coverage (of OH furnace flame)

да́ль·ност·ь на·сти́ль·ност·и (*f*) (gunn) point blank range

на·сти́ль·н·ый, -ая, -ое, (*a*) level, horizontal

~ траекто́р·и·я (*f*) flat trajectory

на·сти́ч·ь, (*fut 3rd pl*) на·сти́г·нут, (*past masc sing*) на·сти́г·, (*perf*) *see* на·стиг·а́·ть

на·стл·а́ть, (*fut 3rd pl*) на·сте́л·ют, (*perf*) *see* на·стил·а́·ть

на·сто́·й, -я, (*m*) infusion, tincture

на·сто́й·к·а, -и, (*g pl*) -о́·ек·, (*f*) liqueur, alcoholic cordial; (pharm) tincture

~, ла́кмус·ов·ая (chem) litmus tincture

на·сто́й·чивост·ь, -и, (*f*) persistence, insistence

на·сто́й·чив·ый, -ая, -ое, (*a*) persistent, insistent

насто́ль (*adv*) *see* на·сто́лько

на·сто́лько (*adv*) so, so far/much/completely, to such an extent

на·ст**о**ль·н·ый, -ая, -ое, (*a*) table, desk, bench, table/desk/bench-mounted

~ кн**и**г·а (*f*) essential reference work

~ л**а**мп·а (*f*) table/desk lamp

на·стор**а**ж·ива·ть, -ют, (*imp*) alert, put on guard

на·сторож·**е** (*adv*) on the alert/lookout

на·сторож·**и**ть, -**а**т, (*perf*) *see* на·-стор**а**ж·ива·ть

на·сто·**я**ни·е, -я, (*n*) insistence

на·ст**о**·янн·ый, -ая, -ое, (*past part pass*) *of* на·сто·**я**ть; infused, steeped

на·сто·**я**тельн·ый, -ая, -ое, (*a*) persistent, insistent; urgent, pressing

на·сто·**я**ть, -**я**т, (*perf*) *see* на·ст**а**·ива·ть

на·сто·**я**щ·ее, -его (*n decl as adj*) the present, the present time

на·сто·**я**щ·ий, -ая, -ее, (*pres part act*) *of* на·сто·**я**ть; real, genuine, true; the present, current

~ вр**е**м·я (*n*) the present time/day, to-day; (*gram*) present tense

    в на·сто·**я**щ·ее вр**е**м·я (*adv*) at present, at the moment, now, nowadays

~ планкт**о**н· (*m*) (biol) euplankton

~ т**о**ч·к·а (*f*) (navig) present position

на·стр**а**г·ива·ть, -ют, (*imp*) plane, shave (a quantity of)

на·стр**а**·ива·ть, -ют, (*imp*) tune, tune in, adjust, line up; incite, incline, dispose; build (a quantity of)

на·стр**а**ч·ива·ть, -ют, (*imp*) stitch, stitch on

на·стр**е**л·ива·ть, -ют, (*imp*) shoot up/down, shoot, kill (a number of)

на·стрел·**я**·ть, -ют, (*perf*) *see* на·-стр**е**л·ива·ть

н**а**·стриг·, -а, (*m*) clip (of wool); на·стр**и**г·, -а, (*m*) (*v n*) *see* на·-стриг·**а**·ть

на·стриг·**а**·ть, -ют, (*imp*) shear (sheep etc.); cut up, cut up small/fine

на·стр**и**чь, (*fut 3rd sing, pl*) на·-стриж·ёт, на·стриг·**у**т, (*past masc sing*) на·стр**и**г· (*perf*); *see* на·стриг·**а**·ть

на·строг·**а**·ть, -ют, (*perf*) plane, shave (a quantity of)

н**а**·строг·о (*adv*) strictly

на·стро·**е**ни·е, -я, (*n*) mood, frame of mind, attitude

на·стр**о**·енн·ый, -ая, -ое, (*past part pass*) *of* на·стр**о**·ить; tuned, adjusted; built-on

~ к**о**нтур· (*m*) (rad) resonant/tuned circuit

на·стр**о**·ечн·ый, -ая, -ое, (*a*) tuning, adjustment

~ табл**и**ц·а (*f*) (TV) tuning picture

на·стр**о**·ить, -ят, (*perf*) *see* на·стр**а**·и·ва·ть

на·стр**о**й·к·а, -и, (*f*) tuning, adjustment

~, бес·ш**у**м·н·ая quiet tuning

    бл**о**к· на·стр**о**й·к·и (*m*) (rad) tuner

~ велич·ин·**ы** давл·**е**ни·я pressure adjustment

~ вы·д**е**рж·к·и вр**е**мен·и timing

~, гр**у**б·ая coarse adjustment; (rad) rough tuning

~ диапаз**о**н·а band/range adjustment

~, др**о**ссель·н·ая (rad) choke tuning

~, кн**о**п·очн·ая push-button tuning

~ лоп**а**т·очк·ой (rad) slug tuning

~, магн**и**т·н·ая permeability tuning механ**и**зм· на·стр**о**й·к·и пр**е**·дел·ов пропорци**а**ль·ност·и (*m*) (autom) throttling range adjuster mechanism

~, одно·р**у**ч·ечн·ая (rad) single tuning, single-knob tuning

~, **о**стр·ая fine tuning/adjustment

~, пол**о**с·ов·**а**я (rad) band-pass tuning

~, по·след·ов**а**тельн·ая stagger tuning

~ серд·**е**чник·ом (rad) slug tuning сигн**а**л· на·стр**о**й·к·и (*m*) tuning signal

~ со·пряж·**е**ни·я (rad) gang tuning шкал·**а** на·стр**о**й·к·и (*f*) tuning dial

на·строч·**и**ть, -**а**т, (*perf*) stitch, stitch on; scribble

на·струг·**а**·ть, -ют, (*perf*) = на·-строг·**а**·ть

на·студ·**и**ть, -ят, (*perf*) chill, cool

на·стуж·**а**·ть, -ют, (*imp*) chill, cool

на·ст**у**ж·ива·ть, -ют, (*imp*) chill, cool

на·ступ·**а**тельн·ый, -ая, -ое, (*a*) offensive

на·ступ·**а**·ть, -ют, (*imp*) step/tread on; come on, begin, set in, ensue (of time, events etc.); (*imp only*) advance, carry out an offensive

на·ступ·**а**ющ·ий, -ая, -ее, (*pres part act*) *of* на·ступ·**а**·ть; oncoming, advancing

на·ступ·**и**ть, -ят, (*perf*) *see* на·ступ·**а**·ть

на·ступл·**е**ни·е, -я, (*n*) (mil) offensive, attack; encroachment (of the sea); onset, advent, coming

на·ступл·éни·е
　температу́р·а на·ступл·éни·я (*f*)
　threshold temperature
настура́н·, -а, (*m*) (min) nasturan,
　uraninite
на·сты́л·ь, -и, (*f*) crust; (met) slag
　incrustation (on furnace refractories)
на·сты́·н·уть, -ут, (*perf*) *see* **на·-
　сты́·ть**
на·сты́·ть, (*fut 3rd pl*) **на·сты́·нут,**
　(*perf*) freeze/congeal onto/over, form
　a crust
на́·сух·о (*adv*) dry, till dry; in a dry
　state
на·су́ч·ива·ть, -ют, (*imp*) (text) twist,
　spin (a quantity of); throw (of silk)
на·суч·и́ть, -а́т, (*perf*) *see* **на·су́ч·и·-
　ва·ть**
на·су́ш·ива·ть, -ют, (*imp*) dry, des-
　sicate (a quantity of)
на·суш·и́ть, ⁼ат, (*perf*) *see* **на·су́ш·и·-
　ва·ть**
на·су́щ·н·ый, -ая, -ое, (*a*) essential,
　vital
на·с·чёт· (*prep + gen*) as regards, con-
　cerning, about
на·с·чит·а́·ть, -ют, (*perf*) count,
　count out (a number of); count on
на·с·чи́т·ыва·ть, -ют, (*imp*) count,
　count out (a number of); count on;
　number, consist of, contain, have
　(a number)
на·сыл·а́·ть, -ют, (*imp*) send (in sev-
　eral stages)
на·сып·а́·ть, -ют, (*imp*) **на·сы́п·ать,
　-лют,** (*perf*) pour over (of solids);
　fill, pour in (of solids)
на·сып·н·óй, -а́я, -óе, (*a*) free-flowing,
　bulk (of cargo etc.); filled, piled
　(with free-flowing solids)
~ вес· (*m*) bulk weight (of free-flow-
　ing material)
~ груз· (*m*) bulk freight (free flowing)
~ плот·и́н·а (*f*) rock-fill dam
на́·сып·ь, -и, (*f*) embankment; fill,
　filling
на́·сып·ью (*adv*) in bulk, bulk
　цемéнт· на́·сып·ью (*m*) bulk
　cement
на·сы́т·им·ый, -ая, -ое, (*pres part
　pass*) of **на·сы́т·ить,** *see* **на·сы-
　щ·а́ем·ый**
на·сы́т·ить, -ят, (*perf*) *see* **на·сы-
　щ·а́·ть**
на·сых·а́·ть, -ют, (*imp*) dry on;
　shrivel, dry out/up (of a quantity of)

на·сыщ·а́емост·ь, -и, (*f*) saturability,
　saturation capacity
на·сыщ·а́ем·ый, -ая, -ое, (*pres part
　pass*) of **на·сыщ·а́·ть;** saturable
на·сыщ·а́·ть, -ют, (*imp*) satiate, satisfy;
　saturate
~ водо·ро́д·ом (chem) hydrogenate
~ угле·кисл·от·óй carbonate
на·сыщ·а́ющ·ий, -ая, -ее, (*pres part
　act*) **на·сыщ·а́·ть;** (*m as a*) saturant
на·сыщ·éни·е, -я, (*n*) saturation,
　satiation; fitting out, installation fit-
　tings, fixtures and installations
　давл·éни·е на·сыщ·éни·я (*n*)
　saturation pressure
~, до·изоляцио́н·н·ое (shipb) pre-
　insulation installations
~, кóрпус·н·ое (*n*) (shipb) hull system
　installations
　коэффициéнт· на·сыщ·éни·я (*m*)
　saturation factor
~, магни́т·н·ое (elec) (*m*) satura-
　tion induction
~ от·нос·и́тельн·о вод·ы́ (*n*) (me-
　teor) water saturation
~ чёрн·ым (*n*) (TV) black saturation
на·сы́щ·енност·ь, -и, (*f*) (TV) sa-
　turation, state of saturation; colour
　saturation
на·сы́щ·енн·ый, -ая, -ое, (*past part
　pass*) *see* **на·сыщ·а́·ть**
~ пар· (*m*) saturated vapour/vapor;
　(st eng) saturated steam
~ ток· (*m*) (elec) saturation current
~ угле·водоро́д· (*m*) (chem) saturated
　hydrocarbon
на·ся́д·ут (*fut 3rd pl*) of **на·сéст·ь**
на·та́лк·ива·ть, -ют, (*imp*) push/shove/
　jog on/onto; persuade to adopt, get
　to understand; **-ся** run into/onto,
　meet
на·та́пл·ива·ть, -ют, (*imp*) heat well;
　melt (a quantity of); (food) warm
на·таск·а́·ть, -ют, (*perf*) *see* **на·та́ск·
　·ива·ть**
на·та́ск·ива·ть, -ют, (*imp*) drag/pull/
　carry/bring up/to (in several stages);
　pull/drag out (in several stages); train,
　teach, cram
на·та́ч·ива·ть, -ют, (*imp*) sharpen
на·тащ·и́ть, ⁼ат, (*perf*) drag/pull/carry
　up/to, bring
на·тёк·, -а, (*m*) (*v n*) *see* **на·тек·а́·ть;**
　leakage fluid, infiltrate; (min) sinter,
　geyserite
на·тек·а́ни·е, -я, (*n*) infiltration, in-
　leakage leakage

на·тек·а́тел·ь, -я, (*m*) flow regulator, infiltrometer

на·тек·а́·ть, -ют, (*imp*) leak in, infiltrate, accumulate, form (of liquids)

на·те́ль·н·ый, -ая, -ое, (*a*) worn next to the skin, under, body (of clothing)

~ бел·ь·ё (*n*) underwear

на·тер·е́ть, (*fut 3rd pl*) на·тр·у́т, (*past masc sing*) на·тёр·, (*perf*); *see* на·тир·а́·ть

на·те́ч·к·а, -и, (*g pl*) -ч·ек·, (*f*) inleakage, infiltration; fine-calibrated orifice

на·те́ч·ь, (*fut 3rd sing, pl*) на·теч·ёт, на·тек·ут, (*past masc sing*) на·тёк·, (*perf*); *see* на·тек·а́·ть

нати́в·н·ый, -ая, -ое, (*a*) (min) native

на·тир·а́ни·е, -я, (*n*) (*v n*) *see* на·тир·а́·ть; (pharm) liniment, ointment, embrocation

на·тир·а́·ть, -ют, (*imp*) rub on/in; polish; grate, grind; rub sore; wipe (a ship in degaussing treatment)

на·ти́р·очн·ый, -ая, -ое, (*a*) rubbing, grating, polishing

~ маши́н·а (*f*) sanding machine (for wood block floors)

на·ти́р·к·а, -и, (*f*) *see* на·тир·а́ни·е

на́·тиск·, -а, (*m*) onslaught, onset; (print) impression

на·ти́ск·а·ть, -ют, (*perf*) *see* на·ти́ск·ива·ть

на·ти́ск·ива·ть, -ют, (*imp*) cram, cram in; (print) impress

на·тк·а́ть, -у́т, (*perf*) weave (a quantity of)

на·тк·у́ть, -у́т, (*perf*) *see* на·тык·а́·ть

на·толк·н·у́ть, -у́т, (*perf*) *see* на·та́лк·ива·ть

на·толо́ч·ь, (*fut 3rd sing, pl*) на·толч·ёт, на·толк·у́т, (*past masc sing*) на·толо́к·, (*perf*); pound, crush, grind (a quantity of)

на·топ·и́ть, -́ят, (*perf*) *see* на·та́пл·ива·ть

на·точ·и́ть, -́ат, (*perf*) sharpen

на·тощ·а́к (*adv*) on an empty stomach

на́тр·, -а, (*m*) soda, sodium oxide

~, е́дк·ий caustic soda

на·трав·и́ть, -́ят, (*perf*) *see* на·тра́вл·ива·ть

на·тра́вл·ива·ть, -ют, (*imp*) egg on, incite, instigate, urge to attack, unleash; poison (a number of); etch

на·трен·ир·ова́ть, -у́ют, (*perf*) train

на·тр·ёт (*fut 3rd sing*) of на·тер·е́ть

на́тр·иев·ый, -ая, -ое, (*a*) (*and see* на́тр·ов·ый *and* натро́н·н·ый) sodium, soda

~ сели́тр·а (*f*) sodium nitrate, Chile saltpetre/saltpeter

на́тр·и·й, -я, (*m*) (chem) sodium, Na

~, борно·ки́сл·ый sodium borate

~, дву·угле·ки́сл·ый sodium bicarbonate

~, йо́д·ист·ый sodium iodide

~, металл·и́ческ·ий metallic sodium

~, угле·ки́сл·ый sodium carbonate

~, хло́р·ист·ый sodium chloride

~ хлорновато·ки́сл·ый sodium chlorate

натрий·ами́д·, -а, (*m*) sodamide, sodium amide

натрий·ами́н·, -а, (*m*) sodamide

натрий·дивини́л·ов·ый, -ая, -ое, (*a*) sodium divinyl, sodium butadiene (rubber)

натрий·орган·и́ческ·ий, -ая, -ое, (*a*) organosodium

на́тр·ов·ый, -ая, -ое, (*a*) of натр·

~ и́звест·ь (*f*) soda lime

на́·тро·е (*adv*) in three

натро·кальци́т·, -а, (*m*) (chem) soda-calcite

натро·ли́т·, -а, (*m*) (min) natrolite

натро́н·н·ый, -ая, -ое, (*a*) soda

~ ва́р·к·а (*f*) (paper) soda pulping

~ мы́л·о (*n*) hard/soda soap

~ стекл·о́ (*n*) soda glass

натроярози́т·, -а, (*m*) (min) natrojarosite

на·тр·у́т (*fut 3rd pl*) of на·тер·е́ть

на·ту́г·а, -и, (*f*) effort, strain

на́·туг·о (*adv*) tightly

нату́р·а, -ы, (*f*) nature; model (art); goods supplied in payment; (cinema) location

в нату́р·е naked, in the natural state; in kind; in the open air

~ зерн·а́ (agr) grain unit

ис·пыт·а́ни·е в нату́р·е (*n*) (met) testing on an unnotched piece

натурализ·а́ци·я, -и, (*f*) (bot etc.) naturalization

натурализ·ова́ть, -у́ют, (*imp or perf*) naturalize

натурали́ст·, -а, (*m*) naturalist

натура́ль·н·ый, -ая, -ое, (*a*) natural; in kind (of payments)

~ велич·ин·а́ (*f*) life size

нату́р·н·ый, -ая, -ое, (*a*) full-scale, life-size; (cinema) location; (econ) commodity; nature (art)

нату́р·н·ый, -ая, -ое

~ съ·ём·к·а (*f*) (cinema) filming on location

натур·о·пла́т·а, -ы, (*f*) payment in goods/kind

на·тык·а́·ть, -ют, (*imp*) **на·ты́к·а·ть, -ют,** (*perf*) put on, mount; **-ся** stumble on, come across

натюрмо́рт·, -а, (*m*) (cinema) still

на·тя́г·, -а, (*m*) tightness (of fit); (mech eng) interference, negative allowance

~, пред·вар·и́тельн·ый preload (bearings etc.)

на·тя́г·ивани·е, -я, (*n*) (*v n*) *see* **на·тя́г·ива·ть** *and* **на·тяж·éни·е**

на·тя́г·ива·ть, -ют, (*imp*) pull tight/taut, tighten, tauten, tension, stretch; pull on/on top/over

~, пред·вар·и́тельн·о preload (bearings), pretension (concrete)

на·тя́г·иваю́щ·ий, -ая, -ее, (*pres part act*) of **на·тя́г·ива·ть;** tensile

на·тяж·éни·е, -я, (*n*) (*v n*) *see* **на·тя́г·ива·ть;** tension, tensioning; tightening, stretching; (eng) shrinkage allowance

~, за́д·н·ее (met roll) back tension

~, меж·фа́з·н·ое (phys) interfacial tension/energy

~, пере́д·н·ее (met roll) front tension

~, по·ве́рх·ностн·ое (phys) surface tension

~, по·сле́д·ующ·ее post-tensioning

~, пред·вар·и́тельн·ое pretensioning

на·тяж·к·а, -и, (*f*) (*v n*) *see* **на·тя́г·и·ва·ть**

на·тяж·н·о́й, -а́я, -о́е, (*a*) tightening, tensioning, strain, tension

~ при·способл·éни·е (*n*) tensioning device, tensioner

~ ро́л·ик· (*m*) idler pulley, idler (belt drive)

на·тя́·нут·ый, -ая, -ое, (*past part pass*) of **на·тя·ну́ть;** tight, taught, strained

на·тя·ну́ть, -у́т, (*perf*) *see* **на·тя́г·и·ва·ть**

на·у·га́д (*adv*) at a guess, by guess-work, at random

на·угле·ро́ж·енн·ый, -ая, -ое, (*a*) (met) carburized, carbon case-hardened; carbonized

на·угле·ро́ж·ивани·е, -я, (*n*) (met) carburizing, case hardening, carbon case hardening; addition of carbon, carbonizing

на·угле·ро́ж·иватель·ь, -я, (*m*) carburizer

на·угле·ро́ж·иваю́щ·ий, -ая, -ее, (*pres part act*) (chem) carburizing

~ пла́м·я (*n*) (weld) carburizing/reducing flame

на·уго́ль·ник·, -а, (*m*) try square (tool); corner, corner plate; (astron) the Rule, Norma

на·у·да́·ч·у (*adv*) on chance, for luck

на·у́к·а, -и, (*f*) science, knowledge

Акаде́м·и·я на·ук· (*f*) Academy of Sciences

~ о -logy

~, то́ч·н·ая exact science

науманни́т·, -а, (*m*) (min) naumannite

науплиус·, -а, (*m*) (zool) nauplius

наутил·и́д·ы (*pl*) (pal, zool) nautilids, *Nautilidae*

наутил·оид·é·я, -и, (*f*) (pal) nautiloid, (*pl*) *Nautiloidea*

наути́лус·, -а, (*m*) (zool) nautilus

науто·фо́н·, -а, (*m*) nautophone, electric foghorn

на·у́тро (*adv*) the next morning

на·утюж·ить, -ат, (*perf*) iron/press well

на·уч·и́ть, -ат, (*perf*) teach; **-ся** learn

на·у́ч·н·о-ис·сле́д·свательск·ий, -ая, -ое, (*a*) research, scientific research

на·у́ч·н·о-популя́р·н·ый, -ая, -ое, (*a*) popular-scientific, popular-science

на·у́ч·н·о-техн·и́ческ·ий, -ая, -ое, (*a*) scientific and technical; applied scientific

на·у́ч·н·о-фантаст·и́ческ·ий, -ая, -ое, (*a*) science-fiction

на·у́ч·н·ость·ь, -и, (*f*) scientific character/nature

на·у́ч·н·ый, -ая, -ое, (*a*) scientific

на·у́ш·ник·, -а, (*m*) headphone; ear cap/flap

на́ф·та, -ы, (*f*) (chem) naphtha

нафтала́н·, -а, (*m*) (chem) naphthalene

нафтала́т·, -а, (*m*) naphthalate

нафтализо́л·, -а, (*m*) naphthalysol (a disinfectant)

нафтали́н·, -а, (*m*) naphthalene

нафталин·дисульфо·кисл·от·а́, -ы́, (*f*) naphthalene-2-sulphonic acid

нафтали́н·н·ый, -ая, -ое, (*a*) naphthalene

~ из·ло́м· (*m*) (met) intercrystalline fracture

нафталин·сульфо·кисл·от·а́, -ы́, (*f*) naphthalene-*n*-sulphonic acid

нафтеи́н·, -а, (*m*) (geol) naphthein

нафте́н·, -а, (*m*) (chem) naphthene

нафтена́т·, -а, (*m*) (chem) naphthenate
~ **ци́нк·а** zinc naphthenate
нафте́н·ов·ый, -ая, -ое, (*a*) naphthene, naphthenic
~ **кисл·от·а́** (*f*) naphthenic acid
нафтен·о́л·ь, -я, (*m*) naphthenol (a drying oil)
нафти́л·, -а, (*m*) (chem) naphthyl
нафтилами́н·, -а, (*m*) (chem) naphthylamine
нафтио́н·ов·ый, -ая, -ое, (*a*) (chem) naphthionic
~ **кисл·от·а́** (*f*) naphthionic acid, 1-naphthylamine-4-sulfonic acid
нафто́й·н·ый, -ая, -ое, (*a*) (chem) naphthoic
~ **альдеги́д·** (*m*) naphthaldehyde
~ **кисл·от·а́** (*f*) naphthoic acid
нафт·о́л·, -а, (*m*) (chem) naphthol
нафто·ли́т·, -а, (*m*) (geol) naphtholith
нафто́л·ов·ый, -ая, -ое, (*a*) naphthol
~ **жёлт·ый** naphthol yellow (a dye)
нафто·ло́г·и·я, -и, (*f*) (geol) naphthology
нафтол·сульфо·кисл·от·а́, -ы́, (*f*) naphthol sulfonic acid
нафтоля́т·, -а, (*m*) (chem) naphtholate
нафто·хино́н·, -а, (*m*) naphthoquinone
на·хвал·и́ть, -ят, (*perf*) extol
на·хват·а́·ть, -ют, (*perf*) see **на·хва́·т·ыва·ть**
на·хва́т·ыва·ть, -ют, (*imp*) snatch, grab, seize (a quantity of)
на·хлест·а́·ть, -ют, or **на·хлёщ·ут,** (*perf*) see **на·хлёст·ыва·ть**
на·хлёст·к·а, -и, (*g pl*) **-т·ок·,** (*f*) overlapping, overlap, lap, jointing
с·ва́р·ива·ть в на·хлёст·к·у lap weld
на·хлёст·очн·ый, -ая, -ое, (*a*) overlapped
на·хлёст·ыва·ть, -ют, (*imp*) whip, beat
на·хлы́·н·уть, -ут, (*perf*) sweep, rush (of liquids, gas etc.)
на·хму́р·ить, -ят, (*perf*) frown
на·ход·и́ть, -ят, (*imp*) find, discover; come, come up/over/across/upon; walk/bump into; cover, obscure; get in, penetrate; (min) strike; **-ся** (*imp only*) be, be located/situated, occur
на·хо́д·к·а, -и, (*g pl*) **-д·ок·,** (*f*) find, discovery; (geol) occurrence
на·хо́д·чив·ый, -ая, -ое, (*a*) resourceful
на·ход·я́щ·ийся, -аяся, -ееся, (*pres part act*) of **на·ход·и́ться;** being, finding oneself (somewhere); (geol) occurring

на·ход·я́щ·ийся, -аяся, -ееся
~ **в бес·коне́ч·ност·и** infinitely distant, at an infinite distance
~ **на борт·у** on board (ship, aircraft)
на·хожд·е́ни·е, -я, (*n*) finding; being (of place)
врем·я на·хожд·е́ни·е в у·ста·но́в·к·е (*n*) (phys, chem) holdup time
ме́ст·о на·хожд·е́ни·я (*n*) location, whereabouts
на·хож·у́ (*pres 1st sing*) of **на·ход·и́ть**
на·цара́п·а·ть, -ют, (*perf*) see **на·цара́п·ыва·ть**
на·цара́п·ыва·ть, -ют, (*imp*) scratch on; scribble
на·цед·и́ть, -ят, (*perf*) see **на·цёж·ива·ть**
на·цёж·ива·ть, -ют, (*imp*) strain, filter, decant (a quantity of)
на·цёл·ива·ть, -ют, (*imp*) aim, aim at
на·цёл·ить, -ят, (*perf*) see **на·цёл·ива·ть**
на·цён·ива·ть, -ют, (*imp*) raise the price of, mark up
на·цен·и́ть, -ят, (*perf*) see **на·цён·ива·ть**
на·цён·к·а, -и, (*g pl*) **-н·ок·,** (*f*) price increase, mark up
~**, това́р·н·ая** trade profit, trader's profit
на·цеп·и́ть, -ят, (*perf*) see **на·цеп·л·я́·ть**
на·цепл·я́·ть, -ют, (*imp*) fasten/hook/pin on
национализ·и́р·овать, -уют, (*imp or perf*) nationalize
национа́ль·ност·ь, -и, (*f*) nationality
национа́ль·н·ый, -ая, -ое, (*a*) national
на́ци·я, -и, (*f*) nation
нач- (*component*) (*abbr*) of (1) **на·ча́ль·-ник·,** (2) **на·ча́ль·ствующ·ий**
на·ча́·вш·ий, -ая, -ее, (*past part act*) of **на·ча́·ть** (*perf*); see **на·чин·а́·ть** (*imp*)
на·чад·и́ть, -ят, (*perf*) smoke, fume
на·ча́·л·о, -а, (*n*) beginning, start, commencement; starting point; origin, source; principle, basis; (*pl*) elements, principles, rudiments
бы·ть под· на·ча́л·ом be/work under (a person), be subordinate to
~ **кип·е́ни·я** (chem) initial boiling point
~**, коллекти́в·н·ое** collective basis
~ **координа́т·** (math) origin of coordinates, (surv) grid origin
~ **крив·о́й** (math) point of curvature

на·ча́·л·о
~ от·с·чёт·а reference point
~ от·с·чёт·а вертика́ль·н·ых угл·о́в vertical reference point
~ от·с·чёт·а вре́мен·и zero time
~ от·с·чёт·а угл·а́ крен·а (rocket) roll datum
~ пере·да́·ч·и (telecom) commence
сигна́л· на·ча́·л·а пере·да́·ч·и (*m*) commencing signal'
~ рабо́т·ы (telecom) clear forward; (rocket) cut-in
сигна́л· на·ча́·л·а рабо́т·ы (*m*) (telecom) clear-forward signal
~ с за·де́рж·к·ой delayed start
~ тек·у́чест·и (met) upper yield point
~ термо·дина́м·ик·и, пе́рв·ое first law of thermodynamics
на·ча́ль·ник·, -а, (*m*) head, chief, superior, foreman, man in charge
~ диста́нц·и·и пут·и́ (rail) permanent way inspector
~ карау́л·а (mil) officer of the guard, guard commander
~ мая́к·а́ lighthouse keeper
~ по́рт·а harbourmaster
~ пу́нкт·а у·правл·е́ни·я по·са́д·к·и (air) approach controller
~ с·вя́з·и (mil) chief signals officer, senior communications officer
~ ста́нц·и·и (rail) station master
~ телегра́ф·н·ой ста́нц·и·и telegraph supervisor
~ цех·а (mech eng etc.) shop superintendent/manager; foreman
на·ча́ль·н·ый, -ая, -ое, (*a*) elementary; initial, first, original; primary (schools)
~ вес· (*m*) initial weight
~ меридиа́н· (*m*) (navig) first meridian
~ про·ниц·а́емост·ь (*f*) (phys) initial permeability
~ то́ч·к·а (*f*) starting point; origin
на·ча́ль·ств·о, -а, (*m*) management, superiors, authorities, (mil) command
на·ча́ль·ств·овать, -уют, (*imp*) be in command/charge, be the head of
на·ча́ль·ств·ующий, -ая, -ее, (*pres part act*) of **на·ча́ль·ств·овать**
~ со·ста́·в· (*m*) executive staff; (mil) specialist officers, nonexecutive officers (e.g. doctors, paymasters etc.)
на·ча́т·к·и, (*m pl*) rudiments, elements
на́·ча·т·ый, -ая, -ое, (*past part pass*) of **на·ча́·ть;** started, begun, commenced
на·ча́·ть, (*fut 3rd pl*) **на·чн·у́т,** (*perf*); see **на·чин·а́·ть**

на·чека́н·ива·ть, -ют, (*imp*) coin, strike (coins)
на·чека́н·ить, -ят, (*perf*) see **на·чека́·н·ива·ть**
на·чек·у́ (*adv*) on the alert/look-out
на·черн·и́ть, -ят, (*perf*) see **на·черн·я́·ть**
на́·черн·о (*adv*) in rough, rough, roughly; in draft form
~, об·раба́т·ыва·ть (mech eng) rough out
~, шлиф·ова́ть rough-grind
на·черн·я́·ть, -ют, (*imp*) black, blacken, paint black
на·че́рп·а·ть, -ют, (*perf*) see **на·че́р·п·ыва·ть**
на·че́рп·ыва·ть, -ют, (*imp*) scoop, scoop up (a quantity); fill, scoop full
на·черт·а́ни·е, -я, (*n*) description; outline, sketch
~ шрифт·ов (print) type face
на·черт·а́тельн·ый, -ая, -ое, (*a*) graphic
~ гео·ме́тр·и·я (*f*) projective geometry
на·черт·и́ть, -ят, (*perf*) see **на·че́рч·ива·ть**
на·че́рч·ива·ть, -ют, (*imp*) draw, sketch, plot
на·чёс·, -а, (*m*) (text) nap
на·чёс·анн·ый, -ая, -ое, (*past part pass*) of **на·чес·а́ть**
на·чес·а́ть, (*fut 3rd pl*) **на·чёш·ут,** (*perf*) see **на·чёс·ыва·ть**
на·чёс·ыва·ть, -ют, (*imp*) comb, comb out/onto, (text) nap
на·чёст·ь, (*fut 3rd sing, pl*) **на·чт·ёт, на·чт·у́т,** (*past masc sing*) **на·чёл,** (*perf*); see **на·чи́т·ыва·ть**
на·чёт·, -а, (*m*) deduction for unauthorized expenditure
на·чёт·ист·ый, -ая, -ое, (*a*) too dear/expensive, not profitable
на·чёт·ническ·ий, -ая, -ое, (*a*) dogmatic
на·чёш·ут (*fut 3rd pl*) of **на·чес·а́ть**
на·чин·а́ни·е, -я, (*n*) enterprise, foundation
на·чин·а́тел·ь, -я, (*m*) initiator, founder
на·чин·а́тельн·ый, -ая, -ое, (*a*) (gram) inceptive
на·чин·а́·ть, -ют, (*imp*) begin, start, commence, initiate
на·чин·а́ющ·ий, -ая, -ее, (*pres part act*) of **на·чин·а́·ть;** (*m decl as a*) beginner, novice

на·чин·а́ющ·ийся, -аяся, -ееся, (*pres part pass*) of **на·чин·а́·ться;** incipient

на·чи́н·ива·ть, -ют, (*imp*) mend, repair; sharpen, point

на·чин·и́ть, -я́т, (*perf*) mend, repair; sharpen, point; stuff, fill

на·чи́н·к·а, -и, (*g pl*) **-н·ок·,** (*f*) (food) stuffing, filling

на·чи́н·очн·ый, -ая, -ое, (*a*) of **на·чи́н·к·а**

на·чин·я́·ть, -ют, (*imp*) stuff, fill

на·числ·е́ни·е, -я, (*n*) (*v n*) see **на··числ·я́·ть;** credit, payment

на·чи́сл·ить, -ят, (*perf*) see **на·чи·сл·я́·ть**

на·числ·я́·ть, -ют, (*imp*) reckon in, add on (an additional sum); credit (an account)

на·чи́ст·ить, -ят, (*perf*) see **на·чи·щ·а́·ть**

на́·чист·о (*adv*) till fine, fine, to a finish; finish; absolutely, completely; open, above-board

~, ков·а́ть finish-forge

~, об·раба́т·ыва·ть (tech) finish

~, строг·а́·ть finish-plane

~, фрезер·ова́ть fine-mill

на·чист·от·у́ (*adv*) frankly, openly

на·чи́т·анн·ый, -ая, -ое, (*a*) well-read

на·чи́т·ыва·ть, -ют, (*imp*) deduct for unauthorized expenditure

на·чищ·а́·ть, -ют, (*imp*) peel, shell, scale (a quantity of); clean/polish well

на·чн·ёт (*fut 3rd sing*) of **на·ча́·ть** (*perf*); see **на·чин·а́·ть** (*imp*)

на·чн·у́т (*fut 3rd pl*) of **на·ча́·ть** (*perf*); see **на·чин·а́·ть** (*imp*)

наш· (*pron, nom sing m, acc sing m inanim*) our; (telecom) letter N

на́ш·а (*pron, nom sing f*) of **наш·** our

нашаты́р·н·ый, -ая, -ое, (*a*) of **на·шаты́р·ь**

~ спирт· (*m*) ammonium hydroxide

нашаты́р·ь, -я, (*m*) sal ammoniac, ammonium chloride

на́ш·е (*pron, nom/acc sing n*) of **наш·** our

на́ш·его (*pron, gen sing m/n*) of **наш·** our; (*pron, acc sing m anim*) of **наш·** our

на́ш·ей (*pron, gen/dat/instr/prep f*) of **наш·** our

на·шёл (*past masc sing*) of **на·йти́** (*perf*), see **на·ход·и́ть** (*imp*)

на́ш·ем (*pron, prep sing m/n*) of **наш·** our

на́ш·ему (*pron, dat sing m/n*) of **наш·** our

на·ше́ств·и·е, -я, (*n*) invasion, inroad

на́ш·и (*pron, nom plur, acc plur inanim*) of **наш·** our

на·шив·а́·ть, -ют, (*imp*) sew on; sew (a quantity of)

на·ши́в·к·а, -и, (*g pl*) **-в·ок·,** (*f*) (*v n*) see **на·шив·а́·ть;** (mil) stripe tab, chevron

на́ш·им (*pron, instr sing m/n, dat plur m/f/n*) of **наш·** our

на́ш·ими (*pron, instr plur*) of **наш·** our

на·ши́·ть, (*fut 3rd pl*) **на·шь·ю́т,** (*perf*) see **на·шив·а́·ть**

на́ш·их (*pron, gen/prep/acc anim*) of **наш·** our

на·шла́ (*past fem sing*) of **на·йти́** (*perf*), see **на·ход·и́ть** (*imp*)

на·шл·ёт (*fut 3rd sing*) of **на·сл·а́ть**

на·шпиг·ова́ть, -у́ют, (*perf*) (food) lard

на·шпи́л·ива·ть, -ют, (*imp*) pin, spike, impale

на·шпи́л·ить, -ят, (*perf*) see **на·шпи́л·ива·ть**

на·штамп·ова́ть, -у́ют, (*perf*) see **на·штамп·о́выва·ть**

на·штамп·о́выва·ть, -ют, (*imp*) stamp, stamp out (in quantity)

на·што́п·а·ть, -ют, (*perf*) darn (a quantity of)

на́ш·у (*pron, acc sing f*) of **наш·** our

на·шум·е́ть, -я́т, (*perf*) make a noise

на·шь·ю́т (*fut 3rd pl*) of **на·ши́·ть**

на·щёлк·а·ть, -ют, (*perf*) crack

на·щип·а́ть, -лют, (*perf*) see **на··щи́п·ыва·ть**

на·щи́п·ыва·ть, -ют, (*imp*) pluck, pick (a quantity of); pinch

на·щу́п·а·ть, -ют, (*perf*) see **на·щу́·п·ыва·ть**

на·щу́п·ыватель·, -я, (*m*) feeler

на·щу́п·ыва·ть, -ют, (*imp*) feel, find by feeling

наэги́т·, -а, (*m*) (min) naegite

на·электриз·ова́ть, -у́ют, (*perf*) electrify

не (*particle*) not; no (expressing lack of degree etc.)

~ под·да·ю́щ·ийся not subject to, proof against

~ при·ним·а́·ть во в·ним·а́ни·е neglect, ignore

не

~ **раз** more than once, many times

не- (*prefix*) *see also words without the prefix and modify the sense as follows:* non-, un-, in-, im-, mis-, dis-, not very, not well

не·аддити́в·н·ый, -ая, -ое, (*a*) non-additive

не·адиабат·и́ческ·ий, -ая, -ое, (*a*) non-adiabatic

не·аккура́т·н·ый, -ая, -ое, (*a*) inaccurate; unpunctual; careless; untidy

не·акти́в·н·ый, -ая, -ое, (*a*) inactive, inert

не·акти́н·ичн·ый, -ая, -ое, (*a*) non-actinic

неандерта́л·ец·, -ль·ц·а, (*i*) **ль·ц·ем,** (*m*) (zool) Neanderthal man

неан·и́ческ·ий, -ая, -ое, (*a*) (zool) neanic

Неа́ркт·ик·а, -и, (*f*) (zool) *Nearctica*

не·аркт·и́ческ·ий, -ая, -ое, (*a*) (zool) Nearctic

не·бала́нс·н·ый, -ая, -ое, (*a*) out-of-balance

не·без·о·па́с·н·ый, -ая, -ое, (*a*) quite dangerous, insecure, unsafe

не·без·основ·а́тельн·ый, -ая, -ое, (*a*) not unfounded, not without reason

не·без·у·спе́ш·н·ый, -ая, -ое, (*a*) not unsuccessful

не·без·ынтере́с·н·ый, -ая, -ое, (*a*) quite interesting, not without interest

не·бел·ён·ый, -ая, -ое, (*a*) unbleached

неб·ес·а́ (*pl*) of **не́б·о**

не·бес·коры́ст·н·ый, -ая, -ое, (*a*) not disinterested

неб·е́сн·ый, -ая, -ое, (*a*) celestial, heavenly

~ **меха́н·ик·а** (*f*) celestial mechanics

~ **паралле́л·ь** (*f*) parallel of celestial latitude

~ **свет·и́л·о** (*n*) heavenly body

~ **с·вод·** (*m*) celestial concave, sphere of the stars, sphere of the heavens

~ **эква́тор·** (*m*) equinoctial, celestial equator

не·благо·да́р·н·ый, -ая, -ое, (*a*) ungrateful

не·благо·по·лу́чн·ый, -ая, -ое, (*a*) troubled, unfortunate

не·благо·при·я́т·н·ый, -ая, -ое, (*a*) unfavourable, adverse, poor, bad

не·благо·ро́д·н·ый, -ая, -ое, (*a*) ignoble, base

не·благо·раз·у́м·н·ый, -ая, -ое, (*a*) imprudent

нёб·н·о- (*component*) palat-, palato-

нёб·н·ый, -ая, -ое, (*a*) (anat) palatal, palatine

нёб·о, -а, (*nom pl*) **неб·ес·а́,** (*g pl*) **не·бе́с·,** (*n*) sky; (*pl*) heavens; *see also* **нёб·о** = palate

~, **вод·ян·о́е** water sky (over ice fields)

~, **лед·ян·о́е** ice blink

нёб·о, -а, (*n*) (anat) palate; *see also* **нёб·о** = sky

~, **мя́гк·ое** (anat) soft palate

не·больш·о́й, -а́я, -о́е, (*a*) small, not very big, slight

~ **выс·от·а́** (*f*) low altitude

~ **у·велич·е́ни·е** (*n*) slight increase

небо·с·во́д·, -а, (*m*) (astron) celestial concave, sphere of the stars/heavens

~, **иску́с·ствен·н·ый** artificial sky

небо·скрёб·, -а, (*m*) (arch) sky scraper

не·бре́ж·ност·ь, -и, (*f*) carelessness, (law etc.) negligence

небуля́р·н·ый, -ая, -ое, (*a*) (astron) nebular

~ **гипоте́з·а** (*f*) nebular hypothesis

не·быв·а́л·ый, -ая, -ое, (*a*) unprecedented

не·быт·и·е́, -я́, (*n*) non-existence

не·бь·ю́щ·ийся, -аяся, -ееся, (*a*) unbreakable

~ **стекл·о́** (*n*) safety glass

не·ва́жн·о it is not important, it doesn't matter

не·ва́жн·ый, -ая, -ое, (*a*) unimportant, insignificant; indifferent, poor

не·в·дал·ек·е́ (*adv*) not far off

не·ве́д·ени·е, -я, (*n*) ignorance

не·ве́д·ом·ый, -ая, -ое, (*a*) unfamiliar

не·вели́к·ий, -ая, -ое, (*a*) small, slight, unimportant

не·ве́р·н·ый, -ая, -ое, (*a*) untrue, incorrect, wrong, mis-; unfaithful; uncertain, faltering

~ **в·ключ·е́ни·е** (*n*) (telecom) misconnection, wrong connection

~ **нул·ь** (*m*) (instr) false zero

~ **от·с·чёт·** (*m*) (instr) misreading

не·вероя́т·н·ый, -ая, -ое, (*a*) improbable, incredible, unlikely; impossible

не·ве́р·ующ·ий, -ая, -ее, (*a*) disbelieving, incredulous

не·вес·о́мост·ь, -и, (*f*) weightlessness, zero-*g*

не·вес·о́м·ый, -ая, -ое, (*a*) weightless; imponderable

не·взаимо·за·мест·и́мост·ь, -и,   (*f*) non-interchangeability

не·в·зир·а́·я на   in spite of, regardless of

не·вз·нос·, -а,   (*m*) non-payment

не·взо́·рв·анн·ый, -ая, -ое,   (*a*) (expl) unfired, unexploded, misfired, dud

не·вз·рыв·чат·ый, -ая, -ое,   (*a*) non-explosive

не·вз·ыск·а́тельн·ый, -ая, -ое,   (*a*) unexacting

не·ви́д·анн·ый, -ая, -оё,   (*a*) unprecedented

не·ви́д·им·ый, -ая, -ое,   (*a*) invisible

не·ви́д·н·ый, -ая, -ое,   (*a*) unseen; unimportant; plain, ugly

не·вибр·и́р·ующ·ий, -ая, -ее,   (*a*) non-vibrating, anti-rattle

не·ви́н·н·ый, -ая, -ое,   (*a*) innocent; ingenuous; virgin

не·вихр·ев·о́й, -а́я, -о́е,   (*a*) irrotational, non-circuit, non-cyclical

не·в·ключ·ённ·ый, -ая, -ое,   (*a*) disconnected, off, not switched on, switched off, off-duty

не·в·мен·я́ем·ый, -ая, -ое,   (*a*) irresponsible

не·в·меш·а́тельств·о, -а,   (*n*) non-interference/intervention

не·в·ня́т·н·ый, -ая, -ое,   (*a*) unintelligible

не́вод·, -а,   (*nom pl*) -а́, (*g pl*) -о́в,   (*m*) (fish) seine net

~, за·ки́д·н·ый   (fish) seine, seine net

~, за·мёт·н·ый   trammel net

~, кошель·ко́в·ый   purse seine

~, много·сте́н·н·ый   trammel net

~, став·н·о́й   (fish) stationary net

не́вод·н·ый, -ая, -ое,   (*a*) of не́·вод·; не·во́д·н·ый, -ая, -ое,   (*a*) non-aqueous, anhydrous

не·водо·сто́й·к·ий, -ая, -ое,   (*a*) hydrolabile

не·воз·бужд·ённ·ый, -ая, -ое,   (*a*) (phys) unexcited

не·возвратно·за·по́р·н·ый   кла́пан·   (*m*) stop-and-check valve

не·воз·вра́т·н·ый, -ая, -ое,   (*a*) irrevocable, irretrievable; non-return, irreversible

~ кла́пан·   (*m*) non-return valve, check valve

~ проце́сс·   (*m*) irreversible process

не·воз·де́л·анн·ый, -ая, -ое,   (*a*) uncultivated, unworked, untilled, untouched

не·воз·мо́ж·н·о   it is impossible, it is not possible

не·воз·мо́ж·ност·ь, -и,   (*f*) impossibility

знак· не·воз·мо́ж·ност·и при·ём·а   (*m*) (telecom) illegibility sign

не·воз·мо́ж·н·ый, -ая, -ое,   (*a*) impossible

не·воз·мущ·ённ·ый, -ая, -ое,   (*a*) undisturbed, (phys) unperturbed

не·во́ль·н·ый, -ая, -ое,   (*a*) involuntary; unintentional; forced

не·во·оруж·ённ·ый, -ая, -ое,   (*a*) unarmed, unequipped

не·во·оруж·ённ·ым   гла́з·ом   (*adv*) with the naked eye

не·вос·пламен·я́ем·ый, -ая, -ое,   (*a*) non-inflammable, safety (film)

не·вос·пламен·я́ющ·ийся,   -аяся, -ееся,   (*a*) non-inflammable, incombustible, safety (film etc.)

не·вос·при·и́м·чив·ый, -ая, -ое,   (*a*) unreceptive, non-susceptible; (med) immune

не·во·ю́ющ·ий, -ая, -ее,   (*a*) non-belligerent

не·в·по·па́д·   (*adv*) untimely, out of place

не·в·раз·ум·и́тельн·ый, -ая, -ое,   (*a*) unintelligible, incomprehensible

невр·алг·и́·я, -и,   (*f*) (med) neuralgia

невр·астен·и́·я, -и,   (*f*) (med) neurasthenia

невр·и́т·, -а,   (*m*) (med) neuritis

невро·лог·и́·я, -и,   (*f*) neurology

невр·о́н·, -а,   (*m*) (anat) neuron

невро·пте́рис·, -а,   (*m*) (pal bot) neuropteris

не·в·руч·е́ни·е, -я,   (*n*) non-delivery (of mail etc.), (law) failure to serve

не·вс·хо́ж·ест·ь, -и,   (*f*) (bot) blindness

невьянски́т·, -а,   (*m*) (min) nevyanskite

не·вы·вер·ённост·ь, -и,   (*f*) misalignment

не·вы·вод·и́м·ый, -ая, -ое,   (*a*) indelible (ink), irremovable (marks)

не·вы́·год·н·ый, -ая, -ое,   (*a*) disadvantageous, unprofitable, unremunerative

не·вы́·дел·анн·ый, -ая, -ое,   (*a*) unfinished; undressed (of leather etc.)

не·вы·зыв·а́ющ·ий, -ая, -ее,   (*a*) not causing

~ пя́тен·   non-staining

не·вы·куп·а́ем·ый, -ая, -ое,   (*a*) (fin) irredeemable

не·вы·мыв·а́ющ·ийся, -аяся, -ееся, (*a*) fast, fixed, indelible

не·вы·нос·и́м·ый, -ая, -ое, (*a*) intolerable, unbearable

не·вы·полн·е́ни·е, -я, (*n*) non-fulfilment/execution, failure to perform

~ **обяз·а́тельств·а** (law) non-feasance

не·вы·полн·и́м·ый, -ая, -ое, (*a*) impracticable, unfeasible

не·вы·прямл·я́ющ·ийся, -аяся, -ееся, (*a*) (elec) non-rectifying, ohmic

не·вы́·раж·енн·ый, -ая, -ое, (*a*) not clearly expressed, faint, indistinct

не·вы́·рожд·енн·ый, -ая, -ое, (*a*) non-degenerate, undegenerate; non-singular

не·выс·о́к·ий, -ая, -ое, (*a*) low, not very high

не·вы́·суш·енн·ый, -ая, -ое, (*a*) undried

не·вы́·ход·, -а, (*m*) absence, failure to appear (at work etc.); (print) non-issue

не·вы·цвет·а́ющ·ий, -ая, -ее, (*a*) non-blooming (of rubber)

не·вы́·чет·, -а, (*m*) (math) non-residue

~, **квадра́т·ичн·ый** (math) non-residue of the second order

~ **сте́пен·и** (*n*) non-residue of the *n*th power, (math) not an *n*th power remainder

не·вы́·ясн·енн·ый, -ая, -ое, (*a*) unclarified, unexplained, obscure

не·вя́з·к·а, -и, (*g pl*) **-з·ок·,** (*f*) discrepancy; (surv) closing error, error of closure, misclosure

не·вя́з·к·ий, -ая, -ое, (*a*) inviscid, non-viscous

невянски́т·, -а, (*m*) (min) nevyanskite

не·габари́т·, -а, (*m*) (min) oversize

не·габари́т·н·ый, -ая, -ое, (*a*) oversize, outsize

не·га́з·ов·ый, -ая, -ое, (*a*) gas-free; non-gaseous

негати́в·, -а, (*m*) (phot) negative

~, **втор·и́чн·ый** counter negative

~, **грав·и́рован·ый** line negative

~, **монт·и́руем·ый** cut negative

~, **основ·н·о́й** master negative

~, **ра́стр·ов·ый** broken-tone negative

~, **сла́б·ый** thin negative

негативно·се́т·чат·ый, -ая, -ое, (*a*) areolate

негати́в·н·ый, -ая, -ое, (*a*) of **негати́в·;** negative

негатро́н·, -а, (*m*) (phys) negatron

не·гаш·ён·ый, -ая, -ое, (*a*) unslaked (of lime)

неггера́т·иев·ые, -ых, (*pl decl as adj*) (pal bot) *Noeggerathiales*

не́·где (*adv*) there is nowhere, nowhere

не·герметиз·и́рованн·ый, -ая -ое, (*a*) not sealed, unsealed

не·гермет·и́чн·ый, -ая, -ое, (*a*) non-hermetic, not tightly sealed, leaky

не·ги́б·к·ий, -ая, -ое, (*a*) inflexible

не·гидрат·и́рованн·ый, -ая, -ое, (*a*) unhydrated

не·гла́д·к·ий, -ая, -ое, (*a*) uneven, rough

не·глуб·о́к·ий, -ая, -ое, (*a*) shallow, not very deep

негно́й-де́рев·о, -а, (*n*) (bot) yew tree, *Taxus baccata*

не·го́д·н·ый, -ая, -ое, (*a*) unfit, unsuitable, unserviceable, faulty; written off

не·голоно́м·н·ый, -ая, -ое, (*a*) (math) non-holonomic

не·гомоге́н·н·ый, -ая, -ое, (*a*) non-homogeneous, inhomogeneous

не·горизонта́ль·ност·ь, -и, (*f*) dislevelment, non-horizontality

не·гор·ю́ч·ий, -ая, -ее, (*a*) incombustible, non-combustible

не·гото́в·ый, -ая, -ое, (*a*) unprepared; unfinished

негоциа́нт·, -а, (*m*) merchant, wholesaler

не·градие́нт·н·ый, -ая, -ое, (*a*) (phys) direct, non-derivative

не·гук·о́в·ый, -ая, -ое, (*a*) (mech) non-Hookian

не·да́в·н·о (*adv*) recently, lately, not long ago

не·да́вн·ий, -яя, -ее, (*a*) recent, late

до не·да́вн·ого вре́мен·и until recently

не·дал·ёк·ий, -ая, -ое, (*a*) nearby, close, not very far; short (of journeys etc.); limited (understanding, intelligence)

не·дал·ек·о́ (*adv*) not far away/from, quite near to

не·да́р·ом (*adv*) not for nothing, not without reason; no wonder

не·движ·и́мост·ь, -и, (*f*) real estate

не·движ·и́м·ый, -ая, -ое, (*a*) of **не·движ·и́мост·ь;** immovable, fixed

не·дее·спосо́б·ност·ь, -и, (*f*) (law etc.) incapacity

не·дей·стви́тельн·ый, -ая, -ое, (*a*) invalid, void, null; RD (on worthless cheques)

не·де́й·ствующ·ий, -ая, -ее,   (*a*) non-operating, inoperative

не·дел·и́м·ый, -ая, -ое, (*a*) indivisible
~ числ·о́ (*n*) (math) prime number

не·дел·ов·о́й, -а́я, -о́е, (*a*) unbusiness-like

неде́ль·н·ый, -ая, -ое,   (*a*) weekly; week's

неде́л·я, -и, (*f*) week

не·де́л·ящ·ийся, -аяся, -ееся,   (*a*) (nucl) non-fissionable

не·держ·а́ни·е, -я, (*n*) non-retention; (med) incontinence

не·де́·ятельн·ый, -ая, -ое, (*a*) (phys) inert; (chem) nulvalent; (opt) inactive

не·диспе́рс·н·ый, -ая, -ое, (*a*) non-dispersive

не·диссоци·и́рованн·ый, -ая, -ое, (*a*) undissociated

не·дифференц·и́рованн·ый, -ая, -ое, (*a*) undifferentiated

недо- (*prefix*) under-, short, incomplete

недо·бе́г·, -а, (*m*) undershoot

недо·бир·а́·ть, -ют, (*imp*) collect/get less than expected/required

недо·бо́р·, -а, (*m*) incomplete set, incomplete collection; shortage; (fin) arrears

недо·бр·а́ть, (*fut 3rd pl*) недо·бер·у́т, (*perf*) *see* недо·бир·а́·ть

не·добро·ка́честв·енн·ый, -ая, -ое, (*a*) poor-quality, low-grade

недо·ва́р·, -а, (*m*) (chem) undercook; (weld) underfilled/sunk weld

недо·ва́р·енн·ый, -ая, -ое, (*a*) undercooked, parboiled; (weld) underfilled, sunk

не·до·ве́р·и·е, -я,   (*n*) distrust; no confidence

не·до·ве́р·чив·ый, -ая, -ое, (*a*) suspicious, mistrustful, distrustful

недо·ве́с·, -а,   (*m*) short weight, light weight, underweight

недо·ве́с·ить, -ят, (*perf*) *see* недо·ве́ш·ива·ть

недо·ве́ш·ива·ть, -ют,   (*imp*) give short weight of

не·до·во́ль·н·ый, -ая, -ое, (*a*) dissatisfied

недо·вы·полн·е́ни·е, -я, (*n*) underfulfilment

недо·гляд·е́ть, -я́т, (*perf*) overlook, neglect, be careless

недо·говор·и́ть, -я́т, (*perf*) be reticent

недо·гре́·в·, -а, (*m*) underheating

недо·гре́·т·ый, -ая, -ое, (*a*) underheated, insufficiently heated

недо·груж·а́·ть, -ют, (*imp*) underload; dispatch/send too little; not fully employ, not work to capacity, under-employ

недо·гру́з·, -а, (*n*) (*v n*) *see* недо·груж·а́·ть; shortfall, under-dispatch

недо·груз·и́ть, -я́т, (*perf*) *see* недо·груж·а́·ть

недо·гру́з·к·а, -и, (*g pl*) -з·ок·, (*f*) (*v n*) *see* недо·груж·а́·ть; under-loading; under-operation (of machines)

недо·да́в·, -а, (*m*) too low a pressure

недо·да·ва́ть, -ют, (*imp*) give/produce too little

недо·да́·ть,   (*fut 3rd pl*) недо·дад·у́т, (*perf*) *see* недо·да·ва́ть

недо·да́·ч·а, -и, (*i*) -ей, (*f*) (*v n*) *see* недо·да·ва́ть; deficiency, short supply, short delivery

недо·де́л·а·ть, -ют,   (*perf*) not make enough, make too little/few

недо·де́л·к·а, -и,   (*g pl*) -л·ок·, (*f*) incompleteness; constructional defect, faulty work

недо·держ·а́ть, -ат, (*perf*) *see* недо·де́рж·ива·ть

недо·де́рж·ива·ть, -ют, (*imp*) not hold long enough; underprocess, (phot) underexpose

недо·де́рж·к·а, -и,   (*g pl*) -ж·ек·, (*f*) (phot) underexposure

недо·же́ч·ь, (*fut 3rd pl*) недо·жг·у́т, (*perf*) *see* недо·жиг·а́·ть

недо·жж·ённ·ый, -ая, -ое, (*past part pass*) of недо·же́ч·ь; underbaked, underroasted

недо·жиг·а́·ть, -ют,   (*imp*) underbake (of bricks etc.), under-roast (of ore etc.)

недо·жо́г·, -а, (*m*) underfiring, under-baking, under-roasting

недо·зр·е́л·ый, -ая, -ое,   (*a*) unripe, immature

недо·и́м·к·а, -и,   (*g pl*) -м·ок·, (*f*) tax arrears

недо·ис·по́льз·овани·е, -я,   (*n*) incomplete utilization, underexploitation

не·до·ка́з·анн·ый, -ая, -ое, (*a*) not proved, unproven

не·до·каз·а́тельн·ый, -ая, -ое,   (*a*) inadequate (of proof), not convincing, unconvincing

не·до·каз·у́ем·ый, -ая, -ое, (*a*) not proveable

недо·ка́л·, -а, (*m*) underheating

недо·ка́т·, -а, (*m*) (met roll) underfill, off-gauge product

недо·ка́ч·ива·ть, -ют, (*imp*) under-inflate

не́до·кис·ь, -и, (*f*) (chem) suboxide ~ свин·ц·а́ lead suboxide

недо·ко́нч·енн·ый, -ая, -ое, (*a*) unfinished, incomplete

недо·ко́рм·, -а, (*m*) underfeeding

не·до́лг·ий, -ая, -ое, (*a*) brief

не·до́лг·о (*adv*) not long, not for long

не·долго·ве́ч·н·ый, -ая, -ое, (*a*) short-lived; temporary, short, transient

недо·лёт·, -а, (*m*) undershot, short (of fall); (air) undershoot

недо·ли́в·, -а, (*m*) (met) short-run casting, short-run, misrun; (plast) short-run moulding, unfilled stock

недо·лив·а́·ть, -ют, (*imp*) underfill, pour in too little

недо·ли́·т·ый, -ая, -ое, (*a*) under/short-filled; (met cast) short-run

недо·ме́р·, -а, (*m*) short measure

недо·ме́р·ива·ть, -ют, (*imp*) give short measure

недо·ме́р·ить, -ят, (*perf*) *see* **недо·ме́р·ива·ть**

недо·ме́р·ок·, -р·к·а, (*m*) undersize

недо·ме́с·, -а, (*m*) under-mixing

недо·мог·а́ни·е, -я, (*n*) poor health, indisposition, (med) dysphoria

недо·мо́л·, -а, (*m*) (food) undermilling, undergrinding; unfinished/underground oversize

недо·мы́·в·, -а, (*m*) (civ eng) sluicing residue

недо·наш·ива́ни·е, -я, (*n*) (med) premature birth

недо·нос·е́ни·е, -я, (*n*) (law) failure to report, withholding information

не·до·нос·и́тельств·о, -а, (*n*) (law) non-information

недо·но́с·ок·, -с·к·а, (*m*) (med) premature baby; miscarriage

недо·но́ш·енност·ь, -и, (*f*) premature birth

недо·но́ш·енн·ый, -ая, -ое, (*a*) premature (of birth)

недо·о·кисл·ённ·ый, -ая, -ое, (*a*) insufficiently oxidised

недо·о·цен·ённ·ый, -ая, -ое, (*past part pass*) of **недо·о·цен·и́ть**

недо·о·це́н·ива·ть, -ют, (*imp*) underestimate, underrate, undervalue

недо·о·цен·и́ть, -ят, (*perf*) *see* **недо·о·це́н·ива·ть**

недо·пла́т·а, -ы, (*f*) underpayment

недо·плат·и́ть, -ят, (*perf*) *see* **недо·пла́ч·ива·ть**

недо·пла́ч·ива·ть, -ют, (*imp*) underpay, pay short

недо·по·луч·а́·ть, -ют, (*imp*) receive/get too little

недо·по·луч·и́ть, -ат, (*perf*) *see* **недо·по·луч·а́·ть**

недо·пресс·о́вк·а, -и, (*g pl*) **-в·ок·,** (*f*) (plast) short/faulty/incomplete moulding

недо·про·бе́г·, -а, (*m*) failure to last specified lifetime, premature failure (of materials, machines etc.)

недо·про·из·во́д·ств·о, -а, (*n*) underproduction

не·до·пуст·и́м·ый, -ая, -ое, (*a*) inadmissible

не·до·пущ·е́ни·е, -я, (*n*) non-admission, prohibition

недо·рабо́т·анн·ый, -ая, -ое, (*past part pass*) of **недо·рабо́т·а·ть;** unfinished, half-finished

недо·раба́т·ыва·ть, -ют, (*imp*) work too little, work short time; leave unfinished, omit to finish; develop inadequately

недо·рабо́т·а·ть, -ют, (*perf*) *see* **недо·раба́т·ыва·ть**

недо·раз·ви́·т·ый, -ая, -ое, (*a*) underdeveloped; (biol) rudimentary

недо·раз·ум·е́ни·е, -я, (*n*) misunderstanding; lack of understanding

недо·ре́з·, -а, (*m*) under-cutting (sound recording)

не·дорог·о́й, -а́я, -о́е, (*a*) inexpensive

недо·ро́д·, -а, (*m*) (agr) crop failure, poor crop

недо́·сл·анн·ый, -ая, -ое, (*a*) shipped/supplied short

недо·смо́тр·, -а, (*m*) oversight, slip; careless inspection

недо·смотр·е́ть, -я́т, (*perf*) overlook, miss

недо·со́л·, -а, (*m*) insufficient salt, salt deficiency

недо·сп·а́ть, -я́т, (*perf*) not have enough sleep

недо·ста·ва́ть, -ю́т, (*imp*) be lacking/insufficient/short; fall short, lack

не·до·ста́вл·енн·ый, -ая, -ое, (*a*) undelivered, not delivered

недо·ста́т·ок·, -т·к·а, (*m*) shortage, deficiency, lack; fault, defect, shortcoming, disadvantage, drawback им·е́ть недо·ста́т·ок· have the disadvantage (that), the drawback is (that)

недо·ста́т·ок

~ нейтро́н·ов (nucl) neutron deficiency

недо·ста́т·очност·ь, -и, (f) lack, inadequacy, imperfection, defectiveness; (med) deficiency

недо·ста́т·очн·о (adv) insufficiently, inadequately, under-

недо·ста́т·очн·ый, -ая, -ое, (a) insufficient, deficient, inadequate; (gram) defective

недо·ста́·ть, (fut 3rd sing) **недо·ста́·-нет,** (perf) see **недо·ста·ва́ть**

недо·ста·ю́щ·ий, -ая, -ее, (pres part act) of **недо·ста·ва́ть;** missing, lacking

не·до·стиж·и́м·ый, -ая, -ое, (a) unattainable, inaccessible

не·досто·ве́р·н·ый, -ая, -ое, (a) unreliable, uncertain

не·до·сто́й·н·ый, -ая, -ое, (a) unworthy

не·до·стро́·енн·ый, -ая, -ое, (a) unfinished (of constructions); (nucl) unfilled, partly filled, incomplete

~ сло́·й (m) (nucl) unfilled level

не·до·сту́п·ност·ь, -и, (f) inaccessibility

по́люс не·до·сту́п·ност·и (m) (geog) pole of inaccessibility

не·до·сту́п·н·ый, -ая, -ое, (a) inaccessible, unapproachable; unavailable, unobtainable; unsuitable, difficult (for understanding)

недо·с·чит·а́·ться, -ются, (perf) see **недо·с·чи́т·ыва·ться**

недо·с·чи́т·ыва·ться, -ются, (imp) find something is missing, find a deficiency

недо·сып·а́·ть, -ют, (imp) pour/tip in too little; not have enough sleep, suffer from lack of sleep; **недо·сы́·п·ать, -лют,** (perf) pour/tip in too little

не·досяг·а́ем·ый, -ая, -ое, (a) unattainable

недо·ум·е́ни·е, -я, (n) perplexity, doubt, quandary

недо·у·равно·ве́ш·енн·ый, -ая, -ое, (a) inadequately balanced, not sufficiently balanced, in an insufficient state of equilibrium

недо·уч·е́ст·ь, -и, (f) not make enough allowance for

недо·хва́т·, -а, (m) shortage

не·до·хо́д·н·ый, -ая, -ое, (a) unprofitable

недо·чёт·, -а, (m) deficit, shortage; defect

не́др·а (nom pl), (g pl) **недр·,** earth's interior; mineral resources

не·дур·н·о́й, -а́я, -о́е, (a) not bad, not a bad

Нее́л·я, температу́р·а (f) (met phys) Neel temperature, antiferromagnetic point

не·есте́ств·енн·ый, -ая, -ое, (a) unnatural

не·жел·а́тельн·ый, -ая, -ое, (a) undesirable

не́жели (conj) than

не·жёстк·ий, -ая, -ое, (a) not rigid, flexible; not tight, loose (of joints); soft (of water)

не·жи·в·о́й, -а́я, -о́е, (a) inanimate, lifeless, inorganic

не·жизне·спосо́б·н·ый, -ая, -ое, (a) non-viable

не·жи·л·о́й, -а́я, -о́е, (a) uninhabited; non-residential

не́жн·ый, -ая, -ое, (a) fond, loving; delicate

не·за·ве́р·енн·ый, -ая, -ое, (a) uncertified

не·за·верш·ённ·ый, -ая, -ое, (a) incomplete, uncompleted, unfinished

~ про·из·во́д·ств·о (n) work-on-hand

не·за·ви́с·им·о (adv) independently, irrespective (of), regardless (of)

не·за·ви́с·имост·ь, -и, (f) independence, freedom

не·за·ви́с·им·ый, -ая, -ое, (a) independent, not dependent, self-contained

~ пере·ме́н·н·ая (f) (math) independent variable

не·за·гру́ж·енн·ый, -ая, -ое, (a) idle (of machines); not charged, uncharged

не·за·гряз·нённ·ый, -ая, -ое, (a) uncontaminated

не·за·до́лг·о до (adv) shortly/before, not long ago

не·за·ко́н·н·ый, -ая, -ое, (a) illegal, illicit

не·законо·ме́р·н·ый, -ая, -ое, (a) irregular, erratic

не·за·ко́нч·енн·ый, -ая, -ое, (a) unfinished, incomplete

не·за·ме́дл·енн·ый, -ая, -ое, (a) (nucl) unmoderated

не·за·медл·и́тельн·ый, -ая, -ое, (a) immediate

не·за·медл·я́ющ·ее вещ·еств·о (*n*) (nucl) non-moderator

не·за·мен·и́м·ый, -ая, -ое, (*a*) irreplaceable: indispensable

не·за·мерз·а́ющ·ий, -ая, -ее, (*a*) non-freezing, ice-free; antifreeze

~ жи́дк·ост·ь (*f*) (M/T) antifreeze

не·за·ме́т·н·ый, -ая, -ое, (*a*) imperceptible; inconspicuous; concealed, treacherous

не·за́·мк·нут·ый, -ая, -ое, (*a*) open

не·за́·нят·ый, -ая, -ое, (*a*) unoccupied, vacant, unfilled, empty; unemployed, not engaged, free

не·за·па́зд·ывающ·ий, -ая, -ее, (*a*) undelayed, (nucl) prompt

не·за·по́лн·енн·ый, -ая, -ое, (*a*) unfilled, blank (of forms), vacant

~ у·ро́вен·ь (*m*) (phys) unfilled/vacant level

не·за·ра́з·н·ый, -ая, -ое, (*a*) non-contagious

не·за·секре́ч·енн·ый, -ая, -ое, (*a*) unclassified (documents etc.)

не·за·стро́·енн·ый, -ая, -ое, (*a*) undeveloped, vacant· (of sites); (nucl) unfilled, vacant

~ оболо́ч·к·а (*f*) (nucl) unfilled shell

не·за·тен·ённ·ый, -ая, -ое, (*a*) unshaded

не·за·то́пл·енн·ый, -ая, -ое, (*a*) not submerged; (hydro) free, free-fall

не·за·тро́·нут·ый, -ая, -ое, (*a*) untouched, unimpaired

не·за·тух·а́ющ·ий, -ая, -ее, (*a*) (phys) undamped, sustained, continuous

~ колеб·а́ни·я (*pl*) sustained vibrations/oscillations

~ тон· (*m*) (teleph) sustained note

не·за·у·ря́д·н·ый, -ая, -ое, (*a*) out of the ordinary, outstanding, unusual, exceptional

не́·за·чем (*adv*) not necessarily; there is no need

не·здоро́в·ост·ь, -и, (*f*) unwholesomeness, insanitary state

не·здоро́в·ый, -ая, -ое, (*a*) sickly, poorly, not strong; unhealthy, bad, unwholesome

не·здоро́в·ь·е, -я, (*n*) ill health

не·зеркаль·н·ый, -ая, -ое, (*a*) non-specular

не·зерн·и́ст·ый, -ая, -ое, (*a*) non-granular

~ лейко·ци́т·, (*m*) (med) agranulocyte

не·знак·о́м·ый, -ая, -ое, (*a*) unknown, unfamiliar; unacquainted, unconversant

не·зн·а́ни·е, -я, (*n*) ignorance

~, кон·е́чн·ое (cyb) final ignorance

не·зна́ч·ащ·ий, -ая, -ее, (*a*) without significance, unimportant; (math) non-significant

не·знач·и́тельн·ый, -ая, -ое, (*a*) insignificant, unimportant, trivial, negligible; small, minute, very little

не·зр·е́л·ый, -ая, -ое, (*a*) unripe, immature

не·идеа́ль·н·ый, -ая, -ое, (*a*) non-ideal, imperfect

не·из·бе́ж·н·ый, -ая, -ое, (*a*) unavoidable, inescapable, inevitable

не·из·ве́ст·н·ый, -ая, -ое, (*a*) unknown; **не·из·ве́ст·н·ое, -ого,** (*n decl as adj*), (math) unknown quantity

не·из·глад·и́м·ый, -ая, -ое, (*a*) indelible

не·и́з·да·нн·ый, -ая, -ое, (*a*) unpublished

не·изентроп·и́ческ·ий, -ая, -ое, (*a*) anisentropic

не·из·ме́н·н·ый, -ая, -ое, (*a*) unalterable, invariable, unchangeable, inconvertible, constant; unchanged, unvaried

не·из·мен·я́ем·ый, -ая, -ое, (*a*) invariable, unchangeable, immutable, unalterable

не·из·мер·и́м·ый, -ая, -ое, (*a*) immeasurable

не·изоме́р·н·ый, -ая, -ое, (*a*) (chem) anisomeric

не·изотро́п·ност·ь, -и, (*f*) anisotropy

не·изотро́п·н·ый, -ая, -ое, (*a*) anisotropic

не·им·е́ни·е, -я, (*n*) lack

не·индукти́в·н·ый, -ая, -ое, (*a*) (elec) non-inductive

не·иониз·у́ющ·ий, -ая, -ее, (*a*) non-ionizing

не·ионоге́н·н·ый, -ая, -ое, (*a*) non-ionogenous/ionic

не·ис·каж·ённ·ый, -ая, -ое, (*a*) undistorted

не·и́скр·ящ·ийся, -аяся, -ееся, (*a*) (elec) non-arcing; non-sparking

не·иску́с·н·ый, -ая, -ое, (*a*) unskilled, inexpert

не·ис·полн·е́ни·е, -я, (*n*) non-fulfilment; non-compliance (with conditions of a contract), non-execution

не·ис·по́лн·енн·ый, -ая, -ое,   (*a*) unfulfilled, unrealized

не·ис·полн·и́м·ый, -ая, -ое,   (*a*) impracticable, unrealizable

не·ис·прав·и́м·ый, -ая, -ое,   (*a*) irreparable; incorrigible

не·ис·пра́в·ност·ь, -и,   (*f*) inaccuracy; fault, trouble, something wrong; state of disrepair

~ дви́г·ател·я   engine trouble, something wrong with the engine

не·ис·пра́в·н·ый, -ая, -ое,   (*a*) faulty, defective, out of order; wrong, incorrect

не·ис·пы́т·анн·ый, -ая, -ое,   (*a*) untested, untried

не·ис·пы́т·ываем·ый, -ая, -ое,   (*a*) not subject to, not exposed to

не·ис·сле́д·ованн·ый, -ая, -ое,   (*a*) unexplored, uninvestigated

не·иссяк·а́ем·ый, -ая, -ое,   (*a*) inexhaustible

не·исчез·а́ющ·ий, -ая, -ее,   (*a*) non-vanishing, non-zero; (chem) persistent

не·ис·черп·а́ем·ый, -ая, -ое,   (*a*) inexhaustible

не·ис·числ·и́м·ый, -ая, -ое,   (*a*) innumerable; indeterminable, incalculable

ней   (*pron, prep*) of **он·а́** (about etc.) her, it

нейзи́льбер·, -а,   (*m*) nickel silver, German silver

нейкокси́н·, -а,   (*m*) (phot chem) new coccine

нейло́н·, -а,   (*m*) nylon

Не́йман·а, за·да́·ч·а   (*f*) (math) Neumann problem, second boundary value problem of potential theory

~, фу́нкци·я   (*f*) (math) Neumann function, Bessel function of the 2nd kind

не́йман·ов·ы ли́н·и·и   (*f pl*) (met) Neumann bands/lamellae

нейри́н·, -а,   (*m*) (chem) neurine

нейри́т·, -а,   (*m*) (med) neuritis

нейро-   (*int component*) (*and see also* **невро**) neuro-

нейро·гипофи́з·, -а,   (*m*) (med) neurohypophysis

нейро·гли́·я, -и,   (*f*) (biol) neuroglia

нейро·керати́н·, -а,   (*m*) (chem) neurokeratin

нейро·лимфоматоз·, -а,   (*m*) (vet) neurolymphomatosis

нейро·эпители·й, -я,   (*m*) (biol) neuroepithelium

нейси́льбер·, -а,   (*m*) = **нейзи́льбер·**

нейтралер·, -а,   (*m*) (elec) neutralator, earthing reactor

нейтрализ·а́тор·, -а,   (*m*) neutralizer, (chem) neutralizing tank

нейтрализ·а́ция, -и,   (*f*) neutralization

нейтрализ·о́ванн·ый, -ая, -ое,   (*past part pass*) of **нейтрализ·ова́ть;** neutralized

нейтрализ·ова́ть, -у́ют,   (*imp or perf*) neutralize

нейтрал·ите́т·, -а,   (*m*) neutrality

нейтра́л·ь, -и,   (*f*) (elec) neutral, neutral point, neutral line, neutral conductor

нейтра́ль·н·ый, -ая, -ое,   (*a*) neutral

~ зо́н·а   (*f*) neutral zone

~ сло́·й   (*m*) (mech) neutral axis (of a beam in bend test)

нейтри́н·н·ый, -ая, -ое,   (*a*) neutrino

нейтри́н·о, -а,   (*n*) (nucl) neutrino

нейтроди́н·, -а,   (*m*) (rad) neutrodyne

нейтроди́н·н·ый, -ая, -ое,   (*a*) (rad) neutrodyne

~ конденс·а́тор·   (*m*) (rad) neutrodyne

нейтро́н·, -а,   (*m*) (phys) neutron

~, бы́стр·ый   fast neutron

~ дел·е́ни·я   fission neutron

~, за·ме́дл·енн·ый   delayed/moderated neutron

~ ис·пы́т·авш·ий с·толк·нове́ни·е   degraded neutron

~, ме́дл·енн·ый   slow neutron

~, перв·и́чн·ый   virgin neutron

~, тепл·ов·о́й   thermal neutron

~, эпи·тепл·ов·о́й   epithermal neutron

нейтро́н·н·ый, -ая, -ое,   (*a*) of **нейтро́н·**

~ ис·то́ч·ник·   (*m*) neutron source

~ каротта́ж·   (*m*) (min) neutron-bombardment well logging

нейтроно·гра́мм·а, -ы,   (*f*) (phys instr) neutron diffraction pattern

нейтроно·гра́ф·, -а,   (*m*) neutron-diffraction apparatus/camera

нейтроно·граф·и́ческ·ий, -ая, -ое,   (*a*) neutron-diffraction

нейтроно·гра́ф·и·я, -и,   (*f*) (phys) neutron diffraction analysis/study

нейтро·фи́л·, -а,   (*m*) (biol) neutrophyl

цека́л·ь, -я,   (*m*) (chem) neckal, sodium dibutylnaphthalene sulphonate

не·капт·и́рованн·ый, -ая, -ое,   (*a*) wild (of oil wells)

не·квалифиц·и́рованн·ый, -ая, -ое,   (*a*) unqualified, without qualifications; unskilled

не·кван́т·ованн·ый, -ая, -ое, (*a*) (phys) unquantized

нéк·ий, -ая, -ое, (*pron*) *see* **нé·котор·ый**

некк·, -а, (*m*) (geol) neck

неклён·, -а, (*m*) (bot) box-elder, *Acer negundo*

не·ков́·к·ий, -ая, -ое, (*a*) (met) unmalleable

нé·когда (*adv*) no time, no spare time; sometime, formerly

не·когерéнт·ност·ь, -и, (*f*) (phys) incoherence

не·когерéнт·н·ый, -ая, -ое, (*a*) incoherent, non-coherent

не·компетéнт·н·ый, -ая, -ое, (*a*) inefficient, incompetent; not competent

не·комплéкт·н·ый, -ая, -ое, (*a*) incomplete, having an incomplete set/outfit; odd, not belonging to a complete set/outfit, extra

не·кондицион́·н·ый, -ая, -ое, (*a*) off-grade, substandard; dirty (of ores etc.)

не·консеквéнт·н·ый, -ая, -ое, (*a*) inconsequent

~ **рек·á** (*f*) (geog) inconsequent river

не·контáкт·н·ый, -ая, -ое, (*a*) non-contact, influence (of firing mechanisms)

~ **вз·рыв·áтел·ь** (*m*) (gunn) influence fuze

не·копт·я́щ·ий, -ая, -ее, (*a*) smokeless

не·короди́й·н·ый, -ая, -ое, (*a*) non-corroding

нé·котор·ый, -ая, -ое, (*pron*) some, a certain

не·кристалл·и́ческ·ий, -ая, -ое, (*a*) non-crystalline, amorphous

не·крити́ч·еск·ий, -ая, -ое, (*a*) non-critical

не·крити́ч·ност·ь, -и, (*f*) non-criticality

некро·био́з·, -а, (*m*) (cyt) necrobiosis

некр·óз·, -а, (*m*) (biol) necrosis

некро·лóг·, -а, (*m*) obituary

некро·ти́ческ·ий, -ая, -ое, (*a*) necrotic

не·круг·ов·óй, -áя, -óе, (*a*) non-circular

не·кстáти (*adv*) inopportunely, untimely; irrelevantly

нектáр·, -а, (*m*) (bot) nectar

нектáр·ник·, -а, (*m*) (bot) nectary

нé·кто (*pron*) someone, a certain

нект·óн·, -а, (*m*) (biol) nekton, necton

некто·фóр·, -а, (*m*) (zool) nectophore, nectocalyx

нé·куда (*adv*) nowhere

не·культýр·н·ый, -ая, -ое, (*a*) uncultured; rude, coarse; (bot) wild

не·кур·я́щ·ий, -ая, -ее, (*a*) non-smoking; (*as noun*) non-smoker

не·латери́т·н·ый, -ая, -ое, (*a*) non-laterited (of soils)

не·легáль·н·ый, -ая, -ое, (*a*) illegal

не·леги́р·ованн·ый, -ая, -ое, (*a*) unalloyed

не·лёгк·ий, -ая, -ое, (*a*) quite a heavy, quite heavy

не·лёт·н·ый, -ая, -ое, (*a*) unsuitable for flying

~ **погóд·а** (*f*) non-flying weather

не·лет·ýч·ий, -ая, -ее, (*a*) non-volatile

не·ликви́д·, -а, (*m*) unconvertible stock

не·ликви́д·н·ый, -ая, -ое, (*a*) (fin) inconvertible, unconvertible

не·лин·éйност·ь, -и, (*f*) non-linearity; (telecom) non-linear distortion

коэфициéнт· не·лин·éйност·и (*m*) (telecom) non-linear distortion factor

не·лин·éйн·ый, -ая, -ое, (*a*) non-linear

~ **ис·каж·éни·е** (*n*) (elec) non-linear distortion

коэфициéнт· не·лин·éйн·ых ис·каж·éни·й (*m*) (elec) non-linear distortion factor

не·лóв·к·ий, -ая, -ое, (*a*) awkward, clumsy; inconvenient, uncomfortable

не·лог·и́чн·ый, -ая, -ое, (*a*) illogical

нельзя́ it is impossible, one cannot; one must/should/ought not; not allowed, prohibited

нём (*pron, prep*) of **он·** (about etc.) him/it

не·магни́т·н·ый, -ая, -ое, (*a*) non-magnetic

не·мáл·о (*adv*) quite a lot, much, greatly, not a little

не·мало·вáж·н·ый, -ая, -ое, (*a*) quite important, not unimportant

не·мáл·ый, -ая, -ое, (*a*) quite a lot, quite a; no little, not a little; quite important

не·мáр·очн·ый, -ая, -ое, (*a*) off-grade, not graded

немато·блáст·и́ческ·ий, -ая, -ое, (*a*) (zool) nematoblastic

немат·о́д·а, -ы, (*f*) (zool) nematode, (*pl*) *Nematoda*

~, ци́трус·ов·ая *Tylenchulus semipenetrans*

немато·мо́рф·н·ый, -ая, -ое, (*a*) nematomorphic

не·ме́дл·енн·о (*adv*) immediately, at once

не·ме́дл·енн·ый, -ая, -ое, (*a*) immediate

немерти́н·ы (*pl*) ribbon worms, (zool) *Nemertinea*

не·мета́лл·, -а, (*m*) (chem) non-metal

не·металл·и́ческ·ий, -ая, -ое, (*a*) non-metallic

~ в·ключ·е́ни·е (*n*) (met) non-metallic inclusion

нем·е́·ть, -ют, (*imp*) become dumb; become numb

не́м·ец, -м·ц·а, (*i*) **-м·ц·ем,** (*m*) German

нем·е́цк·ий, -ая, -ое, (*a*) German

не·мно́г·ие, -их, (*pl*) few, a few, not much/many

не·мно́г·о (*pron*) a little/few, some; (*adv*) slightly, somewhat

не·много·сло́в·н·ый, -ая, -ое, (*a*) laconic, short

не·много·чи́сл·енн·ый, -ая, -ое, (*a*) small, consisting of only a few; scarce

не·мк·у́щ·ийся, -аяся, -ееся, (*a*) (text) uncrushable, crease-resisting

нем·о́й, -а́я, -о́е, (*a*) dumb, mute; (mech eng) blank; outline (of drawings, plans, maps)

не·моното́н·н·ый, -ая, -ое, (*a*) non-monotonic

нем·от·а́, -ы́, (*f*) dumbness, (med) mutism

не́·моч·ь, -и, (*f*) illness, sickness, infirmity, poor health

не·мощ·ён·ый, -ая, -ое, (*a*) unpaved

не́·мощ·н·ый, -ая, -ое, (*a*) poorly, unwell, not strong

не́·мощ·ь, -и, (*f*) = **не́·моч·ь**

не·мы́сл·им·ый, -ая, -ое, (*a*) unthinkable, senseless

не·на·ви́д·еть, -ят, (*imp*) hate, detest

не·на·ви́ст·н·ый, -ая, -ое, (*a*) hated, detested

не·на·гру́ж·енн·ый, -ая, -ое, (*a*) idle, light (of machines)

не·на·дёж·н·ый, -ая, -ое, (*a*) unreliable; insecure

не·над·леж·а́щ·ий, -ая, -ее, (*a*) improper, unfitting, inappropriate

не·на́·доб·н·ый, -ая, -ое, (*a*) useless

не·на·до́лг·о (*adv*) for a short time, not for long

не·на·по́лн·енн·ый, -ая, -ое, (*a*) unfilled, unadulterated, pure

~ каучу́к· (*m*) unfilled/uncompounded rubber

не·на·пра́вл·енн·ый, -ая, -ое, (*a*) non-directional, non-directive

~ радио·ма́як· (*m*) non-directive radio beacon

не·на·ру́ш·енн·ый, -ая, -ое, (*a*) undisturbed

нена́ст·н·ый, -ая, -ое, (*a*) (meteor) bad, inclement

не·на·стро́·енн·ый, -ая, -ое, (*a*) untuned, not adjusted, unadjusted

нена́ст·ь·е, -я, (*n*) inclement/bad/foul weather

не·на·сы́т·н·ый, -ая, -ое, (*a*) insatiable

не·на·сы́щ·енн·ый, -ая, -ое, (*a*) non-saturated, (chem) unsaturated

не·на·тя́·нут·ый, -ая, -ое, (*a*) non-tension, slack

не·на·у́ч·н·ый, -ая, -ое, (*a*) unscientific

не·нес·у́щ·ий, -ая, -ее, (*a*) (civ eng) non-bearing, non load-bearing

~ ба́л·к·а (*f*) (civ eng) false beam/bearing

не·норма́ль·н·ый, -ая, -ое, (*a*) abnormal

не·ну́ж·н·о (*adv*) unnecessarily, not necessarily; it is not necessary

не·ну́ж·н·ый, -ая, -ое, (*a*) unnecessary, needless

не·нул·ев·о́й, -а́я, -о́е, (*a*) (math) non-zero, non-trivial

не·об·де́л·анн·ый, -ая, -ое, (*a*) (tech) unfinished, untrimmed; unmounted (of jewelry)

не·об·ду́м·анн·ый, -ая, -ое, (*a*) ill-considered

не·обеспе́ч·енн·ый, -ая, -ое, (*a*) without sufficient means

не·об·жи·т·о́й, -а́я, -о́е, (*a*) unoccupied, empty (of dwellings)

не·обит·а́ем·ый, -ая, -ое, (*a*) uninhabited

не·о·богащ·ённ·ый, -ая, -ое, (*a*) unenriched; (min) unconcentrated, run-of-the-mine

не·об·осно́в·анн·ый, -ая, -ое, (*a*) unsubstantiated

не·об·рабо́т·анн·ый, -ая, -ое, (*a*) untreated, unworked, undressed, raw, rough

не·об·раст·а́ющ·ий, -ая, -ее, (*a*) (civ eng) anti-fouling

~ **кра́с·к·а** (*f*) anti-fouling paint/composition

не·обрат·и́мост·ь, -и, (*f*) irreversibility; (fin) inconvertibility

не·обрат·и́м·ый, -ая, -ое, (*a*) irreversible; (fin) inconvertible; (autom) non-conservative

не·об·резин·ённ·ое ме́ст·о (*n*) holiday (on rubberized goods)

не·об·рез·н·о́й, -а́я, -о́е, (*a*) untrimmed

не·об·са́ж·енн·ый, -ая, -ое, (*a*) (oil) not cased

~ **интерва́л·** (*m*) barefoot interval (in well)

не·об·слу́ж·иваем·ый, -ая, -ое, (*a*) unattended, unwatched, non-attendant

не·об·сто·я́тельн·ый, -ая, -ое, (*a*) superficial

не·об·тек·а́ем·ый, -ая, -ое, (*a*) (dynam) bluff, not streamlined, un-streamlined

~ **лы́ж·а** (*f*) (a/c) flat ski

не·об·ход·и́м·о (*adv*) necessarily; it is necessary, one must, it must be

для э́то не·об·ход·и́м·о ... this requires/needs ...

~ **ко́нч·ить рабо́т·у** the work must be finished

не·об·ход·и́мост·ь, -и, (*f*) necessity, need

в слу́ч·а·е не·об·ход·и́мост·и if necessary

нет не·об·ход·и́мост·и there is no need

не·об·ход·и́м·ый, -ая, -ое, (*a*) necessary, unavoidable, indispensible

~ **и до·ста́т·очн·ое у·сло́в·и·е** (*n*) (math) necessary and sufficient condition

не·объ·ясн·и́м·ый, -ая, -ое, (*a*) inexplicable, unaccountable

не·обык·нове́нн·ый, -ая, -ое, (*a*) unusual, extraordinary

не·обыч·а́йн·ый, -ая, -ое, (*a*) extraordinary, exceptional

не·обы́ч·н·ый, -ая, -ое, (*a*) unusual

не·обяз·а́тельн·ый, -ая, -ое, (*a*) not obligatory, optional; insincere

неогекса́н·, -а, (*m*) (chem) neohexane, 2,2-dimethylbutane

нео·ге́н·, -а, (*m*) (geol) Neogene

нео·ге́·я, -и, (*f*) (zool) Neogaean realm, *Neogaea*

нео·гнат·и́ческ·ий, -ая, -ое, (*a*) (zool) neognathous

не·о·гран·и́ченн·о (*adv*) unreservedly; indefinitely, endlessly

не·о·гран·и́ченн·ый, -ая, -ое, (*a*) unbounded, unlimited; absolute

нео·дарвин·и́зм·, -а, (*m*) (zool) Neo-Darwinism

неоди́м·, -а, (*m*) (chem) neodymium, Nd

не·один·а́ков·ый, -ая, -ое, (*a*) unlike, dissimilar

не·одно·вре́мен·н·ый, -ая, -ое, (*a*) (phys) noncontemporary

не·одно·зна́ч·ност·ь, -и, (*f*) ambiguity

не·одно·зна́ч·н·ый, -ая, -ое, (*a*) ambiguous

не·одно·кра́т·н·ый, -ая, -ое, (*a*) multiple, repeated, reiterated

не·одно·ро́д·ност·ь, -и, (*f*) non-homogeneity, non-uniformity, heterogeneity, inhomogeneity, irregularity, discontinuity; (cryst) imperfection

~, **акуст·и́ческ·ая** acoustic discontinuity

коэффицие́нт· не·одно·ро́д·ност·и (*m*) irregularity factor

~, **то́ч·ечн·ая** (cryst) point imperfection

~, **штри́х·ов·ая** (cryst) line imperfection

~, **электро·хим·и́ческ·ая** electrochemical contrast

не·одно·ро́д·н·ый, -ая, -ое, (*a*) non-homogeneous, non-uniform, heterogeneous, inhomogeneous

не·одно·с·вя́з·н·ый, -ая, -ое, (*a*) (math) not simply connected, multiply connected

не·о·живл·ённ·ый, -ая, -ое, (*a*) inanimate

не·о·жид·а́нн·ый, -ая, -ое, (*a*) unexpected

нео·зо́·й, -я, (*m*) (geol) Neozoic, Tertiary group

неозо́н·, -а, (*m*) (chem) neozone

не·о·кисл·ённ·ый, -ая, -ое, (*a*) unoxidized

не·о·кисл·я́ем·ый, -ая, -ое, (*a*) unoxidizable, inoxidable, non-oxidizing

неоко́м·, -а, (*m*) (geol) Neocomian

нео·ламарк·и́зм·, -а, (*m*) (genet) Neo-Lamarckism

неолейкори́т·, -а, (*m*) (plast) neoleukorite (phenol-form-aldehyde resin)

нео·ли́т·, -а, (*m*) (geol) Neolithic

нео·лог·и́зм·, -а, (*m*) (gram) neologism

нео́н·, -а, (*m*) (chem) neon, Ne

нео́н·ов·ый, -ая, -ое, (*a*) of **нео́н·**

нео·паллиу́м, -а, (*m*) (zool) neopallium

нео·пента́н·, -а, (*m*) (chem) neopentane, tetramethyl-methane

нео·пи́р·ов·ый, -ая, -ое, (*a*) (geol) neopyric

нео·пла́зи·я, -и, (*f*) (med) neoplasia, neoplasm

не·о·по́·зна·нн·ый, -ая, -ое, (*a*) unidentified

не·о·пре·дел·ённост·ь, -и, (*f*) uncertainty, indeterminancy, indefiniteness; (math) indeterminate form

~, нул·ев·а́я zero-point uncertainty

принци́п не·о·пре·дел·ённост·и (*m*) uncertainty principle

то́ч·к·а не·о·пре·дел·ённост·и, о·со́б·ая (*f*) (math) point of indefiniteness

не·о·пре·дел·ённ·ый, -ая, -ое, (*a*) indeterminate, undetermined, indefinite, diophantine; vague, uncertain, undefined

~ вы·раж·éни·е (*n*) (math) indeterminate form

~ интегра́л· (*m*) (math) indefinite integral

~ коэфициéнт· (*m*) (math) undetermined coefficient

~ на·клон·éни·е (*n*) (gram) infinitive

~ у·равн·éни·е (*n*) (math) diophantine/indeterminate equation

~ фо́рм·а (*f*) (math) positive semi-definite form; (gram) infinitive

не·о·пре·дел·и́м·ый, -ая, -ое, (*a*) indeterminate, indefinable

неопре́н·, -а, (*m*) (plast) neoprene

не·о·про·верж·и́м·ый, -ая, -ое, (*a*) irrefutable

не·о·пущ·ённ·ый, -ая, -ое, (*a*) (bot) hairless, glabrous

не·о́·пыт·н·ый, -ая, -ое, (*a*) inexperienced

не·организ·о́ванность·ь, -и, (*f*) lack of organization

не·организ·о́ванн·ый, -ая, -ое, (*a*) unorganized, badly organized; unmethodical

не·орган·и́ческ·ий, -ая, -ое, (*a*) inorganic

неосальварса́н·, -а, (*m*) (pharm) neosalvarsan

не·о·с·вед·омлённ·ый, -ая, -ое, (*a*) uninformed, ignorant

не·ос·ев·о́й, -а́я, -о́е, (*a*) off-axial

не·о·сла́б·н·ый, -ая, -ое, (*a*) unremitting, constant

не·о·мы́сл·енн·ый, -ая, -ое, (*a*) ill-considered

не·основ·а́тельн·ый, -ая, -ое, (*a*) groundless, unfounded

не·основ·н·о́й, -а́я, -о́е, (*a*) non-basic, non-dominant, (phys) minority

~ нос·и́тел·ь (*m*) minority carrier

не·о·со́б·енн·ый, -ая, -ое, (*a*) (math) non-singular

нео·спори́д·и, -й, (*f pl*) (zool) *Neosporidia*

не·о·спор·и́м·ый, -ая, -ое, (*a*) indisputable

не·о·сторо́ж·ност·ь, -и, (*f*) carelessness, (law) negligence

не·о·сторо́ж·н·ый, -ая, -ое, (*a*) careless, negligent

не·о·сущ·естви́м·ый, -ая, -ое, (*a*) impracticable, unrealizable, unfeasible, not feasible

не·от·вéт·ственн·ый, -ая, -ое, (*a*) unimportant, minor, low-grade (of material)

не·от·дел·и́м·ый, -ая, -ое, (*a*) inseparable

нео·тéни·я, -и, (*f*) (zool) neoteny

нео·ти́п·, -а, (*m*) (biol) neotype

не·от·клон·ённ·ый, -ая, -ое, (*a*) undeflected

нé·от·куда (*adv*) from nowhere

не·от·ло́ж·н·ый, -ая, -ое, (*a*) urgent, pressing

не·от·лу́ч·н·ый, -ая, -ое, (*a*) ever-present, always there, unfailing (in attendance)

не·от·пра́в·к·а, -и, (*g pl*) **-в·ок·,** (*f*) failure to despatch/ship

нео·тро́н·, -а, (*m*) (rad) neotron

нео·троп·и́ческ·ий, -ая -ое, (*a*) (biol) neotropical

не·от·раз·и́м·ый, -ая, -ое, (*a*) irrestistable; immeasurably great

не·от·ры́в·н·ый, -ая, -ое, (*a*) continuous, without a break

не·от·сту́п·н·ый, -ая, -ое, (*a*) unremitting, persistent

не·от·чёт·лив·ый, -ая, -ое, (*a*) indistinct, vague

не·отъ·éмл·ем·ый, -ая, -ое, (*a*) inalienable; integral

неоформали́т·, -а, (*m*) (plast) neoformalite

нео·цéн·, -а, (*m*) (geol) Neocene

не·о·чи́щ·енн·ый, -ая, -ое, (*a*) unrefined, unpurified, impure, untreated; (oil) crude

не·ощут·и́м·ый, -ая, -ое, (*a*) imperceptible

не·па́р·ник·, -а, (*m*) (ent) *Oeneria dispar*

не·парно·копы́т·н·ый, -ая, -ое, (*a*) odd-toed, perissodactyl; (*pl as noun*) (zool) perissodactyls, *Perissodactyla*

не·па́р·н·ый, -ая, -ое, (*a*) unpaired, unmatched, odd, azygous, non-pair

не́пер·, -а, (*m*) (elec) neper (unit of attenuation)

не·пере·крыв·а́ющ·ийся, -аяся, -ееся, (*a*) non-overlapping

не·пере·вод·и́м·ый, -ая, -ое, (*a*) untranslatable, difficult to translate

не·пере·дав·а́ем·ый, -ая, -ое, (*a*) indescribable

не·пере·сек·а́ющ·ий, -ая, -ее, (*a*) non-intersecting, (math) of zero intersection

не·пере·хо́д·н·ый, -ая, -ое, (*a*) non-transition, non-transitive, intransitive, (gram) intransitive

не·период·и́ческ·ий, -ая, -ое, (*a*) aperiodic, non-periodic, acyclic

не·период·и́чност·ь, -и, (*f*) aperiodicity

не́пер·ов·о числ·о́ (*n*) (math) natural logarithm base

не́пер·ов·ы анало́г·и·и (*pl*) (math) Napier's analogies

не·пи́к·ов·ый, -ая, -ое, (*a*) off-peak

не·пла́в·к·ий, -ая, -ое, (*a*) infusible

не·пластиц·и́рованн·ый, -ая, -ое, (*a*) unplasticized

не·плат·ёж·, -а́, (*i*) -е́м, (*m*) non-payment, failure to pay

не·плодо·ро́д·н·ый, -ая, -ое, (*a*) infertile, barren

не·плод·у́щ·ий, -ая, -ее, (*a*) infertile

не·пло́т·н·ый, -ая, -ое, (*a*) uncompacted, low-density; leaky; loose (of fitting parts); (math) non-dense

~ мно́ж·ества (*n*) (math) non-dense set, nowhere dense set

не·по·воро́т·лив·ый, -ая, -ое, (*a*) sluggish, slow; clumsy, awkward, unwieldy, unhandy

не·плох·о́й, -а́я, -о́е, (*a*) not a bad

не·по·врежд·ённ·ый, -ая, -ое, (*a*) unimpaired, undamaged, intact

не·по·глощ·а́ющ·ий, -ая, -ее, (*a*) non-absorptive

не·пого́д·а, -ы, (*f*) bad weather

не·по·дал·ёк·у (*adv*) not far off, near

не·по·да́т·лив·ый, -ая, -ое, (*a*) unyielding, rigid

не·по·дат·н·о́й, -а́я, -о́е, (*a*) tax/duty-free

не·под·верг·а́ющ·ийся, -аяся, -ееся, (*a*) not subject/liable to

не·по·дви́ж·н·ый, -ая, -ое, (*a*) fixed, stationary, immovable, immobile, (zool) non-motile

~ блок· (*m*) fixed pulley

~ инде́кс· (*m*) index mark

~ то́ч·к·а (*f*) fixed point

не·под·де́л·анн·ый, -ая, -ое, (*a*) unadulterated

не·под·де́ль·н·ый, -ая, -ое, (*a*) genuine

не·по·дел·ённ·ый, -ая, -ое, (*a*) not shared, unshared

~ па́р·а (*f*) (phys) unshared pair

не·под·леж·а́щ·ий, -ая, -ее, (*a*) not subject to, exempt from

~ о·глаш·е́ни·ю not for publication

~ пере·пла́в·к·е not remeltable

не·по·до́б·н·ый, -ая, -ое, (*a*) dissimilar, unlike

не·под·пру́ж·енн·ый, -ая, -ое, (*a*) unsprung

не·под·ход·я́щ·ий, -ая, -ее, (*a*) inapplicable, unsuitable, inappropriate

~ для ры́н·к·а unmarketable

не·позицио́н·н·ый, -ая, -ое, (*a*) non-positional

не·по·ла́д·к·а, -и, (*g pl*) -д·ок·, (*f*) trouble, something wrong, breakdown

~, эксплуатацио́н·н·ая operating trouble

не·по́лн·о (*adv*) incompletely, partially, not fully; under, hypo-

не·полно·зу́б·ый, -ая, -ое, (*a*) edentate, (*pl as noun*) (zool) edentates, *Edentata*

не·полно·кристалл·и́ческ·ий, -ая, -ое, (*a*) hypocrystalline, hemicrystalline

не·полно·ме́р·н·ый, -ая, -ое, (*a*) undersize(d), short

не·полно·стекл·ова́т·ый, -ая, -ое, (*a*) hypohyaline

не·полно·це́н·н·ый, бел·о́к·, (*m*) incomplete protein

не·по́лн·ый, -ая, -ое, (*a*) partial, incomplete; (med) subtotal

зо́н·а не·по́лн·ой мо́щ·ност·и (*f*) (nucl) sub-power region

не·по́лн·ый, -ая, -ое
~ инду́кци·я　(*f*) (math) incomplete induction
~ ча́ст·н·ое　(*n*) (math) partial quotient
не·поляри́з·ованн·ый, -ая, -ое,　(*a*) non-polarized
не·поля́р·н·ый, -ая, -ое,　(*a*) non-polar
не·по·ме́р·н·ый, -ая, -ое,　(*a*) excessive; immoderate
не·по·ним·а́ни·е, -я,　(*n*) lack of understanding
не·по·ня́т·лив·ый, -ая, -ое,　(*a*) incomprehensible, unintelligible
не·по·ня́т·н·ый, -ая, -ое,　(*a*) incomprehensible; unintelligible
не·по·пад·а́ни·е, -я,　(*n*) miss
не·по·прав·и́м·ый, -ая, -ое,　(*a*) irreparable, irretrievable
не·по·раж·а́ем·ый, -ая, -ое,　(*a*) not dangerous, harmless; (mil) dead
~ про·стра́н·ств·о　(*n*) (mil) dead ground
не·по·раж·ённ·ый, -ая, -ое,　(*a*) unhit; unaffected, untouched
не·по·ря́д·ок·, -д·к·а,　(*m*) disorder, chaos, confusion
пере·хо́д· не·по·ря́д·к·а-по·ря́-д·ок·　(*m*) (phys) disorder-order transition
не·по·ря́д·оченн·ый, -ая, -ое,　(*a*) disordered, unordered
не·по·ря́д·очн·ый, -ая, -ое,　(*a*) disorderly, dishonourable, bad
не·по·си́ль·н·ый, -ая, -ое,　(*a*) too difficult/hard
не·по·след·ова́тельн·ый, -ая, -ое,　(*a*) inconsequent, inconsistent; out of order/turn
не·по·сре́д·ственн·о　(*adv*) immediately, directly, at once
~ на　on the ... itself, on the actual ...
не·по·сре́д·ственн·ый, -ая, -ое,　(*a*) immediate, direct, first-hand; (mil) close
~ в·пры́ск·　(*m*) direct injection (diesel)
~ пере·да́·ч·а　(*f*) direct drive/transmission
~ под·де́рж·к·а　(*f*) (mil) close support
~ со·се́д·ств·о　(*n*) immediate neighbourhood/vicinity
не·по·сто·я́нн·ый, -ая, -ое,　(*a*) inconstant, variable, unsettled; (geol) staggered; nonuniform (of rates, speeds etc.)

не·по·сто·я́нств·о, -а,　(*n*) inconstancy
не·по·топл·я́емост·ь, -и,　(*f*) unsinkability
та́блиц·ы не·по·топл·я́емост·и　(*pl*) (naut) counter-flooding tables
не·по·топл·я́ем·ый, -ая, -ое,　(*a*) unsinkable
не·по·хо́ж·ий, -ая, -ее,　(*a*) dissimilar, unlike
не·по·ча́·т·ый, -ая, -ое,　(*a*) unbroached, not started/opened
не·правдо·по·до́б·н·ый, -ая, -ое,　(*a*) unlikely, improbable, extraordinary
не·пра́в·ильн·о　(*adv*) wrong, incorrectly; it is not true/correct, it is incorrect/wrong
не·пра́в·ильн·ый, -ая, -ое,　(*a*) incorrect, wrong, untrue, false, mis-; improper, irregular
~ глаго́л·　(*m*) (gram) irregular verb
~ дроб·ь　(*f*) (math) improper fraction
~ от·с·чёт·　(*m*) (instr) misreading
не·право·спосо́б·ност·ь, -и,　(*f*) legal incapacity
не·право·су́д·н·ый, -ая, -ое,　(*a*) (law) wrong, improper, unjust
не·пре·взо·йд·ённ·ый, -ая, -ое,　(*a*) unsurpassed, unexcelled
не·пред·ви́д·енн·ый, -ая, -ое,　(*a*) unforeseen
не·пре·де́ль·ност·ь, -и,　(*f*) (chem) unsaturation
не·пре·де́ль·н·ый, -ая, -ое,　(*a*) unlimited, unbounded; (chem) unsaturated
не·пред·у·смо́тр·енн·ый, -ая, -ое,　(*a*) not envisaged, unprovided for, unforeseen
не·пред·у·смотр·и́тельн·ый, -ая, -ое,　(*a*) unpredictable
не·пре·кло́н·н·ый, -ая, -ое,　(*a*) unswerving, undeviating
не·пре·ме́н·н·о　(*adv*) inevitably, without fail; it is imperative
не·пре·ме́н·н·ый, -ая, -ое,　(*a*) necessary, indispensable; permanent, certain
не·пре·одол·и́м·ый, -ая, -ое,　(*a*) insuperable
не·пре·рыв·а́ем·ый, -ая, -ое,　(*a*) uninterrupted; unchopped (of rays, beams etc.)
не·пре·ры́в·н·о　(*adv*) continuously
~ -дискре́т·н·ый пре·обра́з·ова́тел·ь　analog-to-digital computer, digitizer
не·пре·ры́в·ност·ь, -и,　(*f*) continuity

не·пре·ры́в·н·ый, -ая, -ое, (*a*) continuous, uninterrupted, unbroken, continuous

~ **гру́пп·а** (*f*) (math) topological group

~ **де́й·стви·е** (*n*) continuous action, continuous output; (*in gen as adj*) continuous

~ **дроб·ь** (*f*) (math) continued fraction

~ **про·ка́т·н·ый стан·** (*m*) (met roll) continuous mill

сигна́л не·пре·ры́в·н·ого хара́ктер·а (*m*) (autom) analog signal

~ **фу́нкци·я** (*f*) (math) continuous function

не·пре·ста́н·н·ый, -ая, -ое, (*a*) unceasing, ceaseless

не·при·вод·и́м·ый, -ая, -ое, (*a*) (math) irreducible

не·при·вы́ч·н·ый, -ая, -ое, (*a*) unaccustomed

не·при·го́д·н·ый, -ая, -се, (*a*) unfit, unserviceable, unsuitable

не·при·е́мл·ем·ый, -ая, -ое, (*a*) unacceptable, objectionable

не·при·кос·нове́нност·ь, -и, (*f*) inviolability, (law) immunity

~, **диплома́т·и́ческ·ая** diplomatic immunity

не·при·кос·нове́нн·ый, -ая, -ое, (*a*) (law) protected, immune; reserve

~ **за·па́с·** (*m*) reserve funds, reserve stores; emergency rations

не·при·кры́·т·ый, -ая, -ое, (*a*) not quite closed, ajar (of doors); exposed, uncovered

не·при·мен·и́м·ый, -ая, -ое, (*a*) inapplicable; unsuitable; unusable

не·при·мир·и́м·ый, -ая, -ое, (*a*) irreconcilable, uncompromising

не·примити́в·н·ый, -ая, -ое, (*a*) (autom) finite

~ **схе́м·а** (*f*) finite automaton

не·при·ня́т·и·е, -я, (*n*) rejection

не·при·спосо́бл·енн·ый, -ая, -ое, (*a*) unadapted, not adapted

не·при·сту́п·н·ый, -ая, -ое, (*a*) inaccessible, unapproachable

не·при·су́т·ственн·ый, -ая, -ое, (*a*) absent; free (of work etc.)

~ **ден·ь** (*m*) holiday, official holiday, Bank holiday

не·при·я́т·ел·ь, -я, (*n*) enemy

не·при·я́т·ельск·ий, -ая, -ое, (*a*) enemy, hostile

не·при·я́т·н·ый, -ая, -ое, (*a*) unpleasant

не·про·бив·а́ем·ый, -ая, -ое, (*a*) impenetrable; (elec) puncture-proof

не·про·бод·е́ни·е, -я, (*n*) imperforation

не·про·бод·ённ·ый, -ая, -ое, (*a*) imperforate

не·про·ва́р·, -а, (*m*) (weld) poor/faulty/incomplete fusion; (paper) uncooked pulp

не·про·вод·ни́к·, -а́, (*m*) (elec) non-conductor, insulator

не·про·вод·я́щ·ий, -ая, -ее, (*a*) non-conducting, non-conductive

не·про́·волоч·н·ый, -ая, -ое, (*a*) non-wire

не·про́·да·нн·ый, -ая, -ое, (*a*) unplaced (of goods)

не·про·долж·и́тельност·ь, -и, (*f*) short duration, a short period of time

не·про·долж·и́тельн·ый, -ая, -ое, (*a*) short, brief, of short duration

не·продукти́в·н·ый, -ая, -ое, (*a*) unproductive; barren (of oil wells)

не·про·е́зж·ий, -ая, -ее, (*a*) impassable

не·про·зр·а́чност·ь, -и, (*f*) opacity

не·про·зр·а́чн·ый, -ая, -ое, (*a*) opaque

не·про·из·вод·и́тельн·ый, -ая, -ое, (*a*) unproductive, non-productive; (oil) barren

не·про·из·во́ль·н·ый, -ая, -ое, (*a*) involuntary

не·про·ка́л·ываем·ый, -ая, -ое, (*a*) puncture-proof

не·про·кле́·й, -я, (*m*) bonding/adhesion failure

не·про·мок·а́ем·ый, -ая, -ое, (*a*) waterproof, impermeable, impervious, -proof

~ **ткан·ь** (*f*) proofed cloth

не·про·мы́шл·енн·ый, -ая, -ое, (*a*) unprofitable, uneconomic; non-industrial/commercial; dead (of rocks)

не·про·ниц·а́ем·ый, -ая, -ое, (*a*) impenetrable, impermeable, -tight

~ **для га́з·а** gas-tight

не·про·па́·янн·ый, -ая, -ое, (*a*) solderless, dry (joints etc.)

не·про·пи́т·анн·ый, -ая, -ое, (*a*) unimpregnated, undipped

не·про·пря́д·к·а, -и, (*g pl*) -д·ок·, (*f*) (text) slub

не·пропорциона́ль·н·ый, -ая, -ое, (*a*) disproportionate

не·про·пуск·а́ющ·ий, -ая, -ее, (*a*) (opt) non-transparent, non-transmitting; suppressed (frequency band)

не·про·све́ч·иваем·ый, -ая, -ое, (*a*) opaque

не·про·тк·ну́т·ый, -ая, -ое, (*a*) unpierced, unpenetrated

не·про·хо́д·, -а, (*m*) (mech eng) no-go, not go-in/on (of gauges)

не·про·ход·и́мост·ь, -и, (*f*) impassability; blockage

не·про·ход·и́м·ый, -ая, -ое, (*a*) impassable, blocked

не·про·хо́д·н·ый, -ая, -ое, (*a*) impassable, blocked; (mech eng) not-go (of gauges)

~ сторон·а́ (*f*) not-go plug; not-go stop

не·про́ч·н·ый, -ая, -ое, (*a*) delicate, fragile, flimsy; unstable, labile; fugitive, not fast (colours); insecure, loose (of joints); perishable (of food); incompetent (of rock)

не·пруж·и́нн·ый, -ая, -ое, (*a*) unsprung, non-sprung

не·прям·о́й, -а́я, -о́е, (*a*) indirect

непту́н·и·й, -я, (*m*) (chem) neptunium, Np

нептун·и́ческ·ий, -ая, -ое, (*a*) (geol) neptunic

не·пуст·о́й, -а́я, -о́е, (*a*) (math) non-empty

не·рабо́т·авш·ий, -ая, -ее, (*a*) (nucl) clean

не·рабо́т·ающ·ий, -ая, -ее, (*a*) idle, standing

не·рабо́ч·ий, -ая, -ее, (*a*) inoperative, non-driving, off

~ по·лож·е́ни·е (*n*) off position

~ со·сто·я́ни·е (*n*) idle state/condition

не·ра́вен·ств·о, -а, (*n*) inequality, disparity

~, не·стро́г·ое conditional inequality

~, параллакт·и́ческ·ое anomalistic inequality (tides)

~, стро́г·ое absolute inequality

не·ра́вн·о- (*component*) unequal-, aniso-

не·равно·бо́к·ий, -ая, -ое, (*a*) unequal-sided, unequal, scalene (of triangles)

~ уго́ль·ник· (*m*) (met) unequal-angle bar

не·равно·ве́с·и·е, -я, (*n*) disequilibrium, non-equilibrium, unbalance; bias

не·равно·ве́с·ност·ь, -и, (*f*) *see* **не·-равно·ве́с·и·е**

не·равно·ве́с·н·ый, -ая, -ое, (*a*) unbalanced, not in equilibrium, non-equilibrium

не·равно·гра́н·ник·, -а, (*m*) (math) scalenohedron

не·равно·зерн·и́ст·ый, -ая, -ое, (*a*) inequigranular

не·равно·ло́паст·н·ый, -ая, -ое, (*a*) inequilobate

не·равно·ме́р·ност·ь, -и, (*f*) irregularity, non-uniformity; unevenness

~, вре́мен·н·ая (autom) temporary deviation

коэффицие́нт· не·равно·ме́р·ност·ей (*m*) peaking factor, cyclic variation factor

~, о·ста́т·очн·ая (autom) offset, droop, drift

~ регулир·у́ем·ого пара́метр·а (autom) controlled variable range

~ регуля́тор·а (autom) controller throttling range, controller proportional band

не·равно·ме́р·н·ый, -ая, -ое, (*a*) irregular, non-uniform; uneven, unequal; (opt) non-isochronous; (hydr) unsteady

не·равно·сво́й·ственн·ый, -ая, -ое, (*a*) anisotropic

не·равно·с·тво́р·чат·ый, -ая, -ое, (*a*) (biol) inequivalve

не·равно·сто́рон·н·ий, -яя, -ее, (*a*) scalene, inequilateral, unequal-sided

не·равно·сту́п·н·ый, -ая, -ое, (*a*) unsymmetrical

~ пуч·о́к· (*m*) (elec) unsymmetrical grading

не·равно·чле́н·н·ый, -ая, -ое, (*a*) (biol) anisomerous

не·ра́вн·ый, -ая, -ое, (*a*) unequal

~ ну́л·ю (*dat*) non-zero, non-vanishing

не·радио·акти́в·н·ый, -ая, -ое, (*a*) non-radioactive

не·радио·ге́н·н·ый, -ая, -ое, (*a*) (chem) non-radiogenic

не·раз·ба́вл·енн·ый, -ая, -ое, (*a*) undiluted, neat, straight

не·раз·бо́р·н·ый, -ая, -ое, (*a*) fixed, integral

~ свеч·а́ (*f*) (ICE) integral spark plug

не·раз·бо́р·чив·ый, -ая, -ое, (*a*) undecipherable, indecipherable, illegible; indiscriminating, uncritical

не·раз·ветвл·ённ·ый, -ая, -ое, (*a*) unbranched, non-branching; (bot) excurrent не·раз·ви·т·о́й, -а́я, -о́е, (*a*) underdeveloped; undeveloped

не·раз·дел·ённ·ый, -ая, -ое, (*a*) undivided

~ ка́мер·а с·гор·а́ни·я (*f*) direct-injection chamber, open combustion chamber (diesel)

не·раз·дел·и́м·ый, -ая, -ое, (*a*) *see* не·раз·де́ль·н·ый

не·раздельно·кип·я́щ·ий, -ая, -ее, (*a*) (chem) azeotropic

~ с·мес·ь (*f*) (chem) azeotropic mixture

не·раз·де́ль·н·ый, -ая, -ое, (*a*) inseparable, indivisible

~ им·у́ществ·о (*n*) (law) common estate

не·раз·дел·я́ющ·ийся, -аяся, -ееся, (*a*) inextricable

не·раз·лич·и́м·ый, -ая, -ое, (*a*) indistinguishable

не·раз·лож·и́м·ый, -ая, -ое, (*a*) undecomposable

не·раз·ло́ж·енн·ый, -ая, -ое, (*a*) undecomposed

не·раз·лу́ч·н·ый, -ая, -ое, (*a*) inseparable (of people)

не·раз·ме́щ·енн·ый, -ая, -ое, (*a*) (fin) unplaced

не·раз·множ·а́ющ·ий, -ая, -ее, (*a*) (nucl) non-breeding

не·раз·рез·н·о́й, -а́я, -о́е, (*a*) uncut; continuous, not split up, solid

~ ба́л·к·а (*f*) (civ eng) continuous beam

не·раз·реш·ённ·ый, -ая, -ое, (*a*) unauthorized; forbidden, prohibited; unresolved, unsolved

не·раз·реш·и́м·ый, -ая, -ое, (*a*) insoluble, unresolvable

не·раз·руш·а́ющ·ийся, -аяся, -ееся, (*a*) indestructible; non-volatile (of magnetic recordings)

не·раз·руш·и́м·ый, -ая, -ое, (*a*) indestructible

не·раз·ры́в·ност·ь, -и, (*f*) continuity у·равн·е́ни·е не·раз·ры́в·ност·и (*n*) continuity/conservation equation (fluid flow)

не·раз·ры́в·н·ый, -ая, -ое, (*a*) inseparable, continuous, uninterrupted

не·разъ·ед·а́ем·ый, -ая, -ое, (*a*) incorrodible, corrosion-resistant

не·разъ·ём·н·ый, -ая, -ое, (*a*) non-detachable, non-strippable, fixed, solid, permanent (of joints etc.)

не·рас·ка́л·ывающ·ийся, -аяся, -ееся, (*a*) non-splitting, non-fissile

не·рас·па́·вш·ийся, -аяся, -ееся, (*a*) (chem) undecomposed; (nucl) undecayed

не·рас·плыв·а́ющ·ийся, -аяся, -ееся, (*a*) (chem) non-deliquescent; (phys) non-blurring

не·рас·по·знав·а́ем·ый, -ая, -ое, (*a*) unrecognizable, indiscernible; adiagnostic

не·рас·твор·ённ·ый, -ая, -ое, (*a*) undissolved

не·рас·твор·и́мост·ь, -и, (*f*) insolubility

не·рас·твор·и́м·ый, -ая, -ое, (*a*) indissoluble, (chem) insoluble

не·рас·чёт·лив·ый, -ая, -ое, (*a*) uneconomic, extravagant

не·рас·чёт·н·ый, -ая, -ое, (*a*) off-design

не·рациона́ль·н·ый, -ая, -ое, (*a*) irrational

не́рв·, -а, (*m*) (biol) nerve

нерв·а́ци·я, -и, (*f*) (biol) nervation, venation

не́рв·н·ый, -ая, -ое, (*a*) (anat) nerve, nervous, neural

~ пласт·и́нк·а (*f*) medullary plate

~ систе́м·а (*f*) nervous system

~ ткан·ь (*f*) nerve tissue

нервю́р·а, -ы, (*f*) (a/c) rib; (arch) nerve, nervure; (bot) nerve, midrib

~, вс·по·мог·а́тельн·ая (a/c) false rib

~, до·полн·и́тельн·ая (a/c) former/nose/false rib

~, коро́б·чат·ая box rib

~, ло́ж·н·ая (a/c) false rib

~, норма́ль·н·ая (a/c) plain rib

~, нос·ов·а́я (a/c) cap rib

~, у·сил·ённ·ая (a/c) compression rib, bulkhead rib

не·реаги́р·ующ·ий, -ая, -ее, (*a*) unreactive, nonreacting; (instr) unresponsive

не·реве́рс·и́вн·ый, -ая, -ое, (*a*) non-reversing

не·регул́яр·н·ый, -ая, -ое, (*a*) irregular; imperfect

не·ре́д·к·о (*adv*) frequently, often, not infrequently

не·ре́д·к·ий, -ая, -ое, (*a*) frequent

не·резерв·и́рованн·ый, -ая, -ое, (*a*) without reserves/spares; (text chem) unresisted

не·резе́рв·ованн·ый, -ая, -ое, (*a*) *see* не·резерв·и́рованн·ый

не·ре́з·к·ий, -ая, -ое, (*a*) blurred, indistinct, poorly defined, soft (sound)

не·резона́нс·н·ый, -ая, -ое, (*a*) non-resonant, non-resonance

не·релятиви́ст·ск·ий, -ая, -ое, (*a*) non-relativistic

не́рест·, -а, (*m*) (fish) spawning

нерест·и́лищ·е, -а, (*n*) (fish) spawning bed

нерест·и́ться, (*fut 3rd sing*) **нерести́т-ся,** (*imp*) spawn

не·рефле́ктор·н·ый, -ая, -ое, (*a*) non-reflecting; (autom) non-feedback

не·реш·а́ющ·ий, -ая, -ее, (*a*) inconclusive

не·реш·и́тельн·ый, -ая, -ое, (*a*) indecisive; irresolute

не·реш·и́тельност·ь, -и, (*f*) indecision

не·ржав·е́ющ·ий, -ая, -ее, (*a*) rust-resisting, corrosion-resistant, stainless

~ стал·ь (*f*) stainless steel

нери́т·ов·ый, -ая, -ое, (*a*) (geol) neritic

не́рк·а, -и, (*f*) (fish) sockeye, red salmon, blueback salmon, *Oncorhynchus nerka*

Не́рист·а, ла́мп·а (*f*) Nernst lamp/glower

~, теоре́м·а (*f*) Nernst heat theorem

~, у·равн·е́ни·е (*n*) Nernst equation

~, штифт· (*m*) = **Не́рист·а, ла́мп·а**

~, явл·е́ни·е (*n*) Nernst effect

Не́рнст·а-Эттинга́узен·а, эфф́е́кт· (*m*) Nernst effect

не·ро́вн·о (*adv*) unevenly, raggedly

не·ро́вн·ост·ь (*f*) unevenness, roughness, irregularity, raggedness

~, волно·обра́з·н·ая undulation, corrugation

не·ровн·от·а́, -ы́, (*f*) *see* **не·ро́вн·ост·ь**

не·ро́вн·ый, -ая, -ое, (*a*) uneven, rough, irregular, ragged

неро́л·, -а, (*m*) (chem) nerol

неро́ли·ев·ый, -ая, -ое, (*a*) (chem) nerol, neroli

не́рп·а, -ы, (*f*) seal

~, ко́ль·чат·ая (*f*) (zool) bristly seal, *Phoca hispida*

~, обык·нове́нн·ая *see* **не́рп·а, ко́ль·-чат·ая**

не·ру́д·н·ый, -ая, -ое, (*a*) (min) gangue, non-metallic

нёс·, -а, (*m*) floeberg (ice)

нёс· (*past masc sing*) of **нес·ти́**

не·само·вос·пламен·я́ющ·ийся, -аяся, -ееся, (*a*) non-self-igniting, non-inflammable, nonergolic

не·само·гас·я́щ·ий, -ая, -ее, (*a*) (elec) externally quenched

не·само·со·пряж·енност·ь, -и, (*f*) (math) non-self-adjacency

не·само·со·пряж·ённ·ый, -ая, -ое, (*a*) (math) non-self-adjoint

не·само·хо́д·н·ый, -ая, -ое, (*a*) non-self-propelled

~ ба́рж·а (*f*) dumb barge

не·с·баланс·и́рованн·ый, -ая, -ое, (*a*) unbalanced

~ ра́зн·иц·а (*f*) unbalance

не·с·бы́т·очн·ый, -ая, -ое, (*a*) unrealizable, hopeless

не·с·вар·е́ни·е, -я, (*n*) (med) indigestion

не·с·ве́д·ущ·ий, -ая, -ее, (*a*) inexpert, unskilled, untrained

не·све́ж·ий, -ая, -ее, (*a*) (food) stale, old

не·све́рт·ываем·ый, -ая, -ое, (*a*) incoagulable

не·с·ве́р·нут·ый, -ая, -ое, (*a*) uncoiled; uncoagulated, unset

не·све́т·ящ·ий, -ая, -ее, (*a*) (opt) non-luminous, dark

не·свобо́д·н·ый, -ая, -ое, (*a*) restricted, constrained; (chem) combined

~ систе́м·а (*f*) (dynam) rigid body

не·свое·вре́мен·н·ый, -ая, -ое, (*a*) inopportune

не·сво́й·ственн·ый, -ая, -ое, (*a*) unnatural, non-typical, uncharacteristic

не·с·вя́з·анн·ый, -ая, -ое, (*a*) unconnected; incoherent; (chem) free, not combined, uncombined, (nucl) unbound

~ со·сто·я́ни·е (*n*) (phys) unbound state

~ угле·ро́д· (*m*) (chem) free/uncombined carbon

~ част·и́ц·а (*f*) unbound particle

не·с·вя́з·ност·, -и, (*f*) incoherence, incohesion

не·с·вя́з·н·ый, -ая, -ое, (*a*) incoherent

~ рас·се́·яни·е (*n*) (phys) incoherent scattering

не·с·гиб·а́ем·ый, -ая, -ое, (*a*) inflexible

не·с·гор·а́ем·ый, -ая, -ое, (*a*) fireproof, incombustible

не·с·гор·е́вш·ий, -ая, -ее, (*a*) unburnt

не·с·гущ·а́ем·ый, -ая, -ое, (*a*) non-condensible, incondensible

не·с·да́·ч·а, -и, (*f*) nondelivery

не·с·двой·ни́ков·ый, -ая, -ое, (*a*) (cryst) untwinned, nontwinned

нес·е́ни·е, -я, (*n*) performance, performing, carrying out

~ обяз·а́нност·и performance of duties

нес·ённ·ый, -ая, -ое, (*past part pass*) of нес·ти

нес·ёт (*pres 3rd sing*) of нес·ти

не·с·жим·аем·ый, -ая, -ое, (*a*) incompressible, incoercible (of gases)

не·симметр·ическ·ий, -ая, -ое, (*a*) asymmetrical

не·симметр·ичн·ый, -ая, -ое, (*a*) asymmetrical, (elec) unbalanced

не·синхрон·н·ый, -ая, -ое, (*a*) asynchronous, non-synchronous

нес·йте (*plur imper*) of нес·ти

не·сквоз·н·ой, -ая, -ое, (*a*) blind (of openings, holes etc.)

не·с·клад·чат·ый, -ая, -ое, (*a*) (geol) unfolded

не·с·клон·яем·ый, -ая, -ое, (*a*) (gram) undeclinable

не·скольз·ящ·ий, -ая, -ее, (*a*) non-slip, non-skid, antiskid

несколько several, some, a few; (*adv*) slightly, rather, somewhat

не·с·конч·аем·ый, -ая, -ое, (*a*) unending, interminable

не·с·крепл·ённ·ый, -ая, -ое, (*a*) not fastened/fixed

~ ствол· (*m*) monoblock gun body

нес·ла (*past fem sing*) of нес·ти

не·слеп·ящ·ий, -ая, -ее, (*a*) anti-dazzle

нес·ли (*past plur*) of нес·ти

не·с·лит·ин·ы, -ов, (*m pl*) (met cast) cold shut, teeming arrest

не·с·ли·яни·е, -я, (*n*) (met) cold shut, teeming arrest

нес·ло (*past neut sing*) of нес·ти

не·с·лож·н·ый, -ая, -ое, (*a*) simple, uncomplicated

не·слоисто·по·кров·н·ый, -ая, -ое, (*a*) intectate

не·слых·анн·ый, -ая, -ое, (*a*) unprecedented, unheard of

не·слыш·н·ый, -ая, -ое, (*a*) inaudible

не·с·мач·иваем·ый, -ая, -ое, (*a*) non-wettable

не·с·мач·ивани·е, -я, (*n*) (phys) non-wetting

не·с·мсн·яем·ый, -ая, -ое, (*a*) irremovable, nonremovable

не·с·меш·анн·ый, -ая, -ое, (*a*) unmixed, unblended

не·с·меш·иваем·ый, -ая, -ое, (*a*) immiscible

не·с·меш·ивающ·ийся, -аяся, -ееся, (*a*) immiscible

не·с·мещ·ённ·ый, -ая, -ое, (*a*) unbiased, empirical

не·с·мещ·ённ·ый, -ая, -ое

~ о·цен·к·а (*f*) (math) empirical assessment

не·с·мин·аем·ый, -ая, -ое, (*a*) non-creasing, crease-proof/resisting

~ от·дел·к·а (*f*) (text) non-crease finish

не·смол·ён·ый, -ая, -ое, (*a*) not tarred, untarred, white (of rope)

не·смотр·я на in spite of, despite, notwithstanding

не·с·мыв·аем·ый, -ая, -ое, (*a*) solvent-proof; washable (paint); indelible (ink)

не·с·мык·ани·е, -я, (*n*) (met) cold shut

не·с·нос·н·ый, -ая, -ое, (*a*) unbearable, intolerable

не·со·блюд·ени·е, -я, (*n*) non-observance, infringement

не·соб·ственн·ый, -ая, -ое, (*a*) (math) improper

не·совершенно·лет·и·е, -я, (*n*) (law) minority

не·совершенно·лет·н·ий, -яя, -ее, (*a*) under-age, not having attained legal age; (*n decl as adj*) minor, infant

не·со·верш·ённ·ый, -ая, -ое, (*a*) imperfect, incomplete; (gram) imperfective

не·со·верш·енств·о, -а, (*n*) imperfection

~ решёт·к·и (cryst) lattice irregularity/imperfection

не·со·в·мест·имост·ь, -и, (*f*) incompatibility, conflict

не·со·в·мест·им·ый, -ая, -ое, (*a*) incompatible, inconsistent, conflicting, mutually exclusive

не·со·в·мест·н·ый, -ая, -ое, (*a*) = не·со·в·мест·им·ый

не·со·в·пад·ени·е, -я, (*n*) discrepancy, disagreement; non-alignment, misalignment; anticoincidence

не·со·глас·и·е, -я, (*n*) disagreement, variance, discordance; (geol) unconformity

не·со·глас·н·о (*adv*) at variance (with), in disagreement (with)

не·со·глас·н·ый, -ая, -ое, (*a*) disagreeing, dissenting, differing, discordant; inconsistent; (geol) unconformable

~ за·лег·ани·е (*n*) (geol) discordance, unconformability, unconformity

не·со·глас·ованн·ый, -ая, -ое, (*a*) uncoordinated, maladjusted, mismatched не·со·держ·а́щ·ий, -ая, -ее, (*a*) not containing, -free

~ кисло·ро́д·а (*gen*) oxygen-free

не·со·зн·а́тельн·ый, -ая, -ое, (*a*) irresponsible, unreasonable

не·со·из·мер·и́м·ый, -ая, -ое, (*a*) (math) incommensurable

не·со·крат·и́м·ый, -ая, -ое, (*a*) (math) irreducible

~ дроб·ь (*f*) (math) unit fraction

не·со·мн·е́нн·о (*adv*) undoubtedly, unquestionably, no doubt, without doubt

не·со·мн·е́нност·ь, -и, (*f*) certitude, obviousness

не·со·мн·е́нн·ый, -ая, -ое, (*a*) undoubted, unquestioned, undisputed, undubitable

нес·о́м·ый, -ая, -ое, (*pres part pass*) of нес·ти́

не·со·обра́з·н·ый, -ая, -ое, (*a*) incompatible; foolish, absurd

не·со·о́с·ност·ь, -и, (*f*) malalignment, misalignment

не·со·о́с·н·ый, -ая, -ое, (*a*) misaligned, out of alignment

не·со·от·ве́т·ственн·ый, -ая, -ое, (*a*) inappropriate, incongruous, unrelated; conflicting, not in agreement

не·со·от·ве́т·стви·е, -я, (*n*) non-conformity, disparity, incongruity; misfit, mismatch

не·со·пряж·ённ·ый, -ая, -ое, (*a*) disconnected, non-conjugate

не·со·раз·ме́р·н·ый, -ая, -ое, (*a*) disproportionate, incommensurate

не·сорт·иро́ванн·ый, -ая, -ое, (*a*) unsorted, ungraded, run-of-the-mill; (min) run-of-the-mine

не·со́рт·н·ый, -ая, -ое, (*a*) = не·-сорт·ов·о́й

не·сорт·ов·о́й, -а́я, -о́е, (*a*) low grade/quality

не·со·сто·я́тельн·ый, -ая, -ое, (*a*) (fin) insolvent; unsound, without basis; untenable (of contracts etc.)

не·со·сто·я́тельност·ь, -и, (*f*) (fin) insolvency; unsoundness

не·со·хран·е́ни·е, -я, (*n*) non-conservation

не·со·член·ённ·ый, -ая, -ое, (*a*) (anat) inarticulate; non-jointed, rigid

не·с·пас·а́ем·ый, -ая, -ое, (*a*) non-recoverable (of missiles etc.)

не·с·па́р·енн·ый, -ая, -ое, (*a*) unpaired

не·с·пек·а́ющ·ийся, -аяся, -ееся, (*f*) noncaking (coal)

не·специали́ст·, -а, (*m*) non-specialist, layman

не·спи́н·ов·ый, -ая, -ое, (*a*) (phys) extra-spin

не·сплош·н·о́й, -ая, -ое, (*a*) incomplete; broken, patchy (of coverage)

не·спосо́б·ност·ь, -и, (*f*) inability; incapacity; incompetence

не·спосо́б·н·ый, -ая, -ое, (*a*) incapable, incompetent

не·с·правед·ли́в·ый, -ая, -ое, (*a*) unfair, unjust; unjustifiable, wrong

не·с·раба́т·ывани·е, -я, (*n*) failure to operate, faulty working

не·с·рабо́т·авш·ий, -ая, -ее, (*a*) (instr) nonoperated, undischarged

не·с·равн·е́нн·ый, -ая, -ое, (*a*) incomparable, unmatched, matchless

не·с·равн·и́м·ый, -ая, -ое, (*a*) incomparable, matchless

не·с·ро́д·н·ый, -ая, -ое, (*a*) unnatural; uncongenial

Не́сслер·а, реакти́в·, (*m*) (chem) Nessler's solution

не·стаби́ль·ност·ь, -и, (*f*) instability

не·стаби́ль·н·ый, -ая, -ое, (*a*) unstable, astable

не·станда́рт·н·ый, -ая, -ое, (*a*) off-grade, non-standard; not controlled by standard specification; substandard; non-stock, not usually stocked

не·стар·е́ющ·ий, -ая, -ее, (*a*) non-ageing

не·стациона́р·н·ый, -ая, -ое, (*a*) non-stationary, transient, non-steady, unsteady

нес·ти́, -у́т, (*past masc sing*) нёс·, (*imp deter*) carry, bear; wear (clothes), fly (a flag); carry out, perform; lay (eggs)

не·с·тир·а́ющ·ийся, -аяся, -ееся, (*a*) non-erasible; indelible (of inks etc.)

не·сто́й·к·ий, -ая, -ое, (*a*) (chem) unstable, non-persistent

~ газ· (*m*) non-persistent gas

не·сто́й·кост·ь, -и, (*f*) instability

не·стро́г·ий, -ая, -ое, (*a*) (math) conditional

не·стро·ев·о́й, -а́я, -о́е, (*a*) (mil) non-operational, non-combatant; non-constructional, not suitable for construction (of materials)

не·суд·и́мост·ь, -и, (*f*) (law) absence of previous conviction

нес·у́т (*pres plur*) of нес·ти́
не·стро́й·н·ый, -ая, -ое, (*a*) untidy; out of time, irregular (sound)
нес·у́шк·а, -и, (*g pl*) -шек·, (*f*) laying hen
не·сущ·е́ственн·ый, -ая, -ое, (*a*) unessential, unimportant; minor
нес·у́щ·ий, -ая, -ее, (*pres part act*) of нес·ти́; (telecom) carrier
~ винт· (*m*) lifter airscrew/propeller; main rotor (of helicopter)
~ волн·а́ (*f*) (rad) carrier wave
~ кана́т· (*m*) fixed cable (of cableway)
~ крыл·о́ (*n*) (a/c) lifting wing
~ по·ве́рх·ност·ь (*f*) (a/c) supporting surface; lifting surface
~ спосо́б·ност·ь (*f*) load-carrying capacity
~ част·от·а́ (*f*) (telecom) carrier frequency
не·с·ход·и́мост·ь, -и, (*f*) non-convergence, divergence
не·с·ход·и́м·ый, -ая, -ое, (*a*) non-converging
не·с·хо́д·н·ый, -ая, -ое, (*a*) dissimilar, unlike
не·с·хо́д·ств·о, -а, (*n*) dissimilarity
не·с·ход·я́щ·ий, -ая, -ее, (*a*) non-converging, divergent
не·с·цемент·и́рованн·ый, -ая, -ое, (*a*) (geol) uncemented, loose
не·сча́ст·н·ый, -ая, -ое, (*a*) unfortunate, unlucky, unhappy
~ с·луч·а́·й (*m*) accident
не·с·чёт·н·ый, -ая, -ое, (*a*) innumerable, (math) non-denumerable
нёс·ш·ий, -ая, -ее, (*past part act*) of нес·ти́
не·съ·ед·о́бн·ый, -ая, -ое, (*a*) inedible, uneatable
нес·я́ (*pres gerund*) of нес·ти́
нес·я́к·, -а́, (*m*) floeberg (sea ice)
нет no; there is/are no; is/are not
~ ещё not yet
со·вс·е́м нет not at all
~ - пере·хо́д·, -а, (*m*) (phys) no-transition
нетанни́д·, -а, (*m*) (chem) non-tanning substance, non-tan
не·твёрд·ый, -ая, -ое, (*a*) not very hard, rather soft; unsteady, not very firm; soft (currencies)
не·тек·у́ч·ий, -ая, -ее, (*a*) non-flowing, stagnant
не́тер·ов·ы кольц·а́ (*n pl*) (math) Noetherian rings
не·терп·е́ни·е, -я, (*m*) impatience

не·тип·и́чн·ый, -ая, -ое, (*a*) untypical, (biol) atypical
не·токо·вед·у́щ·ий, -ая, -ое, (*a*) (elec) non-current carrying, dead
не·то́н·ов·ый, -ая, -ое, (*a*) (acous) unpitched
нетопы́р·ь, -я́, (*m*) insectivorous bat, (*pl as adj*) *Vespertilionidae*
не·то́ч·ност·ь, -и, (*f*) inaccuracy
~ с·бо́р·к·и (mech) misalignment
не·то́ч·н·ый, -ая, -ое, (*a*) inaccurate, inexact; untrue (of alignments etc.)
не·тре́б·овательн·ый, -ая, -ое, (*a*) unexacting, undemanding, not very demanding/exacting
не·тривиа́ль·н·ый, -ая, -ое, (*a*) (math) non-trivial
не·тро́·нут·ый, -ая, -ое, (*a*) whole, intact, untouched; maiden, virgin
~ у́гол·ь (*m*) unmined/virgin coal
не·тру́д·н·о (*adv*) easily, without difficulty; it is easy, it is not hard
не·тру́д·н·ый, -ая, -ое, (*a*) not very difficult, easy
не·трудо·спосо́б·н·ый, -ая, -ое, (*a*) disabled
не́тто (*indecl*) (*see also* чи́ст·ый) net (of weights etc.)
~ -реги́стр·ов·ая в·мест·и́мост·ь (*f*) net tonnage
не́т·чик·, -а, (*m*) absentee
не·у·бед·и́тельн·ый, -ая, -ое, (*a*) unconvincing
не·у·бир·а́ющ·ийся, -аяся, -ееся, (*a*) unretractable
~ шасси́ (*n*) (a/c) fixed undercarriage
не·у·быв·а́ющ·ий, -ая, -ее, (*a*) non-decreasing/diminishing
не·у·ве́р·енн·ый, -ая, -ое, (*a*) uncertain, diffident
не·у·вя́з·к·а, -и, (*g pl*) -з·ок·, (*f*) discrepancy
не·у·да́·ч·а, -и, (*i*) -ей, (*f*) failure; misfortune
не·у·да́ч·н·ый, -ая, -ое, (*a*) unsuccessful, unfortunate
не·у·диферент·о́ванн·ый, -ая, -ое, (*a*) (naut) out-of-trim
не·у·до́б·н·ый, -ая, -оё, (*a*) uncomfortable; inconvenient
не·у·до́б·о- (*component*) not easily/lightly
неудобо·по́·нят·н·ый, -ая, -ое, (*a*) not easy to understand, not readily understood, obscure
неудобо·про·ход·и́м·ый, -ая, -ое, (*a*) rather bad (roads etc.)

не·удовлетвор·éни·е, -я,   (*n*) dissatisfaction; non-compliance; rejection (of legal action)

не·удовлетвор·ённ·ый, -ая, -ое,   (*a*) unsatisfied; dissatisfied (of people)

не·удовлетвор·и́тельн·ый, -ая, -ое,   (*a*) unsatisfactory

неужéли   is it possible?, surely not?

не·у·зна·вáем·ый, -ая, -ое,   (*a*) unrecognizable

не·у·клóн·н·ый, -ая, -ое,   (*a*) undeviating

не·уклю́ж·ий, -ая, -ее,   (*a*) clumsy, awkward

не·у·лов·и́м·ый, -ая, -ое,   (*a*) elusive; subtle, imperceptible

не·ум·éл·ый, -ая, -ое,   (*a*) clumsy, unskilful

не·у·мéр·енн·ый, -ая, -ое,   (*a*) excessive, immoderate

не·у·мéст·н·ый, -ая, -ое,   (*a*) misplaced, uncalled for

не·у·молк·áем·ый, -ая, -ое,   (*a*) incessant, unceasing

не·у·мы́шл·енн·ый, -ая, -ое,   (*a*) unpremeditated, inadvertent

не·у·плáт·а, -ы,   (*f*) non-payment

не·у·по·ря́д·оченн·ый, -ая, -ое,   (*a*) disordered, non-ordered, unordered

~ со·сто·я́ни·е   (*n*) (phys) disordered state

не·у·по·треб·и́тельн·ый, -ая, -ое,   (*a*) not in use, not used, obsolete

не·у·правл·я́ем·ый, -ая, -ое,   (*a*) not controlled, uncontrolled, random; uncontrollable; (GW) unguided, free-flight

~ рас·шир·éни·е   (*n*) uncontrolled/random expansion

не·у·пру́г·ий, -ая, -ое,   (*a*) inelastic, non-elastic, anelastic

не·у·равно·вéш·енн·ый, -ая, -ое,   (*a*) unbalanced, out-of-balance, without equilibrium

не·у·рож·á·й, -я,   (*m*) (agr) crop failure, poor crop

не·урóч·н·ый, -ая, -ое,   (*a*) inopportune, untimely (British sense only); unspecified

не·у·спéх·, -а,   (*m*) failure

не·у·спéш·н·ый, -ая, -ое,   (*a*) unsuccessful

не·у·стáн·н·ый, -ая, -ое,   (*a*) relentless, unflagging

не·у·станов·и́вш·ийся, -аяся, -ееся,   (*a*) unsteady, non-steady, non-stationary, transient

не·у·станов·и́вш·ийся, -аяся, -ееся

~ движ·éни·е   (*n*) (dynam) unsteady flow

~ процéсс·   (*m*) transient process

~ режи́м·   (*m*) non-stationary state

не·у·станóвл·енн·ый, -ая, -ое,   (*a*) unestablished, unstated

не·у·стóй·к·а, -и,   (*g pl*) **-ó·ек·,**   (*f*) forfeit, penalty

не·у·стóй·чивост·ь, -и,   (*f*) instability, unsteadiness, lability; (phot) flutter, fluttering, unsteadiness; (cinema) dancing, flutter

~ по крéн·у   (air) roll/rolling instability

~ по тангáж·у   (air) pitch/pitching instability

~ пут·ев·áя   (a/c) directional instability

не·у·стóй·чив·ый, -ая, -ое,   (*a*) unstable, unsteady; fluctuating, changeable; labile

~ равно·вéс·и·е   (*n*) lability, mobile equilibrium

~, терм·и́ческ·и   thermolabile

не·у·стран·и́м·ый, -ая, -ое,   (*a*) irremovable, uncorrectable, ineradicable

не·у·стрó·енн·ый, -ая, -ое,   (*a*) unsettled, disordered, disorganized

не·у·стрóй·ств·о, -а,   (*n*) disorganization, muddle, lack of organization

не·у·том·и́м·ый, -ая, -ое,   (*a*) untiring, tireless

не·у·томл·я́ем·ый, -ая, -ое,   (*a*) (met) fatigue-resistant

не·у·чи́т·ываем·ый, -ая, -ое,   (*a*) negligible

не·у·чт·ённ·ый, -ая, -ое,   (*a*) undetermined, indeterminable, indeterminate

нéф·, -а,   (*m*) (arch) nave

нефел·и́н·, -а,   (*m*) (min) nepheline

нефел·и́нов·ый, -ая, -ое,   (*a*) of **нефел·и́н·**

~ сиени́т·   (*m*) nepheline-syenite

нефелó·метр·, -а,   (*m*) (chem) nephelometer, turbidimeter

нефело·мéтр·и·я, -и,   (*a*) (chem) nephelometric/turbidimetric analysis

нефо·грáф·, -а,   (*m*) (phot, met) nephograph

нефо·лóг·и·я, -и,   (*f*) (meteor) nephology

не·фóрм·ов·óй, -áя, -óе,   (*a*) non-moulded/molded

нефо·скóп·, -а,   (*m*) (meteor) nephoscope

нефр·ектом·и́·я, -и, (*a*) (med) nephrectomy

нефр·идиа́льн·ый, -ая, -ое, (*a*) (anat) nephridial

нефр·и́ди·й, -я, (*m*) (zool) nephridium

нефр·и́т·, -а, (*m*) (min) nephrite; (med) nephritis

нефр·о́з·, -а, (*m*) (med) nephrosis

нефро·литиа́з·, -а, (*m*) (med) nephrolithiasis

нефро·па́т·и·я, -и, (*m*) (med) nephropathy

не́фте- (*root*) oil, petroleum; oil fuel

нефте·ба́з·а, -ы, (*f*) (oil) tank farm, bulk plant, oil installation

нефте·биту́м·, -а, (*m*) bitumen, asphaltic bitumen, asphalt (U.S.)

нефте·вод·ян·о́й, -а́я, -о́е, (*a*) petroleum-water, oil-water

нефте·во́з·, -а, (*m*) (naut) tanker, oil tanker

нефте·га́з·, -а, (*m*) petroleum/oil gas

нефте·га́з·ов·ый, -ая, -ое, (*a*) of **нефте·га́з·**; oil-and-gas

нефте·до·бы́·ч·а, -и, (*f*) petroleum production, oil production/recovery

нефте·за·во́д·, -а, (*m*) oil refinery

нефте·за́·леж·ь, -и, (*f*) (geol) oil pool

нефте·лов·у́шк·а, -и, (*g pl*) **-шек·,** (*f*) oil trap (in sewage system)

нефте·матер·и́нск·ий, -ая, -ое, (*a*) (geol) oil-source

~ по·ро́д·а (*f*) oil-source bed

нефте·на·лив·н·о́й, -а́я, -о́е, (*a*) oil/petrol-carrying/transport

~ су́д·н·о (*n*) (naut) tanker, oiler

~ цисте́рн·а (*f*) (rail) oil tank wagon

~ у·стро́й·ств·о (*n*) oil transportation system

нефте·не·про·ниц·а́ем·ый, -ая, -ое, (*a*) oil-tight

нефте·но́с·ност·ь, -и, (*f*) (geol) oil resources

нефте·но́с·н·ый, -ая, -ое, (*a*) (geol) oil-bearing, petroliferous

~ горизо́нт· (*m*) oil horizon

~ по·ро́д·а (*f*) oil reservoir rock

нефте·пере·ка́ч·ывающ·ий, -ая, -ее, (*a*) oil pumping

~ на·со́с· (*m*) pipeline pump; (mar eng) oil-fuel transfer pump

~ ста́нци·я (*f*) pipeline pumping station

нефте·пере·рабо́т·к·а, -и, (*f*) petroleum refining

нефте·пере·го́н·к·а, -и, (*f*) petroleum distillation

нефте·подо·грев·а́тел·ь, -я, (*m*) fuel oil heater

нефте·про·во́д·, -а, (*m*) oil/petroleum pipeline; fuel oil line (in boiler systems)

~, в·са́с·ывающ·ий fuel oil suction line, fuel collecting pipe (boilers)

~, магистра́ль·н·ый (oil) main/trunk pipeline

~, при·ём·н·ый fuel collecting pipe, fuel-oil suction pipe

~, про·мысл·о́в·ый (oil) gathering line

нефте·проду́кт·, -а, (*m*) petroleum product

~, све́тл·ый white products, light products (*pl*)

~, тёмн·ый black products, black oils (*pl*)

нефте·про́·мысел·, -сл·а, (*m*) oil field

нефте·про·мысл·о́в·ый, -ая, -ое, (*a*) oil-field

нефте·про·мы́шл·енн·ый, -ая, -ое, (*a*) oil-industry

нефте·про·ниц·а́ем·ый, -ая, -ое, (*a*) not oil-tight; permeable to oil, oil-permeable

нефте·про·явл·е́ни·е, -я, (*n*) (oil) oil show

нефте·с·бо́р·, -а, (*m*) (oil) gathering, gathering system

нефте·с·бо́р·ник·, -а, (*m*) (oil) initial storage tank

нефте·с·бо́р·н·ый, -ая, -он, (*a*) oil-gathering

нефте·хим·и́ческ·ий, -ая, -ое, (*a*) petrochemical

нефте·хран·и́лищ·е, -я, (*n*) oil storage tank/facility/reservoir

~, металл·и́ческ·ое (oil) metal storage tank

~, земл·ян·о́е (oil) earthen sump

~, железо·бето́н·н·ое (oil) concrete-lined sump, concrete reservoir

нефт·ь, -и, (*f*) oil, petroleum, mineral oil; crude petroleum, crude oil, crude

~, асфа́льт·ов·ая asphaltic-base crude

~, бензи́н·ов·ая paraffinous crude

~, газ·иро́ванн·ая thinned oil

~, де·газ·и́рованн·ая dead/degassed oil

~, за·гущ·ённ·ая gelled crude

~, лёгк·ая light crude, high-gravity oil

~, ма́сл·ян·ая naphthenic crude

~, нафте́н·ов·ая naphthenic crude

~, не·о·чи́щ·енн·ая crude oil, petroleum

нефт·ь

~, об·вод·нённ·ая   water-cut oil

~, парафи́н·ов·ая   paraffinous crude

~, про·мы́в·очн·ая   flush oil

~, сер·ни́ст·ая   sour crude

~, с·ме́ш·анн·ая   mixed-base crude

~, сорт·ов·а́я   high grade crude

~, сыр·а́я   crude oil, petroleum

~, това́р·н·ая   stock tank oil

~, тяжёл·ая   heavy crude, low-gravity oil

нефт·я́нк·а, -и,   (g pl) **-нок·,** (f) semi-diesel engine

нефт·ян·о́й, -а́я, -о́е, (a) of **нефт·ь;** oil-fired

~ вагра́н·к·а   (f) oil-fired cupola

~ дви́г·ател·ь   (m) petrol/gasoline engine

~ кисл·от·а́   (f) naphthenic acid

~ на·со́с·   (m) (eng) fuel oil pump; (oil) oil pump

~ эфи́р·   (m) petroleum ether

не·хва́т·к·а, -и, (g pl) **-т·ок·,** (f) shortage, deficiency

не·ход·ов·о́й, -а́я, -о́е, (a) unsaleable, unmarketable; not running/working

не·хоро́ш·ий, -ая, -ее, (a) bad, not very good

не́·хот·я (adv) unwillingly, reluctantly

не·целе·со·обра́з·н·о (predic adj) it is not good

не·целе·со·обра́з·н·ый, -ая, -ое, (a) inexpedient, unsuitable; inadvisable

не·центр·и́рованн·ый, -ая, -ое, (a) eccentric

не·цикл·и́чн·ый, -ая, -ое, (a) acyclic

не́·чего·   there is nothing; it/there is no need/use/good

не́·чет·, -а, (m) odd number

не·чёт·к·ий, -ая, -ое, (a) indistinct, imprecise; illegible

не·чёт·н·о-чёт·н·ый, -ая, -ое, (a) odd-even

не·чёт·н·ый, -ая, -ое, (a) (math) odd

~ со·сто·я́ни·е   (n) odd parity

не·чист·от·а́, -ы́, (f) dirtiness; impurity; (pl) sewage

не·чи́ст·ый, -ая, -ое, (a) dirty, unclean; adulterated, impure; careless; (naut) foul

не́·что (pron nom, acc only) somewhat, something of a

не·чувств·и́тельност·ь, -и, (f) insensitivity, lack of response

врем·я́ не·чувств·и́тельност·и (f) (instr) insensitive time, paralysis time

не·чувств·и́тельност·ь

зо́н·а не·чувств·и́тельност·и   (f) (autom) dead/neutral zone, differential gap

сте́пен·ь не·чувств·и́тельност·и (f) (autom) neutral zone ratio

не·чувств·и́тельн·ый, -ая, -ое, (a) insensitive

не·штат·н·ый, -ая, -ое, (a) supernumerary, not belonging to regular staff; employed on a temporary basis, temporary

не·эвкли́д·ов·, -а, -о, (predic adj) (math) non-Euclidean

не·эквивале́нт·н·ый, -ая, -ое, (a) non-equivalent

не·эквипотенциа́ль·ност·ь, -и, (f) non-equipotentiality

не·эконо́м·н·ый, -ая, -ое, (a) uneconomical

не·экран·и́рованн·ый, -ая, -ое, (a) unshielded

не·экспоненциа́ль·н·ый, -ая, -ое, (a) non-exponential

не·экстраги́р·уем·ый, -ая, -ое, (a) non-extractable

не·я́в·к·а, -и, (g pl) **-в·ок·,** (f) non-appearance, absence, failure to report

не·явно·по́люс·н·ый, -ая, -ое, (a) (elec) non-synchronous

не·я́в·н·ый, -ая, -ое, (a) implicit, tacit

не·я́сн·о (predic adj) it is not clear, it is obscure

не·ясно·ра́к·овист·ый, -ая, -ое, (a) (zool) subconchoidal

не·я́сн·ый, -ая, -ое, (a) unclear, indistinct, vague

ни   nu, ν (Greek); not a, not

~ в как·о́м с·лу́ч·а·е   in no case, by no means

~ ... ни   neither ... nor

~ о ком   (prepositional) of **ни·кто́**

~ о чём   (prepositional) of **ни·что́**

н·-и. (abbr) = на·у́ч·н·о-ис·сле́д·ова·тельск·ий   research, scientific research

ниаци́н·, -а, (m) (biochem) niacin, nicotinic acid

нив·а́льн·ый, -ая, -ое, (a) nival

нива́л·ов·ый, -ая, -ое, (a) (chem) nivalic

~ кисл·от·а́   (f) (chem) nivalic acid

нив·а́ци·я, -и, (f) (geol) nivation

нивели́р·, -а, (m) (surv) level, levelling instrument

~, глух·о́й   dumpy level

~, прецизио́н·н·ый   precise level

нивели́р

~, **техн·и́ческ·ий** engineer's level

нивели́р·овани·е, -я, (*n*) (surv) levelling

~, **баро·метр·и́ческ·ое** barometric levelling

~, **гео·метр·и́ческ·ое** differential levelling

~, **геодез·и́ческ·ое** trigonometric levelling

нивелир·о́вк·а, -и, (*f*) (surv) levelling

нивелир·о́вочн·ый, -ая, -ое, (*a*) of **нивелир·о́вк·а**

~ **кни́г·а** (*f*) level book

нивели́р·овать, -уют, (*imp and perf*) level, (surv) take levels

нивени́т·, -а, (*m*) (min) nivenite

ни·где́ (*adv*) nowhere

нигре́йт·, -а, (*m*) nigrate (a bitumen)

нигрози́н·, -а, (*m*) (chem) nigrosine, nigrosine base

нигро́л·, -а, (*m*) residual lubricating oil

~, **авто·тра́ктор·н·ый** gear and axle oil

нидерла́нд·ск·ий, -ая, -ое, (*a*) Dutch, Netherlands'

ниепсо·ти́п·и·я, -и, (*f*) (phot) niepceotype

ни́ж·е (*comp*) lower, shorter; (*prep + gen*) below

    из·ло́ж·енн·ый ни́ж·е given/set out below

~ **нул·я́** below zero

~ **по теч·е́ни·ю** downstream

ниже·из·ло́ж·енн·ый, -ая, -ое, (*a*) given below, below

ниже·леж·а́щ·ий, -ая, -ее, (*a*) underlying, subjacent

ниже·лими́т·н·ый, -ая, -ое, (*a*) within the prescribed limits

ниже·под·пис·а́вш·ийся, -егося, (*m decl as adj*) the undersigned

ниже·при·вед·ённ·ый, -ая, -ое, (*a*) given/stated/mentioned below

ни́ж·ет (*pres 3rd sing*) of **низ·а́ть**

ниж·н·е- (*component*) low, lower, infer-, infra

нижне·бок·ов·о́й, -а́я, -о́е, (*a*) (anat) inferolateral

нижне·за́д·н·ий, -яя, -ее, (*a*) (anat) inferoposterior

нижне·ре́бер·н·ый, -ая, -ое, (*a*) (anat) infracostal

ни́ж·н·ий, -яя, -ее, (*a*) lower, inferior, bottom, base; downstream; (*m decl as adj*) (geol) Lower, Early

~ **бъеф·** (*m*) (hydro elec) tail-bay

~ **за́д·н·ий, -яя, -ее,** (*a*) bottom-rear, inferoposterior

ни́ж·н·ий, -яя, -ее

~ **мёртв·ая то́ч·к·а** (*f*) bottom dead centre

~ **плане́т·а** (*f*) (astron) inferior planet

~ **стро·е́ни·е** (*n*) substructure

~ **теч·е́ни·е** (*n*) lower reaches (of river)

~ **част·ь** (*f*) lower part, bottom

~ **эта́ж·** (*m*) (build) ground floor

ни́ж·ник·, -а, (*m*) bottom, bottom part

~, **кру́гл·ый** dolly (riveting)

ни́ж·ущ·ий, -ая, -ее, (*pres part act*) of **низ·а́ть**

низ·, -а, (*nom pl*) **-ы́,** (*g pl*) **-о́в,** (*m*) bottom, bottom part; lower reaches (of river); (*pl*) low notes (music); (geol) lowest strata

~, **фреат·и́ческ·ий** (geol) phreatic low

низ·а́льн·ый, -ая, -ое, (*a*) threading

низ·а́ть, (*pres 3rd pl*) **ни́жут,** (*imp*) string, thread

низ·бег·а́ющ·ий, -ая, -ее, (*a*) (bot) decurrent, decursive

низ·вед·у́т (*fut 3rd pl*) of **низ·вес·ти́**

низ·вёл (*past masc sing*) of **низ·вес·ти́**

низ·верг·а́·ть, -ют, (*imp*) overthrow, throw down

низ·ве́рг·нуть, -нут, (*perf*) see **низ··верг·а́·ть**

низ·вес·ти́, (*fut 3rd pl*) **низ·вед·у́т,** (*past masc sing*) **низ·вёл,** (*perf*); see **низ·вод·и́ть**

низ·вод·и́ть, -́ят, (*imp*) bring down, reduce

низ·и́н·а, -ы, (*f*) lowland, flat; **низ·и·н·а́, -ы́,** (*f*) low place/situation

низ·и́нн·ый, -ая, -ое, (*a*) low-lying

~ **боло́т·о** (*n*) flat/valley bog

ни́з·к·ий, -ая, -ое, (*a*) low; short (stature); deep (voice); low-temperature (of heat treatments)

ни́з·к·о- (*component*) low-

низко·бо́рт·н·ый, -ая, -ое, (*a*) low-side, low sided

низко·бриза́нт·н·ый, -ая, -ое, (*a*) (evpl) low-velocity/strength

низко·во́льт·н·ый, -ая, -ое, (*a*) low-voltage/tension

низко·вя́з·к·ий, -ая, -ое, (*a*) low-viscosity

низко·глино·зём·ист·ый, -ая, -ое, (*a*) low-alumina

низко·за·мещ·ённ·ый, -ая, -ое, (*a*) (chem) low-molecular

низко·кип·я́щ·ий, -ая, -ее, (*a*) low-boiling-point, low-boiling

низко·леги́р·ованн·ый, -ая, -ое, (*a*) low-alloy

низко·молекуля́р·н·ый, -ая, -ое, (*a*) low-molecular, low molecular-weight

~ со·един·ёни·е (*n*) (chem) low molecular-weight compound

низко·на·по́р·н·ый, -ая, -ое, (*a*) low-head, low-pressure

низко·о́м·н·ый, -ая, -ое, (*a*) (elec) low-resistance

низко·пла́н·, -а, (*m*) low-wing aircraft/monoplane

низко·по·тёр·н·ый, -ая, -ое, (*a*) low-loss

низко·про́б·н·ый, -ая, -ое, (*a*) (met) base-alloy, base

низко·про·хо́д·н·ый, -ая, -ое, (*a*) low-pass

низко ра́м·н·ый, -ая, -ое, (*a*) (eng) low-bed; drop-frame

низко·рас·по·ло́ж·енн·ый, -ая, -ое, (*a*) low-slung/mounted/placed

низко·ро́сл·ый, -ая, -ое, (*a*) short (of persons), low (of plants)

низко·со́рт·н·ый, -ая, -ое, (*a*) low-grade

низко·ство́ль·ник·, -а, (*m*) coppice (forestry)

низко·температу́р·н·ый, -ая, -ое, (*a*) low-temperature

~ об·рабо́т·к·а (*f*) (met) sub-zero treatment

~ полимериз·а́ци·я (*f*) low-temperature polymerization

низко·угле·ро́д·ист·ый, -ая, -ое, (*a*) low-carbon; mild (of steel)

низко·част·о́тн·ый, -ая, -ое, (*a*) low-frequency

ни́змен·ност·ь, -и, (*f*) lowland

ни́змен·н·ый, -ая, -ое, (*a*) (geog) low, low-lying

низ·ов·о́й, -а́я, -о́е, (*a*) of **низ·** and **низ·о́вь·е**

~ вод·а́ (*f*) (hydroelec) tail water

~ мете́л·ь (*f*) (meteor) blowing storm

низ·о́вь·е, -я, (*g pl*) -вь·ев, (*n*) lower reaches (of river)

ни́з·ш·ий, -ая, -ее, (*a*) lower, lowest, inferior; primary, elementary, initial

~ у́·ровен·ь (*m*) lower level; low-lying level

НИИ (*abbr*) = **на·у́ч·н·о-ис·сле́д·о·вательск·ий институ́т·** Research Institute, Scientific Research Institute

ни·ка́к (*adv*) in no way, by no means

ни·как·о́й, -а́я, -о́е, (*pron*) no, none, not any

ни́кел·ев·ый, -ая, -ое, (*a*) nickel

~ блеск· (*m*) (min) sulfarsenide of nickel, gersdorffite

~ изумру́д· (*m*) (min) emerald nickel, zaratite

~ с·пла́в· (*m*) nickel-base alloy, nickel alloy

~ цвет·а́ (*pl*) (min) nickel bloom, annabergite

никеле·но́с·н·ый, -ая, -ое, (*a*) nickeliferous

никеле·со·держ·а́щ·ий, -ая, -ее, (*a*) nickeliferous

ни́кел·иев·ый, -ая, -ое, (*a*) *see* **ни́кел·ев·ый**

никели́н·, -а, (*m*) (met) nickelin; (min) niccolite

никел·и́ровани·е, -я, (*n*) nickelizing, nickel plating

никел·иро́ванн·ый, -ая, -ое, (*a*) nickel plated, nickelized

ни́кел·ь, -я, (*m*) (chem) nickel, Ni

~, кремн·и́ст·ый (met) silicon-containing nickel, silicon nickel

~, серно·ки́сл·ый nickel sulphate

~, сурьм·яни́ст·ый nickel antimonide

ни·ке́м (*instr*) of **ни·кто́**

ни·когда́ (*adv*) never

ни·кого́ (*acc/gen*) of **ни·кто́**

николайт·, -а, (*m*) (min) nicolayite

нико́л·ь, -я, (*m*) (opt) Nicol prism, nicol

Нико́льсон·а, гидро́·метр· (*m*) (phys) Nicholson hydrometer

Нико́ля, при́зм·а (*f*) *see* **нико́л·ь**

ни·кому́ (*dat*) of **ни·кто́**

никоти́н·, -а, (*m*) (chem) nicotine

никотин·ами́д·, -а, (*m*) (chem) nicotinamide

никоти́н·ов·ый, -ая, -ое, (*a*) nicotine, nicotinic

~ кисл·от·а́ (*f*) nicotinic acid

никросила́л·, -а, (*m*) (met) Nicrosilal (Ni-Cr-Si-Al cast iron)

никт·алги́·я, -и, (*f*) (med) nyctalgia

никт·алопи́·я, -и, (*f*) (med) nyctalopia, night blindness

никти·на́сти·я, -и, (*f*) (bot) nyctinasty

никти·тропи́зм·, -а, (*m*) (bot) nyctitropism

ни·кто́, -кого́, (*pron*) nobody, no one

никто·фо́б·и·я, -и, (*f*) (med) nyctophobia

никт·ури́·я, -и, (*f*) (med) nycturia

ни·куда́ (*adv*) nowhere, not in any direction, not anywhere

нилас·, -а, (*m*) (meteor ocean) ice rind

нило́н·, -а, *(m)* = **найло́н·** (text)
nylon

ниль·поте́нт·н·ый, -ая, -ое, *(a)*
(math) nilpotent

ниль·ря́д·, -а, *(m)* (math) nil/null-
series

ни·ма́ло *see* **ни·ско́лько**

ни́мф·а, -ы, *(f)* (ent) nymph; (zool)
nympha (*pl* -ae)

нингидри́н·, -а, *(m)* (chem) ninhydrin

ниоба́т·, -а, *(m)* (min, chem) niobate

нио́б·иев·ый, -ая, -ое, *(a)* niobium,
(chem) niobic

~ кисл·от·а́ *(f)* (chem) niobic acid

нио́б·и·й, -я, *(m)* (chem) niobium, Nb,
columbium

ниоб·и́т·, -а, *(m)* (min) niobite, columbite

ни·от·ку́да *(adv)* from nowhere

Ни́пков·а, диск· *(m)* (TV) Nipkow/
apertured disk, disk prism/scanner

ни́ппел·ь, -я, *(nom pl)* **-я́,** *(g pl)* **-е́й,**
(m) nipple (pipe fitting)

~, двой·н·о́й barrel nipple

нира́л·, -а, *(m)* (naut) downhaul

нирези́ст·, -а, *(m)* (met) Ni-resist
(corrosion-resistant cast iron)

ни·ско́лько· *(adv)* not at all, not in the
least

ниста́гм·, -а, *(m)* (med) nystagmus

нис·ход·и́ть, -́ят, *(imp)* descend, come
down

нис·ход·я́щ·ий, -ая, -ее, *(pres part
act)* of **нис·ход·и́ть;** descending,
downwards sloping, downward;
(meteor) katabatic

~ ветв·ь *(f)* descending/downward
branch (of a curve)

~ ветвь траекто́р·и·и post-vortex
trajectory

~ с·брос· *(m)* (geol) downcast/down-
throw fault

нис·хожд·е́ни·е, -я, *(n)* descent,
descending

ни́т·, -а, *(m)* nit (unit of luminance);
(autom) nit (unit of information)

нитба́нк·, -а, *(m)* (horol) riveting/stak-
ing tongue

ните·ви́д·н·ый, -ая, -ое, *(a)* thread-
like, filar, filiform, filamentous, filamentary; (plast) straight-chain

ните·во́д·, -а, *(m)* thread guide (knitting
machine)

ните·вод·и́тел·ь, -я, *(m)* thread carrier
(knitting machine)

ните·на·тяж·н·о́е у·стро́й·ств·о *(n)*
thread-tensioning device

ните·ло́в·к·а, -и, *(g pl)* **-в·ок·,** *(f)*
(text) thread picker/extractor

ните·но́с·н·ый, -ая, -ое, *(a)* (bot)
filiferous

ните·при·тя́г·иватель·, -я, *(m)* take-
up spring (of sewing machine)

ни́т·к·а, -и, *(g pl)* **-т·ок·,** *(f)* *(see also*
нит·ь) thread, twine; (met roll)
strand

~, лик·ов·ая roping twine

~, цвет·н·а́я seam line (canvas)

нитко·швей·н·ая маши́н·а *(f)*
(print) stitching machine

НИТО *(abbr)* = **на·у́ч·н·ое инже-
не́р·н·о-техн·и́ческ·ое общ·еств·о**
Scientific Engineering and Technical
Society

нито́н·, -а, *(m)* (chem) niton, radon

ни́т·очк·а, -и, *(g pl)* **-чек·,** *(f)* *(dim)*
of **нит·ь, ни́т·к·а;** filament

нитраги́н·, -а, *(m)* (agr chem)
nitragin, inoculants

нитралло́·й, -я, *(m)* (met) nitralloy

нитрами́д·, -а, *(m)* (chem) nitramide

нитрами́н·, -а, *(m)* (chem) nitramine

нитрано́л·, -а, *(m)* (pharm) nitranol

нитра́т·, -а, *(m)* (chem) nitrate

~ ка́льц·и·я (chem) calcium nitrate

~ урани́л·а uranyl nitrate

нитра́т·н·ый, -ая, -ое, *(a)* nitrate;
(bot) nitrophilous

~ бакте́р·и·й *(pl)* nitrate bacteria,
(biochem) nitrite-oxidizing bacteria,
Nitrobacter

нитр·а́ци·я, -и, *(f)* (chem) nitration,
nitrating

нитр·и́д·, -а, *(m)* (chem) nitride

~ бери́лл·и·я beryllium nitride

~ то́ри·я thorium nitride

нитри́л·, -а, *(m)* (chem) nitrile

~ акри́л·ов·ой кисл·от·ы́ acrylonitrile

нитрило́н·, -а, *(m)* (plast) Nitrilon

нитри́ль·н·ый, -ая, -ое, *(a)* (chem)
nitrile

нитр·и́ровани·е, -я, *(n)* (met)
nitriding, nitrogen case hardening

нитр·и́рованн·ый, -ая, -ое, *(a)*
(met) nitrided

нитри́т·, -а, *(m)* (chem) nitrite

~ на́тр·и·я sodium nitrite

нитрифик·а́ци·я, -и, *(f)* (biochem)
nitrification

нитрифиц·и́рук·щ·ий, -ая, -ее, *(a)*
nitrifying

~ бакте́р·и·и *(pl)* (biochem) nitrifying bacteria

нитро·амин·, -а, (*m*) (chem) nitramine
нитро·анилин·, -а, (*m*) (chem) nitro-aniline
нитро·барит·, -а, (*m*) (min) nitro-barite
нитро·бензол·, -а, (*m*) (chem) nitro-benzene
нитр·овальн·ый, -ая, -ое, (*a*) nitrating
нитр·овани·е, -я, (*n*) (chem) nitration, nitrating
нитро·гликол·ь, -я, (*m*) (chem) glycol dinitrate
нитро·глицерин·, -а, (*m*) (chem) nitroglycerine
нитро·групп·а, -ы, (*f*) (chem) nitro-group, nitroxyl
нитро·диэтилен·гликол·ь, -я, (*f*) glycol dinitrate
нитр·оз·а, -ы, (*f*) (chem) nitrose
нитро·за·мещ·ени·е, -я, (*n*) nitro-substitution
нитро·за·мещ·ённ·ый, -ая, -ое, (*a*) nitro-
~ **кисл·от·а** (*f*) nitro-acid
нитрозил·, -а, (*m*) (chem) nitrosyl
нитрозил·сер·н·ая кисл·от·а (*f*) nitrosyl sulphuric acid, nitrososulphuric acid
нитроз·ировани·е, -я, (*n*) (biochem) nitrosation
нитроз·н·ый, -ая, -ое, (*a*) nitrose; nitrous
~ **бактер·и·и** (*pl*) (biochem) nitroso-bacteria, ammonia-oxidizing bacteria
~ **газ·ы** (*pl*) nitrous gases
нитрозо·групп·а, -ы, (*f*) nitroso group
нитрозо·крас·ител·ь, -я, (*m*) nitroso-dyestuff
нитрозо·нафтол·, -а, (*m*) nitroso-naphthene
нитро·крас·ител·ь, -я, (*m*) nitro-dyestuff
нитро·крас·к·а, -и, (*g pl*) **-с·ок·,** (*f*) nitrocellulose paint/enamel
нитроксилол·, -а, (*m*) nitroxyl
нитрол·, -а, (*m*) nitrol
нитро·лак·, -а, (*m*) nitrolac, nitrocellulose varnish
нитролло·й, -я, (*m*) (met) nitralloy
нитрол·ов·ый, -ая, -ое, (*a*) nitrol, nitrolic
~ **кисл·от·а** (*f*) nitrolic acid
нитро·метан·, -а, (*m*) nitromethane
нитро·метр·, -а, (*m*) (chem) nitrometer, Lunge nitrometer

нитрон·, -а, (*m*) (chem) nitron, 1,4-diphenyl-3,5-endanilo-hydrotriazole
нитронафталин·, -а, (*m*) nitronaphthalene
нитрон·ов·ый, -ая, -ое, (*a*) nitronic (acid), (plast) polyacrylonitrile
нитро·парафин·, -а, (*m*) nitroparaffin
нитро·про·из·вод·н·ый, -ого, (*m decl as adj*) (chem) nitro-derivative, nitrocompound
нитро·пруссид·, -а, (*m*) (chem) nitroprusside, nitrosoferricyanide
~ **натр·и·я** sodium nitroprusside, sodium nitrosoferricyanide
нитро·основ·а, -ы, (*f*) (phot) nitrate base
нитро·со·един·ени·е, -я, (*n*) (chem) nitrocompound
~, **аромат·ическ·ое** aromatic nitro-compound
~ **жир·н·ого ряд·а** aliphatic nitro-compound
нитро·толуол·, -а, (*m*) nitrotoluene
нитро·фенол·, -а, (*m*) nitrophenol
нитро·филь·н·ый, -ая, -ое, (*a*) (bot) nitrophilous
нитро·фос·к·а, -и, (*f*) (agr) nitrophosphate fertilizer
нитро·хлор·бензол·, -а, (*m*) nitrochlorobenzene
нитро·целлюлоз·а, -ы, (*f*) nitrocellulose
нитро·цемент·аци·я, -и, (*f*) (met) nitrogen case-hardening, nitriding
нитро·шёлк·, -а, (*m*) nitrocellulose fibre/rayon
нитро·эмал·ь, -я, (*m*) nitrocellulose enamel
нитр·ующ·ий, -ая, -ее, (*a*) nitrating
нит·чатк·а, -и, (*g pl*) **-ток·,** (*f*) (zool, med) filaroid, hairlike nematode; (*pl*) *Filarioidea*
нит·чат·ый, -ая, -ое, (*a*) filamentous, filamentary
~ **бактер·и·и** (*pl*) iron bacteria, *Chlamydobacteriales*
нит·ь, -и, (*f*) thread, filament; yarn (of rope); hair, hairline, wire, cross-wire (of instrument)
~, **вихр·ев·ая** vortex line
~, **гриб·н·ая** (bot) hypha
~, **длин·н·ая** (anat) filum terminale
~, **ланс·ированн·ая** float (weaving)
~ **на·кал·а** (elec) valve filament, filament
~, **основ·н·ая** sinker thread (knitting machine)

нит·ь

~, **семен·н·áя**   (genet) spermatozoon

~, **фýтер·н·ая**   needle thread (knitting machine)

нит·ян·ый, -ая, -ое,   (*a*) thread, hair

~ **крест·**   (*m*) (opt) cross hairs

них   (*pron, prep*) of **они** (about etc.) them

ни·хрóм·, -а,   (*m*) (met) Nichrome

нихросилáл·, -а,   (*m*) Nicrosilal (alloy cast iron)

ни·чегó   (*gen*) of **ни·чтó;** (*adv*) not badly, tolerably; it can't be helped, it does not matter, never mind, too bad (as reply)

ни·чéй, -чья́, -чьё,   (*pron*) nobody's, no one's

ни·чемý   (*dat*) of **ни·чтó**

ничкóм   (*adv*) prone, face downwards

ни·чтó   (*pron*) nothing

ни·чтó·жно   (*adv*) negligibly

~ **мáл·ый**   (*a*) negligible

ни·чтó·жн·ый, -ая, -ое,   (*a*) insignificant, negligible, infinitesimal, trace

~ **концентр·áци·я**   (*f*) trace concentration

ни·чь·я́, -éй,   (*f of* **ни·чéй** *as noun*) draw (sport result)

нйш·а, -и,   (*i*) **-ей,** (*f*) niche, recess, bay, slot

нищ·ет·á, -ы́,   (*f*) poverty

нк   (*abbr*) = **натурáль·н·ый каучýк·,** NK, natural rubber; = **на·рóд·н·ый комиссариáт·,** Peoples' Commissariat (*obsolete*)

нитрозо·со·един·éни·е, -я,   (*n*) nitroso compound

нитро·клет·чáтк·а, -и,   (*g pl*) **-ток·,** (*f*) nitrocellulose, nitrocotton

но   (*conj*) but

нóбел·евск·ая прéм·и·я   (*f*) Nobel prize

нóбел·и·й, -я,   (*m*) (chem) nobelium, No

н-обрáз·н·ый, -ая, -ое,   (*a*) H-shaped, H-, I-shaped, I-

~ **мотóр·**   (*m*) (a/c) H engine

новакулит·, -а,   (*m*) (geol) hovaculite

новарсенóл·, -а,   (*m*) (pharm) neosalvarsan

новасекúт·, -а,   (*m*) (min) novasekite

нов·áтор·, -а,   (*m*) innovator, inventor, person with new ideas

нов·éйш·ий, -ая, -ее,   (*a*) newest, latest

новéлла·, -ы,   (*f*) short story

нов·изн·á, -ы́,   (*f*) novelty

нов·ичóк·, -чк·á,   (*m*) novice, beginning

ново-   (*root*) new, novo-, neo-

ново·в·вед·éни·е, -я,   (*n*) innovation

ново·гóд·н·ый, -яя, -ее,   (*a*) new-year's

ново·грéч·еск·ий, -ая, -ое,   (*a*) modern Greek

ново·из·обрет·ённ·ый, -ая, -ое,   (*a*) new, newly/recently invented

новокайн·, -а,   (*m*) (pharm) novocaine

ново·крыл·ые, -ых,   (*pl decl as adj*) (zool) neopterans, *Neoptera*

новолáк·, -а,   (*m*) (plast) novolak, novolac resin

новолáч·н·ый, -ая, -ое,   (*a*) novolac

~ **смол·á**   (*f*) novolac resin

ново·лед·никóвь·е, -я,   (*n*) (geol) New Glacial period

ново·лýн·и·е, -я,   (*n*) new moon

ново·меж·лед·никóвь·е, -я,   (*n*) (geol) New Interglacial period

ново·образ·овáни·е, -я,   (*n*) new formation; (med) neoformation, neoplasm; (ling) neologism; (biol) neogenesis

ново·образ·óванн·ый, -ая, -ое,   (*a*) newly formed; (biol) neogenic, hysterogenetic; (med) neoplastic

ново·сáд·к·а, -и,   (*g pl*) **-д·ок·,** (*f*) seasonal salt deposition (in lake)

ново·стрóй·к·а, -и,   (*g pl*) **-ó·ек·,** (*f*) building site/project

нóв·ост·ь, -и,   (*f*) news; novelty

нóв·шеств·о, -а,   (*n*) novelty, innovation

нóв·ыва·ть, -ют,   (*imp*) (biol) found (species)

нóв·ый, -ая, -ое,   (*a*) new, recent

нóв·ь, -и,   (*f*) virgin soil

ног·á, -и́,   (*acc sing*) **-у,** (*nom pl*) **-и,** (*g pl*) **ног,** (*d pl*) **-áм,** (*f*) foot; leg; raker (bridge pier)

~, **абдоминáль·н·ая**   (zool) uropod

~, **перéд·н·яя**   forefoot, hand (of quadruped)

~ **шассú**   (a/c) undercarriage leg

нóг·ие, -их,   (*pl component decl as adj*) (zool) -poda

ного·плóд·ник·, -а,   (*m*) (bot) *Podocarpus*

ногот·óк·, -т·к·á,   (*m*) (*dim*) of **нóгот·ь;** (bot) claw; (*pl*) (bot) pot-marigold, *Calendula*

ноготко·вúд·н·ый, -ая, -ое,   (*a*) claw-like, (bot) unguiculate

нóгот·ь, -гт·я,   (*m*) (anat) nail, unguis

ного·хвост·ки   (*pl*) (ent) collemboles, collembolans, *Collembola*

ного·челюст·ь, -и,   (*f*) (zool) maxilliped

ногт·ев·ой, -ая, -ое,   (*a*) (anat) nail, ungual

но·ет   (*pres 3rd sing*) of **ны·ть** aches

нож·, -а,   (*i*) **-ом,** (*m*) knife, blade; coulter (of plough), sickle (of grain harvesting machine)

~, **диск·ов·ый**   rotary knife; disk coulter (of plough)

~, **контакт·н·ый**   (elec) switch blade, contact blade

~, **коро·об·дир·очн·ый**   (agr) spudder

~, **момент·н·ый**   (elec) flicker blade

~, **на·клон·н·ый**   guillotine

~, **про·вод·ков·ый**   (met roll) hanging/stripper guide

~, **про·маз·очн·ый**   (paper) doctor, doctor blade/knife

~, **черен·ков·ый**   knife coulter (of plough)

нож·ев·ой, -ая, -ое,   (*a*) of **нож·**

~ **барабан·**   (*m*) awner (of thresher)

~ **голов·к·а**   (*f*) cutter block (on woodworking machines)

~ **из·дел·и·я**   (*pl*) cutlery

~ **о·пор·а**   (*f*) knife-edge support

ноже·держ·атель·, -я,   (*m*) blade holder

нож·ик·, -а,   (*m*) (*dim*) of **нож·**

нож·к·а, -и,   (*g pl*) **-ж·ек·,** (*f*) (*dim*) of **ног·а;** leg (furniture, mining timbers); stem (of glass); (bot) stalk, peduncle; (anat) pedicle, podium; (print) shank, stem, type body

~, **гребеж·ков·ая**   filament structure (of electric lamp)

~ **зуб·а**   addendum (of gear tooth); root of tooth

~ **лопат·к·и**   blade root (of turbines)

~ **сомит·а**   (anat) nephrostome

~, **черн·ая**   black-leg (potato disease)

нож·ниц·ы   (*f pl*) shear, shears, shearing machine(s); scissors, cutters; (econ) discrepancy; (oil) jar

~, **аллигатор·н·ые**   alligator shears

~, **гильотин·н·ые**   guillotine shear(s)

~ **гор·яч·ей рез·к·и**   hot shear(s)

~, **двух·криво·шип·н·ые**   guillotine shears

~, **диск·ов·ые**   rotary shear(s)

~ **за·кры·т·ого тип·а**   closed shear, double standard shear

~, **колен·чат·ые**   angle shears

нож·ниц·ы

~, **кромко·крош·ительн·ые**   side scrap cutter(s)

~, **лет·уч·ие**   flying shears

~, **лист·ов·ые**   plate shears

~, **маят·ников·ые**   swing shear(s)

~, **много·рез·н·ые**   gang shears

~, **об·рез·н·ые**   end/trimming shears

~ **от·кры·т·ого тип·а**   open-sided shear, single standard shear

~, **параллель·н·ые**   guillotine shears

~, **рыча·ж·н·ые**   alligator/lever shears

~ **с верх·н·им рез·ом**   downcutting shear

~ **с ниж·н·им рез·ом**   upcutting shear

~, **сад·ов·ые**   (hortic) secateurs

~, **стол·ов·ые**   bench shears

~, **стул·ов·ые**   small bench shears

нож·н·ой, -ая, -ое,   (*a*) of **ног·а** (biol) pedal, podal, peduncular

нож·н·ы, -ж·ен·,   (*f pl*) sheath, scabbard

нож·овк·а, -и,   (*g pl*) **-вок·,** (*f*) handsaw, hacksaw

~, **слесар·н·ая**   hacksaw

~, **стол·ярн·ая**   carpenter's saw

~, **уз·к·ая**   keyhole/inlaying saw

нож·овочн·ый, -ая, -ое,   (*a*) of **нож·овк·а**

~ **пил·а**   (*f*) power hacksaw

~ **полотн·о**   (*n*) saw blade, hacksaw blade

~ **стан·ок·**   (*m*) power hacksaw

ноздр·еват·ый, -ая, -ое,   (*a*) spongy, porous, honeycombed

ноздр·я, -и,   (*nom pl*) **ноздр·и,** (*g pl*) **-ей,** (*f*) (anat) nostril, naris; (*pl*) hole (in cheese etc.)

~, **внутр·енн·ая**   cavity

нозеан·, -а,   (*m*) (min) nosean, noselite

нозем·а, -ы,   (*f*) (zool) nosema

~, **пчел·ин·ая**   (zool) *Nosema apis*

нозо·лог·и·я, -и,   (*f*) (med) nosology

нозо·токсин·, -а,   (*m*) (med) nosotoxin

нок·, -а,   (*m*) derrick head; (naut) yardarm

нокаут·, -а,   (*m*) knock-out (sport)

ноктал·, -а,   (*m*) (pharm) noctal

нол·ев·ой, -ая, -ое,   (*a*) *see* **нул·ев·ой**

нолит·, -а,   (*m*) (min) nohlite

нол·ь, -я,   (*m*) *see* **нул·ь;** (*imperative*) now! (moment for instrument reading to be taken)

нóм·а, -ы,   (*m*) (med) noma, cancrum oris

номенклатýр·а, -ы,   (*f*) nomenclature; list

~, бинáр·н·ая   (biol) binomial nomenclature

~, хим·ѝческ·ая   chemical nomenclature, chemical symbols and formulae

нóмер·, -а, (*nom pl*) **номер·á,** (*g pl*) **-óв,** (*m*) number; size; (text) count (of yarns); hotel room; item (programme); (print) issue

~, áтом·н·ый   (phys) atomic number

~ зерн·á   grain size (number)

~ на·коп·ѝтел·я   address (computers)

~, нѝз·к·ий   (text) coarse count

~ оболóч·к·и   (phys) shell number

~ позѝц·и·и   (rockets) station number

~, по·рѧ́д·ков·ый   number, ordinal number, serial number

~ рас·чёт·а   crewman, team-man

у·каз·áтел·ь нóмер·а   (*m*) call indicator (telephones)

номер·áтор·, -а,   (*m*) bell-push indicator (board)

номер·нѝк·, -а,   (*m*) (teleph) small local battery switchboard

номер·н·óй, -áя, -óе, (*a*) of **нóмер·**

номеро·на·бир·áтел·ь, -я,   (*m*) automatic telephone dial

номинáл·, -а,   (*m*) (fin) face value, par value; (autom) control point; rating (of a machine)

~, дл·ѝтельн·ый   continuous-duty rating

~, кратко·врéмен·н·ый   short-time rating

~, пре·рѝв·ист·ый   intermittent duty rating

номинáль·н·ый, -ая, -ое, (*a*) nominal, rated

~ дá·нн·ые   (*pl*) rating (of machine)

~ мóщ·ност·ь (*f*) rated power/output, rating

~ на·грýз·к·а   (*f*) rated load

~ част·от·á   (*f*) rated frequency

номо·грáмм·а, -ы,   (*f*) (math) nomogram, nomograph, alignment chart

~, механ·ѝческ·ая   computer board

номо·граф·ѝческ·ий, -ая, -ое, (*a*) (math) nomographic

номо·грáф·и·я, -и,   (*f*) (math) nomography

нон-, нона-   (*int root*) (chem) non-, nona- (= nine)

нона·декáн·, -а,   (*m*) (chem) nonadecane

нона·декáн·ов·ый, -ая, -ое, (*a*) nonadecanoic, nonadecylic

нона·декéн·, -а,   (*m*) nonadecene

нона·декéн·ов·ый, -ая, -ое, (*a*) nonadecene, nonadecenoic

~ кисл·от·á   (*f*) nonadecenoic acid

нон·áн·, -а,   (*m*) (chem) nonane

нонáн·ов·ый, -ая, -ое, (*a*) nonan, nonanoic

~ альдегѝд·   (*m*) nonaldehyde

~ кисл·от·á   (*f*) nonanoic acid

нон·éн·, -а,   (*m*) (chem) nonene

нон·ѝн·, -а,   (*m*) (chem) nonine

нон·ѝл·, -а,   (*m*) (chem) nonyl

нóниус·, -а,   (*m*) vernier

~, круг·ов·óй   dial

нóниус·н·ый, -ая, -ое, (*a*) vernier

~ конденсáтор·   (*m*) (elec) vernier capacitor

нонóд·, -а,   (*m*) (rad) nonode

нонпарéл·ь, -и,   (*f*) (print) nonpareil

нонтронѝт·, -а,   (*m*) (min) nontronite

нор·á, -ы́, (*nom pl*) **-ы,** (*f*) burrow, lair

норвéж·ск·ий, -ая, -ое, (*a*) Norwegian

нóрд·, -а,   (*m*) North; North wind

норденшельдѝт·, -а,   (*m*) (min) nordenskioldine

нордмаркѝт·, -а,   (*m*) (min) nordmarkite

нóрд·ов·ый, -ая, -ое, (*a*) (naut) north, northern

норѝт·, -а,   (*m*) (geol) norite

норѝчник·ов·ые, -ых, (*pl decl as adj*) (bot) figwort family, *Scrophulariaceae*

нóри·я, -и,   (*f*) bucket elevator/chain

~, водо·от·лѝв·н·ая   chain pump

нóр·к·а, -и, (*g pl*) **-р·ок·,** (*f*) (*dim*) of **нор·á;** (zool) mink, *Lutreola mustelidae*

нóрм·а, -ы,   (*f*) norm, performance standard, standard, rate

~ врéмен·и   job time, time allowance (for a job)

~ вы·рабóт·к·и   output norm/standard

~ вы́·сев·а   (agr) weight of seed per unit area (for sowing)

~ от·бóр·а   (oil) rate of production

~, по·лѝв·н·ая   rate of irrigation/watering

~ процéнт·а   (fin) rate of interest

~ рас·хóд·а   consumption standard/rate, planned rate of consumption

~ с·тóк·а   (geog) normal annual runoff

~, техн·ѝческ·ая   performance standard

нормализ·áтор·, -а,   (*m*) normalizer

нормализ·а́ци·я, -и,   (*f*) normalization; standardization, introduction of standards; (met) normalizing

~ во́з·дух·а   air conditioning

~, све́тл·ая   (met) bright annealing

нормализ·о́ванн·ый, -ая, -ое,   (*a*) normalized; standardized, specified by a standard; conditioned (air)

нормáл·ь, -и,   (*f*) (math) normal, normal/perpendicular line; (tech) standard, standard specification

нормáль·н·о   (*adv*) normally; (math) perpendicular, at right angles, normal to

нормáль·ност·ь, -и,   (*f*) normality

нормáль·н·ый, -ая, -ое,   (*a*) normal; standard, according to standard; perpendicular

~ во́з·дух·   (*m*) standard air

~ на·пряж·éни·е   (*n*) (mech) direct/normal stress, stress in tension or compression

~ плóск·ост·ь   (*f*) (math) normal plane, plane perpendicular

~ рас·твóр·   (*m*) (chem) normal solution

~ свеч·á   (*f*) standard candle (source of light standard)

~ сеч·éни·е   (*n*) (math) normal/perpendicular section

~ толщ·ин·á зу́б·а   (*f*) circular thickness, arc thickness (gear wheel)

~ фóрм·а матри́ц·   (*f*) (math) Jordan form of matrix, simple classical matrix

~ шов·   (*m*) theoretical weld

~ электрóд·   (*m*) (elec) normal cell, standard cell

~ элемéнт·   (*m*) (elec) normal/standard cell

~ элемéнт·, не·на·сы́щ·енн·ый   (*m*) unsaturated standard cell

норматúв·, -а,   (*m*) norm, standard (time, performance quality etc.); (mil) establishment of equipment, ration

норматúв·н·ый, -ая, -ое,   (*a*) of **норматúв·**

норм·ировáни·е, -я,   (*n*) (math) normalizing, norming; standardization, rate-fixing, rating, establishment of performance/consumption standards

~ по·требл·éни·я   rationing; establishment of consumption standards

~ труд·á   time/rate fixing

норм·ирóванн·ый, -ая, -ое,   (*past part pass*) normalized; normed; standard, standardized, fixed, rate-fixed

~ кольц·ó   (*n*) (math) normed vector ring

норм·ирóвк·а, -и,   (*f*) (*v n*) *see* **норм·ировáни·е**

норм·ирóвочн·ый, -ая, -ое,   (*a*) normalizing, standardizing, rate-fixing

норм·ирóвщик·, -а,   (*m*) rate setter/fixer

нормо·блáст·, -а,   (*m*) (zool) normoblast

нормо·грáф·, -а,   (*m*) standard type stencil

нормо·цúт·, -а,   (*m*) (zool) normocyte

нормо·час·, -а,   (*m*) rated output per man-hour

норсульфазóл·, -а,   (*m*) (chem) sulphathiazole, sulfathiazole

нóс·, -а,   (*nom pl*) **-ы́,** (*g pl*) **-óв,** (*m*) nose; (geog) headland; (naut) bow, bows head, prow; point (of tools etc.)

~, антиклинáль·н·ый   (geol) anticlinal nose

~, туп·óй   blunt nose/end

~, ут·úн·ый   duckbill (conveyor loading attachment)

нос·áт·ый, -ая, -ое,   (*a*) big-nosed

нóс·ик·, -а,   (*m*) (*dim*) of **нос·;** spout; (bot) rostellum

нос·úлк·и, -лок·,   (*f pl*) (med) stretcher, litter

нос·úльщик·, -а,   (*m*) porter, carrier; stretcher-bearer

нос·úтел·ь, -я,   (*m*) carrier, bearer; (biol) vector; tape transport (tape recorder)

~ за·рáз·ы   (biol) vector

~ из·ображ·éни·я   (phot) image support

~ катализáтор·а   (chem) catalyst support

~, не·оснóв·н·óй   (chem) minority carrier

~, о·сажд·ённ·ый   carrier precipitate

~ слó·я   (phot) backing

~ тóк·а   (elec) carrier

~, у·дéрж·ивающ·ий   (phys) hold-back carrier

~ цéп·и   (chem) chain carrier

нос·úть, ´-ят,   (*imp indet*) carry, bear; wear (clothes etc.); fly (a flag); **-ся** be carried along, be driven (by wind etc.); be carried/borne; rush, rush about, fuss over; wear (of cloth etc.)

нóс·к·а, -и,   (*f*) carrying, bearing; wearing; laying (eggs)

нóс·к·ий, -ая, -ое,   (*a*) good-wearing, durable (of cloth); heavy-laying (of hen)

нос·кóв·ый, -ая, -ое,   (*a*) of **нос·óк·**

нос·кост·ь, -и, (*f*) durability, wearing qualities (of cloth); laying performance (of hen)

нос·н·ый, -ая, -ое, (*adj component*) -bearing, -iferous

нос·ов·ой, -ая, -ое, (*a*) nose, nasal; (naut) fore, forward, bow

∼ колес·о (*n*) (a/c) nosewheel

∼ част·ь (*f*) bows (of ship); nose (of a/c)

носо·глот·к·а, -и, (*g pl*) **-т·ок·,** (*f*) (anat) nasopharynx

носо·губ·н·ой, -ая, -ое, (*a*) (anat) nasolabial

нос·ок·, -с·к·а, (*m*) sock; toe, toe cap (of shoe); point, tip; spout, lip; leading edge (of aerofoil); (naut) bill, pea (of anchor, hook)

∼, раз·лив·очн·ый pouring lip

∼, телеметр·ическ·ий telemetering head

носо·лог·и·я, -и, (*f*) *see* **нозо·лог·и·я**

носо·рог·, -а, (*m*) (zool) rhinoceros

нос·ух·а, -и, (*f*) (zool) coati, *Nasua rufa*

нот·а, -ы, (*f*) note; (*pl*) music (on paper)

нотабен·а, -ы, (*f*) NB, *nota bene*

нотабене (*n indecl*) *see* **нотабен·а**

нота·печат·ани·е, -я, (*n*) (print) music printing

нотариаль·н·ый, -ая, -ое, (*a*) (law) notarial

нотариус·, -а, (*m*) (law) notary, solicitor

нот·аци·я, -и, (*f*) notation

нотис·, -а, (*m*) notice of readiness (of ship to work cargo)

нотифиц·ир·овать, -уют, (*imp or perf*) (com) notify; (dipl) deliver a note

нот·н·ый, -ая -ое, (*a*) music, printed music

∼ письм·о (*n*) (mus) notation

ното·ге·я, -и, (*f*) (zool) Notogaean realm, *Notogaea*

ното·завр·, -а, (*m*) (pal) Nothosaur, (*pl*) *Nothosauria*

ното·тер·и·й, -я, (*m*) (pal) notothere, Nototherium

ното·унгулят·ы (*pl*) (pal) notoungulates, *Notoungulata*

ното·хорд·, -а, (*m*) (zool) notochord

ноч·евать, -уют, (*imp or perf*) spend the night

ноче·свет·к·а, -и, (*g pl*) **-т·ок·,** (*f*) (zool) *Noctiluca*

ноче·цвет·н·ый, -ая, -ое, (*a*) (bot) noctiflorous

ноч·лег·, -а, (*m*) overnight stopping place; night's lodging/rest

ноч·ниц·а, -ы, (*f*) (zool) little brown bat, *Myotis*; (ent) cutworm moth, *Noctuida*

ноч·н·ой, -ая, -ое, (*a*) night, night's, nocturnal

∼ бинокл·ь (*m*) night glasses/binoculars

∼ о·шиб·к·а (*f*) (rad) night error/effect

∼ схем·а (*f*) (telecom) night-service connection

ноч·ь, -и, (*nom pl*) **-и,** (*g pl*) **-ей,** (*f*) night

∼, бел·ая (*pl*) Northern summer nights, polar day

на́ ноч·ь before going to bed/sleep

ноч·ью (*adv*) at/by night

нош·ени·е, -я, (*n*) (*v n*) of **нос·ить**

но́·ющ·ий, -ая, -ее, (*pres part act*) of **ны·ть**

но·я (*pres gerund*) of **ны·ть** aching

ноябр·ь, -я, (*m*) November

нрав·, -а, (*m*) disposition, temper, character

нрав·иться, -ятся, (*imp*) please

ему́ нрав·ится he likes

нрав·ственн·ый, -ая, -ое, (*a*) moral

ну well, now, what (*exclamations*)

нубук·, -а, (*m*) nubuck (leather), new buck

нувистор·, -а, (*m*) (elec) nuvistor

нуг·, -а, (*m*) (bot) quizotia, *Guizotia abyssinica*

нуг·а, -и, (*f*) (food) nougat

нудн·ый, -ая, -ое, (*a*) tedious

нужд·а, -ы, (*nom pl*) **-ы,** (*f*) need

нужд·а·ться, -ются, (*imp*) need, want, require, be in need of

нужн·о (*predic adj*) it is necessary; one should/ought/must

ему́ нуж·н·о he needs/wants

нуж·н·ый, -ая, -ое, (*a*) necessary, wanted

∼ число́ по·втор·ени·й цикл·ов (*n*) (autom) cycle criterion

нукле·аз·а, -ы, (*f*) (chem) nuclease

нукле·ин·, -а, (*m*) (chem) nuclein, nucleoprotein

нуклеин·ов·ый, -ая, -ое, (*a*) nuclein, nucleinic, nucleic

∼ кисл·от·а (*f*) (chem) nucleic acid

∼ об·мен· (*m*) nuclein metabolism

нукле·озид·, -а, (*m*) (chem) nucleoside

нуклё·озидаз·а, -ы, (*f*) (biochem) nucleosidase

нукле·ол·ь, -я,   (*m*) (cyt) nucleolus, plasmosome

нукле·он·, -а,   (*m*) (phys) nucleon

нуклео·протейд·, -а,   (*m*) (chem) nucleoprotein

нуклео·тид·, -а,   (*m*) (chem) nucleotide

нуклео·тидаз·а, -ы,   (*f*) (chem) nucleotidase

нуклеус·, -а,   (*m*) nucleus; queen cell (of bees)

нукл·ид·, -а,   (*m*) (nucl) nuclide

нукл·он·, -а,   (*m*) (phys) nucleon

~, бомбард·ирующ·ий incident nucleon

~ от·да·ч·и recoil nucleon

нуклон·н·ый, -ая, -ое,   (*a*) nucleonic

нул·ев·ой, -ая, -ое,   (*a*) zero, null

~ вероят·ност·ь (*f*) (math) zero probability

~ гипотез·а (*f*) (math) null hypothesis

~ колеб·ани·е (*n*) zero-point oscillation

~ матриц·а (*f*) (math) null/zero matrix

~ метод· (*m*) zero/balance/null method (of measurement)

~ при·бор· (*m*) zero-deflection instrument

~ систем·а (*f*) (autom) null-seeking system

~ точ·к·а (*f*) (elec) neutral point; datum point

~ у·станов·к·а (*f*) zeroing, zero adjustment/setting

нулли·пор·ы,   (*f pl*) (bot) red algae, *Nullipores*

нул·ь, -я,   (*m*) nought, nil, zero, null

~, абсолют·н·ый absolute zero (−273·16°C, −459·69°F)

~ -валент·ност·ь, -и,   (*f*) (chem) zero valency

~ выс·от· (surv) datum

~ глуб·ин· (ocean) chart datum

~ градус·ов zero degrees

~ граф·ик·а (hydr) gauge datum

~, лож·н·ый false null

~, машин·н·ый round-off (computers)

   не рав·н·ый нул·ю (math) non-vanishing, non-zero

   обращ·а·ться в нул·ь vanish

   от·лич·н·ый от нул·я non-zero, non-vanishing

   при·равн·ива·ть к нул·ю set to zero

нуль·детектор·, -а,   (*m*) null detector

нуль·индикатор·, -а,   (*m*) (elec) null indicator, zero adjuster

~ по·лёт·а (a/c) flight director

нуль·корректор·, -а,   (*m*) (autom) null-seeking device

нуль·у·каз·ател·ь, -я,   (*m*) (elec) null indicator, zero adjuster

нумейт·, -а,   (*m*) (min) noumeite

нумер·атор·, -а,   (*m*) numbering machine; (elec) annunciator

~, свет·ов·ой (elec) annunciator

нумер·аци·я, -и,   (*f*) numbering, giving numbers, enumeration

~, футаж·н·ая (cinema) edge numbering

нумер·овать, -уют,   (*imp*) number

нут·, -а,   (*m*) (bot) chick pea, garbanzo, *Cicer arietinum*

нут·аци·я, -и,   (*f*) (astron, bot) nutation

нутриант·, -а,   (*m*) nutrient

нутрицизм·, -а,   (*m*) (biol) nutricism

нутри·я, -и,   (*f*) (zool) nutria, coypu, *Myocaster*

нутро·мер·, -а,   (*m*) inside calliper/gauge/gage

~, индикатор·н·ый telescope internal gauge

~, микро·метр·ическ·ий internal micrometer

нутч-фильтр·, -а,   (*m*) (chem) Nutsche/Nutsch/suction filter

нуцеллюс·, -а,   (*m*) (bot) nucellus

нч (*abbr*) = **низко·част·отн·ый** low-frequency

ны·вш·ий (*past part act*) of **ны·ть**

ныр·н·уть, -ут,   (*perf*) dive, plunge

ныр·ял·о, -а,   (*n*) plunger, ram

ныр·я·ть, -ют,   (*imp*) dive, plunge

ны·ть, (*pres 3rd pl*) **но·ют,**   (*imp*) ache

ньюберрит·, -а,   (*m*) (min) newberrite

ньютон·, -а,   (*m*) newton (unit of force)

ньютон·а, бином·   (*m*) Newtonian binomial

ньютон·ов·а механ·ик·а   (*f*) Newtonian mechanics

ньютон·овск·ий, -ая, -ое,   (*a*) Newtonian

ню (*indecl*) nu, *v* (Greek)

нюх·, -а,   (*m*) scent, smell

нюх·а·ть, -ют,   (*imp*) smell

О

о (*prep*) (*with prepositional case*) concerning, about, on, of; (*with acc*) up against, against, upon, on

о- (*verbal prefix*) = **об-, обо-, объ-** *signifies*: (1) around, past; (2) all over; (3) becoming, making into; (4) excess; (5) perfective aspect

оа́зис·, -а, (*m*) oasis

об (*prep*) *see* **о**

об. $^0/_0$ (*abbr*) = **объ·ём·н·ый проце́нт·** $^0/_0$ by volume, vol. $^0/_0$

о́ба (*qualifies m, n*) both; the two

~ ... не neither

о·банкро́т·иться, -ятся, (*perf*) go/become bankrupt

оба́б·ок·, -б·к·а, *see* **берёз·овик·**

обапол·, -а, (*m*) offcut (timber)

об·бо́р·к·а, -и, (*f*) (min) cover/overhead caving

об.в. (*abbr*) = **объ·ём·н·ый вес·** bulk weight

об·ва́л·, -а, (*m*) collapse, cave-in, caving, caving-in

~, го́р·н·ый rock fall

~, сне́ж·н·ый avalanche, snow slip/slide

об·ва́л·ива·ть, -ют, (*imp*) make fall, dislodge, bring down; heap round; **-ся** collapse, cave, cave in; (*perf*) **об·вал·я́·ть** roll in, coat/cover with/by rolling

об·вал·и́ть, ⸗ят, (*perf*) *see* **об·ва́л·ива·ть**

об·ва́л·к·а, -и, (*f*) boning (meat)

об·вал·ова́ни·е, -я, (*n*) (*v n*) of **об·ва·л·ова́ть;** banking, damming; bank, protective bank, embankment

об·вал·о́ванн·ый, -ая, -ое, (*past part pass*) of **об·вал·ова́ть**

об·вал·ова́ть, -у́ют, (*perf*) bank up, protect/surround with a bank/embankment/rampart; dam up, build protective dykes/dams etc.

об·вал·я́·ть, -ют, (*perf*) roll in (loose material), coat/cover with (by rolling)

об·ва́р·ива·ть, -ют, (*imp*) scald, blanch

об·вар·и́ть, -я́т, (*perf*) *see* **об·ва́р·ива·ть**

об·ва́р·к·а, -и, (*f*) (*v n*) *see* **об·ва́р·ива·ть;** welding round/up; shoulder, collar (on shaft etc.)

об·вев·а́·ть, -ют, (*imp*) fan, (agr) winnow

об·вед·е́ни·е, -я, (*n*) (*v n*) of **об-·вес·ти об·вод·и́ть**

об·ве́·ива·ть, -ют, (*imp*) (agr) winnow

об·вер·н·у́ть, -у́т, (*perf*) *see* **об·вёр·т·ыва·ть**

об·верт·е́ть, -я́т, (*perf*) *see* **об·вёр·т·ыва·ть**

об·вёрт·к·а, -и, (*g pl*) **-т·ок·,** (*f*) (*v n*) *see* **об·вёрт·ыва·ть;** = **обёрт·-к·а;** (pal) husk

об·вёрт·ыва·ть, -ют, (*imp*) wrap up

об·вёрч·енн·ый, -ая, -ое, (*past part pass*) of **об·верт·е́ть**

об·ве́с·, -а, (*m*) screen (canvas etc.); short weight

об·ве́с·ить, ⸗ят, (*perf*) *see* **об·ве́ш·ива·ть**

об·вес·ти́ (*fut 3rd pl*) **об·вед·у́т,** (*past masc sing*) **об·вёл,** (*perf*); *see* **об·вод·и́ть**

об·ве́тр·енн·ый, -ая, -ое, (*past part pass*) weathered, weather-beaten; chapped (of skin); dried off

об·ветш·а́л·ый, -ая, -ое, (*a*) decrepit, decayed

об·вех·овани·е, -я, (*n*) (naut) buoyage; danning, danlaying

об·ве́ш·а·ть, -ют, (*perf*) hang round, put up round

об·ве́ш·ивани·е, -я, (*n*) (*v n*) of **об-·ве́ш·ива·ть**

об·ве́ш·ива·ть, -ют, (*imp*) give short weight; hang round; mark out with posts/buoys/beacons

об·веш·и́ть, -а́т, (*perf*) mark out with posts (e.g. roads); (naut) mark out with buoys

об·ве́·ять, -ют, (*perf*) fan, blow over, (agr) winnow

об·вив·а́·ть, -ют, (*imp*) wind round, entwine

об·вин·е́ни·е, -я, (*n*) charge, accusation, indictment

об·вин·и́тел·ь, -я, (*m*) (law) prosecutor

об·вин·и́ть, -я́т, (*perf*) *see* **об·вин·я́·ть**

об·вин·я́·ть, -ют, (*imp*) accuse, blame, charge with; prosecute, indict

об·вис·а́·ть, -ют, (*imp*) droop, sag, hang down

об·ви́с·л·ый, -ая, -ое, (*a*) drooping, sagging

об·вис·нуть, -нут, (*past masc sing*) **об·вис·,** (*perf*); *see* **об·вис·а·ть**

об·ви·ть, (*fut 3rd pl*) **об·вь·ют,** (*perf*); *see* **об·вив·а·ть**

об·вод·, -а, (*m*) (*v n*) *see* **об·вод·ить;** by-pass; circle, surround; (shipb) line (lines drawing)

об·вод·ить, -ят, (*imp*) lead/bring round; outline, circle, encircle, surround, enclose

об·вод·к·а, -и, (*g pl*) **-д·ок·,** (*f*) (*v n*) of **об·вод·ить;** (met roll) repeater, guide, trough

об·вод·нéни·е, -я, (*m*) (*v n*) *see* **об·вод·ня·ть;** (agr) irrigation, irrigation system; (geol) inundation, flooding; (min oil) flooding; water contamination (of oil etc.)

об·вод·нённ·ый, -ая, -ое, (*past part pass*) *see* **об·вод·ня·ть;** watery (of oil etc.)

об·вод·нительн·ый, -ая, -ое, (*a*) irrigation; irrigating; flooding; watering

об·вод·н·ить, -ят, (*perf*) *see* **об·вод·ня·ть**

об·вод·н·ый, -ая, -ое, (*a*) by-pass, encircling, leading round

~ **регул·ирова́ни·е** (*n*) by-pass governing (turbines)

об·вод·ня·ть, -ют, (*imp*) irrigate, flood, water; let water into; raise the water level in

об·вод·ня́ющ·ий, -ая, -ее, (*pres part act*) of **об·вод·ня·ть**

~ **вод·á** (*f*) (geol) invading water

об·волак·ива·ть, -ют, (*imp*) envelop, cover, wrap in

об·волак·ивани·е, -я, (*n*) (*v n*) of **об·волак·ива·ть;** (rad, elec) dipping

об·волочь, (*fut 3rd sing, pl*) **об·воло·ч·ёт, об·волок·ут,** (*past masc sing*) **об·волок·** (*perf*); *see* **об·волак·ива·ть**

об·вь·ют (*fut 3rd pl*) of **об·ви·ть**

об·вяз·а́ть, (*fut 3rd pl*) **об·вяж·ут,** (*perf*) *see* **об·вяз·ыва·ть**

об·вяз·к·а, -и, (*g pl*) **-з·ок·,** (*f*) (*v n*) *see* **об·вяз·ыва·ть;** band, binding tie, connection; (build) framework

~ **на·сос·ов** pump manifold

об·вяз·ыва·ть, -ют, (*imp*) tie round, bind, bandage; edge in chain stitch

обгалдер·, -а, (*m*) (naut) hook rope

обгалдыр·ь, -я, (*m*) hook rope

об·глад·ить, -ят, (*perf*) smooth out

об·глад·ыва·ть, -ют, (*imp*) gnaw round, gnaw; browse

об·глод·а́ть, (*fut 3rd pl*) **об·гло́ж·ут,** (*perf*) *see* **об·глад·ыва·ть**

об·гон·, -а, (*m*) (*v n*) *see* **об·гон·я·ть**

об·го́н·н·ый, -ая, -ое, (*a*) passing, overtaking

~ **пункт·** (*m*) (rail) passing place

об·го́н·ят (*fut 3rd pl*) of **обо·гн·а́ть** (*perf*); *see* **об·гон·я́·ть** (*imp*)

об·гон·я́·ть, -ют, (*imp*) overtake, overhaul, pass, get ahead of

об·гор·а́ни·е, -я, (*n*) (*v n*) of **об·гор·а́·ть;** spark wear (of electr. contacts)

   ис·пыт·а́ни·е на об·гор·а́ни·е (*n*) charring test

об·гор·а́·ть, -ют, (*imp*) burn, burn round/at the ends, be partially burnt

об·гор·е́л·ый, -ая, -ое, (*a*) charred, burnt

об·гор·е́ть, -ят, (*perf*) *see* **об·гор·а́·ть**

об·да·ва́ть, -ют, (*imp*) douse, pour over, drench; suffuse

об·да́·ть, (*fut 3rd sing, pl*) **об·да́·ст, об·дад·у́т,** (*perf*); *see* **об·да·ва́ть**

об·дéл·а·ть, -ют, (*perf*) *see* **об·дéл·ыва·ть**

об·дéл·к·а, -и, (*g pl*) **-л·ок·,** (*f*) (*v n*) *see* **об·дéл·ыва·ть;** (civ eng) lining, tunnel lining

об·дéл·очн·ый, -ая, -ое, (*a*) of **об·дéл·к·а**

об·дéл·ыва·ть, -ют, (*imp*) finish, fashion; line, jacket; fix, arrange

об·дер·у́т (*fut 3rd pl*) of **обо·др·а́ть** (*perf*); *see* **об·дир·а́·ть** (*imp*)

об·дир·а́·ть, -ют, (*imp*) flay, strip, skin, bark, peel, hull, shell; (fish) flench; (tech) rough out

~ **на токáр·н·ом стан·к·é** rough-turn

об·дир·к·а, -и, (*g pl*) **-р·ок·,** (*f*) (*v n*) *see* **об·дир·а́·ть;** (met cast) dressing-off

об·ди́р·н·ый, -ая, -ое, (*a*) of **об·ди́р·к·а**

~ **круп·а́** (*f*) hulled grain

об·ди́р·очн·о-токáр·н·ый стан·óк· (*m*) roughing lathe

об·ди́р·очн·о-шлиф·овáльн·ый ста·н·óк· (*m*) rough grinding machine

об·ди́р·очн·ый, -ая, -ое, (*a*) of **об·ди́р·к·а**

об·дув·а́ни·е, -я, (*n*) *see* **об·ду́в·к·а**

об·дув·а́·ть, -ют, (*imp*) blow off/out; blast (cleaning); cheat, dupe

об·ду́в·к·а, -и, (*g pl*) -в·ок·, (*f*) (*v n*) of об·дув·а́·ть; blow-out; blasting, blast cleaning

~ дро́б·ью (met) shot-peening

~ песк·о́м· (met) grit blasting

об·ду́в·очн·ый, -ая, -ое, (*a*) of об·-ду́в·к·а

~ у·стро́й·ств·о (*n*) (st eng) soot blower

об·ду́м·анн·ый, -ая, -ое, (*a*) delibe-rate, considered

об·ду́м·а·ть, -ют, (*perf*) consider, think over

об·ду́·ть, -ют, (*perf*) *see* об·дув·а́·ть

обе (*qualifies* (*f*)) both; the two

о·бе́г·а·ть, -ют, (*imp*) run round/past/about

о·бе́г·аю́щ·ий, -ая, -ее, (*pres part act*) of о·бе́г·а·ть; (autom) scan, scanning

~ контро́л·ь (*m*) (autom) scan check-ing, multi-point scan checking

~ у·стро́й·ств·о (*n*) (autom) scanning device

об·е́д·, -а, (*m*) dinner

об·е́д·а·ть, -ют, (*imp*) dine, have dinner

об·е́д·енн·ый, -ая, -ое, (*a*) dinner, dining

о·бедн·е́ни·е, -я, (*n*) impoverishment, depletion, denudation

    коэффицие́нт· о·бедн·е́ни·я (*m*) depletion factor

~, равно·ве́с·н·ое equilibrium deple-tion

о·бедн·ённ·ый, -ая, -ое, (*past part pass*) *see* о·бедн·я́·ть; depleted, im-poverished, denuded, stripped

о·бедн·и́тел·ь, -я, (*m*) (chem) stripper

о·бедн·и́ть, -я́т, (*perf*) *see* о·бедн·я́·ть

о·бедн·я́·ть, -ют, (*imp*) deplete, im-poverish

о·бедн·я́ющ·ая ча́ст·ь (*f*) (nucl) depleting/stripping section (of a co-lumn)

о·беж·а́ть, (*fut 3rd sing, pl*) о·беж·и́т, о·бег·у́т, (*perf*); *see* о·бег·а́·ть

обез- (*prefix*) = обес- *signifies* deprive/free/rid of; de-, dis-, un-

обез·бо́л·ива·ть, -ют, (*imp*) anaes-thetize, anesthetize

обез·бо́л·ить, -я́т, (*perf*) *see* обез·-бо́л·ива·ть

обез·во́д·е·ть, -ют, (*perf*) dry up

обез·во́д·ить, -я́т, (*perf*) *see* обез·-во́ж·ива·ть

обез·водоро́д·ить, -я́т, (*perf*) de-hydrogenate

обез·во́ж·енн·ый, -ая, -ое, (*past part pass*) of обез·во́д·ить

обез·во́ж·ивани·е, -я, (*n*) dehydra-tion, dessication, (geog) exsiccation; (oil) water separation/removal, (min) dewatering

обез·во́ж·ива·ть, -ют, (*imp*) de-hydrate, dewater

обез·возду́ш·енн·ый, -ая, -ое, (*past part pass*) deaerated

обез·вола́ш·ивани·е, -я, (*n*) depila-tion, unhairing (leather), fellmongering (sheepskin)

обез·вре́д·ить, -я́т, (*perf*) *see* обез·-вре́ж·ива·ть

обез·вре́ж·ивани·е, -я, (*n*) (*v n*) of обез·вре́ж·ива·ть

обез·вре́ж·ива·ть, -ют, (*imp*) render harmless

обез·га́ж·ивани·е, -я, (*n*) degassing, degasification, gas separation/remo-val; (elec) evacuation (of valves), get-tering

обез·гла́в·ить, -я́т, (*perf*) *see* обез·-гла́вл·ива·ть

обез·гла́вл·ива·ть, -ют, (*imp*) deca-pitate, behead

обез·жи́р·енн·ый, -ая, -ое, (*past part pass*) degreased, defatted

обез·жи́р·ивани·е, -я, (*n*) degreasing, fat extraction

обез·за·ра́ж·ива·ть, -ют, (*imp*) disin-fect, decontaminate

обез·за·ра́з·ить, -я́т, (*perf*) *see* обез·-за·ра́ж·ива·ть

обез·зо́л·ивани·е, -я, (*n*) deliming, decalcification; de-ashing

обез·и́л·ивани·е, -я, (*n*) desliming, desilting

обез·ле́с·ени·е, -я, (*n*) deforestation, clearing forest land

обез·ли́ст·влени·е, -я, (*n*) defoliation

обез·ли́ч·енн·ый, -ая, -ое, (*past part pass*), depersonalized; general-ized, reduced to a standard/norm; mechanical (of work performed by people)

~ черт·ёж· (*m*) standard component drawing

обез·лю́д·евш·ий, -ая, -ее, (*past part act*); depopulated, deserted

обез·ма́сл·енн·ый, -ая, -ое, (*a*) de-oiled, oil-free

обез·ма́сл·ивани·е, -я, (*n*) de-oiling, oil separation/removal обез·ме́ж·ивани·е, -я,   (*n*) (met) drossing

обез·обра́ж·енн·ый, -ая, -ое, (*past part pass*) disfigured, mutilated

обез·обра́ж·ени·е, -я, (*n*) disfiguration, mutilation

обез·ору́ж·ивани·е, -я, (*n*) disarmament

обез·репе́·ивающ·ий, -ая, -ее, (*a*) (agr) burr-removing/extracting

обез·ро́ж·ива·ть, -ют, (*imp*) dehorn, pollard

обез·углеро́ж·ивани·е, -я, (*n*) decarburization, decarbonization

обез·ы́л·ивани·е, -я,   (*n*) desilting, desliming

обезья́н·а, -ы, (*f*) monkey, ape

обе́их (*gen/prep*) of обе

о·бе́й·те (*imperative pl*) of о·би́·ть

обели́ск·, -а, (*m*) obelisk

о·бел·е́ни·е, -я, (*n*) (*v n*) of о·бел·я́·ть

о·бел·и́ть, -я́т, (*perf*) *see* о·бел·я́·ть

о·бел·я́·ть, -ю́т,   (*imp*) clear (of suspicion), rehabilitate (a reputation)

о·берег·а́·ть, -ю́т, (*imp*) guard, protect, defend

о·бер·ёт (*fut 3rd sing*) of обо·бр·а́ть

о·бере́ч·ь, (*fut 3rd sing, pl*) о·береж·ёт, о·берег·у́т, (*past masc sing*) о·берёг·, (*perf*); *see* о·берег·а́·ть

обёр·нут·ый, -ая, -ое, (*past part pass*) of обер·н·у́ть; wrapped, enveloped

обер·н·у́ть, -у́т, (*perf*) *see* обёрт·ыва·ть

обёрт·к·а, -и, (*g pl*) -т·ок·, (*f*) (*v n*) *see* обёрт·ыва·ть; wrapper, envelope, paper cover, cover; dust-jacket (of book); (bot) spathe, sheath; shuck, husk (of maize)

~ со·цвет·и·я (bot) involucre

обертóн·, -а, (*m*) (telecom) harmonic; overtone (music)

обёрт·очн·ый, -ая, -ое, (*a*) wrapping, packing

обёрт·ыва·ть, -ют, (*imp*) wind round; wrap up, envelop; turn round (*trans*)

о·бер·у́т (*fut 3rd pl*) of обо·бр·а́ть

обес- (*prefix*) = обез- *signifies* deprive/ free/rid of; de-, dis-, un-

обес·кислоро́д·ить, -ят, (*perf*) de-oxidize

обес·кисло́ч·ить, -ат, (*perf*) de-acidify

обес·кле́·ивани·е, -я, (*n*) boiling-off, degumming (silk)

обес·клювл·ивани·е, -я, (*n*) debeaking (of poultry)

обес·кре́мн·ивани·е, -я, (*n*) desiliconization, silica removal, demineralization (of water)

обес·пе́ч·ени·е, -я, (*n*) (*v n*) *see* обес·пе́ч·ива·ть; provision; guarantee, guaranty, security, warranty; (*gen as adj*) providing for, for

~ без·о·па́с·ност·и safeguarding

~, бо·ев·о́е (mil) security

для обес·пе́ч·ени·я ... to provide/give, for

систе́м·а обес·пе́ч·ени·я по·са́д·к·и само·лёт·ов (*f*) (air) blind landing system

~, социа́ль·н·ое social security

обес·пе́ч·енност·ь, -и, (*f*) extent of provision (for), coverage; frequency (of a desired event)

обес·пе́ч·енн·ый, -ая, -ое, (*past part pass*) of обес·пе́ч·ить; provided-for, secured, ensured, guaranteed; (+ *instr*) provided with, fitted with

~ мо́щ·ност·ь (*f*) firm output (of a hydroelectric station)

обес·пе́ч·ива·ть, -ют, (*imp*) ensure, assure, see to; provide, supply, give; provide for, safeguard, make provisions for; (mil) protect

~ воз·мо́ж·ност·ь provide the opportunity

~ до·каз·а́тельств·а (law) provide/bring forward proof

~ плат·ёж· provide for payment, make provisions/arrangements for payment, guarantee payment

обес·пе́ч·ить, -ат, (*perf*) *see* обес·пе́ч·ива·ть

обес·пло́д·ить, -ят, (*perf*) *see* обес·пло́ж·ива·ть

обес·пло́ж·ива·ть, -ют, (*imp*) sterilize, make barren

обес·поко́·енн·ый, -ая, -ое, (*past part pass*) of обес·поко́·ить; anxious

обес·поко́·ить, -ят, (*perf*) disturb, perturb; -ся, be concerned/worried/anxious

обес·пы́л·ивани·е, -я, (*n*) dust removal/elimination

обес·са́хар·ивани·е, -я, (*n*) sugar extraction

обес·серебр·е́ни·е, -я, (*n*) desilverization, silver extraction

обес·се́р·ивани·е, -я, (*n*) desulphurization, desulfurization

~, гидрогенизацио́н·н·ое (oil) hydro-desulphurization

обес·сла́в·ить, -ят, (*perf*) defame

обес·смо́л·ивани·е, -я, (*n*) tar extraction, detarring

обес·со́л·ивани·е, -я, (*n*) desalification, desalting

обес·то́ч·енн·ый, -ая, -ое, (*past part pass*) of обес·то́ч·ить; (elec) dead, idle, switched off, off

обес·то́ч·ивани·е, -я, (*n*) (elec) disconnection, switching off, de-energizing

обес·то́ч·ить, -ат, (*perf*) (elec) disconnect, turn off current, switch off, de-energize

обес·фо́сфор·ивани·е, -я, (*n*) *see* обес·фо́сфор·изаци·я

обес·фо́сфор·изаци·я, -и, (*f*) (met) dephosphorization, removal of phosphorus

обес·цве́ч·ивани·е, -я, (*n*) decolorization, bleaching; deblooming

обес·цве́ч·ивающ·ий, -ая, -ее, (*pres part act*) decolorizing, bleaching

~ сре́д·ств·о (*n*) decolorant, decolorizer, decolorizing/bleaching agent

обес·це́н·ивани·е, -я, (*n*) devaluation; depreciation

обес·це́н·ива·ть, -ют, (*imp*) (fin) devalue, write down; -ся (*pass*) depreciate

обес·це́н·ить, -ят, (*perf*) *see* обес·це́н·ива·ть

обес·цинк·о́вывани·е, -я, (*n*) zinc removal, dezincification

обес·шла́мл·ивани·е, -я, (*n*) (min) desliming

обеча́·йк·а, -и, (*g pl*) -а́·ек·, (*f*) course (of boiler, large diam. pipe etc.); rim, edge

обещ·а́·ть, -ют, (*imp or perf*) promise

об·жа́л·овани·е, -я, (*n*) (law) appeal

~, при·нос·и́ть lodge an appeal

об·жа́л·овать, -уют, (*perf*) (law) appeal against

об·жа́р·ива·ть, -ют, (*imp*) grill; brown (by frying)

об·жа́р·ить, -ят, (*perf*) *see* об·жа́р·ива·ть

об·жа́т·и·е, -я, (*n*) (*v n*) *see* об·жим·а́·ть; (met roll) reduction, compression, draft

~, абсолю́т·н·ое (met roll) absolute draft

~, двух·кра́т·н·ое (met roll) twofold reduction

коэффицие́нт· об·жа́т·и·я (*m*) (met roll) reduction factor

об·жа́т·и·е

~, лин·е́йн·ое (met roll) absolute draft, linear reduction

мо́дул·ь объ·ём·н·ого об·жа́т·и·я (*f*) (mech) volumetric modulus of elasticity, modulus of cubic compressibility

~, объ·ём·н·ое volumetric/cubic compression

~, от·нос·и́тельн·ое (met roll) reduction of area, draft

об·жа́т·ь, (*fut 3rd sing, pl*) обо·жм·ёт, обо·жм·у́т, (*past masc sing*) о́б·жа́л, (*perf*); *see* об·жим·а́·ть

об·же́ч·ь, (*fut 3rd sing, pl*) обо·ж·ж·ёт, обо·жг·у́т, (*past masc sing*) об·жёг·, (*perf*); *see* об·жиг·а́·ть

о́б·жиг·, -а, (*m*) (*v n*) *see* об·жиг·а́·ть

~, вос·станов·и́тельн·ый (met) magnetic roasting

~, дестиляцио́н·н·ый (met) distillation

~, о·кисл·и́тельн·ый (met) oxidation roast

~, с·пек·а́ющ·ий (met) roast-sintering, blast-roasting

об·жиг·а́тельн·ый, -ая, -ое, (*a*) roasting, firing, baking

~ печ·ь (*f*) kiln, roasting furnace

об·жиг·а́·ть, -ют, (*imp*) burn, scorch; kiln; roast (ore), bake (bricks); fire (ceramics); bream (ship); light up (electronic valves)

о́б·жиг·ов·ый, -ая, -ое, (*a*) of о́б·жиг·, *and see* об·жиг·а́тельн·ый

об·жи́м·, -а, (*m*) (*v n*) *see* об·жим·а́·ть; (met roll) cogging, roughing; (met) shingling, nobbing, nobbling (puddling process)

об·жим·а́·ть, -ют, (*imp*) wring/squeeze/press out

об·жи́м·к·а, -и, (*g pl*) -м·ок·, (*f*) (*v n*) *see* об·жим·а́·ть; (met) reduction; (met forge) swage, swage block, rivet header

об·жи́м·н·о́й, -а́я, -о́е, (*a*) of об·жи́м·

~ стан· (*m*) (met roll) primary/cogging mill

об·жи́м·н·ый, -ая, -ое, (*a*) of об·жи́м·; *see* об·жи́м·н·о́й

об·жи́м·ок·, -м·к·а, (*m*) *see* вы́·жим·к·и

об·жи́м·очн·ый, -ая, -ое, (*a*) of об·жи́м·к·а

обзо·й, -я, (*m*) *see* об·зол·

об·зо́л·, -а, (*m*) rough edge (on a wooden plank etc.)

об·зо́л·ива·ть, -ют, (*imp*) incinerate

об·зо́р·, -а, (*m*) review, survey, summary, outline, synopsis; field of view; scanning, scan, coverage (radar)

~ печа́т·и press review

~ про·стра́н·ств·а scanning (radar)
у́гол· об·зо́р·а (*m*) (opt) arc of vision; aspect angle (of radar target)

об·зо́р·ност·ь, -и, (*f*) vision (e.g. from a motor car, aeroplane etc.)

об·зо́р·н·ый, -ая, -ое, (*a*) of об·зо́р·

~ до·кла́д· (*m*) review, summary

~ индика́тор· (*m*) (rad) plan position indicator

о·бив·а́·ть, -ют, (*imp*) beat/knock off; upholster, line; bind (with hoops etc.)

о·би́в·к·а, -и, (*g pl*) -в·ок·, (*f*) (*v n*)
see о·бив·а́·ть; upholstery, lining

~, ко́ж·ан·ая leather upholstery

о·би́д·н·ый, -ая, -ое, (*a*) offensive, vexing, annoying

о·би́·л (*past masc sing*) of о·би́·ть

оби́л·и·е, -я, (*n*) abundance, plenty

оби́ль·ност·ь, -и, (*f*) copiousness, amount

~ ис·то́ч·ник·а по́л·я (math) field source strength

оби́ль·н·ый, -ая, -ое, (*a*) copious, abundant, plentiful

о·бир·а́·ть, -ют, (*imp*) gather, pick (fruit etc.)

обит·а́ем·ый, -ая, -ое, (*pres part pass*)
of о·бит·а́·ть; inhabited; habitable

обит·а́тел·ь, -я, (*m*) inhabitant, dweller; inmate

~, ис·ко́н·н·ый indigenous inhabitant

о·бит·а́·ть, -ют, (*imp*) inhabit, dwell, live in

о·би́·т·ый, -ая, -ое, (*past part pass*)
of о·би́·ть; upholstered, padded

о·би́·ть, (*fut 3rd sing, pl*) обо·бь·ёт, обо·бь·ю́т, (*past masc sing*) о·би́·л, (*perf*); *see* о·бив·а́·ть

обихо́д·, -а, (*m*) custom, habit
в·ход·и́ть в обихо́д· become usual/customary
вы·йти́ из обихо́д·а go out of fashion/use, become obsolete

об·ка́п·а·ть, -ют, (*perf*) sprinkle/cover with drops, spot with

об·ка́п·ыва·ть, -ют, (*imp*) sprinkle/cover with drops, spot with; dig around

об·ка́рмл·ива·ть, -ют, (*imp*) overfeed

об·ка́т·, -а, (*m*) (*v n*) *see* об·ка́т·ыва·ть; (mech) gear generation

об·кат·а́·ть, -ют, (*perf*) *see* об·ка́т·ыва·ть

об·ка́т·к·а, -и, (*g pl*) -т·ок·, (*f*) (*v n*)
see об·ка́т·ыва·ть; (met) rounding off (forging blanks); burnishing; (rail) running test; generating (gears)
механ·и́зм· об·ка́т·к·и (*m*) generating mechanism (gears)

об·ка́т·н·ый, -ая, -ое, (*a*) *see* об·кат·н·о́й

об·кат·н·о́й, -а́я, -о́е, (*a*) of об·ка́т·к·а

~ стан· (*m*) reeling mill/machine (tube rolling)

об·ка́т·очн·ый, -ая, -ое, (*a*) of об·ка́т·к·а

~ стан·о́к· (*m*) gear-measuring machine

об·ка́т·ывани·е, -я, (*n*) *see* об·ка́т·к·а

об·ка́т·ыва·ть, -ют, (*imp*) roll in, coat with (by rolling); roll/wear smooth; run in (new machinery)

об·ка́ш·ива·ть, -ют, (*imp*) mow round

об·кла́д·к·а, -и, (*g pl*) -д·ок·, (*f*) (*v n*)
see об·кла́д·ыва·ть; facing, lining, casing

~ конденса́тор·а (elec) capacitor plate

об·кла́д·оч·н·ый, -ая, -ое, (*a*) (*v n*) of об·кла́д·к·а; coating

об·кла́д·ыва·ть, -ют, (*imp*) put/place/lay around; face, cover; lay siege to, surround

об·кле́·ива·ть, -ют, (*imp*) glue/stick/paste over; paper (walls)

об·кле́·ить, -ят, (*perf*) *see* об·кле́·ива·ть

об·ко́м·, -а, (*m*) (*abbr*) of о́блас·т·н·о́й комите́т· regional committee

об·коп·а́·ть, -ют, (*perf*) dig around

об·корм·и́ть, -ят, (*perf*) overfeed

об·кос·и́ть, -ят, (*perf*) mow round

об·ко́ш·енн·ый, -ая, -ое, (*past part pass*) of об·кос·и́ть

об·кромс·а́·ть, -ют, (*perf*) cut round, snip off, whittle down

об·ку́р·ива·ть, -ют, (*imp*) smoke, season; smoke in (pipe etc.)

об·кур·и́ть, -ят, (*perf*) *see* об·ку́р·ива·ть

обл- (*component*) (*abbr*) of о́блас·т·н·о́й regional

об·лаг·а́·ть, -ют, (*imp*) impose, levy, exact (tax, fine etc.)

о·благо·ра́ж·ивани·е, -я, (*n*) (*v n*) of о·благо·ра́ж·ива·ть; (met) ageing, age hardening, dispersion hardening; processing (wood)

о·благо·ра́ж·ива·ть, (*imp*) improve; (text) dress, finish; (geol) ennoble о·благо·ро́д·ить, ⸗ят, (*perf*) *see* о·бла·го·ра́ж·ива·ть

облад·а́·ть, -ют, (*imp*+*instr*) possess, own, have, hold

облад·а́ющ·ий, -ая, -ее, (*pres part act*) of облад·а́·ть; having, with

о́блак·о, -а, (*nom pl*) -а́, (*g pl*) -о́в, (*n*) cloud

~ вертика́ль·н·ого раз·ви́т·и·я (*pl*) heap clouds

~ ве́рх·н·его я́рус·а (*pl*) high clouds

~, внутри·ма́сс·ов·ое (meteor) air-mass cloud

~, высоко·куч·ев·ы́е (*pl*) Altocumulus

~, высоко·сло·и́ст·ые (*pl*) Altostratus

~ Ко́ттрелл·а (phys) Cottrell atmosphere

~, кучево·дожд·ев·ы́е (*pl*) Cumulo-nimbus

~, куч·ев·ы́е (*pl*) Cumulus

~ ни́ж·н·его я́рус·а (*pl*) low clouds

~, пер·и́ст·о-куч·ев·ы́е (*pl*) (meteor) Cirrocumulus

~, пер·и́ст·о-сло·и́ст·ые (*pl*) (meteor) Cirrostratus

~, пер·и́ст·ые (*pl*) (meteor) Cirrus

~ препя́ств·и·й (meteor) crest cloud

~, разо́·рв·анн·о-дожд·ев·ы́е (*pl*) (meteor) Fractonimbus

~, разо́·рв·анн·о-куч·ев·ы́е (*pl*) (meteor) Fractocumulus

~, разо́·рванн·о-сло·и́ст·ые (*pl*) (meteor) Fractostratus

~, сло·и́ст·о-дожд·ев·ы́е (*pl*) Nimbostratus

~, сло·и́ст·о-куч·ев·ы́е (*pl*) Stratocumulus

~, сло·и́ст·ые (*pl*) Stratus

~ электро́н·ов (phys) electron cloud

облако·ме́р·, -а, (*m*) (meteor) cloud indicator

об·ла́м·ыва·ть, -ют, (*imp*) break off; break in/down (of will etc.); calve (of ice)

област·н·о́й, -а́я, -о́е, (*a*) of о́бласт·ь

о́бласт·ь, -и, (*f*) province, district, area, region, zone; range, field, sphere; (math) domain

~, бли́ж·н·яя инфра·кра́сн·ая (opt) near infra-red region

~ больш·и́х энерг·и·й (phys) high-energy region/range

~, да́ль·н·яя инфра·кра́сн·ая (opt) far infra-red region

~ де́й·стви·й field of activity; range

~, за·па́ль·н·ая (nucl) seed region

~ кавита́ц·и·и (dynam cavity

о́бласт·ь

~ кон·и́ческого теч·е́ни·я (dynam) conical field

~, много·с·вя́з·н·ая (math) multiple-connected domain

~, нео·троп·и́ческ·ая (biol) neotropical biogeographic region

~ о·пре·дел·е́ни·я фу́нкц·и·и (math) domain of a function, range of an independent function

~ по·ниж·ённ·ого давл·е́ни·я (meteor) depression, area of low pressure, low

~ по·вы́ш·енн·ого давл·е́ни·я (meteor) area of high pressure, high

~ при·мен·е́ни·я field of application

~ само·про·из·во́ль·н·ой на·маг·ни́ч·енност·и domain (ferromagnetism)

~ с·ход·и́мост·и (math) integral domain

~ то́ч·ечн·ого из·луч·е́ни·я (rad) Fraunhofer region (of an antenna)

~, ферро·магни́т·н·ая (phys) ferro-magnetic domain

~ чи́сел· (math) number domain

~, цилиндр·и́ческ·ая (math) tube domain

обла́т·к·а, -и, (*g pl*) -т·ок·, (*f*) wafer

о́блач·к·о, -а, (*nom pl*) -а́, (*g pl*) -о́в, (*n*) (*dim*) of о́блак·о

о́блач·ност·ь, -и, (*f*) cloudiness, cloud(s), cloud amount

    журна́л· о́блач·ност·и (*m*) (meteor) cloud journal

о́блач·н·ый, -ая, -ое, (*a*) cloud, cloudy, overcast

~ ден·ь (*m*) cloudy day

~ по·кро́в· (*m*) cloud cover/deck

об·лег·а́·ть, -ют, (*imp*) fit closely; cover, surround

об·легч·а́·ть, -ют, (*imp*) lighten, ease, alleviate, facilitate, make easier

об·легч·е́ни·е, -я, (*n*) (*v n*) *see* об·легч·а́·ть; alleviation; relief

об·легч·ённ·ый, -ая, -ое, (*past part pass*) of об·легч·и́ть; light-weight, light; (oil) light-fraction; easy, favoured, favored

~ пере·хо́д· (*m*) (phys) favoured transition

об·легч·и́ть, -а́т, (*perf*) *see* об·легч·а́·ть

об·лед·енева́·ть, -ют, (*imp*) ice up, become covered with ice

об·лед·ене́л·ый, -ая, -ое, (*a*) ice-covered

об·лед·снéн·ие, -я, (*n*) icing-up, icing

об·лед·енé·ть, -ют, (*perf*) *see* об·лед·е-невá·ть

об·лез·á·ть, -ют, (*imp*) peel/come off (of paint); wear off, grow bare (of furs etc.)

об·лéз·л·ый, -ая, -ое, (*a*) shabby, bare, patchy

об·лéз·ть, -ут, (*past masc sing*) об·лéз·, (*perf*) *see* об·лез·á·ть

об·лéй·те (*pl imperative*) of об·лй·ть

об·лек·á·ть, -ют, (*imp*) clothe, present, formulate (thoughts etc.); invest/ supply with

об·леп·йть, -ят, (*perf*) *see* об·лепл·я́·ть

об·леп·йх·а, -и, (*f*) (bot) sea buckthorn, *Hippophae Rhamnoides*

об·лепл·я́·ть, -ют, (*imp*) stick/adhere all over/round

об·лес·éни·е, -я, (*n*) afforestation

об·лёт·, -а, (*m*) (*v n*) *see* об·лет·á·ть; test/trial flight; round flight

~, при·ём·н·ый acceptance flight

об·лет·á·ть, -ют, (*imp*) fly round/ around/all over; overtake, get ahead of (flying); fall, drop (of leaves etc.); (*perf*) carry out test flight(s)

об·лет·é·ть, -áт, (*perf*) *see* об·лет·á·ть (*imp*)

об·лёт·ыва·ть, -ют, (*imp*) carry out test flight(s)

об·леч·ённ·ый, -ая, -ое, (*past part pass*) of об·лéч·ь (1); invested/en-trusted with

об·лéч·ь (1), (*fut 3rd sing, pl*) об·леч·ёт, об·лек·ýт, (*past masc sing*) об·лёк·, (*perf*); *see* об·лек·á·ть; об·лéч·ь (2), (*fut 3rd sing, pl*) об·ля́ж·ет, об·ля́г·-ут, (*past masc sing*) об·лёг·, (*perf*); *see* об·лег·á·ть

об·леч·ý (*fut 1st sing*) of об·лет·éть

об·лив·á·ть, -ют, (*imp*) pour/sluice/spill over/on; (ceram) glaze; -ся (*pass*); be covered with (sweat etc.)

об·ливíн·а, -ы, (*f*) *see* об·зóл·

об·лíв·к·а, -и, (*g pl*) -в·ок·, (*f*) (*v n*) *see* об·лив·á·ть; (ceram) glaze

об·лив·н·óй, -áя, -óе, (*a*) glazed

облигáт·н·ый, -ая, -ое, (*a*) (biol) obligate

облиг·áци·я, -и, (*f*) (fin) bond, deben-ture

óб·лик·, -а, (*m*) appearance, look, aspect, shape, configuration, (cryst) habit

~, пласт·íнчат·ый lamellar habit

об·лíст·веннос·т·ь, -и, (*f*) (bot) leaf coverage, leafiness

об·лíст·венн·ый, -ая, -ое, (*a*) folia-ceous, foliated, leafy

об·лй·ть, (*fut 3rd pl*) об·ль·ю́т, (*perf*) *see* об·лив·á·ть

об·лиц·евá·ть, -ýют, (*perf*) face, line

об·лиц·óванн·ый, -ая, -ое, (*past part pass*) of об·лиц·евá·ть; faced, lined

об·лиц·óвк·а, -и, (*g pl*) -вок·, (*f*) (civ eng) facing; lining

~ порóг·а (geol) sill apron

об·лиц·óвочн·ый, -ая, -ое, (*a*) facing, lining

~ кирпíч· (*m*) facing brick

~ с·мес·ь (*f*) (met cast) facing sand

об·лóг·, -а, (*m*) (agr) layland, ley, lea

об·лож·éни·е, -я, (*n*) (*v n*) of об·лож·йть; taxation

об·лóж·енн·ый, -ая, -ое, (*past part pass*) of об·лож·йть; (med) furred

об·лóж·ечн·ый, -ая, -ое, (*a*) of об·-лóж·к·а

об·лож·йть, -ат, (*perf*) put/place/lay around; face, cover; lay siege to, surround; impose, levy, exact (tax, fine etc.)

об·лóж·к·а, -и, (*g pl*) -ж·ек·, (*f*) wrapper, cover, folder, holder

об·лож·н·óй, -áя, -óе, (*a*)

~ дожд·ь (*m*) continuous/steady rain

облóй, -я, (*m*) (met) flash (drop forging or extrusion defect)

об·локот·йться, -ятся, (*perf*) put one's elbow(s) on, lean on

об·лóм·, -а, (*m*) (*v n*) *see* об·лáм·ыва·ть; (arch) molding

об·лóм·анн·ый, -ая, -ое, (*past part pass*) of об·лом·á·ть; broken-off, broken

об·лом·á·ть, -ют, (*perf*) *see* об·лáм·ыва·ть

об·лóм·к·а, -и, (*f*) (*v n*) *see* об·лáм·ыва·ть; (hort) pinching out (excessive shoots)

об·лóм·ок·, -м·к·а, (*m*) fragment; (*pl*) wreckage, debris; ejecta (from volcano)

~ пóл·я floe (ice)

об·лóм·очн·ый, -ая, -ое, (*a*) of об·лó-м·ок·; fragment, fragmentary; (geol) detrital, clastic

~ по·рóд·а (*f*) fragmental/clastic rock, scree

об·лопáч·ивани·е, -я, (*n*) blading, fitting turbine blades

об·луж·éни·е, -я, (*n*) tinning

об·луж·ённ·ый, -ая, -ое, (*a*) tinned, dipped in a tin bath, tin-plated

об·луп·и́ть, -́ят, (*perf*) *see* **об·лу́п·л·ива·ть**

об·лу́пл·ива·ть, -ют, (*imp*) peel, shell, husk, hull

об·луч·а́тел·ь, -я, (*m*) irradiator; feed, exciter (of antenna)

~, анте́нн·ый antenna feed

~, волно·во́д·н·ый waveguide feed

~, о·с·ко́л·оч·н·ый (nucl) fission-fragment irradiator

~, рентге́н·овск·ий X-ray machine

об·луч·а́ем·ый, -ая, -ое, (*a*) (*pres part pass*) of **об·луч·а́·ть**

об·луч·а́·ть, -ют, (*imp*) irradiate, expose to; illuminate (with radar); bombard (with nuclear particles)

об·луч·е́ни·е, -я, (*n*) (*v n*) *see* **об·лу-ч·а́·ть;** irradiation, exposure; (rad) illumination; stimulation (with infra-red rays)

~, вс·его́ те́л·а (phys) whole-body irradiation

~, корпускуля́р·н·ое (nucl) particle bombardment

~ ма́л·ой до́з·ы light irradiation

~, профессиона́ль·н·ое occupational exposure (to radiation)

~ сем·я́н· seed irradiation

~, сумма́р·н·ое total/cumulative irradiation

об·луч·и́ть, -а́т, (*perf*) *see* **об·луч·а́·ть**

об·ль·ю́т (*fut 3rd pl*) of **об·ли́·ть**

об·ля́г·ут (*fut 3rd pl*) of **об·ле́ч·ь** (2)

об·ля́ж·ет (*fut 3rd sing*) of **об·ле́ч·ь** (2)

об·ма́з·ать, (*fut 3rd pl*) **об·ма́ж·ут,** (*perf*); *see* **об·ма́з·ыва·ть**

об·ма́з·к·а, -и, (*g pl*) **-з·ок·,** (*f*) (*v n*) of **об·ма́з·ыва·ть;** coating, covering (electrodes); (build) rendering

об·ма́з·ыва·ть, -ют, (*imp*) grease, smear, coat; (build) render

об·ма́к·ива·ть, -ют, (*imp*) dip

об·мак·н·у́ть, -у́т, (*perf*) *see* **об·ма́-к·ива·ть**

об·ма́н·, -а, (*m*) deception, fraud

~ зр·е́ни·я optical illusion

об·ма́н·к·а, -и, (*g pl*) **-н·ок·,** (*f*) (geol) blende, glance

~, ви́смут·ов·ая (min) bismuth silicate

~, ка́дм·иев·ая cadmium blende, greenockite

~, рог·ов·а́я hornblende

~, ура́н·ов·ая смол·ян·а́я (min) pitchblende, nasturan

об·ма́н·к·а

~, ци́нк·ов·ая (min) sphalerite, zinc blende

об·ман·у́ть, -́ут, (*perf*) *see* **об·ма́н·ы-ва·ть**

об·ма́н·ыва·ть, -ют, (*imp*) deceive, defraud

об·ма́н·чив·ый, -ая, -ое, (*a*) deceptive, illusory

об·мар·а́·ть, -ют, (*perf*) soil/dirty all over

об·марга́нц·ованн·ый, -ая, -ое, (*a*) manganese-enriched

об·ма́т·ыва·ть, -ют, (*imp*) coil/wrap/wind round with, strap, tape, bind, lag pipes

об·ма́х·ива·ть, -ют, (*imp*) fan; dust/sweep off

об·мах·н·у́ть, -у́т, (*perf*) (single action) dust/sweep off

об·ма́ч·ива·ть, -ют, (*imp*) moisten, damp, wet

об·мёл (*past masc sing*) of **об·мес·ти́**

об·мел·а́·ть, -ют, (*perf*) become shallow/shallower; shoal, silt up

об·ме́н·, -а, (*m*) exchange, interchange, swop; barter; (biol) metabolism; (telecom) traffic

~ вещ·е́ств· (biol) metabolism

~, ио́н·н·ый (chem) ion exchange

~, катио́н·н·ый (chem) cation exchange

~, ко́с·венн·ый (phys) indirect exchange

~, меж·цеп·н·о́й (chem) interchain transfer

ли́н·и·я об·ме́н·а (*f*) (elec) tie line

~, основ·н·о́й (biol) basal metabolism

~, про·меж·у́точн·ый (biol) intermediary metabolism

~, хромосо́м·н·ый (genet) chromosome interchange

об·ме́н·ива·ть, -ют, (*imp*) exchange, interchange, swop; barter

об·мен·и́ть, -́ят, (*perf*) *see* **об·ме́н·и-ва·ть**

об·ме́н·ник·, -а, (*m*) exchanger

об·ме́н·н·ый, -ая, -ое, (*a*) exchange; (biol) metabolic

~ взаимо·де́й·стви·е (*n*) (phys) exchange interaction

~ си́л·а (*f*) (phys) exchange force

об·мен·о·спосо́б·н·ый, -ая, -ое, (*a*) (chem) labile

об·мен·я́·ть, -ют, (*perf*) *see* **об·ме́-н·ива·ть**

об·ме́р·, -а, (*m*) (*v n*) *see* **об·ме́р·и·ва·ть;** measurements; (shipb) measurement of tonnage

об·ме́р·енн·ый, -ая, -ое, (*past part pass*) of **об·ме́р·ить**

об·мерз·а́·ть, -ют, (*imp*) ice up/over, be covered with ice; get frozen

об·мёрз·нуть, -нут, (*past masc sing*) **об·мёрз·,** (*perf*); *see* **об·мерз·а́·ть**

об·ме́р·ива·ть, -ют, (*imp*) measure, measure up, take the measurements of; cheat in measuring, give false measure

об·ме́р·ить, -ят, (*perf*) *see* **об·ме́р·и·ва·ть**

об·ме́р·ок·, -р·к·а, (*m*) underweight purchase

об·мер·я́·ть, -ют, (*imp*) *see* **об·ме́·р·ива·ть**

об·ме́р·шик·, -а, (*m*) bulker, measurer

об·мес·ти́, (*fut 3rd pl*) **об·мет·у́т,** (*past masc sing*) **об·мёл** (*perf*) sweep/dust off

об·мет·а́·ть, -ют, (*imp*) sweep/dust off; **об·мет·а́ть, -а́ют,** or **об·ме́·ч·ут,** (*perf*) (text) whip

об·мёт·к·а, -и, (*g pl*) **-т·ок·,** (*f*) (text) whipping

об·мёт·ыва·ть, -ют, (*imp*) (text) whip

об·мин (*abbr*) of **оборо́т·ов в ми·ну́т·у** rpm, revolutions/rotations per minute

обми́н·, -а, (*m*) (*v n*) of **об·мин·а́·ть**

об·мин·а́·ть, -ют, (*imp*) push into shape; wear in (of clothes etc.)

об·мо́·ет (*fut 3rd sing*) of **об·мы́·ть**

об·мола́ч·ива·ть, -ют, (*imp*) (agr) thresh, thrash

об·моло́т·, -а, (*m*) (agr) thrashing, threshing; thrashing yield

об·молот·и́ть, -́ят, (*perf*) *see* **об·мо·ла́ч·ива·ть**

об·моло́ч·енн·ый, -ая, -ое, (*past part pass*) of **об·молот·и́ть**

об·морож·е́ни·е, -я, (*n*) frost bite

об·мора́ж·ива·ть, -ют, (*imp*) get frost bite

об·моро́ж·енн·ый, -ая, -ое, (*past part pass*) of **об·моро́з·ить**

об·моро́з·ить, -ят, (*perf*) *see* **об·мо·ра́ж·ива·ть**

о́б·морок·, -а, (*m*) faint, (med) syncope

об·мо́т·анн·ый, -ая, -ое, (*past part pass*) *see* **об·ма́т·ыва·ть**

об·мот·а́·ть, -ют, (*perf*) *see* **об·ма́·т·ыва·ть**

об·мо́т·к·а, -и, (*g pl*) **-т·ок·,** (*f*) (elec) winding; lagging (pipes), serving (ropes); covering, taping (insulation); armature (gyro compass); coil (degaussing); (*pl*) puttees

~, бараба́н·н·ая (elec) drum winding

~, бифиля́р·н·ая (elec) bifilar winding

~ воз·бужд·е́ни·я (elec) field/excitation winding

~, волн·ов·а́я (elec) wave winding

~, втор·и́чн·ая (elec) secondary winding

~, за́·мк·нут·ая (elec) closed-coil winding

~, компа́унд·н·ая (elec) compound winding

~, компенсацио́н·н·ая (elec) compensating winding

~, коротко·за́·мк·нут·ая (elec) squirrel-cage winding

~, обра́т·н·ая (elec) return winding

~, основ·н·а́я main coil (degaussing)

~, перв·и́чн·ая (elec) primary winding

~, паралле́ль·н·ая (elec) lap winding

~, петл·ев·а́я (elec) lap winding

~ под·магни́ч·ивани·я (elec) bias winding

~, под·раз·дел·ённ·ая (elec) split/tapped winding

~, по́люс·н·ая (elec) pole-face winding

~, по·сле́д·ователь·н·ая (elec) wave/series winding

~, проти́во-компа́унд·н·ая (elec) differential compound winding

~, пуск·ов·а́я (elec) starting winding

~ синхрониз·а́ци·и secondary winding (of a synchro)

~, со·средо·то́ч·енн·ая (elec) lap winding

~, стерж·нев·а́я (elec) bar winding

~, с·чит·а́ющ·ая read drive wire, read driver (computers)

~, у·с·поко·и́тельн·ая damper/amortisseur winding

~, шабло́н·н·ая (elec) form winding (the action)

~ я́кор·я (elec) armature winding

об·мо́т·очн·ый, -ая, -ое, (*a*) of **об·-мо́т·к·а**

~ коэффицие́нт· (*m*) (elec) distribution/breadth factor

~ ле́нт·а (*f*) binding tape

об·моч·и́ть, -́ат, (*perf*) moisten, damp, wet

об·мо́·ют (*fut 3rd pl*) of **об·мы́·ть**

об·мундир·ова́ни·е, -я, (*n*) clothing, kitting out; uniform, kit

об·мур·о́вк·а, -и, (*g pl*) -в·ок·, (*f*) lining (of furnace); setting (of a boiler)

об·мы́·в·, -а, (*m*) (*v n*) *see* об·мы́·в·а́·ть; (naut) washing down

об·мыв·а́ни·е, -я, (*n*) washing, bathing, ablution

об·мыв·а́·ть, -ю́т, (*imp*) wash, bathe

об·мы́·ть, (*fut 3rd pl*) об·мо́·ют, (*perf*); *see* об·мыв·а́·ть

об·мяк·а́·ть, -ю́т, (*imp*) soften, become soft/flabby

об·мя́к·нуть, -нут, (*past masc sing*) об·мя́к·, (*perf*); *see* об·мяк·а́·ть

об·мя́·ть, (*fut 3rd pl*) обо·мн·у́т, (*perf*); *see* об·мин·а́·ть

об·надёж·ить, -ат, (*perf*) give/inspire hope(s)

обнаж·а́·ть, -ю́т, (*imp*) bare, expose, uncover; -ся (*pass*); crop out, lie bare/exposed

обнаж·е́ни·е, -я, (*n*) (*v n*) *see* обнаж·а́·ть; (geol) outcrop, exposure

обнаж·и́ть, -а́т, (*perf*) *see* обнаж·а́·ть

об·найто́в·ить, -ят, (*perf*) *see* об·найто́вл·ива·ть

об·найто́вл·ива·ть, -ю́т, (*imp*) (naut) lash, seize, make fast

об·наро́д·овать, -уют, (*perf*) promulgate, publish

об·наруж·е́ни·е, -я, (*n*) (*v n*) *see* об·наруж·ива·ть; discovery, detection, disclosure; (rad) locating, warning

~, да́ль·н·ее (mil rad) early warning

~ из·луч·е́ни·я radiation detection ста́нци·я об·наруж·е́ни·я воз·ду́ш·н·ых цел·ей (*f*) (mil) air-warning radar

ста́нци·я об·наруж·е́ни·я и на·вед·е́ни·я (*f*) (mil) radar reporting and control centre

ста́нци·я об·наруж·е́ни·я над·во́д·н·ых цел·е́й (*f*) (mil) sea-search radar, air-surface vessel radar

~ фотограф·и́ческ·им мето́д·ом photographic detection

об·нару́ж·енн·ый, -ая, -ое, (*past part pass*) *see* об·нару́ж·ива·ть

об·нару́ж·ива·ть, -ю́т, (*imp*) display, show, reveal; detect, locate, discover, find, strike (oil, minerals); (rad) pick up

об·наруж·и́тел·ь, -я, (*m*) (*see also* детéктор·) detector, finder

об·нару́ж·ить, -ат, (*perf*) *see* об·нару́ж·ива·ть

об·наруж·и́тельн·ый, -ая, -ое, (*a*) detecting, discovering; (rad) search

об·нес·е́ни·е, -я, (*n*) (*v n*) *see* об·нос·и́ть

об·нес·ти́, -у́т, (*past masc sing*) об·нёс·, (*perf*); *see* об·нос·и́ть

об·ним·а́·ть, -ю́т, (*imp*) embrace, grasp/clasp around; grasp, take in (ideas etc.)

об·нищ·а́·ть, -ю́т, (*perf*) become impoverished/poor

об·нов·и́ть, -я́т, (*perf*) *see* об·новл·я́·ть

об·новл·я́·ть, -ю́т, (*imp*) renew, renovate; refresh, reinvigorate, rejuvenate; use for the first time

об·но́с·, -а, (*m*) (naut) overhang

об·нос·и́ть, -я́т, (*imp*) carry around/round; enclose, surround; (naut) stop (tie up temporarily)

об·но́с·к·а, -и, (*f*) (build) pegging/marking-out arrangement

об·но́с·н·ый, -ая, -ое, (*a*) enclosing, embracing, encompassing

~ сéзен·ь (*m*) (naut) sea gasket, furling line

об·но́ш·енн·ый, -ая, -ое, (*past part pass*) of об·нос·и́ть

об·ня́·ть, (*fut 3rd sing*) об·ни́м·ут, (*perf*); *see* об·ним·а́·ть

обо- (*prefix*) *see* о- (*prefix*)

обо·бр·а́·ть, (*fut 3rd pl*) о·бер·у́т, (*perf*) gather, pick (fruit etc.)

об·общ·а́·ть, -ю́т, (*imp*) generalize, make general/common

об·общ·е́ни·е, -я, (*n*) (*v n*) *see* об·общ·а́·ть; generalization

об·общ·ённ·ый, -ая, -ое, (*past part pass*) of об·общ·и́ть; generalized; unified

~ модéл·ь (*f*) unified model; collective model (of a nucleus)

~ нéрв·н·ая схéм·а (*f*) finite automaton

~ си́л·а (*f*) (mech) generalized force

об·обществ·и́ть, -я́т, (*perf*) *see* об·обществл·я́·ть

об·обществл·е́ни·е, -я, (*n*) (*v n*) *see* об·обществл·я́·ть

об·обществл·я́·ть, -ю́т, (*imp*) collectivize, socialize

об·общ·и́ть, -а́т, (*perf*) *see* об·общ·а́·ть

обо·бь·ю́т (*fut 3rd pl*) of о·би́·ть

о·богат·и́мост·ь, -и, (*f*) enrichability; dressability (of ores)

о·богат·и́тел·ь, -я, (*m*) enricher; (min) cullman, dresser, ore-dresser

о·богат·и́тельн·ый, -ая, -ое, (*a*) enriching

~ фа́брик·а (*f*) ore dressing/concentration plant works

о·богат·и́ть, -я́т, (*perf*) *see* о·богащ·а́·ть

о·богащ·а́·ть, -ют, (*imp*) enrich; (min) dress, concentrate

о·богащ·е́ни·е, -я, (*n*) enrichment, concentration, beneficiation, ore dressing

~, гравитаци́он·н·ое gravity concentration

~ изото́п·ов (nucl) isotope enrichment

~, ко́с·венн·ое (geol) indirect enrichment

~, магни́т·н·ое magnetic separation

~, прям·о́е (geol) positive enrichment

~ руд· ore dressing/concentration

~ с·ме́с·и (ICE) mixture enrichment

~, электро·стат·и́ческ·ое electrostatic separation

о·богащ·ённ·ый, -ая, -ое, (*past part pass*) *see* о·богащ·а́·ть; enriched; dressed, concentrated

~ угле·ро́д·ом enriched with carbon, carbon-enriched

обб·гн·анн·ый, -ая, -ое, (*a*) (*past part pass*) of обо·гн·а́ть

обо·гн·а́ть, (*fut 3rd pl*) об·го́н·ят, (*perf*) overtake, overhaul, get ahead of

обо·гн·у́ть, -у́т, (*perf*) go round, skirt; adapt, fit (a curve to points); (naut) double

обо·гре́·в·, -а, (*m*) warming, heating

обо·грев·а́тел·ь, -я, (*m*) heater

обо·грев·а́·ть, -ют, (*imp*) warm, heat; -ся become warm/hot

обо·гре́·ть, -ют, (*perf*) *see* обо·грев·а́·ть

о́бод·, -а, (*m*) rim, hoop, ring; tread (of railway/railroad tyre)

~ молот·о́вищ·а (mech) helve ring

~, раз·двó·енн·ый (mech) split ring

~ шкив·а pulley ring

обод·о́к·, -д·к·а́, (*m*) (*dim*) of о́бод·; ferrule; (bot) cingulum

обод·о́чн·ый, -ая, -ое, (*a*) of о́бод·, обод·о́к·; (anat) colonic

~ киш·к·а (*f*) (anat) colon

обо·др·а́ть, (*fut 3rd pl*) об·дер·у́т, (*perf*); *see* об·дир·а́·ть

о·бодр·и́ть, -я́т, (*perf*) *see* о·бодр·я́·ть

о·бодр·я́·ть, -ют, (*imp*) encourage, hearten, reassure

обое·по́л·ый, -ая, -ое, (*a*) (biol) bisexual, hermaphroditic, (bot) monoecious, androgynous

о·бо́·ечн·ый, -ая, -ое, (*a*) of о·бо́й·к·а

~ маши́н·а (*f*) receiving separator (milling machinery)

обо·жг·у́т (*fut 3rd pl*) of об·же́ч·ь

обо·жд·а́ть, -у́т, (*perf*) wait a bit/while

обо·жж·ённ·ый, -ая, -ое, (*past part pass*) of об·же́ч·ь; calcined, burnt, roasted

обо·жм·ёт (*fut 3rd sing*) of об·жа́т·ь

обо́з·, -а, (*m*) train (of pack animals etc.), (M/T) convoy, string; horse-drawn transport

обо·знач·а́·ть, -ют, (*imp*) designate, indicate, mark; mean

обо·знач·е́ни·е, -я, (*n*) (*v n*) of обо·знач·ить; designation, marking, mark; notation; legend (to drawing etc.)

~, десят·и́чн·ое decimal notation

~ за·по·мин·а́ющего· реги́стр·а absolute address (computers)

~, ма́трич·н·ое (math) matrix notation

~ реги́стр·а address (computers)

систе́м·а обо·знач·е́ни·й (*f*) notation, legend, nomenclature

~, со·кращ·ённ·ое contracted notation (computers)

~, спектроскоп·и́ческ·ое spectroscopic notation

~, у·сло́в·н·ое symbol, conventional sign

обо·знач·енн·ый, -ая, -ое, (*past part pass*) of обо·знач·ить

обо·знач·ить, -ат, (*perf*) designate, indicate, mark, denote

обо́з·н·ый, -ая, -ое, (*a*) of обо́з·

~ ло́шад·ь (*f*) dray, draught horse

обозо·стро·е́ни·е, -я, (*n*) cartbuilding

обо·зр·ева́тел·ь, -я, (*m*) reviewer, critic

обо·зр·ева́·ть, -ют, (*imp*) review; look around, survey

обо·зр·е́ни·е, -я, (*n*) review, sketch

обо·зр·е́ть, -я́т, (*perf*) *see* обо·зр·е·ва́·ть

обо́·и, -ев, (*m pl*) wallpaper

обо́·их (*gen*/*prep*) of об·а

о·бо́й·, -я, (*m*) windfalls, windfall fruit

обо́·йд·енн·ый, -ая, -ое, (*past part pass*) of обо·йти́; by-passed, circumvented

о·бо́й·к·а, -и, (*g pl*) -б·ек·, (*f*) (*v n*) *see* о·бив·а́·ть; scourer, scouring machine (flour milling); (*pl*) flax/hemp waste

обо́йм·а, -ы, (*f*) (mech) band, girdle, ring, yoke, clip; çlip, charger (of ammunition etc.)

~, па́ч·ечн·ая   clip (of ammunition etc.)

~ са́ль·ник·а   gland housing

~, ша́р·иков·ая   (mech) ball cage/race

~, эксцентри́к·ов·ая   eccentric strap

обо́йм·иц·а, -ы,   (*f*) block, pulley block

обо́й·н·ый, -ая, -ое,   (*a*) wallpaper; upholstery

обо·йти́, (*fut 3rd pl*) обо·йд·ут, (*past masc sing*) обо·шёл, (*perf*); *see* об·-ход·и́ть

о·бо́й·щик·, -а, (*m*) upholsterer

о́болон·ь, -и,   (*f*) (bot) alburnum, sapwood

оболо́ч·ечн·ый, -ая, -ое,   (*a*) of оболо́ч·к·а

~ моде́л·ь   (*f*) (nucl) shell model

оболо́ч·к·а, -и,   (*g pl*) -ч·ек·, (*f*) cover, case, casing, coat, coating, envelope, jacket, shell, sheath; (anat) membrane; (shipb) skin plating; (arch) shell; (nucl eng) jacket, can, tank, tamper; serving (of electric cable); (zool) tunic, tunica

~, алюми́н·иев·ая   (nucl) aluminium can

~ а́том·а   atom shell

~, бе́л·очн·ая   (anat) sclera, sclerotic coat

~, вне́шн·яя   (nucl) outer shell

~ выс·о́к·ого давл·е́ни·я   (phys) high-pressure shell

~, га́з·ов·ая   gas blanket

~, герметиз·и́рующ·ая   containment (a vessel)

~, за·мк·ну́т·ая   closed shell

~, за·щи́т·н·ая   (phys) protective envelope

~, ио́н·н·ая   (phys) ion sheath

~, кле́т·очн·ая   (cyt) cell membrane/wall

~, мяк·и́нн·ая   (bot) hull

~, не·до·стро́·енн·ая   (nucl) unfilled/partly-filled shell

оболо́ч·к·а

~, не·за·стро́·енн·ая   (nucl) unfilled/partly-filled shell

~, около·серд·е́чн·ая   (anat) pericardium

~, о·са́д·очн·ая   (geol) lithosphere

~, про·вод·я́щ·ая   (phys) conducting shell

~, ра́д·ужн·ая   (anat) iris

~ реа́ктор·а   (nucl) reactor shell

~, рог·ов·а́я   cornea (of eye)

~, свобо́д·н·ая   (nucl) vacant shell

~, семен·н·а́я   (bot) seed coat/cover/hull

~, се́т·чат·ая   retina (of eye)

~, с·крепл·я́ющ·ая   outer tube (of gun)

~, сли́з·ист·ая   (med) mucous membrane

~ тепло·вы·дел·я́ющ·его элеме́нт·а (nucl) fuel-element jacket/can

~ ша́р·а   balloon envelope

~, электро́н·н·ая   (phys) electron shell (of atom); electronic shell (= one principal quantum number)

оболо́ч·ков·ый, -ая, -ое,   (*a*) of оболо́ч·к·а

оболо́ч·ник·и, -ов, (*m pl*) (zool) tunicates, *Tunicata*

обо·мн·у́т (*fut 3rd pl*) of об·мя́·ть

обон·я́ни·е, -я, (*n*) smell, sense of smell, olfaction

обон·я́тельн·ый, -ая, -ое,   (*a*) olfactory

обо·пр·у́т (*fut 3rd pl*) of о·пер·е́ть

обора́ч·иваемост·ь, -и,   (*f*) (econ) turnover/turnround factor

обора́ч·ива·ть, -ют,   (*imp*) turn; -ся turn round; (fin) turn over

оборач·ива́ющ·ий, -ая, -ее,   (*pres part act*) of обора́ч·ива·ть; (opt) erecting

~ систе́м·а   (*f*) (opt) erecting system

обо·рв·а́ть, -ут,   (*perf*) *see* об·ры-в·а́·ть

о·бо́р·к·а, -и, (*g pl*) -р·ок·, (*f*) (text) frill, flounce

оборо́н·а, -ы,   (*f*) defence; defensive position, defences

~, противо·воз·ду́ш·н·ая   air defence

оборон·и́тельн·ый, -ая, -ое, (*a*) defensive

оборон·и́ть, -я́т, (*perf*) defend

оборо́н·н·ый, -ая, -ое,   (*a*) defence, defensive

оборон·я́·ть, -ют,   (*imp*) defend

оборо́т·, -а, (*m*) revolution, rotation, turn; reverse side, back; (fin) turnover; turn round (transport); (zool) whorl

~ **в мину́т·у** (*pl*) revolutions per minute, r.p.m.

~ **дв·ух фу́нкц·и·й** (math) convolution of two functions

~ **ре́ч·и** (gram) phrase, turn of speech

оборо́т·иться, -ятся, (*perf*) *see* **обо-ра́ч·ива·ться**

оборо́т·н·ый, -ая, -ое, (*a*) of **оборо́т·;** (fin) circulating, working; turn-round; reverse

~ **вре́м·я** (*n*) turn-round time (of transport)

~ **капита́л·** (*m*) (fin) working/circulating capital

~ **котёл·** (*m*) return-tube boiler

~ **круг·** (*m*) (rail) turntable

~ **сре́д·ств·а** (*n pl*) (fin) floating assets

~ **сторон·а́** (*f*) reverse side, back

об·ору́д·овани·е, -я, (*n*) equipment, plant, machinery; equipping, outfitting

~ **контро́ль·н·о-из·мер·и́тельн·ыми при·бо́р·ами** instrumentation

об·ору́д·овать, -уют, (*imp or perf*) equip, fit out

об·основ·а́ни·е, -я, (*n*) (*v n*) *see* **об·осн·ова́ть;** substantiation, evidence, basis (of an argument)

~ **ви́д·а** (biol) foundation of a species

~**, стро́г·ое** rigorous substantiation

об·осно́в·анност·ь, -и, (*f*) validity

об·осно́в·анн·ый, -ая, -ое, (*past part pass*) *see* **об·осно́в·ыва·ть;** well-founded

об·осн·ова́ть, -у́ют, (*perf*) *see* **об·осно́в·ыва·ть**

об·осно́в·ыва·ть, -ют, (*imp*) base, ground (proof etc.), substantiate

обо·со́б·ить (*perf*) *see* **обо·собл·я́·ть**

обо·собл·е́ни·е, -я, (*n*) (*v n*) *see* **обо·-собл·я́·ть;** isolation

обо·со́бл·енн·ый, -ая, -ое, (*pass part pass*) *see* **обо·собл·я́·ть;** single, individual

~ **цех·** (*m*) (eng) one-off shop

обо·собл·я́·ть, -ют, (*imp*) isolate, set apart

об·остр·е́ни·е, -я, (*n*) sharpening, aggravation, exacerbation

~ **и́мпульс·ов** (elec) peaking

об·остр·и́тел·ь, -я, (*m*) peaker, shaper (of pulses)

об·остр·и́ть, -ят, (*perf*) *see* **об·остр·я́·ть**

об·остр·я́·ть, -ют, (*imp*) intensify, sharpen; aggravate, exacerbate

об·остр·я́ющ·ий, -ая, -ее, (*pres part act*) of **об·остр·я́·ть;** (elec) peaking

~ **схе́м·а** (*f*) (elec) peaking circuit

обо·тр·у́т (*fut 3rd pl*) of **об·тер·е́ть**

обо́чин·а, -ы, (*f*) verge (of road)

обо·шёл (*past masc sing*) of **обо·йти́**

обо·шь·ю́т (*fut 3rd pl*) of **об·ши́·ть**

обою́д·н·ый, -ая, -ое, (*a*) mutual

обою́до- (*component*) mutually, to one another

обою́до-во́·гн·ут·ый, -ая, -ое, (*a*) concavo-concave

обою́до-вы́·пук·л·ый, -ая, -ое, (*a*) convexo-convex

обою́до-о́стр·ый, -ая, -ое, (*a*) double-edged

об·раба́т·ываемост·ь, -и, (*f*) workability; (met) machinability

ОБП (*abbr*) of **од·н·а́ бок·ов·а́я полос·а́** single side band

об·раба́т·ываем·ый, -ая, -ое, (*pres part pass*) of **об·раба́т·ыва·ть;** machineable, workable

~ **из·де́л·и·е** (*n*) workpiece

об·раба́т·ыва·ть, -ют, (*imp*) process, treat, machine, work; (agr) cultivate, till

~ **на·чёрн·о** rough out

~ **на·чи́ст·о** finish

~ **по раз·ме́р·у** finish to size

~ **пред·вар·и́тельн·о** pre-treat/process

об·раба́т·ывающ·ий, -ая, -ее, (*pres part act*) of **об·раба́т·ыва·ть**

~ **про·мы́шл·енност·ь** (*f*) manufacturing industry

об·рабо́т·анн·ый, -ая, -ое, (*past part pass*); *see* **об·раба́т·ыва·ть**

об·рабо́т·а·ть, -ют, (*perf*) *see* **об·раба́т·ыва·ть**

об·рабо́т·к·а, -и, (*g pl*) **-т·ок·,** (*f*) (*v n*) *see* **об·раба́т·ыва·ть;** treatment

~**, анали́т·ическ·ая** analysis

~**, гор·ю́ч·ая** (met) hot working

~ **давл·е́ни·ем** (met) shaping, non-cutting shaping, mechanical working

~ **да́·нн·ых** data processing/reduction/handling

~**, до·вс·хо́д·ов·ая** (agr) pre-emergence treatment

~ **древес·и́н·ы** woodworking, wood machining

~ **из·луч·е́ни·ем** radiation treatment

~ **льн·а́** (text) flax processing

~**, механ·и́ческ·ая** machining

об·рабо́т·к·а
~, **мо́кр·ая** wet processing
~ **на вальц·а́х** milling (rubber)
~, **на·ча́ль·н·ая** initial treatment; (nucl) head-end treatment
~, **один·а́рн·ая** (met) single quench
~, **по·сле́д·ующ·ая** aftertreatment
~ **почв·ы́** soil cultivation/tillage
~, **пред·вар·и́тельн·ая** pretreatment
~ **рез·а́ни·ем** (met) machining
~ **санита́р·н·ая** personal decontamination
~ **телегра́мм·** (telecom) traffic handling
~, **терм·и́ческ·ая** (met) heat treatment
~, **тока́р·н·ая** turning (lathe)
~, **хим·и́ческ·ая** chemical processing
~ **хо́лод·ом** (met) sub-zero treatment
о́браз·, -а, (*nom pl*) **-ы,** (*g pl*) **-ов,** (*m*) form, shape; manner; way; image, picture; (math) image; **о́браз, -а,** (*nom pl*) **-а́,** (*g pl*) **-о́в,** (*m*) icon
   гла́в·н·ым о́браз·ом mainly
   как·и́м о́браз·ом how
   не·ко́тор·ым о́браз·ом somehow
   ни·ко́им о́браз·ом not at all, by no means, not in the least
   по·до́б·н·ым о́браз·ом similarly, in the same way
   та́к·им о́браз·ом thus, in this way
образ·е́ц·, -з·ц·а, (*i*) **-з·ц·о́м,** (*m*) specimen, example, sample; test piece; model, pattern, original
~, **аналит·и́ческ·ий** analysis/assay sample
~, **гру́б·ый** bulk sample
~, **ис·пыт·у́ем·ый** test sample/specimen/piece
~, **ис·сле́д·уем·ый** test specimen/sample/piece
~, **лист·ов·о́й** (met) flat test piece, test strip
~, **масси́в·н·ый** (met) massive specimen
~, **над·ре́з·анн·ый** (met) notched specimen
~, **норма́ль·н·ый** (met) standard test piece
~, **об·луч·а́ем·ый** sample being irradiated, exposed sample
~, **о́·пыт·н·ый** experimental model, prototype
~, **по·ро́д·ы** (oil) formation sample
~, **пропорцiона́ль·н·ый** (met) proportional test piece
~, **пяти·кра́т·н·ый** (met) test piece of gauge length 5

образ·е́ц
~, **этало́н·н·ый** standard specimen, standard
о́браз·н·ый, -ая, -ое, (*a*) figurative, graphic; (*as adj component*) -shaped, -form, -oid
образ·ова́вш·ийся, -аяся, -ееся, (*past part act*) of **образ·ова́ться;** (*see also* **образ·о́ванн·ый**) product
~ **полиме́р·** (*m*) (chem) product polymer
~ **ядр·о́** (*n*) (nucl) product nucleus
образ·ова́ни·е, -я, (*n*) formation; production; education
~ **ви́д·ов** (biol) speciation
~, **вы́с·ш·ее** higher education
~ **за·ро́д·ыш·ей** (phys) nucleation
~ **звёзд·** (phys) star production
~ **кана́л·ов** (nucl) channelling
~ **ло́паст·ей** lobation
~ **па́р·ы** (nucl) pair production
~ **по·пере́ч·н·ых с·вяз·е́й** (phys, chem) crosslinking
~ **с·бро́с·ов** (geol) faulting
~ **с·гу́ст·к·ов** (phys) clustering, bunching
   тепл·о́ образ·ова́ни·я (*n*) (chem) heat of formation/combination
~ **тре́щ·ин·** cracking, formation of cracks
~ **ше́·йк·и** (met) necking, necking down
образ·о́ванност·ь, -и, (*f*) education
образ·о́ванн·ый, -ая, -ое, (*past part pass*) of **образ·ова́ть**
образ·ова́тельн·ый, -ая, -ое, (*a*) educational
образ·ова́ть, -у́ют, (*perf or imp*) see **образ·о́выва·ть**
образ·о́выва·ть, -ют, (*imp*) form, make up, constitute; generate, produce; **-ся** (*pass*)
   образ·о́выва·ться в результа́т·е result (from)
образ·у́ющ·ая, -ей, (*f decl us adj*) (math) generatrix, generating line
~ **ко́нус·а воз·мущ·е́ни·я** (aerodynam) edge of Mach cone
образ·у́ющ·ий, -ая, -ее, (*pres part act*) see **образ·о́выва·ть;** (*see also* **образ·у́ющ·ая**) (*as noun*); generating
обра́з·чик·, -а, (*m*) specimen, example, sample
образ·цо́в·ый, -ая, -ое, (*a*) of **образ·е́ц·;** master (of gauges, meters, measures etc.), reference
об·рамл·е́ни·е, -я, (*n*) frame, framing; setting; encompassing, spanning

об·раст·а́ни·е, -я, (*n*) (*v n*) *see* **об·раст·а́·ть;** (zool) epiboly

об·раст·а́·ть, -ют, (*imp*) be overgrown/ covered; be fouled (of ship's bottom etc.)

об·раст·и́, -у́т, (*past masc sing*) **об·ро́с·,** (*perf*); *see* **об·раст·а́·ть**

обра́т·, -а, (*m*) skimmed milk

~, сух·о́й dried skim milk

обра́т·и́мост·ь, -и, (*f*) reversibility, invertibility

обра́т·и́м·ый, -ая, -ое, (*a*) reversible; (chem) inversive; (fin) convertible

обра́т·и́ть, -я́т, (*perf*) *see* **обращ·а́·ть**

обра́т·н·о (*adv*) return, back; inversely
мен·я́·ться обра́т·н·о vary inversely (with); vary as the reciprocal (of)

обратно·за·ви́с·им·ый, -ая, -ое, (*a*) (instr) inverse-time, inverse-relation

обратно·кон·и́ческ·ий, -ая, -ое, (*a*) inversely conical, obconical, obconic

обратно·ланцето·ви́д·н·ый, -ая, -ое, (*a*) (bot) oblanceolate, inversely lanceolate

обра́т·н·о-по·са́д·очн·ый курс· (*m*) (air) runway reciprocal heading

обратно·сердце·ви́д·н·ый, -ая, -ое, (*a*) (bot) obcordate

обратно·тек·у́щ·ий, -ая, -ее, (*a*) returning, reflux

обратно·яйце·ви́д·н·ый, -ая, -ое, (*a*) (bot) obovate

обра́т·н·ый, -ая, -ое, (*a*) reverse, return, back; counter, opposite; (math) inverse, reciprocal

~ велич·ин·а́ (*f*) reciprocal

~ де́й·стви·е (*n*) reversal

~ диффу́з·и·я (*f*) (phys) back-diffusion

~ длин·а́ (*f*) reciprocal/inverse length

~ за·да́·ч·а (*f*) (math) inverse problem
ли́н·и·я обра́т·н·ого рас·шир·е́ни·я (*f*) re-expansion line
ли́н·и·я обра́т·н·ого хо́д·а (*f*) return line; retrace line (teleph)

~ пере·лёт· (*m*) (TV) retrace

~ поля́р·ност·ь (*f*) reversed polarity

~ рас·шир·е́ни·е (*n*) re-expansion

~ с·вяз·ь (*f*) (elec) feedback; (rad) reaction/regenerative coupling

~ с·вяз·ь, ги́б·к·ая (*f*) (autom) floating control mechanism

~ с·вяз·ь, двой·н·а́я double feedback

~ с·вяз·ь, жёстк·ая proportional control mechanism

~ с·вяз·ь, комбин·и́рованн·ая combined feedback

обра́т·н·ый, -ая, -ое

~ с·вяз·ь, коррект·и́рующ·ая correcting feedback

~ с·вяз·ь, лог·и́ческ·ая logical feedback (computers)

~ с·вяз·ь по давл·е́ни·ю pressure feedback

~ с·вяз·ь по о·гиб·а́ющ·ей envelope feedback

~ с·вяз·ь по· по·лож·е́ни·ю position feedback

~ с·вяз·ь по скор·ост·и rate feedback

~ с·вяз·ь, по·лож·и́тельн·ая sustained feedback

~ с·вяз·ь, у·пру́г·ая floating control mechanism

~ теоре́м·а (*f*) (math) converse of a theorem

~ теч·е́ни·е (*n*) refluence, reflux

~ ход· (*m*) return; retrace (telephones); (TV) flyback

~ цикл· (*m*) reflux

обращ·а́·ть, -ют, (*imp*) (*trans*) turn; turn into, change, convert, reduce; **-ся** (*pass*); handle, manage, manipulate, treat (persons); address, appeal/apply to; (*imp only*) (econ) circulate

~ в·ним·а́ни·е (*n*) pay attention (to), give consideration (to); draw attention (to), emphasize

~ в бес·коне́ч·ност·ь (math) become infinite

~ в нул·ь (math) vanish

обращ·а́ющ·ий, -ая, -ее, (*pres part act*) of **обращ·а́·ть;** turning; **-ся** rotating, circulating

обращ·е́ни·е, -я, (*n*) (*v n*) *see* **обращ·а́·ть(ся);** conversion, reduction, reversal, inversion, reversion; rotation, revolution, circulation, manipulation; treatment, behaviour (towards); reference (computers)

~ време́н·и time reversion/reversal
вре́м·я обращ·е́ни·я (*n*) access time (computer storage)

~, де́неж·н·ое (fin) currency, monetary circulation

~, до·пуст·и́м·ые (*pl*) (autom) read-around ratio

~ спектра́ль·н·ых ли́н·и·й spectrum line reversal

~, че́к·ов·ое (fin) cheque currency

обращ·ённ·ый, -ая, -ое, (*past part pass*) of **обрат·и́ть;** (*perf*); *see* **обращ·а́·ть** (*imp*)

о·бр·е́ет (*fut 3rd sing*) of **о·бр·и́ть**

об·ре́ж·ут, (*fut 3rd pl*) of об·ре́з·ать

об·ре́з·, -а, (*m*) edge, cut; sawn off gun (short barrel); wash-tub, tub; verge (of road)

~, ве́рх·н·ий head, top (of book)

~, ни́ж·н·ий foot, tail, lower edge (of book)

~, пере́д·н·ий front edge (of book)

об·рез·а́ни·е, -я, (*n*) (*v n*) *see* об·ре́-з·ыва·ть

~ и́мпульс·ов (elec) clipping

~ кро́м·ок· side trimming, edging

об·ре́з·анн·ый, -ая, -ое, (*past part pass*) of об·ре́з·ать, (*perf*); *see* об·ре́-з·ыва·ть (*imp*)

~ в кра́·й (print) bled

об·ре́з·ать, (*fut 3rd pl*) об·ре́ж·ут, (*perf*), об·рез·а́·ть, -ют, (*imp*), *see* об·ре́з·ыва·ть

об·резин·ённ·ый, -ая, -ое, (*past part pass*) *see* об·резин·ива·ть; rubber-ized

об·рези́н·ива·ть, -ют, (*imp*) rubberize

об·ре́з·к·а, -и, (*g pl*) -з·ок·, (*f*) (*v n*) *see* об·ре́з·ыва·ть; trimming (of forgings), shaving (sheet metal blanks); pruning (trees)

об·рез·н·о́й, -а́я, -о́е, (*a*) trimmed, edged, pruned; trimming, edging, pruning

~ пресс· (*m*) (mech) trimming press

~ стан·о́к· (*m*) (mech) trimmer; (phot) trimmer, guillotine; trimming machine (carpentry)

об·ре́з·ок·, -з·к·а, (*m*) cut-off piece, piece, length; trimming, paring, crop, discard, scrap; (hort) cutting

об·ре́з·ыва·ть, -ют, (*imp*) cut off, cut round, cut, trim, edge, pare, prune

обр·ёл (*past masc sing*) of обрес·ти́

о·бремен·и́тельн·ый, -ая, -ое, (*a*) burdensome, onerous

о·бремен·и́ть, -я́т, (*perf*) *see* о·бремен·я́·ть

о·бремен·я́·ть, -ют, (*imp*) burden

обрес·ти́, (*fut 3rd pl*) обрет·у́т, (*past masc sing*) обрёл·, (*perf*); *see* обре-т·а́·ть

обрет·а́·ть, -ют, (*imp*) find, seek out

обрет·ённ·ый, -ая, -ое, (*past part pass*) of обрес·ти́

обрёт·ш·ий, -ая, -ее, (*past part act*) of обрес·ти́

об·реше́т·ин·а, -ы, (*f*) (build) joist

об·реше́т·к·а, -и, (*g pl*) -т·ок·, (*f*) bracing; (build) lathwork, lathing

об·реше́т·ник·, -а, (*m*) grating

о·бр·е́ют (*fut 3rd pl*) of о·бр·и́ть

Обри́, при·бо́р· (*m*) gyro-angling gear (in a torpedo)

о·бр·ива́·ть, -ют, (*imp*) shave

об·рис·о́ванн·ый, -ая, -ое, (*past part pass*) of об·рис·ова́ть

об·рис·ова́ть, -у́ют, (*perf*) outline, delineate, sketch

об·рис·о́вк·а, -и, (*g pl*) -вок·, (*f*) sketch, outline

об·рис·о́выва·ть, -ют, (*imp*) outline, delineate

обри́т·, -а, (*m*) (min) aubrite

о·бр·и́ть, -е́ют, (*perf*) shave

об·рон·и́ть, -я́т, (*perf*) drop, let fall

об·ро́с·ш·ий, -ая, -ее, (*past part act*) of об·раст·и́; overgrown; (naut) foul

об·ру́б·, -а, (*m*) lopping, clearing (lumbering); stump (of a tree)

об·руб·а́·ть, -ют, (*imp*) chop/hack/ slash off, lop (branches), dock (tails), chip off

об·руб·и́ть, -я́т, (*perf*) *see* об·руб·а́·ть

об·ру́б·к·а, -и, (*f*) (*v n*) of об·ру-б·а́·ть; (met cast) fettling, dress-ing-off

об·руб·н·о́й, -а́я, -о́е, (*a*) trimming, (met) fettling

об·ру́б·ок·, -б·к·а, (*m*) stump; (*pl*) trimmings, scraps, chips

об·ру́б·очн·ый, -ая, -ое, (*a*) chip-ping, trimming, dressing, fettling

~ цех· (*m*) dressing/fettling shop

о́бруч·, -а, (*i*) -ем, (*m*) hoop, ring, band, collar

обруче·о·са́д·очн·ый стан·о́к· (*m*) hooping-up machine (cooperage)

обру́ч·к·а, -и, (*pl g*) -ч·ек·, (*f*) (met roll) narrow band, hoop iron

о́бруч·н·ый, -ая, -ое, (*a*) of о́бруч·

~ желе́з·о (*n*) hoop iron

~ стал·ь (*f*) hoop iron

об·ру́ш·ени·е, -я, (*n*) (*v n*) *see* об·-ру́ш·ива·ть; (civ eng) sliding; (min) caving

ли́н·и·я об·ру́ш·ени·я (*f*) (min) caving line

пло́ск·ост·ь об·ру́ш·ени·я (*f*) (civ eng) sliding plane

~, под·эта́ж·н·ое sublevel caving

при́зм·а об·ру́ш·ени·я (*f*) (civ eng) sliding triangle

~, сло·ев·о́е (min) topslicing, cover caving

об·ру́ш·ива·ть, -ют, *(imp)* bring down, make fall; hull, scour (flour milling); (min) cave/stope in; **-ся** collapse, cave, come down; fall upon, attack

об·ру́ш·ить, -ат, *(perf) see* **об·ру́·ш·ива·ть**

об·ры́в·, -а, *(m) (v n) see* **об·рыв·а́·ть;** (geol) precipice, scarp; break, break-off, sudden termination; twist-off (of drill etc.)

~, берег·ов·о́й cliff

~ ро́ст·а це́п·и (chem) chain termination

об·рыв·а́ни·е, -я, *(n) (v n)* of **об··рыв·а́·ть**

об·рыв·а́·ть, -ют, . *(imp)* tear off; break (of rope etc.) pick, gather (fruit etc.); cease abruptly, break off, cut short; **-ся** lose hold, fall off

об·ры́в·ист·ый, -ая, -ое, *(a)* precipitous, steep; intermittent

об·ры́в·ок·, -в·к·а, *(m)* torn-off portion, scrap, bit, piece

о·бры́зг·, -а, *(n) (v n)* of **о·бры́зг·и·ва·ть**

о·бры́зг·а·ть, -ют, *(perf) see* **о·бры́з·г·ива·ть**

о·бры́зг·ива·ть, -ют, *(imp)* sprinkle, spatter, splash

об·ря́д·, -а, *(m)* ceremony, ritual

об·ряд·и́ть, -я́т, *(perf) see* **об·ряж·а́·ть**

об·ря́д·ов·ый, -ая, -ое, *(a)* ceremonial, ritual

об·ряж·а́·ть, -ют, *(imp)* draw, clean, clean out (carcases, flayed hides etc.); dress, get into line (obs)

об·сад·и́ть, -ят, *(perf) see* **об·са́ж·и·ва·ть**

об·са́д·к·а, -и, *(g pl)* **-д·ок·,** *(f) (v n) see* **об·са́ж·ива·ть;** planting round/out

об·са́д·н·ый, -ая, -ое, *(a)* planted, planting, planted out; casing (wells)

~ коло́нн·а *(f)* (oil) casing string

~ труб·а́ *(f)* casing, drive pipe (of well, borehole etc.)

об·са́д·очн·ый, -ая, -ое, *(a)* of **об··са́д·к·а**

~ пресс· *(m)* drive press, tube-fitting press (fits tubes into manifolds, headers etc.)

об·са́ж·ива·ть, -ют, *(imp)* plant round, plant; case (a well)

об·са́хар·ива·ть, -ют, *(imp)* (chem) saccharify; candy, sugar, preserve in sugar

об·са́хар·ить, -ят, *(perf) see* **об·са́·хар·ива·ть**

об·сев·а́·ть, -ют, *(imp)* (agr) sow

об·сек *(abbr)* = **оборо́т·ов в се·ку́нд·е** r.p.s., revolutions per second

об·сек·а́·ть, -ют, *(imp)* cut round, chip, trim, dress (stone)

об·семен·я́·ть, -ют, *(imp) see* **о·семен·я́·ть**

обсерв·ато́ри·я, -и, *(f)* observatory

обсерв·а́ци·я, -и, *(f)* observation

обсе́рв·ованн·ый, -ая, -ое, *(a)* (navig) observed, by observation

~ ме́ст·о *(n)* (navig) fix

~ шир·от·а́ *(f)* latitude by observation

об·сеч·н·о́й, -а́я, -о́е, *(a)* dressing, trimming

~ пресс· *(m)* nibbling machine (sheet metal working)

об·се́ч·ь, *(fut 3rd sing, pl)* **об·сеч·ёт, об·сек·у́т,** *(past masc sing)* **об·се́к·,** *(perf); see* **об·сек·а́·ть**

об·се́·янн·ый, -ая, -ое, sown

обсидиа́н·, -а, *(m)* (geol) obsidian

обскура́нт·, -а, *(m)* obscurantist

об·сле́д·овани·е, -я, *(n) (v n) see* **об··след·овать;** examination, inspection, investigation

~, дозиметр·и́ческ·ое (nucl) radiation survey

об·сле́д·овать, -уют, *(imp and perf)* investigate, examine, inspect

об·слу́ж·иваем·ый, -ая, -ое, *(pres part pass)* of **об·слу́ж·ива·ть;** service, servicing, serviced

об·слу́ж·ивани·е, -я, *(n) (v n) see* **об·слу́ж·ива·ть;** maintenance, service

~, бы́т·ов·ое public utilities and social services

~ с номеро·у·каз·а́тел·ем (telecom) call-indicator working

~, тек·у́щ·ее routine maintenance

об·слу́ж·ива·ть, -ют, *(imp)* serve; maintain, service; work (telephones); operate, handle, attend (machines); man (guns etc.)

об·слу́ж·ивающ·ий, -ая, -ее, *(pres part act)* of **об·слу́ж·ива·ть;** serving, servicing; being used (of equipment); *(m decl as adj)* attendant

~ организ·а́ци·я *(f)* service organization

~ персона́л· *(m)* staff, assistants, attendants

об·служ·и́ть, -ат, (*perf*) serve; be used (as) (of equipment)

об·со́с·к·а, -и, (*g pl*) **-с·ок·,** (*f*) (met) anode scrap

об·со́х·нуть, -нут, (*past masc sing*) **об·со́х·,** (*perf*) dry, dry out/through

об·ста́в·ить, -ят, (*perf*) *see* **об·ставл·я́·ть**

об·ставл·я́·ть, -ют, (*imp*) put/arrange round, surround; furnish (rooms etc.)

об·стано́в·к·а, -и, (*g pl*) **-в·ок·,** (*f*) conditions, situation; arrangements; furniture; (theat) set

~ **пут·и́** navigation signs (inland waterways)

об·сто·я́тельн·ый, -ая, -ое, (*a*) detailed, thorough, circumstantial; thorough, reliable (of people)

об·сто·я́тельственн·ый, ая, -ое, (*a*) adverbial

об·сто·я́тельств·о, -а, (*n*) circumstance, circumstances, conditions; (gram) adverb

об·сто·я́ть, -я́т, (*imp*) be

об·стра́г·ива·ть, -ют, (*imp*) plane, shave, shave smooth

об·стра́·ива·ть, -ют, (*imp*) build round; build, lay out (towns)

об·стре́л·, -а, (*m*) (gunn) fire, firing; field of fire

об·стре́л·ива·ть, -ют, (*imp*) (gunn) shell, bombard, machine-gun, shoot up

об·стрел·я́·ть, -ют, (*perf*) *see* **об·стре́л·ива·ть**

об·стриг·а́·ть, -ют, (*imp*) shear, clip

об·стри́ж·енн·ый, -ая, -ое, (*past part pass*) of **об·стри́ч·ь**

об·стри́ж·к·а, -и, (*f*) cleaning rags

об·стри́ч·ь, (*fut 3rd sing, pl*) **об·стриж·ёт, об·стриг·у́т,** (*past masc sing*) **об·стри́г·,** (*perf*); *see* **об·стриг·а́·ть**

об·строг·а́·ть, -ют, (*perf*) *see* **об·стра́г·ива·ть**

об·стро́·ить, -ят, (*perf*) *see* **об·стра́·ива·ть**

обстру́кци·я, -и, (*f*) obstruction

об·суд·и́ть, -ят, (*perf*) *see* **об·сужд·а́·ть**

об·сужд·а́·ть, -ют, (*imp*) discuss, go into (a problem)

об·сужд·е́ни·е, -я, (*n*) discussion
    пред·ме́т· об·сужд·е́ни·я (*m*) topic, issue

об·с·чёт·, -а, (*m*) miscalculation; falsification (of figures)

об·сы́п·ать, -лют, (*perf*) *see* **об·сы·п·а́·ть**

об·сып·а́·ть, -ют, (*imp*) strew, sprinkle

об·сых·а́·ть, -ют, (*imp*) dry, dry out/through

об·та́·ива·ть, -ют, (*imp*) melt away, melt on the outside

об·та́ч·ивани·е, -я, (*n*) *see* **об·то́ч·к·а**

об·та́ч·ива·ть, -ют, (*imp*) turn (on lathe); grind, polish

~ **в·тор·е́ц·** (mech eng) face

об·та́·ять, -ют, (*perf*) *see* **об·та́·ива·ть**

об·тек·а́емост·ь, -и, (*f*) extent of streamlining, streamlining

об·тек·а́ем·ый, -ая, -ое, (*pres part pass*) of **об·тек·а́·ть**; streamlined, streamline

~ **по·ве́рх·ност·ь** (*f*) (a/c) fairing

~ **со скольж·е́ни·ем,** (dynam) in yaw

~ **те́л·о** (*n*) (dynam) streamlined body; body in a flow

об·тек·а́ни·е, -я, (*n*) (dynam) flow, flow around, flow past a body, streamline flow

~, **без·вихр·ев·о́е** smooth flow

~, **без·от·ры́в·н·о́е** flow without separation

~ **крыл·а́** wing flow, flow for wing

~ **кли́н·а** wedge flow

~ **ло́паст·ей** flow through blades

~, **не·у·станов·и́вш·ееся** unsteady flow

~, **от·ры́в·н·о́е** separated flow

~, **пла́в·н·ое** streamline flow

~ **пласт·и́нк·и** flat-plate flow

~ **решёт·к·и** (air) cascade flow

~ **те́л·а** flow over/past a body, flow outside a boundary

~ **угл·а́** flow in a corner

~, **у·сто́й·чив·ое** steady flow

об·тек·а́тел·ь, -я, (*m*) fairing, streamlined housing, cowling

~ **анте́нн·ы** (rad) radome

~ **винт·а́** (a/c) propeller spinner; (shipb) propeller boss

~ **двиг·ател·я** engine cowling

~, **нос·ов·о́й** (a/c) nose fairing

~, **от·кид·н·о́й** hinged cowling

~, **ради·о·про·зр·а́чн·ый** (rad) radome

~, **хвост·ов·о́й** (a/c) tail cone

об·тек·а́·ть, -ют, (*imp*) flow round

об·тек·а́ющ·ий, -ая, -ее, (*pres part act*) of **об·тек·а́·ть**; circumfluent

~ **ток·** (*m*) slip flow

об·тер·е́ть, (*fut 3rd pl*) **обо·тр·у́т,** (*past masc sing*) **об·тёр·,** (*perf*); *see* **об·тир·а́·ть**

об·тёс·анн·ый, -ая, -ое, (*past part pass*) of **об·тес·а́ть**; rough trimmed/dressed

об·тес·а́ть, (*fut 3rd pl*) об·те́ш·ут, (*perf*); *see* об·тёс·ыва·ть

об·тёс·ыва·ть, -ют, (*imp*) adze (wood), dress, trim (stone etc.)

об·те́ч·ь, (*fut 3rd sing, pl*) об·теч·ёт, об·тек·у́т, (*past masc sing*) об·тёк·, (*perf*); flow round

об·те́ш·ут (*fut 3rd pl*) of об·тес·а́ть

об·тир·а́·ть, -ют, (*imp*) wipe, wipe dry/clean; rub, rub smooth

об·ти́р·очн·ый, -ая, -ое, (*a*) wiping, rubbing; cleaning, polishing

об·то́ч·енн·ый, -ая, -ое, (*past part pass*) of об·точ·и́ть; turned; ground, polished

об·точ·и́ть, ‑ат, (*perf*) *see* об·та́ч·ива·ть

об·то́ч·к·а, -и, (*f*) (*v n*) *see* об·та́ч·ива·ть; turning (on lathe for external machining)

~, кон·и́ческ·ая   taper turning

~ на· ко́нус·   taper turning

~, про·до́ль·н·ая   straight/plain/longitudinal turning

~ тор·ц·а́   radial turning

~, чист·ов·а́я   finish turning

об·треп·а́ть, -лют, (*perf*) fray

обтур·а́тор·, -а, (*m*) (med) obturator; (gunn etc.) obturator, obturator pad, seal; (cinema) shutter

обтюр·а́тор·, -а, (*m*) *see* обтур·а́тор·

обтюр·а́ци·я, -и, (*f*) obturation

об·тя́г·ива·ть, -ют, (*imp*) cover, stretch over; fit, fit close

об·тя́ж·к·а, -и, (*f*) (*v n*) *see* об·тя́г·ива·ть; (met) stretch-wrap forming; cover, covering, skin, jacketing; lasting (shoes)

в об·тя́ж·к·у   tight, close-fitting

об·тяж·н·о́й,-ая,-ое, (*a*) of об·тя́ж·к·а

об·тя·н·у́ть, ‑ут, (*perf*) *see* об·тя́г·ива·ть

обув·н·о́й, -а́я, -о́е, (*a*) footwear, boot-and-shoe

~ про·мы́шл·енность (*f*) boot-and-shoe industry

о́був·ь, -и, (*f*) footwear

об·угле·ро́ж·ивани·е, -я, (*n*) carburization, addition of carbon, carbonization

об·у́гл·ива·ть, -ют, (*imp*) carbonize, carburize; char

об·у́гл·ить, -ят, (*perf*) *see* об·у́гл·ива·ть

об·у·сло́в·ить, -ят, (*perf*) *see* об·у·сло́в·лива·ть

об·у·сло́вл·енност·ь, -и, (*f*) conditionality

об·у·сло́вл·енн·ый, -ая, -ое, (*past part pass*) of об·у·сло́в·ить; stipulated, dictated/conditioned/determined by, dependent on, due to

об·у·сло́вл·ива·ть, -ют, (*imp*) stipulate, lay down conditions; determine, condition; -ся (*pass*); result from, be due to

о́бух·, -а, (*m*) butt, back edge (of cutting instrument); (naut) eye bolt; eye (of hook)

~, вертлю́ж·н·ый   swivel eye

об·уч·а́·ть, -ют, (*imp*) teach, instruct; -ся (*pass*); learn

об·уч·е́ни·е, -я, (*n*) instruction, training

об·уч·и́ть, ‑ат, (*perf*) *see* об·уч·а́ть

обуш·о́к·, -ш·к·а́, (*m*) (*dim*) of о́бух·; (min) pick

об·ха́ж·ива·ть, -ют, (*imp*) *see* об·хо·д·и́ть

об·хва́т·, -а, (*m*) *see* о·хва́т· у́гол· об·хва́т·а (*m*) (horol) span angle; wrapping angle (pulleys and belts); grip-hold angle (of brake band)

об·хват·и́ть, ‑ят, (*perf*) *see* о·хва́т·ыва·ть

об·хва́т·ыва·ть, -ют, (*imp*) *see* о·хва́т·ыва·ть

об·хо́д·, -а, (*m*) by-pass, circumvention; round, rounds (duty), patrol; evasion

~, вертика́ль·н·ый   vertical missing (e.g. of an electrode)

об·ход·и́ть, ‑ят, (*imp*) go/walk round, circle; by pass, go/walk round, circumvent; go/walk all over/round, patrol; -ся get on (with), do (with), manage; cost, come to (of expense)

об·хо́д·н·ый, -ая, -ое, (*a*) roundabout; (mil) flanking, turning; bypass

об·хо́д·чик·, -а, (*m*) patrolman

~, пут·ев·о́й   (rail) lengthman

об·чи́ст·ить, -ят, (*perf*) *see* об·чищ·а́ть

об·чищ·а́·ть, -ют, (*imp*) clean; peel

об·шив·а́ни·е, -я, (*n*) *see* об·ши́в·к·а

об·шив·а́·ть, -ют, (*imp*) cover with; edge, border, trim (by sewing)

~ де́рев·ом   plank

~ лист·а́ми   plate

~ ме́д·ью   plate with copper

об·ши́в·к·а, -и, (*g pl*) -в·ок·, (*f*) covering, casing, sheathing; (a/c) skin; edging, bordering (by sewing); tank liner (of circuit breaker)

об·ши́в·к·а

~, вну́тр·енн·яя   (shipb) inner bottom plating

~, дерев·я́нн·ая   planking, boarding

~, лед·о́в·ая   (shipb) ice belt

~, нару́х·н·ая   (shipb) shell plating

~, нес·у́щ·ая   load-carrying skin

~, сталь·н·а́я   plating

об·шив·н·о́й, -а́я, -о́е,   (a) of об·ши́в·-к·а; covered, cased, sheathed; cover

~ по́яс·   (m) (shipb) strake

об·ши́в·очн·ый, -ая, -ое,   (a) of об·ши́в·к·а; cover

об·ши́р·ност·ь, -и,   (f) expanse; vastness; magnitude, extent

об·ши́р·н·ый, -ая, -ое,   (a) extensive, spacious, vast

об·ши́·ть,   (fut 3rd pl) обо·шь·ю́т, (perf); see об·шив·а́ть

обшла́г·, -а́,   (m) cuff

обще-   (root) general, all-

обще·до·сту́п·н·ый, -ая, -ое,   (a) available to all; cheap, at popular prices, popular

обще·жи́т·и·е, -я,   (n) hostel; society, community

обще·за·во́д·ск·ий, -ая, -ое,   (a) relating to/concerning the whole factory, interdepartmental, general

обще·из·ве́ст·н·о   (predic adj) it is well known, it is common knowledge

обще·из·ве́ст·н·ый, -ая, -ое,   (a) well/generally/widely-known

обще·на·ро́д·н·ый, -ая, -ое,   (a) public

общ·е́ни·е, -я,   (n) intercourse, contact (between people)

обще·по·ня́т·н·ый, -ая, -ое,   (a) obvious

обще·при́·нят·ый, -ая, -ое,   (a) conventional, generally accepted, usual

обще·с·плав·н·о́й, -а́я, -о́е,   (a)

~ канал·иза́ци·я   (f) combined sewerage system

обще·со·ю́з·н·ый, -ая, -ое,   (a) all-union, nation-wide

общ·е́ственн·ый, -ая, -ое,   (a) social, public

о́бщ·еств·о, -а,   (n) society, association; (com) company

~, акционе́р·н·ое   joint stock company

~, добро·во́ль·н·ое   voluntary society

~, классификацио́н·н·ое   registry of shipping

~, страх·ов·о́е   insurance company

обще·у·по·треб·и́тельн·ый, -ая, -ое,   (a) usual

о́бщ·ий, -ая, -ее,   (a) general, common; total, overall; combined

~ ава́р·и·я   (f) (mar ins) general average

в о́бщ·ем   generally, in general, on the whole

~ вес·   (m) total weight

~ вид·   (m) general/overall view

~ ме́р·а   (f) (math) common measure

~ на·знач·е́ни·е   (n) general purpose

~ наи·ме́ньш·ее крат·н·ое   (n) (math) least common multiple

~ пра́в·о   (n) common law

~ реш·е́ни·е   (n) (math) general solution (of a differential equation)

ру́ч·к·а о́бщ·его ша́г·а   (f) pitch control (of helicopters)

~ су́мм·а   (f) sum total

общ·и́н·а, -ы,   (f) community, society

общ·и́нн·ый, -ая, -ое,   (a) of общ·и́н·а; common, public

об·щип·а́ть, ‑лют,   (perf) see об·щи́п·ыва·ть

об·щи́п·ыва·ть, -ют,   (imp) pluck

о́бщ·ност·ь, -и,   (f) generality; community

объ·е́вш·ий, -ая, -ее,   (past part act) of объ·е́ст·ь, (perf); see объ·ед·а́ть (imp)

объ·ед·а́·ть, -ют,   (imp) eat round, bite round/off

объ·ед·е́нн·ый, -ая, -ое,   (past part pass) of объ·е́ст·ь, (perf); see объ·-ед·а́·ть (imp)

объ·е́д·ет   (fut 3rd sing) of объ·е́·х·ать

объ·един·е́ни·е, -я,   (n) (v n) see объ·-един·я́·ть; unification, amalgamation; (com) corporation; union, society, association; (mil) unit, formation

~ кома́нд·   packing of orders (computers)

~ чи́сел·   packing (computers)

объ·един·ённ·ый, -ая, -ое,   (past part pass) of объ·един·и́ть; unified, united, amalgamated, joint

~ инстит́ут·   (m) Joint Institute

объ·един·и́тельн·ый, -ая, -ое,   (a) unifying, uniting

объ·един·и́ть, -я́т,   (perf) see объ·-един·я́·ть

объ·един·я́·ть, -ют,   (perf) unite, unify, amalgamate; combine

объ·е́д·ут   (fut 3rd pl) of объ·е́х·ать

объ·ед·я́т   (fut 3rd pl) of объ·е́ст·ь

объ·е́зд·, -а,   (m) circuit, circuitous route, detour

объ·е́зд·ить, -ят, *(perf)* drive/travel round/all over; break, break in (horses)

объ·езд·н·о́й, -а́я, -о́е, *(a)* of **объ·е́зд·**; circuitous

объ·е́зд·чик·, -а, *(m)* mounted patrol-man; horse breaker/tamer

объ·езж·а́·ть, -ют, *(imp)* drive/travel round/around, by-pass; drive/travel about/all over; break, break in (horses)

объ·е́зж·ий, -ая, -ее, *(a) see* **объ·езд·н·о́й**

объе́кт·, -а, *(m)* object, objective

~ -микро́·метр·, -а, *(m)* photo-micrography micrometer

объекти́в·, -а, *(m)* (opt) objective, objective lens, lens

~, ис·пра́вл·енн·ый (phot) corrected objective/lens

~, ландша́фт·н·ый (phot) landscape/achromatic lens

~, на·бо́р·н·ый convertible lens

~ па́р·н·ый matched lens

~, свето·си́ль·н·ый (phot) rapid objective

~, с·ло́ж·н·ый compound lens/objective

объекти́в·н·ый, -ая, -ое, *(a)* objective

объе́кт·н·ый, -ая, -ое, *(a)* of **объе́кт·**,

объекто·держ·а́тел·ь, -я, *(m)* specimen/object holder (microscope); substrate (electron microscope)

объ·е́л *(past masc sing)* of **объ·е́ст·ь, (perf); see объ·ед·а́·ть** *(imp)*

объ·ём·, -а, *(m)* volume, cubic capacity, bulk, size, space, interior; (rad) cavity; **объ·е́м** *(fut 1st sing)* of **объ·е́ст·ь**

~, акти́в·н·ый (nucl) sensitive volume

в объ·ём·е *(adv)* inside

в проце́нт·ах объ·ём·а %/o by volume, vol %

~ ка́мер·ы с·жа́т·и·я (ICE) clearance volume

~, поло́з·н·ый useful volume

~, про·тек·а́ющ·ий (phys) volume flow

~ пуст·о́т· void content

~ рабо́т·ы amount/volume of work

~, рабо́ч·ий (ICE) swept volume, displacement volume

~, у·ле́ль·н·ый specific volume

объ·ём·ист·ый, -ая, -ое, *(a)* voluminous, bulky

объёмно·вращ·а́тельн·ый, -ая, -ое, *(a)* positive-displacement (pumps etc.)

объёмно·центр·и́рованн·ый, -ая, -ое, *(a)* (cryst) body-centred

объ·ём·н·ый, -ая, -ое, *(a)* volumetric, volume, three-dimensional, bulk, space; second, dilatational (of viscosity)

~ анали́з· *(m)* (chem) volumetric analysis

~ вя́з·кост·ь *(f)* bulk viscosity

за·ко́н объ·ём·н·ых от·нош·е́·ни·й *(m)* (phys) Gay-Lussac's law of volumes

~ за·ря́д· *(m)* (elec) space charge

коэффицие́нт· объ·ём·н·ого рас·-шир·е́ни·я *(m)* coefficient of cubical expansion

~ рас·шир·е́ни·е *(n)* cubical expansion; dilatation

~ резона́тор· *(m)* (elec) cavity resonator, cavity

~ сво́й·ств·а *(pl)* bulk properties

~ элеме́нт· *(m)* element of volume

~ эффе́кт· *(m)* volume change

объемо́·метр·, -а, *(m)* volumenometer

объ·е́ст·ь, *(fut 3rd sing, pl)* **объ·е́ст·, объ·ед·я́т,** *(past masc sing)* **объ·е́л, (perf); see объ·ед·а́·ть**

объ·е́х·ать, *(fut 3rd pl)* **объ·е́д·ут, (perf)** drive/travel round/around, by-pass; drive/travel about/all over

объ·яв·и́тел·ь, -я, *(m)* announcer; advertiser; author (of a statement)

объ·яв·и́ть, -ят, *(perf) see* **объ·яв·л·я́·ть**

объ·явл·е́ни·е, -я, *(n)* declaration, announcement, notice, advertisement

объ·явл·я́·ть, -ют, *(imp)* announce, declare, state

~ благо·да́р·ност·ь thank officially

объ·ясн·е́ни·е, -я, *(n)* explanation; reason; legend (on illustrations)

объ·ясн·и́м·ый, -ая, -ое, *(pres part pass)* of **объ·ясн·и́ть;** explicable, explainable

объ·ясн·и́тельн·ый, -ая, -ое, *(a)* explanatory

объ·ясн·и́ть, -ят, *(perf) see* **объ·-ясн·я́·ть**

объ·ясн·я́·ть, -ют, *(imp)* explain, elucidate, account for; find the reason for, find out; attribute to; **-ся** *(pass);* be attributed/due to

обы́ден·н·ый, -ая, -ое, *(a)* commonplace

об·ызвествл·е́ни·е, -я, *(n)* (geol) calcification

об·ызвест·ков·а́ни·е, -я,   (*n*) (geol) calcitization

обык·нове́ни·е, -я, (*n*) habit, custom

обык·нове́нн·о (*adv*) usually, in the ordinary way

    как обык·нове́нн·о as usual

обык·нове́нн·ый, -ая, -ое, (*a*) ordinary, usual, normal; (biol) common; (*f decl as adj*) routine (telegrams)

~ **то́ч·к·а** (*f*) (math) regular point

о́б·ыск·, -а, (*m*) search

об·ыск·а́ть, (*fut 3rd pl*) **об·ы́щ·ут,** (*perf*); see **об·ы́ск·ива·ть**

об·ы́ск·ива·ть, -ют, (*imp*) search

об·ы́скр·ивани·е, -я, (*n*) (met) sparking, spark-testing

обы́ч·а·й, -я, (*m*) custom, habit

обы́ч·н·о (*adv*) usually, generally

обы́ч·н·ый, -ая, -ое,   (*a*) usual, ordinary, habitual, conventional, routine, normal, customary; (law) common

~ **о́·черед·ь** (*f*) regular turn

~ **явл·е́ни·е** (*n*) (*as adv*) commonplace

об·ы́щ·ут (*fut 3rd pl*) of **об·ыск·а́ть**

обя́ж·ут (*fut 3rd pl*) of **обяз·а́ть**

обя́з·анност·ь, -и,   (*f*) duty, responsibility, liability

~**, войск·ов·а́я** liability for military service

    ис·полн·я́ющ·ий обя́з·анност·и (*m*) acting

обя́з·анн·ый, -ая, -ое, (*a*) obliged, bound, liable; indebted

обяз·а́тельн·ый, -ая, -ое, (*a*) obligatory, compulsory, mandatory

обяз·а́тельств·о, -а, (*n*) obligation, commitment; liability, responsibility

~**, у·сло́в·н·ое** (fin) contingent liability

обяз·а́ть, (*fut 3rd pl*) **обя́з·ут,** (*perf*); see **обя́з·ыва·ть**

обя́з·ыва·ть, -ют, (*imp*) oblige; commit, make liable; make (somebody do something), see that (somebody does something)

ОВ (*abbr*) = **о·травл·я́ющ·ее ве·щ·еств·о́** toxic agent, poison

ова́л·, -а, (*m*) oval

~ **Касси́ни** (*pl*) (math) ovals of Cassini

овало·гу́б·ц·ы, -ев, (*m pl*) oval-nose pliers

овало·тока́р·н·ый стан·о́к· (*m*) oval lathe, oval-turning lathe

овально·зве́н·н·ый, -ая, -ое, (*a*) (mech) oval-link

овально·о·кру́гл·ый, -ая, -ое, (*a*) (bot) ovate-orbicular

овально·ланцето·ви́д·н·ый, -ая, -ое, (*a*) (bot) ovate-lanceolate

ова́ль·ност·ь, -и, (*f*) ovality

ова́ль·н·ый, -ая, -ое, (*a*) oval

~ **профи́л·ь** (*m*) (met) oval, oval shape

ов·ариа́льн·ый, -ая, -ое, (*a*) (anat) ovarian

овари́н·, -а, (*m*) (pharm) ovarin

оварикри́н·, -а, (*m*) see **овари́н·**

ова́ци·я, -и, (*f*) ovation

Ове́н·, -а, (*m*) (astron) Aries, the Ram

ов·енхи́м·а, -ы, (*f*) (anat) ovenchyma

о́вершот·, -а, (*m*) (oil) overshot

ове́с·, (*g sing*) **овс·а́,** (*m*) (bot) oats, *Avena sativa*

ов·е́ч·ий, -ья, -ье, (*a*) sheep, ovine

ов·е́чк·а, -и, (*g pl*) **-чек·,** (*f*) ewe lamb; (bot) piper mushroom, *Boletus piperatus*

ови́н·, -а, (*m*) drying barn/house, corn-kiln

ови·це́лл·а, -ы, (*f*) (zool) ovicell

ови·ци́д·, -а, (*m*) (chem) ovicide

о·влад·ева́·ть, -ют, (*imp*) (+ *instr*) seize, take possession of; master, become proficient in

о·влад·е́·ть, -ют, (*perf*) (+ *instr*) see **о·влад·ева́·ть**

ово·альбуми́н·, -а, (*m*) (biochem) ovalbumin

ово·вителли́н·, -а, (*m*) (biochem) ovovitellin

ово·гене́з·, -а, (*m*) (zool) ovulation

о́вод·, -а, (*m*) gadfly

~**, желу́д·оч·н·ый** (ent) horse bot, *Gastrophilidae*

~**, ко́ж·н·ый** (ent) warble fly, (*pl*) *Hypodermatidae*

~**, носо·гло́т·оч·н·ый** (ent) bot fly, (*pl*) *Oestridae*

~**, ов·е́ч·ий** (ent) sheep bot fly, *Oestrus ovis*

овод·о́в·ый, -ая, -ое, (*a*) of **о́вод·**

овоид·а́льн·ый, -ая, -ое, (*a*) ovoid

ово·муко́ид·, -а, (*m*) (biochem) ovomucoid

ово·ско́п·, -а, (*m*) egg tester

ово·ци́т·, -а, (*m*) (zool) ovocyte

о́вощ·, -а, (*m*) vegetable

овоще·во́д·ств·о, -а, (*m*) vegetable growing/gardening, olericulture

овощ·н·о́й, -а́я, -о́е, (*a*) vegetable, oleraceous

овощ·н·о́й, -а́я, -о́е
~ консе́рв·ы (*pl*) canned/tinned vegetables

овра́г·, -а, (*m*) ravine, gorge, gully

овра́ж·ист·ый, -ая, -ое, (*a*) (geog) gullied

овс·е́ц·, -а́, (*i*) **-о́м·,** (*m*) (bot) perennial oat, oat grass, *Avenastrum*

овсо·дроб·и́лк·а, -и, (*g pl*) **-лок·,** (*f*) oat grinder

овсо·за·во́д·, -а, (*m*) oats mill

овсо·ру́ш·к·а, -и, (*g pl*) **-ш·ек·,** (*f*) oats milling machine/unit

овсю́г·, -а, (*m*) (bot) wild oats, *Avena fatua*

овс·я́ниц·а, -ы, (*f*) (bot) fescue, *Festuca*

овс·я́нк·а, -и, (*g pl*) **-н·ок·,** (*f*) oatmeal; (orn) bunting, *Emberiza*; (fish) *see* **верховка**

овс·ян·о́й, -а́я, -о́е, (*a*) oat

овс·я́н·ый, -ая, -ое, (*a*) oat; oaten
~ круп·а́ (*f*) oatmeal, porridge oats

овул·я́ци·я, -и, (*f*) (zool) ovulation

ов·ц·а́, -ы́, (*nom pl*) **ов·ц·ы,** (*g pl*) **ов·е́ц·,** (*f*) (zool) sheep, ewe, *Ovis*

овце·бы́к·, -а, (*m*) (zool) musk ox, *Ovibos moschatus*

овце·во́д·, -а, (*m*) sheep farmer/breeder, grazier

овце·во́д·ств·о, -а, (*n*) sheep breeding/farming

ов·ча́р·, -а, (*m*) shepherd

ов·ча́рк·а, -и, (*f*) sheepdog

ов·чи́н·а, -ы, (*f*) sheepskin

о·га́р·ок·, -р·к·а, (*m*) (met) matte; candle stub; (*pl*) cinder, ash

~, колчеда́н·н·ый roasted pyrite residue

огдо·пирами́д·а, -ы, (*f*) (cryst) ogdopyramid

огдо·эдр·и́·я, -и, (*f*) (cryst) ogdohedry

о·гиб·а́·ть, -ют, (*imp*) bend round; go round, skirt; (naut) double; (math) fit (a circle to points)

о·гиб·а́ющ·ая, -ей, (*f decl as a*) (math, rad) envelope

~ и́мпульс·а (phys) pulse envelope

~ част·о́тн·ого от·клон·е́ни·я (rad) frequency deviation envelope

о·гиб·а́ющ·ий, -ая, -ее, (*pres part act*) of **о·гиб·а́·ть**; (*see also* **о·гиб·а́ющ·ая**) (*as noun*)

о·главл·е́ни·е, -я, (*n*) table of contents, contents

о·глас·и́ть, -я́т, (*perf*) *see* **о·глаш·а́·ть**

о·глаш·а́·ть, -ют, (*imp*) proclaim, announce, publish

о·глаш·е́ни·е, -я, (*n*) (*v n*) *see* **о·глаш·а́·ть**
не под·леж·и́т о·глаш·е́ни·ю for restricted circulation, not to be made public, not for publication

о·гле·е́ни·е, -я, (*n*) (geol) gleying, gleization

о·глин·е́ни·е, -я, (*n*) (geol) argillization

огло́бл·я, -и, (*f*) shaft (of cart etc.); (shipb) dagger plate

о·гло́х·нуть, -нут, (*past masc sing*) **о·гло́х·,** (*perf*) become deaf

о·глуш·а́·ть, -ют, (*imp*) deafen; stun

о·глуш·и́ть, -ат, (*perf*) *see* **о·глу·ш·а́·ть**

о·гляд·е́ть, -ят, (*perf*) *see* **о·гляд·ы·ва·ть**

о·гля́д·ыва·ть, -ют, (*imp*) look over, examine; **-ся** look around, get acquainted

о·гляж·у́ (*fut 1st sing*) of **о·гляд·е́ть**

о·гля·н·у́ть, -у́т, (*perf single action*) *see* **о·гля́д·ыва·ть**

огн·еви́к·, -а́, (*m*) (mil) gunner

огн·ёвк·а, -и, (*g pl*) **-вок·,** (*f*) (ent) pyralid, snout moth, (*pl*) *Pyralidae*

огн·ев·о́й, -а́я, -о́е, (*a*) of **ого́н·ь**
~ за·ве́с·а (*f*) (gunn) curtain of fire
~ за·чи́ст·к·а (*f*) (met) flame scarfing, hot deseaming
~ ка́мер·а (*f*) combustion chamber (boiler)
~ кольц·о́ (*n*) fire ring (of stuffing box)
~ коро́б·к·а (*f*) firebox (of a boiler)
~ пози́ци·я (*f*) (gunn) firing position
~ пре·гра́д·а (*f*) (build) fire stop
~ рафин·и́ровани·е (*n*) (met) fire refining, pyrometallurgical refining
~ тепло·те́хн·ик·а (*f*) combustion technology
~ то́ч·к·а (*f*) (mil) strong point

огне·воз·ду́ш·н·ое о·топл·е́ни·е (*n*) solid-fuel hot-air heating

огне·гас·и́тельн·ый, -ая, -ое, (*a*) fire-extinguishing

огне·ды́ш·ащ·ий, -ая, -ее, (*a*) (geol) volcanous

огне·за·щи́т·а, -ы, (*f*) fireproofing (materials), fire protection (buildings etc.)

огне·за·щи́т·н·ый, -ая, -ое, (*a*) fire-proof, fire-resisting, fire-retarding

огне·мёт·, -а, (*m*) flame-thrower/gun

огненно·жи́дк·ий, -ая, -ое, (*a*) (geol) liquid-molten

óгн·енн·ый, -ая, -ое, (*a*) fire

~ шар· (*m*) (mil nucl) fireball

огне·о·пáс·ност·ь, -и, (*f*) fire hazard, risk of fire

огне·о·пáс·н·ый, -ая, -ое, (*a*) inflammable

~ груз· (*m*) hazardous cargo

огне·при·пáс·ы, -ов, (*pl*) (mil) ammunition

огне·про·вóд·н·ый шнур· (*m*) (expl) safety fuse, miners' fuse

огне·стóй·к·ий, -ая, -ое, (*a*) fireproof, fire-resisting

огне·стóй·кост·ь, -и, (*f*) fire resistance

огне·стрéль·н·ое ору́ж·и·е (*n*) firearms

огне·тру́б·н·ый котёл· (*m*) fire-tube boiler

огне·ту́ш·и́тел·ь, -я, (*m*) fire extinguisher

~, пéн·н·ый foam fire extinguisher

~ тетра·хлор·ов·óй tetrachloride fire extinguisher

огне·у·пóр·, -а, (*m*) refractory (material)

~, на·би́в·н·óй lining/hearth refractory; rammed refractory

огне·у·пóр·ност·ь, -и, (*f*) refractoriness

огне·у·пóр·н·ый, -ая, -ое, (*a*) refractory

~ кирпи́ч· (*m*) fire brick (domestic); refractory brick

огн·и́ (*pl*) of огóн·ь

о·говáр·ива·ть, -ют, (*imp*) stipulate, specify; -ся make a slip (in speaking)

о·говор·ённ·ый, -ая, -ое, (*past part pass*) *see* о·говáр·ива·ть; stipulated, specified

о·говор·и́ть, -я́т, (*perf*) *see* о·говáр·ива·ть

о·говóр·к·а, -и, (*g pl*) -р·ок·, (*f*) stipulation

о·гол·ённ·ый, -ая, -ое, (*past part pass*) of о·гол·и́ть; naked, bare, exposed

о·гол·и́ть, -я́т, (*perf*) *see* о·гол·я́·ть

о·голóв·ь·е, -я, (*n*) bridle; headband

о·гол·я́·ть, -ют, (*imp*) bare, denude, strip

ó·гон·, -а, (*m*) eye splice (in rope)

~, подково·ви́д·н·ый horseshoe splice

~, прост·óй soft eye (cordage)

~, раз·ру́б·н·óй cut splice

огон·ёк·, -нь·к·á, (*m*) (*dim*) of огóн·ь

~, блужд·áющ·ий (geol) ignis fatuus

огóн·ь, -гн·я́, (*m*) fire; light (signalling, navigation)

~, авар·и́йн·ый (naut) not under command light, breakdown light

~, артиллер·и́йск·ий gun fire

~, бéг·л·ый (gunn) rapid fire

~, бли́нкер·н·ый (air) blinker light

~, бок·ов·óй (rail) side light

~, борт·ов·óй (naut) side light; (air) navigation light

~, верт·я́щ·ийся (nav) revolving light

~, вéрх·н·ий тóп·ов·ый (naut) after steaming light

~, встрéч·н·ый backfire (firefighting)

~, втор·óй тóп·ов·ый (naut) after steaming light

~, гакобóрт·н·ый (naut) overtaking light

~, гру́пп·о-за·тм·евáющ·ийся (nav) group occulting light

~, гру́пп·о-про·блéск·ов·ый (nav) group flashing light

~, дей·сви́тельн·ый (mil) effective fire

~, до·бáв·очн·ый тóп·ов·ый (naut) after steaming light

~, за·град·и́тельн·ый (gunn) barrage fire; (air) obstruction light

~, зáлп·ов·ый volley fire (small arms), salvo fire (larger guns)

~ за·мык·áни·я маршру́т·а (rail) route-locking light

~ за·нят·ност·и, у·каз·áтельн·ый (rail) occupancy light

~, за·тм·евáющ·ийся (nav) occulting light

~, кинжáль·н·ый (mil) short-range concentrated fire

~, контрóль·н·ый (rail) back light

~ контрóл·я за·мык·áни·я (rail) lock light

~, корм·ов·óй я́кор·н·ый (naut) stern light

~, крыль·ев·óй (a/c) wing-tip light

~, лóж·н·ый phantom light

~, лу́н·н·о-бéл·ый lunar light

~ на раз·ру́ш·éни·е (gunn) destructive fire

~, не·об·слу́ж·иваем·ый (navig) unwatched light

~, ни́ж·н·ий тóп·ов·ый (naut) fore steaming light

~, не·пре·ры́в·н·о гор·я́щ·ий steady light

~, низ·ов·óй ground fire (firefighting)

~, нормáль·н·о-по·гáш·енн·ый normally dark light

ого́н·ь

~ о·гражд·е́ни·я   (nav) obstruction light

~, один·о́чн·ый   (gunn) single-shot fire

~, о·станов·очн·о-раз·реш·и́тель·н·ый   (rail) stop-and-go light

~, от·кры́·т·ый   naked light

~, от·лич·и́тельн·ый   (naut) side light; (rail) marker light

~, пере·ме́н·н·ый   (nav) alternating light

~, по·гран·и́чн·ый   (air) range light

~, под·хо́д·н·ый   (air) approach light

~, по·са́д·очн·ый   (air) landing light

~, по·сто·я́нн·ый   (nav) fixed light

~, пред·у·пред·и́тельн·ый   warning light

~ при·ближ·е́ни·я   approach lighting/light

~, про·бле́ск·ов·ый   (nav) flashing light

~, про·до́ль·н·ый   (gunn) raking fire

~, про·чёс·ыва́ющ·ий   (gunn) sweeping fire

~, пут·ев·о́й   navigation light

~, раз·реш·а́ющ·ий   (rail) clear light

~, с·тво́р·н·ый   leading-in light (ports etc.)

~, то́п·ов·ый   (naut) fore steaming light, masthead light

~, у·каз·а́тельн·ый   indicating light

~, у·сло́в·н·о-раз·реш·а́ющ·ий   (rail) permissive light

~, цвет·ов·о́й   colour light

~, ход·ов·о́й   (naut) navigation light

~, шта́г·ов·ый   riding light, anchor light

~, я́кор·н·ый   (naut) anchor light, riding light

о·гора́ж·ива·ть, -ют,   (imp) fence in, enclose

о·горо́д·, -а,   (m) kitchen/vegetable garden

~, про·мы́шл·енн·ый   market garden (Brit), truck garden (U.S.)

о·город·и́ть, -я́т,   (perf) see о·гора́ж·ива·ть

о·горо́д·ничеств·о, -а,   (m) vegetable growing, olericulture

о·горо́ж·енн·ый, -ая, -ое,   (past part pass) of о·город·и́ть

о·гра́д·а, -ы,   (f) fence, enclosure, wall

о·град·и́тель·ь, -я,   (m) guard, protector

о·град·и́тельн·ый, -ая, -ое,   (a) fencing, guard

~ со·оруж·е́ни·е   (n) (civ eng) breakwater

о·град·и́ть, -я́т,   (perf) see о·гражд·а́·ть

о·гражд·а́·ть, -ют,   (imp) protect/guard against; enclose, fence in; (naut) buoy (a hazard)

о·гражд·е́ни·е, -я,   (n) (v n) see о·гражд·а́·ть; protection; barrier, guard, railing; enclosure; (naut) buoyage

~, ле́ер·н·ое   (shipb) guard rails/chain

~ по́·езд·а   (rail) signalling arrangements

~ ру́б·к·и   conning tower superstructure (submarine)

о·гранич·е́ни·е, -я,   (n) (v n) see о·гранч·ива·ть; restriction, restraint, limitation(s); boundary

    вре́м·я о·гранич·е́ни·я   (n) (elec) clipping time

~ и́мпульс·ов   (phys) pulse clipping

~, ква́нт·ов·ое   (phys) quantum restriction

~, мо́щ·ност·и,   при·нуд·и́тельн·ое   saturation constraint (cybernetics)

~, обою́д·н·ое   (cryst) mutual boundary

~ по моме́нт·у   torque limitation

~ по по·лож·е́ни·ю   (autom) position limitation

~ раст·у́щ·их цеп·е́й   (chem) chain termination

о·грани́ч·енност·ь, -и,   (f) boundedness, limitedness

о·грани́ч·енн·ый, -ая, -ое,   (a) restricted, confined, limited, (math) finite, bounded; (rad) clipped

~ мно́ж·еств·о   (n) (math) finite set

о·грани́ч·ивани·е, -я,   (n) (v n) of о·грани́ч·ива·ть; containment

о·грани́ч·ива·ть, -ют,   (imp) restrict, confine, limit; bound

о·гранич·и́тел·ь, -я,   (m) limiter, stop, gate; (rad) clipper

~, дву·сторо́н·н·ий   (rad) amplitude gate

~, дио́д·н·ый   diode clipper

~ дро́ссел·я   (ICE) throttle gate

~ за·про́с·ов   (telecom) demand limiter

~ и́мпульс·ов   (elec) clipping circuit, clipper

~ максима́ль·н·ых оборо́т·ов   engine speed governor

~ на·пряж·е́ни·я   voltage limiter

~ с·верх·у́   (rad) peak clipper

~ то́к·а   (elec) current limiter

~ толк·а́тел·я   stop pin (diesel unit injector)

~ хо́д·а   (mech eng) stop

о·гранич·и́тельн·ый, -ая, -ое, (*a*) restrictive, restricting, limiting, stopping

~ вы·ключ·а́тел·ь (*m*) limit/end switch, stop

о·грани́ч·ить, -ат, (*perf*) *see* **о·грани́ч·ива·ть**

о·гра́н·к·а, -и, (*g pl*) **-н·ок·,** (*f*) cut, cutting, facetting (gems); (cryst) facet, face, faces·

о·гре́х·, -а, (*m*) (agr) lapse

о·гро́м·н·ый, -ая, -ое, (*a*) huge, enormous, vast

о·губл·ённ·ый, -ая, -ое, (*a*) labial

огу́з·ок·, -з·к·а, (*m*) rump (meat)

огур·е́ц·, -р·ц·а́, (*i*) **-р·ц·о́м,** (*m*) (bot) cucumber, *Cucumis sativus*

~, мор·ск·о́й (zool) sea cucumber, *Thyone briareus*

огур·е́чник·, -а, (*m*) (bot) borage, *Borago officinalis*

огур·е́чн·ый, -ая, -ое, (*a*) cucumber

~ трав·а́ (*f*) *see* **огур·е́чник·**

о·дар·ённост·ь, -и, (*f*) talent

о·дар·ённ·ый, -ая, -ое, (*past part pass*) gifted, talented

о·дев·а́·ть, -ют, (*imp*) dress, clothe; cover

о·де́жд·а, -ы, (*f*) clothes, clothing; surfacing; jacket; revetment

~, доро́ж·н·ая road surfacing/surface

~, за·щи́т·н·ая protective clothing

одеколо́н·, -а, (*m*) eau-de-Cologne

о·де́·нут (*fut 3rd pl*) of **о·де́·ть**

о·дерев·се́ни·е, -я, (*n*) *see* **о·дре́вес·не́ни·е**

о·держ·а́ть, -ат, (*perf*) *see* **о·де́рж·ива·ть**

о·де́рж·ива·ть, -ют, (*imp*) gain, win

~ верх· get the upper hand, overcome

о·держ·и́м·ый, -ая, -ое, (*pres part pass*) of **о·держ·а́ть;** obsessed, possessed; seized with, overcome by

о·дерн·о́вк·а, -и, (*f*) turfing

о·де́·т·ый, -ая, -ое, (*past part pass*) of **о·де́·ть;** clothed, clad, coated

о·де́·ть, -нут, (*perf*) *see* **о·дев·а́·ть**

о·де·я́л·о, -а, (*n*) blanket

оди́н·, (*m numeral*), **одн·а́** (*fem*), **одн·о́** (*neut*), one; the same; alone; only, nothing but; a, an, a certain

~ друг·о́го one another, each other

~ за друг·и́м one by one, one after another

  моде́л·ь одн·о́го те́л·а (*f*) (phys) one-body model

оди́н

~ на· оди́н· tête-à-tête, in private

по одн·ому́ singly, one by one

один·а́ков·о (*adv*) equally; identically

один·а́ков·ый, -ая, -ое, (*a*) identical, the same

~ велич·ин·а́ (*f*) the same value/amount

один·а́рн·ый, -ая, -ое, (*a*) single

оди́н·на·дцат·и- (*component*) eleven-, hendeca-

одиннадцати·уго́ль·ник·, -а, (*m*) (math) hendecagon

одиннадцати·уго́ль·н·ый, -ая, -ое, (*a*) (math) hendecagonal

оди́н·на·дцат·ый, -ая, -ое, (*a*) eleventh

оди́н·на·дцат·ь, -и, (*m*) eleven

один·о́к·ий, -ая, -ое, (*a*) solitary, single

один·о́чник·, -а, (*m*) singlet, singleton

один·о́чн·ый, -ая, -ое, (*a*) single, solitary, solo, one-man

~ вы́·стрел· (*m*) single shot

~ то́ч·к·а (*f*) isolated point

о·ди́ч·ал·ый, -ая, -ое, (*a*) (biol) wild, feral

о·дич·а́·ть, -ют, (*perf*) grow/run wild/feral

одн·а́, (*nom sing fem*) of **оди́н·**

одн·а́жд·ы (*adv*) once

одн·а́к·о (*conj*) however, but; yet, nevertheless

одн·и́ (*nom plur, acc plur inanim*) of **оди́н·** some, certain, certain ones; the same; alone; only, nothing but

одн·и́м (*instr sing m n see* **оди́н·**), (*dat plur m f n see* **одн·и́**)

одн·о́, (*n*) of **оди́н·;** (*as component*) single, one, uni,- mono-, homo-

одно·а́дрес·н·ая систе́м·а (*f*) one-address system (computers)

одно·а́том·н·ый, -ая, -ое, (*a*) monoatomic, monatomic; (chem) monoacid, monobasic; monohydric (alcohol)

~ сло·й (*m*) (phys) monolayer

~ сол·ь (*f*) monoacid salt

~ спирт· (*m*) monohydric alcohol

одно·ба́л·очн·ый, -ая, -ое, (*a*) (civ eng) single-beam/rail

~ кран· (*m*) single-rail overhead traveller

одно·борозд·н·о́й, -а́я, -о́е, (*a*) monocolpate

одно·бо́рт·н·ый, -ая, -ое, (*a*) single-sided; single-breasted (clothing)

одно·бра́т·ственн·ый, -ая, -ое, (*a*) (biol) monadelphous

одно·бра́ч·н·ый, -ая, -ое, (*a*) monogamous

одно·бро́м·ист·ый, -ая, -ое, (*a*) (chem) monobromide (of)

одно·вале́нт·ност·ь, -и, (*f*) (chem) univalence, monovalence

одно·вале́нт·н·ый, -ая, -ое, (*a*) (chem) univalent, monovalent

одно·вариа́нт·н·ый, -ая, -ое, (*a*) monovariant

одно·ветв·и́ст·ый, -ая, -ое, (*a*) single-branched, uniramose, uniramous

одно·вибра́тор·, -а, (*m*) (elec) univibrator, flip-flop/Kipp oscillator

одно·винт·ов·о́й, -а́я, -о́е, (*a*) single-screw

одно·вит·ко́в·ый, -ая, -ое, (*a*) single-turn

одно·во́д·н·ый, -ая, -ое, (*a*) (chem) monohydrate (of)

одно·во́до·росл·ев·ый, -ая, -ое, (*a*) (bot) unialgal

одно·воз·раст·н·о́й, -а́я, -о́е, (*a*) (biol) coeval, coetaneous

одно·вре́мен·н·о (*adv*) simultaneously, at the same time; at a time
　　сущ·ést·овать одно·вре́мен·н·о coexist

одно·време́н·ност·ь, -и, (*f*) simultaneity, simultaneousness, synchronism, isochronism, synchroneity
　　коэфицие́нт· одно·вре́мен·ност·и (*m*) (elec) demand factor

одно·вре́мен·н·ый, -ая, -ое, (*a*) simultaneous, synchronous, isochronous; contemporary, contemporaneous; unit-distance (of computer codes)

~ **флюи́д·** (*m*) (geol) connate fluid

одно·гла́в·ый, -ая, -ое, (*a*) one-headed, monocephalous

одно·гла́з·к·а, -и, (*g pl*) **-з·ок·,** (*f*) (zool) cyclops, monocule

одно·гла́з·н·ый, -ая, -ое, (*a*) (zool) monocular, monophthalmic, one-eyed

одно·гнёзд·н·ый, -ая, -ое, (*a*) unilocular, single-celled; (zool) monothalamous

одно·год·и́чн·ый, -ая, -ое, (*a*) one-year, yearling

одно·го́рб·ый, -ая, -ое, (*a*) single-humped/hump

одно·горл·ов·о́й, -а́я, -о́е, (*a*) single-funnel (vessel); single-die (extruder etc.)

одно·групп·ов·о́й, -а́я, -о́е, (*a*) one-group

~ **тео́р·и·я** (*f*) (phys) one-group theory

одно·дере́в·ая ма́чт·а (*f*) (naut) pole-mast

одно·дере́в·к·а, -и, (*g pl*) **-в·ок·,** (*f*) dugout (boat)

одно·диапазо́н·н·ый, -ая, -ое, (*a*) single-range; (rad) single-waveband/band

одно·дн·е́вк·а, -и, (*g pl*) **-вок·,** (*f*) (ent) mayfly

одно·дн·е́вн·ый, -ая, -ое, (*a*) one-day, one day's

одно·до́ль·н·ый, -ая, -ое, (*a*) (bot) monocotyledonous; (bot) (*pl as noun*) monocotyledons, *Monocotyledoneae*

одно·до́м·н·ый, -ая, -ое, (*a*) (bot) monoecious

одно·дуа́нт·н·ый, -ая, -ое, (*a*) (nucl) one-dee

одно·ду́ж·н·ый, -ая, -ое, (*a*) (zool) synapsid, (*pl as noun*) (pal) synapsids, *Synapsida*

одно·ём·костн·ый, -ая, -ое, (*a*) single-capacity

~ **систе́м·а** (*f*) (autom) single-capacity system

одно·жа́бер·н·ый, -ая, -ое, (*a*) monobranchiate

одно·жгу́т·иков·ый, -ая, -ое, (*a*) (zool) monociliated, uniflagellate

одно·жи́ль·н·ый, -ая, -ое, (*a*) (elec) single-core

~ **ка́бел·ь** (*m*) single-core cable

одно·за·зо́р·н·ый резона́тор· (*m*) (elec) single-gap cavity

одно·за·мещ·ённ·ый, -ая, -ое, (*a*) (chem) monosubstituted

одно·за·ря́д·н·ый, -ая, -ое, (*a*) single-/singly charged

одно·за·хо́д·н·ый, -ая, -ое, (*a*) single-start, single-thread (screws etc.)

одно·зве́н·н·ый, -ая, -ое, (*a*) (elec) single-stage

~ **фильтр·** (*m*) (elec) single-mesh filter

одно·зерн·я́нк·а, -и, (*g pl*) **-нок·,** (*f*) (bot) one-grained wheat, *Triticum monococcum*

одно·зна́ч·ащ·ий, -ая, -ее, (*a*) synonymous

одно·зна́ч·н·о (*adv*) unequivocally, definitely

одно·зна́ч·ност·ь, -и, (*f*) (math) identity
　　сигна́л одно·зна́ч·ност·и (*m*) (rad navig) sense signal

одно·знáч·н·ый, -ая, -ое, (*a*) synonymous; unambiguous, single-valued; one-digit
　взаи́м·о одно·знáч·н·ый one-to-one
~ фýнкци·я (*f*) (math) single-valued/one-valued function
одно·зýб·чат·ый, -ая, -ое, (*a*) (bot) unidentate
одно·зýб·ый, -ая, -óе, (*a*) (zool) monodont
одно·имён·н·ый, -ая, -ое, (*a*) like-named/titled, similarly named; mononomical
одно·и́мпульс·н·ый, -ая, -ое, (*a*) single-pulse, one-pulse; (autom) single-factor
одн·óй (*gen/dat/instr/prep f*) of оди́н·
одно·кали́бер·н·ый, -ая, -ое, (*a*) of the same calibre
одно·кáмер·н·ый, -ая, -ое, (*a*) single-chambered, (biol) monothalamous, unilocular; single/one-stage (of machinery)
~ редýктор· (*m*) one-step reducing valve
~ шлюз· (*m*) (civ eng) single lock
одно·канáль·н·ый, -ая, -ое, (*a*) (rad) single-channel; single-duct (conduits etc.)
одно·канáт·н·ый, -ая, -ое, (*a*) (civ eng) monocable
одно·кáчеств·енн·ый, -ая, -ое, (*a*) of the same quality; (math) isomorphic
одно·ке́рн·ов·ый, -ая, -ое, (*a*) (instr) unipivot
одно·клéт·очн·ый, -ая, -ое, (*a*) unicellular, single-celled, one-celled
одно·клет·ев·óй, -áя, -óе, (*a*) (met roll) single-stage
одно·клино·мéр·н·ый, -ая, -ое, (*a*) (geol) monoclinic
одно·ковш·óв·ый, -ая, -ое, (*a*) one/single-bucket/grab
~ зем·снаря́д· (*m*) dipper dredger
одно·колéй·н·ый, -ая, -ое, (*a*) single-track/line/guage/gage
одно·колéн·чат·ый, -ая, -ое, (*a*) (mech) single-throw
одно·колóнн·ый, -ая, -ое, (*a*) single-column; open-front (of machine tools)
одно·компонéнт·н·ый, -ая, -ое, (*a*) single-component, unicomponent
~ систéм·а (*f*) (phys) monovariant system
одно·кóнтур·н·ый, -ая, -ое, (*a*) (elec) single-loop/circuit

одно·копы́т·н·ый, -ая, -ое, (*a*) (zool) soliped, solidungulate, whole-hoofed
одно·кóрпус·н·ый, -ая, -ое, (*a*) single-stage, single-housing, one-body; (shipb) single-hull; single-furrow (ploughs)
одно·крáт·н·ый, -ая, -ое, (*a*) single, once-through, one-pass, simplex, uni-
~ дéй·стви·е (*n*) single action, (*g as adj*) single-acting
~ пере·гóн·к·а (*f*) flash distillation
~ про·хожд·éни·е (*n*) (phys) single transit
~ цикл· (*m*) once-through cycle/operation
одно·крýж·н·ый, -ая, -ое, (*a*) single/one-circle
одно·кры́л·ый, -ая, -ое, (*a*) one-winged; single-leaf; single-blade
одно·лепéст·ков·ый, -ая, -ое, (*a*) monopetalous
одно·лéт·н·ий, -яя, -ее, (*a*) one-year, one-year-old; (bot) annual
одно·лéт·ник·, -а, (*m*) (bot) annual plant, annual
одно·лин·éйн·ый, -ая, -ое, (*a*) unilinear
одно·лист·н·ый, -ая, -ое, (*a*) single-sheet/leaf; (bot) monophyllous, unifoliate; (math) one-sheeted
одно·луч·ев·óй, -áя, -óе, (*a*) single-beam; one-pronged
~ со·бы́т·и·е (*n*) (phys) one-pronged event
одно·мéр·н·ый, -ая, -ое, (*a*) unidimensional, one-dimensional
одно·мéст·н·ый, -ая, -ое, (*a*) single-seater, single-seat/berth/place; single-station (assembly lines etc.)
одно·минерáль·н·ый, -ая, -ое, (*a*) monomineralic
одно·модáль·н·ый, -ая, -ое, (*a*) (math) unimodular
одно·молекуля́р·н·ый, -ая, -ое, (*a*) monomolecular, unimolecular
~ сло·й (*m*) monolayer
одно·мотóр·н·ый, -ая, -ое, (*a*) single-engined; (elec) single-motor
одно·мýскуль·н·ый, -ая, -ое, (*a*) (zool) monomyaric, monomyarian
одно·на·прáвл·енн·ый, -ая, -ое, (*a*) monodirectional, unidirectional; (bot) homodromous; (math) unicursal
одно·нáтр·иев·ый, -ая, -ое, (*a*) (chem) monosodium
одно·ни́т·н·ый, -ая, -ое, (*a*) unifilar

одно·ни́т·очн·ый, -ая, -ое, (*a*) unifilar

одно·об·мо́т·очн·ый, -ая, -ое, (*a*) single-turn; single-winding

одно·обра́з·и·е, -я, (*n*) uniformity, monotony

одно·обра́з·н·ый, -ая, -ое, (*a*) monotonous, unchanging, uniform, monotone

одно·о́·кис·ь, -и, (*f*) monoxide

одно·осно́в·н·ый, -ая, -ое, (*a*) monobasic

~ **кисл·от·а́** (*f*) (chem) monobasic acid

одно·о́с·ност·ь, -и, (*f*) uniaxiality; alignment

одно·о́с·н·ый, -ая, -ое, (*a*) uniaxial; (bot) haplocaulescent

одно·отве́рст·н·ый, -ая, -ое, (*a*) single-hole/entry

одно·па́л·ый, -ая, -ое, (*a*) (zool) monodactylous

одно·пере·ки́д·н·ый, -ая, -ое, (*a*) (elec) single-flip-flop

одно·петл·ев·о́й, -а́я, -о́е, (*a*) single-loop

одно·по·воро́т·н·ый, -ая, -ое, (*a*) one-turn

одно·подъ·ём·н·ый, -ая, -ое, (*a*) single-lift

одно·поло́с·н·ый, -ая, -ое, (*a*) single-band

одно·полост·н·о́й, -ая, -ое, (*a*) single-cavity

~ **гиперболо́ид·** (*m*) (math) hyperboloid of one sheet

одно·полу·перио́д·н·ый, -ая, -ое, (*a*) (rad) half-wave

~ **вы·прямл·е́ни·е** (*n*) half-wave rectification

одно·по́л·ый, -ая, -ое, (*a*) unisexual

одно·по́ль·н·ый, -ая, -ое, (*a*) single-leaf, single

~ **двер·ь** (*f*) single door

одно·по́люс·ност·ь, -и, (*f*) unipolarity, monopolarity

одно·по́люс·н·ый, -ая, -ое, (*a*) (phys) unipolar, homopolar, monopolar, single-pole

~ **систе́м·а** (*f*) single-current system (telephony)

одно·пост·ов·о́й, -а́я, -о́е, (*a*) single-operator (of machines)

одно·по·то́ч·н·ый, -ая, -ое, (*a*) single-flow

одно·пре·де́ль·н·ый, -ая, -ое, (*a*) (instr) single-range

одно·пре·ломл·я́ющ·ий, -ая, -ее, (*a*) singly-refracting

одно·про́вод·н·ый, -ая, -ое, (*a*) (elec) single-wire

одно·про́·волоч·н·ый, -ая, -ое, (*a*) monofilar

одно·про·хо́д·н·ый, -ая, -ое, (*a*) single-pass (welding, rolling, forging); (*pl as noun*) (zool) monotremes, *Monotremata*

одно·пря́д·н·ый, -ая, -ое, (*a*) single-strand (of rope)

одно·пу́т·н·ый, -ая, -ое, (*a*) (rail) single-track/line, one-way, single-lane/line (road traffic)

одно·ра́з·н·ый, -ая, -ое, (*a*) single

одно·ра́з·ов·ый, -ая, -ое, (*a*) single, single-shot, one-kick

одно·раз·ры́в·н·ый, -ая, -ое, (*a*) single-break (switches)

одно·раз·ря́д·н·ый, -ая, -ое, (*a*) (autom) single/one-column

~ **сумма́тор·** (*m*) one-column adder

одно·рёбер·н·ый, -ая, -ое, (*a*) unicostate

одно·реда́н·н·ый, -ая, -ое, (*a*) (shipb) single-step (of pulls)

одно·ре́льс·ов·ый, -ая, -ое, (*a*) monorail

одно·ресни́т·чат·ый, -ая, -ое, (*a*) monociliated

одно·ро́г·ий, -ая, -ое, (*a*) (zool) monocerous

одно·ро́д·ност·ь, -и, (*f*) uniformity, homogeneity

коэффицие́нт· одно·ро́д·ност·и (*m*) uniformity coefficient

одно·ро́д·н·ый, -ая, -ое, (*a*) uniform; homogeneous, homogenous

~ **деформ·а́ци·я** (*f*) homogeneous deformation

~ **по́л·е** (*n*) (elec) uniform field

~ **у·равн·е́ни·е** (*n*) (math) homogeneous equation

одно·ру́ч·ечн·ый, -ая, -ое, (*a*) single, single-knob, one-spot (tuning etc.)

одно·ручь·ев·о́й, -а́я, -о́е, (*a*) single-drop (forgings etc.); single-groove (rolling)

одно·ря́д·н·ый, -ая, -ое, (*a*) single-row, unilinear; uniserial, uniseriate; (bot) monostichous

одно·с·вя́з·н·ый, -ая, -ое, (*a*) (math) simply connected

одно·семено·до́ль·н·ый, -ая, -ое, (*a*) = **одно·семя·до́ль·н·ый**

одно·семя·до́ль·н·ый, -ая, -ое, (*a*) (bot) monocotyledonous; (*pl as noun*) monocotyledons, *Monocotyledoneae*

одно·сем·я́нн·ый, -ая, -ое, (*a*) (bot) one-seeded, monospermous

одно·семяно·до́ль·н·ый, -ая, -ое, (*a*) = **одно·семя·до́ль·н·ый**

одно·сер·ни́ст·ый, -ая, -ое, (*a*) (chem) monosulphide (of)

одно·с·ка́т·н·ый, -ая, -ое, (*a*) single-slope, single-ended, single

~ **дискримина́тор·** (*m*) (elec) single-ended discriminator

~ **колес·о́** (*n*) single wheel

~ **кры́·ш·а** (*f*) single-slope roof (on major structure); lean-to roof (on subordinate structure)

одно·ско́р·ост·н·ый, -ая, -ое, (*a*) single-speed, one-velocity

~ **тео́ри·я** (*f*) (nucl) one-group theory

одно·с·ло́ж·н·ый, -ая, -ое, (*a*) monosyllabic

одно·сло́й·н·ый, -ая, -ое, (*a*) single-layer, single, one-ply, one-course (paving etc.)

~ **шёлк·ов·ая об·мо́т·к·а** (*f*) single-silk insulation (cables)

одно·стаби́ль·н·ый, -ая, -ое, (*a*) monostable

одно·стан·и́нн·ый, -ая, -ое, (*a*) overhanging (machine tools)

одно·с·тво́рчат·ый, -ая, -ое, (*a*) single-leaf (of doors, screens etc.); single (of doors, gates etc.); (zool) univalve

одно·сте́н·н·ый, -ая, -ое, (*a*) single-walled

одно·сто́·ечн·ый, -ая, -ое, (*a*) open-side, overhanging (of machine tools); (a/c) single-bay

одно·сторо́н·н·ий, -яя, -ее, (*a*) unilateral, one-sided, one-way; open-side (of machine tools); (bot) secund

~ **де́й·стви·е** (*n*) single action; (*gen as adj*) single-acting/action

одно·ступе́н·чат·ый, -ая, -ое, (*a*) single-stage; simple

одно·суста́в·н·ый, -ая, -ое, (*a*) single-jointed

одно·та́кт·н·ый, -ая, -ое, (*a*) single-action, single-cycle, single-stroke, one-stroke

одно·тари́ф·н·ый, -ая, -ое, (*a*) single-fee/rate, fixed-fare

одно·тёс·, -а, (*m*) plank nail

одно·ти́п·н·ый, -ая, -ое, (*a*) of the same type; monotype, single-type; sister (ship)

одно·то́м·ник·, -а, (*m*) single-volume edition

одно·тру́б·н·ый, -ая, -ое, (*a*) tubeless (tire)

одно·тычи́н·ков·ый, -ая, -ое, (*a*) (bot) monandrous

одно·узл·ов·о́й, -а́я, -о́е, (*a*) uni-nodal

одно·утро́б·н·ый, -ая, -ое, (*a*) (zool) monodelphic

одно·фа́з·н·ый, -ая, -ое, (*a*) single-phase, monophase

~ **режи́м·** (*m*) single phasing

одно·фо́кус·н·ый, -ая, -ое, (*a*) single-focusing; confocal

одно·хозя́й·ственн·ый, -ая, -ое, (*a*) (bot) autoecious, autoxenous

одно·цве́т·н·ый, -ая, -ое, (*a*) monochromatic

одно·част·и́чн·ый, -ая, -ое, (*a*) (nucl) single/independent-particle

одно·част·о́тн·ый, -ая, -ое, (*a*) single-frequency

одно челн·о́чн·ый, -ая, -ое, (*a*) plain, one shuttle, single-box (of looms etc.)

одно·черп·а́ков·ый, -ая, -ое, (*a*) dipper (of dredgers etc.)

одно·чле́н·, -а, (*m*) (math) monomial

одно·чле́н·ист·ый, -ая, -ое, (*a*) (zool) uniarticulate

одно·чле́н·н·ый, -ая, -ое, (*a*) monomial

одно·электро́н·н·ый, -ая, -ое, (*a*) one/single-electron

одно·элеме́нт·н·ый, -ая, -ое, (*a*) (instr) single-unit

одно·эта́ж·н·ый, -ая, -ое, (*a*) single-storeyed/deck

одно·я́дер·н·ый, -ая, -ое, (*a*) uni-nuclear; (bot) uninucleate

одно·я́кор·н·ый, -ая, -ое, (*a*) (elec) with one armature, rotary

одно·я́рус·н·ый, -ая, -ое, (*a*) single-tier/deck/stage

одно·ячёй·ков·ый, -ая, -ое, (*a*) unicellular

одн·у́ (*acc sing f*) of **оди́н·**

одобр·е́ни·е, -я, (*n*) (*v n*) see **о·добр·я́·ть;** approval

о·до́бр·енн·ый, -ая, -ое, (*past part pass*) of **о·добр·и́ть**

о·до́бр·ить, -ят, (*perf*) see **о·добр·я́·ть**

о·добр·я́·ть, -ют, (*imp*) approve

одо́·граф·, -а, (*m*) (ocean) odograph, course recorder

одол·ева́·ть, -ют, (*imp*) overcome, surmount

о·дом·а́шнени·е, -я, (*n*) taming, domestication

одо́·метр·, -а, (*m*) odometer; soil compression tester

одонто·глоссу́м·, -а, (*m*) (bot) *Odontoglossum*

одонто́·метр·, -а, (*m*) (mech) odontometer

одонто·ли́т·, -а, (*m*) (min) odontolite

одно·ход·ов·о́й, -а́я, -о́е, (*a*) single-pass; one-way; single-thread (screws etc.)

одонто·ло́г·и·я, -и, (*f*) (med) odontology

одор·а́нт·, -а, (*m*) odorant

одор·а́тор·, -а, (*m*) odorant

одори·ме́тр·, -а, (*m*) (chem) odorimeter

одори·фо́р·, -а, (*m*) (chem) odoriphore

о·древес·не́ни·е, -я, (*n*) (bot) lignification

о·древес·не́·ть, -ют, (*perf*) lignify

о·дря́хл·е́·ть, -ют, (*perf*) become decrepit

о·ду́б·ин·а, -ы, (*f*) tanning waste

о·дув·а́нчик·, -а, (*m*) (bot) dandelion, *Taraxacum*

о·ду́м·а·ться, -ются, (*perf*) change one's mind, reconsider

о́·дур·ь, -и, (*f*) stupor

~, со́н·н·ая (bot) belladonna, *Atropa belladonna*

о·дур·я́·ть, -ют, (*imp*) stupefy

о·душ·ев·и́ть, -я́т, (*perf*) see **о·ду·ш·евля́·ть**

о·душ·евля́·ть, -ют, (*imp*) animate

о·ды́ш·к·а, -и, (*f*) panting, laboured breathing, (med) dyspnoea

о·жг·у́т (*fut 3rd pl*) of **о·же́ч·ь**

Оже́, ли́в·ен·ь (*f*) (nucl) Auger shower

о·жёг· (*past masc sing*) of **о·же́ч·ь**

ожеле́д·ь, -и, (*f*) glazed frost, ice crust

о·желез·не́ни·е, -я, (*n*) (geol) ferruginization

о·жереб·и́ться, -я́тся, (*perf*) (zool) foal

ожере́ль·е, -я, (*n*) necklace

о·жесточ·ённ·ый, -ая, -ое, (*past part pass*) embittered, fierce, desperate

о·же́ч·ь, (*fut 3rd sing, pl*) **о·жж·ёт, о·жг·у́т,** (*past masc sing*) **о·жёг·,** (*perf*); see **об·жиг·а́·ть**

о·жж·ёт (*fut 3rd sing*) of **о·же́ч·ь**

ожива́л·, -а, (*m*) ogive

о·жива́ль·н·ый, -ая, -ое, (*a*) ogival, ogeval, ogee

~ част·ь (*f*) (air) ogive

о·жив·а́·ть, -ют, (*imp*) come to life, revive

о·жив·и́ть, -я́т, (*perf*) see **о·жи·вл·я́·ть**

о·живл·е́ни·е, -я, (*m*) (*v n*) see **о·жи·вл·я́·ть**; resuscitation, revival; (cinema) animation

о·живл·я́·ть, -ют, (*imp*) revive, resuscitate, enliven

о·жиг·а́·ть, -ют, (*imp*) see **об·жиг·а́·ть**

о·жид·а́ем·ый, -ая, -ое, (*past part pass*) of **о·жид·а́·ть**; expected

~ про·долж·и́тельность·ь (*f*) expectancy (of life)

о·жид·а́ни·е, -я, (*n*) (*v n*) of **о·жид·а́·ть**; expectation, anticipation; wait, waiting

вре́м·я о·жид·а́ни·я (*n*) access time (computer)

знак· о·жид·а́ни·я (*m*) (telecom) break-off sign

зо́н·а о·жид·а́ни·я (*f*) waiting area; (air) holding zone

~, математ·и́ческ·ое mathematical expectation, expectation value

о·жид·а́·ть, -ют, (*imp*) (+ *gen*) wait for, expect, anticipate

о·жиж·а́·ть, -ют, (*imp*) liquefy

о·жиж·е́ни·е, -я, (*n*) (*v n*) see **о·жи·ж·а́·ть**; liquefaction

~ угл·е́й destructive hydrogenation, Bergius processing (of coal)

о·жиж·и́тел·ь, -я, (*m*) liquefier

о·жир·е́ни·е, -я, (*n*) obesity, (med) adiposis

о·жи́·ть, (*fut 3rd pl*) **о·жив·у́т,** (*perf*) come to life, revive

о·жо́г·, -а, (*m*) (med) burn; (bot) blight, scorch

~, луч·ев·о́й radiation burn

о́з·, -а, (*m*) (geol) esker

о·забо́т·ить, -я́т, (*perf*) see **о·забо́·ч·ива·ть**

о·забо́ч·ива·ть, -ют, (*imp*) cause/give anxiety/concern, worry, preoccupy; **-ся** concern oneself with, see/attend to

о·за·гла́в·ить, -я́т, (*perf*) give a title/heading, entitle, head

о·за·да́ч·енн·ый, -ая, -ое, (*past part pass*); perplexed, puzzled

озазо́н·, -а, (*m*) (chem) bisphenylhydrazone

озафа́н·, -а, (*m*) (chem) ozaphane

о·зву́ч·ени·е, -я, (*n*) *see* **о·зву́ч·и·вани·е**

о·зву́ч·енн·ый, -ая, -ое (*past part pass*) of **о·зву́ч·ить;** exposed to sonic radiation; (cinema) scored, sound (film)

о·зву́ч·ивани·е, -я, (*n*) (*v n*) *see* **о·зву́ч·ива·ть;** sonic irradiation; (cinema) scoring

~, по·след·ующ·ее (cinema) scoring, post-synchronization

о·зву́ч·ива·ть, -ют, (*imp*) score for sound; expose to sonic radiation/vibration

о·зву́ч·ить, -ат, (*perf*) *see* **о·зву́ч·и·ва·ть**

о·здоров·и́ть, -я́т, (*perf*) *see* **о·здо·ровл·я́·ть**

о·здоровл·я́·ть, -ют, (*imp*) improve the health of, make better/well; make sanitary, sanitate; normalize, put right, clean up (figurative)

озд·а, -ы, (*f*) (shipb) deck beam (wood)

о·зелен·ёни·е, -я, (*n*) creating/planting out gardens/parks

озер·к·о́, -а́, (*nom pl*) **-й,** (*g pl*) **-ов,** (*dim*) of **о́зер·о;** pool; freshwater pool (in sea ice)

озёр·н·ый, -ая, -ое, (*a*) lake, (biol etc.) lacustrine; limnetic

~ руд·а́ (*f*) (min) limonite

о́зер·о, -а, (*nom pl*) **озёр·а,** (*g pl*) **озёр·,** (*n*) lake

~, альп·и́йск·ое tarn

~, ла́в·ов·ое lava cauldron

озеро·ве́д·ени·е, -я, (*a*) limnology

о·зи́м·ый, -ая, -ое, (*a*) (bot) winter, winter-flowering; (*pl as noun*) winter crops

о·знако́м·ить, -ят, (*perf*) *see* **о·зна·комл·я́·ть**

о·знакомл·я́·ть, -ют, (*imp*) acquaint, familiarize

о·знамен·ова́ть, -у́ют, (*perf*) mark, celebrate

о·знач·а́·ть, -ют, (*imp*) mean, indicate; denote

о·знач·ить, -ат, (*perf*) *see* **о·зна·ч·а́·ть**

о·зно́б·, -а, (*m*) (med) rigor, shivering

озобро́м·, -а, (*m*) (phot chem) ozobrome

озокери́т·, -а, (*m*) (min) ozokerite, mineral wax

о·зол·ёни·е, -я, (*n*) ashing, ash formation; combustion

о·зол·ённ·ый, -ая, -ое, (*past part pass*) ashed

озо́н·, -а, (*m*) (chem) ozone

озон·а́т·, -а, (*m*) (chem) ozonide

озон·а́тор·, -а, (*m*) ozonizer

озон·и́д·, -а, (*m*) ozonide

озон·и́р·ова·ть, -уют, (*imp or perf*) ozonize

озоно·ли́з·, -а, (*m*) ozonolysis

озо·ти́п·и·я, -и, (*f*) (phot) ozotype

о·зу́бл·ени·е, -я, (*n*) dentition

о·зу́бл·енн·ый, -ая, -ое, (*a*) denticulate

ойди·и, (*pl*) (bot) oidium

ойко·криста́лл·, -а, (*m*) (cryst) oiko-crystal

ойко·ло́г·и·я, -и, (*f*) (obs bot) ecology

ОКА (*abbr*) = **о·живл·я́ющ·ий кис·ло·ро́д·н·ый аппара́т·** (*m*) artificial respiration unit

о·каз·а́ни·е, -я, (*n*) (*v n*) *see* **о·ка́з·ы·ва·ть**

о·каз·а́ть, (*fut 3rd pl*) **о·ка́ж·ут,** (*perf*); *see* **о·ка́з·ыва·ть; -ся** occur, be, fall (in a range of)

ока́зи·я, -и, (*f*) occasion, opportunity

о·ка́з·ыва·ть, -ют, (*imp*) render, give, show, have, do; exert, exercise (influence, pressure), offer (resistance to) **-ся** (+ *instr*) prove to be, turn out to be

~ в·ли·я́ни·е have/exert an influence on, affect, influence

о·кайм·и́ть, -я́т, (*perf*) *see* **о·кайм·л·я́·ть**

о·каймл·е́ни·е, -я, (*n*) (*v n*) *see* **о·кайм·л·я́·ть;** border, fringe, edge, rim, (bot) margin

~, све́т·л·ое (opt) halo, halation

о·каймл·ённ·ый, -ая, -ое, (*past part pass*) *see* **о·каймл·я́·ть;** (bot) marginate, bordered

о·каймл·я́·ть, -ют, (*imp*) edge, border, fringe, rim

о·ка́л·ин·а, -ы, (*f*) (met) scale

окалино·лом·а́тел·ь, -я, (*m*) (met roll) scale breaker

окалино·образ·ова́ни·е, -я, (*n*) (met) scaling

окалино·сто́й·кост·ь, -и, (*f*) (met) oxidation/scaling resistance

окалино·сто́й·к·ий, -ая, -ое, (*a*) oxidation/scaling-resistant

о·ка́л·ыва·ть, -ют, (*imp*) prick round/all over; split/break round

о·камен·ева́·ть, -ют, (*imp*) fossilize, petrify, lithify

о·камен·е́лост·ь, -и, (*f*) (geol) fossil

о·камен·е́л·ый, -ая, -ое, (a) petrified, fossilized, fossil

о·камен·е́·ть, -ют, (perf) see о·камен·ева́·ть

о·кант·ова́ть, -у́ют, (perf) see о·кант·о́выва·ть

о·кант·о́вк·а, -и, (g pl) -вок·, (f) (text) edge piping; (phot) cardboard mount

о·кант·о́выва·ть, -ют, (imp) (text) pipe; mount (photographs etc.)

о·ка́нч·ива·ть, -ют, (imp) finish, end, complete; graduate (academically)

о·ка́п·ывани·е, -я, (n) (mil) entrenching, digging in

о·ка́п·ыва·ть, -ют, (imp) dig round/in

о·кат·а́·ть, -ют, (perf) see об·ка́т·ыва·ть

о·кат·и́ть, -ят, (perf) see о·ка́ч·ива·ть

о·ка́т·ыва·ть, -ют, (imp) = об·ка́т·ыва·ть

о·ка́ч·ива·ть, -ют, (imp) douse, drench, sluice

о·ка́ш·ива·ть, -ют, (imp) mow round

о·ква́рц·евани·е, -я, (n) (geol) silicification

о·кварц·и́ровани·е, -я, (n) (geol) silicification

океа́н·, -а, (m) ocean

~, Се́вер·н·ый Лед·ови́т·ый (geog) Arctic Ocean

~, Ти́х·ий (geog) Pacific Ocean

океан·и́т·, -а, (m) (min) oceanite

океано́·граф·, -а, (m) oceanographer

океано·гра́ф·и·я, -и, (f) oceanography

океа́н·ск·ий, -ая, -ое, (a) ocean, oceanic; ocean-going (of boats etc.)

о·ки́л·ени·е, -я, (n) (biol) carina

о·ки́л·енн·ый, -ая, -ое, (a) carinate, keeled, keel-shaped

о́·кисел·, -сл·а (m) (chem) oxide

о·кисл·е́ни·е, -я, (n) oxidation, oxidizing

~, ано́д·н·ое anodic oxidation, anodizing

~ -вос·становл·е́ни·е -я, (n) (chem) oxidation-reduction, oxido-reduction, redox

о·кисл·ённ·ый, -ая, -ое, (past part pass) of о·кисл·и́ть; oxidized, blown

~ биту́м· (m) blown asphalt/bitumen

о·кисл·и́тел·ь, -я, (m) oxidant, oxidizing agent, oxidizer

о·кисл·и́тельн·о - вос · станов · и́тель - н·ый, -ая, -ое, (a) (chem) redox

о·кисл·и́тельн·ый, -ая, -ое, (a) oxidizing, oxidative

~ дезамин·и́ровани·е (n) oxidative deamination

о·кисл·и́ть, -я́т, (perf) see о·кисл·я́·ть

о·кисл·я́емост·ь, -и, (f) oxidability; (oil) oxidation test; oxygen consumed (in water)

о·кисл·я́·ть, -ют, (imp) oxidize; -ся (pass); turn sour

о́·кис·н·ый, -ая, -ое, (a) oxide, oxidic

о́·кис·ь, -и, (f) (chem) oxide (higher, or -ic)

~, азотно·ки́сл·ая nitrate

~, без·во́д·н·ая anhydride

~ бери́лл·и·я beryllium oxide

~, во́д·н·ая hydroxide

~ желе́з·а ferric oxide

~ желе́з·а, азотно·ки́сл·ая ferric nitrate

~ -за́·кис·ь ура́н·а uranyl-uranate, uranous-uranic oxide

~, полу́тор·н·ая (chem) sesquioxide

~, серно·ки́сл·ая sulphate, sulfate

~ сульфи́д·ов sulphoxide, sulfoxide

~ угле·ро́д·а carbon monoxide

~ ура́н·а-ба́р·и·я uranium-barium oxide

~ хро́м·а chromic oxide

окклюд·и́р·овать, -уют, (imp and perf) occlude

окклю́з·и·я, -и, (f) occlusion

оккуп·а́нт·, -а, (m) (mil) occupying power/force

оккуп·а́ци·я, -и, (f) (mil) occupation

оккуп·и́р·овать, -уют, (imp or perf) occupy

о·кла́д·, -а, (m) rate of pay/salary, pay, salary; setting, set, framework

~, двер·н·о́й (min) set, frame set

~, квадра́т·н·ый (min) square set

~, контро́ль·н·ый junction set

о·кла́д·к·а, -и, (g pl) -д·ок·, (f) (shipb) carling; deck beam (wood)

о·кле́·ива·ть, -ют, (imp) glue/stick/ paste over; paper (walls)

о·кле́·ить, -ят, (perf) see о·кле́·ива·ть

окн·о́, -а́, (n pl) о́кн·а, (g pl) о́кон·, (n) window, opening, port (of engine)

~, ветр·ов·о́е (M/T) windscreen

~, в·пуск·н·о́е (ICE) inlet port

~ вы́·да·ч·и discharge door (of furnaces)

~, вы·движ·н·о́е sliding window

~, вы·пуск·н·о́е (ICE) exhaust port

~, вы·ход·н·о́е window (of X-ray tube)

~, за·ва́л·очн·ое charging door (furnaces)

~, за·щи́т·н·ое (nucl) radiation/shielding window

~, пере·пуск·н·о́е (ICE) by-pass port

окн·о́
~ **по·са́д·к·и** charging door (furnaces)
~, **про·ду́в·очн·ое** (ICE) scavenging port
~, **рабо́ч·ее** charging door (furnaces)
~, **раз·движ·н·о́е** sliding window
~, **са́д·очн·ое** charging door (furnaces)
~, **экспозицио́н·н·ое** (cinema) aperture gate
о·ко́в·, -а, (*m*) (met roll) collar, collaring
окова́л·ок·, -л·к·а, (*m*) loin (of meat)
о·ков·а́ть, (*fut 3rd pl*) **о·ку·ю́т,** (*perf*); *see* **о·ко́в·ыва·ть**
о·ко́в·к·а, -и, (*g pl*) **-в·ок·,** (*f*) (*v n*) *see* **о·ко́в·ыва·ть**
~ **ло́паст·и** blade tipping
о·ко́в·ыва·ть, -ют, (*imp*) (met) bind; (met roll) collar
о·ко́л·к·а, -и, (*f*) (*v n*) *see* **о·ка́л·ыва·ть**
~ **су́д·н·а** freeing a ship beset in ice
о́коло (*prep* + *gen*) (*adv*) near, nearby, by, around, about, approximately
  вращ·е́ни·е о́коло то́ч·к·и (*n*) (math) rotation about a point
о́коло- (*prefix*) peri-, circum-
о́коло·жа́бер·н·ый, -ая, -ое, (*a*) (zool) peribranchial
о́коло·звук·ов·о́й, -а́я, -о́е, (*a*) trans-sonic
о́коло·зенита́ль·н·ый, -ая, -ое, (*a*) (surv) circumzenithal
о́коло·крит·и́ческ·ий, -ая, -ое, (*a*) near-critical
о́коло·лепе́ст·н·ый, -ая, -ое, (*a*) (bot) peripetalous
о́коло·пе́ст·ичн·ый, -ая, -ое, (*a*) (bot) perigynous
о́коло·пло́д·ник·, -а, (*m*) (bot) pericarp
о́коло·по́люс·н·ый, -ая, -ое, (*a*) (astron) circumpolar
о́коло·рот·ов·о́й, -а́я, -о́е, (*a*) circumoral, adoral
о́коло·серд·е́чн·ый, -ая, -ое, (*a*) (anat) pericardial
о́коло·стволь·н·ый, двор· (*m*) (min) shaft bottom
о·кол·о́ть, -ю́т, (*perf*) *see* **о·ка́л·ыва·ть**
о́коло·у́ст·ь·е, -я, (*n*) (biol) peristome
о́коло·у́ш·ник·, -а, (*m*) (anat) paratoid gland
о́коло·у́ш·н·ый, -ая, -ое, (*a*) (zool) parotid
о́коло·цве́т·ник·, -а, (*a*) (bot) perianth
око́л·ыш·, -а, (*m*) cap-band
око́ль·н·ый, -ая, -ое, (*a*) roundabout, devious

о·кон·е́чност·ь, -и, (*f*) extremity, tip, tail, end
о·кон·е́чн·ый, -ая, -ое, (*a*) final, terminal; output, end
~ **каска́д·** (*m*) (elec) output stage
~ **ста́нци·я** (*f*) terminal station
~ **ста́нци·я, втор·а́я** (rail, telecom) down station
око́н·ниц·а, -ы, (*i*) **-ей,** (*f*) window frame; (shipb) port frame
око́н·ничн·ый, -ая, -ое, (*a*) of **око́н·ниц·а**
оконно·ра́м·н·ый, -ая, -ое, (*a*) window-sash
око́н·н·ый, -ая, -ое, (*a*) of **окн·о́**
~ **пере·плёт·** (*m*) (build) window sash
~ **ра́м·а** (*f*) window sash
о·ко́нтур·енн·ый, -ая, -ое, (*past part pass*); contoured (of maps etc.)
о·конту́р·ивани·е, -я, (*n*) (*v n*) *see* **о·конту́р·ива·ть**
о·конту́р·ива·ть, -ют, (*imp*) outline, map, mark in contours
о·ко́нтур·ирующ·ие шпу́р·ы (*pl*) (expl) peripheral holes
о·конч·а́ни·е, -я, (*n*) termination, conclusion, end, expiry; graduation (academic)
  знак· о·конч·а́ни·я (*m*) (telecom) ending sign
о·конч·а́тельн·ый, -ая, -ое, (*a*) final, ultimate; definitive
о·ко́нч·ить, -ат, (*perf*) *see* **о·ка́нч·ива·ть**
о·ко́п·, -а, (*m*) trench, hole, pit
о·коп·а́ть, -ют, (*perf*) dig round/in
о·ко́п·ник·, -а, (*m*) (bot) comfrey, *Symphytum*
о·кора́ч·ива·ть, -ют, (*imp*) shorten, curtail
о·ко́р·енн·ый, -ая, -ое, (*past part pass*) *see* **о·кор·я́·ть**
о·кор·и́ть, -я́т, (*perf*) *see* **о·кор·я́·ть**
о·ко́р·ива·ть, -ют, (*imp*) = **о·кор·я́·ть**
о·ко́р·к·а, -и, (*f*) (*v n*) *see* **о·кор·я́·ть**; debarking, stripping
о·корм·и́ть, -я́т, (*perf*) overfeed; poison (with food)
о́корок·, -а, (*nom pl*) **-а́, -о́в,** (*m*) (food) ham; leg (of lamb etc.)
о·корот·и́ть, -я́т, (*perf*) *see* **о·кора́ч·ива·ть**
о·ко́р·очн·ый, -ая, -ое, (*a*) of **о·ко́р·к·а**
~ **стан·о́к·** (*m*) debarking machine (for timber)

о·корч·ёвк·а, -и, (*f*) uprooting, grubbing, stubbing

о·кόр·щик·, -а, (*m*) barker, peeler

о·кор·я́·ть, -ют, (*imp*) bark, peel

о·кос·и́ть, ´ят, (*perf*) mow round

о·кост·енева́·ть, -ют, (*imp*) ossify; stiffen, become stiff/numb; harden

о·кост·енέл·ый, -ая, -ое, (*a*) ossified; stiff, numb

о·кост·енέни·е, -я, (*n*) ossification

о·кост·енέ·ть, -ют, (*perf*) *see* о·кост·енева́·ть

окόт·, -а, (*m*) (agr) lambing; lambing time

о·коч·енέни·е, -я, (*n*) stiffening; numbness

~, тру́п·н·ое (med) rigor mortis

окόш·ечк·о, -а, (*g pl*) -чек·, (*n*) (*dim*) of окн·ό; (biol) foramen

окόш·к·о, -а, (*g pl*) -ш·ек·, (*n*) (*dim*) of окн·ό

~, резона́нс·н·ое (elec) resonant window

о·кра́·ин·а, -ы, (*f*) outskirts; frontier/remote area; (bot) peristome

~ гео·синклина́л·и (geol) geosynclinal margin

о·кра́·инн·ый, -ая, -ое, (*a*) bordered

~ мόр·е (*n*) land locked sea

о·кра́с·ить, ´ят, (*perf*) *see* о·кра́ш·ива·ть

о·кра́с·к·а, -и, (*g pl*) -с·ок·, (*f*) (*v n*); *see* о·кра́ш·ива·ть; colour, color, coloration

~, дефор́м·и́рующ·ая dazzle painting

~, за·щи́т·н·ая protective painting; (zool) protective colouring

~, крипт·и́ческ·ая (zool) cryptic coloration

~, лже·предо·стерег·а́ющ·ая (zool) mimicry

~, маск·ирόвочн·ая camouflage painting

~, предо·стерег·а́ющ·ая (zool) warning/aposematic coloration

о·кра́с·очн·ый, -ая, -ое, (*a*) of о·кра́с·к·а; painting

о·кра́ш·енн·ый, -ая, -ое, (*past part pass*) of о·кра́с·ить; coloured, colored, dyed, painted

о·кра́ш·иваемост·ь, -и, (*f*) colorability, dyeability

о·кра́ш·ивани·е, -я, (*n*) (*v n*) of о·кра́ш·ива·ть; colouring, coloring, coloration; (phot) toning

~, не·равно·мέр·н·ое patchy coloration/dyeing

о·кра́ш·ивани·е

~, при·жи́зн·енн·ое (microbiol) intravitam staining

о·кра́ш·ива·ть, -ют, (*imp*) paint, colour/color, dye, stain; (phot) tone

о·кремн·ева́·ть, -ют, (*imp*) (min) silicify

о·кремн·έни·е, -я, (*n*) silicification, silicifying

о·кремн·έ·ть, -ют, (*perf*) *see* о·кремн·ева́·ть

о·кремн·и́ть, -я́т, (*perf*) *see* о·кремн·я́·ть

о·кремн·я́·ть, -ют, (*imp*) silicify

о·крέп·нуть, -нут, (*perf*), (*past masc sing*) о·крέп·; become stronger, get strong again

о·крέст·ност·ь, -и, (*f*) neighbourhood, region, environs vicinity

о·крέст·н·ый, -ая, -ое, (*a*) neighbouring

о·кров·а́вленн·ый, -ая, -ое, (*past part pass*) bloody, bloodstained

о·кроп·и́ть, -я́т, (*perf*) *see* о·кроп·л·я́·ть

о·кропл·я́·ть, -ют, (*imp*) sprinkle

ό·круг·, -а, (*nom pl*) -а́, (*g pl*) -όв, (*m*) district; circuit

о·кругл·έни·е, -я, (*n*) (*v n*) *see* о·кругл·я́·ть; round-off

о·ши́б·к·а о·кругл·έни·я (*f*) (math) round-off error

тόч·к·а о·кругл·έни·я (*f*) (math) umbilical point

о·кругл·ённ·о-όстр·ый, -ая, -ое, (*a*) subacuate

о·кругл·ённ·о-угл·ова́т·ый, -ая, -ое, (*a*) subangular

о·кругл·ённ·ый, -ая, -ое, (*past part pass*) of о·кругл·и́ть; rounded-off, round

о·кругл·и́ть, -я́т, (*perf*) *see* о·кругл·я́·ть

о·кру́гл·ый, -ая, -ое, (*a*) rounded, round; (biol) orbicular

о·кругл·я́·ть, -ют, (*imp*) round off, make round

о·круж·а́·ть, -ют, (*imp*) surround, encircle, ring, enclose

о·круж·а́ющ·ий, -ая, -ее, (*pres part act*) of о·круж·а́·ть; surrounding, ambient, peripheral

~ вόз·дух· (*m*) ambient air, surrounding air

~ сред·а́ (*f*) surrounding medium, environment

~ у·слόв·и·я (*pl*) environmental conditions

о·круж·éни·е, -я, (*n*) (*v n*) *see* **о·кру·ж·áть;** environment, environs, surroundings; encirclement; circle (of persons)

о·круж·и́ть, -áт, (*perf*) *see* **о·кру·ж·á·ть**

о·круж·н·óй, -áя, -óе, (*a*) of **ó·круг·** district

~ комитéт· (*m*) district committee

о·кру́ж·ност·ь, -и, (*f*) circumference, circle

~ в·пáд·ин dedendum circle (of gear wheel)

~, в·пи́с·анн·ая (math) incircle, inscribed circle

~ вы́·ступ·ов tip/addendum circle (of gear wheel)

~ голóв·ок· tip/addendum circle (gear wheel)

~, дел·и́тельн·ая pitch circle (gear wheel)

~ зéркал·а (mech eng) chamfer circle

~, нару́ж·н·ая addendum circle (of gear wheel)

~, на·чáль·н·ая pitch circle (of gear wheel)

~ нóж·ек· dedendum circle (gear wheel)

~, образ·у́ющ·ая (mech) rolling circle

~, о·пи́с·анн·ая circumscribed circle

~, ос·ев·áя (math) circular axis

~, основ·н·áя base circle (of gear wheel)

о·кру́ж·н·ый, -ая, -ое, (*a*) of **о·кру́ж·-ност·ь**

~ скóр·ост·ь (*f*) circular/circumferential velocity, peripheral speed

~ скóр·ост·ь кóнц·ов лóпаст·ей blade-tip velocity/speed (of helicopter)

о·крут·и́ть, -ят, (*perf*) *see* **о·кру́·ч·ива·ть**

о·кру́ч·ива·ть, -ют, (*imp*) wind round

окс·ази́н·, -а, (*m*) (chem) oxazine

окс·азóл·, -а, (*m*) oxazole

оксал·áт·, -а, (*m*) (chem) oxalate

~ урани́л·а uranium oxalate

оксал·и́л·, -а, (*m*) (chem) oxalyl

окс·ами́д·, -а, (*m*) oxamide

оксами́н·ов·ый, -ая, -ое, (*a*) oxamic

~ кисл·от·á (*f*) oxamic acid

окси- (*int component*) (chem) hydroxy-, oxy-

окси·акри́л·ов·ая кисл·от·á (*f*) hydroxy-acrylic acid

окси·альдеги́д·, -а, (*m*) hydroxyaldehyde

окси·ами́н·, -а, (*m*) hydroxamine

окси·антрахинóн·, -а, (*m*) (chem) oxyanthraquinone

окси·биóз·, -а, (*m*) (zool) oxybiosis

окси·гемо·глоби́н·, -а, (*m*) (biochem) oxyhaemoglobin

окси·гемó·метр·, -а, (*m*) (med) oximeter

окси·гемо·мéтр·и·я, -и, (*f*) (med) oximetry

окси·ген·áтор·, -а, (*m*) (med) oxygenizer

окс·и́д·, -а, (*m*) (*see also* **ó·кис·ь**) (chem) oxide

оксид·áз·а, -ы, (*f*) (biochem) oxidase

оксид·áци·я, -и, (*f*) *see* **о·кисл·éни·е;** oxidation

окси·дéндрон·, -а, (*m*) (bot) sourwood, sorrel tree, *Oxydendrum arboreum*

оксиди·мéтр·и·я, -и, (*f*) (chem) oxidimetry

оксид·и́ровани·е, -я, (*n*) oxidation, oxide coating

~, анóд·н·ое (met) anodic oxidation, anodizing

оксид·и́р·овать, -уют, (*imp and perf*) oxidize

окси́д·н·ый, -ая, -ое, (*a*) oxide; oxide-coated

оксидо·редукт·áз·а, -ы, (*f*) (chem) oxidoreductase

окси·ими́д·, -а, (*m*) hydroxyl imide

окси·кетóн·, -а, (*m*) hydroxy-ketone

окси·кисл·от·á, -ы́, (*f*) hydroxy-acid

~, четырёх·основ·н·ая tetrabasic hydroxy-acid

окси·ликви́д·, -а, (*m*) liquid oxygen explosive

окси·ликви́т·, -а, (*m*) = **окси·ликви́д·**

окси·люцифери́н·, -а, (*m*) (biochem) oxyluciferin

окс·и́м·, -а, (*m*) (chem) oxime

окси·мáсл·ян·ая кисл·от·á (*f*) hydroxy butyric acid

окси·мети́л·, -а, (*m*) methylol, hydroxymethyl

окси·метил·целлюлóз·а, -ы, (*f*) (phot) colourcoll, tilosa

окси́·метр·, -а, (*m*) (med) oximeter

окс·ими́д·, -а, (*m*) oximide

оксим·и́ровани·е, -я, (*n*) (chem) oximation

окс·и́н·, -а, (*m*) (chem) oxine, oxyn

окс·индóл·, -а, (*m*) (chem) oxindole

окси·нитри́д·, -а, *(m)* (chem) hydroxy nitrile

окси·олеи́н·ов·ая кисл·от·а́ *(f)* hydroxyoleic/ricinoleic acid

окси·проли́н·, -а, *(m)* (chem) hydroxyproline

окси·тетра·цикли́н·, -а, *(m)* (pharm) oxytetracycline

окси·угле·ро́д·ист·ый, -ая, -ое, *(a)* carboxyl, carboxy-

~ гемо·глоби́н· *(m)* carboxy-haemoglobin, carbon monoxide-haemoglobin

окси·у́ксус·н·ая кисл·от·а́ *(f)* hydroxy-acetic acid

окси·фе́р·, -а, *(m)* (met) ferrite

окси·фи́ль·н·ый, -ая, -ое, *(a)* (biol) oxyphylic

окси·хиноли́н·, -а, *(m)* hydroxyquinoline

окси·хлори́д·, -а, *(m)* (chem) oxychloride, chloride

~ цирко́н·и·я zirconyl chloride

окси·хромати́н·, -а, *(m)* (cyt) oxychromatin

окси·целлюло́з·а, -ы, *(m)* oxycellulose

окси·янта́р·н·ая кисл·от·а́ *(f)* malic acid

оксо- *(int component)* (chem) охо-

оксо́н·иев·ый, -ая, -ое, *(a)* (chem) oxonium

~ со·един·е́ни·е *(n)* (chem) oxonium compound

оксо́н·и·й, -я, *(m)* (chem) oxonium

оксо·со·един·е́ни·е, -я, *(n)* (chem) oxo-compound

о́ксфорд·ск·ий, -ая, -ое, *(a)* Oxford

~ един·и́ц·а *(f)* (pharm) Oxford unit

окс·эти́л·, -а, *(m)* ethoxy

окта́в·а, -ы, *(f)* octave

окта·дека́н·ов·ая кисл·от·а́ *(f)* octadecanoic acid

окта·деканои́л·, -а, *(m)* octadecanoyl

окта·деце́н·ов·ая кисл·от·а́ *(f)* octadecenic acid

окта·деци́л·, -а, *(m)* octadecyl

окта·дие́н·ов·ая кисл·от·а́ *(f)* octadienoic acid

окта́ль·н·ый, -ая, -ое, *(a)* octal

окта·мети́л·, -а, *(m)* (chem) octamethyl

окт·а́н·, -а, *(m)* (chem) octane

окта·нафте́н·, -а, *(m)* (chem) cyclooctane

окта́н·ов·ый, -ая, -ое, *(a)* of **окт·а́н·**

~ числ·о́ *(n)* (oil) octane number

окт·а́нт·, -а, *(m)* (biol) octant

окта́·эдр·, -а, *(m)* (math) octahedron

окта·эдр·и́т·, -а, *(m)* (min) octahedrite

окта·эдр·и́ческ·ий, -ая, -ое, *(a)* octahedral

окт·е́т·, -а, *(m)* octet

окт·и́л·, -а, *(m)* (chem) octyl

окт·о́д·, -а, *(m)* (elec) octode

окт·о́з·а, -ы, *(f)* (biochem) octose

окто·пло́ид·, -а, *(m)* (gen) octoploid

октупо́л·ь, -я, *(m)* (phys) octupole, octopole

октя́бр·ь, -я́, *(m)* October

октя́брь·ск·ий, -ая, -ое, *(a)* October

о·ку·ёт *(fut 3rd sing)* of **о·ков·а́ть**

о·ку́кл·иван·и·е, -я, *(n)* (ent) pupation

окул·иро́вк·а, -и, *(g pl)* **-вок·,** *(f)* (bot) inoculation, grafting, budding

о·культив·и́р·овать, -уют, *(perf)* see **о·культу́р·ива·ть**

о·культу́р·ива·ть, -ют, *(imp)* (agr) cultivate, till; tame, domesticate

о·культу́р·ить, -ят, *(perf)* see **о·культу́р·ива·ть**

окуля́р·, -а, *(m)* (opt) eyepiece, ocular

окуля́р·н·ый, -ая, -ое, *(a)* of **окуля́р·**

~ ка́мер·а *(f)* eyepiece camera, camera attachment (to a telescope, microscope etc.)

~ кольц·о́ *(n)* (opt) eyepiece ring, lens ring

окун·а́ни·е, -я, *(n)* dipping, dip

окун·а́·ть, -ют, *(imp)* dip, plunge, immerse

окуне·о́браз·н·ые, -ых, *(pl decl as adj)* (zool) *Perciformes*

окун·у́ть, -у́т, *(perf)* see **окун·а́·ть**

о́кун·ь, -я, *(nom pl)* **-и,** *(g pl)* **-е́й,** *(m)* (fish) perch, *Perca fluviatilis*

~, мор·ск·о́й (fish) rosefish, *Sebastes marinus*

о·куп·а́емост·ь, -и, *(f)* (fin) cover of expenditure

о·куп·а́·ть, -ют, *(imp)* compensate, repay, pay the expenses/cost of; **-ся** pay, be worthwhile

о·куп·и́ть, -ят, *(perf)* see **о·куп·а́·ть**

о·ку́р·ива·ть, -ют, *(imp)* fumigate; cure, smoke

о·ку́р·енн·ый, -ая, -ое, *(past part pass)* of **о·кур·и́ть**

о·кур·и́ть, -ят, *(perf)* see **о·ку́р·ива·ть**

о·ку́р·ок·, -р·к·а, *(m)* cigarette-end; cigar-stub

о·ку́ск·ован·и·е, -я, *(n)* pelletizing

о·кýч·ивани·е, -я,   (*n*) ridging, earthing-up

о·кýч·ник·, -а,   (*m*) (agr) ridger, ridging plough

о·ку·ю́т,   (*fut 3rd pl*) of о·ков·áть

окципитáль·н·ый, -ая, -ое,   (*a*) (zool) occipital

олáдь·я, -и,   (*g pl*) -ди·й, (*f*) pancake

олеáндр·, -а,   (*m*) (bot) oleander, *Nerium oleander*

олеандр·и́н·, -а,   (*m*) (chem) oleandrin

олеáт·, -а,   (*m*) (chem) oleate

~ свин·ц·á   (chem) lead oleate

о·лед·енéл·ый, -ая, -ое,   (*a*) *see* о·лед·енённ·ый

о·лед·енéни·е, -я,   (*n*) (*v n*) *see* о·лед·енé·ть; (geol) glaciation

о·лед·енённ·ый, -ая, -ое,   (*past part pass*) of о·лед·ен·и́ть; frozen over, covered with ice; (geol) glaciated

о·лед·енé·ть, -ют,   (*perf*) freeze, be covered with ice

о·лед·ен·и́ть, -я́т,   (*perf*) freeze, freeze over, cover with ice

олейн·, -а,   (*m*) (chem) olein

олеиново·ки́сл·ый, -ая, -ое,   (*a*) oleate (of)

олейн·ов·ый, -ая, -ое,   (*a*) oleic

~ кисл·от·á   (*f*) oleic acid

олене·вóд·ств·о, -а,   (*n*) deer/reindeer husbandry/breeding

олéн·ий, -ья, -ье,   (*a*) deer, cervine

~ кóж·а   (*f*) buckskin

олéн·ин·а, -ы,   (*f*) venison; reindeer meat; buckskin, deerskin

олéн·ь, -я,   (*m*) (zool) deer, reindeer; (*pl*) *Cervidae*

~, америкáн·ск·ий   (zool) caribou, *Rangifer*

~, блáго·рóд·н·ый   red deer, *Cervus elaphus*

~, канáд·ск·ий   (zool) elk, wapiti, *Cervus canadensis*

~, сéвер·н·ый   reindeer, *Cervus tarandus*

олео·грáф·и·я, -и,   (*f*) (print) oleography; oleograph

олео·тóракс·, -а,   (*m*) (med) oleothorax

óле·ум·, -а,   (*m*) (chem) oleum, fuming sulphuric acid

оле·фи́н·, -а,   (*m*) (chem) olefin, olefine

оле·фи́н·ов·ый, -ая, -ое,   (*a*) olefin, olefinic

оли́в·а, -ы,   (*f*) (bot) olive

оливени́т·, -а,   (*m*) (min) olivenite

оливи́н·, -а,   (*m*) (min) olivine

оли́в·к·а, -и,   (*g pl*) -в·ок·, (*f*) olive

оли́в·ков·ый, -ая, -ое,   (*a*) olive, olive-coloured, olive-green

олиг·еми́·я, -и,   (*f*) (med) oligaemia, oligemia

олиго·динам·и́ческ·ий, -ая, -ое, (*a*) oligodynamic

олиго·клáз·, -а,   (*m*) (min) oligoclase

олиго·мéр·н·ый, -ая, -ое,   (*a*) (bot) oligomerous

олиго·ми́кт·ов·ый, -ая, -ое,   (*a*) oligomictic

олиго·трóф·н·ый, -ая, -ое,   (*a*) (geog) oligotrophic

олиго·цéн·, -а,   (*m*) (geol) Oligocene

оли́ф·а, -ы,   (*f*) drying oil

óлов·о, -а,   (*n*) tin, stannum, Sn

~, дву·сер·ни́ст·ое   stannic sulphide, tin disulphide

~, дву·хлóр·ист·ое   stannous chloride, tin bichloride

  зá·кис·ь óлов·а   stannous oxide, tin oxide

  ó·кис·ь óлов·а   stannic oxide, tin oxide

~, рóс·сып·н·ое   (min) stream tin

~, сер·ни́ст·ое   stannous sulfide, tin monosulphide

~, хлóр·ист·ое   stannous chloride, tin bichloride

~, хлóр·н·ое   stannic chloride, tin tetrachloride

~, четырёх·хлóр·ист·ое   stannic chloride, tin tetrachloride

олово·алки́л·, -а,   (*m*) alkyl tin

олово·диэти́л·, -а,   (*m*) tin diethyl/ethide

олово·нóс·ност·ь, -и,   (*f*) tin content

олово·нóс·н·ый, -ая, -ое,   (*a*) (geol) stanniferous, tin-bearing; (met) tin-containing

олово·орган·и́ческ·ий, -ая, -ое,   (*a*) (chem) organostannic, organotin

олово·тетраметри́л·, -а,   (*m*) tetramethyl tin

олово·тетраэти́л·, -а,   (*m*) tin tetramethide/tetraethyl

олово·триэти́л·, -а,   (*m*) tin triethyl

олово·эти́л·, хлóр·ист·ый   (*m*) ethyl-tin-chloride

олов·я́нист·ый, -ая, -ое,   (*a*) tin, stannous

~ брóнз·а   (*f*) (met) tin bronze

~ кисл·от·á   (*f*) stannous acid

оловянно·ки́сл·ый, -ая, -ое,   (*a*) stannate (of)

~ нáтр·и·й   (*m*) sodium stannate

оловянно·фтор·ист·ый, -ая, -ое, (*a*) fluostannate (of)

оловянно·хлор·ист·ый, -ая, -ое, (*a*) chlorostannate

олов·янн·ый, -ая, -ое, (*a*) tin, (chem) stannic

~ **ангидрид·** (*m*) tin peroxide/anhydride

~ **ка́мен·ь** (*m*) (min) tin stone, cassiterite

~ **кисл·от·а́** (*f*) (chem) stannic acid

~ **колчеда́н·** (*m*) (min) stannite

~ **про·тра́в·а** (*f*) tin mordant

~ **со́л·ь** (*f*) hydrated stannous chloride, tin salt

~ **фо́льг·а** (*f*) tin foil

~ **чум·а́** (*f*) (met) tin disease/pest

~ **я́зв·а** (*f*) (met) tin disease/pest

олу́ш·а, -и, (*f*) (orn) gannet, *Sulidae*

ольдгами́т·, -а, (*m*) (min) oldhamite

Ольдге́йм·а, му́фт·а (*f*) (mech) Oldham's coupling

ольх·а́, -и́, (*nom pl*) **о́льх·и, (*f*)** (bot) alder, *Alnus*

~, **кра́с·н·ая** (bot) red alder, *Alnus rubra*

~, **се́р·ая** (bot) speckled alder, *Alnus incana*

оля́пк·а, -и, (*f*) (orn) dipper, *Cinclus cinclus*

ольфактори·ме́тр·и·я, -и, (*f*) odorimetry

ом; (*g sing*) о́м·а, (*g pl*) о́м·ов, (*m*) ohm

~, **акуст·и́ческ·ий** acoustical ohm

О́м·а, за·ко́н· (*m*) (elec) Ohm's law

ома́р·, -а, (*m*) (zool) lobster, *Homarus vulgaris*

омбил·и́ческ·ий, -ая, -ое, (*a*) umbilical

~ **то́ч·к·а** (*f*) (math) umbilical point

омбро́·метр·, -а, (*m*) (meteor) ombrometer, rain-gauge

оме́г·а, -и, (*f*) omega Ω, ω (Greek)

омега·тро́н·, -а, (*m*) (nucl) omegatron

о·мед·не́ни·е, -я, (*n*) coppering, copper-plating/coating

о·мед·нённ·ый, -ая, -ое, (*n*) copper-plated

оме́л·а, бе́л·ая (*f*) (bot) European mistletoe, *Viscum album*

о·мертв·е́л·ый, -ая, -ое, (*a*) deadened, (med) necrotic

о·мертв·е́ни·е, -я, (*n*) (med) necrosis, gangrene

о·мертв·е́·ть, -ют, (*perf*) grow numb/deadened

о·мертв·и́ть, -я́т, (*perf*) (med) necrotize; (fin) immobilize

о·мёт·, -а, (*m*) (agr) straw stack

омикро́н·, -а, (*m*) omicron, ο (Greek)

ом·и́ческ·ий, -ая, -ое, (*a*) (elec) ohmic, resistive

омлёт·, -а, (*m*) omlette

ом·ме́тр·, -а, (*m*) ohmmeter

о́мни·бус·, -а, (*m*) bus, omnibus

омни·ме́тр·, -а, (*m*) (surv) omnimeter

омнопо́н·, -а, (*m*) (pharm) omnopon

о́м·овск·ий, -ая, -ое, (*a*) ohmic

о·мо́·ет (*fut 3rd sing*) of **о·мы́·ть**

о·мозол·е́л·ый, -ая, -ое, (*a*) callous

о·мола́ж·ива·ть, -ют, (*imp*) rejuvenate

о·молод·и́ть, -я́т, (*perf*) see **о·мола́·ж·ива·ть**

о·молож·е́ни·е, -я, (*n*) rejuvenation

омо́ним·, -а, (*m*) (ling) homonym

о·моноли́ч·ивани·е, -я, (*n*) *in situ* casting (concrete)

о·мо́·ют (*fut 3rd pl*) of **о·мы́·ть**

о́мул·ь, -я, (*g pl*) -е́й, (*m*) (fish) omul, *Salmo omul*

о́мут·, -а, (*m*) pool, whirlpool

омфал·и́т·, -а, (*m*) (med) omphalitis

омфал·эктом·и́·я, -и, (*f*) (med) omphalectomy

омфало·том·и́·я, -и, (*f*) (med) omphalotomy

омфаци́т·, -а, (*m*) (min) omphacite

о·мыв·а́ни·е, -я, (*n*) (*v n*) see **о·мыв·а́·ть**

о·мыв·а́·ть, -ют, (*imp*) wash, bathe; flow around/over

о·мыл·е́ни·е, -я, (*n*) (chem) saponification

числ·о́ о·мыл·е́ни·я (*n*) (chem) saponification number

о·мыл·я́ющ·ее вещ·еств·о́ (*n*) saponifying agent

о·мы́·ть, (*perf*) (*fut 3rd pl*) о·мо́·ют, see **о·мыв·а́·ть**

он (*pron*) he; it (*of nouns masculine in Russian and neuter in English*); (telecom) letter O

он·а́ (*pron*) she; it (*of nouns feminine in Russian and neuter in English*)

она́грик·ов·ые, -ых, (*pl decl as adj*) (bot) *Onagraceae, Oenotheraceae*

о́нгстрем·, -а, (*m*) (phys) angström, Å

онда́тр·а, -ы, (*f*) (zool) muskrat, *Ondatra/Fiber zibethicus*

ондо́·граф·, -а, (*n*) (elec) ondograph

ондо·метр·и́ческ·ий, -ая, -ое,   (*a*) ondometric

ондул·я́тор·, -а,   (*m*) (telecom) undulator; automatic tape recorder

о·нем·е́·ть, -ют,   (*perf*) grow/become dumb

он·и́   (*pron*) they

они·ев·ый, -ая, -ое,   (*n*) (chem) onium

о́никс·, -а,   (*m*) (min) onyx

оних·атроф·и́·я, -и,   (*f*) (med) onychatrophia

оних·и́т·, -а,   (*m*) (med) onychitis; (min) onychite

онихо·дин·и́·я, -и,   (*f*) (med) onychodynia

онко́·лог·, -а,   (*m*) oncologist

онко·ло́г·и·я, -и,   (*f*) (med) oncology

онко́ль·н·ый, -ая, -ое,   (*a*) (fin) on-call

~ с·чёт·   (*m*) (fin) on-call account

онко·сфе́р·а, -ы,   (*f*) (zool) oncosphere, onchosphere

онк·оти́ческ·ий, -ая, -ое,   (*a*) (med) oncotic

он·о́   (*pron*) it

ономати́п·, -а,   (*m*) (biol) onomatype

ономато·пе́·я, -и,   (*f*) (ling) onomatopoeia

Он сагер·а, со·от·нош·е́ни·я взаи́м·ост·и   (*pl*) (phys) Onsager reciprocity theorem

Онсагер·а, теоре́м·а   (*f*) (phys) Onsager theorem, Onsager's reciprocity theorem

Онсагер·а, у·равн·е́ни·е   (*n*) (phys) Onsager's formula

онто·гене́з·, -а,   (*m*) (biol) ontogenesis

онто·ло́г·и·я, -и,   (*f*) (biol) ontology

оо·гене́з·, -а,   (*m*) (gen) oogenesis

оо·го́н·и·й, -я,   (*m*) (bot) oogonium

оо·ли́т·, -а,   (*m*) (geol) oolite, oolith

оо·ли́т·ов·ый, -ая, -ое,   (*a*) oolitic

оо·мице́т·, -а,   (*m*) (bot) oomycete

оо·спо́р·а, -ы,   (*f*) (biol) oospore

о·па́·вш·ий, -ая, -ее,   (*past part act*) of **о·па́с·ть,** (*perf*); *see* **о·пад·а́·ть** (*imp*)

о·пад·а́·ть, -ют,   (*imp*) shed (leaves etc.); fall off; subside

о·пад·а́ющ·ий, -ая, -ее,   (*pres part act*) of **о·пад·а́·ть**; (bot) fugacious

о·пад·е́ни·е, -я,   (*n*) shedding; subsidence; falling/dropping off

~ ли́ст·ьев   (bot) defoliation

о·па́зд·ывани·е, -я,   (*n*) (*v n*) of **о·па́зд·ыва·ть;** delay, retardation

о·па́зд·ыва·ть, -ют,   (*imp*) be late; be slow (of clocks etc.)

о·па́·ива·ть, -ют,   (*imp*) solder

опак·иллюмина́тор·, -а,   (*m*) vertical illuminator (microscope)

опа́к·ов·ый, -ая, -ое,   (*a*) opaque

о·па́·л   (*past masc sing*) of **о·па́с·ть** (*perf*), *see* **о·пад·а́·ть** (*imp*)

опа́л·, -а,   '(*m*) (min) opal

о·пал·ённ·ый, -ая, -ое,   (*past part pass*) *see* **о·па́л·ива·ть**

опал·есце́нци·я, -и,   (*f*) (phys) opalescence

~, звёзд·н·ая   stellate opalescence

опал·есци́р·овать, -уют,   (*imp and perf*) opalesce

о·па́л·ивани·е, -я,   (*n*) (*v n*) *see* **о·па́л·ива·ть**

о·па́л·ива·ть, -ют,   (*imp*) singe, scorch; bream (a ship)

опали́з·овать, -уют,   (*imp and perf*) opalize

о·пал·и́ть, -я́т,   (*perf*) *see* **о·па́л·ива·ть**

опа́л·ов·ый, -ая, -ое,   (*a*) opal, opaline

о·па́л·очн·ый, -ая, -ое,   (*a*) singeing, scorching

~ печ·ь   (*f*) singer (meat processing)

о·па́луб·ить, -ят,   (*perf*) (build) put up shuttering/formwork/casing; deck

о·па́луб·к·а, -и,   (*f*) shuttering, formwork, casing (for concrete); decking

~, термо·акти́в·н·ая   electrically heated shuttering

о·па́луб·очн·ый, -ая, -ое,   (*a*) of **о·па́луб·к·а**

о·па́л·ыва·ть, -ют,   (*imp*) weed round

о·пал·я́·ть, -ют,   (*imp*) = **о·па́л·ива·ть**

опане́р   (*indecl*) up and down (of anchor cable), a'peak (of anchor)

о·па́р·а, -ы,   (*f*) leavened dough

о·пас·а́·ться, -ются   (*imp*) (+ *gen*) be apprehensive, fear; avoid

о·пас·е́ни·е, -я,   (*n*) apprehension, misgiving, fear

о·па́с·ност·ь, -и,   (*f*) danger, peril, hazard, risk

о·па́с·н·ый, -ая, -ое,   (*a*) dangerous, perilous, hazardous

~ рас·сто·я́ни·е   (*n*) danger area

~ сеч·е́ни·е   (*a*) (civ eng) section liable to greatest stress

о·па́с·ть,   (*fut 3rd pl*) **о·пад·у́т,** (*perf*), *see* **о·пад·а́·ть**

о·пах·а́л·о, -а,   (*n*) whisk, fan; vane (of feather)

о‧пах‧а́ть, (*fut 3rd pl*) о‧па́ш‧ут, (*perf*) plough round

о‧па́х‧ива‧ть, -ют, (*imp*) plough round; swirl round, envelop

о‧пах‧н‧у́ть, -у́т, (*perf*) swirl round, envelop

о‧па́ш‧ут (*fut 3rd pl*) of о‧пах‧а́ть

о́пенер‧, -а, (*m*) (text) opener

опёнок‧, на‧сто‧я́щ‧ий (*m*) (biol) *Armillaria mellea*

о‧пёр‧ (*past masc sing*) of о‧пер‧е́ть

о́пер‧а, -ы, (*f*) opera

операти́в‧н‧ый, -ая, -ое, (*a*) operational; operative, operating

~ в‧меш‧а́тельств‧о (*n*) surgical intervention

~ дежу́р‧н‧ый (*m*) (mil) duty operations officer

~ от‧де́л‧ (*m*) operations section/department

~ план‧и́ровани‧е (*n*) scheduling

~ о‧во́д‧к‧а (*f*) situation report

опера́тор‧, -а, (*m*) operator; (cinema) cameraman; (met roll) the roller; instruction (computer)

~ воз‧мущ‧е́ни‧я (phys) perturbation operator

~ Гами́льто́н‧а (math) Hamiltonian, Hamiltonian operator

~, дифференциа́ль‧н‧ый (math) differential operator

~, маши́н‧н‧ый (autom) machine operator

~, обра́т‧н‧ый (math) inverse operator

~ пере‧стано́в‧к‧и координа́т‧ (nucl) space-exchange operator

~, ра́з‧ностн‧ый (math) difference operator

~, со‧пряж‧ённ‧ый (math) adjoint operator

~ спи́н‧а (nucl) spin operator

опера́тор‧н‧ый, -ая, -ое, (*a*) of опера́тор‧

~ за́‧пис‧ь (*f*) (math) differential operator form

опера́тор‧ск‧ий, -ая, -ое, (*a*) (cinema) cameraman's, camera

~ кран‧ (*m*) (cinema) camera boom/crane

~ теле́ж‧к‧а (*f*) (cinema) perambulator

операцио́н‧н‧ая, -ой, (*f decl as adj*) operations room; (med) operating theatre

операцио́н‧н‧ый, -ая, -ое, (*a*) (*see also* операцио́н‧н‧ая) (*as noun*); operations, operational

операцио́н‧н‧ый, -ая, -ое

~ контро́л‧ь (*m*) working inspection

~ рас‧хо́д‧ы (*pl*) running costs/expenses

опера́ци‧я, -и, (*f*) operation

о‧перед‧и́ть, -я́т, (*perf*) *see* о‧переж‧а́ть

о‧переж‧а́ть, -ют, (*imp*) outstrip, get ahead, lead

о‧переж‧е́ни‧е, -я, (*n*) (*v n*) *see* о‧переж‧а́ть; (tech) advance; (met roll) forward slip/creep, extrusion effect, speed gain

~ в‧пры́ск‧а (ICE) injection advance

~ за‧жиг‧а́ни‧я (ICE) ignition advance

~ за‧кры́т‧и‧я кла́пан‧а (ICE) valve lead

~ от‧кры́т‧и‧я кла́пан‧а (ICE) valve lead

~ по‧да́‧ч‧и то́пл‧ив‧а (ICE) injection advance

у́гол‧ о‧переж‧е́ни‧я (*m*) angle of advance/lead

у́гол‧ о‧переж‧е́ни‧я за‧жиг‧а́‧ни‧я (ICE) advance of ignition

о‧пер‧е́ни‧е, -я, (*n*) (zool) plumage, feathers; (a/c) tail, tail unit; (M/T) body accessories; (fish) fins and rudders

~, вертика́ль‧н‧ое (a/c) fin-rudder unit

~, горизонта́ль‧н‧ое (a/c) stabilizer-elevator unit

~, трёх‧кил‧ев‧о́е (a/c) three-fins tail

~ у‧скор‧и́тел‧я (rocket) booster fins

~, хвост‧ов‧о́е (a/c) tail, tail unit

~, цельно‧по‧воро́т‧н‧ое (a/c) all-flying tail

о‧пер‧е́ть, (*fut 3rd pl*) обо‧пр‧у́т, (*past masc sing*) о‧пёр‧, (*perf*); *see* о‧пир‧а́ть

о‧пер‧и́вш‧ийся, -аяся, -ееся, (*past part act*) (orn) plumose, fledged

опер‧и́р‧овать, -уют, (*imp and perf*) operate

о́пер‧н‧ый, -ая, -ое, (*a*) of о́пер‧а

о‧пёр‧т‧ый, -ая, -ое, (*past part pass*) of о‧пер‧е́ть

о‧пёр‧ш‧ий, -ая, -ее, (*past part act*) of о‧пер‧е́ть

о‧пер‧я́‧ться, -ются, (*imp*) fledge, feather

о‧песко‧стру́‧ивани‧е, -я, (*n*) sand blasting

о‧печа́т‧а‧ть, -ют, (*perf*) *see* о‧печа́‧т‧ыва‧ть

о·печа́т·к·а, -и,   (g pl) -т·ок·,   (f) (print) misprint, erratum

о·печа́т·ыва·ть, -ют, (imp) seal up/off

о́п·и·й, -я,   (m) (pharm) opium

о·пи́л·ивани·е, -я,   (n) (v n) see о·пи́·л·ива·ть

о·пи́л·ива·ть, -ют,   (imp) saw off, shorten, trim (by sawing); file

о·пил·и́ть, -ят, (perf) see о·пи́л·ива·ть

о·пи́л·к·и, -л·ок·, (f pl) filings, turnings; sawdust

о·пил·бето́н·, -а, (m) sawdust concrete

о·пил·о́вк·а, -и, (f) (v n) see о·пи́л·и·ва·ть

о·пил·о́вочн·о-за·чист·н·о́й стан·о́к· (m) (mech eng) filing machine

о·пир·а́·ть, -ют, (imp) lean; -ся lean on, be supported by

о·пис·а́ни·е, -я,   (n) description; list; definition

~ огн·е́й (nav) list of lights

о·пи́с·анн·ый, -ая, -ое, (past part pass) of о·пис·а́ть

о·пис·а́тельн·ый, -ая, -ое, (a) descriptive

о·пис·а́ть,   (fut 3rd pl) о·пи́ш·ут, (perf), see о·пи́с·ыва·ть

о·пи́с·к·а, -и, (g pl) -с·ок·, (f) slip of the pen

о́·пис·н·ый, -ая, -ое, (a) of о́·пис·ь

~ су́д·н·о (n) surveying vessel

описто·со́м·а, -ы, (f) (zool) opisthosoma

описто·де́т·н·ый, -ая, -ое, (a) (zool) opisthodetic

описто·це́ль·н·ый, -ая, -ое, (a) (zool) opisthocoelous

о·пи́с·ыва·ть, -ют,   (imp) describe, depict, portray; list, make (an inventory etc.); (math) circumscribe, describe, draw round; -ся (pass); make a slip (in writing)

о́·пис·ь, -и, (f) inventory, list, schedule

о·пи́ш·ут (fut 3rd pl) of о·пис·а́ть

о·плавл·е́ни·е, -я, (n) (weld) flashing off; sweating (refractories); polishing (of metal surfaces by heat)

о·пла́т·а, -ы, (f) (v n) see о·пла́ч·и·ва·ть; pay, payment, remuneration

~, за́·работ·н·ая wages, salary

~, по·време́н·н·ая time-work payment

~, премиа́ль·н·ая premium-bonus payment

~, с·де́ль·н·ая piecework payment

о·плат·и́ть, -ят, (perf) see о·пла́ч·и·ва·ть

о·пла́ч·ива·ть, -ют, (imp) pay, pay for; repay, reimburse; settle (accounts etc.)

о·плес·ти́,   (fut 3rd pl) о·плет·у́т, (past masc sing) о·плё·л, (perf); see о·плет·а́·ть

о·плет·а́·ть, -ют, (imp) weave around, braid with; entwine

о·плёт·к·а, -и, (g pl) -т·ок·, (f) braid, braiding; (naut) pointing (rope)

~, тата́р·ск·ая point (cordage)

о·плёт·очн·ый, -ая, -ое, (a) of о·п·лёт·к·а

~ стан·о́к· (m) braider (for electric cable)

о·плодо·твор·е́ни·е, -я, (n) (v n) see о·плодо·твор·я́·ть; fertilization

~, близко·ро́д·ственн·ое (gen) inbreeding, inzucht

~, пере·крёст·н·ое (gen) amphimixis

о·плодо·твор·и́ть, -я́т,   (perf) see о·плодо·твор·я́·ть

о·плодо·твор·я́·ть, -ют, (imp) fertilize, make fertile, engender

о·пломб·ир·ова́ть, -у́ют, (perf) seal

о·пло́т·, -а, (m) stronghold, bulwark

о·пло́т·ник·, -а, (m) raftman

оплош·ност·ь, -и, (f) error, mistake

о·плыв·а́·ть, -ют,   (imp) swim/drift/float/sail round/around; melt, crumble, be swept away; splutter (of flames etc.)

о·плыв·и́н·а, -ы, (f) (geol) mud stream

о·плы́·ть,   (fut 3rd pl) о·плыв·у́т, (perf); see о·плыв·а́·ть

о·плюс·о́ванн·ый, -ая, -ое,   (a) padded

о·по·вест·и́тель·, -я, (m) (instr) notifier, announcer

~, по·езд·н·о́й (rail) train describer

~, централи́з·ованн·ый (rail) train describer

о·по·вест·и́ть, -я́т, (perf) see о·по·вещ·а́·ть

о·по·вещ·а́·ть, -ют, (imp) announce, notify

о·по·вещ·е́ни·е, -я, (n) (v n), see о·по·-вещ·а́·ть; announcement, notification; (mil) early warning (radar etc.)

оподельдо́к·, -а, (m) opodeldoc (ointment)

о·под·зо́л·ивани·е, -я,   (n) podsol development (of soil)

опо́·ек·, -о·йк·а,   (m) calf skin; calf leather, calf

о·позд·а́ни·е, -я, (n) (v n) of о·позд·а́·ть; delay, lateness, retardation

~ на́ час· an hour late

о·позд·а́·ть, -ют, (*perf*) be late

о·по·знав·а́ни·е, -я, (*n*) identification, recognition

~ св·о́й-чуж·о́й (radar) IFF **ста́нци·я о·по·знав·а́ни·я** (*f*) IFF radar

о·по·знав·а́тел·ь, -я, (*m*) identifier

о·по·знав·а́тельн·ый, -ая, -ое, (*a*) identification

~ знак· (*m*) identification mark

о·по·зна·ва́ть, -ют, (*imp*) identify, recognize

о·по·зна́·ть, -ют, (*perf*) *see* **о·по·зна·-ва́ть**

о·по·и́ть, -я́т, (*perf*) give too much to drink

о·по́й·ков·ый, -ая, -ое, (*a*) calf-skin, calf-leather

опо́к·а, -и, (*f*) (geol) opoka, gaize; (met cast) moulding box/flask

~, ве́рх·н·яя (met cast) cope

~, ни́ж·н·яя (met cast) drag

о·пола́ск·ива·ть, -ют, (*imp*) rinse, swill out

о·полз·а́·ть, -ют, (*imp*) crawl round/ around; slump, slip, slide, settle

о́·полз·ен·ь, -з·н·я, (*m*) landslide, landslip, creep

~, под·во́д·н·ый (geol) subaqueous slumping

о́·полз·нев·ый, -ая, -ое, (*a*) of **о́·полз·ен·ь;** (geol) landslide, slip, slide

о·полз·ти́, -у́т, (*perf*) *see* **о·полз·а́·ть**

о·полос·н·у́ть, -у́т, (*perf*) rinse, swill out

о·пол·о́ть, -ют, (*perf*) weed round

о·полч·е́ни·е, -я, (*n*) (mil) home guard

о·по́·мн·иться, -ятся, (*perf*) come to one's senses, pull oneself together

о·по́р·, во весь at full speed, at top speed

о·по́р·а, -ы, (*f*) support, prop; (mech) foot; pier (of bridge); pylon, tower, pole; (rad) stub

~, а́нкер·н·ая anchor support/tower (for overhead lines)

~, бере́г·ов·а́я abutment (of bridges)

~, жёстк·ая (mech) fixed foot (of machinery)

~, кач·а́ющ·аяся swing support

~, кле́т·очн·ая crib pier (of bridge)

~, ко́зл·ов·ая scaffold trestle

~, коло́н·ков·ая pillar support

~, конц·ев·а́я abutment (of bridge); dead-end tower (for overhead lines)

~, мёртв·ая (mech) fixed foot

о·по́р·а

~, не·пло́ск·остн·ая locating pin/ stop (of machine tools)

~, не·по·дви́ж·н·ая (mech) fixed foot

~, нож·ев·а́я knife-edge support

~ о·пра́в·к·и arbor support (of milling machine)

~, пло́ск·остн·ая plane-locating pad/ button (of machine tools)

~, по·дви́ж·н·ая (mech) sliding foot

~, полу·жёстк·ая (mech) sliding foot

~, пруж·и́нн·ая sprung support

~, ра́м·н·ая single-bent pier (of bridge)

~, ро́л·иков·ая supporting idler (of conveyors)

~, само·у·стана́вл·ивающ·аяся self-centring locating jig

~, сва́й·н·ая pile pier (of bridge) **то́ч·к·а о·по́р·ы** (*f*) fulcrum (of lever)

~, угл·ов·а́я про·меж·у́точн·ая angle-suspension tower (for overhead lines)

~, четверть·волн·ов·а́я (rad) quarter-wave stub

о·пора́жн·ивани·е, -я, (*n*) (*v n*) of **о·пора́жн·ива·ть кла́пан· опора́жн·ивани·я** (*m*) running-down valve

о·пора́жн·ива·ть, -ют, (*imp*) empty, evacuate; discharge

о·по́р·н·ый, -ая, -ое, (*a*) of **о·по́р·а;** bearing, bearing on; reference

~ велич·ин·а́ (*f*) (autom) reference input

~ диск· (*m*) turntable (of gramophone)

~ кат·о́к· (*m*) track/bogey wheel/ roller (of a tracked tractor)

~ плит·а́ (*f*) bearing plate

~ пло́ск·ост·ь (*f*) (math) plane of support

~ по·ве́рх·ност·ь кла́пан·а (*f*) valve face

~ пункт· (*m*) (surv) reference point, datum; (mil) strong point, defended post

~ реа́кци·я (*f*) reaction, reactive force

~ сто́й·к·а (*f*) stanchion

о·поро́жн·енн·ый, -ая, -ое, (*past part pass*) of **о·порожн·и́ть;** emptied, evacuated

о·порожн·и́ть, -ят, (*perf*) *see* **о·по-ра́жн·ива·ть**

опо́ссум·, -а, (*m*) (zool) opossum

опо·терап·и́·я, -и, (*f*) (med) opotherapy

опо́ч·н·ый, -ая, -ое, (*a*) of **опо́к·а**

о·пошл·ени·е, -я,   (*n*) vulgarization, debasing

о·пояс·ать,   (*fut 3rd pl*) о·пояш·ут, (*perf*); *see* о·пояс·ыва·ть

о·пояс·к·а, -и,   (*g pl*) -с·ок·, (*f*) girdle; (civ eng) strengthened canal wall

о·пояс·н·ый, -ая, -ое,   (*a*) encircling, encompassing

о·пояс·ыва·ть, -ют,   (*imp*) girdle, ring, surround

о·пояш·ут (*fut 3rd pl*) of о·пояс·ать

оппозиц·и·я, -и,   (*f*) opposition

оппанол·, -а,   (*m*) (chem) oppanol

о·прав·а, -ы,   (*f*) setting, mounting, mount; rim (of spectacles etc.)

~, вы·движ·н·ая   (phot) slip-on mount

~, вы·ступ·ающ·ая   (phot) front mount

~ стекл·а   bezel (of a watch)

~, штык·ов·ая   (phot) locking ring mount

о·правд·ани·е, -я,   (*n*) (*v n*) *see* о·правд·ыва·ть; excuse; justification; (law) acquittal

о·правд·а·ть, -ют,   (*perf*) *see* о·правд·ыва·ть

о·правд·ыва·ть, -ют,   (*imp*) justify; excuse, plead; acquit (accused person), find for (a plaintiff)

о·прав·ить, -ят,   (*perf.*) *see* о·правл·я·ть

о·прав·к·а, -и,   (*g pl*) -в·ок·, (*v n*); *see* о·правл·я·ть; mandrel (tube working); arbor (machine tools); drift (riveting)

~, гиб·очн·ая   bending mandrel

~, одно·сторон·н·яя   stump mandrel

~, рас·точ·н·ая   boring bar

~, фрезер·н·ая   cutter arbor

о·правл·я·ть, -ют,   (*imp*) put/set right/in order

оправо·из·влек·ател·ь, -я,   (*m*) mandrel stripper (tube rolling)

о·прав·очн·ый, -ая, -ое,   (*a*) of о·п·рав·к·а

о·праш·ива·ть, -ют,   (*imp*) interrogate, examine

о·праш·ивающ·ий, -ая, -ее,   (*pres part act*) of о·праш·ива·ть

~ у·строй·ств·о   (*n*) (instr) reading device

о·пре·дел·ени·е, -я,   (*n*) (*v n*) of о·пре·дел·я·ть; determination; analysis, assay, test (for), definition, identification; (gram) attribute; (law) decision

~, вал·ов·ое   bulk analysis

~, вес·ов·ое   gravimetric analysis/determination

о·пре·дел·ени·е

~ вс·пыш·к·и   (oil) flash-point test

~ качеств·а   (agr) approbation; qualitative determination/analysis

~ мест·а   location; (nav) fix

~ мест·а по·врежд·ени·я   fault location

~, объ·ём·н·ое   volumetric determination/analysis

по о·пре·дел·ени·ю   by definition

~ радио·девиаци·и   (nav) D/F calibration

~ сорт·а   grade assessment

о·пре·дел·ённ·о   (*adv*) definitely

о·пре·дел·ённост·ь, -и,   (*f*) determinacy, definiteness

о·пре·дел·ённ·ый, -ая, -ое,   (*past part pass*) *see* о·пре·дел·я·ть; definite, specified, a certain/particular

~ интеграл·   (*m*) (math) definite integral

~ член,   (*m*) (gram) definite article

о·пре·дел·им·ый, -ая, -ое,   (*pres part pass*) *see* о·пре·дел·я·ть; determinable, determinate

о·пре·дел·ител·ь, -я,   (*n*) definer, fixer, (math) determinant; identification handbook/manual

~ на·правл·ени·я   direction finder

о·пре·дел·ить, -ят,   (*perf*) *see* о·пре·дел·я·ть

о·пре·дел·я·ть, -ют,   (*imp*) define, determine, fix, assign, diagnose (illness); -ся (*pass*) depend on

~ максимум· функци·и   (math) maximize a function

о·пре·дел·яющ·ий, -ая, -ее,   (*pres part act*) of о·пре·дел·я·ть; decisive

~ точ·к·а   (*f*) (nav) estimated position

о·пресн·ени·е, -я,   (*n*) (*v n*) *see* о·пресн·я·ть

о·пресн·ённ·ый, -ая, -ое,   (*past part pass*) *see* о·пресн·я·ть

~ вод·а   (*f*) distilled water

о·пресн·ител·ь, -я,   (*m*) water distilling plant

о·пресн·ить, -ят,   (*perf*) *see* о·пресн·я·ть

о·пресн·я·ть, -ют,   (*imp*) distil water

о·пресс·овани·е, -я,   (*n*) *see* о·пресс·овк·а

о·пресс·овк·а, -и,   (*g pl*) -вок·, (*f*) pressure testing (pipes, cylinders etc.); compression moulding/molding (plastics)

~ оболоч·к·ой   sheathing

~, пред·вар·ительн·ая   pre-moulding

о·пресс·о́вывани·е, -я, (*n*) pressurization

о·при·хо́д·овани·е, -я, (*n*) temporary storage (perishable stores); filing (of papers)

о·пробк·ове́ни·е, -я, (*n*) (bot) suberization

о·про́б·овани·е, -я, (*n*) (*v n*) *see* о·про́б·овать

~, борозд·ов·о́е (min) trenching

~, по·вто́р·н·ое resampling

~, то́ч·ечн·ое (min) pit sampling

~, шпур·ов·о́е (min) pit sampling

о·про́б·ова́тел·ь (*m*) tester, sampler

о·про́б·овать, -уют, (*imp and perf*) try out, test (before regular use); (min) assay, sample

о·про·верг·а́·ть, -ют, (*imp*) refute, disprove

о·про·ве́рг·н·уть, -ут, (*perf*) *see* о·про·верг·а́·ть

о·про·верж·е́ни·е, -я, (*n*) (*v n*) *see* о·про·верг·а́·ть; refutation; disavowal, denial

о·про·кид·н·о́й, -а́я, -о́е, (*a*) tipping

~ бадь·я́ (*f*) tipping bucket

о·про·ки́д·ывани·е, -я, (*n*) tipping; capsizing, inversion; reversal

~ фаз· phase inversion/reversal

о·про·ки́д·ыватель·ь, -я, (*m*) tipple, tippler, tilter; reverser, inverter

~, ваго́н·н·ый (rail) wagon tippler

~, лоб·ов·о́й end tipper

~ сигна́л·ов (rad) signal inverter

~ с·ли́т·к·ов (met) ingot tilter

о·про·ки́д·ыва·ть, -ют, (*imp*) overturn, upset, tip, topple; capsize, turn upside down, invert, reverse; (mil) overrun

о·про·ки́д·ывающ·ий, -ая, -ее, (*pres part act*) of о·про·ки́д·ыва·ть

~ моме́нт· (*m*) (mech) upsetting moment; driving moment; stalling torque (of elec. motor)

~ схе́м·а (*f*) (elec) phase inverter

о·про·ки́·нут·ый, -ая, -ое, (*past part pass*) of о·про·ки́·нут·ь; inverted, upset, overturned

о·про·ки́·нуть, -ут, (*perf*) *see* о·про·ки́д·ыва·ть

о·про·ме́т·чив·ый, -ая, -ое, (*a*) precipitate, inconsiderate, rash

о·про́с·, -а, (*m*) interrogation

о·прос·и́ть, -я́т, (*perf*) *see* о·пра́ш·ива·ть

о·про́с·н·ый, -ая, -ое, (*a*) interrogatory; answering (telephone)

о·прост·и́ть, -я́т, (*perf*) simplify

о·про́с·чик·, -а, (*m*) (*and see* за·про́с·чик·) interrogator

~ -от·ве́т·чик·, -а, (*m*) (rad) interrogator-responder

о·протест·ова́ть, -у́ют, (*perf*) lodge/enter/make a protest/appeal; (fin) protest (a bill)

о·прощ·а́·ть, -ют, (*imp*) simplify

о·пры́ск·а·ть, -ют, (*perf*) *see* о·пры́ск·ива·ть

о·пры́ск·иватель·ь, -я, (*m*) sprayer, sprinkler

о·пры́ск·ива·ть, -ют, (*imp*) spray, sprinkle

о·пря́т·н·ый, -ая, -ое, (*a*) neat, tidy

опсон·и́н·, -а, (*m*) (bact) opsonin, bacteriotropin

опсон·и́ческ·ий, -ая, -ое, (*a*) (bact) opsonic, bacteriotropic

о́пт·ик·, -а, (*m*) optician

о́пт·ик·а, -и, (*f*) optics; (instr) optical system

~, волн·ов·а́я physical optics

~, звуко·вос·про·из·вод·я́щ·ая (cinema) sound-head lens

~, луч·ев·а́я geometrical optics

~, про·светл·ённ·ая non-reflection optics

~, электро́н·н·ая electron optics

оптима́ль·н·о (*adv*) at an optimum

оптима́ль·н·ый, -ая, -ое, (*a*) optimum

~ крив·а́я (*f*) best performance curve

~ с·вяз·ь (*f*) (elec) optimum/critical coupling

~ систе́м·а (*f*) (autom) optimalizing/adaptive system

опти́·метр·, -а, (*m*) (instr) optimeter

оптимиз·а́ци·я, -и, (*f*) optimization

оптимист·и́ческ·ий, -ая, -ое, (*a*) optimistic

о́птимум·, -а, (*m*) optimum

~ вулканиз·а́ци·и optimum cure (of rubber)

опт·и́ческ·и (*adv*) optically

~ акти́в·н·ый (*a*) optically active

опт·и́ческ·ий, -ая, -ое, (*a*) optic, optical, visual

~ бел·я́щ·ее сре́д·ств·о (*n*) optical whitening agent

~ двой·н·ы́е звёзд·ы (*pl*) (astron) optical double

~ длин·а́ пут·и́ (*f*) optical path

~ об·ма́н· (*m*) optical illusion

~ центр· (*m*) optical centre

опт·и́чност·ь, -и,   (*f*) optical activity

опт·ови́к·, -а́,   (*m*) wholesale dealer/merchant

опт·о́в·ый, -ая, -ое,   (*a*) wholesale

о́пт·ом   (*adv*) wholesale

о·публик·ова́ни·е, -я,   (*n*) publication, promulgation

о·пу́др·ивани·е, -я,   (*n*) powdering, dusting

о·пу́нци·я, -и,   (*f*) (bot) opuntio, *Opuntio*

о·пуск·а́ни·е, -я,   (*n*) (*v n*) *see* о·пуск·а́·ть; (math) drop; (geol) subsidence, sinking, settling

о·пуск·а́·ть, -ют,   (*perf*) lower, let/put/pull down; insert (coin in slot etc.), mail, post (letters etc.); omit, leave out; -ся go down, fall, sink, subside; slacken (of ropes etc.)

~ в во́д·у   immerse in water

о·пуск·н·о́й, -а́я, -о́е,   (*a*) drop, vertically adjustable

~ двер·ь   (*f*) trapdoor

~ окн·о́   (*n*) (build) drop light, drop slide window

~ труб·а́   (*f*) downcomer (of boiler etc.)

о·пуст·е́ни·е, -я,   (*n*) (*v n*) *see* о·пуст·е́·ть; (hydrodyn) cavitation

о·пуст·е́·ть, -ют,   (*perf*) become empty/void

о·пуст·е́л·ый, -ая, -ое,   (*a*) empty

о·пуст·и́·ть, -ят,   (*perf*) *see* о·пуск·а́·ть

о·пуст·оша́·ть, -ют,   (*imp*) devastate

о·пуст·оше́ни·е, -я,   (*n*) (*v n*) *see* о·пуст·оша́·ть; devastation, havoc

о·пуст·ош·и́·ть, -а́т,   (*perf*) *see* о·пуст·оша́·ть

о·пу́т·а·ть, -ют,   (*perf*) *see* о·пу́т·ыва·ть

о·пу́т·ыва·ть, -ют,   (*imp*) wind/tie round, entangle, enmesh

о·пух·а́ни·е, -я,   (*n*) (med) swelling, intumescence

о·пух·а́·ть, -ют,   (*imp*) swell

о·пу́х·лост·ь, -и,   (*f*) intumescence

о·пу́х·л·ый, -ая, -ое,   (*a*) swollen, intumescent

о·пу́х·нуть, -нут,   (*past masc sing*) о·пу́х·, (*perf*); *see* о·пух·а́·ть

о́·пух·олев·ый, -ая, -ое,   (*a*) swelling; (med) tumoral

о́·пух·ол·ь, -и,   (*f*) swelling; (med) tumour

~, жир·ов·а́я   (med) lipome, lipoma

~, зло·ка́честв·енн·ая   malignant tumour

о·пуш·а́·ть, -ют,   (*imp*) cover with fluff/down, cover with fine hairs; cover (lightly with snow etc.); edge/trim (with fur etc.)

о·пуш·ённ·ый, -ая, -ое,   (*past part pass*) of о·пуш·а́·ть; (biol) pubescent, tomentose

о·пуш·и́·ть, -а́т,   (*perf*) *see* о·пуш·а́·ть

о·пу́ш·к·а, -и,   (*g pl*) -ш·ек·, (*f*) edge of forest/wood

о·пущ·е́ни·е, -я,   (*n*) (*v n*) *see* о·пуск·а́·ть; (med) prolapse

~ пе́чен·и   (med) hepatoptosis

о·пыл·е́ни·е, -я,   (*n*) (*v n*) of о·пыл·и́·ть; (bot) pollination

о·пы́л·ивани·е, -я,   (*n*) dusting, powdering

о·пы́л·иватель·ь, -я,   (*m*) (agr) duster

о·пы́л·ива·ть, -ют,   (*imp*) dust (crops etc.)

о·пыл·и́тел·ь, -я,   (*m*) (bot) pollinator, pollinizer

о·пыл·и́·ть, -я́т,   (*perf*) pollinate; dust (crops etc.)

о·пыл·я́·ть, -ют,   (*imp*) (bot) pollinate

о́·пыт·, -а,   (*m*) experiment, test, trial; experience

о́·пыт·ничеств·о, -а,   (*n*) (agr) experimental work

о́·пыт·ност·ь, -и,   (*f*) experience, proficiency

о́·пыт·н·ый, -ая, -ое,   (*a*) of о́·пыт·

~ бассе́йн·   (*m*) experimental tank

~ образ·е́ц·   (*m*) prototype, experimental model

~ рабо́т·ник·   (*m*) experienced technician

~ у·стано́в·к·а   (*f*) pilot plant

о́·пыт·ов·ый, -ая, -ое,   (*a*) (*and see* о́·пыт·н·ый) experimental

о·пьян·е́ни·е, -я,   (*n*) intoxication, inebriation

опя́ть   (*adv*) again, once more

ора́нж·ев·ый, -ая, -ое,   (*a*) orange, orange-coloured/colored

~, мети́л·ов·ый   (chem) methyl orange

~, я́р·к о   brilliant orange

оранжере́·я, -и,   (*f*) (hortic) winter garden; (arch) conservatory

ора́тор·, -а,   (*m*) speaker, public speaker; orator

ор·а́·ть, -у́т,   (*imp*) yell

орбикуля́р·н·ый, -ая, -ое,   (*a*) (bot, geol) orbicular

орби́т·а, -ы,   (*f*) orbit; (air) tracking circle

орби́т·а

~, воз·мущ·ённ·ая   perturbed/disturbed orbit

~ Гоманн·а   (rocket) Hohmann orbit

~, за́·да·нн·ая   (rocket) intended orbit

~, ке́плер·ов·а   (astron) Keplerian orbit

~, молекуля́р·н·ая   (phys) molecular orbital

~, не·за́·мк·нут·ая   open orbit

~, перв·и́чн·ая   primitive orbit

~, перво·на·ча́ль·н·ая   primary orbit

~, у·стой·чив·ая   stationary orbit

орбита́л·ь, -и,   (*f*) (phys) orbital

орбита́ль·н·ый, -ая, -ое,   (*a*) orbital

орг-   (*component*) (*abbr*) of **органи·зацио́н·н·ый**

о́рган·, -а,   (*m*) (*see also under associated adjectives*) organ, organization, unit, department, body; tool, instrument; (instr) element, unit; **орга́н·, -а,** (*m*) (music) organ

~ вы·дел·е́ни·я   (biol) secretory/excretory organ

~, за ча́т·очн·ый   (biol) rudiment

~ из·мер·е́ни·я   (autom) measuring element

~, ис·полн·и́тельн·ый   executive; agency

~ регул·и́ровани·я   (autom) control element

органе́лл·а, -ы,   (*f*) (biol) organelle

организ·а́торск·ий, -ая, -ое,   (*a*) organizer's, organizational, organizing

организ·ацио́нн·ый, -ая, -ое,   (*a*) organization, organizational

организ·а́ци·я, -и,   (*f*) organization

органи́зм·, -а,   (*m*) organism, body

~, много·кле́т·очн·ый   (bot) metaphyte; (zool) metazoon

организ·о́ванност·ь, -и,   (*f*) organization, discipline, self-discipline

организ·о́ван·ный, -ая, -ое,   (*past part pass*) of **организ·ова́ть;** organized, arranged; disciplined

организ·ова́ть, -у́ют,   (*imp and perf*) organize, arrange

орган·и́ческ·ий, -ая, -ое,   (*a*) organic

орга́н·н·ый, -ая, -ое,   (*a*) of **орга́н·**

органо·гене́з·, -а,   (*m*) (biol) organogeny, organogenesis

органо·ге́н·ный, -ая, -ое,   (*a*) organogenous; (geol) biogenetic, biogenic

органо·гли́н·а, -ы,   (*f*) organic clay

органо·зо́л·ь, -я,   (*m*) (chem) organosol

орган·о́ид·, -а ,   (*m*) (biol) organoid

органо·лепт·и́ческ·ий, -ая, -ое,   (*a*) (biol) organoleptic

~ анали́з·   (*m*) (food) organoleptic analysis

органо·ли́т·, -а,   (*m*) (min) organolith, biolith

органо·терап·и́·я, -и,   (*f*) (med) organotherapy

органо·фосфа́т·, -а,   (*m*) (agr) organophosphate

орга·те́хник·а, -и,   (*f*) accounting and business machinery

о́рден·, -а,   (*nom pl*) **-а́,** (*g pl*) **-о́в,** (*m*) order, decoration, medal; **о́рден·, -а,** (*nom pl*) **-ы,** (*g pl*) **-ов,** (*m*) order (organization)

о́рдер·, -а,   (*nom pl*) **-а́,** (*g pl*) **-о́в,** (*m*) warrant, order; (law) writ; (mil) formation, disposition

~, архитекту́р·н·ый   (arch) order

~, дори́ческ·ий   (arch) Doric order

~, иони́ческ·ий   (arch) Ionic order

~, кори́нф·ск·ий   (arch) Corinthian order

~, по·гру́з·очн·ый   consignment note

ордина́ль·н·ый, -ая, -ое,   (*a*) ordinal

ордина́р·, -а,   (*m*) mean/normal level

ордина́р·н·ый, -ая, -ое,   (*a*) ordinary; single

~ с·вяз·ь   (*f*) (phys, chem) single bond

~ у·по́р·н·ый под·ши́п·ник·   (*m*) single-row thrust bearing

ордина́т·а, -ы,   (*f*) (math) ordinate; Y-coordinate

ос·ь ордина́т·   (*f*) Y-axis

ординато·ме́р·, -а,   (*m*) ordinate gauge

ордина́тор·, -а,   (*m*) (med) assistant physician/surgeon, intern

ордови́к·ск·ий, -ая, -ое,   (*a*) (geol) Ordovician

о·ребр·е́ни·е, -я,   (*n*) ribbing, finning; air cooling fins (ICE, refrig. etc.)

пло́щ·ад·ь о·ребр·е́ни·я   (*f*) cooling fin area (of car radiators etc.)

орёл·,   (*g sing*) **орл·а́,** (*m*) (orn) eagle, *Acquila*, (astron) Aquila, the Eagle

орео́л·, -а,   (*m*) (astron etc.) halo, (geol etc.) aureole, (phot) halation

~, конта́кт·н·ый,   (geol) contact aureole

~ от·раж·е́ни·я   (phot) reflex halation, halation by reflexion

рас·се́·яни·я   (*phot*) diffuse halation, halation by diffusion

орео·пите́к·, -а,   (*m*) (pal) *Oreopithecus*

оре́х·, -а,   (*m*) nut

~, гре́цк·ий   walnut

~, земл·ян·о́й   ground nut, pea nut, *Arachis hypogaea*

~, кита́й·ск·ий = **орех·, земл·ян·о́й**

~ ко́л·а   cola nut

~, лес·н·о́й   hazel, hazelnut

~, минда́ль·н·ый   almond

~, муска́т·н·ый   nutmeg

~, фиста́ш·ков·ый   pistachio

оре́х·ов·ый, -ая, -ое,   (*a*) of **оре́х·;** (*pl as noun*) (bot) *Juglandaceae*

~ де́рев·о   (*n*) walnut (wood)

орехо·тво́р·к·а, -и,   (*f*) (zool) gall fly, *Cynipoidea*

оре́ш·ек·, -ш·к·а,   (*m*) (*dim*) of **оре́х·**

~, черн·и́льн·ый   nut gall, Aleppo gall (excrescence on oak)

оре́ш·ков·ый, -ая, -ое,   (*a*) of **оре́·ш·ек·**

оре́ш·ник·, -а,   (*m*) (bot) filbert, *Corylus*

оригина́л·, -а,   (*m*) original; type/authentic specimen; top copy (of typescript); eccentric person, eccentric

~, втор·о́й   metal mother (sound recording disc manufacture)

~, пе́рв·ый   metal master (sound recording disc manufacture)

~, тре́т·ий   metal stamper (sound recording disc manufacture)

ориент·а́ци·я, -и,   (*f*) orientation

~, не·ребр·о́в·ая   (cryst) chaotic orientation

~, обра́т·н·ая   (cryst) inverse/negative/reverse orientation

~, пре·им·у́щественн·ая   (phys) preferred orientation

~, пре·облад·а́ющ·ая   (phys) preferred orientation

~, ребр·о́в·ая   (cryst) preferred orientation

ориенти́р·, -а,   (*m*) landmark, feature; (mil) reference point

~, астроном·и́ческ·ий   (nav) celestial reference

~ -бусс́о́л·ь, -и,   (*f*) (surv) declinator

~, контро́ль·н·ый   check feature/point

~, лин·е́йн·ый   (air) line feature

~, на·зе́м·н·ый   landmark

~, основ·н·о́й   basic reference point

~, свет·ов·о́й   (air) landmark beacon, aerial lighthouse

~, то́ч·ечн·ый   pin-point feature

ориент·и́ровани·е, -я,   (*n*) *see* **ориен·т·иро́вк·а**

ориент·и́рованн·ый, -ая, -ое,   (*past part pass*) of **ориент·и́р·овать;** oriented, directed (towards)

ориент·и́р·овать, -уют,   (*imp and perf*) orientate, orient; **-ся** get one's bearings

ориент·иро́вк·а, -и,   (*g pl*) **-вок·,** (*f*) (*v n*) *see* **ориент·и́р·овать;** orientation, (nav) fix

~, астроном·и́ческ·ий   (nav) star fix

ориент·иро́вочн·ый, -ая, -ое,   (*a*) of **ориент·иро́вк·а;** tentative, rough

~ залп·   (*m*) (gunn) ranging salvo

~ рас·чёт·   (*m*) rough calculation

орикто·цено́з·, -а,   (*m*) (pal) oryctocoenosis

орио́н·, -а,   (*m*) (astron) Orion

орке́стр·, -а,   (*m*) orchestra, band

орлеа́н·ск·ое де́рев·о   (*n*) (bot) arnotto, *Bixa orellana*

орле́ц·, -а́, (*i*) **-о́м,**   (*m*) (min) rhodonite

орл·и́к·, -а,   (*m*) (bot) aquilegia

орл·и́н·ый, -ая, -ое,   (*a*) eagle's, eagle, acquiline

~ де́рев·о   (*n*) (bot) eagleswood, aloeswood

орл·я́к·, -а́,   (*m*) (bot) bracken, *Pteridium*

~, обык·нове́нн·ый   (fish) eagle ray, *Myliobatis aquila*

орля́н·к·а, -а,   (*f*) = **орлеа́н·ск·ое де́рев·о**

орна́мент·, -а,   (*m*) decoration, pattern, design, ornamentation

орнит·и́н·, -а,   (*m*) (biochem) ornithine

орнити́н·ов·ый цикл·   (*m*) (biochem) Krebs cycle, citric acid cycle, tricarboxylic cycle

орнито·ло́г·и·я, -и,   (*f*) ornithology

орнито·пте́р·, -а,   (*m*) (a/c) ornithopter

орнито·фил·и́·я, -и,   (*f*) (bot) ornithophily

оро·гене́з·, -а,   (*m*) (geol) orogenesis, orogeny

оро·ген·и́·я, -и,   (*f*) = **оро·гене́з·**

о·рог·ове́ни·е, -я,   (*n*) (zool) keratinization

о·рог·ове́·ть, -ют,   (*perf*) keratinize

оро·граф·и́ческ·ий, -ая, -ое,   (*a*) orographic

оро·ло́г·и·я, -и,   (*f*) (geol) orology

оро·метр·и́·я, -и,   (*f*) (geol) orometry

о·рос·и́тел·ь, -я,   (*m*) sprinkler; irrigation ditch/canal

о·рос·и́тельн·ый, -ая, -ое,   (*a*) irrigation; sprinkling, spraying

~ но́рм·а   standard irrigation requirement (total amount of water required by crops in the growing season)

о·рос·и́тельн·ый, -ая, -ое
~ **холод·и́льник·**   (*m*) atmospheric evaporative cooler
о·рос·и́ть, -я́т, (*perf*) *see* **о·рош·а́·ть**
о·рош·а́·ть, -ют, (*imp*) water, spray, sprinkle; irrigate
о·рош·éни·е, -я, (*n*) (*v n*) *see* **о·рош·а́·ть;** irrigation; (min) wet dusting; (chem) reflux, refluxing
~ **дожд·еви́ни·ем** sprinkler irrigation
~ **за·топл·éни·ем** flood irrigation
~, **ка́рт·ов·ое** basin irrigation
~ **коло́нн·ы** (chem) reflux
~, **па́·вод·ков·ое** inundation irrigation
~ **по борозд·а́м** furrow irrigation
~, **под·по́чв·енн·ое** subirrigation
пол·я́ о·рош·éни·я (*pl*) sewage farm
Орса, газо·анализа́тор (*m*) Orsat gas analyser
орси́н·, -а, (*m*) (chem) orcinol, dihydroxytoluene
о́рт·, -а, (*m*) (math) unit vector; (min) crosscut, cross drift
~, **един·и́чн·ый** (math) unit vector
орт·ико́н·, -а, (*m*) (TV) orthicon, orthiconoscope
орт·и́т·, -а, (*m*) (min) orthite, allanite
орто- (*int component*) ortho-
орто·алюми́н·иев·ая кисл·от·а́ (*f*) (chem) orthoaluminic acid
орто·бо́р·н·ая кисл·от·а́ (*f*) orthoboric acid, boric acid
орто·водо·ро́д·, -а, (*m*) (chem) orthohydrogen
орто·гéл·и·й, -я, (*m*) (chem) orthohelium
орто·генéз·, -а, (*m*) (biol) orthogenesis
орто·гнéйс·, -а, (*m*) (geol) orthogneiss
о́рт·ов·ый, -ая, -ое, (*a*) of **орт·**
орто·гона́ль·ност·ь, -и, (*f*) orthogonality, perpendicularity, rectangularity
орто·гона́ль·н·ый, -ая, -ое, (*a*) orthogonal
~ **много·члéн·ы** (*pl*) (math) orthogonal polynomials
~ **проéкци·я** (*f*) orthogonal projection
орто·граф·и́ческ·ий, -ая, -ое, (*a*) orthographic, orthogonal
орто·до́м·а, -ы, (*f*) (cryst) orthodome
орт·одо́нт·и́·я, -и, (*f*) (med) orthodontia, dentistry
орто·дро́м·и·я, -и, (*f*) (nav) great circle; (surv) orthodromic line
орто·ио́д·н·ая кисл·от·а́ (*f*) periodic acid, per-iodic acid
орто·иодно·ки́сл·ый, -ая, -ое, (*a*) periodate of

орто·кла́з·, -а, (*m*) (min) orthoclase
орто·мышь·яко́в·ая кисл·от·а́ (*f*) orthoarsenic acid, arsenic acid
орто·метр·и́ческ·ий, -ая, -ое, (*a*) orthometric
орто·норм·иро́ванност·ь, -и, (*f*) (math) degree of orthonormal completeness
орто·норм·иро́ванн·ый, -ая, -ое, (*a*) (math) orthonormal, orthonormalized
о́рто-о́с·ь, -и, (*f*) (cryst) ortho-axis
орто·пед·и́ческ·ий, -ая, -ое, (*a*) (med) orthopaedic, orthopedic
орто·пéд·и·я, -и, (*f*) (med) orthopaedics, orthopedics
орто·ромб·и́ческ·ий, -ая, -ое, (*a*) (cryst) orthorhombic
орто·с·бро́с·, -а, (*m*) (cryst) orthokink
орто·ско́п·, -а, (*m*) orthoscope, orthoscopic polarizing microscope
орто·скоп·и́ческ·ий, -ая, -ое, (*a*) (opt) orthoscopic
орто·сла́н·ец·, -н·ц·а, (*m*) (geol) orthoschist
орто·со·един·éни·е, -я, (*n*) (chem) ortho-compound
орто·сти́л·, -а, (*m*) (arch) orthostyle
орто·сти́х·а, -и, (*f*) (bot) orthostichy
орто·тéст·, -а, (*m*) (eng, instr) compact comparator
орто·тро́п·н·ый, -ая, -ое, (*a*) (bot) orthotropic, orthotropous
орто·у́голь·н·ый, -ая, -ое, (*a*) (chem) orthocarbonic
орто·у́ксус·н·ая кисл·от·а́ (*f*) (chem) orthoacetic acid
орто·фенилендиами́н·, -а, (*m*) (chem) *o*-phenylenediamine
орто·фи́р·, -а, (*m*) (geol) orthophyre
орто·фосфа́т·, -а, (*m*) (chem) orthophosphate
орто·фо́сфор·н·ая кисл·от·а́ (*f*) (chem) orthophosphoric acid
орто·фот·и́ческ·ий, -ая, -ое, (*a*) (phot) orthophotic
орто·хромат·и́чност·ь, -и, (*f*) (phot) orthochromaticity
орто·хромат·и́ческ·ий, -ая, -ое, (*a*) orthochromatic
орто·цéнтр·, -а, (*m*) (math) orthocentre, orthocenter
орто·цимо́л·, -а, (*m*) (chem) orthocymene
орто·эфи́р·, -а, (*m*) (chem) orthoester
ортштéйн·, -а, (*m*) (geol) hardpan, iron pan

о·руд·ене́л·ый, -ая, -ое, (a) mineralized

о·руд·ене́ни·е, -я, (n) mineralization

ору́д·и·е, -я, (n) implement, tool; gun, piece, ordnance

~, зени́т·н·ое anti-aircraft gun

~, карту́з·н·ое breech-loading gun

~ ло́в·а fishing gear

~, патро́н·н·ое quick-firing gun

~, рыбо·ло́в·н·ое fishing gear

~, универса́ль·н·ое dual purpose gun (surface and anti-aircraft)

оруд·и́йн·ый, -ая, -ое, (a) gun

~ ого́н·ь (n) gun fire

~ ста́л·ь (f) gun-barrel steel

~ стан·о́к· (m) gun mounting

оруж·е́йн·ый, -ая, -ое, (a) of ору́ж·и·е

~ ма́стер· (m) gunsmith

ору́ж·и·е, -я, (n) arms, weapon

~, огне·стре́ль·н·ое firearms

~, руч·н·о́е small arms

~, стрел·ко́в·ое infantry weapon

~, термо·я́дер·н·ое thermonuclear weapon

~, у·правл·я́ем·ое guided weapon

~, штат·н·ое standard weapon, weapon in service

ор·у́т (pres 3rd pl) of ор·а́ть

орфо·гра́ф·и·я, -и, (f) orthography

орхид·е́·я, -и, (f) (bot) orchid

орхи́д·н·ые, -ых, (pl decl as adj) (bot) orchids, Orchidaceae

ос·а́, -ы́, (f) (ent) wasp

~, на·сто·я́щ·ая true wasp, (pl) Vespidae

~, общ·е́ственн·ая social wasp

~, один·о́чн·ая solitary wasp, (pl) Sphecidae

~, ро́·ющ·ая solitary wasp, (pl) Sphecidae

о·са́д·а, -ы, (f) siege

о·сад·и́тел·ь, -я, (m) precipitator, precipitant, precipitating agent; settling/precipitating tank

~, электро·стат·и́ческ·ий electrostatic precipitator

о·сад·и́тельн·ый, -ая, -ое, (a) precipitating, precipitation, settling

о·сад·и́ть, -ят, (perf) precipitate; sink (a pile, post etc.); lay siege to, besiege; check, pull up (e.g. a horse)

о·са́д·к·а, -и, (g pl) -д·ок·, (f) subsidence, settling, settlement; sedimentation; slump (of concrete); (met) upset forging, upsetting; (naut) draught; (nucl) pressurizing

о·са́д·к·а

~ в груз·у́ (naut) load draught

~, гидравл·и́ческ·ая hydraulic pressurizing (of nuclear fuel jacket)

~ порожн·ём (naut) light draught

~, рас·чёт·н·ая (naut) load draught

о·са́д·к·и (pl) see о·са́д·ок·, о·са́д·к·а

осадко·ме́р·, -а, (m) (meteor) rain gauge, precipitation gauge, pluviometer

осадко·на·копл·е́ни·е, -я, (n) (geol) sedimentation; (elec) sludging

осадко·образ·ова́ни·е, -я, (n) (geol) lithogenesis; (elec) sludging

о·са́д·н·ый, -ая, -ое, (a) of о·са́д·а

о·са́д·ок·, -д·к·а, (nom pl) о·са́д·к·и, (m) precipitate, precipitation, deposit, sediment, silt, sludge; (pl) (meteor) precipitation

~, атмосфе́р·н·ые (pl) (meteor) precipitation

~, внутри·ма́сс·ов·ые (pl) (meteor) convectional precipitation

~, во́д·н·ые (pl) (geol) water-borne sediments

вы·пад·е́ни·е о·са́д·к·ов (n) (meteor) precipitation; (nucl) fallout

~, до́н·н·ый (chem) bottoms, bottom sediment

~, лед·нико́в·ые (pl) (geol) glacial deposits

~, ли́в·нев·ые (pl) (meteor) heavy shower

~, оро·граф·и́ческ·ие (pl) (meteor) orographic precipitation

~, радио·акти́в·н·ые (pl) radioactive deposit

~, фронта́ль·н·ые (pl) (meteor) frontal precipitation

о·са́д·очн·ый, -ая, -ое, (a) precipitation; (geol) sedimentary

~ оболо́ч·к·а (f) (geol) lithosphere

~ по·ро́д·а (f) sedimentary rock

о·сажд·а́·ть, -ют, (imp) precipitate, settle, deposit; besiege, lay siege to; -ся (pass) separate/precipitate out, fall out

о·сажд·а́ющ·ее сре́д·ств·о (n) precipitating agent, precipitant

о·сажд·е́ни·е, -я, (n) precipitation, settling; sedimentation, deposition

~, втор·и́чн·ое (geol) reprecipitation

~, конта́кт·н·ое (met) electrodeless plating

~ нос·и́тел·ем (phys) carrier precipitation

о·сажд·е́ни·е
~, одно·вре́мен·н·ое (geol) coprecipitation
~, электро·стат·и́ческ·ое electrostatic precipitation
о·сажд·ённ·ый, -ая, -ое, (*past part pass*) *see* о·сажд·а́·ть; precipitated, deposited, settled
о·саж·ённ·ый, -ая, -ое, (*past part pass*) *see* о·са́ж·ива·ть
о·са́ж·ивани·е, -я, (*n*) joggling, jolting (plates); (met) upsetting, jumping up
  ис·пыт·а́ни·е о·са́ж·ивани·ем (met) jumping-up test, knock-down test, slug test
  манёвр· о·са́ж·ивани·ем (*m*) (rail) flat shunting
о·са́ж·ива·ть, -ют, (*imp*) check, pull up, press back; (met) jump up, upset; joggle (plates), sink (posts, piles)
о·са́л·ивани·е, -я, (*n*) (food) becoming rancid
осарсо́л·, -а, (*m*) (pharm) Osarsol, acetarsol
о·са́хар·ивани·е, -я, (*n*) saccharification
~ древе́с·и́н·ы (chem) destructive distillation of wood
о·са́хар·ивател·ь, -я, (*m*) saccharifier
о·са́хар·енн·ый, -ая, -о́е, (*past part pass*) saccharified
о·са́хар·ива·ть, -ют, (*imp*) *see* об·са́хар·ива·ть
о·сва́·ивани·е, -я, (*n*) (*v n*) of о·сва́·ива·ть; familiarization, assimilation
о·сва́·ива·ть, -ют, (*imp*) master, take over; assimilate, digest (knowledge etc.); -ся master, get to know, become familiar with, get used/accustomed to, feel at home
о·с·ве́дом·ить, -ят, (*perf*) *see* о·с·ве-домл·я́·ть
о·с·ведомл·ённ·ый, -ая, -ое, (*past part pass*) *see* о·сведомл·я́·ть; conversant with, versed in, well informed
о·с·ведомл·я́·ть, -ют, (*imp*) inform, notify; -ся inquire, ask about
о·свеж·а́·ть, -ют, (*imp*) refresh, freshen up
о·свеж·ева́ть, -ую́т, (*perf*) skin, dress (carcasses, hide etc.)
о·свеж·е́ни·е, -я, (*n*) (*v n*) *see* о·свеж·а́·ть; refreshment
о·свеж·ённ·ый, -ая, -ое, (*past part pass*) *see* о·свеж·а́·ть

о·свеж·и́ть, -а́т, (*perf*) *see* о·свеж·а́·ть
о·свет·и́тел·ь, -я, (*m*) light, lamp, illuminator, light source; light electrician, lampman, illumination engineer
о·свет·и́тельн·ый, -ая, -ое, (*a*) illuminating, lighting
~ авиа·бо́мб·а (*f*) (air) flare bomb
~ армату́р·а (*f*) (build) light fittings
~ ма́сл·о (*m*) illuminating/lamp oil/kerosene
~ с·на·ря́д· (*m*) (mil) star shell
о·свет·и́ть, -я́т, (*perf*) *see* о·свещ·а́·ть
о·светл·е́ни·е, -я, (*n*) (chem) clarification, clarifying, clearing; thinning (of trees); filtering (water); clearing (a photogr. negative); defecation (of sugar)
~ на·грев·а́ни·ем (chem) hot breakdown
о·светл·и́тел·ь, -я, (*m*) clarifier, clarifying agent, clearing agent; water filtration plant
о·светл·и́ть, -я́т, (*perf*) *see* о·светл·я́·ть
о·светл·я́·ть, -ют, (*imp*) (chem) clarify, clear; defecate (sugar etc.); thin (trees)
о·све́ч·ивани·е, -я, (*n*) luminous emittance
о·свещ·а́·ть, -ют, (*imp*) light up, illuminate; throw light on (figurative)
о·свещ·е́ни·е, -я, (*n*) lighting, illumination; elucidation; interpretation
~, бок·ов·о́е dark-field illumination (of microscope); (phot) side/horizontal lighting
~, ве́рх·н·ее ceiling lighting (of interiors); upper lighting (of exteriors)
~, нату́р·н·ое (cinema) outdoor lighting
~, основ·н·о́е (phot) general lighting
~, прям·о́е direct/broad lighting
о·свещ·ённост·ь, -и, (*f*) illuminance, illumination intensity
  коэффицие́нт· есте́ств·енн·ой о·свещ·ённост·и (*m*) (elec) daylight factor
о·свещ·ённ·ый, -ая, -ое, (*past part pass*) *see* о·свещ·а́·ть
~ по́л·е (*n*) bright field (microscopy)
о·с·вид·е́тельствовани·е, -я, (*n*) examination, survey
~, техн·и́ческ·ое (naut) certificate of seaworthiness
о·свинц·о́ванн·ый, -ая, -ое, (*a*) lead-covered/coated/plated, leaded; lead-sheathed

о·свобод·и́ть, -я́т, (*perf*) *see* о·свобожд·а́·ть

о·свобожд·а́·ть, -ю́т, (*imp*) set free, release, liberate; relieve (of duties); vacate (premises)

о·свобожд·е́ни·е, -я, (*n*) (*v n*) *see* о·свобожд·а́·ть; liberation, release

~ от бел·к·о́в (chem) deproteinization

~ от оболо́ч·к·и (nucl) de-canning/jacketing

о·сво·е́ни·е, -я, (*n*) (*v n*) *see* о·сва́·ива·ть; assimilation; mastering (a new technique)

о·сво́·ить, -ят, (*perf*) *see* о·сва́·ива·ть

ос·ев·о́й, -а́я, -о́е, (*a*) of ос·ь; axle, axial; axial-flow (of pumps etc.)

~ скеле́т· (*m*) (zool) axial skeleton

~ с·ма́з·к·а (*f*) axle grease

~ цили́ндр· (*m*) (zool) axon, axis cylinder

о·се́·вш·ий, -ая, -ее, (*past part act*) of о·се́ст·ь

о·сед·а́ни·е, -я, (*n*) subsidence, settling; settlement, settling (of people); (nucl) fallout

о·сед·а́·ть, -ю́т, (*imp*) subside, settle

о·седл·а́·ть, -ю́т, (*perf*) saddle; straddle

ос·е́й (*gen pl*) of ос·ь

осёл·, -сл·а́, (*m*) (zool) donkey, ass, burro, *Equus asinus*; о·сё·л (*past masc sing*) of о·се́ст·ь

осел·о́к·, -л·к·а́, (*m*) hone, whetstone

о·семен·е́ни·е, -я, (*n*) (*v n*) *see* о·семен·я́·ть; insemination

~, иску́с·ственн·ое (agr) artificial insemination

о·семен·и́ть, -я́т, (*perf*) *see* о·семен·я́·ть

о·семен·я́·ть, -ю́т, (*imp*) inseminate, fertilize; pollinate; seed, sow

осе́н·н·ий, -яя, -ее, (*a*) autumnal, fall

о́сен·ь, -и, (*f*) autumn, fall (U.S.)

о́сен·ью (*adv*) in the autumn/fall

о·серебр·и́ть, -я́т, (*imp*) silver-plate, coat with silver

о·сер·не́ни·е, -я, (*n*) sulphuration, sulfuration

о·сер·н·ённый, -ая, -ое, (*past part pass*) of о·сер·н·и́ть; sulphurated, sulfurated, sulphuretted

о·сер·н·и́ть, -я́т, (*perf*) *see* о·сер·ня́·ть

о·сер·ня́·ть, -ю́т, (*imp*) sulphurate, sulfurate

осе·симетр·и́чн·ый, -ая, -ое, (*a*) axiosymmetrical, axisymmetric(al)

о·се́ст·ь, (*fut 3rd pl*) о·ся́д·ут, (*past masc sing*) о·се́·л, (*perf*); *see* о·сед·а́·ть

осе·тока́р·н·ый, стан·о́к· (*m*) (mech) axle lathe

осётр·, (*g sing*) о·сетр·а́, (*m*) (fish) sturgeon, *Acipenser*

о·сетр·и́н·а, -ы, (*f*) sturgeon (as food)

о·се́ч·к·а, -и, (*g pl*) -ч·ек·, (*f*) (gunn) misfire

о́с·и (*gen, dat, prep, sing, nom pl*) of ос·ь

о·си́л·ива·ть, -ю́т, (*imp*) overcome

оси́н·а, -ы, (*f*) (bot) aspen, *Populus tremula*

оси́н·овик·, -а, (*m*) (bot) *Boletus versipellis*

оси́н·ов·ый, -ая, -ое, (*a*) aspen

ос·и́н·ый, -ая, -ое, (*a*) of ос·а́; wasp, wasp's

о·с·ко́л·ок·, -д·к·а, (*m*) splinter, fragment

~, бы́стр·ый (nucl) fast fragment

~ дел·е́ни·я (nucl) fission fragment

о·с·ко́л·очн·ый, -ая, -ое, (*a*) of о·с·ко́л·ок·

~ де́й·стви·е (*n*) fragmentation effect

о·с·копл·е́ни·е, -я, (*n*) (vet) castration, gelding

о·скорб·и́тельн·ый, -ая, -ое, (*a*) insulting, offensive

о·скорбл·е́ни·е, -я, (*n*) (*v n*) *see* о·скорбл·я́·ть; (law) libel

о·скорбл·я́·ть (*imp*) insult, offend

о·скуд·е́·ть, -ю́т, (*perf*) become scarce/impoverished

оскулю́м·, -а, (*m*) (zool) osculum, oscule

о·слаб·ева́ни·е, -я, (*n*) (*v n*) of о·слаб·ева́·ть; lessening, slackening, deterioration

о·слаб·ева́·ть, -ю́т, (*imp*) become weaker/fainter, weaken, decline, diminish

о·слаб·е́·ть, -ю́т, (*perf*) *see* о·слаб·ева́·ть

о·слаб·и́тел·ь, -я, (*m*) (elec) attenuation; (phot) reducer, reducing agent

~, по·ве́рх·ностн·ый (phot) cutting reducer

~, по·глот·и́тельн·ый (elec) dissipative attenuator

~, пре·де́ль·н·ый (elec) cut-off attenuator, dissipative attenuator

о·слаб·ить, -ят, (*perf*) *see* о·слабл·я́·ть

о·слабл·е́ни·е, -я, (*n*) (*v n*) *see* о·слабл·я́·ть; weakening, slackening, lessening, attenuation; abatement, attenuation; dilution; reduction

длин·а́ о·слабл·е́ни·я (*f*) (phys) attenuation length

~ из·луч·е́ни·я attenuation of radiation

~ интенси́в·ност·и (phys) decrease in intensity

коэффицие́нт· о·слабл·е́ни·я (*m*) attenuation factor; (nucl) total absorption coefficient

~, ме́ст·н·ое (phot) local reduction

~ по·то́к·а flux attenuation/depression; flow decay

сло́·й двух·кра́т·н·ого о·слабл·е́ни·я (*m*) (nucl) half-value layer

~, хим·и́ческ·ое (phot) chemical reduction

о·слабл·енн·ый, -ая, -ое, (*past part pass*) *see* о·слабл·я́·ть; relaxed, deteriorated

о·слабл·я́·ть, -ют, (*imp*) weaken, slacken, attenuate, loosen; reduce (photogr. negatives)

о·сла́б·нуть, -нут, (*perf*) *see* о·слаб·ева́·ть

о·сланц·ева́ни·е, -я, (*n*) (min) rock dusting

о·слеп·и́ть, -я́т, (*perf*) *see* о·слепл·я́·ть

о·слеп·и́тельн·ый, -ая, -ое, (*a*) blinding, dazzling, obscuring

о·слепл·ённ·ый, -ая, -ое, (*past part pass*) *see* о·слепл·я́·ть

о·слепл·я́·ть, -ют, (*imp*) blind; dazzle; obscure, blanket

о·слём·нуть, -нут, (*perf*) become blind/dazzled

о·слиз·не́вш·ий, -ая, -ее, (*past part act*); slimy, slippery; mucid

о·сли́з·нуть, -нут, (*perf*) become slimy/slippery

осл·и́нник·, -а, (*m*) (bot) evening primrose

осл·и́нников·ые, -ых, (*pl decl as adj*) (bot) *Onagraceae, Oenotheraceae*

осл·и́н·ый, -ая, -ое, (*a*) donkey, ass, assinine

осл·и́ц·а, -ы, (*i*) -ей, (*f*) she-ass

осло·во́д·ств·о, -а, (*n*) donkey breeding

о·с·лож·не́ни·е, -я, (*n*) complication

о·с·лож·нённ·ый, -ая, -ое, (*a*) complicated

о·с·лож·н·и́ть, -я́т, (*perf*) *see* о·с·лож·ня́·ть

о·с·лож·ня́·ть, -ют, (*imp*) complicate; -ся become complicated/ aggravated

о·слу́ш·а·ть, -ют, (*perf*) (med) sound; -ся disobey

о·слы́ш·аться, -атся, (*perf*) mishear, hear wrongly

о·слюд·ене́ни·е, -я, (*n*) (geol) micatization

о·сма́л·ива·ть, -ют, (*imp*) tar, coat with tar; resinify

о·сма́тр·ива·ть, -ют, (*imp*) examine, inspect, survey, look over/round, scan; (mil) muster

о·сме́·ива·ть, -ют, (*imp*) ridicule

о·сме·я́ть, -ют, (*perf*) *see* о·сме́·ива·ть

о́см·иев·ый, -ая, -ое, (*a*) of о́см·и·й

о́см·и·й, -я, (*m*) (chem) osmium, Os

о·смо́л·, -а, (*m*) resinous wood; tarry residue

о·смол·е́ни·е, -я, (*n*) (*v n*) *see* о·сма́л·ива·ть; resinification

о·смол·и́ть, -я́т, (*perf*) *see* о·сма́л·ива·ть

осмо́·метр·, -а, (*m*) (chem) osmometer

~, обращ·ённ·ый inverted osmometer

осмо·регул·я́ци·я, -и, (*f*) (bot) osmoregulation

о́смо·с·, -а, (*m*) (phys) osmosis

осм·оти́ческ·ий, -ая, -ое, (*a*) osmotic

о·смо́тр·, -а, (*m*) (*v n*), *see* о·сма́тр·ива̀·ть; inspection, examination

акт· о·смо́тр·а (*m*) inspection certificate

~, медици́н·ск·ий medical examination

~, период·и́ческ·ий routine inspection

~, ста́рт·ов·ый (a/c) between-flight inspection

о·смотр·е́ть, ⸗я́т, (*perf*) *see* о·сма́тр·ива·ть

о·смотр·и́тельност·ь, -и, (*f*) circumspection, discretion, prudence

о·смотр·и́тельн·ый, -ая, -ое, (*a*) circumspect, prudent

о·смо́тр·щик·, -а, (*m*) inspector

осмо·фо́р·, -а, (*m*) (chem) osmophore

о·с·мы́сл·енн·ый, -ая, -ое, (*past part pass*) *see* о·с·мысл·я́·ть; intelligent, sensible

о·с·мы́сл·ива·ть, -ют, (*imp*) *see* о·с·мысл·я́·ть

о·с·мы́сл·ить, -ят, (*perf*) *see* **о·с·мы-сл·я́·ть**

о·с·мысл·я́·ть, -ют, (*imp*) make sense of

о·снаст·и́ть, -я́т, (*perf*) *see* **о·сна-щ·а́·ть**

о·сна́ст·к·а, -и, (*g pl*) **-т·ок·,** (*f*) equipment, equipping, tooling, tooling-up; (naut) rigging

о·снащ·а́·ть, -ют, (*imp*) equip, fit out, tool up; (naut) rig

о·снащ·е́ни·е, -я, (*n*) = **о·сна́ст·к·а**

осно́в·а, -ы, (*f*) base, basis; (text) warp; (phot) film support, backing; (gram) stem; (math) base, radix; (*pl*) principles, fundamentals

~, клуб·ко́в·ая (text) ball warp
    леж·а́ть в осно́в·е form the basis/foundation
    на осно́в·е ме́д·ью (met) on copper base

основ·а́ни·е, -я, (*n*) base, basis; foundation, bed; root (of tooth, wing etc.); reason, grounds, substantion; founding (of institutions); (*pl*) (math) elements
    да·ва́ть основ·а́ни·е give grounds (to suppose), suggest

~ кры·л·а́ (a/c) wing root

~ логари́фм·ов (math) base of logarithms, radix
    на основ·а́ни·и on the basis (of), based on, from

~ на́·двиг·а (geol) fault sole

~ о́блака (meteor) cloud base

~ перпендикуля́р·а (math) foot of a perpendicular

~, пури́н·ов·ое (chem) purine base

~, сва́й·н·ое piled foundation

~ тре·уго́ль·ник·а base of a triangle

~, штыр·ев·о́е central pivot

основ·а́тел·ь, -я, (*m*) founder

основ·а́тельн·ый, -ая, -ое, (*a*) well founded, well grounded, substantiated, sound

осн·ова́ть, -у́ют, (*perf*) *see* **осно́-в·ыва·ть**

основ·н·о́й, -а́я, -о́е, (*a*) (*see also* **осно́в·н·ый**) basic, fundamental, principal, main, major, basal; master (of inspection gauges etc.)
    в основ·н·о́м (*adv*) essentially, mainly; much as (e.g. before)

~ ве́ктор· (*m*) basis vector

~ за·ко́н· (*m*) fundamental law

~ коли́честв·о (*n*) bulk, main bulk, most of

осно́в·н·о́й, -а́я, -о́е

~ ко́мпас· (*m*) master compass

~ ли́н·и·я (*f*) base line

~ материа́л·ы (*pl*) direct materials (in manufacturing)

~ мета́лл· (*m*) matrix metal, matrix (metallography); parent metal, base metal (welding, metal plating etc.)

~ об·ору́д·овани·е (*n*) capital equipment

~ ос·ь (*f*) (cryst, opt) principal axis; (bot) primary axis

~ по·то́к· (*m*) main flow

~ при·бо́р· (*m*) master inspection instrument

~ рас·хо́д·ы (*pl*) (fin) primage

~ со·сто·я́ни·е (*n*) (nucl) ground state

~ сре́д·ств·а (*pl*) fixed assets

~ твёрд·ый рас·тво́р· (*m*) matrix (metallography)

~ теоре́м·а а́лгебр·ы (*f*) fundamental theorem of algebra

~ цвет·а́ (*pl*) primary colours

~ цех· (*m*) (mech ind) production shop

осно́в·ност·ь, -и, (*f*) (chem) basicity

осно́в·н·ый, -ая, -ое, (*a*) (*see also* **о снов·н·о́й**) (chem) basic, sub-; (text) warp, warping

~ аз отно·ки́сл·ый (chem) subnitrate of

~ атла́с· (*m*) (text) warp satin

~ карбона́т· свин·ц·а́ (*m*) (chem) basic lead carbonate

~ сол·ь (*f*) (chem) basic salt

~ спо́соб· (*m*) basic process (of steelmaking)

~ футер·о́вк·а (*f*) basic lining (of metallurgical furnace)

~ шлак· (*m*) (met) basic slag

осново·вяз·а́льн·ый, -ая, -ое, (*a*) (text) warp-knitting

~ маши́н·а (*f*) (text) warp-knitting machine

осново·вя́з·ан·ый, -ая, -ое, (*a*) (text) warp knit, tricot

осново·на·блюд·а́тел·ь, -я, (*m*) (text) warp stop motion

осново·по·ло́ж·ник·, -а, (*m*) founder, initiator

осно́в·ыва·ть, -ют, (*imp*) found, base, establish; **-ся** (*pass*); settle down (of people); be based on

о·со́б·а, -ы, (*f*) person, personage

о·со́б·енн·о (*adv*) particularly, especially; unusually

о·со́б·енност·ь, -и,   (*f*) peculiarity, feature, (math) singularity
    в о·со́б·енност·и   in particular

о·со́б·енн·ый, -ая, -ое,   (*a*) special, particular, peculiar

о·соб·ня́к·, -а́,   (*m*) detached town house

о·со́б·ый, -ая, -ое,   (*a*) special; particular, peculiar, (math) singular

~ реш·е́ни·е,   (*n*) (math) singular solution

~ то́ч·к·а   (*f*) (math) singular point

о́·соб·ь, -и,   (*f*) (biol) individual

о·со·зна·ва́ть, -ют,   (*imp*) realize

о·со·зна́·ть, -ют,   (*perf*) *see* о·со·зна··ва́ть

осо́к·а, -и,   (*f*) (bot) sedge, *Carex*

осо́к·ов·ые, -ых,   (*pl decl as adj*) (bot) sedges, *Cyperaceae*

осоко́р·ь, -я,   (*m*) (bot) black poplar, *Populus nigra*

о·сола́ж·ивани·е, -я,   (*n*) (biochem) malting

о·солод·е́ни·е, -я,   (*n*) solothization (of soil)

о·солон·е́ни·е, -я,   (*n*) salinization

о·солон·цева́ни·е, -я,   (*n*) solonetz formation, alkalization (of soil)

о·солон·чакова́ни·е, -я,   (*n*) solonchak formation (of soil)

о́сп·а, -ы,   (*f*) (med) smallpox, variola

~, ве́тр·ян·ая   (med) chickenpox, varicella

~, натура́ль·н·ая   smallpox

оспадистери́т птиц·   (*m*) (vet) fowlpox, avian diphtheria

о·спа́р·ива·ть, -ют,   (*imp*) dispute, call in question, question (validity)

о́сп·енн·ый, -ая, -ое,   (*a*) smallpox, variolous

о́сп·ин·а, -ы,   (*f*) pock-mark; (met) pit

оспо·ви́д·н·ый, -ая, -ое,   (*a*) pockmarked; (met) pitted

оспо·при·вив·а́ни·е, -я,   (*n*) (med) vaccination

о·спо́р·ить, -ят,   (*perf*) *see* о·спа́р·и·ва·ть

о·сред·не́ни·е, -я,   (*n*) averaging

о·сред·не́нн·ый, -ая, -ое,   (*a*) averaged, faired

~ крив·а́я   (*f*) faired curve

оссе·и́н·, -а,   (*m*) (biochem) ossein

ост·, -а,   (*m*) (naut) east

ОСТ   (*abbr*) of обще·со·ю́з·н·ый станда́рт·   (*m*) OST (All-Union Standard Specification)

о·ста·ва́ться, -ю́тся,   (*imp*) (+ *instr as part of compound predicate*) remain, stay, be left (over)

о·ста́в·ить, -ят,   (*perf*) *see* о·ставл·я́·ть

о·ставл·е́ни·е, -я,   (*n*) (*v n*) *see* о·ставл·я́·ть

о·ставл·енн·ый, -ая, -ое,   (*past part pass*) *see* о·ставл·я́·ть; left, remaining

о·ставл·я́·ть, -ют,   (*imp*) leave, abandon

~ за соб·о́й   retain, reserve

о·ста·ётся   (*pres 3rd sing*) of о·ста·ва́ть·ся

о·ста́л·ивани·е, -я,   (*n*) steel coating

о·ста́·лся   (*past masc sing*) of о·ста́·ться

о·ста́·льн·о́й, -а́я, -о́е,   (*a*) the remaining, the rest of; (*as noun*) the rest, the others
    в о·ста·льн·о́м   in other respects
    вс·ё о·ста·льн·о́е   everything else, all the rest

о·стана́вл·ива·ть, -ют,   (*imp*) stop, halt, pull up, shut down (a machine)

о·стана́вл·ивающ·ий, -ая, -ее,   (*pres part act*) of о·стана́вл·ива·ть; shutoff, shutdown, stop
    о·стана́вл·ива·ться на во·про́с·е   dwell on a question

~ механи́зм·   (*m*) shutoff mechanism, arrester

о·ста́·нется   (*fut 3rd sing*) of о·ста́·ться

о·ста́н·ец·, -н·ц·а,   (*m*) (geol) monadnock

о·стано́в·, -а,   (*m*) (mech) stop, stop block, check, arrester, detent

о·станов·и́вш·ийся, -аяся, -ееся,   (*past part act*) of о·станов·и́ться; dead (motionless)

о·станов·и́ть, -ят,   (*perf*) *see* о·ста-на́вл·ива·ть

о·стано́в·к·а, -и,   (*g pl*) -в·ок·, (*f*) stoppage, stopping; stop, halt; shutdown/off

~, бы́стр·ая   scram (of nuclear reactors)
    врем·я о·стано́в·к·и   (*n*) shutdown time

~, кон·е́чн·ая   terminal (of transport)
    у·правл·е́ни·е о·стано́в·к·ой   (*n*) decking control (of lift)

о·стано́вл·енн·ый, -ая, -ое,   (*past part pass*) of о·станов·и́ть; stopped; standing (of machines)

о·стано́в·очн·ый, -ая, -ое,   (*a*) stopping, checking, halting

о·ста́т·ок·, -т·к·а,   (*m*) residue, remainder, rest, remnant, residuum; (chem) radical; (fin) balance; (*pl*) waste, bottoms, scraps, remains

о·ста́т·ок
~, куб·ов·ы́е (*pl*) (chem) bottoms
~ ряд·а (math) remainder of a series
~, твёрд·ый (ICE) carbon residue, coke
о·ста́т·очн·ый, -ая, -ое, (*a*) residual, remaining; (met) permanent, plastic
~ с·ро́д·ств·о (*n*) (chem) residual affinity
~ член· (*m*) (math) remainder
о·ста́·ться, -нутся, (*perf*) *see* о·ста·-ва́ться
о·ста·ю́тся (*pres 3rd pl*) of о·ста·ва́ть-ся
О́ствальд·а, за·ко́н· раз·ба́вл·е́ни·я (*m*) (chem) Ostwald's dilution law
о·стекл·е́ни·е, -я, (*n*) (*v n*) *see* о·стек-л·я́·ть; (build) glazing; vitrification
~, двой·н·о́е (build) double glazing
о·стекл·и́ть, -я́т, (*perf*) *see* о·стек-л·я́·ть
о·стекл·о́ванн·ый, -ая, -ое, (*a*) vitreous; vitrified; (plast) glasslike
о·стекл·ова́ть, -у́ют, (*perf*) *see* о·стек-л·я́·ть
о·стекл·я́·ть, -ют, (*imp*) glaze, fit glass into/round; -ся (*pass*); vitrify, become glassy/glasslike
остео·бла́ст·, -а, (*m*) (zool) osteoblast
остео·кла́ст·, -а, (*m*) (zool) osteoclast
остео·ло́г·и·я, -и, (*f*) (zool) osteology
осте·о́м·а, -ы, (*f*) (med) osteoma
остео·маля́ци·я, -и, (*f*) (med) osteo-malacia
остео·миэли́т·, -а, (*m*) (med) osteo-myelitis
остео·сарко́м·а, -ы, (*f*) (med) osteo-sarcoma
остео·синте́з·, -а, (*m*) osteosynthesis
остео·пла́ст·ик·а, -и, (*f*) (med) osteo-plasty
остео·том·и́·я, -и, (*f*) (med) osteotomy
остео·фи́л·, -а, (*m*) (biol) bone-seeker
остео·фи́ль·н·ый, -ая, -ое, (*a*) (bot) bone-seeking
остео·фи́т·, -а, (*m*) (med) osteophyte
остео·фо́н·, -а, (*m*) osophone, bone-conduction microphone
о·стерег·а́ни·е, -я, (*n*) (*v n*) of о·сте-рег·а́·ть; warning, caution, admoni-tion
о·стерег·а́·ть, -ют, (*imp*) warn; -ся (+ *gen*) beware of, be careful of
о·стере́ч·ь, (*fut 3rd sing, pl*) о·сте-реж·ёт, о·стерег·у́т, (*past masc sing*) о·стерёг·, (*perf*); *see* о·стере-г·а́·ть

осте·стриг·а́льн·ая маши́н·а (*f*) guard-hair cutting machine (fur pro-cessing)
ост·и́ст·ый, -ая, -ое, (*a*) (bot) barbate, awned, aristate; (anat) spinal, spinous
ост·и́т·, -а, (*m*) (med) osteitis
о́стов·, -а, (*m*) frame, framework, skele-ton, carcass, shell
~, а́том·н·ый (nucl) atom shell
~ под·ши́п·ник·а (mech) bearing shell
~ ро́тор·а (elec) rotor core
~ само·лёт·а (a/c) airframe
~ ядр·а́ core of a nucleus
~ я́кор·а (elec) armature core, hub
о·сто́й·чивост·ь, -и, (*f*) (mech) sta-bility, equilibrium
~, нул·ев·а́я на·ча́ль·н·ая neutral equilibrium
~, от·риц·а́тельн·ая на·ча́ль·н·ая unstable equilibrium
~, по·лож·и́тельн·ая на·ча́ль·н·ая stable equilibrium
~, по·пере́ч·н·ая transverse stability
~, про·до́ль·н·ая longitudinal sta-bility
~, стат·и́ческ·ая statical stability
о·сто́й·чив·ый, -ая, -ое, (*a*) stable
о·сторо́ж·н·о (*adv*) carefully, cautious-ly; (*imperative*) , look/mind out! beware! with care (on packet etc.)
о·сторо́ж·н·ый, -ая, -ое, (*a*) careful, cautious
о·сто·я́вш·ийся, -аяся, -ееся, (*past part act*) (chem) supernatant
о·стра́г·ива·ть, -ют, (*imp*) plane, shave, shave smooth
острак·о́д·а, -ы, (*f*) (zool) ostracod, (*pl*) ostracods, *Ostracoda*
острако·де́рм·ы (*f pl*) (pal) ostraco-derms, *Ostracoderma*
острако·ли́т·, -а, (*m*) (min) ostracolite
о́стр·ее (*comp*) of о́стр·ый
остр·е́ц·, -а́, (*i*) -о́м, (*m*) (bot) sedge, *Carex*
о·стриг·а́·ть, -ют, (*imp*) shear, clip, crop
остр·и·ё, -я́, (*n*) point, spike, cusp; blade, edge, cutting edge; stylus
~, на·правл·я́ющ·ее guide pivot
~ стре́л·к·и (instr) pointer tip
о·стри́ж·енн·ый, -ая, -ое, (*past part pass*) of о·стри́чч·ь, (*perf*); *see* о·стри-г·а́·ть (*imp*)
остр·и́йн·ый, -ая, -ое, (*a*) of остр·и·ё
о·стри́ж·енн·ый, -ая, -ое, (*past part pass*) *see* о·стриг·а́·ть

остр·и́ть, -я́т, (*imp*) sharpen

остр·и́ц·а, -ы, (*i*) -ей, (*f*) (zool) pin-worm, oxyuris, *Enterobius vermicularis*

о·стри́ч·ь, (*fut 3rd sing, pl*) о·стриж·ёт, о·стриг·у́т, (*past masc sing*) о·стри́г·, (*perf*); *see* о·стриг·а́·ть

о́стр·о (*adv*) sharply, keenly; (*as prefix*) oxy-, keenly, highly, sharply

о́стров·, -а, (*nom pl*) -а́, (*g pl*) -о́в, (*m*) island

остро·ве́рх·ий, -ая, -ое, (*a*) peaked

остров·итя́нин·, -а, (*nom pl*) в·итя́н·е, (*m*) islander

остров·н·о́й, -а́я, -о́е, (*a*) insular, island

~ о́т·мел·ь (*f*) (ocean) insular shelf

остров·о́к·, -в·к·а́, (*m*) (*dim*) of о́стров·; islet

острог·а́, -и́, (*f*) harpoon, fish spear, gaff

о·строг·а́·ть, -ют, (*perf*) *see* о·стра́·г·ива·ть

остро·гу́б·ц·ы, -ев, (*pl*) cutting pliers

остро·за·ра́з·н·ый, -ая, -ое, (*a*) highly infectious/contagious

остро·ки́ль·н·ый, -ая, -ое, (*a*) (shipb) sharp-bottomed

остро·кон·е́чн·ый, -ая, -ое, (*a*) pointed, sharp, peaked, (biol) cuspidate, acuminate; sharp-edged, mucronate

остро·ли́ст·, -а, (*m*) (bot) holly

остро·луч·ев·о́й, -а́я, -о́е, (*a*) pencil-beam

остро·на·пра́вл·енн·ый, -ая, -ое, (*a*) narrow; pencil (beam etc.)

о·строп·и́ть, -я́т, (*imp and perf*) sling, strop (goods handling)

о·стро́пл·енн·ый, -ая, -ое, (*past part pass*) of о·строп·и́ть; stropped

остро·пло́д·очник·, -а, (*m*) (bot) crazyweed, *Oxytropis*

остр·от·а́, -ы, (*nom pl*) остр·о́т·ы, (*f*) sharpness, acuity, keenness

~ зр·е́ни·я keenness of vision

~ слу́х·а hearing acuity

остро·уго́ль·н·ый, -ая, -ое, (*a*) (math) acute-angled, acute; oxygonal, sharply angled

остро·у́м·и·е, -я, (*n*) wit

остро·фо́кус·н·ый, -ая, -ое, (*a*) focused, focusing, fine-focus

остро·че́люст·н·ый, -ая, -ое, (*a*) (zool) oxygnathous

о·структу́р·ива·ть, -ют, (*imp*) aggregate, structurize

о́стр·ый, -ая, -ое, (*a*) sharp, acute, keen

остр·я́к·, -а́, (*m*) witty person, wit; (rail) switch rail

о·студ·и́ть, -'ят, (*perf*) *see* о·стуж·а́·ть

о·студ·нева́ни·е, -я, (*n*) (*v n*), *see* о·студ·нева́·ть; gelatinization, gelation, gelatination, jellification; gelling

о·студ·нева́·ть, -ют, (*imp*) gel, gelatinize, gelatinate, jell

о·стуж·а́·ть, -ют, (*imp*) cool

о·сту́ж·енн·ый, -ая, -ое, (*past part pass*) of о·студ·и́ть; cooled, chilled

о·сту́к·а·ть, -ют, (*perf*) *see* о·сту́к·ива·ть

о·сту́к·ива·ть, -ют, (*imp*) tap, rap, tap around

о·ступ·а́·ться, -ются, (*imp*) stumble

о·стыв·а́·ть, -ют, (*imp*) become/get cold, cool down/off

о·сты́·вш·ий, -ая, -ее, (*pres part act*) of о·сты́·ть; cold, congealed

о·сты́·н·уть, -ут, (*perf*) *see* о·стыв·а́·ть

о·сты́·ть, -нут, (*perf*) *see* о·стыв·а́·ть

о́ст·ь, -и, (*f*) (bot) beard, barb, awn; (anat) spine, spina; guard hair (of fur)

о·суд·и́ть, -'ят, (*perf*) *see* о·сужд·а́·ть

о·сужд·а́·ть, -ют, (*imp*) condemn, (law) sentence; censure, blame

о·суш·а́ем·ый, -ая, -ое, (*pres part pass*) of о·суш·а́·ть; drainage (of land)

о·суш·а́·ть, -ют, (*imp*) dry, dehumidify, desiccate; drain (land)

о·суш·а́ющ·ий, -ая, -ее, (*pres part act*) of о·суш·а́·ть

~ вещ·еств·о́ (*n*) desiccant, drying agent

о·суш·е́ни·е, -я, (*n*) (*v n*), *see* о·суш·а́·ть; drainage (of land)

о·суш·и́тел·ь, -я, (*m*) dryer, dehumidifier; desiccant

о·суш·и́тельн·ый, -ая, -ое, (*a*) drying, desiccating; drainage

~ кана́л· (*m*) drainage channel

~ на·со́с· (*m*) (shipb) bilge pump

~.систе́м·а (*f*) (shipb) fire and bilge pumping system

о·суш·и́ть, -'ат, (*perf*) *see* о·суш·а́·ть

о·су́ш·к·а, -и, (*g pl*) -ш·ек·, (*f*) (*v n*); *see* о·суш·а́·ть; foreshore, tidal zone

о·сущ·ествӥмост·ь, -и, (*f*) feasibility, practicability

о·сущ·ествѝм·ый, -ая, -ое,   (*a*) realizable, feasible, practicable

о·существ·ѝть, -я́т,   (*perf*) *see* **о·су·ществл·я́·ть**

о·сущ·ествле́ни·е, -я,   (*n*) (*v n*) of **о·существл·я́·ть;** accomplishment, realization, implementation

о·существл·ённ·ый, -ая, -ое,   (*past part pass*) *see* **о·существл·я́·ть**

о·существл·я́·ть, -ют,   (*imp*) effect, carry out, implement, accomplish, realize, achieve; **-ся** (*pass*); be carried on, occur

осфра́ди·й, -я,   (*m*) (zool) osphradium

осцилло·гра́мм·а, -ы,   (*f*) oscillogram, oscillograph trace

осцилло́·граф·, -а,   (*m*) oscillograph

~, без·инерцио́н·н·ый   cathode-ray oscillograph

~, гальвано·метр·ѝческ·ий   galvanometer oscillograph

~, двух·луч·ев·о́й   dual-trace/beam oscillograph

~, инерцио́н·н·ый   electromagnetic oscillograph

~, катод·н·ый   cathode-ray oscillograph

~, много·луч·ев·о́й   multi-channel/beam oscillograph

~, много·шлейф·ов·ый   multi-channel/beam oscillograph

~, стру́н·н·ый   Duddell oscillograph

~, тѐпл·ов·о́й   hot-wire oscillograph

~, шле́йф·н·ый   loop oscillograph

~, электро·магнѝт·н·ый   electromagnetic oscillograph, moving-coil oscillograph

~, электронно·луч·ев·о́й   cathode-ray oscillograph

осцилло·граф·ѝр·овать, -уют,   (*imp*) take an oscillograph trace of

осцилло·граф·ѝческ·ий, -ая, -ое,   (*a*) oscillographic

осцилло·гра́ф·и·я, -и,   (*f*) oscillography

осцилло·скоп·, -а,   (*m*) oscilloscope

~, двух·луч·ев·о́й   double-trace/beam oscilloscope, two-gun oscilloscope

~, ѝмпульс·н·ый   pulsed oscilloscope, synchronoscope

~, строб·ѝрующ·ий   sampling oscilloscope

~, электро́н·н·ый   cathode-ray oscilloscope

осцилл·я́тор·, -а,   (*m*) (elec) oscillator, oscillatory circuit; (weld) HF generator

осцилл·я́ци·я, -и,   (*f*) oscillation

о·сып·а́ни·е, -я,   (*n*) (*v n*) of **о·сы·па́·ть**

о·сып·а́·ть, -ют,   (*imp*) shower, sprinkle, strew, spread; shed (leaves etc.) **ся** (*pass*); fall (of leaves etc.); crumble; **о·сы́п·ать, -лют,**   (*perf*) *see above*

о́·сып·ь, -и,   (*f*) heap, mound; (geol) talus

~, берег·ов·а́я   coastal debris

о́с·ь, -и,   (*nom pl*) **-и,** (*g pl*) **-ей,**   (*f*) axis (*pl* axes); axle, (instr) spindle, shaft

~ абсцѝсс·   (*math*) axis of abscissae, abscissa X-axis

~ аппликáт·   (math) Z-axis

~ бала́нс·а   (horol) balance staff

~, бѐг·унко́в·ая   (rail) front bogie axle

~, бина́р·н·ая   (cryst) binary axis

~, бок·ов·а́я   lateral axis

~ бок·ов·о́й сѝл·ы   side force axis

~, больш·а́я   (opt cryst) major axis

~, вертика́ль·н·ая   (a/c) normal axis; (gunn) longitudinal axis (of dispersion zone)

~, винт·ов·а́я   (cryst) screw axis

~ в·раз·вѝл·к·у   forked axle

~ вращ·éни·я   (math) rotation axis; transit axis (of survey instruments)

~ вращ·ени·я мг·новéнн·ая   (math) instantaneous axis

~ врéмен·и   (elec) time base; (math) timing axis

~, вы́·дел·енн·ая   preferred axis

~, гла́в·н·ая   principal axis

~, глух·а́я   loose axle

~, горизонта́ль·н·ая   (gunn) lateral axis (of dispersion zone)

~ деформ·а́ци·и   (mech) strain axis

~, диагона́ль·н·ая   (cryst) clinoaxis

~, длѝн·н·ая   macroaxis; (min) major axis

за·ко́н· о́с·и   (*m*) (cryst) parallel twin law

~ за·хва́т·ывающ·их собáч·ек·   (telecom) common bar

~, зерка́ль·н·ая   inversion axis

~, зеркально·по·воро́т·н·ая   (cryst) alternating axis, rotation-reflection axis

~, инверсио́н·н·ая   rotoinversion axis

~ инéрци·и, гла́в·н·ая   principal axis of inertia

~, ис·хо́д·н·ая   reference axis

~, кѐрн·ов·ая   epipolar axis (air photography)

о́с·ь

~ коллим·а́ци·и (surv) line of collimation

~ координа́т· coordinate axis

~, координа́т·н·ая reference axis

~ координа́т·н·ой се́т·к·и grid axis

~, коро́т·к·ая (geol) short axis; (min) minor axis

~, кра́й·н·яя end axle

~ кре́н·а (air) roll axis

~, ма́л·ая minor axis

~ му́льд·ы trough axis

~ на·вед·е́ни·я (GW) guidance axis

~ на·кло́н·а axis of tilt

~, на·ча́ль·н·ая basic/reference axis

~, нейтра́ль·н·ая neutral axis

~, не·по·дви́ж·н·ая dead axle

~, о́бщ·ая common axis, coaxis

~, о·по́р·н·ая (zool) virgula

~, опт·и́ческ·ая principal axis (of a lens); optic axis (of a double refracting crystal); *Z*-axis (of piezocrystal)

~ ордина́т· (math) *Y*-axis

~, основ·н·а́я basic/reference axis

~ *Ox* (dynam) drag axis

~ *Oy* (dynam) lift axis

~ *Oz* (dynam) side force axis

~ па́р·ы сил· torque axis

~, перв·и́чн·ая (cryst) protaxis

~, по·бо́ч·н·ая secondary axis

~ по·воро́т·а slewing axis, axis of swivel/skew; axis of revolution

~, по·груж·а́ющ·аяся (geol) pitching axis

~, по·дви́ж·н·ая live axle

~, под·де́рж·ивающ·ая trailing axle

~, под·чин·ённ·ая sub-axis

~ подъ·ём·н·ой си́л·ы (aerodynam) lift axis

~, по·пере́ч·н·ая lateral axis

~ по·то́ч·н·ой систе́м·ы координа́т· wind axis (in wind tunnel)

~ по́яс·а (cryst) zone axis

~, про·до́ль·н·ая longitudinal axis; fore-and-aft axis

~ про·ка́т·к·и (met) axis/direction of rolling

~, рабо́ч·ая (min) running axis

~ раз·вёрт·к·и (elec) time base

~ с·вя́з·и (rad) main communications line

~ сим·метр·и́·й axis of symmetry

~ сим·метр·и́·и, двой·н·а́я second-order symmetry axis

~ сим·метр·и́·и *n*-его по·ря́д·к·а (cryst) *n*-fold rotation axis of symmetry

о́с·ь

~ с·клон·е́ни·я (astron) declination axis

~, скор·остн·а́я wind axis

~ со·про·вожд·е́ни·я (air) tracking axis

~ с·тво́р·а (nav) leading line

~ стру́·ийн·ого те·ч·е́ни·я (meteor) jet axis, jet-stream axis

~ танга́ж·а (a/c) pitching axis

~, тригона́ль·н·ая (cryst) trigonal axis, axis of trigonal symmetry

~, трой·н·а́я (cryst) triad, third-order symmetry axis

~, холост·а́я dead/free axle

~, час·ов·а́я (astron) hour axis

~, четвер·н·а́я (cryst) tetrad, fourth-order symmetry axis

~, шарни́р·н·ая pivotal axis

~, шест·ерн·а́я (cryst) hexad, sixth-order symmetry axis

осьми·но́г·, -а, (*m*) (zool) octopus

о·ся́д·ет (*fut 3rd sing*) of **о·се́ст·ь**

осяз·а́ем·ый, -ая, -ое, (*pres part pass*) of **осяз·а́·ть**; tangible, palpable

осяз·а́ни·е, -я, (*n*) touch (as sense), (med) tactus

осяз·а́тельн·ый, -ая, -ое, (*a*) tactile; noticeable, tangible, palpable

осяз·а́·ть, -ют, (*imp*) feel

от (*prep + gen*) from, away from, off; of (dates, locations, components etc.); for (special cases); versus, vs. (graphs)

~ винт·о́в·! (a/c) "stand clear!"

врем·я от времен·и from time to time

~ и́мен·и on behalf of

оборо́н·а от (*f*) defence against

от- (*as verbal prefix*) *signifies*: (1) departure, dispatch, disposal; (2) separation; (3) completion and cessation; (4) completion of intensive activity; (5) response

ота́в·а, -ы, (*f*) (agr) aftergrowth, aftercrop

о·та́пл·ивани·е, -я, (*n*) *see* **о·топл·е́ни·е**

о·та́пл·ива·ть, -ют, (*imp*) heat (space)

ота́р·а, -ы, (*f*) flock, flock of sheep

от·ба́в·ить, -ят, (*perf*) *see* **от·бавл·я́·ть**

от·ба́в·к·а, -и, (*f*) (*v n*) of **от·бавл·я́·ть**; subtraction, removal

от·бавл·я́·ть, -ют, (*imp*) dispose of some, remove some, take/pour away some от·бег·а́·ть, -ют, (*imp*) run off/away (a certain distance)

от·беж·а́ть, (*fut 3rd sing, pl*) от·бе-ж·и́т, от·бег·у́т, (*perf*); *see* от·бе-г·а́·ть

от·бе́л·, -а, (*m*) (met) chill, chilling

~, обра́т·н·ый inverse chilling

от·бел·ённ·ый, -ая, -ое, (*past part pass*) *see* от·бе́л·ива·ть

~ сло́·й (*m*) (met) chill

от·бе́л·ивани·е, -я, (*n*) (*v n*) of от·-бе́л·ива·ть; (met) chill

от·бе́л·ива·ть, -ют, (*imp*) (foods etc.) blanch, (text etc.) bleach, cure (sugar); (met) chill

от·бе́л·ивающ·ая гли́н·а (*f*) bleaching/decolorizing clay

от·бел·и́ть, -я́т, (*perf*) *see* от·бе́-л·ива·ть

от·бе́л·к·а, -и, (*g pl*) -л·ок·, (*f*) (text) bleaching

от·бе́ль·н·ый, -ая, -ое, (*a*) bleaching; (met) chilling

от·бензин·ённ·ый, -ая, -ое, (*a*) topped (petroleum)

от·бер·ёт, (*fut 3rd sing*) of ото·бр·а́ть

от·бив·а́·ть, .-ют, (*imp*) ward/fend off, repel, knock/beat off/back; remove, take away (sensation etc.); recapture; beat (out) (a signal, the time etc.); chip/hack off; mark out (boundaries)

от·би́в·к·а, -и, (*f*) (*v n*) *see* от·би-в·а́·ть

от·бир·а́·ть, -ют, (*imp*) take away, remove; tap, bleed, draw off; select, sort, pick up/out

~ про́б·ы take samples, sample

от·би́·ть, (*fut 3rd pl*) ото·бь·ю́т, (*past masc sing*) от·би́·л, (*perf*); *see* от·бив·а́·ть

о́т·блеск·, -а, (*f*) reflected light, reflection, gleam, sheen

от·блок·и́рованн·ый, -ая, -ое, (rail) open

от·бо́·й, -я, (*m*) (*v n*), *see* от·бив·а́·ть; (rad) sign-off; (telecom) ringoff, ring off signal, clearing signal; (mech) cushioning, recoil; all clear (air raid); (mil) retreat (signal); (naut) relax action stations, stand down

    да·ва́ть от·бо́·й ring off (telephones); (mil) sound the retreat

    ключ· от·бо́·я (*m*) (telecom) clearing key

от·бо́й·к·а, -и, (*g pl*) -бо́·ек·, (*f*) hacking/chipping/knocking off; (min) getting, cutting

от·бо́й·н·ый, -ая, -ое, (*a*) of от·-бо́·й and от·бо́й·к·а

~ кла́пан· (*m*) (telephones) clearing/ringoff indicator

~ молот·о́к· (*m*) (min) mechanical pick; mechanical chipping hammer

~ при·способл·е́ни·я (*pl*) fenders, fender arrangements (in ports etc.)

~ сва́·я (*f*) (naut) pile fender, rubbing pile

от·бо́й·щик·, -а, (*m*) (min) getter, pickman

от·бо́р·, -а, (*m*) (*v n*), *see* от·бир·а́·ть; selection; extraction, removal; outgoing materials (in isotope separation)

~ мо́щ·ност·и (mech) power take-off

~ образ·ц·о́в sampling

~ па́р·а steam extraction/bleeding

    пра́в·ил·а от·бо́р·а (*pl*) (phys) selection rules

~ проб· sampling

~ проб· на·у·га́д· random sampling

~ тепл·а́ heat extraction/removal

от·бо́р·к·а, -и, (*g pl*) -р·ок·, (*f*) picking, picking out, selecting

~, руч·н·а́я (min) hand picking/sorting

от·бо́р·ник·, -а, (*m*) sampler, sorter, picker

от·бо́р·н·ый, -ая, -ое, (*a*) selected, select, choice, best

~ нефт·ь (*f*) (oil) selected crude

от·бо́р·очн·ый, -ая, -ое, (*a*) selection

от·борт·о́ванн·ый, -ая, -ое, (*a*) flanged

от·борт·о́вк·а, -и, (*f*) flanging, crimping, flange, crimp

от·бо́р·щик·, -а, (*m*) = от·бо́р·ник·

от·брак·ова́ть, -у́ют, (*perf*) reject, scrap, throw out

от·брак·о́вк·а, -и, (*f*) (*v n*), *see* от·-брак·ова́ть; rejection, scrapping

от·брак·о́вщик·, -а, (*m*) quality control checker

от·бра́с·ывани·е, -я, (*n*) (*v n*), *see* от·бра́с·ыва·ть; rejection; repulsion

~ воз·дух·а biting (by air screw)

~ и́мпульс·ов (elec) pulse rejection

от·бра́с·ыва·ть, -ют, (*imp*) throw aside/back/off; throw, cast (shadows etc.); reject, dismiss, discard (ideas etc.); (math) neglect, take no account of

от·бро́с·, -а, (*m*) deflection, throw (of indicator); (*pl*) refuse, garbage; waste, waste matter, spoil

~ про·из·во́д·ств·а (*pl*) industrial waste

от·брóс·ить, -ят, (*perf*) *see* **от·брá-с·ыва·ть**

от·брóс·н·ый, -ая, -ое, (*a*) waste
~ тепл·ó (*n*) waste heat

от·брóш·енн·ый, -ая, -ое, (*past part pass*) *see* **от·брáс·ыва·ть;** discarded, dismissed, rejected; neglected

от·бýд·ет (*fut 3rd sing*) of **от·бы́·ть**

от·буксúр·овать, -уют, (*perf*) tow, tow away

от·быв·á·ть, -ют, (*imp*) depart, leave, set off

от·бы́т·и·е, -я, (*n*) departure

от·бы́·ть (*fut 3rd sing, pl*) **от·бýд·ет, от·бýд·ут,** (*past masc sing*) **от·бы́·л;** (*perf*); *see* **от·быв·á·ть**

отвáг·а, -и, (*f*) bravery, gallantry

от·вáл, -а, (*m*) dump, heap, pile; spoilbank; (chem) tailings, spent material; blade (of grader); mouldboard, breast (of plough etc.); (naut) taffrail; casting off, departure (of ship)

от·вáл·ива·ть, -ют, (*imp*) roll/heave/pull off; dump; (naut) push/bear/cast off; **-ся** (*pass*); fall off

от·вал·и́ть, -́ят, (*perf*) *see* **от·вáл·ива·ть**

отвало·мест·и́лищ·е, -я, (*n*) dump site

отвало·образ·овáтел·ь, -я, (*m*) (min) stacker, spoil stacker/dumper

от·вáль·н·ый, -ая, -ое, (*a*) of **от·вáл·;** banking, grading
~ плуг· (*m*) (min) spreading plough/plow, spreader, blade grader

от·вальц·óванн·ый, -ая, -ое, (*a*) rolled, calendered; milled (rubber)

от·вáр·, -а, (*m*) decoction, tea
~, мяс·н·óй broth
~, ячмéн·н·ый barley water

от·вáр·ива·ть, -ют, (*imp*) boil, decoct

от·вар·и́ть, -́ят, (*perf*) *see* **от·вáр·ива·ть**

от·вед·éни·е, -я, (*n*) (*v n*) *see* **от·вод·и́ть;** (biol) point of contact

от·вез·ти́, -ýт, (*perf*) *see* **от·воз·и́ть**

от·вёл (*past masc sing*) of **от·вес·ти́**

от·верг·á·ть, -ют, (*imp*) reject, repudiate

от·вéрг·нуть, -нут, (*perf*) *see* **от·верг·á·ть**

о·тверд·евáни·е, -я, (*n*) solidification, hardening, congelation, (med) induration
   температýр·а о·тверд·евáни·я (*f*) (met) crystallization point

о·тверд·евá·ть, -ют, (*imp*) solidify, congeal, (med) indurate

о·тверд·éл·ый, -ая, -ое, (*a*) solidified, congealed

о·тверд·é·ть, -ют, (*perf*) *see* **о·твер-д·евá·ть**

о·тверд·и́тел·ь, -я, (*m*) (chem) hardener, solidifier; curing agent (rubber)

о·тверд·и́ть, -я́т, (*perf*) *see* **о·тверж-д·á·ть**

о·твержд·á·ть, -ют, (*imp*) harden; (plastic) set, cure; hydrogenate (fats)

о·твéржд·енн·ый, -ая, -ое, (*past part pass*) *see* **о·твержд·á·ть**
~ жир· (*m*) hydrogenated fat

от·вер·нýть, -ýт, (*perf*) unscrew; turn back/on/away/aside, turn; **-ся** (*pass*); come unscrewed

отвéрст·и·е, -я, (*n*) opening, orifice, aperture, hole, port; slot (coin entrance); mesh (of screen, sieve etc.); (anat) foramen; span, opening (of a bridge)
~, вентиляциóн·н·ое vent hole
~, в·ход·н·óе (ICE) inlet, inlet port
~, вы·пуск·н·óе (ICE) outlet, outlet port; discharge aperture (extruder)
~, вы·ход·н·óе (ICE) outlet, outlet port
~, за·бóр·н·ое intake
~, за·грýз·очн·ое fire door (boiler); loading aperture/hole (extruder)
~, зр·и́тельн·ое (anat) optic foramen
~, кольц·ев·óе annulus
~, креп·ёжн·ое pin hole (for fixing)
~, луче·ис·пуск·áющ·ее (opt) beam orifice
~, о·гран·и́чивающ·ее (opt) collimating aperture
~, от·нос·и́тельн·ое (phot) relative aperture, F-valve
~, раз·би́·т·ое oversize hole
~, резонáнс·н·ое (rad) resonant aperture
~, слеп·óе blind opening, thimble
~, смотр·ов·óе inspection/peep/sight hole

от·верт·é·ть, -я́т, (*perf*) unscrew; twist off; **-ся** (*pass*); avoid, get out of (difficulties etc.)

от·вёрт·к·а, -и, (*g pl*) **-т·ок·,** (*f*) screwdriver

от·вёрт·ыва·ть, -ют, (*imp*) unscrew; turn back/on/away/aside, turn; (naut) let go (ropes etc.); **-ся** (*pass*); come unscrewed

от·вéрч·енн·ый, -ая, -ое, (*past part pass*) of **от·верт·é·ть**

от·вес·, **-а**, (*m*) plumb, plummet; vertical drop, sheer drop/wall

от·вес·ить, **-ят**, (*perf*) weigh out

от·вес·н·ый, **-ая**, **-ое**, (*a*) steep, very steep; perpendicular, vertical, plumb, sheer

~ **лин·и·я** (*f*) plumb/vertical line

~ **пик·и́рование** (*n*) (air) nosedive

от·вес·ти́, (*fut 3rd sing, pl*) **от·вед·ёт**, **от·вед·у́т**, (*past masc sing*) **от·вёл·**, (*perf*); *see* **от·вод·и́ть**

от·ве́т·, **-а**, (*m*) answer, reply, response

знак· от·ве́т·а (*m*) (telecom) acknowledged signal, answering sign

от·ветв·и́тел·ь, **-я**, (*m*) branch; (rad) coupler

~, **дву·на·пра́вл·енн·ый** bidirectional coupler

~, **на·пра́вл·енн·ый** directional coupler

от·ветв·и́тельн·ый, **-ая**, **-ое**, (*a*) branching, dividing

~ **му́фт·а́** (*f*) (elec) dividing box

от·ветв·и́ть, **-ят**, (*perf*) *see* **от·ветв·л·я́·ть**

от·ветвл·е́ни·е, **-я**, (*n*) branch, branching, off-shoot, branch pipe, tap joint; ramification; (elec) distributing main; (biochem) lateral substituent

~, **бок·ов·о́е** side branch

от·ветвл·ённ·ый, **-ая**, **-ое**, (*past part pass*) *see* **от·ветвл·я́·ть**; branched, branch; (elec) derived, branch, shunt

от·ветвл·я́·ть, **-ют**, (*imp*) branch; (elec etc.) tap; **-ся** branch, branch off

от·ве́т·ить, **-ят**, (*perf*) answer, reply to

от·ве́т·н·ый, **-ая**, **-ое**, (*a*) answering, return, reciprocal

~ **визи́т·** (*m*) return visit

~ **знак·** (*m*) (telecom) answering sign

от·ве́т·ственност·ь, **-и**, (*f*) responsibility

от·ве́т·ственн·ый, **-ая**, **-ое**, (*a*) responsible, in charge; very important, important, key (*as adjective*)

~ **реда́ктор·** (*m*) editor-in-chief

от·ве́т·чик·, **-а**, (*m*) (law) defendant, respondent; (rad) responder

от·веч·а́·ть, **-ют**, (*imp*) answer, reply to, respond; answer for, be responsible for; meet (requirements etc.)

от·ве́ш·енн·ый, **-ая**, **-ое**, (*past part pass*) *see* **от·ве́ш·ива·ть**; weighed

от·ве́ш·ива·ть, **-ют**, (*imp*) weigh out/off

от·винт·и́ть, **-ят**, (*perf*) *see* **от·вин·ч·ива·ть**

от·вин·ч·ива·ть, **-ют**, (*imp*) unscrew

от·вис·а́·ть, **-ют**, (*imp*) sag, droop, hang down

от·вис·л·ый, **-ая**, **-ое**, (*a*) pendulous

от·вис·нуть, **-нут**, (*perf*) *see* **от·вис·а́·ть**

от·влек·а́·ть, **-ют**, (*imp*) distract, divert; abstract; **-ся** (*pass*); digress

от·влеч·ённ·ый, **-ая**, **-ое**, (*past part pass*) *see* **от·влек·а́·ть**; abstract

~ **числ·о́** (*n*) (math) abstract number

от·вле́ч·ь, (*fut 3rd sing, pl*) **от·влеч·ёт**, **от·влек·у́т**, (*past masc sing*) **от·влёк·**, (*perf*); *see* **от·влек·а́·ть**

от·во́д·, **-а**, (*m*) (*v n*) of **от·вод·и́ть**; removal, withdrawal, extraction; bend, branch (piping); outlet, tap; objection (raised in a court of law)

~ **из трубо·про·во́д·а** (autom) pipe tap

~, **корм·ов·о́й** (shipb) propeller guard

~, **прям·о́й** elbow/right-angle bend

~ **тепл·а́** heat removal/withdrawal/extraction

от·вод·и́м·ый, **-ая**, **-ое**, (*pres part pass*) of **от·вод·и́ть**; outgoing

от·вод·и́ть, **-ят**, (*imp*) lead, take; lead away, take aside, withdraw; divert, move/put aside; reject, turn down; allot, allocate; tap, bleed, drain (liquids, gases); (naut) ease the helm

от·во́д·к·а, **-и**, (*g pl*) **-д·ок·**, (*f*) branch, diversion; shifting device

~ **му́фт·ы** clutch shift collar

~, **ремён·н·ая** (mech) belt shifter

от·во́д·ник·, **-а**, (*m*) baffle, baffle plate

от·вод·н·о́й, **-а́я**, **-о́е**, (*a*) *see* **от·во́д·н·ый**

от·во́д·н·ый, **-ая**, **-ое**, (*a*) branch; outlet, drain

~ **труб·а́** (*f*) drain/waste pipe; breathing pipe (of fuel tank)

от·во́д·ок·, **-д·к·а**, (*m*) (bot) layer, cutting

от·вод·я́щ·ий, **-ая**, **-ое**, (*pres part act*) of **от·вод·и́ть**; take-off, outlet; (anat) abductant

~ **кана́л·** (*m*) (hydroelec) take-off, take-off channel

~ **нерв·** (*m*) (zool) abductor

от·воз·и́ть, **-ят**, (*imp*) take, deliver; take away, remove

от·вора́ч·ива·ть, **-ют**, (*imp*) turn back/away/aside; (obs) *see* **от·вер·н·у́ть**

о·твор·и́ть, **-ят**, (*perf*) *see* **о·твор·я́·ть**

от·воро́т·, -а, (*m*) (*v n*) *see* от·вора́-ч·ива·ть; lapel, flap

~ от маршру́т·а (air) dog-leg

от·ворот·и́ть, -́ят, (*perf*) *see* от·вора́-ч·ива·ть

о·твор·я́·ть, -ют, (*imp*) open, swing open

от·врат·и́тельн·ый, -ая, -ое, (*a*) disgusting, repulsive

от·врат·и́ть, -я́т, (*perf*) *see* от·вра-щ·а́·ть

от·вращ·а́·ть, -ют, (*imp*) avert

от·вращ·е́ни·е, -я, (*n*) aversion, repugnance

от·вык·а́·ть, -ют, (*imp*) give up (of habits); become unaccustomed to

от·вы́к·нуть, -нут, (*perf*) *see* от·вы-к·а́·ть

от·вяз·а́ть, (*fut 3rd sing, pl*) от·вя́-ж·ет, от·вя́ж·ут, (*perf*); *see* от·вя́-з·ыва·ть

от·вя́з·ыва·ть, -ют, (*imp*) untie; loose, set free

от·гад·а́·ть, -ют, (*perf*) guess correctly

от·ги́б·, -а, (*m*) bend, crease, fold

от·гиб·а́ни·е, -я, (*n*) (*v n*) *see* от·ги-б·а́·ть; diffraction

от·гиб·а́·ть, -ют, (*imp*) unbend; bend/turn back

от·гиб·н·о́й, -а́я, -о́е, (*a*) folding, bending

~ ша́йб·а (*f*) tab washer

от·глаго́ль·н·ый, -ая, -ое, (*a*) (gram) verbal

от·гла́д·ить, -́ят, (*perf*) *see* от·гла́-ж·ива·ть

от·гла́ж·ива·ть, -ют, (*imp*) iron, press, smooth

от·гни́·ть, -ю́т, (*perf*) rot off

от·гова́р·ива·ть, -ют, (*imp*) dissuade; -ся excuse oneself

от·говор·и́ть, -я́т, (*perf*) *see* от·гова́-р·ива·ть

от·гово́р·к·а, -и, (*g pl*) -р·ок·, (*f*) excuse

от·голо́с·ок·, -с·к·а, (*m*) echo

от·го́н·, -а, (*m*) (*v n*) *see* от·гон·я́·ть; distillation; distillate

~ лёгк·их фра́кц·и·й (chem) prefractionation

от·го́н·к·а, -и, (*g pl*) -н·ок·, (*f*) (*v n*) of от·гон·я́·ть; distillation

от·го́н·очн·ый, -ая, -ое, (*a*) of от·-го́н·к·а

~ коло́нн·а (*f*) (oil) stripping column

от·гон·я́нн·ый, -ая, -ое, (*past part pass*) of от·гон·я́·ть

от·гон·я́нн·ый, -ая, -ое

~ фра́кц·и·и (*f*) (oil) strippings

от·гон·я́·ть, -ют, (*imp*) drive off/away; (chem) distill off, (oil) strip

~ па́р·ом steam out

от·гора́ж·ива·ть, -ют, (*imp*) fence/partition off, separate/divide by

от·город·и́ть, -я́т, (*perf*) *see* от·гора́-ж·ива·ть

от·горо́ж·енн·ый, -ая, -ое, (*a*) (*past part pass*) *see* от·гора́ж·ива·ть; enclosed

от·грани́ч·ива·ть, -ют, (*imp*) delimit, demarcate, draw a line between

от·грани́ч·ить, -ат, (*perf*) *see* от·грани́ч·ива·ть

от·греб·а́·ть, -ют, (*imp*) rake away/off; row off/away

от·грес·ти́ (*fut 3rd pl*) от·греб·у́т, (*past masc sing*) от·грёб·, (*perf*); *see* от·греб·а́·ть

от·груж·а́·ть, -ют, (*imp*) dispatch, ship, send off

от·груз·и́ть, -я́т, (*perf*) *see* от·гру-ж·а́·ть

от·гру́з·к·а, -и, (*g pl*) -з·ок·, (*f*) dispatch, shipment

от·да·ва́·ть, -ют, (*imp*) give back, return; hand over, give up, let go, part with, let have, give away; (*imp only*) give off, smell/taste of; (naut) cast off, slip, drop (anchor etc.); (gunn) recoil; -ся (*pass*); reverberate, resound

~ под стра́ж·у give into custody

~ под суд· prosecute

~ при·ка́з· issue an order

~ чест·ь salute

от·дав·и́ть, -я́т, (*perf*) crush, squash, damage by squeezing

от·дал·е́ни·е, -я, (*n*) (*v n*), *see* от·да-л·я́·ть; distance

в от·дал·е́ни·и at a distance

от·дал·ённ·ый, -ая, -ое, (*past part pass*) *see* от·дал·я́·ть; remote, distant

от·дал·и́ть, -я́т, (*perf*) *see* от·дал·я́·ть

от·дал·я́·ть, -ют, (*imp*) move/draw back/away; postpone; estrange

о́т·да·нн·ый, -ая, -ое, (*a*) (*past part pass*) of от·да́·ть; given, given off

от·да́·ть, (*fut 3rd sing, pl*) от·да́ст, от·дад·у́т, (*past masc sing*) от·да·л, (*perf*); *see* от·да·ва́·ть

от·да́·ч·а, -и, (*f*) (*v n*) *see* от·да·ва́·ть; delivery, yield, output, efficiency; recoil, rebound

~, вес·ов·а́я load ratio

от·да́·ч·а
~, ква́нт·ов·ая quantum efficiency
~, механ·и́ческ·ая mechanical efficiency
~, ни́з·к·ая low efficiency
~, о́бщ·ая overall efficiency
~, поле́з·н·ая efficiency
~, про·мы́шл·енн·ая actual efficiency
~ прото́н·а (nucl) proton recoil
~, прям·а́я surface radiation (of heat)
 коэффицие́нт· прям·о́й от·да́·ч·и (*m*) coefficient of surface radiation (of heat in a boiler)
~, то́к·ов·ая (elec) current efficiency
~, эффекти́в·н·ая effective output
от·дежу́р·ить, -ят, (*perf*) spend time on duty; finish duty
от·де́л·, -а, (*m*) (*and see under associated words*) department, section; (geol) division, formation
~, пере́д·н·ий (zool) prosoma
~, пояс·ни́чн·ый (zool) lumbar spine
от·де́л·анн·ый, -ая, -ое, (*past part pass*) of **от·де́л·а·ть;** finished, dressed
от·де́л·а·ть, -ют, (*perf*) *see* **от·де́л·ыва·ть**
от·дел·е́ни·е, -я, (*n*) (*and see under associated adjectives*) (*v n*), *see* **от·де·л·я́·ть;** separation, isolation, detachment; compartment; department branch; (mil) squad, section
~, бо·ев·о́е gunhouse (ship's turret); fighting cab/compartment (of tank)
~, бага́ж·н·ое luggage compartment; (a/c) luggage hold
~, коли́честв·енн·ое (math) quantitative isolation
~ сло·ёв exfoliation; removal of layers
~ стру́ж·к·и chip formation (machining metal)
~ сту́п·ен·и (rocket) stage separation
~ у·скор·и́тел·я (rocket) booster jettisoning/separation
от·дел·ённ·ый, -ая, -ое, (*past part pass*) *see* **от·дел·я́·ть**
от·дел·и́мост·ь, -и, (*f*) separability; separateness
от·дел·и́м·ый, -ая, -ое, (*a*) separable
от·дел·и́тел·ь, -я, (*m*) separator, divider
от·дел·и́тельн·ый, -ая, -ое, (*a*) dividing, separating
от·дел·и́ть, -ят, (*perf*) *see* **от·дел·я́·ть**
от·де́л·к·а, -и, (*g pl*) -л·ок·, (*f*) (*v n*) of **от·де́л·ыва·ть;** finish, trimmings, trim, dressing
~, ма́т·ов·ая dull finish

от·де́л·к·а
~ мо́лот·ом (build) hammer dressing
~ черн·е́ни·ем (met) black finish
от·де́л·очн·о-рас·то́ч·н·ый стан·о́к· (*m*) fine-boring machine
от·де́л·очн·ый, -ая, -ое, (*a*) of **от·де́л··к·а;** finishing; decoration, decorative
от·де́л·ыва·ть, -ют, (*imp*) finish, trim, dress; **-ся** be/have finished/done with; get rid of
от·де́ль·ност·ь, -и, (*f*) separateness; (geol) jointing, joint; (min) cleavage
 в от·де́ль·ност·и separately
 пло́ск·ост·ь от·де́ль·ност·и (*f*) (geol) joint plane
~ с·жа́т·и·я (geol) compression jointing; (min) compression cleavage
от·де́ль·н·ый, -ая, -ое, (*a*) separate, detached, individual, discrete
~ у́·ров·ен·ь (*m*) (math, phys) discrete level
~ ча́ст·ь (*f*) (mil) detached unit; individual part, (*pl*) different/some parts (of a whole)
от·дел·я́·ть, -ют, (*imp*) separate, detach, divide/cut off, segregate, isolate
от·дёрг·ива·ть, -ют, (*imp*) pull/tug/ jerk back/aside
от·дер·ёт, (*fut 3rd sing*) of **ото·др·а́ть**
от·дёр·н·уть, -ут, (*perf*) *see* **от·дёрг·ива·ть**
от·дир·, -а, (*m*) (*v n*) of **от·дир·а́ть**
 со·противл·е́ни·е от·ди́р·у (*n*) pull-off resistance (of adhesive joints)
от·дир·а́·ть, -ют, (*imp*) tear/pull off
отдор·, -а, (*m*) (meteor) land breeze
от·дох·н·у́ть, -у́т, (*perf*) *see* **от·ды·х·а́·ть**
от·ду́б·ин·а, -ы, (*f*) tan waste
от·дув·а́·ть, -ют, (*imp*) blow back/ away/aside; **-ся** (*pass*); breathe heavily
от·ду́л·ин·а, -ы, (*f*) bulge
от·ду́·ть, -ют, (*perf*) *see* **от·дув·а́·ть**
от·ду́ш·ин·а, -ы, (*f*) vent, vent hole
от·ду́ш·ник·, -а, (*m*) vent, vent hole
о́т·дых·, -а, (*m*) rest; recovery (of materials); relief annealing, lonealing
от·дых·а́·ть, -ют, (*imp*) rest, relax, have/take a rest/holiday/vacation
о·тёк·, -а, (*m*) (med) oedema, edema, dropsy
~ лёгк·их (med) pulmonary oedema
о·тек·а́·ть, -ют, (*imp*) (med) swell, become dropsical
о·тёл·, -а, (*m*) (agr) calving
~ лед·ник·а́ glacier calving

отел·ь, -я, (*m*) hotel
отенит·, -а, (*m*) (min) autunite
о·тен·ить, -ят, (*perf*) *see* о·тен·я·ть
о·тен·я·ть, -ют, (*imp*) shade
о·тепл·ённ·ый, -ая, -ое, (*past part pass*) of о·тепл·ить; heat-insulated
о·тепл·ить, -ят, (*perf*) *see* о·тепл·я·ть
о·тепл·я·ть, -ют, (*imp*) make warm/cold-proof, heat-insulate
о·тер·еть, (*fut 3rd sing, pl*) о·тр·ёт, о·тр·ут, (*past masc sing*) о·тёр·, (*perf*); *see* об·тир·а·ть
о·тес·ать, (*fut 3rd pl*) о·теш·ут, (*perf*); *see* об·тёс·ыва·ть
отец·, -тц·а, (*m*) father
отеч·еск·ий, -ая, -ое, (*a*) paternal, father's
отеч·ественн·ый, -ая, -ое, (*a*) of отеч·еств·о; domestic, home (of industry etc.); patriotic (wars)
отеч·еств·о, -а, (*n*) fatherland, native country, home, homeland
отёч·н·ый, -ая, -ое, (*a*) (med) edematic, oedematic, dropsical
о·теч·ь, (*fut 3rd sing, pl*) о·теч·ёт, о·тек·ут, (*past masc sing*) о·тёк·, (*perf*); *see* о·тек·а·ть
о·теш·ет (*fut 3rd sing, pl*) of о·тес·ать
от·жат·и·е, -я, (*n*) (*v n*), *see* от·жим·а·ть
сигнал· от·жат·и·я (*m*) (telecom) spacing signal
от·жат·ый, -ая, -ое, (*past part pass*) of от·жат·ь
от·жат·ь, (*fut 3rd sing, pl*) ото·жм·ёт, ото·жм·ут, (*past masc sing*) от·жал, (*perf*); *see* от·жим·а·ть; (*fut 3rd sing, pl*) ото·жн·ёт, ото·жн·ут, (*past masc sing*) ото·жа·л, (*perf*); *see* от·жин·а·ть
от·жеч·ь, (*fut 3rd sing, pl*) ото·жж·ёт, ото·жг·ут, (*past masc sing*) от·жёг·, (*perf*); *see* от·жиг·а·ть
от·жив·а·ть, -ют, (*imp*) become obsolete; have had one's day
от·жиг·, -а, (*m*) anneal, annealing, annealing treatment
~ втор·ого род·а annealing for phase recrystallization
~, выс·ок·ий homogenizing
~, диффузион·н·ый homogenizing
~, изо·терм·ическ·ий isothermal annealing
~ на мелк·ое зерн·о grain refining, grain refining annealing
~, не·полн·ый partial anneal

от· жиг
~, низ·к·ий stabilizing anneal
~, нормаль·н·ый полн·ый full anneal
~ перв·ого род·а annealing to improve equilibrium
~, полн·ый full anneal
~, пред·вар·ительн·ый (rad) pre-firing
~, радиацион·н·ый (nucl) radiation annealing
~, раз·мельч·ающ·ий grain refining, grain-refining annealing
~, рекристаллизацион·н·ый recrystallization anneal
~, светл·ый bright annealing
~, ступ·енчат·ый step annealing
~, сфероидиз·ирующ·ий spheroidizing
~, черн·ый black anneal
от·жиг·ательн·ый, -ая, -ое, (*a*) annealing
от·жиг·а·ть, -ют, (*imp*) anneal
от·жим·ани·е, -я, (*n*) (*v n*) of от·-жим·а·ть
от·жим·а·ть, -ют, (*imp*) wring/squeeze out; open, open up (by pressure); (mech) disengage; check away, ease (screws etc.)
от·жим·к·а, -и, (*g pl*) -м·ок·, (*f*) (*v n*), *see* от·жим·а·ть; (*pl*) marc, residue, pressed fruit residue, refuse
~, рыб·н·ые (*pl*) fish manure
от·жим·н·ый, -ая, -ое, (*a*) squeezing, wringing
от·жин·а·ть, -ют, (*imp*) finish reaping/harvesting
от·жи·ть, (*fut 3rd pl*) от·жив·ут, (*perf*); *see* от·жив·а·ть
от·звон·ить, -ят, (*perf*) ring (a signal); finish ringing
от·звук·, -а, (*m*) sound reflection, echo; (autom) response; repercussion
от·звуч·ать, -ат, (*perf*) die away (of sound)
от·зов·ёт, (*fut 3rd sing*) of ото·зв·ать, (*perf*); *see* от·зыв·а·ть (*imp*)
от·зыв·, -а, (*m*) opinion, reference, testimonial; review, report, appraisal, evaluation; reply, password; от·зыв·, -а, (*m*) (*v n*), *see* от·зыв·а·ть; (dipl) recall
от·зыв·а·ть, -ют, (*imp*) recall (from port etc.); take aside; -ся (*pass*); answer, respond to; evoke, call forth; give an opinion

от·зы́в·н·о́й, -а́я, -о́е, (*a*) (dipl) recall

от·зы́в·чивост·ь, -и, (*f*) responsiveness, response

от·зы́в·чив·ый, -ая, -ое, (*a*) responsive, sympathetic

оти́т·, -а, (*m*) (med) otitis

ОТК (*abbr*) of **от·де́л· техн·и́ческ·ого контро́л·я** quality control department

от·ка́ж·ет, (*fut 3rd sing*) of **от·каз·а́ть**

от·ка́з·, -а, (*m*) refusal, repudiation, renunciation; (law) rejection, nonsuit; natural (music); (tech) failure

~ за·ря́д·а (gunn) misfire, charge failure

   по́лн·ый до от·ка́з·а cram full, full to capacity

   рабо́т·а·ть без от·ка́з·а run smoothly/faultlessly

от·каз·а́ть, (*fut 3rd pl*) **от·ка́ж·ут,** (*perf*), *see* **от·ка́з·ыва·ть**

от·ка́з·ыва·ть, -ют, (*imp*) refuse, deny; break down, fail (of mechanism); **-ся** refuse, decline, repudiate, renounce, give up, abandon

~ без·о·па́с·н·о fail safe

от·ка́л·ыва·ть, -ют, (*imp*) chop/break/split off; unpin

от·ка́п·ыва·ть, -ют, (*imp*) dig up/out

от·ка́рмл·ива·ть, -ют, (*imp*) (agr) fatten

от·ка́т·, -а, (*m*) (*v n*), *see* **от·ка́т·ыва·ть;** recoil

от·кат·и́ть, -ят, (*perf*) *see* **от·ка́т·ыва·ть**

от·ка́т·к·а, -и, (*g pl*) **-т·ок·,** (*f*) (*v n*), *see* **от·ка́т·ыва·ть;** (min) haulage

~ груз·овик·а́ми trucking

от·кат·н·о́й, -а́я, -о́е, (*a*) of **от·ка́т·**

~ воро́т·а (*pl*) box gate (dry dock)

от·ка́т·очн·ый, -ая, -ое, (*a*) (min) haulage

~ штрек· (*m*) (min) haulage drift

от·ка́т·чик·, -а, (*m*) (min) trammer, haulageman

от·ка́т·ыва·ть, -ют, (*imp*) roll aside/away/off; (min etc.) haul; **-ся** (*pass*); roll aside/away; recoil, rebound, roll back

от·кач·а́нн·ый, -ая, -ое, (*past part pass*) of **от·кач·а́·ть**

от·кач·а́·ть, -ют, (*perf*) *see* **от·ка́ч·ива·ть**

от·кач·енн·ый, -ая, -ое, (*past part pass*) of **от·кат·и́ть**

от·ка́ч·ива·ть, -ют, (*imp*) pump out, evacuate, scavenge; give artificial respiration

от·ка́ч·ивающ·ий на·со́с· (*m*) pump (for raising liquids); (a/c) scavenge pump

от·ка́ч·к·а, -и, (*g pl*) **-ч·ек·,** (*f*) (*v n*) *see* **от·ка́ч·ива·ть;** evacuation, scavenge

от·кач·н·у́ть, -у́т, (*perf*) swing aside/back

от·кид·н·о́й, -а́я, -о́е, (*a*) hinged, flap, folding, collapsible

~ ста́в·н·и (*pl*) (build) folding shutters

~ шлюп·ба́л·к·а (*f*) (naut) luffing/quadrant davit

~ щит· (*m*) flap plate

от·ки́д·ыва·ть, -ют, (*imp*) throw away/back; fold back, turn up/down

от·ки́н·уть, -ут, (*perf*) *see* **от·ки́д·ыва·ть**

от·кла́д·к·а, -и, (*g pl*) **-д·ок·,** (*f*) (*v n*) *see* **от·кла́д·ыва·ть;** (zool) oviposition

от·кла́д·ыва·ть, -ют, (*imp*) put/place/lay/set aside/apart; postpone, put off, defer; adjourn; (zool) oviposit, lay (eggs); (geol) deposit, lay down

от·кле́·ива·ть, -ют, (*imp*) unstick

от·кле́·ить, -ят, (*perf*) *see* **от·кле́·ива·ть**

о́т·клик·, -а, (*m*) response, answer

от·клик·а́·ться, -ются, (*imp*) respond to, answer; comment on

от·кли́к·н·уться, -утся, (*perf*) *see* **от·клик·а́·ться**

от·клон·е́ни·е, -я, (*n*) (*v n*), *see* **от·клон·я́·ть(ся);** deflection, deflexion, departure, deviation

~, больш·о́е wide-angle deflection (cathode ray tube)

~, вероя́т·н·ое probable error; (gunn) 50% zone, 50% probability zone

~, вероя́т·н·ое бок·ов·о́е (gunn) 50% breadth zone

~, вертика́ль·н·ое vertical deflection

~, горизонта́ль·н·ое horizontal deflection

~, до·пуст·и́м·ое tolerance

~, квадрат·и́чн·ое (math) standard deviation, root mean square deviation

   коэффицие́нт· от·клон·е́ни·я (*m*) departure factor

~, круг·ов·о́е circular deflection

~, ле́в·ое backing (of wind)

   ли́н·и·я от·клон·е́ни·я (*f*) deflection line

от·клон·éни·е
~, магнúт·н·ое magnetic deflection/ variation; (elec) electromagnetic deflection
~ óрган·а у·правл·éни·я (autom) control deflection
~, о·стáт·очн·ое (autom) offset, drift, droop
~ от за·кóн·а X non-X behaviour
~ от нóрм·ы deviation
~ по кáдр·у (TV) frame deflection
~ по строк·é (TV) line deflection
~, прáв·ое veering (of wind)
~, срéд·н·ее mean error (departure from correct value)
~, средне·квадрат·úчн·ое root-mean-square deviation, standard deviation
~, стандáрт·н·ое standard deviation
~ стрéл·к·и pointer/needle throw/ deflection
~ температýр·ы, баллист·úческ·ое ballistic temperature
~ траектóр·и·и по·лёт·а flight path deviation
~ тя́г·и thrust deviation
ýгол· от·клон·éни·я (m) angle of deviation/deflection, (opt) diffraction angle, refraction angle (of light signal)
~, част·óтн·ое (rad) frequency deviation/drift
от·клон·úтел·ь, -я, (m) deflector
~ стру·ú jet deflector
от·клон·úть, -я́т, (perf) see **от·клон·я́·ть**
от·клон·я́·ть, -ют, (imp) deflect, divert; decline, reject; **-ся** (pass); diverge, deviate, turn aside; digress
от·клон·я́ющ·ий, -ая, -ее, (pres part act) of **от·клон·я́·ть; -ся** divergent
~ кат·ýшк·а (f) deflecting coil/yoke (cathode ray tube)
~ пласт·úн·а (f) deflecting plate (cathode ray tube)
~ систéм·а (f) deflector, deflecting yoke/system, yoke (of cathode ray tube)
от·ключ·áтел·ь, -я, (m) disconnecting switch/mechanism
~, автомат·úческ·ий automatic closing mechanism
от·ключ·á·ть, -ют, (imp) switch off, disconnect, isolate, cut off
от·ключ·éни·е, -я, (n) (v n), see **от·ключ·á·ть**

от·ключ·ённ·ый, -ая, -ое, (past part pass) see **от·ключ·á·ть**; (elec) dead, disconnected, detached, switched off, off
от·ключ·úть, -áт, (perf); see **от·ключ·á·ть**
от·ков·áть, (fut 3rd sing, pl) **от·ку·ёт, от·ку·ю́т,** (perf); see **от·кóв·ыва·ть**
от·кóв·ыва·ть, -ют, (imp) forge; hammer off
от·ковы́р·ива·ть, -ют, (imp) pick/ chip off
от·ковыр·я́·ть, -ют, (perf) see **от·ковы́р·ива·ть**
от·колáч·ива·ть, -ют, (imp) strike/ knock/beat off
от·колот·úть, ꞌя́т, (perf) see **от·колáч·ива·ть**
от·кóл·от·ый, -ая, -ое, (past part pass) of **от·кол·óть**
от·кол·óть, ꞌют, (perf) see **от·кáл·ыва·ть**
от·колóч·енн·ый, -ая, -ое, (past part pass) of **от·колот·úть**
от·команд·ирóванн·ый, -ая, -ое, (past part pass) of **от·команд·ирó·выва·ть**
от·команд·ир·овáть, -ýют, (perf) see **от·команд·ирó·выва·ть**
от·команд·ирó·выва·ть, -ют, (imp) send on business, send on detached duty, second, post
от·комáнд·овать, -уют, (perf) be in command (a period of time); relinquish command
от·коп·á·ть, -ют, (perf) dig up/out
от·кóрм·, -а, (m) (v n) of **от·корм·úть**
от·корм·úть, ꞌя́т, (perf) (agr) fatten
от·коррект·úр·овать, -уют, (perf) correct
от·кóс·, -а, (m) (civ eng) slope; (geol) scarp; (min) bank, dip; (met) banks (of OH furnace)
ýгол· естéств·енн·ого от·кóс·а (m) (civ eng) angle of repose
ýгол· от·кóс·а (m) slope
от·кóс·н·ый, -ая, -ое, (a) of **от·кóс·**
~ профúл·ы (pl) (civ eng) slope staking
от·креп·úть, -я́т, (perf) see **от·креп·л·я́·ть**
от·крепл·я́·ть, -ют, (imp) unfasten, unfix; strike off (item on list etc.)
от·кров·éни·е, -я, (n) revelation
от·кров·éнност·ь, -и, (f) frankness

от·кров·е́нн·ый, -ая, -ое, (*a*) frank, outspoken

от·кро́·ет (*fut 3rd sing*) of **от·кры́·ть**

от·крут·и́ть, -́ят, (*perf*) *see* **от·кру́ч·ива·ть**

от·кру́ч·ива·ть, -ют, (*imp*) unwind; unscrew, screw off, undo (a screw); unlay (rope etc.), untwist

от·крыв·а́·ть, -ют, (*imp*) open, uncover, reveal, disclose, discover, detect; open up; offer (possibilities of)

от·кры́т·и·е, -я, (*n*) (*v n*) *see* **от·крыв·а́·ть**; opening; discovery

от·кры́т·к·а, -и, (*g pl*) **-т·ок·,** (*f*) postcard

открыто·пло́д·н·ые, -ых, (*pl decl as adj*) (bot) discomycetes

от·кры́·т·ый, -ая, -ое, (*past part pass*) *see* **от·крыв·а́·ть**; open; overt, undisguised

~ вы́·работ·к·а (*f*) (min) trench

~ маши́н·а (*f*) (elec) open machine

~ о·хлажд·е́ни·е (*n*) evaporative cooling

~ рабо́т·ы (*pl*) (min) open-cast/cut working

~ текст· (*m*) (telecom) plain language

от·кры́·ть, (*fut 3rd sing, pl*) **от·кро́··ет, от·кро́·ют,** (*perf*); *see* **от·кры·в·а́·ть**

от·ку́да (*adv*) where ... from, from where, whence

~ ни wherever ... from

~ -либо (*adv*) *see* **от·ку́да-нибудь**

~ -нибудь (*adv*) from somewhere or other

~ -то (*adv*) from somewhere

от·ку·ёт, (*fut 3rd sing*) of **от·ков·а́ть**

о́т·куп·, -а, (*nom pl*) **-а́,** (*g pl*) **-о́в,** (*m*) (*v n*) of **от·куп·а́·ть**; lease

от·куп·а́·ть, -ют, (*imp*) buy, buy up (all); lease

от·куп·и́ть, -́ят, (*perf*) *see* **от·ку·п·а́·ть**

от·ку́пор·енн·ый, -ая, -ое, (*past part pass*) of **от·ку́пор·ить;** uncorked, open (of bottles)

от·ку́пор·ить, -ят, (*perf*) uncork, open (a bottle)

от·кус·и́ть, -́ят, (*perf*) bite off

от·ку́ш·енн·ый, -ая, -ое, (*past part pass*) of **от·кус·и́ть**

от·ку·ю́т, (*fut 3rd pl*) of **от·ков·а́ть**

от·лаг·а́·ть, -ют, (*imp*) postpone, put off, adjourn; (geol) deposit

от·ла́д·ить, -ят, (*perf*) put right, adjust

от·ла́д·к·а, -и, (*g pl*) **-д·ок·,** (*f*) (*v n*) *see* **от·ла́д·ить;** adjustment

от·ла́ж·ива·ть, -ют, (*imp*) put right, adjust

от·ла́м·ыва·ть, -ют, (*imp*) break off

от·леп·и́ть, -я́т, (*perf*) *see* **от·леп·л·я́·ть**

от·лепл·я́·ть, -ют, (*imp*) unstick, take off

от·лёт·, -а, (*m*) (*v n*) *see* **от·лет·а́·ть;** (air) departure

от·лет·а́·ть, -ют, (*imp*) fly away/off; be thrown back/aside/off; come/fall off, be torn off

от·лет·е́ть, -я́т, (*perf*) *see* **от·лет·а́·ть**

от·ли́·в·, -а, (*m*) (*v n*) *see* **от·лив·а́·ть;** ebb tide, ebb; (arch) drip mould

~, квадрату́р·н·ый neap tide

~ при·бо́·я undertow

с от·ли́·в·ом shot with (a colour)

от·лив·а́·ть, -ют, (*imp*) pour off, decant; discharge (liquids), pump/bail out; (met) cast, found; ebb; be shot with (a colour)

от·ли́в·к·а, -и, (*g pl*) **-в·ок·,** (*f*) (*see also* **лит·ь·ё**) founding, casting (action); casting, cast; pouring off, decanting, discharging (liquids)

~, цель·н·ая monoblock casting

от·лив·н·о́й, -а́я, -о́е, (*a*) (*see also* **от·ли́в·н·ый**) cast, casting, founded; discharge (of liquids)

~ аппара́т· (*m*) (print) casting machine

~ шланг· (*m*) discharge hose

от·ли́в·н·ый, -ая, -ое, (*a*) (*see also* **от·лив·н·о́й**) ebb

~ теч·е́ни·е (*n*) ebb stream

от·ли́п·, -а, (*m*) tackiness; (*past masc sing*) of **от·ли́п·нуть**

от·лип·а́ни·е, -я, (*n*) (*v n*) of **от·ли·п·а́·ть;** peeling off

от·лип·а́·ть, -ют, (*imp*) come unstuck/off

от·ли́п·нуть, -нут, (*perf*) *see* **от·ли·п·а́·ть**

от·ли́·ть, (*fut 3rd sing, pl*) **ото·ль·ёт, ото·ль·ю́т,** (*past masc sing*) **от·ли·л,** (*perf*); *see* **от·лив·а́·ть**

от·ли́·т·ый, -ая, -ое, (*past part pass*) of **от·ли́·ть;** cast; cast in

от·лич·а́·ть, -ют, (*imp*) distinguish; **-ся** be distinguished/distinctive, differ from; distinguish oneself, be noted for

от·ли́ч·и·е, -я, (*n*) difference, distinction

в от·ли́ч·и·и от unlike, in contrast to

от·лич·и́тельн·ый, -ая, -ое,   (*a*) distinctive, distinguishing, characteristic
~ глуб·ин·а́   (*f*) (surv) deep patch
~ ого́н·ь   (*m*) (naut) side light; (rail) marker light
~ при́·знак·   (*m*) distinguishing feature
от·лич·и́ть, -а́т, (*perf*) *see* от·лич·а́·ть
от·ли́ч·ник·, -а,   (*m*) Otlichnik (person distinguished for excellent work)
от·ли́ч·н·ый, -ая, -ое,   (*a*) different, distinct (from), other (than); excellent
~ от нул·я́   (math) non-vanishing
от·ло́г·ий, -ая, -ое,   (*a*) sloping, gently sloping
от·лож·е́ни·е, -я,   (*n*) deposit, deposition, sediment; precipitate, precipitation
~, абисса́ль·н·ые   (*pl*) (ocean) abyssal deposits
~, вну́тр·енн·ее   internal deposition
~ на·но́с·ов   (geol) alluviation
~, реч н·о́е   alluvial deposit
~, ио́н·н·ое   ionic deposition
от·ло́ж·енн·ый, -ая, -ое,   (*past part pass*) of от·лож·и́ть; deposited; deferred, put off
от·лож·и́ть, -а́т, (*perf*) *see* от·кла́д·ыва·ть, от·лаг·а́·ть
от·лож·н·о́й, -а́я, -о́е,   (*a*) turn-down, lay-down (of collars)
от·лом·а́·ть, -ют, (*perf*) *see* от·ла́м·ыва·ть
от·лом·и́ть, -я́т, (*perf*) *see* от·ла́м·ыва·ть
от·луч·а́·ться, -ются, (*imp*) be absent, absent oneself
от·луч·и́ться, -а́тся, (*perf*) *see* от·лу·ч·а́·ться
от·лу́ч·к·а, -и, (*g pl*) -ч·ек·, (*f*) absence, temporary absence
~, само·во́ль·н·ая   absence without leave
от·магни́ч·ивани·е, -я,   (*n*) magnetic separation
от·ма́т·ыва·ть, -ют, (*imp*) wind/reel off
от·мах·а́ть, (*fut 3rd pl*) от·ма́ш·ут, (*perf*) (naut) give a visual signal (e.g. semaphore, light etc.); cover (a distance); wave away/off
от·ма́ч·ива·ть, -ют,   (*imp*) soak off; soak, soak out, wet (flax)
от·ма́ш·ет, (*fut 3rd sing*) of от·мах·а́ть
от·ма́ш·к·а, -и,   (*g pl*) -ш·ек·, (*f*) (*v n*) of от·мах·а́ть
от·меж·ева́ть, -у́ют, (*perf*) mark off (a boundary)
от·мёл, (*past masc sing*) of от·мес·ти́

о́т·мел·ь, -и, (*f*) shelving bottom (of sea. etc.), shelf; shoal
~, матер·ико́в·ая   (ocean) continental shelf
~, над·во́д·н·ая   emerged shoal
~, остров·н·а́я   (ocean) insular shelf
~, песч·а́н·ая   sandbank
~, усть·ев·а́я   bar
от·ме́н·а, -ы,   (*f*) (*v n*) of от·мен·я́·ть; repeal, cancellation
от·мен·и́ть, -я́т, (*perf*) *see* от·мен·я́·ть
от·ме́н·н·ый, -ая, -ое,   (*a*) excellent
от·мен·я́·ть, -ют, (*imp*) abolish, repeal, cancel, countermand
от·мер·е́ть, (*fut 3rd sing, pl*) ото·мр·ёт, ото·мр·у́т, (*past masc sing*) о́т·мер·, (*perf*); die off/out
от·мерз·а́·ть, -ют, (*imp*) (bot) killed by the cold
от·мёрз·нуть, -нут, (*perf*) *see* от·мерз·а́·ть
от·ме́р·ива·ть, -ют,   (*imp*) measure off
от·ме́р·ить, -ят, (*perf*) *see* от·ме́р·ива·ть
от·мер·я́·ть, -ют,   (*imp and perf*) *see* от·ме́р·ива·ть
от·мес·ти́, (*fut 3rd sing, pl*) от·мет·ёт, от·мет·у́т, (*past masc sing*) от·мёл, (*perf*); *see* от·мет·а́·ть
от·мет·а́·ть, -ют,   (*imp*) sweep off/away/aside
от·ме́т·ин·а, -ы,   (*f*) mark, blaze
от·ме́т·ить, -ят, (*perf*) *see* от·меч·а́·ть
от·ме́т·к·а, -и, (*g pl*) -т·ок·, (*f*) mark, indication; (elec) pip, blip; (surv) datum, level, level mark
~, абсолю́т·н·ая   (surv) height above sea level
~ вре́мен·и   timing pip, time mark
~ да́ль·ност·и   (rad) range marker
~, калибр·о́вочн·ая   (rad) marker pip
~, контро́ль·н·ая   reference mark
~, масшта́б·н·ая   (rad) calibration pip
~ на раккорд·е   (cinema) cue
~, нул·ев·а́я   (surv) datum
~, телеметр·и́ческ·ая   telemetry reading
~, хрон·и́рующ·ая   timing pip
~, цел·и   (rad) target echo
от·ме́т·чик·, -а,   (*m*) marker, recorder; (rad) display, indicator (*see also* инди·ка́тор·)
~ вре́мен·и   (rad) timer, time marker
~ круг·ов·о́го об·зо́р·а   (rad) plan position indicator
от·мёт·ш·ий, -ая, -ее,   (*past part act*) of от·мес·ти́

отмеч·а́·ть, -ют,   (*imp*) mark; note, observe, notice, remark upon
   след·ует от·ме́т·ить (*perf*) it is worth noting, it is emphasized
от·меч·а́ющ·ий о́рган· (*m*) marking element (automatic inspection)
от·мир·а́·ть, -ют, (*imp*) die off/out
от·мо́·ет, (*fut 3rd sing*) of **от·мы́·ть**
от·мок·а́·ть, -ют,   (*imp*) become wet through, become soaked/wet/damp; come unstuck/off
от·мо́к·нуть, -нут, (*perf*) *see* **от·мо·ка́·ть**
от·мола́ч·ива·ть, -ют,   (*imp*) (agr) thresh/thrash (for a time)
от·молот·и́ть, -я́т, (*perf*) (agr) thresh/thrash (for a time); finish threshing/thrashing
от·мора́ж·ивани·е, -я, (*n*) (med) frostbite
от·мора́ж·ива·ть, -ют, (*imp*) get frostbitten/frozen
от·моро́з·ить, -я́т, (*perf*) *see* **от·мо·ра́ж·ива·ть**
от·мот·а́·ть, -ют, (*perf*) *see* **от·ма́т·ыва·ть**
от·мо́т·к·а, -и, (*g pl*) **-т·ок·,** (*f*) (*v n*) *see* **от·ма́т·ыва·ть;** (text) backing-off
от·моч·и́ть, -а́т, (*perf*) *see* **от·ма́ч·и·ва·ть**
от·мо́·ют, (*fut 3rd pl*) of **от·мы́·ть**
от·мут·и́ть, -я́т, (*perf*) *see* **от·му́ч·и·ва·ть**
от·му́ч·енн·ый, -ая, -ое, (*past part pass*) *see* **от·му́ч·ива·ть;** elutriated, washed
  ~ гли́н·а (*f*) (chem) washed clay
от·му́ч·ивани·е, -я, (*n*) elutriating, elutriation
от·му́ч·ива·ть, -ют, (*imp*) elutriate
от·мыв·а́·ть, -ют, (*imp*) wash away/off, wash
от·мы́в·к·а, -и, (*g pl*) **-в·ок·,** (*f*) (*v n*) *see* **от·мыв·а́·ть**
от·мык·а́·ть, -ют, (*imp*) unlock, unbolt; unfix (bayonets etc.)
от·мы́·ть, (*fut 3rd sing, pl*) **от·мо́·ет, от·мо́·ют,** (*past masc sing*) **от·мы́·л,** (*perf*); *see* **от·мыв·а́·ть**
от·мы́т·ый, -ая, -ое, (*past part pass*) *see* **от·мыв·а́·ть**
от·мы́ч·к·а, -и, (*g pl*) **-ч·ек·,** (*f*) master/skeleton key
от·мяк·а́·ть, -ют, (*imp*) become soft/softer, soften
от·мя́к·нуть, -нут, (*perf*) *see* **от·мя·ка́·ть**

от·несе́·ни·е, -я, (*n*) (*v n*) *see* **от·но·си́ть;** reference
от·нес·ти́, -у́т, (*perf*) *see* **от·нос·и́ть**
от·нес·ённ·ый, -ая, -ое, (*past part pass*) *see* **от·нос·и́ть;** related/relative/referred to
от·ним·а́·ть, -ют,   (*imp*) take, take from/away/off, subtract; **-ся** (*pass*); be paralysed, lose the use of
от·но́г·а, -и, (*f*) (bot) aerial root
от·но́с·, -а,   (*m*) forward travel (of falling body); (mech) lead (advanced position)
от·нос·и́мост·ь, -и, (*f*) reference
  у́гол· от·нос·и́мост·и (*m*) reference angle
от·нос·и́тельн·о (*adv*) relatively, comparatively; (*prep* + *gen*) concerning, about; relative to, about (a position)
  вращ·е́ни·е от·нос·и́тельн·о о́с·и (*n*) rotation about an axis
  на·сыщ·е́ни·е от·нос·и́тельн·о льд·ы́ (*n*) (meteor) ice saturation
от·нос·и́тельност·ь, -и, (*f*) relativity
  тео́р·и·я от·нос·и́тельност·и (*f*) theory of relativity
от·нос·и́тельн·ый, -ая, -ое, (*a*) relative, ratio, percentage
  ~ велич·ин·а́ крыл·а́ (*f*) (a/c) thickness/chord ratio
  ~ вла́ж·ност·ь (*f*) (meteor) relative humidity
  ~ вы́·ход· (*m*) relative yield
  ~ со·держ·а́ни·е изото́п·ов (*n*) (nucl) abundance ratio
  ~ с·уж·е́ни·е (*n*) (met) percentage reduction of area; (dynam) contraction ratio
  ~ у·длин·е́ни·е (*n*) (met) percentage elongation, elongation
  ~ у·длин·е́ни·е, равно·ме́р·н·ое (*n*) (met) elongation prior to fracture
от·нос·и́ть, -я́т,   (*imp*) carry/take to/away/off; move back, resite; count as, include among; attribute, ascribe; postpone; **-ся** (*pass*); treat, behave/react to; (*imp only*) concern, have to do with, have bearing on; apply/pertain/belong to; be, be in relation to
от·нош·е́ни·е, -я,   (*n*) relation, relationship, ratio; attitude, regard; memorandum, official letter
  в от·нош·е́ни·и with regard to, as to
  ~ ветв·е́й рас·па́д·а (nucl) branching ratio
  ~, гиро·магни́т·н·ое (phys) gyromagnetic ratio

от·нош·éни·е

~, **двой·н·óе**   (math) anharmonic/cross ratio

за·кóн по·сто·я́нн·ых от·нош·é·ни·й (*m*) (chem) law of proportions, law of constant proportions, law of equivalent/reciprocal proportions

~, **магнито·механ·и́ческ·ое**   gyromagnetic ratio

~ **ма́ксимум·а к ми́нимум·а** (math) peak-to-valley ratio

~ **масс·**   mass ratio; (aerodynam) coefficient of dilution

~ **на óc·и, прóбочн·ое** (nucl) axial mirror ratio

~ **площад·éй**   area ratio

по·от·нош·éни·е к   with reference to, relative to, in respect of; about (a position)

~, **по·сто·я́нн·ое**   constant proportion/ratio/relation

~ **по·тóк·ов мáсс·ы**   mass-flow ratio

~, **прóбочн·ое**   (nucl) mirror ratio

~ **секу́нд·н·ых масс·**   mass-flow ratio

~ **сигнáл·/по·мéх·а**   signal-to-noise ratio

~, **с·лóж·н·ое**   double/cross ratio

~ **сторóн· фото·с·ни́м·к·а** (phot) aspect ratio

~, **шаг·ов·ое**   pitch ratio (of screw propeller)

отню́дь не   (*adv*) not in the least, not at all, by no means

от·ня́т·и·е, -я, (*n*) (*v n*), *see* **от·ним·á·ть;** removal

óт·ня·т·ый, -ая, -ое, (*past part pass*) *see* **от·ним·á·ть**

от·ня́·ть, (*fut 3rd sing, pl*) **от·ни́м·ет, от·ни́м·ут,** (*past masc sing*) **óт·ня·л,** (*perf*); *see* **от·ним·á·ть**

ото (*prep or prefix*) *see* **от**

от·ображ·á·ть, -ют, (*imp*) show (of inanimate objects), depict, represent, reflect

от·ображ·áющ·ий, -ая, -ее, (*pres part act*) of **от·ображ·á·ть**

~ **у·стрóй·ств·о**   (*n*) (instr) display, presentation, indication

~ **фу́нкци·я**   (*f*) (math) mapping function

от·ображ·éни·е, -я, (*n*) reflection, image; representation; (math) mapping, transformation, map, correspondence

~, **взаймо·одно·знáч·н·ое**   one-to-one correspondence

от·ображ·éни·е

~, **зеркáль·н·ое**   mirror transformation

~, **конфóрм·н·ое**   (math) conformal/isogonal transformation

~, **тополог·и́ческ·ое**   (math) topological transformation

от·ображ·ённ·ый, -ая, -ое, (*past part pass*) of **от·образ·и́ть**

от·образ·и́ть, -я́т, (*perf*) *see* **от·обра·ж·á·ть**

отó·бр·анн·ый, -ая, -ое, (*past part pass*) *see* **от·бир·á·ть;** selected; bled (of steam, heat etc.)

ото·бр·áть, (*fut 3rd sing, pl*) **от·бер·ёт, от·бер·у́т,** (*past masc sing*) **ото·-бр·áл,** (*perf*); *see* **от·бир·á·ть**

ото·бь·ёт, (*fut 3rd sing*) of **от·би́·ть**

о·товáр·енн·ый, -ая, -ое, (*past part pass*) of **о·товáр·ить;** supplied (by)

о·товáр·ива·ть, -ют, (*imp*) supply with goods under contract, have the contract for

о·товáр·ить, -ят, (*perf*) *see* **о·товá·р·ива·ть**

отó·гн·анн·ый, -ая, -ое, (*past part pass*) *see* **от·гон·я́·ть;** (chem) distilled off, overhead; driven off

~ **продýкт·**   (*m*) (chem) overhead product

ото·гн·áть, (*fut 3rd sing, pl*) **от·гóн·ит, от·гóн·ят,** (*past masc sing*) **ото·-гн·áл,** (*perf*); *see* **от·гон·я́·ть**

отó·гн·ут·ый, -ая, -ое, (*past part pass*) of **ото·гн·у́ть;** (bot) recurved, recurvate

ото·гн·у́ть, -у́т, (*perf*) bend back, unbend

ото·грев·áни·е, -я, (*n*) (*v n*) of **ото·-грев·á·ть**

ото·грев·á·ть, -ют, (*imp*) heat/warm up again, reheat, rewarm

ото·гре́·ть, -ют, (*perf*) *see* **ото·гре-в·á·ть**

ото·двиг·á·ть, -ют, (*imp*) move aside/back/away; postpone, put off

ото·дви́·н·уть, -ут, (*perf*) *see* **ото·двиг·á·ть**

ото·др·áть, (*fut 3rd sing, pl*) **от·дер·ёт, от·дер·у́т,** (*past masc sing*) **ото·-др·áл,** (*perf*); tear/pull off

ото·жг·у́т, (*fut 3rd pl*) of **от·жéч·ь,** (*perf*); *see* **от·жиг·á·ть** (*imp*)

о·тождеств·и́ть, -я́т, (*perf*); *see* **о·тож-дествл·я́·ть**

о·тождествл·éни·е, -я, (*n*) identification (with)

о·тождествл·я́·ть, -ют, (*imp*) identify (with)

о·тожеств·и́ть, -я́т, (*perf*) *see* **о·тождеств·и́ть**

ото·жж·ённ·ый, -ая, -ое, (*past part pass*) *see* **от·жиг·а́·ть;** annealed

ото·жж·ёт, (*fut 3rd sing*) of **от·жёч·ь,** (*perf*); *see* **от·жиг·а́·ть** (*imp*)

ото·жм·ёт, (*fut 3rd sing*) of **от·жа́т·ь,** (*perf*); *see* **от·жим·а́·ть** (*imp*)

ото·жн·ёт, (*fut 3rd sing*) of **от·жа́т·ь,** (*perf*); *see* **от·жин·а́·ть** (*imp*)

ото·зв·а́ть, (*fut 3rd sing, pl*) **от·зов·ёт, от·зов·у́т,** (*past masc sing*); **ото·зв·а́л,** (*perf*); *see* **от·зыв·а́·ть**

ото·йти́, (*fut 3rd sing, pl*) **ото·йд·ёт, ото·йд·у́т,** (*past masc sing*) **ото·шё·л,** (*perf*); *see* **от·ход·и́ть**

ото·ли́т·, -а, (*m*) (anat) otolith, ear bone

ото·ло́г·и·я, -и, (*f*) (med) otology

ото·ль·ёт, (*fut 3rd sing*) of **от·ли́·ть**

ото·мк·н·у́ть, -у́т, (*perf*) *see* **от·мы·к·а́·ть**

ото·мр·ёт, (*fut 3rd sing*) of **от·мер·е́ть**

о·топ·и́тельн·ый, -ая, -ое, (*a*) heating, space-heating

~ **котёл·** (*m*) central-heating boiler

о·топ·и́ть, ◌я́т, (*perf*) heat (space)

о·топл·е́ни·е, -я, (*n*) heating, space heating, central heating

~**, вод·о-вод·ян·о́е** district/area water central heating

~**, вод·ян·о́е** water central heating

~**, мазу́т·н·ое** oil-fired heating; oil firing (of furnaces)

~**, пар·ов·о́е** steam heating

ото·пр·ёт, (*fut 3rd sing*) of **от·пер·е́ть**

о·тора́ч·ива·ть, -ют, (*imp*) (text) edge, trim

ото́·рв·анност·ь, -и, (*f*) isolation

ото·рв·а́ть, -у́т, (*perf*) tear/break off/ away; isolate, cut off, separate; **-ся** (*pass*); (air) take off, become airborne

ото·рино·ларинго·ло́г·и·я, -и, (*f*) (med) otorhinolaryngology

о·тороч·и́ть, -а́т, (*perf*) *see* **о·тора́ч·ива·ть**

о·торо́ч·к·а, -и, (*g pl*) **-ч·ек·,** (*f*) edge, edging, fringe; (geol) margin, fringe

~**, гли́н·ист·ая** (geol) gouge

о·торф·ова́ни·е, -я, (*n*) (geol) peat formation

о·тор·цо́вк·а, -и, (*g pl*) **-вок·,** (*f*) cross-cut (timber)

ото·склеро́з·, -а, (*m*) (med) otosclerosis

ото·ско́п·, -а, (*m*) (med) otoscope

ото·сл·а́ть, (*fut 3rd sing, pl*) **ото·шл·ёт, ото·шл·ю́т,** (*past masc sing*) **ото·сл·а́л,** (*perf*); *see* **от·сыл·а́ть**

ото·тр·ёт, (*fut 3rd sing*) of **от·тер·е́ть,** (*perf*); *see* **от·тир·а́ть** (*imp*)

ото·ше́д·ш·ий, -ая, -ее, (*past part act*) of **ото·йти́**

ото·шл·ёт, (*fut 3rd sing*) of **ото·сл·а́ть**

ото·шь·ёт, (*fut 3rd sing*) of **от·ши́·ть,** (*perf*); *see* **от·шив·а́ть** (*imp*)

о·тощ·а́·ть, -ют, (*perf*) become emaciated/lean

от·пад·а́·ть, -ют, (*imp*) fall off/away; pass, lose point/significance

от·пад·а́ющ·ий, -ая, -ее, (*pres part act*) *see* **от·пад·а́·ть;** (bot) caducous, fugacious

от·па́·ива·ть, -ют, (*imp*) unsolder; seal, seal off (glass ware); feed/treat with (drink)

от·па́й·к·а, -и, (*g pl*) **-а́·ек·,** (*f*) (*v n*) of **от·па́·ива·ть**

от·па́р·ива·ть, -ют, (*imp*) steam-press (clothes etc.); steam off

от·па́р·ить, -ят, (*perf*) *see* **от·па́р·и·ва·ть**

от·па́р·к·а, -и, (*f*) (*v n*), *see* **от·па́р·и·ва·ть;** (oil) stripping

от·па́р·н·ый, -ая, -ое, (*a*) steaming; (oil) stripping

~ **коло́нн·а** (*f*) (oil) stripper

~ **коло́нн·а, вы·нос·н·а́я** (*f*) side-stripping column, side stream stripper

~ **се́кци·я** (*f*) (oil) stripper

от·па́р·ыва·ть, -ют, (*imp*) open, rip open (a sewn seam)

от·па́с·ть, (*fut 3rd sing, pl*) **от·пад·ёт, от·пад·у́т,** (*past masc sing*) **от·па́л,** (*perf*); *see* **от·пад·а́·ть**

от·па́·янн·ый, -ая, -ое, (*past part pass*) of **от·па·я́·ть**

~ **тру́б·к·а** (*f*) (chem) sealed tube

от·па·я́·ть, -ют, (*perf*) unsolder; seal, seal off (glass ware)

от·пер·е́ть, (*fut 3rd sing, pl*) **ото·пр·ёт, ото·пр·у́т,** (*past masc sing*) **о́т·пер·,** (*perf*); *see* **от·пир·а́ть**

о́т·пер·т·ый, -ая, -ое, (*past part pass*) of **от·пер·е́ть**

от·печа́т·ани·е, -я, (*n*) (*v n*), *see* **от·печа́т·ыва·ть**

от·печа́т·а·ть, -ют, (*perf*) *see* **от·печа́т·ыва·ть**

от·печа́т·ок·, -т·к·а, (*m*) print, imprint, mark; impression, indentation; replica (electron microscopy)

~**, конта́кт·н·ый** (phot) contact print

от·печа́т·ок

~, проекцио́н·н·ый (phot) projection print

~, угле·ро́д·ист·ый carbon replica

от·печа́т·ыва·ть, -ют, (imp) print, imprint; type (with typewriter); unseal

от·пи́л·ива·ть, -ют, (imp) saw off

от·пил·и́ть, ⸗ят, (perf) see от·пи́л·ива·ть

от·пир·а́·ть, -ют, (imp) unlock, unbolt, open; (elec) trigger off

от·пир·а́ющ·ая схе́м·а (f) (elec) trigger circuit

от·пла́т·а, -ы, (f) repayment

от·плат·и́ть, ⸗ят, (perf) see от·пла́ч·ива·ть

от·пла́ч·ива·ть, -ют, (imp) repay, pay back

от·плыв·а́·ть, -ют, (imp) swim/drift/float/sail off/out/away

от·плы́т·и·е, -я, (n) departure, sailing (of ship)

от·плы́·ть, (fut 3rd sing, pl) от·плыв·ёт, от·плыв·у́т, (past masc sing) от·плы́·л, (perf); see от·плыв·а́·ть

от·по·и́ть, -я́т, (perf) feed/treat with (drink)

от·полз·а́·ть, -ют, (imp) crawl away/back

от·полз·ти́, -у́т, (perf) see от·полз·а́·ть

от·полир·ова́·ть, -у́ют, (perf) polish

от·по́р·н·ый, -ая, -ое, (a) repulsing, fending off

~ крюк· (m) boathook

от·пор·о́ть, ⸗ют, (perf) see от·па́р·ыва·ть

от·пот·ева́ни·е, -я, (n) (v n) of от·пот·ева́·ть; condensation, sweating, misting (of glass)

от·пот·ева́·ть, -ют, (imp) become damp/moist, mist, mist up/over, sweat

от·пот·е́·ть, -ют, (perf) see от·пот·ева́·ть

от·почк·ова́·ться, -ются, (perf) see от·почк·о́выва·ться

от·почк·о́выва·ться, -ются, (imp) (biol) gemmate, propagate by gemmation

от·прав·и́тел·ь, -я, (m) sender, consigner; (telecom) originator, transmitter (autom)

от·пра́в·ить, -ят, (perf) see от·правл·я́·ть

от·пра́в·к·а, -и, (g pl) -в·ок·, (f) (v n) see от·правл·я́·ть; consignment, shipment

от·пра́в·к·а

~, по·ваго́н·н·ая (rail) wagon-load consignment

от·правл·е́ни·е, -я, (n) (v n), see от·правл·я́·ть

ме́ст·о от·правл·е́ни·я (n) origin, place of dispatch/departure

от·правл·я́·ть, -ют, (imp) dispatch, consign, send, forward; (imp only) administer, perform, exercise, function; -ся (pass); depart, leave, set off/out

от·прав·н·о́й, -а́я, -о́е, (a) departure, starting

~ пункт· (m) (nav) departure point

от·пра́в·очн·ый, -ая, -ое, (a) dispatch, shipping, forwarding, consigning

от·пресс·о́ванн·ый, -ая, -ое, (past part pass) of от·пресс·ова́ть; (plast) moulded, molded

от·пресс·ова́·ть, -у́ют, (perf) press, press out

от·пры́г·ива·ть, -ют, (imp) jump/leap/spring back/aside

от·пры́г·н·уть, -ут, (perf) see от·пры́г·ива·ть

о́т·прыск·, -а, (m) (bot) shoot, sprout, sucker

~, корн·еви́щн·ый stem sucker

~, корн·ев·о́й root sucker

от·пряг·а́·ть, -ют, (imp) unharness, unhitch

от·пряж·ённ·ый, -ая, -ое, (past part pass) of от·пряч·ь

от·пря́·н·уть, -ут, (perf) recoil, jump/start back

от·пря́ч·ь, (fut 3rd sing, pl) от·пряж·ёт, от·пряг·у́т, (past masc sing) от·пря́г·, (perf); see от·пряг·а́·ть

от·пу́г·ива·ть, -ют, (imp) frighten/scare off/away

о́т·пуск·, -а, (m) (v n), see от·пуск·а́·ть; holiday, leave, vacation; (met) tempering, temper, drawing, tempering treatment; issue, distribution; duplicate copy

~, выс·о́к·ий high-temperature tempering, sub-critical annealing

~, много·кра́т·н·ый repeated tempering

~, ни́з·к·ий low-temperature tempering

~, сре́д·н·ий medium-temperature tempering

~, стабилиз·и́рующ·ий stabilizing treatment

от·пуск·а́ни·е, -я, (*n*) release, releasing; (elec) drop-out

от·пуск·а́·ть, -ют, (*imp*) let off/go, release; slack/check off/away, slacken; allot, allocate, assign; issue, distribute, make available; grow, let grow (hair etc.); (met) temper

от·пуск·ни́к·, -а́, (*m*) holidaymaker, vacationer; (naut) libertyman

от·пуск·н·о́й, -а́я, -о́е, (*a*) holiday, vacation, leave; (met) temper, tempering; release; (com) wholesale

~ кла́пан· (*m*) (rail) release valve

~ хру́п·кост·ь (*f*) (met) temper brittleness

~ цен·а́ (*f*) wholesale price

от·пуст·и́ть, -ят, (*perf*) *see* от·пуск·а́·ть

от·пу́щ·енн·ый, -ая, -ое, (*past part pass*) of от·пуст·и́ть

от·рабат·ыва·ть, -ют, (*imp*) work, work off/out/for (a time); clear (by working); finish (a piece of work); run through (a routine)

от·работ·анн·ый, -ая, -ое, (*past part pass*) *see* от·рабат·ыва·ть; used up, finished, exhausted, waste, spent

~ пар· (*m*) exhaust steam

~ щёлок· (*m*) (chem) waste liquor

от·работ·а·ть, -ют, (*perf*) *see* от·ра·бат·ыва·ть

от·работ·к·а, -и, (*g pl*) -т·ок·, (*f*) (*v n*), *see* от·рабат·ыва·ть

о·тра́в·а, -ы, (*f*) poison, toxin

о·трав·и́ть, -ят, (*perf*) *see* о·травл·я́·ть

о·травл·е́ни·е, -я, (*n*) poisoning

~, мг·нове́нн·ое (nucl) prompt poisoning

~, не·стациона́р·н·ое (nucl) transient poisoning

о·травл·я́·ть, -ют, (*imp*) poison

о·травл·я́ющ·ий, -ая, -ее, (*pres part act*) of о·травл·я́·ть; poisonous

~ вещ·еств·о́ (*n*) poisonous chemical agent, chemical warfare agent

~ вещ·еств·о́, бо·ев·о́е war gas

~ вещ·еств·о́, обще·яд·ови́т·ое nerve gas

~ вещ·еств·о́, раз·драж·а́ющ·ее nose gas

от·раж·а́емост·ь, -и, (*f*) reflectivity, reflecting power

от·раж·а́тел·ь, -я, (*m*) reflector; deflector; repeller (of a klystron); tamper (in atomic bomb)

от·раж·а́тел·ь

~, в·движ·н·о́й (nucl) move-in reflector

~, парабол·и́ческ·ий (opt, rad) parabolic reflector

~, прожéктор·н·ый searchlight reflector

~ то́п·очн·ой двéр·к·и deflector shield (of boiler)

~, угол·ко́в·ый (rad) corner reflector

~, шар·ов·о́й spherical reflector

от·раж·а́тельн·ый, -ая, -ое, (*a*) reflective, reflecting; reverberatory; deflecting

~ втýл·к·а (*f*) spill deflector (diesel

~ клистро́н· (*m*) reflex klystron

~ печ·ь (*f*) reverberatory furnace

~ способ·ност·ь (*f*) reflecting power, albedo

от·раж·а́·ть, -ют, (*imp*) reflect, reverberate; repel, repulse, beat off/back; refute

от·раж·а́ющ·ий, -ая, -ее, (*pres part act*) of от·раж·а́·ть, *see* от·ра·ж·а́тельн·ый

от·раж·е́ни·е, -я, (*n*) (*v n*), *see* от·раж·а́·ть; reflection, reverberation, (rad) echo

~, ба́зис·н·ое (cryst) basal reflection

~, времен·н·о́е (phys) time reflection

~, зерка́ль·н·ое mirror image, (opt) specular/regular reflection

коэффициéнт· от·раж·éни·я (*m*) reflection factor

~, на·пра́вл·енн·ое regular reflection

~ норма́ль·н·ое regular reflection

~, обра́т·ное back reflection

~, одно·кра́т·н·ое (rad) single reflection; (*in gen as adj*) one-hop

~ от земл·и́ (rad) ground echo/reflection

~, по́л·н·ое total reflection

~, пра́в·ильн·ое normal/regular reflection

~, про·стра́н·ственн·ое space reflection, inversion

~, пульс·и́рующ·ее (rad) fluctuating echo

~, рас·сé·янн·ое scattered/diffuse reflection

~, скольз·я́щ·ее glancing reflection

от·раж·ённ·ый, -ая, -ое, (*past part pass*) *see* от·раж·а́·ть

~ звук· (*m*) echo

~ сигна́л· (*m*) reflected signal

от·раз·и́ть, -я́т, (*perf*) *see* от·ра·ж·а́·ть

от·рапорт·ова́ть, -у́ют, (*perf*) report
от·расл·ев·о́й, -а́я, -о́е, (*a*) of **от·-расл·ь**
о́т·расл·ь, -и, (*f*) branch, department, field, sphere (of knowledge etc.)
от·раст·а́ни·е, -я, (*n*) (*v n*), see **от·-раст·а́·ть;** (agr) aftergrowing
от·раст·а́·ть, -ют, (*imp*) grow; grow again, sprout
от·раст·и́, -у́т, (*past masc sing*) **от·-ро́с·,** (*perf*) see **от·раст·а́·ть**
от·раст·и́ть, -я́т, (*perf*) see **от·ра́-щ·ива·ть**
от·ра́щ·ива·ть, -ют, (*imp*) (*trans*) grow, let grow.
от·регул·и́р·овать, -уют, (*perf*) adjust, regulate
от·редакт·и́р·овать, -уют, (*perf*) edit
от·ре́ж·ет, (*fut 3rd sing*) of **от·ре́з·ать**
от·ре́з·, -а, (*m*) (*v n*) of **от·рез·а́ть;** piece, length; section, (math) intersection
~ на плат·ь·е (text) dress length
от·рез·а́·ть, -ют, (*imp*) **от·ре́з·ать,** (*fut 3rd pl*) **от·ре́ж·ут,** (*perf*) cut off
от·ре́з·к·а, -и, (*g pl*) **-з·ок·,** (*f*) (*v n*) of **от·ре́з·ать,** (*see also* **от·ре́з·ок·**); cropping (ends of ingots); blanking (sheet metal); cutting up (into lengths)
от·ре́з·к·и (*nom pl*) of **от·ре́з·ок·;** (*g sing and nom pl*) of **от·ре́з·к·а**
от·рез·н·о́й, -а́я, -о́е, (*a*) cut-off; cutting; detachable
~ стан·о́к· (*m*) cutting-off machine
~ штамп· (*m*) (mech) cut-off, shear die
от·ре́з·овк·а, -и, (*f*) (build) pointing trowel
от·ре́з·ок·, -з·к·а, (*nom pl*) **-з·к·и,** (*m*) (*see also* **от·ре́з·к·а**); piece, length, section; (math) segment, interval; intercept; (rad) quarter-wave section; (text) cut; (*pl*) remnants, scrap
~ абсци́сс·ы (math) *X*-intercept
~ апплика́т·ы (math) *Z*-intercept
~ кас·а́тельн·ой (math) length of a tangent
~ ордина́т·ы (math) *Y*-intercept
пра́в·ил·о от·ре́з·к·ов (*n*) (mech) lever principle
~ прям·о́й (math) line segment, intercept of a straight line
~ след·а (nucl) track segment
от·ре́з·ыва·ть, -ют, (*imp*) cut off
от·репет·и́р·овать, -уют, (*perf*) repeat back, run through again

от·репет·ова́ть, -уют, (*perf*) see **от·-репет·и́р·овать;** (naut) acknowledge (a signal)
о·тр·ёт (*fut 3rd sing*) of **о·тер·е́ть**
от·ретуш·и́р·овать, -уют, (*perf*) (phot) retouch
от·риц·а́ни·е, -я, (*n*) (*v n*) of **от·ри-ц·а́·ть;** negation; denial
от·риц·а́тельн·ый, -ая, -ое, (*a*) negative, adverse
~ велич·ин·а́ (*f*) (math) negative quantity
~ за·ря́д· (*m*) (elec) negative charge
~ на·ча́ль·н·ая о·сто́й·чивост·ь (*f*) (shipb) unstable equilibrium
от·риц·а́·ть, -ют, (*imp*) deny, disclaim, contradict
от·ро́г·, -а, (*m*) spur, branch; (*pl*) outskirts
~, бар·и́ческ·ий (meteor) ridge of high pressure
~ о́зер·а (geog) arm of a lake
~ по·вы́ш·енн·ого давл·е́ни·я (meteor) ridge of high pressure
от·ро́д·ь·е, -я, (*n*) (agr) local breed
от·ро́·ет, (*fut 3rd sing*) of **от·ры́·ть**
от·рожд·е́ни·е, -я, (*n*) (ent) hatching
от·ро́с·, (*past masc sing*) of **от·раст·и́**
от·ро́ст·ок, -т·к·а, (*m*) branch, spur; (bot) sprout, shoot; (anat) process, processus
~, черве·обра́з·н·ый (anat) appendix, vermiform process
от·ро́с·ш·ий, -ая, -ее, (*past part act*) of **от·раст·и́**
от·ро́·ют, (*fut 3rd pl*) of **от·ры́·ть**
от·ру́б·, -а, (*m*) (*see also* **о́т·руб·и**) top log, top (of timber); (*pl*) scraps
от·руб·а́ть (*imp*) chop/hack/slash off
о́т·руб·и, -ей, (*pl*) bran, sharps, middlings (flour milling)
от·руб·и́ть, -я́т, (*perf*) see **от·ру-б·а́·ть**
от·руб·н·о́й, -а́я, -о́е, (*a*) of **от·-руб·;** see also **о́т·руб·н·ый**
~ зуб·и́л·о (*n*) caulking chisel
о́т·руб·н·ый, -ая, -ое, (*a*) of **о́т·ру-б·и;** see also **от·руб·н·о́й**
от·ру́б·ок·, -б·к·а, (*m*) bolt (cooperage)
о·тр·у́т, (*fut 3rd pl*) of **о·тер·е́ть**
от·ры́в·, -а, (*m*) (*v n*), see **от·ры-в·а́·ть;** break, pause; breakaway, detachment, separation; (met) brittle rupture/fracture
ис·пыт·а́ни·е на от·ры́в· (weld) tear-apart test; (plast) direct-pull test

от·ры́в

~ **от Земл·и́** (rocket) escape from the Earth

~ **пла́зм·ы** (phys) detachment/breakaway of a plasma

~ **по·гран·и́чн·ого сло́·я** boundary-layer separation

~ **по·то́к·а** (dynam) flow separation/breakaway

про́ч·н·ост·ь на отры́в· (*f*) (met) technical cohesive strength; (rubber) resistance to direct pull

со·противл·е́ни·е от·ры́в·у (*n*) (met) cohesive strength; (rubber) resistance to direct pull

ско́рост·ь от·ры́в·а (*f*) (air) take-off speed; (chem) rate of separation; (phys) escape velocity

температу́р·а от·ры́в·а (*f*) break-off temperature

то́ч·к·а от·ры́в·а (*f*) (dynam) point of separation

от·рыв·а́·ть, -ют, (*imp*) tear/break off/away; isolate, cut off, separate, detach; dig out/up, excavate; **-ся** (*pass*); (air) take off

от·ры́в·ист·ый, -ая, -ое, (*a*) intermittent, jerky

от·ры́в·н·о́й, -а́я, -о́е, (*a*) of **от·-ры́в·**; detachable, tear-off; detached, separated; separating, detaching

~ **календа́р·ь** (*m*) tear-off calendar

~ **систе́рн·а** (*f*) safety tank

~ **ско́р·ост·ь** (*f*) (rocket) escape speed

~ **теч·е́ни·е** (*n*) separated flow

от·ры́в·ок·, -в·к·а, (*m*) fragment

от·ры́в·очн·ый, -ая, -ое, (*a*) fragmentary

от·ры́г·ива·ть, -ют, (*imp*) regurgitate, ruminate, belch

от·ры́г·н·у́ть, -у́т, (*perf*) see **от·-ры́г·ива·ть**

от·ры́·ть, (*fut 3rd sing, pl*) **от·ро́·ет, от·ро́·ют,** (*past masc sing*) **от·ры́·л,** (*perf*); dig out/up, excavate

от·ря́д·, -а, (*m*) detachment, force, squadron, team; (biol) order

от·ряд·и́ть, -я́т, (*perf*) see **от·ря-ж·а́·ть**

от·ряж·а́·ть, -ют, (*imp*) send (on mission)

о·тряс·а́·ть, -ют, (*imp*) shake off

о·тряс·ти́, -у́т, (*perf*) see **о·тряс·а́·ть**

о·тря́х·ива·ть, -ют, (*imp*) = **о·тря-с·а́·ть**

о·тря́х·н·у́ть, -у́т, (*perf*) = **о·тряс·-ти́**

от·сад·и́ть, -ят, (*perf*) see **от·са́ж·и-ва·ть**

от·са́д·к·а, -и, (*g pl*) **-д·ок·,** (*f*) (*v n*) see **от·са́ж·ива·ть**

от·сад·очн·ый, -ая, -ое, (*a*) of **от·-са́д·к·а**

~ **маши́н·а** (*f*) (min) jig, jigger

~ **маши́н·а, пульс·и́рующ·ая** pulsator jig

от·са́ж·ива·ть, -ют, (*imp*) put/seat separately, separate; (hortic) plant out; jig (ore dressing)

от·са́л·ива·ть, -ют, (*imp*) salt out

от·салют·ова́ть, -у́ют, (*perf*) salute

от·са́с·ывани·е, -я, (*n*) (*v n*), see **от·-са́с·ыва·ть;** suction (out)

от·са́с·ыва·ть, -ют, (*imp*) suck/draw/pump out, exhaust

от·са́с·ывающ·ий, -ая, -ее, (*pres part act*) of **от·са́с·ыва·ть;** suction, exhausting, extracting

о́т·свет·, -а, (*m*) reflected light, sheen, shine

от·свет·и́ть, -ят, (*perf*) stop/cease shining

от·све́ч·ивани·е, -я, (*n*) (*v n*) of **от·-све́ч·ива·ть;** glare (e.g. from snow)

от·све́ч·ива·ть, -ют, (*imp*) reflect (light); shine, gleam (in the sun's light); be reflected in (of light sources)

от·се́в·, -а, (*m*) (*v n*), see **от·се́·и-ва·ть;** siftings, screenings

от·се́·ива·ть, -ют, (*imp*) sift, screen; select, choose; eliminate

от·се́к·, -а, (*m*) compartment, cubicle; (agr) corn bin; (*past masc sing*) of **от·се́ч·ь**

~**, бага́ж·н·ый** (a/c) luggage hold

~**, ба́к·ов·ый** (rocket) tank compartment

~**, бо́мб·ов·ый** (a/c) bomb bay

~**, груз·ов·о́й** cargo compartment

~ **пу́льт·а у·правл·е́ни·я** control cubicle

от·сек·а́тел·ь, -я, (*m*) (elec) clipper, clipping circuit

от·сек·а́тельн·ый, -ая, -ое, (*a*) cutting, clipping; intercepting; cut-off

от·сек·а́·ть, -ют, (*imp*) cut/chop off; (math) intercept

от·сеч·ённ·ый, -ая, -ое, (*past part pass*) see **от·сек·а́·ть;** severed

от·се́ч·к·а, -и, (*f*) (*v n*), see **от·се-к·а́·ть;** cut-off

~ **нейтро́н·ов** (nucl) neutron cutoff

~ **пар·о́в** (st eng) cutoff

от·се́ч·н·ый, -ая, -ое, (*a*) cutoff
~ **при·во́д·** (*m*) fuel control (diesel)
от·се́ч·ь, (*fut 3rd sing, pl*) **от·сеч·ёт, от·сек·у́т,** (*past masc sing*) **от·се́к·,** (*perf*); *see* **от·сек·а́·ть**
от·се́·ять, -ют, (*perf*) *see* **от·се́·ива·ть**
от·ска́бл·ива·ть, -ют, (*imp*) scrape/shave off
от·скак·а́ть, (*fut 3rd sing, pl*) **от·ска́·ч·ет, от·ска́ч·ут,** (*perf*) gallop off/away
от·ска́к·ивани·е, -я, (*n*) (*v n*) *see* **от·ска́к·ива·ть**
от·ска́к·ива·ть, -ют, (*imp*) recoil, rebound, jump back/aside/away; come/fly off; gallop off/away
от·ска́ч·ет, (*fut 3rd sing*) of **от·ска·к·а́ть**
от·скобл·и́ть, -я́т, (*perf*) *see* **от·ска́бл·ива·ть**
от·ско́к·, -а, (*m*) (*v n*), *see* **от·скоч·и́ть;** recoil, rebound, jump, bounce
от·скоч·и́ть, -͡ат, (*perf*) recoil, rebound, bounce, jump back/way/aside; come/fly off
от·скреб·а́·ть, -ют, (*imp*) scrape/scratch off
от·скрес·ти́, (*fut 3rd pl*) **от·скреб·у́т,** (*past masc sing*) **от·скрёб·,** (*perf*); *see* **от·скреб·а́·ть**
от·сла́·ивани·е, -я, (*n*) (*v n*), *see* **от·-сла́·ива·ть;** scaling; (rubber) ply/lay separation, delamination
от·сла́·ива·ть, -ют, (*imp*) peel/strip off; **-ся** peel/flake off, exfoliate, desquamate
от·сло·е́ни·е, -я, (*n*) (*v n*), *see* **от·сла́·-ива·ть;** flake, strip, layer; (rubber) ply/lay separation
от·сло·ённ·ый, -ая, -ое, (*past part pass*) *see* **от·сла́·ива·ть**
от·сло·и́ть, -я́т, (*perf*) *see* **от·сла́·ива·ть**
от·служ·и́ть, -͡ат, (*perf*) serve (for a period of time), complete service; be worn out
от·с·ня́·ть, (*fut 3rd pl*) **от·с·ним·ут,** (*perf*); (cinema) turn out, kill; make a film
от·со·ве́т·овать, -уют, (*perf*) dissuade
от·со·един·и́ть, -я́т, (*perf*) *see* **от·со·е-дин·я́·ть**
от·со·един·я́·ть, -ют, (*imp*) disconnect, (elec) isolate
от·сол·и́ть, -я́т, (*perf*) salt out
от·со́л·к·а, -и, (*f*) (chem etc.) salting out

от·сорт·иро́ванн·ый, -ая, -ое, (*past part pass*) *see* **от·сорт·иро́выва·ть**
~ **по·ро́д·н·ая мел·оч·ь** (*f*) (min) rock screenings
от·сорт·ир·ова́ть, -у́ют, (*perf*) *see* **от·сорт·иро́выва·ть**
от·сорт·иро́выва·ть, -ют, (*imp*) sort, sort out, grade, size, separate, screen; select, pick; reject
от·со́с·, -а, (*m*) (*v n*) *see* **от·са́с·ыва·ть;** extract, extract ventilator; blowing (boundary layer control)
~**, мест·н·ый** (build) local vent
от·со́с·анн·ый, -ая, -ое, (*past part pass*) *see* **от·са́с·ыва·ть**
от·сос·а́ть, -у́т, (*perf*) *see* **от·са́с·ыва·ть**
от·со́с·н·ый, -ая, -ое, (*a*) of **от·со́с·**
от·со́х·нуть, -нут, (*perf*) shrivel, wither
от·сро́ч·енн·ый, -ая, -ое, (*past part pass*) *see* **от·сро́ч·ива·ть**
~ **де́й·стви·е** (*n*) delayed action
от·сро́ч·ива·ть, -ют, (*imp*) postpone, defer; adjourn (a meeting); renew (permits etc.)
от·сро́ч·ить, -ат, (*perf*) *see* **от·сро́ч·и-ва·ть**
от·сро́ч·к·а, -и, (*g pl*) **-ч·ек·,** (*f*) postponement, deferment, adjournment; extension, renewal (of permits)
от·став·а́ни·е, -я, (*n*) (*v n*), *see* **от·-ста·ва́ть;** delay, lag, lagging (behind); arrears; backwardness; (met roll) backward slip
~**, времен·н·о́е** time lag
у́гол· от·став·а́ни·я (*m*) (elec) angle of lag; (air) trail angle
у́гол· от·став·а́ни·я ло́паст·и dragging angle (helicopters)
~ **фаз·** phase lag
от·ста·ва́ть, -ю́т, (*imp*) lag, lag/fall/drop behind; be backward/behindhand, be behind schedule; (horol) be slow; come off/unstuck; lose touch/contact; leave/let alone
от·ста́в·ить, -ят, (*perf*) *see* **от·став-л·я́·ть;** (*imperative*) as you were!
от·ста́в·к·а, -и, (*g pl*) **-в·ок·,** (*f*) retirement, resignation; dismissal, discharge (of staff)
в от·ста́в·к·е retired
от·ставл·я́·ть, -ют, (*imp*) put/set/move aside/out
от·ста́в·н·о́й, -а́я, -о́е, (*a*) retired
от·ста·ёт, (*pres 3rd sing*) of **от·ста·ва́ть**

от·ста́·ивани·е, -я, (*n*) (*v n*), *see* **от·-ста́·ива·ть;** settling, sedimentation; (rubber) creaming (latex)

от·ста́·ива·ть, -ют, (*imp*) defend, (mil) hold; uphold; **-ся** (*pass*); settle, become settled; (naut) ride out (a gale)

от·ста́л·ост·ь, -и, (*f*) backwardness

от·ста́·л·ый, -ая, -ое, (*a*) backward, retarded

от·ста́·ть, (*fut 3rd pl*) **от·ста́·нет, от·ста́·нут,** (*past masc sing*) **от·ста́·л,** (*perf*); *see* **от·ста·ва́ть**

от·ста·ю́щ·ий, -ая, -ее, (*pres part act*) *see* **от·ста·ва́ть;** backward

~ гироско́п· (*m*) displacement gyroscope

от·стёг·ива·ть, -ют, (*imp*) undo, unfasten, unbutton, unhook

от·стег·н·у́ть, -у́т, (*perf*) *see* **от·стёг·ива·ть**

от·с·тир·а́·ть, -ют, (*perf*) *see* **от·с·ти́·р·ыва·ть**

от·с·ти́р·ыва·ть, -ют, (*imp*) wash off, launder out

от·сто́·й, -я, (*m*) sludge, sediment, dregs

от·сто́й·ник·, -а, (*m*) (ICE) sump; settling/sedimentation tank/basin

от·сто́й·н·ый, -ая, -ое, (*a*) settling

от·сто́пор·ива·ть, -ют, (*imp*) uncork, open; unlock

от·сто·я́ни·е, -я, (*n*) distance apart, separation distance, spacing

от·сто·я́ть, -ят, (*perf*) *see* **от·ста́·ива·ть;** (*imp*) be, be located (at a distance from)

от·стра́·ива·ть, -ют, (*imp*) complete (building etc.); (rad) tune out

от·стран·е́ни·е, -я, (*n*) (*v n*), *see* **от·-стран·я́·ть;** removal, elimination; dismissal

от·стран·и́ть, -я́т, (*perf*) *see* **от·стран·я́·ть**

от·стран·я́·ть, -ют, (*imp*) supersede, dismiss; remove, get rid of; move aside/back/away

от·стра́ч·ива·ть, -ют, (*imp*) stitch

от·стре́л·, -а, (*m*) (*v n*), *see* **от·стре́·л·ива·ть**

~ ору́ж·и·я (gunn) proving

от·стре́л·ива·ть, -ют, (*imp*) shoot, kill (a hunting quota); expend, use up (ammunition); finish shooting/firing; **-ся** beat off/back (by shooting)

от·стрел·я́·ть, -ют, (*perf*) *see* **от·стре́·л·ива·ть**

от·стриг·а́·ть, -ют, (*imp*) cut/clip/snip off

от·стри́ч·ь, (*fut 3rd sing, pl*) **от·стри́ж·ёт, от·стриг·у́т,** (*past masc sing*) **от·стри́г·,** (*perf*); *see* **от·стриг·а́·ть**

от·стро́·ить, -ят, (*perf*) *see* **от·стра́·ива·ть**

от·стро́й·к·а, -и, (*g pl*) **-о́·ек·,** (*f*) (*v n*), *see* **от·стра́·ива·ть;** (rad) wave selection

от·строч·и́ть, -а́т, (*perf*) stitch

о́т·ступ·, -а, (*m*) paragraph indent/indentation

о́т·ступ·ани·е, -я, (*n*) (*v n*), *see* **от·сту·п·а́·ть;** recession, regression, retreat; slip (of screw propeller)

от·ступ·а́·ть, -ют, (*imp*) step back/away; recede, regress; retreat, draw back; deviate/depart/digress from; **-ся** renounce, give up

~ в бок· (eng) offset

от·ступ·и́ть, -ят, (*perf*) *see* **от·сту·п·а́·ть**

от·ступл·е́ни·е, -я, (*n*) departure, deviation; digression; recession; (mil) retreat. withdrawal

~ част·от·ы́ (rad) frequency departure

от·су́т·стви·е, -я, (*n*) absence; lack

от·су́т·ств·овать, -уют, (*imp*) be absent; be lacking

от·с·чёт·, -а, (*m*) (*v n*), *see* **от·с·чи́·т·ыва·ть;** (instr) reading, count

~, груб·о́й coarse reading

~, дей·стви́тельн·ый real count

~, дистанцио́н·н·ый remote reading

~, мно́го·кра́т·н·ый multiple metering (telephones)

~, нул·ев·о́й zero reading

~, пра́в·ильн·ый true reading

~, про·из·вод·и́ть от·с·чёт· take a reading

~, с·луч·а́йн·ый random count
то́ч·к·а от·с·чёт·а (*f*) point of reference, datum point

~, то́ч·н·ый fine reading

~ фон·ов (phys) background count

~, цифр·ов·о́й digital read-out

от·с·чёт·н·ый, -ая, -ое, (*a*) of **от·-с·чёт;** reference

от·с·чи́т·анн·ый, -ая, -ое, (*past part pass*) *see* **от·с·чи́т·ыва·ть**

от·с·чит·а́·ть, -ют, (*perf*) *see* **от·с·чи́·т·ыва·ть**

от·с·чи́т·ыва·ть, -ют, (*imp*) read/count out/off, take a reading

от·сыл·а́·ть, -ют, (*imp*) send, send off/away, dispatch; refer

от·сы́л·к·а, -и, (*g pl*) -л·ок·, (*f*) (*v n*), *see* от·сыл·а́·ть; reference (in text)

от·сы́п·ать, -лют, (*perf*) *see* от·сы·п·а́·ть, (*first meaning*)

от·сып·а́·ть, -ют, (*imp*) pour out some, measure out (of free-flowing solids); -ся (*pass*); sleep off/out/ enough

от·сы́п·к·а, -и, (*f*) (*v n*), *see* от·сы-п·а́·ть; = от·сы́п·ь

от·сы́п·ь, -и, (*f*) (*v n*), *see* от·сып·а́·ть

~, ка́мен·н·ая (civ eng) rubble mound

от·сыр·ева́·ть, -ют, (*imp*) become damp

от·сыр·е́л·ый, -ая, -ое, (*a*) damp, dampened

от·сыр·е́·ть, -ют, (*perf*) *see* от·сыр·е-ва́·ть

от·сых·а́·ть, -ют, (*imp*) dry up, shrivel, wither

от·сю́да (*adv*) from here, hence

от·та́·ет, (*fut 3rd sing*) of от·та́·ять

от·та́·ивани·е, -я, (*n*) (*v n*), of от·та́·и-ва·ть

от·та́·ива·ть, -ют, (*imp*) thaw, thaw out

от·та́лк·ивани·е, -я, (*n*) (*v n*), *see* от·-та́лк·ива·ть; repulsion

~, взаи́м·н·ое (phys) mutual repulsion, repulsive interaction

   си́л·а от·та́лк·ива́ни·я (*f*) repul-sive force

от·та́лк·ива·ть, -ют, (*imp*) push off/ away, fend off, repel; alienate, antagonize

от·та́лк·ивающ·ий, -ая, -ее, (*pres part act*) of от·та́лк·ива·ть; repul-sive, repellent

от·та́пт·ыва·ть, -ют, (*imp*) tread/ trample on

от·тарт·а́·ть, -ют, (*perf*) *see* от·та́рт-т·ыва·ть

от·та́рт·ыва·ть, -ют, (*imp*) bail down/ out

от·та́ск·ива·ть, -ют, (*imp*) drag/pull back/away/aside

от·та́ч·ива·ть, -ют, (*imp*) sharpen, sharpen to a point; finish sharpening

от·тащ·и́ть, -ат, (*perf*) *see* от·та́ск·и-ва·ть

от·та́·ять, -ют, (*perf*) thaw, thaw out, defrost

от·тёк·, -а, (*m*) syrup (sugar-making); (*past masc sing*) of от·те́ч·ь

от·тек·а́·ть, -ют, (*imp*) flow back

от·тен·ённ·ый, -ая, -ое, (*past part pass*) *see* от·тен·я́·ть; preshadowed (electron microscopy)

от·тен·и́ть, -я́т, (*perf*) *see* от·тен·я́·ть

от·те́н·ок·, -н·к·а, (*m*) tint, hue, tinge, shade; timbre, nuance

от·тен·я́·ть, -ют, (*imp*) shade, shade in (by drawing etc.); tint; set off (con-trast), stress, emphasize

о́т·тепел·ь, -и, (*f*) (meteor) thaw

от·тер·е́ть, (*fut 3rd sing, pl*) ото·тр·ёт, ото·тр·у́т, (*past masc sing*) от·тёр·, (*perf*); *see* от·тир·а́·ть

от·тес·а́ть (*fut 3rd sing, pl*) от·теш·ет, от·теш·ут, (*past masc sing*) от·тес·а́л, (*perf*); *see* от·тёс·ыва·ть

от·тесн·ённ·ый, -ая, -ое, (*past part pass*) *see* от·тесн·я́·ть

от·тесн·и́ть, -я́т, (*perf*) *see* от·тес-н·я́·ть

от·тесн·я́·ть, -ют, (*imp*) force/push back/aside

от·тёс·ыва·ть, -ют, (*imp*) trim (with axe etc.)

от·те́ч·ь, (*fut 3rd sing, pl*) от·теч·ёт, от·тек·у́т, (*past masc sing*) от·тёк·, (*perf*); flow back

от·те́ш·ет, (*fut 3rd sing*) of от·тес·а́·ть,

от·тир·а́·ть, -ют, (*imp*) rub/wipe off; polish

о́т·тиск·, -а, (*m*) imprint; (print etc.) impression, print; reprint (of journal article)

~, корректу́р·н·ый (print) proof im-pression/print

от·ти́ск·ива·ть, -ют, (*imp*) make an impression; (print) impress; press back

от·ти́с·н·уть, -ут, (*perf*) *see* от·ти́с-к·ива·ть

от·титр·о́выва·ть, -ют, (*imp*) (chem) titrate

от·того́ (*adv*) for this reason, therefore

Отто, цикл· (*m*) (ICE) Otto cycle

от·то́к·, -а, (*m*) (*v n*), *see* от·тек·а́·ть; drain/waste pipe; outflow, egress

от·толк·н·у́ть, -у́т, (*perf*) *see* от·та́л-к·ива·ть

от·топт·а́·ть, (*fut 3rd pl*) от·то́пч·ут, (*perf*); *see* от·та́пт·ыва·ть

от·топы́р·енн·ый, -ая, -ое, (*past part pass*); protruding, protruberant, jutt-ing/sticking out

от·торж·е́ни·е, -я, (*n*) (med) disen-gagement

от·то́ч·енн·ый, -ая, -ое, (*past part pass*) *see* от·та́ч·ива·ть

от·то́ч·и·е, -я, (*n*) (print) dots, sepa-rative dots

от·точ·и́ть, -ат, (*perf*) sharpen to a point; finish sharpening

от·ту́да (*adv*) from there

от·туш·ева́ть, -у́ют, (*perf*) shade (by drawing)

от·тя́г·ива·ть, -ют, (*imp*) pull/draw/haul back; (met) draw down; turn inside out, turn out; delay, procrastinate

от·тя́ж·к·а, -и, (*f*) (*v n*), *see* **от·тя́г·ива·ть;** stay, guy, guide rope; (met) drawing down

от·тяж·н·о́й, -а́я, -о́е, (*a*) (met) drawing-down; tensioning, guy, stay, span, spanning; delaying

от·тя́·нут·ый, -ая, -ое, (*past part pass*) of **от·тя́·н·уть;** drawn, drawn down

от·тя́·н·уть, -у́т, (*perf*) *see* **от·тя́г·ива·ть**

о·тума́н·ить, -ят, (*perf*) fog, blur; confuse

отуни́т·, -а, (*m*) (min) autunite

от·уч·а́·ть, -ют, (*imp*) break the habit of

от·уч·и́ть, -ат, (*perf*) *see* **от·уч·а́·ть**

от·фильтр·ова́ть, -у́ют, (*perf*) filter off/out

от·фильтр·о́выва·ть, -ют, (*imp*) filter off/out

от·форм·и́р·овать, -уют, (*perf*) *see* **от·форм·о́выва·ть**

от·форм·о́ванн·ый, -ая, -ое, (*past part pass*) of **от·форм·ова́ть**

от·форм·ова́ть, -у́ют, (*perf*) *see* **от·форм·о́выва·ть**

от·форм·о́выва·ть, -ют, (*imp*) form, shape, mould, mold

от·ха́рк·ивани·е, -я, (*n*) expectoration, spitting

от·ха́рк·ивающ·ее вещ·еств·о́ (*n*) (pharm) expectorant

от·хлы́·н·уть, -ут, (*perf*) flood/rush back

от·хо́д·, -а, (*m*) departure, leaving, going/moving away/back/off, withdrawal, removal; outlet; deviation; (*pl*) waste, refuse, screenings, tailings; waste material/product, by-product

~, высоко·акти́в·н·ые (*pl*) (nucl) high-level/hot waste

~, газо·обра́з·ные (*pl*) (nucl) off-gas

~, жи́дк·ие (*pl*) liquid waste, effluent

~ кокс·ова́ни·я (*pl*) coking by-product

~, мало·акти́в·н·ые (*pl*) (nucl) low-level/cold waste

~, оборо́т·н·ые (*pl*) reverts

порт· от·хо́д·а (*m*) port of departure

~, с·луч·а́йн·ые (*pl*) (nucl) non-recurrent waste

~ хме́л·я (*pl*) hop manure

от·ход·и́ть, -ят, (*imp*) go/move/walk away/back/off; leave, depart; withdraw, fall back; come loose/unstuck/off; diverge, digress, deviate; be taken over, come into the possession of

от·ход·я́щ·ий, -ая, -ее, (*pres part act*) *see* **от·ход·и́ть;** outgoing; exhaust, waste

**~ газ· (*m*) exhaust/burnt/flue gas, off-gas

отц·а́, (*acc/gen sing*) of **оте́ц·**

от·цвест·и́, (*fut 3rd sing, pl*) **от·цвет·ёт, от·цвет·у́т,** (*past masc sing*) **от·цвёл·,** (*perf*); *see* **от·цвет·а́·ть**

от·цвет·а́·ть, -ют, (*imp*) finish flowering/blossoming

от·цве́т·ш·ий, -ая, -ее, (*past part act*) *see* **от·цвет·а́·ть;** (bot) deflorate

от·цед·и́ть, -ят, (*perf*) *see* **от·це́ж·ива·ть**

от·це́ж·ива·ть, -ют, (*imp*) strain off, filter out, decant

от·це́п·, -а, (*m*) (rail) uncoupled/shunted wagon/coach/car

от·цеп·и́ть, -ят, (*perf*) *see* **от·цеп·ля́·ть**

от·це́п·к·а, -и, (*g pl*) **-п·ок·,** (*f*) (*v n*) *see* **от·цепл·я́·ть**

от·цепл·я́·ть, -ют, (*imp*) unhook, uncouple, unfasten, disengage

отц·о́вск·ий, -ая, -ое, (*a*) paternal, father's

от·ча́·ива·ться, -ются, (*imp*) despair

от·ча́л·ива·ть, -ют, (*imp*) cast off, leave (of ships)

от·ча́л·ить, -ят, (*perf*) *see* **от·ча́л·ива·ть**

от·ча́ст·и (*adv*) partly

от·ча́·янн·ый, -ая, -ое, (*a*) desperate

от·ча́·яться, -ются, (*perf*) despair

от·чего́ (*adv*) why

от·чего́-либо (*adv*) for some reason or other

от·чего́-то (*adv*) for some reason

от·чека́н·ива·ть, -ют, (*imp*) mint, strike (coin); pronounce/enunciate clearly

от·чека́н·ить, -ят, (*perf*) *see* **от·чека́н·ива·ть**

от·чёрк·ива·ть, -ют, (*imp*) mark off

от·черк·н·у́ть, -у́т, (*perf*) *see* **от·чёрк·ива·ть**

от·че́рп·а·ть, -ют, (*perf*) *see* **от·че́рп·ыва·ть**

от·черп·н·у́ть, -у́т, (*perf*) *see* **от·чёрп·ыва·ть**

от·че́рп·ыва·ть, -ют, (*imp*) scoop/ladle/bail out

о́тч·еств·о, -а, (*n*) patronymic

от·чёт·, -а, (*m*) account, report, returns

от·чёт·ливост·ь, -и, (*f*) clarity, precision; (phot) definition

от·чёт·лив·ый, -ая, -ое, (*a*) distinct, precise; clearly defined, sharp

от·чёт·ност·ь, -и, (*f*) accountability; book-keeping, accounts, accounting

     контрол·ёр· от·чёт·ност·и (*m*) (fin) auditor

от·чёт·н·ый, -ая, -ое, (*a*) of от·чёт·; financial; under discussion/review

~ год· (*m*) financial year

от·числ·е́ни·е, -я, (*n*) (*v n*) of от·числ·я́·ть

от·числ·и́ть, ⸗ят, (*perf*) *see* от·числ·я́·ть

от·числ·я́·ть, -ют, (*imp*) deduct; discharge, transfer (personnel); expel, dismiss (members of organization)

от·чи́ст·ить, -ят, (*perf*) *see* от·чищ·а́·ть

от·чит·а́·ть, -ют, (*perf*) finish reading; lecture, rebuke; -ся report

от·чищ·а́·ть, -ют, (*imp*) clean, clean off

от·член·ени·е, -я, (*n*) (*v n*), *see* от·член·я́·ть; (ent) limbing

от·член·и́ть, -ят, (*perf*) *see* от·член·я́·ть

от·член·я́·ть, -ют, (*imp*) separate, take part of, give up part of

от·чужд·е́ни·е, -я, (*n*) alienation, estrangement

     полос·а́ от·чужд·е́ни·я (*f*) (rail) right of way

от·шварт·ова́ть, -у́ют, (*imp*) (naut) cast off

от·шед·ш·ий, -ая, -ее, (*pres part act*) of ото·йти́; departing, withdrawing, receding

от·ше́ль·ник·, -а, (*m*) hermit

от·ше́ств·и·е, -я, (*n*) (nav) departure

от·шив·а́·ть, -ют, (*imp*) unnail, remove the nails, open up (a box, crate etc.)

от·ши́·ть, (*fut 3rd sing, pl*) ото·шь·ёт, ото·шь·ю́т, (*past masc sing*) от·ши́·л, (*perf*); *see* от·шив·а́·ть

от·шлиф·ова́ть, -у́ют, (*perf*) *see* от·шлиф·о́выва·ть

от·шлиф·о́выва·ть, -ют, (*imp*) (mech eng) grind, grind off; smooth, polish

от·шнур·ова́ть, -у́ют, (*perf*) unthread, unlace, untie, undo

от·шпи́л·ива·ть, -ют, (*imp*) unpin, unfasten

от·шпи́л·ить, -ят, (*perf*) *see* от·шпи́л·ива·ть

от·ще́п·, -а, (*m*) flake, chip

от·щеп·и́ть, -я́т, (*perf*) *see* от·щепл·я́·ть

от·щепл·е́ни·е, -я, (*n*) (*v n*), *see* от·щепл·я́·ть; detaching, detachment, separation, abstraction

~ водо·ро́д·а (chem) dehydrogenation

~ ио́д·а (chem) deiodination

~ хло́р·а (chem) dechlorination

~ эти́ль·н·ых групп· (chem) deethylation

от·щепл·я́·ть, -ют, (*imp*) chip/split off; remove, separate

от·щип·а́·ть, -лют, (*perf*) *see* от·щи́п·ыва·ть

от·щип·н·у́ть, у́т, (*perf single action*) *see* от·щи́п·ыва·ть

от·щи́п·ыва·ть, -ют, (*imp*) nip/pluck off, pinch out

отъ- (*prefix*) *see* от-

отъ·е́зд·, -а, (*m*) (*v n*), *see* отъ·езж·а́·ть; departure; (cinema) dolling out

отъ·езж·а́·ть, -ют, (*imp*) depart, drive/ride off/away

отъ·ём·, -а, (*m*) (*v n*), *see* от·ним·а́·ть; takeoff, bleeding; (zool) weaning

отъ·ём·н·ый, -ая, -ое, (*a*) of отъ·ём; take-off; detachable

отъ·е́х·ать, (*fut 3rd sing, pl*) отъ·-; е́д·ет, отъ·е́д·ут, (*past masc sing*) отъ·е́х·ал, (*perf*); *see* отъ·езж·а́·ть

от·ыгр·а́·ть, -ют, (*perf*) *see* от·ы́гр·ыва·ть

от·ы́гр·ыва·ть, -ют, (*imp*) win back

от·ымён·н·ый, -ая, -ое, (*a*) (gram) denominative

от·ыск·а́ть, (*fut 3rd pl*) от·ы́щ·ут, (*perf*); *see* от·ы́ск·ива·ть

от·ы́ск·ива·ть, -ют, (*perf*) find, search out

от·ы́щ·ет, (*fut 3rd sing*) of от·ыск·а́ть

о·тягот·и́ть, -я́т, (*perf*) *see* о·тяго·щ·а́·ть

о·тягощ·а́·ть, -ют, (*imp*) weight, weight down, burden, make heavy/heavier

о·тяг·ча́·ть, -ют, (*imp*) = о·тяго·щ·а́·ть; aggravate

о·тяг·ч·и́ть, -а́т, (*perf*) *see* о·тяг·ча́·ть

о·тяжел·ённ·ый, -ая, -ое, (*past part pass*) of о·тяжел·и́ть; weighted, weighed down, heavy

о·тяжел·е́·ть, -ют,   (*perf*) become heavy/heavier

о·тяжел·и́ть, -я́т,   (*perf*) *see* о·тя·жел·я́·ть

о·тяжел·я́·ть, -ют,   (*imp*) make heavy/heavier, weight, weight down; make cumbersome

оф́ерт·а, -ы, (*f*) = офф́ерт·а (com) offer

офи·кальц́ит·, -а,   (*m*) (min) ophicalcite

офио·плутеус·, -а,   (*m*) (zool) ophiopluteus larva

офи́т·, -а, (*m*) (min) ophite

офи·ур·ы   (*f pl*) (zool) ophiuroids, brittle stars, *Ophiuroidea*

офиц́ер·, -а, (*m*) officer

офиц́ер·ск·ий, -ая, -ое,   (*a*) officer, officer's, officers'

офиц́ер·ств·о, -а,   (*n*) officers; commissioned/officer's rank

офици́аль·н·ый, -ая, -ое,   (*a*) official

официа́нт·, -а, (*m*) waiter

офици́оз·, -а, (*m*) semi-official journal

офици́оз·н·ый, -ая, -ое,   (*a*) semi-official

о·флюс·о́ванн·ый, -ая, -ое,   (*past part pass*) of о·флюс·ова́·ть; fluxed

о·флюс·ова́·ть, -у́ют,   (*perf*) flux

о·форм·и́тел·ь, -я,   (*m*) *see* о·форм·л·я́·ть; (print) layout man

о·фо́рм·ить, -ят,   (*perf*) *see* о·форм·л·я́·ть

о·формл·е́ни·е, -я,   (*n*) (*v n*), *see* о·формл·я́·ть; appearance, setting; (theat) décor

о·формл·я́·ть, -ют,   (*imp*) put into shape/order, (print) lay out; draw up (documents), complete formalities, sign, sign on; design, (theat) mount, stage; -ся (*pass*); take shape

о·формл·я́ющ·ий, -ая, -ее,   (*pres part act*) of о·формл·я́·ть; (plast) moulding

офо́рт·, -а, (*m*) (print) etching

о·форт·и́ст·, -а, (*m*) etcher

офс́ет·н·ый, -ая, -ое,   (*a*) (print) offset

~ печа́т·ь (*f*) offset printing

~ репроду́кци·я (*f*) litho-offset

офт (*abbr*) = относи́тельная фа́зовая телегра́фия

офтальм·и́ческ·ий, -ая, -ое,   (*a*) ophthalmic

офтальм·и́·я, -и, (*f*) (med) ophthalmia

офтальмо·ло́г·и·я, -и,   (*f*) (med) ophthalmology

офтальмо·плег·и́·я, -и,   (*f*) (med) ophthalmoplegia

офтальмо·ско́п·, -а,   (*m*) (med) ophthalmoscope

офф́ерт·а, -ы, (*f*) (com) offer

о·ха́п·к·а, -и, (*g pl*) -п·ок·, (*f*) armful, armload

о·хва́т·, -а,   (*m*) (*v n*), *see* о·хва́т·ыва·ть; girth; scope

у́гол· о·хва́т·а (*m*) inclusion angle; angle of contact (gear wheels etc.); (text) angle of lap; (phot) acceptance angle (of exposure meter)

о·хват·и́ть, -́ят,   (*perf*) *see* о·хва́т·ыва·ть

о·хва́т·ываем·ый, -ая, -ое,   (*pres part pass*) of о·хва́т·ыва·ть; male (of engineering parts)

о·хва́т·ыва·ть, -ют,   (*imp*) embrace, surround, encircle, cover; envelop, include, take in/hold of; (mil) outflank

о·хват·ывающ·ий, -ая, -ее,   (*pres part act*) of о·хва́т·ыва·ть; female (of engineering parts)

о·хва́ч·енн·ый, -ая, -ое,   (*past part pass*) of о·хват·и́ть

о·хво́ст·ь·е, -я,   (*n*) (agr) chaff, tailings

о·хлад·и́тел·ь, -я,   (*m*) cooling agent/medium, coolant, refrigerant; cooler, cooling installation/plant

~, ба́шен·н·ый cooling tower

~, до·полн·и́тельн·ый after-cooler

~, о·рос·и́тельн·ый drip cooler, surface-spray cooler

~, про·меж·у́точн·ый intercooler, intercycle cooler

~, циркул·и́рующ·ий recirculation cooler

о·хлад·и́тельн·ый, -ая, -ое,   (*a*) *and see* о·хлажд·а́ющ·ий; cooling

~ коло́н·к·а (*f*) cooling tower

о·хлад·и́ть, -я́т,   (*perf*) *see* о·хлажд·а́·ть

о·хлажд·а́·ть, -ют,   (*imp*) cool, chill, make cold/colder; refrigerate, freeze

о·хлажд·а́ющ·ий, -ая, -ее,   (*pres part act*) of о·хлажд·а́·ть

~ жи́дк·ост·ь coolant

~ на·со́с· (*m*) cooling-water pump

~ пруд· (*m*) cooling pond

~ с·мес·ь (*f*) refrigerating compound

~ стол· (*m*) (met roll) cooling bed/rack

~ у·стро́й·ств·о (*n*) cooling system/arrangements

о·хлажд·е́ни·е, -я,   (*n*) cooling, chilling; refrigerating, freezing

о·хлажд·éни·е
~ **анóд·а** (elec) plate cooling
~, **брызг·áльн·ое** spray cooling
~, **вод·ян·óе** water cooling
~ **вóз·дух·а** air-cooling, cooling the air
~, **воз·дýш·н·ое** air cooling
~, **втор·ѝчн·ое** secondary cooling
~ **вы·пот·евáни·ем** sweat cooling
~, **глуб·óк·ое** deep freezing
~, **ём·кост·н·ое** capacitative cooling
~, **естéств·енн·ое** self-cooling
~, **жѝдк·ост·н·ое** liquid cooling; (in gen) liquid-cooled (adj)
     за·кóн· о·хлажд·éни·я (m) (phys) Newton's law of cooling
~, **за·крѝ·т·ое** closed-circuit cooling
     зóн·а втор·ѝчн·ого о·хлажд·éния· (f) secondary cooling zone (continuous casting)
~ **ис·пар·éни·ем** evaporative cooling
     крив·áя о·хлажд·éни·я (f) (phys) cooling curve
~, **лед·ян·óе** ice cooling
~ **луче·ис·пуск·áни·ем** radiation cooling
~ **массо·об·мéн·ом** mass transfer cooling
~, **от·крѝ·т·ое** evaporative cooling
~, **плён·очн·ое** film cooling
~, **пóр·ист·ое** transpiration cooling
~, **при·нуд·ѝтельн·ое** pump circulation of water; forced cooling
~, **прямо·тóч·н·ое** direct-flow/once-through cooling
~ **рас·пыл·éни·ем** spray/shower cooling, soft spray cooling
~ **рас·шир·éни·ем** cooling by expansion, expansion cooling
~, **регенератѝв·н·ое** regenerative cooling
~, **спрéйер·н·ое** (met) spray cooling
~, **струй·н·ое** hard-spray cooling, water-jet cooling
~, **термо·сифóн·н·ое** thermo-syphon cooling
~, **туннéль·н·ое** ducted cooling
о·хлажд·ённ·ый, -ая, -ое, (past part pass) see **о·хлажд·á·ть**; cooled, refrigerated
о·хлóп·ок·, -п·к·а, (m) combings, tow
о·хóт·а, -ы, (f) hunting, shooting; desire, inclination, wish; (zool) season, oestrus, oestrum, rut
о·хóт·ник·, -а, (m) hunter, trapper; sportsman, gun; enthusiast, lover (person who is keen on something), volunteer; (naut) submarine-chaser

о·хóт·нич·ий, -ья, -ье, (a) of **о·хóт·-ник·**
о·хóт·н·о (adv) willingly, gladly
óхр·а, -ы, (f) ochre, ocher
~, **крáсн·ая** raddle, red ochre
~, **чёрн·ая** wad, black ochre
о·хрáн·а, -ы, (f) guard; guarding, protection; conservation, preservation
~, **воен·изѝрованн·ая** armed security guard
~ **нéдр·ы** (geol) conservation of mineral resources
~ **пóчв·ы** soil conservation
~ **при·рóд·ы** nature preservation
~ **рéйд·ов** (mil) harbour defence
     от·дéл· о·хрáн·ы (m) security department
~ **труд·á** industrial safety measures
о·хран·éни·е, -я, (n) safeguarding; (mil) protection, guard; (naut) screen, picket
о·хран·ённ·ый, -ая, -ое, (past part pass) see **о·хран·я́·ть**
о·хран·ѝтел·ь, -я, (m) protector, guardian; (naut) paravane
о·хран·ѝтельн·ый, -ая, -ое, (a) protective
о·хран·ѝть, -я́т, (perf) see **о·хран·я́·ть**
о·хрáн·н·ый, -ая, -ое, (a) of **о·хрáн·а**
~ **кольц·ó** (n) guard ring
~ **рáм·а** (f) (naut) camel, bumper, fender
~ **у·станóв·к·а** (f) (instr) intrusion detector
о·хран·я́·ть, -ют, (imp) guard, protect; preserve
о·хрѝп·нуть, -нут, (perf) become hoarse
óхр·ист·ый, -ая, -ое, (a) ocherous, ochreous, ochery
о·хром·é·ть, -ют, (perf) become/grow lame
охро·дерм·ѝ·я, -и, (f) (med) ochrodermia
охро·лѝт·, -а, (m) (min) ochrolite
о·хрýп·чивани·е, -я, (n) embrittlement
охр·я́н·ый, -ая, -ое, (a) ochre, ochreous
о·хрящ·евáни·е, -я, (n) (zool) chondrification
о·царáп·а·ть, -ют, (perf) see **о·царá·п·ыва·ть**
о·царáп·ыва·ть, -ют, (imp) scratch, graze

оцелля́р·н·ый, -ая, -ое, (*a*) (zool) ocellar

оцело́т·, -а, (*m*) (zool) ocelot, *Felis pardalis*

о·це́н·ива·ть, -ют, (*imp*) estimate, evaluate, assess, value, appraise; fix the price at, value at

~ ко́рен·ь у·равн·е́ни·я (math) isolate the root of an equation

о·цен·и́ть, -́ят, (*perf*) *see* **о·це́н·ива·ть**

о·це́н·к·а, -и, (*g pl*) **-н·ок·,** (*f*) (*v n*) *see* **о·це́н·ива·ть;** estimation, estimate, evaluation, appraisal, valuation

~, техн·и́ческ·ая engineering estimate

о·це́н·очн·ый, -ая, -ое, (*a*) of **о·це́н·-к·а**

о·це́н·щик·, -а, (*m*) valuer, estimator

о·цепен·е́·ть, -ют, (*perf*) become numb/torpid

о·цепен·е́л·ый, -ая, -ое, (*a*) torpid, numbed

о·цепен·и́ть, -я́т, (*perf*) *see* **о·цепен·я́·ть**

о·цепен·я́·ть, -ют, (*imp*) make numb/torpid, torpidify

о·цеп·и́ть, -́ят, (*perf*) *see* **о·цепл·я́·ть**

о·цепл·я́·ть, -ют, (*imp*) surround, cordon off

о·цинк·ова́ни·е, -я, (*n*) galvanizing, galvanization (ferrous metals); zincing, zinc plating/coating

~, сух·о́е sherardizing

о·цинк·о́ванн·ый, -ая, -ое, (*past part pass*) of **о·цинк·ова́ть;** galvanized, zinc coated/plated

~ желе́з·о (*n*) galvanized iron

о·цинк·ова́ть, -у́ют, (*perf*) galvanize, coat/plate with zinc

о·цинк·о́вк·а, -и, (*f*) (met) galvanization

о·цинк·о́вывани·е, -я, (*n*) *see* **о·цинк·ова́ни·е**

о·цифр·ова́ть, -у́ют, (*perf*) *see* **о·цифр·о́выва·ть**

о·цифр·о́выва·ть, -ют, (*imp*) figure, put in figures, mark (scales etc.)

оча́г·, -а́, (*m*) hearth; source, centre, focus, nidus, seat

~ земле·тряс е́ни·я focus of earthquake

~ на·руш·е́ни·я seat of disturbance

~ со·противл·е́ни·я (mil) pocket of resistance

~ то́п·к·и furnace hearth

~, фу́рмен·н·ый tuyère hearth

оча́г·о́в·ый, -ая, -ое, (*a*) of **оча́г·**

~ о·ста́т·к·и (*pl*) ash, ashes, cinders (from boiler furnace)

~ симпто́м· (*m*) (med) focal symptom

оча́ж·н·ый, -ая, -ое, (*a*) hearth

оче·ви́д·ец·, -д·ц·а, (*m*) eye-witness

оче·ви́д·н·о (*adv*) obviously, of course; it is obvious

оче·ви́д·ност·ь, -и, (*f*) obviousness

оче·ви́д·н·ый, -ая, -ое, (*a*) evident, obvious

о́чень (*adv*) very, very much, greatly

~ выс·о́к·ая част·от·а́ (*f*) very high frequency

~ кре́п·к·ий ве́тер· (*m*) (meteor) fresh gale

о·черед·н·о́й, -а́я, -о́е, (*a*) next, next in turn; routine, regular, usual; (bot) alternate

о·черёд·ност·ь, -и, (*f*) sequence, regular succession, order; priority, precedence

о́·черед·ь, -и, (*nom pl*) **-и,** (*g pl*) **-е́й,** (*f*) turn, succession, sequence, rota, precedence; queue; burst (of gunfire)

вне вся́·к·ой о́·черед·и right out of turn; (telecom) emergency

~, пе́рв·ая first instance/place; first stage/phase, stage/phase one; (telecom) first degree of precedence

по о́·черед·и in turn

~, по·сле́д·н·яя last, last/final stage; (telecom) lowest degree of precedence

о́·черк·, -а, (*m*) outline, sketch, synopsis; essay, feature-story

о·чёрк·ива·ть, -ют, (*imp*) outline, mark round

о·черк·н·у́ть, -у́т, (*perf*) *see* **о·чёрк·ива·ть**

о·черств·е́л·ый, -ая, -ое, (*a*) hardened, callous

о·черт·а́ни·е, -я, (*n*) outline, contour

о·черт·и́ть, -́ят, (*perf*) *see* **о·чёрч·ива·ть**

о·чёрч·ива·ть, -ют, (*imp*) outline, draw/mark round; characterize, depict, describe

о·чёс·, -а, (*m*) combings; (*pl*) tow, flock

о·чес·а́ть, (*fut 3rd pl*) **о·чёш·ут,** (*perf*) *see* **о·чёс·ыва·ть**

о·чёс·к·и, -ов, (*m pl*) combings; tow, flock

о·чёс·ыва·ть, -ют, (*imp*) comb out/flat; hackle (flax)

о·чехл·о́вк·а, -и, (*f*) jacketing, canning (a piece of equipment)

о·чёш·ет, (*fut 3rd sing*) of **о·чес·а́ть**

о·чи́н·ива·ть, -ют, (*imp*) sharpen, point

о·чин·и́ть, - я́т, (*perf*) *see* **о·чи́н·ива·ть**
о·чин·я́·ть, -ют, (*imp*) = **о·чи́н·ива·ть**
о·чист·и́тел·ь, -я, (*m*) cleaner, purifier, cleanser
~ га́з·а scrubber, gas scrubber
о·чист·и́тельн·ый, -ая, -ое, (*a*) cleaning, cleansing, purifying, purification, decontamination, refining
о·чи́ст·ить, -ят, (*perf*) *see* **о·чищ·а́·ть**
о·чи́ст·к·а, -и, (*g pl*) **-т·ок·,** (*f*) (*v n*), *see* **о·чищ·а́·ть;** (met) fettling (castings); clearance (of ships by customs); (*pl*) peelings
~ вод·ы́ water treatment
~ во́з·дух·а air purifying/purification
~ га́з·а gas scrubbing/sweetening/purifying
~, гидравл·и́ческ·ая (met cast) hydraulic fettling, hydroblast
~, из·бир·а́тельн·ая (oil) selective refining, solvent extraction refining
коэффице́нт о·чи́ст·к·и (*m*) decontamination factor
~, огн·ев·а́я (met) flame scarfing, hot deseaming
~, песко·стру́й·н·ая (met) sand blasting
~ по от·хо́д·у clearance outward (customs)
~ по при·хо́д·у clearance inward (customs)
о·чист·н·о́й, -а́я, -о́е, (*a*) cleaning, scrubbing (gas), refining (oil)
~ вы́·ем·к·а (*f*) (min) stoping
~ за·бо́·й (*m*) (min) stope
~ пере·го́н·к·а (*f*) (oil) refinery distillation
~ про·стра́н·ств·о (*n*) (min) goaf
~ рабо́т·ы (*pl*) (min) stoping
~ у·стано́в·к·а (*f*) (chem) purifying/decontamination plant
очи́т·ок·, -т·к·а, (*m*) (bot) stonecrop, *Sedum*
о·чищ·а́·ть, -ют, (*imp*) clean, cleanse, scrub, clean up, clarify, purify, refine; fettle (castings); peel, shell, scale; clear, free; scrub (gas)
о·чищ·а́ющ·ий, -ая, -ее, (*pres part act*) of **о·чищ·а́·ть**
~ по́л·е (*n*) (instr) clearing field
~ сре́д·ств·о (*n*) purifying/cleaning agent
о·чищ·ённ·ый, -ая, -ое, (*past part pass*) *see* **о·чищ·а́·ть**
~ рас·тво́р· (*m*) raffinate
~ сло́·й (*m*) (chem) raffinate layer

очк·и́, -о́в, (*m pl*) (*see also* **очк·о́**) (opt) spectacles, glasses
~, за·щи́т·н·ые goggles
очк·о́, -а́, (*see also* **очк·и́**); pip (marking), point (for scoring); hole; eye (in a rope); (print) type face
~, смотр·ов·о́е peephole, sighthole
очк·о́в·ый, -ая, -ое, (*a*) of **очк·и́**
~ зме·я́ (*f*) (zool) cobra, *Naja naja*
~ о·пра́в·а (*f*) spectacle frame
оч·н·у́ться, -у́тся, (*perf*) regain consciousness
о·чувств·и́тел·ь, -я, (*m*) sensitizer; activator
о·чувствл·ён·и·е, -я, (*n*) sensitization; activation (of thermionic cathodes)
очут·и́ться, -ятся, (*perf*) find oneself, happen to be
о·шварт·ова́ть, -у́ют, (*perf*) (naut) come/secure alongside, tie up to, make fast to
о·шварт·о́ванн·ый, -ая, -ое, (*past part pass*) of **о·шварт·ова́ть**
о·ше́й·ник·, -а, (*m*) collar
ошелом·и́тельн·ый, -ая, -ое, (*a*) astounding
о·шерст·не́ни·е, -я, (*n*) (text) woollenization
о·шиб·а́·ться, -ются, (*imp*) make a mistake, be wrong, be in error, err
о·шиб·и́ться, -у́тся, (*perf*) *see* **о·шиб·а́·ться**
о·ши́б·к·а, -и, (*g pl*) **-б·ок·,** (*f*) *see also* **про·гре́ш·ност·ь;** error, mistake; (autom) deviation
~, аэро·динам·и́ческ·ая position error (of a/c instruments)
~, гру́б·ая gross error
~, динам·и́ческ·ая lag (of instruments), dynamic error (a/c instruments)
за·ко́н· о·ши́б·ок· (*m*) law of errors
~ за·мык·а́ни·я (surv) closing error
знак· о·ши́б·к·и (*m*) (telecom) erase sign
~, инструмента́ль·н·ая instrument/index error
~ на·с·жим·а́емост·ь compressibility error
~ на·блюд·а́тел·я individual error
~ на·лож·е́ни·я (cryst) stacking fault
~ нивели́р·а (surv) level error
~ о·кругл·е́ни·я round-off error
~ от·сек·а́ни·я truncation error
~, по·сто·я́нн·ая systematic/assignable error

о·ши́б·к·а
~, пре·де́ль·н·ая limiting error
~ при·бо́р·а instrumental/index error
~, с·луч·а́йн·ая random/accidental error
~, средне·квадрат·и́чн·ая (math) standard deviation/error, root mean square deviation/error
~ с·чёт·а miscount
~, у·станов·и́вш·аяся steady-state error
~, у·стано́в·очн·ая installation error
~ эксперимéнт·а experimental error
о·ши́б·очн·ый, -ая, -ое, (*a*) erroneous, mistaken
о·шлак·ова́ни·е, -я, (*n*) (*v n*), *see* **о·шлак·о́выва·ть**; slagging
о·шлак·ова́ть, -у́ют, (*perf*) *see* **о·шлак·о́выва·ть**
о·шлак·о́выва·ть, -ют, (*imp*) slag, slag up, form slag; put on slag
о·шпа́р·ива·ть, -ют, (*imp*) scald
о·шпа́р·ить, -ят, (*perf*) *see* **о·шпа́р·ива·ть**
о·шпа́р·к·а, -и, (*f*) scalding
о·штраф·ова́ть, -у́ют, (*perf*) fine
о·штукату́р·ить, -ят, (*perf*) plaster
о·щела́ч·ивани·е, -я, (*n*) alkalization, alkalizing

о·щен·и́ться, -я́тся, (*perf*) pup, whelp, cub
о·щип·а́ть, -лют, (*perf*) *see* **о·щи́п·ыва·ть**
о·щи́п·ыва·ть, -ют, (*imp*) pluck
о·щу́п·а·ть, -ют, (*perf*) *see* **о·щу́п·ыва·ть**
о·щу́п·ыва·ть, -ют, (*imp*) feel, feel over/around
о·щу́п·ывающ·ий, -ая, -ее, (*pres part act*) of **о·щу́п·ыва·ть**
~ у·стро́й·ств·о, электро·механ·и́ческ·ое (*n*) mechanical reader (for punched tapes and cards)
о́·щупь на о́·щупь to the touch
о́·щуп·ью (*adv*) by touch/feeling/groping
о·щут·и́м·ый, -ая, -ое, *see* **о·щут·и́тельн·ый**
о·щут·и́тельн·ый, -ая, -ое, (*a*) sensible, palpable, tangible; appreciable, perceptible
о·щут·и́ть, -я́т, (*perf*) *see* **о·щущ·а́·ть**
о·щущ·а́·ть, -ют, (*imp*) feel, sense
о·щущ·е́ни·е, -я, (*n*) (*v n*), *see* **о·щущ·а́·ть**; sensation, perception
~, такти́ль·н·ое tactile perception
о·ягн·и́ться, -я́тся, (*perf*) lamb

a	adjective
abbr	abbreviation, etc.
a/c	aircraft, aircraft industry
acc	accusative case
acous	acoustics
act	active
adj	adjective
adv	adverb
aerodynam	aerodynamics
agr	agriculture
air	aeronautics, air transport
anat	anatomy
arch	architecture
astron	astronomy
attrib	attributive
autom	automation, automatics, control engineering
bact	bacteriology
biol	biology
bot	botany
Brit	British
build	building
cast	casting, foundry work
ceram	ceramics
chem	chemistry
cinema	cinematography
civ eng	civil engineering
coll	colloquial
com	commerce
comp	comparative
component	word component, complex combining form
conj	conjunction
cryst	crystallography
cyt	cytology
d	dative case
deter	determinate
dim	diminutive
dipl	diplomatic
dynam	dynamics
econ	economics, economy
elec	electrical, electronics
eng	engineering
ent	entomology
esp	especially
expl	explosives
f	feminine noun
fem	feminine, feminine gender
fin	finance, financial
fish	fish, fisheries, fishing ichthyology
food	food, food technology
g	genitive case
g as adj	used in genitive case as an adjective
gen	genetics
geog	geography, geographical
geol	geology
ger	gerund
gram	grammar, grammatical
gunn	gunnery, guns, ordnance, ammunition
GW	guided weapons
horol.	horology
hortic.	horticulture
hydr.	hydraulics hydrology
hydr eng	hydraulic engineering
hydrog	hydrography
i	instrumental case
ICE	internal combustion engine
imp	imperfective verb
imper	imperative
indecl	indeclinable
indet	indeterminate
ins	insurance
instr.	instrumentation, experimental equipment
int	international
intrans	intransitive
law	law, legal
ling	linguistics
m	masculine noun
mar	marine
mar eng	marine engineering
mar ins	marine insurance
masc	masculine, masculine gender
math	mathematics, mathematical
mech.	mechanical, mechanics

RUSSIAN ENGLISH

SCIENTIFIC AND TECHNICAL DICTIONARY

IN TWO VOLUMES

Volume 2

П-Я

RUSSIAN–ENGLISH
SCIENTIFIC
AND
TECHNICAL DICTIONARY

M. H. T. and V. L. ALFORD

Volume 2

П–Я

PERGAMON PRESS

OXFORD · NEW YORK · BEIJING · FRANKFURT
SÃO PAULO · SYDNEY · TOKYO · TORONTO

U.K.	Pergamon Press, Headington Hill Hall, Oxford OX3 0BW, England
U.S.A.	Pergamon Press, Maxwell House, Fairview Park, Elmsford, New York 10523, U.S.A.
PEOPLE'S REPUBLIC OF CHINA	Pergamon Press, Room 4037, Qianmen Hotel, Beijing, People's Republic of China
FEDERAL REPUBLIC OF GERMANY	Pergamon Press, Hammerweg 6, D-6242 Kronberg, Federal Republic of Germany
BRAZIL	Pergamon Editora, Rua Eça de Queiros, 346, CEP 04011, Paraiso, São Paulo, Brazil
AUSTRALIA	Pergamon Press Australia, P.O. Box 544, Potts Point, N.S.W. 2011, Australia
JAPAN	Pergamon Press, 8th Floor, Matsuoka Central Building, 1-7-1 Nishishinjuku, Shinjuku-ku, Tokyo 160, Japan
CANADA	Pergamon Press Canada, Suite No 271, 253 College Street, Toronto, Ontario, Canada M5T 1R5

First edition 1970

Reprinted 1974; 1981; 1984, 1988

Library of Congress Catalog Card No. 73—88348

Printed in Great Britain by A. Wheaton & Co. Ltd., Exeter

ISBN 0-08-012227-2

CONTENTS

RUSSIAN–ENGLISH DICTIONARY П-Я

П

п-обра́з·н·ый, -ая, -ое,   (*a*) U-, U-shaped; pi, π (Greek symbol); square

~ и́мпульс·   (*m*) (elec) square pulse

павиа́н·, -а,   (*m*) (zool) baboon, *Cynocephalus*

павильо́н·, -а,   (*m*) pavilion, hall; (cinema) film studio, set

павино́л·, -а,   (*m*) Pavinol (a leather substitute)

павли́н·, -а,   (*m*) (orn) peacock, peafowl, *Pavo cristatus*

павли́н·ий, -ья, -ье,   (*a*) of **павли́н·**

~ глаз·   (*m*) (ent) thistle butterfly, *Vanessa*

павли́н·ый, -ая, -ое, (*a*) *see* **павли́н·ий**

павли́н·ов·ый, -ая, -ое, (*a*) *see* **павли́н·ий**

па́·вод·ков·ый, -ая, -ое, (*a*) of **па́·вод·ок·**

па́·вод·ок·, -д·к·а, (*m*) flood (of rivers), high water

па́·вш·ий, -ая, -ее, (*past part act*) of **пас·ть,** *see* **па́д·а·ть;** fallen; killed, killed in action

пагин·а́ци·я, -и, (*f*) (print) pagination

пагоди́т·, -а, (*m*) (min) pagodite

па́·губ·н·ый, -ая, -ое, (*a*) ruinous, pernicious

падале·я́д·ы, -ов, (*m pl*) (zool) scavengers, carrion eaters

па́д·алиц·а, -ы, (*i*) **-ей,** (*f*) windfall

па́д·ал·ь, -и, (*f*) carrion

па́д·а·ть, -ют, (*imp*) fall, go/come down, drop, sink; drop/fall/come out (of hair etc.); die (of animals); (geol) dip, hade

па́д·аю·щ·ий, -ая, -ее, (*pres part act*) of **па́д·а·ть;** incident (of rays); (nucl) incoming (of particles)

~ звезд·а́ (*f*) shooting star, meteor

~ мо́лот· (*m*) (met forge) drop hammer

~ характери́стик·а (*f*) (phys) drooping/falling characteristic

пад·е́ж·, -а́, (*i*) **-о́м,** (*m*) (gram) case

пад·ёж·, -а́, (*i*) **-о́м,** (*m*) (vet) epizooty, murrain

пад·е́ни·е, -я, (*n*) (*v n*), *see* **па́д·а·ть;** drop, fall; incidence (of rays); impact (of rockets); (gunn) graze; (geol) dip; gradient (of rivers etc.)

~, на·кло́н·н·ое oblique incidence (of rays); (geol) gentle dip

пад·е́ни·е

~ на·пряж·е́ни·я (elec) potential/voltage drop

~ на·пряж·е́ни·я, ано́д·н·ое anode/plate drop

~ на·пряж·е́ни·я, ом·и́ческ·ое (elec) IR drop

~ по кас·а́тельн·ой grazing incidence

~ с·на·ря́д·а (gunn) fall of shot

~, со·гла́с·н·ое (geol) hade with the dip

то́ч·к·а пад·е́ни·я (*f*) (gunn) point of arrival/impact

угол· пад·е́ни·я (*m*) (geol) dip angle, angle of incidence; angle of incidence, striking angle (rays, beams etc.); (gunn) angle of descent/arrival

пад·ёт, (*fut 3rd sing*) of **пас·ть,** (*perf*) *see* **па́д·а·ть** (*imp*)

па́дуб·, -а, (*m*) (bot) holly, *Ilex*

пад·у́н·, -а, (*m*) drifting ice

пад·у́т, (*fut 3rd pl*) of **пас·ть,** (*perf*); *see* **па́д·а·ть** (*imp*)

пад·у́ч·ий, -ая, -ее, (*a*) falling

~ бол·е́зн·ь (*f*) "the falling sickness", (med) epilepsy

па́д·ш·ий, -ая, -ее, (*past part act*) of **пас·ть,** (*perf*); *see* **па́д·а·ть** (*imp*)

па·ев·о́й, -а́я, -о́е, (*a*) of **па·й**

па·ёк·, (*g sing*) **па́й·к·а,** (*m*) (mil) rations; **па́·ек·** (*g pl*) of **па́й·к·а**

~, сух·о́й dry provisions, hard tack

паёл·, -а, (*m*) *see* **пайо́л·**

па́жит·ник·, -а, (*m*) (bot) trigonella

паз·, -а, (*nom pl*) **-ы́,** (*g pl*) **-о́в,** (*m*) slot, groove, mortise, channel, rabbet; (shipb) seam (hull plating)

~, шпо́н·очн·ый keyway, keyseat

паза́ни·я, -и, (*f*) (bot) tan oak, *Pasania*

паз·и́ть, -ят, (*perf*) flute, groove, mortise, channel

паз·ни́к·, -а́, (*m*) grooving/tongue plane (carpenter's tool)

паз·ова́льн·ый, -ая, -ое, (*a*) slotting, grooving, mortising

~ стан·о́к· (*m*) boring and slot-mortising machine

паз·ова́ть, -у́ют, (*imp and perf*) groove, channel

паз·овик·, -а, (*m*) *see* **паз·ни́к·**

па́з·ов·ый, -ая, -ое, (*a*) of **паз·**

~ сверл·о́ (*n*) slot drill

~ фре́з·а (*f*) grooving cutter

па́з·ух·а, -и, (*f*) (anat) sinus; (bot) axil

~, вено́з·н·ая (zool) sinus venosus

па́з·ушн·ый, -ая, -ое,   (*a*) (anat) sinusal; (bot) axillary

па·й,   (*g sing*) па́·я, (*nom pl*) па·й, (*g pl*) -ёв, (*m*) part, portion, share

па́йлер·, -а,   (*m*) (met roll) piler

па́йз·а, -ы,   (*f*) (bot) Japanese millet, *Echinochloa frumentacea*

па́йдз·а, -ы,   (*f*) *see* па́йз·а

Па́йерлс·а, си́л·а   (*f*) Peierls force

па́й·к·а, -и,   (*g pl*) па́·ек·, (*f*) soldering (under 427°C); brazing (over 427°C); seal (of glass objects, e.g. ampoules, vacuum tubes etc.); пай·к·а́ (*g sing*) of па·ёк·

~, ва́куум·н·ая   vacuum seal

~, га́з·ов·ая   torch brazing

~, индукцио́н·н·ая   induction brazing

~, конта́кт·н·ая   resistance brazing

~ мя́гк·им при·по́·ем   soldering (generally under 300°C)

~ по·груж·ени·ем   dip brazing/soldering

~ твёрд·ым при·по́·ем   brazing (over 600°C)

па́й·ко́в·ый, -ая, -ое,   (*a*) ration, rationed

па́йол·, -а,   (*m*) ceiling (cargo ships)

па́й·щик·, -а,   (*m*) shareholder, member (of cooperative etc.); solderer

па́к·, -а,   (*m*) pack; (ocean) pack ice

~ пит·а́ни·я   (rad) power pack

~, у·плот·нённ·ый   consolidated pack ice

па́к·а, -и,   (*f*) (zool) alpaca

пакга́уз·, -а,   (*m*) warehouse

пакеля́ж·, -а,   (*m*) stone pitching (road-making)

па́кер·, -а,   (*m*) (oil) packer

паке́т·, -а,   (*m*) packet, package, parcel; pack; bale; fagot (of scrap metal); roadbearer, composite roadbearer (on bridges)

~, волн·ов·о́й   (phys) wave packet

~ и́мпульс·ов   pulse packet

~, индивидуа́ль·н·ый пере·вя́з·оч·н·ый   (mil) field dressing

~ лист·о́в   (met) mill pack, pack (for rolling)

~ фоно́н·ов   (nucl) phonon packet

паке́т·и́ровани·е, -я,   (*n*) *see* пак·о·ва́ни·е

~ ло́м·а   (met) scrap baling, fagoting

пакети́р-пресс·, -а,   (*m*) (met) baling/fagoting press

паке́т·н·ый, -ая, -ое,   (*a*) of паке́т·

~ вы·ключ·а́тел·ь   (*m*) (elec) rotary packet switch

паке́т·н·ый, -ая, -ое

~ про·ка́т·к·а   (*f*) (met) ply/pack rolling

~ с·вя́з·к·а   (*f*) (met) fagot

па́ккер·, -а,   (*m*) *see* па́кер·

па́кл·я, -и,   (*f*) tow; oakum

пак·ова́ни·е, -я,   (*n*) (*v n*) of пак·о·ва́ть

пак·ова́ть, -у́ют,   (*imp*) pack, stack, bale, make into packs/bales/stacks

пак·о́вк·а, -и,   (*g pl*) -вок·, (*f*) (*v n*), *see* пак·ова́ть; (text) package

пак·о́вочн·ый, -ая, -ое,   (*a*) of пак·о́в·к·а́

~ пресс·   (*m*) baling press

па́к·ов·ый, -ая, -ое,   (*a*) of па́к·

~ лёд·   (*m*) (ocean) pack ice

па́кост·ить, -ят,   (*imp*) soil, dirty, foul; spoil

пакси́лл·а, -ы,   (*f*) (zool) paxilla, paxillus

па́кт·, -а,   (*m*) (dipl) pact

пал·, -а,   (*nom pl*) -ы́, (*m*) fire clearing (in a wood); ridge (for cultivation); (naut) bollard, pawl; (*pl*) pile mooring, dolphin

палаба́жник·, -а,   (*m*) ice rind

пала·со́м·а, -ы,   (*f*) (biol) palasome, host

пала́т·а, -ы,   (*f*) chamber, House (parliamentary); ward (in hospital)

~ мер·и ве́с·ов   weights and measures department

~ о́бщ·ин·   House of Commons

~, торг·о́в·ая   chamber of commerce

палатализ·ова́ть, -у́ют,   (*imp and perf*) (ling) palatalize

палат·а́льн·ый, -ая, -ое,   (*a*) palatal

палат·и́нов·ый, -ая, -ое,   (*a*) (zool) palatine

~ кра́сн·ый   (*m*) (chem) palatine red

пала́т·к·а, -и,   (*g pl*) -т·ок·, (*f*) tent, awning; stall, booth

пала́т·н·ый, -ая, -ое,   (*a*) (med) ward

палато·пла́ст·ик·а, -и,   (*f*) (med) palatoplasty

пала́т·очн·ый, -ая, -ое,   (*a*) of пала́т·к·а

пала́ш·, -а́,   (*i*) -о́м, (*m*) cutlass, broadsword

палгуп·, -а,   (*m*) (naut) pawl ring

пале·а́рктик·а, -и,   (*f*) (zool) Palaearctic/Palearctic region

па́л·ев·ый, -ая, -ое,   (*a*) pale pinkish yellow, straw-coloured

пал·е́ни·е, -я,   (*n*) (*v n*), *see* пал·и́ть; singeing, scorching

пал·ён·ый, -ая, -ое, (*a*) scorched, singed

палео·анодо́нт·ы, -ов, (*m pl*) (pal) palaeanodonts, *Palaeanodonta*

палео·бота́н·ик·а, -и, (*f*) paleobotany, palaeobotany, fossil botany

палео·ге́н·, -а, (*m*) (geol) Palaeogene, Paleogene

палео·геогра́ф·и·я, -и, (*f*) palaeogeography, paleogeography

палео·гна́т·ы, -ов, (*m pl*) (pal) palaeognaths, *Palaeognathae*

палео·зо́·й, -я, (*m*) (geol) Paleozoic, Palaeozoic

палео·климато·ло́г·и·я, -и, (*f*) (geol) paleoclimatology

палео·ко́нх·, -а, (*m*) (pal) palaeoconch, (*pl*) *Palaeoconcha*

палео·ли́т·, -а, (*m*) (geol) Palaeolithic, Paleolithic

палео·лит·и́ческ·ий, -ая, -ое, (*a*), (geol) palaeolithic, paleolithic

пале·онто́лог·и·я, -и, (*f*) paleontology, palaeontology

палео·си́м·и·я, -и, (*f*) (pal) Palaeosimia

палео·ти́п·н·ый, -ая, -ое, (*a*) (geol) palaeotypic(al), paleotypic(al)

палео·троп·и́ческ·ий, -ая, -ое, (*a*) (bot) palaeotropical, paleotropical

∼ о́бласт·ь (*f*) (bot) palaeotropical/ paleotropical region

палео·флюмено·ло́г·и·я, (*f*) (geol) paleoflumenology

палео·це́н·, -а, (*m*) (geol) Palaeocene, Paleocene

палео·эко́лог·и·я, -и, (*f*) palaeoecology, paleoecology

палео·эхиноиде́·и, -й, (*f pl*) (pal) Palaechinoida

пале́т·а, -ы, (*f*) *see* **пале́тт·а**

пале́т·к·а, -и, (*g pl*) **-т·ок·,** (*f*) (surv) measuring grid, reticulation

пале́тт·а, -ы, (*f*) (horol) pallet, pallet stone

∼, в·ход·н·а́я (horol) entering pallet

∼, вы·ход·н·а́я (horol) exit pallet

па́л·ец·, -ль·ц·а, (*i*) **-ль·ц·ем,** (*m*) finger, digitus, dactylus; pin, peg, stud; cog, catch; (anat) digit

∼, больш·о́й thumb; big toe

∼, пла́в·ающ·ий floating gudgeon pin

∼, порш·нев·о́й piston/wrist/gudgeon pin

∼, у·каз·а́тельн·ый forefinger, index finger

па́·ли (*fut pl*) of **пас·ть** (*perf*), *see* **па́д·а·ть** (*imp*)

пал·и́лк·а, -и, (*f*) (text) singer

пал·и́льн·ый, -ая, -ое, (*a*) singeing; shooting, shot-firing

палим·псе́ст·, -а, (*m*) palimpsest

палин·гене́зис·, -а, (*m*) (geol, biol) palingenesis

палино·ло́г·и·я, -и, (*f*) (bot) palynology, pollen analysis

палиса́д·, -а, (*m*) palissade

палиса́д·ник·, -а, (*m*) front garden

палиса́д·н·ый, -ая, -ое, (*a*) of **палиса́д·**

∼ ткан·ь (*f*) (bot) palisade tissue

палиса́ндр·, -а, (*m*) (bot) rosewood, *Jacaranda*

пал·и́ть, -я́т, (*imp*) singe, scorch; shoot, fire

па́ли·я, -и, (*f*) (fish) char, *Salvelinus*

па́л·к·а, -и, (*g pl*) **-л·ок·,** (*f*) stick

палла́д·а, -ы, (*f*) (astron) Pallas

палла́д·иев·ый, -ая, -ое, (*a*) of **палла́д·и·й** (chem) palladic

палла́д·и·й, -я, (*m*) (chem) palladium, Pd

паллад·и́рованн·ый, -ая, -ое, (*a*) palladinized; palladium-coated

палла́д·ист·ый, -ая, -ое, (*a*) palladous, palladious

палласи́т·, -а, (*m*) (min) pallasite

палл·естези́·я, -и, (*f*) (zool) pallaesthesia

паллетро́н·, -а, (*m*) (phys) palletron

палли·ати́в·, -а, (*m*) (pharm) palliative

палли·ати́вн·ый, -ая, -ое, (*a*) (med) palliative

паллио·эссекси́т·, -а, (*m*) (min) pallioessexite

палло·гра́ф·, -а, (*m*) (shipb) vibration recorder

пало·де́л·ател·ь, -я, (*m*) ridge/dam forming attachment (irrigation tool)

па́л·очк·а, -и, (*g pl*) **-чек·,** (*f*) (*dim*) of **па́л·к·а** stick, baton, rod; (bact) bacillus

палочко·ви́д·н·ый, -ая, -ое, (*a*) rod-like, rod-shaped; bacillary, bacilliform

палочко·обра́з·н·ый, -ая, -ое, (*a*) rod-shaped; bacilliform, bacillary, baculiform

палочко·я́дер·н·ый, -ая, -ое, (*a*) (biol, nucl) stabnuclear

па́л·очн·ый, -ая, -ое, (*a*) of **па́л·к·а**

па́лтус·, -а, (*m*) (fish) halibut

па́лтус·ов·ый, -ая, -ое, (*a*) of па́л·тус·

па́луб·а, -ы, (*f*) (shipb) deck

~, взлёт·н·ая flight deck

~, жил·а́я messdeck, crew space

~, карапа́с·н·ая turtle deck/back, whaleback

~, коммуна́ль·н·ая recreation space

~, на·вес·н·а́я shade deck

~ над·стро́й·к·и bridge deck

~, по·лёт·н·ая flight deck

~, про·гу́л·оч·н·ая promenade deck

~, угл·ов·а́я angled deck

~, шлюп·оч·н·ая boat deck

~ ю́т·а poop deck

па́луб·н·ый, -ая, -ое, (*a*) of па́·луб·а

-па́л·ый, -ая, -ое, (*component adj*) -fingered, -dactylous

палыгорски́т·, -а, (*m*) (min) mountain leather/cork

па́льм·а, -ы, (*f*) (bot) palm, palm tree, *Palmae*

~, коко́с·ов·ая common coconut palm, *Cocos nucifera*

~, фи́ник·ов·ая common date palm, *Phoenix dactylifera*

пальм·ати́н·, -а, (*m*) (chem) palmatine

пальм·итами́д·, -а, (*a*) (chem) palmitamide

пальм·ити́н·, -а, (*m*) (chem) palmatin

пальмитиново·ки́сл·ый, -ая, -ое, (*a*) palmate (of)

~ кисл·от·а́ (*f*) (chem) palmitic acid

па́льм·ов·ый, -ая, -ое, (*a*) of па́льм·а; (*pl as noun*) *Palmae*

пальмо·я́дер·н·ое ма́сл·о (*n*) palm-kernel oil

пальп·а́ци·я, -и, (*f*) (med) palpation

пальпебр·а́ль·н·ый, -ая, -ое, (*a*) palpebral

пальп·и́р·овать, -уют, (*imp and perf*) palpate

пальпит·а́ци·я, -и, (*f*) (physiol) palpitation

пальто́ (*n indecl*) overcoat

па́ль·ц·а, (*g sing*) of па́л·ец·

пальце·ви́д·н·ый, -ая, -ое, (*a*) digitate, finger-like/shaped

пальц·ев·о́й, -а́я, -о́е, (*a*) digital, finger; finger-action (of mechanisms)

~ брус· (*m*) finger bar (of harvesting machine)

~ пласт·и́нк·а (*f*) ledger plate/strip (carpentry)

пальце·обра́з·н·ый, -ая, -ое, (*a*) digitiform, finger-shaped

пальце·ход·я́щ·ий, -ая, -ее, (*a*) (zool) digitigrade

пальчато·ви́д·н·ый, -ая, -ое, (*a*) (bot) digitaliform

пальчато·над·рез·н·о́й, -а́я, -о́е, (*a*) (bot) palmatifid

пальчато·раз·де́ль·н·ый, -ая, -ое, (*a*) palmatipartite

пальчато·рас·сеч·ённ·ый, -ая, -ое, (*a*) (bot) palmatisect, palmatisected

па́ль·чат·ый, -ая, -ое, (*a*) finger, pin, tooth; (bot) palmate

па́ль·чик·, -а, (*m*) (*dim*) of па́л·ец·

па́ль·чиков·ый, -ая, -ое, (*a*) miniature, mini-, bantam

па́ль·щик·, -а, (*m*) (min) shot-firer

палюд·о́зн·ый, -ая, -ое, (*a*) (bot, geol) paludose, paludal, paludine

палюд·ри́н·, -а, (*m*) (pharm) paludrine, paludrin

памахи́н·, -а, (*m*) (pharm) pamaquine

па́мпельмус·, -а, (*m*) (bot) pummelo, shaddock, grapefruit, *Citrus grandis*

па́мят·к·а, -и, (*g pl*) -т·ок·, (*f*) reminder list; rules, book of rules

па́мят·ник·, -а, (*m*) monument, memorial

па́мят·н·ый, -ая, -ое, (*a*) of па́·мят·ь; memorable, unforgetable; reminder

~ за·пи́с·к·а (*f*) memorandum

~ кни́ж·к·а (*f*) notebook

па́мят·ь, -и, (*f*) memory; recollection; memory, store (of computer)

в па́мят·ь (*adv*) in commemoration/ memory

на па́мят·ь (*adv*) as a memento/ keepsake

~, операти́в·н·ая (autom) working/ internal memory/store

~, по·сле́д·ователь·н·ая (autom) serial store/memory

панакси́н·, -а, (*m*) (chem) panaxin

пана́м·а, -ы, (*f*) (text) hopsack; (bot) Panama bark

панари́ци·й, -я, (*m*) (zool) paronychia, whitlow

пан·аце́·я, -и, (*f*) panacea

пан·ге́·я, -и, (*f*) (zool) *Pangaea*

панголи́н·, -а, (*m*) (zool) pangolin, (*pl*) *Pholidota*

пан·дем·и́ческ·ий, -ая, -ое, (*a*) (med) pandemic

пан·дем·и́·я, -и, (*f*) (med) pandemic disease

панéл·ь, -и, (*f*) panel; (elec) panel mounting; pavement, sidewalk

∼, граф·и́ческ·ая graphic control panel

∼, за·щи́т·н·ая (elec) protective panel

∼, консóль·н·ая (autom) console

∼, при·бóр·н·ая instrument panel

∼, смотр·ов·áя access panel

панéль·к·а, -и, (*f*) (*dim*) of **панéл·ь**

∼, октáль·н·ая (rad) octal socket

панéр· (*indecl*) (naut) up-and-down (of anchor)

∼ бóрт·а корабл·я́ ships side vertical projection lines

панидио·мóрф·н·ый, -ая, -ое, (*a*) panidiomorphic

пáник·а, -и, (*f*) panic

пан·ирóвк·а, -и, (*g pl*) **-вок·,** (*a*) (food) coating with egg and breadcrumbs

пан·и́ческ·ий, -ая, -ое, (*a*) panic

пан·карди́т·, -а, (*m*) (med) pancarditis

панкреас·, -а, (*m*) (anat) pancreas

пан·креат·и́н·, -а, (*m*) (biochem) pancreatin

пан·креат·и́ческ·ий, -ая, -ое, (*a*) pancreatic

пан·орáм·а, -ы, (*f*) panorama; (gunn) panoramic sight

пан·орам·и́ровани·е, -я, (*n*) (instr) scanning; (TV, cinema) panning

пан·орáм·н·ый, -ая, -ое, (*a*) panoramic, panorama

пан·офтальми́т·, -а, (*m*) (med) panophthalmitis

пансиóн·, -а, (*m*) boarding school; boarding house, pension; board and lodging, pension

пансион·éр·, -а, (*m*) boarder; lodger

пант·алги́·я, -и, (*f*) (med) pantalgia

панта·трóн·, -а, (*m*) (elec) pantatron

пант·ахромат·и́ческ·ий, -ая, -ое, (*a*) pantachromatic

пан·телегрáф·, -а, (*m*) (telecom) pantelegraph

пантеллер·и́т·, -а, (*m*) (min) pantellerite

пантéр·а, -ы, (*f*) (zool) panther, *Felis panthera*

пантети́н·, -а, (*m*) (biochem) panthetine

пáнт·ов·ый, -ая, -ое, (*a*) of **пáнт·ы**

панто·грав·ёр·, -а, (*m*) (print) pantograver

панто·грáф·, -а, (*m*) (instr) pantograph

пант·одóнт·, -а, (*m*) (pal) pantodont

пантокарен·ы, -ов, (*m pl*) (shipb) cross curves of stability

панто·кри́н·, -а, (*m*) *see* **пáнт·ы** pantocrin (a drug obtained from reindeer horns)

пантó·метр·, -а, (*m*) (surv) pantometer

пант·опóн·, -а, (*m*) (pharm) pantopon

панто·тéн·ов·ая кисл·от·á (*f*) (biochem) pantothenic acid

панто·тéр·и·и, -ев, (*m pl*) (pal) pantotheres, *Pantotheria*

панто·ци́д·, -а, (*m*) pantocide (disinfectant)

пантус·, -а, (*m*) ramp

пáнт·ы, -ов, (*m pl*) horns of young reindeer

пан·хромат·и́ческ·ий, -ая, -ое, (*a*) panchromatic

панци́р·н·о- (*component*) *see* **пан·цы́р·н·о**

панцирно·голóв·ые, -ых, (*pl decl as adj*) (zool) stegocephalians, *Stegocephalia*

панцырно·жгýт·иков·ые, -ых, (*pl decl as adj*) (zool) *Dinoflagellata*

пáнцыр·н·ый, -ая, -ое, (*a*) of **пáн·цыр·ь** armoured, iron-clad, reinforced; (*pl as noun*) (zool) loricates, *Loricata, Polyplacophora*

∼ дино·зáвр·ы (*pl*) (pal) ankylosaurs, armoured/armored dinosaurs, *Ankylosauria*

∼ моллю́ск·и (*pl*) (zool) chitons, *Polyplacophora, Loricata*

∼ сéт·к·а (*f*) reinforcement mesh (concrete)

пáнцыр·ь, -я, (*m*) armour, armor; shell; (zool) shell, test; lorica (of rotifers etc.); iron-clad tube (in electric batteries)

папавери́н·, -а, (*m*) (pharm) papaverine

папаи́н·, -а, (*m*) (biochem) papain

папáх·а, -и, (*f*) (mil) fur cap (senior officers)

пáперт·ь, -и, (*f*) (arch) parvis

папилл·и́т·, -а, (*m*) (med) papillitis

папилл·óм·а, -ы, (*f*) (med) papilloma

папильонáж·, -а, (*m*) cross dredging

папиро·лóг·и·я, -и, (*f*) papyrology

папирóс·а, -ы, (*f*) cigarette, Russian cigarette

папирóс·н·ый, -ая, -ое, (*a*) cigarette, Russian-cigarette папи́рус·, -а, (*m*) (bot) papyrus, *Cyperus papyrus*

па́п·к·а, -и, (*g pl*) -п·ок·, (*f*) file, file cover (for office papers), portfolio

~, ма́тр·ичн·ая (print) flong

пап·маши́н·а, -ы, (*f*) *see* па́п·оч·н·ая маши́н·а

па́поротник·, -а, (*m*) (bot) fern

па́поротник·ов·ый, -ая, -ое, (*a*) fern; (*pl as noun*) true ferns, *Filicales*

папоротнико·обра́з·н·ые, -ых, (*pl as adj*) (bot) pteridophytes, *Pteridophyta*

папоротнико·семя́н·н·ые, -ых, (*pl decl as adj*) (pal) seed ferns, *Pteridospermae*

па́п·оч·н·ый, -ая, -ое, (*a*) of па́п·к·а

~ маши́н·а (*f*) presse pâte (paper-making)

па́пул·а, -ы, (*f*) (med) papula

папше́р·, -а, (*m*) guillotine, paper cutter

папье́-маше́ (*n indecl*) papier mâché

па́р·, -а, (*nom pl*) -ы́, (*g pl*) -о́в, (*m*) steam; vapour, vapor; (agr) fallow, fallow land/soil

~, вод·ян·о́й steam, water vapour/vapor

~ выс·о́к·их параме́тр·ов high-temperature/pressure steam

~, кондицио́н·н·ый standard steam

~, мя́·т·ый dead/exhaust steam

~, на·сы́щ·енн·ый saturated steam

~, о́стр·ый live steam

~, от·рабо́т·авш·ий exhaust steam

~, пере·гре́·т·ый superheated steam

~, пере·сы́щ·енн·ый supersaturated steam

~, рабо́ч·ий live steam

~, све́ж·ий live steam

~, сух·о́й dry steam

~, сыр·о́й wet steam

~, техно·лог·и́ческ·ий process steam

па́р·а, -ы, (*f*) pair, couple; suit (of clothes); (rubber) para

без па́р·ы odd, unpaired

в па́р·е (phys) paired with; coupled with

~, зуб·ча́т·ая gear wheel and pinion

~, ио́н·н·ая (phys) ion pair

моме́нт· па́р·ы (*m*) (mech) moment of couple

~, от·кры́·т·ая (mech) unclosed pair

~, по·ступ·а́тельн·ая (mech) sliding pair

~ сил· (mech) couple, force couple

~, с·крещ·ённ·ая (telecom) transposed pair

па́р·а

~, со·пряж·ённ·ая (phys) conjugate pair

~, черв·я́чн·ая (mech) worm-and-wormwheel

пара- (*int component*) (chem) para-, *p*-

пара·амидофено́л·, -а, (*m*) (chem, phot) para-amidophenol, para-amidooxybenzene

пара·аминосалици́л·ов·ая кисл·от·а́ (*f*) (pharm) para-aminosalicylic acid, PAS

парабан·ов·ая кисл·от·а́ (*f*) (chem) parabanic acid, oxalyl urea

пара́·бол·а, -ы, (*f*) (math) parabola

~, куб·и́ческ·ая cubical parabola

пара·бол·и́ческ·ий, -ая, -ое, (*a*) parabolic

пара·боло́ид·, -а, (*m*) (math) paraboloid

пара·болоида́ль·н·ый, -ая, -ое, (*a*) paraboloidal

пара·ва́н·, -а, (*m*) (naut) paravane

пара·ва́н·н·ый, -ая, -ое, (*a*) of пара·ва́н·

пара·водо·ро́д·, -а, (*m*) parahydrogen

пара·ге́л·и·й, -я, (*n*) (nucl) para-helium

пара·ген·е́з·,-а, (*m*) (geol) paragenesis

пара·гне́йс·, -а, (*m*) (geol) paragneiss

парагони́т·, -а, (*m*) (min) paragonite

пара́·граф·, -а, (*m*) paragraph

пара́д·, -а, (*m*) parade, review

~, воз·ду́ш·н·ый flypast

пара·дейте́р·и·й, -я, (*m*) (nucl) para-deuterium

пара·диоксибензо́л·, -а, (*m*) (phot, chem) parahydroxybenzene, hydroquinone

пара·дихлорбензо́л·, -а, (*m*) (chem) paradichlorbenzene

пара́д·н·ый, -ая, -ое, (*a*) parade, review; gala, festive, ceremonial; (arch) front, main

~ фо́рм·а (*f*) (mil) full dress

парадо́кс·, -а, (*m*) paradox

~ вре́мен·и (phys) clock paradox

~ с час·а́ми (phys) clock paradox

~ час·о́в (phys) clock paradox

парадокса́ль·н·ый, -ая, -ое, (*a*) paradoxical

пара·за·мещ·ённ·ый, -ая, -ое, (*a*) (chem) para-

парази́т·, -а, (*m*) parasite; (rad) spurious effect, bug

паразит·а́рн·ый, -ая, -ое, (*a*) (med) parasitic, transmitted by parasites

парази́т·и́ческ·ий, -ая, -ое,   (*a*) = **парази́т·н·ый**

парази́т·н·ый, -ая, -ое, (*a*) parasitic; spurious; (mech) idler, idle

~ **колес·о́**   (*n*) idler wheel

~ **по·те́р·я**   (*f*) (elec) parasitic loss

~ **с·вяз·ь**   (*f*) (phys) spurious coupling

~ **ток·**   (*m*) parasitic/spurious/stray current

~ **явл·е́ни·я**   (*pl*) (elec) parasitic phenomena, spurious effects, parasitics

паразито·ло́г·и·я, -и,   (*f*) parasitology

па́ра-каучу́к·, -а,   (*m*) para-rubber

па́ра·кла́з·, -а,   (*m*) (geol) paraclase

пар·аксиа́ль·н·ый, -ая, -ое, (*a*) (opt) paraxial

парализ·о́ванн·ый, -ая, -ое,   (*past part pass*) of **парализ·ова́ть**; paralysed, paralytic

парализ·ова́ть, -у́ют,   (*imp and perf*) paralyse

парали́т·и́ческ·ий, -ая, -ое, (*a*) paralytic

парали́ч·, -а́, (*i*) **-о́м,** (*m*) (med) paralysis, palsy

парали́ч·еск·ий, -ая, -ое, (*a*) (geol) paralic

парали́ч·н·ый, -ая, -ое, (*a*) = **парали́ч·еск·ий**

~ **трав·а́**   (*f*) (bot) bryony

паралла́кс·, -а,   (*m*) parallax

~, **век·ов·о́й**   (astron) secular parallax

~, **год·и́чн·ый** (astron) annual parallax; (surv) heliocentric parallax

~, **от·риц·а́тельн·ый**   negative parallax

~, **про·стра́н·ственн·ый**   (opt) local parallax

~, **со́лн·ечн·ый** (astron) solar parallax

~ **со́лнц·а** (surv) horizontal parallax

~, **су́т·очн·ый** (astron) diurnal parallax; (surv) geocentric parallax

параллакт·и́ческ·ий, -ая, -ое,   (*a*) parallactic, parallax

~ **не·ра́вен·ство** (*n*) (astron) parallactic inequality; anomalistic inequality (tides)

параллелепи́пед·, -а,   (*m*) (math) parallelepiped

~, **прям·о́й** right parallelepiped

параллело·гра́мм·, -а,   (*m*) parallelogram; (mech) parallel motion

~ **сил·** (mech) parallelogram of forces

~ **скор·ост·е́й**   parallelogram of velocities

~, **шарни́р·н·ый** (mech) parallel-crank mechanism

паралле́л·ь, -и,   (*f*) parallel; (*pl*) guides (engine)

в паралле́л·ь   (adv) parallel to, in parallel

~, **неб·е́сн·ая**   parallel of celestial latitude

~, **су́т·очн·ая**   (astron) daily parallel

паралле́ль·н·о   (*adv*) parallel (to), in parallel

~, **в·ключ·ённ·ый**   (elec) parallel connected, connected in parallel

паралле́ль·ност·ь, -и,   (*f*) parallelism, being parallel

паралле́ль·н·ый, -ая, -ое, (*a*) parallel, parallel-connected, parallel-operation

~ **со·един·е́ни·е**   (*n*) (elec) connection in parallel, parallel connection

пар·альдеги́д·, -а,   (*m*) (chem) paraldehyde

па́ра·магнети́зм·, -а,   (*m*) (phys) paramagnetism

па́ра·магне́тик·, -а,   (*m*) paramagnet, paramagnetic, paramagnetic material

па́ра·магни́т·н·ый, -ая, -ое,   (*a*) paramagnetic

пара́·метр·, -а, (*m*) parameter, constant, variable, characteristic, (*pl*) conditions

~, **волн·ов·о́й**   wave-length constant

~ **вре́мен·и** (*pl*) (phys) time transients

~ **групп·иров́а́ни·я** (phys) bunching parameter

~ **диффу́з·и·и** (phys) diffusion constant

~ **за·ви́с·им·ый от част·от·ы́** (elec) frequency response

~ **крив·изн·ы́** (nucl) buckling, size-shape factor

~ **на·стро́й·к·и** (instr) setting

~ **па́р·а** (*pl*) steam conditions

~ **раз·бро́с·а** (math) straggling parameter

~ **раз·лож·е́ни·я**   (math) expansion parameter

~, **рас·пре·дел·ённ·ые** (*pl*) distributed parameters

~, **регул·и́руем·ый** controlled variable

~ **решёт·к·и** (phys) lattice parameter/constant

~, **со·средо·то́ч·енн·ые**   (*pl*) (autom) lumped parameters, lumped capacities

~ **с·раба́т·ывани·я** (autom) set point

~ **уда́р·а** (nucl) collision parameter

~ **фо́рм·ы**   shape parameter

параметри́т·, -а,   (*m*) (med) parametritis

параметр·и́ческ·ий, -ая, -ое,   (*a*) parametric, parametral

параметрóн·, -а, (*m*) (autom) parametron, parametric exciter

пáр·ами, (*i pl*) of **пáр·а;** (*adv*) in pairs

пара·молóч·н·ый, -ая, -ое, (*a*) (chem) paralactic

пара·морф·и́зм·, -а, (*m*) (min) paramorphism

пара·морф·óз·а, -ы, (*f*) (min) paramorphism

парани́т·, -а, (*m*) paranite (elec insulating material)

пара·нитроанали́н·, -а, (*m*) (chem) para-nitroaniline, *p*-nitroaniline

~, крáсн·ый (chem) para-red

пара·нóй·я, -и, (*f*) (med) paranoia

парант·гéл·и·й, -я, (*m*) (astron) paranthelion

парантéз·, -а, (*m*) parenthesis

пар·анти·селéн·а, -ы, (*f*) (astron) paratiselene

парапéт·, -а, (*m*) parapet

пара·питéк·, -а, (*m*) (pal) Parapithecus

пара·плег·и́·я, -и, (*m*) (med) paraplegia

пара·пóд·и·й, -я, (*m*) (zool) parapodium

пара·по·лож·éни·е, -я, (*n*) (chem) para-position

пара·процéсс·, -а, (*m*) (phys) para-process, true magnetization

пара·селéн·а, -ы, (*f*) (astron) paraselene

пара·сóл·ь, -я, (*m*) (a/c) parasol

пара·со·сто·я́ни·е, -я, (*n*) (phys) para-state

пара·сти́х·а, -и, (*f*) (bot) parastichy

пара·ти́п·, -а, (*m*) (biol) paratype

пара·тиреоид·и́н·, -а, (*m*) (pharm) parathyroidin

пара·тирео́ид·н·ый, -ая, -ое, (*a*) parathyroid

~ гормóн· (*m*) (biochem) parathormone

пара·ти́ф·, -а, (*m*) (med) paratyphoid

паратóн·, -а, (*m*) (oil) Paratone (additive)

пара·тон·и́ческ·ий, -ая, -ое, (*a*) paratonic

~ движ·éни·я (*pl*) (bot) paratonic movements

пара·фенилендиами́н·, -а, (*m*) (phot, chem) paraphenylenediamine, para-diaminobenzene

пара·фи́з·, -а, (*m*) (bot) paraphysis

парафи́н·, -а, (*m*) paraffin wax (Brit), paraffin (U.S.); (chem) paraffin

~, дейтер·изóванн·ый (nucl) D-paraffin

{, жи́дк·ий (pharm) liquid paraffin, medicinal oil

парафин·ёр·, -а, (*m*) (food) cheese coating/dipping machine

парафин·и́ровани·е, -я, (*n*) waxing, paraffin bedding

парафин·и́рованн·ый, -ая, -ое, (*past part pass*) paraffinized, waxed

парафин·и́стост·ь, -и, (*f*) (chem) paraffin content, paraffinicity

парафи́н·ист·ый, -ая, -óе, (*a*) paraffinous

парафи́н·ов·ый, -ая, -ое, (*a*) wax, paraffin, paraffin wax; (chem) paraffinous, paraffin-base

~ гач· (*m*) (oil) slack wax

~ мáсл·о (*n*) (oil) press oil

~ нефт·ь (*f*) (oil) paraffinous crude, paraffin-base crude

~ о·тёк· (*m*) scale wax

парафи́р·овать, -уют, (*imp and perf*) initial (agreements etc.)

парафлóу (*indecl*) Paraflow (pour-point depressant)

пара·фóрм·, -а, (*m*) (chem) paraform, paraformaldehyde

пара·формальдеги́д·, -а, (*m*) para-formaldehyde

пара·хинóн·, -а, (*m*) paraquinone

пара·хóр·, -а, (*m*) (phys) parachor

пара·цент·éз·, -а, (*m*) (med) paracentesis

пара·циáн·, -а, (*m*) (chem) paracyanogen

парашю́т·, -а, (*m*) parachute; (min) safety catch, keps, bearing-up stops; (naut) cargo tray

~, вы·тяж·н·óй pilot parachute, pilot chute

~, по·сáд·очн·ый (a/c) landing parachute

парашю́т·ик·, -а, (*m*) (*dim*) of **парашю́т·**

парашют·и́ровани·е, -я, (*n*) (a/c) pancaking

парашют·и́рующ·ий, -ая, -ее, (*pres part act*)

~ по·сáд·к·а (*f*) pancake landing

~ с·пуск· (*m*) landing approach at minimum air speed

парашю́т·ист·, -а, (*m*) parachutist

парашю́т·н·о-десáнт·н·ый, (*a*) parachute, parachute-landing

пар·гéл·и·й, -я, (*m*) parhelion, mock sun (meteor)

парéз·, -а, (*m*) (med) paresis

парейа·зáвр·, -а, (*m*) (pal) Pareiasaurus

па́р·ени·е, -я, (*n*) (*v n*) of па́р·ить steaming; пар·ёни·е, -я, (*n*) (*v n*) of пар·и́ть soaring, hovering

пар·енхи́м·а, -ы, (*f*) (bot) parenchyma

па́р·ен·ый, -ая, -ое, (*a*) steamed

па́рен·ь, -рн·я, (*g pl*) -рн·е́й, (*m*) lad, fellow

пар·естези́·я, -и, (*f*) (med) paraesthesia

париет·а́льн·ый, -ая, -ое, (*a*) (biol) parietal

париет·и́н·, -а, (*m*) (chem) parietin

париет·и́т·, -а, (*m*) (med) parietitis

пари́ж·ск·ий, -ая, -ое, (*a*) Parisian, Paris

~ зе́лен·ь (*f*) (chem) Paris green

паризи́т·, -а, (*m*) (min) parisite

парикма́хер·, -а, (*m*) hairdresser, barber

пари́р·овать, -у́ют, (*imp and perf*) parry, ward off

пар·ите́т·, -а, (*m*) (fin) parity

па́р·ить, -ят, (*imp*) steam; пар·и́ть, -я́т, (*imp*) soar, hover

па́рк·, -а, (*m*) park; stock, fleet; depot; bed (of oysters)

~, автомоби́ль·н·ый (M/T) fleet of motor vehicles

~, ваго́н·н·ый (rail) rolling stock

~ культу́р·ы и о́т·дых·а recreation park, park of rest and culture (in U.S.S.R.)

~, мост·ов·о́й bridge/bridging train

~, при·ём·н·ый (rail) reception siding

~, сорт·иро́вочн·ый (rail) marshalling yard

~, ста́н·очн·ый stock of machine tools

~, трамва́й·н·ый tram depot

паркериз·а́ци·я, -и, (*f*) (met) parkerizing

парке́т·, -а, (*m*) (build) parquet

парке́т·н·ый, -ая, -ое, (*a*) of парке́т·

~ дощ·е́чк·а (*f*) parquet block

паркинсон·и́зм·, -а, (*m*) (med) parkinsonism

па́рк·ов·ый, -ая, -ое, (*a*) of па́рк·

парко·хозя́й·ственн·ый ден·ь (*m*) (M/T) servicing day

парла́мент·, -а, (*m*) parliament

парламент·ёр·, -а, (*m*) truce negotiator

парламент·ёрск·ий флаг· (*m*) flag of truce, white flag

парла́мент·ск·ий, -ая, -ое, (*a*) of парла́мент·

пармели·я, -и, (*f*) (bot) parmelia

пар·ни́к·, -а́, (*m*) (hortic) hotbed, forcing frame/bed

~, холо́д·н·ый (hortic) cold frame

парно·борозд·н·ый, -ая, -ое, (*a*) geminicolpate

пар·н·о́й, -а́я, -о́е, (*a*) fresh, new, green

~ молок·о́ (*n*) fresh milk

~ шку́р·а (*f*) green hide

парно·копы́т·н·ый, -ая, -ое, (*a*) (zool) artiodactyl, (*pl as noun*) artiodactyls, *Artiodactyla*

парно·лист·нико́в·ые, -ых, (*pl decl as adj*) (bot) bean-caper family, *Zygophyllaceae*

парно·па́р·н·ый, -ая, -ое, (*a*) (zool) artiodactyl

парно·перисто·сло́ж·н·ый, -ая, -ое, (*a*) (bot) paripinnate

парно·пёр·ист·ый, -ая, -ое, (*a*) (bot) paripinnate

парно·по́р·ов·ые, -ых, (*pl decl as adj*) (pal) diplopores, *Diploporita*

парно·у́с·ые, -ых, (*pl decl as adj*) (zool) *Mandibulata, Antennata*

па́р·н·ый, -ая, -ое, (*a*) *see also* пар·н·о́й; twin (one of pair); double (two together), paired; (bot) bijugate, conjugate; *see* пар·ов·о́й

~ весл·о́ (*n*) dinghy oar; scull

~ игр·а́ (*f*) doubles, two a side (sport)

~ ка́бел·ь (*m*) (elec) paired cable

~ конве́рс·и·я (*f*) (phys) pair conversion

~ эффе́кт· (*m*) (nucl) pair effect

пар·ова́ни·е, -я, (*n*) (agr) fallowing

пар·ови́к·, -а́, (*m*) boiler; steam engine

паро·вод·ян·о́й, -а́я, -о́е, (*a*) steam-and-water, steam-to-water, vapour-water

паро·во́з·, -а, (*m*) (*see also* локомоти́в·); (rail) steam locomotive/engine

паро·воз·ду́ш·н·ый, -ая, -ое, (*a*) vapour-air, steam-and-air

паро·во́з·ник·, -а, (*m*) (rail) engine driver, engineer (U.S.)

паро·во́з·н·ый, -ая, -ое, (*a*) of паро·во́з·

паро·возо·стро·е́ни·е, -я, (*n*) steam locomotive engineering

пар·ов·о́й, -а́я, -о́е, (*a*) steam; vapour, vapor; (agr) fallow; (hortic) forced; steam-heated/driven

~ бараба́н· (*m*) steam drum (of boiler)

~ маши́н·а (*f*) steam engine

~ на·со́с (*m*) steam-driven pump

~ порш·нев·а́я маши́н·а (*f*) steam reciprocating engine

пар·ов·о́й, -а́я, -о́е
~ **фа́з·а** (f) (chem) vapour/vapor phase
паро·в·пу́ск·, -а, (m) steam admission
паро·в·пуск·н·о́й, -а́я, -о́е, (a) inlet, steam-inlet, admission
паро·вы·пуск·н·о́й, -а́я, -о́е, (a) exhaust, steam-exhaust
паро·га́з·ов·ый, -ая, -ое, (a) vapour-gas; steam-gas
~ **турби́н·а** (f) (shipb) HTP turbine
паро·газо·генера́тор·, -а, (m) combustion chamber (in HTP system); (rocket) steam generator
паро·генера́тор·, -а, (m) steam generator
паро·генера́тор·н·ый, -ая, -ое, (a) of **паро·генера́тор·**
~ **цикл·** (m) steam power cycle
паро·гидравл·и́ческ·ий, -ая, -ое, (a) steam-hydraulic
паро·дина́мо, (n indecl) steam-driven dynamo
паро·за·по́р·н·ый, -ая, -ое, (a) steam-cutoff
паро·кисло·ро́д·, -а, (m) steam-oxygen; (shipb) HTP steam
паро·ко́н·н·ый, -ая, -ое, (a) two-horse, pair
~ **ход·** (m) two-horse cart
паро·ли́фт·, -а, (m) steam lift
паро́л·ь, -я, (m) password
паро́м·, -а, (m) ferry, ferry boat; **па́р·ом** (i) of **пар·**
паро·ме́р·, -а, (m) steam flow meter
паро·мото́р·, -а, (m) steam prime mover
паро·не·про·ниц·а́ем·ый, -ая, -ое, (a) steam-tight, vapour-tight
паро·об·рабо́т·к·а, -и, (f) (agr) fallow tillage
паро·обра́з·н·ый, -ая, -ое, (a) steam-like, vaporous
паро·образ·ова́ни·е,-я, (n) (phys) vaporization, evaporation; steam generation
тепл·от·а́ паро·образ·ова́ни·е (f) latent heat of evaporation
паро·образ·ова́тел·ь, -я, (m) steamer, steam generator
паро·о·суш·и́тел·ь, -я, (m) steam dryer/separator
паро·от·бо́й·ник·, -а, (m) steam separator
паро·от·во́д·, -а, (m) steam discharge pipe
паро·от·во́д·н·ый, -ая, -ое, (a) exhaust, steam-exhaust

паро·от·дел·и́тел·ь, -я, (m) steam separator
паро·о·хлад·и́тел·ь, -я, (m) de-superheater
паро·о·чист·и́тел·ь, -я, (m) (agr) field cultivator
паро·пере·грев·а́тел·ь, -я, (m) superheater
паро·пре·образ·ова́тел·ь, -я, (m) evaporative feed-water preheater
паро·про·во́д·, -а, (m) steam pipeline/line/main
паро·про·во́д·н·ый, -ая, -ое, (a) steam supply/feed
паро·про·из·вод·и́тельност·ь, -и, (f) steam capacity, rating (of boiler)
паро·про·ниц·а́емост·ь, -и, (f) permeability to steam
паро·рас·пре·дел·е́ни·е, -я, (n) steam distribution, valve motion/gear (of steam power plant)
паро·рас·пре·дел·и́тел·ь, -я, (m) steam header/distributor
паро·рас·пре·дел·и́тельн·ый, -ая, -ое, (a) steam-distributing, steam
~ **коро́б·к·а** (f) steam box
паро·с·бо́р·ник·, -а, (m) steam collector
паро·сепара́тор·, -а, (m) steam separator
паро·сил·ов·о́й, -а́я, -о́е, (a) steam power
~ **у·стано́в·к·а** (f) steam propulsion plant, steam power plant
па́рос·к·ий, -ая, -ое, (a) (ceram) Parian
паро·сто́й·к·ий, -ая, -ое, (a) steam-proof, vapourproof; steam-resistant, vapour-resistant
паро·стру́й·н·ый, -ая, -ое, (a) steam-jet
~ **эже́ктор·** (m) steam-jet ejector (jet pump)
паро·суш·е́ни·е, -я, (n) steam drying, steam separation
паро·тепло·хо́д·, -а, (m) (shipb) power-driven vessel
паро·те́рм·а, -ы, (f) (geol) steam jet
пар·оти́т·, -а, (m) (med) parotitis
паро·турби́н·а, -ы, (f) and see **турби́н·а** steam turbine
паро·фа́з·н·ый, -ая, -ое, (a) vapour-phase
паро·турбо·во́з·, -а, (m) steam-turbine locomotive
паро·хо́д·, -а, (m) steamship, steamer
~, **груз·ов·о́й** cargo steamer

паро·хо́д·
~, колёс·н·ый paddle steamer
~, ледо·ко́ль·н·ый ice-breaking steamship
паро·хо́д·н·ый, -ая, -ое, (*a*) of **паро·хо́д·**
паро·хо́д·ств·о, -а, (*n*) shipping; shipping line/agency; steam navigation
парсе́к·, -а, (*m*) (astron) parsec
парсони́т·, -а, (*m*) (min) parsonite
парт- (*component*) (*abbr*) of **парт·и́й·н·ый**
па́рт·а, -ы, (*f*) school desk
партено·гене́з·, -а, (*m*) (biol) parthenogenesis
партено·ка́рп·и·й, -я, (*m*) (bot) parthenocarp
партено·карп·и́·я, -и, (*f*) (bot) parthenocarpy
парте́р·, -а, (*m*) (theat) stalls, pit
парт·и́йн·ый, -ая, -ое, (*a*) party; Communist party
партикуля́р·н·ый, -ая, -ое, (*a*) particular
партикул·я́ци·я, -и, (*f*) (biol) particulation
партиту́р·а, -ы, (*f*) score (music)
па́рт·и·я, -и, (*f*) party; batch, lot; game (of chess etc.); part (of musical score)
партнёр·, -а, (*m*) partner
па́рус·, -а, (*nom pl*) **-а́,** (*g pl*) **-о́в,** (*m*) sail; (bot) vane
~, га́фель·н·ый gaff sail
~, гро́т·овн·ый mainsail
~ гуари́ gunter sail
~, кита́й·ск·ий slatted lugsail
~, лати́н·ск·ий lateen sail
~, лю́гер·н·ый lugsail
~, прям·о́й square sail
~, ре́й·ков·ый lugsail
парус·и́н·а, -ы, (*f*) canvas, sail cloth
парус·и́нк·а, -и, (*g pl*) **-нок·,** (*f*) folding/collapsible boat
парус·и́нн·ый, -ая, -ое, (*a*) of **парус·и́н·а**
парус·и́нов·ый, -ая, -ое, (*a*) of **парус·и́н·а**
~ брю́к·и (*pl*) duck trousers
~ строп· (*m*) canvas sling
па́рус·ник·, -а, (*m*) sailmaker; sailing vessel; sailor experienced under sail; yachtsman; (zool) sailfish, *Histiophorus*
па́рус·ност·ь, -и, (*f*) set, suit (of sails); sail area
  центр· па́рус·ност·и (*m*) centre/center of effort of sail area

па́рус·н·ый, -ая, -ое, (*a*) sail; sailing
~ во·оруж·е́ни·е (*n*) (naut) rig, sails and rigging
~ игл·а́ (*f*) sailmaker's needle
~ кора́бл·ь (*m*) full rigged ship
парфюме́р·и·я, -и, (*f*) perfumes and cosmetics; cosmetics manufacture
парце́лл·а, -ы, (*f*) (agr) smallholding
парцелл·и́р·овать, -уют, (*imp and perf*) (agr) divide up into small holdings
парциа́ль·н·ый, -ая, -ое, (*a*) partial
~ давл·е́ни·е (*n*) (phys) partial pressure
парч·а́, -и́, (*i*) **-о́й,** (*f*) (text) brocade
~, була́в·чат·ая uncut brocade
парш·а́, -и́, (*i*) **-о́й,** (*f*) (med) favus; (vet) mange, scab; (bot) scab
~, чёрн·ая (hortic) black scurf, *Rhizostonia*
парш·и́в·ый, -ая, -ое, (*a*) mangy, scabby
пар·я́щ·ий, -ая, -ее, (*pres part act*) of **пар·и́ть**
~ по·лёт· (*m*) soaring/hovering flight
пас·, -а, (*m*) or (*indecl*) pass (in games); (*past masc sing*) of **пас·ти́**
па́сек·а, -и, (*f*) apiary, beehive
пас·е́ни·е, -я, (*n*) (*v n*) of **пас·ти́**
пас·ённ·ый, -ая, -ое, (*past part pass*) of **пас·ти́**
пас·ёт, (*pres 3rd sing*) of **пас·ти́**
па́сеч·н·ый, -ая, -ое, (*a*) of **па́сек·а** beekeeper's, beekeeping
~ инвента́р·ь (*m*) beekeeping equipment
пасик·, -а, (*m*) (instr) small drive belt
пас·и́те, (*pl imper*) of **пас·ти́**
ПАСК (*abbr*) = **пара·амино·сали·цил·ов·ая кисл·от·а́** (pharm) PAS, para-aminosalicylic acid
Паска́л·я, за·ко́н· Pascal's law (of hydrostatics)
паскои́т·, -а, (*m*) (min) pascoite
пас·ла́ (*past f sing*) of **пас·ти́**
паслён·ов·ый, -ых, (*pl decl as adj*) (bot) nightshade family, *Solanaceae*
па́с·ли (*past pl*) of **пас·ти́**
па́см·о, -а, (*n*) (text) cut
па́смурн·ый, -ая, -ое, (*a*) overcast, cloudy, dull
па́сок·а, -и, (*f*) (bot) bleeding sap
пас·о́м·ый, -ая, -ое, (*pres part pass*) of **пас·ти́**
паспарту́ (*n indecl*) passe-partout

па́спорт·, -а, (*nom pl*) -а́, (*g pl*) -о́в, (*m*) passport; data sheet (for machines etc.); certificate

~ за·лёт·ы (air) mission data/brief

~ объекти́в·а (phot) lens formulary

~, свет·ов·о́й (cinema) light-change/printing card/band, control strip

~, суд·ов·о́й sea pass/brief/letter, passport

па́спорт·н·ый, -ая, -ое, (*a*) of па́спорт·; rated

~ да́·нн·ые (*pl*) nominal data, rating (of machines)

~ ско́р·ост·ь хо́д·а (*f*) rated/service/normal speed

па́сс·, -а, (*m*) pass; drive belt

пасса́ж·, -а, (*i*) -ем, (*m*) passage; (build) shopping arcade, arcade of shops/stalls

пассажи́р·, -а, (*m*) passenger

пассажи́р·ск·ий, -ая, -ое, (*a*) passenger

пасса́ж·н·ый, -ая, -ое, (*a*) of пасса́ж·

~ инструме́нт· (*m*) (astron) transit

пасса́т·, -а, (*m*) (meteor) trade wind

пассати́ж·и, -ей, (*pl*) nipping pliers (tool)

пасси́в·, -а, (*m*) (fin) liabilities, debit; (gram) passive voice

пассив·и́рован·и·е, -я, (*n*) (chem, met) passivation, passivating

пассивно·пла́в·ающ·ий, -ая, -ее, (*a*) free/passive-floating

пасси́в·ност·ь, -и, (*f*) passiveness, passivity; (met) passivity

пасси́в·н·ый, -ая, -ое, (*a*) passive; (fin) debit, adverse, liability

~ бала́нс· (*m*) (fin) debit/adverse/unfavourable balance

~ ма́сс·а (*f*) (naut) salvage, salvaged cargo

~ опера́ц·и·и (*pl*) (fin) liability operations

~ у·ча́ст·ок· траекто́р·и·и (*m*) (rocket) coasting trajectory

пасси́·метр·, -а, (*m*) passimeter, inside caliper

па́ст·а, -ы, (*f*) paste

~ ГОИ (mech eng) polishing/buffing paste

па́ст·бищ·е, -а, (*i*) -ем, (*n*) pasture

пасте́л·ь, -и, (*f*) crayon, pastel

пастерелл·ёз·, -а, (*m*) (vet) pasteurellosis

пастериз·а́тор·, -а, (*m*) pasteurizer

~, моло́ч·н·ый milk pasteurizer

пастериз·ацио́нн·ый, -ая, -ое, (*a*) pasteurizing

пастериз·а́ци·я, -и, (*f*) pasteurization

~, луч·ев·а́я radiation pasteurization

пастериз·ова́ть, -у́ют, (*imp and perf*) pasteurize

пастерна́к·, -а, (*m*) (bot) parsnip, *Pastinaca*

пасте́р·овск·ий, -ая, -ое, (*a*) (med) Pasteur

~ ста́нц·и·я (*f*) (med) anti-rabies unit

пас·ти́, -у́т, (*imp*) graze, pasture, feed, tend (grazing animals)

паст·и́рованн·ый, -ая, -ое, (*past part pass*) of паст·и́р·овать

паст·и́р·овать, -уют, (*imp*) paste

паст·у́х·, -а́, (*m*) (agr) herdsman, herd, cowherd, stockman

~, электр·и́ческ·ий electric fence

па́ст·ь, -и, (*f*) mouth, jaws (of animal); trap

пас·ть, (*fut 3rd sing, pl*) пад·ёт, пад·у́т, (*past masc sing*) па·л (*perf*) see па́д·а·ть

пасть·б·а́, -ы́, (*f*) pasturage

пас·у́щ·ий, -ая, -ее, (*pres part act*) of пас·ти́

па́с·ш·ий, -ая, -ее, (*past part act*) of пас·ти́

па·сын·к·ова́ть, -у́ют, (*imp and perf*) prune, disbud, remove/pinch out side shoots

па́·сын·ок·, -н·к·а, (*m*) stepson; (bot) side/lateral shoot; (build) scarfed foundation beam; (*pl*) tabernacle (of mast)

пасю́к·, -а́, (*m*) (zool) brown rat, *Rattus norvegicus*

пас·я́ (*pres gerund*) of пас·ти́

па́т·, -а, (*m*) stalemate (at chess)

пателл·эктом·и́·я, -и, (*f*) (med) patellectomy

пателл·я́рн·ый, -ая, -ое, (*a*) (anat) patellar, patellate

пате́нт·, -а, (*m*) patent; licence; (naut med) bill of health; anti-fouling paint

взя́ть пате́нт· take out a patent

патент·и́рован·и·е, -я, (*n*) (met) lead patenting

пате́нт·н·ый, -ая, -ое, (*a*) of пате́нт·

патент·ова́ни·е, -я, (*n*) (*v n*) of патент·ова́ть

патент·о́ванн·ый, -ая, -ое, (*past part pass*) of патент·ова́ть; patented; patent

~ голуб·о́й (*m*) (chem) patent blue

патент·ова́ть, -у́ют, (*imp and perf*) patent, take out a patent; grant/issue a patent

пате́нто·облад·а́тел·ь, -я, (*m*) patentee, patent holder

патефо́н·, -а, (*m*) gramophone, phonograph

пати́н·а, -ы, (*f*) patina (surface film)

пато·ген·е́з·, -а, (*m*) (biol) pathogenesis

пато·ге́н·ност·ь, -и, (*f*) (med) pathogenicity

пато·ге́н·н·ый, -ая, -ое, (*a*) (med) pathogenic

па́ток·а, -и, (*f*) syrup, treacle, molasses

~, корм·ов·а́я (agr) molasses

~, меж·криста́ль·н·ая high green syrup (sugarmaking)

~, све́тл·ая golden syrup

~,.чёрн·ая molasses

пато́·лог·, -а, pathologist

пато·ло́г·и·я, -и, (*f*) pathology

па́точ·н·ый, -ая, -ое, (*a*) of **па́ток·а**

патро́н·, -а, (*m*) chuck, holder (of machine tools); cartridge, case; (elec) lamp socket/holder; (text) pattern, design; (oil) drillpipe, threadcutter; chief, boss, master

~, бо·ев·о́й (expl) primer

~, дав·и́льн·ый form block (metal spinning)

~, дву·кулач·ко́в·ый box chuck

~, за·па́ль·н·ый (a/c) cartridge starter, torch igniter; combustion torch (diesels); detonator (of explosive devices)

~, кулач·ко́в·ый jaw chuck

~, мембра́н·н·ый diaphragm chuck

~, пла́в·к·ий (elec) cartridge fuse

~, про·ве́р·очн·ый dummy round (ammunition)

~, про·во́д·ков·ый centre chuck

~, пруж·и́нн·ый draw-in collet chuck

~, раке́т·н·ый signal cartridge

~, само·центр·и́рующ·ий concentric/self-centring chuck

~, унита́р·н·ый fixed ammunition

~, уч·е́бн·о-контро́ль·н·ый drill round (ammunition)

~ фи́льтр·а filter cartridge

~, холост·о́й blank cartridge

~, ца́нг·ов·ый collet

патро́н·и́рованн·ый, -ая, -ое, (*past part pass*) of **патро́н·и́р·овать;** (text) drafted; cartridge-packed

патро́н·и́р·овать, -уют, (*imp*) (text) draft a pattern; pack in cartridges/cases

патрони́т·, -а, (*m*) (min) patronite

патро́н·к·а, -и, (*g pl*) **-н·ок·,** (*f*) (text) pattern, design

патро́н·ник·, -а, (*m*) chamber (QF gun)

патро́н·н·ый, -ая, -ое, (*a*) of **патро́н·**

~ за·хва́т· (*m*) socket tool (of manipulator)

~ ору́д·и·е (*n*) QF gun

~ по́·греб· (*m*) magazine

~ пред·о·хран·и́тел·ь (*m*) (elec) cartridge fuse

патро́н·чик·, -а, (*m*) (*dim*) of **патро́н·** (q. v.)

па́·труб·ок·, -б·к·а, (*m*) short pipe, pipe, nozzle, stub; branch, branch pipe; (shipb) sea chest

~, в·са́с·ывающ·ий suction nozzle, suction/intake pipe

~, вы·хлоп·н·о́й exhaust pipe

~, гряз·ев·о́й mud nozzle

~, на·гнет·а́тельн·ый discharge nozzle

~, на·по́р·н·ый discharge nozzle

~, от·вод·я́щ·ий outlet branch (pipe)

~, пере·лив·н·о́й branch nozzle

~, пере·хо́д·н·ый taper/tapered pipe

~ с фла́н·ц·ем flanged nozzle/branch

патру́л·ь, -я, (*m*) (mil) patrol

~, метео́р·н·ый (phot, astr) meteor camera

патули́н·, -а, (*m*) (pharm) patulin

па́уз·а, -ы, (*f*) pause, interval, break, rest; spacing (between pulses); resting time

част·от·а́ па́уз·ы (*f*) (telecom) spacing frequency

пау́к·, -а́, (*m*) (ent) spider; compressed air distributor

пауко·обра́з·н·ый, -ая, -ое, (*a*) spidery, spider-like; (*pl decl as adj*) (ent) arachnids, *Arachnida*

Па́ули, при́нцип· (*m*) (phys) Pauli exclusion principle

пауро·по́д·ы, -ов, (*m pl*) (zool) pauropods, *Pauropoda*

паути́н·а, -ы, (*f*) cobweb, spider's web; spider's silk, web fluid

паути́н·н·ый, -ая, -ое, (*a*) of **паути́·н·а**

~ же́лез·ы (*pl*) silk-producing glands (of spiders)

~ оболо́ч·к·а (*f*) arachnoid (of brain)

пауци·литиони́т·, -а, (*m*) (min) paucilithionite

пауч·о́к·, -ч·к·а́, (*m*) (*dim*) cf **пау́к**

па́х·, -а, (*m*) (anat) groin, inguen

па́х·ар·ь, -я, (*m*) ploughman

пах·а́ть, (*pres 3rd pl*) па́ш·ут; (*imp*) plough, plow, till, cultivate

пахи·дисци́д·а, -ы, (*f*) (pal) pachydiscid

пахи·карпи́н·, -а, (*m*) (biochem) pachycarpine

пахи·не́м·а, -ы, (*f*) (cyt) pachynema

пахи·те́н·а, -ы, (*f*) (cyt) pachytene

па́х·нуть, -нут, (*imp*) smell/reek of; пах·ну́ть, -ну́т, (*perf*) puff, gust, blow back (of fires)

пах·ов·о́й, -а́я, -о́е, (*a*) (anat) groin, inguinal

па́х·от·а, -ы, (*f*) ploughing, tillage; ploughed land/field

па́х·отн·ый, -ая, -ое, (*a*) arable, cultivable

па́хт·а, -ы, (*f*) buttermilk

па́хт·ань·е, -я, (*n*) churning; buttermilk

па́хт·а·ть, -ют, (*imp*) churn

пах·у́ч·ий, -ая, -ее, (*a*) odorous, strong smelling

пах·у́чк·а, -и, (*g pl*) -чек·, (*f*) savory (spice)

пацие́нт·, -а, (*m*) (med) patient

пацю́к·, -а, (*m*) *see* пасю́к·

па́ч·ечн·ый, -ая, -ое, (*a*) of па́ч·к·а

~ маши́н·а (*f*) cigarette packaging machine

~ обо́йм·а (*f*) clip (of ammunition)

па́ч·к·а, -и, (*g pl*) -ч·ек·, (*f*) bundle, sheaf, batch, pile; pack, packet, package; (min) bench (of seam)

~ фольг· foil stack

па́чк·а·ть, -ют, (*imp*) soil, dirty, stain, smear; daub

пачко·вяз·а́льн·ая маши́н·а (*f*) letter-bundling machine

Па́шен·а, за·ко́н· (*m*) (elec) Paschen law

Па́шен·а-Ба́к·а, явл·е́ни·е (*n*) (phys) Paschen–Back effect

Па́шен·а-Ри́тц·а, се́р·и·я (*f*) (spectr) Paschen series

па́ш·енн·ый, -ая, -ое, (*past part pass*) of пах·а́ть

па́ш·ет (*pres 3rd sing*) of пах·а́ть

па́ш·н·я, -и, (*nom pl*) па́ш·ен·, (*f*) ploughed land/field

паште́т·, -а, (*m*) (food) pâté, meat/fish loaf

~, печён·очн·ый liver pâté

па́ш·ущ·ий, -ая, -ее, (*pres part act*) of пах·а́ть

па́юс·н·ая икр·а́ (*f*) (food) pressed caviar

па·я́льник·, -а, (*m*) soldering iron, solderer, copper bit

па·я́льн·ый, -ая, -ое, (*a*) of па́й·к·а

~ гор·е́лк·а (*f*) laboratory-glass blowpipe

~ ла́мп·а (*f*) blowlamp, blowtorch, soldering torch

~ флюс· (*m*) brazing/soldering flux

па·я́льщик·, -а, (*m*) solderer

па·я́ни·е, -я, (*m*) *see* па́й·к·а

па·я́·ть, -ют, (*imp*) solder (below 427°C); braze (above 427°C); seal (glass)

ПВД (*abbr*) = прямо·то́ч·н·ый воз·душно·реакти́в·н·ый дви́г·ател·ь ramjet engine

ПВО (*abbr*) = проти́во·воз·ду́ш·н·ая оборо́н·а air defence

пев·е́ц·, -в·ц·а́, (*i*) -в·ц·о́м, (*m*) singer

пев·у́ч·ий, -ая, -ее, (*a*) melodious

пе́в·ч·ий, -ая, -ее, (*a*) singing

~ пти́ц·а (*f*) song-bird

пе́·вш·ий, -ая, -ее, (*past part act*) of пе·ть; which was sung, sung

Пега́с·, -а, (*m*) (astron) Pegasus, the Square

Пега́с·а, квадра́т· (*m*) = пега́с·

пе́г·ий, -ая, -ое, (*a*) spotted, mottled; piebald (of horses), brindle (of cattle)

пегмат·и́т·, -а, (*m*) (min) pegmatite

пегмат·о́ид·, -а, (*m*) (geol) pegmatoid

пегмато·фи́р·, -а, (*m*) (geol) pegmatophyre

пед·аго́гик·а, -и, (*f*) education, educational science, pedagogy

  на·у́ч·н·о-ис·сле́д·овательск·ий институ́т· пед·аго́гик·и education research institute

пед·агоги́ческ·ий, -ая, -ое, (*a*) teacher's; educational

~ институ́т· (*m*) secondary school teachers' training college, College of Education

~ уч·и́лищ·е (*n*) primary school teachers' training college

педа́л·ь, -и, (*f*) pedal, treadle

~ нож·н·о́го у·правл·е́ни·я (a/c) rudder pedal

педа́ль·н·ый, -ая, -ое, (*a*) pedal, treadle

~ тка́ц·к·ий стан·о́к· (*m*) (text) treadle/tread loom

педи·а́льн·ый, -ая, -ое, (*a*) (cryst) pedial

пед·иа́тр·, -а, (*m*) (med) pediatrician, paediatrician

пед·иатр·и́·я, -и, (*f*) (med) pediatrics, paediatrics

педио́н·, -а, (*m*) (cryst) pedion

педи·па́льп·ы (*pl*) (ent) whip scorpions, *Pedipalpi*

педи·плен·, -а, (*m*) (geol) pediplain

педи·плен·иза́ци·я, -и, (*f*) (geol) pediplanation

педицелл·я́ри·я, -и, (*f*) (zool) pedicellaria, (*pl*) *Pedicellariae*

пе́дл·ь, -я, (*m*) paddle (canoe)

педо·гене́з·, -а, (*m*) (ent) paedogenesis, pedogenesis; (geol) soil formation

педо·ло́г·и·я, -и, (*f*) soil science, pedology

педо́·метр·, -а, (*m*) (surv) pedometer

педо·морф·о́з·, -а, (*m*) (zool) paedomorphosis

пейза́ж·, -а, (*m*) landscape

пей·те, (*pl imper*) of **пи·ть**

пек·, -а, (*m*) pitch (substance)

~, нефт·ян·о́й petroleum pitch

пёк· (*past masc sing*) of **печ·ь** (*imp*)

пека́н·, -а, (*m*) (bot) pecan, *Carya pecan/olivaeformis*; pecan nut

пе́кар·и (*indecl*) (zool) peccary, wild pig

пек·а́рн·я, -и, (*g pl*) **-рен·,** (*f*) bakery

пе́к·ар·ь, -я, (*nom pl*) **-я́,** or **и,** (*g pl*) **-éй** or **-ей,** (*m*) baker

Пе́кле, числ·о́ (*n*) (dynam) Péclet number

пеклева́н·к·а, -и, (*g pl*) **-н·ок·,** (*f*) fine quality rye flour

пек·ли (*past pl*) of **печ·ь** (*imp*)

пе́к·ов·ый, -ая, -ое, (*a*) pitch (substance)

~ я́м·а (*f*) pitch bay

пеко·туш·и́тел·ь, -я, (*m*) pitch cooler (coal tar distillation)

пект·иза́ци·я, -и, (*f*) (chem) pectisation

пект·и́н·, -а, (*m*) (biochem) pectin

пект·ина́з·а, -ы, (*f*) pectinase

пект·и́нов·ый, -ая, -ое, (*a*) pectic

~ вещ·еств·а́ (*pl*) (biochem) pectic substances, pectins

~ кисл·от·а́ (*f*) pectic acid

пект·о́з·а, -ы, (*f*) (chem) pectose

пекто·ли́т·, -а, (*m*) (min) pectolite

пекто·целлюло́з·а, -ы, (*f*) (biochem) pectocellulose

пекуля́р·н·ый, -ая, -ое, (*a*) peculiar

~ движ·е́ни·е (*n*) peculiar motion (of stars)

пек·у́щ·ий, -ая, -ее, (*pres part act*) of **печ·ь;** baking

пёк·ш·ий, -ая, -ее, (*past part act*) of **печ·ь**

пе·л (*past masc sing*) of **пе·ть**

пелаг·иа́л·ь, -я, (*m*) (ocean) pelagic zone

пелаг·и́ческ·ий, -ая, -ое, (*a*) (ocean) pelagic

~ от·лож·е́ни·я (*pl*) (ocean) pelagic sediments

пелами́д·а, -ы, (*f*) (fish) bonito, *Sarda sarda*

пеларго́н·, -а, (*m*) (biochem) pelargone

пеларгон·иди́н·, -а, (*m*) pelargonidin

пеларго́н·и·я, -и, (*f*) (bot) pelargonium

пеларго́н·ов·ая кисл·от·а́ (*f*) (biochem) pelargonic acid

пеле́й·ск·ий, -ая, -ое, (*a*) (geol) Pelian

пелен·а́, -ы́, (*f*) shroud, pall; (bot) hymenium

~, вихр·ев·а́я (dynam) vortex sheet

пе́ленг·, -а, (*m*) bearing, direction; echelon (a/c formation)

~, взаи́м·н·ый reciprocal bearing

~, и́ст·инн·ый true bearing

~, обра́т·н·ый reciprocal bearing

~, от·нос·и́тельн·ый relative bearing

~, пут·ев·о́й (nav) bearing relative to course, relative bearing

пеленг·а́тор·, -а, (*m*) direction finder (rad); azimuth circle (compass)

~, девиацио́н·н·ый (a/c) landing compass

~, одно·зна́ч·н·ый sense direction finder

~, ра́м·оч·н·ый loop direction finder

пеленг·а́ци·я, -и, (*f*) direction finding

пеленг·ова́ни·е, -я, (*n*) direction finding

~, одно·зна́ч·н·ое sense finding

пеленг·ова́ть, -у́ют, (*imp*) take a bearing, take bearings

пелери́н·а, -ы, (*f*) cape, cloak (garment)

пелеци·по́д·ы, -ов, (*m pl*) (pal, geol) clams, pelecypods, *Pelecypoda*

пе́·ли (*past pl*) of **пе·ть**

пелико·за́вр·ы, -ов, (*m pl*) (pal) pelycosaurs, *Pelycosauria*

пелла́гр·а, -ы, (*f*) (med) pellagra

пеллета́йзер·, -а, (*m*) pelletizer

пелли́кул·а, -ы, (*f*) (zool) pellicle, pellicule

пело́р·и·й, -я, (*m*) (bot) peloria

пелор·и́ческ·ий, -ая, -ое, (*a*) (bot) peloric

~ цвет·о́к· (*m*) (bot) pelory

пело́рус·, -а, (*m*) (naut) pelorous

пелоте́з·а, -ы, (*f*) plodder (soapmaking)

пель·ко́мпас·, -а, (*m*) (surv, nav) bearing compass, azimuth finder

пельмато·зо́·й, -ев, (*m pl*) (pal) pelmatozoans, *Pelmatozoa*

пельме́н·и, -ей, (*m pl*) meat dumplings

Пельтье́, явл·е́ни·е (*n*) (elec) Peltier effect

пе́мз·а, -ы, (*f*) pumice, pumice-stone

пемзо·бето́н·, -а, (*m*) pumecrete, pumice concrete

пемз·ова́ть, -у́ют, (*imp*) polish/clean/rub with pumice

пемзо·ви́д·н·ый, -ая, -ое, (*a*) (geol) pumiceous

пемз·о́вк·а, -и, (*f*) (*v n*) of **пемз·ова́ть**; pumicing

пе́мз·ов·ый, -ая, -ое, (*a*) of **пе́мз·а**

пе́н·а, -ы, (*f*) foam, froth, lather

~, во́д·н·ая water-base foam

~, твёрд·ая foam-type insulating material

пена́л·, -а, (*m*) cylindrical case; pencil case

пендул·я́рн·ый, -ая, -ое, (*a*) pendular

пендельфе́дер·, -а, (*m*) (horol) pendulum spring

пен·ёк·, -нь·к·а́, (*m*) (*dim*) of **пен·ь**

пене·пле́н·, -а, (*m*) (geog) peneplain, peneplane

пенетр·а́ци·я, -и, (*f*) penetration

пенетро́·метр, -а, (*m*) penetrometer, penetration tester

пе́·ни·е, -я, (*n*) singing, song

пене·сейсм·и́ческ·ий, -ая, -ое, (*a*) (geol) peneseismic

пе́нис·, -а, (*m*) (anat) penis

пе́н·ист·ый, -ая, -ое, (*a*) foam, foamy, foam-covered, foaming, frothing; lathering

пе́н·ить, -ят, (*imp*) foam, froth, lather

пеницилли́н·, -а, (*m*) (pharm) penicillin

пе́н·к·а, -и, (*g pl*) **-н·ок·,** (*f*) skim, scum, skin (on liquid which has boiled and frothed)

пенни́н·, -а, (*m*) (min) pennine, penninite

пенни·се́тум·, -а, (*m*) (bot) pearl millet, *Pennisetum*

пе́н·н·ый, -ая, -ое, (*a*) foam, froth

~ огне·туш·и́тел·ь (*m*) foam/fire extinguisher

пе́н·о- (*component*) foam, foaming, frothing; (plast) foam, cellular, expanded

пено·бето́н·, -а, (*m*) aerated/cellular/foam concrete

пено·бетон·ёрк·а, -и, (*g pl*) **-рок·,** (*f*) foam-concrete mixer

пено·гас·и́тел·ь, -я, (*m*) antifoaming/antifrothing agent

пено·генера́тор·, -а, (*m*) foam generator

пено·ги́пс·, -а, (*m*) foam plaster (heat insulator)

пено·образ·ова́ни·е, -я, (*n*) foaming

пено·образ·ова́тел·ь, -я, (*m*) frothing agent; foam compound (for fire-fighting equipment)

пено·от·дел·и́тел·ь, -я, (*m*) skimmer

пено·пла́ст·, -а, (*m*) cellular/foam plastic, plastic foam

~, целлюло́з·н·ый cellulose plastic

пено·пласт·ма́сс·а, -ы, (*f*) expanded plastic

пено·поли·стиро́л·, -а, (*m*) (chem) expanded polystyrene

пено·поли·урета́н·, -а, (*m*) (chem) polyurethane foam

пено·с·бо́р·ник·, -а, (*m*) froth trap

пено·стекл·о́, -а́, (*n*) foamglass (heat insulation)

пено·структу́р·а, -ы, (*f*) foam structure

пе́нс·, -а, (*m*) (fin) penny

пенса́к·, -а, (*m*) hulled barley

пе́нси·я, -и, (*f*) pension

пента- (*int component*) penta-, five-

пента·бора́т·, -а, (*m*) (chem) pentaborate

пента·го́н·, -а, (*m*) pentagon

пентагона́ль·н·ый, -ая, -ое, (*a*) pentagonal

пента·го́н·-додека́·эдр·, -а, (*m*) (cryst) pentagonal dodecahedron, pyritohedron

пента·го́н·-три·окта́·эдр·, -а, (*m*) (cryst) pentagonal icositetrahedron, plagiohedron

пента·го́н·-три·тетра́·эдр·, -а, (*m*) (cryst) tetrahedral pentagon, tetrahedron

пента·гра́мм·а, -ы, (*f*) pentagram

пента·гри́д·, -а, (*m*) (rad) heptode, pentagrid tube

пента·дека́н·ов·ый, -ая, -ое, (*a*) (chem) pentadecanoic

пента·деце́н·ов·ый, -ая, -ое, (*a*) pentadecenoic

пента·дие́н·ов·ый, -ая, -ое, (*a*) pentadienoic

пента·деци́л·, -а, (*m*) pentadecyl

пента·коза́н·, -а, (*m*) pentacosane

пента·ли́н·, -а, (*m*) (chem) pentaline, pentachlorethane

пента·метиле́н·, -а, (*m*) (chem) pentamethylene

пент·а́н·, -а, (*m*) (chem) pentane

пента́ндио́л·, -а, *(m)* pentanediol

пента·пло́ид·, -а, *(m)* (gen) pentaploid

пента·при́зм·а, -ы, *(f)* (opt) penta prism

пента·хлор·эти́л·ов·ый, эфи́р· *(m)* (chem) perchlor-ether

пента́·эдр·, -а, *(m)* (math) pentahedron

пента·эдр·и́ческ·ий, -ая, -ое, *(a)* pentahedral

пента·эритри́т·, -а, *(m)* (chem) pentaerythritol

пент·е́н·, -а, *(m)* pentene

пенте́н·ов·ая кисл·от·а́ *(f)* pentenoic acid

пе́нтер-га́к·, -а, *(m)* (naut) trip hook

пент·и́л·, -а, *(m)* pentyl

пенти́н·ов·ая кисл·от·а́ *(f)* pentinoic acid

пентланди́т·, -а, *(m)* (min) pentlandite

пент·о́д·, -а, *(m)* (elec) pentode

~, вы·ход·н·о́й output pentode

~, луч·ев·о́й beam pentode

пенто́д·н·ый, -ая, -ое, *(a)* of **пент·о́д·**

~ сет·к·а *(f)* (rad) suppressor grid

пент·о́з·а, -ы, *(f)* (biochem) pentose

пент·оза́н·, -а, *(m)* (biochem) pentosan

пент·окси́л·, -а, *(m)* (chem) pentoxyl

пент·о́н·, -а, *(m)* pentone

пенто́н·ов·ая кисл·от·а́ *(f)* pentonic acid

пентри́т·, -а, *(m)* (chem) pentaerythritol tetranitrate

пен·ь, *(g sing)* **пн·я,** *(nom pl)* **пн·и,** *(g pl)* **пн·ей,** *(m)* stump, stub

пеньк·а́, -и́, *(f)* hemp fibre, hemp

~, мани́ль·ск·ая manilla hemp

~, мя́т·ая broken/beetled hemp

~, трёп·ан·ая scutched hemp

~, чёс·ан·ая hackled hemp

пеньк·о́в·ый, -ая, -ое, *(a)* hemp, hempen

пеньк·о·пряд·и́льн·ый, -ая, -ое, *(a)* hemp-spinning

пеньк·о·треп·а́льн·ый, -ая, -ое, *(a)* hemp-hackling

пеньк·о·чес·а́льн·ый, -ая, -ое, *(a)* hemp-scutching

пе́н·я, -и, *(f)* fine, penalty

пео́н·, -а, *(m)* (bot) peony, *Paeonia*

пе́пел·, -пл·а, *(m)* ash, ashes

пепел·и́ть, -я́т, *(imp)* reduce to ashes

пе́пель·ниц·а, -ы, *(i)* **-ей,** *(f)* ash-tray

пе́пель·н·ый, -ая, -ое, *(a)* ash; ash-grey

пеперти́п·и·я, -и, *(f)* (phot) pepper-type

пе́пл·ов·ый, -ая, -ое, *(a)* ash, ashy, cinereous

пепло·па́д·, -а, *(m)* ashfall (from volcano)

пепс·и́н·, -а, *(m)* (biochem) pepsin

пепс·и́нов·ый, -ая, -ое, *(a)* peptic, pepsin

пепт·и́д·, -а, *(m)* (biochem) peptide

пепт·ида́з·а, -ы, *(f)* (biochem) peptidase

пептиз·а́ци·я, -и, *(f)* (chem) peptization

пептиз·и́р·овать, -уют, *(imp)* (chem) peptize

пепт·о́н·, -а, *(m)* (biochem) peptone

пептониз·а́ци·я, -и, *(f)* (biochem) peptonization

пепто́н·н·ый, -ая, -ое, *(a)* peptone

пёр· *(past masc sing)* of **пер·е́ть**

пе́рв·енец·, -нц·а, *(i)* **-нц·ем,** *(m)* first-born, firstling

пе́рв·енств·о, -а, *(n)* superiority; championship (sport)

перв·енств·ова́ть, -у́ют, *(imp)* excel, be first

перв·енству́ющ·ий, -ая, -ее, *(pres part act)* of **перв·енств·ова́ть;** pre-eminent

первити́н·, -а, *(m)* (pharm) pervitin, methyl benzedrine

перв·и́чн·о *(adv)* primarily, originally, initially

первично·бес·кры́л·ые, -ых, *(pl decl as adj)* (ent) *Apterygota*

первично·мона́д·н·ые, -ых, *(pl decl as adj)* (zool) *Protomonadidae*

первично·по·кро́в·н·ые, -ых, *(pl decl as adj)* (bot) *Archichlamydeae*

первично·ро́т·ые, -ых, *(pl decl as adj)* (zool) protostomes, *Protostomia*

перв·и́чност·ь, -и, *(f)* primariness

первично·трахе́й·н·ые, -ых, *(pl decl as adj)* (zool) *Prototracheata*

перв·и́чн·ый, -ая, -ое, *(a)* primary, prime, primordial, original, virgin

~ дви́г·ател·ь, *(m)* prime mover

~ ки́ш·к·а *(f)* (zool) gastrocoel, archenteron

~ мета́лл· *(m)* virgin metal

~ нефт·ь *(f)* protopetroleum

~ поло́с·к·а *(f)* (zool) primitive streak

~ по́л·ост·ь тел·а *(f)* (zool) primary body cavity

пéрв·о- (*root*) first, primary; prime, top; newly, just; (biol) archi-, archaeo-, proto-

перво·быт·н·ый, -ая, -ое, (*a*) primitive, primeval

перво·звéр·и, -éй, (*m pl*) (zool) protheres, *Prototheria*

перво·здá·нн·ый, -ая, -ое, (*a*) protogenic

перво·ис·тóч·ник·, -а, (*m*) primary/original source, origin

перво·клáсс·н·ый, -ая, -ое, (*a*) first-class

перво·нач·áльн·ый, -ая, -ое, (*a*) original, initial; primary, prime; elementary

∼ **числ·ó** (*n*) (math) prime number

перво·óбраз·, -а, (*m*) prototype

перво·обрáз·н·ый, -ая, -ое, (*a*) prototype, original; (math) primitive

∼ **кóрен·ь** (*m*) (math) primitive root

∼ **фýнкци·я** (*f*) (math) integrand

перво·о·черёд·н·ый, -ая, -ое, (*a*) first, immediate, most important

перво·печáт·н·ый, -ая, -ое, (*a*) first-edition; incunabular

перво·при·чи́н·а, -ы, (*f*) origin

перво·пти́ц·а, -ы, (*f*) (pal) archaeopteryx, archeopteryx

перво·пуск·áем·ый, -ая, -ое, (*a*) just started up, newly started, virgin (of nuclear reactors)

перво·рóд·н·ый, -ая, -ое, (*a*) first-born

перво·рóд·ств·о, -а, (*n*) primogeniture

перво·род·я́щ·ая, -ей, (*f decl as adj*) (med) primapara, primipara

перво·рóт, -а, (*m*) (zool) protostome

перво·сóрт·н·ый, -ая, -ое, (*a*) first/top grade/quality

перво·степéн·н·ый, -ая, -ое, (*a*) first, prime

перво·цвéт·н·ые, -ых, (*pl decl as adj*) (bot) *Primulales*

пéрв·ый, -ая, -ое, (*a*) first; the former/first-mentioned

в пéрв·ых first of all, in the first place, to begin with

∼ **вертикáл·** (*m*) (nav) prime vertical

∼ **зóн·а молч·áни·я** (rad) primary skip zone

∼ **пó·мощ·ь** (*f*) first aid

пергáмен·, -а, (*m*) *see* **пергáмент·**

пергáмент·, -а, (*m*) parchment

пергами́н·, -а, (*m*) pergamyn, greaseproof paper

пер·гидрáт, -а, (*m*) (chem) perhydrate

пер·гидридáз·а, -ы, (*f*) (chem) perhydridase

пер·гидрóл·ь, -я, (*m*) (chem) perhydrol

пере- (*verbal prefix*) *signifies*: (1) direction/movement across/through/over, trans-; (2) repetition (with or without change), re-; (3) to excess, over-; (4) advantage in action, out-; (5) extension of action to a number, action embracing a number; (6) division into two or into parts; (7) spend time; (8) action to limited degree or for short time; (9) **-ся** mutual action

пере·адрес·áци·я, -и, (*f*) address substitution/modification (computers)

пере·адрес·овáть, -ýют, (*perf*) forward (mail), re-address; substitute an address (computers)

пере·алк·и́ровани·е, -я, (*n*) (chem) transalkylation

пере·амид·и́ровани·е, -я, (*n*) (biochem) transamidation

пере·амин·и́ровани·е, -я, (*n*) (biochem) transamination

пере·баз·и́р·овать, -уют, (*perf*) move (factory, organization etc.), shift base, relocate

пере·баланс·и́р·овать, -уют, (*imp and perf*) rebalance; (autom) retrim

пере·бéг·, -а, (*m*) (*v n*) of **пере·бе·г·á·ть**; (mech eng) overtravel; desertion, deflection

пере·бег·á·ть, -ют, (*imp*) run across, cross; desert, defect

пере·беж·á·ть, (*fut 3rd sing, pl*) **пере·беж·и́т, пере·бег·ýт,** (*perf*); *see* **пере·бег·á·ть**

пере·бéл·ива·ть, -ют, (*imp*) whiten/whitewash again/afresh; overbleach; make a fair copy, copy out

пере·бел·и́ть, -я́т, (*perf*) *see* **пере·бéл·ива·ть**

пере·бер·ёт, (*fut 3rd sing*) of **пере·бр·áть**

пере·бив·á·ть, -ют, (*imp*) interrupt, break in, break; beat/knock in again, shift, move (nails etc.); (mil) knock out, kill (all, many); outbid; reupholster

пере·би́в·к·а, -и, (*f*) (*v n*), *see* **пере·бив·á·ть**; interruption; upholstery; (cinema) cut

пере·бинт·овáть, -ýют, (*perf*) dress (wounds), bandage, bandage again

пере·бир·а́·ть, -ют, (*imp*) sort, sort out, look/go through/over; finger, touch upon, run over (with the hands); overhaul (a machine); recall, call to mind; (print) reset; **-ся** (*pass*); get/cross over; move to

пере·би́·ть, (*fut 3rd sing, pl*) **пере·бь·ёт, пере·бь·ю́т,** (*past masc sing*) **пере·би́·л** (*perf*); *see* **пере·бив·а́·ть**

пере·блок·и́р·овать, -ую́т, (*imp and perf*) (elec) interlock

пере·бо́·й, -я, (*m*) interruption, stoppage (in work); (ICE) miss, misfire; intermission (in pulse); (telecom) erasure

  ~, да·ва́ть (ICE) miss

  ~ пит·а́ни·я (elec) power failure/breakdown

  сигна́л· пере·бо́·я (*m*) (telecom) erasure signal

пере·бо́й·н·ый, -ая, -ое, (*a*) of **пере·бо́·й**

  ~ сигна́л· (*m*) (telecom) erasure sign/signal

пере·бол·е́ть, -ят, (*perf*) have had (an illness); suffer, endure

пере·бо́р·, -а, (*m*) running over, fingering; gearing (of machine tools); excess, surplus

пере·бо́р·к·а, -и, (*g pl*) **-р·ок·,** (*f*) (*v n*), *see* **пере·бир·а́·ть**; partition; (shipb) bulkhead; (text) reblading

  ~, водо·про·ниц·а́ем·ая partition/screen bulk-head

  ~, диаметра́ль·н·ая centreline bulkhead

  ~, корм·ов·а́я afterpeak bulkhead

  ~, тара́н·н·ая collision bulkhead

пере·бо́р·очн·ый, -ая, -ое, (*a*) of **пере·бо́р·к·а**

  ~ стака́н· (*m*) (shipb) bulkhead piece

пере·бо́р·щик·, -а, (*m*) sorter

пере·бра́с·ыва·ть, -ют, (*imp*) throw over/across, throw; transfer; throw too far, overthrow; **-ся** (*pass*); rush across; throw oneself across; throw to and fro; exchange (words etc.)

пере·бр·а́ть, (*fut 3rd sing, pl*) **пере·бер·ёт, пере·бер·у́т,** (*perf*); *see* **пере·бир·а́·ть**

пере·брод·и́ть, -я́т, (*perf*) overferment

пере·бро́с·, -а, (*m*) throwing over; transfer; (elec) overthrow, throwover, flipping; (geol) overthrust fault

пере·брос·а́·ть, -ют, (*perf*) throw (move a pile by throwing)

пере·бро́с·ить, -я́т, (*perf*) *see* **пере·бра́с·ыва·ть**

пере·бро́с·к·а, -и, (*g pl*) **-с·ок·,** (*f*) transfer

пере·бро́ш·енн·ый, -ая, -ое, (*past part pass*) of **пере·бро́с·ить**

пере·бу́р·, -а, (*m*) (min) overdrill, overdrilling

пере·бу́ч·енн·ый, -ая, -ое, (*a*) over-plumped (leather)

пере·бь·ёт (*fut 3rd sing*) of **пере·би́·ть**

пере·ва́л·, -а, (*m*) crossing; (geog) pass; (min) caving

  ме́тод· пере·ва́л·а (*m*) (math) steepest descent method

пере·ва́л·ива·ть, -ют, (*imp*) roll/heave over/across, tranship, transfer; cross (obstacles); be beyond/over/past; **-ся** tumble/fall/roll over

пере·вал·и́ть, -я́т, (*perf*) *see* **пере·ва́л·ива·ть**

пере·ва́л·к·а, -и, (*g pl*) **-л·ок·,** (*f*) transhipment, transfer (of freight); (met) roll change

пере·ва́л·очн·ый, -ая, -ое, (*a*) of **пере·ва́л·к·а**

пере·ва́ль·н·ый, -ая, -ое, (*a*) transhipment; pass, mountain-pass

  ~ знак· (*m*) cross-over sign (sign indicating that shipping must cross to the other side of a river)

  ~ стен·к·а (*f*) baffle plate (in furnaces etc.)

  ~ то́ч·к·а (*f*) hurdle (graphs)

пере·вальц·о́ванн·ый, -ая, -ое, (*a*) dead-milled (of rubber)

пере·ва́р·, -а, (*m*) broth (vaccine culture)

пере·ва́р·енн·ый, -ая, -ое, (*past part pass*) of **пере·вар·и́ть**

пере·вар·и́м·ый, -ая, -ое, (*a*) digestible

пере·ва́р·ива·ть, -ют, (*imp*) boil again; boil too long, overboil; digest; overcure (rubber)

пере·вар·и́ть, -я́т, (*perf*) *see* **пере·ва́р·ива·ть**

пере·вев·а́·ть, -ют, (*imp*) = **пере·ве́·ива·ть**

пере·вез·ти́, -у́т, (*perf*) *see* **пере·во·з·и́ть**

пере·ве́·ива·ть, -ют, (*imp*) winnow (all); winnow again

пере·вёр·нут·ый, -ая, -ое, (*past part pass*) of **пере·вер·н·у́ть**

  ~ по·лёт (*m*) (air) inverted flight

пере·вер·н·у́ть, -у́т, (*perf*) turn, turn over/up; invert, upset

пере·верст·а́ть, -ю́т, (*perf*) *see* **пере·-вёрст·ыва·ть**

пере·вёрст·ыва·ть, -ют, (*imp*) (print) reimpose

пере·верт·е́ть, -́ят, (*perf*) overwind, screw up too far

пере·вёрт·ывани·е, -я, (*n*) (*v n*) of **пере·вёрт·ыва·ть**

~ спин·а (nucl) spin flipping

пере·вёрт·ыва·ть, -ют, (*imp*) turn, turn over/up; invert, upset; overwind, screw up too far

пере·вёрч·енн·ый, -ая, -ое, (*past part pass*) of **пере·верт·е́ть**

пере·ве́с·, -а, (*m*) over-weight, over-balance; preponderance; advantage

~, чи́сл·енн·ый majority, numerical superiority

пере·ве́с·ить, -ят, (*perf*) outweigh, outbalance; resite, move (a suspended object); weigh again; **-ся** lean right out/over

пере·вес·ти́, (*fut 3rd sing, pl*) **пере·ве·д·ёт, пере·вед·у́т,** (*past masc sing*) **пере·вёл,** (*perf*); *see* **пере·вод·и́ть; -ся** disappear

пере·ве́ш·анн·ый, -ая, -ое, (*past part pass*) of **пере·ве́ш·а·ть**

пере·ве́ш·а·ть, -ют (*perf*) hang, hang up (all, much); weigh (all, much)

пере·ве́ш·енн·ый, -ая, -ое, (*past part pass*) of **пере·ве́с·ить**

пере·ве́ш·ива·ть, -ют, (*imp*) *see* **пере·ве́с·ить** (*perf*), **пере·ве́ш·а·ть** (*perf*)

пере·ве́·ять, -ют, (*perf*) *see* **пере·ве́·и·ва·ть**

пере·вибр·и́р·овать, -уют, (*imp*) over-vibrate (concrete)

пере·вив·а́·ть, -ют, (*imp*) wind round, entwine; intertwine, interweave; re-wind, lay up again (rope); wind, twist (all, much)

~ осно́в·у (text) rebeam the warp

пере·винт·и́ть, -ят, (*perf*) *see* **пере·-ви́нч·ива·ть**

пере·ви́нч·ива·ть, -ют, (*imp*) screw up again; resite, move (a screw); damage (by screwing)

пере·ви́т·ый, -ая, -ое, (*past part pass*) *see* **пере·вив·а́·ть**

пере·ви́·ть, (*fut 3rd sing, pl*) **пере·вь·ёт, пере·вь·ю́т,** (*past masc sing*) **пере·-ви́·л** (*perf*); *see* **пере·вив·а́·ть**

пере·во́д·, -а, (*m*) (*v n*), *see* **пере·во·д·и́ть;** translation; (fin) remittance; operation (of switches, levers); (math) conversion

~, маши́н·н·ый (ling) machine/mechanical/automatic translation, MT

~, почт·о́в·ый money order

~, стре́л·очн·ый (rail) turnout, switch

~, телегра́ф·н·ый (fin) cable transfer, telegraph money order

пере·во·д·и́ть, -́ят, (*imp*) lead through/across; transfer, move/direct to, move, shift; convert, reorganize; translate, interpret; remit (money); switch (trains)

пере·во́д·ник·, -а, (*m*) nipple, adapter (pipes)

пере·вод·н·о́й, -а́я, -о́е, (*a*) *see* **пере·-во́д·н·ый**

~, бума́г·а (*f*) carbon paper

пере·во́д·н·ый, -ая, -ое, (*a*) transfer; shift; switch, switching; conversion; (fin) remittance; translated

~ ве́ксел·ь (*m*) bill of exchange

~ коэффицие́нт· (*m*) (math) conversion factor

~ руко·я́т·к·а (*f*) control lever, switch

~ механи́зм· (*m*) switching/reversing gear; (telecom) shift mechanism

~ множ·и́тел·ь (*m*) (math) conversion factor

~ рыча́г· (*m*) switch/reversing lever

пере·во́д·ческ·ий, -ая, -ое, (*a*) translator's, interpreter's

пере·во́д·чик·, -а, (*m*) translator, interpreter

пере·вож·у́, (*pres 1st sing*) of **пере·во·д·и́ть**

пере·во́з·, -а, (*m*) (*v n*), *see* **пере·-во·з·и́ть,** ferry, crossing place

пере·воз·бужд·е́ни·е, -я, (*n*) over-excitation; overdrive, overswing (of vacuum tubes etc.)

пере·воз·и́ть, -́ят, (*imp*) carry/take across/over, transport

пере·во́з·к·а, -и, (*g pl*) **-з·ок·,** (*f*) (*v n*), *see* **пере·воз·и́ть;** transportation, transport, haulage, carriage

в пере·во́з·к·е in transit

~ гру́з·ов freightage, carriage

пере·во́з·н·ый, -ая, -ое, (*a*) of **пере·-во́з·**

пере·во́з·очн·ый, -ая, -ое, (*a*) of **пере·во́з·к·а**

пере·во́з·чик·, -а, (*m*) ferryman

пере·во·оруж·е́ни·е, -я, (*n*) rearma-ment, rearming

пере·ворáч·ивател·ь, -я, (*m*) turner, inverter; (agr) tedder

пере·ворáч·ива·ть, -ют, (*imp*) = **пере·вёрт·ыва·ть**

пере·ворóт·, -а, (*m*) revolution; (air) half-roll; (geol) cataclysm

~, двой·н·óй (air) barrel roll

пере·вос·пит·áни·е, -я, (*n*) re-education

пере·вулканиз·áци·я, -и, (*f*) over-cure, overcuring (of rubber)

пере·вы́·бр·ать, (*fut 3rd pl*) **пере·вы́·бер·ут** (*perf*) re-elect

пере·вы́·пас·, -а, (*m*) (agr) overgrazing

пере·вы·полн·éни·е, -я, (*n*) overful-filment

пере·вы́·полн·ить, -ят, (*perf*) *see* **пере·вы·полн·я́·ть**

пере·вы·полн·я́·ть, -ют, (*imp*) over-fulfil, exceed

пере·вы́·пуск·, -а, (*m*) (min) overdraw-ing

пере·вяз·áть, (*fut 3rd pl*) **пере·вя́ж·ут,** (*perf*) *see* **пере·вя́з·ыва·ть**

пере·вя́з·к·а, -и, (*g pl*) **-з·ок·,** (*f*) bandage, dressing; bond (brick-work); (zool) tiger polecat

~ карт· (text) relacing

~, креп·остн·áя (build) raking bond

~, крест·óв·ая (build) crossbond

~, лóж·ков·ая (build) stretcher bond

~, по·перéч·н·ая (build) heading bond

~, про·дóль·н·ая (build) stretching bond

~, тыч·кóв·ая (build) header bond

пере·вя́з·очн·ый, -ая, -ое, (*a*) (med) dressing

~ материáл· (*m*) (med) dressings

~ пункт· (*m*) (med) dressing station

пере·вя́з·ыва·ть, -ют, ·(*imp*) tie up, lash; tie up again; bandage, dress; apply a new dressing; knit again, re-knit

пére·вяз·ь, -и, (*f*) (med) sling; cross-belt, shoulder-belt

пере·вя́сл·о, -а, (*n*) binding

пере·гáж·енн·ый, -ая, -ое, (*a*) super-saturated with gas

пере·газ·óванност·ь, -и, (*f*) super-saturation with gas

пере·гáр·, -а, (*m*) smell (of something burning or burnt)

пере·гúб·, -а, (*m*) (*v n*), *see* **пере·ги·б·á·ть;** bend, inflection, kink; reverse bending, flexure; (mech) knuckle; hogging (of ship)

пере·гúб

ис·пыт·áни·е на пере·гúб· (*n*) (met) reverse bend test

крив·áя пере·гúб·ов (*f*) curve through inflection points

~ с·клóн·а slope discontinuity

тóч·к·а пере·гúб·а (*f*) (math) point of inflection

~ раз·ря́д·а (nucl) discharge kink

пере·гиб·á·ть, -ют, (*imp*) bend/fold over

пере·глáд·ить, -ят, (*perf*) *see* **пере·глáж·ива·ть**

пере·глáж·ива·ть, -ют, (*imp*) iron/press/smooth again

пере·глас·óвк·а, -и, (*g pl*) **-вок·,** (*f*) (ling) mutation

перé·гн·анн·ый, -ая, -ое, (*past part pass*) *see* **пере·гон·я́·ть**

~, до кóкс·а (oil) distilled to dryness

пере·гн·áть, (*fut 3rd pl*) **пере·гóн·ят** (*perf*), *see* **пере·гон·я́·ть**

пере·гнив·áтел·ь, -я, (*m*) digestion/septic tank (sewerage)

пере·гнив·á·ть, -ют, (*imp*) rot through

пере·гни́·ть, -ют, (*perf*) *see* **пере·гни·в·á·ть**

пере·гнó·й, -я, (*m*) humus

пере·гн·у́ть, -у́т, (*perf*) bend, fold

пере·говáр·ива·ть, -ют, (*imp*) discuss, have a talk; talk down

пере·говóр·, -а, (*m*) (telecom) con-versation, call; (*pl*) negotiations, dis-cussions, talks

~ ключ·óм (telecom) key conversation

~, плáт·н·ый charged/fee call (tele-phones)

пере·говор·и́ть, -я́т, (*perf*) *see* **пере·говáр·ива·ть**

пере·говóр·н·ый, -ая, -ое, (*a*) of **пере·говóр·**

~ бýд·к·а (*f*) call box/booth

~ таблúц·а (*f*) (telecom) standard prearranged message code

~ труб·á (*f*) (telecom) voice pipe

~ у·стрóй·ств·о (*n*) intercommunica-tion system

пере·гóн·, -а, (*m*) driving (animals etc.); stage, span (between stops, stations)

пере·гóн·к·а, -и, (*g pl*) **-н·ок·,** (*f*) (*v n*), *see* **пере·гон·я́·ть;** distillation (of liquids and solids); sublimation (of solids only); (text) dressing (the warp); ferry (aircraft)

~, дрóб·н·ая fractional distillation

~, не·пре·ры́в·н·ая continuous dis-tillation

пере·гóн·к·а
~, **периодúческ·ая** batch distillation
~, **протúво·тóч·н·ая** counter-current distillation
~, **селектúв·н·ая** extractive distillation
~, **сух·áя** dry/destructive distillation
~, **цикл·úческ·ая** batch distillation
пере·гóн·н·ый, -ая, -ое, (*a*) of **пере·гóн·** and **пере·гóн·к·а;** *see also* **пере·гóн·оч·н·ый**
~ **куб·** (*m*) (chem) still
~ **машúн·а** (*f*) (text) warp dressing machine
пере·гóн·оч·н·ый, -ая, -ое, (*a*) of **пере·гóн·к·а;** *see also* **пере·гóн·н·ый;** (air) ferry, ferrying
~ **пере·лёт·** (*m*) ferry flight
пере·гон·я́·ть, -ют, (*imp*) drive (animals etc.); ferry (aircraft); outstrip, outdistance, outrun, get ahead of; (chem) distil (liquid or solid); sublimate (solid only)
пере·горáж·ива·ть, -ют, (*imp*) partition, partition off
пере·гор·á·ть, -ют, (*imp*) burn through/out; fuse, blow (of fuse)
пере·гор·éть, -я́т, (*perf*) *see* **пере·гор·á·ть**
пере·город·úть, -я́т, (*perf*) *see* **пере·горáж·ива·ть**
пере·горóд·к·а, -и, (*g pl*) **-д·ок·,** (*f*) partition; baffle, barrier; (anat) septum; membrane, diaphragm, iris; (hydrodynam) closure; slag skimmer (of furnace)
~, **вентиляциóн·н·ая** (min) air partition
~, **диффузиóн·н·ая** (phys) diffusion barrier
~, **ём·костн·ая** capacitive iris (of waveguide)
~, **периклинáль·н·ая** (bot) periclinal wall
~, **по·глот·úтельн·ая** (elec) absorbing septum
~, **пóр·ист·ая** porous diaphragm
~, **полу·про·ниц·áем·ая** semipermeable membrane
~, **протúво·по·жáр·н·ая** fire wall, fire bridge, baffle (boiler)
~, **тепл·ов·áя** thermal baffle
пере·горóд·очк·а, -и, (*g pl*) **-чек·,** (*f*) (*dim*) of **пере·горóд·к·а;** (anat) septulum
пере·горóд·очн·ый, -ая, -ое, (*a*) of **пере·горóд·очк·а;** (anat) septate, septulate

пере·горóд·чат·ый, -ая, -ое, (*a*) of **пере·горóд·к·а;** (bot etc.) murate
пере·граду·úр·овать, -уют, (*perf*) recalibrate
пере·греб·á·ть, -ют, (*imp*) rake over, stir/mix up
пере·греб·áющ·ая рукоя́т·к·а (*f*) rabble arm (of furnace)
пере·гре́·в·, -а, (*m*) overheating, overheat; superheating; overrun (of vacuum tubes)
~, **мéст·н·ый** thermal spike, hot spot
~ **метáлл·а** (met) overheating (a defect condition)
тепл·от·á пере·гре́·в·а (*f*) superheat
пере·грев·áни·е, -я, (*n*) (*v n*), *see* **пере·грев·á·ть**
пере·грев·áтел·ь, -я, (*m*) overheater
пере·грев·á·ть, -ют, (*imp*) overheat; (st eng) superheat
пере·грес·тú, (*fut 3rd sing, pl*) **пере·греб·ёт, пере·греб·ýт,** (*past masc sing*) **пере·грёб·** (*perf*); *see* **пере·греб·á·ть**
пере·гре́·т·ый, -ая, -ое, (*past part pass*) of **пере·гре́·ть;** overheated; superheated
~ **мéст·о** (*n*) hot spot
~ **пар·** (*m*) superheated steam
пере·гре́·ть, -ют, (*perf*) *see* **пере·грев·á·ть**
пере·груж·áемост·ь, -и, (*f*) overload capacity
пере·груж·áтел·ь, -я, (*m*) conveyor, cargo handling equipment
~, **мост·ов·óй** cantilever-type gantry crane
пере·груж·á·ть, -ют, (*imp*) tranship, handle (cargo); surcharge; overload, overcharge, overdrive
пере·груз·úть, -я́т, (*perf*) *see* **пере·груж·á·ть**
пере·грýз·к·а, -и, (*g pl*) **-з·ок·,** (*f*) (*v n*), *see* **пере·груж·á·ть;** recharging; overriding (a measuring instrument); (air) acceleration g, load/acceleration factor
~, **больш·áя** (air) heavy g-load, high g-force
~, **дл·úтельн·ая** sustained overload
~, **трёх·крáт·н·ая** (air) g-force of three g's
пере·грýз·очн·ый, -ая, -ое, (*a*) of **пере·грýз·к·а**
~ **ис·каж·éни·е** (*n*) (rad) blasting
~ **мост·** (*m*) cantilever-type gantry crane

пере·гру́з·очн·ый, -ая, -ое
~ **ста́нци·я** (*f*) (rail) transhipping station
~ **характери́стик·а** (*f*) (elec) overload characteristic
пере·грунт·ова́ть, -у́ют, (*perf*) apply a new prime/priming coat (for paint)
пере·групп·иро́вк·а, -и, (*g pl*) **-вок·,** (*f*) regrouping, rearrangement, reshuffling; (chem, phys) rearrangement, molecular rearrangement, transformation
~, **бензиди́н·ов·ая** (chem) benzidine transformation/rearrangement
~, **меж·молекуля́р·н·ая** molecular rearrangement
~, **пинаколи́н·ов·ая** pinacoline transformation/rearrangement
пере·грыз·а́·ть, -ют, (*imp*) gnaw through
пере·гры́з·ть, -у́т, (*perf*) *see* **пере·гры·з·а́·ть**
пе́ред (*prep* + *instr*) in front of, before
перёд·, (*g sing*) **пе́ред·а,** (*nom pl*) **пе·ред·а́,** (*g pl*) **перед·о́в,** (*m*) fore-part, front
пере·да·ва́ть, -ю́т, (*imp*) pass, transfer, transmit, send, hand over; communicate, relate; depict, reproduce
~ **то́ч·к·и** (telecom) transmit/send dots
пере·дад·у́т, (*fut 3rd pl*) of **пере·да́·ть**
пе́ре·да·нн·ый, -ая, -ое, (*past part pass*) *see* **пере·да·ва́ть**
пере·да́·ст, (*fut 3rd sing*) of **пере·да́·ть**
пере·да́т·очн·ый, -ая, -ое, (*a*) of **пере·да́·ч·а**
~ **ключ·** (*m*) (telecom) morse/transmitting key
~ **коро́б·к·а** (*f*) (M/T) transfer case
~ **ли́н·и·я** (*f*) (telecom) transfer/lending circuit; interposition trunk (telephones)
~ **механ·и́зм·** (*m*) transmitting/driving mechanism; (horol) motion work
~ **от·нош·е́ни·е** (*n*) (mech) gear ratio
~ **по·те́р·и** (*pl*) (telecom) transmission losses
~ **фу́нкци·я** (*f*) (phys) transfer function
~ **числ·о́** (*n*) (mech) gear ratio
пере·да́т·чик·, -а, (*m*) (telecom) transmitter, sender
~, **вс·по·мог·а́тельн·ый** satellite transmitter
~, **за·да·ю́щ·ий** master transmitter
~ **звук·ов·о́го со·про·вожд·е́ни·я** (TV) aural transmitter
~, **искр·ов·о́й** (rad) spark transmitter

пере·да́т·чик
~, **ла́мп·ов·ый** valve sender, vacuum-tube transmitter
~, **маши́н·н·ый** (rad) alternator transmitter
~, **меш·а́ющ·ий** jamming transmitter
~, **штрих·ов·о́й** facsimile transmitter
пере·да́·ть, (*fut 3rd sing, pl*) **пере·да́·ст, пере·дад·у́т,** (*past masc sing*) **пе́ре·да·л,** (*perf*); *see* **пере·да·ва́ть**
пере·да́·ч·а, -и, (*i*) **-ч·ей,** (*f*) (*v n*), *see* **пере·да·ва́ть**; transmission, drive; transfer; gear, gearing
~, **бес·ступ·е́нчат·ая** (mech) variable-speed gear
~, **борт·ов·а́я** final drive (of a tracked vehicle)
~, **гидравл·и́ческ·ая** hydraulic transmission
~, **гидро·динам·и́ческ·ая** hydrokinetic transmission
~, **гидро·механ·и́ческ·ая** hydraulic and mechanical transmission
~, **гидро·стат·и́ческ·ая** hydrostatic transmission
~, **гла́в·н·ая** (autom) gearing
~, **глобойда́ль·н·ая** double-throated worm gear, cone-drive worm gear
~ **да́·нн·ых** data transmission/transfer (computers)
~, **двой·н·а́я зуб·ча́т·ая** double reduction gear
~, **дел·и́тельн·ая** index drive
~ **Дже́нни** variable speed gear, Williams–Janney gear
~, **дистанцио́н·н·ая** remote indicating/control transmission
~, **зуб·ча́т·ая** gearing, tooth gearing
~ **и́мпульс·а** (phys) momentum transfer
~, **кана́т·н·ая** rope drive
~, **кард·а́н·н·ая** (M/T) propeller shaft and universal joint
~, **клино·ремён·н·ая** V-belt transmission
~, **кон·и́ческ·ая** bevel drive
~, **круг·ов·а́я кана́т·н·ая** continuous rope drive
~, **лёгк·ая син·хро́н·н·ая** indicator magslip transmission
ли́н·и·я пере·да́·ч·и (*f*) (elec) transmission line
~, **лоб·ов·а́я** plate clutch, faceplate coupling
~, **не·по·сре́д·ственн·ая** direct drive; (telecom) direct transmission
~, **остро·на·пра́вл·енн·ая** (TV) beam transmission

пере·да́·ч·а

~, **планета́р·н·ая** planetary gear, sun-and-planet gear

~, **по·выш·а́ющ·ая** (M/T) overdrive

~, **прям·а́я** direct drive

~, **ремён·н·ая** belt drive

~, **рул·ев·а́я** (shipb) steering gear

~, **само·синхрониз·и́рующ·ая син·хро́н·н·ая** synchro transmission

~, **само·тормоз·я́щ·аяся** self-sustaining gear

~ **сил·ов·а́я син·хро́н·н·ая** power magslip transmission

~, **син·хро́н·н·ая** synchronous transmission

~ **со·обш·е́ни·я** information transfer
тео́р·и·я пере·да́·ч·и со·обш·е́·ни·я (f) (autom) information theory

~, **стерж·нев·а́я** rod gear

~, **ступенчато·шки́в·н·ая** cone-pulley drive

~ **тепл·а́** heat transfer

~, **трубо·зуб·ча́т·ая** geared turbine transmission

~, **угл·ов·а́я** angle drive

~, **у·скор·я́ющ·ая** (M/T) overdrive

~, **фрикцио́н·н·ая** friction gearing

~, **цеп·н·а́я** chain drive

~, **черв·я́чн·ая** worm-and-wheel, worm gear/gearing

пере·да·ющ·е-при·ём·н·ый, -ая, -ое, (a) (telecom) transmit–receive

пере·да·ю́щ·ий, -ая, -ее, (pres part act) of **пере·да·ва́ть;** transmission, transfer

~ **у·стро́й·ств·о** (n) = **да́т·чик·**

~ **ста́нци·я** (f) (rad) transmitting station/unit

~ **тру́б·к·а** (f) (TV) camera/pick-up tube

пере·дви́г·ател·ь, -я, (m) manipulator

пере·двиг·а́·ть, -ют, (imp) move, shift, transfer; **-ся** travel, migrate

пере·движ·е́ни·е, -я, (n) (v n), see **пере·двиг·а́·ть;** movement, migration

~ **контине́нт·ов** (geol) continental drift

пере·дви́ж·к·а, -и, (g pl) **-ж·ек·,** (f) (v n) of **пере·двиг·а́·ть;** mobile unit

пере·движ·н·о́й, -а́я, -о́е, (a) mobile, itinerant; movable, transportable

пере·дви́·н·уть, -ут, (perf) see **пере··двиг·а́·ть**

пере·де́л·, -а, (m) (v n), see **пере··дел·я́·ть;** redivision, repartition; (met) conversion

пере·де́л·анн·ый, -ая, -ое, (past part pass) see **пере·де́л·ыва·ть**

пере·де́л·а·ть, -ют, (perf) see **пере··де́л·ыва·ть**

пере·дел·ённ·ый, -ая, -ое, (past part pass) see **пере·дел·я́·ть**

пере·дел·и́ть, ⸰ят, (perf) see **пере··дел·я́·ть**

пере·де́л·к·а, -и, (g pl) **-л·ок·,** (f) (v n), see **пере·де́л·ыва·ть;** alteration, adaption, conversion, remodelling, remaking

пере·де́л·очн·ый, -ая, -ое, (a) of **пере·де́л·к·а**

пере·де́л·ыва·ть, -ют, (imp) remake, alter, convert, adapt

пере·де́ль·н·ый, -ая, -ое, (a) (met) conversion; hot-working, hot-reduction

~ **чугу́н·** (m) conversion iron

пере·дел·я́·ть, -ют, (imp) divide up again, redivide

пере·демпф·и́рованн·ый, -ая, -ое, (past part pass) overdamped

пере·дёрг·ива·ть, -ют, (imp) pull/tug through; jerk, twitch

пере·де́рж·анн·ый, -ая, -ое, (past part pass) see **пере·де́рж·ива·ть**

пере·держ·а́ть, ⸰ат, (perf) see **пере··де́рж·ива·ть**

пере·де́рж·ива·ть, -ют, (imp) hold/do too long; (phot) over-expose

пере·де́рж·к·а, -и, (g pl) **-ж·ек·,** (f) (v n), see **пере·де́рж·ива·ть;** (phot) over-exposure

~ **в мо́ч·к·е** overretting (flax)

пере·дёр·н·уть, -ут, (perf) see **пере··дёрг·ива·ть**

перед·ко́в·ый, -ая, -ое, (a) of **пе·ред·о́к·**

пере́д·н·е- (component) pro-, antero-

передне·бок·ов·о́й, -а́я, -о́е, (a) (anat) antero-lateral

передне·груд·ь, -и, (f) (ent) prothorax

передне·жа́бер·н·ые, -ых, (pl decl as adj) (zool) prosobranchs, *Prosobranchia*

передне·спи́н·к·а, -и, (g pl) **-н·ок·,** (f) (ent) pronotum

передне·шо́в·н·ый, -ая, -ое, (a) (pal) proparian; (pl as noun) proparians, *Proparia*

пере́д·н·ий, -яя, -ее, (*a*) front, fore, leading; (anat) anterior

~ **гран·ь** (*m*) front, face

~ **кро́м·к·а** (*f*) (a/c) leading edge

~ **спин·н·о́й** (*a*) (zool) anterodorsal

~ **сторон·а́** (*f*) front

~ **фронт·** (*m*) leading edge (of a pulse)

~ **ход·** (*m*) forward travel, headway, passage forward

пе́редо (*prep*) *see* **пе́ред**

пере·до·ве́р·енн·ый, -ая, -ое, (*past part pass*) *see* **пере·до·вер·я́·ть**

~ **до·гово́р·** (*m*) subcontract

пере·до·ве́р·ить, -ят, (*perf*) *see* **пе·ре·до·вер·я́·ть**

пере·до·вер·я́·ть, -ют, (*imp*) delegate, transfer trust

перед·ов·и́к·, -а́, (*m*) leader, most successful person

перед·ови́ц·а, -ы, (*i*) **-ей,** (*f*) leading article, leader

перед·ов·о́й, -а́я, -о́е, (*a*) forward, advanced, leading, progressive; foremost; vamp (of shoes)

пере·до́·й, -я, (*m*) (zool) superlactation

перед·о́к·, -д·к·а́, (*m*) front (of cart or sledge); forecarriage (of plough); limber, trailer (of gun carriage); vamp (of shoe)

пере·ду́в·к·а, -и, (*g pl*) **-в·ок·,** (*f*) (met) afterblow

пере·ду́м·а·ть, -ют, (*perf*) change one's mind

пере·е́д·ут, (*fut 3rd pl*) of **пере·е́х·ать**

пере·ды́ш·к·а, -и, (*g pl*) **-ш·ек·,** (*f*) rest break/pause, respite

пере·ед·а́·ть, -ют, (*imp*) corrode, eat through/away; overeat

пере·е́зд·, -а, (*m*) (*v n*), *see* **пере·ез·ж·а́·ть**; crossing, passage; removal (of residence); (rail) level crossing

пере·езж·а́·ть, -ют, (*imp*) drive/ride across/through, cross; move (residence); run over

пере·е́ст·ь, (*fut 3rd sing, pl*) **пере·е́ст, пере·ед·я́т,** (*past masc sing*) **пере·е́л,** (*perf*); *see* **пере·ед·а́·ть**

пере·е́х·ать, (*fut 3rd pl*) **пере·е́д·ут,** (*perf*) *see* **пере·езж·а́·ть**

пере·жа́р·ива·ть, -ют, (*imp*) fry/roast/grill too much

пере·жа́р·ить, -ят, (*perf*) *see* **пере·жа́р·ива·ть**

пере·жг·у́т, (*fut 3rd pl*) of **пере·жéч·ь,** (*perf*) *see* **пере·жиг·а́·ть**

пере·жд·а́·ть, -у́т, (*perf*) *see* **пере·жид·а́·ть**

пере·жев·а́·ть, (*fut 3rd sing, pl*) **пере·ж·у́ёт, пере·ж·у́ют,** (*perf*) masticate, chew, chew up

пере·жéч·ь, (*fut 3rd sing, pl*) **пере·жж·ёт, пере·жг·у́т,** (*past masc sing*) **пере·жёг·,** (*perf*); *see* **пере·жиг·а́·ть**

пере·жж·ённ·ый, -ая, -ое, (*past part pass*) *see* **пере·жиг·а́·ть**; burned, burnt

пере·жив·а́·ть, -ют, (*imp*) live through, experience; survive, outlive

пере·жи́г·, -а, (*m*) (*v n*) of **пере·жиг·а́·ть**; overconsumption, overfiring (of fuel)

пере·жиг·а́·ть, -ют, (*imp*) burn (spoil by burning); burn through; (elec) blow (a fuse); burn up (too much fuel)

пере·жид·а́·ть, -ют, (*imp*) wait till/for; wait for the end of, spend the time till the end of

пере·жи́м·, -а, (*m*) narrowing, contraction, pinch, squeeze, nip; knuckle (of OH furnace)

пере·жи́т·ок·, -т·к·а, (*m*) survival, vestige

пере·жи́·ть, (*fut 3rd pl*) **пере·жи·в·у́т,** (*perf*) *see* **пере·жив·а́·ть**

пере·жо́г·, -а, (*m*) (*v n*), *see* **пере·жиг·а́·ть**; overfire, overfiring (fuel); (met) burning; burnout (of cables etc.)

пере·ж·у́ёт, (*fut 3rd sing*) of **пере·жев·а́ть**

пере·за·ключ·а́·ть, -ют, (*imp*) renew (agreements etc.)

пере·за·ли́·т·ый, -ая, -ое, (*a*) (ICE) overprimed

пере·за·пис·а́ть, (*fut 3rd pl*) **пере·за·пи́ш·ут,** (*perf*) *see* **пере·за·пи́с·ыва·ть**

пере·за·пи́с·ыва·ть, -ют, (*imp*) re-record

пере·за́·пис·ь, -и, (*f*) re-recording; (TV, cinema) dubbing

пере·за·пи́ш·ет, (*fut 3rd sing*) of **пере·за·пис·а́ть**

пере·за́·пуск·, -а, (*m*) restarting

пере·за·ря́д·, -а, (*m*) (elec) overcharge; recharge

пере·за·ряд·и́ть, -я́т, (*perf*) *see* **пе·ре·за·ряж·а́·ть**

пере·за·ря́д·к·а, -и, (*g pl*) **-д·ок·,** (*f*) reloading, recharging; (elec) overcharge

пере·за·ряж·а́·ть, -ют, (*imp*) reload, recharge; (elec) overcharge

пере·за·та́ч·ива·ть, -ют, (*imp*) re-grind (tools)

пере·за·хва́т·, -а, (*m*) recapture

пере·зва́н·ива·ть, -ют, (*imp*) chime; ring chimes

пере·зво́н·, -а, (*m*) chime

пере·звон·и́ть, -я́т, (*perf*) ring/call again/back (of telephone)

пере·зим·ова́ть, -у́ют, (*perf*) *see* **пере·зим·о́выва·ть**

пере·зим·о́вк·а, -и, (*g pl*) -вок·, (*f*) (*v n*), *see* **пере·зим·о́выва·ть**; hibernation

пере·зим·о́выва·ть, -ют, (*imp*) winter, spend the winter; hibernate

пере·зол·ённ·ый, -ая, -ое, (*past part pass*) overlimed

пере·зрев·а́·ть, -ют, (*imp*) overripen, become overripe/overmature

пере·зре́·л·ый, -ая, -ое, (*a*) overripe, overmature

пере·зре́·ть, -ют, (*perf*) *see* **пере·зрев·а́·ть**

пере·игр·а́·ть, -ют, (*perf*) *see* **пере·йгр·ыва·ть**

пере·йгр·ыва·ть, -ют, (*imp*) replay; play (all/much); overplay

пере·из·бр·а́ни·е, -я, (*n*) re-election

пере·из·да·ва́ть, -ю́т, (*imp*) republish, repromulgate

пере·из·да́·ни·е, -я, (*n*) (*v n*), *see* **пере·из·да·ва́ть**; new edition

пере·из·да́·ть, (*fut 3rd sing, pl*) **пере·из·да́·ст, пере·из·дад·у́т**, (*past masc sing*) **пере·из·да́·л**, (*perf*); *see* **пере·из·да·ва́ть**

пере·имен·ова́ни·е, -я, (*n*) renaming

пере·йд·ённ·ый, -ая, -ое, (*past part pass*) of **пере·йти́**, (*perf*); *see* **пере·ход·и́ть** (*imp*)

пере·йд·ёт, (*fut 3rd sing*) of **пере·й·ти́** (*perf*), *see* **пере·ход·и́ть** (*imp*)

перейо·по́д·ы, -ов, (*m pl*) (zool) pereiopods

пере·йти́, (*fut 3rd sing, pl*) **пере·й·д·ёт, пере·йд·у́т**, (*past masc sing*) **пере·шёл**, (*perf*); *see* **пере·ход·и́ть**

пере·ка́л·, -а, (*m*) (*v n*), *see* **пере·ка́л·ива·ть**; over-running, hot shot (of vacuum tubes)

пере·ка́л·ива·ть, -ют, (*imp*) overheat; (met) over-harden; over-run (vacuum tubes)

пере·кал·и́ть, -я́т, (*perf*) *see* **пере·ка́л·ива·ть**

пере·ка́л·ыва·ть, -ют, (*imp*) pin elsewhere; prick (all, much); prick all over

пере·ка́п·ыва·ть, -ют, (*imp*) dig up/across, unearth; dig up again

пере·ка́рмл·ива·ть, -ют, (*imp*) over-feed, overfatten

пере·ка́т·, -а, (*m*) (*v n*), *see* **пере·кат·и́ть**; shoal, shallow; rumble, rumbling; (mil) leapfrog

пере·кат·а́·ть, -ют, (*perf*) roll (all, many); mangle again, mangle (all, many)

пере·кат·и́ть, -я́т, (*perf*) roll, move (by rolling)

пере·ка́т·к·а, -и, (*g pl*) -т·ок·, (*f*) (*v n*), *see* **пере·кат·а́·ть**

пере·ка́т·ыва·ть, -ют, (*imp*) *see* **пере·кат·а́·ть, пере·кат·и́ть**

пере·ка́ч·анн·ый, -ая, -ое, (*past part pass*) *see* **пере·ка́ч·ива·ть**

пере·кач·а́·ть, -ют, (*perf*) *see* **пере·ка́ч·ива·ть**

пере·ка́ч·енн·ый, -ая, -ое, (*past part pass*) of **пере·кат·и́ть**

пере·ка́ч·ивани·е, -я, (*n*) (*v n*), *see* **пере·ка́ч·ива·ть**

~, **опт·и́ческ·ое** (phys) optical pumping

пере·ка́ч·ива·ть, -ют, (*imp*) pump, pump over/across; overinflate (tyres)

пере·ка́ч·к·а, -и, (*g pl*) -ч·ек·, (*f*) (*v n*), *see* **пере·ка́ч·ива·ть**; transfer by pumping; overinflation

пере·ка́ш·ива·ть, -ют, (*imp*) mow/scythe (all, much); distort, twist

пере·квалифиц·и́рованн·ый, -ая, -ое, (*past part pass*) of **пере·квалифи·ц·и́р·овать**

пере·квалифиц·и́р·овать, -уют, (*imp and perf*) train/study for new qualifications

пере·кид·а́·ть, -ют, (*perf*) throw (all, many), move (by throwing)

пере·ки́д·к·а, -и, (*g pl*) -д·ок·, (*f*) (elec) jumper, repair cable; reversal (of valves etc.)

пере·ки́д·н·о́й, -а́я, -о́е, (*a*) throw-over, reversing; collapsible, portable

~ **вы·ключ·а́тел·ь** (*m*) tumbler switch

~ **схе́м·а** (*f*) (elec) flip-flop circuit

~ **мост·** (*m*) (min) crossing

пере·ки́д·ыва·ть, -ют, (*imp*) *see* **пере·ки́·н·уть, пере·кид·а́·ть**

пере·ки́·н·уть, -ут, (*perf*) throw through/across/over; flick over; throw too far; **-ся** (*pass and reflex*); throw to and fro

пере·кип·а́·ть, -ют, (*imp*) boil too long

пере·кип·е́ть, -я́т, (*perf*) *see* **пере·-кип·а́·ть**

пере·кисл·е́ни·е, -я, (*n*) (chem) peroxidation

пе́ре·кис·ь, -и, (*f*) peroxide, superoxide

~ **водо·ро́д·а** hydrogen peroxide

~ **водо·ро́д·а, высоко·со́рт·н·ая** high-test hydrogen peroxide, HTP

~ **водо·ро́д·а, мало·во́д·н·ая** high-test hydrogen peroxide, HTP

~ **на́тр·и·я** sodium peroxide

пере·кла́д·, -а, (*m*) (min) stull, cross-bar

пере·кла́д·ин·а, -ы, (*f*) cross-beam, cross-piece, rung, transom, spar

пере·кла́д·ыва·ть, -ют, (*imp*) shift, move, put/place/lay/set elsewhere/again/too many, add too much; interlay, pack into layers of

пере·кла́д·чик·, -а, (*m*) log transporter (in timber mills)

пере·кле́·ива·ть, -ют, (*imp*) glue/stick/paste elsewhere/again

пере·кле́·ить, -ят, (*perf*) *see* **пере·-кле́·ива·ть**

пере·кле́й·к·а, -и, (*g pl*) **-е́·ек·,** (*f*) (*v n*), *see* **пере·кле́·ива·ть;** plywood

пере·клеп·а́·ть, -ют, (*perf*) *see* **пере·-клёп·ыва·ть**

пере·клёп·ыва·ть, -ют, (*imp*) rivet again/elsewhere; rivet (all, many)

пере·клик·а́·ть, -ют, (*imp*) call over, make a roll call; **-ся** call to one another; have something in common

пере·кли́к·н·уть, -ут, (*perf*) *see* **пере·-клик·а́·ть**

пере·кли́ч·к·а, -и, (*g pl*) **-ч·ек·,** (*f*) (*v n*), *see* **пере·клик·а́·ться;** roll call

пере·ключ·а́тел·ь, -я, (*m*) changeover switch, switch, changeover lever, selector, changer, shift knob/lever

~**, анте́нн·ый** (rad) T-R switch

~**, бараба́н·н·ый** (elec) drum/barrel switch

~ **блок·иро́вк·и магнетро́н·а** ATR switch (radar)

~**, вы·зыв·н·о́й** challenge switch (telephones)

~ **выс·от·ы́** (a/c) altitude selector

~ **диапазо́н·ов** (rad) band selector/switch

~ **масшта́б·а да́ль·ност·и** range-scale switch (radar)

~ **на·правл·е́ни·я** (elec) reverser; direction control switch (radar)

пере·ключ·а́тел·ь

~ **от·ветв·е́ни·я** (elec) tap changer/switch

~**, паке́т·н·ый** packet rotary changeover switch

~ **све́т·а фа́р·а** (M/T) dipper switch

~ **со звезд·о́й на тре·уго́ль·ник·е** (elec) star-delta switch

пере·ключ·а́·ть, -ют, (*imp*) switch, change, switch/change over

пере·ключ·е́ни·е, -я, (*n*) (*v n*), *see* **пере·ключ·а́·ть**

~ **лепест·к·о́в** (rad) lobing

~ **пере·да́·ч·** gear change, changing gear

~ **пит·а́ни·я** (elec) power transfer

пере·ключ·и́ть, -а́т, (*perf*) *see* **пере·-ключ·а́·ть**

пере·ков·а́ть, (*fut 3rd pl*) **пере·ку·ю́т** (*perf*) *see* **пере·ко́в·ыва·ть**

пере·ко́в·ыва·ть, -ют, (*imp*) reforge, forge again, hammer out again; shoe (all, many), reshoe (horses etc.)

пере·кол·о́ть, -ют, (*perf*) *see* **пере·-ка́л·ыва·ть**

пере·компаунд·и́р·овать, -уют, (*imp*) (elec) overcompound

пере·компенса́ци·я, -и, (*f*) overcompensation

пере·констру·и́ровани·е, -я, (*n*) redesigning

пере·ко́п·, -а, (*m*) cross ditch

пере·коп·а́·ть, -ют, (*perf*) *see* **пере·ка́-п·ыва·ть**

пере·ко́рм·, -а, (*m*) overfeeding

пере·ко́с·, -а, (*m*) distortion, twist, skew, (mech) angular deflection, lack of alignment

    угол· пере·ко́с·а (*m*) angularity; angle of obliquity (gears)

пере·кос·и́ть, -я́т, (*imp*) *see* **пере·ка́-ш·ива·ть**

пере·коч·ева́ть, -у́ют, (*perf*) migrate

пере·ко́ш·енн·ый, -ая, -ое, (*past part pass*) of **пере·кос·и́ть;** warped, distorted, twisted

пере·кра́·ива·ть, -ют, (*imp*) recut, cut out again

пере·кра́с·ить, -я́т, (*perf*) *see* **пере·-кра́ш·ива·ть**

пере·кра́ш·енн·ый, -ая, -ое, (*past part pass*) of **пере·кра́с·ить;** recoloured, repainted, redyed

пере·кра́ш·ива·ть, -ют, (*imp*) recolour, recolor, repaint, dye/stain again; colour, paint, **dye**, stain (all, many)

пере·креп·и́ть, -я́т, (*perf*) *see* **пере·-крепл·я́·ть**

пере·крепл·я́·ть, -ют, (*imp*) refasten, refix

пере·крёст·, -а, (*m*) (gen) crossing over, crossing; (anat) chiasma

пере·крёст·и·е, -я, (*n*) (instr) cross lines; (opt) graticule

пере·крест·и́ть, -́ят, (*perf*) cross

перекрёстно·сло·и́ст·ый, -ая, -ое, (*a*) (geol) cross-bedded

пере·крёст·н·ый, -ая, -ое, (*a*) cross, crossed, crossover, decussate; (elec) lattice

~ звен·о́ (*n*) (elec) lattice network

~ на·во́д·к·а (*f*) (telecom) crosstalk, crossfire

пере·крёст·ок·, -ст·к·а, (*m*) crossroads, crossing

пере·крёст·ь·е, -я, (*n*) cross; (instr) cross-hair

~, электро́н·н·ое electronic crosshair

пере·крещ·ённ·ый, -ая, -ое, (*past part pass*) *see* **пере·крещ·ива·ть;** crossed

пере·крещ·ивани·е, -я, (*n*) (*v n*), *see* **пере·крещ·ива·ть;** intersection, decussation; transposition (of wires)

пере·крещ·ива·ть, -ют, (*imp*) cross, cross over, intersect

пере·крещ·ивающ·ий, -ая, -ее, (*pres part act*) of **пере·крещ·ива·ть;** cross

~ штóпор·, (*m*) (air) cross spin

пере·кристаллиз·а́ци·я, -и, (*f*) re-crystallization

пере·кро·и́ть, -я́т, (*perf*) *see* **пере·кра́·ива·ть**

пере·кру́т·, -а, (*m*) (text) overtwist; (med) torsion

пере·крут·и́ть, -́ят, (*perf*) *see* **пере·круч·ива·ть**

пере·круч·ени·е, -я, (*n*) (*v n*), *see* **пере·круч·ива·ть;** (med) torsion

пере·круч·енн·ый, -ая, -ое, (*past part pass*) of **пере·крут·и́ть;** twisted, twisted together, intertwined

пере·круч·ивани·е, -я, (*n*) (*v n*), *see* **пере·круч·ива·ть**

пере·круч·ива·ть, -ют, (*imp*) intertwine; twist too much, overwind; distort (by twisting)

пере·крыв·а́ни·е, -я, (*n*) (*v n*) of **пере·крыв·а́·ть;** overlap

пере·крыв·а́·ть, -ют, (*imp*) cover; cover over/again; roof, bridge, span, overlap; shut off (a valve)

пере·кры́т·и·е, -я, (*n*) (*v n*), *see* **пере·-крыв·а́·ть;** cover; (build) ceiling floor; overlap; (elec) arcing over, flashing over, flashover; (nucl) stopping, shuttering

~, без·ба́л·очн·ое (build) single/girder-less floor

врём·я пере·кры́т·и·я (*n*) time overlap

~ и́мпульс·ов pulse overlap

~ и́мпульс·ом (elec) flashover

~, искр·ов·о́е (elec) sparkover

коэффицие́нт· пере·кры́т·и·я (*m*) engagement factor (of gears)

~ нейтро́н·н·ого пу́ч·к·а (nucl) neutron-beam shuttering

~, не·со·гла́с·н·ое overlap

~, по·пере́ч·н·ое (air phot) lateral overlap

~, про·до́ль·н·ое (air phot) line overlap

~, регресси́в·н·ое offlap

пере·кры́·ть, (*fut 3rd pl*) **пере·кро́·ют,** (*perf*) *see* **пере·крыв·а́·ть**

пере·кры́·ш·а, -и, (*f*) (st eng) lap

~, паро·в·пуск·н·а́я steam lap

~, паро·вы·пуск·н·а́я exhaust lap

пере·кры́ш·к·а, -и, (*f*) (*v n*), *see* **пере·-крыв·а́·ть**

пере·ку·ёт, (*fut 3rd sing*) of **пере·ко·в·а́ть**

пере·лаг·а́·ть, -ют, (*imp*) shift, delegate (work); reset (music etc.)

пере·лак·ирова́ть, -́уют, (*perf*) re-varnish; varnish (all or many)

пере·ла́м·ыва·ть, -ют, (*imp*) break, break in two, fracture, break up; change suddenly

пере·лез·а́·ть, -ют, (*imp*) climb/clamber over/across

пере·ле́з·ть, -ут, (*perf*) *see* **пере·ле-з·а́·ть**

пере·ле́с·ок·, -с·к·а, (*m*) coppice

пере·лёт·, -а, (*m*) (*v n*), *see* **пере·ле-т·а́·ть;** (ornith) passage, transmigration, (a/c) cross-country flight, flight; (a/c) overshoot (on landing), over (of missile fall)

~, бес·по·са́д·очн·ый non-stop flight

~, пере·го́н·очн·ый (air) ferry flight

пере·лет·а́·ть, -ют, (*imp*) fly, fly across/over; overshoot

пере·лет·е́ть, -я́т, (*perf*) *see* **пере·ле-т·а́·ть**

пере·лёт·н·ый, -ая, -ое, (*a*) (orn) migratory

пере·ли́в·, -а, (*m*) (*v n*), *see* **пере·ли·в·а́·ть;** tint, tinge; (hydr) overflow, overfall

пере·лив·а́ни·е, -я, (*n*) (*v n*) of **пере··лив·а́·ть;** transfusion

~ **кро́в·и** blood transfusion

пере·лив·а́·ть, -ют, (*imp*) pour, pour over, decant, (med) transfuse; pour too much; (met) re-cast; play (of colours), be iridescent; **-ся** flow, flow across, overflow, overtop (dams etc.)

пере·ли́в·ист·ый, -ая, -ое, (*a*) iridescent

пере·лив·н·о́й, -а́я, -о́е, (*f*) overflow; pouring; recasting

~ **труб·а́** (*f*) overflow pipe

пере·ли́в·чатост·ь, -и, (*f*) (min) chatoyancy

пере·ли́в·чат·ый, -ая, -ое, (*a*) iridescent, (min) chatoyant

пере·лист·а́·ть, -ют, (*perf*) *see* **пере··ли́ст·ыва·ть**

пере·ли́ст·ыва·ть, -ют, (*imp*) turn (pages); thumb through (book)

пере·ли́·т·ый, -ая, -ое, (*past part pass*) *see* **пере·лив·а́·ть**

пере·ли́·ть, (*fut 3rd pl*) **пере·ль·ю́т,** (*perf*) *see* **пере·лив·а́·ть**

пере·лиц·ева́·ть, -у́ют, (*perf*) *see* **пере·лиц·о́выва·ть**

пере·лиц·о́выва·ть, -ют, (*imp*) turn (a garment); present in a new form, give a new look to

пере·ло́в·, -а, (*m*) exceeding the quota (in fishing or trapping); overfishing

пере·лов·и́ть, -ят, (*perf*) catch (all, many, too many)

пере·ло́г·, -а, (*m*) (agr) layland, ley, lea

пере·лож·и́ть, -ат, (*perf*) *see* **пере··кла́д·ыва·ть, пере·лаг·а́·ть**

пере·ло́м·, -а, (*m*) (*v n*), *see* **пере·ла́·м·ыва·ть;** fracture, break; sudden change, turning point, discontinuity, salient point (of a curve)

пере·лом·а́·ть, -ют, (*perf*) break (all, many)

пере·лом·и́ть, -ят, (*perf*) *see* **пере·ла́·м·ыва·ть**

пере·лопа́ч·ивани·е, -я, (*n*) turn (with shovel); move (with shovel); reblading (turbines etc.)

пере·ль·ёт, (*fut 3rd sing*) of **пере·ли́·ть**

пере·магни́ч·ивани·е, -я, (*n*) magnetic reversal; remagnetization

пере·ма́з·ать, (*fut 3rd pl*) **пере·ма́·ж·ут,** (*perf*); grease/smear/wax/coat again; grease, smear, wax, coat (all, many)

пере·ма́л·ыва·ть, -ют, (*imp*) grind, mill (all, much); grind/mill again

пере·ман·и́ть, -ят, (*perf*) entice

пере·ма́т·ыва·ть, -ют, (*imp*) transfer, wind, reel (from–to); wind/reel (all, much); wind/reel again, rewind

пере·ма́ч·ива·ть, -ют, (*imp*) soak too long, over-ret (flax); soak, damp, moisten (all, many)

пере·ма́щ·ива·ть, -ют, (*imp*) repave, resurface

пере·меж·а́емост·ь, -и, (*f*) intermittance; alternation

пере·меж·а́·ть, -ют, (*imp*) alternate; (rad) interleave (signals); **-ся** be intermittent

пере·меж·а́ющ·ийся, -аяся, -ееся, (*pres part act*) of **пере·меж·а́·ться;** alternating; intermittent; interleaved

~ **по·врежд·е́ни·е** (*n*) intermittent fault

~ **сигна́л·** (*m*) (elec, rad) interleaved signal

пере·меж·ева́ть, -у́ют, (*perf*) re-survey

пере·ме́л·ет, (*fut 3rd sing*) of **пере·мо·л·о́ть**

пере·ме́н·а, -ы, (*f*) alteration, change; exchange; break (between lessons)

~ **зна́к·ов** reversal, alternation

пере·мен·и́ть, -ят, (*perf*) *see* **пере··мен·я́·ть**

пере·ме́н·н·ая, -ой, (*f decl as adj*) (math) variable

~ **автомоде́ль·н·ая** similarity variable

~, **за·ви́с·им·ая** dependant variable

~, **за́·да·нн·ая** fixed variable

~, **регул·и́руем·ая** controlled variable

пере·ме́н·ност·ь, -и, (*f*) variability, changeability

пере·ме́н·н·ый, -ая, -ое, (*a*) *see also* **пере·ме́н·н·ая** (*as noun*); variable, alternating

~ **конденса́тор·** (*m*) (elec) variable capacitor

~ **по́л·е** (*n*) (elec) variable field

~ **ток·** (*m*) alternating current, A.C., ac

~ **шаг·** (*m*) variable pitch

пере·ме́н·чив·ый, -ая, -ое, (*a*) changeable

пере·мен·я́·ть, -ют, (*imp*) alter, change; exchange, reverse

пере·мерз·а́·ть, -ют, (*imp*) freeze over; be killed by frost

пере·мёрз·нуть, -нут, (*perf*) *see* **пере··мерз·а́·ть**

пере·мер·ива·ть, -ют, (*imp*) *see* **пере··мер·я́·ть**

пере·мер·ить, -ят, (*perf*) *see* **пере··мер·я́·ть**

пере·мер·я́·ть, -ют, (*imp*) measure again, re-measure; try on, have another fitting; measure (all, many)

пере·ме́с·, -а, (*m*) overmixing

пере·мест·и́тельност·ь, -и, (*f*) (math) commutativity

пере·мест·и́тельн·ый, -ая, -ое, (*a*) commutative

~ **за·ко́н·** (*m*) (math) commutative law

пере·мест·и́ть, -я́т, (*perf*) *see* **пере··мещ·а́·ть**

пере·мёт·, -а, (*m*) (fish) seine net

пере·мет·а́·ть, -ют, (*perf*) *see* **пере·мё·т·ыва·ть**

пере·метил·и́ровани·е, -я, (*n*) (chem) transmethylation

пере·ме́т·ить, -ят, (*perf*) *see* **пере··меч·а́·ть**

пере·мёт·ыва·ть, -ют, (*imp*) throw on again; (text) baste again, baste under again

пере·меч·а́·ть, -ют, (*imp*) mark again; mark (all, many)

пере·меш·а́·ть, -ют, (*perf*) *see* **пере··ме́ш·ива·ть**

пере·меш·анн·ый, -ая, -ое, (*past part pass*) *see* **пере·ме́ш·ива·ть**

пере·ме́ш·ивани·е, -я, (*n*) (*v n*) of **пере·ме́ш·ива·ть;** agitation

~ **то́нк·ое** fine-scale mixing

пере·меш·ива́тел·ь, -я, (*m*) mixer, stirrer, (telecom) scrambler

пере·ме́ш·ива·ть, -ют, (*imp*) mix, mingle; stir, churn; (telecom) scramble

пере·мещ·а́·ть, -ют, (*imp*) move, transfer, displace, shift, commute; **-ся** (*pass*) move, slide, slip, travel

пере·мещ·éни·е, -я, (*n*) (*v n*), *see* **пере·мещ·а́·ть;** travel, migration, transference, transposition, displacement, shift, translation; (geol) dislocation, displacement; (mech) deflection

~ **а́том·а** displacement of an atom

~, **винт·ов·о́е** (phys) screw translation

~, **воз·мо́ж·н·ые** (*pl*) (mech) virtual displacements

пере·мещ·éни·е

~ **гра́н·иц·** boundary migration

~, **ды́р·очн·ое** hole migration (semiconductors)

~ **континéнт·ов** (geol) continental drift

~, **лин·éйн·ое** (phys) linear translation **ско́рост·ь ·пере·мещ·éни·я** (*f*) (phys) translational velocity

~, **угл·ов·о́е** angular rotation

~ **хромо·со́м·ы** (biol) chromosome translocation

~ **цéл·и, угл·ов·о́е** (gunn) crossing speed

пере·ми́р·и·е, -я, (*n*) truce, armistice

пере·множ·а́·ть, -ют, (*imp*) multiply

пере·множ·éни·е, -я, (*n*) multiplication

пере·мно́ж·ить, -ат, (*perf*) *see* **пере··множ·а́·ть**

пере·модул·я́ци·я, -и, (*f*) (rad) overmodulation

пере·мо́·ет, (*fut 3rd sing*) of **пере·мы́·ть**

пере·мок·а́·ть, -ют, (*imp*) get wet through, become soaked/drenched

пере·мо́к·нуть, -нут, (*perf*) *see* **пере··мок·а́·ть**

пере·мо́л·, -а, (*m*) (*v n*), *see* **пере·ма́·л·ыва·ть;** grist, grinding

пере·мола́ч·ива·ть, -ют, (*imp*) thresh, thrash (all, much); thresh/thrash again

пере·молот·и́ть, ·я́т, (*perf*) *see* **пере··мола́ч·ива·ть**

пере·мол·о́ть, (*fut 3rd pl*) **пере·мé·л·ют,** (*perf*) *see* **пере·ма́л·ыва·ть**

пере·моло́ч·енн·ый, -ая, -ое, (*past part pass*) of **пере·молот·и́ть**

пере·монт·а́ж·, -а, (*m*) re-erection, re-mounting, re-installation

пере·монт·и́р·овать, -уют, (*perf*) erect/mount/install/fit again

пере·мора́ж·ива·ть, -ют, (*imp*) freeze (all, much); over-freeze

пере·моро́з·ить, -ят, (*perf*) *see* **пере··мора́ж·ива·ть**

пере·мост·и́ть, -я́т, (*perf*) repave, resurface

пере·мот·а́·ть, -ют, (*perf*) *see* **пере··ма́т·ыва·ть**

пере·мо́т·к·а, -и, (*f*) (*v n*), *see* **пере··ма́т·ыва·ть;** rewinding; (text) winding

пере·моч·и́ть, ·ат, (*perf*) *see* **пере··ма́ч·ива·ть**

пере·мо́ч·к·а, -и, (*f*) over-retting (of flax fibres etc.)

пере·мощ·ённ·ый, -ая, -ое, (*past part pass*) of **пере·мост·и́ть**

пере·мо́·ют, (*fut 3rd pl*) of **пере·мы́·ть,** (*perf*) *see* **пере·мыв·а́·ть**

пере·мыв·а́·ть, -ют, (*imp*) wash again, rewash; wash (all, many)

пере·мык·а́·ть, -ют, (*imp*) bridge over, connect, rejoin; (elec) place a jumper/tie

пере·мы́·ть, (*fut 3rd pl*) **пере·мо́·ют,** (*perf*) *see* **пере·мыв·а́·ть**

пере·мы́ч·к·а, -и, (*g pl*) **-ч·ек·,** (*f*) tie-plate, connecting piece, bridge, neck, commissure; (civ eng) cofferdam, dam; (arch) straight arch, lintel; (elec) bonding jumper, bridle, connector; (meteor) ridge; (min) fire stopping; (ocean) neck of ice; (anat) lacertus

~, вентиляцио́н·н·ая (min) brattice

~, двер·н·а́я lintel

~, ря́ж·ев·ая crib cofferdam

~, с·пуск·а́тельн·ая (min) chute gate

пере·на·грев·а́ни·е, -я, (*n*) superheating

пере·на·жр·а́нн·ый, -ая, -ое, (*a*) overplumped (leather)

пере·на·лад·и·ть, -ят, (*perf*) reset

пере·на·ла́ж·иваем·ый, -ая, -ое, (*pres part pass*) of **пере·на·ла́ж·ива·ть;** resetable

пере·на·ло́ж·енн·ый, -ая, -ое, (*a*) superimposed

пере·на·пряг·а́·ть, -ют, (*imp*) overstrain, over-exert

пере·на·пряж·е́ни·е, -я, (*n*) (*v n*), *see* **пере·на·пряг·а́·ть;** (elec) excess voltage, surge; (electrochem) overvoltage; (mech) overstress

волн·а́ пере·на·пряж·е́ни·я (*f*) (elec) surge

пере·на·пряч·ь, (*fut 3rd sing, pl*) **пере·на·пряж·ёт, пере·на·пряг·у́т,** (*past masc sing*) **пере·на·пряг·,** (*perf*); *see* **пере·на·пряг·а́·ть**

пере·на·сел·е́ни·е, -я, (*n*) overpopulation

пере·на·сел·ённ·ый, -ая, -ое, (*past part pass*) overpopulated, congested

пере·на·сы́т·и·ть, -ят, (*perf*) *see* **пере·на·сыщ·а́·ть**

пере·на·сыщ·а́·ть, -ют, (*imp*) oversaturate, supersaturate

пере·на·сыщ·е́ни·е, -я, (*n*) supersaturation

пере·на·сы́щ·енн·ый, -ая, -ое, (*past part pass*) of **пере·на·сы́т·и·ть;** supersaturated, oversaturated

пере·нес·е́ни·е, -я, (*n*) (*v n*), *see* **пере·нос·и́ть**

пере·нес·ти́, -у́т, (*perf*) *see* **пере·нос·и́ть**

пере·норм·иро́вк·а, -и, (*f*) renormalization

~ за·ря́д·а (nucl) charge renormalization

пере·но́с·, -а, (*m*) (*v n*), *see* **пере·нос·и́ть;** (phys) transport, transfer; migration (of ions); (print) carryover; carry, carry input (computers)

~ вещ·еств·а́ (phys) mass transfer/transport

гру́пп·а пере·но́с·ов (*f*) (cryst) translation group

длин·а́ пере·но́с·а (*f*) (phys) transport length

~, за·де́рж·анн·ый delayed carry (computers)

~, круг·ов·о́й end-around carry (computers)

~ ли́н·и·и (elec) shifting a line, line shift

~, меж·раз·ря́д·н·ый intercolumnar carry (of computer)

~ нейтро́н·ов (phys) neutron transport/transfer

~ огн·я́ (gunn) shift fire

~, паралле́ль·н·ый parallel displacement

~, сквоз·н·о́й ripple-through carry (computers)

тео́р·и·я пере·но́с·а (*f*) (phys) transport theory

у·равн·е́ни·е пере·но́с·а (*n*) transport equation, diffusivity equation

~, цикл·и́ческ·ий end-around carry (computers)

числ·о́ пере·но́с·а (*n*) (chem) transport number; (math) carry number

~, чи́ст·ый (nucl) net transport

~ электро́н·а electron transfer

пере·нос·и́ть, -я́т, (*imp*) carry, take, carry/take across/over; transfer, shift, switch; re-locate; postpone, adjourn; endure, suffer; **-ся** (*pass*); drift

пере·нос·и́м·ый, -ая, -ое, (*pres part pass*) of **пере·нос·и́ть**

пере·но́с·иц·а, -ы, (*f*) bridge (of nose)

пере·но́с·н·ый, -ая, -ое, (*a*) of **пере·но́с·;** portable; translatory

~ движ·е́ни·е (*n*) (phys) translatory movement

пере·но́с·чик·, -а, (*m*) porter; carrier

пере·но́ч·ева́ть, -у́ют, (*perf*) spend the night

пере·нумер·ова́ть, -у́ют, (*perf*) re-number; number (all, many); (math) relabel

~ **координа́т·** relabel coordinates

пере·об·луч·éни·е, -я, (*n*) over-irradiation/exposure

пере·об·оруд·овать, -уют, (*imp and perf*) re-equip

пере·о·дев·а́·ть, -ют, (*imp*) change (clothes)

пере·о·де́·ть, (*fut 3rd pl*) **пере·о·де́·-нут** (*perf*), *see* **пере·о·дев·а́·ть**

пере·о·пре·дел·éни·е, -я, (*n*) redefinition, redefining

пере·ориент·áци·я, -и, (*f*) reorientation; (phys) flipping (spins)

пере·о·сажд·éни·е, -я, (*n*) reprecipitation

пере·от·клон·éни·е, -я, (*n*) (instr) overshoot, overswing

пере·о·хлажд·éни·е, -я, (*n*) super-cooling, undercooling

пере·о·цéн·ива·ть, -ют, (*imp*) re-value, re-estimate, reappraise; over-estimate, overvalue, overrate

пере·о·цен·и́ть, -ят, (*perf*) *see* **пере·-о·цéн·ива·ть**

пере·пáд·, -а, (*m*) fall (water level); (phys) gradient; water drop (irrigation)

~ **давл·éни·я** pressure gradient; sur-plus pressure

~ **потенциáл·а** potential gradient

~, **ступ·éнчат·ый** (hydroelec) stepped fall

~ **температу́р·ы** temperature gradient

~ **энéрг·и·и** (phys) energy gap

пере·пá·ива·ть, -ют, (*imp*) solder/braze again; solder, braze (all, many); seal again (glass objects)

пере·пáр·ива·ть, -ют, (*imp*) steam too long

пере·пáр·ить, -ят, (*perf*) *see* **пере·-пáр·ива·ть**

пере·пах·áть, (*fut 3rd pl*) **пере·пá-ш·ут** (*perf*), *see* **пере·пáх·ива·ть**

пере·пáх·ива·ть, -ют, (*imp*) plough again/over/across

пере·па·я́·ть, -ют, (*perf*) *see* **пере·пá·и-ва·ть**

пере·пек·á·ть, -ют, (*imp*) overbake; bake (all, many)

пéрепел·, -а, (*nom pl*) **-á,** (*g pl*) **-óв,** (*m*) (orn) quail

пере·печáт·а·ть, -ют, (*perf*) reprint; retype

пере·печáт·к·а, -и, (*g pl*) **-т·ок·,** (*f*) (*v n*), *see* **пере·печáт·ыва·ть;** (print) cancel, cancelling sheet, reprint

пере·печáт·ыва·ть, -ют, (*imp*) re-print; retype

пере·пéч·ь, (*fut 3rd sing, pl*) **пере·-печ·ёт, пере·пек·у́т,** (*past masc sing*) **пере·пёк·,** (*perf*); *see* **пере·пе-к·á·ть**

пере·пи́л·ива·ть, -ют, (*imp*) saw in two; saw, saw up (all, many)

пере·пил·и́ть, -ят, (*perf*) *see* **пере·-пи́л·ива·ть**

пере·пис·áть, (*fut 3rd pl*) **пере·пи́-ш·ут** (*perf*), *see* **пере·пи́с·ыва·ть**

пере·пи́с·к·а, -и, (*g pl*) **-с·ок·,** (*f*) rewriting; retyping, copying; corre-spondence

пере·пи́с·чик·, -а, (*m*) copyist; copy typist

пере·пи́с·ыва·ть, -ют, (*imp*) rewrite, transcribe; retype; list; **-ся** (*pass*); (*imp only*) correspond, conduct a cor-respondence, exchange letters

пéре·пис·ь, -и, (*f*) census; inventory

пере·пи́ш·ет, (*fut 3rd sing*) of **пере·-пис·áть**

пере·плáв·ить, -ят, (*perf*) *see* **пере·-плавл·я́·ть**

пере·плáвл·енн·ый, -ая, -ое, (*past part pass*) *see* **пере·плавл·я́·ть**

пере·плáвл·я́·ть, -ют, (*imp*) melt, smelt (all, some); melt again, remelt; (trans) float, raft (timber etc.)

пере·план·и́р·овать, -уют, (*perf*) re-plan

пере·пластик·áци·я, -и, (*f*) (rubber) overmastication

пере·плат·и́ть, -ят, (*perf*) *see* **пере·-плáч·ива·ть**

пере·плáч·ива·ть, -ют, (*imp*) over-pay, pay too much

пере·плес·ти́, (*fut 3rd pl*) **пере·-плет·у́т,** (*past masc sing*) **пере·-плё·л** (*perf*), *see* **пере·плет·á·ть**

пере·плёт·, -а, (*m*) casing, binding (of book); (build) casement, transom

~, **окóн·н·ый** window casing, sash

пере·плет·á·ть, -ют, (*imp*) bind (books); (text etc.) interweave, inter-lace, intertwine, intermesh; cane (a chair)

пере·плет·áющ·ий, -ая, -ее, (*pres part act*) of **пере·плет·á·ть;** (tele-com) interlock

пере·плет·а́юш·ийся, -аяся, -ееся, (*pres part act*) *see* **пере·плет·а́·ть;** (telecom) interlocked

пере·плет·е́ни·е, -я, (*n*) (*v n*), *see* **пере·плет·а́·ть;** (text) weave; (telecom) interlock; (print) bookbinding

~, атла́с·н·ое satin weave

~, вя́з·ан·ое interlooping, knitting

~, митка́л·ёв·ое calico/plain weave

~ пе́тел·ь (text) meshing

~, полотн·я́н·ое plain/calico weave

~, са́рж·ев·ое twill weave

~, тка́ц·к·ое weave

пере·плёт·н·ый, -ая, -ое, (*a*) binding, bookbinding

пере·плыв·а́·ть, -ют, (*imp*) swim/drift/float/sail across

пере·плы́·ть, (*fut 3rd pl*) **пере·плыв·у́т,** (*perf*), *see* **пере·плыв·а́·ть**

пере·под·гота́вл·ива·ть, -ют, (*imp*) give further training, give a refresher/extension course to

пере·под·гото́в·ить, -ят, (*perf*) *see* **пере·под·гота́вл·ива·ть**

пере·подъ·ём·, -а, (*m*) (min) overwind, overwinding

пере·полз·а́ни·е, -я, (*n*) (*v n*), *see* **пере·полз·а́·ть;** (cryst) climb

пере·полз·а́·ть, -ют, (*imp*) crawl, creep, crawl/creep across

пере·полз·ти́, -у́т, (*perf*) *see* **пере·полз·а́·ть**

пере·полн·е́ни·е, -я, (*n*) (*v n*), *see* **пере·полн·я́·ть;** overflow (computers)

~ раз·ря́д·н·ой се́т·к·и overflow (computers)

пере·полн·ить, -ят, (*perf*) *see* **пере·полн·я́·ть**

пере·полн·я́·ть, -ют, (*imp*) overfill, overcrowd

пере·по́люс·овани·е, -я, (*n*) (phys) polarity reversal

пере·по́н·к·а, -и, (*g pl*) **-н·ок·,** (*f*) (*see also* **пере·горо́д·к·а**); membrane, diaphragm, septum, web

пере·пон·чато·кры́л·ый, -ая, -ое, (*a*) (ent) hymenopterous, (*pl as noun*) *Hymenoptera*

пере·по́н·чат·ый, -ая, -ое, (*a*) membranous, webbed

пере·по́рт·ить, -ят, (*perf*) spoil, ruin (all, many)

пере·по·руч·а́·ть, -ют, (*imp*) hand/turn over (duties etc.)

пере·поя́с·ать, (*fut 3rd pl*) **пере·поя́ш·ут,** (*perf*) girdle, tie round, ring, strap; strap up again, put on a belt again

пере·пра́в·а, -ы, (*f*) crossing, passage; ford; ferry

~, лед·ян·а́я railway track laid on ice

пере·пра́в·ить, -ят, (*perf*) *see* **пере·правл·я́·ть**

пере·правл·я́·ть, -ют, (*imp*) take across/through; consign, send, forward; (telecom) relay; correct; **-ся** (*pass*); cross

пере·прев·а́·ть, -ют, (*imp*) rot; overstew

пере·пре́·л·ый, -ая, -ое, (*a*) rotten

пере·пре́·ть, -ют, (*perf*) *see* **пере·прев·а́·ть**

пере·при·ём·, -а, (*m*) (telecom) re-transmission

пере·при·ём·ник·, -а, (*m*) (elec) transducer

пере·про́б·овать, -уют, (*perf*) try, taste (all, many)

пере·про·ве́р·ить, -ят, (*perf*) recheck

пере·про·да·ва́·ть, -ют, (*imp*) resell

пере·про·да́·ть, (*fut 3rd sing, pl*) **пере·про·да́·ст, пере·про·дад·у́т,** (*past masc sing*) **пере·про́·да·л** (*perf*), *see* **пере·про·да·ва́·ть**

пере·про·из·во́д·ств·о, -а, (*n*) overproduction

пере·про·явл·я́·ть -ют, (*imp*) overdo, overdevelop

пере·пры́г·ива·ть, -ют, (*imp*) jump, leap, spring, jump/leap/spring over/across

пере·пры́г·н·уть, -ут, (*perf*) *see* **пере·пры́г·ива·ть**

пере·пряг·а́·ть, -ют, (*imp*) reharness; change (horses etc.)

пере·пря́ч·ь, (*fut 3rd sing, pl*) **пере·пряж·ёт, пере·пряг·у́т,** (*past masc sing*) **пере·пря́г·** (*perf*); *see* **пере·пряг·а́·ть**

пере·пуг·а́·ть, -ют, (*perf*) frighten

пере·пу́ск·, -а, (*m*) (*v n*) of **пере·пуск·а́·ть;** admission

пере·пуск·а́ни·е, -я, (*n*) (*v n*) of **пере·пуск·а́·ть;** underrunning (of submarine cable)

пере·пуск·а́·ть, -ют, (*imp*) decant, pour out/into; let out; make way for

пере·пуск·н·о́й, -а́я, -о́е, (*a*) of **пере·пу́ск·**

~ ве́нтил·ь by-pass valve (turbine); (hydr) unloading valve

~ лот·о́к· (*m*) overflow chute

пере·пуст·и́ть, -ят, (*perf*) *see* **пере·-пуск·а́·ть**

пере·пу́т·а·ть, -ют, (*perf*) *see* **пере·-пу́т·ыва·ть**

пере·пу́т·ыва·ть, -ют, (*imp*) confuse, muddle, entangle

пере·пу́т·ь·е, -я, (*n*) crossroads

пере·раба́т·ыва·ть, -ют, (*imp*) re-work, remake; process; (oil) refine; convert, transform; (*intrans*) work overtime; overwork

пере·рабо́т·а·ть, -ют, (*perf*) *see* **пере·рабá́т·ыва·ть**

пере·рабо́т·к·а, -и, (*g pl*) **-т·ок·,** (*f*) (*v n*), *see* **пере·рабá́т·ыва·ть;** con-version; (oil) refining; overtime

~, деструкти́в·н·ая (oil) cracking

пере·рас·пре·дел·е́ни·е, -я, (*n*) re-distribution, rearrangement

пере·рас·пре·дел·и́ть, -я́т, (*perf*) *see* **пере·рас·пре·дел·я́·ть**

пере·рас·пре·дел·я́·ть, -ют, (*imp*) redistribute

пере·раст·а́·ть, -ют, (*imp*) outgrow, grow taller than; pass the age limit, grow too old for; grow/develop into

пере·раст·и́, -у́т, (*past masc sing*) **пе-ре·ро́с·,** (*perf*); *see* **пере·раст·а́·ть**

пере·рас·хо́д·, -а, (*m*) overexpendi-ture

пере·рас·хо́д·овать, -уют, (*imp and perf*) overexpend

пере·рас·чёт·, -а, (*m*) re-calculation

пере·рас·шир·е́ни·е, -я, (*n*) over-expansion

пере·рв·а́ть, -у́т, (*perf*) tear, tear up/apart, tear in two

пере·регистр·и́р·овать, -уют, (*imp and perf*) re-register; register (all, many)

пере·регул·и́рования·е, -я, (*n*) re-adjustment; over-adjustment; (autom) overshoot, overshooting, overswing

пере·ре́з·ать, (*fut 3rd pl*) **пере·ре́·ж·ут,** (*perf*) *see* **пере·рез·а́·ть**

пере·рез·а́·ть, -ют, (*imp*) cut, shear, cut up/apart, cut in two; recut, rethread (a screw); block, bar, cut off (access etc.); overcut (a sound record-ing)

пере·ре́з·ыва·ть, -ют, (*imp*) *see* **пе-ре·рез·а́·ть** (*imp*)

пере·ре́з·ывающ·ая си́л·а (*f*) shear-ing force

пере·ректифик·а́ци·я, -и, (*f*) (chem) re-rectification

пере·реш·а́·ть, -ют, (*imp*) reconsider (a decision), change one's mind; solve a lot (of problems)

пере·реш·и́ть, -а́т, (*perf*) *see* **пере·-реш·а́·ть**

пере·рис·ова́ть, -у́ют, (*perf*) *see* **пе-ре·рис·о́выва·ть**

пере·рис·о́выва·ть, -ют, (*imp*) draw (all, many); redraw, draw again; copy (by drawing)

пере·род·и́ть, -я́т, (*perf*) *see* **пере·-рожд·а́·ть**

пере·ро́·ет, (*fut 3rd sing*) of **пере·ры́·ть**

пере·рожд·а́·ть, -ют, (*imp*) transform; **-ся** (*pass*); degenerate

пере·рожд·е́ни·е, -я, (*n*) (*v n*), *see* **пере·рожд·а́·ть;** degeneration

~, жир·ов·о́е fatty degeneration

~, мы́ш·ечн·ое myodegeneration

пере·рос·, (*past masc sing*) of **пере·-раст·и́** (*perf*), *see* **пере·раст·а́·ть** (*imp*)

пере·ру́б·, -а, (*m*) (*v n*), *see* **пере·ру-б·а́·ть;** superfelling (forestry)

пере·руб·а́·ть, -ют, (*imp*) chop/hack up, chop/hack in two

пере·руб·и́ть, -я́т, (*perf*) *see* **пере·-руб·а́·ть**

пере·ры́в·, -а, (*m*) (*v n*), *see* **пере·-рыв·а́·ть;** break, intermission, pause, recess, discontinuity; interruption

пере·рыв·а́·ть, -ют, (*imp*) *see* **пере·-ры́·ть, пере·рв·а́ть**

пере·ры́·ть, (*fut 3rd pl*) **пере·ро́·ют,** (*perf*), dig, dig up/across

пере·сад·и́ть, -я́т, (*perf*) *see* **пере·-са́ж·ива·ть**

пере·са́д·к·а, -и, (*g pl*) **-д·ок·,** (*f*) (*v n*), *see* **пере·са́ж·ива·ть;** change (of conveyance); (fish) planting

пере·сад·о́чн·ый, -ая, -ое, (*a*) of **пере·са́д·к·а**

~ лопа́т·к·а (*f*) (hortic) planting trowel, trowel

~ ста́нци·я (*f*) (rail) connecting station

пере·са́ж·енн·ый, -ая, -ое, (*past part pass*) *see* **пере·са́ж·ива·ть**

пере·са́ж·ива·ть, -ют, (*imp*) reseat, resite, put, move (elsewhere); trans-plant, replant; graft; **-ся** (*pass*); change (conveyance), change places

пере·с·да·ва́·ть, -ю́т, (*imp*) relet; retake (exams); deal again (at cards)

пере·с·да́·ть, (*fut 3rd sing, pl*) **пере·с·-да́·ст, пере·с·дад·у́т,** (*perf*); *see* **пе-ре·с·да·ва́·ть**

пере·сев·а́·ть, -ют, (*imp*) sow again; sieve/screen/sift again

пере·се́·вш·ий, -ая, -ее, (*past part act*) of **пере·се́ст·ь**

пере·седл·а́·ть, -ют, (*perf*) *see* **пере·сёдл·ыва·ть**

пере·сёдл·ыва·ть, -ют, (*imp*) re-saddle

пере·се́·ива·ть, -ют, (*imp*) = **пере·сев·а́·ть**

пере·сек·а́·ть, -ют, (*imp*) cross, cross over, intersect, cut; re-cut (files etc.), re-nick

пере·сек·а́ющ·ий, -ая, -ее, (*pres part act*) of **пере·сек·а́·ть**; crossing; recutting

пере·сек·а́ющ·ийся, -аяся, -ееся, (*pres part pass*) *see* **пере·сек·а́·ть**; (*pl*) intersecting

~ ли́н·и·и intersecting lines

пере·се́л, (*past masc sing*) of **пере·се́ст·ь**

пере·сел·е́нец·, -нц·а, (*m*) migrant, immigrant, emigrant

пере·сел·е́ни·е, -я, (*n*) (*v n*), *see* **пере·сел·я́·ть**; moving; migration; resettlement

пере·сел·и́ть, -я́т, (*perf*) *see* **пере·сел·я́·ть**

пере·сел·я́·ть, -ют, (*imp*) move (residence), resettle; **-ся** (*pass*); migrate

пере·се́ст·ь, (*fut 3rd pl*) **пере·ся́д·ут,** (*perf*) change places, move (sitting place); change (conveyance)

пере·сеч·е́ни·е, -я, (*n*) (*v n*), *see* **пере·сек·а́·ть**; intersection, crossing; crossover (vacuum tubes); traverse

~, глух·о́е (rail) frog

~ мно́ж·еств· (math) intersection/product/meet of sets

~ по·то́к·а (nucl) flux traverse

~ пу́ч·к·а (nucl) beam traverse

то́ч·к·а пере·сеч·е́ни·я (*f*) crossover/crossing/intersection point; (math) point of intersection; junction point (computers)

пере·сеч·ённ·ый, -ая, -ое, (*past part pass*) *see* **пере·сек·а́·ть**; (geog) broken

~ ме́ст·ност·ь (*f*) rugged terrain/country

пере·се́ч·ь, (*fut 3rd sing, pl*) **пере·се·ч·ёт, пере·сек·у́т,** (*past masc sing*) **пере·се́к·,** (*perf*); *see* **пере·сек·а́·ть**

пере·се́·ять, -ют, (*perf*) *see* **пере·сев·а́·ть**

пере·с·жа́т·и·е, -я, (*n*) supercompression

пере·с·каз·а́ть, (*fut 3rd pl*) **пере·с·ка́ж·ут,** (*perf*) retell, repeat (a story etc.)

пере·ска́к·ива·ть, -ют, (*imp*) jump, jump over/across

пере́ски·я, кол·ю́ч·ая (*f*) (bot) *Pereskia aculeata*

пере·ско́к·, -а, (*m*) (*v n*), *see* **пере·ска́к·ива·ть**; (phys) skip, skip effect

пере·скоч·и́ть, ⸗ат, (*perf*) *see* **пере·ска́к·ива·ть**

пере·сла́·ивани·е, -я, (*n*) (*v n*) of **пере·сла́·ива·ть**; interstratification, intercalation

пере·сла́·ива·ть, -ют, (*imp*) put in layers, put in alternate layers; **-ся** alternate in layers, interstratify, intercalate

пере·сла́·ивающ·ийся, -аяся, -ееся, (*pres part act*) of **пере·сла́·ива·ться**; interstratified, interbanded, interleaved; (zool) intercalary

пере·сл·а́ть, (*fut 3rd pl*) **пере·шл·ю́т,** (*perf*) *see* **пере·сыл·а́·ть**

пере·сло·и́ть, -я́т, (*perf*) *see* **пере·сла́·ива·ть**

пере·сма́тр·ива·ть, -ют, (*imp*) revise, review, look over/through

пере·сме́ш·ник·, -а, (*m*) (orn) mockingbird, *Mimus polyglottus*

пере·смо́тр·, -а, (*m*) (*v n*), *see* **пере·сма́тр·ива·ть**; revision, review

пере·смо́тр·енн·ый, -ая, -ое, (*past part pass*) *see* **пере·сма́тр·ива·ть**

пере·смотр·е́ть, ⸗ят, (*perf*) *see* **пере·сма́тр·ива·ть**

пере·с·ним·а́·ть, -ют, (*imp*) rephotograph, retake; copy (a photograph etc.)

пере·с·ня́·ть, (*fut 3rd pl*) **пере·с·ни́м·ут,** (*perf*) *see* **пере·с·ним·а́·ть**

пере·со·един·и́ть, -я́т, (*perf*) *see* **пере·со·един·я́·ть**

пере·со·един·я́·ть, -ют, (*imp*) change round a connection; reconnect

пере·со·зда·ва́ть, -ю́т, (*perf*) recreate

пере·со·зда́·ть, (*fut 3rd sing, pl*) **пере·со·зда́·ст, пере·со·здад·у́т,** (*perf*) *see* **пере·со·зда·ва́ть**

пере·сорт·ир·ова́ть, -у́ют, (*perf*) resort

пере·со́х·нуть, -нут, (*perf*) dry up

пере·со́х·ш·ий, -ая, -ее, (*past part act*) of **пере·со́х·нуть**; desiccated, dried-up

пере·спев·а́·ть, -ют, (*imp*) overripen, get too ripe

пере·спе́·л·ый, -ая, -ое, overripe

пере·спе́·ть, -ют, (*perf*) *see* **пере·-спев·а́·ть**

пере·с·пра́ш·ива·ть, -ют, (*imp*) ask again; ask to repeat; ask (all, many)

пере·с·прос·и́ть, ⸰ят, (*perf*) *see* **пере·с·пра́ш·ива·ть**

пере·ста·ва́·ть, -ют, (*imp*) stop, cease

пере·ста́в·ить, -ят, (*perf*) *see* **пере·-ставл·я́·ть**

пере·ста́вл·енн·ый, -ая, -ое, (*past part pass*) *see* **пере·ставл·я́·ть**

пере·ставл·я́·ть, -ют, (*imp*) move, shift, transfer; (math) permute, commute; transpose, rearrange

пере·став·н·о́й, -а́я, -о́е, (*a*) adjustable

пере·ста́·вш·ий, -ая, -ее, (*past part act*) of **пере·ста́·ть,** (*perf*); *see* **пере·ста·ва́ть** (*imp*)

пере·ста́·ива·ть, -ют, (*imp*) stand too long, spoil; wait for the end of, hold out for

пере·стана́вл·ива·ть, (*imp*) *see* **пере·ста́в·ить**

пере·ста́·нет, (*fut 3rd sing*) of **пере·-ста́·ть**

пере·станов·и́ть, ⸰ят, (*perf*) = **пере·ста́в·ить**

пере·стано́в·к·а, -и, (*g pl*) -вок·, (*f*) transposition, rearrangement; (math) permutation, commutation; (autom) adjusting mechanism

~ в·хо́д·ов (autom) input rearrangement

~ и́ндекс·ов (math) permutation of subscripts

~, круг·ов·а́я cyclic permutation

~, обра́т·н·ая (autom) floating control mechanism

~, тожде́ств·енн·ая (math) identity permutation

~, цикл·и́ческ·ая cyclic permutation/interchange

пере·стано́в·оч·н·ый, -ая, -ое, (*a*) of **пере·стано́в·к·а**

~ со·от·нош·е́ни·е (*n*) commutation relation

пере·ста́·ть, -нут, (*perf*) *see* **пере·-ста·ва́ть**

пере·ствол·е́ни·е, -я, (*n*) relining (gun barrel)

пере·сте́л·ет, (*fut 3rd sing*) of **пере·-стл·а́ть**

пере·стил·а́·ть, -ют, (*imp*) relay, re-floor, replank, repave; make (a bed)

пере·стл·а́ть, (*fut 3rd pl*) **пере·сте́-л·ют,** (*perf*) *see* **пере·стил·а́·ть**

пере·сто́·й, -я, (*m*) overripeness

пере·сто́й·н·ый, -ая, -ое, (*a*) over-mature, overripe, old, kept too long

пере·сто·я́вш·ийся, -аяся, -ееся, (*past part act*) *see* **пере·ста́·ива·ть;** overproof (alcohol); overmature, overripe

пере·сто·я́ть, -я́т, (*perf*) *see* **пере·-ста́·ива·ть**

пере·стра́·ива·ть, -ют, (*imp*) rebuild, reconstruct; reorganize, reform, rearrange; (rad) retune, switch over; replot (a curve); (instr) reset

пере·страх·ова́ни·е, -я, (*n*) reinsurance

пере·стро·е́ни·е, -я, (*n*) (*v n*) *see* **пере·стра́·ива·ть;** (mil) reforming

пере·стро́·енн·ый, -ая, -ое, (*past part pass*) *see* **пере·стра́·ива·ть**

пере·стро́·ить, -ят, (*perf*) *see* **пере·-стра́·ива·ть**

пере·стро́й·к·а, -и, (*g pl*) -о́ек·, (*f*) (*v n*), *see* **пере·стра́·ива·ть;** reconstruction, reorganization, rearrangement

пере·ступ·а́·ть, -ют, (*imp*) step across/over; move, walk; exceed, overstep

пере·ступ·е́н·ь, -я, (*m*) (bot) bryony, *Brionia*

пере·ступ·и́ть, ⸰ят, (*perf*) *see* **пере·-ступ·а́·ть**

пере·сульф·и́рован·и·е, -я, (*n*) (biochem) sulphur transfer

пере·су́ш·ива·ть, -ют, (*imp*) dry again; overdry, dry too much; dry (all, many)

пере·суш·и́ть, ⸰ат, (*perf*) *see* **пере·-су́ш·ива·ть**

пере·с·чёт·, -а, (*m*) (*v n*), *see* **пере·-с·чи́т·ыва·ть;** (autom) translation; (math) scaling; recalculation

~, дво·и́чн·ый binary scaling

~, десят·и́чн·ый decade scaling

коэффицие́нт· пере·с·чёт·а (*m*) (math) conversion factor; (instr) scaling factor

~ на два scale of two

~ но́мер·а register, translator (telephones)

пере·с·чёт·к·а, -и, (*g pl*) -т·ок·, (*f*) scaling circuit/unit

пере·с·чёт·н·ый, -ая, -ое, (*a*) of **пере·с·чёт·**

~ при·бо́р· (*m*) (elec) scaler, counter

пере·с·чёт·н·ый, -ая, -ое
~ **систéм·а** (*f*) (instr, phys) scaler
~ **схéм·а** (*f*) (elec) scaler, scaling circuit
~ **схéм·а, бинáр·н·ая** binary scaler
~ **схéм·а, кольц·ев·áя** ring scaler
~ **схéм·а на со·в·пад·éни·ях** coincidence scaler
~ **схéм·а, универсáль·н·ая** multiscaler
пере·с·чёт·чик·, -а, (*m*) (elec) scaler; register, translator (telephones)
пере·с·чит·á·ть, -ют, (*perf*) *see* **пере·с·чи́т·ыва·ть**
пере·с·чи́т·анн·ый, -ая, -ое, (*past part pass*) *see* **пере·с·чи́т·ыва·ть**
пере·с·чи́т·ыва·ть, -ют, (*imp*) count out/over; count again; scale, convert; recalculate
пере·съ·ём·к·а, -и, (*g pl*) **-м·ок·,** (*f*) (*v n*), *see* **пере·с·ним·á·ть;** (cinema) retake
пере·сыл·á·ть, -ют, (*imp*) send; forward
пере·сы́п·, -а, (*m*) (*v n*) of **пере·сы-п·á·ть**
пере·сы́п·ать, -лют, (*perf*) *see* **пере·-сып·á·ть**
пере·сып·á·ть, -ют, (*imp*) pour (of solids); sprinkle with (in between), intersperse
пéре·сып·ь, -и, (*f*) (geog) bar, barrier beach
пере·сы́т·ить, -ят, (*perf*) *see* **пере·-сыщ·á·ть**
пере·сых·á·ть, -ют, (*imp*) dry up
пере·сыщ·á·ть, -ют, (*imp*) supersaturate, oversaturate
пере·сы́щ·енн·ый, -ая, -ое, (*past part pass*) *see* **пере·сыщ·á·ть**
пере·сы́щ·ени·е, -я, (*n*) supersaturation
пере·ся́д·ет, (*fut 3rd pl*) of **пере·-сéст·ь**
пере·тáпл·ива·ть, -ют, (*imp*) stoke (all, many), heat with (all, many); stoke/heat again; melt, render (fat)
пере·тáск·ива·ть, -ют, (*imp*) drag/pull across/over; drag
пере·тас·овá·ть, -у́ют, (*perf*) shuffle; reshuffle
пере·тáч·ива·ть, -ют, (*imp*) resharpen, regrind (tools), dress (tools); sharpen (all, many)
пере·тащ·и́ть, -ат, (*perf*) *see* **пере·-тáск·ива·ть**
пере·тек·áни·е, -я, (*n*) overflowing
пере·тек·á·ть, -ют, (*imp*) overflow

пере·тер·éть, (*fut 3rd pl*) **пере·тр·у́т,** (*perf*) *see* **пере·тир·á·ть**
пере·тéч·ь, (*fut 3rd sing, pl*) **пере·-теч·ёт, пере·тек·у́т,** (*past masc sing*) **пере·тёк·,** (*perf*); *see* **пере·те-к·á·ть**
пере·тир·áни·е, -я, (*n*) (*v n*), of **пе-ре·тир·á·ть**
пере·тир·á·ть, -ют, (*imp*) wear to pieces, wear out; grate; wipe (all, many)
пере·ти́ш·ь, -и, (*m*) temporary calm
пере·тóк·, -а, (*m*) overflow; (elec) overcurrent
~, обрáт·н·ый reflux
пере·топ·и́ть, -ят, (*perf*) *see* **пере·-тáпл·ива·ть**
пере·тóп·к·а, -и, (*g pl*) **-п·ок·,** (*f*) (*v n*), *see* **пере·тáпл·ива·ть**
пере·тóп·очн·ый, -ая, -ое, (*a*) of **пере·тóп·к·а**
~ котёл· (*m*) fat-rendering vessel
пере·тóч·енн·ый, -ая, -ое, (*past part pass*) *see* **пере·тáч·ива·ть**
пере·тóч·ечн·ый, -ая, -ое, (*a*) of **пере·тóч·к·а**
пере·точ·и́ть, -ат, (*perf*) *see* **пере·-тáч·ива·ть**
пере·тóч·к·а, -и, (*g pl*) **-ч·ек·,** (*f*) (*v n*), *see* **пере·тáч·ива·ть**
~ вáл·к·а (met) roll dressing/regrinding
пере·тóч·н·ый, -ая, -ое, (*a*) of **пере·тóк·**
пере·тр·ёт, (*fut 3rd sing*) of **пере·те-р·éть**
пере·тряс·á·ть, -ют, (*imp*) shake out
пере·тряс·ти́, -у́т, (*perf*) *see* **пере·-тряс·á·ть**
пер·éть, (*fut 3rd pl*) **пр·ут,** (*perf*) press on, put pressure on
пере·тя́г·ива·ть, -ют, (*imp*) pull, draw, haul; struggle over/across; win over, enlist; tie/bind tightly, tighten; tighten/tauten again; pull too tight, overtighten; outbalance, overbalance
пере·тяжел·ённ·ый, -ая, -ое, (*past part pass*); overweighted
~ на корм·у́ stern-heavy
~ на нос· (shipb) bow-heavy, (a/c) nose-heavy
~ на хвост· (a/c) tail-heavy
пере·тя́ж·к·а, -и, (*g pl*) **-ж·ек·,** (*f*) (*v n*), *see* **пере·тя́г·ива·ть;** necking off (screws); overwind (of springs); redrawing (rolls); neck, sausage-like instability (in a plasma column)

пере·тя́·нут·ый, -ая, -ое, (*past part pass*) *see* **пере·тя́г·ива·ть;** constricted

пере·тя·н·у́ть, ́ут, (*perf*) *see* **пере·тя́г·ива·ть**

пере·у·влаж·нéни·е, -я, (*n*) (text) overconditioning

пере·угле·ро́ж·ивани·е, -я, (*n*) overcarburization

пере·у́л·ок·, -л·к·а, (*m*) lane, alley

пере·у·плот·нéни·е, -я, (*n*) overcrowding

пере·у·с·поко·éни·е, -я, (*n*) overdamping

пере·у·станóв·к·а, -и, (*g pl*) **-в·ок·,** (*f*) resetting

пере·у·стрá·ива·ть, -ют, (*imp*) rearrange, reorganize

пере·у·стрó·ить, -ят, (*perf*) *see* **пере·у·стрá·ива·ть**

пере·у·стрóй·ств·о, -а, (*n*) rearrangement, reorganization; modification

пере·у·ступ·á·ть, -ют, (*imp*) cede, give up

пере·у·ступ·и́ть, ́ят, (*perf*) *see* **пере·у·ступ·á·ть**

пере·у·том·и́ть, -я́т, (*perf*) *see* **пере·у·томл·я́·ть**

пере·у·томл·éни·е, -я, (*n*) (*v n*), *see* **пере·у·томл·я́·ть;** exhaustion (physical)

пере·у·томл·я́·ть, -ют, (*imp*) overwork, overstrain

пере·у·чéст·ь, (*fut 3rd pl*) **пере·у·ч·т·у́т,** (*past masc sing*) **пере·у·чёл,** (*perf*); *see* **пере·у·чи́т·ыва·ть**

пере·у·чёт·, -а, (*m*) stocktaking; (fin) discount

пере·у́ч·ивани·е, -я, (*n*) (*v n*), *see* **пере·у́ч·ива·ть;** conversion training

пере·у́ч·ива·ть, -ют, (*imp*) learn/ teach again; **-ся** learn again

пере·у·чи́т·ыва·ть, -ют, (*imp*) take stock of; (fin) discount

пере·уч·и́ть, ́ат, (*perf*) *see* **пере·у́ч·ива·ть**

пере·у·чт·ёт, (*fut 3rd sing*) of **пере·у·чéст·ь**

пере·форм·ир·овá·ть, -у́ют, (*perf*) reform; (rail) re-marshal

пере·фраз·и́р·овать, -уют, (*imp and perf*) paraphrase

пере·хвá́т·, -а, (*m*) interception; waist; (met cast) secondary pipe; capture, piracy, beheading (rivers)

пере·хват·и́ть, ́ят, (*perf*) *see* **пере·хвá́т·ыва·ть**

пере·хвá́т·чик·, -а, (*m*) interceptor

пере·хвá́т·ывани·е, -я, (*n*) (*v n*), *see* **пере·хвá́т·ыва·ть;** interception

пере·хвá́т·ыва·ть, -ют, (*imp*) intercept; behead, capture (rivers)

пере·хвá́ч·енн·ый, -ая, -ое, (*past part pass*) of **пере·хват·и́ть**

пере·хитр·и́ть, -я́т, (*perf*) outwit

пере·хлёст·ывани·е, -я, (*n*) overlapping

пере·хóд·, -а, (*m*) (*v n*), *see* **пере·ход·и́ть;** passage, crossing; transition, transfer, transformation; junction (of semiconductors); (mil) march; reducer (pipe joint)

~ **без из·мен·éни·я чёт·ност·и** (phys) "no" transition

~ **в генер·áци·ю** (rad) spillover

~ **в основ·н·óе со·сто·я́ни·е** (phys) ground-state transition

~, **валéнт·н·ый** (chem) valency transfer

~, **в·плáвл·енн·ый** alloy junction

~, **вы́·ращ·енн·ый** grown junction

~, **дипóль·н·о раз·реш·ённ·ый** (phys) allowed dipole transition

~ **за пре·дéл·** overrun, overshoot, overstroke

~, **каскáд·н·ый** (phys) cascade/successive/stopover transition

~, **квáнт·ов·ый** (phys) quantum transition

~, **колеб·áтельн·ый** (phys) vibrational transition

~, **прям·óй** crossover transition

~, **с·ращ·ённ·ый** grown junction

температу́р·а пере·хóд·а (*f*) (phys) transformation point

тóч·к·а пере·хóд·а (*f*) transition point/temperature; entry point (computers)

~, **у·слóв·н·ый** (autom) discrimination, conditional transfer of control

~, **фáз·ов·ый** (phys) phase transition/ transformation

~, **электронно·ды́р·очн·ый** *n-p* junction (semiconductors)

пере·ход·и́ть, ́ят, (*imp*) cross, go across; pass, pass on/to/into, spread to; go over to, change to

пере·хóд·ник·, -а, (*m*) adapter

пере·хóд·н·ый, -ая, -ое, (*a*) transition(al), passing, transient, transitory; connecting, communicating; (mech) adapter, reducer; (gram) transitive

врéм·я пере·хóд·н·ого процéсс·а (*n*) (autom) duration of transient

пере·хо́д·н·ый, -ая, -ое
~ **движ·е́ни·е** (*n*) transient motion
~ **комплéкт·** (*m*) (telecom) hybrid set
~ **коэффициéнт·** (*m*) (math) conversion factor
~ **мýфт·а** (*f*) reducer (pipe joint); (elec) adapter coupling/joint
~ **периóд·** (*m*) (autom) transfer lag
~ **по·мéх·а** (*f*) (telecom) crosstalk
~ **процéсс·** (*m*) (phys) transient process, transient
~ **режи́м·** (*m*) transient behaviour/conditions
~ **с·вя́з·ь** (*f*) (telecom) cross coupling
~ **сло́·й** (*m*) transition layer
~ **у·стро́й·ств·о** (*n*) (rad) matching device (of antennas); radio-to-line transfer unit
пéре·ход·я́щ·ий, -ая, -ее, (*pres part act*) see **пере·ход·и́ть;** transitory, transient; prize (in periodic competition); overheat (of distillates)
~ **о·ста́т·ок·** (*m*) (fin) carry-over
пéрец·, -рц·а, (*i*) **-рц·ем** (*m*) pepper
~**, вод·ян·о́й** (bot, pharm) water pepper, *Polygonum hydropiper*
~**, испа́н·ск·ий** common garden pepper, capsicum, *Capsicum annuum*
~**, чёрн·ый** (bot) pepper, black pepper, *Piper nigrum*
пере·чёл, (*past masc sing*) of **пере·чéст·ь**
пéречен·ь, -чн·я, (*m*) list, schedule; enumeration; seam (planking)
пере·чёрк·ивать, -ют, (*imp*) cross out; cancel
пере·черк·н·у́ть, -у́т, (*perf*) see **пере·чёрк·ива·ть**
пере·черт·и́ть, -́ят, (*perf*) see **пере·чéрч·ива·ть**
пере·чéрч·ива·ть, -ют, (*imp*) redraft, sketch/plot again; copy, take/make a copy
пере·чéст·ь, (*fut 3rd pl*) **пере·чт·у́т,** (*past masc sing*) **пере·чёл,** (*perf*); see **пере·чи́т·ыва·ть**
пере·чи́н·ива·ть, -ют, (*imp*) mend, repair (all, many); resharpen, repoint
пере·чин·и́ть, -́ят, (*perf*) see **пере·чи́н·ива·ть**
пере·чи́сл·ить, -ят, (*perf*) see **пере·числ·я́·ть**
пере·чи́сл·енн·ый, -ая, -ое, (*past part pass*) see **пере·числ·я́·ть**
пере·числ·я́·ть, -ют, (*imp*) enumerate; (fin) transfer

пере·чит·а́·ть, -ют, (*perf*) see **пере·чи́т·ыва·ть**
пере·чи́т·ыва·ть, -ют, (*imp*) read (all, much); re-read
пéреч·н·ые, -ых, (*pl decl as adj*) (bot) pepper family, *Piperaceae*
пере·чт·ёт, (*fut 3rd sing*) of **пере·чéст·ь**
пере·шé·ек·, -éй·к·а, (*m*) isthmus, neck of land
~ **аóрт·ы** (anat) aortic isthmus
пере·шёл, (*past masc sing*) of **пере·йти́** (*perf*) see **пере·ход·и́ть** (*imp*)
пере·шив·а́·ть, -ют, (*imp*) sew again, alter; re-cover, re-face
пере·ши́·ть, (*fut 3rd pl*) **пере·шь·ю́т,** (*past masc sing*) **пере·ши·л,** (*perf*); see **пере·шив·а́·ть**
пере·шл·ёт, (*fut 3rd sing*) of **пере·сл·а́ть**
пере·шлиф·ова́ть, -у́ют, (*perf*) see **пере·шлиф·о́выва·ть**
пере·шлиф·о́выва·ть, -ют, (*imp*) regrind; reseat (valves)
пере·шь·ёт, (*fut 3rd sing*) of **пере·ши́·ть**
пере·этерифик·а́ци·я, -и, (*f*) (chem) transesterification
перея́р·ок·, -р·к·а, (*m*) (agr) yearling
пери·а́нци·й, -я, (*m*) (bot) perianthium, perianth
пери·артери́т·, -а, (*m*) (med) periarteritis
пери·а́стр·, -а, (*m*) (astron) periastron
пери·блéм·а, -ы, (*f*) (bot) periblem
пери·гé·й, -я, (*m*) (astron) perigee
пери·гéл·и·й, -я, (*m*) (astron) perihelion
пери·ги́н·и·й, -я, (*m*) (bot) perigynium
пери·ги́н·ов·ый, -ая, -ое, (*a*) (bot) perigynous
пери·дéрм·а, -ы, (*f*) (bot) periderm, periderma
пери́ди·й, -я, (*m*) (bot) peridium
пери·динé·и, -й, (*f pl*) (zool) peridinians, *Peridineae*
перидóт·, -а, (*m*) (min) peridot
перидоти́т·, -а, (*m*) (geol) peridotite
пери·ка́мби·й, -я, (*m*) (bot) pericycle
пери·ка́рд·, -а, (*m*) (anat) pericardium
пери·ка́рд·и·й, -я, (*m*) (anat) pericardium
пери·кард·и́т·, -а, (*m*) (med) pericarditis
пери·ка́рп·и·й, -я, (*m*) (bot) pericarpium, pericarp
пери·кла́з·, -а, (*m*) (min) periclase

пери·клн·а, -ы, (*f*) (bot) periclinal wall

пери·клина́л·ь, -я, (*m*) (geol) periclinal ending

пери·клина́ль·н·ый, -ая, -ое, (*a*) periclinal

пери·клн·ов·ый, -ая, -ое, (*a*) pericline

~ **за·ко́н·** (*m*) (cryst) pericline law

пер·и́л·а, -ы, (*f*) rail, handrail

пери́лл·а, -ы, (*f*) (bot) perilla

пери́лл·ов·ый, -ая, -ое, (*a*) perilla, perillic

~ **ма́сл·о** (*n*) perilla oil

пери·медуля́р·н·ый, -ая, -ое, (*a*) (bot) perimedullary

пери́·метр·, -а, (*m*) perimeter

~, **с·мо́ч·енн·ый** (phys) wetted perimeter

периметри́т·, -а, (*m*) (med) perimetritis

пери·метр·и́·я, -и, (*f*) perimetry

пери́од·, -а, (*m*) period, cycle

~, **больш·о́й** long/slow period

~ **жи́зн·и** life cycle

~ **за·де́рж·к·и** delay, time delay

~ **за·па́зд·ывани·я** lag

~ **идент·и́чност·и** (cryst) identity period, lattice parameter

~ **и́мпульс·ов** pulse interval/period

~ **инду́кци·и** (chem) induction period

~, **инкубацио́н·н·ый** incubation period; (met) cooling period

~ **колеб·а́ни·я** (phys) period of oscillation

~, **коро́т·к·ий** short/fast period

~ **на·гре́·в·ы** heating time

~ **на·раст·а́ни·я** (phys) rising/build-up period

~ **не·чувств·и́тельност·и** paralysis/insensitive period/time

~ **поко́·я** quiescent period

~ **полу·об·ме́н·а** (chem) half-time of exchange

~ **полу·рас·па́д·а** (nucl) half-life

~ **прецесс·и·и** (phys) precession time

~ **решёт·к·и** (cryst) lattice parameter/spacing; lattice pitch (nuclear reactors)

~ **то́ч·к·и** (telecom) dot cycle

~ **цикли́ческого на·гружё́ни·я** (mech) period stress cycle

~ **ци́фр·ы** digit time (computers)

~ **числ·а́** word time (computers)

периода́т·, -а, (*m*) (chem) periodate

период·иза́ци·я, -и, (*f*) division into periods

перио́д·ик·а, -и, (*f*) (print) periodical

период·и́ческ·ий, -ая, -ое, (*a*) periodic(al); recurrent; (math) recurring; batch (of process machines)

~ **де́й·стви·е** (*n*) periodic/recurrent action; batch output; (*in gen as adj*) batch, batch-type

~ **на·гру́з·к·а** (*f*) (mech) repeated loading/stress

~ **опера́ци·я** (*f*) batch operation

~ **про́фил·ь** (*m*) (met) repetitive section, (*gen as adj*) indented

стёрж·ен·ь период·и́ческ·ого про́фил·я indented bar (for concrete)

~ **реа́кци·и** (*pl*) (chem) periodic precipitation

~ **систе́м·а** (*f*) (phys, chem) periodic system

~ **фу́нкци·я** (*f*) periodic function

~ **явл·е́ни·е** (*n*) recurrent phenomenon

период·и́чност·ь, -и, (*f*) periodicity, rate

~ **за·гру́з·к·и** feed rate

~ **решёт·к·и** lattice pitch (in a nuclear reactor); (cryst) lattice spacing

период·и́чн·ый, -ая, -ое, (*a*) *see* **период·и́ческ·ий**

периодонти́т·, -а, (*m*) (med) periodontitis

пери·о́ст·, -а, (*m*) periosteum

пери·оста́ль·н·ый, -ая, -ое, (*a*) (anat) periosteal

пери·ости́т·, -а, (*m*) (med) periostitis

пери·остра́кум·, -а, (*m*) (zool) periostracum

пери·пета́ль·н·ый, -ая, -ое, (*a*) (bot) peripetalous

пери·пла́зм·а, -ы, (*f*) (bot) periplasm

пери·плазмо́д·и·й, -я, (*m*) (bot) periplasmodium

пери·про́кт·, -а, (*m*) (zool) periproct

пери·са́рк·, -а, (*m*) (zool) perisarc

пери·ско́п·, -а, (*m*) periscope

~, **коле́н·чат·ый** corner periscope

пери·скоп·и́ческ·ий, -ая, -ое, (*a*) (bot) periscopic

пери·ско́п·н·ый, -ая, -ое, (*a*) periscope

пери·спора́нг·и·й, -я, (*m*) (bot) perisporangium

пери·ста́льт·ик·а, -и, (*f*) (zool) peristalsis

пери·стальт·и́ческ·ий, -ая, -ое, (*a*) peristaltic

пери·стил·ь, -я, (*m*) (arch) peristyle

пёр·ист·о- (*int component*) (biol) pinna-, pinni-

перисто·доль·чат·ый, -ая, -ое, (*a*) pinnatilobate

перисто·жабер·н·ые, -ых, (*pl decl as adj*) (zool) pterobranchs, *Pterobranchia*

перисто·крыл·к·и, -л·ок·, (*f pl*) (ent) caddis flies, *Trichoptera*

пёр·ист·с-куч·ев·ые облак·а (*pl*) cirro-cumulus clouds

перисто·лопаст·н·ый, -ая, -ое, (*a*) pinnatilobate

пёр·ист·о-сло·ист·ые облак·а (*pl*) cirrostratus clouds

пери·стом·, -а, (*m*) (biol) peristome, peristoma

перисто·над·рез·н·ой, -ая, -ое, (*a*) (bot) pinnatifid

перисто·нёрв·н·ый, -ая, -ое, (*a*) (bot) pinninervate

перисто·раз·дель·н·ый, -ая, -ое, (*a*) (bot) pinnatipartite

перисто·рас·сеч·ённ·ый, -ая, -ое, (*a*) (bot) pinnatisect

пёр·ист·ый, -ая, -ое, (*a*) feathery, feathered; (biol) plumose, pinnate

~ облак·а (*pl*) (meteor) cirrus clouds

~ руд·а (*f*) (min) feather ore, jamesonite

перитёкт·ик·а, -и, (*f*) (chem) incongruent saturated solution; incongruent melting point

перитект·ическ·ий, -ая, -ое, (*a*) peritectic

пери·тёц·и·й, -я, (*m*) (bot) perithecium

перитонит·, -а, (*m*) (med) peritonitis

пери·троф·ическ·ий, -ая, -ое, (*a*) (zool) peritrophic

пери·физ·, -а, (*m*) (bot) periphysis

перифер·ийн·ый, -ая, -ое, (*a*) periphery, peripheral

перифер·ическ·ий, -ая, -ое, (*a*) peripheral

перифер·и·я, -и, (*f*) periphery

пери·флоэм·а, -ы, (*f*) (bot) periphloem, pericambium

перифраз·, -а, (*m*) paraphrase

перифраз·а, -ы, (*f*) paraphrase

пери·хёц·и·й, -я, (*m*) (bot) perichaetium

пери·цикл·, -а, (*m*) (bot) pericycle

перициклоид·а, -ы, (*m*) (math) pericycloid

пёр·к·а, -и, (*g pl*) -р·ок·, (*f*) drill/common bit; centre bit (carpenter's tool); flat drill (lathe tool)

~, лож·ечн·ая auger

перкал·ь, -и, (*f*) or -я, (*m*) (text) percale

Пёркин·а, синтез· (*m*) (chem) Perkin's synthesis

пер·кисл·от·а, -ы, (*f*) (chem) peracid

перкнит·, -а, (*m*) (geol) perknite

перкол·ятор·, -а, (*m*) (met) percolator

перкол·яци·я, -и, (*f*) (met) percolation

перкусси·я, -и, (*f*) (med) percussion

пёрл·, -а, (*m*) pearl; (print) pearl; (chem, met) bead

пёр·ла, (*past fem sing*) of пер·ёть

перламутр·, -а, (*m*) mother-of-pearl, nacre

перламутр·ов·ый, -ая, -ое, (*a*) mother-of-pearl, nacreous; pearl (as colour)

пёрлин·ь, -я, (*m*) (naut) hawser

перлит·, -а, (*m*) (met) pearlite; (geol) perlite

перлит·н·ый, -ая, -ое, (*a*) pearlitic

перлит·ов·ый, -ая, -ое, (*a*) (min) perlitic

перл·овиц·а, -ы, (*i*) -ей, (*f*) (zool) pearl oyster

перл·овник·, -а, (*m*) (bot) onion grass, *Melica*

перл·ов·ый -ая, -ое, (*a*) pearl

~ круп·а (*f*) pearl barley

перлон·, -а (*m*) (plast, text) perlon

прелюв·и·й, -я, (*m*) (geol) perluvium

пермалло·й, -я, (*m*) (met) permalloy

пер·манганат·, -а, (*m*) (chem) permanganate

~ кали·я potassium permanganate

перманганато-метр·и·я, -и, (*f*) (chem) permanganate titration

перманёнт·н·ый, -ая, -ое, (*a*) permanent

перматрон·, -а, (*m*) (elec) permatron

пермеа·метр·, -а, (*m*) (elec) permeameter, flux density meter

~, баллист·ическ·ий (elec) Drysdale/ballistic permeameter

~, от·рыв·н·ой traction permeameter

пермендур·, -а, (*m*) (met) Permendur

пермендюр·, -а, (*m*) = пермендур·

перменорм·, -а, (*m*) (met) permenorm

перминвар·, -а, (*m*) (met) perminvar

пёрм·ск·ий, -ая, -ое, (*a*) (geol) Permian

пермутит·, -а, (*m*) (chem) permutite

пе́рм·ь, -и, (*f*) (geol) Permian
пер·на́т·ый, -ая, -ое, (*a*) feathered, plumigerous
пер·о́, -а́, (*nom pl*) **пе́р·ья,** (*g pl*) **-ьев,** (*n*) feather, plume, pinna; pen, stylus, nib; blade, fin; after-piece (of rudder); tongue (of a rail)
~ **вентил·я́тор·а** fan blade
~, **ве́ч·н·ое** fountain pen
~, **пла́в·ательн·ое** (fish) fin
перовски́т·, -а, (*m*) (min) perovskite
пер·ов·о́й, -а́я, -о́е, (*a*) of **пер·о́**
~ **кист·ь** (*f*) liner, fitch (paintbrush)
пер·окси́д·, -а, (*m*) (chem) peroxide
~ **на́три·я** sodium peroxide
пер·оксид·а́з·а, -ы, (*f*) (chem) peroxidase
перо·обра́з·н·ый, -ая, -ое, (*a*) feather-like, pinnate
перо·пух·о́в·ый, -ая, -ое, (*a*) down-and-feather
пер·ора́ль·н·ый, -ая, -ое, (*a*) (pharm) peroral
перпендикуля́р·, -а, (*m*) (math) perpendicular, normal
перпендикуля́р·н·о (*adv*) perpendicular (to)
перпендикуля́р·н·ый, -ая, -ое, (*a*) perpendicular
перрена́т·, -а, (*m*) (chem) perrhenate
перро́н·, -а, (*m*) (rail) platform; (air) concourse; (build) perron
пе́рс·, -а, (*m*) Iranian, Persian
персевер·а́ци·я, -и, (*f*) (med) perseveration
персе́йд·ы, -ов, (*m pl*) (astron) Perseids
Персе́·й, -я, (*m*) (astron) Perseus
персе́·я, -и, (*f*) (bot) persea
перси́д·ск·ий, -ая, -ое, (*a*) Persian, Iranian
пе́рсик·, -а, (*m*) (bot) peach, *Prunus persica*
персона́л·, -а, (*m*) personnel, staff
~, **мла́д·ш·ий об·слу́ж·ивающ·ий** operatives
~, **техн·и́ческ·ий** engineering personnel
персона́ль·н·ый, -ая, -ое, (*a*) personal
перспе́кс·, -а, (*m*) (plast) perspex
перспекти́в·а, -ы, (*f*) perspective; (phot) aerial perspective; (*pl*) prospects, possibilities
перспекти́в·ност·ь, -и, (*f*) promising future, prospects; perspectivity

перспекти́в·н·ый, -ая, -ое, (*a*) perspective; promising; long-term
~ **аэрофото·с·ни́м·ок·** (*m*) oblique aerial photograph
пе́рстен·ь, -тн·я, (*m*) ring (worn on the finger)
перстне·ви́д·н·ый, -ая, -ое, (*a*) cricoid
персто·ви́д·н·ый, -ая, -ое, (*a*) finger-like/shaped, digitiform, digitate
пер·сульфа́т·, -а, (*m*) (chem) persulphate
пе́рт·, -а, (*m*) footrope
пертина́кс·, -а, (*m*) (plast) pertinax
перти́т·, (*m*) (min) perthite
пертози́т·, -а, (*m*) (geol) perthosite
пертурб·а́ци·я, -и, (*f*) perturbation
перуа́н·ск·ий, -ая, -ое, (*a*) Peruvian
~ **бальза́м·** (*m*) (chem) balsam of Peru
перфо·ка́рт·а, -ы, (*f*) punched card
~, **ито́г·ов·ая** summary punched card
перфо·ка́рт·очк·а, -и, (*g pl*) **-чек·,** (*f*) punched card
перфо·ле́нт·а, -ы, (*f*) punched/perforated tape
перфора́тор·, -а, (*m*) (telecom) perforator; punch (punched card systems); (min) reciprocating rock drill, hammer drill; gun perforator (oil wells)
~, **вос·про·из·вод·я́щ·ий** reproducing punch
~, **вы·числ·и́тельн·ый** calculating punch
~, **двух·перио́д·н·ый** punch-die-type punch
~, **дубл·и́рующ·ий** duplicating punch
~, **ито́г·ов·ый** summary punch
~, **клавиату́р·н·ый** (telecom) keyboard perforator
~, **коло́н·ков·ый** (min) drifter
~, **руч·н·о́й** (min) hand-held hammer drill
~, **телеско́п·н·ый** (min) stoper, stoper hammer drill
перфора́тор·и·й, -я, (*m*) (zool) perforatorium, acrosome
перфора́тор·н·ый, -ая, -ое, (*a*) of **перфора́тор·**
перфорацио́н·н·ый, -ая, -ое, (*a*) of **перфора́ци·я**
~ **маши́н·а** (*f*) punched card/tape machine
перфора́ци·я, -и, (*f*) perforation, punching, holing, drilling
перфори́рованно·по·кро́в·н·ый, -ая, -ое, (*a*) punctitegillate

перфор·и́р·овать, -у́ют, (*imp and perf*) perforate, punch, make holes

пер·фто́р·о- (*int component*) (chem) perfluo-

пер·хлора́т·, -а, (*m*) (chem) perchlorate

пе́рхот·ь, -и, (*f*) (med) dandruff, scurf

перце·я́д·ы, -ов, (*m pl*) (zool) toucans, *Rhamphastidae*

перц·о́вк·а, -и, (*g pl*) **-во́к·,** (*f*) pepper vodka, pertsovka

перча́т·к·а, -и, (*g pl*) **-т·ок·,** (*f*) glove; sleeve, joint (cables etc.)

перча́т·очн·ый, -ая, -ое, (*a*) of **перча́т·к·а**

пёр·ш·ий, -ая, -ее, (*past part act*) of **пер·е́ть**

пёр·ышк·о, -а, (*g pl*) **-шек·,** (*f*) (*dim*) of **пер·о́**

пер·ья, (*nom pl*) of **пер·о́**

пёс·, (*g sing*) **пс·а, (*m*) dog

Пёс·, Больш·о́й (astron) Canis Major

Пёс·, Ма́л·ый (astron) Canis Minor

пес·е́ц·, -с·ц·а́, (*i*) **-с·ц·о́м, (*m*) (zool) polar fox, *Felis arcticus*

пёс·ий, -ья, -ье, (*a*) dog's

пескар·ь, -я́, (*m*) (fish) gudgeon, *Gobis gobis*

песк·и́ (*nom pl*) of **песо́к·;** (geol) sands

песк·ова́т·ый, -ая, -ое, (*a*) sandy, sand-like

песко·ду́в·к·а, -и, (*g pl*) **-в·ок·,** (*f*) sand blowing; sand blasting

песко·ду́в·н·ый, -ая, -ое, (*a*) sand-blowing/blasting

~ маши́н·а (*f*) (met cast) compressed air sand slinger

песко·ло́в·к·а, -и, (*g pl*) **-в·ок·,** (*f*) sand trap

песко·лю́б·, -а, (*m*) (bot) beach-grass, *Ammophila arenaria*

песко·мёт·, -а, (*m*) (met cast) sand-slinger

песко·мо́й·к·а, -и, (*g pl*) **-мо́·ек·,** (*f*) (min) sandwasher

песко·раз·бра́с·ыватель·, -я, (*m*) (civ eng) sander, sand spreader

песко·ро́й·к·а, -и, (*g pl*) **-о́·ек·,** (*f*) (zool) sand eel, *Ammocoetes*

песко·стру́й·н·ый, -ая, -ое, (*a*) (met) sand-blast/blasting

~ аппара́т· (*m*) (met) sand blaster

~ об·рабо́т·к·а (*f*) (met) sand blasting, sandblast cleaning

песко·у·креп·и́тельн·ое раст·е́ни·е (bot) sand binder

пе́сн·я, -и, (*g pl*) **пе́сен·,** (*f*) song

песо́к·, -ск·а́, (*m*) sand; (med) gravel

~, зыб·у́ч·ий shifting sand, quicksand

~, по́р·ист·ый open sand

~, са́хар·н·ый granulated sugar

~, сып·у́ч·ий shifting sand, quicksand

~, форм·о́вочн·ый moulding/foundry sand

песо́ч·ин·а, -ы, (*f*) sand mark (on rolled metal); sand buckle (on a casting); sand hole (in a casting)

песо́ч·ник·, -а, (*m*) (orn) sandpiper, *Actitis*

песо́ч·ниц·а, -ы, (*i*) **-ей,** (*m*) sand box (of locomotive); sand trap, riffer (papermaking)

песо́ч·н·ый, -ая, -ое, (*a*) sand, sandy; short (of pastry)

~ час·ы́ (*pl*) sand glass

песса́ри·й, -я, (*m*) (med) pessary

пест·, -а́, (*m*) pestle, stamp

пе́ст·ик·, -а, (*m*) pestle, pounder; (bot) pistil; pin (for cable insulators)

пе́ст·иков·ый, -ая, -ое, (*a*) pistillate

пе́ст·ичн·ые, -ых, (*pl decl as adj*) (bot) pistillate plants

пестро·ли́ст·н·ый, -ая, -ое, (*a*) mottled/variegated-leaved

пестро·тка́н·ь, -и, (*f*) tapestry

пестро·тка́ц·к·ий, -ая, -ое, (*a*) tapestry

~ стан·о́к· (*m*) tapestry loom

пестр·у́шк·и, -шек·, (*f pl*) (zool) *Microtinae* (voles and lemmings)

пёстр·ый, -ая, -ое, (*a*) mottled, variegated

~ мед·н·ая руд·а́ (*f*) (min) variegated copper ore, bornite

~ свинц·о́в·ая руд·а́ (*f*) (min) pyromorphite

песч·а́ник·, -а, (*m*) (min) sandstone; (zool) yellow souslik, *Citellus fulvus*

песч·а́нк·а, -и, (*g pl*) **-нок·,** (*f*) (zool) gerbil, *Gerbillus*; (fish) sand eel, *Ammodytes*; (bot) sandwort, *Arenaria*; (orn) sandpiper, *Crocethia alba*

песч·а́н·ый, -ая, -ое, (*a*) sand, sandy, arenaceous, arenicolous

~ культу́р·а (*f*) (bot) sand culture

~ осо́к·а (*f*) (bot) sea bent, *Carex arenaria*

~ раст·е́ни·е (*n*) (bot) psammophile, psammophyte, arenaceous/arenicolous plant

~ холм· (*m*) sand dune

песч·и́нк·а, -и, (*g pl*) **-нок·,** (*f*) grain of sand

петал·и́т·, -а, (*m*) (min) petalite
петал·о́ид·н·ый, -ая, -ое, (*a*) petaloid
пета́рд·а, -ы, (*m*) detonating cartridge, fire cracker; black powder pellet
пе́тел·ь, (*g pl*) of пе́тл·я
пе́тель·к·а, -и, (*g pl*) -л·ек·, (*f*) small loop, eyelet
пе́тель·н·ый, -ая, -ое, (*a*) of пе́тл·я
~ маши́н·а (*f*) buttonholing machine
пе́тель·чат·ый, -ая, -ое, (*a*) looped, loop-shaped
пети́т·, -а, (*m*) (print) brevier
петл·ева́ни·е, -я, (*n*) (met etc.) looping; (elec) looping-in
петле·ви́д·н·ый, -ая, -ое, (*a*) loop-shaped, looped
петл·ев·о́й, -а́я, -о́е, (*a*) of пе́тл·я
~ стан· (*m*) (met) looping mill
петле·держ·а́тел·ь, -я, (*m*) (met roll) looper, looping channel, repeater
петле·обра́з·н·ый, -ая, -ое, (*a*) loop, loop-shaped
петле·образ·ова́тел·ь, -я, (*m*) (met roll) looper
петле·об·мо́т·очн·ая маши́н·а (*f*) (text) buttonholing machine
петле·регул·и́рова·ни·е, -я, (*n*) (met roll) loop control
петле·регул·я́тор·, -а, (*m*) loop controller
петле·скоп·, -а, (*m*) (instr) hysteresigraph
петл·и́стост·ь, -и, (*f*) looping (defect in cloth)
петл·и́тел·ь, -я, (*m*) (text) looper
петл·и́ц·а, -ы, (*i*) -ей, (*f*) buttonhole; tab; frog (of sword)
пе́тл·я, -и, (*g pl*) пе́тел·ь (*f*) loop, kink; noose, eye; buttonhole; stitch (in knitting); hinge
~, гистере́зис·н·ая (phys) hysteresis loop
~, за́·мк·нут·ая closed loop
~, кос·а́я (air) oblique loop
~ Не́сторов·а (air) wingover
об·мёт·к·а пе́тел·ь (*f*) buttonholing, buttonhole edging, blanket stitching
~, рул·ев·а́я gudgeon (of rudder)
~ с·вя́з·и (phys) coupling loop
~, шарни́р·н·ая hinge
Пе́три, ча́ш·к·а (*f*) Petri dish
петри·фик·а́ци·я, -и, (*f*) petrification
Петро́в·а, жи́дк·ост·ь (*f*) Petrov's fluid (blood substitute)

петро·гене́з·, -а, (*m*) (geol) petrogenesis
петро·гли́ф·, -а, (*m*) petroglyph (anthropology)
петро́·граф·, -а, (*m*) petrographer
петро·граф·и́ческ·ий, -ая, -ое, (*a*) petrographic
петро·граф·и·я, -и, (*f*) petrography
петрола́т·, -а, (*m*) petrolatum
петрола́тум·, -а, (*m*) petrolatum, petroleum jelly
петроле́·йн·ый, -ая, -ое, (*a*) petroleum
~ эфи́р· (*m*) petroleum ether
петро́ле·ум·, -а, (*m*) petroleum jelly
петро·ло́г·и·я, -и, (*f*) petrology
Петрося́н·а, аппара́т· (*m*) (oil) Petrosyan's acid-bottle inclinometer
петро·фи́т·, -а, (*m*) (bot) petrophyte
пету́х·, -а́, (*m*) (orn) cock
петуш·и́н·ый, -ая, -ое, (*a*) of пету́х·
петци́т·, -а, (*m*) (min) petzite
пе́·т·ый, -ая, -ое, (*past part pass*) of пе·ть; sung
пе·ть, (*pres 3rd pl*) по·ю́т, (*past masc sing*) пе·л (*imp*), sing
пех·о́т·а, -ы, (*f*) infantry
~, мор·ск·а́я marines, naval infantry
пех·оти́нец·, -нц·а, (*i*) -ц·ем, (*m*) infantryman
пехште́йн·, -а, (*m*) (min) pitchstone
печа́ль·н·ый, -ая, -ое, (*a*) sad, dismal; unfortunate, regrettable
печа́т·ани·е, -я, (*n*) (*v n*) of печа́·т·а·ть
печа́т·анн·ый, -ая, -ое, (*past part pass*) of печа́т·а·ть; printed
печа́т·а·ть, -ют, (*imp*) print, type; be/have printed/published
печа́т·ающ·ий, -ая, -ее, (*pres part act*) of печа́т·а·ть
~ лин·е́йк·а (*f*) (telecom) print/printer bar
~ у·стро́й·ств·о (*n*) printer (of computers etc.)
печа́т·ник·, -а, (*m*) printer
печа́т·н·ый, -ая, -ое, (*a*) printing, printer's, printed
~ аппара́т· (*m*) printer
~ де́л·о (*n*) typography
~ кра́с·к·а (*f*) printer's ink
~ маши́н·а (*f*) printing press; (text) printing machine
~ маши́н·а, лист·ов·а́я sheet-fed press
~ маши́н·а, пло́ск·ая flat press
~ маши́н·а, ти́гель·н·ая platen press

печа́т·н·ый, -ая, -ое
~ **пла́н·к·а** (*f*) (elec) printed-circuit board
~ **схе́м·а** (*f*) (elec) printed circuit
~ **цили́ндр·** (*m*) (print) impression cylinder
печа́т·ь, -и, (*f*) seal, stamp; press, printing; print
   аге́нт·ств·о печа́т·и (*n*) press agency
   в печа́т·и (print) printing
~, **выс·о́к·ая** relief/typographic/letter-press print
~, **гидро·ти́п·н·ая** (phot) hydrotype printing
~, **глуб·о́к·ая** intaglio
~, **лист·ов·а́я** (print) flat printing
~, **пло́ск·ая** surface/planographic print
~, **опт·и́ческ·ая** (cinema) optical printing
~, **ра́кель·н·ая** screen printing
~, **типо·гра́ф·ск·ая** typographic/relief/letter-press print
~, **трюк·ов·а́я** (cinema) combination printing
~, **у·бо́р·ист·ая** (print) closed spacing
печ·е́ни·е, -я, (*n*) baking; biscuit, cookie
печён·к·а, -и, (*g pl*) **-н·ок·,** (*f*) liver (as food)
печ·ённ·ый, -ая, -ое, (*past part pass*) of **печ·ь**
печён·очник·и, -ов, (*m pl*) (bot) liverworts, *Hepaticae*
печён·очниц·а, -ы, (*f*) (bot) *Hepatica*
печён·очн·ый, -ая, -ое, (*a*) liver, hepatic
~ **про·то́к·** (*m*) (anat) ductus hepaticus
печ·ён·ый, -ая, -ое, (*a*) baked
пе́чен·ь, -и, (*f*) (anat) liver
~, **извест·ко́в·ая се́р·н·ая** (chem) lime hepar
~, **се́р·н·ая** (chem) liver of sulphur
печ·е́нь·е, -я, (*n*) *see* **печ·е́ни·е**
печер·и́ц·а, -ы, (*i*) **-ей,** (*f*) mushroom
печ·ёт, (*pres 3rd sing*) of **печ·ь**
пе́ч·к·а, -и, (*g pl*) **-ч·ек·,** (*f*) stove
печ·н·о́й, -а́я, -о́е, (*a*) of **печ·ь**
печ·ь, (*pres 3rd sing, pl*) **печ·ёт, пе-к·у́т,** (*past masc sing*) **пёк·** (*perf*); bake
печ·ь, -и, (*f*) stove, oven, furnace; heater; (min) bord
~, **бес·ша́хт·н·ая** low-shaft furnace
~, **ва́куум·н·ая** (elec) vacuum oven
~, **ва́нн·ая** tank furnace (glassmaking)
~, **ватержаке́т·н·ая** water-jacketed blast furnace

пе́ч·ь
~, **вос·станов·и́тельн·ая** reducing furnace
~, **глазу́р·н·ая** (ceram) glost oven
~, **го́фман·ск·ая** Hoffman kiln (brickmaking)
~, **диагона́ль·н·ая** (min) crosscut
~, **дистилляцио́н·н·ая** (met) retort
~ **дл·и́тельн·ого гор·е́ни·я** long-burning stove
~, **до́мен·н·ая** (met) blast furnace
~, **дуг·ов·а́я** arc furnace
~, **индукцио́н·н·ая** induction furnace
~, **карби́д·н·ая** (met) bottom-electrode arc furnace
~, **ка́мер·н·ая** reverberatory furnace
~, **карусе́ль·н·ая** rotary-hearth furnace
~, **ко́кс·ов·ая** coke oven
~, **коло́д·цев·ая** (met) soaking pit
~, **колоко́ль·н·ая** bell furnace
~, **колпак·о́в·ая** bell furnace; stack annealing furnace
~, **кольц·ев·а́я** Hoffman ring kiln (brickmaking); (met) rotary furnace
~, **конве́йер·н·ая** conveyor furnace
~, **конта́кт·н·ая** (chem) catalytic reaction furnace
~, **кузн·е́чн·ая** smith's hearth
~, **люл·ечн·о-под·о́в·ая** rocking-bottom oven (bakery)
~, **мазу́т·н·ая** oil-fired furnace
~, **марте́н·овск·ая** (met) open-hearth furnace, OH furnace
~, **метод·и́ческ·ая** continuous furnace
~ **-ми́ксер·, -а,** (*m*) (met) holding furnace
~, **много·ка́мер·н·ый** multiple-hearth furnace
~, **му́фель·н·ая** muffle furnace
~, **на·грев·а́тельн·ая** (met) reheating furnace
~, **об·жиг·ов·ая** roasting furnace, kiln
~, **от·жиг·ов·ая** annealing furnace
~, **от·раж·а́тельн·ая** reverberatory furnace
~, **период·и́ческ·ая** intermittent kiln
~, **плав·и́льн·ая** melting unit/furnace
~ **пла́мен·н·ая** furnace in which charge and fuel are in contact
~, **подо·грев·а́тельн·ая** preheating furnace
~, **по́л·очн·ая** multiple-hearth roasting furnace
~, **про·хо́д·н·ая** continuous furnace

пе́ч·ь

~, рас·хо́д·н·ая holding furnace

~, регенерати́в·н·ая regenerative furnace

~, регенера́тор·н·ая reverberatory furnace

~, руч·н·а́я hand-rabbled furnace

~, сад·о́чн·ая (met) batch furnace

~, секцио́н·н·ая multistage furnace

~ со·противл·е́ни·я resistance furnace

~, терм·и́ческ·ая heat-treatment furnace

~, ти́гель·н·ая crucible furnace

~, толк·а́тельн·ая pusher furna·e

~, тру́б·чат·ая tubular furnace, kiln; (oil) pipe still

~, ша́хт·н·ая shaft furnace

~, электро́н·н·о-луч·ев·а́я electron-beam furnace

пеше·хо́д·, -а, (m) pedestrian

пеше·хо́д·н·ый, -ая, -ое, (a) pedestrian, foot

пе́ш·к·а, -и, (g pl) **-ш·ек·,** (f) pawn (at chess)

пеш·к·о́м (adv) on foot, walking

пешн·я́, -й, (nom pl) **пе́шн·и,** (g pl) **-шен·,** (f) ice spike (for breaking ice)

пеще́р·а, -ы, (f) cave, cavern

пеще́р·ист·ый, -ая, -ое, (a) (anat) cavernous, cavernosus, honeycombed

пеще́р·к·а, -и, (g pl) **-р·ок·,** (f) (dim) of **пеще́р·а**

пеще́р·н·ый, -ая, -ое, (a) cave, spelean

пещеро·ве́д·ени·е, -я, (n) speleology

пз (abbr) = **пуа́з·** poise (unit of viscosity)

пи pi, π (Greek)

пив·н·о́й, -а́я, -о́е, (a) beer, brewing, brewer's

~ дроб·и́н·а (f) brewer's grains

~ су́сл·о (n) brewer's wort

пи́в·о, -а, (n) beer

пиво·ва́р·, -а, (m) brewer

пиво·вар·е́ни·е, -я, (n) brewing

пиво·ва́р·енн·ый, -ая, -ое, (a) brewing

~ за·во́д· (m) brewery

~ дро́жж·и (pl) brewer's yeast

пи́·вш·ий, -ая, -ее, (past part act) of **пи·ть**

пиг·и́ди·й, -ая, (m) (zool) pygidium

пигме́·й, -я, (m) pigmy

пигме́нт·, -а, (m) pigment

пигмент·а́ци·я, -и, (f) pigmentation

пиго·сти́л·ь, -я, (m) (orn) pygostyle

пид·жа́к·, -а́, (m) (text) coat, jacket

~, дву·бо́рт·н·ый double-breasted coat

пие́ло- (int component) see **пиэ́ло-**

пи·еми́·я, -и, (f) (med) pyaemia, pyemia

пижони́т·, -а, (m) (min) pigeonite

пизо·ли́т·, -а, (m) (min) pisolite

пизо·ли́т·ов·ый, -ая, -ое, (a) (min) pisolitic

пи́к·, -а, (m) peak

~ вну́тр·енн·его тр·е́ни·я (phys) damping peak

~ с·мещ·е́ни·я (phys) displacement spike

с·рез·а́ни·е пи́к·ов (n) (rad) peak clipping

пи́к·а, -и, (f) lance; (hortic) dibber; (pl) spades (at cards)

пика́п·, -а, (m) (M/T) pickup van/truck; (agr) pickup attachment

Пика́р·а, ме́тод· (m) (math) Picard method

пике́ (n indecl) (text) piqué; (air) dive

пикел·ева́ни·е, -я, (n) pickling (leather)

пикел·ева́ть, -юют, (imp and perf) = **пикел·и́р·овать**

пикел·и́р·овать, -уют, (imp and perf) pickle (leather)

пике́т·, -а, (m) picket; (surv) stake, peg, bench mark; (rail) 100-metre mark/stake

пикет·а́ж·, -а, (m) picketing, stationing; (surv) pegging out

пик·и́ровани·е, -я, (n) (v n) of **пик·и́р·овать;** (air) dive

~, от·ве́с·н·ое nose dive

у́гол· пик·и́ровани·я (m) (air) angle of dive

пик·и́р·овать, -уют, (imp and perf) (air) dive; (hortic) prick out (seedlings)

пик·и́ро́вщик·, -а, (m) (a/c) dive bomber

пи́кк·а, -и, (f) unfrozen seawater on ice/snow

пи́ккер·, -а, (m) (agr) picker

пикн·и́д·а, -ы, (f) (bot) pycnidium

пикнидио·спо́р·а, -ы, (f) (bot) pycnidiospore

пикн·и́т·, -а, (m) (min) pycnite

пикно́·метр·, -а, (m) pycnometer, pyknometer, density bottle

пикно·спо́р·а, -ы, (f) (bot) pycnospore

пико- (int component) pico-, micro-micro-, 10^{-12}

пико·ампе́р·, -а, (m) (elec) micro-microampere

пи́к·ов·ый, -ая, -ое, (*a*) of **пи́к·**
~ **знач·е́ни·е** (*n*) peak/crest value
~ **на·гру́з·к·а** (*f*) peak load
пик·оли́н·, -а, (*m*) (chem) picoline, methyl pyridine
пиколи́н·ов·ый, -ая, -ое, (*a*) picoline, picolinic
~ **кисл·от·а́** (*f*) (chem) picolinic acid
пико·о·гранич·и́тел·ь, -я, (*m*) (rad) peak limiter/clipper
пикоти́т·, -а, (*m*) (min) picotite
пико·фара́д·а, -ы, (*f*) (elec) pico-farad, micromicrofarad (10^{-12} farad)
пикр·ами́н·ов·ая кисл·от·а́ (*f*) (chem) picramic acid
пикр·а́т·, -а, (*m*) (chem) picrate
пикр·и́нов·ый, -ая, -ое, (*a*) (chem) picric
~ **кисл·от·а́** (*f*) picric acid
пикр·и́л·, -а, (*m*) (chem) picryl
пикр·и́т·, -а, (*m*) (geol) picrite
пикр·олона́т·, -а, (*m*) (chem) picro-lonate
пикр·оло́н·ов·ая кисл·от·а́ (*f*) (chem) picrolonic acid
пикро·мери́т·, -а, (*m*) (min) picromerite
пикро·токси́н·, -а, (*m*) (pharm) picro-toxin
пикте́стер·, -а, (*m*) (paper) pick tester
пик-трансформ·а́тор·, -а, (*m*) (elec) peaking transformer
пи́кул·и, -ей, (*pl*) (food) pickles
пикфа́ктор·, -а, (*m*) (elec) peak factor
пи́кш·а, -и, (*i*) **-ей,** (*f*) (fish) had-dock, *Melanogrammus aeglefinus*
пи·л, (*past masc sing*) of **пи·ть**
пил·а́, -ы́, (*nom pl*) **пи́л·ы,** (*g pl*) **пил·,** (*f*) saw; **пи·ла́** (*past fem sing*) of **пи·ть**
~, **баланси́р·н·ая** travelling-head saw
~, **бу́гель·н·ая** bow saw
~ **гор·я́ч·его рез·а́ни·я** (met) hot saw
~, **ди́ск·ов·ая** circular saw
~, **дол·ев·а́я** pit/rip saw
~, **кант·ова́льн·ая** edging saw
~, **конце·рез·н·а́я** cutoff saw
~, **кругл·а́я** circular saw
~, **ле́нт·очн·ая** bandsaw
~, **луч·ко́в·ая** bow/frame saw
~, **ма́ят·ников·ая** pendulum cross-cut saw
~, **нож·о́вочн·ая** hacksaw
~, **по·пере́ч·н·ая** cross-cut saw
~, **про·до́ль·н·ая** rip/pit saw

пил·а́
~, **сала́з·ков·ая** sliding-frame saw
~ **тр·е́ни·я** friction saw
~ **холо́д·н·ого рез·а́ни·я** (met) cold saw
~, **цеп·н·а́я** chain saw
~, **шли́ц·ев·ая** sash saw
пила́в·, -а, (*m*) (food) pilaff, risotto
пилбари́т·, -а, (*m*) (min) pilbarite
пил·е́ни·е, -я, (*n*) sawing; filing
пи́л·енн·ый, -ая, -ое, (*past part pass*) of **пил·и́ть**
пил·ён·ый, -ая, -ое, (*a*) sawn; filed
~ **лес·** (*m*) lumber, sawn timber
~ **са́хар·** (*n*) lump sugar
пи́·ли, (*past pl*) of **пи·ть**
пилигри́м·ов·ый стан· (*m*) (met roll) Pilger mill
пил·и́ди·й, -я, (*m*) (biol) pilidium
пил·и́льн·ый, -ая, -ое, (*a*) sawing; filing
пил·и́льщик·, -а, (*m*) (ent) sawfly, (*pl*) *Tenthredinoidae*
пил·и́ть, -ят, (*imp*) saw; file
пи́л·к·а, -и, (*g pl*) **-л·ок·,** (*f*) (*v n*), see **пил·и́ть**; nail file
пи́ллерс·, -а, (*m*) (shipb) deck stan-chion, pillar
пи́·ло, (*past neut sing*) of **пи·ть**
пило·ви́д·н·ый, -ая, -ое, (*a*) sawtooth, serrated, serrate, serratiform
пил·о́вочник·, -а, (*m*) log (timber)
пило·зу́б·н·ый, -ая, -ое, (*a*) saw-tooth, serrated, serrate
пило·карпи́н·, -а, (*m*) (chem) pilo-carpine
пило·материа́л·ы, -ов, (*m pl*) lumber, converted/sawn timber
пило·мото́р·н·ый, -ая, -ое, (*a*) (med) pilomotor
пило́н·, -а, (*m*) tower, pylon; pier (of suspension bridge)
пило́н·н·ый, -ая, -ое, (*a*) of **пило́н·**
пило·обра́з·н·ый, -ая, -ое, (*a*) saw-tooth/toothed, serrated, serrate
~ **волн·а́** (*f*) (phys) saw-tooth wave
пило·пра́в·щик·, -а, (*m*) saw-doctor
пилор·и́т·, -а, (*m*) (med) pyloritis
пилор·и́ческ·ий, -ая, -ое, (*a*) pyloric
пилор·спа́зм·, -а, (*m*) (med) pyloro-spasm
пилоро·стено́з·, -а, (*m*) (med) pyloro-stenosis
пил·о́рус·, -а, (*m*) (anat) pylorus
пило́т·, -а, (*m*) (*see also* **лёт·чик·**); (air) pilot
~ **-сква́ж·ин·а** (*f*) (oil) pilot borehole

пилот·а́ж·, -а,   (*m*) (air) piloting, flying, flight

∼, вы́с·ш·ий aerobatics

∼, пере·вёр·нут·ый inverted flight

∼, фигу́р·н·ый aerobatics

пилот·а́жн·ый, -ая, -ое,   (*a*) piloting, pilot, flying, flight

пилот·и́р·овать, -уют,   (*imp*) (air) pilot, fly

пило́т·к·а, -и,   (*g pl*) **-т·ок·,** (*f*) forage cap, fore-and-aft cap

пильбари́т·, -а,   (*m*) (min) pilbarite

пи́льгер·н·ый стан·   (*m*) (met roll) Pilger mill

пи́ль·н·ый, -ая, -ое,   (*a*) sawing; filing

∼ ле́нт·а   (*f*) band saw

∼ ра́м·а   (*f*) saw frame

пильчато·зу́б·чат·ый, -ая, -ое,   (*a*) serrate-dentate

пи́ль·чат·ый, -ая, -ое,   (*a*) sawtooth, sawtoothed, serrate

пилю́л·я, -и,   (*f*) (pharm) pill

пиля́стр·а, -ы,   (*f*) (arch) pilaster

пи-мезо·а́том·, -а,   (*m*) (nucl) pi-mesic atom

пи·ме́зон·, -а, (*m*) pi-mezon, pion

пимели́н·ов·ая кисл·от·а́   (*f*) (chem) pimelic acid

пиме́нт·, -а, (*m*) pimento

пиме́нт·ов·ый, -ая, -ое, (*a*) pimento

пинаго́р·, -а,   (*m*) (fish) lumpsucker, lump fish, *Cyclopterus lumpus*

пинакамфо́н·, -а,   (*m*) (chem) pin-camphone

пинак·о́ид·, -а, (*m*) (cryst) pinacoid

пинак·оли́н·, -а, (*m*) (chem) pina-coline

пинак·о́н·, -а, (*m*) (chem) pinacone

пинако·ци́т·, -а, (*m*) (zool) pinacocyte

пинакрипто́л·ов·ый, -ая, -ое,   (*a*) pinacryptol (dyes)

пинациано́л·, -а,   (*m*) (chem) pina-cyanol

пингви́н·, -а, (*m*) (orn) penguin

пине·а́льн·ая желез·а́   (*f*) (zool) pineal gland, epiphysis cerebri

пин·е́н·, -а, (*m*) (chem) pinene

пини́т·, -а, (*m*) (min) pinite

пинио́л·а, -ы, (*f*) pine nut

пи́ни·я, -и, (*f*) (bot) stone pine, *Pinus pinea*

пин·о́л·, -а, (*m*) pinol, pine camphor

пино́л·ь, -и, (*f*) tail spindle (of a lathe)

пино·пикри́н·, -а, (*m*) (pharm) pino-picrine

пи́нт·а, -ы, (*f*) pint (measure)

пинтадои́т·, -а, (*m*) (min) pintadoite

пинце́т·, -а, (*m*) tweezers, pincers

∼, ла́мп·ов·ый, (elec) lamp extractor

пинц·иро́вк·а, -и,   (*f*) pinching out (the growing tips of plants)

пи́нч·, -а, (*m*) (phys) pinch

∼ -эффе́кт·, -а, (*m*) (phys) pinch effect

пио·ге́н·н·ый, -ая, -ое,   (*a*) (med) pyogenic

пионе́р·, -а,   (*m*) pioneer; young pioneer (member of Soviet youth organization)

пио·нефро́з·, -а,   (*m*) (med) pyone-phrosis

пио·певмотора́кс·, -а,   (*m*) (med) pyopneumothorax

пио·рре́·я, -и,   (*f*) (med) pyorrhoea, pyorrhea

пио·сальпи́нкс·, -а,   (*m*) (med) pyo-salpinx

пио·циани́н·, -а,   (*m*) (microbiol) pyocyanin

пипер·ази́н·, -а,   (*m*) (pharm) pipe-razine

пипер·иди́н·, -а,   (*m*) (biochem) pipe-ridine

пипер·и́н·, -а, (*m*) (chem) piperine

пипер·она́л·, -а, (*m*) (chem) piperonal

пипер·они́л·, -а, (*m*) (chem) piperonyl

пипе́т·к·а, -и, (*g pl*) **-т·ок·,** (*f*) (chem) pipette, pipet

пир·ази́н·, -а, (*m*) (chem) pyrasine

пир·азо́л·, -а, (*m*) (chem) pyrazole

пир·азо́л·ов·ый, -ая, -ое,   (*a*) py-razole

пир·азоло́н·, -а,   (*m*) (chem) pyra-zolone

пир·азоло́н·ов·ый, -ая, -ое,   (*a*) pyrazolone

пир·аме́н·, -а, (*m*) (chem) pyramen

пирами́д·а, -ы,   (*f*) pyramid; arms/ rifle rack

пирамида́ль·н·ый, -ая, -ое,   (*a*) py-ramidal

пир·амидо́н·, -а, (*m*) (chem) pyrami-done, amidopyrin

пир·а́н·, -а, (*m*) (chem) pyran

Пирани, мано́·метр· (*m*) (phys) hot-wire gauge/gage, Pirani gauge/gage

пирано́·граф·, -а, (*m*) recording pyranometer/solarimeter

пирано́·метр·, -а, (*m*) (meteor) sola-rimeter, pyranometer

пир·антро́н·, -а, (*m*) (chem) pyran-throne

пир·аргири́т·, -а, (*m*) (min) pyrar-gyrite

пира́т·ство, -а, *(n)* (law) piracy

пир·гелио́·метр·, -а, *(m)* (meteor) pyrheliometer

пир·гео́·метр·, -а, *(m)* (meteor) pyrgeometer

пир·е́н·, -а, *(m)* (chem) pyrene

пирен·о́ид·, -а, *(m)* (bot) pyrenoid

пирено·ка́рп·и·й, -я, *(m)* (bot) pyrenocarp

пирен·о́л·, -а, *(m)* (chem) pyrenol

пирено·мице́т·ы, -ов, *(m pl)* (bot) *Pyrenomycetes*

пир·етри́н·, -а, *(m)* (chem) pyrethrin

пир·е́трум·, -а, *(m)* (bot, chem) pyrethrum

пир·иди́л·, -а, *(m)* (chem) pyridyl

пир·идилиде́н·, -а, *(m)* (chem) pyridylidene

пир·иди́н·, -а, *(m)* (chem) pyridine, pyridin

пир·идокса́л·, -а, *(m)* (pharm) pyridoxal

пир·идокси́н·, -а, *(m)* (chem) pyridoxine, pyridoxin

пир·имиди́н·, -а *(m)* (chem) pyrimidine

пир·и́т·, -а, *(m)* (min) pyrite

~, волос·и́ст·ый capillary pyrite

~, почко·ви́д·н·ый nodular pyrite

пири́т·ов·ый, -ая, -ое, *(a)* pyrite, pyritic

~ дете́ктор· *(m)* (rad) pyron detector

пиро·белони́т·, -а, *(m)* (min) pyrobelonite

пиро·бензо́л·, -а, *(m)* (chem) benzex

пиро·биту́м·, -а, *(m)* pyrobitumen

пиро·бора́т·, -а, *(m)* (chem) pyroborate, borate

пиро·бо́р·н·ая кисл·от·а́ *(f)* pyroboric acid

пиро·борно·ки́сл·ый, -ая, -ое, *(a)* pyroborate of

~ на́три·й *(m)* sodium pyroborate

пиро·ви́н·н·ая кисл·от·а́ *(f)* (chem) pyrovinnic acid, uvic acid

пиро·вино·гра́д·н·ый, -ая, -ое, *(a)* pyruvic

~ альдеги́д· *(m)* methyl glyoxal

~ кисл·от·а́ *(f)* (biochem) pyruvic/ pyroracemic acid

пиро́г·, -а́, *(m)* pie; clinker (in a coal-fired furnace)

пиро·галло́л·, -а, *(m)* (chem) pyrogallol, pyro, pyrogallic acid

пиро·гелио́·метр·, -а, *(m)* (meteor) pyrheliometer

пиро·генет·и́ческ·ий, -ая, -ое, *(a)* (chem) pyrogenic

пиро·ге́н·н·ый, -ая, -ое, *(a)* (chem) pyrogenic, pyrogenous, pyrolytic

пиро·гео́·метр·, -а, *(m)* (meteor) pyrgeometer

пиро·голо́в·к·а, -и, *(g pl)* **-в·ок·,** *(f)* (a/c) firing head (of ejection seat parachute); fire-extinguisher head

пиро·грани́т·н·ый, -ая, -ое, *(a)* vitreous clay

пиро́ж·н·ое, -ого, *(n decl as noun)* pastry

пиро·за·мо́к·, -мк·а, *(m)* (rocket) separation charge

пиро·за·па́л·, -а, *(m)* (rocket) powder squib

пиро·за·ря́д·, -а, *(m)* explosive charge

пиро·зо́м·ы *(f pl)* (zool) *Pyrosomatidae*

пиро·катехи́н·, -а, *(m)* (chem) pyrocatechin

пиро·кисл·от·а́, -ы́, *(f)* (chem) pyroacid

пиро·кла́пан·, -а, *(m)* (rocket) explosive valve

пиро·класт·и́ческ·ий, -ая, -ое, *(a)* (geol) pyroclastic

пиро·колло́ди·й, -я, *(m)* (chem, expl) pyrocollodion, nitrocellulose, guncotton

пир·оксе́н·, -а, *(m)* (chem) pyroxene

пир·оксени́т·, -а, *(m)* (geol) pyroxenite

пир·оксили́н·, -а, *(m)* (chem, expl) pyroxilin, guncotton, nitrocellulose

пиро·ли́з·, -а, *(m)* pyrolysis, destructive distillation (solids); cracking, pyrolytic aromatization (oils)

пиро·люзи́т·, -а, *(m)* (min) pyrolusite

пиро·металлу́рг·и·я, -и, *(f)* pyrometallurgy

пиро́·метр·, -а, *(m)* pyrometer

~, термо·электр·и́ческ·ий thermocouple pyrometer

~, цвет·ов·о́й colour comparison pyrometer

пиро·метр·и́ческ·ий, -ая, -ое, *(a)* pyrometric

пиро·морф·и́т·, -а, *(m)* (min) pyromorphite

пир·о́н·, -а, *(m)* (chem) pyrone

пиро·на́фт·, -а, *(m)* pyronaphtha, heavy kerosine

пир·о́п·, -а, *(m)* (min) pyrope

пиро·патро́н·, -а, *(m)* cartridge, explosive charge

пиро·пла́зм·а, -ы, *(f)* (zool) pyroplasma

пиро·плазм·о́з·, -а, (*m*) (vet) pyroplasmosis

пиро·при·во́д·, -а, (*m*) (elec) cartridge fuse

пирос·, -а, (*m*) pyros (heat-resisting alloy)

пиро·свеч·а́, -и́, (*f*) cartridge igniter

пиро·се́р·н·ый, -ая, -ое, (*a*) (chem) pyrosulphuric, pyrosulfuric

пиро·ско́п·, -а, (*m*) pyrometric/Seger/ fusion cones (used in furnaces)

пиро·сульф·а́т·, -а, (*m*) (chem) pyrosulphate, pyrosulfate

пиро·сульф·и́т·, -а, (*m*) (chem) pyrosulphite, pyrosulfite

пиро·те́хн·ик·а, -и, (*f*) pyrotechnics

пиро·ура́н·ов·ая кисл·от·а́ (*f*) uranic acid

пиро·филли́т·, -а, (*m*) (min) pyrophyllite

пиро·фо́р·, -а, (*m*) (chem) pyrophorus

пиро·фо́р·н·ый, -ая, -ое, (*a*) (chem) pyrophoric, pyrophorous

~ мета́лл· (*m*) pyrophoric alloy

пиро·фосфа́т·, -а, (*m*) (chem) pyrophosphate

пиро·фосфо́р·н·ая кисл·от·а́ (*f*) (chem) pyrophosphoric acid

пиро·хи́м·и·я, -и, (*f*) pyrochemistry

пиро·хло́р·, -а, (*m*) (min) pyrochlore

пиро·электр·и́ческ·ий, -ая, -ое, (*a*) pyroelectric

пиро·электр·и́честв·о, -а, (*n*) pyroelectricity

пирро- (*int component*) pyrro-, pyrrho-

пирр·о́л·, -а, (*m*) (chem) pyrrole

пирр·олиди́н·, -а, (*m*) pyrrolidine

пирр·оли́н·, -а, (*m*) (chem) pyrroline

пиррот·и́н·, -а, (*m*) (min) pyrrhotine, pyrrhotite

пи́рс·, -а, (*m*) pier

Пи́рсон·а, коэффицие́нт· (*m*) (math) Pearson's coefficient

пирува́т·, -а, (*m*) (chem) pyruvate

пи́с·анн·ый, -ая, -ое, (*past part pass*) of **пис·а́ть**

пи́с·ар·ь, -я, (*nom pl*) **-я́,** (*g pl*) **-е́й,** (*m*) clerk

пис·а́тел·ь, -я, (*m*) writer, author

пис·а́ть, (*pres 3rd sing*) **пи́ш·ут,** (*imp*) write; paint (pictures)

пи́ск·, -а, (*m*) squeak

писк·ли́в·ый, -ая, -ое, (*a*) squeaky

писсуа́р·, -а, (*m*) urinal

пистоле́т·, -а, (*m*) pistol, gun

~, дыро·про·би́в·н·о́й bolt driving and punching gun

пистоле́т

~, конта́кт·н·ый spot-welding gun

~, металлизацио́н·н·ый (met) metal-spraying pistol

~ -пуле·мёт·, -а, (*m*) machine carbine, submachine gun

~ -раке́т·ниц·а, -ы, (*f*) signal pistol

~ -рас·пыл·и́тел·ь, -я, (*m*) paint spray gun

~, сигна́ль·н·ый signal pistol

~, с·ва́р·очн·ый welding gun

писто́н·, -а, (*m*) percussion cap; hollow/tubular rivet

писто́н·н·ый, -ая, -ое, (*a*) of **писто́н·**

~ вы́·вод· (*m*) (elec) stack-mounting terminal

писто́н·чик·, -а, (*m*) eyelet (of vacuum tubes)

писче·бума́ж·н·ый, -ая, -ое, (*a*) writing-paper, stationery

пи́с·ч·ий, -ая, -ее, (*a*) writing

пи́с·чик·, -а, (*m*) pen recorder

пи́сьм·енн·о (*adv*) in writing

пи́сьм·енн·ый, -ая, -ое, (*a*) writing; written; (geol) graphic

~ графи́т· (*m*) (geol) graphic granite

письм·о́, -а́, (*nom pl*) **пи́сьм·о,** (*g pl*) **пи́сем·,** (*n*) letter; script

~, за·ка́з·н·о́е registered letter

письмо·вод·и́тел·ь, -я, (*m*) clerk, correspondence clerk

письмо·но́с·ец·, -с·ц·а, (*i*) **-с·ц·ем,** (*m*) postman

письмо·сорт·иро́вочн·ая маши́н·а (*f*) letter-sorting machine

пит·а́ни·е, -я, (*n*) feed, feeding, supply; (elec) power supply; nourishment, nutrition

~, ано́д·н·ое (elec) plate supply

блок· пит·а́ни·я (*m*) (elec) power pack/unit

~, взаи́м·н·ое cross feed

~, за·кры́·т·ое closed feed (system)

кла́пан· пит·а́ни·я (*m*) main valve (boiler feed regulator)

~ на·ка́л·а (elec) filament power supply, filament supply

~, объ·един·ённ·ое (a/c) crossfeed

регуля́тор· пит·а́ни·я (*m*) automatic feed regulator; (ICE) fuel control governor

~ с·мещ·е́ни·я (elec) grid voltage supply

пит·а́тел·ь, -я, (*m*) feeder, conveyer; (gunn) loading tray

~, голов·н·о́й (text) hopper opener

пит·а́тел·ь
~, **ле́нт·очн·ый** belt conveyer
~, **ли́т·ников·ый** (met cast) feeder
~, **лот·ко́в·ый** pan-type feeder/conveyer
~, **пласт·и́нчат·ый** plate feeder/conveyer
~, **цеп·н·о́й** gravity chain conveyer, chain feeder
пит·а́тельност·ь, -и, (f) food value, nourishment
пит·а́тельн·ый, -ая, -ое, (a) feed, feeding, feeder, supply; nourishing, nutritious, nutrient, nutritive; (st eng) feed-water
~ **систе́м·а** (f) (st eng) feed-water system
~ **с·ме́с·ь** (f) (bot) nutrient mixture
~ **сред·а́** (f) (bact) culture medium; (biol) nutrient medium
пит·а́·ть, -ют, (imp) feed, supply; nourish; **-ся** (pass); live on
пит·а́ющ·ий, -ая, -ее, (pres part act) see **пит·а́·ть;** feed, feeder, feeding
~ **ли́н·и·я** (f) feed line; feeder (of antenna)
пит ек·а́нтроп·, -а, (m) (pal) Pithecanthropus
пи́тинг·, -а, (m) pit, pitting (corrosion)
Пито́, тру́б·к·а (f) Pitot tube
пит·о́мец·, -мц·а, (i) **-мц·ем,** (m) foster-child, nursling
пит·о́мник·, -а, (m) (hortic) nursery
пи́ттинг·, -а, (m) see **пи́тинг·**
пи́ттини́т·, -а, (m) (min) pittinite
пи́тч·, -а, (m) pitch (of gears)
~, **диаметра́ль·н·ый** diametral pitch
~, **о·круж·н·о́й** circular pitch
питч·ев·о́й, -а́я, -о́е, (a) of **питч·**
пи́·т·ый, -ая, -ое, (past part pass) of **пи·ть;** drunk (consumed)
пи·ть, (pres 3rd pl) **пь·ют,** (imp) drink
пит·ь·ё, -я́, (n) (v n), see **пи·ть;** drink, beverage
пить·ев·о́й, -а́я, -о́е, (a) potable, drinkable, drinking
Пифаго́р·а, теоре́м·а (f) (math) Pythagorean theorem
пифаго́р·ов·, -а, (predic adj) Pythagorean
~ **теоре́м·а** (f) (math) Pythagorean theorem
~ **стро́·й** (m) (acous) Pythagorean tuning
пи·фото·мезо́н·, -а, (m) (nucl) photopion

пи́хт·а, -ы, (f) (bot) fir, Abies
~, **кана́д·ск·ая** balsam fir, Abies balsamea Mill
пи́хт·ов·ый, -ая, -ое, (a) (bot) fir, abietic
~ **бальза́м·** (m) (chem) balsam of fir, Canada balsam
~ **кисл·от·а́** (f) (chem) abietic acid
~ **ма́сл·о** (n) (chem) oil of fir
пице́н·, -а, (m) (chem) picene
пи́ш·ет, (pres 3rd sing) of **пис·а́ть**
пи́ш·ущ·ий, -ая, -ее, (pres part act) of **пис·а́ть**
~ **колёс·ик·о** (n) (instr) inking wheel
~ **маши́н·к·а** (f) typewriter
~ **о́рган·** (m) (instr) inking element
пи́щ·а, -и, (i) **-ей,** (f) food
пищ·а́лк·а, -и, (g pl) **-лок·,** (f) (rad) tweeter, high-audiofrequency loudspeaker
пищ·а́ть, -а́т, (imp) squeak, peep, tweet
пище·вар·е́ни·е, -я, (n) digestion
пище·вар·и́тельн·ый, -ая, -ое, (a) digestion, digestive, peptic
~ **тракт·** (m) (zool) alimentary tract/canal, digestive tract, gut
~ **ферме́нт·** (m) (biochem) pepsin
пищ·еви́к·, -а́, (m) food-industry worker
пище·во́д·, -а, (m) (anat) oesophagus, esophagus, gullet
пище·во́д·н·ый, -ая, -ое, (a) oesophageal, esophageal
пищ·ев·о́й, -а́я, -о́е, (a) food, alimentary
пи́щ·ик·, -а, (m) (elec) hummer, buzzer
пищ·у́х·а, -и, (f) (zool) pica (pl) Ochotonidae
пиэл·и́т·, -а, (m) (med) pyelitis
пиэло·гра́ф·и·я, -и, (f) (med) pyelography
пиэло·нефр·и́т·, -а, (m) (med) pyelonephritis
пи·эми́·я, -и, (f) (med) pyaemia, pyemia
пи·я, (pres gerund) of **пи·ть** drinking
пия́в·к·а, -и, (g pl) **-в·ок·,** (f) (zool) leech, (pl) Hirudineae
пиявко·обра́з·н·ые, -ых, (pl decl as adj) (zool) Bdelloida
плав·, -а, (m) (v n) of **пла́в·а·ть;** (met) dross; drossing agent (lead refining)
на плав·у́ (adv) afloat

пла́в·ани·е, -я, (*n*) (*v n*) of **пла́в·а·ть;** sailing, navigation

  в пла́в·ани·и (*adv*) (naut) at sea

~, вну́тр·енн·ее inland waterways navigation

~, генера́ль·н·ое (nav) great circle sailing

~, да́ль·н·ее ocean navigation

~, ди́к·ое (naut) tramping

~, ма́л·ое coastal and inland seas navigation

~ по локсо·дро́м·и·и (nav) plane sailing

~ по орто·дро́м·и·и (nav) great circle sailing

~ по с·числ·éни·ю (nav) sailing by dead reckoning

~, су́т·очн·ое (nav) day's run

~ то́н·а (acoust) wow

пла́в·ательн·ый, -ая, -ое, (*a*) swimming, floating, natatory, natatorial

пла́в·а·ть, -ют, (*imp indet*) swim; float, drift; run, sail (of ship)

пла́в·ающ·ий, -ая, -ее, (*pres part act*) of **пла́в·а·ть**

~ автомоби́ль (*m*) amphibious vehicle, wheeled amphibian

~ вал·о́к· (*m*) (met) middle/friction roll (of a Lauth rolling mill)

~ вту́л·к·а (*f*) (mech) floating bush

~ па́л·ец· (*m*) (ICE) floating gudgeon pin

~ реа́ктор· (*m*) (nucl) swimming-pool reactor

~ регуля́тор· (*m*) (autom) floating controller

~ систе́м·а а́дрес·ов (*f*) (autom) floating/symbolic address system

пла́в·ен·ь, -в·н·я, (*m*) (met) flux

плав·ико́в·ый, -ая, -ое, (*a*) fluoric

~ кисл·от·а́ (*f*) (chem) hydrofluoric acid

~ шпат· (*m*) (chem) fluorspar, fluorite

плав·и́льник·, -а, (*m*) melting pot

плав·и́льн·ый, -ая, -ое, (*a*) melting, smelting

~ за·во́д· (*m*) (met) smelting works

пла́в·ить, -ят, (*imp*) melt, fuse, smelt; **-ся** (*pass*); melt; fuse, blow (of fuses)

пла́в·к·а, -и, (*g pl*) **-в·ок·,** (*f*) (*v n*), *see* **пла́в·ить;** (met) melt, heat (operation and product); fusion (welding); smelt; blow (of Bessemer convertor)

~, бала́нс·ов·ая (met) adjustment heat/melt

пла́в·к·а

~, брак·о́в·ая (met) off-heat/melt

~, бро́с·ов·ая (met) off-heat/melt

~, на·ла́д·очн·ая (met) running-in heat/melt

~, о·сад·и́тельн·ая (met) precipitation refining

~ пири́т·н·ая pyritic smelting

~, у·стано́в·очн·ая (met) adjustment melt/heat

~, эталóн·н·ая reference heat/melt

пла́в·к·ий, -ая, -ое, (*a*) fusible, meltable; smeltable

~ в·ста́в·к·а (*f*) (elec) fuse element/link

~ пред·о·хран·и́тел·ь (*m*) (elec) fuse

пла́в·кост·ь, -и, (*f*) fusibility, meltability

~ диагра́мм·а пла́в·кост·и (*f*) (met) phase/constitution diagram

плав·кра́н·, -а, (*m*) floating crane

плавл·éни·е, -я, (*n*) (*v n*) of **пла́в·ить**

  температу́р·а плавл·éни·я (*f*) melting point; fusion point

  тепл·от·а́ плавл·éни·я (*f*) heat of fusion, heat of melting

пла́вл·ен·ый, -ая, -ое, (*a*) melted, fused

~ сыр· (*m*) processed cheese

пла́в·н·и, -ей, (*pl*) flood plain, seasonally flooded land

плав·ни́к·, -а́, (*m*) fin (of fish); wood cast ashore, drift wood

~, жир·ов·о́й (zool) fleshy fin

плав·нико́в·ый, -ая, -ое, (*a*) of **плав·ни́к·**

плавно·регул·и́руем·ый, -ая, -ое, (*a*) continuously adjustable/variable

пла́в·ност·ь, -и, (*f*) smoothness, evenness (of action)

плавно·у·правл·я́ем·ый, -ая, -ое, (*a*) continuously controlled

пла́в·н·ый, -ая, -ое, (*a*) even, flowing, fluent, smooth

~ о·черт·а́ни·е (*pl*) flowing lines

~ регул·иро́вк·а (*f*) continuous control/regulation

~ ход· (*m*) smooth working/running (of mechanisms); smooth course (of a curve)

плав·сре́д·ств·а, -ств·, (*pl*) (naut) floating craft, craft

плав·у́н·, -а́, (*m*) sinking sand, quicksand; (zool) water opossum, *Chironectes minimus*

плав·унéц·, -нц·á, (*i*) **-нц·óм,** (*m*) (ent) water beetle, (*pl*) *Dytiscidae*

плав·ýчест·ь, -и, (*f*) buoyancy

~, нул·ев·áя neutral buoyancy

~, от·риц·áтельн·ая negative buoyancy

~, по·лож·и́тельн·ая positive buoyancy

си́л·а плав·ýчест·и (*f*) buoyancy

плав·ýч·ий, -ая, -ее, (*a*) floating

~ бáз·а (*f*) depot ship

~ бáз·а-фáбр·ик·а (*f*) factory ship, floating factory

~ мáстер·ск·áя (*f*) repair ship

плáв·щин·а, -ы, (*f*) flotsam, floating object

плагиáт·, -а, (*m*) plagiarizing, plagiarism

плагио·гедр·и́ческ·ий, -ая, -ое, (*a*) (cryst) plagiohedral

плагио·клáз·, -а, (*m*) (min) plagioclase

плагио·клаз·и́т·, -а, (*m*) (min) plagioclasite

плагио·трóп·н·ый, -ая, -ое, (*a*) (bot) plagiotropic

плáз·, -а, (*m*) (shipb) mould loft; (a/c) loft floor

плáзм·а, -ы, (*f*) plasma

~, за·рóд·ышев·ая (biol) germ/germinal plasm

~, ламинáр·н·ая (phys) quiescent plasma

~, не·стационáр·н·ая (phys) transient plasma

~, о·грани́ч·енн·ая (phys) confined plasma

~, с·жáт·ая (phys) pinched plasma

~, у·дéрж·анн·ая (phys) confined plasma

плазма·гéн·, -а, (*m*) (gen) plasmagene

плáзм·енн·ый, -ая, -ое, (*a*) plasma

плазмо·дéсм·а, -ы, (*f*) (cyt) plasmodesma

плазм·óди·й, -я, (*m*) (biol) plasmodium

плазмо·динáм·ик·а, -и, (*f*) plasma dynamics

плазмодио·трофоблáст·, -а, (*m*) (biol) plasmoditrophoblast

плазм·óид·, -а, (*m*) (phys) plasmoid

плазмо·ли́з·, -а, (*m*) (biol) plasmolysis

плазмо·нéм·а, -ы, (*f*) (bot) plasmonema

плазмо·ци́т·, -а, (*m*) (cyt) plasmocyte

плáз·ов·ый, -ая, -ое, (*a*) of **плáз·**

~ кни́г·а (*f*) (shipb) table of mould offsets

плáз·ов·ый, -ая, -ое

~ раз·би́в·к·а (*f*) (shipb) laying out; (a/c) lofting

плак·антиклинáл·ь, -и, (*f*) (geol) placanticline

плакáт·, -а, (*m*) placard, poster, wall sheet

плáк·ать, (*pres 3rd sing*) **плáч·ут,** (*imp*) weep, cry

плак·ировáни·е, -я, (*n*) (*v n*) of **плак·и́р·овáть**

плак·и́р·овáть, -ýют, (*imp and perf*) (met) clad; turf (for protection)

плако·дéрм·ы, (*f pl*) (pal) placoderms, *Placodermi*

плак·одóнт·ы, -ов, (*m pl*) (pal) placodonts, *Placodontia*

плак·óид·н·ый, -ая, -ое, (*a*) placoid, plate-shaped

плак·ýч·ий, -ая, -ее, (*a*) weeping

пламе·гас·и́тел·ь, -я, (*m*) (ICE) flame damper; (gunn) flash eliminator; flash reducing charge

плáмен·и, (*gen, dat, prep sing*) of **плáм·я**

плáмен·ник·, -а, (*m*) (bot) phlox

плáмен·н·ый, -ая, -ое, (*a*) flame, flaming; fiery, ardent

~ за·кáл·к·а (*f*) (met) flame hardening

~ лáмп·а (*f*) flame lamp

~ труб·á (*f*) liner, flame tube (gas turbine)

пламе·стóй·к·ий, -ая, -ое, (*a*) flameproof

пламе·чуств·и́тельн·ый автомáт· (*m*) (autom) burner-flame controller

плáм·я, (*gen, dat, prep sing*) **плáмен·и** (*n*) flame; fire

~, бýнзен·овск·ое Bunsen flame

~, вос·станов·и́тельн·ое (weld) reducing flame

~, дýль·н·ое flash (of gun)

~, на·угле·рóж·ивающ·ее (weld) carbonizing flame

~, о·кисл·и́тельн·ое (weld) oxidizing flame

~, подо·грев·áтельн·ое (weld) preheating flame

~, чувств·и́тельн·ое (phys) sensitive flame

~, холóд·н·ое cool flame

плáн·, -а, (*m*) plan, scheme; plan, large-scale map; plane, surface; (cinema) plot

~, генерáль·н·ый layout

~, зáд·н·ий background

пла́н

~ ме́ст·ност·и   ground plan

~, пере́д·н·ий   foreground

~, темат·и́ческ·ий   syllabus

пла́н·а, -ы,   (*f*) (cryst) symmetry plane

план·аксиа́ль·н·ый, -ая, -ое,   (*a*) plane-axial

планари·и, -й,   (*f pl*) (zool) flat worms, planarians, *Planaria*

план·ёр·, -а,   (*m*) glider; airframe

~ -пар·и́тел·ь, -я,   (*m*) sailplane

~ с·на·ря́д·а   (*m*) missile airframe

планер·и́зм·, -а,   (*m*) gliding

плане́т·а, -ы,   (*f*) (astron) planet

~, больш·а́я   major planet

~, вну́тр·енн·яя   inner planet, inferior planet

~, за·хва́т·ывающ·ая   capturing planet

~, ма́л·ая   minor planet, asteroid

~, ни́ж·н·яя   inferior planet

~, транс·непту́н·ов·ая   trans-Neptunian planet

планет·а́ри·й, -я,   (*m*) planetarium

планет·а́рн·ый, -ая, -ое,   (*a*) planetary, epicyclic

~ механи́зм· по·воро́т·а   (*m*) (М/Т) epicyclic steering gear

~ пере·да́·ч·а   (*f*) (mech) planetary gear

~ тума́н·ност·ь   (*f*) (astron) planetary nebula

планетезима́л·ь, -и,   (*f*) (astron) planetesimal

плани·ме́тр·, -а,   (*m*) (math) planimeter

плани·ме́тр·и·я, -и,   (*f*) plane geometry, planimetry

плани́р·, -а,   (*m*) levelling bar (in coke ovens)

плани́р·н·ый, -ая, -ое,   (*a*) levelling

план·и́ровани·е, -я,   (*n*) *see also* план·иро́вк·а   planning, plan; (air) gliding, glide

~, календа́р·н·ое   scheduling

~, операти́в·н·ое   schedule planning

~, рас·чёт·н·ое   (air) approach glide

~, с·во́д·н·ое   coordinating

     угол· план·и́ровани·я   (*m*) (air) angle of glide

пл ан·и́р·овать, -уют,   (*imp*) (air) glide; plan, draft a plan; **план·ир·о·ва́ть, -у́ют,**   (*imp*) locate, lay out (according to plan)

план·иро́вк·а, -и,   (*f*) (*v n*) of **план·и́р·овать** *see also* план·и́рова·ни·е;** layout, design; location drawing

план·иро́вк·а

~, вертика́ль·н·ая   (civ eng) profile/ subservice location drawing

план·и́рующ·ий, -ая, -ее,   (*pres part act*) *see* план·и́р·овать

плани·сфе́р·а, -ы,   (*f*) (astron) planisphere

Пла́нк·а, по·сто·я́нн·ая   Planck's constant

пла́н·к·а, -и,   (*g pl*) **-н·ок·,** (*f*) plank, strip (of wood etc.), pad, gib; slat, lath; strap (in welding or riveting); cleat, clamp

~, ки́п·ов·ая   (shipb) fairlead

~, на·жи́м·н·ая   clip

~, па́з·ов·ая   seam strap

~, под·кла́д·н·а́я   backing strap

~, при·жи́м·н·ая   clamping strap

~, при·це́ль·н·ая   backsight leaf (of rifle)

~, с·бра́с·ывающ·ая   (text) work/ facing bar

~, сто́пор·н·ая   stop plate

~, с·ты́к·ов·а́я   butt strap

~, щит·ов·а́я   (rail) snow-fence board

планкто́н·, -а,   (*m*) (biol) plankton

планкто́н·н·ый, -ая, -ое,   (*a*) plankton, planktonic

план·ови́к·, -а́,   (*m*) planner

пла́н·ов·о-организацио́н·н·ый, -ая, -ое,   (*a*) organizational planning; planned

пла́н·ов·о-пред·у·пред·и́тельн·ый, -ая, -ое,   (*a*) periodic preventive

~ ремонт·   (*m*) periodic preventive maintenance, maintenance, service

пла́н·ов·ый, -ая, -ое,   (*a*) of **пла́н·**

~ воз·ду́ш·н·ое фотограф·и́рова·ни·е   (*n*) vertical aerial photography

~ за·да́н·и·е   (*n*) (civ eng) specification

плано·гаме́т·а, -ы,   (*f*) (biol) planogamete

плано·зиго́т·а, -ы,   (*f*) (bot) planozygote

плано·ме́р·ност·ь, -и,   (*f*) systematization; development according to plan; planning

плано·ме́р·н·ый, -ая, -ое,   (*a*) planned, according to plan, systematic

плант·а́ж·, -а,   (*m*) (agr) deep ploughing, trenching

плант·а́ци·я, -и,   (*f*) plantation

~, виногра́д·н·ая   vineyard

пла́н·ул·а, -ы,   (*f*) (zool) planula

план·ша́йб·а, -ы,   (*f*) (mech) face chuck/plate

планшѐт·, -а, (*m*) (surv) plane table; (mil) plot, plotting board; map case, chart board

~, мѐнзуль·н·ый plane-table sheet

~, огн·ев·о́й (gunn) firing chart

планшет·и́ст·, -а, (*m*) (rad) plotter

планшѐт·н·ый, -ая, -ое, (*a*) of **план-шѐт·**

планши́р·, -а, (*m*) (shipb) gunwale

пласко́н·, -а, (*m*) (plast) plascon

пласт·, -а́, (*m*) layer, (geol) stratum, bed; (agr) furrow

вы́·ход· пласт·а́ (*m*) (geol) outcrop

~, у́голь·н·ый coal seam/bed

пласт·и́д·а, -ы, (*f*) (cyt) plastid

пла́ст·ик·, -а, (*m*) plastic

~, сло·и́ст·ый laminated plastic

пла́ст·ик·а, -и, (*f*) sculpture, plastic art

пласт·ика́т·, -а, (*m*) plastic compound; plasticized rubber

пласт·ика́тор·, -а, (*m*) (chem) plasticator

пласт·ика́ци·я, -и, (*f*) plastication, plasticization; smearing (in extruding machine)

~, механ·и́ческ·ая mastication

~ на вальц·а́х mill plasticization; mastication

пласт·иле́н·, -а, (*m*) plastilene

пласт·или́н·, -а, (*m*) plasticine

пласт·и́н·а, -ы, (*f*) *see also* **пласт·и́нк·а;** plate, flake, membrane, lamina, platelet; (elec) commutator segment; fin (engine radiator); flitch (timber); link plate (roller chain); record, disc (gramophones)

~, вертика́ль·н·о-от·клон·я́ющ·ая Y-plate (cathode ray tube)

~, горизонта́ль·н·о-от·клон·я́ющ·ая X-plate (cathode ray tube)

~, долго·игр·а́ющ·ая long-playing record/disc

~, за·па́ль·н·ая (nucl) burnup/seed-fuel plate

~, ква́рц·ев·ая piezoelectric crystal plate

~, ма́сс·ов·ая (elec) mass-type accumulator plate

~, паст·и́рованн·ая pasted plate (electric batteries)

~, по·ве́рх·ностн·ая formed plate (electric batteries)

~ прям·о́го вос·про·из·вед·е́ни·я instantaneous record/disc

~, техн·и́ческ·ая industrial rubber sheeting

пласт·и́н·а

~ -утюж·о́к·, -ж·к·а́, (*m*) smoothing plate

~ фо́рм·н·ая (print) form plate

пласт·и́нк·а, -и, (*g pl*) **-нок·,** (*f*) (*dim*) of **пласт·и́н·а** (q. v.); tool tip; vane, blade, reed; (bot) blade; loose tongue (carpentry)

~ вы́·бег·а́ющ·ая sliding vane (of pump)

~ за·вихр·и́тел·я swirl plate (jet eng)

~ ка́пел·ьн·ая (chem) spotting tile

~, паль·цев·а́я ledger plate/strip/board (carpentry)

~, сигна́ль·н·ая signal plate (of cathode ray tube)

~ твёрд·ого с·пла́в·а cemented carbide tip

~, у·пру́г·ая vibrating reed

пласт·и́нник·, -а, (*m*) offcut (timber)

пластино·жа́бер·н·ые, -ых, (*pl decl as adj*) (fish) elasmobranchs, *Elasmobranchii*

пластино·ко́ж·ий, -ая, -ее, (*a*) placodermal, (*pl as noun*) (pal) placoderms, *Placodermi*

пластино·обра́з·н·ый, -ая, -ое, (*a*) platelike, lamellar

пластинчато·жа́бер·н·ые, -ых, (*pl decl as adj*) (zool) lamellibranchs, *Lamellibranchiata*

пластинчато·зу́б·ая крыс·а (*f*) (zool) bandicoot rat, (*pl*) *Nesokia*

пласт·и́нчатост·ь, -и, (*f*) lamination, foliation

пластинчато·у́с·ые, -ых, (*pl decl as adj*) (ent) lamellicorn beetles, *Scarabaeidae*

пласт·и́нчат·ый, -ая, -ое, (*a*) laminated, laminar, lamellar, flaked, flaky, scaly; plate, plate-like, tabular, sheet

~ транспортёр· (*m*) plate conveyer, pallet train

пластифик·а́тор·, -а, (*m*) (chem) plasticizer

пластифик·а́ци·я, -и, (*f*) (chem) plasticization

пластифиц·и́ровани·е, -я, (*n*) (*v n*) of **пластифиц·и́р·овать** plasticizing

пластифиц·и́р·овать, -уют, (*imp and perf*) plasticize

пластиц·и́р·овать, -уют, (*imp and perf*) plasticize

пласт·и́ческ·ий, -ая, -ое, (*a*) plastic

~ деформ·а́ци·я (*f*) (met) plastic deformation

пласт·и́ческ·ий, -ая, -ое
~ **ма́сс·а** (*f*) (*see also* **пласт·ма́сс·а**) plastic
пласт·и́чност·ь, -и, (*f*) plasticity, ductility; plastic deformation
~, **кас·а́тельн·ая** ductility in shear **мо́дул·ь кас·а́тельн·ой пласт·и́чност·и** (*f*) tangent modulus in shear **мо́дул·ь пласт·и́чност·и** (*f*) tangent modulus, modulus of flexibility
пласт·и́чн·ый, -ая, -ое, (*a*) plastic, ductile
пласт·ко́ж·а, -ы, (*f*) fibre/fiber leather
пласт·ма́сс·а, -ы, (*f*) plastic
~, **термо·пласт·и́ческ·ая** thermoplastic plastic
~, **термо·реакти́вн·ая** thermosetting plastic
пласт·ма́сс·ов·ый, -ая, -ое, (*a*) of **пласт·ма́сс·а**
пласто·бето́н·, -а, (*m*) plastic concrete
пласт·ова́ни·е, -я, (*n*) (geol) stratification, bedding
пласт·ов·о́й, -а́я, -о́е, (*a*) of **пласт·**; stratified
~ **давл·е́ни·е** (*n*) (geol) formation/ rock pressure
~ **культу́р·а** (*f*) (agr) furrow-slice/ sod crop
пласто·ме́р·, -а, (*m*) (chem) plastomer
пласто́·метр·, -а, (*m*) plastometer
пласто·со́м·а, -ы, (*f*) (cyt) plastosome
пласто·эласт·и́ческ·ий, -ая, -ое, (*a*) plasto-elastic
пластро́н·, -а, (*m*) (zool) plastron
пла́стыр·ь, -я, (*m*) (med) plaster; (naut) collision mat
пла́т·а, -ы, (*f*) pay, fee; (fin) charge; board (printed circuit); (telecom) mounting plate; (autom) plane (of a core matrix)
~, **за́·работ·н·ая** wages, salary
~, **о·со́б·ая** extra charge
плата́н·, -а, (*f*) (bot) plane tree, *Platanus*, sailmaker's palm
плат·бло́к·, -а, (*m*) (naut) snatch block
плат·ёж·, -еж·а́, (*i*) **-еж·о́м,** (*m*) payment
~ **в рас·сро́ч·к·у** payment by instalments
~, **на·ли́ч·н·ый** cash payment
~, **на·ло́ж·енн·ый** payment forward, cash on delivery
платёже·спосо́б·ност·ь, -и, (*f*) (fin) solvency

платёже·спосо́б·н·ый, -ая, -ое, (*a*) (fin) solvent
плат·ёжн·ый, -ая, -ое, (*a*) pay, payment
~ **бала́нс·** (*m*) (fin) balance of payments
~ **ве́д·омост·ь** (*f*) pay roll
плат·е́льщик·, -а, (*m*) payer
платео·за́вр·, -а, (*m*) (pal) *Plateosaurus*
плати·белодо́н·, -а, (*m*) (pal) *Platybelodon*
пла́тин·а, -ы, (*f*) (chem) platinum, Pt; sinker (of a knitting machine)
   гидро·о́·кис·ь пла́тин·ы (*f*) platinic hydroxide
   дву·о́·кис·ь пла́тин·ы (*f*) platinum dioxide, platinic oxide
~, **дву·хло́р·ист·ая** platinum dichloride, platinous chloride
~, **хло́р·н·ая** (chem) platinum tetrachloride
платин·а́т·, -а, (*m*) (chem) platinate
платин·и́ровани·е, -я, (*n*) (chem) platinizing
платин·и́т·, -а, (*m*) (met) platinite
пла́тин·н·ый, -ая, -ое, (*a*) (text) sinker
пла́тин·ов·ый, -ая, -ое, (*a*) platinum, platinic
платин·о́ид·, -а, (*m*) (met) platinoid
платино·ро́д·нев·ый, -ая, -ое, (*a*) platinum rhodium
платино·хлористо·водо·ро́д·н·ая кис·л·от·а́ (*f*) chloroplatinic acid, platinum chloride (commercial name)
платино·циа́н·ист·ый ба́р·и·й (*m*) barium platinocyanide
плат·и́ть, -ят, (*imp*) pay
плати·филли́н·, -а, (*m*) (pharm) platyphylline
плати·цефа́л·и·я, -и, (*f*) (anat) platycephaly
плати·цера́тид·, -а, (*m*) (pal) platyceratid
пла́т·н·ый, -ая, -ое, (*a*) payable, chargeable; paid; paying
~ **на·гру́з·к·а** (*f*) (a/c) pay load
~ **служ·е́бн·ая радио·телегра́мм·а** (*f*) (telecom) paid service message
плато́ (*n indecl*) plateau
~ **с·чёт·чик·а** (nucl) counter plateau
плат·о́к·, -т·к·а́, (*m*) (text) square
~, **голов·н·о́й** shawl, head-square/ scarf
~, **нос·ов·о́й** handkerchief

платфо́рм·а, -ы, (*f*) platform; (rail) flatcar, flat wagon; throat plate (of sewing machine)

∼, берег·ов·а́я (geol) bench, off-shore bench

∼, лет·а́ющ·ая (a/c) flying bedstead

∼, плав·у́ч·ая подъ·ём·н·ая (oil) drilling barge/platform

платфо́рминг·, -а, (*m*) (oil) platforming (catalytic reforming with a platinum catalyst)

пла́т·ь·е, -я, (*n*) clothes, clothing

плау́н·, -а́, (*m*) club moss, *Lycopodium*, (*pl*) *Lycopodiaceae*

плауно·ви́д·н·ые, -ых, (*pl decl as adj*) (bot) lycopsids, *Lycopsida*

плау́н·ов·ые, -ых, (*pl decl as adj*) (bot) *Lycopodiales*

плафо́н·, -а, (*m*) (arch) plafond; ceiling fitting (for a light); decorated ceiling

плацда́рм·, -а, (*m*) (mil) bridgehead, beachhead; battle area

плац·е́нт·а, -ы, (*f*) (biol) placenta

плац·ента́рн·ый, -ая, -ое, (*a*) placental, placentate, placentiferous, placentigerous, (*pl as noun*) (zool) *Eutheria, Placentaria*

плац·ка́рт·а, -ы, (*f*) seat reservation ticket

пла́ч·ет, (*pres 3rd sing*) of пла́к·ать

пла́ч·ущ·ий, -ая, -ее, (*pres part act*) of пла́к·ать; weeping

пла́ш·к·а, -и, (*g pl*) -ш·ек·, (*f*) die, screw thread die; (oil) slip

плашко́ут·, -а, (*m*) (naut) scow

плашмя́ (*adv*) flat, flatways, flatwise, on the flat side, flat side down; prone, supine

∼ -на·пра́вл·енн·ый, -ая, -ое, (*a*) broadside

плащ·, -а́, (*i*) -о́м, (*m*) cloak; raincoat; (geol) mantle

плев·а́, -ы́, (*f*) membrane, pellicle, film

∼, де́в·ственн·ая (anat) hymen

∼, под·ко́ж·н·ая flesh side (of hides)

плев·а́ть, (*pres 3rd pl*) плю·ю́т, (*imp*) spit, expectorate

плев·а́льниц·а, -ы, (*i*) -ей, (*f*) spittoon

пле́вел·, -а, (*m*) (bot) rye grass, *Lolium*

плев·о́к·, -в·к·а́, (*m*) spit, spittle, sputum

пле́вр·а, -ы, (*f*) (zool) pleura

плевр·ака́нт·н·ый, -ая, -ое, (*a*) pleuracanth

плевр·а́льн·ый, -ая, -ое, (*a*) (zool) pleural

плевр·и́т·, -а, (*m*) (med) pleurisy; (zool) pleurite

плевр·один·и́·я, -и, (*f*) (med) pleurodynia

плевр·ом·е́·и, -й, (*f pl*) (pal) *Pleuromeia*

плевро́н·, -а, (*m*) (zool) pleuron

плевро·пневмон·и́·я, -и, (*f*) (med, vet) pleuropneumonia

плевро·спо́р·а, -ы, (*f*) (bot) pleurospore

плед·, -а, (*m*) rug

плези·адапи́д·ы (*pl*) (pal) plesiadapids, *Plesiadapidae*

плези·а́нтроп·, -а, (*m*) (pal) *Plesianthropus*

плезио·за́вр·, -а, (*m*) (pal) *Plesiosaurus*

плезио·мо́рф·н·ый, -ая, -ое, (*a*) plesiomorphic, plesiomorphous

плезио·ти́п·, -а, (*m*) plesiotype

плейо- (*int component*) *see also* плио-

плейо·ха́з·н·й, -я, (*m*) (bot) pleiochasium

плейсто·фи́р·ов·ый, -ая, -ое, (*a*) (geol) pleistophyric

плейсто·це́н·, -а, (*m*) (geol) Pleistocene

плексигла́с·, -а, (*m*) (plast) plexiglass

плекс·и́т·, -а, (*m*) (med) plexitis

плект·енхи́м·а, -ы, (*f*) (bot) plectenchyma

плекто·сте́л·а, -ы, (*f*) (bot) plectostele

плё·л (*past masc sing*) of плес·ти́

племен·н·о́й, -а́я, -о́е, (*a*) tribal; (agr) pedigree, pure-strain

∼ де́л·о (*n*) pedigree stockbreeding

∼ кни́г·а (*f*) (agr) herd book

∼ рас·са́д·ник· (*m*) pedigree stock selection and breeding centre

пле́м·я, (*gen, dat, prep sing*) пле́мен·и, (*nom pl*) племен·а́, (*n*) tribe; (agr) breed

плем·я́нник·, -а, (*m*) nephew

плен·, -а, (*n*) captivity

плен·а́, -ы́, (*f*) (met) scab

плена́р·н·ый, -ая, -ое, (*a*) plenary

плён·к·а, -и, (*g pl*) -н·ок·, (*f*) film, pellicle; tape (magnetic sound recorder); husk, hull (of cereal grains etc.)

∼ -конденса́т·, -а, (*m*) condensate film

пленко·образ·ова́тел·ь, -я, (*m*) film-forming material

пленко·образ·у́ющ·ий, -ая, -ее, (*a*) film-forming, skinning

плен·ник·, ·а, (*m*) prisoner, captive

плен·н·ый, -ая, -ое, (*a*) captive

плён·оч·ный, -ая, -ое, (*a*) of **плён·-к·а;** film, filmy, pellicular

~ **вод·а́** (*f*) (geol) pellicular water

~ **шунт·** (*m*) film cutout (vacuum tubes)

пле́нум·, -а, (*m*) plenary session

плён·чат·ый, -ая, -ое, (*a*) filmy; scaly; (bot) glumaceous

плео·кристалл·и́ческ·ий, -ая, -ое, (*a*) (cryst) pleocrystalline

плео·морф·и́зм·, -а, (*m*) (biol) pleomorphism

плео·на́ст·, -а, (*m*) (min) pleonaste

плео·хро·и́зм·, -а, (*m*) (cryst) pleochroism

плео·хро·и́ческ·ий, -ая, -ое, (*a*) pleochroic

плео·хро·и́чн·ые орео́л·ы (*pl*) (min) pleochroic haloes

плеро́м·а, -ы, (*f*) (bot) plerome

плеро·церко́ид·, -а, (*m*) (zool) plerocercoid

плёс·, -а, (*m*) reach, stretch of water (river)

пле́сен·ь, -и, (*f*) (biol) mould, mold, must, fungus, mildew

пле́ск·, -а, (*m*) splash

плеск·а́ть, (*pres 3rd pl*) **пле́щ·ут,** (*imp*) splash

плесн·еве́лост·ь, -и, (*f*) mustiness, mouldiness, moldiness

плесн·еве́л·ый, -ая, -ое, (*a*) mouldy, moldy, musty

пле́сн·еве·ть, -ют, (*imp*) grow mouldy/musty

плесн·ев·о́й, -а́я, -о́е, (*a*) of **пле́сен·ь**

~ **гриб·** (*m*) saprophytic fungus

плесне·у·сто́й·чив·ый, -ая, -ое, (*a*) mould-resistant, mold-resistant

плес·н·у́ть, -у́т, (*perf*) splash

плесси́·метр·, -а, (*m*) (med) pleximeter

плесси́т·, -а, (*m*) (min) plessite

плес·ти́, (*pres 3rd pl*) **плет·у́т,** (*imp*) braid, plait, weave

плет·е́льн·ый, -ая, -ое, (*a*) braiding, braided, wicker

~ **маши́н·а** (*f*) (text) braiding machine

плет·е́ни·е, -я, (*n*) (*v n*), *see* **плес·ти́;** braiding, basketwork, wickerwork; network, plexus

плет·ёнк·а, -и, (*g pl*) **-нок·,** (*f*) (text) braiding; wicker basket, hamper; wickerwork, basketwork

плет·ённ·ый, -ая, -ое, (*a*) (*past part pass*) of **плес·ти́**

плет·ён·ый, -ая, -ое, (*a*) braided, plaited

~ **корзи́н·а** (*f*) wicker basket

~ **се́т·к·а** (*f*) (elec) mesh grid

плет·е́н·ь, -т·н·я́, (*m*) wattle, wattling

плет·ёт, (*pres 3rd sing*) of **плес·ти́** (*imp*)

плетизмо·гра́мм·а, -ы, (*f*) plethysmograph recording/chart/trace

плетизмо́·граф·, -а, (*m*) (med, instr) plethysmograph

плето́р·а, -ы, (*f*) (med) plethora

плетор·и́т·, -а, (*m*) (min) plethorite

плет·у́щ·ий, -ая, -ее, (*pres part act*) of **плес·ти́**

плёт·ш·ий, -ая, -ее, (*past part act*) of **плес·ти́**

плеч·ев·о́й, -а́я, -о́е, (*a*) of **плеч·о́;** humeral, brachial

~ **кост·ь** (*f*) (zool) humerus

плече·локт·ев·о́й, -а́я, -о́е, (*a*) (anat) humero-ulnar

плече·луч·ев·о́й, -а́я, -о́е, (*a*) (anat) brachioradial

плече·но́г·ие, -их, (*pl decl as adj*) (zool) brachiopods, *Brachiopoda*

пле́ч·ик·о, -а, (*nom pl*) **-и,** (*g pl*) **-ов,** (*n*) (*dim*) of **плеч·о́;** (text) shoulder strap

плеч·о́, -а́, (*i*) **-ч·о́м,** (*nom pl*) **пле́ч·и,** (*g pl*) **плеч·,** (*n*) shoulder; (anat) brachium, upper arm; (tech) arm, leg

~, **за·да·ю́щ·ее** master arm (of manipulator)

~, **ис·полн·и́тельн·ый,** slave arm (of manipulator)

~ **криво·ши́п·а** crank web

~ **па́р·ы сил·** (mech) arm of couple, couple arm

~ **си́л·ы** moment arm

плеш·и́вост·ь, -и, (*f*) baldness, (med) alopecia

плеш·и́в·ый, -ая, -ое, (*a*) bald, balding

пле́щ·ет, (*pres 3rd sing*) of **плеск·а́ть**

плея́д·а, -ы, (*f*) pleiad, pleyad; (astron) (*pl*) Pleiades

~ **изото́п·ов** (nucl) isotope group/pleyade

пли (*imper*) fire!

пликат·и́вн·ый, -ая, -ое, (*a*) plicated, plicate

плиниа́н·ск·ий, -ая, -ое, (*a*) (geol) Plinian

плинт·, -а, (*m*) plinth; (telecom) terminal/distributing block

плинти́т·, -а, (*m*) (min) plinthite

пли́нтус·, -а, (*m*) plinth; skirting board

плио·питёк·, -а, (*m*) (pal) Pliopithecus

плиотро́н·, -а, (*m*) (rad) pliotron

плио·цён·, -а, (*m*) (geol) Pliocene, Pleiocene

плис·, -а, (*m*) (text) velvet, plush

плиссе́ (*n indecl*) (text) plissé

плисс·ир·ова́ть, -у́ют, (*imp*) pleat

плит·а́, -ы́, (*nom pl*) пли́т·ы, (*g pl*) плит·, (*f*) plate, slab, panel; platen (of press); (geol) platform

~, а́нкер·н·ая anchor plate

~, арм·и́рующ·ая (plast) backing plate

~, бес·конёч·н·ая (phys) infinite slab

~, брон·ев·а́я armour plate

~, вы·тяж·н·а́я (met cast) stripping plate

~, га́з·ов·ая gas cooker, gas stove (domestic)

~, за·жи́м·н·ая (mech) bolster plate

~, клин·ов·а́я (mech) key plate

~, кондýктор·н·ая (mech) bush/bushing plate

~, кузн·ечн·ая (met forge) swage block

~, магни́т·н·ая (mech) magnetic chuck

~, модёль·н·ая (met cast) pattern plate

~, о·по́р·н·ая (mech) bearing plate; (plast) backing plate

~, по·вёр·очн·ая (mech) check faceplate

~, по́д·ов·ая bed plate (cupola)

~, прав·и́льн·ая (mech) dressing/adjustment plate

~, про́бк·ов·ая corkboard

~, раз·мёт·очн·ая (mech) surface reference plate, tracing plate

~, реверси́в·н·ая (met cast) match plate

~, рихтова́ль·н·ая (mech) check faceplate

~, стол·я́рн·ая coreboard

пли́т·к·а, -и, (*g pl*) -т·ок·, (*f*) (*dim*) of плит·а́; tile, thin slab; bar (of chocolate etc.); (elec) hot plate; (mech) slip gauge/gage, gauge/gage block

плит·ня́к·, -а́, (*m*) (build) flagstone, flag; plate coal

плит·ови́н·а, -ы, (*f*) (met roll) mill bedplate, housing shoe

плито·обра́з·н·ый, -ая, -ое, (*a*) plate/slab-shaped, tabular

пли́т·очн·ый, -ая, -ое, (*a*) of плит·а́, пли́т·к·а

~ рабо́т·а (*f*) (build) tiling

пли́ц·а, -ы, (*i*) -ей, (*f*) bailer; paddle (of paddle wheel)

плов·ёц·, -в·ц·а, (*i*) -в·ц·о́м, (*m*) swimmer

плов·у́чест·ь, -и, (*f*) *see* плав·у́чест·ь

плов·у́ч·ий, -ая, -ее, (*a*) *see* плав·у́ч·ий floating

плод·, -а́, (*m*) fruit; foetus, fetus

~, о·камен·ёл·ый (geol) carpolite, carpolith

плод·и́ть, -я́т, (*imp*) procreate, produce

пло́д·ник·, -а, (*m*) (bot) pistil

пло́д·н·ый, -ая, -ое, (*a*) foetal, fetal

плод·ови́тост·ь, -и, (*f*) fertility, fecundity, fruitfulness

плодо·во́д·ств·о, -а, (*n*) fruit farming/culture

плод·о́в·ый, -ая, -ое, (*a*) fruit

~ мýш·к·а (*f*) drosophila, fruit fly

~ тёл·о (*n*) fruit body

плодо·жо́р·к·а, -и, (*g pl*) -р·ок·, (*f*) (ent) codling moth, *Carpocapsa*

плодо·ли́ст·ик·, -а, (*m*) (bot) carpel

плодо·но́ж·к·а, -и, (*g pl*) -ж·ек·, (*f*) (bot) fruit stem, carpophore

плодо·нос·и́ть, -я́т, (*imp*) fruit, bear fruit

плодо·но́с·н·ый, -ая, -ое, (*a*) fertile, fruitful

плодо·овощ·н·о́й, -а́я, -о́е, (*a*) fruit-and-vegetable

плодо·пит·о́мник·, -а, (*m*) (hortic) nursery

плодо·пит·о́мск·ий, -ая, -ое, (*a*) nursery

плодо·ро́д·н·ый, -ая, -ое, (*a*) fertile, productive

плодо·с·мён·, -а, (*m*) crop rotation

плодо·тво́р·н·ый, -ая, -ое, (*a*) fruitful

плодо·я́д·н·ый, -ая, -ое, (*a*) (zool) frugiverous

плод·у́щ·ий, -ая, -ее, (*a*) fruiting, fruit-bearing

пло́й·чат·ый, -ая, -ое, (*a*) plicated, puckered

пломатро́н·, -а, (*m*) (rad) plomatron

пло́мб·а, -ы, (*f*) seal, lead seal; filling, stopping (in tooth)

пломби́р·, -а, (*m*) scaling tool; (food) sundae

пломб·ирова́ни·е, -я, (*n*) (*v n*) of **пломб·ир·ова́ть**

пломб·ир·ова́ть, -у́ют, (*imp*) seal, seal with lead; fill, stop (teeth)

плóск·ий, -ая, -ое, (*a*) flat, plane, two-dimensional

~ **волн·а́** (*f*) (phys) plane wave

~ **движ·éни·е** (*n*) (dynam) translation, translational motion

~ **за·да́·ч·а** (*f*) (math) two-dimensional problem

~ **по·вéрх·ност·ь** (*f*) plane surface

~ **сéт·к·а** (*f*) (cryst) two-dimensional lattice

~ **схéм·а** (*f*) (elec) planar network

~ **характер·и́стик·а** (*f*) (phys) flat response, flat response curve

~ **чéрв·и** (*pl*) (zool) flatworms, *Platyhelminthes*

плóск·о- (*root*) flat-, plane-, plano-, platy-

плоско·вéрх·н·ий, -яя, -ее, (*a*) flat-top

плоско·вó·гн·ут·ый, -ая, -ое, (*a*) planoconcave

плоско·вы́·пукл·ый, -ая, -ое, (*a*) planoconvex

плоско·вяз·а́льн·ая маши́н·а (*f*) (text) flat-bed machine, flat-knitting machine

плоско·голóв·ост·ь, -и, (*f*) (anat) platycephaly

плоско·гóр·ь·е, -я, (*nom pl*) **-р·и·й,** (*n*) (geog) plateau

плоско·гра́н·ност·ь, -и, (*f*) (cryst) plane-face feature

плоско·гу́б·ц·ы, -ев, (*pl*) pliers

~, **комбин·и́рованн·ые** cutting pliers

плоско·дóн·к·а, -и, (*g pl*) **-н·ок·,** (*f*) punt, flat-bottomed boat

плоско·дóн·н·ый, -ая, -ое, (*a*) flat-bottomed

плоско·компа́унд·ированн·ый, -ая, -ое, (*a*) (elec) flat-compounded, plane-compounded

плоско·паралле́ль·н·ый, -ая, -ое, (*a*) (math) plane-parallel; parallel-plate

~ **по·тóк·** (*m*) (dynam) plane flow, two-dimensional flow

плоско·печа́т·н·ая маши́н·а (*f*) flat-bed printing press

плоско·по·крóв·н·ый, -ая, -ое, (*a*) (bot) planitegillate

плоско·поляриз·óванн·ый, -ая, -ое, (*a*) plane-polarized

плоско·пруж·и́нп·ый, -ая, -ое, (*a*) flat-spring, strap

плоск·остн·óй, -а́я, -óе, (*a*) plane, planar, flat; junction, junction-type (of semiconductors)

~ **транзи́стор·** (*m*) junction transistor

плóск·остност·ь, -и, (*f*) flatness

плóск·ост·ь, -и, (*f*) flatness; plane; face (of a pulse); (a/c) main plane

~, **ба́зис·н·ая** (cryst) basal plane

~, **ба́з·ов·ая** (surv) datum plane

~, **гла́в·н·ая** (opt) principal plane

~, **гран·и́чн·ая** (geol) parting plane

~, **двóй·никóв·ая** (cryst) twinning plane

~, **диаметра́ль·н·ая** (shipb) fore-and-aft midship line, middle line, **centre** line

~, **за́·да·нн·ая** (surv) reference plane

~, **ис·хóд·н·ая** (surv) datum plane; (air phot) assumed ground plane

~, **о·пóр·н·ая** (surv) plane of support

~, **оснóв·н·а́я** (cryst) basal plane

~, **от·раж·а́ющ·ая** (cryst) reflection plane

~, **у·рóвен·н·ая** (air phot) reference/ground plane

плоско·шлиф·ова́льн·ая маши́н·а (*f*) (mech) flat surfacing machine

плот·, -а́, (*m*) raft, float

~, **с·пас·а́тельн·ый** life raft

плотв·а́, -ы́, (*f*) (fish) roach, *Rutilus rutilus*

плóт·ик·, -а, (*m*) (geol) bedrock

плот·и́н·а, -ы, (*f*) (civ eng) dam

~, **а́р·очн·ая** horizontal-arch dam

~, **водо·с·ли́в·н·ая** spillway dam; weir

~, **гравитациóн·н·ая** gravity dam

~, **земл·ян·а́я** earth/earth-fill dam

~, **контрфóрс·н·ая** multiple-arch dam

~, **на·брóс·н·ая** rockfill dam

~, **от·вóд·н·ая** diversion dam

~, **раз·бóр·чат·ая** low-head dam

плот·нéйш·ий, -ая, -ее, (*a*) (*superl*) of **плóт·н·ый**

плóт·ник·, -а, (*m*) carpenter

плóт·ничн·ый, -ая, -ое, (*a*) carpentry

~ **дéл·о** (*n*) carpentry

~ **цех·** (*m*) carpenter's shop, (shipb) shipwright's shop

плóт·н·о (*adv*) tightly, closely; densely, compactly

плотно·зерн·и́ст·ый, -ая, -ое, (*a*) close-grained

плотно·мéр·, -а, (*m*) densitometer

плотносте·ме́р·, -а, (*m*) = **плотно-·ме́р·**

пло́т·ност·ь, -и, (*f*) density, compactness, solidity; tightness, closeness; soundness (of a casting)

~ **вероя́т·ност·и** probability density/function

~ **за́·пис·и** packing density (magnetic tape etc.)

~ **об·луч·е́ни·я** intensity of irradiation, irradiance; intensity of illumination (radar)

~**, объ·ём·н·ая** volume density

~ **решёт·к·и** (cryst) lattice tightness

~ **со·сто·я́ни·я** (phys) density of states

~ **то́к·а** (elec) current density, specific current

~ **энерго·вы·дел·е́ни·я** power density

плотно·у·пак·о́ванн·ый, -ая, -ое, (*a*) close-packed

~ **структу́р·а, гексагона́ль·н·ая** (*f*) (cryst) hexagonal close-packed structure

пло́т·н·ый, -ая, -ое, (*a*) compact, dense, close, thick; solid, strong, tight; (geol) massive

~ **мно́ж·еств·о** (*n*) (math) dense set

~ **сред·а́** (*f*) dense/heavy medium; (bact) solid culture medium

плот·ов·о́й, -а́я, -о́е, (*a*) of **плот·**

~ **шум·** (*m*) thump (telephones)

плото·фи́т·, -а, (*m*) (bot) photophyte

плото·хо́д·, -а, (*m*) (civ eng) raft chute

плото·я́д·н·ый, -ая, -ое, (*a*) carnivorous, flesh-eating; (*pl as noun*) (zool) carnivores, *Carnivora*

пло́т·ь, -и, (*f*) flesh

~**, кра́й·н·яя** (anat) prepuce

пло́х·о (*adv*) badly; (*n indecl*) bad mark, bottom marking

плох·о́й, -а́я, -о́е, (*a*) bad, poor

площа́д·к·а, -и, (*g pl*) **-д·ок·,** (*f*) ground, area, space; platform, staging, stage, stand; landing (stairs); plateau (on graph)

~**, груз·ов·а́я** (naut) tea board

~**, за·гру́з·очн·ая** (met) charging table/platform; charging area (nuclear reactor)

~**, по·са́д·очн·ая** airfield

~ **при·земл·е́ни·я** (air) landing zone

~**, пуск·ов·а́я** (rocket) launching platform

площа́д·к·а

~**, рабо́ч·ая** (met) furnace platform/stage/level; working/operations area; furnace platform (of OH furnace)

~**, раз·ли́в·очн·ая** (met) casting platform

~**, стро·и́тельн·ая** building site

~ **тек·у́чест·и** (met) yield point elongation/plateau

пло́щад·ь, -и, (*nom pl*) **-и,** (*g pl*) **-е́й,** (*f*) area; square (in towns)

~ **бассе́йн·а** (geol) catchment/drainage area

~**, де́й·ствующ·ая** effective area

~**, конта́кт·н·ая** contact area

~ **по·ве́рх·ност·и** (math) square measure

~ **по·глащ·е́ни·я** (chem, phys) absorption cross section

за·ко́н· площад·е́й (*m*) (dynam) area rule

пло́щ·иц·а, -ы, (*f*) (ent) crab louse, *Phthirius pubis*

плу́г·, -а, (*nom pl*) **-и́,** (*g pl*) **-о́в,** (*m*) plough, plow

~**, баланси́р·н·ый** balance plough

~**, вис·я́ч·ий** swing plough

~**, виногра́д·ников·ый** vineyard plough

~**, двух·ко́рпус·н·ый** 2-furrow plough

~**, ди́ск·ов·ый** disc plough

~**, крото·дре́н·н·ый** mole plough

~**, леме́ш·н·ый** breast plough

~**, оборо́т·н·ый** one-way plough

~**, от·ва́ль·н·ый** mouldboard plough, moldboard plow

~**, планта́ж·н·ый** deep-draught plough

~**, при·цеп·н·о́й** trailed plough

плуг·а́р·ь, -я́, (*m*) ploughman, plowman

плу́ж·н·ый, -ая, -ое, (*a*) of **плу́г·**

плумб·а́т·, -а, (*m*) (chem) plumbate

плумб·и́д·, -а, (*m*) (chem) plumbide

плумб·и́т·, -а, (*m*) (min) plumbite

плу́нжер·, -а, (*m*) (mech) plunger

плу́нжер·н·ый на·со́с· (*m*) plunger/piston pump

плуте́ус·, -а, (*m*) (zool) pluteus

плуто́н·, -а, (*m*) (astron) Pluto; (geol) pluton, plutonic intrusion

плуто́н·иев·ый, -ая, -ое, (*a*) plutonium

плуто́н·и·й, -я, (*m*) (chem) plutonium, Pu

плутон·и́т·, -а, (*m*) (geol) plutonite

плутон·и́ческ·ий, -ая, -ое, (*a*) (geol) plutonic

плыв·у́н·, -а, *(m)* (geol) quick ground, quicksand; light floe (ice)

плыв·у́ч·ий, -ая, -ее, *(a)* flowing, deliquescent; (geol) quick, running

плыв·у́щ·ий, -ая, -ее, *(pres part act)* of **плы·ть** *(imp deter)*, *see* **пла́в·а·ть** *(imp indet)*

плы́·вш·ий, -ая, -ее, *(past part act)* of **плы·ть** *(imp deter)*, *see* **пла́в·а·ть** *(imp indet)*

плы·ть, *(pres 3rd pl)* **плыв·у́т,** *(imp deter) see* **пла́в·а·ть**

плюви·а́льн·ый, -ая, -ое, *(a)* pluvial

плювио́·граф·, -а, *(m)* (meteor) recording rain-gauge, hyetograph

плювио́·метр·, -а, *(m)* (meteor) pluviometer, rain gauge

плювио·метр·и́ческ·ий, -ая, -ое, *(a)* of **плювио́·метр·**

~ **коэффицие́нт·** *(m)* (meteor) hyetal coefficient

плю·ёт, *(pres 3rd sing)* of **плев·а́ть**

плюмази́т·, -а, *(m)* (geol) plumasite

плюмб·а́т·, -а, *(m) see* **плумб·а́т·**

плюмбо·станни́т·, -а, *(m)* (min) plumbostannite

плюмбо·ярози́т·, -а, *(m)* (min) plumbojarosite

плю́·н·уть, -ут, *(perf)* spit, expectorate

плюр·, -а, *(m)* chromo paper

плюри·гландуля́р·н·ый, -ая, -ое, *(a)* (zool) pluriglandular

плюс·, -а, *(m)* plus

плюс·к·а, -и, *(f)* (bot) cupule

плюсн·а́, -ы́, *(f)* (anat) metatarsus

плюсн·ев·о́й, -а́я, -о́е, *(a)* (anat) metatarsal

плю́с·ов·ый, -ая, -ое, *(a)* of **плю́с·**

плюш·, -а, *(m)* (text) plush

плющ·, -а́, *(i)* **-щ·о́м,** *(m)* (bot) English ivy, *Hedera helix*

плющ·ени·е, -я, *(n)* flattening

плющ·и́лк·а, -и, *(g pl)* **-лок·,** *(f)* (agr) fodder crusher

плющ·и́льн·ый, -ая, -ое, *(a)* flattening

~ **стан·** *(m)* (met) wire flattening mill

~ **стан·о́к·** *(m)* (mech) flatter

плю́щ·ить, -ат, *(imp)* flatten

плю·ю́т, *(pres 3rd pl)* of **плев·а́ть**

пляж·, -а, *(m)* beach

пляс·к·а, -и, *(g pl)* **-с·ок·,** *(f)* dance, folk dance

~, **ви́тт·ов·а** (med) St. Vitus' dance

пневма́т·ик·, -а, *(m)* pneumatic tyre

пневма́т·ик·а, -и, *(f)* pneumatics

пневмат·и́ческ·ий, -ая, -ое, *(a)* pneumatic, compressed-air, air-powered, air

~ **винт·о́вк·а** *(f)* air gun

~ **двиг·ател·ь** *(m)* air motor

~ **жилет·** *(m)* inflatable life jacket

~ **рас·пы́л·ивани·е** *(n)* (ICE) air/blast injection, air blast

~ **тра́нспорт·** *(m)* pneumatic conveyer

пневмат·о́д·а, -ы, *(f)* (bot) pneumatode, pneumathode

пневмато·кард·и́·я, -и, *(f)* (med) pneumatocardia

пневмато́·лиз·, -а, *(m)* (geol) pneumatolysis

пневмато·лит·и́ческ·ий, -ая, -ое, *(a)* (geol) pneumatolithic

пневмато·фо́р·, -а, *(m)* (bot) pneumatophore

пневмат·ури́·я, -и, *(f)* (med) pneumaturia

пневмеркá́тор·, -а, *(m)* pneumatic liquid level gauge

пневмо·баци́лл·а, -ы, *(f)* (bact) pneumobacillus

пневмо·бето́н·, -а, *(m)* gunite

пневмо·га́з·ов·ый вы·ключ·а́тел·ь *(m)* (elec) air-blast breaker

пневмо·да́т·чик·, -а, *(m)* pneumatic/pressure pick-up

пневмо·кат·о́к·, -т·к·а́, *(m)* pneumatic-tyred roller

пневмо·ко́кк·, -а, *(m)* (bact) pneumococcus

пневмо·конио́з·, -а, *(m)* (med) pneumoconiosis

пневмо·костю́м·, -а, *(m)* pneumatic suit, ventilated protective suit

пневмо·метр·и́ческ·ая тру́б·к·а *(f)* (dynam) head meter, head-type flow meter

пневмо·мико́з·, -а, *(m)* (med) pneumonomycosis, pneumomycosis

пневмо·мото́р·, -а, *(m)* air motor

пневмон·и́т·, -а, *(m)* (med) pneumonitis

пневмон·и́ческ·ий, -ая, -ое, *(a)* (anat) pneumonic

пневмон·и́·я, *(f)* (med) pneumonia

пневмоно·гра́ф·и·я, -и, *(f)* (med) pneumonography

пневмоно·конио́з·, -а, *(m)* (med) pneumonokoniosis, pneumonoconiosis

пневмо·по́чт·а, -ы, *(f)* pneumatic tube conveyor, (nucl) pneumatic rabbit

пневмо·про·во́д·, -а, (*m*) compressed-air line

пневмо·та́ксис·, -а, (*m*) (biol) pneumotaxis

пне вмо·транспортёр·, -а, (*m*) pneumatic conveyer

пне ймат- (*int component*) = **пневма́т-** (q. v.)

пнеймон- (*int component*) = **пневмон-** (q. v.)

пне·корч·ева́тел·ь, -я, (*m*) (agr) root grubber

пн·и, (*nom pl*) of **пен·ь**

по (*prep + dat*) along, on, about, over; by means of, by; in accordance with, according to, by; as a consequence of, through, by; in, at, on, by (of time); *see also* **по** (*prep + prepositional case*)

~ **ве́тр·у** down-wind

~ **воз·ду́ш·н·ой по́чт·е** by air-mail

~ **вс·ему́** all over, throughout

~ **длин·е́** along

~ **ка́пл·е** dropwise, drop-by-drop

~ **но́с·у** (naut) on the bow
   о·пре·дел·ённ·ый по вес·у calculated from the weight

~ **о·ши́б·к·е** by mistake, in error

~ **при·чи́н·е** owing/due to

~ **про·ис·хожд·ени·ю** in/by origin
   суд·я́ по judging from

~ **хим·и́ческ·ой шкал·е́** on chemical scale

по (*prep + prepositional case*) on

~ **рас·смотр·е́ни·и** on examination

по (*prep + acc*) each, apiece (with numerals); up to, to, until

~ **два** two each/apiece; in twos

~ **по́яс·** up to the waist

~ **ту сто́рон·у** on the far/other side of

по- (*verbal prefix*): (1) forms the perfective of certain verbs; (2) acquiring a characteristic, become, become more; (3) completing action at one go; (4) completing action in a short time; (5) action in a reduced degree, do a bit; (6) action embracing all, many; (7) with verbs which already have a prefix; action gradual or in several stages

по- (*adjectival prefix*) (1) after, post-; (2) relating to each/every; (3) along, near; (4) with comparatives: —a little, rather

по·бе́г·, -а, (*m*) flight, escape; (bot) shoot, sprout

по·бе́г·а·ть, -ют, (*perf*) run, run a little

по·бег·у́шк·а, -и, (*g pl*) **-ш·ек·,** (*f*) (elec) brush (of distributor)

по·ба́лт·ыва·ть, -ют, (*imp*) shake

по·бе́д·а, -ы, (*f*) victory

победи́т·, -а, (*m*) pobedite (a tungsten-carbide alloy)

победо·но́с·н·ый, -ая, -ое, (*a*) victorious, triumphant

по·беж·а́лост·ь, -и, (*f*) tarnish
   цвет·а́ по·беж·а́лост·и (*pl*) (met) temper colours

по·беж·а́ть, (*fut 3rd sing, pl*) **по·беж·и́т, по·бег·у́т,** (*perf*) begin to run/move

по·бе́й·те, (*pl imper*) of **по·би́·ть**

по·бел·е́·ть, -ют, (*perf*) turn/become white

по·бел·и́ть, -я́т, (*perf*) whiten, blanch, whitewash

по·бе́л·к·а, -и, (*g pl*) **-л·ок·,** (*f*) whitewash, whitening

по·бере́ж·ь·е, -я, (*g pl*) **-ж·и·й,** (*n*) shore; coastland, littoral

по·бере́ч·ь, (*fut 3rd sing, pl*) **по·береж·ёт, по·берег·у́т,** (*past masc sing*) **по·берёг,** (*perf*); keep, look after

по·бесе́д·овать, -уют, (*perf*) have a talk with

по·би́·ть, (*fut 3rd pl*) **по·бь·ю́т,** (*perf*) beat, beat/knock down; **-ся** be bruised (of fruit)

по·блёк·л·ый, -ая, -ое, (*a*) faded; withered

по·блёк·нуть, -нут, (*perf*) fade, become dull/dim; (bot) wither

по·блёск·ива·ть, -ют, (*imp*) gleam

по·бли́з·ост·и (*adv*) near, nearby

по·бо́·й, -ев, (*m pl*) beating

по·бо́льш·е (*comp*) a little/rather larger/bigger; a little/rather more

по·бо́р·, -а, (*m*) exaction, extortion

по·бо́ч·и·е, -я, (*n*) (elec) leak to earth/ground

по·бо́ч·н·ый, -ая, -ое, (*a*) side, secondary, by-, accessory, extraneous, collateral; spurious, false, ghost; natural, illegitimate (of birth)

~ **вид·** (*m*) ghost mode (oscillations)

~ **из·ображ·е́ни·е** (*n*) (opt) ghost image

~ **о·бес·пе́ч·ени·е** (*n*) (fin) collateral security

~ **пред·при·я́т·и·е** (*n*) subsidiary undertaking

~ **проду́кт·** (*m*) by-product

~ **рабо́т·а** (*f*) side line (activity)

по·бо́ч·н·ый, -ая, -ое
~ **реа́кци·я** (*f*) (chem) secondary reaction
~ **эффе́кт·** (*m*) (phys) side effect; secondary effect
по·бр·и́ть, -е́ют, (*perf*) shave
по·бу́д·ет, (*fut 3rd sing*) of **по·бы́·ть**
по·буд·и́тел·ь, -я, (*m*) stimulus, stimulator, motivating agent; (elec) booster
~ **рас·хо́д·а** sampling aspirator (of gas analyser)
по·буд·и́тельн·ый, -ая, -ое, (*a*) impelling, inducing, exciting, motivating
по·буд·и́ть, -я́т, (*perf*) *see* **по·бужд·а́·ть**
по·бужд·а́·ть, -ют, (*imp*) impel, induce, rouse, excite, stimulate, motivate
по·бужд·е́ни·е, -я, (*n*) (*v n*), *see* **по·бужд·а́·ть**; stimulation, impulse, boosting
по·бур·е́·ть, -ют, (*perf*) become/grow brown
по·быв·а́·ть, -ют, (*perf*) be in/at (for a while), visit
по·бы́в·к·а, -и, (*g pl*) **-в·ок·,** (*f*) (mil) leave
по·бы́·ть, (*fut 3rd pl*) **по·бу́д·ут,** (*perf*) be, stay, remain (e.g. on a visit)
по·бь·ёт, (*fut 3rd sing*) of **по·би́·ть**
по·ваго́н·н·ый, -ая, -ое, (*a*) (rail) car, wagon
~ **от·пра́в·к·а** (*f*) (rail) wagon-load, consignment
по·ва́д·к·а, -и, (*g pl*) **-д·ок·,** (*f*) custom, usage, habit
по·ва́л·енн·ый, -ая, -ое, (*past part pass*) of **по·вал·и́ть**
по·вал·и́ть, -ят, (*perf*) fell, knock/throw down/over, upset, overturn
по·ва́ль·н·о (*adv*) without exception
по·ва́ль·н·ый, -ая, -ое, (*a*) all-embracing, widespread; (med) epidemic
по·вал·я́·ть, -ют, (*perf*) (*trans*) roll
по́·вар·, -а, (*nom pl*) **-а́,** (*g pl*) **-о́в,** (*m*) cook
по·ва́р·енн·ый, -ая, -ое, (*a*) cookery, culinary, cooking
~ **сол·ь** (*f*) table/cooking salt, sodium chloride
по·ва́ш·ему (*adv*) according to you, in your opinion
по·вед·е́ни·е, -я, (*n*) conduct, behaviour, behavior; (instr) response

по·вед·е́ни·е
~ **во вре́мен·и** (phys) time behaviour/response
~, **пре·де́ль·н·ое** (math) asymptotic behaviour
по·вед·ёт, (*fut 3rd sing*) of **по·вес·ти́** (*perf*) *see* **по·вод·и́ть** (*imp*)
по·вез·ти́, -у́т, (*perf*) take, convey
по·вел·и́тельн·ый, -ая, -ое, (*a*) (gram etc.) imperative
повелли́т·, -а, (*m*) (min) powellite
по·ве́р·енн·ый, -ая, -ое, (*past part pass*) of **по·ве́р·ить**; (*m decl as adj*) agent, authorized representative
~ **в дел·а́х** (dipl) chargé d'affaires
~ **в суд·е́** attorney-at-law
по·ве́р·ить, -ят, (*perf*) *see* **по·ве·р·я́·ть**
по·ве́р·к·а, -и, (*g pl*) **-р·ок·,** (*f*) (*v n*), *see* **по·вер·я́·ть**; check, verification; roll-call
по·вёр·нут·ый, -ая, -ое, (*past part pass*) *see* **по·вора́ч·ива·ть**
по·вер·н·у́ть, -у́т, (*perf*) *see* **по·вора́ч·ива·ть**
по·ве́р·очн·ый, -ая, -ое, (*a*) check, test, checking
по·ве́рх· (*prep + gen*) (*adv*) over, above
по·ве́рх·ностн·о (*adv*) on the surface; superficially
по·верх·ностн·о·акти́в·н·ый, -ая, -ое, (*a*) surface-acting/active, surfactant
~ **вещ·еств·о́** (*n*) (chem) surfactant
по·верх·ностн·о·слоисто·по·кро́в·н·ый, -ая, -ое, (*a*) (geol) supratectal
по·ве́рх·ностн·ый, -ая, -ое, (*a*) surface; superficial
    коэффицие́нт· по·ве́рх·ностн·ого рас·шир·е́ни·я (*m*) coefficient of flat expansion
~ **эффе́кт·** (*m*) surface effect; (elec) skin effect
по·ве́рх·ност·ь, -и, (*f*) surface
~, **аэро·динам·и́ческ·ая** (dynam) airfoil, aerodynamic surface
~, **винт·ов·а́я** helical surface
~ **воз·мущ·е́нн·й** (dynam) Mach front
~, **волн·и́ст·ая** corrugated surface, (met) surface folding
~ **вращ·е́ни·я** (math) surface of revolution
~ **втор·о́го по·ря́д·к·а** (phys) quadric surface
~, **глисс·и́рующ·ая** (dynam) hydrofoil

по·ве́рх·ност·ь

~, дел·и́тельн·ая   primary surface (of gear wheel)

~, изо·бар·и́ческ·ие   (*pl*) (meteor) areas of equal pressure

~ кас·а́ни·я   tangential surface

~ кат·а́ни·я   (rail) tread area/surface

~, минима́ль·н·ая   (math) minimal surface

~, на·гнет·а́ющ·ая   driving face (of propellers)

~ на·гре́в·а   heating surface

~, не·нес·у́щ·ая   (dynam) non-useful surface

~, нес·у́щ·ая   (a/c) lifting surface

~, о·по́р·н·ая   seating/supporting surface; base (of cutting tool); face (of valve)

~, пере́д·н·яя   top cutting surface (machining)

~ пере·но́с·а   (phys) translation surface

~ поко́·я   (horol) dead face

~ раз·де́л·а   (phys) interface

~ раз·де́л·а, конта́кт·н·ая   contact interface

~ раз·де́л·а фаз·   interphase interface, phase contacting surface

~ раз·реж·е́ни·я   (air) suction wing surface

~ ска́ч·к·а у·плот·не́ни·я   (aerodynam) shock wave envelope

~ с·ко́с·а   (met) edged surface

~, с·мо́ч·енн·ая   (phys) wetted surface

~ со·при·кос·нове́ни·я   contact area

~ с·ры́в·а   (dynam) separation surface

~ то́к·а   (air) stream surface

~ тр·е́ни·я   rubbing/friction surface

~ у́·ровн·я   equipotential surface

~, фа́з·ов·ая   (autom) phase plane

~, фронта́ль·н·ая   (dynam) front, frontal area; (meteor) frontal surface

~, хоро́ш·ая   clean surface

по·вер·я́·ть, -ют,   (*imp*) verify, check; entrust, trust

по·ве́с·ить, -ят,   (*perf*) hang, hang up; replace (telephone receiver)

по·вес·ти́,   (*fut 3rd sing, pl*) по·ве·д·ёт, по·вед·у́т,   (*past masc sing*) по·вёл, (*perf*); *see* по·вод·и́ть

по·ве́ст·к·а, -и,   (*g pl*) -т·ок·,   (*f*) notice; (law) summons, subpoena; (*pl*) call-up papers

~ дн·я   agenda

по́·вест·ь, -и,   (*f*) story, short story

по·ве́тер·, -тр·а,   (*m*) following wind

по·ве́тр·и·е, -я,   (*n*) epidemic

по·ве́ш·ени·е, -я,   (*n*) hanging

по·ве́ш·енн·ый, -ая, -ое,   (*past part pass*) of по·ве́с·ить

по·ви́·в·, -а,   (*m*) lay (of cables)

по·вив·а́·ть, -ют,   (*perf*) twist, twine

по·ви́д·им·ому   (*adv*) apparently; it seems that

по·вин·и́ться, -я́тся,   (*perf*) confess

по·ви́н·ност·ь, -и,   (*f*) duty, obligation, compulsory service

~, во́ин·ск·ая   compulsory military service, conscription

по·вин·ова́ться, -у́ются,   (*imp and perf*) obey

по·вин·ове́ни·е, -я,   (*n*) obedience, compliance

по·ви́с·л·ый, -ая, -ое,   (*a*) hanging-down, drooping, nutant

по·вис·а́·ть, -ют,   (*imp*) hang, hang down, droop

по·ви́с·нуть, -нут,   (*perf*) *see* по·вис·а́·ть

по·вле́ч·ь,   (*fut 3rd sing, pl*) по·влеч·ёт, по·влек·у́т,   (*past masc sing*) по·влёк·, (*perf*); entail

~ за соб·о́й   involve, entail

по·вли·я́·ть, -ют,   (*perf*) influence, affect

по́·вод·, -а,   (*nom pl*) -ы, (*g pl*) -ов,   (*m*) occasion, ground, cause; (nucl) event

по́·вод·, -а,   (*nom pl*) -а́, or ᷓья, (*g pl*) -о́в, or ᷓьев,   (*m*) rein, reins (of harness)

по·вод·ен·ь, -д·н·я,   (*m*) lower bank (on river side)

по·вод·и́ть, ᷓят,   (*imp*) take, lead; move

по·во́д·к·а, -и,   (*g pl*) -д·ок·, (*f*) (met) distortion

по·вод·ко́в·ый, -ая, -ое,   (*a*) of по·во́д·к·а, по·вод·о́к·

~ патро́н·   (*m*) carrier/catch plate (of lathe)

по·вод·о́к·, -д·к·а,   (*m*) reins, short rein (harness); lead, leash (for dogs etc.); carrier (of lathe); tang (of tool)

по·во́д·ья,   (*nom pl*), (*g pl*) -ьев,   (*m*) reins (of harness)

по·во́з·к·а, -и,   (*g pl*) -з·ок·, (*f*) vehicle, carriage, cart, waggon, wagon

по·во́з·очн·ый, -ая, -ое,   (*a*) of по·во́з·к·а; (*m decl as adj*) driver

по·вора́ч·ивани·е, -я,   (*n*) (*v n*) of по·вора́ч·ива·ть

~ ковш·а́   tilting the ladle (continuous casting)

по·вора́ч·ива·ть, -ют, (*imp*) turn, swivel; tilt, swing over; (elec) reverse; deflect, divert, turn aside

по·вора́ч·иваю́щ·ая при́зм·а (*f*) (opt) deflecting prism

по·воро́т·, -а, (*m*) (*v n*), *see* **по·вора́·ч·ива·ть;** turn; turning point, change; bend, curve (river)

  механи́зм· по·воро́т·а (*m*) (M/T) steering unit

  ско́р·ост·ь по·воро́т·а (*f*) (air) rate of turn; slewing/swing speed (of excavator)

  у́гол· по·воро́т·а (*m*) angle of rotation; angle of pitch (cams etc.); crank angle (crankshaft); locking angle (of lever etc.)

по·воро́т·ливост·ь, -и, (*f*) manoeuvrability

  ис·пыт·а́ни·е по·воро́т·ливост·и (*n*) turning/manoeuvring trial(s)

по·воро́т·лив·ый, -ая, -ое, (*a*) manoeuvrable, (naut) handy

по·воро́т·н·ый, -ая, -ое, (*a*) turning, revolving, rotating, rotary, rotational; swing, tilting; (gunn) training, traversing

~ **кран·** (*m*) revolving crane
~ **круг·** (*m*) turntable, turnplate
~ **мост·** (*m*) swing bridge
~ **пункт·** (*m*) turning point

по·вред·и́ть, -я́т, (*perf*) *see* **по·вреж·д·а́·ть**

по·врежд·а́емост·ь, -и, (*f*) damageability; (elec) occurrence of faults, fault rate; probability of faults

по·врежд·а́ем·ый, -ая, -ое, (*a*) (*pres part pass*) *see* **по·врежд·а́·ть;** liable/susceptible to injury/damage

по·врежд·а́·ть, -ют, (*imp*) damage, impair; injure, hurt

по·врежд·е́ни·е, -я, (*n*) damage, fault, trouble; injury

~ **из·луч·е́ни·ем** (nucl) radiation damage

~ **изоля́ци·и** (elec) insulation fault

~, **луч·ев·о́е** (biol) radiation injury

по·време́н·н·о (*adv*) periodically, at times, from time to time

по·време́н·н·ый, -ая, -ое, (*a*) periodical; time

~ **рабо́т·а** (*f*) timework

по·все·дн·е́вн·ый, -ая, -ое, (*a*) everyday, daily

~ **обя́з·анност·и** (*pl*) routine duties

по·все·ме́ст·н·о (*adv*) everywhere, universally

по·все·ме́ст·н·ый, -ая, -ое, (*a*) ubiquitous, universal

по·всю́ду (*adv*) everywhere

по·вто́р·, -а, (*m*) (TV) double image

по·втор·е́ни·е, -я, (*n*) repetition, repeating

  знак· по·втор·е́ни·я (*m*) (telecom) repeat sign

по·втор·и́тел·ь, -я, (*m*) repeater; (nav) repeater compass, compass repeater

~, **и́мпульс·н·ый** transponder

~ **и́мпульс·ов** pulse repeater

~, **като́д·н·ый** (elec) cathode follower

~, **регенерати́в·н·ый** (telecom) interpolator

~ **-синхрониза́тор·, регенерати́в·-н·ый** (telecom) synterpolator

по·втор·и́ть, -я́т, (*perf*) *see* **по·вто-р·я́·ть**

по·вто́р·н·о (*adv*) once more, again; repeatedly

~ **вы·зыв·а́·ть** call again (telephones)

~ **за·ряж·а́·ть** re-charge, charge again

по·вто́р·н·ый, -ая, -ое, (*a*) re-, repeated, iterated, iterative

~ **в·ключ·е́ни·е** (*n*) reclosure (of a circuit breaker)

~ **за·мора́ж·ивани·е** (*n*) refreezing

~ **импеда́нс·** (*m*) (elec) iterative impedance

~ **интегра́л·** (*m*) (math) iterated integral

~ **пере·рабо́т·к·а** (*f*) reprocessing

по·втор·я́емост·ь, -и, (*f*) repetition; recurrence, repetition rate

по·втор·я́ем·ый, -ая, -ое, (*a*) (*pres part pass*) of **по·втор·я́·ть;** repetitive

~ **черт·ёж·** (*m*) standard component drawing

по·втор·я́·ть, -ют, (*imp*) repeat

по·выс·и́тел·ь, -я, (*m*) augmenter; (autom) augment

по·выс·и́тельн·ый, -ая, -ое, (*a*) increasing, augmenting; (elec) step-up

~ **под·ста́нци·я** (*f*) (elec) step-up substation

по·вы́с·ить, -ят, (*perf*) *see* **по·вы-ш·а́·ть**

по·выш·а́·ть, -ют, (*imp*) raise, elevate, heighten

~ **в·дв·о́е** double

по·вы́ш·е (*comp*) a little/rather higher/taller; (*adv*) a little higher up

по·выш·е́ни·е, -я, (*n*) (*v n*), *see* **по·-выш·а́·ть;** rise, elevation; increase, improvement; promotion, advancement; (elec) stepping-up по·вы́ш·енн·ый, -ая, -ое, (*past part pass*) *see* **по·выш·а́·ть**

~ сво́й·ств·а (*pl*) improved properties

~ температу́р·а (*f*) high/elevated temperature

по·вяз·а́ть, (*fut 3rd pl*) **по·вя́ж·ут**, (*perf*) *see* **по·вя́з·ыва·ть**

по·вя́з·к·а, -и, (*g pl*) -з·ок·, (*f*) bandage, band

по·вя́з·ыва·ть, -ют, (*imp*) tie, bandage; knit

пога́н·к·а, -и, (*g pl*) -н·ок·, (*f*) poisonous/inedible mushroom; (orn) loon, *Gavia*

~, бле́д·н·ая (bot) death cup, *Amanita phalloides*

по·гас·а́ни·е, -я, (*n*) (*v n*), *see* **по·гас·а́·ть;** extinction

   потенциа́л· по·гас·а́ни·я (*m*) extinction potential

по·гас·а́·ть, -ют, (*imp*) go out, become extinguished

по·гас·и́ть, ⸗ят, (*perf*) extinguish, put out, quench; damp (vibrations), deaden (sound); liquidate (debts), repay (loans etc.); cancel (validity)

по·га́с·нуть, -нут, (*perf*) *see* **по·гас·а́·ть**

по·гаш·а́·ть, -ют, (*imp*) liquidate (debts), repay (loans etc.); cancel (validity)

по·гаш·е́ни·е, -я, (*n*) (*v n*), *see* **по·гас·и́ть;** extinction; (fin) payment, liquidation, cancellation

по·га́ш·енн·ый, -ая, -ое, (*past part pass*) of **по·гас·и́ть**

по·гекта́р·н·ый, -ая, -ое, (*a*) per hectare

по́·гиб·, -а, (*m*) camber

по·гиб·а́·ть, -ют, (*imp*) perish, be killed, be lost (of ship)

по·ги́б·ельн·ый, -ая, -ое, (*a*) fatal; ruinous

по·ги́б·нуть, -нут, (*perf*) *see* **по·ги·б·а́·ть**

по·глот·и́тел·ь, -я, (*m*) absorber (apparatus); absorbent (substances)

~ нейтро́н·ов, вре́д·н·ый (nucl) neutron poison

~, и́мпульс·н·ый digit absorber, digit-absorbing selector (telephones)

~ пере·на·пряж·е́ни·й (elec) surge absorber

~ тепл·а́ heat absorbent

~ уда́р·а shock absorber

по·глот·и́тельн·ый, -ая, -ое, (*a*) absorptive, absorption

~ спосо́б·ност·ь (*f*) absorptive power/capacity, absorptivity; (chem) coefficient of absorption

~ схе́м·а (*f*) (elec) absorption network, absorber

по·глот·и́ть, ⸗ят, (*perf*) *see* **по·гло·щ·а́·ть**

по·глощ·а́емост·ь, -и, (*f*) absorptivity

по·глощ·а́·ть, -ют, (*imp*) absorb, take in/up

по·глощ·а́ющ·ий, -ая, -ое, (*pres part act*) of **по·глощ·а́·ть;** absorbing absorbent, absorptive

~ вещ·еств·о́ (*n*) absorbent

~ сре́д·ств·о (*n*) absorbent

по·глощ·е́ни·е, -я, (*n*) (*v n*), *see* **по·глощ·а́·ть;** absorption, sorption; (oil) loss (of drilling fluid)

~, дискре́т·н·ое (phys) line absorption

~, косм·и́ческ·ое interstellar absorption

   коэффицие́нт· по·глощ·е́ни·я (*m*) coefficient of absorption, absorption factor, absorptivity

   коэффицие́нт· по·глощ·е́ни·я, по́лн·ый (*m*) total absorption coefficient

~, объ·ём·н·ое volume absorption

~, от·нос·и́тельн·ое fractional absorption

~, по·вто́р·н·ое re-absorption

   ско́р·ост·ь по·глощ·е́ни·я (*f*) absorption rate; gettering rate (vacuum tubes)

по·глу́б·же (comp) rather deeper, a little deeper

по·гляд·е́ть, -я́т, (*perf*) look at, take a look at

по·гляж·у́ (*fut 1st sing*) of **по·гля·д·е́ть**

по·гн·а́ть, (*fut 3rd pl*) **по·го́н·ят**, (*perf*) drive, herd; -ся run/chase after

по·гни́·ть, -ю́т, (*perf*) rot, decay

по·гн·у́ть, -у́т, (*perf*) bend

по·говор·и́ть, -я́т, (*perf*) talk, have a talk

по·гово́р·к·а, -и, (*g pl*) -р·ок·, (*f*) saying, proverb

по·го́д·а, -ы, (*f*) weather

~, не·лёт·н·ая non-flying weather

по·го́д·н·ый, -ая, -ое, (*a*) yearly; weather

погодо·сто́й·к·ий, -ая, -ое, (*a*) weather-proof

по·голо́в·н·ый, -ая, -ое, (*a*) per capita; general

по·голо́в·ь·е (*n*) number (of livestock)

по·го́н·, -а, (*g pl*) **по·го́н·** (*m*) fraction, cut (distillation); race, ball/roller path (of turntables etc.); (mil) shoulder strap

~, голо́в·н·ы́е (*pl*) heads (spirit distilling)

ли́н·и·я по·го́н·ы (*f*) (math) tractrix

~, хвост·ов·ы́е (*pl*) tails (spirit distilling)

~, ша́р·ов·о́й (mech eng) roller path

по·го́н·ит, (*fut 3rd sing*) of **по·гн·а́ть**

по·го́н·н·ый, -ая, -ое, (*a*) linear, lineal; per unit of length

~ вес· (*m*) weight per meter (of tubes, rails etc.)

~ ём·кост·ь (*f*) line capacitance, mutual capacitance per 1 km of transmission line

~ метр· (*m*) linear meter

по·го́н·щик·, -а, (*m*) driver, teamster

по·го́н·ыш·, -а, (*i*) **-ем,** (*m*) (orn) rail, *Porzana*

по·го́н·я, -и, (*f*) pursuit

крив·а́я по·го́н·и (*f*) (air) curve of pursuit, pursuit path

ли́н·и·я по·го́н·и (*f*) (math) tractrix

по·гран·и́чн·ый, -ая, -ое, (*a*) frontier, border, boundary

~ сло́·й (*m*) (mech, phys) boundary layer

по́·греб·, -а, (*nom pl*) **-а́,** (*g pl*) **-о́в,** (*m*) cellar, vault; (naut) magazine, shellroom;

по·грёб· (*past masc sing*) of **по·грес·ти́**

по·греб·а́·ть, -ют, (*imp*) bury, inter

по·греб·е́ни·е, -я, (*n*) burial, interment

по·греб·ёт, (*fut 3rd sing*) of **по·грес·ти́**

по·грем·у́шк·а, -и, (*g pl*) **-шек·,** (*f*) rattler, rattle

по·грес·ти́, (*fut 3rd pl*) **по·греб·у́т,** (*past masc sing*) **по·грёб·,** (*perf*) bury; row

по·гре́·ть, -ют, (*perf*) heat/warm a little, warm (for a time)

по·греш·а́·ть, -ют, (*imp*) err, transgress, make mistakes

по·греш·и́ть, -а́т, (*perf*) *see* **по·гре·ш·а́·ть**

по·гре́ш·ност·ь, -и, (*f*) *see also* **о·ши́б·к·а;** error, mistake; degree of error; defect (in mechanism)

~, абсолю́т·н·ая accuracy (of an instrumental reading)

~ граду·иро́вк·и calibration error

~ из·мер·е́ни·я error in measurement

~ из·мер·и́тельн·ого при·бо́р·а meter error, instrumental error

~, ме́ст·н·ая site/local error

~ место·по·лож·е́ни·я (nav) position error

~, от·нос·и́тельн·ая relative error

~ по·каз·а́ни·я indication error

~, при·вед·ённ·ая percentage/fractional error

по·груж·а́ем·ый, -ая, -ое, (*pres part pass*) of **по·груж·а́·ть;** submersible, dipping

по·груж·а́·ть, -ют, (*imp*) (1) submerge, immerse, dip, plunge, dive; (2) sink (into), drive (piles etc.), embed, imbed; (3) load, ship, embark; **-ся** (*pass and reflex*); settle, sink, subside

по·груж·е́ни·е, -я, (*n*) (*v n*), *see* **по·груж·а́·ть** (1) (2); immersion, dive (submarines); penetration, insertion

глуб·ин·а́ по·груж·е́ни·я (*f*) submerged depth; depth of indentation/penetration (hardness tests)

~, сро́ч·н·ое crash dive (submarine)

~ сте́рж·н·я rod insertion (in nuclear reactor)

по·груж·ённ·ый, -ая, -ое, (*past part pass*) *see* **по·груж·а́·ть;** drowned; swimming-pool (nucl reactor)

по·груж·н·о́й, -а́я, -о́е, (*a*) submersible, immersible, immersion

по·груз·и́ть, -я́т, (*perf*) *see* **по·груж·а́·ть, груз·и́ть**

по·гру́з·к·а, -и, (*g pl*) **-з·ок·,** (*f*) (*v n*), *see* **по·груж·а́·ть** (3); loading; shipment; embarkation, entrainment, embussing

по·гру́з·очн·ый, -ая, -ое, (*a*) of **по·гру́з·к·а**

~ маши́н·а (*f*) handling machine

~ о́рдер· (*m*) consignment note (shipping)

~ при·ча́л· (*m*) loading berth (shipping)

по·гру́з·чик·, -а, (*m*) loader; lift truck

~, ви́л·очн·ый fork-lift truck

под   (*prep* + *acc*) under (movement); for use as, for; to the neighbourhood of; shortly before, towards; to, to the sound of; like, in imitation of

~ вéчер·   towards evening

~ гóр·у   (*adv*) downhill

под   (*prep* + *instr*) under (location); used as, for; in the neighbourhood of

под-   (*verbal prefix*) *signifies:* (1) down, downward; (2) under, underneath; (3) approach; (4) addition; (5) secrecy; (6) repetition or accompaniment; (7) reduced degree

под-   (*noun/adj prefix*) *signifies:* (1) location below/lower; (2) location near; (3) sub-, under, in the sphere of; (4) like, similar

пóд·, -а,   (*m*) hearth (of furnace)

по·дав·áльщик·, -а,   (*m*) waiter; delivery man/boy; server (tennis), pitcher (baseball)

по·дав·áтел·ь, -я,   (*m*) feeder, feed mechanism; magazine platform (of a rifle)

по·да·вáть, -ют,   (*imp*) give, serve; (tech) feed, supply; bring in/up/to (vehicles etc.); submit, forward, hand in (requests etc.); (theat) depict, present, portray; **-ся** (*pass*); give way, yield, move

~ букси́р·   (naut) pass a tow

~ в суд·   (law) bring an action

~ жáл·об·у   lodge/make a complaint

по·дав·и́тел·ь, -я,   (*m*) suppressor; (chem) depressant

~ обрáт·н·ой с·вяз·и   (rad) reaction suppressor, (elec) feedback suppressor

~ от·раж·ённ·ых сигнáл·ов   (rad) echo suppressor

~ по·мéх·   interference suppressor

~ с·вяз·и   (telecom) jammer

по·дав·и́ть, -ят,   (*perf*) *see* **по·давл·я́·ть**

по·давл·éни·е, -я,   (*m*) (*v n*), *see* **по·давл·я́·ть;** suppression, repression

по·давл·я́·ть, -ют,   (*imp*) suppress, repress; overwhelm, crush; (mil) neutralize

по·дáвл·енн·ый, -ая, -ое,   (*past part pass*) *see* **по·давл·я́·ть;** inhibited

по·дáвн·о   (*adv*) all the more

под·áгр·а, -ы,   (*f*) (med) podagra, gout

по·дад·у́т,   (*fut 3rd pl*) of **по·дá·ть,** (*perf*); *see* **по·да·вáть** (*imp*)

по·да·ёт,   (*pres 3rd sing*) of **по·да·вáть**

по·дáль·ш·е   (*adv*) further, somewhat further

по·дáм,   (*fut 1st sing*) of **по·дá·ть,** (*perf*) *see* **по·да·вáть** (*imp*)

по·дар·и́ть, -ят,   (*perf*) give a present, make a gift of, present, give

по·дáр·ок·, -р·к·а,   (*m*) present, gift

по·дá·ст,   (*fut 3rd sing*) of **по·дá·ть,** (*perf*); *see* **по·да·вáть**

по·дá·тел·ь, -я,   (*m*) bearer (of document etc.); petitioner

по·дáт·ливост·ь, -и,   (*f*) compliance; pliability, capacity of yielding

по·дáт·лив·ый, -ая, -ое,   (*a*) easily workable (of materials), pliable, yielding; yielding, pliant, compliant, easily influenced (of people)

по·дáт·очн·ый, -ая, -ое,   (*a*) of **по·дá·ч·а**

по·дá·ть,   (*fut 3rd sing, pl*) **по·дá·ст, по·дад·у́т,** (*past masc sing*) **пó·да·л,** (*perf*); *see* **по·да·вáть**

по·дá·ч·а, -и,   (*i*) **-ей,** (*f*) (*v n*), *see* **по·да·вáть;** feed, feeding, charging, supply, supply system; handing-in (of messages etc.); (M/T) fuel feed system

~, в·рез·н·áя   straight-in feed (machine tools)

глуб·ин·á по·дá·ч·и   (*f*) depth of cut (metal working)

~, двой·н·áя   dual supply/feed

~ и́мпульс·ов   (elec) pulsing

~, комбин·и́рованн·ая   (M/T) gravity and exhaust-pressure fuel feed system

~ о·переж·áющ·их и́мпульс·ов   (elec) prepulsing

~, по·перéч·н·ая   cross feed (machine tools)

~ раз·реж·éни·ем   (M/T) vacuum fuel feed system

~, револьвéр·н·ая   dial feed (machine tools)

~, регул·и́руем·ая   variable delivery (of pumps etc.)

~, само·тёч·н·ая   gravity feed

скóр·ост·ь по·дá·ч·и   (*f*) rate of feed, feed rate

~ с·мещ·éни·я   (elec) biassing

~ с·ни́з·у   underfeed

~, строч·ечн·ая   pick feed (machine tools)

по·дáч·н·ый, -ая, -ое,   (*a*) of **по·дá·ч·а**

по·да·ю́щ·ий, -ая, -ее,   (*pres part act*) of **по·да·вáть**

~ аппарáт·   (*m*) feeder; pusher (tube rolling)

по·да·ю́щ·ий, -ая, -ее
~ кассе́т·а   (*f*) feed reel (of tape recorder etc.)
~ част·ь   (*f*) feed (of a machine)
под·ба́в·ить, -ят,   (*perf*) *see* под·-бавл·я́·ть
под·бавл·я́·ть, -ют,   (*imp*) add a little
под·ба́л·к·а, -и,   (*g pl*) -л·ок·,   (*f*) (eng) bolster; (shipb) stringer
под·ба́лт·ыва·ть, -ют,   (*imp*) stir/beat/mix in
под·бараба́н·ь·е, -я,   (*n*) concave (machine part)
под·барье́р·н·ый, -ая, -ое,   (*a*) (phys) sub-barrier
под·бег·а́·ть, -ют,   (*imp*) run up to; run under/underneath
под·беж·а́ть,   (*fut 3rd sing, pl*) под·-беж·и́т, под·бег·у́т, (*perf*); *see* под·-бег·а́·ть
под·бе́й·те,   (*pl imperative*) of под·-би́·ть (*perf*); *see* под·бив·а́·ть
под·берёз·овик·, -а,   (*m*) (bot) rough-stalked boletus, *Boletus scaber*
под·бер·ёт,   (*fut 3rd sing*) of подо·-бр·а́ть (*perf*); *see* под·бир·а́·ть (*imp*)
под·бив·а́·ть, -ют,   (*imp*) fit/attach underneath, resole, reshoe; (text) line, line with
~ ва́т·ой   (text) wad
~ под·кла́д·к·у   (text) line up
под·бир·а́ни·е, -я,   (*n*) (*v n*) of под·-бир·а́·ть; selection
под·бир·а́·ть, -ют,   (*imp*) pick/take up; select; match; pull/draw in/tighter; tuck up/in; glean
под·би́·ть,   (*fut 3rd pl*) подо·бь·ю́т, (*perf*) *see* под·бив·а́·ть
под·би́ч·ник·, -а,   (*m*) cahir, beater bed (of a threshing machine)
под·бо́·й, -я,   (*m*) (*v n*) of под·би́·ть
под·бо́й·к·а, -и,   (*g pl*) -бо́·ек·,   (*f*) lining; (rail) ballast hammer; packer (machine part)
под·болт·а́·ть, -ют,   (*perf*) *see* под·-ба́лт·ыва·ть
под·бо́р·, -а,   (*m*) (*v n*), *see* под·би-р·а́·ть; collection, selection
в под·бо́р·   (print) run on
~ импеда́нс·ов   (elec) impedance matching
    ме́тод· под·бо́р·а   (*m*) trial-and-error
~ фра́кци·й   grading (in sizes)
~ цвет·о́в   colour/color matching
под·бо́р·к·а, -и,   (*g pl*) -р·ок·,   (*f*) selection

под·бо́р·к·а
~ труд·о́в   (print) selected works
под·бород·ок·, -д·к·а,   (*m*) (anat) chin, mentum
под·бо́р·щик·, -а,   (*m*) (agr) pick-up attachment
~, пресс-   (agr) pick-up baler
под·бра́с·ыва·ть, -ют,   (*imp*) throw/toss up/under; throw/put more (on); leave/deposit secretly
под·бре́кер·н·ый, -ая, -ое,   (*a*) bottom-cushion (of tyres)
под·брос·и́ть, ⸗ят,   (*perf*) *see* под·-бра́с·ыва·ть
под·бро́ш·енн·ый, -ая, -ое,   (*past part pass*) *see* под·бра́с·ыва·ть
~ ребён·ок·   (*m*) foundling
под·бру́ствер·н·ый, -ая, -ое,   (*a*) trench-type
под·брю́ш·ник·, -а,   (*m*) belly band (of harness); (shipb) poppet boards; launching cradle; support straps (boilers etc.)
под·бу́р·ок·, -р·к·а,   (*m*) (min) plug hole
под·ва́л·, -а,   (*m*) basement, cellar
~, зол·ов·ый   ash basement (of boiler plant)
под·ва́л·ива·ть, -ют,   (*imp*) heap up
под·вал·и́ть, ⸗ят,   (*perf*) *see* под·ва́-л·ива·ть
под·ва́л·к·а, -и,   (*f*) (*v n*), *see* под·-ва́л·ива·ть
под·ва́ль·н·ый, -ая, -ое,   (*a*) of под·-ва́л·
под·ва́р·ива·ть, -ют,   (*imp*) reboil, boil again; reweld, weld again; weld to
под·вар·и́ть, ⸗ят,   (*perf*) *see* под·ва́-р·ива·ть
под·ва́хт·енн·ый, -ая, -ое,   (*a*) (naut) off-watch
под·вед·е́ни·е, -я,   (*n*) *see* по·да́·ч·а
под·вед·ёт,   (*fut 3rd sing*) of под·-вес·ти́ (*perf*); *see* под·вод·и́ть (*imp*)
под·ве́д·омственн·ый, -ая, -ое,   (*a*) within/under the jurisdiction of
под·вез·ти́, -у́т,   (*perf*) *see* под·во-з·и́ть
под·вёл·,   (*past masc sing*) of под·-вес·ти́
под·верг·а́ем·ый, -ая, -ое,   (*past part pass*) of под·верг·а́·ть
~ ис·пыт·а́ни·ю   under test
~ из·луч·е́ни·ю   exposed
под·верг·а́·ть, -ют,   (*imp*) subject/expose to, put through (a process); -ся (*pass*); undergo

под·верг·а́·ть
~ де́й·стви·ю expose to
~ ис·пыт·а́ни·ю test, experiment with
~ о́т·жиг·ом (met) anneal, give an annealing treatment to
под·ве́рг·нуть, -нут, (*perf*) *see* **под·-верг·а́·ть**
под·ве́рг·нут·ый, -ая, -ое, (*past part pass*) *see* **под·верг·а́·ть**
~ де́й·стви·ю и́мпульс·ов pulsed
под·ве́рх·еннос·т·ь, -и, (*f*) susceptibility (to)
под·ве́рх·енн·ый, -ая, -ое, (*past part pass*) *see* **под·верг·а́·ть**; having undergone, having been
под·вер·н·у́ть, -у́т, (*perf*) *see* **под·-вёрт·ыва·ть**
под·вёрт·ыва·ть, -ют, (*imp*) turn/bend/tuck back/under/down; screw up/tighter, tighten (a screw)
под·вёрт·ывани·е, -я, (*n*) (*v n*) of **под·вёрт·ыва·ть**
под·ве́с·, -а, (*m*) (*v n*), *see* **под·ве́-ш·ива·ть**; suspension, suspension device, suspender, hanger; ceiling light
~, бифиля́р·н·ый (phys) bifilar suspension
~ма́ят·ник·а (horol) suspension spring
~, универса́ль·н·ый шарни́р·н·ый gimbals
под·ве́с·ить, -ят, (*perf*) *see* **под·ве́-ш·ива·ть**
под·ве́с·к·а, -и, (*g pl*) **-с·ок·,** (*f*) (*v n*), *see* **под·ве́ш·ива·ть**; hanger, suspender, suspension, pendant, (civ eng) sling
~, баланси́р·на·я (М/Т) bogie suspension
~, двой·н·а́я цеп·н·а́я double catenary suspension (elec overhead lines)
под·вес·н·о́й, -а́я, -о́е, (*a*) suspension, suspended; overhead, swinging, hanging, undersling
~ доро́г·а (*f*) overhead/aerial way
~ доро́г·а, жёстк·ая monorail
~ доро́г·а, кана́т·н·ая aerial ropeway
~ доро́г·а, ре́льс·ов·ая overhead trolley conveyor
~ изоля́тор· (*m*) (elec) suspension insulator
~ конве́йер· (*m*) overhead conveyor
~ мото́р· (*m*) (naut) outboard motor
~ систе́м·а парашу́т·а (*f*) parachute harness

под·ве́с·ок·, -с·к·а, (*m*) pendant, pendent; (zool) hyomandibular; (chem) side chain
под·вес·ти́, (*fut 3rd pl*) **под·вед·у́т,** (*past masc sing*) **под·вёл,** (*perf*) *see* **под·вод·и́ть**
под·ве́тр·енн·ый, -ая, -ое, (*a*) lee, leeward
под·ве́ш·енн·ый, -ая, -ое, (*past part pass*) *see* **под·ве́ш·ива·ть**
под·ве́ш·ива·ть, -ют, (*imp*) hang up, suspend, sling
подвздо́шно- (*component*) (zool) ile-, ileo-, ilio-
подвздо́ш·н·ый, -ая, -ое, (*a*) (zool) iliac
~ кост·ь (*f*) (zool) ilium
под·вив·а́·ть, -ют, (*imp*) curl
по́·двиг·, -а, (*m*) feat, exploit, deed
по·дви́г·а·ть, -ют, (*perf*) move a little; make several movements; **по·-двиг·а́·ть, -ют,** (*imp*) move, shift; **-ся** (*pass*); make some progress, move on
под·ви́д·, -а, (*m*) (biol) subspecies
по·дви́ж·к·а, -и, (*g pl*) **-ж·ек·,** (*f*) (geol) movement, shift; adjustment (of ore)
~ льд·а débâcle (of ice), ice motion
по·движ·н·о́й, -а́я, -о́е, (*a*) mobile, moving, travelling
~ катализа́тор· (*m*) (chem) moving-bed catalyst
~ о·по́р·а (*f*) sliding foot (boiler)
~ по·са́д·к·а (*f*) clearance/running fit
~ по́ч·к·а (*f*) (med) floating kidney
~ равно·ве́с·и·е (*n*) (chem) mobile equilibrium
~ со·ста́в· (*m*) (rail) rolling stock
по·дви́ж·нос·т·ь, -и, (*f*) mobility; movement, motion
~, диффузио́н·н·ая (phys) diffusion movement
~, ды́р·очн·ая (phys) hole mobility
~ ио́н·ов (phys) ionic mobility
по·каз·а́тел·ь по·дви́ж·ност·и (*m*) consistency factor (e.g. of cement)
температу́р·а по·те́р·и по·дви́ж·-ност·и (*f*) (oil) pour point
по·дви́ж·н·ый, -ая, -ое, (*a*) *see* **по·движ·н·о́й**
под·винт·и́ть, -я́т, (*perf*) *see* **под·-ви́нч·ива·ть**
под·ви́нч·ива·ть, -ют, (*imp*) screw up/in more, tighten (a screw); screw on underneath

под·вис·а́ни·е, -я, (*n*) hanging, sticking (in vertical passage); hold up (in distillation tower)

~ шйхт·ы hanging (of charge in blast furnace)

под·ви́·ть, (*fut 3rd pl*) подо·вь·ю́т, (*perf*) curl

под·вла́ст·н·ый, -ая, -ое, (*a*) subject

под·во́д·, -а, (*m*) (*v n*), *see* под·во·д·и́ть; intake, admission, feed, supply

под·во́д·а, -ы, (*f*) cart, waggon, wagon, dray

под·вод·и́м·ый, -ая, -ое, (*pres part pass*) *see* под·вод·и́ть

~ мо́щ·ност·ь (*f*) (elec) power input

под·вод·и́ть, -́ят, (*imp*) lead up to, bring; put/place/pass under; (tech) feed, supply

~ бала́нс· (fin) balance

~ ито́г·и sum up

~ фунда́мент· (civ eng) underpin

под·во́д·к·а, -и, (*f*) (*v n*), *see* под·-вод·и́ть

под·во́д·ник·, -а, (*m*) (naut) submariner

под·во́д·н·о́й, -а́я, -о́е, (*a*) *and see* под·во́д·н·ый; *see* под·вод·я́щ·ий

под·во́д·н·ый, -ая, -ое, (*a*) underwater, submarine, submerged

     в под·во́д·н·ом по·лож·е́ни·и submerged

~ ка́бел·ь (*m*) submarine cable

~ ка́мен·ь (*m*) reef

~ кры·л·о́ (*n*) (shipb) hydrofoil

~ ло́д·к·а (*f*) *see also* под·ло́д·к·а; submarine

~ ско́р·ост·ь (*f*) submerged speed

~ та́нкер· (*m*) submarine tanker

под·вод·я́щ·ий, -ая, -ее, (*pres part act*) of под·вод·и́ть

~ кана́л· (*m*) (hydroelec) aqueduct, conduit, diversion canal

~ по́л·ост·ь (*f*) suction/sump chamber (of pump)

под·вож·у́ (*pres 1st sing*) of под·во·д·и́ть; (*pres 1st sing*) of под·воз·и́ть

под·во́з·, -а, (*m*) transport carriage, carting

под·воз·и́ть, -́ят, (*imp*) bring, bring up/more, cart; give a lift to

под·во́·й, -я, (*m*) stock (grafting plants)

под·воло́к·а, -и, (*f*) ceiling; (shipb) deckhead

под·вулканиз·а́ци·я, -и, (*f*) scorching (of rubber)

~, пре́жде·вре́мен·н·ая scorching (of rubber)

под·вулканиз·ова́ть, -у́ют, (*perf*) scorch (rubber)

под·вяз·а́ть, (*fut 3rd pl*) под·вя́-ж·ут, (*perf*) *see* под·вя́з·ыва·ть

под·вя́з·ыва·ть, -ют, (*imp*) tie on underneath, tie up/round; knit on/ longer

под·вя́з·ь, аркат·н·ая (*f*) (text) harness

под·гиб·а́·ть, -ют, (*imp*) bend/fold/ tuck under/back

под·гиб·н·о́й, -а́я, -о́е, (*a*) bending, flap, tuck

под·гнив·а́·ть, -ют, (*imp*) rot underneath; become slightly rotten

под·гни́·ть, -ю́т, (*perf*) *see* под·гнив·а́·ть

под·гно·ённ·ый, -ая, -ое, (*past part pass*) *see* под·гнив·а́·ть

под·голо́в·ок·, -в·к·а, (*m*) neck

~, кон·и́ческ·ий tapered neck (of rivets etc.)

под·го́н·, -а, (*m*) (*v n*), *see* под·го-н·я́·ть; afterspring (timber); (hortic) second growth

под·го́н·к·а, -и, (*f*) (*v n*) of под·го-н·я́·ть; adjustment, trimming, matching, aligning, retouching

~ ко́нтур·ов (elec) alignment of tuned circuits

~ спо́соб·ом наи·ме́ньш·их квадра́-т·ов (math) least-square fit

~ част·от·ы́ ква́рц·ев·ых пласт·и́н· (elec) close etching (of crystals)

под·го́н·очн·ый, -ая, -ое, (*a*) of под·го́н·к·а

~ конденса́тор· (*m*) (elec) trimmer

под·гон·я́·ть, -ют, (*imp*) drive/urge on/under; drive up to; fit, adjust (a fit); -ся (*pass*); (math) draw in

под·горизо́нт·, -а, (*m*) subhorizon

под·гото́вл·ива·ть, -ют, (*imp*) prepare, prepare for, make ready; train, instruct

под·гото́в·и́тельн·ый, -ая, -ое, (*a*) preparatory

~ вы́·работ·к·и (*pl*) (min) development workings

под·гото́в·ить, -ят, (*perf*) *see* под·-гото́вл·ива·ть

под·гото́в·к·а, -и, (*g pl*) -в·ок·, (*f*) (*v n*), *see* под·гото́вл·ива·ть; preparation; training

~, бо·ев·а́я (mil) training

~, вне·во́йск·ов·а́я (mil) pre-conscription training

~, пред·по·лёт·н·ая (air) preflight planning/preparation

под·гото́в·к·а
~ режи́м·а   pre-conditioning (an apparatus)

под·готовл·я́·ть, -ют, (*imp*) = под·-готáвл·ива·ть

под·гру́пп·а, -ы, (*f*) subgroup

под·да·ва́ть, -ю́т, (*imp*) knock/kick up/away; give away (at chess etc.); -ся (*pass*); be amenable to, lend itself/oneself to

по́д·да·нн·ый, -ого, (*m decl as adj*) (polit) subject; (*past part pass*) of под·да́·ть

под·да́·ть, (*fut 3rd sing, pl*) под·-да́·ст, под·дад·у́т, (*past masc sing*) под·да́·л, (*perf*); see под·да·ва́ть

под·да·ю́щ·ийся, -аяся, -ееся, (*pres part pass*) of под·да·ва́ться

~ обрабóт·к·е вод·óй   washable

~ управл·éни·ю   controllable

под·дви́г·, -а, (*m*) (geol) underthrust

под·дёв·, -а, (*m*) (fish) spinning

под·дев·á·ть, -ют, (*imp*) lever/prise up; put on underneath (clothes)

под·дé·л, (*past masc sing*) of под·дé·ть; (*perf*); see под·дев·á·ть (*imp*)

под·дéл·а·ть, -ют, (*perf*) see под·-дéл·ыва·ть

под·дéл·ыва·ть, -ют, (*imp*) counterfeit, falsify, forge (documents)

под·дéль·н·ый, -ая, -ое, (*a*) counterfeit, forged, spurious, false, artificial

под·дé·нет, (*fut 3rd sing*) of под·дé·ть, (*perf*); see под·дев·á·ть

под·держ·áни·е, -я, (*n*) (*v n*), see под·дéрж·ива·ть; support; maintenance, keeping

под·держ·á·ть, -ат, (*perf*) see под·-дéрж·ива·ть

под·дéрж·иваем·ый, -ая, -ое, (*pres part pass*) of под·дéрж·ива·ть; sustained

~ процéсс· (*m*) sustained process

под·дéрж·ива·ть, -ют, (*imp*) support, back up; maintain, keep up

под·дéрж·иваю̇щ·ий, -ая, -ее, (*pres part act*) of под·дéрж·ива·ть; holding, carrying; (*m decl as adj*) carrier

~ си́л·а (*f*) buoyancy

под·дéрж·к·а, -и, (*g pl*) -ж·ек·, (*f*) (*v n*), see под·дéрж·ива·ть; support; prop; supporting, backing; holder-on (of rivets)

под·дé·ть, -нут, (*perf*) see под·де-в·á·ть

под·диапазóн·, -а, (*m*) sub-range/band

под·ди́р·, -а, (*m*) moil (mining tool)

под·дóн·, -а, (*m*) drip-pan, sump, tray; (met cast) bottom plate/place/board; pallet (for cranes etc.)

под·дув·áл·о, -а, (*n*) ashpit (of boiler)

под·дув·áни·е, -я, (*n*) (*v n*), see под·-дув·á·ть; (geol) swelling, heaving, creep

под·дув·á·ть, -ют, (*imp*) blow from below/beneath; be a slight draught

под·ду́·ть, -ют, (*perf*) see под·ду-в·á·ть

по·дéй·ств·овать, -уют, (*perf*) see дéй·ств·овать

по·декáд·н·о (*adv*) every ten days

по·декáд·н·ый, -ая, -ое, (*a*) decade; ten-day, ten-daily

~ иск·áни·е (*n*) (autom) decade line-finding

по·дел·и́ть, -ят, (*perf*) see дел·и́ть; divide; share

по·дéл·к·а, -и, (*g pl*) -л·ок·, (*f*) making, fashioning (by hand); employment, odd job, way (of employment); article; (*pl*) sundries

по·дéл·очн·ый, -ая, -ое, (*a*) of по·-дéл·к·а; repair; sundry-purpose

по·дён·к·и, -н·ок·, (*f pl*) (ent) mayflies, *Ephemeroptera*

по·дён·н·о (*adv*) by the day, daily

по·дён·н·ый, -ая, -ое, (*a*) daily; (*m decl as adj*) casual labourer

по·дён·щик·, -а, (*m*) casual labourer

по·дёрг·а·ть, -ют, (*perf*) see по·дёр-г·ива·ть

по·дёрг·ива·ть, -ют, (*imp*) pluck, pull; pluck/pull out/off; twitch; cover with a thin film of, veil in

по·держ·áни·е, -я, (*n*) (*v n*) of по·-держ·áть
на по·держ·áни·е   on loan

по·держ·анн·ый, -ая, -ое, (*past part pass*) see по·держ·á·ть; used, second-hand

по·держ·áть, -ат, (*perf*) see дер-ж·áть; keep/hold for a while

по·дёр·н·уть, -ут, (*perf*) see по·дёр-г·ива·ть

по·дешев·é·ть, -ют, (*perf*) become cheaper

под·жá·вш·ий, -ая, -ее, (*past part act*) of под·жá·ть

под·жáр·ый, -ая, -ое, (*a*) lean, stringy

под·жáт·и·е, -я, (*n*) (*v n*), see под·-жим·á·ть
коэффициент· под·жáт·и·я (*m*) contraction ratio, compression ratio

под·жа́т·ь, (*fut 3rd pl*) **подо·жм·у́т,** (*past masc sing*) **под·жа́·л,** (*perf*); *see* **под·жим·а́·ть**

под·желу́д·очн·ый, -ая, -ое, (*a*) (anat) pancreatic

под·же́ч·ь, (*fut 3rd sing, pl*) **подо·ж·ж·ёт, подо·жг·у́т,** (*past masc sing*) **под·жёг·,** (*perf*); *see* **под·жиг·а́·ть**

под·жи́г·, -а, (*m*) ignition

под·жиг·а́ни·е, -я, (*n*) (*v n*), *see* **под·жиг·а́·ть;** (nucl) preionization

под·жиг·а́тел·ь, -я, (*m*) ignitor; incendiary; instigator

под·жиг·а́·ть, -ют, (*imp*) set fire to, set on fire, ignite, fire

под·жим·а́·ть, -ют, (*imp*) draw in/up, contract, constrict

под·жо́г·, -а, (*m*) arson

под·за·голо́в·ок·, -в·к·а, (*m*) subheading, subtitle

под·за·ря́д·, -а, (*m*) (elec) boost charge

~, доз·ов·о́й (elec) floating/trickle charge

под·за·ря́д·к·а, -и, (*f*) recharging

под·зву́к·ов·ый, -ая, -ое, (*a*) subsonic

под·зе́м·н·ый, -ая, -ое, (*a*) underground, subterranean

под·зи́м·н·ий, -яя, -ее, (*a*) late-autumn/fall, early-winter

~ по·се́в (*m*) (agr) late-autumn sowing

под·зов·ёт, (*fut 3rd sing*) of **подо·зв·а́ть**

под·зо́л·, -а, (*m*) podzol, podzolic soil, podsol

под·зол·иза́ци·я, -и, (*f*) podzol formation

под·зо́л·ист·ый, -ая, -ое, (*a*) podzol, podzolic, podsolic

под·зо́н·а, -ы, (*f*) zonule, sub-zone

под·зо́р·, -а, (*m*) (arch) cornice; (shipb) counter

под·зо́р·н·ая труб·а́ telescope, spy-glass

под·зыв·а́·ть, -ют, (*imp*) summon, call up

под·и́н·а, -ы, (*f*) bottom (furnace)

под·интегра́ль·н·ый, -ая, -ое, (*a*) (math) sub-integral, under the integral sign

~ фу́нкци·я (*f*) (math) integrand

под·интерва́л·, -а, (*m*) subinterval, subrange

под·кали́бер·н·ый, -ая, -ое, (*a*) (gunn) sub-calibre; skirted (of ammunition)

под·ка́л·ыва·ть, -ют, (*imp*) pin up/back/on; split up

под·кана́в·очн·ый, -ая, -ое, (*a*) undertread, sub-tread (of tyres)

~ сло́·й (*m*) undertread (of tyre)

под·кана́л·, -а, (*m*) (telecom) subcarrier channel

под·ка́п·ыва·ть, -ют, (*imp*) undermine, dig under

под·ка́рмл·ива·ть, -ют, (*imp*) feed up, fatten; dress with fertilizer

под·кас·а́тельн·ая, -ой, (*f decl as adj*) (math) subtangent

под·ка́т·, -а, (*m*) (met) starting strip (for cold rolling)

под·кат·и́ть, -я́т, (*perf*) *see* **под·ка́т·ыва·ть**

под·ка́т·к·а, -и, (*g pl*) **-т·ок·,** (*f*) (*v n*), *see* **под·ка́т·ыва·ть;** (met forge) fullering·

под·ка́т·ыва·ть, -ют, (*imp*) roll/put under; roll/drive/bring up to

под·ка́ч·анн·ый, -ая, -ое, (*past part pass*) of **под·кач·а́·ть**

под·кач·а́·ть, -ют, (*perf*) *see* **под·ка́ч·ива·ть**

под·ка́ч·енн·ый, -ая, -ое, (*past part pass*) of **под·кат·и́ть**

под·ка́ч·ива·ть, -ют, (*imp*) pump, pump more, boost

под·ка́ч·ивающ·ий, -ая, -ее, (*pres part act*) of **под·ка́ч·ива·ть;** auxiliary, booster (of pumps)

под·ка́ш·ива·ть, -ют, (*imp*) mow, mow down/more

под·ки́д·ыва·ть, -ют, (*imp*) = **под·бра́с·ыва·ть**

под·ки́ль·н·ый, -ая, -ое, (*a*) (naut) below the keel

~ кон·е́ц· (*m*) (naut) bottom line

под·ки́н·уть, -ут, (*perf*) *see* **под·бра́с·ыва·ть**

под·кисл·е́ни·е, -я, (*n*) acidification, acidulation

под·кисл·и́ть, -я́т, (*perf*) *see* **под·кисл·я́·ть**

под·кисл·я́·ть, -ют, (*imp*) acidify, acidulate

под·кла́д·к·а, -и, (*g pl*) **-д·ок·,** (*f*) lining, backing, liner; washer; block, chock; mudsill (of bridge); (rail) rail chair, bearing plate, sole plate

~, во́йлоч·н·ая (mech) felt washer

~, у·пру́г·ая cushioning

под·кла́д·н·о́й, -а́я, -о́е, (*a*) underneath, backing, lining; for putting/laying under, underlay

под·клад·н·о́й, -а́я, -о́е
~ **пласт·и́н·а** (*f*) (M/T) press plate
под·кла́д·очн·ый, -ая, -ое, (*a*) (text) lining
~ **ткан·ь** (*f*) (text) lining
под·кла́д·ыва·ть, -ют, (*imp*) put/place/lay under/more; (text) line
под·кла́сс·, -а, (*m*) subclass, subcategory; (math) sub-set
под·кле́·ивани·е, -я, (*n*) (*v n*) of **под·кле́·ива·ть**
под·кле́·ива·ть, -ют, (*imp*) glue/stick/paste under/more; glue up/together
под·кле́·ить, -ят, (*perf*) see **под·-кле́·ива·ть**
под·клеп·а́·ть, -ют, (*perf*) see **под·-клёп·ыва·ть**
под·клёп·ыва·ть, -ют, (*imp*) rivet up; rivet again
под·клин·ива·ть, -ют, (*imp*) wedge/chock/block up, shim
под·клин·ить, -ят, (*perf*) see **под·-клин·ива·ть**
под·ключ·а́·ть, -ют, (*imp*) switch on/in
под·ключ·ённ·ый, -ая, -ое, (*past part pass*) see **под·ключ·а́·ть**; on, live
под·ключ·и́ть, -а́т, (*perf*) see **под·-ключ·а́·ть**
под·ключ·и́чн·ый, -ая, -ое, (*a*) (anat) subclavian
~ **арте́ри·я** (*f*) (anat) arteria subclavia
под·ко́в·а, -ы, (*f*) horseshoe; metal reinforcement (under footwear)
под·ков·а́ть, (*fut 3rd pl*) **под·ку·ю́т,** (*perf*) see **под·ко́в·ыва·ть**
подково·ви́д·н·ый, -ая, -ое, (*a*) horseshoe, horseshoe-shaped
подково·обра́з·н·ый, -ая, -ое, (*a*) horseshoe, horseshoe-shaped
под·ко́в·ыва·ть, -ют, (*imp*) shoe (horses etc.); fit (runners etc.)
под·ко́ж·н·ый, -ая, -ое, (*a*) subcutaneous, hypodermic
под·кол·о́ть, -ют, (*perf*) see **под·-ка́л·ыва·ть**
под·комите́т·, -а, (*m*) sub-committee
под·ко́н·ник·, -а, (*m*) (shipb) deck girder, carling
под·ко́п·, -а, (*m*) (*v n*), see **под·ка́-п·ыва·ть**; underground passageway
под·коп·а́·ть, -ют, (*perf*) see **под·-ка́п·ыва·ть**
под·корен·н·о́й, -а́я, -о́е, (*a*) (math) subradical, under the radical sign
~ **числ·о́** (*n*) (math) radicand

под·ко́р·к·а, -и, (*g pl*) **-р·ок·,** (*f*) (zool) subcortex
под·ко́р·ков·ый, -ая, -ое, (*a*) subcortical
под·корм·и́ть, -ят, (*perf*) see **под·-ка́рмл·ива·ть**
под·ко́рм·к·а, -и, (*g pl*) **-м·ок·,** (*f*) (*v n*), see **под·ка́рмл·ива·ть**; (agr) top dressing
под·кор·ов·о́й, -а́я, -о́е, (*a*) (geol) subcrustal
под·ко́с·, -а, (*m*) (*v n*), see **под·ко-с·и́ть**; strut, brace, bracing strut; (build) knee brace, knee rafter, raking shore/prop
~, **лом·а́ющ·ийся** (a/c) folding strut
под·ко́с·и́ть, -ят, (*perf*) mow, mow down/more
под·ко́с·н·ый, -ая, -ое, (*a*) of **под·-ко́с·**
~ **мост·** (*m*) frame bridge
под·ко́ш·енн·ый, -ая, -ое, (*past part pass*) of **под·кос·и́ть**
под·кра́с·ить, -ят, (*perf*) see **под·-кра́ш·ива·ть**
под·кра́ш·ива·ть, -ют, (*imp*) tint, tincture; touch up (paintwork)
под·креп·и́ть, -я́т, (*perf*) see **под·-крепл·я́·ть**
под·крепл·éни·е, -я, (*n*) (*v n*), see **под·крепл·я́·ть**; reinforcement, support; corroboration, confirmation
под·крепл·я́·ть, -ют, (*imp*) support, reinforce, strengthen, brace; corroborate, confirm
под·крит· и́ческ·ий, -ая, -ое, (*a*) sub-critical
под·крит·и́чност·ь, -и, (*f*) sub-criticality, subcritical state
под·крут·и́ть, -ят, (*perf*) twist, twirl, spin, turn
под·кру́ч·енн·ый, -ая, -ое, (*past part pass*) of **под·крут·и́ть**; twisted
под·кры́л·ок, -л·к·а, (*m*) (a/c) flap, extension flap
под·ку·ёт, (*fut 3rd sing*) of **под·ко-в·а́ть**
по́д·куп·, -а, (*m*) bribe; bribery
под·куп·а́·ть, -ют, (*imp*) bribe
под·ла́м·ыва·ть, -ют, (*imp*) break below, break low down
по́дле (*prep* + *gen*) beside, by the side of
под·ле́гар·ец·, -р·ц·а, (*m*) (shipb) rising
под·лед·ник·о́в·ый, -ая, -ое, (*a*) (geol) subglacial

под·леж·а́ть, -а́т, (*imp*) be subject/liable to

под·леж·а́щ·ий, -ая, -ее, (*pres part act*) of под·леж·а́ть; subject/liable to; (*n decl as adj*) (gram) subject

не под·леж·а́щий о·глаш·е́ни·ю not for publication, confidential

~, об·лож·е́ни·ю по́шлин·ой dutiable

~ со·мн·е́ни·ю open to doubt, doubtful

под·лез·а́·ть, -ют, (*imp*) creep under

под·ле́с·ок·, -с·к·а, (*m*) (agr) undergrowth, brush, bush

под·лёт·, -а, (*m*) (air) hover

под·лив·а́·ть, -ют, (*imp*) add (by pouring), pour more

под·ли́в·к·а, -и, (*g pl*) -в·ок·, (*f*) (*v n*), *see* под·лив·а́·ть; (food) sauce, dressing; (build) mortar

под·лив·н·о́й, -а́я, -о́е, (*a*) of под·-ли́в·к·а

~ колес·о́ (*n*) water wheel

по́длин·ник·, -а, (*m*) original; genuine article/thing

по́длин·н·ый, -ая, -ое, (*a*) original; authentic, genuine; real

с по́длин·н·ым ве́р·н·о certified copy, authenticated (as endorsement)

под·ли́·ть, (*fut 3rd pl*) подо·лlь·ю́т, (*perf*) *see* под·лив·а́·ть

под·ло́д·к·а, -и, (*g pl*) -д·ок·, (*f*) submarine

~ радиолокацио́н·н·ого до·зо́р·а radar picket submarine

**~ ракето·но́с·ец·, -с·ц·а, (*m*) (rocket) missile-carrying submarine

~, торпе́д·н·ая attack submarine

под·ло́д·очн·ый, -ая, -ое, (*a*) of под·ло́д·к·а

под·ло́ж·ечн·ый, -ая, -ое, (*a*) (anat) substernal

под·лож·и́ть, ⌐ат, (*perf*) *see* под·-кла́д·ыва·ть

под·ло́ж·к·а, -и, (*g pl*) -ж·ек·, (*f*) support; backing, base, support, substrate (for condensations); (phot) film support

~ мише́н·и (nucl) target support

под·ло́ж·ечн·ый, -ая, -ое, (*a*) of под·ло́ж·к·а

под·ло́ж·н·ый, -ая, -ое, (*a*) spurious, counterfeit, false

под·локо́т·ник·, -а, (*m*) elbow-rest, arm (of chair etc.)

под·лом·и́ть, ⌐ят, (*perf*) *see* под·ла́-м·ыва·ть

под·лопа́т·очн·ый, -ая, -ое, (*a*) (anat) subscapular

по́дл·ый, -ая, -ое, (*a*) base, ignoble

под·магни́ч·ивани·е, -я, (*n*) magnetic biassing (magnetic recorder); premagnetization, field magnetization, excitation

без под·магни́ч·ивани·я unbiassed

под·ма́з·ать, (*fut 3rd pl*) под·ма́-ж·ут, (*perf*) *see* под·ма́з·ыва·ть

под·ма́з·ыва·ть, -ют, (*imp*) grease/smear/coat lightly/again

под·мал·ева́ть, -ю́ют, (*imp*) touch up with paint, repaint; make an underpainting (art)

под·мал·ёвк·а, -и, (*f*) (*v n*) of под·-мал·ева́ть

под·мал·ёвок·, -вк·а, (*m*) underpainting (art)

под·ма́сл·ива·ть, -ют, (*imp*) oil/butter lightly; add oil/butter

под·ма́сл·ить, -ят, (*perf*) *see* под·-ма́сл·ива·ть

под·масте́р·ь·е, -я, (*g pl*) -ев, (*m*) journeyman

под·ма́т·ыва·ть, -ют, (*imp*) wind/reel/coil more; wind round, bind with

под·ма́ч·ива·ть, -ют, (*imp*) moisten/damp/wet slightly

под·ме́здр·ивани·е, -я, (*n*) fleshing (leather)

под·мёл, (*past masc sing*) of под·-мес·ти́

под·ме́н·ива·ть, -ют, (*imp*) = под·-мен·я́·ть

под·мен·и́ть, ⌐ят, (*perf*) *see* под·-мен·я́·ть

под·мен·я́·ть, -ют, (*imp*) substitute, substitute temporarily; substitute secretly

под·мерз·а́·ть, -ют, (*imp*) freeze slightly

под·мерзл·о́тн·ый, -ая, -ое, (*a*) (geog) infrapermafrost, subpermafrost

под·мёрз·л·ый, -ая, -ое, (*a*) rather frozen

под·мёрз·нуть, -нут, (*perf*) *see* под·-мерз·а́·ть

под·мес·и́ть, ⌐ят, (*perf*) knead/mix in

под·мес·ти́, (*fut 3rd pl*) под·мет·у́т, (*past masc sing*) под·мёл, (*perf*); *see* под·мет·а́·ть

под·мет·а́льн·о-у·бо́р·очн·ая маши́н·а (*f*) road sweeper-collector

под·мет·а́льн·ый, -ая, -ое,   (*a*) sweeping

под·мет·а́·ть, -ют,   (*imp*) sweep, sweep up

под·мет·ёт, (*fut 3rd sing*) of **под·мес·-ти́** (*perf*); *see* **под·мет·а́·ть** (*imp*)

под·ме́т·ить, ⸗ят,   (*perf*) *see* **под·-меч·а́·ть**

под·мёт·к·а, -и,   (*g pl*) **-т·ок·,** (*f*) sole (of shoe)

под·меч·а́·ть, -ют, (*imp*) note

под·меш·а́·ть, -ют, (*perf*) mix/stir in

под·ме́ш·анн·ый, -ая, -ое, (*past part pass*) of **под·меш·а́·ть**

под·ме́ш·енн·ый, -ая, -ое, (*past part pass*) of **под·мес·и́ть**

под·ме́ш·ива·ть, -ют,   (*imp*) mix/stir/knead in; adulterate, doctor

под·мно́ж·еств·о, -а,   (*n*) (math) subset

под·моде́ль·н·ая плит·а́   (*f*) (met cast) pattern plate

под·мо́·ет, (*fut 3rd sing*) of **под·мы́·ть** (*perf*); *see* **под·мыв·а́·ть** (*imp*)

под·мок·а́·ть, -ют, (*imp*) become/get slightly wet

под·мо́к·нуть, -нут, (*perf*) *see* **под·-мок·а́·ть**

под·мора́ж·ива·ть, -ют, (*imp*) (food) chill; (meteor) freeze

под·мороз·и́ть, ⸗ят, (*perf*) *see* **под·-мора́ж·ива·ть**

по́д·мост·и, -ей,   (*pl*) scaffolding, scaffold, trestle, staging

под·мо́ст·к·и, -ов, (*m pl*) stage, staging, boards

под·мот·а́·ть, -ют, (*perf*) *see* **под·-ма́т·ыва·ть**

под·моч·и́ть, ⸗ат, (*perf*) *see* **под·-ма́ч·ива·ть**

под·мы́·в·, -а,   (*m*) erosion, under-mining (of river banks etc.); jetting (pile-driving)

под·мыв·а́·ть, -ют, (*imp*) wash under-neath/lightly; wash away, undermine, erode

под·мы́в·н·ый, -ая, -ое, (*a*) eroding, undermining (by water); jetting (pile-driving)

под·мы́ль·н·ый щёлок· (*m*) spent lye from soap boiling

под·мы́·ть, (*fut 3rd pl*) **под·мо́·ют,** (*perf*) *see* **под·мыв·а́·ть**

под·мы́ш·к·а, -и, (*g pl*) **-ш·ек·,** (*f*) (anat) armpit, axilla

под·мы́ш·ечн·ый, -ая, -ое,   (*a*) (anat) axillary

под·мы́ш·ник·, -а,   (*m*) (text) dress protector

под·на́·двиг·, -а,   (*m*) (geol) under-thrust

под·над·зо́р·н·ый, -ая, -ое, (*a*) under observation, suspect

под·на·ла́д·к·а, -и, (*g pl*) **-д·ок·,** (*f*) re-setting (machine tools)

под·на·ря́д·, -а, (*m*) insert sole (for footwear)

под·нес·ти́, -у́т, (*perf*) *see* **под·но-с·и́ть**

под·нес·у́щ·ая, -ей,   (*f decl as adj*) (rad) subcarrier

под·ним·а́·ть, -ют, (*imp*) raise, lift, hoist, elevate; plough, plow, break, break up (soil); pick up; (*imp only*) be able to carry, have a carrying capacity of; **-ся** (*pass*); rise, ascend, get up

под·ни́т·очник·, -а, (*m*) (text) coun-ter faller wire

под·нов·и́ть, -я́т,   (*perf*) *see* **под·-новл·я́·ть**

под·новл·я́·ть, -ют, (*imp*) touch up (paintwork etc.), freshen up, make some repairs to

под·ногт·ев·о́й, -а́я, -о́е,   (*a*) (zool) hyponychial

под·но́ж·и·е, -я, (*n*) foot (of hill etc.); pedestal

под·но́ж·к·а, -и, (*g pl*) **-ж·ек·,** (*f*) footboard, foot rest; treadle; back-heel (sport)

под·но́ж·н·ый, -ая, -ое, (*a*) under-foot, low-growing

под·нормáл·ь, -и,   (*f*) (math) sub-normal

под·но́с·, -а, (*m*) (*v n*), *see* **под·но-с·и́ть;** tray

под·нос·и́ть, ⸗я́т, (*imp*) bring, carry to; present

под·но́с·ок·, -с·к·а, (*m*) box toe (of shoes)

под·но́с·чик·, -а,   (*m*) carrier

под·нош·у́ (*pres 1st sing*) of **под·но-с·и́ть**

под·ня́т·и·е, -я,   (*n*) (*v n*), *see* **под·-ним·а́·ть;** uplift, rise

под·ня́·ть, (*fut 3rd pl*) **под·ни́м·ут,** (*perf*) *see* **под·ним·а́·ть**

подо (*prep*) *see* **под**

по·до́б·и·е, -я, (*n*) similarity, likeness, similitude

~, динам·и́ческ·ое dynamic similitude

~, кинемат·и́ческ·ое kinematic similitude

по·до́б·и·е
~, по́лн·ое absolute/complete similarity
под·о́бласт·ь, -и, (*f*) subregion, subarea
по·до́б·н·ый, -ая, -ое, (*a*) similar, like, the same kind of
~ пре·образ·ова́ни·е (*n*) (math) transformation of similitude, homothetic transformation
~ сво́й·ств·а (*pl*) affinity
под·оболо́ч·к·а, -и, (*g pl*) -чек·, (*f*) (nucl) subshell
подо́·бр·анност·ь, -и, (*f*) matched condition, match
~ импеда́нс·ов (elec) impedance match
подо́·бр·анн·ый, -ая, -ое, (*past part pass*) of **подо·бр·а́ть;** matched; gathered, picked; selected
подо·бр·а́ть, (*fut 3rd pl*) под·бер·у́т, (*perf*) see **под·бир·а́ть**
подо·бь·ёт, (*fut 3rd sing*) of **под·би́·ть,** (*perf*); see **под·бив·а́·ть** (*imp*)
под·о́в·ый, -ая, -ое, (*a*) hearth
~ ба́л·оч·к·а (*f*) firing floor beam (furnace)
подо·вь·ёт, (*fut 3rd sing*) of **под·ви́·ть**
подо́·гн·анн·ый, -ая, -ое, (*past part pass*) of **подо·гн·а́ть;** adjusted, fitted
подо·гн·а́ть, (*fut 3rd pl*) под·го́·н·ят, (*perf*) see **под·гон·я́·ть**
подо·гн·у́ть, -у́т, (*perf*) see **под·ги·б·а́·ть**
подо·гре́в·, -а, (*m*) preheating, warming up
~, ме́ст·н·ый (mech) hot-spotting
подо·грев·а́тел·ь, -я, (*m*) heater, preheater; primer (of engines)
~, воз·ду́ш·н·ый пуск·ов·о́й flame primer (diesel)
~, изо́·гн·ут·ый folded heater (of vacuum tubes)
~ нефт·и fuel-oil heater (boiler systems)
~, с·ме́ш·ивающ·ий (st eng) open feed-water heater
~, электро·фа́кель·н·ый flame primer (diesel)
подо·грев·а́тельн·ый, -ая, -ое, (*a*) heating, preheating
~ пла́м·я (*n*) (weld) preheating flame
подо·грев·а́·ть, -ют, (*imp*) heat, preheat, warm up
подо·гре́в·н·ый, -ая, -ое, (*a*) heater, heater-type

подо·гре́·ть, -ют, (*perf*) see **подо··грев·а́·ть**
подо·двиг·а́·ть, -ют, (*imp*) move up/closer
подо·дви́·н·уть, -ут, (*perf*) see **подо··двиг·а́·ть**
подо́·енн·ый, -ая, -ое, (*past part pass*) of **по·до·и́ть**
подо·жг·у́т, (*fut 3rd pl*) of **под··же́ч·ь** (*perf*); see **под·жиг·а́·ть** (*imp*)
подо·жд·а́ть, -у́т, (*perf*) wait, wait for
подо·жж·ёт, (*fut 3rd sing*) of **под··же́ч·ь,** (*perf*); see **под·жиг·а́·ть** (*imp*)
подо·жм·ёт, (*fut 3rd sing*) of **под··жа́т·ь,** (*perf*); see **под·жим·а́·ть** (*imp*)
подо·зв·а́ть, (*fut 3rd pl*) под·зов·у́т, summon, call up
подо·зр·ева́·ть, -ют, (*imp*) suspect
подо·зр·е́ни·е, -я, (*n*) suspicion
подо·зр·и́тельн·ый, -ая, -ое, (*a*) suspicious, suspect
по·до·и́ть, -я́т, (*perf*) milk
подо·йд·ёт, (*fut 3rd sing*) of **подо·й··ти́,** (*perf*); see **под·ход·и́ть** (*imp*)
по·до́й·ник·, -а, (*m*) milk-pail/bucket
подо·йти́, (*fut 3rd sing, pl*) подо·й··д·ёт, подо·йд·у́т, (*past masc sing*) подо·шёл, (*perf*); see **под·ход·и́ть**
подо·ка́рп·, -а, (*m*) (bot) *Podocarpus*
под·око́н·ник·, -а, (*m*) (build) window sill/seat, elbow-board
по·до́л·, -а, (*m*) hem (of garment), foot, bottom edge
по·до́лг·у (*adv*) long, for a long time
подоли́т·, -а, (*m*) (min) podolite
подо·ль·ёт, (*fut 3rd sing*) of **под··ли́·ть,** (*perf*); see **под·лив·а́·ть** (*imp*)
по·до́н·к·и, -ов, (*m pl*) dregs
подо·прев·а́·ть, -ют, (*imp*) rot slightly/underneath
подо·пр·ёт, (*fut 3rd sing*) of **под·пе·р·е́ть,** (*perf*); see **под·пир·а́·ть** (*imp*)
подо·пре́·ть, -ют, (*perf*) see **подо··прев·а́·ть**
под·о́·пыт·н·ый, -ая, -ое, (*a*) experimental, test
подо·рв·а́ть, -у́т, (*perf*) blow up, blast, explode
под·оре́ш·ник·, -а, (*m*) *Lactarius volemus* (edible fungus)
по·доро́ж·ник·, -а, (*m*) (bot) plantain, *Plantago*
по·доро́ж·ников·ый, -ая, -ое, (*a*) (bot) plantain, plantago; (*pl as noun*) *Plantaginaceae*

по·доро́ж·н·ый, -ая, -ое, (*a*) on the road; roadside

под·оси́н·овик·, -а, (*m*) (bot) rough stalked boletus, *Boletus versipellis*

подо·сл·а́ть, (*fut 3rd pl*) подо·шл·ю́т, (*perf*) send on a secret mission

подо·ста́т·очн·ый, -ая, -ое, (*a*) subpermanent

подо·сте́л·ет, (*fut 3rd sing*) of подо·стл·а́ть

подо·стл·а́ть, (*fut 3rd pl*) подо·сте́л·ют, (*perf*) spread/lay under

под·от·де́л·, -а (*m*) subdivision, subsection, subdepartment, sub-branch

подо·тр·ёт, (*fut 3rd sing*) of под·тер·ёть, (*perf*); *see* под·тир·а́ть (*imp*)

под·от·ря́д·, -а, (*m*) (biol) suborder

под·от-чёт·ност·ь, -и, (*f*) accountability

под·от·чёт·н·ый, -ая, -ое, (*a*) (fin) accountable

подо·фи́лл·, -а, (*m*) (bot) mayapple, *Podophyllum*

~, щито·ви́д·н·ый (bot) *Podophyllum peltatum*

подо·филл·и́н·, -а, (*m*) (chem) podophyllin

подо·филло·токси́н·, -а, (*m*) (chem) podophyllotoxin

по·до·хо́д·н·ый, -ая, -ое, (*a*) according to income

~ на·ло́г· (*m*) income tax

подо́·шв·а, -ы, (*f*) sole, (anat) pelma, planta; foot (of mountain etc.); floor (of mine); trough (of wave); base, footing, bottom, underside

~ льд·а (ocean) ice foot

~, сла́б·ая (geol) soft ground

подо́·шв·енн·ый, -ая, -ое, (*a*) of подо́·шв·а

подо·шёл, (*past masc sing*) of подо·й·ти́, (*perf*); *see* под·ход·и́ть (*imp*)

подо·шл·ёт, (*fut 3rd sing*) of подо·сл·а́ть

подо·шь·ёт, (*fut 3rd sing*) of под·ши́·ть, (*perf*); *see* под·шив·а́·ть (*imp*)

под·пад·а́·ть, -ют, (*imp*) come under, come under the rule of

под·па́лз·ыва·ть, -ют, (*imp*) creep under; creep up to

под·па́л·ива·ть, -ют, (*imp*) singe, scorch

под·пал·и́ть, -ят, (*perf*) *see* под·па́л·ива·ть

под·па́луб·н·ый, -ая, -ое, (*a*) (shipb) under/below the deck, supporting the deck, deck

под·па́р·, -а, (*m*) pulled/skin wool

под·па́р·ыва·ть, -ют, (*imp*) partly unstitch/unpick/rip open

под·па́с·ть, (*fut 3rd pl*) под·пад·у́т, (*past masc sing*) под·па́·л, (*perf*); *see* под·пад·а́·ть

под·пек·а́·ть, -ют, (*imp*) bake longer/more

под·пер·е́ть, (*fut 3rd pl*) подо·пр·у́т, (*past masc sing*) под·пёр·, (*perf*); *see* под·пир·а́·ть

под·пе́ст·ичн·ый, -ая, -ое, (*a*) (bot) hypogynous, hypopetalous

под·пе́ч·ь, (*fut 3rd sing, pl*) под·печ·ёт, под·пек·у́т, (*past masc sing*) под·пёк·, (*perf*); *see* под·пек·а́·ть

под·пи́л·ива·ть, -ют, (*imp*) saw/file underneath; shorten (by sawing), file down

под·пил·и́ть, -ят, (*perf*) *see* под·пи́л·ива·ть

под·пи́л·ок·, -л·к·а, (*m*) *see* на·пи́ль·ник·; file

под·пир·а́ни·е, -я, (*n*) (*v n*) of под·пир·а́·ть

под·пир·а́·ть, -ют, (*imp*) prop up, chock up, support, underpin

под·пис·а́вш·ийся, -егося, (*m decl as adj*) signatory

под·пис·а́ни·е, -я, (*n*) signing

под·пи́с·анн·ый, -ая, -ое, (*past part pass*) *see* под·пи́с·ыва·ть

под·пис·а́ть, (*fut 3rd pl*) под·пи́ш·ут, (*perf*) *see* под·пи́с·ыва·ть

под·пи́с·к·а, -и, (*g pl*) -с·ок, (*f*) signing; subscription; written undertaking, bond

~, авар·и́йн·ая average bond

под·пис·н·о́й, -а́я, -о́е, (*a*) subscription

~ лист· (*m*) subscription list

под·пи́с·чик·, -а, (*m*) subscriber

под·пи́с·ывани·е, -я, (*n*) signing

под·пи́с·ыва·ть, -ют, (*imp*) sign; subscribe to; write under

по́д·пис·ь, -и, (*f*) signature; (print) caption; (*pl*) lettering (on illustrations etc.)

под·пи́т·к·а, -и, (*g pl*) -т·ок·, (*f*) adding; make-up; trickle feeding

~ ура́н·ом (nucl) uranium make-up

под·пи́т·очн·ый, -ая, -ое, (*a*) feed, make-up

~ вод·а́ (*f*) make-up water

под·пи́ш·ет, (*fut 3rd sing*) of под·пис·а́ть, (*perf*); *see* под·пи́с·ыва·ть (*imp*)

под·пла́ншир·н·ый брус· (*m*) (shipb) gunwale

под·пла́ст·ок·, -т·к·а, (*m*) substratum

под·пле́т·ин·а, -ы, (*f*) (text) tangle, bore

под·плыв·а́·ть, -ют, (*imp*) swim/drift/ float/sail up to/under

под·плы́·ть, (*fut 3rd pl*) **под·плы-в·у́т,** (*perf*) *see* **под·плыв·а́·ть**

под·по·ве́рх·ностн·ый, -ая, -ое, (*a*) subsurface

по́д·пол·, -а, (*m*) *see* **под·по́л·ь·е**

под·по́л·е, -я, (*n*) (phys) sub-field

под·полз·а́·ть, -ют, (*imp*) crawl/creep under; crawl/creep up to

под·полз·ти́, -у́т, (*perf*) *see* **под·полз·а́·ть**

под·полк·о́вник·, -а, (*m*) (mil) Lieutenant-Colonel

под·по́л·ь·е, -я, (*n*) cellar; (mil, polit) underground, resistance

под·по́ль·н·ый, -ая, -ое, (*a*) underground, illegal; underfloor, subfloor, floor

~ кно́п·к·а (*f*) floor pushbutton

~ с·вёрт·ывающ·ая маши́н·а (*f*) (met roll) downcoiler

под·по́р·, -а, (*m*) (civ eng) backwater effect, damming

под·по́р·а, -ы, (*f*) prop, buttress, shore, stay, strut, support, abutment, balk

под·по́р·к·а, -и, (*g pl*) **-р·ок·,** (*f*) = **под·по́р·а** (q.v.); toe bearing; (min) prop

под·по́р·н·ый, -ая, -ое, (*a*) of **под·по́р·а**

~ резервуа́р· (*m*) priming/suction tank

~ сте́н·к·а (*f*) breastwall, retaining wall

под·по́ро·мер·, -а, (*m*) plenum gauge

под·пор·о́ть, -ют, (*perf*) *see* **под·па́р·ыва·ть**

под·по·ря́д·ок·, -д·к·а, (*m*) (bot) suborder

под·по·сле́д·овательност·ь, -и, (*f*) (math) subsequence

под·по·сто·я́нн·ый, -ая, -ое, (*a*) subpermanent

под·по́чв·а, -ы, (*f*) subsoil, subsurface, underground

под·по́чв·енн·ый, -ая, -ое, (*a*) subsoil; underground, subterranean

под·по́яс·ать, (*fut 3rd pl*) **под·по-я́ш·ут,** (*perf*) *see* **под·по́яс·ыва·ть**

под·по́яс·ыва·ть, -ют, (*imp*) girdle, strap up, put on a belt

по·пра́в·ить, -ят, (*perf*) *see* **под·правл·я́·ть**

под·правл·я́·ть, -ют, (*imp*) repair/ mend partially; put in order, put right

под·пресс·о́вк·а, -и, (*g pl*) **-вок·,** (*f*) (rubber) premoulding, premolding

под·програ́мм·а, -ы, (*f*) subprogram, subroutine (computers)

~, библиоте́ч·н·ая library subprogram

под·про·стра́н·ств·о, -а, (*n*) (math) subspace, subregion

под·про·те́ктор·, -а, (*m*) undertread (of tyre)

под·про·те́ктор·н·ый, -ая, -ое, (*a*) of **под·проте́ктор·**; cushion

под·пру́г·а, -и, (*f*) girth, bellyband (of harness)

под·пруж·и́нн·ый, -ая, -ое, (*a*) spring-opposed

под·пры́г·ива·ть, -ют, (*imp*) hop, skip, jump

под·пры́г·н·уть, -ут, (*perf*) *see* **под·-пры́г·ива·ть**

по́д·пуск·, -а, (*m*) trailed multiple-hook fishing line

под·пуск·а́·ть, -ют, (*imp*) let approach

под·пуст·и́ть, -ят, (*perf*) *see* **под·-пуск·а́·ть**

под·пу́ш·ек·, -ш·к·а, (*m*) fluff, down, fuzz; (text) linters, cotton linters

под·пу́щ·енн·ый, -ая, -ое, (*past part pass*) of **под·пуст·и́ть**

под·пя́т·ник·, -а, (*m*) footstep/step bearing; heel (of derrick etc.)

~, сапфи́р·н·ый (instr) sapphire point

под·рабо́т·а·ть, -ют, (*imp*) go/work over again

под·рабо́т·к·а, -и, (*g pl*) **-т·ок·,** (*f*) (min) tapping, robbing (pillars)

~ подо́·шв·ы (min) dinting

под·ра́вн·ива·ть, -ют, (*imp*) level/ even/line up; make more even/level, make straighter

под·раж·а́ни·е, -я, (*n*) (*v n*) of **под·-раж·а́·ть**; imitation

под·раж·а́тельн·ый, -ая, -ое, (*a*) imitative

под·раж·а́·ть, -ют, (*imp* + *dat*) imitate

под·раз·де́л·, -а, (*m*) subsection

под·раз·дел·е́ни·е, -я, (*n*) (*v n*), *see* **под·раз·дел·я́·ть**; subdivision, unit, section

под·раз·дел·и́ть, -ят, (*perf*) *see* **под·-раз·дел·я́·ть**

под·раз·дел·я́·ть, -ют, (*imp*) subdivide, graduate

под·раз·ум·ева́·ть, -ют, (*imp*) imply, mean; **-ся** (*pass*); be implicit in

под·ра́м·ник·, -а, (*m*) sub-frame; frame (for artist's canvas)

под·ра́м·ок·, -м·к·а, (*m*) = **под·-ра́м·ник·**

под·раст·а́·ть, -ют, (*imp*) grow, grow up

под·раст·и́, -у́т, (*past masc sing*) **под·-ро́с·,** (*perf*); *see* **под·раст·а́·ть**

под·раст·и́ть, -ят, (*perf*) *see* **под·-ра́щ·ива·ть**

под·ра́щ·ива·ть, -ют, (*imp*) grow (*transitive*, e.g. flowers)

под·рёбер·н·ый, -ая, -ое, (*a*) (anat) subcostal

под·регул·и́р·овать, -уют, (*imp*) re-adjust

под·ре́з·, -а, (*m*) (*v n*), *see* **под·ре́-з·а́·ть;** undercut (of gear teeth)

под·ре́з·ать, (*fut 3rd pl*) **под·ре́-ж·ут,** (*perf*) *see* **под·рез·а́·ть**

под·рез·а́·ть, -ют, (*imp*) trim, shorten; (mech) face (lathework); undercut (gear teeth etc.); (hortic) clip, prune

под·резин·енн·ый, -ая, -ое, (*a*) rubber mounted/footed

под·ре́з·к·а, -и, (*g pl*) **-з·ок·,** (*f*) (*v n*), *see* **под·рез·а́·ть**

~ **тор·ц·о́в** facing (lathe operation)

~ **у·сту́п·ов** shoulder turning (lathe operation)

под·рез·н·о́й, -а́я, -о́е, (*a*) (mech) facing, undercutting

под·ре́з·ыва·ть, -ют, (*imp*) = **под·-рез·а́·ть**

под·рессо́р·ива·ть, -ют, (*imp*) spring, mount on springs; (hydr) spring-load

под·рессо́р·ивани·е, -я, (*n*) (*v n*), *see* **под·рессо́р·ива·ть**

под·рессо́р·н·ый, -ая, -ое, (*a*) spring-mounted, spring; (hydr) spring-loaded

под·решёт·ин·а, -ы, (*f*) (build) purlin

под·решёт·к·а, -и, (*g pl*) **-т·ок·,** (*f*) (phys) sub-lattice

под·рис·ова́ть, -у́ют, (*perf*) touch up, draw/paint in (photos, drawings)

по·дро́б·н·о (*adv*) in detail

по·дро́б·ность·, -и, (*f*) detail, particularity; (*pl*) minutiae

по·дро́б·н·ый, -ая, -ое, (*a*) detailed, thorough

под·ровн·я́·ть, -ют, (*perf*) *see* **под·-ра́вн·ива·ть**

под·ро́д·, -а, (*m*) (biol) subgenus

под·ро́·ет, (*fut 3rd sing*) of **под·ры́·ть**

под·ро́с·, (*past masc sing*) of **под·-раст·и́,** (*perf*); *see* **под·раст·а́·ть** (*imp*)

под·ро́ст·, -а, (*m*) seedlings, low cover (timber)

под·ро́ст·ок·, -т·к·а, (*m*) juvenile (12–16)

под·руб·а́·ть, -ют, (*imp*) (text) hem; undercut, chop/hack/slash under/off

под·руб·и́ть, -ят, (*perf*) *see* **под·ру-б·а́·ть**

под·ру́б·к·а, -и, (*f*) (*v n*), *see* **под·-руб·а́·ть;** (min etc.) undercut, undercutting

под·ру́л·ива·ть, -ют, (*imp*) steer up to; (air) taxi up to

под·рул·и́ть, -я́т, (*perf*) *see* **под·ру́-л·ива·ть**

под·ру́сл·ов·ый, -ая, -ое, (*a*) (geol) infrabed

под·ру́ч·ник·, -а, (*m*) tool rest (on a lathe)

под·ру́ч·н·ый, -ая, -ое, (*a*) handy; (*m decl as adj*) helper, hand

под·рыв·а́·ть, -ют, (*imp*) undermine, sap; blow up, blast, explode

под·рыв·н·о́й, -а́я, -о́е, (*a*) demolition, blasting; subversive

~ **ста́нци·я** (*f*) (min) firing station

под·ры́·ть, (*fut 3rd pl*) **под·ро́·ют,** (*perf*) undermine, sap

под·ря́д· (*adv*) in succession, running

под·ря́д·, -а, (*m*) (com) contract

под·ря́д·н·ый, -ая, -ое, (*a*) (com) contract

под·ря́д·чик·, -а, (*m*) (com) contractor

под·сад·и́ть, -ят, (*perf*) *see* **под·са́-ж·ива·ть**

под·са́ж·ива·ть, -ют, (*imp*) plant more/near; help to mount; plant (e.g. an agent)

под·са́с·ыва·ть, -ют, (*imp*) suck under

под·са́хар·ива·ть, -ют, (*imp*) sugar, add sugar

под·са́ч·ива·ть, -ют, (*imp*) tap (trees etc.)

под·с·бо́р·к·а, -и, (*g pl*) **-р·ок·,** (*f*) sub-assembly

под·свёж·к·а, -и, (*g pl*) **-ж·ек·,** (*f*) fresh river water in salt water

под·све́т·, -а, (*m*) lighting from below; (instr) illumination

под·свет·и́ть, ⸗ят, (*perf*) *see* **под·-свеч·ива·ть**

под·свет·к·а, -и, (*g pl*) **-т·ок·,** (*f*) (*v n*), *see* **под·свеч·ива·ть;** bias lighting, brightening (of a light); (TV) intensity gate

под·свеч·ивани·е, -я, (*n*) (*v n*), *see* **под·свеч·ива·ть;** = **под·свет·к·а** (q.v.)

под·свеч·ивátел·ь, -я, (*m*) (TV) sweep magnifier/intensifier

под·свеч·ива·ть, -ют, (*imp*) light from below, light up, intensify (light)

под·свеч·ник·, -а, (*m*) candlestick

под·свин·ок·, -н·к·а, (*m*) (agr) gilt, yelt (4–9 months)

под·сев·á·ть, -ют, (*imp*) undersow; sow more

под·се́д·, -а, (*m*) (bot) sprout; fur residues (on hide)

под·сед·е́льник·, -а, (*m*) girth, belly-band (of harness)

под·се́·ива·ть, -ют, (*imp*) = **под·сев·á·ть**

под·сек·á·ть, -ют, (*imp*) cut from under, cut off; clear (land for cultivation); (fish) strike

под·се́кци·я, -и, (*f*) subsection

под·семе́й·ств·о, -а, (*n*) (biol) sub-family

под·семя·до́ль·н·ое колéн·о (*n*) (bot) hypocotyl

под·сер·ни́ст·ая кисл·от·á (*f*) hyposulphurous acid

под·се́ч·к·а, -и, (*g pl*) **-ч·ек·,** (*f*) (*v n*), *see* **под·сек·á·ть;** hardie (forging tool), anvil chisel

под·се́ч·ь, (*fut 3rd sing, pl*) **под·сеч·ёт, под·сек·ýт,** (*past masc sing*) **под·сéк·,** (*perf*); *see* **под·сек·á·ть**

под·сé·ять, -ют, (*perf*) *see* **под·сев·á·ть**

под·син·гóн·и·я, -и, (*f*) (cryst) subsystem

под·си́н·ива·ть, -ют, (*imp*) blue

под·син·и́ть, -я́т, (*perf*) *see* **под·си́н·ива·ть**

под·си́ньк·а, -и, (*f*) (*v n*), *see* **под·си́н·ива·ть**

под·систе́м·а, -ы, (*f*) subsystem

под·скáбл·ива·ть, -ют, (*imp*) scrape/shave/plane lightly

под·с·каз·á·ть, (*fut 3rd pl*) **под·с·кá·ж·ут,** (*perf*); *see* **под·с·кáз·ыва·ть**

под·с·кáз·к·а, -и, (*g pl*) **-з·ок·,** (*f*) (*v n*), *see* **под·с·кáз·ыва·ть;** suggestion

под·с·кáз·ыва·ть, -ют, (*imp*) prompt

под·скак·á·ть, (*fut 3rd pl*) **под·скá·ч·ут,** (*perf*) gallop up to

под·скáк·ива·ть, -ют, (*perf*) jump; gallop up to

подскáльник·, -а, (*m*) (text) warp beam rod

под·скáч·ет, (*fut 3rd pl*) of **под·ска·к·á·ть**

под·скобл·и́ть, -я́т, (*perf*) *see* **под·-скáбл·ива·ть**

под·скоч·и́ть, ⸗ат, (*perf*) jump

под·слá·ива·ть, -ют, (*imp*) make/form a layer under

под·сло́·й, -я, (*m*) sublayer, underlayer, substratum

под·слýш·ивани·е, -я, (*n*) (*v n*), *see* **под·слýш·ива·ть**

под·слýш·ива·ть, -ют, (*imp*) listen in (to), eavesdrop; intercept (radio); tap (line communications)

под·смол·и́ть, -я́т, (*perf*) tar underneath

под·смо́л·ок·, -л·к·а, (*m*) (chem) tar water

под·смо́ль·н·ый, -ая, -ое, (*a*) (chem) pyroligneous

под·со́б·н·ый, -ая, -ое, (*a*) auxiliary, subsidiary

~ **хозя́й·ств·о** (*n*) farm or market garden attached to a works/factory, works farm

под·сов·á·ть, (*fut 3rd pl*) **под·су·ю́т,** (*perf*) *see* **под·со́в·ыва·ть**

под·со́в·ы (*pl*) layered ice (formed by rafting)

под·со́в·ыва·ть, -ют, (*imp*) shove/thrust/poke under

под·со·един·е́ни·е, -я, (*n*) connection

под·со·един·и́ть, -я́т, (*perf*) *see* **под·-со·един·я́·ть**

под·со·един·я́·ть, -ют, (*imp*) connect up (to)

под·со·зná·ни·е, -я, (*n*) subconscious

под·со·зná·тельн·ый, -ая, -ое, (*a*) subconscious

под·сол·ев·о́й, -áя, -о́е, (*a*) (geol) subsalt

под·со́лн·ечник·, -а, (*m*) (bot) sunflower, *Helianthus annus*

под·со́лн·ечн·ый, -ая, -ое, (*a*) of **под·со́лн·ечник·**

~ **жмых·** (*m*) (agr) sunflower cake

под·со́с·, -а, (*m*) inleakage, inflow, insuction

под·сос·áть, -ýт, (*perf*) suck under/in

под·со́с·н·ый, -ая, -ое,   (*a*) (zool) suckling

под·со·сто·я́ни·е, -я,   (*n*) sub-state

под·со́х·нуть, -нут,   (*perf*) *see* **под·-сых·а́·ть**

под·соч·и́ть, -а́т,   (*perf*) tap (trees etc.)

под·со́ч·к·а, -и,   (*f*) (*v n*) of **под·со·ч·и́ть**

под·ста́·в·а, -ы,   (*f*) tray, trough; (shipb) stopping up

под·ста́в·ить, -ят,   (*perf*) *see* **под·-ставл·я́·ть**

под·ста́в·к·а, -и,   (*g pl*) **-в·ок·,** (*f*) support, prop, rest, stand

под·ставл·я́·ть, -ют,   (*imp*) put under/ up to/to; prop, support; substitute, insert

под·став·н·о́й, -а́я, -о́е,   (*a*) support-ing, bracing; substitute, understudy

под·ста́в·очн·ый, -ая, -ое,   (*a*) of **под·ста́в·к·а**

под·стано́в·к·а, -и,   (*g pl*) **-в·ок·,** (*f*) (math) substitution, permutation

~, лине́ар·н·ая   linear substitution

под·ста́нци·я, -и,   (*f*) substation; sub/ branch exchange, satellite office (tele-phones)

~, за·кры́·т·ая (elec) indoor substation

~, ма́чт·ов·ая   (elec) pole-mounted substation

~, от·кры́·т·ая   (elec) outdoor sub-station

~, по·выс·и́тельн·ая   (elec) step-up transformer substation

~, по·низ·и́тельн·ая   (elec) step-down transformer substation

~, пре·обра́з·ова́тельн·ая   (elec) con-verting substation

~, рту́т·н·о-пре·обра́з·ова́тельн·ая (elec) mercury-arc converting station

~, трансформа́тор·н·ая   (elec) trans-former substation

~, тя́г·ов·ая   (elec) railway substation

~, фи́дер·н·ая   (elec) distribution sub-station

под·ство́л·ок·, -л·к·а,   (*m*) (min) draining pit, sump

под·стела́ж·н·ый изоля́тор·   (*m*) (elec) stand insulator

под·стил·а́·ть, -ют,   (*imp*) spread/lay under

под·стил·а́ющ·ий, -ая, -ее, (*pres part act*) of **под·стил·а́·ть;** subjacent, underlying; basement (rocks etc.)

под·сти́л·к·а, -и,   (*g pl*) **-л·ок·,** (*f*) (agr) bedding, litter; (naut) dunnage; floor (of forest)

под·сти́л·очн·ый, -ая, -ое,   (*a*) of **под·сти́л·к·а**

под·стра́·ива·ть, -ют, (*imp*) build on/ against/near; tune, tune in; arrange secretly, contrive

под·стрек·а́·ть, -ют,   (*imp*) incite, instigate; excite

под·стрек·н·у́ть, -у́т, (*perf*) *see* **под·-стрек·а́·ть**

под·стре́л·ива·ть, -ют,   (*imp*) hit, wound, wing (with shot)

под·стрел·и́ть, -́ят, (*perf*) *see* **под·-стре́л·ива·ть**

под·стриг·а́·ть, -ют,   (*imp*) trim, clip, crop, shear; prune, lop

под·стри́ж·к·а, -и,   (*f*) (*v n*), *see* **под·стриг·а́·ть**

под·стри́ч·ь, (*fut 3rd sing, pl*) **под·-стриж·ёт, под·стриг·у́т,** (*past masc sing*) **под·стриг·,** (*perf*); *see* **под·-стриг·а́·ть**

под·стро́·енн·ый, -ая, -ое, (*past part pass*) *see* **под·стра́·ива·ть**

под·стро́·ечн·ый, -ая, -ое,   (*a*) of **под·строй·к·а**

~ конденса́тор·   (*m*) (elec) trimmer, padder

под·стро́·ить, -ят,   (*perf*) *see* **под·-стра́·ива·ть**

под·строй·к·а, -и,   (*f*) (*v n*), *see* **под·-стра́·ива·ть;** (rad) alignment, tuning

~ част·от·у́ frequency control/trim

под·стро́ч·н·ый, -ая, -ое, (*a*) (print etc.) interlinear, below the line; literal, word for word

~ и́ндекс· (*m*) (print) subscript

~ при·меч·а́ни·е, (*n*) (print) footnote

под·структу́р·а, -ы,   (*f*) substructure

по́д·ступ·, -а,   (*m*) (*v n*), *see* **под·-ступ·а́·ть;** approach; (min) bench, intermediate bench

под·ступ·а́·ть, -ют,   (*imp*) approach, come up to/into

под·ступ·ёнок·, -нк·а,   (*m*) (build) height of step/stair

под·ступ·и́ть, -́ят, (*perf*) *see* **под·-ступ·а́·ть**

под·суд·и́м·ый, -ого, (*m decl as adj*) (law) the accused

под·су́д·ност·ь, -и, (*f*) (law) jurisdic-tion, competence

под·су́д·н·ый, -ая, -ое, (*a*) (law) under the jurisdiction/competence of

под·су·ёт, (*fut 3rd sing*) of **под·со·в·а́ть**

под·су́·н·уть, -ут, (*perf*) *see* **под·со́-в·ыва·ть**

под·су́ш·ива·ть, -ют, (*imp*) dry a little/slightly; pre-dry

под·суш·и́ть, -ат, (*perf*) *see* под·су́ш·ива·ть

под·с·чёт·, -а, (*m*) (*v n*), *see* под·-с·чи́т·ыва·ть; estimated total, estimate; count

~ кро́в·и (med) blood count

под·с·чит·а́·ть, -ют, (*perf*) *see* под·-с·чи́т·ыва·ть

под·с·чи́т·ыва·ть, -ют, (*imp*) total/ count/tot up; estimate, reckon

под·сы́п·ать, -лют, (*perf*) *see* под·-сып·а́·ть

под·сып·а́·ть, -ют, (*imp*) pour/heap/ pile/put more on; pile/heap up

под·сых·а́·ть, -ют, (*imp*) dry slightly, become drier, dry

под·та́·ива·ть, -ют, (*imp*) thaw/melt slightly/a little, be affected by thaw

. под·та́лк·ивател·ь, -я, (*m*) pusher, ram

под·та́лк·ива·ть, -ют, (*imp*) jog, jolt, nudge, shove; push/shove towards

под·та́нгенс·, -а, (*m*) (math) subtangent

под·та́пл·ива·ть, -ют, (*imp*) heat/ stoke a little; heat more/again; melt again

под·та́ск·ива·ть, -ют, (*imp*) drag/ pull up to

под·та́ч·ива·ть, -ют, (*imp*) sharpen, resharpen; eat away; (cinema) dub

под·тащ·и́ть, -ат, (*perf*) *see* под·-та́ск·ива·ть

под·та́·ять, -ют, (*perf*) *see* под·-та́·ива·ть

под·тверд·и́ть, -ят, (*perf*) *see* под·-твержд·а́·ть

под·твержд·а́·ть, -ют, (*imp*) confirm, corroborate; assert, state; (telecom) acknowledge; -ся (*pass*); be borne out (by)

под·твержд·е́ни·е, -я, (*n*) confirmation, corroboration; assertion; (telecom) acknowledgement

под·тёк·, -а, (*m*) (*v n*), *see* под·те-к·а́·ть; inflow, seepage; (med) bruise

под·тек·а́ни·е, -я, (*n*) (*v n*), *see* под·-тек·а́·ть; leak, dribble; seepage

под·тек·а́·ть, -ют, (*imp*) flow under; leak, dribble, seep

под·те́кст·, -а, (*m*) concealed meaning

под·тепл·ов·о́й, -а́я, -о́е, (*a*) subthermal

под·тер·е́ть, (*fut 3rd pl*) подо·тр·у́т, (*perf*) *see* под·тир·а́·ть

под·те́ч·ь, (*fut 3rd sing, pl*) под·те-ч·ёт, под·тек·у́т, (*past masc sing*) под·тёк·, (*perf*); *see* под·тек·а́·ть

под·ти́п·, -а, (*m*) subtype

под·тир·а́·ть, -ют, (*imp*) wipe up/off

под·това́р·ник·, -а, (*m*) (naut) dunnage; light poles (timber)

под·толк·н·у́ть, -у́т, (*perf*) *see* под·-та́лк·ива·ть

под·то́к·, -а, (*m*) (*v n*), *see* под·тек·а́·ть

под·тона́ль·н·ый, -ая, -ое, (*a*) sub-audio, infrasonic

под·топ·и́ть, -ят, (*perf*) *see* под·-та́пл·ива·ть, под·топл·я́·ть

под·топл·е́ни·е, -я, (*n*) (*v n*), *see* под·топл·я́·ть

под·топл·я́·ть, -ют, (*imp*) submerge, flood, saturate with ground water

под·точ·и́ть, -ат, (*perf*) *see* под·-та́ч·ива·ть

под·тя́г·ивани·е, -я, (*n*) (*v n*), *see* под·тя́г·ива·ть

под·тя́г·ива·ть, -ют, (*imp*) pull/draw/ haul up to/under/tighter, tighten up; (mil) bring/move up

под·тя́ж·к·а, -и, (*g pl*) -ж·ек·, (*f*) tightening up; tautening; (*pl*) braces, suspenders

под·тя·н·у́ть, -у́т, (*perf*) *see* под·-тя́г·ива·ть

по·дув·а́·ть, -ют, (*imp*) blow

под·узел·ко́в·ый, -ая, -ое, (*a*) sub-nodal

под·уклю́чин·а, -ы, (*f*) crutch/rowlock socket (in a rowing boat)

по·ду́м·а·ть, -ют, (*perf*) think

под·у́·ровен·ь, -я, (*m*) (phys) sub-level/-state

по·ду́·ть, -ют, (*perf*) blow

поду́ш·ечк·а, -и, (*g pl*) -чек·, (*f*) (*dim*) of поду́ш·к·а

поду́ш·к·а, -и, (*g pl*) -ш·ек·, (*f*) pillow; cushion, pad, chock; bedding (elec cable); (rail) rail chair

~, воз·ду́ш·н·ая (aerodynam) air cushioning/cushion/blanket, cushion of air

~, зем·н·а́я (air) ground cushion

под·фа́р·ник·, -а, (*m*) (M/T) side/ park light

под·фюзеля́ж·н·ый, -ая, -ое, (*a*) (a/c) ventral

под·хва́т·, -а, (*m*) (*v n*), *see* под·-хва́т·ыва·ть; carrier arm, cross support; (nucl) pickup

под·хват·и́ть, -ят, (*perf*) *see* под·-хва́т·ыва·ть

под·хва́т·чик·, -а, (*m*) pickup

под·хва́т·ыва·ть, -ют, (*imp*) catch/ snatch, seize; catch/snatch/seize up

под·хва́ч·енн·ый, -ая, -ое, (*past part pass*) see под·хва́т·ыва·ть

под·хо́д·, -а, (*m*) approach; (math) mode of treatment; (air) initial approach

   у́гол· под·хо́д·а (*m*) (air) angle of approach

под·ход·и́ть, -ят, (*imp*) approach, come/go/walk up to; suit, fit

под·хо́д·н·ый, -ая, -ое, (*a*) of под·-хо́д·

~ марке́р· (*m*) approach marker

под·ход·я́щ·ий, -ая, -ее, (*pres part act*) of под·ход·и́ть; suitable, appropriate; (*pl decl as adj*) (math) convergents

под·хож·у́, (*pres 1st sing*) of под·-ход·и́ть

под·ца́пф·енник·, -а, (*m*) trunnion bearing

под·ца́р·ств·о, -а, (*m*) (biol) sub-kingdom

под·цвет·и́ть, -ят, (*perf*) see под·-цве́ч·ива·ть

под·цве́ч·ива·ть, -ют, (*imp*) tint, dye (the same colour but more intense)

под·цеп·и́ть, -ят, (*perf*) see под·-цепл·я́·ть

под·цепл·я́·ть, -ют, (*imp*) hook/ pick up

под·ча́с· (*adv*) occasionally, at times

под·чека́н·к·а, -и, (*g pl*) -н·ок·, (*f*) recaulking

под·челюст·н·о́й, -а́я, -о́е, (*a*) (anat) submaxillary

под·чёрк·ивани·е, -я, (*n*) (*v n*) of под·чёрк·ива·ть; emphasis

под·чёрк·ива·ть, -ют, (*imp*) under-line, emphasize; (rad) accentuate, pre-emphasize (certain frequencies)

под·черк·н·у́ть, -у́т, (*perf*) see под·-чёрк·ива·ть

под·чин·е́ни·е, -я, (*n*) (*v n*), see под·-чин·я́·ть; subordination

под·чин·ённ·ый, -ая, -ое, (*past part pass*) see под·чин·я́·ть; subordinate, under the command of

~ ста́нци·я (*f*) out-station

под·чин·и́ть, -ят, (*perf*) see под·-чин·я́·ть

под·чин·я́·ть, -ют, (*imp*) subordinate, place under (figurative); -ся (*pass*); come under; submit to; obey, satisfy, follow (natural laws etc.)

под·чи́ст·ить, -ят, (*perf*) see под·-чищ·а́·ть

под·чит·а́·ть, -ют, (*perf*) read out/ aloud

под·чищ·а́·ть, -ют, (*imp*) clean up; erase, rub out

под·ша́ш·к·а, -и, (*g pl*) -ш·ек·, (*f*) (min) chock (inserted into coal seam cut to prevent premature caving)

под·шив·а́·ть, -ют, (*imp*) sew in, line; sew on underneath, sole

под·ши́п·ник·, -а, (*m*) bearing

~, бес·сепера́тор·н·ый cageless bear-ing

~, бочко·обра́з·н·о-кон·и́ческ·ий angular-contact roller bearing

~, бу́кс·ов·ый (rail) axlebox bearing

~, втул·очн·ый bush bearing

~, голов·н·о́й crosshead bearing; (ICE) small-end bearing

~, греб·е́нчат·ый collar bearing

~, двой·н·о́й у·по́р·н·ый double thrust bearing

~, двух·ря́д·н·ый double-row bearing

~ жи́дк·остн·ого тр·е́ни·я hydro-dynamic bearing, fluid-film bearing

~, иг о́ль·чат·ый needle bearing

~, кач·а́ющ·ийся tumbler bearing

~ кач·е́ни·я rolling-contact bearing, rolling-element bearing

~, кон·и́ческ·ий tapered roller bearing

~, корен·н·о́й (ICE) main bearing

~ Миче́лл·я Michell thrust bearing

~, моты́л·ев·ый large-end bearing

~, на·гре́в·ш·ийся hot bearing

~, не·разъ·ём·н·ый solid bearing

~, один·а́рн·ый single-row bearing

·~, одно·ря́д·н·ый single-row bearing

~, о·по́р·н·ый plummer bearing

~, ордина́р·н·ый single-row bearing

~, пере·движ·н·о́й у·по́р·н·ый (met roll) thrust-bearing carriage (Man-nesmann piercer)

~, про·меж·у́точн·ый block/inner bearing

~, радиа́ль·н·о-сфер·и́ческ·ий angu-lar-contact bearing, spherical-race radial bearing

~, радиа́ль·н·о-у·по́р·н·ый radial-thrust bearing

~, радиа́ль·н·ый radial bearing

~, разъ·ём·н·ый two-part/split bear-ing

~, ра́м·ов·ый (ICE) main bearing

~, рези́н·ов·ый rubber bearing, (shipb) cutless bearing

~, ро́л·иков·ый roller bearing

под·ши́п·ник

~, ро́лико·кон·и́ческ·ий tapered roller bearing

~, само·с·ма́з·ывающ·ийся self-lubricating bearing

~, само·у·стана́вл·ивающ·ийся self-aligning bearing

~, с·дво́·енн·ый double-/dual-purpose bearing

~, сегме́нт·н·ый segmental bearing

~ скольж·е́ни·я slide/plain bearing

~, сфер·и́ческ·ий spherical-race thrust bearing

~, сферо·кон·и́ческ·ий spherangular bearing

~, у·по́р·н·ый thrust bearing

~, четырёх·ря́д·н·ый four-row bearing

~, ша́р·иков·ый ball bearing

~, шату́н·н·ый large-end bearing

под·ши́п·ников·ый, -ая, -ое, (a) of под·ши́п·ник·

~ с·плав· (m) bearing alloy

под·ши́·т·ый, -ая, -ое, (past part pass) see под·шив·а́·ть

под·ши́·ть, (fut 3rd pl) подо·шь·ю́т, (perf) see под·шив·а́·ть

под·шлиф·о́ванн·ый, -ая, -ое, (past part pass) sueded (of leathers)

под·штукату́р·ить, -ят, (perf) (build) repair plaster, touch up plaster

под·щела́ч·ивани·е, -я, (n) alkalization

под·щела́ч·ива·ть, -ют, (imp) (chem) alkalize

под·щело́ч·енн·ый, -ая, -ое, (past part pass) alkalized

подъ·е́д·ет, (fut 3rd sing) of подъ·е́х·ать, (perf); see подъ·езж·а́·ть (imp)

подъ·е́зд·, -а, (m) (v n), see подъ·езж·а́·ть; approach driveway, drive; porch, entrance

подъ·езд·н·о́й, -а́я, -о́е, (a) see also подъ·е́зд·н·ый; approach, access

~ доро́г·а (f) access road

~ пут·ь (m) drive; (rail) belt line

подъ·е́зд·н·ый, -ая, -ое, (a) see also подъ·езд·н·о́й; entrance

~ двер·ь (f) front door

подъ·езж·а́·ть, -ют, (imp) drive/ride/go up to, approach

подъ·ём·, -а, (m) (v n), see под·ним·а́·ть; lifting, hoisting, raising; lift, rise, ascent; (air) climb; (econ) development, upsurge; (geog) slope, upgrade; enthusiasm, animation; instep (of foot)

подъ·ём

вы́с·от·а́ подъ·ём·а (f) lift

~ дн·и́щ·а (shipb) rise of bottom, dead rise

~ зя́б·и autumn ploughing

~ пар·о́в ploughing the fallow

ско́р·ост·ь подъ·ём·а (f) (air) climbing speed, rate of ascent/climb; hoisting/lifting speed/rate (e.g. of crane); (min) winding speed

у́гол· подъ·ём·а (m) (air) angle of ascent; lifting angle; lead/helix angle (of screw threads)

подъ·ём·ник·, -а, (m) lift, elevator, hoist, jack

~, гидравл·и́ческ·ий hydraulic jack

~, ма́чт·ов·ый (build) platform hoist

~, много·кле́т·очн·ый paternoster lift

~, порш·нев·о́й ram lift

подъ·ём·н·ый, -ая, -ое, (a) lifting; hoisting; elevating; (pl decl as adj) travelling expenses

~ движ·е́ни·е (n) lifting/vertical motion

коэффицие́нт· подъ·ём·н·ой си́л·ы (m) (air) lift coefficient

~ маши́н·а (f) (min) winding engine, hoisting engine (U.S.)

~ механ·и́зм· (m) lifting/hoisting gear; (gunn) elevating gear

~ площа́д·к·а (f) cargo tray

~ си́л·а (f) (air) buoyancy; (rail) carrying capacity

подъ·е́х·ать, (fut 3rd pl) под·е́д·ут, (perf) see подъ·езж·а́·ть

подъ·язы́ч·н·ый, -ая, -ое, (a) (anat) sublingual, hypoglossal

подъ·я́рус·, -а, (m) (geol) substage

под·ынтегра́ль·н·ое вы·раж·е́ни·е (n) (math) integrand

под·ыск·а́ть, (fut 3rd pl) под·ы́щ·ут, (perf) find, seek out (something suitable)

под·ытож·ени·е, -я, (n) (v n) of под·ытож·ить; summation

под·ыто́ж·ить, -ат, (perf) total/count/tot up; sum up

по·дыш·а́ть, ‑ат, (perf) breathe

под·ы́щ·ет, (fut 3rd sing) of под·ыс·к·а́ть

под·эта́ж·, -а, (m) (min) sublevel

по·е́вш·ий, -ая, -ее, (pres part act) of по·е́ст·ь, (perf); see по·ед·а́·ть (imp)

по·ед·а́·ть, -ют, (imp) eat

по·е́д·ет, (fut 3rd sing) of по·е́х·ать

по·ед·я́т, (*fut 3rd pl*) of по·е́ст·ь, (*perf*); *see* по·ед·а́·ть (*imp*)

по́·езд·, -а, (*nom pl*) по·езд·а́, (*g pl*) -о́в, (*m*) train

~, автомоби́ль·н·ый   tractor-trailer unit

~, групп·ов·о́й   assorted train

~, маршру́т·н·ый   through freight train

~, с·бо́р·н·ый   pickup goods train

~, хозя́й·ственн·ый   service/maintenance train

по·е́зд·ить, -ят, (*perf*) make a journey, go for a ride

по·е́зд·к·а, -и, (*g pl*) -д·ок·, (*f*) journey, short journey, trip

по·езд·н·о́й, -а́я, -о́е, (*a*) train

~ ваго́н·н·ый ма́стер·   (*m*) train examiner

поездо́·граф·, -а,   (*m*) (rail) train recorder/graph

по·е́зж·у, (*fut 1st sing*) of по·е́зд·ить

по·ём·н·ый, -ая, -ое, (*a*) of по́·йм·а

~ луг·а́ (*pl*) water meadows

по·е́ни·е, -я, (*n*) (*v n*) of по·и́ть

по·е́ст·ь, (*fut 3rd sing, pl*) по·е́ст·, по·ед·я́т, (*past masc sing*) по·е́л (*perf*); eat

по·ёт, (*pres 3rd sing*) of пе·ть

по·е́х·ать, (*fut 3rd pl*) по·е́д·ут, (*perf*) drive, ride, go; set off, depart

по·жа́л·овани·е, -я, (*n*) (*v n*) of по·жа́л·овать; award, grant

по·жа́л·овать, -уют, (*perf*) grant, bestow, confer; -ся complain

по·жа́л·уй (*adv*) probably, maybe; it is likely/probable

пожа́луйста (*indecl*) please

по·жа́р·, -а, (*m*) fire, conflagration, blaze

по·жа́р·ищ·е, -а, (*n*) site of fire

по·жа́р·н·ый, -ая, -ое, (*a*) fire, firefighting; (*m decl as adj*) fireman, firefighter

~ вы́ш·к·а (*f*) firewatching tower

~ кома́нд·а (*f*) fire brigade

~ маши́н·а (*f*) fire engine

~ на·со́с· (*m*) fire pump; fire engine

~ о·хра́н·а (*f*) fire service

~ рож·о́к· (*m*) fire hydrant

~ систе́м·а (*f*) (shipb) fire protection system

~ су́д·н·о (*n*) fire float, firefighting craft

~ трево́г·а (*f*) fire alarm

по·жа́т·и·е, -я, (*n*) handshake

по·жа́т·ь, (1) (*fut 3rd pl*) по·жм·у́т, (*perf*) squeeze, press; shake (hand as greeting); shrug (shoulders); по·жа́т·ь, (2) (*fut 3rd pl*) по·жн·у́т, (*perf*) reap, finish reaping; reap, earn (rewards etc.)

по·жев·а́ть, (*fut 3rd pl*) по·жу·ю́т, (*perf*) chew, masticate

по·жел·а́·ть, -ют, (*perf*) (+ *gen*) wish

по·желт·е́·ть, -ют, (*perf*) become/turn yellow

по·желт·и́ть, -ят, (*perf*) make/colour yellow, yellow

по·желч·ённ·ый, -ая, -ое, (*past part pass*) of по·желт·и́ть; yellowed

по·жив·ёт, (*fut 3rd sing*) по·жи́·ть

по·жи́зн·енн·ый, -ая, -ое, (*a*) life, lifelong

~ ре́нт·а (*f*) (ins) life annuity

по·жи·л·о́й, -а́я, -о́е, (*a*) middle-aged

по·жим·а́·ть, -ют, (*imp*) squeeze, press; shake (hands); shrug (shoulders)

по·жин·а́·ть, -ют, (*imp*) reap, finish reaping; reap, earn, receive (rewards etc.)

по·жир·а́·ть, -ют, (*imp*) devour

по·жи́·ть, (*fut 3rd pl*) по·жив·у́т, (*perf*) live, stay

по·жм·ёт, (*fut 3rd sing*) of по·жа́т·ь (1)

по·жн·ёт, (*fut 3rd sing*) of по·жа́т·ь (2)

по́·жн·ивн·ый, -ая, -ое, (*a*) post-harvest

по·жр·а́ть, -у́т, (*perf*) devour

по·жу·ёт, (*fut 3rd sing*) of по·жев·а́ть

по́з·а, -ы, (*f*) pose, posture, attitude

по·забо́т·иться, -ятся, (*perf*) look after, take care of; see that, make sure that, take care to

по·за·вчера́ (*adv*) the day before yesterday

по·зад·и́ (*adv*) (*prep*) behind, (naut) astern of, abaft

позади·ло́б·н·ый, -ая, -ое, (*a*) (anat) retropubic

позади·серд·е́чн·ый, -ая, -ое, (*a*) (anat) retrocardiac

по·звол·е́ни·е, -я, (*n*) (*v n*), *see* по·звол·я́·ть; permission

по·зво́л·ить, -ят, (*perf*) *see* по·звол·я́·ть

по·звол·я́·ть, -ют, (*imp*) allow, permit, enable, make possible

   по·звол·я́·ет о·пре·дел·и́ть   can be used to find/determine по·звон·и́ть, -ят, (*perf*) ring, clang

по·звон·о́к, -н·к·а́, (*m*) (anat) vertebra, spondyl

позвон·о́чник·, -а, (*m*) (anat) spine, backbone, spinal column

позвон·о́чн·ый, -ая, -ое, (*a*) vertebrate, vertebral; (*pl as noun*) vertebrates, *Vertebrata*

по·звуч·а́ть, -а́т, (*perf*) sound, resound, ring, be heard

позд·не́йш·ий, -ая, -ее, (*superlative*) of по́зд·н·ий

поздне·спе́·л·ый, -ая, -ое, (*a*) (agr) late-ripening

поздне·тре́т·и́чн·ый, -ого, (*n decl as adj*) (geol) Late Tertiary

по́зд·н·ий, -яя, -ее, (*a*) late; (geol) recent, young

~ голо·це́н· (*m*) (geol) Neoholocene

по·здра́в·ить, -ят, (*perf*) *see* по·-здравл·я́·ть

по·здравл·е́ни·е, -я, (*n*) (*v n*), *see* по·-здравл·я́·ть; congratulations, good wishes, greetings

по·здравл·я́·ть, -ют, (*imp*) congratulate, greet

по·зелен·е́·ть, -ют, (*perf*) become/ turn green

по·зём·, -а, (*m*) manure, dung

по·земе́ль·н·ый, -ая, -ое, (*a*) land, land-tenure/ownership

по·зём·к·а, -и, (*g pl*) -м·ок·, (*f*) *see* по·зём·ок·

по·зём·н·ый, -ая, -ое, (*a*) ground

~ тума́н· (*m*) ground fog

по·зём·ок·, -м·к·а, (*m*) drifting snow

~, сла́б·ый slight or moderate drifting snow

~, си́ль·н·ый heavy drifting snow

по́зж·е (*comp*) (*adv*) later

позити́в·, -а, (*m*) (phot) positive

позити́в·н·ый, -ая, -ое, (*a*) positive; empirical

позитро́н·, -а, (*m*) (phys) positron

позитро́н·и·й, -я, (*m*) (chem) positronium

позитро́н·н·ый, -ая, -ое, (*a*) of позитро́н·

позиционе́р·, -а, (*m*) (weld etc.) positioner

позицио́н·н·ый, -ая, -ое, (*a*) positional, position; positioning, stepping; (mil) static

~ по·лож·е́ни·е (*n*) (math) local value; trimmed-down position (of submarines)

~ у́гол· (*m*) position angle

пози́ци·я, -и, (*f*) position; site; item

~, ис·хо́д·н·ая original/starting position; (mil) forming-up place

~ радио·локацио́н·н·ой ста́нци·и radar site

~, ста́рт·ов·ая (rocket) launching area

по·зна·ва́ем·ый, -ая, -ое, (*pres part pass*) of по·зна·ва́ть; comprehensible

по·зна·ва́ть, -ю́т, (*imp*) comprehend, understand; grasp (ideas)

по·знако́м·ить, -ят, (*perf*) acquaint, introduce

по·зна́·ни·е, -я, (*n*) (*v n*), *see* по·-зна·ва́ть; knowledge, experience

по·зна́·ть, -ю́т, (*perf*) *see* по·зна·-ва́ть

по·золо́т·а, -ы, (*f*) gilt, gilding

по·золот·и́ть, -я́т, (*perf*) gild

по·золо́ч·енн·ый, -ая, -ое, (*past part pass*) of по·золот·и́ть

по·зо́н·н·ый, -ая, -ое, (*a*) zoned, staged, compartmented

~ регул·и́рова́ни·е дут·ь·я́ compartmented air control (of furnace)

по·зо́р·, -а, (*m*) shame, disgrace

по·зо́р·н·ый, -ая, -ое, (*a*) shameful, disgraceful

позуме́нт·, -а, (*m*) braid, galloon

по·зыв·н·о́й, -а́я, -о́е, (*a*) call; (*pl as noun*) call sign; identification letters; ship's number

~, между·на·ро́д·н·ый international code letters (identification)

~ сигна́л· (*m*) (telecom) call-sign

по·игр·а́·ть, -ю́т, (*perf*) play

по·и́лк·а, -и, (*g pl*) -лок·, (*f*) (agr) drinking bowl/trough

по·име́н·н·о (*adv*) by name

по·име́н·н·ый, -ая, -ое, (*a*) nominal

по·и́м·к·а, -и, (*g pl*) -м·ок·, (*f*) (*v n*), *see* по·йм·а́·ть; catching, capture

~ це́л·и target acquisition (range-finder etc.); (rad) target locking

по·им·у́ществен·н·ый, -ая, -ое, (*a*) property

по́йн·, -а, (*m*) (pharm) poine

по·ин·о́му (*adv*) otherwise, differently, in a different way/manner

по·интерес·ова́ться, -у́ются, (*perf*) show interest in

по́·иск·, -а, (*m*) search, scan, scanning, sweep; (rad) search; (mil) raid; (min) prospecting, exploration

~, круг·ов·о́й circular scanning (radar)

~ по горизо́нт·у horizon search

по·иск
~, **радио·локацио́н·н·ый** radar sweep
~, **руч·н·о́й** manual search
по·иск·а́ть (*fut 3rd pl*) **по·йщ·ут,** (*perf*) search/look for, seek
по́·иск·овик·, -а, (*m*) (min) examiner
по́·иск·ов·ый, -ая, -ое, (*a*) of **по́·иск·**
~ **брига́д·а** (*f*) prospecting party
~ **при·ём·ник·** (*m*) (rad) search receiver
по·и́ст·ин·е (*adv*) actually, indeed
по·и́ть, -я́т, (*imp*) water, give to drink
по·и́щ·ет, (*fut 3rd sing*) of **по·иск·а́ть**
по·йд·ёт, (*fut 3rd sing*) of **по·йти́,** (*perf*); *see* **ид·ти́** (*imp*)
пойдо́·метр·, -а, (*m*) poidometer
пойкил·и́тов·ый, -ая, -ое, (*a*) (geol) poikilitic
пойкило·бласт·и́ческ·ий, -ая, -ое, (*a*) (geol) poikiloblastic
пойкило·те́рм·н·ый, -ая, -ое, (*a*) (zool) poikilothermal, poikilothermic, cold-blooded
пойкило·цит·о́з·, -а, (*m*) (med) poikilocytosis
по́й·л·о, -а, (*m*) swill, mash
по́·йм·а, -ы, (*f*) flood plain
по́·йм·анн·ый, -ая, -ое, (*past part pass*) of **по·йм·а́ть**
по·йм·а́·ть, -ют, (*imp*) catch, capture; lock on (radar) '
по́·йм·енн·ый, -ая, -ое, (*a*) of **по́·йм·а**
~ **луг·а́** (*pl*) water meadows
по·йм·ёт, (*fut 3rd sing*) of **по́·нят·ь,** (*perf*); *see* **по·ним·а́·ть** (*imp*)
По́йнтинг·а, теоре́м·а (*f*) (phys) Poynting theorem
по́·йте, (*pl imper*) of **пе·ть**
по·йти́, (*fut 3rd sing, pl*) **по·йд·ёт, по·йд·у́т,** (*past masc sing*) **по·шёл,** (*perf*); *see* **ид·ти́, ход·и́ть**
пока́ (*adv*) for a time/while; (*conj*) while, as/so long as; until, till
~ **не** until
~ **что** for the time being; while
по·ка́др·ов·ый, -ая, -ое, (*a*) (cinema) one-shot, intermittent, time-lapse
~ **кино·съ·ём·к·а** (*f*) time-lapse filming
по·ка́ж·ет, (*fut 3rd sing*) of **по·каз·а́ть,** (*perf*); *see* **по·ка́з·ыва·ть** (*imp*)
по·ка́з·, -а, (*m*) show, showing, demonstration

по·каз·а́ни·е, -я, (*n*) indication, evidence; (instr) reading; (law) testimony; aspect (of railway signals)
~, **дистанцио́н·н·ое** (instr) remote reading
~, **капита́н·ск·ое** manifest (of cargo)
по·ка́з·анн·ый, -ая, -ое, (*past part pass*), *see* **по·ка́з·ыва·ть**
по·каз·а́тел·ь, -я, (*m*) index, indicator, pointer; indication, characteristic, figure; (math) exponent, index
~ **адиаба́т·ы** adiabatic exponent
~ **добр·о́тност·и** figure of merit
~, **ка́честв·енн·ый** performance figure
~ **ко́рн·я** (math) radical sign
~ **Ми́ллер·а** (cryst) Miller index
~ **поли·тро́п·ы** polytropic index
~ **пре·ломл·е́ни·я** (opt) refractive index
~ **сте́пен·и** (math) exponent
~ **сте́пен·и со зна́к·ом** signed exponent
~ **твёрд·ост·и** hardness number
~, **тепл·ов·о́й** heat index
по·каз·а́тельн·ый, -ая, -ое, (*a*) demonstration, show; model, exemplary; significant, characteristic; (math) exponential
~ **строй·к·а** (*f*) (build) demonstration site
~ **уро́к·** (*m*) demonstration lesson
~ **фу́нкци·я** (*f*) (math) exponential function
по·каз·а́ть, (*fut 3rd pl*) **по·ка́ж·ут,** (*perf*) *see* **по·ка́з·ыва·ть**
по·каз·н·о́й, -а́я, -о́е, (*a*) show, demonstration, display
по·ка́з·ывани·е, -я, (*n*) (*v n*) of **по·ка́з·ыва·ть**
по·ка́з·ыва·ть, -ют, (*imp*) show, demonstrate; indicate, point to; (law) testify, give evidence; **-ся** (*pass*); appear, be found
по·ка́з·ывающ·ий, -ая, -ее, (*pres part act*) of **по·ка́з·ыва·ть**
~ **при·бо́р·** (*m*) indicating instrument
по·кат·а́·ть, -ют, (*perf*) roll; take for a drive '
по·кат·и́ть, -́ят, (*perf*) roll; **-ся** set/move off
по·ка́т·ост·ь, -и, (*f*) slope, pitch, declivity
~ **и́мпульс·ов** pulse droop/tilt
по·ка́т·ый, -ая, -ое, (*a*) sloping, declining, drooping
по·ка́ч·анн·ый, -ая, -ое, (*past part pass*) of **по·кач·а́·ть**

по·кач·а́·ть, -ют, (*perf*) *see* **кач·а́·ть**

по·ка́ч·енн·ый, -ая, -ое, (*past part pass*) of **по·кат·и́ть**

по·ка́ч·ива·ть, -ют, (*imp*) rock/shake/vibrate/wobble intermittently

по·кач·н·у́ть, -у́т, (*perf*) shake; **-ся** sway, lurch

по·квартал́ь·н·о (*adv*) by the quarter, per quarter

по·кид·а́·ть, -ют, (*imp*) leave, abandon, desert

по·ки́·н·уть, -ут, (*perf*) *see* **по·кид·а́·ть**

по·клон́·, -а, (*m*) bow (as greeting etc.)

по·ков·а́ть, (*fut 3rd pl*) **по·ку·ю́т,** (*perf*) forge

по·ко́в·к·а, -и, (*g pl*) **-в·ок·,** (*f*) (*v n*) of **по·ков·а́ть;** (met) forging, forged piece

поко́·й, -я, (*m*) rest, repose, stillness, quiet, quiescence, dormancy; (telecom) letter P; small ward (in hospital)

в со·сто·я́ни·и поко́·я (*adv*) at rest; in a dormant state

ток· поко́·я (*m*) (elec) quiescent current

то́ч·к·а поко́·я (*f*) stationary point

у́гол· поко́·я (*m*) angle of repose/rest; (horol) locking angle

покой·н·ый, -ая, -ое, (*a*) restful, peaceful, calm; (phys) rest, quiescent; late, deceased

~ по·лож·е́ни·е (*n*) rest/home/quiescent position

покол·е́ни·е, -я, (*n*) (gen) generation

~, втор·и́чн·ое (gen) secondary generation

~, молод·о́е younger generation

~ я́дер· generation of nuclei

по·колеб·а́·ть, -лют, (*perf*) *see* **колеб·а́ть; -ся** hesitate

по·ко́нч·ить, -ат, (*perf*) finish with, have done with

по·кор·и́ть, -я́т, (*perf*) *see* **по·кор·я́·ть**

по·корм·и́ть, -ят, (*perf*) feed

по·ко́р·н·ый, -ая, -ое, (*a*) submissive, obedient

по·короб·ить, -ят, (*perf*) warp, buckle

по·коро́ч·е (*adv*) somewhat shorter

по·кор·я́·ть, -ют, (*imp*) subjugate, subdue; **-ся** submit to

по·ко́с·, -а, (*m*) haymaking; hay field

по·кос·и́ться, -я́тся, (*perf*) become slanting/askew/distorted, heel over

по·ко́ш·енн·ый, -ая, -ое, (*past part pass*) *see* **по·кос·и́ться;** slanted, sloping

поко́·ящ·ийся, -аяся, -ееся, (*pres part act*); resting, at rest, quiescent

по·кра́п·а·ть, -ют, (*perf*) spot, speckle

по·кра́п·ыва·ть, -ют, (*imp*) spot/speckle intermittently

по·кра́с·ить, -ят, (*perf*) *see* **кра́с·ить**

по·красн·е́·ть, -ют, (*perf*) redden, go/become red, flush

по·кра́ш·енн·ый, -ая, -ое, (*past part pass*) of **по·кра́с·ить,** (*perf*); *see* **кра́с·ить**

по·крив·и́ть, -я́т, (*perf*) bend, distort

по·кро́в·, -а, (*m*) cover, coating, coat; (anat) tegmen, tectum; (geol) sheet, cap, mantle

~, ко́ж·н·ый, (biol) integument

~, на·пыл·ённ·ый spray coating

~ цвет·к·а́ (bot) perianth, floral envelope

по·кров·и́тел·ь, -я, (*m*) patron, sponsor

по·кров·и́тельственн·ый, -ая, -ое, (*a*) of **покров·и́тел·ь;** protective

~ о·кра́с·к·а (*f*) (zool) protective colouring

по·кро́в·н·ый, -ая, -ое, (*a*) of **по·кро́в·;** (anat) tegmental, tectorial

~ раст·е́ни·е (*n*) (agr) cover crop

~ стекл·о́ (*n*) coverglass (in microscopy)

~ тка́н·ь (*f*) (biol) integument

покрово·пло́д·н·ый, -ая, -ое, (*a*) (bot) chlamydocarpous

по·кро́·ет, (*fut 3rd sing*) of **по·кры́·ть,** (*perf*); *see* **по·крыв·а́·ть** (*imp*)

по·кро́·й, -я, (*m*) cut (of clothes); cutting out; direction of cut

по·кро́м·к·а, -и, (*g pl*) **-м·ок·,** (*f*) (text) selvage

по·крош·и́ть, -ат, (*perf*) crumble

по·крыв·а́л·о, -а, (*n*) cover, cover-cloth

по·крыв·а́льц·е, -а, (*n*) (bot) indusium

по·крыв·а́·ть, -ют, (*imp*) cover, put a cover on; coat, face; drown (a noise)

по·крыв·а́ющ·ий, -ая, -ее, (*pres part act*) *see* **по·крыв·а́·ть;** overlying, superincumbent

по·кры́т·и·е, -я, (*n*) covering, coating; surface (of roads); roof (of house); (agr) mating

~, за·щи́т·н·ое protective coating

по·крыт·и·е
~, **катод·н·ое** (elec) cathode coating
коэффициéнт· по·крыт·и·я (*m*) coverage factor
~ **метáлл·ом** metal coating, plating
~ **оболóч·к·ой** jacketing, canning (instruments etc.)
~, **пере·нóс·н·ое** (civ eng) road mat
покрыто·голóв·ый, -ая, -ое, (*a*) (zool) stegocephalous
покрыто·сем·я́нн·ый, -ая, -ое, (*a*) (bot) angiospermous; (*pl as noun*) angiosperms, *Angiospermae*
по·кры́·т·ый, -ая, -ое, (*past part pass*) of **по·кры́·ть**; coated, plated, clad; paid (of debts)
~ **бóр·ом** borated
~ **мéд·ью** copper plated
~ **оболóч·к·ой** sheathed
по·кры́·ть, (*fut 3rd pl*) **по·крó·ют,** (*perf*) see **по·крыв·á·ть**
по·кры́ш·к·а, -и, (*g pl*) **-ш·ек·,** (*f*) covering, cover, casing; (oil, geol) cap rock; (M/T) outer casing, tyre, tire
~, **мнóго·слóй·н·ая** multiple-cord tyre
по·ку·ёт, (*fut 3rd sing*) of **по·ков·áть**
по·куп·áтел·ь, -я, (*m*) buyer, purchaser
по·куп·áтельн·ый, -ая, -ое, (*a*) purchasing, buying
~ **спосóб·ност·ь,** (*f*) purchasing power
по·куп·áтельск·ий, -ая, -ое, (*a*) buyer's, purchaser's
по·куп·á·ть, -ют, (*imp*) buy, purchase; bathe, bath
по·кýп·к·а, -и, (*g pl*) **-п·ок·,** (*f*) buying, purchasing; purchase, buy
по·кýп·н·ой, -áя, -óе, (*a*) of **по··кýп·к·а**
по·кур·и́ть, ⸗ят, (*perf*) smoke (e.g. cigarette)
пóл·, -а, (*nom pl*) **-ы́,** (*g pl*) **-óв,** (*m*) floor; **пóл·, -а,** (*nom pl*) **-ы́,** (*g pl*) **-óв,** (*m*) sex
~, **фри́з·ов·ый** (build) framed floor
пол- (*root*) half, semi-
пол·á, -ы́, (*nom pl*) **пóл·ы,** (*g pl*) **пóл·,** (*f*) skirt, flap; belly (leather)
по·лаг·á·ть, -ют, (*imp*) suppose, assume, believe; **-ся** (*pass*); rely/depend upon
по·лáд·ить, -ят, (*perf*) come to an agreement

поларитóн·, -а, (*m*) (phys) polariton
пóлб·а, -ы, (*f*) (bot) emmer, *Triticum dicoccum*
~, **на·сто·я́щ·ая** spelt, dinkel wheat, *Triticum spelta*
пол·вéк·а, (*m*) half a century
пол·гóд·а, (*m*) half-year
пóл·ден·ь, (*g sing*) **полý·дн·я,** or **пóл·дн·я,** (*m*) midday, noon
~, **срéд·н·ий** (astron) mean solar day
пол·дн·éвн·ый, -ая, -ое, (*a*) noon, midday
пол·дн·я́, (*m*) half-day; **пóл·дн·я,** see **пóл·ден·ь**
пóл·е, -я, (*nom pl*) **-я́,** (*g pl*) **-éй,** (*n*) field, ground, area; (print) margin, edge; brim (hat); land (of rifling)
~, **анáль·н·ое** (zool) anal area, periproct
~, **вéктор·н·ое** vector field
~, **вéрх·н·ее** (print) top, head (of page)
~, **вихр·ев·óе** vortex/rotational field
~ **воз·бужд·éни·я** exciting field
~, **вращ·áющ·ееся** rotating/rotary field
~, **вы·тя́г·ивающ·ее** drawing/pulling field
~, **гнезд·ов·óе** jack panel (telephony)
~, **гравитациóн·н·ое** gravitational/gravity field
~, **деформациóн·н·ое** strain/deformation field
~, **за·медл·я́ющ·ее** (dynam) decelerating field; drift field (of vacuum tubes)
~ **зр·éни·я** field of vision; coverage (of radio antenna)
~ **из·ображ·éни·я** (phot) useful field, circle of good definition
~, **кóд·ов·ый комбинáц·и·й** code pattern (of analogue digital converter)
~, **консервати́в·н·ое** conservative field
~, **контáкт·н·ое** bank (telephones)
~, **кореш·кóв·ое** (print) inside margin
~, **кра·ев·óе** (math) fringing field
~, **лед·ян·óе** ice field
~, **лёт·н·ое** (air) landing area
~, **магни́т·н·ое** magnetic field
~, **ми́н·н·ое** (mil) minefield
~, **мнóго·крáт·н·ое** multiple (telephones)
~, **мýскуль·н·ое** (anat) muscle area
~ **на вы́·ход·е** exit field
~ **на·правл·éни·я** (math) directional field, unit vector field

по́л·е

~, на·правл·я́ющ·ее guiding/guide field

~, не·ви́хр·ев·о́е irrotational/non-circuital field

~, не·одно·ро́д·н·ое non-uniform/inhomogeneous field

~, не·стациона́р·н·ое non-steady/unsteady field

~ об·зо́р·а field of view

~, одно·ро́д·н·ое uniform field

~, о·чищ·а́ющ·ее clearing field

~, пере·ме́н·н·ое variable field

~, пере·мещ·а́ющ·ееся travelling field

~ перфо·ка́рт·ы (autom) card field

~, по·пере́ч·н·ое cross/transverse field

~, по·сторо́н·н·ее extraneous field

~ по·то́к·а flow field

~, про·меж·у́точн·ое jumper field (telephones)

~, проти́во·де́й·ствующ·ий controlling/choking field

~ прям·о́го из·луч·е́ни·я (rad) direct field

~ рас·пре·дел·и́тел·я вы́·зов·ов traffic distributor bank (telephones)

~ рас·се́·яни·я stray field

~, рас·тя́·нут·ый distended field

~, рот·о́в·о́е (biol) peristome

~, с·вя́з·анн·ые (pl) coupled fields

~ сил· от·та́лк·ивани·я (phys) repulsion field

~, сил·ов·о́е field of force; vectorial/vector field

~, си́ль·н·ое strong field

~, скаля́р·н·ое scalar field

~ с·ре́з·а cutoff field

~, стру́н·н·ое piano (autom, telephones)

~, тормоз·я́щ·ее drift field (of vacuum tubes)

у́гол· по́л·я (m) (phot) lens angle

~, у·де́рж·ивающ·ее confining field

~, у·стано́в·ивш·ееся (autom) terminal field

~, фа́з·ов·ое (autom) field of behaviour

~, чёрн·ое black area (photofax)

пол·еви́ц·а, -ы, (i) -ей, (f) (bot) bent grass, red top, *Agrostis*

поле·во́д·, -а, (m) agronomist

поле·во́д·ств·о, -а, (n) agronomy, crop cultivation

пол·ев·о́й, -а́я, -о́е, (a) field

~ а́брис· (m) (surv) field sketch

~ рабо́т·ы (pl) (surv) field work

пол·ев·о́й, -а́я, -о́е

~ слу́ж·б·а (f) (mil) field service

~ шпат· (m) (min) feldspar, felspar

полево·шпа́т·н·ый, -ая, -ое, (a) (min) feldspar, feldspathic, feldspathose

по·лег·а́ни·е, -я, (n) falling (of plant under own weight)

по·ле́гч·е (comp) lighter; easier

по·леж·а́л·ое, -ого, (m decl as adj) charge for luggage or goods, left over the stated period

по·леж·а́ть, -а́т, (perf) lie down for a while

поле·за·щи́т·н·ый, -ая, -ое, (a) windbreak, protective (for fields)

~ полос·а́ (f) (agr) protective afforestation strip

по·ле́з·к·а, -и, (g pl) -з·ок·, (f) (zool) vole

поле́з·н·о (adv) usefully; it is useful (to), it is sound practice (to)

поле́з·н·ый, -ая, -ое, (a) useful, helpful; healthy, wholesome; net, effective

~ выс·от·а́ (f) (air) useful height

~ де́й·стви·е (n) (and see коэффи·ци́ент· поле́з·н·ого де́й·стви·я) efficiency; propulsive efficiency (of jet engine)

~ ис·коп·а́ем·ое (n) mineral, commercial mineral

на·гру́з·к·а (f) payload; (a/c) disposable load; net/useful load (on machines)

~ от·да́·ч·а тепл·а́ (f) thermal efficiency

~ при·мен·е́ни·е (n) utilization

~ рабо́т·а (f) useful work

по·ле́з·ть, -ут; (perf) climb, clamber

по·ле́й·те, (pl imperative) of **по·ли́·ть, (perf)**; see **по·лив·а́·ть (imp)**

полектро́н·, -а, (m) (plast) polectron

поле·ме́р·, -а, (m) field strength meter

~, магни́т·н·ый magnetic field strength meter

полем·изи́р·овать, -ую́т, (imp) dispute, argue about

поле́м·ик·а, -и, (f) controversy, dispute, polemics

полемониу́м·, -а, (m) (bot, pharm) Greek valerian, *Polemonium*

по·ле́с·ь·е, -я, (n) wooded boggy area; (geog) polessie (alluvial plain)

по́л·ет, (pres 3rd sing) of **пол·о́ть**; (for noun see below)

по·лёт·, -а, (m) flight, flying

~, акти́в·н·ый powered flight

по·лёт

~, аэродро́м·н·ый   local flight

~, бр·е́ющ·ий   ground level flying, low-level flight, hedge hopping

вре́м·я по·лёт·а   (*n*) time of flight

~ выс·о́тн·ый   high-altitude flight

~, горизонта́ль·н·ый   horizontal flight, straight-and-level flight

~ зигза́г·ом   (air) dog leg

~, контро́л·н·о-по·ка́з·н·о́й   instructional demonstration flight

~, контро́ль·н·ый   pre-solo test flight

~, ма́рш·ев·ый   (rocket) cruising flight

~, маршру́т·н·ый   cross-country flight

~, модел·и́руем·ый   simulated flight

~ на спин·е́   inverted flight

~ не·у·станов·и́вш·ийся   non-uniform/unsteady flight

~, пар·я́щ·ий   soaring, soaring flight

~, пере·вёр·нут·ый   inverted flight

~, план·и́рующ·ий   gliding, gliding flight

~ по ине́рц·и·и   (rocket) coasting

~ по орто·дро́м·и·и   great circle flying

~ по при·бо́р·ам   instrument/blind flying

~ по се́т·к·е   grid flying

~, ре́йс·ов·ый   scheduled flight

~ с конве́йер·а   roller landing, touch-and-go-landing

~, само·сто·я́тельн·ый   solo flight

~, свобо́д·н·ый   free flight

~, слеп·о́й   instrument/blind flying

~, трасс·ов·о́й   trunk route flight

по·лет·а́·ть, -ют,   (*perf*) fly

по·лет·е́ть, -ят,   (*perf*) fly, fly/take off, take flight; begin to fly

по·лёт·н·ый, -ая, -ое,   (*a*) flying, flight

~ вес·   (*m*) (a/c) all-up weight

по·леч·и́ть, -ат,   (*perf*) (med) treat

по·леч·у́   (*fut 1st sing*) of **по·лет·е́ть;**   (*fut 1st sing*) of **по·леч·и́ть**

полз·,   (*past masc sing*) of **полз·ти́**

по́лз·ани·е, -я,   (*n*) (*v n*), *see* **по́лз·а·ть**

по́лз·а·ть, -ют,   (*imp indet*) crawl, creep

полз·к·о́м   (*adv*) crawling

полз·ти́, -у́т,   (*imp deter*) crawl, creep

полз·у́н·, -а,   (*m*) slide/link bar, slide block, slider, crosshead shoe, slipper; (M/T) selector rod (of gear box)

полз·уно́к·, -нк·а́,   (*m*) (*dim*) of **полз·у́н·,** cursor, sliding contact

~ за·кора́ч·ивающ·ий   (elec) shorting bar

полз·у́чест·ь, -и,   (*f*) creep (of materials)

втор·о́й эта́п· полз·у́чест·и   (*m*) (met) secondary/steady-state stage of creep

деформа́ци·я полз·у́чест·и   (*f*) (met) total creep

~, за·тух·а́ющ·ая   (met) microcreep, transient creep

коэффицие́нт· полз·у́чест·и   (*m*) creep coefficient (concrete)

крив·а́я полз·у́чест·и   (*f*) creep curve

~, на·раст·а́ющ·ая   (met) tertiary creep

~, не·у·станов·и́вш·аяся   (met) transient creep

пе́рв·ый эта́п· полз·у́чест·и   (*m*) (met) primary/transient stage of creep

пре·де́л· полз·у́чест·и   (*m*) (met) limiting creep stress

ско́р·ост·ь полз·у́чест·и   (*f*) (met) creep rate

~ со́л·и   (electrochem) creepage

тре́т·ий эта́п· полз·у́чест·и   (*m*) (met) tertiary stage of creep

~, у·станов·и́вш·аяся   (met) steady-state creep, macrocreep, secondary creep

~, фа́з·н·ая   (telecom) phase creep

полз·у́ч·ий, -ая, -ее,   (*a*) crawling, creeping, climbing (of plants)

~ кран·   (*m*) crawler crane/derrick

~ раст·е́ни·е   (*n*) (bot) creeping plant

полз·у́щ·ий, -ая, -ее,   (*pres part act*) of **полз·ти́**

по́лз·ш·ий, -ая, -ее,   (*pres part act*) of **полз·ти́**

поли-   (*int component*) poly-, multi-

поли·азо·крас·и́тел·ь, -я,   (*m*) poly-azo dye

поли·акрил·ами́д·, -а,   (*m*) poly-acrylamide

поли·акрила́т·, -а,   (*m*) polyacrylate

поли·акрила́т·н·ый, -ая, -ое,   (*a*) polyacrylate

поли·акри́л·ов·ый, -ая, -ое,   (*a*) polyacrylic

поли·акри́ль·н·ый, -ая, -ое,   (*a*) polyalkylated

поли·алли́л·ов·ый, -ая, -ое,   (*a*) polyallyl

поли·ами́д·, -а,   (*m*) polyamide

поли·амид·и́ров·ани·е, -я,   (*n*) poly-amidation

поли·ами́д·н·ый, -ая, -ое,   (*a*) of **поли·ами́д·**

поли·аминотриазо́л·, -а, (*m*) polyaminotriazole, PAT

поли·а́ндр·и·я, -и, (*f*) (biol) polyandry

полианит·, -а, (*m*) (min) polianite

поли·ари́ль·н·ый, -ая, -ое, (*a*) polyarylated

поли·артри́т·, -а, (*m*) (med) polyarthritis

поли·бази́т·, -а, (*m*) (min) polybasite

поли·бенз·оксазо́л, -а, (*m*) polybenzoxazole

поли·бла́ст·, -а, (*m*) (zool) polyblast

поли·бутадиен·, -а, (*m*) (chem) polybutadiene

поли·буте́н·, -а, (*m*) polybutene

по·ли́·в·, -а, (*m*) (*v n*), *see* по·ли·в·а́·ть; (agr) watering, irrigation; (phot) coating

~ за·топл·е́ни·ем flooding (for irrigation)

по·ли́·в·а, -ы, (*f*) glaze

поли·вакси́н·а, -ы, (*f*) (med) polyvaccine

по·лив·а́ни·е, -я, (*n*) (*v n*), *see* по··лив·а́·ть; flood

~, оби́ль·н·ое copious flood (lubrication/cooling)

поли·вариа́нт·н·ый, -ая, -ое, (*a*) polyvariant

по·лив·а́·ть, -ют, (*imp*) pour on/over, sluice, douse; (agr) water

по·ли́в·енн·ый, -ая, -ое, (*a*) (ceram) glazed; (phot) coated

поли·вини́л·, -а, (*m*) polyvinyl

поли·винил·аце́та́л·ь, -я, (*m*) polyvinylacetal

поли·винил·аце́та́т·, -а, (*m*) polyvinylacetate

поли·вини́л·ов·ый, -ая, -ое, (*a*) polyvinyl

поли·винил·пироллидо́н·, -а, (*m*) polyvinylpyrollidon

поли·винил·хлори́д·, -а, (*m*) polyvinylchloride, PVC

по·ли́в·к·а, -и, (*g pl*) -в·ок·, (*f*) (*v n*), *see* по·лив·а́·ть; (*see also* по·ли́·в·); watering, irrigation; flushing

по·лив·н·о́й, -а́я, -о́е, (*a*) of по··ли́·в·, по·ли́в·к·а

по·ли́в·н·ый, -ая, -ое, (*a*) (ceram) glazed

поли́вочно·мо́·ечн·ая маши́н·а (*f*) street-sprinkling and washing vehicle

по·ли́в·очн·ый, -ая, -ое, (*a*) flushing, watering

поли·гали́т·, -а, (*m*) (min) polyhalite

поли·галогени́д·, -а, (*m*) (chem) polyhalide

поли·гало́ид·н·ый, -ая, -ое, (*a*) (chem) polyhalide

поли·гамари́н·, -а, (*m*) (pharm) polygamarin

поли·гармон·и́ческ·ий, -ая, -ое, (*a*) polyharmonic

поли·ге́н·, -а, (*m*) (gen) polygene

поли·ге́н·н·ый, -ая, -ое, (*a*) polygenetic, polygenic

поли·ги́р·н·ый, -ая, -ое, (*a*) (cryst) polygyral, polygyric

поли·глюки́н·, -а, (*m*) (pharm) polyglucin

поли·го́н·, -а, (*m*) polygon; (surv) traverse; casting yard (concrete); (mil) proving ground, practice area, range (missiles)

~, дра́ж·н·ый (min) placer area

~, не·за́·мк·нут·ый (surv) open traverse

~, нивели́р·н·ый (surv) circuit с·мык·а́·ть поли·го́н·, (surv) adjust a traverse

~, теодоли́т·н·ый (surv) theodolite polygon/traverse

полигон·иза́ци·я, -и, (*f*) polygonization

полигоно·метр·и́ческ·ий, -ая, -ое, (*a*) (surv) polygonometric, traverse

~ сет·ь (*f*) (surv) traverse network

~ ход· (*m*) (surv) traverse

полигоно·ме́тр·и·я, -и, (*f*) (surv) polygon measurements, survey/field traverse

поли·граф·и́ст·, -а, (*m*) printing-industry worker, printer

поли·граф·и́ческ·ий, -ая, -ое, (*a*) printing

~ про·мы́шл·енност·ь (*f*) printing-industry

поли·граф·и́·я, -и, printing

поли·гу́б·к·а, -и, (*g pl*) -б·ок·, (*f*) (plast) polyfoam

поли·диспе́рс·ност·ь, -и, (*f*) polydispersity

поли·диспе́рс·н·ый, -ая, -ое, (*a*) polydisperse

полидо́р·овск·ое желе́з·о (*n*) (elec) ferrocart core

по·лиз·а́ть, (*fut 3rd pl*) по·ли́ж·ут, (*perf*) lick

поли·изобутиле́н·, -а, (*m*) (plast) polyisobutylene, oppanol

поли·изопре́н·, -а, (*m*) polyisoprene

по́ли·капроами́д·, -а,   (*m*) (plast) polycapramide, polycaprolactam, nylon 6

по́ли·карбона́т·, -а,   (*m*) polycarbonate

по́ли·карбо́н·ов·ая кисл·от·а́   (*f*) polybasic carboxylic acid

по́ли·ка́рп·и·й, -я,   (*m*) (bot) polycarp

по́ли·ка́рп·и́ческ·ий, -ая, -ое,   (*a*) (bot) polycarpic

по́ли·кисл·от·а́, -ы́,   (*f*) (chem) polyacid

по́ли·кли́ник·а, -и,   (*f*) (med) dispensary, medical post, clinic

по́ли·конденсацио́н·н·ый, -ая, -ое, (*a*) of **по́ли·конденс·а́ци·я**

по́ли·конденс·а́ци·я, -и,   (*f*) (chem) polycondensation, condensation polymerization

~ меж·фаз·н·ая, interfacial polycondensation

по́ли·кра́з, -а, (*m*) (min) polycrase

по́ли·кремн·ёв·ый, -ая, -ое,   (*a*) (chem) polysilicic

по́ли·криста́лл·, -а, (*m*) polycrystal
    ме́тод· по́ли·криста́лл·а   (*m*) (spectr) powder/Debye-Scherrer method

по́ли·кристалл·и́ческ·ий, -ая, -ое, (*a*) polycrystalline

по·ли́·л, (*past masc sing*) of **по·ли́·ть,** (*perf*); *see* **по·лив·а́·ть**

по́ли·линеа́р·н·ый, -ая, -ое,   (*a*) (math) multilinear

полиме́нт·, -а, (*m*) gilding size

полиме́р·, -а, (*m*) (chem) polymer

~, аддити́в·н·ый addition polymer

~, бло́ч·н·ый block copolymer

~, высо́к·ого давл·е́ни·я high-pressure polymer, high polymer

~, высоко·молекуля́р·н·ый high polymer

~, ди́к·ий popcorn polymer

~, дли́нно·цеп·н·о́й long-chain polymer

~, лин·е́йн·ый linear polymer

~, ма́л·ого давл·е́ни·я low-pressure polymer, low polymer

~, при·ви́·т·ый graft copolymer

~, про·стра́н·ственн·ый three-dimensional polymer

~, се́т·очн·ый network polymer, three-dimensional crosslinked polymer

~, с·ме́ш·анн·ый inter-polymer

~, со·в·ме́ст·н·ый copolymer

полимер·иза́т·, -а, (*m*) polymerizate, polimerization product

полимер·иза́тор·, -а, (*m*) polymerizer, polymerization reactor

полимер·иза́ци·я, -и, (*f*) polymerization

~, би́сер·н·ая bead polymerization

~ в ма́сс·е bulk polymerization

~ в рас·тво́р·е solution polymerization

~ вз·ры́в·н·а́я explosive polymerization

~, со·в·ме́ст·н·ая copolymerization

~, стерео·регуля́р·н·ая stereospecific polymerization

~, стерео·специф·и́ческ·ая stereospecific polymerization

~, ступ·е́нчат·ая addition polymerization

~, эмульсио́н·н·ая emulsion polymerization

полимери́з·овать, -у́ют, (*imp*) polymerize

полимер·и́·я, -и,   (*f*) (chem) polymerism; (bot) polimery

полиме́р·н·ый, -ая, -ое, (*a*) polymer, polymeric

по́ли·металл·и́ческ·ий, -ая, -ое, (*a*) polymetallic; complex (of ores)

по́ли·мети́л·, -а, (*m*) (chem) polymethyl

по́ли·метиле́н·, -а, (*m*) (chem) polymethylene

по́ли·метилен·ов·ый, -ая, -ое, (*a*) polymethylene

по́ли·мети́н·, -а, (*m*) (chem) polymethine

по́ли·мети́н·ов·ый, -ая, -ое, (*a*) polymethine

по́ли·микси́н·, -а, (*m*) (pharm) polymyxin

по́ли·ми́кт·н·ый, -ая, -ое, (*a*) (geol) polymict, polymictic

по́ли·морфи́зм·, -а, (*m*) polymorphism, polymorphy

по́ли·мо́рф·н·ый, -ая, -ое, (*a*) polymorphous, polymorphic

по́ли·неври́т·, -а, (*m*) (med) polyneuritis

по́ли·нитро·этиле́н·, -а, (*m*) polynitroethylene

по́ли·но́м·, -а, (*m*) (*and see* **мно́го··чле́н·**) (math) polynomial

~ не·вя́з·к·и residual polynomial

по́ли·нуклеоти́д·, -а, (*m*) (chem) nucleic acid

по·лин·я́л·ый, -ая, -ое, (*a*) discoloured, discolored, faded

по·лин·я́·ть, -ют,   (*perf*) (text) fade, run, lose colour; (zool) moult

поли·о́з·а, -ы,   (*f*) (chem) polyose, polysaccharose

поли·окси-   (*int component*) (chem) polyhydroxy-, polyoxy-

поли·олефи́н·, -а,   (*m*) polyolefin

полио·миели́т·, -а,   (*m*) (med) poliomyelitis

поли·органо·силокса́н·, -а,   (*m*) (chem) silicone

поли́п·, -а,   (*m*) (zool) polyp, polypus

~, прост·е́йш·ий (zool) hydrula

поли·пепти́д·, -а,   (*m*) (chem) polypeptide

поли·пер·фторо·бутади́н·, -а,   (*m*) (plast) polyfluorobutadiene

полип·ня́к·, -а,   (*m*) (zool) polypary, polyparium

полипо·ви́д·н·ый, -ая, -ое,   (*a*) (zool) polypoid

поли·пре́н·, -а,   (*m*) (rubber) polyprene

поли·пропиле́н·, -а, (*m*) polypropylene

поли·рекомбин·а́ци·я, -и,   (*f*) (plast) polyrecombination

поли·решёт·к·а, -и,   (*g pl*) **-т·ок·,** (*f*) (phys) composite lattice

полир·ова́льн·ый, -ая, -ое,   (*a*) polishing

~ бума́г·а (*f*) sandpaper

~ круг· (*m*) buff, polishing buff

~ пыл·ь (*f*) buffing

полир·ова́ни·е, -я,   (*n*) polishing

полир·ова́ть·, -уют,   (*imp*) polish

полир·о́вк·а, -и,   (*f*) polishing, polish

~ окун·а́ни·ем dipping polish

~, рельеф·н·ая (met) relief polishing

по́лис·, -а,   (*m*) (ins) policy

поли·сапро́б·н·ый, -ая, -ое,   (*a*) (biol) polysaprobe

поли·сахар·и́д·, -а,   (*m*) (chem) polysaccharide, polysaccharose, polyose

поли·сер·ни́ст·ый, -ая, -ое,   (*a*) polysulphide (of)

поли·силика́т·н·ый, -ая, -ое,   (*a*) polysilicate

полисилокса́н·, -а,   (*m*) (chem) silicone

поли·с·ме́с·ь, -и,   (*f*) polyblend

по́лис·н·ый, -ая, -ое, (*a*) of **по́лис·**

полисо·держ·а́тел·ь, -я,   (*m*) (ins) policy holder

полиспа́ст·, -а,   (*m*) block-and-tackle, purchase

поли·сперм·и́·я, -и,   (*f*) (biol) polyspermy

поли·стелл·и́ческ·ий, -ая, -ое,   (*a*) (bot) polystelic

поли·стиро́л·, -а,   (*m*) (chem) polystyrene

поли·стиро́л·ов·ый, -ая, -ое,   (*a*) polystyrene

~ плён·к·а (*f*) plasticon

поли·стиро́ль·н·ый, -ая, -ое,   (*a*) polystyrene

по·ли́ст·н·ый, -ая, -ое, (*a*) per sheet

поли·сульфи́д·, -а,   (*m*) (chem) polysulphide

полит- (*abbr component*) = **полит·и́ческ·ий** political

пол·и́те (*pl imperative*) of **пол·о́ть**

полите́н·, -а,   (*m*) (plast) polythene

поли·терпе́н·, -а,   (*m*) (chem) polyterpene

поли·тетра·фтор·этиле́н·, -а,   (*m*) (chem) polytetrafluorethylene, teflon

поли·техн·и́зм·, -а, (*m*) technological/technical/polytechnical training, training on a broad technological front

поли·те́хникум·, -а,   (*m*) polytechnic school, polytechnic

поли·техн·и́ческ·ий, -ая, -се,   (*a*) polytechnic(al)

поли́т·ик·, -а,   (*m*) politician

поли́т·ик·а, -и,   (*f*) policy; politics

поли·тио́н·ов·ый, -ая, -ое, (*a*) (chem) polythionic

полит·и́ческ·ий, -ая, -ое,   (*a*) political

поли·тро́п·а, -ы,   (*f*) polytropic curve

поли·троп·и́ческ·ий, -ая, -ое,   (*a*) polytropic

поли·тро́п·н·ый, -ая, -ое, (*a*) polytropic

поли·ту́р·, -ы,   (*f*) French polish

по·ли́·т·ый ~ая, -ое, (*past part pass*) *see* **по·лив·а́·ть**

по·ли́·ть, (*fut 3rd pl*) **по·ль·ю́т,** (*perf*) *see* **по·лив·а́·ть**

поли·урета́н·, -а,   (*m*) (chem) polyurethane

поли·ури́·я, -и,   (*f*) (med) polyuria

поли·фа́г·и·я, -и,   (*f*) (biol) polyphagy

поли·филет·и́ческ·ий, -ая, -ое,   (*a*) polyphyletic, polyphylous

поли·фил·и́·я, -и,   (*f*) (biol) polyphyly

поли·фо́рминг·, -а,   (*m*) (oil) polyforming

поли·ха́з·и·й, **-я,** *(m)* (bot) polychasium

поли·хе́т·ы *(f pl)* (zool) polychaetes, *Polychaeta*

поли·хлор·вини́л·, **-а,** *(m)* polyvinyl chloride, PVC

поли·хлоро·пре́н·, **-а,** *(m)* polychloroprene

поли·хлор·пине́н·, **-а,** *(m)* polychloropinene

поли·хро·и́зм·, **-а,** *(m)* polychroism, pleochroism

поли·хромази́·я, **-и,** *(f)* *(med)* polychromasia, polychromatophilia

поли·хромат·и́ческ·ий, **-ая, -ое,** *(a)* polychromatic

полиц·е́йск·ий, **-ая, -ое,** *(a)* police

поли·цикл·и́ческ·ий, **-ая, -ое,** *(a)* polycyclic

поли·цитеми́·я, **-и,** *(f)* (med) polycythemia

поли́ци·я, **-и,** *(f)* police

поли·э́др·, **-а,** *(m)* (math) polyhedron

поли·эдр·и́ческ·ий, **-ая, -ое,** *(a)* polyhedral

поли·эте́н·, **-а,** *(m)* polyethylene

поли·этиле́н·, **-а,** *(m)* (plast) polyethylene

поли·этилен·глико́л·ев·ый, **-ая, -ое,** *(a)* polyethyleneglycol

поли·этиле́н·ов·ый, **-ая, -ое,** *(a)* polyethylene

поли·эфи́р·, **-а,** *(m)* (chem) polyester; polyether

~, прост·о́й polyether

полк·, **-а́,** *(m)* (mil) regiment; cf. **пол·к·а, пол·о́к·**

пол·к·а, **-и,** *(g pl)* **-л·ок·,** *(f)* shelf, ledge, rack; flange, boom (of I-beam); leg (of angle bar); (rail) berth, bunk; (hortic) weeding; pan (of primitive firearms); cf. **полк·, пол·о́к·**

~ лонжеро́н·а (a/c) spar flange

полк·о́вник·, **-а,** *(m)* colonel

полк·о́внич·ий, **-ья, -ье,** *(a)* colonel's

полк·ов·о́й, **-а́я, -о́е,** *(a)* regimental

полли́н·и·й, **-я,** *(m)* (bot) pollinium

по́ллукс·, **-а,** *(m)* (astron) Pollux; (min) pollucite

поллуци́т·, **-а,** *(m)* (min) pollucite

поллю́ци·я, **-и,** *(f)* (med) spermatorrhea, pollution

поллярд·и́рованн·ый, **-ая, -ое,** *(past part pass)*; (agr) pollarded, polled

пол·миллио́н·а, *(g sing)* **полу·милли́он·а** *(m)* half a million

полн·е́йш·ий, **-ая, -ее,** *(a)* fullest, most complete

полн·е́·ть, **-ют,** *(imp)* become fuller/fatter

по́лн·о- *(root)* full-, holo-, completely

полно·ве́с·н·ый, **-ая, -ое,** *(a)* full-weight

полно·вла́ст·н·ый, **-ая, -ое,** *(a)* sovereign

полно·во́д·н·ый, **-ая, -ое,** *(a)* deep (of water); high-water

полно·во́д·ь·е, **-я,** *(n)* high water

полно·гра́н·ник·, **-а,** *(m)* (cryst) holohedron

полно·гра́н·н·ый, **-ая, -ое,** *(a)* (cryst) holohedral

полно·кристалл·и́ческ·ий, **-ая, -ое,** *(a)* holocrystalline

полно·кро́в·и·е, **-я,** *(n)* (med) plethora

полно·кро́в·н·ый, **-ая, -ое,** *(a)* full-blooded, (med) plethoric

полно·лу́н·и·е, **-я,** *(n)* (astron) full moon

полно·ме́р·н·ый, **-ая, -ое,** *(a)* full-sized

полно·метр·а́жн·ый, **-ая, -ое,** *(a)* (phot) full-length

полно·мо́ч·и·е, **-я,** *(n)* authority, powers, plenary powers; proxy

полно·мо́ч·н·ый, **-ая, -ое,** *(a)* authorized, (dipl) plenipotentiary

полно·о́с·н·ый, **-ая, -ое,** *(a)* (cryst) holoaxial

полно·пра́в·и·е, **-я,** *(n)* equality of rights

полно·пра́в·н·ый, **-ая, -ое,** *(a)* having equal rights

полно·раз·ме́р·н·ый, **-ая, -ое,** *(a)* full-size

по́лн·ост·ь, **-и,** *(f)* fullness, completeness; stoutness

по́лн·ост·ью *(adv)* fully, completely

полн·от·а́, **-ы́,** *(f)* fullness, completeness; stoutness; height (of a quality); volume (of sound)

коэффицие́нт· вертика́ль·н·ой полн·от·ы́ *(m)* (shipb) vertical prismatic coefficient

коэффицие́нт· полн·от·ы́ о́бщ·ий *(m)* (shipb) block coefficient

коэффициент· полн·от·ы́ пло́ща·д·и *(m)* (shipb) waterplane coefficient

коэффицие́нт· полн·от·ы́ пло́ща·д·и ми́дель-шпангоу́т·а *(m)* (shipb) midship section coefficient

коэффициент· про·до́ль·н·ой полн·от·ы́ *(m)* longitudinal prismatic coefficient

полн·от·а́

~ тормож·е́ни·я (dynam) stagnation density

полно·це́н·н·ый, -ая, -ое, (*a*) full/face-value

полно·цикл·и́ческ·ий, -ая, -ое, (*a*) (biol) holocyclic

полно·ча́ст·постн·ый, -ая, -ое, (*a*) full-frequency, high-fidelity (sound recording)

пол·но́ч·н·ый, -ая, -ое, (*a*) midnight

по́л·ноч·ь, -и or **полу́·ноч·и,** (*f*) midnight

~, сре́д·н·яя гри́нвич·ск·ая (astron) Greenwich mean midnight

по́лн·ый, -ая, -ое, (*a*) full, complete, total; gross, overall; stout, corpulent (of persons)

~ вод·а́ (*f*) (ocean) high tide/water; flood, spate (rivers)

~ выс·от·а́ подъ·ём·а (*f*) total head (pumps etc.)

~ газ· (*m*) (ICE) full power

крив·а́я по́лн·ых давл·е́ни·й (*f*) (hydr) Pitot curve

~ про·из·во́д·н·ая (*f*) (math) total derivative

~ у·сил·е́ни·е (*n*) overall amplification/gain

пол·оборо́т·а (*m indecl*) half-revolution/turn

на пол·оборо́т·а half way round

пол·о́в·а, -ы, (*f*) chaff

половин·а, -ы, (*f*) half; middle (of month); leaf (of door)

в половин·у half as much, half as

~ втор·о́го half past one (o'clock)

~, о·па́с·н·ая dangerous semicircle (of cyclone)

~ при·ли́·в·а half-tide

половин·е́ни·е, -я, (*n*) halving

половин·ник·, -а, (*m*) half-and-half (solder)

половин·н·ый, -ая, -ое, (*a*) half

половин·чат·ый, -ая, -ое, (*a*) half-and-half

пол·ови́ц·а, -ы, (*i*) **-ей,** (*f*) floor board

пол·овня́к·, -а́, (*m*) (build) bat, half-brick

поло·во́д·ь·е, -я, (*n*) flood, spate (of rivers); high water/tide

полов·о́й, -а́я, -о́е, (*a*) floor; sex, sexual, genital

пол·о́вш·ий, -ая, -ее, (*past part act*) of **пол·о́ть**

по́·лог·, -а, (*m*) canopy, cover

по·ло́г·ий, -ая, -ое, (*a*) gently sloping, shallow (of slope), flat (of curves etc.)

полого·па́д·ающ·ий, -ая, -ее, (*a*) gently sloping, gradually falling

полого·с·кла́д·чат·ый, -ая, -ое, (*a*) gently undulating

по·ло́г·ост·ь, -и, (*f*) declivity, slope

поло·ду́·в·, -а, (*m*) chaff blower

по·лож·е́ни·е, -я, (*n*) position; attitude, posture; status, state, footing; statute

~, бо·ев·о́е "on" position; armed position (of ammunition); periscope depth (of submarine)

~ в по·лёт·е flight attitude

~ в про·стра́н·ств·е attitude (of missile etc.)

~ вещ·е́й state of affairs

~ в·ключ·е́ни·я "on" position

~, воен·н·ое war footing

ли́н·и·я по·лож·е́ни·я (*f*) (naut, surv, astron) line of position, circle of position

~ матема́т·ик·и (*pl*) the positions of mathematics

~, материа́ль·н·ое financial position

~ о вы́·бор·ах electoral law/statute

~ равно·ве́с·и·я equilibrium position; (nucl) equilibrium configuration

~, семе́й·н·ое family status

~, системат·и́ческ·ое status

~, с·ме́ж·н·ое adjacent position

~, таксоном·и́ческ·ое (biol) taxonomic status

по·ло́ж·им let us assume/suppose, let it be assumed

по·лож·и́тельн·о (*adv*) positively

по·лож·и́тельн·о-о·пре·дел·ённ·ая фо́рм·а (*f*) (math) positive-definite quadratic form

по·лож·и́тельн·ый, -ая, -ое, (*a*) positive, plus; favourable, good

~ воз·де́й·стви·е (*n*) positive/good effect

~ за·ря́д· (*m*) (elec) positive charge

~ знак· (*m*) plus sign

~ от·ве́т· (*m*) affirmative reply; favourable reply

по·лож·и́ть, -ат, (*perf*) see **кла́ст·ь,** and **по·лаг·а́·ть**

по́лоз·, -а, (*nom pl*) **поло́з·ья,** (*g pl*) **-ьев,** (*m*) (mech) runner, slide, skid, skate

~, конта́кт·н·ый (elec rail) contact skate

~, под·де́рж·ивающ·ий track skid (of tracked vehicle)

~, с·пуск·ов·о́й (shipb) sliding way

полоз·о́к·, -з·к·а, (*dim*) of **по́лоз·** (*pl*) (cinema) pads

поло́з·ья, (*nom/acc*) of **по́лоз·**

полоида́ль·н·ый, -ая, -ое, (*a*) poloidal

пол·о́к·, -л·к·а́, (*m*) (*see also* **полк·, пол·к·а**) suspended platform; flat cart

~, ле́стнич·н·ый (build) landing; (min) sollar, soller

~, про·хо́д·ческ·ий (min) sinking platform

пол·о́л, (*past masc sing*) of **пол·о́ть**

пол·о́льник·, -а, (*m*) (agr) scratcher cultivator

пол·о́льн·ый, -ая, -ое, (*a*) weeding

пол·о́льщик·, -а, (*m*) weeder

по·лом·а́·ть, -ют, (*perf*) break; **-ся** (*pass*); break down

по·ло́м·к·а, -и, (*g pl*) **-м·ок·,** (*f*) (*v n*), *see* **по·лом·а́·ть**; break; breakage; (mach) failure, breakdown

поло·мо́·ечн·ая маши́н·а (*f*) floor washer

поло́н·иев·ый, -ая, -ое, (*a*) polonium

поло́н·и·й, -я, (*m*) polonium, Po

поло́н·ист·ый, -ая, -ое, (*a*) (chem) polonous

полоно·ки́сл·ый, -ая, -ое, (*a*) (chem) polonite (of)

поло·ро́г·ие, -их, (*pl decl as adj*) (zool) *Bovidae, Cavicornia*

полос·а́, -ы́, (*nom pl*) **по́лос·ы,** (*g pl*) **поло́с·,** (*f*) strip, band; stripe; region, belt, area, zone; spell, period; (met) strip, skelp; (met roll) piece, stock; (print) type page

~ без·о·па́с·ност·и safety strip (on airfields)

~, вале́нт·н·ая (phys) valence band

~, ве́рх·н·яя горизонта́ль·н·ая (shipb) rider plate

~, вз·лёт·н·о-по·са́д·очн·ая (air) runway

~, вращ·а́тельн·ая (spectr) rotational band

~, горяче·ка́т·ан·ая (met) hot-rolled strip

~ деформ·а́ци·и (met) strain/slip lines, slip band

~, за·прещ·ённ·ая (phys) forbidden band/zone

~, за·щи́т·н·ая (cinema) barrier

~ льд·а ice-stream

~ не·про·зр·а́чност·и фи́льтр·а (elec) filter stop band

~, оборо́н·ы defensive zone

~ от пе́н·ы (*pl*) (paper) snailing

~ от·во́д·а overall width (roads)

полос·а́

~ по·глощ·е́ни·я (phys) absorption band

~ под·хо́д·ов (air) approach funnel/ zone

~ по·ниж·ённ·ого давл·е́ни·я (meteor) belt of low pressure

~ препя́тств·и·й (mil) assault course

~ про·вод·и́мост·и (phys) conduction band

~ про·пуск·а́ни·я (rad) band width/ pass, pass/transmission band (of a filter)

~, раз·мы́·т·ая (spectr) broadened/ smeared band

~, раз·реш·ённ·ая (phys) permitted band

~, рас·тя́·нут·ая (rad) spread band

~, рул·ёжн·ая (air) taxi strip

~ скольж·е́ни·я (cryst) slip band

~ тума́н·а (meteor) fog bank

~ част·о́т· (rad) frequency band, bandwidth

~ штил·ев·а́я (meteor) calm belt

~, штурм·ов·а́я (mil) assault course

полос·а́тик·, ма́л·ый (*m*) (zool) lesser rorqual, *Balaenoptera acuto-rostrata*

полос·а́тост·ь, -и, (*f*) striation, banding

полос·а́т·ый, -ая, -ое, (*a*) striped, stripy, striated, banded; fasciated

поло́с·к·а, -и, (*g pl*) **-с·ок·,** (*f*) (*dim*) of **полос·а́** (q.v.); streak

~, перв·и́чн·ая (zool) primitive streak

полоск·а́ни·е, -я, (*n*) (*v n*) of **полоск·а́ть**; mouthwash

полоск·а́ть, (*pres 3rd pl*) **поло́щ·ут,** (*imp*) rinse; gargle; **-ся** (*pass*); splash, splash about; flap (of flags etc.)

полос·н·о-за·гражд·а́ющ·ий, -ая, -ее, (*a*) (elec) band-stop/-elimination /-exclusion

полос·н·о-про·пуск·а́ющ·ий, -ая, -ее, (*a*) (elec) band-pass

полос·н·о́й, -а́я, -о́е, (*a*) of **полос·а́,** *see* **полос·ов·о́й**

полос·н·ый, -ая, -ое, (*a*) *see* **полос·о·в·о́й**

полосо·бу́льб·ов·ый, -ая, -ое, (*a*) (met) bulb-plate

полос·ов·о́й, -а́я, -о́е, (*a*) of **полос·а́**

~ на·стро́й·к·а (*f*) (rad) band-pass tuning

~ ста́л·ь (*f*) band/strap/strip iron

~ стан· (*m*) (met) narrow strip mill, skelp mill

~ фи́льтр· (*m*) (elec) band-pass filter

по́л·ост·ь, -и, (*f*) cavity, cavum, hollow, void; gallery, chamber (of mechanism)

~, атриа́ль·н·ая (zool) atrium

~, на·стро́·енн·ая (rad) resonant cavity

~ пор· pore space

пол·остн·о́й, -а́я, -о́е, (*a*) cavity, cavity-type

полос·ча́тост·ь, -и, (*f*) striation, banding; streak (plant disease)

полос·ча́т·ый, -ая, -ое, (*a*) banded, striped, striated

~ структу́р·а (*f*) banded structure

полоте́н·ечн·ый, -ая, -ое, (*a*) towel

~ ткан·ь (*f*) (text) towelling

полоте́н·ц·е, -а, (*i*) **-ем,** (*g pl*) **-н·ец·,** (*n*) towel; sling (pipelaying); (shipb) sling plate

поло·тёр·, -а, (*m*) floor cleaner; (build) floor sander/grinder

поло·тёр·н·ый, -ая, -ое, (*a*) of **поло·тёр·**

полотн·и́щ·е, -а, (*i*) **-ем,** (*n*) width (of cloth), bolt (of canvas); section (of tent), gore (of parachute); (shipb) plating (of bulkhead); ground panel (for display to aircraft)

полотн·о́, -а́, (*nom pl*) **полотн·а,** (*g pl*) **-тен·,** (*n*) linen, cloth; (civ eng) road bed; blade (of saw); (met) barrel, body (of rolling mill roll); (agr) canvas (of binder); band (of conveyer)

~, анте́нн·ое (rad) antenna array

~, гу́сен·ичн·ое track (of tracked vehicle)

~, из·луч·а́ющ·ее (rad) radiating curtain

~, льн·ян·о́е (text) flax linen

полотн·я́н·ый, -ая, -ое, (*a*) linen

~ пере·плет·е́ни·е (text) plain/calico weave

пол·о́ть, -ют, (*imp*) (hortic) weed

по́л·очк·а, -и, (*g pl*) **-чек·,** (*f*) (*dim*) of **по́л·к·а**

по́л·очн·ый, -ая, -ое, (*a*) of **по́л·к·а**; shelved, multishelf, multihearth (furnaces etc.); **пол·о́чн·ый, -ая, -ое,** (*a*) of **пол·о́к·**

~ реа́ктор· (*m*) multihearth reactor

~ суш·и́лк·а (*f*) shelf dryer

поло́щ·ет, (*pres 3rd sing*) of **полоск·а́ть**

пол·пу́нкт·а, (*m indecl*) (print) half a point

пол·пут·и́ (*m indecl*) half-way

полтора́ (*gen/dat/instr/prep*) **полу́тора** (*m, n*) one and a half

полтора́ста, (*gen/dat/instr/prep*) **полу́тораста** one hundred and fifty

полторы́ (*gen/dat/instr/prep*) **полу́тора** (*f*) one and a half

полу- (*component*) half-, semi-, demi-, hemi-

полу·автома́т·, -а, (*m*) semi-automatic machine tool

~, ток·а́рн·ый semi-automatic lathe

полу·автома́т·ик·а, -и, (*f*) semi-automatic gear/mechanism/system

полу·автомат·и́ческ·ий, -ая, -ое, (*a*) semi-automatic, aided

полу·альдеги́д·, -а, (*m*) half-aldehyde

полу·антраци́т·, -а, (*m*) semi-anthracite

полу·ба́к·, -а, (*m*) (shipb) raised forecastle

полу·ба́рхат·, -а, (*m*) (text) velveteen

полу·бата́н·чик·, -а, (*m*) reed case, rack (of loom)

полу·бе́л·ый, -ая, -ое, (*a*) off-white; pale-green (of glass)

полу·бёрд·о, -а, (*a*) half-reed (of a loom)

полу·бес·кон·е́чн·ый, -ая, -ое, (*a*) (phys) semi-infinite

полу·би́мс·, -а, (*m*) (shipb) half-beam

полу·бо́ч·к·а, -и, (*g pl*) **-ч·ек·,** (*f*) (air) half-roll

полу·ваго́н·, -а, (*m*) (rail) open wagon/truck, gondola

полу·ва́л·, -а, (*m*) (arch) half-hip; bottom leather

полу·ва́тман·, -а, (*m*) Whatman imitation paper

полу·век·ов·о́й, -а́я, -о́е, (*a*) 50 year(s)', half-a-century's

полу·волн·ов·о́й, -а́я, -о́е, (*a*) (rad, opt) half-wave

полу·вы·вед·е́ни·е, -я, (*n*) (phys) half-life

    вре́м·я полу·вы·вед·е́ни·я (*n*) (nucl) biological half-life

полу·вы·ём·к·а, -и, (*g pl*) **-м·ок·,** (*f*) (civ eng) cut-and-fill

полу·га́й·к·а, -и, (*g pl*) **-а·ёк·,** (*f*) half-coupling (e.g. of hose)

полу·гидра́т·, -а, (*m*) (chem) hemihydrate

полу·глиссёр·, -а, (*m*) (shipb) hydroplane

полу·го́д·и·е, -я, (*n*) half-year, six months

полу·год·и́чн·ый, -ая, -ое, (*a*) half-a-year long, half-a-year's, six months'

полу·год·ова́льн·ый, -ая, -ое, (*a*) six-months-old

полу·год·ов·о́й, -а́я, -о́е, (*a*) half-yearly, twice-a-year, semiannually

полу·голо́в·ый, -ая, -ое, (*a*) (zool) hemicephalous

полу·гор·я́ч·ий, -ая, -ее, (*a*) half-boiled/-boiling

~ ва́р·к·а (*f*) half-boiling (soup)

полу·гра́н·ник·, а, (*m*) hemihedron

полу·гра́н·н·ый, -ая, -о́е, (*a*) hemihedral

по́лу·гудро́н·, -а, (*m*) unrefined residual axle grease

полу·гу́сен·ичн·ый, -ая, -ое, (*a*) (M/T) half-track

по·лу́д·а, -ы, (*f*) tinning; tin, tin-coat

полу́·ден·н·ый, -ая, -ое, (*a*) midday; meridional, meridian

~ ли́н·и·я (*f*) (nav) meridian line

полу·ди́зел·ь, -я, (*m*) (ICE) semi-diesel

полу·дистанцио́н·н·ый, -ая, -ое, (*a*) (instr) semiremote

по·луд·и́ть, -я́т, (*perf*) tin

полу·дн·е́вн·ый, -ая, -ое, (*a*) half-day, half-a-day's

полу·до́рн·, -а, (*m*) shoulder drum (tyre building)

полу·драго·це́н·н·ый, -ая, -ое, (*a*) semi-precious

полу·ду́ж·ь·е, -я, (*n*) (arch) semi-intrados

полу·дю́рен·, -а, (*m*) semi-splint coal

полу·жа́бр·а, -ы, (*f*) (zool) hemibranch

полу·жёстк·ий, -ая, -ое, (*a*) semi-rigid; (text) semi-stiff

полу·жестко·кры́л·ые, -ых, (*pl decl as adj*) (zool) *Hemiptera*

полу·жи́дк·остн·ый, -ая, -ое, (*a*) semi-fluid

полу·жи́р·н·ый, -ая, -ое, (*a*) half-dark (of type etc.)

полу·за·вод·ск·о́й, -а́я, -о́е, (*a*) pilot, pilot-plant

полу·за·кры́·т·ый, -ая, -ое, (*a*) (elec) semi-/half-enclosed; semi-closed

полу·за́·пис·ь, -и, (*f*) (instr) half-write

и́мпульс· полу·за́·пис·и (*m*) half-write pulse (sound recording)

полу·за·су́ш·лив·ый, -ая, -ое, (*a*) semi-arid

полу·за·то́пл·енн·ый, -ая, -ое, (*a*) waterlogged

полу·за·щищ·ённ·ый, -ая, -ое, (*a*) semi-protected

полу·зо́нт·ичн·ый, -ая, -ое, (*a*) (bot) cymose

полу·ка́др·, -а, (*m*) (TV) frame field

полу·киломе́тр·о́вк·а, -и, (*g pl*) **-вок·,** (*f*) 1/50000-scale map

полу·клю́з·, -а, (*m*) (naut) bull ring, mooring pipe

полу·ко́кс·, -а, (*m*) semicoke, coalite, low-temperature coke

полу·кокс·ова́ни·е, -я, (*n*) semicoking, low temperature carbonization, low temperature distillation (of coal)

полу·колло́ид·, -а, (*m*) (chem) hemicolloid

полу·кольц·о́, -а́, (*i*) **-о́м,** (*n*) half-ring

полу·ко́нус·, -а, (*m*) half-angle cone

полу·копы́т·н·ый, -ая, -ое, (*a*) (zool) subungulate; (*pl as noun*) subungulates, *Subungulata*

полу·кро́в·н·ый, -ая, -ое, (*a*) half-breed, half-bred

полу·кру́г·, -а, (*m*) semicircle

~, вис·я́ч·ий (min instr) protractor-type inclinometer

полу·кру́гл·ый, -ая, -ое, (*a*) semicircular, half-round

~ голо́в·к·а (*f*) button head (of rivets etc.)

~ стал·ь (*f*) segmental bars, half-round steel bars

полу·кру́г·ов·ый, -ая, -ое, (*a*) *see* **полу·кру́гл·ый**

полу·куб·и́ческ·ий, -ая, -ое, (*a*) semicubical

полу·леж·а́ть, -а́т, (*imp*) recline

полу·литр·о́в·ый, -ая, -ое, (*a*) half-litre/liter

полу·логари́фм·и́ческ·ий, -ая, -ое, (*a*) (math) semilogarithmic, semilog

полу·лу́н·н·ый, -ая, -ое, (*a*) crescent-shaped, half-moon

полу·ма́сс·а, -ы, (*f*) (paper) half-stock/-stuff

полу·ма́т·ов·ый, -ая, -ое, (*a*) (paper) semi-matte, flat-finish

полу·ме́р·а, -ы, (*f*) half-measure

полу·ме́сяц·, -а, (*i*) **-ем,** (*m*) half moon; crescent

полу·ме́сяч·н·ый, -ая, -ое, (*a*) fortnightly

полу·метр·о́в·ый, -ая, -ое, (*a*) half-metre/meter

полу·микро·ана́лиз·, -а, (*m*) semi-microanalysis

полу·мра́к·, -а, (*m*) gloom

полу·му́фт·а, -ы, (*f*) half-coupling

полу·му́шкел·в, -я, (*m*) serving mallet

полу·мя́гк·ий, -ая, -ое, (*a*) semi-upholstered

полу·на·вес·н·о́й, -а́я, -о́е, (*a*) semi-mounted

~ с·бо́р·к·а (*f*) (build) semicanti-levering

полу́ндра (*imper*) (naut) stand from under! mind out!

полу·не·пре·ры́в·н·ый, -ая, -ое, (*a*) semicontinuous, semi-batch

полу·не·про·ниц·а́ем·ый, -ая, -ое, (*a*) semitight

полу·но́ч·н·ый, -ая, -ое, (*a*) *see* **пол·но́ч·н·ый**

полу·обезья́н·ы (*f pl*) (zool) Lemuroidea

полу·о·бес·кле́·ен·ый, -ая, -ое, (*a*) (text) souple (of silk)

полу·об·ме́н·, -а,
перио́д· полу·об·ме́н·а (*m*) (chem) half-time of exchange

полу·оборо́т·, -а, (*m*) half-revolution/turn

полу·об·рабо́т·анн·ый, -ая, -ое, (*a*) half-worked, half-blank, semifinished

полу·о·ка́т·анн·ый, -ая, -ое, (*a*) subrounded

полу·о́стров·, -а, (*nom pl*) **-а́,** (*g pl*) **-о́в,** (*m*) peninsular

полу·о́с·ь, -и, (*f*) semi-axis; half/semi-axle

~, больш·а́я major semi-axis

~, дифференциа́ль·н·ая (М/Т) differential axle

~, ма́л·ая minor semi-axis

~, полу·раз·гру́ж·енн·ая (М/Т) semi-floating axle

~, раз·груж·ённ·ая (М/Т) floating axle/half-axle

полу·от·бе́л·к·а, -и, (*f*) (text) half-bleaching

полу·от·раж·ённ·ый, -ая, -ое, (*a*) semi-reflected, semi-direct, semi-indirect

полу·парабол·и́ческ·ий, -ая, -ое, (*a*) semiparabolic

полу·парази́т·, -а, (*m*) (bot) facultative saprophyte, hemiparasite, hemisaprophyte

полу·парали́ч·, -а, (*m*) (med) hemiplegia

полу·пере·воро́т·, -а, (*m*) (air) half-roll

полу·пере·ме́н·н·ый, -ая, -ое, (*a*) semivariable

полу·перио́д·, -а, (*m*) half-period; (elec) half-cycle, half-wave

~ рас·па́д·а (nucl) half-life

полу·пе́тл·я, -и, (*g pl*) **-пе́тел·ь,** (*f*) (air) half-loop; (text) half-stitch

полу·пике́т·, -а, (*m*) 50-metres mark

полу·пло́д·, -а, (*m*) (bot) hemicarp, mericarp

полу·пло́ск·ост·ь, -и, (*f*) half-plane

~, ве́рх·н·яя (dynam) higher half-plane

~ воз·мущ·е́ни·й, пере́д·н·яя (dynam) forward-leaning Mach plane

полу·по́лб·а, -ы, (*f*) = **по́лб·а** (q.v.)

полу·порта́ль·н·ый, -ая, -ое, (*a*) semi-portal, semi-gantry (of cranes etc.)

полу·по·та́й·н·ый, -ая, -ое, (*a*)

~ голо́в·к·а (*f*) raised countersunk head (of rivet)

полу·при·це́п·, -а, (*m*) (М/Т) semi-trailer

полу·про·вод·ни́к·, -а́, (*m*) (phys) semiconductor

~, вы́·рожд·енн·ый degenerate semiconductor

~, ды́р·оч·н·ый *p*-type semiconductor

~ с при́·мес·ами extrinsic semiconductor

~, чи́ст·ый intrinsic semiconductor

~ электро́н·н·ый *n*-type semiconductor

полу·про·вод·нико́в·ый, -ая, -ое, (*a*) semiconductor, semiconducting

~ у·сил·и́тел·ь (*m*) transistor

полу·проду́кт·, -а, (*m*) semifinished product, half-product, half-blank, semi; (chem) intermediate

полу·про·зр·а́чн·ый, -ая, -ое, (*a*) semi-transparent, translucent

полу·про·ниц·а́ем·ый, -ая, -ое, (*a*) semi-permeable

полу·про·све́ч·ивающ·ий, -ая, -ое, (*a*) semi-translucent

полу·про·стра́н·ств·о, -а, (*m*) (phys) half-space

~ твёрд·ое (acous) semi-infinite solid

полу·про́фил·ь, -я, (*m*) (met, civ eng) half-section

полу·пуст·ы́нн·ый, -ая, -ое, (*a*) (geog) semi-arid, semi-desert

полу·равн·и́н·а, -ы, (*f*) (geol) peneplain

полу·раз·ма́х·, -а, (*m*) amplitude, semispan

полу·ра́к·овист·ый, -ая, -ое, (*a*) subconchoidal

полу·рас·ко́с·, -а, (*m*) (civ eng) semi-diagonal, knee brace

полу·рас·па́д·, -а,   (*m*) half-decay/-disintegration

    пери́од· полу·рас·па́д·а (*m*) (nucl) half-life

полу·рас·тво́р·, -а,   (*m*) half-opening/span

    у́гол· полу·рас·тво́р·а   (*m*) (dynam) half-angle

полу·регуля́р·н·ый, -ая, -ое,   (*a*) semiregular

полу·реми́з·к·а, -и,   (*g pl*) **-з·ок·,** (*f*) (text) doup

полу·ры́л·ы   (*pl*) (fish) half-beaks, *Hemirhamphidae*

полу·с·ка́т·, -а,   (*m*) single-axle bogie (of mine cars)

полу·сли́пер·, -а,   (*m*) half-sleeper (timber)

полу·с·мерт·а́льн·ый, -ая, -ое,   (*a*) half-lethal

полу·с·поко́й·н·ый, -ая, -ое,   (*a*) semi-killed, balanced (of steel)

полу·ста́л·ь, -и,   (*f*) (met) semisteel, high-duty iron

полу·стекл·ова́т·ый, -ая, -ое,   (*a*) semivitreous, subvitreous

полу·стопо·ход·я́щ·ий, -ая, -ее,   (*a*) (zool) semiplantigrade

полу·стре́ль·чат·ая а́рк·а   (*f*) (arch) semigothic arch

полу·стро́б·, -а,   (*m*) (elec) half-gate, gate

~, за́д·н·ий   late gate

~, пере́д·н·ий   early gate

полу·строп·и́льник·, -а,   (*m*) square (timber)

полу·с·ты́к·, -а,   (*m*) half-joint

полу·сумма́тор·, -а,   (*m*) half-adder (computers)

полу·су́т·очн·ый, -ая, -ое, (*a*) semidiurnal, 12-hour

полу·сфе́р·а, -ы,   (*f*) hemisphere

полу·тари́ф·н·ый, -ая, -ое,   (*a*) half-rate

полу·твёрд·ый, -ая, -ое,   (*a*) semisolid

полу·те́л·о, -а,   (*n*) (dynam) half-body

полу·тен·ев·о́й, -а́я, -о́е,   (*a*) half-shade

полу·те́н·ь, -и,   (*f*) penumbra

полу·толщ·ин·а́, -ы́,   (*f*) half-depth/thickness

полу·томпа́к·, -а,   (*m*) (met) 80% red brass

полу·то́н·, -а,   (*m*) half-tone, semitone

полу·то́н·ов·ый, -ая, -ое,   (*a*) half-tone, semitone

полу́тора   (*gen/dat/instr/prep*) of **полтора́**

полу́тора-   (*component*) one-and-a-half, sesqui-

полутора·год·ов·о́й, -а́я, -о́е,   (*a*) one-and-a-half yearly, eighteen monthly

полутора·ко́рпус·н·ая под·ло́д·к·а   (*f*) saddle-tank submarine

полутора·ма́чт·ов·ый, -ая, -ое,   (*a*) (naut) main-and-mizzen rigged

полутора·о́·кис·ь, -и,   (*f*) (chem) sesquioxide

полутора·хло́р·ист·ый, -ая, -ое,   (*a*) sesquichloride (of)

полу́тор·н·ый, -ая, -ое,   (*a*) one-and-a-half, sesqui-

~ о́·кис·ь   (*f*) (chem) sesquioxide

полу·угл·ова́т·ый, -ая, -ое,   (*a*) subangular

полу·у·то́пл·енн·ый, -ая, -ое,   (*a*) semirecessed, half-flush

полу·фабрика́т·, -а,   (*m*) semifinished product, semi-product, semi

полу·фарфо́р·, -а,   (*m*) semiporcelain

полу·фо́рм·а, -ы,   (*f*) half-mould/mold

полу·фуга́н·ок·, -н·к·а,   (*m*) long plane (carpenter's tool)

полу·хо́рд·ов·ые, -ых,   (*pl decl as adj*) (zool) hemichordates, *Hemichordata*

полу·хроно́·метр·, -а,   (*m*) chronometer watch

полу·целлюло́з·а, -ы,   (*f*) (chem) hemicellulose

полу·це́л·ый, -ая, -ое,   (*a*) (math) half-integral

~ числ·о́   (*n*) half-integer

полу·цикл·и́ческ·ий, -ая, -ое,   (*a*) semicyclic

полу·ци́ркуль·н·ый, -ая, -ое,   (*a*) (arch) semicircular

полу·час·ов·о́й, -а́я, -о́е,   (*a*) half-hour, half-hourly

по·луч·а́тел·ь, -я,   (*m*) recipient, consignee

по·луч·а́·ть, -ют, (*imp*) receive, obtain, get; **-ся** (*pass*); be, turn out to be, be produced/prepared

по·луч·е́ни·е, -я,   (*n*) (*v n*), see **по·луч·а́·ть;** acquisition, production, preparation

~ да́·нн·ых   data acquisition

~ изото́п·ов   isotope production/preparation

по·луч·éни·е

~ сополимéр·а   preparation of a co-polymer

по·лýч·енн·ый, -ая, -ое, (*past part pass*) *see* по·луч·á·ть; product

полу·чúст·ый, -ая, -ое, (*a*) semi-finished (of machined articles), half-bright

по·луч·úть, ‑ат, (*perf*) *see* по·лу·ч·á·ть

полу·шáг·, -а, (*m*) half-step

полу·шáр·и·е, -я, (*n*) hemisphere

полу·шир·ин·á, -ы́, (*f*) half-width

полу·шир·от·á, -ы́, (*nom pl*) шир·о́т·ы, (*g pl*) шир·о́т·, (*f*) (shipb) half-breadth plan

полу·шпангóут·, -а, (*m*) (shipb, a/c) half-frame

полу·эдр·úческ·ий, -ая, -ое, (*a*) hemihedral

полу·электрóд·, -а, (*m*) (elec) single-electrode system, half-cell

полу·элемéнт·, -а, (*m*) (elec) half-cell

полу·ю́т·, -а, (*m*) (shipb) after super-structure

полу·ядр·óв·ый, -ая, -ое, (*a*) half-grained (soap)

пол·цен·ы́ (*f indecl*) half-price

пол·час·á, (*g sing*) полу·чáс·а (*m*) half an hour

пол·шпáц·и·я, -и, (*f*) (shipb) half-frame spacing

пóл·ый, -ая, -ое, (*a*) hollow; cored (of casting)

полы́н·н·ый, -ая, -ое, (*a*) of по-лы́н·ь

~ вóд·к·а (*f*) absinthe (drink)

~ мáсл·о (*n*) (pharm) wormwood oil

полы́н·ь, -и, (*f*) (bot) wormwood, *Artemisia*

~, гóрь·к·ая (bot) absinthe, *Artemisia absinthium*

~, цитвáр·к·ая Levant wormseed, *Artemisia cina*

пол·ынь·я́, -й, (*g pl*) -н·éй, (*f*) (ocean) polynia (open water in ice)

~, при·берéж·н·ая shore lead (in ice)

пóльдер·, -а, (*m*) (geog) polder

Пóльди, спóсоб· (*m*) Poldi hardness testing

по·ль·ёт, (*fut 3rd sing*) of по·лú·ть, (*perf*); *see* по·лив·á·ть (*imp*)

пóльз·а, -ы, (*f*) benefit, advantage, use в пóльз·у in favour of, in support of, on behalf of

пóльз·овани·е, -я, (*n*) use

пóльз·оваться, -уют̣ся, (*imp*) (+ *instr*) use, make use of; have, possess, enjoy, be

пóлюс·, -а, (*m*) pole

~ воз·бужд·éни·я (elec) field pole

~, вы·ступ·áющ·ий (elec) salient pole

~, гео·магнúт·н·ый geomagnetic pole

~, до·бáв·очн·ый (elec) interpole, commutating pole

~, магнúт·н·ый magnetic pole

~, одно·имён·н·ый analogous pole

~, от·риц·áтельн·ый negative pole

~, разно·имён·н·ый opposite pole

~, сéвер·н·ый North Pole

~ фýнкци·и (math) pole of a function

~ хóлод·а (geog) cold pole

~, ю́ж·н·ый South Pole

~, я́в·н·ый (elec) salient pole

пóлюс·н·ый, -ая, -ое, (*a*) pole

~ башмáк· (*m*) (elec) pole shoe

~ луч· (*m*) (civ eng) funicular line

~ на·кон·éчник· (*m*) (elec) pole piece

~ отвéрст·и·е (*n*) vent (of parachute)

~ пре·облад·áни·е (*n*) (telecom) bias

~ фигýр·а (*f*) (spectr) pole figure

пóл·ют (*pres 3rd pl*) of пол·óть

пóл·ющ·ий, -ая, -ее, (*pres part act*) of пол·óть

пол·я́ (*nom pl*) of пóл·е; (*pres gerund*) of пол·óть

пол·я́н·а, -ы, (*f*) glade

Поля́ни, динамó·метр· (*m*) Polyani dynamometer

поля́р·, -а, (*m*) (*see also* поля́р·а) polar (dye)

поля́р·а, -ы, (*f*) (math) polar, polar-line/curve

~ крыл·á (a/c) drag polar

~, суммáр·н·ая resultant polar diagram

~, удáр·н·ая (dynam) shock polar

поляренциáль·н·ый, -ая, -ое, (*a*) polar differential

поляриз·áтор·, -а, (*m*) polarizer

поляриз·ациóнн·ый, -ая, -ое, (*a*) polarization, polarizing

поляриз·áци·я, -и, (*f*) (phys) polarization

~, круг·ов·áя circular polarization

~, объ·ём·н·ая volume polarization

~, плóск·ая plane polarization

~, по·перéч·н·ая cross polarization

~, пре·дéль·н·ая critical polarization явл·éни·е поляриз·áци·и (*n*) (opt) polarization effect поляриз·ова́ть, -у́ют, (*imp and perf*) (phys) polarize

поляриз·у́емост·ь, -и, (*f*) polarizability, polarization capacity

поляриз·у́ем·ый, -ая, -ое, (*pres part pass*) of поляриз·ова́ть; polarizable

поляри́·метр·, -а, (*m*) (opt) polarimeter

поляри·метр·и́ческ·ий, -ая, -ое, (*a*) polarimetric

поляри·ме́тр·и·я, -и, (*f*) (chem) polarimetry

поляри·ско́п·, -а, (*m*) (opt) polariscope

поля́р·ност·ь, -и, (*f*) polarity

~, обра́т·н·ая (elec) reversed polarity

~, обращ·ённ·ая reversed polarity

~ от·клон·е́ни·я deflection polarity

~, противо·по·ло́ж·н·ая opposite polarity

поля́р·н·ый, -ая, -ое, (*a*) polar, arctic

~ во́з·дух· (*m*) (meteor) polar air

~ жи́дк·ост·ь (*f*) polar liquid

~ звезд·а́ (*f*) (astron) Polaris, Pole star, North star

~ координа́т·ы (*pl*) (math) polar coordinates

~ круг·, се́вер·н·ый (*m*) Arctic circle

~ круг·, ю́ж·н·ый Antarctic circle

~ по́яс· (*m*) (geog) frigid zone

~ с·вя́з·ь (*f*) (chem) polar bond

~ си·я́ни·е (*n*) (astron) Aurora Polaris, Polar light

~ си·я́ни·е, се́вер·н·ое Aurora Borealis, Northern lights

~ си·я́ни·е, ю́ж·н·ое Aurora Australis, Southern lights

поляро·гра́мм·а, -ы, (*f*) polarogram

поляро́·граф·, -а, (*m*) (chem, elec) polarograph

поляро·граф·и́ров·ани·е, -я, (*n*) polarographic analysis

поляро·гра́ф·и·я, -и, (*f*) (chem) polarography

поляр·о́ид·, -а, (*m*) (opt) polaroid, polarizing sheet

поляро́н·, -а, (*m*) (phys) polaron

пом- (*component*) (*abbr*) of по·мо́щ·ник· assistant

пома́д·а, -ы, (*f*) pommade, cream

пома́д·к·а, -и, (*g pl*) -д·ок·, (*f*) fondant

по·ма́з·ать, (*fut 3rd pl*) по·ма́ж·ут, (*perf*) grease, smear, cost

по·маз·о́к·, -з·к·а́, (*m*) brush; shaving brush; tar brush

по·ма́р·к·а, -и, (*g pl*) -р·ок·, (*f*) correction, amendment (in manuscript)

по·мат·ове́ни·е, -я, (*n*) dulling, dimming, tarnishing

по·мед·ненн·ый, -ая, -ое, (*past part pass*) coppered, copper-plated

по·мел·е́·ть, -ют, (*perf*) become shallow

по·мень·ш·е (*comp*) rather/a little less/ smaller

померан·ец·, -н·ц·а, (*m*) (bot) bitter orange, *Citrus aurantium*

померан·цев·ый, -ая, -ое, (*a*) of померан·ец· (*pl*) aurantiaceae

по·мёрзл·ый, -ая, -ое, (*past part pass*) of по·мёрз·нуть

по·мёрз·нуть, -нут, (*perf*) die of cold, be killed by frost

по·ме́р·ить, -ят, (*perf*) measure

по·мест·и́тельн·ый, -ая, -ое, (*a*) spacious, capacious, roomy

по·мест·и́ть, -ят, (*perf*) *see* по·ме·щ·а́·ть

по·ме́ст·н·ый, -ая, -ое, (*a*) position, according to position, positional

~ нумер·а́ци·я (*f*) (math) position notation

по́·мес·ь, -и, (*f*) crossbreed, hybrid

по·ме́сяч·н·ый, -ая, -ое, (*a*) monthly

по·мёт·, -а, (*m*) droppings, dung, castings; offspring, litter

по·ме́т·а, -ы, (*f*) mark, note

по·мет·а́·ть, -ют, (*perf*) *see* мет·а́ть

по·ме́т·ить, -ят, (*perf*) *see* по·меч·а́·ть

по·ме́т·к·а, -и, (*g pl*) -т·ок·, (*f*) (*dim*) of по·ме́т·а

по·ме́х·а, -и, (*g pl*) по·ме́х· (*f*) hindrance, obstacle; (*sing or pl*) (rad) interference, noise; jamming (when intentional)

~, альтиметр·о́в·ые (*pl*) clutter (radar)

~, атмосфе́р·н·ые (*pl*) (rad) atmospheric interference

~, бе́л·ая (TV) white noise

~, диполь·н·ые (*pl*) (rad) window

~, за·град·и́тельн·ые, (*pl*) barrage jamming

~, зеркаль·н·ая (rad) image interference

~, косм·и́ческ·ие (*pl*) (rad) star static, Jansky noise

коэффицие́нт· телефо́н·н·ых по·-ме́х· (*m*) telephone influence factor

~, культу́р·н·ые (*pl*) (rad) man-made interference

~ от дожд·я́ (*pl*) (rad) rain clutter

~ от за·жиг·а́ни·я (*pl*) (rad) ignition noise

~, шум·ов·а́я noise jamming

помехо·глуш·и́тел·ь, -я,   *(m)* (rad) noise silencer

помехо·сто́й·к·ий, -ая, -ое,   *(a)* (rad) noiseproof, antistatic, anti-interference

помехо·у·сто́й·чив·ый, -ая, -ое,   *(a)* (rad) noise proof; anti-interference, anti-static

по·меч·а́·ть, -ют,   *(imp)* mark; date; (nucl) tag

по·ме́ш·анн·ый, -ая, -ое,   *(past part pass)* of **по·меш·а́·ть;** disturbed, confused, deranged, mentally sick

по·меш·а́·ть, -ют, *(perf) see* **меш·а́·ть;** stir/mix a little; **-ся** become deranged, become mentally sick

по·ме́ш·ивани·е, -я, *(n)* *(v n)* of **по·ме́·ш·ива·ть;** agitation

по·ме́ш·ива·ть, -ют,   *(imp)* stir/mix intermittently/gently, agitate

по·мещ·а́·ть, -ют,   *(imp)* place, locate, put; plant (an order; computers); (fin) invest; accommodate, lodge

по·мещ·е́ни·е, -я, *(n)* *(v n), see* **по·ме·щ·а́·ть;** accommodation, apartment, room, flat, compartment, premises; (fin) investment; position, place, station

~, жи·л·о́е living accommodation; (mil) quarters

~ газо·ре́з·к·и (met) cut-off station

~ у·правл·е́ни·я control room

помидо́р·, -а,   *(m)* (bot) tomato, *Lycopersicon esculentum*

по·ми́мо   *(prep + gen)* besides, apart from

по·мину́т·н·о   *(adv)* a minute, every minute, minute by minute

по·мину́т·н·ый, -ая, -ое,   *(a)* per minute, by the minute; incessant

~ тари́ф·   *(m)* charges per minute (e.g. telephones)

по·мн·ёт, *(fut 3rd sing)* of **по·мя́·ть**

по́·мн·ить, -ят,   *(imp)* remember, keep/bear in mind

по·множ·а́·ть, -ют,   *(imp)* multiply

по·мно́ж·ить, -ат, *(perf) see* **по·множ·а́·ть**

по·мн·у́т, *(fut 3rd pl)* of **по·мя́·ть**

по́·мн·ят, *(pres 3rd pl)* of **по́·мн·ить**

по·мог·а́·ть, -ют, *(imp)* help, assist, aid

по-мо́·ему *(conj)* in my opinion; in my way

по·мо́ж·ет, *(fut 3rd sing)* of **по·мо́ч·ь,** *(perf); see* **по·мог·а́·ть** *(imp)*

по·мо́·и, -ев, *(m pl)* slops, swill, dishwater

по·мо́й·н·ый, -ая, -ое, *(a)* of **по·мо́·и**

~ я́м·а *(f)* cesspit

по·мо́л·, -а,   *(m)* grinding, milling; grist, grind

~, выс·о́к·ий high grinding (flour)

~, круп·и́тчат·ый high grinding (flour)

~, прост·о́й plain grinding, single-passage grinding

~, с·ло́ж·н·ый high grinding

помо·ло́г·и·я, -и, *(f)* (hortic) pomology

по·молч·а́·ть, -а́т,   *(perf)* be silent (for a while)

по·мо́р·ник·, -а, *(m)* (orn) skua

по·мо́р·ск·ий, -ая, -ое,   *(a)* coastal, maritime

по·мо́р·ь·е, -я,   *(n)* coastal area, coast, littoral

по·мо́ст·, -а, *(m)* stage, dais, platform

помо́х·а, -и,   *(f)* (geog) pomokha (dust haze over chernozem areas)

по·моч·и́ть, -а́т, *(perf)* moisten/damp slightly

по·мо́ч·ь, *(fut 3rd sing, pl)* **по·мо́ж·ет, по·мо́г·ут,** *(past masc sing)* **по·мо́г·,** *(perf); see* **по·мог·а́·ть**

по·мо́щ·ник·, -а, *(m)* assistant

~ капита́н·а (naut) mate

~ капита́н·а, ста́р·ш·ий (naut) chief officer

~ капита́н·а, тре́т·ий (naut) second officer

~ команди́р·а корабл·я́ first lieutenant (naval)

по́·мощ·ь, -и, *(f)* help, assistance, aid **при по́·мощ·и** with the aid of, with; by means of, by

с по́·мощ·ью *see* **при по́·мощ·и**

~, ско́р·ая first aid

по́мп·а, -ы, *(f)* *(see also* **на·со́с·);** pump; pomp

~ под·ка́ч·к·и (a/c) booster pump

~, топл·и́вн·ая fuel injection pump

помп·а́ж·, -а,   *(m)* (dynam) surge, surging; (autom) pumping, hunting, flutter

~ компре́ссор·а compressor surge **крив·а́я помп·а́ж·а** *(f)* (air) stall line

по·мут·и́ть, -я́т, *(perf)* make cloudy/muddy/turbid

по·мут·не́ни·е, -я,   *(n)* cloudiness, muddiness, turbidity, blooming (of varnish etc.)

температу́р·а по·мут·не́ни·я *(f)* (oil) cloud point

по·мут·не́·ть, -ют, *(perf)* become cloudy/muddy/turbid

по·мч·а́ть, -а́т, *(perf)* rush, race, hurtle

по·мы́л·ить, -ят, *(perf)* soap lightly

по·мя́·т·ый, -ая, -ое, *(past part pass)* of **по·мя́·ть**

по·мя́·ть, *(fut 3rd pl)* **по·мн·у́т,** *(perf)* crumple, crease; knead (dough)

по-на·сто·я́щ·ему *(adv)* properly

по-на́ш·ему *(conj)* in our opinion; in our way

пондеро·мото́р·н·ый, -ая, -ое, *(a)* ponderomotive

по·неде́ль·ник·, -а, *(m)* Monday

по·неде́ль·н·ый, -ая, -ое, *(a)* weekly

по·не·мно́г·у *(adv)* gradually, little at a time, little-by-little

по·нес·ти́, -у́т, *(perf)* carry, take, bear; rush, race, take/convey quickly

по́ни *(m indecl)* (zool) pony

по·ниж·а́тел·ь, -я, *(m)* *(v n)*, *see* **по·ни·ж·а́·ть;** depressor, reducer

по·ниж·а́·ть, -ют, *(imp)* lower, reduce, depress; **-ся** *(pass)*; fall, drop, go down, decrease, diminish

по·ниж·а́·ть в 3 ра́з·а reduce to a third

по·ниж·а́ющ·ий, -ая, -ее, *(pres part act)* of **по·ниж·а́·ть;** (elec) step-down

~ трансформ·а́тор· *(m)* (elec) step-down transformer

по·ни́ж·е *(comp)* a little/rather lower/shorter

по·ниж·е́ни·е, -я, *(n)* *(v n)*, *see* **по·ни·ж·а́·ть;** reduction, drop, decline (in values); degradation; (geol) subsidence, depression, low; (surv) depression

~ горизо́нт·а (surv) apparent depression of the horizon

у́гол по·ниж·е́ни·я *(m)* (surv) angle of depression

по·ни́ж·енн·ый, -ая, -ое, *(a)* *(past part pass)* *see* **по·ниж·а́·ть;** decreased, hypo-

~ кисл·о́тност·ь *(f)* (med) hypo–acidity

по·низ·и́тел·ь, -я, *(m)* depressor, reducer

~ вы́·брос·ов (elec) despiker

~ част·от·ы́ (elec) scaler, scaling circuit

по·низ·и́тельн·ый, -ая, -ое, *(a)* reducing, depressing; (elec) step-down

по·ни́з·ить, -ят, *(perf)* *see* **по·ни·ж·а́·ть**

по́·низ·у *(adv)* low, down below; along the ground/bottom

по·ник·а́·ть, -ют, *(imp)* droop, wilt, flag

по·ни́к·нуть, -нут, *(perf)* *see* **по·ни·к·а́·ть**

по·ним·а́ни·е, -я, *(n)* understanding, comprehension; conception, sense

по·ним·а́·ть, -ют, *(imp)* understand, comprehend

по-но́в·ому *(adv)* anew, afresh, in a new way/manner/fashion

поно́р·, -а, *(m)* (geog) ponor, sinkhole

по·но́с·, -а, *(m)* (med) diarrhoea

по·нос·и́·ть, -ят, *(perf)* carry, wear; *(imp)* abuse

по·но́ш·енн·ый, -ая, -ое, *(past part pass)* of **по·нос·и́·ть;** old (of clothes)

понтаци́л·, -а, *(m)* pontacyl (dye)

по́нти·я, -и, *(f)* pontie, pontil, punty (glass)

понто́н·, -а, *(m)* pontoon; pontoon bridge

~, жёстк·ий camel (pontoon)

понто́н·н·ый, -ая, -ое, *(a)* of **понто́н·**

~ мост· *(m)* pontoon bridge

по·нуд·и́тельн·ый, -ая, -ое, *(a)* coercive, compelling; forced

по·ну́д·ить, -ят, *(perf)* *see* **по·нужд·а́·ть**

по·нужд·а́·ть, -ют, *(imp)* compel, force

пону́р·, -а, *(m)* (civ eng) upstream puddle blanket (of dam etc.)

по·ню́х·а·ть, -ют, *(perf)* sniff

по́·ня·л, *(past masc sing)* of **по·ня́т·ь,** *(perf)*; *see* **по·ним·а́·ть** *(imp)*

по·ня́т·и·е, -я, *(n)* idea, concept

по·ня́т·ливост·ь, -и, *(f)* comprehension

по·ня́т·лив·ый, -ая, -ое, *(a)* quick (to understand), perspicacious

по·ня́т·ност·ь, -и, *(f)* intelligibility, clarity

по·ня́т·н·ый, -ая, -ое, *(a)* comprehensible, intelligible, clear; understandable

по·нят·о́й, -о́го, *(m decl as adj)* (law) authorized/official witness

по́·нят·ый, -ая, -ое, *(a)* understood

по·ня́т·ь, *(fut 3rd pl)* **по·йм·у́т,** *(past masc sing)* **по́·ня·л,** *(perf)*; *see* **по·ним·а́·ть**

по·о́·дал·ь *(adv)* at a distance

по·один·о́чк·е *(adv)* one at a time, one by one

по·о·черёд·н·о *(adv)* in turn, successively

по·о·черёд·н·ый, -ая, -ое, *(a)* successive

по·ощр·и́ть, -я́т, (*perf*) *see* **по·ощр·я́·ть**

по·ощр·я́·ть, -ю́т, (*imp*) encourage

по·пад·а́ни·е, -я, (*n*) (*v n*), *see* **по·па́·д·а·ть** and **по·пад·а́·ть;** hit, impact (of missile); reaching

~, прям·о́е direct hit

по·пад·а́·ть, -ю́т, (*imp*) hit, strike (of missile); find oneself, be, get to, arrive, enter; come upon, find; step/stumble on; get into; **-ся** be/get caught/taken (*and see below*)

по·па́·д·а·ть, -ют, (*perf*) fall (one after the other)

по·па́·л, (*fut 3rd sing*) of **по·па́с·ть,** (*perf*); *see* **по·пад·а́·ть** (*imp*)

по·па́р·н·о (*adv*) in pairs/twos

~ не·пере·сек·а́ющ·иеся (math) mutually disjunct, disjoint

по·па́с·ть, (*fut 3rd pl*) **по·пад·у́т,** (*past masc sing*) **по·па́·л,** (*perf*); *see* **по·пад·а́·ть**

по·перёк· (*adv, prep*) across, crosswise; (naut) athwart, athwartships

~, в·ключ·ённ·ый cross-connected

по·пере·ме́н·н·о (*adv*) alternately, in turn

по·переч·ин·а, -ы, (*f*) cross beam/girder/bar/piece/tie/arm/member; crosshead (of engine)

по·переч·ник·, -а, (*m*) diameter; cross section

поперечно·вяз·а́льн·ая маши́н·а (*f*) (text) framework knitting machine

поперечно·ост·и́ст·ый, -ая, -ое, (*a*) (anat) transversospinal

поперечно·полос·а́т·ый, -ая, -ое, (*a*) transversostriated

поперечно·ро́т·ые, -ых, (*pl decl as adj*) (fish) plagiostomes, *Plagiostomi*

поперечно·строг·а́льн·ый стан·о́к· (*m*) (eng) shaping machine

по·переч·н·ый, -ая, -ое, (*a*) diametrical; transverse, cross, lateral; (naut) athwartships

~ V (*n*) (a/c) dihedral

~ V, обра́т·н·ое (*n*) anhedral

~ V, от·риц·а́тельн·ое (*n*) anhedral, negative dihedral

~ V, по·лож·и́тельн·ое (*n*) positive dihedral

~ масшта́б· (*m*) transversal scale

~ про́фил·ь (*m*) (arch) end elevation; transverse profile (river)

~ раз·ре́з· (*m*) cross section

у́гол· по·переч·н·ого V (*m*) dihedral angle

по·печ·и́тел·ь, -я, (*m*) guardian, trustee

по·пир·а́·ть, -ю́т, (*imp*) trample upon, transgress

по·плав·ко́в·ый, -ая, -ое, (*a*) float, float-type, float-operated

~ ка́мер·а (*f*) (ICE) float chamber/bowl

~ кла́пан· (*m*) float valve

по·плав·о́к·, -в·к·а́, (*m*) float

~ гидро·стат·и́ческ·их вес·ов hydrostatic sink

~ -интегра́тор·, -а, (*m*) float, float gauge (for rate of flow)

поплин·, -а, (*m*) (text) poplin

по·по́зж·е (*comp*) a little/bit later

по·пол·а́м (*adv*) in two; half-and-half

по·полз·у́шк·а, -и, (*g pl*) **-шек·,** (*f*) radius link

по·полз·ти́; -у́т, (*perf*) crawl, creep

по·полн·е́ни·е, -я, (*n*) (*v n*), *see* **по·полн·я́·ть;** replenishment; (mil) reinforcement, draft

~ гор·я́ч·им refuelling, topping up with fuel

по·полн·и́тел·ь, -я, (*m*) (phot) booster/replenishing solution

по·по́лн·ить, -ят, (*perf*) *see* **по·полн·я́·ть**

по·полн·я́·ть, -ю́т, (*imp*) replenish, supplement, add, fill up, top up; (mil) reinforce

по·полоск·а́·ть, -ю́т, or **по·поло́щ·ут,** (*perf*) rinse; gargle

по·полу́·дн·и (*adv*) p.m. (of time)

по·полу́·ноч·и (*adv*) a.m. (of time)

попо́н·а, -ы, (*f*) cloth, blanket (for animal)

по·прав·и́м·ый, -ая, -ое, (*pres part pass*) of **по·пра́в·ить;** reparable, improvable, remediable

по·пра́в·ить, -ят, (*perf*) *see* **по·прав·л·я́·ть**

по·пра́в·к·а, -и, (*g pl*) **-в·ок·,** (*f*) (*see also under associated words*); correction, amendment

в·нес·ти́ по·пра́в·к·у на correct for, make a correction for

~, выс·о́тн·ая (air) altitude correction

~ за про·вис·а́ни·е (surv) catenary correction

~, инструмента́ль·н·ая index correction/error

~ ко́мпас·а compass error

~, кон·цев·а́я end correction

~ на за·па́зд·ывани·е delay/lag correction

по·пра́в·к·а
~ на эффе́кт· сте́н·к·и (dynam) wall correction
~, от·нос·и́тельн·ая relative error
    фо́рмул·а по·пра́в·ок· (f) correlation formula
по·правл·е́ни·е, -я, (n) (v n), see **по·правл·я́·ть**
по·правл·я́·ть, -ют, (imp) repair, put right; correct, amend; tidy, tidy up, put in order, straighten; recover, improve; **-ся** (pass); improve, get better/well, recover; correct oneself
по·пра́в·очн·ый, -ая, -ое, (a) correction
~ коэффицие́нт· (m) correction factor
по·пр·а́ть, -у́т, (perf) see **по·пир·а́·ть**
по·пре́ж·н·ему (adv) as before/ previously
по́·пр·ищ·е, -а, (i) -ем, (n) field (of endeavour); career
по·про́б·овать, -уют, (perf) see **про́б·овать**
по·прос·и́ть, ·я́т, (perf) (+ gen) see **прос·и́ть**
по·прош·у́, (fut 1st sing) of **по·про·с·и́ть**
по·прощ·а́·ться, -ются, (perf) say goodbye, take leave
по·пры́г·а·ть, -ют, (perf) hop, skip, jump; hop/skip/jump one after another
попуга́·й, -я, (m) (orn) parrot, (pl) Psittacidae
попул·яриза́ци·я, -и, (f) popularization
попул·я́рн·ый, -ая, -ое, (a) popular
попул·я́ци·я, -и, (f) (biol) population
по·пуст·и́тельств·овать, -уют, (imp) connive (at)
по·пу́т·н·о (adv) in passing; incidentally
по·пу́т·н·ый, -ая, -ое, (a) accompanying, passing, following; parallel, cocurrent; byproduct
~ ве́тер· (m) following wind, tail wind
~ по·то́к· cocurrent/parallel flow
по·пы́т·к·а, -и, (g pl) **-т·ок·,** (f) attempt, endeavour
по́р·а, -ы, (f) (and see **пор·а́**); pore, cavity
~, пищ·ев·а́я (biol) gastropore
пор·а́, -ы́, (f) time
    до сих пор· till now, so far, until the present

пор·а́
    до тех пор· as long as, until
    с котор·ых пор· since when
    с тех пор· since then
    с тех са́м·ых пор· since when
~, стра́д·н·ая harvest time
по·рабо́т·а·ть, -ют, (perf) work (for a certain time), do some work
по·равн·я́·ться, -ются, (perf) come up level with, draw alongside
по·ра́д·овать, -уют, (perf) make glad/happy; be happy with; **-ся** (+ dat) be glad of, be happy about; be happy
по·раж·а́емост·ь, -и, (f) vulnerability
по·раж·а́ем·ый, -ая, -ое, (pres part pass) of **по·раж·а́·ть**; danger, endangered, vulnerable
~ пло́щад·ь (f) (gunn) beaten zone
~ про·стра́н·ств·о (n) (gunn) dangerous space
~ про·стра́н·ств·о, ме́ст·н·ое (gunn) true dangerous space
~ про·стра́н·ств·о, при·це́ль·н·ое (gunn) theoretical dangerous space
по·раж·а́·ть, -ют, (imp) hit (of missiles); strike, impress; affect (of illness), injure, damage; **-ся** (pass); (+ dat) be struck/impressed by
по·раж·е́ни·е, -я, (n) (v n), see **по·раж·а́·ть**; (med) affection, injury, damage; (gunn) effective/holding/ damaging fire; hitting (of missiles)
~, луч·ев·о́е (med, nucl) radiation injury
по·раз·и́тельн·ый, -ая, -ое, (a) striking, startling
по·раз·и́ть, -я́т, (perf) see **по·раж·а́·ть**
по·ра́зн·ому (adv) differently, in different ways
по·раз·ря́д·н·ый, -ая, -ое, (a) digit-by-digit (computers operations)
по·ран·е́ни·е, -я, (n) (med) wound, injury
по·ра́н·ить, -ят, (perf) wound, injure
по·ра́нь·ш·е (comp) rather/somewhat earlier
по·раст·а́·ть, -ют, (imp) grow over; become overgrown
по·раст·и́, -у́т, (perf) grow, grow a little; grow over; become overgrown
по́·рв·анн·ый, -ая, -ое, (past part pass) of **по·рв·а́ть**
по·рв·а́ть, -у́т, (perf) tear, tear up; disrupt (communications), break (wires etc.); break, break off, interrupt

по·ред·é·ть, -ют, (*perf*) become rarer/thinner

по·réж·ут, (*fut 3rd pl*) of по·ré·з·ать

по·réз·, -а, (*m*) cut, cutting

по·réз·ать, (*fut 3rd pl*) по·réж·ут, (*perf*) cut

порé·й, -я, (*m*) (bot) leek, *Allium porrum*

пóр·ет, (*pres 3rd sing*) of пор·óть (*imp*)

по·ржáв·е·ть, -ют, (*perf*) go rusty

пóр·истост·ь, -и, (*f*) porosity, void content

~, гáз·ов·ая (met cast) porosity due to blowholes, blowhole porosity/content

коэффициент· пóр·истост·и (*m*) voids ratio, degree of porosity

~, тóч·ечн·ая (met) pinhole porosity/content

~, у·сáд·очн·ая (met cast) porosity due to shrinkage cavities

пóр·ист·ый, -ая, -ое, (*a*) porous

по·риц·áни·е, -я, (*n*) censure

поркупáйнопенер·, -а, (*m*) (text) porcupine opener

пó·ровн·у (*adv*) equally, into equal parts/quantities

пор·óвш·ий, -ая, -ее, (*past part act*) of пор·óть

порóг·, -а, (*m*) sill; (phys) threshold, cutoff, point; (*pl*) rapids (in rivers); (ocean) rise

~ за·пир·áни·я (elec) cutoff level

~ из·луч·éни·я radiation cutoff

~ раз·бóр·чивост·и threshold of intelligibility

~ раз·лич·ймост·и threshold of intelligibility

~ рас·сé·яни·я scattering threshold

~ ре·кристалл·изáци·и recrystallization point/threshold

~ слыш·ймост·и audibility threshold

~ спéктр·а spectrum cutoff

~, тóп·очн·ый baffle plate (in furnaces)

~ у·лучш·éни·я (rad) sputter point

~ чувств·йтельност·и discrimination (of an instrument)

поро·гáм·и·я, -и, (*f*) (bot) porogamy

порóг·ов·ый, -ая, -ое, (*a*) of порóг·

~ знач·éни·е (*n*) threshold value

~ у·стрóй·ств·о (*n*) gate (radar)

по·рóд·а, -ы, (*f*) breed, sort, kind; (geol) rock

по·рóд·а

~, гóр·н·ая rock

~, корен·н·áя primary rock

~, лист·венн·ая hardwood

~, матер·йков·ая bed-rock

~, о·сáд·очн·ая sedimentary rock

~, пуст·áя country rock, (min) dirt

~, хвóй·н·ая softwood

по·рóд·ист·ый, -ая, -ое, (*a*) pedigree, thoroughbred

по·рóд·йть, -ят, (*perf*) *see* по·рожд·á·ть

по·рóд·ност·ь, -и, (*f*) (agr) breed, stock

по·рóд·н·ый, -ая, -ое, (*a*) of по·рóд·а

~ от·вáл· (*m*) (min) slag heap

~ скот· (*m*) pedigree cattle

по·рожд·á·ть, -ют, (*imp*) engender, give birth to, generate

по·рожд·éни·е, -я, (*n*) (*v n*), *see* по·рожд·á·ть; creation, product, outcome, result

порóж·ист·ый, -ая, -ое, (*a*) with rapids (of rivers)

порожн·ём (*adv*) (*see also* порожн·як·óм); empty, without load

~, о·сáд·к·а (*f*) (naut) light draught

порóжн·ий, -яя, -ее, (*a*) empty

порóж·н·ый, -ая, -ое, (*a*) of порóг·

порожн·я́к·, -á, (*m*) (rail) empty train

порожн·якóв·ый, -ая, -ое, (*a*) empty

порожн·як·óм (*adv*) (*see also* порожн·ём); empty, without load

вес· порожн·як·óм (*m*) weight empty, tare weight

порозй·метр·, -а, (*m*) porosity meter

пó·розн·ост·ь, -и, (*f*) separateness

пó·розн·ь (*adv*) separately, apart

пор·óй (*adv*) at times, sometimes; (*i*) of пор·á

порóк·, -а, (*m*) vice; defect, flaw, blemish

~, по·вéрх·ностн·ый surface defect

пор·óл, (*past masc sing*) of пор·óть

поролóн·, -а, (*m*) (plast) porolon

порó·метр·, -а, (*m*) (bot) porometer

поро·образ·овáни·е, -я, (*n*) pore-formation; blowing (plastics)

поро·образ·овáтел·ь, -я, (*m*) blowing agent (for plastics)

поро·плáст·, -а, (*m*) (plast) expanded plastic

по·рóс·, (*past masc sing*) of по·раст·й

порос·ёнок·, -нк·а, (*nom pl*) -я́та, (*g pl*) -я́т, (*m*) piglet

порос·и́ться, -я́тся, (*imp*) farrow (of pigs)

по́·росл·ь, -и, (*f*) undergrowth, bush

по·ро́с·ш·ий, -ая, -ее, (*past part act*) of **по·раст·и́**

порос·я́та (*nom pl*) of **порос·ёнок·**

пор·о́ть, ⸗ют, (*imp*) undo, unpick, unstitch; rip open (a seam); thrash, flog

поро·фо́р·, -а, (*m*) (plast) blowing agent

по́рох·, -а, (*nom pl*) **-а́,** (*g pl*) **-о́в,** (*m*) (expl) powder, (rocket) solid propellant

~, без·ды́м·н·ый (expl) cordite

~, бес·пла́мен·н·ый flashless powder/ propellant

~, быстро·гор·я́щ·ий quick-burning powder

~, ды́м·н·ый gunpowder, blackpowder

~, ле́нт·очн·ый (expl) strip grain powder

~, пласт·и́нчат·ый (expl) powder grain

~, призмат·и́ческ·ий (expl) prismatic powder

~, тру́б·очн·ый time-fuse powder

~, тру́б·чат·ый single-perforated powder, cord-grain powder

~, цилиндр·и́ческ·ий rod-/cord-grain powder

~, чёрн·ый gunpowder

порох·ов·о́й, -а́я, -о́е, (*a*) of **по́рох·**

~ за·ря́д· (*m*) (gunn) propellant charge

~ раке́т·н·ый дви́г·ател·ь (*m*) solid-fuel rocket engine

поро́ч·н·ый, -ая, -ое, (*a*) vicious, depraved; faulty, fallacious

поро́ш·ечн·ый, -ая, -ое, (*a*) (pharm) powder, in powder form

порош·и́нк·а, -и, (*g pl*) **-нок·,** (*f*) grain, speck

порош·кова́ни·е, -я, (*n*) pulverization, comminution, (chem etc.) trituration

порош·кова́т·ый, -ая, -ое, (*a*) powdery

порошко·ви́д·н·ый, -ая, -ое, (*a*) powder, powdered

порош·ко́в·ый, -ая, -ое, (*a*) powder; powdery, powdered

~ желе́з·о (*f*) ferrocart

~ металл·у́рг·и·я (*f*) powder metallurgy

~ му́фт·а (*f*) magnetic powder clutch

~ пред·о·хран·и́тел·ь (*m*) (elec) powder-filled fuse

порош·ко́в·ый, -ая, -ое

~ серд·е́чник· (*m*) (elec) powdered-iron core

~ те́л·о (*n*) (met) powder compact

~ фигу́р·а (*f*) (spectr) powder figure, powder pattern

порошко·гра́мм·а, -ы, (*f*) (spectr) powder pattern

порош·о́к·, -ш·к·а́, (*m*) powder

~, пресс·о́ванн·ый compacted powder

~, пресс·о́вочн·ый (plast) moulding/ molding powder

~, тек·у́ч·ий fluidized powder

порпеци́т·, -а, (*m*) (min) porpezite

Порро, призма́ (*f*) (opt) Porro prism

порт·, -а, (*nom pl*) **-ы,** (*g pl*) **-о́в,** (*m*) port, harbour; (shipb) port, porthole

~, вое́н·н·ый naval port

~, во́ль·н·ый free port

~, груз·ов·о́й (shipb) cargo/side port

~ за·хо́д·а port of call

порта́л·, -а, (*m*) (arch) portal; gantry (of crane)

порта́ль·н·ый, -ая, -ое, (*a*) of **порта́л·**

~ кран· (*m*) gantry crane

портати́в·н·ый, -ая, -ое, (*a*) portable

портве́йн·, -а, (*m*) port-type (fortified) wine

по́ртер·, -а, (*m*) porter (beverage)

~, кре́п·к·ий stout (beverage)

порт·иза́ци·я, -и, (*f*) port-winemaking, making (or faking) port wine

по́рт·ик·, -а, (*m*) (arch) portico; (shipb) porthole, port

по́рт·ить, ⸗ят, (*imp*) spoil, damage, harm, put out of order; **-ся** (*pass*); get damaged, be out of order; go bad, rot, deteriorate

портла́нд·цеме́нт, -а, (*m*) Portland cement

~, металл·и́ческ·ий iron cement

портн·о́й, -о́го, (*m decl as adj*) tailor; dressmaker

порт·ови́к·, -а́, (*m*) port employee

порт·о́в·ый, -ая, -ое, (*a*) port, harbour

~ рабо́ч·ий (*m*) docker

порто·фо́н·, -а, (*m*) portable radio-telephone

портре́т·, -а, (*m*) portrait

~, фа́з·ов·ый (autom) phase-plane diagram

портфе́л·ь, -я, (*m*) briefcase; portfolio

по·руб·и́ть, ⸗ят, (*perf*) chop, hack, hew slash, chop up/down по·ру́б·к·а, -и, (g pl) -б·ок·, (f) (v n) of по·руб·и́ть; clearing

по·ру́к·а, -и, (f) pledge, surety

по·рулён·н·ый, -ая, -ое, (a) reel, in reels

по-ру́с·ск·и (adv) in Russian, Russian

по·руч·а́·ть, -ют, (imp) entrust with, instruct/commission to; -ся (pass); stand as guarantor (for)

по·руч·е́нец·, -н·ц·а, (i) -н·ц·ем, (m) (mil) aide-de-camp

по·руч·е́ни·е, -я, (n) assignment, commission

по·ру́ч·енн·ый, -ая, -ое, (past part pass) see по·руч·а́·ть

по́·руч·ен·ь, -ч·н·я, (m) handrail, rail; (civ eng) capping beam

~, скул·ов·о́й (shipb) bilge rail

по·руч·и́тел·ь, -я, (m) guarantor

по·руч·и́тельств·о, -я, (n) guarantee

по·руч·и́ть, -ат, (perf) see по·ру·ч·а́·ть

порфи́р·, -а, (m) (geol) porphyry

порфир·и́н·, -а, (m) (biochem) porphyrin

порфир·и́т·, -а, (m) (geol) porphyrite

порфиро·бласт·и́ческ·ий, -ая, -ое, (a) (geol) porphyroblastic

порфиро·ви́д·н·ый, -ая, -ое, (a) porphyraceous

порфи́р·ов·ый, -ая, -ое, (a) porphyry; (geol) porphyritic

порх·а́·ть, -ют, (imp) flutter, flit

порци·о́нн·ый, -ая, -ое, (a) of по́р·ци·я; batch-output (of processing machines); (food) à la carte

по́рци·я, -и, (f) portion; batch

по́рч·а, -и, (i) -ей, (f) (v n) see по́р·т·ить; damage, harm

по́рч·енн·ый, -ая, -ое, (past part pass) see по́рт·ить

по́рч·ен·ый, -ая, -ое, (a) spoilt, damaged

порш·енёк·, -ньк·а, (m) (dim) of по́рш·ен·ь

по́рш·ен·ь, -ш·н·я, (m) piston, plunger

~, встречно·дви́ж·ущ·иеся (pl) opposed pistons

~, ди́ск·ов·ый slipper piston

~, за·кора́ч·иваюш·ий shorting plunger (waveguides)

~ на·стро́й·к·и (instr) tuning plunger

~, противо·положно·дви́ж·ущ·иеся (pl) opposed pistons

~ с полз·у́н·ом slipper piston

по́рш·ен·ь

~, ска́ль·чат·ый (elec instr) plunger

~, ступ·е́нчат·ый stepped/differential piston

~, тронк·о́в·ый trunk piston

~, уравно·ве́ш·иваюш·ий dummy balance piston (turbine)

порш·нева́ни·е, -я, (n) swabbing (oilwell exploitation)

порш·нев·о́й, -а́я, -о́е, (a) piston, plunger; reciprocating (of machines)

~ кольц·о́ (n) piston ring

~ па́л·ец· (m) gudgeon pin

~ па́л·ец·, пла́в·аюш·ий fully floated gudgeon pin

~ сте́рж·ен·ь (m) piston rod

по·ры́в·, -а, (m) (v n) of по·рыв·а́·ть

по·рыв·а́ни·е, -я, (n) (v n) of по·-рыв·а́·ть; endeavour(s)

по·рыв·а́·ть, -ют, (imp) break, break off, interrupt; -ся endeavour

по·ры́в·ист·ый, -ая, -ое, (a) gusty; rushing (sound); jerky; impetuous

по́р·ют, (pres 3rd pl) of пор·о́ть (imp)

по́р·юш·ий, -ая, -ее, (pres part act) of пор·о́ть

пор·я́, (pres gerund) of пор·о́ть

по·ря́д·ков·ый, -ая, -ое, (a) serial; (math) ordinal

~ но́мер· (m) serial number

~ числ·и́тельн·ое (n) ordinal, ordinal number

~ числ·о́ (n) ordinal number

по·ря́д·овк·а, -и, (g pl) -вок·, (f) bricklayer's gauge

по·ря́д·ок·, -д·к·а, (m) order; sequence; procedure (e.g. for starting · up); order of magnitude; (biol) order; (pl) usage, custom

~, алфави́т·н·ый alphabetical order

~, бли́ж·н·ий (phys) short-range order в по·ря́д·к·е in order, all right; (naut) under control

~ велич·ин·ы́ order of magnitude

~ в·ключ·е́ни·я (elec) switching sequence

~ воз·муш·е́ни·я (phys) order of perturbation

~ воз·раст·а́ни·я ascending order

~, втор·о́й second order; (in gen) second-order, secondary

~, вы́с·ш·ий higher order

~, да́ль·н·ий (phys) long-range order

~ де́й·стви·я order/sequence of operation

~ дн·я agenda; order of the day к по·ря́д·к·у (imper) order!

по·ря́д·ок

~, ма́рш·ев·о́й (mil) march order

моме́нт· вы́с·ш·его по·ря́д·к·а (m) (phys) multipole moment

на дв·а по·ря́д·к·а by two orders of magnitude

не в по·ря́д·к·е out of order, irregular

~, нул·ев·о́й (math) zeroth order

по по·ря́д·к·у in order, in succession, by rote

~, по·сле́д·ователь·н·ый consecutive order

~ пу́ск·а start-up procedure

~ рас·пре·дел·е́ни·я в·ним·а́ни·я (air) sequence of instrument coverage

~ реа́кци·и (chem) order of a reaction

~, тре́т·ий third order; (in gen) third-order, tertiary

числ·о́ с у·чёт·ом по·ря́д·к·ов (n) floating number (computers)

~, ша́хтмат·н·ый staggered order

по·ря́д·очн·ый, -ая, -ое, (a) considerable, sizable, respectable, decent

по·сад·и́ть, ⸗ят, (perf) see **сад·и́ть, саж·а́ть**

по·са́д·к·а, -и, (g pl) **-д·ок·,** (f) (a/c) landing; (agr) planting; boarding (trains, ships etc.), embarkation; (mech) fit; charging (a furnace); (min) artificial caving; seat, carriage (attitude); dip, drop

~, вы́·нужд·енн·ая (a/c) forced landing

~, глух·а́я force fit

~ движ·е́ни·я slide fit

~, легко·ход·ов·а́я free/free-running fit

~ на·пряж·е́ни·я (elec) voltage dip

~, на·пряж·ённ·ая push fit

~, не·по·дви́ж·н·ая close/interference fit

~ оборо́т·ов drop of revs /revolutions

~, парашют·и́рующ·ая (a/c) pancake landing

~, пере·хо́д·н·ая transition fit

~, плот·н·ая drive/tunking/wringing fit

~, пре́сс·ов·ая heavy-drive/press fit

~ по при·бо́р·ам (a/c) instrument landing

~, по·дви́ж·н·ая running/clearance fit

~, пре́сс·ов·ая shrink/interference fit

~ с прям·о́й (a/c) straight-in approach

~, свобо́д·н·ая easy/loose fit

по·са́д·к·а

~, скольз·я́щ·ая sliding/snug fit

~, слеп·а́я (a/c) blind approach

систе́м·а слеп·о́й по·са́д·к·и само·лёт·ов (f) (a/c) blind approach system

тип· по·са́д·к·и (m) class of fit

~, трёх·то́ч·ечн·ая (a/c) three-point landing

~, туг·а́я tight fit

~, ход·ов·а́я medium fit

~, широко·ход·ов·а́я loose/loose-running/coarse-clearance fit

~, экстре́н·н·ая (a/c) forced landing

по·са́д·очн·ый, -ая, -ое, (a) of **по·-са́д·к·а**

~ маши́н·а (f) charging machine (for furnaces); (agr) planting machine

~ площа́д·к·а (f) airfield, landing ground

~ фа́р·а (f) (a/c) landing light

~ щит·о́к· (m) (a/c) landing flap

по·са́ж·енн·ый, -ая, -ое, (past part pass) of **по·сад·и́ть**

по·свеж·е́·ть, -ют, (perf) freshen, freshen up, become cooler

по·свет·и́ть, ⸗ят, (perf) shine, light (for a time)

по·светл·е́·ть, -ют, (perf) become light/lighter

по·свист·а́ть, (fut 3rd pl) **по·сви́щ·ут,** (perf) whistle, give a whistle

по-сво́·ему (adv) in one's own way

по·свят·и́ть, -я́т, (perf) see **по·свя-щ·а́ть**

по·свящ·а́·ть, -ют, (imp) (+ dat) devote, dedicate; initiate/let into

по·се́в·, -а, (m) sowing; crop; (microbiol) inoculation, planting

по·сев·н·о́й, -а́я, -о́е, (a) sowing

по·сед·е́л·ый, -ая, -ое, (past part pass) of **по·сед·е́·ть;** grey, gray

по·сед·е́·ть, -ют, (perf) go/become grey (of hair)

по·се́·ет, (fut 3rd sing) of **по·се́·ять**

по·сел·е́нец·, -нц·а, (i) **-нц·ем,** (m) settler

по·сел·е́ни·е, -я, (n) settlement, colony

по·сел·и́ть, -я́т, (perf) see **по·сел·я́·ть**

по·сёл·ок, -л·к·а, (m) settlement, estate, camp

по·сел·я́·ть, -ют, (imp) settle, accommodate; engender, inspire in

по·серебр·е́ни·е, -я, (n) silver plating

по·серебр·ённ·ый, -ая, -ое, (past part pass) of **по·серебр·и́ть;** silver-plated

по·серед·и́ (*adv*) *see* **по·сред·и́**

по·серед·и́н·е (*adv*) in the middle, half way, half way along

по·сер·е́·ть, -ют, (*perf*) turn/grow/become grey/greyer

посет·и́тел·ь, -я, (*m*) visitor

посет·и́ть, -я́т, (*perf*) visit, call

посещ·а́емост·ь, -и, (*f*) attendance

посещ·а́·ть, -ют, (*imp*) attend (lectures); frequent, pay frequent visits

посещ·е́ни·е, -я, (*n*) (*v n*) *see* **посещ·а́·ть**; visit; attendance

по·се́·ять, -ют, (*perf*) sow

по·сид·е́ть, -я́т, (*perf*) sit (for a time)

по·сиж·у́, (*fut 1st sing*) of **по·сид·е́ть**

по·си́ль·н·ый, -ая, -ое, (*a*) appropriate, feasible, comensurate with the power/strength (available)

по·син·е́·ть, -ют, (*perf*) turn/become blue/dark blue

по·скак·а́ть, (*fut 3rd pl*) **по·ска́ч·ут,** (*perf*) hop; gallop

по·скольз·н·у́ться, -у́тся, (*perf*) slip, slip up

по·ско́льк·у (*conj*) so/as far as; so long as, since

поско́н·ник·, -а, (*m*) (bot) Eupatorium

~, конопле·ви́д·н·ый (bot) hemp agrimony, *Eupatorium cannabinum*

по·скор·е́е (*comp*) rather/somewhat quicker/faster

по·ско́т·ин·а, -ы, (*f*) (agr) pasture, keep

по·скрёб·к·и, -ов, (*m pl*) scrapings, scraps

по·сл·а́ни·е, -я, (*n*) message

по·сл·а́нник·, -а, (*m*) envoy, minister

по́·сл·анн·ый, -ая, -ое, (*past part pass*) of **по·сл·а́ть**

по·сл·а́ть (*fut 3rd pl*) **по·шл·ю́т,** (*perf*) send, dispatch

по́сле (*prep + gen*) after, since; (*adv*) afterwards, later on; (*prefix*) post-, after-

~ того́ как after

по·сле́д·, -а, (*m*) (zool) placenta, afterbirth

после·де́й·стви·е, -я, (*n*) after-effect; reaction

~, магни́т·н·ое magnetic hysteresis

~, у·пру́г·ое (met) elastic after-effect

по·след·и́ть, -я́т, (*perf*) *see* **след·и́ть**

по·сле́д·н·ий, -яя, -ее, (*a*) last; latter; latest, new, recent; lowest, worst (of quality); (*n decl as adj*) the latter

в по·сле́д·н·ее вре́м·я recently

до по·сле́д·н·его време́н·и until recently

по·сле́д·ователь, -я, (*m*) follower

по·сле́д·овательн·о (*adv*) in succession, one after another, in turn; (elec) in series

по·сле́д·овательн·о-воз·вра́т·н·ый, -ая, -ое, (*a*) (met roll) cross-country

по·сле́д·овательн·о-параллéль·н·ый, -ая, -ое, (*a*) (elec) series-parallel

по·сле́д·овательност·ь, -и, (*f*) succession, sequence, order; consistence, continuity; (cryst) stacking sequence, sequence, configuration

~, вес·ов·а́я (math) weighting sequence

~ волн· (phys) wave train

~, и́мпульс·н·ая (elec) pulse train

~ спи́н·ов (nucl) spin sequence

~, числ·ов·а́я number sequence

по·сле́д·овательн·ый, -ая, -ое, (*a*) successive, consecutive, serial, sequential; consistent, consequent; (elec) series, series-connected, in series

~ вы́·бор·к·а (*f*) (math) sequential sampling

~ вы́·бор·к·а кома́нд· control sequence (computers)

~ де́й·стви·е (*n*) serial/series action; (*gen as adj*) serial, successive

ме́тод· по·сле́д·овательн·ых при·бли́ж·éни·й (*m*) (math) iteration method

~ при·бли́ж·éни·я (*pl*) (math) successive approximations

~ со·един·éни·е (*n*) (elec) series connection

по·сле́д·овать, -уют, (*perf*) follow

по·сле́д·ов·ые, -ых, (*pl as adj*) (zool) *Eutheria, Placentaria*

по·сле́д·стви·е, -я, (*n*) consequence, effect, corollary

~ об·луч·éни·я effect/effects of radiation

по·сле́д·ующ·ий, -ая, -ее, (*pres part act*) of **по·сле́д·овать**; following, subsequent, next

после·за́втра (*adv*) the day after tomorrow

после·за·ро́д·ышев·ое раз·ви́т·и·е (*n*) (biol) postfertilization stages

после·и́мпульс·, -а, (*m*) (phys) afterpulse

после·лед·нико́в·ая эпо́х·а (*f*) (geol) post-Glacial

после·операцио́н·н·ый, -ая, -ое, (*a*) (med) postoperative

после·по·лёт·н·ый, -ая, -ое, (*a*) (air) after-flight

по·сле·род·ов·о́й, -а́я, -о́е,   *(a)* (zool) post-natal

~ пери́од· *(m)* (med) puerperium

после·свеч·е́ни·е, -я,   *(n)* (phys) after-glow, persistence

после·тепл·от·а́, -ы́,   *(f)* afterheat

после·трет·и́чн·ый, -ая, -ое, *(a)* (geol) post-Tertiary

после·у·сил·и́тел·ь, -я,   *(m)* post-amplifier

после·у·скор·я́ющ·ий, -ая, -ее,   *(a)* after-/post-acceleration

после·фокус·иро́вк·а, -и,   *(f)* secondary focussing

по·сло́в·иц·а, -ы,   *(i)* **-ей,** *(f)* proverb, saying

по·сло́в·н·ый, -ая, -ое,   *(a)* literal, word-for-word, verbatim

по·сло́й·н·о   *(adv)* layer-by-layer, in layers, layerwise

по·служ·и́ть, ⌐ат, *(perf)* *see* **служ·и́ть**

по·слуш·а́·ть, -ют,   *(perf)* *see* **слу-ш·а́·ть**

по·слу́ш·н·ый, -ая, -ое, *(a)* obedient, dutiful; manageable, responsive (of mechanisms)

по·сма́тр·ива·ть, -ют   *(imp)* keep looking at (at frequent intervals); keep an eye on

по·с·ме́рт·н·ый, -ая, -ое,   *(a)* post-humous

по·сме·я́ться, -ю́тся, *(perf)* laugh

по·смотр·е́ть, ⌐ят, *(perf)* *see* **смо-тр·е́ть**

по·со́б·и·е, -я,   *(n)* grant, allowance, benefit, relief; textbook, manual, guide

по·со́б·ник·, -а, *(m)* accomplice

по·со́б·н·ый, -ая, -ое, *(a)* subsidiary, auxiliary

по·со·ве́т·овать, -уют, *(perf)* *see* **со·ве́-т·овать**

по·со́л·, -а, *(m)* *(v n)* *see* **по·сол·и́ть;** (dipl) ambassador

~, мо́кр·ый   brine salting

~, тузлу́ч·н·ый   brine salting

по·сол·и́ть, -я́т, *(perf)* salt, pickle

по·со́ль·ств·о, -а, *(n)* embassy

по·со́т·енн·о   *(adv)* in hundreds, by the hundred

по·со́х·нуть, -нут, *(perf)* dry up, wither

по·сп·а́ть, -я́т, *(perf)* have a sleep

по·спев·а́·ть, -ют,   *(imp)* ripen; have time, be in time

по·спе́·ть, -ют, *(perf)* *see* **по·спев·а́·ть**

по·спеш·и́ть, -а́т, *(perf)* *see* **спеш·и́ть**

по·спе́ш·н·ый, -ая, -ое, *(a)* hurried, hasty

по·спосо́б·ств·овать, -уют,   *(perf)* *see* **спосо́б·ств·овать**

по·срамл·я́·ть, -ют,   *(imp)* disgrace

по·сред·и́ *(adv)* in the middle (of), among

по·сред·и́н·е *(adv)* = **по·серед·и́н·е**

по·сре́д·ник·, -а,   *(m)* mediator, inter-mediary, negotiator; (mil) umpire

по·сре́д·ническ·ий, -ая, -ое,   *(a)* of **по·сре́д·ник·**

~ опера́ц·и·и   *(pl)* negotiations

по·сре́д·ственн·ый, -ая, -ое,   *(a)* mediocre, fair; satisfactory (as marking)

по·сре́д·ств·ом   *(prep + gen)* by means of, by

пост·, -а́,   *(m)* *(and see under associated adjs.)*; (1) post, position, station, pulpit; (2) fast, abstention from meat

~, бо·ев·о́й   action station

~ пе́рв·ой по́·мощ·и   first aid post

~ у·правл·е́ни·я   control room/position/pulpit

по·ста́·в·, -а, *(g pl)* **-о́в,** *(m)* posture, attitude, carriage; (agr) mill

~, шелуш·и́льн·ый   sheller, husker

по·ста́в·ить, -ят,   *(perf)* *see* **ста́в·ить, по·ставл·я́·ть**

по·ста́в·к·а, -и, *(g pl)* **-в·ок·,** *(f)* delivery, supplying; consignment, supply

по·ста́вл·енн·ый, -ая, -ое, *(past part pass)* of **по·ста́в·ить,** *(perf)* *see* **по·-ставл·я́·ть** *(imp)*; posed, set, formulated

по·ставл·я́·ть, -ют,   *(imp)* supply, furnish

по·став·щи́к·, -а́, *(m)* supplier, caterer, outfitter, chandler

постаме́нт·, -а,   *(m)* pedestal, base, shoe

по·стана́вл·ива·ть, -ют, *(imp)* = **по·-становл·я́·ть**

по·станов·и́ть, ⌐ят, *(perf)* *see* **по·ста-новл·я́·ть**

по·стано́в·к·а, -и,   *(g pl)* **-в·ок·,** *(f)* putting, placing, positioning; arrangement, organization; (civ eng) erection, raising; setting, setting up, setup; posing, putting (a question); stating, statement, formulation; (theat) staging, production

~ за·да́·ч·и   statement of problem

~, корабл·я́ на я́кор·ь   (naut) anchoring

~ мин·   (naut) minelaying

~ эксперим́ент·а   experimental setup; setting up/mounting an experiment по·станов·лéни·е, -я,   (*n*) decree, resolution

по·станов·ля·ть, -ют,   (*imp*) decree, decide, resolve, enact

по·стар·á·ться, -ются,   (*perf*) try, endeavour

по·стар·é·ть, -ют,   (*perf*) become old/older

пост·ганглион·áрн·ый, -ая, -ое,   (*a*) (anat) postganglionic

~ волокн·á   (*pl*) (zool) postganglionic fibre

по·стéл·ет,   (*fut 3rd sing*) по·стл·áть (*perf*); *see* по·стил·á·ть

по·стéл·ист·ый плит·нáк·   (*m*) self-faced flagstone

по·стéл·ь, -и,   (*f*) bed; (geol) substratum, floor; hearth layer/bed (of furnace)

по·степéн·н·о   (*adv*) gradually

по·степéн·н·ый, -ая, -ое,   (*a*) gradual

по·стиг·á·ть, -ют,   (*imp*) overtake, befall, strike (of misfortune etc.); comprehend, perceive

по·стúг·нуть, -нут,   (*perf*) *see* по·стиг·á·ть

по·стил·á·ть, -ют,   (*imp*) spread, lay out

по·стúч·ь,   (*fut 3rd pl*) по·стúг·нут, (*perf*) *see* по·стиг·á·ть

по·стл·áть,   (*fut 3rd pl*) по·стéл·ют, (*perf*) *see* по·стил·á·ть

пóст·н·ый, -ая, -ое,   (*a*) of пост· (2); (food) vegetable

~ мáсл·о   (*n*) vegetable oil

по·столб·цóв·ый, -ая, -ое,   (*a*) column, up the column, for the column

по·стóльку   (*conj*) in so far as, inasmuch as, as

по·сторон·úться, ‑ятся,   (*perf*) move/draw to one side

по·сторóн·н·ий, -яя, -ее,   (*a*) foreign, strange, extraneous, outside, alien; ectogenic, exotic, exogenous; (*m decl as adj*) unauthorized person, outsider

~ в·ключ·éни·е   (*n*) (met) exogenous inclusion

по·сто·я́нн·ая, -ой,   (*f decl as adj*) (*for named constants see under names*); (math) constant

~ врéмен·и   time constant

~ врéмен·и изодрóм·а   integral time constant

~ врéмен·и, мáл·ая   fast-time constant

~, гáз·ов·ая   gas constant

~ на·вед·éни·я   navigation constant

~ раз·реш·éни·я врéмен·и   time-resolution constant

по·сто·я́нн·ая

~ рас·пáд·а   (nucl) decay constant

~ решёт·к·и   (cryst) lattice parameter/constant

по·сто·я́нн·ый, -ая, -ое,   (*a*) *see also* по·сто·я́нн·ая (*as noun*); constant, invariable, permanent

~ велич·ин·á   (*f*) (math) constant

~ во·врéмен·и   time-independent

~ крут·изн·á   (*f*) uniform twist

~ магнúт·   (*m*) permanent magnet

~ на·грýз·к·а   (*f*) dead load

~ с·вяз·ь   (*f*) fixed communications схéм·а по·сто·я́нн·ого тóк·а (*f*) (telecom) closed-circuit system

~ ток·   (*m*) (elec) direct current, d.c., (*gen as a*) d.c.

по·сто·я́нств·о, -а,   (*n*) constancy, permanence, persistence, continuity за·кóн· по·сто·я́нств·а вес·ов·ы́х от·нош·éни·й (*m*) law of constant/reciprocal proportions за·кóн· по·сто·я́нств·а со·стáв·ов (*m*) law of constant/reciprocal proportions

~ мáсс·ов·óго рас·хóд·а   mass flow continuity

~ объ·ём·а   soundness (of cement etc.)

по·сто·я́ть, -ят,   (*perf*) stand (for a short time); up/out for, make a stand

пост·плиоцéн·, -а,   (*m*) (geol) post-Pliocene

по·страд·áвш·ий, -его,   (*m decl as adj*) victim

по·страд·á·ть, -ют,   (*perf*) suffer

по·стран·úчн·ый, -ая, -ое,   (*a*) page (by the page)

по·стрéл·ива·ть, -ют,   (*imp*) shoot, fire (from time to time)

по·стрéл·к·а, -и,   (*g pl*) -л·ок·, (*f*) sporadic firing

по·стрел·я́·ть, -ют,   (*perf*) shoot, fire (for a time)

по·стро·éни·е, -я,   (*n*) (*v n*) *see* стро··úть; construction, building; structure, formation; drawing, plotting, designing

~ калúбр·ов   (met roll) pass designing/drawing

~ крив·ы́х   plotting (graphs)

~ по·тóк·а   (elec) flux plot

по·стрó·ечн·ый, -ая, -ое,   (*a*) building, construction

по·стро·úтел·ь, -я,   (*m*) plotter, plotting device

~, сúнус·н·о-ко·сúнус·н·ый   sine-cosine resolver

по·стро·и́тельн·ый, -ая, -ое,   (*a*) building, constructive, plotting

~ вращ·а́ющ·ийся трансформ·а́тор· (*m*) (autom) plotting resolver

по·стро·и́ть, -́ят, (*perf*) *see* стро·и́ть

по·стро́й·к·а, -и, (*g pl*) -о́·ек·, (*f*) building, construction; building/construction site

по·стро́ч·н·о (*adv*) line-a-time, line-by-line

по·стро́ч·н·ый, -ая, -ое, (*a*) row, line, line-by-line, row-by-row, along-the-line/row

~ раз·лож·е́ни·е (*n*) line scanning (radar)

постскри́птум·, -а, (*m*) postscript, P.S.

по·сту́к·ивани·е, -я, (*n*) tapping

по·сту́к·ива·ть, -ют, (*imp*) tap, tap occasionally

постула́т·, -а, (*m*) (math) postulate

постул·и́р·овать, -уют, (*imp and perf*) postulate

по·ступ·а́тельн·о-воз·вра́т·н·ый, -ая, -ое, (*a*) reciprocating

~ движ·е́ни·е (*n*) reciprocating motion

по·ступ·а́тельн·ый, -ая, -ое, (*a*) progressive, forward-moving; translational, translatory

~ движ·е́ни·е (*n*) forward/progressive movement, progression; (phys) translational/translatory motion, translation

по·ступ·а́·ть, -ют, (*imp*) behave, act; join, start work at/with; be brought/delivered to; reach, enter, pass to/into/along

по·ступ·а́ющ·ий, -ая; -ее, (*pres part act*) of по·ступ·а́·ть; behaving; incoming, entering, supply, supplied

~ во́з·дух· (*m*) supply air

по·ступ·и́ть, -́ят, (*perf*) *see* по·ступ·а́·ть

по·ступл·е́ни·е, -я, (*n*) entrance, admission, joining; receipts (of money); entry (book-keeping); arrival, intake (of air), inflow (of fluid)

~ и́мпульс·ов pulse arrival

ско́р·ост·ь по·ступл·е́ни·я (*f*) rate of arrival (pulse counting)

по·сту́п·ок·, -п·к·а, (*m*) act, deed, action

по́·ступ·ь, -и, (*f*) step, tread

~ винт·а́ advance per revolution of propeller

~ винт·а́, от·нос·и́тельн·ая propeller modulus

по·стуч·а́ть, -а́т, (*perf*) knock, rap

по·су́д·а, -ы, (*f*) ware, dishes; (chem) laboratory ware

по·су́д·н·ый, -ая, -ое, (*a*) of по·су́д·а

посудо·мо́·ечн·ая машн́н·а (*f*) dish-washing machine

по·су́т·очн·ый, -ая, -ое, (*a*) 24-hour, daily (by the day)

по·с·чит·а́ть, -ют, (*perf*) count

по·сыл·а́·ть, -ют, (*imp*) send, dispatch

по·сы́л·к·а, -и, (*g pl*) -л·ок·, (*f*) (*v n*) *see* по·сыл·а́·ть; parcel; premise; (elec) impulse

~ -бандеро́л·ь, -и, (*f*) small packet (international postal classification)

~ вы́·зов·а

сигна́л· по·сы́л·к·и вы́·зов·а (*m*) ringing tone (telephones)

~ и́мпульс·ов pulsing

част·от·а́ по·сы́л·ок· (*f*) train/repetition rate/frequency

частот·а́ рабо́ч·ей по·сы́л·к·и (telecom) marking frequency

по·сы́ль·н·ый, -ая, -ое, (*a*) dispatch; (*m decl as adj*) messenger; (mil) orderly

~ ка́тер· (*m*) despatch boat

по·сы́п·ать, -лют, (*perf*) *see* по·сы·п·а́·ть

по·сы́п·а́·ть, -ют, (*imp*) pour, tip (of small solids); strew, sprinkle

по́т·, -а, (*nom pl*) -ы́, (*g pl*) -о́в, (*m*) sweat, perspiration, sudor

~, шёрст·н·ый suint

пот·азо́т·, -а, (*m*) (agr) potash fertilizer

по·та́·й, -я, (*m*) countersink (for head of screw or bolt)

по·та́й·ник·, -а, (*m*) sunken rock

по·тай·н·о́й, -а́я, -о́е, (*a*) secret, hidden; sunk, countersunk, flush

~ голо́в·к·а (*f*) flat countersunk head (rivets, screws etc.)

потамо·ге́н·н·ый, -ая, -ое, (*a*) (geol) potamogenic

потамо·ло́г·и·я (*f*) (hydr) potamology

потамо·планкто́н·, -а, (*m*) (biol) potamoplankton

по·та́пл·ива·ть, -ют, (*imp*) heat a little; heat up intermittently

пота́ш·, -а, (*i*) -о́м, (*m*) (chem) potash

по·тек·у́т, (*fut 3rd pl*) of по·те́ч·ь

по·темн·е́ни·е, -я, (*n*) darkening

коэффицие́нт· по·темн·е́ни·я (*m*) darkening coefficient

по·темн·é·ть, -ют, (*perf*) become dark/darker

пот·éн·и·е, -я, (*n*) sweating, perspiring

потéнт·ност·ь, -и, (*f*) potency

потенци·áл·, -а, (*m*) potential; voltage

~, бутылко·обрáз·н·ый (phys) wine-bottle potential

~ дальне·дéй·ствующ·их сил· long-range potential

~, дей·ствúтельн·ый real potential

~ дéй·стви·я action potential

~, за·пáзд·ывающ·ий (rad) retarded potential

~, кинет·úческ·ий (mech) kinetic potential, Lagrangian function

~ мúшен·и (TV) target voltage

~ на·сыщ·éни·я saturation voltage

~, о·кисл·úтельн·ый (chem) oxidation potential

~, прямо·угóль·н·ый (nucl) square-well potential

~ раз·дел·éни·я (phys) separation potential

~, стандáрт·н·ый standard electrode potential

потенциалó·метр·, магнúт·н·ый (*m*) magnetic potentiometer

потенциало·скóп·, -а, (*m*) (elec) storage tube, electrostatic memory tube

потенциáль·н·ый, -ая, -ое, (*a*) potential, irrotational (flow etc.)

потенциó·метр·, -а, (*m*) (elec) potentiometer

~, автомат·úческ·ий self-balancing potentiometer

~, коменс·úрующ·ий balance-type potentiometer

~, реохóрд·н·ый slide-wire potentiometer

~ с·мещ·éни·я bias bleeder

~, струн·н·ый slide-wire potentiometer

потенцио·метр·úческ·ий, -ая, -ое, (*a*) potentiometric

~ систéм·а (*f*) (autom) null-seeking system

потенц·úровани·е, -я, (*n*) log-to-linearization

потенц·úрующ·ий, -ая, -ее, (*a*) logarithmic-to-linear, antilogarithmic (computers)

потéнци·я, -и, (*f*) potential

по·тепл·é·ть, -ют, (*perf*) become warm/warmer

по·тер·é·ть, (*fut 3rd pl*) **по·тр·ýт,** (*perf*) rub

по·тéр·и (*nom pl*) *see* **по·тéр·я**

по·терп·éвш·ий, -ая, -ое, (*past part act*) of **по·терп·éть;** (*m decl as adj*) victim

по·терп·éть, -ят, (*perf*) endure; suffer, undergo

по·тёр·т·ый, -ая, -ое, (*past part pass*) of **по·тер·éть;** (text) worn, shabby

по·тёр·ш·ий, -ая, -ее, (*past part act*) of **по·тер·éть**

по·тéр·я, -и, (*f*) loss; (*pl*) losses, loss

~ в плáзм·е (*pl*) (phys) plasma loss

~ в серд·éчник·е (*pl*) (elec) core loss

~ в·слéд·стви·я рас·сé·яни·я (phys) scattering loss

~ давл·éни·я loss of pressure, drop in pressure

~, до·бáв·очн·ые (*pl*) (elec) stray losses

~, кáж·ущ·иеся (*pl*) apparent losses

~, кон·цев·áя (phys) end loss

коэффициéнт· по·тéр·ь (*m*) loss factor

~ на вентиляц·и·ю blade ventilation/windage loss (of steam turbine)

~ на из·луч·éни·е loss by radiation, radiation loss

~ на тóк·и фýко (elec) eddy current loss

~ на·пóр·а (hydr eng) loss of head, drop in pressure

~ от по·глощ·éни·я (*pl*) absorption loss

~, от·нос·úтельн·ая fractional loss

~, паразúт·н·ая (elec) parasitic loss

~ при пере·гóн·к·е (*pl*) (chem) distillation loss

ýгол· по·тéр·ь (*m*) (elec) loss angle

ýгол· по·тéр·и скóр·ост·и (*m*) (air) stall angle

~, у·дéль·н·ая specific loss; (elec) core loss

по·тер·я́·й, -я, (*m*) (shipb) stealer plate

по·тéр·янн·ый, -ая, -ое, (*past part pass*) of **по·тер·я́·ть;** lost, wasted

по·тер·я́·ть, -ют, (*perf*) lose

по·тес·áть, (*fut 3rd pl*) **по·тéш·ут,** (*perf*) adze, trim

по·тесн·úть, -я́т, (*perf*) restrict, confine, cramp; **-ся** make room for, move over

пот·é·ть, -ют, (*imp*) sweat, perspire

по·те́ч·ь, (*fut 3rd sing, pl*) **по·теч·ёт, по·тек·у́т,** (*past masc sing*) **по·тёк·,** (*perf*) flow

по·те́ш·ет, (*fut 3rd sing*) of **по·те·с·а́ть**

по·тир·а́·ть, -ют, (*imp*) rub

по́т·н·ый, -ая, -ое, (*a*) sweaty, perspiring

пот·ов·о́й, -а́я, -о́е, (*a*) sweaty, (anat) sudoral, sudoriferous

пото·го́н·н·ый, -ая, -ое, (*a*) sweat-inducing, sudorific, diaphoretic

по·то́к·, -а, (*m*) stream, flow, current, flux

~, **аксиа́ль·н·ый** axial flux

~, **бок·ов·о́й** side stream

~, **бу́р·н·ый** (dynam) turbulent flow; (geog) torrent

~ **ве́ктор·а** vector flux, flux of a vector

~, **вес·ов·о́й** weight flow

~, **вис·я́ч·ий** (geog) perched stream

~ **во́з·дух·а** air flow/stream/current

~, **воз·мущ·ённ·ый** disturbed/bad flow

~, **встре́ч·н·ый** head-on flow

~, **в·тек·а́ющ·ий** inward flow

~, **вя́з·к·ий** (dynam) frictional flow

~ **га́з·а** gas flow/stream

~ **жи́дк·ост·и** fluid flow

~, **за·ви́с·им·ый** (autom) secondary flow

~, **за·вихр·ённ·ый** eddying/bumpy flow

~ **из·луч·е́ни·я** radiation flow

~ **ио́н·ов** ion beam

~, **кра·ев·о́й** edge flux

~, **ламина́р·н·ый** laminar flow

~, **лив·нев·о́й** (geog) torrent

ли́н·и·я по·то́к·а (*f*) (dynam) stream/flow line; production line

~, **луч·ев·о́й** radiant flux

~, **луч·и́ст·ый** radiant flux

~, **магни́т·н·ый** magnetic flux

~ **ма́л·ой глубин·ы́** (dynam) thin-sheet flow

~, **молекуля́р·н·ый** molecular beam/flow

~ **мо́щ·ност·и** power flux

~, **на·бег·а́ющ·ий** incoming/forward flow, free-stream flow

~, **не·воз·мущённ·ый** (dynam) free stream

ско́р·ост·ь не·воз·мущён·н·ого по·то́к·а (*f*) free-stream velocity

~, **не·вя́з·к·ий** inviscid flow

по·то́к

~, **не·за·ви́с·им·ый** (autom) reference flow

~ **нейтро́н·ов** neutron flux/current

~, **не·пре·ры́в·н·ый** (dynam) continuum flow

~, **не·равно·ме́р·н·ый** pulsating flow

~, **низ·ход·я́щ·ий** (dynam) incident flow

~, **нул·ев·о́й** (elec) zero flux

~, **об·ме́н·н·ый** (nucl) exchange current

~, **обра́т·н·ый** reflux

~, **об·тек·а́ющ·ий** washing flow

~, **объ·ём·н·ый** volumetric flow

~, **ос·ев·о́й** axial flow

~, **основ·н·о́й** reference flow

~, **от·клон·ённ·ый** deviated flow

~, **по·вёр·нут·ый** deviated flow

~, **про·стра́н·ственн·ый** three-dimensional flow

~, **рас·сеч·ённ·ый** split flow

~ **рас·се́·яни·я** (phys) stray/dispersion flux

~, **свет·ов·о́й** (phys) light flux

~, **скольз·я́щ·ий** (air) slip flow

~, **со́·рв·анн·ый** stalled flow

~, **с·поко́й·н·ый** laminar flow

~, **с·пу́т·н·ый** (air) slipstream

~, **с·у́ж·енн·ый** reduced flow

~, **тепл·ов·о́й** heat flow/flux

~, **технолог·и́ческ·ий** production line

~, **электр·и́ческ·ий** electric flux/stream

потоко·с·цепл·е́ни·е, -я, (*n*) (elec) flux linkage

потолко·у·сту́п·н·ый, -ая, -ое, (*a*) overhead, overhand

~ **систе́м·а** (*f*) (min) overhand stoping

потол·о́к·, -л·к·а́, (*m*) ceiling, roof (of mine etc.), crown (of furnace)

~, **баллист·и́ческ·ий** (air) energy height

~, **динам·и́ческ·ий** (air) full-throttle height

~, **практ·и́ческ·ий** (air) service ceiling

~, **стат·и́ческ·ий** (air) hovering ceiling

~, **теорет·и́ческ·ий** (air) absolute ceiling

~, **эксплуатацио́н·н·ый** service ceiling

потол·о́чн·ый, -ая, -ое, (*a*) of **по·тол·о́к·**

~ **кана́л·** (*m*) (elec) ceiling duct

~ **с·ва́р·к·а** (*f*) overhead welding

по·то́м (*adv*) afterwards, later on, then, and then; (*conj*) moreover, besides

пото́·метр·, **-а**, (*m*) (bot instr) potometer

по·то́м·ок·, **-м·к·а**, (*m*) descendant

по·то́м·ственн·ый, **-ая**, **-ое**, (*a*) hereditary

по·то́м·ств·о, **-а**, (*n*) posterity, progeny, descendants

по·тому́ (*adv*) for this reason, consequently, therefore

~ **что** (*conj*) because, for, as

по·тон·у́ть, **-ут**, (*perf*) (*intrans*) sink, submerge, go down; drown

по·то́п·, **-а**, (*m*) (geog) deluge, flood, inundation

по·топ·и́ть, **-ят**, (*perf*) *see* **по·та́пл·ива·ть** *and* **по·топл·я́·ть**

по·топл·е́ни·е, **-я**, (*n*) sinking, drowning; submersion, immersion

по·то́пл·енн·ый, **-ая**, **-ое**, (*past part pass*) of **по·топ·и́ть**, (*perf*); *see* **по·топл·я́·ть** (*imp*)

по·топл·я́·ть, **-ют**, (*imp*) sink, immerse, submerge; drown

по·топт·а́ть, (*fut 3rd pl*) **по·то́пч·ут**, (*perf*) trample, trample/tread down

по·тороп·и́ть, **-ят**, (*perf*) *see* **торо·п·и́ть**

потору́ (*m indecl*) (zool) potoroo, rat kangaroo, (*pl*) *Potoroinae*

по·то́ч·ечн·о (*adv*) pointwise, point-by-point

по·то́ч·ечн·ый, **-ая**, **-ое**, (*a*) point

по·точ·и́ть, **-ат**, (*perf*) sharpen, point (a little)

по·точно·конве́йер·н·ое про·из·во́д·ств·о (*n*) line production

поточно·ма́сс·ов·ое про·из·во́д·ств·о (*n*) large-lot production

по·то́ч·ност·ь, **-и**, (*f*) state of flow-line production; suitability for flow-line production

по·то́ч·н·ый, **-ая**, **-ое**, (*a*) of **по·то́к·**; continuous-flow, continuous; in-line; production-line

~ **ли́н·и·я** (*f*) flow/production line

ли́н·и·я по·то́ч·н·ой об·рабо́т·к·и (*f*) processing line

~ **об·рабо́т·к·а** (*f*) line processing

~ **про·из·во́д·ств·о** (*n*) line production, flow-line production

~ **с·мещ·е́ни·е** (*n*) (oil) in-line blending

по·трав·и́ть, **-ят**, (*perf*) poison, exterminate (pests etc.); (naut) pay out a little, veer a little

по·тра́т·ить, **-ят**, (*perf*) spend, expend, use up

по·тра́ч·енн·ый, **-ая**, **-ое**, (*past part pass*) of **по·тра́т·ить**

по·треб·и́тел·ь, **-я**, (*m*) consumer

по·треб·и́тельск·ий, **-ая**, **-ое**, (*a*) consumer's, consumers', consumer

~ **о́бщ·еств·о** (*n*) (com) consumers' cooperative society, cooperative society

по·треб·и́ть, **-ят**, (*perf*) *see* **по·требл·я́·ть**

по·требл·е́ни·е, **-я**, (*n*) consumption

по·требл·я́ем·ый, **-ая**, **-ое**, (*pres part pass*) of **по·требл·я́·ть**; consumable

по·требл·я́·ть, **-ют**, (*imp*) consume, use up

по·тре́б·ност·ь, **-и**, (*f*) requirements, need, demand

по·тре́б·н·ый, **-ая**, **-ое**, (*a*) required, necessary

~ **коли́чест·во** (*n*) required amount, requirement

~ **мо́щ·ност·ь** (*f*) (elec) demand

по·тре́б·овать, **-уют**, (*perf*) demand; need, require, call for

по·треп·а́ть, **-лют**, (*perf*) *see* **треп·а́ть**

по·тре́ск·а·ться, **-ются**, (*perf*) be covered with cracks

по·тре́ск·ивани·е, **-я**, (*n*) crackling

по·тре́ск·ива·ть, **-ют**, (*imp*) crackle

по·тр·ёт, (*fut 3rd sing*) of **по·тер·е́ть**

по·тро́г·а·ть, **-ют**, (*perf*) touch, finger, keep touching/feeling

потрох·а́, **-о́в**, (*m pl*) guts, bowels, viscera, entrails (of animals)

потрош·и́ть, **-а́т**, (*imp*) gut, eviscerate, disembowel

по·труд·н·е́е (*comp*) rather/somewhat more difficult

по·тр·у́т, (*fut 3rd pl*) of **по·тер·е́ть**

по·тряс·а́·ть, **-ют**, (*imp*) shake; shock, astound

по·тряс·е́ни·е, **-я**, (*n*) shock; upheaval, shaking, convulsion

по·тряс·ти́, **-у́т**, (*perf*) *see* **по·тряс·а́·ть**

по·тря́х·ива·ть, **-ют**, (*imp*) shake occasionally/lightly, shake up

по·туск·не́л·ый, **-ая**, **-ое**, (*a*) tarnished, dull

по·туск·не́·ть, **-ют**, (*perf*) become dull/lustreless/tarnished

по·тух·а́ни·е, **-я**, (*n*) (*v n*) of **по·ту·х·а́·ть**; extinction

по·тух·а́·ть, **-ют**, (*imp*) go/die out, become extinguished по·ту́х·нуть, -нут, (*perf*) *see* по·ту-х·а́·ть

по·ту́х·ш·ий, -ая, -ее, (*past part act*) *see* по·тух·а́·ть; extinct, out (of lights etc.)

по·туш·и́ть, ⸗ат, (*perf*) put out, extinguish; simmer (in cooking)

по·тя́г·ивани·е, -я, (*n*) tugging

по·тя́г·ива·ть, -ют, (*imp*) keep tugging at

по·тя·н·у́ть, -у́т, (*perf*) *see* тя·н·у́ть

по·уро́ч·н·о (*adv*) by/for the job

по·уро́ч·н·ый, -ая, -ое, (*a*) piece-work, per-job

по·уч·а́·ть, -ют, (*imp*) teach, instruct

по·уч·и́тельн·ый, -ая, -ое, (*a*) instructive

по·уч·и́ть, ⸗ат, (*perf*) teach, give a lesson in

по·хвал·а́, -ы́, (*f*) praise, commendation

по·хвал·и́ть, ⸗ят, (*perf*) praise, commend

по·хва́ль·н·ый, -ая, -ое, (*a*) laudatory; praiseworthy, commendable, laudable

~ гра́мот·а (*f*) certificate of commendation, testimonial

по·хищ·е́ни·е, -я, (*n*) theft, abduction; capture, piracy (of rivers)

по·хло́п·а·ть, -ют, (*perf*) slap, flap, clap (for a while)

по·хло́п·ыва·ть, -ют, (*imp*) pat, slap, flap, clap, pat/slap/flap/clap occasionally

по·хо́д·, -а, (*m*) campaign; march, hike, tour; cruise, voyage, trip; a little over (weight etc.)

кило́ с по·хо́д·ом (*n*) a little over a kilo

по·ход·и́ть, ⸗ят, (*perf*) go, walk (a short distance); (*imp*) resemble, be like/similar

по·хо́д·к·а, -и, (*g pl*) -д·ок·, (*f*) walk, gait, step (manner of walking)

по·хо́д·н·ый, -ая, -ое, (*a*) of по·-хо́д·; field, mobile; marching; cruising

~ крова́т·ь (*f*) camp-bed

~ ку́х·н·я (*f*) mobile/field kitchen

~ ла́гер·ь (*m*) bivouac

~ о́рдер· (*m*) (naut) tactical formation

по·хо́ж·ий, -ая, -ее, (*a*) resembling, like

по·холод·а́ни·е, -я, (*n*) (*v n*) of по·-холод·а́·ть; drop in temperature, cold spell

по·холод·а́·ть, -ют, (*perf*) get colder

по·холод·е́·ть, -ют, (*perf*) become cold

по·хорон·и́ть, ⸗ят, (*perf*) bury, inter

по·хоро́н·н·ый, -ая, -ое, (*a*) burial, funeral

по·худ·е́·ть, -ют, (*perf*) become thin/thinner

по·цара́п·а·ть, -ют, (*perf*) scratch lightly; scratch, make a scratch

по·ча́с·н·о (*adv*) by the hour, hourly

початко·ви́д·н·ый, -ая, -ое, (*a*) (bot) spadiciform, spadicose, spadiceous

початко·с·рыв·а́тел·ь, -я, (*n*) (agr) corn picker

по·ча́т·ок·, -т·к·а, (*m*) (bot) cob, spadix; (text) cop

по·ча́т·очн·ый, -ая, -ое, (*a*) of по·-ча́т·ок·; spadiceous

по́чв·а, -ы, (*g pl*) почв· (*f*) soil; ground; floor (of seam etc.)

по́чв·енн·ый, -ая, -ое, (*a*) soil, edaphic; ground, (biol) edaphonic

почво·ве́д·, -а, (*m*) soil scientist, pedologist

почво·ве́д·ени·е, -я, (*n*) soil science, pedology, edaphology

почво·ве́д·ческ·ий, -ая, -ое, (*a*) pedological

почво·за·крепл·я́ющ·ий, -ая, -ее, (*a*) soil-conserving

почво·за·це́п·, -а, (*m*) (M/T) grouser; cleat (of tyre)

почво·об·раба́т·ывающ·ий, -ая, -ее, (*a*) (agr) cultivating

почво·у·глуб·и́тел·ь, -я, (*m*) deep plough/plow

почво·у·сту́п·н·ый, -ая, -ое, (*a*) (min) underhand

почво·у·томл·е́ни·е, -я, (*n*) soil depletion

почво·фре́з·а, -ы, (*f*) rotary hoe

по·чему́ (*adv*) why, for what reason

по·чему́-то (*adv*) for some reason

по́·черк·, -а, (*m*) handwriting; mannerism

по·черн·е́л·ый, -ая, -ое, (*a*) darkened, blackened

по·черн·е́ни·е, -я, (*n*) (*v n*) of по·-черн·е́·ть; blackening, darkening

за·ко́н· по·черн·е́ни·я (*m*) (phot) density law

по·черн·е́·ть, -ют, (*perf*) become black/dark

по·черп·а́·ть, -ют, (*perf*) scoop; get, draw (information)

по·черп·н·у́ть, -у́т, (*perf*) *see* по··черп·а́·ть

по·черств·е́·ть, -ют, (*perf*) become stale/staler; become harsh/harsher

по·чес·а́ть, (*fut 3rd pl*) по·че́ш·ут, (*perf*) comb; scratch

по·чес·у́х·а, -и, (*f*) (med) prorigo

по·чёт·, -а, (*m*) honour, honor, respect

по·чёт·н·ый, -ая, -ое, (*a*) honoured, honored, esteemed; honour, honor; honorary

~ ме́ст·о (*n*) seat of honour/honor

~ член· (*m*) honorary member

по́ч·ечк·а, -и, (*g pl*) -чек·, (*f*) (biol) gemmule

почечно·двенадцати·пе́рст·н·ый,-ая, -ое, (*a*) (anat) renoduodenal

по́ч·ечн·ый, -ая, -ое, (*a*) kidney, nephritic, nephric, renal; (bot) gemmaceous

~ ка́мен·ь (*m*) (min) nephrite, kidney stone; (med) calculus

почечу́·й, -я, (*m*) (med) piles, haemorrhoids

по·че́ш·ут, (*fut 3rd pl*) of по·чес·а́ть

по·чи́н, -а, (*m*) initiative

по·чин·и́ть, -ят, (*perf*) *see* по·чин·я́·ть

по·чи́н·к·а, -и, (*g pl*) -н·ок·, (*f*) (*v n*) *see* по·чин·я́·ть; repair, mend

по·чи́н·очн·ый, -ая, -ое, (*a*) repair

по·чин·я́·ть, -ют, (*imp*) repair, mend

по·чи́ст·ить, -ят, (*perf*) *see* чи́ст·ить

по·чит·а́тел·ь, -я, (*m*) admirer

по·чит·а́·ть, -ют, (*perf*) read (for a short time); (*imp*) admire

по·чи́щ·енн·ый, -ая, -ое, (*past part pass*) of по·чи́ст·ить, (*perf*) *see* чи́ст·ить (*imp*)

по́ч·к·а, -и, (*g pl*) -ч·ек·, (*f*) (anat) kidney, nephros; (bot) bud, gemma; (geol) ball

~, голов·н·а́я (zool) head-/fore-kidney, promephros

~, ту́ловищ·н·ая mesonephros

поч·ков·а́ни·е, -я, (*n*) (bot) budding, gemmation

почко·ви́д·н·ый, -ая, -ое, (*a*) kidney-shaped, reniform, nephroid; gemmiform, gemmaceous

~ руд·а́ (*f*) (min) kidney ore

почко·но́с·н·ый, -ая, -ое, (*a*) (bot) gemmiferous

почко·с·лож·е́ни·е, -я, (*n*) (bot) aestivation, estivation

почко·с·мык·а́ни·е, -я, (*n*) (bot) aestivation, estivation

по·чле́н·н·о (*adv*) (math) term-wise, term-by-term

по́чт·а, -ы, (*f*) post, mail; post-office

~, пневмат·и́ческ·ая pneumatic tube conveyer, pneumatic rabbit (nucl engineering jargon)

почтальо́н·, -а, (*m*) postman

почта́мт·, -а, (*m*) central/main post office

~, центра́ль·н·ый general/head post office

по·чт·е́ни·е, -я, (*n*) respect, esteem

почти́ (*adv*) almost, nearly

~ -период·и́ческ·ая фу́нкци·я (*f*) (math) almost periodic function

~ равно·поля́р·н·ый, -ая, -ое, (*a*) subisopolar

по·чт·и́тельн·ый, -ая, -ое, (*a*) respectful, deferential

почт·о́в·ый, -ая, -ое, (*a*) post, postal, mail

по·чу́вств·овать, -уют, (*perf*) feel, have a sensation of

по·шат·н·у́ть, -у́т, (*perf*) shake, shake loose; -ся (*pass*); lean over, heel; loose balance, reel, stagger

по·ше́д·ш·ий, -ая, -ее, (*past part act*) of по·йти́, (*perf*); *see* ид·ти́ (*imp*)

по·шёл, (*past masc sing*) of по·йти́, (*perf*); *see* ид·ти́

по·ши́·ть, (*fut 3rd pl*) по·шь·ю́т, (*perf*) sew

по·шл·ёт, (*fut 3rd sing*) of по·сл·а́ть

по·шли́, (*past pl*) of по·йти́, (*perf*); *see* ид·ти́ (*imp*)

по́шлин·а, -ы, (*f*) (fin) duty

~, в·во́з·н·ая import duty

~, тамо́ж·енн·ая customs duty, customs

~, у·равн·и́тельн·ая countervailing duty

по·шту́ч·н·о (*adv*) by the piece, piece by piece

по·шту́ч·н·ый, -ая, -ое, (*a*) piece (by the piece)

~ о·пла́т·а (*f*) pay at piecework rates

по·шь·ёт, (*fut 3rd sing*) of по·ши́·ть

по·щад·и́ть, -я́т, (*perf*) spare, show kindness/mercy

по·щёлк·а·ть, -ют, (*perf*) *see* по··щёлк·ива·ть

по·щёлк·ива·ть, -ют, (*imp*) click, snap, crack, pop

по·щип·а́·ть, -ют, or -йпл·ют, (*perf*) *see* по·щи́п·ыва·ть

по·щи́п·ыва·ть, -ют, (*imp*) pluck, nip, pinch

по·щу́п·а·ть, -ют, (*perf*) *see* **щу́-п·а·ть**

по·э́т·ому (*adv*) therefore, accordingly, consequently, for this reason

по·ю́ч·и, (*pres gerund*) of **пе·ть;** singing

по·ю́щ·ий, -ая, -ее, (*pres part act*) of **пе·ть;** (elec) singing, humming

по·яв·и́ться, -́ятся, (*perf*) *see* **по-·явл·я́·ться**

по·явл·е́ни·е, -я, (*n*) (*v n*) *see* **по-·явл·я́·ть;** appearance, manifestation, occurrence, emergence
 потенциа́л по·явл·е́ни·я (*m*) (spectr) appearance potential

~ **ше́й·к·и** (met) necking

по·явл·я́·ться, -ются, (*imp*) appear, show up, be manifest, emerge, occur

пояр·о́к·, -р·к·а, (*m*) lamb's wool

по́яс·, -а, (*nom pl*) **-а́,** (*g pl*) **-о́в,** (*m*) belt, girdle; zone; boom, flange, chord (of girders); (shipb) strake

~, **афот·и́ческ·ий** (ocean) aphotic zone

~, **брюш·н·о́й** (anat) pelvic girdle

~, **за́д·н·ий** (anat) pelvic girdle

~, **зуб·ов·о́й** cutting plank (carpentry)

~, **кил·ев·о́й** (shipb) garboard strake

~ **кон·е́чност·ей** (anat) girdle

~, **лед·о́в·ый** (shipb) icebelt

~, **ни́ж·н·ий** (civ eng) lower chord; pedestal tie bar (of bogie frame)

~, **об·шив·н·о́й** (shipb) strake

~, **пере́д·н·ий** (anat) pectoral/shoulder girdle

~, **поля́р·н·ый** (geog) frigid zone

~, **с·пас·а́тельн·ый** life belt

~, **та́з·ов·ый** (anat) pelvic girdle

~, **час·ов·о́й** time zone

~, **шпунт·ов·о́й** (shipb) garboard strake

пояс·и́н·а, -ы, (*f*) (geog) ice-stream

пояс·ко́в·ый, -ая, -ое, (*a*) of **по-яс·о́к·;** (zool) clitellar

~ **че́рв·и** (*pl*) (zool) *Clitellata*

по·ясн·е́ни·е, -я, (*n*) explanation, elucidation; (civ eng) strapping (of tanks etc.)

по·ясн·и́тельн·ый, -ая, -ое, (*a*) explanatory, elucidatory

по·ясн·и́ть, -я́т, (*perf*) *see* **по·яс-н·я́·ть**

пояс·ни́ц·а, -ы, (*i*) **-ей,** (*f*) (anat) lumbar region, lumbus

поясни́чно·груд·н·о́й, -а́я, -о́е, (*a*) (anat) lumbothoracic

поясни́чно·крест·цо́в·ый, -ая, -ое, (*a*) (anat) lumbosacral

пояс·ни́чн·ый, -ая, -ое, (*a*) (anat) lumbar

~ **част·ь** (*f*) loin (of meat)

пояс·н·о́й, -а́я, -о́е, (*a*) of **по́яс·;** waist, waist-length

~ **вре́м·я** (*n*) zone time

~ **нож·** (*m*) sheath knife

~ **угол·о́к·** (*m*) (met) angle flange

по·ясн·я́·ть, -ют, (*imp*) explain, elucidate

пояс·о́к·, -с·к·а́, (*m*) (*dim*) of **по́яс·;** zonule, zonula; (zool) clitellum

~, **вед·у́щ·ий** driving band (of projectile)

по·я́т, (*pres 3rd pl*) of **по·и́ть**

п/п, (*abbr*) = **по по·ря́д·к·у** serial number, No. (used in tables)

пра- (*prefix*) primitive, original, parent; great- (of relationships)

пра́вд·а, -ы, (*f*) truth

правд·и́в·ый, -ая, -ое, (*a*) true, truthful

правдо·по·до́б·и·е, -я, (*n*) plausibility, likelihood

правдо·по·до́б·н·ый, -ая, -ое, (*a*) probable, likely, plausible

пра́в·ил·о, -а, (*n*) (*see also under associated words*); rule, law, principle; (*pl*) regulations; **прав·и́л·о, -а,** (*n*) straight-edge, dresser, moon-bard (tool)

~ **без·о·па́с·ност·и** (*pl*) safety regulations

~, **гру́б·ое** rough-and-ready rule

~ **зна́к·ов** rule of signs

~ **у́л·ичн·ого движ·е́ни·я** traffic regulations, rules of the road

~ **фаз·** (chem) phase rule

правильно·ги́б·очн·ая маши́н·а (*f*) (met) gag press, straightening machine

правильно·рас·тяж·н·а́я маши́н·а (*f*) (met) stretch-straightening/-levelling machine

пра́в·ильност·ь, -и, (*f*) accuracy

пра́в·ильн·ый, -ая, -ое, (*a*) right, correct, proper; regular (gram, geom); normal (of X-rays etc.); **прав·и́ль-н·ый, -ая, -ое,** (*a*) straightening, levelling

~ **дроб·ь** (*f*) (math) proper fraction

~ **маши́н·а** (*f*) (met) straightening machine (for bars, sections, tubes etc.); (met) levelling/flattening machine (for sheet and strip)

~ **маши́н·а, ро́л·иков·ая** (*f*) roller levelling/flattening machine; roller straightening machine

пра́в·ильн·ый, -ая, -ое
~ **мно́го·уго́ль·ник·** (*m*) (math) regular polygon
~ **от·раж·е́ни·е** (*n*) regular reflection
~ **пресс·** (*m*) straightening press, flattening press
прав·и́тельственн·ый, -ая, -ое, (*a*) government, governmental; (telecom) fourth priority
прав·и́тельств·о, -а, (*n*) government
пра́в·ить, -ят, . (*imp*) steer, drive; correct, put/set right; true, straighten, level, trim; sharpen (blades); block (hats)
пра́в·к·а, -и, (*f*) (*v n*) *see* **пра́в·ить;** truing up
~, **гор·я́ч·ая** hot trimming (of drop forgings)
~ **по шир·ин·е́** (met roll) cross levelling
~, **холо́д·н·ая** cold trimming (of drop forgings)
правл·е́ни·е, -я, (*n*) board (of management); government
пра́в·о, -а, (*g pl*) **прав·** (*n*) right; (naut) starboard; license; law; (*adv*) truly; (*predic adj*) it is true
~, **ме́жду·на·ро́д·н·ое** international law
~ **на борт·** (*imper*) hard a'starboard
~ **на труд·** the right to work
~, **объекти́в·н·ое** objective law
~ **свобо́д·н·ой пра́кт·ик·и** (naut) pratique
~, **у·голо́в·н·ое** criminal law
право·бере́ж·н·ый, -ая, -ое, (*a*) right-bank
право·борт·ов·о́й, -а́я, -о́е, (*a*) starboard
право·ве́д·, -а, (*m*) lawyer, jurist
право·ве́д·ени·е, -я, (*n*) jurisprudence
право·ве́р·н·ый, -ая, -ое, (*a*) orthodox, dogmatic
право·винт·ов·о́й, -а́я, -о́е, (*a*) (math) right-handed, dextrorotatory
прав·ов·о́й, -а́я, -о́е, (*a*) law, legal; lawful, rightful
право·вращ·а́ющ·ий, -ая, -ее, (*a*) clockwise-rotating, dextrorotatory, (math) right-handed
право·ме́р·н·ый, -ая, -ое, (*a*) rightful, lawful
право·мо́ч·и·е, -я, (*n*) (law) competence
право·мо́ч·н·ый, -ая, -ое, (*a*) (law) competent
право·на·руш·е́ни·е, -я, (*n*) infringement of the law

право·пис·а́ни·е, -я, (*n*) spelling, orthography
право·по·ря́д·ок·, -д·к·а, (*m*) law and order
право·спосо́б·н·ый, -ая, -ое, (*a*) (law) capable, competent
право·сторо́н·н·ый, -ая, -ое, (*a*) right-side, starboard
право·су́д·и·е, -я, (*n*) justice
прав·от·а́, -ы́, (*f*) rightness
пра́в·оч·н·ый, -ая, -ое, (*a*) of **пра́в·-к·а**
пра́в·щик·, -а, (*m*) straightener, leveller, straightening/levelling machine operator
пра́в·ый, -ая, -ое, (*a*) right, right-hand, right-handed; (naut, air) starboard; right, just; (polit) right-wing
~ **борт·** (*m*) (naut, air) starboard side
~ **вращ·е́ни·е** (*n*) clockwise rotation
~ **рез·ьб·а́** (*f*) right-handed thread
прав·и́л·о пра́в·ой рук·и́ (*n*) (elec) right-hand rule, Fleming's right-hand rule
~ **со·един·е́ни·е** (*n*) (chem) dextrorotatory compound
пра́в·ящ·ий, -ая, -ее, (*pres part act*) of **пра́в·ить;** ruling, governing
пра·га́усс·, -а, (*m*) (elec) pragauss, 10^{-10} gauss
пра́зд·ник·, -а, (*m*) holiday; celebration
пра́зд·ничн·ый, -ая, -ое, (*a*) holiday, festive, celebration
пра́зд·н·ый, -ая, -ое, (*a*) idle
празем·, -а, (*m*) (min) prase
празео·ди́м·, -а, (*m*) (chem) praseodymium, Pr
празео·ди́м·и·й, -я, (*m*) = **празео·ди́м·**
празео·ли́т·, -а, (*m*) (min) praseolite
пра́кт·ик·, -а, (*m*) practitioner; practical man, handyman
пра́кт·ик·а, -и, (*f*) practice, practice experience/training; (naut) pratique
~, **мор·ск·а́я** seamanship
практ·ика́нт·, -а, (*m*) trainee, probationer
~, **про·из·во́д·ственн·ый** trainee, apprentice (for executive position in industry)
практик·ова́ть, -у́ют, (*imp*) put into practice, apply, practice
пра́кт·икум·, -а, (*m*) study with experimental work; laboratory (in which such study is conducted)
практ·и́ческ·и (*adv*) in practice; practically

практ·и́ческ·ий, -ая, -ое, (*a*) practice, practical, applied

~ **медици́н·а** (*f*) applied medicine

~ **с·на·ря́д·** (*m*) practice projectile

практ·и́чност·ь, -и, (*f*) practicalness, practicality; practical ability

практ·и́чн·ый, -ая, -ое, (*a*) practical

Пра́ндтл·я, тру́б·к·а (*f*) (dynam) Prandtl tube

~, числ·о́ (*n*) (dynam) Prandtl number

пра·род·и́тел·ь, -я, (*m*) ancestor, progenitor

пра́ч·счн·ая, -ой, (*f decl as adj*) laundry, wash house

пращ·а́, -и, (*i*) **-о́й,** (*f*) sling

праще·ви́д·н·ый, -ая, -ое, (*a*) fundiform

пра·язы́к·, -а́, (*m*) (ling) parent language

пре- (*prefix*) *signifies:* — (1) change of condition; (2) excessive action; (3) = **пере** across, from — to; (4) division into parts, into two; (5) *superlative adjective, adverb;* (6) Latin pre = before

пре·а́мбул·а, -ы, (*f*) preamble

пре·бы́в·а́ни·е, -я, (*n*) (*v n*) of **пре·бы́в·а́·ть;** stay; sojourn, residence, tenure (of office)

вре́м·я пре·бы́в·а́ни·я (*n*) stay, time there, residence time (of catalyst)

пре·бы́в·а́·ть, -ют, (*imp*) be, remain, stay

превал·и́ровани·е, -я, (*n*) prevalence, predominance

превал·и́р·овать, -уют, (*imp*) predominate

преве́нтор·, -а, (*m*) (oil) blow-out preventer

пре·взо·йд·ённ·ый, -ая, -ое, (*past part pass*) of **пре·взо·йти́,** (*perf*); *see* **пре·вос·ход·и́ть** (*imp*)

пре·взо·йти́, (*fut 3rd sing, pl*) **пре·взо·йд·ёт, пре·взо·йд·у́т,** (*past masc sing*) **пре·взо·шёл,** (*perf*); *see* **пре·вос·ход·и́ть**

пре·взо·шёл, (*past masc sing*) of **пре·взо·йти́**

пре·воз·мог·а́·ть, -ют, (*imp*) overcome

пре·воз·мо́ч·ь, (*fut 3rd sing, pl*) **пре·воз·мо́ж·ет, пре·воз·мо́г·ут,** (*past masc sing*) **пре·воз·мо́г·,** (*perf*); *see* **пре·воз·мог·а́·ть**

пре·вос·ход·и́тельств·о, -а, (*n*) (dipl) excellency

пре·вос·ход·и́ть, -ят, (*imp*) exceed, surpass, excel

пре·вос·хо́д·н·ый, -ая, -ое, (*a*) excellent, superb

пре·вос·хо́д·ств·о, -а, (*n*) superiority, excellence

пре·вос·ход·я́щ·ий, -ая, -ее, (*pres part act*) of **пре·вос·ход·и́ть;** superior

пре·врат·и́ть, -ят, (*perf*) *see* **пре··вращ·а́·ть**

пре·вращ·а́·ть, -ют, (*imp*) transform, convert, transmute, turn into/to; **-ся** (*pass*); change/turn into/to

пре·вращ·е́ни·е, -я, (*n*) (*v n*) *see* **пре·вращ·а́·ть;** transformation, conversion, transmutation

~ **в газ·** gasification

~, обрат·и́м·ое reversible transformation

~ **по·ря́д·ок·–бес·по·ря́д·ок·** (phys) order-disorder transformation

то́ч·к·а пре·вращ·е́ни·я (*f*) transformation/transition/critical point

~, фа́з·ов·ое phase transformation

~ **элеме́нт·ов** transmutation of the elements

~, я́дер·н·ое nuclear transformation/ transmutation

пре·вы́с·ить, -ят, (*perf*) *see* **пре·вы·ш·а́·ть**

пре·выш·а́·ть, -ют, (*imp*) exceed, be greater than

пре·выш·е́ни·е, -я, (*n*) (*v n*) *see* **пре··выш·а́·ть;** exceeding, excess; overshoot; (air) height separation above, vertical interval; (geog) relative altitude, height difference

прегни́н·, -а, (*m*) (pharm) pregnine

пре·гра́д·а, -ы, (*f*) barrier, obstacle, obstruction

пре·град·и́ть, -ят, (*perf*) *see* **пре··гражд·а́·ть**

пре·гражд·а́·ть, -ют, (*imp*) bar, block, obstruct

пре·гражд·е́ни·е, -я, (*n*) (*v n*) *see* **пре·гражд·а́·ть;** obstruction

пред- (*prefix*) pre-

пре·да·ва́·ть, -ют, (*imp*) betray; commit/consign/bring to

предаззи́т·, -а, (*m*) (min) predazzite

пре́·да·нн·ый, -ая, -ое, (*past part pass*) *see* **пре·да·ва́·ть;** devoted; faithfully, truly (in formal correspondence)

пре·да́·ть, (*fut 3rd sing, pl*) **пре·да́·ст, пре·дад·у́т,** (*past masc sing*) **пре́·да·л** (*perf*); *see* **пре·да·ва́·ть**

пре·да́т·ельств·о, -а,   (*n*) treachery

пред·вар·е́ни·е, -я,   (*n*) (*v n*) *see* пред·-вар·я́·ть; anticipation, advance, lead
> вре́м·я пред·вар·е́ни·я (*n*) (autom) rate time

~, обра́т·н·ое (autom) feedback anticipation

~ от·кры́т·и·я кла́пан·а   (st eng) slide-valve lead

~ по·да́·ч·и то́пл·ив·а   (ICE) injection advance

пред·вар·и́тельн·о   (*adv*) beforehand, as a preliminary, preliminarily, pre-, first of all

~ -на·пряж·ённ·ый, -ая, -ое,   (*a*) prestressed

~ -на·тяж·ённ·ый, -ая, -ое,   (*a*) pretensioned, preloaded

пред·вар·и́тельн·ый, -ая, -ое,   (*a*) preliminary, pre-, prior

~ на·гре́·в· (*m*) preheating

~ на·пряж·е́ни·е (*n*) prestressing

~ на·тя́г· (*m*) preload (bearings)

~ на·тяж·е́ни·е (*n*) pretension, pretensioning

~ об·рабо́т·к·а (*f*) preconditioning, pretreatment

~ о·хлажд·е́ни·е (*n*) precooling

~ пред·у·прежд·е́ни·е   (*n*) advance notice

~ у·сил·е́ни·е (*n*) (elec) pre-amplification

пред·вар·и́ть, -я́т, (*perf*) *see* пред·вар·я́·ть

пред·вар·я́·ть, -ют,   (*imp*) forestal, anticipate

пред·ве́ст·ник·, -а,   (*m*) forerunner, precursor; sign, indication

пред·вещ·а́·ть, -ют, (*imp*) foreshadow, presage, betoken, indicate

пред·взя́·т·ый, -ая, -ое,   (*a*) preconceived

пред·ви́д·ени·е, -я,   (*n*) (*v n*) of пред·-ви́д·еть; foresight

пред·ви́д·еть, -ят,   (*imp*) foresee, envisage

пред·в·ключ·ённ·ый,   -ая, -ое, (*a*)
~ турби́н·а (*f*) superposition turbine

пред·вод·и́тельств·о, -а, (*n*) leadership

пред·вое́н·н·ый, -ая, -ое, (*a*) prewar; preliminary to war

пред·вы́·бор·н·ый, -ая, -ое, (*a*) pre-election

пред·вы·числ·е́ни·е, -я, (*n*) pre-computation

пред·го́р·н·ый, -ая, -ое, (*a*) (geog) piedmont

пред·го́р·ь·е, -я, (*n*) foothills, piedmont

пред·груд·и́нн·ый, -ая, -ое,   (*a*) (anat) presternal

пред·две́р·и·е, -я,   (*n*) vestibule

пред·деформ·а́ци·я, -и,   (*f*) previous deformation, predeformation

пред·диссоци·а́ци·я, -и, (*f*) predissociation

пре·де́л·, -а,   (*m*) (*see also under associated words*); limit; (*pl*) range
> в пре·де́л·ах   in/within the range, within
> ...в пре·де́л·ах ±2%, within ±2%
> ...в пре·де́л·ах 95°–200°C, in the range 95°–200°C, from 95° to 200°C

~ вз·рыв·а́емост·и explosion limit

~ дл·и́тельн·ой про́ч·ност·и (met) rupture stress (in a creep test)

~ дроссел·и́ровани·я   (*pl*) (autom) throttling range, proportional band
> за пре·де́л·ы outside/beyond the range; outside/beyond the scope

~ из·мер·е́ни·я (*pl*) range of measurement

~ об·наруж·е́ни·я detection limit/threshold
> основ·н·ы́е теоре́м·ы о пре·де́-л·ах (*f pl*) (math) fundamental theorems on limits

~ полз·у́чест·и (met) limiting creep stress

~ по·сле́д·овательност·и (math) limit of a sequence

~ пропорциона́ль·ност·и (met) limit of proportionality, proportional elastic limit; (autom) throttling range, proportional band

~ пропорциона́ль·ност·и, у·сло́в·-н·ый (mech) limit of proportionality (at given percentage reduction in the elastic modulus)

~ про́ч·ност·и maximum stress, ultimate strength (in general); tensile strength, ultimate tensile stress (in particular)

~ про́ч·ност·и, дей·стви́тельн·ый true tensile stress

~ про́ч·ност·и, и́ст·инн·ый   true tensile stress

~ про́ч·ност·и на́ с·рез· maximum stress in shear, shear strength

~ про́ч·ност·и при и́з·гиб·е (met) transverse strength, transverse modulus of rupture; flexural strength, (plast) cross-breaking strength

~ про́ч·ност·и при круч·е́ни·и torsional stress

пре·де́л

~ про́ч·ност·и при круч·е́ни·и, и́с·т·инн·ый maximum stress in shear

~ про́ч·ност·и при круч·е́ни·и, у·сло́в·н·ый modulus of rupture in torsion

~ про́ч·ност·и при рас·тяж·е́ни·и maximum stress, tensile strength, ultimate tensile stress

~ про́ч·ност·и при с·жа́т·и·и compressive strength, maximum stress in compression

~ ско́р·ост·и speed limit
пре·де́л·ы скор·ост·е́й (*pl*) range of speeds, speed range

~ тек·у́чест·и yield point, proof stress

~ тек·у́чест·и, ве́рх·н·ий upper yield point

~ тек·у́чест·и, техн·и́ческ·ий engineering stress

~, тек·у́чест·и, у·сло́в·н·ый proof stress

~ тек·у́чест·и, физ·и́ческ·ий yield stress, yield point

~, тепл·ов·о́й (elec) thermal limit

~ у·пру́г·ого со·противл·е́ни·я (gunn) maximum safe pressure

~ фу́нкци·и (math) limit of a variable

пре·дель·н·ый, -ая, -ое, (*a*) limiting, limit, overall, maximum, ultimate, critical; (chem) saturated

~ по·гре́ш·ност·ь (*f*) maximum error

~ с·лу́ч·а·й (*m*) extreme/limiting case

~ со·сто·я́ни·е (*n*) critical state

~ столб· (*m*) (rail) boundary post

~ угле·водо·ро́д·ы (*pl*) (chem) saturated hydrocarbons

пред·за·жиг·а́ни·е, -я, (*n*) (elec) prestrike (of an arc); keep-alive discharge

пред·за·жиг·а́тел·ь, -я, (*m*) trigger/excitation/keep-alive electrode

пред·из·бир·а́тел·ь, -я, (*m*) preselector

пред·иониз·а́тор·, -а, (*m*)
цеп·ь пред·иониз·а́тор·а (*f*) (elec) keep-alive circuit

преди·сло́в·и·е, -я, (*n*) foreword, preface

пред·ка́мер·а, -ы, (*f*) (ICE) pre-combustion chamber, antechamber

пре́д·ков·ый, -ая, -ое, (*a*) of пре́д·ок·, ancestral

пред·корен·н·о́й, -а́я, -о́е, (*a*) (zool) premolar

пред·кры́л·ок·, -л·к·а, (*m*) (a/c) leading-edge flap

пред·лаг·а́·ть, -ют, (*imp*) propose, suggest, offer

пред·леж·а́ни·е, -я, (*n*) presentation (obstetrics)

пред·ли́ст·, -а, (*m*) (bot) prophyll, prophyllum

пред·ло́г·, -а, (*m*) pretext, excuse; (gram) preposition

пред·лож·е́ни·е, -а, (*n*) (*v n*) *see* **пред·лаг·а́·ть;** proposal, suggestion, offer, proposition; (econ) supply; (gram) sentence, clause

~, в·во́д·н·ое (gram) parenthesis

~, вс·по·мог·а́тельн·ое (math) lemma

~, при·да́т·очн·ое (gram) subordinate clause

пред·лож·и́тельн·ый, -ая, -ое, (*a*) suggested, propositional, tentative

пред·лож·и́ть, -ат, (*perf*) *see* **пред·лаг·а́·ть**

пред·ло́ж·н·ый, -ая, -ое, (*a*) (gram) prepositional

пред·маршру́т·н·ый, -ая, -ое, (*a*) (rail) approach

пред·ме́ст·ь·е, -я, (*n*) suburb

пред·ме́т·, -а, (*m*) object, article; topic, theme, subject

~, ме́ст·н·ый (surv) feature, detail

пред·ме́т·н·ый, -ая, -ое, (*a*) of пред·ме́т·

~ катало́г· (*m*) subject catalogue (of books)

~ ми́шен·ь (*f*) object vane (of compass azimuth circle)

~ стекл·о́ (*n*) slide (of microscope)

~ стол· (*m*) stage (of microscope)

~ сто́л·ик· (*m*) stage (of microscope)

~ у·каз·а́тел·ь (*m*) (print) index, subject index

пред·модуля́тор·, -а, (*m*) (elec) sub-modulator

пред·мо́ст·н·ый, -ая, -ое, (*a*) bridge-approach

~ у·крепл·е́ни·е (*n*) (mil) bridgehead

пред·на·знач·а́·ть, -ют, (*imp*) intend/mean/destine/earmark for

пред·на·знач·е́ни·е, -я, (*n*) (*v n*) *see* **пред·на·знач·а́·ть;** destination (of trains etc.); purpose (of machines etc.)
ста́нци·я пред·на·знач·е́ни·я (*f*) destination

пред·на·зна́ч·енн·ый, -ая, -ое, (*past part pass*) *see* **пред·на·знач·а́·ть;** to, bound for (of trains etc.); designed (for), for use (as/with), used (for) (of machines)

пред·на·знач·ить, -ат, (*perf*) *see* пред·на·знач·а·ть

пред·на·мер·енн·ый, -ая, -ое, (*a*) premeditated, deliberate

пред·на·пряж·ённ·ый, -ая, -ое, (*a*) prestressed

предо- (*prefix*) *see* пред-

пред·ок·, -д·к·а, (*m*) ancestor, predecessor

пред·о·кон·ечн·ый, -ая, -ое, (*a*) penultimate

предомин·и́рующ·ий, -ая, -ое, (*a*) predominant

пред·о·пла́т·а, -ы, (*f*) prepayment

предо·пре·дел·и́ть, -я́т, (*perf*) *see* предо·пре·дел·я́·ть

предо·пре·дел·я́·ть, -ют, (*imp*) predetermine, fix/define beforehand

предо·ста́в·ить, -ят, (*perf*) *see* предо·ставл·я́·ть

предо·ставл·е́ни·е, -я, (*n*) (*v n*) *see* предо·ставл·я́·ть

предо·ставл·я́·ть, -ют, (*imp*) assign, allot, place at the disposal of, grant, give, make available; allow, let

пред·о·стерег·а́тельн·ый, -ая, -ое, (*a*) warning, cautionary

пред·о·стерег·а́·ть, -ют, (*imp*) warn, caution

пред·о·стереж·е́ни·е, -я, (*n*) warning, caution

пред·о·сторо́ж·ност·ь, -и, (*f*) precaution, safeguard

  мер·а пред·о·сторо́ж·ност·и (*f*) precaution

пред·от·врат·, -и́ть, -я́т, (*perf*) *see* пред·от·вращ·а́·ть

пред·от·вращ·а́·ть, -ют, (*imp*) avert, prevent

пред·от·вращ·е́ни·е, -я, (*n*) (*v n*) *see* пред·от·вращ·а́·ть; prevention

пред·о·хран·е́ни·е, -я, (*n*) (*v n*) *see* пред·о·хран·я́·ть; protection, preservation, safeguard

пред·о·хран·и́тел·ь, -я, (*m*) safety device/stop/cap/catch, protecting device, guard; (elec) fuse

~, в·ста́в·н·о́й (elec) push-in fuse

~ выс·о́к·ого на·пряж·е́ни·я (elec) high-capacity fuse

~, жи́дк·ост·н·ый (elec) liquid-quenched fuse

~, за·кры́·т·ый (elec) cartridge fuse

~, инерцио́н·н·ый inertia switch

~, ма́сл·ян·ый (elec) oil-quenched fuse

~ на·ка́л·а (elec) filament fuse

пред·о·хран·и́тел·ь

~, от·кры́·т·ый (elec) open fuse

~, патро́н·н·ый (elec) cartridge fuse

~, пла́в·к·ий (elec) fuse, fusible cutout

~, пла́в·к·ий терм·и́ческ·ий (elec) thermal cutout

~, пласт·и́нчат·ый (elec) plate fuse

~, полу·от·кры́·т·ый (elec) semi-enclosed fuse

~, порош·ко́в·ый (elec) powder-filled fuse

~, про́б·очн·ый (elec) screw-plug fuse

~, рог·ов·о́й (elec) horn-break fuse

~, руж·е́йн·ый safety catch (on rifle)

~, стрел·я́ющ·ий (elec) expulsion fuse

~, токо·о·грани́ч·ивающ·ий (elec) current-limiting fuse

~, тру́б·чат·ый (elec) cartridge fuse

пред·о·хран·и́тельн·ый, -ая, -ое, (*a*) of пред·о·хран·и́тел·ь; protective, safety; (expl) permitted, permissible

~ колпа́к· (*m*) safety cap

~ коро́б·к·а (*f*) (elec) fuse box

~ при·бо́р· (*m*) safety device

пред·о·хран·и́ть, -я́т, (*perf*) *see* пред·о·хран·я́·ть

пред·о·хран·я́·ть, -ют, (*imp*) protect, preserve, safeguard

пред·пис·а́ни·е, -я, (*n*) (*v n*) *see* пред·пи́с·ыва·ть; directions, instruction, injunction, order, (med) prescription

~ суд·а́ court order

пред·пис·а́ть, (*fut 3rd pl*) пред·пи́·ш·ут, (*perf*) *see* пред·пи́с·ыва·ть

пред·пи́с·ыва·ть, -ют, (*imp*) prescribe, order; ascribe

пред·пи́ш·ет, (*fut 3rd sing*) of пред··пис·а́ть

пред·плеч·ь·е, -я, (*g pl*) -ч·и·й, (*n*) (anat) forearm, antebrachium

пред·плу́ж·ник·, -а, (*m*) (agr) skim coulter, jointer

пред·плюсн·а́ -ы́, (*nom pl*) -плюсн·ы, (*g pl*) -сен·, (*f*) (anat) tarsus

пред·плюсн·ев·о́й, -а́я, -о́е, (*a*) (anat) tarsal

пред·по·лаг·а́ем·ый, -ая, -ое, (*pres part pass*) of пред·по·лаг·а́·ть; supposed, conjectural

пред·по·лаг·а́·ть, -ют, (*imp*) suppose, conjecture; assume; intend, propose; presuppose

пред·по·лёт·н·ый, -ая, -ое, (*a*) preflight

пред·по·лож·е́ни·е, -я, (*n*) supposition, assumption, hypothesis; proposal, proposition
вы́·сказ·ат·ь пред·по·лож·е́ни·е suggest
пред·по·ло́ж·енн·ый, -ая, -ое, (*past part pass*) of **пред·по·лож·и́ть**
пред·по·лож·и́тельн·ый, -ая, -ое, (*a*) presumable, tentative, conjectural
пред·по·лож·и́ть, ⁻ат, (perf) suppose, conjecture; assume
пред·по́л·ь·е, -я, (*n*) (mil) forward defence area
пред·поля́р·н·ый, -ая, -ое, (*a*) subpolar
предпорт·о́в·ая ста́нци·я (*f*) (rail) harbour station
пред·по·сле́д·н·ий, -яя, -ее, (*a*) penultimate
пред·по·сы́л·к·а, -и, (*g pl*) **-л·ок·,** (*f*) precondition, prerequisite; premise
пред·по·че́ст·ь, (*fut 3rd pl*) **пред·по·чт·у́т,** (perf) prefer
пред·по·чит·а́·ть, -ют, (imp) prefer
пред·по́ч·к·а, -и, (*g pl*) **-ч·ек·,** (*f*) (zool) pronephros, fore/head kidney
пред·по·чт·ёт, (fut 3rd sing) of **пред·по·че́сть**
пред·по·чт·и́тельн·ый, -ая, -ое, (*a*) preferable, preferred
~ **ориент·а́ци·я** (*f*) (cryst) preferred orientation
пред·при·и́м·чив·ый, -ая, -ое, (*a*) enterprising
пред·пр·и́м·ет, (fut 3rd sing) of **пред·при·ня́·ть**
пред·при·ним·а́тел·ь, -я, private company owner/chairman; entrepreneur
пред·при·ним·а́·ть, -ют, (imp) undertake
пред·при·ня́·ть, (fut 3rd pl) **пред·пр·и́м·ут,** (perf) see **пред·при·ним·а́·ть**
пред·при·я́т·и·е, -я, (*n*) (com) undertaking, enterprise, concern
пред·прямо·крыл·ые, -ых, (*pl decl as adj*) (pal) Protorthoptera
пред·рас·по·лаг·а́·ть, -ют, (imp) predispose
пред·рас·по·лож·е́ни·е, -я, (*n*) (*v n*) see **пред·рас·по·лаг·а́·ть**; predisposition, tendency
пред·рас·по·лож·и́ть, ⁻ат, (perf) see **пред·рас·по·лаг·а́·ть**
пред·рас·све́т·н·ый, -ая, -ое, (*a*) pre-dawn

пред·рас·су́д·ок·, -д·к·а, (*m*) prejudice
пред·реш·а́·ть, -ют, (imp) predetermine, decide beforehand
пред·реш·и́ть, -а́т, (perf) see **пред·реш·а́·ть**
пред·род·ов·о́й, -а́я, -о́е, (*a*) (zool) prenatal
пред·ро́ст·ок·, -т·к·а, (*m*) (bot) protonema
пред·рот·ов·о́й, -а́я, -о́е, (*a*) preoral
пред·сед·а́тел·ь, -я, (*m*) chairman, president
пред·сед·а́тельств·овать, -уют, (imp) preside, be in the chair, be chairman
пред·се́рд·и·е, -я, (*n*) (anat) auricle of the heart, atrium cordis
пред·серд·е́чн·ый, -ая, -ое, (*a*) (anat) precardiac, precordial
пред·с·каз·а́ни·е, -я, (*n*) (*v n*) see **пред·с·ка́з·ыва·ть**; forecast, prediction
пред·с·каз·а́ть, (*fut 3rd pl*) **пред·с·ка́ж·ут,** (perf) see **пред·с·ка́з·ыва·ть**
пред·с·ка́з·ыва·ть, -ют, (imp) forecast, predict
пред·с·ме́рт·н·ый, -ая, -ое, (*a*) (med) premortal
пред·став·и́тел·ь, -я, (*m*) representative, agent; example
пред·став·и́тельств·о, -а, (*n*) representation
пред·ста́в·ить, -ят, (perf) see **пред·ставл·я́·ть**
пред·ставл·е́ни·е, -я, (*n*) representation, picture, idea, concept; show (theatre)
пред·ставл·я́·ть, -ют, (imp) present; represent; introduce (people); submit (papers); **-ся** be
~ **себе́** imagine
~ **соб·о́й** be
пред·ста́т·ельн·ый, -ая, -ое, (*a*) (zool) prostate, prostatic
пред·сто·я́ть, -я́т, (imp) be due/coming, be in prospect
пред·сто·я́щ·ий, -ая, -ее, (*pres part act*) of **пред·сто·я́ть**; forthcoming, impending
пред·сущ·еств·у́ющ·ий, -ая, -ее, (*a*) pre-existing
пред·у·бежд·е́ни·е, -я, (*n*) prejudice, bias
пред·узл·ов·о́й, -а́я, -о́е, (*a*) antenodal
пред·у·пред·и́тельн·ый, -ая, -ое, (*a*) warning, preventive, precautionary

пред·у·пред·и́ть, -я́т, (*perf*) *see* пре-д·у·прежд·а́·ть

пред·у·прежд·а́·ть, -ют, (*imp*) give advance notice, let know beforehand, warn; prevent, avert; anticipate, forestall

пред·у·прежд·а́ющ·ий, -ая, -ее, (*pres part act*) of пред·у·прежд·а́·ть; warning, precautionary, preventive

~ о·кра́с·к·а (*f*) (zool) warning/aposematic coloration

пред·у·прежд·е́ни·е, -я, (*n*) (*v n*) *see* пред·у·прежд·а́·ть; prevention; warning

~, за·благо·вре́мен·н·ое (rad) early warning

~, шторм·ов·о́е (meteor) gale warning

пред·у·сил·и́тел·ь, -я, (*m*) (elec) pre-amplifier

пред·у·сма́тр·ива·ть, -ют, (*imp*) envisage, foresee

пред·у·смо́тр·енн·ый, -ая, -ое, (*past part pass*) of пред·у·смотр·е́ть; envisaged, foreseen

пред·у·смотр·е́ть, -я́т, (*perf*) *see* пред·у·сма́тр·ива·ть

пред·у·смотр·и́тельн·ый, -ая, -ое, (*a*) farsighted, prudent

пред·челюст·н·о́й, -а́я, -о́е, (*a*) (anat) premaxillary

пред·чист·ов·о́й, -а́я, -о́е, (*a*) (met roll) leader

пред·ше́ств·енник·, -а, (*m*) predecessor, forerunner, precursor

пред·ше́ств·енн·ый, -ая, -ое, (*a*) preceding, foregoing, prior

пред·ше́ств·и·е, -я, (*n*) (*v n*) *see* пред·ше́ств·овать

пред·ше́ств·овать, -уют, (*imp*) precede

пред·ше́ств·ующ·ий, -ая, -ое, (*pres part act*) of пред·ше́ств·овать; preceding, previous, earlier

предъ·иониз·а́тор·, -а, (*m*) (elec) trigger/starter electrode

предъ·яв·и́тел·ь, -я, (*m*) (fin) bearer

предъ·яв·и́ть, -я́т, (*perf*) *see* предъ·явл·я́·ть

предъ·явл·я́·ть, -ют, (*imp*) present, produce, show, exhibit (documents)

~ прете́нз·и·ю (law) claim

пред·ыд·у́щ·ий, -ая, -ее, (*a*) previous, earlier; (*m deci as adj*) the foregoing

пред·ынтегра́ль·н·ый, -ая, -ое, (*a*) (math) pre-integral

пред·ыониз·а́ци·я, -и, (*f*) pre-ionization

пред·ыск·а́ни·е, -я, (*n*) preselection

пред·ыск·а́тел·ь, -я, (*m*) preselector, line switch (telephony)

пред·ысто́р·и·я, -и, (*f*) prehistory; previous history

пре·е́м·ник·, -а, (*m*) successor

пре·е́м·ственност·ь, -и, (*f*) succession, continuity

пре·е́м·ственн·ый, -ая, -ое, (*a*) successive

пре́жд·е (*adv*) before, first; formerly; (*prep + gen*) before

~ вс·его́ first of all, to begin with

прежде·вре́мен·н·ый, -ая, -ое, (*a*) premature

пре́ж·н·ий, -яя, -ее, (*a*) previous, prior, former, earlier

по пре́ж·н·ему as before

презервати́в·, -а, (*m*) (med) condom, french letter

президе́нт·, -а, (*m*) president

прези́диум·, -а, (*m*) presidium

пре·зр·е́ни·е, -я, (*n*) contempt, scorn

презу́мпц·и·я, -и, (*f*) presumption

пре·им·у́щественн·о (*adv*) preferable; mainly, largely

пре·им·у́щественн·ый, -ая, -ое, (*a*) preferential, preferred

~ ориент·а́ци·я (*f*) (phys) preferred orientation

~ по·глощ·е́ни·е (*n*) (phys) preferential absorption

пре·им·у́ществ·о, -а, (*n*) advantage; preference

пре·ис·по́лн·ить, -ят, (*perf*) *see* пре··ис·полн·я́·ть

пре·ис·полн·я́·ть, -ют, (*imp*) fill with

прейскура́нт·, -а, (*m*) price list

пре·капилля́р·, -а, (*m*) (biol) pre-capillary

пре·кардиа́ль·н·ый, -ая, -ое, (*a*) (anat) precardiac

преколла́ген·ов·ое волокн·о́ (*n*) (zool) reticular fibre

пре·кра́с·н·ый, -ая, -ое, (*a*) beautiful, fine; excellent

пре·крат·и́ть, -я́т, (*perf*) *see* пре··кращ·а́·ть

пре·кращ·а́·ть, -ют, (*imp*) cease, stop, discontinue; -ся (*pass*); come to a halt

пре·кращ·е́ни·е, -я, (*n*) (*v n*) *see* пре·кращ·а́·ть; stop, discontinuation, cessation

пре·кращ·éни·е
~ пит·áни·я (*elec*) power failure
тóч·к·а пре·кращ·éни·я (*f*) (math) point of discontinuity
прелиминáр·н·ый, -ая, -ое, (*a*) preliminary
прéл·ин·а, -ы, (*f*) rotten spot (leather)
пре·лом·úм·ый, -ая, -ое, (*pres part pass*) see **пре·ломл·я́·ть**; refrangible, refractive
пре·лом·úть, ⸗ят, (*perf*) see **пре·-ломл·я́·ть**
пре·ломл·éни·е, -я, (*n*) refraction
~, двой·н·óе birefringence
коэффициéнт· пре·ломл·éни·я (*m*) refractive index
по·каз·áтел·ь пре·ломл·éни·я (*m*) refractive index
пре·ломл·я́емост·ь, -и, (*f*) refractivity, refrangibility; (chem) specific refraction
пре·ломл·я́ем·ый, -ая, -ое, (*a*) see **пре·лом·úм·ый**
пре·ломл·я́тел·ь, -я, (*m*) refractor
пре·ломл·я́·ть, -ют, (*imp*) refract
пре·ломл·я́ющ·ий, -ая, -ее, (*pres part act*) of **пре·ломл·я́·ть**; refractive, refracting
прé·л·ый, -ая, -ое, (*a*) rotten, fusty
прé·л·ь, -и, (*f*) rot
премиáль·н·ый, -ая, -ое, (*a*) of **прéм·и·я**
прем·ир·овáть, -уют, (*imp and perf*) award a prize/bonus (to); reward
прéм·и·я, -и, (*f*) prize; bonus, premium, gratuity
премьéр·, -а, (*m*) prime minister; (theat) lead, star
премьéр·а, -ы, (*f*) (theat, cinema) premiere
пре·не·брег·á·ть, -ют, (*imp*) (+ *instr*) disregard, neglect; scorn
пре·не·бреж·éни·е, -я, (*n*) (*v n*) see **пре·не·брег·á·ть**; disregard, neglect
пре·не·бреж·úм·о (*adv*) negligibly
~ мáл·ый, -ая, -ое, (*a*) negligible
пре·не·бреж·úтельн·ый, -ая, -ое, (*a*) careless, neglectful
пре·не·брéч·ь, (*fut 3rd sing, pl*) **пре·-не·бреж·ёт, пре·не·брег·у́т,** (*past masc sing*) **пре·не·брёг·,** (*perf*); see **пре·не·брег·á·ть**
прé·ни·е, -я, (*n*) (*v n*) (see **прé·ть**) **прéн·и·я, -й,** (*n pl*) debate
пренúт·, -а, (*m*) (*min*) prehnite
пренúт·ов·ый, -ая, -ое, (*a*) prehnite, prehnitic

пре·облад·áни·е, -я, (*n*) predominance, prevalence; (telecom) bias
пре·облад·á·ть, -ют, (*imp*) predominate, prevail
пре·облад·áющ·ий, -ая, -ее, (*pres part act*) of **пре·облад·á·ть**; predominating, predominant, prevalent, prevailing
пре·ображ·á·ть, -ют, (*imp*) transform, change
пре·образ·úть, -я́т, (*perf*) see **пре·-ображ·á·ть**
пре·образ·овáни·е, -я, (*n*) conversion, transformation; (math) transformation, transform, map, mapping
~, аналóго-цифр·óв·ое digitizing, analog-to-digital conversion (computers)
~ дá·нн·ых data conversion/reduction (computers)
~ информ·áци·и information processing (computers)
~ канон·úческ·ое (mech) canonical transformation
~, конфóрм·н·ое (math) conformal/ isogonal/equiangular transformation, conformal mapping/map
~ координáт· (math) transformation of coordinates
~, коррелятúв·н·ое (math) correlation
~, круг·ов·óе (math) cylindrical map
крут·изн·á пре·образ·овáни·я (*f*) (elec) conversion conductance
~ Лóренц·а (mech) Lorentz transformation
~, обрáт·н·ое inverse transformation, reconversion
~ от·ображ·éни·я (math) mapping
~, ротациóн·н·ое (phys) rotational transformation
~, тóч·ечн·ое (math) transform
~ Фурьé (math) Fourier transform
~ част·от·ы́ (rad) frequency conversion
~ энéрг·и·и (elec) power conversion
пре·образ·óванн·ый, -ая, -ое, (*past part pass*) see **пре·образ·óвыва·ть**
пре·образ·овáтел·ь, -я, (*m*) (elec) converter, transducer; reorganizer, processor
~, амплитýд·н·о-времен·н·óй (elec) pulse-height pulse-width converter
~, аналóго-цифр·ов·óй digitizer, analog-to-digital converter (computers)
~, балáнс·ов·ый (elec) balanced converter
~, "вал·цúфра" shaft-position digitizer (computers)

пре·образ·ова́тел·ь
~, **венти́ль·н·ый** inverted rectifier, inverter
~, **вибрацио́н·н·ый** vibrator, chopper
~, **вращ·а́ющ·ийся** rotary/electro-mechanical converter
~ **давл·е́ни·я** pressure transducer
~, **дво·и́чн·ый** binary quantizer (computers)
~, **ём·костн·ый** (elec) capacitance pick-off
~ **из·обра́ж·е́ни·й** (TV) image translator, image converter tube
~ **и́мпульс·н·ых сигна́л·ов** (rad) wave sampler
~ **и́мпульс·ов** pulse converter
~, **индукти́в·н·ый** induction-type transducer
~, **индукцио́н·н·ый** induction-type transducer
~, **каска́д·н·ый** (elec) cascade/motor converter
~ **ко́д·а** code translator (computers)
~, **колле́ктор·н·ый** (elec) frequency changer
~, **лин·е́йн·ый** linear quantizer (computers)
~, **лин·е́йн·о-логарифм·и́ческ·ий** linear-to-log converter (computers)
~, **магни́т·н·ый** magnetic transducer
~, **магни́то·индукцио́н·н·ый** magneto-induction transducer
~, **магни́то·стрикцио́н·н·ый** magnetostriction oscillator
~ **мо́щ·ност·и** transducer
~ **нейтро́н·н·ого по·то́к·а** (nucl) neutron flux converter
~, **не·пре·ры́в·н·о-дискре́т·н·ый** analog-to-digital converter
~, **обра́т·н·ый** inverter
~, **одно·я́кор·н·ый** (elec) rotary/synchronous converter
~, **ом·и́ческ·ий** variable-resistance transducer
~, **при·ём·н·ый** receiving transducer
~, **пьезо·электр·и́ческ·ий** piezoelectric transducer
~, **радио·акти́в·н·ый** radioactive transducer
~ **с фикс·и́рующ·ей звёзд·очк·ой** (elec) star-wheel digitizer
~, **стат·и́ческ·ий** (elec) rectifier
~ **то́к·а** (elec) converter
~ **фаз·** (elec) phase inverter
~, **ферро·дина́м·и́ческ·ий** (elec) moving-iron transducer

пре·образ·ова́тел·ь
~, **фото·электр·и́ческ·ий** (elec) photoelectric pickoff
~, **функциона́ль·н·ый** (autom) function generator
~, **цифр·ов·о́й** digitizer (computers)
~ **част·от·ы́** (rad) frequency converter
~ **част·от·ы́, ла́мп·ов·ый** (rad) frequency converter, mixer, first detector
~ **част·от·ы́, электро·маши́н·н·ый** frequency changer
~, **шаг·ов·ый** step-switch converter (computers)
~, **электро·люминисце́нт·н·ый** electroluminescent transducer
~, **электро·маши́н·н·ый** (elec) electromechanical/rotary converter (current converter); (rad) frequency changer
~, **электронно·опт·и́ческ·ий** (TV) image converter
~, **электро́н·н·ый функциона́ль·н·ый** cathode-ray function generator, photoformer
пре·образ·ова́тельн·ый, -ая, -ое, (a) converting
~ **под·ста́нци·я** (f) (elec) converting substation
пре·образ·ова́ть, -у́ют, (perf) see **пре·образ·о́выва·ть**
пре·образ·о́выва·ть, -ют, (imp) transform, change, reorganize, convert
пре·образ·у́ющ·ий, -ая, -ее, (pres part act) see **пре·образ·о́выва·ть**
~ **фу́нкци·я** (f) transfer function (cybernetics)
пре·одол·ева́·ть, -ют, (imp) overcome, surmount
пре·одол·е́·ть, -ют, (perf) see **пре·одол·ева́·ть**
препара́т·, -а, (m) (chem) preparation; specimen (for biological analysis etc.)
препара́т·н·ый, -ая, -ое, (a) of **препара́т·**
препарато·вод·и́тел·ь, -я, (m) mechanical stage (of microscope)
препар·а́тор·, -а, (m) assistant (in laboratory); demonstrator (at lecture)
препар·и́р·овать, -уют, (imp and perf) prepare
пре·по·дав·а́тел·ь, -я, (m) teacher, instructor
пре·под·нес·ти́, -у́т, (perf) see **пре·под·нос·и́ть**
пре·под·нос·и́ть, -я́т, (imp) present
пре·по́н·а, -ы, (f) obstacle, impediment

пре·про·вод·и́ть, -я́т, (*perf*) *see* **пре·-про·вожд·а́ть**

пре·про·вожд·а́ть, -ют, (*imp*) forward; escort

пре·пя́т·стви·е, -я, (*n*) obstacle, hindrance, impediment

∼, стер·и́ческ·ое (chem) steric hindrance

пре·пя́т·ств·овать, -уют, (*imp*) (+ *dat*) hinder, impede, hamper, inhibit

пре́·рв·анност·ь, -и, (*f*) discontinuity

пре́·рв·анн·ый, -ая, -ое, (*past part pass*) *see* **пре·рыв·а́·ть**

пре·рв·а́ть, -у́т, (*perf*) *see* **пре·ры·в·а́·ть**

пре́ри·я, -и, (*f*) (geog) prairie

пре·рыв·а́ни·е, -я, (*n*) (*v n*) of **пре·-рыв·а́·ть,** interruption, break, discontinuity

пре·рыв·а́тел·ь, -я, (*m*) interrupter, (elec) contact breaker; chopper, ticker (of rays of any kind)

∼, игнитро́н·н·ый (weld) ignitron control; (M/T) ignition interrupter

∼ из·луч·éни·я radiation chopper

∼, камерто́н·н·ый tuning-fork interrupter

∼, конта́кт·н·ый (elec) contact breaker

∼ луч·а́ beam chopper

∼, молот·о́чков·ый (elec) hammer break

∼ пуч·к·а́ (phys) beam chopper

∼, сверх·бы́стр·ый (elec) superchopper

∼, тиратро́н·н·ый (weld) thyratron control

пре·рыв·а́·ть, -ют, (*imp*) interrupt, break off, discontinue; (elec) chop, break; **-ся** (*pass*)

пре·рыв·а́ющ·ийся, -аяся, -ееся, (*pres part act*) of **пре·рыв·а́·ться;** intermittent, discontinuous

прерывисто·пере·горо́д·чат·ый, -ая, -ое, (*a*) fragmentimurate

прерывисто·пе́р·ист·ый, -ая, -ое, (*a*) (bot) interruptedly pinnate

пре·ры́в·истост·ь, -и, (*f*) ragged/ irregular state/condition, state of being full of breaks

пре·ры́в·ист·ый, -ая, -ое, (*a*) intermittent, discontinuous, interrupted

∼ генер·а́ци·я (*f*) (elec) squegging, squagging, squitter

∼ огóн·ь (*m*) (nav) intermittent light

∼ спектр· (*m*) discontinuous spectrum

пре·ры́в·ист·ый, -ая, -ое

∼ ток· (*m*) (elec) intermittent current

пре·ры́в·ност·ь, -и, (*f*) lack of continuity; intermittence

пре·ры́в·н·ый, -ая, -ое, (*a*) discontinuous; interrupting, chopping

пресби·опи́·я, -и, (*f*) (med) presbyopia

пре·сек·а́·ть, -ют, (*imp*) suppress, stop

пре·селéктор·, -а, (*m*) preselector

пресéрв·ы, -ов, (*m pl*) (food) preserves, preserved foods

пре·сéч·ь, (*fut 3rd sing, pl*) **пере·-сеч·ёт, пере·сек·у́т,** (*past masc sing*) **пере·сéк·,** (*perf*); *see* **пре·се·к·а́·ть**

пре·слéд·овани·е, -я, (*n*) pursuit, pursuing, following up; persecution, victimization

крив·а́я пре·слéд·овани·я (*f*) curve of pursuit (air tactics)

∼, суд·éбн·ое (law) prosecution

пре·с·мык·а́·ться, -ются, (*imp*) creep, crawl

пре·с·мык·а́ющ·ееся, -егося, (*n decl as adj*) (zool) reptile, (*pl*) *Reptilia*

пресно·во́д·ност·ь, -и, (*f*) freshness (of water)

пресно·во́д·н·ый, -ая, -ое, (*a*) freshwater

пре́сн·ост·ь, -и, (*f*) lack of taste, flatness; lack of point, uninterestingness

пре́сн·ый, -ая, -ое, (*a*) fresh, sweet (water); unleavened (bread); unflavoured, insipid (food)

пре́сс·, -а, (*m*) (*see also under associated words*) (mech) press

∼, двух·стó·ечн·ый double-arm press

∼, ис·пыт·а́тельн·ый compression testing machine

∼, карусéль·н·ый rotary press

∼, колéн·н·ый toggle press

∼, криво·ши́п·н·ый eccentric press

∼, лéнт·очн·ый wire-cut brickmaking press

∼, много·эта́ж·н·ый (rubber) multiplaten press, multidaylight press

∼, пли́т·очн·ый platen press

∼, рé·ечн·ый bench arbor press

∼, рыча́ж·н·ый bench arbor press

∼, черв·я́чн·ый extruding machine, extruder

∼, эта́ж·н·ый (rubber) platen press

∼, 3-эта́ж·н·ый (rubber) 3-daylight press

пресс·а, -ы, (*f*) the press (newspapers and periodicals)

пресс-дестиллат·, -а, (*m*) (oil) pressure distillate

пресс-камер·а, -ы, (*f*) (rubber) shaping bag

пресс-котёл·, -тл·а́, (*m*) (rubber) press heater

пресс-кро́ш·к·а, -и, (*f*) (plast) coarse moulding powder

пресс·лит·ь·ё, -я́, (*n*) (plast) transfer moulding/molding

пресс-масл·ёнк·а, -и, (*g pl*) **-нок·,** (*f*) pressure lubricator

пресс·материа́л·, -а, (*m*) (plast) moulding/molding compound

пресс-но́ж·ниц·ы (*f pl*) power-driven shears

пресс·ова́льн·ый, -ая, -ое, (*a*) press, pressing

пресс·ова́ни·е, -я, (*n*) (*v n*) of **пресс·ова́ть;** (met) drop/pressure forging; (plast) compression moulding, molding

~, лить·ев·о́е (plast) transfer/flow moulding/molding

~, прям·о́е (plast) compression moulding

 у·си́л·и·е пресс·ова́ни·я (*n*) (plast) compressing/moulding/molding force

пресс·ова́нн·ый, -ая, -ое, (*past part pass*) of **пресс·ова́ть**

~ порош·о́к· (*m*) compacted powder, powder compact

пресс·ова́ть, -у́ют, (*imp*) press; compress; (powder met) compact; (plast) press, mould

пресс·о́вк·а, -и, (*g pl*) **-вок·,** (*f*) see **пресс·ова́ни·е**

пресс·о́вочн·ый, -ая, -ое, (*a*) press, pressing, compacting

~ материа́л· (*m*) (plast) moulding/molding material

пре́сс·ов·ый, -ая, -ое, (*a*) of **пресс·**

~ маши́н·а (*f*) (met cast) squeezer, squeezing machine

~ оригина́л· (*m*) master (sound record)

прессо·ста́т·, -а, (*m*) barostat

пресс-о·ста́т·ок·, -т·к·а, (*m*) press waste

пресс-под·бо́р·щик·, -а, (*m*) (agr) pickup bailer

пресс-порош·о́к·, -ш·к·а́, (*m*) moulding/molding powder; (met) compacting powder

пресс-с·тек·а́тел·ь, -я, (*m*) run-off press (grapes)

пресс·то́ч·к·а, -и, (*g pl*) **-ч·ек·,** (*f*) pressure point (in press)

пресс·фо́рм·а, -ы, (*f*) (*see also* **фо́рм·а**) (plast) compression mould/mold

пресс-шпа́н·, -а, (*m*) (paper) press-board

пресс-штёмпел·ь, -я, (*m*) press plunger

пре·сто́л·, -а, (*m*) throne

престо́·метр·, -а, (*m*) fluid gauge/gage

пре·ступл·ёни·е, -я, (*n*) crime, offence

пре·сыщ·ёни·е, -я, (*n*) satiety, super-saturation

пре·сы́щ·енн·ый, -ая, -ое, (*past part pass*); supersaturated

пр·ёт, (*fut 3rd sing*) of **пер·е́ть**

пре·твор·ёни·е, -я, (*n*) conversion; realization

пре·твор·и́ть, -я́т, (*perf*) see **пре·-твор·я́·ть**

пре·твор·я́·ть, -ют, (*imp*) change, transform

~ в жизн·ь put into practice, realize

претенд·ова́ть, -у́ют, (*imp*) claim

претенз·ио́нн·ый, -ая, -ое, (*a*) of **претёнз·и·я**

претёнз·и·я, -и, (*f*) claim; complaint, grievance

пре·терп·ева́·ть, -ют, (*imp*) suffer, endure, undergo

пре·терп·е́ть, -ят, (*perf*) see **пре·-терп·ева́·ть**

пре́·ть, -ют, (*imp*) rot; simmer

пре·у·велич·ёни·е, -я, (*n*) exaggeration

пре·у·вели́ч·енн·ый, -ая, -ое, (*past part pass*) see **пре·у·вели́ч·ива·ть**

пре·у·вели́ч·ива·ть, -ют, (*imp*) exaggerate, overstate

пре·у·вели́ч·ить, -ат, (*perf*) see **пре·-у·вели́ч·ива·ть**

пре·у·меньш·а́·ть, -ют, (*imp*) underestimate, understate, belittle, minimize

пре·у·меньш·ёни·е, -я, (*n*) (*v n*) see **пре·у·меньш·а́·ть;** understatement

пре·у·меньш·и́ть, -а́т, (*perf*) see **пре·-у·меньш·а́·ть**

пре·у·по·ря́д·оченн·ый, -ая, -ое, (*a*) preordered

пре·у·спев·а́·ть, -ют, (*imp*) succeed, be successful

пре·у·спе́·ть, -ют, (*perf*) see **пре·-у·спев·а́·ть**

префикс·, -а, (*m*) (gram) prefix

пре·ход·я́щ·ий, -ая, -ее, (*a*) transient

прецесс·ио́н·н·ый, -ая, -ое, (*a*) precessional

~ **момент·** (*m*) precession

прецесс·и́р·овать, -уют, (*imp and perf*) precess

прецесс·и·я, -и, (*f*) precession

прецизио́н·н·ый, -ая, -ое, (*a*) precision

преципит·а́т·, -а, (*m*) precipitate; (agr) dicalcium phosphate, precipitated phosphate

~ **фосфа́т·а** (agr chem) precipitated phosphate

преципит·а́ци·я, -и, (*f*) precipitation

преципит·и́н·, -а, (*m*) (biol) precipitin

при (*prep + prepositional case*) in the presence of, at, with, near; as a result of, thanks to, due to; in/at the time of, when, on, during; attached to (organizationally)

~ **из·уч·е́ни·и** when studying, in the study of

~ **на·ли́ч·и·и** in the presence of, if/where there is, with

~ **об·луч·е́ни·и** during/on irradiation/exposure

~ **по́·мощ·и** by means of, by, with

~ **ско́р·ост·и** at a rate/speed/velocity of

~ **у·сло́в·и·и** under/on the condition, in conditions of

~ **э́т·ом** in this case, here

при- (*as verbal prefix*) *signifies:* — (1) to the destination, up to; (2) to a result; (3) to contact, to form an extension/addition; (4) pressing downwards; (5) to/towards oneself; (6) incomplete/limited action (*often in conjunction with verbs which already have another prefix*)

при- (*as noun or adj prefix*) *when not of verbal origin signifies:* — located near, adjoining

при·ба́в·ить, -ят, (*perf*) *see* **при·-бавл·я́·ть**

при·ба́в·к·а, -и, (*g pl*) **-в·ок·,** (*f*) (*v n*) *see* **при·бавл·я́·ть;** addition

при·бавл·е́ни·е, -я, (*n*) (*v n*) *see* **при·-бавл·я́·ть;** increase; (print) appendix

при·бавл·я́·ть, -ют, (*imp*) add; increase; **-ся** (*pass*); rise (of water), wax (of moon), get longer (of daylight)

при·ба́в·очн·ый, -ая, -ое, (*a*) additional; (polit) surplus; (zool) accessory

при·байка́ль·ск·ий, -ая, -ое, (*a*) (geog) Baikal

при·балт·и́йск·ий, -ая, -ое, (*a*) Baltic

при·бег·а́·ть, -ют, (*imp*) resort to, have recourse to; run up to, arrive at the run

при·бе́г·нуть, -нут, (*perf*) resort to, have recourse to

при·беж·а́·ть, (*fut 3rd pl*) **при·бе-г·у́т,** (*perf*) run up to, arrive at the run

при·бе́ж·ищ·е, -а, (*i*) **-ем,** (*n*) refuge

при·берег·а́·ть, -ют, (*imp*) save up, keep/reserve for

при·бер·ёт, (*fut 3rd sing*) of **при·-бр·а́ть;** (*perf*), *see* **при·бир·а́·ть** (*imp*)

при·бере́ч·ь, (*fut 3rd sing, pl*) **при·-береж·ёт, при·берег·у́т,** (*past masc sing*) **при·берёг·;** (*perf*), *see* **при·-берег·а́·ть**

при·бив·а́·ть, -ют, (*imp*) nail on/up; lay (by rain); carry to (by wind or water); (text) beat up

при·би́в·к·а, -и, (*f*) (*v n*) *see* **при·-бив·а́·ть**

при·бир·а́·ть, -ют, (*imp*) tidy, clear up

при·би́·т·ый, -ая, -ое, (*past part pass*) of **при·би́·ть**

при·би́·ть, (*fut 3rd pl*) **при·бь·ю́т,** (*perf*) *see* **при·бив·а́·ть**

при·ближ·а́·ть, -ют, (*imp*) bring/draw nearer, shorten (time), advance (date); **-ся** (*pass*); approach, come nearer; approximate (*imp only*)

при·ближ·е́ни·е, -я, (*n*) (*v n*) *see* **при·ближ·а́·ть;** approach; (math) approximation

~, гру́б·ое rough approximation

~, диофа́нт·ов·о (math) Diophantine approximation

по·пра́в·к·а пе́рв·ого при·бли-ж·е́ни·я (*f*) first-order correction

~, по·сле́д·овательн·ые (*pl*) successive approximations

ско́р·ост·ь при·ближ·е́ни·я (*f*) approach speed

~, то́ч·ечн·ое point approximation

при·ближ·ённост·ь, -и, (*f*) approximation, approximate nature; degree of approximation; proximity

при·ближ·ённ·ый, -ая, -ое, (*past part pass*) *see* **при·ближ·а́·ть;** approximate, rough; approximated

при·близ·и́тельн·о   (*adv*) approximately, roughly

при·близ·и́тельн·ый, -ая, -ое,   (*a*) approximate, rough

при·бли́з·ить, -ят,   (*perf*) *see* **при·-ближ·а́·ть**

при·бо́·й, -я,   (*m*) surf, breakers; increase (in volume of broken parts compared with whole)

    зна́к·и при·бо́·я   (*pl*) (geol) wave marks

    ли́н·и·я при·бо́·я   (*f*) (ocean) wave front

при·бо́й·ник·, -а,   (*m*) rammer (of gun)

при·бо́й·н·ый, -ая, -ое,   (*a*) of **при·-бо́·й**

при·бо́р·, -а,   (*m*) (*and see under associated adj*); instrument, apparatus, device; set (of articles)

~ **быт·ов·о́го об·слу́ж·ивани·я** domestic appliance, labour-saving device

~, **водо·ме́р·н·ый** water gauge

~, **да·ю́щ·ий** transmitter

~, **из·мер·и́тельн·ый** measuring instrument, meter

~, **контро́ль·н·о-из·мер·и́тель·н·ые** (*pl*) instrumentation

~, **контро́ль·н·ый** check/reference instrument; (nucl) monitoring instrument; repeat receiver

~, **нул·ев·о́й** zero-deflection instrument

~, **образ·цо́в·ый** standard instrument

~, **пред·о·хран·и́тельн·ый** safety device

~, **регистр·и́рующ·ий** recording instrument

~, **реш·а́ющ·ий** (autom) resolver

~, **само·пи́ш·ущ·ий** recording instrument

~, **стре́л·очн·ый** pointer-type instrument

~, **с·чёт·н·о-реш·а́ющ·ий** *see* **вы·-числ·и́тельн·ая маши́н·а,** computer

~, **то́ч·н·ый** precision instrument

~, **универса́ль·н·ый из·мер·и́тель·н·ый** multimeter

~, **у·правл·е́ни·я** control/regulating instrument

~ **у·правл·е́ни·я артиллер·и́йск·им огн·ём** (mil) fire control instrument predictor

~, **эталóн·н·ый** standard instrument

при·бор·и́ст·, -а,   (*m*) instrument operator

при·бо́р·н·ый, -ая, -ое,   (*a*) of **при·-бо́р·**

~ **доск·а́**   (*f*) (a/c) instrument panel

прибоpo·стро·е́ни·е, -я,   (*n*) instrument making/manufacture/engineering

прибоpo·стро·и́тел·ь, -я,   (*m*) instrument maker

при́·бр·анн·ый, -ая, -ое,   (*past part pass*) *see* **при·бир·а́ть**

при·бр·а́ть,   (*fut 3rd pl*) **при·бер·у́т,** (*perf*) *see* **при·бир·а́·ть**

при·бре́ж·н·ый, -ая, -ое,   (*a*) coastal, (naut) inshore, (geog) off-shore; riverside, lakeside, riparian, riparious, riparial

при·бу́д·ет,   (*fut 3rd sing*) of **при·-бы́·ть,** (*perf*); *see* **при·быв·а́·ть** (*imp*)

при·быв·а́ни·е, -я,   (*n*) increase, rise

при·быв·а́·ть, -ют,   (*imp*) arrive, come in; increase, grow, rise (of water), wax (of moon)

при·быв·а́ющ·ий, -ая, -ее,   (*pres part act*) of **при·быв·а́·ть**; incoming

при́·бы·л·ь, -и,   (*f*) profit; increase, rise; (met) feeder head, ingot top; riser (of casting)

~, **за·кры́·т·ая** (met cast) blind riser

~, **от·во́д·н·ая** (met cast) blind riser

~, **от·кры́·т·ая** (met cast) open riser

~, **чи́ст·ая** (fin) net profit

при́·быль·н·ый, -ая, -ое,   (*a*) of **при́·бы·л·ь** (q.v.); profitable

~ **над·ста́в·к·а**   (*f*) (met cast) refractory head, hot top, shrinkhead, sinkhead, superimposed feeder head

~ **над·ста́в·к·а, обо·грев·а́ем·ая** exothermic feeder head (continuous casting)

~ **част·ь**   (*f*) (met cast) top, feeder head

при·бы́т·и·е, -я,   (*n*) arrival

    пункт· при·бы́т·и·я   (*m*) (nav) destination point, point of arrival

при·бы́·ть,   (*fut 3rd pl*) **при·бу́д·ут,** (*perf*) *see* **при·быв·а́·ть**

при·бъ·ёт,   (*fut 3rd sing*) of **при·-би́·ть,** (*perf*); *see* **при·бив·а́·ть** (*imp*)

при·ва́л·, -а,   (*m*) halt; (naut) coming/going alongside

при·ва́л·ива·ть, -ют,   (*imp*) roll/push up against; (naut) come alongside

при·вал·и́ть, -́ят,   (*perf*) *see* **при·ва́·л·ива·ть**

при·ва́ль·н·ый, -ая, -ое,   (*a*) of **при·ва́л·**

~ **брус·**   (*m*) (shipb) rubbing strip/strake

при·ва́р·енн·ый, -ая, -ое, (*past part pass*) *see* **при·ва́р·ива·ть**

при·ва́р·ива·ть, -ют, (*imp*) weld on, burn on to; (met) burn in (a furnace)

при·вар·и́ть, -я́т, (*perf*) *see* **при·ва́р·ива·ть**

при·ва́р·к·а, -и, (*f*) (*v n*) *see* **при··ва́р·ива·ть**

~ **шпи́л·ек·** stud welding

при·вар·н·о́й, -а́я, -о́е, (*a*) welding (of joints etc.)

~ **флан·ец·** (*m*) welding flange, flange for fixing by welding

при·ва́р·ок·, -р·к·а, (*m*) victuals

при·ва́р·очн·ый, -ая, -ое, (*a*) of **при·ва́р·ок·**

~ **до·во́льн·стви·е** (*n*) rations

при·вед·ёни·е, -я, (*n*) (*v n*) *see* **при··вод·и́ть**; deduction, (math) reduction

~ **до·каз·а́тельств·** production of proof

~ **глуб·и́н·** (hydr) reduction of soundings

~ **к абсу́рд·у** (math) reductio ad absurdum

~ **к сре́д·н·ему у́·ровн·ю мо́р·я** (nav) correction for height above mean sea level

~ **к шир·от·е́** (nav) reduction for latitude

~ **по·до́б·н·ых член·ов** (math) cancelling of the terms
у́·ровен·ь при·вед·е́ни·я (*m*) datum level

при·вед·ённ·ый, -ая, -ое, (*past part pass*) *see* **при·вод·и́ть**; brought; deduced, derived, (math) reduced, normalized; recurring

~ **ма́сс·а** (*f*) (phys) reduced mass

~ **на·пряж·е́ни·е** (*n*) (elec) lumped/reduced voltage

~ **характер·и́стик·а** (*f*) total characteristics (of vacuum tube)

~ **чувсв·и́тельност·ь** (*f*) (instr) factor/figure of merit

при·вез·ти́, -у́т, (*perf*) bring, deliver

при·вёл, (*past masc sing*) of **при··вес·ти́,** (*perf*); *see* **при·вод·и́ть** (*imp*)

при·ве́рж·енец·, -нц·а, (*m*) adherent, follower

при·ве́рж·енност·ь, -и, (*f*) adherence

при·вер·н·у́ть, -у́т, (*perf*) screw up/in/on; turn down (lamps etc.)

при·верт·е́ть, -я́т, (*perf*) screw up/on/in; bind round, bind/tie up

при·вёрт·ыва·ть, -ют, (*imp*) *see* **при·вер·н·у́ть, при·верт·е́ть**

при·ве́с·, -а, (*m*) increase/gain in weight, weight increment

при·ве́с·ить, -ят, (*perf*) *see* **при·ве́·ш·ива·ть**

при·ве́с·ок, -ка (*m*) side chain (polymers)

при·вес·ти́, (*fut 3rd pl*) **при·вед·у́т,** (*past masc sing*) **при·вёл,** (*perf*); *see* **при·вод·и́ть**

при·ве́т·, -а, (*m*) greeting

при·ве́т·ств·овать, -уют, (*imp*) greet

при·ве́ш·ива·ть, -ют, (*imp*) hang up, suspend

при·вив·а́·ть, -ют, (*imp*) (med) inoculate, vaccinate; (bot etc.) graft; (bot) inoculate; splice; inculcate, implant; **-ся** (*pass*); grow together; take (of vaccination etc.)

при·ви́в·к·а, -и, (*g pl*) **-в·ок·,** (*f*) (*v n*) *see* **при·вив·а́·ть**; (med) inoculation, vaccination; (hortic etc.) graft, grafting

~ **аблакт·иро́вк·ой** grafting by approach

~ **в кро́н·у** (hortic) top-working

~ **в·рас·ще́п·** cleft grafting

~ **глаз·к·о́м** (hortic) inoculation

~ **с·ближ·е́ни·ем** grafting by approach

при·ви́в·ок·, -в·к·а, (*m*) grafting, graft

при·ви́в·очн·ый, -ая, -ое, (*a*) (med) inoculating; (hortic) grafting, graft; (bot) inoculation

привиде́ньев·ые, -ых, (*pl*) (ent) stick insects

привиле́г·и·я, -и, (*f*) privilege

при·винт·и́ть, -я́т, (*perf*) *see* **при··ви́нч·ива·ть**

при·ви́нч·ива·ть, -ют, (*imp*) screw on

при·ви́·т·ый, -ая, -ое, (*past part pass*) *see* **при·вив·а́·ть**; graft (polymers)

при·ви́·ть, (*fut 3rd pl*) **при·вь·ю́т,** (*perf*) *see* **при·вив·а́·ть**

при́·в·кус·, -а, (*m*) flavour, taste

при·влек·а́тельн·ый, -ая, -ое, (*a*) attractive

при·влек·а́·ть, -ют, (*imp*) attract, draw, enlist

~ **к от·ве́т·ственност·и** call to account, (law) institute proceedings against

при·вле́ч·ь, (*fut 3rd sing, pl*) **при··влеч·ёт, при·влек·у́т,** (*past masc sing*) **при·влёк·,** (*perf*); *see* **при·влек·а́·ть**

при·в·нес·ти́, -у́т, (*perf*), *see* **при··в·нос·и́ть**

при·в·нос·, -а,   (*m*) addition, introduction

при·в·нос·ить, ⸗ят,   (*imp*) introduce

при·вод·, -а, (*m*) (*and see under associated adj*); drive, driving gear, operating gear, actuator; (nav) homing

~ **агрегат·ов** (a/c) accessory drive

~, **груз·ов·ой** gravity operating gear

~, **грушп·ов·ой** common drive, group drive

~ **диаграмм·ы** (instr) chart drive

~, **зад·н·ий** (M/T) rear drive, rear-wheel drive

~, **криво·шип·н·ый** crank drive

~, **кулач·ков·ый** cam drive

~ **на обод·е** rim drive

~, **один·очн·ый** independent drive, (elec) individual drive

~, **перед·н·ий** (M/T) front drive, front-wheel drive

~, **прям·ой** direct drive

~, **ремён·н·ый** belt drive

~, **реверсив·н·ый** reversing gear/mechanism

~, **рул·ев·ой** steering gear

~, **руч·н·ой** hand operation, manual control

~, **рыча́ж·н·ый** lever operating gear

~ **с земл·и** (air) ground-controlled approach, GCA

~, **сил·ов·ой** motor operating mechanism

~, **след·ящ·ий** servomechanism

~, **трансмиссион·н·ый** transmission drive; (elec) group drive

~, **храп·ов·ой** ratchet drive

~, **цеп·н·ой** chain drive

при·вод·имост·ь, -и, (*f*) (math) reducibility

при·вод·им·ый, -ая, -ое, (*pres part pass*) of **при·вод·ить**; driven; cited, quoted; reducible

при·вод·ить, ⸗ят, (*imp*) (+ *prep* **в** + *noun, verb expresses meaning of noun*) bring, lead; quote, cite, give; (math) reduce

~ **в готов·ност·ь** get/make ready

~ **в движ·ени·е** set in motion, start

~ **в дей·стви·е** actuate, start up

~ **в ис·полн·ени·е** execute, carry out

~ **в по·ряд·ок·** put in order

~ **в равно·вес·и·е** equilibrate

~ **к масштаб·у** reduce to scale, scale, scale down

~ **к пере·грев·у** cause overheating, result in overheating

при·вод·к·а, -и, (*g pl*) **-д·ок·,** (*f*) (print) register, registration, (phot) register work

при·вод·нéни·е, -я, (*n*) entry into water, water entry (of diving or falling objects)

при·вод·н·иться, -ятся, (*perf*) enter the water

при·вод·н·ой, -ая, -ое, (*a*) driving; power, power driven; (nav) homing

~ **пресс·** (*m*) power press

~ **радио·луч·** (*m*) (air nav) approach beam

~ **радио·маяк·** (*m*) (air nav) homing beacon

~ **рем·éн·ь** (*m*) driving belt

~ **станци·я** (*f*) (nav) homing beacon, homer

~ **станци·я, ближ·н·яя** (air) inner marker beacon

~ **станци·я, даль·н·яя** (air) outer marker beacon

при·воз·, -а, (*m*) (*v n*) *see* **при·во·з·ить**; delivery, consignment

при·воз·ить, ⸗ят, (*imp*) bring, deliver

при·воз·н·ой, -ая, -ое, (*a*) = **при·воз·н·ый**

при·воз·н·ый, -ая, -ое, (*a*) delivery; delivered; imported

при·во·й, -я, (*m*) scion (in grafting plants), graft

при·врат·ник·, -а, (*m*) doorkeeper, gatekeeper; (anat) pylorus

при·вык·а·ть, -ют, (*imp*) get/be accustomed/used to

при·вык·нуть, -нут, (*perf*) *see* **при·-вык·а·ть**

при·выч·к·а, -и, (*g pl*) **-ч·ек·,** (*f*) habit

при·выч·ный, -ая, -ое, (*a*) usual, habitual

при·вь·ёт, (*fut 3rd sing*) of **при·ви·ть,** (*perf*); *see* **при·вив·а·ть** (*imp*)

при·вяж·ет, (*fut 3rd sing*) of **при·-вяз·ать,** (*perf*); *see* **при·вяз·ыва·ть** (*imp*)

при·вяз·анн·ый, -ая, -ое, (*past part pass*) *see* **при·вяз·ыва·ть**; attached

при·вяз·ать, (*fut 3rd pl*) **при·вя·ж·ут,** (*perf*) *see* **при·вяз·ыва·ть**

при·вяз·к·а, -и, (*g pl*) **-з·ок·,** (*f*) (*v n*) *see* **при·вяз·ыва·ть**; (surv) closure, conjunction, tie, connection; fixation, fixing, referencing; (TV) clamping

у́·ровен·ь при·вяз·к·и (*m*) (TV) clamping level

~ **у́·ровн·я** (TV) clamping

при·вяз·н·ой, -ая, -ое, (*a*) tied, secured, fastened; securing, fastening
~ **аэростат·** (*m*) captive balloon
при·вяз·ыва·ть, -ют, (*imp*) tie, fasten, bind, attach, tether; (surv) locate, fix, key, tie, reference (to a map); **-ся** lock on (radar)
при·вязь, -и, (*f*) tie, binding, tether
при·гар·, -а, (*m*) burnt spot, burn; (met cast) burnt-on sand
при·гиб·а·ть, -ют, (*imp*) bend towards/downwards
при·глад·ить, -ят, (*perf*) *see* **при·глаж·ива·ть**
при·глаж·ива·ть, -ют, (*imp*) smooth
при·глас·ительн·ый, -ая, -ое, (*a*) inviting; (rail) call-on, calling-on
~ **с·вязь** (*f*) paging system
~ **сигнал·** (*m*) (rail) call-on signal/ aspect
при·глас·ить, -ят, (*perf*) *see* **при·глаш·а·ть**
при·глаш·а·ть, -ют, (*imp*) invite; call, call in; engage (for work)
при·глаш·ени·е, -я, (*n*) invitation
при·глуш·а·ть, -ют, (*imp*) damp down, muffle, deaden, mute
при·глуш·ить, -ат, (*perf*) *see* **при·глуш·а·ть**
при·гн·ать, (*fut 3rd pl*) **при·гон·ят,** (*perf*) *see* **при·гон·я·ть**
при·гн·уть, -ут, (*perf*) *see* **при·гиб·а·ть**
при·говар·ива·ть, -ют, (*imp*) (law) sentence
при·говор·, -а, (*m*) (law) sentence; (law) verdict, finding
~**, об·вин·ительн·ый** verdict of guilty; finding for the plaintiff
~**, о·правд·ательн·ый** verdict of not guilty; finding for the defendant
при·говор·ить, -ят, (*perf*) *see* **при·говар·ива·ть**
при·год·иться, -ятся, (*perf*) be/prove useful
при·год·ность·, -и, (*f*) suitability, usefulness, fitness, soundness
при·год·н·ый, -ая, -ое, (*a*) suitable, fit
при·гон·к·а, -и, (*g pl*) **-н·ок·,** (*f*) fitting, fitting together, matching; reseating (valve); (surv) adjustment
при·гон·очн·ый, -ая, -ое, (*a*) of **при·гон·к·а**
~ **в·став·к·а** (*f*) (mech eng) making-up strip

при·гон·я·ть, -ют, (*imp*) fit, fit on, fit in, work in; adjust, adapt; bring in (herds etc.)
при·гор·а·ть, -ют, (*imp*) burn, catch (in cooking)
при·гор·ел·ый, -ая, -ое, (*a*) burnt, caught (in cooking)
при·гор·еть, -ят, (*perf*) *see* **при·гор·а·ть**
при·город·, -а, (*m*) suburb
при·город·н·ый, -ая, -ое, (*a*) suburban; toll (of telephone systems)
при·гор·ок·, -р·к·а, (*m*) hill, knoll
при·горш·н·я, -и, (*g pl*) **-ей,** (*f*) handful
при·готавл·ива·ть, -ют, (*imp*) = **при·готовл·я·ть**
при·готов·ительн·ый, -ая, -ое, (*a*) preparatory
при·готов·ить, -ят, (*perf*) *see* **при·готовл·я·ть**
при·готовл·ени·е, -я, (*n*) (*v n*) *see* **при·готовл·я·ть;** preparation
при·готовл·енн·ый, -ая, -ое, (*past part pass*) *see* **при·готовл·я·ть**
~ **на** made of/with/from
~ **к** ready for
при·готовл·я·ть, -ют, (*imp*) prepare, prepare for, make ready
при·грев·а·ть, -ют, (*imp*) heat, warm
при·гре·ть, -ют, (*perf*) *see* **при·грев·а·ть**
при·да·ва·ть, -ют, (*imp*) give, impart; attach, ascribe
~ **важ·ность·** attach importance
~ **водо·не·про·ниц·аемость·** waterproof, make waterproof
~ **жёстк·ость·** stiffen
при·дав·ить, -ят, (*perf*) *see* **при·давл·ива·ть**
при·давл·енн·ый, -ая, -ое, (*past part pass*) *see* **при·давл·ива·ть**
при·давл·ива·ть, -ют, (*imp*) press, press down/against
при·дад·ут, (*fut 3rd pl*) of **при·да·ть,** (*perf*); *see* **при·да·ва·ть** (*imp*)
при·да·ни·е, -я, (*n*) (*v n*) *see* **при·да·ва·ть**
~ **об·тек·аемость·** streamlining
при·дат·ок·, -т·к·а, (*m*) appendage, adjunct, accessory, gadget; (anat) appendix, (*pl*) adnexa
~ **яич·к·а** (anat) epididymis
при·дат·очн·ый, -ая, -ое, (*a*) accessory, additional, (bot) adventitious
~ **пред·лож·ени·е** (*n*) (gram) subordinate clause

при·да́·ть, (*fut 3rd sing, pl*) **при··да́ст, при·дад·у́т,** (*past masc sing*) **при·да́·л,** (*perf*); *see* **при·да·ва́ть**

при·двиг·а́·ть, -ют, (*imp*) move up/closer; **-ся** (*intrans*); approach

при·дви́н·уть, -ут, (*perf*) *see* **при··двиг·а́·ть**

при·де́л·а·ть, -ют, (*perf*) *see* **при··де́л·ыва·ть**

при·де́л·ыва·ть, -ют, (*imp*) fix/fasten on, attach

при·дер·ётся, (*fut 3rd sing*) of **при··др·а́ться,** (*perf*); *see* **при·дир·а́·ться** (*imp*)

при·держ·а́ть, -ат, (*perf*) *see* **при··де́рж·ива·ть**

при·де́рж·ива·ть, -ют, (*imp*) hold back/up; **-ся** (*pass*); (+ *gen*) hold/keep/adhere to

пр·ид·ёт, (*fut 3rd sing*) of **при·йти́,** (*perf*); *see* **при·ход·и́ть** (*imp*)

при·дир·а́·ться, -ются, (*imp*) find fault with, cavil at

при·до́н·н·ый, -ая, -ое, (*a*) bottom, (ocean) benthonic, (biol) benthic

~ **трал·** (*m*) (naut) bottom sweep

при·доро́ж·н·ый, -ая, -ое, (*a*) roadside, road

при·др·а́ться, (*fut 3rd pl*) **при·дер·у́тся,** (*perf*) *see* **при·дир·а́·ться**

при·ду́м·а·ть, -ют, (*perf*) *see* **при··ду́м·ыва·ть**

при·ду́м·ыва·ть, -ют, (*imp*) think of/up, invent, devise

при·дых·а́ни·е, -я, (*n*) (ling) aspiration

при·дых·а́тельн·ый, -ая, -ое, (*a*) (ling) aspirate

при·е́д·ет, (*fut 3rd sing*) of **при·е́х·ать**

при·е́зд·, -а, (*m*) arrival, coming

при·езж·а́·ть, -ют, (*imp*) arrive, come

при·е́зж·ий, -ая, -ее, (*a*) newly-arrived; (as noun) newcomer, visitor

при·ём·, -а, (*m*) (*v n*) *see* **при·ни··ма́·ть;** reception, receipt; method, procedure; movement, stage (of process, activity); dose; admittance, enrolment; (*pl*) (mil) drill

   в три при·ём·а in three stages/movements

~ **во́з·дух·а** air induction

~, **гетероди́н·н·ый,** (rad) heterodyne reception, beat reception

~ **за·ка́з·ов** booking of calls (telephony)

~ **на раз·нес·ённ·ые анте́нн·ы** (rad) aerial diversity reception

при·ём

~, **не·у·сто́й·чив·ый** (rad) freak reception

~, **по́лн·ый** (surv) reciprocal observation

~, **руж·е́йн·ые** (*pl*) rifle drill

при·ем·и́стост·ь, -и, (*f*) responsiveness, susceptibility; (ICE) pickup, acceleration

   вре́м·я при·ем·и́стост·и (*n*) (autom) time constant

   по́мп·а при·ем·и́стост·и (*f*) (ICE) accelerator pump

при·ём·к·а, -и, (*g pl*) **-м·ок·,** (*f*) (*v n*) *see* **при·ним·а́·ть;** reception, receipt; acceptance

~ **про·до·во́ль·стви·я** provisioning

~ **то́пл·ив·а** fuelling

при·ёмл·ем·ый, -ая, -ое, (*a*) acceptable, admissible

при·ём·н·ая, -ой, (*f decl as adj*) reception room; (rad) receiving room

при·ём·ник·, -а, (*m*) receiver; bulb (of thermometer)

~ **воз·ду́ш·н·ых давл·е́ни·й** pilot static head (of air speed indicator)

~, **детéктор·н·ый** (rad) crystal receiver

~ **из·луч·е́ни·я** (nucl) radiation detector

~, **контро́ль·н·ый** (rad) monitoring receiver

~, **нейтроди́н·н·ый** (rad) neutrosonic receiver

~, **об·наруж·и́тельн·ый** warning receiver (radar)

~ **от·ве́т·ов** (nav) responser

~, **по·са́д·очн·ый** (nav) glide-path receiver

~ **прям·о́го у·сил·е́ни·я** (rad) straight receiver

~, **пере·хва́т·н·ый** intercept receiver

~, **по́·иск·ов·ый** (rad) search receiver

~ **стат·и́ческ·ого давл·е́ни·я** static tube (of air speed indicator)

~ **электро́н·ов** (phys, chem) electron sink

при·ём·н·о-пере·да·ю́щ·ая радио··ста́нци·я (*f*) (rad) transceiver

при·ём·н·ый, -ая, -ое, (*a*) *see also* **при·ём·н·ая** (*as noun*); receiving, reception; pickup; acceptance; inlet, suction

~ **акт·** (*m*) acceptance certificate

~ **анте́нн·а** (*f*) (rad) receiving aerial/antenna

при·ём·н·ый, -ая, -ое
~ **голо́в·к·а** (*f*) (instr) pickup/feeler head
~ **кассе́т·а** (*f*) take-up reel (e.g. of tape recorder)
~ **коми́сс·и·я** (*f*) selection committee; acceptance commission
~ **отве́рст·и·е** (*n*) (ICE) inlet port
~ **при·спосо́бл·е́ни·я** (*pl*) acceptance-inspection instruments
~ **про·стра́н·ств·о** (*n*) suction gallery/chamber (pump/pumping systems)
~ **ста́нци·я** (*f*) (rad) receiving set; receiving station
~ **тру́б·к·а** (*f*) (TV) picture tube, kinescope
~ **трубо·про·во́д·** (*m*) suction line
~ **у·стро́й·ств·о** (*n*) (autom) receiver; pouring system (continuous casting)
~ **час·ы́** (*pl*) hours of consultation; office hours
приёмо·от·ве́т·чик·, -а, (*m*) (rad) transponder
приемо·пере·да́т·чик·, -а, (*m*) (rad) transceiver
при·ём·оч·н·ый, -ая, -ое, (*a*) reception; acceptance
при·ём·щик·, -а, (*m*) acceptance inspector, examiner (in factory)
при·е́х·ать, (*fut 3rd pl*) **при·е́д·ут,** (*perf*) arrive, come
при·жа́т·ый, -ая, -ое, (*past part pass*) of **при·жа́т·ь**
~ **луч·** (*m*) (rad) low-altitude beam
при·жа́т·ь, (*fut 3rd pl*) **при·жм·у́т,** (*perf*) see **при·жим·а́·ть**
при·же́ч·ь, (*fut 3rd sing, pl*) **при·ж·ж·ёт, при·жг·у́т,** (*past masc sing*) **при·жёг·,** (*perf*); see **при·жиг·а́·ть**
при·жиг·а́ни·е, -я, (*n*) (med) cauterization
~, **электр·и́ческ·ое** (med) electrocautery
при·жиг·а́тельн·ый, -ая, -ое, (*a*) cauterant
при·жиг·а́·ть, -ют, (*imp*) cauterize
при·жи́зн·енн·ый, -ая, -ое, (*a*) contemporary, during one's life, while alive; intravital
при·жи́м·, -а, (*m*) clamp, clip, holder
при·жим·а́·ть, -ют, (*imp*) press, squeeze; tighten, clamp, screw
при·жи́м·н·ый, -ая, -ое, (*a*) clamping
~ **пла́н·к·а** (*f*) clamping strip

при·жм·ёт, (*fut 3rd sing*) of **при·-жа́т·ь,** (*perf*); see **при·жим·а́·ть** (*imp*)
приз·, -а, (*nom pl*) **-ы́,** (*g pl*) **-о́в,** (*m*) prize
~, **мор·ск·о́й** prize of war
при·за·бо́й·н·ый, -ая, -ое, (*a*) (min) face, at the face
при·зв·а́т·ь, (*fut 3rd pl*) **при·зов·у́т,** (*perf*) see **при·зыв·а́·ть**
при·зе́м·ист·ый, -ая, -ое, (*a*) squat
при·земл·е́ни·е, -я, (*n*) (air) touchdown, landing
~ **на три то́ч·к·и** (air) three-point landing
при·зе́м·н·о́й, -а́я, -о́е, (*a*) surface, ground
при́зм·а, -ы, (*f*) prism; vee block (lathework); gag (of press)
~ **вы·пир·а́ни·я** (civ eng) overturning/tilting triangle (retaining wall calculations)
~, **дисперси́он·н·ая** (opt) dispersion prism
~, **за·медл·я́ющ·ая** (opt) moderating prism
~, **из·мер·и́тельн·ая** (opt) wedge prism
~, **крыше·обра́з·н·ая** (opt) Amici/roof prism
~, **на·кло́н·н·ая** (math) oblique prism
~ **Ни́кол·я** (opt) Nicol prism, Nicol
~, **обора́ч·ивающ·ая** (opt) Amici/inverting/erecting prism
~ **об·ру́ш·е́ни·я** (civ eng) sliding triangle (retaining wall calculations)
~, **объекти́в·н·ая** (astron) objective prism
~, **по·вора́ч·ивающ·ая** (opt) deflecting prism
~, **пра́в·ильн·ая** (math) regular prism
~, **прям·а́я** (math) right prism
~ **прям·о́го зр·е́ни·я** direct-vision prism
~, **пяти·уго́ль·н·ая** penta prism
~, **спектра́ль·н·ая** (opt) dispersing prism
~, **тепл·ов·а́я** (nucl) thermal prism, sigma pile
~, **трёх·гра́н·н·ая** (math) triangular prism
~, **у·сту́п·чат·ая** (TV) prism drum, echelon prism
~, **четырёх·гра́н·н·ая** (math) quadrangular prism
призмат·и́ческ·ий, -ая, -ое, (*a*) prismatic

призмат·о́ид·, -а, (*m*) (math) prismatoid

при́зм·енн·ый, -ая, -ое, (*a*) prism, prismatic

~ ка́мер·а (*f*) (astron) slitless spectroscope, prismatic camera

призм·о́ид·, -а, (*m*) (math) prismoid

при·зна·ва́ть, -ю́т, (*imp*) recognize, acknowledge, admit

при́·знак·, -а, (*m*) sign, indication, symptom, trait; attribute (cybernetics); criterion, test

~ дел·и́мост·и (math) criterion of divisibility

при·зна́·ни·е, -я, (*n*) acknowledgement, recognition; admission

при·зна́·тельност·ь, -и, (*f*) gratitude

при·зна́·тельн·ый, -ая, -ое, (*a*) grateful

при·зна́·ть, -ю́т, (*perf*) *see* при·зна·ва́ть

при·зов·ёт, (*fut 3rd sing*) of при·зв·а́ть,(*perf*); *see* при·зыв·а́·ть (*imp*)

приз·ов·о́й, -а́я, -о́е, (*a*) prize

~ суд· (*m*) (law) prize court

при́·зр·ак, -а, (*m*) illusion, phantom

при́·зр·ачн·ый, -ая, -ое, (*a*) illusory

при·зы́в·, -а, (*m*) call, appeal; (mil) call up, conscription; slogan

при·зыв·а́·ть, -ю́т, (*imp*) call to/for/ up, summon

при·зыв·ни́к, -а́, (*m*) conscript

при·зыв·н·о́й, -а́я, -о́е, (*a*) conscription, call-up; (*m decl as adj*) conscript, draftee

~ во́з·раст· (*m*) call up age, military age

при·зы́в·н·ый, -ая, -ое, (*a*) of при·зы́в·

при́·иск·, -а, (*m*) alluvial/placer mine

при·иск·а́ни·е, -я, (*n*) (*v n*) of при·иск·а́ть

при·иск·а́ть, (*fut 3rd pl*) при·и́щ·ут, (*perf*) find

при·йти́, (*fut 3rd pl*) пр·ид·у́т, (*past masc sing*) при·шёл, (*perf*); *see* при·ход·и́ть

при·кабол·и́ть, -я́т, (*perf*) (naut) seize, mouse

при·ка́ж·ет, (*fut 3rd sing*) of при·каз·а́ть, (*perf*); *see* при·ка́з·ыва·ть (*imp*)

при·ка́з·, -а, (*m*) order, command

при·каз·а́ни·е, -я, (*n*) order, injunction, instruction

при·каз·а́ть, (*fut 3rd pl*) при·ка́ж·ут, (*perf*); *see* при·ка́з·ыва·ть

при·ка́з·ыва·ть, -ют, (*imp*) order, command, direct

при·ка́л·ыва·ть, -ют, (*imp*) pin/ fasten on; kill, slaughter

при·ка́п·ыва·ть, -ют, (*imp*) drip in, add dropwise

при·ка́рмл·ива·ть, -ют, (*imp*) lure (with bait); tame (by feeding)

при·кас·а́тельн·ый, -ая, -ое, (*a*) contact, touching

при·кас·а́·ться, -ются, (*imp*) touch, be tangent to/upon

при·кат·и́ть, -́ят, (*perf*) *see* при·ка́т·ыва·ть

при·ка́т·к·а, -и, (*f*) (*v n*) *see* при·ка́т·ыва·ть

при·ка́т·очн·ый, -ая, -ое, (*a*) of при·ка́т·к·а

~ стан·о́к· (*m*) tyre stitching/ grooving machine

при·ка́т·чик·, -а, (*m*) (rubber) applicator roller

при·ка́т·ывани·е, -я, (*n*) (*v n*) of при·ка́т·ыва·ть; (agr) packing/ compacting the soil

при·ка́т·ыва·ть, -ют, (*imp*) roll up, bring (rolling); roll down/in, pack; (rubber) stitch, roll on

при·ка́ч·енн·ый, -ая, -ое, (*past part pass*) *see* при·ка́т·ыва·ть

при·ки́д·к·а, -и, (*g pl*) -д·ок·, (*f*) (*v n*) *see* при·ки́д·ыва·ть; rough estimate

при·ки́д·ыва·ть, -ют, (*imp*) throw in/on, add; calculate roughly, reckon

при·ки́·н·уть, -ут, (*perf*) *see* при·ки́д·ыва·ть

при·кла́д·, -а, (*m*) butt, butt-stock (of rifle); trimmings (in and on clothes and shoes)

при·клад·н·о́й, -а́я, -о́е, (*a*) applied

~ на·у́к·и (*pl*) applied sciences

~ час· по́рт·а (*m*) establishment of port (tide)

~ час·, сре́д·н·ий mean high water lunitidal interval

при·кла́д·ыва·ть, -ют, (*imp*) put to, place against, apply; append, attach, enclose (in letter); -ся (*pass*); take aim

при·кле́·ен·ый, -ая, -ое, (*a*) glued, stuck, pasted

при·кле́·ива·ть, -ют, (*imp*) glue/stick/ paste on

при·кле́·ить, -ят, (*perf*) *see* при·кле́·ива·ть

при·кле́·й, -я, (*m*) (text) sizing (weight⁰/₀)

при·кле́й·к·а, -и, (*g pl*) -é·ек·, (*f*) (*v n*) *see* при·кле́·ива·ть

при·клеп·а́·ть -ют, (*perf*) *see* при·-клёп·ыва·ть

при·клёп·ыва·ть, -ют, (*imp*) rivet, rivet on; (naut) shackle (join shackles)

при·ключ·а́·ть, -ют, (*imp*) connect up with; -ся happen, occur

при·ключ·éни·е, -я, (*n*) occurrence, adventure

при·ключ·и́ть, -а́т, (*perf*) *see* при·-ключ·а́·ть

при·ков·а́ть, (*fut 3rd pl*) при·ку·ю́т, (*perf*) *see* при·ко́в·ыва·ть

при·ко́в·ыва·ть, -ют, (*imp*) forge, forge on, fix

при·ко́л·, -а, (*m*) (naut) mooring pile/stake; (agr) tethering stake

~, ста́в·ить на lay up (a ship)

при·кола́ч·ива·ть, -ют, (*imp*) nail/tack on

при·колот·и́ть, -ят, (*perf*) *see* при·-кола́ч·ива·ть

при·ко́л·от·ый, -ая, -ое, (*past part pass*) of при·кол·о́ть, (*perf*); *see* при·-ка́л·ыва·ть (*imp*)

при·кол·о́ть, -ют, (*perf*) *see* при·ка́-л·ыва·ть

при·коло́ч·енн·ый, -ая, -ое, (*past part pass*) of при·колот·и́ть, (*perf*); *see* при·кола́ч·ива·ть (*imp*)

при·команд·ир·ова́ть, -у́ют, (*perf*) *see* при·команд·иро́выва·ть

при·команд·иро́выва·ть, -ют, (*imp*) attach (organizationally)

при·конта́кт·н·ый, -ая, -ое, (*a*) contact

при·корм·и́ть, -ят, (*perf*) *see* при·-ка́рмл·ива·ть

при·корн·ев·о́й, -ая, -о́е, (*a*) radical

при·кос·нове́ни·е, -я, (*n*) touch, touching, contact, tangency

при·кос·нове́нн·ый, -ая, -ое, (*a*) touching upon; having to do with, concerned in, connected with, involved in

при·кос·н·у́ться, -у́тся, (*perf*) touch

при·креп·и́ть, -ят, (*perf*) *see* при·-крепл·я́·ть

при·крепл·éни·е, -я, (*n*) (*v n*) *see* при·крепл·я́·ть; registration

при·крепл·ённ·ый, -ая, -ое, (*past part pass*) *see* при·крепл·я́·ть; (biol) sessile

при·крепл·я́·ть, -ют, (*imp*) fasten, attach, fix; register

при·кро́·ет, (*fut 3rd sing*) of при·-кры́·ть, (*perf*) *see* при·крыв·а́·ть (*imp*)

при·крут·и́ть, -ят, (*perf*) *see* при·-круч·ива·ть

при·круч·ива·ть, -ют, (*imp*) twist/turn/wind round/down

при·крыв·а́·ть, -ют, (*imp*) screen, shield, cover; partly shut/close

при·кры́т·и·е, -я, (*n*) (*v n*) *see* при·-крыв·а́·ть; cover, shield, screen, escort

сигна́л· при·кры́т·и·я (*m*) (rail) holding signal

~, щит·ов·о́е shield

при·кры́·ть, (*fut 3rd pl*) при·кро́·ют, (*perf*) *see* при·крыв·а́·ть

при·ку·ёт, (*fut 3rd sing*) of при·ко·в·а́ть, (*perf*) *see* при·ко́в·ыва·ть (*imp*)

при·куп·а́·ть, -ют, (*imp*) buy/purchase some more

при·куп·и́ть, -ят, (*perf*) *see* при·ку-п·а́·ть

при·ку́с·, -а, (*m*) (anat) bite

при·ку́с·к·а, -и, (*g pl*) -с·ок·, (*f*) (med) aerophagy

при·ла́в·ок·, -в·к·а, (*m*) counter (in shop); (geol) rock bench

~ -витри́н·а, -ы, (*f*) display counter

при·лаг·а́тельн·ое, -ого, (*n decl as adj*) (gram) adjective

при·лаг·а́·ть, -ют, (*imp*) append, attach, enclose (in letter); apply

при·ла́д·ить, -ят, (*perf*) *see* при·ла́-ж·ива·ть

при·ла́ж·ива·ть, -ют, (*imp*) fit, adapt, adjust

при·лёг·, (*past masc sing*) of при·ле́ч·ь

при·лег·а́ни·е, -я, (*n*) (*v n*) *see* при·-лег·а́·ть

~, пло́т·н·ое fair bearing

у́гол· при·лег·а́ни·я (*m*) fishing angle (of fish-plates)

при·лег·а́·ть, -ют, (*imp*) lie down; be adjacent, adjoin, border; fit closely

при·лег·а́ющ·ий, -ая, -ее, (*pres part act*) of при·лег·а́·ть; adjoining, adjacent, contiguous, neighbouring; faying (of engineering surfaces); close fitting (of clothes)

~ лист· об·ши́в·к·и (*m*) (shipb) inner strake

~ сторон·а́ (*f*) (eng) faying surface/face

при·леж·а́ни·е, -я, (*n*) diligence, application

при·леж·а́ть, -а́т, (*imp*) adjoin, abut upon, lie next to

при·леж·а́щ·ий, -ая, -ее, (*pres part act*) *see* **при·леж·а́ть;** (math) adjacent

~ **у́гол·** (*m*) (math) adjacent supplementary angle

при·ле́ж·н·ый, -ая, -ое, (*a*) diligent, assiduous, industrious

при·леп·и́ть, -́ят, (*perf*) *see* **при·лепл·я́·ть**

при·лепл·я́·ть, -ют, (*imp*) stick/adhere to

при·лёт·, -а, (*m*) arrival (by air)

при·лет·а́·ть, -ют, (*imp*) fly in, arrive/come by air

при·лет·е́ть, -я́т, (*perf*) *see* **при·лет·а́·ть**

при·лёт·н·ый, -ая, -ое, (*a*) of **при·лёт·;** migratory (of birds)

при·ле́ч·ь, (*fut 3rd sing, pl*) **при·ля́ж·ет, при·ля́г·ут,** (*past masc sing*) **при·лёг·,** (*perf*) lie down (for a time); fall down flat (e.g. grass); lie/be close to

при·ли́в, (*m*) flood tide, rising tide; surge, influx; (mech) lug, tongue, boss, rib, feather, piece; (med) congestion

~**, квадрату́р·н·ый** neap tide

~**, рас·по́р·н·ый** (mech) distance piece

~**, сизиг·и́йн·ый** spring tide

~**, у·по́р·н·ый** (mech) abutment, joggle

при·лив·а́·ть, -ют, (*imp*) add (liquid), pour more; flow to

при·ли́в·н·ый, -ая, -ое, (*a*) tidal, flood-tide

~ **гре́б·ен·ь** (*m*) ice ridge (formed by tide)

~ **полос·а́** (*f*) tide lands

при·ли́в·о-от·ли́в·н·ый, -ая, -ое, (*a*) tidal

~ **де́й·стви·е** (*n*) tide

~ **теч·е́ни·е** (*n*) tidal current

~ **тре́щ·ин·а** (*f*) tide crack (ice)

~ **явл·е́ни·е** (*n*) tidal effect

при·лип·а́емост·ь, -и, (*f*) adherence

при·лип·а́л·а, -ы, (*m*) or (*f*) (fish) remora, *Echeneida*

при·лип·а́ни·е, -я, (*n*) (*v n*) of **при·лип·а́·ть;** adhesion, attachment, (phys) capture

коэффицие́нт· при·лип·а́ни·я (*m*) (phys) sticking coefficient

у́·ровен·ь при·лип·а́ни·я (*m*) (phys) adhesion level

при·лип·а́·ть, -ют, (*imp*) stick/adhere to; press up against

при·ли́п·нуть, -нут, (*perf*) *see* **при·лип·а́·ть**

при·ли́п·чив·ый, -ая, -ое, (*a*) sticky, adhesive; catching, infectious, contagious

при·ли́ст·ник·, -а, (*m*) (bot) stipule

при·ли́ст·ников·ый, -ая, -ое, (*a*) (bot) stipular

при·ли́ст·ничн·ый, -ая, -ое, (*a*) (bot) stipulate, stipulose

при·ли́·ть, (*fut 3rd pl*) **при·ль·ю́т,** (*perf*) *see* **при·лив·а́·ть**

при·ли́ч·н·ый, -ая, -ое, (*a*) decent, proper

при·ло́ж·ени·е, -я, (*n*) (*v n*) *see* **при·кла́д·ыва·ть,** appendix (in book), supplement (to newspaper), enclosure (to letter); application (e.g. of load, stress etc.); (gram) apposition

при·ло́ж·енн·ый, -ая, -ое, (*past part pass*) of **при·лож·и́ть;** applied; appended, enclosed

~ **на·пряж·е́ни·е** (*n*) (mech) applied stress; (elec) applied voltage

при·лож·и́ть, -́ат, (*perf*) *see* **при·кла́д·ыва·ть**

при·ль·ёт, (*fut 3rd sing*) of **при·ли́·ть,** (*perf*); *see* **при·лив·а́·ть** (*imp*)

при·льн·у́ть, -у́т, (*perf*) press up against; stick to

при·ля́г·ут, (*fut 3rd pl*) of **при·ле́ч·ь**

при·ля́ж·ет, (*fut 3rd sing*) of **при·ле́ч·ь**

~**, гли́н·ист·ая** (*f*) (geol, min) flookan, flucan, gouge, salvage

при·ма́н·к·а, -и, (*g pl*) **-н·ок·,** (*f*) bait, lure, enticement

прима́р·н·ый, -ая, -ое, (*a*) primary, prime

прима́т·, -а, (*m*) first place, pride of place; (zool) primate

при·ма́т·ыва·ть, -ют, (*imp*) wind/reel in/on; wind/reel in/on some more; tie/hitch round

примахи́н·, -а, (*m*) (pharm) primaquine

при·ма́ч·ива·ть, -ют, (*imp*) bathe, wash (an injury etc.), apply (lotions etc.)

при·мельк·а́·ться, -ются, (*imp*) become familiar/unnoticeable (due to constant encounter/occurrence/repetition)

при·мен·е́ни·е, -я, (*n*) (*v n*) *see* **при·мен·я́·ть;** application, employment, use; adaptation

при·мен·ённ·ый, -ая, -ое, (*past part pass*) *see* при·мен·я́·ть

при·мен·и́мост·ь, -и, (*f*) applicability, possibility of using/applying

при·мен·и́м·ый, -ая, -ое, (*a*) applicable

при·мен·и́тельн·о (*adv*) in connection with, about

при·мен·и́ть, -ят, (*perf*) *see* при·мен·я́·ть

при·мен·я́емост·ь, -и, (*f*) scope, range of use/application; frequency of use

при·мен·я́·ть, -ют, (*imp*) apply, employ, use; adapt; -ся adapt oneself

при·ме́р·, -а, (*m*) example, instance

при·мерз·а́·ть, -ют, (*imp*) freeze on/to

при·мёрз·л·ый, -ая, -ое, (*past part pass*) of при·мерз·а́·ть

при·мёрз·нуть, -нут, (*perf*) *see* при·мерз·а́·ть

при·ме́р·ива·ть, -ют, (*imp*) = при·мер·я́·ть

при·мер·и́ть, -ят, (*perf*) *see* при·мер·я́·ть

при·ме́р·н·о (*adv*) roughly, approximately; as an example, e.g.

при·ме́р·н·ый, -ая, -ое, (*a*) approximate, rough

при·мер·я́·ть, -ют, (*imp*) try on, fit

при·ме́с·н·ый, -ая, -ое, (*a*) of при́·мес·ь; extrinsic

~ про·вод·и́мост·ь (*f*) extrinsic conductivity

~ свой·ств·а (*pl*) extrinsic properties (of a semiconductor)

~ структу́р·а (*f*) (phys) impurity structure

при·мес·ти́, (*fut 3rd pl*) при·ме·т·у́т, (*perf*) sweep to

при́·мес·ь, -и, (*f*) admixture; impurity; (met) alloying element, impurity/trace element

~, акце́птор·н·ая (phys) acceptor impurity

~, вре́д·н·ая deleterious/undesirable impurity

до·бавл·е́ни·е при́·мес·и (*n*) doping (a semiconductor)

~, до́нор·ск·ая (phys) donor impurity

~, механ·и́ческ·ая mechanical impurity; (oil) (*pl*) sediment insoluble in benzene

~, с·луч·а́йн·ая (met) impurity

~ со·сто·я́ни·я (phys) state admixture

при́·мес·ь

~, специа́ль·н·ая (met) alloying element

пр·и́м·ет, (*fut 3rd sing*) of при·ня́·ть, (*perf*); *see* при·ним·а́·ть (*imp*)

при·ме́т·а, -ы, (*f*) distinguishing mark

при·мет·а́·ть, -ют, (*imp*) sweep to; (*perf*) (text) baste, tack

при·мет·ёт, (*fut 3rd sing*) of при·мес·ти́

при·ме́т·ить, -ят, (*perf*) *see* при·меч·а́·ть

при·мёт·к·а, -и, (*f*) (text) tacking

при·ме́т·н·ый, -ая, -ое, (*a*) discernible, noticeable; distinguishing

~ знак· (*m*) distinguishing mark/sign

~ ме́ст·о (*n*) (naut) buoyed area/position

при·мёт·ыва·ть, -ют, (*imp*) (text) baste

при·меч·а́ни·е, -я, (*n*) note, remark, comment

при·меч·а́·ть, -ют, (*imp*) note, draw attention to, remark, observe

при·меш·а́·ть, -ют, (*perf*) *see* при·ме́ш·ива·ть

при·ме́ш·ива·ть, -ют, (*imp*) admix, mix in, add; adulterate; alloy

при·мин·а́·ть, -ют, (*imp*) crush, squash

при·мир·е́ни·е, -я, (*n*) reconciliation; conciliation

при·мир·и́м·ый, -ая, -ое, (*a*) reconcilable

при·мир·и́тел·ь, -я, (*m*) conciliator, peacemaker

при·мир·и́тельн·ый, -ая, -ое, (*a*) conciliatory

при·мир·и́ть, -я́т, (*perf*) *see* при·мир·я́·ть

при·мир·я́·ть, -ют, (*imp*) reconcile; conciliate

пр·им·и́те, (*pl imper*) of при·ня́·ть, (*perf*); *see* при·ним·а́·ть (*imp*)

примити́в·н·ый, -ая, -ое, (*a*) primitive

при·мк·н·у́ть, -у́т, (*perf*) *see* при·мык·а́·ть

при·мн·ёт, (*fut 3rd sing*) of при·мя́·ть, (*perf*); *see* при·ним·а́·ть (*imp*)

при·мо́·ин·а, -ы, (*f*) alluvial deposits (river)

прим·ордиа́ль·н·ый, -ая, -ое, (*a*) (biol) primordial

при·мо́р·ск·ий, -ая, -ое, (*a*) seaside; maritime

~ кли́мат· (*m*) maritime climate

при·мо́р·ь·е, -я, (*n*) seaside, coast
при·мот·а́·ть, -ют, (*perf*) *see* при·-мат·ыва·ть
при·моч·и́ть, -ат, (*perf*) *see* при·ма́-чива·ть
при·мо́ч·к·а, -и, (*g pl*) -ч·ек·, (*f*) (pharm) wash, lotion, fomentation
примули́н·, -а, (*m*) primuline (dye)
при́мус·, -а, (*m*) primus stove
пр·и́м·ут, (*fut 3rd pl*) of при·ня́·ть, (*perf*) *see* при·ним·а́·ть (*imp*)
при·мык·а́·ть, (*imp*) join, side with, support; put/move against/up to; fix (bayonets); (*imp only*) adjoin, abut, border
при·мык·а́ющ·ий, -ая, -ее, (*pres part act*) of при·мык·а́·ть; adjoining, adjacent, abutting
при·мя́·ть, (*fut 3rd pl*) при·мн·у́т, (*perf*) *see* при·мин·а́·ть
при·над·леж·а́ть, -а́т, (*imp*) belong, pertain to
при·над·ле́ж·ность·, -и, (*f*) appurtenance; (*pl*) belongings; requisites, gear
~, пи́сьм·енн·ые (*pl*) writing materials
~, рыбо·ло́в·н·ые (*pl*) fishing tackle/gear
принайто́в·ить, -ят, (*perf*) (naut) lash, seize
при·нес·ти́, -у́т, (*perf*) *see* при·нос·и́ть
при·ниж·а́·ть, -ют, (*imp*) disparage, belittle, write down, ascribe less significance to; humiliate
при·ниж·е́ни·е, -я, (*n*) (*v n*) *see* при·-ниж·а́·ть; disparagement, depreciation; (air) height separation below, vertical interval downwards
при·ни́з·ить, -ят, (*perf*) *see* при·ниж·а́·ть
при·ник·а́·ть, -ют, (*imp*) press to/against/up to
при·ни́к·нуть, -нут, (*perf*) *see* при·-ник·а́·ть
при·ним·а́·ть, -ют, (*imp*) receive, take, accept, assume; -ся (*pass*); begin, start, set to/about; (bot) take, take root
~ внутр·ь (med) ingest
~ во в·ним·а́ни·е bear in mind, consider
при·ним·а́ющ·ий, -ая, -ее, (*pres part act*) of при·ним·а́·ть
~ при·бо́р· (*m*) receiver, receiving instrument
при·норов·и́ть, -ят, (*perf*) adapt, adjust

при·нос·и́ть, -ят, (*imp*) bring; bring in, yield
~ по́льз·у be of use/benefit/advantage
при́нтер·, -а, (*m*) printer
~, на·бир·а́ющ·ий (autom) gang printer
при·нуд·и́тельн·ый, -ая, -ое, (*a*) compulsory, forced
~ циркул·я́ци·я (*f*) forced circulation
при·ну́д·ить, -ят, (*perf*) *see* при·-нужд·а́·ть
при·нужд·а́·ть, -ют, (*imp*) compel, coerce, constrain, force
при·нужд·е́ни·е, -я, (*n*) compulsion, coercion, constraint
при·нужд·ённость·, -и, (*f*) constraint
при́нцип·, -а, (*m*) (*and see under associated words*); principle
в при́нцип·е principally
из при́нцип·а on principle
~ наи·ме́ньш·его при·нужд·е́ни·я principle of least constraint/curvature
принципиа́ль·н·о (*adv*) on principle; in principle; in the main
принципиа́ль·н·ый, -ая, -ое, (*a*) based/founded on principle(s); basic, fundamental
~ во·про́с· (*m*) question of principle
~ знач·е́ни·е (*n*) fundamental significance
~ схе́м·а (*f*) basic diagram, diagram; flow sheet
при·ня́т·и·е, -я, (*n*) (*v n*) *see* при·-ним·а́·ть; reception, acceptance; assumption
при́·нят·о (*predic of* при́·ня·т·ый) it is assumed/accepted, it is customary
при́·ня·т·ый, -ая, -ое, (*past part pass*) of при·ня́·ть
при·ня́·ть, (*fut 3rd pl*) пр·и́м·ут, (*perf*) *see* при·ним·а́·ть
при·обрес·ти́, (*fut 3rd pl*) при·об·рет·у́т, (*perf*) *see* при·обрет·а́·ть
при·обрет·а́·ть, -ют, (*imp*) acquire, obtain, gain
при·обрет·е́ни·е, -я, (*n*) (*v n*) *see* при·обрет·а́·ть; acquisition
при·общ·а́·ть, -ют, (*imp*) acquaint/familiarize with, introduce to; attach, append
при·общ·и́ть, -а́т, (*perf*) *see* при·-общ·а́·ть
при·озёр·н·ый, -ая, -ое, (*a*) lakeside, lake-land
приори́т·, -а, (*m*) (min) priorite
приорите́т·, -а, (*m*) priority

при·о·стана́вл·ива·ть, -ют, (*imp*) halt, suspend, stop/cease for a time; -ся (*pass*); pause

при·о·станов·и́ть, ´-ят, (*perf*) *see* при·о·стана́вл·ива·ть

при·о·становл·е́ни·е, -я, (*n*) suspension, stay

~ ис·полн·е́ни·я при·гово́р·а (law) stay of execution

при·о·стано́в·к·а, -и, (*g pl*) -в·ок·, (*f*) (*v n*) *see* при·о·стана́вл·ива·ть; halt, suspension

при·о·твор·и́ть, ´-ят, (*perf*) *see* при·о·твор·я́·ть

при·о·твор·я́·ть, -ют, (*imp*) open slightly/a little, set ajar

при·от·крыв·а́·ть, -ют, (*imp*) open slightly/a little

при·от·кры́·ть, (*fut 3rd pl*) при·от·кро́·ют, (*perf*) *see* при·от·крыв·а́·ть

при·пад·а́·ть, -ют, (*imp*) fall down

при·па́д·ок·, -д·к·а, (*m*) (med) fit, seizure, paroxysm

при·па́·ечн·ый, -ая, -ое, (*a*) of при·па́й·к·а

при·па́·ива·ть, -ют, (*imp*) solder, solder on (below 427°C); braze, braze on (above 427°C)

при·па́·й, -я, (*m*) (ocean) landfast ice

при·па́й·к·а, -и, (*g pl*) -а́·ек·, (*f*) soldering/brazing on; soldered/brazed-on part

при·пал·ённ·ый, -ая, -ое, (*past part pass*) *see* при·па́л·ива·ть

при·па́л·ива·ть, -ют, (*imp*) singe

при·пал·и́ть, -я́т, (*perf*) singe

при·па́р·ива·ть, -ют, (*imp*) (med) foment; sweat (leather)

при·пар·и́ть, -я́т, (*perf*) *see* при·па́р·ива·ть

при·па́р·к·а, -и, (*g pl*) -р·ок·, (*f*) (med) fomentation, cataplasm; sweating (leather)

при·па́с·, -а, (*m*) (*pl*) stores, supplies

~, бо·ев·ы́е (*pl*) ammunition

~, съ·ест·н·ые (*pl*) victuals, provisions

при·пас·а́·ть, -ют, (*imp*) store, lay up, save

при·пас·о́ванн·ый, -ая, -ое, (*past part pass*) of при·пас·ова́ть

при·пас·ова́ть, -у́ют, (*perf*) (mech eng) align, match

при·пас·о́вк·а, -и, (*f*) (*v n*) of при·пас·о́выва·ть; alignment; (print) blending (of lines)

при·пас·о́выва·ть, -ют, (*imp*) (mech eng) match, align

при·пас·ти́, -у́т, (*perf*) *see* при·па·с·а́·ть

при·па́с·ть, (*fut 3rd pl*) при·пад·у́т, (*perf*) *see* при·пад·а́·ть

при·па·я́·ть, -ют, (*perf*) *see* при·па́·ива·ть

при·пёк·, -а, (*m*) heat (from the sun); sunbaked spot/area/place; weight increment of loaf after baking

при·пек·а́ни·е, -я, (*n*) (*v n*) *see* при·пек·а́·ть

при·пек·а́·ть, -ют, (*imp*) overbake; get very hot in the sun; (met) sinter

при·пе́ч·ь, (*fut 3rd sing, pl*) при·пе·ч·ёт, при·пек·у́т, (*past masc sing*) при·пёк *see* при·пек·а́·ть

при·пис·а́ть, (*fut 3rd pl*) при·пи́ш·ут, (*perf*) *see* при·пи́с·ыва·ть

при·пи́с·к·а, -и, (*g pl*) -с·ок·, (*f*) addition, postscript (to letter), codicil (to will etc.); registration

порт· при·пи́с·к·и (*m*) (naut) port of registry, port of documentation

при·пи́с·ыва·ть, -ют, (*imp*) add (in writing); register, attach; ascribe, attribute, assign

при·пи́ш·ет, (*fut 3rd sing*) of при·пис·а́ть

при·пла́т·а, -ы, (*f*) (*v n*) *see* при·пла́ч·ива·ть; supplementary payment

при·плат·и́ть, ´-ят, (*perf*) *see* при·пла́ч·ива·ть

при·пла́ч·ива·ть, -ют, (*imp*) pay a supplement, pay additionally

при·плес·ти́, (*fut 3rd pl*) при·плет·у́т, (*perf*) *see* при·плет·а́·ть

при·плет·а́·ть, -ют, (*imp*) weave/plait into/onto; (print) bind in

при·пло́д·, -а, (*m*) litter (young at one birth)

при·плыв·а́·ть, -ют, (*imp*) arrive (on or in the water), swim/drift/float/sail to/up to

при·плы́·ть, (*fut 3rd pl*) при·плыв·у́т, (*perf*) *see* при·плыв·а́·ть

при·плю́с·н·уть, -ут, (*perf*) *see* при·плю́щ·ива·ть

при·плю́щ·ива·ть, -ют, (*imp*) flatten

при·по·ве́рх·ностн·ый, -ая, -ое, (*a*) near-surface, surface-layer, surface

при·под·ним·а́·ть, -ют, (*imp*) raise/lift/elevate/hoist slightly/a little

при·под·ня́·т·ый, -ая, -ое, (*past part pass*) *see* при·под·ним·а́·ть; elevated, elated, uplifted; (civ eng) superelevated

при·под·ня́·ть, (*fut 3rd pl*) при·-
под·ни́м·ут, (*perf*) *see* при·под·-
ним·а́·ть

при·по́·й, -я, (*m*) solder

~, мя́гк·ий soft solder

~, твёрд·ый hard/brazing solder, brazing alloy

при·полз·а́·ть, -ют, (*imp*) crawl/
creep up to

при·полз·ти́, -у́т, (*perf*) *see* при·-
полз·а́·ть

при·по́люс·н·ый, -ая, -ое, (*a*) (geog)
Polar

при·поля́р·н·ый, -ая, -ое, (*a*) Arctic; Antarctic

при·по·мин·а́·ть, -ют, (*imp*) remember, recollect, recall

при·по́·мн·ить, -ят, (*perf*) *see* при·-
по·мин·а́·ть

при·пра́в·а, -ы, (*f*) (food) seasoning,
flavouring, condiment

при·пра́в·ить, -ят, (*perf*) *see* при·-
правл·я́·ть

при·пра́в·к·а, -и, (*g pl*) -в·ок·, (*f*)
(print) make-ready, making-ready

при·правл·я́·ть, -ют, (*imp*) season,
flavour; (print) make ready

при·пу́др·ива·ть, -ют, (*imp*) dust/
powder lightly

при·пудр·и́ть, -я́т, (*perf*) *see* при·-
пу́др·ива·ть

при́·пуск·, -а, (*m*) allowance (in metal
working); overmeasure (cybernetics)

~ на об·рабо́т·к·у machining allowance, machine finish allowance

~ на о·са́д·к·у (weld) push-up allowance

~ на при·го́н·к·у (mech) fitting allowance

~ на шлиф·о́вк·у (mech) grinding
allowance

при·пуск·а́·ть, -ют, (*imp*) let in,
admit, allow to approach; make allowance for (in metal working)

при·пуст·и́ть, -я́т, (*perf*) *see* при·-
пуск·а́·ть

при·пух·а́·ть, -ют, (*imp*) swell slightly, tumefy, intumesce

при·пу́х·лост·ь, -и, (*f*) swelling,
intumescence, tumescence

при·пу́х·л·ый, -ая, -ое, (*a*) swollen,
tumescent

при·пу́х·нуть, -нут, (*perf*) *see* при·-
пух·а́·ть

при·пу́щ·енн·ый, -ая, -ое, (*past
part pass*) of при·пуст·и́ть, (*perf*)
see при·пуск·а́·ть

при·пы́л·, -а, (*m*) (met cast) facing

при·пы́л·ива·ть фо́рм·у (met) under-
face a mould/mold

при·раба́т·ыва·ть, -ют, (*imp*) run
in (of machinery); earn extra

при·рабо́т·а·ть, -ют, (*perf*) *see* при·-
раба́т·ыва·ть

при·рабо́т·к·а, -и, (*g pl*) -т·ок·, (*f*)
(mech eng) running in; running-in
period, wear-in; green test

при́·работ·ок·, -т·к·а, (*m*) extra
earnings

при·ра́вн·ива·ть, -ют, (*imp*) equate;
make level, level up

при·равн·я́·ть, -ют, (*perf*) equate

при·раст·а́·ть, -ют, (*imp*) grow
together/into; increase, grow

при·раст·и́, -у́т, (*past masc sing*)
при·ро́с·, (*perf*) *see* при·раст·а́·ть

при·раст·и́ть, -я́т, (*perf*) *see* при·-
ра́щ·ива·ть

при·ращ·е́ни·е, -я, (*n*) (*v n*) *see* при·-
ра́щ·ива·ть; increment, increase,
accretion

при·ра́щ·ива·ть, -ют, (*imp*) (*trans*)
grow/graft/join on

при·ре́з·, -а, (*m*) (*v n*) *see* при·ре-
з·а́·ть

при·рез·а́·ть, -ют, (*imp*) при·ре́з·ать,
(*fut 3rd pl*) при·ре́ж·ут, (*perf*)
carve up; hack into pieces; divide up

при·ре́з·н·ый, -ая, -ое, (*a*) of при·-
ре́з·

~ стан·о́к· (*m*) rip-saw frame, ripping saw frame

при·ре́з·ь, -и, (*f*) lump (leather)

при·ре́ч·н·ый, -ая, -ое, (*a*) river,
riparian

при·ровн·я́·ть, -ют, (*perf*) make level,
level up

при·ро́д·а, -ы, (*f*) nature

при·ро́д·н·ый, -ая, -ое, (*a*) natural;
innate, inborn; born, by/from birth

при·родо·ве́д·, -а, (*m*) naturalist

при·родо·ве́д·ени·е, -я, (*n*) nature
study, natural history

при·родо·ве́д·ческ·ий, -ая, -ое, (*a*)
naturalists'

при·рожд·ённ·ый, -ая, -ое, (*a*)
innate, inborn, born

при·ро́с·, (*past masc sing*) of при·-
раст·и́, (*perf*); *see* при·раст·а́·ть

при·ро́ст·, -а, (*m*) increase, increment,
accretion, growth, gain

~ су́ш·и (geog) land emergence

~ сто́·имост·и (fin) increment value

при·руб·е́жн·ый, -ая, -ое, (*a*) border, frontier

при·ру́сл·ов·ый, -ая, -ое, (*a*) (geol) channel

при·руч·а́·ть, -ют, (*imp*) tame, domesticate

при·руч·и́ть, -а́т, (*perf*) see **при·руч·а́·ть**

при·сад·и́ть, ⁼ят, (*perf*) see **при·са́·ж·ива·ть**

при·са́д·к·а, -и, (*g pl*) **-д·ок·,** (*f*) (*v n*) of **при·сад·и́ть;** additive, dope (to fuels and lubricants); (met) addition agent/element; (weld) filler, filler metal

∼, вя́з·костн·ая (oil) viscosity-index improver

при·са́д·очн·ый, -ая, -ое, (*a*) of **при·са́д·к·а**

∼ мета́лл· (*m*) (weld) filler metal

∼ прут·о́к· (*m*) welding rod

при·са́ж·ива·ть, -ют, (*imp*) (met) add to the charge, charge addition agents

при·са́с·ыва·ть, -ют, (*imp*) suck in/up; suck to/onto, draw to (by suction)

при·сва́·ива·ть, -ют, (*imp*) award, confer; appropriate, assume

при·сво·е́ни·е, -я, (*n*) appropriation; award, conferment

∼ а́дрес·ов address allocation (computers)

при·сво́·ить, -ят, (*perf*) see **при·сва́·ива·ть**

при·се́мен·ник·, -а, (*m*) = **при·семя́н·ник·**

при·семя́н·ник·, -а, (*m*) (bot) aril

при·скак·а́ть, (*fut 3rd pl*) **при·ска́·ч·ут,** (*perf*) see **при·ска́к·ива·ть**

при·ска́к·ива·ть, -ют, (*imp*) gallop up to; hop/skip up to/nearer

при·ско́рб·н·ый, -ая, -ое, (*a*) sad, sorrowful

при·сл·а́ть, (*fut 3rd pl*) **при·шл·ю́т,** (*perf*) send

при·сло́н·, -а, (*m*) river gorge

при·слон·и́ть, -я́т, (*perf*) see **при·слон·я́·ть**

при·слон·я́·ть, -ют, (*imp*) lean against/on

при·слу́г·а, -и, (*f*) (obs) crew, gun crew; servant, servants

при·слу́ш·а·ться, -ются, (*perf*) see **при·слу́ш·ива·ться**

при·слу́ш·ива·ться, -ются, (*imp*) listen carefully; pay attention

при·сма́тр·ива·ть, -ют, (*imp*) watch over, supervise, look after; **-ся** look closely; get accustomed to

при·смотр·е́ть, ⁼ят, (*perf*) see **при·сма́тр·ива·ть**

при·совокуп·и́ть, -я́т, (*perf*) see **при·совокупл·я́·ть**

при·совокупл·я́·ть, -ют, (*imp*) add

при·со·един·е́ни·е, -я, (*n*) (*v n*) see **при·со·един·я́·ть;** (math) adjunction; (chem) combination, addition, affixture

∼ кисло·ро́д·а (chem) oxygenation продукт· **при·со·един·е́ни·я** (chem) addition compound

при·со·един·ённ·ый, -ая, -ое, (*past part pass*) see **при·со·един·я́·ть;** adjoint, adjunct to; (chem) added

∼ ма́сс·а (*f*) (mech) apparent additional mass, associated/virtual mass

∼ ма́триц·а (*f*) (math) adjoint matrix

при·со·един·и́тельн·ый, -ая, -ое, (*a*) connecting, joining; (elec) terminal

∼ за·жи́м· (*m*) (elec) terminal screw

при·со·един·и́ть, -я́т, (*perf*) see **при·со·един·я́·ть**

при·со·един·я́·ть, -ют, (*imp*) join, (elec etc.) connect; add; (chem) combine, affix, (math) annex

при·со́л·енн·ый, -ая, -ое, (*a*) half-green (leather)

при·со́с·, -а, (*m*) (*v n*) see **при·са́с·ыва·ть;** suction; sucker

при·сос·а́ть, -у́т, (*perf*) see **при·са́с·ыва·ть**

при·со́с·к·а, -и, (*g pl*) **-с·ок·,** (*f*) sucker

при·со́с·н·ый, -ая, -ое, (*a*) sucker

при·со́с·ок·, -с·к·а, (*m*) sucker

при·со́х·нуть, -нут, (*perf*) see **при·сых·а́·ть**

при·спосо́бл·ива·ть, -ют, (*imp*) see **при·способл·я́·ть**

при·способ·и́тельн·ый, -ая, -ое, (*a*) adaptive

∼ реа́кци·я (*f*) (biol) adaptive reaction

при·спосо́б·ить, -ят, (*perf*) see **при·способл·я́·ть**

при·способл·е́ни·е, -я, (*n*) (*v n*) see **при·способл·я́·ть;** adaptation; device, contrivance, gear; attachment (machine tools)

∼, за·жи́м·н·ое clamping fixture, jig зо́н·а при·способл·е́ни·я (*f*) (phys) accommodation zone

∼, на·правл·я́ющ·ее jig

∼, регул·иро́вочн·ое adjusting gear

при·способл·ени·е

~, с·бо́р·очн·ое assembly jig

~, тормоз·н·о́е arrester gear

~, у·стано́в·очн·ое locating jig/fixture

~, фикс·и́рующ·ее fixture

при·спосо́бл·енн·ый, -ая, -ое, (*past part pass*) of при·спосо́б·ить

при·способл·я́емост·ь, -и, (*f*) adaptability

при·способл·я́ем·ый, -ая, -ое, (*pres part pass*) of при·способл·я́·ть; adaptable

при·способл·я́·ть, -ют, (*imp*) adapt, adjust, rearrange, reorganize; -ся (*pass*); be suitable for, be suited to

при·с·пуск·а́·ть, -ют, (*imp*) lower a little, let down a little; half mast (of flags); -ся (*pass*); (naut) pay off (from wind), bear up (tiller)

при·с·пуст·и́ть, -ят, (*perf*) see при·с·пуск·а́·ть

при·став·а́ни·е, -я, (*n*) (*v n*) see при·ста·ва́ть

при·ста·ва́ть, -ю́т, (*imp*) stick/adhere to; (naut) come alongside

при·ста́в·ить, -ят, (*perf*) see при·ставл·я́·ть

при·ста́в·к·а, -и, (*g pl*) -в·ок·, (*f*) lengthening, extending, extension, adding to; extension piece, attachment; (gram) prefix

~, съ·ём·очн·ая (phot) attachment lens

при·ставл·ени·е, -я, (*n*) (*v n*) see при·ставл·я́·ть

при·ставл·я́·ть, -ют, (*imp*) put/set/move to/up to; extend/lengthen with; add on

при·став·н·о́й, -а́я, -о́е, (*a*) attached, attachment

при·ста́в·очн·ый, -ая, -ое, (*a*) attachable, detachable; prefixed

при·ста́·вш·ий, -ая, -ее, (*past part act*) of при·ста́·ть, (*perf*); see при·ста·ва́ть (*imp*)

при·ста́ль·н·ый, -ая, -ое, (*a*) intent, fixed

при·ста́·нет, (*fut 3rd sing*) of при·ста́·ть, (*perf*); see при·ста·ва́ть (*imp*)

при́·стан·н·ый, -ая, -ое, (*a*) of при́·стан·ь

при·станцио́н·н·ый, -ая, -ое, (*a*) station

при́·стан·ь, -и, (*nom pl*) -и, (*g pl*) -е́й, (*f*) wharf, jetty, quay, landing stage

~ снабж·е́ни·я (mil) navigation head

при·ста́·ть, -нут, (*perf*) see при·ста·ва́ть

при·стег·а́·ть, -ют, (*perf*) (text) baste

при·стёг·ива·ть, -ют, (*imp*) fasten/button/hook on; (text) baste

при·стег·н·у́ть, -у́т, (*perf*) fasten/button/hook on

при·сте́н·н·ый, -ая, -ое, (*a*) wall

~ сло́·й (*m*) wall/boundary layer (fluid flow)

при·сте́н·очн·ый, -ая, -ое, (*a*) wall; (biol) parietal

при·стра́·ива·ть, -ют, (*imp*) build on/against; form up in line with, take station alongside/abeam

при·стра́ст·н·ый, -ая, -ое, (*a*) biased, partial, prejudiced

при·стра́ч·ива·ть, -ют, (*imp*) stitch/sew on

при·стре́л·ива·ть, -ют, (*imp*) range on (by observation of shots); carry out test firing; shoot down, kill; -ся range, find the target (by observation of shots)

при·стрел·и́ть, -ят, (*perf*) shoot down, kill

при·стре́л·к·а, -и, (*g pl*) -л·ок·, (*f*) (gunn) finding fire, ranging; harmonizing (a/c guns); test firing

~ воз·ду́ш·н·ого репе́р·а air-burst ranging

~, холо́д·н·ая harmonizing (of a/c guns)

при·стре́л·очн·ый, -ая, -ое, (*a*) ranging (of shot etc.)

при·стре́ль·н·ый = при·стре́л·очн·ый

при·стрел·я́·ть, -ют, (*perf*) range on (by observation of shots); carry out test firing; -ся range, find the target (by observation of shots)

при·стро́·ить, -ят, (*perf*) see при·стра́·ива·ть

при·стро́й·к·а, -и, (*g pl*) -о́·ек·, (*f*) building on; (arch) annexe, extension

при·строч·и́ть, -а́т, (*perf*) see при·стра́ч·ива·ть

при́·ступ·, -а, (*m*) (med) fit, attack, bout; (mil) assault, attack; access

при·ступ·а́·ть, -ют, (*imp*) begin, start, set about, proceed to; -ся (*pass*); approach

при·ступ·и́ть, -ят, (*perf*) see при·ступ·а́·ть

при·суд·и́ть, -ят, (*perf*) see при·сужд·а́·ть

при·сужд·а́·ть, -ют, *(imp)* sentence to, impose (penalty); award, confer

при·сужд·е́ни·е, -я, *(n) (v n) see* при·сужд·а́·ть; award

при·су́т·стви·е, -я, *(n)* presence

при·су́т·ств·овать, -уют, *(imp)* be present, attend

при·су́т·ствующ·ий, -ая, -ее, *(pres part act)* of при·су́т·ств·овать; present, attending; *(pl)* those present

при·су́х·а, -и, *(f)* (min) sticky/frozen coal

при·су́ч·ива·ть, -ют, *(imp)* join by twisting together

при·суч·и́ть, -а́т, *(perf) see* при·су́ч·ива·ть

при·су́ч·к·а, -и, *(f)* (text) warp tying

при·су́ш·енн·ый, -ая, -ое, *(past part pass)* of при·суш·и́ть; rather dry, rather a dry

при·су́ш·ива·ть, -ют, *(imp)* dry slightly/a little

при·суш·и́ть, -а́т, *(perf) see* при·су́·ш·ива·ть

при·су́щ·ий, -ая, -ее, *(a)* inherent, intrinsic, eigen-

при·с·чит·а́·ть, -ют, *(perf) see* при·с·чи́т·ыва·ть

при·с·чи́т·ыва·ть, -ют, *(imp)* add on

при·сыл·а́·ть, -ют, *(imp)* send

при·сы́п·ать, -лют, *(perf) see* при·сып·а́·ть

при·сып·а́·ть, -ют, *(imp)* pour/sprinkle some more; pour/sprinkle up against; sprinkle/powder/dust lightly

при·сы́п·к·а, -и, *(g pl)* -п·ок·, *(f)* *(v n) see* при·сып·а́·ть; powder, dusting powder

при·сых·а́·ть, -ют, *(imp)* dry on/onto, stick fast

прися́г·а, -и, *(f)* oath

присяг·а́·ть, -ют, *(imp)* (law) swear, take the oath

присяг·н·у́ть, -у́т, *(perf) see* при·сяг·а́·ть

присяж·н·ый, -ого, *(m decl as adj)* (law) juror, juryman (not in U.S.S.R.), *(pl)* jury

при·та́пт·ыва·ть, -ют, *(imp)* trample, trample/tread down

при·та́ск·ива·ть, -ют, *(imp)* drag/pull to/from, bring (by pulling)

при·тач·а́·ть, -ют, *(perf)* stitch on

при·та́ч·ива·ть, -ют, *(imp)* fit (by grinding/turning); stitch on

при·тащ·и́ть, -а́т, *(perf) see* при·та́ск·ива·ть

при·твор·и́ть, -я́т, *(perf) see* при·твор·я́·ть

при·тво́р·н·ый, -ая, -ое, *(a)* pretended, simulated

при·твор·я́·ть, -ют, *(imp)* close/shut slightly/a little; -ся *(reflex and pass)*; pretend, simulate

при·тек·а́ни·е, -я, *(n) (v n)* of при·тек·а́·ть; indraft, inflow

при·тек·а́·ть, -ют, *(imp)* flow to

при·тек·а́ющ·ий, -ая, -ее, *(pres part act)* of при·тек·а́·ть; inflowing, incoming

при·тер·е́ть, *(fut 3rd pl)* при·тр·у́т, *(perf) see* при·тир·а́·ть

при·тёр·т·ый, -ая, -ое, *(past part pass) see* при·тир·а́·ть; ground; ground-in; lapped

~ стекл·о́ *(n)* ground glass

при·те́ч·ь, *(fut 3rd sing, pl)* при·те·ч·ёт, при·тек·у́т, *(past masc sing)* при·тёк·, *(perf)* flow to

при·ти́р·, -а, *(m)* (met) lap

~, алма́з·н·ый (met) diamond lap

при·тир·а́·ть, -ют, *(imp)* rub in; lap, grind; grind in (valves)

при·ти́р·к·а, -и, *(f) (v n) see* при·тир·а́·ть; lapping; grinding-in

при·ти́р·очн·ый, -ая, -ое, *(a)* of при·ти́р·к·а

~ стан·о́к· *(m)* (eng) lapping machine

при·ти́ск·ива·ть, -ют, *(imp)* squeeze/press against

при·ти́с·н·уть, -ут, *(perf) see* при·ти́ск·ива·ть

при·тих·а́·ть, -ют, *(imp)* become quiet/quieter, quieten down

при·ти́х·нуть, -нут, *(perf) see* при·тих·а́·ть

при·то́к·, -а, *(m)* (geog) tributary; influx, flow, inflow

при́толок·а, -и, *(f)* lintel

при·то́м *(conj)* besides, besides this, moreover

при·топт·а́·ть, *(fut 3rd pl)* при·то́п·ч·ут, *(perf) see* при·та́пт·ыва·ть

при·то́ч·енн·ый, -ая, -ое, *(past part pass)* of при·точ·и́ть

при·точ·и́ть, -а́т, *(perf)* fit (by grinding/turning), grind in/on

при·то́ч·н·ый, -ая, -ое, *(a)* of при·то́к·

~ вентил·я́ци·я *(f)* convective ventilation

при·тр·ёт, *(fut 3rd sing)* of при·тер·е́ть, *(perf) see* при·тир·а́·ть *(imp)*

при·туп·и́ть, -я́т, (*perf*) *see* **при·тупл·я́·ть**

при·тупл·я́ни·е, -я, (*n*) (*v n*) *see* **при·тупл·я́·ть;** blu ntness, blunt part, dullness

~ **кро́м·к·и** (weld) root face

при·ту́пл·енн·ый, -ая, -ое, (*past part pass*) *see* **при·тупл·я́·ть;** truncated; blunt; dull

при·тупл·я́·ть, -ют, (*imp*) blunt, dull; (elec) despike; truncate

при·тупл·я́ющ·ий, -ая, -ее, (*pres part act*) of **при·тупл·я́·ть**

~ **у·стро́й·ств·о** (*n*) (elec) despiker

при·тяг·а́тельн·ый, -ая, -ое, (*a*) attractive

при·тя́г·ива·ть, -ют, (*imp*) attract; pull/draw/haul to

при·тяж·е́ни·е, -я, (*n*) attraction, gravitation

~, **зем·н·о́е** terrestrial gravitation

~, **капилля́р·н·ое** capillary attraction

~, **молекуля́р·н·ое** (phys) molecular cohesion

си́л·а при·тяж·е́ни·я (*f*) attractive/attraction force, force of attraction

~, **я́дер·н·ое** (phys) nuclear attraction

при·тяз·а́ни·е, -я, (*n*) claim, pretence, pretensions

при·тя́·нут·ый, -ая, -ое, (*past part pass*) *see* **при·тя́г·ива·ть**

при·тя·н·у́ть, -у́т, (*perf*) *see* **при·тя́г·ива·ть**

при·у·кра́с·ить, -я́т, (*perf*) *see* **при·у·кра́ш·ива·ть**

при·у·краш·а́·ть, -ют, (*imp*) = **при·у·кра́ш·ива·ть**

при·у·кра́ш·ива·ть, -ют, (*imp*) touch up (decoration)

при·у·меньш·а́·ть, -ют, (*imp*) reduce, diminish

при·у·меньш·и́ть, -а́т, (*perf*) *see* **при·у·меньш·а́·ть**

при·у·множ·а́·ть, -ют, (*imp*) increase/multiply still more

при·у·мно́ж·ить, -ат, (*perf*) *see* **при·у·множ·а́·ть**

при·уро́ч·енн·ый, -ая, -ое, (*past part pass*) *see* **при·уро́ч·ива·ть;** (geol) confined to (a named period)

при·уро́ч·ива·ть, -ют, (*imp*) time to coincide with, arrange for the same time as

при·уро́ч·ить, -ат, (*perf*) *see* **при·уро́ч·ива·ть**

при·уч·а́·ть, -ют, (*imp*) train/school to

при·уч·е́ни·е, -я, (*n*) training, schooling

при·уч·и́ть, -ат, (*perf*) *see* **при·у·ч·а́·ть**

при·фуг·о́вк·а, -и, (*f*) jointing (woodworking)

при·фуг·о́выва·ть, -ют, (*imp*) joint (woodwork)

при·хва́т·, -а, (*m*) *see* **при·хва́т·к·а**

~, **от·кид·н·о́й** leaf clamp (engineering jig)

при·хват·и́ть, -я́т, (*perf*) *see* **при·хва́т·ыва·ть**

при·хва́т·к·а, -и, (*g pl*) **-т·ок·,** (*f*) (*v n*) *see* **при·хва́т·ыва·ть;** clamp; tack weld

при·хва́т·ыва·ть, -ют, (*imp*) fasten, clamp; (weld) tack; freeze slightly, touch (with frost)

при·хва́ч·енн·ый, -ая, -ое, (*past part pass*) *see* **при·хва́т·ыва·ть**

при·хло́п·н·уть, -ут, (*perf*) *see* **при·хло́п·ыва·ть**

при·хло́п·ыва·ть, -ют, (*imp*) slap, swat, slam

при·хо́д·, -а, (*m*) arrival, coming; (fin) receipts, income

зо́н·а при·хо́д·а (*f*) (rad) impact zone

у́гол· при·хо́д·а (*m*) (air) arrival angle

при·ход·и́ть, -я́т, (*imp*) come, arrive; **-ся** fit, suit; (*impersonal*) be necessary/essential, have to

ему́ при·хо́д·ится he has to, he must

при·хо́д·н·ый, -ая, -ое, (*a*) (fin) receipt

~ **о́рдер·** (*m*) notification of receipt

при·хо́д·овать, -уют, (*imp*) (fin) credit

при·хо́д·о-рас·хо́д·н·ый, -ая, -ое, (*a*) (fin) account

при·ход·я́щ·ий, -ая, -ее, (*pres part act*) of **при·ход·и́ть;** visiting, non-resident, day, daily (of people)

~ **боль·н·о́й** (*m*) (med) out-patient

~ **по·то́к·** (*m*) inward flux

~ **уч·ени́к·** (*m*) day pupil

при·ход·я́щ·ийся, -аяся, -ееся, (*pres part act*) of **при·ход·и́ться;** necessary, required

~ **на** (math) taken over

при·хо́ж·ая, -ей, (*f*) entrance hall, vestibule

при·хож·у́, (*pres 1st sing*) of при·ход·и́ть

при·цвет·ник, -а, (*m*) (bot) bract, bracteole

при·це́л·, -а, (*m*) (gunn etc.) sight; back/rear-sight (of rifle)

~, диоптр·и́ческ·ий peep sight

~, опт·и́ческ·ий telescopic sight

~, радио·локацио́н·н·ый radar pre-dictor

~, раккурс·н·ый predictor sight

при·це́л·ивани·е, -я, (*n*) (*v n*) of при·це́л·ива·ться

   ли́н·и·я при·це́л·ивани·я (*f*) (surv, nav) sight line, line of sight

   пло́ск·ост·ь при·це́ливани·я (*f*) sighting plane

   то́ч·к·а при·це́л·ивани·я (*f*) point of aim

   у́гол· при·це́л·ивани·я (*m*) (gunn) tangent elevation

при·це́л·ива·ться, -ются, (*imp*) aim, sight, take aim

при·це́л·иться, -ятся, (*perf*) *see* при·це́л·ива·ться

при·це́ль·н·ый, -ая, -ое, (*a*) sight, sighting, aiming; (*m decl as adj*) sightsetter

~ на·во́д·к·а (*f*) (naut) quarters firing

~ по·ме́х·а (*f*) (rad) spot jamming

~ рас·сто·я́ни·е (*n*) (phys) impact parameter

при·це́п·, -а, (*m*) trailer

~ -ро́с·пуск·, -а, (*m*) pole trailer

~, сед·е́льн·ый semi-trailer

при·цеп·и́ть, ят, (*perf*) *see* при·цепл·я́·ть

при·цепл·е́ни·е, -я, (*n*) (*v n*) *see* при·цепл·я́·ть

при·це́пл·енн·ый, -ая, -ое, (*past part pass*) of при·цеп·и́ть

при·цепл·я́·ть, -ют, (*imp*) hook on, couple up, attach, fasten

при·цеп·н·о́й, -а́я, -о́е, (*a*) towed, trailed, trailer; towing

~ комба́йн· (*m*) (agr) tractor-drawn combine-harvester

~ плуг· (*m*) trailed plough

~ при·способл·е́ни·е (*n*) towing attachment

~ шат·у́н· (*m*) (eng) link rod, articulated rod

при·ча́л·, -а, (*m*) (*v n*) *see* при·ча́л·ива·ть; (naut) berth, mooring; mooring rope

~, по·гру́з·очн·ый loading berth

при·ча́л·ива·ть, -ют, (*imp*) (naut) berth, come/go alongside

при·ча́л·ить, -ят, (*perf*) *see* при·ча́л·ива·ть

при·ча́л·к·а, -и, (*f*) alignment rope (for bricklaying)

при·ча́ль·н·ый, -ая, -ое, (*a*) berthing, berth

~ фронт· по́рт·а (*m*) harbour wharves

при·ча́ст·и·е, -я, (*n*) (gram) participle

при·ча́ст·ност·ь, -и, (*f*) participation

при·ча́ст·н·ый, -ая, -ое, (*a*) (gram) participial; participating

при·чём (*conj*) in addition, as well; and here, moreover, besides this (or do not translate, but render succeeding phrase "... it being ...", or "... being ...")

при·чес·а́ть, (*fut 3rd pl*) при·че́ш·ут, (*perf*) dress hair

при·чёс·к·а, -и, (*g pl*) -с·ок·, (*f*) hair style, coiffure

при·че́ш·ет, (*fut 3rd sing*) of при·чес·а́ть

при·чи́н·а, -ы, (*f*) cause, reason

при·чин·и́ть, -я́т, (*perf*) *see* при·чин·я́·ть

при·чи́н·ност·ь, -и, (*f*) causality

при·чи́н·н·ый, -ая, -ое, (*a*) causal, causative

при·чин·я́·ть, -ют, (*imp*) cause, occasion

при·числ·е́ни·е, -я, (*n*) (*v n*) *see* при·числ·я́·ть

при·чи́сл·енн·ый, -ая, -ое, (*past part pass*) of при·числ·ить

при·чи́сл·ить, -ят, (*perf*) *see* при·числ·я́·ть

при·числ·я́·ть, -ют, (*imp*) add on; appoint, attach; reckon/count as

при·ша́бр·ива·ть, -ют, (*imp*) scour

при·ша́бр·ить, -ят, (*perf*) *see* при·ша́бр·ива·ть

при·шварт·ова́ть, -у́ют, (*perf*) *see* при·шварт·о́выва·ть

при·шварт·о́выва·ть, -ют, (*imp*) (naut) berth, moor alongside

при·ше́д·ш·ий, -ая, -ее, (*past part act*) of при·йти́, (*perf*); *see* при·ход·и́ть (*imp*)

~ пункт· (*m*) point of destination

~ шир·от·а́ (*f*) latitude inwards

при·шёл, (*past masc sing*) of при·йти́, (*perf*); *see* при·ход·и́ть (*imp*)

при·ше́ств·и·е, -я, (*n*) advent, arrival, coming

при·ши́б·, -а, (*m*) surf; steep bank (river)

при·шив·а́·ть, -ют, (*imp*) sew on, sew

при·шив·н·о́й, -а́я, -о́е, (*a*) sewn-on/in; attached (of collars etc.)

при·ши́·ть, (*fut 3rd pl*) при·шь·ю́т, (*perf*) *see* при·шив·а́·ть

при·шла́, (*past fem sing*) of при·йти́, (*perf*); *see* при·ход·и́ть (*imp*)

при·шл·ёт, (*fut 3rd sing*) of при·сл·а́ть

при·шлиф·о́ванн·ый, -ая, -ое, (*past part pass*) *see* при·шлиф·о́выва·ть; (mech) ground, ground-in

∼ про́б·к·а (*f*) ground glass stopper

при·шлиф·ова́·ть, -у́ют, (*perf*) *see* при·шлиф·о́выва·ть

при·шлиф·о́выва·ть, -ют, (*imp*) grind, grind in/on

при́·шл·ый, -ая, -ое, (*a*) newcomer, new, from elsewhere; (zool) immigrant

при·шоссе́й·н·ый, -ая, -ое, (*a*) highway

при·шпи́л·ива·ть, -ют, (*imp*) pin on, spike, stake

при·шпи́л·ить, -ят, (*perf*) *see* при·шпи́л·ива·ть

при·шь·ёт, (*fut 3rd sing*) of при·ши́·ть (*perf*) *see* при·шив·а́·ть (*imp*)

при·щем·и́ть, -я́т, (*perf*) *see* при·щемл·я́·ть

при·щемл·я́·ть, -ют, (*imp*) pinch

при·ще́п·, -а, (*m*) (*v n*) *see* при·щеп·л·я́·ть; graft (as plant); cleft (in grafting)

при·щеп·и́ть, -я́т, (*perf*) *see* при·щепл·я́·ть

при·щепл·я́·ть, -ют, (*imp*) (hortic) graft, graft on

при·щи́п·к·а, -и, (*f*) (*v n*) *see* при·щи́п·ыва·ть

при·щи́п·н·уть, -ут, (*perf*) *see* при·щи́п·ыва·ть

при·щи́п·ыва·ть, -ют, (*imp*) pinch, pinch out

при·ю́т·, -а, (*m*) refuge, shelter

при·я́м·ок·, -м·к·а, (*m*) basement, cellar, area, areaway; (met) pit

при·я́т·ел·ь, -я, (*m*) friend

при·я́т·ельск·ий, -ая, -ое, (*a*) friendly, amicable

при·я́т·н·ый, -ая, -ое, (*a*) pleasant, pleasing

про (*prep + acc*) about, concerning

про- (*noun and adj prefix*) pro- (for, in favour of)

про- (*verbal prefix*) *signifies*: — (1) action through, between, across; (2) action resulting in hole, indentation; track; (3) movement past; (4) finish, completion, to completion; (5) completion of action lasting certain time; (6) action causing harm; (7) -ся removal of causes bringing about the action

про·анализ·и́р·овать, -уют, (*perf*) analyse

про́б·а, -ы, (*f*) specimen, sample; test, trial, try-out; (met) assay; hallmark; (a/c) running-up

∼, 24-ая 24 carat, pure (of precious metals)

∼, 96-ая pure (of precious metals, old Russian system)

∼, 1000-ая 1000 fine, pure (of precious metals)

∼, до́ктор·ск·ая (oil) doctor test

∼, мо́кр·ая (min) wet assay

∼ на ме́д·н·ую пласт·и́нк·у (oil) copper corrosion test

∼, ти́гель·н·ая fire assay

∼, шёбер·н·ая (met, chem) scorification

про·бе́г·, -а, (*m*) run, race; (rail, M/T) mileage, kilometerage; (air) landing run; (nucl) track, range, path length; rundown (of oscillator)

∼, балла́ст·н·ый (naut) voyage in ballast

∼, ви́д·им·ый visible range

вре́м·я про·бе́г·а (*n*) (phys) transit time

длин·а́ про·бе́г·а (*f*) mean free path (of atomic particles)

из·мер·е́ни·е про·бе́г·ов (*n*) (autom) range measurement

∼, меж·ремо́нт·н·ый kilometres/miles run between overhauls

∼ нейтро́н·а neutron path

∼, о·ста́т·очн·ый residual range

∼, пре·де́ль·н·ый ultimate range

∼, прям·о́й linear range

∼, свобо́д·н·ый (nucl) free path

∼, сре́д·н·ий свобо́д·н·ый (phys) mean free path

про·бег·а́·ть, -ют, (*imp*) run past/by/through/over, pass; cover (distance)

про·беж·а́·ть, (*fut 3rd sing, pl*) про·беж·и́т, про·бег·у́т, (*past masc sing*) про·беж·а́л, (*perf*); *see* про·бег·а́·ть

про·бе́й·те, (*pl imper*) of про·би́·ть, (*perf*); *see* пробив·а́·ть (*imp*) про·бе́л, -а, (*m*) blank, gap, omission, hiatus; space, spacing

про·бел·ённ·ый, -ая, -ое, (*past part pass*) *see* про·бе́л·ива·ть

про·бе́л·ива·ть, -ют, (*imp*) paint/make white; cure (sugar)

про·бел·и́ть, -ят, (*perf*) *see* про·бе́л·ива·ть

про·бе́л·к·а, -и, (*f*) (*v n*) *see* про··бе́л·ива·ть

~ утфел·я centrifuging (sugar)

про·бе́л·очн·ый, -ая, -ое, (*a*) of про·бе́л·к·а

~ клерс· (*m*) washing liquor (sugar)

про́·бел·ь, -и, (*f*) white patch/streak/tint

про·бе́ль·н·ый, -ая, -ое, (*a*) of про·бе́л·; whitening; curing (sugar); (*f decl as adj*) curing room (sugar)

~ и́мпульс· (*m*) (telecom) spacing pulse

~ материа́л· (*m*) (print) spacing material, furniture

~ част·ь (*f*) (print) space

про·бер·ёт, (*fut 3rd sing*) of про··бр·а́ть, (*perf*); *see* про·бир·а́·ть (*imp*)

про·бив·а́ни·е, -я, (*n*) (*v n*) of про··бив·а́·ть

про·бив·а́·ть, -ют, (*imp*) pierce, punch, hole, perforate, puncture, force/push through; lay out, build through (roads etc.); break down (insulation etc.); (telecom) break through; beat, beat/rap out, sound (signals etc.); caulk, stuff

про·би́в·к·а, -и, (*g pl*) -в·ок·, (*f*) (*v n*) *see* про·бив·а́·ть; (met) punching, die-punching

~ дыр· (mech eng) drifting

~ свет·ов·о́й ли́н·и·и (mech eng) optical alignment (shafting etc.)

про·бив·н·о́й, -а́я, -о́е, (*a*) penetrating, penetrative, piercing; (elec) breakdown, disruptive

~ ис·пыт·а́ни·е (*n*) (elec) breakdown test

~ на·пряж·е́ни·е (*n*) (elec) breakdown voltage

~ способ·ност·ь (*f*) penetrating power, penetration

про·би́в·очн·ый, -ая, -ое, (*a*) piercing, perforating, punching

про·бир·а́·ть, -ют, (*imp*) part, separate, make a parting; (hortic) weed, hoe; -ся get through (to), slip through, steal into

про·би́р·к·а, -и, (*g pl*) -р·ок·, (*f*) test tube

про·би́р·н·ый, -ая, -ое, (*a*) assay

~ ка́мен·ь (*m*) (min) lydite, touchstone, Lydian stone

~ клейм·о (*n*) hallmark, mark of assay

~ ма́стер· (*m*) assay-master, assayer

~ пала́т·а (*f*) assay office

про·би́р·щик·, -а, (*m*) assayer, assay-master

про·би́·ть, (*fut 3rd pl*) про·бь·ю́т, (*perf*) *see* про·бив·а́·ть

про́б·к·а, -и, (*g pl*) -б·ок·, (*f*) (bot) cork, phellem; stopper, bung, plug; (eng) plug gauge/gage; (elec) plug-type fuse; filler cap (petrol tank); piercing mandrel (tube drawing); rotor (of hydraulic selector valve); traffic jam; stopper-end (of steel teeming ladle)

~, га́з·ов·ая gas lock

~, ду́ль·н·ая (gunn) tampion

~, магни́т·н·ая (phys) magnetic mirror

~, на·правл·я́ющ·ая (oil) floating plug

~, пре·де́ль·н·ая двух·сторо́н·н·яя go-and-not-go plug gauge/gage

~, про·би́в·очн·ая (oil) cement plug

проб·ковени·е, -я, (*n*) (bot) suberization

пробко·ви́д·н·ый, -ая, -ое, (*a*) cork-like, subereous, suberose

про́б·ков·ый, -ая, -ое, (*a*) cork, subereous, suberous, suberic

~ де́рев·о (*n*) (bot) Amur cork tree, *Phellodendron amurense*

~ доск·а́ (*f*) corkboard

~ дуб· (*m*) (bot) cork oak, *Quercus suber*

~ ка́мб·и·й (*m*) (bot) cork cambium, phellogen

~ кисл·от·а́ (*f*) (chem) suberic acid

~ плит·а́ (*f*) corkboard

~ по́яс· (*m*) lifebelt

пробле́м·а, -ы, (*f*) problem

проблема́т·ик·а, -и, (*f*) problems

проблемат·и́ческ·ий, -ая, -ое, (*a*) problematic

проблемат·и́чн·ый, -ая, -ое, (*a*) problematic

про́·блеск·, -а, (*m*) flash, gleam, ray

~, дли́н·н·ый dash, long (light signalling)

~, коро́т·к·ий dot, short (light signalling)

про́·блеск·ов·ый, -ая, -ое, (*a*) flashing

~ аппара́т· (*m*) flasher (for lighthouse/advertising lights etc.)

проб·ник·, -а, (*m*) (elec) probe; (met) sampler

проб·н·ый, -ая, -ое, (*a*) test, testing, trial; hall-marked

~ за́втрак· (*m*) (med) test meal

~ кран· (*m*) (st eng) pet/test/gauge cock

~ пуск· (*m*) trial run

~ реле́ (*n*) (elec) testing relay

~ шар· (*m*) (dipl) *ballon d'essai*

проб·овани·е, -я, (*n*) trying out, testing

проб·овать, -уют, (*imp*) try/attempt/endeavour to; try, try out, test; taste, feel

про·бод·е́ни·е, -я, (*n*) perforation; puncture

~ оболо́ч·к·и (nucl) shell puncture

про·бод·н·о́й, -а́я, -о́е, (*a*) perforating

про·бо́·ин·а, -ы, (*f*) hole, perforation; bullet-hole

про·бо́·й, -я, (*m*) (elec) breakdown, breaking down (of insulation); clamp, hasp, holdfast

~ га́з·а (elec) gas breakdown

~ диэле́ктрик·а dielectric breakdown **про́ч·ност·ь на про·бо́·й** (*f*) (elec) breakdown strength

про·бо́й·ник·, -а, (*m*) punch, blacksmith's punch; drift (tool)

пробо·от·бир·а́тел·ь, -я, (*m*) sampler

пробо·от·бо́р·ник·, -а, (*m*) sampler, (min) sample cutter

про·бо́р·к·а, -и, (*g pl*) **-р·ок·,** (*f*) (*v n*) *see* **про·бир·а́·ть**

~ осно́в·ы (text) looming, healding

про·бо́р·щик·, -а, (*m*) (text) healder

проб·очник·, -а, (*m*) bottle stopper

проб·очн·ый, -ая, -ое, (*a*) of **про́б·к·а**

про·бр·а́ть, (*fut 3rd pl*) **про·бер·у́т,** (*perf*) *see* **про·бир·а́·ть**

про·бу́д·ет, (*fut 3rd sing*) of **про·бы́·ть**

про·буд·и́ть, -я́т, (*perf*) *see* **про·бужд·а́·ть**

про·бужд·а́·ть, -ют, (*imp*) rouse; **-ся** (*pass*); wake up, awaken

про·бук·со́вк·а, -и, (*g pl*) **-вок·,** (*f*) (rail) skid, skidding

про·бура́в·ить, -ят, (*perf*) *see* **про·бура́вл·ива·ть**

про·бура́вл·ива·ть, -ют, (*imp*) bore through

про·бы́·ть, (*fut 3rd pl*) **про·бу́д·ут'** (*perf*) spend (time), stay

про·бь·ёт, (*fut 3rd sing*) of **про·бй·ть,** (*perf*); *see* **про·бив·а́·ть** (*imp*)

про·ва́л·, -а, (*m*) (*v n*) *see* **про·ва́·л·ива·ться;** subsidence, collapse, failure; valley, dip, trough; sink, slump, depression, hollow; unburnt fuel fallen through fire grate

~, ка́рст·ов·ый (geol) karst hole

про·ва́л·ива·ть, -ют, (*imp*) fall in/into; fail, cause to fail, spoil; **-ся** fall in/into; collapse, fall down, cave in, sink, subside; fail; become sunken (of features etc.)

про·вал·и́ть, -я́т, (*perf*) *see* **про·ва́·л·ива·ть**

про·ва́ль·н·ый, -ая, -ое, (*a*) subsidence, collapse; slack (of loading time etc.)

прова́нск·ое ма́сл·о (*n*) olive oil

про·ва́р·, -а, (*m*) (*v n*) *see* **про·ва́·р·ива·ть;** (weld) fusion, penetration; (paper) degree of cooking

про·ва́р·ива·ть, -ют, (*imp*) cook, boil, cook/boil through

про·вар·и́ть, -я́т, (*perf*) *see* **про··ва́р·ива·ть**

про·ве́д·а·ть, -ют, (*perf*) *see* **про·ве́·д·ыва·ть**

про·вед·е́ни·е, -я, (*n*) (*v n*) *see* **про··вод·и́ть** (*imp*)

~ вы́·работ·ок· (min) development

~ горизонта́ль·н·ых вы́·работ·ок· (min) tunnel/gallery driving

про·вед·ённ·ый, -ая, -ое, (*past part pass*) *see* **про·вод·и́ть** (*imp*)

про·ве́д·ыва·ть, -ют, (*imp*) find out, trace, learn

про·ве́·ет, (*fut 3rd sing*) of **про·ве́··ять,** (*perf*); *see* **про·ве́·ива·ть** (*imp*)

про·вез·ти́, -у́т, (*perf*) transport, convey, take

про·ве́·ива·ть, -ют, (*imp*) winnow

про·вёл, (*past masc sing*) of **про··; вес·ти́,** (*perf*); *see* **про·вод·и́ть** (*imp*)

про·вентил·и́р·овать, -уют, (*perf*) ventilate

про·ве́р·ить, -ят, (*perf*) *see* **про··вер·я́·ть**

про·ве́р·к·а, -и, (*g pl*) **-р·ок·,** (*f*) (*v n*) *see* **про·вер·я́·ть;** verification, check, check-up, test; inspection (quality control)

~ бо́·я (gunn) accuracy test

~, вы́·бор·очн·ая spot check

~ гермет·и́чност·и leak test

про·вер·к·а
~ на дефе́кт·ност·ь marginal checking
~ на ме́ст·е spot check
~ на·ли́ч·и·я stock-taking
~ пе́ре·пис·и (elec) rewrite check (storage tube)
~, пред·ста́рт·ов·ый (rocket) pre-firing check
~ при пере·ме́н·н·ом ко́д·е leapfrog test (computers)
~, сплош·н·а́я complete check; 100% inspection (quality control)
~ с·чёт·ов audit
про·вер·н·у́ть, -у́т, (perf) see **про·-вёрт·ыва·ть**
про·ве́р·очн·ый, -ая, -ое, (a) check, checking, test
~ ис·пыт·а́ни·е, (n) examination (scholastic)
про·верт·е́ть, -я́т, (perf) drill, bore; rotate, spin (for a time)
про·вёрт·ыва·ть, -ют, (imp) drill, bore; mince; rotate, spin
про·ве́рч·енн·ый, -ая, -ое, (past part pass) of **про·верт·е́ть**
про·вер·я́·ть, -ют, (imp) check, verify, control; test
про·ве́с·, -а, (m) sag, dip, bending deflection; (com) underweight
про·ве́с·ить, -ят, (perf) see **про·ве́-ш·ива·ть**
про·вес·ти́, (fut 3rd pl) **про·вед·у́т,** (past masc sing) **про·вё·л,** (perf) see **про·вод·и́ть**
про·ве́тр·ива·ть, -ют, (imp) air, ventilate
про·ве́тр·ить, -ят, (perf) see **про·-ве́тр·ива·ть**
про·ве́ш·енн·ый, -ая, -ое, (past part pass) of **про·ве́с·ить,** (perf); see **про·-ве́ш·ива·ть** (imp)
про·веш·ённ·ый, -ая, -ое, (past part pass) of **про·веш·и́ть**
про·ве́ш·ивани·е, -я, (n) (v n) of **про·ве́ш·ива·ть;** (surv) chain surveying, marking/staking out (a line), tracing (a line)
~ ли́н·и·й (surv) ranging
про·ве́ш·ива·ть, -ют, (imp) give short weight; (food) dry in the open air, hang up to dry; check for alignment/verticality; plumb
~ ли́н·и·ю (surv) range out a line, stake out, trace
про·веш·и́ть, -а́т, (perf) see **про·-ве́ш·ива·ть ли́н·и·ю**

про·ве́·ять, -ют, (perf) see **про·ве́-·ива·ть**
провиа́нт·, -а, (m) provisions, victuals
про·ви́д·ени·е, -я, (n) foresight
прови́з·и·я, -и, (f) (food) provisions, victuals
прови́зор·, -а, (m) pharmacist
провизо́р·н·ый, -ая, -ое, (a) provisional, preliminary; temporary
прови́нц·и·я, -и, (f) province
~, га́з·ов·ая (geol) gas province
~, петро·граф·и́ческ·ая (geol) petrographic province
про·вис·а́ни·е, -я, (n) (v n) see **про·-вис·а́·ть;** creep (of roof)
  по·пра́в·к·а за про·вис·а́ни·е (f) (surv) catenary correction
про·вис·а́·ть, -ют, (imp) sag, dip
про·вис·е́ть, -ят, (perf) hang for a time
про·ви́с·нуть, -нут, (perf) see **про·-вис·а́·ть**
про·витами́н·, -а, (m) (biochem) provitamin
про·во́д·, -а, (m) (v n) see **про·во-д·и́ть; про́·вод·, -а,** (nom pl) **-а́,** (g pl) **-о́в,** (m) (elec) conductor, conducting wire, wire, lead; conduit, piping, line
~, би·металл·и́ческ·ий composite conductor, bimetallic wire
~, ви́·т·ый stranded conductor
~, воз·ду́ш·н·ый open/overhead conductor/wire
~, го́л·ый (elec) uninsulated/bare conductor
~, за·земл·я́ющ·ий grounding/earthing conductor
~, компенсаци́он·н·ый compensating lead/wire
~, конденсаци́он·н·ый condensate return pipe/main
~, конта́кт·н·ый contact/trolley wire/conductor
~, контро́ль·н·ый pilot wire (in remote measuring systems)
~, кра́й·н·ий outer conductor (of elec cable)
~, кро́сс·ов·ый jumper wire (telephones)
~, много·про́·волоч·н·ый stranded conductor
~, монта́ж·н·ый electrical installation wire, wiring conductor
~, нейтра́ль·н·ый neutral/middle conductor (of elec cable)

про·во́д

~, **нул·ев·о́й** neutral/middle conductor (of elec cable)

~, **о·бес·то́ч·енн·ый** dead wire

~, **об·мо́т·анн·ый** covered wire

~, **об·мо́т·очн·ый** winding wire

~, **от·ветвл·ённ·ый** branch conductor

~, **пуч·ко́в·ый** bunched/bundled conductor

~, **стале·алюми́н·иев·ый** aluminium-steel conductor, steel-cored aluminium conductor

~, **стале·ме́д·н·ый** copper-clad steel conductor, steel-cored copper conductor

~ **то́пл·ив·а** fuel pipe

~, **троллей·н·ый** trolley wire

~, **у·стано́в·очн·ый** electrical installation wire, wiring conductor

~, **шла́нг·ов·ый** rubber sheathed/tube conductor

про·вод·и́мостн·ый, -ая, -ое, (a) of **про·вод·и́мост·ь**

про·вод·и́мост·ь, -и, (f) conductivity; (elec) conductance, admittance

~, **акти́в·н·ая** (elec) conductance

~, **ано́д·н·ая** (elec) anode/plate conductance

~, **ве́ктор·н·ая** vector admittance

~, **волн·ов·а́я** characteristic/surge admittance

~, **в·ход·н·а́я ко́мплекс·н·ая** driving-point admittance

~, **ды́р·очн·ая** hole-/p-type conductivity

~, **ём·кост·н·ая** (elec) capacitive susceptance

~ **изоля́ци·и** (elec) leakance, leakage conductance

~, **одно·ро́д·н·ая** isotropic conductivity

~, **одно·сторо́н·н·яя** unilateral conductivity

~, **по́лн·ая** (elec) admittance

~ **по·сто·я́нн·ого то́к·а** (elec) direct-current conductance

~, **при·ме́с·н·ая** impurity/extrinsic conductivity

~, **про·ход·н·а́я** (elec) transfer admittance

~, **реакти́в·н·ая** (elec) susceptance

~ **се́т·к·и, акти́в·н·ая** (rad) grid conductance

~, **сквоз·н·а́я** bulk conductivity

~, **температу́р·н·ая** thermal diffusivity

~, **тепл·ов·а́я** thermal conductivity

про·вод·и́мост·ь

~, **у·де́ль·н·ая** conductivity

~, **щел·ев·а́я** slot conductance

~, **электро́н·н·ая** electron-/n-type conductivity

про·вод·и́м·ый, -ая, -ое, (pres part pass) of **про·вод·и́ть**; in process/progress, under way

~ **о́·пыт·** (m) test in progress, test being conducted

про·вод·и́ть, -ят, (imp) lead, take, pilot; draw/slide/pass across/along/over; carry out, conduct, perform; spend, pass (time); draw (lines etc.); lay, lay out, build (roads etc.); drive/construct (tunnels), sink (shafts); introduce, propose; (imp only) (phys) conduct; **-ся** (pass); be in progress

~ **раз·ли́ч·и·е** draw a distinction, distinguish (between)

про·вод·и́ть, -ят, (perf) see **про·вож·а́ть**

про·во́д·к·а, -и, (g pl) **-д·ок·,** (f) (v n) see **про·вод·и́ть** (imp); (elec) wiring; run, line (of pipes); (naut) piloting; (met roll) guide

~, **в·во́д·н·ая** (met roll) entering/receiving guide

~, **винт·ов·а́я** (met roll) twist guide

~, **вы·вод·н·а́я** (met roll) delivery/stripper guide

~, **геликоида́ль·н·ая** (met roll) twist guide

~, **кант·у́ющ·ая** (met roll) twist guide

~, **на·правл·я́ющ·ая** (met roll) guard

~, **об·вод·н·а́я** (met roll) repeater

~, **ро́л·иков·ая** (met roll) roller guide, tensioning bridle

~, **ро́л·иков·ая кант·у́ющ·ая,** roller twist guide, twist rolls

про·во́д·ков·ый, -ая, -ое, (a) of **про·во́д·к·а**

~ **брус·** (m) (met roll) guide/cramp/rest bar

~ **нож·, ве́рх·н·ий** (m) (met roll) hanging guide, top-stripper guide

~ **нож·, ни́ж·н·ий** (m) (met roll) resting guide, bottom stripper guide

про·вод·ни́к·, -а́, (m) guide, (phys, build) conductor, (geol, build, hortic) leader; messenger, guide-rope

~, **плох·о́й** poor conductor

про·вод·нико́в·ый, -ая, -ое, (a) of **про·вод·ни́к·;** conducting

про·вод·н·о́й, -а́я, -о́е, (a) of **про́·вод·;** wire, line

про·вод·н·о́й, -а́я, -о́е
~ с·вя́зь (*f*) (telecom) line communications; (elec) wired connection
пр·во́д·нᴈст·ь, -и, (*f*) conductance, conductivity
про·вод·я́щ·ий, -ая, -ее, (*pres part act*) of про·вод·и́ть; conductive
~ пут·ь (*m*) (biol) conducting/vascular strand
~ ткан·ь (*f*) (bot) vascular tissue
про·вож·а́т·ый, -ого, (*m decl as adj*) guide
про·вож·а́·ть, -ют, (*imp*) accompany; see off; track (with radar)
про·вож·у́, (*pres 1st sing*) of про·вод·и́ть, (*imp*); (*pres 1st sing*) of про·воз·и́ть, (*fut 1st sing*) of про·вод·и́ть, (*perf*); see про·вож·а́·ть (*imp*)
про·во́з·, -а, (*m*) (*v n*) see про·воз·и́ть; transport, conveyance
про·воз·глас·и́ть, -я́т, (*perf*) see про·воз·глаш·а́·ть
про·воз·глаш·а́·ть, -ют, (*imp*) proclaim; propose (a toast)
про·воз·и́ть, ⁻ят, (*imp*) transport, convey, carry, cart, take
про·воз·н·о́й, -а́я, -о́е, (*a*) transport, carriage, transit (of goods)
провозо·спосо́б·ност·ь, -и, (*f*) carrying capacity, capacity (of transport vehicles)
провока́ци·я, -и, (*f*) provocation
про·вола́к·ива·ть, -ют, (*imp*) drag, draw; draw (wire)
про·воло́к·, (*past masc sing*) of про·воло́ч·ь
про́·волок·а, -и, (*f*) wire
~, була́в·очн·ая pin wire
~, волн·и́ст·ая crimped wire
~, кана́т·н·ая rope wire
~, ка́т·ан·ая wire rod
~, кол·ю́ч·ая barbed wire
~, от·бо́й·н·ая guard wire
~, при·са́д·очн·ая filling/filler/welding wire
~ прямо·уго́ль·н·ого сеч·е́ни·я flat/rectangular wire
~, роя́ль·н·ая piano wire
~, с·ва́р·очн·ая welding/filler/filling wire
~, стру́н·н·ая piano wire
~, то́лст·ая heavy-gauge wire
~, тя́·нут·ая drawn wire
~, электро́д·н·ая electrode wire
про·волок·у́т, (*fut 3rd pl*) of про·воло́ч·ь

про·волоч·и́ть, -а́т, (*perf*) see про·вола́к·ива·ть
про́·волоч·к·а, -и, (*g pl*) -ч·ек·, (*f*) (*dim*) of про́·волок·а (q.v.); про·воло́ч·к·а, -и, (*g pl*) -ч·ек·, (*f*) delay, procrastination
~, из·мер·и́тельн·ые (*pl*) external screw-thread gauge
про́·волоч·ник·, -а, (*m*) (ent) wire worm, click beetle larva; wiremaker
про́·волоч·н·ый, -ая, -ое, (*a*) wire
~ кана́т· (*m*) wire rope
~ на·грев·а́тел·ь стекл·а́ (*m*) electric windscreen heater
~ стан· (*m*) (met) rod mill
про·воло́ч·ь, (*fut 3rd sing, pl*) про·воло́ч·ёт, про·волок·у́т, (*past masc sing*) про·воло́к·, (*perf*) drag, draw
про·вора́ч·ива·ть, -ют, (*imp*) run through/up, do quickly; (ICE) turn over, run; -ся (*pass*); twist/turn/unscrew spontaneously
прово́р·н·ый, -ая, -ое, (*a*) agile, adroit; quick, swift
провоц·и́р·овать, -уют, (*imp and perf*) provoke
про·вощ·ённ·ый, -ая, -ое, (*past part pass*) waxed, wax-impregnated
про·вя́л·ива·ть, -ют, (*imp*) dry-cure, dry, sun-dry
про·вя́л·ить, -ят, (*perf*) see про·вя́л·ива·ть
прога́л·ин·а, -ы, (*f*) open space, gap; current channel (through ice); glade (in woods)
прогано́л·, -а, (*m*) (pharm) proganol
про·га́р·, -а, (*m*) burn, burnt place, burn hole
про·га́р·ин·а, -ы, (*f*) = про·га́р·
прогестеро́н·, -а, (*m*) (biochem) progesterone
про·ги́б·, -а, (*m*) (*v n*) see про·ги·б·а́·ть; sag, sagging, trough; deflection (in bend test); bowing
ли́н·и·я про·ги́б·а (*f*) deflection line
~, обра́т·н·ый sagitta, versed sine, rise
~, с·бро́с·ов·ый (geol) fault trough
стрел·а́ про·ги́б·а (*f*) deflection, sag
про·гиб·а́ни·е, -я, (*n*) (*v n*) see про·гиб·а́·ть; (geol) warping, bowing
про·гиб·а́·ть, -ют, (*imp*) bend, flex, make sag; -ся (*pass*); sag
про·гибо·ме́р·, -а, (*m*) deflectometer
прогино́н· -а, (*m*) (pharm) progynon

про·гла́д·ить, -ят, (*perf*) *see* про·-гла́ж·ива·ть

про·гла́ж·ива·ть, -ют, (*imp*) iron, press, smooth

про·гла́т·ыва·ть, -ют, (*imp*) swallow, ingest

про·глот·и́ть, -ят, (*perf*) *see* про·-гла́т·ыва·ть

про·гнат·и́зм·, -а, (*m*) (zool) prognathism

про·гн·а́ть, (*fut 3rd pl*) про·го́н·ят, (*perf*) *see* про·гон·я́·ть

про·гнив·а́·ть, -ют, (*imp*) rot through

про·гни́·ть, -ют, (*perf*) *see* про·гни-в·а́·ть

про·гно·ённ·ый, -ая, -ое, (*past part pass*) *see* про·гнив·а́·ть

про·гно́з·, -а, (*m*) forecast, prediction, (med etc.) prognosis

~ по·го́д·ы   weather forecast

про·гноз·и́р·овать, -уют, (*imp and perf*) forecast, predict

про·гност·и́ческ·ий, -ая, -ое, (*a*) prognostic, predicted

про·гн·у́ть, -у́т, (*perf*) *see* про·гиб·а́·ть

про·голос·ова́ть, -у́ют, (*perf*) vote

про·го́н·, -а, (*m*) (*v n*) *see* про·го-н·я́·ть; well (between floors); joist, bearer (floor support), purlin (roof support); roadbearer (of bridge); (*pl*) (obs) travelling expenses

~ пере·кры́т·и·я   (build) floor joist

~, сре́д·н·ий   (chem) middlings

про·го́н·к·а, -и, (*g pl*) -н·ок, (*f*) screw die

~, тру́б·чат·ая   spring die

про·гон·я́·ть, -ют, (*imp*) drive along/past/away/through

про·гор·а́ни·е, -я, (*n*) (*v n*) of про·-гор·а́·ть

про·гор·а́·ть, -ют, (*imp trans*) burn through/up

про·гор·е́л·ый, -ая, -ое, (*a*) burnt

про·гор·е́ть, -я́т, (*perf*) burn through/up; burn (for a time)

про·го́рк·л·ый, -ая, -ое, (*a*) rancid, rank

про·го́рк·нуть, -нут, (*perf*) become rancid/rank

про·град·а́ци·я, -и, (*f*) progradation

про·граду·и́рованн·ый, -ая, -ое, (*past part pass*); (instr) graduated, calibrated

програ́мм·а, -ы, (*f*) programme, routine

~, вед·у́щ·ая   master programme (computers)

програ́мм·а

~, вс·по·мог·а́тельн·ая   utility programme (computers)

~, ис·пыт·а́тельн·ая   trials programme, test routine

~, программ·и́рующ·ая   compiler, compiling routine (computers)

~, само·у·стана́вл·ивающ·аяся   optimalizing programme (computers)

~ с·чи́т·ывани·я   read routine (computers)

программ·и́ровани·е, -я, (*n*) (autom) programming

~, блок·схе́м·н·ое   macro-programming

программ·и́р·овать, -уют, (*imp*) programme

программ·и́рующ·ий, -ая, -ое, (*pres part act*) of программ·и́р·овать

~ програ́мм·а   (*f*) compiler, compiling routine

программ·и́ст·, -а, (*m*) programmer

програ́мм·н·ый, -ая, -ое, (*a*) programme, programmed

~ у·стро́й·ств·о   (*n*) programme unit

про·графич·ённ·ый, -ая, ое, (*past part pass*); graphite-impregnated

про·грев·а́·ть, -ют, (*imp*) warm/heat up/through

про·гре́·ве, -а, (*m*) (*v n*) *see* про·-грев·а́·ть; warming up (of motors etc.)

про·грем·е́ть, -я́т, (*perf*) thunder

прогре́сс·, -а, (*m*) progress

прогресси́в·ност·ь, -и, (*f*) progressiveness; rate of progression

~ гор·е́ни·я   (expl) regular rate of burning

прогресси́в·н·ый, -ая, -ое, (*a*) progressive

прогресс·и́р·овать, -уют, (*imp*) progress, make progress

прогре́сс·и·я, -и, (*f*) progression

~, арифмет·и́ческ·ая   (math) arithmetical progression

про·гре́·ть, -ют, (*perf*) *see* про·гре-в·а́·ть

про·грохот·а́ть, (*fut 3rd pl*) про·-грохо́ч·ут, (*perf*) screen

про·грыз·а́·ть, -ют, (*imp*) gnaw, gnaw a hole in

про·гры́з·ть, -у́т, (*perf*) *see* про·-грыз·а́·ть

про·гуд·е́ть, -я́т, (*perf*) hum, buzz, drone; sound (of hooters etc.)

про·гу́л·, -а, (*m*) unauthorized absence, (*pl*) absenteeism про·гу́л·к·а, -и, (g pl) -л·ок·, (f) walk, stroll; outing

про·гу́л·очн·ый, -ая, -ое, (a) excursion, pleasure; promenading, walking

про·гу́ль·щик·, -а, (m) absentee

прод- (abbr component) = про·до·воль·ствени·ый (q.v.)

про·да·ва́ть, -ют, (imp) sell, market
~ в ро́зн·иц·у retail, sell retail
~ в·на·ви́л·к·у sell in bulk
~ на́·сып·ью sell in bulk
~ иопт·ом sell wholesale, sell in bulk, wholesale

про·да́в·ец·, -в·ц·а́, (m) seller; shopassistant, salesman

про·дав·и́ть, -ят, (perf) see про·давл·ива·ть

про·да́вл·ивани·е, -я, (n) (v n) of про·да́вл·ива·ть
ис·пыт·а́ни·е на про·да́вл·ивани·е (n) (rubber) impact puncture test

про·да́вл·ива·ть, -ют, (imp) crush, break, stave in, puncture; squeeze/press through

про·дад·у́т, (fut 3rd pl) of про·да́·ть

про·да́·ж·а, -и, (f) (v n) see про·да·ва́ть; sale, selling
в про·да́·ж·е on sale
~ в ро́зн·иц·у retail
~, опт·о́в·ая wholesale

про·да́ж·н·ый, -ая, -ое, (a) selling, sale, for sale; corrupt, venal

про·да́бл·ива·ть, -ют, (imp) chip, chip/chisel away/through

про·да́·ть, (fut 1st, 3rd sing, 3rd pl) про·да́м, про·да́·ст, про·дад·у́т, (past masc sing) про́·да·л, (perf); see про·да·ва́ть

про·двиг·а́тел·ь ткан·и (m) feed dog (of sewing machine)

про·двиг·а́·ть, -ют, (imp) move/shift/transfer on/onwards/forwards, further, advance, forward; про·дви́г·а·ть, -ют, (perf) move/shift/transfer for a period; -ся (intrans)

про·движ·е́ни·е, -я, (n) (v n) see про·двиг·а́ть; advance, advancement

про·дви́·н·уть, -ут, (perf) see про·двиг·а́·ть

про·дев·а́·ть, -ют, (imp) put/pass through, thread, (naut) reeve; (text) run in

про·дежу́р·ить, -ят, (perf) be on duty

про·де́л·, -а, (m) ground buckwheat

про·де́л·а·ть, -ют, (perf) see про·де́л·ыва·ть

про·дел·ыва́ем·ый, -ая, -ое, (pres part pass) of про·де́л·ыва·ть; in progress

про·де́л·ыва·ть, -ют, (imp) make, do; knock through, make a hole in

про·демонстр·и́р·овать (perf) demonstrate

про·де́·нет, (fut 3rd sing), of про·де́·ть (perf), see про·дев·а́·ть (imp)

про·дёрг·а·ть, -ют, (perf) thin, thin out; pull, tug (for a time)

про·дёрг·ива·ть, -ют, (imp), see про·дёрг·а·ть, (perf), про·дёр·н·уть (perf)

про·дер·ёт, (fut 3rd sing) of про·др·а́ть, (perf); see про·дир·а́·ть (imp)

про·держ·а́ть, -ат, (perf) hold, keep

про·дёр·н·уть, -ут, (perf) put/pass through, thread, (naut) reeve

про·де́·ть, -нут, (perf) see про·дев·а́·ть

про·дефил·и́р·овать, -уют, (perf) march past

про·дешев·и́ть, -ят, (perf) see про·дешевл·я́·ть

про·дешевл·я́·ть, -ют, (imp) sell too cheaply, sell below cost

про·дикт·ова́ть, -у́ют, (perf) dictate

про·дир·а́·ть, -ют, (imp) tear across, tear/wear through; -ся (pass); force one's way through

про·дифференц·и́р·овать, -уют, (perf) differentiate

про·диффунд·и́р·овать, -уют, (perf) diffuse, be diffused

про·дл·ева́·ть, -ют, (imp) prolong, extend (of time)

про·дл·е́ни·е, -я, (n) prolongation, extension

про·дл·и́ть, -я́т, (perf) prolong, extend (of time); -ся last (for a time)

про·до·во́ль·ствени·ый, -ая, -ое, (a) food, provision; (mil) ration

про·до·во́ль·стви·е, -я, (n) foodstuffs, provisions; (mil) victuals, rations

про·долб·и́ть, -я́т, (perf) see про·да́бл·ива·ть

про·долг·ова́т·ый, -ая, -ое, (a) elongated, prolate; oblong, oblongate
~ мозг· (m) (anat) medulla oblongata
~ фигу́р·а (f) oblong

про·должж·а́тел·ь, -я, (m) continuer, successor

про·долж·а́·ть, -ют, (imp) continue, go/keep on; prolong, extend

про·долж·éни·е, -я, (*n*) duration, course; continuation, prolongation, extension
в про·долж·éни·е (*prep + gen*) during, throughout, in the course of
про·долж·ительност·ь, -и, (*f*) *see also* **врéм·я** *and associated words*; duration, period, time (length)
∼ **вулканиз·áци·и** (rubber) vulcanization time, time of cure
∼ **ймпульс·а** (phys) pulse length
ис·пыт·áни·е на про·долж·ительность (*n*) endurance test
∼ **óт·жиг·а** annealing time
∼ **слýж·б·ы** campaign (of a furnace)
∼ **с·чёт·а** counting interval
про·долж·ительн·ый, -ая, -ое, (*a*) long, prolonged
про·дóлж·ить, -ат, (*perf*) *see* **про·долж·á·ть**
про·дóль·н·о (*adv*) lengthwise, longitudinally
продольно·волокн·йст·ый áсбест· (*m*) slip-fibre asbestos
продольно·по·перéч·н·ый автомáт· (*m*) (a/c) bank-and-climb control unit
продольно·рас·пил·óвочн·ый ста·н·óк (*m*) mechanical rip saw, rip-saw frame/machine
продольно·строг·áльн·ый стан·óк· (*m*) planing machine
продольно·фрéзер·н·ый стан·óк (*m*) straight-line milling machine, planer
про·дóль·н·ый, -ая, -ое, (*a*) longitudinal, lengthwise, (naut) fore-and-aft
∼ **бáл·к·а** (*f*) side member (of a truss, chassis etc.)
∼ **кáч·к·а** (*f*) (naut) pitching
∼ **мóдул·ь у·прýг·ост·и** (*m*) modulus of elasticity in tension
∼ **с·вяз·ь** (*f*) through stay (of boiler)
∼ **шов·** (*m*) (weld) longitudinal seam
про·дорож·ённ·ый, -ая, -ое, (*past part pass*); grooved, nicked
про·дорóж·ива·ть, -ют, (*imp*) rifle, groove
прод·от·чёт·, -а, (*m*) state/return of victualling stocks
про·др·áть, (*fut 3rd pl*) **про·дер·ýт,** (*perf*) *see* **про·дир·á·ть**
про·дрейф·овáть, -ýют, (*perf*) drift; drag (of vessel at anchor)
про·дрóг·нуть, -нут, (*perf*) shiver; be chilled
про·дув·áем·ый, -ая, -ое, (*pres part pass*) of **про·дув·á·ть**; air-blast

про·дув·áни·е, -я, (*f*) = **про·дýв·к·а** (q.v.); (med) perflation
∼, **вéрх·н·ее** surface blow-off (of boiler)
про·дув·á·ть, -ют, (*imp*) blow through/off; scavenge (in diesels); blow (for a time)
про·дýв·к·а, -и, (*g pl*) **-в·ок·,** (*f*) (*v n*) *see* **про·дув·á·ть**
∼, **клáпан·н·о-щел·ев·áя** combined valve-and-cylinder-port scavenging
∼, **петле·обрáз·н·ая** loop scavenging
∼, **прямо·тóч·н·ая** uniflow scavenging
∼, **щел·ев·áя** valveless scavenging
про·дув·н·óй, -áя, -óе, (*a*) = **про·дýв·очн·ый** (q.v.) blow/blowing-off; scavenge, scavenging
∼ **кран·** (*m*) blow-off cock (boiler)
про·дýв·очн·ый, -ая, -ое, (*a*) = **про·дув·н·óй** (q.v.) scavenge, scavenging; blow/blowing-off
∼ **вóз·дух·** (*m*) scavenge air
∼ **на·сóс·** (*m*) scavenging pump
продýкт·, -а, (*m*) product; (*esp pl*) produce, food-stuffs
∼ **воз·гóн·к·и** (chem) sublimate
∼, **врéд·н·ый** harmful product
∼ **гор·éни·я** (*pl*) waste gas
∼, **дочéр·н·ий** (nucl) daughter product
∼, **ис·хóд·н·ый** primary product
∼, **кон·éч·н·ый** end product
∼, **по·бóч·н·ый** by-product
∼ **стран·ы́, вал·ов·óй** (econ) gross national product
продуктúв·ност·ь, -и, (*f*) productivity, productiveness; efficiency
продуктúв·н·ый, -ая, -ое, (*a*) productive, producing, efficient
продукт·óв·ый, -ая, -ое, (*a*) produce, food, provision
продукто·про·вóд·, -а, (*m*) (oil) all-product line
продýкци·я, -и, (*f*) production, output; product
∼, **вал·ов·áя** (com) gross output
∼, **товáр·н·ая** (com) output for sale
∼, **чúст·ая** (com) pure output
про·дýм·ать, -ют, (*perf*) *see* **про·дý·м·ыва·ть**
про·дýм·ыва·ть, -ют, (*imp*) think over/out
про·дý·т·ый, -ая, -ое, (*past part pass*) *see* **про·дув·á·ть**
∼ **мáсл·о** (*n*) (chem, oil) blown oil
про·дý·ть, -ют, (*perf*) *see* **про·ду·в·á·ть**
прó·дух·, -а, (*m*) (build) air hole

продуцент·, -а, (*m*) (econ) producer

про·душ·ин·а, -ы, (*f*) air hole; hole in ice

про·дыряв·ить, -ят, (*perf*) *see* **про·-дырявл·ива·ть**

про·дырявл·ива·ть, -ют, (*imp*) hole, make a hole, perforate

про·ед·а·ть, -ют, (*imp*) eat through

про·ед·ут, (*fut 3rd pl*) of **про·éх·ать,** (*perf*); *see* **про·езж·а·ть** (*imp*)

про·ед·ят, (*fut 3rd pl*) of **про·éст·ь,** (*perf*); *see* **про·ед·а·ть**

про·езд·, -а, (*m*) (*v n*) *see* **про·езж·а·ть;** thoroughfare, through driveway

про·езд·ной, -ая, -ое, (*a*) of **про·-езд**

~ билет· (*m*) ticket (for a journey)

про·езж·а·ть, -ют, (*imp*) pass, drive/ride/travel/go past/through/by/for/at; get carried on/past, overshoot

про·езж·ий, -ая, -ее, (*a*) suitable for vehicles; (*as noun*) traveller, passer-by

~ часть дорог·и (*f*) carriageway

проект·, -а, (*m*) project, design, draft

~ до·говор·а (dipl) draft treaty; (com) draft/proposed contract

~, рабоч·ий (civ/mech eng) working specification

~, техн·ическ·ий (civ eng) specification

~, тип·ов·ой series/standard design

проект·ант·, -а, (*m*) project/projecting engineer

проектив·ност·ь, -и, (*f*) (math) projectivity

проектив·н·ый, -ая, -ое, (*a*) projective, projection

~ геометр·и·я (*f*) (math) projective geometry

~ плоск·ост·ь (*f*) (math) projective plane

проект·ировани·е, -я, (*n*) projecting, design, designing

~, одно·вес·ов·ое plumbing with one load (gravimetric survey)

проект·ир·овать, -уют, (*imp*) project, plan, design; project (onto screen)

проект·ировк·а, -и, (*f*) *see* **проект·ировани·е**

проект·ировочн·ый, -ая, -ое, (*a*) designing, planning, project, projecting

проект·ировщик·, -а, (*m*) project engineer, planner, designer

проект·н·ый, -ая, -ое, (*a*) project; designed, rated, planned

~ выс·от·а (*f*) planned/designed height

~ дел·о (*n*) planning, project planning

~ за·да·ни·е (*n*) (civ eng) project requirement

~ институт· (*m*) project planning institute

~ мощ·ност·ь (*f*) planned capacity/output (of factory etc.), rated capacity/power (of machines)

~ скор·ост·ь ход·а (*f*) designed speed

проектор·, -а, (*m*) projector; flood light

~, из·мер·ительн·ый measuring projector, (enlarges small components by projecting their silhouettes onto a screen)

~, ион·н·ый field-ion microscope

~, по·сад·очн·ый (air) landing floodlight

~, электрон·н·ый field-electron microscope

проекцион·н·ый, -ая, -ое, (*a*) projection, projecting

~ фонар·ь (*m*) (opt) projector

проекци·я, -и, (*f*) projection

~, азимут·аль·н·ая zenithal projection

~, аксонометр·ическ·ая clinographic projection

~, вертикаль·н·ая vertical projection, front elevation, front view

~, гномон·ическ·ая reciprocal/gnomonic projection

~, горизонталь·н·ая horizontal projection, plan view

~, зенит·аль·н·ая zenithal projection

~, кон·ическ·ая conical projection, conic projection

~, конформ·н·ая conformal projection, conformable projection, orthomorphic projection

~, меркатор·ск·ая Mercator's projection

~ на сек·ущ·ем конус·е conical projection with standard parallel

~, над·рез·анн·ая interrupted projection

~, ортограф·ическ·ая orthographic projection

~, перспектив·н·ая perspective/geometrical projection

~, по·вер·нут·ая меркатор·н·ая (surv) transverse Mercator projection

~, поли·кон·ическ·ая polyconic projection

прое́кци·я

~, **равно·вели́к·ая** equal-area projection, homolographic projection

~, **равно·площад·н·а́я** equal-area projection

~, **равно·про·меж·у́точн·ая** equidistant projection

~, **равно·уго́ль·н·ая** conformal/orthomorphic projection

~, **синусоида́ль·н·ая** sinusoidal projection

~, **стереограф·и́ческ·ая** stereographic projection

~, **центра́ль·н·ая** gnomonic/central projection

~, **эквивале́нт·н·ая** equivalent projection, equal-area projection

про·е́л, (*past masc sing*) of **про·е́ст·ь,** (*perf*); *see* **про·ед·а́·ть** (*imp*)

про·ём·, -а, (*m*) opening, aperture; **про·е́м,** (*fut 1st sing*) of **про·е́ст·ь**

~, **двер·н·о́й** doorway

~ **стан·и́н·ы** (met roll) housing window

про·е́ст·ь, (*fut 3rd sing, pl*) **проее́ст, про·ед·я́т,** (*past masc sing*) **про·е́л,** (*perf*); eat through

про·е́х·ать, (*fut 3rd pl*) **про·е́д·ут,** (*perf*) *see* **про·езж·а́·ть**

проец·и́р·овать, -уют, (*imp*) (opt) project

про·жа́р·енн·ый, -ая, -ое, (*past part pass*) *see* **про·жа́р·ива·ть;** well-cooked

про·жа́р·ива·ть, -ют, (*imp*) fry/roast/grill thoroughly/well

про·жа́р·ить, -ят, (*perf*) *see* **про··жа́р·ива·ть**

про·жа́т·ь, (*fut 3rd pl*) **про·жм·у́т·,** (*perf*) *see*·**про·жим·а́·ть; про·жа́т·ь,** (*fut 3rd pl*) **по·жн·у́т,** (*perf*) *see* **про·жин·а́·ть**

про·жг·ла́, (*past fem sing*) of **про··же́ч·ь,** (*perf*), *see* **про·жиг·а́·ть** (*imp*)

про·жд·а́ть, -у́т, (*perf*) wait

про·жев·а́ть, (*fut 3rd pl*) **про·жу·ю́т,** (*perf*) *see* **про·жёв·ыва·ть**

про·жёв·ыва·ть, -ют, (*imp*) masticate/chew thoroughly

про·жёг·, (*past masc sing*) of **про··же́ч·ь,** (*perf*) *see* **про·жиг·а́·ть** (*imp*)

прожёкт·, -а, (*m*) (obs) *see* **прое́кт·**

прожект·ёр·, -а, (*m*) obsessional planner, one with a mania for planning

прожект·ёрств·о, -а, (*n*) obsession with plans, indulgence in futile promotional activities, mania for planning, plan mania

прожёктор·, -а, (*m*) projector; searchlight; headlight, headlamp; (phot) spotlight

~ **за·лив·а́ющ·его све́т·а** floodlight

~, **ио́н·н·ый** (nucl) ion gun

~, **ли́нз·ов·ый** (phot) Fresnel-lensed spotlight

~, **пла́зм·енн·ый** (phys) plasma gun

~, **по·са́д·очн·ый** (air) landing light

~, **электро́н·н·ый** electron gun

прожектор·и́ст·, -а, (*m*) projector/searchlight operator

прожёктор·н·ый, -ая, -ое, (*a*) of **прожёктор·**

~ **о·свещ·е́ни·е** (*n*) floodlighting

~ **от·раж·а́тел·ь** (*m*) searchlight reflector

про́·желт·ь, -и, (*f*) yellow tinge, yellow

про·же́ч·ь, (*fut 3rd sing, pl*) **про··жж·ёт, про·жг·у́т,** (*past masc sing*) **про·жёг·,** (*perf*); *see* **про··жиг·а́·ть**

про·жж·ёт, (*fut 3rd sing*) of **про··же́ч·ь**

про·жив·а́ни·е, -я, (*n*) (*v n*) *see* **про·жив·а́·ть;** residence

про·жив·а́·ть, -ют, (*imp*) reside, live, spend (time), stay (at); spend (on living), run through (money)

про·жиг·а́ни·е, -я, (*n*) (*v n*) of **про··жиг·а́·ть;** arc-over

про·жиг·а́·ть, -ют, (*imp*) burn, burn through; burn out (for cleaning)

про·жи́л·к·а, -а, (*g pl*) **-л·ок·,** (*f*) vein

про·жи́л·ков·ый, -ая, -ое, (*a*) veined, streaky

про·жи́л·ок·, -л·к·а, (*m*) (geol) vein, streak

про·жим·а́·ть, -ют, (*imp*) press, squeeze; press/squeeze through

про·жин·а́·ть, -ют, (*imp*) reap, harvest; cut (a path etc. reaping)

про·жи́т·и·е, -я, (*n*) subsistence

про·жи́т·очн·ый, -ая, -ое, (*a*) subsistence

~ **сре́д·ств·а** (*pl*) means of subsistence

про·жи́·ть, (*fut 3rd pl*) **про·жив·у́т,** (*perf*) *see* **про·жив·а́·ть**

про·жм·ёт, (*fut 3rd sing*) of **про··жа́т·ь,** (*perf*); *see* **про·жим·а́·ть**

про·жн·ёт, *(fut 3rd sing)* of **про··жа́т·ь,** *(perf)*; *see* **про·жин·а́·ть** *(imp)*

про·жо́р·лив·ый, -ая, -ое, *(a)* voracious

про·жу·ёт, *(fut 3rd sing)* of **про··жев·а́ть,** *(perf)*; *see* **про·жёв·ы·ва·ть** *(imp)*

про·за·ро́д·ыш·, -а, *(m)* (phot) progerm

про·зв·а́ть, *(fut 3rd pl)* **про·зов·у́т,** *(perf)* name, nickname

про·звен·е́ть, -я́т, *(perf)* tinkle, ring, sound

про́·зв·ищ·е, -а, *(n)* nickname

про·звон·и́ть, -я́т, *(perf)* ring, sound

про·звон·к·а, -и, *(g pl)* **-н·ок·,** *(f)* ringing out, signalling test (telephones)

про·звуч·а́ть, -а́т, *(perf)* sound

про·зев·а́·ть, -ют, *(perf)* miss (opportunities etc.); yawn

прозе́ктор·, -а, *(m)* (anat) prosector, dissector

про́·зелен·ь, -и, *(f)* green tinge, green

проз·енхи́м·а, -ы, *(f)* (bot) prosenchyma

проз·енхи́м·н·ый, -ая, -ое, *(a)* (bot) prosenchymatous

прозери́н·, -а, *(m)* (pharm) proserine, neostygmine

про·зим·ова́ть, -у́ют, *(perf)* *see* **зим·ова́ть**

про·зов·ёт, *(fut 3rd sing)* of **про··зв·а́ть**

проз·о·де́жд·а, -ы, *(f)* work/working clothes

про·зр·а́чност·ь, -и, *(f)* transparency; (phys) transmittance, transmissivity

про·зр·а́чн·ый, -ая, -ое, *(a)* transparent

про·зыв·а́·ть, -ют, *(imp)* name, nickname

про·игр·а́·ть, -ют, *(perf)* *see* **про··йгр·ыва·ть**

про·йгр·ыватель·, -я, *(m)* record-player

про·йгр·ыва·ть, -ют, *(imp)* lose; play, play through/over

про́·игр·ыш·, -а, *(m)* loss, defeat **коэффицие́нт· про́·игр·ыш·а** *(m)* disadvantage factor

про·из·вед·ени·е, -я, *(n)* (v n) *see* **про·из·вод·и́ть;** production, work, composition; (math) product

~ рас·твор·и́мост·и (chem) solubility product

про·из·вед·ённ·ый, -ая, -ое, *(past part pass)* *see* **про·из·вод·и́ть**

про·из·вё·л, *(past masc sing)* of **про··из·вес·ти́,** *(perf)*; *see* **про·из·вод·и́ть**

про·из·вес·ти́, *(fut 3rd pl)* **про·из··вед·у́т,** *(perf)*; *see* **про·из·вод·и́ть**

про·из·вод·и́тел·ь, -я, *(m)* producer; (genet) sire

про·из·вод·и́тельност·ь, -и, *(f)* productivity, output; capacity, performance, output (of machines), discharge (of pump); (elec) rating

про·из·вод·и́тельн·ый, -ая, -ое, *(a)* productive

про·из·вод·и́ть, ⸍ят, *(imp)* produce, make; carry out, make, effect, perform, do; promote; *(imp only)* derive

про·из·во́д·н·ая, -ой, *(f decl as adj)* (math) derivative

~, полн·ая total derivative

~, част·н·ая partial derivative

про·из·во́д·н·ое, -ого, *(n decl as adj)* (chem) derivative

про·из·во́д·н·ый, -ая, -ое, *(a)* (*see also* **про·из·во́д·н·ая, про·из·во́д··н·ое** *as nouns*); derived, derivative

~ фу́нкци·я *(f)* (math) derivative

про·из·во́д·ственник·, -а, *(m)* production worker

про·из·во́д·ственн·ый, -ая, -ое, *(a)* production, industrial, commercial

~ квалифика́ци·я industrial qualification

~ мо́щ·ност·ь *(f)* productive capacity

про·из·во́д·ств·о, -а, *(n)* production, manufacture; (elec) generation; carrying out (work etc.), effecting (payment etc.); promotion

~, не·за·верш·ённ·ое work-in-progress

~ па́рт·и·ями batch production

~, по·то́ч·н·ое line production, flow-line production

про·из·вод·я́щ·ий, -ая, -ее, *(pres part act)* of **про·из·вод·и́ть;** *(f as noun)* (math) ·generating line, generatrix

про·из·во́ль·н·ый, -ая, -ое, *(a)* arbitrary; voluntary; random

про·из·нес·ти́, -у́т, *(perf)* *see* **про··из·нос·и́ть**

про·из·нос·и́тельн·ый, -ая, -ое, *(a)* articulatory

про·из·нос·и́ть, ⸍ят, *(imp)* pronounce, articulate; deliver (speeches etc.)

про·из·нош·éни·е, -я,   (*n*) pronunciation, articulation

про·изо·йти́,   (*fut 3rd pl*) **про·изо·йд·ýт,** (*past masc sing*) **про·изо·шёл,** (*perf*); *see* **про́·ис·ход·и́ть**

про·из·раст·á·ть, -ют,   (*imp*) grow, spring up, sprout; be established (of plants)

про·из·раст·и́, -ýт,   (*perf*) *see* **про··из·раст·á·ть**

про·из·рóс·, (*past masc sing*) of **про··из·раст·и́**

про·иллюстр·и́р·овать, -уют,   (*perf*) illustrate

про·инструкт·и́р·овать, -уют,   (*perf*) instruct

про·интегр·и́р·овать, -уют,   (*perf*) integrate

про·интервью·и́р·овать, -уют, (*perf*) interview, have an interview with

про·информ·и́р·овать, -уют,   (*perf*) inform

про·ис·тек·á·ть, -ют,   (*imp*) result/ arise from

про·ис·тéч·ь,   (*fut 3rd sing, pl*) **про··ис·теч·ёт, про·ис·тек·ýт,** (*past masc sing*) **про·ис·тёк·,** (*perf*); *see* **про··ис·тек·á·ть**

про·ис·ход·и́ть, ⸗ят,   (*imp*) occur, happen, take place; result/arise from, be the result of, originate from; be descended from

про·ис·хожд·éни·е, -я,   (*n*) origin; parentage, descent, genesis

~ ви́д·ов (zool) origin of the species

~ элемéнт·ов (nucl) nucleogenesis

про·ис·шéд·ш·ий, -ая, -ее,   (*past part act*) of **про·изо·йти́,** (*perf*); *see* **про·ис·ход·и́ть**

про·ис·шéств·и·е, -я, (*n*) incident, event, accident

проиц·и́р·овать, -уют, (*perf*) (cinema) project

про́·йд·енн·ый, -ая, -ое, (*past part pass*) of **про·йти́;** (math) passed, traversed

про́·йм·а, -ы,   (*f*) arm-hole (in garment)

про·йти́,   (*fut 3rd pl*) **про·йд·ут,** (*past masc sing*) **про·шёл,** (*perf*); *see* **про·ход·и́ть**

про·каж·ённ·ый, -ая, -ое, (*a*) leprous

про·кáз·а, -ы, (*f*) (med) leprosy

про·кал·ённ·ый, -ая, -ое, (*past part pass*) *see* **про·кáл·ива·ть**

про·кáл·иваемост·ь, -и, (*f*) hardenability (of steel)

про·кáл·ивани·е, -я,   (*n*) (*v n*) *see* **про·кáл·ива·ть;** calcination

по·тéр·я от про·кáл·ивани·я (*f*) loss of ignition (of cements)

про·кáл·ива·ть, -ют,   (*imp*) calcine, calcinate; bring to white heat; **-ся** become hardened (by heating)

про·кал·и́ть, -я́т,   (*perf*) *see* **про··кáл·ива·ть**

про·кáл·ыва·ть, -ют,   (*imp*) pierce through, perforate

про·кáмб·и·й, -я,   (*m*) (bot) procambium

про·каменно·угóль·н·ый, -ая, -ое, procarboniferous

про·канониз·и́р·овать, -уют, (*perf*) (math) reduce to canonical form

про·кáп·ыва·ть, -ют, (*imp*) dig, dig through

про·кáрмл·ива·ть, -ют, (*imp*) feed

про·кáрп·и·й, -я, (*m*) (bot) procarp

про·кáт·, -а, (*m*) rolled metal; rolling; hire; (cinema) release

~, сорт·ов·óй (met roll) rolled metal

~, специáль·н·ый (met) tyres, wheels and axles

про·кáт·анн·ый, -ая, -ое, (*past part pass*) of **про·кат·á·ть**

про·кат·á·ть, -ют, (*perf*) roll, mangle; ride, drive (for a time), take for a ride/ drive

про·кат·и́ть, ⸗ят,   (*perf*) roll, roll along; take for a ride; **-ся** (*pass*); go for a ride

про·кáт·к·а, -и,   (*g pl*) **-т·ок·,** (*f*) (met) rolling

~ бутéрбрóд·ом sandwich rolling

~ в·дол·ь straight-away rolling

~, винт·ов·áя rotary rolling

~, гор·я́ч·ая hot rolling

~, кос·áя helical rolling

лин·и·я про·кáт·к·и (*f*) rolling/ pass line

~ на кóнус· taper rolling

~ на плит·ý plate rolling

~ на ýгол· cross rolling

~, по·перéч·н·ая rotary rolling

~, по·перéч·н·о-винт·ов·áя helical rolling

~, тёпл·ая warm rolling

~ труб· tube rolling

~, холóд·н·ая cold rolling

про·кáт·н·ый, -ая, -ое, (*a*) rolling, rolled; hire, hired, hiring

~ стал·ь (*f*) rolled steel

про·ка́т·н·ый, -ая, -ое
~ стан· (*m*) rolling mill
про·ка́т·чик·, -а, (*m*) rolling mill worker, roller, operator; rolling specialist
про·ка́т·ыва·ть, -ют, (*imp*) *see* про·кат·а́·ть, (*perf*) про·кат·и́ть (*perf*)
про·ка́ч·енн·ый, -ая, -ое, (*past part pass*) of про·кат·и́ть
про·ка́ч·ива·ть, -ют, (*imp*) pump (through/along)
про·ка́ч·к·а, -и, (*g pl*) -ч·ек·, (*f*) through-rate
про·квант·ова́ть, -у́ют, (*perf*) (math) quantize
про·ква́ш·енн·ый, -ая, -ое, (*past part pass*); soured, sour
про·кид·а́·ть, -ют, (*imp*) throw through; throw all
про·ки́д·к·а, -и, (*f*) throw, throwing
~ челн·ок·о́в (text) picking, pick
~ челн·ок·о́в, по·след·ователь·н·ая pick-and-pick
~ челн·ок·о́в, не·регуля́р·н·ая pick-at-will
про·ки́·н·уть, -ут, (*perf*) *see* про·кид·а́·ть
про·кип·а́·ть, -ют, ·(*imp*) (*intrans*) boil well/thoroughly, boil long enough
про·кип·е́ть, -я́т, (*perf*) (*intrans*) boil well/thoroughly; boil (for a while)
про·кипят·и́ть, -я́т, (*perf*) (*trans*) boil (for a while), boil long enough
про·кис·а́·ть, -ют, (*imp*) turn sour
про·ки́с·нуть, -нут, (*perf*) *see* про·кис·а́·ть
про·кла́д·к·а, -и, (*g pl*) -д·ок·, (*f*) (*v n*) *see* про·кла́д·ыва·ть; packing, packing piece, gasket, washer, strip, liner, seating (for windows, doors etc.), ply (e.g. of conveyer belt)
~ голо́в·к·и блок·а (ICE) cylinder head gasket
~, зуб·ча́т·ая corrugated gasket
~, ме́д·н·ая copper jointing ring (in engine cylinder)
~ об·ши́в·к·и liner (in ship's plating)
~, регул·иро́вочн·ая (elec) shim
~, рифл·ён·ая corrugated gasket
про·кла́д·н·о́й, -а́я, -о́е, (*a*) = про·кла́д·очн·ый
про·кла́д·очн·ый, -ая, -ое, (*a*) packing, lining
~ кольц·о́ (*n*) gasket ring
~ материа́л· (*m*) packing material, dunnage

про·кла́д·ыва·ть, -ют, (*imp*) lay, lay out, build (something long); plot, chart; interlay, pack with layers of, interleave
про·кле́·енн·ый, -ая, -ое, (*past part pass*) *see* про·кле́·ива·ть
про·кле́·ива·ть, -ют, (*imp*) glue, stick, paste (for a time); cover with glue; impregnate with glue; (paper) size
про·кле́·ить, -ят, (*perf*) *see* про·кле́·ива·ть
про·клей·к·а, -и, (*g pl*) -е́·ек·, (*f*) (*v n*) *see* про·кле́·ива·ть; (paper) size; glue, paste; glued article
про·клю́ч·ива·ть, -ют, (*imp*) (telecom) switch through
про·ко́в·анн·ый, -ая, -ое, (*past part pass*) *see* про·ко́в·ыва·ть
про·ков·а́ть, (*fut 3rd pl*) про·ку·ю́т, (*perf*) *see* про·ко́в·ыва·ть
про·ко́в·ыва·ть, -ют, (*imp*) (met) forge; beat (gold leaf etc.)
про·ковы́р·ива·ть, -ют, (*imp*) pick a hole in
про·ковыр·я́·ть, -ют, (*perf*) *see* про·ковы́р·ива·ть
про·ко́л·, -а, (*m*) (*v n*) *see* про·ко·л·о́ть; puncture, pinhole, pierced hole
про·ко́л·к·а, -и, (*f*) (*v n*) *see* про·кол·о́ть
проколлаге́н·, -а, (*m*) (pharm) procollagen
про·кол·о́ть, -ю́т, (*perf*) pierce, pierce through, puncture
про·кома́нд·овать, -уют, (*perf*) be in command
про·коммент·и́р·овать, -уют, (*perf*) provide/make a commentary, comment upon
про·компост·и́р·овать, -уют, (*perf*) punch (tickets etc.)
про·конопа́т·ить, -ят, (*perf*) caulk
про·конопа́ч·ива·ть, -ют, (*imp*) caulk
про·конспект·и́р·овать, -уют, (*perf*) summarize, make an abstract/synopsis
про·консульт·и́р·овать, -уют, (*perf*) give a consultation
про·контрол·и́р·овать, -уют, (*perf*) *see* контрол·и́р·овать
про·коп·а́·ть, -ют, (*perf*) dig, dig through
про·копт·и́ть, -я́т, (*perf*) make sooty/smoky; (food) smoke
про·копч·ённ·ый, -ая, -ое, (*past part pass*) of про·копт·и́ть; (food) smoked

про·корм·и́ть, ·ят, (*perf*) feed, provide for, sustain

про·коррект·и́р·овать, -уют, (*perf*) *see* **коррект·и́р·овать**

про·кра́с·ить, -ят, (*perf*) *see* **про·-кра́ш·ива·ть**

про·кра́ш·ива·ть, -ют, (*imp*) colour, color, paint, dye, stain

про·кро́·ет, (*fut 3rd sing*) of **про·-кры́·ть,** (*perf*); *see* **про·крыв·а́·ть** (*imp*)

про·кру́т·к·а, -и, (*f*) motoring (of engine turned by other source)

про·кру́ч·иваем·ый дви́г·ател·ь (*m*) (ICE) motored engine

про·крыв·а́·ть, -ют, (*imp*) cover, cover over/completely; apply a coat/primer

про·кры́·ть, (*fut 3rd pl*) **про·кро́·ют,** (*perf*) *see* **про·крыв·а́·ть**

проксима́ль·н·ый, -ая, -ое, (*a*) (biol) proximal

прокт·алг·и́·я, -и, (*f*) (med) proctalgia

прокт·и́т·, -а, (*m*) (med) proctitis

прокто·рре́·я, -и, (*f*) (med) proctor-rhea

прокто·ско́п·, -а, (*m*) (med) procto-scope

про·ку·ёт, (*fut 3rd sing*) of **про·ко-в·а́ть**

прокурату́р·а, -ы, (*f*) (law) Procurator General's office, Public Prosecutor's office

прокуро́р·, -а, (*m*) (law) procurator, public prosecutor, prosecuting attorney

про·кус·и́ть, ·ят, (*perf*) *see* **про·-ку́с·ыва·ть**

про·ку́с·ыва·ть, -ют, (*imp*) bite through

про·лаг·а́·ть, -ют, (*imp*) lay, lay out, build (something long)

про·лакти́н·, -а, (*m*) (biochem) prolactin

пролами́н·, -а, (*m*) (chem) prolamine

про·ла́м·ыва·ть, -ют, (*imp*) *see* **про·-лом·а́·ть,** (*perf*) **про·лом·и́ть** (*perf*)

прола́н·, -а, (*m*) (biochem) prolan

про·лег·а́·ть, -ют, (*imp*) lead, run lie (of roads etc.)

про·леж·а́ть, -а́т, (*perf*) *see* **про·лё-ж·ива·ть**

про́·леж·ен·ь, -ж·н·я, (*m*) bedsore, (med) decubitus ulcer

про·лёж·ива·ть, -ют, (*imp*) lie (for a time); lie unused/unnoticed; (phys etc.) age

про·лез·а́·ть, -ют, (*imp*) climb/clamber through

про·ле́з·ть, -ут, (*perf*) *see* **про·ле-з·а́·ть**

про·ле́й·те, (*pl imperative*) of **про·-ли́·ть** (*perf*), *see* **про·лив·а́·ть** (*imp*)

про·лёт·, -а, (*m*) flight; (orn) seasonal migration; gap, opening, spacing; bay (in building); span (between supports); well (of stairs etc.)

~, а́нкер·н·ый anchor span (electric transmission lines)

~, воз·ду́ш·н·ый air port (of open-hearth furnace)

вре́м·я про·лёт·а (*n*) transit time (electrons etc.)

~, га́з·ов·ый gas port (of open hearth furnace)

~, монта́ж·н·ый erecting bay (of workshop)

~, мост·ов·о́й bridge span

~ по·прям·о́й crow flight, bee-line, distance as the crow flies

~, под·вес·н·о́й suspended span

~, раз·ли́в·очн·ый (met) casting bay; pitside, pouring side (of open heart furnace)

~, свобо́д·н·ый path (of an electron)

у́гол пролёт·а (*m*) transit angle (of electron)

про·лет·а́·ть, -ют, (*imp*) fly past/by/through

про·лет·е́ть, -ят, (*perf*) *see* **про·ле-т·а́·ть**

про·лёт·н·ый, -ая, -ое, (*a*) of **про·-лёт·**

~ ба́з·а (*f*) (instr) path length

~ вре́м·я (*n*) (phys) transit time

~ стро·е́ни·е (*n*) bridge superstructure

~ щит· (*m*) duckboard

про·леч·у́, (*fut 1st sing*) of **про·ле-т·е́ть**

про·ли́·в·, -а, (*m*) (geog) straits, strait; narrows (on lakes etc.)

про·лив·а́·ть, -ют, (*imp*) spill

про·лив·н·о́й, -а́я, -о́е, (*a*) pouring, very heavy (of rain)

проли́н·, -а, (*m*) (chem) proline

про·ли́·ть, (*fut 3rd pl*) **про·ль·ю́т,** (*perf*); *see* **про·лив·а́·ть**

проли·фера́ци·я, -и, (*f*) (biol) proliferation

проли·фика́ци·я, -и, (*f*) (biol) prolification

про·логарифм·и́р·овать, -уют, (*perf*) *see* **логарифм·и́р·овать**

про·ло́ж·енн·ый, -ая, -ое, (*past part pass*) *see* про·кла́д·ыва·ть; sandwiched

про·лож·и́ть, ⹀ат, (*perf*) *see* про··кла́д·ыва·ть

про·ло́м·, -а, (*m*) (*v n*) *see* про·лом·и́ть; breach, break, rupture, fracture

про·лом·а́·ть, -ют, (*perf*) batter/pound through; break, break through

про·лом·и́ть, ⹀ят, (*perf*) break through, stave in

пролонг·и́р·овать, -уют, (*imp and perf*) prolong, extend (of time)

про·ль·ёт, (*fut 3rd sing*) of про·ли́·ть (*perf*), *see* про·лив·а́·ть (*imp*)

про·лю́в·и·й, -я, (*m*) (geol) proluvium

пром- (*abbr*) of: (1) про·мы́шл·ен·н·ый = industrial; (2) про·мы́·шл·енност·ь = industry; (3) про··мысл·о́в·ый = trade (industrial); (4) про·меж·у́точн·ый = intermediate

про·ма́ж·ет, (*fut 3rd sing*) of про··ма́з·ать

про·ма́з·анн·ый, -ая, -ое, (*past part pass*) of про·ма́з·ать

про·ма́з·ать, (*fut 3rd pl*) про·ма́·ж·ут, (*perf*) *see* про·ма́з·ыва·ть

про·ма́з·к·а, -и, (*g pl*) -з·ок·, (*f*) (*v n*) *see* про·ма́з·ыва·ть

про·ма́з·очн·ый, -ая, -ое, (*a*) spreading, coating, smearing

~ маши́н·а (*f*) (rubber) spreading machine

~ нож· (*m*) (rubber) doctor knife

про·ма́з·ывани·е, -я, (*n*) (*v n*) of про·ма́з·ыва·ть

про·ма́з·ыва·ть, -ют, (*imp*) coat, grease, smear, spread, grease/smear all over; stuff/putty up (cracks etc.)

про·ма́л·ыва·ть, -ют, (*imp*) put through grinder/mill; grind, mill (for a time)

про·ма́сл·енн·ый, -ая, -ое, (*past part pass*) *see* про·ма́сл·ива·ть; oil-soaked; oiled (of paper etc.)

про·ма́сл·ива·ть, -ют, (*imp*) impregnate with oil/grease, oil, grease

про·ма́сл·ить, -ят, (*perf*) *see* про··ма́сл·ива·ть

про·ма́т·ыва·ть, -ют, (*imp*) wind, reel, coil (for a time); squander, dissipate, run through (money)

про́·мах·, -а, (*m*) miss (with projectile); slip, blunder, gross error

про·ма́х·ива·ться, -ются, (*imp*) miss, fail to hit (with projectile)

про·мах·н·у́ться, -у́тся, (*perf*) *see* про·ма́х·ива·ться

про·ма́ч·ива·ть, -ют, (*imp*) soak/wet through, dampen/moisten thoroughly

про·медл·е́ни·е, -я, (*n*) delay, time loss; (autom) dead time

про·ме́дл·ить, -ят, (*perf*) delay

промедо́л·, -а, (*m*) promedole (an analgesic)

про·ме́ж·ност·ь, -и, (*f*) (anat) perineum

про·меж·у́ток·, -тк·а, (*m*) interval, gap, interstice

~, воз·ду́ш·н·ый air gap

~, деио́н·н·ый (elec) deion gap

~, за·щи́т·н·ый (elec) protective gap

~, иск·ов·о́й (elec) spark gap

~, лу́н·н·ый lunitidal interval

~, меж·и́мпульс·н·ый, (elec) pulse separation

~, меж·рёбер·н·ый (anat) intercostal space

~, мёртв·ый (gunn) dead interval

~, про·би́в·н·о́й (elec) disruptive distance

~, раз·ря́д·н·ый (elec) discharge gap

~ ря́д·а (cryst) primitive translation

~, стерж·нев·о́й (elec) rod gap

промежуточно·част·о́тн·ый, -ая, -ое, (*a*) (rad) intermediate-frequency, i-f

про·меж·у́точн·ый, -ая, -ое, (*a*) intermediate, interstitial

~ культу́р·а (*f*) (hortic) catch crop

~ интегра́л· (*m*) (math) intermediate differential

~ проду́кт· (*m*) (chem) intermediate; middling (ore-dressing)

~ со·сто·я́ни·е (*n*) intermediate state; (nucl) compound state

~ ста́нци·я (*f*) (telecom) retransmitting/through station; (rail) wayside station; booster station (pipelines)

~ хозя́·ин· (*m*) (zool) intermediate host

про·ме́л·ет, (*fut 3rd pl*) of про·мо·л·о́ть (*perf*), *see* про·ма́л·ыва·ть (*imp*)

про·ме́н·ива·ть, -ют, (*imp*) exchange, swap, barter

про·мен·я́·ть, -ют, (*perf*) *see* про··ме́н·ива·ть

про·ме́р·, -а, (*m*) (*v n*) *see* про·ме́·р·ива·ть; measurement, survey; sounding (of depths); error in measurement

про·мерз·а́·ть, -ют, (*imp*) freeze, freeze through

про·мёрзл·ый, -ая -ое, (*a*) frozen, frozen-through

про·мёрз·нуть, -нут, (*perf*) *see* **про·-мерз·а́·ть**

про·ме́р·ива·ть, -ют, (*imp*) measure, survey; sound (depths); measure wrongly, make an error in measurement

про·меристе́м·а, -ы, (*f*) (bot) promeristem

про·ме́р·ить, -ят, (*perf*) *see* **про·-ме́р·ива·ть**

про·ме́р·н·ый, -ая, -ое, (*a*) measuring, surveying; sounding (depths)

~ **су́д·н·о** (*n*) survey ship

про·мер·я́·ть, -ют, (*imp*) = **про·-ме́р·ива·ть**

про·мес·и́ть, -ят, (*perf*) knead, mix (for a time); knead/mix well/thoroughly

про·метафа́з·а, -ы, (*f*) (genet) prometaphase

проме́т·и·й, -я, (*m*) (chem) promethium, Pm

про·меш·а́·ть, -ют, (*perf*) mix well/thoroughly

про·ме́ш·ива·ть, -ют. (*imp*) knead/mix well/thoroughly

проми́лле (*n indecl*) a thousandth part

про·мице́л·и·й, -я, (*m*) (bot) promycelium

про·миэлоци́т·, -а, (*m*) (cyt) promyelocyte

про·мо́·ет, (*fut 3rd sing*) of **про·-мы́·ть** (*perf*), *see* **про·мыв·а́·ть**

промо́згл·ый, -ая, -ое, (*a*) dank

про·мо́·ин·а, -ы, (*f*) rain rill/channel, gully; current hole (in ice)

про·мок·а́тельн·ая бума́г·а (*f*) blotting-paper

про·мок·а́·ть, -ют, (*imp*) become wet through, become soaked; blot, mop up

про·мок·а́шк·а, -и, (*g pl*) -шек·, (*f*) blotter, blotting paper

про·мо́к·нуть, -нут, (*perf*) become wet through, become soaked; **про·-мок·н·у́ть, -у́т,** (*perf*) blot, mop up

про·мола́ч·ива·ть, -ют, (*imp*) (agr) thresh, thrash

про·молот·и́ть, -я́т, (*perf*) *see* **про·-мола́ч·ива·ть**

про·мол·о́ть, (*fut 3rd pl*) **про·ме́л·ют,** (*perf*) *see* **про·ма́л·ыва·ть;** grind, mill (for a period)

про·мора́ж·ива·ть, -ют, (*imp*) freeze, freeze through/completely

про·моро́з·ить, -ят, (*perf*) *see* **про·-мора́ж·ива·ть**

про·мот·а́ни·е, -я, (*n*) (*v n*) of **про·мо·т·а́·ть;** (mil) misuse of equipment

про·мот·а́·ть, -ют, (*perf*) *see* **про·ма́·т·ыва·ть;** wind , reel, coil (for a period)

промот·и́р·овать, -уют, (*imp*) promote

промо́тор·, -а, (*m*) (chem) promoter

про·моч·и́ть, -ат, (*perf*) *see* **про·ма́·ч·ива·ть**

про·мо́·ют, (*fut 3rd pl*) of **про·мы́·ть,** (*perf*); *see* **про·мыв·а́·ть** (*imp*)

пром·проду́кт·, -а, (*m*) (chem) intermediate

пром·това́р·н·ый, -ая, -ое, (*a*) of **пром·това́р·ы**

пром·това́р·ы, -ов, (*m pl*) manufactured/consumer goods

пром·фин·пла́н·, -а, (*m*) = **про·-мы́шл·енн·о-фина́нс·ов·ый план·** industrial and financial plan

пром·щи́т, -а, (*m*) intermediate distributing frame (telephones)

про·мы́·в·, -а, (*m*) (geol) channel

про·мыв·а́лк·а, -и, (*g pl*) -лок·, (*f*) (chem) wash bottle, elutriator, washer

про·мыв·а́ни·е, -я, (*m*) (*v n*) *see* **про·-мыв·а́·ть;** (med) irrigation, lavage (internal), bathing (external)

про·мыв·а́тел·ь, -я, (*m*) (min) washer

про·мыв·а́·ть, -ют, (*imp*) wash, wash out/through, flush, rinse

про·мы́в·к·а, -и, (*g pl*) -в·ок·, (*f*) (*v n*) *see* **про·мыв·а́·ть;** wash

~ **суро́в·ь·я** (text) grey wash

~ **тепл·о́м** (chem) heat flush

про·мыв·н·о́й, -а́я, -о́е, (*a*) wash, washing

~ **аппара́т·** (*m*) washer (for films etc.)

~ **маши́н·а** (*f*) (text, phot) washer

про·мы́в·очн·ый, -ая, -ое, (*a*) of **про·мы́в·к·а**

~ **маши́н·а** (*f*) (min) classifier washer

про́·мысел·, -сл·а, (*m*) trade, industry; trading organization; field, ground, place (where trade etc. is carried on)

~, **зверо·ло́в·н·ый** trapping; trapping grounds

~, **кито·бо́й·н·ый** whaling; whale fishery

~, **ко́т·иков·ый** fur-seal hunting; sealing grounds

~, **куста́р·н·ый** cottage industry

~, **нефт·ян·о́й** oil field

про́·мысел
~, о·хо́т·нич·ий hunting; trapping; hunting/trapping grounds
~, ры́б·н·ый fishery; catch (of fish)
~, сол·ян·о́й salt works
~, у́стрич·н·ый oyster farm; oyster farming
про·мысл·ови́к·, -а́, (*m*) professional hunter/fisherman
про·мысл·о́в·ый, -ая, -ое, (*a*) of **про́·мысел·**
~ газ· (*m*) (oil) casing head/natural gas
~ гео·фи́з·ик·а (*f*) (min) geophysical exploration
~ коопер·а́ци·я (*f*) producers' co-operative
~ на·ло́г· (*m*) industrial rate/rating
~ ры́б·а (*f*) food fish
~ с·вид·е́тельств·о (*n*) licence (to shoot/hunt/fish)
~ су́д·н·о (*n*) fishing boat·
про·мы́·т·ый, -ая, -ое, (*past part pass*) see **про·мыв·а́·ть**
про·мы́·ть, (*fut 3rd pl*) **про·мо́·ют,** (*perf*) see **про·мыв·а́·ть**
про·мы́шл·енник·, -а, (*m*) industrialist
про·мы́шл·енност·ь, -и, (*f*) industry
~, до·быв·а́ющ·ая extractive industry
~, об·раба́т·ывающ·ая manufacturing industry
про·мы́шл·енн·ый, -ая, -ое, (*a*) industrial; (econ) commercial, pay
~ масшта́б· (*m*) commercial scale
~ место·рожд·е́ни·е (*n*) (min) commercial deposit, pay lode
пронарко́н·, -а, (*m*) (pharm) pronarkon
про·нес·е́ни·е, -я, (*n*) (*v n*) see **про·нос·и́ть;** (math) transfer
про·нес·ти́, -у́т, (*perf*) see **про·нос·и́ть**
про·нефро́с·, -а, (*m*) (zool) pronephros, fore/head kidney
про·нз·а́·ть, -ют, (*imp*) pierce
про·нз·и́тельн·ый, -ая, -ое, (*a*) piercing
про·нз·и́ть, -я́т, (*perf*) see **про·нз·а́·ть**
про·низ·а́·ть, -ют, or **про·ни́ж·ут,** (*perf*) see **про·ни́з·ыва·ть**
про·ни́з·ыва·ть, -ют, (*imp*) string, thread; bead; pierce; run through, permeate (of ideas)
про·ник·а́ни·е, -я, (*n*) (*v n*) of **про·ник·а́·ть**
про·ник·а́·ть, -ют, (*imp*) penetrate, permeate
про·ник·а́ющ·ая ради·а́ци·я (*f*) penetrating radiation

про·ник·нове́ни·е, -я, (*n*) (*v n*) see **про·ник·а́·ть;** penetration, permeation
~, взаи́м·н·ое interpenetration
глуб·ин·а́ про·ник·нове́ни·я (*f*) depth of penetration
про·ник·нове́нност·ь, -и, (*f*) penetration, insight, astuteness
про·ни́к·нуть, -нут, (*perf*) see **про·ник·а́·ть**
про·ниц·а́емост·ь, -и, (*f*) penetrability, permeability; pellucidity (light); penetration factor (of vacuum tube)
~ диффу́зи·и permeability to diffusion
~, ди·электр·и́ческ·ая permittivity, absolute dielectric constant
~, ко́мплекс·н·ая complex permeability
~, магни́т·н·ая magnetic permeability, permeability
~, на·ча́ль·н·ая initial permeability
~ потенциа́ль·н·ого барье́р·а (phys) potential barrier penetrance, barrier factor
про·ниц·а́ем·ый, -ая, -ое, (*a*) permeable, penetrable, pervious; pellucid (to light)
~ для из·луч·е́ни·й radiotranslucent
про·ниц·а́тельност·ь, -и, (*f*) penetration, insight, astuteness
про·ниц·а́тельн·ый, -ая, -ое, (*a*) astute, observant, penetrating
про·норм·ир·ова́ть, -у́ют, (*perf*) see **норм·ир·ова́ть**
про·но́с·, -а, (*m*) (*v n*) of **про·нос·и́ть;** gap (part of river bed left until last when constructing dam)
про·нос·и́ть, ―ят, (*imp*) carry, carry past/through/by/to
про·нумер·ова́ть, -у́ют, (*perf*) enumerate
про·о́браз·, -а, (*m*) (math) inverse image
про·оли́ф·анн·ый, -ая, -ое, (*a*) oiled, impregnated with linseed oil
~ ткан·ь (*f*) oilskin
про·па́·вш·ий, -ая, -ее, (*past part act*) of **про·па́с·ть,** (*perf*); see **про·пад·а́·ть**
про·пад·а́ни·е, -я, (*n*) (*v n*) of **про·пад·а́·ть;** dropout (magnetic tape); (phys) extinction
про·пад·а́·ть, -ют, (*imp*) be missing/lost/absent/wasted, disappear, vanish, fade (of sound)
про·па́·ива·ть, -ют, (*imp*) solder thoroughly/well; solder (for a period)

про·па́·л, (*past masc sing*) of **про·па́с·ть**, (*perf*); *see* **про·пад·а́·ть** (*imp*)

про·па́л·ыва·ть, -ют, (*imp*) weed

проп·а́н·, -а, (*m*) (chem) propane

пропано́л·, -а, (*m*) propanol, propyl alcohol

пропарги́л·, -а, (*m*) (chem) propargyl

пропарги́л·ов·ый, -ая, -ое, (*a*) propargylic, propargyl

~ **кисл·от·а́** (*f*) propargylic acid

~ **спирт·** (*m*) propargyl alcohol

про·па́р·енн·ый, -ая, -ое, (*past part pass*) *see* **про·па́р·ива·ть**; steamed, steam-cured

про·па́р·ива·ть, -ют, (*imp*) steam out, steam, steam cure; **-ся** take a steam bath

про·па́р·ить, -ят, (*perf*) *see* **про·па́р·ива·ть**

про·па́р·к·а, -и, (*f*) (*v n*) *see* **про·па́р·ива·ть**

про·па́р·очн·ый, -ая, -ое, (*a*) steaming, steam-curing

про́·паст·ь, -и, (*f*) precipice, chasm

про·па́с·ть, (*fut 3rd pl*) **про·пад·у́т**, (*perf*) *see* **про·пад·а́·ть**

про·пах·а́ть, (*fut 3rd pl*) **про·па́ш·ут**, (*perf*) *see* **про·па́х·ива·ть**

про·па́х·ива·ть, -ют, (*imp*) plough through/in/under

про·па́ш·ник·, -а, (*m*) (agr) cultivator

про·паш·н·о́й, -а́я, -о́е, (*a*) inter-row cultivation

~ **культу́р·ы** (*pl*) inter-row/intertilled crops

~ **тра́ктор·** (*m*) tractor-cultivator

про·па́ш·ут, (*fut 3rd pl*) of **про·пах·а́ть**

про·па́·янн·ый, -ая, -ое, (*past part pass*) of **про·па·я́·ть**; well-soldered

про·па·я́·ть, -ют, (*perf*) solder thoroughly/well; solder (for a period)

про·пек·а́·ть, -ют, (*imp*) bake, bake through

пропе́ллер·, -а, (*m*) propeller, airscrew

пропе́ллер·н·ый, -ая, -ое, (*a*) of **пропе́ллер·**

~ **крыл·а́кт·а** (*f*) impeller (of fan etc.)

~ **на·со́с·** (*m*) propeller pump

пропе́н·, -а, (*m*) propene, propylene

про·пепто́н·, -а, (*m*) (biochem) propeptone

Пропе́рции, систе́м·а (*f*) (met) Properzi method

про·печа́т·к·а, -и, (*g pl*) **-т·ок·**, (*f*) magnetic printing/transfer

про·печа́т·ывани·е, -я, (*n*) magnetic printing/transfer

про·печ·ённ·ый, -ая, -ое, (*past part pass*) *see* **про·пек·а́·ть**; well-baked

про·пе́ч·ь, (*fut 3rd sing, pl*) **про·печ·ёт, про·пек·у́т**, (*past masc sing*) **про·пёк·**, (*perf*); *see* **про·пек·а́·ть**

про·пи́л·, -а, (*m*) (*v n*) *see* **про·пи́л·ива·ть**; kerf, saw cut; **про·пи́л·, -а**, (*m*) (chem) propyl

проп·ила́т·, -а, (*m*) (chem) propylate

проп·иле́н·, -а, (*m*) propylene, propene

про·пи́л·ива·ть, -ют, (*imp*) saw, file, saw/file through/out

про·пил·и́т·, -а, (*m*) (min) propylite

пропил·итиза́ци·я, -и, (*f*) (geol) propylitization, propylitic alteration

пропил·ито́в·ый, -ая, -ое, (*a*) propylitic

про·пил·и́ть, -ят, (*perf*) *see* **про·пи́л·ива·ть**; saw/file (for a period)

проп·и́лов·ый спирт· (*m*) propyl alcohol, propanol-1

пропи́н·, -а, (*m*) (chem) propine, allylene

пропио́л·ов·ая кисл·от·а́ (*f*) (chem) propiolic acid, propargylic acid, propine acid

пропио́н·, -а, (*m*) (chem) propione

пропиони́л·, -а, (*m*) propionyl

пропио́н·ов·ый, -ая, -ое, (*a*) propionic

~ **кисл·от·а́** (*f*) propionic acid

про·пи́с·анн·ый, -ая, -ое, (*past part pass*) of **про·пис·а́ть**

про·пис·а́ть, (*fut 3rd pl*) **про·пи́ш·ут**, (*perf*) *see* **про·пи́с·ыва·ть**

про·пи́с·к·а, -и, (*g pl*) **-с·ок·**, (*f*) official entry, registration, inscription

про·пис·н·о́й, -а́я, -о́е, (*a*) registration; capital (of letters)

про·пи́с·ыва·ть, -ют, (*imp*) (med) prescribe; register, record, enter

про́·пис·ью (*adv*) in words (writing out figures)

про·пи́т·анн·ый, -ая, -ое, (*past part pass*) *see* **про·пи́т·ыва·ть**

~ **вод·о́й** waterlogged

про·пит·а́·ть, -ют, (*perf*) *see* **про·пи́т·ыва·ть**

про·пи́т·к·а, -и, (*g pl*) **-т·ок·**, (*f*) (*v n*) *see* **про·пи́т·ыва·ть**; impregnation

~**, твёрд·ая** (elec) jelly impregnation

про·пи́т·очн·ый, -ая, -ое, (*a*) impregnating, impregnation; dipping, coating

~ **компа́унд·** (*m*) impregnation/impregnating compound

~ **лак·** (*m*) impregnating varnish

про·пи́т·очн·ый, -ая, -ое
~ **маши́н·а** (*f*) (text) impregnating/coating machine; (rubber) dipping machine
про·пи́т·ыва·ть, -ют, (*imp*) impregnate; coat (textiles), dip (in rubber)
про·пи́ш·еı, (*fut 3rd sing*) of **про·пис·а́ть,** (*perf*); *see* **про·пи́с·ыва·ть** (*imp*)
про·пла́в·а·ть, -ют, (*perf*) swim, sail, drift, float (for a time)
про·пла́в·ить, -ят, (*perf*) melt, smelt; melt, smelt (for a period)
про·пла́вл·ива·ть, -ют, (*imp*) melt, smelt
про·плавл·я́·ть, -ют, (*imp*) melt, smelt
про·пла́ст·ок·, -т·к·а, (*m*) (geol) parting, intercalation
проплиопите́к·, -а, (*m*) (pal) Propliopithecus
про·плыв·а́·ть, -ют, (*imp*) swim, sail, drift, float; swim/sail/drift/float past/by/through
про·плы́·ть, (*fut 3rd pl*) **про·плыв·у́т,** (*perf*) *see* **про·плыв·а́·ть**
про·по·ве́д·овать, -уют, (*imp*) preach, propagate (a teaching, theory or philosophy)
про·пола́ск·ива·ть, -ют, (*imp*) rinse, swill (for a time); rinse/swill out; gargle
про·полз·а́·ть, -ют, (*imp*) crawl, creep; crawl/creep through/to/into
про·полз·ти́, -у́т, (*perf*) *see* **про·пол·з·а́·ть**
про·по́л·к·а, -и, (*f*) (agr) weeding
про·полоск·а́ть, (*fut 3rd pl*) **про·поло́щ·ут,** (*perf*) *see* **про·пола́ск·и·ва·ть**
про·пол·о́ть, ˉют, (*perf*) weed
про·поло́щ·ет, (*fut 3rd sing*) of **про·по·ло́ск·а́ть,** (*perf*); *see* **про·пола́ск·и·ва·ть**
пропорциона́ль·н·о (*adv*) proportionally, proportionately, in proportion to
пропорциона́ль·ност·ь, -и, (*f*) proportionality, proportion
 коэффицие́нт· пропорциона́ль·ност·и (*m*) (math) proportionality factor
пропорциона́ль·н·ый, -ая, -ое, (*a*) proportional; (*n decl as adj*) proportional
 обра́т·н·о пропорциона́ль·н·ый inversely proportional
 пря́м·о пропорциона́ль·н·ый directly proportional

пропорциона́ль·н·ый, -ая, -ое
~ **со·ставл·я́щ·ая** (*f*) proportional component
~ **у·велич·е́ни·е** (*n*) proportional increase, (math) scaling up
пропо́рци·я, -и, (*f*) proportion
~**, не·пре·ры́в·н·ая** (math) continuous proportion
~**, про·из·во́д·н·ая** (math) proportion by addition and subtraction
про·пот·ева́ни·е, -я, (*n*) heavy sweating, (med) transudation
про·пу́др·ива·ть, -ют, (*imp*) dust, powder
пропульси́в·н·ый, -ая, -ое, (*a*) propulsive
~ **коэффицие́нт·** (*m*) propulsive coefficient
про́·пуск·, -а, (*m*) (*v n*) of **про·пуск·а́·ть;** (*nom pl*) **-а́,** (*g pl*) **-о́в,** admission, passage into; pass (through rolls); pass, permit (document); (*nom pl*) **-и,** (*g pl*) **-ов,** omission, lapse, miss (of engines); blank, gap; absence, non-attendance
~ **за·жиг·а́ни·я** (ICE) misfire
~ **и́мпульс·а** misfiring (in a magnetron); count-down (of a transponder beacon)
~ **на·тонк·у́ю** (rubber) refining
про·пуск·а́емост·ь, -и, (*f*) (phys) transmittance, transmissivity
про·пуск·а́ем·ый, -ая, -ое, (*pres part pass*) of **про·пуск·а́·ть;** transmitted; transmissible
~ **полос·а́** (*f*) (elec) band-pass, pass band
про·пуск·а́ни·е, -я, (*n*) (*v n*) of **про·пуск·а́·ть;** transmission (of light); gating (computers)
 зо́н·а про·пуск·а́ни·я (*f*) transmission band
 коэфицие́нт· про·пуск·а́ни·я (*m*) transmission factor
 крив·а́я про·пуск·а́ни·я (*f*) transmission curve
 полос·а́ про·пуск·а́ни·я (*f*) (elec) passband
 -схе́м·а про·пуск·а́ни·я (*f*) (elec) gating circuit, gate
про·пуск·а́тел·ь, -я, (*m*) (elec) band-pass filter, acceptor
про·пуск·а́·ть, -ют, (*imp*) let through/past/by/in; pass/put through/in; miss out, omit

про·пуск·а́ющ·ий, -ая, -ее, (*pres part act*) of **про·пуск·а́·ть;** transmitting, passing, translucent (to); gate, gating (computers)

~ **из·луч·е́ни·я** radiotranslucent

~ **ла́мп·а** (*f*) (elec) gate valve/tube

про·пуск·н·о́й, -а́я, -о́е, (*a*) of **про́·-пуск·**

~ **бума́г·а** (*f*) blotting paper

~ **свет·** (*m*) transmitted light

~ **спосо́б·ност·ь** (*f*) throughout; traffic capacity, turnround capacity; (telecom) carrying capacity; (dynam) discharge capacity; current-carrying capacity (cables)

про·пуст·и́ть, -ят, (*perf*) see **про·-пуск·а́·ть**

про·пу́щ·енный, -ая, -ое, (*past part pass*) of **про·пуст·и́ть**

про·ра́б·, -а, (*m*) (*abbrev*) = **про·из·-вод·и́тел·ь рабо́т·** foreman, overseer, site engineer; (shipb) ship manager

про·раба́т·ыва·ть, -ют, (*imp*) work, work up/out

про·рабо́т·а·ть, -ют, (*perf*) see **про·-раба́т·ыва·ть**

прора́н·, -а, (*m*) gap (part of river bed left until last when constructing dam)

про·раст·а́ни·е, -я, (*n*) (*v n*) of **про·-раст·а́·ть;** germination; (geol) penetration intergrowth

двой·ни́к· про·раст·а́ни·я (*m*) (cryst) penetration/interpenetration twin

про·раст·а́·ть, -ют, (*imp*) germinate; sprout, shoot; intergrow, penetrate

про·раст·и́, -ут, (*past masc sing*) **про·-ро́с·,** (*perf*) see **про·раст·а́·ть**

про·раст·и́ть, -я́т, (*perf*) see **про·ра́-щ·ива·ть**

про·ра́щ·ивани·е, -я, (*n*) (*v n*) of **про·ра́щ·ива·ть;** germination

про·ра́щ·ива·ть, -ют, (*imp*) make/let germinate; let sprout/shoot

про·рв·а, -ы, (*f*) narrow channel through sand (river), cutoff; water-filled pit; a mass/lot of

про́·рв·анн·ый, -ая, -ое, (*past part pass*) of **про·рв·а́ть**

про·рв·а́ть, -у́т, (*perf*) tear, tear a hole in; burst, break

про·реаги́р·овать, -уют, (*perf*) see **реаги́р·овать**

про·ред·и́ть, -я́т, (*perf*) see **про·-ре́ж·ива·ть**

про·реж·ённ·ый, -ая, -ое, (*past part pass*) of **про·ред·и́ть,** (*perf*) see **про·-ре́ж·ива·ть** (*imp*)

про·ре́ж·ет, (*fut 3rd sing*) of **про·-ре́з·ать** (*perf*), see **про·ре́з·ыва·ть** (*imp*)

про·ре́ж·ивани·е, -я, (*n*) (*v n*) of **про·ре́ж·ива·ть**

про·ре́ж·ива·ть, -ют, (*imp*) thin out, thin

про·ре́з·, -а, (*m*) (*v n*) see **про·ре́з·ы-ва·ть;** cut, furrow, slit, groove, slot; gate (turbine rotor)

про·ре́з·анн·ый, -ая, -ое, (*past part pass*) see **про·ре́з·ыва·ть**

про·рез·а́ть, -ют, (*imp*), **про·ре́-з·ать,** (*fut 3rd pl*) **про·ре́ж·ут,** (*perf*); see **про·ре́з·ыва·ть**

про·рези́н·енн·ый, -ая, - ое, (*past part pass*) see **про·рези́н·ива·ть;** rubber-proofed, rubber-coated, rubberized

про·рези́н·ива·ть, -ют, (*imp*) rubberize, coat with rubber

про·рези́н·ить, -ят, (*perf*) see **про·-рези́н·ива·ть**

про·рези́н·к·а, -и, (*f*) (*v n*) see **про·-ре́зи́н·ива·ть**

про·рез·н·о́й, -а́я, -о́е, (*a*) slit, cut, furrow, groove

~ **карма́н·** (*m*) slit pocket

~ **ло́д·к·а** (*f*) fishhold boat, packer

про·ре́з·ывани·е, -я, (*n*) (*v n*) of **про·ре́з·ыва·ть;** dentition (of teeth)

про·ре́з·ыва·ть, -ют, (*imp*) cut, cut through/out

про́·рез·ь, -и, (*f*) cut, slit; fishhold boat, packer; gate (of a turbine rotor)

проре́ктор·, -а, (*m*) pro-/vice-rector (of university)

про·репет·и́р·овать, -уют, (*perf*) rehearse

про·ржа́в·е·ть, -ют, (*perf*) rust, rust through

про·рис·ова́ть, -у́ют, (*perf*) draw, touch up, touch up the lines/outlines/contours

про·ро́·ет, (*fut 3rd sing*) of **про·ры́·ть**

про·ро́с·, (*past masc sing*) of **про·-раст·и́,** (*perf*); see **про·раст·а́·ть** (*imp*)

про·ро́ст·ок·, -т·к·а, (*m*) (bot) sprout, shoot, germinant

про́·рост·ь, -и, (*f*) deadwood cavity (defect in timber)

про·руб·а́·ть, -ют, (*imp*) chop/hack/slash through

про·руб·и́ть, ⸚ят, (*perf*) see про·-руб·а́·ть

про́·руб·ь, -и, (*f*) ice-hole, hole hacked through ice

про·ры́в·, -а, (*m*) break, breakthrough, burst; outburst; (met cast) breakout; hitch, hold up (in work); (gunn) blow-through

про·рыв·а́·ть, -ют, (*imp*) tear, tear a hole in; burst, break through; dig through

про·ры́в·к·а, -и, (*f*) (*v n*) see про·-рыв·а́·ть; (hortic) thinning out

про·ры́·ть, (*fut 3rd pl*) про·ро́·ют, (*perf*) dig through

про·са́д·к·а, -и, (*g pl*) -д·ок·, (*f*) subsidence

просак·, -а, (*m*) rope walk

про·са́л·енн·ый, -ая, -ое, (*past part pass*) of про·са́л·ить

про·са́ливать (*imp*) grease, impregnate with grease; (food) salt, pickle (with salt); corn (meat)

про·са́л·ить, -ят, (*perf*) grease, impregnate with grease

про·са́ч·ивани·е, -я, (*n*) (*v n*) of про·са́ч·ива·ться; seepage, percolation, permeation, infiltration

про·са́ч·ива·ться, -ются, (*imp*) seep, percolate, infiltrate, trickle through gradually/slowly

про·свёрл·ивани·е, -я, (*n*) (*v n*) of про·свёрл·ива·ть; perforation

про·свёрл·ива·ть, -ют, (*imp*) drill, bore, drill/bore through, perforate

про·сверл·и́ть, -я́т, (*perf*) see про·-свёрл·ива·ть

про·све́т·, -а, (*m*) beam/ray/strip/patch of light; gap, space, clearance; (arch) bay, aperture, opening; (*v n*) see про·све́ч·ива·ть

   контро́л·ь на про·све́т· (*m*) radiographic inspection; (med) trans-illumination

~ лю́к·а hatch opening

про·свет·и́тельн·ый, -ая, -ое, (*a*) enlightening

про·свет·и́ть, ⸚ят, (*perf*) see про·-све́ч·ива·ть; про·свет·и́ть, -я́т, (*perf*) see про·свещ·а́·ть

~ рентге́н·ов·ыми луч·а́ми X-ray

про·светл·е́ни·е, -я, (*n*) (*v n*) see про·светл·я́·ть; coating (a lens)

про·светл·ённ·ый объекти́в· (*m*) (phot) coated lens

про·светл·е́·ть, -ют, (*perf*) (meteor) clear up, brighten

про·светл·и́ть, -ят, (*perf*) see про·-светл·я́·ть

про·светл·я́·ть, -ют, (*imp*) clarify

про·свеч·енн·ый, -ая, -ое, (*past part pass*) see про·свеч·ива·ть

про·свеч·иваемост·ь, -и, (*f*) translucency

про·свеч·ивани·е, -я, (*n*) (*v n*) of про·свеч·ива·ть; fluoroscopy; radiographic inspection, X-ray flaw detection; transillumination

про·свеч·ива·ть, -ют, (*imp*) shine (for a time); shine/show/appear through; examine (by gamma-/X-rays/transmitted light)

про·свеч·ивающ·ий, -ая, -ее, (*pres part act*) of про·свеч·ива·ть; translucent, transmitting (light)

~ экра́н· (*m*) fluoroscopic screen, fluoroscope

про·свещ·а́·ть, -ют, (*imp*) enlighten

про·свещ·е́ни·е, -я, (*n*) enlightenment; education

про·свещ·ённ·ый, -ая, -ое, (*past part pass*) see про·свещ·а́·ть

просви́р·ник·, -а, (*m*) (bot) mallow, *Malva*

про·свист·е́ть, -я́т, or про·свищ·ут, (*perf*) whistle

про·се́в· (*v n*) see про·сев·а́·ть; sievings; gap (in sowing)

про·сев·а́тел·ь, -я, (*m*) sifter, sifting machine

про·сев·а́·ть, -ют, (*imp*) sieve, screen, sift, bolt

про·се́·вш·ий, -ая, -ее, (*past part act*) of про·се́ст·ь

про·сед·а́ни·е, -я, (*n*) subsiding, subsidence

про·сед·а́·ть, -ют, (*imp*) subside

про·се́·ива·ть, -ют, (*imp*) sieve, screen, sift, bolt

про́·сек·, -а, (*m*) ride, cleared strip (in forest); (min) crosscut, through-cut; про·се́к· (*past masc sing*) of про·се́ч·ь

~, кос·ова́т·ый (min) crosscut

~, противо·по·жа́р·н·ый fire lane (in forests etc.)

про́·сек·а, -и, (*f*) = про·се́к· (*m*) ride, cleared strip (in forest)

про·сек·а́·ть, -ют, (*imp*) cut through

про·се́л·, (*past masc sing*) of про·се́ст·ь

про·сёл·ок·, -л·к·а, (*m*) earth road, cart track, side road

про·се́ст·ь, (*fut 3rd pl*) про·ся́д·ут, (*perf*) subside

про·сеч·к·а, -и, (*f*) (*v n*) *see* **про·сек·а·ть;** punching (sheet metal); hot punching, piercing (a solid piece of metal)

про·сечь, (*fut 3rd sing, pl*) **про·се·чёт, про·сек·у́т,** (*past masc sing*) **про·сек·,** (*perf*) *see* **про·сек·а·ть**

про·се́·ять, -ют, (*perf*) *see* **про·се́·ива·ть;** sow (for a time)

про·сид·е́ть, -я́т, (*perf*) *see* **про·си́·ж·ива·ть**

про·си́ж·ива·ть, -ют, (*imp*) sit (for a time)

про·си́кул·а, -ы, (*f*) (zool) prosicula

про́·син·ь, -и, (*f*) blue tinge/tint/ streaks

прос·и́ть, -я́т, (*imp*) (+ *gen*) ask for, ask; invite

про·сифо́н·, -а, (*m*) (zool) prosiphon

про·ска́бл·ива·ть, -ют, (*imp*) scrape, shave, plane; scrape/shave/plane a hole through

про·скак·а́ть, (*fut 3rd pl*) **про·ска́·ч·ут,** (*perf*) gallop, gallop past/through; hop

про·ска́к·ива·ть, -ют, (*imp*) race/ rush across/past/through; gallop, gallop past; hop

про·ска́льз·ыва·ть, -ют, (*imp*) slip/ slide past/through

про·ска́ч·ет, (*fut 3rd sing*) of **про·скак·а́ть**

про·с·клон·я́·ть, -ют, (*perf*) (gram) decline

про·скобл·и́ть, -я́т, (*perf*) *see* **про·ска́бл·ива·ть**

про·скок·, -а, (*m*) (*v n*) *see* **про·ско·ч·и́ть;** overswing (of pointer); racing (of rotating shaft etc.)

~ пла́мен·и flashback

~ стре́л·к·и (instr) overshoot, overswing

про·скольз·н·у́ть, -у́т, (*perf*) *see* **про·ска́льз·ыва·ть**

про·скоч·и́ть, -ат, (*perf*) race/rush across/past/through, overshoot, shoot past

про·скреб·а́·ть, -ют, (*imp*) scrape/ scratch a hole/through

про·скрес·ти́, (*fut 3rd pl*) **про·скре·б·у́т,** (*past masc sing*) **про·скрёб·,** (*perf*); *see* **про·скреб·а́·ть**

про·сла́в·ить, -ят, (*perf*) *see* **про·славл·я́·ть**

про·сла́вл·енн·ый, -ая, -ое, (*past part pass*) *see* **про·славл·я́·ть;** celebrated, famous

про·славл·я́·ть, -ют, (*imp*) make famous; **-ся** become famous

про·сла́·ивани·е, -я, (*n*) (*v n*) of **про·сла́·ива·ть**

~, част·о́тн·ое (TV) frequency interlacing

про·сла́·ива·ть, -ют, (*imp*) place/put/ arrange in layers, place/put/arrange one layer on another, interlay; **-ся** (*pass*); have layers, be in layers

про·след·и́ть, -я́т, (*perf*) *see* **про·слё·ж·ива·ть**

про·слёж·ивани·е, -я, (*n*) (*v n*) of **про·слёж·ива·ть**

про·слёж·ива·ть, -ют, (*imp*) trace, follow up

про·сло́·ек·, -о́й·к·а, (*m*) = **про·сло́·й** (q.v.)

~, воз·ду́ш·н·ый (build) air cavity

про·сло·ённ·ый, -ая, -ое, (*past part pass*) *see* **про·сла́·ива·ть;** interstratified, interlayed

про·сло·и́ть, -я́т, (*perf*) *see* **про·сла́·ива·ть**

про·сло́·й, -я, (*m*) interlayer, intermediate layer, band, layer, (geol) intercalation, parting

про·сло́й·к·а, -и, (*g pl*) **-о́·ек·,** (*f*) (*v n*) *see* **про·сла́·ива·ть(ся);** **про·сло́·й** (q.v.)

~, двой·ников·ая (cryst) twin lamella

~, за·клин·и́вш·аяся (cryst) wedgeshaped twin lamella

про·служ·и́ть, -ат, (*perf*) serve, be in service

про·слу́ш·а·ть, -ют, (*perf*) *see* **про·слу́ш·ива·ть**

про·слу́ш·ивани·е, -я, (*n*) (*v n*) of **про·слу́ш·ива·ть;** audition; (telecom) crosstalk, hum

~, контро́ль·н·ое (telecom) monitoring

про·слу́ш·ива·ть, -ют, (*imp*) listen to; complete, attend (a course etc.)

про·сма́л·ива·ть, -ют, (*imp*) tar, cover/impregnate with tar

про·сма́тр·ива·ть, -ют, (*imp*) look through/over, scan; overlook, miss out

про·смол·и́ть, -я́т, (*perf*) *see* **про·сма́·л·ива·ть**

про·смо́тр·, -а, (*m*) (*v n*) *see* **про·сма́·тр·ива·ть;** scanning; oversight, mistake

про·смо́тр·енн·ый, -ая, -ое, (*past part pass*) *see* **про·сма́тр·ива·ть**

~ пло́щадь (*f*) (rad) scanned/swept area

про·смотр·е́ть, -ят, (*perf*) see **про·-смáтр·ива·ть**

про·смóтр·ов·ый, -ая, -ое, (*a*) viewing, scanning

~ **зал·** (*m*) viewing room

про·сн·у́ться, -у́тся, (*perf*) wake up

прóс·о, -а, (*n*) (bot) common millet, proso, *Panicum miliaceum*; panic grass, *Panicum* (in general)

~, **венгéр·ск·ое** Hungarian millet, *Setaria italica* var. *moharica*

~, **гермáн·ск·ое** German millet, *Panicum germanicum*

~, **итальян·ск·ое** Italian millet, *Setaria italica*

про·сóв·ыва·ть, -ют, (*imp*) shove/thrust/poke through/into

прóс·ов·ый, -ая, -ое, (*a*) of **прóс·о**; (*pl*) *Paniceae*

про·сол·и́ть, -ят, (*perf*) salt; corn (meat)

про·сóм·а, -ы, (*f*) (zool) prosoma

про·сóх·нуть, -нут, (*perf*) dry, dry through/out

про·соч·и́ться, -áтся, (*perf*) see **про·-сáч·ива·ться**

про·сп·áть, -ят, (*perf*) sleep (for a time); oversleep

проспéкт·, -а, (*m*) boulevard; prospect (with Soviet street names); prospectus

про·спирт·овáть, -у́ют, (*perf*) see **про·спирт·óвыва·ть**

про·спирт·óвыва·ть, -ют, (*imp*) alcoholize

про·с·пряг·á·ть, -ют, (*perf*) (gram) conjugate

про·срóч·енн·ый, -ая, -ое, (*past part pass*) see **про·срóч·ива·ть**; overdue

про·срóч·ива·ть, -ют, (*imp*) be overdue, be overdue in /from, be in arrears

про·срóч·ить, -ат, (*perf*) see **про·срó·ч·ива·ть**

про·срóч·к·а, -и, (*g pl*) -ч·ек·, (*f*) overstaying (leave etc.), overrunning (delivery date), failure to complete (a project)

про·стáв·ить, -ят, (*perf*) see **про·стáв·л·я́·ть**

про·стáвл·я́·ть, -ют, (*imp*) write/put/down

про·стáв·ок·, -в·к·а, (*m*) spacer

про·стá·ива·ть, -ют, (*imp*) stand (for a time), stay; be/remain idle/inactive, be/remain out of action

простáт·а, -ы, (*f*) (anat) prostate

про·стег·á·ть, -ют, (*perf*) see **про·стё·г·ива·ть**

про·стёг·ива·ть, -ют, (*imp*) quilt

прост·éйш·ий, -ая, -ее, (*superl adj*) the simplest; (*pl decl as adj*) (zool) protozoans, *Protozoa*

про·стéн·ок·, -н·к·а, (*m*) partition

про·стерили́з·ова·ть, -уют, (*perf*) see **стерили́з·овать**

про·с·тер·éть, (*fut 3rd pl*) **про·с·-тр·у́т**, (*perf*) stretch, extend

простет·и́ческ·ая грýпп·а (*f*) (org chem) prosthetic group

про·сти́л·к·а, -и, (*g pl*) -л·ок·, (*f*) midsole, bottom filler (footwear)

про·с·тир·áни·е, -я, (*n*) (*v n*) of **про·-с·тир·á·ть**; expanse, stretch; extent; (geol) strike, course

ли́н·и·я про·с·тир·áни·я (*f*) (min) line of strike/bearing

про·с·тир·á·ть, -ют, (*imp*) stretch, extend; -ся stretch, extend, run; (geol) strike; (*perf*) wash, launder

про·с·ти́р·ыва·ть, -ют, (*imp*) wash, launder

прост·и́тельн·ый, -ая, -ое, (*a*) excusable, pardonable

прост·и́ть, -я́т, (*perf*) forgive, excuse; -ся say good-bye, take leave

прóст·о (*adv*) simply, merely; (*predic adj*) see **прост·óй**

про·стó·й, -я, (*m*) idle time/period, downtime, outage, demurrage

~ **пéч·и, тек·у́щ·ий** time off-blast (of furnace)

прост·óй, -áя, -óе, (*a*) ordinary, common, simple

~ **глаз·** (*m*) naked eye

~ **дроб·ь** (*f*) simple/common/vulgar fraction

~ **множ·и́тел·ь** (*m*) (math) prime factor

~ **тéл·о** (*n*) (chem) elementary substance

~ **числ·ó** (*n*) (math) prime number

про·стóй·н·ый, -ая, -ое, (*a*) idle, inactive (of machinery etc.)

просто·квáш·а, -и, (*i*) -ей, (*f*) thickened sour milk

про·стóм·а, -ы, (*f*) (zool) prostomium

про·с·тóр·, -а, (*m*) space, expanse; scope

просто·рéч·и·е, -я, (*n*) (ling) popular speech, slang

про·с·тóр·н·ый, -ая, -ое, (*a*) spacious

прост·от·á, -ы́, (*f*) simplicity

про·сто·я́ть, -я́т, (*perf*) see **про·стá·-ива·ть**

про·стран·ственн·о·времен·н·óй, -áя, -óе, (*a*) space-time

про·стран·ственн·о-лин·ейн·ый,
　-ая, -ое, (*a*) linear-dimensional
про·стран·ственн·о-плоск·остн·ой,
　-ая, -ое, (*a*) planar-dimensional
про·стран·ственн·о-по·доб·н·ый,
　-ая, -ое, (*a*) space-like
про·стран·ственн·о-раз·нес·ённ·ый,
　при·ём· (*m*) space-diversity reception
про·стран·ственн·ый, -ая, -ое, (*a*)
　spatial, space, stereo-, steric, dimen-
　sional, solid (angle), three-dimensional
~ в·ли·яни·е (*n*) (plastic) steric inter-
　ference
~ груп·а (*f*) space group
~ за·да·ч·а (*f*) three-dimensional
　problem
~ изомер· (*m*) (plast) stereoisomer
~ крив·ая (*f*) (math) space curve
~ пере·мен·н·ая (*f*) (math) space
　variable
~ про·из·вод·н·ая (*f*) space derivative
~ структур·а (*f*) (plastic) three-
　dimensional structure
~ эффект· (*m*) three-dimensional effect
про·стран·ств·о, -а, (*n*) space
~, вод·ян·ое water space (boiler)
~, вред·н·ое waste/dead space, clear-
　ance
~ груп·ир·овк·и (elec) buncher cavity
~ дрейф·а (elec) drift space
~ импульс·ов (phys) momentum space
~, косм·ическ·ое outer space
~, лин·ейн·ое (math) linear/vector
　space
~, между·желез·н·ое (elec) air gap,
　clearance
~, над·утфель·н·ое space above the
　massecuite (sugar making)
~, не·по·раж·аем·ое (mil) dead
　ground
~, о·чист·н·ое (min) goaf
~, по·раж·аем·ое danger space
~ пре·образ·овáни·я (elec) buncher
　cavity
~, рабоч·ее furnace chamber (of
　furnace); reaction chamber (of reac-
　tor)
~, тёмн·ое (phys) dark space
~, топ·очн·ое furnace, firebox
~, фаз·ов·ое (autom) phase space
~, фараде·ев·ое (phys) Faraday's dark
　space
~ четыр·ёх из·мер·éни·й space time
про·страч·ива·ть, -ют, (*imp*) stitch,
　back-stitch

про·стрел·, -а, (*m*) (*v n*) see про·стре-
　л·ива·ть, (min) see про·стрел·к·а;
　(med) lumbago; (bot) anemone,
　Pulsatilla
~ тиз·луч·éни·я (nucl) radiation leak-
　age/streaming
про·стрел·ива·ть, -ют, (*imp*) shoot
　through; sweep, rake (with gunfire);
　carry out test firing; shoot (wells)
про·стрел·ить, -ят, (*perf*) shoot
　through; shoot (wells)
про·стрел·к·а, -и, (*f*) (*v n*) see про·-
　стрел·ива·ть; (min) springing, cham-
　bering, bulling; shooting (wells)
~ вз·рыв·н·ой скваж·ин·ы blast-
　hole springing
про·стрел·я·ть, -ют, (*perf*) carry out
　test firing; shoot (for a time)
про·с·тр·ёт, (*fut 3rd sing*) of про·с·те-
　р·éть
про·строч·ить, -ат, (*perf*) stitch,
　back-stitch
про·студ·а, -ы, (*f*) (med) common cold
про·студ·иться, -ятся, (*perf*) see про·-
　стуж·áться
про·стуж·á·ться, -ются, (*imp*) (med)
　catch a cold
про·стук·а·ть, -ют, (*perf*) see про·сту·-
　к·ива·ть
про·стук·ивани·е, -я, (*n*) (*v n*) of
　про·стук·ива·ть
про·стук·ива·ть, -ют, (*imp*) tap out
　(a message); sound (by tapping, e.g.
　med); tap away (for a time)
　ис·пыт·áни·е про·стук·ивани·ем
　(*n*) tap test
про·ступ·áни·е, -я, (*n*) (*v n*) see про·-
　ступ·á·ть; show-through, coming
　through
про·ступ·á·ть, -ют, (*imp*) seep through;
　come/show through
~ сквозь пор·ы (biol) exude
про·ступ·ить, -ят, (*perf*) see про·сту-
　п·á·ть
про·ступ·ь, -и, (*f*) (build) width of
　stair/step
простыня, -й, (*nom pl*) простын·и,
　(*g pl*) простын·ь, (*f*) sheet (for a bed)
про·сумм·ир·овать, -уют, (*perf*) see
　сумм·ир·овать
про·су·н·уть, -ут, (*perf*) see про·со·
　в·ыва·ть
про·суш·енн·ый, -ая, -ое, (*past part
　pass*) see про·суш·ива·ть; dried
про·суш·ива·ть, -ют, (*imp*) (*trans*)
　dry, dry through/out, dehumidify,
　desiccate

про·суш·и́ть, -ат, (*perf*) see про·су́-ш·ива·ть

про·су́ш·к·а, -и, (*f*) (*v n*) see про·су́-ш·ива·ть; desiccation

про·сущ·еств·ова́ть, -у́ют, (*perf*) exist

про·с·чёт·, -а, (*m*) (*v n*) see про·с·чи́-т·ыва·ть; count-down; counting loss; error, miscalculation

~, двой·н·о́й   duplication check (computers)

~ со·в·пад·ени·й   coincidence loss

про·с·чит·а́·ть, -ют, (*perf*) see про·-с·чи́т·ыва·ть

про·с·чи́т·ыва·ть, -ют, (*imp*) count, check; miscalculate

про·сып·а́·ть, -ют, (*imp*) spill; oversleep; -ся be spilt; wake up

про·сы́п·ать, -лют, (*perf*) spill

про·сых·а́·ть, -ют, (*imp*) dry, dry through/out

прос·ьб·а, -ы, (*f*) request

про·ся́д·ет, (*fut 3rd sing*) of про·се́ст·ь

прос·ян·о́й, -а́я, -о́е, (*a*) millet

прос·я́щ·ий, -ая, -ее, (*pres part act*) of прос·и́ть

про·табул·и́р·овать, -уют, (*perf*) tabulate

прот·актин·и́д·, -а, (*m*) (chem) protactinide

прот·акти́н·и·й, -я, (*m*) (chem) protactinium, Pa

прота́лин·а, -ы, (*f*) thaw hole (in ice), thawed patch (in snow)

про·та́лк·ивани·е, -я, (*n*) (*v n*) of про·та́лк·ива·ть

про·та́лк·ива·ть, -ют, (*imp*) push/shove through

прот·ами́н·, -а, (*m*) (biochem) protamine

прот·андр·и́ческ·ий, -ая, -ое, (*a*) (biol) protandrous

прот·а́ндр·и·я, -и, (*f*) (biol) protandry

про·та́пл·ива·ть, -ют, (*imp*) heat, stoke, heat/stoke well

про·та́пт·ыва·ть, -ют, (*imp*) tread, tread out, wear (a path etc.)

прот·арго́л·, -а, (*m*) (pharm) protargol

про·тар·и́р·овать, -уют, (*perf*) see тар·и́р·овать

прота́рс·, -а, (*m*) Protars (a fungicide)

про·та́ск·а, -и, (*f*) (naut) seizing, lashing

про·та́ск·ива·ть, -ют, (*imp*) drag/pull through

прот·а́спис·, -а, (*m*) (zool) protaspis

про·тач·а́·ть, -ют, (*perf*) stitch (for a time); stitch/sew through

про·та́ч·ива·ть, -ют, (*imp*) turn (on lathe); wear through, wash out (by water); eat/gnaw through; stitch/sew through

про·та́щ·енн·ый, -ая, -ое, (*past part pass*) see про·та́ск·ива·ть

про·тащ·и́ть, -ат, (*perf*) see про·та́ск·ива·ть

проте·а́з·а, -ы, (*f*) (biochem) protease

про·те́гулум·, -а, (*m*) (zool) protegulum

проте́з·, -а, (*m*) (med) prosthesis, prosthetic device, artificial aid

~, глаз·н·о́й   artificial/glass eye

~, зуб·н·о́й   denture, false teeth

~, мосто·ви́д·н·ый зуб·н·о́й   dental bridge

~, слух·ов·о́й   hearing aid

протез·и́р·овать, -уют, (*imp and perf*) fit/apply a prosthesis

проте́з·н·ый, -ая, -ое, (*a*) of проте́з· prosthetic

проте·и́н·, -а, (*m*) (biochem) protein

проте·ина́з·а, -ы, (*f*) (biochem) proteinase

проте·и́нов·ый, -ая, -ое, (*a*) of проте·и́н· (chem) proteic

проте·ино́ид·, -а, (*m*) albuminoid, scleroprotein

проте́·й, -я, (*m*) (microbiol) Proteus

проте́й·н·ый, -ая, -ое, (*a*) of проте́·й; (*pl as noun*) (bot) *Proteaceae*

про·те́к·а, -ы, (*f*) (zool) protheca

про·тек·а́ни·е, -я, (*n*) (*v n*) of про·-тек·а́·ть; course, progress, passage у́гол· про·тек·а́ни·я анод·н·ого то́к·а (*m*) plate-current operating angle (vacuum tube)

про·тек·а́·ть, -ют, (*imp*) flow; leak; pass, elapse; proceed (of phys/chem reactions)

проте́ктор·, -а, (*m*) (met) protector, zinc protector, sacrificial anode; tire tread; protective rubber coating; (oil) drill pipe protector

проте́ктор·н·ый, -ая, -ое, (*a*) of проте́ктор·

проте·о́з·а, -ы, (*f*) (chem) proteose

протео·лит·и́ческ·ий, -ая, -ое, (*a*) (biochem) proteolytic, proteoclastic

протер·а́ндр·и·я, -и, (*f*) (biol) proterandry, protandry

про·тер·е́ть, (*fut 3rd pl*) про·тр·у́т, (*perf*) see про·тир·а́ть

протеро·ги́н·и·я, -и, (*f*) (genet) proterogyny, protogyny

протеро·зо́·й, -я, (*m*) (geol) Proterozoic

протеро·зо́й·ск·ий, -ая, -ое, (*a*) (geol) Proterozoic

протеро·ти́п·, -а, (*m*) (biol) proterotype

про·тес·а́ть, (*fut 3rd pl*) **про·те́ш·ут,** (*perf*); see **про·тёс·ыва·ть**

протест·ова́ть, -у́ют, (*imp*) protest

про·тёс·ыва·ть, -ют, (*imp*) adze

про·те́ч·к·а, -и, (*g pl*) **-ч·ек·,** (*f*) flow, small flow

про·те́ч·ь, (*fut 3rd sing, pl*) **про·теч·ёт, про·тек·у́т,** (*past masc sing*) **про·-тёк·,** (*perf*); see **про·тек·а́·ть**

про·те́ш·ет, (*fut 3rd sing*) of **про·те·с·а́ть** (*perf*), see **про·тёс·ыва·ть** (*imp*)

про́тив (*prep+gen*) against; opposite; contrary to; to, as against (in comparisons etc.)

~ ве́тр·а into the wind

~ час·ов·о́й стре́л·к·и counter-clockwise

про́тив·ен·ь, -в·н·я, (*m*) baking tin/sheet; pan (in phys/chem experiments etc.)

проти́в·иться, -ятся, (*imp*) (*+dat*) oppose, object to

проти́в·ник·, -а, (*m*) opponent; (mil) enemy

проти́в·н·ый, -ая, -ое, (*a*) opposite; contrary; adverse; offensive, unpleasant, repugnant, disgusting

в проти́в·н·ом с·лу́ч·а·е otherwise

~ ве́тер· (*m*) head wind

до·каз·а́тельств·о от проти́в·н·о·го (*n*) proof by contradiction

противо- (*root*) counter-, anti-, contra-

противо·а́том·н·ый, -ая, -ое, (*a*) anti-atomic, (mil) atomic-defence

противо·брод·и́льн·ый, -ая, -ое, (*a*) (chem) antifermentative

противо·ве́с·, -а, (*m*) counterweight, counterpoise, counterbalance; (rad) artificial/capacity earth, counterpoise/balancing antenna

противо·ветр·ов·о́й, -а́я, -о́е, (*a*) storm-guyed (of masts, poles etc.)

противо·ви́рус·н·ый, -ая, -ое, (*a*) antiviral

противо·в·ключ·е́ни·е, -а, (*n*) opposition/feedback circuit (of an electric battery)

противо·воз·ду́ш·н·ый, -ая, -ое, (*a*) air defence, anti-aircraft

противо·вуал·и́рующ·ее ве́щ·еств·о (*n*) (phot) anti-foggant

противо·га́з·, -а, (*m*) gas mask, respirator, breathing apparatus

~, изол·и́рующ·ий self-contained breathing apparatus

~, фильтр·у́ющ·ий respirator

~, шла́нг·ов·ый само·в·са́с·ываю-щ·ий equalizer-tube breathing apparatus

противо·га́з·н·ый, -ая, -ое, (*a*) of **противо·га́з·**

~ коро́б·к·а (*f*) filtering canister, filter (of respirator)

противо·га́з·ов·ый, -ая, -ое, (*a*) antigas

противо·гли́ст·н·ый, -ая, -ое, (*a*) (pharm) vermifugous, anthelmintic

~ препара́т· (*m*) vermifuge

противо·глуш·и́тел·ь, -я, (*m*) (rad) anti-jamming device

противо·гни́л·ост·н·ый, -ая, -ое, (*a*) (pharm) antiputrefractive

противо·гриб·ко́в·ый, -ая, -ое, (*a*) fungusized, treated against fungus

противо·давл·е́ни·е, -я, (*n*) back pressure, counter pressure

противо·де́й·стви·е, -я, (*n*) counter-action, reaction, opposition

противо·де́й·ств·овать, -уют, (*imp*) oppose, counteract, react against

противо·де́й·ствующ·ий, -ая, -ее, (*past part act*) of **противо·де́й·ств·овать;** opposing, counter, antagonistic; (*+ dative*) anti-

~ кре́н·у antiroll

~ на·пряж·е́ни·е (*n*) (elec) counter/bucking voltage

~ по́л·е (*n*) (phys) controlling/choking field

~ си́л·а (*f*) (phys) opposing/controlling force

~ тр·е́ни·ю antifrictional

противо·деса́нт·н·ый, -ая, -ое, (*a*) (mil) anti-landing

противо·динатро́н·н·ая се́т·к·а (*f*) (elec) suppressor grid

противо·есте́ств·енн·ый, -ая, -ое, (*a*) unnatural

противо·за·ди́р·н·ый, -ая, -ое, (*a*) antiscuffing

противо·за·ко́н·н·ый, -ая, -ое, (*a*) illegal

противо·за·пот·ева́ющ·ий, -ая, -ее, (*a*) anti-dim (of goggles etc.)

противо·за·ча́т·очн·ый, -ая, -ое, (*a*) contraceptive

~ колпач·о́к· (*m*) contraceptive cap

~ сре́д·ств·о (*n*) contraceptive

противо·зени́т·н·ый, -ая, -ое,   (*a*) (mil) anti-aircraft, AA

~ **манёвр·**   (*m*) (air, mil) avoiding action, action to avoid anti-aircraft fire

противо·из·луч·éни·е, -я,   (*n*) counter-radiation, back radiation

противо·ис·треб·и́тельн·ый манёвр· (*m*) (air, mil) avoiding action, action to avoid enemy fighters

противо·ка́тер·н·ый, -ая, -ое,   (*a*) (naut) anti-MTB

противо·ка́ч·ечн·ый, -ая, -ое,   (*a*) (shipb) stabilizing

противо·кисл·óтн·ый, -ая, -ое,   (*a*) (pharm) antacid

противо·колеб·а́тельн·ый, -ая, -ое, (*a*) (instr) antihunt

противо·компáунд·н·ая об·мóт·к·а (*f*) (elec) differential compound winding

противо·кра́ж·н·ый, -ая, -ое,   (*a*) burglar, burglar-/theft-proof

противо·крéн·н·ая систéм·а   (*f*) (shipb) counter-flooding arrangements

противо·леж·а́щ·ий, -ая, -ее,   (*a*) opposite

~ **у́гол·** (*m*) (math) alternate angle

противо·лихора́д·очн·ый, -ая, -ое, (*a*) antipyretic, antifebrile

противо·лóд·очн·ый, -ая, -ое,   (*a*) (naut) antisubmarine

противо·локациóн·н·ый, -ая, -ое, (*a*) radar-jamming, anti-radar

~ **меро·при·я́т·и·я** (*n pl*) (mil) radar countermeasures

противо·лун·а́, -ы́,   (*f*) (astron) anti-selena

~, **по·бóч·н·ая** parantiselena

противо·маляр·и́йн·ый, -ая, -ое, (*a*) (med) antimalarial

противо·мед·ни́тел·ь, -я, (*m*) (gunn) decoppering charge

противо·мéст·н·ый, -ая, -ое,   (*a*) anti-sidetone (telephones)

противо·ми́н·н·ый, -ая, -ое,   (*a*) secondary-armament (naval); mine-countermeasures (naval)

~ **у·толщ·éни·я** (*pl*) (shipb) bulges

противо·мóль·н·ый, -ая, -ое,   (*a*) moth-proofing, anti-moth

противо·об·лед·ени́тел·ь, -я,   (*m*) anti-icer, de-icer, de-icing equipment

~, **жи́дк·остн·ый,** fluid de-icing equipment

~, **механ·и́ческ·ий** de-icer

противо·об·раст·а́ющ·ий, -ая, -ее, (*a*) (civ eng, shipb) anti-fouling

~ **кра́с·к·а** (*f*) anti-fouling paint

противо·о·жóг·ов·ый шлем·   (*m*) anti-flash helmet

противо·о·кисл·и́тел·ь, -я, (*m*) anti-oxidant

противо·ореóль·ност·ь, -и, (*f*) (phot) antihalation

противо·ореóль·н·ый, -ая, -ое,   (*a*) (phot) anti-halo, anti-halation

противо·о·с·кóл·очн·ый, -ая, -ое, (*a*) splinter-proof

противо·от·ка́т·н·ый, -ая, -ое,   (*a*) (gunn) recoil

~ **у·стрóй·ств·о** (*n*) (gunn) recoil arrangement

противо·от·печа́т·ок·, -т·к·а,   (*m*) counterpart

противо·пере·хóд·н·ый, -ая, -ое, (*a*) (telecom) anti-induction, anti-crossfire

противо·по·втóр·ност·ь, -и,   (*f*) **у·правл·éни·е по систéм·е противо·по·втóр·ност·и** (*n*) (autom) one-movement-only control

противо·под·лóд·очн·ый, -ая, -ое, (*a*) anti-submarine

противо·по·жа́р·н·ый, -ая, -ое,   (*a*) fire-fighting, fire, anti-fire

~ **пере·бóр·к·а** (*f*) (shipb) fire bulkhead, (a/c) firewall

~ **пре·гра́д·а** (build) fire stop

~ **у·стрóй·ств·а** (*pl*) firefighting appliances

противо·по·каз·а́ни·е, -я,   (*n*) (law) contrary evidence; (med) contra-indication

противо·по·лож·éни·е, -я,   (*n*) (*v n*) of **противо·по·лож·и́ть;** confrontation; contrast

противо·по·лож·и́ть, -áт,   (*perf*) set against, oppose; confront, contrast

противо·по·лóж·н·о (*adv*) contrariwise, in the opposite way, oppositely

противо·по·лóж·н·о-движ·ущ·иеся пóрш·н·и (*pl*) opposed pistons

противо·по·лóж·ност·ь, -и,   (*f*) contrast, opposition, opposite, antithesis **в противо·по·лóж·ност·ь** in contrast to, by contrast with

противо·по·лóж·н·ый, -ая, -ое,   (*a*) opposite; contrary, opposed, anti-counter

противо·по·мéх·ов·ый, -ая, -ое,   (*a*) (rad) antijamming

противо·по·ста́в·ить, -ят,   (*perf*) *see* **противо·по·ставл·я́·ть**

противо·по·ставл·éни·е, -я, (*n*) (*v n*)
see **противо·по·ставл·я́·ть;** confrontation, contrast; opposition

противо·по·ставл·я́·ть, -ют, (*imp*)
contrast, confront; oppose, put up against, put in opposition to, set against

противо·при·гáр·н·ый материáл· (*m*)
(met cast) foundry facing

противо·пýль·н·ый, -ая, -ое, (*a*)
bullet-proof

∼ брон·я́ (*f*) light armour

противо·ради·áци·я, -и, (*f*) counter/ back radiation

противо·радио·лок·áци·я, -и, (*f*)
radar countermeasure(s)

противо·ракéт·а, -ы, (*f*) (mil) anti-missile

противо·ракéт·н·ый, -ая, -ое, (*a*)
of **противо·ракéт·а**

∼ оборóн·а (*f*) antimissile defence

противо·рв·óтн·ый, -ая, -ое, (*a*)
anti-emetic

противо·рéж·ущ·ая пласт·и́нк·а (*f*)
ledger plate (of harvesting or mowing machine)

противо·реч·áщ·ий, -ая, -ее, (*a*)
contradictory, conflicting

противо·реч·и́в·ый, -ая, -ое (*a*)
contradictory, conflicting

противо·рéч·и·е, -я, (*n*) contradiction, conflict (of ideas)

противо·рéч·ить, -ат, (*imp*) contradict

противо·с·вёрт·ывающ·ее вещ·е·ств·ó (*n*) anticoagulant

противо·с·войлáч·ивающ·ий, -ая, -ее, (*a*) (text) antifelting

противо·с·вя́з·ь, -и, (*f*) *see* **обрáт·н·ая с·вяз·ь;** (autom) feedback

противо·си·я́ни·е, -я, (*n*) (astron) Gegenschein, counterglow

противо·с·мин·áем·ый, -ая, -ое, (*a*)
anticreasing, anticrease

противо·со·в·пад·éни·е, -я, (*n*) anti-coincidence

противо·сóлн·ц·е, -а, (*n*) (astron) anthelion

∼, по·бóч·н·ое paranthelion

противо·срéд·ств·о, -а, (*m*) antidote, remedy

противо·с·рыв·н·óй, -áя, -óе, (*a*)
(dynam) antistall

∼ характери́ст·ик·а (*f*) (dynam)
antistall characteristic

противо·ставл·éни·е, -я, (*n*) contra-position

противо·стар·и́тел·ь, -я, (*m*) age-resistor, preservative; (rubber) anti-oxidant, anti-ager

противо·сто·я́ни·е, -я, (*n*) (*v n*) of
противо·сто·я́ть; resistance, opposition; (astron) opposition

противо·сто·я́ть, -я́т (*imp*) with-stand, resist, oppose

противо·сýдорож·н·ый, -ая, -ое, (*a*) antispasmodic

противо·тáнк·ов·ый, -ая, -ое, (*a*)
(mil) anti-tank, anti-armour/armor, anti-mechanized

противо·тéл·о, -а, (*m*) (bact) antibody

противо·тен·ев·óй, -áя, -óе, (*a*)
antishadow

противо·теч·éни·е, -я, (*n*) counter-current, counterflow

противо·тóк·, -а, (*m*) counterflow, countercurrent, reflux

коэффициéнт· противо·тóк·а (*m*)
reflux ratio

противо·торпéд·н·ая оборóн·а (*f*)
torpedo countermeasures

противо·тóч·н·ый, -ая, -ое, (*a*) of
противо·тóк·

противо·у·сáд·очн·ый, -ая, -ое, (*a*)
shrink-resistant/proofing

противо·у·том·и́тел·ь, -я, (*m*) (plast)
anti-fatigue agent

противо·фáз·а, -ы, (*f*) antiphase, opposite phase

противо·фóн·ов·ый, -ая, -ое, (*a*)
antibackground; (rad) anti-hum

противо·хим·и́ческ·ий, -ая, -ое, (*a*)
(mil) antigas, anti-C.W. (chemical warfare)

противо·электро·дви́ж·ущ·ая си́л·а
(*f*) counter/back e.m.f.

противо·элемéнт·, -а, (*m*) (elec)
counter-emf cell

противо·э́х·ов·ый, -ая, -ое, (*a*) (rad)
anti-reflection

противо·я́д·и·е, -я, (*n*) antidote

против·у·гóн·, -а, (*m*) (rail) anti-creeper

протиг·, -а, (*m*) (shipb) continuous hull planking

прóт·и·й, -я, (*m*) (chem) protium, H^1

про·тир·á·ть, -ют, (*imp*) rub, rub/ wear through; wipe, wipe across/clean

про·ти́р·к·а, -и, (*f*) (*v n*) *see* **про··тир·á·ть**

про·ти́р·очн·ый, -ая, -ое, (*a*) of
про·ти́р·к·а

∼ маши́н·а (*f*) (food) pulping machine

протисто·лог·и·я, -и,   (*f*) (biol) protistology

протист·ы,   (*pl*) (biol) Protista

про·тк·н·у́ть, -у́т,   (*perf*) *see* **про·- тык·а́·ть**

прото·акти́н·и·й, -я,   (*m*) (chem) protoactinium, protactinium, Pa

прото·басти́т·, -а,   (*m*) (min) protobastite

прото·ге́н·н·ый, -ая, -ое,   (*a*) (chem) protogenic

прото·ги́н·, -а,   (*m*) (min) protogine

прото·гин·и́·я, -и,   (*f*) (genet) protogyny

прото·гне́йс·, -а,   (*m*) (geol) protogneiss

прото·де́рм·а, -ы,   (*f*) (bot) protoderm, dermatogen

прото·зиго́т·а, -ы,   (*f*) (genet) protozygote

прото·зо́е·а, -ы,   (*f*) (zool) protozoea

прото·зо́й·н·ый, -ая, -ое,   (*a*) (zool) protozoan

прото·зоо·ло́г·и·я, -и,   (*f*) protozoology

прото·исто́р·и·я, -и,   (*f*) prehistory

про·то́к·, -а,   (*m*) channel, canal; watercourse; branch (of river); (anat) duct, ductus

~**, пузы́р·н·ый** (anat) ductus cysticus

~**, с·лёз·н·ый** (anat) tear duct

протокатехо́л·, -а,   (*m*) (chem) protocatechol

протоко́л·, -а,   (*m*) proceedings, minutes (of meeting etc.), record, record sheet; (dipl) protocol

~ **ис·пыт·а́ни·я**   trials/test record sheet

протоко́ль·н·ый, -ая, -ое,   (*a*) of **протоко́л·**

~ **да́·нн·ые**   (*pl*) performance data/figures

прото·ко́н·, -а,   (*m*) (zool) protocone

прото·кони́д·, -а,   (*m*) (zool) protoconid

прото·ко́нх·, -а,   (*m*) (zool) protoconch

прото·коню́л·ь, -я,   (*m*) (zool) protoconule

прото·кристалл·и́ческ·ий, -ая, -ое,   (*a*) protocrystalline

прот·окси́д·, -а,   (*m*) (chem) protoxide

прото·ксиле́м·а, -ы,   (*f*) (bot) protoxylem

про·толк·н·у́ть, -у́т,   (*perf*) push/shove through

прото·ло́ф·, -а,   (*m*) (anat) protoloph

прото·мито́з·, -а,   (*m*) (genet) protomitosis

прот·о́н·, -а́,   (*m*) (phys) proton

~ **от·да́·ч·и** recoil proton

прото·не́м·а, -ы,   (*f*) (bot) protonema

прото·не́фт·ь, -и,   (*f*) (geol) protopetroleum

протон·н·ый, -ая, -ое,   (*a*) of **прот·о́н·**

протопат·и́ческ·ий, -ая, -ое,   (*a*) (med) protopathic

прото·пекти́н·, -а,   (*m*) (biochem) protopectin

про·топ·и́ть, -я́т,   (*perf*) *see* **про·-та́пл·ива·ть**

прото·пла́зм·а, -ы,   (*f*) (biol) protoplasm

прото·пла́ст·, -а,   (*m*) (biol) protoplast

прото·поди́т·, -а,   (*m*) (zool) protopodite

прото·порфири́н·, -а,   (*m*) (biochem) protoporphyrin

про·топт·а́ть,   (*fut 3rd pl*) **про·то́п·ч·ут,** (*perf*) *see* **про·та́пт·ыва·ть**

про·то́ракс·, -а,   (*m*) (zool) prothorax

про·тор·е́ни·е, -я,   (*n*) (med) itineration

проторо·за́вр·, -а,   (*m*) (pal) Protorosaur

прото·сте́л·а, -ы,   (*f*) (bot) protostele

прото·сте́л·ь, -и,   (*f*) = **прото·-сте́л·а**

прото·те́р·и·и,   (*m pl*) (zool) prototheres, *Prototheria*

прото·те́тик·а, -и,   (*f*) protothetics

прото·ти́п·, -а,   (*m*) prototype; primary standard

прото·тро́п·и·я, -и,   (*f*) (chem) prototropy

прото·флоэ́м·а, -ы,   (*f*) (bot) protophloem

прото·хлорофи́лл·, -а,   (*m*) (biochem) protochlorophyll

про·то́ч·енн·ый, -ая, -ое,   (*past part pass*) of **про·точ·и́ть**

про·точ·и́ть, -а́т,   (*perf*) bore, grind out (on lathe); sharpen (for a time); wear through, wash out (by water); eat/gnaw through

про·то́ч·к·а, -и,   (*f*) (*v n*) of **про·-точ·и́ть**

~**, кольц·ев·а́я** (mech) scroll

~ **плу́нжер·а** plunger scroll (of pump)

про·то́ч·н·ый, -ая, -ое,   (*a*) flowing, running (of water); non-circulating, through-

~ **вод·а́** (*f*) running water

про·то́ч·н·ый, -ая, -ое
~ **с·ма́з·к·а** (*f*) non-circulating lubrication
~ **с·чёт·чик·** (*m*) flow counter
прото·эпифи́т·, -а, (*m*) (bot) proto-epiphyte
прото·э́циум·, -а, (*m*) (zool) proto-oecium
про·тра́в·а, -ы, (*f*) (*v n*) *see* **про·тра́·вл·ива·ть;** (chem) mordant; (met) pickle
про·трав·и́тел·ь, -я, (*m*) (agr chem) seed dressing preparation
про·трав·и́ть, -́ят, (*perf*) *see* **про·тра́вл·ива·ть**
про·тра́в·к·а, -и, (*g pl*) **-в·ок·,** (*f*) (*v n*) *see* **про·тра́вл·ива·ть**
~, **гру́б·ая** mass etch
про·травл·е́ни·е, -я, (*n*) (*v n*) *see* **про·тра́вл·ива·ть**
про·тра́вл·енн·ый, -ая, -ое, (*past part pass*) *see* **про·тра́вл·ива·ть**
про·тра́вл·иватєл·ь, -я, (*m*) seed dresser, (agr) seed-dressing machine
про·тра́вл·ива·ть, -ют, (*imp*) (met) pickle; etch (engraving etc.); (text) mordant; treat (seeds etc.)
про·травл·я́·ть, -ют, (*imp*) = **про·тра́вл·ива·ть**
про·трав·н·о́й, -а́я, -о́е, (*a*) mordant
~ **крас·и́тел·ь** (*m*) mordant dye
про·тра́в·очн·ый, -ая, -ое, (*a*) etching; pickling; mordanting
протра́ктор·, -а, (*m*) (navig) station pointer
про·тра́л·енн·ый, -ая, -ое, (*past part pass*) *see* **про·тра́л·ива·ть;** (navig) swept
про·тра́л·ива·ть, -ют, (*imp*) (naut) sweep (for mines etc.)
про·тра́л·ить, -ят, (*perf*) *see* **про·тра́л·ива·ть**
про·тр·ёт, (*fut 3rd sing*) of **про·те·р·е́ть,** (*perf*); *see* **про·тир·а́·ть** (*imp*)
про·трещ·а́ть, -а́т, (*perf*) crackle; creak; chirp (of insects)
протромби́н·, -а, (*m*) (biochem) prothrombin
протубера́н·ец·, -н·ц·а, (*m*) *see* **про·тубера́нц·**
протубера́нц·, -а, (*m*) (astron) prominence, solar flare
~ **-спектро·ско́п·, -а,** (*m*) (astron) spectro-helioscope
про·тух·а́·ть, -ют, (*imp*) become rotten, go bad, putrefy

про·ту́х·л·ый, -ая, -ое, (*a*) rotten, bad, putrefied, putrid
про·ту́х·нуть, -нут, (*perf*) *see* **про·ту·х·а́·ть**
про·тык·а́·ть, -ют, (*imp*) pierce, pierce through, pierce a hole in
про·тя́г·ива·ть, -ют, (*imp*) stretch, stretch out, extend; (met) broach (a hole); draw (hollow forging); draw down (a forged bar)
про·тяж·е́ни·е, -я, (*n*) extent, expanse, span, length, stretch (of distances), duration (of periods of time)
на про·тяж·е́ни·и for (of time or distance)
про·тяж·ённост·ь, -и, (*f*) = **про·тяж·е́ни·е**
про·тяж·ённ·ый, -ая, -ое, (*a*) extended; extensive, lengthy
~ **ис·то́ч·ник·** (*m*) (nucl) extended source
про·тя́ж·к·а, -и, (*f*) (*v n*) *see* **про·тя́г·ива·ть;** broach (tool)
~ **толк·а́ющ·его ти́п·а** push broach
~ **тя́·нущ·его ти́п·а** pull broach
~, **шпо́н·очн·ая** keyway tool
про·тяж·н·о́й, -а́я, -о́е, (*a*) drawing; drawing-down; broaching
~ **маши́н·а** (*f*) (met cast) mould/pattern drawing machine
~ **механи́зм·** (*m*) (instr) transport mechanism, winder
~ **ручє́·й** (*m*) (met forge) drawing die
~ **стан·о́к·** (*m*) broaching machine
про·тя́ж·н·ый, -ая, -ое, (*a*) = **про·тяж·ённ·ый,** *see also* **про·тяж·н·о́й;** drawn-out; drawling (of speech)
про·тя·н·у́ть, -́ут, (*perf*) *see* **про·тя́г·ива·ть**
про·у́ш·ин·а, -ы, (*f*) (eng) lug, ear, eye
проф- (*component*) (*abbrev*) of (1) **профессиона́ль·н·ый** = occupational, professional trade, (2) **проф·сою́з·н·ый** = trade-union
про·фа́з·а, -ы, (*f*) (cyt) prophase
проф·бол·е́зн·ь, -и, (*f*) occupational disease
про·ферме́нт·, -а, (*m*) (biochem) zymogen, proferment
проф·движ·е́ни·е, -я, (*n*) trade-union movement (in U.S.S.R.)
профессиона́ль·н·ый, -ая, -ое, (*a*) occupational, professional, trade
профе́сс·и·я, -и (*f*) occupation, profession, trade
профе́ссор·, -а, (*nom pl*) **-а́,** (*g pl*) **-о́в,** (*m*) professor

профе́ссор·ск·ий, -ая, -ое, (*a*) professorial

профе́ссор·ств·о, -а, (*n*) professorship

профе́ссор·ств·овать, -уют, (*imp*) be a professor, have a professorship/chair

професс·у́р·а, -ы, (*f*) professorship; professors, professoriate

про·фила́кт·ик·а, -и, (*f*) (med) prophylaxis, prophylactic treatment; (mech) preventive measures, routine maintenance/inspection, trouble shooting

про·филакт·и́ческ·ий, -ая, -ое, (*a*) preventive; (med) prophylactic

~ **о·смо́тр·** (*m*) routine inspection

~ **ремо́нт·** (*m*) preventive/routine maintenance

~ **сре́д·ств·о** (*n*) (pharm, med) prophylactic

про·филакт·о́ри·й, -я, (*m*) health centre, preventorium; (mech) maintenance shop

профил·и́рованн·ый, -ая, -ое, (*past part pass*) of **профил·и́р·овать**

профил·и́р·овать, -уют, (*imp and perf*) (met) shape, form; grind a profile, profile (on machine tools); camber (roads etc.); extrude (rubber)

профил·иро́вк·а, -и, (*g pl*) **-в·ок·,** (*f*) (*v n*) *see* **профил·и́р·овать**

профил·иро́вочн·ый, -ая, -ое, (*a*) of **профил·иро́вк·а**; shaping, forming; cambering

~ **стан·** (*m*) (met) continuous roll-forming machine

~ **стан·о́к·** (*m*) (mech eng) profile-grinding machine

профило́·граф·, -а, (*m*) (mech eng) profilograph, profilcorder, contorograph, surface analyser; river-bed sounder

~, **пьезо·электр·и́ческ·ий** crystal pick-up profilograph, brush surface-analyser

профило́·метр·, -а, (*m*) profilometer; (civ eng) roughness integrator, corrugmeter

про́фил·ь, -я, (*m*) profile, side view/elevation, section; outline, contour; type, character; (met) rolled/structural shape/section/bar; (arch) moulding; (air dynam) airfoil

~, **акти́в·н·ый** active profile (cf gear tooth)

~, **гидро·дина́м·и́ческ·ий** (dynam) hydrofoil profile

про́фил·ь

~ **кры·л·а́** (dynam) airfoil profile

~, **ламина́р·н·ый** laminar-flow airfoil

~, **ова́ль·н·ый** (met) oval

~ **пло́т·ност·ей** (dynam) density profile

~ **по·лёт·а** (nav) flight profile

~, **по·переч·н·ый** cross-section, transverse section

~, **по́чв·енн·ый** soil profile

~, **про·до́лт·н·ый** longitudinal section/prcfile

~, **про·ка́т·н·ый** rolled metal/section/shape/bar

~, **прост·о́й** (met) geometrical shape, regular section (i.e. round, square, hexagon or rectangle)

~ **равно·ве́с·и·я** base level (of a stream)

~, **ромбо·ви́д·н·ый** (dynam) double-wedge profile

~ **скор·ост·е́й** velocity profile

~, **скор·остн·о́й** (dynam) high-speed profile

~, **сорт·ов·о́й** (met roll) sections, shapes

~, **стро·и́тельн·ый** (met) structural shape

тео́ри·я про́фил·я (*f*) (dynam) airfoil theory

~, **фасо́н·н·ый** (met) section, shape, special section

~, **шпунт·ов·о́й** rolled sheet piling

про́филь·н·ый, -ая, -ое, (*a*) of **про́фил·ь**

~ **со·противл·е́ни·е** (*n*) (dynam) profile drag

~ **стал·ь** (*f*) steel shapes/sections

про·фильтр·о́ванн·ый, -ая, -ое, (*past part pass*) of **про·фильтр·ова́ть;** filtered

про·фильтр·ова́ть, -у́ют, (*perf*) filter

проф·о́рг·, -а, (*m*) = **проф·сою́з·-н·ый организ·а́тор·** local trade-union official

проф·организ·а́ци·я, -и, (*f*) local trade-union organization

про·фо́рм·а, -ы, (*m*) proforma

проф·рабо́т·ник·, -а, (*m*) trade-union official

проф·сою́з·, -а, (*m*) trade-union (esp in U.S.S.R.)

проф·тех·шко́л·а, -ы, (*f*) engineering trades school, technical school

про·хлад·и́тельн·ый, -ая, -ое, (*a*) refreshing, cooling

про·хла́д·н·ый, -ая, -ое, (*a*) cool, fresh

про·хо́д·, -а, (*m*) passage, pass, way; alley, aisle (between machines); (naut) fairway, channel; (anat) passage, meatus; pass (welding, met rolling etc.)

~, за́д·н·ий (anat) anus, back passage

~, нару́ж·н·ый слух·ов·о́й (anat) external auditory meatus

~, с·лёз·н·ый (anat) lachrymal duct

~, у·сло́в·н·ый internal diameter, nominal inside diameter, bore

про·ход·и́мост·ь, -и, (*f*) passability (of roads etc.); (М/Т) rough-road performance; permeability

про·ход·и́м·ый, -ая, -ое, (*pres part pass*) of **про·ход·и́ть**; passable (of roads etc.); permeable

про·ход·и́ть, -ят, (*imp*) pass, go, go/walk past/by/through; pass, elapse, be over (of time); circulate (of rumours etc.); (min) sink (shafts etc.), drive, drift (tunnels etc.)

~ со·един·е́ни·е advance a call (telephony)

~ через знач·е́ни·е (math) assume a value

про·хо́д·к·а, -и, (*g pl*) **-д·ок·,** (*f*) (*v n*) *see* **про·ход·и́ть**; (min) tunnelling, driving, drifting; sinking

~ вос·ста·ю́щ·его (min) raising, driving a raise

~ кана́в· trenching

ско́р·ост·ь про·хо́д·к·и (*f*) (min) advance rate (stoping etc.), sinking rate (vertical and inclined workings), tunnelling rate (horizontal); (oil) rate of drilling/penetration

~ шахт· shaft sinking

про·ход·н·о́й, -а́я, -о́е, (*a*) through, communicating; continuous (of machines etc.); straight, straight-through; wall, partition (of bushings, glands etc.); transfer (of electrical characteristics); check (of places where passes are shown); (fish) anadromous

~ аппара́т· (*m*) (text) continuous machine

~ бу́д·к·а (*f*) entrance gate, check post

~ ём·кост·ь (*f*) (elec) transfer capacitance

~ изоля́тор· (*m*) (elec) lead-in insulator, wall-entrance insulator, bushing

~ пе́ч·ь (*f*) continuous furnace

~ сторон·а́ (*f*) go-plug, go-step (of a limit gauge)

про·хо́д·ческ·ий, -ая, -ое, (*a*) (min) tunnelling, tunneller's; sinking, sinker's; driving, driver's

~ цикл· (*m*) development stint

про·хо́д·чик·, -а, (*m*) (min) tunneller, sinker, driver, hewer

про·ход·я́щ·ий, -ая, -ее, (*past part act*) of **про·ход·и́ть**; (telecom) through

~ свет· (*m*) transmitted light

про·хожд·е́ни·е, -я, (*n*) (*v n*) of **про·ход·и́ть**; passage

~ по·то́к·а flow-through

~ сигна́л·а passage/transmission of signal

про·хо́ж·ий, -его, (*m decl as adj*) passer-by

про·хож·у́, (*pres 1st sing*) of **про·ход·и́ть**

про·цве·сти́, (*fut 3rd pl*) **про·цвет·у́т,** (*past masc sing*) **про·цвё·л,** (*perf*), *see* **про·цвет·а́·ть**

про·цвет·а́·ть, -ют, (*imp*) prosper, flourish

про·цед·и́ть, -ят, (*perf*) *see* **про·це́·ж·ива·ть**

процед·у́р·а, -ы, (*f*) procedure; (med) prescribed treatment, treatment

про·це́ж·ива·ть, -ют, (*imp*) strain, filter

процент·, -а, (*m*) per cent, percentage; (*pl*) (fin) interest

~, а́том·н·ый atomic percent, at. %

~, больш·и́е (*pl*) large percentages; (fin) high interest

~, вес·ов·о́й weight percent, percent by weight, wt. %

~, на·ро́с·ш·ие (*pl*) accrued interest

~, объ·ём·н·ый volume percent, percent by volume, vol. %

~, с·ло́ж·н·ые (*pl*) compound interest

процентил·ь, -я, (*m*) (math) percentile

процент·н·ый, -ая, -ое, (*a*) of **про·цент·**

~ бума́г·и (*f*) (fin) interest-bearing securities

~ до·хо́д· (*m*) (fin) interest

~ от·нош·е́ни·е (*n*) percentage

~ ста́в·к·а (*f*) rate of interest

процентуа́ль·н·ый, -ая, -ое, (*a*) percentage

проце́сс·, -а, (*m*) (*see also under associated adjectives*); process; (law) trial

~, бе́рег·ов·ы́е (*pl*) (geol) littoral processes

~, гражд·а́нск·ий suit, lawsuit, civil proceedings

~, обрат·и́м·ый reversible process

процесс·и·я, -и, (*f*) procession

про·цио́н·, -а, (*m*) (astron) Procyon

про·чека́н·ива·ть, -ют, (*imp*) (eng) chase; pierce

про·чекан·и́ть, -ят, (*perf*) *see* про·чека́н·ива·ть

про·чёл, (*past masc sing*) of про·че́ст·ь

про·чёрк·ива·ть, -ют, (*imp*) draw a line (to signify omission)

про·черк·н·у́ть, -у́т, (*perf*) *see* про··чёрк·ива·ть

про·черт·и́ть, -ят, (*perf*) *see* про·чёр·ч·ива·ть

про·чёрч·ива·ть, -ют, (*imp*) delineate, draw

про·чёс·, -а, (*m*) (*v n*) *see* про·чёс·ыва·ть

про·чёс·анн·ый, -ая, -ое, (*past part pass*) *see* про·чёс·ыва·ть

про·чес·а́ть, (*fut 3rd pl*) про·че́ш·ут, (*perf*) *see* про·чёс·ыва·ть

про·че́ст·ь, (*fut 3rd pl*) про·чт·у́т, (*past masc sing*) про·чёл, (*perf*); read, read through

про·чёс·ывани·е, -я, (*n*) (*v n*) *see* про··чёс·ыва·ть; (air) modified sector search

про·чёс·ыва·ть, -ют, (*imp*) (text) comb, card; ripple (flax etc.); comb (search)

про·чёш·ет, (*fut 3rd sing*) of про·чес·а́ть (*perf*), *see* про·чёс·ыва·ть

про́ч·ий, -ая, -ое, (*a*) other; (*n or pl decl as adj*) the rest/remainder, the others

  и про́ч·ее etcetera, and so on/forth

  ме́жду про́ч·им incidentally, by the way, among other things

  при про́ч·их ра́вн·ых у·сло́в·и·ях other conditions being equal, other conditions being the same

про·чи́ст·ить, -ят, (*perf*) *see* про·чи·щ·а́·ть

про·чи́ст·к·а, -и, (*g pl*) -т·ок·, (*f*) (*v n*) *see* про·чищ·а́·ть; scavenging; thinning out (a forest)

про·чит·а́·ть, -ют, (*perf*) read, read through, read (for a time)

про́ч·ить, -ат, (*imp*) intend for, designate

про·чищ·а́·ть, -ют, (*imp*) clean, clean out/through; scavenge (engines); clear (a path etc.); thin (woods etc.); weed

про·чла́, (*past fem sing*) of про·че́ст·ь

проч·ност·н·о́й, -а́я, -о́е, (*a*) strength

проч·ност·ь, -и, (*f*) (*see also under associated words*); strength (of materials); fastness (of dyes etc.)

про́ч·ност·ь

~, вла́ж·н·ая   wet tensile strength (of paper)

~, гидравл·и́ческ·ая bursting strength (e.g. of hose)

~, динам·и́ческ·ая dynamic strength, impact strength

~, диэлектр·и́ческ·ая (phys) dielectric strength

~, дл·и́тельн·ая durability, long-term strength; (met) creep limit, long-time strength

  за·па́с· проч·ност·и (*m*) (eng) factor of safety, safety factor/margin

~ крепл·е́ни·я bond strength, adhesion

~ на из·ги́б· flexural/bending strength; transverse strength (bricks etc.), flexing strength (rubber)

~ на раз·ры́в· tensile strength

~ на рас·тяж·е́ни·е tensile strength

~ на с·двиг· shear strength

~ на с·жа́т·и·е (met) compressive strength; crushing strength (bricks etc.)

~ на с·рез· shear strength

~ на стат·и́ческ·ий из·ги́б· static bending strength

~ на уда́р· impact strength, dynamic strength

~, объ·ём·н·ая (met) saturation pressure (cavitation)

  пре·де́л· про́ч·ност·и (*m*) (*see under* преде́л·) maximum stress, ultimate strength, tensile strength, ultimate tensile stress

~ при... *see* про́ч·ност·ь на...

~, раз·рыв·н·а́я tensile strength

~ с·вя́з·и bond strength (of rubber etc.)

~ с·вя́з·и рези́н·ы с ткан·ью rubber-to-fabric bond strength

~, уда́р·н·ая impact strength, dynamic strength

~, у·ста́л·ост·н·ая fatigue strength

~, хру́п·к·ая technical cohesive strength

~, цикл·и́ческ·ая fatigue strength

~, электр·и́ческ·ая dielectric strength

про́ч·н·ый, -ая, -ое, (*a*) strong (of materials); fast (of dyes etc.)

~ зна́·ни·я (*pl*) sound knowledge

про·чт·е́ни·е, -я, (*n*) reading

про·чт·ёт, (*fut 3rd sing*) of про·че́ст·ь

про·чу́вств·овать, -уют, (*perf*) experience, feel

прочь (*adv*) away, off; (*imper*) out of the way! mind out! stand clear

про·шéд·ш·ий, -ая, -ее, (*past part act*) *see* **про·ход·и́ть;** past, previous; last (of time); (rad) transmitted; (*n decl as adj*) the past

~ **врéм·я** (*n*) (gram) past tense

про·шёл, (*past masc sing*) of **про·й-ти́** (*perf*), *see* **про·ход·и́ть** (*imp*)

про·шéств·и·е, -я, (*n*) passage, lapse (of time)

про·шив·а́·ть, -ют, (*imp*) sew, stitch, sew across/around; (met) pierce

про·ши́в·к·а, -и, (*g pl*) **-в·ок·,** (*f*) (*v n*) *see* **про·шив·а́·ть;** (met) piercing; (met) piercing mandrel; (text) insertion, inserting

про·шив·н·о́й, -а́я, -о́е, (*a*) (met) piercing; (text) inserting

~ **пресс·** (*m*) piercing/punching press

~ **стан·** (*m*) (met roll) piercing mill

про·ши́в·очн·ый, -ая, -ое, (*a*) = **про·шив·н·о́й**

про·ши́·ть, (*fut 3rd pl*) **про·шь·ю́т,** (*perf*) *see* **про·шив·а́·ть**

про·шла́, (*past fem sing*) of **про·йти́** (*perf*), *see* **про·ход·и́ть** (*imp*)

прошло·го́д·н·ий, -яя, -ее, (*a*) last year's

про́шл·ый, -ая, -ое, (*a*) past, last (of time); (*n decl as adj*) the past

про·шнур·ова́ть, -у́ют, (*perf*) thread, string

про·шпакл·ева́ть, -ю́ют, (*perf*) putty

про·шпакл·ёвыва·ть, -ют, (*imp*) putty

про·шпатл·ева́ть, -ю́ют, (*perf*) putty

про·штуди́р·овать, -уют, (*perf*) study

про·штукату́р·ить, -ят, (*perf*) plaster

прош·у́, (*pres 1st sing*) of **прос·и́ть**

про·шь·ёт, (*fut 3rd sing*) of **про·-ши́·ть,** (*perf*); *see* **про·шив·а́·ть**

прощ·а́·йте, (*pl imperative*) good-bye!

прощ·а́·ть, -ют, (*imp*) forgive, excuse; **-ся** say good-bye, take leave

про́щ·е (*comp*) simpler, easier, more ordinary

прощ·éни·е, -я, (*n*) pardon, forgiveness

про·щу́п·а·ть, -ют, (*perf*) *see* **про·-щу́п·ыва·ть**

про·щу́п·ыва·ть, -ют, (*imp*) feel/grope along/through; locate (by feeling)

про·энзи́м·, -а, (*m*) (biochem) pro-enzyme

про·эритроци́т·, -а, (*m*) (cyt) pro-erythrocyte

про·эстру́с·, -а, (*m*) (biol) proestrus

про·яв·и́тел·ь, -я, (*m*) (phot) developer

про·яв·и́тел·ь

~, **метол·гидрохинóн·ов·ый** (phot) metol-hydroquinone developer

про·яв·и́ть, ˮят, (*perf*) *see* **про·явл·я́·ть**

про·явл·éни·е, -я, (*n*) (*v n*) *see* **про·-явл·я́·ть;** manifestation, display, appearance, show; (phot) development

врéм·я про·явл·éни·я (*n*) (phot) developing time; epoch (of a signal)

~ **нéфт·и** (geol) oil show

про·явл·енн·ый, -ая, -ое, (*past part pass*) *see* **про·явл·я́·ть**

про·явл·я́·ть, -ют, (*imp*) show, display, give evidence of; (phot) develop; **-ся** (*pass*); become apparent/manifest, appear, be revealed

про·я́в·очн·ый, -ая, -ое, (*a*) (phot) developing

про·ясн·éни·е, -я, (*n*) clearing, clearing up, clarification

про·ясн·é·ть, -ют, (*perf*) clear, become clear

про·ясн·и́ть, -ят, (*perf*) *see* **про·-ясн·я́·ть**

про·ясн·я́·ть, -ют, (*imp*) make clear/distinct, clarify; **-ся** clear, become clear

пруд·, -а́, (*m*) pond

~, **маль·кóв·ый** (fish) hatchery

~, **сол·ян·óй** (geog) salt pan

пруд·и́ть, -я́т, (*imp*) dam, dam up

пруд·ов·о́й, -а́я, -о́е, (*a*) of **пруд·**

пруж·и́н·а, -ы, (*f*) spring

~ **Бéльвилл·я** Belleville spring washer

~, **бо·ев·а́я** mainspring; striker spring (of grenade)

~, **вед·у́щ·ая** master/lever spring

~, **винт·ов·а́я** coil/coiled/helical spring

~, **ви́·т·ая** coil/coiled/helical spring

~, **воз·вращ·а́ющ·ая** homing/reset spring

~, **гидро·стат·и́ческ·ая** hydrostatic valve spring

~, **дви́ж·ущ·ая** pressure spring (of relays etc.)

~ **из·ги́б·а** bow spring

~ **круч·éни·я** torsion spring

~, **на·жи́м·н·ая** pressure/compression spring

~, **от·тя́г·ивающ·ая** antagonistic/restoring spring

~, **про·дóль·н·ая** axial spring

~ **рас·тяж·éни·я** tension spring

~ **с·жа́т·и·я** compression spring

~, **спира́ль·н·ая** spiral spring

пруж·и́н·а
~, таре́ль·чат·ая　Belleville spring washer
~, уравно·ве́ш·ивающ·ая　compensating/balancing spring
пруж·и́нист·ый, -ая, -ое, (*a*) springy
пружи́н·ить, -ят, (*imp*) be springy, work (of springs etc.)
пруж·и́нк·а, -и, (*g pl*) **-нок·,** (*f*) (*dim*) of **пруж·и́н·а;** (bot) elater
пруж·и́нн·ый, -ая, -ое, (*a*) spring; spring-loaded
~ за·стёж·к·а　(*f*) spring catch
~ кольц·о́　(*n*) snap ring; gimbal ring (of compass)
~ регул·я́тор·　(*m*) (autom) spring-loaded governor
прус·, -а, (*m*) (ent) grasshopper, locust, *Calliptamus*
прусти́т·, -а, (*m*) (min) proustite, ruby silver ore
прут·, -а, (*nom pl*) **пру́т·ья,** (*g pl*) **-ьев,** (*m*) rod, stick; (met) rod; **пр·ут,** (*fut 3rd pl*) of **пер·е́ть**
~, и́л·ов·ый　osier, withe
пру́т·ик·, -а, (*m*) rod, stick, twig
прут·ко́в·ый, -ая, -ое, (*a*) of **пру-т·о́к·**
прут·о́к·, -т·к·а́, (*m*) (met) rod, bar; billet (continuous casting)
~, кру́г·л·ый　(met) round, round bar
пру́т·ья, (*nom/acc pl*) of **прут·**
прут·ян·о́й, -а́я, -о́е, (*a*) wicker
пры́г·а·ть, -ют, (*imp*) leap, jump, spring
пры́г·ающ·ий, -ая, -ее, (*pres part act*) of **пры́г·а·ть;** (zool) saltatorial, saltatory
пры́г·н·уть, -ут, (*perf*) see **пры́г·а·ть**
прыг·у́н·, -а́, (*m*) (zool) spring buck, *Antidorcas marsupialis*; (fish) mud-skipper, mud-springer, *Periophthalmidae*
прыг·у́нчик·и, (*g pl*) **-ов,** (*m pl*) (zool) *Macroscelididae*
прыж·о́к·, -ж·к·а́, (*m*) leap, jump, spring
пры́ск·а·ть, -ют, (*imp*) sprinkle; spurt
пры́с·н·уть, -ут, (*perf*) see **пры́с-к·а·ть**
пры́т·к·ий, -ая, -ое, (*a*) lively, alert
прыщ·, -а́, (*i*) **-о́м,** (*m*) pimple, (med) pustule
пряд·е́ни·е, -я, (*n*) (text) spinning
~, аппара́т·н·ое　condenser spinning (synthetics); flyer spinning (coarse thread)

пряд·е́ни·е
~, греб·ённ·ое　worsted manufacture
~, грубо·греб·ённ·ое　Bradford spinning
~, камво́ль·н·ое　worsted
~, суко́н·н·ое　flyer spinning
~, та́з·ов·ое　pot spinning
~, тонко·греб·ённ·ое　French spinning
~, филье́р·н·ое　funnel spinning
пря́д·енн·ый, -ая, -ое, (*past part pass*) of **пря́·сть**
пря́д·ен·ый, -ая, -ое, (*a*) (text) spun
~ асбе́ст·　(*m*) asbestos thread, spun asbestos
пряд·ёт, (*pres 3rd sing*) of **пря́·сть**
пряд·и́льн·ый, -ая, -ое, (*a*) (text) spinning
~ маши́н·а　(*f*) (text) spinning machine/frame
~ машина период·и́ческ·ого ле́й·-стви·я　(*f*) (text) mule
~ на·со́с·ник·　(*m*) spinning pump (synthetic fibres)
пряд·и́льщик·, -а, (*m*) (text) spinner
пряд·у́щ·ий, -ая, -ее, (*pres part act*) of **пря́·сть**
пря́д·ш·ий, -ая, -ее, (*past part act*) of **пря́·сть**
пря́д·ь, -и, (*f*) strand (of rope etc.)
пряд·я́, (*pres gerund*) of **пря́·сть**
пря́ж·а, -и, (*i*) **-е́й,** (*f*) (text) yarn
~, асбе́ст·ов·ая　spun asbestos
пря́ж·к·а, -и, (*g pl*) **-ж·ек·,** (*f*) clasp, buckle; stage (horse transport)
~, D-обра́з·н·ая　D-ring
пря·л, (*past masc sing*) of **пря́·с·ть**
пря́л·к·а, -и, (*g pl*) **-л·ок·,** (*f*) (text) spinning wheel, distaff
прям·а́я, -о́й, (*f decl as adj*) straight line; leg (of airfield circuit)
на одн·о́й прям·о́й с　in line with
~, образ·у́ющ·ая　(math) rectilinear generator
~, пере·сек·а́ющ·ая　(math) transversal
~ ра́вн·ых выс·о́т·　line of equal altitudes
прям·е́йш·ий, -ая, -ее, (*superlative adj*) straightest, most direct
при́нцип· прям·е́йш·его пут·и́　(*m*) principle of least curvature
прям·изн·а́, -ы́, (*f*) straightness
прям·и́л·о, -а, (*n*) (mech) parallel-motion mechanism, straight-line mechanism

пря́м·о (*adv*) straight, in a straight line, directly; frankly, openly; (*predic adj*) *see* **прям·о́й**

~ **по корм·е** (naut) right astern

прямо·борт·ов·о́й, -а́я, -о́е, (*a*) beadless, straight-sided (of tires)

прямо·воз·бужд·ённ·ый, -ая, -ое, (*a*) directly excited

прямо·волн·ов·о́й, -а́я, -о́е, (*a*) (rad) straight-line-wavelength

прямо·го́н·н·ый, -ая, -ое, (*a*) straight-run (of distilled products)

'прямо·де́й·ствующ·ий, -ая, -ее, (*a*) direct-acting; direct-release (air brakes)

прямо·ём·кости·ый, -ая, -ое, (*a*) (elec) straight-line capacitance

~ **конденс·а́тор·** (*m*) (elec) straight-line capacitor

прямо·за·ви́с·ящ·ий, -ая, -ее, (*a*) directly depending, (instr) direct-relation

прямо·зу́б·ое колес·о́ (*n*) spur-gear wheel

прям·о́й, -а́я, -о́е, (*a*) *see also* **прям·а́я** (*as noun*); straight, direct; (math) right

~ **вос·хожд·е́ни·е** (*n*) (astron) right ascension

~ **движ·е́ни·е** (*n*) (astron) direct motion

~ **на·ка́л·** (*m*) (elec) direct heating; (*gen as adj*) directly heated

~ **со·общ·е́ни·е** (*n*) (rail) through service; (*in gen as adj*) through

~ **со·от·ве́т·стви·е** (*n*) direct correspondence

~ **с·пуск·** (*m*) righthand lay (of rope)

~ **у́гол·** (*m*) right angle

прямо·кры́л·ые, -ых, (*pl decl as adj*) (ent) *Orthoptera*

прямо·лин·е́йн·ый, -ая, -ое, (*a*) straight, straight-line, rectilinear

~ **образ·у́ющ·ая** (*f*) (math) rectilinear generator

~ **шкал·а́** (*f*) slide-rule scale

прямо·на·ка́ль·н·ый, -ая, -ое, (*a*) (elec) directly heated, filament

прямо·пере·горо́д·чат·ый, -ая, -ое, (*a*) rectimurate

прямо·пропорциона́ль·ност·ь, -и, (*f*) direct proportionality

прямо·ре́бер·ност·ь, -и, (*f*) (cryst) straight-line edge feature

прямо·сло́й·н·ый, -ая, -ое, (*a*) straight-grained (timber)

прямо·то́к·, -а, (*m*) (dynam) forward flow

прямо·то́ч·н·ый, -ая, -ое, (*a*) single-flow, uniflow, once through, straight-through, single-pass

~ **воз·ду́ш·н·о-реакти́в·н·ый дви́г·ател·ь** (*m*) ramjet engine

~ **котёл·** (*m*) once-through boiler

прямо·уго́ль·ник·, -а, (*m*) rectangle

прямо·уго́ль·н·ый, -ая, -ое, (*a*) rectangular, right-angled, orthogonal; (rad) square, square-wave

~ **и́мпульс·** (*m*) square pulse

~ **тре·уго́ль·ник·** (*m*) right-angled triangle

прямо·ход·я́щ·ий, -ая, -ое, (*a*) (zool) orthograde

прямо·част·о́тн·ый, -ая, -ое, (*a*) (elec) straight-line-frequency, SLF

пря́н·ик, -а, (*m*) spiced biscuit

пря́н·ост·ь, -и, (*f*) spice

пря́н·ый, -ая, -ое, (*a*) spiced

пряс·ть, (*pres 3rd pl*) **пряд·у́т,** (*past masc sing*) **пря·л,** (*imp*) (text) spin

пря́т·ать, (*pres 3rd pl*) **пря́ч·ут,** (*imp*) conceal, hide, put away

пря́ч·а, (*pres gerund*) of **пря́т·ать**

пря́ч·ущ·ий, -ая, -ее, (*pres part act*) of **пря́т·ать**

псамм·и́т·, -а, (*m*) (geol) psammite, arenite

псамм·и́тов·ый, -ая, -ое, (*a*) (geol) psammitic

псамм·о́м·а, -ы, (*f*) (med) psammoma

псаммо·терап·и́·я, -и, (*f*) (med) psammotherapy

псаммо·фи́т·, -а, (*m*) (bot) psammophyte

псевдо·ве́ктор·, -а, (*m*) (math) pseudovector

псевдо·гла́в·н·ый, -ая, -ое, (*a*) (math) pseudoprincipal

псевдо·глобули́н·, -а, (*m*) (biochem) pseudoglobulin

псевдо·двой·ни́к·, -а́, (*m*) (cryst) pseudotwin

псевдо·зу́хи·и, (*pl*) (pal) pseudosuchians

псевдо·кисл·от·а́, -ы́, (*f*) (chem) pseudo-acid

псевдо·кристалл·и́ческ·ий, -ая, -ое, (*a*) pseudocrystalline

псевдо·ме́р·а, -ы, (*f*) (math) pseudomeasure

псевдо·мер·и́·я, -и, (*f*) (chem) pseudomerism

псевдо·морф·и́зм·, -а,　(*m*) pseudomorphism

псевдо·морф·о́з·а, -ы,　(*f*) (min) pseudomorph

псевдо·ни́м·, -а, (*m*) pseudonym. penname, stage-name

псевдо·о·жиж·ённ·ый, -ая, -ое, (*a*) fluidized

~ сло́·й (*m*) (chem) fluidized bed

псевдо·основ·а́ни·е, -я,　(*n*) (chem) pseudo-base

псевдо·по́д·и·й, -я, (*m*) (zool) pseudopodium

псевдо·по·до́б·н·ый, -ая, -ое,　(*a*) pseudo-similar

псевдо·рас·тво́р·, -а,　(*m*) (chem) pseudosolution

псевдо·скаля́р·, -а, (*m*) (math) pseudoscalar

псевдо·скаля́р·н·ый, -ая, -ое,　(*a*) (math) pseudoscalar

псевдо·сфе́р·а, -ы, (*f*) (math) pseudosphere; pseudospherical surface

псевдо·тахили́т·, -а, (*m*) (min) pseudotachylite

псеф·и́т·, -а, (*m*) (min) psephite

псеф·и́тов·ый, -ая, -ое, (*a*) (geol) psephitic

пси (*indecl*) psi, Ψ (Greek)

псило·мела́н·, -а,　(*m*) (min) psilomelane

псило́т·, -а, (*m*) (bot) psilotum

псилото·ви́д·н·ые, -ых, (*pl decl as adj*) (bot) *Psilopsida*

псило́т·ов·ые, -ых, (*pl decl as adj*) (bot) *Psilotales*

псило·фи́т·, -а, (*m*) (bot) psilophyte

псило·фи́т·ов·ые, -ых, (*pl decl as adj*) (bot) *Psilophytineae*

псило·фито́н·, -а, (*m*) (pl bot) psilophyton

пситтако́·з·, -а, (*m*) (med) psittacosis

пситтако·за́вр·, -а, (*m*) (pal) *Psittacosaurus*

псих·иа́тр·, -а, (*m*) psychiatrist

псих·иатр·и́ческ·ий, -ая, -ое, psychiatric

псих·иатр·и́·я, -и, (*f*) (med) psychiatry

пси́х·ик·а, -и, (*f*) (med) psychics

псих·и́ческ·ий, -ая, -ое, (*a*) (med) psychic

психо·ана́лиз·, -а, (*m*) psychoanalysis

психо·гальван·и́ческ·ий, -ая, -се, (*a*) (zool) psychogalvanic

~ рефле́кс· (*m*) psychogalvanic response

псих·о́з·, -а, (*m*) (med) psychosis

психо́·лог·, -а, (*m*) psychologist

психо·метр·и́ческ·ий, -ая, -ое, (*a*) psychometric

психо·мото́р·н·ый, -ая, -ое, (*a*) (zool) psychomotor

психо·сома́т·ик·а, -и, (*f*) psychosomatic medicine

психо·те́хн·ик·а, -и, (*f*) psychotechnics

психо·фармако·ло́г-и·я, -и, (*f*) psychopharmacology

психо·физ·и́ческ·ий, -ая, -ое, (*a*) psychophysical

психро́·метр·, -а, (*m*) (meteor) psychrometer, wet-and-dry-bulb hygrometer

~, аспирацио́н·н·ый (instr) aspirated psychrometer, Assmann, psychrometer

~, пращ·ев·о́й sling psychrometer

психро·фи́л·, -а, (*m*) (bot) psychrophil

психро·фи́т·, -а, (*m*) (bot) psychrophyte

псор·иа́з·, -а, (*m*) (med) psoriasis

псор·о́зис·, -а, (*m*) psorosis (plant disease)

псофо́·метр·, -а, (*m*) (telecom) psophometer

пт·ене́ц·, -нц·а́, (*i*) -нц·о́м, (*m*) (orn) nestling

птер·анодо́н·, -а, (*m*) (pal) *Pteranodon*

птеридо·спе́рм·ы, (*f pl*) (bot) *Pteridospermae*

птеридо·фи́т·, -а, (*m*) (bot) pteridophyte

птеро·да́ктил·ь, -я, (*m*) (pal) *Pterodactyl*

птеро·за́вр·, -а, (*m*) (pal) pterosaur, *Pterosaurus*

птеро·по́д·, -а, (*m*) (zool) pteropod, pteropodan

птеро·по́д·ов·ый, -ая, -ое, (*a*) (zool) pteropod; (*pl as noun*) pteropods, *Pteropodae*

птер·опси́д·а, -ы, (*f*) (bot) *Pteropsida*, (*pl*) *Pteropsidae*

птиал·и́зм·, -а, (*m*) (med) ptyalism

птиал·и́н·, -а, (*m*) (biochem) ptyalin

птигмат·и́т·, -а, (*m*) (min) ptymatite

птило·ли́т·, -а, (*m*) (min) ptilolite

птихи·ге́н·н·ый, -ая, -ое, (*a*) ptychigenic

птихо·пари́д·н·ый, -ая, -ое, (*a*) ptychoparioid

пти́ц·а, -ы, (*i*) -ей, (*f*) (orn) bird, *Aves*; poultry

птице·бо́й·н·я, -и, (*f*) poultry dressing plant

птице·во́д·, -а,   (*m*) poultry farmer/breeder

птице·во́д·ств·о, -а,   (*n*) poultry farming/breeding; aviculture

птице·е́д·ы, -ов,   (*pl*) (ent) *Mygalomorphae*

птице·зве́р·и, -е́й,   (*m pl*) (zool) monotremes, *Monotremata*

птице·но́г·ие, -их,   (*pl decl as adj*) (pal) ornithopods, *Ornithopoda*

птице·обра́з·н·ый, -ая, -ое,   (*a*) bird-like

птице·та́з·ов·ые, -ых,   (*pl decl as adj*) (pal) *Ornithischia*

пти́ч·ий, -ая, -ее,   (*a*) bird, avian; poultry

пти́ч·к·а, -и,   (*g pl*) **-ч·ек·,** (*f*) (*dim*) of **пти́ц·а;** (print) tick

пто́з·, -а,   (*m*) (med) ptosis

пто́зис·, -а,   (*m*) (med) ptosis

птома́·ин·, -а,   (*m*) (chem) ptomaine

пуа́з·, -а,   (*m*) poise (unit of absolute viscosity)

Пуазёйл·я, за·ко́н·   (*m*) Poiseuille's equation (of viscosity)

пуазо́,   (*abbrev*) of **при·бо́р· у·правл·е́ни·я зени́т·н·ым огн·ём** anti-aircraft director

Пуанкаре́, теоре́м·а   (*f*) (math) Poincaré theorem

пуансо́н·, -а,   (*m*) die, punch (of stamps, presses etc.)

Пуассо́н·а, коэффицие́нт·   (*m*) (phys) Poisson's ratio

~, по·сто·я́нн·ая   (*f*) (phys) Poisson's ratio

пу́блик·а, -и,   (*f*) public; audience

публик·ова́ть, -у́ют,   (*imp*) publicize, publish, make public

публи́ч·н·ый, -ая, -ое,   (*a*) public

пу́г·ал·о, -а,   (*n*) scarecrow, bird guard

пуг·а́·ть, -ют,   (*imp*) frighten, scare; threaten; **-ся** (+ *gen*) be frightened/scared of

пуг·н·у́ть, -у́т,   (*perf*) frighten, scare, give a fright/scare

пу́гов·иц·а, -ы,   (*i*) **-ей,** (*f*)   button

пу́гов·ичн·ый, -ая, -ое, (*a*) of **пу́гов·иц·а**

~ маши́н·а   (*f*) button-sewing machine

пу́гов·к·а, -и,   (*g pl*) **-в·ок·,** (*f*) button, stud

пуд·, -а,   (*nom pl*) **-ы́,** (*g pl*) **-о́в,** (*m*) pood (16·38 kg, 36 lb)

пу́дди́нг·, -а,   (*m*) (geol) pudding stone, conglomerate

пу́динг·, -а,   (*m*) (food) pudding

пудлинг·ова́ни·е, -я,   (*n*) (met) puddling

пу́длинг·ов·ый, -ая, -ое,   (*a*) (met) puddling, puddled

~ печ·ь   (*f*) puddling furnace

~ желе́з·о   (*n*) puddled iron

пуд·ов·о́й, -а́я, -о́е,   (*a*) of **пуд·**

пу́др·а, -ы,   (*f*) powder

~, мра́мор·н·ая   marble flour/dust

пудр·и́лк·а, -и,   (*g pl*) **-лок·,** (*f*) duster, powdering/dusting unit

пу́др·ить, -ят,   (*imp*) powder

пуз·ано́к·, -нк·а́,   (*m*) (fish) shad, *Alosa*

пуз·а́т·ый, -ая, -ое,   (*a*) (biol) ventricose

пузыре·ви́д·н·ый, -ая, -ое, (*a*) bubble-like; bladder-like, vesicular

пузыр·ёк·, -рь·к·а́,   (*m*) (*dim*) of **пузы́р·ь;** bleb (in glass); (biol) vesicle; phial, vial

пузыре·но́г·ие, -их,   (*pl decl as adj*) (ent) bean thrips, *Thysanoptera*

пузы́р·ист·ый -ая, -ое,   (*a*) = **пузы́р·чат·ый**

пузы́р·ник·, -а,   (*m*) (bot) bladder senna, *Colutea*

пузыр·ча́тк·а, -и,   (*g pl*) **-ток·,** (*f*) (bot) bladderwort, *Utricularia*; (med) pemphigus

пузыр·ча́тков·ые, -ых,   (*pl decl as adj*) (bot) *Utriculariaceae*

пузы́р·чатост·ь, -и,   (*f*) vesiculation, blistered/blebby condition; (med) pustule

пузы́р·чат·ый, -ая, -ое,   (*a*) blistered, bubbly, vesicular, blebby

пузы́р·ь, -я́,   (*m*) bubble, blister, bleb; (anat) bl..dder, vesica

~, га́з·ов·ый   (met) blowhole, gas cavity

~, мор·ск·о́й   (pal) cystoid, (*pl*) cystoids, *Cystoidea*

~, моч·ев·о́й   (anat) urinary bladder

~, под·ко́р·ков·ый   (met) subcutaneous blowhole

пузырь·ко́в·ый, -ая, -ое,   (*a*) of **пузыр·ёк·, пузы́р·ь**

~ ка́мер·а   (*f*) (nucl) bubble chamber

пук·, -а,   (*nom pl*) **-и́,** (*g pl*) **-о́в,** (*m*) bunch, bundle

пул·ев·о́й, -а́я, -о́е,   (*a*) of **пу́л·я**

~ шрапне́ль   (*f*) shrapnel

пуле·мёт·, -а,   (*m*) machine gun

пуле·мёт·н·ый, -ая, -ое,   (*a*) of **пуле·мёт·**

~ ле́нт·а   (*f*) cartridge belt (of machine gun)

пуллороз·, -а, (*m*) (vet) pullorum infection, bacillary white diarrhoea

пульвериз·átор·, -а, (*m*) spray, spray gun

пульвериз·áци·я, -и, (*f*) spray-painting

пульвериз·úр·овать, -уют, (*imp and perf*) spray

пульмон·úт·, -а, (*m*) (med) pulmonitis

пульмотóр·, -а, (*m*) pulmotor

пýль·н·ый, -ая, -ое, (*a*) of **пýл·я**

~ стрел·ьб·á (*f*) (mil) aiming rifle fire

пýльп·а, -ы, (*f*) pulp

пульпúт·, -а, (*m*) (med) pulpitis

пульпо·вúд·н·ый, -ая, -ое, (*a*) pulpy

пульпо·вóд·, -а, (*m*) pulp line; spoil/ delivery pipeline (dredges etc.)

пульпо·мéр·, -а, (*m*) pulp density meter

пýльс·, -а, (*m*) pulse

~, о·пóр·н·ый clock pulse (computers)

~ -схéм·а (*f*) (autom) relay-type pulse-generating circuit

пульс·átor·, -а, (*m*) pulsator

пульсátor·н·ый, -ая, -ое, (*a*) of **пульс·átor·**

~ машúн·а (*f*) (met) push-pull fatigue testing machine

пульс·áци·я, -и, (*f*) pulsation, pulse; (elec) ripple

~, коллéктор·н·ая (elec) commutator ripple

коэффициéнт пульс·áци·и (*m*) ripple ratio

~ пере·нóс·а carry ripple (computers)

част·от·á пульс·áци·й (*f*) ripple frequency

пульс·úр·овать, -уют, (*imp*) pulse, pulsate

пульс·úрующ·ий, -ая, -ее, (*pres part act*) of **пульс·úр·овать;** pulsed, pulse; (elec) ripple, rippling; (autom) fluttering

~ воз·дýш·н·о-реактúв·н·ый двú·г·атель (*m*) pulse jet engine

~ режúм· (*m*) (autom) fluttering state

~ ток· (*m*) (elec) ripple current

пýльс·ов·ый, -ая, -ое, (*a*) of **пýльс·**

~ волн·á (*f*) (med) pulse

пульсó·метр·, -а, (*m*) pulsometer pump; engine indicator

пýльт·, -а, (*m*) desk, stand, panel

~, да·ющ·ий data panel

~ у·правл·éни·я control desk, remote control panel, instrumentation panel, console

пýл·я, -и, (*f*) bullet

пумúт·, -а, (*m*) (min) pumice

пумицúт·, -а, (*m*) (min) pumicite

пýнкт·, -а, (*m*) *see also under associated words*; point, place; post, station; item, paragraph; (print) point

~ маршрýт·а, ис·хóд·н·ый (nav) departure point

~ маршрýт·а, кон·éчн·ый (nav) destination point

~ маршрýт·а, крит·úческ·ий point of no return, critical point

~ сýд·н·а ship's position (on chart)

пунктáт·, -а, (*m*) (med) puncture

пунктúр·, -а, (*m*) (print) dashed/dotted line

пунктúр·н·ый, -ая, -ое, (*a*) (print) dotted, dashed

пунктúр·овать, -уют, (*imp*) dot; punctuate

пунктуáль·н·ый, -ая, -ое, (*a*) punctilious

пункту·áци·я, -и, (*f*) (gram) punctuation

пýнкци·я, -и, (*f*) (med) puncture

пунц·óв·ый, -ая, -ое, (*a*) bright red

пýнш·, -а, (*i*) **-ем,** (*m*) cup, wine cup (a drink)

пунш·úр·овать, -уют, (*imp and perf*) punch (telegraph tape)

пуп·, -á, (*m*) navel, (anat) umbilicus

пуп·áвк·а, -и, (*g pl*) **-вок·,** (*f*) (bot) camomile, *Anthemis*

пупилло·скоп·ú·я, -и, (*f*) (med) pupilloscopy

пупилло·тон·ú·я, -и, (*f*) (med) pupillotonia

пупиниз·áци·я, -и, (*f*) (elec) coil loading

пупиниз·úрованн·ый, -ая, -ое, (*a*) (elec) coil-loaded

пупúн·овск·ий, -ая, -ое, (*a*) (elec) Pupin loading

пуп·овúн·а, -ы, (*f*) (anat) umbilical cord

пуп·óк·, -п·к·á, (*m*) *see* **пуп·**

пупочно·кúш·ечн·ый, -ая, -ое, (*a*) (anat) omphaloenteric

пуп·óчн·ый, -ая, -ое, (*a*) umbilical

пург·á, -ú, (*f*) snowstorm

пурúн·, -а, (*m*) (biochem) purine

пурúн·ов·ый, -ая, -ое, (*a*) of **пурúн·**

пýрпур·, -а, (*m*) purple

пýрпур·а, -ы, (*f*) (med) purpura

пурпур·úн·, -а, (*m*) purpurine (dye)

пурпýр·н·ый, -ая, -ое, (*a*) purple

пурпýр·ов·ый, -ая, -ое, (*a*) purple; (chem) purpuric

~ кисл·от·á (*f*) purpuric acid

пýск·, -а, (*m*) (*v n*) *see* **пуск·á·ть;** start-up; (rocket) launching, firing

пу́ск

~, подо·гре́в·н·ый heater starting (fluorescent lamp)

пуск·а́тел·ь, -я, (*m*) starter

пуск·а́·ть, -ют, (*imp*) let, allow, have (something done); release, start, start up, set (in motion), put (into operation/service), turn on (gas, water etc.)

пуск·ов·о́й, -а́я, -о́е, (*a*) starting, actuating; (rocket) launching; (rad) trigger, triggering

~ кла́пан· (*m*) (ICE) starting valve

~ моме́нт· (*m*) (elec) starting torque

~ на·со́с· (*m*) priming pump

~ площа́д·к·а (*f*) (rocket) launching platform

~ схе́м·а (*f*) (elec) trigger circuit

~ характер·и́стик·а (*f*) control characteristic

~ элеме́нт· (*m*) (autom) operating element

пуссбе́р·а, -ы, (*f*) *see* **пусье́р·а**

пуст·е́·ть, -ют, (*imp*) empty, become empty

пуст·и́ть, -ят, .(*perf*) *see* **пуск·а́·ть**

пуст·о́й, -а́я, -о́е, (*a*) empty; trivial; (geol) barren

пуст·от·а́, -ы́, (*f*) emptiness, void, hollow, hollowness; vacuum

~ вы·мыв·а́ни·я (geol) wash-out cavity

~ пере·кристалл·иза́ци·и (cryst) recrystallization interstice

~, супер·капилля́р·н·ая (cryst) supercapillary interstice

~, у·са́д·очн·ая (met) shrinkage cavity, piping

пусто·те́л·ост·ь, -и, (*f*) hollowness, void **коэффицие́нт· пусто·те́л·ост·и** (*m*) (nucl) void coefficient

пусто·те́л·ый, -ая, -ое, (*a*) hollow

пуст·о́тност·ь, -и, (*f*) *see* **пусто·те́л·ост·ь**

пуст·о́тн·ый, -ая, -ое, (*a*) vacuum

~ при·бо́р· (*m*) (elec) vacuum tube

~ фото·элеме́нт· (*m*) vacuum cell

пусто·цве́т·, -а, (*m*) (bot) unfertile flower; unfertilized flower

пу́ст·ош·ь, -и, (*f*) waste land/ground

пу́стул·а, -ы, (*f*) (biol) pustule

пустул·ёзн·ый, -ая, -ое, (*a*) (biol) pustular

пуст·ы́нн·ый, -ая, -ое, (*a*) desert, waste

пуст·ы́н·я, -и, (*f*) desert, waste

пуст·ы́р·ь, -я́, (*m*) waste land/ground

пусть (*particle*) let; (math etc.) let, suppose

пусье́р·а, -ы, (*m*) (met) blue metal/ powder, zinc dust/fume

пу́т·аниц·а, -ы, (*f*) confusion

пу́т·анн·о-волокн·и́ст·ый, -ая, -ое, (*a*) matted-fibre

пу́т·а·ть, -ют, (*imp*) confuse, muddle; tangle; hobble (horses etc.)

пут·ёвк·а, -и, (*g pl*) **-вок·,** (*f*) pass (document); (M/T) work ticket, routing instructions; catalogue slip (libraries)

путе·вод·и́тел·ь, -я, (*m*) guide, guidebook; itinerary, made-up route

пут·ев·о́й, -а́я, -о́е, (*a*) travel; (rail) track, (*m decl as adj*) (rail) trackman **ве́ктор· пут·ев·о́й ско́р·ост·и** (*m*) (air) flight-path vector

~ ка́рт·а (*f*) (air) track chart, navigational chart

~ ко́мпас· (*m*) (air) steering compass

~ лист· (*m*) (M/T) work ticket

~ об·хо́д·чик· (*m*) (rail) lengthman

~ ско́р·ост·ь (*f*) (air) ground speed

~ струг· (*m*) (civ eng) grader

~ хозя́й·ств·о (*n*) (rail) track facilities

путе·из·мер·и́тел·ь, -я, (*m*) (rail) track recording machine

пут·ём (*adv*) *see also* **пут·ь;** by means of, by (*followed by verbal noun*); via

путе·пере·двиг·а́тел·ь, -я, (*m*) (rail) trackshifter

путе·подъ·ём·ник·, -а, (*m*) (rail) track lifter

путе·про·во́д·, -а, (*m*) (rail) overbridge, fly-over

путе·у·кла́д·чик·, -а, (*m*) (rail) rail/ track layer

путе·ше́ств·и·е, -я, (*n*) journey, voyage

путе·ше́ств·овать, -уют, (*imp*) travel; explore

-пут·н·ый, -ая, -ое, (*adj component*) -track

путресци́н·, -а, (*m*) (chem) putrescine

путре·фа́кци·я, -и, (*f*) (med) putrefaction

пут·ь, -и́, (*i*) **-ём,** (*m*) way, track, path, route; way, method; journey, voyage; (anat) tract, duct; (rail) permanent way, track

~, во́д·н·ый inland waterway **ли́н·и·я пут·и́** (*f*) (naut) course made good; (air) track

~ мане́вр·а (*mil*) tactical route

~, об·го́н·н·ый (rail) passing track/ siding/loop

~, обра́т·н·ый (air) reverse track

пут·ь
~, океа́н·ск·ий shipping lane/route
~ орби́т·а orbital path
~ пере·мещ·éни·я (dynam) mixing/ mixture length
~, подъ·éзд·н·ый (rail) belt line
~, по·пере́ч·н·ый lateral road
~, по·тéр·янн·ый (horol) dead path
~ раз·ря́д·а discharge path
~, ре́льс·ов·ый rail track, rails
~, рока́д·н·ый lateral road
  тео́ри·я пут·и́ пере·мéш·ивани·я (f) (dynam) mixture-length theory
~ тормож·éни·я braking distance
~, траве́рс·н·ый (shipb) transfer track
~ у·множ·éни·я by multiplying
пух·, -а, (m) down (feathers), fuzz, fluff
~, хлóп·ков·ый linter
пухери́т·, -а, (m) (min) pucherite
пух·л·ый, -ая, -ое, (a) bulky, bulgy
пух·нуть, -нут, (imp) swell
пухо·от·дел·éни·е, -я, (n) linting (of cotton)
пухо·от·дел·и́тел·ь, -я, (m) (text) linting machine
пуццола́н·, -а, (m) (geol) pozzuolana, puzzolana
пуче·гла́з·и·е, -я, (n) (med) exophthalmos
пуч·ени·е, -я, (n) distention, swelling
пуч·и́н·а, -ы, (f) abyss, chasm; lift, swell (of roads etc. due to frost)
пуч·ить, -ат, (imp) distend; -ся (pass); swell, rise, heave
пуч·ко́вани·е, -я, (n) bundling, bunching, clustering
пуч·ко́ванн·ый, -ая, -ое, (a) bunched
пуч·кова́т·ый, -ая, -ое, (a) bunched, bundled, clustered; (bot) fasciculate, (zool) nodular
пуч·ко́в·ый, -ая, -ое, (a) of пуч·ок· (q.v.); fascicular
~ армату́р·а (f) wire-rod bundle reinforcement (of concrete)
~ ка́бел·ь (m) bunched cable
пу́ч·ност·ь, -и, (f) (phys) antinode, loop
~ волн·ы́ (rad) oscillation loop
~ давл·éни·я pressure antinode
пуч·ók·, -ч·к·а́, (m) bundle, bunch, cluster; (math) pencil, family, bundle; (phys) beam; (telecom) grading; (zool) fascile, fasciculus, bundle, nodule; (bot) fascile
~, вы́·вед·енн·ый (phys) extracted/ external beam
~, вы́·вод·и́ть reject/extract a beam
~, зá·пер·т·ый cutoff beam

пуч·ók·
~, коллатера́ль·н·ый (bot) collateral bundle
~, концентр·и́ческ·ий (bot) concentric vascular bundle
~ луч·éй (phys) beam, pencil of rays
~, молекуля́р·н·ый molecular beam
~, не·равно·до·сту́п·н·ый (telecom) unsymmetrical grading
~, одно·ро́д·н·ый uniform beam
~ о·кру́ж·ност·ей (math) pencil of circles
~, пло́ск·ий flat beam
~ пло́ск·ост·ей (math) pencil of planes
~, про·вод·я́щ·ий (bot) conducting bundle
~ прям·ы́х (math) family/pencil of straight lines
~, равно·сту́п·н·ый (telecom) symmetrical grading
~, радиа́ль·н·ый (bot) radial vascular bundle
~, свет·ов·о́й light beam, pencil/beam of light
~ с·во́д·а (build) crown/apex of roof
~, со·су́д·ист·о·волокн·и́ст·ый (bot) vascular bundle
~, со·су́д·н·ый (bot) conducting bundle
~ сфер· (math) pencil of spheres
~ труб· tube bundle/bank/nest
~, у́з·к·ий (phys) pencil, pencil beam
пуш·ён·ый, -ая, -ое, (a) fluffed up, fiberized
~ асбéст· (m) fiberized asbestos
пу́ш·ечн·ый, -ая, -ое, (a) gun, cannon
~ брóнз·а (f) gunmetal
пуш·и́нк·а, -и, (g pl) -нок·, (f) flock, fluff
пуш·и́ст·ый, -ая, -ое, (a) downy, fluffy, plumose
пу́ш·к·а, -и, (g pl) -ш·ек·, (f) gun, cannon
~, авиацио́н·н·ая aircraft cannon
~, гарпу́н·н·ая harpoon gun
~, лине·мет·а́тельн·ая line-throwing gun
~, цемéнт·н·ая cement gun
~, электро́н·н·о·луч·ев·а́я electron gun, electron-beam gun (of CRT)
пуш·ни́н·а, -ы, (f) furs, fur-skins, pelts
пуш·н·о́й, -а́я, -о́е, (a) fur; fur-bearing
пуш·о́к·, -ш·к·а́, (m) fluff, flue, down; (bot) bloom, pubescence
пуш·о́нк·а, -и, (g pl) -нок·, (f) slaked lime

пушпу́ль·н·ый, -ая, -ое, (*a*) (elec) push–pull

~ у·сил·и́тел·ь (*m*) (telecom) push–pull amplifier

пущ·а, -и, (*i*) -е́й, (*f*) dense forest

пу́щ·енн·ый, -ая, -ое, (*past part pass*) of пуст·и́ть, (*perf*); *see* пуск·а́·ть (*imp*)

пф, (*abbrev*) = пико·фара́д·а micro-microfarad

пфа́фф·ов·а фо́рм·а (*f*) (math) Pfaffian differential form

пчел·а́, -ы́, (*nom pl*) пчёл·ы, (*g pl*) пчёл· (*f*) (ent) bee, honeybee

пчел·и́н·ый, -ая, -ое, (*a*) bees', bee; (*pl decl as adj*) (ent) honeybee family, *Apidae*

пчело·во́д·, -а, (*m*) beekeeper, apiarist

пчело·во́д·ств·о, -а, (*n*) apiculture, bee-keeping

пчело·е́д·, -а, (*m*) (zool) honey buzzard, *Pernis apivorus*

пчёль·ник·, -а, (*m*) (agr) apiary

пшен·и́ц·а, -ы, (*i*) -е́й, (*f*) (bot) wheat, *Triticum*

~, англ·и́йск·ая turgid wheat, *Triticum turgidum*

~, ветв·и́ст·ая ramified wheat

~, мя́гк·ая soft wheat, *Triticum vulgare*

~, по́ль·ск·ая Polish wheat, *Triticum polonium*

~ Спе́льта Spelt wheat, *Triticum Spelta*

~, твёрд·ая hard wheat, *Triticum durum*

пшен·и́чн·ый, -ая, -ое, (*a*) wheat

пшен·о́, -а́, (*n*) hulled millet grain

~ -дра́н·ец·, -н·ц·а, (*m*) whole millet grain

~, дробл·ён·ое ground millet grain

пыж·, -а́, (*i*) -о́м, (*m*) wad (in gun); bow (of river vessel)

пыж·а́ми (*adv*) double-line-ahead, in double line ahead

пыл·а́·ть, -ют, (*imp*) blaze, flame, be on fire

пыле·ви́д·н·ый, -ая, -ое, (*a*) powdered, pulverized, pulverulant

~ то́пл·ив·о (*n*) pulverized fuel/coal

пыле·за·щищ·ённ·ый, -ая, -ое, (*a*) dustproof

пыле·ло́в·к·а, -и, (*g pl*) -в·ок·, (*f*) dust catcher/trap

пыле·ме́р·, -а, (*m*) (meteor) dust meter/counter

пыле·не·про·ниц·а́ем·ый, -ая, -ое, (*a*) dustproof, dust-tight

пыле·о·са́д·очн·ая ка́мер·а (*f*) dust collector

пыле·от·дел·и́тел·ь, -я, (*m*) dust collector/separator

~, инерцио́н·н·ый inertial dust collector

пыле·со́с·, -а, (*m*) vacuum cleaner

пыле·у́голь·н·ый, -ая, -ое, (*a*) pulverized-coal

пыле·у·лов·и́тел·ь, -я, (*m*) dust-catcher, dust trap

~, жалюзи́й·н·ый baffle separator (dust catcher)

пыл·ец·, -лц·а, (*m*) dust particle

пыл·и́нк·а, -и, (*g pl*) -нок·, (*f*) speck of dust, dust particle; (bot) microspore

пыл·и́ть, -я́т, (*imp*) raise the dust, leave a trail of dust

пы́л·к·ий, -ая, -ое, (*a*) ardent

пы́л·ь, -и, (*f*) dust; spray

~, радио·акти́в·н·ая radioactive dust

~, цвет·о́чн·ая (bot) pollen farina

пыль·ник·, -а, (*m*) (bot) anther

пыльнико·но́с·н·ый, -ая, -ое, (*a*) (bot) antheriferous

пы́ль·н·ый, -ая, -ое, (*a*) dust, dusty

пыль·ц·а́, -ы́, (*i*) -о́й, (*f*) (bot) pollen

~, без·борозд·н·а́я acolpate pollen

~, жизне·спосо́б·н·ая vital pollen

пыль·цев·о́й, -а́я, -о́е, (*a*) (bot) pollen

пыльце·в·хо́д·, -а, (*m*) (bot) pollen tube

пыльце·но́с·н·ый, -ая, -ое, (*a*) (bot) polliniferous, pollen-bearing

пыльце·у·лов·и́тел·ь, -я, (*m*) (bot) pollen trap

пыре́·й, -я, (*m*) (bot) quack grass, *Agropyrum*

пыт·а́вш·ий, -ая, -ое, (*past part act*) of пыт·а́·ть

~ от·жиг·ом annealed, after annealing

пыт·а́·ть, -ют, (*imp*) subject to, try, try out; torture; -ся endeavour, attempt, try

пыт·ли́в·ый, -ая, -ое, (*a*) inquisitive, enquiring

пыхт·е́ть, -я́т, (*imp*) puff

пы́ш·н·ый, -ая, -ое, (*a*) luxuriant; fluffy

пьедеста́л·, -а, (*m*) pedestal

пьедмо́нт·, -а, (*m*) (geol) piedmont, piedmont plain

пьезо·дви́г·ател·ь, -я, (*m*) piezo-electric motor

пьезо·диффу́зи·я, -и, *(f)* pressure diffusion

пьезо́ид·, -а, *(m)* (elec, cryst) piezoid

~, пре·образ·у́ющ·ий (elec) transducing piezoid

пьезо·изоба́т·а, -ы, *(f)* (geophys) piezoisobath

пьезо·ква́рц·, -а, *(i)* **-ем,** *(m)* piezoelectric crystal

пьезо·ква́рц·ев·ый, -ая, -ое, *(a)* (geol) piezoquartz; (elec) piezoelectric crystal

пьезо́·метр·, -а, *(m)* (phys) piezometer

пьезо·электр·и́честв·о, -а, *(m)* piezoelectricity

пьезо·элеме́нт·, -а, *(m)* piezoelectric element; pressure-sensitive element

пьезо·хи́м·и·я, -и, *(f)* piezochemistry

пьезо·яче́й·к·а, -и, *(g pl)* **-чеек,** *(f)* piezo cell

пьекс·ы *(nom pl)*, *(g pl)* **пьекс·,** *(f)* ski-boots

пьес·а, -ы, *(f)* (theat) play; piece (of music)

пьемонти́т·, -а, *(m)* (min) piedmontite

пь·ёт, *(pres 3rd sing)* of **пи·ть** *(imp)*

пь·ю́щ·ий, -ая, -ее, *(pres part act)* of **пи·ть**

пья́н·иц·а, -ы, *(i)* **-ей,** *(m and f)* (med) alcoholic, drunkard; (bot) blueberry, *Vaccinium uliginosum*

пья́н·ый, -ая, -ое, *(a)* drunk, intoxicated

пюй *(m indecl)* (geol) pug

пюре́ *(n indecl)* (food) puree

~, карто́фель·н·ое mashed potatoes

пяде́ниц·ы, *(nom pl)* *(f)* (ent) measuring worm moths, *Geometridae*

пя́льц·ы *(nom pl)*, *(g pl)* **-лец·,** *(f)* embroidery/lace frame

пя́ртнерс·, -а, *(m)* (shipb) partner plate

пяст·ь, -и, *(f)* shank, (anat) metacarpus

пят·а́, -ы́, *(nom pl)* **пя́т·ы,** *(g pl)* **пят·** *(f)* heel; (mech etc.) heel, foot, pivot, vertical journal; springer, springing (of arch)

~, гидравл·и́ческ·ая hydraulic-balancing disc

~ рул·я́ (shipb) pintle bearing

~, сфер·и́ческ·ая footstep pivot

~, шар·ов·а́я ball pivot

~ я́кор·я (naut) anchor crown

пят·ери́к·, -а́, *(m)* fiver (containing five); five stranded thread; packet of five

пят·ери́чн·ый, -ая, -ое, *(a)* (math) quinary

пят·ёрк·а, -и, *(g pl)* **-рок·,** *(f)* a five, figure five, number five, set of five; excellent (as marking)

пят·ерни́к·, -а, *(m)* quintuplet, fiveling

пя́т·ер·о, -ы́х, *(d)* **-ы́м,** *(n)* five, five together

пят·ер·ы́х, *(gen/prep/acc anim)* of **пя́т·ер·о**

пяти- *(root)* five, penta-

пяти·а́том·н·ый, -ая, -ое, *(a)* pentatomic

~ основ·а́ни·е *(n)* (chem) pentacid base

пяти·вале́нт·н·ый, -ая, -ое, *(a)* (chem) pentavalent

пяти·во́д·н·ый, -ая, -ое, *(a)* pentahydrate of

пяти·гра́н·ник·, -а, *(m)* (math) pentahedron

пяти·гра́н·н·ый, -ая, -ое, *(a)* pentahedral

пяти·десяти- *(component)* fifty-

пяти́·десят·и *(gen/dat/prep)* of **пять·-десят·**

пяти·десяти·ле́т·и·е, -я, *(n)* fifty years; fiftieth anniversary

пяти·десяти·ле́т·н·ий, -яя, -ее, *(a)* fifty-year; fifty-year-old

пяти·деся́т·ый, -ая, -ое, *(a)* fiftieth

~ го́д·ы *(pl)* the fifties (decade)

пяти·дн·е́вк·а, -и, *(g pl)* **-вок·,** *(f)* pentad, five-day period

пяти·дн·е́вн·ый, -ая, -ое, *(a)* five-day, pentad

пяти·зве́н·н·ый, -ая, -ое, *(a)* five-member/unit

~ полиме́р· *(m)* (chem) pentamer

пяти·зна́ч·н·ый, -ая, -ое, *(a)* (math) five-digit/-unit

пяти·километр·о́вк·а, -и, *(g pl)* **-вок·,** *(f)* 1/5000000 scale map

пяти·кон·е́чн·ый, -ая, -ое, *(a)* five-pointed, pentagonal

пяти·кра́т·н·ый, -ая, -ое, *(a)* five-fold, quintuple

пяти·лепест·ко́в·ый, -ая, -ое, *(a)* (bot) pentapetalous, five-petalled

пяти·ле́т·к·а, -и, *(g pl)* **-ток·,** *(f)* five-year plan

пяти·ле́т·и·е, -я, *(n)* five years, five-year period

пяти·ле́т·н·ий, -яя, -ее, *(a)* five-year; five-year-old

пяти·ли́ст·н·ый, -ая, -ое, *(a)* five-leaved, (bot) quinquefoliate

пяти·лóпаст·н·ый, -ая, -ое, (*a*) five-lobed, pentalobate; five-bladed/vaned

пяти·луч·úст·ый, -ая, -ое, (*a*) five-rayed/branched, pentactinal

пяти·мéсяч·н·ый, -ая, -ое, (*a*) five-month; five-month-old

пяти·недéль·н·ый, -ая, -ое, (*a*) five-week; five-week-old

пяти·ó·кис·ь, -и, (*f*) (chem) pentoxide

пяти·оснóв·н·ый, -ая, -ое, (*a*) (chem) pentahydric

пяти·пáл·ый, -ая, -ое, (*a*) (zool) pentadactyl

пяти·пéст·иков·ый, -ая, -ое, (*a*) (bot) pentagynous

пяти·пóл·ь·е, -я, (*n*) (agr) five-crop rotation

пяти·рúчн·ый, -ая, -ое, (*a*) *see* **пят·ерúчн·ый**

пяти·сер·нúст·ый, -ая, -ое, (*a*) (chem) pentasulphide (of)

пяти·сéт·очн·ая лáмп·а (*f*) pentode

пяти·слóй·н·ый, -ая, -ое, (*a*) five-layer/ply

пяти·сóт·, (*gen*) of **пять·сóт·**

пяти·сот·лéт·и·е, -я, (*n*) five hundredth anniversary, quincentenary

пяти·сóт·ый, -ая, -ое, (*a*) five-hundredth

пяти·стáм, (*dat*) of **пяти·сóт·**

пяти·стáх, (*prep*) of **пять·сóт·**

пяти·сторóн·н·ий, -яя, -ее, (*a*) five-sided

пяти·тóн·н·ый, -ая, -ое, (*a*) five-ton

пяти·тýсяч·н·ый, -ая, -ое, (*a*) five-thousandth

пяти·угóль·ник·, -а, (*m*) (math) pentagon

пяти·угóль·н·ый, -ая, -ое, (*a*) pentagonal, five-cornered, (bot) pentangular

пяти·ýст·к·и, (*g pl*) -т·ок·, (*f pl*) (zool) *Pentastomida*

пяти·хлóр·úст·ый, -ая, -ое, (*a*) (chem) pentachloride

пяти·час·óв·óй, -áя, -óе, (*a*) five-hour; five o'clock

пя́т·к·а, -и, (*g pl*) -т·ок·, (*f*) (*dim*) of **пят·á;** (shipb) sole piece (of stern frame); (text) upper butt (of latch needle)

пят·на·дцати- (*component*) fifteen-

пят·нá·дцат·ый, -ая, -ое, (*a*) fifteenth

пят·нá·дцат·ь, -и, (*i*) -ью, (*f*) fifteen

пятн·á·ть, -ют, (*imp*) spot, cover with spots, leave a spot

пятн·úстост·ь, -и, (*f*) spottiness, spotting; spot, blight (plant disease)

пятн·úст·ый, -ая, -ое, (*a*) spotty, spotted, maculate, maculose

~ **маск·ирóвк·а** (*f*) partial camouflage

пятн·ó, -á, (*nom pl*) **пя́тн·а,** (*g pl*) **пя́тен·,** (*n*) spot, patch, blot, blotch, macula

~, **бéл·ые** (*pl*) (met) fish eyes, snowflakes

~, **вед·ýщ·ее** (astron) preceding spot

~, **иóн·н·ое** (rad) ion burn

~, **катóд·н·ое** (rad) cathode spot

~ **свет·овó·е** light/luminous spot, highlight

~, **сóлн·ечн·ое** (astron) sun spot

~, **фóкус·н·ое** focal spot (of an X-ray tube)

~, **чёрн·ое** (spectr) dark spot

пятно·вы·вод·úтел·ь, -я, (*m*) cleaner, stain remover

пя́тн·ышк·о, -а, (*g pl*) шек·, (*n*) speck

пят·ов·óй, -áя, -óе, (*a*) of **пят·á**

~ **бáл·к·а** (*f*) (civ eng) skewback

пя́т·очк·а, -и, (*g pl*) -чек·, (*f*) (*dim*) of **пя́т·к·а;** butt (of knitting machine latch needle)

пя́т·ый, -ая, -ое, (*a*) fifth

половúн·а пя́т·ого half past four

пят·ь, -ú, (*i*) -ью́, (*f*) five

пять·десят, (*g*) **пятú·десят·и,** (*i*) **пятью́·десят·ью** fifty

пять·сóт·, (*g*) **пяти·сóт·** five hundred

пят·ью́ (*adv*) five times

пятью́·десят·ью (*instr*) of **пять·-десят·**

пятью́·стáми, (*instr*) of **пять·сóт·**

Р

р. (*abbrev*) = **рентге́н·** roentgen
рабд·и́т·, -а, (*m*) (zool) rhabdite
рабдо·ли́т·, -а, (*m*) (zool) rhabdolith
рабдо·мио́м·а, -ы, (*f*) (med) rhabdomyoma
рабдо·сарко́м·а, -ы, (*f*) (med) rhabdosarcoma
рабдо·со́м·а, -ы, (*f*) (zool) rhabdosome
рабитти́т·, -а, (*m*) (min) rabbitite
Ра́битц·а, се́т·к·а (*f*) Rabitz wire netting
раб·ко́р·, -а, (*m*) (print) special industrial correspondent
раб·ко́р·овск·ий, -ая, -ое, (*a*) of **раб·ко́р·**
рабо́т·а, -ы, (*f*) work; labour, job; operation, duty, running (of machines); (print) article, paper, study; (*pl*) works, operations
~ **в реа́ль·н·ом масшта́б·е вре́мен·и** (autom) real-time operation
~ **волн·** wave action
~ **вы·лёт·а** = **р. вы́·ход·а**
~ **вы́·ход·а** work function (of an electron)
~, **го́р·н·ые** (*pl*) mining, underground mining
~, **ка́бель·н·ая** cable-laid rope
~ **ключ·о́м** (rad) keying
коэффицие́нт· рабо́т·ы (*m*) (elec rail) operating factor
~, **лёгк·ая** light work; light duty (of machines)
~, **от·кры́·т·ые** (*pl*) (min) opencast/open-cut/pit/open-pit mining
~ **на не·больш·о́й мо́щ·ност·и** low-power operation
~, **по·вре́мен·н·ая** timework
~ **с реа́ль·н·ой аппара́т·у́р·ой, со·в·ме́ст·н·ая** (autom) on-line operation
~, **сверх·уро́ч·н·ая** overtime
~, **с·де́ль·н·ая** piecework
рабо́т·а́·ть, -ют, (*imp*) work; run, operate, work (machines); be open (of institutions etc.)
рабо́т·ник·, -а, (*m*) worker (at superior level), staff member, official, executive, officer (e.g. scientific officer)
~, **инжене́р·н·о-техн·и́ческ·ие** (*pl*) senior technical personnel, technical officers

рабо́то·ме́р·, -а, (*m*) dynamometer
рабо́то·спссо́б·ност·ь, -и, (*f*) fitness/capacity for work; efficiency (of machines); durability, resilience (of materials)
рабо́то·спосо́б·н·ый, -ая, -ое, (*a*) hardworking; ablebodied
рабо́ч·ий, -ая, -ее, (*a*) work, working, operating, service; running; effective; worker's; (*as noun*) worker, workman, operative, labourer
~ **вещ·еств·о́** (*n*) working substance
~ **вре́м·я** (*n*) running time, operating time
~ **груз·** (*m*) useful/working load
~ **давл·éни·е** (*n*) working/effective pressure
~ **ис·каж·éни·е** (*n*) (telecom) service distortion
~ **ка́мер·а** (*f*) working chamber (e.g. of a caisson)
~ **костю́м·** (*m*) overall, overalls
~ **лист·о́к·** (*m*) time sheet/card
~ **ме́ст·о** (*n*) work place/stations
~ **с·мес·ь** (*f*) (ICE) mixture, airfuel mixture
от·де́л· рабо́ч·его снабж·éни·я (*m*) staff shop (shop run by factory etc. to supply personal wants of staff)
~ **площа́д·к·а** (*f*) working/work area; furnace platform; (phys) effective area
~ **по·лож·éни·е** (*n*) "on" position (of machines etc.)
~ **полос·а́** (*f*) (rad) service band
~ **про·стра́н·ств·о** (*n*) main chamber (of a furnace)
~, **пут·ев·о́й** (*m*) (rail) permanent way man
~ **си́л·а** (*f*) manpower
~ **сторон·а́** (*f*) driving face (e.g. of propellers)
схе́м·а рабо́ч·его то́к·а (*f*) (telecom) open-circuit system
~ **то́ч·к·а** (*f*) quiescent point (of a vacuum tube)
~ **характер·и́стик·а** (*f*) performance (of a machine)
~ **част·ь** (*f*) (met) test length (of test piece); (instr) measurement range
равенду́к·, -а, (*m*) (text) duck
ра́вен·ств·о, -а, (*n*) equality, parity
знак· ра́вен·ств·а (*m*) (print) equality/equals sign

ра́вен·ств·о
~, обра́тное
    центр· обра́т·н·ого ра́вен·ств·а
    (*m*) (cryst) symmetry centre/center
~, стро́г·ое absolute equality
~, тождеств·енн·ое identity
равенту́х·, -а, (*m*) = равенду́к· (q.v.)
равн·е́ни·е, -я, (*n*) alignment, (mil) dressing
равн·и́н·а, -ы, (*f*) (geog) plain
равн·и́нн·ый, -ая, -ое, (*a*) (geog) plain
равн·и́тел·ь, -я, (*m*) (paper) dandy roll
равн·о́ (*adv*) equally, the same, alike; is equal to, is
~ как и... as are...
ра́вн·о- (*root*) equi-, iso-, homo-
равно·бег·у́щ·ий, -ая, -ее, (*a*) isodromic
равно·бе́др·енн·ый, -ая, -ое, (*a*) (math) isosceles
равно·бо́к·ий, -ая, -ое, (*a*) equilateral
~ уго́ль·ник· (*m*) (met) equal-angle bar
равно·бо́ч·н·ый, -ая, -ое, (*a*) (math) equilateral; isosceles
равно·вели́к·ий, -ая, -ое, (*a*) isometric, (math) equigraphic, equivalent, of equal magnitude
равно·вероя́т·ност·ь, -и, (*f*) equal probability
равно·вероя́т·н·ый, -ая, -ое, (*a*) equally probable, equiprobable
равно·ве́с·ие, -я, (*n*) equilibrium; balance, equipoise
~, без·раз·ли́ч·н·ое (mech) neutral equilibrium
~ дв·ух фаз· biphase equilibrium
~ момéнт·ов (dynam) balance of couples/moments, couple/moment balance
~ момéнт·ов танга́ж·а (aerodyn) pitching-moment balance
~, по·движ·н·о́е mobile/dynamic balance/equilibrium
~ сил· (phys) force balance
    со·сто·я́ни·е равно·ве́с·и·я (*n*) state of equilibrium
~, у·сто́й·чив·ое true/stable balance/equilibrium
    центр· равно·ве́с·и·я (*m*) (mech) centre/center of equilibrium
равно·ве́с·н·ый, -ая, -ое, (*a*) equilibrium; equiponderant
~ орби́т·а (*f*) stable/equilibrium orbit
~ пло́т·ност·ь (*f*) (chem) equilibrium density
~ рабо́т·а (*f*) (mech) equilibrium operation

равно·ве́с·н·ый, -ая, -ое
    лин·и·я равно·ве́с·н·ой рабо́т·
    (*f*) (mech) equilibrium operating line
~ со·сто·я́ни·е (*n*) state of equilibrium
равно·воз·мо́ж·н·ый, -ая, -ое, (*a*) equally possible
равно·времен·ност·ь, -и, (*f*) (phys) isochronism
равно·времен·н·ый, -ая, -ое, (*a*) isochronous, tautochronous
равно·де́й·ствующ·ий, -ая, -ее, (*a*) acting equally, resultant; (*f decl as adj*) resultant, resultant force
~ ве́ктор· (*m*) resultant
равно·дел·е́ни·е, -я, (*n*) equipartition
равно·де́н·ственн·ый, -ая, -ое, (*a*) (astron) equinoctial, equidiurnal
равно·де́н·стви·е, -я, (*n*) (astron) equinox
~, весе́н·н·ее vernal equinox
~, осе́н·н·ее autumnal equinox
    то́ч·к·а весе́н·н·его равно·де́н·-стви·я (*f*) vernal equinoctial point, first point of Aries
равно·до·сту́п·н·ый, -ая, -ое, (*a*) (elec) random-access, equally accessible
равно·ду́ш·н·ый, -ая, -ое, (*a*) indifferent
равно·зна́ч·ащ·ий, -ая, -ее, (*a*) equivalent; synonymous
равно·зна́ч·ност·ь, -и, (*f*) equivalence
равно·зна́ч·н·ый, -ая, -ое, (*a*) equivalent
равно·из·мен·ённ·ый, -ая, -ое, (*a*) equal-variance, homoscedastic
равно·кры́·л·ый, -ая, -ое, (*a*) (ent) homopterous; (*pl as adj*) Homoptera
равно·ло́паст·н·ый, -ая, -ое, (*a*) (zool) isocercal
равно·ме́р·н·о (*adv*) uniformly, evenly
равномерно·с·кла́д·чат·ый, -ая, -ое, (*a*) uniformly folded
равно·ме́р·ност·ь, -и, (*f*) uniformity, evenness
равно·ме́р·н·ый, -ая, -ое, (*a*) uniform, even
~ движ·е́ни·е (*n*) uniform motion
~ не·пре·ры́в·ност·ь, (*f*) (math) uniform continuity
~ с·хо́д·и́мост·ь (*f*) (math) uniform convergence
равно·мо́щ·ност·ь, -и, (*f*) equivalence
равно·мо́щ·н·ый, -ая, -ое, (*a*) equally strong, equivalent (in strength), of the same power, equipotent

равно·му́скуль·н·ый, -ая, -ое, (*a*) (anat) isomyaric, isomyarian

равно·не·пре·ры́в·н·ый, -ая, -ое, (*a*) equicontinuous

равно·но́г·ий, -ая, -ое, (*a*) (zool) isopodous; (*pl as noun*) isopods, *Isopoda*

равно·о́с·н·ый, -ая, -ое, (*a*) equiaxial, equiaxed

равно·от·сто·я́щ·ий, -ая, -ее, (*a*) equidistant, equally/evenly spaced

равно·па́д·ающ·ий, -ая, -ее, (*a*) like-falling

равно·пе́р·ист·ый, -ая, -ое, (*a*) (bot) paripinnate

равно·плеч·ий, -ая, -ее, (*a*) (instr) equal-arm

равно·площа́д·н·ый, -ая, -ое, (*a*) equal-area

равно·поля́р·н·ый, -ая, -ое, (*a*) isopolar

равно·потенциа́ль·н·ый, -ая, -ое, (*a*) (elec) equipotential

равно·пра́в·и·е, -я, (*n*) (law) equality of rights, equality

равно·пра́в·ност·ь, -и, (*f*) equality of rights

равно·пра́в·н·ый, -ая, -ое, (*a*) of **равно·пра́в·и·е**

равно·при·ли́в·н·ый, -ая, -ое, (*a*) (ocean) cotidal

равно·про·меж·у́точн·ый, -ая, -ое, (*a*) equidistant, equally spaced

равно·про́ч·н·ый, -ая, -ое, (*a*) equally strong, equivalent in strength, of equal strength

равно·раз·ме́р·н·ый, -ая, -ое, (*a*) isometric, equidimensional, same-sized

равно·рас·пре·дел·е́ни·е, -я, (*n*) equal distribution, equipartition

равно·ресни́ч·н·ый, -ая, -ое, (*a*) (zool) holotrichous

~ **инфузо́р·и·и** (*pl*) (zool) Holotricha

равно·сво́й·ственн·ый, -ая, -ое, (*a*) isotropic

равно·сигна́ль·н·ый, -ая, -ое, (*a*) equisignal

~ **зон·** (*m*) (rad) equisignal zone

~ **ме́тод·** (*m*) (rad) lobe-switching method

~ **мая́к·** (*m*) (nav) radio range, radio-range beacon

равно·си́ль·ност·ь, -и, (*f*) equivalence,

равно·си́ль·н·ый, -ая, -ое, (*a*) equal, equally strong, equivalent, idempotent, equipotent

равно·си́ль·н·ый, -ая, -ое

~ **у·равн·е́ни·я** (*pl*) (math) equivalent equations

равно·спо́р·ов·ый, -ая, -ое, (*a*) (bot) homosporous

равно·сторо́н·н·ий, -яя, -ее, (*a*) (math) equilateral

равно·сту́п·н·ый, пуч·о́к· (*m*) (tele-com) symmetrical grading

равно·уго́ль·ник·, -а, (*m*) (math) isogon

равно·уго́ль·н·ый, -ая, -ое, (*a*) isogonal, equiangular

равно·це́н·ност·ь, -и, (*f*) equivalence, equal value

равно·це́н·н·ый, -ая, -ое, (*a*) equivalent, equally valuable, equal (in value); of the same price, equally dear

равно·экце́сс·н·ый, -ая, -ое, (*a*) homokurtic

равн·ый, -ая, -ое, (*a*) equal, the same

    ли́н·и·я ра́вн·ого за·па́зд·ывани·я (*f*) (rad) line of equal time difference

    ли́н·и·я ра́вн·ых а́зимут·ов (*f*) (surv) isoazimuth; (navig) curve of equal bearing

    ли́н·и·я ра́вн·ых выс·о́т· (*f*) (navig) circle of altitude

    ли́н·и·я ра́вн·ых радио·пе́ленг·ов (*f*) (nav) curve of equal bearing

~ **у·сло́в·и·я** (*pl*) equal conditions; **при проч·их ра́вн·ых у·сло́в·и·ях** all other conditions being equal

равн·я́·ть, -ют, (*imp*) equalize, equate, make equal to, make level with; consider equal, compare with, put on the same level as; **-ся** (*pass*); be equal to, amount to; (mil) dress

рагиокри́н·ов·ая клет·к·а (*f*) (zool) histocyte

рагу́ (*n indecl*) (food) ragout

рада́р·, -а, (*m*) (*and see* **радио·лок·а́·ци·я**) radar

радар·ско́п·, -а, (*m*) radarscope

ра́ди (*prep + gen*) for the sake of

радиал·триангул·я́ци·я, -и, (*f*) (surv) radial line control/plotting/triangulation

радиа́ль·н·о (*adv*) radially

радиа́ль·н·о-луч·и́ст·ый, -ая, -ое, (*a*) (cryst) divergent

радиа́ль·н·о-пере·ме́н·н·ый, шаг· (*m*) radially increasing pitch

радиально·сверл·и́льн·ый стан·о́к· (*m*) radial drill

радиально·сто́лб·чат·ый, -ая, -ое, (*a*) (cryst) divergent-columnar

радиа́ль·н·о-сфер·и́ческ·ий под·-ши́п·ник· (*m*) spherical-race radial bearing

радиа́ль·н·ый, -ая, -ое, (*a*) radial

ради·а́н·, -а, (*m*) (math) radian

ради·а́нн·ый, -ая, -ое, (*a*) radian

ради·а́нт·, -а, (*m*) (astron) radiant

ради·а́тор·, -а, (*m*) *see also* **из·луч·а́·тел·ь**; radiator, emitter; cooling fin (of X-ray tube)

~, пласт·и́нчат·ый finned/ribbed/gilled radiator

~, пли́нтус·н·ый (build) base-board radiator

~, со́т·ов·ый (M/T) cellular/honeycomb radiator

~, тру́б·чат·ый tubular radiator

радиацио́н·н·ый, -ая, -ое, (*a*) *see also* **луч·и́ст·ый**; radiation, radiative

ради·а́ци·я, -и, (*f*) *see also* **из·луч·е́·ни·е** radiation

ра́ди·ев·ый, -ая, -ое, (*a*) radium

ра́ди·й, -я, (*m*) (chem) radium, Ra

ради́й·со·держ·а́щ·ий, -ая, -ее, (*a*) radium-bearing

радика́л·, -а, (*m*) radical; (math) radical sign

~, кисл·о́тн·ый (chem) acid radical

~, свобо́д·н·ый (chem) radical

радика́ль·н·ый -ая, -ое, (*a*) radical

~ ос·ь (*f*) (math) radical axis

радик·у́л·а, -ы, (*f*) (bot) radicle, radicule, rootlet, primary root

радикули́т·, -а, (*m*) (med) radiculitis

радикуля́р·н·ый, -ая, -ое, (*a*) radicular

ра́дио (*n indecl*) (rad) radio, wireless

~, напра́вленное directional radio

радио- (*int component*) radio, operated by radio, radio-controlled, radar; (chem) radio-, radioactive

радио·а́вто·гра́мм·а, -ы, (*f*) autoradiogram, autoradiograph, radioautograph

радио·авто́·граф·, -а, (*m*) autoradiogram, autoradiograph, radioautograph

радио·авто·гра́ф·и·я, -и, (*f*) autoradiography

радио·акти́в·ност·ь, -и, (*f*) radioactivity

радио·акти́в·н·ый, -ая, -ое, (*a*) radioactive

~ ряд· (*m*) (phys) radioactive disintegration series

радио·акти́н·и·й, -я, (*m*) (nucl chem) radioactinium

радио·аппара́т·, -а, (*m*) radio set

радио·аппара́т·н·ая, -ой, (*f decl as adj*) (rad) control room, equipment room

радио·аппарат·у́р·а, -ы, (*f*) radio equipment

радио·астро·но́м·и·я, -и, (*f*) radio astronomy

радио·био́·лог·, -а, (*m*) radiobiologist

радио·бу́·й, -я, (*m*) (nav) sonobuoy

радио·бюро́ (*n indecl*) radio office, radiotelegraph office/station

радио·ва́хт·а, -ы, (*f*) radio watch

радио·в·во́д·, -а, (*m*) inlet/bushing/insulator for radio leads

радио·вещ·а́ни·е, -я, (*n*) broadcasting

радио·вещ·а́тельн·ый, -ая, -ое, (*a*) broadcasting

радио·вз·рыв·а́тел·ь, -я, (*m*) (mil) proximity fuze, radar pulse fuze, radio fuze

радио·волн·а́, -ы́, (*f*) radio wave

радио·вс·по·мог·а́тельн·ые сре́д·ств·а (*pl*) (nav) radio aids

радио·высото·ме́р·, -а, (*m*) (air) radio altimeter, radar altimeter, terrain clearance indicator

радио·гальвано́·метр·, -а, (*m*) radio-micrometer

радио·ге́н·н·ый, -ая, -ое, (*a*) (chem) radiogenic

радио·гидро·акуст·и́ческ·ий бу́·й (*m*) (naut) radiosonobuoy

радио·гонио́·метр·, -а, (*m*) radiogoniometer

радио·гра́мм·а, -ы, (*f*) radiogram, radio message/telegram; (phot etc.) radiogram, radiograph, crystallogram, X-ray photograph; X-ray diffraction pattern

радио·граммо·фо́н·, -а, (*m*) radio-gramophone, record player

радио́·граф·, -а, (*m*) radiograph, roentgenogram

радио·гра́ф·и·я, -и, (*f*) radiography, X-ray photography

радио·даль·номе́р·, -а, (*m*) radio range finder, range-only radar

радио·да́·нн·ые, -ых, (*pl decl as adj*) wavelengths and callsigns, radio service particulars

радио·деви·а́ци·я, -и, (*f*) (navig) radio deviation, local D/F errors

~, четвёрт·н·ая quadrantal D/F errors **о·пре·дел·и́ть радио·деви·а́ци·и** carry out D/F calibration

радио·де́л·о, -а, (*n*) radio, radio art

радио·дефекто·скоп·и́·я, -и, (*f*) X-ray flaw detection, radiographic inspection
радио·за·глуш·е́ни·е, -я, (*n*) radio jamming
радио·за·про́с·чик·, -а, (*m*) interrogator (radar)
радио·за·се́ч·к·а, -и, (*g pl*) **-ч·ек·,** (*f*) (nav) radio fix
радио·звезд·а́, -ы́, (*f*) (astron) radio star
радио·зе́ркал·о, -а, (*n*) antenna reflector
радио·зо́н·а, -ы, (*f*) (law) radio territorial waters
радио·зо́нд·, -а, (*m*) (meteor) radio-sonde, radio sounding balloon
радио·зонд·и́ровани·е, -я, (*n*) radio-sonde observations
радио·из·луч·е́ни·е, -я, (*n*) (astron) radio waves/emission
радио·изото́п·, -а, (*m*) (chem) radio-isotope
радио·интерферо́·метр·, -а, (*m*) (astron) radio-interferometer
радио·ис·то́ч·ник·, -а, (*m*) radioactive source; (astron) radio source
радио·кана́л·, -а, (*m*) radio channel
радио·кера́м·ик·а, -и, (*f*) radio ceramics, radio engineering ceramics
радио·комба́йн·, -а, (*m*) radiogram-recorder
радио·коммента́р·и·й, -я, (*m*) radio/broadcast commentary
радио·компар·а́тор·, -а, (*m*) (rad) comparative field intensity/strength meter
радио·ко́мпас·, -а, (*m*) radio compass
радио·контро́л·ь, -я, (*m*) remote checking by radio
радио·круж·о́к·, -ж·к·а, (*m*) radio amateur's/ham's group
радио·курс·ов·о́й у́гол· (*m*) (navig) relative bearing by D/F
радио́л·а, -ы, (*f*) radio-gramophone, radiogram, radio-phonograph
радио·ла́мп·а, -ы, (*f*) (elec) valve, vacuum tube, tube
ради·ола́р·и·й, -я, (*m*) (zool) radio-larian, (*pl*) *Radiolaria*
радио́·лиз·, -а, (*m*) radiolysis, radio-lytic decomposition
радио·ли́н·и·я, -и, (*f*) microwave link; wire-broadcasting line
радио·лит·и́ческ·ий, -ая, -ое, (*a*) radiolytic
радио́·лог·, -а, (*m*) radiologist
радио·ло́г·и·я, -и, (*f*) (med) radiology

радио·лог·и́ческ·ий, -ая, -ое, (*a*) radiological
~ вой·н·а́ (*f*) radiological warfare
радио·лок·а́тор·, -а, (*m*) radar set/equipment, radar
~, бес·по́·иск·ов·ый non-rotating radar
~ бомбо·мет·а́ни·я bomb-aiming radar
~ да́ль·н·его об·наруж·е́ни·я early-warning radar
~, даль·номе́р·н·ый ranging radar
~, диспе́тчер·ск·ий airfield control radar
~ круг·ов·о́го об·зо́р·а plan position indicator, PPI
~, мор·ск·о́й shipborne radar
~ на·вед·е́ни·я ис·треб·и́тел·ей ground-controlled interception radar, GCI
~ об·зо́р·а воз·ду́ш·н·ого движ·е́·ни·я airport surveillance radar
~, об·зо́р·н·ый (air) surveillance radar
~ об·наруж·е́ни·я warning radar
~ об·наруж·е́ни·я воз·ду́ш·н·ых цел·е́й air-warning set, early-warning radar
~ об·наруж·е́ни·я над·во́д·н·ых це·л·е́й sea-search radar, air-to-surface vessel radar, ASV radar
~ об·наруж·е́ни·я облак·о́в cloud-detection radar
~ о·по·зна·ва́ни·я IFF radar
~ оруд·и́йн·ой на·во́д·к·и (mil) fire-control radar, gunlaying radar
~ пере·хва́т·а interception radar
~ пере·хва́т·а, берег·ов·о́й (mil) coastal interception radar
~ по́·иск·а search radar
~, по·са́д·очн·ый (air) precision-approach radar
~ пред·у·прежд·е́ни·я столк·нове́·ни·й anticollision radar
~, при·вод·н·о́й (nav) approach radar
~ слеж·е́ни·я tracking radar
~ целе·у·каз·а́ни·я acquisition radar
радио·лока́тор·щик·, -а, (*m*) radar operator
радио·локацио́н·н·ый, -ая, -ое, (*a*) radar
~ о·гражд·е́ни·е, (*n*) (mil) radar fence
~ со·де́й·стви·е (*n*) (navig) radar aids
~ сре́д·ств·а (*pl*) (navig) radar aids
~ ста́нци·я (*f*) *see also* **радио·лок·а́тор·**; radar set, radar, radar equipment

радио·локацио́н·н·ый, -ая, -ое
~ **у·равн·е́ни·е** (*n*) radar equation
радио·лок·а́ци·я, -и, (*f*) radar (system)
~, **выс·о́тн·ая** height-finding radar
~, **и́мпульс·н·ая** pulse-modulation radar
~, **карто·граф·и́ческ·ая** ground-mapping radar
~ **о·по·зна·ва́ни·я** radar recognition
~ **о·по·зна·ва́ни·я автома́т·н·ая** separate band radar recognition
~, **равно·сигна́ль·н·ая** lobing radar
радио·ло́т·, -а, (*m*) radar/radio altimeter
радио·лу́ч·, -а, (*m*) radio beam, radio-frequency beam
радио·люб·и́тел·ь, -я, (*m*) radio ham/amateur
радио·люмин·есце́нци·я, -и, (*f*) radioluminescence
радиоля́р·иев·ый, -ая, -ое, (*a*) radiolarian
радиоля́р·и·и, -й, (*f pl*) (zool) Radiolaria
радиоляр·и́т, -а, (*m*) radiolarite
радио·ма́ркер·, -а, (*m*) (nav) radio-marker
радио·маск·иро́вк·а, -и, (*f*) (mil) signals security and deception
радио·ма́стер·, -а, (*m*) radio technician
радио·мастер·ск·а́я, -о́й, (*f decl as adj*) radio workshop/repair shop
радио·ма́чт·а, -ы, (*f*) radio mast
радио·мая́к·, -а́, (*m*) radio beacon
~, **глисса́д·н·ый** glide path transmitter (air fields)
~, **зона́ль·н·ый** radio-range beacon, equi-signal beacon
~ **круг·ов·о́го де́й·стви·я** non-directive beacon
~, **курс·ов·о́й** (air) localizer beacon, localizer transmitter
~, **ма́ркер·н·ый** marker beacon
~, **много·зона́ль·н·ый** (nav) Consol/Sonne radio navigation system
~ **на·пра́вл·енн·ого из·луч·е́ни·я** radio-range beacon
~, **на·правл·я́ющ·ий** landing approach beacon
~, **не·на·пра́вл·енн·ый** non-directive beacon
~ **-от·ве́т·чик·, -а,** (*m*) transponder, responder beacon
~, **пе́ленг·ов·ый** rotating beacon
~, **при·вод·н·о́й** (navig) homing beacon
~, **равно·сигна́ль·н·ый** (navig) radio range, radio-range beacon

радио·мая́к
~, **се́ктор·н·ый** Sonne-type long-range navigational aid
~, **с·тво́р·н·ый** radio range beacon
~, **трёх·кана́ль·н·ый** triple modulation beacon
радио·метео·гра́ф·, -а, (*m*) radio-meteograph, radiosonde
радио́·метр·, -а, (*m*) radiometer, radiation meter, radioactivity meter
~, **звук·ов·о́й** acoustic radiometer
радио·метр·и́ст·, -а, (*m*) radar operator
радио·ме́тр·и·я, -и, (*f*) radiometry, measurement of radioactivity
радио·молч·а́ни·е, -я, (*n*) radio silence
радио·монт·а́ж·, -а́, (*i*) **-о́м,** (*m*) radio review; radio assembly work
радио·монт·ёр·, -а, (*m*) radio mechanic
радио·наблюд·е́ни·е, -я, (*n*) (mil) radio observation, radio reconnaissance
радио·навигацио́н·н·ый, -ая, -ое, (*a*) radionavigation, radionavigational
~ **по·со́б·и·я** (*pl*) radionavigational aids
~ **сре́д·ств·а** (*pl*) radionavigational aids
радио·навиг·а́ци·я, -и, (*f*) radio navigation
~, **угло·ме́р·н·о-даль·номе́р·н·ая** Rho Theta navigation system
~, **фа́з·ов·ая да́ль·н·яя** Decca-type long-range radio navigation
радио·на·правл·е́ни·е, -я, (*n*) radio link
радио·не·про·ниц·а́ем·ый, -ая, -ое, (*a*) radiopaque, impenetrable by radio
радио·нивел·и́ровани·е, -я, (*n*) radio measurement/altimetry/surveying
радио·об·ме́н·, -а, (*m*) radio traffic
радио·об·наруж·е́ни·е, -я, (*n*) radar detection
радио·опера́тор·, -а, (*m*) radio/wireless operator
радио·ориент·иро́вк·а, -и, (*f*) (nav) fixing by D/F, radio position finding
радио·от·ме́т·чик·, -а, (*m*) radio marker beacon
радио·пе́ленг·, -а, (*m*) *see also* **пе́ленг·;** D/F bearing, radio bearing
радио·пеленг·а́тор·, -а, (*m*) radio direction finder, D/F set/station
радио·пеленг·а́ци·я, -и, (*f*) radio direction finding

радио·пере·да́т·чик·, -а, (*m*) *and see* **пере·да́т·чик·;** radio transmitter

радио·пере·да́·ч·а, -и, (*f*) radio transmission

радио·по·го́д·а, -ы, (*f*) radio-wave propagation conditions

радио·полиго́н·, -а, (*m*) (mil) signals training area

радио·полу·ко́мпас·, -а, (*m*) fixed-loop radio compass

радио·по·ме́х·а, -и, (*f*) radio interference, (*pl*) radio interference

радио·при·во́д· авто·пило́т·а (*m*) radio-autopilot coupler

радио·при·ём·, -а, (*m*) radio reception

радио·при·ём·ник·, -а, (*m*) *and see* **при·ём·ник·;** radio receiver

~, ма́ркер·н·ый (a/c) marker-beacon receiver

радио·при·ём·н·ый, -ая, -ое, (*a*) of **радио·при·ём·**

~ то́ч·к·а (*f*) radio reproduction point

радио·про·вод·н·о́й, -а́я, -о́е, (*a*) radio-to-line

~ вещ·а́ни·е (*n*) radio rediffusion

~ пере·хо́д·н·ое у·стро́й·ств·о (*n*) radio-to-line transfer unit

радио·прогно́з·, -а, (*m*) radio forecast

радио·прожёктор·, -а, (*m*) beamed radio beacon, directional radio beacon; radio-controlled searchlight

радио·про·ниц·а́ем·ый, -ая, -ое, (*a*) radio-transparent, penetrable by radio

радио·противо·де́й·стви·е, -я, (*n*) radio countermeasures

радио·раз·ве́д·к·а, -и, (*g pl*) **-д·ок·,** (*f*) (mil) signals intelligence

радио·рас·сто·я́ни·е, -я, (*n*) radio range

радио·резисте́нт·ност·ь, -и, (*f*) (nucl, biol) radioresistance

радио·ре́й·к·а, -и, (*g pl*) **-е́·ек·,** (*f*) radio water-gauge

радио·реле́й·н·ый, -ая, -ое, (*a*) radio-relay, microwave

~ ли́н·и·я с·вя́з·и (*f*) microwave link

~ свя́з·ь (*f*) radio-relay/microwave communications

радио·ре·трансляцио́н·н·ая сет·ь (*f*) broadcast relay/rediffusion network

радио·ру́б·к·а, -и, (*g pl*) **-б·ок·,** (*f*) (naut) radio office, (air) radio operator compartment

радио·свеч·ени·е, -я, (*n*) radio-luminescence

радио·с·вя́з·ь, -и, (*f*) radio communication; radio link

радио·с·вя́з·ь

~, дву·сторо́н·н·яя two-way radio link **ли́н·и·я радио·с·вя́з·и** (*f*) radio circuit

радио·секста́нт·, -а, (*m*) radio sextant

радио·се́т·ь, -и, (*f*) antenna/aerial system/array, antenna system; (mil) radio net

~, зо́нт·ичн·ая umbrella aerial/antenna

радио·сигна́л·, -а, (*m*) radio signal

радио·ско́п·, -а, (*m*) (nucl) radioscope

радио·скоп·и́·я, -и, (*f*) radioscopy

радио·с·ни́м·ок·, -мка, (*m*) *see* **ра·дио́·граф·**

радио·спектро·скоп·и́·я, -и, (*f*) (phys) radio-frequency spectroscopy

~, микро·волн·ов·а́я (rad) micro-wave spectroscopy

радио·сре́дств·а, -ств·, (*n pl*) radio facilities

радио·ста́нци·я, -и, (*f*) radio set; radio station

~, вед·о́м·ая (nav) slave station

~, вед·у́щ·ая (nav) master station

~, при·ём·н·о·пере·да·ю́щ·ая transceiver

радио·сто́й·кост·ь, -и, (*f*) (nucl biol) radioresistance

радио·сту́д·и·я, -и, (*f*) broadcasting studio

радио·схе́м·а, -ы, (*f*) radio/electronic circuit

радио·телегра́мм·а, -ы, (*f*) radio telegram, radio message, radiogram

~, прав·и́тельственн·ая government message

~, служ·е́бн·ая unpaid service message

радио·телегра́ф·, -а, (*m*) radio telegraph

радио·телеграф·и́р·овать, -ую́т, (*imp and perf*) send a radio telegram/message

радио·телеграф·и́·я, -и, (*f*) radio telegraphy

радио·теле·из·мер·е́ни·е, -я, (*n*) radio telemetry, radiotelemetering

радио·теле·ме́тр·и·я, -и, (*f*) radio-telemetry, radiotelemetering

радио·теле·меха́н·ик·а, -и, (*f*) radio-telecontrol

радио·теле·механ·и́ческ·ий, -ая, -ое, (*a*) radio-telecontrolled, radio-telecontrol, radio controlled

~ автома́т· (*m*) radio robot

~ реле́ (*n indecl*) radio-control relay

радио·теле·скóп·, -а, (*m*) radio telescope

радио·теле·у·правл·éни·е, -я, (*n*) radio remote control

радио·телефóн·, -а, (*m*) radio telephone

радио·телефон·ú·я, -и, (*f*) radio telephony, radiophony

радио·тéн·ь, -и, (*f*) (rad) screened zone

радио·терап·ú·я, -и, (*f*) (med) radiotherapy, radiation therapy

радио·тéхн·ик·, -а, (*m*) radio mechanic

радио·тéхн·ик·а, -и, (*f*) radio engineering/technology

радио·техн·úческ·ий, -ая, -ое, (*a*) radio, radio engineering, electronic

~ **даль·номéр·н·ая систéм·а** (*f*) radar range measuring equipment

радио·тóр·и·й, -я, (*m*) (chem) radiothorium, RaTh

радио·трансляциóн·н·ый, -ая, -ое, (*a*) of **радио·трансл·яци·я**

~ **с·вязь·** (*f*) relayed radio communication, radio relay, broadcasting rediffusion

~ **сет·ь** (*f*) broadcasting rediffusion network/system

~ **ýзел·** (*m*) broadcast relay centre/exchange; sound reproduction system/equipment, internal broadcast system/equipment

радио·трансл·яци·я, -и, (*f*) broadcasting rediffusion; internal broadcast system

радио·угле·рóд·, -а, (*m*) radiocarbon, radioactive carbon

радио·ýзел·, -зл·á, (*m*) broadcast relay exchange, rediffusion station; (mil) radio section; internal broadcast equipment

~, **при·ём·н·ый** central receiving room

радио·у·правл·éни·е, -я, (*n*) radio control

радио·у·правл·яем·ый, -ая, -ое, (*a*) radio-controlled

радио·уровно·мéр·, -а, (*m*) radio water-level gauge/recorder

радио·у·станóв·к·а, -и, (*g pl*) -в·ок·, (*f*) radio set

радио·фарфóр·, -а, (*m*) radio engineering porcelain

радио·фик·áци·я, -и, (*f*) installation of radio equipment, setting up rediffusion services, extension of broadcast facilities

радио·фото·с·ним·ок·, -м·к·а, (*m*) radio-photogram, photo-radiogram

радио·фото·телеграф·ú·я, -и, (*m*) facsimile radiotelegraphy

радио·хúм·и·я, -и, (*f*) radiochemistry

радио·цéнтр·, -а, (*m*) radio communicating centre

радио·част·от·á, -ы́, (*nom pl*) -óт·ы, (*g pl*) -óт, (*f*) radio frequency

радио·част·óтн·ый, -ая, -ое, (*a*) (rad) radio-frequency

радио·электрóн·ик·а, -и, (*f*) radio-electronics

радио·электрóн·н·ый, -ая, -ое, (*a*) radio-electronic

радио·эх·о, -а, (*n*) radio echo, radar echo

рад·úр·овать, -уют, (*imp and perf*) radio

рад·úст·, -а, (*m*) radio operator; (naut) radio officer; telegraphist (naval)

рáдиус·, -а, (*m*) radius

~ **áтом·а** atomic radius

~ **-вéктор·**, -а, (*m*) (phys) radius vector

~ **дéй·стви·я** radius of action, range

~ **криво·шúп·а** (mech) crank throw

~ **по·раж·éни·я** (gunn) effective radius

радиусо·граф·úческ·ий, -ая, -ое, (*a*) radiusographic

рáд·овать, -уют, (*imp*) make glad/happy; **-ся** (+ *dat*) be glad of, be happy about

радóн·, -а, (*m*) (chem) radon

радóн·ов·ый, -ая, -ое, (*a*) of **радóн·**

рáд·остн·ый, -ая, -ое, (*a*) glad, joyful

рáд·ост·ь, -и, (*f*) joy, gladness

рáд·уг·а, -и, (*f*) rainbow

рáд·ужин·а, -ы, (*f*) (anat) iris, iris diaphragm

рáд·ужк·а, -и, (*g pl*) -жек·, (*f*) (anat) iris, iris diaphragm

рáд·ужност·ь, -и, (*f*) iridescence

рáд·ужн·ый, -ая, -ое, (*a*) iris, iridescent, rainbow

~ **оболóч·к·а** (*f*) (anat) iris diaphragm, iris

~ **цвет·á** (*pl*) prismatic/iridescent colours/colors

рáдул·а, -ы, (*f*) (zool) radula

радýш·н·ый, -ая, -ое, (*a*) cordial

раж·ý, (*pres 1st sing*) of **раз·úть**

раз·, -а, (*nom pl*) -ы́, (*g pl*) **раз·** (*m*) time, occasion

раз (*adv*) one, once, one time, time; (*conj*) since, as

раз
>в дв·а ра́з·а twice as; by a half, half
>в два ра́з·а ме́ньш·е half as big/great/much
>вс·е ра́з·ом all at once, all together
>два ра́з·а twice, two times
>ещё раз· once more
>как раз· just as
>не раз· not infrequently, frequently, often

раз- *as verbal prefix signifies:* — (1) breaking to pieces, dividing; (2) disconnecting, separating, undoing; (3) distributing; (4) action in several directions, dispersal; (5) action over entire surface; (6) achievement of result in course of the action; (7) cancellation of result before the action; (8) intensity of action; (9) **-ся** development of action and achievement of high degree of excess

раз- *as noun or adj prefix signifies:* — highest degree of manifestation of a quality

раз·аррет·и́р·овать, -уют, *(imp or perf)* uncage (a gyro)

раз·бав·и́тел·ь, -я, *(m)* diluent

раз·ба́в·ить, -ят, *(perf) see* **раз·-бавл·я́·ть**

раз·бавл·е́ни·е, -я, *(m) (v n) see* **раз·-бавл·я́·ть;** dilution; rarefication (of gases)

раз·ба́вл·енн·ый, -ая, -ое, *(past part pass) see* **раз·бавл·я́·ть;** dilute; thin, weak (solution); rare, thin (gas)

раз·бавл·я́·ть, -ют, *(imp)* dilute, thin, cut, rarefy (a gas)

раз·бала́нс·, -а, *(m)* unbalance

раз·бала́нс·ов·ый, -ая, -ое, *(a)* unbalanced, out-of-balance

раз·ба́лт·ыва·ть, -ют, *(imp)* shake/stir up; **-ся** work/get loose; be mixed

раз·бе́г·, -а, *(m)* picking up speed, gaining speed; (air etc) take-off run

раз·бег·а́·ться, -ются, *(imp)* scatter, run in all directions; pick up speed, gain speed; take a run at; (a/c) make a take-off run

раз·беж·а́ться, *(fut 3rd pl)* **раз·-бег·у́тся,** *(perf); see* **раз·бсг·а́·ться**

раз·бер·ёт, *(fut 3rd sing)* of **разо·-бр·а́ть,** *(perf); see* **раз·бир·а́·ть** *(imp)*

раз·бив·а́·ть, -ют, *(imp)* smash, break up, break; divide up, break up/down; lay/mark/peg out; pitch, set up (tents etc.); cripple; **-ся** *(pass);* crash

раз·бив·а́·ть
>~ ли́н·и·ю (surv) stake/peg out a line
>~ на пла́з·е (shipb) lay off; (a/c) loft
>~ на шпа́ц·и·и (print) space

раз·би́в·к·а, -а, *(g pl)* **-в·ок·,** *(f) (v n) see* **раз·бив·а́·ть;** (shipb) laying off; (a/c) lofting; (print) spacing; (build) setting out, pegging out, location; break down (of data etc.)
>~ тра́сс·ы location survey
>~ тра́сс·ы доро́г·и (surv) highway/road location

раз·бив·н·о́й, -а́я, -о́е, *(a)* separable

раз·би́в·очн·ый, -ая, -ое, *(a)* of **раз·би́в·к·а**

раз·би·е́ни·е, -я, *(n)* break-up, separation; partitioning, subdivision; fragmentation, decomposition

раз·бинт·ова́·ть, -у́ют, *(perf) see* **раз·бинт·о́выва·ть**

раз·бинт·о́выва·ть, -ют, *(imp)* (med) unbind, remove a bandage/dressing; unwrap

раз·бир·а́ем·ый, -ая, -ое, *(pres part pass)* of **раз·бир·а́·ть;** under review/discussion, in question

раз·бир·а́тельств·о, -а, *(n)* examination, enquiry

раз·бир·а́·ть, -ют, *(imp)* take (all separately); sort out; take to pieces, dismantle, strip; investigate, look into, examine, analyse; distinguish, make out; (gram) parse; (print) distribute; **-ся** *(pass);* make sense of, get an understanding of

раз·би́·т·ый, -ая, -ое, *(past part pass) see* **раз·бив·а́·ть**

раз·би́·ть, *(fut 3rd pl)* **разо·бь·ю́т,** *(perf) see* **раз·бив·а́·ть**

раз·бло́к·, -а, *(m)* (telecom) replacement device

раз·блок·и́р·овать, -уют, *(perf)* unblock, release

раз·богат·е́·ть, -ют, *(perf)* get/become rich

раз·бо́й·нича·ть, -ют, *(imp)* rob, plunder

раз·болт·а́·ть, -ют, *(perf) see* **раз·ба́л·т·ыва·ть**

раз·бомб·и́ть, -я́т, *(perf)* bomb to pieces, destroy (by bombing)

раз·бо́р·, -а, *(m) (v n) see* **раз·бир·а́·ть;** examination, analysis, review, criticism; choice, selection; distribution (of type face); grade
>без раз·бо́р·а indiscriminately

раз·бо́р·к·а, -и,   (*f*) (*v n*) *see* **раз·би́-р·а́ть;** sorting; dismantling, stripping, taking apart, disassembling

~ буке́т·ов (agr) singling

разборно·пере·но́с·н·ый, -ая, -ое, (*a*) sectional

~ по·кры́т·и·е   (*n*) road mat

раз·бо́р·н·ый, -ая, -ое, (*a*) demountable, dismountable, sectional

~ ла́мп·а (*f*) (elec) demountable valve/tube

раз·бо́р·очн·ый, -ая, -ое, (*a*) stripping, disassembling, dismounting

раз·борт·о́вк·а, -и, (*f*) flanging (tubes)

раз·бо́р·чивост·ь, -и, (*f*) discrimination; intelligibility (of sounds); legibility (of script etc.)

раз·бо́р·чив·ый, -ая, -ое, (*a*) legible; intelligible; exacting, choosy

раз·брак·о́ванн·ый, -ая, -ое, (*past part pass*) of **раз·брак·ова́ть;** reject, scrap

раз·брак·ова́ть, -у́ют, (*perf*) reject, scrap

раз·бра́с·ывани·е, -я, (*n*) (*v n*) of **раз·бра́с·ыва·ть;** dispersion

раз·бра́с·ывател·ь, -я, (*m*) (agr etc.) spreader

раз·бра́с·ыва·ть, -ют, (*imp*) throw about, scatter, strew, spread, broadcast; **-ся** (*pass*); dissipate one's energies, flit from one thing to another

раз·бро́с·, -а, (*m*) (*v n*) *see* **раз·бра́-с·ыва·ть;** scatter, spread, dispersal

~, аппарату́р·н·ый (instr) instrumental spread/scatter

~ вре́мен·и time jitter/spread

~ по угл·а́м angular spread/straggle

~, статист·и́ческ·ий statistical spread/scatter

~ то́ч·ек· spread/scatter of points (e.g. on a graph)

~, част·о́тн·ый (rad) frequency diversity; frequency splitting (radar)

раз·бро́с·анн·ый, -ая, -ое, (*past part pass*) *see* **раз·бра́с·ыва·ть;** sparse, scattered, dispersed

раз·брос·а́ть, -ют, (*perf*) *see* **раз·бра́-с·ыва·ть**

раз·брос·н·о́й, -а́я, -о́е, (*a*) (agr) spreading

раз·бро́с·н·ый, -ая, -ое, (*a*) straggling

раз·бры́зг·а·ть, -ют, (*perf*) *see* **раз·-бры́зг·ива·ть**

раз·бры́зг·ивани·е, -я, (*n*) (*v n*) of **раз·бры́зг·ива·ть**

раз·бры́зг·ивател·ь, -я, (*m*) sparger; spray, atomizer

раз·бры́зг·ива·ть, -ют, (*imp*) sprinkle, splash, spray; sparge

раз·бры́зг·ивающ·ий, -ая, -ее, (*pres part act*) of **раз·бры́зг·ива·ть;** spray

раз·буд·и́ть, -́ят, (*perf*) wake, awaken, arouse

раз·бу́ж·енн·ый, -ая, -ое, (*past part pass*) of **раз·буд·и́ть**

раз·бура́в·ить, -ят, (*perf*) *see* **раз·бу-ра́вл·ива·ть**

раз·бура́вл·ива·ть, -ют, (*imp*) open up a hole

раз·бур·ённ·ый, -ая, -ое, (*past part pass*) *see* **раз·бу́р·ива·ть**

раз·бу́р·ива·ть, -ют, (*imp*) drill, bore, drill/bore through

~ не·габари́т· (min) block-hole

раз·бур·и́ть, -́ят, (*perf*) *see* **раз·бу́-р·ива·ть**

раз·бух·а́ни·е, -я, (*n*) (*v n*) of **раз·бу-х·а́·ть;** inflation, distention

раз·бух·а́·ть, -ют, (*imp*) swell, puff up, bulge, distend

раз·бу́х·нуть, -нут, (*perf*) *see* **раз·бу-х·а́·ть**

раз·бу́х·ш·ий, -ая, -ее, (*past part act*) *see* **раз·бух·а́·ть;** swollen, bulged

раз·ва́л·, -а, (*m*) (*v n*), *see* **раз·ва́л·и-ва·ть;** disintegration, collapse; (nucl) breakup, split; mess; (shipb) flare

    у́гол· ра·-ва́л·а (*m*) camber angle (of wheel)

раз·ва́л·ива·ть, -ют, (*imp*) upset, knock over, break up, pull down (a pile or heap); (M/T) set wheels at an angle; **-ся** (*pass*); fall to pieces, collapse, disintegrate

раз·ва́л·енн·ый, -ая, -ое, (*past part pass*) *see* **раз·ва́л·ива·ть;** ruined

раз·ва́л·ин·ы (*f pl*) ruins

раз·вал·и́ть, -́ят, (*perf*) *see* **раз·ва́-л·ива·ть**

раз·вальц·ева́ть, -у́ют, (*imp*) open up a hole, expand a pipe, flare

раз·ва́льц·о́вк·а, -и, (*f*) flaring, pipe expanding

раз·ва́р·ивани·е, -я, (*n*) (*v n*) of **раз·-ва́р·ива·ть**

раз·ва́р·ива·ть, -ют, (*imp*) boil well; boil soft, boil to pulp/shreds; part (metal assay)

раз·вар·и́ть, -́ят, (*perf*) *see* **раз·ва́р·и-ва·ть**

ра́зве (*conj*) unless, except; (*particle*) really

раз·вев·а́·ть, -ют, (*imp*) = **раз·ве́·и·ва·ть**

раз·ве́д·аннос·т·ь, -и, (*f*) extent of exploration; advancement of prospecting

раз·ве́д·анн·ый, -ая, -ое, (*past part pass*) *see* **раз·ве́д·ыва·ть**

раз·ве́д·а·ть, -ют, (*perf*) *see* **раз·ве́·д·ыва·ть**

раз·вед·е́ни·е, -я, (*n*) (*v n*), *see* **раз·во·д·и́ть;** dilution, thinning; breeding, rearing; cultivation, propagation; (*bact*) culture

~ **в себ·е́** autosexing

~, **чи́ст·ое** (agr) stock breeding, bloodstock breeding

раз·вед·ённ·ый, -ая, -ое, (*past part pass*) *see* **раз·вод·и́ть**

раз·вед·ёт, (*fut 3rd sing*) of **раз·ве·сти́,** (*perf*); *see* **раз·вод·и́ть** (*imp*)

раз·ве́д·к·а, -и, (*f*) (*v n*) *see* **раз·ве́д·ы·ва·ть;** prospecting, exploration; reconnaissance, scouting; intelligence, intelligence service

~, **аге́нт·у́рн·ая** espionage, covert intelligence

~, **бо·ев·а́я** battle reconnaissance

~ **бо́·ем** (mil) reconnaissance in force

~, **воз·ду́ш·н·ая** (mil) aerial reconnaissance; (geol) airborne prospecting

~, **звук·ов·а́я** (mil) sound ranging

~, **на·зе́м·н·ая** ground reconnaissance

~, **опт·и́ческ·ая** (mil) flash spotting

~ **по·го́д·ы** weather reconnaissance

~, **радиацио́н·н·ая** (mil) radiation reconnaissance

~, **радиоакти́в·н·ая** radioactive prospecting

раз·ве́д·очн·ый, -ая, -ое, (*a*) of **раз··ве́д·к·а;** prospecting, exploring, exploratory; reconnaissance, reconnoitring; intelligence

раз·ве́д·чик·, -а, (*m*) scout; reconnaissance aircraft; prospector; spy

раз·ве́д·ывательн·ый, -ая, -ое, (*a*) prospecting, exploring, exploratory; reconnaissance, scout; intelligence

раз·ве́д·ыва·ть, -ют, (*imp*) prospect, explore; reconnoitre, reconnoiter, scout

раз·вез·ённ·ый, -ая, -ое, (*past part pass*) *see* **раз·воз·и́ть**

раз·вез·ти́, -у́т, (*perf*) *see* **раз·воз·и́ть**

раз·ве́·ива·ть, -ют, (*imp*) blow about, disperse, scatter

раз·вё·л, (*past masc sing*) of **раз·вес·ти́** (*perf*), *see* **раз·вод·и́ть** (*imp*)

разверз·а́·ть, -ют, (*imp*) open up/wide

раз·ве́рз·нуть, -нут, (*perf*) *see* **разверз·а́·ть**

раз·вё́р·нут·ый, -ая, -ое, (*past part pass*) *see* **раз·вё́рт·ыва·ть;** developed, (math) expanded; extensive, large-scale; broken (of flags)

~ **спо́соб·** (*m*) (met roll) butterfly method

~ **схе́м·а** (*f*) (elec) series circuit

~ **у́гол·** (*m*) (math) straight angle

раз·вер·н·у́ть, -у́т, (*perf*) *see* **раз·вё́р·т·ыва·ть**

раз·верст·а́·ть, -ют, (*perf*) *see* **раз··вёрст·ыва·ть**

раз·вёрст·к·а, -и, (*g pl*) **-т·ок·,** (*f*) (*v n*) *see* **раз·вёрст·ыва·ть;** allotment, allocation

раз·вёрст·ыва·ть, -ют, (*imp*) allot, allocate, apportion

раз·верт·е́ть, -я́т, (*perf*) *see* **раз·вёр·ч·ива·ть**

раз·вёрт·к·а, -и, (*g pl*) **-т·ок·,** (*f*) (*v n*), *see* **раз·вёрт·ыва·ть;** reamer, broach; scan, sweep, trace (on cathode ray tube); (math) developed surface; evolute

~ **бег·у́щ·им луч·о́м** indirect scanning, flying-spot scanning

~, **времен·н·а́я** time base

~, **жд·у́щ·ая** triggered sweep, ratchet time-base

~, **кон·и́ческ·ая** (mech) taper/conical reamer; (elec) conical seam

ли́н·и·я раз·вёрт·к·и (*f*) sweep trace, scanning line

~, **мало·стро́ч·н·ая** coarse scanning

~, **маши́н·н·ая** machine/chucking reamer

~, **много·поло́с·н·ая** multiple scanning

~, **на·са́д·н·ая** shell reamer

~, **пер·ов·а́я** flat reamer

~, **по·стро́ч·н·ая** sequential/line-by-line scanning

~ **пуч·к·о́м** beam scanning

~, **радиа́ль·н·ая** circular trace

~, **раз·жи́м·н·ая** expanding/adjustable reamer

~, **ра́стр·ов·ая** (TV) frame time base

~, **рас·тя́·нут·ая** expanded sweep/time-base

~, **свет·ов·а́я** light scan

ско́р·ост·ь раз·вёрт·к·и (*f*) scanning rate (radar)

раз·вёрт·к·а
~, **стро́ч·н·ая** line scanning/scan
~ **ти́п·а Б** B-scan (radar)
~, **це́ль·н·ая** solid reamer
~, **черес·стро́ч·н·ая** line-jump scanning, interlaced scanning
~, **чёрн·ов·ая** pilot reamer
~, **эллипт·и́ческ·ая** elliptical time-base
~ **эпю́р·ы** exploded diagram
раз·вёрт·оч·ный, **-ая**, **-ое**, (*a*) of **раз·вёрт·к·а**
раз·вёрт·ывани·е, **-я**, (*n*) (*v n*) of **раз·вёрт·ыва·ть**; *see also* **раз·вёрт·к·а**
~ **про·стра́н·ств·а** scanning, scansion (TV, radar etc.)
раз·вёрт·ыватель, **-я**, (*m*) (rad) scanner
~ **перё́д·н·его се́ктор·а** forward-looking scanner
раз·вёрт·ыва·ть, **-ют**, (*imp*) unfold, unwrap, unroll, unwind, unfurl, straighten (shoulders etc.); (mil) deploy; develop, expand; establish, set up; turn; unfold, display, expound; (met etc.) ream, broach; (elec) scan
раз·вёрт·ывающ·ий, **-ая**, **-ее**, (*pres part act*) of **раз·вёрт·ыва·ть**
~ **мото́р·** (*m*) (elec) sweep motor
~ **пятн·о́** (*n*) (TV) scanning spot
~ **у·стро́й·ств·о** (*n*) (TV) scanner
раз·вёрт·ывающ·ийся, **-аяся**, **-ееся**, (*pres part act*) *see* **раз·вёрт·ыва·ть**; (math) developable
~ **по·ве́рх·ност·ь** (*f*) (math) developable surface
раз·ве́рч·ива·ть, **-ют**, (*imp*) unscrew; ream, broach; spin (wheels etc.)
раз·ве́с·, **-а**, (*m*) weighing, weighing out; (text) number
на раз·ве́с· by weight
раз·ве́с·ист·ый, **-ая**, **-ое**, (*a*) (bot) spreading
раз·ве́с·ить, **-ят**, (*perf*) *see* **раз·ве́ш·ива·ть**
раз·ве́с·к·а, **-и**, (*f*) weighing out
раз·вес·н·о́й, **-а́я**, **-о́е**, (*a*) sold by weight
раз·ве́с·очн·ая, **-ой**, (*f decl as a*) (rubber) compounding room
раз·ве́с·очн·ый, **-ая**, **-ое**, (*a*) of **раз·ве́с·к·а**; *see also* **раз·ве́с·очн·ая** (*as noun*)
раз·вес·ти́, (*fut 3rd pl*) **раз·вед·у́т**, (*perf*) *see* **раз·вод·и́ть**
раз·ветв·и́тел·ь, **-я**, (*m*) (elec) splitter

раз·ветв·и́тел·ь
~, **волно·во́д·н·ый** (rad) waveguide splitter
раз·ветв·и́ть, **-я́т**, (*perf*) *see* **раз·-ветвл·я́·ть**
раз·ветв·и́тельн·ый, **-ая**, **-ое**, (*a*) branched, branching; splitting, dividing
~ **коро́б·к·а** (*f*) (elec) splitter box
раз·ветвл·е́ни·е, **-я**, (*n*) (*v n*), *see* **раз·ветвл·я́·ть**; branch; fork, bifurcation; ramification; dividing breeching (of fire hose)
раз·ветвл·я́·ть, **-ют**, (*imp*) branch, ramify, divide, subdivide; fork, bifurcate
раз·ве́ш·ива·ть, **-ют**, (*imp*) weigh, weigh out (portions); hang out; spread (of branches etc.)
раз·ве́·ять, **-ют**, (*perf*) *see* **раз·ве́·ива·ть**
раз·вив·а́ем·ый, **-ая**, **-ое**, (*pres part pass*) of **раз·вив·а́·ть**; developed
раз·вив·а́·ть, **-ют**, (*imp*) develop, work up; untwist, unlay (ropes etc.), unplait, uncurl
~ **ско́р·ост·ь** work/get up speed, gather speed
~ **ход·** (naut) get under way
раз·ви́л·ин·а, **-ы**, (*f*) branch, fork, furcation, (bot) bifurcation, crotch
раз·ви́л·ист·ый, **-ая**, **-ое**, (*a*) branched, branching, furcate, bifurcate, forked
раз·ви́л·к·а, **-и**, (*g pl*) **-л·ок·**, (*f*) = **раз·ви́л·ин·а**
раз·винт·и́ть, **-я́т**, (*perf*) *see* **раз·ви́нч·ива·ть**
раз·ви́нч·ивани·е, **-я**, (*n*) (*v n*) of **раз·ви́нч·ива·ть**; loosening, working loose (of screws)
раз·ви́нч·ива·ть, **-ют**, (*imp*) unscrew; **-ся** come unscrewed, work loose (of screws)
раз·ви́т·и·е, **-я**, (*n*) development, evolution
~ **цеп·н·о́й реа́кци·и** (chem) auto-catalysis
раз·ви·т·о́й, **-а́я**, **-о́е**, (*a*) developed
раз·ви́·т·ый, **-ая**, **-ое**, (*past part pass*) of **раз·ви́·ть**; *see also* **раз·ви·т·о́й**
раз·ви́·ть, (*fut 3rd pl*) **разо·вь·ю́т**, (*perf*) *see* **раз·вив·а́·ть**
раз·влек·а́·ть, **-ют**, (*imp*) entertain, amuse
раз·влеч·ь, (*fut 3rd sing, pl*) **раз·вле·ч·ёт**, **раз·влек·у́т**, (*past masc sing*) **раз·влёк·**, (*perf*); *see* **раз·влек·а́·ть**

раз·во́д·, -а, (*m*) posting (sentries etc.); separating, opening, parting; set (of a saw); (law) divorce; (mil) inspection of the guard; (*pl*) pattern, design

~, волн·и́ст·ый wavy set (of a saw blade)

раз·вод·и́ть, -ят, (*imp*) conduct, take (each to appropriate place), post (sentries etc.); separate, open, part; (law) divorce; dilute (liquids), dissolve (solids); get up, make get up (of waves etc.); raise, get up (steam); make up (fire etc.); breed, rear, raise; cultivate, grow, propagate; set (saw teeth); **-ся** breed

раз·во́д·к·а, -и, (*f*) separating, opening, dividing; saw set, tooth setter; cable laying/layout; (text) ratch; setting-out (leather)

раз·вод·н·о́й, -а́я, -о́е, (*a*) *see also* **раз·во́д·н·ый**; separating, opening

~ ключ· (*m*) adjustable spanner

~ кольц·о́ (*n*) split ring

~ маши́н·а (*f*) setting-out machine (leather)

~ мост· (*m*) swing/draw/bascule bridge

~ шплинт· (*m*) split pin, cutter

раз·во́д·н·ый, -ая, -ое, (*a*) *see also* **раз·вод·н·о́й**; divorce

раз·во́д·ь·е, -я, (*n*) opening, open water/patch (between ice)

раз·вод·я́щ·ий, -ая, -ее, (*pres part act*) of **раз·вод·и́ть**; distributing; (*m as noun*) corporal of the guard

~ сет·ь (*f*) distribution system (water supply)

раз·во́з·, -а, (*m*) delivery, conveyance, transport (to several places)

раз·воз·и́ть, -ят, (*imp*) distribute (using transport), deliver (to several places)

раз·воз·н·о́й, -ая, -ое, (*a*) of **раз·во́з·**; delivery, distribution; home-delivery, delivered to the home

раз·волокн·е́ни·е, -я, (*n*) (biol) tissue splitting/fibering

раз·вора́ч·ивани·е, -я, (*n*) (*v n*) of **раз·вора́ч·ива·ть**

раз·вора́ч·ива·ть, -ют, (*imp*) = **раз·вёрт·ыва·ть** (q.v.); scatter, shatter, make a mess of; turn; **-ся** (*passive*); swing round

раз·воро́т·, -а, (*m*) turn; opening (of book); inside (of folded sheet etc.)

~, бо·ев·о́й (air) stall turn, chandelle

~, рас·чёт·н·ый (air) procedure turn

раз·воро́т

~, станда́рт·н·ый (air) standard-rate turn

у́гол· раз·воро́т·а (*m*) angle of turn

раз·ворот·и́ть, -ят, (*perf*) scatter, shatter, make a mess of

раз·вра́т·, -а, (*m*) depravity

раз·врат·и́ть, -я́т, (*perf*) *see* **раз·вра·щ·а́·ть**

раз·вращ·а́·ть, -ют, (*imp*) deprave, corrupt

раз·вьюч·ива·ть, -ют, (*imp*) unload, unsaddle (a pack animal)

раз·вью́ч·ить, -ат, (*perf*) *see* **раз·вью́ч·ива·ть**

раз·вяз·а́ть, (*fut 3rd pl*) **раз·вя́ж·ут,** (*perf*) *see* **раз·вя́з·ыва·ть**

раз·вя́з·к·а, -и, (*g pl*) **-з·ок·,** (*f*) outcome, issue; (elec) decoupling, uncoupling, bypassing; (theat) denouement

~ на зе́мл·ю (elec) bypass to ground/earth

раз·вя́з·ывани·е, -я, (*n*) (*v n*) of **раз·вя́з·ыва·ть**; (chem) lyolysis; (naut) marking the leadline

раз·вя́з·ыва·ть, -ют, (*imp*) untie, unfasten, undo, unbind; (elec) uncouple, decouple

раз·вя́з·ывающ·ий, -ая, -ее, (*pres part act*) of **раз·вя́з·ыва·ть**

~ на·пряж·е́ни·е (*n*) (elec) stopper, uncoupling resistor

~ у·сил·и́тел·ь (*m*) (elec) isolating amplifier

~ фильтр· (*m*) (elec) decoupling filter

~ цеп·ь (*f*) (rad) decoupling circuit

раз·гад·а́·ть, -ют, (*perf*) guess correctly

раз·га́р·, -а, (*m*) (*v n*), *see* **раз·гор·а́·ться**; climax, height; erosion (of gunbarrel); wear (of refractories)

сет·к·а раз·га́р·а (*f*) (met) crazing (of moulds)

раз·герметиз·а́ци·я, -и, (*f*) depressurization

раз·ги́б·, -а, (*m*) (*v n*), *see* **раз·ги·б·а́·ть**; crease, fold

раз·гиб·а́ни·е, -я, (*n*) (*v n*) of **раз·гиб·а́·ть**; (anat) extension

раз·гиб·а́тел·ь, -я, (*m*) (anat) extensor

раз·гиб·а́·ть, -ют, (*imp*) unbend, straighten, straighten out; (anat) extend

раз·глад·ить, -ят, (*perf*) *see* **раз-·глаж·ива·ть**

раз·глаж·ива·ть, -ют, (*imp*) smooth/iron out

раз·гляд·е́ть, -я́т, (*perf*) discern, make out

раз·гля́д·ыва·ть, -ют, (*imp*) examine, look over

раз·гова́р·ива·ть, -ют, (*imp*) converse with, talk, speak, talk/speak with/to

раз·гово́р·, -а, (*m*) conversation, talk

~, пере·хо́д·н·ый crosstalk (telephones)

раз·гово́р·ник·, -а, (*m*) phrase book

раз·гово́р·н·ый, -ая, -ое, (*a*) colloquial; conversation

~ бу́д·к·а (*f*) telephone box/booth

~ полос·а́ (*f*) (telecom) voice-frequency band

раз·го́н·, -а, (*m*) (*v n*), *see* **раз·го-н·я́·ть;** dispersal; acceleration, boost; racing (of unconnected engine); riding/going up (**of nuclear reactor**), runaway (of uncontrolled reactor); (print) spacing

   вре́м·я раз·го́н·а (*n*) (autom) time constant (of the controlled object)

   длин·а́ раз·го́н·а (*f*) starting length

   пери́од· раз·го́н·а (*m*) (rocket) duration of boost

   режи́м· раз·го́н·а (*m*) period range (of a nuclear reactor)

~ с ме́ст·а acceleration from rest

   ско́р·ост·ь раз·го́н·а (*f*) sensitivity, accuracy (of automatic controller)

~ сту́п·ен·и (rocket) stage boost

раз·го́н·к·а, -и, (*f*) fractional distillation; flattening (leather), spreading; (met) drawing down, fullering, spreading

раз·го́н·н·ый, -ая, -ое, (*a*) of **раз-·го́н·**

~ дви́г·ател·ь (*m*) (elec) starting motor; barring motor

раз·го́н·очн·ый, -ая, -ое, (*a*) of **раз·го́н·к·а**

~ на·со́с· (*m*) fractionating pump

раз·гон·я́·ть, -ют, (*imp*) drive/chase away, disperse, dispell; (chem) distill off; accelerate, race; (print) space, space out

раз·гора́ж·ива·ть, -ют, (*imp*) partition; **-ся** partition off

раз·гор·а́ни·е, -я, (*n*) (*v n*), *see* **раз-·гор·а́·ться;** flare-up, build-up

раз·гор·а́·ться, -ются, (*imp*) flare/flame up; begin to burn

раз·гор·е́·ться, -я́тся, (*perf*) *see* **раз-·гор·а́·ться**

раз·город·и́ть, -я́т, (*perf*) *see* **раз-·гора́ж·ива·ть**

раз·гра́б·ить, -ят, (*perf*) plunder, pillage

раз·град·и́ть, -я́т, (*perf*) *see* **раз-·гражд·а́·ть**

раз·гражд·а́·ть, -ют, (*imp*) (mil) take down/remove obstacles

раз·гранич·е́ни·е, -я, (*n*) (*v n*), *see* **раз·грани́ч·ива·ть;** demarcation

раз·грани́ч·ива·ть, -ют, (*imp*) delimit, demarcate; differentiate, discriminate

раз·грани́ч·ить, -ат, (*perf*) *see* **раз-·грани́ч·ива·ть**

раз·граф·и́ть, -я́т, (*perf*) *see* **раз-·графл·я́·ть**

раз·гра́ф·к·а, -и, (*f*) (*v n*) of **раз-·граф·и́ть**

~ лист·о́в sheet numbering/division (of a map)

раз·графл·ённ·ый, -ая, -ое, (*past part pass*) *see* **раз·графл·я́·ть**

раз·графл·я́·ть, -ют, (*imp*); rule columns on, divide into columns

раз·греб·а́·ть, -ют, (*imp*) rake/shovel away/back

раз·грес·ти́, (*fut 3rd pl*) **раз·греб·у́т,** (*past masc sing*) **раз·грёб·,** (*perf*); *see* **раз·греб·а́·ть**

раз·гро́м·, -а, (*m*) rout, defeat; havoc, destruction

раз·груж·а́тел·ь, -я, (*m*) unloader; kicker (on a conveyor)

раз·груж·а́·ть, -ют, (*imp*) unload, discharge (load etc.); relieve/remove (the load from)

~ под·ши́п·ник· relieve pressure on bearing

раз·гр./з·и́ть, -я́т, (*perf*) *see* **раз-·груж·а́·ть**

раз·гру́з·к·а, -и, (*f*) (*v n*), *see* **раз-·груж·а́·ть;** relief (of load, stress etc.)

   клапан· раз·гру́з·к·и (*m*) relief/unloading valve

раз·гру́з·очн·ый, -ая, -ое, (*a*) unloading, discharging, load-removing, relief, relieving

раз·гру́з·очн·ый, -ая, -ое
~ **диск·** (*m*) hydraulic balancing disc
~ **кла́пан·** (*m*) (hydr) relief/unloading valve
~ **маши́н·а** (*m*) unloading machine
раз·групп·ирова́ни·е, -я, (*n*) (*v n*), *see* **раз·групп·ир·ова́ть**
раз·групп·ир·ова́ть, -у́ют, (*perf*) group, divide into groups, allocate to groups; debunch
раз·дав·а́ни·е, -я, (*n*) (*v n*), of **раз·да·ва́ть**
раз·да·ва́ть, -ют, (*imp*) distribute, dispense; give/hand/serve out; deal (cards), spread, open out; **-ся** (*pass*); sound, be heard; make way, stand back
раз·дав·и́ть, -ят, (*perf*) *see* **раз·да́·вл·ива·ть**
раз·да́вл·енн·ый, -ая, -ое, (*past part pass*) *see* **раз·да́вл·ива·ть**
раз·да́вл·ивани·е, -я, (*n*) (*v n*) of **раз·да́вл·ива·ть**
~, внеза́п·н·ое (min) goth
раз·да́вл·ива·ть, -ют, (*imp*) crush, squash, run over
раз·дад·у́т, (*fut 3rd pl*) of **раз·да́·ть** (*perf*), *see* **раз·да·ва́ть**
раз·да́лбл·ива·ть, -ют, (*imp*) chip/gouge/chisel out, enlarge (holes etc.)
раз·да́·ст, (*fut 3rd sing*) of **раз·да́·ть,** (*perf*), *see* **раз·да·ва́ть**
раз·да́т·очн·ый, -ая, -ое, (*a*) distributing, dispensing
~ **коро́б·к·а** (*f*) (M/T) distributor box
~ **на·кон·е́чник·** (*m*) delivery nozzle
раз·да́·ть, (*fut 3rd sing, pl*) **раз·да́ст, раз·дад·у́т,** (*past masc sing*) **ро́з·да·л,** (*past fem sing*) **раз·да·ла́,** (*perf*); *see* **раз·да·ва́ть**
раз·да́·ч·а, -и, (*f*) (*v n*), *see* **раз·да·ва́ть**; distribution; (met forge) drawing, spreading, enlarging a hole by drawing down the piece
раз·два́·ива·ть, -ют, (*imp*) divide into two, bisect; **-ся** (*pass*); fork, bifurcate
раз·дви́г·, -а, (*m*) width apart/separation
раз·двиг·а́ни·е, -я, (*n*) (*v n*) of **раз·двиг·а́·ть,** *and see* **раз·движ·е́·ни·е**; separation
раз·двиг·а́·ть, -ют, (*imp*) move apart, open (of two movables), part
раз·движ·е́ни·е, -я, (*n*) (*v n*) of **раз·двиг·а́·ть**

раз·движ·е́ни·е
~ **вы́·брос·ов** pip spacing (radar)
регул·я́тор· раз·движ·е́ни·я (*m*) spread control (radar)
раз·дви́ж·к·а, -и, (*g pl*) **-ж·ек·,** (*f*) opening, parting, separation (of rolls)
раз·движ·н·о́й, -а́я, -о́е, (*a*) extensible, expanding, telescopic; opening, openable (of two movables)
~ **га́·ечн·ый клю́ч·** (*m*) adjustable spanner/wrench
~ **двер·ь** (*f*) sliding door
~ **за́·на·вес·** (*m*) (theat) tableau curtains
~ **мост·** (*m*) opening/movable bridge
раз·дви́·нут·ый, -ая, -ое, (*past part pass*) of **раз·дви́·н·уть**
раз·дви́·н·уть, -ут, (*perf*) *see* **раз·двиг·а́·ть**
раз·дво·е́ни·е, -я, (*n*) (*v n*), *see* **раз·два́·ива·ть**; dividing/splitting into two, dichotomy, division, split; bifurcation, forking
раз·дво́·енн·ый, -ая, -ое, (*past part pass*) *see* **раз·два́·ива·ть**; split; bifurcated, bifurcate, bifid, forked
раз·дво·и́ть, -я́т, (*perf*) *see* **раз·два́·ива·ть**
раз·двой·никова́ни·е, -я, (*n*) (cryst) detwinning
раз·дев·а́лк·а, -и, (*g pl*) **-лок·,** (*f*) cloak-room, changing room
раз·дев·а́·ть, -ют, (*imp*) undress, strip
раз·де́л·, -а, (*m*) division; (print etc.) section; (telecom) break sign
гран·и́ц·а раз·де́л·а (*f*) separation boundary; (phys, chem) interface
знак· раз·де́л·а (*m*) (telecom) long break sign, space sign
ли́н·и·я раз·де́л·а (*f*) boundary line, parting line
по·ве́рх·ност·ь раз·де́л·а (*f*) interface
раз·де́л·анн·ый, -ая, -ое, (*past part pass*) *see* **раз·де́л·ыва·ть**
раз·де́л·а·ть, -ют, (*perf*) *see* **раз·де́·л·ыва·ть**
раз·дел·е́ни·е, -я, (*n*) (*v n*), *see* **раз·дел·я́·ть**; division, separation, partition; fractionation (of gases); (chem, opt) resolution
~, амплиту́д·н·ое (telecom) amplitude separation
~ **во вре́мен·и** (telecom) time division

раз·дел·éни·е

~, **гравитаци́он·н·ое** (min) gravity separation

знак· раз·дел·éни·я (*m*) division mark (inland navig)

~ **кана́л·ов** (telecom) multiplexing

~ **кана́л·ов, времен·н·о́е** (telecom) time-division multiplexing

~ **кана́л·ов, част·о́тн·ое** (telecom) frequency-division multiplexing

~ **пере·ме́н·н·ых** (math) separation of variables

раз·дел·ённ·ый, -ая, -ое, (*past part pass*) *see* **раз·дел·я́·ть**; divided, split, separated

~ **ка́мер·а** (*f*) (ICE) divided combustion chamber

~ **цеп·ь** (*f*) (telecom) multiplexed/ divided circuit

раз·дел·и́вш·ийся, -аяся, -ееся, (*past part act*) *see* **раз·дел·я́·ть**; divided, separated; (nucl) fissioned; (chem) parted

раз·дел·и́м·ый, -ая, -ое, (*pres part pass*) *see* **раз·дел·я́·ть**; divisible, separable; distinguishable

раз·дел·и́тел·ь, -я, (*m*) separator, divider; (zool) disjunctor; guide card (card index); separator vessel (pressure gauge)

~, **воз·ду́ш·н·ый** air testing pump

раз·дел·и́тельн·ый, -ая, -ое, (*a*) dividing, separating, (gram) partitive; dichotomous, disjunctive

~ **аппара́т·** (*m*) (chem) liquid-air fractionating tower

~ **пла́в·к·а** (*f*) Orford process/processing (nickel and copper matte)

~ **ток·** (*m*) (telecom) spacing current

~ **цеп·ь** (*f*) (telecom) separation circuit

раз·дел·и́ть, ⸗я́т, (*perf*) *see* **раз·дел·я́·ть**

раз·де́л·к·а, -и, (*g pl*) **-л·ок·,** (*f*) (*v n*), *see* **раз·де́л·ыва·ть, раз·де·л·я́·ть**; jointing (meat); farming out (elec cable)

у́гол раз·де́л·к·и кро́м·ок· (*m*) groove angle

раз·де́л·ыва·ть, -ют, (*imp*) dress, prepare; joint (meat), flense (whales); finish in imitation of, finish; **-ся** have done with

раздельно·лепе́ст·н·ый, -ая, -ое, (bot) choripetalous, dialypetalous

раздельно·ли́ст·н·ый, -ая, -ое, (*a*) (bot) choriphyllous

раз·дельно·по́л·ый, -ая, -ое, (*a*) (bot) dioecious

раз·де́ль·н·ый, -ая, -ое, (*a*) of **раз··де́л·;** separate, distinct

~ **о·пре·де·ле́ни·е,** (*n*) **про·из·во·д·и́ть** distinguish

~ **пункт·** (*m*) (rail) blockpost

раз·дел·я́·ть, -ют, (*imp*) divide, separate, part; share; lease (weaving)

раз·дел·я́ющ·ий, -ая, -ее, (*pres part act* of **раз·дел·я́·ть**; dividing, separating; **-ся** divisible, (nucl) fissionable

~ **прут·к·и́** (*pl*) (text) lease rods

раз·дёрг·ива·ть, -ют, (*imp*) pull/tug apart; pay out (ropes etc.), (naut) check, slack away

раз·дер·ёт, (*fut 3rd sing*) of **разо··др·а́ть** (*perf*)

раз·дёр·н·уть, -ут, (*perf*) *see* **раз··дёрг·ива·ть**

раз·де́·т·ый, -ая, -ое, (*past part pass*) of **раз·де́·ть**; undressed, stripped

раз·де́·ть, -нут, (*perf*) undress, strip

раз·ди́р·, -а, (*m*) tear

раз·дир·а́ни·е, -я, (*n*) (*v n*) of **раз··дир·а́·ть**; (rubber) tear growth, tearing

раз·дир·а́·ть, -ют, (*imp*) tear up, tear, shred

раз·ди́р·очн·ый, -ая, -ое, (*a*) tearing, shredding

~ **маши́н·а** (*f*) (met roll) pack opener

раз·до́·й, -я, (*m*) (agr) increase in milk yield; increasing the milk yield

раз·долб·и́ть, -я́т, (*perf*) *see* **раз··да́лбл·ива·ть**

раз·до́л·ь·е, -я, (*n*) expanse; freedom

раз·до́р·, -а, (*m*) discord, dissension; (geog) branching (streams, rivers)

раз·драж·а́·ть, -ют, (*imp*) irritate; stimulate

раз·драж·а́ющ·ий, -ая, -ее, (*pres part act*) of **раз·драж·а́·ть**; irritant; irritating

раз·драж·е́ни·е, -я, (*n*) (*v n*), *see* **раз·драж·а́·ть**; irritation; stimulation

раз·драж·и́мост·ь, -и, (*f*) (biol) irritability

раз·драж·и́тел·ь, -я, (*m*) irritant; stimulant, stimulus

раз·драж·и́тельн·ый, -ая, -ое, (*a*) irritable; irritating

раз·драж·и́ть, -а́т, (*perf*) *see* **раз··драж·а́·ть**

раз·древес·не́ни·е, -я, (*n*) (bot) delignification

раз·дроб·и́ть, -я́т, (*perf*) *see* **раз··дробл·я́·ть**

раз·дробл·е́ни·е, -я, (*n*) (*v n*), *see* **раз·дробл·я́·ть;** fragmentation

раз·дро́бл·енн·ый, -ая, -ое, (*past part pass*) *see* **раз·дробл·я́·ть**

раз·дробл·я́·ть, -ют, (*imp*) break up, disintegrate, shatter, crush; (math) turn into, express in terms of

раз·ду́в·, -а, (*m*) (*v n*), *see* **раз·дув·а́·ть;** (geol, min) bulge swell

раз·дув·а́ни·е, -я, (*n*) (*v n*) of **раз··дув·а́·ть;** inflation

~ в фо́рм·е (*n*) (plast) blow moulding/molding

раз·дув·а́·ть, -ют, (*imp*) blow apart, scatter; blow up, inflate, swell; blow fan (fires); **-ся** (*pass*); bulge, swell

раз·ду́в·к·а, -и, (*f*) inflation

раз·ду́м·а·ть, -ют, (*perf*) change one's mind

раз·ду́м·ыва·ть, -ют, (*imp*) be in doubt, be unable to make up one's mind, waver, hesitate

раз·ду́·т·ый, -ая, -ое, (*past part pass*) *see* **раз·дув·а́·ть;** bulging

раз·ду́·ть, -ют, (*perf*) *see* **раз·дув·а́·ть**

раз·жа́т·ь, (*fut 3rd pl*) **разо·жм·у́т,** (*perf*) *see* **раз·жим·а́·ть**

раз·жев·а́ть, (*fut 3rd pl*) **раз·жу·ю́т,** (*perf*) *see* **раз·жёв·ыва·ть**

раз·жёв·ыва·ть, -ют, (*imp*) chew, masticate

раз·жёг·ш·ий, -ая, -ее, (*past part act*) of **раз·жéч·ь** (*perf*), *see* **раз··жиг·а́·ть**

раз·желоб·к·а, -и, (*g pl*) **-б·ок·,** (*f*) (build) valley

раз·жéч·ь, (*fut 3rd sing, pl*) **разо·ж·ж·ёт, разо·жг·у́т,** (*past masc sing*) **раз·жёг·,** (*perf*); *see* **раз·жиг·а́·ть**

раз·жиг·а́ни·е, -я, (*n*) (*v n*) of **раз··жиг·а́·ть**

раз·жиг·а́·ть, -ют, (*imp*) light (fuel), kindle, ignite; stir up, arouse

раз·жид·и́ть, -я́т, (*perf*) *see* **раз··жиж·а́·ть**

раз·жиж·а́·ть, -ют, (*imp*) dilute, thin out, rarefy; liquify, liquidize

раз·жиж·éни·е, -я, (*n*) (*v n*), *see* **раз·жиж·а́·ть;** dilution, rarefaction; liquefaction

раз·жиж·и́тел·ь, -я, (*m*) thinner (paints)

раз·жи́м·, -а, (*m*) spreader (e.g. for tires)

раз·жим·а́·ть, -ют, (*imp*) open (vice etc.), unclench (hand etc.), expand (of springs), release, undo; **-ся** (*pass*); spring away

раз·жи́м·н·ый, -ая, -ое (*a*) of **раз··жи́м·;** expanding, spreading, opening, releasing

~ о·пра́в·к·а (*f*) expanding mandrel

раз·жир·é·ть, -ют, (*perf*) get/grow/become fat

раз·жу·ёт, (*fut 3rd sing*) of **раз·жев·а́ть** (*perf*), *see* **раз·жёв·ыва·ть**

раз·зенк·ова́ть, -у́ют, (*perf*) *see* **зенк·ова́ть**

раз·зенк·о́вк·а, -и, (*g pl*) **-вок·,** (*f*) counter-bore

раз·и́тельн·ый, -ая, -ое, (*a*) striking, notable

раз·и́ть, -я́т, (*imp*) strike, strike down

раз·лаг·а́тел·ь, -я, (*m*) (TV) dissector, analyser

раз·лаг·а́·ть, -ют, (*imp*) (chem) decompose, dissociate; (math) expand (into a series etc.); demoralize; **-ся** (*pass*); disintegrate

раз·ла́д·, -а, (*m*) *see* **раз·ла́д·к·а**

раз·ла́д·ить, -ят, (*perf*) *see* **раз·ла́·ж·ива·ть**

раз·ла́д·к·а, -и, (*f*) discord, disharmony

раз·ла́ж·енност·ь, -и, (*f*) failure; maladjustment

раз·ла́ж·ива·ть, -ют, (*imp*) disarrange, disorder, upset; **-ся** (*pass*); go wrong, get out of order

раз·ла́м·ывани·е, -я, (*n*) (*v n*) of **раз··ла́м·ыва·ть;** fracture

раз·ла́м·ыва·ть, -ют, (*imp*) break, break up, fracture

раз·лé·йте, (*pl imperative*) of **раз·ли́·ть**

раз·лет·а́·ться, -ются, (*imp*) fly apart, disperse, scatter, spread

раз·лет·éться, -я́тся, (*perf*) *see* **раз··лет·а́·ться**

раз·лёт·, -а, (*m*) (*v n*), *see* **раз·лет·а́·ться;** dispersion, scatter, spread, divergence (of flying particles)

у́гол· раз·лёт·а (*m*) angle of divergence (of nuclear particles)

раз·ли́в·, -а, (*m*) *see also* **раз·ли́в·к·а;** pouring out (into several receptacles); spreading/covering capacity (of paints etc.); flood, flooding (of rivers), freshet (of streams)

раз·лив·а́ни·е, -я, (*n*) (*v n*) of **раз·ли́в·а́·ть**

раз·лив·а́·ть, -ют, (*imp*) pour out (into several receptacles), can, bottle (etc.); rack (wine, beer etc.); (met) teem; spill; **-ся** (*pass and reflex*); spread

раз·ли́в·к·а, -и, (*g pl*) **-в·ок·,** (*f*) *see also* **раз·ли́·в·;** pouring out (into several receptacles); (met) teeming, casting

~, не·пре·ры́в·н·ая (met) continuous casting

~ с·ве́рх·у (met) top teeming, direct pouring

~, сифо́н·н·ая (met) bottom pouring
ско́р·ост·ь раз·ли́в·к·и (*f*) (met) casting speed, rate of casting

раз·лив·н·о́й, -а́я, -о́е, (*a*) on tap, draught (of beverages)

раз·ли́в·очн·ый, -ая, -ое, (*a*) of **раз·ли́в·к·а**

~ ковш· (*m*) (met) ladle, casting ladle

~ маши́н·а (*f*) (met) casting machine; bottling/filling machine

~ у·стро́й·ств·о (*n*) (met) tundish, trough, bakie

раз·лин·ева́ть, -ю́ют, (*perf*) *see* **раз·лин·о́выва·ть**

раз·лин·ова́ть, -у́ют, (*perf*) *see* **раз·лин·о́выва·ть**

раз·лин·о́выва·ть, -ют, (*imp*) rule/line, make lines

раз·лист·ова́ть, -у́ют, (*imp and perf*) foliate

раз·ли́т·и·е, -я, (*n*) (*v n*), *see* **раз·лив·а́·ть**

раз·ли·т·о́й, -а́я, -о́е, (*a*) diffuse, widespread

~ тума́н· (*m*) widespread fog

раз·ли́·т·ый, -ая, -ое, (*past part pass*) *see* **раз·лив·а́·ть**

раз·ли́·ть, (*fut 3rd pl*) **разо·ль·ю́т,** (*perf*) *see* **раз·лив·а́·ть**

раз·лич·а́·ть, -ют, (*imp*) distinguish, discern; discriminate; **-ся** differ, be different; be distinguished/distinctive/discernible

раз·лич·е́ни·е, -я, (*n*) (*v n*), *see* **раз·лич·и́ть;** discrimination

раз·ли́ч·и·е, -я, (*n*) distinction, difference
знак· раз·ли́ч·и·я (*m*) (mil) insignia, badge (of rank)

раз·лич·и́м·ый, -ая, -ое, (*a*) distinguishable, discernible

раз·лич·и́тел·ь, -я, (*m*) discriminator

раз·лич·и́тельн·ый, -ая, -ое, (*a*) distinctive

раз·лич·и́ть, -а́т, (*perf*) distinguish, discern; (elec etc.) discriminate; **-ся** (*pass*); differ

раз·ли́ч·н·ый, -ая, -ое, (*a*) different, distinct; various, diverse

раз·лож·е́ни·е, -я, (*n*) (*v n*) of **раз·ло·ж·и́ть;** decomposition, dissociation; (plast) degradation; splitting, splitting up, breaking down (facts, data); resolution, analysis; (math) expansion (into a series); demoralization

~ бино́м·а (math) binomial expansion

~ в ряд· Фурье́ (math) expansion into a Fourier series, Fourier analysis/expansion, Fourier-series expansion, harmonic expansion

~, двой·н·о́е (chem) double decomposition

~ из·ображ·е́ни·я (TV) resolution

~, катали́т·и́ческ·ое (chem) catalytic decomposition

~ на множ·и́тел·и (math) factorization

~, об·ме́н·н·ое (chem) double decomposition

по́л·е раз·лож·е́ни·я (*n*) (math) splitting field

~, по·стро́ч·н·ое (TV) line scanning

~ си́л·ы (mech) resolution of a force

~, спектра́ль·н·ое (opt) spectral decomposition

~, стро́к·ов·ое (TV) line scanning

~, терм·и́ческ·ое (chem) thermal dissociation

ка́мер·а раз·лож·е́ни·я decomposition chamber

раз·ло́ж·енн·ый, -ая, -ое, (*past part pass*) of **раз·лож·и́ть**

раз·лож·и́ть, -ат, (*perf*) lay/set out, distribute, place, put; spread out (sheets etc.); make, build (fires); demoralize; (chem) decompose, dissociate; (math) expand (into a series etc.)

раз·ло́м·, -а, (*m*) breaking up; break, rupture, fracture; (geol) fault

раз·лом·а́·ть, -ют, (*perf*) *see* **раз·ла́м·ыва·ть**

раз·лом·и́ть, -ят, (*perf*) *see* **раз·ла́м·ыва·ть**

раз·луч·а́·ть, -ют, (*imp*) separate

раз·луч·е́ни·е, -я, (*n*) separation, separating

раз·луч·и́ть, -а́т, (*perf*) *see* **раз·лу·ч·а́·ть**

раз·магни́т·ить, -ят, (*perf*) *see* **раз·магни́ч·ива·ть**

раз·магни́ч·ивани·е, -я, (*a*) demagnetization; (naut) degaussing

раз·магни́ч·ива·ть, -ют, (*imp*) demagnetize; (naut) degauss; **-ся** be/become/get demagnetized

раз·маз·а́ть, (*fut 3rd pl*) **раз·ма́ж·ут,** (*perf*) *see* **раз·ма́з·ыва·ть**

раз·ма́з·ывани·е, -я, (*n*) (*v n*) of **раз·ма́з·ыва·ть**

~ **пла́зм·ы** (phys) disassembly of a plasma

~ **потенциа́ль·н·ой я́м·ы** (phys) smearing of a potential well

раз·ма́з·ыва·ть, -ют, (*imp*) spread, smear, grease, spread/smear/grease over, coat; blur

раз·ма́л·ываемост·ь, -и, (*f*) grindability; (paper) beatability

раз·ма́л·ыва·ть, -ют, (*imp*) grind, mill (flour etc.) pulverize; (paper) beat; (rubber) crush

раз·ма́т·ывани·е, -я, (*n*) uncoiling, decoiling

раз·ма́т·ывател·ь, -я, (*m*) (met) uncoiler, decoiler, uncoiling reel

раз·ма́т·ыва·ть, -ют, (*imp*) unwind, uncoil, unreel

раз·ма́х·, -а, (*m*) swing, sweep; peak-to-peak value, scope, range, double, amplitude; wing span

~ **кры·л·а́** (a/c) wing span

~ **кры·л·а́, от·нос·и́тельн·ый** (a/c) aspect ratio

~ **раз·вёрт·к·и** (rad) scan width

~ **рельеф·а** (geol) scale/scope of relief

~ **с·брос·а** (geol) amount of throw

раз·ма́х·ива·ть, -ют, (*imp*) wave about, brandish, gesticulate

раз·мах·н·у́ть, -у́т, (*perf*) make a movement/stroke, give a wave, wave, sweep, swing

раз·ма́ч·ива·ть, -ют, (*imp*) soak, soften (by soaking), macerate; **-ся** become soft/swollen (by soaking)

раз·меж·ева́ни·е, -я, (*n*) (*v n*), *see* **раз·меж·ёвыва·ть;** delimitation, demarcation

раз·меж·ева́ть, -у́ют, (*perf*) *see* **раз·меж·ёвыва·ть**

раз·меж·ёвыва·ть, -ют, (*imp*) delimit, demarcate, fix the boundaries/limits of, mark out

раз·мё·л, (*past masc sing*) of **раз·мес·ти́**

раз·мёл·ет, (*fut 3rd sing*) of **раз·мо·л·о́ть** (*perf*), *see* **раз·ма́л·ыва·ть**

раз·мельч·а́·ть, -ют, (*imp*) break up fine/small, crush (rocks etc.), pulverize; **-ся** (*pass*); become/get smaller

раз·мельч·а́ющ·ий, -ая, -ее, (*pres part act*) of **раз·мельч·а́·ть**

~ **о́т·жиг·** (*m*) (met) grain refining anneal

раз·мельч·е́ни·е, -я, (*n*) (*v n*), *see* **раз··мельч·а́·ть**

~ **зерн·а́** (met) grain refining

раз·мельч·и́ть, -а́т, (*perf*) *see* **раз··мельч·а́·ть**

раз·мен·, -а, (*m*) (*v n*), *see* **раз·мен·ива·ть**

раз·мен·ива·ть, -ют, (*imp*) change, get change for (high value money)

раз·мен·я́·ть, -ют, (*perf*) *see* **раз··мен·ива·ть**

раз·мер·, -а, (*m*) dimension, size; metre, measure (music etc.); (*pl*) dimensions, size

~, **габари́т·н·ый** overall dimension

~, **груз·ов·о́й** (naut) displacement curve

~ **зёрен·** grain size

~, **наи·бо́льш·ий пре·де́ль·н·ый** high/H limit, maximum size

~, **най·ме́ньш·ий пре·де́ль·н·ый** low/L limit, minimum size

~, **номина́ль·н·ый** nominal/basic size

~, **пре·де́ль·н·ые** (*pl*) limits

раз·мер·е́ни·е, -я, (*n*) (*v n*) of **раз··мер·ить;** dimensions, measurement

раз·мер·ить, -ят, (*perf*) measure out/off

раз·мер·ност·ь, -и, (*f*) (math) dimensionality, number of dimensions; dimension, unit of measurement (e.g. in tabulated data); comparability

ана́лиз·раз·ме́р·ност·ей (*m*) (math) dimensional analysis, method of dimensions

~ **вре́мен·и** units of time

~, **им·е́ющ·ий** which is measured in...

теори·я раз·ме́р·ност·и (*f*) dimensional theory

фо́рмул·а раз·ме́р·ност·и (*f*) dimensional formula

раз·ме́р·н·ый, -ая, -ое, (*a*) size, dimension, dimensional

~ **фа́ктор·** (*m*) size factor

~ **эффе́кт·.** (*m*) scale/size effect

раз·мер·я́·ть, -ют, (*imp*) measure out/off

раз·мес·и́ть, -ят, (*perf*) knead (dough etc.), puddle (clay etc.), churn (butter)

раз·ме·сти́, *(fut 3rd pl)* **раз·мет·у́т,** *(perf)* sweep, sweep up

раз·мест·и́ть, -а́т, *(perf) see* **раз·-мещ·а́ть**

раз·мёт·анн·ый, -ая, -ое, *(past part pass)* of **раз·мет·а́ть** *(perf)*; scattered

раз·мет·а́·ть, -ют, *(imp)* sweep, sweep up; **раз·мет·а́ть,** *(fut 3rd pl)* **раз·-меч·ут,** *(perf)* throw about, scatter

раз·мет·ённ·ый, -ая, -ое, *(past part pass)* of **раз·ме·сти́**

раз·мет·ёт, *(fut 3rd sing)* of **раз·ме·сти́**

раз·ме́т·ить, -ят, *(perf) see* **раз·-меч·а́·ть**

раз·ме́т·к·а, -и, *(f) (v n),* see **раз·ме·ч·а́·ть**

раз·мёт·к·а, -и, *(f) (v n) see* **раз·-мет·а́ть**

раз·мет·очн·ый, -ая, -ое, *(a)* marking-out

~ **ци́ркул·и** *(pl)* dividers

раз·ме́т·чик·, -а, *(m)* marker, marker-out; (shipb) plater, marking-off plater (Brit), marker-off (U.S.A.)

раз·мёт·ыва·ть, -ют, *(imp)* throw about/around, scatter

раз·меч·а́·ть, -ют, *(imp)* mark off/out; peg out

раз·ме́ч·енн·ый, -ая, -ое, *(past part pass) see* **раз·меч·а́·ть**

раз·ме́ч·ет, *(fut 3rd sing)* of **раз·ме·т·а́ть** *(perf)*

раз·ме́ш·анн·ый, -ая, -ое, *(past part pass)* of **раз·меш·а́ть**; well-mixed

раз·меш·а́·ть, -ют, *(perf)* mix, stir, mix evenly/well

раз·ме́ш·енн·ый, -ая, -ое, *(past part pass)* of **раз·мес·и́ть**; kneaded, puddled, churned

раз·ме́ш·ивани·е, -я, *(n) (v n)* of **раз·-ме́ш·ива·ть**

раз·ме́ш·ива·ть, -ют, *(imp)* mix, stir, mix evenly/well; knead (dough etc.), puddle (clay etc.), churn (butter etc.)

раз·мещ·а́·ть, -ют, *(imp)* place (several), distribute, dispose; find places for, accommodate, house; **-ся** *(pass)*; take a seat

раз·мещ·е́ни·е, -я, *(n) (v n),* see **раз·-мещ·а́·ть**; arrangement, disposition, distribution; (math) *(pl)* combinations

раз·мещ·ённ·ый, -ая, -ое, *(past part pass) see* **раз·мещ·а́·ть**

раз·мин·а́·ть, -ют, *(imp)* soften (by working), knead, puddle

раз·мин·и́р·овать, -уют, *(imp and perf)* (mil) clear of mines

раз·множ·а́·ть, -ют, *(imp)* multiply, reproduce, (biol) propagate; (nucl) breed (fuel), multiply (particles)

раз·множ·а́ющ·ий, -ая, -ее, *(pres part act)* of **раз·множ·а́·ть**; fert·e

~ **с·бо́р·к·а** *(f)* (nucl) fertile-material assembly

~ **спосо́б·ност·ь** *(f)* multiplication capacity, regenerative efficiency

~ **сред·а́** *(f)* (nucl) neutron-multiplying medium; fertile medium

раз·множ·е́ни·е, -я, *(n) (v n) see* **раз·-множ·а́·ть**; multiplication, (biol etc.) reproduction; (print) duplication

коэффицие́нт· **раз·множ·е́ни·я** *(m)* multiplication factor

~, **пол·ов·о́е** sexual reproduction

раз·множ·и́тел·ь, -я, *(m)* (nucl) breeder reactor

раз·мно́ж·ить, -ат, *(perf) see* **раз·-множ·а́·ть**

раз·мо́·ет, *(fut 3rd sing)* of **раз·мы́·ть** *(perf), see* **раз·мыв·а́·ть** *(imp)*

раз·мок·а́·ть, -ют, *(imp)* be/become/ get sodden/soaked/soft

раз·мо́к·нуть, -нут, *(perf) see* **раз·-мок·а́·ть**

раз·мо́л·, -а, *(m)* grind, grist; grinding, milling, pulverizing; (paper) beating; (rubber) crushing

мук·а́ кру́п·н·ого **раз·мо́л·а** *(f)* coarse-ground flour

мук·а́ ме́ле·ого **раз·мо́л·а** *(f)* fine-ground flour

раз·молод·и́ть, -я́т, *(perf)* unkink (rope etc.)

раз·молож·е́ни·е, -я, *(n)* unkinking (rope etc.)

раз·мол·о́ть, *(fut 3rd pl)* **раз·ме́л·ют,** *(perf) see* **раз·ма́л·ыва·ть**

раз·мора́ж·ива·ть, -ют, *(imp)* unfreeze, defrost, thaw out

раз·моро́ж·енн·ый, -ая, -ое, *(past part pass) see* **раз·мора́ж·ива·ть**; thawed, unfrozen, defrosted

раз·моро́з·ить, -ят, *(perf) see* **раз·мо-ра́ж·ива·ть**

раз·мот·а́·ть, -ют, *(perf) see* **раз·ма́-т·ыва·ть**

раз·мо́ч·енн·ый, -ая, -ое, *(past part pass) see* **раз·ма́ч·ива·ть**

раз·моч·и́ть, -ат, *(perf) see* **раз·ма́-ч·ива·ть**

раз·мы́в·, -а, *(m) (v n), see* **раз·мы-в·а́·ть**; (geol) erosion, scour; (spectr) broadening, blurring, smearing (of lines, spots etc.)

раз·мыв·а́ни·е, -я, (*n*) *see* **раз·мы́·в·** and **раз·мы́т·и·е**

раз·мыв·а́·ть, -ют, (*imp*) wash away/out, erode (by water)

раз·мык·а́ни·е, -я, (*n*) (*v n*) of **раз··мык·а́·ть**

раз·мык·а́тел·ь, -я, (*m*) (elec) breaker, release

раз·мык·а́·ть, -ют, (*imp*) disconnect, open, (elec) break (a circuit); **-ся** (*pass*) open up

раз·мык·а́ющ·ий, -ая, -ее, (*pres part act*) of **раз·мык·а́·ть**

~ ключ· (*m*) (elec) interruption/splitting key

~ конта́кт· (*m*) break contact

~ механ·и́зм· (*m*) tripping mechanism

раз·мы́сл·ить, -ят, (*perf*) *see* **раз··мышл·я́·ть**

раз·мы́т·и·е, -я, (*n*) (*v n*) of **раз··мы́·ть;** (spectr) broadening, smearing, blurring; (phys) blow-up (of a beam); disassembly (of plasma)

~ лин·и·и (spectr) line broadening/smearing

раз·мы́·т·ый, -ая, -ое, (*a*) (*past part pass*) *see* **раз·мыв·а́·ть;** broad, diffuse

раз·мы́·ть, (*fut 3rd pl*) **раз·мо́·ют,** (*perf*) *see* **раз·мыв·а́·ть**

раз·мышл·я́·ть, -ют, (*imp*) meditate, reflect

раз·мягч·а́·ть, -ют, (*imp*) soften, make soft; **-ся** (*pass*); become soft, soften

раз·мягч·а́ющ·ий, -ая, -ее, (*pres part act*) of **раз·мягч·а́·ть;** softening, emolient

раз·мягч·е́ни·е, -я, (*n*) (*v n*) of **раз··мягч·а́·ть;** softening, (med) malacia **температу́р·а раз·мягч·е́ни·я** (*f*) softening point

раз·мягч·ённ·ый, -ая, -ое, (*past part pass*) *see* **раз·мягч·а́·ть**

раз·мягч·и́·ть, -а́т, (*perf*) *see* **раз··мягч·а́·ть**

раз·мяк·а́·ть, -ют, (*imp*) become soft, soften

раз·мя́к·л·ый, -ая, -ое, (*past part pass*) of **раз·мяк·а́·ть;** soft

раз·мя́к·нуть, -нут, (*perf*) *see* **раз··мяк·а́·ть**

раз·мя́·ть, (*fut 3rd pl*) **разо·мн·у́т,** (*perf*) *see* **раз·мин·а́·ть**

раз·найто́в·ить, -ят, (*perf*) *see* **раз··найто́вл·ива·ть**

раз·найто́вл·ива·ть, -ют, (*imp*) (naut) unlash, take off a seizing

раз·на·ря́д·к·а, -и, (*f*) (com) multiple purchase order

раз·на́ш·ивани·е, -я, (*n*) broadening through wear (of shoes); drift, creep

~, динам·и́ческ·ое dynamic drift

раз·нес·е́ни·е, -я, (*n*) (*v n*), *see* **раз··нос·и́ть;** (rad) diversity spacing

раз·нес·ённост·ь, -и, (*f*) *see also* **раз·но́с·;** spacing, separation, diversity

раз·нес·ённ·ый, -ая, -ое, (*past part pass*) *see* **раз·нос·и́ть;** (rad) spaced, diversity spaced

~ анте́нн·а (*f*) diversity antenna **при·ём· на раз·нес·ённ·ые ан·те́нн·ы** (*m*) diversity reception

раз·нес·ти́, -у́т, (*perf*) *see* **раз·нос·и́ть**

раз·ним·а́·ть, -ют, (*imp*) take apart, take to pieces, dismantle, strip

ра́зн·иться, -ятся, (*imp*) differ

ра́зн·иц·а, -ы, (*f*) difference

ра́зн·о- (*root*) different, hetero-

разно·бо́·й, -я, (*m*) lack of coordination, irregularity

~ ору́д·и·й (gunn) spread

разно·вероя́т·ност·ь, -и, (*f*) unequal probability

разно·ве́с·, -а, (*m*) set of small weights

разно·ви́д·ност·ь, -и, (*f*) variety; (geol) facies

разно·ви́д·н·ый, -ая, -ое, (*a*) a different kind of, variant; (*pl*) various, diverse, different

разно·време́н·н·о (*adv*) at different times

разно·време́н·ност·ь, -и, (*f*) anachronism; lack of synchronization; (rad) diversity **коэффицие́нт· разно·време́н·ност·и** (*m*) (rad) diversity factor

разно·време́н·н·ый, -ая, -ое, (*a*) non-contemporary, belonging to a different time; non-simultaneous, taking place at different times; unsynchronized

разно·гла́с·и·е, -я, (*n*) disagreement, difference (of opinion); discrepancy

ра́зн·ое, -ого, (*n decl as adj*) miscellany

разно·зерн·и́ст·ый, -ая, -ое, (*a*) inequigranular, of different grain size

разно·знач·а́ющ·ий, -ая, -ее, (*a*) *see* **разно·зна́ч·н·ый**

разно·зна́ч·н·ый, -ая, -ое, (*a*) with a different meaning

разно·зу́б·ый, -ая, -ое, (*a*) (zool) heterodont

разно·имён·н·ый, -ая, -ое, (*a*) unlike, opposite; different-named

разно·ли́ст·ый, -ая, -ое, (*a*) (bot) heterophyllous

разно·ма́ст·н·ый, -ая, -ое, (*a*) different-coloured/colored

разно·му́скуль·н·ые, -ых, (*pl decl as adj*) (zool) anisomyarians, Anisomyaria

разно·мы́сл·и·е, -я, (*n*) difference of opinion

разно·но́г·ие, -их, (*pl as adj*) (zool) amphipods, Amphipoda

разно·обра́з·и·е, -я, (*n*) variety, diversity

разно·обра́з·ить, -ят, (*imp*) diversify, make varied

разно·обра́з·ност·ь, -и, (*f*) variety, diversity

разно·обра́з·н·ый, -ая, -ое, different kind of; (*pl*) various, diverse, miscellaneous

разно·па́р·н·ый, -ая, -ое, (*a*) with pairs of different kinds, with unlike pairs, composite, compound (of elec cable)

разно·племён·н·ый, -ая, -ое, (*a*) of mixed race/pedigree/stock, hybrid

разно·пло́д·н·ый, -ая, -ое, (*a*) (bot) heterocarpous

разно·по́л·ый, -ая, -ое, (*a*) heterosexual; of opposite sex

разно·поля́р·н·ый, -ая, -ое, (*a*) heteropolar

разно·рабо́ч·ий, -его, (*m decl as adj*) odd-job man

разно·ре́ч·ивост·ь, -и, (*f*) inconsistency, contradiction

разно·ре́ч·ив·ый, -ая, -ое, (*a*) contradictory

разнородно·по·кро́в·н·ый, -ая, -ое, (*a*) (bot) heterochlamydous

разно·ро́д·ност·ь, -и, (*f*) heterogeneity

разно·ро́д·н·ый, -ая, -ое, (*a*) heterogeneous, heterogenous; unlike, dissimilar hybrid

раз·но́с·, -а, (*m*) (*v n*), *see* **раз·но́с·и́ть**; separation; (ICE etc.) racing, overspeed; (min) box-cut working, open pit; (nucl) runaway (of reactor); (naut) rake (of rigging)

~ **ме́жду пи́к·ами** peak separation

~ **по част·от·е́** frequency separation

~, **част·о́тн·ый** (rad) frequency diversity

раз·нос·и́ть, ⸗ят, (*imp*) distribute, deliver, serve, carry round; disperse, scatter, carry in all directions; enter, plot (on paper in several places); explode, blow up; **-ся** (*pass*); sound, resound

раз·но́с·н·ый, -ая, -ое, (*a*) delivery, distribution; racing, runaway (of machinery)

~ **кни́г·а** (*f*) delivery register; (telecom) receipt book

~ **ско́р·ост·ь** (*f*) runaway/racing speed, excess speed

разно·со́л·, -а, (*m*) (food) mixed pickles

разно·спо́р·овост·ь, -и, (*f*) (bot) heterospory

разно·спо́р·ов·ый, -ая, -ое, (*a*) heterosporous

ра́зн·ост·н·ый, -ая, -ое, (*a*) difference

~ **ана́лог·** (*m*) (math) difference analogue/analog

~ **сигна́л·** (*m*) difference signal

~ **тон·** (*m*) difference tone

разно·сто́лб·чатост·ь, -и, (*f*) (bot) heterostyly

разно·сторо́н·н·ий, -яя, -ее, (*a*) unequal-sided, inequilateral, (math) scalene; versatile, varied

ра́зн·ост·ь, -и, (*f*) difference

~ **вре́мен·и, нул·ев·а́я** zero-time difference

~, **кон·е́чн·ая** finite difference

~ **потенциа́л·ов** (elec) potential, potential difference

~ **потенциа́л·ов, конта́кт·н·ая** (elec, chem) contact potential

у́гол· ра́зн·ост·и пе́лент·ов (*m*) (navig) conversion angle

~ **хо́д·а** (phys) propagation difference (of beams)

раз·но́с·чик·, -а, (*m*) deliveryman; salesman (carrying goods); (biol) vector

разно·ти́п·ност·ь, -и, (*f*) different types, diversity

разно·ти́п·н·ый, -ая, -ое, (*a*) of a different type; (*pl*) different types of

разно·то́лщ·и́нност·ь, -и, (*f*) difference in thickness, variable thickness

разно·тра́в·ь·е, -я, (*n*) herbage

разно·фо́рм·енн·ый, -ая, -ое, (*a*) heteromorphous

разно·хозя́й·ственн·ый, -ая, -ое, (*a*) (biol) heteroecious

разно·цвет·н·ый, -ая, -ое, *(a)* multi-coloured, variegated, polychromatic; different-coloured/colored

разно·цен·н·ый, -ая, -ое, *(a)* indifferent; of/at varying prices

раз·нош·у, *(pres 1st sing)* of **раз·нос·ить**

разно·щит·ков·ые, -ых, *(pl decl as adj)* (pal) heterostracans, Heterostraci

разно·экцесс·н·ый, -ая, -ое, *(a)* heterokurtic

разн·ый, -ая, -ое, *(a)* different, various, miscellaneous, various different

раз·ня·ть, *(fut 3rd pl)* **раз·ним·ут,** *(perf) see* **раз·ним·а·ть**

раз·облач·ени·е, -я, *(n)* exposure, unmasking (of deceit etc.)

разоб·бр·анн·ый, -ая, -ое, *(past part pass) see* **раз·бир·а·ть;** sorted, picked over

разо·бр·а́ть, *(fut 3rd pl)* **раз·бер·ут,** *(perf) see* **раз·бир·а·ть**

раз·общ·а·ть, -ют, *(imp)* separate, isolate; disconnect, disengage, uncouple

раз·общ·ени·е, -я, *(n)* *(v n), see* **раз·общ·а·ть;** separation, isolation; disconnection, disengagement, uncoupling

раз·общ·ённ·ый, -ая, -ое, *(past part pass) see* **раз·общ·а·ть**

раз·общ·ительн·ый, -ая, -ое, *(a)* disconnecting, separating, uncoupling

~ **кла́пан·** *(m)* stop valve

~ **муфт·а** *(f)* clutch

раз·общ·ить, -а́т, *(perf) see* **раз·общ·а·ть**

разо·бь·ёт, *(fut 3rd sing)* of **раз·би·ть** *(perf), see* **раз·бив·а·ть** *(imp)*

раз·ов·ой, -а́я, -ое, *(a)* single, one-way, one-time

ра́з·ов·ый, -ая, -ое, *(a) see* **раз·ов·ой**

разо·вь·ёт, *(fut 3rd sing)* of **раз·ви́·ть** *(perf), see* **раз·вив·а·ть** *(imp)*

разо́·гн·анн·ый, -ая, -ое, *(past part pass) see* **раз·гон·я·ть**

разо·гн·а́ть, *(fut 3rd pl)* **раз·го́н·ят,** *(perf) see* **раз·гон·я·ть**

разо·гн·у́ть, -у́т, *(perf) see* **раз·ги·б·а́·ть**

разо·гре́·в·, -а, *(m)* *(v n), see* **разо·грев·а·ть;** warming-up

разо·грев·а́тел·ь, -я, *(m)* heater, warmer

~, **доро́ж·н·ый** (civ eng) devil

разо·грев·а́·ть, -ют, *(imp)* heat, warm, heat/warm up

разо·гре́·ть, -ют, *(perf) see* **разо·-грев·а́·ть**

разо́·др·анн·ый, -ая, -ое, *(past part pass)* of **разо·др·а́ть;** torn, shredded, tattered

разо·др·а́ть, *(fut 3rd pl)* **раз·дер·у́т,** *(perf)* tear up, tear, shred

разо·жг·у́т, *(fut 3rd pl)* of **раз·жёч·ь** *(perf), see* **раз·жиг·а́·ть**

разо·жж·ёт, *(fut 3rd sing)* of **раз·-жёч·ь** *(perf), see* **раз·жиг·а́·ть**

разо·жм·ёт, *(fut 3ra sing)* of **раз·-жа́т·ь** *(perf), see* **раз·жим·а́·ть**

разо·йд·ётся, *(fut 3rd sing)* of **разо·йтсь** *(perf), see* **рас·ход·и́ться** *(imp)*

разо·йти́сь, *(fut 3rd pl)* **разо·й-д·у́тся,** *(perf), see* **рас·ход·и́ться**

разо·ль·ёт, *(fut 3rd sing)* of **раз·-ли́·ть** *(perf), see* **раз·лив·а́·ть** *(imp)*

ра́з·ом *(adv)* at one stroke/go

разо·мк·нутност·ь, -и, *(f)* openness; (elec) open condition/state

разо́·мк·нут·ый, -ая, -ое, *(past part pass)* of **разо·мк·н·у́ть;** open

~ **гнезд·о́** *(n)* (elec) open-circuit jack

~ **стро·й** *(m)* (mil) open order

~ **цепь** *(f)* (elec) open circuit

разо·мк·н·у́ть, -у́т, *(perf) see* **раз·-мык·а́·ть**

разо·мн·ёт, *(fut 3rd sing)* of **раз·-мя́·ть** *(perf), see* **раз·мин·а́·ть** *(imp)*

разо·пр·ёт, *(fut 3rd sing)* of **рас·-пер·е́ть** *(perf), see* **рас·пир·и́·ть** *(imp)*

разо́·рв·анн·о- *(component)* (meteor) fracto-

~ **-дожд·ев·ы́е облак·а́** *(pl)* (meteor) fracto-nimbus

~ **-куч·ев·ы́е облак·а́** *(pl)* fracto-cumulus

~ **-сло·и́ст·ые облак·а́** *(pl)* fracto-stratus

разо́·рв·анн·ый, -ая, -ое, *(past part pass) see* **раз·рыв·а́·ть;** (geol) faulted, ruptured; (med) lacerated

разо·рв·а́ть, -у́т, *(perf) see* **раз·ры-в·а́·ть**

разор·ени·е, -я, *(n)* *(v n), see* **раз-ор·я́·ть;** ruination, destruction

разор·ённ·ый, -ая, -ое, *(past part pass) see* **разор·я́·ть**

разор·и́тельн·ый, -ая, -ое, *(a)* destructive, ruinous

разор·и́ть, -я́т, *(perf) see* **разор·я́·ть**

раз·оруж·а́·ть, -ют, *(imp)* disarm; dismantle; (naut) unrig

раз·оруж·и́ть, -а́т, (*perf*) *see* **раз·-оруж·а́·ть**

разор·я́·ть, -ют, (*imp*) ruin, destroy

разо·сл·а́ть, (*fut 3rd pl*) **разо·шл·ю́т**, (*perf*) *see* **рас·сыл·а́·ть**

разо·стл·а́ть, (*fut 3rd pl*) **рас·сте́·л·ют**, (*perf*) *see* **рас·стил·а́·ть**

раз·о·травл·е́ни·е, -я, (*n*) (nucl) depoisoning

разо·тр·ёт, (*fut 3rd sing*) of **рас·те·р·е́ть**, (*perf*) *see* **рас·тир·а́·ть** (*imp*)

раз·о·чар·ова́ни·е, -я, (*n*) disappointment

разо·чл·а́, (*past fem sing*) of **рас·-че́ст·ь** (*perf*)

разо·чт·ёт, (*fut 3rd sing*) of **рас·-че́ст·ь** (*perf*)

разо·шёлся, (*past masc sing*) of **разо·йти́сь** (*perf*), *see* **рас·ход·и́ться**

разо·шл·ёт, (*fut 3rd sing*) of **разо·сл·а́ть** (*perf*), *see* **рас·сыл·а́·ть** (*imp*)

разо·шь·ёт, (*fut 3rd sing*) of **рас·-ши́·ть** (*perf*), *see* **рас·шив·а́·ть** (*imp*)

раз·раба́т·ываем·ый, -ая, -ое, (*pres part pass*) of **раз·раба́т·ыва·ть**; developable, exploitable; under development, under exploitation, under cultivation

~ **у·ча́ст·ок·** (*m*) (min) field

раз·раба́т·ыва·ть, -ют, (*imp*) work out/up, develop, devise; (min) exploit, mine, work (deposits etc.); (agr) cultivate

раз·рабо́т·анност·ь, -и, (*f*) stage/degree of development

раз·рабо́т·анн·ый, -ая, -ое, (*past part pass*) *see* **раз·раба́т·ыва·ть**

раз·рабо́т·а·ть, -ют, (*perf*) *see* **раз·-раба́т·ыва·ть**

раз·рабо́т·к·а, -и, (*g pl*) -т·ок·, (*f*) (*v n*), *see* **раз·раба́т·ыва·ть**; development, elaboration; (min) (*pl*) field, workings

~, **алма́з·н·ые** diamond field/workings

~ **нефт·ян·о́го место·рожд·е́ни·я** oil-field development

~, **перво·на·ча́ль·н·ая** (min) early development

раз·ра́вн·ива·ть, -ют, (*imp*) level, make even/smooth

раз·раж·а́·ться, -ются, (*imp*) break/burst out (start violently)

раз·раз·и́ться, -ятся, (*perf*) *see* **раз·-раж·а́·ться**

раз·раст·а́ни·е, -я, (*n*) (*v n*) of **раз·-раст·а́·ться**; growth, expansion; (bot) proliferation

раз·раст·а́·ться, -ются, (*imp*) grow thickly, spread, expand, proliferate

раз·раст·и́сь, -у́тся, (*past masc sing*) **раз·ро́с·ся**, (*perf*) *see* **раз·раст·а́·ться**

раз·ращ·е́ни·е, -я, (*n*) excessive growth, proliferation

раз·регул·иро́ванност·ь (*f*) misalignment, maladjustment

раз·регул·и́р·овать, -уют, (*imp*) put/get out of adjustment/alignment, adjust boldly

раз·ред·и́ть, -я́т, (*perf*) *see* **раз·реж·а́·ть**

раз·реж·а́·ть, -ют, (*imp*) rarefy (gases); thin out, thin; evacuate

раз·реж·а́ющ·ий, -ая, -ее, (*pres part act*) of **раз·реж·а́·ть**; vacuum

~ **на·со́с·** (*m*) vacuum pump

раз·реж·е́ни·е, -я, (*n*) (*v n*), *see* **раз·-реж·а́·ть**; rarefaction; thinning (of trees etc.); negative pressure; vacuum (*see also* **ва́куум·**)

волн·а раз·реж·ени·я (*f*) low-pressure wave, rarefication wave

раз·реж·ённост·ь, -и, (*f*) rarefaction, rarity; thinness (lacking density), openness; vacuity; vacuum (*see also* **ва́куум·**)

раз·реж·ённ·ый, -ая, -ое, (*past part pass*) *see* **раз·реж·а́·ть**; thin; low density (of gases); open (of plantations etc.); evacuated, exhausted (of vessels etc.)

~ **лёд·** (*m*) (ocean) open pack ice

раз·ре́ж·ёт, (*fut 3rd sing*) of **раз·-ре́з·ать** (*perf*), *see* **раз·ре́з·ыва·ть** (*imp*)

раз·ре́з·, -а, (*m*) (*v n*), *see* **раз·ре́·з·ыва·ть**; cut, slit, (med) section, incision; section, sectional view; (min) open-cast mine, open working/cut; log (trace from logged well)

~, **бур·ов·о́й** (min) drill log

в раз·ре́з·е (*adv*) cutaway (of drawing etc.); in the opinion (of someone); **моде́л·ь в раз·ре́з·е** cutaway model

~, **геолог·и́ческ·ий** geologic cross-section

~, **по́лн·ый** full section

~, **по·пере́ч·н·ый** cross section (view), body plan; (med) transverse section

~, **про·до́ль·н·ый** longitudinal section/plan

~, **с·ло́ж·н·ый** offset section

~, **част·и́чн·ый** broken section

раз·рез·а́емост·ь, -и, (*f*) sectility

раз·рез·а́ем·ый, -ая, -ое, (*pres part pass*) *see* **раз·ре́з·ыва·ть**; sectile

раз·рез·а́ни·е, -я, (*n*) (*v n*), *see* **раз·ре́з·ыва·ть**

раз·ре́з·анн·ый, -ая, -ое, (*past part pass*) *see* **раз·ре́з·ыва·ть**

раз·рез·а́·ть, -ют, (*imp*) *see* **раз·ре́з·ыва·ть, раз·ре́з·ать,** (*fut 3rd pl*) **раз·ре́ж·ут,** (*perf*) *see* **раз·ре́з·ыва·ть,**

раз·ре́з·к·а, -и, (*f*) = **раз·рез·а́ни·е**

раз·рез·н·о́й, -а́я, -о́е, (*a*) cut-up; slit, split, slotted; slitting, ripping; (bot) laciniated

~ **ано́д·** (*m*) (elec) split anode

~ **кры·л·о́** (*n*) (a/c) slotted aerofoil

~ **но́ж·ик·** (*m*) paper-knife

~ **пил·а́** (*f*) rip saw

раз·ре́з·ывани·е, -я, (*n*) (*v n*) of **раз·ре́з·ыва·ть**

раз·ре́з·ыва·ть, -ют, (*imp*) cut, slit, (med) incise; cut up, cut into pieces; intersect, cut in two

~ **по·пере́ч·н·о** cut into lengths, cut across

~ **про·до́ль·н·о** slit, cut lengthwise

раз·реш·а́·ть, -ют, (*imp*) allow, permit, authorize, clear (formalities); solve, resolve

раз·реш·а́ющ·ий, -ая, -ее, (*pres part act*) of **раз·реш·а́·ть**

~ **спосо́б·ност·ь** (*f*) (instr) resolving power, resolution; (rad) discrimination

~ **спосо́б·ност·ь по а́зимут·у** (rad) bearing discrimination, azimuth resolution

~ **ядр·о́** (*n*) (math) resolvent kernel

раз·реш·е́ни·е, -я, (*n*) permission, authorization; licence, permit, clearance; (math etc.) solution; (instr) resolution

~ **на в·ход·** (naut) clearance/permission to enter

~ **на провиа́нт·** (naut) victualling bill, shipping bill

~ **на радио·у·стано́в·к·у** radio licence

~ **пере·но́с·а**
сигна́л· раз·реш·е́ни·я пере·но́с·а (autom) carry-clear signal

~, **энергет·и́ческ·ое** (phys) energy resolution

раз·реш·ённ·ый, -ая, -ое, (*past part pass*) *see* **раз·реш·а́·ть**

~ **пере·хо́д·** (*m*) (phys) allowed transition

раз·реш·ённ·ый, -ая, -ое

~ **спектр·** (*m*) allowed spectrum

~ **част·от·а́** (*f*) authorized/assigned frequency

раз·реш·и́тельн·ый, -ая, -ое, (*a*) (rail) permissive

~ **сигна́л·** (*m*) (rail) permissive signal

раз·реш·и́ть, -а́т, (*perf*) *see* **раз·ре·ш·а́·ть**

раз·ровн·я́·ть, -ют, (*perf*) level, make even/smooth

раз·ро́·ет, (*fut 3rd sing*) of **раз·ры́·ть**

раз·ро́зн·енн·ый, -ая, -ое, (*past part pass*) *see* **раз·ро́зн·ива·ть**; broken, incomplete (of sets etc.); separate (of parts of set); disunited

раз·ро́зн·ива·ть, -ют, (*imp*) break, break up (sets etc.); **-ся** (*pass*); become/be incomplete

раз·ро́зн·ить, -ят, (*perf*) *see* **раз·ро́з·н·ива·ть**

раз·ро́с·ся, (*past masc sing*) of **раз·раст·и́сь** (*perf*), *see* **раз·раст·а́·ться** (*imp*)

раз·ро́с·т·ийся, -аяся, -ееся, (*past part act*) of **раз·раст·и́сь** (*perf*), *see* **раз·раст·а́·ться** (*imp*)

раз·руб·а́·ть, -ют, (*imp*) cut/chop up, cut/chop in two, cut into pieces

раз·руб·и́ть, -ят, (*perf*) *see* **раз·ру·б·а́·ть**

раз·руш·а́·ть, -ют, (*imp*) destroy, demolish, ruin; **-ся** (*pass*); fail (of materials), break down

раз·руш·а́ющ·ий, -ая, -ее, (*pres part act*) of **раз·руш·а́·ть**; (mech) breaking; destructive; **-ся** (*see next entry*)

~ **моме́нт·** (*m*) (mech) breaking moment

~ **на·гру́з·к·а** (*f*) (mech) breaking load

~ **на·пряж·е́ни·е** (*n*) (met) rupture stress

раз·руш·а́ющ·ийся, -аяся, -ееся, (*pres part act*) of **раз·руш·а́·ться**; self-destructive

~ **за·по·мин·а́ющ·ее у·стро́й·ств·о** (*n*) volatile store (computers)

раз·руш·е́ни·е, -я, (*n*) (*v n*) *see* **раз·руш·а́·ть**; destruction, demolition; (mech) rupture, failure, collapse, fracture; mechanical breakdown (polymers)

~, **вя́з·к·ое** (met) tough fracture

~, **дл·и́тельн·ое** (met) delayed fracture

~, **за·ме́дл·енн·ое** (met) delayed fracture

~, **кавитацио́н·н·ое** (met) cavitation failure

раз·руш·е́ни·е
~ **от у·ста́л·ост·и** fatigue failure
~ **по крепл·е́ни·ю** (rubber) bond failure
~, **по·пере́ч·н·ое** transverse contamination (in a wind tunnel)
~, **хру́п·к·ое** brittle rupture; (rubber) brittle break/breakage
~ **эму́льс·и·и** demulsification; **ско́р·ост·ь раз·руш·е́ни·я эму́льс·и·и** (f) (oil) demulsification number
раз·ру́ш·енн·ый, -ая, -ое, (past part pass) see **раз·руш·а́·ть**
раз·руш·и́тельн·ый, -ая, -ое, (a) destructive, destroying
раз·ру́ш·ить, -ат, (perf) see **раз·ру·ш·а́·ть**
раз·ры́в·, -а, (m) break, gap, discontinuity; tear, tearing; (mech) tensile rupture/failure/fracture; (geol) fault, rupture; (cyt) disconjunction; central break (in a drawing); burst, explosion; **раз·ры́·в·,** (pres gerund) of **раз·ры́·ть**
~, **воз·ду́ш·н·ый** (expl) air burst
~, **дл·и́тельн·ый** (met) delayed fracture
~ **желу́д·к·а** (med) gastrorrhexis
  ис·пыт·а́ни·е до раз·ры́в·а (n) test to failure/rupture, destructive test, testing
  ис·пыт·а́ни·е на раз·ры́в· (n) tensile test
  мо́дул·ь раз·ры́в·а (m) modulus of rupture
~ **не·пре·ры́в·ност·и** discontinuity
~ **по·ро́д·** (geol, min) rock burst
~, **про·из·во́д·н·ой** (math) derivative discontinuity
  про́ч·ност·ь на раз·ры́в· (f) tensile strength
~, **рас·се́·ян·ый** (geol) dispersed fault
~ **с·вя́з·ей** (phys, chem) bond breaking
~ **се́рд·ц·а** heart failure
~, **сла́б·ый** (math) removable/first-order discontinuity
~ **сле́д·а** (dynam) track gap/break
  со·противл·е́ни·е раз·ры́в·у tensile strength
~ **спло́ш·ност·и** discontinuity
  то́ч·к·а раз·ры́в·а (f) (math) discontinuity, point of discontinuity
~ **-трав·а́, -ы́,** (f) (bot) saxifrage, *Saxifraga*
~, **уда́р·н·ый** (dynam) shock-wave discontinuity
~ **фу́нкци·и** (math) discontinuity of a function

раз·рыв·а́·ть, -ют, (imp) tear, tear up/apart; (med) lacerate; break, interrupt; blow up (by explosion); dig, dig up; **-ся** (pass); explode, blow up, burst; break, fail, rupture (of materials)
раз·рыв·н·о́й, -а́я, -о́е, (a) bursting, breaking; tearing, tensile; discontinuous
~ **длин·а́** (f) (text) breaking length
~ **за·ря́д·** (m) explosive charge
~ **за·ря́д·, ли·т·о́й** (m) cast/poured filling
~ **маши́н·а** (f) tensile testing machine
~ **мо́щ·ност·ь** (f) (elec) breaking capacity
~ **на·пряж·е́ни·е** (n) (mech) breaking stress
~ **фу́нкци·я** (f) (math) discontinuous function
раз·ры́в·ност·ь, -и, (f) discontinuity
раз·ры́в·вш·ий, -ая, -ее, (past part act) of **раз·ры́·ть**
раз·ры́·ть, (fut 3rd pl) **раз·ро́·ют,** (perf) dig up
раз·рыхл·е́ни·е, -я, (n) (v n), see **раз··рыхл·я́·ть**
раз·ры́хл·енност·ь, -и, (f) looseness, friability, openness
раз·рыхл·и́тел·ь, -я, (m) (agr, civ eng) scarifier; cutterhead (suction dredger); (text) opener
раз·рыхл·и́тельн·ый, -ая, -ое, (a) loosening, scarifying; opening (bales)
~ **агрега́т·** (m) (text) opener
раз·рыхл·и́ть, -я́т, (perf) see **раз·рых·л·я́·ть**
раз·рыхл·я́·ть, -ют, (imp) scarify, stir/break up, loosen (solids), make crumbly/friable; open (bales)
раз·рыхл·я́ющ·ий, -ая, -ее, (pres part act) of **раз·рыхл·я́·ть;** (phys) loosening, bond-loosening
раз·ря́д·, -а, (m) discharge; category, grade, class, rating, order, rank; (math) place, digit, bit (computers)
~, **а·реоиод·и́ческ·ий** overdamping (computers)
~, **выс·о́к·ий** (math) high order
~, **дво·и́чн·ый** (math) binary digit; binary digit bit (computers)
~, **дежу́р·н·ый** keep-alive discharge (of vacuum tubes)
~ **деся́т·к·ов** tens digit (computers)
~, **дуг·ов·о́й** (elec) arc discharge
~, **дуго·обра́з·н·ый** quasi-/semi-arc discharge

раз·ря́д
~, знáк·ов·ый   sign digit/place (computers)
~, из·гиб·áющ·ийся   (elec) wriggling discharge
~, и́мпульс·н·ый   (elec) pulsed discharge
~, кист·ев·óй   (elec) brush discharge
~, контрóль·н·ый   check bit (computers)
~, корóн·н·ый   (elec) corona discharge
~, кра·ев·óй   marginal discharge
~, один·áков·ый   (math) equal order
~, оснóв·н·óй   (mil) primary charge
~, о·стáт·очн·ый   (elec) residual discharge, soaking out
~, от·шнур·óванн·ый   (phys) pinched discharge
~, пéрв·ый   first class, (*gen as adj*) first-class
~ пере·полн·éни·я   extra order
~, порох·ов·óй   (mil) propellant charge
~, про·бив·н·óй   (elec) disruptive discharge
~ с крá·я   (phys) marginal discharge
~, сло·и́ст·ый   (elec) striated discharge
~, стáр·ш·ий   top digit, high order (computers)
~, ти́х·ий   (phys) silent/glow discharge
~, тл·éющ·ий   (elec) glow discharge
~ числ·á   digit order number (computers)
раз·ряд·и́ть, -я́т,   (*perf*) *see* раз·ря·ж·á·ть
раз·ря́д·к·а, -и,   (*g pl*) -д·ок·, (*f*) discharge, discharging; unloading; (print) spacing out, spacing
раз·ря́д·ник·, -а,   (*m*) discharger; (elec) spark gap, surge diverter, lightening arrester
~, вращ·áющ·ийся   rotary spark-gap/discharger
~, вы·хлоп·н·óй   expulsion gap
~, гáз·ов·ый   spark-gap tube
~, деиóн·н·ый   dion gap
~, ди́ск·ов·ый   (elec) disc discharger
~, искр·ов·óй   (rad) spark gap; (elec) spark-gap arrester
~, искро·гас·я́щ·ий   (rad) quenched spark gap
~, мнóго·крáт·н·ый   multigap discharger
~, пóр·ист·ый   autovalve lightning arrester
~, резонáнт·н·ый   (rad) gas-discharge switch

раз·ря́д·ник
~, рог·ов·óй   (elec) arcing horn, horn arrester
~, трýб·чат·ый   expulsion gap
~, ýголь·н·ый   carbon protector, carbon lightening protector
раз·ря́д·н·ый, -ая, -ое,   (*a*) discharge
~ рас·сто·я́ни·е   (*n*) (elec) arcing distance
~ сéт·к·а   (*f*) word (computers)
раз·ряж·á·ть, -ют,   (*imp*) (elec) discharge, run down; unload; relieve; (print) space out
раз·у·бед·и́ть, -я́т,   (*perf*) *see* раз·у·бéжд·á·ть
раз·у·бежд·á·ть, -ют,   (*imp*) dissuade
раз·у·бóж·ивани·е, -я,   (*n*) impoverishment, depletion
раз·у·крупн·éни·е, -я,   (*n*) (*v n*), *see* раз·у·крупн·я́·ть
раз·у·крупн·и́ть, -я́т,   (*perf*) *see* раз·у·крупн·я́·ть
раз·у·крупн·я́·ть, -ют,   (*imp*) make smaller, make less big, reorganize in smaller units
рáз·ум·, -а,   (*m*) mind, intellect; reason, intelligence
раз·ум·é·ться, -ются,   (*imp*) be understood
    раз·ум·é·ется   is meant/understood; naturally, of course, it is clear/understandable
    сáм·о соб·óй раз·ум·é·ется   it stands to reason, it is obvious/clear
раз·ýм·н·ый, -ая, -ое,   (*a*) reasonable, sensible, judicious
раз·у·плот·нéни·е, -я,   (*n*) gland/seal failure
раз·у·по·ря́д·очени·е, -я,   (*n*) disorder, disordering; randomization
раз·у·по·ря́д·очива·ть, -ют,   (*imp*) disorder, randomize
раз·у·проч·нéни·е, -я,   (*n*) (met) softening, weakening
раз·ýч·ива·ть, -ют,   (*imp*) study, practice, learn by heart; -ся forget, lose the art of
раз·уч·и́ть, -ат,   (*perf*) *see* раз·ýч·ива·ть
разъ·ед·áемост·ь   (*f*) corrodibility
разъ·ед·áни·е, -я,   (*n*) (*v n*) of разъ·ед·á·ть; corrosion, attack
разъ·ед·á·ть, -ют,   (*imp*) eat away, corrode
разъ·éд·енн·ый, -ая, -ое,   (*past part pass*) *see* разъ·ед·á·ть

разъ·един·е́ни·е, -я, (*n*) (*v n*), *see* **разъ·един·я́·ть;** separation, disconnection; (elec) isolation; release (telephony)

разъ·един·ённ·ый, -ая, -ое, (*past part pass*) *see* **разъ·един·я́·ть**

разъ·един·и́тел·ь, -я, (*m*) (elec) isolator, isolating/disconnecting switch

~, кач·а́ющ·ийся rocker-type disconnecting switch

~, про·ход·н·о́й (elec) bush-mounted isolating switch

~, секцио́н·н·ый шин·н·ый busbar sectionalizing switch, bus-section switch

~, штéпсель·н·ый plug-type isolating switch

разъ·един·и́тельн·ый, -ая, -ое, (*a*) (elec) disconnecting, isolating

~ в·ста́в·к·а (*f*) (elec) isolating/disconnecting link

разъ·един·и́ть, -я́т, (*perf*) *see* **разъ·-един·я́·ть**

разъ·един·я́·ть, -ют, (*imp*) disconnect, separate, part

разъ·е́д·утся, (*fut 3rd pl*) of **разъ·-е́х·аться**

разъ·ед·я́т, (*fut 3rd pl*) of **разъ·е́ст·ь** (*perf*), *see* **разъ·ед·а́·ть**

разъ·е́зд·, -а, (*m*) departure; (rail) passing loop, siding; (mil) patrol

разъ·езд·н·о́й, -а́я, -о́е, (*a*) of **разъ·-е́зд·;** travelling, journeying

~ ка́тер· (*m*) (naut) harbour launch

разъ·езж·а́·ть, -ют, (*imp*) travel, travel about; **-ся** separate, drive off; drive past

разъ·е́л, (*past masc sing*) of **разъ·е́ст·ь** (*perf*), *see* **разъ·ед·а́·ть** (*imp*)

разъ·ём·, -а, (*m*) (*v n*), *see* **раз·ним·а́·ть;** parting, joint

~ ва́л·к·а roll parting/joint
ли́н·и·я разъ·ём·а (*f*) (met) parting/die/flash line

~, штéпсель·н·ый (elec) plug connection

разъ·ём·н·ый, -ая, -ое, (*a*) partable, split; dismantlable, sectional; detachable, dismountable

~ ка́ртер· (*m*) (M/T) split crankcase

~ под·ши́п·ник· (*m*) split bearing

разъ·е́ст·ь, (*fut 3rd sing, pl*) **раз·е́ст, разъ·я́т,** (*past masc sing*) **разъ·е́л,** (*perf*); *see* **разъ·ед·а́·ть**

разъ·е́х·аться, (*fut 3rd pl*) **разъ·-е́д·утся,** (*perf*) separate, drive off; drive past

разъ·ясн·е́ни·е, -я, (*n*) elucidation

разъ·ясн·и́тельн·ый, -ая, -ое, (*a*) elucidatory, explanatory

разъ·ясн·и́ть, -я́т, (*perf*) *see* **разъ·-ясн·я́·ть**

разъ·ясн·я́·ть, -ют, (*imp*) make clear, elucidate; **-ся** become clear

раз·ыск·а́ть, (*fut 3rd pl*) **раз·ы́щ·ут,** (*perf*) *see* **раз·ы́ск·ива·ть**

раз·ы́ск·ива·ть, -ют, (*imp*) seek out, find (by searching)

раз·ы́щ·ет, (*fut 3rd sing*) of **раз·ыс-к·а́ть**

ра́·й, -я, (*m*) paradise

рай- (*abbr component*) = **райо́н·н·ый**

райгра́с·, -а, (*m*) (bot) rye grass, *Lolium*

райм·о́вк·а, -и, (*g pl*) **-вок·,** (*f*) (met) zinc retort slag

райо́н·, -а, (*m*) district, region, zone, area; "rayon" (Soviet administrative area)

~ воз·ду́ш·н·ых под·хо́д·ов (air) approach funnel

~, во́ль·н·ый free zone
~ дéй·стви·я radius of action
~ рас·по·лож·éни·я (mil) assembly area
~ с·бо́р·а (mil) rallying area
~ с·ва́л·к·и гру́нт·а (naut) spoil ground

райо́н·и́ровани·е, -я, (*n*) (*v n*), *see* **райо́н·и́р·овать;** regionalization

райо́н·и́рованн·ый, -ая, -ое, (*past part pass*) of **райо́н·и́р·овать;** zoned, regionalized; multi-exchange (telephones)

~ сет·ь (*f*) multi-exchange area (telephones)

райо́н·и́р·овать, -уют, (*imp and perf*) zone, divide into regions/districts

райо́н·н·ый, -ая, -ое, (*a*) of **райо́н·;** regional, district

ра́й·ск·ий, -ая, -ое, (*a*) paradise

ра́к·, -а, (*m*) (zool) crayfish; (med) cancer, carcinoma; (bot) cancer; (astron) Cancer, the Crab

~, реч·н·о́й (zool) crayfish, *Astacus leptodactylus*

ра́кел·ь, -я, (*m*) (print) doctor

ракéт·а, -ы, (*f*) *see also* **с·на·ря́д·;** rocket; racket (sport)

~, зени́т·н·ая у·правл·я́ем·ая surface-to-air missile

~, лине·мет·а́тельн·ая (naut) life-saving rocket

раке́т·а
~, по́рох·ов·а́я со·ста́в·н·а́я all solid-propellant rocket
~, по·са́д·очн·ая (air) landing rocket
~, сверх·да́ль·н·ая баллист·и́ческ·ая intercontinental ballistic missile
~, сигна́ль·н·ая signal rocket/flare
~, со·ста́в·н·а́я staged rocket
~ -спу́т·ник·, -а, (m) orbital rocket
~, ста́рт·ов·ая (air) jet-assisted take-off unit
~, тормоз·н·а́я retrorocket

ракет·ниц·а, -ы, (i) -ей, (f) signal pistol

раке́т·н·ый, -ая, -ое, (a) rocket, rocket-propelled
~ дви́г·ател·ь (m) rocket engine
~ дви́г·ател·ь, жи́дк·остн·ый (m) liquid fuel/propellant rocket engine
~ патро́н· (m) signal cartridge
~ стан·о́к· (m) (naut) rocket life-saving gear
~ старт· (m) (a/c) rocket-assisted takeoff

ракето·дро́м·, -а, (m) rocket launching base

ракето·но́с·ец·, -с·ц·а, (m) rocket carrier

ракето·пла́н·, -а, (m) rocket aircraft

раки́т·а, -ы, (f) (bot) goat willow, *Salix caprea*

раки́т·ник·, -а, (m) (bot) broom, *Cytisus*

ракко́рд·, -а, (m) see **рако́рд·**
ракку́рс·, -а, (m) see **раку́рс·**
ракли́ст·, -а, (m) (text) printer
ра́кл·я, -и, (f) (text) doctor

рако·ви́д·н·ый, -ая, -ое, (a) cancerous

ра́ковин·а, -ы, (f) shell, (pal) test; (met) hole, pipe, cavity; (anat) concha; (naut) quarter; (telecom) earphone, earpiece; (build) sink, kitchen sink; open-air stage/bandstand; cissing (paintwork)
~, втор·и́чн·ая у·са́д·очн·ая (met) secondary pipe
~, га́з·ов·ая (met) gas cavity/hole, blowhole
~, звёзд·чат·ая (pal) stellate test
~, земл·ян·а́я (met) sand hole
~, ле́в·ая (naut) port quarter
~, о·кал·и́нн·ая (met) scale pit
~ на по·ве́рх·ност·ь (met) skin hole
~, пес·о́чн·ая (met) sand hole
~, пра́в·ая (naut) starboard quarter
~, трёх·ря́д·н·ая (pal) triserial test

ра́ковин·а
~, у·са́д·очн·ая (met) pipe, tube, contraction/shrinkage cavity

раковин·к·а, -и, (g pl) -н·ок·, (f) (met cast) skin hole

ра́ковин·н·ый, -ая, -ое, (a) of **ра́ковин·а**
~ амёб·ы (pl) (zool) *Thecamoebina*

ра́ковин·чат·ые, -ых, (pl decl as adj) (zool, pal) ostracods, *Ostracoda*

ра́к·овист·ый, -ая, -ое, (a) shelly, conchoidal

ра́к·ов·ый, -ая, -ое, (a) (zool) crayfish; (med) cancerous, carcinomatous
~ шей·к·а (f) (bot) bistorta, *Polygonum bistorta*

рако·обра́з·н·ый, -ая, -ое, (a) cancroid; crustaceous; (pl as noun) crustaceans, *Crustacea*; (see also associated adjectives used as nouns)

рако́рд·, -а, (m) (cinema) leader, trailer

рако·скорпио́н·ы, -ов, (m pl) (zool) merostomes

рако·у·сто́й·чив·ый, -ая, -ое, (a) (agr) cancer-resistant

ракс·, -а, (m) (naut) hank, parrel; jaw rope

ракс-бу́гел·ь, -я, (m) (naut) traveller ring

раку́рс·, -а, (m) (opt) foreshortening
~ це́л·и (air) target aspect

раку́рс·н·ый, -ая, -ое, (a) of **раку́рс·**
~ при·це́л· (m) (gunn) predictor sight

рак·у́шечник·, -а, (m) (geol) coquina, shellrock

рак·у́шечн·ый, -ая, -ое, (a) of **рак·у́шк·а**
~ корро́з·и·я (f) (met) pitting

рак·у́шк·а, -и, (g pl) -шек·, (f) mussel, cockle shell, (pl) barnacles

рак·у́шков·ые, -ых, (pl decl as adj) (zool) *Ostracoda*

ра́л·о, -а, (m) (anat) vomer

ра́м·а, -ы, (f) frame; foot (of boiler); mounting (of engine); ladder (of dredger); carrier (of gun); (print) chase; (geol) enclosing rock
~, гусен·и́чн·ая track frame (of a tractor)
~, двер·н·а́я door jamb
~, креп·ёжн·ая (min) set, frame set
~, лесо·пи́ль·н·ая frame saw
~, о·хра́н·н·ая (naut) bumping fender, camel
~, фунда́мент·н·ая bedplate, bearer

рами (*n indecl*) (bot) rhamie, *Boehmeria*

рами·се́кци·я, -и, (*f*) (med) ramisection

ра́м·к·а, -и, (*g pl*) **-м·ок·,** (*f*) frame; (rad) frame aerial, loop antenna; (shipb) floor, frame; margin (of map etc.); (*pl*) scope, limits

~, **браке́т·н·ая** (shipb) bracket floor

~, **вращ·а́ющ·аяся** (rad) rotating frame aerial

~, **об·де́л·очн·ая** (shipb) boundary angle

~ **с гнёзд·ами** jack strip (telephony)

~, **сплош·н·а́я** (shipb) non-watertight frame

~, **флор·н·ая** (shipb) solid floor

~, **холост·а́я** jack spacer, end panel (telephones)

рамн·ази́н·, -а, (*m*) (chem) rhamnazin

рамн·ети́н·, -а, (*m*) (chem) rhamnetin

рамн·о́з·а, -ы, (*f*) (chem) rhamnose

ра́м·н·ый, -ая, -ое, (*a*) of **ра́м·а**

~ **констру́кци·я** (*f*) (civ eng) framework

~ **стан·о́к·** (*m*) frame saw

~ **шпангоут·** (*m*) (shipb) web frame

ра́м·ов·ый, -ая, -ое, (*a*) of **ра́м·а,** *see also* **ра́м·н·ый**

ра́м·очн·ый, -ая, -ое, (*a*) of **ра́м·к·а**

~ **анте́нн·а** (*f*) (rad) loop/coil/frame antenna/aerial

ра́мп·а, -ы, (*f*) ramp; (theat) footlights; wharf (of coke oven)

ра́м·щик·, -а, (*m*) saw frame operator

ра́н·а, -ы, (*f*) (med) wound, vulnus

ранверсма́н·, -а, (*m*) (air) stall turn

ра́нг·, -а, (*m*) rank

капита́н· втор·о́го ра́нг·а (*m*) Captain Second Rank (equiv. Commander R.N./U.S.N.)

капита́н· перв·о́го ра́нг·а (*m*) Captain 1st Rank (equiv. Captain R.N./U.S.N.)

капта́н· тре́ть·его ра́нг·а Captain Third Rank (equiv. Lieutenant-Commander R.N./U.S.N.)

ранго́ут·, -а, (*m*) (naut) masts and spars

ранго́ут·н·ое де́рев·о (*n*) (naut) spar

ранд·ба́л·к·а, -и, (*g pl*) **-л·ок·,** (*f*) (build) wall beam

ран·ев·о́й, -а́я, -о́е, (*a*) wound, vulnerary

ра́н·ее (*comp adj*) = **ра́нь·ш·е** earlier

ран·е́ни·е, -я, (*n*) wounding, vulneration; wound

ра́н·енн·ый, -ая, -ое, (*past part pass*) of **ра́н·ить;** wounded

ра́н·ен·ый, -ого, (*m or f decl as adj*) wounded/injured person

ра́н·ец·, -н·ц·а, (*m*) rucksack, knapsack, haversack, (mil) pack

ран·и́мост·ь, -и, (*f*) vulnerability

ра́н·ить, -ят, (*imp and perf*) wound

ранне·зре́·л·ый, -ая, -ое, (*a*) (hortic) early-maturing, early

ранне·спе́·л·ый, -ая, -ое, (*a*) (hortic) early-ripening, early

ранне·трет·и́чн·ый, -ая, -ое, (*a*) (geol) Early Tertiary

ра́нн·ий, -яя, -ее, (*a*) early

~ **голоце́н·** (*m*) (geol) Eoholocene

ра́н·о (*adv*) early

рано·о·пад·а́ющ·ий, -ая, -ее, (*a*) (bot) caducous

ра́нт·, -а, (*m*) edge, ledge; welt (of footwear)

рант·ов·о́й, -а́я, -о́е, (*a*) of **рант·**

ра́нт·ов·ый, -ая, -ое, (*a*) of **рант·**

ра́н·цев·ый, -ая, -ое, (*a*) of **ра́н·ец·**

ра́нь·ш·е (*comp*) earlier; (*prep + gen*) before, previously, formerly

~ **сро́к·а** ahead of schedule

рап·а́, -ы́, (*f*) (geol) salt water, brine

рапиди́н·, -а, (*m*) (elec) cascade exciter

рапи́д·ы, -ов, (*m pl*) rapides, rapid dyes

ра́п·н·ый, -ая, -ое, (*a*) brine, saltwater

ра́порт·, -а, (*m*) report

рапорт·ова́ть, -у́ют, (*imp and perf*) report, make/deliver a report

ра́ппер·, -а, (*m*) (met) wrapper (for coiling strip)

раппо́рт·, -а, (*m*) (text) repeat (of pattern)

ра́пс·, -а, (*m*) (bot) rape

ра́пс·ов·ый, -ая, -ое, (*a*) of **рапс·**

рас- (*prefix*) see **раз-**

ра́с·а, -ы, (*f*) (genet) race

рас·кал·ённост·ь, -и, (*f*) incandescence, glow

рас·кал·ённ·ый, -ая, -ое, (*past part pass*) see **рас·кал·я́·ть;** very hot, incandescent

~ **до·бел·а́** white-hot

~ **до·красн·а́** red-hot

рас·кал·и́ть, -я́т, (*perf*) see **рас·кал·я́·ть**

рас·ка́л·ывани·е, -я,   (*n*) (*v n*) of **рас·ка́л·ыва·ть**

рас·ка́л·ыва·ть, -ют,   (*imp*) split, split up, cleave; chop/break up; **-ся** (*pass*); crack, break into pieces

рас·кал·я́·ть, -ют,   (*imp*) heat, make very hot; **-ся** become very hot/ incandescent

~ **до·бел·а́**   bring to white heat

~ **до·красн·а́**   bring to red heat

рас·ка́п·ыва·ть, -ют,   (*imp*) dig, enlarge (by digging); dig up, unearth; excavate, carry out excavations

рас·ка́рмл·ива·ть, -ют, (*imp*) fatten

рас·ка́т·, -а,   (*m*) (*v n*), *see* **рас·кат·а́·ть** and **рас·кат·и́ть**; speed (rolling or sliding downhill); slide, slope; (met) slab; (print) ink distribution; peal (of thunder)

рас·ка́т·анн·ый, -ая, -ое, (*past part pass*) of **рас·кат·а́·ть**

рас·кат·а́·ть, -ют, (*perf*) unroll, unreel, unwind; roll, roll flat/thin, iron (clothes, linen); break out, unreef (a sail), pay out (a rope etc.); (print) distribute ink (with a roller)

рас·ка́т·ист·ый, -ая, -ое, (*a*) rolling, booming (of sound)

рас·кат·и́ть, -́ят, (*perf*) set/get rolling; roll apart; (met) flare, expand by rolling, reel (tubes)

рас·ка́т·к·а, -и,   (*f*) (*v n*), *see* **рас·кат·а́·ть**; (*v n*), *see* **рас·кат·и́ть**; (met) reeling (tubes); flaring, expanding (flat pieces with holes)

ис·пыт·а́ни·е на рас·ка́т·к·у   (*n*) (met) hammering test

рас·ка́т·н·ой, -а́я, -о́е, (*a*) of **рас·ка́т·** and **рас·ка́т·к·а**

~ **стан·**   (*m*) reeler, reeling machine (for furnishing and expanding tubes)

рас·ка́т·очн·ый, -ая, -ое,   (*a*) of **рас·ка́т·к·а**

~ **ва́л·ик·**   (*m*) (print) distributor roller

рас·ка́т·ыва·ть, -ют,   (*imp*) *see* (1) **рас·кат·а́·ть** (*perf*), (2) **рас·кат·и́ть** (*perf*)

рас·ка́ч·анн·ый, -ая, -ое, (*past part pass*) *see* **рас·ка́ч·ива·ть**

рас·кач·а́·ть, -ют, (*perf*) *see* **рас·ка́ч·ива·ть**

рас·ка́ч·енн·ый, -ая, -ое, (*past part pass*) of **рас·кат·и́ть**

рас·ка́ч·ивани·е, -я,   (*n*) (*v n*) of **рас·ка́ч·ива·ть**

~, **про·до́ль·н·ое** (a/c) phugoid

рас·ка́ч·иватель·, -я,   (*m*) (rad) driver

рас·ка́ч·ива·ть, -ют,   (*imp*) rock, shake, wobble; oscillate, swing; drive (vacuum tubes)

рас·ка́ч·к·а, -и,   (*f*) surging (of a/c propeller); drive (of vacuum tubes); build up (of an oscillation)

рас·ка́ш·ива·ть, -ют,   (*imp*) brace; put/make aslant; mow all/across

рас·квартир·ова́ни·е, -я,   (*n*) (mil) billeting, quartering

рас·ква́с·ить, -ят, (*perf*) *see* **рас··кваш·ива·ть**

рас·ква́ш·ива·ть, -ют,   (*imp*) reduce to mush, make mushy

рас·кид·а́·ть, -ют,   (*perf*) scatter, spread, distribute (by throwing)

рас·кид·и́ст·ый, -ая, -ое,   (*a*) scattered, spreading (e.g. of trees etc.)

рас·кид·н·о́й, -а́я, -о́е,   (*a*) folding, collapsible

рас·кид·ыватель·, -я,   (*m*) (agr) spreader

рас·кид·ыва·ть, -ют,   (*imp*) *see* (1) **рас·кид·а́·ть** (*perf*), (2) **рас·ки́·н·уть** (*perf*)

рас·ки́·н·уть, -ут,   (*perf*) throw/spread/ lay out; pitch (a tent)

рас·кис·а́·ть, -ют,   (*imp*) rise (of dough); become soft/boggy

рас·кисл·е́ни·е, -я,   (*n*) (*v n*), *see* **рас·кисл·я́·ть**; (chem) deoxidation; killing (of steel)

рас·ки́сл·енн·ый, -ая, -ое, (*past part pass*) *see* **рас·кисл·я́·ть**; killed (of steel)

рас·кисл·и́тел·ь, -я,   (*m*) deoxidizer, deoxidant; (met) scavenger

рас·кисл·и́ть, -я́т,   (*perf*) *see* **рас··кисл·я́·ть**

рас·кисл·я́·ть, -ют,   (*imp*) deoxidize

рас·ки́с·нуть, -нут,   (*perf*) *see* **рас··кис·а́·ть**

рас·кла́д·к·а, -и,   (*f*) (*v n*), *see* **рас··кла́д·ыва·ть**

рас·клад·н·о́й, -а́я, -о́е,   (*a*) folding, collapsible

рас·кла́д·очн·ый, -ая, -ое, (*a*) distributing, spreading

~ **маши́н·а** (*f*) (text) spreadboard

рас·кла́д·чик·, -а, (*m*) folder

рас·кла́д·ыватель·, -я,   (*m*) (agr) spreader

рас·кла́д·ыва·ть, -ют,   (*imp*) lay/set out, distribute, put, place; spread out, spread; (math) expand; make, build (fires)

рас·клѐ·ива·ть, -ют, (*imp*) unstick; stick up (in several places), put up/out posters/bills (etc.)

рас·клѐ·ить, -ят, (*perf*) *see* рас·клѐ·ива·ть

рас·клеп·а́·ть, -ют, (*perf*) *see* рас·клёп·ыва·ть

рас·клёп·ыва·ть, -ют, (*imp*) unrivet, remove rivets from; unshackle

рас·клѝн·ива·ть, -ют, (*imp*) knock away wedge(s)/chock(s), unwedge, unchock; split, open (with wedge); expand, open up (a crack)

рас·клѝн·ивающ·ий, -ая, -ее, (*pres part act*) of рас·клѝн·ива·ть

~ давл·ѐни·е (*n*) disjoining pressure

~ момѐнт· (*m*) wedging moment

рас·клин·ѝть, -я́т, (*perf*) *see* рас·клѝн·ива·ть

рас·клин·цо̂вк·а, -и, (*g pl*) -вок·, (*f*) (build) garreting, galleting, blinding

рас·ко́в·анн·ый, -ая, -ое, (*past part pass*) *see* рас·ко́в·ыва·ть; beaten (of gold leaf etc.)

рас·ков·а́ть, (*fut 3rd pl*) рас·ку·ю́т, (*perf*) *see* рас·ко́в·ыва·ть

рас·ко́в·к·а, -и, (*f*) (*v n*), *see* рас·ко́в·ыва·ть

рас·ко́в·ыва·ть, -ют, (*imp*) forge/hammer out, flatten (by hammering), beat flat, beat (metal); unshoe (horses); unshackle

рас·ковы́р·ива·ть, -ют, (*imp*) pick at a hole, enlarge a hole by picking

рас·ковыр·я́ть, -ют, (*perf*) *see* рас·ковы́р·ива·ть

рас·код·и́р·овать, -уют, (*imp and perf*) decode

рас·ко́л·, -а, (*m*) (*v n*) *see* рас·ка́л·ыва·ть; split, cleavage

рас·кола́ч·ива·ть, -ют, (*imp*) flatten (by beating); knock to pieces

рас·коло́т·, -а, (*m*) (min) divider

рас·колот·и́ть, -я́т, (*perf*) *see* рас·кола́ч·ива·ть

рас·ко́л·от·ый, -ая, -ое, (*past part pass*) *see* рас·ка́л·ыва·ть

рас·кол·о́ть, -ют, (*perf*) *see* рас·ка́л·ыва·ть

рас·коло́ч·енн·ый, -ая, -ое, (*past part pass*) *see* рас·кола́ч·ива·ть

рас·компенс·а́ци·я, -и, (*f*) decompensation

рас·консерв·а́ци·я, -и, (*f*) bringing out of reserve; demothballing; degreasing (equipment in reserve)

рас·ко́п·анн·ый, -ая, -ое, (*past par pass*) *see* рас·ка́п·ыва·ть

рас·копа́·ть, -ют, (*perf*) *see* рас·ка́п·ыва·ть

рас·ко́п·к·а, -и, (*g pl*) -п·ок·, (*f*) (*v n*), *see* рас·ка́п·ыва·ть; (*pl*) excavation(s) dig, digging

рас·корач·ива·ть, -ют, (*imp*) (elec) unshort

рас·корм·и́ть, -я́т, (*perf*) *see* рас·ка́рмл·ива·ть

рас·корот·и́ть, -я́т, (*perf*) *see* рас·кора́ч·ива·ть

рас·корч·ева́ть, -у́ют, (*perf*) *see* рас·корч·ёвыва·ть

рас·корч·ёвыва·ть, -ют, (*imp*) uproot, grub/stub up/out, clear (area of roots)

рас·ко́с·, -а, (*m*) cross stay/strut, angle brack; (naut) sprit

рас·кос·и́н·ы (*f pl*) (naut) riders

рас·кос·и́ть, -я́т, (*perf*) brace; put/make aslant; рас·кос·и́ть, -я́т, (*perf*); mow all/across

рас·ко́с·ый, -ая, -ое, (*a*) slanting

рас·кра́·ива·ть, -ют, (*imp*) cut out

рас·крас·и́ть, -я́т, (*perf*) *see* рас·кра́ш·ива·ть

рас·кра́с·к·а, -и, (*g pl*) -с·ок·, (*f*) (*v n*), *see* рас·кра́ш·ива·ть; coloured/colored pattern, colouration, coloration

рас·кра́ш·ива·ть, -ют, (*imp*) colour, paint, dye, stain (with several colours); tint (a photograph); crumble

рас·креп·и́ть, -я́т, (*perf*) *see* рас·крепл·я́·ть

рас·крепл·я́·ть, -ют, (*imp*) unfasten, unlash

рас·критик·ова́ть, -у́ют, (*perf*) criticize severely/sharply

рас·кро·и́ть, -я́т, (*perf*) *see* рас·кра́·ива·ть

рас·кро́·й, -я, (*m*) (*v n*), *see* рас·кра́·ива·ть; cutout, cut-out portion/part

~, лѐнт·очн·ый (text) band knife cutting

~ на·сти́л·а (text) cutting out the lay

рас·кро́·ет, (*fut 3rd sing*) of рас·кры́·ть (*perf*), *see* рас·крыв·а́·ть (*imp*)

рас·кро́ш·енн·ый, -ая, -ое, (*past part pass*) of рас·крош·и́·ть

рас·крош·и́ть, -а́т, (*perf*) crumble

рас·крут·и́ть, -я́т, (*perf*) *see* рас·кру́ч·ива·ть

рас·кру́т·к·а, -и,   (*f*) (*v n*), *see* **рас·-кру́ч·ива·ть;** acceleration (of a turbine); over-revving (of a propeller)

рас·кру́ч·енн·ый, -ая, -ое, (*past part pass*) *see* **рас·кру́ч·ива·ть**

рас·кру́ч·ива·ть, -ют, (*imp*) untwist, unravel; rev up, accelerate (a rotary engine); **-ся** (*pass*); gather speed (of rotary engines)

рас·кры́·в·, -а, (*m*) aperture, flare

~ **антённ·ы** (rad) antenna aperture
   у́гол· рас·кры́·в·а (*m*) flare angle (of horn antenna)

рас·крыв·а́·ть, -ют, (*imp*) open, open up; uncover, disclose; discover; **-ся** (*pass and reflex*); (bot) dehisce

рас·крыв·а́ющ·ийся, -аяся, -ееся, (*pres part act*) *see* **рас·крыв·а́·ть;** (bot) dehiscent

рас·кры́т·и·е, -я, (*n*) (*v n*), *see* **рас·-крыв·а́·ть;** (math) solution, solving

~ **кольц·а́** (chem) ring opening

~ **не·о·пре·дел·ённост·и** (math) evaluation of an indeterminate form
   у́гол· рас·кры́т·и·я шв·а (*m*) (weld) included angle

рас·кры́·ть, (*fut 3rd pl*) **рас·кро́·ют,** (*perf*) *see* **рас·крыв·а́·ть**

рас·кряж·ёвк·а, -и, (*g pl*) **-вок·,** (*f*) cross-cutting (wood), sawing up (a felled tree trunk)

рас·ку·ёт, (*fut 3rd sing*) of **рас·ко·в·а́ть** (*perf*), *see* **рас·ко́в·ыва·ть** (*imp*)

рас·куп·а́·ть, -ют, (*imp*) buy up

рас·куп·и́ть, -́ят, (*perf*) *see* **рас·ку-п·а́·ть**

рас·ку́пор·ива·ть, -ют, (*imp*) uncork, unstop; open

рас·ку́пор·ить, -ят, (*perf*) *see* **рас·-ку́пор·ива·ть**

рас·ку́пор·к·а, -и, (*f*) (*v n*), *see* **рас·-ку́пор·ива·ть**

рас·кус·и́ть, -́ят, (*perf*) *see* **рас·ку́-с·ыва·ть**

рас·ку́с·ыва·ть, -ют, (*imp*) bite to pieces

рас·ку́т·а·ть, -ют, (*perf*) *see* **рас·-ку́т·ыва·ть**

рас·ку́т·ыва·ть, -ют, (*imp*) unwrap

ра́с·ов·ый, -ая, -ое, (*a*) racial

рас·па́·вш·ийся, -аяся, -ееся, (*past part act*) *see* **рас·пад·а́·ться**

рас·па́д·, -а, (*m*) (*v n*), *see* **рас·па-д·а́·ться;** disintegration, decomposition, dissociation, decay, breakdown

~**, а́том·н·ый** atomic decay

рас·па́д

~**, вз·рыв·н·о́й** (nucl) explosive disintegration

~ **зерн·а́** (phot) grain disintegration
   ме́ст·о рас·па́д·а (*n*) breakdown point (polymers)
   ме́ст·о рас·па́д·а электро́д·а pitch circle (elec arc furnace)
   мо́дул·ь рас·па́д·а (*m*) (nucl) decay modulus

~ **на две част·и́ц·ы** (nucl) two-particle decay
   тепл·от·а́ рас·па́д·а (*f*) heat of dissociation

~**, терм·и́ческ·ий** thermal decomposition; thermal breakdown (polymers)

~ **ядр·а́** nuclear disintegration

рас·пад·а́·ться, -ются, (*imp*) disintegrate, decompose, dissociate, crumble, fall apart; (nucl) decay, disintegrate; fall into (groups, categories etc.), be split/divided up

рас·пад·е́ни·е, -я, (*n*) (*v n*), *see* **рас·-пад·а́·ться;** fall, collapse (of an organization)

~ **ли́н·и·й** (math) decomposition of curves

рас·па́·ива·ть, -ют, (*imp*) unsolder

рас·па́й·к·а, -и, (*f*) (*v n*), *see* **рас·-па́·ива·ть**

рас·пакер·о́вк·а, -и, (*f*) (oil) packer removal

рас·пак·о́ванн·ый, -ая, -ое, (*past part pass*) *see* **рас·пак·о́выва·ть**

рас·пак·ова́ть, -у́ют, (*perf*) *see* **рас·-пак·о́выва·ть**

рас·пак·о́вк·а, -и, (*f*) (*v n*), *see* **рас·-пак·о́выва·ть**

рас·пак·о́выва·ть, -ют, (*imp*) unpack; undo (a parcel or package)

рас·па́·лся, (*past masc sing*) of **рас·-па́с·ться** (*perf*), *see* **рас·пад·а́·ться** (*imp*)

рас·па́луб·ить, -ят, (*perf*) dismantle (concrete), remove the shuttering/formwork

рас·па́луб·к·а, -и, (*f*) dismantling (concrete)

рас·па́р·, -а, (*m*) cylinder, belly, bosh parallel (of blast furnace)

рас·па́р·енный, -ая, -ое, (*past part pass*) *see* **рас·па́р·ива·ть;** steam-softened

рас·па́р·ива·ть, -ют, (*imp*) steam (to soften)

рас·па́р·ить, -ят, (*perf*) *see* **рас·па́-р·ива·ть**

рас·па́р·к·а, -и, (*f*) steam softening

рас·па́р·ыва·ть, -ют, (*imp*) unrip, rip open, undo (sewn seams), unstitch

рас·па́с·ться, (*fut 3rd pl*) **рас·па·д·у́тся,** (*perf*) *see* **рас·пад·а́·ться**

рас·пах·а́ть, (*fut 3rd pl*) **рас·па́ш·ут,** (*perf*) plough/plow up, till

рас·па́х·ива·ть, -ют, (*imp*) plough/ plow up, till; throw open, open wide

рас·пах·н·у́ть, -у́т, (*perf*) throw open, open wide

рас·па́ш·к·а, -и, (*f*) ploughing/plowing up

рас·паш·н·о́й, -а́я, -о́е, (*a*) double (of doors, gates), two/double-leaf (of doors), double-sash (of windows); arable, ploughed, plowed

~ **весл·о́** (*n*) (naut) single-banked oar

рас·па́·янн·ый, -ая, -ое, (*past part pass*) *see* **рас·па́·ива·ть;** unsoldered

рас·па·я́·ть, -ют, (*perf*) *see* **рас·па́·и·ва·ть**

рас·пер·е́ть, (*fut 3rd pl*) **разо·пр·у́т,** (*perf*) *see* **рас·пир·а́·ть**

рас·печа́т·а·ть, -ют, (*perf*) *see* **рас··печа́т·ыва·ть**

рас·печа́т·ыва·ть, -ют, (*imp*) unseal, open

рас·пи́л·, -а, (*m*) cut (in sawing)

~, **по·пере́ч·н·ый** cross cut/sawing

рас·пи́л·енн·ый, -ая, -ое, (*past part pass*) *see* **рас·пи́л·ива·ть;** resawn, resawed

рас·пи́л·ива·ть, -ют, (*imp*) saw, saw up; resaw

рас·пил·и́ть, -́ят, (*perf*) *see* **рас·пи́·л·ива·ть**

рас·пил·о́вк·а, -и, (*g pl*) **-вок·,** (*f*) resawing

рас·пил·о́вочн·ый, -ая, -ое, (*a*) saw; resawing

рас·пир·а́·ть, -ют, (*imp*) burst (from pressure inside or below)

рас·пис·а́ни·е, -я, (*n*) time-table, schedule

~, **бо·ев·о́е** (nav) watch and station bill

~, **корабе́ль·н·ое** (naut) station bill

~ **по за·ве́д·ывани·ем** maintenance routine

~ **по·езд·о́в** (rail) timetable

рас·пис·а́ть, (*fut 3rd pl*) **рас·пи́ш·ут,** (*perf*) *see* **рас·пи́с·ыва·ть**

рас·пи́с·к·а; -и, (*g pl*) **-с·ок·,** (*f*) (*v n*) *see* **рас·пи́с·ыва·ть;** receipt

~, **долг·ов·а́я** (fin) I.O.U.

~, **обра́т·н·ая** receipt, voucher

рас·пис·н·о́й, -а́я, -о́е, (*a*) decorated

рас·пи́с·ыва·ть, -ют, (*imp*) copy/ write out, transcribe (excerpts); register, enter; schedule, assign; decorate; describe, depict; **-ся** (*pass*); sign, sign for, authenticate (by signing)

рас·пи́ш·ет, (*fut 3rd sing*) of **рас·пи·с·а́ть** (*perf*), *see* **рас·пи́с·ыва·ть** (*imp*)

рас·пла́в·, -а, (*m*) melt, heat; fusion й́ндекс· **рас·пла́в·а** (*m*) melt index, melt flow index (rubber)

рас·пла́в·ить, -ят, (*perf*) *see* **рас·· плавл·я́·ть**

рас·плавл·е́ние, -я, (*n*) (*v n*), *see* **рас·· плавл·я́·ть;** fusion

рас·пла́вл·енн·ый, -ая, -ое, (*past part pass*) *see* **рас·плавл·я́·ть;** molten

рас·плавл·я́·ть, -ют, (*imp*) melt, melt down, fuse

рас·план·ир·ова́ть, -у́ют, (*perf*) *see* **рас·план·иро́выва·ть**

рас·план·иро́вк·а, -и, (*f*) (*v n*), *see* **рас·план·иро́выва·ть**

рас·план·иро́выва·ть, -ют, (*imp*) schedule, plan; plan, lay out to plan, lay out; level, clear (an area)

рас·пла́ст·анн·ый, -ая, -ое, (*past part pass*) *see* **рас·пла́ст·ыва·ть**

рас·пласт·а́·ть, -ют, (*perf*) *see* **рас·· пла́ст·ыва·ть**

рас·пла́ст·ыва·ть, -ют, (*imp*) stratify, laminate; split, split/divide into layers; sprawl; stretch (of wings)

рас·пла́т·а, -ы, (*f*) (*v n*), *see* **рас·· пла́ч·ива·ть;** (fin) settlement

рас·плат·и́ться, -́ятся, (*perf*) *see* **рас·· пла́ч·ива·ться**

рас·пла́ч·ива·ться, -ются, (*imp*) (fin) settle with, pay off/for

рас·плё·л, (*past masc sing*) of **рас·· плес·ти́** (*perf*), *see* **рас·плет·а́·ть**

рас·плеск·а́ть, (*fut 3rd pl*) **рас·плё·щ·ут,** (*perf*) *see* **рас·плёск·ива·ть**

рас·плёск·ива·ть, -ют, (*imp*) splash, splash over, spill

рас·плес·ти́, (*fut 3rd pl*) **рас·плет·у́т,** (*perf*) *see* **рас·плет·а́·ть**

рас·плет·а́·ть, -ют, (*imp*) untwine, unplait, unplat, unlay (rope etc.)

рас·плёщ·ет, (*fut 3rd sing*) of **рас·· плеск·а́ть** (*perf*), *see* **рас·плёск·и·ва·ть**

рас·плод·и́ть, -я́т, (*perf*) breed, propagate

рас·плож·а́·ть, -ют, (*imp*) breed, propagate

рас·плыв·а́ни·е, -я, (n) (v n) of рас·‑
плыв·а́·ться

рас·плыв·а́·ться, -ются, (imp) run
(on ink), spread (of liquids, gases),
deliquesce (of solids); be/become in‑
distinct/unclear/blurred; swim/drift/
float/sail apart

рас·плыв·а́ющ·ийся, -аяся, -ееся,
(pres part act) of рас·плыв·а́·ться;
deliquescent

рас·плы́в·чат·ый, -ая, -ое, (a) in‑
distinct, unclear, blurred

рас·плы́·ться, (fut 3rd pl) рас·плы‑
в·у́тся, (perf) see рас·плыв·а́·ться

рас·плю́щ·ивани·е, -я, (n) (v n) of
рас·плю́щ·ива·ть; (met) spreading,
fullering

 ис·пыт·а́ни·е на рас·плю́щ·и‑
вани·е (n) rivet test (for steel bars)

рас·плю́щ·ива·ть, -ют, (imp) flatten
out, hammer out, draw out (by forg‑
ing); -ся (pass); spread

рас·плю́щ·ить, -ат, (perf) see рас·‑
плю́щ·ива·ть

рас·по·зна·ва́ем·ый, -ая, -ое, (pres part
pass) of рас·по·зна·ва́ть; recogniz‑
able, distinguishable, identifiable

рас·по·зна·ва́ни·е, -я, (n) recognition,
discrimination, identification

~ зна́к·ов character recognition (com‑
puters)

~ из·ображ·е́ни·я pattern recognition
(computers)

рас·по·зна·ва́ть, -ют, (imp) recognize,
discern, diagnose

рас·по·зна́·ть, -ют, (perf) see рас·‑
по·зна·ва́ть

рас·по·лаг·а́ем·ый, -ая, -ое, (pres
part pass) of рас·по·лаг·а́·ть; (mech)
available

~ мо́щ·ност·ь (f) available power

рас·по·лаг·а́·ть, -ют, (imp) dispose,
arrange, place, situate, locate; win
over; (imp only) have at one's disposal,
have available, possess, make use of;
(imp only) make one inclined/disposed/
prone/liable to

рас·по·лаг·а́ющ·ий, -ая, -ее, (pres
part act) of рас·по·лаг·а́·ть; con‑
ducive to

рас·полз·а́·ться, -ются, (imp) crawl/
creep apart; open (of seams etc.)

рас·полз·ти́сь, -у́тся, (perf) see рас·‑
полз·а́·ться

рас·по́л·иться, -ятся, (perf) become
free of ice (rivers)

рас·по·лож·е́ни·е, -я, (n) disposition,
arrangement; situation, location; in‑
clination, tendency

~ гусь·к·о́м tandem arrangement

~, о́бщ·ее general arrangement/view,
layout; (build) situation, siting

~, по·сле́д·овательн·ое tandem ar‑
rangement

~, эшело́н·н·ое unit system (engine)

рас·по·ло́ж·енн·ый, -ая, -ое, (past
part pass) see рас·по·лаг·а́·ть

~ гусь·к·о́м in tandem

рас·по·лож·и́ть, ‑ат, (perf) see рас·‑
по·лаг·а́·ть

рас·по́р·, -а, (m) (mech) thrust; (min)
setting (of pit props)

 у·си́л·и·е рас·по́р·а (n) thrust force

рас·по́р·к·а, -и, (g pl) -р·ок·, (f) (v n),
see рас·пир·а́·ть; distance piece/bar,
spacer, spreader (between wires);
(a/c) strut; shore (of dry dock); waling
board

рас·по́р·н·ый, -ая, -ое, (a) thrust,
brace, distance

~ вту́л·к·а (f) distance piece; gear
retainer (of diesel)

~ кольц·о́ (n) distance collar

~ тру́б·к·а (f) distance piece

~ тя́г·а (f) supporting rod, distance bar

рас·пор·о́ть, ‑ют, (perf) see рас·па́‑
р·ыва·ть

рас·по́р·очн·ый, -ая, -ое, (a) of рас·‑
по́р·к·а

~ изол·я́тор· (m) standoff insulator

рас·по·ряд·и́тел·ь, -я, (m) man/person
in charge

рас·по·ряд·и́тельн·ый, -ая, -ое, (a)
control, controlling; efficient, effective,
active (of persons)

~ пост· (m) (rail) traffic control office

~ ста́нци·я (f) control exchange/office
(telephones); (mil) regulating station,
railhead regulating station

рас·по·ряд·и́ть, -ятся, (perf) see рас·‑
по·ряж·а́·ться

рас·по·ря́д·ок·, -д·к·а, (m) established
order, routine, regulations

~, неде́ль·н·ый week's routine, duties
for the week

рас·по·ряж·а́·ться, -ются, (imp +
instr) order, instruct, direct; be in
charge, take charge of

рас·по·ряж·е́ни·е, -я, (n) (v n) see
рас·по·ряж·а́·ться; order, instruc‑
tion, directive, direction, ordinance

 в рас·по·ряж·е́ни·и at the disposal of

рас·поя́с·ать (*fut 3rd pl*) **рас·поя́-ш·ут,** (*perf*) *see* **рас·поя́с·ыва·ть**

рас·поя́с·ыва·ть, -ют, (*imp*) ungirdle, unstrap, take off a belt

рас·пра́в·ить, -ят, (*perf*) *see* **рас·-правл·я́·ть**

рас·правл·е́ни·е, -я, (*n*) (*v n*), *see* **рас·правл·я́·ть**

рас·правл·я́·ть, -ют, (*imp*) straighten, smooth out, put right/straight; **-ся** (*pass*); take reprisals against

рас·пре·дел·е́ни·е, -я, (*n*) (*v n*), *see* **рас·пре·дел·я́·ть;** distribution, allocation; arrangement

~ **амплиту́д·ы и́мпульс·ов** (phys) pulse amplitude distribution

~ **в·ним·а́ни·я** (a/c) instrument coverage

~ **во вре́мен·и** (phys) time distribution

~ **вы́·ход·а по ма́сс·ам** (phys) yield-mass distribution

~, **га́усс·ов·о** (phys) Gaussian/normal distribution

диагра́мм·а рас·пре·дел·е́ни·я (*f*) (st eng) valve diagram

коэффицие́нт· рас·пре·дел·е́ни·я (*m*) (chem) distribution/partition coefficient

~ **по и́мпульс·ам** (phys) momentum distribution

~ **по угл·а́м** angular distribution

~ **у·добр·е́ни·й** (agr) fertilizer spreading

рас·пре·дел·ённ·ый, -ая, -ое, (*past part pass*) *see* **рас·пре·дел·я́·ть**

~ **по Га́усс·у** normally distributed

~ **по·сто·я́нн·ая** (*f*) (elec) distributed constant

рас·пре·дел·и́тел·ь, -я, (*m*) distributer; (a/c) selector valve

~ **во́з·дух·а, пуск·ов·а́я** starting-air distributor (diesel)

~, **вращ·а́ющ·ийся** rotary distributor

~ **вы́·зов·ов** call distributor (telephones)

~, **золот·ниќов·ый** (a/c) piston-type selector valve

~ **и́мпульс·ов** (elec) command pulse distributor

~, **кла́пан·н·ый** (a/c) poppet-type selector valve

~, **кран·ов·о́й** (a/c) rotor-type selector valve

~ **Мак-Ки** McKee revolving top (blast furnace)

~ **на·во́з·а** (agr) muck/manure spreader

рас·пре·дел·и́тел·ь

~, **ша́г·ов·ый** (elec) step-by-step selector

рас·пре·дел·и́тельн·ый, -ая, -ое, (*a*) distributing, distributive, distribution

~ **вал·** (*m*) (mech) camshaft

~ **за·ко́н·** (*m*) (math) distributive law

~ **ка́мер·а** (*f*) distribution leader (boiler)

~ **коро́б·к·а** (*f*) (st eng) steam chest; (elec) switch box

~ **механ·и́зм·** (*m*) (eng) valve gear; (autom) cycling mechanism

~ **механ·и́зм·, золот·ниќов·ый** slide-valve gear

~ **механ·и́зм·, кла́пан·н·ый** poppet-valve gear

~ **пункт·** (*m*) (elec) distributing centre

~ **у·стро́й·ств·о** (*n*) (elec) switchgear; (autom) controlling unit/means

~ **у·стро́й·ств·о, компле́кт·н·ое** (elec) metal-clad switchgear

~ **у·стро́й·ство, с·бо́р·н·ое** cubicle-type switchgear

~ **щит·** (*m*) (elec) switchboard, distribution board; (telecom) distribution board

~ **щит·, гла́в·н·ый** (elec) main switchboard

~ **я́щик·** (*m*) (elec eng) distribution box

рас·пре·дел·и́ть, -ят, (*perf·*) *see* **рас·-пре·дел·я́·ть**

рас·пре·дел·я́·ть, -ют, (*imp*) distribute, allot, allocate; (agr) spread, distribute; arrange, order

рас·пресс·о́вк·а, -и, (*f*) (*v n*), *see* **рас·-пресс·о́выва·ть**

рас·пресс·о́выва·ть, -ют, (*imp*) press out

рас·про·да·ва́ть, -ют, (*imp*) sell off/out

рас·про·да́·ж·а, -и, (*i*) **-ей,** (*v n*), *see* **рас·про·да·ва́ть;** sale, clearance sale

рас·про·да́·ть, (*fut 3rd sing, pl*) **рас·-про·да́·ст, рас·про·дад·у́т,** (*past masc sing*) **рас·про́·да·л,** (*perf*); *see* **рас·про·да·ва́ть**

рас·про·с·тер·е́ть, (*no future*), (*past masc sing*) **рас·про·с·тёр,** (*perf*); *see* **рас·про·с·тир·а́ть**

рас·про·с·тир·а́·ть, -ют, (*imp*) stretch out, extend, spread

рас·про·стран·е́ни·е, -я, (*n*) propagation, spreading, dissemination, diffusion; (geol) occurrence; (biol) distribution

~, **аном́аль·н·ое** anopropagation

рас·про·стран·éни·е

~ **в свобóд·н·ом про·стрáн·ств·е** free-space propagation

~, **по·вéрх·ностн·ое** ground propagation

по·сто·я́нн·ая рас·про·стран·é·ни·я (*f*) (phys) propagation constant/factor

~ **фýнкци·й** (math) extension of functions

фýнкци·я рас·про·стран·éни·я (*f*) propagation function

рас·про·стран·ённост·ь, -и, (*f*) abundance; extent

~ **в процéнт·ах** percentage abundance

~ **изотóп·ов** (nucl) abundance ratio

рас·про·стран·ённ·ый, -ая, -ое, (*past part pass*) *see* **рас·про·стран·я́·ть;** propagated; abundant; widely used, widespread, common

рас·про·стран·и́тел·ь, -я, (*m*) spreader, disseminator; (biol) vector

рас·про·стран·и́ть, -я́т, (*perf*) *see* **рас·про·стран·я́·ть**

рас·про·стран·я́·ть, -ют, (*imp*) propagate, spread/extend over; distribute, disseminate, give off/out

рас·пры́ск·а·ть, -ют, (*perf*) *see* **рас··пры́ск·ива·ть**

рас·пры́ск·ива·ть, -ют, (*imp*) use up, expend (by spraying), spray away

рас·пряг·á·ть, -ют, (*imp*) unharness

рас·пряж·ёт, (*fut 3rd sing*) of **рас··пря́ч·ь** (*perf*), *see* **рас·пряг·á·ть** (*imp*)

рас·прям·и́ть, -я́т, (*perf*) *see* **рас··прямл·я́·ть**

рас·прямл·éни·е, -я, (*n*) (*v n*), *see* **рас·прямл·я́·ть**

рас·прямл·я́·ть, -ют, (*imp*) straighten out, unbend

рас·пря́ч·ь, (*fut 3rd sing, pl*) **рас·пряж·ёт, рас·пряг·ýт,** (*past masc sing*) **рас·пря́г·,** (*perf*); *see* **рас·пряг·á·ть**

рас·пуск·áни·е, -я, (*n*) (*v n*) of **рас··пуск·á·ть**

~ **лист·в·ы́** (bot) frondescence

рас·пуск·á·ть, -ют, (*imp*) let go/off/out, release, slacken off; disband, dissolve; unreef, set clear (sails), unfurl; **-ся** (*pass*); open (of flowers); put/come out (of leaves); become sticky/soft/boggy, melt, run

рас·пуст·и́ть, -я́т, (*perf*) *see* **рас·пуск·á·ть**

рас·пýст·к·а, -и, (*g pl*) **-т·ок·,** (*f*) (zool) heat, rutting

рас·пýт·а·ть, -ют, (*perf*) *see* **рас·пý·т·ыва·ть**

рас·пýт·иц·а, -ы, (*i*) **-ей,** (*f*) period of bad road conditions

рас·пýт·ыва·ть, -ют, (*imp*) disentangle, unravel, untangle

рас·пýт·ь·е, -я, (*g pl*) **-т·и·й,** (*n*) crossroads, road intersection/parting

рас·пух·áни·е, -я, (*n*) swelling, tumefaction, intumescence

рас·пух·á·ть, -ют, (*imp*) swell, tumefy

рас·пýх·нуть, -нут, (*perf*) *see* **рас··пух·á·ть**

распýщенный, -ая, -ое, (*past part pass*) *see* **рас·пуск·á·ть**

рас·пы́л·, -а, (*m*) *see* **рас·пыл·éни·е**

рас·пыл·éни·е, -я, (*n*) (*v n*), *see* **рас··пыл·я́·ть;** atomization (of liquids); pulverization (of solids); sputtering (thermionics)

~ **в атмосфéр·у** (nucl) dispersion into the air

~, **газо·пламен·н·ое,** (*n*) (plastic) flame spraying

~ **гéттер·а** (elec) getter flash

~, **катóд·н·ое** (elec) sputtering, cathode sputtering

~ **метáлл·а** metal spraying, metallization

~, **механ·и́ческ·ое** solid/mechanical/airless injection (diesels)

~ **ни́т·и** (elec) disintegration of a filament

~, **пневмат·и́ческ·ое** air/blast injection (diesels)

~, **струй·н·ое** direct injection (diesels)

рас·пы́л·ивани·е, -я, (*n*) *see* **рас·пы·л·éни·е**

рас·пы́л·ива·ть, -ют, (*imp*) = **рас··пыл·я́·ть** (q.v.)

рас·пы́л·ивающ·ий стéрж·ен·ь (*m*) atomizer pintle

рас·пыл·и́тел·ь, -я, (*m*) atomizer; pulverizer; spray gun, spray, sprayer; (agr) duster

~, **штифт·ов·óй** pintle nozzle (diesel)

рас·пыл·и́ть, -я́т, (*perf*) *see* **рас·пы·л·я́·ть**

рас·пыл·я́емост·ь, -и, (*f*) dispersibility

рас·пыл·я́·ть, -ют, (*imp*) spray, atomize; pulverize, reduce to dust

рас·пя́л·ива·ть, -ют, (*imp*) stretch (on frame etc.)

рас·пял·и́ть, -ят, (*perf*) *see* **рас·пя́л·ива·ть**

рас·сáд·а, -ы, (*f*) (hortic) seedling

рас·сад·и́ть, [⸗]**ят,** *(perf) see* **рас·са́·ж·ива·ть**

рас·са́д·ник·, -а, *(m)* (hortic) seedbed, hotbed

рас·са́ж·ивани·е, -я, *(n) (v n)* of **рас·са́ж·ива·ть**

рас·са́ж·ива·ть, -ют, *(imp)* set out, seat (an assembly of people), put out, put (a number in several places); (hortic) plant out

рас·са́с·ывани·е, -я, *(n)* resorption, (med) resolution

рас·са́с·ыва·ть, -ют, *(imp)* resorb; **-ся** *(pass)*; (med) resolve

рас·све·ло́, *(past neut sing)* of **рас·свес·ти́** *(perf), see* **рас·свет·а́·ть**

рас·свёрл·ивани·е, -я, *(n) (v n)* of **рас·свёрл·ива·ть**

рас·свёрл·ива·ть, -ют, *(imp)* ream, bore out, enlarge by drilling

рас·сверл·и́ть, -я́т, *(perf) see* **рас·свёрл·ива·ть**

рас·свес·ти́, *(fut 3rd sing)* **рас·све·т·ёт,** *(perf) see* **рас·свет·а́·ть**

рас·свёт·, -а, *(m)* dawn, daybreak

рас·свет·а́·ть, *(pres 3rd sing)* **рас·свет·а́·ет,** *(imp)* get light (at dawn)

рас·се́в·, -а, *(m)* sowing; plansifter (flour milling machine)

рас·сев·а́·ть, -ют, *(imp)* sow

рас·седл·а́·ть, -ют, *(perf) see* **рас·сёдл·ыва·ть**

рас·сёдл·ыва·ть, -ют, *(imp)* unsaddle

рас·се́ет, *(fut 3rd sing)* of **рас·се́·ять** *(perf), see* **рас·се́·ива·ть** *(imp)*

рас·се́·ивани·е, -я, *(n) (v n)* of **рас·се́·ива·ть;** dispersion, diffusion, dissemination
   пло́щад·ь рас·се́·ивани·я *(f)* (gunn) zone of dispersion
   э́ллипс· рас·се́·ивани·я *(m)* (gunn) zone of dispersion

рас·се́·ивател·ь, -я, *(m)* (opt) diffuser; (nucl) scatterer

рас·се́·ива·ть, -ют, *(imp)* sow; scatter, disperse, dissipate, diffuse, disseminate

рас·се́·ивающ·ий, -ая, -ее, *(pres part act)* of **рас·се́·ива·ть;** dissipative, dispersive

~ **вещ·еств·о́** *(n)* (nucl) scatterer, scattering material

~ **ли́нз·а** *(f)* (opt) spreading/diverging lens

~ **мо́щ·ност·ь** *(f)* (phys) dissipative power

рас·се́·ивающ·ий, -ая, -ее

~ **спосо́б·ност·ь** *(f)* (phys) scattering power

рас·сек·а́тел·ь, -я, *(m)* (TV) dissector

рас·сек·а́·ть, -ют, *(imp)* cut up, dissect; cut, slash

~ **волн·у́** (naut) keep head on to the sea

рас·секре́т·ить, -ят, *(perf) see* **рас·секре́ч·ива·ть**

рас·секре́ч·ива·ть, -ют, *(imp)* declassify, release (a secret); withdraw access to secret information

Ра́ссел·, -а, *(m) (see also* **Ре́ссел·)** Russell

рас·се́л·ин·а, -ы, *(f)* (geol) rift, fissure, crevasse

рас·сел·е́ни·е, -я, *(n) (v n) see* **рас·сел·я́·ть;** resettlement; (biol) spread

рас·сел·и́ть, -я́т, *(perf) see* **рас·се·л·я́·ть**

рас·сел·я́·ть, -ют, *(imp)* resettle, move; live apart

рас·се́ч·ь, *(fut 3rd sing, pl)* **рас·се·ч·ёт, рас·сек·у́т,** *(past masc sing)* **рас·сёк·,** *(perf); see* **рас·сек·а́·ть**

рас·се́·яни·е, -я, *(n) (v n) see* **рас·се́·ива·ть;** (phys) scattering, scatter (of reflected radiations); dispersion, diffusion (of transmitted radiations); spread; dissipation (of energy, heat etc.)

~, **бес·по·ря́д·очн·ое** random scattering

~, **диффу́з·н·ое** diffuse scattering

~ **зву́к·а** (phys) acoustic scattering

~, **когере́нт·н·ое** (phys) Bragg/coherent scattering

~, **комбинаци́он·н·ое** (opt) Raman effect/scattering

круж·о́к· рас·се́·яни·я *(m)* (phot) circle of confusion/diffusion

~, **мно́го·кра́т·н·ое** (phys) plural dissipation

~ **мо́щ·ност·и** power dissipation

~ **на бо́льш·ие угл·ы́** (phys) large-angle scattering

~ **на с·вя́з·анн·ых а́том·ах** (nucl) bound scattering

~ **на экра́н·е** (elec) screen dissipation

~, **обра́т·н·ое** back scattering

~ **от стен·** (phys) room scattering

пло́щад·ь рас·се́·яни·я *(f)* (phys) scattering cross-section

по́л·е рас·се́·яни·я *(n)* (phys) stray field

рас·се́·яни·е

~, **рэле́·евск·ое** (opt) Rayleigh scattering

~ **тексту́р·ы** (met) spread (of grain orientations)

ток· рас·се́·яни·я (m) (elec) stray current

~ **у·пру́г·ое** (phys) elastic scattering

рассеянно·ли́ст·н·ый, -ая, -ое, (a) (bot) sparsifolious

рассеянно·не́рв·н·ый, -ая, -ое, (a) (bot) vaginervose

рассеянно·по́р·ист·ый, -ая, -ое, (a) diffuse porous

рас·се́·янн·ый, -ая, -ое, (past part pass) see **рас·се́·ива·ть;** scattered, diffused, diffuse; dissipated; stray

~ **от·раж·е́ни·е** (n) scattered reflection

~ **про·пуск·а́ни·е** (n) diffuse transmission

рас·се́·ять, -ют, (perf) see **рас·се́·ива·ть**

рас·синхрониз·а́ци·я, -и, (f) desynchronization

рас·сказ·а́ть, (fut 3rd pl) **рас·ска́·ж·ут,** (perf) see **рас·ска́з·ыва·ть**

рас·ска́з·ыва·ть, -ют, (imp) expound, relate, narrate, tell

рас·сла́б·ить, -ят, (perf) see **рас··слабл·я́·ть**

рас·слабл·я́·ть, -ют, (imp) weaken

рас·сла́·ивани·е, -я, (n) = **рас··сло·е́ни·е** (q.v.)

рас·сла́·ива·ть, -ют, (imp) separate/divide into layers, layer out (of liquids etc.), stratify, laminate; **-ся** (reflex and pass); (geol) exfoliate, foliate; (chem) demix

рас·сла́нц·евани·е,· -я, (n) (geol) schist formation

рас·сланц·о́ванност·ь, -и, (f) (geol) schistosity

рас·сланц·о́ванн·ый, -ая, -ое, (a) (geol) schistose

рас·сле́д·овани·е, -я, (n) (v n) see **рас·сле́д·овать;** enquiry, investigation

рас·сле́д·овать, -уют, (imp and perf) investigate, examine, look into

рас·сло·е́ни·е, -я, (n) (v n) see **рас··сла́·ива·ть;** stratification, lamination, exfoliation, foliation, spalling; lamination (defect in rolled metal); separation (into layers); ply separation (in tires)

рас·сло·и́ть, -я́т, (perf) see **рас·сла́··ива·ть**

рас·сло́·й, -я, (m) lamination

рас·смотр·ивани·е, -я, (n) (v n) see **рас·смотр·ива·ть;** examination, consideration; (cinema) viewing

рас·сма́тр·ива·ть, -ют, (imp) consider, regard, examine, deal with, cover (a subject)

рас·сма́тр·ыва́ем·ый, -ая, -ое, (pres part pass) of **рас·сма́тр·ива·ть;** under consideration/investigation, in question

рас·смотр·е́ни·е, -я, (n) examination, scrutiny; consideration, treatment

~, **групп·ов·о́е** group treatment

рас·смотр·енн·ый, -ая, -ое, (past part pass) see **рас·сма́тр·ива·ть**

рас·смотр·е́ть, -́ят, (perf) see **рас··сма́тр·ива·ть**

рас·со·глас·ова́ни·е, -я, (n) mismatch, misalignment; error

сигна́л· рас·со·глас·ова́ни·я (m) (telecom) error signal

у́гол· рас·со·глас·ова́ни·я (m) (instr) displacement/error angle

рас·со·глас·о́ванн·ый, -ая, -ое, (a) out-of-step (instruments etc.), unsynchronized

рас·со́л·, -а, (m) brine

~, **матер·и́нск·ий** bittern

рас·со́ль·н·ый, -ая, -ое, (a) of **рас··со́л·**

рас·сорт·ир·ова́ть, -у́ют, (perf) see **рас·сорт·иро́выва·ть**

рас·сорт·иро́вк·а, -и, (g pl) **-вок·,** (f) (v n) see **рас·сорт·иро́выва·ть**

рас·сорт·иро́выва·ть, -ют, (imp) sort, grade, (met etc.) screen

рас·со́с·анн·ый, -ая, -ое, (past part pass) of **рас·сос·а́ть;** resorbed; resolved

рас·сос·а́ть, -у́т, see **рас·са́с·ыва·ть**

рас·со́х·нуться, -нутся, (perf) crack (owing to dryness)

рас·с·пра́ш·ивани·е, -я, (n) (v n) of **рас·с·пра́ш·ива·ть;** interrogation

рас·с·пра́ш·ива·ть, -ют, (imp) question, interrogate

рас·с·прос·и́ть, -́ят, (perf) see **рас··с·пра́ш·ива·ть**

рас·средо·то́ч·ени·е, -я, (n) (v n) see **рас·средо·то́ч·ива·ть;** dispersion, dispersal, spread

рас·средо·то́ч·ива·ть, -ют, (imp) disperse, send to dispersed stations; **-ся** disperse, take up dispersed stations рас·средо·то́ч·ить, -ат, (*perf*) *see* рас·средо·то́ч·ива·ть

рас·сро́ч·ива·ть, -ют, (*imp*) set in-stalments/stages for (payment or com-pletion)

рас·сро́ч·ить, -ат, (*perf*) *see* рас·-сро́ч·ива·ть

рас·сро́ч·к·а, -и, (*f*) (*v n*) *see* рас·-сро́ч·ива·ть
в рас·сро́ч·к·у by instalments

рас·став·а́ни·е, -я, (*n*) (*v n*) *see* рас·-ста·ва́ться

рас·ста·ва́ться, -ются, (*imp*) part (of people), abandon (ideas etc.)

рас·ста́в·ить, -ят, (*perf*) *see* рас·-ставл·я́·ть

рас·ста́вл·енн·ый, -ая, -ое, (*past part pass*) *see* рас·ставл·я́·ть

рас·ставл·я́·ть, -ют, (*imp*) arrange, set/put out, put, place; dispose, station, post; set/put/move apart, open (for a distance); let out (a gar-ment)

рас·станавл·ива·ть, -ют, (*imp*) ar-range, set/put out, put, place; dispose, station, post

рас·станов·и́ть, ⸹ят, (*perf*) *see* рас·-станавл·ива·ть

рас·стан·о́вк·а, -и, (*g pl*) -вок·, (*f*) (*v n*) *see* рас·станавл·ива·ть; ar-rangement, order; (M/T) parking

рас·ста́·ться, -нутся, (*perf*) *see* рас·-ста·ва́ться

рас·стег·а́·ть, -ют, (*imp*) make a pat-terned quilt

рас·стёг·ива·ть, -ют, (*imp*) unfasten, unbutton, unbuckle, unhook; (*imp only*) make a patterned quilt

рас·стег·н·у́ть, -у́т, (*perf*) *see* рас·-стёг·ива·ть

рас·стекл·ова́ни·е, -я, (*n*) devitri-fication

рас·стекл·о́ванн·ый, -ая, -ое, (*a*) devitrified

рас·сте́л·ет, (*fut 3rd sing*) of разо·-стл·а́ть, (*perf*) *see* рас·стил·а́·ть (*imp*)

рас·сти́л·, -а, (*m*) (*v n*) *see* рас·сти-л·а́·ть

рас·стил·а́·ть, -ют, (*imp*) spread, spread/lay out

рас·сто́·ечн·ая доск·а́ (*f*) pastry board

рас·стопор·ивани·е, -я, (*n*) (*v n*) *see* рас·стопор·и́ть

рас·стопор·и́ть, -ят, (*perf*) release, unlock, unlatch; uncage (a gyro)

рас·сто·я́ни·е, -я, (*n*) distance, range, interval, spacing

~, диффузио́н·н·ое (rad) diffusion length

~ до маяк·а́ (nav) omnidistance

~ до объе́кт·а (opt) object distance

~, зени́тн·ое (astron) zenith distance

~, меж·а́том·н·ое (phys) atomic spac-ing

~, меж·пло́ск·остн·ое (cryst) inter-planar spacing
на рас·сто·я́ни·и at a distance

~ скач·к·а́ (rad) skip distance

~, угл·ов·о́е angular distance

~, фо́кус·н·ое (opt) focal length

рас·стра́·ива·ть, -ют, (*imp*) disorder, disarrange, upset; put out of order/alignment/tune; (rad) detune

рас·стре́л·, -а, (*m*) execution by a firing squad; (min) buntons

рас·стре́л·ива·ть, -ют, (*imp*) expend (ammunition); wear (barrel by shoot-ing); shoot (execute); shoot to pieces

рас·стрел·я́·ть, -ют, (*perf*) *see* рас·-стре́л·ива·ть

рас·стро́·енност·ь, -и, (*f*) maladjust-ment; state of disorder

рас·стро́·енн·ый, -ая, -ое, (*past part pass*) *see* рас·стра́·ива·ть; dis-turbed, maladjusted

рас·стро́·ить, -ят, (*perf*) *see* рас·-стра́·ива·ть

рас·стро́й·к·а, -и, (*f*) (*v n*) *see* рас·-стра́·ива·ть; maladjustment; detun-ing

~ по част·от·е́ frequency separation

рас·стро́й·ств·о, -а, (*m*) (*v n*) *see* рас·стра́·ива·ть; disorder, disar-rangement

рас·суд·и́тельн·ый, -ая, -ое, (*a*) reasonable, sensible

рас·суд·и́ть, ⸹ят, (*perf*) judge

рас·су́д·ок·, -д·к·а, (*m*) reason

рас·су́д·очн·ый, -ая, -ое, (*a*) rational

рас·сужд·а́·ть, -ют, (*imp*) reason, argue, discuss, debate

рас·сужд·е́ни·е, -я, (*n*) (*v n*) *see* рас·сужд·а́·ть; reasoning, argument, discussion

рас·су́ч·ива·ть, -ют, (*imp*) untwist

рас·суч·и́ть, ⸹ат, (*perf*) *see* рас·су́-ч·ива·ть

рас·с·чи́т·анн·ый, -ая, -ое, (*past part pass*) *see* рас·с·чи́т·ыва·ть

рас·с·чит·а́·ть, -ют, (*perf*) *see* рас·-с·чи́т·ыва·ть

рас·с·чи́т·ыва·ть, -ют, (*imp*) calculate, reckon; design; dismiss, discharge; (mil) number; **-ся** (*pass*); (fin) settle

рас·сыл·а́·ть, -ют, (*imp*) send (to various places), send round/about, distribute, circulate (papers etc.)

рас·сы́ль·н·ый, -ого, (*m decl as adj*) messenger, courier, delivery man

рас·сып·а́ни·е, -я, (*n*) (*v n*) see **рас·-сып·а́·ть**

рас·сы́п·ать, -лют, (*perf*) see **рас·-сып·а́·ть**

рас·сып·а́·ть, -ют, (*imp*) sprinkle, strew, scatter; pour (solids); **-ся** (*pass*); crumble, fall to pieces; shower (of solids)

рас·сы́п·к·а, -и, (*f*) (*v n*) see **рас·сы́-п·а́·ть;** weight loss in free-flowing materials

рас·сып·н·о́й, -а́я, -о́е, (*a*) (*and see* **ро́с·сып·н·ый**); free-flowing, suitable for sprinkling/strewing/scattering; granular; loose (of goods)

рас·сы́п·чат·ый, -ая, -ое, (*a*) crumbly, friable

рас·сых·а́·ться, -ются, (*imp*) crack (owing to dryness)

рас·та́·ива·ть, -ют, (*imp*) melt

рас·та́лк·ивани·е, -я, (*n*) (*v n*) of **рас·та́лк·ива·ть;** repulsion

~ электро́н·ов electron repulsion

рас·та́лк·ива·ть, -ют, (*imp*) push/shove apart; shake, rouse

рас·та́лк·иваю·щ·ий, -ая, -ее, (*pres part act*) of **рас·та́лк·ива·ть;** repulsing, repelling

рас·та́пл·ивани·е, -я, (*n*) (*v n*) of **рас·та́пл·ива·ть**

рас·та́пл·ива·ть, -ют, (*imp*) light, light up (furnaces etc.); melt

рас·та́пт·ыва·ть, -ют, (*imp*) trample, crush

рас·та́ск·ива·ть, -ют, (*imp*) drag/pull apart

рас·та́ч·ивани·е, -я, (*n*) see **рас·-то́ч·к·а**

рас·та́ч·ива·ть, -ют, (*imp*) bore, bore out

рас·тащ·и́ть, -ат, (*perf*) see **рас·-та́ск·ива·ть**

рас·та́·ять, -ют, (*perf*) see **рас·та́·ива·ть**

рас·тво́р·, -а, (*m*) (chem) solution; (build) mortar; opening, aperture, flare

~, биту́м·н·ый bituminous grout

рас·тво́р

~, бур·ов·о́й (oil) drilling fluid, mud, slush

~ ва́л·ик·ов roll opening (of rolling mills etc.)

~ в·недр·е́ни·я intersticial solution

~, во́д·н·ый aqueous solution

~, гли́н·ист·ый, (oil) drilling fluid, mud, slush

~ за·мещ·е́ни·я substitutional solution

~ зна́к·ов (naut) position with open transit

~, ма́т·очн·ый (chem) mother liquor

~, от·сто·я́вш·ийся (chem) supernatant liquor/liquid

~, регуля́р·н·ый ordered solution

~, стро·и́тельн·ый building mortar

~, твёрд·ый solid solution

~, титр·о́ванн·ый standard solution, titrant

у́гол· рас·тво́р·а (*m*) cone/divergence angle, flare; (rad) aperture angle, antenna beam width, beam angle

~, у·по·ря́д·оченн·ый ordered/super-latticed solid solution

рас·твор·е́ни·е, -я, (*n*) dissolution, dissolving, solution

~, по·вто́р·н·ое re-solution

тепл·от·а́ рас·твор·е́ни·я (*f*) heat of solution

рас·тво́р·енн·ый, -ая, -ое, (*past part pass*) opened, open

рас·твор·ённ·ый, -ая, -ое, (*past part pass*) dissolved

рас·твор·и́мост·ь, -и, (*f*) solubility

рас·твор·и́м·ый, -ая, -ое, (*pres part pass*) of **рас·твор·я́·ть;** solute, soluble

~ вещ·еств·о́ (*n*) solute substance, solute

~, легк·о́ readily soluble

рас·твор·и́тел·ь, -я, (*m*) solvent; solution mixer; thinner (for paint)

рас·твор·и́ть, -я́т, (*perf*) dissolve; (perf) open (something hinged)

раствор·о·меш·а́лк·а, -и, (*g pl*) **-лок·,** (*f*) mortar mixer

раствор·о·на·со́с·, -а, (*m*) mortar pump

рас·твор·я́·ть, -ют, (*imp*) dissolve; open (something hinged); **-ся** (*pass*) be soluble

рас·твор·я́ющ·ий, -ая, -ее, (*pres part act*) of **рас·твор·я́·ть;** solvent, dissolving

рас·твор·я́ющ·ийся, -аяся, -ееся, (*pres part act*) of **рас·твор·я́·ться;** solute

рас·тек·а́ни·е, -я, (*n*) (*v n*) of **рас·тек·а́·ться**

рас·тек·а́·ться, -ются, (*imp*) run, spread (of liquids on solids); flow (of viscous materials)

рас·тёл·, -а, (*m*) (agr) calving

раст·е́ни·е, -я, (*n*) (bot) plant; (*pl*) flora

~, влаго·люб·и́в·ое moisture-loving plant

~, ни́з·ш·ее lower plant

~, со́р·н·ое weed

растение·во́д·, -а, (*m*) (agr) grower; plant breeder

растение·во́д·ство, -а, (*n*) plant/crop growing/husbandry

рас·тер·е́ть, (*fut 3rd pl*) **разо·тр·у́т,** (*perf*) see **рас·тир·а́ть**

рас·тёр·т·ый, -ая, -ое, (*past part pass*) see **рас·тир·а́·ть;** ground

рас·тер·я́·ть, -ют, (*perf*) lose (in stages); **-ся** (*pass*); become confused (of people), be at a loss

рас·тес·а́ть, (*fut 3rd pl*) **рас·теш·у́т,** (*perf*) see **рас·тёс·ыва·ть**

рас·тёс·ыва·ть, -ют, (*imp*) hack out

раст·ёт, (*pres 3rd sing*) of **раст·и́**

рас·те́ч·ься, (*fut 3rd sing, pl*) **рас·те·ч·ётся, рас·тек·у́тся,** (*past masc sing*) **рас·тёк·ся,** (*perf*); see **рас·тек·а́·ться**

рас·те́ш·ет, (*fut 3rd sing*) of **рас·тес·а́ть,** (*perf*); see **рас·тёс·ыва·ть** (*imp*)

раст·и́, -у́т, (*past masc sing*) **рос·,** (*imp*) grow, increase

раст·и́льн·я, -и, (*f*) (agr) germinator, sprouter

рас·тир·а́тел·ь, -я, (*m*) grinder; (min) muller

рас·тир·а́·ть, -ют, (*imp*) grind, rub to powder/pieces, triturate; rub about, spread (by rubbing); rub, massage

рас·ти́р·к·а, -и, (*g pl*) **-р·ок·,** (*f*) (*v n*) see **рас·тир·а́·ть;** rubber (sailmaking)

раст·и́тельност·ь, -и, (*f*) vegetation

раст·и́тельн·ый, -ая, -ое, (*a*) plant; vegetable, vegetative, vegetal

~ ма́сл·о (*n*) vegetable oil

~ форм·а́ци·я (*f*) (biol) phytoformation

раст·и́ть, -я́т, (*imp*) grow, let grow, raise, rear

рас·то́лк·анн·ый, -ая, -ое, (*past part pass*) of **рас·толк·а́·ть**

рас·толк·а́·ть, -ют, (*perf*) push/shove apart; shake, rouse

рас·толк·о́ванн·ый, -ая, -ое, (*past part pass*) see **рас·толк·о́выва·ть**

рас·толк·ова́ть, -у́ют, (*perf*) see **рас·толк·о́выва·ть**

рас·толк·о́выва·ть, -ют, (*imp*) make clear, explain

рас·толк·у́т, (*fut 3rd pl*) of **рас·толо́ч·ь**

рас·толк·у́ют, (*fut 3rd pl*) of **рас·толк·ова́ть**

рас·толо́ч·ь, (*fut 3rd sing, pl*) **рас·толч·ёт, рас·толк·у́т,** (*past masc sing*) **рас·толо́к·** (*perf*) pound, crush, grind

рас·топ·и́ть, ⸚ят, (*perf*) see **рас·та́пл·ива·ть**

рас·то́п·к·а, -и, (*g pl*) **-п·ок·,** (*f*) lighting up (a furnace); kindling (material)

рас·топл·я́·ть, -ют, (*imp*) = **рас·та́пл·ива·ть**

рас·топт·а́ть, (*fut 3rd pl*) **рас·то́пч·ут,** (*perf*) trample, crush

рас·торг·а́·ть, -ют, (*imp*) annul, abrogate

рас·то́рг·нуть, -нут, (*perf*) see **рас·торг·а́·ть**

рас·торг·ова́ть, -у́ют, (*perf*) sell out/off

рас·торж·е́ни·е, -я, (*n*) (*v n*) see **рас·торг·а́·ть**

рас·тормо́ж·ивани·е, -я, (*n*) brake release, releasing the brake; (med) disinhibition

рас·точ·а́·ть, -ют, (*imp*) squander, lavish

рас·точ·е́ни·е, -я, (*n*) squandering, wasting

рас·то́ч·енн·ый, -ая, -ое, (*past part pass*) of **рас·точ·и́ть;** bored, turned

рас·точ·ённ·ый, -ая, -ое, (*past part pass*) of **рас·точ·и́ть;** squandered, lavished

рас·точ·и́тельн·ый, -ая, -ое, (*a*) wasteful, extravagant

рас·точ·и́ть, ⸚ат, (*perf*) bore, bore out; **рас·точ·ить, -а́т,** (*perf*) squander, lavish

рас·то́ч·к·а, -и, (*f*) boring, boring/turning out (on lathe)

~, кон·и́ческ·ая taper boring

~ со·о́с·н·ая line boring

~, толк·а́ющ·ая push boring

~, тя́·нущ·ая pull boring

рас·то́ч·н·ый, -ая, -ое, (*a*) boring

~ голо́в·к·а (*f*) boring head

~ о·пра́в·к·а (*f*) boring bar

рас·то́ч·н·ый, -ая, -ое
~ **рез·е́ц·** (*m*) inside-boring tool
~ **стан·о́к·** (*m*) boring machine
ра́стр·, -а, (*m*) (opt) raster, grating; (print) screen; (TV) scanning field/pattern
~ **из·ображ·е́ни·я** (TV) picture pattern
~, **лин·е́йн·ый** (print) lined screen
~, **лин·е́йчат·ый** (TV) striated pattern
~, **много·уго́ль·н·ый** (print) polygap screen
~, **от·раж·а́тельн·ый** (opt) reflection grating
~, **плён·очн·ый** (print) acetate screen
~, **полу·тон·ов·о́й** half-tone screen
~, **про·пуск·а́ющ·ий** (opt) transmission grating
рас·тра́т·а, -ы, (*f*) waste; embezzlement
рас·тра́т·ить, -ят, (*perf*) see **рас·тра́·ч·ива·ть**
рас·тра́ч·ива·ть, -ют, (*imp*) embezzle; squander
рас·трево́ж·ива·ть, -ют, (*imp*) alarm
рас·трево́ж·ить, -ат, (*perf*) see **рас··трево́ж·ива·ть**
рас·треп·а́ть, -лют, (*perf*) see **рас··трёп·ыва·ть**
рас·трёп·ыва·ть, -ют, (*imp*) fray, tatter, tousle
рас·тре́ск·а·ться, -ются, (*perf*) see **рас·тре́ск·ива·ться**
рас·тре́ск·ивани·е, -я, (*n*) (*v n*) see **рас·тре́ск·ива·ться**; checking, crazing, cracking, spalling; (bot) dehiscence; (nucl) spallation
~, **коррозио́н·н·ое** (met) corrosion/season cracking
~, **сезо́н·н·ое** (met) season cracking
рас·тре́ск·ива·ться, -ются, (*imp*) become covered with cracks, become checked, decrepitate, crack up
рас·тре́ск·ивающ·ийся, -аяся, -ееся, (*pres part act*) of **рас·тре́ск·ива·ться**; (bot) dehiscent
ра́стр·ов·ый, -ая, -ое, (*a*) of **растр·**
~ **микро·ско́п·** (*m*) grating/scanning electron microscope
~ **о́пт·ик·а** (*f*) diffraction grating; diffraction grating optics
рас·тру́б·, -а, (*m*) bell mouth, bell, socket; (acous) flare; (shipb) sea chest; (bot) ocrea, ochrea
~ **вентил·я́тор·а** ventilator cowl, cowl ventilator

рас·тру́б
  со·един·е́ни·е рас·тру́б·ом (*n*) bell-and-spigot joint
рас·трус·и́ть, -я́т, (*perf*) see **рас·тру́·ш·ива·ть**
рас·тру́ш·ива·ть, -ют, (*imp*) spill, lose (small solids); strew, spread (a layer)
рас·тряс·а́ть, -ют, (*imp*) shake out (hay etc.)
рас·тряс·ти́, -у́т, (*perf*) see **рас·тряс·а́·ть**
рас·тря́с·ыва·ть, -ют, (*imp*) = **рас··тряс·а́·ть**
рас·туш·ёвк·а, -и, (*g pl*) **-вок·,** (*f*) shading (on drawing); stump (for shading)
раст·у́щ·ий, -ая, -ее, (*pres part act*) of **раст·и́**
рас·ты́к·а·ть, -ют, (*perf*) see **рас·ты́·к·ива·ть**
рас·тык·а́·ть, -ют, (*imp*) see **рас·ты́·к·ива·ть**
рас·ты́к·ива·ть, -ют, (*imp*) put, shove in/on
рас·тюк·ова́ть, -у́ют, (*perf*) unpack, unbale
раст·я́, (*pres gerund*) of **раст·и́**
рас·тя́г·ивани·е, -я, (*n*) (*v n*) of **рас··тя́г·ива·ть**; see also **рас·тяж·е́·ни·е**; protraction, expansion, extension
~ **диапазо́н·а** (rad) bandspreading
~ **до́з·ы** (nucl) dose protraction
~ **раз·вёрт·к·и** sweep expansion (radar)
рас·тя́г·иватель·, -я, (*m*) stretcher, extender
~ **и́мпульс·а** pulse stretcher
рас·тя́г·ива·ть, -ют, (*imp*) stretch, stretch out, extend, pull/draw out; prolong, protract; (med) pull, strain, sprain
рас·тяж·е́ни·е, -я, (*n*) (*v n*) see **рас··тя́г·ива·ть**; extent, range; (met, mech) extension, tension, elongation; (med) stretching, dilation
  диагра́мм·а рас·тяж·е́ни·я (*f*) (mech) load-extension/-elongation curve
  ис·пыт·а́ни·е на рас·тяж·е́ни·е (*n*) (met) tensile test
  крив·а́я рас·тяж·е́ни·я (*f*) (mech) load-elongation curve
~, **лин·е́йн·ое** extension
  мо́дул·ь лин·е́йн·ого рас·тяж·е́·ни·я (*m*) (mech) modulus of elasticity, Young's modulus

рас·тяж·éни·е
мóдул·ь объ·ём·н·ого рас·тя·ж·éни·я (*m*) modulus of contraction
~, объ·ём·н·ое contraction
рас·тяж·ймост·ь, -и, (*f*) stretch-ability, tensility, extensibility
рас·тяж·йм·ый, -ая, -ое, (*a*) tensile, extensible, stretchable
рас·тяж·к·а, -и, (*g pl*) -ж·ок·, (*f*) (*v n*) see **рас·тяг·ива·ть;** *see also* **рас·тяж·éни·е;** (build) bracing wire/bolt
~ диапазóн·а (rad) band spread
рас·тяж·н·óй, -áя, -óе, (*a*) extension, expansion, stretch
~ листо·прáв·ильн·ая машйн·а (*f*) (met roll) stretcher levelling/flattening machine
рас·тя́·нут·ый, -ая, -ое, (*past part pass*) of **рас·тя·н·у́ть;** extended, distended, expanded, spread, stretched; prolonged, protracted
рас·тя·н·у́ть, -ут, (*perf*) see **рас·тя·г·ива·ть**
рас·фаз·йровани·е, -я, (*n*) dephasing; misphasing
рас·фас·óванн·ый, -ая, -ое, (*past part pass*) of **рас·фас·овáть**
рас·фас·овáть, -уют, (*perf*) weigh and package (goods for sale); package
рас·фас·óвк·а, -и, (*g pl*) -вок·, (*f*) packaging
рас·фокус·йровани·е, -я, (*n*) = **рас··фокус·ирóвк·а**
рас·фокус·йрованн·ый, -ая, -ое, (*past part pass*); out-of-focus, defocussed
рас·фокус·ирóвк·а, -и, (*f*) defocus-sing; debunching (electrons)
рас·форм·ир·овáть, -у́ют, (*perf*) see **рас·форм·ирóвыва·ть**
рас·форм·ирóвыва·ть, -ют, (*imp*) disband (organizations); reform (trains)
рас·хвáл·ива·ть, -ют, (*imp*) praise highly
рас·хвал·йть, -ят, (*perf*) see **рас·хвá·л·ива·ть**
рас·хóд·, -а, (*m*) expenditure, con-sumption; discharge (of rivers and pumps), delivery (of pumps), flow rate (of pipelines etc.); gap, span; (*pl*) expense, expenses
~, вес·ов·óй weight/mass flow-rate, weight-to-mass flow rate
~, вне·про·из·вóд·ственн·ые (*pl*) selling and distribution expenses
~ вод·ы́ water consumption; water flow rate (in pipes); water discharge (of rivers)

рас·хóд
~ вóз·дух·а air flow rate
~ вóз·дух·а, мáсс·ов·ый mass air flow
~ жйдк·ост·и flow rate; fluid velocity
~, километр·óв·ый consumption per kilometer
~, кóс·венн·ые (*pl*) (fin) overheads
~, мáсс·ов·ый mass rate
~, на·клад·н·ы́е (*pl*) oncost
~, объ·ём·н·ый volumetric flow rate
~, основ·н·ы́е (*pl*) manufacturing expenses
~, по·сто·я́нн·ые (*pl*) fixed expenses
~, прям·ы́е (*pl*) prime cost
~ тóпл·ив·а fuel consumption; (rocket) propellant consumption
~, у·дéль·н·ый specific consumption
~, час·ов·óй consumption per hour, hourly consumption
рас·ход·ймост·ь, -и, (*f*) divergence
тóч·к·а рас·ход·ймост·и (*f*) (meteor) centre of anti-cyclone
рас·ход·йться, -ятся, (*imp*) disperse, separate, go different ways, part, be separated; diverge, fork; diverge, differ; be used up, be sold out; come on heavily (of rain, snow); gather speed; (*imp only*) walk/pace up and down
рас·хóд·к·а, -и, (*f*) (zool) rutting
рас·хóд·н·ый, -ая, -ое, (*a*) of **рас··хóд·**
~ бак· (*m*) service tank; (a/c) collector tank
~ кнйг·а (*f*) expenses book
~ характер·ӥстик·а (*f*) (hydr) dis-charge characteristic
рас·хóд·овани·е, -я, (*n*) (*v n*) of **рас··хóд·овать;** expenditure, consump-tion
рас·хóд·овать, -уют, (*imp*) spend, expend; consume
рас·ходó·граф·, -а, (*m*) recording flowmeter
рас·ходо·мéр·, -а, (*m*) flowmeter
~, кры́ль·чат·ый turbine flowmeter
~, масс·ов·óй mass flowmeter
~, объ·ём·н·ый displacement flow-meter
~, порш·нев·óй piston-type flow-meter
~, скор·остн·óй inferential flowmeter
~, тахо·метр·ӥческ·ий tachometric flowmeter
~, термо·анемо·метр·ӥческ·ий hot-wire anemometer flowmeter

рас·ходо·мéр·н·ый, -ая, -ое, (*a*) of рас·ходо·мéр·

~ сопл·ó (*n*) flow nozzle

рас·хóд·уем·ый, -ая, -ое, (*pres part pass*) of рас·хóд·овать; consumable

рас·ход·я́щ·ийся, -аяся, -ееся, (*pres part act*) of рас·ход·и́ться; divergent, diverging

~ ряд· (*m*) (math) divergent series

рас·хожд·éни·е, -я, (*n*) (*v n*) of рас··ход·и́ть; divergence, discrepancy, difference (of opinion); separation; widening (of cracks)

~ луч·á beam divergence

~ по фáз·е (elec) phase split

прáв·ил·а рас·хожд·éни·я суд·óв в мóр·е (*pl*) (naut) rule of the road

~ с·ты́к·а separation of a joint

рас·холáж·ивани·е, -я, (*n*) (*v n*) of рас·холáж·ива·ть; (nucl) cooling shut-down

рас·холáж·ива·ть, -ют, (*imp*) cool down

рас·холод·и́ть, -я́т, (*perf*) *see* рас··холáж·ива·ть

рас·царáп·а·ть, -ют, (*perf*) *see* рас··царáп·ыва·ть

рас·царáп·ыва·ть, -ют, (*imp*) scratch, cover with scratches

рас·цвест·и́, (*fut 3rd pl*) рас·цвет·у́т, (*past masc sing*) рас·цвё·л, (*perf*); *see* рас·цвет·á·ть

рас·цвéт·, -а, (*m*) bloom, blossoming; flourishing

рас·цвет·á·ть, -ют, (*imp*) bloom, blossom; flourish, prosper

рас·цвет·и́ть, -я́т, (*perf*) *see* рас··цвéч·ива·ть

рас·цвéт·к·а, -и, (*g pl*) -т·ок·, (*f*) (*v n*) *see* рас·цвéч·ива·ть; colouring, coloring, coloration; color/colour coding (e.g. of cables etc.)

~, пёстр·ая (bot) mottling, chlorosis

рас·цвéч·енн·ый, -ая, -ое, (*past part pass*) *see* рас·цвéч·ива·ть

рас·цвéч·ива·ть, -ют, (*imp*) colour, color, tint, colour/decorate brightly; (naut) dress ship

рас·цен·ённ·ый, -ая, -ое, (*past part pass*) *see* рас·цéн·ива·ть

рас·цéн·ива·ть, -ют, (*imp*) price, value

рас·цен·и́ть, -я́т, (*perf*) *see* рас·цéн·ива·ть

рас·цéн·к·а, -и, (*g pl*) -н·ок·, (*f*) (*v n*) *see* рас·цéн·ива·ть; price, established/fixed price; fixed rate (of remuneration)

рас·цéн·очн·ый, -ая, -ое, (*a*) rate/price-fixing

рас·цéн·щик·, -а, (*m*) rate-fixer

рас·цéп·, -а, (*m*) trip, release

рас·цеп·и́ть, -ят, (*perf*) *see* рас·цеп·л·я́·ть

рас·цéп·к·а, -и, (*f*) (*v n*) *see* рас·цеп·л·я́·ть; disengagement, separation

~ ступ·ен·и (rocket) stage separation

рас·цепл·éни·е, -я, (*n*) (*v n*) *see* рас··цепл·я́·ть; *see also* рас·цéп·к·а; breaking (a chain of molecules or equations)

рас·цéпл·енн·ый, -ая, -ое, (*past part pass*) *see* рас·цепл·я́·ть

рас·цепл·я́ющ·ий, -ая, -ее, (*pres part act*) of рас·цепл·я́·ть

~ механ·и́зм· (*m*) release gear, tripping mechanism

рас·цепл·я́·ть, -ют, (*imp*) unhook, unlink, uncouple, unfasten; trip, disengage, release, separate

рас·цех·óвк·а, -и, (*f*) shop circularization (industrial organization)

рас·чáл·енн·ый, -ая, -ое, (*past part pass*) *see* рас·чáл·ива·ть; braced

рас·чáл·ива·ть, -ют, (*imp*) brace

рас·чáл·ить, -я́т, (*perf*) *see* рас·чá·л·ива·ть

рас·чáл·к·а, -и, (*g pl*) -л·ок·, (*f*) bracing; brace, wire-brace, bracing wire

~, инерцирóн·н·ая (a/c) anti-drag wire

~, лоб·ов·áя (a/c) drag wire

~, нес·у́щ·ая (a/c) lift wire, flying wire

~, под·дéрж·ивающ·ая (a/c) anti-lift wire, landing wire

рас·чекáн·ива·ть, -ют, (*imp*) (met etc.) caulk

рас·чекáн·ить, -я́т, (*perf*) *see* рас··чекáн·ива·ть

рас·чес·á·ть, (*fut 3rd pl*) рас·чéш·ут, (*perf*); *see* рас·чёс·ыва·ть

рас·чёс·ыва·ть, -ют, (*imp*) comb, comb out/smooth; scratch

рас·чёт·, -а, (*m*) calculation, computation; design; (fin) settlement, account; (mil) crew, team

~, аэро·динам·и́ческ·ий aerodynamic analysis

~, без·на·ли́ч·н·ый (fin) non-cash settlement, paper settlement, clearing

в рас·чёт·е на referred to, calculated per

из рас·чёт·а на calculated on

~ на по·сáд·к·у (air) initial approach

~, не·вéр·н·ый miscalculation

рас·чёт

~ **объектив·а** (phot) formula (of a lens)

~ **по·вод·к·ов** (civ eng) flood estimate

~, **полн·ый** (fin) full settlement, settlement in full

при·ня́ть в рас·чёт· take into account, allow for

~ **реактор·а** (nucl) reactor design

~ **решёт·к·и профил·ей** (a/c) cascade design/designing

с так·им рас·чёт·ом so (that)

~, **старт·ов·ый** (rocket) launching crew

рас·чёт·лив·ый, -ая, -ое, (a) economical; calculating, careful

рас·чёт·н·ый, -ая, -ое, (a) calculated, calculating; rated, designed, design; (fin) settlement, pay, accounting

~ **выс·от·а́** (f) (air) rated altitude; throat (of a weld)

~ **год·** (m) financial year

~ **да́·нн·ые** (pl) theoretical figures, calculated data

~ **длин·а́** (f) (met) gauge length (of test piece)

~ **книж·к·а** (f) pay book

~ **курс·** (m) (fin) settlement rate

~ **лист·** (m) (mech) design sheet

~ **мест·о** (n) (nav) dead-reckoning position

~ **модел·ь сет·и** (m) (elec) network analyser

~ **мощ·ност·ь** (f) rated power/capacity

~ **на·пряж·ени·е** (n) (elec) voltage rating

~ **палат·а** (f) (fin) clearing house

~ **период·** (m) (fin) accounting period, account

~ **план·ировани·е** (n) (air) approach glide

~ **стол·** (m) (elec) network analyser

~ **схем·а** (f) (civ eng) structural simulator/model/analyser

~ **с·чёт·** (m) (fin) clearing/settlement account

~ **формул·а** (f) design formula

рас·чехл·ить, -ят, (perf) see **рас·чехл·я́·ть**

рас·чехл·я́·ть, -ют, (imp) uncover, remove the cover from

рас·чеш·ет, (fut 3rd sing) of **рас·чес·ать,** (perf); see **рас·чёс·ыва·ть** (imp)

рас·чист·ить, -ят, (perf) see **рас·чищ·а́·ть**

рас·чищ·а́·ть, -ют, (imp) clear, free (from obstructions)

рас·чищ·енн·ый, -ая, -ое, (past part pass) see **рас·чищ·а́·ть;** free, clear

рас·член·ени·е, -я, (n) (v n) see **рас·член·я́·ть;** (math) differentiation; (print) scanning

рас·член·ённ·ый, -ая, -ое, (past part pass) see **рас·член·я́·ть;** separated, differentiated, disjoint; (print) scanned

рас·член·ить, -ят, (perf) see **рас·член·я́·ть**

рас·член·я́·ть, -ют, (imp) dismember, break up

рас·шат·анн·ый, -ая, -ое, (past part pass) see **рас·шат·ыва·ть;** loose, rickety, shaky, wobbly

рас·шат·а́·ть, -ют, (perf) see **рас·шат·ыва·ть**

рас·шат·ыва·ть, -ют, (imp) shake loose, make unsteady/shaky; **-ся** be/get/become unsteady/shaky/loose

рас·шив·а́·ть, -ют, (imp) tear, rip, tear/rip open (something sewn); embroider; point (brickwork etc.), ding (walls), fan out (a stranded wire/cable)

рас·шив·к·а, -и, (g pl) **-в·ок·,** (f) (v n) see **рас·шив·а́·ть;** fan, fanning out (e.g. an electric cable)

рас·шир·ени·е, -я, (n) (v n) see **рас·шир·я́·ть;** enlargement, expansion; (spectr) broadening; (med) dilation, ectasis, ectasia

~, **двой·н·ое** double expansion; (st eng) compound expansion

~, **дву·крат·н·ое** double expansion; (st eng) compound expansion

~ **импульс·ов** (phys) pulse stretching/lengthening/widening

~ **кольц·а́** (chem) ring enlargement

~, **кон·ическ·ое** (dynam) taper expansion

коэффициент· лин·ейн·ого рас·шир·ени·я (m) coefficient of linear expansion

коэффициент· рас·шир·ени·я coefficient of expansion

коэффициент· тепл·ов·ого линеа́р·н·ого рас·шир·ени·я thermal coefficient of linear expansion

~ **лин·и·й** (spectr) line broadening

лин·и·я рас·шир·ени·я (f) expansion line/curve (on an indicator diagram)

рас·шир·е́ни·е
~, мно́го·кра́т·н·ое (st eng) multiple expansion
~, объ·ём·н·ое cubical/volumetric expansion, dilation; мо́дул·ь объ·ём··н·ого рас·шир·е́ни·я (m) modulus of cubical expansion
~, от·нос·и́тельн·ое expansion ratio; area ratio (of expanding nozzles etc.)
~, по·ве́рх·ностн·ое flat/surface expansion
~, прост·о́е (st eng) single expansion
~, тепл·ов·о́е thermal expansion
~, тро́й·н·о́е (st eng) triple expansion у́гол· рас·шир·е́ни·я (m) (dynam) angle of convergence
~, четвер·н·о́е (st eng) quadruple expansion; (gen as adj) quadruple-expansion
рас·шир·ённ·ый, -ая, -ое, (past part pass) see рас·шир·я́·ть
рас·шир·и́тел·ь, -я, (m) expander; (build) expansion tank; oil conservator (of transformers); expansion pipe/trunk; (med) dilator
~ и́мпульс·ов (phys) pulse widener/stretcher
рас·шир·и́тельн·ый, -ая, -ое, (a) expansion, expanding, widening, broadening
рас·ши́р·ить, -ят, (perf) see рас··шир·я́·ть
рас·шир·я́емост·ь, -и, (f) expansibility, dilatability
рас·шир·я́·ть, -ют, (imp) widen, broaden; enlarge, expand, dilate; increase
рас·шир·я́ющ·ийся, -аяся, -ееся, (pres part act) see рас·шир·я́·ть; expansible, extensible; expanded (of concretes etc.); (plastic) expansible
рас·ши́·ть, (fut 3rd pl) разо·шь·ю́т, (perf) see рас·шив·а́·ть
рас·шифр·ова́ть, -у́ют, (perf) see рас·шифр·о́выва·ть
рас·шифр·о́вк·а, -и, (g pl) -вок·, (f) (v n) see рас·шифр·о́выва·ть; expansion (of an abbreviation); interpretation
рас·шифр·о́вочн·ая маши́н·а (f) interpreter (punched cards etc.)
рас·шифр·о́вщик·, -а, (m) coder, interpreter
~ зву́к·а (cinema) sound reader/reproducer

рас·цифр·о́выва·ть, -ют, (imp) decypher, decode; interpret, make out, read
рас·шлиф·о́выва·ть, -ют, (imp) grind out
рас·шлихт·о́вк·а, -и, (f) (text) desizing, destarching
~, бактериа́ль·н·ая rot-steep desizing
~, ки́сл·ая sour desizing
рас·шнур·ова́ть, -у́ют, (perf) unlace, unthread
рас·ще́л·ин·а, -ы, (f) (geol) cleft, fissure; (naut) lead (through ice)
рас·ще́п·, -а, (m) split, scissure
рас·щеп·и́ть, -я́т, (perf) see рас··щепл·я́·ть
рас·щепл·е́ни·е, -я, (n) (v n) see рас·щепл·я́·ть; fission, disintegration; (chem) decomposition, -lysis, cleavage; (nucl) spallation, splitting
~ аммиа́к·ом (chem) ammonolysis
~ бел·к·а́ (chem) albuminolysis
~ ли́н·и·й спе́ктр·а (spectr) line splitting, spectroscopic splitting
~ ма́гм·ы (geol) magma splintering
~, магни́т·н·ое; фа́ктор· магни́т··н·ого рас·щепл·е́ни·я (m) (phys) g factor
~ позвон·к·а́ (med) rachischisis, spina bifida
~ при́·знак·ов (biol) splitting
~ у́·ровн·ей (phys) splitting of levels
~ фа́з·ы phase splitting
~ хромо·со́м·ы chromosome break
~, шта́рк·овск·ое (phys) Stark effect
~ ядр·а́ (nucl) nuclear splitting/disintegration
рас·щепле·но́г·ие, -их, (pl decl as adj) (zool) schizopods, Schizopoda
рас·щеплённо·но́г·ие, -их, (pl decl as adj) = рас·щепле·но́г·ие
рас·щепл·ённ·ый, -ая, -ое, (past part pass) see рас·щепл·я́·ть
~ на две ча́ст·и (biol) bifid, bifidate
рас·щепл·я́емост·ь, -и, (f) fissility
рас·щепл·я́ем·ый, -ая, -ое, (pres part pass) of рас·щепл·я́·ть; fissile, fissionable
рас·щепл·я́·ть, -ют, (imp) split, splinter, chip, split up; -ся (pass); disintegrate (of a nucleus); (chem) decompose
рати́н·, -а, (m) (text) ratine
ратифиц·и́р·овать, -уют, (imp and perf) ratify

ра́тлер·, -а, (*m*) ratler (brick testing machine)

ратовки́т·, -а, (*m*) (min) ratovkite, rattoffkite

Рау́л·я, за·ко́н· (*m*) Raoult's law (vapour pressure)

раухтопа́з·, -а, (*m*) (min) cairngorm, smoky quartz

раф·ано́л·, -а, (*m*) (biochem) raphanol

раф·и́д·, -а, (*m*) (biochem) raphide

рафина́д·, -а, (*m*) lump sugar

рафина́д·н·ый, -ая, -ое, (*a*) lump-sugar

~ пресс· (*m*) lump-sugar press

~ за·во́д· (*m*) sugar factory/refinery

рафин·ацио́нн·ый, -ая, -ое, (*a*) refining

рафин·а́ци·я, -и, (*f*) refining

рафин·ёр·, -а, (*m*) (paper) refiner, perfecting engine

рафин·и́ровани·е, -я, (*n*) refining

~, пе́рв·ое improving, softening (of lead)

~, пиро·металлург·и́ческ·ое (met) fire refining

~ шла́к·ом (met) slag refining

рафин·и́р·овать, -уют, (*imp and perf*) refine

раффино́з·а, -ы, (*f*) (chem) raffinose

ра́ффи·я, -и, (*f*) raffia, raffia bast; (bot) raffia palm, *Raphia ruffia*

рахио·плеги́·я, -и, (*f*) (med) rachioplegia

рахио·том·и́·я, -и, (*f*) (med) rachiotomy

ра́хис·, -а, (*m*) (bot) rachis

рах·и́т·, -а, (*m*) (med) rachitis, rickets

рах·ити́чн·ый, -ая, -ое, (*a*) rachitic

рахи·том·и́·я, -и, (*f*) (med) rachitomy

рахи·центе́з·, -а, (*m*) (med) rachicentesis

рацем·а́т·, -а, (*m*) (biochem) racemate

рацем·иза́ци·я, -и, (*f*) (chem) racemization

рацем·и́ческ·ий, -ая, -ое, (*a*) racemic

~ ви́н·н·ая кисл·от·а́ (*f*) (chem) racemic acid

~ со·един·е́ни·е (*n*) (chem) racemic compound

рацем·о́зн·ый, -ая, -ое, (*a*) (bot) racemose

рацио́н·, -а, (*m*) ration

рационализ·а́тор·, -а, (*m*) efficiency expert; innovator

рационализ·а́тор·ск·ий, -ая, -ое, (*a*) efficiency, improving

~ пред·лож·е́ни·е efficiency suggestion

рационализ·а́ци·я, -и, (*f*) rationalization, improvement

рационализ·и́р·овать, -уют, (*imp and perf*) improve, make more efficient

рациона́ль·ност·ь, -и, (*f*) rationality

за·ко́н· рациона́ль·ност·и пара·ме́тр·ов (*m*) (cryst) law of rational indices, law of rationality

рациона́ль·н·ый, -ая, -ое, (*a*) rational; efficient

ра́ци·я, -и, (*f*) (*abbr*) = **радио·ста́нци·я** (q.v.) radio set/station

ра́ч·ий, -ья, -ье, (*a*) of **рак·** crayfish

ра́ч·к·а, -и, (*g pl*) **-ч·ек·,** (*f*) (eng) ratchet

Ра́шиг·а, кольц·о́ (*n*) (chem) Raschig ring

ра́шкул·ь, -я, (*m*) charcoal pencil

ра́шпил·ь, -я, (*m*) rasp (tool)

ращ·е́ни·е, -я, (*n*) (*v n*) see **раст·и́ть**

ращ·у́, (*pres 1st sing*) of **раст·и́ть**

рв·а (*g sing*) of **ров·**

рв·а́вш·ий, -ая, -ее, (*past part act*) of **рв·ать**

рв·ани́н·а, -ы, (*f*) fissure, crack, flaw, transverse crack, tear

рва·н·у́ть, -у́т, (*perf*) jerk, tug; **-ся** rush at

рв·а́н·ый, -ая, -ое, (*a*)

~ ка́мен·ь (*m*) rubble

рв·а́н·ь, -и, (*f*) rag, rags, ragged garment

рв·ань·ё, -я́, (*n*) tearing, laceration; rags, ragged clothing

рв·а́тельн·ый, -ая, -ое, (*a*) extractor

~ кольц·о́ (*n*) (min) core lifter

рв·ать, -ут, (*imp*) tear, tear up; pick, pluck, gather; vomit; blow up (with explosive); extract (teeth); break off (relations)

рв·о́т·а, -ы, (*f*) emesis; vomit, vomiting

рв·о́тн·ый, -ая, -ое, (*a*) emetic, vomitory

~ ка́мен·ь (*m*) tartar emetic

~ ко́рен·ь (*m*) (pharm) ipecacuanha

~ сре́д·ств·о (*n*) emetic

рв·у́шк·а, -и, (*g pl*) **-шек·,** (*f*) soluble parting piece

рв·у́щ·ий, -ая, -ее, (*pres part act*) of **рв·ать**

рв·ы, (*nom/acc pl*) of **ров·**

рде́ст·, -а, (*m*) (bot) pondweed, *Potamogeton*

ре (*n indecl*) D (in music)

реаге́нт·, -а, (*m*) *see also* **реакти́в·;** (chem) reactant

реаг·и́ровани·е, -я, *(n) (v n)* of **реаг·и́р·овать;** reaction; (instr) response
реаг·и́р·овать, -уют, *(imp)* react; (instr) respond
реакта́нс·, -а, *(m)* (elec) reactance
реакта́нс·н·ый, -ая, -ое, *(a)* (elec) reactance
~ схе́м·а *(f)* reaction circuit
реакта́нц·, -а, *(m) see* **реакта́нс·**
реакти́в·, -а, *(m)* (chem) reagent
реакти́вно·спосо́б·н·ый, -ая, -ое, *(a)* reactive
реакти́в·ност·ь, -и, *(f)* (phys) reactivity, reaction rate; (elec) reactance
~, до́л·ев·а́я reactance per unit, per-unit reactance
за·па́с· реакти́в·ност·и *(m)* (nucl) excess reactivity
коэффицие́нт· реакти́в·ност·и *(m)* (phys) coefficient of reactivity; (elec) reactive factor
~, о·ста́т·очн·ая (nucl) shut-down reactivity
~, уда́р·н·ая subtransient reactance
реакти́в·н·ый, -ая, -ое, *(a)* reaction, reactive, reagent; (elec) reactive, reactance, repulsion; (a/c) rocket, jet, jet-propelled
~ бума́ж·к·а *(f)* (chem, phot) test paper
~ во·оруж·е́ни·е *(n)* (mil) rocket armament
~ дви́г·ател·ь *(m)* reaction engine (in general), jet engine (specifically)
~ дви́г·ател·ь, воз·ду́ш·н·о- thermal jet engine
~ дви́г·ател·ь, порох·ов·о́й solid propellant engine
~ дви́г·ател·ь, прямо·то́ч·н·ый воз·ду́ш·н·о- ram jet, ram-jet engine
~ дви́г·ател·ь, пульс·и́рующ·ий pulse jet, resonant jet
~ движ·е́ни·е *(n)* reaction propulsion
~ движ·и́тел·ь *(m)* (a/c) reaction nozzle
~ кат·у́шк·а *(f)* (elec) reactance/choking coil, inductor, reactor
~ мо́щ·ност·ь *(f)* (elec) reactive power
~ синхро́н·н·ый мото́р· *(m)* (elec) repulsion motor
~ сопл·о́ *(n)* (a/c) reaction nozzle
~ сопл·о́, кон·цев·о́е blade-tip reaction nozzle (of helicopter)
~ со·противл·е́ни·е *(n)* (elec) reactance
~ со·ставл·я́ющ·ая *(f)* (elec) reactive/quadrature/wattless/idle component

реакти́в·н·ый, -ая, -ое
~ ток· *(m)* (elec) reactive current
~ то́пл·ив·о *(n)* (a/c) jet fuel
~ турби́н·а *(f)* (st eng) reaction turbine
~ у·стано́в·к·а *(f)* (mil) rocket-launching frame
реакти́·метр·, -а, *(m)* reactivity meter
реакто·пла́ст·, -а, *(m)* thermosetting plastic
реа́ктор·, -а, *(m)* reactor
~, бассе́йн·ов·ый (nucl) swimming-pool reactor, pool reactor
~, бето́н·н·ый (elec) concrete-base reactor
~, бло́ч·н·ый (nucl) lumped reactor
~, бы́стр·ый (nucl) fast reactor
вне реа́ктор·а (nucl) out-of-pile/-reactor
внутр·и́ реа́ктор·а (nucl) in-pile/-reactor
~, во́д·н·ый (nucl) aqueous reactor
~ -дви́г·ател·ь, -я, *(m)* (nucl) propulsion reactor
~, дейто́н·н·ый (nucl) heavy water reactor
~, кип·я́щ·ий (nucl) boiling-water/water-boiler reactor
~, ма́сл·ян·ый (elec) liquid-immersed reactor
~, мн·и́м·ый (nucl) image reactor
~ на бы́стр·ых нейтро́н·ах (nucl) fast reactor
~, по·груж·ённ·ый (nucl) swimming-pool reactor
~, про·меж·у́точн·ый (nucl) intermediate reactor
~ -раз·множ·и́тел·ь, -я, *(m)* (nucl) breeder reactor
~, сух·о́й (elec) dry-type reactor
~, тепл·ов·о́й (nucl) thermal reactor
~, теплофика́цион·н·ый (nucl) heat reactor
~, техно·лог·и́ческ·ий (nucl) breeder
~, ши́н·н·ый (elec) bus reactor
~, энергет·и́ческ·ий (nucl) power reactor
~, я́дер·н·ый nuclear reactor
реа́ктор·н·ый, -ая, -ое, *(a)* of **реа́ктор·**
реакторо·стро·е́ни·е, -я, *(n)* reactor engineering
реакцио́нно·спосо́б·н·ый, -ая, -ое, *(a)* reactive
реакцио́н·н·ый, -ая, -ое, *(a)* reaction, reactional
~ ка́мер·а *(f)* (chem) reaction chamber
~ ине́рт·ност·ь *(f)* reactionlessness
спосо́б·ност·ь *(f)* reactivity

реа́кци·я, -и, (*f*) reaction; response; pH value (of soil)

~, групп·ов·áя (chem) group reaction

~, до·полн·и́тельн·ая side reaction

~, индукт·иру́ем·ая (chem) induced reaction

~ на и́мпульс·н·ое воз·мущ·éни·е (autom) impulse response

~ об·мéн·а exchange reaction

~, обрат·и́м·ая reversible reaction

~, обра́т·н·ая back/reverse reaction

~ пере·да́ч·и цеп·и (chem) chain transfer reaction

~ по крéн·у, обра́т·н·ая (a/c) aileron reversal

~, прям·а́я direct/straight/simple reaction

~ рóст·а цéп·и (chem) growth

~ со·един·éни·я (chem) combination reaction

~, цеп·н·а́я chain reaction

~, я́дер·н·ая nuclear reaction

реа́л·, -а, (*m*) (print) composing frame, rack

реализ·а́ци·я, -и, (*f*) (fin) realization; execution, achievement

~ пла́н·а execution of a plan

~ проду́кци·и sale of output

~ фóнд·ов receipt of allocations

реализ·óванн·ый, -ая, -ое, (*past part pass*) of **реализ·ова́ть**

реализ·ова́ть, -у́ют, (*imp and perf*) (fin) realize, convert into money; achieve, put into practice; **-ся** (*pass*); materialize

реальга́р·, -а, (*m*) (min) realgar

реа́ль·н·ый, -ая, -ое, (*a*) real, actual; imperfect; modern (as opposed to classical: of education); (autom) real-time

~ газ· (*m*) (chem) imperfect gas

~ масшта́б· врéмен·и real time

реб·ёнок·, -нк·а, (*nom pl*) **реб·я́т·а,** (*g pl*) **реб·я́т·,** (*m*) child

~, груд·н·óй infant

рéбер·н·ый, -ая, -ое, (*a*) costal, rib; edge

ребóрд·а, -ы, (*f*) rim, flange (on wheel)

ребр·и́ст·ый, -ая, -ое, (*a*) ribbed, costate, gilled, finned, corrugated

~ труб·á (*f*) gilled/finned/radiator piping

ребр·ó, -á, (*nom pl*) **рёбр·а,** (*g pl*) **рёбер·,** (*n*) (anat) rib, costa; rib, fin; edge (of a geometrical shape); arm (of spoked wheels etc.)

~ ата́к·и (a/c) leading edge

ребр·ó

~ воз·вра́т·а (math) line of striction (of a ruled surface), cuspidal edge, edge of regression

~, горизонта́ль·н·ое (shipb) horizontal stiffener

длин·á ребр·á (*f*) (cryst) spacing, primitive translation

~ криста́лл·а crystal edge

~, на·правл·я́ющ·ее guide vane

~ об·тек·а́ни·я (a/c) trailing edge

~, о·пóр·н·ый knife-edge

~, о·хлажд·а́ющ·ее (mech) cooling fin

~ рул·я́ rudder arm

~ цили́ндр·а (ICE) cylinder fin

ребр·óв·ый, -ая, -ое, (*a*) of **ребр·ó**

~ стан·óк·, кругло·пи́ль·н·ый roller-feed circular deep resawing machine

ребр·óм (*adv*) edgewise, edge-on; bluntly, directly

~ -на·пра́вл·енн·ая антéнн·а (*f*) (rad) end-fire antenna array

рёбр·ышк·о, -а, (*g pl*) **-шек·,** (*n*) (*dim*) of **ребр·ó;** riblet

реб·я́т·а, (*pl*) of **реб·ёнок·** children; lads

рёв·, -а, (*m*) roar

рева́нт·, -а, (*m*) (naut) roband, roving

ревдински́т·, -а, (*m*) (min) revdinskite

рев·éвш·ий, -ая, -ее, (*past part act*) of **рев·éть**

ревéн·н·ый, -ая, -ое, (*a*) rhubarb

ревéн·ь, -я́, (*m*) (bot) rhubarb, *Rheum*

реверберациóн·н·ый, -ая, -ое, (*a*) reverberation

~ ка́мер·а (*f*) (acous) reverberation chamber

ревербер·а́ци·я, -и, (*f*) reverberation

ревербирó·метр·, -а, (*m*) (acous) reverberation-time meter

рéверс·, -а, (*m*) reverse, reversal; reversing gear; rundown (of nuclear reactor)

~ -реду́ктор·, -а, (*m*) reverse-reduction gear

~ тя́г·и (ICE) thrust reversal

ревéрсер·, -а, (*m*) reverser

реверси́в·ност·ь, -и, (*f*) reversibility

реверси́в·н·ый, -ая, -ое, (*a*) reversing, reversible, reverse

~ дви́г·ател·ь (*m*) direct-reversing engine

~ стан· (*m*) (met roll) reversing mill

реверсиóн·н·ый, -ая, -ое, (*a*) (biol) reversionary, atavistic

реверс·и́ровани·е, -я, (*n*) reversal, reversing

реверс·и́р·овать, -уют, (*imp and perf*) (*intrans*) reverse

реве́рси·я, -и, (*f*) (biol) reversion, atavism

реве́рсор·, -а, (*m*) (elec) reverser

рев·е́ть, -у́т, (*imp*) roar, howl

ревизио́н·н·ый, -ая, -ое, (*a*) of **реви́зи·я**

реви́зи·я, -и, (*f*) inspection; examination; audit (of accounts); cleanout pipe

ревизо́р·, -а, (*m*) inspector; examiner; auditor

ревмати́зм·, -а, (*m*) (med) rheumatism

ревн·ова́ть, -у́ют, (*imp*) be jealous of

ревок·а́ци·я, -и, (*f*) (dipl) recall; (com) revocation, cancellation; stopping (a cheque)

револьве́р·, -а, (*m*) (gunn) revolver; revolving nosepiece (of microscope)

револьве́р·н·ый, -ая, -ое, (*a*) revolving; revolver; capstan (of machine tools)

~ голо́в·к·а (*f*) turret/capstan head; revolving head

~ стан·о́к· (*m*) turret/capstan lathe

револьве́р·щик·, -а, (*m*) capstan lathe operator

революцио́н·н·ый, -ая, -ое, (*a*) revolutionary

револю́ци·я, -и, (*f*) revolution

рев·у́н·, -а́, (*m*) howler; (naut) loud-sounding buzzer, whistling fog signal; (zool) howler, (*pl*) *Alouattinae, Mycetinae*

рев·у́щ·ий, -ая, -ее, (*pres part act*) of **рев·е́ть**

рега́т·а, -ы, (*f*) regatta

регенера́т·, -а, (*m*) reclaimed rubber

регенерати́в·н·ый, -ая, -ое, (*a*) regenerative; (rad) regenerative, retroactive, reaction, reactive

~ при·ём· (*m*) (rad) reactive/retroactive/regenerative reception

регенер·а́тор·, -а, (*m*) regenerator, regenerative heater, (met) regenerative chamber; (rad) regenerative amplifier

~, одно·оборо́т·н·ый (met) single-pass regenerator

регенер·а́ци·я, -и, (*f*) regeneration, reclaiming (rubber), recovery (oil), purification (air); reprocessing (of atomic fuel); (rad) regeneration, retroaction, reaction, self-oscillation

регенер·а́ци·я

~ гор·е́л·ой земл·и́ (met cast) sand reclamation

~ от·хо́д·ов waste recovery

регенер·ацио́нн·ый, -ая, -ое, (*a*) reclaiming, reclamation, recovering

регенер·и́рованн·ый, -ая, -ое, (*past part pass*); reclaimed, recovered, regenerated

региона́ль·н·ый, -ая, -ое, (*a*) regional

реги́стр·, -а, (*m*) register, recorder; coder, register (telephones); register, accumulator (computers); damper, register (in heating systems); pitch, register (music); case, register (typewriters)

~, ве́рх·н·ий (print) upper case

~, дым·ов·о́й carbon dioxide recorder

~ множ·и́м·ого multiplicand register (computers)

~, мор·ск·о́й registry of shipping

~, с·дви́г·ов·ый shift register (typewriters)

регистр·а́тор·, -а, (*m*) registrar; recorder, recording apparatus; (nucl) monitor

~ интепси́в·ност·и пуч·к·а́ (nucl) beam monitor

~, много·то́ч·ечн·ый scanning recorder

~ пу́льс·а (med) sphygmograph

~ со·в·пад·е́ни·й coincidence counter

~ толч·к·о́в impact recorder

~, цифр·ов·о́й digital recorder

~, электро́н·н·о-луч·ев·о́й cathode-ray recorder

регистрацио́н·н·ый, -ая, -ое, (*a*) registration, recording

регистр·а́ци·я, -и, (*f*) registration, recording; detection (of radiations); logging (computers)

~, цифр·ов·а́я (autom) digital logging

регистр·и́р·овать, -уют, (*imp and perf*) register, record

регистр·и́рующ·ий, -ая, -ее, (*pres part act*) of **регистр·и́р·овать**; recording, registering

реги́стр·ов·ый, -ая, -ое, (*a*) of **реги́стр·;** registered, register

~ в·мест·и́мост·ь (*f*) (naut) register tonnage

~ выс·от·а́ бо́рт·а (naut) registered depth

регла́мент·, -а, (*m*) regulations; procedure (at conferences etc.); maintenance routine

регламент·а́ци·я, -и,   (*f*) regulation, regulating

регламент·и́р·овать, -уют,   (*imp*) stipulate, regulate, lay down rules for

регла́мент·н·ый, -ая, -ое,   (*a*) of **регла́мент·**

~ рабо́т·ы (*pl*) (a/c) periodic servicing

регле́т·, -а,   (*m*) (print) clump

ре́гм·а, -ы,   (*f*) (bot) regma

регресси́в·н·ый, -ая, -ое,   (*a*) regressive, retrogressive

регре́сси·я, -и,   (*f*) regression; retreat (of the sea)

регуле́кс·, -а,   (*m*) (elec) regulex exciter

регул·и́ровани·е, -я,   (*n*) (*v n*) see **регул·и́р·овать**; regulation, adjustment, control

~, автомат·и́ческ·ое automatic control

~, автоно́м·н·ое autonomous automatic control

~, а·период·и́ческ·ое aperiodic damping control

~, а·стат·и́ческ·ое floating control

~, байпа́сс·н·ое by-pass governing (of turbine)

~, бес·ступ·е́нчат·ое stepless control **врем·я регул·и́ровани·я**   (*n*) (autom) time constant

~, гармон·и́ческ·и колеб·а́тельн·ое steady oscillation control

~ гро́м·кост·и (elec) volume control

~ движ·е́ни·я traffic control; (mil) movement control

~, дро́ссель·н·ое throttle governing (turbine)

~, за·тух·а́ющ·ее damped oscillation control

~, изодро́м·н·ое automatic reset control, proportional-speed floating control, integral control

~, ка́честв·енн·ое throttle governing (of turbine)

~, коли́честв·енн·ое nozzle-control governing (turbine)

~ контра́ст·ност·и (TV) contrast control

~, об·во́д·н·ое by-pass governing (turbine)

~, пла́в·н·ое slide control

~ по у·скор·е́ни·ю second-derivative control

~, пре·ры́в·ист·ое (autom) sampled-data control, discontinuous control

~, програ́мм·н·ое programme control

~ раз·ме́р·ов dimensional control

регул·и́ровани·е

~, рас·ход·я́щ·ееся increased oscillation control

~ ру́сл·а (civ eng) stream regulation

~, с·вя́з·анн·ое cascade control, multiple automatic control

~, с·ло́ж·н·ое simultaneous automatic control

~, сопл·ов·о́е nozzle control governing (turbine)

~, с·ход·я́щ·ееся steady oscillation control

~ у·сил·е́ни·я (rad) gain control

~, экстрема́ль·н·ое (autom) extremum adaptive control

регул·и́рованн·ый, -ая, -ое,   (*past part pass*) of **регул·и́р·овать**

регул·и́р·овать, -уют,   (*imp and perf*) regulate, adjust; control, govern

регул·иро́вк·а, -и,   (*f*) see **регул·и́·ровани·е**

регул·иро́вочн·ый, -ая, -ое, (*a*) *and see* **регул·и́рующ·ий**; regulating, controlling, control, adjusting

~ реоста́т· (*m*) rheostatic controller

~ кольц·о́ (*n*) adjusting ring

~ про·кла́д·к·а (*f*) (elec) shim

регул·и́руем·ый, -ая, -ое, (*pres part pass*) of **регул·и́р·овать**; adjustable; controlled

~ велич·ин·а́ control quantity

~ проце́сс· (*m*) (autom) controlled system

~ сред·а́ (*f*) controlled medium

регул·и́рующ·ий, -ая, -ее, (*pres part act*) of **регул·и́р·овать**; *see also* **регул·иро́вочн·ый**

~ автомат·и́ческ·ое у·стро́й·ств·о (*n*) automatic controller

~ механ·и́зм· (*m*) control mechanism

~ при·бо́р· (*m*) controller, regulator

~ систе́м·а (*f*) control system; controlling unit

регул·и́рующ·ийся, -аяся, -ееся, (*pres part act*) *see* **регул·и́р·овать**; adjustable; controlled

регуля́р·н·ый, -ая, -ое,   (*a*) regular

~ вой·ск·а́ (*pl*) (mil) regular forces, regulars

регул·я́тор·, -а,   (*m*) regulator, regulating mechanism, controller, control mechanism; speed governor, governor (of prime movers); reducing valve (pressure systems)

~, а·стат·и́ческ·ий floating controller

~, анти·помп·а́жн·ый surge regulator (of compressor)

регул·я́тор
~, баро·метр·и́ческ·ий (a/c) barometric pressure controller
~, ва́куум·н·ый pneumatic governor
~, вибрацио́н·н·ый (elec) on-off regulator
~ воз·бужд·е́ни·я (elec) field rheostat
~, все·режи́м·н·ый variable-speed governor
~, гидравл·и́ческ·ий hydraulic controller/governor
~ двой·н·о́го де́й·стви·я double-response controller
~, двух·и́мпульс·н·ый two-factor regulator/controller
~, двух·позицио́н·н·ый on-off/two-position controller
~, изодро́м·н·ый automatic reset control, proportional speed floating control mechanism
~, двух·режи́м·н·ый maximum-and-minimum speed governor
~ коли́честв·а flow controller/regulator
~, колоко́ль·н·ый inverted-bell pressure controller
~, комбин·и́рованн·ый rate-action controller, derivative controller
~, максима́ль·н·ый overspeed governor
~, много·позицио́н·н·ый multiposition/multistep controller
~ на·пряж·е́ни·я voltage regulator
~ на·пряж·е́ни·я, ступ·е́нчат·ый step voltage regulator
~ на·пряж·е́ни·я, у́голь·н·ый (a/c) carbon-pile voltage regulator
~, пла́в·ающ·ий floating controller
~ полимериз·а́ци·и (chem) polymerization modifier/retarder
~ по·пере́ч·н·ого по́л·я, электро··маши́н·н·ый (elec) amplidyne
~ по·сто·я́нн·ого у́ровн·я, closed-feed regulator (boiler)
~ по·сто·я́нн·ого числ·а́ оборо́т·ов (a/c) constant speed unit
~ по·то́к·а flow controller
~ по·то́к·а, стабил·и́рующ·ий constant-pressure flow controller
~, пре·де́ль·н·ый overspeed governor
~ пре·ры́в·н·ого де́й·стви·я discontinuous-action controller
~, прецизио́н·н·ый constant-speed governor
~ про·движ·е́ни·я (a/c elec) advance control

регул·я́тор
~ про·до́ль·н·ого по́л·я, электро··маши́н·н·ый (elec) rototrol amplifier
~, пропорциона́ль·н·ый proportional controller
~, пруж·и́нн·ый spring-loaded governor
~ рас·хо́д·а flow controller
~ рас·хо́д·а то́пл·ив·а fuel consumption controller
~ с пред·вар·е́ни·ем anticipatory/pre-act/derivative controller
~, с·вя́з·анн·ый cascade-controlled controller
~ скольж·е́ни·я (elec) slip regulator
~ ско́р·ост·и governor, speed governor
~, с·пуск·ов·о́й (horol) escapement; proportional controller
~, стат·и́ческ·ий proportional controller
~, у́голь·н·ый (elec) carbon-pile regulator
~ у·сил·е́ни·я (rad) gain control
~, центро·бе́ж·н·ый pendulum governor
~ числ·а́ оборо́т·ов governor, speed governor
~, шар·ов·о́й pendulum governor
~, электро·маши́н·н·ый (elec) amplidyne control
регуля́тор·н·ый, -ая, -ое, (a) of **регул·я́тор·**
регуляцио́н·н·ый, -ая, -ое, (a) regulating
~ со·оруж·е́ни·я (pl) regulating works (rivers etc.)
ред- (component) (abbr) = **редакци·о́н·н·ый**
редакт·и́р·овать, -уют, (imp) edit; word (own composition)
реда́кт·ор·, -а, (m) editor
~, гла́в·н·ый editor-in-chief
   от·реда́кт·ор·а editor's foreword; editorial note
~, от·ве́т·ственн·ый sub-editor
реда́ктор·ск·ий, -ая, -ое, (a) editor's, editorial
редакцио́н·н·ый, -ая, -ое, (a) editorial, editing
реда́кци·я, -и, (f) editing; editorship; editorial staff; editorial office; version
реда́н·, -а, (m) step (on hull bottom)
реда́н·н·ый, -ая, -ое, (a) step (on hull bottom)
~ ко́рпус· (m) (shipb) stepped hull

Ре́двуд·а, визкози́·метр· (*m*) Redwood viscometer

ред·е́·ть, -ют, (*imp*) become sparse, thin out

ред·ин·а́, -ы́, (*f*) sacking, hessian

реди́с·, -а, (*m*) (bot) radish

ре́ди·я, -и, (*f*) (zool) redia

ре́д·к·ий, -ая, -ое, (*a*) rare, unusual, uncommon; sparse, thin (of density); infrequent

~ **земл·я́** (*f*) (chem) rare earth

~ **мета́лл·** (*m*) rare metal; less common metal

~ **ого́н·ь** (*m*) (mil) slow rate of fire

ре́д·к·о (*adv*) rarely, seldom, infrequently

редко·ды́р·чат·ый, -ая, -ое, (*a*) (biol) oligoforate

редко·земе́ль·н·ый, -ая, -ое, (*a*) rare-earth

~ **мета́лл·** (*m*) rare-earth metal

~ **элеме́нт·** (*m*) (chem) rare-earth element

редко·ли́ст·н·ый, -ая, -ое, (*a*) sparsely leaved

ред·колле́г·и·я, -и, (*f*) editorial board

редко·на·сел·ённ·ый, -ая, -ое, (*a*) sparsely/thinly populated

редко·ребр·и́ст·ый, -ая, -ое, (*a*) rare-costate

редко·сто́лб·чат·ый, -ая, -ое, (*a*) oligobaculate

ре́д·кост·ь, -и, (*f*) rarity; sparseness

редко·уда́р·н·ый, -ая, -ое, (*a*) (mech) slow-hitting

редко·яче́·ист·ый, -ая, -ое, (*a*) oligobrochate

редокси·потенциа́л·, -а, (*m*) (chem) redox potential

реду́кто·р, -а, (*m*) reduction gear; reducing valve, reducer, regulator (air and liquid systems)

~, **борт·ов·о́й** (mot ind) final drive

~, **двух·мото́р·н·ый** double-reduction gear

~, **кисло·ро́д·н·ый** oxygen regulator

~, **кон·и́ческ·ий** bevel-gear reduction gear

~, **одно·скор·остн·о́й** single-speed reduction gear

редукцио́н·н·о-вы·тяж·н·о́й стан· (*m*) (met roll) stretch mill

редукцио́н·н·ый, -ая, -ое, (*a*) reducing

~ **кла́пан·** (*m*) reducing valve, regulator, check valve

реду́кци·я, -и, (*f*) reduction

редуц·и́р·овать, -уют, (*imp and perf*) reduce

ре́дьк·а, -и, (*g pl*) **ре́дек·,** (*f*) (bot) radish; pointing (of a rope)

реёк·, -а, (*m*) (naut) spar, yard; **ре·ёк·,** (*g pl*) of **рей·к·а**

рее́стр·, -а, (*m*) register, list

ре́·ет, (*pres 3rd sing*) of **ре́·ять**

ре́·ечн·о- (*component*) (mech) rack

реечно·рыча́ж·н·ый, -ая, -ое, (*a*) lever-and-rack-operated

реечно·фре́зер·н·ый стан·о́к· (*m*) rack milling machine

ре́·ечн·ый, -ая, -ое, (*a*) of **рей·к·а** rake, rack

~ **классифик·а́тор·** (*m*) (min) rake classifier

ре́ж·а, (*pres gerund*) of **ре́з·ать**

ре́ж·е, (*comp*) of **ре́д·к·ий**

режёктор·н·ый, -ая, -ое, (*a*) (elec) band-elimination

режел·я́ци·я, -и, (*f*) regelation

ре́ж·ет, (*pres 3rd sing*) of **ре́з·ать** (*imp*)

режи (*indecl*) roller for hauling ships on to a beach

режи́м·, -а, (*m*) regime; condition, state; conditions, operating conditions, duty (of a machine); behaviour, routine; hydrologic cycle, regimen (rivers); (med) regime, regimen

~, **авто·колеб·а́тельн·ый** oscillatory state

~, **времен·н·о́й** time behaviour

~, **дл·и́тельн·ый** long-term operating conditions

   мо́щ·ност·ь дл·и́тельн·ого режи́·м·а (*f*) long-term rating

~ **круг·ов·о́го об·зо́р·а** continuous scan (radar)

~ **на·гру́з·к·и** duty (of a machine)

~ **наи·бо́льш·его благо·при·я́т·-ствовани·я** most-favoured-nation agreement/conditions

~, **не·пре·ры́в·н·ый** continuous duty

~, **не·у·станов·и́вш·ийся** transient process; transient behaviour; non-stationary conditions

~, **не·у·сто́й·чив·ый** unstable state

~, **номина́ль·н·ый** rating, duty (of machines)

~ **нул·ев·о́й тя́г·и** (air) zero-thrust conditions

~, **пере·хо́д·н·ый** (instr) transient behaviour/response; transient conditions

режи́м

~ по·вто́р·н·о-кратко·вре́мен·н·ый intermittent duty

~ по·лёт·а, втор·о́й   (air) stalled flight

~ рабо́т·ы operating conditions, routine, duty (of machines)

~ радио·лока́тор·а radar performance

~ ре́верс·а (a/c) negative-thrust conditions

~, регуля́р·н·ый по·вто́р·н·о-кратко·вре́мен·н·ый periodic duty

~, се́ктор·н·ого по́·иск·а sector scan (radar)

~, тепл·ов·о́й thermal conditions, thermal operating conditions (of furnace)

~, у·станов·и́вш·ийся steady state, steady-state conditions

~ фикс·а́ци·и реш·е́ни·я hold condition (computers)

~ форм·ова́ни·я moulding/molding conditions

~, час·ов·о́й hourly operating conditions;
мо́щ·ност·ь час·ов·о́го режи́м·а (f) hourly rating

режиссёр·, -а, (m) (theat) producer
помо́щ·ник· режиссёр·а (m) stage manager

ре́ж·ущ·ий, -ая, -ее, (pres part act) of ре́з·ать; cutting; -ся sectile

реж·ьте, (pl imper) of ре́з·ать

ре́з·, -а, (m) cut

рез·а́к·, -а́, (m) cutter; chopping knife, chopper; (weld) cutting torch, torch, lance

~, га́з·ов·ый (weld) cutter, cutting torch

~, огн·ев·о́й flame cutter

ре́з·альн·ый, -ая, -ое, (a) cutting

~ маши́н·а (f) cutting machine, guillotine

ре́з·ани·е, -я, (n) = ре́з·к·а (q.v.); machining (metals)
глуб·ин·а́ ре́з·ани·я (f) depth of cut (metal machining)

ре́з·анн·ый, -ая, -ое, (past part pass) of ре́з·ать

ре́з·ательн·ый, -ая, -ое, (a) cutting

ре́з·ать, (pres 3rd pl) ре́ж·ут, (imp) cut, incise; cut up, slice; carve, engrave; kill (with knife)

ре́зв·ый, -ая, -ое, (a) frisky, frolicsome

резед·а́, -ы́, (f) (bot) mignonette, *Reseda*

ре·зе́кци·я, -и, (f) (med) resection

резе́н·, -а, (m) (chem) resene

ре́зен·кил·ь, -я, (m) (shipb) hog (of keel)

резен·ли́н·и·я, -и, (f) (naut) rising line

резе́рв·, -а, (m) reserve, reserves; (text) resist

резерв·а́т·, -а, (m) reservation, reserve (territory for a particular ethnic group)

резерв·а́ци·я, -и, (f) reservation; reservation, reserve (territory for a particular ethnic group)

резерв·и́р·овать, -уют, (imp and perf) reserve

резе́рв·н·ый, -ая, -ое, (a) reserve

резервуа́р·, -а, (m) storage tank, tank, container; liquid container (of gyrocompass)

~, водо·на·по́р·н·ый (civ eng) pressure-equalizing tank

~ вы́·держ·к·и (nucl) decay tank

~, нефт·ян·о́й oil/petroleum storage tank

~, от·сто́й·н·ый (nucl) retention pond
печ·ь-резервуа́р·, -а, (m) holding furnace

~, у·равн·и́тельн·ый (hydroelec) surge tank

ре́зерфорд·, -а, (m) (nucl) rutherford, rd

ре́зерфорд·овск·ий, -ая, -ое, (a) (phys) Rutherford

рез·е́ц·, -з·ц·а́, (i) -з·ц·о́м, (m) cutter, cutting tool; (anat) incisor; stylus (disc recording)

~, в·став·н·о́й bit

~, га́льтель·н·ый corner cutter/tool

~, грав·ирова́льн·ый (print) engraving stylus

~, керам·и́ческ·ий ceramic tool tip

~, лев·ый left-cut tool

~, на·вар·н·о́й welded tipped tool

~, на·па́й·н·ый soldered tipped tool

~, от·кид·н·о́й swing tool

~, от·рез·н·о́й parting/cut-off tool

~, плу́ж·н·ый (agr) plough/plow cou·ter/cutter

~, под·рез·н·о́й рас·то́ч·н·ый bottom-boring cutter

~, пра́в·ый right-cut tool

~, фасо́н·н·ый form cutter

~, черн·ов·о́й roughing tool

резили·у́м·, -а, (m) (zool) resilium

резили·фе́р·, -а, (m) (zool) resilifer

рези́н·а, -ы, (f) rubber, vulcanized/processed rubber; rubber pad/packing/piece/band

~, бег·ов·а́я tread rubber

~, бес·са́ж·ев·ая unfilled rubber

рези́н·а
~, **гор·е́л·ая**   burnt rubber
~, **гу́б·чат·ая**   cellular/sponge rubber
~, **жев·а́тельн·ая**   chewing gum
~, **лёгк·ая**   low-density rubber
~, **лист·ов·а́я**   sheet rubber
~, **микро·по́р·ист·ая**   microporous rubber
~, **на·по́лн·енн·ая**   filled rubber
~, **по́р·ист·ая**   porous/expanded rubber
~, **рог·ов·а́я**   ebonite
~, **са́ж·ев·ая**   black-filled rubber
~, **сыр·а́я**   uncured rubber
резина́т·, -а,   (*m*) (chem) resinate
резин·и́стост·ь, -и,   (*f*) rubberiness
резини́т·, -а,   (*m*) (plast) rezinit, thiokol
рези́н·к·а, -и,   (*g pl*) **-н·ок·,**   (*f*) eraser, india-rubber, rubber; (text) elastic
рези́н·ов·ый, -ая, -ое,   (*a*) rubber
~ **под·ши́п·ник·**   (*m*) rubber bearing; (shipb) cutless bearing
~ **про·кла́д·к·а**   (*f*) rubber gasket, rubber seating
~ **техн·и́ческ·ие из·де́л·и·я**   (*pl*) rubber engineering components, rubber mechanicals
резинози́с·, -а,   (*m*) (bot) resinosis, resin flow
резино·обув·н·о́й, -а́я, -о́е,   (*a*) rubber-footwear
резино·по·до́б·н·ый, -ая, -ое,   (*a*) rubbery
резино·с·мес·и́тел·ь, -я,   (*m*) (rubber) mixer
~, **за·кры́·т·ый**   internal mixer
рези́но·техн·и́ческ·ие из·де́л·и·я (*pl*) rubber mechanicals, rubber engineering components
резиста́нс·, -а,   (*m*) (*and see* **со·про·тивл·е́ни·е**) (elec) resistance
резиста́нц·н·ый, -ая, -ое,   (*a*) resistance, variable-resistance, rheostatic, rheostat
резисти́в·н·ый, -ая, -ое,   (*a*) (elec) resistor, resistance, resistance-wire
рези́стор·, -а,   (*m*) *see also* **со·противл·е́ни·е;** non-ohmic resistor (semiconductors)
рези́т·, -а,   (*m*) (plast) resite, C-stage resin
резито́л·, -а,   (*m*) (plast) resitol, B-stage resin
ре́з·к·а, -и,   (*g pl*) **-з·ок·,**   (*f*) cutting
~, **га́з·ов·ая**   gas/oxy-acetylene cutting
~, **кисло·ро́д·н·ая**   oxygen/flame cutting
~, **огн·ев·а́я**   oxygen/flame cutting

ре́з·к·а
~, **паке́т·н·ая**   (weld) stack cutting
~, **по·пере́ч·н·ая**   cross cutting
~, **пре·ры́в·ист·ая**   (weld) skip cutting
ре́з·к·ий, -ая, -ое,   (*a*) sharp, abrupt, sudden; drastic; shrill, harsh (sound), pungent (smell); pronounced, definite, crisp (outlines)
ре́з·к·о   (*adv*) sharply, abruptly, suddenly; drastically; clearly
~ **воз·раст·а́ющ·ий**   suddenly rising; runaway, breakaway
~ **вы́·раж·енн·ый**   pronounced, marked
резко·о·черн·ённ·ый, -ая, -ое,   (*a*) (phot) sharp
ре́з·кост·ь, -и,   (*f*) sharpness, abruptness; shrillness; (phot) definition
~, **выс·о́к·ая**   (phot) good definition
резнатро́н·, -а,   (*m*) (elec) resnatron
рез·н·о́й, -а́я, -о́е,   (*a*) carved, fretted
резо́л·, -а,   (*m*) (plast) resol, A-stage resin
резольве́нт·а, -ы,   (*f*) (math) resolvent
резольво́·метр·, -а,   (*m*) (phot) resolution tester
резо́ль·н·ый, -ая, -ое,   (*a*) resol
~ **смол·а́**   (*f*) (plast) bakelite
резолю́ци·я, -и,   (*f*) resolution, decision; signature to a resolution
резона́нс·, -а,   (*m*) resonance
~ **на·пряж·е́ни·я** (elec) voltage/series resonance
~, **по·бо́ч·н·ый**   spurious resonance
~, **пол·остн·о́й**   cavity resonance
~, **по·сле́д·овательн·ый**   series resonance
~, **раз·ма́з·анн·ый**   (phys) smeared-out resonance
~, **раз·реш·ённ·ый** resolved resonance
~, **рас·пре·дел·и́тельн·ый** space resonance
~ **то́к·ов**   (elec) current/parallel resonance, antiresonance
~, **то́к·ов·ый**   (elec) antiresonance
~, **я́дер·н·ый магни́т·н·ый** nuclear magnetic resonance
резона́нс·н·ый·, -ая, -ое,   (*a*) resonance (of properties); tuned, resonant (of components)
~ **из·луч·е́ни·е** (*n*) resonance radiation
~ **ко́нтур·**   (*m*) resonant circuit
~ **крив·а́я**   (*f*) resonance curve
~ **око́ш·к·о**   (*n*) (rad) resonant window (of waveguide)
~ **по·глощ·е́ни·е**   (*n*) resonance absorption

резона́нс·н·ый, -ая, -ое
~ по́л·ост·ь (*f*) resonant cavity
~ част·от·а́ (*f*) resonant frequency
~ щел·ь (*f*) resonant gap
резон·а́тор·, -а, (*m*) resonator; cavity, resonant cavity (of vacuum tubes)
~, в·ход·н·о́й buncher, input resonator
~, вы·ход·н·о́й output resonator, catcher
~, ква́рц·ев·ый quartz resonator
~, магнито·стрикцио́н·н·ый magnetostrictive resonator
~, объ·ём·н·ый cavity, cavity resonator
~, по́л·ый cavity resonator
~, прямо·уго́ль·н·ый rectangular cavity
~, реакти́в·н·ый reaction cavity
~ -у·скор·и́тел·ь, -я, (*m*) (nucl) accelerator cavity
резон·и́р·овать, -уют, (*imp*) resonate
резон·и́рующ·ий, -ая, -ее, (*pres part act*) of **резон·и́р·овать**; resonant
резо́рбци·я, -и, (*f*) resorption
резорци́н·, -а, (*m*) (chem) resorcinol, (pharm) resorcin
резорци́н-формальдеги́д·н·ая смол·а́ (*f*) resorcinol-formaldehyde resin
результа́нт·, -а, (*m*) (math) resultant
результа́т·, -а, (*m*) result
   в результа́т·е as a result
результа́т·н·ый, -ая, -ое, (*a*) fruitful, successful
результ·и́рующ·ий, -ая, -ее, (*a*) resultant, product, net
~ амплиту́д·а (*f*) net amplitude
рез·у́н·, -а́, (*m*) (ocean) young ice
резурге́нт·н·ый, -ая, -ое, (*a*) resurgent
ре́зус·, -а, (*m*) (zool) rhesus monkey, *Macacus rhesus*
ре́з·ц·е- (*component*) cutter, tool
резце·держ·а́тел·ь, -я, (*m*) tool holder (of cutting machine)
резц·о́в·ый, -ая, -ое, (*a*) tool, cutter
рез·ц·ы́, (*pl*) of **рез·е́ц·**
ре́з·ч·е, (*comp*) of **ре́з·к·ий** and **ре́з·-к·о**
ре́з·чик·, -а, (*m*) engraver; carver; (met roll) sawman
ре́з·чицк·ий, -ая, -ое, (*a*) of **ре́з·чик·**
ре́з·ь, -и, (*f*) (med) colic
рез·ьб·а́, -ы́, (*f*) screw thread, thread; carving
~, вне́ш·н·яя male thread
~, вну́тр·енн·яя female thread

рез·ьб·а́
~, га́·ечн·ая female thread
~, га́з·ов·ая gas/pipe thread
~, двух·за·хо́д·н·ая double thread
~, дюйм·о́в·ая inch/British thread
~, за·кругл·ённ·ая rounded/radiused thread
~, креп·ёжн·ая securing thread
~, ле́в·ая left-handed thread
~, много·за·хо́д·н·ая multi-entry thread
~, много·ход·ов·а́я multiple thread
~, нару́ж·н·ая male thread
~, не·по́лн·ая imperfect/faulty thread
~, одно·за·хо́д·н·ая single thread
~, по·ло́г·ая slow-pitch thread
~, при·ту́пл·енн·ая flattened thread
~, прямо·уго́ль·н·ая square thread
~, тре·уго́ль·н·ая angular thread
~, трёх·ни́т·очн·ая triple thread
~, у·по́р·н·ая buttress thread
рез·ьбов·о́й, -а́я, -о́е, (*a*) thread, threading, threaded; screw
~ греб·е́ц· (*m*) thread chaser
~ кали́бр· (*m*) screw gauge/gage
резьбо·из·мер·и́тельн·ый инстру-ме́нт· (*m*) thread gauges/gages (*pl*)
резьбо·ме́р·, -а, (*m*) screw-pitch gauge/gage, thread gauge/gage
резьбо·на·ка́т·н·ый, -ая, -ое, (*a*) thread-rolling
~ голо́в·к·а (*f*) thread-rolling attachment
~ ро́л·ик· (*m*) thread-rolling tool
~ стан·о́к· (*m*) thread-rolling machine
резьбо·на·рез·н·о́й, -а́я, -о́е, (*a*) thread/screw-cutting, screwing, threading
~ голо́в·к·а (*f*) screwing attachment
~ стан·о́к· (*m*) thread-cutting/screwing machine
резьбо·ток·а́рн·ый стан·о́к· (*m*) screw-/thread-cutting lathe
резьбо·фре́зер·н·ый стан·о́к· (*m*) thread-milling machine
резьбо·шлиф·ова́льн·ый стан·о́к· (*m*) thread-grinding machine
резюме́ (*n indecl*) summary, résumé
резюм·и́р·овать, -уют, (*imp*) give a résumé, review
ре́йн·, -а, (*m*) (chem) rhein
ре́·й, -я, (*m*) (naut) yard
ре́йбал·, -а, (*m*) reamer
ре́йд·, -а, (*m*) (naut) roadstead, road; raid
~, вну́тр·енн·ый inner harbour/harbor
~, лес·н·о́й timber trap

рейд
~, от·кры́·т·ый   open roadstead
   о·хра́н·а рейд·ов (*f*) (mil) harbour
   defence
ре́йдер·, -а, (*m*) raider
рейд·ов·ый, -ая, -ое, (*a*) roadstead,
   harbour, harbor, coastal
~ сто·я́нк·а (*f*) anchorage
рей·к·а, -и, (*g pl*) **ре́·ек·,** (*f*) lath,
   batten; graduated pole/rod/staff;
   (mech) rack; rake
~, ба́зис·н·ая (surv) base-measuring
   bar
~, водо·ме́р·н·ая   water board-gauge/
   gage
~, вы́·вер·очн·ая   infinity bar (on
   rangefinder)
~, гла́д·к·ая (mech) flat-faced rack
~, дально·ме́р·н·ая (surv) range rod
~, зуб·ча́т·ая   rack, toothed rack
~, ко́рпус·н·ая (shipb) mould-loft
   batten
~, ме́р·н·ая   surveyor's rod/pole
~, не·по·дви́ж·н·ая (mech) fixed rack
~, нивели́р·н·ая (surv) levelling staff/
   rod
~, па́з·ов·ая (shipb) seam strap
~, по·движ·н·а́я   floating rack
~, реж·ущ·ая   rack-type gear-planing
   cutter
рей·ков·ый, -ая, -ое, (*a*) (naut) lug
   (of rigging)
рейко·на·рез·н·о́й стан·о́к· (*m*) rack-
   cutting machine
рейко·фре́зер·н·ый стан·о́к· (*m*)
   rake-milling machine
ре́йс·, -а, (*m*) trip, run, round; (naut)
   voyage, passage; (air) scheduled flight
~, челн·о́чн·ый (air) feeder service
ре́йсмас·, -а, (*m*) *see* **ре́йсмус·**
ре́йсмус·, -а, (*m*) scribing block,
   surface/marking gauge/gage; thicknesser, panel planer (woodworking)
~, четырёх·сторон·н·ий   four-cutter
   planer (woodworking)
рейсмус·ов·ый строг·а́льн·ый ста-
   н·о́к· (*m*) thicknesser, panel planer,
   planer and thicknesser (woodworking)
ре́йсмасс· -а, (*m*) *see* **ре́йсмус·**
ре́йс·ов·ый, -ая, -ое, (*a*) of **ре́йс·**;
   scheduled, regular (of transport)
~ пассажи́р·ск·ий само·лёт· (*m*)
   airliner
ре́йстрек·, -а, (*m*) (nucl) race track
рейсфе́дер·, -а, (*m*) ruling pen; pencil
   holder

рейсфе́дер
~, двой·н·о́й   road/railroad pen (used
   in cartography)
~, криво·лин·е́йн·ый   contour pen
рейсши́н·а, -ы, (*f*) tee-square (drawing)
ре́йтер·, -а, (*m*) rider (e.g. on a chemical balance)
рек·а́, -и́, (*nom pl*) **ре́к·и,** (*g pl*) **рек·,**
   (*f*) river
~, в·ло́ж·енн·ая   engrafted river
~, исчез·а́ющ·ая   intermittent river
~, лед·ян·а́я   ice stream
~, ма́л·ая   shallow navigable river
~, о·бед·нённ·ая   misfit/underfed river
~, под·чин·ённ·ая   tributary
~, ста́р·ая   mature river
рекапитул·я́ци·я, -и, (*f*) (biol) reca-
   pitulation
рекарбониз·а́тор·, -а, (*m*) (chem)
   recarbonizer
реквиз·и́р·овать, -уют, (*imp and perf*)
   requisition, commandeer
реквизи́т·, -а, (*m*) (theat) properties,
   props; (*pl*) particulars
реквизи́тор·, -а, (*m*) (theat) properties
   man
реквизи́ци·я, -и, (*f*) requisition, requi-
   sitioning
рекла́м·а, -ы, (*f*) advertisement; pub-
   licity
реклам·а́ци·я, -и, (*f*) complaint,
   protest; notification of defects
реклам·и́р·овать, -уют, (*imp and
   perf*) advertise; notify defects; put
   in a complaint
рекла́м·н·ый, -ая, -ое, (*a*) advertising
~ вещ·а́ни·е (*n*) commercial broadcasting
рекламо·да́т·ел·ь, -я, (*m*) advertiser;
   sponsor (of commercial broadcasting)
рекогносц·и́р·овать, -уют, (*imp and
   perf*) reconnoitre
рекогносц·иро́вк·а, -и, (*g pl*) **-вок·,**
   (*f*) reconnaissance
рекомбин·а́тор·, -а, (*m*) (nucl)
   recombiner, recombining system
рекомбинацио́н·н·ый, -ая, -ое, (*a*)
   recombination
рекомбин·а́ци·я, -и, (*f*) recombination
рекомбин·и́р·овать, -уют, (*imp*)
   recombine
рекоменд·а́тельн·ый, -ая, -ое, (*a*)
   of **рекоменд·а́ци·я**; recommendary
~ письм·о́ (*n*) letter of recommen-
   dation/introduction
рекоменд·а́ци·я, -и, (*f*) recommendation; reference, testimonial

реко́менд·ова́ть, -у́ют, *(imp)* recommend

реконстру·и́ровани·е, -я, *(n)* *(v n)* of **реконстру·и́р·овать;** reorganization; reconstruction, restoration

реконстру·и́р·овать, -уют, *(imp and perf)* reconstruct; reorganize

реконстру́кци·я, -и, *(f)* reconstruction; reorganization

реко́рд·, -а, *(m)* record (highest achievement)

∼, мир·ов·о́й world record

реко́рд·ер·, -а, *(m)* *see also* **само·пи́с·ец·;** recorder; recording head

реко́рд·и́ст·, -а, *(m)* record breaker/holder, champion

бык-рекорди́ст·, -а, *(m)* *(agr)* champion bull

рекордсме́н·, -а, *(m)* = **реко́рд·и́ст·** (q.v.)

ре·кристалл·иза́ци·я, -и, *(f)* recrystallization

∼, без·за·ро́д·ышев·ая recrystallization *in situ*

∼ об·рабо́т·к·и primary recrystallization

∼, со·бир·а́тельн·ая secondary recrystallization

рексиген·н·ая по́л·ост·ь *(f)* *(biol)* lysigenous space

рект·а́льн·ый, -ая, -ое, *(a)* *(anat)* rectal

ректи·град·а́ци·я, -и, *(f)* *(gen)* rectigradation

рект·и́т·, -а, *(m)* *(med)* rectitis

ректи·фикацио́н·н·ый, -ая, -ое, *(a)* rectification, rectifying

∼ коло́нн·а *(f)* *(chem)* rectification column

ректи·фик·а́ци·я, -и, *(f)* *(chem)* rectification, redistillation

ректи·фиц·и́р·овать, -уют, *(imp and perf)* *(chem)* rectify

ре́ктор·, -а, *(m)* rector, head of university

ректо·ско́п·, -а, *(m)* *(med)* rectoscope

ректо·кретра́ль·н·ый *(anat)* recto-urethral

рекуперати́в·н·ый, -ая, -ое, *(a)* recuperative; regenerative

∼ тормож·е́ни·е *(n)* *(elec)* regenerative braking

рекупер·а́тор·, -а, *(m)* recuperator (of a furnace); recuperative air-heater

рекупер·а́ци·я, -и, *(f)* recuperation, regeneration

рекурре́нт·н·ый, -ая, -ое, *(a)* recurrent, recurrence, recurring, repeating

рекурси́в·н·ый, -ая, -ое, *(a)* recursive, recursion

реку́рс·и·я, -и, *(f)* *see also* **воз·вра·щ·е́ни·е;** recursion, recurrence

релакс·а́тор·, -а, *(m)* *(phys)* relaxation oscillator

релакс·ацио́нн·ый, -ая, -ое, *(a)* relaxation

∼ генер·а́тор· *(m)* relaxation oscillator

∼ звен·о́ *(n)* *(autom)* aperiodic stage

релакс·а́ци·я, -и, *(f)* relaxation

∼, вращ·а́тельн·ая *(phys)* rotational relaxation

релаксо́·метр·, -а, *(m)* relaxometer

реле́ *(n indecl)* relay

∼ бд·и́тельност·и *(rail)* acknowledging relay

∼, без·инерцио́н·н·ое extra high-speed relay

∼, блок·иро́вочн·ое locking relay

∼, быстро·де́й·ствующ·ее fast-acting relay

∼, в·ключ·а́ющ·ее closure/cut-in relay

∼, вод·ян·о́е priming relay (of pumps)

∼, време́н·н·о́е time-delay relay, timing relay, timer

∼, вс·по·мог·а́тельн·ое auxiliary relay

∼ в·ступл·е́ни·я time-delay relay, timing relay, timer

∼, втор·и́чн·ое *(autom)* pilot relay

∼, вы·ключ·а́ющ·ее shutdown relay

∼, выше·сто·я́щ·ее back-up relay

∼, га́з·ов·ое gas-pressure relay, gas actuated relay, gas detector

∼, гальвано·метр·и́ческ·ое galvanometer relay

∼ двой·н·о́го де́й·стви·я double-acting relay

∼, дифференци·а́ль·н·ое balanced/differential relay

∼, ём·костн·ое capacitive relay

∼, жд·у́щ·ее time-delay relay, slow-acting relay

∼, за·де́рж·ивающ·ее guard relay (telephones)

∼, за·щи́т·н·ое protective relay

∼, импеда́нс·н·ое impedance relay

∼, и́мпульс·н·ое impulse-accepting relay (telephony)

∼, инерцио́н·н·ое slug relay

∼ ис·хо́д·н·ого режи́м·а reset relay

реле́

~, **максима́ль·н·ое** maximum/over-current relay, current limiter

~, **мембра́н·н·ое** diaphragm relay

~ **минима́ль·н·ого то́к·а** minimum/undercurrent relay

~, **минима́ль·н·ое** minimum/undercurrent relay

~, **на·пра́в·л·енн·ое** directional relay

~ **на·пряж·е́ни·я** voltage relay

~, **нейтра́ль·н·ое** neutral relay

~, **нул·ев·о́е** zero-phase-sequence relay; no-voltage relay

~ **обра́т·н·ого то́к·а** reverse-current relay

~, **от·ключ·а́ющ·ее** trip-free relay

~, **перв·и́чн·ое** (autom) controlling unit/means

~, **позицио́н·н·ое** positioning relay; air relay/amplifier (of pneumatic controller)

~ **по́лн·ого со·противл·е́ни·я** impedance relay

~, **про́б·н·ое** testing relay

~, **про·меж·у́точн·ое** auxiliary relay

~ **раз·гру́з·к·и** discharge measurement relay (of pumps)

~ **реакти́в·н·ого со·противл·е́ни·я** reactance relay

~, **регул·и́рующ·ее** regulating relay

~ **-регул·я́тор·, -а,** (m) current-and-voltage regulator

~ **режи́м·а реш·е́ни·я** computer relay

~, **само·в·ключ·а́ющ·ее** reclosing relay

~, **сильфо́н·н·ое** bellows-type pneumatic relay

~ **со·един·и́тельн·ых ли́н·и·й** outgoing junction relay (telephony)

~, **стру́й·н·ое** flow-meter relay (for closed conduits); jet relay (of hydraulic jet-type controller)

~, **тепл·ов·о́е** thermal relay

~, **у·каз·а́тельн·ое** flag indicating relay

~ **у·те́ч·к·и** leakage relay

~, **фа́з·ов·ое** phase-sequence relay

~, **част·о́тн·ое** frequency relay

~, **чувст·и́тельн·ое** sensitive relay

~ **ша́г·ов·ое** (elec) stepping relay

~, **электро·магни́т·н·ое бала́нс·н·ое** balanced armature relay

~, **электро·механ·и́ческ·ое резо-на́нс·н·ое** reed relay

~, **электро́н·н·ое** electronic relay; electron relay (in vacuum tubes)

Ре́ле·й, -я, (m) see also **Рэ́ле·й** (phys) Rayleigh

ре́ле·евск·ий, -ая, -ое, (a) see **рэле́·-евск·ий;** Rayleigh

реле́йно·конта́кт·н·ая схе́м·а (f) relay circuit diagram, network diagram

реле́й·н·ый, -ая, -ое, (a) relay, relaying, repeating

~ **за·щи́т·а** (f) relay protection, protective relay

~ **ли́н·и·я с·вя́з·и** (f) radio relay link

~ **се́кци·я** (f) relay unit

~ **систе́м·а** (f) (autom) two-position system; all-relay system (telephones); (rad) radio-relay link/system

~ **централиз·а́ци·я** (f) (rail) all-relay interlocking

ре́ле·я, (g) of **ре́ле·й**

рели́кви·я, -и, (f) relic

рели́кт·, -а, (m) relic; (biol) relict; (geog) relict mountain

рели́кт·ов·ый, -ая, -ое, (a) (biol) relict

~ **о́зер·о** (n) relict lake

релье́ф·, -а, (m) relief, topography

~, **потенциа́ль·н·ый** (elec) charge image

релье́ф·ност·ь, -и, (f) relief; (TV) plasticity; depth (of a photograph)

релье́ф·н·ый, -ая, -ое, (a) relief, raised, projecting

~ **с·ва́р·к·а** (f) projection welding

ре́льс·, -а, (m) rail

~ **Ви́ньол·я** flat-bottomed rail

~, **двух·голо́в·ый** bull-head rail

~, **жело́б·чат·ый** tram rail

~, **конта́кт·н·ый** (rail) third/contact/conductor rail

~, **ра́м·н·ый** (rail) stock rail

~, **трамва́й·н·ый** tram rail

~, **широко·подо́·шв·енн·ый** (rail) flat-bottomed rail

~ **-форм·, -а,** (m) (civ eng) road forms

рельсо·ба́л·очн·ый стан· (m) (met roll) rail mill

ре́льс·ов·ый, -ая, -ое, (a) rail, track

~ **пут·ь** (m) rails, track

~ **со·един·е́ни·е** (n) (elec rail) track rail-bond

~ **цеп·ь** (f) (elec rail) track circuit

рельсо·про·ка́т·н·ый, -ая, -ое, (a) rail-rolling

~ **стан·** (m) rail mill

реля́кс·а́тор·, **-а**, (*m*) *see* **релакс·а́-тор·**

реляти́в·ность·, **-и**, (*f*) relativity

реляти́в·н·ый, **-ая**, **-ое**, (*a*) relative

релятиви́ст·ск·ий, **-ая**, **-ое**, (*a*) relativist, relativistic

ремалло́·й, **-я**, (*m*) (met) Remalloy

рем·ённ·ый, **-ая**, **-ое**, (*a*) belt

∼ при·во́д· (*m*) belt, belt drive

рем·е́н·ь, **-м·н·я́**, (*m*) belt, strap, thong

∼, вентиля́тор·н·ый fan belt

∼, клино·ви́д·н·ый V/Vee-belt

∼, при·вяз·н·о́й (a/c) safety belt

∼, руж·е́йн·ый rifle sling

ремёсл·енник·, **-а**, (*m*) craftsman, artisan, tradesman (qualified industrial worker)

ремёсл·енн·ый (*a*) trade, craft

ремесл·о́, **-а́**, (*nom pl*) **ремёсл·а** (*n*) trade, craft

рем·ешко́в·ый, **-ая**, **-ое**, (*a*) of **рем·ешо́к·**

рем·ешо́к·, **-шк·а́**, (*m*) (*dim*) of **рем·е́н·ь**

реми́з·, **-а**, (*m*) (fin) remittance

реми́з·а, **-ы**, (*f*) (text) healds (*pl*)

реми́з·к·а, **-и**, (*g pl*) **-з·ок·**, (*f*) heald, (text) heald shaft

реми́зо·подъ·ём·н·ая каре́т·к·а (*f*) (text) dobby

реми́сси·я, **-и**, (*f*) (med) remission

ремите́нт·, **-а**, (*m*) (fin) remitter

ремне·ви́д·н·ый, **-ая**, **-ое**, (*a*) strap/ribbon shaped, lorate

ремн·е́ц·, **-а́**, (*i*) **-о́м**, (*m*) (zool) Ligula intestinalis

ремне·цве́т·ник·, **-а**, (*m*) (bot) loranth

ремне·цве́т·н·ые, **-ых**, (*pl decl as adj*) (bot) mistletoe family, *Loranthaceae*

ремо́нт·, **-а**, (*m*) repair, repairs, overhaul, maintenance; (shipb) refit; (mil) remount (cavalry)

∼, авар·и́йн·ый emergency repairs **в ремо́нт·е** under repair; (shipb) refitting

∼, вос·станов·и́тельн·ый renovation; (shipb) rehabilitation refit; retreading (tyres)

∼, капита́ль·н·ый major/capital/heavy repairs; (shipb) major refit

∼, навигацио́н·н·ый (naut) maintenance at sea

∼, плано-пред·у·пред·и́тельн·ый periodic maintenance

ремо́нт

∼, пред·у·пред·и́тельн·ый maintenance, trouble-shooting

∼, профилакт·и́ческ·ий maintenance, trouble-shooting

∼, сре́д·н·ий intermediate overhaul/refit

∼, тек·у́щ·ий routine repairs; annual refit

ремонта́нт·н·ый, **-ая**, **-ое**, (*a*) (bot) remontant

ремонт·и́р·овать, **-у́ют**, (*imp and perf*) repair, refit; remount (of cavalry)

ремо́нт·ник·, **-а**, (*m*) maintenance man, repair man

ремо́нт·н·о-стро·и́тельн·ый цех· (*m*) works-and-buildings department

ремо́нт·н·ый, **-ая**, **-ое**, (*a*) of **ремо́нт·**

∼ цех· (*m*) repair shop

ремонтуа́р·, **-а**, (*m*) stem winding watch

рему (*indecl*) (ocean) up-and-down currents; disturbed air

ренарди́т·, **-а**, (*m*) (min) renardite

рен·а́т·, **-а**, (*m*) (chem) rhenate

рендзи́н·а, **-ы**, (*f*) rendzina (soil)

ре́н·иев·ый, **-ая**, **-ое**, (*a*) (chem) rhenium, perrhenic

∼ ангидри́д· (*m*) rhenium heptoxide

∼ кисл·от·а́ (*f*) perrhenic acid

ре́н·и·й, **-я**, (*m*) (chem) rhenium, Re

∼, сер·ни́ст·ый rhenium disulphide; rhenium heptasulphide

∼, фто́р·ист·ый rhenium hexafluoride

∼, хло́р·ист·ый rhenium pentachloride; rhenium tetrachloride; rhenium trichloride

ре́н·ист·ый, **-ая**, **-ое**, (*a*) rhenium, rhenic

∼ ангидри́д· (*m*) rhenium trioxide

∼ кисл·от·а́ (*f*) rhenic acid

рен·и́т·, **-а**, (*m*) rhenite

Ре́нкин·а, **цикл·** (*m*) (eng) Rankine cycle

ренкло́д·, **-а**, (*m*) (bot) greengage

ренн·а́з·а, **-ы**, (*f*) (biochem) rennase

ренн·и́н·, **-а**, (*m*) (biol) rennin

ре́нт·а, **-ы**, (*f*) rent; investment income; government bonds, undated government stock

∼, еже·го́д·н·ая annuity

∼, земе́ль·н·ая ground rent

рента́бель·ность·, **-и**, (*f*) profitability, commercial value

рента́бель·н·ый, -ая, -ое, (*a*) paying, showing a profit, profitable, commercial, economic

рентге́н·, -а, (*g pl*) **рентге́н·,** (*m*) roentgen, röntgen (unit of exposure dose of X or gamma radiation)

~ **-эквивале́нт·, био·лог·и́ческ·ий** roentgen-equivalent man, rem

~ **-эквивале́нт·, физ·и́ческ·ий** roentgen-equivalent physical, rep

рентген·изи́р·овать, -уют, (*imp and perf*) X-ray

рентген·ме́тр·, -а, (*m*) roentgenometer, roentgen meter, ionometer

~**, интегра́ль·н·ый** condenser roentgenometer

рентге́н·ов, -а, -о, (*posses adj*) Roentgen, X-

~ **луч·и́** (*pl*) X-rays

рентге́н·овск·ий, -ая, -ое, (*a*) X-ray, Roentgen

~ **луч·и́** (*pl*) X-rays

~ **про·све́ч·ивани·е** (*n*) fluoroscopy, X-ray flaw detection

~ **спектра́ль·н·ый ана́лиз·** (*m*) X-ray spectral/spectrum/spectroscopic analysis

~ **структу́р·н·ый ана́лиз·** X-ray crystallography

~ **труб·а́** (*f*) X-ray tube

~ **труб·а́, ио́н·н·ая** (*f*) gas-filled X-ray tube

рентгено·гра́мм·а, -ы, (*f*) roentgenogram, radiograph, X-ray photograph; X-ray diffraction pattern

~ **враще́·ни·я** (spectr) rotating-crystal pattern

~ **колеб·а́ни·я** oscillating-crystal pattern

ренгено́·граф·, -а, (*m*) (spectr) X-ray diffraction apparatus

рентгено·гра́ф·и·я, -и, (*f*) radiography, (obs) roentgenography; (spectr) X-ray diffraction analysis

~**, техн·и́ческ·ая** industrial radiography

рентгено·дефекто·ско́п·и·я, -и, (*f*) X-ray flaw detection

рентгено·диагно́ст·ик·а, -и, (*f*) X-ray diagnostic radiology, X-ray diagnostics

рентгено·контра́ст·н·ый, -ая -ое, (*a*) X-ray contrast

рентгено́·лог·, -а, (*m*) radiologist

рентгено·ло́г·и·я, -и, (*f*) radiology

рентгено·люмин·исце́нци·я, -и, (*f*) roentgenoluminescence

рентгено·плён·к·а, -и, (*g pl*) **-н·ок·,** (*f*) X-ray film

рентгено·про·зр·а́чн·ый, -ая, -ое, (*a*) radioparent, radio-transparent, X-ray-transparent

рентгено·про·ниц·а́ем·ый, -ая, -ое, (*a*) radiolucent

рентгено·про·све́ч·ивани·е, -я, (*n*) fluoroscopy, X-ray flaw detection

рентгено·про·све́ч·иваюш·ий, -ая, -ее, (*a*) radiolucent

рентгено·ско́п·, -а, (*m*) roentgenoscope, fluoroscope

рентгено·скоп·и́·я, -и, (*f*) X-ray examination, fluoroscopy

рентгено·с·ни́м·ок·, -м·к·а, (*m*) X-ray photograph, roentgenogram, radiograph, skiagraph

рентгено·спектро·скоп·и́·я, -и, (*f*) X-ray spectroscopy, X-ray spectral/spectrum analysis

рентгено·структу́р·н·ый ана́лиз· (*m*) X-ray crystallography

рентгено·съ·ём·к·а, -и, (*f*) X-ray photography

рентгено·терап·и́·я, -и, (*f*) (med) X-ray therapy, radiotherapy

рентгено·те́хн·ик·, -а, (*m*) X-ray technician/engineer

рентгено·флюоро·гра́ф·и·я, -и, (*f*) fluorography

ре́нт·н·ый, -ая, -ое, (*a*) of **ре́нт·а**

Реньо́, калори́·метр· (*m*) Regnault calorimeter

рео·ба́з·а, -ы, (*f*) (physiol) rheobase

рео·вискози·ме́тр·, -а, (*m*) rheoviscosimeter

рео·жо́лоб·, -а, (*m*) (min) rheolaveur

рео·лог·и́ческ·ий, -ая, -ое, (*a*) rheological

рео·ло́г·и·я, -и, (*f*) (phys) rheology

рео́·метр·, -а, (*m*) rheometer, flowmeter

рео·мо́й·к·а, -и, (*g pl*) **-мо́·ек·,** (*f*) (min) rheolaveur

рео·морф·и́ческ·ий, -ая, -ое, (*a*) rheomorphic

Реомю́р·а, шкал·а́ (*f*) (phys) Reaumur scale

рео·пиро́·метр·, -а, (*m*) resistance pyrometer

рео·планкто́н·, -а, (*m*) (biol) rheoplankton

ре·организ·а́ци·я, -и, (*f*) reorganization

ре·организ·ова́ть, -у́ют, (*imp and perf*) reorganize

рео·ста́т·, -а, (*m*) (elec) rheostat, variable resistor

~, балла́ст·н·ый ballast resistor, thermistor

~ воз·бужд·е́ни·я field rheostat

~ выс·от·ы́ (gunn) height control rheostat

~, жи́дк·остн·ый (elec) liquid controller/rheostat

~, ла́мп·ов·ый (elec) lamp resistance

~, на·гру́з·очн·ый loading rheostat

~ на·ка́л·а filament rheostat/potentiometer

~, полз·унко́в·ый sliding-contact rheostat

~, пуск·ов·о́й starting rheostat

~, регул·иро́вочн·ый rheostatic controller

~ -темн·и́тел·ь, пере·движ·н·о́й (*m*) (cinema) dimmer bank

~, у́голь·н·ый carbon rheostat

~, фигу́р·н·ый functional rheostat

~, ши́бер·н·ый sliding-contact rheostat

рео·ста́т·н·ый, -ая, -ое, (*a*) rheostat, rheostat-type, rheostatic, potentiometric

~ да́т·чик· (*m*) potentiometer transmitter

~ да́т·чик·, петл·ев·о́й loop-wound potentiometer transmitter

~ тормож·е́ни·е (*n*) (elec) rheostatic braking

рео·стри́кци·я, -и, (*f*) (phys) pinch effect, rheostriction

рео·та́ксис·, -а, (*m*) (biol) rheotaxis

рео·такт·и́ческ·ий, -ая, -ое, (*a*) rheotactic

рео·троп·и́зм·, -а, (*m*) (biol) rheotropism

рео·фи́ль·н·ый, -ая, -ое, (*a*) (biol) rheophylic

рео·фо́р·, -а, (*m*) rheophore, connection, terminal

рео·хо́рд·, -а, (*m*) (elec) slide-wire resistor, slide resistance

ре́п·а, -ы, (*f*) (bot) turnip, *Brassica rapa*

реп·е́·й, (*g sing*) реп·ь·я́, (*nom pl*) реп·ь·и́, (*g pl*) реп·ь·ёв, (*m*) (bot) burdock, *Lappa major*

реп·е́йник·, -а, (*m*) = **реп·е́йничек·**

реп·е́йничек·, -чк·а, (*m*) (bot) agrimony, *Agrimonia*

реп·е́йн·ый, -ая, -ое, (*a*) burdock

~ ма́сл·о (*n*) (pharm) burdock oil

репе́р·, -а, (*m*) datum mark, datum/reference point, (surv) bench mark; (gunn) aiming point

репе́р·н·ый, -ая, -ое, (*a*) of **репе́р·**; reference, datum

репети́р·, -а, (*m*) (horol) repeater

репет·и́р·овать, -уют, (*imp*) rehearse, coach

репети́тор·, -а, (*m*) tutor, coach

репет·и́ци·я, -и, (*f*) (theat) rehearsal; (horol) striking work

репет·и́чн·ый, -ая, -ое, (*a*) repeating

репет·ова́ть, -у́ют, (*imp*) repeat, (naut) acknowledge (a signal)

репи́тер·, -а, (*m*) repeater (in remote indicating systems)

ре́п·к·а, -и, (*g pl*) -п·ок·, (*f*) crown knot

ре́плик·а, -и, (*f*) replica

репорт·а́ж·, -а, (*m*) reporting (for newspapers etc.)

репорт·ёр·, -а, (*m*) reporter, correspondent (newspapers)

репрезентати́в·ност·ь, -и, (*f*) (math) representation

репрезентати́в·н·ый, -ая, -ое, (*a*) representative

репресса́ли·и, -й, (*f pl*) reprisals

репре́сси·я, -и, (*f*) repression, repressive punishment

репроду́ктор·, -а, (*m*) reproducer; loudspeaker

~, перфораци́он·н·ый (autom) reproducing puncher, printing punch

репродукци́он·н·ый, -ая, -ое, (*a*) of **репроду́кци·я**

~ у·станов·к·а (*f*) (phot) copying apparatus

репроду́кци·я, -и, (*f*) reproduction; (phot) copying

репродуц·е́нт·, -а, (*m*) reproducer

репродуц·и́р·овать, -уют, (*imp and perf*) (print etc.) reproduce; (phot) copy

ре́пс·, -а, (*m*) (text) repp

ре́пс·ов·ый, -ая, -ое, (*a*) repp

репти́л·и·я, -и, (*f*) (zool) reptile, (*pl*) reptiles, *Reptilia*

репульси́он·н·ый, -ая, -ое, (*a*) repulsion

~ дви́г·ател·ь (*m*) (elec) repulsion motor

репут·а́ци·я, -и, (*f*) reputation

ре́п·чат·ый, -ая, -ое, (*a*) turnip-shaped/like

~ лук· (*m*) onion

реп·ь·и́, (*nom pl*) of **реп·е́·й**

ресекве́нт·н·ый, -ая, -ое, (*a*) resequent

реси́вер·, -а, (*m*) receiver (steam and compressed air); scavenging air receiver (diesel); turbine manifold; receiver (telegraphy)

ресинхрониз·а́ци·я, -и, (*f*) resynchronization

ресму́с·, -а, (*m*) *see* **рейсму́с·**

ресни́т·чат·ый, -ая, -ое, (*a*) (biol) ciliate, ciliated, ciliary

ресни́ц·а, -ы, (*i*) **-ей,** (*f*) eyelash, (zool) cilium

ресни́чат·ый, -ая, -ое, (*a*) = **ресни́т·· чат·ый**

ресни́ч·к·а, -и, (*g pl*) **-ч·ек·,** (*f*) (*dim*) of **ресни́ц·а;** (biol) cilium

ресни́ч·н·ый, -ая, -ое, (*a*) eyelash, (biol) ciliary

~ чёрв·и (*pl*) (zool) flatworms, *Turbellaria*

ресо́рбци·я, -и, (*f*) resorption

респира́тор·, -а, (*m*) *see also* **противо·га́з·** respirator, breathing apparatus

респира́тор·н·ый, -ая, -ое, (*a*) respiratory

респиро́·метр·, -а, (*m*) (med) respiration apparatus, respirometer

респу́блик·а, -и, (*f*) republic

Ре́ссел·я-Са́ундерс·а, с·вяз·ь (*f*) (phys) Russel–Saunders coupling

рессо́р·а, -ы, (*f*) leaf spring, spring

рессо́р·н·ый, -ая, -ое, (*a*) spring

рессу́рс·, -а, (*m*) *see* **ресу́рс·**

реставр·а́ци·я, -и, (*f*) restoration

реставр·и́р·овать, -уют, (*imp*) restore

реститу́ци·я, -и, (*f*) restitution

рестора́н·, -а, (*m*) restaurant

рестри́кци·я, -и, (*f*) restriction

ресупина́т·н·ый, -ая, -ое, (*a*) resupinate

ресурге́нт·н·ый, -ая, -ое, (*a*) resurgent

ресу́рс·, -а, (*m*) safe life (of engine etc.); (*pl*) resources

ретардёр·, -а, (*m*) (*and see* **за·мед·· л·и́тел·ь**) retarder

рети́в·ый, -ая, -ое, (*a*) zealous; ardent

ретикул·и́н·, -а, (*m*) (biochem) reticulin

ретикуло·ци́т·, -а, (*m*) (biol) reticulocyte

ретикуло·эндотелиа́ль·н·ый, -ая, -ое, (*a*) (physiol) reticuloendothelial

ретикул·я́рн·ый, -ая, -ое, (*a*) reticular, reticulate, retiform

рети́н·а, -ы, (*f*) (anat) retina

ретина́ль·н·ый, -ая, -ое, (*a*) (anat) retinal

ретин·е́н·, -а, (*m*) retinene (vitamin A product)

ретин·и́т·, -а, (*m*) (med) retinitis; (min) retinite

ретино·бласт·о́м·а, -ы, (*f*) (med) retinoblastoma

ретин·о́л·, -а, (*m*) retinol (vitamin A product)

ретино·ско́п·и·я, -и, (*a*) (physiol) retinoscopy

рето́рт·а, -ы, (*f*) (chem, met) retort, still

~, глух·а́я stop-end retort

~, сквоз·н·а́я through retort

рето́рт·н·ый, -ая, -ое, (*a*) of **рето́рт·а**

ретрансляцио́н·н·ый, -ая, -ое, (*a*) of **ретранс·ля́ци·я**

ретранс·ля́ци·я, -и, (*f*) (*and see* **транс· л·я́ци·я**) repeating (telephones); (radio) relaying, retransmitting

ре·трансми́сс·и·я, -и, (*f*) radio relay, radio-relay equipment

ре·трансми́ттер·, -а, (*m*) retransmitter (telegraphy)

ретре́дер·, -а, (*m*) retreader (tyres)

ретро·акти́в·н·ый, -ая, -ое, (*a*) retroactive

ретро·бульба́р·н·ый, -ая, -ое, (*a*) (opt) retrobulbar

ретро·вакцин·а́ци·я, -и, (*f*) (med) retrovaccination

ретро·град·а́ци·я, -и, (*f*) retrogradation

ретро·грессив·н·ый, -ая, -ое, (*a*) retrogressive

ретро·морф·о́з·, -а, (*m*) (biol) retromorphosis

ретурбе́нт·, -а, (*m*) return bend/pipe

ре·туш·ева́льн·ый, -ая, -ое, (*a*) (phot) retouching, finishing

~ стан·о́к· (*m*) retouching stand

ре·туш·ёвк·а, -и, (*f*) (*v n*) *see* **ре·ту·· ш·и́р·овать**

ре·туш·ёр·, -а, (*m*) (phot) retoucher-and-finisher

ре·туш·ёрн·ый, -ая, -ое, (*a*) = **ре·· туш·ева́льн·ый**

ре·туш·и́р·овать, -уют, (*imp and perf*) (phot) retouch (negatives); finish (prints)

ре́·туш·ь, -и, (*f*) (phot) retouching (negatives); finishing (print)

рефера́т·, -а, (*m*) abstract; paper (academic)

реферат·и́вн·ый, -ая, -ое, (*a*) abstract

~ журна́л· (*m*) journal of abstracts, abstracts

референдум·, -а, (*m*) referendum

референт·, -а, (*m*) abstractor; reader, speaker (delivering academic paper); consultant, adviser

рефери (*m indecl*) referee (sport)

референци·я, -и, (*f*) reference, testimonial

рефер·и́р·овать, -уют, (*imp and perf*) abstract; referee, umpire

рефле́кс·, -а, (*m*) (physiol) reflex, reflex action; (phot) ghost

рефле́кс·и·я, -и, (*f*) reflection

рефле́кс·н·ый, -ая, -ое, (*a*) reflection; reflex; feedback

~ схе́м·а (*f*) (rad) reflex/dual-amplification circuit

~ у·стро́й·ств·о (*n*) (autom) feedback control system

рефлект·и́р·овать, -уют, (*imp*) react; reflect

рефлекто́·метр·, -а, (*m*) (opt) reflectometer, reflecting power measuring instrument

рефле́ктор·, -а, (*m*) (*and see* **от·ра·ж·а́тел·ь**) reflector; (astron) reflecting telescope

~, парабол·и́ческ·ий parabolic reflection

~, угол·ко́в·ый angled/corner reflector

рефле́ктор·н·ый, -ая, -ое, (*a*) reflector, reflecting; **рефлекто́р·н·ый, -ая, -ое,** (*a*) reflex, feedback

рефо́рм·а, -ы, (*f*) reform

рефо́рминг·, -а, (*m*) (oil chem) reforming

рефра́ктер·н·ый пери́од· (*m*) (zool) refractory period

рефракто́·метр·, -а, (*m*) (opt) refractometer

~, по·груж·н·о́й dipping refractometer

рефракто·метр·и́ческ·ий, -ая, -ое, (*a*) refractometric

рефра́ктор·, -а, (*m*) (astron etc.) refractor, refracting telescope

рефракцио́н·н·ый, -ая, -ое, (*a*) refraction

рефра́кци·я, -и, (*f*) (*and see* **пре·лом·л·е́ни·е**) (phys) refraction

рефрижера́тор·, -а, (*m*) refrigerator; refrigerator ship/car/wagon

рефрижера́тор·н·ый, -ая, -ое, (*a*) refrigerator, refrigerated

рефуле́р·, -а, (*m*) shore/barge-delivery suction dredger; delivery pipeline (of above)

рефуле́р·н·ый, -ая, -ое, (*a*) of **рефуле́р·**

~ земле·со́с· (*m*) shore/barge-delivery suction dredger

рехиге́н·н·ый, -ая, -ое, (*a*) (biol) rhexigenous

ре́ценз·е́нт·, -а, (*m*) reviewer, critic

реце́нз·и·я, -и, (*f*) review

реце́пт·, -а, (*m*) (food) recipe; (med) prescription; (rubber) formula, mix formula

реце́птор·, -а, (*m*) (biol) receptor

рецепт·у́р·а, -ы, (*f*) (pharm) prescriptions; recipes, formulas, formulation

рецепт·у́рн·ая, -ой, (*f decl as adj*) (rubber) compounding department

~ с·пра́в·очник· (*m*) formulary

рецесси́в·н·ый, -ая, -ое, (*a*) recessive

рециди́в·, -а, (*m*) (med) relapse, recurrence

рециркуляцио́н·н·ый, -ая, -ое, (*a*) of **рециркул·я́ци·я**

рециркул·я́ци·я, -и, (*f*) recirculation, recycling

реч·ев·о́й, -а́я, -о́е, (*a*) vocal, speech

ре́ч·к·а, -и, (*g pl*) **-ч·ек·,** (*f*) (*dim*) of **рек·а́**

реч·н·о́й, -а́я, -о́е, (*a*) river, fluvial

ре́ч·ь, -и, (*nom pl*) **-и,** (*g pl*) **-е́й,** (*f*) speech; conversation, talk

реш·а́·ть, -ют, (*imp*) decide; solve

реш·а́ющ·ий, -ая, -ее, (*pres part act*) of **реш·а́·ть**; decisive; (autom) computing, compute, computer

~ фа́ктор· (*m*) decisive factor

~ элеме́нт· (*m*) computing element

реш·е́ни·е, -я, (*n*) (*v n*) *see* **реш·а́·ть**; decision; solution

~ под·бо́р·ом trial-and-error solution

реш·ённ·ый, -ая, -ое, (*past part pass*) *see* **реш·а́·ть**

решёт·а, (*nom pl*) of **решет·о́**

решёт·к·а, -и, (*g pl*) **-т·ок·,** (*f*) grating, grate, grid; lattice; railing; (dynam) cascade, blading; strainer (piping)

~, акти́в·н·ая (phys) active lattice; impulse cascade (of turbine)

~, анте́нн·ая (rad) antenna array

~, а́том·н·ая (phys) atomic lattice

~ Браве́ (cryst) Bravais lattice

~, гране·центр·и́рованн·ая куб·и́ческ·ая (cryst) face-centred cubic lattice

~, групп·ов·а́я (phys) cluster lattice

~, дифракцио́н·н·ая (spectr) diffraction grating

решёт·к·а

~, **жалюзи́й·н·ая**   baffle separator (dust catcher)

~, **за·щи́т·н·ая**   (a/c) air-intake screen

~, **иго́ль·чат·ая**   spiked lattice conveyor (of cotton bale breaker)

~, **колосни́к·ов·ая**   furnace grate; (rocket) grate; (pl) grids (of cotton bale breaker)

~, **компа́кт·н·ая**   (cryst) close-packed lattice, tight lattice

~, **кристалл·и́ческ·ая**   (phys) crystal/space lattice

~, **ку́б·и́ческ·ая**   (cryst) cubic lattice

~ **лопа́т·ок·**   vane cascade, blading (of turbine)

~, **мно́го·рас·ко́с·н·ая**   (civ eng) multiple lattice

~, **мно́го·ря́д·н·ая**   (rad) mattress array

~, **на·бух·а́ющ·ая**   expanding lattice

~, **на·кло́н·н·ая**   (met roll) gravity-feed bed

~, **на·ру́ш·енн·ая**   (phys) perturbed lattice

~, **обра́т·н·ая**   (cryst) reciprocal lattice

~, **объёмно·центр·и́рованн·ая ку́б·и́ческ·ая**   (cryst) body-centred cubic lattice

~, **пит·а́ющ·ая**   feed lattice (of bale breaker)

~, **пласт·и́нчат·ая**   (phys) slab lattice

~, **плотно·у·пак·о́ванн·ая**   (phys) close-packed lattice

~, **пра́в·ильн·ая**   (phys) perfect/regular lattice

~, **про·стра́н·ственн·ая**   (phys) space lattice

~ **про́фил·ей**   (aerodynam) cascade, airfoil cascade

~ **про́фил·ей, кольц·ев·а́я**   (aerodynam) annular cascade

~, **пря́м·а́я**   (phys) direct lattice

~, **радиолокацио́н·н·ая**   radar antenna array

~ **рыбо·хо́д·а**   fish screen (on weirs etc.)

~, **с·прямл·я́ющ·ая**   (dynam) diffuser grid, gate

     тео́ри·я решёт·ок·   (f) (dynam) cascade theory

~, **те́сн·ая**   (phys) close-spaced lattice, tight lattice

~, **тес·о́в·ая**   crate

~, **тру́б·н·ая**   tube plate (of boilers etc.), tube support

~, **центр·и́рованн·ая**   body-centred lattice

решёт·к·а

~, **цеп·н·а́я**   chain-grate stoker (of furnace)

решёт·ник·, -а,   (m) (build) lathing, laths

решёт·н·ый, -ая, -ое,   (a) of **ре-шёт·к·а**

решет·о́, -а́,   (nom pl) **решёт·а,** (g pl) **решёт·,** (m) sieve, screen

решето·ви́д·н·ый, -ая, -ое,   (a) sieve-like, ethmoid, cribriform

решётчато·ви́д·н·ый, -ая, -ое,   (a) sieve-like, ethmoid, cribriform

решётчато·нос·ов·о́й, -а́я, -о́е,   (a) (anat) ethmonasal

решёт·чат·ый, -ая, -ое,   (a) sieve-like, grated, grate, grid; lattice, latticed; ethmoid, clathrate

реш·и́тельн·ый, -ая, -ое,   (a) decisive; resolute, determined

реш·и́ть, -а́т,   (perf) see **реш·а́·ть**

решта́к·, -а,   (m) (min) plank chute

ре́·юшк·а, -и,   (g pl) **-шек·,** (f) (naut) dory

ре́·ющ·ий, -ая, -ее,   (pres part act) of **ре́·ять**

ре́·ять, -ют,   (imp) glide, flow

рж·а, -и,   (i) **-о́й,** (f) rust

ржа́в·е·ть, -ют,   (imp) rust

ржа́вл·ени·е, -я,   (n) rust, rusting

ржа́в·чин·а, -ы,   (f) (met, bot) rust

~, **пузы́р·чат·ая**   resin vein (defect in timber)

ржа́в·чинн·ые гриб·ы́   (pl) (bot) rust fungi, *Uredinales*

ржа́в·ый, -ая, -ое,   (a) rusty

рж·ан·о́й, -а́я, -о́е,   (a) rye

рж·а́ть, -ут,   (imp) neigh

рз. (abbr) = **ре́д·к·ие земл·и́** rare earths

риако·ли́т·, -а,   (m) (min) rhyacolite

ри́ас·, -а,   (m) (geol) ria

рибеки́т·, -а,   (m) (min) riebeckite

риб·о́з·а, -ы,   (f) (chem) ribose

рибо́йлер·, -а,   (m) (oil chem) reboiler

рибон·ов·ый, -ая, -ое,   (a) (biochem) ribonic

~ **кисл·от·а́**   (f) ribonic acid

рибо·нукле·а́з·а, -ы,   (f) ribonuclease

рибо·нуклейн·ов·ая кисл·от·а́   (f) ribonucleonic acid

рибо·флави́н·, -а,   (m) riboflavin, vitamin B_2

ривано́л·, -а,   (m) rivanol, acricide

риверсайди́т·, -а,   (m) (min) riversideite

ри́г·а, -и,   (f) (agr) threshing/thrashing shed/barn

ри́гел·ь, -я, *(m)* stayblock, collar beam; don catch; (geol) rock bar, riegel

Ри́ги-Ледук·а, эффе́кт· *(m)* (phys) Righi–Leduc effect

риг·оле́н·, -а, *(m)* (chem) rhigolene

ри́дберг·, -а, *(m)* rydberg (spectroscopic unit)

Ри́дберг·а, по·сто·я́нн·ая *(f)* (phys) Rydberg constant

ри́дерс·, -а, *(m)* (shipb) rider

риз·и́н·, -а, *(m)* (bot) rhizine

риз·о́ид·, -а, *(m)* (bot) rhizoid

ризо·карпи́н·ов·ый, -ая, -ое, *(a)* (bot) rhizocarpic, rhizocarpous; (*pl as noun*) rhizocarps, *Rhizocarpeae*

риз·о́м·, -а, *(m)* = **ризо́м·а**

риз·о́м·а, -ы, *(f)* (bot) rhizome

ризо·мо́рф·, -а, *(m)* (bot) rhizomorph

ризо·по́д·ы, -ов, *(m pl)* (zool) rhizopods, rhizopodans, *Rhizopoda, Rhizopodea*

ризо·сфе́р·а, -ы, *(f)* (bot) rhizosphere

рикка́ти, у·равн·е́ни·е *(n)* (math) Rikkati equation

риккёттс·и·и, -й, *(f pl)* (med) rickettsia

риккетс·ио́з·, -а, *(m)* (med) ricketsiosis

рикове́ри *(n indecl)* (rubber) recovery

рикоше́т·, -а, *(m)* ricochet

ри́ллинг·, -а, *(m)* (met roll) reeler, reeling machine/mill

ри́ман·ов, -а, -о, *(possess adj)* (math) Riemann, Riemannian

~ геоме́тр·и·я *(f)* Riemannian geometry

~ по·ве́рх·ност·ь *(f)* Riemann surface

~ про·стра́н·ств·о *(n)* Riemannian space

ри́м·ск·ий, -ая, -ое, *(a)* Roman

~ цифр· *(m)* Roman number/numeral

рингел-аппара́т·, -а, *(m)* (text) striping mechanism

рин·и́т·, -а, *(m)* (med) rhinitis

рино·ло́г·и·я, -и, *(f)* (med) rhinology

рино·склер·о́м·а, -ы, *(f)* (med) rhinoscleroma

ринсо́н·, -а, *(m)* (shipb) deadwood, rising wood

ри́н·уться, -утся, *(perf)* rush, dash

ринхо·не́лл·ы *(pl)* (pal) *Rhynchonellaceae*

ринхо·спо́р·а, -ы, *(f)* (bot) beak rush/sedge, *Rhynchospora*

ринхо·цефа́л·ы, -ов, *(m pl)* (zool) rhynchocephalians, *Rhynchocephalia*

рио·ли́т·, -а, *(m)* (geol) rhyolite

рио·та́ксис·, -а, *(m)* (biol) rhyotaxis

ри́ппер·, -а, *(m)* (civ eng) ripper

рип·ша́йб·а, -ы, *(f)* pattern drum (circular knitting machine)

рирпрое́кци·я, -и, *(f)* (cinema) rear projection

рис· *(abbr)* = **рис·у́нок·** (print) illustration, figure, fig.

рис·, -а, *(m)* (bot) rice, *Oryza sativa*

~, дробл·ён·ый ground rice

~, не·об·ру́ш·енн·ый unhulled/unhusked rice, rough/paddy rice

~ -сыр·е́ц·, -р·ц·а́, *(m)* unhulled/unhusked rice, rough/paddy rice

~, шлиф·о́ванн·ый polished rice

риса́йкл·, -а, *(m)* (chem) recycle

рисбе́рм·а, -ы, *(f)* (civ eng) dam bucket

ри́ск·, -а, *(m)* risk

ри́ск·а, -и, *(f)* line, mark, graduation mark; rib (of tyre)

~, у·стано́в·очн·ая (mech eng) index mark

риск·н·у́ть, -у́т, *(perf)* take a risk/chance

риск·о́ванн·ый, -ая, -ое, *(past part pass)* of **риск·ова́ть;** risky

риск·ова́ть, -у́ют, *(imp)* risk, chance, take a risk/chance

риск·о́в·ый, -ая, -ое, *(a)* risky

рис·ова́льн·ый, -ая, -ое, *(a)* drawing

~ бума́г·а *(f)* drawing paper

~ при·способл·е́ни·е *(n)* projection attachment (of microscope)

рис·ова́ни·е, -я, *(n)* (*v n*) see **рис·ова́ть**

рис·ова́ть, -у́ют, *(imp)* draw, depict, paint

рис·о́вк·а, -и, *(g pl)* **-вок·,** *(f)* drawing; (orn) Java sparrow, *Munia oryzivora*

рисо·ви́д·к·а, -и, *(g pl)* **-д·ок·,** *(f)* (bot) ricegrass, *Oryzopsis*

рисо·во́д·ств·о, -а, *(n)* rice growing/cultivation

ри́с·ов·ый, -ая, -ое, *(a)* rice; (*pl as noun*) *Oryzeae*

~ за·во́д· *(m)* rice mill

~ по́чв·а *(f)* paddy soil

~ се́ч·к·а *(f)* rice semolina

рисо·ру́ш·к·а, -и, *(g pl)* **-ш·ек·,** *(f)* rice scourer, debranner

рисс·вю́рм·ск·ий, -ая, -ое, *(a)* (geol) Riss-Würm

рис·у́нок·, -нк·а, *(m)* drawing, illustration, figure; figure (in wood)

~ проте́ктор·а tread pattern (of tyre)

ритм·, **-а**, (*m*) rhythm

ритм·и́ческ·ий, **-ая**, **-ое**, (*a*) rhythmic

ритм·и́чност·ь, **-и**, (*f*) rhythmicity, rhythm

ритм·и́чн·ый, **-ая**, **-ое**, (*a*) rhythmic

ритурне́л·ь, **-и**, (*f*) (mus) ritornello

ритурбе́нт·, **-а**, (*m*) return bend (piping)

ри́тц·а, **при́нцип·** (*m*) (phys) Ritz combination principle

риф·, **-а**, (*m*) (geol, naut) reef; sole pattern (footwear)

∼, кора́лл·ов·ый coral reef

рифа́йнер·, **-а**, (*m*) (rubber) refiner

риф·ить, **-ат**, (*imp*) reef (a sail)

рифл·ён·ый, **-ая**, **-ое**, (*a*) serrated, knurled, fluted, grooved; chequered (of metal plate); reefed-down (of sail)

рифл·ить, **-ят**, (*perf*) riffle, flute, groove, make flutes/grooves/serrations

рифлуа́р·, **-а**, (*m*) riffler, riffler file

ри́фл·я, **-и**, (*f*) riffle, flute, groove; (*pl*) serrations, fluting

риф·ов·ый, **-ая**, **-ое**, (*a*) of **риф·**

риформинг·, **-а**, (*m*) (oil) reforming

риформинг-бензи́н·, **-а**, (*m*) reformed gasoline

рифта́л·и (*pl*) double whip tackle

рифт·ова́ть, **- уют**, (*imp*) riffle, flute, groove, make flutes/grooves/serrations

рихт·ова́льн·ый, **-ая**, **-ое**, (*a*) straightening

рихт·о́ванн·ый, **-ая**, **-ое**, (*past part pass*) of **рихт·ова́ть**; straightened

рихт·ова́ть, **-у́ют**, (*imp*) straighten

рихт·о́вк·а, **-и**, (*f*) (met) straightening

рици́н·, **-а**, (*m*) (chem) ricin

рици́ник·, **-а**, (*m*) (pharm) ricinus

рицини́н·, **-а**, (*m*) ricinine

рицино́л·ев·ый **-ая**, **-ое**, (*a*) ricinoleic

∼ кисл·от·а́ (*f*) ricinoleic acid

рицинолейн·ов·ый, **-ая**, **-ое**, (*a*) = **рицино́л·ев·ый**

рицину́с·, **-а**, (*m*) (bot) castor, *Ricinus*

Ри́чардсон·а, **фо́рмул·а** (*f*) (phys) Richardson's equation

ри́шт·а, **-ы**, (*f*) (zool) Guinea worm, *Dracunculus medinensis*

ро (*indecl*) rho, ϱ (Greek)

ро́б·к·ий, **-ая**, **-ое**, (*a*) timid

ров·, (*g sing*) **рв·а**, (*nom pl*) **рв·ы**, (*g pl*) **рв·ов**, (*m*) ditch, trench

рови́т·, **-а**, (*m*) (min) rauvite

ровн·и́тел·ь, **-я**, (*m*) (paper) ass, drainage horn

ро́вн·иц·а, **-ы**, (*i*) **-ей**, (*f*) (text) roving, rove

ро́вн·ичн·ый, **-ая**, **-ое**, (*a*) of **ро́вн·иц·а**

∼ каре́т·к·а (*f*) (text) flyer

∼ маши́н·а (*f*) *see also* **банкабро́ш·**; (text) carder (wool); flyer frame (cotton)

ро́вн·о (*adv*) smoothly, steadily, evenly; just, exactly, (math) precisely

ро́вн·ый, **-ая**, **-ое**, (*a*) even, level, flat, smooth, equal

ровн·я́·ть, **-ют**, (*imp*) align, line up, (mil) dress; level, make even

рог·, **-а**, (*nom pl*) **рог·а́**, (*g pl*) **ро-г·о́в**, (*m*) horn; (anat) horn, cornu; (geog) point, cape; bick, beak, horn (of anvil)

∼, оле́н·ий antler

рога́лит·, **-а**, (*m*) (food) croissant, crescent roll

рог·а́т·ый, **-ая**, **-ое**, (*a*) horned

∼ скот·, кру́п·н·ый (*m*) cattle

рог·а́ч·, **-а**, (*i*) **-о́м**, (*m*) (ent) stag-beetle, *Lucanus cervus*; (zool) horned adder, *Cerastes cornutus*; (bot) cerato-carpus

рого·ви́д·н·ый, **-ая**, **-ое**, (*a*) horn-like/shaped, horned, corniform, cera-toid, keratoid

рог·ови́к·, **-а́**, (*m*) (geol) hornfels, hornstone; (bot) rescue-grass

рог·ови́н·а, **-ы**, (*f*) wart; healed scar (in leather)

рог·ови́ц·а, **-ы**, (*i*) **-ей**, (*f*) (anat) cornea

рог·ови́чн·ый, **-ая**, **-ое**, (*a*) (anat) corneal

∼ част·ь (*f*) cornea

рог·ов·о́й, **-а́я**, **-о́е**, (*a*) horn, horny, keratic

∼ об·ма́н·к·а (*f*) (min) hornblende, amphibole

∼ оболо́ч·ка (*f*) (zool) cornea

∼ пред·о·хран·и́тел·ь (*m*) (elec) horn-break fuse

∼ серебр·о́ (*n*) (min) horn silver, cerargyrite

рого·гла́в·ник·, **-а**, (*m*) (bot) *Cerato-cephalus*

рого́ж·а, **-и**, (*i*) **-ей**, (*f*) bast matting

рого́ж·к·а, **-и**, (*g pl*) **-ж·ек·**, (*f*) (text) hopsack

рого·обра́з·н·ый, **-ая**, **-ое**, (*a*) horn-shaped, horny, corniform

рого·образ·ова́ни·е, **-я**, (*n*) (zool) keratogenesis, ceratogenesis

рого·хвост·, -а,   (*m*) (ent) horntail, *Sirex*

рогу́ль·к·а, -и,   (*g pl*) **-л·ек·,**   (*f*) (text) flyer

род·, -а,   (*nom pl*) **-ы́,** (*g pl*) **-о́в,**   (*m*) family, stock, race; (biol) genus; generation; kind, sort, type, species; (gram) gender
     ро́д·ы (*nom pl*), (*g pl*) **ро́д·ов** labour, labor (in childbirth)

~ **а́льгебра·и́ческ·ой крив·о́й** order of an algebraic curve
     втор·о́й род·   second type; (*in gen as adj*) second-order/-type

род·ами́н·, -а,   (*m*) (chem) rhodamine

род·а́н·, -а,   (*m*) (chem) thiocyanogen

род·ана́т·, -а,   (*m*) rhodanate, thiocyanate

родани́д·, -а,   (*m*) = **родани́н·**

род·ани́н·, -а,   (*m*) rhodanine, thiocyanate, sulphocyanide

роданисто·водо·ро́д·н·ый, -ая, -ое,   (*a*) thiocyanic

~ **кисл·от·а́** (*f*) thiocyanic acid
     эфи́р· роданисто·водо·ро́д·н·ой кисл·от·ы́ thiocyanate

рода́н·ист·ый, -ая, -ое,   (*a*) thiocyanate (of), sulfocyanate (of)

~ **ка́ли·й**   (*m*) potassium thiocyanate

рода́н·ов·ый, -ая, -ое,   (*a*) of **род·а́н·**

~ **числ·о́**   (*n*) thiocyanogen number

родано·ме́тр·и·я, -и,   (*f*) thiocyanometry

род·а́т·, -а,   (*m*) rhodate

ро́д·иев·ый, -ая, -ое,   (*a*) (chem) rhodium

род·изи́т·, -а,   (*m*) (min) rhodizite

ро́д·и·й, -я,   (*m*) (chem) rhodium, Rh

~**, серно·ки́сл·ый** rhodium sulphate/ sulfate

род·и́льниц·а, -ы,   (*f*) (med) parturient

род·и́льн·ый, -ая, -ое,   (*a*) maternity, puerperal, parturient

ро́д·ин·а, -ы,   (*f*) native land, homeland

родинчок·, -чк·а,   (*m*) (zool) fontanelle

род·и́тел·ь, -я,   (*m*) parent

род·и́тельн·ый, -ая, -ое,   (*a*) (gram) genitive

род·и́тельск·ий, -ая, -ое,   (*a*) parental, paternal

род·и́ть, -я́т,   (*perf*) give birth to, bear; **-ся** be born

род·ни́к·, -а́,   (*m*) spring, source (of water)

род·н·о́й, -а́я, -о́е,   (*a*) native, natal; own

род·ов·о́й, -а́я, -о́е,   (*a*) birth; family; (biol) generic; (gram) gender

родо·вс·по·мог·а́тельн·ый, -ая, -ое,   (*a*) (med) obstetric, maternity

родо·вс·по·мож·е́ни·е, -я,   (*n*) (med) maternity service

родо·де́ндрон·, -а,   (*m*) (bot) rhododendron

род·оксанти́н·, -а,   (*m*) rhodoxanthin (pigment)

родо·ли́т·, -а,   (*m*) (min) rhodolite

родо·на·ча́ль·ник·, -а,   (*m*) founder, father, originator; forefather

род·они́т·, -а,   (*m*) (min) rhodonite

род·опси́н·, -а,   (*m*) (biochem) rhodopsin

родо·сло́в·и·е, -я,   (*n*) genealogy, pedigree

родо·сло́в·н·ый, -ая, -ое,   (*a*) genealogical; (*f as noun*) pedigree, genealogy

родо·хрози́т·, -а,   (*m*) (min) rhodochrosite

родри́г·а, фо́рмул·а   (*f*)   (math) Rodrigues formula

ро́д·ственн·ый, -ая, -ое,   (*a*) related

ро́д·ств·о, -а,   (*n*) relationship; affinity

ро́д·ы, -ов,   (*m pl*) labour, labor (in childbirth)

ро·ёвн·я, -и,   (*g pl*) **-вен·,**   (*f*) swarm box (for bees)

ро·ев·о́й, -а́я, -о́е,   (*a*) (ent) swarm, swarming

ро·е́ни·е, -я,   (*n*) (ent) swarming

ро·ет,   (*pres 3rd sing*) of **ры·ть** (*imp*)

ро́ж·а, -и,   (*i*) **-ей,** (*f*) (med) erysipelas

~ **свин·е́й**   (vet) rhusiopathia suum

рожд·а́емост·ь, -и,   (*f*) birth rate

рожд·а́·ть, -ют,   (*imp*) give birth to, bear; **-ся** be born

рожд·а́ющ·ийся, -аяся, -ееся, (*pres part act*) of **рожд·а́·ться;** nascent

рожд·е́ни·е, -я,   (*n*) (*v n*) *see* **рожд·а́·ть;** birth, genesis, creation, production, formation

~**, кра́т·н·ое** (nucl) multiple production

~**, мно́ж·ественн·ое** (nucl) multiple production

     опера́тор· рожд·е́ни·я (*m*) (nucl) creation operator

~ **па́р·** (nucl) pair production

~**, с·вя́з·анн·ое** (nucl) associated production

~**, со·в·ме́ст·н·ое**   (nucl) associated production

~ **ядр·а́** (nucl) nucleogenesis

рожд·ённ·ый, -ая, -ое, (*past part pass*) *see* **рожд·а́·ть;** born

рож·ист·ый, -ая, -ое, (*a*) (med) erysipelatous

рож·о́к·, -ж·к·а́, (*m*) (*dim*) of **рог·;** hydrant; (*pl*) ergot (plant disease); (*pl*) shoe-horn

~, по·жа́р·н·ый fire hydrant

рож·ь, (*g sing*) **рж·и,** (*f*) (bot) rye, *Secale cereale*

ро́з·а, -ы, (*f*) rose; rose-tree/bush; (arch) rose window

~ ветр·о́в (meteor) wind rose

~, ко́мпас·н·ая compass rose

роз·анили́н·, -а, (*m*) (chem) rosaniline

ро́з·анн·ые, -ых, (*pl decl as adj*) (bot) rose family, *Rosaceae*

ро́зг·а, -и, (*g pl*) **-зог·,** (*f*) rod, switch

ро́з·да·л, (*past masc sing*) of **раз·да́·ть,** (*perf*); *see* **раз·да·ва́ть** (*imp*)

Ро́зебом·а, пра́в·ил·о (*n*) (phys chem) Roozeboom phase law

розе́т·к·а, -и, (*g pl*) **-т·ок·,** (*f*) rosette; (arch) rosace, rose, rosette; (elec) socket outlet; deflector (of hose); spider/ratchet wheel

~, крест·о́в·ая (elec) ceiling rose

~, от·ветв·и́тельн·ая (elec) connection box, ceiling rose

~, ште́псель·н·ая (elec) socket outlet

розе́т·очн·ый, -ая, -ое, (*a*) of **розе́т·к·а**

~ шнур· (*m*) instrument cord (telephones)

роз·жи́г·, -а, (*m*) (*v n*) *see* **раз·жиг·а́·ть**

ро́з·ли·в·, -а, (*m*) (*v n*) *see* **раз·лив·а́·ть**

розмари́н·, -а, (*m*) (bot) rosemary, *Rosmarinus officinalis*

ро́з·мах·, -а, (*m*) handle

ро́зн·иц·а, -ы, (*i*) **-ей,** (*f*) (com) retail; retail goods

    про·да·ва́ть в ро́зн·иц·у retail, sell retail

ро́зн·ичн·ый, -ая, -ое, (*a*) (com) retail

ро́з·ня·л, (*past masc sing*) of **раз·ня́·ть,** (*perf*); *see* **раз·ним·а́·ть** (*imp*)

ро́з·ов·ый, -ая, -ое, (*a*) rose; pink (as colour), rose-coloured; (*pl as noun*) *Rosaceae*

~ лиша́·й (*m*) (med) acne rosacea

розол·ов·ая кисл·от·а́ (*f*) (chem) rosolic acid

розо·цве́т·н·ые, -ых, (*pl decl as adj*) (bot) *Rosacea*

ро́з·ыгр·ыш·, -а, (*m*) draw, drawn game

~ по·лёт·а (air) grope, dry swim

ро́з·ыск·, -а, (*m*) search, inquiry, investigation

ро·и́ть, -я́т, (*imp*) gather (bees); **-ся** swarm (of insects)

ро́·й, -я, (*nom pl*) **ро·и́,** (*g pl*) **ро·ёв,** (*m*) (ent) swarm; (nucl) cluster

ро́й·ер·, -а, (*m*) (met cast) sand-mixing machine

рой·те, (*pl imper*) of **ры·ть**

рока́д·а, -ы, (*f*) (mil) lateral road

рока́д·н·ый пут·ь (*m*) (mil) lateral road

Ро́квелл·у, твёрд·ост·ь по (met) Rockwell hardness

рок·иро́вк·а, -и, (*g pl*) **-вок·,** (*f*) castling (at chess); (mil) lateral move

рококо́ (*n indecl*) (arch) rococo

рокот·а́ни·е, -я, (*n*) (*v n*) of **роко·т·а́ть;** rumble

рокот·а́ть, (*pres 3rd sing*) **рокоч·ет,** (*imp*) rumble, roar

рокси́т·, -а, (*m*) (plast) roxite

рокфо́р·, -а, (*m*) Roquefort-type cheese, Roquefort cheese (rare)

ро́л·, -а, (*m*) roll

рол·ев·о́й, -а́я, -о́е, (*a*) rolled (in a roll)

ро́л·ик·, -а, (*m*) roller, roll

~, ви·т·о́й wound/spiral roller

~, за·жим·н·о́й pressure roller

~, за·ка́т·ывающ·ий (met) seaming roll

~, на·ка́т·н·ый thread roller; (eng) knurling roller/wheel

~, на·ма́з·ывающ·ий (print) inking roller

~, на·правл·я́ющ·ий guide/guiding roller/pulley, idler (belt transmission etc.)

~, на·тяж·н·о́й idler, tension pulley

~ обра́т·н·ой по·да́·ч·и (met roll) stripper roll

~, от·вод·я́щ·ий discharging roller/roll

~, от·жим·н·о́й squeeze roller/roll

~, от·тяж·н·о́й floating pulley/sheave

~, по·да́·ющ·ий pinch roll

~, под·де́рж·ивающ·ий bearing/supporting roll; travelling roll (of a levelling machine)

~, прн·ём·н·ый feed roller, take-up roller

~, при·жим·н·о́й contact roller/roll

~, рабо́ч·ий bending roll (levelling and straightening)

~, с·ва́р·очн·ый (weld) contact/seam-welding roller

~, холост·о́й loose roller

рол·иков·ый, -ая, -ое, (*a*) of **рол·ик·**

~ машин·а (*f*) seam-welding machine

~ стан·ок· (*m*) (met) roll-forming machine, forming rolls

~ транспорт·ёр· (*m*) roller conveyer

ролико·гиб·очн·ая машин·а (*f*) roll-forming machine, forming rolls

ролико·о·пор·а, -ы, (*f*) supporting roller

ролико·под·шип·ник·, -а, (*m*) *see also* **под·шип·ник·;** roller bearing

ролл·, -а, (*m*) (paper) beater, beater roll

~, масс·н·ый (paper) pulp beater

роллер·, -а, (*m*) roller, tealeaf roller

рол·ь, -и, (*nom pl*) **рол·и,** (*g pl*) **рол·ей,** (*f*) role, part; roll, list; coil (of sheet metal)

рольганг·, -а, (*m*) roller conveyer; (met roll) roller table

роль·н·ый, -ая, -ое, (*a*) rolled, coiled (into a roll)

ром·, -а, (*m*) rum

роман·, -а, (*m*) novel, romance; (shipb) driving-wedge ram

роман·цемент·, -а, (*m*) roman/Parker's cement

ромаш·к·а, -и, (*g pl*) **-ш·ек·,** (*f*) (bot) camomile, *Matricaria*

ромаш·ник·, -а, (*m*) (bot) pyrethrum

ромб·, -а, (*m*) rhomb, rhombus; diamond (-shape)

ромб·ическ·ий, -ая, -ое, (*a*) rhombic, (cryst) orthorhombic

~ антенн·а (*f*) (rad) rhombic antenna

~ систем·а (*f*) (cryst) orthorhombic system

ромбо·вид·н·ый, -ая, -ое, (*a*) rhomboid, diamond-shaped

ромб·ов·ый, -ая, -ое, (*a*) rhombic

ромбо·додека·эдр·, -а, (*m*) (cryst) rhombic dodecahedron

ромб·оид·, -а, (*m*) rhomboid

ромб·оидаль·н·ый, -ая, -ое, (*a*) rhomboid, rhomboidal

ромб·оидн·ый, -ая, -ое, (*a*) rhomboid

ромбо·нос·н·ые, -ых, (*pl decl as adj*) (pal) rhombifers, *Rhombifera*

ромбо·образ·н·ый, -ая, -ое, (*a*) diamond-shaped

ромбо·эдр·, -а, (*m*) (cryst) rhombohedron

ромбо·эдр·ическ·ий, -ая, -ое, (*a*) rhombohedral

~ син·гон·и·я (*f*) (cryst) rhombohedral class

ро·мезон·, -а, (*m*) (nucl) rho meson

ромейт·, -а, (*m*) (min) romeite

ро·металл·, -а, (*m*) Rhometal (Fe–Ni–Cr–Si alloy)

ром·ов·ый, -ая, -ое, (*a*) rum

ронгалит·, -а, (*m*) (chem) rongalite

рон·я·ть, -ют, (*imp*) drop, let fall; shed

ропак·, -а, (*m*) small hummock, "ropak" (ice)

ропас·ить, -ят, (*perf*) *see* **ропач·ить**

ропач·ить, -ат, (*imp*) hummock, form hummocks; **-ся** be beset by ice

ропсокет·, -а, (*m*) (oil) rope socket

рос·, (*past masc sing*) of **раст·и**

рос·а, -ы, (*nom pl*) **рос·ы,** (*g pl*) **рос·,** (*f*) dew; (bot) downy mildew

  точ·к·а рос·ы (*f*) (phys etc.) dew point

рос·ени·е, -я, (*n*) retting (flax)

рос·ист·ый, -ая, -ое, (*a*) dewy

роскош·н·ый luxurious, luxuriant

роскоэлит·, -а, (*m*) (min) roscoelite

рос·ла, (*past fem sing*) of **раст·и**

рос·ли, (*past pl*) of **раст·и**

рос·ло; (*past neut sing*) of **раст·и**

росл·ост·ь, -и, (*f*) lankiness; tendency to gain growth

росл·ый, -ая, -ое, (*a*) tall; full-grown

~ стал·ь (*f*) wild steel

~ ладан· (*m*) (chem) benzoin gum

росо·граф·, -а, (*m*) (meteor) recording drosometer

росо·мер·, -а, (*m*) (meteor) drosometer

рос·плыв·ь, -и, (*f*) log drive (not in a raft)

рос·пуск·, -а, (*m*) (*v n*) *see* **рас·пуск·а·ть;** (rail) sorting

  руб·ёж· рос·пуск·а (*m*) (a/c) formation break-off line

рос·сып·н·ый, -ая, -ое, (*a*) of **рос·сып·ь**

рос·сып·ь, -и, (*f*) (*v n*) *see* **рас·сы·п·а·ть;** log drive; (geol) placer, alluvial deposit; weight loss in transit (of free-flowing materials); (ocean) bar, shallow

  в рос·сып·ью (*adv*) in bulk

рост·, -а, (*m*) growth, increase; height (of person etc.); (print) type height

~, в·став·очн·ый (bot) intercalary growth

~ чугун·а (met) cast iron growth, growth

ростбиф·, -а, (*m*) roast beef

рост-блок·, -а, (*m*) boat's crutches

ростверк·, -а, (*m*) grillage, grating

рост·ец·, -т·ц·а, (*m*) (bot) thallus

рост·ов·ой, -ая, -ое, (*a*) of **рост·**

~ мишен·ь (*f*) full-height figure target

рост·о́к·, -т·к·а́, (*m*) (bot) sprout, shoot
росто·ме́р·, -а, (*m*) (bot) auxanometer
ро́стр·, -а, (*m*) (zool) rostrum
ро́стр·а, -ы, (*f*) (build) rostrum; (*pl*) (shipb) booms, skids
ростр·а́льн·ый, -ая, -ое, (*a*) rostral, rostrate
ростр·у́м·, -а, (*m*) (zool) rostrum
ро́с·чин·, -а, (*m*) leaven
росшевр·о́, -а́, (*n*) glazed horse leather
ро́с·ш·ий, -ая, -ее, (*past part act*) of **раст·и́**
ротено́н·, -а, (*m*) (agr chem) rotanone, derris
рот·, рт·а, (*nom pl*) **рт·ы,** (*g pl*) **рт·ов,** (*m*) (biol) mouth, stoma
~, перв·и́чн·ый (zool) blastopore
ро́т·а, -ы, (*f*) (mil) company
рота́·метр·, -а, (*m*) rotameter, direct-reading flow meter
ротати́вн·ый, -ая, -ое, (*a*) rotary
рота́тор·, -а, (*m*) rotator; (print) rotary press, rotary reproducer/duplicator
~ бума́г·а (*f*) duplicating paper
ротацио́н·ный, -ая, -ое, (*a*) rotary; (phys) rotational, rotation
~ компре́ссор· (*m*) rotary compressor
~ на·со́с· (*m*) rotary pump
~ полос·а́ (*f*) (phys) rotational band
рота́ци·я, -и, (*f*) rotation; rotary printing press
ро́т·н·ый, -ая, -ое, (*a*) of **ро́т·а**
рот·ов·о́й, -а́я, -о́е, (*a*) oral, mouth, (biol) stomate, stomatic, stomatose, stomatous
рото·гло́т·к·а, -и, (*g pl*) **-т·ок·,** (*f*) (anat) oropharynx
рото·лёт·, -а, (*m*) (air) annular-wing guided missile
рото́н·, -а, (*m*) (nucl) roton
рото́нд·а, -ы, (*f*) (arch) rotunda
рото·но́г·ие, рако·обра́з·н·ые (*pl*) (zool) *Stomatopoda*
ро́тор·, -а, (*m*) rotor; revolving table (of drilling rig); (math) curl (of a vector function)
~ Бу́шеро (elec) Boucherot rotor
~, коротко·за́·мк·нут·ый (elec) short-circuited rotor, squirrel-cage rotor
~, не·явно·по́люс·н·ый (elec) cylindrical rotor
~ -пит·а́тел·ь, -я, (*m*) feed rotor (of snowplough)
~ с глуб·о́к·им па́з·ом (elec) closed-slot rotor
~ с конта́кт·н·ыми ко́льц·ами (elec) slip-ring rotor, wound rotor

ро́тор
~, явно·по́люс·н·ый (elec) salient-pole rotor
ро́тор·н·о-ро́л·иков·ый на·со́с· (*m*) rotary roller pump
ро́тор·н·ый, -ая, -ое, (*a*) rotor, rotor-type, rotary
рототро́л·, -а, (*m*) (elec) rototrol amplifier
ро́ульс·, -а, (*m*) (naut) roller
ротте́йг·, -а, (*m*) first matte (nickel produ‹c›tion); coarse matte (copper produ‹c›tion)
ро́щ·а, -·‹и›, (*i*) **-ей,** (*f*) grove, coppice
Ро́уленд·а, круг· (*m*) (opt) Rowland circle
ро́щ·ин·а, -ы, (*f*) leaven (breadmaking)
ро́·ющ·ий, -ая, -ее, (*pres part act*) of **ры·т·ь;** digging; (zool) fossorial
ро́·я, (*pres gerund*) of **ры·ть;** digging
роя́л·ь, -я, (*m*) piano, grand piano
роя́ль·н·ый, -ая, -ое, (*a*) of **роя́л·ь**
~ про́·волок·а (*f*) (met) piano wire
р-р (*abbr*) of **рас·т·во́р·** (chem) solution
рт-ст, (*abbr*) of **рту́т·н·ый столб·** mercury column
мм. рт. ст. mm Hg (pressure measurement)
рту́т·ник·, -а, (*m*) mercury-arc rectifier
рту́т·н·о-пре·образ·ова́тельн·ая под·ста́нци·я (*f*) (elec) mercury-arc conversion station
рту́т·н·ый, -ая, -ое, (*a*) mercury, mercurial; (chem) mercuric, mercurous; (elec) mercury-arc, mercury-vapour/vapor
~ баро́·метр· (*m*) mercury barometer
~ вы·прям·и́тел·ь (*m*) (elec) mercury-arc rectifier
~ като́д· (*m*) mercury-pool cathode
~ ма́з·ь (*f*) (pharm) ointment of mercury
~ на·со́с· (*m*) mercury-vapour pump
~ столб· (*m*) mercury column
рту́т·ь, -и, (*f*) (chem) mercury, Hg
~, грем·у́ч·ая (chem, expl) fulminate of mercury
~, дву·хло́р·ист·ая mercuric chloride
~, ио́д·ист·ая mercurous iodide, iodide of mercury
~, ио́д·н·ая mercuric iodide
~, сер·ни́ст·ая mercuric sulphide
~, хло́р·ист·ая mercurous chloride
~, хло́р·н·ая mercuric chloride
рту́ть·диаки́л·, -а, (*m*) mercuride

ртуть-орган·и́ческ·ий, -ая, -ое,   (*a*) (chem) organomercuric

руба́н·, -а,   (*m*) edge of land-fast ice

руба́н·ок·, -н·к·а,   (*m*) plane (tool)

~, шипо·ре́з·н·ый   dovetailing plane

руб·а́х·а, -и,   (*f*) shirt

~, водо·ла́з·н·ая   diving dress

руб·а́шк·а, -и,   (*g pl*) **-шек·,**   (*f*) shirt; (mech eng) jacket; liner (of a reflector); (math) lateral area (of a cone, cylinder etc.)

~, вод·ян·а́я   water jacket/passage

~, воз·ду́ш·н·ая   air shroud (jet engines)

~, пар·ов·а́я   steam jacket

~ цили́ндр·а   cylinder jacket

рубелли́т·, -а,   (*m*) (min) rubellite

руб·е́ж·, -а́,   (*i*) **-ом,**   (*m*) boundary, border; (mil) line, position(s)

~ об·наруж·е́ни·я   detection perimeter (of air warning system)

~ раз·мык·а́ни·я   (a/c) break-off line

руб·е́жн·ый, -ая, -ое,   (*a*) of **ру-б·е́ж·**

рубербо́ид·, -а,   (*m*) *see* **рубербо́йд·**

рубербо́йд·, -а,   (*m*) ruberoid (a roofing material)

руб·е́ц·, -б·ц·а́,   (*i*) **-б·ц·о́м,**   (*m*) (med, bot) scar, cicatrix; (zool) rumen

руби́д·иев·ый, -ая, -ое,   (*a*) rubidium

руби́д·и·й, -я,   (*m*) (chem) rubidium, Rb

руб·и́льник·, -а,   (*m*) (elec) knife switch

~, дву·по́люс·н·ый   tandem knife-switch

руби́н·, -а,   (*m*) (min) ruby, true ruby; ruby glass

руби́н·ов·ый, -ая, -ое,   (*a*) ruby

~ шпине́л·ь   (*f*) (min) ruby spinel

руб·и́ть, ·ят,   (*imp*) chop, hack, slash, hew; fell, cut down; (naut) **furl** and stow (sails etc.)

ру́б·к·а, -и,   (*g pl*) **-б·ок·,**   (*f*) (*v n*) *see* **руб·и́ть;** (shipb) deckhouse, house, room

~, бо·ев·а́я   conning tower (warships)

~, котл·ови́нн·ая   gap felling (of timber)

~, рул·ев·а́я   (shipb) wheelhouse

~, у·равн·и́тельн·ая   selective cutting/felling (of timber)

~, ход·ов·а́я   (shipb) wheelhouse

~, шту́рман·ск·ая   (shipb) chart house, chart room

рубл·ёв·ый, -ая, -ое,   (*a*) rouble

ру́бл·енн·ый, -ая, -ое,   (*past part pass*) of **руб·и́ть**

ру́бл·ен·ый, -ая, -ое,   (*a*) chopped; (build) log

ру́бл·ь, -я́,   (*m*) (fin) rouble

рубра́кс·, -а,   (*m*) rubrax (a blown asphalt)

ру́брик·а, -и,   (*f*) (print) heading, column heading; column

рубрик·а́ци·я, -и,   (*f*) (print) arrangement/lay-out

руб·у́х·а, -и,   (*f*) wildfire (plant disease)

руб·цева́ни·е, -я,   (*n*) (biol) cicatrization

руб·ц·ева́ться, -у́ются,   (*imp*) cicatrize

руб·цева́т·ый, -ая, -ое,   (*a*) scarred

руб·цо́в·ый, -ая, -ое,   (*a*) (biol) cicatricial, (min) scar, gash

ру́б·чат·ый, -ая, -ое,   (*a*) (geol) ribbed, corrugated; cicatricose

ру́б·чик·, -а,   (*m*) (bot) scar; groove, joint (of book)

ру́б·щик·, -а,   (*m*) chipper (man)

ру́г·а, -и,   (*f*) (biol) ruga

руг·а́тельн·ый, -ая, -ое,   (*a*) abusive

руд·а́, -ы́,   (*nom pl*) **ру́д·ы,**   (*g pl*) **руд·,**   (*f*) (*see also under associated words*); ore

~, агрегат·и́вн·ая   aggregate ore

~, бе́д·н·ая   lean/low-grade ore

~, бе́л·ая свин·цо́в·ая   white lead ore, cerussite

~, боб·о́в·ая   pisolitic/pea ore

~, боло́т·н·ая   bog/marsh ore

~, бурунду́ч·н·ая   banded ore (zincblende in association with galenite)

~, в·кра́пл·енн·ая   impregnated ore

~, вс·кры́·т·ая   developed ore

~, вы́·ветр·ивш·аяся   efflorescent/dust ore

~, высоко·проце́нт·н·ая   high-grade ore

~, грубо·пятн·и́ст·ая   mottled ore

~, дерн·о́в·ая   meadow ore

~, желвак·о́в·ая   nodular ore

~, кус·ков·а́я   lump ore

~, мелко·в·кра́пл·енн·ая   chatty ore

~, не·сорт·иро́ванн·ая   run-of-the-mine ore

~, не·тро́·нут·ая   solid ore

о·богащ·е́ни·е руд·   (*n*) ore dressing/beneficiation

~, поли·металл·и́ческ·ая   complex ore

~, про·мы́шл·енн·ая   pay ore

~, с·кры́·т·ая   hidden ore

руд·а́

~, **сплош·н·а́я** massive ore

~, **ячé·ист·ая** alveolar ore

рудера́ль·н·ый, -ая, -ое,   (*a*) (bot) ruderal

ру́дерпис·, -а,   (*m*) (shipb) rudder stock, vertical shaft

ру́дерпост·, -а,   (*m*) (shipb) rudder post

рудимéнт·, -а,   (*m*) rudiment

рудимент·а́рн·ый, -ая, -ое,   (*a*) (biol) rudimentary, vestigial

руди́ст·ы   (*pl*) (geol) Rudistes

руд·ни́к·, -а́,   (*m*) (min) mine, pit

руд·нико́в·ый, -ая, -ое,   (*a*) (min) mine

руд·ни́чн·ый, -ая, -ое,   (*a*) mine, mining

~ **газ·**   (*m*) fire-damp

~ **лес·**   (*m*) mining timber

ру́д·н·ый, -ая, -ое,   (*a*) ore; mining

~ **дéл·о**   (*n*) mining, ore/metalliferous mining

~ **процéсс·**   (*m*) pig-and-ore process (of open-hearth steelmaking)

~ **тéл·о**   (*n*) ore body

~ **столб·**   (*m*) ore shoot/column

рудо·во́з·, -а,   (*m*) ore carrier (ship)

рудо·до·ста́в·очн·ый, -ая, -ое,   (*a*) (min) cutting-and-loading

рудо·дроб·и́лк·а, -и, (*g pl*) **-лок·,** (*f*) (min) ore crusher

рудо·ко́п·, -а,   (*m*) miner

рудо·нос·и́тел·ь, -я,   (*m*) (geol) ore bringer

рудо·но́с·ност·ь, -и,   (*f*) ore content; presence of ore

рудо·но́с·н·ый, -ая, -ое,   (*a*) ore-bearing, metalliferous

рудо·про·вод·я́щ·ий, -ая, -ее,   (*a*) mineralizing

рудо·про·мы́в·очн·ый, -ая, -ое,   (*a*) ore-washing

рудо·про·явл·éни·е, -я,   (*n*) show of ore

рудо·раз·бо́р·к·а, -и,   (*f*) (min) hand picking/sorting

рудо·сорт·иро́вк·а, -и,   (*f*) (min) screening

рудо·с·пу́ск·, -а,   (*m*) ore-chute

рудо·сто́й·к·а, -и, (*g pl*) **-о·ек·,** (*f*) (min) pit prop

рудо·со·держ·а́щ·ий, -ая, -ее,   (*a*) ore-bearing, metalliferous

рудс·у·правл·éни·е, -я,   (*n*) mine management

руж·éйн·ый, -ая, -ое, (*a*) rifle, gun

руж·ь·ё, -я, (*nom pl*) **ру́ж·ь·я,** (*n*) rifle, gun

руи́н·а, -ы, (*f*) ruin, ruins

рук·а́, -и́, (*acc*) **ру́к·у,** (*nom pl*) **ру́·к·и,** (*g pl*) **рук·,** (*f*) (anat) hand, manus; arm

    от рук·и́ (*adv*) by hand, manually

    под рук·о́й (*adv*) to hand

рук·а́в·, -а́, (*m*) sleeve; hose, flexible tube; branch (of a river), arm (of the sea); sock, bag

~, **авто·ка́мер·н·ый** inner tubing (for tires)

~, **в·са́с·ывающ·ий** suction hose

~, **вы·кид·н·о́й** delivery hose

~, **кругло·тк·а́нн·ый** fabric-reinforced hose

~, **на·гнет·а́тельн·ый** pressure hose

~, **на·по́р·н·ый** pressure hose

~, **по·жа́р·н·ый** fire hose

~, **про·кла́д·очн·ый** mandrel-wrapped hose

~, **спира́ль·н·ый** wire-reinforced hose

рук·ави́ц·а, -ы, (*i*) **-ей,** (*f*) mitten

рук·ави́чн·ый, -ая, -ое, (*a*) mitten

рук·а́вн·ый, -ая, -ое, (*a*) of **рук·а́в·**

рукаво·обра́з·н·ый, -ая, -ое, (*a*) (geol) channel (of deposits etc.)

руко·вод·и́тел·ь, -я, (*m*) head, leader, director, manager

~ **по·са́д·к·и** (air) runway controller

руко·вод·и́ть, -я́т, (*imp*) (+ *instr*) direct, run (an organization); manage, lead (people)

руко·во́д·ств·о, -а, (*n*) leadership, management, direction, supervision; instructions, directions; handbook, textbook, manual

~ **для пла́в·ани·я** (naut) pilot (book)

~, **операти́в·н·ое** operating instructions/manual, directions for using

руко·во́д·ств·оваться, -уются, (*imp*) be guided by

руко·вод·я́щ·ий, -ая, -ее, (*pres part act*) of **руко·вод·и́ть**

~ **ис·коп·а́ем·ое** (*n*) (geol) index/guide fossil

~ **персона́л·** (*m*) managerial/supervisory personnel

руко·дéль·н·ый, -ая, -ое, (*a*) hand-made

руко·держ·а́тел·ь, -я, (*m*) arm-holder

руко·кры́·л·ое, -ого, (*n decl as adj*) (zool) bat; (*pl*) *Cheiroptera, Chiroptera*

руко·па́ш·н·ый, бо́·й (*m*) hand-to-hand fighting

руко́·пе́р·ые, -ых, (*pl decl as adj*) (fish) *Lophiiformes*

руко́·пис·н·ый, -ая, -ое, (*a*) manuscript

ру́ко·пис·ь, -и, (*f*) manuscript

руко·плеск·а́ни·е, -я, (*n*) applause, clapping

руко·по·жа́т·и·е, -я, (*n*) handshake

руко·я́т·к·а, -и, (*g pl*) **-т·ок·,** (*f*) handle, haft, helve, hilt, grip, hand-hold; lever, hand lever/control; (anat) manubrium; (ent) stipes

∼, с·пуск·ов·а́я trigger, release; firing lever

∼ у·правл·е́ни·я control lever; (a/c) control column, joystick

руко·я́т·чик·, -а, (*m*) (min) banks-man

руко·я́т·ь, -и, (*f*) *see* **руко·я́т·к·а**

рул·ев·о́й, -а́я, -о́е, (*a*) of **рул·ь;** steering; (*m as noun*) helmsman

∼ автома́т· (*m*) (naut) automatic/ gyro pilot/steerer

∼ коло́н·к·а (*f*) (M/T) steering column

∼ крюк· (*m*) (shipb) rudder pintle

∼ маши́н·а (*f*) (a/c) control actuator; steering engine (of ship)

∼ маши́н·к·а (*f*) (a/c) servo unit

∼ маши́н·к·а выс·от·ы́ elevator servo unit

∼ маши́н·к·а кре́н·а aileron servo unit

∼ маши́н·к·а ку́рс·а rudder servo unit

∼ пере·да́·ч·а (*f*) (M/T) steering gear

∼ пер·о́ (*n*) (zool) rectrix

∼ при·во́д· (*m*) steering gear

∼ при·во́д·, винт·ов·о́й (shipb) screw steering gear

∼ при·во́д·, ру́мпель·н·ый rod-and-chain steering gear

∼ при·во́д·, се́ктор·н·ый quadrant steering gear

∼ при·во́д· с пруж·и́нн·ыми амор-тиза́тор·ами, се́ктор·н·ый spring tiller, elastic quadrant steering gear

∼ у·каз·а́тел·ь (*m*) (shipb) helm indicator

∼ у·правл·е́ни·е (*n*) (M/T) steering

∼ у·стро́й·ств·о (*n*) steering gear/ system

рул·ёжк·а, -и, (*f*) (*v n*) *see* **рул·и́ть**

рул·ёжн·ый, -ая, -ое, (*a*) of **ру-л·ёжк·а**

∼ доро́ж·к·а (*f*) (air) taxi-way

рул·е́ни·е, -я, (*n*) (*v n*) of **рул·и́ть**

рулёт·, -а, (*m*) (food) roulade, beef olive

рулёт·к·а, -и, (*g pl*) **-т·ок·,** (*f*) tape-measure; roulette

рулётт·а, -ы, (*f*) (math) trochoid

рул·и́ть, -я́т, (*imp*) steer; (air) taxi; scull (with one oar)

руло́н·, -а, (*m*) roll (of paper etc.); coil (of strip metal); slug (of rubber)

руло́н·н·ый, -ая, -ое, (*a*) of **руло́н·**

∼ материа́л (*m*) material sold in rolls (e.g. paper, linoleum, roofing etc.)

∼ старт·сто́п·н·ый аппара́т· (*m*) (telecom) page-print teletype

рул·ь, -я́, (*m*) rudder; steering wheel (of automobile); handlebar (of bi-cycle); fin (of ship stabilizer); (a/c) control, control vane, rudder

∼, акти́в·н·ый· (shipb) active rudder

∼, аэро·дина́м·и́ческ·ий (rocket) external rudder, air vane

∼, баланси́р·н·ый (shipb) balanced rudder

∼ -ба́н·к·а, -и, (*f*) (naut) cable roller

∼, бок·ов·о́й (shipb) fin stabilizer

∼, борт·ов·о́й (shipb) fin stabilizer

∼, вертика́ль·н·ый rudder (of sub-marine)

∼, воз·ду́ш·н·ый (rocket) air vane, external rudder

∼ выс·от·ы́ (a/c) elevator

∼, га́з·ов·ый (rocket) jet vane, gas control-vane; deflected thrust con-trol (in V.T.O.L. aircraft)

∼ глуб·ин·ы́ (a/c) elevator

∼, горизонта́ль·н·ый hydroplane (of submarine)

∼, двух·сло́й·н·ый (shipb) double-plate rudder

∼, до·ба́в·очн·ый (shipb) supplemen-tary rudder

∼ Ки́чен·а (shipb) Kitchen rudder

∼ кре́н·а (a/c) aileron

∼ на·правл·е́ни·я (a/c) rudder

∼, обык·нове́нн·ый (shipb) unbal-anced rudder

∼, одно·сло́й·н·ый (shipb) single-plate rudder

∼ по·воро́т·а (a/c) underhung rudder

∼, под·вес·н·о́й (a/c) underhung rud-der

∼, прям·о́й про·до́ль·н·ый (shipb) single-arm tiller

∼, пусто·те́л·ый (shipb) double-plate rudder, side-plate rudder

рул·ь
~, **сплош·н·о́й**   (shipb)   single-plate rudder
~, **стру́й·н·ый**   (a/c) ducted-fan control
~, **-та́ли**   (pl)   (naut) relieving tackle
ру́мб·, **-а**, (m) point (of the compass)
~, **гла́в·н·ый**   cardinal point
~, **обра́т·н·ый**   opposite point
~, **трёх·бу́кв·енн·ый**   three-letter point
~, **четверт·н·о́й**   semi-cardinal point
ру́мб·ов·ый, **-ая**, **-ое**,   (a) of **румб·**
Ру́мкорф·а, **кат·у́шк·а**   (f)   (elec) Ruhmkorff coil
ру́мпел·ь, **-я**, (m) (shipb) helm, tiller
~, **по·пере́ч·н·ый**   yoke, rudder crosshead
~, **про·до́ль·н·ый**   single-arm tiller
~, **се́ктор·н·ый**   sectional tiller, quadrant tiller
~ **-штёрты**   (pl) yoke lines
ру́мпель·н·ый, **-ая**, **-ое**,   (a) (shipb) tiller
~ **от·дел·е́ни·е**   (n) tiller compartment, steering gear compartment, tiller flat
румы́н·ск·ий, **-ая**, **-ое**,   (a) Rumanian, Romanian, Roumanian
румя́н·а, **-ы**,   (f) rouge, paint
румя́н·ец·, **-н·ц·а**,   (i) **-н·ц·ем**, (m) blush, (bot) erubescence
румя́н·к·а, **-и**,   (g pl) **-н·ок·**, (f) (bot) viper's bugloss, *Echium*
рунди́ст·, **-а**, (m) uncut stone (gem)
рунду́к·, **-а́**,   (m) locker, bin; settee berth (in a ship)
рун·е́ц, **-н·ц·а́**, (i) **-н·ц·о́м**, (m) (zool) sheep ked, *Melophagus ovinus*
ру́н·н·ый, **-ая**, **-ое**,   (a) of **рун·о́**
рун·о́, **-а́**,   (nom pl) **ру́н·а**, (g pl) **рун·**, (n) fleece
~, **от·бо́р·н·ое**   blue fleece
~, **с·ва́л·янн·ое**   cotted fleece
рупач·и́ть, **-а́т**,   (imp) see **ропа·ч·и́ть**
ру́пор·, **-а**,   (m) speaking trumpet, megaphone; (rad) horn
~, **квадра́т·н·ый**   pyramidal horn
~, **кру́гл·ый**   conical horn
~, **прямо·уго́ль·н·ый**   rectangular horn
~, **се́ктор·н·ый**   sectoral horn
ру́пор·н·ый, **-ая**, **-ое**,   (a) of **ру́пор·**
~ **анте́нн·а**   (f) (rad) horn antenna/aerial
~ **из·луч·а́тел·ь**   (m) (rad) horn radiator
руптур·а, **-ы**,   (f) (med) rupture
руса́к·, **-а́**,   (m) (zool) winter/snow hare

руслен·и   (pl) (naut) chains
ру́сл·о, **-а**,   (n) river bed, channel
русл·ов·о́й, **-а́я**, **-о́е**,   (a) of **ру́сл·о**; fluvial
русл·о́в·ый, **-ая**, **-ое**,   (a) see **русл·ов·о́й**
ру́с·ск·ий, **-ая**, **-ое**,   (a) Russian; (m or f decl as adj) Russian
ру́с·ск·о-   (component) Russian-, Russo-
ру́стер·ы   (pl) (naut) hatch grating
ру́ст·ик·а, **-и**,   (f) (build) rustic, rustic work
русто́в·, **-а**,   (m) (naut) shank painter
руст·ова́ни·е, **-я**,   (n) (build) rustication
ру́т·а, **-ы**,   (f) (bot) rue, *Ruta*
руте́н·иев·ый, **-ая**, **-ое**,   (a) ruthenium
руте́н·и·й, **-я**,   (m) (chem) ruthenium, Ru, ruthenic
рути́л·, **-а**,   (m) (min) rutile
рути́н·, **-а**,   (m) (pharm) rutin
рути́н·а, **-ы**,   (f) routine
рути́н·н·ый, **-ая**, **-ое**,   (a) routine
ру́т·ов·ые, **-ых**,   (pl decl as adj) (bot) rue family, *Rutaceae*
рухл·я́к·, **-а́**,   (m) (geol) marl
рухл·яко́в·ый, **-ая**, **-ое**,   (a) marlaceous
ру́х·н·уть, **-ут**,   (perf) collapse
руч·а́тельств·о, **-а**,   (n) guaranty, warranty, guarantee
руч·а́·ться, **-ются**,   (imp) warrant, guarantee, certify
руче·ёк·, **-ей·к·а́**,   (m) (dim) of **ручé·й**
ручé·й, **-чь·я́**,   (nom pl) **-чь·и́**, (g pl) **-чь·ёв**, (m) stream, brook, rivulet; (met) groove (of a roll); impression (of a die); channel (in machinery)
~, **за·гото́в·и́тельн·ый**   (forge) roughing/dummying die/impression; (roll) roughing groove/pass
~, **за·кры́·т·ый**   (roll) box groove, dead hole
ос·ь ручь·я́   (f) (roll) pass line
~, **от·кры́·т·ый**   (roll) live hole
~, **пра́в·очн·ый**   (forge) trimming die
~, **про·тя́ж·н·о́й**   (forge) drawing die/impression
~ **раз·ря́д·а**   (nucl) streamer
~, **фо́рм·о́вочн·ый**   (forge) roughing impression
ручéй·ник·, **-а**,   (m) (ent) caddis fly, Trichoptera
руч·éй·н·ый, **-ая**, **-ое**,   (a) of **ручé·й**
ру́ч·к·а, **-и**,   (g pl) **-ч·ек·**, (f) (dim) of **рук·а́**; handle, haft, grip, knob

ру́ч·к·а
~, автома́т·и́ческ·ая　fountain pen
~ у·правл·е́ни·я　(a/c) control column/
　stick
руч·и́йк·, -а́,　(m) hand tool; hammer;
　sash **tool** (painting)
~, кру́гл·ый　sash tool
~, трафаре́т·н·ый　stencil brush
руч·н·о́й, -а́я, -о́е,　(a) hand, arm,
　hand-operated, manual; tame
~ перфор·а́тор·　(m) (min) hand-held
　hammer drill
~ при·во́д·　(m) hand-operating gear,
　manual control
~ то́п·к·а　(f) hand-fired furnace
ручь·ев·о́й, -а́я, -о́е,　(a) of **ручё·й**
ручь·и́,　(pl) of **ручё·й**
ру́ш·ени·е, -я,　(n) (v n) of **ру́ш·ить;**
　collapse
ру́ш·ить, -ат,　(imp) knock/tear/pull
　down; husk, hull, shell; **-ся** fall
　down, collapse
рцы　(indecl) letter "R"
ры́б·а, -ы,　(f) (see also associated adjec-
　tives); fish, (pl) (astron) Pisces, the
　Fishes
~, кост·и́ст·ая　bony fish, (pl) Teleo-
　stomi
~, лет·у́ч·ая　flying fish
~, потрош·ён·ая　(food) eviscerated
　fish
　　соб·ственн·о ры́б·ы　(fish) Pisces
рыб·а́к·, -а́,　(m) fisherman, fisher
рыб·а́цк·ий, -ая, -ое,　(a) fisherman's,
　fishing, piscatory
рыб·а́ч·ий, -ья, -ье,　(a) = **рыб·а́ц·
　к·ий**
рыб·а́ч·ить, -ая,　(imp) fish
рыб·ёшк·а, -и, (g pl) ⁻шек·,　(f) (fish)
　fry
ры́б·ий, -ья, -ье,　(a) fish, piscine
~ кле́·й　(m) fish glue, isinglass
ры́бин·а, -ы,　(f) (naut) ribband;
　(shipb) diagonal; (pl) cargo battens
ры́бинс·ы, -ов,　(m pl) (naut) cargo
　battens, spar ceiling
ры́б·ниц·а, -ы, (i) -ей,　(f) fishing
　depot ship
ры́б·н·ый, -ая, -ое,　(a) fish, piscine
~ консе́рв·ы　(pl) canned fish
~ мук·а́　(f) (agr) fish meal
~ полу·фабрика́т·ы　(pl) dried and
　frozen fish
~ про́·мысел·　(m) fishery, fishing field
~ про·мы́шл·енность·　(f) (food)
　fish industry
~ хозя́й·ств·о　(n) fishing industry

рыбо·ви́д·н·ый, -ая, -ое,　(a) fish-like;
　fish-bellied (beams, girders etc.)
рыбо·во́д·, -а,　(m) fish farmer, pisci-
　culturist
рыбо·во́д·н·ый, -ая, -ое,　(a) fish-
　farming, piscicultural
~ за·во́д·　(m) fish farm
рыбо·во́д·ств·о, -а,　(n) fish farming/
　breeding/culture, pisciculture
рыбо·во́д·ческ·ий, -ая, -ое,　(a) fish-
　farming, piscicultural
рыбо·за·во́д·, -а,　(m) fish factory
рыбо·иск·а́тел·ь, -я,　(m) fish finder
　(echosounding set)
рыбо·консе́рв·н·ый, -ая, -ое,　(a)
　fish-canning
рыбо·ло́в·, -а,　(m) fisherman; (orn)
　osprey, Pandion haliactus
рыбо·лов·е́цк·ий, -ая, -ое,　(a) fisher-
　man's, fishing, fishery
рыбо·ло́в·н·ый, -ая, -ое,　(a) fishing
~ биле́т·　(m) fishing licence
~ ору́д·и·е　(n) fishing gear
~ снаст·ь　(f) fishing tackle (sport)
рыбо·ло́в·ств·о, -а,　(n) fishing, fishery
рыбо·на·со́с·, -а,　(m) fish pump
рыбо·обра́з·н·ый, -ая, -ое,　(a) fish-
　shaped, pisciform, (zool) ichthyoid
рыбо·о·хра́н·а, -ы,　(f) fishery pro-
　tection
рыбо·пит·о́мник·, -а,　(m) fish farm
рыбо·плав·нико́в·ые, -ых,　(pl decl
　as adj) (zool) ichthyopterygians, Ichtyo-
　pterygia
рыбо·по·до́б·н·ый, -ая, -ое,　(a)
　fish-like, ichthyoid
рыбо·подъ·ём·ник·, -а,　(m) (civ eng)
　Borland-type fish pass
рыбо·про́·мысел·, -сл·а,　(m) fishery,
　fishing field
рыбо·про·мысл·о́в·ый, -ая, -ое,　(a)
　fishing, fishery
рыбо·про·мы́шл·енност·ь, -и,　(f)
　(food) fish industry
рыбо·про·мы́шл·енн·ый, -ая, -ое,
　(a) of **рыбо·про·мы́шл·енност·ь**
рыбо·хо́д·, -а,　(m) (civ eng) fish pass
рыбо·хо́д·н·ая ле́стниц·а　(f) (civ eng)
　fish ladder
рыбо·я́д·н·ый, -ая, -ое,　(a) (zool)
　fish-eating, piscivorous, ichthyo-
　phagous
рыв·о́к·, -в·к·а́,　(m) jerk, jolt
ры́·вш·ий, -ая, -ее,　(past part act) of
　ры·ть
рыж·ева́т·ый, -ая, -ое,　(a) gingerish,
　yellowish, rusty

рыж·ий, -ая, -ее, (*a*) ginger (colour); chestnut (of horses); red (of squirrels)

рыж·ик·, -а, (*m*) (bot) saffron milk cap, *Lactarius deliciosus*; falso flax, *Camelina sativa*

ры·л, (*past masc sing*) of **ры·ть**

ры́·л·о, -а, (*n*) snout; **ры́·ло** (*past neut sing*) of **ры·ть**

ры́ль·ц·е, -а, (*i*) **-ем,** (*n*) (*dim*) of **ры́·л·о**; (bot) stigma, snout

рым·, -а, (*m*) ring bolt, lifting eye

~, шварто́в·н·ый (naut) mooring ring

ры́нд·а, -ы, (*f*) ship's bell

ры́нда-бу́лин·ь, -я, (*m*) bell rope

ры́н·ок·, -н·к·а, (*m*) market

ры́н·очн·ый, -ая, -ое, (*a*) market

рыс·а́к·, -а́, (*m*) trotter (horse)

ры́ск·ани·е, -я, (*n*) (*v n*) of **ры́ск·а·ть**; (instr) hunting

    моме́нт· ры́ск·ани·я (*m*) (air) yawing moment

    у́гол· ры́ск·ани·я (*m*) (air) angle of yaw

ры́ск·а·ть, -ют, *or* **ры́щ·ут,** (*imp*) (naut) yaw, (air) yaw, swing; rove, roam

ры́ск·ли́вост·ь, -и, (*f*) tendency to yaw/swing

ры́ск·н·уть, -ут, (*perf single action*) *see* **ры́ск·а·ть**

рыск·ов·о́й я́кор·ь (*m*) kedge anchor

ры́с·ь, -и, (*f*) (zool) lynx, *Felis lynx*; trot (of horses etc.)

ры́т·вин·а, -ы, (*f*) rut, groove

ры·ть, (*pres 3rd pl*) **ро́·ют,** (*imp*) dig

ры́·т·ый, -ая, -ое, (*a*) (*past part pass*) of **ры·ть**

рыт·ь·ё, -я, (*n*) digging

рыхл·е́ни·е·, -я, (*n*) (*v n*) of **рыхл·и́ть**; (agr) cultivation

    глуб·ин·а́ рыхл·е́ни·я (*f*) (met) loose depth

рыхл·е́·ть, -ют, (*imp, intrans*) become loose/friable/crumbly/porous

рыхл·и́тел·ь, -я, (*m*) (civ eng) scarifier; (agr) cultivator

~, под·по́чв·енн·ый (agr) subsoiler

рыхл·и́ть, -я́т, (*imp*) loosen (solids), make crumbly/friable/porous, (agr) mellow, cultivate

рыхло·зерн·и́ст·ый, -ая, -ое, (*a*) loose-grained

рыхло·ком·кова́т·ый, -ая, -ое, (*a*) loose-lumpy/cloddy

рыхло·на·сы́п·н·ый, -ая, -ое, (*a*) loose (of free-flowing materials)

рыхл·ост·ь, -и, (*f*) friability, looseness, porousness, porosity

рыхл·ый, -ая, -ое, (*a*) friable, crumbly, loose, porous

рыхл·от·а́, -ы́, (*f*) friability, looseness, porousness, porosity

рыча́г·, -а́, (*m*) lever

~, за·да·ю́щ·ий master arm (of manipulator)

    за·ко́н· рычаг·а́ (*m*) (mech) lever principle/rule

~, ис·полн·и́тельн·ый slave arm (of manipulator)

~, кач·а́ющ·ийся rocking lever

~, кла́виш·н·ый, (telecom) key level

~, кла́пан·н·ый valve rocker (diesels)

~, коле́н·чат·ый crank

~ пе́рв·ого ро́д·а (mech) lever of the first kind

~ по·воро́т·н·ой ца́пф·ы (М/Т) steering arm

~, про·меж·у́точн·ый connecting rod, intermediate lever

~, тормоз·н·о́й brake lever

    пра́в·ил·о рычаг·а́ (*n*) (phys, chem) lever rule/principle

рыча́ж·н·ый, -ая, -ое, (*a*) lever, lever-operated

~ при·во́д· (*m*) lever operating gear

рыча́ж·ок·, -ж·к·а́, (*m*) (*dim*) of **рыча́г·**

рыч·а́ть, -а́т, (*imp*) growl, snarl

ры́ш·ет, (*pres 3rd sing*) of **ры́ск·а·ть**

рья́н·ый, -ая, -ое, (*a*) zealous

рэле́·евск·ий, -ая, -ое, (*a*) (phys) Rayleigh

Рэле·я, диск· (*m*) (phys) Rayleigh disk

~, за·ко́н· (*m*) (phys) Rayleigh law

~, теоре́м·а (*f*) (math) Rayleigh's theorem

Рэле·я-Джи́нс·а· за·ко́н· (*m*) Rayleigh–Jeans law (of radiation)

рюкза́к·, -а, (*m*) rucksack

рю́м·к·а, -и, (*g pl*) **-м·ок·,** (*f*) wine glass

рю́м·очк·а, -и, (*g pl*) **-чек·,** (*f*) (*dim*) of **рю́м·к·а**

рю́ттель-маши́н·а, -ы, (*f*) (met cast) jar-/jolt-rammer

ряб·и́н·а, -ы, (*f*) (bot) rowan, *Sorbus*

ряб·и́нник·, -а, (*m*) (bot) spirea, *Sorbaria*

ряб·и́ть, -я́т, (*imp*) ripple

ряб·о́й, -а́я, -о́е, (*a*) speckled; pitted, pock-marked

ря́б·чик·, -а, (*m*) (bot) fritillary, *Fritillaria*

ря́б·ь, -и, (*f*) ripples, rippling, undulation

ря́д·, -а, (*nom pl*) **-ы́,** (*g pl*) **-о́в,** (*m*) row, rank, line; series; course (brickwork); bank (of machines, instruments)

~, арифмет·и́ческ·ий arithmetic series

~, аромат·и́ческ·ий (chem) aromatic series

~, бес·кон·е́чн·ый infinite series
в ряд·у́ among

~, гармон·и́ческ·ий (math) harmonic progression

~, геометр·и́ческ·ий geometrical progression

~, дискре́т·н·ый (math) discrete range

~, знако·пере·ме́н·н·ый (math) alternating series

~ и́мпульс·ов (phys) pulse train

~, кон·е́чн·ый finite series

~ на·пряж·е́ни·я (chem) displacement/electromotive series

~ пе́рв·ого по·ря́д·к·а, арифмет·и́ческ·ий arithmetic progression

~, пе́тель·н·ый (text) wale

~, пре·вос·ход·я́щ·ий (math) majorant, majorant series

~ пре·вращ·е́ни·й (phys) transmutation series

~, рас·ход·я́щ·ийся (math) divergent series

ря́д

~ решёт·к·и (cryst) lattice row

~ со·сто·я́ни·й (phys) sequence of states

~, степе́н·н·ый (math) power series

~, с·ход·я́щ·ийся (math) convergent series

~ с·чёт·чик·ов (nucl) counter tray

~, це́л·ый a number (of), a whole lot (of), several, a wide variety of

ря́д·а́ми (*adv*) in rows/banks

ря́д·ко́в·ая жа́т·к·а (*f*) (agr) side delivery reaper

ря́д·н·ый, -ая, -ое, (*a*) in-line, lined-up; (*as component*) -row, -rank, -line, -series

~ мото́р· (*m*) (a/c) in-line engine

ря́д·ов·о́й, -а́я, -о́е, (*a*) ordinary, common; (*m as noun*) private (mil), rating (naval), airman

~ со·ста́·в· (*m*) (mil) other ranks, ratings, rank-and-file

~ у́гол·ь (*m*) run-of-the-mine coal

рядо·зу́б·ые, -ых, (*pl decl as adj*) (zool) taxodonts, *Taxodonta*

ря́д·о́к·, -д·к·а́, (*m*) (*dim*) of **ря́д·**

ря́д·ом (*adv*) beside, alongside, next to

ряж·, -а, (*i*) **-ем,** (*m*) (civ eng) crib, underwater crib, cribwork

ря́ск·а, -и, (*f*) (bot) duckweed, *Lemna*

С

с (*prep* + *instr*) with; (*prep* + *gen*) from, off, down from, since; (*prep* + *acc*) the size of, about

~ одн·о́й сторон·ы́ on the one hand

с- (*as verbal prefix*) *signifies*: (1) movement (a) from, away from, (b) to one side, (c) down, down from, from off, (d) there and back, (e) concentrating in one place; (2) joining, fastening; (3) accompaniment, participation, concert; (4) comparison; (5) copying; (6) using up, expending; (7) **-ся** mutuality; (8) perfective aspect

сабади́лл·а, -ы, .(*f*) (pharm, bot) sabadilla, *Sabadilla officinalis*

саба́л·ь, -я, (*f*) (bot) palmetto, *Sabal*

саба́н·ы, -ов, (*m pl*) scaffolding, staging

Сабатье́, ката́·лиз· (*m*) (chem) Sabatier reaction

са́бель·ник·, -а, (*m*) (bot, pharm) potentilla, *Comarum*

сабине́н·, -а, (*m*) (chem) sabinene

сабино́л·, -а, (*m*) (chem) sabinol, sabinene

сабле·ви́д·н·ый, -ая, -ое, (*a*) sabre-shaped; (air) swept-back

сабле·зу́б·ый, -ая, -ое, (*a*) (pal) sabre-toothed, saber-toothed, machairdont, (*pl as noun*) machairodonts, *Machairodontidea*

са́бл·я, -и, (*g pl*) **са́бел·ь** (*f*) sabre, saber, cutlass

сабота́ж·, -а, (*i*) **-ем,** (*m*) sabotage

сабота́ж·ник·, -а, (*m*) saboteur

сабу́р·, -а, (*m*) (bot, pharm) aloe

сабу́р·ов·ый, -ая, -ое, (*a*) aloe

сава́нн·а, -ы, (*f*) (geobot) savanna

саво́·й, -я, (*m*) = **саво́й·ск·ая капу́ст·а**

саво́й·ск·ая капу́ст·а (*f*) (bot) Savoy cabbage, *Brassica oleracea sabauda*

саги́тт·а, -ы, (*f*) (zool) sagitta

сагитта́ль·н·ый, -ая, -ое, (*a*) sagittal

~ пло́ск·ост·ь (*f*) (opt, zool) sagittal plane

са́го (*n indecl*) (food) sago

са́г·овник·, -а, (*m*) (bot) cycad, (*pl*) *Cycadales*

са́г·овников·ые, -ых, (*pl decl as adj*) (bot) cycadophytes, *Cycadophytae*

сагово·обра́з·н·ые, -ых, (*pl decl as adj*) (pal, bot) *Cycadeoidales*

са́д·, -а, (*nom pl*) **-ы́,** (*g pl*) **-о́в,** (*m*) garden

са́д

~, де́т·ск·ий kindergarten

~, плод·о́в·ый orchard

са́д·ик·, -а, (*m*) (*dim*) of **сад·**

сад·и́льник·, -а, (*m*) (hortic) dibber, dibble

садист·и́ческ·ий, -ая, -ое, (*a*) (med) sadistic

сад·и́ть, ⸗ят, (*imp*) plant, place, put; (naut) haul down (on sheets); **-ся** set, go down (of sun etc.); sit down; take, board (public vehicles), get in (to automobiles); (air) land; settle, subside; shrink (of cloth)

~ на мел·ь (naut) strand, beach, put ashore;

сад·и́ться на мел·ь go aground

~ под аре́ст· place under arrest

са́д·к·а, -и, (*g pl*) **-д·ок·,** (*f*) (*v n*) of **сад·и́ть;** putting, placing, planting; charge (of furnace); capacity (of OH furnace); thickening, settling (of liquids, gases); shrinking (of cloth); snaring, trapping (animals); **сад·к·а́** (*g sing*) of **сад·о́к·**

сад·о́вник·, -а, (*m*) gardener; (ent) blastophag beetle

сад·о́внич·ий, -ья, -ье, (*a*) gardener's

садо·во́д·, -а, (*m*) horticulturist, gardener

садо·во́д·ств·о, -а, (*n*) horticulture, gardening

~, пейза́ж·н·ое landscape gardening

садо·во́д·ческ·ий, -ая, -ое, (*a*) horticultural, gardening

сад·о́в·ый, -ая, -ое, (*a*) garden

~ культив·а́тор· (*m*) orchard cultivator

сад·о́к·, -д·к·а́, (*m*) breeding place; fish tank; run (for small animals); bird trap/snare; pigeon shooting stand; **са́д·ок·,** (*g pl*) of **са́д·к·а**

са́д·очн·ый, -ая, -ое, (*a*) charging; trapping, snaring, shooting (sport)

~ окн·о́ (*n*) (met) charging door (of furnace)

са́·ек·, (*g pl*) of **са́й·к·а**

са́ж·а, -и, (*i*) **-ей,** (*f*) carbon black, soot

~, ацетиле́н·ов·ая acetylene black

~, бе́л·ая white carbon black, (rubber) silica filler

~, га́з·ов·ая gas black

~, кана́ль·н·ая channel black

~, ла́мп·ов·ая lamp black

са́ж·а
~, **печ·н·а́я** furnace black
~, **терма́том·н·ая** thermatomic carbon
~, **швё́ллер·н·ая** channel black
саж·а́лк·а, -и, (*g pl*) **-лок·,** (*f*) fish breeding pond; (agr) planter
саж·а́льник·, -а, (*m*) (hortic) dibber, dibble
саж·а́льн·ый, -ая, -ое, (*a*) planting
саж·а́·ть, -ют, (*imp*) put, place, seat, set, plant; (naut) haul down (on sheets)
са́ж·ев·ый, -ая, -ое, (*a*) of **са́ж·а**
~ **с·мес·ь** (*f*) carbon black and rubber mix
саже·ма́сл·ян·ый, -ая, -ое, (*a*) oil-black (rubber)
саже·на·полн·е́ни·е, -я, (*n*) carbon black content
саже·на·по́лн·енн·ый, -ая, -ое, (*a*) (rubber) black, black-filled/reinforced
са́ж·енец·, -нц·а, (*i*) **-нц·ем,** (*a*) nursery tree, sapling; seedling
са́ж·енн·ый, -ая, -ое, (*past part pass*) of **сад·и́ть**
са́жен·ь, -и, (*g pl*) **сажен·** or **са-жён·ей** (*f*) sagene (213 cm); (naut) fathom (183 cm)
~, **мор·ск·а́я** fathom (183 cm)
саже·об·ду́в·очн·ое у·стро́й·ств·о (*n*) soot-blower (boilers)
са́ж·ист·ый, -ая, -ое, (*a*) sooty
са́ж·к·а, -и, (*f*) smut (plant disease)
са́ж·н·ый, -ая, -ое, (*a*) of **са́ж·а**
са́з·а, -ы, (*f*) (bot) sasa, *Saasa*
саза́н·, -а, (*m*) (fish) carp, *Cyprinus carpio*
сайг·а́, -й, (*f*) (zool) saiga, *Saiga tatarica*
сайга́к·, -а, (*m*) = **сайг·а́**
са́йд·а, -ы, (*f*) (fish) pollack, *Pollachius vireus*
са́й·к·а, -и, (*g pl*) **са́·ек·,** (*f*) bread roll
сайлентбло́к·, -а, (*m*) silent-block (bushings)
сайоди́н·, -а, (*m*) (pharm) sajodin, calcium iodobehenate
са́к·, -а, (*m*) sack, bag
саквоя́ж·, -а, (*i*) **-ем,** (*m*) travelling-bag
са́ки (*indecl*) (zool) saki, *Pithecia*
сакр·а́льн·ый, -ая, -ое, (*a*) (anat) sacral
сакро·люмба́ль·н·ый, -ая, -ое, (*a*) (anat) sacrolumbar
Са́ксмит·а, вес·ы́ (*pl*) Sucksmith balance

саксони́т·, -а, (*m*) (geol) saxonite
сала́з·к·и, -з·ок·, (*f pl*) toboggan, sledge; skids; slide; (shipb) sliding ways; sliding carriage, carriage, slide rest, slide (of machine tools)
~, **паралла́кс·ов** parallax slide (air photography)
~, **по·воро́т·н·ые** slide rest (of machine tool)
~, **при·жи́м·н·ые** pressure guides
сала́к·а, -и, (*f*) (fish) sprat, *Clupea sprattus*; herring, *Clupea harengus*
салама́ндр·а, -ы, (*f*) (zool) salamander, (*pl*) *Salamandroidea*
сала́т·, -а, (*m*) (bot) lettuce, *Lactuca*
~, **ри́м·ск·ий** cos lettuce, *Lactuca sativa romana*
салив·а́ци·я, -и, (*f*) (physiol) salivation
салигени́н·, -а, (*m*) (chem) saligenin, salicyl alcohol
са́линг·, -а, (*m*) (naut) cross-tree
салипири́н·, -а, (*m*) (pharm) salipyrin
салирга́н·, -а, (*m*) (pharm) salyrgan
сали́т·, -а, (*m*) (min) sahlite
са́л·ить, -ят, (*imp*) grease
салици́л·ов·ый, -ая, -ое, (*a*) salicylic
~ **кисл·от·а́** (*f*) (chem) salicylic acid
~ **спирт·** (*m*) salicylic alcohol, saligenin
салицилово·ки́сл·ый, -ая, -ое, (*a*) salicylate (of)
~ **на́тр·и·й** (*m*) sodium salicylate
салици́н·, -а, (*m*) (chem) salicin
са́лм·а́, -ы, (*f*) (geog) channel
са́л·о, -а, (*n*) fat, grease, suet, lard, tallow; grease ice, sludge
~, **го́р·н·ое** (geol) mountain tallow
~, **иску́с·ственн·ое** hydrogenated fat
~, **лед·ян·о́е** (ocean) grease ice, sludge
сало́л·, -а, (*m*) (pharm) salol
салоли́н·, -а, (*m*) = **салома́с·**
сало·ма́с·, -а, (*m*) (food) hydrogenated fat/oil
сало́н·, -а, (*m*) lounge (in hotel); saloon (in ship etc.)
сало·то́п·к·а, -и, (*g pl*) **-п·ок·,** (*f*) fat-melting plant
сало·то́п·н·ый, -ая, -ое, (*a*) fat-melting
салфе́т·к·а, -и, (*g pl*) **-т·ок·,** (*f*) napkin
~, **ма́рл·ев·ая,** (med) gauze dressing
са́л·ь, -и, (*f*) (geol) sial
сальварса́н·, -а, (*m*) (pharm) salvarsan

са́льви·я, -и,   (*f*) (bot) sage, *Salvia*

са́льдо   (*n indecl*) (fin) balance

са́льд·ов·ый, -ая, -ое,   (*a*) (fin) balance

са́льз·а, -ы,   (*f*) (geog) mud volcano, salse

са́ль·ник·, -а,   (*m*) stuffing box, gland, packing gland; (anat) omentum; water-swivel (oil wells)

~, лабири́нт·ов·ый labyrinth packing gland

са́ль·ников·ый, -ая, -ое,   (*a*) (mech) gland, stuffing-box; (anat) omental

~ кран·   (*m*) screwed-grand cock

~ на·би́в·к·а   (*f*) stuffing-box packing

са́ль·ниц·а, -ы,   (*f*) grease box

са́ль·ност·ь, -и,   (*f*) greasiness

са́ль·н·ый, -ая, -ое,   (*a*) fatty, greasy, sebaceous; tallow; obscene

~ желез·а́   (*f*) (zool) sebaceous gland

сальпинг·и́т·, -а,   (*m*) (med) salpingitis

сальпинго·офори́т·, -а,   (*m*) (med) salpingo-oophoritis

сальпинго·скоп·и́·я, -и,   (*f*) (med) salpingoscopy

сальсолиди́н·, -а,   (*m*) (pharm) salsoline

сальсоли́н, -а,   (*m*) (pharm) salsoline

салю́т·, -а,   (*m*) salute

~, ли́ч·н·ый personal salute

салют·ова́ть, -у́ют,   (*imp*) salute

са́л·ящ·ий, -ая, -ее,   (*pres part act*) of **са́л·ить**

сам·   (*pronoun*) myself, himself, itself

~ по себ·е́   (*adv*) alone by oneself/itself/himself, *per se*

сам·а́   (*pronoun*)   (*f*) myself, herself, itself

сама́н·, -а,   (*m*) adobe, adobe clay

сама́н·н·ый, -ая, -ое,   (*a*) adobe

сама́р·иев·ый, -ая, -ое,   (*a*) (chem) samarium, samaric

сама́р·и·й, -я,   (*m*) (chem) samarium, Sm

самарски́т·, -а,   (*m*) (min) samarskite

сам·е́ц·, -м·ц·а́,   (*i*) **-м·ц·о́м,**   (*m*) male, he-

са́м·и (*pronoun*) (*pl*) yourselves, ourselves, themselves

самире́си́т·, -а,   (*m*) (min) samiresite

са́м·к·а, -и,   (*g pl*) **-м·ок·,**   (*f*) female, she-

сам·о́ (*pronoun*) itself

~ по себ·е́ by itself, *per se*

~ соб·о́й раз·ум·е́·ется   it goes without saying, obviously

само-   (*root*) self, auto-, spontaneous

само·баланс·и́рующ·ийся, -аяся, -ееся,   (*a*) (elec) self-balancing

само·бес·плод·н·ый, -ая, -ое,   (*a*) (biol) self-sterile/-incompatible

само·блок·иро́вк·а, -и,   (*g pl*) **-вок·,**   (*f*) (elec) automatic interlock/interlocking

само·блок·и́рующ·ий, -ая, -ее,   (*a*) (elec) locking-in, holding

само·блок·и́рующ·ийся, -аяся, -ееся,   (*a*) (elec) interlocking; self-locking

само·брож·е́ни·е, -я,   (*n*) (chem) autofermentation

само·бы́т·ност·ь, -и,   (*f*) originality

само·ва́р·, -а,   (*m*) samovar, hot-water urn (for making tea)

само·вес·ы́, -о́в,   (*m pl*) automatic scales

само·вз·мыв·а́ни·е, -я,   (*n*) (a/c) ballooning

само·вз·рыв·н·о́й фуга́с·   (*m*) (mil) chemical-contact mine

само·в·ключ·а́ющ·ий, -ая, -ее,   (*a*) (elec) reclosing; (mech) self-engaging

само·в·ключ·е́ни·е, -я,   (*n*) (elec) reclosure; (mech) self-engaging

само·внуш·е́ни·е, -я,   (*n*) autosuggestion

само·воз·бужд·а́ющ·ийся, -аяся, -ееся,   (*a*) (elec) self-excited/-exciting/-energizing; (rad) self-oscillating

само·воз·бужд·е́ни·е, -я,   (*n*) (elec) self-excitation; (rad) self-oscillation

с само·воз·бужд·е́ни·ем   self-excited

само·воз·вра́т·, -а,   (*m*) self-recovery, self-reset

само·воз·гор·а́ни·е, -я,   (*n*) spontaneous combustion/ignition

само·воз·гор·а́ющ·ийся, -аяся, -ееся   (*a*) (expl) self-propagating

само·воз·ник·а́ющ·ий, -ая, -ее,   (*a*) spontaneous

само·во́ль·н·ый, -ая, -ое,   (*a*) wilful, self-willed; unwarranted

само·вос·пламен·е́ни·е, -я,   (*n*) spontaneous combustion/ignition, auto-ignition

само·вос·пламен·я́ющ·ийся, -аяся, -ееся,   (*a*) hypergolic, pyrophoric, pyrophorous, self-igniting

само·вос·стана́вл·ивающ·ийся, -ая·ся, -ееся,   (*a*) (elec) self-restoring; self-healing

само·вос·становл·е́ни·е, -я,   (*n*) self-recovery

само·вращ·е́ни·е, -я, (*n*) (air) auto-rotation (of helicopter rotor); wind-milling (of propeller)

само·в·са́с·ывающ·ий, -ая, -ее, (*a*) auto-intake; self-priming (of pumps)

само·вы́·воз·, -а, (*m*) (com) collection by purchaser

само·вы·ключ·а́ющ·ийся, -аяся, -ееся, (*a*) automatically disconnecting; self-disengaging

~, стекло·очист·и́тел·ь (*m*) self-parking windscreen wiper

само·вы·ключ·е́ни·е, -я, (*n*) failure (e.g. of an engine); self-disengagement

само·вы·ра́вн·ивани·е, -я, (*n*) self-alignment; (autom) self/inherent regulation

само·гас·я́щ·ийся, -аяся, -ееся, (*a*) self-quenched; self-stifling (of reactions)

~ с·чёт·чик· (*m*) (nucl) self-quenched counter

само·го́н·, -а, (*m*) home-made spirit, poteen

само·гоно·вар·е́ни·е, -я, (*n*) illicit distillation/distilling

само·гру́з·н·ый, -ая, -ое, (*a*) self-weighted

само·дви́ж·н·ый, -ая, -ое, (*a*) automotive

само·дви́ж·ущ·ийся, -аяся, -ееся, (*a*) self-propelled; automotive

само·де́й·ствующ·ийся, -аяся, -ееся, (*a*) automatic, self-acting

само·де́ль·н·ый, -ая, -ое, (*a*) home-made; do-it-yourself kit

само·де́·ятельност·ь, -и, (*f*) independent activity; enterprise; amateur performance/work

само·де́·ятельн·ый, -ая, -ое, (*a*) enterprising; amateur; (econ) gainfully employed

само·диффу́з·и·я, -и, (*f*) (phys) self-diffusion

коэффицие́нт· само·диффу́з·и·и (*m*) self-diffusion coefficient

само·довле́·ющ·ий, -ая, -ее, (*a*) self-sufficient/-supporting

само·до·во́ль·н·ый, -ая, -ое, (*a*) complacent, self-satisfied

само·ду́в·н·ый, -ая, -ое, (*a*) natural-draft (boilers etc.)

сам·оё, (*pron*) (*acc sing f*) of **сам·** herself, itself

само·за·бв·е́нн·ый, -ая, -ое, (*a*) selfless

само·за·вод·я́щ·ийся, -аяся, -ееся, (*a*) self-winding, automatic (of watches)

само·за·груж·а́ющ·ийся, -аяся, -ееся, (*a*) self-loading

само·за·гряз·не́ни·е, -я, (*n*) (chem) intrinsic contamination

само·за·жиг·а́ни·е, -я, (*n*) spontaneous ignition, autoignition

само·за·ка́л·ивани·е, -я, (*n*) (met) air hardening

само·за·ка́л·к·а, -и, (*g pl*) **-л·ок·,** (*f*) (met) air hardening; air-/self-hardening steel

само·за·ква́ш·ивани·е, -я, (*n*) (biol) self-scouring

само·за·кле́·ивающ·ий, -ая, -ее, (*a*) self-sealing

~ со·ста́·в· (*m*) (rubber) sealant

само·за·ко́нтр·ивающ·ийся, -аяся, -ееся, (*a*) self-locking (screws)

само·за·крыв·а́ющ·ийся, -аяся, -ееся, (*a*) self-closing, self-restoring

само·за·ле́ч·ивани·е, -я, (*n*) self-healing (e.g. of concrete)

само·за·лив·а́ющ·ийся, -аяся, -ееся, (*a*) self-priming (pumps)

само·за·мык·а́ющ·ийся, -аяся, -ееся, (*a*) self-closing

само·за·пир·а́ющ·ийся, -аяся, -ееся, (*a*) (elec) self-squegging, squegging (of oscillators); (mech) self-catching/-tripping

само·за·пи́с·ывающ·ий, -ая, -ее, (*a*) (instr) recording

само·за·ряж·е́ни·е, -я, (*n*) (biol, med) autoinfection

само·за·рожд·е́ни·е, -я, (*n*) (biol) autogeny, autogenesis, spontaneous generation

само·за·ря́д·н·ый, -ая, -ое, (*a*) (gunn) self-loading, semi-automatic

само·за·тя́г·ивающ·ийся, -аяся, -ееся, (*a*) self-sealing (e.g. tank)

само·за·хва́т·ывающ·ийся, -аяся, -ееся, (*a*) self-locking

само·за·щи́т·а, -ы, (*f*) self-defence

само·за·щищ·а́ющ·ий, -ая, -ее, (*a*) self-protecting; (nucl) self-shielding

само·за·щищ·ённ·ый, -ая, -ое, (*a*) (elec) self-protecting; (nucl) self-shielding

само·зеркаль·н·ый, -ая, -ое, (*a*) self-mirrored

само·из·луч·е́ни·е, -я, (*n*) self-emission

само·индукцио́н·н·ый, -ая, -ое, (*a*) (elec) self-inductive, self-induction

само·инду́кци·я, -и,   (*f*) (elec) self-induction

само·ион·иза́ци·я, -и,   (*f*) auto-ionization

само·ис·цел·е́ни·е, -я,   (*n*) self-healing (concrete)

само·ка́л·, -а,   (*m*) = **само·за·-ка́л·к·а**

само·кале́ч·ени·е, -я,   (*n*) (zool) autotomy

само·ка́л·к·а, -и,   (*f*) = **само·за·-ка́л·к·а**

само·кас·а́·ться, -ются,   (*imp*) (math) be tangent to itself

само·ка́т·, -а,   (*m*) scooter (toy)

само·ката·лит·и́ческ·ий, -ая, -ое,   (*a*) (chem) autocatalytic

само·ква́с·, -а,   (*m*) (chem) self-precipitated casein

само·колеб·а́ни·е, -я,   (*n*) self-oscillation; (*pl*) (a/c) flutter; shimmy (of wheel)

само·контрол·и́рующ·ийся, -аяся, -ееся,   (*a*) self-supervising; self-contained

само·контро́л·ь, -я,   (*m*) automatic control; self-checking

само·кри́т·ик·а, -и,   (*f*) self-criticism

само·лёт·, -а,   (*m*) *see also* **гидро·-само·лёт·**; aircraft, aeroplane, airplane

~ -амфи́би·я, -и,   (*f*) amphibian aircraft, amphibian

~, борт·ов·о́й carrier aircraft, carrier-borne aircraft; shipborne aircraft

~ -буксир·о́вщик·, -а,   (*m*) towing aircraft

~, вед·о́м·ый No. 2 aircraft in formation

~, вед·у́щ·ий leading aircraft

~ -верто·лёт·, -а,   (*m*) convertiplane

~ -за·пра́в·щик·, -а,   (*m*) tanker aircraft

~, катапу́льт·н·ый catapult aircraft

~ -корект·иро́вщик·, -а, (*m*) (gunn) spotter aircraft

~, ло́д·очн·ый flying boat

~, мало·мо́щ·н·ый light aircraft

~ -ма́т·к·а, -и,   (*f*) mother-aircraft, carrier-aircraft; rocket-launching aircraft

~ -мише́н·ь, -и,   (*f*) (a/c) drone

~ -нос·и́тел·ь, -я,   (*m*) rocket-carrier, rocket-launching aircraft

~, па́луб·н·ый deck-landing aircraft

~, по·плав·ко́в·ый seaplane, float plane

само·лёт

~, раз·ве́д·ывательн·ый (mil) reconnaissance aircraft; prospecting aircraft

~, раке́т·н·ый rocket-propelled aircraft

~, реакти́в·н·ый jet aircraft

~, санита́р·н·ый air ambulance

~ с·вя́з·и (mil) liaison aircraft

~ -с·на·ря́д·, -а,   (*m*) aerodynamic/ cruise missile

~, сухо·пу́т·н·ый landplane

~ -тренажёр·, -а,   (*m*) instructional fuselage

~ -штурм·ови́к·, -а́,   (*m*) (mil) ground-attack aircraft

само·лёт·ик·, -а,   (*m*) aircraft image, miniature aircraft (in a/c instruments)

само·лёт·н·ый, -ая, -ое,   (*a*) aircraft, airborne

~ радио·ста́нци·я   (*f*) airborne radio

~ у·кры́т·и·е   (*n*) aircraft pen

самолёто·вожд·е́ни·е, -я,   (*n*) air navigation

~ по зем·н·ы́м ориенти́р·ам air pilotage

~ по свет·и́л·ам astronavigation

самолёто·вы·лет·, -а,   (*m*) (mil) sortie

самолёто·подъ·ём·ник·, -а,   (*m*) aircraft lift

самолёто·стро·е́ни·е, -я,   (*n*) aircraft construction

самолёто·стро·и́тельн·ый, -ая, -ое, (*a*) aircraft-construction, aircraft

само·леч·е́ни·е, -я,   (*n*) (physiol) autotherapy

само·ле́ч·ивани·е, -я,   (*n*) self-healing; (met) annealing out (of defects)

само·ликвид·а́тор·, -а,   (*m*) self-destruction device

само·ли́ч·н·о (*adv*) personally

само·модул·я́ци·я, -и,   (*f*) (rad) self-pulse modulation

само·на·вед·е́ни·е, -я,   (*n*) (GW) homing, self-guidance, self-reacting

~, акти́в·н·ое active homing

~ на кон·е́чн·ом у·ча́ст·к·е terminal homing

сигна́л· само·на·вед·е́ни·я   (*m*) (GW) homing signal

само·на·вед·ённ·ый, -ая, -ое,   (*a*) self-induced

само·на·вод·я́щ·ийся, -аяся, -ееся, (*a*) (GW) homing, self-guided, self-reacting; (autom) self-directing/ instructed

само·на·грев·а́ни·е, -я,   (*n*) spontaneous heating

само·на·ка́л·, -а,   (*m*) direct heating (vacuum tubes)

само·на·кла́д·, -а,   (*m*) (print) automatic feeding/feeder

само·на·кла́д·чик·, -а,   (*m*) (print) automatic feeder

само·на·правл·я́ем·ый, -ая, -ое,   (*a*) *see* **само·на·вод·я́щ·ийся**

само·на·са́с·ывающ·ийся, -аяся, -ееся,   (*a*) self-saturating

само·на·стра́·ивающ·ийся, -аяся, -ееся,   (*a*) self-adjusting, adaptive; (rad) autotuning

~ систе́м·а   (*f*) (autom) self-adjusting system, adaptive optimalizing system

само·на·стро́й·к·а, -и,   (*f*) (autom) self-adjustment; (rad) autotune

само·обеспе́ч·енн·ый, -ая, -ое,   (*a*) self-sufficient

само·облад·а́ни·е, -я,   (*n*) self-possession

само·образ·ова́ни·е, -я,   (*n*) self-education

само·обращ·е́ни·е, -я,   (*n*) self-reversal

само·об·слу́ж·ивани·е, -я,   (*n*) self-service

само·об·слу́ж·ивающ·ийся, -аяся, -ееся,   (*a*) self-contained (e.g. of workshops)

само·о·кисл·е́ни·е, -я,   (*n*) (chem) auto-oxidation

само·о·куп·а́ем·ый, -ая, -ое,   (*a*) self-supporting

само·о·плодо·твор·е́ни·е, -я,   (*n*) (biol) autogamy, self-fertilization

само·о·пре·дел·е́ни·е, -я,   (*n*) self-determination

само·о·про·ки́д·ывающ·ийся, -аяся, -ееся,   (*a*) self-tipping, tipping, dump, dumping

само·о·пыл·е́ни·е, -я,   (*n*) (bot) self-pollination, selfing

само·организ·у́ющ·ийся, -аяся, -ееся,   (*a*) (autom) adaptive, self-adjusting

само·о·стана́вл·ивающ·ийся, -аяся, -ееся,   (*a*) self-stopping

само·о·стано́в·, -а,   (*m*) automatic stop

само·от·ве́рж·енн·ый, -ая, -ое,   (*a*) selfless

само·от·жиг·, -а,   (*m*) (met) self-annealing

само·от·лив·н·о́й, -а́я, -о́е,   (*a*) self-draining

само·от·пуск·, -а,   (*m*) (met) self-tempering

само·о·травл·е́ни·е, -я,   (*n*) (med) auto-intoxication

само·о·хлажд·а́ем·ый, -ая, -ое,   (*a*) self-cooled

само·оче·ви́д·н·ый, -ая, -ое,   (*a*) self-evident

само·о·чищ·а́ющ·ийся, -аяся, -ееся,   (*a*) self-cleaning; self-wiping (of electric contacts)

само·о·чищ·е́ни·е, -я,   (*n*) self-purification; self-cleansing

само·о·чи́щ·енн·ый, -ая, -ое,   (*a*) (civ eng) self-cleansing

само·пере·ва́р·ивани·е, -я,   (*n*) (chem) autolysis

само·пере·сеч·е́ни·е, -я,   (*n*) (math) crunode

само·пи́с·ец·, -с·ц·а,   (*i*) **-с·ц·ем,** (*m*) recorder, recording instrument

~, много·то́ч·ечн·ый   multipoint recorder

~ мор·ск·и́х теч·е́ни·й   (ocean) automatic current meter

~ пере·гру́з·ок·   accelerograph

~ про́·йд·енн·ого пут·и́   (a/c) distance recorder

~, цифр·ов·о́й   digital recorder

само·пит·а́тел·ь, -я,   (*m*) automatic feeder

само·пи́ш·ущ·ий, -ая, -ее,   (*a*) (instr) recording, registering

~ ру́ч·к·а   (*f*) fountain pen

само·пла́в·к·ий, -ая, -ое,   (*a*) self-fluxing (ores etc.)

само·пло́д·н·ый, -ая, -ое,   (*a*) (bot) self-fertilizing, self-compatible

само·по·глощ·е́ни·е, -я,   (*n*) (phys) self-absorption

само·под·де́рж·ивающ·ийся, -аяся, -ееся,   (*a*) self-sustaining

само·под·жим·н·о́й, -а́я, -о́е,   (*a*) self-sealing (e.g. rubber goods)

само·по·йлк·а, -и,   (*g pl*) **-лок·,** (*f*) (agr) bin-type mash feeder

само·при·кос·нове́ни·е, -я,   (*n*) (math) osculation, self-tangency

то́ч·к·а само·при·кос·нове́ни·я   (*f*) point of osculation, double cusp

само·про·гре́в·, -а,   (*m*) self-heating

само·про·из·во́ль·н·ый, -ая, -ое,   (*a*) spontaneous

само·пря́л·к·а, -и,   (*g pl*) **-л·ок·,** (*f*) spinning wheel

само·пу́ск·, -а,   (*m*) self-starter

~, воз·ду́ш·н·ый   compressed-air starter

само·раз·груж·а́ющ·ийся, -аяся, -ееся, (*a*) self-unloading/discharging/dumping

само·раз·мык·а́ни·е, -я, (*n*) automatic opening

само·разо·гре́·в·, -а, (*m*) spontaneous heating

само·раз·ря́д·, -а, (*m*) (electrochem) self-discharge, local action
   ис·пыт·а́ни·е на само·раз·ря́д· (*n*) shelf test (batteries)

само·рас·про·стран·я́ющ·ийся, -аяся, -ееся, (*a*) self-propagating

само·рас·се́·яни·е, -я, (*n*) self-scattering

само·рас·твор·е́ни·е, -я, (*n*) autolysis

само·рас·це́п·, -а, (*m*) (rail) spontaneous uncoupling

само·рас·цепл·я́ющ·ийся, -аяся, -ееся, (*a*) self-detaching, automatically disconnecting

само·регул·и́ровани·е, -я, (*n*) self-/inherent regulation/adjustment

само·регул·и́рующ·ий, -ая, -ее, (*a*) self-regulating/adjusting

само·ро́д·н·ый, -ая, -ое, (*a*) (min) native
   ~ зо́лот·о (*n*) native gold

само·ро́д·ок·, -д·к·а, (*m*) nugget

само·са́д·к·а, -и, (*g pl*) -д·ок·, (*f*) natural salt

само·с·бро́с·к·а, -и, (*g pl*) -с·ок·, (*f*) (agr) self-rake reaper

само·с·ва́л·, -а, (*m*) (М/Т) tipper-lorry, dump-truck

само·свет·я́щ·ийся, -аяся, -ееся, (*a*) fluorescent, luminous

само·се́·в·, -а, (*m*) (hort) self-sown seedling

само·с·жа́т·и·е, -я, (*n*) (phys) self-constriction, pinching
   ~, лин·е́йчат·ое (nucl) Z-pinch

само·с·жа́·т·ый, -ая, -ое, (*a*) self-constricted

само·си́н·, -а, (*m*) (elec) synchro, selsyn

само·синхрониз·а́ци·я, -и, (*f*) (elec) self-synchronization/-synchronizing

само·синхрониз·и́рующ·аяся переда́·ч·а, одно·об·мо́т·очн·ая delta-wound single-phase synchro

само·синхрониз·и́рующ·ий, -ая, -ее, (*a*) self-synchronizing/-synchronous, selsyn; self-pulsing (radar)

само·с·крепл·е́ни·е, -я, (*n*) (gunn) autofrettage, self-hooping

само·с·лип·а́ни·е, -я, (*n*) (rubber) autohesion

само·слыш·имост·ь, -и, (*f*) sidetone (telephony)

само·с·ма́з·к·а, -и, (*g pl*) -з·ок·, (*f*) automatic lubrication

само·с·ма́з·очн·ый, -ая, -ое, (*a*) self-lubricating

само·с·ма́з·ывающ·ийся, -аяся, -ееся, (*a*) self-lubricating

само·с·мещ·е́ни·е, -я, (*n*) self-bias (vacuum tubes)

само·со·глас·о́ванн·ый, -ая, -ое, (*a*) self-consistent/-congruent
   ~ по́л·е (*n*) (phys) self-consistent field

само·со·пряж·ённ·ый, -ая, -ое, (*a*) (math) self-conjugate/adjoint
   ~ матри́ц·а (*f*) (math) Hermitian conjugate matrix; adjoint matrix
   ~ пре·образ·ова́ни·е (*n*) (math) self-adjoint transformation
   ~ у·равн·е́ни·е (*n*) (math) self-adjoint equation

само·со·хран·е́ни·е, -я, (*n*) self-preservation

само·с·пас·а́тел·ь, -я, (*m*) (min) portable breathing apparatus
   ~ изол·и́рующ·ий self-contained breathing apparatus

само·с·пек·а́ющ·ийся, электро́д· (*m*) (elec, met) self-baking electrode

само·с·пу́ск·, -а, (*f*) self-triggering

само·стери́ль·н·ый, -ая, -ое, (*a*) (bot) self-sterile/-incompatible

само·сто·я́тельност·ь, -и, (*f*) independence

само·сто·я́тельн·ый, -ая, -ое, (*a*) independent; self-contained, self-sustaining

само·стре́ль·н·ый, -ая, -ое, (*a*) automatic (of guns)

само·с·тя́г·ивани·е, -я, (*n*) (phys) self-constriction, pinch

само·та́ск·а, -и, (*g pl*) -сок·, (*f*) drag conveyer

само·тёк·, -а, (*m*) gravity feed/flow, flow; drift (of affairs)

само·тек·а́ющ·ий, -ая, -ее, (*a*) gravity-flow, gravity (of conveyers etc.)

само·тёк·ом (*adv*) by gravity

семо·тёч·н·ый, -ая, -ое, (*a*) gravity
   ~ ли́н·и·я (*f*) gravity pipeline

само·тормоз·я́щ·ийся, -аяся, -ееся, (*a*) self-braking, self-stopping; self-sustaining (gear)

само·то́ч·к·а, -и, (*g pl*) -ч·ек·, (*f*) automatic lathe

само·тя́г·а, -и, (*f*) natural draft

само·у·плот·няющ·ийся,   **-аяся, -ееся,**   (*a*) self-sealing; self-consolidating; self-packing

само·у·правл·éни·е, -я,   (*n*) self-government, autonomy, local government

само·у·правл·яем·ый, -ая, -ое,   (*a*) (GW) self-guided, self-reacting, homing; self-governing, autonomous

само·у·прáв·ств·овать, -уют,   (*imp*) take the law into one's own hands, act illegally

само·уравно·вéш·ивающ·ийся, -аяся, -ееся,   (*a*) self-balancing

само·у·станáвл·ивающ·ийся, -аяся, -ееся,   (*a*) self-adjusting, self-aligning; (autom) optimalizing

∼ шарико·под·шйп·ник·   (*m*) self-aligning ball-bearing

само·у·станóв·к·а, -и,   (*f*) self-alignment/adjustment

само·уч·йтел·ь, -я,   (*m*) self-instruction manual/textbook/pamphlet

само·ýч·к·а, -и,   (*g pl*) **-ч·ек·,**   (*f*) self-educated person

само·фаз·ирóвк·а, -и,   (*f*) autophasing

само·флюс·ýющ·ийся, -аяся, -ееся,   (*a*) *see* **само·плáв·к·ий**

само·форм·óвк·а, -и,   (*f*) (rubber) scaled-up moulding

само·хвáт·, -а,   (*m*) self-operating grab, automatic grab

само·хóд·, -а,   (*m*) (M/T) cross-country vehicle; (elec) shunt running (of a meter)

само·хóд·н·ый, -ая, -ое,   (*a*) self-propelled

само·хрон·йрующ·ий, -ая, -ее,   (*a*) (elec) self-pulsed

само·цвéт·, -а,   (*m*) gem, gem stone

само·центр·йровани·е, -я,   (*n*) self-centring/aligning

само·центр·йрующ·ийся, -аяся, -ееся,   (*a*) self-centring/aligning

само·экран·йровани·е, -я,   (*n*) self-shielding

самсон-пóст·, -а,   (*m*) (naut) Samson post

самýм·, -а,   (*m*) (geog) sand/dust storm

самшйт·, -а,   (*m*) (bot) box, *Buxus*

сáм·ый, -ая, -ое,   (*pronoun*) the very, the same/selfsame, the actual, itself; the most

в **сáм·ом дéл·е** indeed; actually

в **эт·у сáм·ую минýт·у** at this very moment

сáм·ый, -ая, -ое

∼ вéрх·н·ий   (*a*) the highest, the uppermost

∼ мáл·ый   (*a*) the smallest/least/lowest

∼ мáл·ый ход· dead slow

с **сáм·ого на·чá·л·а** from the very beginning, right from the first

у **сáм·ой вод·ы** right beside the water, at the water's edge

у **сáм·ой стен·ы** right by the wall, close into the wall

сан-   (*abbr*) = **санитáр·н·ый** (q.v.)

санатóр·и·й, -я,   (*m*) sanatorium

сан·áци·я, -и,   (*f*) (*v n*), *see* **сан·йр·овать;**   (med) treatment; restoration (of currencies, bankrupt organizations etc.)

сангвинарйн·, -а,   (*m*) (pharm) sanguinarine

сандáл·, -а,   (*m*) (bot) sandal-wood tree

сандалéт·а, -ы,   (*g pl*) **-лéт·,**   (*f*) sandal (footwear)

сандáл·и·я, -и,   (*f*) sandal (footwear)

сандáл·ов·ый, -ая, -ое,   (*a*) of **сандáл·**

∼ дéрев·о   (*n*) sandalwood

∼ мáсл·о   (*n*) sandalwood oil

сандáль·н·ый, -ая, -ое,   (*a*) of **сандáл·и·я** and **сандáл·**

сандарáк·, -а,   (*m*) sandarac, sandarach (resin)

сандарáк·ов·ый, -ая, -ое,   (*a*) of **сандарáк·**

сáн·и, -éй,   (*pl*) sledge, sled, sleigh

санидйн·, -а,   (*m*) (min) sanidine

сан·йровани·е, -я,   (*n*) (*v n*) of **сан·йр·овать**

сан·йр·овать, -уют,   (*imp and perf*) (med) restore to health, treat; (fin) restore to a healthy state

санитáр·, -а,   (*m*) male nurse, (mil) medical orderly

санитар·й·я, -и,   (*f*) hygiene, public health, preventive medicine

∼, про·мышл·енн·ая industrial health/hygiene

санитáр·к·а, -и,   (*g pl*) **-р·ок·,**   (*f*) (med) nurse

санитáр·н·о-ветеринáр·н·ый, -ая, -ое,   (*a*) veterinary, veterinary-service

санитáр·н·о-курóрт·н·ый, -ая, -ое,   (*a*) sanatorium-and-health-resort

санитáр·н·о-техн·йческ·ий, -ая, -ое,   (*a*) sanitary-engineering

∼ из·дéл·и·я   (*pl*) sanitary ware

санитáр·н·о-трáнспорт·н·ое сýд·н·о   (*n*) hospital ship

санита́р·н·о-хим·и́ческ·ая за·щи́т·а (*f*) anti-gas protection

санита́р·н·о-эпидемио·лог·и́ческ·ий (*a*) prophylactic-epidemiological, anti-epidemics, epidemiological

санита́р·н·ый, -ая, -ое, (*a*) public health, health, preventive medicine; medical, ambulance, hospital; sanitary

~ **автомоби́л·ь** (*m*) ambulance

~ **врач·** (*m*) medical officer for health, public health specialist

~ **инспе́кци·я** (*f*) health inspectorate

~ **инспе́ктор·** (*m*) health officer; (mil) junior medical officer

~ **коми́сси·я** (*f*) hygiene committee (e.g. in blocks of flats)

~ **об·рабо́т·к·а** (*f*) personal decontamination

~ **о·хра́н·а** (*f*) public health measures

~ **о·хра́н·а атмосфе́р·н·ого во́з·ду·х·а** (*f*) prevention of air pollution

~ **о·хра́н·а гран·и́ц·** immigration health regulations

~ **пате́нт·** (*m*) (naut) bill of health

~ **пост·** (*m*) medical aid post

~ **при·бо́р·** (*m*) sanitary ware

~ **про·пуск·н·о́й пункт·** (*m*) medical/ health check-point

~ **само·лёт·** (*m*) ambulance plane

~ **те́хн·ик·а** (*f*) sanitary engineering

санкафе́н·, -а, (*m*) (pharm) sancaphene, santonin

санкцион·и́р·овать, -уют, (*imp and perf*) give approval, sanction

са́н·н·ый, -ая, -ое, (*a*) of **са́н·и**

са́н·очн·ый, -ая, -ое, (*a*) of **са́н·и**

санта́л·, -а, (*m*) see **санда́л·**

сан·те́хн·ик·а, -и, (*f*) sanitary engineering

санти·ба́р·, -а, (*m*) centibar (unit of pressure)

санти·бе́л·, -а, (*m*) (telecom) centibel

санти·гра́мм·, -а, (*g pl*) -ов, (*m*) centigramme, centigram

санти·ме́тр·, -а, (*m*) centimetre, centimeter; centimetre rule/tape

санти·метр·о́в·ый, -ая, -ое, (*a*) centimetre, centimeter; centimetric

~ **волн·а́** (*f*) (rad) centimetric wave

санти·пуа́з·, -а, (*m*) centipoise (unit of viscosity)

санти·сто́кс·, -а, (*m*) centistoke (unit of viscosity)

сантони́н·, -а, (*m*) (pharm) santonin

сантори́н·, -а, (*m*) (geol, build) santorin

сантохло́р·, -а, (*m*) (chem) santochlor

са́п·, -а, (*m*) (med) glanders

са́п·а, -ы, (*m*) (mil) sap

сапёр·, -а, (*m*) (mil) sapper, field engineer

сапе́т·к·а, -и, (*g pl*) -т·ок·, (*f*) (agr) crib

сап·и́ндов·ые, -ых, (*pl decl as adj*) (bot) soapberry family, *Sapindaceae*

сапо́г·, -а́, (*nom pl*) -и́, (*m*) boot, high boot

сапогени́н·, -а, (*m*) (chem) sapogenin

сапо́ж·ник·, -а, (*m*) shoemaker; cobbler

сапо́ж·н·ый, -ая, -ое, (*a*) shoe, shoemaking, shoemaker's, cobbler's

сапож·о́к·, -ж·к·а́, (*nom pl*) -о́ж·к·и, (*g pl*) -ж·к·о́в, (*m*) light boot

сапон·а́ри·я, -и, (*f*) (bot) soapwort, *Saponaria*

сапон·и́н·, -а, (*m*) (chem) saponin

сапон·и́т·, -а, (*m*) (min) saponite

сапони·фика́ци·я, -и, (*f*) saponification

сапо́т·ов·ый, -ая, -ое, (*a*) (bot) sapodilla, (*pl as noun*) sapodilla family, *Sapotaceae*

сапр·еми́·я, -и, (*f*) (med) sapraemia, sapremia

сапро·ге́н·н·ый, -ая, -ое, (*a*) (biol) saprogenic, saprogenous

сапро·ли́т·, -а, (*m*) (min) saprolite

сапро·пели́т·, -а, (*m*) (geol) sapropelite

сапро·пели́т·ов·ый, -ая, -ое, (*a*) sapropelic; sapropelite

сапро·пе́л·ь, -я, (*m*) (geol) sapropel

сапро·фа́г·, -а, (*m*) (biol) saprophage

сапро·фи́т·, -а, (*m*) (biol) saprophyte

сапу́н·, -а́, (*m*) (ICE) breather pipe

сапфи́р·, -а, (*m*) sapphire

сапфир·и́н·, -а, (*m*) (min) sapphirine

сапфи́р·н·ый, -ая, -ое, (*a*) of **сапфи́р·**

сапфи́р·ов·ый, -ая, -ое, (*a*) of **сапфи́р·**

сара́·й, -я, (*m*) barn, shed, outhouse

сара́й·н·ый, -ая, -ое, (*a*) of **сара́·й**

сара́н·, -а, (*m*) (plast) saran

саранч·а́, -и́, (*i*) -о́й, (*f*) (ent) locust

~, **пере·лёт·н·ая** migratory locust

саранч·о́в·ый, -ая, -ое, (*a*) locust, grasshopper, (*pl as noun*) grasshoppers, *Acrididae*

сарга́н·, -а, (*m*) (fish) garfish, *Belone belone*, (*pl*) *Belonidae*

саргано·обра́з·н·ые, -ых, (*pl decl as adj*) (fish) *Beloniformes*

сарга́сс·о, -а, (*n*) (bot) sargasso, gulfweed, *Sargassum*

сарга́сс·ов·ые во́до·росл·и (*pl*) (bot) *Sargassaceae*; *Sargassum*

са́рд·, -а, (*m*) (geol) sard

сардéль·к·а, -и,　(*g pl*) **-л·ек·,**　(*f*) small fat sausage; (fish) anchovy
сардúн·а, -ы,　(*f*) (fish) sardine
сардонúкс·, -а,　(*m*) (min) sardonyx
сáрж·а, -и,　(*i*) **-ей,**　(*f*) (text) serge, twill
сáрж·ев·ый, -ая, -ое,　(*a*) of **сáрж·а**
сарк·óд·а, -ы,　(*f*) (zool) sarcode
сарк·óдов·ые, -ых,　(*pl decl as adj*) (zool) *Sarcodina*
сарко·кáрп·и·й, -я,　(*m*) (bot) sarcocarp
сарко·лéмм·а, -ы,　(*f*) (zool) sarcolemma
сарко·лúт·, -а,　(*m*) (zool) sarcolyte; (min) sarcolite
сарк·óм·а, -ы,　(*f*) (med) sarcoma
сарко·мéр·, -а,　(*m*) (cyt) sarcomere
сарко·плáзм·а, -ы,　(*f*) (zool) sarcoplasm
сарк·озúн·, -а,　(*m*) (chem) sarcosine
сарко·спор·úди·и, -ев,　(*m pl*) (zool) *Sarcosporidia*
сарпúн·к·а, -и,　(*f*) printed striped calico
сарц·енхúм·а, -ы,　(*f*) (bot) sarcenchyma
сарцúн·а, -ы,　(*f*) (bot) Sarcina
сары́ч·, -á,　(*i*) **-óм,**　(*m*) (orn) common buzzard, *Buteo buteo*
сассафрáс·, -а,　(*m*) (bot) sassafras, *Sassafras*
сассолúн·, -а,　(*m*) (min) sassolite
сассолúт·, -а,　(*m*) (min) sassolite
сателлúт·, -а,　(*m*) satellite; (mech eng) planetary bevel gear, satellite
сатúн·, -а,　(*m*) (text) sateen
сатин·úр·овать, -уют,　(*imp and perf*) (text) satinize; glaze (paper, leather)
сатúн·ов·ый, -ая, -ое,　(*a*) sateen; satin
~ дéрев·о　(*n*) satin-wood
~ пере·плет·éни·е　(*n*) (text) satin weave
сатур·áтор·, -а,　(*m*) (chem) saturator; carbonator
сатурациóн·н·ый, -ая, -ое,　(*a*) of **сатур·áци·я**
~ котёл·　(*m*) saturation boiler (sugar-making)
сатур·áци·я, -и,　(*f*) saturation; carbonation
Сатýрн·, -а,　(*m*) (astron) Saturn
саурé·я, -и,　(*f*) (bot) savory, *Satureia hortensis*
сафлóр·, -а,　(*m*) (bot) safflower, *Carthamus tinctorius*

сафлóр·ов·ый, -ая, -ое,　(*a*) of **сафлóр·**
~ мáсл·о　(*n*) safflower oil
сафран·áл·, -а,　(*m*) (biochem) safranal
сафран·úн·, -а,　(*m*) (chem) safranine
сафр·óл·, -а,　(*m*) (chem) safrole, safrol
сафрóн·, -а,　(*m*) (food) saffron
саффлорúт·, -а,　(*m*) (min) safflorite
сафьян·, -а,　(*m*) morocco (leather)
сáхар·, -а,　(*nom pl*) **-á,**　(*g pl*) **-óв,**　(*m*) sugar
~, виногрáд·н·ый　grape sugar, glucose
~, голов·н·óй　loaf sugar
~, гриб·н·óй　(biochem) mycose
~, древéс·н·ый　wood sugar, xylose
~, жж·ён·ый　caramel
~, камéд·н·ый　gum sugar, arabinose
~, кристалл·úческ·ий　granulated sugar
~, кукурýз·н·ый　maizena
~, молóч·н·ый　milk-sugar, lactose
~, мы́ш·ечн·ый　(biochem) inositol
~, мяс·н·óй　(biochem) inositol
~, печён·очн·ый　liver sugar, glycogen
~, пил·ён·ый　lump sugar
~, плод·óв·ый　fruit sugar, fructose
~ -сатýрн·, -а,　(*m*) (chem) crystalline lead acetate, lead sugar
~, свекл·овúчн·ый　beet sugar
~, солод·óв·ый　malt sugar, maltose
~, тростник·óв·ый　cane sugar, saccharobiose
~, фрукт·óв·ый　fructose
сахар·áз·а, -ы,　(*f*) (biochem) saccharose
сахар·áт·, -а,　(*m*) saccharate
сахарú·метр·, -а,　(*m*) saccharimeter
сахари·мéтр·и·я, -и,　(*f*) saccharimetry
сахар·úн·, -а,　(*m*) (chem) saccharin
сахарúн·ов·ый, -ая, -ое,　(*a*) saccharin, (chem) saccharinic
сахарин·óз·а, -ы,　(*f*) saccharinose
сáхар·истост·ь, -и,　(*f*) sugar content, saccharinity
сáхар·ист·ый, -ая, -ое,　(*a*) sugary
сáхар·ить, -ят,　(*imp*) sugar, add sugar to
сáхар·н·ый, -ая, -ое,　(*a*) sugar; (chem) saccharic
~ глазýр·ь　(*m*) icing sugar, icing
~ кисл·от·á　(*f*) saccharic acid
~ пес·óк·　(*m*) granulated sugar
~ разъ·един·úтел·ь　(*m*) soluble firing plug (in underwater explosive systems)

са́хар·н·ый
~ **свёкл·а** (*f*) sugar beet
~ **тростни́к·** (*m*) (bot) sugar-cane, *Saccharum*

сахаро·био́з·а, -ы, (*f*) (chem) saccharobiose, sucrose

сахаро·вар·е́ни·е, -я, (*n*) sugar refining

сахаро·ви́д·н·ый, -ая, -ое, (*a*) sugary; (geol) saccharoidal

сахар·о́з·а, -ы, (*f*) (chem) saccharose

сахаро·мице́т·ы, -ов, (*m pl*) (bot) saccharomycetes, *Saccharomycetaceae*

сахаро́·метр·, -а, (*m*) saccharometer

сахаро·но́с·, -а, (*m*) sugar plant

сахаро·образ·ова́ни·е, -я, (*n*) sugar formation, saccharification

сахаро·по·до́б·н·ый, -ая, -ое, (*a*) sugar-like, saccharoid

сахаро·с·ниж·а́ющ·ий, -ая, -ое, (*a*) hypoglycemic, sugar-reducing

сач·о́к·, -ч·к·а́, (*m*) (fish) landing net; butterfly net

сб. (*abbr*) = **с·бо́р·ник·** (q.v.); (print) collection, symposium

сб. (*abbr*) = **стилб·** stilb (unit of brightness)

с·ба́в·ить, -ят, (*perf*) see **с·бавл·я́·ть**

с·ба́в·к·а, -и, (*f*) (*v n*), see **с·бавл·я́·ть**; reduction, cut, deduction

с·ба́вл·енн·ый, -ая, -ое, (*a*) (*past part pass*) see **с·бавл·я́·ть**

с·бавл·я́·ть, -ют, (*imp*) reduce, cut, deduct, take off

с·баланс·и́рованност·ь, -и, (*f*) balance

с·баланс·и́р·овать, -уют, (*perf*) regain balance; balance, strike a balance

с·ба́лт·ыва·ть, -ют, (*imp*) shake/stir up

с·бег·а́ни·е, -я, (*n*) (*v n*) of **с·бег·а́·ть**

с·бег·а́·ть, -ют, (*imp*) run down/out/away; **-ся** (*pass*); run together, converge

с·бег·а́ющ·ий, -ая, -ее, (*pres part act*) of **с·бег·а́·ть**
~ **с·кат·** (*m*) trailing edge (of pulse)

с·беж·а́ть, (*fut 3rd sing, pl*) **с·бе·ж·и́т, с·бег·у́т,** (*perf*); see **с·бег·а́·ть**

с·бе́й·те, (*pl imper*) of **с·би·ть,** (*perf*); see **с·бив·а́·ть** (*imp*)

с·берег·а́тельн·ый, -ая, -ое, (*a*) (fin) savings
~ **ка́сс·а** (*f*) savings bank

с·берег·а́·ть, -ют, (*imp*) save, preserve, protect; save, save up, economize

с·береж·е́ни·е, -я, (*n*) (*v n*) of **с·берег·а́·ть**

с·бере́ч·ь, (*fut 3rd sing, pl*) **с·бере·ж·ёт, с·берег·у́т,** (*past masc sing*) **с·берёг·,** (*perf*); see **с·берег·а́·ть**

сбер·ка́сс·а, -ы, (*f*) (fin) savings bank

с·бив·а́ем·ый, -ая, -ое, (*pres part pass*) of **с·бив·а́·ть**

с·бив·а́емост·ь с·ли́в·ок· (*f*) churnability of cream

с·бив·а́льн·ый, -ая, -ое, (*a*) churning, whipping, whisking, beating

с·бив·а́ни·е, -я, (*n*) (*v n*), see **с·бив·а́·ть**
~ **пе́ленг·а** (mil, rad) meaconing

с·бив·а́·ть, -ют, (*imp*) knock down/off; bring/shoot down; divert, throw/put off/out, confuse; reduce, bring down; (mil) dislodge, throw back; knock together/up; churn, beat, whip
~ **во́йлок·** (text) felt

с·бив·а́·ться, -ются, (*imp*) (*pass*); go astray, lose the way, go off course; gather close together, crowd in; felt up, become matted
~ **в во́йлок·** (text) felt up
~ **в ком·к·и́** (paper) ball, ball up

с·би́в·к·а, -и, (*f*) (*v n*), see **с·бив·а́·ть**
~ **нул·я́** stick-off (of servomechanisms)

с·бив·н·о́й, -а́я, -о́е, (*a*) beating, churning, whipping; beaten, churned, whipped

с·би́в·чивост·ь, -и, (*f*) inconsistency, unevenness; confusedness, state of confusion

с·би́в·чив·ый, -ая, -ое, (*a*) inconsistent, inconsequential; uneven, unrhythmic; confused

с·би·ть, (*fut 3rd pl*) **со·бь·ю́т,** (*perf*) see **с·бив·а́·ть**

с·би́·т·ый, -ая, -ое, (*a*) (*past part pass*) see **с·бив·а́·ть**; stripped (of conveyer belts); (instr) biassed, out-of-alignment; (mil, a/c) shot down

с·ближ·а́·ть, -ют, (*imp*) bring/move/draw together/nearer/closer, (naut) marry (of ropes etc.); **-ся** approach, come nearer/closer; become similar

с·ближ·е́ни·е, -я, (*n*) (*v n*), see **с·бли·ж·а́·ть**; approach; convergence
~ **меридиа́н·ов** (surv) meridian/geodetic convergence
ско́р·ост·ь с·ближ·е́ни·я (*f*) rate of approach/closing/closure
у́гол· с·ближ·е́ни·я меридиа́н·ов (*m*) (nav) grid inclination

с·бли́ж·енност·ь, -и, (*f*) close proximity

с·бли́ж·енн·ый, -ая, -ое, (*past part pass*) *see* с·бли́ж·а́·ть: (bot) connivent; (geol) contiguous; adjacent

с·бли́з·ить, -ят, (*perf*) *see* с·ближ·а́·ть

с·блок·и́р·овать, -уют, (*perf*) interlock; join, join up, assemble

сбо́ечно·бур·и́льн·ая маши́н·а (*f*) (min) crosscut-boring/drilling machine, large-diameter drilling machine

с·бо́·ечн·ый, -ая, -ое, (*a*) of с·бо́й·к·а

с·бой, (*g sing*) с·бо́·я, (*nom pl*) с·бо·́й, (*g pl*) с·бо·ёв, (*m*) jerk, jerking (due to a blow); misfire, misfiring (of engines); residue, remains, husks, discard

с·бо́й·к·а, -и, (*g pl*) с·бо́·ек· (*f*) (min) crosscut, box-hole; connecting up (stopes)

с·бо́й·н·ый, -ая, -ое, (*a*) of с·бо·й

~ теч·е́ни·е (*n*) (ocean) tidal bore

с·бо́к·у (*adv*) from/on/at the side
   вид· с·бо́к·у side view

с·бо́лт·анн·ый, -ая, -ое, (*a*) (*past part pass*) of с·болт·а́·ть

с·болт·а́·ть, -ют, (*perf*) shake/stir up

с·болт·и́ть, -я́т, (*perf*) *see* с·бо́лч·ива·ть

с·бо́лч·енн·ый, -ая, -ое, (*past part pass*) *see* с·бо́лч·ива·ть

с·бо́лч·ива·ть, -ют, (*imp*) bolt together/up, fasten with bolts

с·бо́р·, -а, (*m*) collection, gathering; takings, gate-money; dues, charge, charges, fee, toll; assembly, muster; picking

~, груз·ов·о́й wharfage (in a port)

~, ло́цман·ск·ий (naut) pilotage

~, по́лн·ый (theat, etc.) "sold out"

~, порт·о́в·ый harbour dues

~, при·ча́ль·н·ый berthage (in a port)

с·бо́р·к·а, -и, (*g pl*) -р·ок·, (*f*) (eng) assembly, assembling, putting together; (text) gathering, (*pl*) gathers (dressmaking etc.)

~ зо́н·ы вос·про·из·во́д·ств·а (nucl) blanket assembly

~, про·то́ч·н·ая conveyer-belt assembly, progressive assembly

~, секцио́н·н·ая assembly of prefabricated parts, assembly of sections

с·бо́р·ник·, -а, (*m*) collector, receiver; collection tank, manifold, header; collection, selection, symposium (of written works)

с·бо́р·ник

~ стат·е́й collected articles

с·бо́р·н·ый, -ая, -ое, (*a*) composite, sectional, fabricated; assembly, collecting, rallying; collected, assembled; miscellaneous

~ бето́н· (*m*) precast concrete

~ ваго́н· (rail) mixed-goods wagonload

~ дом· (*m*) prefabricated/sectional house/building

~ домо·стро·е́ни·е (*n*) (build) house prefabrication, sectional construction

~ колле́ктор· (*m*) (a/c) collector tank

~ лист· (*m*) (print) map index

~ пункт· (*m*) (mil) assembly point, rallying point

~ по́·езд· (*m*) local/pick-up goods train

~ стати́в· (*m*) (telecom) mixed-equipment rack

с·бо́р·очн·о-автома́т·и́ческ·ий, -ая, -ое, (*a*) automatic assembly

~ ли́н·и·я (*f*) (eng) automatic assembly line

сбо́рочно·с·ва́р·очн·ый, -ая, -ое, (*a*) (weld) fabricating, fabrication

~ цех· (*m*) fabricating/prefabrication shop

с·бо́р·очн·ый, -ая, -ое, (*a*) assembly

~ ме́т·к·а (*f*) location mark (for assembling equipment)

~ при·способл·е́ни·е (*n*) (eng) assembly jig

~ цех· (*m*) assembly shop; (shipb) assembly berths

с·бо́р·щик·, -а, (*m*) collector; assembler, fitter

с·бра́ж·ива·ть, -ют, (*imp*) ferment

с·бра́с·ываем·ый, -ая, -ое, (*pres part pass*) of с·бра́с·ыва·ть; expendable

~ бак· (*m*) (a/c) drop tank

с·бра́с·ывани·е, -я, (*n*) (*v n*), *see* с·бра́с·ыва·ть; disposal; (geol) faulting; through clearance (telephones)

~ ли́ст·ьев (bot) defoliation
   то́ч·к·а с·бра́с·ывани·я (*f*) point of release (bombs etc.)

с·бра́с·ыватель·, -я, (*m*) ejector, throw-/sweep-off gear; throw-off carriage (of conveyer); kicker, log kicker (in sawmills); (geol) fault fissure; (rail) derailer

с·бра́с·ыва·ть, -ют, (*imp*) throw down/off/back; drop, release, jettison; shed, discard, dump; discharge (into) (water engineering)

~ давл·е́ни·е depressurize

с·брас·ывающ·ий, -ая, -ее, (*pres part act*) of **с·брас·ыва·ть**
~ **башмак·** (*m*) (rail) derailer
~ **план·к·а** (*f*) (text) work bar
~ **стрел·к·а** (*f*) (rail) derailing points/switch
~ **трещ·ин·а** (*f*) (geol) fault fissure
с·бре́·ет, (*fut 3rd sing*) of **с·бри·ть,** (*perf*); *see* **с·брив·а́·ть**
с·брив·а́·ть, -ют, (*imp*) shave off
с·бри·ть, (*fut 3rd pl*) **с·бре́·ют,** (*perf*) *see* **с·брив·а́·ть**
с·брод·и́ть, -ят, (*perf*) ferment
с·брос·, -а, (*m*) (*v n*), *see* **с·брас·ыва·ть**; (geol) fault; (cryst) dislocation
~, **зи·я́ющ·ий** (geol) open fault
и́мпульс· с·брос·а (*m*) (autom) reset pulse
~, **нис·ход·я́щ·ий** (geol) downthrow
~, **обра́т·н·ый** reserve fault
~ **попад·е́ни·ю** dip fault
~ **реги́стр·а** (autom) reset, resetting
~, **ступ·е́нчат·ый** step fault
с·брос·анн·ый, -ая, -ое, (*a*) (*past part pass*) *see* **с·брас·ыва·ть**
с·брос·ить, -ят, (*perf*) *see* **с·брас·ыва·ть**
с·брос·н·о́й, -а́я, -о́е, (*a*) discharge (water engineering); waste
~ **кана́л·** (*m*) discharge canal/channel
~ **рас·тво́р·** (*m*) waste solution
с·брос·н·ый, -ая, -ое, (*a*) *see* **с·брос·-н·о́й**
с·брос·ов·ый, -ая, -ое, (*a*) *see* **с·брос·-н·о́й**; (geol) fault
~ **бре́кчи·я** (*f*) (geol) fault breccia
сбросо·образ·ова́ни·е, -я, (*n*) (geol) faulting; (cryst) dislocation
сбросо·с·дви́г·, -а, (*m*) (geol) strike fault
с·бро́ш·енн·ый, -ая, -ое, (*a*) (*past part pass*) *see* **с·брас·ыва·ть**
с·брошюр·ова́ть, -у́ют, (*perf*) (print) stitch
с·бу́д·ет, (*fut 3rd sing*) of **с·бы·ть,** (*perf*); *see* **с·быв·а́·ть**
с·быв·а́·ть, -ют, (*imp*) market, sell; go down (of water level); **-ся** be realized, come true (e.g. of wishes)
с·бы·т·, -а, (*m*) sale, market
с·быт·ов·о́й, -а́я, -о́е, (*a*) sales, marketing
с·бы·ть, (*fut 3rd pl*) **с·бу́д·ут,** (*perf*) *see* **с·быв·а́·ть**
св. (*abbr*) = **свеч·а́** candle (unit of luminous intensity)
сваби́·рование·, -я, (*n*) (oil) swabbing
свае- (*root*) (civ eng) pile-

сваe·вы·держ·иватель·, -я, (*m*) (civ eng) pile-driver
сва́·ек·, (*g pl*) of **сва́й·к·а**
сва́·и, (*g sing, nom/acc pl*) of **сва́·я**
сва́й·к·а, -и, (*g pl*) **сва́·ек·,** (*f*) (*dim*) of **сва́·я**; (naut) marline spike; (elec) spud (vacuum tubes); (biol) palus
сва́й·н·ый, -ая, -ое, (*a*) of **сва́·я**
~ **башма́к·** (*m*) pile shoe
~ **куст·** (*m*) mooring posts
~ **мо́лот·** (*m*) piling hammer
~ **о·по́р·а** (*f*) pile pier
с·ва́л·енн·ый, -ая, -ое, (*a*) (*past part pass*) *see* **с·ва́л·ива·ть**
с·ва́л·ивани·е, -я, (*n*) (*v n*), *see* **с·ва́л·ива·ть**; stall, stalling; canting (of a rifle)
~, **по·вто́р·н·ое** (air) secondary stall
ско́р·ост·ь с·ва́л·ивани·я (*f*) (air) stalling speed
с·ва́л·ива·ть, -ют, (*imp*) knock down/over; heave/throw off, dump; (air) stall; **-ся** fall off/over/down
~ **ся на кры·л·о́** (air) fall off on one wing
с·вал·и́ть, -ят, (*perf*) *see* **с·ва́л·ива·ть**
с·ва́л·к·а, -и, (*g pl*) **-л·ок·,** (*f*) (*v n*), *see* **с·ва́л·ива·ть**; dump, garbage dump, tip, rubbish tip
с·вал·я́вш·ийся, -аяся, -ееся, (*past part act*) of **с·вал·я́·ть**; felted; (bot) caespitose, cespitose
с·вал·я́·ть, (*perf*) (text) felt; **-ся** (*pass*)
с·ва́р·енн·ый, -ая, -ое, (*past part pass*) *see* **с·ва́р·ива·ть**
с·ва́р·иваемост·ь, -и, (*f*) (met) weldability; tendency to stick (of coiled or stacked sheet)
~, **ис·пыт·а́ни·е на** welding test
с·ва́р·ивани·е, -я, (*n*) (*v n*) of **с·ва́р·и·ва·ть**; sticking (of metal sheets stacked together for rolling or annealing)
с·ва́р·ива·ть, -ют, (*imp*) weld; **-ся** (*pass*); stick together (of sheet stacked for rolling or annealing)
с·вар·и́ть, -ят, (*perf*) weld; boil; cook; melt, melt down
с·вар·и́вш·иеся лист·ы (*pl*) (met) stickers
с·ва́р·к·а, -и, (*g pl*) **-р·ок·,** (*f*) welding
~, **авто·ге́н·н·ая** fusion welding, non-pressure welding; oxy-acetylene welding
~, **аргоно·дуг·ов·а́я** argon-arc welding
~, **автомат·и́ческ·ая то́ч·ечн·ая** stitch welding

с·вáр·к·а

~, **атомно·водо·рóд·н·ая** atomic-hydrogen welding

~ **в пласт·ѝческ·ом со·сто·я́ни·и** solid-phase welding

~ **в сред·é инéрт·н·ого гáз·а, дуг·о·в·áя** inert-gas-shielded-arc welding

~, **вéрх·н·яя** overhead welding

~ **в·на·пýск·** scarf welding

~ **гáз·ов·ая** gas welding

~, **газо·прéсс·ов·ая** oxy-acetylene welding

~, **газо·электр·ѝческ·ая** gas-shielded welding

~, **горн·ов·áя** forge/hammer welding

~ **давл·éни·ем** welding with pressure

~, **двух·дуг·ов·áя** twin-arc welding

~, **двух·ѝмпульс·н·ая** pulsation welding

~, **дуг·ов·áя** arc welding

~, **ѝмпульс·н·ая** stored-energy welding, shot welding

~, **конденсáтор·н·ая** electrostatic welding, electrostatic-percussive welding

~, **контáкт·н·ая** resistance welding

~, **куз·нéчн·ая** smith/forge welding

~, **лéв·ая** forehand welding

~, **лин·éйн·ая** seam welding

~ **металл·ѝческ·им электрóд·ом, дуг·ов·áя** metal/metallic-arc welding

~, **мнóго·слóй·н·ая** multilayer welding

~ **о·плавл·éни·ем, с·тык·ов·áя** flash welding

~ **плавл·éни·ем** fusion welding

~ **под флю́с·ом, автомáт·ѝческ·ая** submerged-arc welding

~, **потол·óчн·ая** overhead welding

~, **прáв·ая** back-hand welding

~, **рельéф·н·ая** projection welding

~, **рóл·иков·ая** seam welding

~ **с·жáт·ой дуг·óй, высоко·темпе·ратýр·н·ая** plasma-jet welding

~, **с·тык·ов·áя** butt welding

~, **с·тык·ов·áя контáкт·н·ая** flash-butt welding

~, **термѝт·н·ая** thermit welding

~, **тóч·ечн·ая** spot welding

~ **тр·éни·ем** friction welding

~ **ýголь·н·ым электрóд·ом, дуг·о·в·áя** carbon-arc welding

~, **холóд·н·ая** cold pressure welding

~ **шв·ом** seam welding

~, **шлáнг·ов·ая** submerged-arc welding

~, **электро·дуг·ов·áя** electric-arc welding

с·вáр·к·а

~, **электрóнно·луч·ев·áя** electron-beam welding

~, **электро·шлáк·ов·ая** electroslag welding

с·вáр·н·óй, -áя, -óе, (*a*) welded

~ **констрýкци·я** (*f*) fabricated structural member

~ **со·един·éни·е** (*n*) weld, welded joint

~ **шов·** (*m*) weld

с·вáр·очн·ый, -ая, -ое, (*a*) welding

~ **аппарáт·** (*m*) welding set

~ **вáнн·а** (*f*) weld pool

~ **голóв·к·а,** (*f*) blowpipe

~ **инструмéнт·** (*m*) electrode holder

~ **машѝн·а** (*f*) welding set, welder

~ **пистолéт·** (*m*) welding gun

~ **регуля́тор· тóк·а** (*m*) welding regulator

~ **редýктор·** (*m*) welding regulator (gas welding)

~ **щит·óк·** (*m*) (weld) hand shield

свáр·щик·, -а, (*m*) welder

свá·я, -и, (*f*) pile, stilt, post; spud (of tractor wheel); anchoring spud (of dredger)

~, **áнкер·н·ая** anchor pile

~, **вис·я́ч·ая** friction pile

~, **за·вѝнч·ивающ·ая** screw pile

~, **за·мык·áющ·ая** key pile

~, **корен·н·áя** foundation pile

~ **Лаковáн·а** Lackawanna sheet pile

~, **на·бив·н·áя** filling/moulded-in-place pile

~, **на·клóн·н·ая** batter pile

~, **от·бóй·н·ая** (naut) pile fender, rubbing pile

~, **от·кóс·н·ая** batter pile

~, **перед·ов·áя** leading pile

~, **у·пóр·н·ая** stay pile

~, **шпунт·ов·áя** sheet pile

свёд·а, -ы, (*f*) (bot) seepweed, *Suaeda*

с·вед·éни·е, -я, (*n*) (*v n*) of **с·вéс·ти, с·вод·ѝть;** (math) reduction, contraction; (med) cramp, contraction

с·вéд·ени·е, -я, (*n*) information, knowledge

ýгол· с·вед·éни·я (*m*) convergence angle

с·вед·ённ·ый, -ая, -ое, (*past part pass*) *see* **с·вод·ѝть;** traced back to, reduced to, reduced

с·вéд·ущ·ий, -ая, -ее, (*a*) well-informed, knowledgeable

свёж·е (*adv*) freshly, while fresh; recently, newly

свеж·ева́ть, -у́ют, (*imp*) flay, skin, dress

свеже·за·моро́ж·енн·ый, -ая, -ое, (*a*) fresh-frozen/chilled; quick-frozen

свеже·ис·печ·ённ·ый, -ая, -ое, (*a*) fresh-/newly-baked, fresh, new (of bread, cakes etc.)

свеже·пере́·гн·анн·ый, -ая, -ое, (*a*) fresh-distilled

свёж·ест·ь, -и, (*f*) freshness

свеж·е́·ть, -ют, (*imp*) freshen

свёж·ий, -ая, -ее, (*a*) fresh, new, latest; live (steam)

с·вез·ти́, -у́т, (*perf*) *see* **с·воз·и́ть**

с·ве́й·те, (*pl imper*) of **с·ви·ть** (*perf*) *see* **с·вив·а́·ть**

свёкл·а, -ы, (*f*) (bot, food) beet, *Beta*

~, корм·ов·а́я mangel wurzel

~, са́хар·н·ая sugar beet

~, стол·о́в·ая beetroot

свекл·ови́ц·а, -ы, (*i*) **-ей,** (*f*) sugarbeet

свекл·ови́чн·ый, -ая, -ое, (*a*) of **свекл·ови́ц·а**

~ му́х·а (*f*) (ent) beet leaf miner, *Pegomia hyosciami*

~ се́·ялк·а (*f*) (agr) sugar-beet sower

~ са́хар· (*m*) beet sugar

свекло·во́д·ств·о, -а, (*m*) beet production/growing

свекло·са́хар·н·ый, -ая, -ое, (*a*) beetsugar, sugar-beet

свекло·мо́й·к·а, -и, (*g pl*) **-о·ек·,** (*f*) (agr) beet washer

свекло·подъ·ём·ник·, -а, (*m*) (agr) beet lifter

с·верг·а́·ть, -ют, (*imp*) overthrow, throw down, detrude

с·ве́рг·нуть, -нут, (*perf*) *see* **с·верг·а́·ть**

с·верж·е́ни·е, -я, (*n*) overthrow, detrusion

с·ве́р·к·а, -и, (*f*) (*v n*), *see* **с·вер·я́·ть;** collation, comparison; (horol) regulation

с·ве́р·ить, -ят, (*perf*) *see* **с·вер·я́·ть**

сверк·а́ни·е, -я, (*n*) (*v n*) of **сверк·а́·ть**

сверк·а́·ть, -ют, (*imp*) sparkle, glitter, twinkle

сверк·н·у́ть, -у́т, (*perf*) show a gleam, flash

сверл·е́ни·е, -я, (*n*) (*v n*), *see* **сверл·и́ть**

сверл·и́лк·а, -и, (*g pl*) **-лок·,** (*f*) drill, drilling machine

сверл·и́л·о, -а, (*n*) borer

~, корабе́ль·н·ое (ent) shipworm, *Teredo*

сверл·и́льн·о-долб·ёжн·ый стан·о́к· (*m*) drilling and mortising machine

~ -на·рез·н·о́й стан·о́к· (*m*) boring-and-threading machine

~ -паз·ова́льн·ый стан·о́к· (*m*) drilling-and-mortising machine

~ -рас·то́ч·н·о-фре́зер·н·ый стан·о́к· (*m*) boring-facing-and-milling machine

сверл·и́льн·ый, -ая, -ое, (*a*) drill, drilling

~ патро́н· (*m*) drill chuck; drillstock (on hand drill)

~ стан·о́к· (*m*) drilling machine

~ стру́ж·к·а (*f*) chips, drillings

сверл·и́мост·ь, -и, (*f*) drillability **ис·пыт·а́ни·е на сверл·и́мост·ь** (*n*) drill test

сверл·и́ть, -я́т, (*imp*) (mech) drill; (zool etc.) bore

сверл·о́, -а́, (*nom pl*) **свёрл·а,** (*g pl*) **свёрл·,** (*n*) drill, drilling bit

~, пер·ов·о́е flat drill

~, пу́ш·ечн·ое gun barrel drill

~, спира́ль·н·ое twist drill

~, центр·о́вочн·ое centre drill

сверл·я́нк·а, -и, (*g pl*) **-нок·,** (*f*) (ent) borer, *Pholas tullus*

сверл·я́щ·ие жив·о́тн·ые (*pl*) (zool) marine borers

с·вёр·нут·ый, -ая, -ое, (*past part pass*) *see* **с·вёрт·ыва·ть;** convolute

~, спира́ль·н·о helical, spiral

~, улитко·обра́з·н·о helicoid

с·вер·н·у́ть, -у́т, (*perf*) *see* **с·вёрт·ыва·ть**

с·вёрст·анн·ый, -ая, -ое, (*past part pass*) of **с·верст·а́ть**

~ на·бо́р· (*m*) (print) imposition

с·верст·а́·ть, -ют, (*perf*) (print) impose

с·вёрст·ыва·ть, -ют, (*imp*) (print) impose

с·вёрт·к·а, -и, (*g pl*) **-т·ок·,** (*f*) (*v n*) *see* **с·вёрт·ыва·ть;** (*g sing*) of **свёрт·ок**

~ фу́нкци·и (math) convolution

с·вёрт·ок, -т·к·а, (*m*) roll (of rolled up material); (*g pl*) of **с·вёрт·к·а**

с·вёрт·ывани·е, -я, (*n*) (*v n*) of **с·вёрт·ыва·ть;** convolution

с·вёрт·ыва·ть, -ют, (*imp*) roll up, roll, furl (sails etc.); curtail, reduce; wind up (organizations etc.); turn, turn aside/off; (mil) deploy; **-ся** (*pass*); coagulated, curdle (of milk); (bot) convolve, convolute

с·вёрт·ывающ·ая машин·а (met roll) roll-type coiler

с·верх· (*prep* + *gen*) over, above, beyond; besides, in addition to, in excess of

~ того moreover, besides this

с·верх- (*component*) super-, supra-, ultra-, extra-, over-, hyper-; in excess of, over-and-above, supplementary to, outside (estimates, quotas etc.)

с·верх·атмосфер·н·ое давл·ени·е (*n*) superpressure

с·верх·баллон·, -а, (*m*) superballoon tyre

с·верх·быстро·ход·н·ый, -ая, -ое, (*a*) ultra-super high-speed

с·верх·быстр·ый, -ая, -ое, (*a*) ultra-rapid, superfast, ultraspeed

с·верх·выс·ок·ий, -ая, -ое, (*a*) ultra-high, super; super-high (of frequencies)

~ на·пряж·ени·е (*n*) (elec) super voltage

с·верх·высоко·част·от·а, ·ы, (*f*) super high frequency

с·верх·высоко·част·отн·ый, -ая, -ое, (*a*) super high-frequency

с·верх·вяз·кост·ь, -и, (*f*) super-viscosity

с·верх·гетеро·дин·, -а, (*m*) (rad) superhet, superheterodyne receiver

с·верх·групп·а, -ы, (*f*) (cryst) super-group

с·верх·давл·ени·е, -я, (*n*) superpressure; overpressure

с·верх·даль·н·ий, -яя, -ее, (*a*) very long, very long-range

~ баллист·ическ·ая ракет·а (*f*) intercontinental ballistic missile

с·верх·дально·бой·н·ый, -ая, -ое, (*a*) (gunn) very long-range

с·верх·жёст·к·ий, -ая, -ое, (*a*) ultrahard

с·верх·за·тух·ани·е, -я, (*n*) over-damping

с·верх·звук·ов·ой, -ая, -ое, (*a*) super-sonic (above the speed of sound)

с·верх·комплект·н·ый, -ая, -ое, (*a*) supernumerary

с·верх·корон·а, -ы, (*f*) (astron) super-corona

с·верх·крит·ическ·ий, -ая, -ое, (*a*) supercritical

с·верх·лёгк·ий, -ая, -ое, (*a*) extra light

с·верх·лимит·н·ый, -ая, -ое, (*a*) (fin) extra-ordinary, extra-budgetary, supplementary (to quota/allowance/budget etc.)

с·верх·мал·ый, -ая, -ое, (*a*) midget

с·верх·мер·н·ый, -ая, -ое, (*a*) over-size, excessively large

с·верх·микро·скоп·, -а, (*m*) super/electron microscope

с·верх·мощ·н·ый, -ая, -ое, (*a*) superpowered, superpower

~ сет·ь (*f*) (elec) supergrid

с·верх·на·правл·енн·ый, -ая, -ое, (*a*) superdirectional; supergain (of antennas)

с·верх·низ·к·ий, -ая, -ое, (*a*) ultralow

~ част·от·а (*f*) ultralow frequency

с·верх·норматив·н·ый, -ая, -ое, (*a*) extra-ordinary, in excess of the standard/rate

с·верх·об·мен·н·ый, -ая, -ое, (*a*) superexchange

с·верх·плав·к·ий, -ая, -ое, (*a*) hyper-fusible

с·верх·план·ов·ый, -ая, -ое, (*a*) (econ) in excess of that planned, over-and-above the plan

с·верх·плоск·ост·ь, -и, (*f*) (math) hyperplane

с·верх·пре·ломл·ени·е, -я, (*n*) super-refraction

с·верх·при·бы·л·ь, -и, (*f*) (fin) excess/excessive profit

с·верх·про·вод·имост·ь, -и, (*f*) (phys) superconductivity

с·верх·про·вод·ник·, -а, (*m*) (phys) superconductor

с·верх·раз·мер·ы, -ов, (*pl*) (min) oversize

с·верх·раз·реш·ённ·ый, -ая, -ое, (*a*) (phys) superallowed

с·верх·ранн·ий, -яя, -ее, (*a*) (hortic) very early, winter

с·верх·рас·кач·к·а, -и, (*g pl*) -ч·ек·, (*f*) (mech eng) overdrive, overswing

с·верх·рас·чёт·н·ый, -ая, -ое, (*a*) in excess of the calculated/rated, overrating

~ на·пряж·ени·е (*n*) overrating voltage

с·верх·регенер·аци·я, -и, (*f*) (rad) super-regeneration/-reaction

с·верх·с·жат·и·е, -я, (*n*) overcompression

с·верх·скор·остн·ой, -ая, -ое, (*a*) ultra high-speed, ultraspeed, superspeed, ultrafast

~ кино·съём·к·а, (*f*) slow-motion cinematography, superspeed slow-motion cinematography

с·верх·скор·остн·ой, -а́я, -о́е
~ фото·гра́ф·и·я (f) ultra-/super-speed photography
с·верх·с·ме́т·н·ый, -ая, -ое, (a) extra-budgetary, supplementary (to estimates)
с·верх·со·пряж·е́ни·е, -я, (n) hyperconjugation
с·верх·сро́ч·ник·, -а, (m) (mil) extended-service man
с·верх·сро́ч·н·ый, -ая, -ое, (a) extended (of period)
~ слу́ж·б·а (f) (mil) extended service
с·верх·срочно·слу́ж·ащ·ий, -его, (m decl as adj) see с·верх·сро́ч·ник·
с·верх·стар·е́ни·е, -я, (n) (met) overaging
с·верх·сто́·имост·ь, -и, (f) excess cost
с·верх·структу́р·а, -ы, (f) (cryst) superstructure, superlattice
с·верх·твёрд·ый, -ая, -ое, (a) extra-hard, ultra hard
с·верх·тек·у́чест·ь, -и, (f) (phys) superfluidity
с·верх·тек·у́ч·ий, -ая, -ее, (a) (phys) superfluid
~ жи́дк·ост·ь (f) superfluid
с·верх·то́к·, -а, (m) (elec) excess current, overcurrent
с·верх·то́нк·ий, -ая, -ое, (a) (spectr, cryst) hyperfine; superfine, ultrafine
~ структу́р·а (f) (spectr) hyperfine structure
~ част·и́ц·ы (pl) ultrafines (of sieved materials); superfines (powder metallurgy)
с·верх·то́ч·н·ый, -ая, -ое, (a) super-precision; exceedingly accurate
с·верх·тяж·ёл·ый, -ая, -ое, (a) superheavy
с·ве́рх·у (adv, prep + gen) from above, from the top; on top, overhead
с·верх·у·пру́г·ий, -ая, -ое, (a) hyperelastic
с·верх·уро́ч·н·ый, -ая, -ое, (a) overtime; (pl as noun) overtime pay
~ рабо́т·а (f) overtime
с·верх·у·сил·е́ни·е, -я, (n) super-amplification
с·верх·фото·регистр·а́ци·я, -и, (f) moving-image photo-recording
с·верх·центри·фу́г·а, -и, (f) (chem) supercentrifuge
с·верх·чи́ст·ов·ый, -ая, -ое, (a) (mech eng) superfinishing, superfinish
~ об·рабо́т·к·а (f) honing

с·верх·чи́ст·ый, -ая, -ое, (a) (chem) ultra-pure; (mech eng) superfinished
с·верх·чувств·и́тельн·ый, -ая, -ое, (a) supersensitive, hypersensitive; (phot) ultraspeed, superspeed
~ регул·я́тор· (m) (instr) super-regulator
с·верх·шта́т·н·ый, -ая, -ое, (a) supernumerary, unestablished, surplus to establishment
сверч·ко́в·ые, -ых, (pl decl as adj) (ent) Grylloidea
сверч·о́к·, -ч·к·а́, (m) (ent) cricket, Gryllus
со́б·ственн·о сверч·к·и́ (pl) (ent) Gryllidae
с·вер·я́·ть, -ют, (imp) compare, check, collate
с·ве́с·, -а, (m) overhang
~, корм·ов·о́й (shipb) counter
у́гол· с·ве́с·а (m) droop angle (of helicopter blades)
с·ве́с·ить, -ят, (perf) see с·ве́ш·ива·ть
с·вес·ти́, (fut 3rd pl) с·вед·у́т, (past masc sing) с·вё·л, (perf) see с·во·д·и́ть
све́т·, -а, (m) light; world
в све́т·е in the light of, from the point of view of, considering
в свет·у́ (adv) inside (a hollow piece)
диа́метр· в свет·у́ (m) internal/inside diameter
~, за·лив·а́ющ·ий floodlight
~, за·полн·я́ющ·ий fill light
~, на·пра́вл·енн·ый parallel/collimated light
~, не·акти́в·н·ый non-actinic light
~, по·да·ю́щ·ий incident light
~, про·пу́щ·енн·ый transmitted/emergent light
~, про·ход·я́щ·ий transmitted/emergent light
свет·а́·ть, -ют, (imp) become/get light/lighter; dawn
свет·и́л·о, -а, (n) (astron) heavenly body; star, luminary (of persons)
свет·и́льн·ик·, -а, (m) lamp, lamp fitting, luminaire
свет·и́льн·ый, -ая, -ое, (a) illuminating, lamp
~ газ· (m) illuminating/town gas
свет·и́мост·ь, -и, (f) (astron) luminosity; brightness, luminance (light)
свет·и́ть, -́ят, (imp) shine, give light
светл·е́·ть, -ют, (imp) brighten, become lighter/light/brighter

све́тл·о-　　(*root*) light (of colours); bright

светло·бу́р·ый, -ая, -ое,　　(*a*) light brown

светл·ова́т·ый, -ая, -ое,　(*a*) lightish

светл·ови́н·а, -ы,　　(*f*) light/bright spot; (met) ghost, ferrite ghost, phantom line

светло·голо́в·ый, -ая, -ое,　(*a*) fair-haired

светло·голуб·о́й, -а́я, -о́е,　(*a*) very pale blue

светло·ка́т·анн·ый, -ая, -ое,　　(*a*) (met) cold/bright-rolled

светло·по́л·ый, -ая, -ое,　(*a*) (opt) bright-field

светло·серд·е́чн·ый чугу́н·　(*m*) whiteheart malleable cast iron

светл·от·а́, -ы́, (*f*) bright light; brightness (e.g. of colours)

светл·о́тн·ый, -ая, -ое,　(*a*) of **све·тл·от·а́**

све́тл·ый, -ая, -ое,　(*a*) light; lucid, clear; bright; white; (print) hollow (of rings, squares etc., as graphical symbols)

~ **вод·а́** (*f*) clear water

~ **нефте·проду́кт·ы** (*pl*) (oil) white oils

~ **о́т·жиг·** (*m*) (met) bright annealing

~ **ромб·и́к·** (*m*) (print) hollow diamond

светл·я́к·, -а, (*m*) (ent) glow-worm, *Lampyris noctiluca*

све́т·ност·ь, -и, (*f*) *see* **свет·и́мост·ь**

све́т·о- (*root*) light-, photo-

свето·бли́к·, -а,　(*m*) light pattern (sound recording)

свето·бо·я́зн·ь, -и, (*f*) (med) photophobia

свето·во́д·, -а, (*m*) (phys) light pipe/guide

свет·ов·о́й, -а́я, -о́е, (*a*) light, luminous, photo-

~ **год·** (*m*) (astron) light year

~ **давл·е́ни·е** (*n*) (phys) radiation pressure

~ **из·луч·е́ни·е** (*n*) radiation of light; (nucl) flash

~ **от·да́·ч·а** (*f*) *see* **свето·от·да́·ч·а**

~ **по·то́к·** (*m*) (phys) light/luminous flux

~ **рекла́м·а** (*f*) illuminated sign

~ **ста́д·и·я** (*f*) (bot) photostage

~ **су́мм·а** (*f*) (phys) light sum

~ **эне́рг·и·я** (*f*) (phys) luminous energy

свето·вы́·ход·, -а,　(*m*) light yield/output

свето·гра́мм·а, -ы,　(*f*) (telecom) light signal, heliogram

свето·дел·и́тел·ь, -я, (*m*) (opt) beam splitter

свето·за·ка́л·к·а, -и,　(*f*) (agr) light hardening

свето·из·мер·и́тельн·ый, -ая, -ое, (*a*) photometric

~ **ла́мп·а** (*f*) comparison lamp

свето·коп·ирова́льн·ый, -ая, -ое, (*a*) photocopying

~ **аппара́т·** (*m*) photocopying equipment, photocopier

свето·ко́п·и·я, -и,　(*f*) (phot) copy, photostat, photocopy

свето·куль́т·у́р·а, -ы,　(*f*) (hortic) artificial-light culture

свето·леч·е́ни·е, -я,　(*n*) (med) phototherapy

свето·люб·и́в·ый, -ая, -ое,　(*a*) (bot) luciphilous, photophilous

свето·маск·иро́вк·а, -и, (*g pl*) **-вок·,** (*f*) blacking out, blackout

свето·маск·иро́вочн·ый, -ая, -ое, (*a*) of **свето·маск·иро́вк·а**

~ **на·са́д·к·а** (*f*) dipper (for headlights)

свето·ма́сс·а, -ы, (*f*) phosphorescent/luminous compound

свето·мая́к·, -а́, (*m*) (air) light beacon, beacon light

свето·ме́р·, -а, (*m*) *and see* **фото́·-метр·**; light meter, photometer

свето·ме́р·н·ый, -ая, -ое, (*a*) photometric

~ **шар·** (*m*) sphere photometer

свето·модул·я́тор·, -а,　(*m*) (TV) light modulator

свето·не·про·ниц·а́ем·ый, -ая, -ое, (*a*) light-proof, opaque, light-tight

свето·но́с·н·ый, -ая, -ое, (*a*) luminous

свето·об·о·знач·е́ни·е, -я, (*n*) marking with lights (routes, roads etc.)

свето·ориент·а́ци·я, -и,　(*f*) (bot) phototropism

свето·от·да́·ч·а, -и, (*f*) luminous efficiency (of radiant energy); luminosity, luminosity factor, light efficiency, light output ratio (of light source)

свето·печа́т·ани·е, -я, (*n*) heliography

свето·по·глощ·е́ни·е, -я, (*n*) luminous absorption

све́то·пис·ь, -и, (*f*) photography

свето·план·, -а,   *(m)* (air nav) plan position indicator

свето·пре·ломл·éни·е, -я,   *(n)* (opt) refringence, refracting, refraction

свето·про·вóд·, -а,   *(m)* (phys) pipe/guide

свето·про·ниц·áем·ый, -ая, -ое, *(a)* transparent

свето·про·пуск·áемост·ь, -и,   *(f)* transparence

свето·про·пуск·áни·е, -я,   *(n)* light transmission

свето·прóч·н·ый, -ая, -ое, *(a)* fast (of colours); (phys) photostable

свето·рас·сé·ивающ·ий, -ая, -ее, *(a)* light-scattering (reflecting); light-diffusing (transmitting)

свето·сигнализ·áци·я, -и,   *(f)* flash signalling, flashing

свето·сигнáль·н·ая лáмп·а *(f)* signalling lamp

свето·сил·а, -ы,   *(f)* luminous power/intensity, light intensity, candlepower (where given in candle units); (phot) transmission; rapidity, speed, focal ratio; aperture ratio (of optical system)

свето·сúль·н·ый, -ая, -ое, *(a)* (phot) rapid

свето·со·бир·áющ·ий, -ая, -ее, *(a)* light-collecting

свето·со·стá·в, -а, *(m)* luminous compound, phosphor, luminophor

свето·стóй·к·ий, -ая, -ое, *(a)* photostable, light-proof, fast (dye)

свето·стóй·кост·ь, -и,   *(f)* light resistance

свето·тéн·ь, -и, *(f)* light-and-shade

свето·тéхн·ик·а, -и,   *(f)* illumination engineering

свето·фúльтр·, -а,   *(m)* light filter, (phot) filter

~, субтрактúв·н·ый (phot) subtractive filter

свето·фóр·, -а, *(m)* traffic light

свето·чувств·úтельност·ь, -и,   *(f)* (phot) sensitivity, speed (of films etc.); (biol) photosensitivity, light-sensitivity

свето·чувств·úтельн·ый, -ая, -ое, *(a)* (phot) photographic, sensitized; photosensitive, photoactive, light-sensitive

~ бумáг·а *(f)* sensitized/photographic paper

~ галóид·н·ая сол·ь *(f)* (chem) photohalide

свето·чувств·úтельн·ый, -ая, -ое

~ слó·й *(m)* (phot) emulsion, photographic emulsion, sensitive emulsion

~ элемéнт· *(m)* photocell

свет·я́щ·ий, -ая, -ее, *(a)* see also **свет·я́щ·ийся**; illuminating, lighting; **свéт·ящ·ий, -ая, -ее,** *(pres part act)* of **свет·úть**

~ бý·й *(m)* (naut) lit buoy

свет·я́щ·ийся, -аяся, -ееся, *(a)* luminous, phosphorescent, fluorescent

~ авиа·бóмб·а *(f)* (air) flare bomb

~ со·стá·в· *(m)* see **свето·со·стá·в·**

~ экрáн· *(m)* fluorescent screen

свеч·á, -й, *(i)* -ой, *(nom pl)* **свéч·и,** *(g pl)* **свеч·,** *or* **свеч·éй,** candle; (ICE) spark plug; (oil) string, stand, drill column; blow-off pipe (of gas producer); bleeder (of OH furnace etc.); (med) suppository

~ вос·пламен·úтел·я igniter sparking plug

~, за·пáль·н·ая (ICE) spark plug

~, между·на·рóд·н·ая (opt) international candle, candle

~ на·кал·ивáни·я heater plug (diesel)

~, пуск·ов·áя igniter plug (gas turbine)

~, слюд·ян·áя mica-insulated spark plug

~, с·мáз·очн·ая plug/candle of grease

свеч·éни·е, -я, *(n)* luminescence, glow; (zool) phosphorescence

~, втор·úчн·ое secondary emission

~, кáтод·н·ое cathode glow

~ кристáлл·ов crystalloluminescence

~ от тр·éни·я triluminescence

свéч·к·а, -и, *(g pl)* **-ч·ек·,** *(f)* = **свеч·á**

свеч·н·óй, -áя, -óе, *(a)* of **свеч·á**

с·вéш·енн·ый, -ая, -ое, *(past part pass)* see **с·вéш·ива·ть**; overhanging

с·вéш·ива·ть, -ют, *(imp)* let down, lower; **-ся** hang/lean down/over, overhang

с·вив·á·ть, -ют, *(imp)* twist, twine, plait, plat; twist (strand into ropes); wind (wires into strands); **-ся** *(pass)*; roll/coil up

с·вúв·к·а, -и, *(g pl)* **-в·ок·,** *(f)* *(v n)* see **с·вив·á·ть**; lay (of rope)

~, альбéрт·овск·ая Lang's/Albert lay (wire rope)

~, двóй·н·áя hawser lay (of rope)

~, крест·ов·áя regular/ordinary lay

~, один·áрн·ая spiral construction (of wire rope)

~, одно·сторóн·н·яя Albert/Lang's lay

~, прáв·ая right-hand lay

с·ви́в·к·а
~, **трой·н·а́я** cable lay (of rope)
с·вид·а́ни·е, -я, (*n*) meeting, appointment; visit
с·вид·е́тел·ь, -я, (*m*) eye-witness; (law) witness; (geol) outlier; blank/control sample
с·вид·е́тельств·о, -а, (*n*) evidence, testimony, statement; certificate, licence
~, **а́втор·ск·ое** (print) copyright, certificate of copyright
~, **до·смо́тр·ов·ое** (naut) survey report
~, **про·би́р·н·ое** (min, met) assay certificate
~, **суд·ов·о́е** (naut) certificate of registry
с·вид·е́тельств·овать, -уют, (*imp*) witness, bear witness, testify, certify; show, indicate, be indicative of
свя́зен·ь, -и, (*f*) *see* **свя́тен·ь**
с·вил·ева́тост·ь, -и, (*f*) wavy grain (in wood)
с·вил·ева́т·ый, -ая, -ое, (*a*) wavygrained, knaggy, knotty (of wood)
с·ви́л·ь, -и, (*f*) curve, knot (in wood); spill (on surface of a casting); cord, striae (in glass)
свин·а́рк·а, -и, (*g pl*) **-рок·,** (*f*) pig/hog keeper (woman)
свин·а́рник·, -а, (*m*) pig/hog sty/sties
свин·а́р·ь, -я, (*m*) pig/hog keeper
свин·е́ц·, -н·ц·а́, (*i*) **-н·ц·о́м,** (*m*) (chem) lead, plumbum, Pb
~, **азотно·ки́сл·ый** (chem) lead nitrate
~, **акти́н·иев·ый** (nucl) actinium lead, actinium D
~, **кра́сн·ый** (chem) red lead, minium
~, **сер·ни́ст·ый** (chem) lead sulphide, (min) galena, lead glance
~, **сурьм·яни́ст·ый** (met) antimonial/hard lead
~, **угле·ки́сл·ый** (chem) lead carbonate
~, **уксусно·ки́сл·ый** (chem) sugar of lead, lead acetate
~, **хромово·ки́сл·ый** (chem) lead chromate
~, **черн·ов·о́й** (met) base bullion, work lead
~, **щавелево·ки́сл·ый** (chem) plumbic oxalate
свинец·со·держ·а́щ·ий, -ая, -ее, (*a*) (min) lead-bearing, plumbiferous
свин·и́н·а, -ы, (*f*) pork
сви́н·к·а, -и, (*g pl*) **-н·ок·,** (*f*) (*dim*) of **свин·ь·я́**; (med) mumps; (met) pig
~, **мор·ск·а́я** guinea pig

свино·во́д·ств·о, -а, (*n*) pig/hog breeding/rearing/farming
свин·о́й, -а́я, -о́е, (*a*) pig, hog, swine; pork
свино·ма́т·к·а, -и, (*g pl*) **-т·ок·,** (*f*) sow (mature female hog)
с·винт·и́ть, -я́т, (*perf*) *see* **с·ви́нч·ива·ть**
свинц·ева́ни·е, -я, (*n*) lead plating/coating
свинц·ева́ть, -у́ют, (*imp*) plate/coat with lead
свинцовисто·ки́сл·ый, -ая, -ое, (*a*) (chem) plumbite (of)
~ **на́три·й** (*m*) sodium plumbite
свинц·о́вист·ый, -ая, -ое, (*a*) leadcontaining, leaded; (chem) plumbous
~ **кисл·от·а́** (*f*) plumbous acid
~ **лату́н·ь** (*f*) (met) leaded brass
свинц·о́вк·а, -и, (*g pl*) **-вок·,** (*f*) (bot) leadwort, *Plumbago*
свинцово·ки́сл·ый, -ая, -ое, (*a*) (chem) plumbic, plumbate (of)
свинцово·плав·и́льн·ый, -ая, -ое, (*a*) lead-smelting
свин·цо́в·ый, -ая, -ое, (*a*) lead, leaden, plumbic, saturine
~ **аккумул·я́тор·** (*m*) (elec) lead-acid accumulator
~ **бел·и́л·а** (*pl*) white lead, lead white
~ **блеск·** (*m*) (min) galena
~ **вод·а́** (*f*) (chem, pharm) diluted solution of sugar of lead
~ **глёт·** (*m*) (chem) litharge, lead monoxide
~ **кисл·от·а́** (*f*) (chem) plumbic acid
~ **крон·** (*m*) lead chromate, chrome yellow (pigment)
~ **от·ве́с·** (*m*) plumb line
~ **су́рик·** (*m*) red lead, minium
~ **эквивале́нт·** (*m*) (nucl) lead equivalent
свин·ча́тков·ые, -ых, (*pl decl as adj*) (bot) *Plumbaginaceae*
с·ви́нч·енн·ый, -ая, -ое, (*past part pass*) *see* **с·ви́нч·ива·ть**
с·ви́нч·ива·ть, -ют, (*imp*) screw, screw together/up; strip (a screw thread); unscrew
свинье·обра́з·н·ый, -ая, -ое, (*a*) pig-like, suiform
свин·ь·я́, -и́, (*nom pl*) **сви́н·ь·и,** (*g pl*) **свин·е́й,** (*f*) pig, hog, swine, (*pl*) pigs, *Suidae*
~, **мор·ск·а́я** (zool) common porpoise, *Phocoena phocoena*
~, **на·сто·я́щ·ая** pig, *Sus*

свин·ь·я́

~, реч·н·а́я   hippopotamus, *Potamochoerus*

~, со́б·ствен·о *(pl)* (zool) pigs, *Suinae*

свире́п·ый, -ая, -ое, *(a)* fierce, ferocious; violent

с·ви́с·, -а, *(m)* droop; **с·ви́с·** *(past masc sing)* of **с·ви́с·нуть**

с·вис·а́·ть, -ют, *(imp)* hang, hang down, dangle, droop

с·вис·а́ющ·ий, -ая, -ее, *(pres part act)* of **с·вис·а́·ть**; pendent

с·ви́с·л·ый, -ая, -ое, *(a)* drooping, pendent

с·ви́с·нуть, -нут, *(perf) see* **с·вис·а́·ть**

свист·, -а, *(m)* whistle (noise)

свист·а́ть, *(pres 3rd pl)* **сви́щ·ут,** *(imp)* whistle, (naut, orn) pipe

свист·е́ть, -я́т, *(imp)* = **свист·а́ть**

сви́ст·н·уть, -ут, *(perf)* give a whistle

свистов·, -а, *(m)* (naut) swifter

свист·о́к·, -т·к·а́, *(m)* whistle

свист·я́щ·ий, -ая, -ее, *(pres part act)* of **свист·е́ть**; whistling; sibilant

свит·а, -ы, *(f)* retinue, entourage; (geol) suite, series, formation

~, аркт·и́ческ·ая   (geol) Arctic suite

~, жи́ль·н·ая   vein series

~, нефте·но́с·н·ая   oil-bearing formation

~, сол·ян·а́я   salt series

сви́тен·ь, -и, *(f)* (naut) tail of rope, tail, lanyard

сви́тер·, -а, *(m)* (text) sweater

с·вит·о́к·, -т·к·а́, *(m)* roll, scroll

сви́т·ск·ий, -ая, -ое, *(a)* of **сви́т·а**

с·ви́·т·ый, -ая, -ое, *(past part pass) see* **с·вив·а́·ть**

с·ви·ть, *(fut 3rd pl)* **со·вь·ю́т,** *(perf) see* **с·вив·а́·ть**

свищ·, -а́, *(i)* **-о́м,** *(m)* air hole, hole; (med) fistula; warble hole (in leather); knot hole (in timber); *(pl)* honeycombing

свищ·ева́т·ый, -ая, -ое, *(a)* (biol) fistulous, fistular

свищ·ев·о́й, -а́я, -о́е, *(a)* of **свищ·**

сви́щ·ет, *(pres 3rd sing)* of **свист·а́ть**

свобо́д·а, -ы, *(f)* freedom, liberty

свобо́д·н·о *(adv)* free, freely

свободно·лепе́ст·н·ый, -ая, -ое, *(a)* (bot) apopetalous

свободно·пе́ст·ичн·ый, -ая, -ое, *(a)* (bot) apocarpous

свободно·пла́в·ающ·ий, -ая, -ее, *(a)* free-floating, (zool) planktonic

свободно·порш·нев·о́й, -а́я, -о́е, *(a)* free-piston

свобо́д·н·ый, -ая, -ое, *(a)* free

~ ём·кост·ь   *(f)* idle capacity (telephones)

~ кисл·от·а́   *(f)* (chem) free acid

~ ме́ст·о   *(n)* vacancy

~ от   free of, devoid of, without, un-, -free

~ от ава́р·и·и   (naut) free of average

~ от ис·каж·е́ни·й   undistorted, not distorted, without distortions

~ от слу́ж·б·ы   off-duty

~ теч·е́ни·е   *(n)* free/unrestricted flow

~ у́·ровен·ь   *(m)* (phys) empty band, conduction band

с·во́д·, -а, *(m)* arch, vault, roof (of interior space), (biol) fornix; code, code book; (rubber) hold-up, hold-up material

~, вееро·обра́з·н·ый   fan vaulting

~, гор·я́щ·ий   (expl) web

~, между·на·ро́д·н·ый   (telecom) international code

~, небе́с·н·ый   (astron) firmament, celestial concave

~ -оболо́ч·к·а, -и, *(f)* (arch) shell

~, распорно·а́роч·н·ый   braced arch roof

~, цилиндр·и́ческ·ий   (build) barrel/wagon/tunnel vault

с·вод·и́мост·ь, -и, *(f)* (math) reducibility

с·вод·и́ть, -я́т, *(imp)* lead/bring/take down; bring together/to, amalgamate; fell (trees); remove; (math) reduce to; numb; **-ся** amount to, come to, result in

~ на нет   (mech) fire away

~ к нул·ю́   reduce to zero, nullify

~ к ми́нимум·у   minimize, keep to a minimum

с·во́д·к·а, -и, *(g pl)* **-д·ок·,** *(f)* *(v n), see* **с·вод·и́ть**; summary, résumé, report; (print) revise

~ ка́рт·ы   (surv) map adjustment

~ лист·о́в   junction (of a map)

с·вод·н·о́й, -а́я, -о́е, *(a) see also* **с·во́д·-н·ый**; joinable, assemblable; joining, assembling

~ ма́л·к·а   *(f)* (shipb) close bevel

с·во́д·н·ый, -ая, -ое, *(a) see also* **с·вод·н·о́й**; summary; composite; step- (of kin)

с·во́д·ов·ый, -ая, -ое, *(a)* arch, vault, roof

с‧вод‧чат‧ый, -ая, -ое, (*a*) vaulted, arched, (biol) fornicate

сво‧ё, (*pron, nom/acc sing neut*) of сво‧й; own, one's own

свое‧воль‧н‧ый, -ая, -ое, (*a*) self-willed

свое‧времен‧н‧о (*adv*) opportunely, in good time, at the proper time

свое‧времен‧н‧ый, -ая, -ое, (*a*) opportune, timely; well/correctly timed

сво‧ей, (*pron, g, d, i, p, f sing*) of сво‧й

свое‧образ‧и‧е, -я, (*n*) peculiarity, originality

свое‧образ‧н‧ый, -ая, -ое, (*a*) peculiar, singular

сво‧его, (*g sing*) of сво‧й

с‧воз‧ить, ⸗ят, (*imp*) bring together/down, take, take down

сво‧й, (*nom pl*) of сво‧й

св‧ой, (*acc/gen sing*) сво‧его (*pron*) own, one's own, my, your, our, its, his, her, their; (*pl as noun*) one's own, one's own side

~ -чуж‧ой, (*not declined*) friend or foe

~ -чуж‧ой, о‧по‧зна‧ва́ни‧е I.F.F. (radar)

с‧войлач‧ивани‧е, -я, (*n*) (text) felting

свой‧ственност‧ь, -и, (*f*) peculiarity, singularity, feature, characteristic

свой‧ственн‧ый, -ая, -ое, (*a*) peculiar, characteristic; inherent, intrinsic; (+ *dat*) peculiar to, characteristic of, inherent in

свой‧ств‧о, -а, (*n*) feature, property, trait, quality, characteristic; свой‧ств‧о́, -а́, (*n*) relationship by marriage

~, внутр‧енн‧ие (*pl*) intrinsic properties

~, рабоч‧ие (*pl*) service properties

с‧волак‧ива‧ть, -ют, (*imp*) rake together

с‧волоч‧ь, (*fut 3rd sing, pl*) с‧волоч‧ёт, с‧волок‧у́т, (*past masc sing*) с‧волок‧, (*perf*); see с‧волак‧ива‧ть

с‧ворач‧ивани‧е, -я, (*n*) (*v n*) of с‧во‧рач‧ива‧ть; turning, swinging

с‧ворач‧ива‧ть, -ют, (*imp*) twist/swing off; knock/smash round

с‧ворот‧ить, ⸗ят, (*perf*) see с‧ворач‧ива‧ть

с‧вороч‧енн‧ый, -ая, -ое, (*past part pass*) see с‧ворач‧ива‧ть

сво‧я, (*pron, nom sing fem*) of сво‧й; own, one's own

с‧вулканиз‧ованн‧ый, -ая, -ое, (*a*) vulcanized

с‧вулканиз‧ова́ть, -у́ют, (*perf*) vulcanize

с‧вык‧а‧ться, -ются, (*imp*) get used/accustomed to

с‧вык‧нуться, -нутся, (*perf*) see с‧вы‧к‧а‧ться

с‧выш‧е (*adv*) from above; (*prep + gen*) beyond, over

с‧вяж‧ет, (*fut 3rd sing*) of с‧вяз‧а́ть, (*perf*); see с‧вяз‧ыва‧ть (*imp*)

с‧вяз‧анност‧ь, -и, (*f*) constriction; connectedness

с‧вяз‧анн‧ый, -ая, -ое, (*past part pass*) see с‧вяз‧ыва‧ть; (chem) fixed, combined; (phys) bound; coupled (*pl*)

~ азот‧ (*m*) (chem) fixed nitrogen, combined nitrogen

~ вод‧а́ (*f*) (biol) bound water

~ вод‧о́й waterbound

~ за‧ря́д‧ (*m*) (phys) bound charge

~ конту́р‧ы (*pl*) (elec) coupled circuits

не с‧вяз‧анн‧ый free

~ с водо‧ро́д‧н‧ой с‧вяз‧ью hydrogen-bonded

~ с ним associated, related

~ систе́м‧ы (*pl*) (rad) coupled circuits

~ со‧сто‧я́ни‧е (*n*) (phys) bound state

~ у‧равн‧е́ни‧я (*pl*) (math) coupled equations

с‧вяз‧а́ть, (*fut 3rd pl*) с‧вяж‧ут, (*perf*) see с‧вяз‧ыва‧ть

связ‧и́ст‧, -а, (*m*) signalman, communications man, communicator

с‧вяз‧к‧а, -и, (*g pl*) -з‧ок‧, (*f*) binding, lashing, bond, tie; bunch, bundle, sheaf; (anat) ligament, chord; strap (of a magnetron); (gram) copula

~, в‧ход‧н‧а́я (autom) input bundle

~, голос‧ов‧а́я (anat) vocal chord

~ о‧кру́ж‧ност‧ей (math) sheaf of circles

~ пло́ск‧ост‧ей (math) sheaf/bundle of planes

связко‧зу́б‧ые, -ых, (*pl decl as adj*) (pal) desmodonts, *Desmodonta*

с‧вяз‧ни́к‧, -а́, (*m*) (bot) connective

с‧вяз‧н‧о́й, -а́я, -о́е, (*a*) see also с‧вяз‧н‧ый; communications, liaison; connecting, fixing; (*m decl as adj*) (mil) runner

~ тру́б‧к‧а (*f*) stay tube (boiler)

с‧вяз‧ност‧ь, -и, (*f*) coherence, cohesion; connectivity, connectedness, connection; cohesiveness

с·вя́з·н·ый, -ая, -ое, (*a*) *see also* **с·вя́з·н·ой;** closely reasoned/argued; cohesive, coherent; (math) connected

~ грунт· (*m*) cohesive soil

~ мно́ж·еств·о (*n*) (math) connected set

~ о́бласт·ь (*f*) (math) connected region

с·вя́з·очн·ый, -ая, -ое, (*a*) of **с·вя́з·к·а;** (anat) ligamentous

~ резона́нс·н·ая систе́м·а (*f*) (phys) strapped resonant system

с·вяз·у́ющ·ий, -ая, -ее, (*pres part act*) *see* **с·вя́з·ыва·ть;** connecting; linking, coupling; binding, bonding; (*n decl as adj*) binder; bonding agent (glueing)

~ вещ·еств·о́ (*n*) binder, binding agent

~ глин·а́ (*f*) (chem) bonding clay

с·вя́з·ывани·е, -я, (*n*) (*v n*) of **с·вя́з·ыва·ть**

с·вя́з·ыва·ть, -ют, (*imp*) tie together, link, couple, connect; tie up, bind; (chem etc.) bind, bond, fix, combine; (text) knit, knit up; connect with, relate/attribute to (of facts); **-ся** (*pass*); communicate, get in touch with; be associated with

~ по·па́р·н·о couple

с·вя́з·ывающ·ий, -ая, -ее, *see* **с·вя́з·у́ющ·ий**

с·вя́з·ь, -и, (*f*) tie, bond, connection, link, relation; communications, telecommunications; liaison, contact; (elec) coupling; (chem, phys) bond; (mech eng) stay, rod, connection; (mech) constraint

~, авар·и́йн·ая emergency communications

~, а́нкер·н·ая tie/connecting rod

~, ано́д·н·ая (elec) anode coupling

~. а́том·н·ая (phys, chem) atomic/covalent bond

в с·вя́з·и́ regarding, reference, in connection with

~, вале́нт·н·ая (phys, chem) valence bond

~, взаи́м·н·ая interconnection; interrelation, interrelationship

~, водо·ро́д·н·ая (chem) hydrogen bond; hydrogen bridge

~, вре́д·н·ая (phys) spurious coupling

~, гальван·и́ческ·ая (elec) direct coupling

~, гетеро·поля́р·н·ая (phys, chem) heteropolar bond, ionic bond

с·вя́з·ь

~, геоме́тр·и́ческ·ая (mech) holonomic constraint

~, гомео·поля́р·н·ая (phys, chem) covalent/homopolar bond

~, да́ль·н·яя long-range communications

~, дати́в·н·ая (phys) inverse coordination bond

~, двой·н·а́я (chem) double bond

~, двух·сторо́н·н·яя two-way/either-way communication; (mech) bilateral constraint

~, диспе́тчер·ск·ая central control communications (railways etc.)

~, дли́н·н·ая through stay (boiler)

~, до·лолн·и́тельн·ая (autom) feed-forward

~, дро́ссель·н·ая (elec) impedance coupling

~, ём·костн·ая (elec) capacitance/capacitive coupling

~, ём·кост·о-ом·и́ческ·ая (elec) RC-coupling

~, жёстк·ая (eng) rigid connection; (autom) rigid path

~, из·бир·а́тельн·ая selective communication

~, индукти́в·н·ая (elec, rad) inductive/inductance/transformer coupling

~, индукти́в·н·о-ём·костн·ая (rad) reactance coupling

~, ио́н·н·ая (phys, chem) ionic bond

~, като́д·н·ая (rad) cathode coupling

~ кле́р·ом plain-language communication

~, кома́нд·н·ая (mil) command link

~, комбин·и́рованн·ая (telecom) composite communication link

~, кондукти́в·н·ая (elec) direct coupling

~, конъюг·и́рованн·ая (chem) conjugate double bond

~, коро́т·к·ая short stay (boiler)

коэффицие́нт· свя́з·и (*m*) (rad) coupling factor/coefficient

~, кра́т·н·ые (*pl*) (chem) multiple bonds

~, криста́лл·и́ческ·ая (cryst) crystal bonds

ли́н·и·я свя́з·и (*f*) communications link

~, магни́т·н·ая (elec) magnetic inductive coupling

~, магнито·стрикцио́н·н·ая (autom) magnetostrictive coupling

с·вя́з·ь

~, **металл·и́ческ·ая**   (phys, chem) metallic bond

~, **механ·и́ческ·ая** (mech) constraint; mechanical linkage/connection

~, **мно́го·кра́т·н·ая**   (phys) multiple bond/linkage/link; (telecom) multiple-channel connection

~, **молекуля́р·н·ая**   (chem) molecular bond

~, **на·пра́вл·енн·ая** (rad) beam communication

~, **не·по·сре́д·ственн·ая** (elec) direct coupling

~, **не·станциона́р·н·ая** (mech) moving constraint

~, **об·ме́н·н·ая** (phys) exchange coupling/interaction

~, **обра́т·н·ая**   (elec) feedback; (rad) reaction, coupling reaction, regenerative coupling

   с обра́т·н·ой с·вя́з·ью   self-saturating

~, **ордина́р·н·ая** (chem, phys) single bond

~, **пере·хо́д·н·ая** (telecom) cross coupling

~, **поля́р·н·ая** (phys, chem) polar bond

~, **по·пере́ч·н·ая** (phys, chem) crosslink; (civ eng) cross-bracing

~, **по·сто·я́нн·ая**   fixed communications

~, **почт·о́в·ая** postal communications, post

~, **про·во́д·н·ая** line communications

~, **про·до́ль·н·ая**   (civ eng) lateral bracing/tie

~, **прям·а́я**   (elec) direct coupling; tie-line (telephones); direct communication; (autom) direct path

~, **про·из·во́д·ственн·ая**   industrial internal communications

~, **прост·а́я**   (chem) single bond; simple connection

~, **про́ч·н·ая** firm/close link/connection; (chem) stable bond

~, **радиа́ль·н·ая**   (civ eng) radial bracing; (mil telecom) rearward communications

~, **ра́дио·про·во́д·н·ая** (rail) inductive communication

~, **рас·пре·дел·ённ·ая** (phys) space coupling

~, **реоста́т·н·ая** (elec) resistance coupling

~, **рока́д·н·ая** (mil) lateral communications

с·вя́з·ь

~ **само·лёт·а с земл·е́й**   air-to-ground communications

~, **само·лёт·н·ая вну́тр·енн·яя**   aircraft intercom

~, **селе́ктор·н·ая** selective telephony

~, **си́ль·н·ая**   close connection/relationship; tight/close/strong coupling, over-coupling; tight bond

~, **си́мплекс·н·ая** simplex communications

~, **синхро́н·н·ая** synchro system

~, **склероно́м·н·ая** (mech) stationary constraint

~, **сла́б·ая**   weak relation, slight connection; (phys) weak/loose coupling, undercoupling

~, **спин-орбита́ль·н·ая** (phys) spin-orbit coupling

~, **стациона́р·н·ая** (mech) stationary constraint

   тео́р·и·я с·вя́з·и (f) theory of communications

~, **трансформ·а́торн·ая** (elec) transform coupling, inductive coupling

~, **у·де́рж·ивающ·ая**   (mech) bilateral constraint

~, **хим·и́ческ·ая** chemical bond/binding

~, **циркуля́р·н·ая**   conference call/communication (telephones)

~, **цифр·ов·а́я** (autom) digital communication

~, **электр·и́ческ·ая** telecommunications

~, **электро·вале́нт·н·ая** (phys, chem) ionic/electrovalent bond

~, **электро́н·н·ая**   electron coupling (of circuits); beam coupling (in a vacuum tube)

   эне́рги·я с·вя́з·и (f) binding/coupling energy

свящ·е́нн·ый, -ая, -ое, (a) sacred

с·га́р·к·и (pl) slag

с·ги́б·, -а, (m) (v n), see **с·гиб·а́·ть;** (geol) flexure, folding

с·гиб·а́ем·ый, -ая, -ое, (pres part pass) of **с·гиб·а́·ть;** flexible, pliable, pliant

с·гиб·а́тел·ь, -я, (m) (biol) flexor

с·гиб·а́ни·е, -я, (n) (v n) of **с·ги·б·а́·ть**

с·гиб·а́·ть, -ют, (imp) bend, fold, bend/fold over/round

с·гиб·а́ющ·ийся, -аяся, -ееся, (pres part act) see **с·гиб·а́·ть;** flexible, pliable, pliant

с·глáд·ить, -ят, (*perf*) *see* с·глáж·ива·ть

с·глáж·енн·ый, -ая, -ое, (*past part pass*) *see* с·глáж·ива·ть

с·глáж·ивани·е, -я, (*n*) (*v n*), *see* с·глáж·ива·ть; (elec) smoothing коэффициéнт· с·глáж·ивани·я (*n*) (elec) smoothing factor

с·глáж·ивател·ь, -я, (*m*) smoother, flattener

с·глáж·ива·ть, -ют, (*imp*) smooth, smooth/iron out

с·глáж·ивающ·ий, -ая, -ее, (*pres part act*) of с·глáж·ива·ть

~ у·стрóй·ств·о (*n*) (elec) smoothing equipment

~ фильтр· (*m*) (elec) smoothing/ripple filter, smoother

с·гнá·ива·ть, -ют, (*imp*) let rot, leave to rot

с·гнив·á·ть, -ют, (*imp*) rot, decay, putrefy, rot to pieces

с·гни·ть, -ют, (*perf*) *see* с·гнив·á·ть

с·гно·ённ·ый, -ая, -ое, (*past part pass*) *see* с·гнá·ива·ть

с·гно·и́ть, -я́т, (*perf*) *see* с·гнá·и·ва·ть

с·гон·, -а, (*m*) (*v n*), *see* с·гон·я́·ть; joint-release pipe thread

с·гóн·ит, (*fut 3rd sing*) of со·гн·áть (*perf*) *see* с·гон·я́·ть (*imp*)

с·гóн·н·ый, -ая, -ое, (*a*) of сгóн·; flotation, floated (timber); (chem) distilled

~ мýфт·а (*f*) screwed pipe union

~ резервуáр· (*m*) rundown tank

с·гóн·щик·, -а, (*m*) log driver

с·гон·я́·ть, -ют, (*imp*) drive/chase off/away; float down (wood on river); round up (herds etc.), drive together

с·гор·áем·ый, -ая, -ое, (*pres part pass*) of с·гор·á·ть; combustible

с·гор·áни·е, -я, (*m*) (*v n*) of с·го·р·á·ть; combustion, burning; burn-up (of nuclear fuel)

~, внýтр·енн·ее internal combustion двúг·ател·ь внýтр·енн·его с·го·р·áни·я (*m*) internal combustion engine

кáмер·а с·гор·áни·я (*f*) combustion chamber

продýкт·ы с·гор·áни·я (*pl*) combustion gases

тепл·от·á с·гор·áни·я (*f*) heat of combustion

с·гор·á·ть, -ют, (*imp*) burn, be burnt out/down/up; (bot) be dried up

с·гóрбл·енн·ый, -ая, -ое, (*past part pass*); hunched, crooked

с·гор·é·ть, -я́т, (*perf*) *see* с·гор·á·ть

сграффúто (*n indecl*) sgraffito, graffito, scratch work

с·греб·á·ть, -ют, (*imp*) rake together/up

с·грес·ти́, (*fut 3rd pl*) с·греб·ýт, (*past masc sing*) с·грёб·, (*perf*); *see* с·греб·á·ть

с·груд·и́ть, -я́т, (*perf*) *see* с·грý·ж·ива·ть

с·груж·á·ть, -ют, (*imp*) unload

с·грýж·енн·ый, -ая, -ое, (*past part pass*) *see* с·грýж·ива·ть and с·гру·ж·á·ть

с·грýж·ива·ть, -ют, (*imp*) bunch, pile up; -ся crowd in

с·груз·и́ть, -я́т, (*perf*) *see* с·гру·ж·á·ть

с·групп·ирóванн·ый, -ая, -ое, (*past part pass*) of с·групп·ир·овáть

с·групп·ировáни·е, -я, (*n*) (*v n*) of с·групп·ир·овáть

с·групп·ир·овáть, -ýют, (*perf*) group, bunch, bank; classify

с·грыз·á·ть, -ют, (*imp*) gnaw/nibble up

с·грыз·ть, -ýт, (*perf*) *see* с·грыз·á·ть

сгс (*abbr*) (phys, elec, eng) cgs, centimetre-gram-second

с·густ·и́тел·ь, -я, (*m*) thickener; (paper) decker, concentrator, thickener

с·густ·и́ть, -я́т, (*perf*) *see* с·гущ·á·ть

с·гýст·ок·, -т·к·а, (*m*) clot, coagulate, coagulum; (phys) cluster (of electrons), bunch (of particles)

~, плáзм·енн·ый (phys) plasmoid

с·гущ·áемост·ь, -и, (*f*) condensability

с·гущ·á·ть, -ют, (*imp*) thicken, condense, concentrate; coagulate, clot

~ контрáст· (*m*) (print) deepen the tone/contrast

с·гущ·éни·е, -я, (*n*) (*v n*), *see* с·гу·щ·á·ть; condensation; crowding (of parallel lines)

~ со·сто·я́ни·й (phys) crowding of levels/states

тóч·к·а с·гущ·éни·я (*f*) (math) condensation point (of a set)

~, фáз·ов·ое (phys) phase bunching

с·гущ·ённ·ый, -ая, -ое, (*past part pass*) *see* с·гущ·á·ть

с·да·вáть, -ю́т, (*imp*) hand over/in, deliver; pass (an exam/test); surrender, yield, give up

с·да·ва́ть
~ в аре́нд·у lease
~ в·на·ём· hire out; lease, let
с·дав·и́ть, ₋ят, (*perf*) *see* с·да́вл·и·ва·ть
с·да́вл·енн·ый, -ая, -ое, (*past part pass*) *see* с·да́вл·ива·ть
с·да́вл·ива·ть, -ют, (*imp*) squeeze, compress, constrict, constrain
с·да́·нн·ый, -ая, -ое, (*past part pass*) *see* с·да·ва́ть
с·да́т·очн·ый, -ая, -ое, (*a*) of с·да́·ч·а
~ ис·пыт·а́ни·е (*n*) delivery test/trials; (shipb) acceptance trials
с·да́т·чик·, -а, (*m*) delivery agent
с·да·ть, (*fut 1st, 3rd sing*) с·да·м, с·да·ст, (*fut 3rd pl*) с·дад·у́т, (*past masc sing*) с·да́·л, (*perf*); *see* с·да·ва́ть
с·да́·ч·а, -и, (*i*) -ей, (*f*) (*v n*), *see* с·да·ва́ть; delivery, handing-over; surrender; change (money)
    срок· с·да́·ч·и (*m*) delivery date, completion date
с·два́·ивани·е, -я, (*n*) (*v n*), *see* с·два́·ива·ть; duplication, repetition
с·два́·ива·ть, -ют, (*imp*) double, double up; connect two; repeat, duplicate, make a second
с·двиг·, -а, (*m*) displacement, shift, (mech) shear, shearing, pure shear, slip; (autom) offset
~ ви́д·а (elec) mode shift
~, генет·и́ческ·ий genetic drift
~, двой·нико́в·ый (met) twinning
~, изото́п·н·ый (nucl) isotope shift
    ис·пыт·а́ни·е на с·двиг· (*n*) (mech) shear test
    мо́дул·ь с·дви́г·а (*m*) (mech) shear modulus, modulus of elasticity in shear, modulus of transverse elasticity, modulus of rigidity
~, нул·ев·о́й zero offset/drift
~, одно·ро́д·н·ый (met) homogeneous shear
~ орби́т·ы (nucl) orbital shift, orbit shift
    про́ч·ност·ь на с·двиг· (*f*) (mech) shearing strength
    реги́стр с·дви́г·а (*m*) shift register (computers)
    со·против·ле́ни·е с·двиг·у (*n*) shear strength
    у́гол· с·дви́г·а (*m*) angle of shift/displacement
    у́гол· сдви́г·а фаз· (*m*) angle of phase difference, phase angle

с·двиг
~ у́·ровн·я (phys) level shift
~ фаз· (phys) phase shift/difference
с·двиг·а́ни·е, -я, (*n*) (*v n*), *see* с·двиг·а́·ть; displacement
с·двиг·а́тел·ь, -я, (*m*) shift unit (computers)
~, в·ход·н·о́й input shift unit
с·двиг·а́·ть, -ют, (*imp*) move, shift, displace, move off; move together/nearer
с·двиг·а́ющ·ий, -ая, -ее, (*pres part act*) of с·двиг·а́·ть
~ си́л·а (*f*) shearing force
с·дви́г·ов·ый, -ая, -ое, (*a*) of с·двиг·
с·движ·н·о́й, -а́я, -о́е, (*a*) movable; folding
с·дви́·нут·ый, -ая, -ое, (*past part pass*) *see* с·двиг·а́·ть; displaced, shifted, offset; out-of-line
~ нул·ь (*m*) (instr) false zero
~ по фаз· out-of-phase
~ со·в·пад·е́ни·е (*n*) delayed coincidence
с·дви́·н·уть, -ут, (*perf*) *see* с·двиг·а́·ть
с·дво·е́ни·е, -я, (*n*) *see* с·два́·ива·ни·е
с·дво́·енн·ый, -ая, -ое, (*past part pass*) *see* с·два́·ива·ть; double, binary, duplex, (cryst) twinned
с·дво·и́ть, -я́т, (*perf*) *see* с·два́·ива·ть
с·де́л·анн·ый, -ая, -ое, (*past part pass*) of с·де́л·а·ть; manufactured
с·де́л·а·ть, -ют, (*perf*) make, do, produce
с·де́л·а·ться, -ются, (*perf*) (+ *instr*) become, be getting; happen, be going on
с·де́л·к·а, -и, (*g pl*) -л·ок·, (*f*) deal, transaction
с·де́л·очн·ый, -ая, -ое, (*a*) of с·де́л·к·а
с·де́ль·н·о (*adv*) by the piece
с·де́ль·н·ый, -ая, -ое, (*a*) piece-work, piece-rate
~ о·пла́т·а труд·а́ (*f*) piece-rate system (of payment)
~ рабо́т·а (*f*) piece work
с·де́ль·щик·, -а, (*m*) piece-rate worker, piece worker
с·де́ль·щин·а, -ы, (*f*) piece work
с·дёрг·ива·ть, -ют, (*imp*) pull off
с·дер·ёт, (*fut 3rd sing*) of со·др·а́ть, (*perf*); *see* с·дир·а́·ть (*imp*)

с·дёрж·анн·ый, -ая, -ое, (*past part pass*) *see* с·дёрж·ива·ть; reserved; moderated

с·держ·а́ть, ⸗ат, (*perf*) *see* с·дёрж·ива·ть

с·дёрж·ива·ть, -ют, (*imp*) restrain, repress, hold back, pull up, check

с·дёр·н·уть, -ут, (*perf*) *see* с·дёрг·ива·ть

с·дир· silk cocoon

с·дир·а́·ть, -ют, (*imp*) strip, peel, bark, skin, flay, tear off

сдо́б·н·ый, -ая, -ое, (*a*) (food) milk-dough

~ бу́л·к·а (*f*) bun

с·дрейф·ова́ть, -у́ют, (*perf*) *see* дрейф·ова́ть

с·дубл·и́рованн·ый, -ая, -ое, (*past part pass*) of с·дубл·и́р·овать; doubled; plied-up (of tires)

с·дубл·и́р·овать, -уют, (*perf*) double; ply up (tires)

с·дув·, -а, (*m*) (*v n*), *see* с·дув·а́·ть; (aerodynam) suction method of boundary layer control

с·дув·а́·ть, -ют, (*imp*) blow away/off; blow together

с·ду́в·к·а, -и, (*g pl*) -в·ок·, (*f*) (*v n*), *see* с·дув·а́·ть; (paper) blower

с·дув·н·о́й, -а́я, -о́е, (*a*) blowoff

~ кран· (*m*) blowoff cock

с·ду́в·очн·ый, -ая, -ое, (*a*) of с·ду́в·к·а

с·ду·ть, ⸗ют, (*perf*) *see* с·дув·а́·ть

сеа́нс·, -а, (*m*) (theat etc.) performance, house; session, sitting

себака́т·, -а, (*m*) (chem) sebacate

себациново·ки́сл·ый, -ая, -ое, (*a*) sebacate (of)

себаци́н·ов·ый, -ая, -ое, (*a*) (chem) sebacic

~ кисл·от·а́ (*f*) sebacic acid
   эфи́р· себаци́н·ов·ой кисл·от·ы́ (*m*) sebacate

себе́ (*pron*), (*dat/prep*) *see* себя́

себе·сто́·имост·ь, -и, (*f*) (econ) cost price, cost of production, cost

себя́ (*pron*) (*acc/gen*) (him, it, your etc.) self

се́·в·, -а, (*m*) sowing

сев·а́лк·а, -и, (*g pl*) -лок·, (*f*) seed cod

се́вер·, -а, (*m*) north

се́вер·н·ее (*adv*) more northerly, further north, to the north of

се́вер·н·ый (*a*) north, northern

се́вер·о-вос·то́ч·н·ый (*a*) north-east, north-eastern, north-easterly

се́вер·о-вос·то́ч·н·ый

~ -за́·пад·н·ый, -ая, -ое, (*a*) north-west, north-western, north-westerly

север·я́нин·, -а, (*nom pl*) север·я́н·е, (*g pl*) север·я́н· (*m*) northerner

се́в·н·ый, -ая, -ое, (*a*) crop

сево·оборо́т·, -а, (*m*) (agr) crop rotation

севрю́г·а, -и, (*f*) (fish) sturgeon (type of), *Acipenser stellatus*

се́·вш·ий, -ая, -ее, (*past part act*) of сест·ь, (*perf*); *see* сад·и́ться (*imp*)

се́гарс·, -а, (*m*) (naut) mast loop

сегме́нт·, -а, (*m*) segment; (zool) merome, metamere, segment

~, жа́бер·н·ый (zool) branchiomere

~ ног·и́ (zool) podomere

~ нож·а́ knife/sickle section (of a harvesting machine)

сегмент·а́ци·я, -и, (*f*) segmentation, metamerism

сегме́нт·н·ый, -ая, -ое, (*a*) segmental, metameric

~ про́фил·ь (*m*) (met) half-round (steel section)

сегнето·акти́в·н·ый, -ая, -ое, (*a*) ferroelectric

сегнет·ов·ая сол·ь (*f*) (chem) Seignette's salt, Rochelle salt

сегнето·кера́м·ик·а, -и, (*f*) ferroelectric ceramic

сегнето·эле́ктр·ик·, -а, (*m*) (phys) ferroelectric (material)

сегнето·электр·и́ческ·ий, -ая, -ое, (*a*) ferroelectric

сегнето·электр·и́честв·о, -а, (*n*) ferroelectrics

сего́·дн·я today

~ у́тр·ом this morning

сего́дня·шн·ий, -яя, -ее, (*a*) today's

сего·ле́т·к·а, -и, (*f*) fish under one year old

сегрег·а́т·, -а, (*m*) segregate

сегрег·а́ци·я, -и, (*f*) (*and see* ликв·а́·ци·я) segregation

сегрег·и́ровани·е, -я, (*n*) segregation

сед·а́лищ·е, -а, (*i*) -ем, (*n*) (anat) buttock; seat, seating

сед·а́лищн·ый, -ая, -ое, (*a*) (anat) sciatic, ischial, ischiadic

~ кост·ь (*f*) (zool) ischium

сед·ёлк·а, -и, (*g pl*) -лок·, (*f*) saddle; collar beam

сед·е́льн·ый, -ая, -ое, (*a*) saddle

~ при·це́п· (*m*) (M/T) semi-trailer

сед·е́·ть, -ют, (*imp*) become/grow grey

седимент·а́ци·я, -и, (*f*) sedimentation

седименто·метр·и́ческ·ий, -ая, -ое, (*a*) sedimentometric

седл·а́·ть, -ют, (*imp*) saddle

сед·л·о́, -а́, (*nom pl*) **сёд·л·а,** (*g pl*) **сёд·ел·** (*n*) saddle; seat (of valve), foot (of boiler); (meteor) col, saddleback

~, до·полн·и́тельн·ое (geol) adventive saddle

~, сифо́н·н·ое (geol) siphonal saddle

седл·ова́тост·ь, -и, (*f*) (shipb) saddle backing

седл·ова́т·ый, -ая, -ое, (*a*) saddleback, saddlebacked

седло·ви́д·н·ый, -ая, -ое, (*a*) saddle-shaped, selliform, sellaeform

седл·ови́н·а, -ы, (*f*) saddleback; (meteor) col, saddleback; (geol) saddle; saddle point, trough (of a surface, curve etc.)

седл·о́в·ый, -ая, -ое, (*a*) of **седл·о́**

~ то́ч·к·а (*f*) trough, saddle (on a graph)

седло·обра́з·н·ый, -ая, -ое, (*a*) saddle-shaped, selliform, sellaeform

сед·о- (*root*) grey, gray

сед·ова́т·ый, -ая, -ое, (*a*) greyish, grayish

сед·о́й, -а́я, -о́е, (*a*) grey, gray

седьм·о́й, -а́я, -о́е, (*a*) seventh

се́·ет, (*pres 3rd sing*) of **се́·ять**

сеза́м·, -а, (*m*) *see* **кунжу́т·;** sesame

се́зен·ь, -и, (*f*) (naut) gasket, becket, furling line

сезо́н·, -а, (*m*) season

сезо́н·ност·ь, -и, (*f*) seasonal fluctuation/occurence/prevalence

сезо́н·н·ый, -ая, -ое, (*a*) season, seasonal

сей (*pron*), (*g*) **сего́** this
    при сём herewith

сейва́л·, -а, (*m*) (zool) sei whale, Rudolph's rorqual, *Balaenoptera borealis*

сеймуриа·мо́рф·ы, -ов, (*m pl*) (pal) seymouriamorphs, *Seymourismorpha*

сейму́ри·я, -и, (*f*) (pal) Seymouria

се́йн·а, -ы, (*f*) (fish) seine net

се́йнер·, -а, (*m*) seiner (fishing vessel)

сейсм·, -а, (*m*) (geophys) seism

сейсм·и́ческ·ий, -ая, -ое, (*a*) (geophys) seismic

~ звук·ов·ы́е колеб·а́ни·я (*pl*) seismic transients

~ про́фил·ь (*m*) seismic section

сейсм·и́чност·ь, -и, (*f*) (geophys) seismicity

сейсмо·гра́мм·а, -ы, (*f*) seismogram, seismograph recording/trace

сейсмо́·граф·, -а, (*m*) (geophys) seismograph

сейсмо·гра́ф·и·я, -и, (*f*) seismography

сейсмо·карота́ж·, -а, (*m*) seismic well-logging

сейсмо·ло́г·и·я, -и, (*f*) seismology

сейсмо́·метр·, -а, (*m*) seismometer

сейсмо·наст·и́·я, -и, (*f*) (bot) seismonasty

сейсмо·при·ём·ник·, -а, (*m*) seismic detector

сеймо·профил·и́ровани·е, -я, (*n*) seismic profile shooting

сейсмо·раз·ве́д·к·а, -и, (*f*) seismic prospecting

сейсмо·сто́й·к·ий, -ая, -ое, (*a*) (civ eng) earthquake-proof/-resistant

сей-та́ли (*pl*) (naut) stay tackle

се́·йте (*plur imper*) of **се́·ять**

сейф·, -а, (*m*) safe, strong box

сей·ча́с·, (*adv*) immediately, now; just now

сейш·а, -и, (*f*) (geophys) seiche, standing wave

~, дву·узл·ов·а́я binodal seiche

сек. (*abbrev*) = **секу́нд·а** sec., second

сек·, (*past masc sing*) of **сеч·ь**

секале́н·, -а, (*m*) (pharm) secalen

се́канс·, -а, (*m*) (math) secant

сека́тор·, -а, (*m*) (hort) secateurs

сек·а́ч·, -а, (*i*) **-о́м,** (*m*) cutter; (zool) young boar; young male fur seal

секве́стр·, -а, (*m*) (med) sequestrum; sequestration

секвестр·а́ци·я, -и, (*f*) sequestration

секво́й·я, -и, (*f*) (bot) sequoia

сек·и́те, (*plur imper*) of **сеч·ь**

се́к·ли, (*past pl*) of **сеч·ь**

сек·о́м·ый, -ая, -ое, (*pres part pass*) of **сеч·ь**

секре́т·, -а, (*m*) secret; (text) carder; (biol) secretion; secrecy device/mechanism, secret lock/spring/catch

секретариа́т·, -а, (*m*) secretariat

секрета́р·ск·ий, -ая, -ое, (*a*) secretarial, secretary's

секрета́р·ь, -я́, (*m*) secretary; (zool) secretary bird

секрети́н·, -а, (*m*) (zool) secretin

секре́т·н·о (*adv*) secretly, covertly; (as classification) secret, confidential

секре́т·ност·ь, -и, (*f*) secrecy

секре́т·н·ый, -ая, -ое, (*a*) secret, covert

~ за·мо́к· (*m*) combination lock

секрето́р·н·ый, -ая, -ое, (*a*) (biol) secretary

~ жи́дк·ост·ь (*f*) (biol) secretion

секрецио́н·н·ый, -ая, -ое, (*a*) secretory

секре́ци·я, -и, (*f*) (biol) secretion

Сексмит·а, вес·ы́ (*pl*) (instr) Sucksmith balance

сексо·ло́г·и·я, -и, (*f*) (med) sexology

се́кст·а, -ы, (*f*) sixth (in music)

секста́н·, -а, (*m*) (nav instr) sextant

~ пузыр·к·о́м bubble sextant

секста́нт·, -а, (*m*) = **секста́н·**

сексто́л·, -а, (*m*) (chem) sextol, cyclohexanol

сексу·а́льн·ый, -ая, -ое, (*a*) sexual

се́ктор·, -а, (*nom pl*) **-ы,** or **-а́,** (*m*) sector (geometrical); segment, quadrant; division (organizational); (naut) log ship; mandrel (for vulcanizing tires)

~ га́з·а (a/c) throttle quadrant

~, зуб·ча́т·ый (mech eng) sector gear

~, калибр·о́ванн·ый (met roll) grooved die (of Rockrite machine)

~ кули́с·ы (mech eng) slotted link

~ подъ·ём·н·ый (surv etc.) elevating arc

~, рас·пре·дел·и́тельн·ый distributor segment (of magnets)

~, рул·ев·о́й (naut) rudder quadrant

~, черв·я́чн·ый worm sector (of gear)

се́ктор·н·ый, -ая, -ое, (*a*) sector, quadrant, sectoral

секуля́р·н·ый, -ая, -ое, (*a*) (math) secular

секу́нд·а, -ы, (*f*) second (of time or arc)

~, угл·ов·а́я second of arc

секунд·а́нт·, -а, (*m*) second (assistant to contestant)

секу́нд·н·ый, -ая, -ое, (*a*) second (of time); per second

~ объ·ём· (*m*) volume per second

секундо·ме́р·, -а, (*m*) stopwatch

~, двух·стре́л·очн·ый split-hand stopwatch

сек·у́щ·ий, -ая, -ее, (*pres part act*) of **сеч·ь;** cutting, intersecting, (geol) cross-cutting; (*f as noun*) (math) secant

секцион·и́ровани·е, -я, (*n*) sectionalization, subdivision, graduation

секцион·и́рованн·ый, -ая, -ое, (*past part pass*) sectioned, sectional, subdivided, sectionalized; graduated; tapped (of vacuum tubes)

~ кат·у́шк·а (*f*) (rad) tapped/sectional coil

секцио́н·н·ый, -ая, -ое, (*a*) sectional, section; (med) autopsy

~ на·со́с· (*m*) ring-type pump

се́кци·я, -и, (*f*) section; (med) section, autopsy

~, блок- (shipb) box-section

~, кольц·ев·а́я объ·ём·н·ая (shipb) hull box-section

~, объ·ём·н·ая (shipb) double-sided section

~, пло́ск·остн·ая (shipb) panel

сек·ш·ий, -ая, -ее, (*past part act*) of **сеч·ь**

се·л, (*past masc sing*) of **сест·ь** (*perf*), *see* **сад·и́ться** (*imp*)

селагине́лл·а, -ы, (*f*) (bot) club moss, *Selaginella*

села́хи·и (*pl*) (zool) selachians, *Selachii*

сел·ев·о́й, -а́я, -о́е, (*a*) of **сел·ь;** (geog) run-off, torrent·

~ во́д·ы (*pl*) (geog) run-off waters

селёд·к·а, -и, (*g pl*) **-д·ок·,** (*f*) = **сельд·ь**

селёд·очн·ый, -ая, -ое, (*a*) = **сельд·я·н·о́й, се́льд·ев·ый**

селезён·к·а, -и, (*g pl*) **-н·ок·,** (*f*) (anat) spleen

селезён·очн·ый, -ая, -ое, (*a*) splenic, splenetic

се́лез·ен·ь, -з·н·я, (*m*) (orn) drake

селекти́в·н·о-о·чищ·ённ·ый, -ая, -ое, (*a*) (oil) selective-refined

селекти́в·ность·, -и, (*f*) selectivity

селекти́в·н·ый, -ая, -ое, (*a*) selective

~ рас·твор·и́тел·ь (*m*) (oil) selective solvent

селект·и́ровани·е, -я, (*n*) (rad) strobing

селе́ктор·, -а, (*m*) selector, selector switch; gate (electronic device); *and see* **иска́тель** (telephones)

~, времен·н·о́й time gate

~ и́мпульс·ов, амплиту́д·н·ый (elec) pulse-height selector

~, криотро́н·н·ый cryotron gate

~ скор·ост·е́й velocity selector

~, спектра́ль·н·ый spectrum selector

селе́ктор·н·ый, -ая, -ое, (*a*) of **селе́ктор·**

~ и́мпульс· (*m*) (elec) gate, gate/strobe pulse

~ с·вяз·ь (*f*) selective telephony

~ у·правл·е́ни·е (*n*) (elec) gate control

селектро́н·, -а, (*m*) (elec) selectron

селекцион·е́р·, -а, (*m*) (agr) selector; breeder

селекцио́н·н·ый, -ая, -ое, (*a*) selection, breeding; selected

селе́кци·я, -и, (*f*) selection; (agr) selection, breeding

селе́н·, -а, *(m)* (chem) selenium, Se

~, сер·ни́ст·ый selenosulphide

~, серно·ки́сл·ый selenosulphate

селен·а́т·, -а, *(m)* (chem) selenate

селен·и́д·, -а, *(m)* (chem) selenide

сел·е́ни·е, -я, *(n)* village, settlement

селен·и́л·, -а, *(m)* (chem) selenyl

селе́н·ист·ый, -ая, -ое, *(a)* selenium, selenous, selenious; selenide (of)

~ водо·ро́д· *(m)* hydrogen selenide

~ ангидри́д· *(m)* selenium dioxide, selenious oxide

~ кисл·от·а́ *(f)* selenious/selenous acid

селен·и́д·, -а, *(m)* selenide

селен·и́т·, -а, *(m)* (min) selenite

селе́н·ов·ый, -ая, -ое, *(a)* selenium; (chem) selenic

~ вы·прям·и́тел·ь *(m)* selenium rectifier

~ кисл·от·а́ *(f)* selenic acid

селено·гра́ф·и·я, -и, *(f)* (astron) selenography

селено·кисл·от·а́, -ы́, *(f)* selenonic acid

селенола́н·, -а, *(m)* selenolane

селено·ли́т·, -а, *(m)* (min) selenolite

селено́н·иев·ый, -ая, -ое, *(a)* selenonium

селенон·и́л·, -а, *(m)* selenonyl

се́· ли *(past plur)* of **сест·ь** *(perf)*, *see* **са·д·и́ться** *(imp)*; *(pl)* of **се·л**

сели́тр·а, -ы, *(f)* (chem) saltpetre, nitre; nitrate

~, аммиа́ч·н·ая ammonium nitrate

~, аммо́н·иев·ая ammonium nitrate

~, извест·ко́в·ая calcium nitrate

~, ка́л·иев·ая potassium nitrate, saltpetre

~, ка́льц·иев·ая calcium nitrate

~, на́тр·ов·ая (min) soda nitre, Chile saltpetre

сели́тр·ян·ый, -ая, -ое, *(a)* of **сели́тр·а**; saltpeterous, nitric

сел·и́ть, -я́т, *(imp)* settle (people)

сел·о́, -а́, *(nom pl)* **сёла,** *(g pl)* **сёл·,** *(n)* village, large village

сёл·ь, -я, *(m)* (geog) torrent

сель- *(root)* rural, village, agricultural

сёльд·ев·ый, -ая, -ое, *(a)* herring, *(pl as noun)* Clupeidae

сельде·обра́з·н·ые, -ых, *(pl decl as adj)* (fish) Clupeiformes

сельдере́·й, -я, *(m)* celery

~, корн·ев·о́й (bot, food) celeraic, knob celery, *Apium graveolens* var. *rapaceum*

~, череш·ко́в·ый celery, *Apium graveolens* var. *dulce*

сёльд·ь, -и, *(nom pl)* **-и,** *(g pl)* **-е́й,** *(f)* (fish) herring

~, атлант·и́ческ·ая (fish) Atlantic herring, *Clupea harengus*

~, копч·ён·ая bloater, smoked herring

сельд·ян·о́й, -а́я, -о́е, *(a)* herring

~ кит· *(m)* (zool) fin whale, razorback, *Balaenoptera physalus*

сельпо́ *(n indecl)* rural/village consumers' co-operative society, country co-op

сельси́н·, -а, *(m)* selsyn, synchro, selsyn/synchro generator/motor

~ гру́б·ого от·с·чёт·а coarse selsyn

~ -да́т·чик·, -а, *(m)* selsyn transmitter, transmitting synchro, synchrotorque transmitter

~ -при·ём·ник· *(m)* synchro-selsyn repeater

~ то́ч·н·ого от·с·чёт·а fine selsyn

сельи́н·н·ый, -ая, -ое, *(a)* of **сельи́н·**

се́ль·ск·ий, -ая, -ое, *(a)* village, rural, country

~ хозя́й·ств·о *(n)* agriculture, farming

сельско·хозя́й·ственн·ый, -ая -ое, *(a)* agricultural, farming, farm

сельфа́ктор·, -а, *(m)* (text) mule

сель·хоз- *(abbr component)* = **сельско·хозя́й·ственн·ый,** agricultural, farming, farmers'

сем- *(abbr component)* = **сем·енн·о́й** seed

сём, *see* **сей**

се́ма́нт·ик·а, -и, *(f)* semantics

сема·фо́р·, -а, *(m)* (rail etc.) semaphore

~, двух·кры́·л·ый double-arm semaphore

сёмг·а, -и, *(f)* (zool) Atlantic salmon, *Salmo salar*

сёмг·ов·ый, -ая, -ое, *(a)* of **сёмг·а**

семе́й·н·ый, -ая, -ое, *(a)* family, familial, married, domestic

семе́й·ств·о, -а, *(n)* family

~ крив·ы́х (math) family of curves

~ по·ве́рх·ност·ей (math) family of surfaces

~ про́фил·ей airfoil family

~ рас·па́д·а (nucl) decay family

~ траекто́р·и·й family of trajectories

~ ура́н·а (min) uranium family

~ характер·и́стик· family of characteristic curves

семен·а́ *(nom/acc pl)* of **сём·я** (q.v.)

семен·и́ст·ый, -ая, -ое, *(a)* full of seeds, having many seeds

семен·ни́к·, -а́, *(m)* seed plant; seed pod; (anat) testicle, testis

семен·н·ой, -а́я, -о́е, (*a*) seed, seminal, spermatic; (*pl as noun*) (bot) spermatophytes, *Spermatophyta*
~ **кана́т·ник·** (*m*) (zool) spermatic cord
~ **нит·ь** (*f*) (zool) spermatozoon
~ **хозя́й·ств·о** (*n*) (hort) seedgrowing
семено·ве́д·ени·е, -я, (*n*) (bot) study of seeds
семено·во́д·ств·о, -а, (*n*) (agr) seed growing/production
семено·до́л·я, -и, (*f*) *see* **семя·до́л·я**
семено·но́с·н·ый, -ая, -ое, (*a*) (bot) seed-bearing, seminiferous
семено·суш·и́лк·а, -и, (*g pl*) **-лок·,** (*f*) seed dryer
сем·ери́чн·ый, -ая, -ое, (*a*) septenary
сем·ёрк·а, -и, (*g pl*) **-рок·,** (*f*) a seven, a number seven
сем·ерн·о́й, -а́я, -о́е, (*a*) seven times as large/big, sevenfold
сем·ер·о, -ы́х, (*collective numeral*) seven, seven together
семе́стр·, -а, (*m*) semester, term
сем·ечк·о, -а, (*g pl*) **-чек·,** (*n*) (*dim*) of **сем·я;** pit, stone (of fruit); pip/ seed coal
сем·ечков·ый, -ая, -ое, (*a*) of **се́-м·ечк·о**
~ **плод·** (*m*) cored/pit fruit, pome
семи- (*root*) seven-, hepta; (*int component*) semi
семи·а́том·н·ый, -ая, -ое, (*a*) (chem) heptabasic, heptatomic, heptahydric
семи·вале́нт·ност·ь, -и, (*f*) (chem) septavalence, heptavalence
~ **спирт·** (*m*) heptahydric alcohol
семи·вале́нт·н·ый, -ая, -ое, (*a*) heptavalent, septavalent
семи·во́д·н·ый, -ая, -ое, (*a*) heptahydrate
семи·гра́н·ник·, -а, (*m*) heptahedron
семи·гра́н·н·ый, -ая, -ое, (*a*) heptahedral
семи́·десят·и, (*g, d, p*) of **се́мь·де·сять** seventy
семи́·десяти- (*component*) seventy-
семи·десяти·ле́т·и·е, -я, (*n*) seventieth birthday/anniversary; seventy year period
семи·деся́т·ый, -ая, -ое, (*a*) seventieth
семи·ди́н·ов·ая пере·групп·иро́вк·а (*f*) (chem) semidine rearrangement
семи·дн·е́вн·ый, -ая, -ое, (*a*) seven-day
семи·зна́ч·н·ый, -ая, -ое, (*a*) (math) seven-place

семи·карбази́д·, -а, (*m*) (chem) semicarbazide
семи·ки́с·ь, -и, (*f*) (chem) heptoxide
семи·колло́ид·, -а, (*m*) semicolloid
семи·кра́т·н·ый, -ая, -ое, (*a*) seven times, sevenfold, septuple
семи·кристалл·и́ческ·ий, -ая, -ое, (*a*) semicrystalline
семи·ле́т·и·е, -я, (*n*) seventh anniversary; seven year period
семи·ле́т·к·а, -и, (*g pl*) **-т·ок·,** (*f*) seven year plan; seven year school
семи·ле́т·н·ий, -яя, -ее, (*a*) seven-year, septennial; seven-year-old
семина́р·, -а, (*m*) seminar
семина́р·и·й, -я, (*m*) seminar
семина́р·и·я, -и, (*f*) seminary
семио·ло́г·и·я, -и, (*f*) (med) semeiology, semiology, symptomatology
семи·по́л·ь·е, -я, (*n*) seven-field system (of crop rotation)
семи·по́ль·н·ый, -ая, -ое, (*a*) of **семи·по́л·ь·е**
семи·поля́р·н·ый, -ая, -ое, (*a*) (phys) semipolar
семи·ри́чн·ый, -ая, -ое, (*a*) septenary
семи·со́т·, (*g*) of **семь·со́т·**
семи·со́т·ый, -ая, -ое, (*a*) seven-hundredth
семи·ст·а́м, (*d*) of **семь·со́т·**
семи·ст·а́х, (*p*) of **семь·со́т·**
семи·ты́сяч·н·ый, -ая, -ое, (*a*) seven-thousandth
семи·тыч·и́нков·ый, -ая, -ое, (*a*) (bot) heptandrous
семи·уго́ль·ник·, -а, (*m*) (math) heptagon
семи·уго́ль·н·ый, -ая, -ое, (*a*) heptagonal
семи·час·ов·о́й, -а́я, -о́е, (*a*) seven-hour; seven o'clock
семи·чле́н·н·ый, -ая, -ое, (*a*) seven-membered/-termed
~ **кольц·о́** (*n*) (chem) heptatomic ring
семи·цикл·и́чн·ый, -ая, -ое, (*a*) heptacyclic
сем·на́·дцат·и- (*prefix*) seventeen
сем·на́·дцат·ый, -ая, -ое, (*a*) seventeenth
сем·на́·дцат·ь, -и, (*f*) seventeen
сем·фо́нд·, -а, (*m*) seed stock
сем·ь, -и́, (*i*) **-ью,** (*numeral*) seven
се́мь·десят·, (*g, d, p*) **семи́·десят·и** (*i*) **семью́·десят·ью** seventy

семь·со́т·, (g) **семи·со́т·,** (*numeral*) seven hundred

се́м·ью (*adv*) seven times

семью́·десят·ью, (*i*) of **се́мь·де·сят·ь** seventy

семью́·ст·а́ми, (*i*) of **семь·со́т·**

семь·я́, -й, (*nom pl*) **се́мь·и,** (*g pl*) **семе́·й,** (*d pl*) **семь·я́м,** (*f*) (zool) family; (ent) colony

се́м·я, (g) **се́мен·и,** (*n*) seed, semen, sperm

семя·в·хо́д·, -а, (*m*) (bot) micropyle

семя·вы·нос·я́щ·ий про·то́к· (*m*) (zool) spermiduct, spermaduct, ductus deferens, ductus spermaticus

семя·до́ль·н·ый, -ая, -ое, (*a*) (bot) cotyledonary

семя·до́л·я, -и, (*f*) (bot) cotyledon

семя·за·ча́т·ок·, -т·к·а, (*m*) (bot) ovule

сем·я́н·, (*g pl*) of **се́м·я**

сем·я́нк·а, -и, (*g pl*) **-нок·,** (*f*) (bot) achene, akene, achaenocarp

семяно·до́ль·н·ый, -ая, -ое, (*a*) (bot) cotyledonous

семя·но́ж·к·а, -и, (*g pl*) **-ж·ек·,** (*f*) (bot) funicle, seed stalk

семя·но́с·ец·, -с·ц·а, (*i*) **-с·ц·ем,** (*m*) (bot) placenta

семяно·но́с·н·ый, -ая, -ое, (*a*) (bot) seed-bearing, seminiferous

семяно·суш·и́лк·а, -и, (*g pl*) **-лок·,** (*f*) seed dryer

семя·по́ч·к·а, -и, (*g pl*) **-ч·ек·,** (*f*) (bot) ovule

семя·про·во́д·, -а, (*m*) (zool) seminal duct; seed tube (of seed drill)

се́м·ячк·о, -а, (*n*) *see* **се́м·ечк·о**

сенармони́т·, -а, (*m*) (min) senarmonite

сен-вена́н·а, при́нцип· (*m*) (phys) Saint-Venant principle

сенда́ст·, -а, (*m*) (met) Alsifer

сене́г·а, -и, (*f*) (bot, pharm) senega, *Polygala senega*

сенеги́н·, -а, (*m*) (pharm) senegin

се́н·и, -е́й, (*pl*) (arch) entrance hall

се́н·к·и, *see* **се́н·и**

се́нн·а, -ы, (*f*) (bot, pharm) senna

сенни́ц·а, -ы, (*f*) *see* **се́н·и**

сен·н·о́й, -а́я, -о́е, (*a*) hay

~ лихора́д·к·а, (*f*) (med) hay fever

~ пресс· (*m*) hay press/baler

се́н·о, -а, (*n*) hay

сено·ва́л·, -а, (*m*) hay loft

сено·волок·у́ш·а, -и, (*f*) hay sweep

сено·ворош·и́лк·а, -и, (*g pl*) **-лок·,** (*f*) (agr) swath-runner/-tedder

сено·е́д·ы, -ов, (*m pl*) (ent) Copeognatha, *Psocoptera*

сено·за·гото́в·к·а, -и, (*f*) haymaking

сено·копн·и́тел·ь, -я, (*m*) hay stacker

сено·ко́с·, -а, (*m*) haymaking, hay-mowing; haymaking/mowing time; hay field/meadow

сено·ко́с·ец·, -с·ц·а, (*i*) **-с·ц·ем,** (*m*) (ent) daddy-long-legs, *Phalangid*

сено·кос·и́лк·а, -и, (*g pl*) **-лок·,** (*f*) mower, mowing machine, hay cutter

~, вал·ко́в·ая windrower

сено·ко́с·н·ый, -ая, -ое, (*a*) of **сено·ко́с·**

сено·кош·е́ни·е, -я, (*n*) hay mowing

сеномá́н·, -а, (*m*) (geol) Cenomanian stage

сеномá́н·ск·ий я́рус· (*m*) (geol) Cenomanian stage

сено́н·, -а, (*m*) (geol) Senonian

сено·на·гру́з·чик·, -а, (*m*) hay-loader

сено·ста́в·к·а, -и, (*g pl*) **-в·ок·,** (*f*) (ent) tree-creeper, *Certhia familiaris*

сено·у·бо́р·к·а, -и, (*f*) hay harvesting/making

сенса́ци·я, -и, (*f*) sensation

сенсибилиз·а́тор·, -а, (*m*) sensitizer, sensitizing agent

сенсибилиз·а́ци·я, -и, (*f*) sensitization, sensitizing

сенси́лл·а, -ы, (*f*) (zool) sensilla

сенсито·гра́мм·а, -ы, (*f*) (phot) gamma/sensitivity strip

сенсито́·метр·, -а, (*m*) (phot) sensitometer; penetrameter (radiography)

сенсито·мéтр·и·я, -и, (*f*) (opt) sensitometry

сенсо́р·н·ый, -ая, -ое, (*a*) (physiol) sensory

~ вос·пит·а́ни·е (*n*) sensory learning

сентя́бр·ь, -я́, (*m*) September

се́п·ь, -и, (*f*) canopy

сепара́т·, -а, (*m*) separated substance (e.g. oil from gas); trapped moisture

сепара́т·н·ый, -ая, -ое, (*a*) separate

сепара́тор·, -а, (*m*) separator; steam trap; cage (of roller bearing); drop eliminator (air conditioning)

~, аккумуля́тор·н·ый cell separator (of electric battery)

~, изото́п·н·ый (nucl) isotope/mass separator

~, лéнт·очн·ый (min) belt separator

~, магни́т·н·ый magnetic separator

сепара́тор
~ под·ши́п·ник·а bearing cage
~, уда́р·н·ый toboggan-type electrostatic separator
~, центрифуга́ль·н·ый centrifuge, centrifugal separator
сепара́тор·н·ый, -ая, -ое, (*a*) of **сепара́тор·**
сепарацио́н·н·ый, -ая, -ое, (*a*) separation, separating
~ материа́л· (*m*) (naut) dunnage
сепар·а́ци·я, -и, (*f*) separation
сепар·и́р·овать, -уют, (*imp and perf*) separate
сепио·ли́т·, -а, (*m*) (min) sepiolite, meerschaum
сепио́н·, -а, (*m*) (zool) sepion, cuttlebone
се́пи·я, -и, (*f*) sepia (colour); (zool) cuttle-fish
сепси́н·, -а, (*m*) (biochem) sepsine
се́псис·, -а, (*m*) (med) sepsis
се́пт·а, -ы, (*f*) (biol) septum
септа́л·и·й, -я, (*m*) (zool) septal neck
септа́ль·н·ый, -ая, -ое, (*a*) (biol) septal
септа́р·иев·ый, -ая, -ое, (*a*) (geol) septarian
септа́р·и·я, -и, (*f*) (geol) septarium, septarian nodule
се́пт·ик·, -а, (*m*) septic tank (sewage)
се́птим·а, -ы, (*f*) seventh (music)
септ·и́ческ·ий, -ая, -ое, (*a*) septic
септиц·еми́·я, -и, (*f*) (med) septicaemia, septicemia
септ·о́л·ь, -я, (*m*) septole (music)
септорио́з·, -а, (*m*) septoria spot (plant disease)
септо·том·и́·я, -и, (*f*) (med) septotomy
септо·флав·и́н·, -а, (*m*) (pharm) Septoflavine, proflavine
се́птул·а, -ы, (*f*) (biol) septulum
се́р·а, -ы, (*f*) (chem) sulphur, sulfur, S
~, га́з·ов·ая, reclaimed sulphur (from industrial smoke)
~, ком·ов·а́я lump/block sulphur
~, о́бщ·ая (chem) total sulphur
~, ромб·и́ческ·ая (min) orthorhombic sulphur
~, само·ро́д·н·ая native/naturally-occurring sulphur
~, хло́р·ист·ая sulphur chloride
~, уш·н·а́я (zool) cerumen
~, черен·ко́в·ая cane brimstone
сераде́лл·а, -ы, (*f*) (bot) seradilla, *Ornithopus sativus*
серва́л·, -а, (*m*) (zool) serval, *Felix serval*; serval (fur)

серванти́т·, -а, (*m*) (min) cervantite
серви́з·, -а, (*m*) service, set (of crockery)
серво·бло́к·, -а, (*m*) servo unit
серво·дви́г·ател·ь, -я, (*m*) servomotor, servoactuator
серво·компенс·а́тор·, -а, (*m*) (a/c) balance tab, servo-tab
серво·ко́нтур·, -а, (*m*) servoloop
серво·механ·и́зм·, -а, (*m*) servomechanism, servo, servo unit
серво·мото́р·, -а, (*m*) servomotor
~, ма́сл·ян·ый hydraulic servomotor
серво·при·во́д·, -а, (*m*) servo mechanism, servo
серво·систе́м·а, -ы, (*f*) servosystem, servo
серво·три́ммер·, -а, (*m*) (a/c) servo-tab
серво·у·правл·е́ни·е, -я, (*n*) servo-control
серво·у·сил·и́тел·ь, -я, (*m*) servo-amplifier
серд·е́чник·, -а, (*m*) core; heart (of rope); (bot) cuckoo flower, *Cardamine*
лист· серд·е́чник·а (*m*) (elec) core disc/plate/lamination
~, магни́т·н·ый magnetic core
~ по́люс·а (elec) pole core/shank
сердечно·дых·а́тельн·ый, -ая, -ое, (*a*) (anat) cardiorespiratory
сердечно·со·су́д·н·ый, -ая, -ое, (*a*) (anat) cardiovascular
серд·е́чн·ый, -ая, -ое, (*a*) heart, cardiac; kindly, cordial
серд·и́ть, ⸗ят, (*imp*) anger; -ся be angry
сердоли́к·, -а, (*m*) (min) carnelian
се́рд·ц·е, -а, (*i*) -ем, (*nom pl*) -а́, (*g pl*) -д·е́ц·, (*n*) (anat) heart, core
сердце·би·е́ни·е, -я, (*n*) palpitation, (med) tachycardia
сердце·ви́д·к·а, -и, (*g pl*) -д·ок·, (*f*) (fish) cockle, *Cardium*
сердце·ви́д·н·ый, -ая, -ое, (*a*) heart-shaped, cordiform, (bot) cordate
серд·цеви́н·а, -ы, (*f*) core, centre, middle, heart; core (of ingot); web (of a drill); (bot) pith, medulla; duramen, heartwood
серд·цеви́нн·ый, -ая, -ое, (*a*) of **серд·цеви́н·а;** (bot) medullary
~ луч·и́ (*pl*) (bot) medullary/vascular rays
~ полос·а́ (*f*) (gunn) 70% zone
~ полос·а по да́ль·ност·и (*f*) 70% length zone

сердце·ли́ст·н·ый, -ая, -ое, *(a)* (bot) cardiophyllous

сердце·обра́з·н·ый, -ая, -ое, *(a)* = **сердце·ви́д·н·ый**

серебр·е́ни·е, -я, *(n)* silvering (glass); silver plating (metal); (plastic) crazing

серебр·и́ст·ый, -ая, -ое, *(a)* silvery, silver; argentous, argentiferous, argentate

серебр·и́ть, -я́т, *(imp)* silver, silver plate, stain silver; **-ся** *(pass)*; become/appear silvery

серебр·о́, -а́, *(n)* (chem) silver, Ag; (plastics) crazing

~, азотно·ки́сл·ое (chem) silver nitrate

~, бли́к·ов·ое unrefined silver obtained in cupellation of lead

~, бро́м·ист·ое (chem, phot) silver bromide, photobromide

~, кош·а́ч·ье cat silver

~, рог·ов·о́е (min) horn silver, ceragyrite

~, суса́ль·н·ое silver-leaf

~, хло́р·ист·ое (chem) silver chloride, (phot) photochloride

серебро·но́с·н·ый, -ая, -ое, *(a)* (geol) argentiferous, silver-bearing

серебро·сто́й·кост·ь, -и, *(f)* (plast) resistance to crazing

серебр·я́нк·а, -и, *(f)* (met) bright bar, bright steel bar

сере́бр·ян·ый, -ая, -ое, *(a)* silver

~ блеск· *(m)* (min) silver glance, argentite

серед·и́н·а, -ы, *(f)* middle, centre, center; mean

серед·и́нн·ый, -ая, -ое, *(a)* centre, center, central, middle, mid, mean

серёж·к·а, -и, *(g pl)* **-ж·ек·,** *(f)* lug, ear, ring; (bot) catkin, amentum

серёжко·цве́т·н·ый, -ая, -ое, *(a)* (bot) amentaceous

серёж·чат·ый, -ая, -ое, *(a)* (bot) amentaceous

сер·е́·ть -ют, *(imp)* turn/become grey/gray; look/appear grey/gray

сержа́нт·, -а, *(m)* sergeant

сериа́ль·н·ый, -ая, -ое, *(a)* (bot) serial

се́рнес·н·ый, -ая, -ое, *(a)* (elec) series, series-characteristic/-wound

~ дви́г·ател·ь *(m)* (elec) series motor, series-characteristic motor

сер·и́йност·ь, -и, *(f)* order, sequence; precedence, priority; series/multiple production system

сер·и́йн·ый, -ая, -ое, *(a)* serial, series

~ про·из·во́д·ств·о *(n)* multiple/series production

сер·и́н·, -а, *(m)* (biochem) serine

сер·ици́н·, -а, *(m)* sericine, silk gum

сериц·и́т·, -а, *(m)* (min) sericite

сер·и·я, -и, *(f)* series, serial; train (of pulses, waves etc.); run (of measurements)

~, вращ·а́тельн·ая (phys) rotational series

~ и́мпульс·ов (phys) pulse train

се́рн·а, -ы, *(f)* (zool) chamois, *Rupicapra*

сернисто·ка́дм·иев·ый, -ая, -ое, *(a)* cadmium-sulphide/sulfide

сернисто·ки́сл·ый, -ая, -ое, *(a)* sulphurous acid; sulphite of

~ на́тр·и·й *(m)* sodium sulphate/sulfite

сернисто·сере́бр·ян·ый фото·эле-ме́нт· *(m)* (elec) silver-sulphur photocell

сер·ни́ст·ый, -ая, -ое, *(a)* sulphur, sulfur, sulphurous, sulfurous; sulphide (of), sulfide (of)

~ ангидри́д· *(m)* (chem) sulphur dioxide

~ водо·ро́д· *(m)* hydrogen sulphide

~ ди·эти́л· *(m)* ethyl thioether

~ кисл·от·а́ *(f)* sulphurous acid

~ на́тр·и·й *(m)* sodium sulphide

~ уран·и́л· *(m)* uranyl sulphide

се́р·н·о- *(component)* sulphate (of)

серновато·ки́сл·ый, -ая, -ое, *(a)* hyposulphate (of), thiosulphate (of)

сер·нова́тист·ый, -ая, -ое, *(a)* hyposulphurous, thiosulphuric

~ кисл·от·а́ *(f)* hyposulphurous/thiosulphuric acid

сер·нова́т·ый, -ая, -ое, *(a)* hyposulphuric, hyposulfuric, dithionic

серно·кисл·о́тн·ый, -ая, -ое, *(a)* sulphuric acid

серно·ки́сл·ый, -ая, -ое, *(a)* sulphate (of)

~ аммо́н·и·й *(m)* ammonium sulphate

~ ка́льц·и·й, двух·во́д·н·ый hydrated calcium sulphate, gypsum

~ ура́н· *(m)* uranium sulphate

се́р·н·ый, -ая, -ое, *(a)* sulphur, sulfur, sulphuric, sulfuric

~ ангидри́д· *(m)* sulphuric anhydride

~ кисл·от·а́ *(f)* sulphuric/sulfuric acid

~ кисл·от·а́ дым·я́щ·ая fuming sulphuric acid

сéр·н·ый, -ая, -ое
~ **колчедáн·** (*m*) (min) pyrites

серо·бактéр·и·и, -й, (*f pl*) (bact) sulphur/sulfur bacteria, *Thiobacteriaceae*

сéро-бýр·ый, -ая, -ое, (*a*) grey/gray-brown

сер·овáт·ый, -ая, -ое, (*a*) greyish, grayish

серо·водо·рóд·, -а, (*m*) hydrogen sulphide/sulfide

серо·водо·рóд·н·ая кисл·от·á (*f*) hydrosulphuric/hydrosulfuric acid

серо·диагнóст·ик·а, -и, (*f*) (med) serodiagnosis

серо·зём·, -а, (*m*) sierozem, grey desert soil

сер·óзн·ый, -ая, -ое, (*a*) (zool) serous
~ **оболóч·к·а** (*f*) (zool) serosa, serous membrane

серо·лóг·и·я, -и, (*f*) (med) serology

серо·ó·кис·ь, -и, (*f*) oxysulphide, oxysulfide

серо·о·чи́ст·к·а, -и, (*f*) removal of sulphur (from gases)

серо·профилáкт·ик·а, -и, (*f*) (med) inoculation

серо·сине·рóд·ист·ый, -ая, -ое, (*a*) thiocyanate (of)

серо·терап·и́·я, -я, (*f*) (med, vet) serotherapy

серо·тони́н·, -а, (*m*) (biochem) serotonin

серо·угле·рóд·, -а, (*m*) (chem) carbon bisulphide/bisulfide/disulfide

серп·, -á, (*m*) sickle, reaping hook; (anat) falx

серпанти́н·, -а, (*m*) paper streamer

серпанти́н·а, -ы, (*f*) *see* **серпенти́н·а**

серпенти́н·, -а, (*f*) (min) serpentine

серпенти́н·а, -ы, (*f*) hairpin bend (in road)

серпентин·изáци·я, -и, (*f*) (geol) serpentinization

серпентин·и́т·, -а, (*m*) (geol) serpentine, serpentine rock

серпéт·к·а, -и, (*g pl*) **-т·ок·,** (*f*) (hortic) pruning knife, billhook

серпо·ви́д·н·ый, -ая, -ое, (*a*) crescent/sickle-shaped, falcate, falciform, drepaniform, drepanoid

серпо·плóд·н·ый, -ая, -ое, (*a*) (bot) sickle-fruited, drepanocarpus

серпули́т·, -а, (*m*) (pal) serpulite

серп·ýх·а, -и, (*f*) (bot) saw-wort, *Serratula*

сертификáт·, -а, (*m*) certificate; (fin) government bond

сéрум·, -а, (*m*) (med) serum

серу·со·держ·á·щий, -ая, -ее, (*a*) sulphur-containing

сéр·ый, -ая, -ое, (*a*) grey, gray
~ **тéл·о** (*n*) (phys) grey/gray body

серьг·á, -и́, (*nom pl*) **сéрьг·и,** (*g pl*) **серёг·,** (*f*) ear-ring; connecting link, clevis, shackle; (naut) slip rope
~, **на·тя́ж·н·áя** tensioning blade-holder (of saw)
~, **с·цеп·н·áя** coupling link

серьёз·н·ый, -ая, -ое, (*a*) serious, grave

сер·я́нк·а, -и, (*g pl*) **-нок·,** (*f*) resin/pitch pocket (defect in pinewood)

сескви·геóл·, -а, (*m*) (chem) sesquigeol

сескви·терпéн·, -а, (*m*) (chem) sesquiterpene

сéсси·я, -и, (*f*) session, (law) term

сестóн·, -а, (*m*) (biol) seston

сестр·á, -ы́, (*nom pl*) **сёстр·ы,** (*g pl*) **сестёр·,** (*f*) sister; (med) nurse

сест·ь, (*fut 3rd pl*) **ся́д·ут,** (*past masc sing*) **сé·л,** (*perf*); *see* **сад·и́ться**

сете·ви́д·н·ый, -ая, -ое, (*a*) net-like, retiform, reteform, reticular

сет·ев·óй, -áя, -óе, (*a*) of **сет·ь**
~ **за·град·и́тель·** (*m*) (naut) netlayer
~ **на·сóс·** (*m*) district-heating supply pump
~ **при·ём·ник·** (*m*) (rad, elec) mains receiver

сете·вяз·áльн·ая маши́н·а (*f*) (text) net-making machine

сете·вяз·áни·е, -я, (*n*) netmaking

сете·подъ·ём·ник·, -а, (*m*) small fishing vessel

сете·подъ·ём·н·ая маши́н·а (*f*) (fish) net/trawl winch

сéт·к·а, -и, (*g pl*) **-т·ок·,** (*f*) (*dim*) of **сет·ь**; netting, network, net; wire/metal cloth/gauze/netting; grid; squares (on ruled paper); (paper) machine wire, wire cloth, wire; (opt) graticule, reticule; (anat) reticulum, plexus; screen, liner (of oil wells); (*pl*) pigeon holes
~, **анти·динатрóн·н·ая** suppressor grid (of vacuum tube)
~ **вариáтор·а** (elec) buncher grid
~, **венти́ль·н·ая** (autom) rectifier network
   гибербол·и́ческ·ая кáрт·а с гибербол·и́ческ·ими сéт·к·ами (surv) lattice chart
~, **груз·ов·áя** cargo net

сет·к·а

~, за·де́рж·ивающ·ая pickup/signal plate (of vacuum tube)

~, за·земл·ённ·ая (elec) earthed grid, suppressor grid

~, за·щи́т·н·ая (elec) shield grid

~, координа́т·н·ая (surv etc.) grid, reference grid

~, логарифм·и́ческ·ая (math) logarithmic grid/coordinate

~, па́нцир·н·ая reinforcement mesh (concrete)

пе́ленг· по сет·к·е (*m*) (navig) grid bearing

пло́ск·ост·ь сет·к·и (*f*) (cryst) net-plane

~, при·ём·н·ая suction strainer (of a pump)

~ при·це́л·а (surv, navig) sight graticule

~, про·стра́н·ственн·ая three-dimensional network

~, раз·ря́д·н·ая word (computers)

~, рас·са́с·ывающ·ая space-charge grid (of vacuum tube)

~, ре́д·к·ая sparse/open network

се́вер· по сет·к·е (nav) grid north

~, со·бир·а́ющ·ая (autom) counting grid

~, тари́ф·н·ая payment schedule/scale

~, у·правл·я́ющ·ая (elec) control grid

~, у·скор·я́ющ·ая (elec) accelerating grid

~ фи́льтр·а filter gauze

~, экра́н·н·ая (elec) screen grid

сетко·держ·а́тел·ь, -я, (*m*) (cinema) diffuser frame, gauze holder

сет·н·о́й, -а́я, -о́е, (*a*) (fish) net

сет·очн·ый, -ая, -ое, (*a*) of **сет·к·а**

~ ко́нтур· (*m*) grid circuit (vacuum tubes)

~ с·мещ·е́ни·е (*n*) (elec) grid bias

~ ток· (*m*) (elec) grid current

~ у·те́ч·к·а (*f*) grid leak

~ характер·и́стик·а (*f*) (elec) grid characteristic, grid current characteristic

сет·ча́тк·а, -и, (*g pl*) **-ток·,** (*f*) (anat) retina

сетчато·кры́·л·ый, -ая, -ое, (*a*) (ent) neuropterous, (*pl as noun*) *Neuroptera*

сет·чатост·ь, -и, (*f*) netting, reticulation

сет·чат·ый, -ая, -ое, (*a*) net, netted, network, reticular, reticulate; three-dimensional

~ структу́р·а (*f*) (phys, chem) reticular/lattice/network structure

сет·ь, -и, (*nom pl*) **сет·и,** (*g pl*) **сет·е́й,** (*f*) *see also* **сет·к·а**; net, network, (zool) rete; mains, supply mains/network/system; (rad) array

~, анте́нн·ая (rad) broadside array

~, водо·про·во́д·н·ая water-supply system

~, дри́фтер·н·ая (fish) drift net

~, за·гражд·е́ни·я (naut) boom-defence net

~, канализацио́н·н·ая sewerage system

~, конта́кт·н·ая (elec rail) overhead contact system

~, коро́т·к·ая lead (elec furnace)

~, ли́в·нев·ая surface water sewer, storm sewer

~ о·по́р·н·ых то́ч·ек· (surv) survey network

~ пла́н·ов·ых о·по́р·н·ых то́ч·ек· (surv) horizontal control network

~, про·мысл·о́в·ая oilfield gathering system

~, раз·вод·н·а́я distribution system (of water supply)

~, реч·н·а́я (geog) drainage

~, со·бир·а́тельн·ая (oil) gathering system

~, тепло·фикацио́н·н·ая district heating system

~, чуд·е́сн·ая (zool) rete mirabile

~ яи́ч·к·а (anat) rete testis

се́ч·а, (*pres gerund*) of **сеч·ь**

сеч·е́ни·е, -я, (*n*) (*v n*) of **сеч·ь;** section, cross-section; revolved section (of engineering drawings)

~, вы́·дел·енн·ое detail section (of engineering drawings)

~ вы́·ход·а (phys) yield cross-section

~, жи·в·о́е cross-section (of river)

~ за·хва́т·а (phys) capture cross-section

крив·а́я сеч·е́ни·я (*f*) cross-section curve

~, крит·и́ческ·ое throat (of nozzle); (met roll) neutral plane/point, no-slip point

~, кру́гл·ое circular section

~ лопа́т·к·и (aerodynam) blade section

~, ми́дел·ев·о́е (shipb) midship section

~, на·ло́ж·енн·ое section revolved in place (eng drawings)

~, нейтра́ль·н·ое (met roll) neutral plane/point

~ -не́тто (eng) effective section

~ пере·но́с·а (phys) transport cross-section

сеч·éни·е
~ **по·глощ·éни·я** (phys) absorption cross-section
~ **по·глощ·éни·я тепл·ов·ы́х ней·трóн·ов** neutron-capture cross-section
~, **пóлн·ое** (phys) total cross-section
~, **по·перéч·н·ое** cross-section
~ **сопл·á крит·и́ческ·ое** (rocket) nozzle throat
~, **срéд·н·ее** mid-span section (of wing)
~ **с·толк·новéни·я** (phys) collision cross-section
~, **эффектúв·н·ое** (phys) nuclear cross-section; (mil) radar cross-section (of target)

сéч·енн·ый, -ая, -ое, (*past part pass*) of **сеч·ь**

сеч·ёт, (*pres 3rd sing*) of **сеч·ь**

сéч·к·а, -и, (*g pl*) **-ч·ек·,** (*f*) (*v n*) of **сеч·ь;** chopper, chopping knife; high-grade hemp; (food) kibbled grain; (agr) chaff

сеч·ь, (*pres 3rd sing, pl*) **сеч·ёт, сек·у́т,** (*past masc sing*) **сек·,** (*imp*) cut up/off, hew; flog, whip

сé·ющ·ий, -ая, -ее, (*pres part act*) of **сé·ять**

сé·я, (*pres gerund*) of **сé·ять**

сé·ялк·а, -и, (*g pl*) **-лок·,** (*f*) (agr) seed drill, sower
~, **квадрáт·н·о-гнезд·ов·áя** check-row planter
~, **овощ·н·áя** vegetable seed drill
~, **тýк·ов·ая** fertilizer spreader

сé·яльн·ый, -ая, -ое, (*a*) sowing, seed; sifting

сé·янец·, -нц·а, (*i*) **-нц·ем,** (*m*) (hortic) seedling

сé·ять, -ют, (*imp*) (agr) sow, drill, plant; sift; (bact) inoculate (a culture); **-ся** (*pass*); shower

с·жá·л, (*past masc sing*) of **с·жат·ь,** *see* **с·жим·á·ть, с·жин·á·ть**

с·жáт·и·е, -я, (*n*) (*v n*), *see* **с·жим·á·ть;** compression, compaction; constriction, pinch; shrinkage, contraction
вос·пламен·éни·е от с·жáт·и·я (*n*) (ICE) compression ignition
~, **все·сторóн·н·ее** cubic compression
мóдул·ь все·сторóн·н·его с·жá·т·и·я (*m*) modulus of cubic compressibility, volumetric modulus of elasticity, bulk modulus
ис·пыт·áни·е на с·жáт·и·е (*n*) (mech) compression test
кáмер·а с·жáт·и·я (*f*) compression chamber

с·жáт·и·е
коэффициéнт· с·жáт·и·я (*m*) coefficient of compressibility/contraction
лúн·и·я с·жáт·и·я (*f*) compression line; (math) line of striction
~, **объ·ём·н·ое** *see also* **все·сторóн·н·ее с·жáт·и·е** (*above*); cubic compression
~, **одно·сторóн·н·ее** unilateral/linear compression
~, **от·нос·и́тельн·ое** compression stiffness
~ **раз·ря́д·а** (nucl) pinching
со·противл·éни·е с·жáт·и·ю (*n*) compressive strength
такт· с·жáт·и·я (*m*) (ICE) compression stroke
~, **удáр·н·ое** shock compression

сжато·спирáль·н·ый, -ая, -ое, (*a*) compact-spiral, close-coiled

с·жáт·ост·ь, -и, (*f*) compression ratio; conciseness, brevity

с·жáт·ый, -ая, -ое, (*past part pass*) *see* (1) **с·жим·á·ть,** (2) **с·жин·á·ть;** concise; pinched

с·жат·ь, (*fut 3rd pl*) **со·жм·ут,** (*past masc sing*) **с·жá·л,** (*perf*) *see* **с·жим·á·ть** (*imp*); **с·жат·ь,** (*fut 3rd pl*) **со·жн·ут,** (*past masc sing*) **с·жа·л,** (*perf*) *see* **с·жин·á·ть** (*imp*)

с·жиг·áни·е, -я, (*n*) (*v n*) of **с·жиг·á·ть;** incineration, combustion
кáмер·а с·жиг·áни·я (*f*) combustion chamber

с·жиг·á·ть, -ют, (*imp*) burn up, incinerate

с·жид·и́ть, -ят, (*perf*) *see* **с·жиж·á·ть**

с·жиж·á·ть, -ют, (*imp*) liquefy

с·жиж·éни·е, -я, (*n*) liquefaction, liquefying

с·жúж·енн·ый, -ая, -ое, (*past part pass*) *see* **с·жиж·á·ть;** liquefied

с·жúм·, -а, (*m*) clamp, grip

с·жим·áемост·ь, -и, (*f*) (mech, phys) compressibility; (dynam) coercibility

с·жим·áем·ый, -ая, -ое, (*pres part pass*) of **с·жим·á·ть;** compressible; coercible, constrictable

с·жим·áтел·ь, -я, (*m*) (anat) constrictor

с·жим·áтел·ь-рас·шир·и́тел·ь, -я, (*m*) (dynam) compandor

с·жим·á·ть, -ют, (*imp*) compress, compact, reduce; squeeze, clench, constrict, grip, nip, pinch; **-ся** (*pass*); shrink, contract

с·жим·а́ющ·ий, -ая, -ее, (*pres part act*) of **с·жим·а́·ть;** compressive, compression

с·жим·а́ющ·ийся, -аяся, -ееся, (*pres part act*) of **с·жим·а́·ться;** contracting, contractile, shrinking

с·жин·а́·ть, -ют, (*imp*) reap, harvest

с·за́д·и (*adv*) from behind, from the back; behind; (*prep* + *gen*) behind

с·зыв·а́ни·е, -я, (*n*) (*v n*) of **с·зыв·а́·ть;** convocation

с·зыв·а́·ть, -ют, (*imp*) summon, call (to meeting); convene, convoke

си (*n indecl*) B (music)

сиал·адени́т·, -а, (*m*) (med) sialadenitis

сиал·и́зм·, -а, (*m*) (med) sialism

сиало·ло́г·и·я, -и, (*f*) (med) sialology

сиало·ррё·я, -и, (*f*) (med) sialorrhoea, sialorrhea

си·а́л·ь, -и, (*f*) (geol) sial

сибани́т·, -а, (*m*) (plast) cibanite

сибире·я́зв·енн·ый, -ая, -ое, (*a*) anthrax, anthracic

~ вакци́н·а (*f*) anthrax vaccine

сиби́р·к·а, -и, (*g pl*) **-р·ок·,** (*f*) (med) anthrax; (bot) sibirea, *Sibirea*

сиби́р·ск·ий, -ая, -ое, (*a*) Siberian

~ я́зв·а (*f*) (med, vet) anthrax

сибо·та́ксис·, -а, (*m*) cybotaxis

сива·питёк·, -а, (*m*) (pal) Sivapithecus

сив·у́х·а, -и, (*f*) a raw vodka (weak); fusel oil

сив·у́ч·, -а́, (*i*) **-о́м·,** (*m*) (zool) Steller's sea lion, *Eumetopias jubata*

сив·у́шн·ый, -ая, -ое, (*a*) of **сив·у́х·а**

~ ма́сл·о (*n*) (chem) fusel oil

си́в·ый, -ая, -ое, (*a*) grey (of horses)

сиг·, -а́, (*m*) (fish) lake whitefish, *Coregonus*

сига́р·а, -ы, (*f*) cigar

~, мор·ск·а́я cigar-shaped raft (logging)

сигаре́т·а, -ы, (*f*) cigarette; cheroot

сигилля́р·иев·ые, -ых, (*pl decl as adj*) (pal bot) Sigillariaceae

сигилля́р·и·и (*pl*) (pal bot) Sigillaria

си́гм·а, -ы, (*f*) sigma, σ (Greek)

сигмо·ви́д·н·ый, -ая, -ое, (*a*) sigmoid, sigmoidal

~ ки́ш·к·а (*f*) (anat) sigmoid flexure of the colon

сигм·о́ид·а, -ы, (*f*) sigmoid

сигмоид·а́льн·ый, -ая, -ое, (*a*) sigmoid

сигна́л·, -а, (*m*) signal

~, авар·и́йн·ый emergency signal; (naut) distress signal, SOS

~, бо·ев·о́й (mil) alarm

~, больш·о́й strong signal

сигна́л

~, в·ход·н·о́й (autom) input signal; (rail) entry/home signal

~, вы·зыв·н·о́й calling signal

~, вы·ход·н·о́й (autom) output signal; (rail) leave/starting signal

~ -вы́ш·к·а, -и, (*f*) (surv) tower, signal tower

~, гас·я́щ·ий blanking signal

~, геодез·и́ческ·ий (surv) tower, survey signal

~, двух·зна́ч·н·ый (rail) two-aspect signal

~, двух·та́кт·н·ый (telecom) push-pull signal

~, дискре́т·н·ый (autom) sampled signal

~, звук·ов·о́й sound/audio signal

~, ис·пыт·а́тельн·ый test/testing signal

~, карли́к·ов·ый (rail) ground signal

~, квит·и́рующ·ий answer signal (telephones)

~, кома́нд·н·ый (autom) control signal

~, коммут·и́рующ·ий keying signal (telegraphy); switching signal

~, компенс·и́рующ·ий (TV) shading signal

~, контро́ль·н·ый (autom) pilot signal; (rail) repeating signal

~, коррект·и́рующ·ий (telecom) correcting signal; phasing signal (photo facsimile, TV)

~, кратко·вре́мен·н·ый fast signal

~, ма́л·ый weak signal

~, маршру́т·н·ый (rail) inner-distant/-home signal

~, много·зна́ч·н·ый (rail) multiple-aspect signal

~, на·ча́ль·н·ый (autom) initiating signal

~, обра́т·н·ый inverse signal; (rail) recall signal

~, о·по·зна·ва́тельн·ый recognition signal, identification signal

~, о·по́р·н·ый reference/comparison signal

~, от·бо́й·н·ый clear/clearing signal (telephones)

~, от·раж·ённ·ый (rad) echo signal

~, пере·го́н·н·ый (rail) intermediate signal

~, пере·да́т·очн·ый transfer signal (telephones)

~, пере·меж·а́ющ·ийся (telecom) interleaved signal

~, по·бо́ч·н·ый ghost/spurious signal

~, по·вто́р·н·ый (rad) multipath signal

сигна́л
~, по·жа́р·н·ый   fire alarm
~, по·зы́в·н·о́й   (rad) call-sign
~, поле́з·н·ый   (TV) useful/wanted signal
~, по́лн·ый цвет·ов·о́й   (TV) composite colour signal
~, пред·у·пред·и́тельн·ый   warning signal, offering signal (telephones); (rail) distant signal
~, прямо·уго́ль·н·ый   (rad) square-wave signal
~, пуск·ов·о́й   start/start-up signal; actuating signal
~, раз·реш·а́ющ·ий   (rail) clear signal
~, результ·и́рующ·ий   net signal
~, симметр·и́ческ·ий   (rad) balanced signal
~, с·ло́ж·н·ый   (telecom) compound/serrated signal
~, ступ·е́нчат·ый   stepped /staircase signal
~, ступ·е́нчат·ый в·ход·н·о́й   (autom) step-input signal
~, та́кт·ов·ый   timing signal; cadence (telephones)
~, тре·уго́ль·н·ый   (rad) sawtooth signal
~, тума́н·н·ый   fog signal
~, у·правл·я́ющ·ий   input/control signal
~, цвет·ов·о́й   (TV) colour signal
~, што́рм·ов·о́й   (naut) storm signal cone

сигна́л-э́х·о, -а,   (*n*) echo signal, echo (radar)

сигнализ·а́тор·, -а,   (*m*) indicator, signalling equipment; alarm, warning light/buzzer/indicator
~, по·жа́р·н·ый   fire alarm

сигнализ·а́ци·я, -и,   (*f*) signalling, indicating, indication, annunciation
~, звук·ов·а́я   sound signalling
~, зр·и́тельн·ая   visual signalling
~, по́·иск·ов·ая   (telecom) paging system
~, свет·ов·а́я   flashing, flash signalling
~, ствол·ов·а́я   mine shaft signalling system

сигнализ·и́р·овать, -уют,   (*imp and perf*) signal; warn

сигнало·образ·у́ющ·ий, -ая, -ее, (*a*) signal-shaping

сигна́ль·н·ый, -ая, -ое, (*a*) signalling, indicating, signal
~ бу́·й   (*m*) indicator buoy

сигна́ль·н·ый, -ая, -ое
~ ла́мп·а   (*f*) signal lamp, indicating light
~ о́рган·   (*m*) (autom) indicating element
~ прожёктор·   (*m*) signalling projector
~ ра́з·ност·ь то́к·а   (*f*) (telecom) current margin
~ со·ста́·в·   (*m*) flaring/flashing compound (pyrotechnics)
~ со·чет·а́ни·е   (*n*) (naut) hoist of signals
~ цеп·ь   (*f*) alarm circuit

сигна́ль·щик·, -а,   (*m*) signalman, signaller

сигнату́р·а, -ы, (*f*) (print) signature, section, signature mark; (pharm) label

сиголо́в·, -а, (*m*) (fish) sigolov, *Coregonus lavaretus beari*

сида́льце·я, -и, (*f*) (bot) prairie mallow, *Sidalcea*

сид·е́ни·е, -я, (*n*) *see also* **сид·е́нь·е;** seating, seat; sitting

сид·е́нь·е, -я, (*n*) seat, seating
~, вы·бра́с·ываем·ое   (a/c) ejection seat
~, катапульт·и́руем·ое   (a/c) ejection seat
~, корм·ов·о́е   (naut) stern sheets
~ -ча́ш·к·а, -и,   (*f*) seat pan, bucket seat

сидер·а́льн·ый, -ая, -ое, (*a*) *see also* **сидер·и́ческ·ий;** (astron) sidereal
~ вре́м·я   (*n*) sidereal time

сидер·а́ци·я, -и, (*f*) (agr) green manuring

сидер·и́т·, -а, (*m*) (geol) siderite; (min) chalybite, siderite, spathic iron

сидери́т·ов·ый, -ая, -ое, (*a*) sideritic

сидер·и́ческ·ий, -ая, -ое, (*a*) *see also* **сидер·а́льн·ый;** (astron) sidereal
~ ме́сяц·   (*m*) (astron) sidereal month
~ пери́од·   (*m*) (astron) sidereal period

сидер·о́з·, -а, (*m*) (med) siderosis

сидеро·мела́н·, -а, (*m*) (min) sideromelane

сидеро·пизоли́т·, -а, (*m*) (min) sideropisolite

сидеро·ста́т·, -а, (*m*) (astron) siderostat

сидеро·филли́т·, -а, (*m*) (min) siderophyllite

сидеро·фи́ль·н·ый, -ая, -ое, (*a*) siderophil, siderophilous
~ элеме́нт·ы   (*pl*) (chem) siderophile elements

сидеро·ци́т·, -а, (*m*) (zool) siderocyte

сид·е́ть, -я́т, (*imp*) sit; be (of location); fit, sit (of clothes)

си́др·, -а, (*m*) cider

сидяче·гла́з·ые, -ых, (*pl decl as adj*) (zool) Basommatophora

сид·я́ч·ий, -ая, -ее, (*a*) sitting; sedentary; sessile

~ дислок·а́ци·я (*f*) (cryst) sessile dislocation

~ фо́рм·а (*f*) (biol) sessile form

сид·я́щ·ий, -ая, -ее, (*pres part act*) of **сид·е́ть**

сиён·а, -ы, (*f*) sienna (pigment)

~, жж·ён·ая burnt sienna

сиени́т·, -а, (*m*) (min) syenite

~, нефели́н·ов·ый (min) nepheline-syenite

сиённ·а, -ы, (*f*) *see* **сиён·а**

сиж·у́, (*pres 1st sing*) of **сид·е́ть**

сиза́л·ь, -я, (*m*) (bot, text) sisal, sisal hemp, *Agave sisalana*

сиз·ева́т·ый, -ая, -ое, (*a*) (bot) glaucous, glaucescent

сизи́г·и·и, -й, (*f pl*) (astron) syzygy

сизиг·и́йн·ый, -ая, -ое, (*a*) syzygial

~ при·ли́в· (*m*) spring tide

си́з·ый, -ая, -ое, (*a*) bluish dark grey; blue (of various animals, e.g. pigeon)

сиккати́в·, -а, (*m*) siccative, paint dryer

сико́з·, -а, (*m*) (med) sycosis

сикоза́к·, -а, (*m*) (ocean) land-fast polar ice

сикомо́р·, -а, (*m*) (bot) sycamore, *Ficus sycomorus*

си́л·а, -ы, (*f*) *see also under associated words*; (mech) force; power; strength, intensity

    в си́л·ах; быть в си́л·ах be in a position to

    в си́л·е in force, valid

    в си́л·у on the strength of, by virtue of

~, вале́нт·н·ые (*pl*) (chem, phys) valence forces

    ве́ктор· си́л·ы (*m*) vector/line of force

~ ве́тр·а wind strength

~, вое́н·н·о-воз·ду́ш·н·ые (*pl*) (mil) air force

~, вое́н·н·о-мор·ск·и́е (*pl*) navy

~, воз·вращ·а́ющ·ая restoring force

    во·йти́ в си́л·у come into force, become effective

~, вращ·а́ющ·ая angular force; (instr) deflecting force

~, вс·плыв·н·ая buoyancy (of submerged object), free lift

си́л·а

~, дви́ж·ущ·ая motive/moving force; (cryst) driving force

~, жи·в·а́я manpower; kinetic energy

~, за·да·ва́ем·ая external force

~ звук·и sound intensity, volume of sound

~ земле·тряс·е́ни·я earthquake intensity

~ из·луч·е́ни·я intensity of radiation

~, кориолис·ов·а (dynam) Coriolis force

~, лошад·и́н·ая (*abbr* **л.с.**) horse-power

~ магни́т·а magnet power

~, магнито·дви́ж·ущ·ая magneto-motive force

~, ма́сс·ов·ая body force

~, меж·а́том·н·ые (*pl*) interatomic forces

~, много·част·и́чн·ые (*pl*) many-body forces

~, на·правл·я́ющ·ая guide/controlling force

    моме́нт· си́л·ы (*m*) moment of a force; moment of force

~, объ·ём·н·ые (*f*) volumetric forces

~, подъ·ём·н·ая carrying capacity, lifting capacity; lift (aerodynamics); buoyancy (of lighter-than-air craft), aerostatic lift

~, противо·де́й·ствующ·ая opposing force; (instr) controlling force

~, равно·де́й·ствующ·ая resultant force

~, результ·и́рующ·ая resultant force

~ све́т·а luminous intensity, candle power

~ с·вя́з·и (phys) binding force; (chem) bonding strength

~, с·де́рж·ивающ·ая holding force

~ сигна́л·а signal strength

~ со·противл·е́ни·я (dynam) drag; (elec) resistance

    со·ставл·я́ющ·ая си́л·ы (*f*) component force

~ с·цепл·е́ни·я adhesive force

~, термо·электро·дви́ж·ущ·ая (*abbr* **тэдс**) thermoelectromotive force, temf

~ то́к·а (elec) current intensity, current

~ тя́г·и tractive force; thrust (of jet engines etc.)

~ тя́ж·ест·и force of gravity

~, у·де́рж·ивающ·ая confining force

~, центро·бе́ж·н·ая centrifugal force

~, электро·дви́ж·ущ·ая (*abbr* **эдс**) electromotive force, emf

сила́л·, -а,   (*m*) (met) Silal

сила́н·, -а,   (*m*) (chem) silane, silicon hydride

си́л·и,   (*nom/acc pl*) of **сил·ь**, *see* **сел·ь**

силикаге́л·ь, -я,   (*m*) (chem) silica gel

силикальци́т·, -а,   (*m*) silicalcite

силик·а́т·, -а,   (*m*) (chem) silicate

~ **бери́лл·и·я**   beryllium silicate

~ **-глы́б·а, -ы,**   (*f*) lump sodium silicate

силикатиз·а́ци·я, -и,   (*f*) silicatization

силика́т·н·ый, -ая, -ое,   (*a*) silicate

силико-   (*int component*) silico-

силик·о́з·, -а,   (*m*) (med) silicosis

силико·ма́рган·ец·, -н·ц·а,   (*m*) silico-manganese

силико·мета́н·, -а,   (*m*) (chem) silico-methane

силико́н·, -а,   (*m*) (chem) silicone

силико́н·ов·ый, -ая, -ое,   (*a*) silicone

силико·терм·и́·я, -и,   (*f*) (met) silico-thermic process

силико·шпи́гел·ь, -я,   (*m*) silicon spiegel iron

сили́т·, -а,   (*m*) silite (electrical insulating material)

силици́д·, -а,   (*m*) (chem) silicide

силициз·а́ци·я, -и,   (*f*) (geol) silicification

сили́ц·и·й, -я,   (*m*) (chem) silicon, Si

силиц·и́ровани·е, -я,   (*n*) (met) siliconizing

силицифик·а́ци·я, -и,   (*f*) (geol) silicification

си́лл·, -а,   (*m*) (geol) sill

силлаб·и́ческ·ий, -ая, -ое,   (*a*) syllabic

силлимани́т·, -а,   (*m*) (met) sillimanite

сил·ов·о́й, -а́я, -о́е,   (*a*) power, force; propulsion, actuating

~ **дета́л·ь**   (*f*) load-bearing part/component

~ **ли́н·и·я**   (*f*) (phys) line of force

~ **ли́н·и·я магни́т·н·ая**   magnetic line of force

~ **ли́н·и·я электр·и́ческ·ая**   electric line of force

~ **пит·а́ни·е**   (*n*) (elec) power supply

~ **по́л·е**   (*n*) (phys) field of force, force field

~ **при·во́д·**   (*m*) power drive

~ **тре·уго́ль·ник·**   (*m*) (mech) triangle of forces

~ **тру́б·к·а**   (*f*) (phys) field tube

~ **у·стано́в·к·а**   (*f*) power plant, propulsion unit, propulsion machinery

~ **фу́нкци·я**   (*f*) (phys) force function

сило·во·оруж·ённост·ь, -и,   (*f*) power-weight ratio

сило·из·мер·и́тел·ь, -я,   (*m*) dynamometer

сило·из·мер·и́тельн·ый, -ая, -ое,   (*a*) force-measuring

сил·о́й   (*adv*) by force

силокса́н·, -а,   (*m*) (chem) siloxane

силокса́н·ов·ая смол·а́   (*f*) siloxane resin

сило·ме́р·, -а,   (*m*) dynamometer

си́лос·, -а,   (*m*) (agr) silo; silage; blunger (for mixing slaked lime and water)

си́лос·н·ый, -ая, -ое,   (*a*) of **си́лос·**

~ **ба́шн·я**   (*f*) silo

~ **культ·у́р·а**   (*f*) silage crop

~ **ма́сс·а**   (*f*) silage

силос·ова́ть, -у́ют,   (*imp*) make in silage

силосо·ре́з·к·а, -и,   (*g pl*) **-з·ок·,** (*f*) silage cutter

силуми́н·, -а,   (*m*) silumin

силу́р·, -а,   (*m*) (geol) Silurian

силур·и́йск·ий, -ая, -ое,   (*a*) (geol) Silurian

силуэ́т·, -а,   (*m*) silhouette

си́л·ы   (*nom/acc pl*) of **си́л·а**

сил·ь, -я,   (*m*) *see* **сел·ь**

сильван·и́т·, -а,   (*m*) (min) sylvanite

сильви́н·, -а,   (*m*) (min) sylvite

сильвини́т·, -а,   (*m*) (min) sylvinite

силь·не́йш·ий, -ая, -ее,   (*superlative*) of **си́ль·н·ый** (q.v.) (sport) best, better; victorious, winning

си́ль·н·о   (*adv*) powerfully, strongly; intensely, vigorously, highly, heavily, closely, extremely

сильно·де́й·ствующ·ий, -ая, -ее,   (*a*) drastic, strong

сильно·леги́р·ованн·ый, -ая, -ое,   (*a*) heavily alloyed; heavily doped (semi-conducting materials etc.)

сильно·магни́т·н·ый, -ая, -ое,   (*a*) highly magnetic

сильно·с·вя́з·анн·ый, -ая, -ое,   (*a*) closely coupled, tightly bound

сильно·то́ч·н·ый, -ая, -ое,   (*a*) (elec) power, heavy-current

-си́ль·н·ый,   (*adj component*) -horsepower

си́ль·н·ый, -ая, -ое,   (*a*) strong, powerful, potent, intense, vigorous, heavy (of rain etc.), hard, severe, (med) sthenic

~ **взаимо·де́й·стви·е**   (*n*) vigorous interaction; (nucl) strong coupling

~ **деформ·а́ци·я**   (*f*) severe strain

~ **с·вяз·ь**   (*f*) strong/close/tight coupling; (chem) strong/tight bond; close connection (figurative)

Си́льсби, гипоте́з·а (*f*) (phys) Silsbee hypothesis

сильфо́н·, -а, (*m*) bellows

~, у·правл·я́ющ·ий (autom) actuating bellows

сильфо́н·н·ый, -ая, -ое, (*a*) bellows, (instr) bellows-type, bellows-operated

~ регул·я́тор· (*m*) bellows-type controller

сильхро́м·, -а, (*m*) (met) silchrome

симбат·но (*adv*) symbatically

симба́т·ност·ь, -и, (*f*) (biol) symbasis; agreement, similarity

сим·био́з·, -а, (*m*) (biol) symbiosis

сим·био́нт·, -а, (*m*) (biol) symbiont

сим·биот·и́ческ·ий, -ая, -ое, (*a*) symbiotic

си́мвол·, -а, (*m*) symbol

~ зна́к·а sign digit (computers)

симв系л·и́зм·, -а, (*m*) symbolism

си́менс·, -а, (*m*) (elec) mho (unit of conductance)

Си́менс·, -а, (*m*) Siemens

симметр·и́р·овать, -уют, (*imp and perf*) balance, match

симметр·и́рующ·ий, -ая, -ее, (*pres part act*) of **симметр·и́р·овать**

~ у·стро́й·ств·о (*n*) (elec) balancer; (telecom) bazooka

симметр·и́ческ·ий, -ая, -ое, (*a*) symmetrical

симметр·и́чност·ь, -и, (*f*) symmetry

симметр·и́чн·ый, -ая, -ое, (*a*) symmetrical, (telecom) balanced

~ цеп·ь (*f*) (telecom) balanced circuit

симметр·и́·я, -и, (*f*) symmetry, symmetrical shape/layout

~ втор·о́го по·ря́д·к·а (cryst) twofold symmetry, binary system

~ вы́с·ш·его по·ря́д·к·а (cryst) sixfold symmetry, hexagonal system

~ движ·е́ни·я (cryst) movement symmetry

~, зерка́ль·н·ая plane symmetry

~ от·нос·и́тельн·о о́с·и axial symmetry

~ по от·нош·е́ни·ю вращ·е́ни·я rotational symmetry

~, то́ч·ечн·ая point symmetry

~ тре́т·ьего по·ря́д·к·а (cryst) threefold symmetry, trigonal system

це́нтр· симме́тр·и·и (*m*) (cryst) centre/center of symmetry

симпат·и́ческ·ий, -ая, -ое, (*a*) sympathetic

~ нерв·н·а́я систе́м·а (*f*) sympathetic nervous system

симпат·и́ческ·ий, -ая, -ое

~ черн·и́л·а (*pl*) sympathetic/invisible ink

симпат·и́чн·ый, -ая, -ое, (*a*) likeable, likable

симпа́т·и·я, -и, (*f*) sympathy

симпато·бла́ст·, -а, (*m*) (zool) sympathoblast

симпла́ст·, -а, (*m*) (biol) symplast

си́мплекс·н·ый, -ая, -ое, (*a*) simplex, single, one-way

~ кана́л· (*m*) (telecom) simplex/oneway channel

~ с·вяз·ь (*f*) (telecom) simplex communication

симподиа́ль·н·ый, -ая, -ое, (*a*) sympodial

Си́мпсон·а, фо́рмул·а (*f*) (math, surv) Simpson's rule

симпто́м·, -а, (*m*) (med) symptom

симптома́т·ик·а, -и, (*f*) (med) symptomatology

симптомат·и́ческ·ий, -ая, -ое, (*a*) (med) symptomatic

симул·и́р·овать, -уют, (*imp and perf*) simulate

симульта́н·н·ый, -ая, -ое, (*a*) simultaneous

симул·я́нт·, -а, (*m*) simulator; malingerer

симул·я́ци·я, -и, (*f*) simulation

син·а́нтроп·, -а, (*m*) (pal) Peking man, Sinanthropus

син·а́нги·й, -я, (*m*) (bot) synangium

син·а́пс·а, -ы, (*f*) (physiol) synapse

син·апси́д·ы (*pl*) (pal) synapsids, Synapsida

син·а́псис·, -а, (*m*) (genet) synapsis

син·апти́кул·а, -ы, (*f*) (zool) synapticula

син·апт·и́ческ·ий, -ая, -ое, (*a*) (physiol) synaptic

синга́м·, -а, (*m*) (zool) Syngamus trachea

син·га́м·и·я, -и, (*f*) (genet) syngamy

син·гам·о́з·, -а, (*m*) (vet) syngamiasis

син·ген·е́з·, -а, (*m*) (genet) syngenesis

син·генет·и́ческ·ий, -ая, -ое, (*a*) syngenetic, contemporaneous

син·гени́т·, -а, (*m*) (min) syngenite

сингле́т·, -а, (*m*) (chem, phys) singlet

сингле́т·н·ый, -ая, -ое, (*a*) (phys) singlet

сингле́т-сингле́т·н·ый, -ая, -ое, (*a*) (phys) singlet-singlet

син·гомо·клина́л·ь, -и, (*f*) (geol) synhomocline

син·гон·и́·я, -и, (*f*) (cryst) syngony, system

~, агир·н·ая triclinic system

~, дигир·н·ая orthorhombic system

~, изо·метр·и́ческ·ая isometric/cubic system

~, квадра́т·н·ая tetragonal system

~, клино·ромб·и́ческ·ая monoclinic system

~, полигир·н·ая cubic/isometric system

~, пра́в·ильн·ая cubic symmetry

~, ромб·и́ческ·ая orthorhombic system

~, шести·член·н·ая hexagonal system

сингуля́р·ност·ь, -и, (*f*) singularity

сингуля́р·н·ый, -ая, -ое, (*a*) singular

~ то́ч·к·а (*f*) (math) singular point

син·дактил·и́·я, -и, (*f*) (zool) syndactyly

синдесмо́·з·, -а, (*m*) (zool) syndesmosis

синдесмо·ло́г·и·я, -и, (*f*) (anat) syndesmology

синдесмо·том·и́·я, -и, (*f*) (med) syndesmotomy

син·детико́н·, -а, (*m*) seccotine

син·дет·и́ческ·ий, -ая, -ое, (*a*) syndetic(al)

синдика́т·, -а, (*m*) syndicate

синдио·такт·и́ческ·ий, -ая, -ое, (*a*) (chem) syndiotactic

синдиц·и́р·овать, -уют, (*imp and perf*) syndicate

син·дро́м·, -а, (*m*) (med) syndrome

сине- (*root*) blue; ' cyano- (*see also* **циано-**); sine-

сине·багр·о́в·ый, -ая, -ое, (*a*) livid

син·ев·а́, -ы́, (*f*) blue, dark blue; blue stain (defect in timber)

син·ева́т·ый, -ая, -ое, (*a*) bluish

сине·голо́в·ник·, -а, (*m*) (bot) eryngo, *Eryngium*

син·еклиз·а, -ы, (*f*) (geol) syneclise

сине·ло́м·к·ий, -ая, -ое, (*a*) (met) blue-short/brittle

сине·ло́м·кост·ь, -и, (*f*) (met) blue brittleness

синемю́рск·ий ярус· (*m*) (geol) Sinemurian stage

син·ерг·и́д·ы (*pl*) (bot) synergids, synergidae

син·ерг·и́зм·, -а, (*m*) (physiol) synergism

син·ерг·и́·я, -и, (*f*) (med) synergy

син·ере́зис·, -а, (*m*) (chem) syneresis

сине·ро́д·, -а, (*m*) *and see* **циа́н·** (chem) cyanogen

сине·ро́д·ист·ый, -ая, -ое, (*a*) cyanide (of) (lower or -ous)

~ мед·ь (*f*) cuprous cyanide

син·е́·ть, -ют, (*imp*) go/become/turn blue; look/appear blue

синигри́н·, -а, (*m*) (chem) sinigrid, potassium myronate

си́н·ий, -яя, -ее, (*a*) blue, dark blue

~, гидро́н·ов·ый (chem) hydron blue

~ ко́п·и·я (*f*) (phot) blue print

син·и́льн·ый, -ая, -ое, (*a*) prussic, hydrocyanic; bluing

~ кисл·от·а́ (*f*) hydrocyanic acid, prussic acid

син·и́ть, -я́т, (*imp*) blue

син·и́ц·а, -ы, (*i*) -ей, (*f*) (orn) titmouse, *Parus*, (*pl*) titmice, *Paridae*

син·кариофи́т·, -а, (*m*) (bot) synkaryophyte, syncaryophyte

син·ка́рп·и·я, -и, (*f*) (bot) syncarpy

син·карп·и́ческ·ий, -ая, -ое, (*a*) (bot) syncarpous

син·кла́з·а, -ы, (*f*) (geol) synclase

син·клина́л·ь, -и, (*f*) (geol) syncline

син·клина́ль·н·ый, -ая, -ое, (*a*) (geol) syncline, synclinal

син·клино́р·и·й, -я, (*m*) (geol) synclinorium

син·мо́рф·н·ый, -ая, -ое, (*a*) synmorphic

синова́ль·н·ый, -ая, -ое, (*a*) of **синови·я**

~ жи́дк·ост·ь (*f*) (zool) synovia

~ оболо́ч·к·а (*f*) (zool) synovial membrane

синови·я, -и, (*f*) (anat) synovia

синод·и́ческ·ий, -ая, -ое, (*a*) synodic

~ ме́сяц· (*m*) (astron) synodic month

сино·ло́г·и·я, -и, (*f*) sinology

син·о́ним·, -а, (*m*) synonym

син·они́м·ик·а, -и, (*f*) synonymy

син·о́псис·, -а, (*m*) synopsis

син·о́пт·ик·, -а, (*m*) weather forecaster, synoptic meteorologist

син·опт·и́ческ·ий, -ая, -ое, (*a*) synoptic

~ ка́рт·а (*f*) (meteor) weather map, synoptic chart

син·оро·ген·и́ческ·ий, -ая, -ое, (*a*) (geol) synorogenic

синосто́з·, -а, (*m*) (zool) synostosis

си́нтаксис·, -а, (*m*) (gram) syntax

си́н·тез·, -а, (*m*) synthesis

~ ядр·а́ nuclear fusion

синтез·а́тор·, -а, (*m*) synthesizer

синтез·и́р·овать, -уют, (*imp and perf*) synthesize

синтер·, -а, (*m*) sinter

синтер·ованн·ый, -ая, -ое, (*past part pass*) sintered

синтет·ическ·ий, -ая, -ое, (*a*) synthetic

синтин·, -а, (*m*) synthole (synthetic gasoline)

син·тип·, -а, (*m*) syntype

синтол·, -а, (*m*) *see* **синтин·**

синто·мицин·, -а, (*m*) (pharm) synthomycin

син·тон·изаци·я, -и, (*f*) (telecom) tuning

син·трон·, -а, (*m*) (elec) syntron

син·узи·я, -и, (*f*) (biol) synusia

синус·, -а, (*m*) (math) sine; (anat) sinus

~ -буссол·ь, -и, (*f*) (elec) sine galvanometer

~ -версус·, -а, (*n*) (math) versed sinus

~ -ко·синус·н·ый, -ая, -ое, (*n*) sine-cosine

синус·ит·, -а, (*m*) (med) sinusitis

синус·н·ый, -ая, -ое, (*a*) sine; (med) sinus

синус·оид·а, -ы, (*f*) (math) sinusoid, sine curve

синусоидаль·н·ый, -ая, -ое, (*a*) (math) sinusoidal

~ за·кон· (*m*) sine law

син·фаз·ировани·е, -я, (*n*) (elec) phase synchronization

син·фаз·ност·ь, -и, (*f*) (elec) phase synchronism

син·фаз·н·ый, -ая, -ое, (*a*) (elec) cophasal, cophased, in-phase

~ антенн·а (*f*) (rad) cophased array

~ работ·а (*f*) in-phase operation

синхон·ин·, -а, (*m*) (chem) cinchonine

синхро·вс·пыш·к·а (TV) burst

сигнал синхро·вс·пыш·к·и (*m*) (TV) burst signal

синхро·генер·атор·, -а, (*m*) selsyn, synchro, magslip, selsyn generator, synchrogenerator, synchronizing-pulse generator

синхро·дин·н·ый, -ая, -ое, (*a*) synchrodyne

синхрон·доз·, -а, (*m*) (zool) synchrondosis

синхрониз·атор·, -а, (*m*) synchronizer, synchronizing gear; (M/T) synchromesh unit

синхронизацион·н·ый, -ая, -ое, (*a*) synchronization, synchronizing

синхрониз·аци·я, -и, (*f*) synchronization; locking-in (frequencies); locking-on (radar)

синхрониз·аци·я

    интервал· синхрониз·аци·и (*m*) locking range (radar)

~, при·нуд·ительн·ая forced synchronization; capture, capture effect, (radar) locking

~ раз·верт·к·и (TV) line synchronization

синхрониз·ированн·ый, -ая, -ое, (*past part pass*) of **синхрониз·ир·овать;** synchronized; (elec) synchronous

~ асинхрон·н·ый двиг·атель· (*m*) (elec) synchronous-asynchronous motor

синхрониз·ир·овать, -уют, (*imp and perf*) synchronize, bring into step/phase, lock in/on

синхрониз·ируемост·ь, -и, (*f*) (elec, rad) locking-in capacity; lock-in feature (radar)

синхрониз·ирующ·ий, -ая, -ее, (*pres part act*) of **синхрониз·ир·овать**

~ генер·атор· (*m*) synchronizing-pulse generator, synchro, selsyn

синхрон·изм·, -а, (*m*) (elec) synchronism

синхрон·ичност·ь, -и, (*f*) synchronism, synchroneity

синхрон·н·о-след·ящ·ая систем·а (*f*) servo-mechanism

синхрон·ност·ь, -и, (*f*) synchronism, synchroneity

синхрон·н·ый, -ая, -ое, (*a*) synchronous

~ генер·атор· (*m*) (elec) synchronous generator

~ машин·а (*f*) (elec) synchronous machine

~ пере·да·ч·а (*f*) synchronous transmission

~ с·вяз·ь (*f*) (elec) synchroconnection, synchro

~ с·вяз·ь, бес·про·вод·н·ая unwired synchroconnection

синхроно·скоп·, -а, (*m*) (instr) synchroscope

синхро·пере·ключ·атель·, -я, (*m*) (elec) synchronous changeover switch

синхро·трон·, -а, (*m*) (nucl) synchrotron, synchrotron accelerator

синхро·фазо·трон·, -а, (*m*) (nucl) proton-synchrotron accelerator

синхро·цикло·трон·, -а, (*m*) (nucl) synchrocyclotron accelerator, f-m cyclotron

синхро·час·ы, -ов, (*pl*) synchroclock

син·ци́т·и·й, -я, (*m*) (cyt) syncytium

си́н·ь, -и, (*f*) blue, dark blue

~, ко́бальт·ов·ая cobalt blue

~, метиле́н·ов·ая alkaline methylene blue

~, молибде́н·ов·ая molybdenum blue

~, пари́ж·ск·ая French blue

си́нь·ка, -и, (*f*) (*and see* **си́н·ь**) (phot, eng) blue-print; blueing; blue

синь·ка́л·и (*n indecl*) (chem) potassium ferrocyanide

синьор·ите́т·, -а, (*m*) seniority

син·эко·ло́г·и·я, -и, (*f*) (bot) syn-ecology

синэстро́л·, -а, (*m*) (pharm) sinestrol, dihydro stilbestrol

син·ю́х·а, -и, (*f*) (bot) Jacob's ladder, *Polemonium*; (med) cyanosis

син·я́к·, -а́, (*m*) bruise, (med) ecchymosis; (bot) Viper's bugloss, *Echium*; blewit, *Gyroporus cyanescens* (edible fungus)

сип·, -а, (*m*) (orn) vulture, *Gyps*; hoarseness; **сип·,** (*past masc sing*) of **си́п·нуть**

сип·е́ть, -я́т, (*imp*) speak hoarsely

си́пл·ый, -ая, -ое, (*a*) hoarse

си́п·нуть, -нут, (*imp*) become hoarse

сип·у́х·а, -ы, (*f*) (orn) owl, *Strix*

сире́н·а, -ы, (*f*) siren; (*pl*) (zool) *Sirenia*

сире́н·ев·ый, -ая, -ое, (*a*) lilac; lilac-coloured; (*pl noun decl as adj*), (zool) *Sirenidae*

сире́н·ь, -и, (*f*) (bot) lilac

сиринго·бульби·я, -и, (*f*) (med) syringobulbia

сиринго·миэл·и́·я, -и, (*f*) (med) syringomyelia

сиринго·том·и́·я, (med) syringotomy

сиро́кко (*n indecl*) (meteor) Sirocco; blower, fan

сиро́п·, -а, (*m*) syrup

сиро́п·н·ый, -ая, -ое, (*a*) syrup

сир·от·а́, -ы́, (*f*) orphan

сирфи́д·ы (*pl*) (ent) hoverflies, *Syrphidae*

си́р·ый, -ая, -ое, (*a*) foresaken

систе́м·а, -ы, (*f*) *see under associated words*; system

систематиз·а́ци·я, -и, (*f*) systematization, classification

систематиз·и́р·овать, -уют, (*imp and perf*) systematize

системáт·ик·, -а, (*m*) (biol) taxonomist

систематик·а, -и, (*f*) (phys) systematics; (biol) systematics, taxonomy, classification

системáт·ик·а

~ жив·о́тн·ых орган·и́зм·ов (zool) animal systematics

~ почв· soil classification

системат·и́ческ·ий, -ая, -ое, (*a*) systematic

системат·и́чн·ый, -ая, -ое, (*a*) systematic

систе́рн·а -ы, (*f*) (*and see* **цисте́рн·а**) (shipb) tank

~, акти́в·н·ая anti-rolling tank

~, балла́ст·н·ая ballast tank

~, борт·ов·а́я side tank

~, гла́в·н·ая main tank

~, глуб·о́к·ая deep tank

~, диффере́нт·н·ая trimming tank (submarine)

~, за·мест·и́тельн·ая quick-acting compensating tank (submarine)

~, кре́н·ов·ая compartment with flooding arrangements

~, на·по́р·н·ая gravity tank

~, противо·с·плы́т·н·ая automatic inboard vent tank (submarine)

~, с·пас·а́тельн·ая safety tank (submarine)

~, с·то́ч·н·ая drain oil tank, settling tank

~, у·равн·и́тельн·ая compensating tank (submarine)

~, у·с·поко·и́тельн·ая anti-rolling tank

си́стол·а, -ы, (*f*) (physiol) systole

ситéг·а, -и, (*f*) sea mist, fine rain

си́т·ец·, -т·ц·а, (*i*) **-т·ц·ем,** (*m*) (text) print, cotton print, chintz

~, на·бив·н·о́й printed calico/cotton

си́т·ник·, -а, (*m*) (bot) rush, *Juncus*; sitnik (an edible fungus); (meteor) sea mist

си́т·ников·ые, -ых, (*pl decl as adj*) (bot) rush family, *Juncaceae*

си́т·о, -а, (*n*) sieve, sifter, screen, riddle

~, бараба́н·н·ое rotary screen

~, вибрацио́н·н·ое jigging screen

~, со·тряс·а́тельн·ое impact screen

сито·ви́д·н·ый, -ая, -ое, (*a*) sieve-like, cribriform, cribrose

сит·ови́н·а, -ы, (*f*) rotted wood

си́т·ов·ый, -ая, -ое, (*a*) of **си́т·о**

~ анáлиз· (*m*) screen analysis (of crushed/ground material)

сито·ло́г·и·я, -и, (*f*) (food) sitology

сито·токси́н·, -а, (*m*) (food) sitotoxin

си́т·очн·ый, -ая, -ое, (*a*) of **си́т·о**

ситуáци·я, -и, (*f*) situation (of events); details (on map)

ситу́г·а, -и, *(f)* see **ситег·а**
ситу́х·а, -и, *(f)* see **ситег·а**
си́т·цев·ый, -ая, -ое, *(a)* (text) print, chintz
ситце·печа́т·ани·е, -я, *(n)* (text) calico/cotton printing
~, вы·трав·н·о́е (text) discharge printing
си́т·чат·ый, -ая, -ое, *(a)* of **си́т·о**
сиф *(indecl)* (comm) cif (cost, insurance, freight)
си́филис·, -а, *(m)* (med) syphilis
сифо́н·, -а, *(m)* siphon; siphon trap (sanitary engineering)
де́й·стви·е сифо́н·а *(n)* siphonage, siphoning
~ -реко́рдер·, -а, *(m)* (telecom) undulator
сифон·и́ровани·е, -я, *(n)* siphoning
сифо́н·н·ый, -ая, -ое, *(a)* of **сифо́н·**
сифоно·зо́ид·, -а, *(m)* (zool) siphonozooid
сифоно·сте́л·а, -ы, *(f)* (bot) siphonostele
сифоно·сте́л·ь, -и, *(f)* = **сифоно·-сте́л·а**
сифоно·сте́ль·н·ый, -ая, -ое, *(a)* (bot) siphonostelic
сифоно·сто́м·н·ый (biol) siphonostomatous
сифоно·фо́р·ы, -ов, *(m pl)* (zool) siphonophores, *Siphonophora*
сифу́нкл·, -а, *(m)* (zool) siphuncle
си·хром·а́л·ь, -я, (met) Sicromal (Si–Cr–Al alloy)
си́штоф·, -а, *(m)* Si-stoff (waste product of aluminium sulphate)
си·я́ни·е, -я, *(n)* radiance
~, поля́р·н·ое (astron) aurora, polar aurora
~, се́вер·н·ое Northern Lights, Aurora Borealis
~, ю́ж·н·ое Aurora Australis
си·я́·ть, -ют, *(imp)* spine, beam
СК *(abbr)* **синт·ет·и́ческ·ий каучу́к·** synthetic rubber
скабио́з·а, -ы, *(f)* (bot) scabiosa, *Scabiosa*
с·ка́ж·ет, *(fut 3rd sing)* of **с·каз·а́ть**
с·каз·а́ть, *(fut 3rd pl)* **с·ка́ж·ут,** *(perf)* say, speak, tell; **-ся** tell on, have an effect, affect
с·каз·у́ем·ое, -ого, *(n decl as adj)* (gram) predicate
с·ка́з·ыва·ться, -ются, *(imp)* tell on, have an effect

скак·а́ть, *(pres 3rd pl)* **ска́ч·ут,** *(imp)* jump, skip, hop; jump/skip/hop along; gallop
скак·н·у́ть, -у́т, *(perf)* make a jump/skip/hop
скак·ов·о́й, -а́я, -о́е, *(a)* racing, steeplechasing
скак·у́н·, -а́, *(m)* skipper, jumper; (ent) *(pl)* tiger beetles, *Cicindelidae*
скал·а́, -ы́, *(nom pl)* **ска́л·ы,** *(f)* rock; scale (graduated)
скалено·э́др·, -а, *(m)* (cryst) scalenohedron
скалено·эдр·и́ческ·ий, -ая, -ое, *(a)* scalenohedral
скал·и́ст·ый, -ая, -ое, *(a)* rocky
ска́лк·а, -и, *(g pl)* **-л·ок·,** *(f)* roller, pin; (food) rolling-pin; ironing roller; (mech eng) plunger, piston, ram; rod, spindle (of slide valve)
с·ка́л·ывани·е, -я, *(n)* *(v n)* see **с·ка́·л·ыва·ть;** (mech) shearing, simple shear; shearing strain/deformation; spallation
пло́ск·ост·ь с·ка́л·ывани·я *(f)* shear plane (metal machining); (cryst) gliding plane
про́ч·н·ость на с·ка́л·ывани·е *(f)* shearing strength, strength in simple shear
у́гол· с·ка́л·ывани·я *(m)* cutting angle (metal machining)
с·ка́л·ыва·ть, -ют, *(imp)* shear/split/chop off; pin/spike together; prick out, mark (with pins etc.)
с·ка́л·ывающ·ий, -ая, -ее, *(pres part act)* of **с·ка́л·ыва·ть**
с·кальк·и́р·овать, -уют, *(perf)* trace, trace a copy of (on paper)
с·калькул·и́р·овать, -уют, *(perf)* calculate
с·ка́ль·н·ый, -ая, -ое, *(a)* of **скал·а́**
ска́льп·, -а, *(m)* (anat) scalp
ска́льпел·ь, -я, *(m)* (med) scalpel
скальп·и́р·овать, -уют, *(imp and perf)* scalp
скаля́р·, -а, *(m)* (math) scalar quantity/field
скаля́р·н·ый, -ая, -ое, *(a)* (math) scalar
скаме́·ечн·ый, -ая, -ое, *(a)* bench
скаме́й·к·а, -и, *(f)* bench
~, про·во́д·ков·ая *(f)* (met roll) guide bench
скамь·я́, -и́, *(nom pl)* **скамь·и́,** *(g pl)* **скаме́·й,** *(f)* bench
~ под·суд·и́м·ых (law) dock

скамь·я́

~, с·пас·а́тельн·ая (naut) life raft

ска́нд·иев·ый, -ая, -ое, (a) (chem) scandium

ска́нд·и·й, -я, (m) (chem) scandium, Sc

сканд·и́ровани·е, -я, (n) scanning

сканд·и́р·овать, -уют, (imp) scan

с·ка́пл·ива·ть, -ют, (imp) save, save up, accumulate; **-ся** (pass); collect, gather, pile up, accumulate

скаполи́т·, -а, (m) (min) scapolite, wernerite

с·капот·и́р·овать, -уют, (perf) (air) nose over

с·ка́п·ыва·ть, -ют, (imp) dig off, level (by digging)

скарифик·а́тор·, -а, (m) (civ eng, agr) scrarifier; (agr) seed-scoring tool

скарифик·а́ци·я, -и, (f) scarification

скарифиц·и́р·овать, -уют, (imp and perf) scarify

скарлати́н·а, -ы, (f) (med) scarlet fever, scarlatina

с·ка́рмл·ива·ть, -ют, (imp) feed, use up (in feeding)

ска́рн·, -а, (m) (min) scarn

с·ка́т·, -а, (m) incline, slope; (min) ore chute; (build) roof slope; (rail) wheel set/assembly; (mech eng) shoulder; (phys) edge (of a pulse); (fish) skate, ray, (pl) *Batoidea*

~, за́д·н·ий trailing edge (of a pulse)

~, на·бег·а́ющ·ий leading edge (of a pulse)

~, с·бег·а́ющ·ий trailing edge (of a pulse)

угол· с·ка́т·а (m) angle of pitch

с·ка́т·анн·ый, -ая, -ое, (past part pass) see **с·ка́т·ыва·ть**

с·кат·а́·ть, -ют, (perf) see **с·ка́т·ы·ва·ть**

скато́л·, -а, (m) (chem) skatole

ска́терт·н·ый, -ая, -ое, (a) of **ска́терт·ь**

с·ка́терт·ь, -и, (f) table-cloth

с·ка́т·н·ый, -ая, -ое, (a) rolling

с·ка́т·ыва·ть, -ют, (imp) roll up, roll (into a shape); roll off/down; (text) felt

скафа́ндр·, -а, (m) one-piece suit

~, авиацио́н·н·ый pressure suit

~, водо·ла́з·н·ый diving equipment, diver's suit

~, выс·о́тн·ый high-altitude pressure suit

~, стратосфе́р·н·ый high-altitude pressure suit

скафо·по́д·, -а, (m) (zool) scaphopod

скач·а́, (pres gerund) of **скак·а́ть**

ска́ч·ет, (pres 3rd sing) of **скак·а́ть**

скач·к·а, -и, (g pl) -ч·ек·, (f) gallop; (pl) races, horse racing; **скач·к·а́;** (gen sing) of **скач·о́к·**

скачко·обра́з·н·ый, -ая, -ое, (a) spasmodic, intermittent, jerky, irregular; sudden, abrupt; (zool) saltatory

~ из·мен·е́ни·е (n) abrupt/sudden change; irregular variation

скач·о́к·, -ч·к·а́, (m) jump, leap, bound; (phys) discontinuity, jump, step, bound, sudden change; (rad) skip

~ давл·е́ни·я pressure jump/discontinuity

~, един·и́чн·ый (autom) unit step

~, за·пазд·ыва́ющ·ий discontinuity

ли́н·и·я скач·к·о́в (f) (math) saltus curve

~ по·глощ·е́ни·я (phys) absorption discontinuity/edge

~ потенциа́л·а (elec) potential jump, jump in potential

~ раз·реж·е́ни·я (aerodyn) expansion shock; low-pressure wave

~ температу́р·ы jump in temperature

~ у·плот·не́ни·я (aerodyn) shock wave, shock

~ у·плот·не́ни·я, при·со·един·ён·н·ое attached shock

~ у·плот·не́ни·я, прям·о́й normal shock

скач·к·а́ми (adv) in jumps/jerks, jerkily, irregularly

скач·к·о́м (adv) suddenly, abruptly

ска́ч·ущ·ий, -ая, -ее, (pres part act) of **скак·а́ть**

с·ка́ш·ивани·е, -я, (n) (v n) of **с·ка́ш·ива·ть**

с·ка́ш·ива·ть, -ют, (imp) mow, mow down, cut, scythe; slant, incline, bevel, slope; twist, turn (of limbs etc.)

скваж·ин·а, -ы, (f) hole, drill hole; borehole, well; chink, slit

~, бур·ов·а́я borehole

~, вз·рыв·н·а́я blast hole

~, в·ру́б·ов·ая (min) cut hole

~ -газо·генер·а́тор·, -а, (m) (min) blind borehole

~, за·мо́ч·н·ая keyhole

~, инжекцио́н·н·ая (oil) injection borehole/well

~, ко́нтур·н·ая (min) line hole

~, котл·о́в·ая (min) sprung/squibbed hole

~, мор·ск·а́я (oil) off-shore oil well

скваж·ин·а
~, **на·гнет·а́тельн·ая** (oil) injection well, water-injection well
~, **на·пра́вл·енн·ая** (min) diverted hole
~, **о·по́р·н·ая** (oil) test well
~, **по·глот·и́тельн·ая** (oil) absorption well; injection well (for nuclear refuse)
~, **под·са́с·ывающ·ая** (oil) offset well
~, **пульс·и́рующ·ая** (oil) surging well
~, **раз·ве́д·очн·ая** (min) exploratory well
~, **раз·рез·н·а́я** (min) empty/uncharged hole
~, **ру́д·н·ая** ore hole
сква́ж·инн·ый, -ая, -ое, (*a*) of **сква́ж·ин·а**
сва́ж·истост·ь, -и, (*f*) porosity, porousness
сква́ж·ист·ый, -ая, -ое, (*a*) porous
сква́ж·ност·ь, -и, (*f*) porosity; (elec) pulse duty factor
сквале́н·, -а, (*m*) squalene
с·ква́нт·овать, -уют, (*perf*) *see* **ква́нт·овать**
скве́р·, -а, (*m*) square, garden square (in town)
скве́р·н·ый, -ая, -ое, (*a*) nasty, repulsive, bad
сквидж·, -а, (*m*) cushion stock (of a tire)
сквоз·и́ть, -я́т, (*imp*) blow through, be draughty; shine/show through
сквоз·ник·, -а, (*m*) passageway
сквоз·н·о́й, -а́я, -о́е, (*a*) through; built-up (telephones)
~ **брига́д·а** (*f*) start-to-finish production team
~ **движ·е́ни·е** (*n*) through traffic
~ **дол·и́н·а** (*f*) (geol) watergap
~ **со·един·е́ни·е** (*n*) through/built-up connection (telephones)
~ **цеп·ь** (*f*) built-up circuit (telephones)
сквоз·ня́к·, -а́, (*m*) draught
сквозь (*prep + acc*) through
сквозь·магмат·и́ческ·ий, -ая, -ое, (*a*) (geol) transmagmatic
сквор·е́ц·, -р·ц·а́, (*i*) **-р·ц·о́м,** (*m*) (orn) starling
сквор·е́чник·, -а, (*m*) roke (defect in rolled metal); nesting box for birds, on a long pole
сквор·е́чниц·а, -ы, (*i*) **-е́й,** (*f*) *see* **сквор·е́чник·**
сквороден·ь, -дн·я, (*m*) dovetail
~, **кос·о́й** splayed dovetail
скеле́т·, -а, (*m*) skeleton, frame, shell

скеле́т·н·ый, -ая, -ое, (*a*) of **скеле́т·**; skeletal
~ **схе́м·а** (*f*) skeleton/simplified diagram
~ **флор·** (*m*) (ship) skeleton floor
скепт·и́ческ·ий, -ая, -ое, (*a*) sceptic(al)
скиа·скоп·и́·я, -и, (*f*) (med) skiascopy, retinoscopy
скиа·тро́н·, -а, (*m*) (elec) skiatron, dark-trace tube
скид·, -а, (*m*) skid, pallet, stillage (for mechanical handling of materials)
скиддер·, -а, (*m*) skidder (logging)
с·ки́д·к·а, -и, (*g pl*) **-д·ок·,** (*f*) (*v n*), *see* **с·ки́д·ыва·ть;** rebate, reduction, discount
с·ки́д·ыва·ть, -ют, (*imp*) throw down/off; deduct, reduce, allow for
с·ки́н·уть, -ут, (*perf*) *see* **с·ки́д·ыва·ть**
скин-сло́й (*m*) skin
  глуб·ин·а́ скин-слоя skin depth
  толщ·ина́ скин-слоя skin depth
скин-эффе́кт·, -а, (*m*) (elec) skin effect
скио·фи́ль·н·ый, -ая, -ое, (*a*) (bot) skiophilous, sciophilous
скип·, -а, (*m*) (min, civ eng) skip
скипида́р·, -а, (*m*) (chem) turpentine, oil of turpentine
скипидар·и́н·, -а, (*m*) (chem) turpentine substitute, white spirit
скип·ов·о́й, -а́я, -о́е, (*a*) of **скип·**
~ **я́м·а** (*f*) (met) skip pit
~ **подъ·ём·ник·** (*m*) skip hoist
скирд·, -а́, (*m*) *see* **скирд·а́**
скирд·а́, -ы́, (*nom pl*) **скӣрд·ы,** (*g pl*) **скирд·** (*f*) (agr) stack, rick
скирд·ова́ни·е, -я, (*n*) stacking, ricking; building stacks/ricks
скирдо·во́з·, -а, (*m*) (agr) stack carrier
с·кис·а́·ть, -ют, (*imp*) become/turn sour
с·ки́с·нуть, -нут, (*perf*) *see* **с·кис·а́·ть**
скиф·, -а, (*m*) (naut) skiff, dinghy
с·кла́д·, -а, (*m*) warehouse, storehouse, store, (mil) dump; build, physique; disposition, mentality
~, **во́ль·н·ый** bonded warehouse
с·клад·а́льн·ый, -ая, -ое, (*a*) folding, pleating, creasing
с·клад·а́льщик·, -а, (*m*) (text) folder
с·клад·и́р·овать, -уют, (*imp and perf*) store
с·кла́д·к·а, -и, (*g pl*) **-д·ок·,** (*f*) fold, crease, crimp, (text) pleat, tuck; (met) lap; (anat) fold, wrinkle, plica; unit

с·кла́д·к·а
~, ба́нт·ов·ая   (text) box pleat
~, в·ло́ж·енн·ая   (geol) nested fold
~, во́·гн·ут·ая   (geol) downfold
~ дв·ух фу́нкци·й   (math) convolution
~ ме́ст·ност·и   (surv) accident of ground
~, син·клина́ль·н·ая   (geol) syncline
то́ч·к·а с·кла́д·к·и   (f) plait point (on solubility curve)

скла́дко·вы·прямл·е́ни·е, -я,   (n) (geol) unfolding

скла́дко·держ·а́тел·ь, -я,   (m) blank holder (on forging press)

скла́дко·образ·ова́ни·е, -я,   (n) (geol) folding

с·кла́д·н·о́й, -а́я, -о́е,   (a) see also с·кла́д·н·ый; folding, collapsible
~ шлю́п·к·а   (f) folding/collapsible boat

с·кла́д·н·ый, -ая, -ое,   (a) see also с·кла́д·н·о́й; well made/ordered, coherent, harmonious

с·кла́д·очник·, -а,   (m) (text) tucker

с·кла́д·очн·ый, -ая, -ое,   (a) store, storage; pool, pooled (resources etc.)

с·кла́д·ск·о́й, -а́я, -о́е,   (a) warehouse, storehouse, store

с·кла́д·чатост·ь, -и,   (f) (geol) folding

с·кла́д·чат·ый, -ая, -ое,   (a) folded; pleated; wrinkled; (bot) plicate

с·кла́д·чин·а, -ы,   (f) pooling (resources)

с·кла́д·ыва·ть, -ют,   (imp) put/place/lay together, gather, make up; (math) add; fold; give up, surrender; -ся (pass); form, take shape, mature, be established

с·кла́д·ывающ·ий, -ая, -ее,   (pres part act) of с·кла́д·ыва·ть
~ при·бо́р·   (m) (instr, math) adder

с·кле́·енн·ый, -ая, -ое,   (past part pass) see с·кле́·ива·ть; glued, cemented; (cinema) mounted; (math) joined, identified

с·кле́·ечн·ый, -ая, -ое,   (a) of с·кле́й·-к·а
~ пресс·   (m) (cinema) splicer

с·кле́·ива·ть, -ют,   (imp) glue/stick/paste together, cement; (cinema) splice; (build) size, dress; (math) join, identify

с·кле́·ить, -ят,   (perf) see с·кле́·ива·ть

с·кле́й·к·а, -и,   (f) (v n), see с·кле́·ива·ть; adhesive joint; patch

скле́п·, -а,   (m) (arch) crypt

с·кле́пп·анн·ый, -ая, -ое,   (past part pass) see с·кле́п·ыва·ть

с·клеп·а́·ть, -ют,   (perf) see с·клёп·ыва·ть

с·клёп·ыва·ть, -ют,   (imp) rivet, fasten, shackle
~ одно·ря́д·н·ым шв·ом rivet single

скле́р·а, -ы,   (f) (zool) sclera

склер·ейд·а, -ы,   (f) (bot) sclereid, sclerid

склер·енхи́м·а, -ы,   (f) (bot) sclerenchyma

скер·и́т·, -а,   (m) (med) scleritis; (zool) sclerite

склери·фик·а́ци·я, -и,   (f) (bot) sclerosis

склеро·дерм·и́·я, -и,   (f) (med) sclerodermia

склеро́·з·, -а,   (m) (med) sclerosis

склеро́·метр·, -а,   (m) sclerometer, hardness tester

склеро·проте́ин·, -а,   (m) (biochem) scleroprotein

склеро·ско́п·, -а,   (m) (met) scleroscope, rebound hardness tester

склеро·ти́ни·я, -и,   (f) (biol) Sclerotinia

склерот·и́ческ·ий, -ая, -ое,   (a) sclerotic

склеро·том·и́·я, -и,   (f) (med) sclerotomy

склеро́·ци·й, -я,   (m) (bot) sclerotium

скли́з·, -а,   (m) slide; (naut) wooden slip/slipway; (text) race board, shuttle race

скли́з·к·ий, -ая, -ое,   (a) slippery, slimy

склодовски́т·, -а,   (m) (min) sklodowskite

с·кло́н·, -а,   (m) slope, incline, descent; side (of a hill or valley)
~, матер·ико́в·ый   (ocean) continental slope
~, остров·н·о́й   (ocean) insular slope
~, перед·ов·о́й   (geol) frontal slope

с·клон·е́ни·е, -я,   (n) (v n), see с·клон·я́·ть; slope, incline; (math) inclination; (astron) declination; variation; declination (of magnetic compass); hade (of ore bodies); (gram) declension
~, вертика́ль·н·ое   (geol) hade
~, вос·то́ч·н·ое   (nav) easterly variation
~, за́·пад·н·ое   (nav) westerly variation
~, магни́т·н·ое   (nav) magnetic variation, (surv) magnetic declination
~, со́лн·ечн·ое   sun's/solar declination

с·клон·ённ·ый, -ая, -ое,   (past part pass) see с·клон·я́·ть

с·клон·и́ть, -я́т,   (perf) see с·клон·я́·ть

с·кло́н·ност·ь, -и,   (f) disposition, tendency, inclination

с·клóн·н·ый, -ая, -ое, (*a*) prone (to), inclined (to), tendency (to)

с·клон·я́·ть, -ют, (*imp*) incline, bow, bend; strike (a flag); (gram) decline; **-ся** (*pass*); droop, bend over/down; tend; turn to

склян·к·а, -и, (*g pl*) **-н·ок·,** (*f*) hour glass; phial, bottle; bell (of ship's time); (ocean) ice rind

скоб·á, -ы́, (*nom pl*) **скóб·ы,** (*g pl*) **скоб·** (*f*) *see also* **скóб·к·а;** cramp, cramp iron, U-shaped piece, staple, clip; fastener, fastening; snap gauge/gage; shackle; (zool) rotula, kneecap

~, áнкер·н·ая dog stay, girder stay (boiler)

~, буксúр·н·ая towing shackle

~, дву·сторóн·н·яя пре·дéл·н·ая go and not-go snap gauge/gage

~, за·жúм·н·ая clip, binding clip

~, из·мер·úтельн·ая gauge, calliper gauge/gage

~, калúбер·н·ая snap gauge

~, под·вес·н·áя suspension clip

~, пре·дéль·н·ая limit gauge/gage

~, со·един·úтельн·ая joining shackle

~, такелáж·н·ая screw shackle

скобел·ь, -я, (*m*) spokeshave, shaving-knife, draw/drawing-knife

скóб·к·а, -и, (*g pl*) **-б·ок·,** (*f*) *see* **скоб·á** (q.v.); (print etc.) bracket; (horol) pin-pallets

~, клеп·áльн·ая yoke riveter

~, квадрáт·н·ые (*pl*) square brackets

~, крýгл·ые (*pl*) parentheses, round brackets

~, с·пуск·ов·áя trigger guard (rifle)

~, фигýр·н·ые (*pl*) (print) braces, brace brackets

скóбл·енн·ый, -ая, -ое, (*past part pass*) of **скобл·úть**

скобл·úть, -я́т, (*imp*) scrape, shave, plane

скóб·очк·а, -и, (*g pl*) **-чек·,** (*f*) wire staple

скóб·очн·ый, -ая, -ое, (*a*) of **скоб·á, скóб·к·а**

~ машúн·а (*f*) stapling machine, stapler; champagne bottle-fastening machine

скоб·ян·óй, -áя, -óе, (*a*) hardware, ironmongery

~ из·дéл·и·я (*pl*) ironmongery, hardware

с·кóв·анн·ый, -ая, -ое, (*past part pass*) *see* **с·кóв·ыва·ть**

с·ков·áть, (*fut 3rd pl*) **с·ку·ю́т,** (*perf*) *see* **с·кóв·ыва·ть**

сковород·á, -ы́, (*nom pl*) **сковорóд·ы,** (*g pl*) **сковорóд·** (*f*) pan, frying-pan

с·кóв·ыва·ть, -ют, (*imp*) forge, hammer; forge/hammer together; fetter, tie/pin down; freeze hard

с·кóв·ывающ·ий, -ая, -ее, (*pres part act*) of **с·кóв·ыва·ть**

~ грýпп·а (*f*) (mil) holding force

с·кóл·, -а, (*m*) (*v n*), *see* **с·кáл·ыва·ть;** split, chip; cleavage face, sheared/shear/chip surface

ýгол· с·кóл·а (*m*) (cryst) angle of shear

с·колáч·ива·ть, -ют, (*imp*) knock up, knock together, run up; organize, form

сколек·одóнт·, -а, (*m*) (pal) scolecodont

сколёкс·, -а, (*m*) (zool) scolex

сколец·úт·, -а, (*m*) (min) scolecite

сколиó·з·, -а, (*m*) (med) scoliosis

сколиó·метр·, -а, (*m*) (med etc.) scoliometer

с·кóл·к·а, -и, (*g pl*) **-л·ок·,** (*f*) (*v n*), *see* **с·кáл·ыва·ть**

с·кóл·ок·, -л·к·а, (*m*) splinter, chip; perforated/pricked pattern; copy

сколоп·úди·й, -я, (*m*) (ent) scolopidium

с·колот·úть, -я́т, (*perf*) *see* **с·колáч·ива·ть**

с·кóл·от·ый, -ая, -ое, (*past part pass*) *see* **с·кáл·ыва·ть**

с·кол·óть, -ю́т, (*perf*) *see* **с·кáл·ыва·ть**

сколо·фóр·, -а, (*m*) (ent) scolophore

с·колóч·енн·ый, -ая, -ое, (*past part pass*) of **с·колот·úть,** (*perf*); *see* **с·колáч·ива·ть** (*imp*)

сколь (*adv*) so, to such extent

~ у·гóд·н·о (*adv*) arbitrarily

скольж·éни·е, -я, (*n*) slide, sliding, slip, slipping, glide, gliding; (air) sideslip, glissade; (elec) slip

~ винт·á (air) propeller slip

лúн·и·я скольж·éни·я (*f*) (civ eng) line of slip; (met) (*pl*) slip lines/planes

~ мнóж·естве·нн·ое multiplet slip

~ на кры·л·ó (air) wing slide, sideslip, glissade

~ на хвост· (air) tail slide

~ плóск·ост·и plane slippage

плóск·ост·ь скольж·éни·я (*f*) slip plane; (geol) slickensides

под·шúп·ник· скольж·éни·я (mech eng) plain bearing

скольж·éни·е
   по·прáв·к·а на скольж·éни·е (*f*) (dynam) slipstream correction
   скóр·ост·ь скольж·éни·я (*f*) slip velocity/speed
   тр·éни·е скольж·éни·я (mech) sliding friction
~, угл·ов·óе (mech) angular slip
   ýгол скольж·éни·я (*m*) (mech) sliding angle; (math) glancing angle; (air) side-slip angle; (cryst) Bragg angle
скольз·úть, -я́т, (*imp*) slide, slip
скóльз·к·ий, -ая, -ое, (*a*) slippery
скольз·н·ýть, -ýт, (*perf*) slide, slip (single action)
скольз·я́щ·ий, -ая, -ее, (*a*) (*pres part act*) of **скольз·úть**
~ контáкт· (*m*) (elec) sliding contact
~ от·раж·éни·е (*n*) (cryst) sliding reflection
~ режúм· (*m*) (autom) fluttering
~ удáр· (*m*) glancing blow/impact
скóлько how much/many
   не стóлько ... скóлько not so much ... as
~ -нибудь (*adv*) any, any amount/number
~ -то (*adv*) a certain amount/number
с·комáнд·овать, -уют, (*perf*) give/issue an order/command
с·комбин·úр·овать, -уют, (*perf*) combine
с·кóм·ка·ть, -ют, (*perf*) crumple; spoil by hurrying
с·компенс·úр·овать, -уют, (*perf*) compensate
с·компил·úр·овать, -уют, (*perf*) compile, make a compilation
с·комплект·овáть, -ýют, (*perf*) make up, complete (sets etc.); bring up to strength (personnel etc.)
с·компон·овáть, -ýют, (*perf*) put together, make up, assemble, arrange
с·констру·úр·овать, -уют, (*perf*) design; form (an organization)
с·конч·á·ться, -ются, (*perf*) die
с·кóп·, -а, (*m*) = **с·копл·éни·е**
с·кóп·анн·ый, -ая, -ое, (*past part pass*) see **с·кáп·ыва·ть**
с·коп·á·ть, -ют, (*perf*) see **с·кáп·ыва·ть**
с·коп·úвш·ийся, -аяся, -ееся, (*past part act*) of **с·коп·úться**; accumulated; occluded
с·коп·úр·овать, -уют, (*perf*) see **ко·п·úр·овать**; copy

с·коп·úть, -я́т, (*imp*) castrate, geld;
с·коп·úть, -я́т, (*perf*) see **с·кáпл·и·ва·ть(ся)**
с·копл·éни·е, -я, (*n*) (*v n*), see **с·кáп·л·ива·ть(ся)**; accumulation, pile-up, cluster, clustering; (astron) cluster; (geol) occlusion
~ áтом·ов (phys) clustering/piling-up of atoms; cluster/aggregate/pileup of atoms
~ вóз·дух·а air lock/pocket
~ дислок·áци·й (cryst) dislocation pile-up
~ изотóп·ов (nucl, med) hot spot
~ кáпел·ь blob of droplets
с·кóпл·енн·ый, -ая, -ое, (*past part pass*) see **с·кáпл·ива·ть**; collected, accumulated
скополамúн·, -а, (*m*) (pharm) scopolamine
скóрб·н·ый, -ая, -ое, (*a*) sad, sorrowful
скорбýт·, -а, (*m*) (med) scurvy, scorbutus
скорбýт·н·ый, -ая, -ое, (*a*) (med) scorbutic
скóрб·ь, -и, (*nom pl*) **-и,** (*g pl*) **-éй,** (*f*) sorrow
скор·ée (*comp*) quicker, faster; rather, preferably
~ вс·егó most probably/likely
скорлуп·á, -ы́, (*nom pl*) **-ýп·ы,** (*f*) shell, peel, hull
скорлупо·вúд·н·ый, -ая, -ое, (*a*) = **скорлупо·обрáз·н·ый**
скорлупо·обрáз·н·ый, -ая, -ое, (*a*) shell-shaped, conchiform; (a/c) monocoque
скорлýп·чат·ые, -ых, (*pl decl as adj*) (pal) ostracods, *Ostracoda*
с·корм·úть, -я́т, (*perf*) see **с·кáрмл·и·ва·ть**
скорн·я́жн·ый, -ая, -ое, (*a*) fur, fur-dressing/processing
~ об·рабóт·к·а (*f*) fur dressing/processing
~ товáр·, (*m*) furriery
скорн·я́к·, -á, (*m*) furrier, fur dresser/processer
скóр·о (*adv*) quickly, fast, rapidly; (*root*) fast-, quick-
скоро·вáр·к·а, -и, (*g pl*) **-р·ок·,** (*f*) (food) pressure cooker
скородúт·, -а, (*m*) (min) scorodite
скоро·мороз·úлк·а, -и, (*g pl*) **-лок·,** (*f*) quick freezer
скоро·пúс·н·ый, -ая, -ое, (*a*) cursive
скóро·пис·ь, -и, (*f*) cursive script

скоро·плод·ност·ь, -и,   (*f*) (hortic) early maturity

скоро·под·ём·ност·ь, -и,   (*f*) (air) rate of climb, vertical speed

скоро·порт·ящ·ийся, -аяся, -ееся, (*a*) perishable

скоро·по·стиж·н·ый, -ая, -ое,   (*a*) sudden, unexpected (death)

скоро·спе·л·ый, -ая, -ое, (*a*) early-/ quick-maturing

скоросте·мер·, -а, (*m*) velocity meter

скор·остн·ой, -ая, -ое, (*a*) fast, high-speed, rapid; speed, velocity, rate

~ **модул·яци·я** (*f*) velocity modulation

~ **по·каз·áтел·ь** (*m*) rate index

~ **по·каз·áтел·ь твёрд·ост·и** (met) rate index of hardness

~ **расходо·мер·,** (*m*) inferential flow-meter

~ **фотогрáф·и·я** (*f*) high-speed photography

~ **энéрг·и·я** (*f*) kinetic energy

скоро·стрéль·ност·ь, -и, (*f*) (gunn) rate of fire

скоро·стрéль·н·ый, -ая, -ое, (*a*) (gunn) quick-firing, QF

скóр·ост·ь, -и, (*f*) speed, velocity (of movement or travel); rate (of a process)

~, **больш·áя** high speed

~ **в руч·ь·é** (met roll) pass-line speed

~, **вес·ов·áя** mass flow rate

~, **вз·лёт·н·ая** (air) take-off speed

~ **вóз·дух·а** air velocity, velocity of air flow

~, **воз·дýш·н·ая** (air) air speed

~ **вращ·éни·я** rotational velocity/speed; rate of rotation; speed (of rotating devices); scanning rate (of antenna)

~ **вращ·éни·я, угл·ов·áя** angular velocity

~, **втор·áя косм·и́ческ·ая** (rocket) escape velocity

~ **вы́·ход·а** exit speed/velocity; (rail) release speed; rate of delivery

~ **вы́·ход·а на редáн·** hump speed (flying boat)

~ **вы́·ход·а от·цéп·а** (rail) cut-release speed

~, **гипер·звук·ов·áя** hypersonic speed/velocity

~ **гор·éни·я** combustion/burning rate

~, **гран·и́чн·ая** cutoff velocity

~, **групп·ов·áя** group of velocity (of a wave)

скóр·ост·ь

~ **движ·éни·я** speed

~, **дей·стви́тельн·ая** actual speed; (naut) speed made good

~, **до·звóл·енн·ая** speed limit

~ **дрéйф·а** (phys) drift velocity, charge-drift velocity

~, **дýль·н·ая** initial velocity (of projectile), muzzle velocity (of gun)

~ **звýк·а** speed/velocity of sound; sonic speed/velocity

~ **звýк·а, мéст·н·ая** local speed of sound

~, **звук·ов·áя** sonic speed/velocity

~, **зем·н·áя** ground speed

~ **из·мен·éни·я** rate of change, rate

~ **из·мен·éни·я áзимут·а** (nav) azimuth rate

~ **из·мен·éни·я пéленг·а** (nav) angular rate

~ **из·мен·éни·я пéленг·а по кóм·пас·у** (nav) compass-bearing rate

~ **изодрóм·а** (autom) floating rate

~, **индикáтор·н·ая** equivalent air speed

~ **ис·теч·éни·я** exhaust/discharge velocity

~ **ис·теч·éни·я гáз·ов** (rocket) exhaust velocity

~, **ис·хóд·н·ая** reference velocity

~ **канáв·к·и** (acous) groove speed

~, **коммéрч·еск·ая** (naut) speed made good

~, **кон·éчн·ая** final speed (of machine); remaining/terminal velocity (of projectile)

~, **контрáкт·н·ая** designed speed

~, **косм·и́ческ·ая** (rocket) escape velocity

~, **крéйсер·ск·ая** (air) best speed for range, (naut) economical cruising speed

~, **крит·и́ческ·ая** stalling/critical velocity/speed

~, **лин·éйн·ая** linear/rectilinear velocity; link speed (of chain drive); (expl) burning rate, rate of flame propagation

~, **мáл·ая** low speed

~, **мáрш·ев·ая** (rocket) cruising speed

~, **маршрýт·н·ая** (rail) average speed for journey

~, **мг·новéнн·ая** instantaneous speed/velocity

~, **мéр·н·ая** (shipb) speed on the measured mile/distance

~ **молекуля́р·н·ого движ·éни·я** molecular velocity

ско́р·ост·ь

~ **на в·хо́д·е** indraft/entrance velocity (jet a/c); (mech eng) input speed

~ **на вы́·ход·е** outlet velocity (e.g. of air from engine etc.); delivery speed (of a machine)

~ **на·бег·а́ющ·его по·то́к·а** (aerodynam) upstream/remote velocity

~ **на·бо́р·а выс·от·ы́** rate of climb

~ **най·вы·год·не́йш·ая** (air) best flying speed

~, **наи·ме́ньш·ая** (naut) steerage way

~, **на·ча́ль·н·ая** initial speed; initial velocity (of projectile), true muzzle velocity (of gun), (rocket) launching speed

~, **норма́ль·н·ая** normal speed; (naut) service speed

~, **нул·ев·а́я** zero speed/velocity

~, **обра́т·н·ая** reciprocal velocity

~, **около·звук·ов·а́я** transonic speed

~, **о·конч·а́тельн·ая** instantaneous/ remaining velocity (of projectile)

~, **о·круж·н·а́я** periferal velocity; tip speed (of rotors); peripheral/ surface speed (metal machining)

~, **ос·ев·а́я** axial velocity; pitch speed (of a screw)

~, **па́спорт·н·ая** normal speed, service speed
пере·йти́ на друг·у́ю ско́р·ост·ь (M/T) change gear

~, **пере·ме́н·н·ая** variable speed

~, **пла́в·н·ая** comfortable speed

~, **по·ве́рх·ности·ая** (air) superficial velocity; surface speed (machining)

~ **по·да́ч·и** feed rate

~ **по·лёт·а** flying/flight speed; rate of travel (projectile)

~ **по·лёт·а в пуст·от·е́** (rocket) drag-free/vacuum velocity

~, **по·са́д·очн·ая** (air) landing speed

~, **по·ступ·а́тельн·ая** (phys) translational velocity; forward speed

~ **по·то́к·а** flow rate; (dynam) flow velocity, velocity

~, **пре·де́ль·н·ая** maximum speed; extreme/limiting speed, limiting transmission speed (synchros); terminal velocity (of a falling body)

~, **при·бо́р·н·ая** (air) indicated air speed

~, **прое́кт·н·ая** designed speed

~ **про·тек·а́ни·я** rate

~, **пульс·и́рующ·ая** fluctuating velocity

~, **пут·ев·а́я** (air) ground speed

ско́р·ост·ь

~ **пуч·к·а́** (phys) beam velocity

~, **рабо́ч·ая** normal speed, service speed; operating speed (of a machine)

~, **рас·по·лаг·а́ем·ая** available velocity

~ **рас·про·стран·е́ни·я** rate of spread/ propagation, velocity

~ **рас·про·стран·е́ни·я волн·ы́** wave speed/velocity

~ **рас·про·стран·е́ни·я звук·а** speed of sound

~ **рас·про·стран·е́ни·я пла́мен·и** flame velocity

~, **рас·чёт·н·ая** (air) design speed; (rail) computed speed

~ **реа́кци·и** (chem) reaction rate, rate of reaction; (instr) rate of response

~ **ре́з·ани·я** (mech eng) rate of cutting

~, **результ·и́рующ·ая** resultant speed / velocity

~, **с·верх·звук·ов·а́я** supersonic speed

~ **све́т·а** velocity of light

~, **с·да́т·очн·ая** (naut) trials speed

~, **синхро́н·н·ая** synchronous speed

~ **с·ко́ш·енн·ого в·низ· по·то́к·а** (dynam) down-wash velocity

~, **техн·и́ческ·ая** air speed (of a/c), sea speed (of ship), scheduled speed (of train)

~, **угл·ов·а́я** angular velocity; (a/c) rate (of pitch, roll etc.); speed (of propeller)
у·каз·а́тел·ь ско́р·ост·и (m) speedometer

~, **у·станов·и́вш·аяся** steady speed; (phys, dynam) steady-state velocity

~, **фа́з·ов·ая** (phys) phase velocity (of a wave)

~, **факт·и́ческ·ая** (air) ground speed

~ **хо́д·а** travelling speed, speed

~ **холост·о́го хо́д·а** no-load speed

~, **эволюти́в·н·ая** safety speed

~, **эконом·и́ческ·ая** economical speed

скоро·с·шив·а́тел·ь, -я, (m) folder, file (for papers)

скоро·те́ч·н·ый, -ая, -ое, (a) short-lived, brief

скорпио́н·, -а, (m) (astron) The Scorpion, Scorpius; (zool) scorpion, (pl) *Scorpionida*

скорпуло·ви́д·н·ый, -ая, -ое, (a) (a/c) monocoque

~ **констру́кци·я** (f) (a/c) monocoque structure/construction

скорц·а, -ы, (f) (geol) scorza, schorza

ско́р·ый, -ая, -ое, (a) quick, fast, rapid; forthcoming, impending, near (of time)

ско́р·ый, -ая, -ое
~ **по́·езд·** (*m*) fast train
~ **по́·мощ·ь** (*f*) (med) first aid
с·ко́с·, -а, (*m*) slant, bevel, chamfer, slope; (naut) sloped deck
~ **по·то́к·а в·верх·** (dynam) upwash
~ **по·то́к·а в·низ·** (dynam) downwash
у́гол· с·ко́с·а (*m*) (mech eng) chamfer/bevel angle, scarfing angle; (air) rake angle; downwash angle (of flow)
с·кос·и́ть, -ят, (*perf*) see **с·ка́ш·ива·ть**
скот·, -а́, (*m*) livestock, stock, cattle
~, **кру́п·н·ый рог·а́т·ый** cattle
~, **моло́ч·н·ый** dairy cows/cattle
~, **мяс·н·о́й** beef cattle
скот·и́н·а, -ы, (*f*) livestock; cattle
ското·бо́й·н·ый, -ая, -ое, (*a*) slaughter
~ **де́л·о** (*n*) slaughtering
ското·бо́й·н·я, -и, (*g pl*) **-о́·ен·,** (*f*) slaughter-house, abattoir
ското·во́д·, -а, (*m*) stockbreeder
ското·во́д·ств·о, -а, (*n*) stockbreeding
ското́м·а, -ы, (*f*) (med) scotoma
ското́·метр·, -а, (*m*) (med) scotometer
ското·планкто́н·, -а, (*m*) (biol) skotoplankton
ското·та́ксис·, -а, (*m*) (bot) skatotaxis
скоци·я, -и, (*f*) (arch) scotia
с·ко́ш·енн·ый, -ая, -ое, (*past part pass*) of **с·кос·и́ть,** (*perf*); see **с·ка́ш·ива·ть** (*imp*); **с·кош·ённ·ый, -ая, -ое,** slanted, inclined, sloped, bevelled
~ **в·низ· по·то́к·** (*m*) (dynam) down wash
с·кра́д·ыва·ть, -ют, (*imp*) conceal, make imperceptible; capture (rivers)
скра́п·, -а, (*m*) scrap, scrap metal
~, **ано́д·н·ый** anode scrap
~ **-проце́сс·, -а,** (*m*) (met) pig-and-scrap process
скрап·н·о́й, -а́я, -о́е, (*a*) of **скрап·**
~ **про·лёт·** (*m*) (met) scrap bay
скрапо·от·дел·и́тел·ь, -я, (*m*) (met) scrap separator
скрёб·, (*past masc sing*) of **скрес·ти́**
скреб·ённ·ый, -ая, -ое, (*past part pass*) of **скрес·ти́**
скреб·ёт, (*pres 3rd sing*) of **скрес·ти́**
скреб·ко́в·о-ковш·о́в·ый конве́йер· (*m*) push-plate conveyer
скреб·ко́в·о-штанг·ов·ый конве́йер· (*m*) drag-link conveyer
скреб·ко́в·ый, -ая, -ое, (*a*) of **скреб·о́к·**
~ **ковш·** (*m*) scraper, flight, drag link, push plate (of conveyer)

скреб·ко́в·ый, -ая, -ое
~ **конве́йер·** (*m*) scraper/flight/drag/ drag-chain/scraper-chain conveyer
скреб·ли́, (*past pl*) of **скрес·ти́**
скребл·о́, -а́, (*n*) doctor knife, doctoι
скреб·о́к·, -б·к·а́, (*m*) scraping/doctor knife/plate; scraper, drag link, push plate, flight (of conveyer); go-devil (for cleaning interior of pipes)
скреб·у́щ·ий, -ая, -ее, (*pres part act*) of **скрес·ти́**
скрёб·ш·ий, -ая, -ее, (*past part act*) of **скрес·ти́**
скрёп·а, -ы, (*f*) (*v n*), see **с·крепл·я́·ть;** clamp, tie, fastener; counter signature, witnessing signature
скре́пер·, -а, (*m*) (civ eng) skimmer, scraper-excavator
скре́пер·н·ая лебёд·к·а (*f*) dragline hoist
с·креп·и́ть, -я́т, (*perf*) see **с·крепл·я́·ть**
с·креп·к·а, -и, (*g pl*) **-п·ок·,** (*f*) fastening, securing, fixing; clip, paper clip; fastener, clamp
с·крепл·е́ни·е, -а, (*n*) (*v n*), see **с·крепл·я́·ть;** (gunn) hooping; fastening, fastener, tie, clamp; countersignature, authentication
~, **ре́льс·ов·ое** (rail) rail fastening/tie
с·крепл·ённ·ый, -ая, -ое, (*past part pass*) see **с·крепл·я́·ть**
с·крепл·я́·ть, -ют, (*imp*) fasten, secure, fix, strengthen, brace; countersign, sign (as witness), authenticate, stamp, witness (a signature)
с·креп·очн·ый, -ая, -ое, (*a*) fastening, securing
с·крес·ти́, (*pres 3rd pl*) **скреб·у́т,** (*imp*) scrape, scratch
с·крест·и́ть, -я́т, (*perf*) see **с·кре́щ·ива·ть**
с·крещ·е́ни·е, -я, (*n*) crossing, criss-crossing; intersection, intersecting
с·крещ·ённ·ый, -ая, -ое, (*past part pass*) see **с·кре́щ·ива·ть**
с·кре́щ·ивани·е, -я, (*n*) (*v n*) of **с·кре́щ·ива·ть;** (telecom) crossing, transposition; (biol) hybridization, cross fertilization, cross breeding, cross
~ **по о·по́р·е** (telecom) point transposition
шаг· с·кре́щ·ивани·я (*m*) (telecom) subdivision transposition
с·кре́щ·ива·ть, -ют, (*imp*) cross, criss-cross, intersect; (gen) cross, interbreed, hybridize; **-ся** (*pass*); form a cross

с·крив·и́ть, -я́т, (*perf*) bend, crook, twist

скрип·, -а, (*m*) squeak, creak

скрип·а́ч·, -а́, (*i*) **-о́м,** (*m*) violinist

скрип·е́ть, -я́т, (*imp*) squeak, creak

скрип·к·а, -и, (*g pl*) **-п·ок·,** (*f*) violin

скрип·н·уть, -ут, (*perf*) give/make/emit a creak/squeak

скрип·у́ч·ий, -ая, -ее, (*a*) squeaky, creaking

с·кро́·енн·ый, -ая, -ое, (*past part pass*) of **с·кро·и́ть**

с·кро́·ет, (*fut 3rd sing*) of **с·кры·ть,** (*perf*); *see* **с·крыв·а́·ть** (*imp*)

с·кро·и́ть, -я́т, (*perf*) cut out; cut (figurative, e.g. a fine figure)

с·кро́м·н·ый, -ая, -ое, (*a*) modest, unassuming

с·кро́·ют, (*fut 3rd pl*) of **с·кры·ть,** (*perf*); *see* **с·крыв·а́·ть** (*imp*)

скру́ббер·, -а, (*m*) (chem) gas scrubber

с·кругл·е́ни·е, -я, (*n*) rounding off

скру́п·ул·, -а, (*m*) scruple (weight)

с·крут·и́ть, -я́т, (*perf*) *see* **с·круч·и·ва·ть**

с·крут·к·а, -и, (*g pl*) **-т·ок·,** (*f*) (*v n*), *see* **с·круч·ива·ть;** twist, lay (of ropes); twist joint (for cables etc.); splice; crossing (of lines)

  ~, ги́льз·ов·ая twisted-sleeve joint (cable)

  коэффицие́нт· с·кру́т·к·и (*m*) lay ratio (cable)

  ~ провод·о́в (telecom) transposition

  у́гол· с·кру́т·к·и (*m*) pitch angle (screw thread etc.)

с·кру́т·очн·ый, -ая, -ое, (*a*) twisting

  ~ стан·о́к· (*m*) (elec) cable-twisting machine

скручен·но·спира́ль·н·ый, -ая, -ое, (*a*) streptospiral

с·кру́ч·енн·ый, -ая, -ое, (*past part pass*) *see* **с·круч·и·ва·ть**

с·круч·ивани·е, -я, (*n*) (*v n*) of **с·кру́·ч·ива·ть;** twist, torque; curling (photographic prints etc.)

  моме́нт· с·круч·ивани·я (*m*) torque

  про́ч·ност·ь на с·круч·ивани·е torsional strength

с·кру́ч·ива·ть, -ют, (*imp*) twist coil; roll (cigarettes etc.); bind, tie up; (text) double (threads)

  ~ па́р·ами pair up (elec cable)

с·круч·ивающ·ий, -ая, -ее, (*pres part act*) of **с·круч·ива·ть**

  ~ шта́нг·а (*f*) torque bar

с·крыв·а́·ть, -ют, (*imp*) hide, conceal

скрытно·е́д·ы, -ов, (*m pl*) (ent) silken fungus beetles, *Cryptophagidae*

с·кры́т·ност·ь, -и, (*f*) secretiveness, reticence, reserve; (telecom) security

скрытно·хо́бот·ник·, -а, (*m*) (ent) snout beetle, *Ceuthorrhynchus*

с·кры́т·н·ый, -ая, -ое, (*a*) secretive, reticent, reserved, cryptic; (telecom) secure

с·кры́т·о- (*component*) crypto-

скрыто·генет·и́ческ·ий, -ая, -ое, (*a*) cryptogenetic

скрыто·гла́в·ы (*f pl*) (ent) *Cryptocephala*

скрыто·жа́бер·ник·, -а, (*m*) (zool) hellbender, *Cryptobranchus*

скрыто·жа́бер·н·ые, -ых, (*pl decl as adj*) (zool) *Cryptobranchidae*

скрыто·под·зо́л·ист·ый, -ая, -ое, (*a*) cryptopodzolic (of soil)

скрыто·се́мян·н·ый, -ая, -ое, (*a*) (bot) angiospermous, (*pl as noun*) angiosperms, *Angiospermae*

скрыто·ше́й·н·ые череп·а́х·и (*pl*) (zool) *Cryptodira*

с·кры́·т·ый, -ая, -ое, (*past part pass*) *see* **с·крыв·а́·ть;** secret; (phys) latent

  ~ из·ображ·е́ни·е (*n*) (phot) latent image

  ~ тепл·от·а́ (*f*) (phys) latent heat

с·кры·ть, (*fut 3rd pl*) **с·кро́·ют,** (*perf*) *see* **с·крыв·а́·ть**

скряб·к·а, -б·ок·, (*f*) = **скреб·о́к·**

скуд·е́·ть, -ют, (*imp*) become scarce; become impoverished/poor

скуд·н·ый, -ая, -ое, (*a*) scarce, scanty, meagre; impoverished, poor

с·ку·ёт, (*fut 3rd sing*) of **с·ков·а́ть,** (*perf*); *see* **с·ко́в·ыва·ть** (*imp*)

скук·а, -и, (*f*) boredom

скул·а́, -ы́, (*nom pl*) **скул·ы,** (*f*) (anat) cheek-bone, zygoma; (naut) bilge

скул·ов·о́й, -а́я, -о́е, (*a*) (anat) zygomatic, malar; (naut) bilge

  ~ блок· (*m*) (naut) bilge block

  ~ дуг·а́ (*f*) (zool) zygoma, zygomatic arch

  ~ кост·ь (*f*) (zool) jugal, zygomatic bone

скуло·ло́б·н·ый, -ая, -ое, (*a*) (anat) zygomaticofrontal

скульпт·у́р·а, -ы, (*f*) sculpture

скульпту́р·н·ый, -ая, -ое, (*a*) sculptural

ску́мбр·иев·ые, -ых, (*pl decl as adj*) (fish) *Scombridae*

ску́мбр·и·я, -и, (*f*) (fish) mackerel, *Scomber scombrus*

ску́мпи·я, -и, (*f*) (bot) sumac, *Cotinus*

скунс·, -а, (*m*) (zool) skunk

с·куп·а́·ть, -ют, (*imp*) (com) buy up, make a corner in

скупи́т·, -а, (*m*) (min) schoepite

с·куп·и́ть, ⸗ят, (*perf*) *see* **с·куп·а́·ть**

с·ку́п·к·а, -и, (*f*) (*v n*) of **с·куп·а́·ть**

с·куп·н·о́й, -а́я, -о́е, (*a*) *see* **с·ку́п·оч-н·ый**

ску́п·о (*adv*) sparingly

скуп·о́й, -а́я, -о́е, (*a*) miserly, stingy

с·ку́п·очн·ый, -ая, -ое, (*a*) (com) cornering

ску́тер·, -а, (*m*) (naut) outboard hydro-plane, skimming boat

скуттеруди́т·, -а, (*m*) (min) skutte-rudite

скуч·а́·ть, -ют, (*imp*) be bored

~ по дом·у be homesick

с·ку́ч·ива·ть, -ют, (*imp*) heap, pile; **-ся** (*pass*); crowd together /up

с·ку́ч·енност·ь, -и, (*f*) density; crowd

с·ку́ч·енн·ый, -ая, -ое, (*past part pass*) *see* **с·ку́ч·ива·ть**; heaped, piled-up; crowded, dense

с·ку́ч·ить, -ат, (*perf*) *see* **с·ку́ч·ива·ть**

ску́ч·н·ый, -ая, -ое, (*a*) tedious, boring, dull

сл, (*abbr*) = **санти·ли́тр·** centilitre, centiliter

слаб·е́·ть, -ют, (*imp*) become weak/weaker/slack/slacker, diminish, slacken off, get fainter

слаб·ин·а́, -ы́, (*f*) slack, bight (of rope); slackness, play

~, рабо́ч·ая (eng) working fit

слаб·и́тельн·ый, -ая, -ое, (*a*) (med) purgative, laxative

сла́блин·ь, -я, (*m*) (naut) lacing line

сла́б·о (*adv*) weakly, feebly; slightly, lightly, mildly

слабо·воз·бужд·ённ·ый, -ая, -ое, (*a*) (elec) feebly excited

~ со·сто·я́ни·е (*n*) (phys) low-lying state

слабо·за·прещ·ённ·ый, -ая, -ое, (*a*) (phys) unfavoured

слабо·кип·я́щ·ий, -ая, -ее, (*a*) light-boiling

слабо·кисл·о́т·н·ый, -ая, -ое, (*a*) (chem) subacid

слабо·ки́сл·ый, -ая, -ое, (*a*) weakly/slightly acidic/acid

слабо·леги́р·ова·нн·ый, -ая, -ое, (*a*) light-alloyed

слабо·лет·у́ч·ий, -ая, -ее, (*a*) low-volatility, non-volatile

слабо·магни́т·н·ый, -ая, -ое, (*a*) weak/low-magnetic

слабо·на·мо́т·анн·ый, -ая, -ое, (*a*) loosely wound/coiled, slack

~ руло́н· (*m*) (paper) soft/slipped roll

слабо·на·тя́·нут·ый, -ая, -ое, (*a*) slack

слабо·не́рв·н·ый, -ая, -ое, (*a*) nervous; (med) neurasthenic

слабо·о·богащ·ённ·ый, -ая, -ое, (*a*) slightly enriched

слабо·осно́в·н·ый, -ая, -ое, (*a*) (chem) slightly/feebly basic

слабо·по·глощ·а́ющ·ий, -ая, -ее, (*a*) (nucl) low-capture

слабо·полимериз·о́ванн·ое масл·о (*n*) short oil

слабо·про·кле́й·н·ый, -ая, -ое, (*a*) slack-/soft-sized

слабо·радио·акти́в·н·ый, -ая, -ое, (*a*) slightly radioactive

слабо·раз·ви́·т·ый, -ая, -ое, (*a*) under-developed

слабо·с·вя́з·анн·ый, -ая, -ое, (*a*) (phys) loosely bound, weakly coupled

слабо·си́ль·н·ый, -ая, -ое, (*a*) weak, feeble, asthenic

слаб·ост·ь, -и, (*f*) weakness, feeble-ness; weakness, failing, weak point, (med) asthenia, debility

слабо·сып·у́ч·ий, -ая, -ее, (*a*) non-free-flowing

слабо·то́ч·н·ый, -ая, -ое, (*a*) (elec) weak-current, communications (cable etc.)

слабо·у́м·и·е, -я, (*n*) (med) dementia

слабо·у́м·н·ый, -ая, -ое, (*a*) imbecile, feeble-minded

слабо·щелоч·н·о́й, -а́я, -о́е, (*a*) weak-alkaline, alkalescent

сла́б·ый, -ая, -ое, (*a*) weak, slight, faint, feeble, mild, dilute (of solutions), low-level (radiation etc.)

~ с·вяз·ь (*f*) (phys) loose/weak coupling

~ ток· (*m*) (elec) weak current, communications current; (*in gen as adj*) low-power/current, communications

сла́в·а, -ы, (*f*) glory, fame, renown

сла́в·к·а, -и, (*g pl*) **-в·ок·,** (*f*) (orn) whitethroat, *Sylvia*

сл·а́вш·ий, -ая, -ее, (*past part act*) of **сл·ать**

слав·яни́н·, -а, (*nom pl*) **-в·я́н·е,** (*g pl*) **-в·я́н·,** (*m*) Slav

с·лаг·а́ем·ое, -ого, (*n decl as adj*) (math) component, term, member

с·лаг·а́ем·ое
~, втор·о́е (math) augend
~, пе́рв·ое (math) addend
с·лаг·а́·ть, -ют, (*imp*) compose, make up, put/lay together; put/lay down, resign, hand over (responsibilities etc.); add, add up
с·лаг·а́ющ·ая, -ей, (*f decl as adj*) (math) component
~ ве́ктор·а (math) vector component
с·лаг·а́ющ·ий, -ая, -ее, (*pres part act*) *see* с·лаг·а́·ть; *see also* с·лаг·а́ем·ое, с·лаг·а́ющ·ая; component
с·ла́д·ить, -ят, (*perf*) put in order
сла́д·к·ий, -ая, -ое, (*a*) sweet
сладко·ко́рен·ь, -рн·я, (*m*) (bot) wall fern polypody, *Polypodium vulgare*
сладко·па́х·нущ·ий, -ая, -ее, (*a*) sweet smelling
сладко·пло́д·н·ый, -ая, -ое, (*a*) sweet-fruited
сладо·стра́ст·и·е, -я, (*n*) (med) libido
сла́д·ост·ь, -и, (*f*) sweetness; (*pl*) sweet food, sweets, candy
с·ла́ж·енн·ый, -ая, -ое, (*past part pass*) of с·ла́д·ить; co-ordinated, harmonious
сл·ал, (*past masc sing*) of сл·ать
с·ла́м·ыва·ть, -ют, (*imp*) break, subdue, overcome
слан·ец·, -н·ц·а, (*i*) -н·ц·ем, (*m*) (geol) shale, slate, schist
~, кедр·о́в·ый (bot) dwarf pine, *Pinus pumila*
~, нефте·но́с·н·ый oil-shale
сл·а́нн·ый, -ая, -ое, (*past part pass*) of сл·ать
слан·цева́тост·ь, -и, (*f*) (geol) schistosity, foliation
слан·цева́т·ый, -ая, -ое, (*a*) shaly, slaty, schistose
сла́н·цев·ый, -ая, -ое, (*a*) shale, schist, schistous
~ гли́н·а (*f*) (geol) shale, schistous clay
~ ма́сл·о (*n*) shale oil
слан·ь, -и, (*f*) (naut) ceiling
сла́ст·и, -е́й, (*f pl*) sweet food, sweets, candy
сл·ать, (*pres 3rd pl*) шл·ют, (*imp*) send, dispatch
сла́щ·е (*comp*) sweeter
с·ле́в·а, (*adv*) on/to/from the left
~ за корм·о́й (naut) fine on the port quarter
~ на тра́верз·е on the port beam
~ на нос·у́ fine on the port bow
с·легк·а́ (*adv*) slightly, somewhat

след·, -а, (*nom pl*) -ы́, (*g pl*) -о́в, (*m*) track, trace; foot-print; in-sole (of shoe); (math) spur
~, вихр·ев·о́й (dynam) trailing vortex
~ звезд·ы́ (astron) star trail
~, инверсио́н·н·ый (air) contrail
~ конденс·и́рованн·ых пар·о́в (air, meteor) condensation trail, contrail
~, метео́р·н·ый (astron) meteor trail
~, по·пу́т·н·ый (naut) wake
~ раз·вёрт·к·и (rad) sweep trace
~, турбуле́нт·н·ый (dynam) wake, turbulent wake
~ част·и́ц·ы (nucl) track
след·и́ть, -я́т, (*imp*) follow, track, keep track of, watch, keep watch over; watch out (for), be careful (that), see (that); leave tracks/traces
~ за тем, что́·бы ... be careful that/to, see that
след·ова́ни·е, -я, (*n*) (*v n*), *see* след·овать
~, да́ль·н·ее (*in gen as adj*) long-distance (of transport)
~ част·от·а́ след·ова́ни·я (*f*) pulse repetition frequency
след·ова́тел·ь, -я, (*m*) (law) examining magistrate, police interrogator; coroner
след·ова́тельн·о (*adv*) consequently, hence, for this reason, therefore
след·овать, -уют, (*imp*) follow, come/go after/next; follow, obey (instructions etc.); proceed/move/go to/towards, be bound for; (impersonal) ought, must, should
как и след·овало о·жид·а́ть as was to be expected, as expected
след·ует за·ме́т·ить it is important to note, it is important
след·ов·о́й, -а́я, -о́е, (*a*) of след·
сле́д·ом (*adv*) immediately after
следо·пы́т·, -а, (*m*) pathfinder (radar)
следо·у·каз·а́тел·ь, -я, (*m*) (agr) marker
сле́д·ственн·ый, -ая, -ое, (*a*) investigation, investigatory, enquiry
сле́д·стви·е, -я, (*n*) consequence, result; inquiry, inquest, examination
~, пред·вар·и́тельн·ое preliminary inquiry
след·ует, (*pres 3rd sing*) of след·овать
след·у́ющ·ий, -ая, -ее, (*pres part act*) *see* след·овать; following, next; (*n decl as adj*) the following
след·я́щ·ий, -ая, -ее, (*pres part act*) *see* след·и́ть; follow-up, servo- (*see also* серво-)

след·я́щ·ий, -ая, -ее
~ **антéнн·а** (*f*) track antenna (radar)
~ **кольц·ó** (*n*) phantom (gyro compass)
~ **механ·и́зм·** (*m*) servomechanism, servo
~ **при·вóд·** (*m*) servomechanism, servo
~ **систéм·а** (*f*) servo, servomechanism, follow-up system
~ **систéм·а, бес·контáкт·н·ая** proportional follow-up system
~ **систéм·а, много·кóнтур·н·ая** multiple-loop servomechanism
~ **систéм·а, одно·кóнтур·н·ая** single-loop servomechanism
~ **систéм·а, пропорционáль·н·ая** proportional follow-up system
~ **систéм·а, релéй·н·ая** step-by-step follow-up system
~ **у·стрóй·ств·о** (*n*) follow-up system; follower, tracker (radar), slave
с·леж·а́ться, -а́тся, (*perf*) *see* **с·лёж·ива·ться**
слеж·éни·е, -я, (*n*) (*v n*) of **след·и́ть;** following, tracking (radar)
~ **за с·на·ря́д·ом** missile tracking
~, **синхрониз·и́рованн·ое** lock following (radar)
с·лёж·енн·ый, -ая, -ое, (*past part pass*) *see* **с·лёж·ива·ться**
слёж·ечн·ый, -ая, -ое, (*a*) track (radar)
~ **на·вед·éни·е** (*n*) track homing (radar)
с·лёж·ива·ться, -ются, (*imp*) become compacted/consolidated, subside, cake
слёж·к·а, -и, (*f*) (*v n*), *see* **след·и́ть**
с·лез·, (*past masc sing*) of **с·лез·ть**
слёз·, (*g pl*) of **слез·á**
с·лез·á, -ы́, (*nom pl*) **слёз·ы,** (*g pl*) **слёз·,** (*f*) (physiol) tear
~, **лáв·ов·ая** (geol) lava tear
с·лез·а́·ть, -ют, (*imp*) climb/get off/down
с·лёз·к·а, -и, (*g pl*) **-з·ок·,** (*f*) (*dim*) of **слез·á;** teardrop; (ocean) ice hummock; (elec) insulating bead
~, **батáв·ск·ая** glass tear
с·лез·ни́к·, -á, (*m*) (build) dripstone, larmier
с·лез·ни́чк·а, -и, (*g pl*) **-чек·,** (*f*) (anat) lachrymal gland
с·лез·ничóк·, -чк·á, (*m*) (anat) lachrymal/lacrimal sac
слезно·нóс·ов·óй, -а́я, -óе, (*a*) (anat) lacrimonasal
слезно·решёт·чат·ый, -ая -ое, (*a*) (anat) lacrimoethmoidal

с·лёз·н·ый, -ая, -ое, (*a*) lachrymal, lacrimal, tear
~ **канáл·ец·** (*m*) lachrymal duct, tear duct
~ **тóч·к·а** (*f*) lachrymal gland, tear gland
слезо·теч·éни·е, -я, (*n*) (physiol) lachrymation, lacrimation
слезо·точ·и́в·ый, -ая, -ое, (*a*) lacrimatory, lachrymatory; running (of eyes)
~ **газ·** (*m*) (chem) lacrimator, lachrymator, tear gas
~ **о·правл·я́ющ·ее вещ·ествó** (*n*) tear gas, lacrimator
с·лéз·ть, -ут, (*perf*) *see* **с·лез·а́·ть**
с·лéй·те, (*pl imper*) of **с·ли́·ть,** (*perf*); *see* **с·лив·á·ть** (*imp*)
слеп·éн·ь, -п·н·я́, (*m*) (ent) horsefly, gadfly, *Tabanus*, (*pl*) *Tabanidae*
слеп·и́ть, -ят, (*imp*) blind, dazzle; **с·леп·и́ть, -ят,** (*perf*) *see* **с·леп·л·я́·ть**
с·лéпл·енн·ый, -ая, -ое, (*past part pass*) *see* **с·лепл·я́·ть**
с·лéпл·ива·ть, -ют, (*imp*) = **с·леп·л·я́·ть**
с·лепл·я́·ть, -ют, (*imp*) model, fashion, shape; stick together; **-ся** (*pass*); stick, adhere
слéп·нуть, -нут, (*imp*) become blind
слеп·óй, -á·я, -ó·е, (*a*) blind
~ **ки́ш·к·а** (*f*) (anat) caecum, cecum
~ **по·лёт·** (*m*) blind flying
~ **по·сáд·к·а** (*f*) (air) blind landing; **вы́·ход· на слеп·ýю по·сáд·к·у** blind approach
с·лéп·ок, -п·к·а, (*m*) copy, moulded copy; (geol) cast
~, **ги́пс·ов·ый** plaster casting
слеп·от·á, -ы́, (*f*) blindness
слеп·ы́ш·, -а, (*m*) (zool) (*pl*) spalacides, *Spalacidae*
слесáр·н·ый, -ая, -ое, (*a*) fitting, fitter's
~ **рабóт·а** (*f*) (mech eng) bench work
~ **инструмéнт·** (*m*) fitter's/fitting tools
слесáр·н·я, -и, (*g pl*) **-р·ен·,** (*f*) (mech eng) fitting/fitter's shop; lock-smithery
слéсар·ь, -я, (*m*) (mech eng) fitter, engine fitter; locksmith
~ **-трубо·про·вóд·чик·, -а,** (*m*) pipe fitter
с·лёт·, -а, (*m*) (*v n*), *see* **с·лет·á·ть;** (text) slub
с·лёт·анност·ь, -и, (*f*) (air) formation flying

с·лет·а́·ть, -ют, (*imp*) fly down/off
с·лет·е́ть, -я́т, (*perf*) *see* с·лет·а́·ть
слёшер·, -а, (*m*) multiblade frame-saw
с·ли́в·, -а, (*m*) overflow; pouring off/out, discharging (liquid cargo), decanting
~, авар·и́йн·ый (air) jettisoning (fuel)
сли́в·а, -ы, (*f*) (bot) plum, plum tree, *Prunus*
с·лив·а́ни·е, -я, (*n*) (*v n*) of с·лив·а́·ть, *and see* с·ли·я́ни·е
с·лив·а́·ть, -ют, (*imp*) pour together, mix, blend; pour off, decant; -ся (*pass*) flow together; blend, mix, merge
с·лив·а́ющ·ийся, -аяся, -ееся, (*pres part act*) *see* с·лив·а·ться; confluent
с·ли́в·к·и, -в·ок·, (*f pl*) cream
~, вз·би́·т·ые whipped cream
сливко·образ·ова́ни·е, -я, (*n*) (rubber) creaming
сливко·от·дел·и́тел·ь, -я, (*m*) cream separator
сливко·под·готов·и́тел·ь, -я, (*m*) cream separator
с·лив·н·о́й, -а́я, -о́е, (*a*) *see also* с·ли́в·н·ый; mixing, blending (of liquids); overflow; pouring, decanting
~ ста́нци·я (*f*) (civ eng) septic tank
сли́в·н·ый, -ая, -ое, (*a*) plum
сли́в·ов·ый, -ая, -ое, (*a*) plum
с·ли́в·очн·ый, -ая, -ое, (*a*) cream
~ ма́сл·о (*n*) butter
слив·я́нк·а, -и, (*g pl*) -нок·, (*f*) slivovitz, plum brandy
с·ли́ж·ет, (*fut 3rd sing*) of с·лиз·а́ть
слиз·, -а, (*m*) (fish) blenny, *Blennius cagnota*
с·лиз·а́ть, (*fut 3rd pl*) с·ли́ж·ут, (*perf*) *see* с·ли́з·ыва·ть
слиз·еви́к·, -а, (*m*) (bot) slime mould/mold, (*pl*) *Myxomycetes*
слиз·ев·о́й, -а́я, -о́е, (*a*) of слиз·ь mucous, slime; mucic
~ кисл·от·а́ (*f*) (chem) mucic acid
слизево·ки́сл·ый, -ая, -ое, (*a*) (chem) mucate (of)
слиз·ен·ь, -з·н·я, (*m*) (zool) slug, *Limax*; (*pl*) (shipb) grease irons
слизе·теч·е́ни·е, -я, (*n*) (med) blennorrhoea; mucilaginous discharge (trees)
слизисто·гно́й·н·ый, -ая, -ое, (*a*) (med) mucopurulent
слизисто·пере·по́н·чат·ый, -ая, -ое, (*a*) (biol) mucomembranous

сли́з·ист·ый, -ая, -ое, (*a*) mucous, mucilaginous, (bot) mucose; slimy
~ гриб· (*m*) (bot) slime mould, (*pl*) *Myxomycetes*
~ желез·а́ (*f*) (zool) mucous gland
~ оболо́ч·к·а (*f*) (anat) mucous membrane, mucosa
~ спор·о́вик·и́ (*pl*) (zool) *Myxosporidia*
слиз·ня́к·, -а́, (*m*) = сли́з·ен·ь
слиз·няко́в·ый, -ая, -ое, (*a*) (zool) limacine
с·ли́з·ыва·ть, -ют, (*imp*) lick off/away
сли́з·ь, -и, (*f*) (biol) mucus, mucilage; slime; (*pl*) (naut) parrels
сли́п·, -а, (*m*) (shipb) slipway, slip, patent slip, marine railway; ramp (of whaling factory ship); slip (of propeller)
~, судо·стро·и́тельн·ый transfer building berth and launching slip
с·лип·а́емост·ь, -и, (*f*) stickiness
с·лип·а́ни·е, -я, (*n*) (*v n*), *see* с·лип·а́·ться; agglutination
с·лип·а́·ться, -ются, (*imp*) stick together, agglutinate
сли́перс·ы, -ов, (*m pl*) (shipb) sleepers
слипинг-док·, -а, (*m*) (shipb) = сли́п· (q.v.)
с·ли́п·нуться, -нутся, (*perf*) *see* с·лип·а́·ться
слитко·во́з·, -а, (*m*) (met) ingot chariot/car/buggy, pot car
слитко·вы·жим·а́тел·ь, -я, (*m*) (met) ingot stripper
с·ли́т·ков·ый, -ая, -ое, (*a*) of с·ли́т·ок
слитно·борозд·н·ый, -ая, -ое, (*a*) syncolpate
слитно·вы·рост·н·ый, -ая, -ое, (*a*) sympilate
с·ли́т·ност·ь, -и, (*f*) coherence
с·ли́т·н·ый, -ая, -ое, (*a*) coherent, connected, continuous
с·ли́т·ок·, -т·к·а, (*m*) ingot (ferrous metals), bar, bullion (gold, silver), block (tin), cake (zinc), brick (copper)
с·ли́т·очн·ый, -ая, -ое, (*a*) of с·ли́т·ок·
с·ли́·т·ый, -ая, -ое, (*past part pass*) *see* с·лив·а́·ть
с·ли·ть, (*fut 3rd pl*) со·ль·ю́т, (*perf*) *see* с·лив·а́·ть
с·лич·а́·ть, -ют, (*imp*) compare, collate
с·лич·е́ни·е, -я, (*n*) (*v n*), *see* с·лич·а́·ть; comparison
с·лич·и́тельн·ый, -ая, -ое, (*a*) comparative; collating

с·лич·и́ть, -а́т, (*perf*) compare, collate

с·ли́ш·к·ом (*adv*) too, more than, too much/many

с·ли·я́ни·е, -я, (*n*) сonfluence, junction, blending; coalescence; fusion, merging

~ я́дер· (phys) nuclear fusion

у́гол· с·ли·я́ни·я (*m*) angle of convergence (roads etc.)

слов·а́рн·ый, -ая, -ое, (*a*) of слов·а́р·ь; lexical, lexicographic

слов·а́р·ь, -я́, (*m*) dictionary, vocabulary, glossary

~, част·о́тн·ый (ling) frequency dictionary

слов·е́сн·ый, -ая, -ое, (*a*) verbal

слов·ник·, -а, (*m*) glossary, word-list

слов·н·о (*conj*) as if

слов·о, -а, (*nom pl*) слов·а́, (*g pl*) слов·, (*n*) word; words, remarks, speech; (telecom) letter "S"

слово·из·мен·е́ни·е, -я, (*n*) (ling) accidence

слов·ом (*conj*) in short

слово·ли́т·н·я, -и, (*f*) (print) type/ letter foundry

слово·образ·ова́ни·е, -я, (*n*) (ling) word-formation, word-derivation

слово·образ·ова́тельн·ый, -ая, -ое, (*a*) word-building/forming

с·ло́г·, -а, (*nom pl*) сло́г·и, (*g pl*) слог·о́в, (*m*) syllable, logatom

сло·ева́тост·ь, -и, (*f*) lamination; (geol) schistosity

сло·еви́щ·е, -я, (*n*) (bot) frond, thallus

сло·ев·о́й, -а́я, -о́е, (*a*) of сло·й

сло·евцо́в·ые, -ых, (*pl decl as adj*) (bot) thallophytes, *Thallophyta*

сло·ёк·, (*g sing*) слой·к·а́, (*m*) (*dim*) of сло·й; lamina, layer, thin layer

сло·е́ни·е, -я, (*n*) (*v n*) of сло·и́ть; lamination; exfoliation

сло·ён·ый, -ая, -ое, (*a*) flaky; (nucl) sandwich

~ ис·то́ч·ник· (*m*) (nucl) sandwich source

~ те́ст·о (*n*) puff pastry

с·лож·е́ни·е, -я, (*n*) (math) addition, summation; build (of body), texture, structure, composition; laying down, handing over (duties etc.)

~, зерн·и́ст·ое (geol) granular structure; (*in gen as adj*) granular

знак· с·лож·е́ни·я (*m*) (math) add/summation sign

и́мпульс· с·лож·е́ни·я (*m*) add pulse (computers)

с·лож·е́ни·е

~, по·раз·ря́д·н·ое digit-by-digit adding

~ си́л·ы (mech) composition of a force

с·ло́ж·енн·ый, -ая, -ое, (*past part pass*) see с·кла́д·ыва·ть; sectionally constructed

с·лож·и́ть (*perf*) see с·кла́д·ыва·ть, с·лаг·а́·ть

сложно·сло·и́ст·ый, -ая, -ое, (*a*) (geol) multiple-bedded

с·ло́ж·ност·ь, -и, (*f*) complexity, intricacy

сложно·сферо·ли́т·ов·ый, -ая, -ое, (*a*) (geol) compound-spherulitic

сложно·цве́т·н·ые, -ых· (*pl decl as adj*) (bot) *Compositae*

сложно·цикл·и́чн·ый, -ая, -ое, (*a*) multicyclic

сложно·эфи́р·н·ый, -ая, -ое, (*a*) ester

с·ло́ж·н·ый, -ая, -ое, (*a*) complex, complicated; compound, composite; (in compound words) -syllabic

~ вещ·еств·о́ (*n*) chemical compound

~ гипоте́з·а (*f*) complex hypothesis

~ крив·а́я рас·па́д·а (*f*) (nucl) composite decay curve

~ пере·плет·е́ни·е (*n*) (text) compound/composed weave

~ фу́нкци·я (*f*) (math) composite function

~ проце́нт·ы (*pl*) (fin) compound interest

~ со·бы́т·и·е (*n*) (math) compound event

сло·и́, (*nom/acc pl*) of сло́·й

сло·и́ст·о-дожд·ев·ы́е (*pl adj*) (meteor) nimbo-stratus

сло·и́ст·о-куч·ев·ы́е (*pl adj*) (meteor) strato-cumulus

слоисто·по·кро́в·н·ый, -ая, -ое, (*a*) tectal

сло·и́стост·ь, -и, (*f*) stratification, lamination, bedding

~, сезо́н·н·ая (geol) seasonal varving

сло·и́ст·ый, -ая, -ое, (*a*) layered, layer, stratified, lamellar, lamellated, lamellate, laminated, laminar; (*as component*) -ply

~ облак·а́ (*pl*) (meteor) Stratus

~ пла́ст·ик· (*m*) laminated plastic

сло·и́ть, -я́т, (*imp*) make layers, make into layers; -ся flake, peel off, exfoliate; layer

сло́·й, -я, (*nom pl*) сло·и́, (*g pl*) сло·ёв, (*m*) layer, lamina, stratum, bed; coat, coating, film; ply (of materials)

сло́·й

~, **адсорбцио́н·н·ый** (phys) adsorbed film

~ **атмосфе́р·ы, ве́рх·н·ий** (meteor) upper atmosphere

~ **атмосфе́р·ы, ни́ж·н·ий** (meteor) substandard layer

~, **вихр·ев·о́й** (dynam) vorticity layer

~, **вулкан·и́ческ·ий** (geol) ash bed

~, **год·и́чн·ый** (bot) annual zone/ring

~, **за·пир·а́ющ·ий** (dynam) barrier layer

~, **кип·я́щ·ий** fluidized bed (ore roasting etc.)

~, **кор·ов·о́й** (bot) cortex

~, **косте·но́с·н·ый** (geol) bone bed

~, **не·до·стро́·енн·ый** (phys) unfilled level

~, **нейтра́ль·н·ый** (mech) neutral axis (of a beam in a bend test)

~, **пласт·и́ческ·ий** plastic zone (in coke oven)

~, **по·гран·и́чн·ый** (dynam) boundary layer

~, **под·стил·а́ющ·ий** underlying layer, underlayer

~ **половин·н·ого с·слабл·е́ни·я** (phys) half-value thickness/layer, half-thickness

~, **про·меж·у́точн·ый** intermediate layer; (dynam) buffer layer

~, **с·ме́ш·анн·ый** (nucl) mixed bed (in a column)

~, **тепл·ов·о́й по·гран·и́чн·ый** (dynam) thermal boundary layer

~ **температу́р·н·ого скач·к·а́** thermocline

~, **то́нк·ий** thin layer; (dynam) shallow layer

~, **то́к·ов·ый** (nucl) current sheath

~, **торф·ян·о́й** (geol) peat bed

сло́й·к·а, -и, (g pl) **сло́·ек·,** (f) puff pastry

сло́й·ност·ь, -и, (f) number of plies (in tires etc.)

-сло́й·н·ый, -ая, -ое, (adj component) -layered; -ply

сло́й·чатост·ь, -и, (f) lamination; bedding (of rock)

с·ло́м·, -а, (m) breakdown, breaking down; break; breaking up, scrapping

с·ло́м·анн·ый, -ая, -ое, (past part pass) of **с·лом·а́·ть**

с·лом·а́·ть, -ют, (perf) see **лом·а́·ть**

с·лом·и́ть, -ят, (perf) break down, subdue, overcome

слон·, -а́, (m) (zool) elephant, *Elephas*; bishop (chess)

слон·ик, -а, (m) (dim) of **слон·;** (ent) weevil, (pl) *Curculionidae*

слон·и́х·а, -и, (f) cow elephant

слон·о́вост·ь, -и, (f) (med) elephantiasis

слон·о́в·ый, -ая, -ое, (a) elephant, elephantine; (pl as noun) *Elephantidae*

~ **бума́г·а** (f) (paper) elephant

~ **кост·ь** (f) ivory; ivory black (pigment)

слуг·а́, -и́, (nom pl) **слу́г·и,** (g pl) **слуг·** (f) servant

слу́д·, -а, (m) (meteor) slood, surface snow on ice

слу́ж·ащ·ий, -ая, -ее, (pres part act) see **служ·и́ть;** (m decl as adj) employee, white-collar/office/professional worker

слу́ж·б·а, -ы, (f) service, duty, employment; (mil) branch, department; (pl) outbuildings, outhouses

~, **вое́н·н·ая** military service

вре́м·я слу́ж·б·ы working life; life (of a machine)

~, **де́й·стви́тельн·ая** (mil) active service

~ **мер· и вес·о́в** Department of Weights and Measures

~ **по·го́д·ы** weather forecasting service

по·ступ·и́ть на слу́ж·б·у take up employment, enter the service

~ **пред·о·стереж·е́ни·я** meteorological warning service

~ **прогно́з·ов** (meteor) forecasting service (weather, flying conditions and hydrology)

~ **пут·и́** (rail) track maintenance

~, **сверх·сро́ч·н·ая** (mil) extended service

~ **с·вя́з·и** communications department; (mil) signals/communications service

служ·е́бн·ый, -ая, -ое, (a) service, official, office; auxiliary

~ **вре́м·я** (n) office/working hours

~ **знак·** (m) (telecom) procedure sign

~ **суд·а́** (pl) (naut) auxiliary craft

~ **та́й·н·а** (f) official secret

~ **телегра́мм·а** (f) (telecom) service message

служ·и́ть, -ат, (imp) serve, be employed (of persons); be used, be in use (of things)

слу́х·, -а, (m) hearing; rumour, rumor

слух·а́ч·, -а́, (i) -о́м, (m) listener; (telecom) sound-reading operator

слух·ов·о́й, -а́я, -о́е,   (*a*) auditory, aural, acoustic
~ аппара́т· (*m*) hearing aid
~ окн·о́ (*n*) (arch) attic window
с·луч·а·й, -я, (*m*) case; occasion, chance, opportunity; event, occurrence, incident
  в кра́й·н·ем с·лу́ч·а·е at the worst, at least; in an emergency
  в о́бщ·ем с·лу́ч·а·е in a general case, in general, generally
  в проти́в·н·ом с·лу́ч·а·е otherwise
  в с·лу́ч·а·е in case (of), in the event (of)
  в э́т·ом с·лу́ч·а·е in this case/instance, under these circumstances
  во вся́·к·ом с·лу́ч·а·е at any rate, in any case
~, кра́й·н·ий extreme case; emergency
~, не·с·ча́ст·н·ый accident
  по с·лу́ч·а·ю on account of, owing to
~, пре·де́ль·н·ый extreme/limiting/boundary case
~, ча́ст·н·ый special/particular case
  при с·лу́ч·а·е occasionally, on a suitable occasion
с·луч·а́йн·о (*adv*) by chance/accident; randomly
с·луч·а́йност·ь, -и, (*f*) chance, hazard; (math) chance/random/stochastic variable
с·луч·а́йн·ый, -ая, -ое, (*a*) chance, random, accidental, fortuitous, casual
~ бло́к·и (*pl*) (math) randomized blocks
~ велич·ин·а́ (*f*) chance/random/stochastic variable
~ о·ши́б·к·а (*f*) random error
~ проце́сс· (*m*) (math) stochastic process
~ со·бы́т·и·е (*n*) (math) random event
с·луч·а́·ть, -ют, (*imp*) pair, couple (for breeding); **-ся** mate; happen, occur
с·луч·и́ть, -а́т, (*perf*) *see* **с·луч·а́·ть**
с·лу́ч·к·а, -и, (*f*) pairing, coupling (for breeding)
слу́ш·ани·е, -я, (*n*) (*v n*) of **слу́ш·а·ть;** audition, hearing; attendance (at lectures)
слу́ш·ател·ь, -я, (*m*) listener, hearer; student (attending lectures); (*pl*) audience
слу́ш·а·ть, -ют, (*imp*) listen to; attend (lectures etc.); (law) hear (a case); **-ся** (+ *gen*) obey, pay attention to, listen to
с·лу́щ·ивани·е, -я, (*n*) (biol) desquamation, exfoliation, defurfuration

слыв·у́щ·ий, -ая, -ее, (*pres part act*) of **слы·ть**
слы́·вш·ий, -ая, -ее, (*past part act*) of **слы·ть**
с·лы·ть, (*pres 3rd pl*) **слыв·у́т,** (*imp*) be reputed
с·лых·а́·ть, -ют, (*imp*) hear
слы́ш·ать, -ат, (*imp*) hear
слы́ш·имост·ь, -и, (*f*) audibility, hearing
слы́ш·им·ый, -ая, -ое, (*pres part pass*) of **слы́ш·ать;** audible
~ генер·а́ци·я (*f*) (telecom) howl, howlback
слы́ш·н·о (*adv*) audibly; they say, it is said
слы́ш·н·ый, -ая, -ое, (*a*) audible
слюд·а́, -ы́, (*nom pl*) **слю́д·ы,** (*g pl*) **слюд·,** (*f*) (min) mica
~, бе́л·ая white mica, muscovite
~, желе́з·ист·ая lepidomelane
~, ко́л·от·ая mica film
~, кори́чнев·ая amber mica, phlogopite
~, ли́т·иев·ая (min) lithia-mica, lepidolite
слюд·иза́ци·я, -и, (*f*) (geol) micatization
слюд·ист·ый, -ая, -ое, (*a*) (min) micaceous
~ песч·а́ник· (*m*) micaceous sandstone
слюд·и́т·, -а, (*m*) (min) micaite, glimmer
слю́д·к·а, -и, (*f*)
~, желе́з·н·ая (min) micaceous iron ore
~, ура́н·ов·ая (min) uranite
слюдо·образ·ова́ни·е, -я, (*n*) (geol) micatization
слюд·ян·о́й, -а́я, -о́е, (*m*) mica, micaceous
~ окн·о́ (*n*) mica window
~ пласт·и́нк·а (*f*) mica sheet
~ сла́н·ец· (*m*) (geol) mica-schist
слюн·а́, -ы́, (*f*) saliva, spittle
слюн·и́ть, -я́т, (*imp*) lick, wet (by licking)
слю́н·н·ый, -ая, -ое, (*a*) salivary
~ ко́лик·а (*f*) (med) sialolith
слюно·ви́д·н·ый, -ая, -ое, (*a*) (biol) sialoid
слюно·го́н·, -а, (*m*) (bot) pellitory of Spain, *Anacyclus pyrethrum*
~, лека́р·ственн·ый common pellitory, *Anacyclus officinalis*
слюно·го́н·н·ый, -ая, -ое, (*a*) (physiol) sialogogic
слюно·от·дел·е́ни·е, -я, (*n*) (anat) salivation

слюно·теч·éни·е, -я, (*n*) (anat) salivation; (med) sialorrhea

сляб·, -а, (*m*) (met) slab

слябинг·, -а, (*m*) (met) slabbing mill

слякот·ь, -и, (*f*) slush

см. (*abbr*) = сантиметр· cm, centimetre, centimeter; (*abbr*) = смотр·úте see, v., *vide*

с·мáж·ет, (*fut 3rd pl*) of с·мáз·ать, (*perf*) see с·мáз·ыва·ть (*imp*)

с·мáз·анн·ый, -ая, -ое, (*a*) (*past part pass*) see с·мáз·ыва·ть; blurred

~ из·обрáж·éни·е (*n*) (opt) blurred image

с·мáз·ать, (*fut 3rd pl*) с·мáж·ут, (*perf*) see с·мáз·ыва·ть

с·мáз·к·а, -и, (*f*) lubrication, greasing; lubricating grease, grease

~, амунúч·н·ая leather grease

~, брикéт·н·ая briquet lubrication

~, гидродинам·úческ·ая film lubrication

~, гран·úчн·ая boundary lubrication

~, графúт·ов·ая graphite grease; graphite lubrication

~, густ·áя grease; grease lubrication, greasing

~, жúдк·остн·ая fluid-film lubrication, fluid lubrication

~, за·щúт·н·ая protective grease; protective greasing

~, кáльц·иев·ая lime-base grease

~, кáпель·н·ая drip-feed lubrication

~, кáртер·н·ая flood lubrication

~, кольц·ев·áя ring lubrication

~, консистéнт·н·ая grease

~, нáтр·ов·ая soda-base grease

~ о·куп·áни·ем flood lubrication

~, ос·ев·áя axle grease

~ по·груж·éни·ем flood lubrication

~ под давл·éни·ем pressure/force-feed/full-pressure lubrication

~, полу·жúдк·ая pudding grease

~, про·тóч·н·ая non-circulating lubrication

~, пýш·ечн·ая gun grease

~, раз·брызг·ивающ·ая splash lubrication

~ рас·пыл·éни·ем oil mist lubrication, atomized lubrication

~, с·на·ряд·н·ая shell grease

~, твёрд·ая solid lubricant

~, технолог·úческ·ая process oil; cutting oil (metal machining); (met roll) rolling oil

~ тóч·ечн·ая point lubrication

~, туго·плáв·к·ая hot-neck grease

с·мáз·к·а

~, фитúль·н·ая capillary lubrication

~, циркуляциóн·н·ая circulating lubrication

с·мáз·очн·ый, -ая, -ое, (*a*) lubricating

~ мáсл·о (*n*) lubricating oil

с·мáз·ыва·ть, -ют, (*imp*) grease, lubricate, put on grease/oil/fat; smear, rub (grease, half-dry paint etc.); blur, make indistinct

смáк·, -а, (*m*) savour, relish; (naut) smack

смáл·ец·, -ль·ц·а, (*i*) -ль·ц·ем, (*m*) lard

смáльт·а, -ы, (*f*) smalt (glass)

смальтúн·, -а, (*m*) (min) smaltine, smaltite

с·мáн·ива·ть, -ют, (*imp*) entice, lure

с·ман·úть, -ят, (*perf*) see с·мáн·и·ва·ть

смарáгд·, -а, (*m*) (min) emerald

смарагд·úт·, -а, (*m*) (min) smaragdite

смарúд·а, -ы, (*f*) (fish) smaride, *Smaris smaris*

с·мастер·úть, -ят, (*perf*) make, run up, knock together

смáтр·ива·ть, -ют, (*imp*) see смотр·éть

с·мáт·ывани·е, -я, (*n*) (*v n*) of с·мáт·ыва·ть

с·мáт·ыва·ть, -ют, (*imp*) uncoil, decoil, unreel, unwind; pour, lay (rods); wind, wind up

с·мáх·ива·ть, -ют, (*imp*) sweep/whisk/brush off; (*imp only*) look like

с·мах·н·ýть, -ýт, (*perf*) see с·мáх·и·ва·ть

с·мáч·иваемост·ь, -и, (*f*) (phys) wettability

с·мáч·ивани·е, -я, (*n*) (*v n*) of с·мáч·ива·ть (phys, chem) wetting ýгол· с·мáч·ивани·я (*m*) (phys) angle of contact, contact angle

с·мáч·иватель·, -я, (*m*) wetting agent, wetting-out agent

с·мáч·ива·ть, -ют, (*imp*) moisten, damp, wet, soak

с·мáч·ивающ·ий, -ая, -ее, (*pres part act*) of с·мáч·ива·ть

~ вещ·еств·ó (*n*) wetting agent

смегматúт·, -а, (*m*) (min) smegmatite

с·меж·ник·, -а, (*m*) sub-contracting factory, sub-contractor

с·меж·ност·ь, -и, (*f*) adjacency, juxtaposition

ассоциáци·я по с·меж·ност·и (*f*) association by contiguity

с·меж·н·ый, -ая, -ое, (*a*) adjacent, adjoining, neighbouring; related (of subjects etc.); (chem) vicinal

∼ по·лож·éни·е (*n*) (chem) vicinal position

∼ ýгол (*m*) (math) supplementary angle

с·мéл·ет, (*fut 3rd sing*) of **с·мол·óть**

смé·л·ый, -ая, -ое, (*a*) bold, daring, courageous

с·мéн·а, -ы, (*f*) (*v n*), *see* **с·мен·я́·ть;** change, replacement, re-, supersession, relief (from duty); shift, session; change of clothing; successors, rising generation; "smena" (tool-resurfacing alloy)

∼ ламп· (rad) revalving, tube replacement

∼, ноч·н·áя night shift, dog shift

∼ при·ли́·в·а turn of the tide

с·мен·и́ть, ⌐ят, (*perf*) *see* **с·мен·я́·ть**

с·мéн·н·ый, -ая, -ое, (*a*) of **с·мéн·а;** change, changeable, removable

∼ детáл·ь (*f*) (mech) replaceable/change part

∼ мáстер· (*m*) shift chargehand/foreman

с·мен·я́емост·ь, -и, (*f*) replaceability; periodic replacement

с·мен·я́ем·ый, -ая, -ое, (*pres part pass*) of **с·мен·я́·ть;** replaceable

с·мен·я́·ть, -ют, (*imp*) change, replace, supersede, relieve (from duty); **-ся** (*pass and reflex*); take turns, work in opposite shifts/watches

с·мерз·áни·е, -я, (*n*) (*v n*), *see* **с·мерз·á·ться**

с·мерз·á·ться, -ются, (*imp*) freeze up/together/hard, congeal

с·мёрз·нуться, -нутся, (*perf*) *see* **с·мерз·á·ться**

с·мéр·ить, -ят, (*perf*) measure, gauge, gage

с·мерт·éльн·ый, -ая, -ое, (*a*) death, dying; fatal, deadly, lethal; mortal

∼ бор·ьб·á (*f*) mortal combat

∼ дóз·а (*f*) lethal dose

∼ рáн·а (*f*) fatal injury

с·мéрт·ност·ь, -и, (*f*) death-rate, mortality

с·мéрт·н·ый, -ая, -ое, (*a*) death, deadly, fatal, mortal; deathly

∼ при·говóр· (*m*) (law) death sentence

с·мерто·éд·, -а, (*m*) (bot) hellebore, *Helleborus*

с·мерто·нóс·н·ый, -ая, -ое, (*a*) lethal, mortal, deadly

с·мéрт·ь, -и, (*nom pl*) **-и,** (*g pl*) **-éй,** (*f*) death, mors

смéрч·, -а, (*i*) **-ем,** (*m*) (meteor) whirlwind, tornado; waterspout; sandstorm

с·месе·образ·овáни·е, -я, (*n*) carburation (petrol engine); mixing (diesels, gas turbines)

с·мес·и́мост·ь, -и, (*f*) miscibility **коэффициéнт· с·мес·и́мост·и** (*m*) mixing ratio

с·мес·и́м·ый, -ая, -ое, (*pres part pass*) of **с·мес·и́ть;** miscible, mixable

с·мес·и́тел·ь, -я, (*m*) mixer, mixing/compounding machine; (elec) mixer, mixer tube/valve; (autom) "or" circuit

∼, винт·ов·óй screw conveyer

∼ и́мпульс·ов (elec) pulse mixer

∼, лáмп·ов·ый (elec) mixer valve/tube

∼ сигнáл·ов (TV) adder

с·мес·и́тельн·ый, -ая, -ое, (*a*) mixing, compounding

с·мес·и́ть, ⌐ят, (*perf*) knead, puddle, mix, compound

с·мéс·к·а, -и, (*f*) blend, mix

с·мес·ти́ (*fut 3rd pl*) **с·мет·ýт,** (*past masc sing*) **с·мё·л,** (*perf*) *see* **с·ме·т·á·ть**

с·мест·и́ть, -я́т, (*perf*) *see* **с·мещ·á·ть**

с·мéс·ь, -и, (*f*) mixture, mix, blend; stock (in processes using mixes)

∼, двой·н·áя (chem) binary mixture

∼ ис·хóд·н·ых вещéств· (chem) stock

∼, рабóч·ая (ICE) the mixture, air-fuel mixture

∼, резúн·ов·ая rubber stock

∼, тóпл·ивн·ая (rocket) composite propellant

с·мéт·а, -ы, (*f*) estimate

с·мет·áн·а, -ы, (*f*) sour cream

∼, желéз·н·ая (min) micaceous hematite

с·мет·áнн·ый, -ая, -ое, (*a*) of **с·мет·áн·а**

с·мёт·анн·ый, -ая, -ое, (*past part pass*) of **с·мет·á·ть**

с·мет·á·ть, -ют, (*imp*) sweep off/away; (*perf*) *see* **с·мёт·ыва·ть**

с·мёт·к·а, -и, (*g pl*) **-т·ок·,** (*f*) (*v n*), *see* (1) **с·мет·á·ть** (*imp*), (2) **с·мёт·ыва·ть;** brush; (*pl*) sweepings

с·мéт·н·ый, -ая, -ое, (*a*) estimate, estimated, budget, planned

с·мёт·ыва·ть, -ют, (*imp*) throw together, stack, throw in a heap/pile; (text) baste, baste under

смé·ть, -ют, (*imp*) dare

смех·, -а, (*m*) laughter, laugh, risus

смешанно·сло·и́ст·ый, -ая, -ое, (*a*) mixed-layered

с·ме́ш·анн·ый, -ая, -ое, (*past part pass*) *see* **с·ме́ш·ива·ть;** composite, compound; (biol) hybrid

~ числ·о́ (*n*) (math) mixed number

с·меш·а́·ть, -ют, (*perf*) *see* **с·ме́ш·ива·ть**

с·меш·е́ни·е, -я, (*n*) (*v n*), *see* **с·ме́ш·ива·ть;** mixture; blend; confusion
тепл·от·а́ с·меш·е́ни·я (*f*) heat of mixing

с·ме́ш·енн·ый, -ая, -ое, (*past part pass*) of **с·мес·и́ть**

с·ме́ш·иваемост·ь, -и, (*f*) miscibility

с·ме́ш·иваем·ый, -ая, -ое, (*pres part pass*) of **с·ме́ш·ива·ть;** miscible

с·ме́ш·ивани·е, -я, (*n*) (*v n*) of **с·ме́ш·ива·ть;** jumpering (telephones), grading (traffic)

с·ме́ш·ива·ть, -ют, (*imp*) mix, mix up, compound; shuffle (order etc.); blend; confuse, mistake for

смеш·н·о́й, -а́я, -о́е, (*a*) funny; laughable, ludicrous, ridiculous

с·мещ·а́·ть, -ют, (*imp*) displace, dislodge, remove, shift

с·мещ·е́ни·е, -я, (*n*) (*v n*), *see* **с·мещ·а́·ть;** displacement, removal, shifting, shift; trail (of falling object); (elec) bias

~, автомат·и́ческ·ое self-bias (vacuum tubes)

~ а́том·ов   atomic displacement; discomposition of atoms

~, бок·ов·о́е (air) cross trail

~, гравитаци́он·н·ое   (astron) Einstein shift

~ гран·и́ц· (phys) boundary movement, wall migration

~, ди·электр·и́ческ·ое electric displacement (dielectrics)

~, като́д·н·ое   cathode bias (vacuum tubes)

~ контине́нт·ов   (geol) continental drift

~, кра́сн·ое (astron) red shift

~ ку́рс·а (nav) squeeze
на·пряж·е́ни·е с·мещ·е́ни·я (*n*) (elec) bias voltage

~, нул·ев·о́е zero bias (vacuum tubes); (instr) zero creep

~, обра́т·н·ое (phys) back bias

~, от·риц·а́тельн·ое (elec) negative bias

с·мещ·е́ни·е

~ по́люса (phys) pole shift
пра́в·ил·о с·мещ·е́ни·я (*n*) (phys) displacement law

~ пуч·к·а́ (phys) beam translation

~, с·бро́с·ов·ое (geol) fault displacement

~, сет·очн·ое (rad) grid bias

~ те́рм·ов (math) term displacement

~, угл·ов·о́е angular displacement
у́гол· с·мещ·е́ни·я (*m*) angle of displacement; (civ eng) angle of pull

~, фа́з·ов·ое phase shift
част·от·а́ с·мещ·е́ни·я (*f*) offset frequency

~ част·от·ы́ frequency drift

~, электр·и́ческ·ое electric displacement (dielectrics)

с·мещ·ённ·ый, -ая, -ое, (*a*) (*past part pass*) *see* **с·мещ·а́·ть;** offset; biased

сме·я́ться, -ются, (*imp*) laugh

с·мин·а́·ть, -ют, (*imp*) crumple, crease, crush

с·ми́р·н·о (*imperative*) quiet; attention

с·ми́р·н·ый, -ая, -ое, (*a*) quiet, mild

смитсони́т·, -а, (*m*) (min) smithsonite

с·мо́г·ш·ий, -ая, -ее, (*past part act*) of **с·моч·ь**

с·мо́·ет, (*fut 3rd pl*) of **с·мы́·ть,** (*perf*); *see* **с·мыв·а́·ть** (*imp*)

с·мо́ж·ет, (*fut 3rd sing*) of **с·мот·ь**

смо́кв·а, -ы, (*f*) (bot) fig

смокед-шит·, -а, (*m*) (rubber) smoked sheet

смок·о́вниц·а, -ы, (*i*) **-ей,** (*f*) (bot) fig tree, *Ficus*

смол·а́, -ы, (*nom pl*) **смо́л·ы,** (*g pl*) **смол·,** (*f*) resin, gum, pitch, tar

~, акри́л·ов·ая acrylic resin

~, акроиди́н·ов·ая botany-bay gum

~, анионо·об·ме́н·н·ая   anion-exchange resin, anionite

~, бензо́й·н·ая (chem) benzoin

~, вон·ю́ч·ая (pharm) asafetida, asafoetida

~, га́з·ов·ая (chem) gas tar

~, генера́тор·н·ая (chem) gas tar

~, го́р·н·ая mineral pitch

~, дама́р·ов·ая dammar

~, ионо·об·ме́н·н·ая (chem) ion-exchange resin

~, ка́мен·н·о-у́голь·н·ая coal tar

~, карбоми́д·н·ая (chem) urea resin

~, конденсаци́он·н·ая (plast) condensation resin

~, кре́мн·и·й-орган·и́ческ·ая (chem) silicone resin

смол·а́

~, **можжевёл·ов·ая**   juniper tar

~, **моч·евинн·ая**   (chem) urea resin

~, **натура́ль·н·ая**   (chem) resin

~, **новола́ч·н·ая**   novolak resin

~, **поли·эфи́р·н·ая**   polyester resin

~, **резо́ль·н·ая**   (plast) bakelite

~, **сад·о́в·ая**   (hortic) grafting wax

~, **синтет·и́ческ·ая**   (plast) synthetic resin, plastic

~, **сосн·о́в·ая**   pine tar, rosin

~, **термо·реакти́в·н·ая** (plast) thermo-setting resin

~, **термо·пласт·и́ческ·ая**   (plast) thermoplastic resin

~, **факт·и́ческ·ая**   gum existent (in fuels under test)

~, **чёрн·ая**   pitch

~, **эпокси́д·н·ая**   epoxy resin

смол·ёвк·а, -и,   (g pl) **-вок·,** (f) (bot) campion, *Silene*

смол·ев·о́й, -а́я, -о́е,   (a) of **смол·а́**

смол·ёни·е, -я,   (n) tarring, resinifying

смол·ён·ый, -ая, -ое,   (a) tarred, pitch-coated, resined

~ **трос·**   (m) tarred rope

смол·и́ст·ый, -ая, -ое,   (a) resinous, resinaceous; tarry

смол·и́ть, -я́т,   (imp) tar, coat with pitch/tar, resinify

смо́л·к·а, -и,   (g pl) **-л·ок·,** (f) (bot) Viscaria

с·молк·а́·ть, -ют,   (imp) become silent/quiet, cease (of noise)

с·мо́лк·нуть, -нут,   (perf) see **с·молк·а́·ть**

смоло·ва́р·н·я, -и,   (g pl) **-р·ен·,** (f) tar-distillery, tar works

смоло·кур·е́ни·е, -я, (n) resin distillation, tar distilling

смоло·на·по́лн·енн·ый, -ая, -ое, (a) resin-filled

смоло·но́с·н·ый, -ая, -ое, (a) resin-iferous

смоло·образ·ова́ни·е, -я,   (n) gum formation

смоло·от·дел·е́ни·е, -я,   (n) tar separation

смоло·семя́н·ник·, -а, (m) (bot) pittosporum

с·молот·и́ть, -я́т, (perf) thresh, thrash

с·мо́л·от·ый, -ая, -ое, (past part pass) of **с·мол·о́ть**

с·мол·о́ть, (fut 3rd pl) **с·ме́л·ют,** (perf) grind, mill (corn etc.); mince (meat etc.)

с·моло́ч·енн·ый, -ая, -ое, (past part pass) of **с·молот·и́ть**

смол·ь·ё, -я́,   (n) resin-impregnated wood

смо́ль·н·ый, -ая, -ое, (a) = **смол·и́ст·ый**

смо́ль·н·я, -и, (f) (shipb) tarring loft

смол·я́к·, -а́, (m) = **смол·ь·ё** (q.v.)

смол·ян·о́й, -а́я, -о́е, (a) resin, rosin, resinous, pitch, tar, tarry, bituminous; tarred, bituminized

~ **ка́мен·ь** (m) (geol) pitch-stone

~ **кисл·от·а́** (f) (chem) resin acid

~ **ле́нт·а** (f) bituminized tape

~ **об·ма́н·к·а** (f) (min) pitchblende

~ **руд·а́** (f) pitchblende

~ **спирт·** (m) (chem) resinol

~ **с·гу́ст·к·и** (pl) (chem) tar fog

~ **ход·** (m) (bot) resin canal/duct

смоляно·ки́сл·ый, -ая, -ое, (a) (chem) resinate (of)

~ **ка́льц·и·й** (m) calcium resinate

с·монт·и́рованн·ый, -ая, -ое, (past part pass) of **с·монт·и́р·овать**

с·монт·и́р·овать, -уют, (perf) erect, install, mount

сморо́д·ин·а, -ы, (f) (bot) currant, currant bush, *Ribes*

сморо́д·инн·ые, -ых, (pl as adj) (bot) currant family, *Grossulariaceae*

сморч·о́к·, -ч·к·а́, (m) (bot) morel, *Morchella esculenta*

с·мо́рщ·енн·ый, -ая, -ое, (past part pass) see **с·мо́рщ·ива·ть**

с·мо́рщ·ива·ть, -ют, (imp) wrinkle, corrugate

с·мо́рщ·ить, -ат, (perf) see **с·мо́рщ·и·ва·ть**

с·мо́т·анн·ый, -ая, -ое, (past part pass) of **с·мот·а́·ть**

с·мот·а́·ть, -ют, (perf) see **с·ма́т·ы·ва·ть**

смотр·, -а, (nom pl) **-ы,** (g pl) **-ов,** (m) review, inspection

смотр·е́ть, ⌐я́т, (imp) look, look at, see; regard, look upon; examine, review, inspect, look over/through/round; (naut) grow (of cable)

~ **за** (+ instr) look after, see to

смотр·и́тел·ь, -я, (m) keeper, man-in-charge; inspector

смотр·ов·о́й, -а́я, -о́е, (a) review; observation, inspection, sight

~ **окн·о́** (n) viewing window, sight glass, peep hole, inspection hole

~ **отве́рст·и·е** (n) sight/inspection/peep hole, eyesight elbow (blast furnace)

~ **стекл·о́** (n) sight glass; facepiece (of protective suit etc.)

смотр·ов·ой, -ая, -ое
~ **фонáр·ь** (*f*) sight glass
~ **щел·ь** (*f*) visor, eye slit
смотр·я́, (*pres gerund*) of **смотр·éть**
~ **на** according to, depending on
  не смотр·я́ на despite
с·мóч·енн·ый, -ая, -ое, (*past part pass*) of **с·моч·и́ть;** wet, wetted; wet-bulb (of thermometers)
с·моч·и́ть, -áт, (*perf*) *see* **с·мáч·и·ва·ть**
с·моч·ь, (*fut 3rd sing, pl*) **с·мóж·ет, с·мóг·ут,** (*past masc sing*) **с·мóг·,** (*perf*) be able
с·мó·ют, (*fut 3rd pl*) of **с·мы·ть** (*perf*); *see* **с·мыв·á·ть** (*imp*)
смрáд·н·ый, -ая, -ое, (*a*) stinking
смýгл·ый, -ая, -ое, (*a*) dark-skinned
смýш·ек·, -ш·к·а, (*m*) = **смýш·к·а**
смýш·к·а, -и, (*g pl*) **-ш·ек·,** (*f*) astrakhan (fur)
~**, каракýль·ск·ая** astrakhan (fur)
~**, перси́д·ск·ая** Persian lamb (fur)
с·мущ·éни·е, -я, (*n*) confusion, embarrassment
с·мы́·в·, -а, (*m*) (*v n*), *see* **с·мыв·á·ть;** erosion (from running water)
с·мыв·áем·ый, -ая, -ое, (*pres part pass*) of **с·мыв·á·ть;** removable, wipe-off
с·мыв·á·ть, -ют, (*imp*) wash off/away
с·мы́в·к·а, -и, (*g pl*) **-в·ок·,** (*f*) (*v n*) of **с·мыв·á·ть;** flushing; paint remover
с·мыв·áльн·ый, -ая, -ое, (*a*) washing, wiping, flushing
с·мыв·áни·е, -я, (*n*) (*v n*) of **с·мы·в·á·ть,** *see* **с·мы́в·к·а** *and* **с·мы·в·**
с·мыв·н·ой, -áя, -óе, (*a*) wash, washing, flushing
~ **бак·** (*m*) flushing tank/cistern (sanitary eng)
с·мык·áни·е, -я, (*n*) (*v n*), *see* **с·мык·á·ть;** closure
  плóск·ост·ь с·мык·áни·я (*f*) line of join (moulds)
с·мык·á·ть, -ют, (*imp*) close, close up, join, connect
~ **полигóн·** (*surv*) adjust a traverse, close a polygon
с·мы́сл·, -а, (*m*) sense, meaning
  в том с·мы́сл·е, что ... in that
  в э́т·ом с·мы́сл·е in this respect
~ **вращ·éни·я** direction of rotation
  нет с·мы́сл·а there is no point/ sense in
  пос·мы́сл·у according to, in the opinion of

с·мысл·ов·ой, -áя, -óе, (*a*) sense, meaning; semantic
с·мы́т·ост·ь, -и, (*f*) erodibility
с·мы́·т·ый, -ая, -ое, (*past part pass*) *see* **с·мыв·á·ть;** eroded
с·мы·ть, (*fut 3rd pl*) **с·мó·ют,** (*perf*) *see* **с·мыв·á·ть**
с·мы́ч·к·а, -и, (*g pl*) **-ч·ек·,** (*f*) (*v n*) of **с·мык·á·ть;** join, connection; clamp, link; shackle (of chain cable); fellowship, union
с·мыч·кóв·ый, -ая, -ое, (*a*) string, stringed (music)
~ **инструмéнт·** (*m*) stringed instrument, string (music)
с·мыч·óк·, -ч·к·á, (*m*) bow (music)
с·мя́·вш·ий, -ая, -ее, (*past part act*) of **с·мя·ть**
с·мягч·á·ть, -ют, (*imp*) soften, mitigate, mollify, alleviate, ease
с·мягч·áющ·ий, -ая, -ее, (*pres part act*) of **с·мягч·á·ть;** extenuating; (med) emollient, softening
~ **об·сто·я́тельств·о** (*n*) extenuating/ mitigating circumstances
~ **срéд·ств·о** (*n*) (pharm) emollient; softening agent
с·мягч·éни·е, -я, (*n*) (*v n*), *see* **с·мяг·ч·á·ть**
~ **спéктр·а** (phys) spectrum softening/ degradation
с·мягч·ённ·ый, -ая, -ое, (*past part pass*) *see* **с·мягч·á·ть;** (gram) palatal
с·мягч·и́ть, -áт, (*perf*) *see* **с·мягч·á·ть**
с·мя́т·и·е, -я, (*n*) (*v n*), *see* **с·мя·ть;** buckling crushing
  прóч·ност·ь на с·мя́т·и·е (*f*) crushing strength
с·мя́·т·ый, -ая, -ое, (*past part pass*) of **с·мя·ть**
с·мя·ть, (*fut 3rd pl*) **со·мн·ýт,** (*perf*) crumple, crease, crush; puddle (clay), crush, break (flax)
сн·а, (*g sing*) of **сон·**
снабд·и́ть -ят, (*perf*) *see* **снабж·á·ть**
снабж·á·ть, -ют, (*imp*) supply, provide
снабж·éни·е, -я, (*n*) (*v n*), *see* **снаб·ж·á·ть;** supply, provision; supplies, provisions; providing, furnishing
~ **про·до·вóль·стви·ем** provisioning, victualling
снабж·ённ·ый, -ая, -ое, (*past part pass*) *see* **снабж·á·ть;** fitted/furnished/supplied/equipped (with), with, complete with
~ **руб·áшк·ой** jacketed
~ **транзи́стор·ами** transistorized

снайпер·, -а, (*m*) (mil) sniper

с·найтóв·ить, -ят, (*perf*) lash (secure with rope)

с·нарýж·и (*adv*) on the outside, from the outside

с·на·ря́д·, -а, (*m*) equipment, gear, apparatus; (gunn) shell, projectile; (GW) missile; (rocket) rocket, vehicle; dredger, dredge

~, авиациóн·н·ый у·правл·я́ем·ый air-launched guided missile, air-to-ground guided missile

~, áтом·н·ый atomic missile; atomic shell

~, броне·про·жиг·áющ·ий hollow-charge shell

~ воз·дýш·н·ого бó·я air-to-air missile

~, выс·óтн·о-зонд·и́рующ·ий high-altitude sounding vehicle

~, дно·у·глуб·и́тельн·ый dredger, dredge

~, жи́дк·остн·ый ракéт·н·ый liquid-fuel rocket

~, за·жиг·áтельн·ый incendiary shell

~, земле·сóс·н·ый suction dredger (Brit), hydraulic dredge (U.S.)

~, земле·черп·áтельн·ый bucket dredger, mechanical dredge

~, зени́т·н·ый у·правл·я́ем·ый ра·кéт·н·ый surface-to-air guided missile

~, зонд·и́рующ·ий sounding rocket

~, крыл·áт·ый aerodynamic/winged missile

~, кумуляти́в·н·ый hollow-charge shell

~, лафето·прóб·н·ый paper shot, proof shot

~, меж·континентáль·н·ый баллист·и́ческ·ий intercontinental ballistic missile

~, мно́го·ковш·óв·ый bucket dredger, multi-bucket dredger, ladder dredger

~, мно́го·черп·акóв·ый bucket/multi-bucket/ladder dredger

~ на·зéм·н·ого бó·я ground-to-ground missile

~ -на·рýш·и́тел·ь, -я, (*m*) (mil) intruder

~ оборóн·ы объéкт·а (mil) point-defence missile

~, одно·ковш·óв·ый dipper dredger/dredge

~, о·свет·и́тельн·ый star shell

~, о·с·кóл·очн·ый fragmentation shell

с·на·ря́д

~ -пере·хвáт·чик·, -а, (*m*) anti-missile missile

~, перфорáтор·н·ый (oil) drilling barge

~, практ·и́ческ·ий practice shell/round

~, реакти́в·н·ый (GW) missile; (gunn) rocket shell

~, само·на·вод·я́щ·ийся self-guided missile

~, скало·дроб·и́тельн·ый rock-breaker dredger/dredge

~, скв·áж·инн·ый (oil) go-devil

~, со·стáв·н·óй staged missile

~, с·пас·áем·ый recoverable missile

~, трасс·и́рующ·ий tracer shell

~, у·правл·я́ем·ый guided missile

~, уч·éбн·ый drill shell, dummy shell

~, фугáс·н·ый H.E. shell

~, шалáнд·ов·ый barge-delivery dredger/dredge

~, штáнг·ов·ый земле·черп·áтель·н·ый dipper dredger/dredge

с·на·ря́д·и́ть, -ят, (*perf*) see **с·на·ря·ж·á·ть**

с·на·ря́д·н·ый, -ая, -ое, (*a*) of **с·на·ря́д·** (*m decl as adj*) loader

с·на·ряж·á·ть, -ют, (*imp*) equip, furnish, fit out, get ready

с·на·ряж·éни·е, -я, (*n*) (*v n*), see **с·на·ряж·á·ть**; equipment, outfit; loading, charging

~, ли́ч·н·ое (mil) accoutrement

~ тóпл·ив·ом (rocket) propellant loading

с·на·ряж·ённ·ый, -ая, -ое, (*past part pass*) see **с·на·ряж·á·ть**

снаст·и́ть, -ят, (*imp*) (naut) rig

снáст·ь, -и, (*nom pl*) **-и,** (*g pl*) **-éй,** (*f*) gear, tackle, equipment; rope, cordage

с·на·чáл·а (*adv*) at first, to start with; from the beginning, again

сн·е (*p sing*) of **сон·**

снег·, -а, (*nom pl*) **-á,** (*g pl*) **-óв,** (*m*) snow

~, мóкр·ый wet snow

~, пере·мещ·áющ·ийся drifting snow

~, порош·ковáт·ый powder snow

~, рас·сы́п·чат·ый loose snow

~ с дожд·ём sleet

~, тá·л·ый slush

снег·овáни·е, -я, (*n*) covering with snow, piling up snow; snow clamping (vegetables); frosting up (of refrigerators)

снег·ов·óй, -áя, -óе, (*a*) snow

~ гран·и́ц·а (*f*) (geog) snow line

снег·ов·ой, -ая, -ое
~ **за·бор·** (*m*) snow fence
~ **на·груз·к·а** (*f*) (civ eng) snow load
снего·за·держ·áни·е, -я, (*n*) (rail) snow fencing/retention; (agr) irrigation by snow
снего·за·щи́т·н·ый, -ая, -ое, (*a*) snow, snow-retaining/-protection
~ **за·гражд·éни·е** (*n*) snow fence/screen/barrier
снего·мéр·, -а, (*m*) snow gauge/gage, snow density gauge/gage
~**, вес·ов·ой** weighing-type snow density gauge
снего·мéр·н·ый, -ая, -ое, (*a*) snow, snow-measuring
снего·о·чист·и́тел·ь, -я, (*m*) snow remover
~**, плуг·ов·ой** (rail) snow-plough wagon attachment
~**, плу́ж·н·ый** (M/T) snow plough/plow
~**, рóтор·н·ый** snow blower
~**, сов·кóв·ый** bulldozer snow remover
~**, тарáн·н·ый** (rail) snow-plough locomotive, snow plough
~**, центро·бéж·н·ый** snow blower
снего·о·чи́ст·к·а, -и, (*f*) snow clearing, clearing the snow
снего·пáд·, -а, (*m*) snow fall
снего·пáх·, -а, (*m*) draw-type snow plough/plow
снего·по·гру́з·чик·, -а, (*m*) snow loader
снего·та·я́лк·а, -и, (*g pl*) **-лок·,** (*f*) snow-melting equipment
снего·у·бóр·очн·ый, -ая, -ое, (*a*) snow-removal
снего·у·бóр·щик·, -а, (*m*) (rail) snow collector
снего·хóд·, -а, (*m*) (M/T) snow cat
снед·óк·, -д·к·á, (*m*) (bot) chervil, *Anthriscus cerefolium*
снеж·и́нк·а, -и, (*g pl*) **-нок·,** (*f*) snow-flake
снеж·и́ть, -ат, (*imp*) (meteor) snow
снеж·ни́ц·а, -ы, (*i*) **-ей,** (*f*) fresh water pool (formed by snow melting on ice)
снежно·я́год·ник·, -а, (*m*) (bot) snowberry, *Symphoricarpus*
снéж·н·ый, -ая, -ое, (*a*) snow, snowy
~ **зёрн·а** (*pl*) (meteor) fine soft hail
~ **круп·á** (*f*) soft hail
~ **по·крó·в·** (*m*) snow cover
снеж·óк·, -ж·к·á, (*m*) snowball; light snow

снеж·ур·á, -ы́, (*f*) slush
с·нес·ённ·ый, -ая, -ое, (*past part pass*) *see* **с·нос·и́ть**
с·нес·ти́, -у́т, (*perf*) *see* **с·нос·и́ть**
снет·óк·, -т·к·á, (*m*) (fish) smelt, *Osmerus eperlanus*
с·ниж·áтел·ь, -я, (*m*) depresser, depressant
с·ниж·á·ть, -ют, (*imp*) lower, bring down, reduce; **-ся** (*pass*); descend, come down, become lower, sink, fall, drop, decrease, abate
с·ниж·éни·е, -я, · (*n*) (*v n*), *see* **с·ниж·á·ть;** reduction, abatement; descent; down lead (of antenna)
~ **давл·éни·я** drop/decrease in pressure; reduction of pressure; depressurizing
~ **скóр·ост·и с·чёт·а** (instr) scaling down
  ýгол· с·ниж·éни·я (*m*) angle of depression; (air) angle of descent
с·ни́ж·енн·ый, -ая, -ое, (*past part pass*) *see* **с·ниж·á·ть**
с·ни́з·ить, -ят, (*perf*) *see* **с·ниж·á·ть**
с·ни́з·у (*adv*) from below, underneath, below
с·ним·áни·е, -я, (*n*) (*v n*) of **с·ним·á·ть;** taking off/away, removing; skimming; gathering, harvesting; stripping (coatings, emulsions etc.)
с·ним·á·ть, -ют, (*imp*) take, take off/from/away, remove; skim, skim off; raise (blockade etc.), free from; gather (harvest etc.), harvest (crops etc.), pick, reap; photograph, take a photograph; make a survey, survey
~ **результáт·** take information (from computer)
с·ни́м·ок·, -м·к·а, (*m*) *see also* **аэро·-с·ни́м·ок·;** photograph, print
~**, рентгéн·овск·ий** X-ray photograph
~**, у·вели́ч·енн·ый** (phot) enlargement print
~**, у·меньш·ённ·ый** (phot) reduction print
~**, фото·грáф·и́ческ·ий** photograph
с·нóв·а (*adv*) anew, again
снов·áльн·ый, -ая, -ое, (*a*) (text) warping
~ **маши́н·а** (*f*) (text) warper, warping machine
~ **маши́н·а, лéнт·очн·ая** sectional warping machine
~ **маши́н·а, партиóн·н·ая** beam warping machine
снов·áни·е, -я, (*n*) (*v n*) of **снов·á·ть**

снов·а́ть, (*pres 3rd pl*) **сну·ю́т,** (*imp*) (text) warp

сно·вид·е́ни·е, -я, (*n*) dream

снóв·к·а, -и, (*f*) (*v n*), *see* **снов·а́ть**

сноп·, -а́, (*m*) sheaf; shaft (of light); (gunn) cone

~ **траект́óр·и·и** (gunn) dispersion cone

снопо·ви́д·н·ый, -ая, -ое, (*a*) sheaf-shaped/like

снопо·вяз·а́лк·а, -и, (*g pl*) **-лок·,** (*f*) (agr) binder

снопо·нóс·, -а, (*m*) (agr) sheaf/bundle carrier, binder attachment

снопо·по·да·ва́тел·ь, -я, (*m*) (agr) sheaf elevator

снор·óвк·а, -и, (*g pl*) **-вок·,** (*f*) knack, skill

с·нóс·, -а, (*m*) (*v n*), *see* **с·нос·и́ть;** drift; (civ eng) demolition; (min) stripping, strip pit

~ **верш·и́н·ы** (geol) decapitation

по·пра́в·к·а на с·нóс· (*f*) (nav) drift correction

у́гол· с·нóс·а (*m*) (air) drift angle

с·нос·и́ть, -ят, (*imp*) carry/take off/away/to/down; tear down and remove; bring together; endure, bear; **-ся** (*pass*); establish contact with, contact, communicate with, enter into relations with; drift

с·нóс·к·а, -и, (*g pl*) **-с·ок·,** (*f*) (print) footnote

с·нóс·н·ый, -ая, -ое, (*a*) tolerable, bearable

сно·твóр·н·ый, -ая, -ое, (*a*) somnifacient, somniferous, soporific, sleep-inducing

с·нош·е́ни·е, -я, (*n*) dealings, relations, intercourse

с·нош·у́, (*pres 1st sing*) of **с·нос·и́ть**

сну·ёт, (*pres 3rd sing*) of **снов·а́ть**

с·ня́т·и·е, -я, (*n*) (*v n*), *see* **с·ним·а́ть;** un-, de-; (phot) stripping; relieving, relief (of pressure, load etc.)

~ **воз·бужд·е́ни·я** de-energizing

~ **корóт·к·ого** (elec) unshorting

с·ня·т·óй, -а́я, -óе, *see also* **с·ня·т·ый**

~ **молок·ó** (*n*) skimmed milk

с·ня·т·ый, -ая, -ое, (*a*) (*past part pass*) *see* **с·ним·а́ть;** *see also* **с·ня·т·óй**

с·ня·ть, (*fut 3rd pl*) **с·ним·ут,** (*perf*) *see* **с·ним·а́ть**

со (*prep*) *see* **с** (*prep*)

со- (*prefix*) *see* **с-** (*prefix*); **со-,** joint

со·а́втор·, -а, (*m*) co-author, joint author

собáк·а, -и, (*f*) (zool) dog, *Canis familiaris*

~ **-блóк·, -а,** (*m*) (naut) jack block

собако·вóд·ств·о, -а, (*n*) dog breeding

собако·зу́б·ые, -ых, (*pl decl as adj*) (pal) cynodonts, *Cynodonta*

собáч·ий, -ья, -ье, (*a*) dog, dog's, canine

~ **ва́хт·а** (*f*) (naut) dog watch

~ **дыр·а́** (*f*) (naut) lubber's hole

~ **петр·у́шк·а** (*f*) (bot) dog parsley, *Aethusa cynapium*

собáч·к·а, -и, (*g pl*) **-ч·ек·,** (*f*) (*dim*) of **собáк·а;** trigger; pawl, ratchet, catch, dog

~, **за·де́рж·ивающ·ая** retaining pawl

со·бер·ёт, (*fut 3rd sing*) of **со·бр·а́ть,** (*perf*); *see* **со·бир·а́ть** (*imp*)

со·бесе́д·ник·, -а, (*m*) interlocutor, party/participant in conversation/discussion, person spoken with

со·бесе́д·овани·е, -я, (*n*) discussion

со·бир·а́ни·е, -я, (*n*) (*v n*) of **со·бир·а́ть**

со·бир·а́тел·ь, -я, (*m*) collector, gatherer

со·бир·а́тельн·ый, -ая, -ое, (*a*) collective, collecting; (opt) condensing, converging; (elec) current-collecting

~ **ли́нз·а** (*f*) (opt) converging/condensing lens; (cinema) collecting lens

~ **кольц·ó** (*n*) (elec) slip ring

~ **систе́м·а** (*f*) (autom) "or" circuit

~ **электрóд·** (*m*) (elec) collecting electrode

со·бир·а́ть, -ют, (*imp*) collect, gather; assemble, put together; **-ся** (*pass*); assemble, gather together; get ready to/for

со·бир·а́ющ·ий, -ая, -ее, (*pres part act*) of **со·бир·а́ть**

со·блюд·а́ть, -ют, (*imp*) observe, keep, maintain, pay attention to

со·блюд·е́ни·е, -я, (*n*) (*v n*), *see* **со·блюд·а́ть;** observance, maintenance

со·блю·сти́, (*fut 3rd pl*) **со·блюд·у́т,** (*perf*) *see* **со·блюд·а́ть**

соб·óй, (*pron, instr*) of **себ·я́** (by/with etc., your/it/him etc.) self

со·болéз·новани·е, -я, (*n*) condolence

соббл·ий, -ья, -ье, (*a*) sable (fur)

сóбол·ь, -я, (*m*) (zool) sable, Russian sable, *Mustela zibellina*

со·бóр·, -а, (*m*) cathedral

соб·óю, *see* **соб·óй**

со·бр·а́ни·е, -я, (*n*) meeting, assembly, conference; collection

со·бр·а́ни·е
~ **со·чин·е́ни·й** collected works
со́·бр·анн·ый, -ая, -ое, (*past part pass*) *see* **со·бир·а́ть**
~ **на магни́стор·ах** (elec) magnistorized
со·бр·а́ть, (*fut 3rd pl*) **со·бер·у́т,** (*perf*) *see* **со·бир·а́ть**
соб·ственник·, -а, (*m*) owner, proprietor
со́б·ственн·о (*adv*) properly, strictly; (particle) proper
~ **говор·я́** strictly/properly speaking
~ **го́род·** (*m*) the town proper
~ **ра́к·и** (*pl*) (zool) eucrustaceans, *Eucrustacea*
~ **таба́к·** (*m*) (bot) tobacco, *Nicotiana tobacum* (distinguishing most important species from named, less important species)
соб·ственност·ь, -и, (*f*) property, estate
~, **ча́ст·н·ая** private property
со́б·ственн·ый, -ая, -ое, (*a*) own; proper; (phys) natural, free, intrinsic; (math) eigen
~ **ве́ктор·** (*m*) (math) eigen vector
~ **движ·е́ни·е** (*n*) (astron) proper motion
~ **знач·е́ни·е** (*n*) (math) eigen value
~ **импеда́нс·** (*m*) (elec) self-impedance
~ **и́м·я** (*n*) (gram) proper name
~ **колеб·а́ни·е** (*n*) (phys) free/natural vibration/oscillation
~ **ну́жд·ы** (*pl*) own requirements; (elec) house supplies; (*gen as adj*) house
со·бы́т·и·е, -я, (*n*) event
~, **взаи́м·н·о ис·ключ·а́ющ·ие** (*pl*) (math) mutually exclusive events
~, **с·ло́ж·н·ое** compound/complex event
~, **с·луч·а́йн·ое** (math) random event
со·бь·ёт, (*fut 3rd sing*) of **с·би·ть,** (*perf*); *see* **с·бив·а́·ть** (*imp*)
сов- (*abbr*) = **со·ве́т·ск·ий** Soviet
сов·а́, -ы́, (*nom pl*) **со́в·ы,** (*g pl*) **сов·,** (*f*) (orn) owl
сов·а́ть, (*pres 3rd pl*) **су·ю́т,** (*imp*) shove, thrust, poke
со·верш·а́ть, -ют, (*imp*) accomplish, perform, make; commit, perpetrate
со·верш·е́ни·е, -я, (*n*) (*v n*), *see* **со·верш·а́ть;** accomplishment, completion, performance
со·верш·е́нн·о (*adv*) completely, absolutely, quite, utterly
~ **секре́т·н·о** Top Secret

совершенно·ле́т·н·ий, -яя, -ее, (*a*) (law) adult
совершенно·ро́т·ые, -ых, (*pl decl as adj*) (pal) teleostomes, *Teleostomi, Actinopterygii*
со·верш·е́нн·ый, -ая, -ое, (*a*) perfect, ideal, absolute; (gram) perfective
со·верш·е́нств·о, -а, (*n*) perfection
со·верш·е́нств·овать, -уют, (*imp*) improve, perfect
со·верш·и́ть, -а́т, (*perf*) *see* **со·верш·а́·ть**
со́вест·лив·ый, -ая, -ое, (*a*) conscientious
со́вест·ь, -и, (*f*) conscience
со·ве́т·, -а, (*m*) council, board; advice, counsel, (law) opinion; Soviet
~ **мини́стр·ов** (*m*) Council of Ministers
~, **педагог·и́ческ·ий** headmasters' council/conference
со·ве́т·ник·, -а, (*m*) adviser, (dipl) counsellor
со·ве́т·овать, -уют, (*imp*) advise, counsel; **-ся** ask/seek advice, consult; confer
со·ве́т·ск·ий, -ая, -ое, Soviet
со·вещ·а́ни·е, -я, (*n*) conference; consultation
со·вещ·а́тельн·ый, -ая, -ое, (*a*) conference; consultative, deliberative
со·вещ·а́·ться, -ются, (*imp*) confer, consult
сов·и́н·ы, -ая, -ое, (*a*) owl, owl's, owlish
со́в·к·а, -и, (*g pl*) **-в·ок·,** (*f*) (ent) cutworm, *Agrotis,* (*pl*) *Noctuidae, Phalaenidae;* (*g sing*) of **сов·о́к·**
~, **луг·ов·а́я** (ent) army worm, *Cirphis unipuncta*
совкаи́н·, -а, (*m*) (pharm) sovcain, percain
сов·ко́в·ый, -ая, -ое, (*a*) of **сов·о́к·**
со·влад·е́лец·, -льц·а, (*i*) **-льц·ем,** (*m*) joint owner
со·в·мест·и́мост·ь, -и, (*f*) compatibility
со·в·мест·и́м·ый, -ая, -ое, (*a*) compatible
со·в·мест·и́ть, -ят, (*perf*) *see* **со·в·мещ·а́·ть**
со·в·ме́ст·н·о (*adv*) in common, jointly; in conjunction/combination (with)
со·в·ме́ст·н·ый, -ая, -ое, (*a*) joint, combined, co-; (math) consistent, compatible
~ **де́й·стви·е** (*n*) joint action, co-operation

со·в·мест·н·ый, -ая, -ое
~ **об·уч·ени·е** (*n*) co-education
~ **опер·аци·я** (*f*) (mil) combined operations
~ **полимериз·аци·я** (*f*) (chem) co-polymerization
~ **работ·а** (*f*) collaboration, co-operation
~ **работ·а с реаль·н·ым объект·ом** (autom) on-line operation
~ **рас·твор·ител·ь** (*m*) (chem) cosolvent
~ **у·равн·ени·я** (*pl*) (math) consistent/compatible equations

со·в·мещ·а·ть, -ют, (*imp*) combine, combine with; match, bring into line, line up; joggle (joints); **-ся** (*pass*); coincide, be congruent

со·в·мещ·а·ние, -я, (*n*) (*v n*), *see* **со·в·мещ·а·ть;** (math) coincidence, congruency
   цеп·ь со·в·мещ·ени·я (*f*) (elec) coincidence circuit/gate

со·в·мещ·ённ·ый, -ая, -ое, (*past part pass*) *see* **со·в·мещ·а·ть;** mixed (of polymers); joggled (of plate joints)

сов·ок·, -в·к·а, (*m*) scoop, shovel; dust pan; *see also* **сов·к·а** (*nom sing*) (*f*)

совокуп·ительн·ый, -ая, -ое, (*a*) copulation, copulatory

совокуп·ить, -ят, (*perf*) *see* **совокупл·я·ть**

совокупл·ени·е, -я, (*n*) (*v n*), *see* **совокупл·я·ть;** (zool) coitus, copulation

совокупл·я·ть, -ют, (*imp*) combine, unite; (zool) copulate

совокуп·ност·ь, -и, (*f*) aggregate, totality, combination; (math) set, ensemble
   в совокуп·ност·и all together, altogether
~ **систем·** system ensemble
~, **у·по·ряд·оченн·ая** (math) ordered set

совокуп·н·ый, -ая, -ое, (*a*) combined, aggregate, total

со·в·пад·а·ть, -ют, (*imp*) coincide (with), be the same (as), be congruent, agree with
~ **по·фаз·е** be in phase (with)

со·в·пад·а·ющ·ий, -ая, -ее, (*pres part act*) of **со·в·пад·а·ть;** coincident, co-
~ **ос·ями** (*pl*) co-axial

со·в·пад·ени·е, -я, (*n*) coincidence, congruence, agreement, match

со·в·пад·ени·е
~, **за·держ·анн·ые** (*pl*) (elec) delayed coincidence
   лин·и·я со·в·пад·ени·я (*f*) (mech) match line

со·в·па·л, (*past masc sing*) of **со·в·пас·ть**

со·в·пас·ть, (*fut 3rd pl*) **со·в·пад·ут,** (*perf*); *see* **со·в·пад·а·ть**

со·времён·ник·, -а, (*m*) contemporary

со·времён·н·ый, -ая, -ое, (*a*) contemporary, current, modern, present-day

со·вс·ем (*adv*) quite, completely

со·вулканиз·аци·я, -и, (*f*) co-vulcanization

сов·хоз·, -а, (*m*) state farm, sovkhoz

со·вь·ёт, (*fut 3rd sing*) of **с·ви·ть,** (*perf*); *see* **с·вив·а·ть** (*imp*)

со·гидролиз·, -а, (*m*) cohydrolysis

со·глас·и·е, -я, (*n*) agreement, consent, assent; accord
   в со·глас·и·и in accordance with; in agreement with

со·глас·иться, -ятся, (*perf*) *see* **со·глаш·а·ться**

со·глас·н·о (*adv*) according to, as per

со·глас·н·ый, -ая, -ое, (*a*) agreeing, concordant, consonant; consistent with, according/conforming to; (*m decl as adj*) (gram) consonant

со·глас·овани·е, -я, (*n*) (*v n*), *see* **со·глас·овыва·ть;** coordination, correlation; (gram) agreement
~ **со·противл·ени·я** (elec) impedance matching

со·глас·ова·ть, -ую, (*perf*) *see* **со·глас·овыва·ть**

со·глас·овыва·ть, -ют, (*imp*) coordinate, match, make agree with; line up, match (pointers etc.)

со·глаш·а·ться, -ются, (*imp*) agree, concur; consent, assent

со·глаш·ени·е, -я, (*n*) agreement

со·гн·ать, (*fut 3rd pl*) **с·гон·ят,** (*perf*) drive off/away

со·гн·уть, -ут, (*perf*) bend, fold, bend/fold over/round

со·грев·а·ть, -ют, (*imp*) warm, heat; **-ся** (*pass*); become warm/hot

со·гре·ть, -ют, (*perf*) *see* **со·грев·а·ть**

сод·а, -ы, (*f*) (chem) soda
~, **дву·угле·кисл·ая** soda bicarbonate, bicarbonate of soda
~, **кальцин·ированн·ая** soda ash
~, **кауст·ическ·ая** caustic soda
~, **кристалл·ическ·ая** soda crystals

со́д·а

~, пить·ев·а́я soda bicarbonate, bicarbonate of soda

содали́т·, -а, (*m*) (min) sodalite

содди́йт·, -а, (*m*) (min) soddyite

со·де́й·стви·е, -я, (*n*) assistance, co-operation

со·де́й·ств·овать, -уют, (*imp and perf*) (+ *dat*) assist, co-operate in

со·де́й·ствующ·ий, -ая, -ее, (*pres part act*) of **со·де́й·ств·овать;** (zool) synergic, synergetic

~ сре́д·ств·о (*n*) (pharm) synergist

Содерберг·а, электро́д· (*m*) (met) Soderberg electrode

со·держ·а́ни·е, -я, (*n*) (*v n*) of **со·держ·а́ть;** content(s); maintenance, upkeep, keeping; allowance(s)

~ зол·ы́ ash content

~, проце́нт·н·ое percentage (of content)

с вы·со́к·им со·держ·а́ни·ем … rich in …, high- …

у·каз·а́тел·ь со·держ·а́ни·я (*m*) (print) contents list

со·держ·а́тельн·ый, -ая, -ое, (*a*) full of interest

со·держ·а́ть, -ат, (*imp*) contain, include; maintain, keep

~ в себ·е́ incorporate

со·держ·и́м·ое, -ого, (*n decl as adj*) contents

со́д·и·й, -я, (*m*) = **на́тр·и·й** (q.v.)

со́д·ов·ый, -ая, -ое, (*a*) soda

~ печ·ь (*f*) (paper) black-ash kiln

со·до·кла́д·, -а, (*m*) supplementary report

содо·пота́ш·н·ый, -ая, -ое, (*a*) soda-and-potash

со́·др·анн·ый, -ая, -ое, (*past part pass*) of **со·др·а́ть**

со·др·а́ть, (*fut 3rd pl*) **с·дер·у́т,** (*perf*) strip, skin, flay, bark

со·дрог·а́ни·е, -я, (*n*) shudder

со·дрог·а́·ться, -ются, (*imp*) shudder

со·дрог·н·у́ться, -у́тся, (*perf*) shudder

со·дру́ж·еств·о, -я, (*m*) (polit) commonwealth

со́·ев·ый, -ая, -ое, (*a*) soya, soy, soya-bean, soybean

со·един·е́ни·е, -я, (*n*) (*v n*) see **со·един·я́ть;** joint, jointing (plates, rails, beams, wood); (elec) connecting, connection, bonding (cables, rails); (chem) compound, combination; (math) combination; (mil) formation, force; (astron) conjunction

со·един·е́ни·е

~ без за·зо́р·а closed joint

~ без с·ко́с·а (weld) plain/square joint

~, бок·ов·о́е edge joint

~ в·ви́л·к·у fork joint

~ в·глад·ь flush joint; (shipb) flush plating

~, ве́рх·н·ее (astron) superior conjunction

~ в·за·мо́к· fish-mouth joint

~ в·ла́ск·у scarf joint

~ в·на·кро́·й lap joint

~ в·на·пу́ск· scarf joint

~ в·на·са́д·к·у constricted-end joint

~, внутри·компле́кс·н·ое (chem) chelate compound, sub-complex salt

~, во́д·н·ое (chem) hydrate

~ в·при·ты́к· Tee joint, edge butt joint

~ в·рас·тру́б· bell-and-spigot joint

~ в·с·ты́к· butt joint

~ высоко·молекуля́р·н·ое (chem) macro-/high-molecular compound

~, гетеро·поля́р·н·ое heteropolar compound

~, двой·н·о́е binary compound

~, дюри́т·ов·ое hose joint (piping)

~ ёл·очк·ой (shipb) clinker plating

~, жёстк·ое rigid connection, fixed joint

~ за·ги́б·ом folded-over joint

~, за·клёп·очн·ое rivet/riveted joint; riveting

~, за·мк·о́в·ое scarf joint

~ звезд·о́й (elec) star connection

~ звезд·а́-зигза́г· (elec) interconnected star connection

~, звёзд·н·ое (elec) star connection

~ зигза́г·ом (elec) zigzag/interstar connection

~, каска́д·н·ое (elec) cascade connection

~, компле́кс·н·ое (chem) complex compound

~ кра·й на кра·й (shipb) clinker plating

~, лоб·ов·о́е (elec) end connection

~, лом·а́ющ·ееся frangible coupling

~ лю́верс·ами grommet joint

~, манёвр·енн·ое task force (naval)

~, меж·элеме́нт·н·ое (elec) intercell connector

мест·о со·един·е́ни·я joint

~, металли́ческое intermetallic compound

~, металло·орга́н·и́ческ·ое (chem) metallo-organic compound, organo-metalic compound

со·един·ён·и·е

~, мно́го·по́люс·н·ое   (elec) multi-terminal connection

~, мост·ов·о́е   (elec) bridge connection

~, му́фт·ов·ое   pipe coupling, sleeve joint

~ на́ ус·   mitre joint

~, не·сквоз·н·ое про́·б·очн·ое   plug weld

~, ни́ж·н·ее   (astron) inferior conjunction

~ оболо́ч·ек·   sheath bonding (of elec cables)

~ обра́т·н·ое   (chem) recombining

~, от·борт·о́ванн·ое бок·ов·о́е   edge lap fillet joint

~ паралле́ль·н·о   (elec) connection in parallel, parallel connection

~, паралле́ль·н·о-по·сле́д·овательн·ое   (elec) multiple-series connection, series-parallel connection

~, пере·крёст·н·ое   (elec) back-to-back connection

~ пе́тл·ей   clasp joint

~, по·переч·н·ое   cross bonding (elec cable)

~, по·сле́д·овательн·ое   (elec) connection in series, cascade/series connection

~, про́ч·н·ое   strength joint

~ с вы·са́д·к·ой   butt strap joint

~ с за·зо́р·ом   (weld) open joint

~ с на·кла́д·к·ой   strap/strapped joint (plating); fished joint (rails or beams)

~, с·ва́р·н·ое   weld

~, сквоз·н·о́е про́б·очн·ое   rivet weld

~, со·в·мещ·ённ·ое   joggled lap joint

~, с·ты́к·ов·о́е   butt joint; (rail) conductor-rail bond

~, тор·цев·о́е   edge joint

~ тре·уго́ль·ник·ом   (elec) delta connection

~, угл·ов·о́е   corner joint; angle joint

~, фла́н·цев·ое   flanged joint

~, цен·ов·ое   (chem) sandwich compound

~, цеп·н·о́е   chain riveting; (chem) chain compound

~, цикл·и́чн·ое   (chem) ring compound

~, шарни́р·н·ое   hinge, hinged joint, articulated joint

~, шар·ов·о́е   ball-and-socket joint (piping)

~, ша́хмат·н·ое   zig-zag riveting

~, шпунт·ов·о́е   matched/joggled joint

~, штёпсель·н·ое   (elec) plug-and-socket connection

со·един·ён·и·е

~, штык·ов·о́е   bayonet joint

~, электро́н·н·ое   (met) electron compound

со·един·ённ·ый, -ая, -ое,   (a) (past part pass) see со·един·я́·ть; united

~ на·кла́д·к·ой   strapped (of plates), fished (of beams or rails)

со·един·и́тел·ь, -я,   (m) connector, connection; (elec rail) bond

~, лин·е́йн·ый   final selector (telephones)

~, между·ваго́н·н·ый   (rail) jumper

~, между·ре́льс·ов·ый   (rail) continuity bond

~, ре́льс·ов·ый   (rail) rail-bond

~, с·ты́к·ов·о́й   (rail) conductor-rail bond

со·един·и́тельн·ый, -ая, -ое,   (a) connecting, connection, connective, joining, junction; bonding (elec cables, rails etc.); (gram) copulative, conjunctive

~ га́й·к·а   (f) hose coupling

~ коро́б·к·а   (f) (elec) connection box

~ ли́н·и·я   (f) junction circuit (telephone)

~ скоб·а́   (f) shackle

~ ткан·ь   (f) (zool) connective tissue

со·един·и́ть, -я́т,   (perf) see со·един·я́·ть

со·един·я́·ть, -ют,   (imp) connect, join; combine

~ на·прям·у́ю   put through (telephones)

со·жал·е́ни·е, -я,   (n) regret

к со·жал·е́ни·ю   (adv) unfortunately

со·жи́т·ельственн·ый, -ая, -ое,   (a) symbiotic

со·жи́т·ельств·о, -а,   (n) (biol) symbiosis

со·жм·ёт,   (fut 3rd sing) of с·жат·ь, (perf); see с·жим·а́·ть

со·жн·ёт,   (fut 3rd sing) of с·жат·ь, (perf); see с·жин·а́·ть (imp)

со·зв·а́ть,   (fut 3rd pl) со·зов·у́т, (perf) summon, call (to meeting); convene, convoke

со·звёзд·и·е, -я,   (n) (astron) constellation

со·зда·ва́тел·ь, -я,   (m) creator, originator, founder

со·зда·ва́ть, -ю́т,   (imp) create, originate, give rise to; found, establish, set up

со·зда́·ни·е, -я,   (n) (v n) see со·зда·ва́ть; creation; creature

со·зда́·ть, *(fut 3rd sing, pl)* **со·зда́·ст, со·здад·у́т,** *(past masc sing)* **со́·зда·л,** *(perf)*; *see* **со·зда·ва́ть**

созерц·а́тельн·ый, -ая, -ое, *(a)* contemplative, meditative

со·зид·а́тельн·ый, -ая, -ое, *(a)* creative, constructive

со·зна·ва́ть, -ют, *(imp)* be aware/conscious, realize; acknowledge

со·зна́·ни·е, -я, *(n)* consciousness

со·зна́·тельност·ь, -и, *(f)* conscientiousness

со·зна́·тельн·ый, -ая, -ое, *(a)* conscious; deliberate; conscientious

со·зна́·ть, -ют, *(perf) see* **со·зна·ва́ть**

со·зов·ёт, *(fut 3rd sing)* of **со·зв·а́ть**

со·зрев·а́ни·е, -я, *(n) (v n)* of **со·зре·в·а́·ть**

со·зрев·а́·ть, -ют, *(imp)* ripen' (fruits etc.), mature (wine, game, cheese etc.), harden, cure (rubber)

со·зре́·ть, -ют, *(perf) see* **со·зрев·а́·ть**

со·зы́в·, -а, *(m) (v n)* of **со·зыв·а́·ть**

со·зыв·а́·ть, -ют, *(imp)* summon, call (to meeting); convene, convoke

со·из·да́·тел·ь, -я, *(m)* co-publisher

со·из·мер·и́м·ый, -ая, -ое, *(a)* commensurate

со·иск·а́ни·е, -я, *(n)* competition

со·йти́, *(fut 3rd pl)* **со·йд·у́т,** *(past masc sing)* **со·шё·л,** *(perf) see* **с·хо·д·и́ть** *(imp)*

со́к·, -а, *(m)* juice, sap

~, желу́д·очн·ый (zool) gastric juice

~, кле́т·очн·ый (bot) cell sap

~, млéч·н·ый (zool) chyle; (bot) latex

соко·ва́р·, -а, *(m)* pre-evaporator

соко·вы·жим·а́тел·ь, -я, *(m)* juicer, juice extractor

со́кол·, -а, *(m)* falcon; wedge-driving tool; trowel, plasterer's trowel

сокол·и́н·ые, -ых, *(pl decl as adj)* (orn) falcon family, *Falconidae*

со·кол·о́к·, -л·к·а́, *(m)* brisket (meat)

соко·со·держ·а́ни·е, -я, *(n)* juiciness, sappiness, juice/sap content

со·крат·и́мост·ь, -и, *(f)* (math) reducibility; (biol) contractibility

со·крат·и́м·ый, -ая, -ое, *(a)* (math) reducible; (biol) contractive, contractile

со·крат·и́тельн·ый, -ая, -ое, *(a)* contracting; (met) concentrating

со·крат·и́ть, -я́т, *(perf) see* **со·кра·щ·а́·ть**

со·кращ·а́·ть, -ют, *(imp)* shorten, abbreviate, abridge, curtail, reduce, cut; (math) reduce, cancel; (phys) contract; (biol) contract

со·кращ·éни·е, -я, *(n) (v n) see* **со·кращ·а́·ть;** abbreviation, reduction, abridgement; (math) cancellation, reduction; (phys) contraction; (biol) contraction

~ про́б·ы (min) splitting/reducing a sample

со·кристаллиз·а́ци·я, -и, *(f)* cocrystallization

со·кров·éнн·ый, -ая, -ое, *(a)* concealed, secret

со·кро́в·ищ·е, -а, *(i)* **-ем,** *(n)* treasure

со·круш·а́·ть, -ют, *(imp)* smash, shatter

со·круш·éни·е, -я, *(n) (v n) see* **со·кру·ш·а́·ть;** demolition, destruction

со·круш·и́ть, -а́т, *(perf) see* **со·кру·ш·а́·ть**

соксле́т·, -а, *(m)* Soxhlet apparatus

солакто́л·, -а, *(m)* (chem) solactol, ethyl lactate

солани́н·, -а, *(m)* (chem) solanine base

солда́т·, -а, *(m)* soldier; serviceman; (zool) soldier (ant); (naut) reef point

соле·вар·éни·е, -я, *(n)* salt making

соле·ва́р·н·я, -и, *(g pl)* **-р·ен·,** *(f)* salt works, saltern

сол·ев·о́й, -ая, -о́е, *(a)* salt

соле·вы·но́с·лив·ый, -ая, -ое, *(a)* (bot) salt-tolerant

соле·мéр·, -а, *(m)* brine/salt gauge/gage, salinometer, salimeter

сол·éни·е, -я, *(n)* (food) salting, pickling in salt, curing

со́л·енн·ый, -ая, -ое, *(a) (past part pass)* of **сол·и́ть;** (food) salt, salted, corned

солен·о́ид·, -а, *(m)* (elec) solenoid

соленоида́ль·н·ый, -ая, -ое, *(a)* (elec) solenoidal

солено·мéр·, -а, *(m)* = **соле·мéр·**

соле·но́с·н·ый, -ая, -ое, *(a)* salt-bearing, saliferous

солено·стéл·а, -ы, *(f)* (bot) solenostele

сол·ёност·ь, -и, *(f)* salinity, saltiness

солено·ци́т·, -а, *(m)* (zool) solenocyte

сол·ён·ый, -ая, -ое, *(a)* salt, salty, salted, saline; (food) pickled, salt, corned

~ капу́ст·а *(f)* sauerkraut

~ огур·éц· *(m)* (food) gherkin

~ о́зер·о *(n)* (geog) salt lake

сол·е́нь·е, -я, (*g pl*) -н·и·й, (*n*) salted/
pickled food
соле·обра́з·н·ый, -ая, -ое, (*a*) (chem)
salt-like
соле·образ·ова́ни·е, -я, (*n*) (geol)
halogenesis, salt formation
соле·рас·твор·и́тел·ь, -я, (*m*) brine
mixer
соле·ро́с·, -а, (*m*) (bot) glasswort,
Salicornia
соле·со·дсрж·а́ни·е, -я, (*n*) salt content
соле·со́с·, -а, (*m*) brine pump
соле·сто́й·к·ий, -ая, -ое, (*a*) (bot)
halophytic
со́л·и, (*g/d/p sing, nom/acc pl*) of со́л·ь
солида́р·ност·ь, -и, (*f*) solidarity
соли́д·н·ый, -ая, -ое, (*a*) solid; reliable, considerable, substantial; reput-
able, well-known, respectable
солидо́л·, -а, (*m*) grease
солидоло·на·гнет·а́тел·ь, -я, (*m*)
pressure lubricator
соли́дус·, -а, (*m*) (chem, phys) solidus
по·ве́рх·ност·ь соли́дус·а (*f*) solidus
солитёр·, -а, (*m*) (zool) tapeworm,
Taenia
солитёр·, -а, (*m*) solitaire (jewellery)
сол·и́ть, -я́т, (*imp*) salt, pickle, corn
(meat)
солифлю́кци·я, -и, (*f*) solifluction
(of soil)
со́л·к·а, -и, (*f*) (*v n*), *see* сол·и́ть
солн·е́чник·и, -ов, (*m pl*) (fish) *Zei-
formes*; (microbiol) *Heliozoa*
со́лн·ечн·ый, -ая, -ое, (*a*) sun, sunny,
solar
~ акти́в·ност·ь (*f*) (phys, astron)
solar activity
~ батар·е́·я (*f*) (elec) solar battery
~ ка́мен·ь (*m*) (min) aventurine feldspar, sunstone
~ пятн·о́ (*n*) (astron) sunspot
~ систе́м·а (*f*) (astron) solar system
~ с·плет·е́ни·е (*n*) (anat) solar plexus
~ су́т·к·и (*pl*) (astron) solar day
~ уда́р· (*m*) (med) sunstroke
~ час·ы́ (*pl*) sundial
со́лн·ц·е, -а, (*i*) -ем, (*n*) sun
~, голуб·о́е blue sun
~, ло́ж·н·ое (astron) parhelion, mock
sun
по·со́лн·ц·у by/with the sun;
clockwise
~, сре́д·н·ее (astron) mean sun
~, экваториа́ль·н·ое сре́д·н·ее second
mean sun

со́лн·ц·е
~, эклипт·и́ческ·ое сре́д·н·ее first
mean sun
солнце·иск·а́тел·ь, -я, (*m*) sunseeker
солнце·леч·е́ни·е, -я, (*n*) (med) heliotherapy
солнце·сто·я́ни·е, -я, (*n*) (astron)
solstice
~, зи́м·н·ее Winter solstice
~, марси́ан·ск·ое Martian solstice
солнце·цве́т·, -а, (*m*) (bot) sunrose,
Helianthemum
солов·е́й, -в·ь·я́, (*m*) (orn) nightingale,
Erithacus
со́лод·, -а, (*m*) malt
солод·е́л·ый, -ая, -ое, (*a*) sweetish
(of damp flour etc.)
солод·и́ть, -я́т, (*imp*) malt
соло́д·к·а, -и, (*g pl*) -д·ок·, (*f*) (bot)
liquorice, licorice, *Glycyrrhiza glabra*
солод·ко́в·ый, -ая, -ое, (*a*) liquorice,
licorice
солод·ова́ни·е, -я, (*n*) malting
солод·о́в·ый, -ая, -ое, (*a*) malt
солодо·дроб·и́лк·а, -и, (*g pl*) -лок·,
(*f*) malt mill
солодо·суш·и́лк·а, -и, (*g pl*) -лок·,
(*f*) maltdryer
со́лод·ь, -и, (*f*) soloth soil
солож·е́ни·е, -я, (*n*) malt conversion,
malting
соло́м·а, -ы, (*f*) straw
соло́м·енн·ый, -ая, -ое, (*a*) straw,
strawy
соло́м·ин·а, -ы, (*f*) (agr) straw (single
stalk), (bot) culm
соло́м·ист·ый, -ая, -ое, (*a*) (bot)
culmiferous
солом·и́т·, -а, (*m*) thatchboard (heat
insulating material)
соло́м·к·а, -и, (*g pl*) -м·ок·, (*f*) (*dim*)
of соло́м·а; straw, haulm
соломо·по·гру́з·чик·, -а, (*m*) (agr)
straw loader
соломо·ре́з·к·а, -и, (*f*) (agr) straw/
chaff cutter
соломо·тря́с·, -а, (*m*) (agr) straw
shaker
солон·е́ц·, -н·ц·а́, (*i*) -н·ц·о́м, (*m*)
solonetz soil
солон·и́н·а, -ы, (*f*) corned/salt beef
солон·ова́т·ый, -ая, -ое, (*a*) brackish,
salty
солон·цева́тост·ь, -и, (*f*) alkalinity
(of soil)
солон·цева́т·ый, -ая, -ое, (*a*) solonetz, alkaline (of soil)

солон·ча́к·, -а,   (*m*) solonchak soil; (geog) salt marsh/bottom

солон·чакова́тост·ь, -и,   (*f*) salinity (of soil)

солон·чако́в·ый, -ая, -ое,   (*a*) solonchak, salty (of soil)

со́л·ь, -и,   (*nom pl*) **-и,** (*g pl*) **-е́й,** (*f*) salt (*and see under associated adj*); (*n indecl*) G (music); (fish) sole, *Solea vulgaris*

~, азотно·ки́сл·ая   nitrate

~, англ·и́йск·ая   (pharm) Epsom salts

~, винно·ки́сл·ая   tartrate

~, во́д·н·ая   hydrous salt

~, втор·и́чн·ая   caustic salt

~, вы́·вар·очн·ая   evaporation salt

~, гла́убер·ов·а   Glauber's salt

~, двой·н·а́я   (chem) double salt

~, ка́мен·н·ая   rock salt

~, ки́сл·ая   super salt

~, не·на·ру́ш·енн·ая   undomed salt

~, по·ва́р·енн·ая   common salt, sodium chloride

~, рас·пла́вл·енн·ая   fused salt

~, само·са́д·очн·ая   solar salt

сольва́т·, -а,   (*m*) (chem) solvate

сольве́й-крас·и́тел·ь, -я,   (*m*) (chem) solway dye

Сольве́·я, спо́соб·   (*m*) (chem) Solvay process

сольве́нт-на́фт·а, -ы,   (*f*) (oil) solvent naphtha

сольво́·лиз·, -а,   (*m*) (chem) solvolysis

со·ль·ёт, (*fut 3rd sing*) of **с·ли́·ть,** (*perf*); *see* **с·лив·а́·ть** (*imp*)

со́ль·н·ый, -ая, -ое,   (*a*) (music) solo

сольфата́р·а, -ы,   (*f*) (geol) solfatara

сол·ян·о́й, -а́я, -о́е,   (*a*) salt, saline

~ кисл·от·а́   (*f*) hydrochloric acid

~ рас·тво́р·   (*m*) brine

соляно·кисл·о́тн·ый, -ая, -ое,   (*a*) (chem) hydrochloric-acid

соляно·ки́сл·ый, -ая, -ое,   (*a*) hydrochloride (of), muriatic

соля́р·, -а,   (*m*) (oil) gas/solar oil

соляриз·а́ци·я, -и,   (*f*) (phot) solarization

соляри́·метр·, -а,   (*m*) solar radiation meter

соля́р·к·а, -и, (*g pl*) **-р·ок·,** (*f*) gas/solar oil

соля́р·ов·ое ма́сл·о   (*n*) gas/solar oil

сом·, -а́,   (*m*) (fish) silurus, *Silurus*

со́м·а, -ы,   (*f*) (zool) soma

сомат·иза́ци·я, -и,   (*f*) somatization

сомат·и́ческ·ий, -ая, -ое,   (*a*) (biol) somatic

сомато·гене́з·, -а,   (*m*) (zool) somatogenesis

сомато·ло́г·и·я, -и,   (*f*) (biol) somatology

сомато·пле́вр·а, -ы,   (*f*) (zool) somatopleura

сомато·фи́т·, -а,   (*m*) (bot) somatophyte

соми́т·, -а,   (*m*) (zool) somite, body segment

со́·мк·нут·ый, -ая, -ое,   (*past part pass*) *see* **с·мык·а́·ть;** closed, close, closed-up

~ стро́·й   (*m*) (mil) close order

со·мк·н·у́ть, -у́т, (*perf*) *see* **с·мык·а́·ть**

сомн·амбули́зм·, -а,   (*m*) (med) somnambulism

со·мн·ева́·ться, -ются,   (*imp*) doubt, have doubts, question (the validity of)

со·мн·е́ни·е, -я,   (*n*) doubt

вне со·мн·е́ни·я   undoubtedly

Со́мнер·а, ли́н·и·я   (*f*) (surv) Sumner line, position line

со·мн·ёт, (*fut 3rd sing*) of **с·мя·ть**

со·мн·и́тельн·ый, -ая, -ое,   (*a*) doubtful, dubious, questionable

~ зо́н·а   (*f*) (nav) dangerous zone

со·множ·и́тел·ь, -я,   (*m*) (math) factor, multiplier; (*pl*) multiplier and multiplicand

сомо·ви́д·н·ые, -ых,   (*pl decl as adj*) (fish) *Siluridae*

сон·, (*g sing*) **сн·а,** (*m*) sleep

сона́р·, -а,   (*m*) sonar (underwater sound navigation and ranging)

сонбута́л·, -а,   (*m*) (pharm) sonbutal, 2-butyl-beta bromallil barbituric acid

сон·ли́в·ый, -ая, -ое,   (*a*) somnolent

со́н·н·ый, -ая, -ое,   (*a*) sleepy, sleeping; (anat) carotid

~ арте́ри·я   (*f*) (anat) carotid artery

~ бол·е́зн·ь   (*f*) (med) sleeping sickness, trypanosomiasis

~ о́дур·ь   (*f*) (bot) belladonna, deadly nightshade

соно́·граф·, -а,   (*m*) sonograph

соно́·метр·, -а,   (*m*) (acous) monochord, sonometer

сонор·н·ый, -ая, -ое,   (*a*) sonorous

сонтохи́н·, -а,   (*m*) (pharm) sontochin, santoquine

со·ображ·а́·ть, -ют,   (*imp*) consider, think over/out, deliberate; understand, grasp (an idea)

со·ображ·е́ни·е, -я,   (*n*) consideration, deliberation; understanding, reason

со·образ·и́тельн·ый, -ая, -ое,   (*a*) quick-witted, bright, shrewd

со·образ·и́ть, -я́т, (*perf*) *see* **со·обра·ж·а́·ть**

со·обра́з·н·о (*adv*) according to, in conformity/compliance with

со·обра́з·н·ый, -ая, -ое, (*a*) compatible, consistent

со·образ·ова́ть, -у́ют, (*imp and perf*) match, make consistent, adjust to fit; **-ся** conform

со·обща́ (*adv*) jointly, together

со·обща́·ть, -ют, (*imp*) communicate, report, inform, advise; connect up (equipment); **-ся** (*pass*); (*imp only*) be joined/connected, give/have access to, communicate

со·обще́ни·е, -я, (*n*) (*v n*) *see* **со·обща́·ть**; communication, report, message; traffic (telephones)

~, бли́ж·н·ее local, suburban traffic (telephones)

~, вое́н·н·ые (mil) lines of communication

~, да́ль·н·ее long-distance communications, (rail) long-distance service; (*gen as adj*) (rail) long-distance; trunk (telephones)

~, земл·ян·о́е (elec) fault to earth, ground fault

~, ме́ст·н·ое (rail) local service

~, прям·о́е direct communications; (rail) through service

~ пут·и́ (*pl*) means of communication, railways

тео́р·и·я со·обще́·ни·я (*f*) (math) information theory

со·обще́нн·ый, -ая, -ое, (*past part pass*) *see* **со·обще́·а·ть**

со·о́бщ·еств·о, -а, (*n*) community, association; (biol) community, biocoenosium; (geol) assemblage

со·обще́·ить, -а́т, (*perf*) *see* **со·обще́·а·ть**

со·оруд·и́ть, -я́т, (*perf*) *see* **со·ору·ж·а́·ть**

со·оруж·а́·ть, -ют, (*imp*) build, construct, erect

со·оруж·е́ни·е, -я, (*n*) (*v n*) *see* **со·о·руж·а́·ть**; building, construction, structure, (*pl*) works

~, гидро·техн·и́ческ·ое hydraulic engineering structure, (*pl*) h.e. works

~, голов·н·ые (*pl*) (hydr eng) headworks

~, иску́с·ственн·ое (rail) civil engineering works

~, кора́лл·ов·ое (ocean) coral reef

~, порт·о́в·ые (*pl*) harbour facilities/works

со·о·сажд·е́ни·е, -я, (*n*) (chem) coprecipitation

со·о́с·ност·ь, -и, (*f*) coaxiality; axial alignment

со·о́с·н·ый, -ая, -ое, (*a*) coaxial, dead in line

~ винт· (*m*) contra-rotating propeller

со·от·ве́т·ственн·о (*adv*) accordingly, correspondingly, according to, in accordance with; respectively (of enumerated data)

со·от·ве́т·ственн·ый, -ая, -ое, (*a*) corresponding, relevant, appropriate

~ со·сто·я́ни·я (*pl*) (phys) corresponding states

со·от·ве́т·стви·е, -я, (*n*) accordance, conformity, correspondence

в со·от·ве́т·стви·и in accordance with

~, взаи́м·н·о одно·зна́ч·н·ое one-to-one correspondence

~, одно-одно·зна́ч·н·ое one-to-one correspondence

при́нцип· со·от·ве́т·стви·я (*m*) (phys) correspondence principle

со·от·ве́т·ств·овать, -уют, (*imp*) correspond to, fit, suit, be appropriate for, be in keeping with, meet (requirements etc.)

со·от·ве́т·ствующ·ий, -ая, -ее, (*pres part act*) of **со·от·ве́т·ств·овать**; appropriate, suitable, pertinent

со·оте́ч·ественник·, -а, (*m*) compatriot, fellow-countryman

со·от·нес·ти́, -у́т, (*perf*) *see* **со·от·-нос·и́ть**

со·от·нос·и́тельн·ый, -ая, -ое, (*a*) correlative

со·от·нос·и́ть, -я́т, (*imp*) establish the relation between, compare, relate

со·от·нош·е́ни·е, -я, (*n*) correlation, relationship, relation, ratio

~ ветв·е́й (nucl) branching ratio

~, лин·е́йн·ое linear relationship

~, пре·де́ль·н·ое asymptotic relation

~, равно·ве́с·н·ое equilibrium

~, сигна́л/шум· signal-to-noise ratio

со·пе́р·ник·, -а, (*m*) rival

со́п·к·а, -и, (*g pl*) **-п·ок·,** (*f*) coniform hill/mountain; mud cone/volcano

~, гряз·ев·а́я mud volcano

сопл·о́, -а́, (*nom pl*) **со́пла,** (*g pl*) **со́пел·,** (*n*) nozzle, jet; throat (of wind tunnel); orifice (of nozzle)

~, в·ход·н·о́е suction nozzle

~, вы·хлоп·н·о́е discharge nozzle; (a/c) exhaust nozzle/cone

сопл·о́

~, калибр·о́вочн·ое metering nozzle

~, кон·цев·о́е реакти́в·н·ое blade-tip reaction nozzle (helicopter)

~ Лава́л·я convergent-divergent nozzle, Laval nozzle

~, на·по́р·н·ое discharge nozzle

~, не·регул·и́руем·ое fixed-area nozzle

~, ото́·гн·ут·ое canted nozzle

~, пере·рас·ши́р·енн·ое overexpanded nozzle

~, подо·грев·а́тельн·ое gas-cutter nozzle

~ по·сто·я́нн·ого сеч·е́ни·я fixed nozzle

~, рас·ходо·ме́р·н·ое flow nozzle

~, рас·ход·я́щ·ееся divergent/diverging nozzle

~, рас·шир·я́ющ·ееся expanding nozzle, effuser

~, реакти́в·н·ое (a/c, rocket) jet/propelling/exhaust nozzle

~, регул·и́руем·ое variable-area nozzle

~, ре́ж·ущ·ее gas-cutter jet

~, сверх·звук·ов·о́е (aerodynam) supersonic nozzle; supersonic throat (wind tunnel)

~, с·у́ж·ивающ·ееся, converging nozzle

сопл·ов·о́й, -а́я, -о́е, (a) = **со́пл·ов·ый**

со́пл·ов·ый, -ая, -ое, (a) of **сопл·о́**

~ аппара́т· (m) nozzle box (jet eng)

~ коро́б·к·а (f) nozzle box (turbines)

~ регул·и́ровани·е, (n) nozzle-control governing (turbines)

со·пло́д·и·е, -я, (n) (bot) fruit system

сопл·я́к·, -а́, (m) (naut) fairlead

со·под·чин·и́ть, -я́т, (perf) see **со·-под·чин·я́ть**

со·под·чин·я́ть, -ют, (imp) make subordinate (a number); (gram) coordinate

со·полиме́р·, -а, (m) (chem) copolymer

~, бло́ч·н·ый block copolymer

~, двой·н·о́й dipolymer

~, при·ви́т·о́й graft copolymer

со·полимериз·а́ци·я, -и, (f) (chem) copolymerization

со·по·ста́в·и́м·ый, -ая, -ое, (a) comparable

со·по·ста́в·ить, -ят, (perf) see **со·по·-ставл·я́·ть**

со·по·ставл·е́ни·е, -я, (n) comparison, confrontation

пра́в·ил·о со·по·ставл·е́ни·я (n) (math) associative law

со·по·ставл·я́·ть, -ют, (imp) compare

со́п·очн·ый, -ая, -ое, (a) of **со́п·к·а**

~ бре́кчи·я (f) (geol) mud brecchia

со·пре·де́ль·н·ый, -ая, -ое, (a) bordering, contiguous, adjacent

со·пре́·ть, -ют, (perf) rot

со·при·кас·а́·ться, -ются, (imp) be touching, be in contact with, be adjacent to, be contiguous upon, border, border on, osculate (of circles)

со·при·кас·а́ющ·ийся, -аяся, -ееся, (pres part act) of **со·при·кас·а́·ться;** touching, contiguous; osculating (of curves)

~ о·кру́ж·ност·ь (f) (math) osculating circle

со·при·кос·нове́ни·е, -я, (n) contact, contiguity; osculation (of curves)

пло́щад·ь со·при·кос·нове́ни·я (f) contact area

со·при·кос·н·у́ться, -у́тся, (perf single action) touch, come into contact with

со·про·вод·и́ть, -я́т, (perf) see **со·-про·вожд·а́·ть**

со·про·вод·и́тел·ь, -я, (m) escort; tracker (radar)

со·про·вожд·а́·ть, -ют, (imp) accompany, escort; track (radar)

со·про·вожд·а́ющ·ий, -ая, -ее, (pres part act) of **со·про·вожд·а́·ть;** associated; (math) moving

~ трёх·гра́н·ник· (m) (math) moving trihedral

со·про·вожд·е́ни·е, -я, (n) (v n) see **со·про·вожд·а́·ть;** accompaniment, escort

~ по да́ль·ност·и range tracking (radar)

~ ис·треб·и́тел·ями (air) fighter escort/protection

~ огн·ём (mil) support fire

~, синхрониз·и́рованн·ое synchronized tracking (radar)

со·противл·е́ни·е, -я, (n) resistance; (elec) resistance (of circuits, conductors), resistivity (of a material), resistor (device); (dynam) drag

~, акти́в·н·ое (elec) resistance, ohmic resistance

~, акуст·и́ческ·ое acoustic/acoustical resistance/impedance

~, анти·парази́т·н·ое (elec) grid suppressor

~, аэро·динам·и́ческ·ое aerodynamic/air-flow resistance

со·противл·éни·е

~, **балла́ст·н·ое** (elec) ballast, ballast resistor; (rail) ballast resistance

~, **без·индукцио́н·н·ое** (elec) non-inductive resistor

~ **в цéп·и** (elec) resistor

~, **вихр·ев·óе** (dynam) eddy resistance, eddy-making resistance, (a/c) vortex drag

~, **в·нос·и́м·ое** (rad) coupled impedance

~, **воз·ду́ш·н·эе** (dynam) air resistance

~, **волн·ов·óе** (elec) wave impedance; (dynam) wave resistance/drag, wave-making resistance

~, **врéд·н·ое** (aerodynam) parasitic drag

~, **врéмен·н·ое** (met) tensile strength

~, **в·ход·н·óе** (elec) input resistance

~ **вы·ступ·áющ·их част·éй** (shipb) resistance of appendages

~, **вя́з·костн·ое** (dynam) viscous drag

~, **гас·я́щ·ее** (elec) dropping resistor

~, **гидро·динам·и́ческ·ое** hydro-dynamic/water resistance

~, **голов·н·óе** (aerodynam) nose drag

~ **движ·éни·ю** resistance (of trains, ships, M/T)

~ **деформ·áци·и** (met) elastic stiffness; (met roll) mean specific roll pressure

~, **до·бáв·очн·ое** (elec) instrument series resistor

~, **долго·врéмен·н·ое** (met) creep limit

~, **дóн·н·ое** (rocket) base drag

~, **до·полн·и́тельн·ое** (elec) additional resistance; booster resistor

~, **ём·костн·ое** (elec) capacitive reactance, condensance

~, **за·земл·я́ющ·ее** (elec) earthing resistor, grounding resistor

~ **из·ги́б·у** transverse/bending strength, flexing resistance

~ **из·лóм·у** (mech) modulus of rupture in bending

~ **из·луч·éни·я** radiation resistance (of antenna)

~, **индукти́в·н·ое** (elec) inductance; (aerodynam) induced drag

~, **кáж·ущ·ееся** apparent resistance, (elec) impedance

~ **корóт·к·ого за·мык·áни·я** (elec) short-circuit impedance

~ **крив·óй** (rail) curve resistance

~, **лоб·ов·óе** (dynam) drag, head resistance

   коэффициéнт· лоб·ов·óго со·про·тивл·éни·я (m) drag coefficient

со·противл·éни·е

   си́л·а лоб·ов·óго со·противл·éни·я (f) drag

~, **магни́т·н·ое** (elec) reluctance

~ **материáл·ов** (mech) strength of materials

~, **механ·и́ческ·ое** (phys) mechanical impedance; (dynam) mechanical resistance

   мóдул·ь со·противл·éни·я сеч·é-ни·я (f) (mech) modulus of section

~, **молекуля́р·н·ое** (phys) molar resistivity

   момéнт· со·противл·éни·я (m) moment of inertia of area, moment of resistance/stiffness

~ **на раз·ры́в·, врéмен·н·ое** tensile strength

~, **на·грев·áтельн·ое** (elec) heating resistor

~ **на·грýз·к·е** (aerodynam) load resistance

~, **на·грýз·очн·ое** (elec) loading resistor

~ **не·по·срéд·ственн·ому раз·ру-ш·éни·ю** (mech) technical cohesive strength

~, **не·линеáр·н·ое** (elec) non-linear resistance; non-ohmic resistor

~, **не·прó·волоч·н·ое** (elec) carbon resistor

~, **о·гран·ичи́тельн·ое** (elec) current-limiting resistor

~, **ом·и́ческ·ое** (elec) direct current resistance, d.c. resistance

~, **основ·н·óе** (rail) net resistance

~, **о·стáт·очн·ое** (elec) residual resistance (of circuit); (phys) residual resistivity (of material); (dynam) residuary resistance

~ **от у·клóн·а** (rail) grade resistance

~, **от·риц·áтельн·ое** (elec) negative resistance

~ **от·ры́в·у** (mech) technical cohesive strength, resistance to direct pull

~, **пере·мéн·н·ое** (elec) variable resistor, rheostat

~ **пере·рéз·ывани·ю** (mech) torsional strength

~, **пере·хóд·н·ое** (elec) contact resistance, electrode-contact resistance; (chem) transfer resistance

~, **пóлн·ое** (elec) impedance; (dynam) total drag

~ **подъ·éм·у** (rail) grade resistance, resistance due to gradient

со·противл·éни·е

~ **полу·про·вод·нúк·а** (elec) spreading resistance

~, **по·сто·я́нн·ое** (elec) fixed resistor

~, **про́·волоч·н·ое** (elec) wire-wound resistor, ww resistor

~ **про·до́ль·н·ому из·гúб·у** (mech) buckling resistance

~, **про́филь·н·ое** (dynam) profile/form drag

~, **пуск·ов·о́е** (elec) starting resistor

~, **рабо́ч·ее** (mech) working load

~ **раз·руш·éни·ю** (mech) real fracture stress, true tensile stress, true breaking stress; resistance to fracture

~, **раз·ры́в·н·о́е** breaking stress (rope)

~, **раз·ры́в·у** (mech) tensile strength; (text) splitting resistance

~ **раз·ры́в·у, úст·инн·ое** (mech) true tensile/breaking stress, real fracture stress

~, **раз·ря́д·н·ое** (elec) field-discharge resistor

~ **рас·тек·áни·ю** (elec) spreading resistance

~, **реактúв·н·ое** (elec) reactance

~, **свето·чувств·úтельн·ое** (elec) photoresistance

~ **с·вя́з·и** (elec) coupling resistance/impedance

~ **сéт·к·и** (elec) grid resistance

~ **с·мещ·éни·ю** (elec) bias resistor

~ **с·мя́т·и·ю** bearing resistance (of concrete etc.)

~, **со́б·ственн·ое** (elec) internal resistance

~, **со·глас·о́ванн·ое** (rad) matched impedance

~, **со·пряж·ённ·ые** (pl) (telecom) conjugate impedances

~ **с·ре́з·у** (mech) torsional strength, maximum shear stress

~ **с·ре́з·ывани·ю** see **со·противл·éни·е с·ре́з·у**

~ **с·цепл·éни·я** (mech) technical cohesive strength

~, **тёмн·ое** dark resistance (photoelectricity)

~, **тепл·ов·о́е** (phys) heat-transfer resistance; (elec) thermal resistance; thermal resistivity (of a dielectric)

~, **тепло·чувств·úтельн·ое** (elec) thermistor

~, **тормоз·н·о́е** (elec) braking resistor; (aerodynam) brake drag; (rail) brake resistance

со·противл·éни·е

~ **тр·éни·я** (dynam) frictional resistance; skin drag (of a/c); skin-friction resistance (of ship)

~, **у́голь·н·ое** (elec) carbon resistor

~ **удáр·у** (mech) impact strength

~, **у·дéль·н·ое** (phys) electrical volume resistivity, volume resistivity, resistivity (of a material); (rail) average resistance (to motion)

~, **у·дéль·н·ое магнúт·н·ое** (phys) reluctivity

~, **у·пру́г·ое** (expl) safe pressure

~ **фо́рм·ы** (dynam) eddy-making resistance, form drag

~, **эквивалéнт·н·ое** (elec) equivalent resistance (of circuit); (phys) equivalent resistivity (of substance)

~, **этало́н·н·ое** (elec) standard resistor

со·противл·я́емост·ь, -и, (f) resistive capacity, resistivity, strength

со·противл·я́·ться, -ются, (imp) resist, oppose

со·противл·я́ющ·ийся, -аяся, -ееся, (pres part act) of **со·противл·я́·ться**; resistant

со·пряг·á·ть, -ют, (imp) conjugate, mate; (elec) gang, track, pad

со·пряг·áющ·ий, -ая, -ее, (pres part act) of **со·пряг·á·ть**

~ **конденс·áтор·** (m) (and see **со·пряж·ённ·ое к.**) (rad) tracking capacitor

со·пряж·éни·е, -я, (n) (v n) see **со·пряг·á·ть;** conjugation

~ **гетероди́н·а** (rad) oscillator padding

~ **ко́нтур·ов** (rad) tracking

регул·иро́вк·а со·пряж·éни·я (f) (TV) gang adjustment

со·пряж·ённ·ый, -ая, -ое, (past part pass) see **со·пряг·á·ть;** (phys, chem, math) conjugate; (rad) gang, tracking, (gunn) cross, bilateral

~ **диáметр·** (m) conjugate diameter

~ **конденс·áтор·ы** (pl) (rad) gang capacitor

~ **на·блюд·éни·е** (n) (gunn) cross observation

~ **про·стрáн·ств·о** (n) conjugate space

~ **с·вя́з·и** (pl) (chem) conjugate double bonds

~ **то́ч·к·и** (pl) (opt) conjugate foci

~ **у·равн·éни·е** (n) adjoint (of a differential equation)

~ **фу́нкци·я** (f) (math) adjoint function; (pl) conjugate harmonic functions

~ **част·о́т·ы** (pl) (rad) tracking frequencies

со·пряж·ённ·ый, -ая, -ое
~ част·ь   (*f*) (eng) mate
со·пряч·ь,   (*fut 3rd sing, pl*) **со·пряж·ёт, со·пряг·ут,**   (*past masc sing*) **со·пряг·,** (*perf*); see **со·пряг·а́·ть**
со·пут·ств·овать, -уют,   (*imp*) accompany
со·пут·ствующ·ий, -ая, -ее,   (*pres part act*) of **со·пут·ств·овать;** accompanying, associated;   (*pl as n*) (chem) impurities
сор·, -а,   (*m*) litter, rubbish; (geol) sor
со·раз·мер·ени·е, -я,   (*n*) (*v n*) *see* **со·раз·мер·я́·ть**
со·раз·мер·енн·ый, -ая, -ое,   (*past part pass*) *see* **со·раз·мер·я́·ть**
со·раз·мер·ить, -ят,   (*perf*) *see* **со··раз·мер·я́·ть**
со·раз·мер·н·ый, -ая, -ое,   (*a*) commensurate, proportionate
со·раз·мер·я́·ть, -ют, (*imp*) proportion, match, harmonize
сорбе́нт·, -а,   (*m*) sorbent
сорби́н·, -а,   (*m*) sorbin
сорби́н·ов·ый, -ая, -ое, (*a*) sorbic
~ кисл·от·а́   (*f*) (chem) sorbic acid
сорб·и́ровани·е, -я,   (*n*) sorbing, sorption
сорб·и́р·овать, -уют,   (*imp and perf*) (chem) sorb
сорби́т·, -а,   (*m*) (met) sorbite; (chem) sorbitol
сорбитиз·а́ци·я, -и,   (*f*) (met) sorbitizing, quenching to sorbite
сорби́т·ный, -ая, -ое, (*a*) sorbitic
сорб·о́з·а, -ы,   (*f*) (chem) sorbose
со́рбци·я, -и,   (*f*) (chem) sorption
~, хим·и́ческ·ая   chemisorption
со́·рв·анн·ый, -ая, -ое, (*past part pass*) of **со·рв·а́ть;** stalled (of fluid flow)
~ рез·ьб·а́   (*f*) stripped thread
со·рв·а́ть, -у́т, (*perf*) strip/tear off
с·организ·ова́ть, -у́ют, (*perf*) organize
со́рго   (*n indecl*) (bot) sorghum, *Sorghum*
~, ком·ов·о́е   *Sorghum cernuum*
~, са́хар·н·ое   sweet sorghum, *Sorghum saccharatum*
со·ревн·ова́ни·е, -я,   (*n*) competition, contest
Соре́л·я, цеме́нт·   (*m*) (build) Sorel's cement
со́рлин·ь, -я,   (*m*) (naut) rudder chains
сорма́йт·, -а,   (*m*) (met) sormite
сор·и́нк·а, -и,   (*g pl*) **-нок·,** (*f*) speck/particle of dirt
со́р·ник·, -а,   (*m*) dust pan

со́р·н·ый, -ая, -ое,   (*a*) of **сор·** littered, spattered, dirty; (agr) weedy
~ трав·а́   (*f*) weed
сор·ня́к·, -а́,   (*m*) weed
соровча́к·, -а,   (*m*) (naut) N.E. wind
со́рок·, -а́,   (*m*) forty; **сорок·** (*gen pl*) of **соро́к·а**
соро́к·а, -и,   (*f*) (orn) magpie, *Pica pica*; **сорок·а́** (*gen/dat/instr/prep*) of **сорок·**
сорока́-   (*as component*) forty-, fortieth
сорока·восьми·гра́н·ник·, -а,   (*m*) (cryst) hexakisoctahedron
сорока·лет·и·е, -я,   (*n*) fortieth anniversary/birthday
сорока·час·ов·о́й, -а́я, -о́е,   (*a*) forty-hour
сорок·ов·о́й, -а́я, -о́е,   (*a*) fortieth
~, бу́й·н·ые   (*pl n*) (meteor) roaring forties
сороко·но́ж·к·а, -и,   (*g pl*) **-ж·ек·,** (*f*) (ent) centipede
сорокопу́т·, -а,   (*m*) (orn) shrike, *Lanius*
соро́ч·к·а, -и,   (*g pl*) **-ч·ек·,** (*f*) shirt
~, жён·ск·ая   chemise
со́рт·, -а,   (*nom pl*) **сорт·а́,** (*g pl*) **сорт·о́в,** (*m*) sort, kind, variety, type, brand, quality
~, пе́рв·ый   first/top quality/grade
~, у·рож·а́йн·ый   (agr) productive variety, high yielder
сорта́мент·, -а,   (*m*) selection, grades, range (of goods), inventory, assortment
сортиме́нт·, -а,   (*m*) = **сорта́мент·**
сорт·ирова́льн·ый, -ая, -ое,   (*a*) = **сорт·иро́вочн·ый**
сорт·ирова́ни·е, -я,   (*n*) (*v n*) of **сорт·ир·ова́ть**
сорт·иро́ванн·ый, -ая, -ое,   (*past part pass*) of **сорт·ир·ова́ть**
сорт·ир·ова́ть, -у́ют,   (*imp*) sort, assort, grade, size, separate, screen, classify
сорт·иро́вк·а, -а,   (*g pl*) **-вок·,** (*f*) (*v n*) *see* **сорт·ир·ова́ть;** grader, sorter, separator, screener; (rail) marshalling, sorting (wagons), classifying (freight)
зда́·ни·е сорт·иро́вк·и   (*n*) (min) screenhouse
сорт·иро́вочн·ый, -ая, -ое,   (*a*) sorting, grading, sizing, separating‘ screening; (rail) classifying, sorting, classification, marshalling, shunting
~ го́р·к·а (*f*) (rail) hump (of classification yard)
~ парк·   (*m*) = **с. ста́нци·я** (q.v.)

сорт·иро́вочн·ый, ая, -ое

~ **пу́т·ь** (*m*) (rail) sorting/shunting siding

~ **ста́нци·я** (*f*) (rail) sorting/marshalling/classification yard

~ **ста́нци·я, го́р·очн·ая** hump marshalling/sorting/shunting/classification yard

сорт·иро́вщик·, -а, (*m*) sorter, grader, screener, separator (man or machine)

со́рт·ност·ь, -и, (*f*) grade, quality, rating; high grade/quality

числ·о́ со́рт·ност·и (*n*) knock rating under take-off conditions (of aviation fuel)

со́рт·н·ый, -ая, -ое, (*a*) high grade/quality; (hortic) varietal; branded, graded (of goods)

сорто·ве́д·ени·е, -я, (*n*) (hortic) crop improvement research, variety investigation

сорто·во́д·ств·о, -а, (*n*) plant breeding

сорт·ов·о́й, -а́я, -о́е, (*a*) graded, sorted; high grade, top quality; (agr) variety, strain, sort; (met) section

~ **про·ка́т·** (*m*) (met) sections, rolled sections, shapes

сортов·ы (*pl*) (naut) lanyard

сорто·об·новл·е́ни·е, -я, (*n*) (gen) strain renovation

сорто·раз·ме́р·, -а, (*m*) grade size

сор·тро́с·, -а, (*m*) (naut) limber rope

со́рус·, -а, (*m*) (bot) sorus

со́рус·ов·ый, -ая, -ое, (*a*) (bot) soral

сос·а́вш·ий, -ая, -ее, (*past part act*) of **сос·а́ть**

сос·а́льц·е, -а, (*i*) **-ем,** (*n*) (zool) sucker

сос·а́льщик·, -а, (*m*) (vet) fluke, (zool) trematoda, (*pl*) *Trematoda*

сос·а́ни·е, -я, (*n*) (*v n*) of **сос·а́ть**

со́с·анн·ый, -ая, -ое, (*past part pass*) of **сос·а́ть**

сос·а́тельн·ый, -ая, -ое, (*a*) sucking, suctorial

сос·а́ть, -у́т, (*imp*) suck, suck/draw in

со·се́д·, -а, (*nom pl*) **-и,** (*g pl*) **-ей,** (*m*) neighbour

со·се́д·н·ий, -яя, -ее, (*a*) neighbouring, adjacent

со·се́д·ств·о, -а, (*n*) neighbourhood, vicinity

сос·ёт, (*pres 3rd sing*) of **сос·а́ть**

сос·е́ц·, -ц·а́, (*i*) **-ц·о́м,** (*m*) nipple

сосис·к·а, -и, (*g pl*) **-с·ок·,** (*f*) sausage, fresh link sausage

сос·и́те, (*pl imper*) of **сос·а́ть**

со́с·к·а, -и, (*g pl*) **-с·ок·,** (*f*) teat (for a baby's bottle); dummy

со·скабл·ива·ть, -ют, (*imp*) scrape/shave/plane off

со·скак·ива·ть, -ют, (*imp*) jump/fall off/down; come off

со·ска́льз·ывание·е, -я, (*n*) (*v n*) see **со·ска́льз·ыва·ть**; (cryst) slip

со·ска́льз·ыва·ть, -ют, (*imp*) slide/slip off/down

со·ско́бл·енн·ый, -ая, -ое, (*past part pass*) see **со·скабл·ива·ть**

со·скобл·и́ть, -я́т, (*perf*) see **со·скаб·л·ива·ть**

соско·ви́д·н·ый, -ая, -ое, (*a*) (biol) mamillar, mamilliform, papilliform

сос·ко́в·ый, -ая, -ое, (*a*) (zool) mamillary

со·скольз·н·у́ть, -у́т, (*perf*) see **со·ска́льз·ыва·ть**

со·скоч·и́ть, -́ат, (*perf*) see **со·ска́к·ива·ть**

со·скреб·а́·ть, -ют, (*imp*) scrape/scratch off

со·скреб·ённ·ый, -ая, -ое, (*past part pass*) see **со·скреб·а́·ть**

со·скрес·ти́, (*fut 3rd pl*) **со·скреб·у́т,** (*perf*) see **со·скреб·а́·ть**

со·с·лаг·а́тельн·ый, -ая, -ое, (*a*) associative; (gram) subjunctive

~ **на·клон·е́ни·е** (*n*) (gram) subjunctive mood

со́·сл·анн·ый, -ая, -ое, (*past part pass*) of **со·сл·ать**; sent away, ...ished, exiled deported

со·сл·а́ть, (*fut 3rd pl*) **со·шл·ю́т,** (*perf*) see **с·сыл·а́·ть**

со·сло́в·н·ый, -ая, -ое, (*a*) class (social)

со·служ·и́вец·, -вц·а, (*i*) **-вц·ем,** (*m*) colleague

сосн·а́, -ы́, (*nom pl*) **со́сн·ы,** (*gl pl*) **со́сен·** (*f*) (bot) pine, pine tree, *Pinus* sp.

~, **австр·и́йск·ая** Austrian pine, *Pinus nigra*

~, **боло́т·н·ая** longleaf pine, *Pinus palustris*

~, **ве́ймут·ов·а** eastern white pine, *Pinus strobus*

~, **ги́б·к·ая** limber pine, *Pinus flexilis*

~, **го́р·н·ая ве́ймут·ов·а** western white pine, *Pinus moniticola*

~, **еж·о́в·ая** shortleaf pine, *Pinus echinata*

~, **жёлт·ая** ponderosa pine, *Pinus ponderosa*

~, **жёстк·ая** pitch pine, *Pinus rigida*

сосн·а́

~, ла́дан·н·ая   loblolly pine, *Pinus taeda*

~, лес·н·а́я   Scotch pine, *Pinus sylvestris*

~, са́хар·н·ая   sugar pine, *Pinus lambertiana*

~, сиби́р·ск·ая   Swiss stone pine, *Pinus cembra* var. *sibirica*

~, смол·и́ст·ая   Norway/red pine, *Pinus resinosa*

~ Ту́нберг·а   Japanese black pine, *Pinus thunbergi*

сосн·о́в·ый, -ая, -ое, (*a*) pine; (*pl decl as adj*) *Pinaceae*

~ верт·у́н·   (*m*) pine twig blight

~ ма́сл·о   (*n*) (chem) turpentine

~ слон·и́к·   (*m*) (zool) pine chafer/weevil, *Hylobius abietis*

~ смол·а́   (*f*) (chem) rosin

~ шелко·пря́д·   (*m*) (zool) pine lappet/moth, *Dendrolimus pini*

сосн·я́к·, -а́, (*m*) pine forest; pine grove, pinetum

сос·о́к·, -с·к·а́, (*m*) (anat) papilla, mamilla

~, груд·н·о́й   nipple

~ зр·и́тельн·ого не́рв·а   (anat) optic papilla

сос·о́м·ый, -ая, -ое, (*a*) (*pres part pass*) of **сос·а́ть**

сос·о́чек·, -чк·а, (*m*) (anat) papilla

сосочко·ви́д·н·ый, -ая, -ое, (*a*) papilliform

сос·о́чков·ый, -ая, -ое, (*a*) (anat) papillary

со́с·очн·ый, -ая, -ое, (*a*) of **со́с·к·а**; **сос·о́чн·ый, -ая, -ое,** (*a*) of **со́с·о́к·**

со·средо·то́ч·ени·е, -я, (*n*) (*v n*) of **со·средо·то́ч·ить**; concentration

со·средо·то́ч·енност·ь, -и, (*f*) concentration (of attention, thought)

со·средо·то́ч·енн·ый, -ая, -ое, (*past part pass*) see **со·средо·то́ч·ива·ть;** (elec) lumped

~ пара́метр·ы   (*pl*) (elec) lumped parameters/constants

со·средо·то́ч·ива·ть, -ют, (*imp*) concentrate, centre

со·средо·то́ч·ить, -ат, (*perf*) see **со·средо·то́ч·ива·ть**

соссюри́т·, -а, (*m*) (min) saussurite

соссюритиз·а́ци·я, -и, (*f*) (geol) saussuritization

со·ста́·в·, -а, (*m*) composition, make-up; compound; staff

в·ход·и́ть в со·ста́·в·   join, become a member (of); be included in

со·ста́·в

~, грануло·метр·и́ческ·ий   size distribution (of free-flowing materials)

~, групп·ов·о́й   (chem) group composition

~, ис·полн·и́тельн·ый   (theat) cast

~, за·ли́в·очн·ый   (elec) filling compound

~, инженер·н·о-техн·и́ческ·ий   technical staff

~, кома́нд·н·ый   (mil) officers

~, лёт·н·ый   air crews

~, ли́ч·н·ый   personnel, staff

~, на·ли́ч·н·ый   effective personnel, strength

~, на·ча́л·ствующ·ий   managerial/senior/directing staff, executives

~, по·дви́ж·н·о́й   (rail) rolling stock

~, про·пи́т·очн·ый   impregnating compound, dipping composition (rubber)

~, социа́ль·н·ый   social structure

~, фракцио́н·н·ый   fractional composition

~, хим·и́ческ·ий   chemical composition

~, чи́сл·енн·ый   numerical strength

со·став·и́тел·ь, -я, (*m*) compiler; originator, author (of project etc.)

~ по·езд·о́в   (rail) yardmaster

~ програ́мм·ы   programmer (e.g. broadcasting, automation etc.)

~ слов·ар·я́   maker/author of a dictionary

со·ста́в·ить, -ят, (*perf*) see **со·ставл·я́·ть**

со·ставл·е́ни·е, -я, (*n*) (*v n*) see **со·ставл·я́·ть;** composition, compilation

~ карт·   (surv) charting

~ ка́рт·ы   (surv) mapping

~ програ́мм·ы   (autom) programming, coding, programme design

~ с·ме́с·и   (rubber) compounding a mix, compounding

со·ста́вл·енн·ый, -ая, -ое, (*past part pass*) see **со·ставл·я́·ть**

со·ставл·я́·ть, -ют, (*imp*) put together, compile, compose, collect, form (from separate parts), make/draw up; **-ся** (*pass*); consist of/in; (math) make, total

со·ставл·я́ющ·ая, -ей, (*f decl as adj*) component, component part, constituent

~ крут·я́щ·его момє́нт·а   (mech) torque component

~, по·пере́ч·н·ая   transverse component

со·ставл·я́ющ·ая
~ **си́л·ы тя́г·и** thrust component
со·ставл·я́ющ·ий, -ая, -ее, (*pres part act*) of **со·ставл·я́·ть;** component; (*see also* **со·ставл·я́ющ·ая** *as noun*)
со·став·н·о́й, -а́я, -о́е, (*a*) compounded, compound, composite, sectional, fabricated; component, constituent, integral
~ **ли́нз·а** (*f*) compound lens
~ **част·ь** (*f*) component part, constituent
~ **числ·о́** (*n*) (math) compound number
со·ста́р·ивани·е, -я, (*n*) ageing, aging; ageing treatment
~, **магни́т·н·ое** (met) magnetic ageing
со·ста́р·ива·ть, -ют, (*imp*) age
со·ста́р·ить, -ят, (*perf*) *see* **со·ста́р·ива·ть**
со·сто·я́ни·е, -я, (*n*) (*see under associated adj*) state, condition, status; (chem) constitution; property, fortune
   бы·ть в со·сто·я́ни·и be in a position to; be in a state of
   диагра́мм·а со·сто·я́ни·я (*f*) (phys) constitution/constitutional/ equilibrium diagram
   индик·а́тор· со·сто·я́ни·я (*m*) positioning indicator (telemechanics)
~, **основ·н·о́е** (phys) ground state
~, **рабо́ч·ее** working order
~ **равно·ве́с·и·я** state of equilibrium; (elec) stable state
   у·равн·е́ни·е со·сто·я́ни·я (*n*) (phys) equation of state
~, **у·станов·и́вш·ееся** steady state
~ **хроно́метр·а** chronometer rate
   шкал·а́ со·сто·я́ни·я мо́р·я (*f*) (meteor) sea disturbance scale, sea state scale
со·сто·я́тельност·ь, -и, (*f*) competence; solvency; validity
со·сто·я́тельн·ый, -ая, -ое, (*a*) valid, well-grounded, sound (of reasoning) well-to-do; (fin) solvent
со·сто·я́ть, -я́т, (*imp*) be; be composed of, consist of, be made up of, comprise; **-ся** take place
со·стра́г·ива·ть, -ют, (*imp*) plane/ shave off
со·страд·а́ни·е, -я, (*n*) compassion, sympathy
со·стрел·я́ть, -я́т, (*perf*) *see* **со·стрел·я́·ть**
со·стре́л·к·а, -и, (*f*) (*v n*) of **со·стрел·я́·ть;** (mil) calibration firing

со·стрел·я́·ть, -ют, (*imp*) calibrate (guns)
со·стриг·а́·ть, -ют, (*imp*) shear/clip/ snip/cut off
со·стри́ж·енн·ый, -ая, -ое, (*past part pass*) *see* **со·стриг·а́·ть**
со·стри́ч·ь, (*fut 3rd sing, pl*) **со·стри·ж·ёт, со·стриг·у́т,** (*past masc sing*) **со·стри́г·,** (*perf*); *see* **со·стриг·а́·ть**
со·строг·а́·ть, -ют, (*perf*) *see* **со·стра́·г·ива·ть**
со·с·ты́к·ованн·ый, -ая, -ое, (*a*) butt-jointed
состяз·а́ни·е, -я, (*n*) contest, competition; controversy
со·су́д·, -а, (*m*) vessel; (bot) trachea, vessel
~, **аккумуля́тор·н·ый** (elec) accumulator jar
~, **волос·н·о́й** (biol) capillary
~, **рту́т·н·ый** mercury box (compass)
со·су́д·ик·, -а, (*m*) vas, (anat) vasculum
со·су́д·ист·о·волокн·и́ст·ый, -ая, -ое, (*a*) vascular, fibro-vascular
~ **пуч·о́к·** (*m*) (bot) vascular bundle
со·су́д·ист·ый, -ая, -ое, (*a*) vascular, (bot) vasculiform, vasiform
~ **раст·е́ни·е** (*n*) vasiform/vasculiform plant
~ **шов·** (*m*) (med) vasoligature
сосудо·волокн·и́ст·ый, -ая, -ое, (*a*) (bot) fibro-vascular, vascular
сосудо·дви́г·ательн·ый, -ая, -ое, (*a*) (anat) vasomotor
сосудо·образ·у́ющ·ий, -ая, -ее, (*a*) (zool) vasoformative, vasifactive
сосудо·рас·шир·е́ни·е, -я, (*n*) (anat) vasodilation
сосудо·рас·шир·я́ющ·ий, -ая, -ее, (*a*) (anat) vasodilator, vasodilatory, vasodilating
сосудо·с·уж·е́ни·е, -я, (*n*) (anat) vasoconstriction
сосудо·с·у́ж·ивающ·ий, -ая, -ее, (*a*) (anat) vasoconstrictor, vasoconstrictive
сос·у́льк·а, -и, (*g pl*) **-лек·,** (*f*) icicle
сос·у́н·, -а́, (*m*) sucker, suckling; suction pipe (of a dredger)
сос·уно́к·, -нк·а́, (*m*) (zool) suckling, sucker
сос·у́т, (*pres 3rd pl*) of **сос·а́ть**
со·сущ·ествова́ни·е, -я, (*n*) co-existence
со·сущ·еств·ова́ть, -у́ют, (*imp*) co-exist

сос·у́щ·ий, -ая, -ее, (*pres part act*) of **сос·а́ть**; sucking; (zool) suctorial

сосце·ви́д·н·ый, -ая, -ое, (*a*) (anat) mastoid

со·с·чи́т·анн·ый, -ая, -ое, (*past part pass*) of **со·с·чит·а́·ть**

~ и́мпульс· (*m*) (nucl) tube count

со·с·чит·а́·ть, -ют, (*perf*) count, compute, reckon

сос·я́, (*pres gerund*) of **сос·а́ть**

сот·, (*m*) *see* **со́т·ы**

-сот (*numeral component*) -hundred

сот·ая, -ой, (*f decl as adj*) (*see also* **со́т·ый**) hundredth, hundredth part

со·твор·ённ·ый, -ая, -ое, (*past part pass*) of **со·твор·и́ть**

со·твор·и́ть, -я́т, (*perf*) create

сотéй·ник·, -а, (*m*) saucepan

со́т·енн·ый, -ая, -ое, (*a*) hundred, centenary, centesimal (graduations etc.)

~ вес·ы́ (*pl*) centesimal scale/weighing machine

~ иск·а́те·ль (*m*) hundreds selector (telephone)

-со́т·енн·ый, -ая, -ое, (*adj component*) -hundred

сотéрн·, -а, (*m*) Sauterne-type wine; Sauterne (rare)

со·тк·а́·ть, -у́т, (*perf*) weave

со́т·н·я, -и, (*g pl*) **со́т·ен·,** (*f*) a hundred

сото·ви́д·н·ый, -ая, -ое, (*a*) honeycomb, cellular

со́т·ов·ый, -ая, -ое, (*a*) honeycomb, cellular

~ кат·у́шк·а (*f*) (rad) honeycomb/lattice coil

~ мёд· (*m*) honey in the comb

сото·обра́з·н·ый, -ая, -ое, (*a*) honeycomb, cellular

со·тр·ёт, (*fut 3rd sing*) of **с·тер·éть**

со·тру́д·ник·, -а, (*m*) collaborator, co-worker, colleague; contributor (to periodical etc.); staff member, official, worker

~, на·у́чн·ый staff member, scientist (working at institute etc.)

со·тру́д·нича·ть, -ют, (*imp*) collaborate, co-operate; contribute

со·тряс·а́·ть, -ют, (*imp*) shake; concuss

со·тряс·éни·е, -я, (*n*) (*v n*) *see* **со·тряс·а́·ть**; tremor, vibration; (med) concussion

~ земл·и́ (seismol) earth tremor

~ мо́зг·а (med) commotio cerebri, concussion of the brain

со·тряс·ённ·ый, -ая, -ое, (*past part pass*) *see* **со·тряс·а́·ть**

со·тряс·ти́, -у́т, (*perf*) *see* **со·тряс·а́·ть**

со́т·ы, со́т·ов, (*m*) (*pl*) honeycomb; bottle rack

со́т·ый, -ая, -ое, (*a*) hundredth

соублок·, -а, (*m*) (met forge) sow block, anvil cap

со·удар·éни·е, -я, (*n*) (*and see* **с·толк·новéни·е**); (nucl) collision, impingement, encounter

со́ус·, -а, (*m*) sauce, gravy

со·у́ст·и·е, -я, (*n*) (anat) anastomosis

со·у·ча́ст·в·овать, -уют, (*imp*) participate in, take part in, share in

со·у·ча́ст·и·е, -я, (*n*) participation, collaboration; (phys) sharing

со·у·ча́ст·ник·, -а, (*m*) participant, collaborator

со·у·ча́ст·н·ый, -ая, -ое, (*a*) participating, collaborating; (phys) sharing

софи́т·, -а, (*m*) (arch) soffit; overhead bank of lights; strip/horizon-type light

со·фо́кус·н·ый, -ая, -ое, (*a*) confocal

софо́р·а, -ы, (*f*) (bot) sophora, *Sophora*

софори́н·, -а, (*m*) (biochem) sophorine

соффи́т·, -а, (*m*) *see* **софи́т·**

сох·, (*past masc sing*) of **со́х·нуть**; (*g pl*) of **сох·а́**

сох·а́, -и́, (*nom pl*) **со́х·и,** (*g pl*) **сох·,** (*f*) wooden plough

со́х·нуть, -нут, (*imp*) dry, dry out/up, become dry; become too dry, wither, become parched

со·хран·éни·е, -я, (*n*) preservation; conservation; retention (of ·qualities etc.)

~ вя́з·кост·и retention of viscosity

~ за·ря́д·а (phys) conservation of charge

~ и́мпульс·а (phys) conservation of momentum

~ коли́честв·а движ·éни·я (phys) conservation of momentum

~ ма́сс·ы (phys) conservation of mass, mass conservation

за·ко́н· со·хран·éни·я ма́сс·ы (*m*) law of conservation of matter, law of mass conservation

~ по́чв·ы (agr) soil conservation

~ прав· (law) reservation of rights

~ чёт·кост·и (phys) conservation of parity

~ энéрг·и·и (phys) conservation of energy

со·хран·ённ·ый, -ая, -ое, (*past part pass*) *see* **со·хран·я́·ть**

со·хран·и́вш·ийся, -аяся, -ееся, (*past part act*) *see* **со·хран·я́·ться;** preserved, surviving, extant

со·хран·и́ть, -я́т, (*perf*) *see* **со·хра·н·я́·ть**

со·хра́н·ност·ь, -и, (*f*) state of preservation; life in state of preservation, shelf life (e.g. of elec. battery); keeping capacity/property (e.g. of fruit)

со·хра́н·н·ый, -ая, -ое, (*a*) of **со·хра·н·я́ющ·ийся**

со·хран·я́·ть, -ют, (*imp*) preserve, keep, maintain, retain; conserve; **-ся** (*pass*); keep well, last; survive

~ си́л·у remain in force, hold (of conditions etc.)

со·хран·я́ющ·ийся, -аяся, -ееся, (*pres part act*) of **со·хран·я́·ться;** preservable, long-storage, keeping, lasting; (autom) non-volatile, permanent

соц- (*component*) (*abbr*) of (1) **социа·лист·и́ческ·ий** = socialist; (2) **со·циа́ль·н·ый** = social

со·цвéт·и·е, -я, (*n*) flower cluster, (bot) raceme, inflorescence

~, цим·о́зн·ое (bot) cyme

социал·исти́ческ·ий, -ая, -ое, (*a*) socialist

социа́ль·н·ый, -ая, -ое, (*a*) social

соц·со·ревн·ова́ни·е, -я, (*n*) socialist competition

соц·стра́х·, -а, (*m*) social insurance

соч·еви́чник·, -а, (*m*) (bot) peavine, *Lathyrus*

со·чет·а́ни·е, -я, (*n*) (*v n*) *see* **со·чет·а́·ть;** combination, blend, set; (*pl*) (math) combinations

~ координа́т· (autom) coordinate multiplexing

реа́кци·я со·чет·а́ни·я (*f*) (chem) coupling reaction

~ сигна́л·ов (naut) signal hoist

со·чет·а́тельн·ый, -ая, -ое, (*a*) combinative, associative

~ за·ко́н· (*m*) (math) associative law

со·чет·а́·ть, -ют, (*imp and perf*) combine, blend, associate; **-ся** (*pass*); match, harmonize

со·чин·éни·е, -я, (*n*) (*v n*) *see* **со·чи·н·я́·ть;** composition, work; (gram) coordination

со·бр·а́ни·е со·чин·éни·й (*n*) complete works (of one author)

со·чин·и́тельн·ый, -ая, -ое, (*a*) (gram) coordinating

со·чин·и́ть, -я́т, (*perf*) *see* **со·чин·я́·ть**

со·чин·я́·ть, -ют, (*imp*) compose, make up, invent

соч·и́ть, -а́т, (*imp*) tap (trees etc.); drip, dribble; **-ся** seep, ooze, exude, trickle, bleed

со·чла́, (*past fem sing*) of **с·честь,** (*perf*); *see* **с·чит·а́·ть** (*imp*)

со·член·éни·е, -я, (*n*) articulation, jointing, joint, (anat) symphysis; articulated/mechanical joint/connection

~, ём·кост·н·ое (rad) capacitative joint

~, колéн·чат·ое knee joint

~, сфер·и́ческ·ое ball joint

со·член·ённ·ый, -ая, -ое, (*past part pass*) *see* **со·член·я́·ть;** articular, linked

со·член·и́ть, -я́т, (*perf*) *see* **со·чле·н·я́·ть**

со·член·о́вн·ый, -ая, -ое, (*a*) (anat) arthroidal

со·член·я́·ть, -ют, (*imp*) joint, join, link up (mechanically); articulate; concatenate, catenate; (rad, elec) gang

со́ч·ност·ь, -и, (*f*) juiciness, succulence, sappiness

со́ч·н·ый, -ая, -ое, (*a*) juicy, succulent, sappy

со·чт·ёт, (*fut 3rd sing*) of **с·честь,** (*perf*); *see* **с·чит·а́·ть** (*imp*)

со·чу́вств·енн·ый, -ая, -ое, (*a*) sympathetic

со·чу́вств·и·е, -я, (*n*) sympathy

со·чу́вств·овать, -уют, (*imp*) (+ *dat*) sympathize

со·шёл, (*past masc sing*) of **со·йти́,** (*perf*); *see* **с·ход·и́ть** (*imp*)

со́ш·к·а, -и, (*g pl*) **-ш·ек·,** (*f*) pitman; bipod, tripod; (mil) gun rest

~, рул·ев·а́я (M/T) drop arm, steering gear arm

со·шл·а́, (*past fem sing*) of **со·йти́,** (*perf*); *see* **с·ход·и́ть** (*imp*)

со·шл·ёт, (*fut 3rd sing*) of **со·сл·а́ть,** (*perf*); *see* **с·сыл·а́·ть** (*imp*)

со·шлиф·о́выва·ть, -ют, (*imp*) grind off

сош·ни́к·, -а́, (*m*) ploughshare, plowshare; coulter; spade (of gun carriage), spike (of machine gun); (anat) vomer

~, а́нкер·н·ый (agr) hoe drill-coulter, hoe furrow-opener

~, ди́ск·ов·ый (agr) disk drill-coulter, disk furrow-opener

сош·ни́к

~, по·ло́з·ков·ый (agr) shoe drill-coulter, shoe furrow-opener

со·шь·ёт, (*fut 3rd sing*) of **с·ши́·ть,** (*perf*); see **с·шив·а́·ть** (*imp*)

сою́з·, -а, (*m*) union, alliance, league; (gram) conjunction

сою́з·к·а, -и, (*g pl*) **-з·ок·,** (*f*) vamp (of footwear)

сою́з·ник·, -а, (*m*) ally

сою́з·ни́ческ·ий, -ая, -ое, (*a*) allied

сою́з·н·ый, -ая, -ое, (*a*) allied; union

~ чи́сл·а (*pl*) (math) amicable numbers

со́·я, -и, (*f*) (bot) soya, soy, soybean, *Glycine*

с·па́·вш·ий, -ая, -ое, (*past part act*) see **с·пад·а́·ть**

с·па́д·, -а, (*n*) (*v n*) see **с·пад·а́·ть;** drop, abatement, fall, decrease

с·пад·а́ни·е, -я, (*n*) (*v n*) of **с·пад·а́·ть;** (phys) decay

врём·я с·пад·а́ни·я и́мпульс·а (*n*) pulse decay time

с·пад·а́·ть, -ют, (*imp*) fall/come off; fall, abate, go down, slacken, decrease, droop; **-ся** fall in, collapse

с·пад·е́ни·е, -я, (*n*) (physiol) collapse (e.g. of veins etc.)

с·па́·ек·, (*g pl*) of **с·па́й·к·а** (q.v.)

спа́зм·, -а, (*m*) spasm

спа́зм·а, -ы, (*f*) = **спазм·**

спазмод·и́ческ·ий, -ая, -ое, (*a*) spasmodic

спазмо·лити́н·, -а, (*m*) (pharm) spasmolytin

спазмо·фили́·я, -и, (*f*) (med) spasmophilia

с·па́·ивани·е, -я, (*n*) (*v n*) of **с·па́·ива·ть**

с·па́·ива·ть, -ют, (*imp*) solder, solder together (below 427°C); braze, braze together (above 427°C); seal, seal together (glass); be a hard drinker

с·па́·й, -я, (*m*) soldered joint/junction/seam; seal (glassware); (geol) joint, juncture; (*pl*) cold shut (defect in cast metal)

~, гор·я́ч·ий hot junction (of thermocouple)

спа́йдер·, -а, (*m*) (oil) spider

с·па́й·к·а, -и, (*g pl*) **с·па́·ек·** (*f*) (*v n*) see **с·па́·ива·ть;** soldered joint/junction/seam; (biol, geol) commissure

спайно·лепе́ст·н·ый, -ая, -ое, (*a*) (bot) sympetalous, gamopetalous, (*pl as noun*) *Sympetalae*

спайно·ли́ст·н·ый, -ая, -ое, (*a*) (bot) gamophyllous

спайно·пло́д·н·ый, -ая, -ое, (*a*) (bot) gamocarpous

с·па́й·ност·ь, -и, (*f*) (cryst, min) cleavage, cleavability

плоск·ост·ь с·па́й·ност·и (*f*) (cryst) cleavage plane, composition plane (twinning)

по́люс· с·па́й·ност·и (*m*) (cryst) cleavage pole

с·па́й·н·ый, -ая, -ое, (*a*) cleavable, cleavage

с·па́й·щик·, -а, (*m*) splicer, jointer (of elec. cables etc.)

сп·ал, (*past masc sing*) of **сп·ать; с·па·л,** (*past masc sing*) of **с·пас·ть** (*perf*) see **с·пад·а́·ть**

с·па́лз·ыва·ть, -ют, (*imp*) creep, creep off/down

с·пал·ённ·ый, -ая, -ое, (*past part pass*) of **с·пал·и́ть;** sunburnt, sunscorched, sunbaked, dried up/killed by the sun

с·пал·и́ть, -я́т, (*perf*) burn up; singe, scorch; **-ся** (*pass*); get sunburnt/sunscorched

сп·а́льн·ый, (*a*) sleeping

сп·а́льн·я, -и, (*g pl*) **-лен·,** (*f*) bedroom; sleeping cabin, cabin, berth; (rail) sleeper

спан·а́ндри·я, -и, (*f*) (biol) spanandry

спано·га́ми·я, -и, (*f*) (biol) spanogamy

спа́нсон·, -а, (*m*) see **спо́нсон·**

спарде́к·, -а, (*m*) (naut) spar deck

спарде́ч·н·ый, -ая, -ое, (*a*) (naut) spar-deck

с·па́р·енн·ый, -ая, -ое, (*past part pass*) see **с·па́р·ива·ть;** coupled, paired, twin; dual, duplex; biconjugate; (rad) gang, ganged

~ конденс·а́тор· (*m*) (rad) gang capacitor

спа́рж·а, -и, (*i*) **-ей,** (*f*) (bot) asparagus, *Asparagus*

спа́рж·ев·ый, -ая, -ое, (*a*) of **спа́р·ж·а** (bot) (*pl as n*) asparagus family, *Asparagoideae*

~ ка́мен·ь (*m*) (min) asparagus stone

~ капу́ст·а (*f*) (bot) broccoli

с·па́р·ивани·е, -я, (*n*) (*v n*) of **с·па́р·ива·ть**

с·па́р·ива·ть, -ют, (*imp*) couple, pair; (gen) mate, couple

с·па́р·ить, -ят, (*perf*) see **с·па́р·ива·ть**

спа́р·ов·ые, -ых, (*pl decl as adj*) (fish) sea bream, *Sparidae* спарти́н·а, -ы, (*f*) (bot) cord/marsh grass, *Spartina*

спа́рт·о, -а, (*n*) (bot) alpha esparto, *Stipa tenacissima*

с·па́р·ыва·ть, -ют, (*imp*) rip off

с·пас·аем·ый, -ая, -ое, (*pres part pass*) *see* с·пас·а́·ть; (rocket) recoverable

с·пас·а́ни·е, -я, (*n*) (*v n*) of с·пас·а́·ть; rescue, life-saving; salvage, recovery

с·пас·а́тел·ь, -я, (*m*) (min) rescue worker; (naut) salvage vessel; submarine rescue ship

с·пас·а́тельн·ый, -ая, -ое, (*a*) rescue, life-saving, life-; salvage; escape (submarines)

~ де́л·о (*n*) (naut) salvage, salvaging, salvage work; (min) rescue work

~ жиле́т· (*m*) life jacket

~ колоко́л· (*m*) escape bell (submarine escape)

~ круг· (*m*) lifebuoy

~ ле́стниц·а (*f*) fire escape

~ лин·ь (*f*) rocket line, life line

~ от·се́к· (*m*) escape department (in submarine)

~ при·бо́р· индивидуа́ль·н·ый (*m*) submerged escape apparatus

~ слу́ж·б·а (*f*) (naut) lifeboat service; (min) rescue service

~, сре́д·ств·а (*pl*) rescue apparatus; life-saving appliances, personal safety equipment

с·пас·а́·ть, -ют (*imp*) rescue, save (lives); salvage, recover (equipment); -ся escape

с·пас·е́ни·е, -я, (*n*) = с·пас·а́ни·е

с·пас·ённ·ый, -ая, -ое, (*past part pass*) *see* с·пас·а́·ть

спаси́бо (*particle*) thank you, thanks

с·пас·и́тел·ь, -я, (*m*) rescuer

с·пас·ти́, -у́т, (*perf*) *see* с·пас·а́·ть

с·пас·ть, (*fut 3rd pl*) с·пад·у́т, (*past masc sing*) с·па́·л, (*perf*) *see* с·пад·а́·ть

с·па́с·ш·ий, -ая, -ее, (*past part act*) of с·пас·ти́ (*perf*); *see* с·пас·а́·ть (*imp*)

с·па́с·ш·ийся, -егося, (*n decl as adj*) survivor

сп·ать, -ят, (*imp*) sleep, be asleep

с·па́·яннос·т·ь, -и, (*f*) cohesion, unity; solidarity

с·па́·янн·ый, -ая, -ое, (*past part pass*) *see* с·па́·ива·ть; united; soldered

с·па·я́·ть, -ют, (*perf*) *see* с·па́·ива·ть

спе́йсер·, -а, (*m*) (mech eng) automatic spacing table, spacer (of punching press)

спейси́стор·, -а, (*m*) (elec) spacistor

с·пек·а́емост·ь, -и, (*f*) tendency to cake/clinker; caking capacity (of coals)

с·пек·а́ни·е, -я, (*n*) (*v n*) *see* с·пек·а́·ть; caking (of coal), sintering (of powder materials), fritting (of ceramics)

с·пек·а́·ть, -ют, (*imp*) frit, sinter, fuse, cause to cake/clinker/sinter; -ся sinter, clinker, cake/coalesce/agglomerate (on heating)

с·пек·а́ющ·ийся, -аяся, -ееся, (*pres part act*) of с·пек·а́·ться

~ у́гол·ь (*m*) caking coal

спекта́кл·ь, -я, (*m*) (theat) performance

спе́ктр·, -а, (*m*) (phys) spectrum

~, вращ·а́тельн·ый rotational spectrum

~ втор·о́го за·прещ·е́ни·я second forbidden spectrum

~, дуг·ов·о́й arc spectrum

~ зву́к·а acoustic/acoustical spectrum

~ ис·пуск·а́ни·я (phys) emission spectrum

~ комбинацио́н·н·ого рас·се́·яни·я Raman spectrum

~, лин·е́йчат·ый line spectrum
ли́н·и·я спе́ктр·а (*f*) spectral line

~ люмин·есце́нци·и (phys) emission spectrum

~ об·тек·а́ни·я (dynam) flow field

~ об·тек·а́ни·я ло́паст·и (aerodynam) airflow over blade

~ по·глощ·е́ни·я (spectr) absorption spectrum

~, полос·а́т·ый band spectrum

~, ротацио́н·н·ый rotational spectrum

~, с·ло́ж·н·ый complex spectrum

~, степе́н·н·ый (math) power-law spectrum

~, энергет·и́ческ·ий (cryst) energy distribution

спектра́ль·н·ый, -ая, -ое, (*a*) spectral, spectrum

~ ана́лиз· (*m*) spectrum/spectral/spectroscope analysis

~ кла́сс·а (*f*) (astron) spectral type

~ ли́н·и·я (*f*) spectral/spectrum line

~ се́р·и·я (*f*) spectral series (of lines)

~ фотогра́ф·и·я (*f*) spectrography

~ чувств·и́тельност·ь (*f*) (anat, phot) spectral sensitivity

спектро·анализ·а́тор·, -а, (*m*) spectrum analyser

спектро·ви́зор·, -а, (*m*) spectrum viewer

спектро·гелио́·граф·, -a, (*m*) (astron) spectroheliograph

спектро·гелио·ско́п·, -a, (*m*) (astron) spectrohelioscope

спектро·гра́мм·a, -ы, (*f*) spectrogram

спектро́·граф·, -a, (*m*) spectrograph

спектро·граф·и́ческ·ий, -ая, -ое, (*a*) spectrographic

спектро·компар·а́тор·, -a, (*m*) (instr) spectrum comparator

спектро́·метр·, -a, (*m*) (phys) spectrometer

∼ пар· (nucl) pair spectrometer

спектро·метр·и́ческ·ий, -ая, -ое, (*a*) spectrometric

спектро·ме́тр·и·я, -и, (*f*) spectrometry

спектро·прое́ктор·, -a, (*m*) (instr) spectrum projector

спектро·сенсито́·метр·, -a, (*m*) (phot) monochromatic sensitometer

спектро·ско́п·, -a, (*m*) spectroscope

спектро·скоп·и́ческ·ий, -ая, -ое, (*a*) spectroscopic

∼ рас·щепл·е́ни·е (*n*) spectroscopic splitting

    фа́ктор· спектро·скоп·и́ческ·ого рас·щепл·е́ни·я (*m*) (phys) Landé g factor

спектро·скоп·и́·я, -и, (*f*) spectroscopy

спектро·фото́·метр·, -a, (*m*) spectrophotometer

спектро·фото·метр·и́ческ·ий, -ая, -ое, (*a*) spectrophotometric

спектро·хим·и́ческ·ий, -ая, -ое, (*a*) spectrochemical

спекул·я́нт·, -a, (*m*) (fin) speculator

спекуляри́т·, -a, (*m*) (min) specularite, specular iron ore

спекуляти́в·н·ый, -ая, -ое, (*a*) speculative

спекул·я́ци·я, -и, (*f*) (fin) speculation

с·пёк·ш·ийся, -аяся, -ееся, (*past part act*) *see* **с·пек·а́·ться**

∼ материа́л· (*m*) sinter

спе́л·ост·ь, -и, (*f*) ripeness, maturity, mellowness

∼ по́чв·ы tilth (of soil)

спе́·л·ый, -ая, -ое, (*a*) ripe, mature, mellow

спел·ь, -я, (*m*) graphite flakes (in molten cast iron)

спе́льт·, -a, (*m*) (bot) spelt wheat, *Triticum Spelta*

с·перв·а́ (*adv*) at first

с·пе́ред·и (*adv, prep* + *gen*) at/from the front, in front, in front of

    вид· с·пе́ред·и (*m*) front view

сперм·a, -ы, (*f*) (gen) sperm

спермат·и́д·, -a, (*m*) (gen) spermatid, spermatoblast

спермато·гене́з·, -a, (*m*) (gen) spermatogenesis

спермато·го́н·и·й, -я, (*m*) (gen) spermatogonium

спермато·зо́ид·, -a, (*m*) (zool) spermatozoon; (bot) spermatozoid

спермато·фи́т·, -a, (*m*) (bot) spermatophyte, spermophyte

спермато·фо́р·, -a, (*m*) (gen) spermatophore

спермато·ци́д·н·ое сре́д·ств·о (*n*) (med) spermatocide, spermicide

срермато·ци́т·, -a, (*m*) (zool) spermatocyte

спермаце́т·, -a, (*m*) (chem) spermaceti

спермаци·й, -я, (*m*) (bot) spermatium

сперм·и́н·, -a, (*m*) (gen) spermine

спермио- (*int component*) *see also* **спермато-**

спермио·ци́т·, -a, (*m*) (gen) spermiocyte

спе́рмо- (*int component*) *see also* **спермато-**

спермо·токси́н·, -a, (*m*) (gen) spermotoxin

сперрили́т·, -a, (*m*) (min) sperrylite

спессарти́н·, -a, (*m*) (min) spessartine, spessartite

спе́·ть, -ют, (*imp*) ripen, become ripe, mature; **с·пе·ть,** (*fut 3rd pl*) **с·по·ю́т,** (*perf*) sing

спе́ц·, -a, (*i*) **-ем,** (*m*) specialist

спец- (*component*) (*abbr*) of **специа́ль·-н·ый** special, special purpose; specialist

спеце·е́д·ств·о, -a, (*n*) anti-specialist attitude, hostility to specialists (on political grounds)

специализ·а́ци·я, -и, (*f*) specialization

специализ·и́рованн·ый, -ая, -ое, (*past part pass*) of **специализ·и́р·о·вать;** trained/adapted for specialized work, adapted/developed for a special purpose

специализ·и́р·овать, -уют, (*imp and perf*) give specialist training/instruction in; **-ся** specialize

специал·и́ст·, -a (*m*) specialist

специа́ль·н·о (*adv*) specially, especially, specifically

специа́ль·ност·ь, -и, (*f*) speciality, specialty, branch, field, line

специа́ль·н·ый, -ая, -ое, (*a*) special, special-purpose, specialist

∼ стал·ь (*f*) special/alloy steel

специ́фик·а, -и, (*f*) distinguishing features, characteristics

специфик·а́ци·я, -и, (*f*) specification, specifications; description; materials list (on eng drawing)

специ́фик·ум·, -а, (*m*) distinguishing features, characteristics

специфиц·и́р·овать, -ую́т, (*imp and perf*) specify

специф·и́ческ·ий, -ая, -ое, (*a*) specific

спе́ци·я, -и, (*f*) (food) spice, condiment

спец·карто́н·, -а, (*m*) spetskarton (a leather substitute)

спец·ко́р·, -а, (*m*) special correspondent

спец·о·де́жд·а, -ы, (*f*) working clothes, protective/industrial clothing; overalls

спец·под·гото́в·к·а, -и, (*f*) specialist training

спец·слу́ж·б·а, -ы, (*f*) special service

с·печ·ённ·ый, -ая, -ое, (*past part pass*) *see* **с·пек·а́·ть**

с·печ·ь, (*fut 3rd sing, pl*) **с·печ·ёт, с·пек·у́т,** (*past masc sing*) **с·пёк·,** (*perf*); *see* **с·пек·а́·ть**

спе́ш·ить, -ат, (*imp*) (*trans and intrans*) hurry, hasten; be fast (of clocks)

спе́ш·н·ый, -ая, -ое, (*a*) hurried; urgent, immediate, express

~ по́чт·а (*f*) express mail/post

спиго·ре́з·к·а, -и, (*g pl*) **-з·ок·,** (*f*) (food) fat chopping machine

спидо́·метр·, -а, (*m*) (M/T) speedometer

спи́к·а, -и, (*f*) (astron) Spica

с·пик·и́р·овать, -ую́т, (*perf*) (air) dive

спи́к·ул·а, -ы, (*f*) spicule, spiculum

с·пи́л·енн·ый, -ая, -ое, (*past part pass*) *see* **с·пи́л·ива·ть**

с·пи́л·ива·ть, -ют, (*imp*) saw, saw off; cut down, fell (with saw); file, file off/down

спил·и́т·, -а, (*m*) (geol) spilite

спил·и́тов·ый, -ая, -ое, (*a*) (geol) spilitic

с·пил·и́ть, -я́т, (*perf*) *see* **с·пи́л·ива·ть**

с·пи́л·к·а, -и, (*f*) (*v n*) *see* **с·пи́л·ива·ть;** (*g sing*) of **с·пи́л·ок·**

с·пи́л·ок·, -л·к·а, (*n*) split (leather)

~, лиц·ев·о́й grain split (leather)

спи́н·, -а, (*m*) (phys) spin; (*g pl*) of **спин·а́**

~, вы́·стро·енн·ые (*pl*) aligned spins

~, це́л·ый integral spin

спин-за́·мк·нут·ый сло́·й (*m*) (nucl) spin-filled layer

спин-спи́н·ов·ый, -ая, -ое, (*a*) (nucl) spin-spin

спин·а́, -ы́, (*nom pl*) **спи́н·ы,** (*g pl*) **спин·** (*f*) back, (anat) dorsum, spina; **спи́н·а** (*g sing*) of **спин·**

спина́кер·, -а, (*m*) = **спи́ннакер·**

спина́ль·н·ый, -ая, -ое, (*a*) spinal

спи́н·к·а, -и, (*g pl*) **-н·ок·,** (*f*) (*dim*) of **спин·а́;** back, spine, back edge; (anat) dorsum

~ нож·а́ knife back

спинна́кер·, -а, (*m*) (naut) spinnaker

спи́ннинг·, -а, (*m*) angler's fishing rod

спинно·брюш·н·о́й, -а́я, -о́е, (*a*) (anat) dorsoventral

спин·н·о́й, -а́я, -о́е, (*a*) dorsal, spinal

~ ао́рт·а (*f*) (zool) dorsal aorta

~ мозг· (*m*) spinal chord

спинно·ло́б·н·ый, -ая, -ое, (*a*) (anat) dorsofrontal

спинно·мозг·ов·о́й, -а́я, -о́е, (*a*) (anat) cerebrospinal

~ жи́дк·ост·ь (*f*) cerebrospinal fluid

спи́н·ов·ый, -ая, -ое, (*a*) (nucl) spin

~ фу́нкци·я (*f*) (phys) spin-wave function

спино́р·, -а, (*m*) spinor (quantum mechanics)

спино́р·н·ый, -ая, -ое, (*a*) spinor

~ част·и́ц·а (*f*) (phys) spinor particle

спино·ро́г·, -а, (*m*) (zool) file-fish, Balister, (*pl*) *Balistoidei*

спин·решёт·очн·ый, -ая, -ое, (*a*) (phys) spin-lattice

спинтари·ско́п·, -а, (*m*) (instr) spinthariscope

спинто́метр·, -а, (*m*) torque meter

спи́р·а, -ы, (*f*) (biol) spire, coil

спира́кул·а, -ы, (*f*) (zool) spiracle

спира́л·ь, -и, (*f*) spiral, coil, helix; spiral filament; (mil) concertina (wire entanglement)

~ винт·ов·а́я helix

~, двойн·а́я (elec) coiled coil

~ корню (math) cornu spiral

~, крут·а́я fast spiral

~, логарифм·и́ческ·ая (math) logarithmic/equiangular spiral

~ одн·ого́ зна́к·а за·кру́ч·ивани·я one-sense spiralization coil (polymers)

~, с·жа́т·ая close spiral

спира́ль·н·о-кон·и́ческ·ий, -ая, -ое, (*a*) trochoid

спира́ль·н·о-плоск·ост·н·о́й, -а́я, -о́е, (*a*) planispiral

спирально·с·вёр·нут·ый, -ая, -ое, (*a*) coiled

спира́ль·ност·ь, -и, (f) helicity
спира́ль·н·ый, -ая, -ое, (a) spiral, volute, coiled, helical
~ ка́мер·а (f) volute chamber (of turbine)
~ дробл·е́ни·е (n) (zool) spiral/alternating cleavage
~ кла́пан· (m) (zool) spiral valve
~ пруж·и́н·а (f) spiral/coiled spring
~ сверл·о́ (n) twist drill
спире́м·а, -ы, (f) (gen) spireme
спири·фери́д·ы (pl) (pal) *Spiriferacea*
спиркетингс·, -а, (m) (shipb) spirketting
спиро·ги́р·а, -ы, (f) (bot) spirogyra
спиро·ги́р·н·ый, -ая, -ое, (a) spirogyrate
спиро́·граф·, -а, (m) (med, instr) spirograph
спиро́·метр·, -а, (m) (med, instr) spirometer
спиро́н·, -а, (m) (eng) ram
спиро·не́м·а, -ы, (f) (biol) spironeme
спиро·хе́т·а, -ы, (f) (bact) spirochaeta, (pl) spirochaetes, spirochetes, *Spirochaetaceae*
~, блед·н·ая (bact) *Spirochaeta/Treponema pallida*
спиро·хет·о́з·, -а, (m) (med) spirochaetosis, spirochetosis
спи́рт·, -а, (nom pl) -ы́, (g pl) -о́в, (m) alcohol, spirit
~, ви́н·н·ый ethyl alcohol
~, древе́с·н·ый methyl alcohol
~, мети́л·ов·ый methyl alcohol
~, нашаты́р·н·ый spirit of hartshorn, ammonia spirit
~, о·чи́щ·енн·ый rectified alcohol
~, поливини́л·ов·ый (chem) polyvinyl alcohol
~ -ректифика́т·, -а, (m) rectified alcohol 95%
~, станда́рт·н·ый proof spirit
~ -сыр·е́ц·, -р·ц·а́, (m) unrectified alcohol 88%
~, хле́б·н·ый ethyl alcohol
~, эти́л·ов·ый ethyl alcohol, ethanol
спирт·а́з·а, -ы, (f) (biochem) alcoholase
спирт·н·о́й, -а́я, -о́е, (a) alcoholic, spirituous
спирт·ова́ни·е, -я, (n) fortifying (wines)
спирт·о́ванн·ый, -ая, -ое, (past part pass) of спирт·ова́ть; fortified
~ на·сто́й (m) alcoholate (liquor production)
спирт·ова́ть, -у́ют, (imp) alcoholize; fortify (wines)

спирт·о́вк·а, -и, (g pl) -вок·, (f) spirit stove/lamp
спирт·ов·о́й, -а́я, -о́е, (a) alcohol, spirit, alcoholic, spirituous
~ лак· (m) spirit varnish
~ у́·ровен·ь (m) spirit level
спирто·кето́н·, -а, (m) (chem) alcohol ketone
спирто·кисл·от·а́, -ы́, (f) alcohol/alcoholic acid, hydroxyacid
спирто·ме́р·, -а, (m) alcoholimeter, alcoholometer
спирто·мор·и́лк·а, -и, (g pl) -лок·, (f) spirit mordant
спирто·рас·твор·и́м·ый, -ая, -ое, (a) spirit (of dyes)
с·пис·а́ни·е, -я, (n) (v n) of с·пис·а́ть; writing off (equipment); discharge (of personnel)
с·пи́с·анн·ый, -ая, -ое, (past part pass) see с·пи́с·ыва·ть
с·пис·а́ть, (fut 3rd pl) с·пи́ш·ут, (perf) see с·пи́с·ыва·ть
с·пи́с·ок·, -с·к·а, (m) list, roll; transcription
~ из·бир·а́тел·ей electoral roll
~, корабе́ль·н·ый (naut) crew list
~ от·печа́т·ок· (print) errata
с·пи́с·ывани·е, -я, (n) (v n) of с·пи́с·ыва·ть
с·пи́с·ыва·ть, -ют, (imp) copy, transcribe, write down; write off (unserviceable equipment); (naut) draft, discharge (personnel); -ся (pass); correspond with
спи́ц·а, -ы, (i) -ей, (f) spoke; knitting needle
спи́ц·ев·ый, -ая, -ое, (a) of спи́ц·а; spoked
спи́ч·, -а, (i) -ем, (m) welcoming remarks; toast
спи́ч·ечниц·а, -ы, (i) -ей, (f) match box
спи́ч·ечн·ый, -ая, -ое, (a) match
спи́ч·к·а, -и, (g pl) -ч·ек·, (f) match, matchstick
с·пи́ш·ет, (fut 3rd sing) of с·пис·а́ть
с·плав·, -а, (m) (v n) see с·плавл·я́·ть; alloy; fusion; melt; float, raft (of timber)
~, антифрикцио́н·н·ый antifriction metal
~, бе́л·ый white metal
~, бина́р·н·ый binary alloy
~ выс·о́к·ого со·противл·е́ни·я electrical resistance alloy
~, высоко·леги́р·ованн·ый high alloy

с·плав

~, композицио́н·н·ый (weld) composite hard-facing rod

~, коше́ль·н·ый boom towing (timber)

~, лёгк·ий light alloy

~, лёгко·пла́в·к·ий fusible alloy

~, ли́·т·ый кус·ков·о́й (weld) cast tungsten carbide hard-facing rod

~, мол·ев·о́й stream driving (timber)

~, мул·ев·о́й stream driving

~, плот·ов·о́й rafting (timber)

~, под·ши́п·ников·ый bearing alloy

~, рас·сы́п·н·о́й stream driving (timber)

~, сверх·твёрд·ый diamond-substitute alloy

~ со́р·ом stream driving (timber)

~, стерж·нев·о́й (weld) composite hard-facing rod

~, твердо·пла́в·к·ий refractory alloy

~, твёрд·ый hard metal, cutting alloy

~, трой·н·о́й ternary alloy

~, труб·чат·о-зерн·и́ст·ый (weld) granular tungsten carbide hard-facing rod

с·пла́в·ин·а, -ы, (f) (bot) floating mat

с·пла́в·ить, -ят, (perf) see с·плавл·я́·ть

с·плавл·е́ни·е, -я, (n) fusion, melting

~ конта́кт·ов (elec) sticking of contacts

с·плавл·я́·ть, -ют, (imp) melt, fuse; float, drive (downstream); -ся (pass); fuse together

с·плав·н·о́й, -а́я, -о́е, (a) timber-flotation, timber-rafting

~ су́д·н·о (n) lumber raft

с·пла́в·щик·, -а, (m) rafter (man)

с·план·и́р·овать, -уют, (perf) plan; (air) glide

спланхно·пле́вр·а, -ы, (f) (zool) splanchnopleure

спланхно·пто́з·, -а, (m) (med) splanchnoptosis

спланхно·то́м·, -а, (m) (zool) splanchnotome

с·пла́ч·ива·ть, -ют, (imp) close up, close the ranks, unite, present a united front; join/fasten/lash together sideways on (e.g. beams, planks); make a raft, lash together to make a raft

с·плёв·ыва·ть, -ют, (imp) spit out, expectorate

сплен·и́т·, -а, (m) (med) splenitis

сплен·о́в·ый, -ая, -ое, (a) (anat) splenetic, splenic

сплено·ло́г·и·я, -и, (f) (med) splenology

сплено·мега́ли·я, -и, (f) (med) splenomegaly

с·пле́с·ен·ь, -с·н·я, (m) (naut) splice

~, раз·го́н·н·ый long splice

с·плес·н·и́ть, -ят, (perf) see с·плес·-ныва·ть

с·плес·ныва·ть, -ют, (imp) splice

с·плес·ти́, (fut 3rd pl) с·плет·у́т, (past masc sing) с·плё·л, (perf); see с·плет·а́·ть

с·плет·а́·ть, -ют, (imp) weave, plait, braid; splice, weave/plait together

с·плет·е́ни·е, -я, (n) intersection, crossing, junction; weave, plait, braid; (anat) plexus

~, со́лн·ечн·ое (anat) solar plexus

с·плет·ённ·ый, -ая, -ое, (past part pass) see с·плет·а́·ть; (anat) plexal, plexiform

с·плёт·к·а, -и, (g pl) -т·ок·, (f) splice

с·плёт·н·я, -и, (g pl) -тен·, (f) gossip, rumour, rumor

сплин·, -а, (m) (anat) spleen

с·плот·и́ть, -ят, (perf) see с·пла́ч·и·ва·ть

с·пло́т·к·а, -и, (f) (v n) see с·пла́ч·и·ва·ть; rafting, lashing together; (g sing) of с·пло́т·ок·

с·пло́т·ок·, -т·к·а, (m) raft

с·пло́т·очн·ый, -ая, -ое, (a) of с·пло́т·к·а

~ маши́н·а (f) raft-making machine

с·плоч·е́ни·е, -я, (n) (v n) see с·пла́·ч·ива·ть; uniting, presenting a solid front

с·плоч·ённост·ь, -и, (f) solidarity, unity; cohesion, density, closeness

с·плоч·ённ·ый, -ая, -ое, (past part pass) see с·пла́ч·ива·ть; solid, tightly packed, close, close-packed

сплош·н·о́й, -а́я, -о́е, (a) continuous, unbroken, complete, solid, compact

~ ли́н·и·я (f) (print) solid line

~ ма́сс·а (f) solid mass

~ про·ве́р·к·а (f) 100% inspection

~ систе́м·а (f) (min) longwall system, longwalling

~ сред·а́ (f) continuum

сплош·ь (adv) completely, entirely

с·плы́в·, -а, (m) (v n) see с·плыв·а́·ть; slump (of cement or concrete)

с·плыв·а́·ть, -ют, (imp) float downstream; overflow; -ся (pass); blend, mix, run together

с·плы́·ть, (fut 3rd pl) с·плыв·у́т, (past masc sing) с·плы́·л, (perf); see с·плыв·а́·ть

с·плю́·н·уть, -ут, (*perf*) *see* **с·плёв·ы·ва·ть**

с·плю́с·нут·ый, -ая, -ое, (*past part pass*); flattened, oblate (of spheroid)

сплю́ш·к·а, -и, (*g pl*) -ш·ек·, (*f*) (orn) owl, *Otus*

с·плю́щ·енн·ый, -ая, -ое, (*past part pass*) *see* **с·плю́щ·ива·ть;** oblate (of spheroids)

с·плю́щ·ива·ть, -ют, (*imp*) flatten; -**ся** be flat/oblate; become flat

с·плю́щ·ить, -ат, (*perf*) *see* **с·плю́щ·ива·ть**

спо́д·и·й, -я, (*m*) spodium, animal charcoal

спод·и́т·, -а, (*m*) (min) spodite

спо́д·умен·, -а, (*m*) (min) spodumene, triphane

с·по·ёт, (*fut 3rd sing*) of **с·пе·ть**

спо́йлер·, -а, (*m*) (a/c) spoiler
~ **тя́г·и** (a/c) thrust spoiler

с·поко́й·н·ый, -ая, -ое, (*a*) calm, peaceful, tranquil, placid, quiescent
~ **во́з·дух·** (*m*) still air
~ **мо́р·е** (*n*) calm sea
~ **стал·ь** (*f*) killed steel

с·поко́й·стви·е, -я, (*n*) calm, tranquillity

с·поласк·ива·ть, -ют, (*imp*) rinse, swill out

с·полз·а́ни·е, -я, (*n*) (v n) of **с·полз·а́·ть**
~ **траекто́ри·и** (gunn) trajectory drop
~ **нул·я́** (instr) zero creep
~ **то́н·а** wow (acoustics)
~ **част·от·ы́** (rad) frequency departure **эффе́кт· с·полз·а́ни·я** (*m*) (phys) reptation effect

с·полз·а́·ть, -ют, (*imp*) crawl/creep down/off; slip down/off

с·полз·ти́, -у́т, (*perf*) *see* **с·полз·а́·ть**

с·полн·а́ (*adv*) in full, completely

с·полос·н·у́ть, -у́т, (*perf*) *see* **с·поласк·ива·ть**

спонг·и́н·, -а, (*m*) (zool) spongin

спонгио·бла́ст·, -а, (*m*) (cyt) spongioblast

спонги·о́зн·ый, -ая, -ое, (*a*) (biol) spongiose

спонгио·ци́т·, -а, (*m*) (cyt) spongiocyte

спонго·ли́т·, -а, (*m*) (geol) spongiolite, spongolite

спондил·артри́т·, -а, (*m*) (med) spondylarthritis

спондил·и́т·, -а, (*m*) (med) spondylitis

спондило·листе́з·, -а, (*m*) (med) spondylolisthesis

спондило·пат·и́·я, -и, (*f*) (med) spondylopathy

спонди́ль·н·ый, -ая, -ое, (*a*) (zool) spondylous

спо́н·к·а, -и, (*g pl*) -н·ок·, (*f*) *see* **шпо́н·к·а**

спо́нсон·, -а, (*m*) (naut) sponson

спонта́н·н·ый, -ая, -ое, (*a*) spontaneous

спо́р·, -а, (*m*) dispute, controversy, argument

спо́р·а, -ы, (*f*) (biol) spore

спорад·и́ческ·ий, -ая, -ое, (*a*) sporadic

спорангие·но́с·н·ый, -ая, -ое, (*a*) (bot) sporangiferous

спор·а́нг·и·й, -я, (*m*) (bot) sporangium

спор·ангио·фо́р·, -а, (*m*) (bot) sporangiophore

спор·и́ди·я, -и, (*f*) (bot) sporidium

спо́р·ить, -ят, (*imp*) dispute, argue

спо́р·н·ый, -ая, -ое, (*a*) of **спор·;** disputable, disputed, controversial, questionable, debatable
~ **вод·а́** (*f*) rapids, race (e.g. on rivers)

споро·болю́с·, -а, (*m*) (bot) dropseed, *Sporobolus*

спор·овик·и́, -о́в, (*m pl*) (zool) sporozoans, *Sporozoa*

спо́р·ов·ый, -ая, -ое, (*a*) (bot) sporophytic, (*pl as noun*) sporophytes, *Sporophyta*

споро·ген·е́з·, -а, (*m*) (bot) sporogenesis

споро·го́н·и·й, -я, (*m*) (bot) sporogonium

споро·го́н·и·я, -и, (*f*) (bot) sporogony; (*g sing*) of **споро·го́н·и·й**

споро·зои́т·, -а, (*m*) (zool) sporozoite

споро·ка́рп·и·й, -я, (*m*) (bot) sporocarp

споро·ку́ч·к·а, -и, (*g pl*) -ч·ек·, (*f*) (bot) sorus

споро·но́с·ец, -с·ц·а, (*m*) (bot) sporophore

споро·но́с·н·ый, -ая, -ое, (*a*) (bot) sporiferous

с·пор·о́ть, -ют, (*perf*) rip off

споро·фи́лл·, -а, (*m*) (bot) sporophyll

споро·фи́т·, -а, (*m*) (bot) sporophyte

споро·ци́ст·а, -ы, (*f*) (zool) sporocyst

спо́рт·, -а, (*m*) sport

спорти́в·н·ый, -ая, -ое, (*a*) sports-, sporting, games, athletic, recreation

спортсме́н·, -а, (*m*) sportsman, athlete

спортсмён·к·а, -и, (*g pl*) **-н·ок·,** (*f*) sportswoman, athlete

спорул·яци·я, -и, (*f*) (bot) sporulation

спорынь·я́, -и́, (*f*) (bot) ergot, *Claviceps purpurea*

спо́соб·, -а, (*m*) (*see under associated words*) method, mode, way

~ **за·да́·ни·я** (math) representation

~ **рабо́т·ы** mode of operation

спосо́б·ност·ь, -и, (*f*) (*and see under qualifying words*) ability, capacity, power

~ **дел·и́ться** (nucl) fissionability

~**, диффузио́н·н·ая** diffusivity

~**, из·бир·а́тельн·ая** selectivity

~ **о·сажд·а́·ться** precipitability

~**, реакцио́н·н·ая** reactivity

спосо́б·н·ый, -ая, -ое, (*a*) able, capable

спосо́б·ств·овать, -уют, (*imp*) (+ *dat*) promote, further, facilitate, help, aid, favour, favor

с·по··тк·у́ться, -у́тся, (*perf*) stumble

с·по·тык·а́·ться, -ются, (*imp*) stumble

с·пра́в·а (*adv*) on/to/from the right, (naut) to starboard

~ **за корм·о́й** (naut) fine on the starboard quarter

~ **на тра́верз·е** on the starboard beam

~ **по но́с·у** fine on the starboard bow

с·правед·ли́в·о (*adv*) fairly, justly; it is true, it holds

с·правед·ли́вост·ь, -и, (*f*) fairness, justice

с·правед·ли́в·ый, -ая, -ое, (*a*) just, fair, right; justified, correct, valid

с·пра́в·иться, -ятся, (*perf*) *see* **с·прав·л·я́·ться**

с·пра́в·к·а, -и, (*g pl*) **-в·ок·,** (*f*) information, reference; certificate

с·правл·я́·ться, -ются, (*imp*) be capable of, be able to, manage, master; find out about, consult, check

с·пра́в·очник·, -а, (*m*) reference book, handbook, manual

с·пра́в·очн·ый, -ая, -ое, (*a*) reference, inquiry

~ **бюро́** (*n*) information/inquiry office

~ **кни́г·а,** (*f*) reference book, handbook, manual

с·пра́ш·ива·ть, -ют, (*imp*) ask, enquire

спрее́р·н·ый, -ая, -ое, (*a*) spray

с·пресс·о́ванн·ый, -ая, -ое, (*past part pass*) of **с·пресс·ова́ть**

с·пресс·ова́ть, -у́ют, (*perf*) press, compress; compact (powder metallurgy)

спри́нклер·, -а, (*m*) sprinkler head, spray nozzle (automatic fire protection system); diffuser branch pipe (of fire hose)

спри́нклер·н·ый, -ая, -ое, (*a*) of **спри́нклер·**

~ **об·ору́д·овани·е** (*n*) sprinkler (fire protection system)

сприн ц·ева́ть, -у́ют, (*imp*) (med) syringe

спринц·о́вк·а, -и, (*g pl*) **-вок·,** (*f*) syringing; syringe

с·провоц·и́р·овать, -уют, (*perf*) provoke

с·про́ект·и́р·овать, -уют, (*perf*) project, plan, design

с·про́с·, -а, (*m*) demand

  без с·про́с·а without permission/asking

  коэффицие́нт· с·про́с·а (*m*) (elec) demand factor

с·прос·и́ть, -я́т, (*perf*) *see* **с·пра́ш·ива·ть**

с·профил·и́р·овать, -уют, (*perf*) (met) shape, form; camber (roads etc.)

с·прош·у́, (*fut 1st sing*) of **с·прос·и́ть,** (*perf*); *see* **с·пра́ш·ива·ть** (*imp*)

спру (*indecl*) (med) sprue

спру́с·, -а, (*m*) spruce (timber)

спру́т·, -а, (*m*) (zool) octopus

с·пры́г·ива·ть, -ют, (*imp*) jump down/off

с·пры́г·н·уть, -ут, (*perf*) *see* **с·пры́г·ива·ть**

с·пры́ск·, -а, (*m*) nozzle, spray nozzle (of hose etc.)

с·пры́ск·ива·ть, -ют, (*imp*) sprinkle, spray

с·пры́с·н·уть, -ут, (*perf*) *see* **с·пры́ск·ива·ть**

с·пряг·а́·ть, -ют, (*imp*) harness together; (gram) conjugate

с·пряж·е́ни·е, -я, (*n*) (*v n*) *see* **с·пряг·а́·ть;** (gram) conjugation

с·прям·и́ть, -я́т, (*perf*) *see* **с·прямл·я́·ть**

с·прямл·е́ни·е, -я, (*n*) (*v n*) *see* **с·прямл·я́·ть;** rectification

с·прямл·я́ем·ый, -ая, -ое, (*pres part pass*) of **с·прямл·я́·ть;** rectifiable

~ **крив·а́я** (*f*) (math) rectifiable curve

с·прямл·я́·ть, -ют, (*imp*) straighten; regulate (rivers); (math) rectify

с·пряч·ь, (*fut 3rd sing, pl*) **с·пряж·ёт, с·пряг·у́т,** (*past masc sing*) **с·пряг·,** (*perf*) harness together

с·пуск·, -а, (*m*) (*v n*), *see* **с·пуск·а́·ть;** descent, slope, slide, chute; (print) imposition; (horol) escapement; (gunn) detent; (pharm) cerate

~, автомат·и́ческ·ий (phot) automatic/timer release

~, а́нкер·н·ый (horol) lever escapement

~, бок·ов·о́й (shipb) broadside launching/launch

~, винт·ов·о́й spiral chute

~, гравитацио́н·н·ый gravity slide/chute; (shipb) conventional launching, inclined launching

~, есте́ств·енн·ый (shipb) floating and towing out

~, иску́с·ственн·ый (shipb) inclined launching

~, крут·о́й steep descent

~ кур·к·а́ trigger release (of rifle)

~ ла́г·ом (shipb) broadside launching

~, обра́т·н·ый left-hand lay (rope)

~, парашют·и́рующ·ий (air) landing approach at minimum air speed

~, план·и́рующ·ий (a/c) gliding descent

~, при·нуд·и́тель·н·ый (*m*) (shipb) inclined launching

~, про́·волоч·н·ый (phot) wire release

~, прям·о́й right-hand lay (rope)

~, ро́л·иков·ый (horol) roller escapement

~ с воз·вра́т·н·ым хо́д·ом (horol) recoil escapement

ско́р·ост·ь с·пу́ск·а (*f*) (rail) coasting speed

~ с воз·ду́ш·н·ым со·противл·е́·ни·ем (horol) fly governor

~ с не·свобо́д·н·ым хо́д·ом frictional rest escapement

~ со свобо́д·н·ым а́нкер·н·ым хо́д·ом lever escapement

~, холост·о́й water-wasting pressure regulator, jet diffuser (water turbine inlet pressure control)

~, цилиндр·и́ческ·ий (horol) cylinder escapement

с·пуск·а́·ть, -ют, (*imp*) let down, lower; let out, drain off, deflate; let go, release, trip; launch (ships etc.); lay (rope); (print) impose; **-ся** (*pass*); descend, go/come down

с·пуск·н·о́й, -а́я, -о́е, (*a*) drain, draining

~ клинке́т· (*m*) sluice valve

~ кран· (*m*) drain cock

с·пуск·ов·о́й, -а́я, -о́е, (*a*) releasing, trigger; launching (ships)

~ доро́ж·к·и (*pl*) (shipb) launching ways

~ крюч·о́к· (*m*) trigger (rifle etc.)

~ регул·я́тор· ско́р·ост·и (*m*) (*see* **с·пуск·**) (horol) escapement

~ руко·я́т·к·а (*f*) firing key (launching)

~ рыча́г· (*m*) (gunn) sear, sear lever, firing key/lever

~ ско́б·к·а (*f*) trigger guard (rifle)

~ у·стро́й·ств·о (*n*) launching gear (ships)

с·пуст·и́ть, ~ят, (*perf*) *see* **с·пуск·а́·ть**

с·пуст·я́ (*prep* + *acc*) after, afterwards

с·пу́т·анн·ый, -ая, -ое, (*past part pass*) *see* **с·пу́т·ыва·ть**

с·пу́т·а·ть, -ют, (*perf*) *see* **с·пу́т·ыва·ть**

с·пу́т·ник·, -а, (*m*) satellite; fellow-traveller; guide-book; (min) associated mineral/rock

~ гла́в·н·ой жи́·л·ы (geol) main reef leader

~ -за·пра́в·щик·, -а, (*m*) (rocket) orbital refuelling base

~ земл·и́, иску́с·ственн·ый artificial Earth-satellite, sputnik

~, иррегуля́р·н·ый irregular satellite

~, иску́с·ственн·ый (rocket) satellite, satellite vehicle, artificial satellite, sputnik

~, не·обит·а́ем·ый unmanned satellite

~, обит·а́ем·ый manned satellite

~, обра́т·н·ый irregular satellite

~, ото·рв·а́вш·ийся escaped satellite

~, пасси́в·н·ый inert satellite

~ плане́т·ы secondary planet

~, прям·о́й regular satellite

~, раз·ве́д·ывательн·ый reconnaissance satellite

~ с·вя́з·и communication satellite

с·пу́т·ывани·е, -я, (*n*) (*v n*) of **с·пу́·т·ыва·ть;** entanglement; confusion

с·пу́т·ыва·ть, -ют, (*imp*) tangle, entangle, muddle, confuse; hobble (animals)

с·пу́щ·енн·ый, -ая, -ое, (*past part pass*) of **с·пуст·и́ть,** (*perf*); *see* **с·пуск·а́·ть** (*imp*)

сп·я́чк·а, -и, (*f*) dormancy, torpor

~, зи́м·н·яя (zool) hibernation

~, ле́т·н·яя (zool) aestivation, estivation

сп·я́щ·ий, -ая, -ее, (*pres part act*) of **сп·а́ть;** sleeping, dormant

ср. (*abbr*) = **сре́д·н·ий** mean; = **с·рав·н·и́** ср., compare

с·рабáт·ывани·е, -я, (*n*) (*v n*) of **с·рабáт·ыва·ть;** actuation, action, operation, operating cycle, cycle of operation; wear, wear-and-tear

    врéм·я с·рабáт·ывани·я (*n*) operating time, time of action (elec relay); response time (counter etc.)

~, двух·позициóн·н·ое (autom) two-position action

~, не·прáв·ильн·ое malfunction, malfunctioning

    скóр·ост·ь с·рабáт·ывани·я (*f*) (instr) speed of response

    ток· с·рабáт·ывани·я (*m*) (elec) operating current

    тóч·к·а с·рабáт·ывани·я (*f*) trip point

~ трýб·к·и (nucl) tube count

    ý·ровен·ь с·рабáт·ывани·я (*n*) (instr) trip level

    част·от·á с·рабáт·ывани·я (*f*) duty classification (of relay)

с·рабáт·ыва·ть, -ют, (*imp*) function, operate, act, work (of mechanism); **-ся** wear out, become worn out; work harmoniously with

с·рабóт·аннос·ть, -и, (*f*) wear, age, worn-out state (of machines); co-operation, understanding, harmony (in working with others)

с·рабóт·анн·ый, -ая, -ое, (*past part pass*) see **с·рабáт·ыва·ть;** worn-out, worn

с·рабóт·а·ть, -ют, (*perf*) see **с·рабáт·ыва·ть**

с·равн·éни·е, -я, (*n*) comparison; congruence (math, number theory); (gram) simile

    вы́с·ш·ий ... по с·равн·éни·ю с higher than

    по с·равн·éни·ю с compared to, in comparison with, than

    электрóд· с·равн·éни·я (*m*) reference electrode

с·рáвн·ивани·е, -я, (*n*) (*v n*) of **с·рáвн·ива·ть;** comparison; levelling

с·рáвн·ива·ть, -ют, (*imp*) compare; equate, equalize; level, make level; **-ся** (*pass*); be equal to; be level with

с·равн·и́мост·ь, -и, (*f*) comparability; congruity

с·равн·и́м·ый, -ая, -ое, (*pres part pass*) see **с·рáвн·ива·ть;** comparable; congruent (number theory)

с·равн·и́тельн·о (*adv*) comparatively; in comparison

с·равн·и́тельн·ый, -ая, -ое, (*a*) comparative

с·равн·и́ть, -я́т, (*perf*) compare

с·равн·я́·ть, -ют, (*perf*) equate, equalize; **-ся** (*pass*); be equal to

с·раж·á·ть, -ют, (*imp*) strike, strike down; **-ся** fight

с·раж·éни·е, -я, (*n*) (mil) battle, operation(s)

с·раз·и́ть, -я́т, (*perf*) see **с·раж·á·ть**

с·рáз·у (*adv*) at once, immediately, directly after

срáм·, -а, (*m*) shame, disgrace

с·раст·áни·е, -я, (*n*) (*v n*) of **с·раст·á·ться;** intergrowth; accretion, concrescence

    двой·ни́к· с·раст·áни·я (*m*) (cryst) contact twin

~, двой·никóв·ое (cryst) twinning

~, клино·ви́д·н·ое (geol) structure peg

    плóск·ост·ь с·раст·áни·я (*f*) (cryst) composition plane

с·раст·á·ться, -ются, (*imp*) grow together, intergrow; accrete

с·раст·и́сь, -ýтся, (*perf*) see **с·раст·á·ться**

с·раст·и́ть, -я́т, (*perf*) see **с·рáщ·ива·ть**

с·ращ·éни·е, -я, (*n*) (*v n*) see **с·рáщ·ива·ть** (biol) accretion, concrescence; (chem etc.) coalescence

с·рáщ·ива·ть, -ют, (*imp*) grow/graft/join together; join/unite/combine end-to-end; lengthen (pipes, beams, rods etc.); splice (ropes etc.)

сред·á, -ы́, (*acc sing*) **-ý,** (*nom pl*) **срé-д·ы,** (*f*) (*see under associated words*) medium, environment, surroundings; agent (when active medium); **сред·á, -ы,** (*acc sing*) **срéд·у,** (*nom pl*) **срé-д·ы,** (*dat pl*) **сред·ам,** (*f*) Wednesday

~, дéй·ствующ·ая agent

~, сплош·н·áя (phys, mech) continuum, continuous medium

~, с·вяз·ýющ·ая binding agent

~, о·круж·áющ·ая environment

сред·и́ (*prep + gen*) among, amongst, in the middle of, amidst

среди·земно·мóр·ск·ий, -ая, -ое, (*a*) Mediterranean

среди·зéм·н·ый, -ая, -ое, (*a*) (geog) inland, landlocked

~ мóр·е (*n*) the Mediterranean; inland sea, landlocked sea

среди·зéм·ь·е, -я, (*n*) (geog) Mediterranean territory/lands

сред·и́н·а, -ы, (*f*) = **серед·и́н·а**

сред·и́нн·ый, -ая, -ое,   (*a*) (*see also* **серед·и́нн·ый**) middle, median, medial

~ **от·клон·е́ни·е**   (*n*) (gunn) 50% probability zone

~ **от·клон·е́ни·е бок·ов·о́е**   (*n*) 50% breadth zone

~ **от·клон·е́ни·е по да́ль·ност·и**   (*m*) 50% breadth zone

сре́д·н·е-   (*component*) mean, average, medium; middle, median; (geog) central; (biol) medi-, medio-, meso-, mes-

средне·арифмет·и́ческ·ий, -ая, -ое, (*a*) arithmetic mean

средне·век·о́в·ый, -ая, -ое,   (*a*) medieval

средне·вес·ов·о́й, -а́я, -о́е, (*a*) weight-average

средне·вз·ве́ш·енн·ое, -ого,   (*n decl as adj*) (math) weighted mean, weighted arithmetic average

средне·влаго·люб·и́в·ый, -ая, -ое, (*a*) (bot) mesophytic

средне·год·ов·о́й, -а́я, -о́е, (*a*) average-annual

средне·груд·ь, -и,   (*f*) (ent) mesothorax

сре́д·н·ее, -его, (*n decl as adj*) (math) mean, average

~**, арифмет·и́ческ·ое** arithmetic mean/average

~**, квадрат·и́чн·ое** root-mean-square

~ **по вре́мен·и** time average

~ **по мно́ж·еств·у** assembly average, mathematical expectation

~**, пропорциона́ль·н·ое**   geometric mean/average

средне·европ·е́йск·ий, -ая, -ое, (*a*) Central European

средне·дол·ев·о́й, -а́я, -о́е, (*a*) middle-lobe

средне·за́д·н·ий, -яя, -ее, (*a*) (biol) medioposterior

средне·зерн·и́ст·ый, -ая, -ое, (*a*) medium-grained

средне·кали́бер·н·ый, -ая, -ое, (*a*) (gunn) medium-calibre

средне·квадрат·и́чн·ый, -ая, -ое, (*a*) (math) root-mean-square, RMS

средне·ключ·и́чн·ый, -ая, -ое, (*a*) (anat) medioclavicular

средне·леги́р·ованн·ый, -ая, -ое, (*a*) (met) medium-alloyed/-alloy

средне·лед·ников́ь·е, -я,   (*n*) (geol) Middle Glacial Period

средне·меж·лед·ников́ь·е, -я,   (*n*) (geol) Middle Interglacial Period

средне·лисг·ов·о́й стан·   (*m*) (met roll) plate mill (4–20 mm × 1000–2000 mm)

средне·ме́сяч·н·ый, -ая, -ое,   (*a*) average monthly

средне·на·вес·н·о́й, -а́я -о́е,   (*a*) centrally suspended/mounted

средне·пла́в·к·ий, -ая, -ое,   (*a*) medium-fusing

средне·пла́н·, -а,   (*m*) (a/c) mid-wing monoplane

средне·порфи́р·ов·ый, -ая, -ое, (*a*) (geol) mediophyric

средне·со́рт·н·ый, -ая, -ое,   (*a*) medium-grade

средне·спи́н·к·а, -и,   (*g pl*) **-н·ок·,** (*f*) (zool) mesotergum

средне·су́т·очн·ый, -ая, -ое,   (*a*) average daily

средне·твёрд·ый, -ая, -ое, (*a*) medium-hard

средне·тяжёл·ый, -ая, -ое, (*a*) medium-weight

средне·угле·ро́д·ист·ый, -ая, -ое, (*a*) (met) medium-carbon

средне·част·о́тн·ый, -ая, -ое,   (*a*) (rad) medium-frequency

средне·чи́сл·енн·ый, -ая, -ое,   (*a*) number-average

средне·шир·о́тн·ый, -ая, -ое,   (*a*) (geog) middle-latitude

средне·яче́·ист·ый, -ая, -ое,   (*a*) mesobrochate

сре́д·н·ий, -яя, -ее,   (*a*) (*see also* **сре́д·н·ее**) (*as noun*); middle, central, median; average, mean, medium; (gram) neuter

в сре́д·н·ем on average

в сре́д·н·ем бу́д·ут да·ва́ть will average

~ **велич·ин·а́** (*f*) mean value

~ **вре́м·я** (*n*) (astron) mean solar time

~ **голоце́н·** (*m*) (geol) Mesopolocene

~ **долго·ве́ч·ност·ь** (*f*) average life

~ **крив·изн·а́** (*f*) (math) average curvature

~ **ли́н·и·я**   (*f*) (math) centre line, midline, median

~ **мозг·** (*m*) (anat) mesencephalon

~ **по·ро́д·а** (*f*) (geol) intermediate rock

~ **ремо́нт·** (*m*) intermediate repairs

~ **род·** (*m*) (gram) neuter

~ **сол·ь** (*f*) (chem) normal salt

~ **член·** (*m*) (math) middle term

~ **шко́л·а** (*f*) secondary/high school

~ **эффекти́в·н·ое знач·е́ни·е**   (*n*) mean effective value

средо·стéн·и·е, -я, (*n*) (anat) mediastinum

средо·тóч·и·е, -я, (*n*) concentrating/assembly/congregating point

~ **прú·мес·ей** (chem, met) impurity centre

срéд·ств·о, -а, (*n*) (*and see under associated words*) means, agent; (pharm) remedy; (*pl*) means, facilities, resources, equipment; (fin) resources

~, мéст·н·ые (*pl*) local resources

~, от·бéл·очн·ое bleaching agent

~, плов·ýч·ие (*pl*) boats, ships, craft

~ **по·требл·éни·я** (*pl*) consumer goods

~ **с·вя́з·и** (*pl*) communications system/facility/equipment

~, с·мáч·ивающ·ее wetting agent

~, тониз·úрующ·ее (pharm) tonic

срéд·ыш·, -а, (*m*) (naut) harbour-entrance rock

с·рéж·ет, (*fut 3rd sing*) of **с·рéз·ать,** (*perf*); *see* **с·рéз·ыва·ть** (*imp*)

с·рéз·, -а, (*m*) (*v n*) *see* **с·рéз·ыва·ть;** cut, slice; shear; section; edge

~, гисто·лог·úческ·ий (biol) histological section

~, дýль·н·ый (gunn) muzzle face

~, зáд·н·ий trailing edge

ис·пыт·áни·е на с·рез· (*n*) (mech) shearing test

~, казён·н·ый breech end (gun barrel)

прóч·ност·ь на с·рез· (*f*) (mech) shear strength

~ **сопл·á** nozzle edge

~ **сопл·á, вы·ход·н·óй** nozzle exit

~, Х- X-cut (crystals)

~, У- Y-cut (crystals)

с·рез·áни·е, -я, (*n*) (*v n*) *see* **с·рéз·ыва·ть;** cutting

~ **маршрýт·а** (air) cutting a corner

с·рез·á·ть, -́ют, (*imp*); **с·рéз·ать,** (*fut 3rd pl*) **с·рéж·ут,** (*perf*); *see* **с·рéз·ыва·ть**

с·рез·áющ·ий, -ая, -ее, (*pres part act*) *see* **с·рéз·ыва·ть;** shear, shearing

с·рез·н·óй, -áя, -óе, (*a*) cut, cut-off, truncated

с·рéз·к·а, -и, (*f*) (*v n*) of **с·рéз·ыва·ть;** (*g sing*) of **с·рéз·ок·**

с·рéз·ок·, -з·к·а, (*m*) cut-off piece/portion/section, piece

с·рéз·очн·ый, -ая, -ое, (*a*) of **с·рéз·к·а**

~ **горизóнт·** (*m*) mean low level (of rivers)

с·рéз·ываем·ый, -ая, -ое, (*pres part pass*) of **с·рéз·ыва·ть;** in shear

с·рéз·ыватель·, -я, (*m*) (elec) clipper

с·рéз·ыва·ть, -ют, (*imp*) cut, cut off, shear, slice, clip

с·рéз·ывающ·ий, -ая, -ее, (*pres part act*) of **с·рéз·ыва·ть**

~ **на·пряж·éни·е** (*n*) shearing stress

с·рис·овáть, -ýют, (*perf*) *see* **с·рис·ó·выва·ть**

с·рис·ó·выва·ть, -ют, (*imp*) copy (by drawing)

с·ровн·я́·ть, -ют, (*perf*) level, make level; **-ся** be level with

с·рóд·н·ый, -ая, -ое, (*a*) kindred, cognate; innate, inherent

с·рóд·ственн·ый, -ая, -ое, (*a*) (gen) kindred, related

с·род·ств·ó, -á, (*n*) affinity, relationship, (zool) consanguinity

~, о·стáт·очн·ое (chem) affinity, residue

~, хим·úческ·ое chemical affinity

с·рó·ет, (*fut 3rd sing*) of **с·ры·ть**

срóк·, -а, (*m*) term, period; time, specified time, date, specified date

~ **вы·полн·éни·я** date of completion

~ **гóд·ност·и** (phot) life

~ **полно·мóч·и·я** term of office

~ **с·дá·ч·и** date of delivery, date specified for delivery, delivery deadline

~ **слýж·б·ы** term/period of service (of personnel); life, useful/service life (of equipment)

~ **хран·éни·я** storage life; shelf life (of elec. battery)

с·рóс·ся, (*past masc sing*), of **с·раст·úсь,** (*perf*); *see* **с·раст·á·ться** (*imp*)

с·рост·úть, -я́т, (*perf*) join together end-to-end; splice (rope)

сростно·голóв·чат·ый, -ая, -ое, (*a*) (bot) syncephalic

сростно·лепéст·н·ый, -ая, -ое, (*a*) (bot) sympetalous, gamopetalous, (*pl as noun*) *Sympetalae*

с·рóст·ок·, -т·к·а, (*m*) growth, concretion, concrement; extension

~, двой·никóв·ый (cryst) twin

с·рóс·ш·ийся, -аяся, -ееся, (*past part act*) *see* **с·раст·á·ться;** (bot) adnate

срóч·н·о (*adv*) immediately, urgently, at once, by return (in correspondence); immediate, urgent (as instruction)

срóч·ност·ь, -и, (*f*) urgency

при·бóр· срóч·ност·и (*m*) arming delay switch (of bombs, mines etc.)

сигнал· срóч·ност·и (*m*) (telecom) emergency priority signal, "PAN" signal, "XXX" signal

сро́ч·н·ый, -ая, -ое, *(a)* immediate; pressing, urgent, rush; timely, punctual, prompt; periodic

~ **за·ка́з·** *(m)* rush order

~ **плат·ёж·** *(m)* instalment (hire purchase payment)

~ **письм·о́** *(n)* express letter

с·ру́б·, -а, *(m) (v n) see* **с·руб·а́·ть;** cut, cut surface; wooden framework

с·руб·а́·ть, -ют, *(imp)* chop/hack/slash/cut down, fell; furl and stow (sails)

с·руб·и́ть, -́ят, *(perf) see* **с·руб·а́·ть**

с·ры́в·, -а, *(m) (v n) see* **с·рыв·а́·ть;** (dynam) breakaway, separation, breakdown, stall; disruption; exposed/stripped surface; failure, setback

~**, волн·ов·о́й** (dynam) shockwave breakdown/separation; (air) shock/compressibility stall

до с·ры́в·а (air) below stall

~**, кон·цев·о́й** (air) tip stall

~ **нейтрон·а** (nucl) neutron stripping

~ **пла́мен·и** (a/c) flame-out (of jet engines)

по́сле с·ры́в·а (air) above stall

~ **по·то́к·а** (dynam) flow separation/breakaway; (aerodynam) stall

~ **по·то́к·а, корн·ев·о́й** (air) root stall

~ **по·то́к·а, раз·ви́·вш·ийся** (air) pronounced/advanced stall

реа́кци·я с·ры́в·а *(f)* (nucl) stripping reaction

ско́р·ост·ь с·ры́в·а *(f)* (air) stalling speed; blowoff velocity (of rocket flare)

~ **-спира́л·ь, -и,** *(f)* (air) inadvertent spin

то́ч·к·а с·ры́в·а *(f)* breakaway point

с·рыв·а́·ть, -ют, *(imp)* tear off/down/away, strip off; pick, pluck; disrupt, upset; overdraw, strip (a screw thread); **-ся** *(pass)*; break/come off/away/loose/adrift; fall off, lose hold; dart off/away

с·рыв·к·а́ *(adv)* by jerking, with a jerk

с·рыв·н·о́й, -а́я, -о́е, *(a)* of **с·рыв·;** **крив·а́я с·рыв·н·ы́х режи́м·ов по·лёт·а** *(f)* (air) stall line

~ **характе́р·и́стик·а** *(f)* (air) stalling characteristic

с·ры́г·ивани·е, -я, *(n)* regurgitation

с·ры·ть, ** *(fut 3rd pl)* **с·ро́·ют, *(perf)* dig away, level

с·са́д·ин·а, -ы, *(f)* scratch, abrasion, excoriation

с·сад·и́ть, -́ят, *(perf) see* **с·са́ж·ива·ть**

с·са́ж·ива·ть, -ют, *(imp)* set/put/help down, drop (passengers); scratch, abrade, excoriate

с·са́с·ыва·ть, -ют, *(imp)* suck/draw off; suck dry

ссо́р·а, -ы, *(f)* quarrel

с·сос·а́ть, -у́т, *(perf) see* **с·са́с·ыва·ть**

с·со́х·нуться, -нутся, *(perf) see* **с·сых·а́·ться**

с·су́д·а, -ы, *(f)* loan

~**, без·воз·вра́т·н·ая** (fin) grant

~ **до вос·тре́б·овани·я** (fin) loan on call, call loan

~**, кратко·сро́ч·н·ая** short-term loan

~**, проце́нт·н·ая** interest-bearing loan

с·суд·и́ть, -́ят, *(perf) see* **с·суж·а́·ть**

с·суж·а́·ть, -ют, *(imp)* lend, loan

с·су́ч·ива·ть, -ют, *(imp)* (text) spin; strand (rope); throw (silk)

с·суч·и́ть, -а́т, *(perf) see* **с·су́ч·ива·ть**

с·сыл·а́·ть, -ют, *(imp)* exile, deport, banish; **-ся** *(pass)*; refer to, cite, quote

с·сы́л·к·а, -и, *(g pl)* **-л·ок·,** *(f)* exile, deportation; reference

с·сы́л·очн·ый, -ая, -ое, *(a)* reference

~ **ка́рт·очк·а** *(f)* cross-reference card

с·сы́п·ать, -лют, *(perf) see* **с·сып·а́·ть** *(imp)*

с·сып·а́·ть, -ют, *(imp)* pour (solids)

с·сып·н·о́й, -а́я, -о́е, *(a)* pouring (of solids)

~ **пункт·** *(m)* (agr) grain-collecting station

с·сых·а́·ться, -ются, *(imp)* shrivel; clot, form hard lumps; become parched/chapped

ст·а *(gen/dat/instr/prep) of* **ст·о**

стабилиз·а́тор·, -а, *(m)* stabilizer; (chem) stabilizer, stabilizing agent; (elec) stabilizer, regulator; (a/c) stabilizer, tail plane; (rocket) fin assembly

~**, вертика́ль·н·ый** (a/c) fin

~ **дуг·и́** (weld) arc stabilizer

~**, коро́б·чат·ый** (a/c) box fin

~ **ку́рс·а** (a/c) directional stabilizer/gyro

~ **на·пряж·е́ни·я** (elec) voltage stabilizer/regulator

~ **пла́мен·и** flame stabilizer

~**, про·до́ль·но-по·пере́ч·н·ый** (a/c) vertical gyro

стабилизацио́н·н·ый, -ая, -ое, *(a)* stabilizing

~ **коло́нн·а** *(f)* (chem) stabilizer column/tower

стабилиз·а́ци·я, -и,   (*f*) stabilization

~, азимута́ль·н·ая (a/c, radar) line-of-sight stabilization

~ вращ·е́ни·ем (rocket) spin stabilization

~ гор·е́ни·я flame holding/stabilization

~, курс·ов·а́я (a/c) yaw/directional stabilization, heading alignment (radar)

~ нул·ев·о́го у́·ровн·я (autom) equalization of high-frequency response

~ пла́мен·и flame stabilization

~ по·лож·е́ни·я (rocket) attitude stabilization

схе́м·а стабилиз·а́ци·и (*f*) anti-hunt circuit (computers)

стабилиз·и́рованн·ый, -ая, -ое, (*past part pass*) of **стабилиз·и́р·овать**

а·стат·и́ческ·ий стабилиз·и́рованн·ый регул·я́тор· (*m*) (autom) proportional-speed floating controller, integral controller

стабилиз·и́р·овать, -уют, (*imp and perf*) stabilize

стабилиз·и́рующ·ий, -ая, -ее, (*pres part act*) of **стабилиз·и́р·овать**

~ сре́д·ств·о (*n*) (autom) antihunt measure/means

стабилиз·о́ванн·ый, -ая, -ое, *see* **стабилиз·и́рованн·ый**

стабилиз·ова́ть, -у́ют, (*imp and perf*) stabilize

стабило·во́льт·, -а, (*m*) (elec) stabilivolt, voltage regulator tube

стабило́·метр· част·от·ы́ (*m*) (rad) frequency departure meter

стаби́ль·ност·ь, -и, (*f*) stability

стаби́ль·н·ый, -ая, -ое, (*a*) stable

~ уч·е́бник· (*m*) standard textbook, set book

ста́в·ен·ь, -в·н·я, (g pl) -в·н·ей, (*m*) shutter (of window)

ста́в·ить, -ят, (*imp*) put, place, stand, set; pose, postulate, raise (problems etc.); stage, produce, put on; stake, bet

~ диагно́з· (med) diagnose, put forward a diagnosis

~ на бо́ч·к·у (naut) secure to buoy

~ на ме́ст·о position, put in position

~ на при·ко́л· lay up (a ship)

~ реко́рд· make/create a record

ста́в·к·а, -и, (g pl) -в·ок·, (*f*) (fin) rate; bet, stake; headquarters

~, порт·о́в·ые (*pl*) port rates

~, у·чёт·н·ая discount rate

ста́в·н·я, -и, (g pl) -в·ен·, (*f*) shutter (of window etc.)

ставроли́т·, -а, (*m*) (min) staurolite

ста́·вш·ий, -ая, -ее, (*past part act*) of **ста·ть,** (*perf*) (q.v.), *and see* **станов·и́ться** (*imp*)

стадиа́ль·н·ый, -ая, -ое, (*a*) stepwise, step by step, phased

стад·и́йност·ь, -и, (*f*) occurrence in stages

стад·и́йн·ый, -ая, -ое, (*a*) = **стадиа́ль·н·ый**

стадио́н·, -а, (*m*) (arch) stadium

ста́д·и·я, -и, (*f*) stage, phase

~, лед·нико́в·ая (geol) glacial stage

ста́д·ност·ь, -и, (*f*) gregariousness, gregarious instinct

ста́д·н·ый, -ая, -ое, (*a*) (zool) gregarious, herd

ста́д·о, -а, (nom pl) -а́, (*n*) herd, flock; (fish) shoal, school

с·та́·ет, (*fut 3rd sing*) of **с·та́·ять**

ста́ж·, -а, (i) -ем, (*m*) service, period of service; apprenticeship, experience

~, ис·пыт·а́тельн·ый probationary period

стаж·ёр·, -а, (*m*) apprentice, trainee, probationer

стаж·и́р·овать, -уют, (*imp*) serve apprenticeship

стаж·иро́вк·а, -и, (g pl) -вок·, (*f*) apprenticeship

ста́з·, -а, (*m*) (med) stasis

с·та́·ива·ть, -ют, (*imp*) melt away, thaw

стака́н·, -а, (*m*) glass, tumbler, beaker; sleeve, liner, cylinder; case, body (of projectile); (met) nozzle (of casting ladle)

~, за·па́ль·н·ый (gunn) primer, gaine, magazine fuse

~, лед·нико́в·ый (geol) ice column

~, на·правл·я́ющ·ий valve cage (diesel)

~, раз·ли́в·очн·ый (met) downtube (continuous casting)

стака́н·чик·, -а, (*m*) (*dim*) of **стака́н·;** beaker (laboratory)

~, про́б·н·ый (met) sampling spoon

ста́кер·, -а, (*m*) stacker

ста́ксел·ь, -я, (*m*) (naut) staysail

ста·л, (*past masc sing*) of **ста·ть** (*perf*) (q.v.), *and see* **станов·и́ться** (*imp*)

сталагм·и́т·, -а, (*m*) (geol) stalagmite

сталагмо́·метр·, -а, (*m*) (phys) stalagmometer

сталакт·и́т·, -а, (*m*) (geol) stalactite; (build) stalactited masonry

сталe·алюмин·иев·ый, -ая, -ое, (*a*) steel-cored aluminium, aluminium-steel, aluminum-steel (of elec conductors)

стале·бетóн·, -а, (*m*) steel-aggregate granolithic concrete

стале·вáр·, -а, (*m*) melter foreman (steelmaking)

стале·лит·éйн·ый, -ая, -ое, (*a*) steel founding/casting

~ за·вóд· (*m*) steel foundry

стале·мéд·н·ый, -ая, -ое, (*a*) (elec) copper-clad steel, steel-cored copper

стале·плав·ильн·ый, -ая, -ое, (*a*) steelmaking

~ за·вóд· (*m*) steelworks

стале·про·кáт·н·ый, -ая, -ое, (*a*) steel rolling, rolling

стá·ли, (*past pl*) of **стá·ть** (*perf*) (q.v.), *and see* **станов·иться** (*imp*)

стал·ийн·ый ден·ь (*m*) (com) lay day

сталинит·, -а, (*m*) (met) stalinit (cutting alloy); stalinit (plate glass)

стáл·ист·ый, -ая, -ое, (*a*) steely

с·тáлк·ивани·е, -я, (*n*) (*v n*) of **с·тáл·к·ива·ть;** ejection

с·тáлк·ивател·ь, -я, (*m*) knock-off arm, pusher, kicker (on a conveyor)

~ об·рéз·к·ов (met roll) crop pusher

с·тáлк·ива·ть, -ют, (*imp*) push/shove/ knock off/away; **-ся** (*pass*); collide, bump into, clash with; impinge

с·тáлк·ивающ·ий, -ая, -ее, (*pres part act*) of **с·тáлк·ива·ть**

с·тáлк·ивающ·ий болт· (*m*) stripper bolt (plastics moulding)

с·тáлк·ивающ·ийся, -аяся, -ееся, (*pres part act*) of **с·тáлк·ива·ть·ся**

с·тáлк·ивающ·ийся нейтрóн· (*m*) impinging neutron

стá·ло, (*past neut sing*) of **стá·ть** (*perf*) (q.v.) *and see* **станов·иться** (*imp*)

~ бы·ть so, consequently, it seems

стáл·ь, -и, (*f*) steel

~, авиациóн·н·ая aircraft steel

~, автомáт·н·ая free-cutting steel, screw-stock steel

~, азот·ирован·ая nitrided steel

~, азот·ируем·ая nitriding steel

~, арматýр·н·ая reinforcement bar (for concrete)

~, аустенит·н·ая austenitic steel

~, бессемéр·овск·ая converter/ Bessemer steel

~, болт·ов·áя screw-stock steel

~, бóр·ист·ая boron steel

~, булáт·н·ая damascus blade

стáл·ь

~, бур·ов·áя drill steel

~, быстро·рéж·ущ·ая high-speed steel

~, ворон·ён·ая blue/blued/burnished steel

~, высоко·кáчеств·енн·ая high-quality steel

~, высоко·кремн·ист·ая silicon steel

~, высоко·угле·рóд·ист·ая high-carbon steel

~, горяче·кáт·анн·ая hot-rolled steel

~, графитиз·ирован·ая graphitized steel

~, дик·ая wild steel

~, динáм·н·ая dynamo steel

~, дисперсиóн·н·о у·прóч·няем·ая dispersion-hardening steel, artificial ageing steel

~, до·эвтектóид·н·ая hypoeutectoid steel

~, дур·н·áя wild steel

~, жаро·прóч·н·ая creep-resisting steel

~, жаро·стóй·к·ая oxidation-resisting steel, scale-resisting steel ·

~, жаро·у·пóр·н·ая heat-resisting steel, steel resistant to creep and scale

~, за·эвтектóид·н·ая hypereutectoid steel

~, инструментáль·н·ая tool steel

~, калибр·óванн·ая bright-drawn steel

~, канáт·н·ая cable steel

~, кáчеств·енн·ая quality steel

~, кип·ящ·ая rimming steel

~, кисл·ая acid steel

~, кислото·у·пóр·н·ая acid-resisting stainless steel

~, конструкциóн·н·ая engineering/ machine/machinery steel

~, корóб·чат·ая channel bars (steel)

~, корыт·н·ая channel bars (steel)

~, котéль·н·ая boiler plate

~, крóв·ельн·ая roofing steel sheets

~, легир·óванн·ая alloy steel

~, ледебурит·н·ая ledeburite steel

~, лист·ов·áя steel plate, sheet, strip

~, ли·т·áя cast steel, foundry steel

~, магнит·н·ая magnet steel

~, магнито·мягк·ая magnetically soft steel

~, магнито·твёрд·ая magnetically hard steel

~, мало·магнит·н·ая low-magnetism steel

~, мало·угле·рóд·ист·ая mild steel

стал·ь

~, **марган·цо́в·ая** manganese steel

~, **марте́н·овск·ая** open-hearth steel

~, **машино·стро·и́тельн·ая** engineering/machine/machinery steel

~, **ме́д·ист·ая** copper steel, copper-bearing steel

~, **мя́гк·ая** mild steel

~, **наргарт·о́ванн·ая** cold-worked steel

~, **не·магни́т·н·ая** non-magnetic steel

~, **не·равно·бо́к·ая угл·ов·а́я** unequal steel angles

~, **не·ржав·е́ющ·ая** stainless steel

~, **машино·по·де́л·очн·ая** engineering/machine/machinery steel

~, **не·стар·е́ющ·ая** non-ageing steel

~, **низко·леги́р·ованн·ая** low-alloy steel

~, **низко·угле·ро́д·ист·ая** mild steel

~, **ни́кел·ев·ая** nickel steel

~, **нож·ев·а́я** cutlery steel

~, **о́бруч·н·ая** band steel

~ **обык·нове́нн·ого ка́честв·а** standard steel

~, **одно·с·ва́р·очн·ая** single shear steel

~, **окалино·сто́й·к·ая** oxidation-resistant steel

~, **оруд·и́йн·ая** gun-barrel steel

~, **осно́в·н·ая** basic steel

~ **о·со́б·ого на·знач·е́ни·я** special/alloy steel

~, **пере·жж·ённ·ая** burnt steel

~, **плак·иро́ванн·ая** clad steel

~, **полу·с·поко́й·н·ая** balanced/semi-killed steel

~, **прецизио́н·н·ая** special/alloy steel

~, **про·ка́т·н·ая** rolled steel

~, **про́филь·н·ая** structural shapes, steel sections

~, **пруж·и́нн·ая** spring steel

~, **рас·ки́сл·енн·ая** killed steel

~, **рессо́р·н·ая** spring steel

~, **ро́сл·ая** wild steel

~, **ряд·ов·а́я** standard/ordinary quality steel

~, **само·за·ка́л·ивающ·аяся** air-hardening steel

~, **с·ва́р·иваем·ая** weldable/welding steel

~, **с·ва́р·очн·ая** shear steel

~, **сегме́нт·н·ая** segmental bar steel

~, **сорт·ов·а́я** structural shapes, structural steel shapes; merchant bar (smaller sizes)

~, **специа́ль·н·ая** special/alloy steel

~, **с·поко́й·н·ая** killed steel

стал·ь

~, **стро·и́тельн·ая** structural steel

~, **судо·стро·и́тельн·ая** ship plate

~, **сыр·а́я** raw steel

~, **тепло·про́ч·н·ая** creep-resisting steel

~, **тепло·у·сто́й·к·ая** creep-resisting steel

~, **ти́гель·н·ая** crucible steel

~, **толсто·лист·ов·а́я** steel plate

~, **тома́с·овск·ая** basic/Thomas steel

~, **томл·ён·ая** blister steel

~, **трансформа́тор·н·ая** transformer steel

~, **угле·ро́д·ист·ая** carbon steel

~, **угл·ов·а́я** steel angles

~, **у·лучш·а́ем·ая** through-hardening steel

~, **универса́ль·н·ая широко·поло́с·н·ая** universe steel plates

~; **у·с·поко́·енн·ая** killed steel

~, **фасо́н·н·ая** structural shapes, sections

~, **фосфори́т·н·ая** high-phosphorus steel

~, **холодно·ка́т·ан·ая** cold-rolled steel

~, **холодно·тя́·нут·ая** cold-drawn steel

~, **хро́м·ист·ая** chromium steel

~, **хромо·ни́кел·ев·ая** nickel-chromium steel

~, **цемент·и́рованн·ая** cemented steel

~, **цемент·и́руем·ая** surface/case-hardening steel

~, **четвер·н·а́я** quaternary steel

~, **шарико·под·ши́п·ников·ая** ball-bearing steel

~, **широко·поло́с·н·ая универса́ль·н·ая** universal steel plates

~, **эле́ктро-** electric steel (made in an electric furnace)

~, **электро·техн·и́ческ·ая** electrical engineering steel, electrical steel

ста́ль·ник·, -а, (*m*) (bot) restharrow, *Ononis*

сталь·н·о́й, -а́я, -о́е, (*a*) steel

~ **про·ка́т·** (*m*) rolled steel products

стаме́с·к·а, -и, (*g pl*) **-с·ок·,** (*f*) firmer chisel (joiner's tool)

стам·ик·, -а, (*m*) (ocean) submerged ice barrier

стамин·о́ди·й, -я, (*m*) (bot) staminodium, staminode

стам·о́й лёд· (*m*) polar ice

стам·у́х·а, -и, (*f*) hummock aground (ice)

ста́н·, -а, *(m)* (met roll) mill; camp, halt, temporary halt; summer fishing settlement (Arctic); stature

~, автома́т·и́ческ·ий (met roll) automatic mill, plug mill

~, банда́ж·н·ый tire mill

~, волоч·и́льн·ый (met) wire draw bench

~, двадцати·вал·ко́в·ый twenty-roll cluster mill

~ до́ппель·ду́о double-duo mill, Dowlais mill

~, дресс·иро́вочн·ый skin-pass mill, temper mill

~, за·гото́в·очн·ый intermediate mill

~, за·гото́в·очн·ый сорт·ов·о́й billet mill

~, калибр·о́вочн·ый sizing mill

~ ква́рто four-high mill

~, колесо·про·ка́т·н·ый wheel mill

~, кольце·про·ка́т·н·ый hoop mill

~, крупно·со́рт·н·ый heavy section mill

~, ленто·про·ка́т·н·ый narrow strip mill

~, ле́нт·очн·ый narrow strip mill

~, лин·е́йн·ый (met roll) open train **ли́н·и·я ста́н·а** *(f)* (met roll) one stand of tandem mill; (met roll) open train

~, лист·ов·о́й sheet mill

~, мелко·со́рт·н·ый light section mill; merchant mill

~, не·реверси́в·н·ый non-reversing mill

~, не·пре·ры́в·н·ый continuous mill

~, об·жим·н·о́й cogging/roughing/primary mill

~, об·ка́т·н·ый (met roll) reeling machine

~, от·де́л·очн·ый finishing mill

~, пере·ка́т·очн·ый rerolling mill

~, пилигри́м·ов·ый Pilger mill

~, планета́р·н·ый planetary mill

~, плющ·и́льн·ый flatting mill

~, полос·ов·о́й strip mill

~, полу·не·пре·ры́в·н·ый semi-continuous mill

~, по·сле́д·овательн·о-воз·вра́т·н·ый cross-country mill

~, по·сле́д·овательн·ый tandem mill

~, про·ка́т·н·ый rolling mill

~, про·ши́в·н·о́й piercing mill

~, рас·ка́т·н·о́й reeling mill, reeling machine

~, реверси́в·н·ый reversing mill

ста́н

~, редукцио́н·н·ый reducing/reduction mill

~, ре́·ечн·ый tube-push bench

~, ре́льсо·ба́л·очн·ый rail-and-section mill

~ рокра́йт (met) Rockrite machine

~, с·ва́р·очн·ый welding mill (tube making)

~, с·дво́·енн·ый (met roll) two-stand tandem train

~, сорт·ов·о́й section mill

~, средне·со́рт·н·ый medium-section mill

~, сутуночно·за·гото́в·очн·ый sheet-bar-mill

~ та́ндем (met roll) tandem train

~, трёх·вал·ко́в·ый three-high mill

~, трёх·вал·ко́в·ый у·длин·и́тель·н·ый Assel mill

~ три́о three-high mill

~, трубо·волоч·и́льн·ый tube draw-bench

~, трубо·с·ва́р·очн·ый pipe-welding mill

~, универса́ль·н·ый universal mill

~, фольго·про·ка́т·н·ый foil rolling mill

~, форм·о́вочн·ый forming mill (tube making)

~, черн·ов·о́й roughing mill

~, шести·вал·ко́в·ый cluster mill, 6-roll cluster mill

~, широко·поло́с·н·ый wide-strip mill

~ шту́ч·н·ой про·ка́т·к·и, тонко··лист·ов·о́й pack mill

станда́рт·, -а, *(m)* standard

~, госуда́р·ственн·ый обще·сою́з·-н·ый State standard, State standard specification; (often written GOST)

~, ленингра́д·ск·ий Leningrad standard hundred (165 cu. ft. of timber)

~ част·от·ы́ (rad) primary frequency standard

стандартиз·а́тор·, -а, *(m)* standardizer

стандартиз·а́ци·я, -и, *(f)* standardization

стандартиз·и́р·овать, -уют, *(imp and perf)* standardize

станда́рт·н·ый, -ая, -ое, *(a)* standard, standardized, conforming to standard specification

~ атмосфе́р·а *(f)* (phys) standard atmosphere

~ водо·из·мещ·е́ни·е *(n)* (shipb) standard displacement

станда́рт·н·ый, -ая, -ое
~ **дом·** (*m*) standardized house, standard-design house
~ **кирпи́ч·** (*m*) standard (size of brick)
ста́·н·ет, (*fut 3rd sing*) of **ста·ть** (*perf*) (q.v.), *and see* **станов·и́ться** (*imp*)
стан·и́н·а, -ы, (*f*) mount, frame, bed; bed (of machine tool), frame (of pile driver), housing (of rolling mill), stand (of cine projector), trail (of gun)
~ **за·кры́·т·ого ти́п·а** (met roll) closed-top housing
~ **ла́мп·ы** (cinema) lampholder
~ **но́ж·ниц·** (met) shear frame/housing
~ **от·кры́·т·ого ти́п·а** (met roll) open-top housing
~, **раз·дви́ж·н·а́я** sliding bed; (gunn) (*pl*) split trail
стани́ол·ь, -я, (*m*) tin foil paper, (pop) silver paper
стандерс·, -а, (*m*) (naut) standard, stand, davit socket
стан·ко́в·ый, -ая, -ое, (*a*) of **стан·о́к·;** mounted, standing, free-standing
станко·стро·е́ни·е, -я, (*n*) machine-tool engineering/building
станко·стро·и́тельн·ый, -ая, -ое, (*a*) machine-tool engineering/construction/building
~ **про·мы́шл·енност·ь** (*f*) machine-tool industry
станн·а́н·, -а, (*m*) (chem) stannane
станн·а́т·, -а, (*m*) (chem) stannate
станн·и́д·, -а, (*m*) (met) stannide, intermetallic tin compound
станн·иле́н·, -а, (*m*) (chem) stannylene
станн·и́н·, -а, (*m*) (min) stannite
станн·и́т·, -а, (*m*) (min, chem) stannite
станн·о́ни·й, -я, (*m*) (chem) stannonium
станн·о́нов·ая кисл·от·а́ (*f*) (chem) stannonic acid
станов·и́ться, -ятся, (*imp*) stand, take up position; stand up; stop, halt; freeze over (of rivers); (+ *instr*) become, get, grow
~ **к ору́д·и·ям** (mil) man the guns, (naut) close up at the guns
~ **на коле́н·и** kneel
~ **на я́кор·ь** (naut) anchor, let go an anchor
~ **уч·и́тел·ем** become a teacher
стан·ов·о́й, -а́я, -о́е, (*a*) of **стан·;** central
~ **бо́ч·к·а** (*f*) mooring buoy

стан·ов·о́й, -а́я, -о́е
~ **я́кор·ь** (*m*) (naut) bower anchor
стан·о́к·, -нка́, (*m*) (*see also under associated adjectives*); machine tool, machine; bench; (gunn) mounting; (rocket) launcher frame; (agr) stall, cradle, crate, crush
~ **-автома́т·, -а,** (*m*) automatic, automatic lathe/loom/press/machine
~, **ве́рх·н·ий** (gunn) saddle
~, **винто·ре́з·н·ый** screw-cutting lathe
~, **-кач·а́лк·а, -и,** (*f*) (oil) pumping jack
~, **ковро·тк·а́цк·ий** (text) carpet loom
~, **монта́ж·н·ый** engine erecting stand
~, **ни́ж·н·ий** (gunn) carriage body
~, **печа́т·н·ый** printing press
~, **пи́ль·н·ый** saw bench, saw
~, **раке́т·н·ый** (naut) rocket apparatus
~, **револьве́р·н·ый** capstan/turret lathe
~, **тк·а́цк·ий** (text) loom
~, **ток·а́рн·ый** lathe
стан·о́чек·, -чк·а, (*m*) (*dim*) of **стан·о́к·;** rack, stand
стан·о́чник·, -а, (*m*) machine-tool operator
стан·о́чн·ый, -ая, -ое, (*a*) of **стан·о́к·**
~ **ли́н·и·я** (*f*) transfer/automation line
ста́·нут, (*fut 3rd pl*) of **ста·ть** (*perf*) (q.v.), *and see* **станов·и́ться** (*imp*)
станцио́н·н·ый, -ая, -ое, (*a*) of **ста́нци·я**
ста́нци·я, -и, (*f*) (*see also under associated words*); station, set, equipment, exchange (telephones)
~ **бы́стр·ой с·вя́з·и** toll exchange, CLR office (telephones)
~, **кон·е́чн·ая** terminal station
~, **радиолокацио́н·н·ая** (*and see under adj*); radar, radar set/equipment
~, **сорт·иро́вочн·ая** (rail) shunting/marshalling yard
~, **телефо́н·н·ая** telephone exchange
~, **узл·ов·а́я** (rail) junction
~, **центра́ль·н·ая с·ма́з·очн·ая** (met roll) oil cellar
~, **шир·о́тн·ая** (meteor) latitude observatory
~, **электр·и́ческ·ая** electric power station; electric generating set
ста́·н·ьте, (*pl imper*) of **ста·ть** (*perf*) (q.v.) *and see* **станов·и́ться** (*imp*)
ста́пел·ь, -я, (*nom pl*) **-я́,** (*g pl*) **-е́й,** (*m*) (shipb) building slip/berth; (a/c) assembly jig
~, **бок·ов·о́й** side-launching berth/slip
~ **-конду́ктор·, -а,** (*m*) ship assembling jig

ста́пел·ь
~, **по·дви́ж·н·о́й** mobile building berth
~ **-по·сте́л·ь, -и,** (*f*) inverted building berth
~, **про·до́ль·н·ый** end-on slip/berth
ста́пель·н·ый, -ая, -ое, (*a*) of **ста́-пел·ь**
~ **ме́ст·о** (*n*) building berth
~ **лес·а́** (*pl*) (shipb) scaffolding
~ **площа́д·к·и** (*pl*) (shipb) grid, gridiron
с·та́пл·ива·ть, -ют, (*imp*) melt together, mix (by melting)
с·та́пт·ыва·ть, -ют, (*imp*) tread/wear down
стар·а́ни·е, -я, (*n*) endeavour, effort
стар·а́тел·ь, -я, (*m*) (min) free miner, private prospector
стар·а́тельн·ый, -ая, -ое, (*a*) assidu-ous, conscientious, keen, painstaking
стар·а́тельск·ий, -ая, -ое, (*a*) of **стар·а́тел·ь**
стар·а́·ться, -ются, (*imp*) endeavour, try, seek to; (phys) tend to/toward
стар·е́йш·ий, -ая, -ее, (*a*) the oldest
стар·е́ни·е, -я, (*n*) ageing, aging; age hardening (non-ferrous metals)
  вре́м·я стар·е́ни·я (*n*) ageing/ aging time
~, **деформацио́н·н·ое** (met) strain ageing
~, **есте́ств·енн·ое** natural ageing
~ **за·ка́л·к·и** (met) quench ageing
~, **иску́с·ственн·ое** artificial ageing
~, **механ·и́ческ·ое** (met) strain ageing
~, **норма́ль·н·ое** (met) quench ageing
~, **свет·ов·о́е** light ageing
~, **тепл·ов·о́е** heat ageing
~, **термо·цикл·и́ческ·ое** (met) ther-mocyclic ageing, thermocycling
стар·е́·ть, -ют· (*imp*) age, grow old
стар·и́к·, -а́, (*m*) old man
стар·и́нн·ый, -ая, -ое, (*a*) ancient, antique; old-fashioned
ста́р·ить, -ят, (*imp*) age, make old, mature
ста́р·иц·а, -ы, (*i*) **-ей,** (*f*) (geog) oxbow-lake
старли́т·, -а, (*m*) (min) starlite
старнкни́ц·а, -ы, (*f*) (naut) sternpost knee
старнпо́ст·, -а, (*m*) (naut) sternpost, propeller post
старнти́мберс·, -а, (*m*) (naut) stern timber
ста́р·о- (*root*) old
старо·да́в·н·ий, -яя, -ее, (*a*) ancient

старо·за́·леж·н·ый, -ая, -ое, (*a*) (agr) long-fallow
старо·мо́д·н·ый, -ая, -ое, (*a*) old-fashioned, outmoded
старо·обра́з·н·ый, -ая, -ое, (*a*) old-looking
ста́р·ост·ь, -и, (*f*) old age
ста́рт·, -а, (*m*) start; (air) takeoff; take-off strip; (rocket) launching; launching platform
  ли́н·и·я ис·полн·и́тельн·ого ста́р-т·а (*f*) (air) runway threshold
  ли́н·и·я пред·вар·и́тельн·ого ста́рт·а (*f*) (air) marshalling line
~, **пред·вар·и́тельн·ый** (air) mar-shalling
ста́ртер·, -а, (*m*) starter, self-starter, starting/starter motor/switch; (air) marshaller; starter (sport)
~, **воз·ду́ш·н·ый** compressed-air starter
~, **инерцио́н·н·ый** inertia starter
~, **патро́н·н·ый** cartridge starter
~, **пневмат·и́ческ·ий** air starting motor
~ **-фи́нишер·, -а,** (*m*) (air) runway controller
старт·ова́ть, -у́ют, (*imp and perf*) start; (air) take off
ста́рт·ов·ый, -ая, -ое, (*a*) start, starting; (air) takeoff; (rocket) launch-ing
~ **дви́г·ател·ь** (*m*) starter motor; (rocket) booster, launching booster
~ **кома́нд·а** (*f*) (rocket) firing crew
~ **ко́мплекс·** (*m*) (rocket) launching station
~ **кома́нд·н·ый пункт·** (*m*) (air) runway control caravan
~ **раке́т·н·ый дви́г·ател·ь** (*m*) (a/c) rocket-assisted take-off unit
~ **у·стано́в·к·а** (*f*) (rocket) launcher
старто·кома́нд·н·ый, -ая, -ое, (*a*) (a/c) takeoff-control
~ **пункт·** (*m*) (air) takeoff control point
старто·финиш·н·ая площа́д·к·а (*f*) (air) take-off and landing area
старт·стоп·н·ый аппара́т· (*m*) (tele-com) start-stop teleprinter
ста́р·ческ·ий, -ая, -ое, (*a*) senile, aged
ста́р·ш·е- (*component*) senior
старше·ку́рс·ник·, -а, (*m*) final year/ term/course student, passing-out/ qualifying classman

ста́р·ш·ий, -ая, -ее, (*a*) older, elder; senior; oldest, eldest; (*m decl as adj*) (mil) senior officer

~ **коэффицие́нт·** (*m*) leading coefficient

~ **по·мо́щ·ник·** (*m*) chief officer (merchant navy), executive officer (navy)

стар·шин·а́, -ы́, (*nom pl*) **стар·ши́н·ы,** (*m*) (mil) sergeant major (Brit), master sergeant (U.S.); petty officer (naval)

~ **втор·о́й стат·ь·и́** leading seaman

~ **пе́рв·ой стат·ь·и́** petty officer

стар·ши́нск·ий, -ая, -ое, (*a*) of **стар·шин·а́**

стар·шинств·о́, -а́, (*n*) seniority; **стар·ши́нств·о, -а,** (*n*) non-commissioned rank; petty-officers' rate

ста́р·ый, -ая, -ое, (*a*) old

с·та́ск·ива·ть, -ют, (*imp*) drag, pull, haul, tow; drag/pull off/away

с·тас·ова́ть, -у́ют, (*perf*) shuffle (cards etc.)

стассфурти́т·, (*m*) (geol) stassfurtite

стат·е́йн·ый, -ая, -ое, (*a*) of **стат·ь·я́**

стати́в·, -а, (*m*) stand, rack

ста́т·ик·а, -и, (*f*) statics

стати́ст·, -а, (*m*) (theat) extra

стати́ст·ик·, -а, (*m*) statistician

стати́ст·ик·а, -и, (*f*) statistics; (*g sing*) of **стати́ст·ик·**

статист·и́ческ·ий, -ая, -ое, (*a*) statistical

ста́тиц·е, -а, (*n*) (bot) thrift, *Statice*

стат·и́ческ·ий, -ая, -ое, (*a*) static, statical

~ **давл·е́ни·е** (*n*) (aerodynam) static pressure

~ **о·сто́й·чивост·ь** (*f*) static/statical stability

~ **тя́г·а** (*f*) (aerodynam) static thrust

стато·бла́ст·, -а, (*m*) (zool) statoblast

стато·ли́т·, -а, (*m*) (biol) statolith

ста́тор·, -а, (*m*) stator

стато·ско́п·, -а, (*m*) (a/c) statoscope, pressure altimeter

стато·спо́р·а, -ы, (*f*) (bot) statospore

стато·ци́ст·а, -ы, (*f*) (zool) statocyst

ста́т·ск·ий, -ая, -ое, (*a*) = **шта́т·ск·ий** (q.v.)

стату́т·, -а, (*m*) statute

ста́ту·я, -и, (*f*) statue

ста́·ть, -нут, (*perf*) *see* **станов·и́ть·ся;** begin to (+ *infinitive*); -ся happen, become

стат·ь·я́, -и́, (*f*) article; clause, item; rate (naval); (agr) (*pl*) points (e.g. of horse)

~, **перед·ов·а́я** leader, leading article

стафило·ко́кк·, -а, (*m*) (bact) Staphylococcus

стафило́м·а, -ы, (*f*) (med) staphyloma

стафило·ррафи·я, -и, (*f*) staphylorrhaphy (surgery)

стациона́р·, -а, (*m*) permanent establishment, permanently located institution; (med) infirmary, hospital; fixed mounting/base

стациона́р·ност·ь, -и, (*f*) stationariness, steady state, steadiness

стациона́р·н·ый, -ая, -ое, (*a*) of **стациона́р·;** stationary, fixed, permanent; (phys) steady-state, steady

~ **боль·н·о́й** (*m*) hospital patient, in-patient

~ **леч·е́ни·е** (*n*) hospitalization, hospital treatment

~ **проце́сс·** (*m*) (phys) steady-state process

~ **реш·е́ни·е** (*n*) steady-state solution

~ **со·сто·я́ни·е** (*n*) (phys) steady state; (math) stationary state

~ **фронт·** (*m*) stationary front

стациони́р·, -а, (*m*) (naut) guard ship

с·тач·а́·ть, -ют, (*perf*) stitch

с·та́ч·ечн·ый, -ая, -ое, (*a*) of **с·та́ч·к·а**

~ **фонд·** (*m*) strike fund

с·та́ч·ива·ть, -ют, (*imp*) stitch; sharpen; grind down (in sharpening etc.)

ста́ч·к·а, -и, (*g pl*) **-ч·ек·,** (*f*) strike (industrial)

~, **ди́к·ая** unofficial strike, wild-cat strike

~, **италь·я́нск·ая** work to rule (industrial dispute); go slow (industrial dispute); overtime ban (industrial dispute)

~, **по́ль·ск·ая** stay-in strike; (min) stay-down strike

~, **сид·я́ч·ая** sit-down strike

~ **солида́р·ност·и** strike in sympathy, supporting strike

с·та́щ·енн·ый, -ая, -ое, (*past part pass*) *see* **с·та́ск·ива·ть**

с·тащ·и́ть, -а́т, (*perf*) *see* **с·та́ск·ива·ть**

с·та́·ют (*fut 3rd pl*) of **с·та́·ять**

ста́·я, -и, (*f*) (orn) flock, flight, covey; (fish) shoal, school; (zool) pack

с·та́·ять, -ют, (*perf*) thaw, melt away

ствол·, -а́, (*m*) stem, stalk, trunk; shaft; shank; branch (of fire hose); (gunn) barrel, body

ствол

~, в·клад·н·о́й   (gunn) sub-calibre tube

~, вс·по·мог·а́тельн·ый   (gunn) sub-calibre tube

~, лафе́т·н·ый   fire-fighting monitor

~, пере·крыв·н·о́й   emergency-type diffuser branch (of fire hose)

~ -рас·пыл·и́тел·ь, -я,   (m) diffuser branch (of fire hose)

~, свобо́д·н·ый   (gunn) loose barrel

~ скваж·ин·ы   borehole

~, с·крепл·ённ·ый   hooped gun body

~, ша́хт·н·ый   (min) shaft, main shaft

ствол·и́ст·ый, -ая, -ое,   (a) stalky; high-branching (of trees)

ствол·ов·о́й, -а́я, -о́е,   (a) of **ствол·;** (m decl as adj) (min) cager, hanger-on, onsetter

~ артилле́р·и·я   (f) conventional artillery

ство́л·ов·ый, -ая, -ое,   (a) = **ство·л·ов·о́й**

стволь·н·ый, -ая, -ое,   (a) **ствол·**

~ коро́б·к·а   (f) breech (of rifle)

с·тво́р·, -а,   (m) = **с·тво́р·к·а;** (surv) transit line, (nav) leading line; section line (of rivers etc.)

~, на·правл·я́ющ·ий   (nav) clearing line

    при·вод·и́ть на с·тво́р·   (m) (surv) bring (two objects) into line, line up (two objects)

с·твора́ж·ива·ть, -ют,   (imp) curdle

с·твор·и́ть, -я́т,   (perf) see **с·твор·я́·ть**

с·тво́р·к·а, -и,   (g pl) -р·ок·, (f) = **створ·;** fold, flap, door, leaf (of door etc.); (bot) glume; (zool) valve; (pl) twin doors/gates/flaps/windows etc.

~ тормоз·н·а́я   (a/c) brake flap

с·тво́р·н·ый, -ая, -ое,   (a) of **створ·**

~ знак·и   (pl) (nav) leading marks

~ радио·мая́к·   (m) radio-range beacon

~ столб·ы́   (pl) leading line posts, transit marks

с·творо́ж·енн·ый, -ая, -ое,   (past part pass) of **с·творо́ж·ить;** curdled

с·творо́ж·ить, -ат,   (perf) curdle

с·тво́р·чат·ый, -ая, -ое,   (a) folding; (biol) valved, valvular

~ воро́т·а   (pl) double-leaf gate

с·твор·я́·ть, -ют,   (imp) get in line/ transit

стеа·пси́н·, -а,   (m) (biochem) steapsin

стеар·а́т·, -а,   (m) (biochem) stearate

стеар·и́н·, -а,   (m) (chem) stearin, stearine, technical stearic acid; glycerol tristearate

стеар·и́нов·ый, -ая, -ое,   (a) (chem) stearin

~ альдеги́д·   (m) (chem) stearaldehyde

~ кисл·от·а́   (f) (chem) stearic acid

стеаро́л·ов·ая кисл·от·а́   (f) stearolic acid

стеат·и́т·, -а,   (m) (min) steatite, soapstone

стеат·о́м·а, -ы,   (f) (med) steatoma

стеато·па́т·и·я, -и,   (f) (med) steatopathy

стеато·рре́·я, -и,   (f) (med) steatorrhoea, steatorrhea

стебел·ёк·, -ль·к·а́,   (m) (dim) of **сте́бел·ь**

сте́бел·ь, -бл·я,   (nom pl) -бл·и, (g pl) -бл·е́й,   (m) stem, stalk, shank, shaft, haulm, culm, blade (e.g. of grass)

стебель·ко́в·ый, -ая, -ое,   (a) stalked, stipitate, caulescent

сте́бель·н·ый, -ая, -ое,   (a) of **сте́бел·ь** (bot) caulescent (having a stem); cauline (appertaining to a stem)

стебельчато·гла́з·ые, -ых,   (pl decl as adj) (zool) *Stylommatophora*

стебе́ль·чат·ый, -ая, -ое,   (a) stalky, stalk-like, stipiform

стебле·ви́д·н·ый, -ая -ое,   (a) stem-shaped, stipiform, cauliform

стебл·ев·о́й, -а́я, -о́е,   (a) stem, stalk, (bot) cauline

стебл·ёв·ый, -ая, -ое,   (a) = **стеб·л·ев·о́й**

стебле·корн·ев·о́й, -а́я, -о́е,   (a) (bot) caulorrhizous

стебле·объ·ёмл·ющ·ий, -ая, -ее,   (a) (bot) amplexicaul

стебле·пло́д·н·ый, -ая, -ое,   (a) (bot) caulocarpic

стебле·ру́б·, -а,   (m) (agr) stalk-cutter

стег·а́·ть, -ют,   (imp) quilt; whip, lash

стег·н·у́ть, -у́т,   (perf) whip, lash

стего·за́вр·, -а,   (m) (pal) stegosaur, *Stegosaurus*, (pl) stegosaurs, stegosaurians, *Stegosauria*

стего·цефа́л·ы, -ов,   (m pl) (pal) stegocephalians, *Stegocephalia*

стёж·к·а, -и,   (g pl) -ж·ек·, (f) (v n) of **стег·а́·ть;** seam

стеж·о́к·, -ж·к·а́,   (m) stitch

с·тек·а́ни·е, -я,   (n) (v n) see **с·те·к·а́·ть;** run-off; leakage

~ за·ря́д·а   (phys) charge leakage

с·тек·а́тел·ь, -я,   (m) deflector

с·тек·а́·ть, -ют,   (imp) flow, run (of liquids etc.); flow down/off; **-ся** flow together

стекл·ене́·ть, -ют, (*imp*) become glassy/ vitrified

стекл·о́, -а́, (*nom pl*) **стёкл·а,** (*g pl*) **стёкол·,** (*n*) glass; glass, piece of glass

~, арм·и́рованн·ое wired glass

~, астроном·и́ческ·ое (a/c) astro-dome

~, без·о·с·ко́л·очн·ое splinterproof glass

~, бе́м·ск·ое Bohemian glass

~, бо́р·н·ое borax glass

~, ветр·ов·о́е (M/T) windscreen, wind-shield

~, водо·ме́р·н·ое water-gauge glass

~, глуш·ённ·ое translucent glass

~, жаро·у·по́р·н·ое heat-resistant glass

~, жи́дк·ое (chem) water glass

~, зерка́ль·н·ое plate glass

~ Кли́нгер·а Klinger glass (water gauge glass)

~, ма́т·ов·ое ground glass

~, на·клад·н·о́е cased glass

~, не·темн·е́ющ·ее non-browning glass

~, опа́л·ев·ое opal glass

~, опт·и́ческ·ое optical glass

~, орган·и́ческ·ое organic glass, transparent plastic, perspex

~, по·кро́в·н·ое cover glass (micro-scopy)

~, пред·ме́т·н·ое microscope slide, object carrier

~, рас·твор·и́м·ое (chem) water glass

~, свин·цо́в·ое lead glass

~, смотр·ов·о́е sight glass

~, стро·и́тельн·ое structural glass

~, то́лст·ое зерка́ль·н·ое plate glass

стекло·бо́·й, -я, (*m*) cullet

стекл·ова́льн·ый, -ая, -ое, (*a*) glass-making

стекл·ова́льщик·, -а, (*m*) glassmaker

стекл·ова́ни·е, -я, (*n*) vitrification; (plastic) glass transition, freezing

    температу́р·а стекл·ова́ни·я (*f*) (plast) glass-transition temperature/ point, glass point

    то́ч·к·а стекл·ова́ни·я (*f*) (plast) glass-transition point, glass point

стекло·ва́р·, -а, (*m*) glassmaker

стекло·вар·е́ни·е, -я, (*n*) glassmaking

стекло·ва́р·енн·ый, -ая, -ое, (*a*) glassmaking

~ печ·ь (*f*) glassmaking furnace

стекло·ва́т·а, -ы, (*f*) glass wool

стекл·ова́т·ый, -ая, -ое, (*a*) glassy, vitreous, hyaline

~ пол·ев·о́й шпат· (*m*) (min) glassy feldspar

стекл·ова́т·ый, -ая, -ое

~, по́лн·ост·ью (*a*) holohyaline

стекл·ова́ться, -у́ются (*imp*) vitrify; (plast) freeze, become glassy, pass the glass transition point

стекло·ви́д·н·ый, -ая -ое, (*a*) glassy, vitreous, hyaline

стекло·волокн·о́, -а́, (*n*) glass fibre/ fiber

стекло́·граф·, -а, (*m*) (print) collotype press

стекло·гра́ф·и·я, -и, (*f*) (print) collo-type

стекло·де́л·, -а, (*m*) glassmaker

стекло·де́л·ательн·ый, -ая, -ое, (*a*) glass-shaping/moulding/making

стекло·ду́·в·, -а, (*m*) glass blower

стекло·жгу́т·, -а, (*m*) glass roving

стекло·лако·тк·а́н·ь, -и, (*f*) (elec) impregnated glass cloth

стекло·ма́сс·а -ы, (*f*) molten glass

стекло·ма́т·, -а, (*m*) glass mat

стекло·микани́т·, -а, (*m*) fibre-glass-backed micanite

стекло·обра́з·н·ый, -ая, -ое, (*a*) glasslike, glassy, vitreous; (ceram) frit

стекло·образ·у́ющ·ий, -ая, -ее, (*a*) glass-forming

стекло·о·чист·и́тел·ь, -я, (*m*) (M/T) windscreen wiper

стекло·пла́в·, -а, (*m*) glassmaker

стекло·плав·и́льн·ый, -ая, -ое, (*a*) glass, glassmaking

~ за·во́д· (*m*) glass works

стекло·подъ·ём·ник·, -а, (*m*) (M/T) window mechanism

стекло·ре́з·, -а, (*m*) glass cutter

стекло·рого́ж·к·а, -и, (*g pl*) **-ж·ек·,** (*f*) glass-fibre mat, fibreglass mat

стекло·слюд·ян·о́й, -а́я, -о́е, (*a*) mica-and-fibreglass

стекло·текстоли́т·, -а, (*m*) bonded glass mat, fabric glass laminate

стекло·тк·а́н·ь, -и, (*f*) glass cloth

стекло·у·проч·ни́тел·ь, -я, (*m*) (plast) glass-fibre filler

стекло·фи́р·н·ый, -ая, -ое, (*a*) (geol) vitrophyric

стекло·форм·о́вочн·ый, -ая, -ое, (*a*) glass-moulding/shaping

~ маши́н·а (*f*) glass moulding/mold-ing/shaping machine

стекло·цеме́нт·, -а, (*m*) prestressed fibreglass-reinforced cement

стёкл·ышк·о, -а, (*g pl*) **-шек·,** (*n*) glass, piece of glass; enamel bead

стекл·я́нн·ый, -ая, -ое, (*a*) glass
~ бо́·й (*m*) cullet
~ кры́ш·к·а (*f*) glass lid/cover; verge glass (compass)
~ шлиф· (*m*) glass grind
стёкол·, (*g pl*) of **стекл·о́**
стеко́ль·н·ый, -ая, -ое, (*a*) glass-making, glass; (build) glazing
~ бо·й (*m*) cullet
стеко́ль·щик·, -а, (*m*) glazier
сте́л·а, -ы, (*f*) (bot) stele; (arch) stele
сте́л·ем·ый, -ая, -ое, (*pres part pass*) of **стл·а́ть**
сте́л·ет, (*pres 3rd sing*) of **стл·а́ть**
сте́л·ечн·ый, -ая, -ое, (*a*) inner-sole (of shoes)
стел·и́те, (*pl imper*) of **стл·а́ть**
стелла́ж·, -а́, (*i*) **-о́м,** (*m*) shelving, bins, stowage, stillage, cradle, rack, racking, grid
стелл·ари́т·, -а, (*m*) (min) stellarite
стелли́т·, -а, (*m*) Stellit (Co-base cutting alloy)
сте́л·ь, -и, (*f*) (bot) stele
сте́ль·к·а, -и, (*g pl*) **-л·ек·,** (*f*) inner sole, sock (of shoe)
с·те́ль·ност·ь, -и, (*f*) pregnancy (of cows), being in calf
стелю́г·а, -и, (*f*) stage (scaffolding)
сте́л·ют, (*pres 3rd pl*) of **стл·а́ть**
сте́л·ющ·ий, -ая, -ее, (*pres part act*) of **стл·а́ть;** *see also* **сте́л·ющ·ийся**
сте́л·ющ·ийся, -аяся, -ееся, (bot) decumbent, creeping
стел·я́, (*pres gerund*) of **стл·а́ть**
стел·я́рн·ый, -ая, -ое, (*a*) of **стел·а́** and **стел·ь**
стём·, -а, (*m*) (shipb) stem
стем·ки́ц·а, -ы, (*f*) (shipb) stemson
стемсо́н·, -а, (*m*) (shipb) stemson
стён·, -а, (*m*) (phys) sthene, sn (unit of force)
стен·а́, -ы́, (*acc sing*) **стён·у,** (*nom pl*) **стён·ы,** (*d pl*) **стен·а́м,** (*f*) wall; *see also* **стён·к·а**
~, бок·ов·а́я side wall
~, за́д·н·яя end wall
~, под·по́р·н·ая retaining wall
стен·газе́т·а, -ы, (*f*) wall newspaper
стёнд·, -а, (*m*) (mech) test bench/stand/range/bed; stand, bench; (print) bank
~, ис·пыт·а́тельн·ый test bed/bench/stand
~, контро́ль·н·ый test bed/bench/stand
~, лет·а́ющ·ий (a/c) jet-powered flat riser, flying bedstead

стёнд
~, по·воро́т·н·ый tilting stand
~, ре́льс·ов·ый (rocket) test sled
~, с·ва́р·очн·ый welding stand
стёндер·, -а, (*m*) stand pipe
стён·к·а, -и, (*g pl*) **-н·ок·,** (*f*) wall; quay wall; web (of rolled section); (anat) wall, paries
~, бок·ов·а́я side casing (of boiler)
~, гимнаст·и́ческ·ая wall bars
~, ло́ж·н·ая (zool) pseudotheca
~, на·кло́н·н·ая flaring wall
~, от·ко́с·н·ая head wall
~, под·око́н·н·ая (build) breast
~, под·по́р·н·ая breast wall, retaining wall
~ фор·ша́хт·ы (min) collar wall
~, шпунт·ов·а́я sheet-piling quay wall
стен·ме́тр·, -а, (*m*) kilojoule (unit of work)
стен·ни́к·, -а́, (*m*) (bot) candituft, *Iberis*
стен·н·о́й, -а́я, -о́е, (*a*) wall, mural; (biol) parietal
стено·ба́т·н·ый, -ая, -ое, (*a*) (zool) stenobathic
стено·био́нт·н·ый, -ая, -ое, (*a*) (zool) stenobiontic
стен·ов·о́й, -а́я, -о́е, (*a*) walling, wall
стено·гали́н·н·ый, -ая, -ое, (*a*) (zool) stenohaline
стено·гра́мм·а, -ы, (*f*) shorthand record
стено́·граф·, -а, (*m*) stenographer; (ent) stenograph bark beetle, *Ipo sexdentatus*
стено·гра́ф·и·я, -и, (*f*) stenography, shorthand
стено́·з·, -а, (*m*) (med) stenosis
стено·мо́рф·н·ый, -ая, -ое, (*a*) (bot) stenomorphic
стен·о́п·, -а, (*m*) (phot) stenopaic/pinhole camera
стено·синклина́л·ь, -и, (*f*) (geol) stenosyncline
стено·те́рм·н·ый, -ая, -ое, (*a*) (biol) stenothermic
стено·тип·и́ст·, -а, (*m*) (print) stenotypist
стено·то́п·н·ый, -ая, -ое, (*a*) (biol) stenotopic
стено·тро́н·, -а, (*m*) (elec) stenotron
стено·фа́г·и·я, -и, (*f*) (zool) stenophagy
стено·фот·и́ческ·ий, -ая, -ое, (*a*) (bot) stenophotic

сте́ньг·а, -и, (*f*) (naut) topmast

сте́ньг·ов·ый, -ая, -ое, (*a*) top-mast

степен·н·о́й, -а́я, -о́е, (*a*) *see* **степе́н·н·ый**

степе́н·н·ый, -ая, -ое, (*a*) of **сте́пен·ь**

~ вы́·чет· (*m*) (math) power residue

~ за·ви́с·имост·ь (*f*) power relationship, exponential dependence

~ за·ко́н· (*m*) power/exponential law

~ крив·а́я (*f*) (math) exponential curve

~ ряд· (*m*) (math) power series

~ фу́нкци·я (*f*) (math) power function

сте́пен·ь, -и, (*g pl*) **-е́й,** (*f*) degree, extent, stage; ratio; (math) power, index (of a radical), order (of an equation)

   в знач·и́тельн·ой сте́пен·и to a considerable extent, in large part, largely, greatly

   в сте́пен·и 2 to the power of 2, squared

   воз·вод·и́ть в сте́пен·ь raise to the power of ...

~ вос·становл·е́ни·я давл·е́ни·я ram efficiency (ram jet)

~ диссоциа́ци·и (phys, chem) degree of dissociation

~ до́ктор·а doctorate

~ за·раж·е́ни·я (nucl) contamination dose

~ за·тух·а́ни·я (phys) degree of damping/attenuation; (autom) damping ratio

~ ко́рн·я (math) radical index

~ на·полн·е́ни·я (dynam) degree of admission; (plast) filler content

~ пере·кры́т·и·я engagement factor (gears)

   по·каз·а́тел·ь сте́пен·и (*m*) (math) exponent

~ по·раж·е́ни·я (nucl) absorbed dose

~, пре·вос·хо́д·н·ая (gram) superlative degree

~ рас·шир·е́ни·я (ICE) expansion ratio, divergence ratio (nozzles etc.)

~ свобо́д·ы (mech) degree of freedom

~ свобо́д·ы, вращ·а́тельн·ая rotational degree of freedom

~ свобо́д·ы, колеб·а́тельн·ая vibrational degree of freedom

~ свобо́д·ы, по·ступ·а́тельн·ая (mech) translatory degree of freedom

~ с·жа́т·и·я (ICE) compression ratio; (plast) compression

~ то́ч·к·и (math) power of a point

~, уч·ён·ая university degree

сте́пен·ь

~ у·ширен·и·я degree of expansion; (ICE) compression ratio

~, четвёрт·ая (math) fourth power

степ·н·о́й, -а́я, -о́е, (*a*) (geog) steppe

сте́пс·, -а, (*m*) (naut) mast step; capstan socket

сте́п·ь, -и, (*g pl*) **-е́й,** (*f*) (geog) steppe

стерадиа́н·, -а, (*m*) (math) steradian

стерег·и́те, (*pl imper*) of **стере́ч·ь**

стерег·у́щ·ий, -ая, -ее, (*pres part act*) of **стере́ч·ь**

стереж·а́, (*pres gerund*) of **стере́ч·ь**

стереж·ённ·ый, -ая, -ое, (*past part pass*) *see* **стере́ч·ь**

стереж·ёт, (*pres 3rd sing*) of **стере́ч·ь**

стере·и́д·, -а, (*m*) (bot) stereid, stone cell

стерео·ба́т·, -а, (*m*) (arch) stereobate

стерео·гра́мм·а, -ы, (*f*) stereogram, stereograph

стерео·граф·и́ческ·ий, -ая, -ое, (*a*) stereographic

стерео·гра́ф·и·я, -и, (*f*) stereography

стерео·даль·номе́р·, -а, (*m*) stereoscopic rangefinder

стерео·зву́к·, -а, (*m*) stereophonic/stereosonic sound

стерео·изоме́р·, -а, (*m*) (chem) stereoisomer

стерео·кин·е́з·, -а, (*m*) (physiol) stereokinesis

стерео·компара́тор·, -а, (*m*) stereocomparator

стерео·ображ·е́ни·е, -я, (*n*) stereoscopic picture

стере·о́м·, -а, (*m*) (bot) stereome

стере́о·метр·, -а, (*m*) (surv) stereometer

стерео·метр·и́ческ·ий, -ая, -ое, (*a*) stereometric

стерео·ме́тр·и·я, -и, (*f*) solid geometry

стерео·моде́л·ь, -и, (*f*) (surv) stereoscopic model

стерео·па́р·а, -ы, (*f*) (opt) stereoscopic image

стерео·плани́·граф·, -а, (*m*) stereoplanigraph

стерео·прое́кци·я, -и, (*f*) stereoscopic projection

стерео·регуля́р·н·ый, -ая, -ое, (*a*) stereospecific, stereoregular

стерео·ско́п·, -а, (*m*) (phot, opt) stereoscope

стерео·скоп·и́ческ·ий, -ая, -ое, (*a*) stereoscopic

стерео·скоп·и́·я, -и, (*f*) stereoscopy

стерео·с·ни́м·ок·, -м·к·а, (*m*) stereoscopic photograph

стерео·специф·и́ческ·ий, -ая, -ое, (*a*) stereospecific

стерео·спонди́л·ов·ый, -ая, -ое, (*a*) (pal) stereospondylous

стерео·теле·ви́д·ени·е, -я, (*n*) stereoscopic television

стерео·ти́п·, -а, (*m*) (print) stereotype

стерео·ти́п·и·я, -и, (*f*) stereotype printing

стерео·труб·а́, -ы́, (*f*) stereoscopic telescope, (mil) battery commander's telescope

стерео·фи́з·ик·а, -и, (*f*) stereophysics

стерео·фон·и́ческ·ий, -ая, -ое, (*a*) stereophonic

стерео·фото·грам·метр·и́ческ·ий, -ая, -ое, (*a*) stereophotogrammetric

~ ка́мер·а (*f*) stereophotogrammetric camera

стерео·фото·на·са́д·к·а, -и, (*f*) stereo-photographic attachment (for microscope etc.)

стерео·хи́м·и·я, -и, (*f*) stereochemistry

с·тер·е́ть, (*fut 3rd pl*) со·тр·у́т, (*perf*) rub out/off, erase; rub sore, abrade; grind to powder, triturate

стере́ч·ь, (*pres 3rd sing, pl*) стере·ж·ёт, стерег·у́т, (*past masc sing*) стерёт·, (*imp*) guard, watch over; watch, watch for

сте́рж·ен·ь, -ж·н·я, (*nom pl*) -и, (*g pl*) -е́й, (*m*) rod; bar (in metal strength tests); (met cast) core; stem (of valve); shank (of bolts, rivets etc.); shaft (of a pillar or column); (nucl) rod; (biol) rachis, shaft; (geol) spine, column

~, авар·и́йн·ый (nucl) safety/scram rod

~, армату́р·н·ый reinforcement bar (concrete)

~, габари́т·н·ый (met cast) cover core

~, грибо·ви́д·н·ый mushroom-valve stem

~, колеб·а́тельн·ый vibrating rod

~, компенс·и́рующ·ий (nucl) shim rod

~, кукуру́з·н·ый (food) corn cob

~, нару́ж·н·ый (met cast) cover core

~, порш·нев·о́й piston rod

~, пред·о·хран·и́тельн·ый detent (ammunition)

~, регул·и́рующ·ий (nucl) regulating rod

~, рас·пы́л·ивающ·ий atomizer

сте́рж·ен·ь

~, стоп- (nucl) safety rod, scram rod

~, у·правл·я́ющ·ий (nucl) control rod

~, централь·н·ый spindle

стержне·ви́д·н·ый, -ая, -ое, (*a*) rod-like/shaped, (biol) rachiform

стерж·нев·о́й, -а́я, -о́е, (*a*) of сте́рж·ен·ь; (biol) rachial, rhachial, rachidal

~ карка́с· (*m*) (met cast) core iron/grid/gagger

~ маши́н·а (*f*) (met cast) core-making machine

~ раз·ря́д·ник· (*m*) (nucl) rod gap

~ с·мес·ь (*f*) core compound

~ цех· (*m*) (met cast) core room

~ я́щик· (*m*) (met cast) core box

стери́гм·а, -ы, (*f*) (bot) sterigma

стерилиз·а́тор·, -а, (*m*) sterilizer

стерилизацио́н·н·ый, -ая, -ое, (*a*) sterilization

стерилиз·а́ци·я, -и, (*f*) sterilization

стерилиз·о́ванн·ый, -ая, -ое, (*past part pass*) of стерилиз·ова́ть; sterilized

стерилиз·ова́ть, -у́ют, (*imp and perf*) sterilize

стери́ль·ност·ь, -и, (*f*) sterility

стери́ль·н·ый, -ая, -ое, (*a*) sterile

стер·и́н·, -а, (*m*) (biochem) sterol

стер·и́ческ·ий, -ая, -ое (*a*) (chem) steric, spacial

стерко·били́н·, -а, (*m*) (biochem) stercobilin

стерко·ли́т·, -а, (*m*) (med) stercolith

стеркор·и́т·, -а, (*m*) (min) stercorite

стерку́ли·я, -и, (*f*) (bot) Sterculia

сте́рлинг·, -а, (*m*) sterling

сте́рляд·ь, -и, (*g pl*) -е́й, (*f*) (fish) sterlet, *Acipenser ruthenus*

стерн·ев·о́й, -а́я, -о́е, (*a*) stubble

стерн·и́т·, -а, (*m*) (zool) sternite

сте́рн·ум·, -а, (*m*) (zool) sternum

сте́рн·ь, -и, (*f*) = стерн·я́

стерн·я́, -и́, (*f*) (agr) stubble field; stubble

стер·о́ид·, -а, (*m*) (biochem) steroid

стер·о́л·, -а, (*m*) (biochem) sterol

стер·о́н·, -а, (*m*) (biochem) sterone

с·терп·е́ть, -я́т, (*perf*) endure, bear

стерр·а́стр·, -а, (*m*) (zool) sterraster

с·тесн·ённ·ый, -ая, -ое, (*past part pass*) of с·тесн·и́ть; constrained; crowded

с·тесн·и́тельн·ый, -ая, -ое, (*a*) restrictive; restraining, constraining; inconvenient; embarrassing с·тесн·и́ть, -я́т, (*perf*) *see* с·тесн·я́·ть

с·тесн·я́·ть, -ю́т, (*imp*) restrict, confine, limit, make narrow/small/tight, pack close; -ся (*pass*); crowd; (*imp only*) (+ *gen*) be shy/embarrassed

стето́·метр·, -а, (*m*) (med) stethometer

стето·ско́п·, -а, (*m*) (med) stethoscope

Сте́фан·а-Бо́льцман·а, по·сто́·ян·н·ая (*f*) (phys) Stefan–Boltzmann constant

стефани́т·, -а, (*m*) (min) stephanite

стехио·ло́г·и·я, -и, (*f*) (physiol) stoichiology, stoicheiology, stoechiology

стехио·метр·и́ческ·ий, -ая, -ое, (*a*) (chem) stoichiometric(al), stoechiometric

стехио·ме́тр·и·я, -и, (*f*) (chem) stoichiometry, stoicheimetry, stoechiometry

с·теч·е́ни·е, -я, (*n*) (*v n*) *see* с·теч·ь; confluence; concourse; coincidence (of events etc.)

с·теч·ь, (*fut 3rd sing, pl*) с·теч·ёт, с·тек·у́т, (*past masc sing*) с·тёк·, (*perf*) flow, run (of liquids etc.); flow down; -ся flow together

стиб·и́д·, -а, (*m*) (chem) stibide, antimonide

стиб·и́л·, -а (*m*) (chem) stibyl, antimonyl

стиб·и́н·, -а, (*m*) (chem) stibine, antimony/antimonous hydride

стиб·и́нов·ый, -ая, -ое, (*a*) (chem) stibonic

~ кисл·от·а́ (*f*) (chem) stibonic acid

стибио·висмути́н·, -а, (*m*) (min) stibiobismuthine

стибио·танта́ли́т·, -а, (*m*) (min) stibiotantalite

стибни́т·, -а, (*m*) (min) stibnite, antimony glance

стиб·о́ни·й, -я, (*m*) (chem) stibonium

сти́гм·а, -ы, (*f*) (biol) stigma, (*pl*) stigmata

стигм·а́ри·й, -я, (*m*) (bot) stigmarium

стигма·стери́н·, -а, (*m*) (biochem) stigmasterol

стигмат·и́ческ·ий, -ая, -ое, (*a*) stigmatic

стигмат·о́з·, -а, (*m*) (med) stigmatosis

стил·е́т·, -а, (*m*) (biol) stylet; (med) stylet

стилиз·о́ванн·ый, -ая, -ое, (*past part pass*) of стилиз·ова́ть; stylized, conventionalized

стилиз·ова́ть, -у́ют, (*imp and perf*) stylize

стило·ба́т·, -а, (*m*) (arch) stylobate

стило·ли́т·, -а, (*m*) (geol) stylolite

стило·ли́т·ов·ый, -ая, -ое, (*a*) (geol) stylolitic

стило́·метр·, -а, (*m*) (met) spectrophotometer

стило́с·, -а, (*m*) stylus

стило·ско́п·, -а, (*m*) (met, cryst) steeloscope, pattern spectroscope

стило·спо́р·а, -ы, (*f*) (bot) stylospore, pycnospore

сти́л·к·а, -и, (*f*) (*v n*) of стл·ать

сти́л·очн·ый, -ая, -ое, (*a*) of сти́л·к·а

~ маши́н·а (*f*) spreader; peat spreader

стил·ь, -я, (*m*) style

~, но́в·ый new style, Gregorian calender (of dates)

~, ста́р·ый old style, Julian calender (of dates)

сти́льб·, -а, (*m*) stilb, sb (unit of brightness)

стильб·е́н·, -а, (*m*) (chem) stilbene

стильб·и́т·, -а, (*m*) (min) stilbite, desmine

стильб·эстро́л·, -а, (*m*) (biochem) stilboestrol, stilbestrol

стильпно·мела́н·, -а, (*m*) (min) stilpnomelane

стильпно·сидери́т·, -а, (*m*) (min) stilpnosiderite

сти́мул·, -а, (*m*) (biol) stimulus

стимул·и́ровани·е, -я, (*n*) stimulation, stimulating

стимул·и́р·овать, -у́ют, (*imp and perf*) stimulate

стимул·я́тор·, -а, (*m*) stimulus, stimulant, stimulator

стипе́нди·я, -и, (*f*) scholarship, allowance (to support student graduate)

стиптици́н·, -а, (*m*) (chem) stypticine

с·тир·а́емост·ь, -и, (*f*) erasability (of magnetic tape)

стира́кс·, -а, (*m*) storax (balsam)

стира́кс·ов·ые, -ых, (*pl decl as adj*) (bot) storax family, *Styracaceae*

с·тир·а́льн·ый, -ая, -ое, (*a*) washing, laundering

~ порош·о́к· (*m*) washing powder, detergent

с·тир·а́ни·е, -я, (*n*) (*v n*) of с·тир·а́·ть; abrasion, detrition, trituration; erasure

с·тир·а́·ть, -ю́т, (*imp*) wipe, wipe off, rub, rub off, erase; abrade, rub sore; rub to powder, triturate; (*imp only*) launder, wash

с·тир·а́ющ·ий, -ая, -ее, (*pres part act*) of с·тир·а́·ть

~ голо́в·к·а (*f*) erase head (magnetic recording)

стир·и́л·, -а, (*m*) (chem) styryl

с·ти́р·к·а, -и, (*g pl*) -р·ок·, (*f*) laundering, laundry

стиро·кри́л·, -а, (*m*) styrocryl

стир·о́л·, -а, (*m*) (chem) styrene

стиро́л·ов·ый, -ая, -ое, (*a*) styrene

стирофле́кс·, -а, (*m*) (plast) styroflex, polystyrene film

с·ти́ск·ива·ть, -ют, (*imp*) squeeze, clench, hug, grip

с·ти́с·н·уть, -ут, (*perf*) *see* с·ти́ск·ива·ть

стифн·а́т·, -а, (*m*) (chem) styphnate, trinitroresorcinate

~ свин·ц·а́ (expl) lead styphnate/trinitroresorcinate

стифн·и́нов·ый, -ая, -ое, (*a*) styphnic

~ кисл·от·а́ (*f*) styphnic acid

с·тих·а́·ть, -ют, (*imp*) become quiet/silent; slacken, ease, fall (of wind etc.), become calmer/calm, abate

стих·и́ди·й, -я, (*m*) (microbiol) stichidium

стих·и́йн·ый, -ая, -ое, (*a*) elemental, natural, spontaneous

~ бе́д·стви·е (*n*) (ins) act of God

стих·и́·я, -и, (*f*) (geophys) element

с·ти́х·нуть, -нут, (*perf*) *see* с·тих·а́·ть

стл·а́вш·ий, -ая, -ее, (*past part act*) of стл·ать

стл·ал, (*past masc sing*) of стл·ать

стл·а́ник·, -а, (*m*) (bot) carpeting bush

стл·а́нн·ый, -ая, -ое, (*past part pass*) of стл·ать

стл·ань·ё, -я́, (*n*) (*v n*) *see* стл·ать; dew retting (of flax)

стл·ать (*pres 3rd pl*) сте́л·ют, (*imp*) spread, spread out, lay; make (beds); -ся (*pass*); cover; trail

сто, (*gen, dat, instr, prep*) ста hundred; (*as component*) hundred-, centi-

стог·, -а, (*nom pl*) -а́, (*g pl*) -о́в, (*m*) stack, rick (hay etc.)

стог·ова́ть, -у́ют, (*imp*) stack, make stacks, pile in stacks

стог·ов·о́й, -а́я, -о́е, (*a*) stacked

стого·мет·а́ни·е, -я, (*n*) making a rick, stacking (hay etc.)

стого·мет·а́тел·ь, -я, (*m*) (agr) stacker, hay stacker

~, пере·кид·н·о́й (agr) overshot stacker

сто·гра́дус·н·ый, -ая, -ое, (*a*) centigrade

сто́·ек·, (*g pl*) of сто́й·к·а

сто́·ечн·ый, -ая, -ое, (*a*) post

сто́ж·ить, -ат, (*imp*) = стог·ова́ть

сто́·имост·ь, -и, (*f*) cost; value

~, курс·ов·а́я market value

~, мен·ов·а́я exchange value

~, при·ба́в·очн·ая surplus value

сто́·ит, (*pres 3rd sing*) of сто́·ить; сто·и́т, (*pres 3rd sing*) of сто·я́ть

сто́·ить, -ят, (*imp*) cost, be worth; (+ *gen*) deserve, be worthy of, be worth

сто́й·к·а, -и, (*g pl*) сто́·ек·, (*f*) post, support, stanchion, strut; (min) prop; web (of rail); (elec) rack; rack frame, bay; stand, pedestal; (a/c) strut; (M/T) side column (of radiator); standard (of plough); shank (of cultivator)

~, амортизацио́н·н·ая (a/c) shock absorber strut

~, вертика́ль·н·ая pillar (of theodolite etc.)

~, конта́кт·н·ая (elec) clip, contact clip (knife-switch)

~, на·кло́н·н·ая (min) raking/battered prop

~ пере·бо́р·к·и (shipb) bulkhead stiffener

~ пере·ключ·е́ни·я patching bay (telephones)

~, по·да́т·лив·ая (min) yielding/adjustable prop

~ про́б·н·ых иск·а́тел·ей test selector rack (telephones)

~, руд·ни́чн·ая (min) pit prop

~ стан·и́н·ы (met roll) roll stand post

сто́й·к·ий, -ая, -ое, (*a*) stable, steady; durable (of materials), persistent (of gases); (*as component*) -resistant, -proof, -fast

сто́й·кост·ь, -и, (*f*) stability, steadiness; durability, resistance (to wear, corrosion etc.); life (of cutting tools)

~, детонацио́н·н·ая (oil) anti-knock value

~, хим·и́ческ·ая (met) resistance to chemical corrosion, chemical corrosion resistance, corrosion resistance

сто́й·л·о, -а, (*n*) (agr) stall; bay (in workshop)

сто́й·м·я (*adv*) upright, standing

с·то́к·, -а, (*m*) (*v n*), *зоо* с·тек·а́·ть; flow, flowing, drainage, run-off; discharge (of river); drain, discharge, sink flow

с·то́к

~, **дожд·ев·о́й** rainfall run-off **коэффицие́нт· с·то́к·а** (*m*) (civ eng) run-off factor

~, **по·ве́рх·ностн·ый** surface run-off

~, **под·зе́м·н·ый** subsurface drainage

~ **эне́рг·и·й** (phys) energy sink

сто́кер·, -а, (*m*) underfeed stoker (for boilers)

сто·километр·о́в·ый, -ая, -ое, (*a*) hundred-kilometer/kilometre

с·то́к·ов·ый, -ая, -ое, (*a*) of **с·ток·**

~ **ста́нци·я** (*f*) river discharge observatory

сто·кра́т·н·ый, -ая, -ое, (*a*) (math) centuple, a hundredfold

сто́кс·, -а, (*m*) stokes (unit of viscosity)

сто́кс·ов·ый коэффицие́нт· (*m*) (phys) Stokes coefficient

стол·, -а́, (*m*) table, desk; department, section; (agr) deck (of binder)

~ **за·ка́з·ов** order department

~, **контро́ль·н·ый** service-observation desk, monitoring desk (telephones)

~, **концентрацио́н·н·ый** (min) concentrating table

~, **ле́нт·очн·ый** (min) vanner

~, **о·пуск·а́ющ·ийся** (met roll) depressing table

~, **подъ·ём·н·о-кач·а́ющ·ийся** (met roll) tilting table; rising table (for shears)

~, **подъ·ём·н·о-о·пуск·а́ющ·ийся** lifting table, elevating platform

~, **подъ·ём·н·ый** (met roll) lifting table

~, **пуск·ов·о́й** (rocket) launching pad

~, **со·тряс·а́тельн·ый** (min) jerking table

столб·, -а́, (*m*) post, pillar, pole, column; (geol) pillar, column, pinnacle

~, **вод·ян·о́й** water column (unit of pressure); **мм водян. столб.** mm Wg

~, **лед·ян·о́й** (geog) ice pinnacle

~, **пере·ва́ль·н·ый** light post (rivers)

~, **позвон·о́чн·ый** (anat) spinal column

~, **рту́т·н·ый** mercury column; **мм ртутн. столб.** mm Hg

~, **термо·электр·и́ческ·ий** (nucl) thermopile

столб·е́ц·, -б·ц·а́, (*m*) (print) column

сто́лб·ик·, -а, (*m*) (*dim*) of **столб·** peg, (bot) stylus, columella; (elec) stub, stack

~, **селе́н·ов·ый** (elec) selenium stack

~, **сифона́ль·н·ый** siphonal column

столб·и́ть, -я́т, (*imp*) peg/mark out, mark with posts

столбл·ённ·ый, -ая, -ое, (*past part pass*) of **столб·и́ть**; pegged/marked-out

столб·ня́к·, -а́, (*m*) (med) tetanus

столб·ов·о́й, -а́я, -о́е, (*a*) of **столб·**

~ **систе́м·а** (*f*) (min) bord-and-pillar

столб·цо́в·ый, -ая, -ое, (*a*) of **столб·е́ц**

сто́лб·чат·ый, -ая, -ое, (*a*) columnar, acicular; (bot) palisade

~ **диагра́мм·а** (*f*) (math) histogram

~ **криста́лл·** (*m*) acicular crystal

~ **ткан·ь** (*f*) (bot) palisade tissue

сто·ле́т·и·е, -я, (*n*) century; centenary

сто·ле́т·н·ий, -яя, -ее, (*a*) centennial, hundred-year(s)

сто́л·ик·, -а, (*m*) (*dim*) of **стол·**

~, **пред·ме́т·н·ый** stage (of microscope)

сто·литр·о́в·ый, -ая, -ое, (*a*) hundred-litre/liter

стол·и́ц·а, -ы, (*i*) **-ей,** (*f*) capital, capital city

стол·и́чн·ый, -ая, -ое, (*a*) of **стол·и́ц·а**

с·толк·нове́ни·е, -я, (*n*) collision; (mil) engagement, encounter

~, **бли́ж·н·ее** (phys) close collision

~, **да́ль·н·ее** (phys) distant collision **то́ч·к·а с·толк·нове́ни·я стру·й** (*f*) point of impingement

с·толк·н·у́ть, -у́т, (*perf*) *see* **с·та́лк·и·ва·ть**

с·толк·у́т, (*fut 3rd pl*) of **с·толо́чь·**

стол·о́в·ая, -ой, (*f decl as adj*) dining room/hall, (mil) mess

стол·о́в·ый, -ая, -ое, (*a*) table, desk; table-mounted; (*see also* **стол·о́в·ая** *as noun*)

~ **вин·о́** (*n*) table wine

~ **гор·а́** (*f*) (geog) mesa, flat-topped mountain

~ **плоско·го́р·ь·е** (*n*) (geol) mesa

стол·о·зо́ид·, -а, (*m*) (zool) stolozooid

с·толо́к·ш·ий, -ая, -ее, (*past part act*) of **с·толо́чь·**

столо́н·, -а, (*m*) (bot) stolon

столо́н·ов·ый, -ая, -ое, (*a*) stolonate

столоно·но́с·н·ый, -ая, -ое, (*a*) (biol) stoloniferous

стол·о·обра́з·н·ый, -ая, -ое, (*a*) tabular, table-shaped

столо·те́к·а, -и, (*f*) (bot) stolotheca

с·толо́чь·, (*fut 3rd sing, pl*) **с·толч·ёт, с·толк·у́т,** (*past masc sing*) **с·толо́к·** (*perf*) pound, crush, grind; pound/crush/grind together

с·толп·и́ться, -я́тся, (*perf*) crowd in/ up/together

с·толч·ённ·ый, -ая, -ое, (*past part pass*) of **с·толо́чь**

столь (*adv*) so

с·то́лько so/as much/many
  ещё с·то́лько же as much/many again
~ **ско́лько** as/so much/many as

стол·я́р·, -а́, (*m*) joiner

стол·я́рк·а, -и, (*f*) joiner's shop

стол·я́рн·ый, -ая, -ое, (*a*) joiner's
~ **вя́з·к·а** (*f*) joiner's joint
~ **де́л·о** (*n*) joinery
~ **из·де́л·и·я** (*pl*) joiner's work, joinery

стомат·и́т·, -а, (*m*) (med) stomatitis

стомат·и́ческ·ий, -ая, -ое, (*a*) stomatic, oral

стомато·ло́г·и·я, -и, (*f*) (med) stomatology

стомато·ско́п·, -а, (*m*) (med) stomatoscope

сто·метр·о́вк·а, -и, (*g pl*) **-вок·,** (*f*) (surv) 1/10000 map

сто́н·, -а, (*m*) moan, groan

сто·но́ж·к·а, -и, (*g pl*) **-ж·ек·,** (*f*) (ent) centipede

стоп (*imper*) stop!

стоп-а́нкер·, -а, (*m*) (naut) kedge anchor

стоп-ва́нн·а, -ы, (*f*) (phot) stop bath

стоп-кра́н·, -а, (*m*) (rail) emergency stop lever, emergency lever; (rocket) cut-off valve

стоп·а́, -ы́, (*nom pl*) **сто́п·ы** (*f*) pile, stack; (anat) foot, pes; (*nom pl*) **стоп·ы́** (paper) ream

стопва́тер·, -а, (*m*) (shipb) stop water

стопи́н·, -а, (*m*) quick match

стоп·и́р·овать, -уют, (*imp and perf*) (paper) put into reams

с·топ·и́ть, -я́т, (*perf*) melt together, mix (by melting)

стоп·к·а, -и, (*g pl*) **-п·ок·,** (*f*) pile, stack

стоплат·, -а, (*m*) (naut) tabling (of sail)

с·то́пл·енн·ый, -ая, -ое, (*past part pass*) of **с·топ·и́ть**

стоп·ов·о́й, -а́я, -о́е, (*a*) stacked, piled

сто́пор·, -а, (*m*) stop, stopper, catch, pawl
~ **вз·во́д·а** (gunn) sear
~**, винт·ов·о́й** (naut) screw slip; compressor (of chain cable)
~**, шла́к·ов·ый** cinder-notch stopper (of blast furnace)

стопо·ре́з·к·а, -и, (*g pl*) **-з·ок·,** (*f*) guillotine (for paper etc.

сто́пор·ить, -ят, (*imp*) stop (engines etc.), (naut) stop engines

сто́пор·к·а, -и, (*f*) (naut) seizing

сто́пор·н·ый, -ая, -ое, (*a*) stop, stopper, locking
~ **винт·** (*m*) set screw
~ **кла́пан·** (*m*) stop valve
~. **пла́н·к·а** (*f*) stop plate; lock washer
~ **сте́рж·ен·ь** (*m*) stopper rod (of metal casting ladle)
~ **ша́р·ик·** (*m*) locking ball

стопо·ход·я́щ·ий, -ая, -ее, (*a*) (zool) plantigrade, (*pl as noun*) plantigrades, *Plantigrada*

сто́ппер·, -а, (*m*) (plast) short-stopper

сто·проце́нт·н·ый, -ая, -ое, (*a*) hundred per cent

с·топт·а́ть, (*fut 3rd pl*) **с·то́пч·ут,** (*perf*) tread/wear down

стоп·цили́ндр·ов·ая маши́н·а (*f*) (print) stop-cylinder machine

с·торг·ова́ться, -у́ются, (*perf*) agree a price, strike a bargain

сторож·, -а, (*nom pl*) **-а́,** (*g pl*) **-е́й,** (*m*) watchman, guard, warden

сторож·ев·о́й, -а́я, -о́е, (*a*) watch, patrol, guard
~ **вы́ш·к·а** (*f*) watch tower
~ **за·ста́в·а** (*f*) picket
~ **ка́тер·** (*m*) (naut) patrol boat
~ **кора́бл·ь** (*m*) (naut) escort vessel
~ **сигнализ·а́ци·я** (*f*) burglar alarm system

сторож·о́к·, -ж·к·а́, (*m*) catch (on measuring instruments); (horol) escapement

сторон·а́, -ы́, (*acc*) **сто́рону,** (*nom pl*) **сто́рон·ы,** (*g pl*) **сторо́н·,** (*f*) side; (law) party; land, place
  в сторон·е́ aside, apart
  в сторон·у aside; in the direction of, towards
~**, вы·зыв·а́ющ·ая** (telecom) originator
~ **давл·е́ни·я** pressure face (of propellers, turbine blades etc.)
~**, до·гова́р·ивающ·аяся** contracting party
~**, за·мык·а́ющ·ая** end line (of a polygon)
~**, лиц·ев·а́я** face, obverse side, right/ top side
~**, оборо́т·н·ая** reverse side, other side, back
~**, при·лег·а́ющ·ая** faying surface (of riveted plate)

сторон·а́

~ **раз·реж·е́ни·я** suction face (of turbine vanes, propeller blades etc.) **с одн·о́й сторон·ы́** on one hand ... **со сторон·ы́** from elsewhere; on the part of

сторон·и́ться **~ятся**, (*imp*) stand aside, make way; (+ *gen*) avoid, shun

сторо́н·к·а, -и, (*g pl*) **-н·ок·,** (*f*) (*dim*) of **сторон·а́**

~, за́д·н·яя (print) back board/cover

~, пере́д·н·яя (print) front board/cover

-сторо́н·ий, -яя, -ее, (*adj component*) -sided, -lateral

сторо́н·ник·, -а, (*m*) supporter, adherent

сторон·о́й (*adv*) sideways

стохаст·и́ческ·ий, -ая, -ое, (*a*) stochastic

~ **проце́сс·** (*m*) (math) stochastic process

с·точ·и́ть, ~ат, (*perf*) sharpen; grind down (in sharpening)

с·то́ч·н·ый, -ая, -ое, (*a*) of **с·ток·**

~ **во́д·ы** (*pl*) sewage

~ **во́д·ы, атмосфе́р·н·ые** storm/surface water

~ **во́д·ы, про·мы́шл·енн·ые** industrial sewage/effluent

~ **во́д·ы, фека́ль·н·ые** soil/sanitary sewage

~ **во́д·ы, хозя́й·ственн·о-быт·ов·ы́е** house/domestic sewage

~ **ма́сл·ян·ая систе́рн·а** (*f*) drain oil tank

~ **труб·а́** (*f*) drain pipe

сто́·я (*adv*) upright, standing

сто·я́к·, -а́, (*m*) post, stanchion; standpipe, stack-pipe

~, дым·ов·о́й chimney stack

~, ли́т·ников·ый (met cast) sprue, downgate

~ **под·ши́п·ник·а** bearing pedestal

сто·я́ни·е, -я, (*n*) standing, stand; stop, halt, wait

~, густ·о́е (agr) close stand

~ **при·ли́в·а** slack water, stand of the tide

~, ре́д·к·ий (agr) thin stand

сто·я́нк·а, -и, (*g pl*) **-нок·,** (*f*) stopping place, parking place (of vehicles); (naut) anchorage, berth; (air) hard standing; stand, stop (action); (pal) camp site

сто·я́ночн·ый, -ая, -ое, (*a*) of **сто·я́н·к·а**

~ **у́гол·** (*n*) (air) stalling angle

сто́·ят, (*pres 3rd pl*) of **сто́·ить; сто·я́т** (*pres 3rd pl*) of **сто·я́ть**

сто·я́ть, -я́т, (*imp*) stand; be; be situated; stop, halt

сто·я́ч·ий, -ая, -ее, (*a*) standing, stand up, stand, standard

~ **вод·а́** (*f*) standing water, stand of the tide

~ **волн·а́** (*f*) (rad) standing wave; (ocean) stationary wave

~ **пар·ов·о́й коте́л·** (*m*) vertical boiler

сто́·ящ·ий, -ая, -ее, (*pres part act*) *see* **сто́·ить;** worthwhile; worthy; **сто·я́щ·ий, -ая, -ее,** (*pres part act*) of **сто·я́ть**

стр. (*abbr*) = **стран·и́ц·а** page

с·трав·и́ть, ~ят, (*perf*) *see* **с·тра́вл·и·ва·ть**

с·травл·е́ни·е, -я, (*n*) *see* **с·тра́вл·и·вани·е**

с·тра́вл·енн·ый, -ая, -ое, (*a*) (*past part pass*) *see* **с·тра́вл·ива·ть**

с·тра́вл·ивани·е, -я, (*n*) (*v n*) of **с·тра́вл·ива·ть;** electrolytic stripping

~, ни́з·к·ое (agr) close grazing

с·тра́вл·ива·ть, -ют, (*imp*) (chem, met) etch, scour, strip/remove by etching; poison; let off, release (gas under pressure); (agr) graze; set on to one another, make fight each other

с·травл·я́·ть, -ют, (*perf*) *see* **с·тра́вл·ива·ть**

страд·а́, -ы́, (*nom pl*) **страд·ы,** (*g pl*) **страд·,** (*f*) harvest time

страд·а́·ть, -ют, (*imp*) suffer

стра́д·н·ый, -ая, -ое, (*a*) of **страд·а́**

стра́ж·а, -и, (*i*) **-ей,** (*f*) guard, watch

~, берег·ов·а́я coast guard

стра́з·, -а, (*m*) strass, paste (jewellery)

стран·а́, -ы́, (*nom pl*) **стра́н·ы,** (*g pl*) **стран·,** (*f*) country, land

~ **све́т·а** cardinal point (of compass)

стран·и́ц·а, -ы, (*i*) **-ей,** (*f*) page

~, у·длин·ённ·ая deep page

~, чёт·н·ая (print) folio verso, left-hand page

стран·и́чн·ый, -ая, -ое, (*a*) page, paginal

стра́н·ническ·ий, -ая, -ое, (*a*) wandering, wanderer's

стра́н·н·о (*adv*) strangely, oddly; it is strange/odd/curious

стра́н·ност·ь, -и, (*f*) strangeness

стра́н·н·ый, -ая, -ое, (*a*) strange, curious

стра́н·ств·овать, -уют, (*imp*) wander, itinerate, travel

стра́н·ствующ·ий, -ая, -ее, (*pres part act*) of **стра́н·ств·овать;** wandering; itinerant

стра́ст·н·ый, -ая, -ое, (*a*) passionate, impassioned, ardent

страсто·цве́т·, -а, (*m*) (bot) passion-flower, *Passiflora*

стра́ст·ь, -и, (*nom pl*) **-и,** (*g pl*) **-е́й,** (*f*) passion; addiction

стра́т·а, -ы, (*f*) (elec) stria, striation, (*pl*) striae

страте́г·, -а, (*m*) strategist

страте́г·и·я, -и, (*f*) strategy

стратег·и́ческ·ий, -ая, -ое, (*a*) strategic

страти́·граф·, -а, (*m*) (geol) stratigrapher

страти·гра́ф·и·я, -и, (*f*) (geol) stratigraphy

страти·гра́ф·и́ческ·ий, -ая, -ое, (*a*) stratigraphic, stratigraphical; stratified

~ горизо́нт· (*m*) (geol) stratigraphical level, geological horizon

~ за́·леж·ь (*f*) (geol) stratified deposit

страти·сфе́р·а, -ы, (*f*) (geog) lithosphere

страти·фик·а́ци·я, -и, (*f*) stratification

страт·о́ид·н·ый, -ая, -ое, (*a*) stratoidal

страто·на́вт·, -а, (*m*) high-altitude pilot; high-altitude flight specialist

страто·па́уз·а, -ы, (*f*) (geophys) stratopause

страто·пла́н·, -а, (*m*) stratosphere aircraft

страто·ста́т·, -а, (*m*) (geophys) stratosphere balloon

страто·сфе́р·а, -ы, (*f*) (geophys) stratosphere

страто·сфе́р·н·ый, -ая, -ое, (*a*) stratospheric, stratosphere

стра́ус·, -а, (*m*) (zool) ostrich, *Struthio*, (*pl*) *Struthioniformes*

~, австрал·и́йск·ий cassowary, *Casuaris*

~, америка́н·ск·ий rhea, *Rhea americana*

стра́ус·ник·, -а, (*m*) (bot) ostrich fern, *Struthiopteris germanica*

стра́х·, -а, (*m*) fear, phobia; (*as prefix*) insurance

    на сво·й страх· и риск· (ins) at own risk

    от·да·ва́ть на страх· obtain insurance/cover for

    при·ним·а́·ть на страх· (ins) accept risk, provide cover

страх·ка́сс·а, -ы, (*f*) insurance office

страх·ова́ни·е, -я, (*n*) insurance

~, мор·ск·о́е marine insurance

страх·о́ванн·ый, -ая, -ое, (*past part pass*) of **страх·ова́ть**

страх·ова́тел·ь, -я, (*m*) insured, insurance policy holder

страх·ова́ть, -у́ют, (*imp*) insure; **-ся** insure, take out an insurance

страх·о́вк·а, -и, (*g pl*) **-вок·,** (*f*) (*v n*) of **страх·ова́ть;** insurance premium

страх·ов·о́й, -а́я, -о́е, (*a*) insurance

~ по́лис· (*m*) insurance policy

страх·о́вщик·, -а, (*m*) insurer, insurance underwriter

стра́ш·н·ый, -ая, -ое, (*a*) terrible, frightful, fearful

Стре́бел·я, котёл· (*m*) Strebel boiler

стре́ж·ен·ь, -ж·н·я, (*m*) path of fastest currents (on river), current line

стре́йнер·, -а, (*m*) (rubber) strainer

стре́йнер-грануля́тор·, -а, (*m*) strainer-pelletizer

стрейн·и́р·овать, -уют, (*imp and perf*) strain

стрек·а́тельн·ый, -ая, -ое, (*a*) (ent) stinging

~ клет·к·а (*f*) stinging cell, cnidoblast

стрек·а́ющ·ие, -их, (*pl decl as adj*) (zool) cnidarians, *Cnidaria*

стрекоз·а́, -ы́, (*nom pl*) **стреко́з·ы,** (*g pl*) **стреко́з·,** (*f*) (ent) dragon-fly, *Libellula*, (*pl*) *Odonata*

стреко́т·, -а, (*m*) clattering, rattling (of machinery)

стрекот·а́ни·е, -я, (*n*) clattering, rattling (of machinery)

стрел·а́, -ы́, (*nom pl*) **стре́л·ы,** (*g pl*) **стрел·,** (*f*) (*see also* **стре́л·к·а**); arrow; pointer, indicator; derrick, boom, jib (of crane); leader (of piling frame); crown (of tire)

~, ажу́р·н·ая lattice derrick

~, вре́мен·н·ые (*pl*) shearlegs

~, груз·ов·а́я derrick

~, о·гран·ичи́тельн·ая cheek plate (on rubber mill)

~ по·ги́б·а camber height at centre

~, под·вес·н·а́я hanging leader (of piling frame)

~ подъ·ём·а (civ eng) rise, versed sine (of an arch)

~ про·ве́с·а sag, dip (of ropeways etc.)

~ про·ги́б·а (mech) sag, deflection, sagitta

~, решёт·чат·ая lattice derrick

стрел·е́ц·, -ль·ц·а́, (*i*) **-ль·ц·о́м,** (*m*) (astron) Sagittarius

стре́л·к·а, -и, (*g pl*) **-л·ок·,** (*f*) (*dim*) of **стрел·а́;** pointer, hand, needle (of compass etc.); (geog) spit; (bot) scape; (rail) switch-point, switch (U.S.), point (Brit); (*pl*) (ent) *Colnagrionidae*

~, воз·ду́ш·н·ая (elec) trolley frog

~, контро́ль·н·ая (autom) control point setting pointer, control index

~ курсо·за·да́т·чик·а (a/c) heading index/setting-pointer

~, от·жи́м·н·ая (rail) spring points/switch

~, по·шёрст·н·ая (rail) trailing point/switch

~, противо·шёрст·н·ая (rail) facing point/switch

~, след·я́щ·ая follow-up pointer

стрел·кова́ни·е, -я, (*n*) shooting, bolting (of vegetable plants)

стрел·ко́в·ый, -ая, -ое, (*a*) arrow; (mil) rifle, shooting; infantry

~ ка́рт·очк·а (*f*) range card

~ от·дел·е́ни·е (*n*) (mil) rifle section

~ ору́ж·и·е (*n*) infantry weapon

стрело·ви́д·ност·ь, -и, (*f*) sweep, sweepback, arrow shape

~, обра́т·н·ая negative/forward sweep, sweepforward (e.g. of a/c wing)

~, прям·а́я positive/backward sweep, sweepback

у́гол· стрело·ви́д·ност·и (*m*) angle of sweep

стрело·ви́д·н·ый, -ая, -ое, (*a*) arrow-shaped, sagittal, swept-back, swept

стрел·ов·о́й, -а́я, -о́е, (*a*) of **стрел·а́;** (*m decl as adj*) derrick arm

стрел·о́к·, -л·к·а́, (*m*) shooter, shot; rifleman; gunner (in tank), air gunner (in aircraft)

~, за́д·н·ий rear gunner

стрело·ли́ст·, -а, (*m*) (bot) arrowhead, *Sagittaria*

стре́л·очник·, -а, (*m*) (rail) pointsman, switchman

стре́л·очн·ый, -ая, -ое, (*a*) of **стре́л··к·а**

~ гальвано́·метр· (*m*) moving magnetic needle galvanometer

~ остр·я́к· (*m*) (rail) switch point

~ пере·во́д· (*m*) (rail) turnout, switch, points

~ поду́ш·к·а (*f*) switching plate

~ у́лица (*f*) (rail) switch track

стрел·ьб·а́, -ы́, (*nom pl*) **стре́л·ьб·ы,** (*g pl*) **стрел·ьб·,** (*f*) shooting, shoot, firing, fire

~, зени́т·н·ая anti-aircraft fire

~, морти́р·н·ая upper-register firing, high-angle firing

~, на·ве́с·н·ая steep-trajectory firing

~, на·сти́ль·н·ая shallow-trajectory firing

пло́ск·ост·ь стрел·ьб·ы́ (*f*) (gunn) vertical gun plane, plane of fire

~, пу́ль·н·ая aiming rifle fire

табл·и́ц·а стрел·ьб·ы́ (*f*) (gunn) range table

~, уч·е́бн·ая practice firing/shoot

стре́л·ьбищ·е, -а, (*i*) **-ем,** (*n*) shooting range/ground

стрель·н·у́ть, -у́т, (*perf*) fire a round/shot

стре́ль·чат·ый, -ая, -ое, (*a*) (arch) lancet; (bot) sagittate, arrow-shaped

стре́л·ян·ый, -ая, -ое, (*a*) (gunn) used; seasoned, veteran (of troops)

стрел·я́·ть, -ют, (*imp*) shoot, fire

стрел·я́щ·ий, -ая, -ее, (*pres part act*) of **стрел·я́·ть;** gun, gun-type (of non-military devices)

~ перфор·а́тор· (*m*) (oil) gun perforator

стрем·гла́в· (*adv*) headlong

стремен·а́, (*nom/acc pl*) of **стре́м·я**

стре́м·ечк·о, -а, (*n*) (*dim*) of **стре́м·я;** (anat) stapes

стрем·и́тельн·ый, -ая, -ое, (*a*) energetic; rushing, dashing, swift

~ марш· (*m*) (mil) forced march

стрем·и́ть, -я́т, (*imp*) impel, drive; **-ся** strive, aspire (of persons); approach, tend, incline (of inanimates); dash, rush

стремл·е́ни·е, -я, (*n*) (*v n*) see **стрем·и́ть(ся);** striving, aspiration; tendency, inclination

стрем·ни́н·а, -ы, (*f*) (geog) rapids, race

стре́м·я, (*gen, dat, prep*) **стре́мен·и,** (*nom pl*) **стремен·а́,** (*g pl*) **стрем·я́н·** (*n*) stirrup; (anat) stapes

стрем·я́нк·а, -и, (*g pl*) **-нок·,** (*f*) step-ladder, steps

стре́нг·а, -и, (*f*) strand (of rope etc.)

стре́нд·а, -ы, (*f*) = **стренд·ь**

стре́нд·ь, -и, (*f*) strand (of rope); rope (component of cable laid rope)

с·тре·но́ж·ить, -ат, (*perf*) hobble (horses etc.)

с·тре́пет·, -а, (*m*) (orn) bastard, *Otis tetrax*

стрептиди́н·, -а, (*m*) (chem) streptidine

стрепто·бактѐр·и·я, -и, (*f*) (bact) streptobacterium

стрепто·баци́лл·а, -ы, (*f*) (bact) Streptobacillus

стрепто·биозами́н·, -а, (*m*) (pharm) streptobiosamine

стрепто·дѐрм·и·я, -и, (*f*) (med) streptodermatitis

стрепто·ко́кк·, -а, (*m*) (bact) Streptococcus

стрепто·мици́н·, -а, (*m*) (pharm) streptomycin

стрептоци́д· бѐл·ый (*m*) (pharm) sulphanilamide

стрептоци́д· кра́сн·ый (*m*) (pharm) prontosil red

~ ~, рас·твор·и́м·ый (pharm) soluble prontosil, neoprontosil

стриг·, (*past masc sing*) of **стрич·ь**

стриг·а́л·ь, -я́, (*m*) shearer

стриг·а́льн·ый, -ая, -ое, (*a*) shearing, clipping (wool etc.)

~ маши́н·а (*f*) (text) shearer, shearing machine

стриг·а́льщик·, -а, (*m*) shearer (person)

стри́г·ли, (*past pl*) of **стрич·ь**

стриговит·, -а, (*m*) (min) strigovite

стриг·у́н·, -а́, (*m*) (ent) pine-shoot beetle, *Myelophilus*

стриг·у́щ·ий, -яя, -ее, (*pres part act*) of **стрич·ь**

стри́г·ш·ий, -ая, -ее, (*past part act*) of **стрич·ь**

стриж·, -а́, (*i*) **-óм,** (*m*) (orn) martin, swift, *Apus*

стри́ж·а, (*pres gerund*) of **стрич·ь**

стри́ж·енн·ый, -ая, -ое, (*past part pass*) of **стрич·ь**

стри́ж·ен·ый, -ая, -ое, (*a*) shorn, clipped, cut

стриж·ёт, (*pres 3rd sing*) of **стрич·ь**

стри́ж·к·а, -и, (*g pl*) **-ж·ек·,** (*f*) (*v n*) *see* **стрич·ь**; shear, clip; haircut, hairstyle

стри́к·, -а, (*m*) by-point (of compass); stripe, streak (plant disease)

стрикт·у́р·а, -ы, (*f*) (med) stricture

стрикцио́н·н·ый, -ая, -ое, (*a*) striction

~ ли́н·и·я (*f*) (math) line of striction

стри́кци·я, -и, (*f*) striction

стри́нгер·, -а, (*m*) (shipb, a/c) longitudinal, stringer

~, бо́рт·ов·о́й (shipb) side stringer

~, гла́в·н·ый (a/c) main longitudinal

~, дн·и́щев·о́й (shipb) intercostal girder

стри́нгер

~, интеркостѐль·н·ый (shipb) intercostal girder

~, па́луб·н·ый (shipb) stringer plate

~, про·ме́ж·у́точн·ый (a/c) secondary longitudinal

~, скул·ов·о́й (shipb) bilge stringer

стри́нгер·н·ый, -ая, -ое, (*a*) of **стри́нгер·**

~ уго́ль·ник· (*m*) (shipb) shell bar, stringer bar

стри́ппер·, -а, (*m*) (met) ingot stripper

стрихниди́н·, -а, (*m*) (pharm) strychnidine

стрихни́н·, -а, (*m*) (chem) strychnine, strychnin

стрич·ь, (*pres 3rd sing, pl*) **стриж·ёт, стриг·у́т,** (*past masc sing*) **стриг·,** (*imp*) shear, clip, crop, cut off; **-ся** (*pass*); have a haircut

стро́б·, -а, (*m*) (elec) gate, strobe

стро́бил·, -а, (*m*) (bot) strobile, strobilus

стробил·а, -ы, (*f*) (zool) strobila

стробил·я́ци·я, -и, (*f*) (zool) strobilization, strobilation

стро́б·и́мпульс·, -а, (*m*) (elec) gating pulse, gate

стро́б·и́ровани·е, -я, (*n*) (rad) gating, strobing

стро́б·и́рующ·ий, -ая, -ее, (*a*) strobe, strobing, gate, gating

стробо·ско́п·, -а, (*m*) (instr) stroboscope

стробо·тро́н·, -а, (*m*) strobotron, strobe, strobe lamp

строг·а́л·ь, -я́, (*m*) shaver, planer

строга́льно·кал·ёвочн·ый стан·о́к· (*m*) combination planer, surfacer and thicknesser (woodworking)

строг·а́льн·ый, -ая, -ое, (*a*) shaving, planing, chipping (soap)

~ коло́д·а (*f*) shaving beam (leather)

~ маши́н·а (*f*) shaving machine (leather)

~ стан·о́к· (*m*) planer, planing machine

~ стан·о́к·, вертика́ль·н·о- (mech eng) vertical shaper, slotter

~ стан·о́к·, по·пере́ч·н·о- (mech eng) shaper

~ стан·о́к·, про·до́ль·н·о- (mech eng) planer

~ стан·о́к·, про·пуск·н·о́- thicknesser, panel planer (woodworking)

строг·а́льщик·, -а, (*m*) shaver, planer

строг·а́ни·е, -я, (*n*) planing; shaving; chipping (soap)

~ по·копи́р·у form planing

строг·а́·ть, -ют,   (*imp*) plane, shave, chip (soap)

стро́г·ий, -ая, -ое,   (*a*) strict, rigorous, stringent, severe, precise

~ **аре́ст·**   (*m*) (law) close arrest

~ **не·ра́вен·ств·о**   (*n*) (math) strict inequality

стро·ев·а́я, -о́й,   (*f decl as adj*)

~ **по ватер·ли́н·и·ям**   (*f*)   (shipb) curve of waterplane areas, curve of areas of immersed sections

~ **по шпанго́ут·ам**   (*f*) (shipb) curve of cross sectional areas

стро·ев·о́й, -а́я, -о́е,   (*a*) building, construction; drill, ceremonial; line, combatant; executive, deck (naval); (*m decl as adj*) seaman rating; (*see also* **стро·ев·а́я** *as noun*)

~ **лес·**   (*m*) building timber

~ **офице́р·**   (*m*) (mil) line officer; executive officer, deck officer (naval)

~ **под·гото́в·к·а**   (*f*) (mil) drill

~ **смотр·**   (*m*) (mil) parade

~ **у·ста́в·**   (*m*) drill regulations

~ **шаг·**   (*m*) (mil) ceremonial march/ step

стро·е́ни·е, -я,   (*n*) structure; (build) building, construction

~, **с·бро́с·ов·ое**   (geol) faulted structure

~, **хим·и́ческ·ое**   chemical constitution

с·тро·ённ·ый, -ая, -ое,   (*past part pass*) of **с·тро·и́ть;** triple, triplex, three-

стро́ж·а, -и,   (*f*) = **строг·а́ни·е**

стро́·и,   (*pl*) of **стро́·й**

стро·и́тел·ь, -я,   (*m*) builder; (shipb) ship manager

~ **ри́ф·ов**   (zool, geol) reef-builder

стро·и́тельн·ый, -ая, -ое,   (*a*) building, construction(al), civil engineering, structural

~ **меха́н·ик·а**   (*f*) structural mechanics

~ **площа́д·к·а**   (*f*) building/construction site

~ **пра́в·ил·а**   (*pl*) building regulations

~ **стал·ь**   (*f*) structural steel

стро·и́тельск·ий, -ая, -ое,   (*a*) of **стро·и́тел·ь**

стро·и́тельств·о, -а,   (*n*) building, construction, civil engineering; building/construction site

~, **зелён·ое**   parks-and-gardens construction, landscaping

стро́·ить, -ят,   (*imp*) build, construct, devise; form, draw up; **с·тро·и́ть, -я́т,** (*perf*) triple

стро́·й, -я,   (*nom pl*) **стро́·и,** or **стро·и́,** (*g pl*) **стро́·ев** or **стро·ёв,** (*m*) (mil, naval) formation; order, system, structure; pitch (music)

строй-   (*component*) building, construction

-стро́·й, -я,   (*component of m*) stroy, Soviet construction company

в стро́·ю   on service; on station, in formation

в·вес·ти́ в стро́·й   bring into action/ use, put into operation; bring into line

в·ста́·ть в стро́·й   take up duties/an appointment; be put in operation (of inanimates)

вы́·йти из стро́·я   get out of order, go wrong, become unserviceable

~, **госуда́рств·енн·ый**   state structure/ system, system of government

~, **гус·и́н·ый**   arrowhead formation

~ **кильва́тер·а**   (naut) column

~ **кли́н·а**   arrowhead formation

~, **прост·о́й**   single-line formation

~ **пе́ленг·а**   line-of-bearing formation

~, **с·ло́ж·н·ый**   multiple-line formation

~ **у·сту́п·а**   staggered-column formation

~ **фро́нт·а**   line abreast

стро́й·к·а, -и,   (*g pl*) **стро́·ек·,** (*f*) building, construction; building/construction site

стро́й·н·ый, -ая, -ое,   (*a*) orderly, well-proportioned/composed/ordered; harmonious (of sounds)

строк·а́, -и́,   (*nom pl*) **стро́к·а,** (*g pl*) **строк·,** (*f*) (print etc.) line

~, **вы·дел·и́тельн·ая**   (print) display, line·

~, **кон·цев·а́я**   (print) break line, turn-over

~, **кра́сн·ая**   (print) centred line, new paragraph

~, **линоти́п·н·ая**   (print) linotype slug

~ **о·пре·дел·и́тел·я**   (math) row of a determinant

~ **раз·лож·е́ни·я**   (TV) scanning line

строко·ме́р·, -а,   (*m*) type scale

строко·от·лив·н·а́я маши́н·а   (*f*) (print) line-casting machine

стро́м·а, -ы,   (*f*) (biol) stroma

стромато·по́р·а, -ы,   (*f*) (pal) stromato-pore, (*pl*) *Stromatopora*

стромато·пороиде́·и   (*pl*) (pal) stromatoporoids, *Stromatoporoidea*

стро́мболиан·ск·ий, -ая, -ое,   (*a*) (geol) Strombolian

стромейери́т·, -а,   (*m*) (min) stromeyer-ite

стронциани́т·, -а, (*m*) (min) strontianite

стро́нц·иев·ый, -ая, -ое, (*a*) (chem) strontium

стро́нц·и·й, -я, (*m*) (chem) strontium, Sr

строп·, -а, (*m*) sling; strop; rigging/suspension/shroud line (of parachute)

~, бо́ч·ков·ый barrel sling

~, кабо́л·очн·ый selvagee strop

~, па́р·н·ые (*pl*) snotters

~, под·вес·н·о́й rigging cord/line (of parachute)

~ подъ·ём·а lifting sling

~, цеп·н·о́й can-hook sling, chain sling

строп·и́л·о, -а, (*n*) (build) rafter

строп·и́льн·ый, -ая, -ое, (*a*) (build) rafter

~ ног·а́ (*f*) (build) rafter

~ с·вяз·ь (*f*) (build) rafter

~ фе́рм·а (*f*) (build) truss

стро́п·к·а, -и, (*f*) (naut) grommet; becket

строп·о́вк·а, -и, (*f*) slinging (cargo etc.)

строф·а́нт·, -а, (*m*) (bot) Strophanthus

строф·ант·и́н·, -а, (*m*) (pharm) strophanthin

строф·о́ид·а, -ы, (*f*) (math) strophoid

строфо·мени́д·а, -ы, (*f*) (pal) strophomenid, (*pl*) *Strophomenidae*

стро́ч·ек·, (*g pl*) of **стро́ч·к·а**

строч·и́ть, -а́т, (*imp*) stitch

стро́ч·к·а, -и, (*g pl*) **-ч·ек·** (*f*) (*dim*) of **строк·а́;** stitch; (met) (*pl*) stringers)

строч·н·о́й, -а́я, -о́е, (*a*) (print, TV) line

~ алфави́т· (*m*) lower-case alphabet, lower case

~ бу́кв·а (*f*) (print) small letter, lower-case letter (of typewriter)

~ ис·каж·е́ни·е (*n*) (TV) line distortion

~ раз·вёрт·к·а (*f*) (TV) line scanning

стро́·ящ·ийся, -аяся, -ееся, (*pres part act*) *see* **стро́·ить;** under construction

струбци́н·а, -ы, (*f*) = **струбци́н·к·а**

струбци́н·к·а, -и, (*g pl*) **-н·ок·,** (*f*) adjustable/screw clamp

Стру́ве, фу́нкци·я (*f*) (math) Struve function

струви́т·, -а, (*m*) (min) struvite

струг·, -а, (*m*) spokeshave, drawknife (carpenter's tool); shaving knife (leather); (civ eng) scraper-loader; flatbottomed river boat

струж·ёк·, -ж·к·а́, (*m*) dugout canoe; **стру́ж·ек·,** (*g pl*) of **стру́ж·к·а**

стру́ж·к·а, -и, (*g pl*) **-ж·ек·,** (*f*) shaving (from wood planing); chip (from metal machining); **струж·к·а́** (*g sing*) of **струж·о́к·**

~, воск·ов·а́я wax chip (sound recording)

~ над·ло́м·а fragmented chip

от·дел·е́ни·е стру́ж·к·и (*n*) chip formation

~ с·ка́л·ывани·я discontinuous chip

~ с·лив·н·а́я continuous chip

струж·о́к·, -ж·к·а́, (*m*) (*dim*) of **струг·;** spokeshave (tool)

стру·и́ться, -я́тся, (*imp*) stream, spurt, jet

стру́й·к·а, -и, (*g pl*) **стру́·ек·,** (*f*) (*dim*) of **стру·я́;** (dynam) filament band

~ во́з·дух·а (dynam) air filament

стру́й·н·ый, -ая, -ое, (*a*) of **стру·я́;** jet, spray; (naut) wake

~ аппара́т· (*m*) (eng) ejector

~ конденс·а́тор· (*m*) (st eng) jet condenser

~ рас·пы́л·ивани·е (*n*) direct injection (diesel)

~ тео́ри·я (*f*) (dynam) free boundaries theory

стру́й·чат·ый, -ая, -ое, (*a*) jet, stream, spray

~ движ·е́ни·е (*n*) (hydrodyn) jet flow

~ за·ка́л·к·а (*f*) (met) spray hardening

структу́р·а, -ы, (*f*) structure; (geol) texture, fabric, structure

~, вращ·а́тельн·ая (phys) rotational structure

~, втор·и́чн·ая (geol) secondary texture

~, гру́б·ая (phys) coarse structure

~, зона́ль·н·ая zonal structure; (met) band structure

~, молекуля́р·н·ая (phys, chem) molecular structure

~, пи́сьм·енн·ая (geol) graphic texture

~, порфи́р·н·ая (geol) porphyritic texture

~, про·стра́н·ственн·ая three-dimensional structure

~, то́нк·ая fine structure

структур·и́ровани·е, -я, (*n*) structurization, structure formation; (chem) cross-linking

структур·и́рованн·ый, -ая, -ое, (*a*) (chem) crosslinked

~ полистиро́л· (*m*) crosslinked polystyrene

структур·и́рующ·ий, -ая, -ее,   (*a*) structure-forming

структу́р·ност·ь, -и,   (*f*) (chem) extent/degree of structure

структурно·чувств·и́тельн·ый, -ая, -ое,   (*a*) structure-sensitive

структу́р·н·ый, -ая, -ое, (*a*) structure, structural

~ табл·и́ц·а   (*f*) table of components (switching algebra)

структуро·образ·ова́ни·е, -я,   (*n*) structure formation, structurization, cross-linking (of polymers)

струн·а́, -ы́, (*nom pl*) стру́н·ы, (*g pl*) струн· (*f*) string, tensed string; cross brace, stay, tie-wire; rung (of ladder); dropper (supporting wire for overhead lines); (anat) cord, chorda

~, спи́н·н·ая   (zool) notochord

струн·е́ц·, -н·ц·а́, (*i*) -н·ц·о́м, (*m*) (zool) nematode

стру́н·к·а, -и, (*g pl*) -н·ок·, (*f*) (*dim*) of струн·а́

стру́н·н·ый, -ая, -ое, (*a*) of струн·а́; stringed, string

~ гальвано́·метр·   (*m*) (elec) string galvanometer/electrometer

~ по́л·е   (*n*) piano (telephones)

~ мост·   (*m*) (elec) slide-wire bridge

струно·бето́н·, -а,   (*m*) long-line tensioned concrete

струп·, -а, (*nom pl*) стру́п·ь·я, (*g pl*) -ь·ев, (*m*) (med) scab, rupia; (bot) scurf, scab

струч·кова́т·ый, -ая, -ое, (*a*) (bot) siliquiform

струч·ко́в·ый, -ая, -ое, (*a*) (bot) siliquose, siliquous

~ пе́р·ец· (*m*) (bot) red pepper, *Capsicum longum*

струч·о́к·, -ч·к·а́, (*m*) (agr) pod

~, у·дли́н·ённ·ый (bot) silique, silicule

струч·о́чек·, -ч·к·а, (*m*) (bot) silicle

Стру́хал·я, числ·о́ (*n*) (phys) Strouhal number

стру·я́, -и́, (*nom pl*) стру́·и, (*g pl*) стру́·й, (*d pl*) стру·я́м, (*f*) jet, stream, spray; (a/c) jet stream; (naut) wake

~, воз·ду́ш·н·ая   air stream

~, вы·хлоп·н·а́я   (a/c) exhaust/exhausted jet stream

~, пла́зм·енн·ая   (phys) plasma jet

~, реакти́в·н·ая   (a/c) jet stream, jet

~, с·пу́т·н·ая   (aerodynam) slipstream, (hydrodyn) wake, backwash

с·тряс·а́·ть, -ют, (*imp*) = со·тряс·а́·ть

с·тряс·ти́, -у́т, (*perf*) = со·тряс·ти́

с·тря́х·ива·ть, -ют, (*imp*) shake off/down

с·тря́х·н·у́ть, -у́т, (*perf*) *see* с·тря́х·ива·ть

студ·ени́ст·ый, -ая, -ое,   (*a*) jelly-like, gelatinous

студе́нт·, -а, (*m*) student

студе́нч·еск·ий, -ая, -ое,   (*a*) student

сту́д·ен·ь, -д·н·я, (*m*) jelly, jel, gel; jellied meat/fish

~, зооглей·н·ый (bact) zoogloeal mass

студ·и́йн·ый, -ая, -ое,   (*a*) studio

студ·и́ть, -ят, (*imp*) chill, cool

сту́д·и·я, -и, (*f*) studio; art school

студне·обра́з·н·ый, -ая, -ое,   (*a*) jelly-like, gelatinous

сту́ж·ен·ый, -ая, -ое,   (*a*) chilled, cooled; jellified

сту́к·, -а, (*m*) knock, rap, tap, clink, thud, bang; (ICE) knocking, knock

сту́к·а·ть, -ют, (*imp*) knock, rap, tap, clink, thud, bang

сту́к·н·уть, -ут, (*perf*) *see* сту́к·а·ть

стуко·ме́р·, -а, (*m*) (M/T) knockmeter

сту́л·, -а, (*nom pl*) сту́л·ь·я, (*m*) chair; support, pedestal, mounting; stock (of anvil); (med) stool

стул·ов·о́й, -а́я, -о́е,   (*a*) mounted, bench/pedestal-mounted

сту́л·ь·я, (*pl*) of стул·

сту́п·а, -ы, (*f*) mortar (for grinding)

ступ·а́·ть, -ют, (*imp*) step, tread; come go

ступ·е́нек·, (*n pl*) of ступ·е́ньк·а

ступенчато·шки́в·н·ый при·во́д· (*m*) belt-pulley drive

ступ·е́нчат·ый, -ая, -ое, (*a*) step, step-shaped/like, (biol) scalariform, echelon; step, stepped, stepwise, staged, staggered, gradual

~ в·ключ·е́ни·е   (*n*) grading (autom telephony)

~ воз·бужд·е́ни·е   (*n*) (nucl) step-by-step excitation

~ диафра́гм·а   (*f*) (spectr) step diaphragm

~ диск·   (*m*) (horol) snail

~ жи́л·а   (*f*) (geol) ladder vein

~ регул·я́тор· на·пряж·е́ни·я   (*m*) (elec) step voltage regulator

~ с·вя́з·к·а   (*f*) echelon strapping (of magnetron)

~ фу́нкци·я   (*f*) (math) step function

~ шкив·   (*m*) stepped pulley/sheave

ступ·éн·ь, **-и**, (*g pl*) **-éй**, step, stair, rung (of ladder); stage, degree; (*g pl*) **-éй**, (*f*) (rocket) stage; (rad) amplifier, stage; (min) bench

~, **бýфер·н·ая**   (rad) buffer amplifier

~, **вы·ход·н·áя**   (rad) output stage

~ **давл·éни·й**   pressure stage (of turbine)

~, **за·бéж·н·ая**   (build) dancing step

~, **мáрш·ев·ая**   (rocket) basic stage/shell

~, **по·слéд·н·яя**   (rocket) final stage

прáв·ил·о ступ·ен·éй   (*n*) (phys) phase law

~, **раз·гóн·н·ая**   (rocket) booster stage

~, **рукоя́т·к·и контрóллер·а**   (elec) controller notch

~ **с воз·врáт·ом**   single-stage recycle

~ **с·брóс·а**   (geol) fault bench

~ **скóр·ост·и**   velocity stage (of turbine)

~ **стабилиз·áтор·а**   (rocket) stabilizer stagger

~ **турбúн·ы**   turbine stage cell

ступ·éньк·а, **-и**, (*g pl*) **-нек·**, (*f*) (*dim*) of **ступ·éн·ь**

ступ·úть, **⸗ят**, (*perf*) *see* **ступ·á·ть**

ступ·úц·а, **-ы**, (*i*) **-ей**, (*f*) hub, boss

стýп·к·а, **-и**, (*g pl*) **-п·ок·**, (*f*) mortar (for grinding)

ступ·н·я́, **-й**, (*f*) foot

стýпп·а, **-ы**, (*f*) (met) stupp

стурúн·, **-а**, (*m*) (chem) sturine

стуртúт·, **-а**, (*m*) (min) sturtite

стуч·áть, **-áт**, (*imp*) make noise, knock, throb, thud, rattle, clink

с·туш·евáть, **-ýют**, (*perf*) *see* **с·туш·ёвыва·ть**

с·туш·ёвыва·ть, **-ют**, (*imp*) make gradually lighter, make outlines less distinct; **-ся** fade out, gradually fade out of the picture; be forgotten

стý·вш·ий, **-ая**, **-ее**, (*past part act*) of **сты·ть**

стыд·, **-á**, (*m*) shame

стыд·úться, **-я́тся**, (*imp*) (+ *gen*) be ashamed of

стыд·лúв·ый, **-ая**, **-ое**, (*a*) diffident, shy

с·тык·, **-а**, (*m*) butt, butt joint, joint, commissure; (rail) bond; cut point (in fractional distillation)

~, **дрóссель·н·ый**   (elec rail) impedance bond

~, **ступ·éнчат·ый**   step/zigzag joint

с·тык·овáни·е, **-я**, (*n*) (*v n*) of **с·тык·овáть**; butt-jointing; (rubber) splicing

~ **кáмер·**   tube sleeving (tyres)

с·тык·овáть, **-ýют**, (*imp*) make a butt joint, join edge-to-edge/end-to-end; join up, put together (sub-assemblies); splice

с·тык·ов·óй, **-áя**, **-óе**, (*a*) butt, joint

~ **плáн·к·а**   (*f*) butt strap

~ **с·вáр·к·а**   (*f*) butt welding

~ **ýзел·**   (*m*) (a/c) wing root fitting/attachment

сты·л, (*past masc sing*) of **сты·ть**

стý·н·уть, **-нут**, (*imp*) = **сты·ть**

стý·ть, **-нут**, (*imp*) become cold/cool, cool down/off; freeze

с·ты́ч·к·а, **-и**, (*g pl*) **-ч·ек·**, (*f*) skirmish

с·ты́ч·н·ый, **-ая**, **-ое**, (*a*) butt-jointed, joined/laid edge to edge

стэн   (*m*) sthene (unit of force)

стэнд·, **-а**, (*m*) *see* **стенд·**

стю́ард·, **-а**, (*m*) steward

с·тя́г·ивани·е, **-я**, (*n*) (*v n*) of **с·тя́г·и·ва·ть**; constriction, contraction

с·тя́г·ива·ть, **-ют**, (*imp*) tighten, draw tight/together, constrict, confine; tie up/together, fasten, secure; gather/draw up/together; (math) subtend; pull/haul off; **-ся** (*pass*); contract, shrink

с·тяж·áни·е, **-я**, (*n*) accretion (of wealth)

с·тя́ж·ек·, (*g pl*) of **с·тя́ж·к·а**

с·тяж·éни·е, **-я**   (*n*) fastening/securing/tying together; mutual attraction; gathering; (cryst) bundle; (min) concretion, nodule; cluster, clustering (phonetics); (geol) pressure, thrust

с·тяж·к·а, **-и**, (*g pl*) **-ж·ек·** (*f*) tie piece, tightening device; drawbar; fastening, securing, tying together/up

с·тяж·н·óй, **-áя**, **-óе**   (*a*) tie, tightening, brace

~ **гáй·к·а**   (*f*) tightening nut, turnbuckle

~ **пресс·**   (*m*) tightening press (cooperage)

~ **тя́г·а**   (*f*) tie rod

с·тя́·нут·ый, **-ая**, **-ое**, (*past part pass*) of **с·тя́·н·уть**, tightened; puckered, drawn, drawn together; constricted, confined

с·тя·н·уть, **⸗ут**, (*perf*) *see* **с·тя́г·и·ва·ть**

су·аурý   (*indecl*) (vet) surra

суб-   (*int component*) sub-

суб·аквáль·н·ый, **-ая**, **-ос**, (*a*) subaqueous

суб·артезиáн·ск·ий, **-ая**, **-ое**,   (*a*) (hydr) subartesian

суб·а́том·н·ый, -ая, -ое, (*a*) (phys) subatomic

суб·бореа́ль·н·ый, -ая, -ое, (*a*) (geophys) subboreal

суббо́т·а, -ы, (*f*) Saturday

субве́нци·я, -и, (*f*) subvention, subsidy

суб·гармон·и́ческ·ий, -ая, -ое, (*a*) (acous) sub-harmonic

суб·гедра́ль·н·ый, -ая, -ое, (*a*) (cryst) subhedral

субер·и́н·, -а, (*m*) (bot) suberin

субери́н·ов·ый, -ая, -ое, (*a*) suberin; (chem) suberic

~ **кисл·от·а́** (*f*) (chem) suberic acid

~ **пласт·и́нк·а** (*f*) (bot) suberin lamellar

~ **сло́·й** (*m*) (bot) suberin lamella

суб·кли́макс·, -а, (*m*) (bot) subclimax

сублим·а́т·, -а, (*m*) (chem, geol) sublimate

сублим·а́ци·я, -и, (*f*) sublimation

сублим·и́р·овать, -уют, (*imp and perf*) sublimate, sublime

суб·литора́л·ь, -и, (*f*) (geog) sublittoral

суб·микро·по́р·истост·ь, -и, (*f*) sub-microporosity

суб·микро·ще́л·ь, -и, (*f*) sub-micro-crack

суб·норма́л·ь, -и, (*f*) (math) subnormal

суб·нукло́н·н·ый, -ая, -ое, (*a*) (phys) sub-nucleonic

суб·под·ря́д·, -а, (*m*) (com) sub-contract

суб·регsiона́ль·н·ый, -ая, -ое, (*a*) (geog) subareal

субсид·и́р·овать, -уют, (*imp and perf*) (fin) subsidize; give aid (in kind)

субси́д·и·я, -и, (*f*) subsidy

субстанти́в·н·ый, -ая, -ое, (*a*) substantive

~ **краш·е́ни·е** (*n*) (chem, text) substantive dyeing

субста́нци·я, -и, (*f*) substation; substance

субстра́т·, -а, (*m*) (cyt) substrata; (geol) substratum, basement

~, **по́чв·енн·ый** (geol) bedrock

суб·стратосфе́р·а, -ы, (*f*) (meteor) tropopause

суб·структу́р·а, -ы, (*f*) substructure

субтили́н·, -а, (*m*) (microbiol) subtilin

субти́ль·н·ый, -ая, -ое, (*a*) frail, slender

субти́тр·, -а, (*m*) (cinema) sub-title

суб·тро́п·ик·и, -ов, (*m pl*) (geog) sub-tropics

суб·троп·и́ческ·ий, -ая, -ое, (*a*) subtropical

субъе́кт·, -а, (*m*) subject

субъекти́в·н·ый, -ая, -ое, (*a*) subjective

суверен·ите́т·, -а, (*m*) (law) sovereignty

сува́льд·а, -ы, (*f*) lever, ward (of a lock)

суво́йк·и (*pl*) (zool) Vorticella

су·гли́н·ист·ый, -ая, -ое, (*a*) loamy

су·гли́н·ок·, -н·к·а, (*m*) (geol) loam, loamy clay

~, **извест·ко́в·ый** marl

су·гро́б·, -а, (*m*) snowdrift

сугу́б·ый, -ая, -ое, (*a*) especial, particular

суд·, -а́, (*m*) ~~(law)~~ law-court, court; opinion, judgement

~, **апелляцио́н·ный** court of appeal

~, **верх·о́вн·ый** supreme court

~, **вое́н·ный** court martial

~ **втор·о́й инста́нци·и** court of appellate jurisdiction

~, **кассацио́н·ный** court of cassation/ appeal

~, **мор·ск·о́й** Admiralty court

~ **пе́рв·ой инста́нци·и** court of original jurisdiction

~, **трет·е́йск·ий** arbitration, arbitration court

суд·а́, (*nom/acc pl*) of **суд·н·о́**

суда́к·, -а́, (*m*) (fish) perch, *Lucioperca*

~, **мор·ск·о́й** (fish) sea perch, *Lucioperca marina*

суда́н·, -а, (*m*) sudan (dye)

суда́н·ск·ая трав·а́ (*f*) (bot) Sudan grass, *Sorghum sudanensis*

судебно·медици́н·ск·ий, -ая, -ое, (*a*) medico-legal

суд·е́бн·ый, -ая, -ое, (*a*) court, judicial; forensic, legal

~ **за·сед·а́ни·е** (*n*) court sitting

~ **из·де́рж·к·и** (*pl*) (law) costs, legal costs

~ **медици́н·а** (*f*) forensic medicine

~ **проце́сс·** (*m*) judicial proceedings

~ **сле́д·стви·е** (*n*) judicial enquiry

суд·е́йск·ий, -ая, -ое, (*a*) judge's; umpire's, referee's

су́д·ен·, (*g pl*) of **суд·н·о́** = bed-pan

суд·и́ть, -ˊят, (*imp*) judge, form a judgement, form an opinion; try (in law-court); umpire, referee

суд·н·о́, -а́, (*nom pl*) **суд·а́, -о́в,** (*n*) (*see also associated adj*); ship, vessel, craft, boat: **суд·н·о́, -а́,** (*nom pl*) **-а́,** (*g pl*) **-д·ен·,** (*n*) (med) bed-pan

~, **барра́ж·н·ое** boom-defence vessel

суд·н·о́

~ **больш·о́го кабота́ж·а** long-range coaster

~, **брод·я́ч·ее** (naut) tramp

~, **гладко·па́луб·н·ое** flush-decked vessel

~, **голов·н·о́е** (shipb) class leader, prototype; leading ship (of formation)

~, **груз·ов·о́е** freighter, cargo vessel

~, **грузо·пассажи́р·ск·ое** cargo liner

~ **да́ль·н·его пла́в·ани·я** ocean-navigation vessel, ocean-going vessel

~ **ди́к·ого пла́в·ани·я** (naut) tramp

~, **кабота́ж·н·ое** coaster, coasting vessel

~, **колёс·н·ое** paddle ship

~, **коло́дез·н·ое** well-deck vessel

~, **комме́рч·еск·ое** merchant vessel, merchantman

~, **легко·па́луб·н·ое** awning-deck vessel

~ **лед·ов·о́го пла́в·ани·я** ship strengthened for navigation in ice

~ **-лов·у́шк·а, -и,** (g pl) **-ш·ек·,** (f) Q-ship

~ **ма́л·ого пла́в·ани·я** inshore and coasting vessel

~ **-мише́н·ь, -и,** (f) floating target, target ship

~, **мото́р·н·ое** motorship

~, **на·ли́в·н·о́е** tanker

~, **нефте·на·ли́в·н·о́е** (naut) oiler

~, **один·о́чн·ое** ship proceeding independently

~, **одно·ти́п·н·ое** sister ship

~, **остро·ки́ль·н·ое** sharp-bottomed vessel

~, **па́рус·н·ое** sailing vessel

~, **полно·на·бо́р·н·ое** deadweight cargo carrier

~, **порт·о́в·ое** port service craft

~, **про·мысл·о́в·ое** fishing vessel

~, **пуст·о́е** light ship, ship travelling light

~, **ре́йд·ов·ое** harbour launch

~, **с·пас·а́тельн·ое** salvage vessel

~, **с·пла́в·н·о́е** lumber raft

~, **торг·о́в·ое** merchant vessel

~, **тра́мп·ов·ое** tramp

~, **экспедицио́н·н·ое** research ship, expeditionary ship

судно·ко́рпус·н·ый, -ая, -ое, (a) (shipb) hull

су́до- (component) = (1) **суд·ов·о́й;** (2) **суд·е́бн·ый**

суд·о́в, (g pl) of **суд·н·о́**

судо·ве́рф·ь, -и, (f) shipyard, ship-building yard

судо·влад·е́лец·, -льц·а, (m) ship-owner

судо·вод·и́тел·ь, -я, (m) mariner, sailor

судо·вожд·е́ни·е, -я, (n) (naut) navigation

суд·ов·о́й, -а́я, -о́е, (a) ship, ship's, shipborne, marine

~ **архитекту́р·а** (f) naval architecture, ship construction

~ **ди́зел·ь** (m) marine diesel

~ **доро́г·а** (f) incline (in canal system)

~ **докуме́нт·ы** (pl) ship's papers

~ **рол·ь** (f) crew list

~ **с·вид·е́тельств·о** (n) (naut) certificate of registry

~ **систе́м·ы** (pl) (shipb) hull systems

~ **у·стро́й·ств·а** (pl) (shipb) hull equipment

~ **ход·** (m) fairway, channel

судо·мо́й·к·а, -и, (g pl) **-о́·ек·,** (f) washer-up (of crockery)

судо·подъ·ём·, -а, (m) ship raising, raising ships

судо·подъ·ём·ник·, -а, (m) ship-lift (e.g. in canal)

~, **на·клон·н·ый** rail-type slipway (in ship repair yard etc.); incline (canal system)

судо·подъ·ём·н·ый, -ая, -ое, (a) (naut) salvage, ship-raising

~ **понто́н·** (m) (naut) camel

~ **рабо́т·ы** (pl) salvage operations, lifting operations

~ **строп·** (m) lifting wire (salvage)

~ **су́д·н·о** (n) salvage vessel

судо·про·из·во́д·ств·о, -а, (n) legal procedure

судо·ремо́нт·, -а, (m) (naut) refit, ship repairing

~ **за·во́д·** (m) ship repair yard

~ **сух·о́й док·** (m) graving dock

су́дорог·а, -и, (f) (physiol) cramp, spasm

су́дорож·н·ый, -ая, -ое, (a) convulsive

судо·с·бо́р·очн·ая верф·ь (f) ship assembly yard

судо·стро·е́ни·е, -я, (n) shipbuilding

судо·стро·и́тел·ь, -я, (m) shipbuilder

судо·стро·и́тельн·ый, -ая, -ое, (a) shipbuilding

~ **верф·ь** (f) shipbuilding yard

~ **за·во́д·** (m) shipbuilding and marine engineering yard

~ **лес·** (m) shipbuilding timbers

судо·стро·и́тельн·ый, -ая, -ое
~ **лист·** (*m*) (met) ship plate
~ **цех·** (*m*) (shipb) hull construction shop
судо·у·стро́й·ств·о, -а, (*n*) judicial system
судо·хо́д·ност·ь, -и, (*f*) navigability
судо·хо́д·н·ый, -ая, -ое, (*a*) navigable
~ **кана́л·** (*m*) ship canal
судо·хо́д·ств·о, -а, (*n*) shipping, navigation
судо·я́м·а, -ы, (*f*) camber, natural building basin
суд·ьб·а́, -ы́, (*f*) fate, destiny
суд·ь·я́, -и́, (*nom pl*) **су́д·ь·и,** (*g pl*) **суд·е́й,** (*m*) judge; umpire, referee
суе·ве́р·и·е, -я, (*n*) superstition
су·ёт, (*pres 3rd sing*) of **сов·а́ть**
с·уж·а́·ть, -ют, (*imp*) make narrow/narrower, constrict, restrict; taper (of an extremity); **-ся** (*pass*); become narrow/narrower, contract, narrow, converge
с·уж·а́ющ·ий, -ая, -ее, (*pres part act*) of **с·уж·а́·ть;** (*see also* **с·у́ж·ивающ·ий**); **-ся** (met) necked-down; contracting, converging (of pipes etc.)
сужд·е́ни·е, -я, (*n*) judgement, opinion
с·уж·е́ни·е, -я, (*n*) (*v n*) (*see* **с·у·ж·а́·ть**); contraction, constriction, convergence; coning; (med) stenosis
~ **и́мпульс·а** pulse narrowing
~ **крыл·а́** (a/c) taper ratio
~, **от·нос·и́тельн·ое** contraction ratio; (met) reduction of area (on rupture)
у́гол· с·уж·е́ни·я (*m*) (dynam) angle of divergence
с·у́ж·енн·ый, -ая, -ое, (*past part pass*) *see* **с·уж·а́·ть;** narrowed, constricted, (med) stenotic
с·у́ж·ива·ть, -ют, (*imp*) = **с·уж·а́·ть**
с·у́ж·ивающ·ий, -ая, -ее, (*pres part act*) *see* **с·уж·а́·ть,** (*see also* **с·уж·а́·ющ·ий**)
~ **част·ь** (*f*) constrictor (fluid flow)
с·у́ж·ивающ·ееся, сопл·о́ (*n*) converging nozzle (of turbine)
с·у́з·ить, -ят, (*perf*) *see* **с·уж·а́·ть**
сук·, -а́, (*nom pl*) **су́ч·ь·я,** (*m*) branch, bough (of tree); knot (in wood)
су́к·а, -и, (*f*) (zool) bitch
суккуле́нт·, -а, (*m*) (bot) succulent
сукн·о́, -а́, (*nom pl*) **су́кн·а,** (*g pl*) **су́кон·,** (*n*) woollen cloth
сукно·ва́л·, -а, (*m*) (text) fuller
сукно·ва́ль·н·ый, -ая, -ое, (*a*) (text) fulling

сукно·ва́ль·н·ый, -ая, -ое
~ **маши́н·а** (*f*) fulling/milling machine
~ **цех·** (*m*) (text) fullery
сукно·вед·у́щ·ий ва́л·ик· (*m*) (paper) felt roll
сук·ова́т·ый, -ая, -ое, (*a*) branchy (of trees); knotted (of wood)
суко́н·к·а, -и, (*g pl*) **-н·ок·,** (*f*) piece of cloth, cloth, rag
суко́н·н·ый, -ая, -ое, (*a*) cloth, woollen-cloth
~ **шерст·ь** (*f*) clothing wool
суко́н·щик·, -а, (*m*) cloth maker
су́кров·иц·а, -ы, (*i*) **-ей,** (*f*) (med) sanies
су́кров·ичн·ый, -ая, -ое, (*a*) (med) sanious
су·кру́т·ин·а, -ы, (*f*) (text) snarl
сукце́сси·я, -и, (*f*) (biol) succession
сукцин·ами́д·, -а, (*m*) (chem) succinamide
сукцин·ами́н·ов·ый, -ая, -ое, (*a*) (chem) succinamic
~ **кисл·от·а́** (*f*) succinamic acid
сукцин·амои́л·, -а, (*m*) (chem) succinamoyl, succinamyl
сукцин·и́л·, -а, (*m*) (chem) succinyl
сукцинил·янта́р·н·ый, -ая, -ое, (*a*) succinylsuccinic
сукцин·и́т·, -а, (*m*) (min) succinite
сукцино·дегидра́з·а, -ы, (*f*) (biochem) succinodehydrogenase
сулем·а́, -ы́, (*f*) mercuric chloride, (chem) mercuric bichloride, sublimate, corrosive sublimate
султа́н·, -а, (*m*) sultan (title); plume, cluster of feathers
султано·ви́д·н·ый, -ая, -ое, (*a*) plume-like
сульвани́т·, -а, (*m*) (min) sulvanite
сульги́н·, -а, (*m*) (pharm) sulgine, para-amino-benzene sulfaguanadine
сульф- (*int component*) sulph- (Brit), sulf- (U.S.)
сульфадимези́н·, -а, (*m*) (pharm) sulfadimezine
сульфази́д·, -а, (*m*) sulphazide, sulfazide
сульфази́н·, -а, (*m*) sulphadiazine, sulfadiazine
сульфазо́л·, -а, (*m*) (pharm) sulphamethylthiazole, sulfamethylthiazole
сульфама́т·, -а, (*m*) sulphamate, sulfamate
сульф·ами́д·, -а, (*m*) (chem) sulphamide, sulfamide; (pharm) sulphonamide, sulfonamide

сульфами́д·н·ые препара́т·ы   (*pl*) sulphonamide/sulfonamide drugs, sulphonamides, sulpha/sulfa drugs

сульфами́но-   (*int component*) sulphamino-, sulfamino-

сульфами́л·, -а,   (*m*) sulphamyl, sulfamyl

сульфани́л·, -а,   (*m*) sulphanilyl, sulfanilyl

сульфаниламид·, -а,   (*m*) sulphanilamide, sulfanilamide

сульфаниламид·н·ые препара́т·ы (*pl*) = **сульфонами́д·н·ые препара́т·ы**

сульфани́л·ов·ый, -ая, -се,   (*a*) sulphanilic, sulfanilic

~ **кисл·от·а́**   (*f*) sulphanilic/sulfanilic acid

сульф·а́т·, -а,   (*m*) sulphate, sulfate

~ **аммо́ни·я**   ammonium sulphate/sulfate, sulphate/sulfate of ammonia

~, **ки́сл·ый**   acid sulphate/sulfate, bisulphate

~ **ура́н·а**   uranium sulphate/sulfate

сульфат·а́з·а, -ы,   (*f*) (chem) sulphatase, sulfatase

сульфат·а́ци·я, -и,   (*f*) (elec) sulphation/sulfation

сульфатиззо́л·, -а,   (*m*) (pharm) sulphathiazol, sulfathiazole, thiazamide

сульфат-канкрини́т·, -а,   (*m*) (min) sulphatic cancrinite, wishnewite

сульфат·и́ровани·е, -я,   (*n*) sulphation, sulfation

сульфатно·сто́й·к·ий, -ая, -ое,   (*a*) sulphate-resistant

сульфа́т·н·ый, -ая, -ое,   (*a*) sulphate, sulfate

~ **спсcоб· ва́р·к·и**   (*m*) (paper) sulphate/sulfate process

~ **стекл·о́**   (*n*) sulphate of soda glass

~ **целлюло́з·а**   (*a*) (paper) sulphate/sulfate pulp

сульфато-   (*int component*) sulphato-, sulfato-

сульфаци́л·, -а,   (*m*) (pharm) sulphacil, sulphacetamide

сульф·гидра́т·, -а,   (*m*) hydrosulphide, hydrosulfide, sulphhydrate, sulfhydrate

сульф·гидри́л·, -а,   (*m*) (biochem) thiol, sulphhydryl, sulfhydryl

сульф·гидри́ль·н·ый, -ая, -ое,   (*a*) of **сульф·гидри́л·**

сульфенами́д·, -а,   (*m*) sulphenamide (rubber accelerator)

сульфи́д·, -а,   (*m*) sulphide, sulfide

~ **на́три·я**   sodium sulphide/sulfide

сульфиди́н·, -а,   (*m*) (pharm) sulphidine/sulfidine, sulphapyridine

сульфид·и́ровани·е, -я,   (*n*) sulphidizing, sulfidizing

сульфи́д·н·ый, -ая, -ое,   (*a*) sulphide, sulphidic, sulfide, sulfidic

~ **сéр·а**   (*f*) sulphidic/sulfidic sulphur

сульфизсксазо́л·, -а,   (*m*) (pharm) sulfisoxazole

сульфими́д·, -а,   (*m*) sulphimide, sulfimide

сульфина́т·, -а,   (*m*) sulphinate, sulfinate

сульфини́д·, -а,   (*m*) sulphimide, sulphinide, sulfimide, sulfinide

~ **бензо́й·н·ой кисл·от·ы́**   benzosulphimide, benzosulfinide

сульфини́л·, -а,   (*m*) sulphinyl, sulfinyl, thionyl

сульфи́но-   (*int component*) sulphino-, sulfino-

сульфи́н·ов·ый, -ая, -ое,   (*a*) sulphinic, sulfinic

~ **кисл·от·а́**   (*f*) sulphinic/sulfinic acid

сульф·и́ровани·е, -я,   (*n*) sulphonation (BrᎥᵗ), sulfonation (U.S.)

сульф·и́рованн·ый, -ая, -ое,   (*past part pass*) sulphonated, sulfonated

сульфи́т·, -а,   (*m*) sulphite, sulfite

~ **на́три·я**   sodium sulphite/sulfite

сульфит·а́тор·, -а,   (*m*) (food) sulphitizer, sulfitizer, sulphiter/sulfitor tower

сульфит·а́ци·я, -и,   (*f*) sulphitation, sulfitation, sulphiting, sulfiting

сульфи́т·н·ый, -ая, -ое,   (*a*) sulphite, sulfite

~ **спосо́б· ва́р·к·и**   (*m*) (paper) sulphite/sulfite process

~ **целлюло́з·а**   (*f*) (paper) sulphite/sulfite wood pulp

~ **щёл·ок·**   (*m*) (paper) sulphite/sulfite liquor, sulphite waste liquor

сульфито́·метр·, -а,   (*m*) sulphite/sulfite dosimeter/feeder

сульфит·целлюло́з·н·ый, -ая, -ое, (*a*) sulphite/sulfite-cellulose

~ **щёл·ок·**   (*m*) sulphite/sulfite waste liquor

~ **экстра́кт·**   (*m*) sulphite/sulfite cellulose extract (leather tanning)

сульфнитра́м·, -а,   (*m*) (agr) sulphur-ammonium nitrate fertilizer

сульфо-   (*int component*) sulpho-, sulfo-

сульфо·аммо·фо́с·, -а,   (*m*) (agr) ammonium phosphate and sulphate fertilizer

сульфо·гру́пп·а, -ы,   (*f*) sulphonic acid group

сульфо·кисл·от·а́, -ы́,   (*f*) sulphonic/sulfonic acid

сульф·окси́д·, -а,   (*m*) sulphoxide, sulfoxide

сульф·оксила́т·, -а,   (*m*) sulphoxylate, sylfoxylate

сульф·окси́л·ов·ая кисл·от·а́   (*f*) sulphoxylic/sulfoxylic acid

сульф·о́н·, -а,   (*m*) (chem) sulphone, sulfone

сульфон·а́л·, -а,   (*m*) sulphonal

сульфон·ами́д·, -а,   (*m*) (pharm) sulphonamide, sulfonamide

сульфон·ами́д·н·ые препара́т·ы   (*pl*) sulpho/sulfo drugs

сульфон·а́т·, -а   (*m*) sulphonate, sulfonate

сульфо́н·иев·ый, -ая, -ое,   (*a*) of **сульфо́н·и·й**

∼ со·един·е́ни·е   (*n*) sulphine, sulfine, sulphonium compound

сульфо́н·и·й, -я,   (*m*) sulphonium, sulfonium

сульфон·и́л·, -а,   (*m*) (biochem) sulphonyl, sulfonyl

сульфон·и́ровани·е, -я,   (*n*) sulphonation, sulfonation

сульфо́н·ов·ый, -ая, -ое,   (*a*) sulphonic, sulfonic

∼ кисл·от·а́   (*f*) sulphonic/sulfonic acid

сульфо·оксо́л·ь, -я,   (*m*) "sulpho-oxole" (a drying oil)

сульфопо́н·, -а,   (*m*) "sulphopone" (a pigment)

сульфо·со·един·е́ни·е, -я,   (*n*) sulpho/sulfo-compound

сульфо·со́л·ь, -и,   (*f*) thio salt

сульфо·у́гол·ь, -гл·я,   (*m*) thiocarbonate

сульфо·хлор·и́ровани·е   (*n*) chlorosulphonation

сульфур·а́тор·, -а,   (*m*) sulphonator, sulfonator

сульфур·и́л·, -а,   (*m*) sulphuryl, sulphonyl, sulfuryl, sulfonyl

сум·а́, -ы́,   (*f*) pouch, bag

сума́х·, -а,   (*m*) (bot) sumac, sumach, *Rhus*

сумах·и́рованн·ый, -ая, -ое,   (*a*) sumac-tanned (leather)

сумас·ше́д·ш·ий, -ая, -ее,   (*a*) insane, mad

сумас·ше́ств·и·е, -я,   (*n*) insanity, madness

сумбу́л·, -а,   (*m*) (bot) sumbul, musk, *Ferula moschata/sumbul*

су́·мер·ечн·ый, -ая, -ое,   (*a*) twilight; (*pl as noun*) (ent) crepusculars, *Crepuscularia*

су·мер·к·и, -р·ек·,   (*f pl*) twilight, dusk

∼, астроном·и́ческ·ие astronomical twilight

∼, гражд·а́нск·ие civil twilight

∼, навигацио́н·н·ый nautical twilight

с·ум·е́·ть, -ют,   (*perf*) be able to, succeed in

су́м·к·а, -и,   (*g pl*) **-м·ок·,** (*f*) bag, pouch, wallet; (anat) sac, bursa, capsule; (zool) pouch; (bot) ascus, sac, bursa

∼, вы́·вод·ков·ая (zool) brood pouch, marsupium

сумко·спо́р·а, -ы,   (*f*) (bot) ascospore

су́мм·а, -ы,   (*f*) sum, total

∼, свет·ов·а́я (phys) light sum

сумм·а́рн·ый, -ая, -ое,   (*a*) total, summated, combined, sum; summary, summarized

∼ де́й·стви·е (*n*) resultant action

∼ тон· (*m*) (acous) summation tone

сумм·а́тор·, -а,   (*m*) adder, summator

∼, арифмет·и́ческ·ий non-algebraic adder

∼, бала́нс·н·ый (autom) balanced summator

∼ вы́с·ш·его раз·ря́д·а left-hand adder

∼, дво·и́чн·ый binary adder

∼, десят·и́чн·ый decimal adder

∼, комбинацио́н·н·ый adder

∼, комбин·и́рованн·ый coincidence adder

∼ ле́в·ого раз·ря́д·а left-hand adder

∼, на·ка́пл·ивающ·ий counter-type adder; accumulator, accumulator register (computers)

∼ ни́з·ш·его раз·ря́д·а right-hand adder

∼, одно·раз·ря́д·н·ый one-column adder

∼, паралле́ль·н·ый parallel adder

∼, по·сле́д·овательн·ый serial adder

сумм·а́ци·я, -и,   (*f*) summation

сумм·и́ровани·е, -я,   (*n*) (*v n*) of **сумм·и́р·овать;** summation, addition

сумм·и́р·овать, -уют,   (*imp and perf*) add up, find the sum/total; sum up, summarize

сумм·и́руем·ый, -ая, -ое,   (*pres part pass*) of **сумм·и́р·овать;** (math) integrable, summable

сумм·и́рующ·ий, -ая, -ее, (*pres part act*) of **сумм·и́р·овать**

~ **маши́н·а** (*f*) algebraic adding machine

~ **у·стро́й·ств·о** (*n*) adder, totalizer

сумм·ов·о́й тон· (*m*) (acous) summation tone

су́м·очк·а, -и, (*f*) (*dim*) of **су́м·к·а;** (biol) utricle

су́м·очн·ый, -ая, -ое, (*a*) of **су́м·оч·к·а;** utricular

су́·мрак·, -а, (*m*) gloom

су́м·чат·ый, -ая, -ое, (*a*) (bot) ascomycetous; (zool) marsupial, (*pl as noun*) marsupials, *Marsupialia*

~ **барсу́к·** (*m*) (zool) bandicoot, (*pl*) *Peramelidae*

~ **гриб·** (*m*) (bot) ascomycete

~, **много·рез·цо́в·ые** (*pl*) (zool) *Polyprotodontia*

сунду́к·, -а́, (*m*) trunk, coffer, box

су́·н·уть, -ут, (*perf*) shove, thrust, poke

су́п·, -а, (*nom pl*) **-ы́,** (*g pl*) **-о́в,** (*m*) soup

супер- (*int component*) super-

супер·аэродина́м·ик·а, -и, (*f*) superaerodynamics

супер·гетероди́н·, -а, (*m*) (rad) superheterodyne

супер·иконоско́п·, -а, (*m*) (TV) super-iconoscope

супер·дуралюми́н·, -а, (*m*) (met) super duralumin

супер·кала́ндр·, -а, (*m*) (paper) super-calander

супер·ка́рго (*n indecl*) (com) supercargo

супермалло́·й, -я, (*m*) (met) Supermalloy

супер·об·ло́ж·к·а, -и, (*g pl*) **-ж·ек·,** (*f*) dust-cover/-jacket (of book)

супер·об·ма́з·к·а, -и, (*g pl*) **-з·ок·,** (*f*) fungicidal paste (for wood)

супер·ортико́н·, -а, (*m*) (TV) image orthicon

супер·пози́ци·я, -и, (*f*) superposition, superimposition; (cryst) pile-up

супер·регенерати́в·н·ый, -ая, -ое, (*a*) (rad) super-regenerative/-reactive

супер·фи́ниш·, -а, (*m*) (mech eng) super-finish

супер·фосфа́т·, -а, (*m*) (agr) superphosphate

супер·фузи́в·н·ый, -ая, -ое, (*a*) (geol) superfusive

супер·центрифу́г·а, -и, (*f*) (chem) supercentrifuge

су́·пес·н·ый, -ая, -ое, (*a*) = **су·пес·ча́н·ый**

су·песч·а́н·ый, -ая, -ое, (*a*) sandy-loam

су́·пес·ь, -и, (*f*) sandy loam (soil)

суп·ов·о́й, -а́я, -о́е, (*a*) soup

су́порт·, -а, (*m*) *see* **су́ппорт·**

су́ппорт·, -а, (*m*) carriage, saddle (of machine tools)

~, **крест·о́в·ый** compound rest

~, **по·воро́т·н·ый** swivel carriage

~, **револьве́р·н·ый** turret carriage

~, **тангенциа́ль·н·ый** tangential-feed carriage

супра- (*int component*) supra-

супра·литора́ль·н·ый, -ая, -ое, (*a*) (geog) supralittoral

супра·рени́н·, -ая, -ое, (*a*) (chem) suprarenin

супра·фосфа́т·, -а, (*m*) (agr) supraphosphate, basic slag

су·проти́в·н·ый, -ая, -ое, (*a*) (bot) opposite, opposed

су·пру́г·, -а, (*m*) husband

су·пру́г·а, -и, (*f*) wife

су·пру́ж·еск·ий, -ая, -ое, (*a*) matrimonial, (med) conjugal

сургу́ч·, -а́, (*i*) **-о́м,** (*m*) sealing-wax

сургу́ч·ниц·а, -ы, (*f*) sealing wax heater

су́рик·, -а, (*m*) red lead, minium

~, **желе́з·н·ый** Indian red

~, **свин·цо́в·ый** red lead

сури́ль·е, -я, (*n*) (zool) skunk, *Conepatus suffocans*

сури́ль·о, -а, (*n*) = **сури́ль·е**

суро́в·ый, -ая, -ое, (*a*) severe, rigorous, stern; (text) unbleached, grey, gray

суров·ь·ё, -я́, (*n*) (text) grey/gray cloth

сур·о́к·, -р·к·а́, (*m*) (zool) marmot, (*pl*) *Marmota*

суррога́т·, -а, (*m*) substitute

сурьм·а́, -ы́, (*f*) (chem) antimony, stibium, Sb

~, **пяти·сер·ни́ст·ая** antimony black

~, **пяти·хло́р·ист·ая** antimonic chloride

~, **трёх·фто́р·ист·ая** antimony trifluoride

~, **черн·ов·а́я** regulus metal

сурьми́н·, -а, (*m*) (pharm) "surmin" para-acetamidobenzene, sodium stibonate

сурьмин·и́рованн·ый, -ая, -ое, (*a*) stibiated

сурьм·яни́ст·ый, -ая, -ое, (*a*) antimonous, stibious, antimonide (of); (met) antimonial

сурьм·яни́ст·ый, -ая, -ое
~ **водо·ро́д·** (*m*) antimony/antimonous hydride, stibine
~ **кисл·от·а́** (*f*) antimonous acid
~ **мета́лл·** (*m*) antimonide
~ **свин·е́ц·** (*m*) (met) antimonial lead
сурьм·я́н·ый, -ая, -ое, (*a*) antimony, antimonic stibic, antimonial
~ **ангидри́д·** (*m*) antimony pentoxide
~ **блеск·** (*m*) (min) stibnite, antimony glance
~ **руд·а́** (*f*) antimony ore
~ **руд·а́, блёк·л·ая** stibnite, grey antimony
суса́к·, -а́, (*m*) (bot) flowering rush, *Butomus*
суса́л·ь, -и, (*f*) gold leaf
сусе́к·, -а, (*m*) corn bin
су́слик·, -а, (*m*) (zool) souslik, suslik, pouched marmot, *Citellus*
су́сл·о, -а, (*m*) must (winemaking); wort (brewing); mash (wood hydrolysis)
~ **-ага́р·, -а,** (*m*) (microbiol) wort agar
суспенд·и́р·овать, -уют, (*imp and perf*) suspend
суспе́нз·и·я, -и, (*f*) (chem) suspension
суспенз·о́ри·й, -я, (*m*) (med) suspensorium
суста́в·, -а, (*m*) joint, articulated joint, articulation, (anat) diarthrosis
~**, шаро·ви́д·н·ый** (anat) arthrodia
сустав·н·о́й, -а́я, -о́е, (*a*) articular, articulated, jointed, (zool) arthrodial
суста́в·чат·ый, -ая, -ое, (*a*) articulate, jointed; (*pl decl as noun*) (zool) articulates, *Articulata*
су́т·к·и (*nom/acc pl*), (*g pl*) **су́т·ок·** 24 hours, day(s)
~**, Бе́ссел·ев·ы** Besselian day
~**, гражд·а́нск·ие** civil day(s)
~**, звёзд·н·ые** sidereal day(s)
~**, и́ст·инн·ые** apparent/true day(s)
~**, кру́гл·ые** round the clock, continuously
~**, лу́н·н·ые** lunar day(s)
~**, марсиа́н·ск·ие** Martian day(s)
~**, при·ли́в·н·ые** tidal day(s)
~**, со́лн·ечн·ые** solar day(s)
~**, сре́д·н·ие со́лн·ечн·ые** mean solar day(s)
~**, юлиа́н·ск·ие** Julian day(s)
су́т·очн·ый, -ая, -ое, (*a*) 24-hour, diurnal, daily, day's; (*pl as noun*) (fin) daily allowance
~ **амплиту́д·а** (*f*) (meteor) daily range

су́т·очн·ый, -ая, -ое
~ **пла́в·ани·е** (*n*) (nav) day's run
~ **при·ли́в·** (*m*) diurnal tide
~ **ход·** (*m*) (horol) daily rate; diurnal variation (of wind)
суту́л·ить, -ят, (*imp*) stoop, hunch, bend (spine etc.)
суту́н·к·а, -и, (*g pl*) **-н·ок·,** (*f*) (met) sheet bar
су́т·ь, -и, (*f*) crux, main point, essence
по су́т·и in fact
су·тяж·ни́ческ·ий, -ая, -ое, (*a*) litigious
суфлёр·, -а, (*m*) (theat) prompter; (min) breather, piper; (geol) fumarole; (eng) blower, au vent
суфля́р·, -а, (*m*) = **суфлёр·**
су́ффикс·, -а, (*m*) (gram) suffix
суффо́зи·я, -и, (*f*) (geol) undermining, suffosion
суха́р·ник·, -а, (*m*) (met cast) gagger
суха́р·ь, -я́, (*m*) dried bread, rusk; (mech eng) insert piece, block; refractory kaolin clay
сухо·адиабат·и́ческ·ий гради́ент· (*m*) (meteor) dry adiabatic lapse rate
сух·ова́т·ый, -ая, -ое, (*a*) dryish
сухо·ве́·й, -я, (*m*) (meteor) dry wind
сухо·воз·ду́ш·н·ый, -ая, -ое, (*a*) dry-air
~ **на·со́с·** (*m*) dry-air pump
сухо·жи́л·и·е, -я, (*n*) (anat) tendon
сухо·жи́ль·н·ый, -ая, -ое, (*a*) (anat) tendinous
сух·о́й, -а́я, -о́е, (*a*) dry, dried, desiccated, arid
~ **молок·о́** (*n*) dried milk
~ **пере·го́н·к·а** (*f*) destructive/dry distillation
~ **термо́·метр·** (*m*) dry-bulb thermometer
сухо·лю́б·, -а, (*m*) (bot) xerophyte
сухо·люб·и́в·ый, -ая, -ое, (*a*) (bot) xerophilous
сухо·па́р·ник·, -а, (*m*) steam dome/receiver
сухо·па́р·ый, -ая, -ое, (*a*) lean, spare, thin
сухо·пло́д·н·ый, -ая, -ое, (*a*) (bot) xerocarpous
сухо·пу́т·н·ый, -ая, -ое, (*a*) land (as opposed to sea etc.), terrestrial
сухо·сто́·й, -я, (*m*) dead wood, dead standing wood
су́х·ост·ь, -и, (*f*) dryness, aridity
сухо́т·к·а, -и, (*g pl*) **-т·ок·,** (*f*) (med) tabes

сухо́т·к·а
~ спин·н·о́го мо́зг·а   (met) tabes dorsalis

сухо́т·очн·ый, -ая, -ое,   (a) (med) tabetic

сухо·фру́кт·ы, -ов,   (m pl) dried fruit(s)

сухо·цвѐт·, -а,   (m) (bot) xeranthemum

суч·е́ни·е, -я,   (n) (v n) of **суч·и́ть;** twist, twisting

суч·ён·ый, -ая, -ое,   (a) (text) twisted

суч·и́ть, -а́т,   (imp) (text) twist, spin; throw (silk)

суч·кова́т·ый, -ая, -ое,   (a) knotted (of wood)

сучко·ре́з·, -а,   (m) (agr) lopping shears; knot cutter/bumper (wood)

суч·о́к·, -ч·к·а́,   (m) (dim) of **сук·;** knot (in wood)

су́ч·ь·я,   (nom/acc pl) of **сук·**

су́ш·а, -и,   (i) -ей,   (f) land, dry land (as opposed to sea)

су́ш·е,   (comp) of **сух·о́й**

суш·е́ни·е, -я,   (n) (v n) see **суш·и́ть;** dried fruit/vegetables

суш·ёнк·а, -и,   (g pl) -нок·,   (f) dried peat

суш·ён·ый, -ая, -ое,   (a) dry, dried, desiccated

суш·и́лк·а, -и,   (g pl) -лок·,   (f) dryer

~, ва́льц·ев·а́я   rotary dryer

~, ка́мер·н·ая   compartment dryer

~, ле́нт·очн·ая   apron dryer

~, плён·очн·ая   film dryer

~, под·о́в·ая   hearth dryer

~, тунне́ль·н·ая   tunnel dryer/kiln

~, цилиндр·и́ческ·ая   rotary dryer; (text) tumble dryer

~, ша́хт·н·ая   shaft dryer

суш·и́л·о, -а,   (m) dryer

суш·и́льн·ый, -ая, -ое,   (a) drying

~ бараба́н·   (m) tumbler/rotary dryer; (text) steam-can dryer

~ плит·а́   (f) (met cast) core/drying plate

~ стан·о́к·   (m) (phot) drying rack

~ шкаф·   (m) drying cabinet, dryer, desiccator

суш·и́льн·я, -и,   (f) drying room

суш·и́ть, -ат,   (imp) dry, desiccate, cure (food)

су́ш·к·а, -и,   (f) (v n) of **суш·и́ть**

сущ·е́ственн·о   (adv) essentially, substantially

~ о·со́б·ая то́ч·к·а   (f) (math) essential singular point

сущ·е́ственност·ь, -и,   (f) significance, essence, basic/fundamental nature/feature

сущ·е́ственн·ый, -ая, -ое,   (a) significant, considerable, basic, fundamental, important, essential

сущ·естви́тельн·ое, -ого,   (n decl as adj) (gram) noun, substantive

сущ·еств·о́, -а́,   (n) main point, essence, substance; creature

по сущ·еств·у́   essentially

сущ·ествова́ни·е, -я,   (n) existence, (geol) occurrence

сущ·еств·ова́ть, -у́ют,   (imp) exist

сущ·еств·у́ющ·ий, -ая, -ее,   (pres part act) of **сущ·еств·ова́ть;** existent, current, present-day

су́щ·ий, -ая, -ее,   (a) real, downright

су́щ·ност·ь, -и,   (f) main point, essence, substance

су·я́гн·ост·ь, -и,   (f) pregnancy, being in lamb (of sheep), being in kid (of goats)

сфагну́м·, -а,   (m) (bot) peat moss, Sphagnum

с·фаз·и́р·овать, -уют,   (perf) see **фа-з·и́р·овать**

сфалер·и́т·, -а,   (m) (min) sphalerite, zincblende

сфацеля́р·иев·ые во́до·росл·и   (pl) (bot) Sphacelariales

сфен·, -а,   (m) (min) sphene, titanite

сфен·одо́н·, -а,   (m) (zool) sphenodont

сфен·о́ид·, -а,   (m) (cryst) sphenoid

сфеноида́ль·н·ый, -ая, -ое,   (a) sphenoidal

сфеноид·и́т·, -а,   (m) (med) sphenoiditis

сфен·оид·о́м·а, -ы,   (f) (cryst) dihedron

сфено·ли́т·, -а,   (m) (geol) sphenolith

сфено·пте́рис·, -а,   (m) (pal, bot) Sphenopteris

сфено·фи́лл·ы, -ов,   (m pl) (pal, bot) Sphenophyllales

сфе́р·а, -ы,   (f) sphere

~, армилля́р·н·ая   armillary sphere

~, неб·е́сн·ая   (astron) celestial sphere

~ сла́б·ост·и   (geol) asthenosphere, tectosphere, zone of weakness

сфер·и́ческ·ий, -ая, -ое,   (a) spherical

~ из·бы́т·ок·   (m) (math, surv) spherical excess

~ ма́ят·ник·   (m) (phys) spherical pendulum

~ тре·уго́ль·ник·   (m) (math) spherical triangle

сфер·и́чност·ь, -и,   (f) sphericity

сфер·о́ид·, -а,　(*m*) (math) spheroid, ellipsoid of revolution

сфероида́ль·н·ый, -ая, -ое,　(*a*) spheroidal, globular

~ **структу́р·а**　(*f*) (geol, min, met) globular/spheroidal structure

сфероидиз·а́ци·я, -и,　(*f*) (met) spheroidize annealing

сфероид·и́ческ·ий, -ая, -ое,　(*a*) spheroidal, globular

~ **тре·уго́ль·ник·**　(*m*) (math) spheroidal/geodetic triangle

сферо·ка́рп·ов·ый, -ая, -ое,　(*a*) (bot) sphaerocarpous

сферо·ли́т·, -а,　(*m*) spherulite

сферо·ли́т·ов·ый, -ая, -ое,　(*a*) spherulite, (geol) spherulitic

сферо́·метр·, -а,　(*m*) (opt) spherometer

сферо·пла́ст·, -а,　(*m*) (bot) spheroplast, sphaeroplast

сферо·сидер·и́т·, -а,　(*m*) (min) spherosiderite

сферо·те́к·а, -и,　(*f*) (bot) *Sphaerotheca*

сфер·у́л·а, -ы,　(*f*) (biol) spherule

сфигмо́·граф·, -а,　(*m*) (med) sphygmograph

сфигмо·ло́г·и·я, -и,　(*f*) (med) sphygmology

сфигмо·ма́но·метр·, -а,　(*m*) (med) sphygmomanometer

сфингози́н·, -а,　(*m*) (chem) sphingosine

сфингмиэли́н·, -а,　(*m*) (biochem) sphyngomyelin

сфи́нкс·, -а,　(*m*) (ent) sphinx moth

сфи́нктер·, -а,　(*m*) (anat) sphincter

с·фокус·и́р·овать, -у:ют,　(*perf*) focus; **-ся** come into focus

с·форм·ир·ова́ть, -у́ют,　(*perf*) form, shape; mould, mold; **-ся** (*pass*); take shape

с·форм·ова́ть, -у́ют,　(*perf*) *see* **форм·ова́ть**

с·формул·и́р·овать, -уют,　(*perf*) formulate

с·фотограф·и́р·овать, -уют,　(*perf*) photograph

с·хва́т·, -а,　(*m*) grab, gripping device, tongs; fastening, grip, clamp

с·хват·и́ть, ·ят,　(*perf*) *see* **с·хва́т·ыва·ть**

с·хват·к·а, -и,　(*g pl*) **-т·ок·,**　(*f*) (*see also* **с·хва́т·**); (mil) engagement; meeting (sport); muscular pain, cramp; fastening, grip, clamp; (naut) stopping, seizing (on rope etc.)

с·хва́т·ывани·е, -я,　(*n*) (*v n*) of **с·хва́т·ыва·ть**; set, setting (concrete); (rubber) setting up

температу́р·а с·хва́т·ывани·я　(*f*) set point (of concrete)

с·хва́т·ыва·ть, -ют,　(*imp*) seize, grab, snatch, catch; secure, fasten, tie, catch in (of clothes); **-ся** (*pass*): set (concrete)

с·хва́ч·енн·ый, -ая, -ое,　(*past part pass*) *see* **с·хва́т·ыва·ть**

схе́м·а, -ы,　(*f*) diagram, schematic diagram, layout diagram; scheme; (elec) circuit, gate, electronic circuit; (aerodynam) configuration

~ **анти·со·в·пад·е́ни·я**　anticoincidence circuit (counters)

~, **аэродинам·и́ческ·ая**　aerodynamic configuration

~, **бло́ч·н·ая**　block diagram

~, **ве́нтиль·н·ая**　gate circuit

~ **в·ключ·е́ни·я**　(elec eng) circuit diagram

~ **вы́·держ·к·и**　(elec) delay network

~ **гаш·е́ни·я**　quench circuit

~, **дво·и́чн·ая пере·с·чёт·н·ая**　binary scaler, scale-of-two circuit (computers)

~, **двух·ме́р·н·ая**　(elec) network analyser

~, **двух·та́кт·н·ая**　push-pull circuit

~, **дека́д·н·ая пере·с·чёт·н·ая**　decade scaler, scale-of-ten circuit (computers)

~, **дел·и́тельн·ая**　scaling circuit

~ **до·полн·е́ни·я и при·бавл·е́ни·я пере·но́с·а**　complement-and-carry-add circuit (computers)

~, **за·по·мин·а́ющ·ая**　memory/storage circuit (computers)

~ **за·прещ·е́ни·я**　"except" gate, inhibitory gate

~ **за́·пуск·а**　trigger circuit

~ **"и"**　"and" gate/circuit

~, **из·бир·а́тельн·ая**　decoder (telephones)

~ **из·мер·е́ни·я**　measuring circuit

~ **или**　"or" gate/circuit (computers)

~ **ис·ключ·е́ни·я**　exclusive circuit (computers)

~, **ключ·ев·а́я**　gate/gating circuit

~, **кольц·ев·а́я десят·и́чн·ая**　ring-of-ten circuit (computers)

~, **кольц·ев·а́я пере·с·чёт·н·ая**　ring scaler (computers)

~, **лин·е́йн·ая параметр·и́ческ·ая**　(elec) linear varying parameter network

схём·а
~, **лог·и́ческ·ая** (eng) functional/logical diagram; (elec) logical/"and" circuit
~ **лог·и́ческ·ого и** "and" circuit
~ **лог·и́ческ·ого не** "not" circuit
~, **металлиз·и́рованн·ая** printed circuit
~, **модел·и́рующ·ая** analogous/simulator circuit
~ **на·гру́з·ок·** (eng) load diagram
~ **не·со·в·пад·е́ни·я** anticoincidence circuit
~ **не** "not" circuit
~, **не·примити́в·н·ая** (elec) finite automaton
~ **нет** "not" circuit
~, **нул·ев·а́я** null circuit
~, **объ·един·е́ни·я** "or" gate/circuit
~, **о·грани́ч·ивающ·ая** clipping circuit
~ **одно·вре́мен·ност·и** time-coincidence circuit
~ **о·про·ки́д·ывани·я** flip-flop circuit
~ **пере·но́с·а, ве́нтиль·н·ая** carry-gating circuit
~, **пере·кид·н·а́я** flip-flop circuit
~, **печа́т·н·ая** printed circuit
~ **по·сле́д·овательност·и** series circuit
~, **потенц·и́рующ·ая** antilogarithmic circuit
~ **по·то́к·а** (dynam) flow pattern
~, **принципиа́ль·н·ая** block diagram; basic diagram
~ **про·из·во́д·ственн·ого проце́сс·а** flow sheet
~, **одно·ва́ль·н·ая** (rocket) single-shaft configuration
~, **про·пуск·а́ющ·ая** gate circuit
~, **раз·вёр·нут·ая** series circuit
~ **раз·дел·е́ни·я** "or" gate/circuit
~ **разно·вре́мен·ност·и** time anti-coincidence circuit
~ **рас·па́д·а** (nucl) decay diagram
~ **рас·по·лож·е́ни·я** layout/location diagram, arrangement
~, **рас·чёт·н·ая** design diagram
~, **реле́й·н·о-конта́кт·н·ая** network diagram (switching algebra)
~, **рефле́кс·н·ая** (rad) reflex/dual-amplification circuit
~ **само·лёт·а** aircraft configuration
~, **само·лёт·н·ая** (rocket) twist and steer layout
~ **с·бро́с·а** reset circuit
~, **скеле́т·н·ая** skeleton diagram

схём·а
~, **со·бир·а́тельн·ая** "or" gate/circuit
~ **со·в·пад·е́ни·я** "and"/coincidence circuit/gate
~ **со·един·е́ни·я** (elec) circuit diagram
~, **с·пуск·ов·а́я** trigger circuit
~, **технолог·и́ческ·ая** flow chart/sheet
~, **три́ггер·н·ая** (elec) trigger/flip-flop circuit
~ **у·стано́в·к·и** installation diagram
~, **фикс·и́рующ·ая** (TV) clamping circuit, clamper; hold circuit (computers)
~ **форм·ирова́ни·я и́мпульс·ов** pulse-shaping circuit
~ **цоколо́вк·и** (rad) basing diagram (vacuum tubes)
~, **электр·и́ческ·ая** circuit diagram
~, **элеме́нт·н·ая** block diagram
схематиз·а́ци·я, -и, (f) (v n) of **схематиз·и́р·овать**
схематиз·и́р·овать, -уют, (imp and perf) outline, give an outline of, present diagrammatically
схемат·и́ческ·ий, -ая, -ое, (a) diagrammatic; outline, schematic
схемат·и́чн·о (adv) in outline
схемат·и́чн·ый, -ая, -ое, (a) outline
схемо·те́хн·ик·а, -и, (f) circuitry (computers)
схиз- (int component) see also **шиз-**
схиз·а́нтус·, -а, (m) (bot) butterfly flower, Schizanthus
схизо- (int component) see also **шизо-**
схизо·ге́н·н·ый, -ая, -ое, (a) (bot) schizogenetic, schizogenic
схизо·гнат·и́ческ·ий, -ая, -ое, (a) (zool) schizognathous
схизо·ка́рп·и·й, -я, (m) (bot) schizocarp, schizocarpium
схизо·ли́т·, -а, (m) (min) schizolite
схизо·мице́т·ы, -ов, (m pl) (microbiol) schizomycetes, Schizomycetes
схизо·мо́рф·н·ый, -ая, -ое, (a) schizomorphic
схизо·фре́н·и·я, -и, (f) (med) schizophrenia
схисто- (int component) see also **шисто-**
схисто·о́ид·, -а, (m) (geol) schistoid
схисто·це́рк·а, -и, (f) (ent) schisto-cercus, (pl) Schistocerca
схисто·ци́т·, -а, (m) (cyt) schistocyte
с·хлёст·ывани·е, -я, (n) (v n) of **с·хлёст·ыва·ть**
~ **про·вод·о́в** whipping of wires (telephone lines)
с·хлёст·ыва·ть, -ют, (imp) whip, lash

с·хло́п·ывани·е, -я,   (n) (v n) see
   хлоп·а́·ть; collapse (of a plasma)

с·хо́д·, -а,   (m) going/coming off/down,
   descent; path down; tailings, tails
   (refuse from· milling); convergence;
   (pl) belly shoulder shoe leather

~ с нул·я́   (instr) zero creep

с·хо́д·имост·ь, -и,   (f) convergence
   то́ч·к·а с·хо́д·имост·и (f) point of
   convergence;. (meteor) centre of
   cyclone

с·ход·и́ть, -ят,   (imp) descend, go/get/
   come off/down, alight; (perf only)
   make a return journey, go to … and
   back; -ся meet, come together,
   gather, converge; agree

с·хо́д·н·ый, -ая, -ое,   (a) similar, like;
   suitable, consistent (with)

с·хо́д·н·я, -и,   (g pl) -ей, (f) (naut)
   gangway, gang plank, brow

с·хо́д·ственн·ый, -ая, -ое,   (a) (math)
   compatible

с·хо́д·ств·о, -а,   (n) similarity, resemb-
   lance, likeness

~ по·ря́д·к·а   (pal) homotaxis

с·ход·я́щ·ийся, -аяся, -ееся,   (pres
   part act) see с·ход·и́ться; convergent,
   converging

с·хожд·е́ни·е, -я,   (n) (v n) see с·хо-
   д·и́ть; convergence

~, угл·ов·о́е   (surv, nav) convergence
   of meridians

с·хож·у́,   (pres 1st sing) of с·ход·и́ть

схоласт·и́ческ·ий, -ая, -ое,   (a)
   scholastic

с·цара́п·а·ть, -ют,   (perf) see с·цара́-
   п·ыва·ть

с·цара́п·ыва·ть, -ют,   (imp) scratch/
   scrape off/away

с·цед·и́ть, -ят,   (perf) see с·це́ж·ива·ть

с·це́ж·енн·ый, -ая, -ое,   (past part
   pass) see с·це́ж·ива·ть

с·це́ж·ива·ть, -ют,   (imp) decant,
   pour off, rack; strain/filter off

с·цемент·и́р·овать, -уют,   (perf)
   cement

сце́н·а, -ы,   (f) scene; (theat) stage

сцена́р·и·й, -я,   (m) scenario, script;
   stage directions

сцен·и́чност·ь, -и,   (f) theatrical suitability

сцен·и́чн·ый, -ая, -ое,   (a) theatrical

с·це́п·, -а,   (m) (v n) see с·цепл·я́·ть;
   (and see с·це́п·к·а)

с·цеп·и́ть, -ят,   (perf) see с·цепл·я́·ть

с·це́п·к·а, -и,   (g pl) -п·ок·, (f) (v n)
   see с·цепл·я́·ть; (agr) hitch; (rail)
   coupling; tow, chain (of linked vehicles); coupler (electronic circuits)

с·цепл·е́ни·е, -я,   (n) (v n) see с·цеп-
   л·я́·ть; linking, linkage; adhesion,
   cohesion, bond, grip; interface strength
   (cement); (rad) gauging; (M/T) clutch
   коэффицие́нт· с·цепл·е́ни·я   (m)
   coefficient of friction/adhesion; coupling factor

~, механ·и́ческ·ое   mechanical bond
   (between concrete and iron); (rad)
   ganging
   си́л·а с·цепл·е́ни·я   (f) (chem,
   phys) cohesive force; (dynam)
   adhesive force
   тя́г·а с·цепл·е́ни·я   (f) (M/T)
   tow bar

с·це́пл·енн·ый, -ая, -ое,   (past part
   pass) see с·цепл·я́·ть

с·цепл·я́емост·ь, -и,   (f) adhesiveness,
   adhesive capacity

с·цепл·я́·ть, -ют,   (imp) couple, link,
   hook up, hitch; clutch, engage, put
   into gear, mesh (of gears); -ся (pass);
   adhere, bond

с·цеп·н·о́й, -а́я, -о́е,   (a) clutch, coupling, linking, linkage; adhesive, cohesive, bonding, binding

~ вес·   (m) adhesive weight

~ кулач·ко́в·ая му́фт·а   (f) square-
   jaw clutch

сци́лл·а, -ы,   (f) (bot) squill, *Scilla*

сцинтигра́мм·а, -ы,   (f) scintigram

сцинтилло́·метр·, -а,   (m) (phys)
   scintillometer, scintillation counter

сцинтилло·ско́п·, -а,   (m) (phys) scin-
   tilloscope

сцинтилл·я́тор·, -а,   (m) (phys) scin-
   tillator

сцинтилляцио́н·н·ый, -ая, -ое,   (a)
   scintillation

~ с·чёт·чик·   (m) scintillation counter

сцинтиля́ци·я, -и,   (f) scintillation

сцио·фи́ль·н·ый, -ая, -ое,   (a) (bot)
   sciophilous, skiophilous, shade-loving

сцио·фи́т·, -а,   (m) (bot) sciophyte,
   skiophyte

сцифи·сто́м·а, -ы,   (f) (zool) scyphis-
   toma, (pl) scyphistomae, scyphistomas

сциф·о́ид·н·ые, -ых,   (pl decl as adj)
   (zool) scyphozoans, *Scyphozoa*

сцифо·меду́з·а, -ы,   (f) (zool) scypho-
   medusa, (pl) *Scyphomedusae*

сцм   (abbr) систе́м·а це́нтр·а масс·
   centre/center-of-mass system

с·ча́л·, -а, (*m*) tow, barge train, pusher-tug train/tow; trot (of moored vessels)

с·част·ли́в·ый, -ая, -ое, (*a*) happy; fortunate, lucky

с·ча́ст·ь·е, -я, (*n*) happiness; good fortune, luck

с·чест·ь, (*fut 3rd pl*) со·чт·у́т, (*past masc sing*) с·чё·л (*perf*) see с·чит·а́·ть

с·чёт·, -а, (*m*) (*v n*) see с·чит·а́·ть; count, counting; metering; account (*pl*) -а́; bill; score (at games); (*pl* -ы́ *also*) abacus

~, блок·и́рованн·ый (fin) blocked/frozen account

в кон·е́чн·ом с·чёт·е in the final analysis, eventually, finally

~ вре́мен·и (rocket) count-down

за с·чёт· at the expense of; on account of, because of

~, за·ключ·и́ть balance an account

~, за·пис·а́ть на carry to the account (of)

~, изото́п·н·ый (phys) isotropic count

~ коло́нн· column count (computers)

~, конто·корре́нт·н·ый (fin) current account (at a bank); (com) account current

~, один·о́чн·ый single-channel counting

~, рас·чёт·н·ый (fin) settlement/clearing account

~, сумма́р·н·ый total/integral count

~, тек·у́щ·ий (fin) current account

~ -факту́р·а (com) invoice

с·чет·а́, (*nom/acc pl*) see с·чёт· (fin)

с·чёт·н·о-аналит·и́ческ·ая маши́н·а (*f*) punched card/tape machine

с·чёт·н·о-импульс·н·ая систе́м·а (*f*) pulse-counting system

с·чёт·н·о-перфораци́он·н·ая маши́н·а (*f*) punched card/tape machine

с·чёт·н·о-расфас·о́вочн·ая маши́н·а (*f*) counting-and-packing machine

с·чёт·н·о-реш·а́ющ·ий, -ая, -ее, (*a*) computing

~ у·стро́й·ств·о (*n*) (*and see* вы·чис·л·и́тельн·ая маши́н·а) computer

с·чёт·ност·ь, -и, (*f*) countability, denumerability

с·чёт·н·ый, -ая, -ое, (*a*) account, accounting, book-keeping; calculating

~ глаз·о́к· (*m*) (text) counting glass

~ кни́г·а (*f*) account book

~ колес·о́ (*n*) (horol) locking plate

~ ла́мп·а (*f*) (phys) counter tube

~ лин·е́йк·а (*f*) slide-rule

с·чёт·н·ый, -ая, -ое

~ маши́н·а (*f*) accounting machine; calculating machine

~ ста́нци·я (*f*) data processing centre

~ схе́м·а (*f*) counting circuit

~ цеп·ь (*f*) counter circuit

с·чето·во́д·, -а, (*m*) (fin) book-keeper, account/ledger clerk

с·чето·во́д·ств·о, -а, (*n*) (fin) book-keeping

с·чёт·чик·, -а, (*m*) counter, integrating meter; register (of ship's log)

~, вибраци́он·н·ый oscillating meter

~, все·во́лн·ов·ый (nucl) long counter

~, га́з·ов·ый gas-filled counter

~, газо·раз·ря́д·н·ый gas-discharge counter, Geiger–Müller counter

~ Ге́йгер·а (phys) Geiger/Geiger-Müller counter

~ групп· batch counter

~, дво·и́чн·ый binary counter

~, десят·и́чн·ый decade counter

~ -зонд·, -а, (*m*) counter-probe

~, индукци́он·н·ый (elec) induction meter, induction motor meter

~, ионизаци́он·н·ый ionization counter

~ компле́кс·н·ого по·требл·е́ни·я (elec) two-part tariff meter

~, кольц·ев·о́й (nucl) ring counter

~, кристалл·и́ческ·ий crystal counter

~, на·ка́пл·ивающ·ий accumulator (computers)

~ оборо́т·ов revolution counter, tachometer

~ о·пла́ч·енн·ой эне́рг·и·и prepayment meter (telephones etc.)

~, полови́н·н·ый scale-of-two counter

~, про·то́ч·н·ый га́з·ов·ый gas-flow counter

~ рабо́ч·его ме́ст·а position meter (telephones)

~ рту́т·н·ый вращ·а́ющ·ийся (elec) mercury motor meter

~, само·гас·я́щ·ийся self-quenched counter

систе́м·а с·чёт·чик·ов (*f*) (nucl) counter array

~ со·в·пад·е́ни·й (nucl) coincidence counter

~, сцинтиляци́он·н·ый scintillation counter

~ та́кт·ов cycle counter (of computers)

~, тор·цо́в·ый end-on counter

~, фото·электро́н·н·ый photoelectric counter

~, цифербла́т·н·ый dial counter

с·чёт·чик
~ Черенко́в·а (phys) Čerenkov counter
~, электро́н·н·ый electronic counter
~ эне́рг·и·й (elec) energy meter, kilowatt-hour meter
с·чёт·ы, -ов, (m pl) abacus
с·числ·е́ни·е, -я, (n) numeration, counting; arithmetic, reckoning
~, пи́сьм·енн·ое (nav) dead reckoning
~, прост·о́е (nav) simple dead reckoning
~ пут·и́, су́т·очн·ое (nav) day's work
систе́м·а с·числ·е́ни·я (f) number system
~, с·ло́ж·н·ое (nav) complex dead reckoning
~, со·ста́в·н·о́е (nav) complex dead reckoning
с·числ·и́тел·ь, -я, (n) reckoner, counter, computer
с·чи́ст·ить, -ят, (perf) see с·чищ·а́·ть
с·чит·а́·ть, -ют, (imp) 'count, reckon; consider, take into account; consider, regard, deem; -ся (pass and reflex); (perf) check (by reading), read off
с·чи́т·ывани·е, -я, (n) (v n) of с·чи́т·ыва·ть; (instr) reading; read-out (computers)
с·чи́т·ыва·ть, -ют, (imp) check (by reading); read off/out
с·чи́т·ывающ·ий, -ая, -ее, (pres part act) of с·чи́т·ыва·ть
~ голо́в·к·а (f) reproduce head (magnetic recorder)
~ об·мо́т·к·а (f) read drive-wire, read driver (magnetic recording)
~ у·стро́й·ств·о (n) reader (magnetic recording)
с·чищ·а́·ть, -ют, (imp) clean/clear off; peel off
с·ше́й·те, (pl imper) of с·ши·ть, (perf); see с·шив·а́·ть
с·шив·а́льн·ый, -ая, -ое, (a) sewing
~ маши́н·а (f) sewing machine
с·шив·а́ни·е, -я, (n) (v n) of с·ши·в·а́·ть; (math) joining; (chem) cross-linking; (med) suture
~ реш·е́ни·й (math) joining of solutions
с·шив·а́тел·ь, -я, (m) stapler, stapling machine
с·шив·а́·ть, -ют, (imp) sew, sew up/together; fasten, join (sheet material ; make (with sheet material); staple (paper etc.)
с·шив·а́ющ·ий мост·и́к (m) (chem) crosslinking bridge

с·ши́в·к·а, -и, (g pl) -в·ок·, (f) (v n) see с·шив·а́·ть; seam, seaming, join, joining; joint (e.g. in belting); (chem) cross-linking
с·ши·л, (past masc sing) of с·ши·ть, (perf); see с·шив·а́·ть
с·ши́·т·ый, -ая, -ое, (past part pass) see с·шив·а́·ть; (chem) cross-linked, built-up
~ полиме́р· (m) (chem) cross-linked/built-up polymer
~ структу́р·а (f) (chem) cross-linked structure
с·ши·ть, (fut 3rd pl) со·шь·ю́т, (perf) see с·шив·а́·ть
съ·ед·а́·ть, -ют, (imp) eat up; corrode, eat away, (elec) kill (voltage etc.)
съ·е́д·ет, (fut 3rd sing) of съ·е́х·ать, (perf); see съ·езж·а́·ть (imp)
съ·ед·о́бн·ый, -ая, -ое, (a) edible, eatable
съ·ёж·ива·ться, -ются, (imp) shrivel up, become wrinkled
съ·ёж·иться, -атся, (perf) see съ·ё·ж·ива·ться
съ·е́зд·, -а, (m) (v n) see съ·езж·а́·ть; congress, convention; sloping/inclined path/track/way/road
съ·е́зд·ить, -ят, (perf) make a visit
съ·езж·а́·ть, -ют, (imp) go down, come down, descend; move (house); -ся assemble, meet
у́гол· съ·е́зд·а (m) (rail) angle of departure
съ·ел (past masc sing) of съ·ест·ь, (perf); see съ·ед·а́·ть (imp)
съ·ём·, -а, (m) (v n) see с·ним·а́·ть; (hortic) picking; throughput (of process); съ·ем (fut 1st sing) of съ·ест·ь
~ тепл·а́ heat removal/extraction
съ·ём·к·а, -и, (g pl) -м·ок·, (f) taking, taking off/from/away, removal; taking, taking a picture, photographing; survey, surveying
~, га́з·ов·ая emanation survey
~, глазо·ме́р·н·ая exploratory survey, eye survey
~, и́мпульс·н·ая flashlight photography
~, кинематограф·и́ческ·ая (phot) shot, shooting, take, taking
~, маркше́йдер·ск·ая mining survey
~, ме́нзуль·н·ая (surv) plane tabling
~ ме́р·н·ой ле́нт·ой chain survey
~, обра́т·н·ая back-reflection X-ray photography
~ с я́кор·я (naut) weighing anchor

съ·ём·к·а
~, чёрн·о-бе́л·ая   black-and-white photography
съ·ём·ник·, -а,   (*m*) (mech eng) puller, remover, lifter, stripper, extractor
~ кла́пан·а   valve lifter/remover
~ под·ши́п·ник·а   bearing puller
съ·ём·н·ый, -ая, -ое,   (*a*) removable, detachable, replaceable; dismantlable; (hortic) ready for picking
съ·ём·очн·ый, -ая, -ое,   (*a*) surveying, survey; (phot, cinema) taking, shooting, filming, photographing, camera
~ труб·к·а   (*f*) (TV) camera tube
съ·ём·щик·, -а,   (*m*) tenant; (hortic) picker; surveyor; (eng) stripper, remover (man)
съ·ест·н·о́й, -а́я, -о́е,   (*a*) food
~ при·па́с·ы   (*pl*) victuals
съ·ест·ь,   (*fut 3rd sing, pl*) **съ·ест·, съ·ед·я́т,** (*past masc sing*) **съ·ел,** (*perf*); see **съ·ед·а́ть**
съ·е́х·ать,   (*fut 3rd pl*) **съ·е́д·ут,** (*perf*); see **съ·езж·а́ть**
сы́ворот·к·а, -и,   (*f*) (med) serum; (food) whey
~, антире́зус·   Rh serum
~, моло́ч·н·ая   whey
~, преципит·и́рующ·ая   precipitinogen
сы́ворот·очн·ый, -ая, -ое,   (*a*) of **сы́ворот·к·а;** serum, serumal; serous
~ бол·е́зн·ь   (*f*) (med) serum sickness
с·ыгр·а́·ть, -ют,   (*perf*) play
сын·, -а,   (*nom pl*) **сын·овь·я́,** (*acc/gen pl*) **сын·ов·е́й,** (*m*) son
сын·овь·я́,   (*nom pl*) of **сын·**
сы́п·анн·ый, -ая, -ое,   (*past part pass*) of **сы́п·ать**
сы́п·ать, -пл·ют,   (*imp*) pour (free-flowing solids), sprinkle, strew, scatter; rain (blows etc.); lay (fishing nets etc.)
сы́пл·ющ·ий, -ая, -ее,   (*pres part act*) of **сы́п·ать**
сы́п·н·о́й, -а́я, -о́е,   (*a*) of **сы́п·ь;** (med) exanthematous, exanthematic
сы́п·у́чест·ь, -и,   (*f*) free-flowing properties, flowability
сы́п·у́ч·ий, -ая, -ее,   (*a*) free-flowing (of bulk solids); granulated
~ песо́к·   (*m*) quicksand; shifting sand
сы́п·ь, -и,   (*f*) (med) rash, eruption, exanthema; (met) rust spot/trace
сы́пь·те,   (*pl imper*) of **сы́п·ать**
сыр·, -а,   (*nom pl*) **-ы́,** (*g pl*) **-о́в,** (*m*) cheese

сы́р
~, пла́вл·ен·ый   processed cheese
сыр·е́ц·, -р·ц·а́,   (*i*) **-р·ц·о́м,** (*m*) raw/unfinished product
~, са́хар-   raw sugar
~, спирт-   unrectified alcohol
~, шёлк-   raw silk
сыр·ко́в·ый, -ая, -ое,   (*a*) of **сыр·о́к·**
~ ма́сс·а   (*f*) cream cheese, processed curds
сы́р·н·ый, -ая, -ое,   (*a*) cheese
сыро-   (*root*) cheese; raw, uncooked
сыро·ва́р·, -а,   (*m*) = **сыро·де́л·**
сыро·ва́р·н·ый, -ая, -ое,   (*a*) = **сыро·де́ль·н·ый**
сыро·ва́р·н·я, -и,   (*g pl*) **-р·ен·,** (*f*) cheese dairy
сыр·ова́т·ый, -ая, -ое,   (*a*) slightly damp
сыро·де́л·, -а,   (*m*) cheese-maker
сыро·дел·а́тельн·ый, -ая, -ое,   (*a*) cheese-making, cheese
сыро·де́л·и·е, -я,   (*n*) cheese-making
сыро·де́ль·н·ый, -ая, -ое,   (*a*) cheese-making, cheese
сыро·ду́·т·ый проце́сс·   (*m*) (met) direct process
сыро·ду́·т·ый горн·   (*m*) (met) bloomery
сыро́еж·к·а, -и,   (*g pl*) **-ж·ек·,** (*f*) (bot) russule, *Russula*
сыр·о́й, -а́я, -о́е,   (*a*) damp; raw, crude, green; uncooked, unboiled, unfinished
~ нефт·ь   (*f*) crude petroleum/oil, crude
сыро·мя́т·н·ый, -ая, -ое,   (*a*) tawing (of equipment), tawed (of hides)
сы́ро·мя́т·ь, -и,   (*f*) tawed leather
сы́р·ост·ь, -и,   (*f*) dampness, damp
сыр·цо́в·ый, -ая, -ое,   (*a*) unfinished, raw
сыр·ь·ё, -я́,   (*n*) raw material
сыр·ьев·о́й, -а́я, -о́е,   (*a*) raw-material
сы́т·ост·ь, -и,   (*f*) satiation
сы́т·ый, -ая, -ое,   (*a*) replete, satisfied
сы́т·ь, -и,   (*f*) cyperus, *Cyperus*
сычу́г·, -а,   (*m*) (zool) abomasum, fourth stomach, reed, rennet
сычуж·о́к·, -ж·к·а́,   (*m*) (biochem) rennet
са́бин·, -а,   (*m*) sabin (unit of sound absorption)
с·эконо́м·ить, -ят,   (*perf*) economize
сэндвиче·обра́з·н·ый, -ая, -ое,   (*a*) sandwich

сюда́ (*adv*) hither, here (movement implied)

сюже́т·, -а, (*m*) subject, topic

сюрпри́з·, -а, (*m*) surprise; (mil) booby-trap

ся́д·ут, (*fut 3rd pl*) of **сест·ь,** (*perf*); *see* **сад·и́ться** (*imp*)

ся́д·ьте (*pl imper*) of **сест·ь,** (*perf*); *see* **сад·и́ться** (*imp*)

ся́ж·ет, (*fut 3rd sing*) of **сест·ь,** (*perf*); *see* **сад·и́ться** (*imp*)

сяж·ко́в·ый, -ая, -ое, (*a*) antennate, horned

сяж·о́к·, -ж·к·а́, (*m*) (ent) antenna, horn feeler

т

т. (*abbr*) = **то́нн·а** t., ton (metric); (*abbr*) = **температу́р·а** temperature; (*abbr*) = **том·** vol.

та (*pron, nom sing f*) of **тот** that; the other; the right/correct

таба́к·, -а́, (*m*) (bot) tobacco, tobacco plant, *Nicotiana*

~, душ·и́ст·ый tobacco, flower-garden tobacco, *Nicotiana affinis*

~, жев·а́тельн·ый chewing tobacco

~ -махо́р·к·а, -и, (*f*) makhorka, *Nicotiana rustica*, (coarse European tobacco)

~, нюх·а́тельн·ый snuff

~, све́тл·ый bright tobacco

~, сига́р·н·ый вну́тр·енн·ий cigar filler tobacco

~, сига́р·н·ый обёрт·очн·ый cigar wrapper tobacco

со́б·ственн·о таба́к· tobacco, *Nicotiana tabacum*

табако·во́д·ств·о, -а, (*n*) tobacco growing/cultivation

табако·суш·и́тельн·ый, -ая, -ое, (*a*) tobacco-curing

табаку́р·, -а, (*m*) tobacco curer

таба́н·ить, -ят, (*imp*) back water (rowing)

таба́ч·ник·, -а, (*m*) tobacco worker

таба́ч·н·о-махо́р·очн·ый, -ая, -ое, (*a*) tobacco

таба́ч·н·ый, -ая, -ое, (*a*) tobacco

~ из·де́л·и·я (*pl*) tobacco products, tobacconist's wares

табашир·, -а, (*m*) (bot) tabashir

та́бел·ь, -я, (*f*) table, list, scale; time-board (recording attendance)

та́бель·н·ый, -ая, -ое, (*a*) of **та́бел·ь**

~ час·ы́ (*pl*) printing clock, time clock (recording attendance)

та́бель·щик·, -а, (*m*) timekeeper (person)

та́бес·, -а, (*m*) (med) tabes dorsalis

таблет·и́ровани·е, -я, (*n*) (*v n*) of **таблет·и́р·овать**; pelleting, tableting, preforming

таблет·и́рованн·ый, -ая, -ое, (*past part pass*) of **таблет·и́р·овать**

~ катализ·а́тор· (*m*) pellet/bead catalyst

~ са́ж·а (*f*) carbon black pellets

таблет·и́р·овать, -уют, (*imp and perf*) (plast) preform; tablet, make into tablets/beads/pellets

таблет·к·а, -и, (*g pl*) **-т·ок·,** (*f*) tablet, pellet, pill, (plast) preform

таблет·очн·ый, -ая, -ое, (*a*) preforming, tableting; tablet, preformed, pellet, bead

~ ге́ттер· (*m*) pellet getter (vacuum tubes)

~ маши́н·а (*f*) tableting/pelleting/preforming machine, tablet-compressing machine

табл·и́ц·а, -ы, (*i*) **-ей,** (*f*) (*see under associated words*); table (of information), chart; (math) table, array

~, по·ве́р·очн·ая check list

~ пре·образ·ова́ни·я conversion table

табл·и́чк·а, -и, (*g pl*) **-чек·,** (*f*) (*dim*) of **табл·и́ц·а**; tablet, plate, notice; (phys) tabular crystal; (anat) plate

~, за·во́д·ск·ая factory/maker's name-plate

табл·и́чн·ый, -ая, -ое, (*a*) table, tabular, tabulated

~ ра́з·ност·ь (*f*) tabular difference

~ да́·нн·ые (*pl*) tabulated data

табло́ (*n indecl*) notice-board; (elec) mimetic diagram, signal/mimic panel; (rail) illuminated track diagram

~, желоб·ко́в·ое (rail) slot/notched track diagram

~, свет·ов·о́е illuminated indicator board

~, пост·ов·о́е (rail) signal box diagram

табльдо́т·, -а, (*m*) (food) table d'hôte

табул·и́р·овать, -уют, (*imp*) tabulate

та́бул·ь, -и, (*f*) (zool) tabula, (*pl*) tabulae

табул·я́тор·, -а, (*m*) tabulator (computers etc.)

табул·я́т·ы, -ов, (*m pl*) (pal) tabulate corals, *Tabulata*

табу́н·, -а́, (*m*) herd of horses; (chem) tabun (nerve gas)

табуре́т·к·а, -и, (*g pl*) **-т·ок·,** (*f*) stool

тависток·и́т·, -а, (*m*) (min) tavistockite

та́волг·а, -и, (*f*) (bot) spirea, *Spirea*

таво́т·, -а, (*m*) grease, tallow

таво́т·ник·, -а, (*m*) grease-cup

таво́т·н·ый, -ая, -ое, (*a*) grease, tallow

таво́т·ниц·а, -ы, (*i*) **-ей,** (*f*) grease cup

тавото·на·гнет·а́тел·ь, -я, (*m*) pressure-greasing equipment

тавр·ён·ый, -ая, -ое, (*a*) branded

тавр·и́ть, -я́т, (*imp*) brand (animals)

тавр·о́, -а́ (*nom pl*) **та́вр·а,** (*g pl*) **тавр·а́м** (*n*) brand; (agr) branding iron; (*as component*) T-, tee-

тавро·би́мс·ов·ый, -ая, -ое, (*a*) (met) T-bulb

тавро·бу́льб·ов·ый, -ая, -ое, (*a*) (met) T-bulb

тавр·о́в·ый, -ая, -ое, (*a*) T, tee, tee-shaped

~ стал·ь (*f*) T-bar, T-beam (of steel)

тавто- (*int component*) *see also* **тауто-**

тавто·ло́г·и·я, -и, (*f*) tautology

тавт·они́м·и·я, -и, (*f*) (biol) tautonymy

таёж·н·ый, -ая, -ое, (*a*) taiga, Northern coniferous forest land

та́·ет, (*pres 3rd sing*) of **та́·ять**

та́з·, -а, (*nom pl*) **-ы́,** (*g pl*) **-о́в,** (*m*) basin, pan; (anat) pelvis

~, ле́нт·оч·н·ый (text) can

та́з·ик·, -а, (*m*) (*dim*) of **таз·** (zool) coxa, (*pl*) coxae

тазо·бе́др·енн·ый, -ая, -ое, (*a*) hip, (anat) coxofemoral

та́з·ов·ый, -ая, -ое, (*a*) of **таз·** (anat) pelvic

~ кост·ь (*f*) (anat) pelvis

~ пояс· (*m*) (anat) pelvic girdle

~ ровн·и́чн·ая маши́н·а (*f*) (text) slubber

та·и́нственн·ый, -ая, -ое, (*a*) mysterious, enigmatic; secret

та·и́ть, -я́т, (*imp*) hide, conceal

тайг·а́, -и́, (*f*) taiga, Northern coniferous forest land

тай·к·о́м (*adv*) secretly, in secret

таймен·ь, -я, (*m*) (fish) huchen, *Salmo hucho*

та́ймер·, -а, (*m*) timer

таймо́·граф·, -а, (*m*) timograph, recording timer

таймотро́л·, -а, (*m*) (elec) thyratron motor control unit

тайм·та́ктор·, -а, (*m*) (elec) timing relay

тайм·-ча́ртер·, -а, (*m*) (com) time charter

тайм·шит·, -а, (*m*) (naut) time sheet

та́й·н·а, -ы, (*f*) secret, secrecy

~, госуда́рств·енн·ая official/state secret(s)

~, служ·е́бн·ая official secret(s)

тайно·бра́ч·и·е, -я, (*n*) (bot) cryptogamy

тайно·бра́ч·н·ый, -ая, -ое, (*a*) cryptogamic, cryptogamous

та́йно·пис·ь, -и, (*f*) cryptography

та́й·н·ый, -ая, -ое, (*a*) secret, confidential

тайри́стор·, -а, (*m*) (elec) thyristor

тайри́т·, -а, (*m*) thyrite

тайфу́н·, -а, (*m*) (meteor) typhoon

тайфу́н-кана́т·, -а, (*m*) (naut) hurricane hawser

так (*adv*) so, thus, like this/that; (*conj*) as, since

~ же как in the same way as, just like

~ же как и just as, as also

~ и есть and so it is

и так да́лее and so on

и так и сяк in every possible way

и́мен·н·о так just so, just like this

как ... так both ... and

~ как as, since, because

~ на·зыв·а́ем·ый, -ая, -ое, (*a*) so-called, ... as it is known

~ что so that

такела́ж·, -а, (*i*) **-ем,** (*m*) (naut) rigging

~, бег·у́ч·ий running rigging

~, сто·я́ч·ий standing rigging

такела́ж·н·ая, -ой, (*f decl as adj*) (shipb) rigging loft

такела́ж·ник·, -а, (*m*) rigger

такела́ж·н·о-па́рус·н·ый цех· (*m*) (shipb) rigging shop

та́к·же (*adv*) also, as well, too

...~ не... ... does not either ..., nor does...

таки́ (*particle*) after all

такнода́ль·н·ый, -ая, -ое, (*a*) tacnodal

так·о́в·, -а́, -о́, (*pron*) such, like

они так·ов·ы́ that is what they are like, they are like that

так·ов·о́й, -а́я, -о́е, (*a*) the same

так·о́й, -а́я, -о́е, (*pron*) such, this/that kind of, so

в так·о́м с·лу́ч·а·е in such a case, in a case like that, in this case, then

~ же как the same as

~ как·о́й such as

так·и́м о́браз·ом in this way, thus

~ -то (*pron*) so and so, such and such; such a person

такон·и́т·, -а, (*m*) (min) taconite

тако́н·ск·ий, -ая, -ое, (*a*) (geol) Taconic

та́кс·а, -ы, (*f*) statutory price, tariff; (zool) dachshund

такс·а́тор·, -а, (*m*) assessor; rate-fixer

~, лес·н·о́й appraiser (sylviculture)

такса·фо́н·, -а, (*m*) = **таксо·фо́н·**

такс·а́ци·я, -и, (*f*) (*v n*) *see* **такс·и́р·овать**

такси́ (*n indecl*) (M/T) taxi

~, груз·ов·о́е removal truck/van

~, легк·ов·о́е taxi

такси·метр·, -а, (*m*) (nav) bearing card

такс·ин·, -а, (*m*) (chem) taxine

такс·и́р·овать, -у́ют, (*imp and perf*) assess, evaluate; fix (rates); cost (expenditure)

та́ксис·, -а, (*m*) (biol) taxis

такси́т·, -а, (*m*) (min) taxite

~, брекчие·ви́д·н·ый ataxite

~, полос·а́т·ый eutaxite

такси́т·ов·ый, -ая, -ое, (*a*) (min) taxitic

та́кс·ов·ый, -ая, -ое, (*a*) tariff; rate-fixing; costing

таксо́д·иев·ые, -ых, (*pl decl as adj*) (bot) taxodium family, *Taxodiaceae*

таксо́д·и·й, -я, (*m*) (bot) deciduous cypress, *Taxodium distichum*

таксо́·метр·, -а, (*m*) taximeter, meter (in taxi)

таксо·мото́р·, -а, (*m*) *see* **такси́**

такс·о́н·, -а, (*m*) (biol) taxon, (*pl*) taxa

таксо·ном·и́ческ·ий, -ая, -ое, (*a*) (biol) taxonomic

таксо·но́м·и·я, -и, (*f*) (biol) taxonomy

таксо·фо́н·, -а, (*m*) coin-operated telephone

такт·, -а, (*m*) time (rhythm etc.); stroke (of engine); tact (behaviour)

~ в·са́с·ывани·я suction stroke

~ вы́·хлоп·а exhaust stroke

~ рабо́ч·его хо́д·а power stroke (of piston)

~ рас·шир·е́ни·я expansion stroke

~ с·жа́т·и·я compression stroke

та́кт·ик·, -а, (*m*) tactician

та́кт·ик·а, -и, (*f*) tactics

та́кт·ик·о-техн·и́ческ·ое за·да́·ни·е (*n*) design specification, staff requirements (of weapons, armaments, warships etc.)

такт·и́льн·ый, -ая, -ое, (*a*) (biol) tactile

такт·и́рующ·ий, -ая, -ее, (*pres part act*); timing, (telecom) cadence

~ и́мпульс· (*m*) clock pulse (computers)

такти́т·, -а, (*m*) (min) tactite

такт·и́ческ·ий, -ая, -ое, (*a*) tactical; (air) ground-support, tactical

такт·и́чн·ый, -ая, -ое, (*a*) tactful

та́кт·ов·ый, -ая, -ое, (*a*) time, timing

~ сигна́л· (*m*) (telecom) cadence

~ и́мпульс· (*m*) (elec) clock pulse

~ у·стро́й·ств·о (*n*) clock (computers)

~ черт·а́ (*f*) bar (music)

~ электро·магни́т· (*m*) (telecom) time tapper

такы́р·, -а, (*m*) takyr, takyr/desert soil

тала́мус·, -а, (*m*) (biol) thalamus

тала́нт·лив·ый, -ая, -ое, (*a*) talented, gifted

таласс·и́ческ·ий, -ая, -ое, (*a*) (ocean) thalassic

талассо·ге́н·н·ый, -ая, -ое, (*a*) (geol) thalassogenic

талассо·гра́ф·и·я, -и, (*f*) (ocean) thalassography

талассо·крат·и́ческ·ий, -ая, -ое, (*a*) (geol) thalassocratic

талассо·терап·и́·я, -и, (*f*) (med) thalassotherapy

талассо·фи́ль·н·ый, -ая, -ое, (*a*) (biol) thalassophile

талатто·за́вр·, -а, (*m*) (pal) thalattosaur

та́л·ев·ый, -ая, -ое, (*a*) of **та́л·ь** and **та́л·и**

талени́т·, -а, (*m*) (min) thalenite

та́лер·, -а, (*m*) (print) type-bed

та́л·и, -ей, (*f pl*) (*see also* **та́л·ь**) (naut) tackle, purchase

~, диференциа́ль·н·ые differential purchase, geared blocks

~, механ·и́ческ·ие geared blocks

риф-тал·и, -ей, (*f pl*) double whip purchase

топена́нт-тал·и, -ей, (*f pl*) topping lift purchase (derrick)

хват-та́л·и, -ей, (*f pl*) luff (purchase)

испа́н·ск·ие хват-та́л·и Spanish burton (purchase)

~, шлюп·очн·ые (naut) boats falls

та́л·иев·ый, -ая, -ое, (*a*) of **та́л·и·я**

та́лия, -и, (*f*) waist

талл·а́т·, -а, (*m*) (chem) thallate

та́лл·иев·ый, -ая, -ое, (*a*) (chem) thallic, thallium

та́лл·и·й, -я, (*m*) (chem) thallium, Tl

~, угле·ки́сл·ый thallous carbonate

~ ма́сл·о (*n*) (biochem) tall oil

талло́л·ь, -я, (*m*) (biochem) tall oil

та́ллом·, -а, (*m*) (bot) thallome

та́ллом·н·ые раст·е́ни·я (*pl*) (bot) *Thallophyta*

талло·спо́р·а, -ы, (*f*) (bot) thallospore

талло·фи́т·, -а, (*m*) (bot) thallophyte, (*pl*) *Thallophyta*

та́ллус·, -а, (*m*) (bot) thallus

тало́н·, -а, (*m*) coupon; stub (in cheque book etc.); duplicate (in order-book etc.)

~, от·ры́в·н·о́й coupon

~, страх·ов·о́й insurance note/slip

талофи́д·н·ый фо́то·элеме́нт·   (*m*) (elec) thalofide cell

та́лреп·, -а, (*m*) (naut) rigging screw

та́л·ь, -и, (*f*) (*see also* **та́л·и**) (*pl*) lifting/ pulley block

~, диференциа́ль·н·ая differential pulley block

~, черв·я́чн·ая worm-gear pulley block

та́льбот·, -а, (*m*) (phys) talbot (= 10^7 lumergs)

Та́льбот-проце́сс·, -а, (*m*) (met) Talbot continuous process

та́львег·, -а, (*m*) (surv, geol) thalweg; (math) line of lowest points, line of minima (on a graph)

та́льк·, -а, (*m*) (min) talc, talcum

та́льк·ов·ый, -ая, -ое, (*a*) of **тальк·**; (min) talcose

талько·обра́з·н·ый, -ая, -ое, (*a*) (min) talcous

там (*adv*) there

тамари́кс·ов·ый, -ая, -ое, (*a*) = **тамари́ск·ов·ый**

тамари́нд·, -а, (*m*) (bot) tamarind, *Tamarindus indica*

тамари́ск·ов·ые, -ых, (*pl decl as adj*) (bot) tamarisk family, *Tamaricaceae*

тамаруги́т·, -а, (*m*) (min) tamarugite

та́мбур·, -а, (*m*) (build) covered passage, vestibule; (rail) covered platform (between carriages); (naut) companion way; (elec) cable drum carriage; (arch) tambour; (text) chain stitch

та́мбур·н·ый, -ая, -ое, (*a*) of **та́мбур·**

~ вы́·шив·а́ни·е (*n*) (text) chain stitching

~ кру́ж·ев·о (*n*) (text) tambour lace

тамо́ж·енн·ый, -ая, -ое, (*a*) customs

~ деклар·а́ци·я (*f*) bill of entry

~ до·смо́тр·щик· (*m*) customs officer

~ за·кла́д·н·ая (*f*) customs bond

~ сою́з· (*m*) customs union

тамо́ж·н·я, -и, (*g pl*) **-ж·ен·,** (*f*) customs house, customs

тампо́н·, -а, (*m*) (med) plug, pad, tampon

тампона́д·а, -ы, (*m*) (med) tamponade

тампон·и́р·овать, -уют, (*imp and perf*) plug, plug up; (civ eng) grout, cement up

танато·ло́г·и·я, -и, (*f*) (med) thanatology

танато·цено́з·, -а, (*m*) (pal) thanatocoenose, thanatocoenosis, (*pl*) thanatocoenoses

танато·фо́б·и·я, -и, (*f*) (med) thanatophobia

танга́ж·, -а, (*m*) (air etc.) pitch, pitching

движ·е́ни·е танга́ж·а (*n*) pitch, pitching, pitching movement, pitch motion

моме́нт· танга́ж·а (*m*) (aerodynam) pitch/pitching moment

от·клон·е́ни·е по танга́ж·у (*n*) pitch displacement

по·лож·е́ни·е по танга́ж·у (*n*) pitch attitude

про·из·во́д·н·ые моме́нт·а танга́ж·а (*pl*) pitching derivatives

угл·ов·о́е у·скор·е́ни·е танга́ж·а (*n*) pitch acceleration

у́гол· танга́ж·а (*m*) (a/c) pitching angle

тангеи́т·, -а, (*m*) (min) tangeite

та́нгенс·, -а, (*m*) (math) tangent

теоре́м·а та́нгенс·ов (*f*) (math) law of tangents, tangent law

~ угл·а́ на·кло́н·а slope tangent

~ угл·а́ по·те́р·и (elec) loss-angle tangent, loss tangent

тангенс·о́ид·а, -ы, (*f*) (math) tangent curve

тангент·а, -ы, (*f*) (telecom) transmitter cut-out switch

тангенциа́ль·н·ый, -ая, -ое, (*a*) tangential

~ у·скор·е́ни·е (*n*) (math) tangential acceleration

та́ндем·, -а, (*m*) tandem, tandem arrangement/layout

~ -маши́н·а, -ы, (*f*) (st eng) tandem engine

~, стан· (met roll) tandem mill

та́н·ец·, -н·ц·а, (*i*) **-н·ц·ем,** (*m*) dance

танжери́н·, -а, (*m*) (bot) tangerine, *Citrus deliciosas*

тани́н·, -а, (*m*) *see* **танни́н·**

та́нк·, -а, (*m*) tank

~, ле́т·н·ий (shipb) summer tank (in tankers)

~ -паро·во́з·, -а,, (*m*) (rail) tank locomotive

~ -тра́ль·щик·, -а, (*m*) (mil) mineclearing tank/vehicle, tank minesweeper

танк·а́ж·, -а, (*m*) tankage (fertilizer)

~, живо́т·н·ый animal tankage

~, ути́ль·н·ый garbage tankage

та́нкер·, -а, (*m*) (shipb) tanker, tank vessel

танке́т·к·а, -и, (*g pl*) **-т·ок·,** (*f*) wedge heel; wedge-heel shoe

танк·и́ст·, -а,   (*m*) tank soldier, tank-crew member

танко·вожд·éни·е, -я,   (*n*) (mil) tank driving, tanking

та́нк·ов·ый, -ая, -ое,   (*a*) tank; (mil) tank, armoured/armored, tank-borne

∼ ли́хтер·   (*m*) (naut) tank lighter

танко·до·сту́п·н·ый, -ая, -ое,   (*a*) (mil) tankable, approachable with tanks

∼ ме́ст·ност·ь   (*f*) tank country

танко·дро́м·, -а,   (*m*) (mil) tank exercise area

танко·стро·éни·е, -я,   (*n*) tank construction

танн·а́з·а, -ы,   (*f*) (biochem) tannase

танн·альби́н·, -а,   (*m*) (chem) tannalbin

танн·и́н·, -а,   (*m*) (chem) tannin, tannic acid, gallo-tannin

танта́л·, -а,   (*m*) (chem) tantalum, Ta

тантал·а́т·, -а,   (*m*) (chem, min) tantalate

тантал·и́т·, -а,   (*m*) (min) tantalite

танта́л·ов·ый, -ая, -ое,   (*a*) tantalum, (chem) tantalic

танц·ева́ть, -у́ют,   (*imp*) dance

тапету́м·, -а,   (*m*) (bot) tapetum

тапио́к·а, -и,   (*f*) (food) tapioca

тапиоли́т·, -а,   (*m*) (min) tapiolite

тапи́р·, -а,   (*m*) (zool) tapir, (*pl*) *Tapiridae*

та́р·а, -ы,   (*f*) package, packaging, packing case/crate; tare

  вес· та́р·ы   (*m*) (com) tare

тарака́н·, -а,   (*m*) (ent) cockroach, (*pl*) *Blattidae*

тарака́н·ов·ые, -ых,   (*pl decl as adj*) (ent) cockroaches, *Blattidae*

тара·ксанти́н·, -а, (*m*) (biochem) taraxanthin

тара́н·, -а,   (*m*) (mech) ram

∼, гидравл·и́ческ·ий   hydraulic ram

тара́н·ить, -ят,   (*imp*) ram

тара́н·н·ый, -ая, -ое, (*a*) ram, ramming

∼ пере·бо́р·к·а   (*f*) (shipb) collision bulkhead

тара́нтул·, -а,   (*m*) (ent) tarantula, (*pl*) *Lycosidae*

тара́н·ь, -и,   (*f*) (fish) sea roach, *Rutilus rutilus heckeli*

тарапаки́т·, -а,   (*m*) (min) tarapacaite

та́рви·я, -и,   (*f*) (civ eng) tarvia

тарденуа́з·, -а,   (*m*) Tardenoisian (anthropology)

таре́л·к·а, -и,   (*g pl*) **-л·ок·,** (*f*) plate, dish, disc, disk, pan; (*pl*) cymbals

∼, бу́фер·н·ая   buffer disc

таре́л·к·а

∼, колпач·ко́в·ая   bubble cap tray (of distillation tower etc.)

∼, ве́рх·н·яя   spring carrier (of valve)

таре́л·очк·а, -и,   (*g pl*) **-чек·,** (*f*) (*dim*) of **таре́л·к·а**; flare, flange (of vacuum tube)

таре́л·очн·ый, -ая, -ое,   (*a*) of **таре́л·к·а**

таре́ль·чат·ый, -ая, -ое,   (*a*) dish-shaped, dished, dish, disc

∼ і ла́цан·   (*m*) disc valve

тарза́ль·н·ый, -ая, -ое,   (*a*) tarsal

∼ кост·ь   (*f*) (anat) tarsus

тарзи·о́ид·ы   (*pl*) (zool) tarsioids, *Tarsioidea*

тарзи·у́с·, -а,   (*m*) (zool) Tarsius

тар·и́ровани·е, -я,   (*n*) (*v n*) of **тар·и́р·овать**; calibration

тар·и́р·овать, -уют,   (*imp and perf*) tare (packaged goods); (instr) calibrate

тар·иро́вк·а, -и,   (*g pl*) **-вок·,** (*f*) (*v n*) of **тар·и́р·овать**; calibration, calibration curve

тари́ф·, -а   (*m*) tariff, rate, (*pl*) scale (of payments, charges)

∼, двой·н·о́й   (elec) two-part/night-and-day tariff

∼, двух·ста́в·очн·ый   (elec) load-rate tariff

∼, еди́н·ый   all-in tariff

∼, льг·о́тн·ый   reduced rate

∼, много·коло́нн·ый   (elec) multi-linear tariff

∼, много·ста́в·очн·ый   multirate tariff

∼, опт·о́в·ый   bulk tariff

∼, прост·о́й   flat rate, flat-rate tariff

∼, с·ло́ж·н·ый   multiple tariff

тарифик·а́ци·я, -и,   (*f*) tariff/rate fixing

тарифиц·и́р·овать, -уют,   (*imp and perf*) tariff, establish scales of charges; fix the rate for

тари́ф·н·ый, -ая, -ое,   (*a*) of **тари́ф·**

∼ зо́н·а   (*f*) exchange area (telephones)

та́р·н·ый, -ая, -ое,   (*a*) of **та́р·а**

∼ груз·   (*m*) package freight/cargo

∼ я́щик·   (*m*) packing case

тарсус·, -а,   (*m*) (anat) tarsus

тарт·а́льн·ый, -ая, -ое,   (*a*) bailing

∼ чан·   (*m*) (min) bailing tub

тарт·а́ни·е, -я,   (*n*) bailing

тартр·ази́н·, -а,   (*m*) (chem) tartrazine

тартр·а́т·, -а,   (*m*) (chem) tartrate

тартыш·и   (*m pl*) growlers (ice)

тары́н·, -а,   (*m*) (geog) ice step/sill

таск·а́·ть, -ют,   (*imp indet*) drag, pul

тасма́н·и́т·, -а, *(m)* (min) tasmanite

тасма́н·о́ид·, -а, *(m)* Tasmanoid (anthropology)

тас·ова́ть, -у́ют, *(imp)* shuffle (cards etc.)

тастату́р·а, -ы, *(f)* (telecom) keyset/sender

тата́р·ск·ий, -ая, -ое, *(a)* Tartar

~ о·плёт·к·а *(f)* point (cordage)

тату·и́р·овать, -у́ют, *(imp and perf)* tattoo

та́убер·ов·а теоре́м·а *(f)* (math) Tauberian theorem

та́у *(indecl)* tau, Greek letter τ

таумаси́т·, -а, *(m)* (min) thaumasite

тау-мезо́н·, -а, *(m)* (nucl) tau-meson, τ-meson, heavy meson

тау·ме́тр·, -а, *(m)* tau-meter (luminescence)

таур·и́н·, -а, *(m)* (biochem) taurine, taurin

тауро·хо́л·ев·ый, -ая, -ое, *(a)* (biochem) taurocholic

та́у-сагы́з·, -а, *(m)* (bot) tau-saghyz, *Scorzonera tau-saghyz*

тауто·мери́з·овать, -у́ют, *(imp and perf)* (chem) tautomerize

тауто·мер·и́·я, -и, *(f)* (chem) tautomerism

~, ко́ль·чат·о-цеп·н·а́я ring-chain tautomerism

тауто·ме́р·н·ый, -ая, -ое, *(a)* tautomeric

тауто·хро́н·, -а, *(m)* (math) tautochrone

тафо·ни́м·и·я, -и, *(f)* (pal) taphonymy

тафо·но́м·и·я, -и, *(f)* (pal) taphonomy

тафо·цено́з·, -а, *(m)* (biol) taphocoenosis

тафро·гене́з·, -а, *(m)* (geol) taphrogenesis

тафро·гео·синклина́л·ь, -и, *(f)* (geol) taphrogeosyncline

тафт·а́, -ы́, *(f)* (text) taffeta

тафт·ян·о́й, -а́я, -о́е, *(a)* taffeta

тах·гидри́т·, -а, *(m)* (min) tachyhydrite, tachydrite

тахео́·метр·, -а, *(m)* (surv) tacheometer, tachymeter

тахео·ме́тр·и·я, -и, *(f)* (surv) tacheometry

тахи кард·и́·я, -и, *(f)* (med) tachycardia

тахи·ли́т·, -а, *(m)* (min) tachylyte, tachylite

тахи́·ме́тр·, -а, *(m)* (surv) tachymeter, tacheometer

тахи́н·а, -ы, *(f)* (ent) tachina fly

тахи·стери́н·, -а, *(m)* (pharm) tachysterol

тахо·генера́тор·, -а, *(m)* A.C. velocity generator

тахо́·граф·, -а, *(m)* recording tachometer

тахо·дина́мо *(n indecl)* D.C. velocity generator

тахо́·метр·, -а, -а, *(m)* tachometer, engine speed indicator

~ вибрацио́н·н·ого ти́п·а tachometer with reed-type frequency meter

~, вольтметр·ов·ого ти́п·а generator-type tachometer, magneto speed indicator

~, гиро·скоп·и́ческ·ий (a/c) rate gyro

~, жи́дк·ост·н·ый fluid-displacement tachometer

тахо·метр·и́ческ·ий, -ая, -ое, *(a)* tachometric

~ электр·и́ческ·ая маши́н·а *(f)* velocity generator

тахо·осцилло·гра́мм·а, -ы, *(f)* tacho-oscillogram

тахо·ско́п·, -а, *(m)* tachoscope, combined revolution counter and stop watch

тахо·ста́т·, -а, *(m)* constant speed unit

тач·а́·ть, -ют, *(imp)* stitch

та́ч·к·а, -и, *(g pl)* **-ч·ок·,** *(f)* stitching; wheelbarrow

тащ·и́ть, -ат, *(imp det)* drag, pull

та́·ющ·ий, -ая, -ее, *(pres part act)* of **та́·ять**

та́·явш·ий, -ая, -ее, *(past part act)* of **та́·ять**

та́·яни·е, -я, *(n)* thaw, thawing; melting

та·я́т, *(pres 3rd pl)* of **та·и́ть**

та́·ять, -ют, *(imp)* thaw; melt

та·я́щ·ий, -ая, -ее, *(pres part act)* of **та·и́ть**

тверд·е́ни·е, -я, *(n)* (v n) of **тверд·е́·ть**

~, дисперсио́н·н·ое (met) dispersion/precipitation hardening

~, радиацио́н·н·ое (nucl) radiation hardening

тверд·е́·ть, -ют, *(imp)* harden, become hard, set; become harder

тверд·и́ть, -я́т, *(imp)* reiterate

твёрд·о *(adv)* thoroughly, firmly, firm, well; (telecom) letter T

твердо·ко́ж·и·е, -я, *(n)* (med) scleroderma

твердо·ме́р·, -а, *(m)* hardometer, hardness tester

твердо·нёб·н·ый, -ая, -ое, *(a)* palatal

твердо·с·пла́в·н·ый, -ая, -ое,   (*a*) hard-metal, cutting-alloy

твёрд·ост·ь, -и,   (*f*) hardness; solidity; firmness

~, акти́в·н·ая   abrasive hardness

~, дл·и́тельн·ая   (met) long-time hardness

~, дл·и́тельн·ая гор·я́ч·ая   (met) long-time hardness

~ по Бринел·ю   Brinell hardness

~ по Ви́ккерс·у   Vickers hardness

~ по Дефо́   Defo value (of rubber)

~ по Кну́п·у   Knoop hardness

~ по Мо́ос·у   Mohs hardness

~ по Ро́квелл·у   Rockwell hardness

~ по склеро·ско́п·у   Shore/scleroscope hardness

~ по Шо́р·у   Shore/scleroscope hardness

твердо·тя́·нут·ый, -ая, -ое,   (*a*) hard-drawn (of copper wire)

твердо·фа́з·н·ый, -ая, -ое,   (*a*) (phys, chem) solid-phase

твёрд·ый, -ая, -ое,   (*a*) hard; solid; firm

~ о·ста́т·ок·   (*m*) (ICE) carbon residue, coke

~ пред·лож·е́ни·е   (*n*) (com) firm offer

~ рас·тво́р·   (*m*) (phys, chem) solid solution

~ рас·тво́р· в·недр·е́ни·я   intersticial solid solution

~ рас·тво́р· вы·чит·а́ни·я   substitutional solid solution

~ рас·тво́р· за·мещ·е́ни·я   substitutional solid solution

~ рас·тво́р·, основ·н·о́й   matrix

~ рас·тво́р·, регуля́р·н·ый   ordered solid solution

~ спирт·   (*m*) chemical fuel block (e.g. methaldehyde)

~ с·пла́в·   (*m*) hard metal, cutting alloy

~ со·сто·я́ни·е   (*n*) solid state

~, тел·о   (*n*) (phys) solid; (mech) rigid body

дина́м·ик·а твёрд·ого тел·а   (*f*) rigid-body dynamics

физ·ик·а твёрд·ого тел·а   (*f*) solid-state physics

~ цен·а́   (*f*) (com) fixed price

твёрж·е,   (*comp*) of **твёрд·ый**

тверж·ённ·ый, -ая, -ое,   (*past part pass*) of **тверд·и́ть**

твинде́к·, -а,   (*m*) (naut) 'tweendeck

тви́стор·, -а,   (*m*) twistor (computer memory element)

тво·его́,   (*g*) of **тво·й, тво·ё**

тво·й,   (*nom pl*) of **тво·й**

тво·и́х,   (*acc anim pl, g pl*) of **тво·й**

тво·й,   (*pron*) **-я́, -ё,** your (familiar)

твор·и́л·о, -а,   (*n*) lime slaker, lime-slaking trough

твор·и́ть, -я́т,   (*imp*) create; do, accomplish; slake (lime)

творо́г·, -а́,   (*m*) (food) curds, cream cheese

творо́ж·ист·ый, -ая, -ое,   (*a*) clotted, curdled, caseous

творо́ж·н·ый, -ая, -ое,   (*a*) of **творо́г·**

тво́р·ческ·ий, -ая, -ое,   (*a*) creative, constructive

тво́р·честв·о, -а,   (*n*) creation, work, works

тво·я́,   (*fem*) of **тво·й**

т.е.   (*abbr*) = **то есть** i.e., that is

те   (*pron nom pl, acc pl inanim*) of **тот** those

теалли́т·, -а,   (*m*) (min) teallite

теа́тр·, -а,   (*m*) theatre

театра́ль·н·ый, -ая, -ое,   (*a*) theatre, theatrical

теба·и́н·, -а,   (*m*) (pharm) thebaine

тебе́   (*pron dat/prep*) of **ты**

тебя́   (*pron acc/gen*) of **ты**

тевя́к·, -а,   (*m*) (zool) grey seal, *Halichoecus grypus*

тегме́н·, -а,   (*m*) (ent) tegmen, (*pl*) tegmina

тегме́нтум·, -а,   (*m*) (biol) tegmentum, tegumentum, (*pl*) tegmenta, tegumenta

теза́н·, -а,   (*m*) (med) thesan

те́зис·, -а,   (*m*) thesis

те·и́н·, -а,   (*m*) (biochem) theine

тейле́р·и·и   (*pl*) (microbiol) Theileria

тейлерио́з·, -а,   (*m*) (med) theileriosis, theileriasis

Те́йлор·а, фо́рмул·а   (*f*) (math) Taylor's formula

тейлори́т·, -а,   (*m*) (min) taylorite

тёк·,   (*past masc sing*) of **теч·ь**

те́к·а, -и,   (*f*) (biol) theca, (*pl*) thecae

те́к·ев·ое де́рев·о   (*n*) teak (wood)

тек·ли́,   (*past pl*) of **теч·ь**

те́к·ов·ое де́рев·о   (*n*) teak (wood)

тек·одо́нт·ы,   (*pl*) (pal) thecodonts, *Thecodontia*

тек·оиде́·и, -й,   (*f pl*) (pal) thecoids, *Thecoidea*

теко́м·а, -ы,   (*f*) (bot) tecoma, *Tecoma*

те́кс·, -а,   (*m*) shoemaker's tack

тексро́п·, -а,   (*m*) (mech eng) texrope, V-belt

текст·, **-а**, (*m*) text; words (music)

~, от·кры́·т·ый (telecom) plain language

тексти́л·ь, **-я**, (*m*) textile(s)

тексти́ль·н·ый, **-ая**, **-ое**, (*a*) textile

тексто·вини́т·, **-а**, (*m*) (plast) textovinit (PVC-coated fabric)

текстоли́т·, **-а**, (*m*) textolite (laminated resin on cotton or glass fabric base)

текстуа́ль·н·ый, **-ая**, **-ое**, (*a*) textual

тексту́р·а, **-ы**, (*f*) (*see under associated words*) texture, fabric

~, ребр·ов·а́я (cryst) cube-on-edge texture

текстур·о́ванн·ый, **-ая**, **-ое**, (*a*) textured

тексту́ро·образ·ова́ни·е, **-я**, (*n*) texture formation, formation of a texture

текти́т·, **-а**, (*m*) (geol) tektite

текто·ге́н·, **-а**, (*m*) (geol) tectogene

текто·генет·и́ческ·ий, **-ая**, **-ое**, (*a*) (geol) tectogenic

текто́н·ик·а, **-и**, (*f*) (geol) tectonics; (arch) tectonics

тектон·и́т·, **-а**, (*m*) (geol) tectonite

тектон·и́ческ·ий, **-ая**, **-ое**, (*a*) (geol) tectonic

тектоно·фи́з·ик·а, **-и**, (*f*) (geol) tectonophysics

тектор·, **-а**, (*m*) cement gun

те́ктум·, **-а**, (*m*) (zool) tectum

тек·у́т, (*pres 3rd pl*) of **теч·ь**

тек·у́чест·ь, **-и**, (*f*) fluidity, flow; (met) yielding, yield

  зуб· тек·у́чест·и (*m*) upper yield point

  и́ндекс· тек·у́чест·и (*m*) melt flow index (rubber)

  на·ча́·л·о тек·у́чест·и (*n*) upper yield point

  площа́д·к·а тек·у́чест·и (*f*) (met) yield-point elongation

~ по хо́лод·у (rubber) cold flow

  пре·де́л· тек·у́чест·и (*m*) (*and see under* **пре·де́л·**) yield point

~ рабо́ч·ей си́л·ы labour/labor turnover

  температу́р·а по·те́р·и тек·у́чест·и (*f*) (oil) pour point

  температу́р·а тек·у́чест·и (*f*) flow temperature/point

  фигу́р·ы тек·у́чест·и (*pl*) (met) stretcher strains, **strain** figures, Lüders/Piobert/Hartmann lines

тек·у́ч·ий, **-ая**, **-ее**, (*a*) fluid, flowing, running

тек·у́щ·ий, **-ая**, **-ее**, (*pres part act*) of **теч·ь**; current, present

~ выс·от·а́ (*f*) (air) present height

~ ремо́нт· (*m*) minor routine repairs, maintenance

~ с·чёт· (*m*) (fin) current account

тёк·ш·ий, **-ая**, **-ее**, (*past part act*) of **теч·ь**

теле·авто́граф·, **-а**, (*m*) (telecom) teleautograph, telewriter

теле·автоматиз·а́ци·я, **-и**, (*f*) remote-controlled automation

теле·ангиэктаз·и́·я, **-и**, (*f*) (med) teleangiectasia

теле·ви́д·ени·е, **-я**, (*n*) television, televiewing

теле·ви́д·еть, **-ят**, (*imp*) teleview, view television

телевизио́н·н·ый, **-ая**, **-ое**, (*a*) television

~ тру́б·к·а, пере·да·ю́щ·ая (*f*) camera tube, telecamera

~ тру́б·к·а, при·ём·н·ая (*f*) kinescope, picture tube

~ центр· (*m*) television broadcasting station, telecasting station

телеви́зор·, **-а**, (*m*) televisor, television set/receiver

телеви́зи·я, **-и**, (*f*) television

теле́г·а, **-и**, (*f*) cart, waggon

теле·га́м·и·я, **-и**, (*f*) (bot) telegamy

телего́н·, **-а**, (*m*) (elec) telegon (a synchro)

теле·го́н·и·я, **-и**, (*f*) (biol) telegony

телегра́мм·а, **-ы**, (*f*) telegram, cable, cablegram; message

~, метеоролог·и́ческ·ая weather report

~, пере·во́д·н·ая telegraph money order

телегра́ф·, **-а**, (*m*) telegraph; telegraph office

~, маши́н·н·ый (shipb) engine-room telegraph

телеграф·и́ровани·е, **-я**, (*n*) (*v n*) of **телеграф·и́р·овать**; telegraphy, telegraph system/service

~, абоне́нт·ск·ое teletypewriter exchange service

~, встре́ч·н·ое duplex telegraph system

~, высоко·част·о́тн·ое carrier telegraphy

~, дву·сторо́н·н·ее duplex telegraph service

~, двух·по́люс·н·ое double-current telegraph system

телеграф·ѝровани·е
~, дуплékс·н·ое   duplex telegraph system
~, много·крáт·н·ое   multiplex telegraph system
~, мост·ов·óе дуплékс·н·ое   bridge duplex telegraph system
~, над·тонáль·н·ое   super-audio frequency telegraphy
~, одно·пóлюс·н·ое   single-current telegraph system
~, под·тонáль·н·ое   infra-/sub-audio telegraphy
~, стрéл·очн·ое   dial telegraphy
~, тонáль·н·ое   voice frequency telegraphy, audio-frequency telegraphy
телеграф·ѝр·овать, -ую̈т,   (*imp and ₁erf*) telegraph, wire, cable
телеграф·ѝст·, -а,   (*m*) telegraphist, telegraph operator
телеграф·ѝ·я, -и,   (*f*) *see* **телеграф·ѝ·ровани·е** telegraphy
телегрáф·н·ый, -ая, -ое,   (*a*) telegraph, telegraphic
~ агéнт·ств·о   (*n*) news agency
~ áдрес·   (*m*) telegraphic address
~ аппарáт·   (*m*) telegraph; teleprinter, teletypewriter
~ аппарáт·, рулóн·н·ый   roll teleprinter
~ аппарáт·, старт·стóп·н·ый   start-stop teleprinter
~ бланк·   (*m*) telegram/message form
~ об·мéн·   (*m*) telegraph traffic
теле·дáт·чик·, -а,   (*m*) remote pickup
телéж·к·а, -и,   (*g pl*) **-ж·ек·,**   (*f*) (*dim*) of **телéг·а**; trolley, truck; crab, carriage (of overhead crane); trolley (of monorail systems); (rail) bogie
~, аккумулѝтор·н·ая   battery/electric truck
~, бег·ункóв·ая   (rail) leading bogie
~, грéйфер·н·ая   grab-type crab (of overhead crane)
~, груз·ов·áя   industrial truck
~, зáд·н·яя   (rail) hind bogie
~, мýльд·ов·ая   charging bogie (met furnace)
~, перéд·н·яя   (rail) leading bogie
~, под·дéрж·ивающ·ая   (rail) hind bogie
~, пояс·н·áя   diamond-frame bogie
~ с подъ·ём·н·ой рáм·ой   elevating platform truck
~ шассѝ   (a/c) bogie

телéж·ник·, -а,   (*m*) carter; trolley/carriage/bogie/cradle driver/operator
теле·зр·ѝтел·ь, -я,   (*m*) (TV) viewer, televiewer
теле·из·мер·éни·е, -я,   (*n*) telemetering
теле·из·мер·ѝтельн·ый, -ая, -ое,   (*a*) telemetering
телейто·спóр·а, -ы,   (*f*) (bot) teletospore
теле·кинó   (*n indecl*) television cinema
телéки·я, -и,   (*f*) (bot) oxeye, *Buphthalamum*
теле·комáнд·овани·е, -я,   (*n*) remote instruction system
теле·контрóл·ь, -я,   (*m*) remote checking
теле·мéтр·, -а,   (*m*) telemeter
~, фотограф·ѝческ·ий   (phot) telemeter, rangefinder, range meter
теле·метр·ѝческ·ий, -ая, -ое,   (*a*) telemetry, telemetering, telemetered
теле·мéтр·и·я, -и,   (*f*) telemetry, telemetering
теле·механ·изáци·я, -и,   (*f*) provision/installation of remote control facilities
теле·механ·ик·а, -и,   (*f*) telemechanics, remote control
теле·мотóр·, -а,   (*m*) telemotor
теле·на·бóр·, -а,   (*m*) telegraphic typesetting
теле·негатѝв·, -а,   (*m*) (phot) telephoto lens back element
тел·ёнок·, -нк·а,   (*nom pl*) **-л·ѝт·а,** (*g pl*) **-л·ѝт·,**   (*m*) calf
теленóмус·, -а,   (*m*) (ent) Telenomus
теле·объектѝв·, -а,   (*m*) (opt) tele-objective, telephoto lens
телео·лóг·и·я, -и,   (*f*) (biol) teleology
телео·фѝль·н·ый, -ая, -ое,   (*a*) (bot) teleophilic
теле·пере·дáт·чик·, -а,   (*m*) telemetering transmitter; television transmitter
теле·позитѝв·, -а,   (*m*) (phot) telephoto lens front element
теле·при·ём·ник·, -а,   (*m*) television receiver
теле·регул·ѝровани·е, -я,   (*n*) remote control/regulation
теле·регул·ѝруем·ый, -ая, -ое,   (*a*) remote-controlled
теле·рáд·иев·ая терап·ѝ·я   (*f*) teleradiography, (med) radium-beam therapy
теле·с·вѝз·ь, -и,   (*f*) telecommunication, telecommunications; remote-control link

теле·сигнализ·а́ци·я, -и, (*f*) remote indication

теле·ско́п·, -а, (*m*) telescope; optical bar (of rangefinder); lift (of gasholder)

~ **больш·о́й свето·си́л·ы** high-powered telescope

~**, вед·у́щ·ий** (astronav) tracking telescope, (rocket) star tracker

~**, двой·н·о́й** doublet telescope

~**, зерка́ль·н·ый** (astron) reflecting telescope

~**, ли́нз·ов·ый** refracting telescope

~**, ло́м·ан·ый** Coudé telescope

~**, мени́ск·ов·ый** meniscus telescope

~**, на·кло́н·н·ый** tilting telescope

~**, про·бе́ж·н·о-ионизацио́н·н·ый** (nucl) range-ionization telescope

~**, раз·дви́ж·н·о́й** draw-tube telescope

~**, след·я́щ·ий** tracking telescope, (rocket) star follower

теле·скоп·и́ческ·ий, -ая, -ое, (*a*) telescopic

теле·ско́п·н·ый, -ая, -ое, (*a*) of **те·ле·ско́п·**

тел·е́сност·ь, -и, (*f*) corporeality

тел·е́сн·ый, -ая, -ое, (*a*) corporal, corporeal

~ **по·врежд·е́ни·е** (*n*) physical injury

~ **у́гол·** (*m*) (math) solid angle

~ **цвет·** (*m*) flesh colour

теле·съ·ём·к·а, -и, (*f*) telephotography

теле·та́йп·, -а, (*m*) teletype

теле·термо·ско́п·, -а, (*m*) telethermoscope

теле·у·каз·а́ни·е, -я, (*n*) remote indication

теле·у·правл·е́ни·е, -я, (*n*) telecontrol

теле·у·правл·я́ем·ый, -ая, -ое, (*a*) remote-controlled; (air) pilotless

теле·фо́н·, -а, (*m*) telephone

~ **-автома́т·, -а,** (*m*) coin-operated telephone

~**, автомат·и́ческ·ий** dial telephone

~**, голов·н·о́й** headphone

~**, моне́т·н·ый** coin-operated telephone

~**, паралле́ль·н·ый** bridging telephone

~**, станцио́н·н·ый** (rail) yard telephone

телефон·и́ровани·е, -я, (*n*) (*v n*) of **телефон·и́р·овать**; telephony, telephone operation/system

~**, высоко·част·о́тн·ое** carrier telephony

телефон·и́р·овать, -уют, (*imp and perf*) telephone; equip with telephone

телефон·и́ст·, -а, (*m*) telephone operator, telephonist

телефон·и́·я, -и, (*f*) telephony

телефо́н·н·ый, -ая, -ое, (*a*) telephone; telephonic

~ **аппара́т·** (*m*) telephone, telephone set

~ **аппара́т·, до·ба́в·очн·ый** extension set, extension

~ **аппарат·, ме́ст·н·ой батаре́·и** local-battery telephone

~ **коммута́тор·** (*m*) switchboard

~ **сет·ь** (*f*) telephone exchange area

~ **ста́нци·я** (*f*) telephone exchange, central office (U.S.)

~ **ста́нци·я, автомат·и́ческ·ая** automatic telephone exchange

~ **ста́нци·я, между·горо́д·н·ая** trunk exchange, long-distance/inter-city office

~ **тру́б·к·а** (*f*) telephone receiver

телефоно·гра́мм·а, -ы, (*f*) phonogram, telephoned message, telephoned telegram

теле·фото·гра́ф·и·я, -и, (*f*) (phot) telephotography, photography with a telephoto lens

теле·фото́·метр·, -а, (*m*) long-range photometer

тел·е́ц·, -ль·ц·а́, (*m*) (astron) Taurus, the Bull

теле·це́нтр·, -а, (*m*) telecasting/television station

теле·яче́й·к·а, -и, (*g pl*) **-е́·ек·,** (*f*) telemetering unit

тел·и́н·, -а, (*m*) (biochem) theelin

тели·пла́зм·а, -ы, (*f*) thelyplasm, female plasm

тели·то́ки·я, -и, (*f*) (biol) thelytocia, thelytoky

тел·и́ться, -я́тся, (*imp*) calve

тёл·к·а, -и, (*g pl*) **-л·ок·,** (*f*) heifer, young cow

теллу́р·, -а, (*m*) (chem) tellurium, Te

теллур·а́т·, -а, (*m*) (chem) tellurate

теллур·и́д·, -а, (*m*) (chem) telluride

теллу́р·и·й, -я, (*m*) (astron) tellurian, tellurion; (chem) = **теллу́р·**

теллур·и́л·, -а, (*m*) (chem) telluryl

теллу́р·ист·ый, -ая, -ое, (*a*) (chem) tellurous

теллур·и́т·, -а, (*m*) (chem, min) tellurite

теллур·и́ческ·ий, -ая, -ое, (*a*) tellurian, terrestrial

теллуро·висмут·и́т·, -а, (*m*) (min) tellurobismuthite

теллу́р·ов·ый, -ая, -ое, (*a*) (chem) telluric

те́л·о, -а, (*nom pl*) **-а́,** (*n*) body; (math) solid (geometry), field (algebra)

тéл·о
~ **вращ·éни·я** (math) solid of revolution
~, **инерциóн·н·ое** (rocket) G weight
~, **ино·рóд·н·ое** foreign body
~ **кач·éни·я** rolling contact
~, **неб·éсн·ое** (astron) celestial/heavenly body
~, **по·сторóн·н·ее** foreign body
~, **сéр·ое** (phys) grey body
~, **твёрд·ое** (phys) solid; (mech) rigid body
~, **чёрн·ое** (phys) black body, complete radiator, Hohlraum
тело·блáст·, -а, (m) (zool) teloblast
тело·гéн·, -а, (m) telogen
тело·движ·éни·е, -я, (n) body movement
тел·одóнт·ы, -ов, (m pl) (pal) thelodonts, *Thelodontidae*
тело·кинéз·, -а, (m) (gen) telekinesis
тел·óл·, -а, (m) (biochem) theelol
тело·лецитáль·н·ый, -ая, -ое, (a) (zool) telolecithal
телó·м·, -а, (m) (bot) telome
тело·мéр·, -а, (m) (gen) telomere
тело·мериз·áци·я, -и, (f) (chem) telomerization
тело·по·дóб·н·ый, -ая, -ое, (a) field-like (algebra)
тело·рéз·, обык·новéнн·ый (m) (bot) water soldier, *Stratiotes aloides*
тело·с·лож·éни·е, -я, (n) build, figure (of person); (med) habitus
тело·спорúди·и (pl) (zool) *Telosporidea*
тело·тáксис·, -а, (m) (zool) telotaxis, goal orientation
тело·фáз·а, -ы, (f) (gen) telophase
тéль·н·ый, -ая, -ое, (a) body, skin, next-to-the-skin
~ **цвет·** (m) skin/flesh colour
тель·нЯшк·а, -и, (g pl) **-шек·,** (f) sailor's striped vest
тéльсон·, -а, (m) (zool) telson
тéльфер·, -а, (m) telfer, telpher
тéль·ц·е, -а, (i) **-ем·** (n) corpuscle; body, small/minute body
тел·энцефалóн·, -а, (m) (anat) telencephalon
тел·Ят·а, (nom pl) of **тел·ёнок·**
тел·Ятин·а, -ы, (f) (food) veal
тел·Яч·ий, -ья, -ье, (a) (zool) calf, (food) veal
тем (adv) so much the …; (pron instr sing m and n, dat pl m, f and n) of **тот**
~ **бóл·ее** the more so
~ **не мéн·ее** nevertheless

тем
~ **сáм·ый** thereby
тéм·а, -ы, (f) subject, theme, topic
темáт·ик·а, -и, (f) subjects
темат·úческ·ий, -ая, -ое, (a) of **тéм·а**
~ **план·** (m) syllabus; advance list of publications
тéмбр·, -а, (m) timbre
тéмен·и, (g, d, p) of **тéм·я**
теменно·вúс·очн·ый, -ая, -ое, (a) (anat) parietotemporal
теменно·за·тЫл·очн·ый, -ая, -ое, (a) (anat) parietooccipital
темен·нóй, -áя, -óе, (a) of **тéм·я;** (anat) parietal
~ **глаз·** (m) (zool) parapineal body
~ **кост·ь** (f) (anat) parietal
тéми, (pron instr pl) of **тот**
темн·é·ть, -ют, (imp) become/grow dark/darker
темн·úтел·ь, -я, (m) dimmer (lighting)
тёмн·о- (component) dark
темн·овáт·ый, -ая, -ое, (a) darkish, rather dark
темн·ов·óй, -áя, -óе, (a) dark
~ **со·противл·éни·е** (n) dark resistance (of photo-electric cells)
тёмн·о-крáсн·ый, -ая, -ое, (a) dark-red
темно·кóж·ий, -ая, -ее, (a) dark-skinned
тёмно·лúст·н·ый, -ая, -ое, (a) dark-leaved
тёмно·о·крáш·енн·ый, -ая, -ое, (a) dark-coloured/colored
темн·от·á, -Ы, (f) darkness, dark
темно·цвéт·н·ый, -ая, -ое, (a) dark
тёмн·ый, -ая, -ое, (a) dark; obscure (meanings); uneducated, uncultured; deep; filled (of printers' symbols)
~ **нефте·продýкт·ы** (m pl) dark/heavy oils, black products
~ **про·стрáн·ств·о** (n) (phys) dark space
~ **рóмб·ик·** (m) filled diamond (symbol on graph etc.)
тéмп·, -а, (m) rate, pace; tempo, time (music)
тéмпер·а, -ы, (f) (build) distemper
температýр·а, -ы, (f) (see also associated nouns); temperature, point (on temperature scale)
~, **абсолЮт·н·ая** absolute/Kelvin temperature
~ **абсолЮт·н·ого чёрн·ого тéл·а** (phys) black-body temperature

температу́р·а

~, **вне́ш·н·яя** external/outside temperature, (geol) surface temperature

~, **выс·о́к·ая** high/elevated temperature

~, **деба́·евск·ая** (phys) Debye temperature

~ **кип·е́ни·я** boiling point

~, **ко́мнат·н·ая** room temperature

~, **кон·е́чн·ая** ˙ final temperature, end point

~, **обра́т·н·ая** inverse temperature

~ **о·круж·а́ющ·ей сред·ы́** ambient temperature

~ **пад·е́ни·я пе́рв·ой ка́пл·и** over-point (in fractionation)

~, **по·вы́ш·енн·ая** elevated temperature; higher temperature

~, **потенциа́ль·н·ая** (meteor) potential temperature

~, **пре·де́ль·н·ая** limiting temperature; maximum/ceiling temperature; minimum temperature

~ **рос·ы́** dew point

~, **спектро·фотограф·и́ческ·ая** (astron) colour temperature

~, **сре́д·н·е-объём·н·ая** volume-average temperature

~, **сре́д·н·яя** average/bulk temperature

~, **цвет·ов·а́я** (phot) colour temperature

~, **эквивале́нт·н·ая** equivalent temperature

~, **эффекти́в·н·ая** (astron) effective temperature

~, **я́р·костн·ая** brightness temperature; (astron) radiation temperature

температу́р·н·ый, -ая, -ое, (a) temperature

~ **интерва́л·** (m) temperature range

~ **труб·а́** (f) thermometer tube

~ **уда́р·** (m) (met) thermal shock

~ **шов·** (m) expansion joint

температуро·про·во́д·ност·ь, -и, (f) thermal diffusivity

температуро·сто́й·кост·ь, -и, (f) temperature/heat-stability

те́мплет·, -а, (m) templet, template; metallographic section

тём·я, (g, d, p) **те́мен·и,** (n) crown (of head etc.), (anat) sinciput, vertex, bregma

тенар ди́т·, -а, (m) (min) thenardite

тенденцио́з·н·ый, -ая, -ое, (a) tendentious

тенде́нци·я, -и, (f) tendency

~, **по·лож·и́тельн·ая** rising tendency (of barometer)

те́ндер·, -а, (m) (rail) tender; (naut) cutter; (naut) rigging screw

тендо·вагини́т·, -а, (m) (med) tenovaginitis

тендо·синови́т, -а, (m) (med) tenosynovitis

те́н·е- (root) shade, scio-, skio-; shadow

тен·ев·о́й, -а́я, -о́е, (a) shadow, shady, shaded

~ **из·об·раж·е́ни·е** (n) (dynam) shadowgraph

~ **при·бо́р·** (m) shadow-indication instrument, instrument with shadow-type indication

~ **фотогра́ф·и·я** (f) (instr) shadowgraph, shadow photograph

тене·вы·но́с·лив·ый, -ая, -ое, (a) (bot) sciophilous, shade-enduring

тене·лю́б·, -а, (m) (bot) sciophyte, skiophyte

тене·люб·и́в·ый, -ая, -ое, (a) (bot) shade-loving, sciophilous, skiophilous, umbraticolous

тенере́с·, -а, (m) (expl) lead trinitro-resorcinate

тенёт·а, -нёт· (n pl) net, snare

тензи́·метр·, (m) = **тенси́·метр·**

тензо·да́т·чик·, -а, (m) (instr) strain gauge/gage, pressure transducer

тензо́·метр·, -а, (m) (instr) extensometer, strain gauge/gage, strainometer

~, **механ·и́ческ·ий** mechanical extensometer

~ **пере·ме́н·н·ого со·противл·е́ни·я** electrical extensometer, resistance strain gauge/gage

~, **про́·волоч·н·ый** wire-type resistance extensometer

~, **стру́н·н·ый** vibrating-wire extensometer, vibrating-wire strain gauge/gage

~, **электр·и́ческ·ий** electrical extensometer, electrical strain gauge

тензо·метр·и́ческ·ий, -ая, -ое, (a) extensometric, extensometer, strain gauge/gage

~ **рас·ходо·ме́р·** (m) strain-gauge flowmeter

тензо·ме́тр·и·я, -и, (f) extensometry, strain measurement

те́нз·ор·, -а, (m) (math) tensor

~ **дефо́рм·а́ци·и** (mech) strain tensor

~ **ине́рц·и·й** (mech) tensor of inertia

~ **на·пряж·е́ни·й** (mech) stress tensor

~, **шар·ов·о́й** isotropic/spherical tensor

те́нзор·н·ый, -ая, -ое, (*a*) tensor, tensorial

~ ис·числ·е́ни·е (*n*) (math) calculus of tensor, tensor calculus

тензо·чувств·и́тельн·ый, -ая, -ое, (*a*) strain-sensitive

тен·и́ди·й, -я, (*m*) (ent) taenidium

тени·одо́нт·ы (*pl*) (pal) taeniodonts, *Taeniodonta, Taeniodontia*

тенио·пте́рис·, -а, (*m*) (pal, bot) taeniopteris, *Taeniopteris*

тен·и́ст·ый, -ая, -ое, (*a*) shady, shaded, shadowy

те́ни·я, -и, (*f*) (zool) taenia, tenia, (*pl*) taeniae, taenias

теннанти́т, -а, (*m*) (min) tennantite

те́ннис·н·ый, -ая, -ое, (*a*) tennis

тенонто·мио·том·и́·я, -и, (*f*) (med) tenomyotomy

тенонто·пла́ст·ик·а, -и, (*f*) (med) tenoplasty

тенори́т·, -а, (*m*) (min) tenorite

тено·рра́фи·я, -и, (*f*) (med) tenorrhaphy

тено·том·и́·я, -и, (*f*) (med) tenotomy

тенси́·метр·, -а, (*m*) (instr) tensimeter, vapour pressure gauge, surface tension apparatus

тенсо́·метр· (*m*) = **тенси́·метр·**

те́нт·, -а, (*m*) (naut) awning

тентакул·и́т·ы, -ов, (*m pl*) (pal) tentaculites, *Tentaculitidae*

те́н·ь, -и, (*g pl*) **-е́й,** (*f*) shade, shadow, umbra; (naut) by-point (compass)

~, аэродинам·и́ческ·ая (aerodynam) wind shadow

~, дожд·ев·а́я (meteor) rain shadow

~, озо́н·н·ая (astron) ozone shadow

тео·бро́м·, -а, (*m*) (pharm) theobromine

теодоли́т·, -а, (*m*) (surv) theodolite

~, ки́не- cine-theodolite

~, маркше́йдер·ск·ий miners' theodolite

~, техн·и́ческ·ий tacheometer, tachymeter

~, универса́ль·н·ый transit (instrument)

~, шаро·пило́т·н·ый (meteor) pilot-balloon theodolite

теоре́м·а, -ы, (*f*) (*see also* **за·ко́н·**) (*for named theorems see under names*); (math) theorem

~, еди́н·ственност·и uniqueness theorem

~, инве́рси·и circle theorem

~, наи·ме́ньш·ей рабо́т·ы least work theorem, Castigliano's theorem

~, обрат·и́мост·и reciprocity theorem

теоре́м·а

~, обра́т·н·ая conserve of a theorem

~, про·меж·у́точн·ая mean-value theorem

~, равно·веро́ят·ност·и equal-probability theorem

~, тепл·ов·а́я Nernst heat theorem

теоре́т·ик·, -а, (*m*) theorist

теорет·и́ческ·ий, -ая, -ое, (*a*) theoretical

теорет·и́чн·ый, -ая, -ое, (*a*) abstract, purely theoretical

тео́ри·я, -и, (*f*) (*and see under associated words*) theory

~, корабл·я́ naval architecture

~, линеариз·и́рованн·ая linear/linearized theory

~, модифиц·и́рованн·ая improved theory

~, об·общ·ённ·ая generalized theory

~, от·нос·и́тельност·и theory of relativity

~, по́л·я, еди́н·ая unified field theory

тео·филли́н·, -а, (*m*) (biochem) theophylline

тео·ци́н·, -а, (*m*) (biochem) theocine, theophylline

тепер·е́шн·ий, -яя, -ее, (*a*) present, present-day

тепе́рь (*adv*) now

тепл·е́·ть, -ют, (*imp*) get/grow/become warm/warmer

те́пл·ить, -ят, (*imp*) make warm/warmer, warm, warm up

тепл·и́ц·а, -ы, (*f*) hothouse, greenhouse

тепл·и́чн·ый, -ая, -ое, (*a*) hothouse, greenhouse

тепл·о́, -а́, (*n*) (*and see under associated nouns*) (phys) heat; warmth

　　гра́дус·ы тепл·а́ (*pl*) degree above freezing/zero

~, ис·по́льз·ованн·ое used/utilized heat

~, о·ста́т·очн·ое residual heat; shutdown heat, afterheat (nuclear reactors)

~, от·вед·ённ·ое rejected heat

~, ото́·бр·анн·ое rejected heat

~, пере·да·нн·ое exchanged heat

~, рас·се́·янн·ое dissipated heat

~, физ·и́ческ·ое sensible heat

те́пл·о- (*root*) heat, thermal, thermo-

тепло·а́мпул·а, -ы, (*f*) bulb (of thermometer)

тепл·ова́т·ый, -ая, -ое, (*a*) warmish, tepid

тепло·влаго·об·ме́н·, -а, (*m*) heat and moisture exchange

тепло·вла́ж·ностн·ый, -ая, -ое, (*a*) hygrothermal

тепло·во́д·н·ый, -ая, -ое, (*a*) warm-water

тепло·во́з·, -а, (*m*) diesel-electric locomotive; diesel locomotive

тепл·ов·о́й, -а́я, -о́е, (*a*) heat, thermal, caloric

~ **бала́нс·** (*m*) heat balance

~ **барье́р·** (*m*) heat barrier

~ **гистере́зис·** (*m*) thermal hysteresis

~ **дви́г·ател·ь** (*m*) heat engine

~ **на·со́с·** (*m*) heat pump

~ **нейтро́н·** (*m*) thermal neutron

~ **об·рабо́т·к·а** (*f*) heat treatment

~, **почти́** (*a*) near-thermal

~ **при·бо́р·** (*m*) (elec) hot-wire instrument

~ **рас·шир·е́ни·е** (*n*) thermal expansion

~ **режи́м·** (*m*) heating/firing cycle/conditions (furnaces); thermodynamics (of soil)

~ **сет·ь** (*f*) (build) central heating system

~ **уда́р·** (*m*) (med) heat stroke; (met) thermal shock; (min) thermoplegia

~ **эффе́кт·** (*m*) heat/thermal effect; calorific value (of a reaction)

тепло·вос·при·я́т·и·е, -я, (*n*) heat reception

тепло·вы·дел·е́ни·е, -я, (*n*) heat liberation/release

    то́ч·к·а тепло·вы·дел·е́ни·я (*f*) (rocket) point of heat release

тепло·вы·дел·я́ющ·ий, -ая, -ее, (*a*) heat-liberating; (rocket) heat-releasing

~ **с·бо́р·к·а** (*f*) (nucl) fuel assembly

~ **элеме́нт·** (*m*) (nucl) fuel element

тепло·гаш·е́ни·е, -я, (*n*) (phys) thermoquenching

тепло·ём·кост·ь, -и, (*f*) thermal/heat/calorific capacity/content, heat

~, **молекуля́р·н·ая** molecular heat

~, **объ·ём·н·ая** molal specific heat, volumetric heat

~, **у·де́ль·н·ая** specific heat

тепло·из·луч·е́ни·е, -я, (*n*) thermal/heat radiation

тепло·изол·и́рующ·ий, -ая, -ее, (*a*) thermal-/heat-insulating

тепло·изол·я́тор·, -а, (*m*) thermal/heat insulator

тепло·изол·я́ци·я, -и, (*f*) thermal/heat insulation

тепло·кро́в·н·ый, -ая, -ое, (*a*) (zool) warm blooded, homothermous, homoiothermal

тепло·леч·е́ни·е, -я, (*n*) (med) heat therapy

тепло·люб·и́в·ый, -ая, -ое, (*a*) (biol) thermophilic, thermophile, thermophilous

тепло·ме́р·, -а, (*m*) heat-meter, calorimeter (measures consumption)

тепло·на·копл·е́ни·е, -я, (*n*) heat build-up

тепло·на·пряж·ённост·ь, -и, (*f*) calorific intensity, heat release rate (of furnace); (met) state of thermal stress

тепло·на·пряж·ённ·ый, -ая, -ое, (*a*) (met) thermally stressed, in a state of thermal stress

тепло·нос·и́тел·ь, -я, (*m*) heat-transfer agent, heat carrier, coolant (nuclear reactors)

~, **жи́дк·ий** heat-exchange/-transfer fluid

~, **жидко·металл·и́ческ·ий** (nucl) liquid-metal coolant

~, **металл·и́ческ·ий,** heat-transfer metal

тепло·об·ме́н·, -а, (*m*) heat exchange

    коэффицие́нт· тепло·об·ме́н·а (*m*) heat transfer coefficient

тепло·об·ме́н·ник·, -а, (*m*) heat-exchanger

~, **воздухо·воз·ду́ш·н·ый** air-to-air heat exchanger

~, **инже́ктор·н·ый** spray-type heat exchanger

~, **кожухо·тру́б·н·ый** tube-and-shell type heat exchanger

~, **противо·то́ч·н·ый** counterflow heat exchanger

~, **с·мес·и́тельн·ый** direct-contact exchanger

тепло·образ·ова́ни·е, -я, (*n*) heat production/generation; heat build-up (in tyres etc.)

тепло·от·бо́р·, -а, (*m*) heat take-off/rejection

тепло·от·во́д·, -а, (*m*) heat removal/extraction/rejection, cooling

тепло·от·да́т·чик·, -а, (*m*) hot body

тепло·от·да́·ч·а, -и, (*f*) output/emission of heat; heat transfer; heat rejection

    коэффицие́нт· тепло·от·да́·ч·и (*m*) heat transfer coefficient

    по·ве́рх·ност·ь тепло·от·да́·ч·и (*f*) heat delivery surface

тепло·от·да·ю́щ·ий, -ая, -ее, (*a*) exothermic

тепло·пеленг·а́тор·, -а, *(m)* thermal D/F set, thermal direction finder
тепло·пере·да́·ч·а, -и, *(f)* heat transfer
   дл·и́тельность·ь тепло·пере·да́··ч·и *(f)* heat transfer retention time
   коэффицие́нт· тепло·пере·да́·ч·и *(m)* heat transfer coefficient
тепло·пере·хо́д·, -а, *(m)* heat transfer
тепло·по·те́р·я, -и, *(f)* heat loss
тепло·про·во́д·, -а, *(m)* heat conductor; hot water pipe, steam pipe
тепло·про·вод·и́мост·ь, -и, *(f)* thermal conductivity
тепло·про·во́д·ност·ь, -и, *(f)* thermal conductivity
   коэффицие́нт· тепло·про·во́д·-ност·и *(m)* coefficient of thermal conductivity
тепло·про·во́д·н·ый, -ая, -ое, *(a)* heat-conducting
тепло·проду́кц·я, -и, *(f)* (zool) thermo-genesis
тепло·про·зр·а́чност·ь, -и, *(f)* dia-thermancy
тепло·про·зр·а́чн·ый, -ая, -ое, *(a)* diathermal, diathermic
тепло·про·из·вод·и́тельност·ь, -и, *(f)* calorific value/power
тепло·про·из·вод·я́щ·ий, -ая, -ее, *(a)* heat-generating
тепло·про·ниц·а́емост·ь, -и, *(f)* dia-thermancy
тепло·сил·ов·о́й, -а́я, -о́е, *(a)* thermal power
~ **у·стано́в·к·а** *(f)* thermal power station/plant
тепло·с·ме́н·а, -ы, *(f)* thermal cycling
тепло·со·держ·а́ни·е, -я, *(n)* (phys) heat content, enthalpy
тепло·сто́й·к·ий, -ая, -ое, *(a)* heat-resistant/-proof, thermotolerant, thermostable
тепло·сто́й·кост·ь, -и, *(f)* heat resis-tance, thermal tolerance/stability
~ **по Ма́ртенс·у** Martens heat-stability
тепло·съ·ём·, -а, *(m)* heat removal/ extraction
тепл·от·а́, -ы, *(f)* *(and see under asso-ciated nouns)* (phys) heat; warmth
~ **гор·е́ни·я** heat of combustion
~ **ис·пар·е́ни·я** heat of evaporation
~**, луч·и́ст·ая** radiant heat
~ **нейтрализ·а́ци·и** (chem) neutraliz-ing heat
~ **образ·ова́ни·я** heat of formation

тепл·от·а́
~ **паро·образ·ова́ни·я** evaporation/ vaporization heat
~**, с·кры́·т·ая** latent heat
~**, у·де́ль·н·ая** specific heat
тепло·тво́р·ност·ь, -и, *(f)* calorific value
тепло·тво́р·н·ый, -ая, -ое, *(a)* calor-ific, heat-producing
~ **спосо́б·ност·ь** *(f)* calorific value
тепло·те́хн·ик·, -а, *(m)* heat engineer
тепло·те́хн·ик·а, -и, *(f)* heat engineer-ing
тепло·техн·и́ческ·ий, -ая, -ое, *(a)* heat-engineering
тепло·у·сто́й·чив·ый, -ая, -ое, *(a)* heat-resisting, thermostable, thermo-tolerant; (met) creep-resisting; (phys) thermoduric
тепло·фикацио́н·н·ый, -ая, -ое, *(a)* of **тепло·фик·а́ци·я**
тепло·фик·а́ци·я, -и, *(f)* combined heat-and-power supply, district heat-ing
тепло·фи́льтр·, -а, *(m)* heat shield
тепло·хо́д·, -а, *(m)* motor ship/vessel
тепло·центра́л·ь, -и, *(f)* district heating plant/station
тепло·чувств·и́тельн·ый, -ая, -ое, *(a)* heat-sensitive
тепло·электр·и́ческ·ий, -ая, -ое, *(a)* thermoelectric; heat-and-power
тепло·электро·генера́тор·н·ая ла́м-п·а *(f)* lamp-type thermoelectric generator
тепло·электро·воздухо·центра́л·ь,-и, *(f)* heat, power and compressed air station
тепло·электро·центра́л·ь, -и, *(f)* heat-and-power station
тепло·энерге́т·ик·а, -и, *(f)* heat-and-power engineering
тепл·у́шк·а, -и, *(g pl)* **-шек·,** *(f)* (rail) heated freight car/wagon; heated shelter (on construction site etc.)
тёпл·ый, -ая, -ое, *(a)* warm; heat, thermal; (elec, instr) hot-wire; (build) heat-insulated
~ **ампер·ме́тр·** *(m)* (elec) hot-wire ammeter, thermal ammeter
~ **люч·ин·а** *(f)* insulated hatch cover
~ **фронт·** *(m)* (meteor) warm front
~ **я́щик·** *(m)* (st eng) hot well
тепл·я́к·, -а́, *(m)* (build) winter shelter
тёр·, *(past masc sing)* of **тер·е́ть**
тера́ли́т·, -а, *(m)* (geol) theralite

терап·е́вт·, -а,   (*m*) (med) therapeutist, therapist

терап·е́вт·и́ческ·ий, -ая, -ое,   (*a*) (med) therapeutic

терап·и́·я, -и,   (*f*) (med) therapy

терапс·и́дн·ые, -ых,   (*pl decl as adj*) (zool) therapsids, *Therapsida*

терато·гене́з·, -а,   (*m*) (biol) teratogenesis

терато·ло́г·и·я, -и,   (*f*) (biol) teratology

терато́·м·а, -ы,   (*f*) (med) teratoma

те́рб·иев·ый, -ая, -ое,   (*a*) terbium

те́рб·и·й, -я,   (*m*) (chem) terbium, Tb

те́рг·и́т·, -а,   (*m*) (zool) tergite

те́рг·у́м·, -а,   (*m*) (zool) tergum

тереб·е́н·, -а,   (*m*) (pharm) terebene

тереб·и́лк·а, -и,   (*g pl*) **-лок·,** (*f*) (agr) puller, plucker

тереб·и́нов·ый, -ая, -ое,   (*a*) (chem) terebinic

~ **кисл·от·а́**   (*f*) terebinic acid

тереб·и́ть, -я́т,   (*imp*) pluck, pull

теребл·е́ни·е, -я,   (*n*) plucking, pulling

~ **льн·а́** flax pulling

теребрат·уло́ид·н·ый   (*a*) (pal) terebratuloid

тереске́н·, -а,   (*m*) (bot) winterfat, *Eurotia*

тер·е́ть,   (*pres 3rd pl*) **тр·у́т,** (*imp*) rub; polish; chafe, abrade; grate, grind

тере·фта́л·ев·ый, -ая, -ое,   (*a*) (chem) terephthalic

~ **кисл·от·а́** (*f*) (chem) terephthalic acid

терз·а́·ть, -ют,   (*imp*) tear to pieces; torment

терилён·, -а,   (*m*) (plast, text) terylene

тёр·к·а, -и,   (*g pl*) **-р·ок·,** (*f*) grater, rasp; (agr) huller, hulling machine; float (plasterer's trowel); (zool) radula

тёр·ли,   (*past pl*) of **тер·е́ть**

терлингуа́йт·, -а, (*m*) (min) terlinguaite

тёр·ло,   (*past n sing*) of **тер·е́ть**

те́рм·, -а,   (*m*) therm (unit of heat); (nucl) term, energy level

те́рм·а, -ы, (*f*) (geol) hot/thermal spring

терма́кс·, -а,   (*m*) thermax (a thermatomic carbon)

термализ·а́ци·я, -и, (*f*) thermalization

термалло́·й, -я, (*m*) (met) thermalloy

терма́ль·н·ый, -ая, -ое,   (*a*) (geol) thermal

те́рмин·, -а, (*m*) (ling) term

термина́ль·н·ый, -ая, -ое, (*a*) terminal

термино·ло́г·и·я, -и, (*f*) terminology

термио́н·, -а, (*m*) (phys) thermion

термио́н·н·ый, -ая, -ое, (*a*) thermionic

терм·и́ст·, -а, (*m*) (met) heat treater

терми́стор·, -а,   (*m*) (phys) thermistor

терми́т·, -а,   (*m*) thermite, thermit (Al–FeO mixture); (ent) termite

терми́т·н·ый, -ая, -ое,   (*a*) (chem) thermite, thermit; (ent) termite

~ **с·ва́р·к·а** (*f*) thermit welding

терм·и́ческ·ий, -ая, -ое,   (*a*) heat, thermal, thermic, (instr) hot-wire

~ **диссоци·а́ци·я** (*f*) thermal dissociation

~ **об·рабо́т·к·а** (*f*) heat treatment

~ **сто́й·кост·ь** (*f*) thermal stability

~ **у·сто́й·чивост·ь** (*f*) (elec) thermal stability

~ **эква́тор·** (*m*) (meteor) thermal/heat equator

те́рмо- (*int component*) thermo-, thermal, heat, (instr) hot-wire

термо·акце́птор·, -а,   (*m*) thermal acceptor

термо·ана́лиз·, -а, (*m*) (chem) thermoanalysis, thermal analysis

термо·анемо́·метр·, -а, (*m*) (meteor) hot-wire anemometer

термо·бар·и́ческ·ое по́л·е (*n*) (meteor) temperature-and-pressure field

термо·баро·ка́мер·а, -ы,   (*f*) altitude chamber

термо·баро́·метр·, -а,   (*m*) (meteor) hypsometer

термо·батере́·я, -и, (*f*) (elec) thermopile

термо·бати·гра́ф·, -а,   (*m*) (ocean) bathythermograph, deep-sea thermograph

термо·би·мета́лл·, -а, (*m*) (met) bimetal

термо·бу́р·, -а, (*m*) (civ eng) heat drill

термо·ве́сы, -ов, (*m pl*) (instr) thermobalance

термо·вульканиз·а́ци·я, -и,   (*f*) (rubber) heat/thermal vulcanization

термо·вы·све́ч·ивани·е, -я, (*n*) thermoluminescence

термо·ге́н·н·ый, -ая, -ое, (*a*) (zool) thermogenetic

термо·гигро́·граф·, -а, (*m*) (meteor) thermohygrograph

термо·гла́з·, -а,   (*m*) (instr) heat-radiation head

термо·глубино·ме́р·, -а, (*m*) (ocean) thermometer-type depth gauge/gage

термо·гра́мм·а, -ы, (*f*) (chem) thermogram; thermograph trace

термо́·граф·, -а, (*m*) thermograph

термо·да́т·чик·, -а, (*m*) temperature-sensitive element

термо·дестру́кци·я, -и, (*f*) (chem) thermal degradation

термо·дина́м·ик·а, -и, (*f*) thermodynamics

    пе́рв·ое на·ча́·л·о термо·дина́м·ик·и (*n*) first law of thermodynamics

~, **техн·и́ческ·ая** engineering thermodynamics

термо·динам·и́ческ·ий, -ая, -ое, (*a*) thermodynamic

термо·диффу́з·и·я, -и, (*f*) thermal diffusion

термо·за·гото́в·к·а, -и, (*g pl*) **-в·ок·,** (*f*) thermometer blank

термози́т·, -а, (*m*) "thermozit" (concrete filler)

термо·зо́нд·, -а, (*m*) (instr) thermoprobe, thermocouple probe; (ocean) deep sea thermograph

термо·из·вещ·а́тел·ь, -я, (*m*) (a/c) flame detector switch

термо·изол·я́ци·я, -и, (*f*) heat insulation

термо·изо·пле́т·а, -ы, (*f*) (meteor) thermoisopleth

термо·ио́н·н·ый, -ая, -ое, (*a*) thermionic

термо·каротта́ж·, -а, (*m*) (geol) thermal well-logging

термо·ка́рст·, -а, (*m*) (geol) thermal karst

термо·като́д·, -а, (*m*) (elec) hot cathode

термо·конста́нт·н·ый, -ая, -ое, (*a*) temperature-constant

термо·лаби́ль·н·ый, -ая, -ое, (*a*) (chem) thermolabile

термо́·лиз·, -а, (*m*) (chem) thermolysis

термо·магни́т·н·ый, -ая, -ое, (*a*) thermomagnetic; temperature-magnetization

~ **крив·а́я** (*f*) temperature-magnetization curve

~ **явл·е́ни·е** (*n*) (phys) thermomagnetic effect

термо́·метр·, -а, (*m*) thermometer

~, **вла́ж·н·ый** wet-bulb thermometer

~, **га́з·ов·ый** gas thermometer

~, **га́з·ов·ый мано·метр·и́ческ·ий** gas-expansion thermometer

~ **-глубино·ме́р·, -а,** (*m*) thermometer-type depth gauge

~, **глубоко·во́д·н·ый** deep-sea thermometer

~, **дистанцио́н·н·ый** distance/distant-reading/transmitting thermometer

термо́·метр

~, **жи́дк·остн·ый мано·метр·и́ческ·ий** liquid-expansion thermometer

~, **максима́ль·н·ый** maximum thermometer

~, **мано·метр·и́ческ·ий** expansion thermometer

~, **пар·ов·о́й мано·метр·и́ческ·ий** vapour-pressure thermometer

~ **рас·шир·е́ни·я** expansion thermometer

~ **со·противл·е́ни·я** resistance thermometer, resistance pyrometer

~, **с·мо́ч·енн·ый** wet-bulb thermometer

~, **сух·о́й** dry-bulb thermometer

термо·метр·и́ческ·ий, -ая, -ое, (*a*) thermometer, thermometric

термо·ме́тр·и·я, -и, (*f*) thermometry, temperature measurement

термо·метро́·граф·, -а, (*m*) recording thermometer, thermograph

термо·механ·и́ческ·ий, -ая, -ое, (*a*) thermomechanical

~ **крив·а́я** (*f*) temperature vs. deformation curve

~ **об·рабо́т·к·а** (*f*) (met) thermomechanic treatment, combined heat treatment and mechanical working

термо·на́ст·и·я, -и, (*f*) (bot) thermonasty

термо·натри́т·, -а, (*m*) (min) thermonatrite

термо·об·рабо́т·к·а, -и, (*g pl*) **-т·ок·,** (*f*) heat treatment

~, **цикл·и́ческ·ая** (met) thermal cycling

термо·осмо́с·, -а, (*m*) (phys) thermoosmosis, fountain effect

термо·па́р·а, -ы, (*f*) thermocouple

~ **платино·ро́д·иев·ая** platinum/rhodium thermocouple

~, **по·грузж·н·а́я** immersion thermocouple

~, **у·равн·и́тельн·ая** booster thermocouple

термо·патро́н·, -а, (*m*) thermobulb

термо·пла́ст·, -а, (*m*) (plast) thermoplast, thermoplastic, thermoplastic material/resin

термо·пла́стик·, -а, (*m*) *see* **термо·-пла́ст·**

термо·пла́стик·а́т·, -а, (*m*) thermally plasticized rubber/material

термопла́стик·а́ци·я, -и, (*f*) thermal plasticization/softening

термо·пласт·ѝческ·ий, -ая, -ое, (*a*) thermoplastic

термопласт·ѝчн·ый, -ая, -ое, (*a*) thermoplastic

термо·полимериз·а́ци·я, -и, (*f*) heat/thermal polymerization

термо·пре́н·, -а, (*m*) (rubber) thermoprene

термо·пре·образ·ова́тел·ь, -я, (*m*) (elec) thermal converter/element, thermocouple converter

термо·при·бо́р·, -а, (*m*) thermocouple instrument

термо·при·ём·ник·, -а, (*m*) thermal pick-up

термо·рас·па́д·, -а, (*m*) (chem) thermal degradation

термо·рас·цеп·ѝтел·ь, -я, (*m*) (elec) thermal trip

термо·реактѝв·н·ый, -ая, -ое, (*a*) thermosetting

~ пласт·ма́сс·а (*f*) thermosetting plastic

термо·регул·я́тор·, -а, (*m*) thermocontroller, thermoregulator, temperature controller

термо·регул·я́ци·я, -и, (*f*) (phys) heat regulation, temperature control

термо·реле́ (*n indecl*) thermal relay

термо·реце́птор·, -а, (*m*) (physiol) thermoreceptor

те́рмос·, -а, (*m*) thermos flask/bottle, vacuum flask

термо·синте́з·, -а, (*m*) (nucl) thermofusion

тепло·сифо́н·н·ый, -ая, -ое, (*a*) natural-circulation (e.g. of cooling systems)

термо·ско́п·, -а, (*m*) (phys) thermoscope

термо·с·ме́н·а, -ы, (*f*) (phys) thermal cycle

термо·с·ме́н·н·ый, -ая, -ое, (*a*) thermal cycling

термо·со·противл·е́ни·е, -я, (*n*) thermistor

термо·с·па́·й, -я, (*m*) thermojunction

термо·стабилиз·ѝрованн·ый, -ая, -ое, (*a*) thermally stabilized

термо·стабѝль·н·ый, -ая, -ое, (*a*) (phys) thermostable

термо·стар·е́ни·е, -я, (*n*) heat-ageing/aging

термо·ста́т·, -а, (*m*) thermostat

~, ква́рц·ев·ый (rad) crystal oven

термо·стат·ѝровани·е, -я, (*n*) thermostatic control

термо·стат·ѝрованн·ый, -ая, -ое, (*a*) thermostatically-controlled, thermostated

термо·ста́т·н·ый, -ая, -ое, (*a*) thermostat, thermostatic

термо·сто́й·к·ий, -ая, -ое, (*a*) heat-resistant, (chem) thermostable, (bot) thermotolerant, (phys) thermoduric

термо·сто́лб·ик·, -а, (*m*) (elec) thermopile

термо·сфе́р·а, -ы, (*f*) (meteor) thermosphere

термо·та́ксис·, -а, (*m*) (biol) thermotaxis

термо·телефо́н·, -а, (*m*) thermophone, thermal telephone receiver, hot-wire telephone

термо·терап·ѝ·я, -и, (*f*) (med) thermotherapy

термо·то́к·, -а, (*m*) heat flow

термо·троп·ѝзм·, -а, (*m*) (biol) thermotropism

термо·у·пру́г·ий, -ая, -ое, (*a*) thermoelastic

термо·фѝль·н·ый, -ая, -ое, (*a*) (biol) thermophilic, thermophilous, thermophile

термо·фо́н·, -а, (*m*) (phys) thermophone

термо·фо́р·, -а, (*m*) (oil) thermofor

термо·фосфа́т·, -а, (*m*) "thermophosphate" (a fertilizer)

термо·хѝм·и·я, -и, (*f*) thermochemistry

термо·хром·ѝровани·е, -я, (*n*) (met) chromizing

термо·чувств·ѝтельн·ый, -ая, -ое, (*a*) thermoscopic, heat-sensitive

термо·шу́м·, -а, (*m*) thermal agitation noise, thermal/Johnson noise (vacuum tubes)

термо·щу́п·, -а, (*m*) thermal detector/probe

термо·эдс (*indecl*) (phys) thermo-emf

термо·электр·ѝческ·ий, -ая, -ое, (*a*) thermoelectric; (instr) thermocouple, thermel

~ батаре́·я (*f*) thermopile

~ эффе́кт· (*m*) thermocouple/thermoelectric effect

термо·электр·ѝчеств·о, -а, (*n*) thermoelectricity

термо·электро·двѝж·ущ·ая сѝл·а (*f*) thermo-emf, thermoelectromotive force

терм·электро́н·, -а, (*m*) thermoelectron

термо·электрон·н·ый, -ая, -ое, (*a*) thermionic

термо·элемéнт·, -а, (*m*) thermocouple

термо·я́дер·н·ый, -ая, -ое, (*a*) thermonuclear

тёрн·, -а, (*m*) (bot) sloe. *Prunus spinosa*

терн·и́ст·ый, -ая, -ое, (*a*) thorny

терн·óвн·ик·, -а, (*m*) = **тёрн·**

терн·óв·ый, -ая, -ое, (*a*) thorn, thorny

терно·сли́в·, -а, (*m*) = **терно·сли́в·а**

терно·сли́в·а, -ы, (*f*) (bot) damson, damson plum, *Prunus insititia*

теро·мóрф·, -а, (*m*) (pal) theromorph

теро·фи́т·, -а, (*m*) (bot) therophyte

тёр·очн·ый, -ая, -ое, (*a*) grinding, grating

терп·ели́в·ый, -ая, -ое, (*a*) patient

терп·éн·, -а, (*m*) (chem) terpene

терп·éни·е, -я, (*n*) patience, sufferance, endurance

терп·éнов·ый, -ая, -ое, (*a*) terpene, terpenic

терпенти́н·, -а, (*m*) turpentine

терпенти́н·ов·ый, -ая, -ое, (*a*) turpentine

~ **дéрев·о** (*n*) (bot) terebinth, *Pistacia terebinus*

~ **мáсл·о** (*n*) turpentine, oil of turpentine

терп·éть, -я́т, (*imp*) put up with, be patient with, have patience; suffer, endure, undergo

~ **бéд·стви·е** be in distress

терп·и́м·ый, -ая, -ое, (*a*) (*pres part pass*) see **терп·éть;** tolerant; tolerable, endurable, bearable

терпи́н·, -а, (*m*) (chem) terpine

терпин·гидрáт·, -а, (*m*) terpine hydrate

терп·инéн·, -а, (*m*) (chem) terpinene

терп·инеóл·, -а, (*m*) (chem) terpineol

терп·инолéн·, -а, (*m*) (chem) terpinolene

тéрп·к·ий, -ая, -ое, (*a*) astringent

тéрп·кост·ь, -и, (*f*) astringency

терпýг·, -á, (*m*) rasp, file; (fish) (*pl*) *Hexagrammidae*

~**, одно·пéр·ый** (fish) atka fish, *Pleurogrammus monopterygius*

терракóт·а, -ы, (*f*) terracotta

терра·мици́н·, -а, (*m*) (pharm) terramycin, oxytetracycline

террáри·й, -я, (*m*) (zool) terrarium; (zool) terrapin

террáри·ум·, -а, (*m*) (zool) terrarium

терра-рóсс·а, -ы, (*f*) terra rossa (soil)

террáс·а, -ы, (*f*) terrace

~**, аллювиáль·н·ая** (geol) alluvial terrace

террас·и́р·овать, -уют, (*imp and perf*) terrace

терри·гéн·н·ый, -ая, -ое, (*a*) (geol) terrigenous

террикóн·, -а, (*m*) *see* **террикóн·ик·**

террикóн·ик·, -а, (*m*) (min) spoil/waste pile/heap

территориáль·н·ый, -ая, -ое, (*a*) territorial, area

~ **вóд·ы** (*pl*) territorial waters

территóри·я, -и, (*f*) territory; area

террóр·, -а, (*m*) terror

террофазори́т·, -а, (*m*) (build) asbestos cement sheeting

тёр·т·ый, -ая, -ое, (*past part pass*) of **тер·éть;** grated, ground

~ **крáс·к·а** (*f*) paste paint, oil paste

терфени́л·, -а, (*m*) terphenyl

тёр·ш·ий, -ая, -ее, (*past part act*) of **тер·éть**

тер·я́·ть, -ют, (*imp*) lose

тёс·, -а, (*m*) ½-inch boards (13–25 mm)

тес·áк·, -á, (*m*) adze; broadsword

тёс·ан·ый, -ая, -ое, (*a*)

~**, кáмен·ь** (*m*) (build) ashlar, ashler

тес·áть, (*pres 3rd pl*) **тéш·ут,** (*imp*) adze, trim (wood), dress (stone); hew, hack

тесём·к·а, -и, (*g pl*) **-м·ок·,** (*f*) = **тесьм·á**

~ **противо·гáз·а** respirator head-band

тесём·очн·ый, -ая, -óе, (*a*) tape, band, braid

тес·и́н·а, -ы, (*f*) half-inch board

Тесла, трансформ·áтор· (*m*) (elec) Tesla transformer/coil

тес·л·ó, -á, (*nom pl*) **тёс·л·а,** (*g pl*) **тёс·ел·,** (*n*) adze

тесн·и́н·а, -ы, (*f*) (geog) narrows; gorge, ravine

тесн·и́ть, -я́т, (*imp*) restrict, confine, limit, pack close, be tight, be too small; **-ся** (*pass*); crowd, cluster

тéсн·о (*adv*) closely, tightly

~ **с·вя́з·анн·ый** closely connected/related

тéсн·ый, -ая, -ое, (*a*) confined, restricted; close, close-packed; tight

тес·óв·ый, -ая, -ое, (*a*) of **тёс·;** board, plank

тессер·áльн·ый, -ая, -ое, (*a*) (cryst) tesseral, cubic

тéст·, -а, (*m*) test

тéст·ер·, -а, (*m*) tester; (elec) tube/valve tester

тест·и́кул·а, -ы, (*f*) (zool) testicle, testis

тест·и́т·, -а, (*m*) (med) testitis
те́ст·о, -а, (*n*) dough; pastry; paste
~, цеме́нт·н·ое slurry
те́сто·мес·и́лк·а, -и, (*g pl*) **-лок·,** (*f*) dough mixer
те́сто·обра́з·н·ый, -ая, -ое, (*a*) pasty; doughy
те́сто·стеро́н·, -а, (*m*) (biochem) testosterone
тесьм·а́, -ы́, (*f*) (text) tape, band, braid
~, верет·ённ·ая banding
~, плет·ён·ая (text) braid
тетан·и́·я, -и, (*f*) (med) tetany
те́тан·ус·, -а, (*m*) (med) tetanus; (zool) tetanus, tetanic contraction
тетарто́·эдр·, -а, (*m*) (cryst) tetartohedron
тетарто·эдр·и́ческ·ий, -ая, -ое, (*a*) (cryst) tetartohedral
те́терев·, -а, (*nom pl*) **-а́,** (*g pl*) **-о́в·** (*m*) (orn) grouse
тетерев·и́н·ые, -ых, (*pl decl as adj*) (orn) grouse, (*pl*) *Tetraonidae*
тетив·а́, -ы́, (*f*) bow string; string, taut string; string board, stringer (stairs)
те́тис·, -а, (*m*) (geol) tethys
тетра- (*int component*) tetra-
тетра·бор·а́т·, -а, (*m*) (chem) tetraborate
тетра·борно·ки́сл·ый, -ая, -ое, (*a*) tetraborate (of)
тетра·бо́р·н·ый, -ая, -ое, (*a*) tetraboric
тетра·вакци́н·а, -ы, (*f*) choleratyphoid-and-paratyphoid vaccine
тетра·бром·мета́н·, -а, (*m*) carbon tetrabromide
тетра·гекса́·эдр·, -а, (*m*) (cryst) tetrahexahedron
тетра·гекса·эдр·и́ческ·ий, -ая, -ое, (*a*) tetrahexahedral
тетра·гидро- (*int component*) tetrahydro-
тетрагидро·фенантре́н·, -а, (*m*) (chem) tetranthrene
тетрагидро·фура́н·, -а, (*m*) (chem) tetrahydrofuran
тетра·ги́р·, -а, (*m*) (cryst) tetragyre, fourth-order symmetry axis
тетра·ги́р·н·ый, -ая, -ое, (*a*) (cryst) tetragyric, tetragonal
тетра·го́н·, -а, (*m*) (math) tetragon
тетра·гона́ль·н·ый, -ая, -ое, (*a*) tetragonal
~ систе́м·а (*f*) (cryst) tetragonal system

тетр·а́д·а, -ы, (*f*) (biol etc.) tetrad
тетрадими́т·, -а, (*m*) (min) tetradymite
тетра́д·к·а, -и, (*g pl*) **-д·ок·,** (*f*) = **тетра́д·ь**
тетрадопенто́з·а, -ы, (*f*) (biochem) tetradopentose
тетра́д·ь, -и, (*f*) notebook, notepaper; section (bookbinding)
тетразе́н·, -а, (*m*) (chem) tetrazene, tetracene
тетрази́н·, -а, (*m*) tetrazine
тетразо·крас·и́тел·ь, -я, (*m*) tetrazodye
тетр·азо́л·, -а, (*m*) tetrazole
тетразо́н·, -а, (*m*) tetrazone
тетра·иод·мета́н·, -а, (*m*) carbon tetra-iodide
тетра·ка́ин·, -а, (*m*) (pharm) tetracaine
тетра·карбони́л· ни́кел·я (*m*) (chem) nickel carbonyl
тетра·карбо́н·ов·ая кисл·от·а́ (*f*) tetrabasic carboxylic acid, tetracarboxylic acid
тетра·коза́н·ов·ая кисл·от·а́ (*f*) tetracosanic acid
тетра·кози́л·, -а, (*m*) tetracosyl
тетракти́н·, -а, (*m*) (zool) tetraxon
тетрали́н·, -а, (*m*) (chem) Tetralin, tetranaphthahydroline
тетраме́р·, -а, (*m*) (chem) tetramer
тетра·нитро·мета́н·, -а, (*m*) (chem) tetranitromethane
тетра·нитро·пент·а·эритри́т·, -а, (*m*) (expl) pentalrythritol tetranitrate, PETN
тетра·пло́ид·, -а, (*m*) (gen) tetraploid
тетра·спо́р·а, -ы, (*f*) (bot) tetraspore
тетра·спора́нги·й, -я, (*m*) (bot) tetrasporangium
тетра·тиона́т·, -а, (*m*) tetrathionate
~ ни́кел·я nickel tetrathionate
тетра·тионово·ки́сл·ый, -ая, -ое, (*a*) tetrathionate (of)
~ кисл·от·а́ (*f*) tetrathionic acid
тетра·фтор·и́д·, -а, (*m*) tetrafluoride
тетра·фтор·мета́н·, -а, (*m*) carbon tetrafluoride
тетра·хлор·бензо́л·, -а, (*m*) tetrachlorobenzene
тетра·хлор·мета́н·, -а, (*m*) tetrachloromethane, carbontetrachloride
тетра·хлор·эта́н·, -а, (*m*) (chem) tetrachloroethane
тетра·хлор·этиле́н·, -а, (*m*) (chem) tetrachloroethylene, perchlorethylene
тетра·цикл·и́н·, -а, (*m*) (chem) tetracycline

тетра·цикл·и́ческ·ий, -ая, -ое,   (*a*) tetracyclic

тетра́·эдр·, -а,   (*m*) (math) tetrahedron

тетра·эдр·и́т·, -а,   (*m*) (min) tetrahedrite

тетра·эдр·и́ческ·ий, -ая, -ое,   (*a*) tetrahedral

тетра·этил·свин·е́ц·, -н·ц·а́·   (*m*) (chem) tetraethyl lead, tetraethyllead

тетр·и́л·, -а,   (*m*) (expl) tetryl

тетр·о́д·, -а,   (*m*) (elec) tetrode

тетро́л·ов·ая кисл·от·а́   (*f*) (chem) tetrolic acid, methyl propionic acid

тефи·гра́мм·а, -ы,   (*f*) (meteor) tephigram

тефло́н·, -а,   (*m*) (plast) teflon

тефр·и́т·, -а,   (*m*) (geol) tephrite

тефро·и́т·, -а,   (*m*) (min) tephroite

тетра·нитро·эфи́р· эритри́т·а   (*m*) tetranitrol

тех-   (*component*) technical

тех,   (*pron, gen/prep/acc anim*) of **тот**

тех·ми́нимум·, -а,   (*m*) minimum qualifications, qualifying standard (of technical knowledge)

техн·е́ци·й, -я,   (*m*) (chem) technetium, Tc, masurium

те́хн·ик·, -а,   (*m*) technician (with technical school qualifications), engineer (not with university degree)

~, зуб·н·о́й   dental mechanic

~ -констру́ктор·, -а,   (*m*) draughtsman

~ -меха́н·ик·, -а,   (*m*) mechanic, mechanical engineer

~ -рад·и́ст·, -а,   (*m*) radio mechanic/technician

~ -эле́ктр·ик·, -а,   (*m*) electrician

те́хн·ик·а, -и,   (*f*) technology, engineering; technique

~ без·о·па́с·ност·и   accident prevention, safety precautions/engineering

~, ва́куум·н·ая   vacuum technology

~, высоко·част·о́тн·ая   high-voltage engineering

~ по·лёт·ов   airmanship

~, раке́т·н·ая   rocket technology, rocketry

~, реа́ктор·н·ая   (nucl) reactor engineering

~ с·вя́з·и   telecommunications/communication engineering

~, холод·и́льн·ая   refrigeration engineering

~, электро́н·н·ая   electronics

те́хн·ик·о-про·из·во́д·ственн·ый, -ая, -ое,   (*a*) production-engineering

те́хн·икум·, -а,   (*m*) technical high school

техн·и́ческ·ий, -ая, -ое,   (*a*) technical; industrial; commercial (of graded products)

~ за·да́·ни·е   (*n*) (civ eng) engineering problem, project requirement

~ ка́мен·ь   (*m*) (build) artificial stone

~ контро́л·ь   (*m*) quality control

от·де́л· техн·и́ческ·ого контро́л·я   (*m*) (*abbr* **ОТК**) quality control department/section

~ культу́р·а   (*f*) (agr) industrial crop

~ ми́нимум·   (*m*) minimum qualifications, qualifying standard (of technical knowledge)

~ персона́л·   (*m*) engineering personnel/grades

~ скор·ост·ь   (*f*) maximum working speed (of machines)

~ с·плав·   (*m*) commercial-/industrial-grade alloy

~ тре́б·овани·я   (*n pl*) (civ eng) specifications

~ у·сло́в·и·е   (*n*) local standard, technical condition/provision/specification

~ флот·   (*m*) dredging vessels

~ чист·от·а́   (*f*) commercial-grade purity

техн·и́чк·а, -и, (*g pl*) **-чек·,** (*f*) cleaner, charwoman; machine

техн·и́чн·ый, -ая, -ое,   (*a*) highly-qualified/skilled

техно·гене́з·, -а,   (*m*) (geochem) technogenesis

техно́·лог·, -а,   (*m*) technologist; production-process engineer

техно·лог·и́ческ·ий, -ая, -ое,   (*a*) technological; production-process, process

~ ба́з·а   (*f*) (mech eng) datum

~ газ·   (*m*) process gas

~ контро́л·ь   (*m*) process control

~ на·ря́д·   (*m*) job sheet

~ проце́сс·   (*m*) production process

~ схе́м·а   (*f*) flow chart/sheet

техно·лог·и́чност·ь, -и,   (*f*) (eng) suitability/adaptability for industrial production (of a design)

техно·лог·и́чн·ый, -ая, -ое,   (*a*) suitable, practicable, commercial (of designs)

техно·ло́г·и·я, -и,   (*f*) technology; production process, production technology

техно·ло́г·о-норм·иро́вочн·ая ка́рт·а   (*f*) planning sheet

техно·ру́к·, -а, (*m*) works/technical manager

тех·персона́л·, -а, (*m*) technical personnel

тех·по́·мощ·ь, -и, (*f*) technical aid/assistance

тех·ре́д·, -а, (*m*) technical editor

те́ц·и·й, -я, (*m*) (bot) thecium

теч·а́, (*pres gerund*) of **теч·ь**

тече·иск·а́тел·ь, -я, (*m*) leak detector (high vacuum systems)

теч·е́ни·е, -я, (*n*) (*v n*) of **теч·ь;** flow, current, stream, course; trend, tendency

~, **без·вихр·ев·о́е** clean/irrotational flow

в теч·е́ни·е in the course of, during; for, in (of periods of time)

~ **в реше́т·к·е** (aerodynam) cascade flow

в·верх· по теч·е́ни·ю (adv) upstream

~, **в·да·ю́щ·ееся** inward flow, indraft

~, **вихр·ев·о́е** vortex flow

~ **вне по·гран·и́чн·ого сло́·я** (aerodynam) main flow

в·низ· по теч·е́ни·ю (adv) downstream

~ **во́з·дух·а** airflow

~, **встре́ч·н·ое** counterflow

~, **вя́з·костн·ое** viscous flow

~, **глуб·и́нн·ое** undercurrent

~, **дре́йф·ов·ое** (ocean) drift current

~, **кавитацио́н·н·ое** cavitation flow

~, **конвекцио́н·н·ое** convection current

~, **корен·н·о́е** (ocean) main stream

крив·а́я теч·е́ни·я (*f*) flow curve

~, **ламина́р·н·ое** laminar flow

~, **ме́дл·енн·ое** tranquil stream/flow

~, **меж·ло́паст·н·ое** (aerodynam) blade-to-blade flow

~, **не·у·станов·и́вш·ееся** unsteady flow

~, **ни́ж·н·ее** (meteor) undercurrent; lower reaches (of river)

~, **обра́т·н·ое** backflow, refluence

~, **от·ли́в·н·ое** (ocean) ebb stream

~, **пере·ме́н·н·ое рас·сло́·енн·ое прост·о́е** (ocean) variable-layered simple current

~, **пере·ме́н·н·ое рас·сло́·енн·ое с·ло́ж·н·ое** (ocean) variable-layered complex current

~, **пере·хо́д·н·ое** transition flow

~, **период·и́ческ·ое** (ocean) periodical current

~, **пласт·и́ческ·ое** plastic flow

теч·е́ни·е

~, **пло́т·ностн·ое** convection flow

~, **по·до́б·н·ое** correspondence flow

~, **по·сто·я́нн·ое** (ocean) permanent current

~, **по·сто·я́нн·ое рас·сло́·енн·ое** (ocean) constant-layered current

~, **по·сто·я́нн·ое сплош·н·о́е** (ocean) constant-set-and-rate current

~, **при·ли́в·н·ое** (ocean) flood stream

~, **при·ли́·в·о-от·ли́в·н·ое** (ocean) tidal current

~, **раз·ры́в·н·о́е** (dynam) non-continuum flow; (ocean) rip current/tide

~, **регуля́р·н·ое** steady-state flow

~, **стру́й·н·ое** (aerodynam) jet stream/flow, jet, stream

~ **по́чв·ы** (geol) solifluction

теч·ёт, (*pres 3rd sing*) of **теч·ь**

те́ч·к·а, -и, (*g pl*) **-ч·ек·,** (*f*) (zool) oestrus, estrus, rut; spout, chute

теч·ь, -и, (*f*) leak, leakage, flow; discharge (from wound etc.)

теч·ь, (*pres 3rd sing, pl*) **теч·ёт, те·к·у́т,** (*past masc sing*) **тёк·** (*imp*) flow, stream, run; leak, be leaky

тешемахери́т·, -а, (*m*) (min) teschemacherite

тёш·ет, (*pres 3rd sing*) of **тес·а́ть**

тёш·ить, -ат, (*imp*) gratify, please

тиази́н·, -а, (*m*) (chem) thiazine

тиази́н·ов·ый, -ая, -ое, (*a*) thiazine

~ **крас·и́тел·ь** (*m*) (chem) thiazine dye

ти·азо́л·, -а, (*m*) (chem) thiazole

ти·азол·и́н·, -а, (*m*) thiazoline

ти·ами́д·, -а, (*m*) (chem) thiamide

ти·ами́н·, -а, (*m*) (biochem) thiamine, vitamin B_1

ти·аминпирофосфа́т·, -а, (*m*) thiamine pyrophosphate

тиа́р·а, -ы, (*f*) tiara

тибе́т·ск·ий, -ая, -ое, (*a*) Tibetan

~ **медици́н·а** (*f*) (med) nature cures

тибо́н·, -а, (*m*) (pharm) tibione

ти́гел·ь, -гл·я, (*m*) (met, chem) crucible; (print) platen

~ **-электро́д·, -а,** (*m*) (spectr) electrode crucible

ти́гель·н·ый, -ая, -ое, (*a*) of **ти́гел·ь**

~ **маши́н·а** (*f*) (print) platen machine

~ **про́б·а** (*f*) fusion (assaying)

~ **проце́сс·** (*m*) (met) crucible/Huntsman process

тигли́н·ов·ая кисл·от·а́ (*f*) isoangelic acid

тигмо·та́ксис·, -а, (*m*) (biol) thigmotaxis

тигмо·такт·и́ческ·ий, -ая, -ое, (*a*) (biol) thigmotactic

тигмо·троп·и́зм·, -а, (*m*) (biol) thigmotropism

ти́гр·, -а, (*m*) (zool) tiger, *Felis tigris*

тигр·о́в·ый, -ая, -ое, (*a*) of **тигр·**

~ глаз· (*m*) (min) tiger's eye

тигр·о́ид·, -а, (*m*) (biol) tigroid

тигро́ид·н·ое вещ·еств·о́ (*n*) (biol) tigroid

тигро́·лиз·, -а, (*m*) (biol) tigrolysis

ти́к·, -а, (*m*) (text) tick, ticking; (med) tic, habit spasm; (bot) teak, *Tectona grandis*

ти́к·ани·е, -я, (*n*) ticking

ти́к·а·ть, -ют, (*imp*) tick (sound)

ти́ккер·, -а, (*m*) (rad) tikker, ticker, chopper

ти́к·ов·ый, -ая, -ое, (*a*) teak

~ де́рево (*n*) (bot) teak, *Tectona grandis*

тико́нд·, -а, (*m*) "Tikond" (TiO_2 cermet dielectric)

тиксо·троп·и́·я, -и, (*f*) (chem etc.) thixotropy

тиксо·тро́п·н·ый, -ая, -ое, (*a*) thixotropic

тилази́т·, -а, (*m*) (min) tilasite

тиле, реа́кци·я (*f*) (chem) Thiele reaction

тилиади́н·, -а, (*m*) (chem) tiliadine

тилиакори́н·, -а, (*m*) (biochem) tiliacorine

тиллейт·, -а, (*m*) (min) tilleyite

ти́лл·ы (*pl*) (bot) tyloses

ти́лл·ь, -я, (*m*) (geol) till, boulder clay

тило́·з·, -а, (*m*) (med) tylosis

ти́льд·а, -ы, (*f*) (print) tilde, swung dash

тимаг·, -а, (*m*) "Timag" (Ti–Mg cermet dielectric)

тиманни́т·, -а, (*m*) (min) tiemannite

тимбер·ова́ть, -у́ют, (*imp*) repair a wooden ship

ти́мберс·, -а, (*m*) (shipb) ship timbers

тимиа́н·, -а, (*m*) (bot) thyme, *Thymus*

тим·идо́л·, -а, (*m*) (chem) thymidol

тим·иди́л·ов·ая кисл·от·а́ (*f*) thymidylic acid

тим·иди́н·, -а, (*m*) (biochem) thymidine

тим·и́л·, -а, (*m*) (biochem) thymyl

тим·и́н·, -а, (*m*) (biochem) thymine

тим·овиди́н·, -а, (*m*) (biochem) thymovidin

тим·о́л·, -а, (*m*) (pharm) thymol, thyme camphor

тимо́л·ов·ый, си́н·ий (*m*) thymol blue

тим·олфтале́ин·, -а, (*m*) (chem) thymolphthalein

тимо·нуклео́н·ов·ая кисл·от·а́ (*f*) thymonucleonic acid

тимофе́евк·а, -и, (*f*) (bot) timothy, *Phleum*

~, луг·ов·а́я timothy, *Phleum pratense*

тимо·ци́т·, -а, (*m*) (cyt) thymocyte

тимпа́н·, -а, (*m*) (anat, arch) tympanum

тимпан·и́т·, -а, (*m*) (med) tympanites

ти́мус·, -а, (*m*) (anat) thymus, thymus gland

тимья́н·, -а, (*m*) = **тимиа́н·**

ти́н·а, -ы, (*f*) pond scum, slime

ти́ндал·евск·ий ко́нус· (*m*) (opt) Tyndall cone

тиндал·иза́ци·я, -и, (*f*) tyndalization (method of sterilization)

тиндало́·метр·, -а, (*m*) tyndallimeter (determination of concentration of matter suspended in a liquid)

ти́ндал·я, эффе́кт· (*m*) (opt) Tyndall effect

ти́н·ист·ый, -ая, -ое, (*a*) slimy, scummy

тинка́л·, -а, (*m*) (min) tincal

тинкту́р·а, -ы, (*f*) (pharm) tincture

тино́л·ь, -я, (*m*) tinol (a soldering paste)

тинто́·метр·, -а, (*m*) (chem) tintometer

тио- (*int component*) (chem) thio-

тио·ами́д·, -а, (*m*) (chem) thiamide

тио·бакте́ри·и, -й, (*f pl*) (bact) *Thiobacteriaceae*

тио·бора́т·, -а, (*m*) (chem) thioborate

тио·ка́ин·, -а, (*m*) (pharm) thiocain

тио·карбами́д·, -а, (*m*) thiocarbamide, thiourea

тио·карбанили́д·, -а, (*m*) (chem) thiocarbanilide

тио·кето́н·, -а, (*m*) (chem) thioketone

тио·кисл·от·а́, -ы́, (*f*) (chem) thioacid

тиоко́л·, -а, (*m*) (plast) thiokol; (pharm) thiocol, orthocoll

ти·окса́н·, -а, (*m*) thioxane

ти·оксе́н·, -а, (*m*) thioxene

ти·о́л·, -а, (*m*) (chem) thiol, thioalcohol

тио́л·ов·ый, -ая, -ое, (*a*) thiolic

~ кисл·от·а́ (*f*) thiolic acid

тио·моч·еви́н·а, -ы, (*f*) (chem) thiourea, thiocarbamide

тио·нафте́н·, -а, (*m*) thionaphthene

тиони́л·, -а, (*m*) thionyl, sulfinyl, sulphinyl

~, хло́р·ист·ый thionyl/sulphinyl chloride

тио́н·ов·ый, -ая, -ое, (*a*) thionic

~ кисл·от·а́ (*f*) thionic acid

тио·олов·я́нн·ая кисл·от·а́ (*f*) thiostannic acid

тио·пентáл·, -а, (*m*) (pharm) thiopental sodium

тиосауйт·, -а, (*m*) (min) thiosauite

тио·семи·карбазйд·, -а, (*m*) thiosemicarbazide

тио·семи·карбазóн·, -а, (*m*) (pharm) thiosemicarbazone

тио·серно·кйсл·ый, -ая, -ое, (*a*) thiosulphate, hyposulphate, thiosulfate, hyposulfate

тио·сéр·н·ый, -ая, -ое, (*a*) thiosulphuric, thiosulfuric, hyposulphurous; hyposulphide (of)

~ кисл·от·á ( ) thiosulphuric/thiosulfuric/hyposulphurous/hyposulfurous acid

тио·сóл·ь, -и, (*f*) thiosalt

тио·спйрт·, -а, (*m*) thio alcohol, thiol

тио·станнáт·, -а, (*m*) thiostannate

тио·сульфáт·, -а, (*m*) thiosulphate, thiosulfate

~ нáтри·я sodium thiosulphate/thiosulfate, (phot) sodium hyposulphate

тио·урацйл·, -а, (*m*) (pharm) thiouracil

тио·фéн·, -а, (*m*) (chem) thiophene

тио·фенóл·, -а, (*m*) (chem) thiophenol

тиофóс·, -а, (*m*) "Thiophos" (insecticide)

тио·фосфáт·, -а, (*m*) thiophosphate

тио·хрóм·, -а, (*m*) (biochem) thiochrome

тио·циáн·ов·ый, -ая, -ое, (*a*) thiocyanic

~ кисл·от·á (*f*) thiocyanic acid

тио·эфйр·, -а, (*m*) thioether

~, диаллйл·ов·ый allyl sulphide/sulfide

тйп·, -а, (*m*) type, pattern, model, design, make, class (of ship etc.); (phys) mode

~ вйд·а (biol) holotype

~ волн· wave mode

~ рас·пáд·а (nucl) mode of decay

~ рóд·а (biol) genotype

типиз·áци·я, -и, (*f*) standardization; classification, typification

типиз·йр·овать, -уют, (*imp and perf*) standardize

тип·йческ·ий, -ая, -ое, (*a*) typical, characteristic

тип·йчн·ый, -ая, -ое, (*a*) typical characteristic, representative

тип·ов·óй, -áя, -óе, (*a*) standard, standardized, model

~ проéкт· (*m*) standard design

типо·генéз·, -а, (*m*) (biol) typogenesis

типó·граф·, -а, (*m*) printer; typographer; (ent) bark beetle, *Ips typographus*

типо·грáф·и·я, -и, (*f*) printing-house press, press

типо·граф·йческ·ий, -ая, -ое, (*a*) *see* **типо·грáф·ск·ий**

типо·грáф·ск·ий, -ая, -ое, (*a*) typographic, typographical; type; printing, printer's

~ машйн·а (*f*) typographic/relief printing machine

~ о·шйб·к·а (*f*) compositor's error

~ с·плав· (*m*) type metal

типо·лито·грáф·и·я, -и, (*f*) relief-and-offset printing house

типо·лóг·и·я, -и, (*f*) (biol) typology

типо·мéтр·и·я, -и, (*f*) printer's system of measurement

тйр·, -а, (*m*) pitch, tar; shooting range, butts

тирáж·, -а, (*m*) (print) circulation, edition, number of copies printed; draw (of lottery etc.)

тираже·у·стóй·чивост·ь, -и, (*f*) number of impressions (which can be taken from a printing forme)

тир·амйн·, -а, (*m*) (biochem) tyramine

тирáн·, -а, (*m*) tyrant

тиранно·зáвр·, -а, (*m*) (pal) tyrannosaur, *Tyrannosaurus*

тира·трóн·, -а, (*m*) (elec) thyratron

~ с холóд·н·ым катóд·ом cold-cathode thyratron

тира·трóн·н·ый, -ая, -ое, (*a*) of **тира·трóн·**

~ пре·рыв·áтел·ь (*m*) (weld) thyratron control

тирé (*n indecl*) (print) dash; (elec) leg (of a filament)

тирео- (*int component*) thyro-, thyreo- = **тиро-** (q.v.)

тирео·глобулйн·, -а, (*m*) (biochem) thyroglobulin

тирео·трóп·н·ый, -ая, -ое, (*a*) thyrotropic

тирйт·, -а, (*m*) (elec) thyrite (silicon carbide material)

тирйт·ов·ый, -ая, -ое, (*a*) thyrite

тиро- (*int component*) thyro- = **тирео-** (q.v.)

тиро·глобулйн·, -а, (*m*) (biochem) thyroglobulin

тироз·йн·, -а, (*m*) (biochem) tyrosine

тироз·инáз·а, -ы, (*f*) (biochem) tyrosinase

тир·óид·, -а, (*m*) (anat) thyroid

тир·оксйн·, -а, (*m*) (biochem) thyroxin

тиролйт·, -а, (*m*) (min) tyrolite

тиро·токсико́н·, -а, *(m)* tyrotoxicon (cheese poison)

тиро·трици́н·, -а, *(m)* (pharm) tyrothricin

ти́ск·а·ть, -ют, *(imp)* squeeze, hug, grip, press; (print) pull

тиск·и́, -о́в, *(m pl)* vice (tool)

~, паралле́ль·н·ые parallel-jaw vice

~, по·воро́т·н·ые swivel vice

~, слеса́р·н·ые bench vice

~, сту́л·ов·ые leg vice

тиск·о́в·ый, -ая, -ое, *(a)* of **тиск·и́**

тисн·е́ни·е, -я, *(n)* embossing

тисн·ённ·ый, -ая, -ое, *(a)* embossed

~ ко́ж·а *(f)* embossed leather

тисн·ён·ый, -ая, -ое, *(a)* = **тисн·ён·н·ый**

тисн·и́ть, -я́т, *(imp)* emboss

ти́сн·уть, -ут, *(perf)* give a hug/squeeze; (print) pull

тисони́т·, -а, *(m)* (min) tisonite

тис·о́чк·и, -чек·, *(f pl)* *(dim)* of **тиск·и́**

тис·о́чн·ый, -ая, -ое, *(a)* of **тиск·и́**

ти́сс·, -а, *(m)* (bot) yew, *Taxus*

~, кана́д·ск·ий Canada yew, ground hemlock, *Taxus Canadiensis*

~, коротко·ли́ст·н·ый Pacific yew, *Taxus brevifolia*

~, остро·кон·е́чн·ый Japanese yew, *Taxus cuspidata*

~, я́год·н·ый English yew, *Taxus baccata*

ти́сс·ов·ые, -ых, *(pl decl as adj)* (bot) yew family, *Taxaceae*

тита́н·, -а, *(m)* (chem) titanium, Ti

~, гу́б·чат·ый titanium sponge

дву·о́·кис·ь тита́н·а *(f)* titanium oxide, titania

~, магниетерм·и́ческ·ий titanium produced by the Kroll method

~, четырёх·хло́р·ист·ый titanium tetrachloride

титан·а́т·, -а, *(m)* (chem) titanate

титан·и́д·, -а, *(m)* titanide

титан·и́л·, -а, *(m)* (chem) titanyl

тита́н·ист·ый, -ая, -ое, *(a)* titanium, titaniferous; (chem) titanous

~ желез·ня́к· *(m)* (min) titaniferous iron ore, ilmenite

титан·и́т·, -а, *(m)* (min) titanite, sphene

тита́н·ов·ый, -ая, -ое, *(a)* titanic, titanium

титано·магнети́т·, -а, *(m)* (min) titanomagnetite

титано·те́ри·и, -ев, *(m pl)* (pal) titanotheres, *Titanotheria*

титано·це́н·, -а, *(m)* (chem) titanocene

ти́тр·, -а, *(m)* (chem) titre, titer; (text) titre; (cinema) caption

титр·а́тор·, -а, *(m)* titrator

титр·ова́льн·ый, -ая, -ое, *(a)* (chem) titration

титр·ова́ни·е, -я, *(n)* (chem) titration, titrimetry

~, амперо·метр·и́ческ·ое amperometric titration

~, кондукто·метр·и́ческ·ое conductimetric titration

~, потенцио·метр·и́ческ·ое potentiometric titration

титр·ова́ть, -у́ют, *(imp and perf)* titrate

титро́·метр·, -а, *(m)* (chem) titrimeter

~, ку·метр·и́ческ·ий *Q*-metric titrimeter

титро·метр·и́ческ·ий, -ая, -ое, *(a)* titrimetric

титро·ме́тр·и·я, -и, *(f)* titrimetry

ти́тул·, -а, *(m)* title; (print) title, head, heading; (civ eng) head, item (in a schedule of works)

титул·ова́ть, -у́ют, *(imp and perf)* title, entitle, call

ти́туль·н·ый, -ая, -ое, *(a)* of **ти́тул·**

~ лист· *(m)* (print) title page

~ с·пи́с·ок· *(m)* (civ eng) schedule of works

~ шрифт· *(m)* (print) headline type

тиура́м·, -а, *(m)* thiuram (rubber accelerator)

ти́ф·, -а, *(m)* (med) typhus

~, брюш·н·о́й typhus abdominalis, typhoid fever

~, воз·вра́т·н·ый recurrent typhus

~, сып·н·о́й exanthematous typhus

ти́фдрук·, -а, *(m)* (print) mezzotint

тифе́н·, -а, *(m)* (pharm) tiphen

тифл·и́т·, -а, *(m)* (med) typhlitis

тифло·коли́т·, -а, *(m)* (med) typhlocolitis

тифло·педаго́г·ик·а, -и, *(f)* education of the blind

тифло·стено́з·, -а, *(m)* (med) typhlostenosis

тиф·о́з·н·ый, -ая, -ое, *(a)* typhous, typhoid

тиф·о́ид·, -а, *(m)* (med) typhoid

тиф·о́ид·н·ый, -ая, -ое, *(a)* typhoid

тифо́н·, -а, *(m)* foghorn

ти́х·ий, -ая, -ое, *(a)* quiet, silent; soft, gentle, faint, calm; slow

~ ве́тер· *(m)* (meteor) light air

~ океа́н· *(m)* Pacific Ocean

~ хо́д· *(m)* slow speed

тих·и́т·, **-а**, (*m*) (min) tychite

тихо·океа́н·ск·ий, **-ая**, **-ое**, (*a*) (geog) Pacific

тихо·хо́д·к·и, (*g pl*) **-д·ок·**, (*f pl*) (zool) tardigrades, *Tardigrada*

тихо·хо́д·н·ый, **-ая**, **-ое**, (*a*) slow-speed/running, slow

ти́ш·е, (*comp*) of **ти́х·ий**

тиш·ин·а́, **-ы́**, (*f*) quiet, silence

т.к. (*abbr*) of **так как** since, because, as

тк·а́вш·ий, **-ая**, **-ее**, (*past part act*) of **тк·ать**

ткан·ев·ый, **-ая**, **-ое**, (*a*) of **ткан·ь**

ткане·на·прав·и́тел·ь, **-я**, (*m*) cloth feeder

ткане·терап·и́·я, **-и**, (*f*) (med) tissue therapy

ткане·эквивале́нт·, **-а**, (*m*) (nucl) tissue-equivalent

тк·а́нн·ый, **-ая**, **-ое**, (*past part pass*) of **тк·ать**

тка́н·ый, **-ая**, **-ое**, (*a*) woven

тка́н·ь, **-и**, (*f*) (text) cloth, fabric, material, textile; (biol) tissue

~, **костю́м·н·ая** (text) suiting

~, **механ·и́ческ·ая** (bot) mechanical tissue

~, **об·резин·ённ·ая** rubber-coated fabric

~, **основ·н·а́я** (bot) parenchyma

~, **пла́т·ельн·ая** (text) dress goods/material

~, **про·резин·ённ·ая** rubberized fabric

~, **со·един·и́тельн·ая** (biol) connective tissue

~, **стекл·я́нн·ая** glass fabric, fibre-glass, fiberglass

~, **техн·и́ческ·ая** industrial textile

ткан·ь·ё, **-я́**, (*n*) (*v n*) see **тк·ать**

тк·ать, **-ут**, (*imp*) weave

тк·а́цк·ий, **-ая**, **-ое**, (*a*) weaving, weaver's

~ пере·плет·е́ни·е (*n*) weave

~ стан·о́к· (*m*) (text) loom

~ стан·о́к·, **вы·шив·а́льн·ый** swivel loom

~ стан·о́к·, **каре́т·очн·ый** dobby loom

тк·а́цк·о-от·де́л·очн·ый, **-ая**, **-ое**, (*a*) weaving-and-finishing

тк·а́цк·о-пряд·и́льн·ый, **-ая**, **-ое**, (*a*) spinning-and-weaving

тк·ач·, **-а́**, (*i*) **-о́м**, (*m*) weaver

тк·а́честв·о, **-а**, (*n*) weaving

тк·ачи́х·а, **-и**, (*f*) weaver (woman)

тк·ёт, (*pres 3rd sing*) of **тк·ать**

ткм (*abbr*) of **тонно·кило́·метр·** t.km, ton-kilometer

тк·н·уть, **-у́т**, (*perf, single action*) poke/prod/stick into

тк·о́м·ый, **-ая**, **-ое**, (*pres part pass*) of **тк·ать**

тк·у́чи, (*pres gerund*) of **тк·ать**

тк·у́щ·ий, **-ая**, **-ее**, (*pres part act*) of **тк·ать**

тл = **те́сла** tesla (unit of magnetic induction)

тл·е́ни·е, **-я**, (*n*) (*v n*) see **тл·е·ть**; putrefaction

тл·е́·ть, **-ют**, (*imp*) decay, rot, putrefy; smoulder, glow

тл·е́ющ·ий, **-ая**, **-ее**, (*pres part act*) see **тл·е́·ть**; (elec) glow, glow-discharge

~ раз·ря́д· (*m*) glow discharge

тл·я́, **-и**, (*f*) (ent) aphid, aphis, plant louse, (*pl*) *Aphidae*

тми́н·, **-а**, (*m*) caraway, caraway seeds

~, **обык·нове́нн·ый** (bot) caraway, *Carum carvi*

т.н. (*abbr*) = **так на·зыв·а́ем·ый** so-called, as it is called, (or simply ("...")

то (*pron, nom/acc sing*) of **тот** that; the other; the right/correct; (*conj*) then; now, to now

    а не то if not, otherwise

~ есть that is, i.e.

~ же the same, ditto

    не то otherwise

    не то ... не то neither ... nor

~ ли ... то ли whether ... or, either ... or

~ ... то ... sometimes/now ... sometimes/now ...

тобо́й, (*pron, instr of* **ты**)

това́р·, **-а**, (*m*) (*esp pl*) merchandise, goods, wares, commodity; (text) cloth; (rail) goods

това́рищ·, **-а**, (*i*) **-ем**, (*m*) comrade

това́рищ·еств·о, **-а**, (*m*) comradeship; association, company

това́р·ност·ь, **-и**, (*f*) (agr) saleable/marketable surplus

това́р·н·ый, **-ая**, **-ое**, (*a*) of **това́р·**

~ би́рж·а (*f*) (fin) commodity exchange

~ за·па́с·ы (*pl*) (com) stock-in-trade

~ знак· (*m*) trade mark

~ конто́р·а (*f*) (rail) goods office

~ по́·езд· (*m*) goods/freight train

~ про·изво́д·ств·о (*n*) commercial production, production for the retail market

~ проду́кци·я (*f*) output for sale

това́р·н·ый, -ая, -ое
~ **регул·я́тор·** (*m*) (text) cloth regulator
~ **со·глаш·е́ни·е** (*n*) (econ) commodity agreement
~ **хозя́й·ств·о** (*n*) commercial economy/economic system
товаро·ве́д·, -а, (*m*) merchandiser, retail trade executive
товаро·ве́д·ени·е, -я, (*n*) merchandising, science of supply and demand
товаро·материа́ль·н·ые це́н·ност·и (*pl*) stock (accounting)
товаро·об·ме́н·, -а, (*m*) barter, commodity exchange
товаро·об·ме́н·н·ый, -ая, -ое, (*a*) barter, commodity-exchange
товаро·обора́ч·иваемост·ь, -и, (*f*) (com) turnover period
товаро·оборо́т·, -а, (*m*) turnover, trade/commodity/merchandise turnover
товаро·от·во́д·, -а, (*m*) (text) draw-off roll
товаро·от·прав·и́тел·ь, -я, (*m*) consigner, consignor
товаро·по·луч·а́тел·ь, -я, (*m*) consignee
тогда́ (*adv*) then
~ **же** at the same time
~ **и то́лько** if, and only if
~ **как** (*conj*) whereas, while
~ **когда́** when
~ **-то** at that time
того́, (*pron, gen sing m, n, acc sing m anim*) of **тот**
тожде́ств·енност·ь, -и, (*f*) identity
тожде́ств·енн·ый, -ая, -ое, (*a*) identical, the same
то́ждеств·о, -а, (*n*) identity
     обращ·а́·ться в то́ждеств·о (math) become identical
то́же (*conj*) also, too, as well
тожёств·енност·ь, -и, (*f*) = **тожде́ств·енност·ь**
този́л·, -а, (*m*) (chem) tosyl
той (*pron, gen/dat/instr/prep sing f*) of **тот**
ток·, -а, (*nom pl*) **-и,** (*g pl*) **-ов,** (*m*) current; (*nom pl*) **-а́,** (*g pl*) **-о́в,** (agr) thrashing/threshing floor; (zool) mating season
~, **акти́в·н·ый** (elec) active/wattful current
~, **акцио́н·н·ый** (bioelec) nerve/action current
     без то́к·а (elec) dead

ток
~, **блужд·а́ющ·ий** (phys) stray current
~, **вихр·ев·о́й** eddy current
~ **воз·бужд·е́ни·я** (elec) field/exciting current
~ **во́з·дух·а** air current, draught
~, **в·ход·я́щ·ий** arriving/incoming current
     крив·а́я в·ход·я́щ·его то́к·а (*f*) (telecom) arrival curve
~ **де́й·стви·я** (bioelec) nerve/action current
~, **ды́р·очн·ый** (phys) hole current
~ **за·мык·а́ни·я на земл·ю́** (elec) earth/ground current
~, **за·ря́д·н·ый** (elec) charging current
~, **зна́ч·н·ый** marking current (telegraphy)
~, **индуц·и́рованн·ый** induced current
~, **ионизацио́н·н·ый** (phys) ionization current
~, **ио́н·н·ый** ionic/ion current
~, **ко́нтур·н·ый** (telecom) loop/mesh current
~ **коро́т·к·ого за·мык·а́ни·я** (elec) short-circuit current
~, **круг·ов·о́й** (elec) circulating/ring current
     ли́н·и·я то́к·а (*f*) line of flow; stream line
~ **ми́нимум·а** valley current
~, **молот·и́льн·ый** (agr) thrashing/threshing floor
     моме́нт· то́к·а (*m*) electric moment
~ **на мише́н·ь** (nucl) target current
~ **на·ка́л·а** filament/heating current (vacuum tubes)
~ **на·сыщ·е́ни·я** saturation current
~, **номина́ль·н·ый** nominal current
~, **обра́т·н·ый** inverse/reversed current
~, **одно·сторо́н·н·ий** (telecom) single current
~ **от·пуск·а** (elec) release current
~, **о·ста́т·очн·ый** (phys) aftercurrent, residual current
~, **парази́т·н·ый** eddy current
~, **перв·и́чн·ый** primary current
~, **пере·жиг·а́ющ·ий** blowing current (fuses)
~, **пере·ме́н·н·ый** (elec) alternating current, A.C., ac
~ **пит·а́ни·я** (elec) feed current
~ **по·врежд·е́ни·я** fault current
     под то́к·ом (elec) current-carrying, alive

ток
~ **подо·грé·в·а** heat current (vacuum tubes)
~ **покó·я** quiescent current, zero-signal current; spacing current (telegraphy)
~ **покó·я, анóд·н·ый** zero-signal plate current (vacuum tubes)
~, **по·сто·я́нн·ый** (elec) direct current, D.C., dc; continuous current/flow
~, **пре·ры́в·ист·ый** intermittent current
~, **про·бéль·н·ый** (telecom) spacing current
~, **про·дóль·н·ый** (telecom) longitudinal current
~, **прям·óй** (elec) forward current
~, **пульс·и́рующ·ий** impulsive/pulsating current
~, **пуск·ов·óй** (elec) starting current
~ **пуч·к·á** scanning-beam current, beam current (of cathode ray tubes)
~, **рабóч·ий** (telecom) marking current
~, **раз·ря́д·н·ый** (elec) discharge current
~, **реактúв·н·ый** (elec) wattless/reactive current
~, **сéт·очн·ый** (elec) grid current
~, **сúль·н·ый** (elec) heavy current
~, **слáб·о пульс·и́рующ·ийся** ripple current
~, **слáб·ый** (elec) light/weak/communications current
~, **с·лóж·н·ый** complex/combination current
~ **с·мещ·éни·я** (phys) displacement current
~ **с·рабáт·ывани·я** (instr) operating current
~, **темн·ов·óй** (photoelec) dark current
~, **тёмн·ый** dark current
~, **термо·электр·úческ·ий** thermo-current, thermionic current
~ **у·тéч·к·и** (elec) leakage current
~ **Фýко** eddy current, Foucault current
~, **электрóн·н·ый** thermionic current
токá·й, -я, (m) Tokay, Tokay-type wine
токарно·винто·рéз·н·ый стан·óк· (m) (mech) screw-cutting lathe
токарно·дав·úльн·ый стан·óк· (m) (met) spinning lathe
токарно·долб·éжн·ый стан·óк· (m) backing-off lathe
токарно·за·ты́л·очн·ый стан·óк· (m) relieving/backing-off lathe
токарно·об·дúр·очн·ый стан·óк· (m) roughing lathe

токарно·от·рéз·н·ый стан·óк· (m) cutting-off lathe
токарно·револьвéр·н·ый стан·óк· (m) turret lathe
токáр·н·ый, -ая, -ое, (a) lathe; turning (machine process)
~ **автомáт·** (m) automatic lathe
~ **об·рабóт·к·а** (f) turning
~ **стан·óк·** (m) lathe
~ **стан·óк·, бес·цéнтр·ов·ый** centreless lathe
~ **стан·óк·, от·дéл·очн·ый** fine-turning lathe
~ **стан·óк·, фасóн·н·о-от·рез·н·óй** forming lathe
~ **стан·óк·, центр·ов·óй** centre lathe
~ **цех·** (m) turning shop, turnery
тóкар·ь, -я, (nom pl) **-я́,** or **-и,** (g pl) **-éй,** or **ей,** (m) turner, lathe operator
тóк·о- (root) current; (int component) toco- (childbirth)
ток·овáни·е, -я, (n) (v n) of **ток·о·вáть;** courting, courtship
ток·овáть, -ýет, (imp) (zool) exhibit courtship behaviour, utter mating calls
токо·вед·ýщ·ий, -ая, -ее, (a) (elec) live, current-carrying
токо·вращ·áтел·ь, -я, (m) D.C. ringer, pole changer (telephones)
тóк·ов·ый, -ая, -ое, (a) of **ток·**
токо·грáмма, -ы, (f) (instr) current record
токо·лóг·и·я, -и, (f) (med etc.) tocology, obstetrics
токо·нес·ýщ·ий, -ая, -ее, (a) (elec) current-carrying, live
токо·о·гран·ичúтел·ь, -я, (m) (elec) current limiter
токо·о·гранúч·ивающ·ий пред·о··хран·úтел·ь (m) current-limiting fuse
токо·от·вóд·, -а, (m) terminal (of accumulator battery)
токо·при·ём·ник·, -а, (m) current consumer (e.g. motor etc.); current collector (rail, tram), trolley
токо·про·вод·я́щ·ий, -ая, -ее, (a) current-carrying/conducting
токо·про·хожд·éни·е, -я, (n) passage of current
токо·рас·пре·дел·úтел·ь, -я, (m) (elec) sequence switch
токо·съ·ём·ник·, -а, (m) (elec) current pick-off/-up; slip-ring
токо·ферóл·, -а, (m) (biochem) tocopherol, vitamin E

токс·емú·я, -и, *(f)* (med) toxaemia, toxemia

токсик·óз·, -а, *(m)* (med) toxicosis

токсико·инфékци·я, -и, *(f)* (med) toxinfection

токсико·лóг·и·я, -и, *(f)* toxicology

токс·úн·, -а, *(m)* (biochem) toxin

токси·пептóн·, -а, *(m)* (chem) toxipeptone

токси·терап·ú·я, -и, *(f)* (med) toxitherapy

токс·úческ·ий, -ая, -ое, *(a)* toxic

токс·úчность·ь, -и, *(f)* toxicity

~ из·луч·éни·я radiotoxicity

токс·úчн·ый, -ая, -ое, *(a)* toxic

токсо·гéн·, -а, *(m)* (med) toxogen

токс·одóнт·ы, -ов, *(m pl)* (pal) toxodonts, *Toxodontia*

токс·óид·, -а, *(m)* (bact) toxoid

токсо·плáзм·а, -ы, *(f)* (zool) toxoplasma

токсо·плазм·óз·, -а, *(m)* (med, vet) toxoplasmosis

токс·эмú·я, -и, *(f)* (med) toxaemia, toxemia

тóл·, -а, *(m)* (expl) TNT

тóл·ев·ый, -ая, -ое, *(a)* of **тол·ь**

толерáнт·н·ая дóз·а *(f)* (biol, nucl) tolerance dose

толúл·, -а, *(m)* (chem) tolyl

толúл·ов·ый, -ая, -ое, *(a)* (chem) tolyl

тóлк·, -а, *(m)* sense; *(esp pl)* rumours, talk

толк·áем·ый, -ая, -ое, *(pres part pass)* of **толк·á·ть**

~ воз· *(m)* pusher-tug train

толк·áни·е, -я, *(n)* *(v n)* of **толк·á·ть;** (naut) push-towing

толк·áтел·ь, -я, *(m)* pusher, push rod, ram; tappet, valve lifter; follower (in cam mechanism)

~, печ·н·óй furnace charger

~, ре́·ечн·ый rack-type pusher

толк·á·ть, -ют, *(imp)* shove, push, jolt, jog, bump, jerk; put (shot in sport)

толк·áч·, -á, *(i)* **-óм,** *(m)* pusher, push rod, rammer; pusher tug; pusher engine, pusher locomotive

~, гидравл·úческ·ий hydraulic jack

толк·áющ·ий, -ая, -ее, *(pres part act)* of **толк·á·ть**

~ винт· *(m)* pusher propeller

толк·úте, *(pl imper)* of **толóч·ь**

толк·лú, *(past pl)* of **толóч·ь**

толк·н·ýть, -ýт, *(perf)* *see* **толк·á·ть**

толк·овáни·е, -я, *(n)* *(v n)* of **толк·о·вáть;** elucidation, interpretation, explanation

толк·овáть, -ýют, *(imp)* make sense of; elucidate, explain, interpret, make clear

толк·óв·ый, -ая, -ое, *(a)* sensible; intelligible, clear; explanatory

~ слов·áр·ь *(m)* dictionary, monolingual dictionary

толк·ýющ·ий, -ая, -ее, *(pres part act)* of **толóч·ь**

тóл·ов·ый, -ая, -ое, *(a)* of **тол·**

толóк·, *(past masc sing)* of **толóч·ь**

толóк·н·ó, -á, *(n)* oat flour

толóк·ш·ий, -ая, -ее, *(past part act)* of **толóч·ь**

толóч·ь, *(pres 3rd sing, pl)* **толч·ёт, толк·ýт,** *(past masc sing)* **толóк·,** *(imp)*; pound, crush, pulverize

толп·á, -ы́, *(nom pl)* **тóлп·ы,** *(f)* crowd

толст·é·ть, -ют, *(imp)* become/grow fat

тóлст·о- *(root)* thick, pachy-, crassi-

толсто·гýб·ый, -ая, -ое, *(a)* (zool) thick-lipped, pachylabial

толсто·кóж·ий, -ая, -ее, *(a)* (zool) thick-skinned, pachydermatous, pachydermic

толсто·кра·ев·óй, -áя, -óе, *(a)* (bot) crassimarginate

толсто·лист·ов·óй, -áя, -óе, *(a)* (met) plate (4-60 mm); (bot) pachyphyllous

~ стан· *(m)* (met) plate mill

толсто·по·крóв·н·ый, -ая, -ое, *(a)* crassitegillate

толсто·стéн·н·ый, -ая, -ое, *(a)* thick-walled

толсто·хóд·н·ый, -ая, -ое, *(a)* thick-sheaved (of pulley block)

тóлст·ый, -ая, -ое, *(a)* thick, heavy; corpulent, fat (of persons)

~ киш·к·á *(f)* (anat) large intestine, intestinum crassum

~ лист· *(m)* (met) plate

толст·я́нк·а, -и, *(g pl)* **-нок·,** *(f)* (bot) houseleek, *Crassula*

толст·я́нков·ые, -ых, *(pl decl as adj)* (bot) houseleek family, *Crassulaceae*

толуáн·ск·ий бальзáм· *(m)* (chem) Tolu balsam

толу·гидрохинóн·, -а, *(m)* (chem) toluhydroquinone

толу·идúн·, -а, *(m)* (chem) toluidine

толуúл·, -а, *(m)* (chem) toluyl, tolyloyl

толуúл·ов·ый, -ая, -ое, *(a)* toluyl, toluic

толуи́л·ов·ый, -ая, -ое
~ **альдеги́д·** (*m*) (chem) tolualdehyde, toluic aldehyde
~ **кисл·от·а́** (*f*) (chem) toluic acid
толу·о́л·, -а, (*m*) (chem) toluene
толч·а́, (*pres gerund*) of **толо́чь**
толч·е́йн·ый, -ая, -ое, (*a*) of **толч·е́·я**
~ **ста·в·** (*m*) (min) stamp battery
толч·е́ни·е, -я, (*n*) (*v n*) of **толо́чь**
толч·ённ·ый, -ая, -ое, (*past part pass*) of **толо́чь**
толч·ёт, (*pres 3rd sing*) of **толо́чь**
толч·е́·я, -й, (*f*) stamp, stamper, crushing stamp; ore/stamping mill; lop (on surface of water)
толч·к·а́, (*gen*) of **толч·о́к·**
толчко·обра́з·н·ый, -ая, -ое, (*a*) impulsive, jerky
толч·к·а́ми (*adv*) jerkily
толч·о́к·, -ч·к·а́, (*m*) push, shove, jolt, jerk, bump; impulse, stimulus; (geol) shock, tremor
то́лщ·а, -и, (*i*) **-ей,** (*f*) mass; thickness; (geol) rock mass, series
то́лщ·е (*comp*) thicker; fatter, stouter (of persons)
толщ·ин·а́, -ы́, (*f*) thickness, depth; (print) set, width
~ **зу́б·а по дел·и́тельн·ому цили́ндр·у** circular/arc thickness (gears)
~ **зу́б·а, хорда́ль·н·ая** chordal thickness (gearwheel)
~, **от·нос·и́тельн·ая** (a/c) thickness ratio
~ **полови́н·н·ого по·глущ·е́ни·я** (nucl) half-thickness
~ **по·те́р·и и́мпульс·а** (dynam) momentum thickness
толщино·ме́р·, -а, (*m*) feeler/finger/ thickness gauge/gage
то́л·ь, -я, (*m*) (build) roofing/saturated felt, sanded roofing felt
~, **ко́ж·а, -и,** (*f*) plain roofing paper
то́лько (*adv*) only
как то́лько as soon as
лишь то́лько no sooner than
~ **что** just
то́м·, -а, (*nom pl*) **-а́,** (*g pl*) **-о́в,** (*m*) volume (book); (*pron, prep sing m n*) of **тот**
томаз·и́т·, -а, (*m*) (min) thaumasite
томас·и́ровани·е, -я, (*n*) (met) basic Bessemer process
тома́с·овск·ий, -ая, -ое, (*a*) (phys) Thomas; (met) basic, basic Bessemer, Thomas
~ **стал·ь** (*f*) basic/Thomas steel

томас·шла́к·, -а, (*m*) (met etc.) basic slag
тома́т·, -а, (*m*) (bot) tomato, *Lycopersicon esculentum*
тома́т·н·ый, -ая, -ое, (*a*) tomato
то́мболо (*m indecl*) (geog) tombolo
томбу́·й, -я, (*m*) (naut) anchor buoy
то́м·ик·, -а, (*m*) (*dim*) of **том·**
том·и́льн·ый, -ая, -ое, (*a*) (met) soaking, malleablizing (cast iron)
~ **зо́н·а** (*f*) (met) soaking zone
~ **печ·ь** (*f*) (met) soaking pit
том·и́ть, -я́т, (*imp*) weary, tire/wear out; (food) steam; (met) malleablize (cast iron), soak; cure (tobacco)
томл·е́ни·е, -я, (*n*) (*v n*) of **том·и́ть**
томо·гра́мм·а, -ы, (*f*) tomogram
томо́·граф·, -а, (*m*) (med) tomograph
томо·гра́ф·и·я, -и, (*f*) tomography, body-layer radiography
томпа́к·, -а́, (*m*) (met) tombac (Cu–Zn alloy)
То́мсон·а, мо́ст·ик· (*m*) (elec) Kelvin bridge
~, **лот·** (*m*) (naut) Kelvin's sounding machine
~, **явл·е́ни·е** (*n*) (elec) Thomson/Kelvin effect
томсон·иза́ци·я, -и, (*f*) (elec) shunt-loading
томсон·о́вск·ий, -ая, -ое, (*a*) (nucl) Thomson
тому́, (*pron, dat sing m n*) of **тот**
то́н·, -а, (*m*) tone; (*nom pl*) **-а́,** (*g pl*) **-о́в,** tone, shade (of colour)
~, **во́·ющ·ий** warble
~, **с·ло́ж·н·ый** complex tone (telephony)
~ **-част·от·а́, -ы́,** (*f*) (rad) audio frequency
тона́ль·ност·ь, -и, (*f*) key (music), tonality
тона́ль·н·ый, -ая, -ое, (*a*) tone, tonal; (telecom) voice, audio
~ **вы́·зов·** (*m*) voice-frequency ringing (telephones)
~ **част·от·а́** (*f*) audio frequency
тона́рм·, -а, (*m*) pickup, tone arm (sound recording)
тонзилл·и́т·, -а, (*m*) (med) tonsillitis
тонзилл·эктом·и́·я, -и, (*f*) (med) tonsillectomy
тонз·и́р·овать, -у́ют, (*imp and perf*) tone up
тонз·и́рующ·ий, -ая, -ее, (*pres part act*) of **тонз·и́р·овать**
~ **сре́д·ств·о** (*n*) (med) tonic

тон·ин·а́, -ы́, (*f*) fineness, thinness

тон·и́ровани·е, -я, (*n*) (phot) toning

тон·и́ческ·ий, -ая, -ое, (*a*) tonic

тон·и́чност·ь, -и, (*f*) (physiol) tomicity

то́нк·ий, -ая, -ое, (*a*) thin, fine; delicate, subtle

~ киш·к·а́ (*f*) (anat) small intestine

~ сло́·й (*m*) thin layer/film

~ структу́р·а (*f*) (phys) fine structure

~ у·стано́в·к·а (*f*) fine adjustment

то́нко- (*component*) thin, fine, (biol) lepto-, tenui-

тонко·волокн·и́ст·ый, -ая, -ое, (*a*) fine-fibred, fine-fibrous, fine-grained (e.g. of wood)

тонко·диспе́рс·н·ый, -ая, -ое, (*a*) finely dispersed/disperse

тонко·зерн·и́ст·ый, -ая, -ое, (*a*) fine-grained, fine

тонко·из·мельч·ённ·ый, -ая, -ое, (*a*) finely ground/divided, fine

тонко·киш·е́чн·ый, -ая, -ое, (*a*) (anat) small-intestine

тонко·ко́ж·ий, -ая, -ее, (*a*) thin-skinned, leptodermatous

тонко·кра·ев·о́й, -а́я, -о́е, (*a*) (bot) tenuimarginate

тонко·лист·ов·о́й, -а́я, -о́е, (*a*) (met) strip, sheet

~ стал·ь (*f*) sheet steel

тонко·луч·ни́к·, -а́, (*m*) (bot) fleabane, *Erigeron annuus*

тонко·но́г·, -а, (*m*) (bot) koeleria, *Koeleria*

тон·контро́л·ь, -я, (*m*) tone control (sound recording etc.)

тонко·па́л·ост·ь, -и, (*f*) (zool) leptodactylis

тонко·па́л·ый, -ая, -ое, (*a*) (zool) leptodactylous

тонко·па́нцыр·н·ые, -ых, (*pl decl as adj*) (zool) Leptostraca, Phyllocarida

тонко·плён·чат·ый, -ая, -ое, (*a*) thin-filmed, (bot) thin-pellicled

тонко·по·кро́в·н·ый, -ая, -ое, (*a*) tenuitegillate

тонко·рас·се́·янн·ый, -ая, -ое, (*a*) diffuse, finely dispersed

тонко·скорлу́п·н·ые, -ых, (*pl decl as adj*) (pal) leptostracans, *Leptostraca*

тонко·сло́й·н·ый, -ая, -ое, (*a*) thin-layered, thin-layer, laminated, lamellar

тонко·сте́н·н·ый, -ая, -ое, (*a*) thin wall/walled, (bot) leptodermous

то́нк·ост·ь, -и, (*f*) thinness, fineness; subtleness, delicacy; (*pl*) finer points, details, refinements

тон·манипул·я́тор·, -а, (*m*) (telecom) tone keyer/manipulator

тон·ме́йстер·, -а, (*m*) (TV etc.) audio operator

то́нн·а, -ы, (*f*) ton

~, англ·и́йск·ая long ton (2240 lbs, 1016·047 kg)

~, англ·и́йск·ая суд·ов·а́я short ton (2000 lbs, 907·185 kg)

~, больш·а́я long ton (2240 lbs, 1016·047 kg)

~, бру́тто-реги́стр·ов·ая gross register ton

~ в·мест·и́мост·и ton burden

~ водо·из·мещ·е́ни·я displacement ton (35 cu ft = 1 m³)

~ гузо·подъ·ём·ност·и ton burden

~, дли́н·н·ая long ton (2240 lbs, 1016·047 kg)

~, испа́н·ск·ая Spanish ton (2271·6 lbs)

~, коро́т·к·ая short ton (2000 lbs, 907·185 kg)

~, ма́л·ая short ton (2000 lbs, 907·185 kg)

~, метр·и́ческ·ая metric ton (2204·6 lbs, 1000 kg)

~, не́тто-реги́стр·ов·ая net register ton

~, об·ме́р·н·ая cargo ton (40 cu ft., 1·12 m³), freight ton

~, реги́стр·ов·ая register ton (2·832 m³)

~, фрахт·о́в·ая cargo ton (40 cu ft., 1·12 m³), freight ton

тонн·а́ж·, -а, (*m*) tonnage

бру́тто-тонн·а́ж· (shipb) gross tonnage, gross register tonnage

~ бру́тто, вал·ов·о́й (shipb) gross tonnage, gross register tonnage

~, вал·ов·о́й реги́стр·ов·ый gross tonnage, gross register tonnage

~, груз·ов·о́й cargo tonnage

~, до́ллар·ов·ый dollar tonnage

~ не́тто, реги́стр·ов·ый net register tonnage

пред·лож·е́ни·е тонн·а́ж·а (*n*) tonnage offering

~, реги́стр·ов·ый register(ed) tonnage

~, сте́рлинг·ов·ый sterling tonnage

~, чи́ст·ый реги́стр·ов·ый net tonnage, net register tonnage

тонна́ж·н·ый, -ая, -ое, (*a*) tonnage

~ с·бор· (*m*) tonnage dues

тоннѐл·ь, -я, (*m*) (*see also* **туннѐл·ь**); tunnel, subway

~ греб·н·о́го ва́л·а (shipb) shaft tunnel

то́нно·киломе́тр·, -а, (*m*) ton/kilo-meter

то́нно·ми́л·я, -и, (*f*) ton-mile

-то́нн·ый, -ая, -ое, (*adj component*) -ton

тон·ов·о́й, -а́я, -о́е, (*a*) of **тон·**

тоно·вос·про·из·вед·е́ни·е, -я, (*n*) (phot) tone separation

то́н·ов·ый, -ая, -ое, (*a*) *see* **тон·ов·о́й**

тоно́·граф·, -а, (*m*) (med) tonograph

тоно́·метр·, -а, (*m*) (med) tonometer

тоно·пла́ст·, -а, (*m*) (cyt) tonoplast

тоно·фибри́лл·ы (*f pl*) (cyt) tono-fibrillae

тон·сигна́л·, -а, (*m*) (rad) audio signal

то́нус·, -а, (*m*) (physiol) tonus, tonicity

тон·у́ть ,·́ут, (*imp*) (*intrans*) sink; drown

тон·фи́льм·, -а, (*m*) sound motion picture, talkie; sound track, tape-recording

тонч·а́йш·ий, -ая, -ее, (*superl adj*) finest, most/very fine, thinnest, ultra-thin, most/very/highly subtle

тонч·а́·ть, -ют, (*imp*) become thinner/finer

то́нь·ше (*comp*) thinner, finer; subtler, more delicate

то́н·я, -и, (*f*) fishery; (fish) haul

топа́з·, -а, (*m*) (min) topaz

~, ды́м·чат·ый (min) smoky quartz, cairngorm

топа́з·ов·ый, -ая, -ое, (*a*) (min) topaz, topazine

топазо·ли́т·, -а, (*m*) (min) topazolite

то́п·а·ть, -ют, (*imp*) stamp (with foot)

топена́нт·, -а, (*m*) (naut) topping lift

~, по·сто·я́нн·ый standing topping lift

топинамбу́р·, -а, (*m*) (bot) Jerusalem artichoke, *Helianthus tuberosus*

топ·и́ть, ·́ят, (*imp*) (trans) sink, flood, immerse; melt, render; stoke, light (stoves etc.), warm, heat (with stove)

топ·и́ческ·ий, -ая, -ое, (*a*) topical, local

то́п·к·а, -и, (*g pl*) **-п·ок·,** (*f*) heating; stoking; melting; fire box/chamber (of boiler etc.), furnace, furnace stoker

~, коте́ль·н·ая boiler furnace/firebox

~, механ·и́ческ·ая automatic/mechanical stoker, stoker-fired furnace

~, на·клён·н·о-пере·та́лк·ивающ·ая inclined overfeed stoker

~, печ·н·а́я furnace chamber

~, полу·механ·и́ческ·ая semi-automatic stoker

то́п·к·а

~, руч·н·а́я hand-fired furnace

~ с ве́рх·н·ей по·да́·ч·ей overfeed stoker

~ с цеп·н·о́й решёт·к·ой chain-grate stoker

~, цикло́н·н·ая cyclone furnace

~, ша́хт·н·ая shaft furnace

то́п·к·ий, -ая, -ое, (*a*) marshy, swampy, boggy

топл·е́ни·е, -я, (*n*) (*v n*) of **топ·и́ть**

то́пл·енн·ый, -ая, -ое, (*past part pass*) of **топ·и́ть**; immersed, sunk, flooded; melted, rendered

топл·ён·ый, -ая, -ое, (*a*) melted, rendered

то́пл·ивник·, -а, (*m*) firebox (of space-heating furnace)

то́пл·ивн·ый, -ая, -ое, (*a*) fuel; fuel-feed/-supply/-injection

~ аппара́т· (*m*) fuel-injection system (of diesel)

~ на·со́с· (*m*) oil pump (of ship); fuel-injection pump (of engine)

то́пл·ив·о, -а, (*n*) fuel; (rocket) propellant

~, бу́нкер·н·ое bunker fuel, fuel oil

~, быт·ов·о́е household/domestic fuel

~, вы·тесн·я́ем·ое (rocket) pressurized propellant

~, двух·компоне́нт·н·ое binary fuel; (rocket) bipropellant

~, двух·осно́в·н·ое (rocket) double-base propellant

~, ди́зель·н·ое high-speed diesel fuel

~, жи́дк·ое liquid fuel; (rocket) liquid propellant

~, жидко·металл·и́ческ·ое (nucl) liquid-metal fuel

~, ис·коп·а́ем·ое (min) fossil fuel

~, колло́йд·н·ое colloidal fuel; (rocket) colloidal propellant

~, коте́ль·н·ое furnace fuel oil

~, лёгк·ое light petroleum fractions

~, металл·и́ческ·ое (rocket) metal-base propellant

~, много·компоне́нт·н·ое (rocket) multipropellant, composite propellant

~, мото́р·н·ое heavy diesel fuel; motor engine fuel

~, не·детон·и́рующ·ее anti-knock fuel

~, одно·компоне́нт·н·ое (rocket) monopropellant

~, одно·осно́з·н·ое (rocket) single-base propellant

~, о·ста́т·очн·ое residual fuel

то́пл·ив·о

~, порошко·обра́з·н·ое   powder fuel; (rocket) solid propellant

~, пуск·ов·о́е   priming/starting fuel

~, пыле·ви́д·н·ое   pulverised/powdered fuel (solid); atomized fuel (liquid)

~, реакти́в·н·ое   (a/c) jet fuel

~, само·вос·пламен·я́ющ·ееся   hypergolic fuel; (rocket) hypergolic propellant

~, с·ме́щ·анн·ое   blended fuel (liquid)

~, спирт·ов·о́е   (ICE) alcoholic fuel

~, твёрд·ое   solid fuel; (rocket) solid propellant

~, технолог·и́ческ·ое   process fuel

~, унита́р·н·ое   monofuel; (rocket) monopropellant

~, у·сло́в·н·ое   fuel unit, theoretical standard fuel (solid)

~, этало́н·н·ое   reference fuel (liquid)

топливо·за·пра́в·очн·ый, -ая, -ое,   (*a*) fueling

топливо·за·пра́в·щик·, -а,   (*m*) (air) bowser

топливо·ме́р·, -а,   (*m*) fuel level gauge, fuel gauge/gage

топливо·по·да́·ч·а, -и,   (*f*) fuel handling plant (solid); fuel feed (of engine)

топливо·под·вод·я́щ·ий, -ая, -ее,   (*a*) fuel-feed (liquid)

~ по́л·ост·ь   (*f*) sump/suction gallery/ chamber (of pump)

~ трубо·про·во́д·   (*m*) fuel feed line

топливо·под·ка́ч·ивающ·ий, -ая, -ое,   (*a*) fuel-feed (of pumps)

топливо·со·держ·а́щ·ий, -ая, -ее,   (*a*) (nucl) fuel-bearing/-carrying

топл·я́к·, -а́,   (*m*) (naut) jetsam; wreckage, wreck; sunken log (timber logging)

то́п·н·уть, -ут,   (*perf*) stamp (with foot)

топо-   (*int component*) topographic, topographical, survey

то́п·ов·ый, -ая, -ое,   (*a*) (naut) top

~ фигу́р·а   (*f*) topmark (of buoy)

топо·вы·числ·и́тельн·ый, -ая, -ое,   (*a*) (mil) survey-computing

топо́·граф·, -а,   (*m*) topographer

топо·граф·и́ческ·ий, -ая, -ое,   (*a*) topographical

~ анато́м·и·я   (*f*) topographic anatomy

топо·гра́ф·и·я, -и,   (*f*) (surv etc.) topography

то́пол·ев·ый, -ая, -ое,   (*a*) of то́пол·ь

топо·лог·иза́ци·я, -и,   (*f*) (math) topologization

топо·лог·и́ческ·ий, -ая, -ое,   (*a*) topological

~ про·стра́н·ств·о   (*n*) (math) topological space

топо·ло́г·и·я, -и,   (*f*) (math) topology

~, комбина́тор·н·ая   combinatorial topology

то́пол·ь, -я,   (*nom pl*) -я́, (*g pl*) -е́й, (*m*) (bot) poplar, *Populus*

~, бальзам·и́ческ·ий   balsam/tacamehac poplar, *Populus balsamifera*

~, волосисто·пло́д·н·ый   black cottonwood, *Populus trichocarpa*

~, осино·обра́з·н·ый   quaking aspen, *Populus tremuloides*

топо́р·, -а́,   (*m*) axe, ax

~, пло́т·ничн·ый   broad axe/ax

топо́р·ик·, -а,   (*m*) hatchet

топор·и́щ·е, -а,   (*n*) axe handle/helve

топо́р·н·ый, -ая, -ое,   (*a*) axe; rough, rough-hewn

топоро·но́г·ие, -их,   (*pl decl as adj*) (zool) pelecypods, *Pelecypoda*

топоро·обра́з·н·ый, -ая, -ое,   (*a*) axe/ax-shaped, dolabriform

топо́рщ·ить, -ая,   (*imp*) bristle (of animals), puff (of feathers)

топот·а́ть,   (*pres 3rd pl*) **топо́ч·ут,** (*imp*) tramp, clatter

топо·ти́п·, -а,   (*m*) (zool) topotype

топо·хи́м·и·я, -и,   (*f*) topochemistry

топо́ч·ет,   (*pres 3rd sing*) of **топот·а́ть**

то́п·очн·ый, -ая, -ое,   (*a*) of **то́п·к·а**

~ двер·ц·а   (*f*) fire door

~ мазу́т·   (*m*) furnace fuel oil, boiler fuel oil

~ про·стра́н·ств·о   (*n*) furnace, furnace chamber

~ с·вод·   (*m*) furnace roof

~ фронт·   (*m*) furnace front plate

топрик·, -а,   (*m*) (naut) davit span

то́псел·ь, -я,   (*m*) (naut) gaff topsail

топт·а́ть,   (*pres 3rd pl*) **то́пч·ут,** (*imp*) trample, trample down; knead (with feed); -ся stamp (with feet)

топти́мберс·, -а,   (*m*) (shipb) top timbers

то́пч·ет,   (*pres 3rd sing*) of **топт·а́ть**

топ·ь, -и,   (*f*) marsh, bog

то́р·, -а,   (*m*) (*see also* **тор·о́ид·**) (math) torus, anchor ring; (elec) toroid; tor, mm Hg (unit of pressure); (bot) torus

~ прямо·уго́ль·н·ого   сеч·ени·я (elec) rectangular-loop toroid

торак·а́лг·и·я, -и,   (*f*) (med) thoracalgia

торако·пла́ст·ик·а, -и, (*f*) (med) thoracoplasty

торако·ско́п·, -а, (*m*) (med) thoracoscope

то́ракс·, -а, (*m*) (zool) thorax

тора·центе́з·, -а, (*m*) (med) thoracentesis, thoracocentesis

торбани́т·, -а, (*m*) (min) torbanite

торберни́т·, -а, (*m*) (min) torbernite

торг·, -а, (*nom pl*) **-и́,** (*g pl*) **-о́в,** (*m*) (*v n*) *see* **торг·ова́ть;** trading organization; (*as component*) (names of Soviet trading organizations, which should be transliterated and not translated)

торг- (*root*) = **торг·о́в·ый** trade, commercial

торг·ова́ть, -у́ют, (*imp*) trade, trade in; be open for business, be open; **-ся** bargain

торг·о́вец·, -в·ц·а, (*m*) trader, dealer, merchant

~, опт·о́в·ый merchant, wholesaler

~, ро́зн·ичн·ый retailer

торг·о́вл·я, -и, (*f*) trade, trading, commerce

~, мен·ов·а́я barter

торг·о́в·ый, -ая, -ое, (*a*) trade, commercial, trading

~ автома́т· (*m*) vending machine

~ аге́нт· (*m*) commercial agent

~ знак· (*m*) trademark

~ организ·а́ци·я (*f*) trading organization

~ пала́т·а (*f*) chamber of commerce

~ пра́в·о (*n*) commercial law

~ пред·ста́в·и́тел·ь (*m*) head of USSR trade delegation

~ пред·ста́в·и́тельств·о (*n*) trade delegation of the USSR

~ со·ве́т·ник· (*m*) **СССР** (dipl) commercial counsellor

~ флот· (*m*) merchant marine/navy

торг·пре́д·, -а, (*m*) head of trade delegation (from USSR resident abroad)

торг·пре́д·ств·о, -а, (*n*) Soviet trade delegation

тордох·, -а, (*m*) reindeer tent

тор·е́ц·, -р·ц·а́, (*i*) **-р·ц·о́м,** (*m*) end, face, end face, butt end; wooden paving-block

~ под·ши́п·ник·а bearing face

торже́ств·енн·ый, -ая, -ое, (*a*) solemn, ceremonial; festive; triumphant

торжеств·о́, -а́, (*n*) festival, celebration; triumph

торжеств·ова́ть, -у́ют, (*imp*) celebrate; triumph, exult

торзио- (*int component*) *see* **торсио-**

торзио́н·н·ый, -ая, -ое, (*a*) torsion, torsional, torque

ториани́т·, -а, (*m*) (min) thorianite

тори́д·, -а, (*m*) (chem) thoride

ториево·окси́д·н·ый, -ая, -ое, (*a*) (elec) thoriated; (chem) thoria

то́р·иев·ый, -ая, -ое, (*a*) thorium

то́р·и·й, -я, (*m*) (chem) thorium, Th

то́рилис·, -а, (*m*) (bot) hedge parsley, *Torilis*

тор·и́рованн·ый, -ая, -ое, (*past part pass*) thoriated

тор·и́т·, -а, (*m*) (min) thorite

тор·и́ть, -я́т, (*imp*) wear, tread (paths etc.)

торриче́лли·ев·, -а, -о, (*predic adj*) *see* **торриче́лли·ев·**

тор·и́ческ·ий, -ая, -ое, (*a*) toric

торкре́т-бето́н·, -а, (*m*) (build) gunite

торкрет·и́р·овать, -уют, (*imp and perf*) (build) gunite

тормож·е́ни·е, -я, (*n*) (*v n*) of **тормоз·и́ть;** braking action, deceleration, moderation, inhibition, inhibiting action; (dynam) stagnation (of a flow)

~, вну́тр·енн·ее (biol) endogenous inhibition

~, вя́з·к·ое (dynam) viscous drag

~, динам·и́ческ·ое (elec) dynamic braking

~ до·звук·ов·о́го по·то́к·а (dynam) subsonic recovery

кла́пан· авар·и́йн·ого тормож·е́ни·я само·лёт·а (*m*) aircraft emergency brake control valve

коэффицие́нт· тормож·е́ни·я (*m*) (dynam) drag coefficient

~ противо·ключ·е́ни·ем (elec) plugging

пут·ь тормож·е́ни·я (*m*) braking distance

~, рекуперати́в·н·ое (elec) regenerative braking

~, реоста́т·н·ое (elec) rheostatic braking

~, ступ·е́нчат·ое graded braking

температу́р·а тормож·е́ни·я (*f*) stagnation temperature (of gas turbines)

тормож·ённ·ый, -ая, -ое, (*past part pass*) of **тормоз·и́ть**

то́рмоз·, -а, (*nom pl*) **-а́,** (*g pl*) **-о́в,** (*m*) brake; braking; hindrance, drag; (biol) inhibition, inhibiting effect

тóрмоз
~, аэродинам·и́ческ·ий (a/c) air brake
~ борт·ов·óй пере·да́·ч·и differential steering brake (of tractor)
~ вало·про·вóд·а shaft brake
~, воз·ду́ш·н·ый air brake
~, гидравл·и́ческ·ий hydraulic brake; (instr) water brake, fluid-friction dynamometer
~, дву·стóрон·н·ий clasp brake
~, динамометр·и́ческ·ий dynamometer brake, absorption dynamometer
~, ди́ск·ов·ый disc brake
~, ду́ль·н·ый muzzle brake (of gun)
~, ка́мер·н·ый (a/c) expander-tube brake
~, колóд·очн·ый block brake
~, лéнт·очн·ый band brake
~, на·вóй·н·ый (text) warp beam brake
~ на·ка́т·а (gunn) run-out speed controller
~, оснóв·н·ый (text) warp-beam brake
~ от·ка́т·а (gunn) recoil absorber
~, парашю́т·н·ый drop brake
~, пневмат·и́ческ·ий air brake
~, рéльс·ов·ый (rail) track brake
~, электро·механ·и́ческ·ий electro-mechanical/Maley brake
тормоз·и́ть, -я́т, (*imp*) brake; slow down, decelerate, hinder, retard, hold up; inhibit (a process)
тормоз·н·óй, -áя, -óе, (*a*) of **тóрмоз·**
~ бараба́н· (*m*) brake drum
~ башма́к· (*m*) brake shoe/block
~ гак· (*m*) (a/c) arrester hook
~ из·луч·éни·е (*n*) (nucl) Bremsstrahlung
~ квант· (*m*) (nucl) Bremsstrahlung quantum
~ колóд·к·а (*f*) brake block/shoe
~ лéнт·а (*f*) brake band
~ мóщ·ност·ь (*f*) brake horse power
~ на·кла́д·к·а (*f*) brake lining
~ пут·ь (*m*) braking distance
~ спосóб·ност·ь (*f*) braking/stopping power
~ трос· (*m*) (air) arrester wire
~ щит·óк· (*m*) (a/c) dive brake
тормоз·я́щ·ий, -ая, -ее, (*pres part act*) of **тормоз·и́ть;** retarding, damping, stopping, inhibiting
~ пóл·е (*n*) retarding field (vacuum tubes)
торна́ри·я, -и (*f*) (zool) tornaria
торо·гумми́т·, -а, (*m*) (min) thoro-gummite

тор·óид·, -а, (*m*) (*see also* **тор·**) (math, elec) toroid
~, ферри́т·ов·ый ferrite toroid
тороида́ль·н·ый, -ая, -ое, (*a*) toroidal
~ кат·у́шк·а (*f*) (elec) toroidal coil
тор·óн·, -а, (*m*) (nucl chem) thoron, Tn, thorium emanation
тороп·и́ть, -́ят, (*imp*) hurry, hasten, expedite
тороп·ли́в·ый, -ая, -ое, (*a*) hurried, hasty
торóс·, -а, (*m*) hummock (ice)
торóс·ист·ый, -ая, -ое, (*a*) hummocky (of ice)
торо·тунгсти́т·, -а, (*m*) (min) thoro-tungstite
торо·уранини́т·, -а, (*m*) (min) thoro-uraninite
торош·éни·е, -я, (*n*) hummocking (of ice)
торпéд·а, -ы, (*f*) torpedo; shaped charge (oil wells)
~, паро·га́з·ов·ая H.T.P. torpedo
~, под·лóд·очн·ая submarine-launched torpedo
~, противо·лóд·очн·ая anti-submarine torpedo
~, само·на·вод·я́щ·аяся homing torpedo
торпед·и́ровани·е, -я, (*n*) (*v n*) of **торпед·и́р·овать;** shaped-charge shooting (oil wells)
торпед·и́р·овать, -ую́т, (*imp and perf*) (naut) torpedo; shoot (oil wells)
торпéд·н·ый, -ая, -ое, (*a*) torpedo
~ аппара́т· (*m*) torpedo tube(s)
~ перфор·а́тор· (*m*) gun perforator (oil wells)
торпедо·держ·а́тел·ь, -я, (*m*) (a/c) torpedo rack·
торпедо·мет·а́ни·е, -я, (*n*) torpedo dropping
торпедо·нóс·ец·, -с·ц·а, (*m*) (a/c) torpedo bomber
торпедо·с·бра́с·ывател·ь, -я, (*m*) (a/c) torpedo release mechanism
торреóл·, -а, (*m*) (chem) torreol
торриче́лли·ев·а пуст·от·а́ (*f*) (phys) Torricellian vacuum
тóрс·, -а, (*m*) (math) torse, trunk, developable surface; (anat) torso
торсиó·граф·, -а, (*m*) (mech eng) torsiograph
торсиó·метр·, -а, (*m*) torsiometer, torsion dynamometer, torque meter
торсиóн·, -а, (*m*) torsion bar
тóрт·, -а, (*m*) (food) cake

торти·ко́н·, -а, (*m*) (zool) torticone
торус·, -а, (*m*) = **тор·**
то́рф·, -а, (*m*) peat, turf
~, верх·ов·о́й highmoor peat
~, жи́р·н·ый bituminous peat
торфо·бето́н·, -а, (*m*) peat-bitumen concrete
торфо·до·бы́·ч·а, -и, (*i*) **-ей,** (*f*) peat winning/digging
торфо·раз·рабо́т·к·а, -и, (*g pl*) **-т·ок·,** (*f*) peat digging
торфо·ре́з·к·а, -и, (*g pl*) **-з·ок·,** (*f*) turf cutter
торфо·ту́к·, -а, (*m*) (agr) peat fertilizer
торфо·убо́р·очн·ая маши́н·а (*f*) (min) peat combine
торф·яни́к·, -а́, (*m*) (geol) peat bog; peat-industry worker
торф·ян·о́й, -а́я, -о́е, (*a*) peat, turf
торц·ева́ть, -у́ют, (*imp*) (mech eng) face, machine the end/face of
торц·ев·о́й, -а́я, -о́е, (*a*) *see* **торц·о́в·ый**
торц·о́вк·а, -и, (*g pl*) **-вок·,** (*f*) face milling; cross-cutting (timber); cross-cut saw
~, ма́ят·ников·ая pendulum cross-cut saw
торц·о́в·ый, -ая, -ое, (*a*) of **тор·е́ц·;** end, face; wood-block (paving)
~ ключ· (*m*) box spanner; winder (of clockwork mechanism)
~ окн·о́ (*n*) (nucl) end window (of counter)
~ мост·ов·а́я (*f*) wood-block paving
~ по·ве́рх·ност·ь (*f*) face, end, end face
~ со·един·е́ни·е (*n*) edge joint
~ фрезер·ова́ни·е (*n*) (mech) face milling
~ шлиф·ова́ни·е (*n*) (mech eng) face grinding
торц·о́вочн·ый, -ая, -ое, (*a*) of **тор·ц·о́вк·а**
~ стан·о́к· (*m*) cross-cut saw
торч·а́ть, -а́т, (*imp*) jut/stand/stick out, protrude
торшо́н·, -а, (*m*) (text) torchon; (paper) linen/chased paper
торшон·и́ровани·е, -я, (*n*) (print) chasing
тоск·а́, -и́, (*f*) melancholy; yearning, longing; boredom
то́ст·, -а, (*m*) toast (drink etc.)
тот (*pron, nom sing m, acc sing m inanim*) that; the other; the right/correct
     без того́ что́·бы unless

тот
     для того́ что́·бы in order that/to
     до того́ to such an extent
~ или ин·о́й something or other
     к тому́ же furthermore
     кроме того́ furthermore, moreover, besides this/that
     по ме́р·е того́ как inasmuch as, as
тота́ль·н·ый, -ая, -ое, (*a*) total
то́т·час· (*adv*) immediately, at once
точ·е́ни·е, -я, (*n*) (*v n*) of **точ·и́ть**
~, фасо́н·н·ое form turning
точ·ён·ый, -ая, -ое, (*a*) sharpened; turned (on lathe)
то́ч·ечн·о (*adv*) pointwise
то́ч·ечн·ый, -ая, -ое, (*a*) point, spot, dot; dotted, spotted; point-like
~ ис·то́ч·ник· (*m*) point source
~ конта́кт· (*m*) point contact
~ корро́зи·я (*f*) pinhole corrosion
~ маши́н·а (*f*) spot welder
~ не·одно·ро́д·ност·ь (*f*) (cryst) point imperfection
~ с·ва́р·к·а (*f*) spot welding
~ транзи́стор· (*m*) point-contact transistor
точ·и́л·о, -а, (*n*) grindstone; sharpener
точ·и́льн·ый, -ая, -ое, (*a*) sharpening
~ ка́мен·ь (*m*) grindstone
~ рем·е́н·ь (*m*) strop
~ сла́н·ец· (*m*) hone
~ стан·о́к· (*m*) sharpener, tool sharpener
точ·и́льщик·, -а, (*m*) sharpener, grinder (person); (ent) furniture beetle, *Anobium*
точ·и́ть, ⸗ат, (*imp*) sharpen, grind (to sharpen); turn (on lathe); eat/gnaw holes in, eat away
то́ч·к·а, -и, (*g pl*) **-ч·ек·,** (*f*) (*v n*) of **точ·и́ть;** point (*see also associated words*); dot; full stop
~, бес·кон·е́ч·о у·дал·ённ·ая (math) ideal point, infinite point
~ воз·вра́т·а (math) cusp
~ воз·вра́т·а втор·о́го ро́д·а cusp of the second kind
~ воз·вра́т·а пе́рв·ого ро́д·а (math) simple cusp, cusp of the first kind
~, во·ображ·а́ем·ая (math) imaginary point
~ вс·пы́ш·к·и flash point
~, двой·н·а́я (math) double point
~, един·и́чн·ая (math) unit point
~ зр·е́ни·я point of view, standpoint
~, ис·хо́д·н·ая starting point, initial position

то́ч·к·а
~ **кас·а́ни·я** point of contact/tangency
~ **кип·е́ни·я** boiling point
~, **кра́т·н·ая** (math) multiple point, K-tuple point
~, **крит·и́ческ·ая** critical point; (dynam) stagnation point
~, **материа́ль·н·ая** material point
~, **мёртв·ая** (mech) dead centre
~, **нул·ев·а́я** datum point; (elec) neutral point
~, **о·со́б·ая** (math) singular point
~, **пре·де́ль·н·ая** (math) accumulation/limit point
~, **прост·а́я** (math) simple point, ordinary point
~ **раз·ветвл·е́ни·я** (math) branch-point
~, **регуля́р·н·ая** (math) regular/ordinary point
~ **рос·ы́** (meteor) dew point
~ **с запя́т·о́й** (print) semi-colon
спо́соб· то́ч·к·а-кольц·о́ circle-dot method (computers)
~, **угл·ов·а́я** (math) point of inflection
~, **узл·ов·а́я** (math) double point, node
то́ч·ност·ь, -и, (f) exactness, precision; accuracy; fidelity (sound reproduction)
~, **дво·и́чн·ая** double precision
~ **по вы́с·от·е́** elevation accuracy (radar)
~, **по·вы́ш·енн·ая** (math) multiple precision
с то́ч·ност·ью до correct to/within, exact up to, within
то́ч·н·ый, -ая, -ое, (a) exact, precise; precision; accurate
~ **при·бо́р·** (m) precision instrument
~ **регул·и́р· овани·е** (n) precise control, accurate regulation/adjustment
тошн·от·а́, -ы, (f) (med) nausea
тошно·тво́р·н·ый, -ая, -ое, (a) nauseating, (med) nauseous
то́щ·ий, -ая, -ее, (a) emaciated, lean, (med) jejunal
~ **сыр·** (m) skim-milk cheese
~ **у́гол·ь** (m) hard/lean/meagre coal
трабе́кул·а, -ы, (f) (biol) trabecula, (pl) trabeculae, trabeculas
трав·а́, -ы́, (nom pl) **тра́в·ы,** (f) (bot) grass; herb
~, **бого·ро́д·ск·ая** lemon thyme, *Thymus serpyllum*
~, **за́яч·ья** sorrel, dock, *Rumex*
~, **ку́х·онн·ая** culinary herb
~, **лека́р·ственн·ая** medicinal herb

трав·а́
~, **медве́ж·ья** lungwort, *Pulmonaria officinalis*
~, **огур·е́чн·ая** borage, *Borago officinalis*
~, **о́ч·н·ая** eyebright, *Euphrasia officinalis*
~, **почечу́й·н·ая** lady's thumb, *Polygonum persicaria*
~, **со́р·н·ая** weed
~, **цинг·о́тн·ая** scurvy grass, *Cochlearia officinalis*
тра́верз·, -а, (m) (see also **тра́верс·**) (naut) beam (direction)
тра́верз·а, -ы, (f) see **тра́верс·а**
тра́верс·, -а, (m) (see also **тра́верс·а**) traverse; crosshead, crossarm; (civ eng) cross dyke, groyne
тра́верс·а, -ы, (f) cross piece/member/head/beam/arm; (a/c) torsion link; head, crosshead (hydraulic press)
~, **ве́рх·н·яя не·по·дви́ж·н·ая** upper fixed head (of hydraulic press)
~, **груз·ов·а́я** load bar (lifting attachment)
~, **кольц·ев·а́я** (elec) brush ring
~, **о·по́р·н·ая** (met roll) carrier
траверс·и́р·овать, -уют, (imp and perf) traverse
тра́верс·н·ый, -ая, -ое, (a) of **тра́верс·а**; transverse; transfer (of assembly lines etc.)
~ **платфо́рм·а** (f) (rail) transfer platform, transborder
~ **пут·ь** (m) (shipb) transfer track
травертин·, -а, (m) (geol) travertine, travertin, calcareous sinter
трав·и́льност·ь, -и, (f) etchability, suitability for etching/pickling
трав·и́льн·ый, -ая, -ое, (a) etching, pickling, scouring
~ **агрега́т·** (m) (met) pickling line
~ **маши́н·а** (f) (print) etching machine; (met) pickling machine
~ **хру́п·кость** (f) (met) acid/pickling brittleness
трав·и́нк·а, -и, (g pl) **-нок·,** (f) blade (of grass)
трав·и́тел·ь, -я, (m) etchant
трав·и́ть, -я́т, (imp) (chem) attack, scour, burn (with acid etc.); (met) etch (to reveal structure); pickle (to remove scale); (phot, print) etch; (text) mordant; blow off, let off (steam, gas); (agr) trample down, damage (crops); pay out, veer (ropes); use up on; hunt, stalk

трáв·к·а, -и, (*g pl*) **-в·ок·,** (*f*) (*v n*) of **трав·и́ть;** (*dim*) of **трав·á**

~, не·пре·ры́в·н·ая (met) continuous pickling

травл·éни·е, -я, (*n*) (*v n*) of **трав·и́ть**

~, анóд·н·ое (met) anodic etching

~, втор·óе (met) white pickling

~, диагност·и́ческ·ое (min) diagnostic etching

~, из·бир·áтельн·ое (met, print) local etching

~ клишé (print) dot etching

~, свéтл·ое (met) white pickling

~, фасóн·н·ое contour etching

фигýр·а травл·éни·я (met) etch figure

трáвл·енн·ый, -ая, -ое, (*past part pass*) of **трав·и́ть**

трáвл·ен·ый, -ая, -ое, (*a*) etched; hunted, stalked

трáвл·я, -и, (*f*) hunt, hunting, stalking

трáвм·а, -ы, (*f*) (med) trauma, injury

травмат·и́зм·, -а, (*m*) (med) traumatism

травмат·и́н·, -а, (*m*) (biochem) traumatin, wound hormone

травмат·и́ческ·ий, -ая, -ое, (*a*) traumatic

травмато·лóг·и·я, -и, (*f*) (med) traumatology

травмато·нáст·и·я, -и, (*f*) (bot) traumatonasty

травмато·троп·и́зм·, -а, (*m*) (bot) traumotropism

травм·и́ровани·е, -я, (*n*) traumatization

травмо·троп·и́зм·, -а, (*m*) (bot) traumotropism

траво·вéд·, -а, (*m*) herbalist

траво·кос·и́лк·а, -и, (*g pl*) **-лок·,** (*f*) grass cutter, lawn mower

траво·пóл·ь·е, -я, (*n*) grass-arable crop rotation

траво·стó·й, -я, (*m*) (agr) grass stand, standing grass crop

траво·я́д·ност·ь, -и, (*f*) (zool) herbivority

траво·я́д·н·ый, -ая, -ое, (*a*) (zool) herbivorous, (*pl as noun*) herbivores, *Herbivora*

трáв·ушк·а, -и, (*g pl*) **-шек·,** (*f*) blade of grass

трав·яни́ст·ый, -ая, -ое, (*a*) grass, grassy, herbaceous

трав·ян·óй, -áя, -óе, (*a*) grass, grassy, herbaceous, herbal

траг·акáнт·, -а, (*m*) tragacanth gum; (bot) tragacanth, *Astragalus*

трагáнт·, -а, (*m*) *see* **трагакáнт·**

траг·и́ческ·ий, -ая, -ое, (*a*) tragic

траг·ули́д·ы (*pl*) (zool) tragulids, *Tragulidae*

традициóн·н·ый, -ая, -ое, (*a*) traditional

трад́ци·я, -и, (*f*) tradition

траектóри·я, -и, (*f*) trajectory; (air) flight path, path

~, баллист·и́ческ·ая ballistic/free trajectory

~ вз·лёт·а (a/c) take-off path

~, воз·мущ·ённ·ая actual path

~ зá·пуск·а (rocket) launching trajectory

~, канон·и́ческ·ая (cyb) canonical path

~, крут·áя steep trajectory

~, морти́р·н·ая (gunn) upper register trajectory

~, на·вéс·н·ая high trajectory

~, на·сти́ль·н·ая flat trajectory; (a/c) flat path

~, от·лóг·ая shallow trajectory

~ пасси́в·н·ого по·лёт·а (rocket) coasting path

~ по·лёт·а flight path

~, по·лóг·ая flat trajectory; (air) flat path

~ пуч·к·á (phys) beam path

~, срéд·н·яя (gunn) trajectory to mean point of impact

~, фáз·ов·ая (autom) phase-plane vector

~ электрóн·а electron path

трáк·, -а, (*m*) track link/shoe/block (of tracked vehicle)

трáкт·, -а, (*m*) highway, route; (mech) duct; (elec) channel; (anat) tract

~, воз·дýш·н·о-гáз·ов·ый air/gas flue/uptake (of furnace)

~ звуко·зá·пис·и sound-recording channel

трактáт·, -а, (*m*) treatise

тракт·овáть, -ýют, (*imp*) treat, discuss, interpret (a problem)

тракт·óвк·а, -и, (*g pl*) **-вок·,** (*f*) treatment, interpretation

трáктор·, -а, (*nom pl*) **-ы** or **á,** (*g pl*) **-ов** or **óв,** (*m*) (M/T) tractor

~ -амфи́би·я, -и, (*f*) (mil) tracked landing vehicle

~, гýсен·ичн·ый crawler/caterpillar tractor

~, колёс·н·ый wheeled tractor

тра́ктор
~ -про·па́ш·ник·, -а, (*m*) row-crop tractor
~, с·ва́р·очн·ый automatic welding head, welding tractor
трактор·иза́ци·я, -и, (*f*) introduction/use of tractors
трактор·и́ст·, -а, (*m*) tractor driver
тра́ктор·н·ый, -ая, -ое, (*a*) tractor; tractor-mounted
~ по́·езд· (*m*) tractor-trailer train
~ комба́йн· (*m*) (agr) tractor-drawn combine
тра́кторо- (*component*) tractor
тракторо·ремо́нт·н·ый, -ая, -ое, (*a*) tractor-repair
тракторо·стро·е́ни·е, -я, (*n*) tractor engineering/production
трактри́с·а, -ы, (*f*) (math) tractrix
тра́л·, -а, (*m*) (fish) beam trawl; (naval) sweep, minesweep; (mil) mine clearing devices
~, гидро·граф·и́ческ·ий (ocean) surveying trawl
~, змей·ко́в·ый (naval) oropesa sweep
тра́л·ени·е, -я, (*n*) (fish) trawling; mine sweeping
тра́л·ить, -ят, (*imp*) (fish) trawl; sweep (mines etc.)
тра́л·ов·ый, -ая, -ое, (*a*) (naut) trawl, trawling; fishing, deep-sea fishing
~ ба́з·а (*f*) fishing port, home port (of trawlers etc.)
~ лебёд·к·а (*f*) trawl winch
~ флот· (*m*) fishing fleet
тра́ль·н·ый, -ая, -ое, (*a*) (naval) sweeping, swept
~ лебёд·к·а (*f*) sweeping winch (of minesweeper)
тра́ль·щик·, -а, (*m*) minesweeper; (obs) trawler
тра́м·а, -ы, (*f*) (bot) trama
трамб·ова́ть, -у́ют, (*imp*) tamp, ram
трамб·о́вк·а, -и, (*g pl*) -вок·, (*f*) (build, civ eng) tamping, ramming, punning; tamper, rammer, ram, punner
~, вз·рыв·н·а́я explosion ram
~, пневмат·и́ческ·ая air/compressed-air tamper/punner; (met cast) air rammer
трамва́·й, -я, (*m*) tramway, street railway; tram, tram/street-car
~, реч·н·о́й river ferry, small river steamer
трампли́н·, -а, (*m*) spring-board; jump, jumping platform (skiing)

тра́н·ец·, -н·ц·а, (*i*) -н·ц·ем, (*m*) (shipb) transom
транзи́стор·, -а, (*m*) (elec) transistor
~, двух·сторо́н·н·ий double-surface transistor
~, кана́л·ев·ый field-effect transistor
~, кре́мн·иев·ый плоск·остн·о́й silicon-junction transistor
на транзи́стор·ах transistorized
~, ните·ви́д·н·ый filamentary transistor
~, одно·конта́кт·н·ый unijunction transistor
~, сло·и́ст·ый junction transistor
~, то́ч·ечн·ый point-contact transistor
транзи́стор·н·ый, -ая, -ое, (*a*) (elec) transistor
транзи́т·, -а, (*m*) transit (of transport), through shipping/routing
транзити́в·н·ый, -ая, -ое, (*a*) transitive
~ от·нош·е́ни·е (*n*) transitive relation
транзи́т·н·ый, -ая, -ое, (*a*) transit; through (of telephone operations); tandem (of telephone equipment)
~ иск·а́тел·ь (*m*) tandem selector
~ по́·езд· (*m*) (rail) through/transit train
~ ста́нци·я (*f*) tandem exchange/office (telephones)
~ со·един·е́ни·е (*n*) through connection/call (telephones)
транзитро́н·, -а, (*m*) (elec) transitron
транс- (*int component*) trans-, (chem) *trans-*
трансбо́рдер·, -а, (*m*) (rail) transborder, transfer platform; (shipb) transfer cradle
трансве́ртер·, -а, (*m*) (elec) transverter
транс·глабелля́р·н·ый, -ая, -ое, (*a*) transglabellar
транс·гре́сс·и·я, -и, (*f*) (geol etc.) transgression; (math) overlapping, intersection
трансду́ктор·, -а, (*m*) (elec) transducer; transductor, saturable reactor
~, двух·ви́д·ов·ый two-mode transducer
транс·звук·ов·о́й, -а́я, -о́е, (*a*) transonic
транс·изоме́р·, -а, (*m*) (chem) *trans-*isomer
транс·кристалл·иза́ци·я, -и, (*f*) transcrystallization; (met) ingotism
транс·кристалл·и́ческ·ий, -ая, -ое, (*a*) transcrystalline
транс·кюри·ев·ый, -ая, -ое, (*a*) (nucl) transcurial

транс·литер·а́ци·я, -и,   (*f*) (ling) transliteration

трансл·и́р·овать, -уют,   (*imp and perf*) (telecom) repeat (line), relay (radio); broadcast live; rediffuse

трансл·и́руемост·ь, -и,   (*f*) repeatability

транс·лок·а́ци·я, -и,   (*f*) translocation

трансл·я́тор·, -а,   (*m*) (telecom) repeater

трансляцио́н·н·ый, -ая, -ое,   (*a*) translational; (telecom) relaying; broadcasting

~ **ра́дио·у́зел·**   (*m*) rediffusion exchange; internal broadcast system

~ **сет·ь**   (*f*) rediffusion net; internal broadcast system

~ **ста́нци·я**   (*f*) (rad) relay station; repeater station/office (line telecom)

трансл·я́ци·я, -и,   (*f*) translation (motion); (met) slip, simple slip, glide; (rad) broadcasting (from live performance); rediffusion (of broadcast over line); (telecom) relay, relaying (of radio communications); repeater, repeating (line communications)

~, **ис·правл·я́ющ·ая**   regenerative repeater

~ **по про·вод·а́м**   line/wire broadcast; line/wire rediffusion

~ **по ра́дио**   live broadcast

~, **регенерати́в·н·ая**   (telecom) regenerative repetition, interpolation

~, **реле́й·н·ая**   telegraph repeater; (rad) relaying

~, **четырёх·про·вод·н·а́я**   four-wire repeater (telephones)

~, **шнур·ов·а́я**   cord-circuit repeater (telephones)

трансмиссио́н·н·ый, -ая, -ое,   (*a*) of **трансми́сс·и·я**

трансми́сс·и·я, -и,   (*f*) (mech eng) transmission, transmission system

трансми́ттер·, -а,   (*m*) (telecom) tape transmitter; (autom) reader (tape or card)

~, **фо́то·электр·и́ческ·ий**   (autom) photoelectric tape reader

~, **электро·механ·и́ческ·ий**   (autom) mechanical reader

транс·мут·а́ци·я, -и,   (*f*) transmutation

транспара́нт·, -а,   (*m*) lined backing-paper; (phot) transparency, diapositive; transparent banner (for propaganda etc.)

транспир·а́ци·я, -и,   (*f*) (biol) transpiration

трансплант·а́т·, -а,   (*m*) (med) transplant, graft

трансплант·а́ци·я, -и,   (*f*) transplantation

транс·плуто́н·иев·ый, -ая, -ое,   (*a*) transplutonium

транспози́ци·я, -и,   (*f*) transposition

транспо́ндер·, -а,   (*m*) (rad) transponder

транспон·и́ровани·е, -я,   (*n*) (math etc.) transposition

транспон·и́р·овать, -уют,   (*imp and perf*) transpose

тра́нспорт·, -а,   (*m*) transport, transportation; consignment (of goods); (naut) fleet auxiliary, supply ship; (mil) train; **транспо́рт·, -а,**   (*m*) (fin) carrying forward

~, **авто·гуж·ев·о́й**   road transport

~, **автомоби́ль·н·ый**   motor transport

~, **войск·ов·о́й**   troopship

~, **результ·и́рующ·ий**   (phys) net transport

транспорт·ёр·, -а,   (*m*) (*see also* **конве́йер·**) conveyer, conveyor, transporter; (rail) well wagon; (met roll) transfer, transfer table

~, **винт·ов·о́й**   screw conveyer

~, **крюк·ов·о́й**   (met roll) hook transfer

~, **ле́нт·очн·ый**   belt conveyer

~, **лот·ко́в·ый**   plate/pan conveyer/feeder

~, **му́фель·н·ый**   (met roll) muffle transfer

~, **от·бо́р·очн·ый**   take-off conveyer

~, **пла́в·ающ·ий**   (mil) load-carrying amphibian

~, **пласт·и́нчат·ый**   plate conveyer, pallet train

~, **при·ём·н·ый**   take-off conveyer

~ **с нес·у́щ·ими це́п·ями**   (met roll) chain-lift transfer, lift-chain cold bed

транспорт·и́р·, -а,   (*m*) (instr) protractor

транспорт·и́ровани·е, -я,   (*n*) transporting, transportation, transport, conveying; movement (ciné cameras etc.)

транспорт·и́рующ·ая маши́н·а   (*f*) handling machine

тра́нспорт·н·ый, -ая, -ое,   (*a*) of **тра́нспорт·**

~ **сеч·е́ни·е**   (*n*) (nucl) transport cross-section

транс·сиби́р·ск·ий, -ая, -ое,   (*a*) Trans-Siberian

транс·суда́т·, -а, (*m*) (med) transudate
транс·урани́д·, -а, (*m*) transuranide
транс·ура́н·ов·ый, -ая, -ое, (*a*) (phys, chem) transuranium
трансфе́р·, -а, (*m*) *see* **трансфе́рт·**
тран сферка́р·, -а, (*m*) (met) transfer car
трансфе́рт·, -а, (*m*) (fin) transfer
трансфер·ши́н·а, -ы, (*f*) (elec) transfer bus, reserve busbar
транс·фигур·а́ци·я, -и, (*f*) tranfisguration
транс·фини́т·н·ый, -ая, -ое, (*a*) transfinite
транс·флю́ксор·, -а, (*m*) (elec) transfluxor
транс-фо́рм·а, -ы, (*f*) (chem) *trans*-configuration
трансформ·а́тор·, -а, (*m*) (elec) transformer
~, анте́нн·ый (rad) aerial transformer
~, бала́нс·н·ый (telecom) repeating coil, balancing transformer
~, брон·ев·о́й shell-type transformer
~, воз·ду́ш·н·ый air-core transformer
~, вольто·до·ба́в·очн·ый booster transformer
~, вращ·а́ющ·ийся rotary converter, dynamotor rotary transformer; resolver (of computer)
~, вы·ход·н·о́й output transformer
~, грозо·у·по́р·н·ый (elec) non-resonating transformer
~ за·тух·а́ющ·их колеб·а́ни·й magnetic-coupling transformer
~ звук·ов·о́й част·от·ы́ audio-frequency transformer
~, из·мер·и́тельн·ый instrument transformer
~, и́мпульс·н·ый (elec) peak/pulse transformer
~ каска́д·н·ого ти́п·а cascade transformer
~ крут·я́щ·его моме́нт·а torque converter (hydraulic gearing)
~, лин·е́йн·ый (telecom) compensating transformer
~, ма́сл·ян·ый oil transformer
~, между·ла́мп·ов·ый inter-valve/interstage transformer
~, много·вит·ко́в·ый wound-primary-type transformer
~ на·ка́л·а filament transformer
~ на·пряж·е́ни·я voltage transformer
~, одно·вит·ко́в·ый through-type transformer
~, пере·хо́д·н·ый (telecom) repeating/repeater coil

трансформ·а́тор
~, пи́к·ов·ый peak transformer
~, по·выш·а́ющ·ий step-up transformer
~, по·ниж·а́ющ·ий step-down transformer
~, по·стро·и́тельн·ый вращ·а́ющ·ийся plotting resolver (computers)
~ про·меж·у́точн·ой част·от·ы́ intermediate-frequency transformer
~, про·ход·н·о́й bushing current transformer
~, пут·ев·о́й (rail) track-feed transformer
~ с коэффицие́нт·ом one-to-one transformer
~, с·ва́р·очн·ый arc welding transformer
~ с·вя́з·и coupling transformer
~, серие́с·н·ый booster transformer
~, сил·ов·о́й power transformer
~, симметр·и́чн·ый balanced transformer
~ си́нус-косинус·, вращ·а́ющ·ийся sine-cosine resolver (computing)
~, со·глас·у́ющ·ий matching transformer
~ со·противл·е́ни·я impedance transformer
~, стабилиз·и́рующ·ий antihunt transformer
~, стерж·нев·о́й bar-type current transformer
~ то́к·а current transformer
~, трёх·об·мо́т·очн·ый three-winding transformer
~ у·правл·е́ни·я variable voltage-control transformer
трансформ·а́торн·ый, -ая, -ое, (*a*) of **трансформ·а́тор·**
~ под·ста́нци·я (*f*) (elec) transformer substation, transforming station
трансформ·и́р·овать, -у́ют, (*imp and perf*) transform, convert; (surv) restitute, translate (aerial photographs)
транс·фу́зи·я, -и, (*f*) (med) transfusion
трансцендента́ль·н·ый, -ая, -ое, (*a*) transcendental
трансценде́нт·н·ый, -ая, -ое, (*a*) transcendental
~ фу́нкци·я (*f*) (math) transcendental function
~ чи́сл·а (*pl*) (math) transcendental numbers
трансценде́нт·ность·, -и, (*f*) transcendence

трансша́льтер·, -а, (*m*) (elec) isolator, isolating switch

тра́н·цев·ый, -ая, -ое, (*a*) of **тра́н·ец·**

~ доск·а́ (*f*) (shipb) transom

~ корм·а́ (*f*) (shipb) transom stern

траншее·коп·а́тел·ь, -я, (*m*) trencher

транш·е́йн·ый, -ая, -ое, (*a*) trench

транш·е́·я, -и, (*f*) trench

тра́п·, -а, (*m*) trap, water seal trap (sanitary eng); interceptor trap (ventilation); (oil) gas and oil separator; (naut) ladder; trap faceting (jewels)

~, за·бо́рт·н·ый (naut) accommodation ladder

~, от·кид·н·о́й ramp (of landing craft)

трапеце·ви́д·ност·ь, -и, (*f*) taper (of a/c wing)

трапеце·ви́д·н·ый, -ая, -ое, (*a*) trapezoid(al)

трапецие·ви́д·н·ый, -ая, -ое, (*a*) trapezoid(al)

трапе́ци·я, -и, (*f*) (math) trapezoid; trapezium; (astron) the Trapezium (in Orion); trapeze (gymnastics)

~, равно·бо́ч·н·ая isosceles trapezoid

трапецоида́ль·н·ый, -ая, -ое, (*a*) trapezoid(al)

трапецоида́ль·ност·ь, -и, (*f*) (elec) keystone effect/distortion

трапец·о́ид·, -а, (*m*) trapezoid

трапецоида́ль·н·ый, -ая, -ое, (*a*) trapezoidal

~ ис·каж·е́ни·е (*n*) (elec) keystone distortion

трапецо́·эдр·, -а, (*m*) (cryst) trapezohedron

тра́пп·, -а, (*m*) (geol) trap, traprock

трапп·о́ид·, -а, (*m*) (geol) trappoid

тра́сс·, -а, (*m*) (min) trass

тра́сс·а, -ы, (*f*) track, route, path, course, line; sketch, draught; (gunn) tracer trail

~, воз·ду́ш·н·ая (air) trunk route

~, по́д·ня·т·ая elevated path (of radio waves)

~ полиго́н·а (rocket) range corridor

трасс·а́нт·, -а, (*m*) (fin) drawer

трасс·а́т·, -а, (*m*) (fin) drawee

трасс·ёр·, -а, (*m*) (gunn) tracer; (air) tracer, flare

трасс·и́р·овать, -уют, (*imp and perf*) mark/map/trace the route/track; (fin) draw a cheque

трасс·иро́вк·а, -и, (*g pl*) **-вок·,** (*f*) (*v n*) *see* **трасс·и́р·овать**

трасс·иро́вочн·ый, -ая, -ое, (*a*) marker, route-marking

трасс·и́рующ·ий, -ая, -ое, (*pres part act*) of **трасс·и́р·овать;** tracer

~ с·на·ря́д· (*m*) (gunn) tracer shell

~ шнур· (*m*) tracing tape

трасс·ов·о́й, -а́я, -о́е, (*a*) of **тра́сс·а;** and of **трасс·**

~ аэродро́м· (*m*) intermediate aerodrome

трассо·ло́г·и·я, -и, (*f*) detection, detective work

тра́т·а, -ы, (*f*) (*v n*) *see* **тра́т·ить;** expenditure

тра́т·ить, -ят, (*imp*) spend, expend, use up

тра́тт·а, -ы, (*f*) (fin) draft

тра́улер·, -а, (*m*) (fish) trawler

траум·и́ровани·е, -я, (*n*) *see* **травм·и́ровани·е**

тра́ур·, -а, (*m*) mourning

тра́ур·ниц·а, -ы, (*f*) (ent) Camberwell Beauty, *Vanessa antiopa*

тра́ур·н·ый, -ая, -ое, (*a*) mourning; funeral; memorial, remembrance

трафаре́т·, -а, (*m*) stencil, template; (min) coarse screen; (ling) cliché

трафаре́т·ить, -ят, (*imp*) stencil

трафаре́т·к·а, -и, (*g pl*) **-т·ок·,** (*f*) stencil brush; title block (eng. drawings)

трафаре́т·н·ый, -ая, -ое, (*a*) stencilled; stereotyped, conventional

тра́ффик·, -а, (*m*) radio traffic

трахе·а́льн·ый, -ая, -ое, (*a*) (biol) tracheal

трахе·и́д·а, -ы, (*f*) (bot) tracheide

трахе·и́т·, -а, (*m*) (med) tracheitis

трахе́й·н·ый, -ая, -ое, (*a*) (biol) tracheal

трахело·рра́ф·и·я, -и, (*f*) (med) trachelorrhaphy

трахело·том·и́·я, -и, (*f*) (med) trachelotomy

трахе·о́л·а, -ы, (*f*) (biol) tracheole

трахео·бронхиа́ль·н·ый, -ая, -ое, (*a*) (anat) tracheobronchial

трахео·том·и́·я, -и, (*f*) (med) tracheotomy

трахе́·я, -и, (*f*) (biol) trachea, (*pl*) tracheae, tracheas

трахи·андези́т·, -а, (*m*) (geol) trachyandesite

трахи·база́льт·, -а, (*m*) (min) trachybasalt

трахи·меду́з·ы (*f pl*) (zool) *Trachymedusae, Trachomedusae*

трах·и́т·, -а, (*m*) (geol) trachyte

трах·и́тов·ый, -ая, -ое, (*a*) (geol) trachytic

трахи·то́ид·, -а, (*m*) (geol) trachytoid

трах·одо́нт·, -а, (*m*) (pal) trachodont, *Trachodon*

трах·о́м·а, -ы, (*f*) (med) trachoma

тра́ч·енн·ый, -ая, -ое, (*past part pass*) of **трат·ить**

тре- (*int component*) three, tri-

тре́б·овани·е, -я, (*n*) (*v n*) of **тре́б·овать;** demand(s), requirement(s)

~, та́кт·ик·о-техн·и́ческ·ое staff requirements (of weapons, armaments etc.)

~, техн·и́ческ·ое technical specifications

тре́б·овательност·ь, -и, (*f*) exacting requirements

тре́б·овательн·ый, -ая, -ое, (*a*) exacting, demanding

тре́б·овать, -уют, (*imp*) (+ *gen*) demand, require, call for (*impers*); **-ся** (*pass*); need, want
что и тре́б·овалось до·каз·а́ть (math) q.e.d., Q.E.D., quod erat demonstrandum

тре́б·уем·ый, -ая, -ое, (*pres part pass*) of **тре́б·овать;** required, specified, requisite

трево́г·а, -и, (*f*) alarm; anxiety
бить трево́г·у sound/give the alarm

~, вод·ян·а́я (naval) collision stations

трево́ж·ить, -ат, (*imp*) alarm, worry, cause anxiety

трево́ж·н·ый, -ая, -ое, (*a*) alarm; alarming, disquieting, disturbing; anxious, worried

трегало́з·а, -ы, (*f*) (biochem) trehalose

тре́гер·, -а, (*m*) (chem) carrier

трёгери́т·, -а, (*m*) (min) troegerite

тред·юнио́н·, -а, (*m*) trade union (outside Soviet bloc)

трезв·е́·ть, -ют, (*imp*) become sober, sober up

тре·звуч·и·е, -я, (*n*) (acous) common chord, triad

тре́зв·ый, -ая, -ое, (*a*) sober

тре·зу́б·ец, -б·ц·а, (*i*) **-б·ц·ем,** (*m*) trident

тре·зу́б·ый, -ая, -ое, (*a*) trident, tridental

тре́йсер·, -а, (*m*) (instr) tracer, guide finger

тре́к·, -а, (*m*) track

тре́к·ов·ый, -ая, -ое, (*a*) track

трел·ева́ть, -ю́ют, (*imp*) skid (logs)

трел·ёвк·а, -и, (*g pl*) **-вок·,** (*f*) skidding (logging)

~, воз·ду́ш·н·ая overhead skidding

~, гуж·ев·а́я animal logging

~, на·зе́м·н·ая ground skidding

трел·ёвочн·ый, -ая, -ое, (*a*) skidding

тре́л·ь, -и, (*f*) skidding trail (logging)

трелья́ж·, -а, (*i*) **-ем,** (*m*) trellis; three-leaved mirror

тр·ём, (*dat*) of **тр·и**

тремат·о́д·а, -ы, (*f*) (zool) trematode, (*pl*) *Trematoda*

тремат·опси́д·, -а, (*m*) (pal) trematopsid

тре́мблер·, -а, (*m*) (instr) trembler

треммато·ло́г·и·я, -и, (*f*) (biol) tremmatology

тремоли́т·, -а, (*m*) (min) tremolite

трём·ст·а́м, (*dat*) of **три́·ст·а**

тр·емя́, (*instr*) of **тр·и**

тремя·ст·а́ми, (*instr*) of **три́·ст·а**

трен·а́ж·, -а, (*m*) *see* **трен·иро́вк·а**

трен·ажёр·, -а, (*m*) training mock-up, training apparatus/machine/device; (air) link trainer

~ -катапу́льт·а, -ы, (*f*) ejector seat test rig

тре́н·ер·, -а, (*m*) trainer (sport etc.), physical culture teacher

тре́нзел·ь, -я, (*m*) bit, snaffle (of harness); (rad) strobe

~, больш·о́й (mech) adjusting/quadrant plate

~, мал·ый (mech) tumbler plate

тр·е́ни·е, -я, (*n*) friction, rubbing

~, вне́ш·н·ее (mech) friction
коэффицие́нт· вне́ш·н·его тр·е́ни·я (*m*) coefficient of friction (metal machining)

~, вну́тр·енн·ее (mech) internal friction, damping capacity (measured)
тепл·от·а́ вну́тр·енн·его тр·е́ни·я (*f*) viscous heat

~, внутри·кристалл·и́ческ·ое (mech) damping capacity, internal friction

~, гран·и́чн·ое limiting friction

~, жи́дк·остн·ое fluid friction

~ кач·е́ни·я rolling friction
ко́нус· тр·е́ни·я (*m*) cone of friction
коэффицие́нт· тр·е́ни·я (*m*) coefficient of friction
мо́дул·ь тр·е́ни·я (*m*) (phys) out-of-phase modulus
моме́нт· тр·е́ни·я (*m*) (mech) moment of friction

~, о·по́р·н·ое static friction
по·ве́рх·ност·и тр·е́ни·я (*pl*) rubbing surfaces

тр·е́ни·е
~, по·ве́рх·ностн·ое   skin/surface friction
~ поко́·я   static friction
~ скольж·е́ни·я   sliding friction
~, сух·о́е   dry friction
~, тормоз·я́щ·ее   drag friction
    у́гол· тр·е́ни·я   (*m*) angle of friction
тре́нинг·, -а,   (*m*) training (horses)
трен·иро́ванн·ый, -ая, -ое,   (*a*) (*past part pass*) of **трен·ир·ова́ть**
трен·ир·ова́ть, -у́ют,   (*imp*) train; (met) hold, soak; age (vacuum tubes)
трен·иро́вк·а, -и,   (*f*) (*v n*) of **трен·и·р·ова́ть**
~ на част·от·у́   (rad) frequency ageing
~, пред·ста́рт·ов·ая   (rocket) dry run
трен·иро́вочн·ый, -ая, -ое,   (*a*) training; (elec) ageing
тре·но́г·а, -и,   (*f*) tripod; hobble (for horse)
тре·но́г·ий, -ая, -ое,   (*a*) tripod
тре·но́ж·ник·, -а,   (*m*) tripod
тренц·ева́ни·е, -я,   (*n*) worming (of rope)
тре́нц·ы   (*pl*) serving (on rope)
треньша́льтер·, -а,   (*m*) (elec) isolating switch
тре·они́н·, -а,   (*m*) (biochem) threonine, threonin
треп·а́лк·а, -и,   (*g pl*) **-лок·,** (*f*) = **тре·п·а́л·о**
треп·а́л·о, -а,   (*n*) (text) beater, scutch
~, руч·н·о́е   (text) brake; swingle (wooden)
треп·а́льн·ый, -ая, -ое,   (*a*) scutching
~ маши́н·а   (*f*) (text) scutcher
трепан·а́ци·я, -и,   (*f*) (med) trepanation
трепа́нг·, -а,   (*m*) (zool) trepang
треп·а́ни·е, -я,   (*n*) (*v n*) see **треп·а́ть**
трепан·и́р·овать, -у́ют,   (*imp and perf*) (med) trepan, trephine
трёп·ан·ый, -ая, -ое,   (*a*) (text) scutched
треп·а́ть, -лют,   (*imp*) pluck, tug, pull, wrench, touse; tease, worry, nag; (text) beat, scutch, swingle, brake (fibres)
треп·а́ч·, -а́,   (*i*) **-о́м,** (*m*) tow beater (ropework)
тре́пел·, -а,   (*m*) (geol) tripoli, diatomite
тре́пет·, -а,   (*m*) trembling, flutter, quiver; trepidation
трепет·а́ни·е, -я,   (*n*) (*v n*) of **трепе·т·а́ть**
трепет·а́ть,   (*pres 3rd pl*) **трепе́щ·ут,**   (*imp*) tremble, quiver, flutter

трёп·к·а, -и,   (*g pl*) **-п·ок·,** (*f*) (*v n*) of **треп·а́ть**
трепл·о́, -а́,   (*m, f or n*) (text) scutching blade
тре́ск·, -а,   (*m*) crackle, crack (sound)
треск·а́, -и́,   (*f*) (fish) cod, codfish, *Gadus*
~, атлант·и́ческ·ая   cod, codfish, *Gadus morhua*
~, балт·и́йск·ая   cod, codfish, *Gadus callarias*
~, тихо·океа́н·ск·ая   Pacific codfish, *Gadus macrocephalus*
тре́ск·ани·е, -я,   (*n*) (*v n*) of **тре́ск·а·-ться**
тре́ск·а·ться, -ются,   (*imp*) crack, split, chap
тре́ск·аю́щ·ийся, -аяся, -ееся,   (*pres part act*) of **тре́ск·а·ться**; crackle, alligatoring (of paint finishes)
треск·о́в·ый, -ая, -ое,   (*a*) cod; (*pl decl as adj*) *Gadidae*
треско·обра́з·н·ые, -ых,   (*pl decl as adj*) (fish) *Gadiformes*
треск·отн·я́, -и́,   (*f*) crackle, crackling, rattle
треск·у́ч·ий, -ая, -ее,   (*a*) crackling, rasping, ringing
тре́с·н·уть, -ут,   (*perf*) crack, split, burst; give a crackle/crack/creak/rattle; (zool) give a chirp
тре́ст·, -а,   (*m*) (econ) trust
трест·а́, -ы́,   (*f*) stock (of flax or hemp)
~, конопл·я́н·ая   hemp stock
трест·и́р·овать, -у́ют,   (*imp*) (econ) form a trust, combine into a trust
трет-   (*root*) (chem) tert-
тр·ёт,   (*pres 3rd sing*) of **тер·е́ть**
трет·е́йск·ий, -ая, -ое,   (*a*) arbitration, arbitrator's
тре́т·ий, -ья, -ье,   (*a*) third
~ лиц·о́   (*n*) third party/person
~ за·ко́н· термо·дина́м·ик·и   third law of thermodynamics
~ рельс·   (*m*) (rail, elec) third rail
трет·и́чн·ый, -ая, -ое,   (*a*) tertiary, third; (geol) Tertiary
~ ами́д·   (*m*) (chem) triacid amide
~ об·мо́т·к·а   (*f*) (elec) tertiary winding
~ основ·а́ни·е   (*n*) (chem) tertiary base
~ пери́од·   (*m*) (geol) Tertiary
~ спирт·   (*m*) (chem) tertiary alcohol
трет·ни́к·, -а́,   (*m*) (met) quick solder
трёт·ь, -и,   (*nom pl*) **-и,** (*g pl*) **-е́й,** (*f*) a/one third
тре́т·ье,   (*nom/acc sing neut*) of **тре́т·ий** (q.v.)

тре́т·ье- (*component*) third

третье·год·и́чн·ый, -ая, -ое, (*a*) concerning the year-before-last, year-before-last's, of /from the year before last

третье·кла́сс·ник·, -а, (*m*) third-form pupil, pupil/student in class 3

третье·со́рт·н·ый, -ая, -ое, (*a*) third class/grade; third-rate (denigration), mediocre

третье·степе́н·н·ый, -ая, -ое, (*a*) insignificant, unimportant

тре́т·ья, (*nom sing f*) of **тре́т·ий** (q.v.)

тре·уго́ль·ник·, -а, (*m*) triangle

~, арифмет·и́ческ·ий arithmetic triangle

~, дву·прямо·уго́ль·н·ый (math) birectangular triangle

~, из·мер·и́тельн·ый (gunn) range triangle

~, кос·о́й scalene triangle

~, ло́ж·н·ый (nav) cocked hat

~ не·вя́з·к·и (surv) triangle of error

~, основ·н·о́й (math) coordinate triangle

~, пло́ск·ий (math) plane triangle

~, само·со·пряж·ённ·ый (math) polar triangle

~ сил· (dynam) force triangle, triangle of forces

~ скор·ост·е́й (dynam) triangle of velocities, velocity triangle

~ со·противл·е́ни·я (elec) impedance triangle

тре·уго́ль·н·ый, -ая, -ое, (*a*) triangular, trigonal

тре́ф·, -а, (*m*) (met roll) wobbler, driving/palm/tenon end

тре́ф·а, -ы, (*f*) (*esp pl*) clubs (at cards)

тре́ф·ов·ый, -ая, -ое, (*a*) of **треф·**, and **тре́ф·а**

~ кон·е́ц· (*m*) = **треф·**

~ му́фт·а (*f*) palm-ended coupling

тре·фо́л·ь, -и, (*f*) *see* **три·фо́л·ь**

тр·ёх, (*gen/prep*) of **тр·и**

тр·ёх- (*component*) (*see also* **тр·и-**) tri-, three-, triple

трёх·а́том·н·ый, -ая, -ое, (*a*) (chem) triatomic, trihydric, trihydroxy

~ спирт· (*m*) trihydroxy/trihydric alcohol

трёх·ба́лль·н·ый, -ая, -ое, (*a*) force/scale-three (of measured meteorological phenomena)

трёх·бараба́н·н·ый, -ая, -ое, (*a*) three-drum

трёх·бро́м·ист·ый, -ая, -ое, (*a*) (chem) tribromide (of)

трёх·буго́р·чат·ый, -ая, -ое, (*a*) (zool) tritubercular, trituberculate, (*pl as noun*) trituberculates, *Trituberculata*

трёх·бу́кв·енн·ый, -ая, -ое, (*a*) three-letter

трёх·вале́нт·ност·ь, -и, (*f*) (chem) trivalence, tervalence

трёх·вале́нт·н·ый, -ая, -ое, (*a*) trivalent, tervalent

трёх·вал·к·а, -и, (*g pl*) **-л·ок·,** (*f*) (met) three-roll bending-machine

трёх·валь·ко́в·ый, -ая, -ое, (*a*) three-roll

~ стан· (*m*) (met) three-high mill

трёх·ветв·и́ст·ый, -ая, -ое, (*a*) three-branched, triramose

трёх·ви́д·ност·ь, -и, (*f*) trimorphism

трёх·ви́д·н·ый, -ая, -ое, (*a*) trimorphic

трёх·винт·ов·о́й, -а́я, -о́е, (*a*) triple-screw, three-screw; three-propeller

трёх·во́д·н·ый, -ая, -ое, (*a*) (chem) trihydric

~ гидра́т· (*m*) (chem) trihydrate

трёх·гиро·ско́п·н·ый, -ая, -ое, (*a*) three-gyro

трёх·гла́в·ый, -ая, -ое, (*a*) (zool) tricephalous, tricipital

~ мы́ш·ц·а (*f*) (anat) triceps

трёх·гнёзд·н·ый, -ая, -ое, (*a*) trilocular

трёх·год·и́чн·ый, -ая, -ое, (*a*) three-year, three-years', (bot) triennial

~ об·уч·е́ни·е (*n*) three-years' training, three-year course

трёх·год·ова́л·ый, -ая, -ое, (*a*) three-year-old

трёх·год·ов·о́й, -а́я, -о́е, (*a*) three-year, three-years', triennial

трёх·го́рл·ый, -ая, -ое, (*a*) three-necked

~ ко́лб·а (*f*) (chem) Wolff bottle

трёх·гра́н·ник·, -а, (*m*) (math) trihedron

трёх·гра́н·н·ый, -ая, -ое, (*a*) trihedral, three-sided; three-edged; (biol) triquetrous

~ на·пи́ль·ник· (*m*) three-cornered/triangular file, three-angular file

~ про́фил·ь (*m*) (met) three-square, triangular shape

~ у́гол· (*m*) (math) trihedral angle

трёх·дн·е́вн·ый, -ая, -ое, (*a*) three-day, three-days'; (med) tertian

трех·до́ль·н·ый, -ая, -ое, (*a*) three-lobe, three-lobed, trilobate, trilobed

трёх·до́ль·чат·ый, -ая, -ое, (*a*) trilobate

трёх·до́м·н·ый, -ая, -ое, (*a*) (bot) trioecious

трёх·жи́ль·н·ый, -ая, -ое, (*a*) (elec) three-core; (bot) triple-nerved, trinervate

трёх·за·мещ·ённ·ый, -ая, -ое, (*a*) (chem) trisubstituted, tertiary

~ **сол·ь фосфо́р·н·ой кисл·от·ы́** (*f*) tertiary orthophosphate

~ **фосфа́т· ка́льци·я** (*m*) tricalcium phosphate

трёх·за·ря́д·н·ый, -ая, -ое, (*a*) (nucl) tricharged

трёх·за·хо́д·н·ый, -ая, -ое, (*a*) three-start (of screw threads etc.)

трёх·зна́ч·н·ый, -ая, -ое, (*a*) three-sign/symbol/letter, three; (math) three-figure/digit; three-place (of decimals); (rail) three-aspect (signalling)

~ **код·** (*m*) (telecom) three-unit code

~ **числ·о́** (*n*) three-figure/-digit number

трёх·зу́б·ый, -ая, -ое, (*a*) three-toothed/pointed/pronged, trident, tridentate

трёх·имен·н·ый, -ая, -ое, (*a*) trinomial

трех·и́мпульс·н·ый, -ая, -ое, (*a*) (autom) three-factor

трех·ио́д·ист·ый, -ая, -ое, (*a*) (chem) triiodide (of)

трёх·ка́мер·н·ый, -ая, -ое, (*a*) trilocular; three-cell

трёх·кисл·о́тн·ый, -ая, -ое, (*a*) triacid

трёх·клино·ме́р·н·ый, -ая, -ое, (*a*) (cryst) triclinic

трех·коле́н·чат·ый, -ая, -ое, (*a*) (mech eng) three-throw

трёх·колёс·н·ый, -ая, -ое, (*a*) three-wheeled, tricycle

трёх·ко́ль·чат·ый, -ая, -ое, (*a*) triannulate

трёх·компоне́нт·н·ый, -ая, -ое, (*a*) three-component, triple, ternary

трёх·конта́кт·н·ый, -ая, -ое, (*a*) three-pin (of elec. plugs etc.)

трёх·коро́б·чат·ый, -ая, -ое, (*a*) (bot) three-capsular, tricapsular

трёх·кра́й·ник·, -а, (*m*) trihedron

трёх·кра́т·н·ый, -ая, -ое, (*a*) triple, three-fold

~ **рас·шир·е́ни·е**, (*n*) (st eng) triple expansion

трёх·кре́мн·иев·ая кисл·от·а́ (*f*) (chem) trisilicic acid

трёх·кулач·ко́в·ый, -ая, -ое, (*a*) three-jawed; three-lobe (pumps etc.)

трёх·лепе́ст·н·ый, -ая, -ое, (*a*) (bot) tripetalous

трёх·ле́т·и·е, -я, (*n*) third anniversary; three years, three-year period

трёх·ле́т·к·а, -и, (*g pl*) -т·ок·, (*f*) three-year plan/period etc.

трёх·ле́т·н·ий, -яя, -ее, (*a*) three-year, triennial; three-year-old

трёх·лин·е́йн·ый, -ая, -ое, (*a*) three-line, trilinear

трёх·ли́ст·н·ый, -ая, -ое, (*a*) three-leaved/sheet, trifoliate, triphyllous

трёх·ло́паст·н·ый, -ая, -ое, (*a*) three-lobed, trilobate

трёх·луч·ев·о́й, -а́я, -о́е, (*a*) three-ray/beam; (biol) triradiate; (TV) three-gun

~ **спи́кула** (*f*) (zool) triaxon

трехмани́т·, -а, (*m*) (min) trechmannite

трёх·ме́р·н·ый, -ая, -ое, (*a*) three-dimensional

трёх·ме́ст·н·ый, -ая, -ое, (*a*) three-berth/seater/place

трёх·меш·ко́в·ый, -ая, -ое, (*a*) trisaccate

трёх·мо́рф·н·ый, -ая, -ое, (*a*) trimorphous, triform, triformed

трёх·о́·кис·ь, -и, (*f*) (chem) trioxide

~ **ура́н·а** uranium trioxide

~ **хро́м·а** chromium trioxide

трёх·осно́в·н·ый, -ая, -ое, (*a*) (chem) tribasic

~ **кисл·от·а́** (*f*) tribasic/trihydric acid

трёх·о́с·н·ый, -ая, -ое, (*a*) triaxial; three-axle

~ **с·жа́т·и·е** (*n*) triaxial compression

~ **теле́ж·к·а** (*f*) (rail) three-axle bogie

трёх·остро́в·н·о́й, -а́я, -о́е, (*a*) (shipb) three-island

трёх·па́л·ый, -ая, -ое, (*a*) three-fingered/toed, tridactyl, tridactylous

~ **ча́й·к·а** (*f*) (orn) kittiwake, *Rissa tridactyla*

трёх·пере·горо́д·чат·ый, -ая, -ое, (*a*) (biol) triseptate

трёх·пе́ст·ичн·ый, -ая, -ое, (*a*) (bot) trigynous

трёх·позицио́н·н·ый, -ая, -ое, (*a*) three-step/position

трёх·по·лож·и́тельн·ый, -ая, -ое, (*a*) tripositive

трёх·поло́с·н·ый, -ая, -ое, (*a*) three-lane (of roads)

трёх·по́ль·н·ый, -ая, -ое, (*a*) (agr) three-field, three-field-rotation

трёх·пол·ь·е, -я, (*n*) (agr) three-field rotation

трёх·полюс·н·ый, -ая, -ое, (*a*) (phys) three-pole, tripolar

трёх·про·вод·н·ой, -а́я, -о́е, (*a*) (elec) three-wire/point/contact

~ **гнезд·о́** (*n*) (elec) three-point jack/socket

~ **систе́м·а** (*f*) (elec) three-wire system

трёх·проце́нт·н·ый, -ая, -ое, (*a*) three-per-cent

трёх·пря́д·н·ый, -ая, -ое, (*a*) three-strand, three-stranded

трёх·пу́т·н·ый, -ая, -ое, (*a*) (civ eng) three-way/-lane

трёх·раз·ветвл·ённ·ый, -ая, -ое, (*a*) trifurcate, trifurcated

трёх·раз·де́ль·н·ый, -ая, -ое, (*a*) tripartite, trifid

трёх·раз·ме́р·н·ый, -ая, -ое, (*a*) three-dimensional

трёх·ра́з·ов·ый, -ая, -ое, (*a*) thrice, triple

трёх·раз·ря́д·н·ый, -ая, -ое, (*a*) (math) three-digit

трёх·ре́бер·н·ый, -ая, -ое, (*a*) three-ribbed/-finned/-edged, (biol) tricostate, tricostal

трёх·резона́нс·н·ый, -ая, -ое, (*a*) (rad) triple-tuned

трёх·ря́д·н·ый, -ая, -ое, (*a*) three-row, three-bank, (biol) triserial, tristichous

трёх·с·вяз·н·о́й, -а́я, -о́е, (*a*) (math) triple-connected

трёх·сер·ни́ст·ый, -ая, -ое, (*a*) (chem) trisulphide/trisulfide (of)

~ **бор·** (*m*) boron sulphide/trisulfide

трёх·с·ло́ж·н·ый, -ая, -ое, (*a*) (gram) three-syllable

трёх·слой·н·ый, -ая, -ое, (*a*) three-layered/ply, triplex

трёх·с·ме́н·н·ый, -ая, -ое, (*a*) three-shift

трёх·со́т·, (*g*) of **три́·ст·а**

трёх·сот·ле́т·и·е, -я, (*n*) tercentenary

трёх·со́т·ый, -ая, -ое, (*a*) three-hundredth

трёч·ст·а́х, (*prep*) of **три́·ст·а**

трёх·с·тво́р·чат·ый, -ая, -ое, (*a*) three-leaved (of doors etc.), triple-sash (of windows); (anat) tricuspid; (biol) three-valved

трёх·степе́н·н·ый, -ая, -ое, (*a*) three-stage

~ **гиро·ско́п·** (*m*) three-frame gyroscope

трёх·сторо́н·н·ий, -яя, -ее, (*a*) three-sided, trilateral

трёх·стре́нд·н·ый, -ая, -ое, (*a*) three-stranded, three-strand

трёх·ступе́н·чат·ый, -ая, -ое, (*a*) three-stage; triple-cascade

трёх·та́кт·н·ый, -ая, -ое, (*a*) three-stroke

трёх·то́м·н·ый, -ая, -ое, (*a*) (print) three-volume

трёх·то́ч·ечн·ый, -ая, -ое, (*a*) three-point

~ **по·са́д·к·а** (*f*) (air) three-point landing

~ **само·пи́с·ец·** (*m*) (instr) three-pen recorder

трёх·ту́бус·н·ый, -ая, -ое, (*a*) three-neck (flasks etc.)

трёх·уго́ль·ник·, -а, (*m*) *see* **тре·-уго́ль·ник·**

трёх·фа́з·н·ый, -ая, -ое, (*a*) three-phase

трёх·фо́рм·енн·ый, -ая, -ое, (*a*) (cryst) trimorphic, trimorphous

трёх·хло́р·ист·ый, -ая, -ое, (*a*) (chem) trichloride (of)

~ **ура́н·** (*m*) uranium trichloride

трёх·ход·ов·о́й, -а́я, -о́е, (*a*) three-way; three-pass

~ **кла́пан·** (*m*) three-way valve

трёх·цвет·к·а, -и, (*g pl*) **-т·ок·,** (*f*) (bot) pansy, *Viola tricolor*; (TV) three-colour/-primary system

трёх·цве́т·н·ый, -ая, -ое, (*a*) three-coloured/-colored, tricolour, tricolor, (phot etc.) trichromatic; three-aspect (railway signalling)

трёх·час·ов·о́й, -а́я, -о́е, (*a*) three-hour; three o'clock

трёх·четверть·волн·ов·ый, -ая, -ое, (*a*) (rad) three-quarter-wave

трёх·чле́н·, -а, (*m*) (math) trinomial

~, **квадра́т·н·ый** trinomial square

трёх·чле́н·ист·ый, -ая, -ое, (*a*) (zool) triarticulate

трёх·чле́н·н·ый, -ая, -ое, (*a*) trinomial; three-membered/member, ternary, trimembral

~ **кольц·о́** (*n*) (chem) three-member ring

трёх·ше́рст·н·ый, -ая, -ое, (*a*) (zool) three-colour/-color

трёх·электро́д·н·ый, -ая, -ое, (*a*) three-electrode

~ **ла́мп·а** (*f*) (elec) triode, three-electrode valve

трёх·элеме́нт·н·ый, -ая, -ое, (*a*) three-element/unit

трёхъ·я́дер·н·ый, -ая, -ое,   (*a*) trinuclear, three-member, tricyclic

~ со·един·е́ни·е (*n*) (chem) tricyclic compound

трёхъ·язы́ч·н·ый, -ая, -ое,   (*a*) trilingual

трёх·эта́ж·н·ый, -ая, -ое, (*a*) (build) three-storey, three-storeyed

трещ·а́ть, -а́т, (*imp*) crackle, crack, rattle, creak; (zool) chirp

трещ·ётк·а, -и, (*f*) *see* **трещ·о́тк·а**

трещ·ёточн·ый, -ая, -ое, (*a*) of **тре·щ·о́тк·а**

~ колес·о́ (*n*) (horol) ratchet-tooth escape wheel

тре́щ·ин·а, -ы, (*f*) crack, split, fissure; (met) crack, check; check, shake, split (timber); (geol) crevasse, crevice, cleft, fissure, fracture, joint; flaw (in gemstone)

~, волос·н·а́я (met) hairline crack, flake, snowflake

~, втор·и́чн·ая (geol) exfoliation joint

~, за·ка́л·очн·ая (met) quenching/hardening crack

~, за·ро́д·ышев·ая (met) nucleating/nuclear crack

~, кольц·ев·а́я (geol) ring fracture; cup shake (timber)

образ·ова́ни·е трещ·и́н· (*n*) crazing, checking, cracking

~, по·ве́рх·ностн·ая surface crack; seasoning split (timber)

~, по·пере́ч·н·ая tear, transverse crack

~, при·ли́в·о-от·ли́в·н·ая tide crack (ice)

~, радиа́ль·н·ая radiating crack; heart shake/check (timber); (geol) radial rift

~ раз·га́р·а (met) thermal crack, hot check

~, тепл·ов·а́я (met) thermal crack, hot check

~, у·са́д·очн·ая (met) shrinkage crack/check

~, у·ста́л·ости·ая fatigue crack

~, шлиф·о́вочн·ая (met) grinding crack/check

тре́щ·инн·ый, -ая, -ое, (*a*) of **тре́·щ·ин·а**

~ из·ли·я́ни·е (*n*) (geol) fissure eruption

трещ·инова́тост·ь, -и, (*f*) cracked/checked state; (geol) fissured state, jointing

трещ·инова́т·ый, -ая, -ое, (*a*) fissured, crazed, checked, full of cracks, covered with cracks

трещино·сто́й·кост·ь, -и, (*f*) crack/cracking resistance

трещ·о́тк·а, -и, (*g pl*) **-ток·,** (*f*) (*see also* **трещ·ётк·а**) (acous) rattle, rattler; ratchet, ratchet stop; ratchet drill

тр·и, (*g*) **тр·ёх** (*numeral*) three

три- (*root*) *see also* **тр·ёх-**

три·а́д·а, -ы, (*f*) triad

триази́н·, -а, (*m*) (chem) triazine

триа́з·о, -а, (*n*) (chem) triazo

триазо́л·, -а, (*m*) (chem) triazole

три·азо·со·един·е́ни·е, -я, (*n*) (chem) triazo-compound

три·аканти́н·, -а, (*m*) (pharm) triacanthine

триакис·окта́·эдр·, -а, (*m*) (cryst) triakisoctahedron

триакис·тетра́·эдр·, -а, (*m*) (cryst) triakistetrahedron

триаконт·и́л·, -а, (*m*) (chem) triacontyl

три·алки́л·, -а, (*m*) (chem) trialkyl

три·алки́ль·н·ый, -ая, -ое, (*a*) (chem) trialkylated

триами́л·, -а, (*m*) (chem) triamyl

триамилами́н·, -а, (*m*) (chem) triamylamine

три·ангул·я́ци·я, -и, (*f*) triangulation

~ I кла́сс·а (surv) first-order triangulation

три·ари́л·, -а, (*m*) (chem) triaryl

три·арил·мета́н·ов·ый, -ая, -ое, (*a*) (chem) triarylmethane

три·ари́ль·н·ый, -ая, -ое, (*a*) (chem) triarylated

триа́с·, -а, (*m*) (geol) Triassic, Trias

триа́с·ов·ый, -ая, -ое, (*a*) (geol) Triassic

три·ацета́т·, -а, (*m*) (chem) triacetate

три·ацетил·глицери́н·, -а, (*m*) triacetin

три·ацетил·целлюло́з·а, -ы, (*f*) cellulose acetate

три·аци́ль·н·ый, -ая, -ое, (*a*) triacylated

три́б·, -а, (*m*) pinion (clockwork)

три́б·а, -ы, (*f*) (biol) tribe

трибестрис·, -а, (*m*) (pharm) tribestris

три́б·к·а, -и, (*g pl*) **-б·ок·,** (*f*) (*dim*) of **триб·**

~, за·вод·н·а́я (horol) winding pinion

трибо- (*int component*) tribo-, friction

трибо·люмин·есце́нци·я, -и, (*f*) (phys) triboluminescence

трибо·ме́тр·и·я, -и, (*f*) tribometry, friction measurement

трибо·электр·и́честв·о, -а, (*n*) triboelectricity, frictional electricity

три·бро́м-　(*int component*) (chem) tribrom-, tribromo-
трибро́м·фено́л·, -а,　(*m*) (chem) tribromophenol
трибу́ль·к·а, -и,　(*g pl*) **-л·ек·,** (*f*) (bot) chives, *Allium schoenoprasum*
трибу́н·а, -ы,　(*f*) platform, rostrum; stand (of stadium etc.)
трибуна́л·, -а, (*m*) (law) tribunal
трибути́л·, -а, (*m*) (chem) tributyl
три·вакци́н·а, -ы,　(*f*) (med) ТАВ vaccine, triple vaccine
тривиа́ль·н·ый, -ая, -ое,　(*a*) trite, hackneyed; (math) trivial
три́виум·, -а, (*m*) (zool) trivium, (*pl*) trivia
три·гармон·и́ческ·ий, -ая, -ое,　(*a*) triharmonic
тригатро́н·, -а,　(*m*) (elec) trigatron
три́ггер·, -а,　(*m*) (elec) trigger circuit/relay; flip-flop (computers)
~, динам·и́ческ·ий　gate
~, не·за·де́рж·анн·ый　dynamic flip-flop (computers)
~ пере·но́с·а　carry flip-flop (computers)
~, реакти́в·н·ый　flip-flop
~, стат·и́ческ·ий　flip-flop
три́ггер·н·ый, -ая, -ое,　(*a*) trigger; flip-flop (computers)
~ ла́мп·а　(*f*) trigger valve/tube
~ ячѐй·к·а　(*f*) (elec) flip-flop
три·ги́р·а, -ы,　(*f*) (cryst) trigyre, third-order symmetry axis
три·ги́р·н·ый, -ая, -ое,　(*a*) (cryst) trigonal
три·гли́ф·, -а,　(*m*) (arch) triglyph
три·глицери́д·, -а,　(*m*) (chem) triglyceride
три́гл·ы　(*pl*) (fish) gurnards, *Triglidae*
тригона́ль·н·ый, -ая, -ое,　(*a*) (cryst) trigonal
тригонелли́н·, -а,　(*m*) (biochem) trigonelline
тригони́д·, -а,　(*m*) (anat) trigonid
тригоно·метр·и́ст·, -а,　(*m*) (surv) triangulator
тригоно·метр·и́ческ·ий, -ая, -ое, (*a*) (math) trigonometric
~ пункт·　(*m*) (surv) trigonometrical station
тригоно·ме́тр·и·я, -и,　(*f*) (math) trigonometry
~, пло́ск·ая　plane trigonometry
~, прямо·лин·е́йн·ая　plane trigonometry

тригоно·ме́тр·и·я
~, сфер·и́ческ·ая　spherical trigonometry
тридек·а́н·, -а,　(*m*) (chem) tridecane
тридека́н·ов·ый, -ая, -ое,　(*a*) (chem) tridecanic
тридеке́н·, -а,　(*m*) (chem) tridecene
тридециле́н·, -а,　(*m*) (chem) tridecylene
тридим·и́т·, -а,　(*m*) (min) tridymite
тридио́н·, -а,　(*m*) (pharm) tridione
три́·дцат·и-　(*component*) thirty
тридцати·гра́н·ник·, -а, (*m*) triacontahedron, polyhedron with thirty sides
тридцати·дн·е́вн·ый, -ая, -ое,　(*a*) thirty-day
тридцати·километр·о́вк·а, -и,　(*g pl*) **-вок·,** (*f*) (surv) 1:3000,000 map
тридцати·ле́т·н·ий, -яя, -ее,　(*a*) thirty-year; thirty-year-old
тридцати·морщ·и́нн·ый, -ая, -ое, (*a*) triacontarugate
три·дца́т·ый, -ая, -ое,　(*a*) thirtieth
три́·дцат·ь, -й,　(*i*) **-ью,** (*f*) thirty
три́ер·, -а,　(*m*) (agr) grain cleaner/grader, screening machine; (agr) cockle separator
триер·ова́ть, -у́ют,　(*imp*) (agr) screen, grade (grain)
три́·жды　(*adv*) three times
трииотра́ст·, -а,　(*m*) (pharm) triiotrast
три·карбо́н·ов·ая кисл·от·а́ (*f*) (chem) tricarboxylic acid
три·клин·и́ческ·ий, -ая, -ое,　(*a*) (cryst) triclinic
три·кли́н·н·ый, -ая, -ое,　(*a*) (cryst) triclinic
трико́　(*n indecl*) (text) tricot; tights (clothing)
трикоза́н·, -а,　(*m*) (chem) tricosane
трикози́л·, -а,　(*m*) (chem) tricosyl
три·кон·одо́нт·ы, -ов,　(*m pl*) (pal) triconodonts, *Triconodonta*
трикота́ж·, -а,　(*m*) knit; knitwear
~, дву·ласт·и́чн·ый　interlock
~, изна́н·очн·ый　reverse knit
~, техн·и́ческ·ий　industrial knit
~, уточно·вя́з·ан·ый　filling knit
трикота́ж·н·ый, -ая, -ое,　(*a*) knitted, knitting
~ маши́н·а　(*f*) knitter
три·латер·а́ци·я, -и,　(*f*) (surv) trilateration
~, радио·геодез·и́ческ·ая　electronic trilateration
три·ли́ст·ник·, -а,　(*m*) trefoil; (bot) clover, *Trifolium*

три·ли́ст·н·ый, -ая, -ое, (*a*) three-leaved, trifoliate, trifoliolate, triphyllous

триллио́н·, -а, (*m*) billion (10^{12}) (Brit); trillion (10^{12}) (U.S.)

три·лоб·и́т·, -а, (*m*) (pal) trilobite, (*pl*) *Trilobita*

тример·, -а, (*m*) (chem) trimer

тримéр·н·ый, -ая, -ое, (*a*) (chem) trimeric

три·мéстр·, -а, (*m*) term (academic)

три·металл·и́ческ·ий, -ая, -ое, (*a*) trimetallic

тримети́л·, -а, (*m*) (chem) trimethyl, trimethide

триметилам́ин·, -а, (*m*) trimethylamine

тримети́н·, -а, (*m*) trimethine

три́ммер·, -а, (*m*) (elec) trimmer capacitor, trimmer; (a/c) trimming tab, trimmer; trimming machine (carpentry), trimmer

три·молекуля́р·н·ый, -ая, -ое, (*a*) trimolecular, termolecular

три·мóрф·н·ый, -ая, -ое, (*a*) (bot) trimorphic, (cryst) trimorphous

три·мýж·еств·о, -а, (*n*) (bot) triandry

три·мýж·н·ый, -ая, -ое, (*a*) (bot) triandrous

три·на́·дцат·и- (*component*) thirteen-

тринадцати·лéт·н·ий, -яя, -ее, (*a*) thirteen-year; thirteen-year-old

три·на́·дцат·ый, -ая, -ое, (*a*) thirteenth

три·на́·дцат·ь, -и, (*f*) thirteen

три·нитро·бензóл·, -а, (*m*) (chem) trinitrobenzene

три·нитр·оксилóл·, -а, (*m*) (expl) trinitroxylene

три·нитро·резорцина́т· свин·ц·а́ (*m*) (expl) lead trinitroresorcinate

три·нитро·толуóл·, -а, (*m*) (expl) trinitrotoluene, TNT

три·нитро·фенóл·, -а, (*m*) trinitrophenol, (chem) picric acid

тринóм·, -а, (*m*) (math) trinomial

три́о (*n indecl*) trio (music); (met roll) three-high

триóд·, -а, (*m*) (elec) triode

~, двой·н·óй twin triode

~, кристалл·и́ческ·ий transistor

~, плоск·остн·óй junction transistor

~, полу·про·вод·никóв·ый transistor

~, тóч·ечн·ый point-contact transistor

~ -гексóд·, -а, (*m*) (elec) triode-hexode

три·óз·а, -ы, (*f*) (biochem) triose

триозо·фосфа́т·, -а, (*m*) (chem) triosephosphate

три·óкси- (*int component*) (chem) trioxy-, trihydroxy-

триокси·метилéн·, -а, (*m*) trioxymethylene

три·окта·эдр·и́ческ·ий, -ая, -ое, (*a*) (cryst) trioctahedral, gyroidal, cubic hemihedral holoaxial

трионáл·, -а, (*m*) (chem) trional, sulfonethylmethane

три·óникс·ы, -ов, (*m pl*) (zool) Trionyx, Trionychidae

три·ортогона́ль·н·ый, -ая, -ое, (*a*) (math) trirectangular

три·óстр·енник·, -а, (*m*) (bot) arrowgrass, *Triglochin*

три·óстр·енн·ый, -ая, -ое, (*a*) tricuspidate

трипа́н·, голуб·óй (*m*) (chem) trypan blue

трипано·сóм·а, -ы, (*f*) (zool) (*pl*) trypanosomes, *Trypanosoma*

трипано·сомóз·, -а, (*m*) (med) trypanosomiasis

трипафлави́н·, -а, (*m*) (pharm) trypaflavine

три·пéрст·ов·ый, -ая, -ое, (*a*) (zool) tridactyle

три́плекс·, -а, (*m*) triplex (glass)

триплéт·, -а, (*m*) triplet; (phot) Cooke lens

триплéт·н·ое со·сто·я́ни·е (*n*) (phys) triplet state

трипл·и́т·, -а, (*m*) (min) triplite

трипл·óид·, -а, (*m*) (gen) triploid

три́ппер·, -а, (*m*) (med) gonorrhea

три́пс·, -а, (*m*) (ent) thrips

трипс·и́н·, -а, (*m*) (biochem) trypsin

трипсино·гéн·, -а, (*m*) (physiol) trypsinogen

три·пта́н·, -а, (*m*) (chem) triptane

триптó·лиз·, -а, (*m*) (physiol) tryptolysis

трипто·фа́н·, -а, (*m*) (biochem) tryptophan, tryptophane

трипуги́т·, -а, (*m*) (min) tripuhyite

три·ромбо·эдр·и́ческ·ий, -ая, -ое, (*a*) trirhombohedral

три́с·, -а, (*m*) (chem) (*abbr*) = **три·этилсили́л·** triethyl-silyl

три·сахар·и́д·, -а, (*m*) (chem) trisaccharide

три·сéкци·я, -и, (*f*) trisection

три́сел·ь, -я, (*m*) (naut) trisail

три́·ст·а,   (*g*) **трёх·со́т·,** (numeral) three hundred

три·стеари́н·, -а,   (*m*) (chem) tristearine

тристе́ц·а, -ы,   (*f*) tristeza (plant disease)

тр·и́те, (*pl imper*) of **тер·е́ть**

три·тетра́·эдр·, -а, (*m*) (cryst) tristetrahedron

три́т·и·й, -я, (*m*) (chem, phys) tritium, H^3, ^3H, T

трити́л·, -а, (*m*) (chem) trityl

тритиона́т·, -а, (*m*) trithionate

тритио́н·ов·ый, -ая, -ое, (*a*) trithionic ~ **кисл·от·а́** (*f*) trithionic acid

тритоми́т·, -а, (*m*) (min) tritomite

трито́н·, -а,   (*m*) (zool) newt, (*pl*) *Triturus*; (nucl) triton

~, **греб·е́нчат·ый** (zool) newt, *Triton cristatus*

трито́ни·я, -и, (*f*) (bot) montbretia, *Tritonia*

три·у́ксус·н·ый, -ая, -ое, (*a*) (chem) triacetate (of)

триумфа́ль·н·ый, -ая, -ое, (*a*) triumphal

трифа́н·, -а, (*m*) (min) triphane, spodumene

трифени́л·, -а, (*m*) (chem) triphenyl

трифенил·азо́т·, -а, (*m*) (chem) nitrogen triphenyl

трифенил·бо́р·, -а, (*m*) triphenyl boron

трифенил·мета́н·, -а, (*m*) triphenyl methane

трифенил·мети́л·, -а, (*m*) trityl

три·фили́н·, -а, (*m*) (min) triphylite, triphyline

трифоли́н·, -а, (*m*) (pharm) trifolin

три·фо́л·ь, -и, (*m*) (bot) buckbean, *Menyanthes trifoliata*

три·фосге́н·, -а, (*m*) (chem) triphosgene, hexachloro-dimethylcarbonate

три·фто́р·- (*component*) (chem) trifluoro-

три·фто́р·и́д·, -а, (*m*) (chem) trifluoride

три·фто́р·мета́н·, -а, (*m*) trifluoromethane

три·хальци́т·, -а, (*m*) (min) trichalcite

трих·иа́з·, -а, (*m*) (med) trichiasis

трих·и́н·а, -ы, (*f*) (zool) trichina

~, **спира́ль·н·ая** *Trichinella spiralis*

трихинелл·ёз·, -а, (*m*) (med) trichinosis, trichiniasis

трих·ино́з·, -а, (*m*) (med) trichinosis, trichiniasis

трих·и́т·, -а, (*m*) (bot) trichite

три·хло́р·- (*component*) (chem) trichlor-, trichloro-

трихлор·ксило́л·, -а, (*m*) xylene trichloride

трихлор·мета́н·, -а, (*m*) (chem) chloroform, trichloromethane

трихлор·у́ксус·н·ый, -ая, -ое, (*a*) trichloroacetic

трихлор·этиле́н·, -а, (*m*) (chem) trichloroethylene, ethylene trichloride

трихо·бакте́ри·и, -й, (*f pl*) *Chlamydobacteriales*

трихо·ге́н·, -а, (*m*) (ent) trichogen

трихо·ги́н·, -а, (*m*) (bot) trichogyne

трих·о́м·а, -ы, (*f*) (bot) trichome

трихо·мици́н·, -а, (*m*) (pharm) trichomycin

трихо·мона́д·а, -ы, (*f*) (zool) trichomonad, *Trichomonas*

трихо·монна́з·, -а, (*m*) (med) trichomoniasis

трихо·цефалёз·, -а, (*m*) (med) trichocephaliasis

трихо·ци́ст·, -а, (*m*) (zool) trichocyst

три·хро·и́чност·ь, -и, (*f*) (cryst) trichroism

три·хро·и́чн·ый, -ая, -ое, (*a*) (cryst) trichroic

три·цера́топс·, -а, (*m*) (pal) Triceratops

три·щет·и́нник·, -а, (bot) trisetum, *Trisetum*

три·э́др·, -а, (*m*) (math) trihedral

три·этанол·ами́н·, -а, (*m*) (chem) triethanolamine

три·этилен·глико́л·ь, -я, (*m*) (chem) triethylene glycol

тр-к (*abbr*) = **тре·уго́ль·ник·** triangle

троака́р·, -а, (*m*) (med) trocar, trochar

тро́г·, -а, (*m*) (geol) trough

тро́г·ани·е, -я, (*n*) (*v n*) *see* **тро́г·а·ть;** pickup (of relay)

   ток· тро́г·ани·я (*m*) pickup current (of a relay)

тро́г·а·ть, -ют, (*imp*) touch; disturb, trouble; **-ся** (*pass*); move/start/set off, start, move

трого́н·, -а, (*m*) (*pl*) (orn) trogons, *Trogoniformes*

тро́·е, (*g*) **тро·и́х,** (*collective numeral*) three, three together

тро́е- (*root*) three

тро́·ек· (*g pl*) of **тро́й·к·а**

трое·кра́т·н·о (*adv*) three times, thrice

трое·кра́т·н·ый, -ая, -ое, (*a*) three-times, threefold, triple

тро́·ечн·ый, -ая, -ое, (*a*) of **тро́й·к·а**

тро·и́ть, -я́т, (*imp*) divide into three; make up into threes; repeat three times

тро́·ичн·ый, -ая, -ое, (*a*) ternary, tripartite, divided into three

тро́й·к·а, -и, (*g pl*) **тро́·ек·,** (*f*) three, a three, a number three; satisfactory (class marking); three-piece suit; troika, three-horse team; triplet, set of three

~, дво·и́чн·ая (math) binary triplet

~ на·правл·е́ни·я (math) directed/ coordinate trihedral

тро́й·ни́к, -а́, (*m*) (cryst) trilling; tee, three-way piece/joint, tee-/T-joint/ junction; (mech eng) shamrock plate, monkey face; three-barrelled gun

тро́й·нико́в·ый, -ая, -ое, (*a*) triple, three-way/part/piece

~ му́фт·а (*f*) (elec) dividing box

тро́й·ни́чн·ый, -ая, -ое, (*a*) (zool) trifacial, trigeminal

тро́й·н·о́й, -а́я, -о́е, (*a*) threefold, triple, triplex, ternary

~ диагра́мм·а (*f*) (chem) ternary diagram

~ рас·шир·е́ни·е (*n*) (st eng) triple expansion

~ с·вя́з·ь (*f*) (chem) triple bond/link

~ то́ч·к·а (*f*) (chem) triple point

тро́й·н·я, -и, (*g pl*) **-о́·ен·,** (*f*) triplets

тро́й·ск·ий, -ая, -ое, (*a*) troy (weight)

~ у́нци·я (*f*) troy ounce

тро́й·ственн·ый, -ая, -ое, (*a*) triple

тройчато·пе́р·ист·ый, -ая, -ое, (*a*) tripinnate, tripinnated

трой·ча́т·ый, -ая, -ое, (*a*) three-part/piece

~ ви́л·ы (*pl*) three-pronged fork/ pitchfork

трокто·ли́т·, -а, (*m*) (geol) troctolite

тролиту́л·, -а, (*m*) (plast) trolitul

тролле́й·бус·, -а, (*m*) trolley bus

троллей·во́з·, -а, (*m*) trolley truck/ lorry

троллей·ка́р·, -а, (*m*) trolley car, industrial trolley truck

тро́мб·, -а, (*m*) (meteor) whirlwind, tornado; (med) clot of blood, thrombus

~, антаркт·и́ческ·ий (meteor) whirly

тромб·и́н·, -а, (*m*) (biochem) thrombin

тромбо·ангии́т·, -а, (*m*) (med) thromboangiitis

тромбо́·з·, -а, (*m*) (med) thrombosis

тромбо·зи́м·, -а, (*m*) (physiol) thrombozyme

тромбо·кина́з·а, -ы, (*f*) (physiol) thrombokinase

тромбо·пласти́н·, -а, (*m*) (biochem) thromboplastin

тромбо·флеби́т·, -а, (*m*) (med) thrombophlebitis

тромбо·ци́т·, -а, (*m*) (cyt) thrombocyte

тромбоцито·пени́·я, -и, (*f*) (med) thrombocytopoenia

трон·, -а, (*m*) throne

тро́н·а, -ы, (*f*) (min) trona

тро́нк·, -а, (*m*) trunk piston

тро́нк·ов·ый, -ая, -ое, (*a*) of **тро́нк·**

~ дви́г·ател·ь (*m*) trunk-piston engine

тро́·н·уть, -ут, (*perf*) *see* **тро́г·а·ть**

троост·и́т·, -а, (*m*) (met) troostite

троп·а́, -ы́, (*g pl*) **тро́п·ы,** (*d pl*) **тро·п·а́м,** (*f*) path

троп·и́зм·, -а, (*m*) (biol) tropism

тро́п·ик·, -а, (*m*) (geog) tropic

~ Козе·ро́г·а Tropic of Capricorn

~ Ра́к·а Tropic of Cancer

тропи́н·, -а, (*m*) (pharm) tropine

троп·и́нк·а, -и, (*g pl*) **-нок·,** (*f*) path, track

троп·и́ческ·ий, -ая, -ое, (*a*) tropical, tropic

~ во́з·дух· (*m*) (meteor) tropical air

~ при·ли́·в· (*m*) (ocean) tropic tide

~ фронт· (*m*) (meteor) intertropical front, intertropical convergence zone

тропа́т·, -а, (*m*) (chem) tropate

тропеоли́н·, -а, (*m*) (chem) tropaeolin (dye)

тро́п·ов·ая кисл·от·а́ (*f*) (chem) tropaic/tropic acid

трополо́н·, -а, (*m*) (pharm) tropolone

тропо·па́уз·а, -ы, (*f*) (geophys) tropopause

тропо·сфе́р·а, -ы, (*f*) (meteor) troposphere

тропо·сфе́р·н·ый, -ая, -ое, (*a*) tropospheric

тропо·та́ксис·, -а, (*m*) (bot) tropotaxis

тро́с·, -а, (*m*) (*and see* **кана́т·**) rope, cable

~, бел·ён·ый white rope

~, бо́род·очн·ый second-grade hemp rope

~, букси́р·н·ый tow rope

~, вы́блен·очн·ый ratline

~, вы·тяж·н·о́й rip-cord (parachute)

~, ги́б·к·ий flexible steel-wire rope

~, за·щи́т·н·ый (elec) overhead earth wire

~, коко́с·ов·ый coir rope, (naut) grass line

трóс

~, **комбин·и́рованн·ый** fibre-and-wire rope

~, **мани́ль·ск·ий** manilla rope

~, **на·дёж·н·ый** endless whip

~, **не·смол·ён·ый** white rope

~, **нес·у́щ·ий** (elec) suspension cable, catenary wire

~ **один·áрн·ой с·ви́в·к·и, авиациóн·н·ый** aircraft strand

~, **пеньк·óв·ый** hemp rope

~, **полу·жёстк·ий** steel-wire rope (19–37 wires each strand)

~, **прó·волоч·н·ый** wire rope

~, **раст·и́тельн·ый** fibre rope

~, **сизáль·ск·ий** sisal rope

~, **смол·ён·ый** tarred rope

~, **сталь·н·óй** steel-wire rope

~, **тормоз·н·óй** (air) arrester wire

~, **трёх·пря́д·н·ый** three-strand hawser-laid rope

~ **у·правл·éни·я** (a/c) control cable

~, **хлопчато·бумáж·н·ый** cotton rope

трóсик·, -а, (m) (dim) of **трос·**; stranded wire

тросо·блóч·н·ая систéм·а (f) rope-and-pulley system

трóс·ов·ый, -ая, -ое, (a) of **трос·**; (f as noun) (naut) cordage room

~ **рабóт·а** (f) hawser lay (of rope)

тросо·мéр·, -а, (m) rope gauge

трост·и́льн·ый, -ая, -ое, (a) (text) doubling

~ **маши́н·а** (f) (text) doubler winding frame

трост·и́нк·а, -и, (g pl) **-нок·,** (f) cane, reed (stem)

трост·и́ть, -я́т, (imp) (text) double

трост·ни́к·, -á, (m) (bot) reed, rush, *Phragmites*

~, **обык·новéнн·ый** common reed, *Phragmites communis*

~, **сáхар·н·ый** (bot) sugar cane, *Saccharum officinarum*

трост·никóв·ый, -ая, -ое, (a) of **трост·ни́к·**

~ **сáхар·** (m) cane sugar

троти́л·, -а, (m) (expl) trotyl, trinitro-toluene

тротуáр·, -а, (m) pavement, side-walk

трофé·й, -я, (m) trophy; captured war material, spoils of war, booty

трóф·ик·а, -и, (f) (bot) trophic system

троф·и́ческ·ий, -ая, -ое, (a) (biol) trophic

трофо·плáзм·а, -ы, (f) (cyt) tropho-plasm

трофо·сóм·а, -ы, (f) (zool) trophosome

трофо·тáксис·, -а, (m) (biol) tropho-taxis

трофо·трóп·изм·, -а, (m) (bot) tropho-tropism

трофо·хроматúн·, -а, (m) (cyt) tropho-chromatin

трохáнтер·, -а, (m) (zool) trochanter

трохантúн·, -а, (m) (zool) trochantin, trochantine

трох·óид·а, -ы, (f) (math) trochoid

трох·оидáль·н·ый, -ая, -ое, (a) trochoid, trochoidal

~ **волн·á** (f) (ocean) trochoid wave

трохо·трóн·, -а, (m) (phys) trochotron

трохо·фóр·а, -ы, (f) (zool) trochophore

трощ·ён·ый, -ая, -ое, (a) doubled

трощ·éни·е, -я, (n) (v n) of **трост·и́ть;** (text) doubling

тро·я́к·ий, -ая, -ое, (a) threefold, triple

тро·я́к·о (adv) in three ways

труб·á, -ы́, (nom pl) **трýб·ы,** (f) (see also **трýб·к·а, трубо·про·вóд·**) (see also associated adjectives); tube, pipe, duct; (elec) conduit, cable conduit; chimney, funnel, smoke-stack; (gunn) barrel; (anat) tube, tuba; trumpet, horn; (geol) pipe, column, shoot; telescope

~, **аэро·динам·и́ческ·ая** (and see under **аэро·динам·и́ческ·ая т.**), wind tunnel

~, **без·мýфт·ов·ая** inserted-joint piping

~, **бес·шóв·н·ая** seamless tube/pipe

~, **бур·и́льн·ая** drill pipe

~, **вед·у́щ·ая** Kelly (rotary boring); (surv, astron) finder telescope

~, **вентиляциóн·н·ая** ventilating pipe; vent, breathing pipe (of fuel tank)

~, **вест·ов·ая** overflow (diesel injector)

~, **визи́р·н·ая** (surv) sighting telescope

~, **водо·подъ·ём·н·ая** rising main

~, **водо·про·пуск·н·áя** water pass (e.g. under railways)

~, **водо·раз·бóр·н·ая** take-off pipe

~, **вос·ход·я́щ·ая** riser, stand pipe, ascending pipe

~, **в·пуск·н·áя** inlet/feed/supply/intake pipe

~, **в·сáс·ывающ·ая** suction intake pipe; (hydroelec) draft tube

~, **вы·брáс·ывающ·ая** ejection pipe

~, **вы·пуск·н·áя** outlet/exit/waste/ blowoff pipe

~ **выс·óк·ого давл·éни·я** pressure pipe/tube

труб·а́

~, вы·тяж·н·áя　exhaust/vent pipe (of ventilation systems)

~, вы·хлоп·н·áя　exhaust pipe; (a/c) tail/jet pipe

~, гельмпóрт·ов·ая　(shipb) rudder trunk/tube/case

~, гидро·динам·и́ческ·ая　(dynam) water tunnel

~, глисс·áжн·ая　skid pipe (furnace)

~ -гри́д·, -а,　(m) (surv, astron) finder telescope

~, дейдву́д·н·ая　(shipb) stern tube

~, дрен·áжн·ая　drain tube

~, ды́м·ов·áя　chimney; (naut) funnel, smoke-stack

~, дымо·гáр·н·ая　heating pipe/tube (of boiler)

~, жар·ов·áя　furnace tube (boiler); flame tube (gas turbine)

~, жёстк·ая телескоп·и́ческ·ая　(a/c) flying boom

~, зр·и́тельн·ая　telescope

~, кавитациóн·н·ая　cavitation tunnel

~, калибр·óванн·ая　quantity calibration tube

~, канализациóн·н·ая　sewage pipe/conduit

~, кипят·и́льн·ая　generating tube (of boiler)

~, колóн·ков·ая　core barrel (of diamond drill)

~, на·гнет·áтельн·ая　delivery pipe (pump)

~, на·пóр·н·ая　discharge pipe (of pump); penstock (of hydro power station)

~, об·вóд·н·ая　by-pass pipe

~ обрáт·н·ой вод·ы́　return tube (boiler)

~, об·сáд·н·ая　casing (of wells)

~, одно·отвéрст·н·ая　(elec) single-duct conduit

~, окуля́р·н·ая　(opt) eye tube

~, о·пуск·н·áя　downcomer

~, от·вод·н·áя　offtake/waste pipe; breathing pipe (of fuel tank); offset/branch pipe

~, от·кáч·ивающ·ая　scavenge pipe

~, пар·ов·áя　steam pipe

~, пере·лив·н·áя　overflow pipe

~ пере·мéн·н·ого сеч·éни·я　variable-section pipe

~, пере·хóд·н·ая　reducing pipe

~, плáмен·н·ая　flame tube; liner (gas turbine)

~, по·движ·н·áя　(instr) draw tube

труб·а́

~, под·вод·я́щ·ая　feed pipe (engine); supply pipe (systems)

~, под·зóр·н·ая　telescope, spy glass

~, подъ·ём·н·ая　riser, uptake

~ по·сто·я́нн·ого сеч·éни·я　uniform-section tube/pipe

~, рабóч·ая　(oil) kelly

~, рас·тру́б·н·ая　bell-and-spigot pipe, inserted-joint pipe

~, рас·ходо·мéр·н·ая　flow tube

~, ребр·и́ст·ая　ribbed/gilled pipe

~, ру́д·н·ая　(geol) ore pipe/shoot

~, само·тёч·н·ая　tubular chute

~, свобóд·н·ая　(gunn) loose barrel

~, с·лив·н·áя　tap/overflow pipe, bleeder (gas systems)

~, с·пуск·н·áя　fall/down/discharge/drain pipe

~, стерео·скоп·и́ческ·ая　stereoscopic telescope

~, с·тóч·н·ая　drain pipe

~, толк·áющ·ая　torque/pinion tube (of universal joint)

~, трáнспорт·н·ая　transfer pipe

~, удáр·н·ая　(aerodynam) shock tube

~, фаллóпи·ев·а　(anat) Fallopian tube

~, флáн·цев·ая　flanged pipe

~, цельно·тя́·нут·ая　solid-drawn tube

~, шлáм·ов·ая　(oil) sludge collector

~, шпигáт·н·ая　(shipb) scupper pipe

~, экрáн·н·ые　water wall (of boiler)

труб·áч, -á,　(m) trumpeter; (orn) trumpeter, *Psophia crepitans*

труб·и́ть, -я́т,　(imp) blow/sound the trumpet/horn, (mil) sound

тру́б·к·а, -и,　(g pl) **-б·ок·,** (f) (dim) of **труб·а́;** short length of pipe/tubing, nozzle; (elec) valve, vacuum tube, tube (*see also* **лáмп·а**) igniting fuze (in ammunition); (biol) tubule; (geol) pipe, canal, chimney

~ áзимут·а　bearing tube (radar)

~ бег·у́щ·его луч·á　(elec) travelling-beam tube

~ Брáун·а　X-ray tube

~, в·вóд·н·ая　(elec) lead-in pipe/bush; entrance/inlet pipe

~, вéктор·н·ая　(math) tube, vector tube

~ Венту́ри　(hydr) Venturi tube/meter

~, водо·грéй·н·ая　generating tube (boiler)

~, вос·пламен·и́тельн·ая　(gunn) vent tube

~, вы·тяж·н·áя　vent pipe

тру́б·к·а

~, газо·све́т·н·ая   (elec) glow-discharge lamp

~, гидро·динам·и́ческ·ая   (hydr) flow nozzle

~ да́ль·ност·и   range scope (radar)

~, динам·и́ческ·ая   (instr) impact tube

~, диссектор·н·ая   (TV) image dissector tube

~, дистанцио́н·н·ая   (expl) time fuse

~, до́н·н·ая   (gunn) base percussion fuse

~, дрен·а́жн·ая   (agr) drain tube; (a/c) vent pipe; trap (radio)

~, дымо·га́р·н·ая   fire tube (boiler)

~, за·жиг·а́тельн·ая   (min) miner's squib

~, за·по·мин·а́ющ·ая   memory/storage tube (computers)

~, индика́тор·н·ая   display, display tube (radar)

~, капилля́р·н·ая   capillary tube

~, като́д·н·ая   (elec) cathode-ray tube

~, код·и́рующ·ая   (elec) coding tube

~, контро́ль·н·ая   overflow pipe (car radiator)

~ круг·ов·о́го об·зо́р·а   plan position indicator, P.P.I. (radar)

~, микро·телефо́н·н·ая   telephone handset

~ модул·я́тор·а   grid sleeve (of vacuum tube)

~, на·коп·и́тельн·ая   storage tube (computers)

~, от·клон·я́ющ·ая   (TV) beam-deflector tube

~, пере·да·ю́щ·ая   (TV) camera tube

~ Пито́   pitot tube

~ по́лн·ого на·по́р·а   pitot tube

~, при·ём·н·ая   (TV) picture tube

~, про·лёт·н·ая   (elec) drift/flight-path tube

~, прост·а́я дымо·га́р·н·ая   plain tube (of fire tube boiler)

~, раз·ря́д·н·ая   (elec) discharge tube

~, рас·по́р·н·ая   (gunn) distance piece

~, рентге́н·овск·ая   X-ray tube

~ сапун·а́   breather pipe

~, с·вяз·н·а́я дымо·га́р·н·ая   stay tube (boiler)

~, сил·ов·а́я   (phys) tube of force, field tube

~, с·чёт·н·ая   counter tube (computers etc.)

~, телефо́н·н·ая   receiver, telephone hand receiver

~, температу́р·н·ая   thermometer tube

тру́б·к·а

~ то́к·а   (dynam) stream tube

~ электро́н·н·ой за́·пис·и   (elec) electron recording tube, E-R tube

~, электронно·луч·ев·а́я   cathode ray tube

трубко·вёрт·ы, -ов,   (*m pl*) (ent) *Attelabidae*

трубко·зу́б·, -а,   (*m*) (zool) aardvark

~, ка́п·ск·ий   aardvark, *Orycteropus capensis*

трубко·зуб·ые, -ых,   (*pl decl as adj*) zool) aardvarks, *Tubulidentata*

трубко·кры́л·ы, -ов,   (*m pl*) (fish) *Solenostomidae*

трубко·цве́т·н·ый, -ая, -ое,   (*a*) (bot) tubiliflorous

трубно·я́йч·ников·ый, -ая, -ое,   (*a*) (anat) tuboovarian

тру́б·н·ый, -ая, -ое,   (*a*) of **труб·а́;** tubal; (*as component*) -funnelled

~ дос·к·а́   (*f*) (met) tube sheet

~ ключ·   (*m*) pipe wrench

~ со·един·е́ни·е   (*n*) pipe joint

тру́бо-   (*root*) tube, pipe; chimney, funnel

трубо·волоч·и́льн·ый, -ая, -ое,   (*a*) (met) tube-drawing

трубо·ги́б·очн·ый, -ая, -ое,   (*a*) pipe-bending

трубо·за·гото́в·очн·ый стан·   (*m*) (met) tube-rounds mill (ingots to tube rounds)

трубо·лит·е́йн·ый, -ая, -ое,   (*a*) tube-casting

трубо·ло́в·к·а, -и,   (*g pl*) **-в·ок·,**   (*f*) (oil) tubing catcher

трубо·ме́д·ницк·ий, -ая, -ое,   plumbing, pipefitting

трубо·на·рез·н·о́й, -а́я, -о́е,   (*a*) (mech eng) pipe-threading

трубо·от·рез·н·о́й, -а́я, -о́е,   (*a*) pipe-cutting

трубо·про·во́д·, -а,   (*m*) (*see also* **тру·б·а́, тру́б·к·а**) piping, tubing, pipeline, conduit, duct, pipe, line

~, в·пуск·н·о́й   (ICE) intake manifold

~, в·са́с·ывающ·ий   (ICE) intake manifold; (a/c) induction manifold; suction line; aspirant tube (air systems)

~, вы·пуск·н·о́й   exhaust manifold/pipe

~, диферент·о́вочн·ый   (mar eng) trimming line

~, на·гнет·а́тельн·ый   pressure piping; (mar eng) fuel service piping

тру́бо·про·во́д

~, на·по́р·н·ый (hydroelec) penstock

~, с·бо́р·н·ый gathering line (oilfields)

~, то́пл·ивн·ый fuel feed; (shipb) fuel oil service piping

~, ~ ма́л·ого га́з·а (a/c) pilot fuel feed

трубо·про·во́д·н·ый, -ая, -ое, (a) of **трубо·про·во́д·**

трубо·про·ка́т·н·ый, -ая, -ое, (a) (met) tube-rolling

~ агрега́т· (m) tube-rolling works, set of tube-rolling mills

~ стан· (m) tube-rolling mill

трубо·прут·ко́в·ый пресс· (m) tube-and-bar press

трубо·ре́з·, -а, (m) pipe-cutter

трубо·ро́г·, -а, (m) (zool) whelk, *Buccinidae*

трубо·с·ва́р·очн·ый, -ая, -ое, (a) (met) pipe-welding

~ агрега́т· (m) pipe-welding machine

трубо·у·кла́д·чик·, -а, (m) pipelaying crane

трубо·чи́ст·, -а, (m) chimney-sweep/cleaner

тру́б·очк·а, -и, (g pl) **-чек·,** (f) (dim) of **тру́б·к·а;** tubule; sleeve (of vacuum tube)

~ А́рльд·а (elec) cable jointing

тру́б·очн·ый, -ая, -ое, (a) of **тру́б·к·а, труб·а́, тру́б·очк·а**

тру́б·чат·ый, -ая, -ое, (a) tubular, tube

~ коло́д·ец· (m) tube well (for water)

~ пред·о·хран·и́тел·ь (m) (elec) cartridge fuse

труб·шта́г·, -а, (m) (naut) funnel guy

труд·, -а́, (m) labour, labor, work; (pl) (print) transactions, proceeding (of society), works (of individual)

труд·и́ться, ⸗ятся, (imp) labour, labor, toil

тру́д·н·о (adv) with difficulty; (predic adj) it is difficult

тру́д·н·о- (component) almost un-/in-, not easily/readily, hardly

трудно·до·сту́п·н·ый, -ая, -ое, (a) difficult/hard to reach, almost inaccessible, not easily/readily available

трудно·за·топл·я́ем·ый, -ая, -ое, (a) almost unsinkable

трудно·об·раба́т·ываем·ый, -ая, -ое, (a) almost unworkable

трудно·о·пре·дел·и́м·ый, -ая, -ое, (a) hard to determine/define, almost indefinable

трудно·пла́в·к·ий, -ая, -ое, (a) hardly fusible, almost infusible; high-melting-point, refractory

трудно·пре·одол·и́м·ый, -ая, -ое, (a) almost insurmountable

трудно·про·ход·и́м·ый, -ая, -ое, (a) almost impassable

трудно·рас·твор·и́м·ый, -ая, -ое, (a) barely/hardly soluble, almost insoluble

тру́д·ност·ь, -и, (f) difficulty

тру́д·н·ый, -ая, -ое, (a) difficult, hard; arduous, hard, heavy

труд·ов·о́й, -а́я, -о́е, (a) labour, labor, labouring, laboring, work, working

~ до·гово́р· (m) contract between employer and employee, terms of employment

~ кни́г·а (f) employment record book, workers' paybook

трудо·де́н·ь, -дн·я́, (m) work-day

трудо·ём·к·ий, -ая, -ое, (a) laborious, time/labour/labor-consuming, heavy (of manual labour)

трудо·люб·и́в·ый, -ая, -ое, (a) industrious, hard-working

трудо·спосо́б·ност·ь, -и, (f) ability to work; capacity for work; (med) fitness for work

трудо·спосо́б·н·ый, -ая, -ое, (a) able-bodied, fit (for work)

трудо·терап·и́·я, -и, (f) (med) occupational therapy

труд·я́щ·ийся, -аяся, -ееся, (pres part act) of **труд·и́ться;** working; (as noun) worker

тру́ж·еник·, -а, (m) toiler; worker

труксин·ов·ая кисл·от·а́ (f) (chem) isotruxillic acid

трула́·й, -я, (m) true-lay rope

тру́п·,-а, (m) dead body, corpse, cadaver

тру́п·н·ый, -ая, -ое, (a) of **труп·**

~ яд· (m) ptomaine poison

тру́пп·а, -ы, (f) (theat) troupe

тру́с·, -а, (m) coward

тру́с·ить, -ят, (imp) be afraid

тру́т·, -а, (m) tinder

тру́т·ен·ь, -т·н·я, (m) (ent) drone

трут·ови́к·, -а́, (m) (bot) shelf fungus, false tinder fungus

трут·о́вк·а, -и, (g pl) **-вок·,** (f) (ent) laying worker bee

тру́т·ов·ые, -ых, (pl decl as adj) (bot) *Polyporaceae*

трух·а́, -и́, (f) (agr) chaff; (pal, bot) debris

трухл·я́в·ый, -ая, -ое, (a) rotten, dry-rotted

тр·у́щ·ийся, -аяся, -ееся, (*a*) (*pres part act*) of **тер·е́ться;** rubbing

~ **конта́кт·** (*m*) wiping contact (sound recording)

~ **по·ве́рх·ност·ь** (*f*) rubbing/friction surface

трущо́б·а, -ы, (*f*) thicket, overgrown place; slum

трюк·, -а, (*m*) (cinema) trick

трюк·ов·ая съ·ём·к·а (*f*) (cinema) trick photography

трюм·, -а, (*m*) (naut) hold; bilge; ballast tanks (of submarine)

~, грунт·ов·о́й hopper (of dredger)

трюм·н·ый, -ая, -ое, (*a*) of **трюм·**

~ **бимс·** (*m*) (shipb) hold beam

~ **на·со́с·** (*m*) bilge pump

трюфел·ь, -я, (*nom pl*) -и, (*g pl*) -е́й, (*m*) (bot) truffle

тряп·и́чн·ый, -ая, -ое, (*a*) rag; ragged

~ **полу·ма́сс·а** (*f*) (paper) rag pulp

тряп·к·а, -и, (*g pl*) -п·ок·, (*f*) rag

тряп·ь·ё, -я́, (*n*) rags

тряс·, (*past masc sing*) of **тряс·ти́**

тряс·е́ни·е, -я, (*n*) (*v n*) of **тряс·ти́**

тряс·ённ·ый, -ая, -ое, (*past part pass*) of **тряс·ти́**

тряс·и́льн·ый, -ая, -ое, (*a*) shaking

тряс·и́н·а, -ы, (*f*) (geog) quagmire

тря́с·к·а, -и, (*f*) (*v n*) of **тряс·ти́;** flutter (airscrew)

тряс·ли́, (*past pl*) of **тряс·ти́**

трясо·гу́з·к·а, -и, (*g pl*) -з·ок·, (*f*) (orn) wagtail, *Motacilla*

тряс·ти́, -у́т, (*imp*) shake, shake off/down/out; **-ся** (*pass*); shake, shiver, tremble

тряс·у́нк·а, -и, (*f*) (bot) quaking grass, *Briza*

тряс·у́щ·ий, -ая, -ее, (*pres part act*) of **тряс·ти́**

тря́с·ш·ий, -ая, -ее, (*past part act*) of **тряс·ти́**

тряс·я́, (*pres gerund*) of **тряс·ти́**

трях·н·у́ть, -у́т, (*perf*) shake

тсу́г·а, -и, (*f*) (bot) hemlock, *Tsuga*

тсумеби́т·, -а, (*m*) (min) tsumebite

тсуна́ми (*indecl*) (ocean) tsunami, earthquake shock waves

тсутсугаму́ши (*indecl*) (med) shimamushi/tsutsugamushi/flood fever, Japanese river fever

ТУ (*abbr*) = **техн·и́ческ·ие у·сло́в·и·я** industrial specifications

ту, (*pron, acc sing f*) of **тот**

туале́т·, -а, (*m*) toilet; dressing-table

туале́т·н·ый, -ая, -ое, (*a*) of **туале́т·**

~ **вод·а́** (*f*) lotion

ту́б·а, -ы, (*f*) tuba (musical instrument); tube, dispenser tube (e.g. for toothpaste)

тубази́д·, -а, (*m*) (pharm) tubazid

тубер·и́ди·и, -й, (*f pl*) (bot) tuberidia

тубе́ркул·, -а, (*m*) (med) tubercle

туберкул·ёз·, -а, (*m*) (med) tuberculosis

туберкул·ёзн·ый, -ая, -ое, (*a*) (med) tuberculous, tubercular, consumptive

туберкул·и́н·, -а, (*m*) (med) tuberculin

туберо́з·а, -ы, (*f*) (bot) tuberose, *Polianthes tuberosa*

тубер·о́зн·ый, -ая, -ое, (*a*) (med etc.) tuberous

тубо·за·жи́м·н·ая маши́н·а (*f*) (pharm) tube-sealing machine

тубо·курари́н·, -а, (*m*) (pharm) tubo-curarine

тубу́с·, -а, (*m*) (opt) viewing hood, **body tube**

ту́г·о (*adv*) tight, tightly; with difficulty

туг·о́й, -а́я, -о́е, (*a*) tight, taut; difficult, hard

туго·пла́в·к·ий, -ая, -ое, (*a*) infusible, refractory, high-melting, high-melting-point; hard (glass); hot-neck (grease)

~ **с·пла́в·** (*m*) refractory alloy

туго·пла́в·кост·ь, -и, (*f*) infusibility, refractoriness; hardness (of glass)

туго·у́зд·ый, -ая, -ое, (*a*) hardmouthed (of horses)

туго·у́х·ий, -ая, -ое, (*a*) rather deaf, hard of hearing

туда́ (*adv*) there, thither

~ **и обра́т·н·о** there and back, return (of tickets)

~ **и сюд·а́** to and fro, hither and thither

ту́·ев·ый, -ая, -ое, (*a*) of **ту́·я;** (chem) thujic, tujic

~ **ма́сл·о** (*n*) thuja/tuja oil

ту́ер·, -а, (*m*) chain ferry

тужу́р·к·а, -и, (*g pl*) -р·ок·, (*f*) doublebreasted jacket; (naut) monkey jacket

туз·, -а́, (*m*) = **туз·ик·;** ace

ту·зе́м·н·ый, -ая, -ое, (*a*) native, local, indigenous

ту́з·ик·, -а, (*m*) dinghy, skiff

тузлу́к·, -а́, (*m*) concentrated brine, brine (for curing hides)

тузлук·ова́ни·е, -я, (*n*) brining (hides)

ту́й·и, (*g sing*) of **ту́й·я**

туй·и́л·, -а, (*m*) (chem) thujyl

туй́ил·ов·ый, -ая, -ое, (*a*) (chem) thujyl
~ спирт· (*m*) thujol, thujyl alcohol
туй·и́н·, -а, (*m*) (chem) thujine
туй·о́л·, -а, (*m*) (biochem) thujol, thuja oil
туй·о́н·, -а,· (*m*) (biochem) thujone
ту́й·я, -и, (*f*) = **ту́·я**
ту́к·, -а, (*m*) (agr) fertiliser, artificial manure, mineral fertilizer; fat, lard
ту́к·ов·ый, -ая, -ое, (*a*) of **тук·**
туко·дроб·и́лк·а, -и, (*g pl*) **-лок·,** (*f*) fertilizer grinder/crusher
туко·с·ме́с·ь, -и, (*f*) mixed fertilizer
ту́л·иев·ый, -ая, -ое, (*a*) (chem) thulium, Tm
ту́л·и·й, -я, (*m*) (chem) thulium, Tm
тули́т·, -а, (*m*) (min) thulite
ту́ловищ·е, -а, (*n*) (anat) trunk, torso
туль·я́, -и́, (*f*) crown (of hat)
туляр·еми́·я, -и, (*f*) (med) tularaemia, tularemia
туляр·и́н·, -а, (*m*) (pharm) tularin
тума́н·, -а, (*m*) fog, mist, haze
~, адвекти́в·н·ый (meteor) advection fog
~, густ·о́й (meteor) thick fog (1 on visibility scale)
~, о́чень густ·о́й (meteor) dense fog (0 on visibility scale)
~, по·зе́м·н·ый (meteor) ground fog
~ пятн·а́ми (meteor) fog patches
~, радиаци́он·н·ый (meteor) radiation fog
~, сплош·н·о́й (nucl) continuous cloud, complete fogging
~ сре́д·н·ей густ·от·ы́ (meteor) fog (2 on visibility scale)
~, сух·о́й (meteor) dry haze
~, у·ме́р·енн·ый (meteor) moderate fog (3 on visibility scale)
тума́н·ить, -ят, (*imp*) (*dim*) make hazy/misty; **-ся** become hazy/misty
тума́н·ност·ь, -и, (*f*) fog, mist, haze; (astron) nebula; impreciseness, obscurity
~, вне·галакт·и́ческ·ая extragalactic nebula
~, га́з·ов·ая gaseous nebula
~, диффу́з·н·ая diffuse nebula
~, крабо·ви́д·н·ая Crab nebula
~, не·пра́в·ильн·ая irregular nebula
~, планета́р·н·ая planetary nebula
~, тёмн·ая dark nebula
тума́н·н·ый, -ая, -ое, (*a*) foggy, misty, hazy; cloudy (liquids)
тумано·у·лов·и́тел·ь, -я, (*m*) spray trap

ту́мб·а, -ы, (*f*) pedestal; (naut) bollard
~, па́р·н·ая (naut) double bollard
~, шварт·о́вн·ая (naut) bollard
ту́мблер·, -а, (*m*) tumbler switch
ту́нг·, -а, (*m*) (bot) tung tree, *Aleurites*
~ Фо́рд·а tung tree, *Aleurites fordii*
~, япо́н·ск·ий mu tree, *Aleurites cordata*
тунга́р·, -а, (*m*) (elec) tungar tube
ту́нг·ов·ый, -ая, -ое, (*a*) tung
~ де́рев·о кита́й·ск·ое tung tree, *Aleurites fordii*
~ ма́сл·о (*n*) tung oil, China wood oil
тунгстен·и́т·, -а, (*m*) (min) tungstenite
тунгст·и́т·, -а, (*m*) (min) tungstite
ту́ндр·а, -ы, (*f*) (geog) tundra
ту́ндр·ов·ый, -ая, -ое, (*a*) tundra
тун·е́ц·, -н·ц·а́, (*m*) (fish) tunny, (*pl*) *Thunnidae*
туни́к·а, -и, (*f*) tunic
туниц·и́н·, -а, (*m*) (biol) tunicine
туннел́·ь, -я, (*m*) (*see also* **тоннел́·ь**) tunnel, subway
туннел́ь·н·ый, -ая, -ое, (*a*) tunnel, tunnel-type
~ эффе́кт· (*m*) (elec) tunnel effect
тунце·ло́в·, -а, (*m*) tunny fishing vessel
тун·цо́в·ый, -ая, -ое, (*a*) of **тун·е́ц·;** (*pl as noun*) *Thunnidae*
туп·е́·ть, -ют, (*imp*) become blunt/dull
туп·и́к·, -а́, (*m*) dead end, impasse, deadlock; blind alley, cul-de-sac; (min) blind room; blunt knife; (rail) railhead, terminal station, switchback; (orn) puffin, *Fratercula*
~, обык·нове́нн·ый Atlantic puffin, *Fratercula arctica*
ста́в·ить в туп·и́к· bring to an impasse
ста·ть в туп·и́к· be at a loss
~, тихо·океа́н·ск·ий horned puffin, *Fratercula corniculata*
туп·и́ть, -я́т, (*imp*) make blunt, blunt
тупо·зу́б·ый, -ая, -ое, (*a*) blunt-toothed
туп·о́й, -а́я, -о́е, (*a*) blunt, obtuse; dull
~ крив·а́я (*f*) broad curve
~ на·стро́й·к·а (*f*) flat/broad tuning
~ у́гол· (*m*) (math) obtuse angle
тупо·кон·е́чн·ый, -ая, -ое, (*a*) blunt-ended, obtuse
тупо·ло́б·ый, -ая, -ое, (*a*) dim-witted
тупо·но́с·н·ый, -ая, -ое, (*a*) blunt-nosed/ended
тупо·уго́ль·н·ый, -ая, -ое, (*a*) obtuse-angled

тупо·у́м·и·е, -я,   (*n*) stupidity, (med) hebetude

ту́р·, -а,   (*m*) stage, round; (geol) cairn; (zool) aurocks, *Bos primigenius*; (zool) mountain goat, *Capra*

тур·а́, -ы́,   (*f*) rook, castle (chess); (bot) fucoid, *Fucus*; (*pl*) *Fucaceae*

тура́нит·, -а,   (*m*) (min) turanite

туран·о́з·а, -ы,   (*f*) (biochem) turanose

тур·а́чк·а, -и,   (*g pl*) **-чок·,** (*f*) (shipb) warping drum

турбе́ри·я, -и,   (*f*) (bot) thurber cotton, *Thurberia thespesioides*

турбиди́·метр·, -а,   (*m*) turbidimeter

турбиди·метр·и́·я, -и,   (*f*) (chem) turbidimetric/nephelometric analysis, turbidimetry

турби́н·а, -ы,   (*f*) turbine

~, акти́в·н·ая   impulse-type steam turbine, impulse turbine

~, вертика́ль·н·ая vertical-shaft turbine

~, вентиля́тор·н·ая turbo-fan

~, винт·ов·а́я propeller-type water turbine, propeller turbine

~, вод·ян·а́я hydraulic/water turbine

~ выс·о́к·ого давл·е́ни·я high-pressure turbine

~, га́з·ов·ая gas turbine

~, гидравл·и́ческ·ая hydraulic/water turbine

~, двух·пер·ов·а́я double-bladed turbine

~, двух·по·то́ч·н·ая double-flow turbine

~ за́д·н·его хо́д·а (mar eng) astern turbine

~ за́·мк·нут·ого ци́кл·а closed-cycle turbine

~, из·бы́т·очн·ая reaction watch turbine

~, ковш·о́в·ая Pelton wheel, impulse water turbine

~, конденсацио́н·н·ая condensing turbine

~, лоб·ов·а́я pipe turbine (water)

~, ло́паст·н·ая feathering-propeller turbine, Kaplan water turbine

~, много·ступ·е́нчат·ая multi-pressure-stage turbine (steam)

~ мя́·т·ого па́р·а exhaust-steam turbine

~, на·по́р·н·ая reaction water turbine

~, напорно·стру́й·н·ая reaction-impulse water turbine

~, не·по́лн·ая impulse water turbine

турби́н·а

~ ни́з·к·ого давл·е́ни·я low-pressure turbine

~, одно·ступ·е́нчат·ая акти́в·н·ая simple-impulse turbine, single-stage impulse turbine (steam)

~, ос·ев·а́я axial-flow turbine

~, пар·ов·а́я steam turbine

~, паро·га́з·ов·ая Н.Т.Р. turbine

~ пере́д·н·его хо́д·а (mar eng) ahead turbine

~, поворотно·ло́паст·н·ая Kaplan water turbine

~, по́лн·ая reaction water turbine

~, пред·в·ключ·ённ·ая superimposed turbine, topping unit (steam)

~, пропе́ллер·н·ая propeller water turbine, propeller turbine

~, радиально·ос·ев·а́я radially axial turbine

~, реакти́в·н·ая reaction-type steam turbine, reaction turbine

~ с зуб·ча́т·ой пере·да́·ч·и geared turbine

~ с от·бо́р·ом па́р·а extraction turbine (steam)

~ с от·бо́р·ами и противо·давл·е́·ни·ем automatic extraction non-condensing turbine

~ с про·меж·у́точн·ым от·бо́р·ом па́р·а bleeder turbine (steam)

~ с противо·давл·е́ни·ем back-pressure turbine

~, свободно·стру́й·н·ая impulse water turbine

~ с·ме́ш·енн·ого давл·е́ни·я mixed-pressure turbine

~ со ступ·е́н·ями давл·е́ни·й mixed-pressure turbine, pressure-compounded turbine

~ со ступ·е́н·ями ско́р·ост·и velocity-compounded turbine

~ со́б·ственн·ых нужд· house turbine (steam)

~ сре́д·н·его давл·е́ни·я medium-pressure turbine

турби́н·к·а, -и, (*g pl*) **-нок·,** (*f*) (*dim*) of **турби́н·а;** impeller

турби́н·н·ый, -ая, -ое, (*a*) turbine

турбо·агрега́т·, -а, (*m*) turbo-set

турбо·бу́р·, -а, (*m*) (oil) turbodrill

турбо·вентил·я́тор·, -а, (*m*) turbo-driven fan, turbofan

турбо·вентиля́тор·н·ый двиг·ател·ь (*m*) (a/c) turbofan, ducted fan

турбо·винт·ов·о́й, -а́я, -о́е, (*a*) (a/c) turboprop

турбо·винт·ов·ой, -а́я, -о́е

~ **дви́г·ател·ь** (*m*) turboprop, propeller-turbine engine

турбо·водо·подъ·ём·ник·, -а, (*m*) (civ eng) turbopump, turbolift

турбо·во́з·, -а, (*m*) turboelectric locomotive

турбо·воздухо·дув·к·а, -и, (*g pl*) **-в·ок·,** (*f*) turbo-blower, turbo-charger

турбо·газо·ду́в·к·а, -и, (*g pl*) **-в·ок·,** (*f*) turbo-blower/charger

турбо·ген·ера́тор·, -а, (*m*) (elec) turbogenerator

турбо·дета́ндер·, -а, (*m*) turbo-expansion engine

турбо·дина́мо (*n indecl*) (elec) turbo-dynamo

турбо·зуб·ча́т·ый, -ая, -ое, (*a*) geared-turbine

~ **агрега́т·** (*m*) geared turbine unit

~ **на·со́с·** (*m*) turbine-lubricating gear pump

~ **пере·да́·ч·а** (*f*) geared turbine transmission

турбо·компре́ссор·, -а, (*m*) turbo-compressor

турбо·компре́ссор·н·ый, -ая, -ое, (*a*) (elec) turbo-compressor; (a/c) turbo-jet

турбо·конденса́т·н·ый на·со́с· (*m*) turbine-driven condensate pump ·

турбо·котёл·, -т·л·а́ (*m*) combined boiler and turbine unit

турбо·ла́мп·а, -ы, (*f*) (min) turbolamp

турбо·ма́сл·ян·ый на·со́с· (*m*) turbine-driven lubricating pump

турбо·на·гнет·а́тел·ь, -я, (*m*) turbo-blower/charger; (a/c) turbosupercharger

турбо·на·со́с·, -а, (*m*) turbopump

турбо·на·со́с·н·ый, -ая, -ое, (*a*) turbopump

турбо·прямо·то́ч·н·ый, -ая, -ое, (*a*) turboram-jet

~ **реакти́в·н·ый дви́г·ател·ь** (*m*) turboram-jet

турбо·реакти́в·н·ый, -ая, -ое, (*a*) (a/c) turbojet

~ **дви́г·ател·ь** (*m*) turbojet, jet-turbine engine

~ **дви́г·ател·ь, двух·ко́нтур·н·ый** ducted fan turbine engine, turbofan

турбо·стро·е́ни·е, -я, (*n*) turbine construction/building/engineering

турбо·циркуляцио́н·н·ый на·со́с· (*m*) turbine-driven circulating pump

турбо·электр·и́ческ·ий, -ая, -ое, (*a*) turbo-electric

турбо·электро·хо́д·, -а, (*m*) turbo-electric motor ship

турбуле́нт·ност·ь, -и, (*f*) turbulence

~, **жесто́к·ая** violent turbulence

~, **си́ль·н·ая** severe turbulence

турбуле́нт·н·ый, -ая, -ое, (*a*) turbulent

~ **теч·е́ни·е** (*n*) (dynam) turbulent flow

турбулиза́тор·, -а, (*m*) swirl vane, vortex generator (fluid flow)

турги́т·, -а, (*m*) (min) turgite, hydro-haematite

туре́л·ь, -и, (*f*) (a/c etc.) gun turret, turret

туре́ль·н·ый, -ая, -ое, (*a*) turret

тур·е́цк·ий, -ая, -ое, (*a*) Turkish

тури́лл·, -а, (*m*) (chem) tourile

турингѝт·, -а, (*m*) (min) thuringite

турист·и́ческ·ий, -ая, -ое, (*a*) tourist, tourist's

тури́ц·а, -ы, (*f*) (bot) stickseed, *Lappula*

туркме́н·ск·ий, -ая, -ое, (*a*) Turkmen

турмали́н·, -а, (*m*), (min) tourmaline, turmaline

турмалин·иза́ци·я, -и, (*f*) (geol) tourmalinization

турмали́н·ов·ый, -ая, -ое, (*a*) tourmal-inic

Ту́рнбул·а, син·ь (*f*) (chem) Turnbull blue

турне́пс·, -а, (*m*) (bot) turnip, swede, *Brassica rapa* var *ropa*, *Brassica campestris* var *ropa*

турнике́т·, -а, (*m*) turnstile; (med) tourniquet

турни́р·, -а, (*m*) tournament

ту́р·ок·, -р·к·а, (*g pl*) **тур·ок·,** (*m*) Turk

турри·ко́н·, -а, (*m*) (pal) turriliticone

тускло·блест·я́щ·ий, -ая, -ее, (*a*) dully gleaming/glistening

ту́ск·лост·ь, -и, (*f*) dullness, dimness, tarnish

ту́ск·л·ый, -ая, -ое, (*a*) dull, dim, lustreless, tarnished

туск·не́·ть, -ют, (*imp*) become dull/dim/tarnished

тут (*adv*) here; (*noun*) = **ту́т·а**

ту́т·а, -ы, (*f*) (bot) mulberry, *Morus*

ту́т·овник·, -а, (*m*) = **ту́т·а**

ту́т·ов·ый, -ая, -ое, (*a*) mulberry; (*pl as noun*) mulberry family, *Moraceae*

~ **де́рев·о** (*n*) = **ту́т·а**

~ **шелко·пря́д·** (*m*) silkworm

тýт·ов·ый, -ая, -ое

~ **ягод·а** (*f*) mulberry

тутокаин·, -а, (*m*) (pharm) tutocaine

туф·, -а, (*m*) (geol) tuff; tufa

~, **вулькан·ическ·ий** (geol) tuff

~, **извест·ков·ый** (geol) tufa

тýфель·к·а, -и, (*g pl*) **-л·ек·,** (*f*) (*dim*) of **тýфл·я;** (zool) slipper, *Paramecium*

тýфл·я, -и, (*g pl*) **тýфел·ь** (*f*) slipper, shoe

туфо·бетóн·, -а, (*m*) tufferete

тýф·ов·ый, -ая, -ое, (*a*) (geol) tuffaceous; (geol) tufaceous

~ **вулькан·ическ·ий** (geol) tuff, tuffaceous

~ **извест·ков·ый** (geol) tufa, tufaceous

тýх·л·ый, -ая, -ое, (*a*) putrid, rotten-smelling

тýх·нуть, -нут, (*imp*) go/die out; become putrid/rotten, go bad

тухолит·, -а, (*m*) (min) thucholite

тýч·а, -и, (*i*) **-ей,** (*f*) cloud

туч·ев·óй, -áя, -óе, (*a*) of **тýч·а**

тýч·н·ый, -ая, -ое, (*a*) obese, adipose; fertile, rich (of soil); (bot) succulent

тýш·а, -и, (*i*) **-ей,** (*f*) carcass

туш·евáни·е, -я, (*n*) (*v n*) of **туш·е·вáть**

туш·ёванн·ый, -ая, -ое, (*past part pass*) of **туш·евáть**

туш·евáть, -ýют, (*imp*) shade

туш·ёвк·а, -и, (*g pl*) **-вок·,** (*f*) shading

туш·éни·е, -я, (*n*) (*v n*) of **туш·ить**

тýш·енн·ый, -ая, -ое, (*past part pass*) of **тýш·ить;** quenched, extinguished, slaked; **туш·ённ·ый, -ая, -ое,** (*past part pass*) of **туш·ить;** stewed, simmered

туш·ён·ый, -ая, -ое, (*a*) stewed, simmered

тýш·и, (*nom/acc pl*) of **тýш·ь;** (*g sing, nom/acc pl*) of **тýш·а**

туш·ильн·ый, -ая, -ое, (*a*) quenching, extinguishing, slaking

~ **бáшн·я** (*f*) quenching tower (coke production)

туш·ить, -ат, (*imp*) put out, extinguish, quench, slake; simmer (in cooking), stew

тушкáн·чик·, -а, (*m*) (zool) jerboa (*pl*) *Dipodinae*

тýш·ь, -и, (*f*) Indian ink

тý·я, -и, (*f*) (bot) arborvitae, *Thuja*

~, **вос·тóч·н·ая** Eastern arborvitae, *Thuja occidentalis*

тý·я

~, **с·клáд·чат·ая** giant arborvitae, *Thuja plicata*

~, **япóн·ск·ая** Japanese arborvitae, *Thuja standishii*

тхмелин·, -а, (*m*) (pharm) thchmelin

т/ч (*abbr*) = **тóнн·ы в час·** t/h, tons per hour

тщáтель·ност·ь, -и, (*f*) care

тщáтель·н·ый, -ая, -ое, (*a*) careful, thorough

тще·слáв·н·ый, -ая, -ое, (*a*) vain

тщéт·н·о (*adv*) in vain

тщéт·н·ый, -ая, -ое, (*a*) vain, futile, unavailing

ты (*pron*) you (familiar address, obligatory to children and animals)

тык·ать, (*pres 3rd pl*) **тыч·ут,** (*imp*) poke/prod/stick into/in

тыкв·а, -ы, (*f*) (bot) gourd, pumpkin, *Cucurbita*

тыкв·енн·ые, -ых, (*pl decl as adj*) (bot) cucurbis family, gourds, *Cucurbitaceae*

тыл·, -а, (*nom pl*) **-ы,** (*g pl*) **-óв,** (*m*) rear, back; (mil) rear services/area, rear

тыл·ов·óй, -áя, -óе, (*a*) (mil) rear, rear services; rear, back

тыл·ов·ый, -ая, -ое, (*a*) = **тыл·ов·óй**

тыль·н·ый, -ая, -ое, (*a*) rear, back

тын·, -а, (*m*) lath fence

тысяч·а, -и, (*i*) **-ей,** or **-ью,** (*f*) thousand

тысяче·килогрáмм·ов·ый, -ая, -ое, (*a*) thousand-kilogram

тысяче·лéт·и·е, -я, (*n*) thousand years; thousandth anniversary

тысяче·лист·ник·, -а, (*m*) (bot) yarrow, *Achillea*

тысяче·нóж·к·а, -и, (*g pl*) **-ж·ек·,** (*f*) (ent) millipede, *Julus*

тысяч·н·ый, -ая, -ое, (*a*) thousandth; thousand(s) strong

~ **дистáнци·я** (*f*) (gunn) mil (3·6′ of arc)

~ **дóл·я** (*f*) a thousandth

тыч·ет, (*pres 3rd sing*) of **тык·ать**

тыч·инк·а, -и, (*g pl*) **-нок·,** (*f*) (bot) stamen

тычинко·нóс·н·ый, -ая, -ое, (*a*) (bot) stamen-bearing, staminiferous, staminate

тыч·иночн·ый, -ая, -ое, (*a*) (bot) staminate

тыч·к·á, (*g sing*) of **тыч·óк·**

тыч·кóв·ый, -ая, -ое, (*a*) of **тыч·óк·**

тыч·о́к·, -ч·к·а́, (*m*) post, stake; (naut) spar; header (brickwork); (elec) prong, peg, plug, pin

ты́ч·ущ·ий, -ая, -ее, (*pres part act*) of ты́к·ать

тьм·а, -ы, (*f*) darkness, dark; a great number

Тью́ринг·а, маши́н·а (*f*) Turing machine (computers)

тэдс (*abbr*) = те́рмо·электро·дви́·ж·ущ·ая си́л·а thermo e.m.f., thermo-electromotive force

тэйлори́т·, -а, (*m*) (min) taylorite

тэнио·ли́т·, -а, (*m*) (min) tainiolite

тэн·и́т·, -а, (*m*) (min) taenite

ТЭС (*abbr*) = тетра·эти́л·свин·е́ц· tetraethyl lead

тэта (*indecl*) theta, θ, ϑ (Greek)

тюа́д·, -а, (*m*) tuad

тю́б·ик·, -а, (*m*) (*see also* ту́б·а) tube, dispenser tube (e.g. for toothpaste)

тю́бинг·, -а, (*m*) (min, civ eng) tubbing frame, (*pl*) tubbing

тю́бинг·ов·ый, -ая, -ое, (*a*) tubbing

~ кре́п·ь (*f*) tubbing

тю́бинго·у·кла́д·чик·, -а, (*m*) erector arm (tunnel driving)

тюк·, -а́, (*m*) bale, package, packet

тюк·ова́льн·ый, -ая, -ое, (*a*) baling

тюк·о́ванн·ый, -ая, -ое, (*past part pass*) of тюк·ова́ть

тюк·ова́ть, -у́ют, (*imp*) bale, package, pack

тю́л·ев·ый, -ая, -ое, (*a*) (text) tulle

тюлене·бо́й·н·ый, -ая, -ое, (*a*) sealing

~ про́·мысел· (*m*) sealing grounds

тюлене·ви́д·н·ый, -ая, -ое, (*a*) (zool) phocoid, phociform, seal-like

тюле́н·ев·ый, -ая, -ое, (*a*) (zool) seal, phocine

тюлен·ёнок·, -нк·а, (*nom pl*) тюле́·н·я́т·а, (*m*) seal pup

тюле́н·ий, -ья, -ье, (*a*) seal(s)

тюле́н·ин·а, -ы, (*f*) seal meat

тюле́н·ь, -я, (*m*) (zool) seal

~, горбо·но́с·н·ый grey seal, *Halichoerus grypus*

~, гренла́нд·ск·ий Greenland seal, *Phoca groenlandia, Pagophoca groenlandia*

~, на·сто·я́щ·ий hair seal, (*pl*) *Phocidae*

~, обык·нове́нн·ый common seal, *Phoca vitulis*

~, сивуче·обра́з·н·ый = тюле́н·ь уш·а́ст·ый

~, уш·а́ст·ый fur seal, (*pl*) *Otariidae*

тюле́н·ь

~ У́эддел·а Weddel's seal, *Leptonichoetes weddelli*

~, шести·рез·цо́в·ый hooded seal, *Cystophora cristata*

тюле́нь·к·а, -и, (*f*) sealer (boat)

тю́л·к·а, -и, (*g pl*) -л·ок·, (*f*) log

тю́л·ь, -я, (*m*)· (text) tulle

тюльпа́н·, -а, (*m*) (bot) tulip, *Tulipa*

тюрбо́ (*n indecl*) (fish) turbot, *Rhombus*

тю́рем·, (*g pl*) of тюрьм·а́

тюре́м·н·ый, -ая, -ое, (*a*) of тюрьм·а́

тюринги́т·, -а, (*m*) (min) thuringite

тюрьм·а́, -ы́, (*nom pl*) тю́рьм·ы, (*g pl*) тю́рем·, (*f*) prison, jail, gaol

тюфя́к·, -а́, (*m*) mattress

~, фаши́н·н·ый (civ eng) fascine mattress

тюямуни́т·, -а, (*m*) (min) tyuyamunite

тя́г·а, -и, (*f*) (*v n*) of тя·н·у́ть; draught, draft, pull; thrust, propulsive thrust; tie/pull/connecting rod, stay

~, а́нкер·н·ая tie rod

~, бес·форса́ж·н·ая dry thrust

~, букси́р·н·ая towing power

~, вз·лёт·н·ая (a/c) takeoff thrust; (rocket) acceleration thrust

~, вы·тя́ж·н·а́я induced draught/draft

~, гусен·и́чн·ая caterpillar traction

~, двой·н·а́я four-wheel drive

~, есте́ств·енн·ая natural draught

~, из·бы́т·очн·ая (a/c) reserve thrust

~, иску́с·ственн·ая mechanical draught/draft

~, ко́с·венн·ая ejector draught (boiler)

~, литр·о́в·ая (rocket) volumetric thrust

~, лоб·ов·а́я thrust per unit cross-section area (of a/c engine)

~, на·гнет·а́тельн·ая blast/forced draught/draft

~, на·правл·я́ющ·ая guide bar

~, не·симметр·и́чн·ая (air) asymmetric power

~, номина́ль·н·ая nominal/rated thrust

~, обра́т·н·ая reverse thrust; back draft/draught

~, объ·ём·н·ая (rocket) density thrust

~, пере·во́д·н·ая (rail) reversing rod

~, по·пере́ч·н·ая (M/T) track rod

~, при·вед·ённ·ая (a/c) standard thrust

~, рас·по́р·н·ая supporting rod

регул·я́тор· тя́г·и (*m*) damper

~, рул·ев·а́я (M/T) drag link, steering rod

~, с·тя́ж·н·а́я tie rod

тя́г·а

~ с·цепл·е́ни·я tow bar

~, тра́ктор·н·ая tractor traction
    на тра́ктор·н·ой тя́г·е tractor-drawn

~, у·де́ль·н·ая specific thrust

~, у·де́ль·н·ая лоб·ов·а́я (a/c) thrust/frontal-area ratio

~ у·правл·е́ни·я control rod/link

~, у·равно·ве́ш·енн·ая balanced draught (boiler)

~, экцентрик·ов·ая eccentric rod

тяг·а́ч·, -а́, (i) **-о́м,** (m) towing vehicle, tractor

~, вое́н·н·ый military tractor

~, тра́нспорт·н·ый cargo tractor

тяго·во·оруж·ённост·ь, -и, (f) (a/c, rocket) thrust-weight ratio

тя́г·ов·ый, -ая, -ое, (a) tractive, traction, draw; draft, draught, drawing

~ кана́т· (m) traction/traversing rope (of cableway)

~ крюк· (m) (rail) drawbar hook

~ рессо́р·а (f) draw spring

~ у·си́л·и·е (n) tractive/towing force/effort, thrust, pull

тяго·ду́ть·ев·ый, -ая, -ое, (a) draught, draft

тяго·ме́р·, -а, (m) draught/draft gauge

~, мембра́н·н·ый diaphragm vacuum gauge/gage

тягот·а́, -ы́, (nom pl) **тя́гот·ы,** (g pl) **тя́гот·,** (f) weight, burden (figurative)

тягот·е́ни·е, -я, (n) (phys) gravitation, gravity; attraction, inclination

~, все·ми́р·н·ое universal gravitation

~, иску́с·ственн·ое artificial gravity
    коэффицие́нт· тяг·оте́ни·я (m) gravity constant
    си́л·а тяг·оте́ни·я (f) attractive/attraction force

тягот·е́·ть, -ют, (imp) gravitate; be attracted; loom/hang over

тягот·и́ть, -я́т, (imp) burden, be a burden to

тяг·ун·, -а, (m) leer, lehr, lear (glass-making)

тяг·у́чест·ь, -и, (f) ductility, elasticity; viscosity, thickness

тяг·у́ч·ий, -ая, -ее, (a) ductile, elastic; viscous, viscid, thick (of liquids)

тяж·, -а́, (i) **-о́м,** (m) driving band/rope/cord; tie rod; (biol) cord, strand

тяжел·е́е, (comp) of **тяжёл·ый**

тяжел·е́·ть, -ют, (imp) become/grow heavy

тяжел·и́ть, -я́т, (imp) make heavy/heavier

тяжело·ве́с·н·ый, -ая, -ое, (a) heavy; ponderous; overloaded

тяжело·во́д·н·ый, -ая, -ое, (a) heavy-water, deuterium oxide

тяжело·во́з·, -а, (m) draft horse, cart horse

тяжело·гру́з·н·ый, -ая, -ое, (a) heavy-cargo/freight/load (of ships, vehicles etc.)

тяжело·хо́д·ы, -ов, (m pl) (zool) gravigrades, *Gravigrada*

тяжёл·ый, -ая, -ое, (a) heavy; difficult, burdensome, hard; serious, grave

~ вод·а́ (f) (chem) heavy water, deuterium oxide

~ водо·ро́д· (m) (chem) heavy hydrogen, deuterium

~ на нос· (a/c) nose-heavy

~ на хвост· (a/c) tail-heavy

~ рабо́т·а (f) hard work, (M/T etc.) heavy duty

тя́ж·ест·ь, -и, (f) weight, heaviness, load; gravity
    центр· тя́ж·ест·и (m) centre of gravity

тя́ж·к·а, -и, (f) staking, scouring, drafting (leather)

тя·ну́льн·ый, -ая, -ое, (a) stretching

~ маши́н·а (f) staking machine (leather)

тя·ну́т·ый, -ая, -ое, (a) (*past part pass*) of **тя·н·у́ть;** (met) drawn

тя·н·у́ть, -ут, (imp) pull, draw, drag, haul; draw, draw out/down, stretch; procrastinate, delay, prolong, drag on/out; **-ся** (*pass*); stretch, extend, range, continue; strive; follow (of springs)

тя́·нущ·ий, -ая, -ее, (*pres part act*) of **тя·н·у́ть**

~ винт· (m) tractor propeller

~ клет·ь (f) (met) withdrawing rolls (continuous casting)
    при·во́д· тя́·нущ·их клет·е́й (m) roll drives (continuous casting)

тя́п·а·ть, -ют, (imp) chop, hack; hoe

тя́п·к·а, -и, (g pl) **-п·ок·,** (f) chopper; hoe

У

у (*prep + gen*) at, by, near, close to, with; of, belonging to

~ **вла́ст·и** in power/office

~ **меня́** I have; at my place/house, with me

у- (*verbal prefix*) *signifies:* — (1) movement away from, right away from; (2) removal of part, reduction of quantity; (3) action embracing whole object; (4) achievement of required result, completion in spite of difficulties; (5) firmness, decisiveness of action; (6) inclusion, disposal within certain limits; (7) acquisition of new or heightened quality as a result of action; (8) perfective aspect of some verbs

уайт-спири́т·, -а, (*m*) (chem) white spirit, turpentine substitute

уа́тт·, -а, (*m*) *see* **ватт·**

у·ба́в·ить, -ят, (*perf*) *see* **у·бавл·я́·ть**

у·бавл·е́ни·е, -я, (*n*) (*v n*) *see* **у·бавл·я́·ть;** reduction, lessening, abatement

у·бавл·я́·ть, -ют, (*imp*) reduce, decrease, diminish, lessen, shorten, narrow

у·бег·а́·ть, -ют, (*imp*) run away; escape; run over (of boiling liquids); rush/fly past (of stationary objects seen by moving viewer)

у·бег·а́юш·ий, -ая, -ее, (*pres part act*) of **у·бег·а́·ть;** outgoing

у·бед·и́тельн·ый, -ая, -ое, (*a*) convincing, persuasive; earnest, pressing, strong

у·бед·и́ть, -ят, (*perf*) *see* **у·бежд·а́·ть**

у·беж·а́·ть, (*fut 3rd sing, pl*) **у·беж·и́т, у·бег·у́т,** (*perf*) *see* **у·бег·а́·ть**

у·бежд·а́·ть, -ют, (*imp*) convince; persuade; **-ся** (*pass*); make sure

у·бежд·е́ни·е, -я, (*n*) (*v n*) *see* **у·бежд·а́·ть:** conviction, belief; persuasion

у·бе́ж·иш·е, -а, (*n*) refuge, shelter; (law) asylum

~, зи́м·н·ее (zool) hibernaculum

у·бе́й·те, (*pl imper*) of **у·би́·ть** (*perf*) *see* **у·бив·а́·ть** (*imp*)

у·бел·ённ·ый, -ая, -ое, (*past part pass*) *see* **у·бел·я́·ть**

у·бел·и́ть, -ят, (*perf*) *see* **у·бел·я́·ть**

у·бел·я́·ть, -ют, (*imp*) make white, whiten (completely), bleach

у·берег·а́·ть, -ют, (*imp*) safeguard, guard against, preserve/protect from

у·бер·ёт, (*fut 3rd sing*) of **у·бр·а́ть** (*perf*) *see* **у·бир·а́·ть** (*imp*)

у·бере́ч·ь, (*fut 3rd sing, pl*) **у·береж·ёт, у·берег·у́т** (*past masc sing*) **у·берёг** (*perf*) *see* **у·берег·а́·ть**

у·бив·а́·ть, -ют, (*imp*) kill, slay, slaughter; murder; assassinate; beat, compact

у·бив·а́юш·ий, -ая, -ее, (*pres part act*) of **у·бив·а́·ть;** (*component*) -cidal

~ вред·и́тел·ей сре́д·ств·о (*n*) pesticide

у·би́й·ств·о, -а, (*n*) (*v n*), *see* **у·бив·а́·ть;** murder; assassination

~, не·пред·у·мы́шл·енн·ое (law) manslaughter

у·би́й·ц·а, -ы, (*i*) **-ей,** (*m or f*) murderer; assassin

у·бир·а́ем·ый, -ая, -ое, (*pres part pass*) of **у·бир·а́·ть;** retractable, retractive

у·бир·а́·ть, -ют, (*imp*) take away, remove; clear/tidy away/up; stow, stow/put away; throttle back (a/c engine); gather (crops), harvest; decorate, adorn; **-ся** (*pass and reflex*); clear off, beat it, scram

у·бир·а́юш·ийся, -аяся, -ееся, (*pres part act*) *see* **у·бир·а́·ть;** retractable, retractive, folding

~ шасси́ (*n*) (a/c) retractable undercarriage

у·би́·т·ый, -ая, -ое, (*past part pass*) *see* **у·бив·а́·ть;** crushed, depressed (of emotions); beaten, compacted; (*m decl as adj*) killed, dead

у·би́·ть, (*fut 3rd pl*) **у·бь·ю́т,** (*perf*) *see* **у·бив·а́·ть**

у·блю́д·ков·ый, -ая, -ое, (*a*) hybrid

у·блю́д·ок·, -д·к·а, (*m*) (biol) hybrid

у·блю́д·очн·ый, -ая, -ое, (*a*) hybrid, mongrel

у·бо́г·ий, -ая, -ое, (*a*) wretched, miserable

у·бо́·й, -я, (*m*) (agr) slaughter, killing

у·бо́й·ност·ь, -и, (*f*) (mil) destructive/ killing power

у·бо́й·н·ый, -ая, -ое, (*a*) slaughter, killing; intended for slaughter; (mil) lethal

~ вес· (*m*) (food) dressed carcass weight

~ интерва́л· (*m*) lethal radius

~ си́л·а (*f*) destructive power, killing power

~ скот· (*m*) cattle for slaughter

у·бо́р·, -а, (*m*) (obs) dress, garment

~, голов·н·о́й   head-dress, headgear, hats and caps, hats

у·бо́р·ист·ый, -ая, -ое, (*a*) (print) close

у·бо́р·к·а, -и, (*g pl*) **-р·ок·,** (*f*) (*v n*) *see* **у·бир·а́·ть;** harvest time, harvest; retraction, retracting (of a/c undercarriage)

~ льн·а flax pulling

~ по по·то́к·у (a/c) retracting backward

~ по·ро́д·ы (min) mucking

~ хло́п·к·а cotton picking

~ ча́·я tea picking

у·бо́р·н·ая, -ой, (*f decl as adj*) lavatory, W.C.; (theat) dressing room

у·бо́р·очн·ый, -ая, -ое, (*a*) of **у·бо́р··к·а;** harvesting

~ маши́н·а (*f*) harvesting machine, harvester, picker, collector

~ маши́н·а, под·мета́ль·н·о- (М/Т) road sweeper-collector

у·бо́р·щик·, -а, (*m*) cleaner; waste remover; harvester, picker

у́·бр·анн·ый, -ая, -ое, (*past part pass*) *see* **у·бир·а́·ть**

у·бр·а́нств·о, -а, (*n*) decoration, furnishing, appointment(s)

у·бр·а́ть, (*fut 3rd pl*) **у·бер·у́т,** (*perf*) *see* **у·бир·а́·ть**

у·бу́д·ет, (*fut 3rd sing*) of **у·бы́·ть** (*perf*) *see* **у·быв·а́·ть**

у·быв·а́ни·е, -я, (*n*) (*v n*) of **у·бы-в·а́·ть;** decrease, reduction

у·быв·а́·ть, -ют, (*imp*) diminish, decrease, lessen, decline, subside, go down; leave, go away, be removed from the strength

у·быв·а́ющ·ий, -ая, -ее, (*pres part act*) of **у·быв·а́·ть;** decaying (oscillations etc.)

~ по·сле́д·овательн·ост·ь (*f*) descending order/sequence

у́·бы·л, (*past masc sing*) of **у·бы́·ть** (*perf*) *see* **у·быв·а́·ть** (*imp*)

у́·бы·л·ь, -и, (*f*) reduction, decrease, subsidence; loss, losses

у·бы́т·ок·, -т·к·а, (*m*) loss

у·бы́т·очн·ый, -ая, -ое, (*a*) unprofitable, losing, loss-incurring

у·бы́·ть, (*fut 3rd pl*) **у·бу́д·ут,** (*perf*) *see* **у·быв·а́·ть**

у·бь·ёт, (*fut 3rd sing*) of **у·би́·ть** (*perf*) *see* **у·бив·а́·ть** (*imp*)

у·важ·а́ем·ый, -ая, -ое, (*pres part pass*) of **у·важ·а́·ть;** respected, esteemed; dear (*in correspondence*)

у·важ·а́·ть, -ют, (*imp*) respect, esteem

у·важ·е́ни·е, -я, (*n*) respect, esteem

у·важ·и́тельн·ый, -ая, -ое, (*a*) worthy; valid

у·ва́л·, -а, (*m*) (text) fulling; (geog) ridge, spur; rut (in road etc.)

у·ва́л·ивани·е, -я, (*n*) (*v n*) of **у·ва́-л·ива·ть** (text) fulling

у·ва́л·ива·ть, -ют, (*imp*) roll over, roll; stream (of people); (text) full

у·ва́л·ива·ться под ве́тер· (naut) drop to leeward

у·вал·и́ть, ⁻ят, (*perf*) roll over, roll; stream (of people)

у·вал·и́ться под ве́тер· (naut) drop to leeward

у·ва́ль·н·ый, -ая, -ое, (*a*) (geog) ridged

у·ва́ль·чивост·ь, -и, (*f*) (naut) lee helm

у·ва́ль·чив·ый, -ая, -ое, (*a*) (naut) carrying lee helm

у·вал·я́·ть, -ют, (*perf*) (text) full; roll (food etc.)

увани́т·, -а, (*m*) (min) uvanite

у·ва́р·, -а, (*n*) (*v n*) *see* **у·ва́р·ива·ть;** loss on boiling

у·ва́р·енн·ый, -ая, -ое, (*past part pass*) *see* **у·ва́р·ива·ть;** concentrated, reduced

у·ва́р·ива·ть, -ют, (*imp*) boil down, concentrate/reduce by boiling; (food) boil

у·вар·и́ть, ⁻ят, (*perf*) *see* **у·ва́р·ива·ть**

уваровит·, -а, (*m*) (min) uvarovite, uwarowite, ouvarovite

у·вед·ёт, (*fut 3rd sing*) of **у·вес·ти́** (*perf*) *see* **у·вод·и́ть** (*imp*)

у·ведом·и́тельн·ый, -ая, -ое, (*a*) informative, notifying, advisatory

~ письм·о́ (*n*) (com) letter of advice

у·ве́дом·ить, -ят, (*perf*) *see* **у·ведом-л·я́·ть**

у·ведомл·е́ни·е, -я, (*n*) (*v n*) of **у·ве-домл·я́·ть;** notification, advice, note of advice

у·ведомл·я́·ть, -ют, (*imp*) inform, notify, advise

у·вез·ти́, -у́т, (*perf*) *see* **у·воз·и́ть**

у·ве́й·те, (*pl imper*) of **у·ви́·ть** (*perf*) *see* **у·вив·а́·ть**

у·веко·ве́ч·ени·е, -я, (*n*) perpetuation; immortalization

у·вё·л, (*past masc sing*) of **у·вес·ти́** (*perf*) *see* **у·вод·и́ть** (*imp*)

у·велич·е́ни·е, -я, (*n*) (*v n*) *see* **у·ве-ли́ч·ива·ть;** increase, extension, augmentation; (phot) enlargement; (opt) magnification

у·вели́ч·енн·ый, -ая, -ое, (*past part pass*) *see* **у·вели́ч·ива·ть**

у·вели́ч·ивател·ь, -я, (*m*) increaser, intensifier, augmented

~ **ли́п·кост·и** (rubber) tackifier

у·вели́ч·ива·ть, -ют, (*imp*) increase, extend, augment, enhance; magnify, enlarge

~ **жизн·ь** extend the life (of equipment)

~ **у́гол· о·переж·éни·я за·жиг·á·ни·я** (ICE) advance the ignition

у·велич·и́тел·ь, -я, (*m*) (phot) enlarger

у·велич·и́тельн·ый, -ая, -ое, (*a*) enlarging, magnifying; augmentative

у·вели́ч·ить, -ат, (*perf*) *see* **у·вели́ч·ива·ть**

у·венч·á·ть, -ют, (*imp*) crown, cap, top

у·венч·á·ться у·спéх·ом succeed, be crowned with success

у·вер·éни·е, -я, (*n*) (*v n*) *see* **у·вер·я́·ть;** assurance

у·вéр·енност·ь, -и, (*f*) certainty, confidence

у·вéр·енн·ый, -ая, -ое (*past part pass*) *see* **у·вер·я́·ть;** confident, sure, assured, firm, unhesitating

у·вéр·ить, -ят, (*perf*) *see* **у·вер·я́·ть**

у·вéрт·к·а, -и, (*g pl*) **-т·ок·,** (*f*) subterfuge, evasion, shift

у·вер·я́·ть, -ют, (*imp*) assure, persuade, convince

у·весел·и́тельн·ый, -ая, -ое, (*a*) pleasure, entertainment

у·весел·и́ть, -ят, (*perf*) *see* **у·весе·л·я́·ть**

у·весел·я́·ть, -ют, (*imp*) cheer up, amuse, entertain

у·вéс·ист·ый, -ая, -ое, (*a*) weighty

у·вес·ти́, (*fut 3rd pl*) **у·вед·у́т,** (*past masc sing*) **у·вё·л,** (*perf*) *see* **у·вод·и́ть**

увéч·ь·е, -я, (*g pl*) **-ч·и·й,** (*n*) mutilation, disabling injury

увéч·ить, -ат, (*perf*) mutilate, maim, cripple

у·вéш·а·ть, -ют, (*perf*) *see* **у·вéш·и·ва·ть**

у·вéш·ива·ть, -ют, (*imp*) hang out, hang/suspend all over, festoon

у·вив·á·ть, -ют, (*imp*) twist/wind round/onto

у·ви́д·еть, -ят, (*perf*) see, catch sight of; become aware of, perceive, see, understand, realize

у·ви́ж·у, (*fut 1st sing*) of **у·ви́д·еть**

увио́л·ев·ый, -ая, -ое, (*a*) uviol, ultra-violet

у·ви́·ть, (*fut 3rd pl*) **у·вь·ю́т,** (*perf*) *see* **у·вив·á·ть**

у·влажн·éни·е, -я, (*n*) (*v n*) *see* **у·-влажн·я́·ть**

у·влажн·ённ·ый, -ая, -ое, (*past part pass*) *see* **у·влажн·я́·ть;** wet-bulb (of thermometers)

у·влажн·и́тел·ь, -я, (*m*) humidifier

у·влажн·и́ть, -ят, (*perf*) *see* **у·влаж-н·я́·ть**

у·влажн·я́·ть, -ют, (*imp*) moisten, damp, wet, humidify; make moister/damper/wetter; condition (by moistening)

у·влек·á·тельн·ый, -ая, -ое, (*a*) absorbing, fascinating, interesting

у·влек·á·ть, -ют, (*imp*) drag, drag/carry away/off, entrain; fascinate, captivate, absorb; **-ся** (+ *instr*) become/be absorbed in, become/be fascinated with/by

у·влеч·éни·е, -я, (*n*) (*v n*) *see* **у·влек·á·ть;** enthusiasm, animation; entrainment, drag

у·влеч·ённ·ый, -ая, -ое, (*past part pass*) *see* **у·влек·á·ть**

у·влéч·ь, (*fut 3rd sing, pl*) **у·влеч·ёт, у·влек·у́т,** (*past masc sing*) **у·влёк·** (*perf*) *see* **у·влек·á·ть**

у·во́д·, -а, (*m*) (*v n*) *see* **у·вод·и́ть;** withdrawal; drift, wandering, displacement; (a/c) tilt (of propeller blades)

у́гол· у·во́д·а (*m*) drift angle (of wheels)

у·вод·и́ть, -я́т, (*imp*) lead/take away/off, withdraw

у·воз·и́ть, -я́т, (*imp*) ship, dispatch, send away/off; carry/take away/off

у·вола́к·ива·ть, -ют, (*imp*) drag/draw away/off

у·во́л·ить, -ят, (*perf*) *see* **у·воль·ня́·ть**

у·волóк·, (*past masc sing*) of **у·волóч·ь** (*perf*) *see* **у·вола́к·ива·ть** (*imp*)

у·волóч·ь, (*fut 3rd sing, pl*) **у·волоч·ёт, у·волок·у́т,** (*past masc sing*) **у·волóк·** *see* **у·вола́к·ива·ть**

у·воль·нéни·е, -я, (*n*) (*v n*) *see* **у·воль·-ня́·ть;** retirement, discharge, dismissal, release (from employment)

у·воль·ни́тельн·ый, -ая, -ое, (*a*) discharge, release

у·воль·ня́·ть, -ют, (*imp*) retire, discharge, dismiss; lay off (from work temporarily); **-ся** (*pass*); retire, leave (employment), (mil) obtain discharge

~ **в за·пáс·** transfer to the reserve

у·воль·ня́·ть
~ **в от·ста́в·к·у**   retire
~ **на бе́рег·**   (naut) give shore leave, give liberty
увч (abbr) = **ультра·выс·о́к·ая час·т·от·а́**   ultra-high frequency, VHF
у·вь·ёт, (fut 3rd sing) of **у·ви́·ть** (perf) see **у·вив·а́·ть** (imp)
у·вяд·а́ни·е, -я, (n) (v n) see **у·вя·д·а́·ть;** wilt (plant disease)
~, **пол·ов·о́е**   (zool) climacteric
у·вяд·а́·ть, -ют, (imp) wither, wilt, fade
у·вяз·а́ть, (fut 3rd pl) **у·вя́ж·ут,** (perf) see **у·вя́з·ыва·ть; у·вяз·а́·ть, -ют,** (imp) bog down, become/get stuck
у·вя́з·к·а, -и, (g pl) **-з·ок·,** (f) (v n) see **у·вя́з·ыва·ть;** correlation, co-ordination, interrelationship
у·вя́з·нуть, -нут, (perf) bog down, become stuck
у·вя́з·ыва·ть, -ют, (imp) tie up, bind, bundle, strap up; correlate, establish the connection/link, connect, link, fit in, tie up
у·вя́·л·ый, -ая, -ое, (a) withered, wilted, faded
у·вя́·нуть, -нут, (perf) see **у·вяд·а́·ть**
у·гад·а́·ть, -ют, (perf) see **у·га́д·ы·ва·ть**
у·га́д·ыва·ть, -ют, (imp) guess, conjecture, divine; make out, distinguish
у·га́р·, -а, (m) loss, waste (in industrial processes); (met) loss on ignition; (ICE) carbon monoxide fumes; carbon monoxide poisoning
у·га́р·н·ый, -ая, -ое, (a) of **у·га́р·**
~ **пряд·е́ни·е** (n) (text) waste/vignogne spinning
у·гас·а́·ть, -ют, (imp) go out, die down/out, become extinct
у·гас·и́ть, ´-ят, (perf) see **у·гаш·а́·ть**
у·га́с·нуть, -нут, (perf) see **у·гас·а́·ть**
у·га́с·ш·ий, -ая, -ее, (pres part pass) see **у·гас·а́·ть;** extinct
у·гаш·а́·ть, -ют, (imp) extinguish, put out
у·га́ш·енн·ый, -ая, -ое, (past part pass) see **у·гаш·а́·ть**
у́гле- (root) carbon; coal
угле·ба́р·иев·ая сол·ь (f) (chem) barium carbonate
угле·во́д·, -а, (m) (chem) carbohydrate
угле·во́д·н·ый, -ая, -ое, (a) carbohydrate
угле·водо·ро́д·, -а, (m) (chem) hydro-carbon

угле·водо·ро́д
~, **жи́р·н·ый** aliphatic hydrocarbon
~, **нафте́н·ов·ый** naphthene
~, **не·на·сы́щ·енн·ый** unsaturated hydrocarbon
~, **не·пре·де́ль·н·ый** unsaturated hydrocarbon
~, **парафи́н·н·ый** paraffin
~, **поли·метиле́н·ов·ый** polymethylene hydrocarbon
угле·водо·ро́д·ист·ый, -ая, -ое, (a) of **угле·водо·ро́д·**
угле·во́з·, -а, (m) (naut) collier, coal ship
угле·вы́·жиг·а́тельн·ая печ·ь (f) charcoal kiln
угле·держ·а́тел·ь, -я, (m) (elec) carbon holder
угле·до·бы́·ч·а, -и, (f) coal winning/extraction/mining
угле·дроб·и́лк·а, -и, (g pl) **-лок·,** (f) coal crusher
угле·жж·е́ни·е, -я, (n) charcoal burning
угле·жо́г·, -а, (m) charcoal burner
угле·кисл·от·а́, -ы́, (f) (chem) carbonic acid
~, **твёрд·ая** dry ice
угле·ки́сл·ый, -ая, -ое, (a) carbonate (of)
~ **аммо́н·и·й,** (m) ammonium carbonate
~ **вод·а́** (f) carbonated water
~ **газ·** (m) carbonic acid gas, carbon dioxide
~ **ка́льц·и·й** (m) calcium carbonate
~ **мед·ь** (f) cuprous carbonate
~ **ртут·ь** (f) mercurous carbonate
угле·мо́й·к·а, -и, (g pl) **-о·ек·,** (f) coal washing plant, coal washery
угле·но́с·ност·ь, -и, (f) presence of coal; concentration (of coalfield)
коэффицие́нт· угле·но́с·ност·и (m) carbon ratio
угле·но́с·н·ый, -ая, -ое, (a) coal-bearing, carboniferous
угле·пло́т·ност·ь, -и, (f) coal density, amount of coal per km^2 (of coal field)
угле·по·да́т·чик·, механ·и́ческ·ий stoker, mechanical stoker
угле·по·да́·ч·а, -и, (f) coal handling
угле·про·мы́шл·енн·ый, -ая, -ое, (a) coal-mining, coal-industry, coal-industry's
угле·раз·мо́ль·н·ая у·стано́в·к·а (f) coal pulverizer
угле·ро́д·, -а, (m) (chem) carbon, C

угле·ро́д

~, дву·сер·ни́ст·ый carbon disulfide/bisulfide

~ карби́д·а (met) combined carbon

~, одно·сер·ни́ст·ый carbon monosulphide

~, свобо́д·н·ый free carbon

~, с·вя́з·анн·ый combined/fixed carbon

~, четырёх·бро́м·ист·ый tetrabromomethane

~, четырёх·хло́р·ист·ый carbon tetrachloride

угле·ро́д·ист·ый, -ая, -ое, (*a*) carbon, carbonic, carbonaceous, carboniferous

~ ста́л·ь (*f*) carbon steel

угле·ро́д·н·ый, -ая, -ое, (*a*) carbon, carbonic

угле·со·держ·а́щ·ий, -ая, -ее, (*a*) carbonaceous; coal-bearing

у́гл·ист·ый, -ая, -ое, (*a*) carbonaceous, coaly

~ желез·ня́к· (*m*) (min) blackband ironstone

у́гло- (*root*) angle, gonio-

угло·бу́льб·, -а, (*m*) (met) bulb angle

угло·бу́льб·ов·ая ста́л·ь (*f*) bulb-angle steel

углова́то·ли́ст·н·ый, -ая, -ое, (*a*) (bot) angular-leaved, goniophyllous

углова́то·не́рв·н·ый, -ая, -ое, (*a*) (bot) angular-nerved, angulinervate

углова́то·пло́д·н·ый, -ая, -ое, (*a*) (bot) angular-fruited, goniocarpous

угл·ова́тост·ь, -и, (*f*) angularity

угл·ова́т·ый, -ая, -ое, (*a*) angular, bumpy

угл·ов·о́й, -а́я, -о́е, (*a*) angular, angled, corner

~ кла́пан· (*m*) angle valve

~ коэффицие́нт· (*m*) gradient, slope

~ лека́л·о (*n*) (build) angle gauge/gage

~ ме́р·а (*f*) (mech eng) angle gauge/gage

~ раз·реш·е́ни·е (*n*) (opt) angular resolution

~ со·един·е́ни·е (*n*) corner joint

~ ста́л·ь (*f*) angle steel, steel angle bars, steel angles

угло·зу́б·ые, -ых, (*pl decl as adj*) (zool) Asiatic salamanders, *Hynobiidae*

угло·ме́р·, -а, (*m*) protractor; (mech eng) protractor, bevel protractor; (surv, astron) azimuth/angular instrument; goniometer (in general sense)

~ -транспорт·и́р·, -а, (*m*) bevel protractor

угло·ме́р·н·ый, -ая, -ое, (*a*) azimuth, azimuthal, goniometric

~ инструме́нт· (*m*) azimuthal/angular instrument, goniometer

~ круг· (*m*) azimuth scale plate

угло·ме́ст·н·ый, -ая, -ое, (*a*) elevation, angle-of-elevation

~ реги́стр· (*m*) elevation register (radar)

угло·с·прямл·е́ни·е, -я, (*n*) squaring

у·глуб·и́ть, -я́т, (*perf*) *see* **у·глубл·я́ть**

у·глу́б·к·а, -и, (*f*) deepening, sinking

у·глубл·е́ни·е, -я, (*n*) (*v n*) *see* **у·глубл·я́ть**; deepening, depression, recess, cavity, hollow, fossa; draught (of ship)

~ модул·я́ци·я ве́рх·н·их част·о́т· (rad) pre-emphasis/pre-accentuation

у·глубл·я́ть, -ют, (*imp*) deepen, make deeper; sink (piles etc.), drive in deeper; intensify, make more profound; **-ся** (*pass*); go deep/deeper, go down; become deeper; become more intense/profound/absorbed

~ дн·о (naut) dredge

у·гн·а́ть, (*fut 3rd pl*) **у·го́н·ят,** (*perf*) drive away/off/to; keep up behind, not lag

у·гнезд·и́ться, -я́тся, (*perf*) be lodged, lodge oneself; make oneself comfortable, settle down

у·гнё·л, (*past masc sing*) of **у·гнес·ти́** (*perf*) *see* **у·гнет·а́·ть**

у·гнес·ти́, (*fut 3rd pl*) **у·гнет·у́т,** (*perf*) *see* **у·гнет·а́·ть**

у·гнет·а́·ть, -ют, (*imp*) oppress, (med) depress

у·гнет·е́ни·е, -я, (*n*) oppression, oppressing; (med) depression

у·гова́р·ива·ть, -ют, (*imp*) persuade; **-ся** agree, arrange

у·говор·и́ть, -я́т, (*perf*) *see* **у·гова́р·ива·ть**

у·год·и́ть, -ят, (*perf*) *see* **у·гожд·а́·ть**

у·го́д·н·ый, -ая, -ое, (*a*) pleasing, satisfactory

у·го́д·ь·е, -я, (*n*) land (usable for agriculture); area (usable for fishing, trapping etc.)

~, сельско·хозя́й·ственн·ое agricultural land

~, о·хо́т·нич·ье hunting ground

~, ры́б·н·ое fishing water

у·гожд·а́·ть, -ют, (*imp*) please, satisfy

у́гол·, (*g sing*) **угл·а́** (*m*) (*see also under associated nouns*) angle; corner

~ ата́к·и (dynam) angle of attack

у́гол

~ ата́к·и, до·крит·и́ческ·ий (aerodynam) angle of incidence below stalling

~ ата́к·и, за·крит·и́ческ·ий (aerodynam) angle of incidence beyond stalling

~ ата́к·и, крит·и́ческ·ий (aerodynam) angle of stall

~ ата́к·и, с·ры́в·н·о́й (dynam) angle of stall

~, баланс·иро́вочн·ый (dynam) trim angle

~, больш·о́й large/wide angle

~, вертика́ль·н·ый vertical angle; (surv) angle of elevation

~, верш·и́нн·ый (math) vertex/vertical angle

~, вне́ш·н·ий exterior angle

~, вну́тр·енн·ий interior angle; included angle

~ воз·мущ·е́ни·й Mach angle

~ в·перёд· (mil) arrowhead

~, вс·по·мог·а́тельн·ый за́д·н·ий end-relief angle (of cutting tool)

~, в·ход·я́щ·ий (math) re-entrant angle

~, гла́в·н·ый за́д·н·ий side-relief angle (of cutting tool)

~, голов·н·о́й cap angle (of crystal)

~, груд·и́нн·ый (anat) sternal angle

~ давл·е́ни·я pressure/flank angle (gears)

~, дирекцио́н·н·ый (nav) grid bearing

~, до·полн·и́тельн·ый (math) complementary angle

~, до·полн·и́тельн·ый до 360° (math) conjugate/explementary angle

~, за́д·н·ий back angle (cutting tools), clearance angle

~, за·щи́т·н·ый angle of cut-off (illumination)

~, координа́т·н·ый quadrant, quadrant angle

~, кра·ев·о́й (phys) angle of contact (surface tension)

~, крит·и́ческ·ий critical angle; (met roll) no-slip angle

~, курс·ов·о́й (nav) relative bearing (of ship); heading (of a/c)

~, лин·е́йн·ый (math) angle between lines

~, мёртв·ый dead pocket (in. water jacket)

~, мно́го·гра́н·н·ый polyhedral angle

~ на·клон·а траекто́ри·и flight path angle

у́гол

~, на·кре́ст·леж·а́щ·ий (math) alternate angle

~, нейтра́ль·н·ый (met roll) no-slip angle

~ нул·ев·о́й подъ·ём·н·ой си́л·ы (dynam) zero-lift angle

~, о́стр·ый acute angle

~, от·нос·и́тельн·ый reference angle

~ о·хва́т·а inclusion angle

~, па́з·уш·н·ый (bot) axil

~, параллакт·и́ческ·ий (nav) parallax in altitude

~, пере́д·н·ий cutting angle (cutting tools)

~, пере·ме́н·н·ый alternate angle

~, пло́ск·ий plane angle

под угл·о́м at an angle, through an angle

~, полови́н·н·ый half-angle

~ по·те́р·и ско́р·ост·и (dynam) stall angle

~, пре·де́ль·н·ый (mech) critical angle, limiting angle

~, при·леж·а́щ·ий supplementary angle

~ при·тек·а́ни·я inflow angle

~, про·стра́н·ственн·ый solid angle

~, противо·леж·а́щ·ий alternate angle

~, прям·о́й right angle

~, пут·ев·о́й (nav) track angle

~, рабо́ч·ий operating angle

~, ради́о·курс·ов·о́й (nav) relative bearing by D/F

~ рас·кры́в·а flare angle

~ с норма́л·ью angle of incidence (ballistics)

~, скольз·я́щ·ий (astron) grazing angle, glancing angle

~, с·ме́ж·н·ый supplementary angle

~ с·но́с·а drift angle

спо́соб· тр·ёх угл·о́в с·но́с·а (m) (air) multiple drift method (of finding wind velocity)

~, с·ре́з·анн·ый truncated angle

~, тел·е́сн·ый (math) solid angle

~, туп·о́й obtuse angle

~, час·ов·о́й (astron) hour angle

угол·ко́в·ый, -ая, -ое, (a) of **угол·о́к·**

~ манже́т·а (f) cup packing

у·голо́в·н·ый, -ая, -ое, (a) criminal

угол·о́к·, -л·к·а́, (m) (dim) of **у́гол·;** lecture and display room

~, не·равно·бо́к·ий (met) unequal angle (bar)

~, равно·бо́к·ий (met) equal angle (bar)

у́гол·ь, (*m*) у́гл·я, (*nom pl*) у́гл·и, or у́гол·ь·я, (*g pl*) у́гл·ей or у́го-л·ь·ев coal; charcoal; carbon

~, актив·и́рованн·ый (chem) activated carbon

~, бес·пла́мен·н·ый blind/anthracite coal

~, битуминóз·н·ый bituminous coal

~, бóгхед·н·ый boghead coal

~, бу́р·ый brown coal, lignite

~, бес·сéр·н·ый sweet coal

~, вон·ю́ч·ий stink coal

~, гáз·ов·ый gas coal

~, длинно·пла́мен·н·ый long-flame coal

~, дóмен·н·ый furnace coal

~, дуг·ов·óй arc carbon

~, живóт·н·ый animal charcoal

~, жи́р·н·ый fat coal

~, ис·коп·áем·ый coal, mineral coal

~, ка́мен·н·ый coal, hard coal

~, кéннел·ев·ый cannel coal

~, кóкс·ов·ый coking coal

~, коротко·пла́мен·н·ый short-flame coal

~, кост·ян·óй bone charcoal/black

~, куз·нéчн·ый forge coal

~, мéл·к·ий small coal

~, от·риц·а́тельн·ый negative carbon (of arc)

~, паровично·жи́р·н·ый fat steam coal

~, пар·ови́чн·ый steam coal

~, ряд·ов·óй run-of-mine coal

~, с·пек·а́ющ·ий caking coal

~, стрел·я́ющ·ий explosive coal

~, тóщ·ий lean coal

~, электро·техн·и́ческ·ий electrical carbon

угóль·ник·, -а, (*m*) angle bar, angle iron, angle; pipe bend, elbow; angle bracket; (build etc.) try-square, set-square (drawing)

~, вéрх·н·ий кил·ев·óй (shipb) continuous top angle

~, вéрх·н·ий флóр·н·ый (shipb) reverse frame

~, дн·и́щев·óй (shipb) double bottom frame

~, не·равно·бóк·ий unequal angle bar

~, ни́ж·н·ий кил·ев·óй (shipb) continuous bottom angle

~, ни́ж·н·ий флóр·н·ый frame (of ship's floor)

~ па́луб·н·ого стри́нгер·а (shipb) gunwhale bar

угóль·ник

~ пере·бóр·к·и, об·дéл·очн·ый (shipb) bulkhead boundary angle

~, равно·бóк·ий equal angle bar

~, стри́нгер·н·ый (shipb) shell/stringer bar

угольно·мети́л·ов·ый эфи́р· (*m*) methyl carbonate

у́голь·н·ый, -ая, -ое, (*a*) coal, carbon, carbonic; у́голь·н·ый angle, angular, corner; (*as component*) -gonal

~ ангидри́д· (*m*) carbonic anhydride, carbon dioxide

~ бассéйн· (*m*) (geol) coalfield

~ кисл·от·á (*f*) (chem) carbonic acid

~ комбáйн· (*m*) (min) combine coal cutter-loader

~ ла́мп·а (*f*) carbon filament lamp

~ мéл·оч·ь (*f*) fine coal, slack

~ регул·я́тор· на·пряж·éни·я (*m*) carbon-pile voltage regulator

~ со·противл·éни·е (*n*) (elec) carbon rheostat

~ струг· (*m*) (min) coal stripper

~ у·плот·нéни·е (*n*) (st eng) carbon gland

у́голь·щик·, -а, (*m*) coal miner; coal heaver; coalship, collier

у·гóн·, -а, (*m*) (*v n*) *see* у·гон·я́·ть; (rail) creep (of rails)

у·гóн·ят, (*fut 3rd pl*) of у·гн·а́ть

у·гон·я́·ть, -ют, (*imp*) drive away/ off/to; steal, rustle

у·гóр·, -а, (*m*) steep river bank

у·гор·а́·ть, -ют, (*imp*) get carbon monoxide poisoning; be reduced

у·гор·éть, -я́т, (*perf*) *see* у·гор·а́·ть

у́гор·ь, (*g*) угр·я́, (*m*) (fish) eel; (med) acne, rosacea

~, обык·новéнн·ый common eel, *Anguilla rostrata*

~, электр·и́ческ·ий electric eel, *Electrophorus electricus*

у·гост·и́ть, -я́т, (*perf*) *see* у·гощ·а́·ть

у·гощ·а́·ть, -ют, (*imp*) entertain/treat/ invite to

угр·ева́т·ый, -ая, -ое, (*a*) pimply

угре·ви́д·н·ый, -ая, -ое, (*a*) eel-like

угр·и́ц·а, -ы, (*f*) (*dim*) of у́гор·ь; (fish) grig; (med) *Anguillula*

у·грож·а́·ть, -ют, (*imp*) threaten, menace

у·грóз·а, -ы, (*f*) threat, menace

угрю́м·ый, -ая, -ое, (*a*) sullen, morose

уд. вес· (*abbr*) = у·дéль·н·ый вес· specific gravity

уд·а́, -ы́, (*nom pl*) **у́д·ы,** (*g pl*) **уд·,** (*f*) fish-hook

у·да́бр·ива·ть, -ют, (*imp*) = **у·добр·я́·ть**

у·да́в·, -а, (*m*) (zool) boa, (*pl*) *Boidae*

у·да·ва́ться, -ются, (*imp*) succeed, be a success, work well

у·дав·и́ть, -ят, (*perf*) *see* **у·да́вл·ива·ть**

у·дав·к·а, -и, (*g pl*) **-в·ок·,** (*f*) (naut) timber-hitch

у·давл·е́ни·е, -я, (*n*) strangulation, strangling

у·да́вл·ива·ть, -ют, (*imp*) strangle

у·дад·у́тся, (*fut 3rd pl*) of **у·да́·ться** (*perf*) *see* **у·да·ва́ться** (*imp*)

у·дал·е́ни·е, -я, (*n*) (*v n*) *see* **у·дал·я́·ть;** removal, disposal, withdrawal, extraction (of teeth)

~ **влаг·и** dehumidification
~ **волос·** hair removal, depilation
~ **зол·ы** ash disposal/removal
~ **луб·а** desortication (timber)
~ **лист·ьев** (bot) defoliation
~ **от·хо́д·ов** waste disposal
~ **пе́чен·и** (med) hepatectomy
~ **ядр·а́** enucleation

у·дал·ённ·ый, -ая, -ое, (*a*) (*past part pass*) *see* **у·дал·я́·ть;** distant, remote, at a distance from

у·дал·и́ть, -я́т, (*perf*) *see* **у·дал·я́·ть**

у·да·ло́сь (*past neut sing*) of **у·да́·ться** it was possible to
ему́ у·да·ло́сь he was able to, he succeeded in

у·дал·я́·ть, -ют, (*imp*) remove, move/ send/clear away; extract (teeth)

уда́р·, -а, (*m*) hit, blow, stroke; (mech) impact; (phys) shock; (mil) assault, attack; (med) stroke

~, **акуст·и́ческ·ий** acoustic shock
~, **бок·ов·о́й** glancing blow
~, **гидравл·и́ческ·ий** water hammer, hydraulic shock
~, **звук·ов·о́й** acoustic shock, sonic boom
~ **земле·тряс·е́ни·я** earthquake shock
зо́на уда́р·а (*f*) shock/impact zone
ис·пыт·а́ни·е на уда́р (*n*) impact test
~, **лоб·ов·о́й** head-on impact
~, **обра́т·н·ый** (ICE) kick-back
~, **прям·о́й** direct impact; (nucl) knock-on impact
тео́ри·я уда́р·а (*f*) (nucl) hit theory
~, **терм·и́ческ·ий** thermal shock

уда́р

~, **то́ч·ечн·ый** localized impact
центр· уда́р·а (*m*) (mech) centre/ center of percussion

удар·е́ни·е, -я, (*n*) (ling) stress, emphasis

уда́р·ить, -ят, (*perf*) *see* **удар·я́·ть**

уда́р·ник·, -а, (*m*) strike, pellet, firing pin; pistol (of torpedo); shock worker

~, **уда́р·н·ый** (gunn) percussion pellet

уда́р·н·о-кана́т·н·ое бур·е́ни·е (*n*) (min) cable-tool drilling

ударно·про́ч·н·ый, -ая, -ое, (*a*) robust, impact/blow/shock-proof, high impact

уда́р·н·о-штанг·ов·ое бур·е́ни·е (*n*) (min) rod-tool drilling

уда́р·н·о-эрозио́н·н·ый, -ая, -ое, (*a*) impact-erosion

уда́р·н·ый, -ая, -ое, (*a*) shock, impact, percussion, percussive, impulsive; (mil) assault

~ **бур·е́ни·е** (*n*) (min) percussion drilling
~ **воз·бужд·е́ни·е** (*n*) (elec) shock excitation
~ **волн·а́** (*f*) shock wave
~ **вя́з·кост·ь** (*f*) impact strength/ toughness (of materials)
~ **вя́з·кост·ь, у·де́ль·н·ая** (met) impact toughness number
~ **механ·и́зм·** (*m*) (gunn) striker mechanism
~ **на·груж·е́ни·е, со·сред·о·точ·н·ое** (*n*) (mech) shock loading
~ **при·способл·е́ни·е** (*n*) rapping gear (of electrostatic filter)
~ **про́ч·ност·ь** (*f*) (mech) impact/ dynamic strength
~ **с·ва́р·к·а** (*f*) percussive welding
~ **с·жа́ти·е** (*n*) shock compression, high-velocity compression
~ **си́л·а** (*f*) impulsive force
~ **твёрд·ост·ь** (*f*) abrasion hardness
~ **труб·а́** (*f*) shock tube
~ **тру́б·к·а** (*f*) (gunn) percussion tube/fuse

удар·я́·ть, -ют, (*imp*) strike, hit, punch, kick, knock

у·да́·стся, (*fut 3rd sing*) of **у·да́·ться,** (*perf*); *see* **у·да·ва́ться** (*imp*)

у·да́·ться, (*fut 3rd sing, pl*) **у·да́·стся, у·дад·у́тся,** (*perf*); *see* **у·да·ва́ться**

у·да́·ч·а, -и, (*i*) **-ей,** (*f*) good luck, success

у·да́чн·ый, -ая, -ое, (*a*) successful; fortunate, happy

у·двá·ивани·е, -я, (*n*) (*v n*) of у·двá·и-
ва·ть

у·двá·ива·ть, -ют, (*imp*) double,
duplicate

у·двá·ивающ·ий, -ая, -ее, (*pres part
act*) of у·двá·ива·ть; doubling, duplicating

~ регúстр· (*m*) doubling register
(computers)

у·дво·éни·е, -я, (*n*) doubling, duplicating
врéм·я у·дво·éни·я (*n*) (autom)
change time; (nucl) doubling time

у·дво·úтел·ь, -я, (*m*) doubler

~ част·от·ы́ (rad) frequency doubler

у·двó·ить, -ят, (*perf*) see у·двá·ива·ть

у·дéл·, -а, (*m*) lot, portion, fate

у·дел·ённ·ый, -ая, -ое, (*past part
pass*) see у·дел·я́·ть

у·дел·úть, -я́т, (*perf*) see у·дел·я́·ть

у·дéль·н·ый, -ая, -ое, (*a*) specific, unit

~ актúв·ност·ь (*f*) specific activity

~ вес· (*m*) specific gravity/weight;
importance, significance, relative im-
portance/share

~ влáж·ност·ь (*f*) specific humidity

~ объ·ём· (*m*) specific volume

~ объ·ём· тóпл·ив·а (rocket) propellant density

~ от·раж·áтельн·ая спосóб·ност·ь
(*f*) reflection factor, reflectivity

~ про·вод·úмост·ь (*f*) (elec) specific
conductance, conductivity

~ рас·хóд· тóпл·ив·а specific fuel
consumption

~ рас·хóд· тóпл·ив·а, индикáтор·-
н·ый specific fuel consumption
(ICE in relation to I.H.P.)

~ рас·хóд· тóпл·ив·а, эффектúв·-
н·ый specific fuel consumption

~ со·противл·éни·е (*n*) (elec) specific
resistance, volume resistivity, resistivity (of a material)

~ тепло·ём·кост·ь (*f*) specific heat

~ тепл·от·á (*f*) specific heat

~ магнúт·н·ое со·противл·éни·е (*n*)
(phys) reluctivity

у·дел·я́·ть, -ют, (*imp*) apportion, allocate, give (part, some), appropriate;
devote, pay (attention etc.)

у·держ·áни·е, -я, (*n*) see у·дéрж·и-
вани·е

у·держ·áть, -ат, (*perf*) see у·дéрж·и-
ва·ть

у·дéрж·иваем·ый, -ая, -ое, (*pres
part pass*) of у·дéрж·ива·ть

~ объ·ём· (*m*) containment volume
(fluid mechanics)

у·дéрж·ивани·е, -я, (*n*) (*v n*) of у·дéр-
ж·ива·ть; retention, confinement,
containment

~ плáзм·ы (phys) plasma confinement

~ тóк·а current containment

у·дéрж·ива·ть, -ют, (*imp*) retain,
keep, hold back, restrain, confine,
keep from, withhold, suppress; (fin)
deduct; -ся retain one's position,
hold out/on; refrain, keep from;
keep/hold to, remain

у·держ·нúк·, -á, (*m*) retaining pawl

у·десят·ерённ·ый, -ая, -ое, (*past
part pass*) increased ten times, tenfold

у·дешев·úть, -я́т, (*perf*) see у·дешев-
л·я́·ть

у·дешевл·ённ·ый, -ая, -ое, (*past part
pass*) see у·дешевл·я́·ть; cheap-rate

у·дешевл·я́·ть, -ют, (*imp*) make cheaper, reduce cost/price; -ся become
cheaper

у·див·úтельн·о (*adv*) astonishingly,
surprisingly; extremely, very

у·див·úть, -я́т, (*perf*) see у·дивл·я́·ть

у·дивл·я́·ть, -ют, (*imp*) astonish,
amaze, surprise; -ся (+ *dat*) be
astonished/amazed/surprised at

удил·á, -úл·, (*n pl*) bit (harness)

уд·úлищ·е, -а, (*n*) (fish) rod

уд·úть, -я́т, (*imp*) fish for (with rod),
angle

у·дифферéнт·овать, -уют, (*perf*)
trim (a ship)

у·длин·éни·е, -я, (*n*) (*v n*) see у·длин-
·я́·ть; elongation, extension, prolongation (of time); (telecom) tailing
(a/c) aspect ratio

~, абсолют·н·ое (met) extension
коэффициéнт· у·длин·éни·я (*m*)
(mech) modulus of elongation

~ крыл·á (a/c) aspect ratio

~, лин·éйн·ое linear extension

~, най·бóльш·ее мéст·н·ое maximum
local elongation

~, о·стáт·очн·ое (mech) permanent
set, residual elongation; (plast) tension
set

~, от·нос·úтельн·ое relative elongation; (met) percentage elongation,
elongation

~, от·нос·úтельн·ое равно·мéр·н·ое
elongation before reduction of area

~, раз·рыв·н·óе breaking elongation,
extension ratio (of rubbers and
plastics)

у·длин·ённ·ый, -ая, -ое, (*past part pass*) *see* **у·длин·я́·ть;** elongated, extended, prolate

~ **цикл·о́ид·а** (*f*) (math) prolate cycloid

у·длин·и́тел·ь, -я, (*m*) elongator, extender; (rad) attenuator, pad

~ **и́мпульс·ов** (rad) pulse lengthener/stretcher/widener

~, **сантиметр·о́в·ый** (rad) microwave attenuator

~ **угл·о́в визи́р·овани·я** (air) collimator handle (of gyro bomb sight)

у·длин·и́ть, -я́т, (*perf*) *see* **у·длин·я́·ть**

у·длин·я́·ть, -ют, (*imp*) lengthen, elongate, make longer, prolong

у·до́б·н·о (*adv*) conveniently; it is convenient, it is a good thing to

у·до́б·н·ый, -ая, -ое, (*a*) convenient, suitable, handy; opportune; comfortable

у·до́б·о- (*component*) readily, easily; -ible, -able

удобо·вар·и́м·ый, -ая, -ое, (*a*) digestible, easily digestible/digested, easy to digest

удобо·об·раба́т·ываем·ый, -ая, -ое, (*a*) workable, easily worked/processed

удобо·об·тек·а́емост·ь, -и, (*f*) (a/c) aerodynamic cleanness

удобо·об·тек·а́ем·ый, -ая, -ое, (*a*) streamlined

удобо·по·ня́т·н·ый, -ая, -ое, (*a*) easily comprehensible/understood, intelligible

удобо·у·кла́д·ываемост·ь, -и, (*f*) placeability (of concrete)

удобо·у·правл·я́ем·ый, -ая, -ое, (*a*) controllable, easily controlled

удобо·чит·а́ем·ый, -ая, -ое, (*a*) legible, easily read

у·добр·е́ни·е, -я, (*n*) (*v n*) *see* **у·добр·я́·ть;** fertilizer, manure; dressing, application (of fertilizer)

~, **азо́т·ист·ое** nitrogenous fertilizer

~, **азо́т·н·ое** nitrogen fertilizer

~, **бактериа́ль·н·ое** bacterial fertilizer

~, **на·во́з·н·ое** manure, farmyard manure

у·добр·ить, -ят, (*perf*) *see* **у·добр·я́·ть**

у·добр·я́·ть, -ют, (*imp*) apply fertilizer, fertilize, manure, topdress

у·до́б·ств·о, -а, (*n*) convenience, ease; comfort

удовлетвор·е́ни·е, -я, (*n*) (*v n*) *see* **удовлетвор·я́·ть;** satisfaction

удовлетвор·и́тельн·ый, -ая, -ое, (*a*) satisfactory

удовлетвор·и́ть, -я́т, (*perf*) *see* **удовлетвор·я́·ть**

удовлетвор·я́·ть, -ют, (*imp*) satisfy, content; answer, meet (needs etc.); supply with, meet the requirements for

у·до·во́ль·стви·е, -я, (*n*) pleasure

удо́д·, -а, (*m*) (orn) hoopoe, (*pl*) *Upupidae*

у·до́·й, -я, (*m*) milk yield; milking

у·дорож·а́·ть, -ют, (*imp*) make more expensive, raise the price/cost of; **-ся** become more expensive, go up in price

у·дорож·и́ть, -а́т, (*perf*) *see* **у·дорож·а́·ть**

у·до·ста́·ива·ть, -ют, (*imp*) award, confer upon, honour with

удосто·вер·е́ни·е, -я, (*n*) (*v n*) *see* **удосто·вер·я́·ть;** certificate, certification, certifying, licence

~, **квалификацио́н·н·ое** certificate of qualifications

~ **ли́ч·ност·и** identity card

~ **о го́д·ност·и к по·лёт·у** certificate of airworthiness

~ **о про·ис·хожд·е́ни·е това́р·а** (com) certificate of origin

удосто·ве́р·ить, (*perf*) *see* **удосто·вер·я́·ть**

удосто·вер·я́·ть, -ют, (*imp*) certify, attest, witness (for certification); **-ся** (*pass*); ascertain, make sure/certain

у·до·сто́·ить, -ят, (*perf*) *see* **у·до·ста́·ива·ть**

у·дочер·е́ни·е, -я, (*n*) (law) adoption (of girl)

у́д·очк·а, -и, (*g pl*) **-чек·,** (*f*) (fish) rod and line

у·душ·а́·ть, -ют, (*imp*) suffocate, asphyxiate

у·душ·и́ть, -́ат, (*perf*) *see* **у·душ·а́·ть**

у·ду́ш·лив·ый, -ая, -ое, (*a*) stifling, suffocating, asphyxiating; asthmatic

у·ду́ш·ь·е, -я, (*n*) (med) asthma, asphyxia

у·е́д·ет, (*fut 3rd sing*) of **у·е́х·ать** (*perf*) *see* **у·езж·а́·ть** (*imp*)

у·един·е́ни·е, -я, (*n*) (*v n*) *see* **у·един·я́·ть;** isolation, seclusion, solitude

у·един·и́ть, -я́т, (*perf*) *see* **у·един·я́·ть**

у·един·я́·ть, -ют, (*imp*) isolate, set apart, seclude

у·езж·а́·ть, -ют, (*imp*) depart, leave, go away (in vehicle) у·е́х·ать, (*fut 3rd pl*) у·е́д·ут, (*perf*) *see* у·езж·а́·ть

уж (*adv*) already, as low/soon/early/ small as, even; (particle) really

уж·, -а́, (*m*) (zool) grass-snake, *Tropidonotus natrix*

у·жа́л·ить, -ят, (*perf*) sting, bite (of insect etc.)

у́жас·, -а, (*m*) terror, horror

ужас·а́·ть, -ют, (*imp*) terrify, horrify

ужас·н·у́ть, -у́т, (*perf*) *see* ужас·а́·ть

ужа́с·н·ый, -ая, -ое, (*a*) terrible, horrible

у·жа́т·ь, (*fut 3rd pl*) у·жм·у́т, (*perf*) *see* у·жим·а́·ть

уже́ (*adv*) already, even, as early as; уж·е (*comp adj*) narrower

~ не no longer

уж·е́ни·е, -я, (*n*) angling, fishing

у·жестч·ённ·ый, -ая, -ое, (*a*) (instr) ruggedized

уж·и́, (*nom pl*) of уж·

у·жим·а́·ть, -ют, (*imp*) pinch, constrict

у·жим·и́н·а, -ы, (*f*) (met) kink, hanger tear (continuous cast)

у́жин·, -а, (*m*) supper

у·жм·ёт, (*fut 3rd sing*) of у·жа́т·ь (*perf*) *see* у·жим·а́·ть (*imp*)

у·за·ко́н·ива·ть, -ют, (*imp*) legalize, legitimate

у·за·ко́н·ить, -ят, (*perf*) *see* у·за·ко́н·ива·ть

у·за·кон·я́·ть, -ют, (*imp*) = у·за·ко́н·ива·ть

узанс·, -а, (*m*) (com) usage; usance (of bills)

узбеки́т·, -а, (*m*) (min) uzbekite

узд·а́, -ы́, (*nom pl*) у́зд·ы, (*f*) bridle; (anat) frenum

узд·е́чк·а, -и, (*g pl*) -чек·, (*f*) bridle; (anat) frenum, frenulum; cork fastening

у́зел·, (*g*) узл·а́, (*m*) knot; node, nodal point, junction, branch point, (anat) ganglion; (math) cusp locus; (eng) unit, subassembly, assembly; multiple joint, junction (in assembling); (telecom) centre, center; node, vertex, junction/branch point (of elec. circuit); (cryst) site, point

~, арифмет·и́ческ·ий arithmetic unit (computer)

~ бала́нс·а (horol) balance wheel assembly

~, вака́нт·н·ый (cryst) vacancy

~, варико́з·н·ый (anat) varix

~, вос·ход·я́щ·ий (astron) ascending node

у́зел

~, гидро·техн·и́ческ·ий (civ eng) hydraulic works system

~, голов·н·о́й (hydroelec) headworks

~, го́р·н·ый (geog) plexus of mountains

~ давл·е́ни·я (acoust) pressure node

~ доро́г· road junction

~, еди́н·ый integral unit (of a machine structure etc.)

~, железно·доро́ж·н·ый railway junction

~, жёстк·ий rigid/fixed joint

~, из·мер·и́тельн·ый measuring means (computers)

~ колеб·а́ни·я (phys) vibration node

~ кущ·е́ни·я (bot) tillering node

~ крепл·е́ни·я attachment fitting, joint (structures)

~, кру́п·н·ый (rocket) subsystem лин·и·я узл·о́в (*f*) (astron) line of nodes

~, мор·ск·о́й (naut) bend, hitch

~, не́рв·н·ый (zool) ganglion, nerve centre

~ оборо́н·ы (mil) strong point, defended area

~ орби́т·ы (astron) orbital node

~, про·меж·у́точн·ый (cryst) interstitial site

~, прям·о́й (naut) reef knot

~, радиа́ль·н·ый (cyb) radial node

~ решёт·к·и (cryst) lattice point/site

~, с·бо́р·очн·ый (eng) subassembly

~ с·вя́з·и communications centre/ center

~, семя·до́ль·н·ый (bot) cotyledonary node

~, серд·е́чн·ый (anat) cardiac ganglion

~, стере́йд·н·ый (bot) stereid bundle

~, сумм·и́рующ·ий adder (computers)

~ то́к·а (elec) current node

~ у·правл·е́ни·я (autom) control assembly

~ фа́з·ов·ой траекто́ри·и (cyb) equilibrium node

~, шарни́р·н·ый articulated/hinged joint

узел·ко́в·ый, -ая, -ое, (*a*) of узел·о́к; nodular

узел·о́к·, -л·к·а́, (*m*) (*dim*) of у́зел·; nodule, nodulus

у́з·к·ий, -ая, -ое, (*a*) narrow; (print) lean, condensed

~ ме́ст·о (*n*) bottleneck

узко·голо́в·ый, -ая, -ое, (*a*) (zool) stenocephalous

узко·го́рл·ый, -ая, -ое, (*a*) narrow-necked

узко·кол·е́йн·ый, -ая, -ое, (*a*) (rail) narrow-gauge/gage

узко·ли́ст·н·ый, -ая, -ое, (*a*) (bot) narrow-leaved, angustifoliate, stenophyllous

узко·ло́б·ый, -ая, -ое, (*a*) (anat) low-browed

узко·над·кры́л·ые, -ых, (*pl decl as adj*) (ent) false blister beetles, *Oedemeridae*

узко·пере·горо́д·чат·ый, -ая, -ое, (*a*) angustimurate

узко·плён·очн·ый, -ая, -ое, (*a*) (phot) 16 mm

узко·поло́с·иц·а, -ы, (*f*) (agr) strip cultivation

узко·поло́с·н·ый, -ая, -ое, (*a*) narrow-band, narrow-strip

узко·ро́т·ый, -ая, -ое, (*a*) narrow-mouthed, (anat) stenostomatous

узко·сифо́н·н·ый, -ая, -ое, (*a*) stenosiphonate

узко·специализ·и́рованн·ый, -ая, -ое, (*a*) highly specialized

узко·те́л·к·и, -л·ок·, (*f pl*) (ent) cylindrical bark beetles, *Colydiidae*

узл·ова́т·ый, -ая, -ое, (*a*) nodular, knotty, (anat) gangliform, ganglionic

узл·ов·о́й, -а́я, -о́е, (*a*) of **у́зел·** (q.v.) nodal, nodular

~ пункт· (*m*) junction; (rad) key point

~ ста́нци·я (*f*) (rail) junction; (telecom) tandem exchange, tandem central office

~ то́ч·к·а (*f*) nodal point; (math) crunode

~ шарни́р· (*m*) multibar joint

узло·вяз·а́тел·ь, -я, (*m*) knotter

узло·лов·и́тел·ь, -я, (*m*) (paper) knotter

узло·образ·ова́ни·е, -я, (*n*) knotting; (telecom) toll switch planning

у·зна·ва́ни·е, -я, (*n*) (*v n*) *see* **у·зна·ва́ть;** recognition

у·зна·ва́ть, -ю́т, (*imp*) recognize; find out, learn

у·зна́·ть, -ю́т, (*perf*) *see* **у·зна·ва́ть**

у·зо́р·, -а, (*m*) pattern, design

у·зо́р·чат·ый, -ая, -ое, (*a*) patterned, covered/decorated with patterns

у́з·ост·ь, -и, (*f*) narrowness

у́зус·, -а, (*m*) (law) usage

у́з·ы, (*g pl*) **уз·,** (*f pl*) bonds, ties

уинта·и́т·, -а, (*m*) (min) uintaite, uintahite

уинта·те́р·и·й, -я, (*m*) (pal) Uintatherium

Уи́тстон·а, мо́ст·ик· (*m*) (elec) Wheatstone bridge

у·йд·ёт, (*fut 3rd sing*) of **у·йти́** (*perf*) *see* **у·ход·и́ть** (*imp*)

у·йти́, (*fut 3rd pl*) **у·йд·у́т,** (*past masc sing*) **у·шёл** (*perf*) *see* **у·ход·и́ть**

укав·и́ст·, -а, (*m*) (rad) ham, VHF amateur

у·ка́ж·ет, (*fut 3rd sing*) of **у·каз·а́ть** (*perf*) *see* **у·ка́з·ыва·ть** (*imp*)

у·ка́з·, -а, (*m*) decree, edict

у·каз·а́ни·е, -я, (*n*) (*v n*) *see* **у·ка́з·ыва·ть;** indication; instructions, advice

у·каз·а́тел·ь, -я, (*m*) indicator, pointer; signpost, road sign, street sign; index, directory; indicating meter/gauge, indicator

~ абоне́нт·ов телефо́н·н·ой се́т·и telephone directory

~, алфави́т·н·ый alphabetical index

~, библио·гра́ф·и́ческ·ий bibliography

~ вертика́ль·н·ой ско́р·ост·и (air) rate-of-climb indicator

~ выс·от·ы́ (a/c) altimeter

~, двух·стре́л·очн·ый crossed-pointer indicator

~, звук·ов·о́й audible/sound indicator

~ ка́честв·а с·ме́с·и (a/c) fuel-mixture indicator

~ ку́рс·а (nav) course indicator, compass

~ на·пряж·е́ни·я по·ме́х· psophometer

~ по·воро́т·а (a/c) turn indicator; (M/T) trafficator

~ по·лож·е́ни·я глисса́д·ы (a/c) glide path indicator

~, свет·ов·о́й illuminated/light indicator

~ скольж·е́ни·я и по·воро́т·а (air) turn-and-bank indicator

~ ско́р·ост·и (air) air-speed indicator

~ узл·о́в колеб·а́ни·й (rad) nodalizer

~ у·клон·а (rail) gradient post

~ у́·ровн·я level/depth indicator

~ у́·ровн·я то́пл·ив·а fuel level indicator

~ у·скор·е́ни·я indicating accelerometer

~, фла́ж·н·ый flag/banner indicator

~ числ·а́ Ма́х·а (a/c) machmeter

~, штык·ов·о́й dipstick

у·каз·а́тельн·ый, -ая, -ое, (*a*) of **у·каз·а́тел·ь;** indicating у·каз·а́тельн·ый, -ая, -ое

~ па́л·ец· (*m*) forefinger

~ стре́л·к·а (*f*) (instr) pointer

у·каз·а́ть, (*fut 3rd pl*) у·ка́ж·ут, (*perf*) *see* у·ка́з·ыва·ть

у·ка́з·к·а, -и, (*g pl*) -з·ок·, (*f*) pointer (stick)

у·ка́з·ыва·ть, -ют, (*imp*) indicate, point to; point out; tell, instruct

у·ка́л·ыва·ть,·-ют, (*imp*) prick, puncture

у·кат·а́·ть, -ют, (*perf*) roll (to flatten), roll/wear down/smooth

у·кат·и́ть, ⸗ят, (*perf*) roll away/off

у·ка́т·к·а, -и, (*g pl*) -т·ок·, (*f*) (*v n*) *see* у·ка́т·ыва·ть; rolling (roads etc.)

~, по·вто́р·н·ая back rolling (of roads)

у·ка́т·ыва·ть, -ют, (*imp*) roll (to flatten), roll/wear down/smooth; roll away/off

у·ка́ч·анн·ый, -ая, -ое, (*past part pass*) of у·кач·а́·ть

у·кач·а́·ть, -ют, (*perf*) *see* у·ка́ч·ива·ть

у·ка́ч·енн·ый, -ая, -ое, (*past part pass*) of у·кат·и́ть

у·ка́ч·ива·ть, -ют, (*imp*) make sick (from motion); rock to sleep

у·ка́ш·ива·ть, -ют, (*imp*) mow, cut (the whole of)

у·кип·а́·ть, -ют, (*imp*) boil down/away; boil till ready

у·кип·е́ть, -я́т, (*perf*) *see* у·кип·а́·ть

у·кла́д·, -а, (*m*) usage

~, реч·ев·о́й (ling) speech usage

у·кла́д·к·а, -и, (*g pl*) -д·ок·, (*f*) (*v n*) *see* у·кла́д·ыва·ть; stowage
деф́ект· у·кла́д·к·и (*m*) (phys) stacking fault

у·кла́д·очн·ый, -ая, -ое, (*a*) of у·кла́д·к·а

~ маши́н·а (*f*) cigarette packing machine

у·кла́д·чик·, -а, (*m*) layer; packer; compacting machine (roadmaking); piler (of sheet metal)

~, ка́бель·н·ый cable-laying equipment

у·кла́д·ывани·е, -я, (*n*) (*v n*) of у·кла́д·ыва·ть

у·кла́д·ыватель·, -я, (*m*) (met roll) piler; (met roll) shuffle bars

~ сляб·ов (met roll) slab piler

у·кла́д·ыва·ть, -ют, (*imp*) lay, lay out; put to bed; pack, stow, stack, pile (flat things); pour, place (concrete)

у·кле́й·к·а, -и, (*g pl*) -е·ек·, (*f*) (fish) bleak, *Alburnus*

у·клон·, -а, (*m*) inclination, slope, gradient, incline; rake (from vertical); bias, deviation

~, гидравл·и́ческ·ий hydraulic gradient

~, конве́йер·н·ый (min) belt incline

~, техн·и́ческ·ий technical bias

у·клон·е́ни·е, -я, (*n*) (*v n*) *see* у·клон·я́·ться; deviation, deflection; evasion, avoidance

у·клон·и́ться, ⸗ятся, (*perf*) *see* у·клон·я́·ться

у·клон·н·ый, -ая; -ое, (*a*) of у·клон·

~ доск·а́ (*f*) (shipb) declivity board

уклоно·ме́р·, -а, (*m*) inclinometer, gradiometer, gradienter

уклоно·у·каз·а́тел·ь, -я, (*m*) gradient indicator/board

у·клон·я́·ться, -ются, (*imp*) deviate, digress; avoid, evade

у·клю́ч·ин·а, -ы, (*f*) (naut) rowlock, crutch

у·ко́в·к·а, -и, (*f*) (met) reduction (in forge drawing)

у·ко́л·, -а, (*m*) (*v n*) *see* у·кол·о́ть; prick, puncture; (med) injection

у·кола́ч·ива·ть, -ют, (*imp*) beat down/smooth; hammer (leather); nail down

у·колот·и́ть, ⸗ят, (*perf*) *see* у·кола́·ч·ива·ть

у·ко́л·от·ый, -ая, -ое, (*past part pass*) of у·кол·о́ть (*perf*)

у·кол·о́ть, ⸗ют, (*perf*) prick, puncture

у·колоч·енн·ый, -ая, -ое, (*past part pass*) of у·колот·и́ть (*perf*) *see* у·ко·ла́ч·ива·ть (*imp*)

у·комплект·ова́ни·е, -я, (*n*) (*v n*) *see* у·комплект·о́выва·ть; outfitting; staffing

у·комплект·о́ванн·ый, -ая, -ое, (*past part pass*) *see* у·комплект·о́выва·ть; complete with; staffed

у·комплект·ова́ть, -у́ют, (*perf*) *see* у·комплект·о́выва·ть

у·комплект·о́выва·ть, -ют, (*imp*) make up a set, complete a set, equip, fit up/out; fill vacancies, make up staff, recruit, man

у·ко́р·, -а, (*m*) reproach

у·кора́ч·ивани·е, -я, (*n*) (*v n*) of у·кора́ч·ива·ть; curtailment

у·кора́ч·ива·ть, -ют, (*imp*) shorten, make shorter, abbreviate, reduce (length); make too short, curtail; (chem) degrade (a chain)

у·корен·е́ни·е, -я, (*n*) (*v n*) *see* у·корен·я́·ть

у·корен·и́ть, -я́т, (*perf*) *see* у·коре-н·я́·ть

у·корен·я́·ть, -ют, (*imp*) allow to take root; implant; -ся (*pass*); take root

у·кор·и́зн·а, -ы, (*f*) reproach

у·кор·и́ть, -я́т, (*perf*) *see* у·кор·я́·ть

у·корот·и́ть, -я́т, (*perf*) *see* у·кора́-ч·ива·ть

у·короч·е́ни·е, -я, (*n*) (*v n*) *see* у·кора́ч·ива·ть; short-cut

у·кор·я́·ть, -ют, (*imp*) reproach

у·ко́с·, -а, (*m*) moving, cutting; hay crop

у·ко́с·ин·а, -ы, (*f*) strut, bracing strut; boom, jib (of crane)

у·кос·и́ть, -я́т, (*perf*) mow, cut (the whole of)

у·кра́д·енн·ый, -ая, -ое, (*past part pass*) of у·кра́с·ть (*perf*) stolen

у·кра́д·к·ой (*adv*) by stealth, stealthily

украи́н·ск·ий, -ая, -ое, (*a*) Ukrainian

у·кра́·л, (*past masc sing*) of у·кра́с·ть

у·кра́с·ить, -ят, (*perf*) *see* у·кра-ш·а́·ть

у·кра́с·ть, (*fut 3rd pl*) у·крад·у́т, (*perf*) steal

у·краш·а́·ть, -ют, (*imp*) decorate, ornament; dress (with flags)

у·краш·е́ни·е, -я, (*n*) (*v n*) *see* у·краш·а́·ть; decoration, ornament

~, леп·н·о́е (build) stucco moulding

у·креп·и́ть, -я́т, (*perf*) *see* у·крепл·я́·ть

у·крепл·е́ни·е (*v n*) *see* у·крепл·я́·ть; fortification; reinforcement

~, пред·мо́ст·н·ое (mil) bridgehead

у·крепл·ённ·ый, -ая, -ое, (*past part pass*) *see* у·крепл·я́·ть

у·крепл·я́·ть, -ют, (*imp*) strengthen, reinforce, consolidate, fortify; fix, lodge

у·крепл·я́ющ·ий, -ая, -ее, (*pres part act*) of у·крепл·я́·ть; (med) tonic, restorative

у·кро́·ет, (*fut 3rd sing*) of у·кры́·ть (*perf*) *see* у·крыв·а́·ть

укро́п·, -а, (*m*) (bot) fennel, dill, *Anethum*

~, о·горо́д·н·ый dill, *Anethum graveolens*

укрот·и́ть, -я́т, (*perf*) tame, curb

укрощ·а́·ть, -ют, (*imp*) tame, curb

у·крупн·е́ни·е, -я, (*n*) (*v n*) *see* у·крупн·я́·ть; enlargement, coarsening; (cryst) growth (of grains); coalescence (of droplets); (chem) cross-linking (of molecules)

у·крупн·ённ·ый, -ая, -ое, (*past part pass*) *see* у·крупн·я́·ть

у·крупн·и́ть, -я́т, (*perf*) *see* у·крупн·я́·ть .

у·крупн·я́·ть, -ю́т, (*imp*) enlarge, make larger; amalgamate, form into larger units; -ся (*pass*); become larger, grow, increase

у·кру́т·к·а, -и, (*g pl*) -т·ок·, (*f*) (text) twist contraction

у·крыв·а́ни·е, -я, (*n*) (*v n*) of у·крыв·а́·ть

у·крыв·а́тельств·о, -а, (*n*) concealment; (law) receiving (stolen goods), aiding and abetting (a crime)

у·крыв·а́·ть, -ют, (*imp*) cover, cover up; conceal, hide; shelter, harbour; (law) aid and abet; -ся (*pass*); hide, take cover

у·крыв·и́стост·ь, -и, (*f*) obscuration, hiding/obliterating power (of paints, pigments etc.)

у·кры́т·и·е, -я, (*n*) (*v n*) *see* у·крыв·а́·ть; shelter, cover, pen (for aircraft)

зо́н·а у·кры́т·и·я (*f*) shelter zone/area

у·кры́·ть, (*fut 3rd pl*) у·кро́·ют, (*perf*) *see* у·крыв·а́·ть

у́ксус·, -а, (*m*) vinegar

~, ви́н·н·ый wine vinegar

~, во́д·очн·ый distilled vinegar

~, пив·н·о́й malt vinegar

~, пло́д·н·ый cider vinegar

~, солод·о́в·ый malt vinegar

уксусно·ки́сл·ый, -ая, -ое, (*a*) (chem) acetate (of)

~ бакте́ри·я (*f*) Bacterium aceti

~ брож·е́ни·е (*n*) (chem) acetic fermentation

~ свин·е́ц· (*m*) lead acetate

уксусно·мышьяково·ки́сл·ый, -ая, -ое, (*a*) acetoarsenide (of)

уксусно·эти́л·ов·ый эфи́р· (*m*) (chem) ethyl acetate

у́ксус·н·ый, -ая, -ое, (*a*) (food) vinegar, vinegary; (chem) acetic

~ ангидри́д· (*m*) acetic anhydride, acetic oxide

~ де́рев·о (*n*) (bot) staghorn sumach, *Rhus typhina*

~ кисл·от·а́ (*f*) acetic acid

~ эссе́нци·я (*f*) (food) vinegar essence, 80% acetic acid

у·ку́пор·енн·ый, -ая, -ое, (*past part pass*) *see* у·ку́пор·ива·ть у·ку́пор·ивани·е, -я   (*n*) (*v n*) of у·ку́-пор·ива·ть

у·ку́пор·ива·ть, -ют,   (*imp*) cork, stopper, stop/bung up, seal

у·ку́пор·ить, -ят,   (*perf*) *see* у·ку́по-р·ива·ть

у·ку́пор·к·а, -и,   (*f*) (*v n*) *see* у·ку́по-р·ива·ть

у·ку́пор·очн·ый, -ая, -ое,   (*a*) corking, stoppering, sealing

у·ку́с·, -а,   (*m*) bite, sting

у·кус·и́ть, -ят,   (*perf*) bite, sting

у·ку́т·а·ть, -ют,   (*perf*) *see* у·ку́т·ы-ва·ть

у·ку́т·ыва·ть, -ют,   (*imp*) wrap/muffle up

у·ку́ш·енн·ый, -ая, -ое,   (*past part pass*) of у·кус·и́ть

у·ла́вл·иватель·, -я,   (*m*) (*see also* у·лов·и́тел·ь) (elec) catcher; (met) wrapper (for strip coiling)

у·ла́вл·ивани·е, -я,   (*n*) (*v n*) of у·ла́в-л·ива·ть; catchment

у·ла́вл·ива·ть, -ют,   (*imp*) catch, trap (particles, etc.) recover (waste)

у·ла́д·ить, -ят,   (*perf*) *see* у·ла́ж·и-ва·ть

у·ла́ж·ива·ть, -ют,   (*imp*) settle (affairs), put in order

у·лёг·ся,   (*past masc sing*) of у·ле́ч·ься

у́·ле·й, (*g*) у́ль·я,   (*m*) hive, beehive

улей·н·ый, -ая, -ое,   (*a*) hive, beehive

улекси́т·, -а,   (*m*) (min) ulexite

у·лет·а́·ть, -ют,   (*imp*) fly away/off, depart, leave (by air); fly past, pass quickly (of time)

у·лет·е́ть, -я́т,   (*perf*) *see* у·лет·а́·ть

у·летуч·ивани·е, -я,   (*n*) (*v n*) *see* у·летуч·ива·ться; volatilization, evaporation

у·летуч·ива·ться, -ются,   (*imp*) volatilize, evaporate

у·летуч·иться, -аяся,   (*perf*) *see* у·ле-туч·ива·ться

у·леч·у́,   (*fut 1st sing*) of у·лет·е́ть

у·ле́ч·ься,   (*fut 3rd sing, pl*) у·ля́ж·ется, у·ля́г·утся, (*past masc sing*) у·лёг·-ся, (*perf*) lie down (to rest); be properly placed; settle (of dust etc.); subside, abate

у·ли́к·а, -и,   (*f*) evidence

ул·и́т·, -а,   (*m*) (med) ulitis

ули́т·к·а, -и,   (*g pl*) -т·ок·, (*f*) (zool) snail, (*pl*) *Helicidae*; helix, limaçon; spiral casing, volute chamber (of centrifugal pump etc.); (anat) cochlea

~ Паска́л·я   (math) Pascal's limaçon

улитко·ви́д·н·ый, -ая, -ое,   (*a*) conchi-form, conchoidal, trochiform, trochi-spiral

ули́т·ков·ый, -ая,   (*a*) snail's; helical, screw

~ пит·а́тел·ь   (*m*) screw feeder

улитко·обра́з·н·ый, -ая, -ое,   (*a*) conchoidal, conchiform, trochiform, trochispiral

у́л·иц·а, -ы,   (*i*) -ей,   (*f*) street

у·лич·а́·ть, -ют,   (*imp*) prove/find guilty, convict

у·лич·и́ть, -а́т,   (*perf*) *see* у·лич·а́·ть

у́л·ичн·ый, -ая, -ое,   (*a*) street

у·ло́в·, -а,   (*m*) (fish etc.) catch

у·лов·и́м·ый, -ая, -ое,   (*pres part pass*) *see* у·ла́вл·ива·ть; audible, percept-ible

у·лов·и́тел·ь, -я,   (*m*) catcher, trap; separator, precipitator (in dust filters)

~, инерцио́н·н·ый   impact-type sepa-rator (dust catcher)

~, ио́н·н·ый   (elec) ion trap

у·лов·и́ть, -ят,   (*perf*) *see* у·ла́вл·ива·ть

у·ло́в·к·а, -и,   (*g pl*) -в·ок·,   (*f*) ruse, trick

у·ло́вл·енн·ый, -ая, -ое,   (*past part pass*) of у·лов·и́ть (*perf*) *see* у·ла́в-л·ива·ть (*imp*)

у·лож·е́ни·е, -я,   (*n*) (*v n*) of у·лож·и́ть

у·ло́ж·енн·ый, -ая, -ое,   (*a*) (*past part pass*) *see* у·кла́д·ыва·ть

у·лож·и́ть, -ат,   (*perf*) *see* у·кла́д·ы-ва·ть

у·луч·а́·ть, -ют,   (*imp*) find, choose (time), seize (opportunity)

у·луч·и́ть, -а́т,   (*perf*) *see* у·луч·а́·ть

у·лучш·а́ем·ый, -ая, -ое,   (*a*) (*pres part pass*) of у·лучш·а́·ть

у·лучш·а́·ть, -ют,   (*imp*) improve, make better, ameliorate

у·луч·а́ющ·ий, -ая, -ее,   (*pres part act*) of у·лучш·а́·ть

у·лучш·е́ни·е, -я,   (*n*) (*v n*) *see* у·луч-ш·а́·ть; improvement; (met) quench-ing and tempering

~ ка́честв·а   refinement, improvement

у·лу́чш·енн·ый, -ая, -ое,   (*past part pass*) *see* у·лучш·а́·ть

у·лу́чш·ить, -ат,   (*perf*) *see* у·луч-ш·а́·ть

улыб·а́·ться, -ются,   (*imp*) smile

улы́б·к·а, -и,   (*g pl*) -б·ок·,   (*f*) smile

улыб·н·у́ться, -у́тся,   (*perf*) *see* улы-б·а́·ться

У́льман·а, реа́кци·я   (*f*) (chem) Ullman reaction

ульман·и́т·, **-а**, (*m*) (min) ullmannite
ульм·и́н·, **-а**, (*m*) (biochem) ulmin, humus
ульми́н·ов·ый, **-ая**, **-ое**, (*a*) ulminic, ulmic, ulmous
~ **кисл·от·а́** (*f*) (chem) humic acid
ультра·аку́ст·ик·а, **-и**, (*f*) (phys) ultrasonics
ультра·ви́рус·, **-а**, (*m*) (med) filter virus, filter-passer
ультра·высоко·част·о́тн·ый, **-ая**, **-ое**, (*a*) ultrahigh frequency, VHF
ультра·зву́к·, **-а**, (*m*) (phys) ultrasonics (above 15,000–20,000 cps); ultrasound
ультра·звук·ов·о́й, **-а́я**, **-о́е**, (*a*) ultrasonic
~ **техно·ло́г·и·я** (*f*) ultrasonics, ultrasonic engineering
ультра·коро́т·к·ий, **-ая**, **-ое**, (*a*) ultra-short
~ **волн·а́** (*f*) (rad) ultra-short wave (less than 10 m)
ультра·коротко·волн·ов·ый, **-ая** **-ое**, (*a*) ultrashort-wave, very high-frequency, VHF (of equipment)
ультра·мари́н·, **-а**, (*m*) ultramarine, ultramarine blue (pigment)
ультра·мари́н·ов·ый, **-ая**, **-ое**, (*a*) ultramarine, ultramarine blue
ультра·маф·и́ческ·ий, **-ая**, **-ое**, (*a*) (geol) ultramafic
ультра·микро́б·, **-а**, (*m*) (biol) ultra-microbe
ультра·микро·вес·ы́, **-о́в**, (*m pl*) (chem) ultramicrobalance
ультра·микро·ско́п·, **-а**, (*m*) ultra-microscope
ультра·микро·скоп·и́·я, **-и**, (*f*) ultra-microscopy
ультра·микро·то́м·, **-а**, (*m*) ultra-microtome
ультра·осно́в·н·ый, **-ая**, **-ое**, (*a*) (geol) ultrabasic
ультра·сфер·и́ческ·ий, **-ая**, **-ое**, (*a*) (math) ultraspherical
ультра·у·скор·и́тел·ь, **-я**, (*m*) (chem) ultra-accelerator
ультра·фи́льтр·, **-а**, (*m*) ultrafilter
ультра·фильтр·а́ци·я, **-и**, (*f*) (chem) ultrafiltration
ультра·фиоле́т·ов·ый, **-ая**, **-ое**, (*a*) ultra-violet
ультра·центри·фу́г·а, **-и**, (*f*) (chem, phys) ultracentrifuge
у·ля́г·утся, (*fut 3rd pl*) of **у·ле́ч·ься**
у·ля́ж·ется, (*fut 3rd sing*) of **у·ле́ч·ься**

ум·, **-а́**, (*m*) mind, intellect
у·мал·е́ни·е, **-я**, (*n*) (*v n*) *see* **у·мал·я́·ть**; belittlement; depreciation
у·мал·и́ть, **-я́т**, (*perf*) *see* **у·мал·я́·ть**
ума·лиш·ённ·ый, **-ая**, **-ое**, (*a*) (med) insane
у·ма́лч·ива·ть, **-ют**, (*imp*) keep quiet about, fail to acknowledge
у·мал·я́·ть, **-ют**, (*imp*) reduce; be little; **-ся** (*pass*); depreciate, diminish
умангит·, **-а**, (*m*) (min) umangite
умбелла́р·ов·ая кисл·от·а́ (*f*) (chem) umbellaric acid
умбелли·феро́н·, **-а**, (*m*) (biochem) umbelliferone
умбе́лл·ов·ая кисл·от·а́ (*f*) (biochem) umbellic acid
умбелл·ул·а, **-ы**, (*f*) (zool) umbellula
у́мбр·а, **-ы**, (*f*) (min) umber; (fish) umbra, *Umbra krameri*
~, **обо·жж·ённ·ая** burnt umber
~, **при·ро́д·н·ая** raw umber
ум·е́л·ый, **-ая**, **-ое**, (*a*) able, capable, skilful
ум·е́ни·е, **-я**, (*n*) ability, skill
у·меньш·а́ем·ое, **-ого**, (*n decl as adj*) (math) minuend
у·меньш·а́·ть, **-ют**, (*imp*) reduce, lessen; **-ся** (*pass*); diminish
у·меньш·е́ни·е, **-я**, (*n*) (*v n*) *see* **у·меньш·а́·ть**; decrease, reduction, diminution
~ **амплиту́д·ы** amplitude contraction
~ **жёстк·ост·и** degradation (of spectrum)
~ **эне́рг·и·и** (phys) energy degradation/attenuation
у·меньш·ённ·ый, **-ая**, **-ое**, (*past part pass*) *see* **у·меньш·а́·ть**; scaled-down (in size)
у·меньш·и́тельн·ый, **-ая**, **-ое**, (*a*) reducing; (gram) diminutive
у·ме́ньш·ить, **-ат**, (*perf*) *see* **у·меньш·а́·ть**
у·мер·е́ни·е, **-я**, (*n*) (*v n*) *see* **у·мер·я́·ть**; moderation
у·ме́р·енност·ь, **-и**, (*f*) moderation, moderateness, temperance
у·ме́р·енн·ый, **-ая**, **-ое**, (*past part pass*) *see* **у·мер·я́·ть**; moderate, temperate
~ **кли́мат·** (*m*) temperate climate
~ **што́рм·** (*m*) (meteor) moderate gale
у́·мер·, (*past masc sing*) of **у·мер·е́ть**
у·мер·е́ть, (*fut 3rd pl*) **у·мр·у́т**, (*perf*) die

у·ме́р·ива·ть, -ют, (*imp*) = у·ме-
р·я́·ть

у·ме́р·ить, -ят, (*perf*) *see* у·мер·я́·ть

у́·мер·ло (*past neut sing*) of у·мер·е́ть

у·мертв·и́ть, -я́т, (*perf*) *see* у·мер-
щвл·я́·ть

у·ме́р·ш·ий, -ая, -ее, (*past part act*)
see у·мер·е́ть; dead, deceased

у·мерщвл·я́·ть, -ют, (*imp*) kill, destroy

у·мер·я́·ть, -ют, (*imp*) moderate

у·мест·и́ть, -я́т, (*perf*) *see* у·мещ·а́·ть

у·ме́ст·н·ый, -ая, -ое, (*a*) appropriate,
suitable, pertinent, relevant, timely

ум·е́·ть, -ют, (*imp*) be able, know how

у·мещ·а́·ть, -ют, (*imp*) put/get in
(all), find/have/be room for; place,
put, lodge; -ся (*pass*); get/go in
(completely)

у·мин·а́·ть, -ют, (*imp*) knead, soften
(by working); squash, tread down,
consolidate

у·мир·а́·ть, -ют, (*imp*) die

у·мир·а́ющ·ий, -ая, -ее, (*pres part
act*) of у·мир·а́·ть; dying; moribund

у·миро·твор·е́ни·е, -я, (*n*) pacification;
appeasement

у·мн·ёт, (*fut 3rd sing*) of у·мя́·ть
(*perf*) *see* у·мин·а́·ть (*imp*)

ум·не́·ть, -ют, (*imp*) get/grow wiser

у·множ·а́·ть, -ют, (*imp*) (math etc.)
multiply; increase

у·множ·а́ющ·ий, -ая, -ее, (*pres part
act*) of у·множ·а́·ть

~ у·стро́й·ств·о (*n*) multiplier

у·множ·е́ни·е, -я, (*n*) (*v n*) *see* у·мно-
ж·а́·ть; multiplication; breeding (of
nuclear fuel)

у·множ·и́тел·ь, -я, (*m*) multiplier;
(math) factor

~, втор·и́чн·о-электро́н·н·ый (elec)
electron multiplier, secondary-emis-
sion valve

~, фото·электро́н·н·ый (elec) photomultiplier, multiplier phototube

~, электро́н·н·ый electron multiplier
(vacuum tubes)

у·мно́ж·ить, -ат, (*perf*) *see* у·мно-
ж·а́·ть

у́м·н·ый, -ая, -ое, (*a*) clever, intelligent, sensible

у·мо́·ет, (*fut 3rd sing*) of у·мы́·ть
(*perf*) *see* у·мыв·а́·ть (*imp*)

умо·за·ключ·а́·ть, -ют, (*imp*) deduce,
conclude, come to a conclusion

умо·за·ключ·и́ть, -а́т, (*perf*) *see*
умо·за·ключ·а́·ть

умо·зр·е́ни·е, -я, (*n*) speculation

умо·зр·и́тельн·ый, -ая, -ое, (*a*)
speculative

у·мола́ч·ива·ть, -ют, (*imp*) thresh,
thrash (all)

у·мол·и́ть, -́ят, (*perf*) persuade, move
(by entreaties)

у·молк·а́·ть, -ют, (*imp*) become/fall
silent, stop, cease (of noise)

у·мо́лк·нуть, -нут, (*perf*) *see* у·мол-
к·а́·ть

у·моло́т·, -а, (*m*) grain yield (after
threshing)

у·молот·и́ть, -́ят, (*perf*) thresh, thrash
(all)

у·мо́лч·енн·ый, -ая, -ое, (*past part
pass*) *see* у·ма́лч·ива·ть

у·молч·а́ть, -а́т, (*perf*) *see* у·ма́лч·и-
ва·ть

у·мол·я́·ть, -ют, (*imp*) entreat, implore

умо·по·мрач·е́ни·е, -я, (*n*) temporary
insanity

у·мор·и́ть, -я́т, (*perf*) kill; tire out

у·мр·ёт, (*fut 3rd sing*) of у·мер·е́ть

у́м·ственн·ый, -ая, -ое, (*a*) mental,
intellectual

умфо́рмер·, -а, (*m*) (elec) transformer;
rotary converter, electromechanical
converter (earlier designation; obsolete)

у·мыв·а́льн·ая, -ой, (*f decl as adj*)
wash-room/place

у·мыв·а́льник·, -а, (*m*) lavatory/wash
basin

у·мыв·а́льн·ый, -ая, -ое, (*a*) (*see
also* у·мыв·а́льн·ая) (*as noun*); wash,
washing, ablutionary, lavatory

у·мыв·а́·ть, -ют, (*imp*) wash, perform
ablutions

у́·мысел·, -сл·а, (*m*) secret design/
intention

у·мы́·ть, (*fut 3rd pl*) у·мо́·ют, (*perf*)
see у·мыв·а́·ть

у·мы́шл·енн·ый, -ая, -ое, (*a*) premeditated, intentional, deliberate

у·мягч·а́·ть, -ют, (*imp*) soften, make
soft/softer, mollify

у·мягч·е́ни·е, -я, (*n*) (*v n*) *see* у·мяг-
ч·а́·ть; mollification

у·мягч·и́ть, -а́т, (*perf*) *see* у·мяг-
ч·а́·ть

у·мя́·ть, (*fut 3rd pl*) у·мн·у́т, (*perf*)
see у·мин·а́·ть

уна́би (*n indecl*) (bot) jujube, *Zizyphus*

у·на·во́ж·ива·ть, -ют, (*imp*) spread
dung, manure

у·на·во́з·ить, -ят, (*perf*) *see* у·на·во́-
ж·ива·ть

у·на·след·овать, -уют, (*perf*) inherit

ундекан·, -а, (*m*) (chem) undecane, hendecane

ундекан·ов·ый, -ая, -ое, (*a*) undecanoic

ундекен·, -а, (*m*) undecene

ундецен·, -а, (*m*) undecene

ундецил·, -а, (*m*) undecyl

ундул·ир·овать, -уют, (*imp*) undulate

ундул·яци·я, -и, (*f*) undulation

у·нес·ти, -ут, (*perf*) *see* **у·нос·ить**

универмаг·, -а, (*m*) (= **универсаль·н·ый магазин·**) department store

универсал·, -а, (*m*) (astron) astronomical transit; (surv) engineer's transit

универсаль·ност·ь, -и, (*f*) universality

универсаль·н·ый, -ая, -ое, (*a*) universal, multi-/all-/general-/dual-purpose

~ **газ·ов·ая по·сто·янн·ая** (*f*) (phys) universal gas constant, gas constant

~ **из·мер·итель·н·ый при·бор·** (*m*) (elec) multimeter

~ **инструмент·** (*m*) = **универсал·** (q.v.)

~ **ис·пыт·ательн·ая машин·а** (*f*) (met) universal testing machine

~ **магазин·** (*m*) department store

~ **пере·с·чёт·н·ая схем·а** (*f*) (elec) multiscaler

~ **раз·движ·н·ой га·ечн·ый ключ·** (*m*) adjustable wrench

~ **стал·ь, широко·полос·н·ая** (*f*) universal steel plates

универс·ант·, -а, (*m*) university graduate; student, undergraduate

универс·итет·, -а, (*m*) university

универс·итет·ск·ий, -ая, -ое, (*a*) university

универс·ум·, -а, (*m*) universe

у·ниж·а·ть, -ют, (*imp*) lower, reduce, degrade; humiliate, disparage

у·ниж·ет, (*fut 3rd sing*) of **у·низ·ать,** (*perf*); *see* **у·низ·ыва·ть** (*imp*)

у·ниж·у, (*fut 1st sing*) of **у·низ·ать;** **у·ниж·у,** (*fut 1st sing*) of **у·низ·ить**

у·низ·ать, (*fut 3rd pl*) **у·ниж·ут,** (*perf*); *see* **у·низ·ыва·ть**

у·низ·ить, -ят, (*perf*) *see* **у·ниж·а·ть**

у·низ·ыва·ть, -ют, (*imp*) cover (as with beads), stud, string, thread (till covered)

уникаль·н·ый, -ая, -ое, (*a*) unique

уник·ум·, -а, (*m*) unique person/object/ example; great rarity

уни·курзаль·н·ый, -ая, -ое, (*a*) (math) unicursal

уни·курсаль·н·ый, -ая, -ое, (*a*) (math) unicursal

уни·модуляр·н·ый, -ая, -ое, (*a*) (math) unimodular

уни·поляр·н·ый, -ая, -ое, (*a*) unipolar, homopolar

~ **машин·а** (*f*) (elec) unipolar/homopolar/acyclic generator

унитаз·, -а, (*m*) WC/water-closet pan

унитар·ност·ь, -и, (*f*) unitarity

унитар·н·ый, -ая, -ое, (*a*) unitary

~ **пре·образ·овани·е** (*n*) (math) unitary transformation

~ **топл·ив·о** (*n*) (rocket) monopropellant

унитал·ь·н·ый, -ая, -ое, (*a*) unital, unitary

уни·фик·аци·я, -и, (*f*) unification

уни·филяр·, -а, (*m*) declinometer

уни·филяр·н·ый, -ая, -ое, (*a*) unifilar

уни·фиц·ир·овать, -уют, (*imp and perf*) unify

униформиз·аци·я, -и, (*f*) (math) uniform convergence

у·ничтож·а·ть, -ют, (*imp*) destroy; exterminate, annihilate; abolish

у·ничтож·ени·е, -я, (*n*) (*v n*) *see* **у·ничтож·а·ть;** destruction; extermination, annihilation; abolition, elimination

~, **взаим·н·ое** (math) cancelling out

~ **девиаци·й** (nav) compass correction/ adjustment

у·ничтож·ить, -ат, (*perf*) *see* **у·ничтож·а·ть**

уни·я, -и, (*f*) union

у·нос·, -а, (*m*) (*v n*) *see* **у·нос·ить;** carry-over; (chem) entrainment, removal; flue dust, soot (in chimneys etc.)

~ **масс·ы** ablation

у·нос·ить, -ят, (*imp*) carry/take away/ off; **-ся** (*pass*); rush/fly away/past

у·нош·у, (*pres 1st sing*) of **у·нос·ить**

унц·ев·ый, -ая, -ое, (*a*) of **унц·и·я**

унц·и·я, -и, (*f*) ounce

~, **аптекар·ск·ая** apothecaries' ounce

уны·л·ый, -ая, -ое, (*a*) despondent, doleful; dismal, depressing

Уокер·а, лаг· (*m*) (naut) Neptune log (Walker's)

Уоррик·а, катализт· (*m*) (plast) Warwick catalyst

у·па·вш·ий, -ая, -ее, (*past part act*) of **у·пас·ть**

у·пад·ёт, (*fut 3rd sing*) of **у·пас·ть**

у·па́д·ок·, -д·к·а, (*m*) decline, depression, decay, decrease

у·па́д·очн·ый, -ая, -ое, (*a*) declining, depressing, depressive, decadent

у·пак·о́ванн·ый, -ая, -ое, (*past part pass*) see у·пак·о́выва·ть

~, не·пло́т·н·о (phys) loose-packed

~, пло́т·н·о (phys) close-packed

у·пак·ова́ть, -у́ют, (*perf*) see у·пак·о́выва·ть

у·пак·о́вк·а, -и, (*g pl*) -вок·, (*f*) packing, packaging; packing material, packaging, pack

~, гексагона́ль·н·ая плот·не́йш·ая (phys) hexagonal tight packing

~, гексагона́ль·н·ая прост·е́йш·ая (phys) hexagonal close packing
    дефе́кт· у·пак·о́вк·и (*m*) (cryst) stacking fault

~, за·кры́·т·ая sealed package

~, пло́т·н·ая (phys) close packing

у·пак·о́вочн·ый, -ая, -ое, (*a*) packing; packaging, wrapping; (*f decl as adj*) packing-room

~ бума́г·а (*f*) packaging/wrapping paper

~ коэффицие́нт· (*m*) (phys) packing fraction

~ пресс· (*m*) baling press

~ цех· (*m*) packaging shop/department/shed

у·пак·о́выва·ть, -ют, (*imp*) pack; package, pack/wrap up, make into a parcel/bale/pack

у·па́·л, (*past masc sing*) of у·па́с·ть

у·пар·е́ни·е, -я, (*n*) (*v n*) see у·па́р·ива·ть; (chem) inspissation

у·па́р·ивани·е, -я, (*n*) (*v n*) see у·па́р·ива·ть, у·пар·е́ни·е

у·па́р·ива·ть, -ют, (*imp*) steam (to process); concentrate by evaporation, boil down

у·па́р·ить, -ят, (*perf*) see у·па́р·ива·ть

у·па́р·к·а, -и, (*f*) (*v n*) see у·па́р·ива·ть

у·па́с·ть, (*fut 3rd pl*) у·пад·у́т, (*past masc sing*) у·па́·л, (*perf*) fall, drop, sink

у·пер·е́ть, (*fut 3rd pl*) у·пр·у́т, (*perf*) see у·пир·а́·ть

у·пир·а́·ть, -ют, (*imp*) prop, set against, (mech) thrust; -ся (*pass*); lean, support; hinge, turn (of argument etc.)

у·пит·а́·ть, -ют, (*perf*) see у·пи́т·ыва·ть

у·пи́т·ыва·ть, -ют, (*imp*) fatten, nourish

у·пла́т·а, -ы, (*f*) payment, paying

~ част·я́ми payment by instalments

у·плат·и́ть, -́ят, (*perf*) see у·пла́ч·ива·ть

у·пла́ч·ива·ть, -ют, (*imp*) pay, discharge (debts)

у·плотн·е́ни·е, -я, (*n*) (*v n*) see у·плотн·я́·ть; compaction, consolidation, condensation, thickening; (mech) seal, packing, gland; (telecom) multiplex, multiplexing

~, ва́куум·н·ое vacuum seal

~, вод·ян·о́е water seal, hydraulic packing

~, времен·н·о́е (telecom) time-division multiplex

~, гермет·и́чн·ое tight seal, vacuum seal

~, конта́кт·н·ое contact/bearing seal

~, лабири́нт·ов·ое (st eng) labyrinth packing

~, по·движ·н·о́е (eng) running/floating packing

~, са́ль·ников·ое packing gland, gland
    с·кач·о́к у·плот·не́ни·я (*m*) (aerodynam) shock wave, shock (*see under* с. у.)

~, у́голь·н·ое (st eng) carbon gland

~, част·о́тн·ое (telecom) frequency-division multiplex

у·плотн·ённост·ь, -и, (*f*) compaction; (telecom) multiplex

у·плотн·ённ·ый, -ая, -ое, (*past part pass*) see у·плотн·я́·ть

~ кана́л· (*m*) (telecom) shared channel

~ по·ро́д·а (*f*) consolidated rock

~ по·се́в· (*m*) (agr) interplanting; companion crop

~ по́чв·а (*f*) compact/compacted/consolidated soil

~ с·вя́з·ь (*f*) (telecom) multichannel line/link

у·плотн·и́тел·ь, -я, (*m*) seal, sealing agent; thickener; compactor

у·плотн·и́тельн·ый, -ая, -ое, (*a*) seal, sealing

у·плотн·и́ть, -я́т, (*perf*) see у·плотн·я́·ть

у·плотн·я́емост·ь, -и, (*f*) compactibility

у·плотн·я́·ть, -ют, (*imp*) compact, consolidate; thicken, condense (liquids, emulsions, loose materials); close up (gaps), make/put/arrange closer, concentrate; tighten, seal (joints etc.), pack (glands etc.); do in less time; do more during

у·плотн·я́ющ·ий, -ая, -ее, (*pres part act*) of у·плотн·я́·ть

у·плотн·я́ющ·ий, -ая, -ее
~ кольц·о́ (*n*) sealing ring, ring seal
у·площ·ённост·ь, -и, (*f*) flatness
у·площ·ённ·ый, -ая, -ое, (*a*) flat
у·плыв·а́·ть, -ют, (*imp*) swim/drift/float/sail away
у·плы́·ть, (*fut 3rd pl*) у·плыв·у́т, (*perf*) *see* у·плыв·а́·ть
у·по·до́б·ить, -ят, (*perf*) *see* у·по·доб·л·я́·ть
у·по·добл·я́·ть, -ют, (*perf*) liken to, compare with; -ся (*pass*); become like/similar
у·полз·а́·ть, -ют, (*imp*) crawl/creep away
у·полз·ти́, -у́т, (*perf*) *see* у·полз·а́·ть
у·полно·мо́ч·енн·ый, -ая, -ое, (*past part pass*) *see* у·полно·мо́ч·ива·ть; (*m decl as adj*) authorized agent, representative; (dipl) plenipotentiary
у·полно·мо́ч·ива·ть, -ют, (*imp*) empower, authorize
у·полно·мо́ч·и·е, -я, (*n*) authority
    по у·полно·мо́ч·и·ю on behalf of
у·полно·мо́ч·ить, -ат, (*perf*) *see* у·полно·мо́ч·ива·ть
у·по·мин·а́ем·ый, -ая, -ое, (*pres part pass*) of у·по·мин·а·ть; mentioned
у·по·мин·а́·ть, -ют, (*imp*) mention, refer to
у·по·мя·н·у́ть, -у́т, (*perf*) *see* у·по··мин·а́·ть
у·по́р·, -а, (*m*) (*v n*) *see* у·пир·а́·ть; (mech) stop, thrust pad/block; support, chock; thrust
~, бу́фер·н·ый (rail) buffers
    в у·по́р· (*adv*) point-blank; face-to-face
~, кольц·ев·о́й (mech eng) clamping-ring stop
~, ме́р·н·ый measuring gauge/gage
~, не·по·дви́ж·н·ый positive stop (on machine tool)
~, пере·движ·н·о́й (met roll) stop-and-measuring gear, shear gauge/gage
~, под·вод·и́м·ый adjustable wedge stop (on a jig)
~, туп·ико́в·ый (rail) fixed buffer
~, у·стана́вл·ивающ·ийся gauge stop (machine tools)
у·по́р·к·а, -и, (*g pl*) -р·ок·, (*f*) (mech eng) counter-stop, sizing stop; stretcher (in rowing boat)
у·по́р·н·ый, -ая, -ое, (*a*) (mech) stop, thrust; stubborn, tenacious, persistent

у·по́р·н·ый, -ая, -ое
~ болт· (*m*) (mech) stop, stop bolt
~ брус· (*m*) (civ eng) shore (support)
~ вал· (*m*) (mech) thrust shaft
~ винт· (*m*) backing-up/locking screw
~ кат·о́к (*m*) bogie wheel
~ кольц·о́ (*n*) (mech) thrust collar
~ поду́ш·к·а (*f*) (mech) thrust pad
~ под·ши́п·ник· (*m*) (mech) thrust bearing
у·по́р·ств·о, -а, (*n*) persistence, tenacity; obstinacy
у·по·ря́д·очени·е, -я, (*n*) (*v n*) *see* у·по·ря́д·очива·ть; (phys) ordering
~, спи́н·ов·ое (nucl) spin ordering
у·по·ря́д·оченност·ь, -и, (*f*) orderliness, degree of order, rank, order
у·по·ря́д·оченн·ый, -ая, -ое, (*past part pass*) *see* у·по·ря́д·очива·ть
~ пере·ме́н·н·ая (*f*) (math) ordered variable
у·по·ря́д·очива·ть, -ют, (*imp*) regulate, order, put in order
у·по·ря́д·оч·ить, -ат, (*perf*) *see* у·по··ря́д·очива·ть
у·по·треб·и́тельн·ый, -ая, -ое, (*a*) commonly/generally used, common, customary
у·по·треб·и́ть, -ят, (*perf*) *see* у·по··требл·я́·ть
у·по·требл·е́ни·е, -я, (*n*) use, usage
у·по·требл·я́·ть, -ют, (*imp*) use, make use of
управ·де́л·, -а, (*m*) = у·прав·дел·а́ми
у·прав·дел·а́ми (*m indecl*) office manager
у·пра́в·ить, -ят, (*perf + instr*) *see* у·правл·я́·ть
у·правл·е́ни·е, -я, (*n*) (*v n*) *see* у·правл·я́·ть; direction, directorate, board; management, administration (organiz); control; steering, handling (vehicles, ships); (GW) guidance
~, авто·колеб·а́тельн·ое direct sensitivity control
~, автоно́м·н·ое (GW) internal control, preset guidance
~, гла́в·н·ое Chief directorate, Main administration (Soviet organization)
~, двой·н·о́е dual control; (M/T) double steering
~, дву·сторо́н·ее dual control
~ дел·а́ми secretarial department; registry
~, дистанцио́н·н·ое remote control
~ дро́ссель·н·ой за·сло́н·к·ой (ICE) throttle control

у·правл· éни·е

~ за·хóд·ом на по·сáд·к·у   (air) approach control

~, золот·никóв·ое (autom) valve operation control

~, инерциóн·н·ое (GW) inertia guidance

~, кнóп·очн·ое   button/push-button control

~ кýрс·ом (rocket) yaw control
лúн·и·я у·правл·éни·я (f) (rail) control line
механ·úзм· у·правл·éни·я (m) control mechanism; (M/T) control system

~ огн·ём (gunn) fire control

~ по крéн·у (rocket) rolling control

~ по про·из·вóд·н·ой derivative control

~ по рáдио radio control, control by radio

~ по·лож·éни·ем (rocket) attitude/position control; positioning

~ по·сáд·к·ой с земл·ú (air) ground control of landing

~, прогрáмм·н·ое programme control; (GW) preset control
пункт· у·правл·éни·я по·сáд·к·ой (m) (air) approach control point

~ пýск·ом start-up control

~ раз·мéр·ами dimension control (machine tools)

~ реверс·úровани·ем плáмен·и flame control reversal

~, рул·ев·óе (a/c) flying controls; (M/T) steering

~, руч·н·óе manual control
сигнáл· у·правл·éни·я (m) (autom) input/control signal; (GW) steering command

~, со·в·мещ·ённ·ое (nucl) integrated control

~ стерж·н·ями control of rod types

~, ступ·éнчат·ое (elec) cascade control

~, тóч·н·ое sensitive control

~ тяг·ой (rocket) control by impulse; throttle control (of jet engine)

~, цифр·ов·óе digital control

~, числ·ов·óе numerical control

~ шáг·ом pitch control (of propeller)

у·правл·яемост·ь, -и, (m) manoeuvrability, maneuvrability; controllability

у·правл·яем·ый, -ая, -ое, (pres part pass) of у·правл·я·ть; controlled, guided

~, дистанциóн·н·о remote-controlled

у·правл·яем·ый, -ая, -ое

~ с·на·ряд· (m) guided missile

у·правл·я·ть, -ют, (imp + instr) govern, rule, administer, direct, control, be in charge of, be in command of; (imp only) operate, control, steer, drive; conduct (music etc.)

у·правл·яющ·ий, -ая, -ее, (pres part act) of у·правл·я·ть; (m decl as adj) manager, director; head, chief, man in charge

~ вы·числ·úтельн·ая машúн·а (f) control computer

~ дел·áми office/business manager

~ огн·ём (gunn) fire-control officer

~ у·стрóй·ств·о (n) (autom) control unit

упражн·éни·е, -я, (n) (v n) see упражн·я·ть; exercise

упражн·я·ть, -ют, (imp) exercise

у·празд·нéни·е, -я, (n) (v n) of у·п·раздн·úть; abolition, abolishment

у·праздн·úть, -ят, (perf) abolish

у·праздн·я·ть, -ют, (imp) abolish

у·прáш·ивани·е, -я, (n) (v n) of у·прáш·ива·ть; persuasion

у·прáш·ива·ть, -ют, (imp) try to persuade; persuade

у·пред·úть, -ят, (perf) see у·прежд·á·ть

у·прежд·á·ть, -ют, (imp) forestall

у·прежд·éни·е, -я, (n) (v n) see у·прежд·á·ть; anticipation, lead, advance; (gunn) aim-off, deflection, lead (anti-tank)

~, лин·éйн·ое (gunn) linear allowance
лúн·и·я у·прежд·éни·я (f) line of flight (ballistics)

~ при·лú·в·а (ocean) priming of the tides
ýгол· у·прежд·éни·я (m) (eng) angle of advance/lead; deflection angle, angular allowance (a/c gunnery)

у·прежд·ённ·ый, -ая, -ое, (past part pass) see у·прежд·á·ть; (gunn) future

~ дáль·ност·ь (f) future range

упрёк·, -а, (m) reproach, rebuke

упрек·á·ть, -ют, (imp) reproach, rebuke

упрек·н·ýть, -ýт, (perf) see упрек·á·ть

у·пр·ёт, (fut 3rd sing) of у·пер·éть (perf) see у·пир·á·ть (imp)

у·прос·úть, -́ят, (perf) persuade

у·прост·úть, -ят, (perf) see у·про·щ·á·ть

у·прóч·ени·е, -я, (n) (v n) see у·прóч·ива·ть, and see у·проч·нéни·е

у·пр́оч·ива·ть, -ют, (*imp*) strengthen; make permanent, confirm

у·пр́оч·ить, -ат, (*perf*) *see* **у·пр́оч·ива·ть**

у·проч·н́ени·е, -я, (*n*) strengthening, toughening; (plast) reinforcement; (met) strain hardening
  коэффици́ент· у·проч·н́ени·я (*m*) (mech) modulus of strain hardening

~, механ·́ическ·ое (met) strain hardening/strengthening
  м́одул·ь у·проч·н́ени·я modulus of strain hardening

~, по·в́ерх·ностн·ое wear hardening

у·пр́ош·енн·ый, -ая, -ое, (*past part pass*) of **у·прос·́ить**

у·прощ·́а·ть, -ют, (*imp*) simplify, make less complicated; **-ся** (*pass*); become simpler

у·прощ·́ени·е, -я, (*n*) simplification, simplifying

у·прощ·ённост·ь, -и, (*f*) simplicity

у·прощ·́енств·о, -а, (*n*) oversimplification

у·прощ·́енск·ий, -ая, -ое, (*a*) oversimplified

у·прощ·́енчеств·о, -а, (*n*) *see* **у·про·щ·́енств·о**

у·пр́уг·ий, -ая, -ое, (*a*) elastic, resilient, flexible

~ деформ·́аци·я (*f*) elastic strain/deformation

~ л́ин·и·я (*f*) elastic line/axis (of beam under test)

~ м́уфт·а (*f*) (mech eng) flexible coupling

~ по·сто·́янн·ая (*f*) (mech) elastic stiffness coefficient

~ ш́айб·а (*f*) spring washer

~ элем́ент· (*m*) (instr) pressure-sensitive element

упруго·в́яз·к·ий, -ая, -ое, (*a*) viscoelastic

упруго·пласт·́ическ·ий, -ая, -ое, (*a*) elastoplastic

упруго·с·в́яз·анн·ый, -ая, -ое, (*a*) flexible/elastically attached

у·пр́уг·ост·ь, -и, (*f*) resilience; elasticity (of a solid); pressure, tension (of a gas)

~ диссоци·́аци·и (phys, chem) dissociation pressure

~, за·п́азд·ывающ·ая delayed elasticity

~, кас·́ательн·ая elasticity in shear
  коэффици́ент· у·пр́уг·ост·и (*m*) coefficient of elasticity

у·пр́уг·ост·ь

~, мг·нов́енн·ая instantaneous elasticity
  м́одул·ь кас·́ательн·ой у·пр́уг·ост·и (*m*) modulus of elasticity in shear, shear modulus
  м́одул·ь мг·нов́енн·ой у·пр́уг·ост·и (*m*) modulus of instantaneous elasticity (rubber)
  м́одул·ь норм́аль·н·ой у·пр́уг·ост·и (*m*) (mech) modulus of elasticity, Young's modulus
  м́одул·ь об·ъ·ём·н·ой у·пр́уг·ост·и (*m*) volumetric modulus of elasticity, modulus of cubic compressibility, bulk modulus
  м́одул·ь по·пер́еч·н·ой у·пр́уг·ост·и (*m*) modulus of transverse elasticity, modulus of elasticity in shear, modulus of rigidity
  м́одул·ь про·д́оль·н·ой у·пр́уг·ост·и (*m*) modulus of elasticity, Young's modulus
  м́одул·ь у·пр́уг·ост·и (*m*) elastic modulus
  м́одул·ь у·пр́уг·ост·и при с·ж́а·т·и·е (*m*) (mech) volumetric modulus of elasticity

~, норм́аль·н·ая elasticity, elasticity in tension

~, об·ъ·ём·н·ая elasticity of bulk; (rubber) bulk modulus

~ п́ар·а vapour tension/pressure
  пре·д́ел· у·пр́уг·ост·и (*m*) elasticity limit (of a material)
  пре·д́ел· у·пр́уг·ост·и физ·́ическ·ий true elastic limit

~ при с·ж́а·т·и·и elasticity of bulk

~ рас·твор·́ени·я (chem) solution pressure

у·пр́уг·о-у·ст́ой·чив·ый, -ая, -ое, (*a*) elastically stable

у·пр·́ут, (*fut 3rd pl*) of **у·пер·́еть** (*perf*) *see* **у·пир·́а·ть** (*imp*)

у·пр́яж·к·а, -и, (*g pl*) **-ж·ек·,** (*f*) team (of draught/draft animals); harness

у·пряж·н·́ой, -́ая, -́ое, (*a*) draught, draft, dray (of animals); (rail) coupling

~ крюк· (*m*) (rail) coupling hook

́у·пряж·ь, -и, (*f*) harness; (rail) coupling

у·пр́ям·ый, -ая, -ое, (*a*) obstinate, stubborn, intractable

у·пуск·́а·ть, -ют, (*imp*) let go/slip, allow to escape; miss (an opportunity); miss out, omit; overlook (e.g. an error, a task)

у·пуст·и́ть, ∠ят, (*perf*) *see* **у·пуск·а́·ть**

у·пущ·е́ни·е, -я, (*n*) omission; dereliction, oversight

у·пу́щ·енн·ый, ‚ая, -ое, (*past part pass*) of **у·пуст·и́ть,** (*perf*); *see* **у·пуск·а́·ть**

упы́р·ь, -я́, (*m*) (zool) vampire, *Vampirus*

у·пятер·ённ·ый, -ая, -ое, (*a*) quintuple, fivefold

у·пятер·и́ть, -я́т, (*perf*) *see* **у·пятер·я́·ть**

у·пятер·я́·ть, -ют, (*imp*) multiply by five, quintuplicate

у́р·, -а, (*m*) (pal) urus, *Bos primigenius*

у·равн·е́ни·е; -я, (*n*) (*v n*) *see* **у·равн·я́·ть** (*see also under associated words*); (math) equation

~ в кон·е́чн·ых ра́з·ност·ях (math) differential equation

~, век·о́в·ое secular equation

~, волн·ов·о́е wave equation

~ вре́мен·и (astron) equation of time

~, вс·по·мог·а́тельн·ое auxiliary/ reduced equation

~ втор·о́й сте́пен·и quadratic equation

~ движ·е́ни·я equation of motion

~, дифференциа́ль·н·ое differential equation

~, канон·и́ческ·ое canonical correlation

~, квази·линеа́р·н·ое quasi-linear equation

~ крив·о́й equation of a curve

~, куб·и́ческ·ое cubic equation

~, лин·е́йн·ое linear equation

~ моме́нт·ов momental equation

~, не·о·пре·дел·ённ·ое indeterminate equation

~ не·пре·ры́в·ност·и (dynam) equation of continuity, continuity equation

~, не·при·вод·и́м·ый irreducible equation

~ не·раз·ры́в·ност·и (dynam) continuity equation, equation of continuity

~, не·со·в·ме́ст·н·ые inconsistent/ incompatible equation

~, норма́ль·н·ое standard equation

~, обык·нове́нн·ое дифферен·циа́ль·н·ое ordinary differential equation

~, одно·ро́д·н·ое homogeneous equation

~, о·пре·дел·я́ющ·ее fundamental equation

~, пре·образ·о́ванн·ое modified equation

у·равн·е́ни·е

~, про·из·во́д·н·ое derived equation

~, равно·ве́с·н·ое balance equation

~, равно·си́ль·н·ое equivalent equation

~, ра́з·ностн·ое difference equation (math)

~, раз·реш·а́ющ·ее resolvent equation

~, само·со·пряж·ённ·ое self-adjoint equation

~, со·в·ме́ст·н·ые (*pl*) compatibility equations

~ со·сто·я́ни·я equation of state

~, у·сло́в·н·ое conditional equation

~ Шре́дингер·а Schroedinger's equation

у·ра́вн·ивани·е, -я, (*n*) (*v n*) of **у·ра́вн·ива·ть;** equalization; (elec) balancing; (surv) adjustment

~ координа́т· adjustment of coordinates

~ се́т·и (surv) station adjustment

у·ра́вн·ива·ть, -ют, (*imp*) equalize, equate; level, smooth, even, make level/smooth/even

у·равн·и́тел·ь, -я, (*m*) equalizer, evener, leveller; (elec) balancer

у·равн·и́тельн·ый, -ая, -ое, (*a*) equalizing, equating, compensating; levelling, leveling

уравно·ве́с·ить, -ят, (*perf*) *see* **уравно·ве́ш·ива·ть**

уравно·ве́ш·енн·ый, -ая, -ое, (*past part pass*) *see* **уравно·ве́ш·ива·ть**

~ тя́г·а (*f*) balanced draught (of boiler)

~ цеп·ь (*f*) (elec) balanced circuit

уравно·ве́ш·ивани·е, -я, (*n*) (*v n*) of **уравно·ве́ш·ива·ть;** compensation; (geol) isostasy, isostacy

~, груз·ов·о́е mechanical balancing

уравно·ве́ш·ива·ть, -ют, (*imp*) balance, equilibrate, counterbalance, compensate

уравно·ве́ш·ивающ·ий, -ая, -ее, (*pres part act*) of **уравно·ве́ш·ива·ть**

~ по́рш·ен·ь (*m*) dummy balance piston (turbine)

у·равн·я́·ть, -ют, (*perf*) equalize, equate

урага́н·, -а, (*m*) (meteor) hurricane (12: Beaufort scale)

урага́н·н·ый, -ая, -ое, (*a*) hurricane

~ што́рм· (*m*) hurricane (12: Beaufort scale)

у·раз·ум·ева́·ть, -ют, (*imp*) understand, make out

у·раз·ум·е́·ть, -ют, (*perf*) *see* **у·раз·у·м·ева́·ть**

уракон*и́*т·, -а, (*m*) (min) uraconite

урали́т·, -а, *(m)* (min) uralite

уралит·иза́ци·я, -и, *(f)* (geol) uralitization

ура́н·, -а, *(m)* (chem) uranium, U; (astron) Uranus

∼ **в блóк·ах** lump/lumped uranium

∼**, пяти·хлóр·и́ст·ый** (chem) uranium pentachloride

уран·а́т·, -а, *(m)* (chem) uranate

уран·и́д·, -а, *(m)* (chem) uranide

уран·и́л·, -а, *(m)* (chem) uranyl

∼**, сер·ни́ст·ый** uranyl sulphate/sulfate

уранил·сульфа́т·, -а, *(m)* uranyl sulphate/sulfate

уранини́т·, -а, *(m)* (min) uraninite

уран·и́т·, -а, *(m)* (min) uranite, pitchblende

∼**, ба́р·иев·ый** barium uranite

уранноби́т·, -а, *(m)* (min) uranniobite

ура́н·ов·ый, -ая, -ое, *(a)* uranium; (chem) uranic

∼ **кисл·от·а́** *(f)* (chem) uranic acid

∼ **об·ма́н·к·а** *(f)* (min) pitchblende

∼ **смóл·к·а** *(f)* (min) pitchblende

∼ **слюд·к·а** *(f)* (min) uranite

урано·графи́т·н·ый, -ая, -ое, *(a)* (nucl) graphite-moderated uranium

урано·гра́ф·и·я, -и, *(f)* (astron) uranography

урано·лепиди́т·, -а, *(m)* (min) uranolepidite

урано·пили́т·, -а, *(m)* (min) uranopilite

урано·пла́ст·ик·а, -и, *(f)* uranoplasty (surgery)

урано·со·держ·а́щ·ий, -ая, -ее, *(a)* uranium-containing; (geol) uranium-bearing, uraniferous

урано·сфери́т·, -а, *(m)* (min) uranosphaerite

урано·талли́т·, -а, *(m)* (min) uranothallite

урано·тори́т·, -а, *(m)* (min) uranothorite

урано·фа́н·, -а, *(m)* (min) uranophane

урано·цирци́т·, -а, *(m)* (min) uranocircite

ураци́л·, -а, *(m)* (chem) uracil

урбан·иза́ци·я, -и, *(f)* (econ) urbanization

уре·а́з·а, -ы, *(f)* (biochem) urease

у·регул·и́р·овать, -уют, *(perf)* regulate, adjust; regularize

уредо·спóр·а, -ы, *(f)* (bot) uredospore

у·ре́ж·ут, *(fut 3rd pl)* of **у·ре́з·ать**

у·ре́з·, -а, *(m)* *(v n)* *see* **у·рез·а́ть;** cut-away portion

∼ **вод·ы́** water line (on shore)

у·рез·а́·ть, -ют, *(imp)*; **у·ре́з·ать,** *(fut 3rd pl)* **у·ре́ж·ут,** *(perf)* cut, cut down, reduce, curtail, shorten (by cutting)

уре·и́д·, -а, *(m)* (biochem) ureid

урем·и́ческ·ий, -ая, -ое, *(a)* (med) uremic

ур·еми́·я, -и, *(f)* (med) uraemia, uremia

ур·ета́н·,. -а, *(m)* (chem) urethane

уретер·алги́·я, -и, *(f)* (med) ureteralgia

уретер·и́т·, -а, *(m)* (med) ureteritis

уретеро·ли́т·, -а, *(m)* (med) ureterolith

уретеро·ррафи́·я, -и, *(f)* (med) ureterorrhaphy

уре́тр·а, -ы, *(f)* (anat) urethra

уретр·а́льн·ый, -ая, -ое, *(a)* (anat) urethral

уретр·и́т·, -а, *(m)* (med) urethritis

уретро·вагина́ль·н·ый, -ая, -ое, *(a)* (anat) urethrovaginal

уретро·скóп·, -а, *(m)* (med) urethroscope

урика́з·а, -ы, *(f)* (biochem) uricase

ури́н·а, -ы, *(f)* (med) urine

у́рн·а, -ы, *(f)* urn; ballot box; litter box/receptacle

уро·баци́лл·а, -ы, *(f)* (bact) urobacillus

уро·били́н·, -а, *(m)* (biochem) urobilin

уро·билиноге́н·, -а, *(m)* (biochem) urobilinogen

у́·ровен·ь, -вн·я, *(m)* level; (build, surv) spirit level, level tube, level

∼**, без·о·па́с·н·ый** safe level

∼ **бе́л·ого** (TV) white/bright level

∼ **ве́ч·н·ой мерзл·от·ы́** (geophys) permafrost table

∼ **вод·ы́** water level, height of water (above mean level)

∼**, вращ·а́тельн·ый** (phys) rotational level

∼ **грóм·кост·и** (telecom, cinema) sensation level

∼ **грунт·ов·ы́х вод·** (geol) water table/plane, ground water level

∼**, жи́зн·енн·ый** (econ) standard of living

∼**, за́·ня·т·ый** (phys) occupied level/band

∼ **за·пир·а́ни·я** (elec) blanking level

∼**, ис·хóд·н·ый** initial level; level of reference, datum level

∼ **квант·ова́ни·я** (phys) quantization level

∼ **конденс·а́ци·и** (meteor) condensation level

ли́н·и·я у́·ровн·я *(f)* (surv) contour line

у́·ровен·ь
~, меж·е́нн·ый mean level (of water)
~ мо́р·я sea level (in USSR, zero on scale at Kronshtat)
~, нул·ев·о́й datum level
~, о·по́р·н·ый base level
~, по·пере́ч·н·ый cross level
~ при·вед·е́ни·я datum level
~ при·ли́·в·а, сре́д·н·ий mean tide level
~ про·ка́т·к·и (met roll) pass line
~, пьезо·метр·и́ческ·ий (hydr) piezometric surface
~, с·вя́з·анн·ый (phys) bound level
~ си́л·ы звук·а (acous) intensity level
~, труб·чат·ый tubular spirit level, level tube
~ черн·е́е чёрн·ого (TV) blacker-than-black level
~ чёрн·ого (TV) black level
~, шта́рк·овск·ий (phys) Stark level
уровне·ме́р·, -а, (m) level gauge/gage
у́·ровн·я́·ть, -ют, (perf) level/smooth/even out, make level/smooth/even
уро·генита́ль·н·ый, -ая, -ое, (a) (anat) urogenital
у·ро́д·, -а, (m) (biol) monster, monstrosity, freak
у·род·и́ть, -я́т, (perf) see у·рожд·а́·ть
у·ро́д·ливост·ь, -и, (imp) deformity
у·ро́д·лив·ый, -ая, -ое, (a) deformed, misshapen, abnormal, distorted
у·ро́д·овать, -уют, (imp) disfigure, mutilate
у·ро́д·ств·о, -а, (n) (biol) monstrosity
у·рож·а́·й, -я, (m) harvest, crop; yield
у·рож·а́йност·ь, -и, (f) (agr) yield, yielding/cropping capacity
у·рожд·а́·ть, -ют, (imp) bear, give a crop
у·рожд·ённ·ая, -ой, (f adj) née, born (of married women)
у·рож·е́нец·, -нц·а, (i) -нц·ем, (m) native, native son
уро́к·, -а, (m) lesson
уро·ло́г·и·я, -и, (f) (med) urology
у·ро́н·, -а, (m) losses, damage
у·рон·и́ть, -я́т, (perf) drop, let fall; shed
уро́н·ов·ая кисл·от·а́ (f) (chem) uronic acid
уро·порфири́н·, -а, (m) (biochem) uroporphyrin
уро·поэ́з·, -а, (m) (physiol) uropoiesis
уро·сти́л·ь, -я, (m) (zool) urostyle
уро·сульфа́н·, -а, (m) (pharm) urosulfane

уро·тропи́н·, -а, (m) (pharm) urotropine
уро·хро́м·, -а, (m) (biochem) urochrome
уро́ч·н·ый, -ая, -ое, (a) fixed, determined
уру́т·ь, -и, (f) (bot) foxtail, *Myriophyllum*
урч·а́ть, -а́т, (imp) rumble
урюк·, -а, (m) dried apricot
у́с·, -а, (nom pl) -ы́, (g pl) -о́в· (m) whisker; (fish) barbel; (ent) feeler, antenna; (bot) tendril, awn, beard; (pl) moustache
~, кит·о́в·ый whalebone
со·един·е́ни·е в ус· (n) mitre/miter joint (carpentry)
у·са́д·ебн·ый, -ая, -ое, (a) of у·са́д·ьб·а
у·сад·и́ть, -я́т, (perf) see у·са́ж·ива·ть
у·са́д·к·а, -и, (g pl) -д·ок·, (f) (v n) see у·са́ж·ива·ть; shrinkage, contraction (of materials), settling (of levels)
~ оборо́т·ов (a/c) rev drop, r.p.m. drop
~ стру́ж·к·и cutting ratio (metal machining)
у·са́д·очн·ый, -ая, -ое, (a) of у·са́д·к·а
~ метр· (m) (met cast) patternmaker's shrink rule, shrinkage/moulder's rule
~ пуст·от·а́ (f) (met) shrinkage/contraction cavity
~ ра́ковин·а (f) (met) pipe, piping
~ шов· (m) (build) keyed expansion joint
у·са́д·ьб·а, -ы, (g pl) -д·еб·, or -д·ьб·, (f) farm, farm buildings; manor house
у·саж·а́·ть, -ют, (perf) plant (over an area); cover (with many)
у·са́ж·енн·ый, -ая, -ое, (past part pass) see у·са́ж·ива·ть
у·са́ж·ива·ть, -ют, (imp) seat, ask to sit down, make sit down; set (seat person for task); place, set down (things); plant out (over an area); cover (with many); -ся (pass and reflex); take a seat, settle, shrink
ус·а́тк·а, -и, (f) (bot) awned wheat, *Triticum vulgare* var. *aristota*
ус·а́т·ый, -ая, -ое, (a) whiskered, moustached; (bot) awned, bearded, barbate
~ син·и́ц·а (orn) bearded titmouse, *Parus biarmicus*
ус·а́ч·, -а́, (i) -о́м, (m) (fish) barbel, *Barbus fluviatilis*; (ent) longhorn beetle, (pl) *Cerambycidae*

у·сва́·ива·ни·е, -я, (*n*) *see* у·сво·е́ни·е

у·сва́·ива·ть, -ют, (*imp*) adopt, appropriate; assimilate; master, learn, become familiar with

у·сво·е́ни·е, -я, (*n*) (*v n*) *see* у·сва́·и·ва·ть; adoption, appropriation; assimilation, familiarization; mastery (of subject)

у·сво́·енн·ый, -ая, -ое, (*past part pass*) *see* у·сва́·ива·ть

у·сво·и́ть, -ят, (*perf*) *see* у·сва́·ива·ть

у·сво·я́емост·ь, -и, (*f*) assimilability; comprehensibility

у·се́·вшийся, -аяся, -ееся, (*past part act*) of у·се́ст·ься

у·се́·ива·ть, -ют, (*imp*) sow (an area); cover, scatter with

у·сек·а́·ть, -ют, (*imp*) cut off, shorten (by cutting), cut short, cut, truncate, (med) excise

у·се́·лся, (*past masc sing*) of у·се́ст·ься

у·семер·и́ть, -ят, (*perf*) *see* у·семер·я́·ть

у·семер·я́·ть, -ют, (*imp*) increase seven times

у·се́рд·н·ый, -ая, -ое, (*a*) zealous, diligent

у·се́ст·ься, (*fut 3rd pl*) у·ся́д·утся (*perf*) take a seat, settle

у·сеч·ён·ный, -ая, -ое, (*past part pass*) *see* у·сек·а́·ть

~ ко́нус· (*m*) (math) frustum of a cone

~ пирами́д·а (*f*) (math) frustum of a pyramid

~ призм· (*m*) (math) truncated prism

у·се́чь, (*fut 3rd sing, pl*) у·сеч·ёт, у·сек·у́т, (*past masc sing*) у·се́к· (*perf*) *see* у·сек·а́·ть

у·се́·янн·ый, -ая, -ое, (*past part pass*) of у·се́·ять

у·се́·ять, -ют, (*perf*) *see* у·се́·ива·ть

у·сид·е́ть, -я́т, (*perf*) *see* у·си́ж·ива·ть

у·си́д·чив·ый, -ая, -ое, (*a*) assiduous

у·си́ж·ива·ть, -ют, (*imp*) remain seated/sitting

у́с·ик·, -а, (*m*) (*dim*) of ус· (q.v.)

у·сил·е́ни·е, -я, (*n*) (*v n*) *see* у·си́л·и·ва·ть; strengthening, reinforcement, intensification, heightening; (elec) amplification; (rad) gain

~, га́з·ов·ое (elec) gas amplification

~ ко́нтур·а регул·и́ровани·я (rad) control-loop gain

коэффицие́нт· у·сил·е́ни·я (*m*) mu-factor, magnification factor; (elec) amplification factor

у·сил·е́ни·е

~, пред·вар·и́тельн·ое (elec) pre-amplification

~ про·меж·у́точн·ой част·от·ы́ (rad) supersonic amplification

у·си́л·енн·ый, -ая, -ое, (*past part pass*) *see* у·си́л·ива·ть; heightened; stricter; aggravated (of undesirable effects)

~ за·ко́н· (*m*) stricter law, more rigorous law

у·си́л·ива·ть, -ют, (*imp*) strengthen, make stronger, reinforce, intensify, amplify (sound etc.), increase, enhance; -ся (*pass*); become stronger, become more intense, grow, increase

у·си́л·ивающ·ий, -ая, -ее, (*pres part act*) of у·си́л·ива·ть

~ экра́н· (*m*) intensifying screen (X-ray etc.)

у·си́л·и·е, -я, (*n*) effort, internal force, force

~, кас·а́тельн·ое crank effort

~ на конта́кт·е contact force

~, при·жи́м·н·ое (plast) normal/clamping force

~ ре́з·ани·я (met) cutting stress/strength

~, с·двиг·ов·ое shearing force

у·сил·и́тел·ь, -я, (*m*) reinforcer; active/reinforcing filler (rubber); intensifier; (elec etc.) amplifier, repeater

~, бло́ч·н·ый plug-in amplifier

~, вертика́ль·н·ый vertical-deflection amplifier (of cathode-ray tube)

~, гидравл·и́ческ·ий hydraulic intensifier/amplifier

~, горизонта́ль·н·ый horizontal-deflection amplifier (of cathode ray tube)

~, да·ю́щ·ий (autom) data repeater

~, двух·каска́д·н·ый (elec) two-stage amplifier

~, двух·та́кт·н·ый push-pull amplifier

~, ди·электр·и́ческ·ий dielectric amplifier

~, ду́плекс·н·ый two-wire repeater (telephones)

~ за́·пис·и (instr) writing amplifier

~, звук·ов·о́й sound intensifier/amplifier (acoustics); (rad) audio amplifier

~, каскад·н·о́й cascade amplifier

~, каско́д·н·ый (elec) cascade amplifier

~, като́д·н·ый cathamplifier

~, квант·ов·ый (phys) maser amplifier

~, клистро́н·н·ый klystron amplifier

у·сил·и́тел·ь

~, компенс·и́рующ·ий   cancellation amplifier

~ -компре́ссор·, -а,   (*m*) compandor (telephones)

~, кристалл·и́ческ·ий   (elec) transistor amplifier

~, ла́мп·ов·ый   thermionic amplifier/magnifier, vacuum-tube amplifier (U.S.)

~, лин·е́йн·ый   line amplifier; (telecom) telephone repeater

~, магни́т·н·ый   magnetic amplifier

~, молекуля́р·н·ый   (phys) maser, maser amplifier

~ на·пряж·е́ни·я   voltage amplifier; magnetic modulator (servos)

~, не·пере·груж·а́ющ·ийся   anti-saturation amplifier

~, пара·метр·и́ческ·ий   parametric/reactance amplifier

~, полос·ов·о́й   band-pass amplifier; band-pass repeater (telephones)

~ по·пере́ч·н·ого по́л·я   amplidyne

~ -пре·рыв·а́тел·ь, -я,   (*m*) (elec) chopping amplifier

~, раз·дел·и́тельн·ый   (elec) buffer amplifier

~, резона́нс·н·ый   (elec) resonance amplifier

~, реоста́т·н·ый   RC coupling, resistance-coupled amplifier

~, реш·а́ющ·ий   computing amplifier

~ с като́д·н·ой на·гру́з·к·ой   (elec) bootstrap amplifier

~ с·като́д·н·ой с·вя́з·ью   cathode-coupled amplifier

~, симметр·и́чн·ый   balanced amplifier

~, стру́й·н·ый   jet amplifier (of hydraulic or pneumatic system)

~ с·чи́т·ывани·я   reading amplifier

~, шнур·ов·о́й   cord-circuit repeater (telephones)

~, электро·маши́н·н·ый   metadyne generator, metadyne

~, электро·маши́н·н·ый реакти́в·-н·ый   amplidyne

~, электро·механ·и́ческ·ий   dynamo-electric amplifier

~, электро́н·н·ый   vacuum/electron-tube amplifier, thermionic amplifier

у·сил·и́тельн·ый, -ая, -ое,   (*a*) amplifying, intensifying, reinforcing

у·си́л·ить, -ят,   (*perf*) *see* у·си́л·ива·ть

у·скак·а́ть,   (*fut 3rd pl*) у·ска́ч·ут, (*perf*) *see* у·ска́к·ива·ть

у·ска́к·ива·ть, -ют,   (*imp*) gallop/hop/skip/bound/jump away

у·ска́льз·ыва·ть,   (*imp*) = у·скольз·а́·ть

у·ска́ч·ет,   (*fut 3rd sing*) of у·скак·а́ть (*perf*) *see* у·ска́к·ива·ть

у·скольз·а́·ть, -ют,   (*imp*) slide/slip away/off

у·скольз·н·у́ть, -у́т,   (*perf*) *see* у·скольз·а́·ть

у·скор·е́ни·е, -я,   (*n*) (*v n*) *see* у·ско·р·я́·ть; acceleration

~ в моме́нт· за́·пуск·а   (rocket) initial acceleration

~, кас·а́тельн·ое   tangential component of acceleration

~, кориоли́с·ов·о   Coriolis acceleration

~ кре́н·а   (aerodynam) acceleration in roll, roll acceleration

~, норма́ль·н·ое   normal acceleration

~, от·риц·а́тельн·ое   (aerodynam) drag, deceleration

~, по·воро́т·н·ое   acceleration of Coriolis, complementary acceleration

~, про·до́ль·н·ое   axial acceleration

~ свобо́д·н·ого пад·е́ни·я   acceleration due to gravity, *g*

~ си́л·ы тя́ж·ест·и   (phys) acceleration due to gravity, *g*

~ танга́ж·а, угл·ов·о́е   acceleration in pitch, pitch acceleration

~, угл·ов·о́е   angular acceleration

~, центро·стрем·и́тельн·ое   centripetal acceleration, centripetal (component) of acceleration

у·ско́р·енн·ый, -ая, -ое,   (*past part pass*) *see* у·скор·я́·ть; fast (e.g. of trains)

у·скор·и́тел·ь, -я,   (*m*) accelerator; (rocket) booster, boost motor; (rail) rapid acting valve, accelerator

~ вращ·е́ни·я   (rocket) roll booster

~ за·ря́ж·енн·ых част·и́ц·   (nucl) charged-particle accelerator

~, и́мпульс·н·ый   (nucl) pulsed accelerator

~, лин·е́йн·ый   (nucl) linear accelerator

~ на·бо́р·а вод·ы́   (st eng) accelerator, accelerator pump

~ пластик·а́ци·и   (plast) peptizer

~, раке́т·н·ый   (a/c) boost motor, booster

~, ста́рт·ов·ый   (a/c) rato, rocket-assisted takeoff motor

у·скор·и́тельн·ый, -ая, -ое,   (*a*) accelerating

у·ско́р·ить, -ят,   (*perf*) *see* у·скор·я́·ть

у·скор·я́·ть, -ют, (*imp*) accelerate, speed up, hasten, expedite, hurry up; **-ся** increase speed, quicken, hurry

у·скор·я́ющ·ий, -ая, -ее, (*pres part act*) of у·скор·я́·ть

~ пере·да́·ч·а (*f*) (M/T) overdrive

~ сту́п·ен·ь (*f*) (M/T) overdrive

~ электро́д· (*m*) accelerator (of klystron)

у·сла́вл·ива·ться, -ются, (*imp*) = у·сло́вл·ива·ться

у·сла́вш·ий, -ая, -ее, (*past part act*) of у·сла́ть

у·сла́·ть, (*fut 3rd pl*) у·шл·ю́т, (*perf*) send, send away

у·след·и́ть, -я́т, (*perf*) *see* у·слёж·и·ва·ть

у·слёж·ива·ть, -ют, (*imp*) look after, watch over; follow (with attention)

у·сло́в·и·е, -я, (*n*) condition(s), term(s), stipulation(s)

   в у·сло́в·и·ях ... under ... conditions

~, гран·и́чн·ое boundary condition

~, до·полн·и́тельн·ое auxiliary condition

~, кра·ев·о́е boundary/limiting condition

~, пре·де́ль·н·ое limiting condition

   при про́ч·их ра́вн·ых у·сло́в·и·ях other conditions being equal

   при у·сло́в·и·и что under the condition that, providing, so long as ...

~ сред·ы́ environment

у·сло́в·иться, -ятся, (*perf*) *see* у·сло́вл·ива·ться

у·сло́вл·енн·ый, -ая, -ое, (*past part pass*) *see* у·сло́вл·ива·ться

у·сло́вл·ива·ться, -ются, (*imp*) arrange, settle, agree

у·сло́в·н·о (*adv*) conditionally; conventionally; arbitrarily

условно·рефлекто́р·н·ый, -ая, -ое, (*a*) conditioned-reflex

у·сло́в·н·ый, -ая, -ое, (*a*) conditional, conditioned; nominal, conventional, hypothetical; agreed, prearranged

~ бомбо·мет·а́ни·е (*n*) simulated bombing

~ знак· (*m*) conventional sign/symbol

~ на·имен·ова́ни·е (*n*) code name, designation

~ на·пряж·е́ни·е (*n*) nominal stress

~ обо·знач·е́ни·е (*n*) conventional sign/symbol; legend

~ при·гово́р· (*m*) (law) suspended sentence

у·сло́в·н·ый, -ая, -ое

~ про·хо́д· (*m*) bore, nominal inside diameter (of pipe)

~ рефле́кс· (*m*) (physiol) conditioned reflex

~ с·ход·и́мост·ь (*f*) (math) conditional convergence

~ то́пл·ив·о (*n*) reference/standard/comparison fuel

у·с·ложн·е́ни·е, -я, (*n*) complication

у·с·ложн·ённ·ый, -ая, -ое, (*past part pass*) *see* у·с·ложн·я́·ть

у·с·ложн·и́ть, -я́т, (*perf*) *see* у·с·ложн·я́·ть

у·с·ложн·я́·ть, -ют, (*imp*) complicate

у·слу́г·а, -и, (*f*) service

у·слу́ж·ива·ть, -ют, (*imp*) serve

у·служ·и́ть, -ат, (*perf*) render a service

у·слых·а́·ть, (*fut 3rd pl*) у·слы́ш·ат, (*perf*) = у·слы́ш·а·ть

у·слы́ш·ать, -ат, (*perf*) hear

у·сма́тр·ива·ть, -ют, (*imp*) look after, watch over; perceive, see, discern; **-ся** (*imp only*) be considered/seen as

у·смех·а́·ться, -ются, (*imp*) smile, grin

у·смех·н·у́ться, -у́тся, (*perf*) *see* у·смех·а́·ться

у·сме́ш·к·а, -и, (*g pl*) -ш·ек·, (*f*) smile, grin

у·с·мир·и́ть, -я́т, (*perf*) *see* у·с·мир·я́·ть

у·с·мир·я́·ть, -ют, (*imp*) pacify, quieten; suppress (mutiny etc.)

у·смотр·е́ни·е, -я, (*n*) discretion, judgement

у·смотр·е́ть, -ят, (*perf*) *see* у·сма́тр·ива·ть

у·снаст·и́ть, -я́т, (*perf*) *see* у·снащ·а́·ть

у·снащ·а́·ть, -ют, (*imp*) equip, cover with (equipment); deck, decorate, garnish

у·сна́щ·ива·ть, -ют, (*imp*) = у·снащ·а́·ть

у·сн·у́ть, -у́т, (*perf*) go to sleep, fall asleep; die (of fish)

у·со·верш·е́нствовани·е, -я, (*n*) (*v n*), *see* у·со·верш·е́нств·овать; improvement; advanced training/courses/studies, refresher courses

   институ́т у·со·верш·е́нствовани·я врач·е́й (*m*) institute of advanced medical studies

у·со·верш·е́нств·овать, -уют, (*perf*) improve; perfect

ус·о́вк·а, -и, (*f*) mitring (carpentry

ус·ов·о́й, -а́я, -о́е, (*a*) of **ус·**

ус·о́вочн·ый, -ая, -ое, (*a*) of **ус·о́вк·а**

~ стан·о́к· (*m*) mitring machine (carpentry)

у·со·мн·и́ться, -я́тся, (*perf*) doubt

усо·но́г·ие, -их, (*pl decl as adj*) (zool) cirripedes, cirripeds, *Cirripedia*

усо·ре́з·, -а, (*m*) mitre cutter/knife (carpenter's tool)

у·со́х·нуть, -нут, (*perf*) dry up, shrivel

у·со́х·ш·ий, -ая, -ее, (*past part act*) of **у·со́х·нуть;** desiccated, dried-up, shrivelled

у·спев·а́емост·ь, -и, (*f*) degree of success, progress (in studies)

у·спев·а́·ть, -ют, (*imp*) be in time, have time; succeed in, be able to

у·спе́·ть, -ют, (*perf*) *see* **у·спев·а́·ть**

у·спе́х·, -а, (*m*) success, (*pl*) successes, advances, progress

у·спе́ш·ност·ь, -и, (*f*) success, successfulness

у·спе́ш·н·ый, -ая, -ое, (*a*) successful

у·с·пока́·ивани·е, -я, (*n*) (*v n*) of **у·с·пока́·ива·ть;** (med) sedation

у·с·пока́·ива·ть, -ют, (*imp*) calm, quiet, soothe

у·с·пока́·иваю·щий, -ая, -ее, calming, soothing, sedative

у·с·поко·е́ни·е, -я, (*n*) calming, (med) sedation; killing (steel)

у·с·поко·и́тел·ь, -я, (*m*) damper, vibration damper; (shipb) stabilizer

~, гироскоп·и́ческ·ий (shipb) gyro stabilizer

~, магни́т·н·ый eddy-current brake, magnetic-coupling torsional vibration damper

~, ма́сл·ян·ый viscous-fluid torsional vibration damper

у·с·поко·и́тельн·ый, -ая, -ое, (*a*) calming, sedative; (shipb) anti-rolling, stabilizing

~ цисте́рн·а (*f*) (shipb) anti-rolling tank

у·с·поко́·ить, -ят, (*perf*) *see* **у·с·по·ка́·ива·ть**

у·средн·е́ни·е, -я, (*n*) average, averaging

~ по гру́пп·е group averaging **фо́рмул·а у·средн·е́ни·я** (*f*) averaging formula

у·средн·ённ·ый, -ая, -ое, (*past part pass*) *see* **у·средн·я́·ть;** average, averaged

у·средн·и́ть, -я́т, (*perf*) *see* **у·сред·н·я́·ть**

у·средн·я́·ть, -ют, (*imp*) average

уст. (*abbr*) = **у·стар·е́л·ый** obsolete

уст·а́, (*g pl*) **уст·,** (*d pl*) **-а́м,** (*pl*) (biol) mouth, stoma

у·ста́·в·, -а, (*m*) regulations; statute, charter

у·ста·ва́·ть, -ют, (*imp*) become/get/be tired

у·ста́в·ить, -ят, (*perf*) *see* **у·ставл·я́·ть**

у·ста́в·к·а, -и, (*f*) (*v n*) *see* **у·ставл·я́·ть;** (autom) setting, set point (of relay); control point (of computer)

у·ставл·я́·ть, -ют, (*imp*) arrange, set, put

у·ста́·ива·ть, -ют, (*imp*) remain standing; remain in being, remain; hold out, hold one's ground; **-ся** become calm, settle (of liquids)

у·ста́л·ост·н·ый, -ая, -ое, (*a*) fatigue

~ маши́н·а (*f*) fatigue-testing machine, fatigue tester

~ раз·руш·е́ни·е (*n*) fatigue failure

~ тре́щ·ин·а (*f*) fatigue crack

у·ста́л·ост·ь, -и, (*f*) fatigue, tiredness, weariness

~, мало·цикл·ов·а́я short-time fatigue **пре·де́л· у·ста́л·ост·и** (*m*) (mech) fatigue limit

~, терм·и́ческ·ая thermal fatigue

у·ста́·л·ый, -ая, -ое, (*a*) tired, fatigued

у·стана́вл·ива·ть, -ют, (*imp*) set up; mount, install, place; (instr) set, adjust; establish; determine, fix, ascertain, find

у·стана́вл·иваю·щ·ийся, -аяся, -ееся, (*pres part pass*) *see* **у·стана́вл·ива·ть;** (phys) transient, unsteady

у·станов·и́вш·ийся, -аяся, -ееся, (*past part pass*) *see* **у·стана́вл·и·ва·ть;** settled, sustained; (phys) steady-stage, steady, stable, stationary

~ движ·е́ни·е (*n*) steady/steady-state movement/flow

~ проце́сс· (*m*) steady/sustained process; (roll) conditions after bite

~ по·го́д·а (*f*) (meteor) settled weather

~ режи́м· (*m*) steady state

у·станов·и́ть, -ят, (*perf*) *see* **у·ста·на́вл·ива·ть**

~ на фо́кус· focus

у·стано́в·к·а, -и, (*g pl*) **-в·ок·,** (*f*) (*v n*), *see* **у·стана́вл·ива·ть** (*see also associated adjectives*); installation, plant, unit; set-up (apparatus); (instr) adjustment

~, артиллер·и́йск·ая gun mounting (including gun)

у·станóв·к·а
~ **в по·лож·éни·е** positioning
~, **грýб·ая** (instr) coarse adjustment/ setting
~, **ис·пыт·áтельн·ая** (chem, phys) experimental set-up
~, **котéль·н·ая** boiler unit
~ **кристáлл·а** crystal derivation
~ **лопáт·к·и** blade setting (turbines etc.)
~, **модел·úрующ·ая** simulator
~ **на нул·ь** (instr) zero adjustment
~, **ó·пыт·н·ая** (chem, phys) experimental set-up; pilot/experimental plant
~, **пуск·ов·áя** (rocket) launcher
стó·имост·ь с у·станóв·к·и (f) installed cost
ýгол у·станóв·к·и (m) (a/c) angle of pitch, rigging angle of incidence (of wing); blade angle (of propeller)
~ **шáг·а** pitch setting (of propeller)
у·становл·éни·е, -я, (n) (v n) see **у·станáвл·ива·ть;** establishment, establishing
врéм·я у·становл·éни·я (n) setting-up time, setting time; transient period
у·станóвл·енн·ый, -ая, -ое, (past part pass) see **у·станáвл·ива·ть**
~, **за·рáн·ее** pre-set
у·станóв·очн·ый, -ая, -ое, (a) adjusting, setting; installation
~ **болт·** (m) adjuster bolt
~ **кольц·ó** (n) collar, distance/setting ring
~ **механ·úзм·** (m) (met roll) screw-down gear
~ **о·шúб·к·а** (f) installation error
~ **прó·вод·** (m) electrical installation wire
~ **рычаж·óк·** (m) (horol) adjusting rod
у·станóв·щик·, -а, (m) adjuster, installer, fitter; (gunn) setter
у·стар·евá·ть, -ют, (imp) become obsolete/old-fashioned
у·стар·éвш·ий, -ая, -ее, (past part act) see **у·стар·евá·ть;** obsolete, old-fashioned
у·стар·éлост·ь, -и, (f) obsoleteness
у·стар·éл·ый, -ая, -ое, (a) obsolete, old-fashioned
у·стар·é·ть, -ют, (perf) see **у·стар·евá·ть**
у·стá·ть, -нут, (perf) see **у·ста·вá·ть**
у·стéл·ет, (fut 3rd sing) of **у·стл·áть** (perf) see **у·стил·á·ть** (imp)
у·стил·á·ть, -ют, (imp) spread over, cover

ýст·ичн·ый, -ая, -ое, (a) (biol) stomatous
у·стл·áть, (fut 3rd pl) **у·стéл·ют,** (perf) see **у·стил·á·ть**
ýст·н·ый, -ая, -ое, (a) oral, verbal
у·стó·й, -я, (m) abutment (of bridge etc.); foundations, support; cream
у·стóй·чивост·ь, -и, (f) stability, steadiness; resistance (to change)
~, **а·период·úческ·ая** dead-beat stability, aperiodic stability
~ **деформ·áци·и** (mech) stiffness
за·пáс· у·стóй·чивост·и (m) stability; (elec) stability factor
~, **коррозиóн·н·ая** (met) corrosion resistance
~, **курс·ов·áя** directional stability
~, **о·гранúч·енн·ая** marginal stability; conditional stability
~ **от·мéт·ок·** positional stability (radar)
~, **по·перéч·н·ая** lateral stability; rolling stability
~ **при хран·éни·и** keeping quality/ qualities
~, **пут·ев·áя** directional stability
~, **пут·ев·áя стат·úческ·ая** weathercock stability
~ **пут·ú** (air) directional stability
~, **сóб·ственн·ая** inherent stability
~, **стат·úческ·ая** static stability; (elec) steady-state stability
~ **тéл·а** (mech) stability
~, **у·слóв·н·ая** conditional stability
~, **флюгер·н·ая** (aerodynam) weathercock stability
~ **част·от·ы** (elec) frequency stability
у·стóй·чив·ый, -ая, -ое, (a) stable, steady; resistant
~ **движ·éни·е** (n) (mech) stable motion
~ **ос·ь** (f) permanent axis (of rotating body)
~ **равно·вéс·и·е** (n) (mech) stable equilibrium
у·сто·я́ть, -я́т, (perf) see **у·стá·ива·ть**
у·стрá·ива·ть, -ют, (imp) arrange, organize, set/fix up
у·стран·éни·е, -я, (n) (v n) see **у·стран·я́·ть;** removal, elimination
у·стран·ú·ть, -я́т, (perf) see **у·стран·я́·ть**
у·стран·ённ·ый, -ая, -ое, (past part pass) see **у·стран·я́·ть**
у·стран·я́·ть, -ют, (imp) remove, eliminate, get rid of
у·страш·á·ть, -ют, (imp) frighten, scare

у·страш·и́ть, -а́т, (*perf*) *see* **у·стра·ш·а́·ть**

у·стрем·и́ть, -я́т, (*perf*) *see* **у·стре·мл·я́·ть**

у·стремл·е́ни·е, -я, (*n*) (*v n*) *see* **у·стре·мл·я́·ть;** aspiration, endeavour; tendency

у·стремл·я́·ть, -ют, (*imp*) direct, point, turn; **-ся** (*pass*); rush

у́стриц·а, -ы, (*f*) (zool) oyster, *Ostrea*

устрице·во́д·, -а, (*m*) oyster culturist; oyster farmer

устрице·во́д·ств·о, -а, (*n*) oyster culture; oyster farming

у́стрич·ник·, -а, (*m*) oyster bed

у́стрич·н·ый, -ая, -ое, (*a*) oyster

~ ба́н·к·а (*f*) oyster bed

~ про́·мысел· (*m*) oyster farming; oyster farm

~ с·тво́р·к·и (*pl*) oystershell

у·стро·и́тел·ь, -я, (*m*) organizer

у·стро́·ить, -ят, (*perf*) *see* **у·стра́·и·ва·ть**

у·стро́й·ств·о, -а, (*n*) (*see also associated adjectives*) equipment, mechanism, device; arrangement, organization, layout, structure

~, вы·числ·и́тельн·ое computer

~, модел·и́рующ·ее simulator

~, пуск·ов·о́е launcher

~, суд·ов·ы́е (*pl*) (shipb) hull equipment

у·сту́п·, -а, (*m*) offset, step, stagger; ledge, shoulder; (arch) projection; (civ eng) bench, berm, terrace, shelf; (geol) scarp, bench, ledge; altar course (of dry dock); (mil) echelon

у·ступ·а́·ть, -ют, (*imp*) give up, let have; concede, cede, yield, give in

у·ступ·и́ть, -я́т, (*perf*) *see* **у·ступ·а́·ть**

у·сту́п·к·а, -и, (*g pl*) **-п·ок·,** (*f*) (*v n*) *see* **у·ступ·а́·ть;** concession; reduction (from set price), rebate

уступо·ме́р·, -а, (*m*) (mech eng) shoulder depth gauge/gage

уступо·обра́з·н·ый, -ая, -ое, (*a*) ledge/step-like, bench-formed

у·сту́п·чат·ый, -ая, -ое, (*a*) stepped, ledged

у·сту́п·чив·ый, -ая, -ое, (*a*) ready to make concessions, yielding

у́ст·ь·е, -я, (*g pl*) **-ев,** (*n*) mouth, orifice, aperture; mouth (of river), outfall (of tributary); (biol) mouth, stoma, (zool) ostium

усть·ев·о́й, -а́я, -о́е, (*a*) of **у́ст·ь·е**

у́сть·иц·е, -а, (*n*) (bot) stoma, (*pl*) stomata

у·су·губ·и́ть, -я́т, (*perf*) *see* **у·су·губ·л·я́·ть**

у·су·губл·я́·ть, -ют, (*imp*) increase, intensify, heighten; aggravate, worsen

у·су́ш·к·а, -и, (*f*) shrinkage

у·сыл·а́·ть, -ют, (*imp*) send, send away

у·сын·овле́ни·е, -я, (*n*) (law) adoption (of boy)

у·сы́п·ать, -лют, (*perf*); **у·сып·а́·ть, -ют,** (*imp*) cover, strew/sprinkle/scatter all over

у·сып·и́тельн·ый, -ая, -ое, (*a*) sleep-inducing, (med) narcotic

у·сып·и́ть, -я́т, (*perf*) *see* **у·сыпл·я́·ть**

у·сыпл·я́·ть, -ют, (*imp*) put to sleep; lull

у·сых·а́ни·е, -я, (*n*) drying up, desiccation

у·сых·а́·ть, -ют, (*imp*) dry up, shrivel

у·ся́д·ется, (*fut 3rd sing*) of **у·се́ст·ься**

у·та́·ивани·е, -я, (*n*) (*v n*) of **у·та́·и·ва·ть;** concealment

у·та́·ива·ть, -ют, (*imp*) keep secret, conceal, hide; secretly appropriate

у·та·и́ть, -я́т, (*perf*) *see* **у·та́·ива·ть**

у·та́пт·ыва·ть, -ют, (*imp*) trample/tread flat; compact, tamp down

у·та́ск·ива·ть, -ют, (*imp*) drag/pull away

у·тащ·и́ть, -а́т, (*perf*) *see* **у·та́ск·и·ва·ть**

у́·твар·ь, -и, (*f*) utensils

у·тверд·и́тельн·ый, -ая, -ое, (*a*) affirmative

у·тверд·и́ть, -я́т, (*perf*) *see* **у·твер·жд·а́·ть**

у·твержд·а́·ть, -ют, (*imp*) confirm; affirm, assert; convince; approve (officially), accept (thesis), ratify (treaty)

у·твержд·е́ни·е, -я, (*n*) (*v n*) *see* **у·твержд·а́·ть;** confirmation; corroboration; affirmation, assertion, statement; approval (official), acceptance (of thesis), ratification (of treaty etc.)

у·тек·а́·ть, -ют, (*imp*) flow/run away

ут·ёнок·, -нк·а, (*nom pl*) **ут·я́т·а,** (*g pl*) **-ут·я́т·,** (*m*) duckling

у·тепл·ённ·ый, -ая, -ое, (*past part pass*) *see* **у·тепл·я́·ть**

у·тепл·и́тел·ь, -я, (*m*) warmer, heater; (met cast) hot top, shrink/feeder head

~, пла́в·ающ·ий (met cast) loose hot top

у·тепл·и́ть, -я́т, (*perf*) *see* **у·тепл·я́·ть**

у·тепл·я́·ть, -ют, (*imp*) make warm/warmer, warm, heat

у·тер·е́ть, (*fut 3rd pl*) у·тр·у́т, (*perf*) *see* у·тир·а́·ть

у·тер·я́·ть, -ют, (*perf*) lose

у·тёс·, -а, (*m*) cliff, steep cliff

у·тёс·ист·ый, -ая, -ое, (*a*) steep (of cliffs etc.), cliffy

у·те́ч·к·а, -и, (*g pl*) -ч·ек·, (*f*) leak, leakage, loss, (com) ullage

~, до·пуст·и́м·ая leak tolerance

~ се́т·к·и (elec) grid leak

  ток у·те́ч·к·и (*m*) (elec) leakage current

у·те́ч·ь, (*fut 3rd sing, pl*) у·теч·ёт, у·тек·у́т, (*past masc sing*) у·тёк· (*perf*) *see* у·тек·а́·ть

у·теш·и́тельн·ый, -ая, -ое, (*a*) consoling

утилиз·а́ци·я, -и, (*f*) utilization; salvage, recovery of waste materials

утилита́р·н·ый, -ая, -ое, (*a*) utilitarian

утѝл·ь, -я, (*m*) salvage, disposable waste, scrap

утиль·рези́н·а, -ы, (*f*) scrap rubber

утиль·сыр·ь·ё, -я́, (*n*) = утѝл·ь

ут·и́н·ый, -ая, -ое, (*a*) duck's, (*pl as noun*) (orn) *Anatidae*

~ нос· (*m*) (min) duckbill

у·тир·а́·ть, -ют, (*imp*) wipe, wipe away

у·тих·а́·ть, -ют, (*imp*) become quiet/quieter, die down/away

у·ти́х·нуть, -нут, (*perf*) *see* у·тих·а́·ть

у́т·к·а, -и, (*g pl*) у́т·ок·, (*f*) (orn) duck; (naut) cleat; false rumour, canard; (*g sing*) of у·то́к·

~, дом·а́шн·яя domesticated barnyard duck, domestic duck, *Anas domestica*

у·тк·а́·ть, -ут, (*perf*) embroider

утко·но́с·, -а, (*m*) (zool) platypus, duckbill, *Ornithorhynchus anatinus*

Утле́гар·ь, -я, (*m*) (naut) jib-boom

у·то́к·, у·тк·а́ (*m*) (text) weft; (*g pl*) of у́т·к·а

~, одно·ни́т·очн·ый single weft

~, у·га́р·н·ый waste weft

утол·и́ть, -ят, (*perf*) *see* утол·я́·ть

у·толст·и́ть, -ят, (*perf*) *see* у·толщ·а́·ть

у·толщ·е́ни·е, -я, (*n*) thickening; bulge, intumescence, swelling

у·толщ·а́·ть, -ют, (*imp*) make thicker

утол·я́·ть, -ют, (*imp*) satisfy, slake, quench; soothe, alleviate

у·том·и́тельност·ь, -и, (*f*) tiringness, tiresomeness, tediousness

у·том·и́тельн·ый, -ая, -ое, (*a*) fatiguing, tiring, tedious

у·том·и́ть, -я́т, (*perf*) *see* у·томл·я́·ть

у·томл·е́ни·е, -я, (*n*) fatigue, tiredness, weariness

у·томл·я́емост·ь, -и, (*f*) fatigability, fatigue

  пре·де́л· у·томл·я́емост·и (*m*) endurance/fatigue limit

у·томл·я́·ть, -ют, (*imp*) fatigue, tire, weary

у·тон·е́ни·е, -я, (*n*) (*v n*) *see* у·тон·ч·а́·ть

у·тон·и́ть, -я́т, (*perf*) = у·тонч·и́ть

у·тон·у́ть, -ут, (*perf*) sink, submerge; drown

у·тонч·а́·ть, -ют, (*imp*) make thinner/finer, thin/fine down; make more sensitive, refine; -ся become thinner/finer; become more sensitive/refined

у·тонч·е́ни·е, -я, (*n*) (*v n*) *see* у·тон·ч·а́·ть; thinning, thinning down; refinement (of a theory etc.)

  длин·а́ у·тонч·е́ни·я (*f*) (nucl) thin-down length

у·тонч·ённост·ь, -и, (*f*) refinement, subtlety

у·тонч·ённ·ый, -ая, -ое, (*past part pass*) *see* у·тонч·а́·ть; finer, thinner; refined, subtle

у·тонч·и́ть, -а́т, (*perf*) *see* у·тонч·а́·ть

у·тон·я́·ть, -ют, (*imp*) = у·тонч·а́·ть

у·топ·а́·ть, -ют, (*imp*) sink, submerge; drown; у·то́п·а·ть, -ют, (*perf*) tread/trample down

у·топ·и́ть, -я́т, (*perf*) sink, submerge; (trans) drown

утоп·и́ческ·ий, -ая, -ое, (*a*) Utopian

у·то́пл·енник·, -а, (*m*) drowned person (male)

у·то́пл·енн·ый, -ая, -ое, (*past part pass*) *see* у·топ·и́ть; countersunk, flush

у·топт·а́ть, (*fut 3rd pl*) у·то́пч·ут, (*perf*) *see* у·та́пт·ыва·ть

утóр·, -а, (*m*) croze, groove (for barrel head)

утóр·н·ый, -ая, -ое, (*a*) crozing

~ стан·о́к· (*m*) crozing machine (cooperage)

~ стан·о́к·, двой·н·о́й croze-cutting machine (cooperage)

у·тóч·ин·а, -ы, (*f*) (text) wefting

у·точн·е́ни·е, -я, (*n*) (*v n*) *see* у·точ·н·я́·ть; refinement, amendment, detailed improvement; explanation, enquiry, detailed explanation/enquiry у·точн·ённ·ый, -ая, -ое, *(past part pass) see* у·точн·я́·ть

у·точн·и́ть, -я́т, *(perf) see* у·точн·я́·ть

уточно·мот·а́льн·ая маши́н·а *(f)* (text) cop winder

уточно·на·сти́л·очн·ый, -ая, -ое, *(a)* (text) weft-faced

у·то́ч·н·ый, -ая, -ое, *(a)* (text) weft

~ атла́с· *(m)* (text) weft satin

~ пря́ж·а *(f)* (text) weft

~ нит·ь *(f)* (text) weft

у·точн·я́·ть, -ют, *(imp)* make more precise/exact/accurate, clarify (an argument etc.); define/explain more precisely/exactly/accurately; elaborate; find out in greater detail, obtain more details

у·тра́·ива·ть, -ют, *(imp)* triple, treble

у·трамб·ова́ть, -у́ют, *(perf) see* у·трамб·о́выва·ть

у·трамб·о́выва·ть, -ют, *(imp)* tamp, ram, ram down/in

у·тра́т·а, -ы, *(f)* loss, deprivation, depletion

у·тра́т·ить, -ят, *(perf) see* у·тра́ч·ива·ть

у·тра́ч·ива·ть, -ют, *(imp)* lose, be deprived of

у́тр·енн·ий, -яя, -ее, *(a)* morning

~ зар·я́ *(f)* (meteor) sunrise colours

у́тр·енник·, -а, *(m)* morning frost; (theat) matinee, morning performance

у·тр·ёт, *(fut 3rd sing)* of у·тер·е́ть *(perf) see* у·тир·а́·ть *(imp)*

утр·и́р·овать, -уют, *(imp)* overemphasize; exaggerate

у́тр·о, -а, *(m)* morning, a.m.

утро́ба·, -ы, *(f)* (anat) uterus

утро́б·н·ый, -ая, -ое, *(a)* (anat) uterine

у·тро·е́ни·е, -я, *(n)* tripling, trebling

утроенн·пе́р·ист·ый, -ая, -ое, *(a)* (bot) tripinnate

утроенно·тро́й·ча́т·ый, -ая, -ое, *(a)* (bot) triternate

у·тро́·енн·ый, -ая, -ое, *(past part pass)* of у·тро́·ить; triple, threefold

у·тро́·ить, -ят, *(perf)* triple, treble

у́тр·ом *(adv)* in the morning, morning

у·трудн·и́ть, -я́т, *(perf) see* у·трудн·я́·ть

у·трудн·я́·ть, -ют, *(imp)* make more difficult/complicated; cause trouble/inconvenience

утру́с·к·а, -и, *(f)* spillage, leakage (of grain etc.)

у·тр·у́т *(fut 3rd pl)* of у·тер·е́ть *see* у·тир·а́·ть

у·тряс·а́·ть, -ют, *(imp)* shake down, compact by shaking

у·тряс·ти́, -у́т, *(perf) see* у·тряс·а́·ть

у́тфел·ь, -я, *(m)* massecuite (sugar production)

утю́г·, -а́, *(m)* iron, flat-iron (domestic); (civ eng) drag

~, доро́ж·н·ый drag, road drag, planer, drag planer

утю́ж·ить, -ат, *(imp)* iron, press (clothes etc.)

утю́ж·к·а, -и, *(g pl)* -ж·ек·, *(f)* pressing, ironing (of clothes); dragging (roadmaking)

у·тя́г·ивани·е, -я, *(n) (v n)* of у·тя́г·ива·ть

у·тя́г·ива·ть, -ют, *(imp)* draw out, reduce by drawing out; drag, drag out (someone against their will); draw/bind/pull tight

у·тяжел·е́ни·е, -я, *(n) (v n) see* у·тяжел·я́·ть; weighting

у·тяжел·и́тел·ь, -я, *(m)* weighting material (e.g. as used in drilling wells)

у·тяжел·и́ть, -я́т, *(perf) see* у·тяжел·я́·ть

у·тяжел·я́·ть, -ют, *(imp)* make heavy/heavier, increase the weight of

у·тяжел·я́ющ·ий, -ая, -ее, *(pres part act)* of у·тяжел·я́·ть; weighting

у·тя́ж·ин·а, -ы, *(f)* (met) pull, pull crack

у·тя́ж·к·а, -и, *(g pl)* -ж·ек·, *(f) (v n) see* у·тя́г·ива·ть; (met roll/forge) drawing, pulling, drawing out, reduction in height/area (due to drawing out of metal)

у·тя́·ну·т·ый, -ая, -ое, *(past part pass) see* у·тя́г·ива·ть

у·тя·н·у́ть, -ут, *(perf) see* у·тя́г·ива·ть

ут·я́т·а, *(nom pl)* of ут·ёнок· ducklings

ут·я́тин·а, -ы, *(f)* duck meat, duck

ут·я́тник·, -а, *(m)* (agr) duck shed

УФ *(abbr)* = ультра·фиоле́т·ов·ый UV, ultra-violet; *(abbr)* = ура́ль·ск·ий филиа́л· Urals Branch (of Academy of Sciences)

ухаб·, -а, *(m)* pothole (in road)

у·ха́ж·ивани·е, -я, *(n) (v n)* of у·ха́ж·ива·ть; care

у·ха́ж·ива·ть, -ют, *(imp)* nurse, look after, tend, take care of; court, curry favour/favor

у·хва́т·, -а, *(m)* grab, gripping device

у·хват·и́ть, ‑ят, (*perf*) *see* у·хва́т·ыва·ть

у·хва́т·ыва·ть, ‑ют, (*imp*) seize, grasp, catch/take firmly

у·хва́т·к·а, ‑и, (*g pl*) ‑т·ок·, (*f*) mannerism, manner, way, trick

у·хва́ч·енн·ый, ‑ая, ‑ое, (*past part pass*) *see* у·хва́т·ыва·ть

у·хво́ст·ь·е, ‑я, (*n*) (naut) downstream end/part

у·хищр·е́ни·е, ‑я, (*n*) contrivance, trick

у́х·о, ‑а, (*nom pl*) у́ш·и, (*g pl*) уш·е́й, (*n*) ear

~, мор·ск·о́е (zool) abalone, (*pl*) *Haliotidae*

~, электр·и́ческ·ое (autom) electric ear

ухо·ве́рт·к·а, ‑и, (*g pl*) ‑т·ок·, (*f*) (ent) earwig, (*pl*) *Dermaptera*

уховёртко·обра́з·н·ые, ‑ых, (*pl decl as adj*) (ent) dermapterans, *Dermaptera*

у·хо́д·, ‑а, (*m*) (*v n*) *see* у·ход·и́ть *and* у·ха́ж·ива·ть; departure, withdrawal; lapse (of time); (phys) drift; care, management, attendance (of machinery)

~ за боль·н·ы́ми care of the sick

~ за культу́р·ами (agr) crop cultivation

~ за по́чв·ой (agr) soil management

~ нес·у́щ·ей (rad) carrier shift

~ част·от·ы́ (rad) frequency drift

у·ход·и́ть, ‑ят, (*imp*) go/come/run away/off/in, leave, depart, quit; elapse, pass, run out (of time); get away from, escape from, out-strip, outdistance; be spent/expended

у·хож·у́ (*pres 1st sing*) of у·ход·и́ть

у·худш·а́·ть, ‑ют, (*imp*) make worse; ‑ся (*pass*); become worse, worsen, deteriorate

у·худш·е́ни·е, ‑я, (*n*) (*v n*) *see* у·худш·а́·ть; deterioration, worsening; (elec) degradation

ско́р·ост·ь у·худш·е́ни·я (*f*) (elec) degradation rate

у·ху́дш·ить, ‑ат, (*perf*) *see* у·худш·а́·ть

у·цел·е́·ть, ‑ют (*perf*) be left whole/intact, remain whole, escape harm, survive unharmed

у·це́н·ива·ть, ‑ют, (*imp*) reduce/cut the price of

у·цен·и́ть, ‑ят, (*perf*) *see* у·це́н·ива·ть

у·цеп·и́ть, ‑ят, (*perf*) *see* у·цепл·я́·ть

у·цепл·я́·ть, ‑ют, (*imp*) hook on, catch hold of

у·ча́л·к·а, ‑и, (*g pl*) ‑л·ок·, (*f*) securings (of pusher tug train)

у·ча́ств·овать, ‑уют, (*imp*) take part in, participate, collaborate; have a share in

у·ча́ств·ующ·ий, ‑его, (*m decl as adj*) participant, participator, collaborator

у·ча́ст·и·е, ‑я, (*n*) participation, collaboration; share, sharing; sympathy, concern

при·ним·а́·ть у·ча́ст·и·е take part in, share in, participate in

~ электро́н·ов (phys) electron sharing

у·част·и́ть, ‑ят, (*perf*) *see* у·чащ·а́·ть

у·ча́ст·ков·ый, ‑ая, ‑ое, (*a*) of у·ча́ст·ок·

у·ча́ст·ник·, ‑а, (*m*) participant, participator, member, sharer; competitor (sport)

у·ча́ст·ок·, ‑т·к·а, (*m*) part, section, sector, division; segment, length; area, region; (agr) plot, lot, parcel, strip

~ по·лёт·а, кон·е́чн·ый (rocket) terminal phase

~ раз·го́н·а (rocket) boost phase

~ траекто́ри·и, акти́в·н·ый (rocket) powered-flight phase

~ тра́сс·ы route stage/section, leg (of course)

у́·част·ь, ‑и, (*f*) lot, fate

у·чащ·а́·ть, ‑ют, (*imp*) make more frequent, increase the frequency, quicken; ‑ся (*pass*); become more frequent, quicken

уч·а́щ·ийся, ‑аяся, ‑ееся, (*pres part act*) *see* уч·и́ться; (*m decl as adj*) student, pupil

уч·ёб·а, ‑ы, (*f*) studies, training

уч·е́бник·, ‑а, (*m*) text-book, manual

уч·е́бн·ый, ‑ая, ‑ое, (*a*) educational, instructional, training; practice (equipment)

~ за·вед·е́ни·е (*n*) training/educational institution/establishment

~ за·вед·е́ни·е, вы́с·ш·ее higher educational establishment

~ план· (*m*) curriculum

~ програ́мм·а (*f*) curriculum

~ стрел·ьб·а́ (*f*) (mil) practice shoot

~ су́д·н·о (*n*) training ship

у·чё·л, (*past masc sing*) of у·че́ст·ь (*perf*) *see* у·чи́т·ыва·ть (*imp*)

уч·е́ни·е, ‑я, (*n*) (*v n*) of уч·и́ть; tuition, instruction; study; training; teaching, doctrine

~ о (+ *noun*) -logy, science

уч·ени́к·, -а́, (*m*) pupil, learner; apprentice; disciple, follower, adherent

уч·ени́ческ·ий, -ая, -ое, (*a*) of **уч·ени́к·**

уч·ени́честв·о, -а, (*n*) school attendance; apprenticeship

у́ч·енн·ый, -ая, -ое, (*past part pass*) of **уч·и́ть**

уч·ёност·ь, -и, (*f*) erudition, learning

уч·ён·ый, -ая, -ое, (*a*) learned, erudite; scientific; performing (of animals); (*m decl as adj*) scholar, savant; noted scientist, scientist (when authoritative)

~ **за·пи́с·к·и** (*pl*) (print) proceedings (of learned societies)

~ **о́бщ·еств·о** (*n*) learned society

~ **секрета́р·ь** (*m*) secretary (of a learned society, institute etc.)

уч·е́нь·е, -я, (*n*) = **уч·е́ни·е**

у·че́ст·ь, (*fut 3rd pl*) **у·чт·у́т,** (*perf*) *see* **у·чи́т·ыва·ть**

у·чёт·, -а, (*m*) (*v n*) *see* **у·чи́т·ыва·ть;** account, report; returns, record, register; (math) allowance; (fin) discount

~**, операти́в·н·ый** progress-of-work report
с у·чёт·ом with regard to, in respect of; with allowance for, taking account of

~ **това́р·ов** stock taking, inventory control

у·четвер·е́ни·е, -я, (*n*) quadrupling

у·четвер·ённ·ый, -ая, -ое, (*past part pass*) *see* **у·четвер·я́·ть;** quadrupled, fourfold

у·четвер·и́ть, -я́т, (*perf*) *see* **у·четвер·я́·ть**

у·четвер·я́·ть, -ют, (*imp*) quadruple

у·чёт·н·ый, -ая, -ое, (*a*) of **у·чёт·**

~ **банк·** (*m*) discount bank

~ **ве́д·омост·ь** (*f*) tally sheet, inventory list

~ **маши́н·а** (*f*) business machine

~ **ста́в·к·а** (*f*) (fin) discount rate

у·чёт·чик·, -а, (*m*) clerk; checkman

уч·и́лищ·е, -а, (*i*) **-ем,** (*f*) college, academy, school

уч·и́тел·ь, -я, (*nom pl*) **-я́,** (*g pl*) **-е́й,** (*m*) teacher, instructor

уч·и́тельск·ий, -ая, -ое, (*a*) teachers', instructors'

у·чи́т·ыва·ть, -ют, (*imp*) account for, take stock of; take into account/consideration, allow for; (fin) discount (bills)

уч·и́ть, -ат, (*imp*) teach, instruct; learn, commit to memory; **-ся** learn, study

у·чла́, (*past fem sing*) of **у·чёст·ь** (*perf*) *see* **у·чи́т·ыва·ть** (*imp*)

учред·и́ть, -я́т, (*perf*) *see* **учрежд·а́·ть**

учрежд·а́·ть, -ют, (*imp*) found, establish, institute, set up

учрежд·е́ни·е, -я, (*n*) (*v n*) *see* **учрежд·а́·ть;** institution, establishment

у·чт·ённ·ый, -ая, -ое, (*past part pass*) *see* **у·чи́т·ыва·ть**

у·чт·ёт, (*fut 3rd sing*) of **у·чёст·ь** (*perf*) *see* **у·чи́т·ыва·ть** (*imp*)

уш·а́н·, -а, (*m*) (zool) bat, *Plecotus*

уш·а́ст·ый, -ая, -ое, (*a*) big-eared

~ **лис·и́ц·а** (*f*) (zool) large-eared fox, *Otocyon megalotis*

у·ше́д·ш·ий, -ая, -ое, (*past part act*) of **у·йти́** (*perf*) *see* **у·ход·и́ть** (*imp*)

у·шёл, (*past masc sing*) of **у·йти́** (*perf*), *see* **у·ход·и́ть** (*imp*)

у́ш·и, (*nom pl*) of **у́х·о** (q.v.)

у·ши́б·, -а, (*m*) (*v n*) *see* **у·шиб·а́·ть;** hurt place, bruise, (med) contusion; (*past masc sing*) of **у·шиб·и́ть**

у·шиб·а́·ть, -ют, (*imp*) hurt, bruise, (med) contuse

у·шиб·и́ть, -у́т, (*perf*) *see* **у·шиб·а́·ть**

у·ши́бл·енн·ый, -ая, -ое, (*past part pass*) *see* **у·шиб·а́·ть;** bruised, contused

у·шив·а́·ть, -ют, (*imp*) take in/up, shorten (by sewing); embroider all over

у·шир·е́ни·е, -я, (*n*) (*v n*) *see* **у·шир·я́·ть;** (spectr) broadening; (met roll) spread; spreading (of liquid over a surface); stretching (of pulses)

~**, аппара́т·н·ое** (spectr) instrumental broadening

~**, лин·е́йн·ое** (met roll) absolute spread

~ **луч·а́** beam spread

~**, не·свобо́д·н·ое** (met roll) restricted spread

по·каз·а́тел·ь у·шир·е́ни·я (*m*) (met roll) proportional spread, spread-to-draught ratio

~**, свобо́д·н·ое** (met roll) free spread, unrestricted spread

~ **сопл·а́** nozzle expansion

у·ши́р·енн·ый, -ая, -ое, (*past part pass*) *see* **у·шир·я́·ть;** broadened (of spectral lines); stretched (of pulses)

у·шир·и́тел·ь, -я, (*m*) widener, broadener

у·ши́р·ить, -ят, (*perf*) *see* **у·шир·я́·ть**

у·шир·я́·ть, -ют, (*imp*) make wider/broader, widen, broaden

у·ши́·ть, (*fut 3rd pl*) у·шь·ю́т, (*perf*)
see у·шив·а́·ть

у́ш·к·о, -а, (*nom pl*) -и, (*g pl*) -ш·ек·,
(*n*) (*dim*) of у́х·о; lug, ear, (biol)
auricle; tab, tag;
уш·к·о́, -а́, (*nom pl*) -й, (*g pl*) -о́в,
(*n*) eye (of needle etc.)

уш·ко́в·ый, -ая, -ое, (*a*) of уш·к·о́
∼ болт· (*m*) eye bolt

у·шла́, (*past fem sing*) of у·йти́ (*perf*)
see у·ход·и́ть (*imp*)

у·шл·ёт, (*fut 3rd sing*) of у·сл·а́ть

у·шли́, (*past pl*) of у·йти́ (*perf*) see у·хо-
д·и́ть (*imp*); у·шл·и́ (*sing imper*) of
у·сл·а́ть

у·шл·и́те, (*pl imper*) of у·сл·а́ть

уш·ни́к·, -а́, (*m*) (med) aurist

уш·н·о́й, -а́я, -о́е, (*a*) ear, aural

у·шь·ёт, (*fut 3rd sing*) of у·ши́·ть
(*perf*) see у·шив·а́·ть (*imp*)

у·щёл·ист·ый, -ая, -ое, (*a*) ravined

у·щёл·ь·е, -я, (*g pl*) -л·и·й, (*n*) gorge,
ravine, canyon

у·щем·и́ть, -я́т, (*perf*) see у·щемл·я́·ть

у·щемл·я́·ть, -ю́т, (*imp*) pinch, jam,
catch, (med) incarcerate, impact

у·щёрб·, -а, (*m*) damage, detriment,
loss, (com) prejudice
воз·мещ·е́ни·е у·щёрб·а (*n*) (law)
damages
на у·щёрб·е   on the decline; on
the wane (of moon)

у·щёрб·н·ый, -ая, -ое, (*a*) waning
(of moon); declining

у·щип·н·у́ть, -у́т, (*perf*) pinch

у·ю́т·н·ый, -ая, -ое, (*a*) comfortable,
cosy

у·язв·и́мост·ь, -и, (*f*) vulnerability

у·язв·и́м·ый, -ая, -ое, (*a*) vulnerable

у·ясн·и́ть, -я́т, (*perf*) see у·ясн·я́·ть

у·ясн·я́·ть, -ю́т, (*imp*) clear/size up
(problems)

Ф

фа (*n indecl*) F (music)

фаб·зав·ко́м·, -а, (*m*) factory trade-union committee

фаб·за́в·уч·, -а, (*m*) factory training school

Фабри́-Перо́, интерферо́·метр· (*m*) (opt) Fabry and Perot interferometer

фа́брик·а, -и, (*f*) (*see also under associated adjectives*); factory, plant, mill

~ **-ку́х·н·я, -и,** (*f*) large-scale catering-establishment

~ **суко́н·н·ая** (text) cloth/woollen

~, **швей·н·ая** (text) clothing factory

фабрик·а́нт·, -а, (*m*) factory owner, industrialist

фабрика́т·, -а, (*m*) finished/manufactured product

фабри́ч·н·о-за·вод·ск·о́й, -а́я, -о́е, (*a*) factory, works

фабри́ч·н·ый, -ая, -ое, (*a*) factory, mill; manufacturing, industrial

~ **ма́р·к·а** (*f*) trademark

~ **ста́р·ост·ь** (*f*) shop steward (in England)

фаг·, -а, (*m*) bacteriophage, phage

фаго́т·, -а, (*m*) bassoon (music)

фаго·терап·и́·я, -и, (*f*) (med) phago-therapy

фаго·ци́т·, -а, (*m*) (biol) phagocyte

фаго·цит·о́з·, -а, (*m*) (biol) phago-cytosis

фа́з·а, -ы, (*f*) phase, stage

~ **в·во́д·а** (rocket) gathering phase

~ **в·недр·е́ни·я** (phys, chem) interstitial phase

~ **во вре́мен·и** time phase

   гран·и́ц·а фаз· (*f*) (phys, chem) interface, phase interface

   диагра́мм·а фаз· газо·рас·пре·дел·е́ни·я (*f*) (ICE) valve diagram, valve acceleration diagram

~, **интерметалли́д·ная** (met) intermediate constituent phase

~, **лу́н·н·ая** (astron) phase of the moon

~ **на·вед·е́ни·я, на·ча́ль·н·ая** (rocket) gathering phase

~, **на·ча́ль·н·ая** initial phase; (rocket) gathering phase

   от·став·а́ни·е фа́з·ы (*n*) (telecom) phase delay

   пра́в·ил·о фаз· (*n*) (phys, chem) phase law

фа́з·а

   с·дви́г· фаз· (*m*) phase shift/displacement/difference

   с·дви́·нут·ый по фа́з·е (*a*) out-of-phase

~ **с·кла́д·чатост·и** (geol) folding stage

~, **твёрд·ая** solid phase

   у́гол· с·дви́г·а фаз· (*m*) phase angle

фаза́н·, -а, (*m*) (orn) pheasant, *Phasianus*

фазеол·и́н·, -а, (*m*) (biochem) phaseoline

фазео·сапони́н·, -а, (*m*) (biochem) phaseosaponine

фаз·ирова́ни·е, -я, (*n*) phasing

фаз·иро́ванн·ый, -ая, -ое, (*past part pass*) of **фаз·ир·ова́ть**; phased

фаз·ир·ова́ть, -у́ют, (*imp and perf*) phase

фаз·иро́вщик·, -а, (*m*) (telecom), phaser

фа́зис·, -а, (*m*) = **фа́з·а**

фазитро́н·, -а, (*m*) (rad) phasitron

фа́з·ност·ь, -и, (*f*) phasal nature

фазо·вращ·а́тел·ь, -я, (*m*) (elec) phase inverter/shifter, phaser

фазо·вращ·а́тельн·ый, -ая, -ое, (*a*) phase-inverting/shifting, phasing

фа́з·ов·ый, -ая, -ое, (*a*) phase

~ **ана́лиз·** (*m*) (chem, phys) phase analysis; (elec) phase-shift analysis

~ **диагра́мм·а** (*f*) (phys, chem) phase/equilibrium diagram

~ **контра́ст·** (*m*) phase contrast (microscope technique)

~ **модул·я́ци·я** (*f*) (rad) phase modulation

~ **пере·хо́д·** (*m*) (phys, chem) phase transformation/transition

~ **пере·хо́д· втор·о́го ро́д·а** (phys) second-order phase transition

~ **портре́т·** (*m*) (autom) phase-plane diagram

~ **по·сто·я́нн·ая** (*f*) (telecom) phase/wavelength constant

~ **про·стра́н·ств·о** (*n*) (math) phase space

~ **равно·ве́с·и·е** (*n*) (phys, chem) phase equilibrium

~ **рас·ка́ч·к·а** (*f*) (TV) phase swinging

~ **систе́м·а слеп·о́й по·са́д·к·и** (*f*) (nav) lateral guidance system

~ **траекто́ри·я** (*f*) (autom) phase-plane vector

~ **у́гол·** (*m*) phase angle

фазо·вы·ра́вн·иватель·, -я, (*m*) (telecom) phase compensator/equalizer

фазо·инеёртор·, -а, (*m*) (elec) phase inverter

фазо·индика́тор·, -а, (*m*) (rad) phase indicator/monitor

фазо·индика́тор·н·ый орга́н· (*m*) phase-indicating element

фазо·компенс·а́тор·, -а, (*m*) (elec) phase advancer

фазо́·метр·, -а, (*m*) (elec) power factor meter, phase indicator/meter

фазо·метр·и́ческ·ий, -ая, -ое, (*a*) phase-meter

фазо·модул·и́рованн·ый, -ая, -ое, (*a*) phase-modulated

фазо·о·про·ки́д·ыватель·, -я, (*m*) phase inverter

фазо·пере·вора́ч·ивающ·ее у·стро́й·-ств·о (*n*) (elec) phase inverter

фазо·пре·образ·ова́тель·, -я, (*m*) (elec) phase adjuster, rotatable phase-adjusting transformer

~, электро·маши́н·н·ый rotary phase converter

фазо·рас·щеп·и́тель·, -я, (*m*) phase splitter

фазо·регул·я́тор·, -а, (*m*) (elec) phase shifter, phase-shifting transformer

фазо·с·двиг·а́ющ·ий, -ая, -ее, (*a*) phase-shifting/shift

фазо·тро́н·, -а, (*m*) (nucl) synchrocyclotron, accelerator, synchrocyclotron, frequency-modulated cyclotron

фазо·у·каз·а́тель·, -я, (*m*) (elec) phase-sequence indicator

фазо·хро́н·н·ый, -ая, -ое, (*a*) phase-chronous

фазо·част·о́тн·ый, -ая, -ое, (*a*) phase-frequency

фазо·чу́вств·и́тельн·ый орган· (*m*) (instr) phase-detecting element

фа́·й, -я, (*m*) (text) faille

файнште́йн·, -а, (*m*) (met) copper-nickel matte

фа́кел·, -а, (*m*) torch, flare, flame, burner jet, tongue of flame

~, вы́·хлоп·н·о́й jet flame (of jet engine)

~ за·жиг·а́ни·я pilot flame

~ пла́мен·и flame jet, tongue of flame

~, пуск·ов·о́й pilot flame; (a/c) starting torch

~ реакти́в·н·ого двиг·ател·я jet-engine flame

~, сигна́ль·н·ый signal flare

фа́кел

~, со́лн·ечн·ый (astron) solar flare

~ то́пл·ив·а fuel spray

факел·е́ни·е, -я, (*n*) torching (jet engine)

фа́кель·н·ый, -ая, -ое, (*a*) of **фа́кел·**

факоида́ль·н·ый, -ая, -ое, (*a*) (geol) phacoidal

фако·ли́т·, -а, (*m*) (geol) phacolith; (min) phacolite

фако·маля́ци·я, -и, (*f*) (med) phacomalacia

фако·ци́ст·, -а, (*m*) (biol) phacocyst

факси́миле (*n indecl*) facsimile; photocopy; signature rubber-stamp

фа́кт·, -а, (*m*) fact

факти́с·, -а, (*m*) factice, factis, vulcanized oil

факт·и́ческ·и (*adv*) actually; in fact

факт·и́ческ·ий, -ая, -ое, (*a*) actual, factual; virtual

~ да́·нн·ые (*pl*) the facts/data

~ пут·ев·о́й у́гол· (*m*) (air) track made good

ли́н·и·я факт·и́ческ·ого пут·и́ (*f*) (air) track made good

~ смо́л·ы (*pl*) gum existent (oil analysis)

фа́ктор·, -а, (*m*) (*see also under associated words*) and *see* **коэффицие́нт;** factor

~, масшта́б·н·ый scale factor

~ масшта́б·н·ого пре·образ·ова́·ни·я scale factor (computers)

~ мно́ж·еств·о, -а, (*n*) (math) factor/quotient set

~ фо́рм·а (*f*) (phys) form factor; shape factor

факториа́л·, -а, (*m*) (math) factorial

факториа́ль·н·ый, -ая, -ое, (*a*) factorial

факто́р·и·я, -и, (*f*) trading post/station (for furs etc.)

факту́р·а, -ы, (*f*) (com) invoice, bill; texture; facing (stonework)

~ -с·чёт·, -а, (*m*) *see* **с·чёт·-факту́-р·а**

факту́р·н·ый, -ая, -ое, (*a*) of **фак-ту́р·а**

~ маши́н·а (*f*) invoicing machine

факультати́в·н·ый, -ая, -ое, (*a*) optional; (biol) facultative

факульте́т·, -а, (*m*) faculty, department, school (in university)

фа́л·, -а, (*m*) (naut) halyard

~, вы́·тяж·н·о́й static line (of parachute)

фала́нг·а, -и, (*f*) (anat) phalanx, (*pl*) phalanges

фала́нг·ов·ый, -ая, -ое, (*a*) (anat) phalangeal

фа́лин·ь, -я, (*m*) boats painter (rope)

фалло́пи·ев·а труб·а́ (*f*) (anat) Fallopian tube

фалло·со́м·а, -ы, (*f*) (zool) phallosome

фа́лреп·, -а, (*m*) (naut) manrope

фальсифика́т·, -а, (*m*) forgery, cheat; adulterated food

фальсифик·а́ци·я, -и, (*f*) falsification; (food) adulteration; counterfeiting, counterfeit (of money etc.)

фальсифиц·и́р·овать, -уют, (*imp and perf*) falsify; adulterate; counterfeit

фальстарнпост·, -а, (*m*) (shipb) inner post (wooden construction)

фальстем·, -а, (*m*) (naut) apron

фа́льц·, -а, (*i*) **-ем,** (*m*) fold, foldover seam (in sheet metal); groove, rebate, rabbet (in wood), fillister (in window sash bar); (print) fold; shaving/ currier's knife (leather)

фальц·бло́к·, -а, (*m*) shaving beam (leather)

фальц·го́бел·ь, -я, (*m*) *see* **фальц·губел·ь**

фальц·губел·ь, -я, (*m*) rabbeting, plane, fillister (carpenter's tool)

фальц·ева́льн·ый, -ая, -ое, (*a*) rebating, rabbeting; folding

фальц·ева́ть, -у́ют, (*imp*) rebate, rabbet, groove; (paper) fold, score; seam (sheet metal); shave (leather)

фа́льц·ев·ый, -ая, -ое, (*a*) rabbeting, grooving; folding

~ кассе́т·н·ая маши́н·а (*f*) (print) plate folding machine

~ нож·ев·а́я маши́н·а (*f*) (print) knife folding machine

фальц·о́вк·а, -и, (*f*) (*v n*) of **фальц·ева́ть**

фальц·о́вочн·ый, -ая, -ое, (*a*) of **фальц·о́вк·а**

фальш·бо́рт·, -а, (*m*) (shipb) bulwark; (mil) skirting plate (of tank)

фальш·и́в·ый, -ая, -ое, (*a*) false, spurious, forged, counterfeit; artificial, imitation, dummy; (naut) jury

~ рул·ь (*m*) (naut) jury rudder

фальш·ки́л·ь, -я, (*m*) (shipb) false keel shoe

фальшфе́йер·, -а, (*m*) (naut) signal flare, blue light

фами́л·и·я, -и, (*f*) surname

фамилья́р·н·ый, -ая, -ое, (*a*) brash, too familiar

фана́л·ев·ый, -ая, -ое, (*a*) (chem) fanal

фана́л·ь, -я, (*m*) (chem) fanal

~ -ро́з·а, -ы, (*f*) fanal pink

фанат·и́ческ·ий, -ая, -ое, (*a*) fanatical

фа́нг·, -а, (*m*) (text) double-rib

фан·гломера́т·, -а, (*m*) (geol) fanglomerate

фа́нг·ов·ый, -ая, -ое, (*a*) of **фанг·**

~ маши́н·а (*f*) (text) double-rib knitter machine

фангсбо́т·, -а, (*m*) sealer (boat)

фане́р·а, -ы, (*f*) veneer; plywood

~, кле·ён·ая plywood

~, лущ·ён·ая rotary-cut veneer

~, металлиз·и́рованн·ая metal-faced plywood

~, много·сло́й·н·ая multi-ply

~, нож·ев·а́я knife-cut veneer

~ -пере·кле́й·к·а, -и, (*f*) plywood

~, пил·ён·ая sawn/sawed veneer

~, ре́з·ан·ая knife-cut veneer

~, стро́г·ан·ая sliced veneer

фанер·и́т·, -а, (*m*) (min) phanerite

фанери́т·ов·ый, -ая, -ое, (*a*) (min) phaneritic

фане́р·н·ый, -ая, -ое, (*a*) veneer, plywood

фанер·ова́ть, -у́ют, (*imp and perf*) veneer

фанеро·га́м·, -а, (*m*) (bot) phanerogam

фанеро·га́м·н·ый, -ая, -ое, (*a*) (bot) phanerogamous

фанеро·пи́ль·н·ый стан·о́к· (*m*) veneer saw frame

фанеро·строг·а́льн·ый, -ая, -ое, (*a*) veneer-cutting

фанеро·фи́т·, -а, (*m*) (bot) phanerophyte

фа́н·ов·ый, -ая, -ое, (*a*) (naut) sanitary, sludge, sewage

~ сист·е́м·а (*f*) (mar eng) sanitary system

фанта́з·и·я, -и, (*f*) fantasy, fancy, imagination

фантаст·и́ческ·ий, -ая, -ое, (*a*) fantastic

фантастро́н·, -а, (*m*) (rad) phantastron

фантастро́н·н·ый, -ая, -ое, (*a*) phantastron

фанто́м·, -а, (*m*) phantom

фанто́м·н·ая цеп·ь (*f*) (telecom) phantom circuit

фаоли́т·, -а, (*m*) (plast) faolite

фао·планктóн·, -а,   (*m*) (microbiol) phaoplankton

фао·сóм·а, -ы,   (*f*) (zool) phaosome

фáр·а, -ы,   (*f*) headlight, headlamp; (a/c) light, landing light

   на·клон·éни·е фар·   (*n*) (M/T) dipping of headlights

~, по·сáд·очн·ая   (a/c) landing light

~, рул·ёжн·ая   (a/c) taxiing light

фарáд·а, -ы,   (*f*) (elec) farad

~, абсолют·н·ая   absolute farad

~, между·на·рóд·н·ая   international farad

фарадé·ев·о тёмн·ое про·стрáн·-ство (*n*) (phys) Faraday dark space

фарадé·ев·ое про·стрáн·ств·о   (*n*) (phys) Faraday dark space

фарадé·й, -я,   (*m*) (phys) faraday, Faraday unit

Фарадé·я, за·кóн·ы   (*pl*) Faraday's laws of electrolysis

~, цилíндр·   (*m*) (nucl) beam catcher, Faraday cylinder

~, числ·ó   (*n*) (elec) Faraday constant

~, эффéкт·   (*m*) Faraday effect (electro-statics)

~, явл·éни·е   (*n*) (phys) Faraday effect

фарадиз·áтор·, -а,   (*m*) (med) faradizer

фарадиз·áци·я, -и,   (*f*) (med) faradiza-tion, faradism

фарад·íческ·ий, -ая, -ое,   (*a*) (elec) faradic

~ ток·   (*m*) (med) faradic current

фарад·мéтр·, -а,   (*m*) faradmeter, microfaradmeter

фарб·лáк·, -а,   (*m*) coloured/colored varnish

фарвáтер·, -а,   (*m*) (naut) fairway, channel

~, за·гражд·ённ·ый   obstructed chan-nel

~, об·слéд·ованн·ый про·мéр·ом   surveyed channel

~, о·гражд·ённ·ый   buoyed channel

Фаренгéйт·, -а,   (*m*) (phys) Fahrenheit

Фаренгéйт·а, шкал·á   (*f*) Fahrenheit scale (temperature)

фарингáль·н·ый, -ая, -ое,   (*a*) (anat) pharyngeal

фарингеáль·н·ый, -ая, -ое,   (*a*) (anat) pharyngeal

фаринг·íт·, -а,   (*m*) (med) pharyngitis

фаринго·плегí·я, -и,   (*f*) (med) pharyngoplegia

фаринго·скóп·, -а,   (*m*) (med) pha-ryngoscope

фарино·тóм·, -а,   (*m*) (bot) farinotom

фармако·гнóз·и·я, -и,   (*f*) (pharm) pharmacognosy

фармако·динáм·ик·а, -и,   (*f*) (pharm) pharmacodynamics

фармако·лíт·, -а,   (*m*) (min) phar-macolite

фармако·лóг·и·я, -и,   (*f*) (pharm) pharmacology

фармако·пé·я, -и,   (*f*) (pharm) phar-macopoeia, pharmacopeia

фармако·сидерíт·, -а,   (*m*) (min) pharmacosiderite

фармако·хím·и·я, -и,   (*f*) pharmaceuti-cal chemistry

фармац·éвт·, -а,   (*m*) pharmacist, pharmaceutist

фармац·éвт·ик·а, -и,   (*f*) pharmaceutics

фармац·евт·íческ·ий, -ая, -ое,   (*a*) pharmaceutical

фармац·í·я, -и,   (*f*) pharmacy, pharma-ceutics

фáртук·, -а,   (*m*) apron

~ сýппорт·а   saddle apron (of machine tool)

фартýч·н·ый, -ая, -ое,   (*a*) apron

~ транспорт·ёр·   (*m*) apron conveyer

фарфóр·, -а,   (*m*) china, porcelain; chinaware

фарфоро·вíд·н·ый, -ая, -ое,   (*a*) porcelanic, porcelaneous

фарфóр·ов·ый, -ая, -ое,   (*a*) of **фар-фóр·**

~ изол·ятор·   (*m*) (elec) porcelain insulator

фáрш·, -а,   (*i*) **-ем,** (*m*) stuffing, force-meat; sausage meat

фарш·ир·овáть, -ýют,   (*imp*) (food) stuff

фáс·, -а,   (*m*) face; (mil) fire bay; (geol) scarp, scarpment; (com) fas, free alongside

~ кристáлл·а   crystal face

фасáд·, -а,   (*m*) (arch) façade, front

фасéт·, -а,   (*m*) facet

фасéт·к·а, -и,   (*g pl*) **-т·ок·,** (*f*) facet

фасéт·очн·ый, -ая, -ое,   (*a*) facetted; (zool) compound (eye)

фáс·к·а, -и,   (*g pl*) **-с·ок·,** (*f*) face, facet; bevel edge; (build) fascia

~ клáпан·а   valve face

фас·овáть, -ýют,   (*imp*) package, bag (weighed quantities)

фас·óвк·а, -и,   (*g pl*) **-вок·,** (*f*) (*v n*) of **фас·овáть**

фас·óвочн·ый, -ая, -ое,   (*a*) packaging

фасо́л·ев·ый, -ая, -ое, (*a*) of фасо́л·ь

фасо́л·ь, -и, (*f*) (bot) bean, *Phaseolus*

~, обык·нове́нн·ая kidney/common bean, *Phaseolus vulgaris*

фасо́н·, -а, (*m*) fashion, style, appearance, pattern, shape, form; pattern (dressmaking)

~, мо́д·н·ый fashionable style/appearance

фасо́н·ирова́ни·е, -я, (*n*) shaping

фасо́н·ист·ый, -ая, -ое, (*a*) = фасо́н·н·ый

фасо́н·н·о-от·ре́з·н·ый автома́т· (*m*) (mech eng) repetition bar-stock lathe

фасо́н·н·о-строга́ль·н·ый стан·о́к· (*m*) (mech eng) shaping machine

фасо́н·н·ый, -ая, -ое, (*a*) of фасо́н·

~ инструме́нт· (*m*) (mech eng) forming tool

~ лит·ь·ё (*n*) (met) shaped casting

~ кирпи́ч· (*m*) (build) moulded/molded brick

~ мета́лл· (*m*) (met) shapes

~ про́фил·ь (*m*) (met) section, shape

~ рез·е́ц· (*m*) forming tool

~ стал·ь (*f*) steel shapes/sections

фасци·а́ци·я, -и, (*f*) (bot) fasciation

фасци·о́л·а, -ы, (*f*) (zool) fasciole

фасци·оле́з·, -а, (*m*) (med) fascioliasis

фасциоли́н·, -а, (*m*) (pharm) fasciolin

фасц·и́т·, -а, (*m*) (med) fascitis

фа́сци·я, -и, (*f*) (anat) fascia

фата́ль·н·ый, -ая, -ое, (*a*) fatal

фа́улер·ов рас·тво́р· (*m*) (pharm) Fowler's solution

фа́ун·а, -ы, (*f*) (zool) fauna

фаун·исти́ческ·ий, -ая, -ое, (*a*) (zool) faunal

фахве́рк·, -а, (*m*) (build) framework

фаце́ли·я, -и, (*f*) (bot) phacelia, *Phacelia*

фаце́лл·а, -ы, (*f*) (zool) phacella

фаце́т·, -а, (*m*) = фасе́т·

фациа́ль·н·ый, -ая, -ое, (*a*) (geol) facies

фа́ци·я, -и, (*f*) (geol) facies

фаше́ппис·, -а, (*m*) (shipb) fashion piece

фаши́н·а, -ы, (*f*) (civ eng) facine

~, по·груж·ённ·ая water fascine

фаши́н·ник·, -а, (*m*) sticks (for fascine making); fascine

Фа́·я, анома́ли·я (*f*) free air effect/reduction

фая́нс·, -а, (*m*) (ceram) faience, fayance, fayence, earthenware, glazed earthenware

фая́нс·ов·ый, -ая, -ое, (*a*) of фая́нс·

февра́л·ь, -я́, (*m*) February

федера́ль·н·ый, -ая, -ое, (*a*) federal

федерати́в·н·ый, -ая, -се, (*a*) federative

федер·а́ци·я, -и, (*f*) federation

фе́динг·, -а, (*m*) (rad) fading

федо́метр·, -а, (*m*) (chem) fadometer

фейерве́рк·, -а, (*m*) firework(s)

фейерве́р·оч·н·ый, -ая, -ое, (*a*) of фейерве́рк·

фейхо́а (*m indecl*) (bot) *Feijoa sellowiana*

фекали́н·, -а, (*m*) (agr) fecalin, faeces, human excrement used as manure

фека́л·и·я, -и, (*f*) (zool) (*pl*) faeces, feces

фека́ль·н·ый, -ая, -ое, (*a*) faecal, fecal

Фе́линг·а, жи́дк·ост·ь (*f*) (chem) Fehling's solution

фелле́м·а, -ы, (*f*) (bot) phellem

фелло·ге́н·, -а, (*m*) (bot) phellogen

фелло·де́рм·а, -ы, (*f*) (bot) phelloderm

фелл·о́ид·, -а, (*m*) (bot) phelloid

фёльген·а, реа́кци·я (*f*) (chem) Feulgen reaction

фельдма́ршал·, -а, (*m*) (mil) Field-Marshal

фе́льдшер·, -а, (*nom pl*) -а́, (*g pl*) -о́в, (*m*) (med) medical assistant, male nurse; (vet) veterinary surgeon's assistant

фельдшпати́д·, -а, (*m*) (min) feldspathoids

фельдшпати́т·ов·ый, -ая, -ое, (*a*) (min, geol) feldspathic

фельето́н·, -а, (*m*) satirical article

фельзи́т·, -а, (*m*) (geol) felsite

фельз·и́ческ·ий, -ая, -ое, (*a*) (geol) felsic

фельзо·фи́р·, -а, (*m*) (min) felsophyre

фем·и́ческ·ий, -ая, -ое, (*a*) (min) femic

фён·, -а, (*m*) (meteor) foehn, foehn wind; hair-dryer

фен·адо́н·, -а, (*m*) (pharm) phenadon

фен·ази́н·, -а, (*m*) (chem) phenazine

фенак·и́т·, -а, (*m*) (min) phenacite, phenakite

фенак·о́дус·, -а, (*m*) (pal) phenacodus, *Phenacodus*

фенако́н·, -а, (*m*) (pharm) phenacon

фенами́н·, -а, (*m*) (pharm) phenamine, benzadrine

фенантре́н·, -а, (*m*) (chem) phenanthrene

фенантр·оли́н·, -а, (*m*) (chem) phenanthroline

фенати́н·, **-а,** (*m*) (pharm) phenatin

фенафа́н·, **-а,** (*m*) (pharm) phenaphan

фенацеми́д·, **-а,** (*m*) (pharm) phenacemide

фенацетами́д·, **-а,** (*m*) phenyl acetamide

фен·ацети́н·, **-а,** (*m*) (pharm) phenacetin

фенаци́т·, **-а,** (*m*) (min) phenacite

фенг·и́т·, **-а,** (*m*) (min) phengite

фен·ето́л·, **-а,** (*m*) (chem) phenetole, phenyl ethyl ether

фенико·хрои́т·, **-а,** (*m*) (min) phoenicochroite, phenicochroite

фен·и́л·, **-а,** (*m*) (chem) phenyl

фенилалани́н·, **-а,** (*m*) (chem) phenyl alanine

фенилгидрази́н·, **-а,** (*m*) (chem) phenylhydrazine

фенилглици́н·, **-а,** (*m*) (pharm) phenylglycine, phenyl amino-acetic acid

фенилдихлорарси́н·, **-а,** (*m*) phenyldichlorarsine

фенилендиами́н·, **-а,** (*m*) phenylenediamine

фенил·у́ксус·н·ая кисл·от·а́ (*f*) phenylacetic acid

фенил·эти́л·ов·ый спирт· (*m*) phenyl ethyl alcohol

фениц·и́т·, **-а,** (*m*) (geol) phoenicite

фено- (*int component*) pheno-, phenolic

фён·ов·ый, **-ая, -ое,** (*a*) (meteor) foehn

фено·кре́ст·, **-а,** (*m*) = **фено·-кристалл·**

фено·криста́лл·, **-а,** (*m*) (cryst) phenocryst

феноксиметилпеницилли́н·, **-а,** (*m*) (pharm) phenoxymethylpenicillin

фен·о́л·, **-а,** (*m*) (chem) phenol

фенола́з·а, **-ы,** (*f*) phenolase, phenoloxydase

фiноли́т·, **-а,** (*m*) (plast) phenolite, phenolic resin

фено·ло́г·и·я, **и-,** (*f*) (biol) phenology

фенолкето́н·, **-а,** (*m*) (chem) phenolic ketone

фенолоксида́з·а, **-ы,** (*f*) (chem) phenoloxydase, phenolase

фенолро́т·, **-а,** (*m*) (chem) phenol red

фенолформальдеги́д·н·ый, **-ая, -ое,** (*a*) (plast) phenol-formaldehyde

фенолфтале́ин·, **-а,** (*m*) (chem) phenolphthalein

фено́ль·н·ый, **-ая, -ое,** (*a*) phenolic

феноля́т·, **-а,** (*m*) (chem) phenolate

фено́мен·, **-а,** (*m*) phenomenon

феномена́ль·н·ый, **-ая, -ое,** (*a*) phenomenal

феномено·лог·и́ческ·ий, **-ая, -ое,** (*a*) phenomenological

фено·пла́ст·, **-а,** (*m*) phenolic plastic/resin

феносфрани́н·, **-а,** (*m*) phenosafranine (dye)

фено·сульфо·кисл·от·а́, **-ы́,** (*f*) sulphocarbolic acid

фенотиази́н·, **-а,** (*m*) phenothiazine

фено·ти́п·, **-а,** (*m*) (biol) phenotype

фено·тип·и́ческ·ий, **-ая, -ое,** (*a*) (biol) phenotypic

фено·фа́з·а, **-ы,** (*f*) (biol) phenophase

фе́нхел·ь, **-я,** (*m*) (bot) fennel, *Foenicum*

феода́ль·н·ый, **-ая, -ое,** (*a*) feudal

фео·фи́лл·, **-а,** (*m*) (biochem) phaeophyll

фео·фити́н·, **-а,** (*m*) (biochem) phaeophytin

фербери́т·, **-а,** (*m*) (min) ferberite

фергани́т·, **-а,** (*m*) (min) ferganite, ferghanite

фергусони́т·, **-а,** (*m*) (min) fergusonite

ферз·ь, **-я́,** (*m*) queen (at chess)

фе́ринг·, **-а,** (*m*) (a/c) fairing

фе́рм·а, **-ы,** (*f*) girder, truss; (agr) farm

~, **ба́л·очн·ая** girder beam

~, **мост·ов·а́я** bridge girder/truss; lead-in girder (of dry dock)

~, **решёт·чат·ая** lattice girder

~, **сквоз·н·а́я** lattice girder

~, **тре·уго́ль·н·ая** (build) English roof truss

~, **хвост·ов·а́я** (a/c) tail boom

Ферма́ (*indecl*) (opt) Fermat

~, **при́нцип·** (*m*) (opt) Fermat's rule

ферме́нт·, **-а,** (*m*) (biochem) ferment, enzyme

~, **брод·и́льн·ый** fermenting agent

фермент·а́ци·я, **-и,** (*f*) (biochem) fermentation

фермент·и́р·овать, **-у́ют,** (*imp*) ferment

ферме́нт·н·ый, **-ая, -ое,** (*a*) ferment

~ **яд·** (*m*) (bot) inhibitor

фе́рмер·, **-а,** (*m*) farmer (in capitalist economy)

фе́рми-дира́к·а, **статист·ик·а** (*f*) (phys) Fermi-Dirac statistics

фе́рми·евск·ий во́з·раст· (*m*) (nucl) Fermi age

~ **опера́тор·** (*m*) (phys) Fermi operator

фе́рми·й, **-я,** (*m*) (nucl, chem) fermium, Fm

фермио́н·, **-а,** (*m*) (nucl) fermion

фермо·во́з·, -а, (*m*) (civ eng) girder transporter

фернамбу́к·ов·ое де́рев·о (*n*) (bot) Brazil wood, *Caesalpina brasiliensis*

фернико́ (*indecl*) (met) fernico (Fe-Ni-Co alloy)

фер·о́н·, -а, (*m*) (biochem) pheron

ферр·а́т·, -а, (*m*) (chem) ferrate

~ **ка́ли·я** potassium ferrate

феррати́н·, -а, (*m*) (pharm) ferratin

ферри·магнет·и́зм·, -а, (*m*) (phys) ferrimagnetism

ферри·магни́т·н·ый, -ая, -ое, (*a*) ferrimagnetic

ферри·молибди́т·, -а, (*m*) (min) ferri-molybdite, molybdite, molybdic ochre

ферри·натри́т·, -а, (*m*) (min) ferri-natrite

ферр·и́т·, -а, (*m*) ferrite (structural component of steel); ferrite (cermet magnetic material)

~ **-грана́т·, -а,** (*m*) (met) garnet-type ferrite, ferrite garnet

ферритиз·а́тор·, -а, (*m*) (met) ferritizer

ферри́т·н·ый, -ая, -ое, (*a*) ferrite, ferritic

ферри́т·ов·ый, -ая, -ое, (*a*) ferrite, ferritic

ферро·бо́р·, -а, (*m*) (met) ferroboron

ферро·граф·и́ческ·ий, -ая, -ое, (*a*) ferrographic

ферро·вана́ди·й, -я, (*m*) ferro-vanadium, (met) ferrovanadium

ферро·вольфра́м·, -а, (*m*) ferro-tungsten, ferrotungsten

ферро·динам·и́ческ·ий, -ая, -ое, (*a*) (instr) ferrodynamic, moving-coil

ферро·зо́нд·, -а, (*m*) (phys) ferroprobe, ferromagnetic probe

ферро·ка́рт·, -а, (*m*) (elec) ferrocart, compressed-iron-core coil

ферро·кри́т·, -а, (*m*) (civ eng) ferro-crete

ферро·ли́т·, -а, (*m*) (build) ferrolithic

ферро·магнет·и́зм·, -а, (*m*) (phys) ferromagnetism

ферро·магне́т·ик·, -а, (*m*) ferro-magnet, ferromagnetic

ферро·магни́т·н·ый, -ая, -ое, (*a*) (phys) ferromagnetic

ферро·магно́н·, -а, (*m*) (phys) ferro-magnon

ферро·ма́рган·ец·, -н·ц·а, (*m*) ferro-manganese, (met) ferromanganese

ферро́·метр·, -а, (*m*) ferrometer

~, **электро́н·н·о-луч·ев·о́й** cathode-ray ferrometer

ферро·молибде́н·, -а, (*m*) ferro-molybdenum, (met) ferromolybdenum

ферро·нио́би·й, -я, (*m*) ferro-niobium, (met) ferroniobium

ферро·нихро́м·, -а, (*m*) (met) ferro-nichrome

ферро·резона́нс·, -а, (*m*) (elec) ferro-resonance, ferromagnetic resonance

ферро·сили́ци·й, -я, (*m*) ferro-silicon, ferrosilicon

ферро·с·пла́в·, -а, (*m*) ferroalloy

ферро·ти́п·и·я, -и, (*f*) (phot) ferrotype

ферро·тита́н·, -а, (*m*) (met) ferro-titanium, ferrotitanium

ферро·тори́т·, -а, (*m*) ferrothorite

ферро·хро́м·, -а, (*m*) ferro-chromium, (met) ferrochrome

~, **угле·ро́д·ист·ый** high-carbon ferro-chromium

ферро·це́н·, -а, (*m*) (chem) ferrocene

ферро·цирко́ни·й, -я, (*m*) ferro-zirconium, (met) ferrozirconium

ферро·шпине́л·ь, -и, (*f*) ferrospinel

ферро·электр·и́ческ·ий, -ая, -ое, (*a*) ferroelectric

ферсмани́т·, -а, (*m*) (min) fersmanite, fersmannite

фе́рт·, -а, (*m*) (telecom) letter F (**ф**)

ферти́ль·ност·ь, -и, (*f*) (biol) fertility

ферто́инг·, -а, (*m*) (naut) mooring swivel

фестме́тр·, -а, (*m*) cubic metre (of timber)

фесто́н·, -а, (*m*) (arch etc.) festoon; (text) scallop

фесто́н·н·ый, -ая, -ое, (*a*) of **фесто́н·**

~ **компенса́тор·** (*m*) (rubber) festoon compensator

фе́тр·, -а, (*m*) (text) felt

~, **пух·о́в·ый** fur felt

фе́тр·ов·ый, -ая, -ое, (*a*) (text) felt

фехра́л·ь, -я, (*m*) (met) Fe-Cr-Al alloy, fecral

фехт·ова́ни·е, -я, (*n*) fencing (sport)

фи phi ϕ (Greek)

фиа́л·к·а, -и, (*g pl*) **-л·ок·,** (*f*) (bot) violet, *Viola*

фиа́л·ков·ые, -ых, (*pl decl as adj*) (bot) violet family, *Violaceae*

~ **де́рев·о** (*n*) (bot) myall wood, *Acacia homalophylla*

фи́бр·а, -ы, (*f*) fibre, fiber; fibreboard, fiberboard

фибр·и́лл·а, -ы, (*f*) (biol) fibril, fibrilla

фибр·и́н·, -а, (*m*) (biochem) fibrin

фибрино·ге́н·, -а, (*m*) (biochem) fibrinogen

фибро·адено́м·а, -ы, (*f*) (med) fibro-adenoma

фибро·бла́ст·, -а, (*m*) (biol) fibroblast

фи́бр·ов·ый, -ая, -ое, (*a*) of **фи́бр·а;** fibrous

фибр·о́з·, -а, (*m*) (med) fibrosis

фибр·оз·и́т·, -а, (*m*) (med) fibrositis

фибр·о́з·н·ый, -ая, -ое, (*a*) fibrous

фибро·и́н·, -а, (*m*) fibroin, double strand (of silk)

фибро·ли́т·, -а, (*m*) (build) fibrolite, fibreboard, fiberboard; (min) fibro-lite, sillimanite

фибр·о́м·а, -ы, (*f*) (med) fibroma

фибро·миози́т·, -а, (*m*) (med) fibro-myositis

фибро·сарко́м·а, -ы, (*f*) (med) fibro-sarcoma

фибро·ферри́т·, -а, (*m*) (min) fibro-ferrite

фиг. (*abbr*) = **фигу́р·а** figure, illustration

фи́г·а, -и, (*f*) (bot) fig tree; (food) fig

фи́г·ов·ый, -ая, -ое, (*a*) (bot) fig

фигу́р·а, -ы, (*f*) figure, configuration, shape; (print) figure, illustration; piece (chessman superior to pawn); court-card (at cards)

~, бес·кон·е́чн·ая (cryst) extended group

~ вы́с·ш·его пилот·а́ж·а (air) aero-batic manoeuvre

~ земл·и́ shape of the earth

~, прямо·уго́ль·н·ая (math) orthogon

~, сигна́ль·н·ая (naut) signalling shape

~ тек·у́чест·и (*pl*) (met) strain figures, stretcher strains

~, то́п·ов·ая (surv) topmark

~ травл·е́ни·я (met) etch/etching figure

~, тупо·уго́ль·н·ая (math) amblygon

фигура́ль·н·ый, -ая, -ое, (*a*) figurative

фигура́нт·, -а, (*m*) (theat) extra, figurant

фигур·и́р·овать, -уют, (*imp*) figure, appear

фигу́р·к·а, -и, (*g pl*) **-р·ок·,** (*f*) (*dim*) of **фигу́р·а**

фигу́р·н·ый, -ая, -ое, (*a*) of **фи-гу́р·а;** figured; shaped, fancy

~ пилот·а́ж· (*m*) (air) aerobatics

~ реоста́т· (*m*) functional rheostat

~ скоб·к·и (*pl*) (print) braces, brace brackets

фи́дер·, -а, (*m*) feeder (cable); feeder (glass production)

~, анте́нн·ый aerial feeder

фи́дер

~ бег·у́щ·их волн· non-resonant feeder

~, дубл·и́рующ·ий duplicate feeder

~, объ·един·я́ющ·ий interconnecting feeder

~, от·ход·я́щ·ий outgoing feeder

~ сто·я́ч·их волн· resonant feeder

~, туп·ико́в·ый dead-ended feeder

физа́лис·, -а, (*m*) (bot) groundcherry, *Physalis*

~, съ·ед·о́бн·ый (*m*) Cape gooseberry, *Physalis angulata*

физ·иатри́·я, -и, (*f*) (med) physiatrics

фи́з·ик·, -а, (*m*) physicist

фи́з·ик·а, -и, (*f*) physics

~ криста́лл·ов crystal physics

~ реа́ктор·ов (nucl) reactor physics

~ твёрд·ого те́л·а solid-state physics

~, я́дер·н·ая nuclear physics

фи́з·ик·о-матема́т·и́ческ·ий, -ая, -ое, (*a*) physicomathematical

фи́з·ик·с-техн·и́ческ·ий, -ая, -ое, (*a*) physico-technical; applied physics

фи́з·ик·о-хим·и́ческ·ий, -ая, -ое, (*a*) physicochemical

физио·гра́ф·и·я, -и, (*f*) (geol) physio-graphy

физио·лог·и́ческ·ий, -ая, -ое, (*a*) physiological

~ аку́ст·ик·а (*f*) physiological accoustics

физио·ло́г·и·я, -и, (*f*) (biol) physiology

физио·терап·и́·я, -и, (*f*) (med) physio-therapy

физ·и́ческ·ий, -ая, -ое, (*a*) physical

~ гео·ло́г·и·я (*f*) physical geology

физ·культу́р·а, -ы, (*f*) athletics, sports, physical-exercise

физ·культу́р·н·ый, -ая, -ое, (*a*) of **физ·культу́р·а**

физо·га́стри·я, -и, (*f*) (zool) physo-gastry

физо·сти́гм·а, -ы, (*f*) (bot) calabar bean, *Physostigma*

физо·стигм·и́н·, -а, (*m*) (biochem) physostigmine

физ·под·гото́в·к·а, -и, (*f*) physical training

Фи́к·а, за·ко́н· (*m*) (nucl) Fick's law

фико·били́н·, -а, (*m*) (biochem) phyco-bilin

фико·ксанти́н·, -а, (*m*) (biochem) phycoxanthin

фико·ло́г·и·я, -и, (*f*) (biol) phycology, algology

фико·мице́т·, -а, (*m*) (bot) phycomycete

фико·фейн·, -а, (*m*) (biochem) phyco-
phaein

фико·хризин·, -а, (*m*) (biochem)
phycochrysin

фико·циан·, -а, (*m*) (biochem) phyco-
cyanin

фико·эритрин·, -а, (*m*) (biochem)
phycoerythrin

фикс·аж·, -а, (*i*) -ем, (*m*) (phot)
fixing; fixer; fixing solution/agent

~, **кисл·ый** (phot) acid fixing

фиксатив·, -а, (*m*) fixative (for drawings etc.)

фикс·атор·, -а, (*m*) fastener, locking
device, lock, catch, clamp; fixer;
(biol) fixator, fixative

~ **шаг·а** (a/c) pitch-change governor

фикс·аци·я, -и, (*f*) fixing; (biol) fixation; immobilization (of atomic
waste); (TV) clamp, clamping

~ **азот·а** nitrogen fixation

~, **у·правл·яем·ая** (TV) keyed clamp

~ **у·ровн·я** clamping (vacuum tubes)

~ **у·ровн·я черн·ого** (TV) black-
level clamping

фикс·ирование, -я, (*m*) (*v n*) of **фик-
с·ир·овать**; (phot) fixing; (TV)
clamping

фикс·ированн·ый, -ая, -ое, (*past
part pass*) of **фикс·ир·овать**

фикс·ир·овать, -уют, (*imp*) fix,
secure, lock; record, register, settle

фикс·ирующ·ий, -ая, -ее, (*pres part
act*) of **фикс·ир·овать**

~ **схем·а** (*f*) (TV) clamping circuit,
clamper; (autom) holding circuit

фиктив·н·ый, -ая, -ое, (*a*) fictitious;
imaginary; dummy; theoretical, hypothetical

~ **на·груз·к·а** (*f*) (elec) fictitious
power

фикус·, -а, (*m*) (bot) ficus

фикци·я, -и, (*f*) fiction

филе (*n indecl*) (food) fillet

~, **рыб·н·ое** (food) fish fillet

филей·н·ый, -ая, -ое, (*a*) fillet

~ **вы·шив·к·а** (*f*) (text) fillet

филён·к·а, -и, (*g pl*) -н·ок·, (*f*) panel;
jack spacer, end panel (telephones)

филён·чат·ый, -ая, -ое, (*a*) panelled

~ **двер·ь** (*f*) panelled/framed door

филер·, -а, (*m*) filler

фил·ец·, -л·ц·а, (*m*) *see* **фильц·**

филиал·, -а, (*m*) branch, affiliated
branch

фили·гран·, -а, (*m*) = **филигран·ь**

фили·гран·ь, -и, (*f*) (paper) watermark;
filigree

филин·, -а, (*m*) (orn) brown/tawny owl,
Bubo

филлипс·ит·, -а, (*m*) (min) phillipsite

филл·ит·, -а, (*m*) (min) phyllite

филл·оди·й, -я, (*m*) (bot) phyllode

филло·карид·ы, -ов, (*m pl*) (zool)
phyllocarids, *Phyllocarida*

филло·клад·н·й, -я, (*m*) (bot) phylloclade

филло·ксер·а, -ы, (*f*) (ent) phylloxera

филл·ом·, -а, (*m*) (bot) phyllome

филло·ман·и·я, -и, (*f*) (bot) phyllo-
mania

филл·онит·, -а, (*m*) (min) phyllonite

филло·под·ы, -ов, (*m pl*) (*pl*) (zool)
phyllopods, *Phyllopoda*

филло·под·и·й, -я, (*m*) (bot) phyllo-
podium

филло·сперм·, -а, (*m*) (bot) phyllo-
sperm

филло·таксис·, -а, (*m*) (bot) phyllotaxis

филло·хинон·, -а, (*m*) phylloquinone,
vitamin K_1

филло·цист·, -а, (*m*) (zool) phyllocyst

фило·генез·, -а, (*m*) (biol) phylo-
genesis

фило·генет·ик·а, -и, (*f*) (biol) phylo-
genetics

фило·генет·ическ·ий, -ая, -ое, (*a*)
phylogenetic

фило·ген·и·я, -и, (*f*) (biol) phylogeny

фило·дендрон·, -а, (*m*) (bot) philodendron, *Philodendron*

фило·лог·и·я, -и, (*f*) (ling) philology

фило·под·и·и, -й, (*f pl*) (zool) filopodia

фило́·соф·, -а, (*m*) philosopher

фило·соф·и·я, -и, (*f*) philosophy

фил·ум·, -а, (*m*) (biol) phylum

фильдекос·, -а, (*m*) (text) lisle, lisle
thread

фильдеперс·, -а, (*m*) (text) Persian
thread, high-grade lisle

фильер·а, -ы, (*m*) (text) spinneret;
(met) draw/drawing die/plate, draw-
hole

фильм·, -а, (*m*) film

фильмо·скоп·, -а, (*m*) (phot) transparency projector

фильмо·стат·, -а, (*m*) (cinema) fire-
proof film safe

фильмо·тек·а, -и, (*f*) (cinema) film
library/store

фильм·пак·, -а, (*m*) (phot) film pack

фи́льтр·, -а, (*m*) filter, strainer; liner (of oil wells)

~, биолог·и́ческ·ий filter bed (sewage)

~ ве́рх·н·их част·о́т· (elec) high-pass filter

~, вы·ра́вн·ивающ·ий (elec) equalizer

~, вы·рез·а́юш·ий (rad) band-elimination filter

~, вы·ход·н·о́й (elec) output filter

~, гидравл·и́ческ·ий viscous-fluid torsional vibration damper

~, греб·е́нчат·ый (elec) comb filter

~, двух·зве́н·н·ый two-section filter

~, двух·по́люс·н·ый приёмо·пере-·да·ю́щ·ий (telecom) directional filter

~, жи́дк·остн·ый viscous fluid torsional vibration damper

~, за·гражд·а́ющ·ий (rad) wave trap, band-elimination filter, band-stop filter

~, раз·вя́з·ывающ·ий (rad) decoupling filter

~, за·щи́т·н·ый (rad) interference filter

~, индукцио́н·н·ый eddy-current brake, magnetic-coupling torsional vibration damper

~, интерференцио́н·н·ый (opt) interference filter

~, ква́рц·ев·ый sand filter (water treatment); (elec) quartz filter

~ кла́сс·а M (telecom) m-derived filter

~, конденса́тор·н·ый (rad) anti-interference capacitor

~, лин·е́йн·ый (elec) line filter

~, лист·ов·о́й leaf filter

~ -лов·у́шк·а, -и, (*g pl*) -шек·, (*f*) (phys) entrainment filter; (telecom) band-elimination filter

~, ма́сл·ян·ый lubricating oil filter

~, ме́дл·енн·ый ква́рц·ев·ый slow sand filter (water treatment)

~, меш·о́чн·ый bag filter

~, микро·по́р·ист·ый micropore filter

~, мост·ов·о́й Т-обра́з·н·ый (elec) bridged T-filter/network

~, на·по́р·н·ый ква́рц·ев·ый pressure sand filter (water treatment)

~, нефт·ян·о́й (ICE) fuel oil filter

~, нижне·про·пуск·н·о́й (telecom) low-pass filter

~ ни́ж·н·их част·о́т· (telecom) low pass filter

~, о·без·пло́ж·ивающ·ий filter bed (sewage)

фи́льтр

~, объ·ём·н·ый (rad) cavity-coupled filter

~ -от·сто́й·ник·, -а, (*m*) (ICE) fuel strainer, fuel filter

~, паралле́ль·н·ый Т-обра́з·н·ый (telecom) parallel T-circuit filter

~, пласт·и́нчат·о-щел·ев·о́й edge-plate filter

~, полос·ов·о́й (telecom) band-pass filter

~ -про́б·к·а, -и, (*f*) (telecom) wave trap

~, про́·волоч·н·о-щел·ев·о́й percolator filter (diesels)

~, раз·вя́з·ывающ·ий (elec) decoupling filter

~, рук·а́вн·ый bag filter

~, с·гла́ж·ивающ·ий (elec) smoothing circuit

~, су́м·чат·ый bag filter

~, толсто·сло́й·н·ый deep-bed filter

~, узко·полос·н·ый ква́рц·ев·ый (elec) narrow band quartz filter

~, цеп·н·о́й (elec) "iterative" impedance filter

~, щел·ев·о́й bar filter

фильтра́т·, -а, (*m*) filtrate, filter liquor

фильтрацио́н·н·ый, -ая, -ое, (*a*) filter

~ давл·е́ни·е (*n*) pore pressure (on dams etc.)

фильтр·а́ци·я, -и, (*f*) (*v n*) *see* **филь-тр·ова́ть;** seepage (of a fluid through a porous medium), percolation; (telecom) filter discrimination

~, капилля́р·н·ая capillary percolation

коэффицие́нг· фильтр·а́ци·и (*m*) permeability, permeability coefficient (of a porous medium); leakage factor

фильтр·ова́льн·ый, -ая, -ое, (*a*) filter, filtering

~ лист· (*m*) filter leaf

~ ткан·ь (*f*) filter cloth

фильтр·ова́ни·е, -я, (*n*) (*v n*) of **фильтр·ова́ть** filtration

~, от·раж·а́тельн·ое momentum separation (of a/c engine)

фильтр·о́ванн·ый, -ая, -ое, (*past part pass*) of **фильтр·ова́ть**

фильтр·ова́ть, -у́ют, (*imp*) filter strain; **-ся** (*pass*), percolate, seep, infiltrate

фи́льтр·ов·ый, -ая, -ое, (*a*) filter

фильтр·пре́сс·, -а, (*m*) filter press

~, лист·ов·о́й enclosed filter press

фильтр·пре́сс

~, ра́м·н·ый   plate-and-frame filter press

~, черв·я́чн·ый   strainer (rubber)

фильтр·пре́сс·н·ый, -ая, -ое,   (*a*) of **фильтр·пре́сс·**

фильтр·у́ющ·ий, -ая, -ее,   (*pres part act*) of **фильтр·ова́ть**; filter

~ со·оруж·е́ни·е   (*n*) filtering basin (for water)

~ цеп·ь   (*f*) (elec) filter circuit

~ элеме́нт·   (*m*) filter element

фильтр·у́ющ·ийся, -аяся, -ееся, (*pres part act*) of **фильтр·ова́ться**; filterable (of liquids); permeable (of porous solids)

~ ви́рус·   (*m*) (bact) filterpasser, filter-passing virus, filterable virus

фи́льц·, -а,   (*m*) (text) felt

~, микрофо́н·н·ый   microphone felt washer

филя́ри·и   (*pl*) (zool) *Filaroidea*

филярио́з·, -а,   (*m*) (med) filariasis

фим·о́з·, -а,   (*m*) (med) phimosis

фин-   (*abbr component*) = **фина́нс·о·в·ый**

фина́л·, -а,   (*m*) final (sport); (theat) finale

финанс·и́р·овать, -уют,   (*imp and perf*) finance

финанс·и́ст·, -а,   (*m*) financier

фина́нс·ов·ый, -ая, -ое,   (*a*) finance; revenue

фина́нс·ы, -ов,   (*m pl*) finance(s)

финва́л·, -а,   (*m*) (zool) fin whale, razorback, *Balaenoptera physalus*

финзен·, -а,   (*m*) (med) finsen

фи́ник·, -а,   (*m*) (*see also* **фи́ник·ов·ая па́льм·а**) (food) date

финик·и́йск·ий, -ая, -ое,   (*a*) Phoenecian

фи́ник·ов·ый, -ая, -ое,   (*a*) (food) date

~ па́льм·а   (*f*) (bot) date palm, *Phoenix*

~ па́льм·а, финико·но́с·н·ая   evergreen date palm, *Phoenix dactylifera*

фини́метр·, -а,   (*m*) pressure gauge (on gas cylinders etc.)

фи́ниш·, -а, (*i*) -ем,   (*m*) finish (sport); (a/c) end of landing run

фи́ниш·ер·, -а,   (*m*) batsman (aircraft carrier); finishing machine (road-making)

фи́ниш·н·ый, -ая, -ое,   (*a*) finish; surface finish

финнемани́т·, -а,   (*m*) (min) finnemanite

фин·от·де́л·, -а,   (*m*) finance department

фин·пла́н·, -а,   (*m*) budget, financial plan

фи́н·ск·ий, -ая, -ое,   (*a*) Finnish

фиоле́т·ов·ый, -ая, -ое,   (*a*) violet (*as noun and/or adj*)

~, мети́л·ов·ый   methyl violet (dye)

~, про́ч·н·ый   solid violet

фио́н·, -а,   (*m*) (biol) phyone

фио́рд·, -а,   (*m*) (geog) fiord, fjord

фиори́т·, -а,   (*m*) (min) fiorite

фи́рм·а, -ы,   (*f*) (com) firm

фи́рм·енн·ый, -ая, -ое,   (*a*) (com) firm's, company, company's

~ ма́рк·а   (*f*) trademark

фирн·, -а,   (*m*) (geog) firn, névé

фиска́ль·н·ый, -ая, -ое,   (*a*) (fin) fiscal

фисси·пе́ди·и   (*pl*) (zool) fissipeds, *Fissipeda*

фиста́ш·к·а, -и, (*g pl*) -ш·ек·,   (*f*) (bot) pistachio, *Pistacia*

фи́стул·а, -ы,   (*f*) (med) fistula

фистулёз·н·ый, -ая, -ое,   (*a*) fistulose, fisulous

фистуля́р·н·ый, -ая, -ое,   (*a*) fistular, fistulous

фит·а́з·а, -ы,   (*f*) (biochem) phytase

фит·а́т·, -а,   (*m*) (chem) phytate

фитера́л·, -а,   (*m*) (pal) phyteral

фитеу́м·а, -ы,   (*f*) (bot) mixed flower, *Phyteuma*

фити́л·ь, -я́,   (*m*) wick; (expl) fuse, fuse match

фит·и́н·, -а,   (*m*) (biochem, pharm) phytin

фи́тинг·, -а,   (*m*) pipe fitting

фито·бе́нтос·, -а,   (*m*) (bot) phyto-benthos, phytobenthon

фито·биоло́г·и·я, -и,   (*f*) (bot) phyto-biology, plant biology

фито·гене́з·, -а,   (*m*) (bot) phyto-genesis

фито·ге́н·н·ый, -ая, -ое,   (*a*) (geol) phytogenic

фито·гормо́н·, -а,   (*m*) (gen) phyto-hormone

фито·за́вр·ы, -ов,   (*m pl*) (pal) phyto-saurs, phytosaurians, *Phytosauria*

фит·о́л·, -а,   (*m*) (biochem) phytol

фитола́кк·а, -и,   (*f*) poke, garget, scove, (bot) *Phytolacca*

фито·ло́г·и·я, -и,   (*f*) phytology, botany

фит·о́м·, -а,   (*m*) (bot) phytome

фитон·ци́д·, -а,   (*m*) (biochem) phyton-cide

фитон·ци́д·н·ый, -ая, -ое,   (*a*) phyton-cidal

фито·пато·лог·и·я, -и, (*f*) (bot) phytopathology

фито·планктон·, -а, (*m*) (microbiol) phytoplankton

фито·стерин·, -а, (*m*) (biochem) phytosterol

фито·фаг·и·я, -и, (*f*) (biol) phytophagy

фито·фтор·а, -ы, (*f*) (bot) *Phytophthora*

фито·ценоз·, -а, (*m*) (bot) phytocoenosis, phytocenosis

Фиттиг·а, реакци·я (*f*) (chem) Fittig's synthesis

фиттинг·, -а, (*m*) *see* **фитинг·**

фихтелит·, -а, (*m*) (min) fichtelite

фиш·к·а, -и, (*g pl*) **-ш·ек·,** (*f*) counter, chip (for scoring); (elec) plug, peg

~, пере·ход·н·ая (elec) insert

фиш·луп·а, -ы, (*f*) (naut) fishfinder, fish echo-sounder

флавакридин·, -а, (*m*) (pharm) acriflavine

флавантрен·, -а, (*m*) flaventhrene, indanthrene yellow (dye)

флав·ин·, -а, (*m*) (biochem) flavine

флаво·ксантин·, -а, (*m*) (chem) flavoxanthin

флав·он·, -а, (*m*) (chem) flavone

флаво·протеин·, -а, (*m*) (biochem) flavoprotein

флаво·пурпурин·, -а, (*m*) (chem) flavopurpurin

флаг·, -а, (*m*) flag, pennant, ensign, colours

~, блужд·ающ·ий (naut) flag of convenience

~, букв·енн·ый alphabet flag

~, вымпель·н·ый pennant

~, корм·ов·ой ensign

~, маяч·н·ый lightship flag

~ от·ход·а blue peter

флагдук·, -а, (*m*) (naut) bunting

флагелл·ят·ы (*pl*) (zool) *Flagellata, Mastigophora*

флагман·, -а, (*m*) Flag Officer (naval); flagship

флагман·ск·ий, -ая, -ое, (*a*) staff, admiral's, fleet, squadron

флагшток·, -а, (*m*) flagstaff, ensign staff

флаж·ков·ый, -ая, -ое, (*a*) of **флаж·ок·**

флаж·н·ый, -ая, -ое, (*a*) flag

флаж·ок·, -ж·к·а, (*m*) of **флаг·** hand flag; flag indicator, padole

флакон·, -а, (*m*) bottle, small bottle, vial, phial

фламанд·ск·ий, -ая, -ое, (*a*) Flemish

фламинго (*m indecl*) (orn) flamingo(es), (*pl*) *Phoenicopteriformes*

фланг·, -а, (*m*) (mil) flank

фланг·ов·ый, -ая, -ое, (*a*) flank; side

фланел·евк·а, -и, (*f*) sailor's flannel

фланел·ев·ый, -ая, -ое, (*a*) (text) flannel

фланел·ь, -и, (*f*) (text) flannel

флан·ец·, -н·ц·а, (*i*) **-н·ц·ем,** (*m*) flange (on pipes); joggle (on plate)

~, глух·ой fixed flange

~, двой·н·ой companion flange

~, жёстк·ий fixed flange (pipeline)

~, за·кры·т·ый dead flange/hole

~ крепл·ени·я mounting flange

~, лож·н·ый counterflange

~, на·кид·н·ой lapped flange

~, от·кры·т·ый live flange/hole

~, при·вар·н·ой welding flange

~, свобод·н·ый slip-on flange

~, с·тяж·н·ой coupling flange

~, угол·ков·ый collar flange

~ цилиндр·а (ICE) cylinder collar

~, цоколь·н·ый base fin (of vacuum tube)

фланж·ировк·а, -и, (*f*) flanging; joggling (plating)

фланж·ировочн·ый, -ая, -ое, (*a*) of **фланж·ировк·а**

~ пресс· (*m*) flanger

~ стан·ок· (*m*) plate joggling machine

фланц·евать, -уют, (*imp*) joggle (a metal plate); **-ся** flange (of pipes)

флан·цев·ый, -ая, -ое, (*a*) flanged, flange

~ муфт·а (*f*) flange coupling, faceplate coupling (pipelines)

фланце·от·гиб·очн·ый стан·ок· (*m*) plate-joggling machine

флаттер·, -а, (*m*) (dynam) flutter

~, с·пар·енн·ый classical flutter

флеб·ит·, -а, (*m*) (med) phlebitis

флебо·грамм·а, -ы, (*f*) (med) phlebogram

флебо·лит·, -а, (*m*) (med) phlebolith

флебо·томус·, -а, (*m*) (ent) *Phlebotomus*

флеб·эктази·я, -и, (*f*) (med) phlebectasia

флегм·а, -ы, (*f*) phlegm; (chem) reflux, bottoms

флегмат·изатор·, -а, (*m*) (expl) phlegmatizer, desensitizer, (chem) retarder

флегмат·ичн·ый, -ая, -ое, (*a*) phlegmatic

флегм·ов·ый, -ая, -ое, (*a*) of **флёгм·а**
~ числ·о (*n*) (chem) reflux ratio
флегмо́н·а, -ы, (*f*) (med) phlegmon
флёйт·а, -ы, (*f*) flute (music)
флёйц·, -а, (*i*) **-ем·** (*m*) flat paint brush
флёкс·и·я, -и, (*f*) (gram) inflection
флексо́·метр·, -а, (*m*) (instr) flexometer
флёкс·ор·, -а, (*m*) (anat) flexor
флекс·у́р·а, -ы, (*f*) (geol) flexure
флект·и́вн·ый, -ая, -ое, (*a*) (ling) inflected
флёп·, -а, (*m*) flap (of tyre)
флёр·, -а, (*m*) (text) veiling
флёр·ниц·а, -ы, (*i*) **-ей,** (*f*) (ent) chrysop, (*pl*) *Chrysopa*
Флеттнер·а, рул·ь (*m*) (shipb) Flettner rudder
флёш·ь, -и, (*f*) tea shoot
флѝгел·ь, -я, (*nom pl*) **-я́,** (*g pl*) **-е́й,** (*m*) (arch) wing
флѝнт·, -а, (*m*) (opt) flint glass
флинтгла́с·, -а, (*m*) (opt) flint glass
флици́д·, -а, (*m*) fly-killer, flit, flycide
флѝш·, -а, (*m*) (geol) flysch
фло·бафе́н·ы, -ов, (*m pl*) (biochem) phlobaphenes
флогопи́т·, -а, (*m*) (min) phlogopite, amber mica
флóкен·ы, -ов, (*m pl*) (met) flakes, lemon spots, hairline cracks (defect in alloyed steel)
~ раз·ры́в·а shatter cracks, snow-flakes (defect in alloyed steel)
флоки́т·, -а, (*m*) (min) flokite
флоккул- (*int component*) *see also* **флокул-**
флóккул·а, -ы, (*a*) *see* **флóккул·ы**
флоккул·и́р·овать, -уют, (*imp and perf*) flocculate
флоккул·и́рующ·ий, -ая, -ее, (*pres part act*) of **флоккул·и́р·овать;** (chem) flocculent
флóккул·ы (*f pl*) (astron) flocculi, plages, chromospheric faculae
флоккул·я́нт·, -а, (*m*) (chem) flocculant
флоккуляцио́н·н·ый, -ая, -ое, (*a*) flocculation
~ с·чи́ст·к·а (*f*) flocculation
флоккул·я́ци·я, -и, (*f*) (chem) flocculation
флóкс·, -а, (*m*) (bot) phlox, *Phlox*
флокс·и́н·, -а, (*m*) (chem) phloxine
флокул- (*int component*) *see also* **флоккул-**
флокул·е́нт·, -а, (*m*) flocculent
флокул·и́р·овать, -уют, (*imp and perf*) flocculate
флокул·я́нт·, -а, (*m*) *see* **флокку-л·я́нт·**
флокул·я́тор·, -а, (*m*) flocculator
флокул·я́ци·я, -и, (*f*) *see* **флокку-л·я́ци·я**
флóр·, -а, (*m*) (shipb) floor plate
~, скеле́т·н·ый (shipb) skeleton floor
флóр·а, -ы, (*f*) (bot) flora
~, жёстк·ая coarse vegetation
флоренси́т·, -а, (*m*) (min) florencite
Флóри-Ре́нер·а, у·равн·е́ни·е (*n*) (chem) Flory–Rehner equation
флори·ге́н·, -а, (*m*) (bot) florigen
флори·ге́н·н·ый, -ая, -ое, (*a*) florigenic
флоридзи́н·, -а, (*m*) phorizein
флор·исти́ческ·ий, -ая, -ое, (*a*) floral, floristic
флóр·н·ый, -ая, -ое, (*a*) of **флóр·**
~ уго́ль·ник·, ве́рх·н·ий (*m*) (shipb) reverse frame
~ уго́ль·ник·, ни́ж·н·ий (*m*) (shipb) frame
флороглюцино́л·, -а, (*m*) (chem) phloroglucinol
флор·ти́мберс·, -а, (*m*) (shipb) floor timbers
флóт·, -а, (*m*) fleet; navy
~, вое́н·н·о-воз·ду́ш·н·ый air force
~, вое́н·н·о-мор·ск·о́й navy
~, воз·ду́ш·н·ый civil aviation
~, мор·ск·о́й merchant marine/navy
~, реч·н·о́й inland water transport
~, рыбо·про·мысл·о́в·ый fishing fleet
~, техн·и́ческ·ий dredging vessels
~, торг·о́в·ый merchant marine/navy
флотацио́н·н·ый, -ая, -ое, (*a*) of **флот·а́ци·я**
флот·а́ци·я, -и, (*f*) (min) floatation, flotation
~, вы́·бор·очн·ая selective flotation
~, коллекти́в·н·ая bulk flotation
~, ма́сл·ян·ая oil flotation
~, пере·чи́ст·н·ая cleaner flotation
~, пневмат·и́ческ·ая air flotation
~, селекти́в·н·ая differential flotation
флоти́ли·я, -и, (*f*) (naut) flotilla
флот·и́руемост·ь, -и, (*f*) (min) floatability
флото·маши́н·а, -ы, (*f*) floatation/flotation machine
флото·ка́мер·а, -ы, (*f*) floatation/flotation cell (ore dressing)
флото·реаге́нт·, -а, (*m*) (min) floatation/flotation reagent

флот·ск·ий, -ая, -ое,   (*a*) of **флот·**; naval

фло·э́м·а, -ы,   (*f*) (bot) phloem, bast, bast-tissue

флукту-   (*int component*) *see also* **флюк-ту-**

флукту·а́ци·я, -и,   (*f*) fluctuation

~, бес·по·ря́д·очн·ая   random fluctuation

~, с·луч·а́йн·ая   random fluctuation

флуор-   (*int component*) *see also* **флюор-**

флуор·е́н·, -а,   (*m*) (chem) fluorene, diphenylenemethane

флуор·есце́йн·, -а,   (*m*) (chem) fluorescein, resorcinol-phthalein

флуор·есце́нт·н·ый, -ая, -ое,   (*a*) fluorescence

~ микроско́п·   (*m*) fluorescence microscope

флуор·есце́нци·я, -и,   (*f*) (phys) fluorescence

~, резона́нс·н·ая   resonance fluorescence

~, уда́р·н·ая   impact fluorescence

флуоресц·и́рующ·ий, -ая, -ее,   (*pres part act*); fluorescent, fluorescing

~ экра́н·   (*m*) fluorescent screen

флуори́·метр·, -а,   (*m*) (phys) fluorimeter, fluorometer

флуори·метр·и́·я, -и,   (*f*) fluorimetry; fluorimetric analysis

флуоро́·метр·, -а,   (*m*) (phys) fluorometer, fluorimeter

флуоро·метр·и́ческ·ий, -ая, -ое,   (*a*) fluorometric

флуоро·скоп·, -а,   (*m*) (phys) fluoroscope

флуоро·скоп·и́·я, -и,   (*f*) (phys) fluoroscopy

флуоро·те́н·, -а,   (*m*) (chem) fluorothene

флуоро·фото·метр·и́ческ·ий, -ая, -ое,   (*a*) fluorophotometric

флювио·гляциа́ль·н·ый, -ая, -ое,   (*a*) (geol) fluvioglacial

флюга́р·к·а, -и,   (*g pl*) **-р·ок·,** (*f*) chimney cowl; weathercock, wind vane; (naut) ship's crest

флю́гер·, -а,   (*nom pl*) **-а́,** (*g pl*) **-о́в,** (*m*) wind vane, vane

флю́гер·н·ый, -ая, -ое,   (*a*) of **флю́гер·**; (a/c) feathering

~ винт·   (*m*) feathering propeller

~ на·со́с·   (*m*) (a/c) feathering pump

~ по·лож·е́ни·е   (*n*) (a/c) full-feathered position

~ у·сто́й·чивост·ь   (*f*) (a/c) weathercock stability

флюг·и́ровани·е, -я,   (*n*) (air) feathering

флюелли́т·, -а,   (*m*) (min) fluellite

флюйд·, -а,   (*m*) fluid

флюида́ль·н·ый, -ая, -ое,   (*a*) fluidal

флюидиз·а́ци·я, -и,   (*f*) fluidization

флюидиз·и́рованн·ый, -ая, -ое,   (*past part pass*); fluidized; fluid

~ катализ·а́тор·   (*m*) (chem) fluid catalyst

флю́кс·и·я, -и,   (*f*) fluxion

флюкс·ме́тр·, -а,   (*m*) (elec) fluxmeter

флюкс·у́ющ·ийся, -аяся, -ееся,   (*pres part act*); fluxing

флюкту-   (*int component*) *see also* **флукту-**

флюктуацио́н·н·ый, -ая, -ое,   (*a*) fluctuational

флюкту·а́ци·я, -и,   (*f*) *see* **флукту·а́ци·я**

флюо·бори́т·, -а,   (*m*) (min) fluoborite

флюо́р-   (*int component*) *see* **флуор-**

флюооридиз·а́ци·я, -и,   (*f*) fluoridization

флюориз·а́ци·я, -и,   (*f*) (med, geol) fluoridation

флюор·есце́йн·, -а,   (*m*) = **флуор·есце́йн·**

флюор·и́т·, -а,   (*m*) (min) fluorite

флюори́т·ов·ый, -ая, -ое,   (*a*) fluorite

флюор·о́з·, -а,   (*m*) (med) fluorosis

флюоро·ско́п·, -а,   (*m*) (phys) fluoroscope

флюо·сидери́т·, -а,   (*m*) (min) fluosiderite

флюо·цери́т·, -а,   (*m*) (min) fluocerite

флю́с·, -а,   (*nom pl*) **-ы́,** (*g pl*) **-о́в,** (*m*) flux; **флю́с·, -а,** (*nom pl*) **-ы,** (*g pl*) **-ов,** (*m*) (med) gumboil

~, с·ва́р·очн·ый   welding flux

флюс·ова́ни·е, -я,   (*n*) fluxing

флюс·ова́ть, -у́ют,   (*imp*) flux

флю́тбет·, -а,   (*m*) (civ eng) sill (of weir etc.)

флюэлли́т·, -а,   (*m*) (min) fluellite

фля́г·а, -и,   (*f*) flask, (mil) waterbottle

фляго·мо́·ечн·ая маши́н·а   (*f*) bottle-washing machine

фля́ж·к·а, -и,   (*g pl*) **-ж·ек·,** (*f*) (*dim*) of **фля́г·а**

фля́н·ец·, -н·ц·а,   (*m*) = **фла́н·ец·**

фля́ттер·, -а,   (*m*) = **фла́ттер·**

фоб   (*int abbr*) (com) f.o.b., free on board

фо́б·и·я, -и,   (*f*) (med) phobia

фобо·та́ксис·, -а,   (*m*) (biol) phobotaxis

фова́ль·н·ый, -ая, -ое,   (*a*) (biol) foveolate, foveolar, foveate

фогли́т·, -а,   (*m*) (min) voglite

фойе́ (*n indecl*) (arch) foyer

фойнико·хройт·, **-а**,   (*m*) (min) phoenicochroite, phenicochroite

фойяйт·, **-а**, (*m*) (min) foyaite

фок·, **-а**, (*m*) (naut) foresail; (*as component*) fore-

фока́ль·н·ый, **-ая**, **-ое**, (*a*) focal

~ **пара́метр·** (*m*) (math) ½ latus rectum

фоки́·метр·, **-а**, (*m*) (phot) focimeter

фок·ма́чт·а, **-ы**, (*f*) (naut) foremast

фоко́·метр·, **-а**, (*m*) (opt) focometer

фо́кус·, **-а**, (*m*) focus; conjuring trick

~, **аэро·динам·и́ческ·ий** aerodynamic focus/centre/center

~, **воз·мо́ж·н·ый** virtual focus

~, **гла́в·н·ый**   principal focus, focal point

~, **не·у·сто́й·чив·ый** (autom) unstable equilibrium focus

~ **хо́рд·ы** (a/c) quarter-chord point

фокус·и́ровани·е, **-я**,   (*n*) (*v n*) of **фокус·и́р·овать**

фокус·и́р·овать, **-уют**,   (*imp*) focus, get in focus, bring into focus

фокус·иро́вк·а, **-и**,   (*g pl*) **-вок·**, (*f*) (*v n*) of **фокус·и́р·овать**

~, **га́з·ов·ая** gas focusing (cathode ray tube)

~, **жёстк·ая**   strong focusing

фо́кус·н·ый, **-ая**, **-ое**, (*a*) of **фо́кус·**; focal

~ **пятн·о́** (*n*) focal point

~ **рас·сто·я́ни·е** (*n*) (opt) focal length

фола́ды (*pl*) (zool) *Pholadidae*

фолиа́нт·, **-а**, (*m*) folio (book)

фо́ли·ев·ый, **-ая**, **-ое**, (*a*) folic

~ **кисл·от·а́** (*f*) (chem) folic acid

фоли́·я, **-и**, (*f*) (phot) gelatine film

фолли́к·ул·, **-а**, (*m*) (anat) follicle

фолликул·и́н·, **-а**, (*m*) (biochem) folliculin, oestrone

фолликул·и́т·, **-а**, (*m*) (med) folliculitis

фолликул·я́рн·ый, **-ая**, **-ое**,   (*a*) follicular

фо́льг·а, **-и**, (*f*) (met) foil, leaf

~, **золот·а́я** (met, print) gold leaf

~, **контро́ль·н·ая**   (nucl) monitoring foil

фомо́з·, **-а**, (*m*) phomosis (plant disease)

фон·, **-а**, (*m*) background; phon (acoustics); background electrolyte/solution

~, **по·гре́в·н·ый** (rad) heater-cathode hum

~, **тепл·ов·о́й** thermal background

~, **шум·ов·о́й** background noise

фона́р·ик·, **-а**, (*m*) (*dim*) of **фона́р·ь**

фона́р·н·ый, **-ая**, **-ое**,   (*a*) of **фона́р·ь**; (*f decl as adj*) lamp room

~ **столб·** (*m*) lamp post

фона́р·щик·, **-а**, (*m*) lamp trimmer

фона́р·ь, **-я**, (*m*) lamp, lantern, light; (arch) abatjour, lantern, clerestory, skylight, lantern light; (a/c) canopy; (instr) gyro frame (of gyroscope)

~, **аристо́тел·ев**   (zool) Aristotle's lantern

~ **каби́н·ы** (a/c) cockpit canopy

~, **карма́н·н·ый** pocket lamp

~, **лаборато́р·н·ый** (phot) safe-light, dark-room lamp

~, **лоб·ов·о́й** (rail) head-lamp

~, **смотр·ов·о́й** sight glass

фо́нгсбот·, **-а**, (*m*) *see* **фа́нгсбот·**

фо́нд·, **-а**, (*m*) fund; standing, stock (of good will etc.); allocation (of materials etc.); (fin) bond, debenture, loan stock

~, **амортизацио́н·н·ый** (econ) sinking fund

~, **основ·н·о́й** (econ) fixed capital

~, **основ·н·о́й** **про·из·во́д·ственн·ый** (econ) productive assets

фонд·и́р·овать, **-уют**,   (*imp*) create a fund, float a loan; allocate (in USSR)

фо́нд·ов·ый, **-ая**, **-ое**, (*a*) of **фонд·**

~ **материа́л·ы** (*pl*) allocated materials

~ **опер·а́ци·и**   (*pl*) (fin) borrowing/lending operations

фоне́м·а, **-ы**, (*f*) (ling) phoneme

фон·ендо·ско́п·, **-а**,   (*m*) (med) phonendoscope

фоне́т·ик·а, **-и**, (*f*) phonetics

фонет·и́ст·, **-а**, (*m*) phonetician

фон·и́ческ·ий, **-ая**, **-ое**, (*a*) phonic; sound-signal

~ **вы́·зов·** (*m*) (telecom) buzzer calling

~ **кольц·о́** (*n*) (telecom) phonic wheel

фо́н·ов·ый, **-ая**, **-ое**, (*a*) background

фоно·гра́мм·а, **-ы**,   (*f*) phonogram, sound recording/record

~, **магни́т·н·ая** magnetic sound-record

фоно́·граф·, **-а**, (*m*) phonograph

фоно·ли́т·, **-а**, (*m*) (min) phonolite

фоно·ли́т·ов·ый, **-ая**, **-ое**, (*a*) (geol) phonolitic

фоно·ло́г·и·я, **-и**, (*f*) (ling) phonology

фоно́·метр·, **-а**, (*m*) (instr) phonometer

фон·о́н·, **-а**, (*m*) (phys) phonon

фонта́н·, **-а**, (*m*) fountain, spout, spurt

~, **мо́щ·н·ый** (oil) gusher

~, **от·кры́·т·ый** (oil) spouter

фонтан·и́р·овать, **-уют**, (*imp*) gush, spout, spurt

фонта́н·н·ый, -ая, -ое, (*a*) of **фонта́н·**
~ **армату́р·а** (*f*) (oil) Christmas tree and tubing head

фонту́р·а, -ы, (*f*) needle bed (circular knitting machine)

фор- (*int component*) fore

фора́мен·, -а, (*m*) foramen, (*pl*) foramina, foramens

форамини́·фер·а, -ы, (*f*) (zool) foraminifer, (*pl*) *Foraminifera*

форамини́·фер·ов·ый, -ая, -ое, (*a*) foraminiferal

форбези́т·, -а, (*m*) (min) forbesite

фор·ва́куум·, -а, (*m*) (phys) fore-vacuum, initial vacuum

форголлев·, -а, (*m*) pusher (tube rolling)

фордак·, -а, (*m*) (naut) wind right aft

фордеви́нд·, -а, (*m*) (naut) wind right aft

форе́з·, -а, (*m*) (elec) phoresis

фор·е́зи·я, -и, (*f*) (biol) phoresia, phoresy

форе́л·ь, -и, (*f*) (fish) trout
~, **европ·е́йск·ая** brown trout, *Salmo trutta*
~, **северо·америка́н·ск·ая** rainbow trout, *Salmo guirdnerii*

фо́рзац·, -а, (*i*) **-ем,** (*m*) fly-leaf, end leaf (of book)

фор·ка́мер·а, -ы, (*f*) (ICE) antechamber, precombustion chamber, (a/c) afterburner chamber, (rocket) precombustion chamber; (hydr) stilling chamber

фор·ки́л·ь, -я, (*m*) (a/c) tail fillet/fairing

фо́рм·а, -ы, (*f*) form, shape; (phys) mode; mould, mold; uniform (dress); (print) forme
~, **бли́з·к·ая** related form
~ **взаимо·де́й·стви·я** (phys) mode of interaction
~ **волн·ы́** wave-form, wave-shape
~, **вы́·суш·енн·ая** (met cast) dry sand mould
~, **вя́л·ен·ая** (met cast) partially dried mould
~, **дво·и́чн·ая** (math) bilinear form
~, **квадрат·и́чн·ая** (math) quadratic form
~, **ко́р·ков·ая** (cast) shell mould/mold
коэффициент· фо́рм·ы (*m*) form factor
~, **лит·е́йн·ая** mould, mold; (plast) injection mould/mold
~, **на́·бр·анн·ая** (print) live forme

фо́рм·а
~, **оболо́ч·ков·ая** (cast) shell mould/mold
~ **о·де́жд·ы** uniform (dress)
~, **от·ли́в·н·а́я** (print) mould, mold
~, **печа́т·н·ая** (print) forme
~ **по·ве́рх·ност·и** surface configuration
~, **ра́з·ов·ая** (met cast) sand mould/mold
~, **разъ·ём·н·ая** (met cast) split mould
~ **ска́ч·к·а у·плотн·е́ни·я** (aerodynam) shock line
~, **стерж·нев·а́я** (met cast) cored mould
~, **сыр·а́я** (met cast) green sand mould
~, **тро́·ичн·ая** (math) trilinear form

формали́зм·, -а, (*m*) formalism
~, **корпускуля́р·н·ый** (phys) particle formalism

форм·али́н·, -а, (*m*) (chem) formalin, formaline

форм·али́н·ов·ый, -ая, -ое, (*a*) (chem) formalin, formaline

формалист·и́ческ·ий, -ая, -ое, (*a*) formalistic

форм·альдеги́д·, -а, (*m*) (chem) formaldehyde

форма́ль·н·ый, -ая, -ое, (*a*) formal

форм·ами́д·, -а, (*m*) (chem) formamide

формани́т·, -а, (*m*) (geol) formanite

форма́нт·, -а, (*m*) (gram) affix, formant

форма́нт·а, -ы, (*f*) (acous) formant

форма́т·, -а, (*m*) (print) format
~ **из·ображ·е́ни·я** (TV) aspect/picture ratio
~ **ка́др·а** (TV) aspect/picture ratio

форма́т·н·ый, -ая, -ое, (*a*) of **форма́т·**
~ **вал·** (*m*) (paper) forming roll
~ **стан·о́к·** (*m*) copy-sawing machine, forming/shaping machine (for plywood)

форм·а́тор·, -а, (*m*) shaper (rubber)
~ **-вулканиз·а́тор·, -а,** (*m*) shaper-vulcanizer (for tyres)

форм·а́ци·я, -и, (*f*) formation

формбло́к·, -а, (*m*) forming block (of rubber die press)

фо́рм·енк·а, -и, (*g pl*) **-нок·,** (*f*) (naut) duck jumper

фо́рм·енн·ый, -ая, -ое, (*a*) uniform (dress)

формиа́т·, -а, (*m*) (chem) formate, formiate

форми́л·, -а, (*m*) (chem) formyl

форм·ирова́ни·е, -я, (*n*) (*v n*) of **форм·ир·ова́ть**

~ и́мпульс·ов (elec) pulse shaping

форм·ирова́тел·ь, -я (*m*) shaper

~ и́мпульс·ов pulse generator/shaper

~, ко́д·ов·ый word pulse shaper

форм·ир·ова́ть, -у́ют, (*imp*) form, shape, give shape to; form, organize, make up, compose; (rail) marshal, sort

форм·иро́вк·а, -и, (*f*) (*v n*) of **форм·ир·ова́ть**

фо́рмиум·, -а, (*m*) (bot) New Zealand flax, *Formium tenax*

фо́рм·н·ый, -ая, -ое, (*a*) (print) forme

~ пласт·и́н·а (*f*) (print) forme plate

форм·образ·ова́ни·е, -я, (*n*) (ling) form-derivation

форм·ова́ни·е, -я, (*n*) (*v n*) of **форм·о·ва́ть** (*and see* **форм·о́вк·а**) (hortic) topiary

~, ва́куум·н·ое vacuum forming (of plastics)

~, пласт·и́ческ·ое plastic forming

~, по·сле́д·ующ·ее (plast) post-forming

форм·о́ванн·ый, -ая, -ое, (*past part pass*) of **форм·ова́ть**

форм·ова́ть, -у́ют, (*imp*) mould, mold; form, shape

форм·о́вк·а, -и, (*f*) (*v n*) of **форм·о·ва́ть;** (elec) forming (e.g. batteries, rectifiers, coils), formation; (met cast) moulding, molding; (forge) roughing; (plast) forming, moulding, molding; (paper) formation; (hortic) topiary

~, без·опо́ч·н·ая (met cast) snap-flask moulding/molding

~ в по́чв·е (cast) pit moulding/molding, floor sand moulding/molding

~ вы·дув·а́ни·ем (plast) blow moulding/molding

~ вы́·тяж·к·ой (plast) stretch forming

~, гор·я́ч·ая (plast) thermoforming

~, негати́в·н·ая ва́куум·н·ая (plast) straight vacuum forming

~, позити́в·н·ая (plast) drape forming

~, по́чв·енн·ая (met) floor moulding/molding

~, руч·н·а́я (met cast) hand ramming

форм·ов·о́й, -а́я, -о́е, (*a*) moulded, molded; moulding, molding

форм·о́вочн·ая, -ой, (*f decl as adj*) forming shop (for electric batteries)

форм·о́вочн·ый, -ая, -ое, (*a*) of **форм·о́вк·а;** (*see also* **форм·о́·вочн·ая**) (*as noun*)

~ глин·а́ (*f*) (met cast) moulding/molding sand/loam

форм·о́вочн·ый, -ая, -ое

~ кисл·от·а́ (*f*) (elec) forming acid

~ маши́н·а (*f*) (met cast) moulding/molding/ramming machine; cake-forming machine (oil expression)

~ маши́н·а, вс·тря́х·ивающ·ая (met cast) jolt/jolting/jarring machine

~ маши́н·а, пресс·ов·ая (met cast) squeezer, squeezing machine

~ с·ме́с·ь (*f*) (met cast) moulding/molding sand mixture; (plast) moulding/molding mix

~ пресс· (*m*) mould/mold press

форм·о́вщик·, -а, (*m*) (met cast) moulder/molder

формо·из·мен·е́ни·е, -я, (*n*) change of shape, deformation; changing the shape, shaping, forming; shape effect

~, гидро·стат·и́ческ·ое (met) hydro-forming

~, тепл·ов·о́е thermal expansion/deformation

формо́л·, -а, (*m*) (pharm) formal

формонитри́л·, -а, (*m*) (chem) formonitrile

формо·образ·ова́тельн·ый, -ая, -ое, (*a*) formative

фо́рмул·а, -ы, (*f*) (*see under associated words*); formula

вы́·вод·ить фо́рмул·у derive a formula

~, зуб·н·а́я (zool) dental formula

формул·и́р·овать, -уют, (*imp and perf*) formulate

формуля́р·, -а, (*m*) log book (for machines etc.); book ticket (in libraries)

~ чит·а́тел·я readers' loans book/card/ticket (in library)

форм-фа́ктор·, -а, (*m*) (phys) shape/form factor

фо́рм·щик·, -а, (*m*) (paper) mould-maker, mold-maker

форо·зо́он·, -а, (*m*) (zool) phorozoon

форо́н·, -а, (*m*) (chem) phorone

форон·и́д·а, -ы, (*f*) (zool) phoronid, (*pl*) phoronids, *Phoronidea, Phoronida*

форора́кус·, -а, (*m*) (pal) *Phororhacos*

фор·пи́к·, -а, (*m*) (shipb) fore-peak

фор·полиме́р·, -а, (*m*) (chem) pre-polymer

форса́ж·, -а, (*m*) (a/c) afterburner, reheat, boost

форса́ж·н·ый, -ая, -ое, (*a*) of **фор·са́ж·**

~ ка́мер·а (*f*) (a/c) afterburner

фо́рсац·, -а, (*m*) *see* **фо́рзац·**

форс·и́ровани·е, -я, (*n*) (*v n*) of **форс·и́р·овать;** (mil) assault crossing (of rivers)

форс·и́рованн·ый, -ая, -ое, (*past part pass*) of **форс·и́р·овать**

форс·и́р·овать, -уют, (*imp and perf*) force, force the pace, speed up; increase the performance/output of, boost (an engine); (mil) force (an obstacle); overcome (an obstacle); spike (a nuclear reactor)

фор-сте́ньг·а, -и, (*f*) (naut) fore-top-mast

форстери́т·, -а, (*m*) (min) forsterite

форс·у́н·, -а́, (*m*) forcer, pusher; steam jet blower (for creating draught)

форс·у́нк·а, -и, (*g pl*) **-нок·,** (*f*) sprayer, injector nozzle, atomizer, fuel nozzle, burner

~, вихр·ев·а́я swirl atomizer/injector

~, воз·ду́ш·н·ая air-jet burner

~, двух·компоне́нт·н·ая (a/c) integral igniter

~, двух·ко́нтур·н·ая duplex burner, duplex nozzle (jet eng)

~, за·кры́·т·ая closed injector nozzle

~, механ·и́ческ·ая pressure jet burner

~, много·ды́р·чат·ая multiple-orifice-type injector (diesel)

~ -на·со́с·, -а, (*m*) unit injector (of diesel)

~, одно·ды́р·чат·ая single-orifice-injector (diesel)

~, одно·ко́нтур·н·ая simplex burner, simplex nozzle (jet eng)

~, от·кры́·т·ая open injector nozzle

~, от·раж·а́ющ·ая (a/c) mix-plate injector

~, пар·ов·а́я steam-jet burner

~, с·дво́·енн·ая duple burner (jet eng)

~, рабо́ч·ая (a/c) main jet

~, раз·бры́зг·ивающ·ая (a/c) splash plate injector

~, центро·бе́ж·н·ая swirl injector

~, щел·ев·а́я (a/c) split jet injector

фо́рт·, -а, (*nom pl*) **-ы́,** (*g pl*) **-о́в,** (*m*) (mil) fort

Фо́ртен·а, баро́·метр· (*m*) Fortin's barometer

форт епья́но (*n indecl*) piano

фортифик·а́ци·я, -и, (*f*) (mil) fortification

фо́рт·очк·а, -и, (*g pl*) **-чек·,** (*f*) cold-weather ventilation-window; (a/c) air vent

фор·ча́н·, -а, (*m*) kettle (in oil extraction)

форша́льтер·, -а, (*m*) trunk-junction switchboard (telephones)

форшальт-турби́н·а, -ы, (*f*) super-imposed turbine, topping unit

фор·штёв·ен·ь, -в·н·я, (*m*) (shipb) stem

фор·што́с·, -а, (*m*) (chem) adapter; (met roll) stop

фос·ге́н·, -а, (*m*) (chem) phosgene

фосгени́т·, -а, (*m*) (min) phosgenite

фос·мук·а́, -и, (*f*) (agr) rock phosphate meal, tricalcic phosphate

фо́сс·а, -ы, (*f*) (zool) fossa, (*pl*) fossae

фоссилиз·а́ци·я, -и, (*f*) (geol) fossilization

фоссилиз·и́р·овать, -уют, (*imp and perf*) (geol) fossilize

фо́ссул·а, -ы, (*f*) (zool) fossula

фосфа·ге́н·, -а, (*m*) (biochem) phosphagen

фосфако́л·, -а, (*m*) (pharm) phosphacol

фосф·а́т·, -а, (*m*) (chem) phosphate

~ аммо́ни·я ammonium phosphate

~, во́д·н·ый hydrous phosphate

~ и́звест·и phosphate of lime

~ ка́льци·я calcium phosphate

~, сыр·о́й rock phosphate

фосфат·а́з·а, -ы, (*f*) (biochem) phosphatase

фосфат·а́ци·я, -и, (*f*) phosphation

фосфат·и́д·, -а, (*m*) (chem) phosphatide

фосфат·и́рование·, -я, (*n*) phosphate water softening; (met) phosphatizing, phosphate coating

фосфа́т·н·ый, -ая, -ое, (*a*) phosphate

фосфат·ури́·я, -и, (*f*) (med) phosphaturia

фосф·и́д·, -а, (*m*) (chem) phosphide

фосф·и́н·, -а, (*m*) (chem) phosphine

фосф·и́н·ов·ый, -ая, -ое, (*a*) phosphinic

~ кис·лот·а (*f*) phosphinic acid

фосф·ин·ово·ки́сл·ый, -ая, -ое, (*a*) phosphonic

фосф·и́т·, -а, (*m*) (chem) phosphite

фосфо·вольфра́м·ов·ая кисл·от·а́ (*f*) phosphotungstic acid

фосфо·глюко́н·ов·ая кисл·от·а́ (*f*) phosphogluconic acid

фосфо·липи́д·, -а, (*m*) (biochem) phospholipid, phospholipide

фосфо·липи́н·, -а, (*m*) (chem) phospholipin, phosphatide

фосфо·молибда́т·, -а, (*m*) (geol) phosphomolybdate

фосфо·протеи́д·, -а, (*m*) (biochem) phosphoprotein

фосфо·протеи́н·, -а,   (m) (biochem) phosphoprotein

фо́с·фор·, -а,   (m) (chem) phosphorus, P; phosphor, luminophor

~, вс·пы́ш·ечн·ый   flash phosphor

~, пяти·хло́р·ист·ый   phosphorus pentachloride

фосфоресслери́т·, -а,   (m) (min) phosphorroesslerite

фосфор·есце́нци·я, -и,   (f) (phys) phosphorescence

фосфоресц·и́р·овать, -уют,   (imp) (phys) phosphoresce

фосфоресц·и́рующ·ий, -ая, -ее, (pres part act) of фосфоресц·ир·ова́ть; (phys) phosphorescent

фосфо·рибо́з·а, -ы,   (f) (biochem) phosphoribose

фосфориз·ова́ть, -у́ют,   (imp and perf) phosphorate, phosphorize

фосфор·ила́з·а, -ы,   (f) (biochem) phosphorylase

фосфор·и́л·, -а,   (m) phosphoryl

~, хло́р·ист·ый   phosphoryl chloride

фосфорил·и́ровани·е, -я,   (n) (chem) phosphorylation

фо́сфор·ист·ый, -ая, -ое,   (a) (chem) phosphorous, phosphide (of)

~ ангидри́д·   (m) phosphorus trioxide, phosphorous oxide

~ бро́нз·а   (f) (met) phosphor-bronze

~ водо·ро́д·   (m) (chem) phosphoretted hydrogen, phosphine

~ кисл·от·а́   (f) phosphorous acid

~ мед·ь   (f) copper/cupric phosphide

фосфор·и́т·, -а,   (m) (min) phosphorite, rock phosphate

~, кост·ян·о́й   (min) osteolite

фосфори́т·н·ый, -ая, -ое,   (a) phosphoritic

~ мук·а́   (f) (agr) tricalcic phosphate

фосфор·и́чност·ь, -и,   (f) phosphorescence

фосфор·новати́ст·ый, -ая, -ое,   (a) hypophosphorous

фосфор·нова́т·ый, -ая, -ое,   (a) hypophosphoric

фосфорно·ки́сл·ый, -ая, -ое,   (a) (chem) phosphate of; (geol) phosphatic

~ аммо́ни·й   (m) ammonium phosphate

фо́сфор·н·ый, -ая, -ое,   (a) (chem) phosphorus

~ ангидри́д·   (m) phosphorus pentoxide

~ кисл·от·а́   (f) (chem) phosphoric acid

фосфоро·бактери́н·, -а,   (m) phosphorobacteria (fertilizer)

фосфоро·ге́н·, -а,   (m) phosphorogen

фосфоро·гра́ф·и·я, -и,   (f) (phys) phosphor printing

фосфоро́·лиз·, -а,   (m) (chem) phosphorolysis

фосфор·орган·и́ческ·ий, -ая, -ое, (a) (chem) organophosphoric

фосфоро·сидери́т·, -а,   (m) (min) phosphorosiderite

фосфоро·ско́п·, -а,   (m) (instr) phosphoroscope

фосфо·ферри́т·, -а,   (m) (min) phosphoferrite

фосфо·филли́т·, -а,   (m) (min) phosphophyllite

фосф·уранили́т·, -а,   (m) (min) phosphuranylite

фо́т·, -а,   (m) (phys) phot, centimetre-candle

фотико́н·, -а,   (m) (TV) photicon

фоти́ни·я, -и,   (f) (bot) photinia, *Photinia*

фо́то   (n indecl) photograph

фото-   (int component) photo; photo-electric

фото·автома́т·, -а,   (m) coin-operated photograph booth

фото·алогра́фи·я, -и,   (f) (print) photoalography

фото·актив·а́ци·я, -и,   (f) photo-activation, photostimulation

фото·анали́тик·, -а,   (m) photoanalyst

фото·аппара́т·, -а,   (m) (phot) camera

~, двух·объекти́в·н·ый зерка́ль·-н·ый   twin-lens reflex camera

~, зерка́ль·н·ый   reflex camera

~, мало·форма́т·н·ый   miniature camera

~, плён·очн·ый   roll-film camera

~, с·клад·н·о́й   folding camera

фото·ателье́   (n indecl) photographer's studio

фото·бач·о́к·, -ч·к·а́,   (m) (phot) developing tank

фото·бума́г·а, -и,   (f) photographic paper

фото·визуа́ль·н·ая велич·ин·а́   (f) (astron) photovisual magnitude

фото·воз·бужд·ённ·ый, -ая, -ое,   (a) photoexcited

фото·вс·пы́ш·к·а, -и,   (g pl) -ш·ек·, (f) photo-flash, flashlight

фото·гало́ид·, -а,   (m) (chem) photo-halide

фото·гальван·и́ческ·ий, -ая, -ое, (a) photovoltaic

~ элеме́нт·   (m) photoelectrolytic/ photovoltaic/photochemical cell

фото·гальван·и́ческ·ий, -ая, -ое
~ **эффе́кт·** (*m*) (phys) photovoltaic effect
фото·гелио́·граф·, -а, (*m*) (astron) coronagraph
фото·ге́н·, -а, (*m*) photogen
фото·ген·и́ческ·ий, -ая, -ое, (*a*) photogenic
фото·грав·ирова́льн·ая маши́н·а (*f*) photo-engraving machine
фото·грав·и́р·овать, -у́ют, (*imp and perf*) (print) photo-engrave
фото·гравю́р·а, -ы, (*f*) (print) photogravure, photoaquatint, photo-intaglio
фото·гра́мм·а, -ы, (*f*) (instr) record, recording; (surv) photogram
фото·грам·метр·и́ческ·ий, -ая, -ое, (*a*) (surv) photogrammetric
фото·грам·ме́тр·и·я, -и, (*f*) (surv) photogrammetry
фото́·граф·, -а, (*m*) photographer
фото·граф·и́́ровани·е, -я, (*n*) photography, photographing
~, **перспекти́в·н·ое воз·ду́ш·н·ое** oblique aerial photography
~, **пла́н·ов·ое воз·ду́ш·н·ое** vertical aerial photography
фото·граф·и́ческ·ий, -ая, -ое, (*a*) (*see also under* **фото-**); photographic; photographer's
~ **аппара́т·** (*m*) (phot) camera
фото·гра́ф·и·я, -и, (*f*) photography; photograph
фото·да́т·чик·, -а, (*m*) photoelectric pick-up
фото·дел·е́ни·е, -я, (*n*) (phys) photofission
фото·дешифр·ова́ни·е, -я, (*n*) aerial photo interpretation
фото·диагно́ст·ик·а, -и, (*f*) (med) photographic diagnosis
фото·дина́м·ик·а, -и, (*f*) (bot) photodynamics
фото·дио́д·, -а, (*m*) (elec) photodiode, photorectifier
фото·дио́д·н·ый, -ая, -ое, (*a*) photodiode
фото·диссоци·а́ци·я, -и, (*f*) (chem) photodissociation
фото·дози·ме́тр·, -а, (*m*) (nucl) film badge
фото·за́·пис·ь, -и, (*f*) photographic record
фото·за·тво́р·, -а, (*m*) (*see under* **за·-тво́р·**) (phot) shutter, camera shutter
фото·и́мпульс·, -а, (*m*) photoimpact

фото·иониз·а́ци·я, -и, (*f*) (phys) photoionization
фото·ка́льк·а, -и, (*g pl*) **-лек·,** (*f*) (phot) paper negative
фото·ка́мер·а, -ы, (*f*) (phot) camera
фото·ка́рт·а, -ы, (*f*) (surv) photomap
фото·ката́лиз·, -а, (*m*) photocatalysis
фото·ката́лиз·а́тор·, -а, (*m*) photocatalyst
фото·като́д·, -а, (*m*) photoelectric cathode, photocathode
фото·кине́з·, -а, (*m*) (zool) photokinesis
фото·ки́но- (*component*) cinematography, cinematographic, cinema, ciné
фотокино·пуле·мёт·, -а, (*m*) camera gun
фото·кле́·й, -я, (*m*) (phot) mounting paste
фото·колори́·метр·, -а, (*m*) (chem) photocolorimeter
фото·контро́л·ь, -я, (*m*) (mil) strike photography
фото·коп·ирова́льн·ая маши́н·а (*f*) (phot) contact printer; photocopying machine, photocopier (office equipment)
фото·ко́п·и·я, -и, (*f*) photocopy
фото·ксило·гра́ф·и·я, -и, (*f*) (print) photoxylography
фото·лаборато́р·и·я, -и, (*f*) dark room
фото·ла́мп·а, -ы, (*f*) (phot) photoflood
фото́·лиз·, -а, (*m*) (chem) photolysis
фото·лит·и́ческ·ий, -ая, -ое, (*a*) (chem) photolytic
фото·лито·гра́ф·и·я, -и, (*f*) (print) photolithography
фото·люмин·есце́нци·я, -и, (*f*) (phys) photoluminescence
фото·люмино·фо́р·, -а, (*m*) photoluminophor
фото·магни́т·н·ый, -ая, -ое, (*a*) photomagnetic, photoelectromagnetic
~ **эффе́кт·** (*m*) photoelectromagnetic effect
фотома́т·, -а, (*m*) (print) line of matrices
фото·материа́л·, -а, (*m*) photographic material
фотомато́н·, -а, (*m*) photomaton, coin-operated photograph booth
фото́·метр·, -а, (*m*) photometer, photometric receiver
~ **вы·ра́вн·ивани·я яр·кост·ей** equality of brightness photometer, luminosity photometer

фото·метр
~, **интегра́ль·н·ый** integrating photometer
~, **клин·ов·о́й** wedge photometer
~, **контра́ст·н·ый** split-field photometer
~, **миг·а́ющ·ий** flicker photometer
~, **пла́мен·н·ый** flame photometer
~, **ту́бус-** tube photometer
~, **шар·ов·о́й** globe photometer
фото·метр·и́ровани·е, -я, (n) photometry, photometric evaluation
фото·метр·и́ческ·ий, -ая, -ое, (a) photometric
~ **ку́б·ик·** (m) photometer head
~ **ла́мп·а** (f) standard lamp (in photometry)
фото·ме́тр·и·я, -и, (f) photometry
~, **объекти́в·н·ая** physical photometry
~, **субъекти́в·н·ая** visual photometry
фото·механ·и́ческ·ий, -ая, -ое, (a) photomechanical
фото·монта́ж·, -а́, (i) **-о́м,** (m) (phot) montage, photo-montage
фот·о́н·, -а, (m) (phys) photon
фото·на·бо́р·, -а, (m) (print) photocomposition, photo-setting
фото·на·бо́р·н·ая маши́н·а (f) (print) photocomposing machine, photosetter
фото·на́ст·и·я, -и, (f) (biol) photonasty
фото·нейтро́н·, -а, (m) (phys) photoneutron
фото·нос·и́тел·ь, -я, (m) carrier of photo effect
фото́н·н·ый, -ая, -ое, (a) photon
фото·об·рабо́т·к·а, -и, (f) (phot) processing
фото·образ·ова́ни·е, -я, (n) (nucl) photoproduction
фото·объекти́в·, -а, (m) (phot) objective, lens
фото·от·печа́т·ок·, -т·к·а, (m) (phot) print, photograph
фото·от·щепл·е́ни·е, -я, (n) (phys) photodetachment
фото·па́т·и·я, -и, (f) (biol) photopathy
фото·перио́д·, -а, (m) (bot) photoperiod
фото·перио́д·и́зм·, -а, (m) (biol) photoperiodism
фото·печа́т·ающ·ий, -ая, -ее, (a) photoprinting, photographic printing
~, **вы·вод·н·о́е у·стро́й·ств·о** (n) photographic output printer
фото·печа́т·ь, -и, (f) photoprinting

фото·пиро́·метр·, -а, (m) photopyrometer, photoelectric pyrometer
фот·опи́·я, -и, (f) (physiol) photopia
фото·пла́н·, -а, (m) large-scale photomap
фото·пласт·и́нк·а, -и, (g pl) **-нок·,** (f) photographic plate
фото·плён·к·а, -и, (g pl) **-н·ок·,** (f) (phot) film
фото·под·ло́ж·к·а, -и, (g pl) **-ж·ек·,** (f) (phot) support
фото·полимер·иза́ци·я, -и, (f) photopolymerization
фото·про·вод·и́мост·ь, -и, (f) (phys) photoconductivity, photoconduction, internal photo effect
фото·про·вод·ни́к·, -а, (m) photoconductor
фото·про·вод·я́щ·ий, -ая, -ее, (a) photoconductive
~ **сло·й** (m) photoconductive target (vacuum tubes)
фото·прото́н·, -а, (m) (phys) photoproton
фото·раз·ве́д·к·а, -и, (g pl) **-д·ок·,** (f) photographic reconnaissance
фото·рас·па́д·, -а, (m) (phys) photodisintegration
фото·рас·щепл·е́ни·е, -я, (n) (phys) photo-disintegration
фото·регистр·а́тор·, -а, (m) photographic recorder, photographic detector
фото·реду́ктор·, -а, (m) (phot) reduction printer
фото·реду́кци·я, -и, (f) (biochem) carbon dioxide reduction; (phot) reduction
фото·редуц·и́ровани·е, -я, (n) (phot) photo-reduction, reduction
фото·реле́ (n indecl) photoelectric relay, photoswitch
фото·рентгено·гра́ф·и·я, -и, (f) photofluography, fluography
фото·репроду́кци·я, -и, (f) photocopy
фото·реце́птор·, -а, (m) (zool) photoreceptor
фото·рожд·е́ни·е, -я, (n) (nucl) photoproduction
фото·си́нтез·, -а, (m) (bot) photosynthesis
фото·синхрониз·а́ци·я, -и, (f) phototiming
фото·скоп·и́ческ·ий, -ая, -ое, (a) photoscopic

фото·сло́·й, -я,   (*m*) photosensitive layer

фот·с·ни́м·ок·, -м·к·а,   (*m*) photograph, photo; (cinema) shot

фото·со·противл·е́ни·е, -я,   (*n*) photoconductive cell/detector; (nucl) photoconductive target

фото·ста́т·, -а,   (*m*) photocopying apparatus; photostat, photostatic copy

фото·сфе́р·а, -ы,   (*f*) (astron) photosphere

фото·схе́м·а, -ы,   (*f*) mosaic (photogrammetry)

фото·съ·ё́м·к·а, -и,   (*g pl*) **-м·ок·,** (*f*) photographing; (cinema) shooting; (surv) photographic survey

фото·та́ксис·, -а,   (*m*) (biol) phototaxis

фото·телегра́мм·а, -ы,   (*f*) (telecom) facsimile telegram

фото·телегра́ф·, -а,   (*m*) (telecom) facsimile telegraph, phototelegraph

фото·телеграф·и́·я, -и,   (*f*) facsimile telegraphy, phototelegraphy

фото·теодоли́т·, -а,   (*m*) (surv) phototheodolite

фото·терапевт·и́ческ·ий, -ая, -ое, (*a*) (med) phototherapeutic

фото·терап·и́·я, -и,   (*f*) (med) phototherapy

фото·ти́п·и·я, -и,   (*f*) (print) phototype, albert-type, aquatone

фото·то́к·, -а,   (*m*) photoelectric current, photocurrent

    сигна́л· фото·то́к·а   (*m*) (TV) photo-signal

фото·топо·гра́ф·и·я, -и,   (*f*) photographic surveying

фото·транзи́стор·, -а,   (*m*) phototransistor

фото·трансми́ттер·, -а,   (*m*) photoelectric reader

фото·трансформ·а́тор·, -а, (*m*) rectifier (aerial photography)

фото·триангул·я́ци·я, -и,   (*f*) (surv) photo-triangulation

фото·трио́д·, сло·и́ст·ый   junction phototransistor

фото·тро́н·, -а,   (*m*) (elec) phototron

фото·троп·и́зм·, -а,   (*m*) (biol) phototropism

фото·у·велич·и́тел·ь, -я,   (*m*) (phot) enlarger

фото·у·множ·и́тел·ь, -я,   (*m*) (elec) photomultiplier, photomultiplier tube

фото·у·пру́г·ий, -ая, -ое,   (*a*) photoelastic

фото·у·пру́г·ост·ь, -и,   (*f*) (phys) photoelasticity

фото·у·стано́в·к·а, -и,   (*g pl*) **-в·ок·,** (*f*) camera mounting/stand

фото·фа́з·а, -ы,   (*f*) (bot) photophase

фото·фи́ль·н·ый, -ая, -ое,   (*a*) (biol) photophilous

фото·фо́б·и·я, -и,   (*f*) (biol) photophobia

фото·фо́б·н·ый, -ая, -ое,   (*a*) (biol) photophobic

фото·фо́р·, -а,   (*m*) (zool) photophore; (med, instr) photophore

фото·форе́з·, -а,   (*m*) (chem) photophoresis

фото·хим·и́ческ·ий, -ая, -ое,   (*a*) photochemical

фото·цинко·гра́ф·и·я, -и,   (*f*) (print) zincography, photoetching

фото·чувств·и́тельн·ый, -ая, -ое, (*a*) photosensitive, photographic

~ **бума́г·а**   (*f*) photographic paper

фото·шир·от·а́, -ы́,   (*f*) (phot) density range

фото·шла́нг·, -а,   (*m*) (cinema) optical sound cable

фото·эдс   (*indecl abbr*) photo e.m.f.

фото·электр·и́ческ·ий, -ая, -ое,   (*a*) photoelectric

~ **эффе́кт·**   (*m*) (*and see* **фото·эф·фе́кт·**) photoelectric effect

~ **явл·е́ни·е**   (*n*) (*and see* **фото·эф·фе́кт·**) photoelectric effect

фото·электро́н·н·ый, -ая, -ое,   (*a*) photoelectron, photoelectric

~ **ток·**   (*m*) (phys) photocurrent

~ **у·множ·и́тел·ь**   (*m*) photomultiplier, photoelectric multiplier, photomultiplier tube

фото·элеме́нт·, -а,   (*m*) photocell, photoelectric cell; phototube

~, **ва́куум·н·ый**   photoemissive cell, vacuum photoelectric cell

~, **ве́нтиль·н·ый**   photovoltaic cell, barrier layer cell

~, **га́зо·на·по́лн·енн·ый**   gas-filled photoelectric cell

~, **кре́мн·иев·ый**   silicon photovoltaic cell

~, **пуст·о́тн·ый**   vacuum photoelectric cell, photoemissive cell

~ **резисти́в·н·ого де́й·стви·я**   photoconductive cell

~, **селе́н·ов·ый**   selenium cell

~, **та́лл·иев·ый**   thalofide cell

~, **эмиссио́н·н·ый**   photoemissive cell, vacuum photoelectric cell

фото·эмисс·и·я, -и, (*f*) (phys) photo-emission

фото·эму́льс·и·я, -и, (*f*) (phot) photogene, photographic emulsion

фото·эффе́кт·, -а, (*m*) photoeffect, photoelectric effect

~, вне́ш·н·ий external/normal photoelectric effect

~, вну́тр·енн·ий photoconduction, internal photoelectric effect

фоулери́т·, -а, (*m*) (min) fowlerite

фошаги́т·, -а, (*m*) (min) foshagite

фрагме́нт·, -а, (*m*) fragment

фрагмента́р·н·ый, -ая, -ое, (*a*) fragmentary

фрагмент·а́ци·я, -и, (*f*) fragmentation

фрагмо·ко́н·, -а, (*m*) (zool) phragmocone

фрагмо·пла́ст·, -а, (*m*) (gen) phragmoplast

фрагмо·спо́р·а, -ы, (*f*) (bot) phragmospore

фра́з·а, -ы, (*f*) phrase

фразео·ло́г·и·я, -и, (*f*) (ling) phraseology

фракто·гра́ф·и·я, -и, (*f*) (met) fractography

фракт·у́р·а, -ы, (*f*) (med) fracture

фракцион·и́ровани·е, -я, (*n*) (chem) fractionation, fractionating, fractional distillation

фракцион·и́рованн·ый, -ая, -ое, (*past part pass*) of **фракцион·и́р·о·вать;** (chem) fractionated

фракцион·и́р·овать, -уют, (*imp and perf*) (chem) fractionate

фракцио́н·н·ый, -ая, -ое, (*a*) fractional, fraction; (polit) factional

~ со·ста́в· (*m*) (oil) distillation fractions

фра́кц·и·я, -и, (*f*) fraction; (oil) fraction, cut; (polit) faction

~, вс·плыв·а́ющ·ая (min) float fraction

~, ис·хо́д·н·ая (nucl) parent fraction

~, лёгк·ая (oil) light fraction; (min) float

~, ме́л·к·ая (min) fines

~, о·бед·нённ·ая (nucl) stripped/depleted/downflow fraction

~, пла́в·ающ·ая (min) float fraction

~, шир·о́к·ая (oil) wide boiling-range fraction

фраму́г·а, -и, (*f*) transom (of window, door)

фра́нк·, -а, (*m*) (fin) franc

Фра́нк·а-Ри́д·а, механ·и́зм· (*m*) (cryst) Frank–Read mechanism

франк·и́р·овать, -уют, (*imp and perf*) stamp (mail), prepay (transit of goods)

франклиниз·а́ци·я, -и, (*f*) (med, elec) franklinization

франклини́т·, -а, (*m*) (min) franklinite

фра́нко (*indecl*) (com) free, franco, paid

~ -алонгса́йд·, -а, (*m*) (com) free alongside

~ -ба́рж·а, -и, (*f*) (com) free into barge

~ -бо́рт·, -а, (*m*) free on board, fob

~ -бу́нкер·, -а, (*m*) free into bunker

~, -ваго́н·, -а, (*m*) free on rail, for

~ -до́к·, -а, (*m*) free-dock, franco-dock

~ -за·во́д·, -а, (*m*) for collection (from factory), to be collected by purchaser

~ -на́·береж·н·ая, -ой, (*f*) free alongside quay

фра́нц·и·й, -я, (*m*) (chem) francium, Fr

францу́з·, -а, (*m*) Frenchman

францу́з·ск·ий, -ая, -ое, (*a*) French

~ ключ· (*m*) monkey wrench

фраунго́фер·ов·ы ли́н·и·и (*pl*) (opt) Fraunhofer lines

фра́хт·, -а, (*m*) freight

~, мёртв·ый dead freight

~, обра́т·н·ый back freight

фрахт·ова́ни·е, -я, (*n*) (naut) chartering, affreightment

фрахт·ова́тел·ь, -я, (*m*) (naut) charterer

фрахт·ова́ть, -у́ют, (*imp*) (naut) charter

фрахт·о́вк·а, -и, (*g pl*) **-вок·,** (*f*) freightage

фрахт·о́в·ый, -ая, -ое, (*a*) freight; (naut) charter

~ с·де́л·к·а (*f*) charter-party

фрахт·о́вщик·, -и, (*m*) carrier, shipowner

фреат·и́ческ·ий, -ая, -ое, (*a*) (geol, bot) phreatic

фреато·фи́т·, -а, (*m*) (bot) phreatophyte

фрега́т·, -а, (*m*) (naut) frigate; (orn) frigate bird

фре́дгольм·ост·ь, -и, (*f*) (math) Fredholm property

фрез·а́, -ы́, (*f*) milling cutter, cutter; cutterhead (of dredger); (civ eng) harrow, scarifier; (agr) scarifier, rotary cultivator

~, вы́·пукл·ая convex milling cutter

~, гальте́ль·н·ая corner cutter

~, доро́ж·н·ая road harrow

~, за·ты́л·о́ванн·ая relieved/backed-off cutter

фрез·á

~, крупно·зу́б·ая   coarse-pitch cutter

~, лоб·ов·áя   end milling cutter

~, мóдуль·н·ая   gear cutter

~, на·бóр·н·ая   gang cutter

~, одно·зуб·ов·áя   fly cutter, single point cutter

~, пáз·ов·ая   groove cutter

~, пóчв·енн·ая   (agr) scarifier, rotary cultivator

~, торц·óв·ая   end milling cutter

~, фасóн·н·ая form cutter, forming tool

~, цек·óвочн·ая   round-up cutter

~, черв·я́чн·ая   hobbing cutter, hob

~, шар·ов·áя   ball cutter

~, шпóн·очн·ая   cotter/slotting mill

фрéзер·, -а,   (m) see **фрез·á;** milling tool (oil wells)

фрéзер·н·о-про·бéль·н·ый стан·óк· (m) (print) routing machine

фрéзер·н·ый, -ая, -ое,   (a) milling (machine operation); scarifier (soil and road surfaces)

~ голóв·к·а   (f) milling head

~ машúн·а   (f) (agr, civ eng) scarifier, rotary cultivator

~ стан·óк·   (m) milling machine; vertical spindle moulder (woodworking)

~ стан·óк·, про·дóль·н·о-   (m) bed-type milling machine

фрезер·овáльн·ый, -ая, -ое,   (a) milling; scarifying

фрезер·овáни·е, -я,   (n) milling, cutting (machining); scarifying

фрезер·овáть, -ýют,   (imp and perf) mill, cut (machining); scarify (soils etc.)

фрезер·óвк·а, -и,   (g pl) -вок·, (f) = **фрезер·овáни·е**

фрезер·óвщик·, -а,   (m) milling-machine operator

фрейалúт·, -а, (m) (min) freyalite

фрейбергúт·, -а, (m) (min) freibergite

Фрейнд·а, реáкци·я (f) (chem) Freund reaction

фрейринúт·, -а, (m) (min) freirinite

фрейслебенúт·, -а, (m) (min) freislebenite

фрейфáл·, -а, (m) (oil) free-fall (boring tool)

френ·áлги·я, -и, (f) (med) phrenalgia

Френé, фóрмул·ы (pl) (math) Frenet equations

Френéл·я, би·зеркал·á (pl) (opt) Fresnel double mirrors, Fresnel's mirrors

Френéл·я

~, зеркал·á   (pl) (opt) Fresnel double mirrors, Fresnel's mirrors

~, зóн·а   (f) (phys) Fresnel zone

~, интегрáл·ы   (pl) (math) Fresnel integrals

~, лúнз·а   (f) (opt) Fresnel lens

френико·костáль·н·ый, -ая, -ое,   (a) (anat) phrenicocostal

френико·том·ú·я, -и, (f) (med) phrenicotomy

Фрéнкел·я, дефéкт·   (m) (cryst) Frenkel defect

Фрéнкел·я-Эйринг·а, у·равн·éни·е (n) Frenckel–Eyring equation

френозúн·, -а, (m) (chem) phrenosin

френо·кáрд·и·я, -и, (f) (med) phrenocardia

френо·пáт·и·я, -и, (f) (med) phrenopathy

фреóн·, -а, (m) (chem) freon

фрéск·а, -и, (g pl) -сок·, (f) fresco (art)

фригорú·метр·, -а, (m) frigorimeter

фриделúт·, -а, (m) (min) friedelite

Фридéл·я-Крáфтс·а, реáкци·я (f) (chem) Friedel–Crafts reaction

фрúз·, -а, (m) (arch) frieze

фрúзер·, -а, (m) (food) freezer, ice-cream freezer

фрúз·ов·ый, -ая, -ое,   (a) frieze; bordered

фрикциóн·, -а, (m) friction clutch, clutch; friction integration (of computer)

~, борт·ов·óй   steering clutch (tractor)

фрикциóн·н·ый, -ая, -ое,   (a) friction, frictional

~ лебёд·к·а   (f) friction-drive winch

~, механ·úзм·   (m) friction integrator (computer)

~ пере·дá·ч·а   (f) friction gear

фрúм·а, -ы, (f) (bot) lopseed, *Phryma*

Фрúс·а, прáв·ил·о (n) (chem) Fries rule

~, пере·группирóв·к·а   (f) (chem) Fries rearrangement

фритерн·, -а, (m) (com) free of turn

фрúтт·а, -ы, (f) (ceram) frit, fritt

фрúттер·, -а,   (m) acoustic shock absorber (telephony)

фритт·овáни·е, -я, (n) (ceram) fritting

фритт·óванн·ый, -ая, -ое, (past part pass) fritted

фритчеúт·, -а, (m) (min) fritscheite

фрóнт·, -а, (nom pl) -ы, (g pl) -óв, (m) front, frontage; line abreast (a/c, ships, tanks); casing plate (of boiler)

~ волн·ы́   wave front

фр о́нт
~ **воз·мущ·éни·й** (aerodynam) Mach front
~, **втор·и́чн·ый** (meteor) secondary front
~, **за́д·н·ий** trailing edge (of a pulse); back casing (boiler)
~ **и́мпульс·а** (phys) pulse front
~, **крут·óй** steep edge (of a pulse)
~, **перéд·н·ий,** leading edge (of a pulse); front casing (boiler)
~ **по·гру́з·к·и-вы́·груз·к·и** (rail) goods depot
~, **поля́р·н·ый** (meteor) polar front
~, **при·ча́ль·н·ый** harbour/harbor wharves
~, **тёпл·ый** (meteor) warm front
~, **тóп·очн·ый** furnace front plate
~, **троп·и́ческ·ий** (meteor) intertropical convergence zone, intertropical front
~, **холóд·н·ый** (meteor) cold front
фронта́ль·н·ый, -ая, -ое, (a) frontal
~ **зон·а** (f) (meteor) frontal surface
~ **на·ступл·éни·е** (n) (mil) frontal attack
~ **по·вéрх·ност·ь** (f) front, front surface; (meteor) frontal surface, surface of discontinuity
фронтиспи́с·, -а, (m) (print) frontispiece
фронт·ови́к·, -а́, (m) (mil) front-line soldier
фронт·ов·óй, -áя, -óе, (a) front, frontal, front-line
~ **атáк·а** (f) frontal attack
фронто·генéз·, -а, (m) (meteor) frontogenesis
фронтó·лиз·, -а, (m) (meteor) frontolysis
фронтóн·, -а, (m) (arch) pediment
Фру́д·а, числ·ó (n) (hydrodynam) Froude number
фру́кт·, -а, (m) fruit
фрукт·óв·ый, -ая, -ое, (a) fruit
~ **сáхар·** (m) (chem) fructose
фрукт·óз·а, -ы, (f) (biochem) fructose
фруктозáн·, -а, (m) (chem) fructosan
фруктозо·дифосфáт·, -а, (m) (pharm) fructose diphosphate
фрукто·пиранóз·а, -ы, (f) (biochem) fructopyranose
фрукто·с·ним·áтел·ь, -я, (n) (hortic) fruit picker
фтал- (int component) (chem) phthal-, phtalo-
фтал·азóл·, -а (m) (pharm) phthalazon

фтал·ами́д·, -а, (m) phthalamide
фтал·амои́л·, -а, (m) phthalamoyl
фтал·áт·, -а, (m) phthalate
фтáл·ев·ый, -ая, -ое, (a) (chem) phthalic
~ **ангидри́д·** (m) (chem) phthalic anhydride
~ **кисл·от·á** (f) (chem) phthalic acid
фталидилидéн·, -а, (m) phthalidylidene
фталилидéн·, -а, (m) phthalylidene
фтал·ими́д·, -а, (m) (chem) phthalimide
фтало·циани́н·, -а, (m) phthalocyanine
фтало·цианин·ов·ый крас·и́тел·ь (m) (chem) phthalocyanine dye
фтивази́д·, -а, (m) (pharm) isoniazid
фтиз·иатри́·я, -и, (f) (med) phthisiology
фтéр·, -а, (m) (chem) fluorine, F; (*prefix*) fluor-, fluoro-
фтор·ацетóн·, -а, (m) fluoroacetone
фтор·бензóл·, -а, (m) (chem) fluorobenzene
фтор·гидро·ó·кис·ь, -и, (f) hydroxyfluoride
фтор·за·мещ·ённ·ая кисл·от·á (f) fluoro acid
фтор·и́д·, -а, (m) (chem) fluoride
~ **бóр·а** (chem) boron fluoride
фторид·и́ровани·е, -я, (n) (met) fluoride treatment/process
фтор·и́ровани·е, -я, (n) fluorination, fluorinating (e.g. of water)
фтор·и́р·овать, -уют, (imp) fluorinate
фтористо·водо·рóд·н·ая кисл·от·á (f) hydrofluoric acid
фтóр·ист·ый, -ая, -ое, (a) (chem) fluoric, (lower) fluoride of
~ **алюми́н·и·й** (m) aluminium/aluminum fluoride
~ **бор·** (m) boron fluoride
~ **водо·рóд·** (m) hydrogen fluoride
~ **мети́л·** (m) fluomethane
~ **угле·рóд·** (m) fluoro-carbon
~ **урáн·** (m) uranium fluoride
~ **урани́л·** (m) uranium oxyfluoride
фтор·нáтр·и·й, -я, (m) sodium fluoride
фтóр·н·ый, -ая, -ое, (a) fluorine, (higher) fluoride (of)
фторо·бор·áт·, -а, (m) borofluoride
фторо·водо·рóд·, -а, (m) hydrogen fluoride
фтор·ó·кис·ь, -и, (f) oxyfluoride
фтор·олефи́н·, -а, (m) (chem) fluorolefine
фторо·пла́ст·, -а, (m) fluoroplastic
~ **-3** (plast) polytrifluorochloroethylene, PTFCE

фторо·пласт
~ -4 (plast) polytetrafluoroethylene, teflon

фторо·полимéр·, -а, (*m*) fluoropolymer

фторо·полиэфúр·, -а, (*m*) fluorinated polyester

фторо·прéн·, -а, (*m*) (plast) fluoroprene

фтор·орган·úческ·ий, -ая, -ое, (*a*) organofluorine

фтор·силикáт·, -а, (*m*) (chem) fluorosilicate

фтор·угле·рóд·, -а, (*m*) (chem) fluorocarbon

фтор·цирконáт·, -а, (*m*) (chem) fluozirconate

фуг·áнок·, -нк·а, (*m*) large plane, jointer plane; overhand planing machine, surfacer

фуг·áночн·ый, -ая, -ое, (*a*) of **фуг·á·нок·**

фугáс·, -а, (*m*) (mil) landmine, fougasse

фугáс·ност·ь, -и, (*f*) (chem) fugacity

фугáс·н·ый, -ая, -ое, (*a*) high-explosive, H.E.
~ дéй·стви·е (*n*) high explosive effect
~ с·на·ряд· (*m*) (mil) H.E. shell

фугитúв·ност·ь, -и, (*f*) (chem) fugacity

фуг·овáльн·ый, -ая, -ое, (*a*) planing, surfacing
~ стан·óк· (*m*) surfacer, overhand planer (woodworking)

фуг·овáни·е, -я, (*n*) (*v n*) of **фуг·овáть**

фуг·овáть, -ýют, (*imp*) plane, surface (with plane)

фузариóз·, -а, (*m*) fusarial wilt (plant disease)

фузáриум·, -а, (*m*) (bot) Fusarium

фузиáн·, -а, (*m*) (geol) fusian

фузо·спирохетóз·, -а, (*m*) (med) fusospirochetosis

фузулúн·а, -ы, (*f*) (pal) fusulina, (*pl*) fusulinas, *Fusulina*

фузулинúд·ы (*pl*) (pal) fusulinids, *Fusulinidae*

Фукó, мáят·ник· (*m*) (astron) Foucault's pendulum
~, ток· (*m*) (elec) eddy/Foucault current

фукозáн·, -а, (*m*) = **фукосáн·**

фук·óид·, -а, (*m*) (bot) fucoid

фуко·ксантúн·. -а, (*m*) (biochem) fucoxanthin

фукос·áн·, -а, (*m*) (biochem) fucosan

фуксúн·, -а, (*m*) (chem) fuchsin, fuchsine (dye)

фуксúт·, -а, (*m*) (min) fuchsite

фýкс·ов·ый, -ая, -ое, (*a*) (math) Fuchsian

фуксшвáнц·, -а, (*m*) hand saw, stair builder's saw

фýкус·, -а, (*m*) (bot) rockweed, *Fucus*

фýкус·ов·ые вóдо·росл·и (*pl*) (bot) fucoid seaweeds, *Fucales*

фульв·áт·ы (*pl*) fulvates (soil)

фульво·кисл·от·á, -ы́, (*f*) fulvic acid, fulvoacid

фульгур·úт·, -а, (*m*) (geol) fulgurite

фульмин·áт·, -а, (*m*) (chem) fulminate

фуляр·, -а, (*m*) (text) foulard

фумáр·ов·ая кисл·от·á (*f*) (chem) fumaric acid

фумар·óл·а, -ы, (*f*) (geol) fumarole

фумиг·áнт·, -а, (*m*) fumigant

фумиг·áци·я, -и, (*f*) fumigation

фунги·сúд·, -а, (*m*) *see* **фунги·цúд·**

фунги·цúд·, -а, (*m*) fungicide

фундáль·н·ая желез·á (*f*) (anat) fundic gland

фундáмент·, -а, (*m*) (build) foundation; bedplate, seating, bearer (of machines etc.)
~, котéль·н·ый boiler seating/bearer
~, с·пуск·ов·óй (shipb) groundways (of slips)

фундаментáль·ност·ь, -и, (*f*) fundamental nature

фундаментáль·н·ый, -ая, -ое, (*a*) fundamental, basic
~ фýнкци·я (*f*) (math) fundamental/characteristic function, eigenfunction
~ числ·ó (*n*) (math) eigen value

фундáмент·н·ый, -ая, -ое, (*a*) of **фундáмент·**
~ болт· (*m*) foundation bolt, holding-down bolt
~ доск·á (*f*) seating plate
~ рáм·а (*f*) bedplate, bed, sole, bearer

фунд·úрованн·ый, -ая, -ое, (*past part pass*) (fin) funded, consolidated

фундýк·, -á, (*m*) (bot) filbert, *Corylus*

фуникулёр·, -а, (*m*) funicular railway

фунúкулус·, -а, (*m*) (zool) funiculus, (*pl*) funiculi

фуникуляр·н·ый, -ая, -ое, (*a*) funicular

функционáл·, -а, (*m*) (math) functional

функциональ·ност·ь (*f*) functionality

функционáль·н·ый, -ая, -ое, (*a*) functional; function
~ у·стрóй·ств·о (*n*) functional/function generator (of computer)

функцион·и́р·овать, -уют,   (*imp*) function

фу́нкци·я, -и, (*f*) function
   в фу́нкци·и   as a function of, against, versus, vs. (graphs etc.)

~, вес·ов·а́я   weighting function

~ в·ли·я́ни·я   influence function, (nucl) kernel

~, воз·мущ·а́ющ·ая   forcing function

~, в·ход·н·а́я   input function

~ вы́·год·ы   utility function

~ -вы·раж·е́ни·е, -я,   (*n*) (math) functional expression

~, вы·ход·н·а́я   output function

~, дей·стви́тельн·ая   real function

~, круг·ов·а́я   inverse trigonometric function

~, не·я́в·н·ая   implicit function

~, норм·ир́ованн·ая   normalized function

~ оборо́т·а   return function

~, обра́т·н·ая   inverse/reciprocal function

~, одно·зна́ч·н·ая   single-valued function

~, одно·ро́д·н·ая   homogeneous function

~, от·обра́ж·а́ющ·ая   mapping function

~ от·риц·а́ни·я   "not" function

~, пере·да́т·оч·н·ая   transfer function

~ по·лож·е́ни·я   position function; (dynam) station function

~ по·те́р·ь   loss function

~, по·каз·а́тельн·ая   exponential function

~, про·из·вод·я́щ·ая   generating function

~ про·из·во́ль·н·ого ви́д·а   general function

~ с·вя́з·и   coupling function

~ скач·к·о́в   jump/step function

~, скачко·обра́з·н·ая   jump function

~, со·пряж·ённ·ая   (math) associate function; (*pl*) adjoint functions

~, степе́н·н·ая   power function

~, ступ·е́нчат·ая   step function
   тео́ри·я фу́нкци·й   (*f*) theory of real functions

~, це́л·ая   (math) integral/entire function

~, це́л·ая рациона́ль·н·ая   (math) rational integral function, polynomial in one variable

~, я́в·н·ая   (math) explicit function

~ ядр·а́   (math) kernel

фу́нт·, -а,   (*m*) pound (453·59 g); (fin) pound

~, англ·и́йск·ий торг·о́в·ый   pound avoirdupois

~, апте́к·арск·ий   apothecaries' pound (5760 grains)

~, мексика́н·ск·ий   (text) libra (460·25 g)

~ стéрлинг·ов   (fin) pound sterling

~, тро́й·ск·ий   pound troy (373·24 g)

фу́р·а, -ы,   (*f*) waggon

фура́ж·, -а́, (*i*) -о́м, (*m*) fodder, forage

фура́ж·к·а, -а, (*g pl*) -ж·ек·, (*f*) (mil) peaked cap

фура́ж·н·ый, -ая, -ое,   (*a*) forage, fodder

фур·а́н·, -а,   (*m*) (chem) furan

фуран·карбо́н·ов·ая кисл·от·а́   (*f*) furoic acid
   хлорангидри́д фуран·карбо́н·о·в·ой кисл·от·ы́   furyl chloride
   эфи́р·фуран·карбо́н·ов·ой кисл·о·т·ы́ (*m*) furoate

фурано́з·а, -ы, (*f*) (chem) furanose

фурго́н·, -а,   (*m*) van (vehicle)

фу́рк·а, -и, (*f*) (zool) furca

фу́рм·а, -ы, (*f*) (met) tuyere (of blast furnace); lance (LD process)

фурмарьери́т·, -а,   (*m*) (min) fourmarierite

фу́рм·енн·ый, -ая, -ое,   (*a*) of **фу́рм·а**

~ по́яс·   (*m*) (met) wind box, air belt

~ тру́б·к·а (*f*) bustle pipe

фурм·о́вщик·, -а,   (*m*) tuyere operator

фу́рм·очк·а, шла́к·ов·ая (*f*) cinder notch, monkey (blast furnace)

фурниту́р·а, -ы, (*f*) accessories, appurtenances

фуроа́т·, -а,   (*m*) furoate

фуру́нкул·, -а,   (*m*) (med) furuncle, boil

фурфур·а́л·, -а,   (*m*) (chem) furfural

фурфур·и́л·, -а,   (*m*) (chem) furfuryl

фурфур·о́л·, -а,   (*m*) (chem) furfural

фурфуро́ль·н·ый, -ая, -ое,   (*a*) furfural

Фурье́, пре·обра́з·ова́ни·е (*n*) (math) Fourier transform

~, ряд·   (*m*) (math) Fourier series

фу́стик·, -а,   (*m*) (bot) fustic, yellow brazil wood

фу́с·ы (*pl*) residual coal tar

фу́т·, -а,   (*m*) foot (30·5 cm)

~, доск·о́в·ый   (build) board foot; board measure (timber)

фут-свеч·а́, -и́, (*f*) (opt) foot candle

футбо́ль·н·ый, -ая, -ое,   (*a*) football, soccer

фу́тер·, -а, (*m*) lining, refractory lining

фу́тер·н·ая нит·ь (*f*) needle thread (knitting machine)

футер·ова́ть, -у́ют, (*imp*) line (with refractories)

футер·о́вк·а, -и, (*f*) lining (with refractories); lining, refractory lining

футля́р·, -а, (*m*) sheath, case, scabbard, box

~, батере́й·н·ый battery cell crate

~ по́рох·а powder case

футля́р·н·ый, -ая, -ое, (*a*) of **футля́р·**

~ с·на·ряж·е́ни·е (*n*) block filling (ammunition)

фу́т·ов·ый, -ая, -ое, (*a*) one-foot (length)

фу́токс·, -а, (*m*) (shipb) futtock

фу́тор·, -а, (*m*) cover, patch; key plate

фу́тор·к·а, -и, (*f*) adapter nut (for piping)

футшто́к·, -а, (*m*) tide gauge/gage/pole

фуфа́·ечн·ый, -ая, -ое, (*a*) of **фуфа́й·к·а**

фуфа́й·к·а, -и, (*g pl*) **-á·ек·,** (*f*) (text) jersey, sweater, pullover

фы́рк·а·ть, -ют, (*imp*) snort, snuffle

фы́рк·н·уть, -ут, (*perf*) *see* **фы́рк·а·ть**

фье́лд·, -а, (*m*) (geog) fjeld

фье́рд·, -а, (*m*) (geog) fjärd, Baltic fiord

фьо́рд·, -а, (*m*) (geog) fiord, fjord

фюзеля́ж·, -а, (*i*) **-ем,** (*m*) (a/c) fuselage, body, (rocket) body

фюзеля́ж·н·ый, -ая, -ое, (*a*) of **фюзеля́ж·**

~ бак· (*m*) (rocket) body tank

фюзе́н·, -а, (*m*) (min) fusain

фюзениз·а́ци·я, -и, (*f*) (min) fusainization

фюзини́т·, -а, (*m*) (min) fusinite

фюзовитри́т·, -а, (*m*) (min) fusovitrite

X

хабази́т·, -а, (*m*) (min) chabasite

хаве́г·, -а, (*m*) (plast) Haveg

Ха́ггинс·а, конста́нт·а (*f*) Huggins constant (polymers)

хадро·за́вр·, -а, (*m*) (pal) hadrosaur, *Hadrosaurus*

хадр·о́м·, -а, (*m*) = **гадр·о́м·** (bot) hadrome, hadrom

хазмато·пла́зм·а, -ы, (*f*) (biol) chasmatoplasm

хазмо·га́м·и·я, -и, (*f*) (bot) chasmogamy

хазмо·фи́т·, -а, (*m*) (bot) chasmophyte

хайда́йн·, -а, (*m*) hydyne (rocket propellant)

ха́ки (*n indecl*) khaki

хала́з·а, -ы, (*f*) (biol) chalaza, (*pl*) chalazea, chalazas

халазо·га́м·и·я, -и, (*f*) (biol) chalazogamy

хала́т·, -а, (*m*) robe, smock, coat; bath-robe, dressing-gown

хала́т·ност·ь, -и, (*f*) slackness, feebleness

халв·а́, -ы́, (*f*) (food) khalva, halva

халиб·и́т·, -а, (*m*) (min) chalybite

халик·о́з·, -а, (*m*) (med) chalicosis

халико·те́ри·и, -ев, (*m pl*) (pal) chalicotheres, *Chalicotherioidea*

халистат·и́ческ·ий, -ая, -ое, (*a*) halistatic

халцедо́н·, -а, (*m*) (min) chalcedony

хальк·анти́т·, -а, (*m*) (min) chalcanthite

халько·алюми́т·, -а, (*m*) (min) chalcoalumite

халько·гени́д·, -а, (*m*) (chem) chalcogenide

халькоз·и́н·, -а, (*m*) chalcosine, (min) chalcocite, copper glance, redruthite

халькоз·и́т·, -а, (*m*) (min) chalcocite, copper glance, redruthite

халько·ли́т·, -а, (*m*) (min) chalcolite, torbernite

халько·мени́т·, -а, (*m*) (min) chalcomenite

халько·пири́т·, -а, (*m*) (min) chalcopyrite, copper pyrite

халько·стиби́т·, -а, (*m*) (min) chalcostibite, wolfsbergite

халько·трихи́т·, -а, (*m*) (min) chalcotrichite

халько·фи́ль·н·ый, -ая, -ое, (*a*) chalcophile

халько·фи́ль·ный, -ая, -ое

~ элеме́нт·ы (*pl*) (chem) chalcophile elements

халькоц·и́т·, -а, (*m*) = **халькози́т·**

хальци́д·ы (*pl*) (ent) *Chalcidoidea*

хаме·лео́н·, -а, (*m*) (zool) chameleon, (*pl*) *Chamaeleontidae*

ха́мел·евск·ое те́л·о (*n*) (math) Hamel body

хаме·ро́пс·, -а, (*m*) (bot) chamaerops palm, *Chamaerops*

хаме·фи́т·, -а, (*m*) (bot) chamaephyte

хаммари́т·, -а, (*m*) (min) hammarite

Ха́ммет·а, констант (*m*) (chem) Hammet constant

хамс·а́, -ы́, (*f*) (fish) anchovy, *Engraulis encrasicholus*

хао́с·, -а, (*m*) chaos

хаот·и́ческ·ий, -ая, -ое, (*a*) chaotic, random

хаот·и́чност·ь, -и, (*f*) chaotic state, randomness

хапло- (*int component*) *see* **гапло-**

ха́р·а, -ы, (*f*) (bot) Chara

хара́ктер·, -а, (*m*) character, nature

~ индик·а́ци·й це́л·и target information (radar)

~ ме́ст·ност·и (geog) relief, terrain

характериз·ова́ть, -у́ют, (*imp and perf*) characterize, describe, define; **-ся** (*pass*); show, exhibit

характер·и́стик·а, -и, (*f*) characteristics, performance, response, nature, properties; characteristic curve; pattern (radar)

~, амплиту́д·н·ая (autom) amplitude response

~, ано́д·н·ая (elec) plate characteristic curve

~, вольт·ампе́р·н·ая (elec) volt-ampere/current–voltage characteristic

~ втор·о́го сем·е́йств·а (aerodynam) minus Mach line

~, динам·и́ческ·ая dynamic response; (a/c) load characteristic

~, за·ви́с·им·ая inverse time characteristic (of elec. relay)

~ из·луч·е́ни·я radiation pattern

~, и́мпульс·н·ая pulse response

~ локомоти́в·а locomotive performance

~ ма́ркер·а marker pattern (radar)

~ на·пра́вл·енност·и (GW) beam pattern

характер·и́стик·а
~, не·за·ви́с·им·ая   definite-time characteristic (of elec. relay)
~, о·грани́ч·енн·о-за·ви́с·им·ая   inverse-definite time characteristic (of elec. relay)
~, оптима́ль·н·ая   best performance curve
~, па́д·ающ·ая   negative characteristic (of vacuum tube)
~ пе́рв·ого сем·е́йств·а   (aerodynam) plus Mach line
~, пере·хо́д·н·ая   transient response
~, при·вед·ённ·ая   total characteristic (of an elec. valve/tube)
~ при·ём·а   (rad) reception diagram
~ при·ём·ник·а   (rad) receiver response/performance
~ про́фил·я   airfoil characteristic curve
~, рабо́ч·ая   performance (of machine)
рас·чёт· характер·и́стик·   (m) performance computation (of a machine)
~, се́т·очн·ая   (elec) grid characteristic
~, служ·е́бн·ая   personal report, testimonial
~, с·чёт·н·ая   counting characteristic curve
~, цвет·ов·а́я   colour response/characteristic
~, част·о́тн·ая   (autom) frequency response; (aerodynam) dynamic response/characteristic
~ чувств·и́тельност·и   response curve
характер·исти́ческ·ий, -ая, -ое, (a) characteristic; performance
~ да́·нн·ые   (pl) performance figures/data
характе́р·н·о (adv) typically
характе́р·н·ый, -ая, -ое, (a) characteristic, typical, distinctive, representative; (theat) character
характеро́·граф·, -а, (m) (instr) voltage-versus-current curve tracer
хараци́н·ов·ые, -ых, (pl decl as adj) (fish) characinid, Characinidae
ха́рз·а, -ы, (f) (zool) yellow-throated marten, Martes flavigula
ха́риус·, -а, (m) (fish) grayling, Thymallus
ха́рк·а·ть, -ют, (imp) spit, expectorate
ха́рк·н·уть, -ут, (perf) see **ха́рк·а·ть**
ха́р·ов·ые, -ых, (pl decl as adj) (bot) charophytes, Charophyta
ха́рти·я, -и, (f) (polit) charter
Ха́рти-фо́к·а, ме́тод· (m) (phys) Hartry-Fock method

хасмо- (int component) see **хазмо**
ха́т·а, -ы, (f) hut, croft
хаульму́гр·ов·ый, -ая, -ое, (a) chaulmoogra, (chem) chaulmoogric
~ кисл·от·а́ (f) chaulmoogric acid
~ ма́сл·о (n) chaulmoogra oil
хвал·е́бн·ый, -ая, -ое, (a) eulogistic, laudatory
хвал·и́ть, -́ят, (perf) praise, commend; -ся (pass); (+ instr) boast
хва́ст·а·ться, -ются, (imp) boast
хва́т-та́л·и, -ей, (f pl) (naut) jigger, luff (tackle)
хват·а́тельн·ый, -ая, -ое, (a) (biol) prehensile
хват·а́·ть, -ют, (imp) seize, snatch, catch, grab, grasp; be sufficient/enough, suffice
хват·и́ть, -́ят, (perf) be sufficient/enough, suffice
хва́т·к·а, -и, (g pl) -т·ок·, (f) grip, grasp; (orn) talon
хва́ч·енн·ый, -ая, -ое, (past part pass) of **хват·и́ть**
хво́й·ник·, -а, (m) (bot) jointfir, Mormon tea, Ephedra
хво́й·ников·ые, -ых, (pl decl as adj) (bot) Gnetales
хво́й·н·ый, -ая, -ое, (a) (bot) coniferous, abietic, (pl as noun) conifers, Coniferae
~ де́рев·о (n) conifer, coniferous tree
~ древес·и́н·а (f) softwood
хво́рост·, -а, (m) brushwood
хвост·, -а́, (m) tail, cauda; queue; (a/c) tail, rear; (met) gripping end of a forging piece; (pl) tailings
~, ла́сточ·кин· dovetail (woodworking etc.), swallowtail
~, раз·дво́·енн·ый swallowtail
~, тёпл·ов·ый (phys) thermal spike
~, экспоненциа́ль·н·ый exponential tail (of a curve)
хвост·а́т·ый, -ая, -ое, (a) tailed, caudate, (pl as noun) (zool) caudates, Caudata
хвост·е́ц·, -т·ц·а́, (i) -т·ц·о́м, (m) (anat) coccyx, urostyle
хво́ст·ик·, -а, (m) (dim) of **хвост·;** tab (of a vacuum tube cathode); (text) lower butt (of latch needle)
хво́ст·ник·, -а, (m) (bot) marestail, Hippurus
хвосто·ви́д·н·ый, -ая, -ое, (a) tail-like, caudiform
хвост·ови́к·, -а́, (m) dovetail; (mech eng) shaft, shank, small stem; tang (of a tool); liner (of oil well)

хвост·ов·о́й, -а́я, -о́е,   (*a*) tail, (zool) caudal; (a/c) tail, rear

~ **антéнн·а**   (*f*) rear-looking antenna/aerial

~ **колес·о́**   (*n*) (a/c) tail wheel

~ **о·пер·éни·е**   (*n*) (a/c) tail unit, empennage

хвосто·ко́л·, -а,   (*m*) (fish) stingray

хвосто·хран·и́лищ·е, -а,   (*n*) (min) tailings dump

хвощ·, -а́,   (*i*) -о́м, (*m*) (bot) horsetail, *Equisetum*

хвоще·ви́д·н·ые, -ых,   (*pl decl as adj*) (bot) *Equisetineae*

хвощ·о́в·ый, -ая, -ое,   (*a*) (bot) equisetic, (*pl as noun*) *Equisetineae*

хво́·я, -и,   (*f*) (bot) coniferous needle(s)

хевé·я, -и,   (*f*) (bot) rubber tree, *Hevea brasiliensis*

Хéвисайд·а, сло́·й   (*m*) (phys) Heaviside layer

хéдер·, -а,   (*m*) header (cutter of combine harvester)

хейл·и́т·, -а,   (*m*) (med) cheilitis, chilitis

хейл·о́з·, -а,   (*m*) (med) cheilosis, chilosis

хейло·пла́стик·а, -и,   (*f*) (med) cheiloplasty, chiloplasty

хейло·том·и́·я, -и,   (*f*) (med) cheilotomy, chilotomy

хейро-   (*int component*) = **хиро-**

хейро·стро́б·ов·ые, -ых,   (*pl decl as adj*) (pal) cheirostrobils, *Cheirostrobila*

хейро·ти́п·, -а,   (*m*) (biol) chirotype

хéл·а, -ы,   (*f*) (zool) chela, (*pl*) chelae

хела́т·н·ый, -ая, -ое,   (*a*) (chem) chelate, chelated

хелато·метр·и́·я, -и,   (*f*) (pal) chelatometry

хели́н·, -а,   (*m*) (pharm) chelin

хели·пéд·а, -ы,   (*f*) (zool) cheliped

хели·цéр·а, -ы,   (*f*) (zool) chelicera, chelicer, (*pl*) chelicerae, chelicers, cheliceres

хели·цéр·ов·ые, -ых,   (*pl decl as adj*) (zool) chelicerates, *Chelicerata*

хеми·люмин·есцéнци·я, -и,   (*f*) (chem, phys) chemiluminescence

хеми·люмино·фо́р·, -а,   (*m*) (chem, phys) chemiluminophor

хеми·со́рбци·я, -и,   (*f*) (chem) chemisorption

хемо·гéн·н·ый, -ая, -ое,   (*a*) (geol) chemogenic, chemogenous

хемо́з·, -а,   (*m*) (med) chemosis

хемо·кинéзис·, -а,   (*m*) (biol) chemokinesis

хемо·на́ст·и·я, -и,   (*f*) (bot) chemonasty

хемо·па́уз·а, -ы,   (*f*) (geophys) chemopause

хемо·рецéптор·, -а,   (*m*) (zool) chemoreceptor

хемо·си́нтез·, -а,   (*m*) (bot) chemosynthesis

хемо·со́рбци·я, -и,   (*f*) (chem) chemisorption

хемо·сфéр·а, -ы,   (*f*) (geophys) chemosphere

хемо·та́ксис·, -а,   (*m*) (biol) chemotaxis

хемо·терап·и́·я, -и,   (*f*) (med) chemotherapy

хемо·троп·и́зм·, -а,   (*m*) (bot) chemotropism

хеномелес·, -а,   (*m*) (bot) dwarf quince, *Chaenomeles*

хéрес·, -а,   (*m*) sherry (rare), sherry-type wine

хéскер·, -а,   (*m*) (agr) husker

~ **-пи́ккер·, -а,**   (*m*) husker-picker

~ **-шрéдер·, -а,**   (*m*) husker-shredder

хéт·а, -ы,   (*f*) (zool) chaeta, seta

хето·гна́т·ы, -ов,   (*m pl*) (zool) chaetognaths, *Chaetognatha*

хи chi, χ (Greek)

хиа́зм·а, -ы,   (*f*) (biol) chiasma

хиазм·о́д·ы   (*pl*) (fish) *Chiasmodontidae*

хиазмо·ли́т·, -а,   (*m*) chiasmolith

хиасто·ли́т·, -а,   (*m*) (min) chiastolite

хиасто·нéвр·и·я, -и,   (*f*) (zool) chiastoneury

хи́жин·а, -ы,   (*f*) hut, hovel

хил·еми́·я, -и,   (*f*) (med) chylaemia, chylemia

хил·и́ди·й, -я,   (*m*) (zool) chilidium

хили·фик·а́ци·я, -и,   (*f*) (physiol) chylification

хило·микро́н·, -а,   (*m*) (zool) chylomicron

хил·ури́·я, -и,   (*f*) (med) chyluria

хи́лус·, -а,   (*m*) (zool) chyle

Хил-Шоу, на·со́с·   (*m*) Hele-Shaw variable delivery pump

хи́л·ый, -ая, -ое,   (*a*) feeble, puny

Хи́льгер·а, при·бо́р·   (*m*) Hilger apparatus

хим-   (*component*) (*abbr*) of **хим·и́ческ·ий** = chemical; chym-

хим·агéнт·, -а,   (*m*) chemical agent

хим·а́з·а, -ы,   (*f*) (biochem) chymase

химéр·а, -ы,   (*f*) chimaera, chimera; (fish) (*pl*) *Chimaeriformes*

химе́р·а

~, **при·ви́в·очн·ая**　plant chimaera, chimaera

химиз·а́ци·я, -и,　(*f*) adoption of chemical methods/agents/media

~ **земле·де́л·и·я**　chemurgy

~ **про·из·во́д·ств·а**　introduction of/conversion to chemical processes

хим·и́зм·, -а,　(*m*) chemism

хи́м·ик·, -а,　(*m*) chemist; chemical industry worker; (mil) chemical warfare officer

~ **-дози·метр·и́ст·, -а,**　(*m*) (nucl) radiac monitor

химика́ли·и, -й,　(*f pl*) (com) chemicals

химика́т·ы, -ов,　(*m pl*) (com) chemicals

химио-　(*int component*) = **хемо-** (q.v.)

химио·профила́кт·ик·а, -и,　(*f*) (med) chemoprophylaxis

химио·терап·и́·я, -и,　(*f*) (med) chemotherapy

химио·троп·и́зм·, -а,　(biol) chemotropism

хим·и́ческ·и　(*adv*) chemically

~ **сто́й·к·ий**　(*a*) corrosion-resistant, stainless (of steels)

хим·и́ческ·ий, -ая, -ое,　(*a*) chemical; (mil) chemical warfare, CW, gas

хи́м·и·я, -и,　(*f*) chemistry

~, **про·стра́н·ственн·ая**　stereochemistry

хим·комбина́т·, -а,　(*m*) chemical combine

химоз·и́н·, -а,　(*m*) (biochem) chymosin, rennin

хим·лес·хо́з·, -а,　(*m*) wood chemistry works/plant

химо·трипси́н·, -а,　(*m*) (biochem) chymotrypsin

химо·трипсино·ге́н·, -а,　(*m*) (biochem) chymotrypsinogen

хим·сыр·ь·ё, -я́,　(*n*) chemical raw materials

хи́м·ус·, -а,　(*m*) (zool) chyme

хин-　(*int component*) (chem) quin-

хин·а, -ы,　(*f*) = **хини́н·**

хин·акри́н·, -а,　(*m*) (pharm) quinacrine

хин·ализари́н·, -а,　(*m*) quinalizanine, 1,2,5,8 tetra-oxy-anthraquinone (dyestuff)

хин·альди́н·, -а,　(*m*) (chem) quinaldine

хин·альди́н·ов·ая кисл·от·а́　(*f*) (chem) quinaldinic acid

хин·ами́н·, -а,　(*m*) (pharm) quinamine

хинафто́л·, -а,　(*m*) (pharm) quinaphthol

хин·гидро́н·, -а,　(*m*) (pharm) quinhydrone

хин·иди́н·, -а,　(*m*) (pharm) quinidine

хинизари́н·, -а,　(*m*) (chem) quinizarine, 1,4 dihydroxyanthraquinone

хин·и́н·, -а,　(*m*) (pharm) quinine

хи́н·н·ый, -ая, -ое,　(*a*) (chem) quinic; quinine, cinchona

~ **де́рев·о**　(*n*) cinchona tree

~ **кисл·от·а́**　(*f*) (chem) quinic acid

~ **кор·а́**　(*f*) (bot) cinchona bark

хинозо́л·, -а,　(*m*) quinosol, 8-oxyquinoline sulphate

хино́ид·, -а,　(*m*) (chem) quinoid

хино́ид·н·ый, -ая, -ое,　(*a*) quinoid

хиноксали́н·, -а,　(*m*) (chem) quinoxaline

хиноли́н·, -а,　(*m*) (chem) quinoline, benzopyridine

хиноли́н·ов·ый жёлт·ый　quinoline yellow

хин·о́н·, -а,　(*m*) (chem) quinone, benzoquinone

хино·токси́н·, -а,　(*m*) (chem) quinotoxin

хио·ли́т·, -а,　(*m*) (min) chiolite

хионо·до́кс·а, -ы,　(*f*) (bot) glory-of-the-snow, chionodoxa

хиро-　(*int component*) = **хейро** (q.v.)

хиро·ти́п·, -а,　(*m*) (biol) chirotype

хиру́рг·, -а,　(*m*) surgeon

хирург·и́ческ·ий, -ая, -ое,　(*a*) surgical

хирург·и́·я, -и,　(*f*) surgery

хит·и́н·, -а,　(*m*) (biol) chitin

хит·ина́з·а, -ы,　(*f*) (biochem) chitinase

хит·и́н·ов·ый, -ая, -ое,　(*a*) (biol) chitinous

хитозами́н·, -а,　(*m*) (chem) chitosamine

хитр·и́ди·й, -я,　(*m*) (bot) chytridium

хитр·и́ть, -я́т,　(*imp*) be/use cunning

хи́тр·ост·ь, -и,　(*f*) cunningness, artfulness; trick

хи́тр·ый, -ая, -ое,　(*a*) cunning, artful

хи́щ·ник·, -а,　(*m*) predator, beast/bird of prey

хи́щ·ническ·ий, -ая, -ое,　(*a*) rapacious, predatory

хи́щ·н·ый, -ая, -ое,　(*a*) predatory; (*pl as noun*) *Carnivora*

хлад·аге́нт·, -а,　(*m*) refrigerant, refrigerating medium, cryogen; (nucl) cooling medium, coolant

хладни́т·, -а,　(*m*) (min) chladnite

хладно·кро́в·н·ый, -ая, -ое,　(*a*) calm, composed (of behaviour); cold-blooded

хладно·ло́м·к·ий, -ая, -ое,　(*a*) (met) cold-short

хладно·ло́м·кост·ь, -и, (*f*) (met) cold shortness

хла́д·н·ый, -ая, -ое, (*a*) *see* **холо́д·н·ый**

хладо·бо́й·н·я, -и, (*g pl*) **-о́·ен·,** (*f*) small slaughterhouse with cold storage

хладо·сто́й·к·ий, -ая, -ое, (*a*) cold-resistant, antifreeze

хладо·тек·у́чест·ь, -и, (*f*) cold flow

хламидо·мона́д·, -а, (*m*) (microbiol) chlamydomonad

хламидо·спо́р·а, -ы, (*f*) (bot) chlamydospore

хлеб·, -а, (*m*) bread; loaf; **хлеб·, -а,** (*nom pl*) **-а́,** (*g pl*) **-о́в,** (*m*) (agr) grain, corn, cereals, corn/grain crop

~, о·зи́м·ый (agr) winter cereals

~, пеклева́н·н·ый wholemeal bread

хлеб·ец·, -б·ц·а, (*m*) small loaf

хле́б·ин·а, -ы, (*f*) flower dust (beekeeping)

хле́б·н·ый, -ая, -ое, (*a*) bread; baker's; grain, corn

~ де́рев·о (*n*) (bot) breadfruit, breadfruit tree, *Artocarpus altibus*

~ зла́к·и (*pl*) bread grain

~ магази́н· (*m*) baker's shop

~ ры́н·ок· (*m*) grain market

хлебо·за·во́д·, -а, (*m*) (food) mechanical bakery

хлебо·за·гото́в·к·а, -и, (*g pl*) **-в·ок·,** (*f*) grain delivery/collection

хлебо·пек·а́рн·ый, -ая, -ое, (*a*) (food) baking, baker's

хлебо·пек·а́рн·я, -и, (*g pl*) **-р·ен·,** (*f*) (food) bakery

хлебо·у·бо́р·к·а, -и, (*g pl*) **-р·ок·,** (*f*) grain harvest

хле́в·, -а, (*nom pl*) **-а́,** (*g pl*) **-о́в,** (*m*) (agr) shed, sty

хлест·а́ни·е, -я, (*n*) (*v n*) of **хлест·а́·ть**

хлест·а́ть, (*pres 3rd pl*) **хле́щ·ут,** (*imp*) lash, whip

хлест·н·у́ть, -у́т, (*perf*) lash, whip, give a lash

хле́щ·ет, (*pres 3rd sing*) of **хлест·а́ть**

хло·анти́т·, -а, (*m*) (min) chloanthite

хло́п·, -а, (*m*) clap, crack, bang, thump, pop, flap, patter; explosion (in flue etc.)

хло́п·ань·е, -я, (*n*) (*v n*) *see* **хло́п·а·ть**; (rad) motorboating, popping

хло́п·а·ть, -ют, (*imp*) bang, clap, thump, crack, pop, flap, patter

хлопини́т·, -а, (*m*) (min) khlopinite

хлопко·во́д·, -а, (*m*) cotton grower

хлопко·во́д·ств·о, -а, (*n*) cotton growing/culture

хло́п·ков·ый, -ая, -ое, (*a*) cotton

~ де́рев·о (*n*) cottonwood

~ ма́сл·о (*n*) cottonseed oil

~ пух· (*m*) (text) linter, cotton linter

хлопко·за·во́д·, -а, (*m*) (text) ginnery

хлопко·за·гото́в·и́тельн·ый, -ая, -ое, (*a*) cotton-collecting

хлопко·о·чист·и́тельн·ый, -ая, -ое, (*a*) (text) gin, ginning

~ маши́н·а (*f*) (text) gin

хлопко·о·чи́ст·к·а, -и, (*f*) ginning (of cotton)

хлопко·пряд·е́ни·е, -я, (*n*) (text) cotton spinning

~, у·га́р·н·ое cotton waste spinning/condensing

хлопко·ро́б·, -а, (*m*) cotton grower

хлопко·у·бо́р·к·а, -и, (*f*) cotton picking

хлопко·у·бо́р·очн·ая маши́н·а (*f*) cotton picker

хло́п·н·уть, -ут, (*perf*) give a bang/clap/thump/crack/pop/flap; go bang/pop

хло́п·ок·, -п·к·а, (*m*) cotton; **хлоп·о́к·, -п·к·а́,** (*m*) clap, crack, bang, thump, pop, flap, patter; explosion (in flue etc.)

~ -волокн·о́, -а́, (*n*) (text) lint

~, звук·ов·о́й (dynam) sonic boom/bang

~, сверх·звук·ов·о́й (dynam) supersonic boom/bang

~ -сыр·е́ц·, -р·ц·а́, (*m*) raw/seed cotton

хлопот·а́ть, (*pres 3rd pl*) **хлопо́ч·ут,** (*imp*) fuss/bustle about/around

хлопо́ч·ет, (*pres 3rd sing*) of **хлопо·т·а́ть**

хлоп·у́шк·а, -и, (*g pl*) **-шек·,** (*f*) fly-swat; (mech) flap; (bot) bladder catch-fly, *Silene venosa*

хлоп·ча́тник·, -а, (*m*) (bot) cotton, cotton plant, *Gossypium*

хлопчато·бума́ж·н·ый, -ая, -ое, (*a*) cotton

хлоп·ча́т·ый, -ая, -ое, (*a*) cotton; flakey, flocculent

хлопье·ви́д·н·ый, -ая, -ое, (*a*) flaky, flocculent, floccular

хлопье·образ·ова́ни·е, -я, (*n*) flocculation

хлоп·ь·я, -ев, (*m pl*) flakes, flocs; flocculate polymers

хлор·, -а, (*m*) (chem) chlorine, Cl; (*as component*) chlor-, chloro-

хлораго·ге́н·н·ая кле́т·к·а (*f*) (zool) chloragen/chloragogen cell
хлоразѝд·, -а, (*m*) chlorine azide
хлоразѝн·, -а, (*m*) (pharm) chlorazine
хлора́л·, -а, (*m*) chloral, trichloracetaldehyde
хлора́л·ь, -и, (*f*) (chem) chloral, trichloracetaldehyde
хлор·альгидра́т·, -а, (*m*) (pharm) chloral hydrate
хлор·алюминѝт·, -а, (*m*) (min) chloraluminite
хлор·амѝд·, -а, (*m*) (chem) chloramide, chloral amide
хлор·амѝн·, -а, (*m*) (chem) chloramine
хлорамфенико́л·, -а, (*m*) (pharm) chloramphenicol
хлорангидрѝд·, -а, (*m*) acid chloride
 -изофта́л·ев·ой кисл·от·ы́ isophthalyldichloride
~ **каприл·ов·ой кисл·от·ы́** capryl chloride
~ **карбамѝн·ов·ой кисл·от·ы́** carbamyl chloride
 -муравь·ѝн·ой кисл·от·ы́ formyl chloride
хлоранѝл·, -а, (*m*) (chem) chloranil, tetrachloro-quinone
хлор·апатѝт·, -а, (*m*) (min) chlorapatite
хлор·аргирѝт·, -а, (*m*) (min) chlorargyrite, cerargyrite
хлор·астролѝт·, -а, (*m*) (min) chlorastrolite
хлор·а́т·, -а, (*m*) (chem) chlorate
~ **ка́ли·я** potassium chlorate
хлор·а́тор·, -а, (*m*) chlorinator
хлораура́т·, -а, (*m*) (chem) chloroaurate
хлорацето́н·, -а, (*m*) (chem) chloroacetone
хлорацетофено́н·, -а, (*m*) (chem) chloroacetophenone, phenacyl chloride
хлорацѝд·, -а, (*m*) chloratsid (a disinfectant)
хлор·бензо́л·, -а, (*m*) (chem) chlorobenzene
хлор·бутано́л·, -а, (*m*) chlorobutanol
хлор·винѝл·, -а, (*m*) vinyl chloride, chlorovinyl
хлор·винил·дихлор·арсѝн·, -а, (*m*) chlorovinyldichlorarsine, lewisite
хлор·водо·ро́д·, -а, (*m*) hydrogen chloride
хлор·гидрѝн·, -а, (*m*) chlorohydrin
хлорда́н·, -а, (*m*) chlordane (insecticide)
хлоре́кс·, -а, (*m*) chlorex, dichlorethyl ether

хлор·е́лл·а, -ы, (*f*) (bot) chlorella, *Chlorella*
хлор·енхѝм·а, -ы, (*f*) (bot) chlorenchyma
хлорефидро́з·, -а, (*m*) (med) chlorephidrosis
хлор·за·мещ·ённ·ый, -ая, -ое, (*a*) (chem) chloro
~ **кисл·от·а́** (*f*) chloroacid, chloracid
~ **эфѝр·** (*m*) chloro ester, chloro ether
хлор·ѝд·, -а, (*m*) (chem) chloride
~ **на́три·я** sodium chloride
хлорѝд·н·ый, -ая, -ое, (*a*) of **хлорѝд·**
~ **спо́соб·** (*m*) (met) chloridizing roasting
хлор·ѝровани·е, -я, (*n*) chlorination, chlorinating; (met) chloridizing roasting
хлор·ѝр·овать, -уют, (*imp and perf*) chlorinate
хлористо·водо·ро́д·н·ый, -ая, -ое, (*a*) (chem) muriatic, hydrochloric, hydrochloride of
~ **анилѝн·** (*m*) aniline hydrochloride
~ **кисл·от·а́** (*f*) hydrochloric/muriatic acid
хлористо·кѝсл·ый, -ая, -ое, (*a*) chlorite (of)
~ **ка́ли·й** (*m*) potassium chlorite
хло́р·ист·ый, -ая, -ое, (*a*) chlorous, chloride (of)
~ **азо́т·** (*m*) nitrous chloride
~ **бензѝл·** (*m*) benzyl chloride
~ **винѝл·** (*m*) vinyl chloride
~ **водо·ро́д·** (*m*) hydrogen chloride
~ **желе́з·о** (*n*) iron chloride (ferrous or ferric)
~ **ка́льци·й** (*m*) calcium chloride
~ **кисл·от·а́** (*f*) (chem) chlorous acid
~ **свин·е́ц·** (*m*) lead chloride
~ **серебр·о́** (*n*) silver chloride
~ **титáн·** (*m*) titanium tetrachloride
хлорѝт·, -а, (*m*) (min) chlorite; (chem) chlorite
хлоритиз·а́ци·я, -и, (*f*) (geol) chloritization
хлорѝт·ов·ый, -ая, -ое, (*a*) chloritic
хлоритóид·, -а, (*m*) (min) chloritoid
хлор·кальцѝт·, -а, (*m*) (min) chlorocalcite, hydrophilite
хлор·каучу́к·, -а, (*m*) chlorinated rubber
хлорманганокалѝт·, -а, (*m*) (min) chlormanganokalite
хлор·метѝл·ов·ый, -ая, -ое, (*a*) (chem) chloromethyl

хлор·моч·евѝн·а, -ы,   (*f*) (chem) chlorourea

хлор·муравь·ѝн·ый эфѝр· (*m*) chloro-formic ester, chloro-formate

хлор·нова́тист·ый, -ая, -ое,   (*a*) hypochlorous; hypochlorite of

~ **ка́льци·й** (*m*) calcium hypochlorite

~ **кисл·от·а́** (*f*) (chem) hypochlorous acid

хлорновато·ки́сл·ый, -ая, -ое,   (*a*) chlorate (of)

~ **ка́ли·й** (*m*) potassium chlorate

~ **свин·е́ц·** (*m*) lead chlorate

хлор·нова́т·ый, -ая, -ое, (*a*) (chem) chloric

~ **кисл·от·а́** (*f*) (chem) chloric acid

хлорно·ки́сл·ый, -ая, -ое, (*a*) per-chlorate (of)

~ **ка́ли·й** (*m*) potassium perchlorate

хло́р·ност·ь, -и,   (*f*) (ocean) chloride content

хло́р·н·ый, -ая, -ое,   (*a*) chlorine, chloric, chloride (of)

~ **ангидри́д·** (*m*) perchloric acid anhydride

~ **вод·а́** (*f*) aqueous solution of chlorine

~ **и́звест·ь** (*f*) chloride of lime

~ **кисл·от·а́** (*f*) perchloric acid

хлоро·бакте́ри·и, -й,   (*pl*) *Chloro-bacteriaceae*, green sulphur bacteria

хлоро·бро́м·ист·ый, -ая, -ое, (*a*) chlorobromide

хлоро·водо·ро́д·, -а,   (*m*) hydrogen chloride

хлоро·ге́н·ов·ая кисл·от·а́ (*f*) chloro-genic acid

хлоро·гидри́н·, -а, (*m*) chlorohydrine

хлоро·да́н·, -а, (*m*) chlordan, chlor-dane (insecticide)

хлоро́·з·, -а, (*m*) (med, bot) chlorosis

хлоро·за·мещ·ённ·ый, -ая, -ое, (*a*) chloro

хлоро·кальци́т·, -а, (*m*) (min) chloro-calcite, hydrophilite

хлор·о́·кис·ь, -и, (*f*) (chem) oxychloride

~ **ви́смут·а** bismuth oxychloride

хлоро·круори́н·, -а,   (*m*) (chem) chlorocruorin

хлоро·ксифи́т·, -а, (*m*) (min) chloro-xiphite

хлоро·лейкеми́·я, -и, (*f*) (med) chloro-leukemia

хлоро·мелани́т·, -а, (*m*) (min) chloro-melanite

хлоро́·метр·, -а, (*m*) chlorimeter

хлоро·мицети́н·, -а, (pharm) chloro-mycetin, chloramphenicol

хлор·опа́л·, -а, (*m*) (min) chloropal

хлоро·паллада́т·, -а,   (*m*) (chem) chloropalladate

хлоро·пла́ст·, -а, (*m*) (cyt) chloroplast

хлоро·платина́т·, -а,   (*m*) platini-chloride, chloroplatinate

~ **на́три·я** sodium platinichloride

~ **ка́ли·я** potassium chloroplatinate

хлоро·платини́т·, -а,   (*m*) platino-chloride

хлоро·пре́н·, -а, (*m*) chloroprene

хлоро·пре́н·ов·ый каучу́к· (*m*) chloro-prene rubber

хлоро·орган·и́ческ·ий, -ая, -ое, (*a*) (chem) organochlorine

хлоро·рас·твор·и́тел·ь, -я, (*m*) chloro-inated solvent

хлоро·сере́бр·ян·ый, -ая, -ое,   (*a*) (phot) silver chloride

хлоро·станна́т·, -а, (*m*) chlorostannate

хлоро·тиони́т·, -а, (*m*) (min) chloro-thionite

хлор·оти́ческ·ий, -ая, -ое, (*a*) (biol) chlorotic

хлоро·тори́т·, -а, (*m*) chlorothorite

хлоро·фа́н·, -а, (*m*) (min) chlorophane

хлоро·феници́т·, -а, (*m*) (min) chloro-phoenicite

хлоро·фи́лл·, -а, (*m*) (biochem) chloro-phyll

хлоро·филла́з·а, -ы, (*f*) (bot) chloro-phyllase

хлоро·филли́т·, -а, (*m*) (min) chloro-phyllite

хлоро·фи́лл·ов·ое зерн·о́ (*n*) (bot) chlorophyll corpuscle

хлоро·фо́р·, -а, (*m*) (bot) chlorophore

хлоро·фо́рм·, -а, (*m*) (chem) chloroform

хлор·пикри́н·, -а, (*m*) (chem) chloro-picrin

хлор·про·из·во́д·н·ый полиме́р· (*m*) chlorine-derivate polymer

хлорпромази́н·, -а, (*m*) (pharm) chlor-promazine

хлор·со·держ·а́щ·ий, -ая, -ее, (*a*) chlorine-containing

хлор·стиро́л·, -а, (*m*) chlorostyrene

хлор·сульфо́н·ов·ая кисл·от·а́ (*f*) chlorosulfonic acid

хлор·тетра·цикли́н·, -а, (*m*) (biochem) chlortetracycline

хлор·толуо́л·, -а, (*m*) (chem) chloro-toluene

хлор·у́голь·н·ый эфѝр· (*m*) chloro-carbon ester, chloro-formate

хлор·у́ксус·н·ый, -ая, -ое, (*a*) (chem) chloroacetic, chloro-acetic

хлор·у́ксус·н·ый, -ая, -ое
~ кисл·от·а́ (*f*) monochloroacetic acid
хлы́·н·уть, -ут, (*perf*) gush, rush, flood
хлыст·, -а́, (*m*) whip; felled tree, full-length log
хмеле·во́д·ств·о, -а, (*n*) (agr) hop growing
хмел·ев·о́й, -а́я, -о́е, (*a*) (agr) hop
хмеле·гра́б·, -а, (*m*) (bot) hophornbeam, *Ostrya*
хмел·ёк·, -ль·к·а́, (*m*) (bot) hop clover, *Trifolium*
хмел·ь, -я, (*m*) (bot) hop, *Humulus*
хмель·ник·, -а, (*m*) hop field
хмель·н·о́й, -а́я, -о́е, (*a*) intoxicating, stupefacient; intoxicated
хму́р·ить, -ят, (*imp*) frown
хму́р·ый, -ая, -ое, (*a*) gloomy
хн·а, -ы, (*f*) henna (reddish-yellow colour)
хоа́н·а, -ы, (*f*) (anat) choana, (*pl*) choanae
хоано·ци́т·, -а, (*m*) (zool) choanocyte
хо́бот·, -а, (*m*) (zool) trunk, proboscis; (mech) projection, nose, snout; trail (of gun carriage)
хо́бот·н·ые, -ых, (*pl decl as adj*) (pal, zool) *Proboscidea*
хобот·о́к·, -т·к·а́, (*m*) (ent) proboscis
ход·, -а, (*nom pl*) **-ы,** *or* **ы́,** (*g pl*) **-ов** *or* **о́в,** (*m*) motion, travel, run, running, stroke; course, progress; speed, pace, rate; shape, trend, variation, dependence (of graphs); (astron, meteor, geol) change, variation, rate; (horol) escapement; (surv) traverse, line; undercarriage, runners; move (chess), play, go (of games); (anat) duct, passage, meatus, use, demand, circulation, fashion; access
в больш·о́м ход·у́ in great demand, in fashion/vogue
в хо́д·е in the course of, during
в ход·у́ at work, in operation
~ в·верх· (mech) upstroke
~, век·ов·о́й secular change/variation
~ вз·ры́в·а (ICE) expansion/explosion/firing stroke
~ в·низ· (mech) downstroke
~, воз·вра́т·н·ый (mech) return stroke, recoil
~ в·пуск·а (ICE) admission stroke
~, времен·н·о́й time dependence (of a curve)
~ вы́·пуск·а (ICE) exhaust stroke
~, год·ов·о́й (astron) annual change

ход
~, да·ть start up/off; (M/T) let in the clutch, start moving; give an opportunity
~, да·ть за́д·н·ий go into reverse; (naut) go astern, put the engines astern
~ дел· course of events; rate things are going
~, дипломат·и́ческ·ий diplomatic move
~, жёлч·н·ый (anat) biliary duct
~, жёстк·ий (ICE) rough running
~, за́д·н·ий backwards motion; backing, reversing; (naut) sternway
~, контро́ль·н·ый (surv) checking traverse
~, ли́т·ников·ый (met, cast) runner, fireclay runner
~, ма́л·ый slow speed/motion
~, мёртв·ый backlash, play (in gears etc.)
на ход·у́ (*adv*) in motion, while in operation, while working/running, in the course of; (naut) under way
~ на·полн·е́ни·я prefill stroke (of power press)
~, нивели́р·н·ый (surv) levelling line
~, обра́т·н·ый return/back stroke; fly-back, return trace (of a beam on a screen, e.g. TV); reverse motion
~, пара́д·н·ый (arch) front door/entrance
~, перёд·н·ий forwards travel/motion; (naut) headway, ahead travel, speed ahead
~, полигоно·метр·и́ческ·ий (surv) traverse
~, по́лн·ый full speed, full strength, all out, full blast
~, по́лн·ый перёд·н·ий full speed ahead
~ по́рш·н·я piston stroke
~, пуст·ь в start up, set in motion, get started
~, рабо́ч·ий power stroke (of engine); forward/cutting stroke, forward motion (of machine tools)
~, рези́н·ов·ый rubber tires; **на рези́н·ов·ом ход·у́** rubber-tired
~ ры́б·ы run of fish
~, со́·мк·нут·ый (surv) closed traverse
~ со·общ·е́ни·я (mil) communication trench
~, сре́д·н·ий half speed

ход
~, су́т·очн·ый (horol) daily rate; diurnal variation (of magnetic compass etc.); (met) daily range; (astron) daily change/variation
~ температу́р·ы temperature curve/variation/dependence
~, ти́х·ий slow speed
~, холост·о́й idle running, idling; return stroke
~, цили́ндр·ов·ый (horol) cylinder escapement
~, чёрн·ый (arch) back door/entrance
ходата́й·ств·овать, -ую́т, (imp) solicit, petition
хо́д·ик·и, -ов, (m pl) (horol) going-train
ход·и́льн·ый, -ая, -ое, (a) (zool) ambulatory
ход·и́мост·ь, -и, (f) running life, life (of tires etc.)
ход·и́ть, -я́т, (imp indet) go, come; walk; (mech etc.) run, work, operate, move, travel; attend, visit; wear (garments etc.); tend, look after
хо́д·к·ий, -ая, -ое, (a) saleable, salable, marketable; fast (for given power)
хо́д·кост·ь, -и, (f) performance coefficient, speed/power ratio (for a vehicle, ship etc.)
ход·ов·о́й, -а́я, -о́е, (a) moving, running, travelling, (naut) steaming, underway, (air) flying, flight; (fish) running, (zool) migrating; (com) popular
~ ис·пыт·а́ни·е (n) (shipb) sea trial(s)
~ колес·о́ (n) (horol) escape wheel
~ кон·е́ц· (m) running part (of rope), hauling part (of pulley block), outboard end (of anchor cable)
~ ско́р·ост·ь (f) travelling speed, rate of travel
~ су́д·н·о (n) boat moving upstream (in rivers)
~ част·ь (f) (mech) working/moving part; (M/T) running gear
ход·о́к·, -а́, (m) walker; (min) footway, manway
хо́д·ом (adv) quickly, at a good pace
ходо·ме́р·, -а, (m) (mech) worm/worm-gear/worm-thread gauge
ход·у́лочник·, -а, (m) (orn) stilt-bird, *Himantopus*
ход·у́льн·ый, -ая, -ое, (a) stilt
ход·у́л·я, -и, (g pl) **-ей,** (f) stilt
ход·ьб·а́, -ы́, (f) walking
ход·я́ч·ий, -ая, -ее, (a) walking; ambulant; widespread, widely used, common, popular

хожд·е́ни·е, -я, (n) walking
хож·у́, (pres 1st sing) of **ход·и́ть**
хоз- (component) (abbr) of **хозя́й·ст·венн·ый** economic, business
хоз·рас·чёт·, -а, (m) (econ) business accountability, non-subsidized/self-supporting operation; cost accounting
хоз·рас·чёт·н·ый, -ая, -ое, (a) self-financing/supporting
хоз·спо́соб·, -а, (m) own resources
хозя́·ин·, -а, (nom pl) **-я́ев·а,** (g pl) **-я́·ев,** (m) owner, master, boss; host
хозя́й·нич·а·ть, -ют, (imp) run (a business etc.), keep house; be bossy
хозя́й·ственник·, -а, (m) manager
хозя́й·ственност·ь, -и, (f) ability to run a business
хозя́й·ственн·ый, -ая, -ое, (a) of **хозя́й·ств·о**; economic, business; household, domestic
~ год· (m) fiscal/business/working year
~ с·то́ч·н·ые во́д·ы domestic sewage
~ цех· (m) domestic services department (shipyard)
хозя́й·ств·о, -а, (n) economy; business, undertaking, organization, setup, show, party, industry; husbandry, farming, farm
~, во́д·н·ое utilization and control of water
~, инструмента́ль·н·ое tool organization (of a works)
~, коммуна́ль·н·ое public service/services
~, компре́ссор·н·ое compressed air service
~, лес·н·о́е forestry
~, на·ро́д·н·ое national economy
~, ры́б·н·ое fishing, fishing industry
~, с·клад·ск·о́е storekeeping
хокке́·й, -я, (m) hockey (sport)
хол·ангио́м·а, -ы, (f) (med) cholangioma
хол·анги́т·, -а, (m) (med) cholangitis
холдени́т·, -а, (m) (min) holdenite
хо́л·ев·ая кисл·от·а́ (f) (chem) cholic acid
холево·ки́сл·ый, -ая, -ое, (a) (chem) cholate of
холейн·ов·ая кисл·от·а́ (f) (biol) choleic acid
холе́р·а, -ы, (f) (med) cholera
холе·стеат·о́м·а, -ы, (f) (med) cholesteatoma
холе·стери́н·, -а, (m) (biochem) cholesterol

холе·стерин·емѝ·я, -и, (f) (med) cholesteraemia, cholesteremia, cholesterolaemia, cholesterolemia

холе·стерѝн·ов·ый, -ая, -ое, (a) (biochem) cholesteric

холе·цистѝт·, -а, (m) (med) cholecystitis

холе·цисто·граф·ѝ·я, -и, (f) (med) cholecystography

холе·цистомѝ·я, -и, (f) (med) cholecystostomy

холе·цистектом·ѝ·я, -и, (f) (med) cholecystectomy

хол·ѝн·, -а, (m) (chem) choline

холин·ерг·ѝческ·ий, -ая, -ое, (a) (anat) cholinergic

холѝн·ов·ый, -ая, -ое, (a) (chem) cholinic

хо́лл·, -а, (m) hall; vestibule, entrance hall

Хо́лл·а, по·двѝж·ност·ь (f) (phys) Hall mobility

~, по·сто·я́нн·ая (f) (phys) Hall constant

~, явл·е́ни·е (n) (elec) Hall effect

холм·, -а́, (m) hill

хо́лм·ик·, -а, (m) hillock, knoll

холм·ѝст·ый, -ая, -ое, (a) hilly

холмквистѝт·, -а, (m) (min) holmquistite

холмо·го́р·ь·е, -я, (g pl) -р·и·й, (n) hilly country

холо·ба́зи·я, -и, (f) (bot) cholobasy

хо́лод·, -а, (nom pl) -а́, (g pl) -о́в, (m) cold, coldness, chill

  волн·а́ хо́лод·а (f) (meteor) cold wave
  гра́дус·хо́лод·а (m) degree below zero
  на холод·у́ (adv) cold, in the cold
  ша́п·к·а хо́лод·а (f) (meteor) cold cap

холод·е́·ть, -ют, (imp) become/grow cold

холод·ѝльник·, -а, (m) (see also кон·денс·а́тор·) (mech) cooler, condenser; refrigerator, cold-storage plant; (met roll) cooling/hot bed, cooling rack; (rocket) heat exchanger, cooler

~, втор·ѝчн·ый (a/c) aftercooler

~, двух·сторо́н·н·ий (met) double-sided cooling bed

~, квартѝр·н·ый domestic refrigerator

~ ма́сл·а lubricating oil cooler

~, одно·сторо́н·н·ий (met) single-sided cooling bed

~, про·меж·у́точн·ый intercooler (of compressor)

~, ре́·ечн·ый (met) carryover-type cooling bed

холод·ѝльник

~, ша́р·иков·ый Allihn condenser

холод·ѝльн·ый, -ая, -ое, (a) cooling; refrigerating, freezing

~ аге́нт· (m) refrigerant, cryogen; coolant

~ агрега́т· (m) refrigeration/refrigerating unit (of refrigerator)

~ коэффицие́нт· (m) coefficient of performance (of a refrigerating machine)

~ машѝн·а (f) refrigerating/refrigeration machine, refrigerator

~ машѝн·а, абсорбцио́н·н·ая absorption refrigerating machine

~ машѝн·а, воз·ду́ш·н·ая cold-air refrigerating machine

~ машѝн·а, компрессио́н·н·ая vapour-compression refrigerating machine

~ тра́нспорт· (m) refrigerated transportation

~ у·стано́в·к·а (f) refrigerating plant

~ шкаф· (m) refrigerator

холод·ѝть, -я́т, (imp) cool, make cold, chill

холодно·вы́·сад·очн·ый, -ая, -ое, (a) (met) cold-heading

холодно·ка́т·ан·ый, -ая, -ое, (a) (met) cold-rolled

холодно·кро́в·н·ый, -ая, -ое, (a) (zool) cold-blooded, poikilothermal; (pl as noun) (zool) poikilotherms

холодно·ло́м·к·ий, -ая, -ое, (a) (met) cold-short

холодно·ло́м·кост·ь, -и, (f) (met) cold shortness

холодно·люб·ѝв·ый, -ая, -ое, (a) (biol) psychrophilic

холодно·об·рабо́т·анн·ый, -ая -ое, (a) cold-worked

холодно·сто́й·к·ий, -ая, -ое, (a) cold/cool-resistant, cold-hardy

холодно·сто́й·кост·ь, -и, (f) cold resistance; (bot) cold hardiness

холодно·тя́·нут·ый, -ая, -ое, (a) (met) bright/cold-drawn

холо́д·н·ый, -ая, -ое, (a) cold, cool

~ про·ка́т·к·а (f) (met) cold rolling/rectifying/reduction

холод·о́к·, -д·к·а́, (m) cold, chill

холодо·про·из·вод·ѝтельност·ь, -и, (f) refrigerating capacity

холож·ён·ый, -ая, -ое, (a) cooled, chilled

холо·лѝт·, -а, (m) (med) chololith

холост·ѝть, -я́т, (imp) castrate, geld

холост·о́й, -а́я, -о́е, (*a*) (tech) idle, free, dead; empty, blank, dummy; (elec) open-circuit; unmarried (persons); (agr) barren, sterile, castrated, gelded

~ **вы́·стрел·** (*m*) blank round/shot

~ **ход·** (*m*) free-idle running, no-load
ис·пыт·а́ни·е на холост·о́м ход·у́ (*n*) no-load test
моме́нт· холост·о́го хо́д·а (*m*) frictional torque
ток· холост·о́го хо́д·а (*m*) (elec) open-circuit current

холост·я́к·, -а́, (*m*) bachelor, unmarried man; (shipb) unused drilled hole

холо·ферме́нт·, -а, (*m*) (physiol) choloferment

холощ·е́ни·е, -я, (*n*) (*v n*) of **холост·и́ть**; castration, gelding

холст·, -а́, (*m*) (text) coarse linen, canvas (artist's), sackcloth; lap (rolled sheet of fibres for carding)

~, **аппрет·и́рованн·ый** buckram

хо́лст·ик·, -а, (*m*) (*dim*) of **холст·**

холст·и́н·а, -ы, (*f*) = **холст·**

холст·и́нк·а, -и, (*g pl*) **-нок·,** (*f*) piece of canvas

холст·о́вщик·, -а, (*m*) (text) card tender

холсто·вы·тяж·н·а́я маши́н·а (*f*) (text) ribbon lap machine

холщ·о́в·ый, -ая, -ое, (*a*) of **холст·**

холя́в·а, -ы, (*f*) cylinder, blown cylinder (glass technology)

хомо·фи́т·, -а, (*m*) (bot) chomophyte

хому́т·, -а́, (*m*) collar, yoke; clamp, clip (for tubes, cable etc.); ferrule (carpentry)

~, **лафе́т·н·ый** (oil) spider

~, **не·по·дви́ж·н·ый** (oil) spider

хому́т·ик·, -а, (*m*) (*dim*) of **хому́т·**

~, **при·це́ль·н·ый** backsight slide (of rifle)

хомя́к·, -а́, (*m*) (zool) hamster, *Cricetus cricetus*, (*pl*) *Cricetinae*

хомяко·обра́з·н·ые, -ых, (*pl decl as adj*) (zool) *Cricetidae*

хо́н·, -а, (*m*) hone (tool)

хо́ндр·а, -ы, (*f*) (geol) chondrule

хондри́лл·а, -ы, (*f*) (bot) skeletonweed, *Chondrilla*

хондрио·зо́м·а, -ы, (*f*) (gen) chondriosome

хондрио·кине́з·, -а, (*m*) (gen) chondriokinesis

хондрио·ко́нт·, -а, (*m*) (gen) chondriocont

хондрио́·м·, -а, (*m*) (cyt) chondriome, chondrioma

хондрио·ми́т·, -а, (*m*) (cyt) chondriomite

хондрио·со́м·а, -ы, (*f*) (gen) chondriosome

хондр·и́т·, -а, (*m*) (geol) chondrite; (med) chondritis

хондрод·и́т·, -а, (*m*) (min) chondrodite

хондрозами́н·, -а, (*m*) (chem) chondrosamine

хондр·о́ид·, -а, (*m*) (vet) chondroid

хондро·ити́н·, -а, (*m*) (biochem) chondroitin

хондр·о́м·а, -ы, (*f*) (med) chondroma

хондро·фо́р·, -а, (*m*) (zool) chondrophore

хондро·ци́т·, -а, (*m*) (cyt) chondrocyte

хонинг·ова́льн·ый, -ая, -ое, (*a*) (mech) honing

~ **голо́в·к·а** (*f*) hone

~ **стан·о́к·** (*m*) honer

хонинг·ова́ни·е, -я, (*n*) honing

~, **жи́дк·остн·ое** liquid honing

хонинг·ова́ть, -у́ют, (*imp*) hone

хоно·ли́т·, -а, (*m*) (geol) chonolith

хо́ппер·, -а, (*m*) (rail) hopper wagon/car

хо́р·, -а, (*m*) choir; chorus; (*pl* arch) choir stall

хорва́т·ск·ий, -ая, -ое, (*a*) Croatian

хо́рд·а, -ы, (*f*) (math, a/c) chord; (anat) chord, chorda, (zool) notochord

~, **кон·цев·а́я** (a/c) tip chord

~, **корн·ев·а́я** (a/c) root chord

~, **ос·ев·а́я** (a/c) centre chord

хорда́ль·н·ый, -ая, -ое, (*a*) chordal

~ **выс·от·а́ голо́в·к·и зу́б·а** (*f*) chordal addendum

хо́рд·ов·ые, -ых, (*pl decl as adj*) (zool) chordates, *Chordata*

хо́рд·ов·ый, -ая, -ое, (*a*) (*see also* **хо́рд·ов·ые**) (*as noun*); chordal

хор·ёк·, -рь·к·а́, (*m*) (zool) polecat, *Putorius*

хорео·гра́ф·и·я, -и, (*f*) (theat) choreography

хоре́·я, -и, (*f*) (med) chorea

хо́рион·, -а, (*m*) (biol) chorion

хорио·ретин·и́т·, -а, (*m*) (med) chorioretinitis

хор·ов·о́й, -а́я, -о́е, (*a*) of **хор·**

хорон·и́ть, -я́т, (*imp*) bury, inter

хоро·пле́т·а, -ы, (*f*) (geol) choropleth

хорош·е́ньк·о (*adv*) properly, soundly

хоро́ш·ий, -ая, -ее, (*a*) good; beautiful, handsome, good-looking

хорош·о́   (*adv*) well; (*neut predic*) of **хоро́ш·ий**

хот·е́вш·ий, -ая, -ее, (*past part act*) of **хот·е́ть**

хот·е́ть,   (*pres 3rd sing, pl*) **хо́ч·ет, хот·я́т,** (*imp*) (+ *gen*) want, wish
ему́ хот·е́лось   he wanted/wished to
ему́ хот·е́лось бы   he would like to
ему́ хот·е́лось пи·ть   he was thirsty

хот·и́м, (*pres 1st pl*) of **хот·е́ть**

хот·и́те, (*pres 2nd pl*) of **хот·е́ть**

хоть (*conj*) = **хотя́**

хотя́ (*conj*) though, although
~ бы   if only, even if
~ бы и так   even if it were so

хот·я́т, (*pres 3rd pl*) of **хот·е́ть**

хот·я́щ·ий, -ая, -ее, (*pres part act*) of **хот·е́ть**

хохл·а́т·ый, -ая, -ое, (*a*) tufted, crested

хохол·, -хл·а́, (*m*) tuft, (orn etc.) crest

хохот·а́ть, (*pres 3rd pl*) **хохо́ч·ут,** (*imp*) laugh, guffaw

хохо́ч·ет, (*pres 3rd sing*) of **хохот·а́ть**

хоху́л·я, -и, (*f*) (zool) Russian desman, *Desman moschata*

хо́ч·ет, (*pres 3rd sing*) of **хот·е́ть**
мне хо́ч·ется   I want/wish to

хра́бр·ост·ь, -и, (*f*) bravery

хра́бр·ый, -ая, -ое, (*a*) brave

хра́м·, -а, (*m*) (arch) temple

хран·е́ни·е, -я, (*n*) (*v n*) of **хран·и́ть**; storage; custody
~, без·гара́ж·н·ое   car park equipped with starting aids
воз·мо́ж·ност·ь   дл·и́тельн·ого хран·е́ни·я   (*f*) long shelf life
~ пере·но́с·а   carry storage (computers)

хран·и́лищ·е, -а, (*i*) **-ем,** (*n*) storehouse, repository, stowage compartment

хран·и́мост·ь, -и, (*f*) shelf life

хран·и́тел·ь, -я, (*m*) keeper, custodian, curator

хран·и́ть, -я́т, (*imp*) keep, preserve, save, store

хра́п·, -а, (*m*) snore, stertor; clip hook, calliper hook

храп·е́ть, -я́т, (*imp*) snore, breathe stertorously

храп·ови́к·, -а́, (*m*) (mech) ratchet, ratchet wheel

храп·ов·о́й, -а́я, -о́е, (*a*) (mech) ratchet

храп·о́к·, -п·к·а́, (*m*) snore (sound); snore piece (of mine pump)

хребе́т·, -бт·а́, (*m*) (geol) ridge, crest, range, (ocean) ridge; (anat) spine, rachis
~, спин·н·о́й   (anat) spinal column

хребе́т·н·ый, -ая, -ое, (*a*) spinal

хребт·о́в·ый, -ая, -ое, (*a*) (geol) ridge, crest, range, (ocean) ridge; (anat) spinal
~ част·ь   (*f*) sirloin (of meat)

хре́н·, -а, (*m*) (bot) horse-radish
~, вод·ян·о́й   water cress, *Nasturtium amphibium*
~, дерев·е́нск·ий   horseradish, *Armoracia rusticana*

хресто·ма́т·и·я, -и, (*f*) (print) reader, selected works/passages, selections

хризали́д·а, -ы, (*f*) (ent) chrysalis

хриз·анте́м·а, -ы, (*f*) (bot) chrysanthemum

хриз·анили́н·, -а, (*m*) (chem) chrysaniline

хризо·бери́лл·, -а, (*m*) (min) chrysoberyl

хриз·оид·и́н·, -а, (*m*) (chem) chrysoidine (dye)

хризо·ко́лл·а, -ы, (*f*) (min) chrysocolla

хризо·ли́т·, -а, (*m*) (min) chrysolite

хризо·мона́д·ы, (*pl*) (microbiol) *Chrysophyceae*

хризо·пра́з·, -а, (*m*) (min) chrysoprase

хризо·ти́л·, -а, (*m*) (min) chrysotile

хризоти́л-асбе́ст·, -а, (*m*) chrysotile asbestos

хризо·фени́н·, -а, (*m*) (chem) chrysophenine

хрип·, -а, (*m*) wheeze, (med) crepitation, rale

хри́пл·ый, -ая, -ое, (*a*) hoarse

хрип·от·а́, -ы́, (*f*) hoarseness

хро́м·, -а, (*m*) (chem) chromium, Cr; box-calf (leather)

хрома́л·ь, -я, (*m*) (met) Chromal, Cr-Al alloy

хроманси́л·ь, -я, (*m*) (met) Chromansil

хрома́т·, -а, (*m*) (chem) chromate

хромат·и́д·а, -ы, (*f*) (biol) chromatid

хромати́д·н·ый об·ме́н· (*m*) (biol) crossing break

хромат·и́н·, -а, (*m*) (cyt) chromatin

хромат·и́ровани·е, -я, (*n*) zinc-yellow painting

хромат·и́ческ·ий, -ая, -ое, (*a*) chromatic
~ аберр·а́ци·я   (*f*) (opt) chromatic aberration
~ адапт·а́ци·я   (*f*) (bot) chromatic adaptation

хромат·и́ческ·ий, -ая, -ое
~ **га́мм·а** (*f*) chromatic scale (music)
~ **ра́з·ность·ь** (*f*) (opt) chromatism
хромат·и́чность·ь, -и, (*f*) (phys) chromaticity
хромат·и́чн·ый, -ая, -ое, (*a*) = **хромат·и́ческ·ий**
хромато·гра́мм·а, -ы, (*f*) (chem) chromatogram, chromatograph
хромато·гра́ф·и·я, -и, (*f*) (chem) chromatography
~, **га́з·ов·ая** gas chromatography
~, **рас·пре·дел·и́тельн·ая** partition chromatography
хромато·метр·и́·я, -и, (*f*) (chem) chromatometric analysis, potassium bichromate titration
хромато·фо́р·, -а, (*m*) (biol) chromatophore
хроматро́н·, -а, (*m*) (TV) chromatron
хром·а́·ть, -ют, (*imp*) limp, be lame
хромаффи́н·н·ый, -ая, -ое, (*a*) (zool) chromaffine
хром·е́л·ь, -я, (*m*) (met) chromel
хром·и́л·, -а, (*m*) (chem) chromyl
~, **хло́р·ист·ый** chromyl chloride
хром·и́ровани·е, -я, (*n*) (met) chromizing, chromium plating; chrome tanning (leather)
~, **диффузио́н·н·ое** chromizing
~, **электро·лит·и́ческ·ое** chromium plating
хром·и́рованн·ый, -ая, -ое, (*past part pass*) of **хром·и́р·овать;** (met) chromium plated, chromized; chrome, chromed (leather)
хром·и́р·овать, -уют, (*imp and perf*) chromium plate, chromize; chrome-tan, chrome (leather)
хром·и́ровочн·ый, -ая, -ое, (*a*) chroming
хро́м·ист·ый, -ая, -ое, (*a*) (chem) chromous, chromite (of); (met) chromium
~ **желез·ня́к·** (*m*) (min) chromite, chrome iron ore
~ **кисл·от·а́** (*f*) chromous acid
~ **стал·ь** (*f*) chromium steel
хром·и́т·, -а, (*m*) (min, chem) chromite
хроми́т·ов·ый, -ая, -ое, (*a*) chromite
хромо- (*int component*) chromium, chromo-, chrome-
хромо·ба́ктер·, -а, (*m*) (bact) chromobacterium
хромо·бласто·мико́з·, -а, (*m*) (med) chromoblastomycosis, chromomycosis

хромово·ки́сл·ый, -ая, -ое, (*a*) (chem) chromate (of)
~ **аммо́ни·й** (*m*) ammonium chromate
~ **серебр·о́** (*n*) silver chromate
хро́м·ов·ый, -ая, -ое, (*a*) chromium, chrome, chromic; chrome, box-calf (leather)
~ **ангидри́д·** (*m*) (chem) chromic anhydride, chromium trioxide
~ **дубл·е́ни·е** (*n*) chrome tanning
~ **зе́лен·ь** (*f*) chrome green
~ **квасц·ы́** (*pl*) chrome alum
~ **кисл·от·а́** (*f*) (chem) chromic acid
~ **краш·е́ни·е** (*n*) chrome dyeing
~ **с·ме́с·ь** (*f*) potassium or sodium bichromate dissolved in sulphuric acid
хромо·ге́н·, -а, (*m*) (cyt) chromogen
хромо·зо́л·ь, -я, (*m*) (chem) chromosol
хром·о́й, -а́я, -о́е, (*a*) lame, limping
хромо·лито·гра́ф·и·я, -и, (*f*) (print) chromolithography; chromolithograph
хромо·магнези́т·ов·ый, -ая, -ое, (*a*) chrome-magnesite (refractory)
хромо·ма́рган·цев·ый, -ая, -ое, (*a*) (met) chromium-manganese
хромо·ме́р·, -а, (*m*) (cyt) chromomere
хромо́·метр·, -а, (*m*) chromometer
хромо·ни́кел·ев·ая стал·ь (*f*) nickel-chromium steel
хромо·пла́ст·, -а, (*m*) (bot) chromoplast
хромо·се́р·н·ая кисл·от·а́ (*f*) chromatosulphuric acid
хромо·ско́п·, -а, (*m*) (phot) chromoscope
хромо·со́м·а, -ы, (*f*) (cyt) chromosome
хромо·сфе́р·а, -ы, (*f*) (astron) chromosphere
хром·от·а́, -ы́, (*f*) lameness
хромо·ти́п·и·я, -и, (*f*) (print) chromotype
хромо·тро́п·, -а, (*m*) (chem) chromotrope
хромо·тро́п·н·ый, -ая, -ое, (*a*) (chem) chromotropic
~ **кисл·от·а́** (*f*) (chem) chromotropic acid
хромо·фо́р·, -а, (*m*) (chem) chromophore
хромо·цисто·скоп·и́·я, -и, (*f*) (med) chromocystoscopy
хро́мпик·, -а, (*m*) (chem) potassium bichromate
хромпикоти́т·, -а, (*m*) (min) chromopicotite
хрон·акси́·я, -и, (*f*) (physiol) chronaxy, chronaxie
хрон·иза́тор·, -а, (*m*) (rad) timer

хро́н·ик·, -а, (*m*) chronic invalid
хро́н·ик·а, -и, (*f*) chronicle; news; (cinema) newsreel
хроника́ль·н·ый, -ая, -ое, (*a*) (cinema) documentary
хроникёр·, -а, (*m*) news editor
хрон·и́ровани·е, -я, (*n*) timing
~, из·бир·а́тельн·ое (rad) selective timing
хрон·и́рованн·ый, -ая, -ое, (*past part pass*) of **хрон·и́р·овать**; timing; keyed (radar)
хрон·и́р·овать, -ують, (*imp*) time
хрон·и́рующ·ий, -ая, -ее, (*pres part act*) of **хрон·и́р·овать**; timing
~ у·стро́й·ств·о (*n*) timer
~ част·от·а́ (*f*) timing frequency
хрон·и́ст·, -а, (*m*) chronicler
хрон·и́ческ·ий, -ая, -ое, (*a*) chronic
хроно·вы·ключ·а́тел·ь, -я, (*m*) telechron timer
хроно·гра́мм·а, -ы, (*f*) time base (radar); chronograph trace
хроно́·граф·, -а, (*m*) (horol) chronograph
~, печа́т·ающ·ий printing chronograph
~, пи́ш·ущ·ий writing chronograph
~, фотограф·и́ческ·ий photographic chronograph
~, электро́н·н·ый electronic-counter chronograph
хроно·изоте́рм·а, -ы, (*f*) chrono-isotherm
хроно·лог·и́ческ·ий, -ая, -ое, (*a*) chronological
хроно·ло́г·и·я, -и, (*f*) chronology
хроно́·метр·, -а, (*m*) (horol) chronometer
~, звёзд·н·ое sidereal chronometer
~, карма́н·н·ый chronometer watch
хроно·метр·а́ж·, -а, (*m*) time-and-motion study; timing
хроно·метра́ж·н·ый, -ая, -ое, (*a*) of **хроно·метр·а́ж·**; time
хроно·метр·и́ческ·ий, -ая, -ое, (*a*) chronometric, chronometer
~ каю́т·а (*f*) (naut) chronometer room
хроно·реле́ (*n indecl*) timing relay
хроно·сигна́л·, -а, (*m*) time signal
хроно·ско́п·, -а, (*m*) (instr) chronoscope
хроно·с·чёт·чик·, -а, (*m*) time meter
хронотро́н·, -а, (*m*) (instr) chronotron
хронофе́р·, -а, (*m*) (telecom) chronopher
хру́п·к·ий, -ая, -ое, (*a*) fragile; brittle

хру́п·к·ий, -ая, -ое
~, раз·руш·ени·е (*n*) (met) brittle failure/rupture
со·противл·ени·е хру́п·к·ому раз·руш·ени·ю (*n*) (met) cohesive strength
хру́п·кост·ь, -и, (*f*) fragility; brittleness, embrittlement
~, водо·ро́д·н·ая (met) hydrogen embrittlement/attack
~, кауст·и́ческ·ая (met) caustic embrittlement/cracking
~, об·рабо́т·очн·ая (met) cutting brittleness
~, от·пуск·н·а́я (met) temper brittleness, Krupp krankheit
~, тепл·ов·а́я (met) temperature embrittlement
хру́ст·, -а, (*m*) crunch, crackle
хруста́л·ик·, -а, (*m*) crystalline lens (of eye)
хруста́л·ь, -я́, (*m*) (min) crystal; crystal glass
~, го́р·н·ый (min) rock crystal
хруста́ль·н·ый, -ая, -ое, (*a*) of **хруста́л·ь**; crystal (non-technical)
хруст·е́ть, -я́т, (*imp*) crunch, crackle
хрущ·, -а́, (*i*) -о́м, (*m*) (ent) cockchafer
хрю́к·а·ть, -ют, (*imp*) grunt
хряк·, -а́, (*m*) (zool) boar
хрящ·, -а́, (*i*) -о́м, (*m*) (anat) cartilage; (geol) gravel, shingle, coarse sand
хрящ·ева́т·ый, -ая, -ое, (*a*) cartilaginous, gristly; gravelly, chondral
~ грунт· (*m*) coarse sand bottom (sea bed)
хряще·ви́д·н·ый, -ая, -ое, (*a*) (anat) chondroid, cartilaginoid
хрящ·ев·о́й, -а́я, -о́е, (*a*) cartilaginous, chondral, gristly
~ гано́ид·ы (*pl*) (fish) *Chondrostei*
~ ры́б·ы (*pl*) (fish) cartilaginous fishes, *Chondrichthyes*
хряще·пёр·ый, -ая, -ое, (*a*) (fish) selachian
хряще·по·до́б·н·ый, -ая, -ое, (*a*) (anat) cartilaginoid, chondroid
худ·е́ть, -ют, (*imp*) become thin/lean
худо́ж·ественн·ый, -ая, -ое, (*a*) artistic
~ литерату́р·а (*f*) belles-lettres, literature
худо́ж·еств·о, -а, (*n*) art
худо́ж·ник·, -а, (*m*) artist, painter, sculptor
худ·о́й, -а́я, -о́е, (*a*) thin, lean; bad, ill; worn

худо·со́ч·и·е, -я,   (*n*) (med) cachexia, cachexy

худо·со́ч·н·ый, -ая, -ое,   (*a*) (med) cachectic

худ·ощав·ый, -ая, -ое,   (*a*) lean, slender, thin

ху́д·ш·ий, -ая, -ее,   (*a*) worse, the worst

ху́ж·е   (*comp*) worse

хулига́н·, -а,   (*m*) hooligan, rowdy

хурм·а́, -ы́,   (*f*) (bot) persimmon, *Diospyros*

ху́тор·, -а,   (*nom pl*) **-а́,** (*g pl*) **-о́в,** (*m*) farmstead; hamlet

хуттони́т·, -а,   (*m*) (min) huttonite

хэк·, -а,   (*m*) (fish) hake, *Merluccius*

Ц

ц. (*abbr*) = **цéнтнер·** metric centner, 100 kg

Ц-систéм·а, -ы, (*f*) (phys) centre/center-of-mass system

цáнг·а, -и, (*f*) (mech eng) collet, collet chuck

цáнг·ов·ый, -ая, -ое, (*a*) of **цáнг·а**

цáп·а·ть, -ют, (*imp*) hoe, hack

цáп·к·а, -и, (*g pl*) -**п·ок·,** (*f*) hand hoe

цáпл·я, -и, (*g pl*) -**пел·ь,** (*f*) (orn) heron, *Ardea*

цапон·лáк·, -а, (*m*) nitrocellulose varnish

цáпф·а, -ы, (*f*) pivot, pin, journal (at end of shaft), neck (as part of shaft); (*pl*) trunnion bearing

~, греб·éнчат·ая collar journal

~, на·правл·я́ющ·ая guide teat (machine tools)

~, по·ворóт·н·ая slewing journal; (M/T) king pin

~ по·перéч·ин·ы crosshead pin

цáпф·енн·ый, -ая, -ое, (*a*) of **цáпф·а**

царáп·ани·е, -я, (*n*) (*v n*) *see* **царá·п·а·ть; scratches**

~ -рéз·ани·е, -я, (*n*) (met) abrasion cutting

царáп·ань·е, -я, (*n*) *see* **царáп·ани·е**

царáп·а·ть, -ют, (*imp*) scratch, make scratches, scrape, abrade

царáп·ин·а, -ы, (*f*) scratch, abrasion

царáп·нут·ый, -ая, -ое, (*past part pass*) *see* **царáп·а·ть**

царáп·н·уть, -ут, (*perf*) *see* **царáп·а·ть**

царатúт·, -а, (*m*) (min) zaratite, emerald nickel

цáрг·а, -и, (*f*) shell section, course (of a sheet metal drum, tank, tower etc.)

цáр·ск·ий, -ая, -ое, (*a*) Czar's, Tsar's, King's, royal, basilic

~ вéн·а (*f*) (anat) basilic vein

~ вóд·к·а (*f*) (chem) aqua regia

цáр·ств·о, -а, (*m*) kingdom, realm; reign

~, живóт·н·ое (biol) animal kingdom

~, раст·úтельн·ое (biol) plant kingdom

цар·ь, -я́, (*m*) (pol) Tsar

ЦБ (*abbr*) = **центрáль·н·ая батарé·я** (telecom) common/central battery

цв. (*abbr*) = **цвет·** colour, color

цвё·л, (*past masc sing*) of **цвест·ú**

цве·лú, (*past pl*) of **цвест·ú**

цвéл·ь, -и, (*f*) mould, green mould

цвест·ú, -у́т, (*imp*) bloom, blossom, flower; flourish, thrive; become green/mouldy/moldy

цвéт·, -а, (*nom pl*) -**á,** (*g pl*) -**óв,** (*m*) colour, color; (*nom pl*) -**ы́,** (*g pl*) -**óв,** (*m*) flower, bloom, blossom

~, до·полн·úтельн·ый complementary colour/color

~, основ·н·óй primary colour/color

~ по·беж·áлост·и (met) temper colour/color

~, фон·ов·óй (print) base colour/color

~, цúнк·ов·ый (min) zinc bloom

цвет·éни·е, -я, (*n*) (*v n*) of **цвест·ú;** florescence

цвéт·ен·ь, -т·н·я, (*m*) (bot) pollen

цвет·ёт, (*pres 3rd sing*) of **цвест·ú**

цвет·кóв·ый, -ая, -ое, (*a*) (bot) phanerogamous, flowering

~ раст·éни·я (*pl*) (bot) *Phanerogamae, Anthophyta*

цвет·нúк·, -á, (*m*) flower garden/bed

цвет·н·óй, -áя, -óе, (*a*) colour, color, coloured, colored, chromatic; variegated, multi-coloured/colored; (met) non-ferrous

~ метáлл· (*m*) non-ferrous metal

~ фильм· (*m*) colour/color film

цвéт·ност·ь, -и, (*f*) chromaticity, chroma; (TV) chrominance

сигнáл· цвéт·ност·и (*m*) (TV) chrominance signal

цвето·вéд·ени·е, -я, (*n*) chromatics

цвето·вóд·ств·о, -а, (*n*) floriculture, flower growing

цвет·ов·óй, -áя, -óе, (*a*) colour, color

~ пирó·метр· (*m*) colour/color-comparison pyrometer

~ тре·угóль·ник· (*m*) (phot) colour/color triangle

цвето·вос·при·я́т·и·е, -я, (*n*) (physiol) colour/color perception

цвето·вос·про·из·вед·éни·е, -я, (*n*) colour/color reproduction

цвето·дел·éни·е, -я, (*n*) (phot) colour/color separation

цвето·дел·ённ·ый, -ая, -ое, (*a*) colour/color-separate

цвето·дел·úтел·ь, -я, (*m*) (print) colour/color-separation scanner

~, электрóн·н·ый (print) electron colour-separation scanner

цвето·éд·, -а, (*m*) (ent) curculio

цве́то·за·мещ·а́ем·ый, -ая, -ое, (*a*) allochromatic

цвет·о́к·, -т·к·а́, (*m*) flower, blossom, bloom

цве́то·ло́ж·е, -а, (*n*) (bot) torus

цве́то·ме́р·, -а, (*m*) (opt) colorimeter

цве́то·ме́тр·и·я, -и, (*f*) colorimetry

цве́то·на·сы́щ·енност·ь, -и, (*f*) colour/color saturation

цве́то·но́ж·к·а, -и, (*g pl*) -ж·ек·, (*f*) (bot) scape, peduncle

цве́то·но́с·н·ый, -ая, -ое, (*a*) flower-bearing, floriferous

цве́то·пере·да́·ч·а, -и, (*f*) (phot) colour/color rendition

цве́то·рас·по·лож·е́ни·е, -я, (*n*) (bot) inflorescence

цве́то·сто́й·к·ий, -ая, -ое, (*a*) colour/color-fast

цве́то·у·сто́й·чив·ый, -ая, -ое, (*a*) colour/color-fast

цвет·о́чник·, -а, (*m*) flowerpot

цвет·о́чн·ый, -ая, -ое, (*a*) flower, floral

~ по·кро́·в· (*m*) (bot) perianth

~ щит·о́к· (*m*) (bot) corymb

цве́то·чувств·и́тельност·ь, -и, (*f*) chromatic/colour/color sensitivity

цве́то·чувств·и́тельн·ый, -ая, -ое, (*a*) colour/color-sensitive

цвет·у́щ·ий, -ая, -ее, (*pres part act*) of цвест·и́

цве́т·ш·ий, -ая, -ее, (*past part act*) of цвест·и́

цвет·ы́, (*pl*) of цвет·о́к· and цвет·

цвет·я́, (*pres gerund*) of цвест·и́

цеаксанти́н·, -а, (*m*) (biol) zeaxanthin

цеа́р·а, -ы, (*f*) (bot) manicoba, ceara rubber

цевардѝн·, -а, (*m*) (biochem) cevadine

цевиа́н·, -а, (*m*) (math) cevian

цеви́н·, -а, (*m*) (biochem) cevine

цёв·к·а, -и, (*g pl*) -в·ок·, (*f*) (zool) tarsometatarsus; (mech) pin (of lantern wheel etc.), pin tooth; bobbin, spool, teat

цёв·очн·ый, -ая, -ое, (*a*) of цёв·к·а; lantern-wheel

~ механ·и́зм· (*m*) lantern-wheel mechanism

~ колес·о́ (*n*) lantern pinion/wheel

цевь·ё, -я́, (*n*) stock (of anchor, rifle etc.)

цед·и́лк·а, -и, (*g pl*) -лок·, (*f*) cullender, colander, strainer

цед·и́льн·ый, -ая, -ое, (*a*) filter, filtering, straining

цед·и́ть, -я́т, (*imp*) strain, filter; decant; -ся (*pass*); percolate

це́др·а, -ы, (*f*) (food) dried peel

цедрари́н·, -а, (*m*) (biochem) cedrarine

цедр·е́н·, -а, (*m*) cedrene

цедр·и́н·, -а, (*m*) cedrin

цедр·о́л·, -а, (*m*) (chem) cedrol

цеж·е́ни·е, -я, (*n*) (*v n*) see цед·и́ть; percolation

це́ж·енн·ый, -ая, -ое, (*past part pass*) of цед·и́ть

цезальпи́н·иев·ые, -ых, (*pl decl as adj*) (bot) senna family, *Caesalpiniaceae*

цезальпи́н·и·я, -и, (*f*) (bot) senna, *Caesalpinia*

це́з·иев·ый, -ая, -ое, (*a*) caesium, cesium, Cs

це́з·и·й, -я, (*m*) (chem) caesium, cesium, Cs

цез·и́рованн·ый, -ая, -ое, (*a*) cesium-coated

цеи́н·, -а, (*m*) (chem) zein

цейлон·и́т·, -а, (*m*) (min) ceylonite, pleonaste

цейло́н·ск·ий, -ая, -ое, (*a*) Ceylon, Singhalese

цейнери́т·, -а, (*m*) (min) zeunerite

Це́йнер·а, диагра́мм·а (*f*) (st eng) Zeuner valve diagram

це́йсс·а, ли́нз·а (*f*) (opt) Zeiss lens

це́йт·лу́п·а, -ы, (*f*) (phot) high-speed slow-motion camera

це́йт·но́т·, -а, (*m*) time trouble (at chess)

цейтра́фер·н·ое у·стро́й·ств·о (*n*) (cinema) time-lapse mechanism

це́йхга́уз·, -а, (*m*) arsenal, armoury, armory

цек·ова́ни·е, -я, (*n*) = цек·о́вк·а (*f*)

цек·о́вк·а, -и, (*f*) (mech eng) rounding up

цел·ака́нт·, -а, (*m*) (fish) coelacanth

цел·е́бн·ый, -ая, -ое, (*a*) salubrious, curative, remedial, healing, health-giving, healthy

цел·ев·о́й, -а́я, -о́е, (*a*) of цел·ь; special-purpose, specific, purposeful

целе·на·пра́вл·енн·ый, -ая, -ое, (*a*) single-minded, purposeful

цел·ентера́т·ы (*pl*) (zool) coelenterates, *Coelenterata*

целе·со·обра́з·н·о (*predic adj*) it **is** a good thing (to), it is best (to), it is advisable (to)

целе·со·обра́з·ност·ь, -и, (*f*) advisability, expediency; (good) sense, **sound practice**

целе·со·образ·н·ый, -ая, -ое, (a) expedient, advisable, good

целестин·, -а, (m) (min) celestite, celestine

целе·у·каз·ани·е, -я, (n) target indication

целе·у·стремл·ённ·ый, -ая, -ое, (a) purposeful

цел·ик·, -á, (m) sighting point, backsight (of optical instruments); (gunn) calculated deflection; (min) pillar

механ·изм· цел·ик·á (m) (gunn) lateral aim-off mechanism

~, о·хран·н·ый (min) shaft pillar

цел·ик·óм (adv) wholly, as á whole, entirely

цел·ин·á, -ы́, (f) virgin soil/land

цел·инн·ый, -ая, -ое, (a) virgin (of soil etc.)

цел·и́тельн·ый, -ая, -ое, (a) healing, curative

цел·ить, -ят, (imp) aim

целеншáльтер·, -а, (m) (elec) battery regulating switch

целло·биóз·а, -ы, (f) (biochem) cellobiose, cellose

целло·зóльв·, -а, (m) (plast) cellosolve

целлóн·, -а, (m) (plast) cellon

целло·фáн·, -а, (m) cellophane

целло·фáн·ов·ый, -ая, -ое, (a) cellophane

целлул·óид·, -а, (m) see целлюл·óид·

целлюл·áз·а, -ы, (biochem) cellulose

целлюл·и́т·, -а, (m) (med) cellulitis; (plast) cellulith

целлюл·óз·а, -ы, (f) (biochem) cellulose; (paper) pulp

~, натрóн·н·ая (paper) soda pulp

целлюлóз·н·о-бумáж·н·ый, -ая, -ое, (a) pulp-and-paper, paper

~ про·из·вóд·ств·о (n) papermaking

целлюлóз·н·ый, -ая, -ое, (a) (biochem) cellulose; (paper) pulp

~ за·вóд· (m) (paper) pulp mill

целлюл·óид·, -а, (m) celluloid

целлюляр·н·ый, -ая, -ое, (a) cellular

целляриýс·, -а, (m) (chem) Cellarius tourill

цело- (int component) coelo-

цело·блáстул·а, -ы, (f) (zool) coeloblastula

цел·овáть, -ýют, (imp) kiss

цéл·ое, -ого, (n decl as adj) (math) integer

целóзи·я, -и, (f) (bot) cockscomb, Celosia

целóм·, -а, (m) (zool) coelom, coelome

целом·и́ческ·ий, -ая, -ое, (a) (zool) coelomic

целóм·н·ые, -ых, (pl decl as adj) (zool) coeloms, Coelomata

целомодýкт·, -а, (m) (zool) coelomoduct

цело·стáт·, -а, (m) (astron) coelostat

цéл·остност·ь, -и, (f) integrity

óбласт·ь цéл·остност·и (f) (math) integral domain

цéл·остн·ый, -ая, -ое, (a) integral

цéл·ост·ь, -и, (f) wholeness, entirety; safety, intact/unharmed/unimpaired condition

в цéл·ост·и (adv) intact; completely

целофáн·, -а, (m) see целлофáн·

цело·тóн·н·ый, -ая, -ое, (a) whole-tone (music)

цело·чи́сл·енн·ый, -ая, -ое, (a) (math) countable, countably infinite, enumerable, enumerative

~ решёт·к·а (f) (math) countable/ enumerable set

статист·ик·а цело·чи́сл·енн·ых велич·и́н· (f) enumerative statistics

цéл·ый, -ая, -ое, (a) whole, entire, integral; intact, safe; (n decl as adj) integer

~ велич·ин·á integral amount/value

~ и не·вред·и́м·ый safe and sound

по цéл·ым недéл·ям for weeks on end, week after week

~ ряд· (m) a large number (of)

~ числ·ó (n) (math) integral number, integer

цéл·ь, -и, (f) target, aim, goal; purpose, object, objective

в цéл·ях with the aim of, with a view to

~ -квадрáт·, -а, (m) (mil, air) target coordinates

ли́н·и·я цéл·и (f) (gunn) line of sight, line of sight to target

~, перспекти́в·н·ая long-term goal/ aim

плóск·ост·ь цéл·и (f) (gunn) plane of position

~, по·дáвл·енн·ая (gunn) neutralized target

с цéл·ью with the aim of, in order to, to

с э́т·ой цéл·ью to this end

стáв·ить себ·é цéл·ью aim to

цельзиáн·, -а, (m) (min) celsian

цéльзи·я, -и, (f) (bot) celsia, Celsia

цéль·н·о- (component) all-, completely, wholly, holo-

цельно·голо́в·ые, -ых, (*pl decl as adj*) (fish) holocephalans, holocephalians, *Holocephali*

цельно·за́·мк·нут·ый, -ая, -ое, (*a*) completely closed

цельно·за·мык·а́ни·е, -я, (*n*) complete closure

цельно·ка́т·ан·ый, -ая, -ое, (*a*) (met) unit/solid/seamless-rolled, one-piece-rolled

~ колес·о́ (*n*) rolled wheel, wrought-steel wheel

цельно·ко́в·ан·ый, -ая, -ое, (*a*) unit-/solid-forged, one-piece forged

цельно·ко́ст·н·ые, -ых, (*pl decl as adj*) (fish) holosteans, *Holostei*

цельно·ко́ст·н·ый, -ая, -ое, (*a*) (zool) holostean, holosteous

цельно·кра́й·н·ий, -яя, -ее, (*a*) entire, intact

цельно·ли́·т·ый, -ая, -ое, (*a*) unit-cast

цельно·металл·и́ческ·ий, -ая, -ое, (*a*) all-metal, solid-metal

цельно·по·воро́т·н·ый, -ая, -ое, (*a*) fully rotatable

~ о·пар·е́ни·е (*n*) (a/c) all-flying tail

цельно·с·ва́р·н·ый, -ая, -ое, (*a*) all-welded

це́ль·ност·ь, -и, (*f*) = **це́л·ост·ь** (q.v.)

цельно·тя́·нут·ый, -ая, -ое, (*a*) (met) whole-/solid-drawn, seamless

~ труб·а́ (*f*) whole-/solid-drawn tube, seamless tube

це́ль·н·ый, -ая, -ое, (*a*) whole, entire, unified, unit; one-piece; whole, undiluted, untouched

~ вин·о́ (*n*) undiluted wine

~ молок·о́ (*n*) (food) whole milk

~ снег· (*m*) virgin/undisturbed/untouched snow

Це́льси·й, -я, (*m*) (phys) centigrade, Celsius

Це́льси·я, шкал·а́ (*f*) (phys) centigrade/Celsius scale

це́льти·й, -я, (*m*) hafnium, Hf

цеме́нт·, -а, (*m*) cement

~, англ·и́йск·ий Keene's cement

~, бокси́т·ов·ый aluminous/alumina cement, ciment fondu

~, быстро·с·хва́т·ывающ·ийся quick-setting cement

~, быстро·тверд·е́ющ·ий rapid-hardening cement, high-early cement

~, волокн·и́ст·ый fibre/fiber cement

~, гидро·фи́ль·н·ый water-retaining cement

цеме́нт

~, гидро·фо́б·н·ый water-repellent cement

~, глино·зём·ист·ый alumina/aluminous cement, ciment fondu

~, зуб·н·о́й (med) cementum

~, зубо·вра́ч·ебн·ый (med) dental cement

~, кла́д·очн·ый masonry cement

~, кли́нкер·н·ый neat cement

~ на́·сып·ью bulk cement

~, низко·терм·и́чн·ый low-heat cement

~, огне·у·по́р·н·ый refractory/fire cement

~, па́ро·ск·ий Parian cement

~ -пу́ш·к·а, -и, (*f*) (civ eng) cement gun

~, рас·шир·я́ющ·ийся expanding cement

~, ру́д·н·ый iron-ore cement

~, тампона́ж·н·ый oil-well cement

~, това́р·н·ый bagged/sacked/finished cement

~, тяжёл·ый weighted cement

цемент·а́ж·, -а, (*m*) (civ eng) cementation, cementing, grouting

цемент·а́ци·я, -и, (*f*) (civ eng) cementing, cementation, grouting; (met) cementation (extractive non-ferrous); cementation, case-hardening, carburizing (ferrous)

~, га́з·ов·ая gas carburizing

~, жи́дк·остн·ая liquid carburizing

~ твёрд·ым карбониз·а́тор·ом solid carburizing

цемент·и́руем·ый, -ая, -ое, (*pres part pass*) of **цемент·и́р·овать**

~ стал·ь (*f*) carburizing/case-hardening steel

цемент·и́рованн·ый, -ая, -ое, (*past part pass*) of **цемент·и́р·овать**

цемент·и́р·овать, -уют, (*imp and perf*) cement, cement in, pour cement, grout; (met) case-harden, carburize, cement

цемент·и́т·, -а, (*m*) (met) cementite; (med) cementitis

цеме́нт·н·ый, -ая, -ое, (*a*) cement

~ молок·о́ (*n*) cement grout; bleeding cement (as defect)

цементо·бето́н·, -а, (*m*) cement concrete

цемент·о́ванн·ый, -ая, -ое, (*a*) = **цемент·и́рованн·ый**

цемент·ова́ть, -у́ют, (*imp and perf*) = **цемент·и́р·овать**

цементо·во́з·, -а, (*m*) cement lorry/truck

цем·я́нк·и, -нок·, (*f pl*) brick flour (hydraulic addition agent for cement)

цен·а́, -ы́, (*nom pl*) це́н·ы, (*f*) price, value; (text) shed rod

~ дел·е́ни·я (instr) multiplying factor, scale division/length/spacing, graduation

~, за·во́д·ск·ая factory price, first cost

~, за·гото́в·и́тельн·ая procurement price

~, за·ку́п·очн·ая purchase price

~, на·риц·а́тельн·ая (fin) par value (of stocks and shares)

~, обо·зна́ч·енн·ая marked price

~, рас·чёт·н·ая factory price

~, твёрд·ая fixed price

~, фабр·и́чн·ая factory price, first cost

цен·енхи́м·а, -ы, (*f*) (biol) coenenchyme, coenenchyma

це́нз·, -а, (*m*) (polit) qualification

це́нзор·, -а, (*m*) censor

цензу́р·а, -ы, (*f*) censorship

цензур·о́ванн·ый, -ая, -ое, (*past part pass*) censored

цен·и́ть, -ят, (*imp*) price, value, appraise, estimate

це́н·ник·, -а, (*m*) price list

це́н·ностн·ый, -ая, -ое, (*a*) of це́н·ност·ь; valuable, of value

це́н·ност·ь, -и, (*f*) value, value function; (*pl*) valuables

~, материа́ль·н·ые (*pl*) (econ) assets, realizable assets

~, товаро·материа́ль·н·ые (*pl*) (econ) stock

це́н·н·ый, -ая, -ое, (*a*) priced; valuable, of worth; registered (mail); (fin) negotiable

~ бума́г·и (*pl*) (fin) securities

цен·ов·о́й, -а́я, -о́е, (*a*) of цен·а́ (text)

~ па́л·очк·и (*pl*) (text) lease rods

цено- (*int component*) coeno-, ceno-, caeno-

цено·гаме́т·а, -ы, (*f*) (gen) coenogamete

цено·гене́зис·, -а, (*m*) (gen) caenogenesis, cenogenesis

цено·генет·и́ческ·ий, -ая, -ое, (*a*) (biol) caenogenetic, cenogenetic

цено́·з·, -а, (*m*) (biol) cocnosis, cenosis

~, живо́т·н·ый (zool) zoocoenosis

~, раст·и́тельн·ый (bot) phytocoenosis

цено·зиго́т·а, -ы, (*f*) (gen) coenozygote

ценози́т·, -а, (*m*) (min) cenosite

цено·са́рк·, -а, (*m*) (biol) coenosarc

цен·остеу́м·, -а, (*m*) (zool) coenosteum

цен·о́ци·й, -я, (*m*) (zool) coenoecium

цено·ци́т·, -а, (*m*) (biol) coenocyte

це́нт·, -а, (*m*) (fin) cent

цента́вр·, -а, (*m*) (astron) Centaurus, the Centaur

центаури́н·, -а, (*m*) (pharm) centaurin

центи·бе́л·, -а, (*m*) (acous) centibel

це́нтнер·, -а, (*m*) 100 kilograms, metric centner (220·46 lb)

~, англ·и́йск·ий hundredweight, cwt. (45·3592 kg)

~, двой·н·о́й double/metric centner (100 kg)

це́нтр·, -а, (*m*) (*and see under associated words*); centre, center; (cryst) nucleus

~, бар·и́ческ·ий (meteor) centre/center of pressure

~ велич·ин·ы́ (shipb) centre/center of buoyancy

крив·а́я це́нтр·ов величи́н·ы́ (*f*) (shipb) curve of centres/centers of buoyancy

~, вну́тр·енн·ий female centre/center (of lathe)

~ в·пи́с·анн·ого кру́г·а (math) incentre/incenter

~ вращ·е́ни·я, мг·нове́нн·ый (mech) instantaneous centre/center

~ гру́пп·ы (math) centre/center of a group

~, за́д·н·ий tail/back centre/center (of lathe)

~ концентр·а́ци·и focus of concentration

ли́н·и·я це́нтр·ов (*f*) line of centres (gearing)

~, на·са́д·н·о́й false centre/center (machine tools)

~ о·пи́с·анн·ого кру́г·а (math) circumcentre, circumcenter

~, пере́д·н·ий live centre/center (of lathe)

~, у·по́р·н·ый fixed centre/center

центр·адени́·я, -и, (*f*) (biol) centradenia

централиз·а́тор·, -а, (*m*) centralizer; (rail) interlocker

централизацио́н·н·ый, -ая, -ое, (*a*) of централиз·а́ци·я

централиз·а́ци·я, -и, (*f*) centralization, centralizing; interlocking (of railway signals)

~, реле́й·н·ая (rail) all-relay interlocking

централиз·о́ванн·ый, -ая, -ое, (*past part pass*) of **централиз·ова́ть**

~ **у·правл·е́ни·е** (*n*) (rail) interlocking control

централиз·ова́ть, -у́ют, (*imp and perf*) centralize; (rail) interlock

центра́л·ь, -и, (*f*) (*see also* **маги-стра́л·ь**) main; (math) central; (elec) central station, station

~ **гру́пп·ы** (math) central of a group

~, **трёх·гироско́п·н·ая** ' three-gyro unit

центра́ль·н·о-ра́з·ностн·ый, -'ая, -ое, (*a*) (math) central-difference

центра́ль·н·ый, -ая, -ое, (*a*) central, centre, center

~ **коммут·а́тор·** (*m*) (telecom) main exchange switchboard

центри·о́л·ь, -я, (*m*) (gen) centriole

центрир·, опт·и́ческ·ий (*m*) (surv) vertical collimator

центр·и́ровани·е, -я, (*n*) (*v n*) *see* **центр·ова́ть**

    ру́ч·к·а центр·и́ровани·я выс·о́-т·ы (*f*) (a/c) pitch control wheel

    ру́ч·к·а центр·и́ровани·я ку́рс·а (a/c) fine turn control

центр·и́р·овать, -уют, (*imp and perf*) = **центр·ова́ть**

центри·фу́г·а, -и, (*f*) (tech) centrifuge

~, **ио́н·н·ая** (phys) ionic centrifuge, magnetron

~, **противо·то́ч·н·ая** counter-current centrifuge

~, **про·то́ч·н·ая** co-current/flow-through centrifuge

центри·фуг·а́т·, -а, (*m*) centrifugal product

центри·фуг·и́ровани·е, -я, (*n*) (*v n*) of **центри·фуг·и́р·овать;** centrifugal separation

центри·фуг·и́р·овать, -уют, (*imp and perf*) centrifuge; cast centrifugally (concrete)

центри·фуг·ова́ть, -у́ют, (*imp and perf*) **центри·фуг·и́р·овать**

центр·и́ческ·ий, -ая, -ое, (*a*) (min) centric

центро·аффи́н·н·ый, -ая, -ое, (*a*) centroaffine

центро·бе́ж·н·ый, -ая, -ое, (*a*) centrifugal; (anat) efferent

центр·ова́льн·ый, -ая, -ое, (*a*) centring, centre, center

~ **стан·о́к·** (*m*) centre/center drill

центр·ова́ни·е, -я, (*n*) = **центр·о́вк·а**

центр·ова́ть, -у́ют, (*imp and perf*) centre, center, align; dot, mark (with centre punch)

центр·о́вк·а, -и, (*f*) (*v n*) of **центр·о·ва́ть;** centering, alignment; dotting, marking (with centre punch)

~ **само·лёт·а** (air) positioning of centre/center of gravity

центр·ов·о́й, -а́я, -о́е, (*a*) centre, center, centrae

центро·ге́н·н·ый, -ая, -ое, (*a*) (physiol) centrogenic

центро·десмо́з·, -а, (*m*) (cyt) centro-desmus, centrodesmose

центро·зо́м·а, -ы, (*f*) = **центро·со́м·а**

центр·о́ид·, -а, (*m*) centroid

центр·о́ид·а, -ы, (*f*) (math) centroid; (mech) polhode

центро·клина́л·ь, -и, (*f*).(geol) centro-cline

центро·лецита́ль·н·ый, -ая, -ое, (*a*) (zool) centrolecithal

центро·ме́р·, -а, (*m*) (gen) centromere

центро·пла́н·, -а, (*m*) (a/c) centre section (of wing)

центро·симметр·и́·я, -и, (*f*) (cryst) centrosymmetry

центро·со́м·а, -ы, (*f*) (cyt) centro-some, astrocentre/astrocenter

центро·стрем·и́тельн·ый, -ая, -ое, (*a*) centripetal; (anat) afferent

центро·сфе́р·а, -ы, (*f*) (cyt, geol) centrosphere

центу́р·и·й, -я, (*m*) centurium, Cn

цену́р·, -а, (*m*) (zool) coenurus

ценур·о́з·, -а, (*m*) (vet) coenurosis

цео·ли́т·, -а, (*m*) (geol) zeolite

цеолитиз·а́ци·я, -и, (*f*) (geol) zeoliti-zation

цео·ли́т·ов·ый, -ая, -ое, (*a*) (geol) zeolitic

це·о́м·а, -ы, (*f*) (bot) Caeoma

цео·филл·и́т·, -а, (*m*) (min) zeophyllite

цеп·, -а́, (*m*) (agr) flail

цепе·вяз·а́льн·ый автома́т· (*m*) chain-bending machine

цепен·е́·ть, -ют, (*imp*) become stiff/rigid/numb

це́п·ен·ь, -п·н·я, (*m*) (zool) tapeworm, *Taenia*

~ **-мозг·ови́к·, -а́,** (*m*) (zool) gid-parasite, *Multiceps multiceps*

цепе·обра́з·н·ый, -ая, -ое, (*a*) chain-like, cateniform, catenoid, (bot) caten-ate, catenulate

це́п·и (*pl*) of **цеп·ь**

цеп·к·ий, -ая, -ое,   *(a)* prehensile, tenacious, clinging; sticky

цепл·я·ться, -ются,   *(imp)* cling, clutch/hang on

цеп·н·ой, -áя, -óе,   *(a)* chain; (elec) ladder, iterative

~ лин·и·я *(f)* (math) catenary

~ пере·дá·ч·а *(f)* chain drive

~ прáв·ил·о *(n)* (math) chain rule

~ реáкци·я *(f)* (chem, nucl) chain reaction

~ схéм·а *(f)* (elec) iterated/recurrent/ ladder network/filter

цепно·фрéзер·н·ый стан·óк·   *(m)* chain-mortising machine (woodworking)

цеп·ов·óй, -áя, -óе,   *(a)* of **цеп·**

цеп·óчечн·ый, -ая, -ое,   *(a)* of **цеп·óчк·а**

цеп·óчк·а, -и,   *(g pl)* **-чек·,** *(f)* *(dim)* of **цеп·ь**; file (of persons); (elec) ladder network/circuit

~ рас·пáд·ов   (phys) decay/disintegration chain

~ стáнци·й на·вед·éни·й   home chain (radar)

цепочко·обрáз·н·ый, -ая, -ое,   *(a)* see **цепе·обрáзн·ый**

цéп·ь, -и,   *(nom pl)* **-и,**   *(g pl)* **-éй,** *(f)* chain, chain-cable; (elec) circuit *(see also* **схéм·а, кóнтур·** *and associated words)*; (mil) line; (autom) network

~, англ·ийск·ая   short-link chain

~, анóд·н·ая   (elec) anode circuit

~, бес·кон·éчн·ая   continuous/endless chain

~, бес·шýм·н·ая   (mech) silent chain, inverted-tooth chain; (rad) quiet tuning circuit

~, бок·ов·áя   (chem) side chain

~, бýфер·н·ая   (elec) buffer circuit

~, воз·дýш·н·ая   (elec) open-wire circuit

~, втýл·очн·о-рóл·иков·ая   (mech) bush-roller chain

~, вы·рáвн·ивающ·ая   (elec) compensator circuit

~, вы·ход·н·áя   output circuit (computers)

~ Гáлл·я   drive chain

~, грýз·ов·áя   hoisting chain

~, двух·пóлюс·н·ая   (autom) two-terminal network

~, двух·ря́д·н·ая   duplex chain

~, демпф·ирующ·ая   (elec) antihunt circuit

цéп·ь

~, зá·мк·нут·ая   closed circuit, loop

~, земле·мéр·н·ая   (surv) pole chain

~, зуб·чáт·ая   inverted-tooth chain

~, из·мер·ительн·ая   (elec) metering circuit

~, искýс·ственн·ая   (telecom) superposed circuit, simplex

~, калибр·óванн·ая   ordinary close chain

~, коменс·ирующ·ая   (elec) compensator circuit

~ контрóл·я перио́д·а   period-control circuit

~, коррект·ирующ·ая   (autom) correction network

~, крюч·кóв·ая   hooklink chain

~, магнéт·н·ая   magnetic circuit

~, мéр·н·ая   (surv) chain table, measuring chain, surveyor's chain

~, много·крáт·н·ая   (elec) multiple

~, много·ря́д·н·ая   multiple chain

~, молекуля́р·н·ая   (phys, chem) molecular chain

~, на·кáл·н·ая   filament circuit

~ на·стрóй·к·и   (autom) adjustment network

~, не·зá·мк·нут·ая   open/incomplete circuit

~, об·остр·я́ющ·ая   (elec) peaking circuit

~, обрáт·н·ая   (autom) inverse network

~ обрáт·н·ой с·вя́з·и   (autom) loop; **сигнáл· цéп·и обрáт·н·ой с·вя́з·и в·ход·н·ой** *(m)* (autom) loop input signal

~, одно·ря́д·н·ая   simple chain

~, основ·н·áя   (chem) main chain; side circuit (telephone)

~, от·ветвл·ённ·ая   (elec) derived circuit

~, пласт·инчат·ая втýл·очн·ая   pin-and-link chain

~ пре·ры́в·ист·ого дéй·стви·я   sampling circuit (computers)

~, прóб·н·ая   testing circuit (telephone)

~, противо·скольж·éни·я   (M/T) non-skid chain, snow chain

~, прям·áя   (autom) forward circuit

~, раз·ветвл·ённ·ая   (elec) forked/tree circuit; (chem) branching chain

~, разó·мк·нут·ая   (elec) broken circuit

~, раст·ýщ·ая   (chem) growing chain

~ регул·ирова́ни·я   (elec) control loop

~, рéльс·ов·ая   (rail) track circuit

~, рóл·иков·ая   roller chain

цеп·ь
~ **сет·к·и** (elec) grid circuit
~, **сигнáль·н·ая** (elec) alarm circuit
~, **симметр·и́чн·ая** (telecom) balanced/bilateral circuit
~, **с·лóж·н·ая** (elec) compound circuit
~, **снег·ов·áя** (M/T) snow/tire chain
~, **с·чёт·н·ая** (autom) counter circuit
~, **тауэр·н·ая** towing/pulling chain
~, **телефóн·н·о-телегрáф·н·ая** (telecom) composite/composited circuit
~, **трёх·ря́д·н·ая** triplex chain
~, **у·длин·я́ющ·ая** (elec) integrating circuit/network, integrator
~, **фантóм·н·ая** (telecom) phantom circuit
~, **фикс·и́рующ·ая** hold/clamping circuit
~, **шарни́р·н·ая** roller chain
~, **шнур·ов·áя** cord circuit (telephone)
~ **Эвáрт·а** Ewart chain, open-hook chain
церази́к·, -а, (m) gouge (leather)
церамици́т·, -а, (m) (min) ceramicite
церáт·, -а, (m) cerate, cerate paste
церат·и́т·, -а, (m) (pal) ceratite
церато·зáвр·, -а, (m) (pal) Ceratosaurus
церат·óпс·, -а, (m) (pal) Ceratops
церат·оптéрис·, -а, (m) (bot) Ceratopteris
церáц·и·й, -я, (m) (zool) Ceratium
цервикáль·н·ый, -ая, -ое, (a) (anat) cervical
цервико·везикáль·н·ый, -ая, -ое, (a) (anat) cervicovesical
цервиц·и́т·, -а, (m) (med) cervicitis
церебрáль·н·ый, -ая, -ое, (a) (anat) cerebral
церебр·и́н·, -а, (m) (biochem) cerebrin
церебр·ози́д·, -а, (m) (biochem) cerebroside
церебро·склерóз·, -а, (m) (med) cerebrosclerosis
церебро·спинáль·н·ый, -ая, -ое, (a) (anat) cerebrospinal
церез·и́н·, -а, (m) (chem) ceresine, mineral wax, ceresine wax
церези́т·, -а, (m) (build) ceresit
церемониáль·н·ый, -ая, -ое, (a) ceremonial
церемóн·и·я, -и, (f) ceremony
церéр·а, -ы, (f) (astron) Ceres
цér·иев·ый, -ая, -ое, (a) cerium, ceric
~ **земл·я́** (f) (min) cerite earth
цéр·и·й, -я, (m) (chem) cerium, Ce

цер·и́л·, -а, (m) (chem) ceryl
церил·éн·, -а, (m) (chem) cerylene
цери́л·ов·ый, -ая, -ое, (a) ceryl
цери·метр·и́·я, -и, (f) (chem) cerium sulphate titration
цер·и́н·, -а, (m) (biochem) cerin; (min) cerine
церио·метр·и́·я, -и, (f) (chem) ceriometry
цер·и́т·, -а, (m) (min) cerite
цéрк·а, -и, (f) (zool) cercus
церк·áри·я, -и, (f) (zool) cercaria
цéрк·ов·ь, -к·в·и, (g pl) **-к·в·éй,** (f) church
церко·пóд·а, -ы, (f) (zool) cercopod
церко·спорóз·, -а, (m) cercosporosis (plant disease)
церо·ксил·и́н·, -а, (m) (biochem) ceroxylin
церо·ли́т·, -а, (m) (min) cerolite
церóл·ь, -я, (m) cerol (dye)
церот·и́н·ов·ая кисл·от·á (f) (chem) cerotic/cerotinic acid
церро·матрикс·, -а, (m) cerromatrix
церро·сéйф·, -а, (m) cerrosafe
церте·пáрт·и·я, -и, (f) (com) charter party
церулéйт·, -а, (m) (min) ceruleite
церусс·и́т·, -а, (m) (min) cerussite, lead carbonate
цесáр·к·а, -и, (g pl) **-р·ок·,** (f) (orn) Guinea fowl, *Numida meleagris*
цет·áн·, -а, (m) (chem) cetane, hexadecane
цетáн·ов·ое числ·ó (n) cetane number (of diesel oils)
цет·éн·, -а, (m) (chem) cetene
цет·и́л·, -а, (m) cetyl
цети́л·ов·ый спирт· (m) (chem) cetyl alcohol, *n*-hexadecyl alcohol
цетрáри·я, -и, (f) (bot) Iceland moss, *Cetraria*
цефал·аспи́д·ы (f pl) (pal) Cephalaspida
цефал·и́н·, -а, (m) (biochem) cephalin
цефал·и́ческ·ий, -ая, -ое, (a) (anat) cephalic
цефал·óн·, -а, (m) (zool) cephalon
цефало·пóд·ы (pl) (zool) cephalopodans, cephalopods, *Cephalopoda*
цефало·тóракс·, -а, (m) (zool) cephalothorax
цефал·я́ри·я, -и, (f) (bot) cephalaria, *Cephalaria*
цефе·и́д·а, -ы, (f) (astron) cepheid variable
цефé·й, -я, (m) (astron) Cepheus

цех·, **-а**, (*nom pl*) **-и**, or **-á**, (*m*) (*and see under associated words*); shop, workshop

цех-автомáт·, **-а**, (*m*) automatic workshop

~, гор·я́ч·ий   hot workshop

~, куз·нéчн·ый   forge

~, лит·éйн·ый   foundry

~, основ·н·óй   production shop

~ -пит·óмник·, **-а**, (*m*) metallurgical shop (of a heavy engineering works)

~, про·кáт·н·ый   rolling mill (everything from soaking pits to finishing mill inclusive)

~, терм·и́ческ·ий   (met) heat-treatment department/shop

цех·ов·óй, **-áя**, **-óе**, (*a*) shop, workshop

цехштéйн·, **-а**, (*m*) (geol) Zechstein

цецé (*f indecl*) (ent) tsetse fly, *Glossina*

цецели́т·, **-а**, (*m*) (min) cecelite

цеци́ди·й, **-я**, (*m*) (bot) cecidium, gall

циáн·, **-а**, (*m*) (chem) cyan; cyanogen

циан·ами́д·, **-а**, (*m*) (chem) cyanamide

~ кáльци·я   calcium cyanamide

циан·áт·, **-а**, (*m*) (chem) cyanate

циан·гидри́н·, **-а**, (*m*) (chem) cyanhydrin, cyanohydrin

циан·é·я, **-и**, (*f*) (zool) Cyanea capillata

циан·и́д·, **-а**, (*m*) (chem) cyanide

циан·иди́н·, **-а**, (*m*) cyanidine

циан·изáци·я, **-и**, (*f*) cyanization

циан·или́д·, **-а**, (*m*) cyanilide

циан·и́н·, **-а**, (*m*) (biochem) cyanin (natural colouring matter); cyanine (dye)

циани́н·ов·ый крас·и́тел·ь (*m*) cyanine dye, cyanine

циан·и́ровани·е, **-я**, (*n*) (met) cyaniding, cyanide hardening (ferrous); cyaniding, cyanidation (non-ferrous)
   твёрд·ой средé· (met) pack cyaniding

~, гáз·ов·ое   (met) gas cyaniding

циан·и́р·овать, **-уют**, (*imp and perf*) cyanide

цианисто·водо·рóд·н·ый, **-ая**, **-ое**, (*a*) (chem) hydrocyanic, hydrocyanide of

~ кисл·от·á   (*f*) hydrocyanic/prussic acid

циáн·ист·ый, **-ая**, **-ое**, (*a*) cyanide of, cyanide

~ водо·рóд·   (*m*) (chem) hydrogen cyanide

~ кáли·й   (*m*) potassium cyanide

циáн·ист·ый, **-ая**, **-ое**

~ мéд·ь   (*f*) cuprous cyanide (1); cupric cyanide (2)

~ плав·   (*m*) (chem) cyanide flux

циан·и́т·, **-а**, (*m*) (min) cyanite, kyanite

циан·кобаламúн·, **-а**, (*m*) (biochem) cyanocobalamin

циан·муравь·и́н·ая кисл·от·á (*f*) cyanoformic acid

цианово·ки́сл·ый, **-ая**, **-ое**, (*a*) cyanate of

~ нáтри·й (*m*) sodium cyanate

циáн·ов·ый, **-ая**, **-ое**, (*a*) cyanic

~ кисл·от·á (*f*) (chem) cyanic acid

циан·óз·, **-а**, (*m*) (med) cyanosis

циано·ти́п·и·я, **-и**, (*f*) (phot) cyanotype, blue-printing

циано·трих·и́т·, **-а**, (*m*) (min) cyanotrichite

циано·у́ксус·н·ый, **-ая**, **-ое**, (*a*) cyanoacetic

~ кисл·от·á (*f*) cyano-acetic acid

циано·филли́н·, **-а**, (*m*) (biochem) cyanophylline

циано·хрóит·, **-а**, (*m*) (min) cyanochroite

циано·этил·и́ровани·е, **-я**, (*n*) (chem) cyanoethylation

циан·плáв·, **-а**, (*m*) = **циáн·ист·ый плав·**

циан·стирóл, **-а**, (*m*) (chem) cyanostyrene

циан·у́р·ов·ая кисл·от·á (*f*) (chem) cyanuric acid

циатé·ев·ые, **-ых**, (*pl decl as adj*) (bot) Cyatheaceae

циато·ли́т·, **-а**, (*m*) (geol) cyatholith

циáц·и·й, **-я**, (*m*) (bot) cyathium

ци́б·а, **-ы**, (*f*) ciba (dye)

цибети́н·, **-а**, (*m*) civetin, (biochem) zibethum

цибетóн·, **-а**, (*m*) (biochem) cycloheptadecene-10-one, civetine

цибо·тáксис·, **-а**, (*m*) = **сибо·тáксис·** (q.v.)

цивéтт·а, **-ы**, (*f*) (zool) civet cat

цивилиз·áци·я, **-и**, (*f*) civilization

циви́ль·н·ый, **-ая**, **-ое**, (*a*) civil

цигадéнус·, **-а**, (*m*) (bot) death camass, *Zygadenus*

цигéй·к·а, **-и**, (*g pl*) **-é·ек·**, (*f*) beaver lamb (fur)

ци́гер·, **-а**, (*m*) cheese albumin

Ци́глер·а, катали́ст· (*m*) (rubber) Ziegler catalyst

цикáд·а, **-ы**, (*f*) (ent) cicada, (*pl*) *Cicadidae*; (bot) cycad, *Cycas*

цика́д·к·а, -и,   (*f*) (ent) leafhopper, *Cicadella*

цика́д·ов·ые, -ых, (*pl decl as adj*) (ent) Auchenorhyncha; (bot) Cycadales

цикад·оид·е́·и, -й,   (*f pl*) (pal bot) Cycadeoidea

цикадо·фи́т·ы, -ов,   (*m pl*) (bot) cycadophytes, *Cycadophytae*

ци́кл·, -а, (*m*) cycle; (chem) ring

~, век·ов·о́й   secular cycle

~, ве́рх·н·ий   (math) cocycle

~, двой·н·о́й   dual/binary cycle

~, за́·мк·нут·ый   closed cycle; (autom) closed-loop feedback

~, ма́л·ый   minor cycle (computers)

~ на бы́стр·ых нейтро́н·ах   (nucl) fast-neutron cycle

~, не·за́·мк·нут·ый   open cycle

~, по·вто́р·н·ый   (chem) recycle; (autom) repetitive cycle

~, по·сто·я́нн·ый   fixed cycle

~, пре·де́ль·н·ый   limit cycle

~, рабо́ч·ий   operating cycle, cycle of operation

~, разо́·мк·нут·ый   open cycle; (autom) open-loop feedback

~, регенерати́в·н·ый   recycle, recovery cycle

~ с тремя́ давл·е́ни·ями   three-pressure steam cycle

~, с·ло́ж·н·ый   compound cycle

~, с·ме́ш·анн·ый   dual cycle (diesel)

~, шести·чле́н·н·ый   (chem) six-membered ring

цикламе́н·, -а,   (*m*) (bot) cyclamen, *Cyclamen*

цикл·ами́н·, -а, (*m*) (chem) cyclamine

цикл·а́н·, -а, (*m*) (chem) cyclane

цикл·ева́льн·ый стан·о́к·   router (woodworking machine)

цикл·ева́ни·е, -я, (*n*) = **цикл·ёвк·а**

цикл·ёвк·а, -и,   (*f*) routing (woodworking)

цикли́д·ы Дюпе́н·а (*pl*) (math) cyclides of Dupin

цикл·иза́ци·я, -и,   (*f*) cyclization, (chem) chelation, ring formation

цикл·и́ческ·ий, -ая, -ое, (*a*) cyclic, cycling, batch (of productive processes or plant)

~ реа́ктор· (*m*) (nucl) batch reactor

~ со·един·е́ни·е (*n*) (chem) cyclo-compound, cyclic compound

цикл·и́чност·ь, -и, (*f*) cyclicity, cyclic recurrence

цикл·и́чн·ый, -ая, -ое, (*a*) = **цикл·и́ческ·ий**

цикло·алкадие́н·, -а, (*m*) cycloalcadiene

цикло·бута·ди·е́н·, -а,   (*m*) (chem) cyclobutadiene

цикло·бута́н·, -а,   (*m*) (chem) cyclobutane, tetramethylene

цикл·ов·о́й -а́я, -о́е, (*a*) cycle, cyclic

цикло·гекса́н·, -а,   (*m*) (chem) cyclohexane

цикло·гексано́л·, -а, (*m*) cyclohexanol

цикло·гексано́н·, -а,   (*m*) cyclohexanone

цикло·гене́з·, -а, (*m*) (meteor) cyclogenesis

цикло·гепта́н·, -а, (*m*) (chem) cycloheptane

цикло·гра́мм·а, -ы, (*f*) (elec) cyclogram

цикло́·граф·, -а,   (*m*) (instr) cyclograph; crane cycle recorder

цикло·диа́лиз·, -а, (*m*) (med) cyclodialysis

цикло·дро́м·, -а, (*m*) cycle race-track

цикло·за·мещ·ённ·ый, -ая, -ое, (*a*) (chem) cyclo-substituted

цикл·о́ид·а, -ы, (*f*) (math) cycloid

циклоида́ль·н·ый, -ая, -ое, (*a*) cycloidal

цикло́ид·н·ый, -ая, -ое, (*a*) cycloid

цикло·каучу́к·, -а, (*m*) (chem) cyclorubber

цикло́·лиз·, -а, (*m*) (meteor) cyclolysis

цикло·ме́р·, -а, (*m*) cycle counter

цикло́·метр·, -а, (*m*) (instr) cyclometer

цикло́н·, -а,   (*m*) (meteor) cyclone; (min etc.) cyclone, cyclone dust separator

~, батаре́й·н·ый   (min) multiple cyclone

ли́н·и·я це́нтр·а цикло́н·а   (*f*) track of cyclone

~, терм·и́ческ·ий   (meteor) thermal depression

~, троп·и́ческ·ий   (meteor) cyclone, tropical revolving storm

циклон·и́ческ·ий, -ая, -ое, (*a*) cyclonic, cyclone

цикло́н·н·ый, -ая, -ое, (*a*) cyclone

цикло·окта́н·, -а, (*m*) (chem) cyclooctane

цикло·олефи́н·, -а, (*m*) (chem) cycloolefine

цикло́п·, -а, (*m*) (zool) Cyclops, (*pl*) cyclopoids, *Cyclopoida*

цикло·парафи́н·, -а,   (*m*) (chem) cycloparaffin

цикло·пента́н·, -а, (*m*) (chem) cyclopentane, pentamethylene

цикло́п·и́ческ·ий, -ая, -ое, (*a*) (build) cyclopean

цикло·пропа́н·, -а, (*m*) (chem) cyclopropane, trimethylene

цикло·симметр·и́·я, -и, (*f*) cyclosymmetry

цикло·сто́м·ов·ый, -ая, -ое, (*a*) (zool) cyclostomate

цикло·те́м·а, -ы, (*f*) (geol) cyclothem

цикло·тим·и́·я, -и, (*f*) (med) cyclothymia

цикло·трие́н·, -а, (*m*) (chem) cyclotriolefin

цикло·тро́н·, -а, (*m*) (phys) cyclotron

~, и́мпульс·н·ый pulsed cyclotron

~, миг·а́ющ·ий pulsed cyclotron

~ с двумя́ режи́м·ами double-mode cyclotron

цикло·тро́н·н·ый, -ая, -ое, (*a*) cyclotron

циклофо́н·, -а, (*m*) (elec) cyclophon

цикло·фора́з·н·ый, -ая, -ое, (*a*) (biol) cyclophorase

цикло·хи́н·, -а, (*m*) (pharm) cycloquine

цикло·холестери́н·, -а, (*m*) (biochem) cyclocholesterol

ци́кл·я, -и, (*f*) spokeshave, scraper (woodworking); slicker, scouring stone (leather)

цико́р·и·й, -я, (*m*) (bot) chicory, *Cichorium*

цику́т·а, -ы, (*f*) (bot) water hemlock, *Cicuta*

цикутокси́н·, -а, (*m*) (pharm) cicutoxin

цилиа́р·н·ый, -ая, -ое, (*a*) (biol) ciliary

цили́ндр·, -а, (*m*) cylinder, drum; (med) cast; top-hat

~ Вене́льт·а (elec) control electrode (of C.R.T.); control grid shield

~, дел·и́тельн·ый pitch cylinder (of gears)

~, коло́н·ков·ый (min) core barrel

~, осно́в·н·о́й base cylinder (of gears)

~ пере·ключ·е́ни·я shifter cylinder (pneumatic systems)

~, сил·ов·о́й actuating cylinder

~, скольз·я́щ·ий yawed cylinder

~ сре́д·н·его давл·е́ни·я intermediate-pressure cylinder (reciprocating engine)

~, тормоз·н·о́й braking cylinder; (M/T) brake drum

цилиндр·и́т·, -а, (*m*) (min) cylindrite

цилиндр·и́ческ·ий, -ая, -ое, (*a*) cylindrical

цилиндр·о́в·ый, -ая, -ое, (*a*) of **цили́ндр·**

цилиндр·о́в·ый, -ая, -ое

~ вту́л·к·а (*f*) cylinder liner

~ кры́ш·к·а (*f*) cylinder cover

цилиндр·о́ид·, -а, (*m*) (math) cylindroid

цимари́н·, -а, (*m*) (biochem) cymarin

цимари́н·ов·ая кисл·от·а́ (*f*) cymaric acid

цимар·о́з·а, -ы, (*f*) cymarose

цима́т·, -а, (*m*) zinc dimethyldithiocarbamate

цимато·ли́т·, -а, (*m*) (min) cymatolite

цимби·морфо́з·, -а, (*m*) (vet) cymbimorphiasis

цимици́н·, -а, (*m*) (pharm) cimicin

цимо·ге́н·, -а, (*m*) (chem) cymogene

цим·о́з·н·ый, -ая, -ое, (*a*) (bot) cymose

цим·о́л·, -а, (*m*) (chem) cymene

цимоли́т·, -а, (*m*) (min) cimolite

цимо·фа́н·, -а, (*m*) (min) cymophane

цинапи́н·, -а, (*m*) (chem) cynapine

цинг·а́, -и́, (*f*) (med) scurvy, scorbutus

цинг·о́тн·ый, -ая, -ое, (*a*) scorbutic, scurvy

цине́б·, -а, (*m*) zineb, zinc ethylene-bis-(dithiocarbamate) (a fungicide)

цинео́л·, -а, (*m*) (chem) cineole

цинера́ри·я, -и, (*f*) (bot) cineraria, *Cineraria*

ци́нк·, -а, (*m*) (chem) zinc, Zn; (*pl*) zinc slabs/protectors, sacrificial zinc anodes (against corrosion)

~, амальгам·и́рованн·ый zinc amalgam

~, като́д·н·ый electrolytic zinc

~, сер·ни́ст·ый zinc sulphide

~, серно·ки́сл·ый zinc sulphate

~, сыр·о́й (met) spelter

~, тимотиново·ки́сл·ый zinc thymotate

~, хромо·ки́сл·ый zinc chromate

цинк·а́т·, -а, (*m*) (chem) zincate

цинк·димети́л·, -а, (*m*) (chem) zinc methide

Ци́нке, ме́тод· (*m*) (chem) Zincke method

цинкени́т·, -а, (*m*) (min) zinkenite

цинки́т·, -а, (*m*) (min) zincite, red oxide of zinc

цинки́т·ов·ый, -ая, -ое, (*a*) zincite

цинк·ова́ни·е, -я, (*n*) (*v n*) of **цинк·ова́ть**

цинк·ова́ть, -у́ют, (*imp*) galvanize, galvanise, zinc, zinc-plate

ци́нк·ов·ый, -ая, -ое, (*a*) zinc, zincic, zincous

~ бел·и́л·а (*pl*) zinc white (pigment)

цинк·ов·ый, -ая, -ое
~ **кольц·о́**   (*n*) zinc protector ring (against corrosion)
~ **об·ма́н·к·а** (*f*) (min) zinc blende, sphalerite
~ **цвет·ы́**   (*pl*) (min) zinc bloom, hydrozincite
цинко·гра́ф·и·я, -и, (*f*) (print) zincography
~, **штрих·ов·а́я** line engraving
цинкози́т·, -а, (*m*) (min) zincosite, zinkosite
цинк·орган·и́ческ·ий, -ая, -ое, (*a*) (chem) organozinc
цинко·со·держ·а́щ·ий, -ая, -ее, (*a*) zinciferous
цинк·сульфи́д·н·ый, -ая, -ое, (*a*) zinc sulphide
ци́нн·а, -ы, (*f*) (bot) wood reed, *Cinna*
циннамо́мум·, -а, (*m*) (bot) cinnamon, *Cinnamomum*
циннвальди́т·, -а, (*m*) (min) zinnwaldite
ци́нни·я, -и, (*f*) (bot) zinnia
цин·о́вк·а, -и, (*g pl*) **-вок·,** (*f*) mat, matting
цино·глосс·и́н·, -а, (*m*) (pharm) cynoglossine
цин·одо́нт·ы, -ов, (*m pl*) (pal) cynodonts, *Cynodontia*
цину́бел·ь, -я, (*m*) tooth plane (woodworking)
цинхо-   (*int component*) (biochem) cincho-
цинхо́н·а, -ы,   (*f*) (bot) cinchona, *Cinchona*
цинхон·и́н·, -а, (*m*) (pharm) cinchonine
цинхони́н·ов·ая кисл·от·а́ (*f*) cinchoninic acid
цинхо́н·ов·ая кисл·от·а́ (*f*) cinchonic acid
циппе́йт·, -а, (*m*) (min) zippeite
циполи́н·, -а, (*m*) (min) cipolin
циприн·, -а, (*m*) (min) cyprine
цира́м·, -а, (*m*) ziram, zinc dimethyldithiocarbamate (fungicide)
ци́рк·, -а, (*m*) (theat) circus; (geog) cirque, corrie
циркалло́·й, -я, (*m*) (met) zircalloy
циркели́т·, -а, (*m*) (min) zirkelite
цирк·ов·о́й, -а́я, -о́е, (*a*) of **цирк·**
цирко́н·, -а, (*m*) (min) zircon
цирконáт·, -а, (*m*) (chem) zirconate
цирко́н·иев·ый, -ая, -ое, (*a*) (chem) zirconic, zirconium; zirconia
~ **ангидри́д·** (*m*) zirconia, zirconium anhydride/oxide/dioxide

цирко́н·иев·ый, -ая, -ое
~ **кисл·от·а́** (*f*) zirconic acid
~ **кирпи́ч·** (*m*) zirconia brick
цирко́н·и·й, -я, (*m*) (chem) zirconium, Zr
~, **четырёх·хло́р·ист·ый** zirconium tetrachloride
циркон·и́л·, -а, (*m*) (chem) zirconyl
цирконил·се́р·н·ая кисл·от·а́ (*f*) sulphozirconylic acid
цирко́н·ов·ый, -ая, -ое, (*a*) (*and see* **цирко́н·иев·ый**) zircon
~ **кирпи́ч·** (*m*) zircon brick, zirconium silicate brick
циркон·о́йд·, -а, (*m*) (geol) zirconoid
циркул·и́р·овать, -уют, (*imp*) circle; circulate
циркул·и́рующ·ий, -ая, -ее, (*pres part act*) of **циркул·и́р·овать**
ци́ркул·ь, -я, (*m*) compass(es) (drawing instrument)
~, **дел·и́тельн·ый** dividers
~, **-за·клёп·очник·, -а,** (*m*) bow compasses
~ **-из·мер·и́тель·, -я,** (*m*) dividers
~, **параллéль·н·ый** scribing compasses
~, **пропорциона́ль·н·ый** bisecting compasses
~, **рыча́ж·н·ый** beam compass(es)
~, **трёх·кон·е́чн·ый** triangular compasses
ци́ркуль·н·ый, -ая, -ое, (*a*) of **ци́ркул·ь**; circular
циркул·я́нт·н·ое, -ого (*n decl as adj*) (math) circulant
циркул·я́р·, -а, (*m*) circular, circular notice
циркул·я́рк·а, -и, (*g pl*) **-рок·,** (*f*) circular saw
циркул·я́рн·о- (*component*) circularly
циркул·я́рн·о-поляриз·о́ванн·ый, -ая, -ое, (*a*) circularly polarized
циркул·я́рн·ый, -ая, -ое, (*a*) circular; (telecom) group conference, round-table, multiple-call
~ **с·вяз·ь** (*f*) conference service (telephones)
циркул·я́торн·ый, -ая, -ое, (*a*) circulatory
циркуляцио́н·н·ый, -ая, -ое, (*a*) circulating, circulation, circulatory
циркул·я́ци·я, -и, (*f*) circulation; turning circle (of ships etc.)
~, **по·вто́р·н·ая** recirculation
~, **при·нуд·и́тельн·ая** forced circulation
~, **свобо́д·н·ая** free/natural circulation

циркум·нут·а́ци·я, -и, (*f*) (bot) circumnutation

циркум·поля́р·н·ый, -ая, -ое, (*a*) circumpolar

цирри·пе́д·иев·ый, -ая, -ое, (*a*) (zool) cirripedial

цирри·пе́д·и·и, -ев, (*m pl*) (zool) cirripeds, *Cirripedia*

цирро́·з·, -а, (*m*) (med) cirrhosis

цирро·ли́т·, -а, (*m*) (min) cirrolite

ци́ррус·, -а, (*m*) (biol) cirrus

цирто·ко́н·, -а, (*m*) (geol) cyrtocone

цирто·ли́т·, -а, (*m*) (min) cyrtolite

цирто·церако́н·, -а, (*m*) (zool) cyrtoceracone

цирце́·я, -и, (*f*) (bot) circaea, *Circaea*

цис- (*int component*) (chem) *cis-*

цис-изоме́р·, -а, (*m*) *cis*-isomer

цис-конфигур·а́ци·я, -и, (*f*) *cis*-configuration

цисс·о́ид·а, -ы, (*f*) (math) cissoid

~ **Диокле́с·а** cissoid of Diocles

ци́ст·а, -ы, (*f*) (biol) cyst

цисте·и́н·, -а, (*m*) (biochem) cysteine

цист·екта́зи·я, -и, (*f*) (med) cystectasia

цисте́рн·а, -ы, (*f*) (*see also* **систе́рн·а**) tank, cistern; (rail) tank wagon/car; (anat) cistern, cisterna

~**, акти́в·н·ая** (shipb) anti-rolling tank

~**, дифферѐнт·н·ая** (shipb) trimming tank

~**, за·пас·н·а́я** main feed tank (of boiler)

~**, изо·терм·и́ческ·ая** insulated tank; (rail) insulated tank-wagon/-car

~**, на·по́р·н·ая** gravity tank

~**, нефте·на·ли́в·н·а́я** (rail) oil tank wagon

~**, нефт·ян·а́я** (naut) fuel oil tank; (rail) fuel oil tank wagon

~**, рас·хо́д·н·ая** service tank

~**, рас·хо́д·н·о-на·по́р·н·ая** gravity service tank

~**, с·то́ч·н·ая** oil drain tank

цист·и́д·, -а, (*m*) (bot) cystid, cystidium

цист·и́н·, -а, (*m*) (chem) cystine

цист·и́т·, -а, (*m*) (med) cystitis

цисти·це́рк·, -а, (*m*) (zool) cysticercus

цисти·церко́з·, -а, (*m*) (med) cysticercosis

цисто·зо́ид·, -а, (*m*) (zool) cystozooid

цист·оиде́·и, -й, (*f pl*) (pal) cystoideans, *Cystoidea*

цисто·ка́рп·и·й, -я, (*m*) (bot) cystocarp

цисто·ли́т·, -а, (*m*) (bot) cystolith

цисто·обра́з·н·ый, -ая, -ое, (*a*) cystic, cystous

цисто·ско́п·, -а, (*m*) (med) cystoscope

цисто·ско́п·и́·я, -и, (*f*) (med) cystoscopy

цисто·сто́м·и́·я, -и, (*f*) (med) cystostomy

цисто·то́м·, -а, (*m*) (med) cystotome

цисто·то́м·и́·я, -и, (*f*) (med) cystotomy

цис-транс- (*int component*) (chem) *cis-trans-*

цис-транс изо·ме́р·и·я, -и, (*f*) *cis-trans* isomerism

цит·а́з·а, -ы, (*f*) (biochem) cytase

цита́т·а, -ы, (*f*) quotation, quote

цитва́р·н·ый, -ая, -ое, (*a*) (bot) wormseed, zedoary

~ **ко́рен·ь** (*m*) (bot) zedoary

~ **полы́н·ь** (*f*) (bot) wormseed, *Artemisia cina*

цитизи́н·, -а, (*m*) (chem) cytisine

цити́зус·, -а, (*m*) (bot) broom, *Cytisus*

цит·иро́ванн·ый, -ая, -ое, (*past part pass*) of **цит·и́р·овать**

цит·и́р·овать, -уют, (*imp*) quote, cite

цито·га́м·и·я, -и, (*f*) (gen) cytogamy

цито·ге́н·, -а, (*m*) (gen) cytogene

цито·ге́н·н·ый, -ая, -ое, (*a*) (gen) cytogenous

цито·кине́з·, -а, (*m*) (gen) cytokinesis

цито́·лиз·, -а, (*m*) (biol) cytolysis

цито·лизи́н·, -а, (*m*) cytolysin, cytotoxin

цито·ли́мф·а, -ы, (*f*) (cyt) cytolymph

цито·ло́г·и·я, -и, (*f*) (biol) cytology

цито·пла́зм·а, -ы, (*f*) (cyt) cytoplasm

цито·плазмат·и́ческ·ий, -ая, -ое, (*a*) (cyt) cytoplasmic

цито·сто́м·, -а, (*m*) (zool) cytostome

цито·хи́м·и·я, -и, (*f*) cytochemistry

цито·хро́м·, -а, (*m*) (biochem) cytochrome

цито·хро́м·оксида́з·а, -ы, (*f*) cytochrome oxidase

цито·хро́м·редукта́з·а, -ы, (*f*) cytochrome, reductase

цитр·ако́н·ов·ая кисл·от·а́ (*f*) (chem) citraconic acid

цитр·а́л·ь, -я, (*m*) (chem) citral

цитр·а́т·, -а, (*m*) (chem) citrate

цитратно·рас·твор·и́м·ый, -ая, -ое, (*a*) (chem) citrate-soluble

цитр·и́н·, -а, (*m*) (min) citrine, false topaz; (biochem) citrin, vitamin P

цитро́н·, -а, (*m*) (bot) citron, *Citrus medica*

цитрон·е́лл·а, -ы, (*f*) (bot) citronella grass, *Andropogon nardus*

цитрон·елл·а́л·ь, -я, (*m*) (chem) citronellal

цитрон·елл·о́л·, -а, (*m*) (chem) citron-ellol

цитр·улли́н·, -а, (*m*) (chem) citrulline

ци́тр·ус·, -а, (*m*) (bot) citrus

ци́тр·ус·ов·ый, -ая, -ое, (*a*) (bot) citrous, citrus

цифербла́т·, -а, (*m*) dial, dial plate, face (of clocks etc.)

цифербла́т·н·ый, -ая, -ое, (*a*) of **цифербла́т·**

цифр·а, -ы, (*f*) digit, figure (number)

~, дво·и́чн·ая bit, binary digit

~, зна́ч·ащ·ая significant digit

~, контро́ль·н·ые (*pl*) planned/projected figures

~ пе́рв·ого раз·ря́д·а first-order digit

~ пе́рв·ого с·лаг·а́ем·ого addend, addend digit

~, ста́р·ш·ая зна́ч·ащ·ая most significant digit (computers)

цифр·а́тор·, -а, (*m*) digitizer, analog-digital converter (computers)

~, матри́ч·н·ый step-matrix digitizer

~ с период·и́ческ·им с·чёт·ом chronometric digitizer, fixed-interval digitizer

~ угл·а́ shaft-position encoder/digitizer, angular-position quantizer

цифр·ова́тел·ь, -я, (*m*) (*and see* **цифр·а́тор·**) digitizer

цифр·ов·о́й, -а́я, -о́е, (*a*) digital, figure; numerical, numeral

~ да́·нн·ые (*pl*) figures, numerical data

~ маши́н·а (*f*) digital computer

ци́церо (*n indecl*) (print) pica

цмин·, -а, (*m*) (bot) everlasting, *Helichrisum*

ЦНИИ (*abbr*) = **центра́ль·н·ый на́-·уч·н·о-ис·след·ователь·ск·ий институ́т·** Central Scientific Research Institute

цоизи́т·, -а, (*m*) (min) zoisite

цо́к·а·ть, -ют, (*imp*) click, clink, clatter

цо́к·н·уть, -ут, (*perf*) click, clink, give a click/clink

цокол·ёвк·а, -и, (*f*) basing (of vacuum tubes)

цокол·ёвочн·ый стан·о́к· (*m*) basing machine (vacuum tubes)

цо́кол·ь, -я, (*m*) (arch) socle, (build) plinth course; (elec) base (of a vacuum tube), cap (of a lamp)

~, четырнадцати·штырь·ко́в·ый diheptal base (vacuum tubes)

~, штык·ов·о́й (elec) bayonet cap

цо́коль·н·ый, -ая, -ое, (*a*) of **цо́кол·ь**

~ вы·ём·к·а (*f*) basing dimple (vacuum tubes)

~ эта́ж· (*m*) (build) ground floor

цо́кор·, -а, (*m*) (zool) zokor, *Myospalax*

ЦОС (*abbr*) = **центра́ль·н·ая о́·пыт·-·н·ая ста́нци·я** Central Experimental Station

цуг·, -а, (*m*) tandem harness; train (of waves)

~ волн· wave train

цука́т·, -а, (*m*) (food) candied peel; candied/glace fruit

цуна́ми (*f indecl*) tsunami (seismology)

цыга́н·ск·ий, -ая, -ое, (*a*) Gipsy

цыпл·ёнок·, -нк·а, (*nom pl*) **-я́т·а,** (*g pl*) **-я́т,** (*m*) (agr) chick, poult

цыпл·я́т·а, (*nom pl*) of **цыпл·ёнок·**

цы́почк·ах, на (*adv*) on tiptoe

Ч

ч. (*abbr*) **час·** hour; **част·ь** part
чабáн·, -á, (*m*) (agr) shepherd
чабёр·, -бр·á, (*m*) (bot) savory, *Satureia*
чабр·éц·, -á, (*i*) **-óм,** (*m*) (bot) thyme, *Thymus*
чáвк·а·ть, -ют, (*imp*) munch, champ
чавы́ч·а, -и, (*i*) **-ей,** (*m*) (fish) king salmon, chinook, spring salmon, *Oncorhynchus tschawytscha*
чагерáк·, -а, (*m*) (bot) camelthorn, *Alhagi*
чáд·, -а, (*m*) fumes, acrid smoke
чае- (*root*) tea-
чае·вóд·, -а, (*m*) tea planter/grower
чае·дроб·и́лк·а, -и, (*g pl*) **-лок·,** (*f*) tea rolling machine
чае·за·вя́л·очн·ая машúн·а (*f*) tea-withering machine
чá·ек·, (*g pl*) of **чáй·к·а**
чае·у·бóр·очн·ый, -ая, -ое, (*a*) tea-picking
ча·úнк·а, -и, (*g pl*) **-нок·,** (*f*) tea-leaf
чá·й, -я, (*m*) tea; (bot) tea plant, *Thea sinensis*
~, луг·ов·óй (bot) moneywort, *Lysimachia nummularia*
чáй·к·а, -и, (*g pl*) **чá·ек·,** (*f*) (orn) gull, *Larus*, (*pl*) *Larinae*; gull (of aircraft wing)
~, оборóт·н·ая inverted gull (of aircraft wing)
~, серебр·и́ст·ая (orn) herring gull, *Laris argentatus*
~, трёх·пáл·ая (orn) kittiwake, *Rissa tridactyla*
чáй·ник·, -а, (*m*) teapot; kettle
чáй·н·ые, -ых, (*pl decl as adj*) (bot) *Theaceae*
чáй·н·ый, -ая, -ое, (*a*) tea
~ дéрев·о (*n*) tea, tea plant, *Thea sinensis*
чайóт·, -а, (*m*) = **чайóт·а**
чайóт·а, -ы, (*f*) (bot) chayote, *Sechium edule*
чáк·, -а, (*m*) (shipb) chock; (build) furring, firring
чáл·, -а, (*m*) = **чáл·к·а**
чáл·ить, -ят, (*imp*) (naut) secure along·side, make fast
чáл·к·а, -и, (*g pl*) **-л·ок·,** (*f*) securing/berting rope
чалм·á, -ы́, (*f*) turban
чалты́к·, -а, (*m*) paddy, unhusked rice
чáл·ый, -ая, -ое, (*a*) roan (colour)

чальмерзи́т·, -а, (*m*) (min) chalmersite
чáн·, -а, (*nom pl*) **-ы́,** (*g pl*) **-óв,** (*m*) vat, tub
~, контáкт·н·ый conditioning tank (ore dressing)
~, от·стóй·н·ый settling/clarifying tank
~, тарт·áльн·ый bailing tub (oil wells)
чáн·н·ый, -ая, -ое, (*a*) vat, tub
чан·ов·óй, -áя, -óе, (*a*) vat, tub
чарнокúт·, -а, (*m*) (min) charnockite
чар·овáть, -у́ют, (*imp*) charm, fascinate
чáртер·, -а, (*m*) (*see also* **чáртер-пáрт·и·я**) (com) charter
чáртер-пáрт·и·я, -и, (*f*) (com) charter party
час·, -а, and (with numbers 2, 3, 4) **-á,** (*nom pl*) **-ы́,** (*g pl*) **-óв,** (*m*) hour; (*pl*) clock, watch, sand-glass; hours
~, академ·и́ческ·ий school period (45 mins in USSR)
~, гáз·ов·ые (*pl*) gas meter, gas consumption meter
~, звёзд·н·ый (astron) sidereal hour
~, квáрц·ев·ые (*pl*) quartz clock, piezo-electric crystal clock
~, обрáт·н·ый (nucl) inverse hour, in hour
~ óт·дых·а rest time
~, песóч·н·ые (*pl*) sand glass, hour glass
~, при·ём·н·ые (*pl*) reception hours, consultation hours
~, при·клад·н·óй establishment, high water full and change (of a harbour)
~, руч·н·ы́е (*pl*) wrist watch
~, с·лич·и́тельн·ые (*pl*) (nav) desk watch
~, служ·éбн·ые (*pl*) office hours
~, сóлн·ечн·ые sun-dial; (astron) solar hours
~, учéб·н·ый school period (45 mins in USSR)
час·óвн·я, -и, (*g pl*) **-вен·,** (*f*) chapel
час·ов·óй, -áя, -óе, (*a*) hour, hour-long, an hour's, hourly; clock, watch; (*as noun*) sentry
~ грáф·ик· (*m*) time chart (industrial organization)
~ дéл·о (*n*) horology, clock/watch making
~ механ·и́зм· (*m*) clockwork
~ пó·езд· (*m*) the one o'clock train
~ пóяс· (*m*) time zone

час·ов·о́й, -а́я, -о́е

~ **рас·хо́д·** (*m*) an hour's consumption; hourly consumption

~ **стре́л·к·а** (*f*) hour-hand
по час·ов·о́й стре́л·к·е clockwise
про́тив· час·ов·о́й стре́л·к·е anticlockwise

~ **у́гол·** (*m*) (astron) hour angle

час·овщи́к·, -а́, (*m*) watch/clock maker/repairer

част·и́к·, -а́, (*m*) fish caught in a fine-mesh net

част·и́ц·а, -ы, (*i*) **-ей,** (*f*) particle; small part

~, **дли́нно·про·бе́ж·н·ая** (nucl) long-range particle

~, **коро́тко·про·бе́ж·н·ая** (nucl) short-range particle

~, **лави́н·н·ая** (nucl) cascade particle

~, **на·лет·а́ющ·ая** (nucl) bombarding/impinging/incident particle, projectile

~, **основ·н·а́я** (nucl) fundamental particle

~, **о·станов·и́вш·аяся** (nucl) arrested particle, particle at rest

~ **от·да́·ч·и** (nucl) recoil particle

~, **перв·и́чн·ая** (nucl) initial/primary particle, precursor

~ **-пред·ше́ств·енник·, -а,** (*m*) (nucl) precursor

~, **про́б·н·ая** (nucl) probe particle

~, **свобо́д·н·ая** (nucl) unbound/uncoupled particle

~, **с·луч·а́йн·ая** (nucl) uncontrolled/stray particle

~, **то́ч·ечн·ая** (nucl) point particle

част·и́чк·а, -и, (*g pl*) **-чек·** (*dim*) of **част·ь** (*f*)

част·и́чн·о (*adv*) partially, partly

част·и́чн·ый, -ая, -ое, (*a*) partial, fractional; particle

~ **рабо́ч·ий** (*m*) free-lance/home worker

~ **раз·ре́з·** (*m*) broken/partial section (engineering drawing)

ча́ст·н·о- (*component*) private-; (math) partial

ча́ст·н·ое, -ого (*n decl as adj*) (math) quotient

частно·пред·при·ним·а́тельск·ий, -ая, -ое, (*a*) (econ) private-enterprise

частно·со́б·ственническ·ий, -ая, -ое, (*a*) private-ownership/property

ча́ст·ност·, -и, (*f*) particular feature, particularity, peculiarity; detail
в ча́ст·ност·и in particular

ча́ст·н·ый, -ая, -ое, (*a*) (see also **ча́ст·н·ое**) (*as noun*); particular; private; (math) particular, partial

~ **про·из·во́д·н·ая** (*f*) (math) partial derivative

~ **реш·е́ни·е** (*n*) (math) particular solution

~ **с·лу́ч·а·й** (*m*) particular case

~ **со́б·ственност·ь** (*f*) (econ) private property

ча́ст·о (*adv*) often, frequently

часто·ко́л·, -а, (*m*) paling, palisade

част·от·а́, -ы́, (*nom pl*) **-о́т·ы,** (*f*) frequency

~, **альфа·гран·и́чн·ая** (TV) alpha-cutoff frequency

~ **би·е́ни·й** beat frequency

~, **бок·ов·а́я** (rad) side frequency

~, **вед·у́щ·ая** control/master frequency

~, **воз·бужд·а́ющ·ая** exciting/master frequency

~, **выс·о́к·ая** radio/high frequency (15–30 Mc/s)

~, **гас·я́щ·ая** quench frequency

~, **за·да·ю́щ·ая** driving/master frequency

~, **за·пир·а́ющ·ая** (elec) critical frequency (of filter)

~, **звук·ов·а́я** voice frequency

~, **зерка́ль·н·ая** image frequency

~, **и́мпульс·н·ая** pulsed frequency

~, **ка́др·ов·ая** (TV) frame frequency

~, **контро́ль·н·ая** pilot frequency

~, **круг·ов·а́я** circular/angular/cyclic frequency

~, **навигацио́н·н·ая** (rad) beacon frequency

~, **на·знач·енн·ая** (rad) allotted frequency

~ **на·руш·е́ни·я** (min) rate of failure
крив·а́я част·от·ы́ на·руш·е́ни·я (*f*) (min) rate of failure curve

~, **не·синхрон·и́рованн·ая** free-running frequency

~, **нес·у́щ·ая** (rad) carrier frequency

~ **нес·у́щ·ей из·ображ·е́ни·я** (TV) visual carrier frequency

~, **ни́з·к·ая** low/audio frequency

~, **о·по́р·н·ая** reference frequency

~, **основ·н·а́я** master/basic frequency

~, **по·бо́ч·н·ая** spurious frequency

~ **по·втор·е́ни·я** repetition frequency

~ **по·втор·е́ни·я и́мпульс·ов** pulse-repetition frequency

~, **под·нес·у́щ·ая** (rad) subcarrier frequency

част·от·á
~ полу·кáдр·ов (TV) field frequency
~, пре·дéль·н·ая limiting/cutoff frequency
~ пýск·ов (rocket) launching rate
~ раз·дел·éни·я crossover frequency
~, рáз·ностн·ая (elec) slip frequency
~ резонáнс·а тóк·ов antiresonance frequency
~, резонáнс·н·ая resonant frequency
~, сóб·ственн·ая free/natural frequency
~, срéд·н·яя medium frequency; resting/centre/center frequency (frequency modulation)
~ с·рéз·а cutoff/critical frequency
~ с·рýв·а колеб·áни·я quench frequency
~, стрóч·н·ая (TV) line frequency
~ тóч·ек· dot frequency
~, угл·ов·áя angular/radian frequency, pulsatance

частотно·времен·н·óй, -áя, -óе, (a) time-and-frequency

частотно·за·вúс·им·ый, -ая, -ое, (a) frequency-dependent

частотно·úмпульс·н·ая систéм·а (f) pulse-frequency system

част·óтн·о-модул·úрованн·ый, -ая, -ое, (a) frequency-modulated

частотно·фáз·ов·ая характер·úстик·а (f) frequency-response characteristic

част·óтност·ь, -и, (f) frequency factor, frequency

част·óтн·ый, -ая, -ое, (a) frequency
~ код· (m) pulse-frequency code
~ раз·дел·éни·е канáл·ов (m) (telecom) frequency-division of channels
~ у·плотн·éни·е (n) (telecom) frequency-division multiplex
~ характер·úстик·а (f) frequency characteristic

частото·мéр·, -а, (m) frequency meter, freq-meter
~, вибрациóн·н·ый vibrating-reed frequency meter
~, цифр·ов·óй digital frequency meter
~, электронно·с·чёт·н·ый electron-counting frequency meter

част·ýх·а, -и, (f) (bot) water plantain, *Alisma*

чáст·ый, -ая, -ое, (a) frequent, repeated, rapid; close, dense, fine
~ грéб·ен·ь (m) fine comb
~ дождь· (m) heavy rain
~ огóн·ь (m) (mil) rapid fire

чáст·ый, -ая, -ое
~ пере·плет·éни·е (n) (text) close weave

чáст·ь, -и, (nom pl) -и, (g pl) -éй, (f) (see also associated adjectives) part, portion, piece, share; section, department, (mil) unit; (anat) part, pars
~, бо·ев·áя warhead (of missile); (mil) operational unit; department (in warship)
~, больш·áя the majority, most of
~, воéн·н·ая military unit
~, голов·н·áя head, head section; (rocket) nose, nose cone
~, дóн·н·ая base, bottom, base section
~, за·пáс·н·ая spare part
~, нос·ов·áя nose, nose section
~, перéд·н·яя (rocket) fore-body
по·част·ям in parts/lots; in instalments
~, пере·хóд·н·ая adapter
~ рéч·и (gram) part of speech
~, хвост·ов·áя tail, tail section

чáст·ью (adv) partly, in part
больш·ей чáст·ью for the most part, mostly

час·ы́, (nom/acc pl) of час·

чатамúт·, -а, (min) chathamite

чатал·, -а, (m) (agr) fruit-tree prop

чаулмýгр·ов·ый, -ая, -ое, (a) chaulmoogric
~ мáсл·о (n) chaulmoogra oil

чáх·л·ый, -ая, -ое, (a) stunted (of plants), wasted, sickly

чáх·нуть, -нут, (imp) wither, become wasted/sickly

чах·óтк·а, -и, (f) wasting away, (med) phthisis
~, лёг·очн·ая pulmonary tuberculosis

чах·óточн·ый, -ая, -ое, (a) wasted, (med) phthisic, phthisical

чáш·а, -и, (i) -ей, (f) cup, bowl, basin, pot, dish, pan
~, бег·ýнн·ая edge-runner mill
~ больш·óго кóнус·а hopper, cone (blast furnace)
~ вес·óв scale pan, load platform (of weighing machine)
~, лúт·ников·ая (met cast) runner cup

чаше·вúд·н·ый, -ая, -ое, (a) cup/bowl-shaped, scyphoid, cupulate

чаше·лúст·ик·, -а, (m) (bot) sepal

чаше·обрáз·н·ый, -ая, -ое, (a) cup/bowl-shaped, scyphiform, cyathiform

чаше·цвéт·ник·, -а, (m) (bot) sweet shrub, *Calycanthus* ча́ш·ечк·а, -и, (*g pl*) -чек·, (*f*) (*dim*) of ча́ш·к·а; cup; pan (of weighing machine); (bot) calyx, scyphus (of lichens)

~ из·ло́м·а (met) fracture cup

чашечко·ви́д·н·ый, -ая, -ое, (*a*) cup-shaped, cupped, (bot) patellate, patelliform

ча́ш·ечн·ый, -ая, -ое, (*a*) of ча́ш·к·а

~ манжёт·а (*f*) cup packing

ча́ш·к·а, -и, (*g pl*) -ш·ек·, (*f*) cup, bowl, pan, dish; scale, pan (of scales); cistern (in mercury holding instruments)

~, коле́н·н·ая (zool) patella

~, над·коле́н·н·ая (zool) patella

~ Пе́три (bact) Petri dish

ча́щ·а, -и, (*i*) -ей, (*f*) thicket

ча́щ·е (*comp*) more often/frequently; more closely/densely/finely

~ вс·его́ most often/frequently, usually

ча́·яни·е, -я, (*n*) expectation, hope

ча́·ять, -ют, (*imp*) hope for, expect

чда (*abbr*) = чи́ст·ый для ана́лиз·а (chem) analytically pure, analytical grade

чего́ (*gen*) of что

чей (*pron, nom sing m, acc sing m inanim*) whose

Чебышёв·а, много·член·ы (*pl*) (math) Tchebycheff's polynomials

чебышёв·ск·ий, -ая, -ое, (*a*) (math) Tchebycheff

че́к·, -а, (*m*) (fin) cheque, check

~, рас·чёт·н·ый crossed cheque

~, тури́ст·ск·ий traveller's cheque/ check

чек·а́, -и́, (*f*) cotter, cotter pin, linchpin (of wheel)

~, на·тяж·н·а́я gib

~, пред·о·хран·и́тельн·ая retaining pin (of hand grenade etc.)

~, раз·вод·н·а́я split pin

чека́н·, -а, (*m*) (*v n*) *see* чека́н·ить; coining/embossing die; caulking/calking iron

чека́н·ить, -ят, (*imp*) (met) coin, emboss, size; caulk (seams); (hortic) pinch out (buds)

чека́н·к·а, -и, (*g pl*) -н·ок·, (*f*) (*v n*) of чека́н·ить

чека́н·очн·ый, -ая, -ое, (*a*) of чека́н·к·а

~ молот·о́к· (*m*) caulking tool, caulker/calker

~ пресс· (*m*) (met forge) sizing press

чекма́р·н·ый, -ая, -ое, (*a*) of чек- ма́р·ь

чекма́р·ь, -я́, (*m*) smoother, smoothing tool; (civ eng) tamper, bumper

че́к·ов·ый, -ая, -ое, (*a*) (fin) cheque, check

чеко·да́т·ел·ь, -я, (*m*) (fin) drawer of cheque/check

чёл, (*past masc sing*) of честь (*imp*)

Челли́ни гало́ (*n*) (phys) Cellini halo

челн·о́к·, -а́, (*m*) cock-boat, dugout canoe; (text) shuttle; (nucl) rabbit

ки́д·к·а челн·ок·а́ (*f*) (text) picking

~, само·лёт·н·ый (text) fly/flying shuttle

челн·о́чн·ый, -ая, -ое, (*a*) of. челн·о́к·

~ коро́б·к·а (*f*) shuttle box (of loom)

челове́к·, -а, (*g pl with numeral*) чело- ве́к·, (*m*) man, person, (*pl*) people, persons, men

~, гейдельбе́рг·ск·ий *Homo heidelbergensis* (anthropology)

челове́к·о-де́н·ь, -дн·я́, (*m*) (econ) man-day

человеко·не·на·ви́ст·ник·, -а, (*m*) (med) misanthrope

человеко·обра́з·н·ый, -ая, -ое, (*a*) anthropoid; (*pl as noun*) (zool) anthropoids

~ обезья́н·ы (*pl*) (zool) anthropoid apes

человеко·по·до́б·н·ый, -ая, -ое, (*a*) anthropoid

челове́к·о-ча́с·, -а, (*nom pl*) -ы́, (*g pl*) -о́в, (*m*) (econ) man-hour

челове́ч·еск·ий, -ая, -ое, (*a*) human; humane, kind

челове́ч·еств·о, -а, (*n*) mankind, humanity

челове́ч·н·ый, -ая, -ое, (*a*) humane, kind

челюст·н·о́й, -а́я, -о́е, (*a*) jaw, gnathic, maxillary

челюстно·ро́т·ые, -ых, (*pl decl as adj*) (zool) gnathostomes, *Gnathostomata*

челюстно·с·лёз·н·ый, -ая, -ое, (*a*) (anat) maxillolacrimal, maxillolacrymal

че́люст·ь, -и, (*f*) jaw

~, ве́рх·н·яя (anat) maxilla

~, ни́ж·н·яя (anat) mandibula

чемери́ц·а, -ы, (*f*) (bot) false hellebore, *Veratrum*

чем (*pron*) (*instr*) of что (q.v.); (*conj*) than; (+ *inf*) instead of

чём (*pron*) (*prep*) of что

чемода́н·, **-а,** (*m*) suitcase, trunk, bag
~, матро́с·ск·ий (naut) kitbag
чемпио́н·, **-а,** (*m*) champion
чему́ (*pron*) (*dat*) of **что**
чепра́к·, **-а́,** (*m*) bends (leather); saddle cloth
чепух·а́, **-и́,** (*f*) nonsense
че́рв·а, **-ы,** (*f*) bee grub
черве·обра́з·н·ый, **-ая, -ое,** (*a*) vermiform, vermicular, worm-like, (*pl as noun*) (zool) *Vermoidea*
~ от·ро́ст·ок· (*m*) (anat) vermiform appendix
черв·ец·ы́, **-о́в,** (*m pl*) (ent) scale insects, coccines, *Coccina*
черве·я́д·н·ый, **-ая, -ое,** (*a*) (zool) vermivorous
че́рв·и (*pl*) of **черв·ь**
червив·е·ть, **-ют,** (*imp*) become wormy/ worm-ridden/eaten
черв·и́в·ый, **-ая, -ое,** (*a*) worm-eaten/ ridden, wormy
черво́н·ец·, **-н·ц·а,** (*m*) (obs) ten-rouble note
черво́н·н·ый, **-ая, -ое,** (*a*) pure (gold)
черво·то́ч·ин·а, **-ы,** (*f*) worm hole(s)
черво·то́ч·н·ый, **-ая, -ое,** (*a*) worm-eaten
черво·я́д·н·ый, **-ая, -ое,** (*a*) (zool) vermivorous
черв·ь, **-я́,** (*nom pl*) че́рв·и, (*g pl*) **-е́й,** (*m*) (zool) worm; (telecom) letter **Ч;** (*f pl*) hearts (at cards)
~, колюче·голо́·вые (*pl*) acantho-cephalans, acanthocephalids, *Acantho-cephala, Acanthocephali*
~, корабе́ль·н·ый shipworm, *Teredo*
~, ко́ль·чат·ые (*pl*) annelids, *Annelida, Annelides*
~, кру́гл·ые (*pl*) nemathelminths, *Nemathelminthes*
~, ле́нт·очн·ые (*pl*) tapeworms, *Cestoda*
~, мало·щет·и́нков·ые (*pl*) oligo-chaetes, *Oligochaeta*
~, мно́го·щет·и́нков·ые (*pl*) poly-chaetes, *Polychaeta*
~, ни́т·чат·ый nematode, *Nematodes*
~, пло́ск·ие (*pl*) flatworms, platy-helminths, *Platyhelminthes*
~, ресни́ч·н·ые (*pl*) turbellarians, *Turbellaria*
~, шелк·ови́чн·ый silkworm, *Bombyx*
~, щетинко·но́г·ие (*pl*) chaetopodans, *Chaetopoda*
~, щетинко·челюст·н·ы́е (*pl*) chaeto-gnaths, *Chaetognatha*

черв·я́г·и (*f pl*) (zool) *Gymnophiona, Caeciliidae*
черв·я́к·, **-а́,** (*m*) worm; screw (of extruder etc.)
~ и се́ктор· (mech) worm-and-sector
~, па́д·аю́щ·ий drop/trip worm
~, по·да·ю́щ·ий conveyer worm/screw
~, про·из·вод·я́щ·ий generating worm (worm gear cutting)
~, само·тормоз·я́щ·ий irreversible worm
черв·я́чн·ый, **-ая, -ое,** (*a*) worm; screw (of extruder etc.)
~ колес·о́ (*n*) (mech) worm wheel
~ па́р·а (*f*) worm-and-wheel, worm-and-worm wheel
~ пресс· (*m*) extruder, tubing/extrud-ing machine (for rubber etc.)
~ транспорт·ёр· (*m*) worm/screw conveyer
чердак·, **-а́,** (*m*) (arch) attic, garret
черда́ч·н·ый, **-ая, -ое,** (*a*) attic, garret
черед·а́, **-ы́,** (*f*) (bot) beggarticks, *Bidens*
черед·ова́ни·е, **-я,** (*n*) (*v n*) of **чере-д·ова́ть;** alternation, cycling inter-change; sequence, order (of events)
~ вс·пы́ш·ек· (ICE) firing sequence/ order
~ о·лед·ене́ни·й (geol) glacial sequence
черед·ова́ть, **-у́ют,** (*imp*) alternate, keep exchanging/changing; **-ся** alter-nate, take turns
че́рез (*prep* + *acc*) across, over; through, (pharm) per; via, by way of; in, after (of time); by, in terms of
~ за́д·н·ий про·хо́д· (pharm) per anum
~ лес· through/via the wood
~ пол·час·а́ in/after half an hour
~ рот· (pharm) per os
~ рек·у́ over/across the river
~ у́л·иц·у across the road
через·зёрн·иц·а, **-ы,** (*f*) unseeded rye ear
через·кра́й·н·ый шов· (*m*) (text) oversewn seam
черёмух·а, **-и,** (*f*) (bot) cherry, bird-cherry, *Padus*
Черенко́в·а-Вави́лов·а эффе́кт· (*m*) (phys) Čerenkov radiation
черен·кова́ни·е, **-я,** (*n*) (*v n*) *see* **че-ренк·ова́ть**
черенк·ова́ть, **-у́ют,** (*imp*) (hortic) graft, take cuttings

черенко́в·ск·ий, -ая, -ое, (*a*) Čerenkov
черен·ко́в·ый, -ая, -ое, (*a*) of **че-
рен·о́к·**; = **черенко́в·ск·ий**
черен·о́к·, -н·к·а́, (*m*) (hortic) scion,
cutting; handle, haft; solen (shellfish)
че́реп·, -а, (*nom pl*) -а́, (*g pl*) -о́в,
(*m*) (anat) skull, cranium
череп·а́х·а, -и, (*f*) (zool) tortoise;
turtle; (*pl*) *Chelonia, Testudinata*
~, ко́ж·ист·ая leatherback turtle,
Dermochelys coriacea
~, кус·а́ющ·аяся snapping turtle,
Chelydra serpentina
~, мя́гк·ая soft-shelled turtle, (*pl*)
Tryonychidae
~, вре́д·н·ая (ent) shell-bug, *Eury-
gaster integriceps*
череп·а́шн·ий, -ья, -ье, (*a*) of **чере-
п·а́х·а**
череп·и́тчат·ый, -ая, -ое, (*a*) (build)
tiled, (biol) tegular
череп·и́ц·а, -ы, (*i*) -ей, (*f*) (build)
tile, tegula; tiling, tiles; (biol) tegula,
(*pl*) tegulae
~, вальм·о́в·ая hip tile
~, голла́нд·ск·ая (build) pantile
~, жело́б·чат·ая Spanish tile
~, конь·ко́в·ая ridge tile
~, пло́ск·ая crown/plane tile
~, ри́м·ск·ая Italian tile
~, шпунт·ов·а́я bent/gutter tile
череп·и́чн·ый, -ая, -ое, (*a*) tile,
tiled; (biol) tegular
черепно·гло́т·очн·ый, -ая, -ое, (*a*)
(anat) craniopharyngeal
череп·н·о́й, -а́я, -о́е, (*a*) cranial;
(*pl as noun*) (zool) *Craniata*
черепно·мозг·ов·о́й, -а́я, -о́е, (*a*)
(anat) craniocerebral
череп·о́в·ый, -ая, -ое, (*a*) *see* **че-
реп·н·о́й**
~ по́яс· (*m*) reinforced bow planking
(wooden ship)
череп·о́к·, -п·к·а́, (*m*) (ceram) sherd,
potsherd
черепо·ко́ж·н·ый, -ая, -ое, (*a*) (biol)
testaceous
чересл·о́, -а́, (*m*) (agr) coulter
черес·поло́с·иц·а, -ы, (*f*) strip holding
(of lands)
черес·сед·е́льник·, -а, (*m*) saddle-
strap
черес·стро́ч·н·ый, -ая, -ое, (*a*)
(TV, radar) line-jump, interlaced
~ раз·вёрт·к·а (*f*) (radar) line-jump/
interlaced scanning

чересчу́р (*adv*) too, too much, excessively, immoderately
череш·ко́в·ый, -ая, -ое, (*a*) of **че-
реш·о́к**; stalked, pedunculate, pedi-
cellate
череш·о́к·, -ш·к·а́, (*m*) peduncle,
pedicle, pedicel, stalk, foot-stalk, stem
череш·о́чек·, -ч·к·а, (*m*) of **чере-
ш·о́к·**
черк·а́·ть, -ют, (*imp*) cross out/off
черк·н·у́ть, -у́т, (*perf*) leave a mark/
furrow/line; scribble
черн·е́ни·е, -я, (*n*) (*v n*) *see* **черн·е́·ть;**
blackening
черн·ённ·ый, -ая, -ое, (*past part
pass*) of **черн·е́·ть;** blackened
черн·е́·ть, -ют, (*imp*) become/turn/
grow black; appear/show black
черн·и́к·а, -и, (*f*) (bot) bilberry,
Vaccinium
Черникеев·а, лаг· (*m*) (naut) Cherni-
keef log
черн·и́л·а (*n pl*) ink
~, не·с·мыв·а́ем·ые indelible ink
черн·и́льниц·а, -ы, (*i*) -ей, (*f*) ink-
pot/well
черн·и́льн·ый, -ая, -ое, (*a*) ink
~ за́·пис·ь (*f*) ink/pen recording
~ каранда́ш· (*m*) indelible pencil
~ меш·о́к· (*m*) (zool) ink sac
~ оре́ш·ек· (*m*) (bot) gall nut
черн·и́ть, -я́т, (*imp*) blacken, make
black
чёрн·о-бе́л·ая съ·ём·к·а (*f*) black-
and-white photography
черн·о-бу́р·ый, -ая, -ое, (*a*) dark-
brown
черно·бы́ль·ник·, -а, (*m*) (bot) mug-
wort, *Artemisia vulgaris*
Черно́в·а, ли́н·и·и (*pl*) (met) Lüders/
Piobert/Hartmann lines, stretcher
strains, strain figures, worms
~, то́ч·к·и (*pl*) (met) critical points
черн·ова́т·ый, -ая, -ое, (*a*) blackish
черн·ов·а́я, -о́й, (*f decl as adj*) = **черн·овы́к·**
черн·овы́к·, -а́, (*m*) rough copy/notes
черн·ов·о́й, -а́я, -о́е, (*a*) (*see also*
черн·ов·а́я) (*as noun*); (mech) rough,
roughing; draft (of documents); (met)
regulus, unrefined, crude, raw
~ мета́лл· (*m*) regulus, raw/unrefined
metal
~ об·рабо́т·к·а (*f*) (mech) roughing
~ сурьм·а́ (*f*) regulus metal
черно·го́р·ск·ий, -ая, -ое, (*a*) Monte-
negrin

черно·зём·, -а, (*m*) chernozem (soil)

черно·зём·н·ый, -ая, -ое, (*a*) chernozem, chernozemic

черно·клён·, -а, (*m*) Tatar maple, *Acer tataricum*

черно·корен·ь, -рн·я, (*m*) (bot) hound's tongue, *Cynoglossum*

черно·лес·ь·е, -я, (*n*) deciduous forest

черно·мор·ск·ий, -ая, -ое, (*a*) Black Sea

черно·пиш·ущ·ий аппарат· (*m*) (telecom) ink writer

черно·рабоч·ий, -ая, -ее, (*a*) unskilled; (as noun) unskilled worker

черно·слив·, -а, (*m*) (food) prunes

черн·от·а, -ы, (*f*) blackness

чёрн·ый, -ая, -ое, (*a*) black; unskilled, manual; (met) ferrous; (mech eng) rough, rough-finished; illegal

~ **вод·а** (*f*) water on frozen river

~ **гидрат· никел·я** (chem) nickelic hydroxide

~ **круж·ок·** (*m*) filled circle (symbol on graphs etc.)

~ **металл·ург·и·я** (*f*) ferrous metallurgy, iron-and-steel industry

~ **пар·ы** (*pl*) (agr) fallow land

~ **порох·** (*m*) (expl) gunpowder

~ **рын·ок·** (*m*) black market

~ **хлеб·** (*m*) black bread, rye bread

черн·ыш·, -а, (*i*) -**ом,** (*m*) (bot) rough-stalked boletus, *Boletus scaper*

чёрн·ь, -и, (*f*) black (pigment etc.)

черп·ак·, -а, (*m*) scoop, bucket, ladle, bailer; gooseneck (of die-caster machine)

черп·алк·а, -и, (*g pl*) -**лок·,** (*f*) = **черп·ак·**

черпало·вид·н·ый, -ая, -ое, (*a*) (zool) arytaenoid

черп·альн·ый, -ая, -ое, (*a*) scooping, ladling, bailing

~ **форм·а** (*f*) (paper) mould, mold

черп·а·ть, -ют, (*imp*) scoop, ladle, bail; derive, draw on (figurative)

~ **борт·ом** (naut) ship at sea

~ **корм·ой** (naut) be pooped

черп·ачк·а, -и, (*g pl*) -**чек·,** (*f*) dipper

черп·н·уть, -ут, (*perf*) *see* **черп·а·ть**

черств·е·ть, -ют, (*imp*) become stale/ hardened

чёрств·ый, -ая, -ое, (*a*) stale; hardened, callous

черт· (*abbr*) = **черт·ёж·** drawing

чёрт·, -а, (*nom pl*) **черт·и,** (*g pl*) -**ей,** (*m*) devil

черт·а, -ы, (*f*) trace, feature, characteristic; line; boundary; (min) streak; (print) vinculum, bar

  в общ·их черт·ах in outline, briefly

~, **курс·ов·ая** (naut) lubber's line; drift line (a/c bombsight)

~, **нул·ев·ая** zero mark

~, **от·лич·ительн·ая** distinctive feature

черт·ёж·, (*g*) **черт·еж·а,** (*i*) -**ом,** (*m*)

~, **габарит·н·ый** outline drawing

  деталь·черт·ёж· (*m*) detail drawing

~, **индивидуаль·н·ый** special-component drawing

~, **ис·полн·ительн·ый** working drawing

~, **компон·овочн·ый** layout drawing, general arrangement

~, **конструктив·н·ый** design drawing

~, **монтаж·н·ый** (mech eng) installation drawing; (civ eng) erection drawing

~, **нем·ой** blank standard drawing

~, **нормализ·ованн·ый** departmental standard drawing

~, **о·без·лич·енн·ый** standard-component drawing

~ **общ·его рас·по·лож·ени·я** general-arrangement drawing

~, **о·пыт·н·ый** prototype drawing

~, **от·чёт·н·ый** alternative-and-additions drawing

~, **по·втор·яем·ый** standard-component drawing

~, **рабоч·ий** working drawing

~, **ремонт·н·ый** repairs drawing; (shipb) refit drawing

~, **с·бор·очн·ый** assembly drawing

~, **сер·ийн·ый** finalized series-drawing

~, **стандарт·н·ый** State standard drawing

~, **табл·ичн·ый** tabular drawing

~, **теорет·ическ·ий** (shipb) lines drawing

~, **тип·ов·ой** standard-project drawing

~, **унифиц·ированн·ые** (*pl*) selected set of standard drawings

черт·ёжн·ая, -ой, (*f decl as adj*) drawing office

черт·ёжник·, -а, (*m*) draughtsman, draftsman

черт·ёжн·ый, -ая, -ое, (*a*) (*see also* **черт·ёжн·ая**) (*as noun*); drawing, draughting, drafting

~ **бумаг·а** (*f*) drawing paper

~ **стан·ок·** (*m*) draughting machine

черт·илк·а, -и, (*f*) (eng etc.) scriber, marking tool

черт·и́ть, -ˈят, (*imp*) draw, plot
~ ка́рт·у map, chart
чертополо́х·, -а, (*m*) (bot) bristle-thistle, *Carduus*
черч·е́ни·е, -я, (*n*) (*v n*) of **черт·и́ть;** drawing, plotting
черчи́т·, -а, (*m*) (min) churchite
чес·а́вш·ий, -ая, -ее, (*past part act*) of **чес·а́ть**
чес·а́лк·а, -и, (*g pl*) **-лок·,** (*f*) (text) comb; hackle (for flax etc.)
чес·а́льн·ый, -ая, -ое, (*a*) (text) combing, carding; hackling (of flax etc.)
~ маши́н·а (*f*) (text) carding engine
~ маши́н·а, ва́л·ичн·ая roller carding engine
~ маши́н·а, шля́п·очн·ая revolving flat carding engine
чес·а́льщик·, -а, (*m*) (text) carder
чес·а́ни·е, -я, (*n*) (*v n*) of **чес·а́ть**
чёс·анн·ый, -ая, -ое, (*a*) (*past part pass*) of **чес·а́ть**
чес·а́ть, (*pres 3rd pl*) **чёш·ут,** (*imp*) comb; (text) comb, card; hackle (flax etc.); scratch (to alleviate irritation)
чесно́к·, -а́, (*m*) (bot) garlic, *Allium sativum*
чесно́ч·н·ый, -ая, -ое, (*a*) (bot) garlic, *Allium sativum*
чес·о́тк·а, -и, (*f*) (med) scabies
чес·о́точн·ый, -ая, -ое, (*a*) scabby
че́ств·овать, -уют, (*imp*) celebrate in honour/honor of, celebrate
че́ст·ност·ь, -и, (*f*) honesty, integrity
че́ст·н·ый, -ая, -ое, (*a*) honest, honourable, honorable
често·люб·и́в·ый, -ая, -ое, (*a*) ambitious
че́ст·ь, -и, (*f*) honour, honor
чест·ь, (*pres 3rd pl*) **чт·ут,** (*imp*) see **чит·а́·ть, с·чит·а́·ть;** (obs) read; (obs) consider, deem
чёт·, -а, (*m*) even number
четве́рг·, -а́, (*m*) Thursday
четвер·е́ньк·и, -н·ек·, (*f pl*); **на четвер·е́ньк·ах**
четвер·и́чн·ый, -ая, -ое, (*a*) quaternary
~ за́·пис·ь (*f*) quaternary/tetral notation
четвёр·к·а, -и, (*g pl*) **-р·ок·,** (*f*) a four, a foursome, a set/team of four, a number four; quad (telephone cable); good (4 marks out of 5 in test etc.); four-oared boat
четвер·ни́к·, -а́, (*m*) four-eye plate; (cryst) fourling

четвер·н·о́й, -а́я, -о́е, (*a*) quadruple, fourfold; (chem) quaternary; tetra-
~ диагра́мм·а (*f*) (chem) quaternary diagram
~ рас·шир·е́ни·е (*n*) (st eng) quadruple expansion
~ систе́м·а (*f*) (chem, met) quaternary system
~ то́ч·к·а (*f*) quadruple point (on a graph)
че́твер·о, -ы́х, (*n*) four, four together/altogether
четверо·но́г·ий, -ая, -ое, (*a*) quadruped, tetrapodal, four-legged/footed, (*pl as noun*) (zool) quadrupeds, tetrapods, *Tetrapoda*
четверо·ру́к·ий, -ая, -ое, (*a*) four-armed, tetrabranchial, (*pl as noun*) (zool) quadrumanes, *Quadrumana*
четвёр·очн·ый, -ая, -ое, (*a*) of **четвёр·к·а**
~ ка́бел·ь (*m*) (elec) quadded cable
четверт·и́н·а, -ы, (*f*) quartering (timber); quarter; quarter-litre bottle
четверт·и́чн·ый, -ая, -ое, (*a*) quaternary, tetral; (geol) Quaternary
~ со·един·е́ни·е (*n*) (chem) quarternary compound
четверт·н·о́й, -а́я, -о́е, (*a*) quarter; quadrantal
~ ра́дио·девиа́ци·я (*f*) (rad) quadrantal D/F error
четверт·у́шк·а, -и, (*g pl*) **-шек·,** (*f*) (print) quarto
четвёрт·ый, -ая, -ое, (*a*) fourth
че́тверт·ь, -и, (*nom pl*) **-и,** (*g pl*) **-е́й,** (*f*) quarter, a fourth, quadrant (of a circle)
~ кирпич·а́ (build) quarter bat
~ лист·а́ (paper) quarto
четверть·волн·ов·ый, -ая, -ое, (*a*) (phys) quarter-wavelength/-wave
~ ли́н·и·я (*f*) (rad) quarter-wavelength line
четверть·километр·о́вк·а, -и, (*f*) (surv) scale 1/250,000 map
четверть·фина́ль·н·ый, -ая, -ое, (*a*) quarter-final (of competition)
чёт·к·ий, -ая, -ое, (*a*) clear, clear-cut; precise, accurate
чёт·кост·ь, -и, (*f*) precision, accuracy; sharpness, definition, contrast
чёт·н·о-не́·чет·н·ый, -ая, -ое, (*a*) even-odd
чёт·ност·ь, -и, (*f*) parity
~, вну́тр·енн·яя intrinsic parity

чёт·н·о-чёт·н·ый, -ая, -ое, (*a*) even-even

чёт·н·ый, -ая, -ое, (*a*) even (of numbers)

чёт·ш·ий, -ая, -ее, (*past part act*) of **чест·ь** (*imp*)

четы́р·е, (*g*) **четыр·ёх** (numeral) four

четыре·деся́т·ник·, -а, (*m*) chronometer watch

четы́ре·жд·ы (*adv*) four times

четыр·ём (*dat*) of **четы́р·е**

четы́ре·ст·а, (*g*) **четырёх·со́т·,** (numeral) four hundred

четырём·ста́м· (*dat*) of **четы́ре·ст·а**

четыр·ёх (*gen/prep*) of **четы́р·е;** four-, tetra-

четырёх·а́том·н·ый, -ая, -ое, (*a*) tetratomic, tetrabasic (of alcohol)

четырёх·бро́м·ист·ый, -ая, -ое, (*a*) (chem) tetrabromide (of)

четырёх·вале́нт·н·ый, -ая, -ое, (*a*) (chem) tetravalent, quadrivalent

чстырёх·вал·ко́в·ый, -ая, -ое, (*a*) four-high (rolling mill); four-bowl (calender)

четырёх·галоге́н·н·ое со·един·е́ни·е (*n*) (chem) tetrahalide

четырёх·глав·ый, -ая, -ое, (*a*) (anat) quadriceps

четырёх·гра́н·ник·, -а, (*m*) (math) tetrahedron

четырёх·гра́н·н·ый, -ая, -ое, (*a*) (math) tetrahedral, four-sided/edged

четырёх·дн·е́вн·ый, -ая, -ое, (*a*) four-day; (med) quartan

четырёх·жа́бер·н·ый, -ая, -ое, (*a*) tetrabranchiate, (*pl as noun*) (zool) tetrabranchiates, tetrabranchs, *Tetrabranchiata*

четырёх·за·мещ·ённ·ый, -ая, -ое, (*a*) (chem) tetra-substituted; (*pl decl as adj*) tetraderivatives

четырёх·за·ря́д·н·ый, -ая, -ое, (*a*) (nucl) quadruply charged

четырёх·звен·ник·, -а, (*m*) four-link chain

~, шарни́р·н·ый (mech) four-bar motion/linkage

четырёх·зна́ч·н·ый, -ая, -ое, (*a*) (math) four-digit/-figure; four-unit/-letter (codes etc.); four-aspect (rail signals)

четы́рёх·йо́д·ист·ый, -ая, -ое, (*a*) (chem) tetraiodide of

четырёх·коле́н·н·ый, -ая, -ое, (*a*) (mech) four-throw (of crankshaft)

четырёх·колёс·н·ый, -ая, -ое, (*a*) four-wheeled, four-wheel

четырёх·кра́т·н·ый, -ая, -ое, (*a*) quadruple, four times, fourfold

~ рас·шир·е́ни·е (*n*) (st eng) quadruple expansion

~ систе́м·а (*f*) (telecom) quadruplex

четырёх·кры́·л·ый, -ая, -ое, (*a*) four-winged, tetrapterous, quadripennate

четырёх·ла́п·н·ый, -ая, -ое, (*a*) four-lugged/-clawed/-fluked

~ я́кор·ь (*m*) (naut) grapnel

четырёх·ле́т·н·ий, -яя, -ее, (*a*) four-year; four-year-old

четырёх·ли́ст·н·ый, -ая, -ое, (*a*) four-leaved, quadrifoliate, tetraphyllous

четырёх·ло́паст·н·ый, -ая, -ое, (*a*) (bot) four-lobed, quadrilobate; (mech) four-bladed/vaned

четырёх·луч·ев·о́й, -а́я, -о́е, (*a*) tetraradiate

четырёх·ме́ст·н·ый, -ая, -ое, (*a*) four place/seater/berth

четырёх·молекуля́р·н·ый, -ая, -ое, (*a*) (chem) quadrimolecular

четырёх·мото́р·н·ый, -ая, -ое, (*a*) (a/c) four-engined

четырёх·но́г·ий, -ая, -ое, (*a*) four-footed/legged, (zool) quadruped

четырёх·о́·кис·ь, -и, (*f*) (chem) tetroxide

~ се́р·ы sulphur/sulfur tetroxide

четырёх·осно́в·н·ый, -ая, -ое, (*a*) (chem) tetrabasic, quadribasic, tetrahydric, tetra-

~ карбо́н·ов·ая кисл·от·а́ (*f*) tetracarboxylic acid

~ кисл·от·а́ (*f*) tetracid, tetrahydric acid

четырёх·па́л·ый, -ая, -ое, (*a*) four-fingered/digited/toed, tetradactyl, tetradactylous

четырёх·па́р·н·ый, -ая, -ое, (*a*) (bot) quadrijugate, quadrijugous, (zool) quadrijugal

четырёх·пёр·ый, -ая, -ое, (*a*) four-blade

четырёх·по́ль·н·ый, -ая, -ое, (*a*) (agr) four-field

четырёх·по́люс·ник·, -а, (*m*) quadripole, four-pole; (telecom) network

~, симметр·и́ческ·ий balanced network, symmetrical four-terminal network

четырёх·по́люс·н·ый, -ая, -ое, (*a*) (elec) four-pole/-terminal/-contact

четырёх·про·вóд·н·ый, -ая, -ое, (*a*)
(elec) four-wire
~ **систéм·а** (*f*) (elec) four-wire system
~ **штéпсел·ь** (*m*) (elec) four-pin plug
четырёх·прáд·н·ый, -ая, -ое, (*a*)
four-stranded, four-strand
четырёх·раз·дéль·н·ый, -ая, -ое, (*a*)
quadrifid
четырёх·рóг·ий, -ая, -ое, (*a*) (zool)
tetracerous, quadricornate
четырёх·рýч·н·ый, -ая, -ое, (*a*)
(zool) quadrumanous
четырёх·рáд·н·ый, -ая, -ое, (*a*)
quadriserial; four-row, (bot) tetra-
stichous
~ **матрúц·а** (*f*) (math) 4-by-4 matrix
четырёх·сер·нúст·ый, -ая, -ое, (*a*)
(chem) tetrasulphide/tetrasulfide (of)
четырёх·с·лóж·н·ый, -ая, -ое, (*a*)
(gram) tetrasyllabic, four-syllable
четырёх·слóй·н·ый, -ая, -ое, (*a*)
four-layer, four-ply
четырёх·сóт· (*gen*) of **четы́ре·ст·а**
четырёх·сóт·ый, -ая, -ое, (*a*) four-
hundredth
четырёх·стáх· (*prep*) of **четы́ре·ст·а**
четырёх·с·твóр·чат·ый, -ая, -ое, (*a*)
(bot) quadrivalve
четырёх·сторóн·н·ий, -яя, -ее, (*a*)
(math) quadrilateral, four-sided;
quadripartite
четырёх·стрéнд·н·ый, -ая, -ое, (*a*)
four-stranded, four-strand
четырёх·ступ·éнчат·ый, -ая, -ое, (*a*)
four-stage
четырёх·тáкт·н·ый, -ая, -ое, (*a*)
(ICE) four-stroke
четырёх·угóль·ник·, -а, (*m*) (math)
tetragon, quadrangle
четырёх·угóль·н·ый, -ая, -ое, (*a*)
tetragonal, quadrangular; (bot) tetra-
gonous
четырёх·фтóр·ист·ый, -ая, -ое, (*a*)
(chem) tetrafluoride (of)
~ **угле·рóд·** (*m*) carbon tetrafluoride
четырёх·хлóр·ист·ый, -ая, -ое, (*a*)
(chem) tetrachloride (of), chloride (of)
четырёх·хóд·ов·ый, -ая, -ое, (*a*)
four-way; four-start/-thread (of
screws)
четырёх·час·ов·óй, -áя, -óе, (*a*)
four-hour; four o'clock
четырёх·част·úчн·ый, -ая, -ое, (*a*)
four-piece
четырёх·чáст·н·ый, -ая, -ое, (*a*)
four-part, tetramerous

четырёх·члéн·н·ый, -ая, -ое, (*a*)
four-membered/part, tetramerous,
(math) four-termed
~ **кольц·ó** (*n*) (chem) tetratomic ring
четырёх·этúль·н·ый, -ая, -ое, (*a*)
(chem) tetraethyl
четырёх·я́дер·н·ый, -ая, -ое, (*a*)
(chem) tetracyclic, tetranuclear
четырнадцати·дн·éвн·ый, -ая, -ое,
(*a*) fourteen-day
четырнадцати·лéт·н·ий, -яя, -ее, (*a*)
fourteen-year; fourteen-year-old
четыр·нá·дцат·ый, -ая, -ое, (*a*)
fourteenth
четыр·нá·дцат·ь, -и, (*f*) fourteen
четыр·ьмя́ (*instr*) of **четы́р·е**
четырьмя·ст·áми (*instr*) of **четы́ре·-
ст·а**
чéфер·, -а, (*m*) chafer
чехард·á, -ы́, (*f*) leap-frog
чéхл·ик·, -а, (*m*) (*dim*) of **чехóл·**;
(bot) cap, moss cap, calyptra
~ **кóрн·я** (bot) root cap
чехл·úть, -я́т, (*imp*) cover, case, put
hood/cover on
чехло·вúд·н·ая бол·éзн·ь (*f*) (bot)
cat-tail fungus infection
чехóл·, -х·л·á, (*m*) hood, cover, case;
(nucl) can, jacket, sheath
чехóль·чик·, -а, (*m*) (*dim*) of **чехóл·**;
(biol) indusium
чехо·словáц·к·ий, -ая, -ое, (*a*)
Czecho-Slovak
чечев·úц·а, -ы, (*i*) -ей, (*f*) (bot)
lentil, Lens; (opt) lens; (horol) lenticular bob
~, **тарéл·очн·ая** (bot) large lentil
чечевице·обрáз·н·ый, -ая, -ое, (*a*)
lens-shaped, lenticular, lentiform
чечев·úчк·а, -и, (*g pl*) -чек·, (*f*) (*dim*)
of **чечев·úц·а**; (bot) lenticel, lenti-
cula
чечев·úчн·ый, -ая, -ое, (*a*) of **чече-
в·úц·а**; pea, bean (of coal, ore etc.)
чеш·á (*pres gerund*) of **чес·áть**
чéш·ет (*pres 3rd sing*) of **чес·áть**
чéш·ск·ий, -ая, -ое, (*a*) Czech
чешуе·кóр·ый, -ая, -ое, (*a*) (bot)
scaly-barked
чешуе·кры́·л·ый, -ая, -ое, (*a*) (ent)
lepidopterous, (*as noun*) lepidopteron,
(*pl*) *Lepidoptera*
чешуе·нóс·н·ый, -ая, -ое, (*a*) (bot)
scale-bearing, squamiferous
чешуе·хвóст·ик·, -а, (*m*) (bot) sickle-
grass, *Pholiurus*

чешу́й·к·а, -и, *(g pl)* **-у́·ек·,** *(f)* (biol) scale, squamule; flake

~, слюд·ян·а́я mica flake

чешу́й·ниц·а, -ы, *(i)* **-ей,** *(f)* (ent) silverfish, *Lepisma saccharina*

чешуй·ча́т·ник·, -а, *(m)* (pal bot) lepidodendrid, *Lepidodendron*

чешуйчато·сосце·ви́д·н·ый, -ая, -ое, *(a)* (med) squamomastoid

чешуйчато·стебе́ль·чат·ый, -ая -ое, *(a)* (bot) scaly stalked

чешуйчато·стволь·н·ые, -ых, *(pl decl as adj)* (pal bot) lepidophytes

чешу́й·чат·ый, -ая, -ое, *(a)* scaly, squamous, squamose, squamate; *(pl as noun)* (zool) squamates, *Squamata*

че́ш·ущ·ий, -ая, -ое, *(a)* *(pres part act)* of **чес·а́ть**

чешу·я́, -й, *(f)* (bot, zool) squama, squame; flake

~, колос·ко́в·ая (bot) glume

чивиати́т·, -а, *(m)* (min) chiviatite

чиж·, -а́, *(i)* **-о́м,** *(m)* (orn) siskin, *Carduelis spinus*

чизел·ева́ни·е, -я, *(n)* (agr) subsoiling

чи́зел·ь, -я, *(m)* (agr) subsoiler

чика́го-2с-кисл·от·а́, -ы́, *(f)* (chem) Chicago 2S acid, SS acid, Chicago acid

чилдрени́т·, -а, *(m)* (min) childrenite

чили́ж·ник·, -а, *(m)* (bot) peashrub, *Caragana*

чилибу́х·а, -и, *(f)* (bot) Strychnos nux-vomica

чил·и́йск·ий, -ая, -ое, *(a)* Chilean, Chile

~ сели́тр·а *(f)* (chem) Chile nitre/salt-petre/saltpeter

чили́м·, -а, *(m)* (bot) water chestnut, *Trapa*

чили́ч·а, -и, *(f)* = **чили́ж·ник·**

чиллаги́т·, -а, *(m)* (min) chillagite

чимини́т·, -а, *(m)* (min) ciminite

чин·, -а, *(nom pl)* **-ы́,** *(g pl)* **-о́в,** *(m)* rank, grade

чин·а, -ы, *(f)* (bot) peavine, beach pea, *Lathyrus*

~, мохна́т·ая rough peavine, wild winter pea, *Lathyrus hirsutus*

чина́р·, -а, *(m)* (bot) plane, *Platanus*

чин·енн·ый, -ая, -ое, *(past part pass)* of **чин·и́ть**

чин·и́ть, -я́т, *(imp)* mend, repair; point, sharpen

чин·к·а, -и, *(g pl)* **-н·ок·,** *(f)* *(v n)* of **чин·и́ть**

чин·о́вник·, -а, *(m)* official; bureaucrat

чиполи́но *(m indecl)* (min) cipolin, cipolino

чир·е́й, -р·ь·я, *(m)* (med) furuncle

чири́к·а·ть, -ют, *(imp)* chirp, twitter

чир·о́к·, -р·к·а́, *(m)* (orn) teal, *Anas*

чи́сл·енн·о *(adv)* numerically

чи́сл·енност·ь, -и, *(f)* number, quantity; strength (of personnel)

чи́сл·енн·ый, -ая, -ое, *(a)* numerical

~ масшта́б· *(m)* representative fraction (scale on a map)

~ ряд· *(m)* (math) numerical series

~ со·ста́·в· *(m)* strength (of personnel)

числ·и́тел·ь, -я, *(m)* (math) numerator

~ от·нош·е́ни·я (math) antecedent

числ·и́тельн·ое, -ого, *(n decl as adj)* (math) numeral

~, коли́честв·енн·ое cardinal number

~, по·ря́д·ков·ое ordinal number

чи́сл·ить, -ят, *(imp)* count, calculate, reckon; **-ся** be reckoned/counted, be among, belong to

числ·о́, -а́, *(nom pl)* **чи́сл·а,** *(g pl)* **-сел·,** *(n)* *(see also associated words)*; number; (math) quantity; date (calendar)

~, ацети́ль·н·ое (chem) acetyl value

~, больш·о́е large number; **за·ко́н· больш·и́х· чи́сел·** (math) law of large numbers

в том числ·е́ including, among the…

в числ·о́ этих… among these …

~, вещ·е́ственн·ое (math) real number

~, дво·и́чн·ое binary number; binary digit, bit (computers)

~, дро́б·н·ое fractional number, fraction

~, еди́н·ственн·ое (gram) singular

~, имен·о́ванн·ое concrete number

~ -и́мпульс·н·ая систе́м·а *(f)* (telecom) pulse-code system

~ и́мпульс·ов pulse repetition rate

~ и́мпульс·ов фо́н·а background counts

~, иррациона́ль·н·ое (math) irrational number

~, йо́д·н·ое (chem) iodine value

~, кавитацио́н·н·ое (dynam) cavitation coefficient

~, кисл·о́тн·ое (chem) acid number

~, ко́мплекс·н·ое complex number

~, ко́мплексно-со·пряж·ённ·ое conjugate complex number, conjugate imaginary

~, координацио́н·н·ое (cryst) coordination number

~, кра́т·н·ое (math) aliquot multiple

~ лейко·ци́т·ов (biol) white count

числ·о́

~, ма́сс·ов·ое (phys) mass number

~, менделе́ев·ск·ое (chem) atomic number

~, мн·и́м·ое imaginary number

~, мно́ж·ественн·ое (gram) plural

~, натура́ль·н·ое natural number, positive integer

~, не·из·ве́ст·н·ое unknown quantity

~, не·чёт·н·ое odd number

~ оборо́т·ов speed (of rotating machines), number of revolutions

~ оборо́т·ов вал·к·о́в roll speed

~ оборо́т·ов, крит·и́ческ·ое critical speed

~ оборо́т·ов, максима́ль·н·ое maximum speed

~ оборо́т·ов, по·сто·я́нн·ое constant speed

~ об·ращ·е́ни·й, до·пуст·и́м·ое read-around ratio (computers)

~, одно·зна́ч·н·ое digit, simple number, one-digit figure

~, о·кругл·ённ·ое roundoff number (computers)

~, окта́н·ов·ое (chem) octane number

~, от·влеч·ённ·ое abstract number

~, пере·да́т·очн·ое gear ratio

~ пере·нос·а ио́н·ов (phys) migration rate

~ по·втор·е́ни·й, ну́ж·н·ое cycle criterion

~, под·корен·н·о́е (math) radicand

~, полу·це́л·ое (math) half-integer

~, по·ря́д·ков·ое (math) ordinal

~, про·ве́р·очн·ое checking number

~, прост·о́е (math) prime number

~, рациона́ль·н·ое (math) rational number

~ след·о́в (phys) track population

~ твёрд·ост·и hardness number

~, трансценде́нт·н·ое (math) transcendental number

~, флёгм·ов·ое reflux ratio

~, це́л·ое integral/whole number, (math) integer; за·ко́н це́л·ых чи́сел· law of rational indices

~, це́л·ое алгебра·и́ческ·ое (math) algebraic integer

~, це́л·ое ко́мплекс·н·ое Gaussian/complex integer

~, цета́н·ов·ое (chem) cetane number

~, чёт·н·ое even number

~, чи́ст·о мн·и́м·ое pure imaginary number

числ·ов·о́й, -а́я, -о́е, (a) numerical

~ знач·е́ни·е (n) numerical value

числ·ов·о́й, -а́я, -о́е

~ у·правл·е́ни·е (n) (autom) numerical control

чист·е́ц·, -т·ц·а́, (m) (bot) betony, *Stachys*

чи́ст·иков·ые, -ых, (*pl decl as adj*) (orn) guillemots, *Alcidae*

чист·и́льн·ый, -ая, -ое, (a) (*and see* чи́ст·ить) cleaning, cleansing, scouring, polishing

~ маши́н·а (f) (text) brushing-and-steaming mill

чи́ст·ить, -ят, (*imp*) clean, polish; clear; peel, shell, scale; purge, purify

чи́ст·к·а, -и, (*g pl*) -т·ок·, (f) (v n) see чи́ст·ить; purge; (met) conditioning, cleaning; (text) brushing and steaming

чи́ст·о (*adv*) cleanly, neatly; purely, completely, all-

~ гипер·звук·ов·о́й, -а́я, -о́е, (a) (aerodynam) all-hypersonic

чист·ови́к·, -а́, (m) fair copy

чист·ов·о́й, -а́я, -о́е, (a) fair, clean; finishing (machining process)

~ об·рабо́т·к·а (f) (mech eng) finishing

~ экземпля́р· (m) fair copy

чисто·кро́в·н·ый, -ая, -ое, (a) (agr) thoroughbred, pedigree

чисто·лин·е́йн·ый, -ая, -ое, (a) (gen) pure-strain

чист·о́л·ь, -я, (m) metal polish

чисто·пис·а́ни·е, -я, (n) calligraphy

чисто·по·ро́д·н·ый, -ая, -ое, (a) purebred

чисто·серд·е́чн·ый, -ая, -ое, (a) candid, frank

чисто·со́рт·н·ый, -ая, -ое, (a) (agr) selected

чист·от·а́, -ы́, (f) cleanness, purity; finish, degree of roughness (of surface); clearness, fidelity (of reproduced sounds)

~ по·ве́рх·ност·и (mech eng) surface finish

чисто·те́л·, -а, (m) (bot) celandine, *Chelidonium*

чисто·у́ст·, -а, (m) (bot) osmunda, *Osmunda*

чи́ст·ый, -ая, -ое, (a) clean; pure; blank, clear; fine-finished; net(t) (of weights, quantities etc.)

~ вес· (m) net weight

~ в·мест·и́мост·ь (f) net capacity

~ для анали́з·а analytically pure

~ ли́н·и·я (f) (zool) pure line

чист·ый, -ая, -ое
~ **от при·мес·ей** clean
~ **при·быль·ь** net profit
~ **стó·имост·ь** (*f*) net cost
~, **техн·ическ·и** commercially pure, commercial grade
~, **хим·ическ·ий** chemically pure
чит·áльн·ый, -ая, -ое, (*a*) reading
чит·áльн·я, -и, (*g pl*) **-лен·,** (*f*) reading-room
чит·áтел·ь, -я, (*m*) reader
чит·á·ть, -ют, (*imp*) read; give (lectures)
чит·áющ·ий, -ая, -ее, (*pres part act*) of **чит·á·ть**; reading; giving (lectures)
~ **при·бóр·** (*m*) reader (instrument)
ч.и.т.д. (*abbr*) = **что и тре́б·овалось до·каз·áть** Q.E.D.
чит·к·а, -и, (*g pl*) **-т·ок·,** (*f*) reading
чих·áни·е, -я, (*n*) (*v n*) of **чих·á·ть**; (med) sternutation
чих·á·ть, -ют, (*imp*) sneeze
чих·н·у́ть, -у́т, (*perf*) sneeze
чищ·е (*comp*) of **чист·ый**
чищ·енн·ый, -ая, -ое, (*past part pass*) of **чист·ить**
чла (*past fem sing*) of **чест·ь** (*imp*)
член·, -а, (*m*) member; limb; part; (math) term; (anat) member, membrum; (gram) article
~, **глáв·н·ый** (math) dominant term
~, **до·полн·ительн·ый** (math) additive term
~, **за·тух·áющ·ий** (math) transient term
~ **-корреспондéнт·, -а,** (*m*) corresponding member (of learned societies in USSR), honorary fellow
~, **стáр·ш·ий** (math) leading term
член·ик·, -а, (*m*) segment (of arthropod body)
членисто·нóг·ий, -ая, -ое, (*a*) (zool) arthropodal, (*as noun*) (*pl*) arthropods, *Arthropoda*
член·ист·о-пéр·ист·ый, -ая, -ое, (*a*) (bot) articulate-pinnate
член·ист·о-слóж·н·ый, -ая, -ое, (*a*) (bot) articulate compound
членисто·стéбель·н·ые, -ых, (*pl decl as adj*) (pal bot) arthrophytes, *Arthrophyta*
член·ист·ый, -ая, -ое, (*a*) (biol) jointed, articulate, segmented
член·и́ть, -я́т, (*imp*) dismember, take apart, take to pieces

член·корреспондéнт·, -а, (*m*) corresponding member (of learned societies in USSR), honorary fellow
члено·раз·дéль·н·ый, -ая, -ое, (*a*) articulate, (*pl as noun*) (zool) articulates, *Articulata*
члено·раз·дéль·ност·ь, -и, (*f*) articulation
члено·раз·дéль·н·ый, -ая, -ое, (*a*) articulate, distinct
член·ск·ий, -ая, -ое, (*a*) member's, membership
чли (*past pl*) of **чест·ь** (*imp*)
чло (*past neut sing*) of **чест·ь** (*imp*)
ЧМ (*abbr*) = **част·óтн·ая модул·я́·ци·я** FM, frequency modulation
чóк·, -а, (*m*) (gunn) choke bore
чóкер·, -а, (*m*) choker (logging)
чорт·, мор·ск·óй (*m*) angler fish, *Lophius piscatores*
чохрáльск·ого, спóсоб· (*m*) Czochralski method (of growing crystals)
чрев·áт·ый, -ая, -ое, (*a*) pregnant
чрéв·о, -а, (*n*) (anat) viscera
чрево·сеч·éни·е, -я, (*n*) (med) laparotomy
чрез- (*prefix*) trans-, over-, extra-
чрез·брюш·и́нн·ый, -ая, -ое, (*a*) (anat) transperitoneal
чрез·выч·áйн·ый, -ая, -ое, (*a*) extraordinary
~ **по·лож·éни·е** (*n*) (pol) state of emergency
~ **по·сóл·** (*m*) (dipl) Ambassador Extraordinary
чрез·лоб·кóв·ый, -ая, -ое, (*a*) (anat) transpubic
чрез·мéр·н·ый, -ая, -ое, (*a*) excessive, immoderate, inordinate, over-
~ **дóз·а** (*f*) overdose
~ **на·пряж·éни·е** (*n*) (mech) overstrain
чрес·плевáль·н·ый, -ая, -ое, (*a*) (zool) transpleural
чрес·пузы́р·н·ый, -ая, -ое, (*a*) (zool) transvesical
чт·éни·е, -я, (*n*) (*v n*) of **чест·ь** reading; interpretation (of photographs); (*pl*) series of lectures
чт·éнн·ый, -ая, -ое, (*past part pass*) of **чест·ь** (*imp*)
чт·ёт (*pres 3rd sing*) of **чест·ь** (*imp*)
чт·ец·, -á, (*i*) **-óм,** (*m*) reader; lecturer
чт·и́те (*pres 2nd pl*) of **чт·ить**; (*pl imper*) of **чест·ь**
чт·ить, -я́т, (*imp*) honour, honor, revere, respect

что (*pron*) what; which, that; (*conj*) that
~ **за** what, what kind/sort of (question); what a (exclamation)
~ **-либо = чтó-нибудь**
 ни за что on no account, in no circumstances
~ **-нибудь** something (indefinite), anything
~ **-то** something (definite)
чтоб = чтóбы
чтóбы (*conj*) that, in order that/to, so that
~ **не** in case
чт·ýщ·ий, -ая, -ее, (*pres part act*) of **чест·ь** (*imp*)
чтя (*pres gerund*) of **чест·ь** (*imp*)
чубáр·ый, -ая, -ое, (*a*) dappled (of horses)
чубýк·, -á, (*m*) (agr) vine slip
чубýш·ник·, -а, (*m*) (bot) mock-orange, *Phyladelphus*
чýвств·енност·ь, -и, (*f*) sensuality
чýвств·енн·ый, -ая, -ое, (*a*) sensual
чувств·ѝтельност·ь, -и, (*f*) sensitiveness, sensitivity, response, sensibility
 коэффициéнт· чувств·ѝтельно-ст·и (*m*) (met) notch sensitivity factor
~ **по мóщ·ност·и** power sensitivity
~**, при·вед·ённ·ая** (elec) factor/figure of merit
~**, тóк·ов·ая** (elec) current response/sensitivity
чувств·ѝтельн·ый, -ая, -ое, (*a*) sensitive; (biol) sensory, sensible
~ **нéрв·н·ые волóкн·а** (*pl*) (anat) sensory nerve fibres
~ **элемéнт·** (*m*) sensitive element (instrument); (autom) input element
чýвств·о, -а, (*n*) sense, feeling, sensation
чýвств·овани·е, -я, (*n*) sensation
чýвств·овать, -уют, (*imp*) feel, have a sensation/feeling of
чугýн·, -á, (*m*) (met) cast iron, iron; pig iron, pig
~**, антифрикциóн·н·ый** antifriction cast iron
~**, бéл·ый** white cast iron, white iron
~ **в чýш·к·ах** pig iron, pigs
~**, высоко·кремн·ѝст·ый** silicoeisen
~**, высоко·марган·цóв·ый** manganese iron
~**, высоко·прóч·н·ый** high-duty cast iron
~**, дóмен·н·ый** pig iron, pig
~**, зеркáль·н·ый** spiegel iron
~ **кóв·к·ий** malleable cast iron, malleable iron

чугýн
~**, легѝр·ованн·ый** alloy cast iron
~**, лит·éйн·ый** foundry iron
~**, мартéн·овск·ий** open-hearth iron
~**, модифиц·ѝрованн·ый** inoculated cast iron
~**, от·бел·ённ·ый** chilled iron
~**, пере·дéль·н·ый** conversion pig iron, conversion pig
~**, половѝн·чат·ый** mottled cast iron, mottled iron
~**, светло·серд·éчн·ый** whiteheart cast iron
~**, светло·серд·éчн·ый кóв·к·ий** whiteheart malleable cast iron
~**, сéр·ый** grey cast iron, grey iron
~**, специáль·н·ой** ferro-alloy
~**, стáл·ист·ый** semisteel
~**, сыр·цóв·ый** pig iron, pig
~**, томáс·овск·ий** basic pig iron
~**; черно·серд·éчн·ый** blackheart cast iron
~**, чýш·ков·ый** pig iron, pig
чугýн·н·ый, -ая, -ое, (*a*) of **чугýн·**
~ **лёт·к·а** (*f*) iron notch (blast furnace)
чугуно·вóз·, -а, (*m*) (met) hot metal car
чугун·óк·, -н·к·á, (*m*) cast iron pot
чугуно·лит·éйн·ый, -ая, -ое, (*a*) foundry, casting, founding
~ **цех·** (*m*) foundry
чуд·áк·, -á, (*m*) eccentric person, crank
чуд·ес·á (*nom/acc pl*) of **чýд·о**
чуд·éсн·ый, -ая, -ое, (*a*) wonderful
чýд·н·ый, -ая, -ое, (*a*) marvellous, miraculous, wonderful
чýд·о, -а, (*nom pl*) **чуд·ес·á,** (*n*) wonder, miracle
чуд·óвищ·е, -а, (*i*) **-ем,** (*n*) monster
чуд·óвищн·ый, -ая, -ое, (*a*) monstrous
чý·ем·ый, -ая, -ое, (*pres part pass*) of **чý·ять**
чуж·бѝн·а, -ы, (*f*) foreign land
чужд·á·ться, -ются, (*imp*) (+ *gen*) shun, avoid
чýжд·ый, -ая, -ое, (*a*) alien, foreign, strange
чуже·зéм·н·ый, -ая, -ое, (*a*) strange, outlandish
чуже·рóд·н·ый, -ая, -ое, (*a*) alien, foreign, (gen) allogenic
чуже·я́д·н·ый, -ая, -ое, (*a*) (bot) parasitic
чуж·óй, -áя, -óе, (*a*) someone/something else's, another's, foreign, strange
~**, свой-** (rad) I.F.F., identification friend or foe

чула́н·, -а,　(*m*) box-room (in house)

чул·о́к·, -лк·а́,　(*f*) (text) stocking, legging

~, котто́н·н·ый　full/fully-fashioned stocking

чул·о́чн·о-нос·о́чн·ый, -ая, -ое,　(*a*) stocking and sock, hosiery

чул·о́чн·ый, -ая, -ое,　(*a*) stocking

чум·а́, -ы́,　(*f*) (med) plague, pest

~, ко́ж·н·ая　(med) cellulocutaneous plague

~, лёг·очн·ая　pneumonic plague

~, олов·я́нн·ая　(met) tin disease

~ плото·я́д·н·ых　canine distemper

~ птиц·　(vet) fowl plague/pest

~ свин·е́й　(vet) swine fever, hog cholera

~, септ·и́ческ·ая　septicemic plague

чуми́з·а, -ы,　(*f*) (bot) Italian/foxtail millet, *Setaria italica*

чу́м·к·а, -и,　(*f*) (vet) distemper

чум·н·о́й, -а́я, -о́е,　(*a*) (med) pestilential

чурба́н·, -а,　(*m*) block, wooden block

чу́р·к·а, -и,　(*g pl*) **-р·ок·,**　(*f*) piece of wood/metal; chock

чу́т·к·ий, -ая, -ое,　(*a*) sensitive, keen; tactful

чуть　(*adv*) hardly, slightly

~ не　within an inch, almost

чут·ь·ё, -я́,　(*n*) scent; sense; sharpness (of senses), sensitivity (of perception)

чу́чел·о, -а,　(*n*) stuffed animal; (mil) dummy enemy

~ попуга́·я　stuffed parrot

чу́ш·к·а, -и,　(*g pl*) **-ш·ек·,**　(*f*) (met) pig

чу́ш·ков·ый, -ая, -ое,　(*a*) (met) pig, in pigs

чушко·ло́м·, -а,　(*m*) (met) pig breaker

чу́·ющ·ий, -ая, -ее,　(*pres part act*) of **чу́·ять**

чу́·ять, -ют,　(*imp*) scent

чьё　(*pron, nom/acc sing neut*) of **чей** whose

чьи　(*pron, nom pl, acc pl inanim*) of **чей** whose

чью　(*pron, acc sing fem*) of **чей** whose

чья　(*pron, nom sing fem*) of **чей** whose

Ш

ш-обра́з·н·ый, -ая, -ое, (*a*) E-shaped
шабази́т·, -а, (*m*) (min) chabasite
шаба́ш (*indecl*) (naut) boat your oars! (order)
ша́бер·, -а, (*m*) scraper; doctor (paper-making)
~, изо́·гн·ут·ый hook-scraper
~, трёх·гра́н·н·ый three-edged scraper, cant scraper
~, четырёх·гра́н·н·ый four-square scraper
шабли́ (*n indecl*) (food) Chablis-type wine, Chablis
шабло́н·, -а, (*m*) pattern, model, mould, mold; template, templet, gauge, gage; (build) full-scale drawing; former (plasterwork); forming/lacing board (elec. cables); former (elec. lamp and valve coils); stencil
~, вращ·а́ющ·ий (met cast) sweep
~, монт·а́жн·ый (mech eng) jig
~, на·правл·я́ющ·ий (mech eng) jig
~, основ·н·о́й master template/form
~, ра́диус·н·ый fillet/radius gauge/gage
~, рез·ьбов·о́й screw-pitch gauge/gage
~, фасо́н·н·ый plate-form gauge/gage
~, центр·ов·о́й centre/center gauge/gage
шабло́н·к·а, -и, (*g pl*) **-н·ок·,** (*f*) (*dim*) of **шабло́н·**
~, слюд·ян·а́я cut mica shape
шабло́н·н·ый, -ая, -ое, (*a*) of **шабло́н·;** commonplace, hackneyed
~ кат·у́шк·а (*f*) (elec) preformed/form-wound coil
шабо́т·, -а, (*m*) anvil block
ша́бр·ени·е, -я, (*n*) (*v n*) of **ша́бр·ить**
ша́бр·енн·ый, -ая, -ое, (*past part pass*) of **ша́бр·ить**
ша́бр·ить, -ят, (*imp*) (mech) scrape, scour
шабр·о́ванн·ый, -ая, -ое, (*past part pass*) *see* **ша́бр·ить**
шабр·ова́ть, -у́ют, (*imp*) *see* **ша́бр·ить**
шабр·о́вк·а, -и, (*g pl*) **-вок·,** (*f*) (*v n*) *see* **ша́бр·ить**
шабр·о́вочн·ый, -ая, -ое, (*a*) of **шабр·о́вк·а**
~ стан·о́к· (*m*) (mech) scraper
ша́г·, -а, (with 2, 3, 4) **-а́,** (*nom pl*) **-и́,** (*g pl*) **-о́в,** (*m*) step, pace; pitch (of coils, threads); spacing; twist (of rifling)
~, аксиа́ль·н·о-пере·ме́н·н·ый axially increasing pitch

ша́г
~, больш·о́й (mech eng) steep pitch; **у́гол· больш·о́го ша́г·а** (a/c) coarse pitch (of propeller)
-газ
ру́ч·к·а шаг-газ· collective pitch control (lever) (of helicopters)
~, геометр·и́ческ·ий (aerodynam) experimental mean pitch
~ за·клёп·ок· rivet spacing
~ за·цепл·е́ни·я circular pitch/spacing (gear wheels)
~, из·мен·я́ем·ый variable pitch
~ и́мпульс·ов pulse spacing
~ ка́др·а (cinema) frame gauge/gage
~, круг·ов·о́й angular/circular pitch
~, кру́п·н·ый coarse pitch
~, ма́л·ый fine/low pitch; close spacing; **с ма́л·ым ша́г·ом** (phys) closely spaced (of lattices); **у́гол· ма́л·ого ша́г·а** fine pitch (a/c propeller)
~ -мото́р·чик·, -а, (*m*) repeater motor (of gyro compass)
~, норма́ль·н·ый full pitch (of elec. windings); normal circular pitch (of gears)
~, о́бщ·ий collective pitch (helicopter rotor)
~, о·круж·н·о́й real/circumferential circular pitch (gear wheels)
~, ос·ев·о́й axial pitch
~, основ·н·о́й base pitch (of gear wheel)
~, от·риц·а́тельн·ый reverse pitch
~, пере·ме́н·н·ый variable pitch
~ по дуг·е́ tooth spacing (gears)
~ по колле́ктор·у (elec) commutator pitch
~, по·сто·я́нн·ый fixed/uniform pitch
~, про·до́ль·н·ый longitudinal spacing
~, радиа́ль·н·о-пере·ме́н·н·ый radially increasing pitch
~, реверси́в·н·ый reverse pitch
~, регул·и́руем·ый adjustable pitch
~ с·кру́т·к·и length of lay (of electric cable)
~, тор·цев·о́й real/circumferential circular pitch (gears)
~, флю́гер·н·ый feathering pitch
~, цикл·и́ческ·ий **ру́ч·к·а цикл·и́ческ·ого ша́г·а** cyclic pitch control (lever) (of helicopters)

шаг·а́·ть, -ют, (*imp*) step, stride, pace

шаг·а́ющ·ий, -ая, -ее, (*pres part act*) of **шаг·а́·ть;** walking (of machines)

шаг·н·у́ть, -у́т, (*perf*) take a pace/step

ша́г·ов·ый, -ая, -ое, (*a*) step, step-by-step; pitch

~ иск·а́ни·е (*n*) (telecom) step-by-step selection

~ систе́м·а (*f*) (telecom) step-by-step system/exchange

ша́г·ом (*adv*) at a walk, at a walking pace

~ марш· (mil) quick march!

шаго·ме́р·, -а, (*m*) (surv) pedometer; (mech eng) worm gauge/gage

шагре́н·ев·ый, -ая, -ое, (*a*) of **шагрен·ь**

шагрен·и́ровани·е, -я, (*n*) shagreen-finishing, pebble-graining (leather)

шагре́н·ь, -и, (*f*) shagreen (leather)

ша́·ечн·ый, -ая, -ое, (*a*) of **ша́й·к·а**

ша́йб·а, -ы, (*f*) washer; ring; die, die-plate (of extruder); puck (sport)

~, ви́·т·ая у·пру́г·ая double coil washer, Thackeray washer

~ Гро́вер·а check/spring washer

~, за·мк·о́в·ая lock washer

~, клино·ви́д·н·ая bevel washer

~, контр·о́воч·н·ая lock washer

~, ко́нус·н·ая angle washer

~, кулач·ко́в·ая cam ring

~, магни́т·н·ая disc magnet

~, на·жи́м·н·ая packing washer

~, об·лиц·о́вочн·ая screw cup

~, от·ги́б·н·ая tab washer

~, пруж·и́нн·ая зуб·ча́т·ая spring lock washer

~, раз·рез·н·а́я open/C-washer

~ с зуб·ц·а́ми, у·пру́г·ая spring lock washer

~, сто́пор·н·ая tab washer

ша́йб·ов·ый, -ая, -ое, (*a*) of **ша́йб·а**

ша́й·к·а, -и, (*g pl*) **ша́·ек·** (*f*) gang, band (e.g. of robbers); wooden baler/scoop

шака́л·, -а, (*m*) (zool) jackal, *Canis aureus*

ша́кш·а, -и, (*f*) grainer (leather)

шала́нд·а, -ы, (*f*) (naut) hopper barge

~, му́сор·н·ая garbage/sullage boat

шала́ш·, -а́, (*i*) **-о́м,** (*m*) hut

шаллери́т·, -а, (*m*) (min) schallerite

шало́т·, -а, (*m*) (bot) shallot, *Allium ascelonicum*

шалфе́·й, -я, (*m*) (bot) sage, salvia, *Salvia*

шалфе́й·н·ый, -ая, -ое, (*a*) sage, salsify

шал·ы́ (*pl*) (food) unhusked rice

шалы́г·а, -и, (*f*) slight silting up (especially around ship aground in river)

ша́л·ь, -и, (*f*) shawl

ша́льтер·, -а, (*m*) switch

шальште́йн·, -а, (*m*) (min) schalstein

шамози́т·, -а, (*m*) (min) chamosite, chamoisite

шамо́т·, -а, (*m*) fireclay

шамо́т·н·ый, -ая, -ое, (*a*) fireclay

шампаниз·а́ци·я, -и, (*f*) secondary fermentation in bottle/bulk

шампа́н·ск·ое, -ого, (*n decl as adj*) champagne, champagne-type wine, sparkling wine

шампиньо́н·, -а, (*m*) (bot) common/field mushroom, *Psalliota campestris*

шампле́н·, -а, (*m*) (geol) Champlainian

шампу́н·ь, -я, (*m*) shampoo

шамуази́т·, -а, (*m*) (min) chamoisite, chamosite

шандо́р·а, -ы, (*f*) shutter panel (for sluices, weirs etc.)

шандо́р·н·ая сте́н·к·а (*f*) shutter wall/gate/dam

шандр·а́, -ы́, (*f*) (bot) hoarhound, horehound, Marrubium

шанжи́р·н·о-шир·и́льн·ая маши́н·а (*f*) (text) jig stenter/tenter

ша́нкр·, -а, (*m*) (med) chancre

~, мя́гк·ий chancroid

~, твёрд·ый chancre

ша́нс·, -а, (*m*) chance

шанта́ж·, -а́, (*i*) **-о́м,** (*m*) (law) blackmail

ша́нц·ев·ый, -ая, -ое, (*a*) (mil) trenching, entrenching

шапбахи́т·, -а, (*m*) (min) schapbachite

ша́п·к·а, -и, (*g pl*) **-п·ок·,** (*f*) cap; (geol) cap-rock; (print) banner head-lines; cape (of parachute)

~, га́з·ов·ая (geol) gas cap

~, лед·ян·а́я (geol) ice cap

~, поля́р·н·ая polar contact face (of gyro compass)

~ -уш·а́нк·а, -и, (*g pl*) **-нок·,** (*f*) fur cap with ear flaps

ша́п·очк·а, -и, (*g pl*) **-чек·,** (*f*) (*dim*) of **ша́п·к·а**

ша́пп·, -а, (*m*) (text) schappe

шар·, -а, or **-а́,** (*nom pl*) **шар·ы́,** (*m*) sphere, globe, ball; balloon; (geol) ball

~, воз·ду́ш·н·ый balloon

~, за·па́ль·н·ый (ICE) hot bulb

~, зем·н·о́й (geog) globe, terrestrial sphere

шар
~ **-зо́нд·, -а,** (*m*) (meteor) sonde, sounding balloon, balloon
~ **-зонд·, автомат·и́ческ·ий** unmanned balloon
~ **-пило́т·, -а,** (*m*) (meteor) pilot balloon
~, **полу́·ден·н·ый** (naut) time ball
~, **у́голь·н·ый** (geol) coal ball/apple
~, **чёрн·ый** (naut) signal ball
шарголи́н·, -а, (*m*) (plast) Shargolin (a PVC leather substitute)
ша́рж·, -а, (*m*) caricature, distorted image
шарж·и́рн·ый, -ая, -ое, (*a*) (met) charging
ша́р·ик·, -а, (*m*) (*dim*) of **шар·**; globule, bead
~, **бузи́н·н·ый** pith ball
~, **моло́ч·н·ый** milk corpuscle
~, **о·по́р·н·ый** (mech) fulcrum ball
~, **сто́пор·н·ый** (mech) locking ball
ша́р·иков·ый, -ая, -ое, (*a*) of **ша́р·ик·**
~ **катализ·а́тор·** (*m*) bead catalyst
~ **под·ши́п·ник·** (*m*) (*see also* **под·-ши́п·ник·**) ball bearing
шарико·под·ши́п·ник·, -а, (*m*) see **под·ши́п·ник·, ша́р·иков·ый**
шарико·под·ши́п·ников·ый, -ая -ое, (*a*) ball bearing
шарла́х·, -а, (*m*) scarlet (dye)
шарло́т·, -а, (*m*) = **шало́т·**
шарло́т·к·а, -и, (*f*) = **шало́т·**
Ша́рл·я, за·ко́н· (*m*) (phys) Charles law
шарни́р·, -а, (*m*) hinge; (geol) bend
~ **вз·ма́х·а** flapping hinge
~, **в·рез·н·о́й** blind hinge
~, **горизонта́ль·н·ый** flapping hinge (of helicopter main rotor)
~ **Гу́к·а** (mech) Hooke's/universal/cardan joint, ball-and-socket joint
~, **иго́ль·чат·ый** needle joint
~, **карда́н·н·ый** cardan/universal/Hooke's joint, ball-and-socket joint
~, **ос·ев·о́й** pitch-changing/feathering hinge (helicopters)
~, **пя́т·ов·о́й** (civ eng) skewback
~, **угл·ов·о́й** elbow joint
~, **универса́ль·н·ый** universal joint
~, **шар·ов·о́й** ball joint
шарни́р·н·ый, -ая, -ое, (*a*) hinged, articulated, jointed, swing-, swivel-, link
~ **бо́лт·** (*m*) link bolt
~ **механ·и́зм·** (*m*) (mech) linkage, link motion

шарни́р·н·ый, -ая, -ое
~ **механ·и́зм·, на·правл·я́ющ·ий** straight-line mechanism
~ **пе́тл·я** (*f*) hinge
~ **со·един·е́ни·е** (*n*) hinge, swing/hinged/pin/rocket joint
~ **со·един·е́ни·е ви́л·к·ой** knuckle joint
~ **цеп·ь** (*f*) roller chain
шаро·ви́д·ност·ь, -и, (*f*) sphericity
шаро·ви́д·н·ый, -ая, -ое, (*a*) spherical, globular
шар·о́вк·а, -и, (*g pl*) **-вок·,** (*f*) (agr) blind cultivation
шар·ов·о́й, -а́я, -о́е, (*a*) sphere, spherical, ball, globe, globular
~ **кла́пан·** (*m*) (mech) ball valve
~ **па́р·а** (*f*) ball and socket
~ **по·го́н·** (*m*) (mech) ball track, roller path
~ **с·вод·** (*m*) = **шар·ов·о́й сегме́нт·**
~ **сегме́нт·** (*m*) (math) spherical segment
~ **сло́·й** (*m*) (math) spherical segment of two bases
~ **фу́нкци·я** (*f*) spherical function
шаро·обра́з·ност·ь, -и, (*f*) sphericity
шаро·обра́з·н·ый, -ая, -ое, (*a*) spherical, spheroidal, globular, globulose, globate
шаро·пило́т·н·ый, -ая, -ое, (*a*) pilot-balloon
шаро·пло́д·к·а, -и, (*f*) (bot) henbane, *Hyoscyamus*
шаро·под·ши́п·ник·, -а, (*m*) see **под·ши́п·ник·, ша́р·иков·ый**
шаро́ш·ечн·ый, -ая, -ое, (*a*) (mech) cutting, milling
шаро́ш·к·а, -и, (*g pl*) **-ш·ек·,** (*f*) (mech) milling cutter, rasp
~, **кон·и́ческ·ая** taper rasp
~, **цилиндр·и́ческ·ая** round-nose rasp
Ша́рпи́, копёр· (*m*) (met) Charpy impact tester
~, **над·ре́з·** (*m*) (met) keyhole notch, Charpy-type B notch
шарпи́т·, -а, (*m*) (min) sharpite
ша́рф·, -а, (*m*) (text) scarf, muffler
шарья́ж·, -а, (*m*) (geol) mass overthrust
шасси́ (*n indecl*) (M/T) chassis; (a/c) undercarriage, landing gear
~, **велосипе́д·н·ое** (a/c) bicycle/tricycle undercarriage
~, **вы́·пущ·енн·ое** undercarriage down/lowered

шасси́

~, двух·колёс·н·ое twin-wheel undercarriage

~, лыж·н·о-колёс·н·ое ski-wheel landing gear

~, лыж·н·ое ski undercarriage, ski landing gear

~, много·колёс·н·ое bogie undercarriage

~, не·у·бир·а́ющ·ееся (a/c) fixed/non-retractable undercarriage

~, по·плав·ко́в·ое (a/c) float alighting gear

~ с пере́д·н·им колес·о́м tricycle undercarriage

~, сала́з·очн·ое skid landing gear

~, теле́ж·ечн·ое bogie undercarriage

~, у·бир·а́ющ·ееся retractable undercarriage

~, у́·бр·анн·ое undercarriage up/raised

ша́ст·алк·а, -и, (*g pl*) **-лок·,** (*f*) (agr) awner

шат·а́ни·е, -я, (*n*) (*v n*) *see* **шат·а́·ть;** (mech) play

шат·а́·ть, -ют, (*imp*) shake, rock, sway; shake loose; **-ся** (*pass*); be/get/work loose; be unsteady, reel, stagger

шательперро́н·, -а, (*m*) (geol) Chatelperronian

шатёр·, -тр·а́, (*m*) marquee, tent

ша́т·к·ий, -ая, -ое, (*a*) shaky, unsteady, precarious

шат·н·у́ть, -у́т, (*perf*) rock, sway (once), lurch

шатон·, -а, (*m*) (horol) bezel

шатр·о́в·ый, -ая, -ое, (*a*) of **шатёр·;** tent-shaped

~ котёл· (*m*) three-drum boiler

~ кры́·ш·а (*f*) hipped roof

шаттуки́т·, -а, (*m*) (min) shattuckite

шат·у́н·, -а́, (*m*) (mech) connecting rod

~, вилко·обра́з·н·ый fork connecting rod

~, гла́в·н·ый master rod

~, при·цеп·н·о́й articulated rod, link rod

~, с·два́·ивающ·ий coupling rod

шат·у́нн·ый, -ая, -ое, (*a*) of **шат·у́н·**

~ болт· (*m*) (M/T) connecting-rod bolt, big-end bolt

~ под·шип·ник· (*m*) (M/T) big-end bearing

шафра́н·, -а, (*m*) (bot) saffron, *Crocus*

шах·, -а, (*m*) check (as chess); shah (title)

шахмат·и́ст·, -а, (*m*) chess player

ша́хмат·н·ый, -ая, -ое, (*a*) chess; staggered (of positions), chess/chequer-board (of patterns)

~ по·ря́д·ок· (*m*) staggered rows, chessboard order

~ шов· (*m*) staggered weld; zig-zag/reeled seam (riveted)

ша́хмат·ы (*f pl*) chess

ша́хт·а, -ы, (*f*) shaft, trunk, well, vertical shaft; (min) mine, pit

~, вентиляцио́н·н·ая ventilating shaft/trunk

~, есте́ств·енн·ая (geol) deep karst channel

~, раз·ве́д·очн·ая (min) test pit

~, у́голь·н·ая coal mine

~ -хран·и́лищ·е, -а, (*n*) (nucl) storage well

шахт·ёр·, -а, (*m*) miner

ша́хт·н·ый, -ая, -ое, (*a*) of **ша́хт·а**

~ печ·ь (*f*) shaft furnace

~ по́л·е (*n*) mine area

~ ствол· (*m*) (min) shaft

ша́шен·ь, мор·ск·о́й (*m*) (zool) ship worm, *Teredo navalis*

ша́ш·ечн·ый, -ая, -ое, (*a*) of **ша́ш·к·а;** staggered-row (formation), checker-board-pattern

ша́ш·к·а, -и, (*g pl*) **-ш·ек·,** (*f*) sabre; (expl) block, cartridge; candle, pot (smokes); (rocket) charge, grain; draughts, checkers (game); (civ eng) paving block

~, брон·иро́ванн·ая inhibited grain

~, дым·ов·а́я smoke candle

~, много·кана́ль·н·ая multiperforated grain

~, под·ры́в·н·а́я (expl) blasting cartridge

~, порох·ов·а́я (rocket) grain

~, стерж·нев·а́я rod grain

~, то́л·ов·ая (expl) TNT block

~, тру́б·чат·ая tubular grain

~, ядовито·ды́м·н·ая toxic smoke candle

шашлы́к·, -а́, (*m*) (food) shashlyk, kebabs (mutton etc. on skewer)

шв·а (*gen*) of **шов·**

шва́бр·а, -ы, (*f*) swab, mop

шва́нн·овск·ая оболо́ч·к·а (*f*) (zool) Schwann's sheath, neurolemma

шва́рт·, -а, (*m*) (naut) sheet anchor; (med) adhesion, induration

шварт·о́в·, -а, (*m*) (naut) berthing hawser

шварт·ова́ть, -у́ют, (*imp*) (naut) moor (alongside), berth; (air) picket

шварт·о́вн·ый, -ая, -ое, (*a*) mooring

шварт·о́в·ый, -ая, -ое, (*a*) mooring, berthing

~ ис·пыт·а́ни·е (*n*) (shipb) basin/dock trials

шварцембергі́т·, -а, (*m*) (min) schwartzembergite

шве́д·ск·ий, -ая, -ое, (*a*) Swedish

~ спи́ч·к·а safety match

швей·ниц·а, -ы, (*i*) -ей, (*f*) seamstress, sempstress, clothing industry worker, sewing-machine operator

швейнфу́рт·ск·ая зе́лен·ь (*f*) Schweinfurth green, Paris green

швей·н·ый, -ая, -ое, (*a*) sewing

~ маши́н·а (*f*) (text) sewing machine; (print) sewer

~ про·мы́шл·енность·ь (*f*) clothing industry

~ ни́т·к·а (*f*) sewing thread, sewing

швейца́р·, -а, (*m*) door-keeper, hall porter, commissionaire

швейца́р·ец·, -р·ц·а, (*m*) Swiss

швейца́р·ск·ий, -ая, -ое, (*a*) Swiss, (*f as noun*) hall porter's lodge/room/office

швейцер·ов·, реакти́в· (*m*) (chem) Schweitzer's reagent

швел·ева́ни·е, -я, (*n*) low-temperature carbonization/distillation (of coal)

шве́ллер·, -а, (*m*) (met) channel, channel bar

шве́ллер·н·ая сталь·ь (*f*) steel channels

швель·га́з·, -а, (*m*) carbonization gas (from low-temperature distillation)

швель·ко́кс·, -а, (*m*) low-temperature coke, semicoke, coalite

швель·ша́хт·а, -ы, (*f*) low-temperature carbonization/distillation shaft (of gas producer)

шверобо́т·, -а, (*m*) centreboard boat

шве́рт·, -а, (*m*) (shipb) centreboard

шверт·бо́т·, -а, (*m*) centreboard boat

шве́рц·, -а, (*m*) (naut) leeboard

шве·я́, -й, (*f*) sempstress, dressmaker, sewing woman

швинг·турби́н·а, -ы, (*f*) (text) scutcher, flax-scutching mill

швиц·ева́ни·е, -я, (*n*) sweating (leather)

швор·ень·, -р·н·я, (*m*) = **шкво́р·ень·ь** (q.v.)

швор·нев·о́й, -а́я, -о́е, (*a*) = **шквор·-нев·о́й**

шв·ы (*nom/acc pl*) of **шов·**

швыр·н·у́ть, -у́т, (*perf*) throw, toss, fling

швыр·о́к·, -р·к·а́, (*m*) a throw/toss; (naut) sand bag (as heaving line weight); firewood; clay pigeon (for shooting practice)

швыр·я́лк·а, -и, (*m*) (agr) grain-cleaning and grading machine

швыр·я́·ть, -ют, (*imp*) fling/hurl/toss about

шевели́н·, -а, (*m*) flax-tow felt (heat-insulator)

шевел·и́ть, -я́т, (*imp*) disturb (something at rest), move slightly, stir; rustle; (agr) turn (for drying)

шевель·н·у́ть, -у́т, (*perf*) make a slight movement, make a stir

Шевенар·а, дилато́·метр· (*m*) (met) Chevenard dilatometer

ше́вер·, -а, (*m*) (mech eng) shaver, shaving cutter/tool

~, ди́ск·ов·ый rotary shaving cutter

~ -ре́й·к·а, -и, (*f*) rack-type shaving tool

~ -шест·ерн·я́, -и, (*f*) rotary shaving tool/cutter

шевинг·ова́ни·е, -я, (*n*) (mech eng) shaving

шевио́т·, -а, (*m*) (text) cheviot

шевре́т·, -а, (*m*) kid (leather from sheepskin)

шевро́ (*n indecl*) kid (leather from goatskin)

шевр·о́в·ый, -ая, -ое, (*a*) of **шевр·о́**

шевро́н·, -а, (*m*) chevron, stripe

шевро́н·н·ый, -ая, -ое, (*a*) chevron; (mech) double-helical, chevron

~ колес·о́ (*n*) double-helical gear-wheel

~ манже́т·а (*f*) chevron packing

ше́динг·, -а, (*m*) shading

ше́д·ов·ый, -ая, -ое, (*a*) (build) north-light, sawtooth

~ кры́·ш·а (*f*) north-/northern-light roof, sawtooth roof

~ фона́р·ь (*m*) (build) north/sawtooth light

ше́д·ш·ий, -ая, -ее, (*a*) (*past part act*) of **ид·ти́**

Ше́еле, зе́лен·ь (*f*) Scheele's green

шеели́т·, -а, (*m*) (min) scheelite

Шези, фо́рмул·а (*f*) (hydrodyn) Chezy formula

ше́й·к·а, -и, (*g pl*) ше́·ек·, (*f*) (*dim*) of **ше́·я**; neck; journal (of shaft); (met) web (of H bar etc.); (anat) neck, cervix

~ бедр·а́ (anat) femoral neck

~ вал·к·а́ roll neck/journal

шéй·к·а
~, вы́·точ·енн·ая   neck journal
~, корен·н·áя   (mech) main journal
~ мáт·к·и   (anat) cervix uteri
~, мотыл·ев·ая   crankpin
  образ·овáни·е шéй·к·и   (n) (met, plast) necking, necking down
~ раз·дел·éни·я   (nucl) scission neck
~ рéльс·а   rail web
~ сомѝт·а   (anat) nephrostome
~, шат·ýнн·ая   (mech) crankpin
шéй·н·ый, -ая, -ое,   (a) jugular; cervical, neck
~ вéн·а   (f) jugular vein
шейно·грýд·н·ой, -áя, -óе,   (a) cervicothoracic
шейно·плеч·ев·óй, -áя, -óе,   (a) (anat) cervicobrachial
шéй·те   (pl imper) of **ши·ть**
шéк·, -а,   (m) (shipb) cutivater
шёл·   (past masc sing) of **ид·тѝ**
шелёв·к·а, -и,   (g pl) **-в·ок·,** (f) ¾-inch board
шéлест·, -а,   (m) rustle, rustling
шелест·éть, -я́т,   (imp) rustle
шёлк·, -а,   (nom pl) **-á,** (g pl) **-óв,** (m) silk
~, ацетáт·н·ый   (text) acetate rayon
~, вискóз·н·ый   viscose rayon
~, за·мáсл·енн·ый   oiled silk
~, искýс·ственн·ый   artificial silk, rayon
~, круч·ён·ый   thrown silk
~, медно·аммиáч·н·ый   copper rayon
~ -сыр·éц·, -р·ц·á,   (m) raw/reeling silk
~, у·тяжел·ённ·ый   loaded silk
шелк·овѝн·а, -ы,   (f) silk fibre/fiber/thread
шелк·овѝнк·а, -и,   (g pl) **-нок·,** (f) silk thread
шелк·овѝст·ый, -ая, -ое,   (a) silky
шелк·овѝц·а, -ы,   (i) **-ей,** (f) (bot) mulberry tree, *Morus*
шелк·овѝчн·ый, -ая, -ое,   (a) (bot) mulberry
~ червь   (m) silkworm, silkworm larva/caterpillar
шелкóв·к·а, -и,   (g pl) **-в·ок·,** (f) (food) lentil meal/flour, ground lentils
шелко·вóд·ств·о, -а,   (n) sericulture
шёлк·ов·ый, -ая, -ое,   (a) silk, silken
~ пря́ж·а   (f) spun silk, filature
шёлко·крут·ѝльщик·, -а,   (m) throwster
шёлко·круч·éни·е, -я,   (n) throwing, silk throwing

шёлко·мот·áльн·ая фáбрик·а   (f) filature
шёлко·пря́д·, -а,   (m) (ent) silkworm, (pl) *Bombycidae*
~, тýт·ов·ый   mulberry silkworm, *Bombyx mori*
шёлко·пря́д·éни·е, -я,   (n) schappe silk spinning
шёлко·трафарéт·н·ая фóрм·а   (f) (print) silk screen forme
шеллáк·, -а,   (m) shellac (resin)
шеллáч·н·ый, -ая, -ое,   (a) shellac
шéллер·, -а,   (m) (agr) sheller, huller, husker
шéлль·ск·ий, -ая, -ое,   (a) (geol) Chellean
шелох·н·ýть, -ýт,   (perf) move slightly, stir
шелух·á, -ѝ,   (f) peel, shell, husk, hull, pod, skin, scale
шелух·ов·ый, -ая, -ое,   (a) scaly, squamate, squamose, squamous, (bot) furfuraceous
шелуш·éни·е, -я,   (n) (v n) see **шелу·ш·ѝть;** (biol) desquamation; furfuration
шелуш·ѝльн·ый, -ая, -ое,   (a) shelling, husking, hulling, podding, descaling
~ машѝн·а   (f) shelling/husking machine
шелуш·ѝть, -áт,   (imp) peel, shell, husk, hull, pod, scale; **-ся** (pass); peel/come off; desquamate, furfurate
шелы́г·а, -и,   (f) (arch) extrados
шéль·е, -я,   (n) gently sloping rock shore
шельтердéк·, -а,   (m) (shipb) shelter deck
шельтердéч·н·ое сýд·н·о   (n) shelter-deck ship
шéльф·, -а,   (m) shelf
~, континентáль·н·ый   (geol) continental shelf
шелю́г·а, -и,   (f) (bot) sharpleaf willow, *Salix acutifolia*
шеневиксѝт·, -а,   (m) (min) chenevixite
шенѝт·, -а,   (m) (min) shoenite
шепеля́в·ост·ь, -и,   (f) lisping, (med) parasigmatism
шéпинг·, -а,   (m) (mech eng) shaper
шёпѝт·, -а,   (m) (min) schoepite
шеп·н·ýть, -ýт,   (perf) whisper
шёпот·, -а,   (m) whisper
шептал·á, -ы́,   (f) dried peach
шепт·áть,   (pres 3rd pl) **шéпч·ут,** (imp) whisper
шéпч·ет   (pres 3rd sing) of **шепт·áть**

шерардиз·а́ци·я, -и, (*f*) (met) sherardizing

ше́рбер·, -а, (*m*) scorifier (assaying)

ше́рбер·н·ая про́б·а (*f*) scorification (assaying)

шерге́н·ь, -и, (*f*) (shipb) declivity board

шере́нг·а, -и, (*f*) rank (men in line)

Ше́ринг·а, мост· (*m*) (elec) Schering bridge

шёрл·, -а, (*m*) (min) schorl, black tourmaline

шерли́т·, -а, (*m*) (min) schorlite

шерломи́т·, -а, (*m*) (min) schorlomite

шерох·ова́льн·ый, -ая, -ое, (*a*) roughening; buffing

~ стан·о́к· (*m*) buffing machine (for tires)

шероховато·ли́ст·н·ый, -ая, -ое, (*a*) rough-leaved, trachyphyllous

шероховато·семя́н·н·ый, -ая, -ое, (*a*) (bot) trachyspermous

шерох·ова́тост·ь, -и, (*f*) roughness (of surface); ragging (for gripping)

коэффицие́нт· шерох·ова́тост·и (*m*) (hydrodyn) roughness coefficient/factor; degree of porosity (of rubber fillers)

шерох·ова́т·ый, -ая, -ое, (*a*) rough (of surfaces); ragged (of surfaces); (bot) asperate

шерох·о́вк·а, -и, (*f*) = **шерох·ова́тост·ь**

шерсте- (*root*) *see also* **шерсто-**

шерсте·мо́·ечн·ый, -ая, -ое, (*a*) wool-washing

шерсте·сте́бель·ник·, -а, (*m*) (bot) pipewort, *Eriocaulon*

шерст·и́нк·а, -и, (*g pl*) -нок·, (*f*) strand of wool

шерстисто·пло́д·н·ый, -ая, -ое, (*a*) (bot) eriocarpous

шерст·и́стост·ь, -и, (*f*) woolliness

шерстисто·цвет·ко́в·ый, -ая, -ое, (*a*) (bot) erianthous

шерст·и́ст·ый, -ая, -ое, (*a*) woolly

шёрсто- (*root*) *see also* **шёрсте-**

шерсто·мо́й·н·ый, -ая, -ое, (*a*) wool-washing

шерсто·пряд·е́ни·е, -я, (*n*) wool spinning

~, греб·е́нн·о́е worsted spinning

шерсто·пряд·и́льн·ый, -ая, -ое, (*a*) wool-spinning

шерсто·тк·а́честв·о, -а, (*n*) wool weaving

шерст·ь, -и, (*nom pl*) -и, (*g pl*) -е́й, (*f*) wool

~, блест·я́щ·ая bright wool

~, во́йлоч·н·ая felting wool

~, греб·енн·а́я worsted wool

~, жёстк·ая harsh/brashy wool

~, регенер·и́рованн·ая shoddy

~, ру́н·н·ая fleece wool

~, сорно·репе́й·н·ая shabby/burry wool

~, с·пу́т·анн·ая cotty wool

~, чум·н·а́я murrain wool

шерст·я́к·, -а́, (*m*) (bot) cupgrass, *Eriochloa*

шерст·я́нк·а, -и, (*g pl*) -нок·, (*f*) bowl paper, calender-roll paper

шерст·ян·о́й, -а́я, -о́е, (*a*) woollen, wool

шёрт·, -а, (*m*) (geol) chert

шерхе́бел·ь, -я, (*m*) incline/jack plane

шершА́в·е·ть, -ют, (*imp*) become rough (of surfaces)

шершА́в·ый, -ая, -ое, (*a*) rough (of surfaces)

шёрш·ен·ь, -ш·н·я, (*m*) (ent) hornet, *Vespa*

~, обык·нове́нн·ый hornet, *Vespa crabro*

шест·, -а́, (*m*) pole, rod, staff, spar

~, от·по́р·н·ый (naut) boathook

шестак·, -а, (*m*) caulking irons, caulker's tool kit

ше́ств·и·е, -я, (*n*) procession

шест·ерённ·ый, -ая, -ое, (*a*) of **шестер·н·я́**

~ на·со́с· (*m*) spur-gear pump

шест·ерёнчат·ый, -ая, -ое, (*a*) = **шест·ерённ·ый**

шест·ери́к·, -а́, (*m*) six-yarn line (rope)

шест·ёрк·а, -и, (*g pl*) -рок·, (*f*) a six, a number six, team/set of six, six-oared boat

шест·ерни́к·, -а́, (*m*) (cryst) sixling

шест·ерн·о́й, -а́я, -о́е, (*a*) six times as big/large/great, six-fold; six-part

шест·ерн·я́, -и́, (*g pl*) -рён·, (*f*) (mech) pinion

~, вед·о́м·ая driven pinion

~, вед·у́щ·ая driving pinion

~, кон·и́ческ·ая bevel wheel

~, ре́ж·ущ·ая gear cutter

~, тор·цев·а́я face gear

~, цилиндр·и́ческ·ая cylindrical pinion, spur-gear wheel

~, шеврон·н·ая double-helical gear-wheel, herringbone gear-wheel

ше́ст·ер·о, -ых, (*collective numeral*) six, six together/altogether шести·а́том·н·ый, -ая, -ое,   (*a*) hexatomic, hexahydric

~ спирт· (*m*) hexahydric alcohol

~ со·един·е́ни·е   (*n*) hexatomic compound

шести·вале́нт·н·ый, -ая, -ое,   (*a*) (chem) hexavalent, sexivalent

шести·вал·ко́в·ый, -ая, -ое,   (*a*) six-roll

~ стан· (*m*) (met roll) cluster mill, six-high mill

шести·гра́н·ник·, -а,   (*m*) (math) hexahedron

шести·гра́н·н·ый, -ая, -ое,   (*a*) (math) hexahedral

~ про́фил·ь (*m*) hexagon (metal shape)

шести́·десят·и   (*gen/dat/prep*) of шесть·деся́т·

шести·десяти·ри́чн·ый, -ая, -ое, (*a*) (math) sexagesimal

шести·деся́т·ый, -ая, -ое, (*a*) sixtieth

шести·зве́н·н·ый, -ая, -ое, (*a*) six-unit/member

~ полиме́р· (*m*) (chem) sexamer

шести·зна́ч·н·ый, -ая, -ое, (*a*) six-figure; six-unit (codes)

шести·кра́т·н·ый, -ая, -ое,   (*a*) sextuple, six-fold

~ за·мещ·е́ни·е   (*n*) (chem) hexa-substitution

~ систе́м·а   (*f*) (telecom) multiplex hexode system

шести·лепе́ст·н·ый, -ая, -ое,   (*a*) (bot) hexapetalous

шести·ле́т·н·ий, -яя, -ее,   (*a*) six-year, six-year-old

шести·ли́ст·н·ый, -ая, -ое, (*a*) (bot) hexaphyllous

шести·луч·ев·о́й, -а́я, -о́е, (*a*) six-rayed, hexactinal; (*pl as noun*) (zool) *Hexacorallia*

шести·ме́сяч·н·ый, -ая, -ое,   (*a*) six-month; six-monthly, half-yearly

шести·му́ж·н·ий, -яя, -ее, (*a*) (bot) hexandrous

шести·над·рез·н·о́й, -а́я, -о́е,   (*a*) (bot) sexfid

шести·но́г·ий, -ая, -ое,   (*a*) six-footed/legged, (zool) hexapod, hexapodous, (*pl as noun*) (zool) hexapods, hexapodans, *Hexapoda*

шести·о́·кис·ь, -и, (*f*) (chem) hexoxide

шести·осно́в·н·ый, -ая, -ое,   (*a*) hexabasic, hexatomic

~ кисл·от·а́ (*f*) hexahydric/hexabasic/hexatomic acid

шести·па́л·ост·ь, -и, (*f*) (med) hexadactylism

шести·па́л·ый, -ая, -ое, (*a*) six-fingered/toed, sex-digitate, hexadactyle

шести·па́р·н·ый, -ая, -ое, (*a*) (bot) sejugous, sejugate

шести·пе́ст·иков·ый, -ая, -ое, (*a*) (bot) hexagynous

шести·по́люс·ник·, -а,   (*m*) (elec) six-terminal network

шести·ри́чн·ый, -ая, -ое, (*a*) (math) senary

шести·сло́й·н·ый, -ая, -ое, (*a*) six-ply/layer

шести·со́т· (*gen*) of шесть·со́т·

шести·со́т·ый, -ая, -ое,   (*a*) six-hundredth

шести·ст·а́м (*dat*) of шесть·со́т·

шести·ст·а́х (*prep*) of шесть·со́т·

шести·сторо́н·н·ий, -яя, -ее,   (*a*) six-sided, hexalateral

шести·тыч·и́ночн·ый, -ая, -ое,   (*a*) (bot) hexastemonous, hexandrous

шести·уго́ль·ник·, -а, (*m*) hexagon

шести·уго́ль·н·ый, -ая, -ое,   (*a*) (math) hexagonal; (bot) sexangular

шести·фто́р·ист·ый, -ая, -ое,   (*a*) (chem) hexafluoride (of)

~ ура́н· (*m*) uranium hexafluoride

шести·час·ов·о́й, -а́я, -о́е, (*a*) six-hour; six o'clock

шести·чле́н·н·ый, -ая, -ое, (*a*) six-membered

шести·электро́д·н·ая ла́мп·а   (*f*) (elec) hexode

шестнадцати·ле́т·н·ий, -яя, -ее, (*a*) sixteen-year, sixteen-year-old

шестнадцати·ри́чн·ый, -ая, -ое, (*a*) (math) sexadecimal, sextodecimo

шест·на́·дцат·ый, -ая, -ое,   (*a*) sixteenth

шест·на́·дцат·ь, -и, (*i*) -ью, (*f*) sixteen

шест·ова́т·ый, -ая, -ое, (*a*) columnar, rod-/pole-like

шест·ови́н·ы (*pl*) pole-marks (leather)

шест·ов·о́й, -а́я, -о́е, (*a*) pole, rod, staff, spar

~ ли́н·и·я (*f*) (telecom) overhead line, air line

шест·о́й, -а́я, -о́е, (*a*) sixth

~ по·ря́д·ок· (*m*) sixth order, (*in gen as adj*) sextic

~ сте́пен·ь (*f*) sixth degree, (*in gen as adj*) sextic

шест·ь, -и́, (*i*) -ью́, (*f*) six

шесть·деся́т· sixty

шесть·со́т·, (*g*) **шести·со́т·**, (*numeral*) six hundred

шест·ью (*adv*) six times; **шест·ью** (*instr*) of **шест·ь**

шестью́·деся́т·ью (*instr*) of **шесть·-деся́т·**

шестью·ст·а́ми (*instr*) of **шесть·со́т·**

шетелиги́т·, **-а**, (*m*) (min) scheteligite

шеф·, **-а**, (*m*) chief; chef; patron, adviser and helper, patron organization

шеф·ств·о, **-а**, (*n*) adoption (of an organization, with the aim of assisting), patronage

шеф·ств·овать, **-уют**, (*imp*) act as patron, give advice and assistance

шефферит·, **-а**, (*m*) (min) schefferite

ше́·я, **-и**, (*f*) neck, collum, cervix

ши́бер·, **-а**, (*m*) plate damper (in flues, chimneys etc.); gate/sluice valve (in water pipes)

ши́бер·н·ый, **-ая**, **-ое**, (*a*) of **ши́бер·**

~ на·со́с· (*m*) sliding-vane rotary pump

ши́вер·а, **-ы**, (*f*) rapids (of river)

ши́·вш·ий, **-ая**, **-ее**, (*past part act*) of **ши·ть**

шиз·а́ндр·а, **-ы**, (*f*) (bot) Schizandra, magnolia vine

шизо- (*int component*) schizo-

шизо·га́м·и·я, **-и**, (*f*) (zool) schizogamy

шизо·гене́з·, **-а**, (*m*) (gen) schizogenesis

шизо·гон·и́·я, **-и**, (*f*) (zool) schizogony

шиз·о́ид·, **-а**, (*m*) (med) schizoid

шизо·ка́рп·и·й, **-я**, (*m*) (bot) schizocarp

шизо·ли́т·, **-а**, (*m*) (min) schizolite

шиз·о́нт·, **-а**, (*m*) (gen) schizont

шизо·сте́л·а, **-ы**, (*f*) (bot) schizostele

шизо·фи́т·, **-а**, (*m*) (bot) schizophyte

шизо·френ·и́·я, **-и**, (*f*) (med) schizophrenia

шика́р·н·ый, **-ая**, **-ое**, (*a*) smart, stylish

Ши́к·а, реа́кци·я (*f*) (med) Schick's reaction/test

ши́кш·а, **-и**, (*f*) (bot) crowberry, *Empetrum*

ши·л (*past masc sing*) of **ши·ть**

ши́·ли (*past pl*) of **ши·ть**

шиллериз·а́ци·я, **-и**, (*f*) (geol) schillerization

ши́ллер-шпа́т·, **-а**, (*m*) (min) schiller spar

ши́·л·о, **-а**, (*nom pl*) **ши́·л·ья**, (*g pl*) **ши́·л·ьев** (*n*) awl, pricker, spike; **ши́·ло** (*past neut sing*) of **ши·ть**

шило·ви́д·н·ый, **-ая**, **-ое**, (*a*) awl-shaped, (bot) subulate; (anat) styloid

шило·гло́т·очн·ый, **-ая**, **-ое**, (*a*) (anat) stylopharyngeal

шило·клю́в·к·а, **-и**, (*g pl*) **-в·о·к**, (*f*) (orn) avocet, *Recurvirostra*

шило·подъ·язы́ч·н·ый, **-ая**, **-ое**, (*a*) (anat) stylohyoid

шило·хво́ст·ь, **-и**, (*f*) (orn) pintail duck; *Anas acuta*

ши́ль·ник·, **-а**, (*m*) (bot) subularia, awl-wort, *Subularia*

ши́ль·ц·е, **-а**, (*i*) **-ем**, (*n*) (bot) plumule

ши́·л·ья (*nom/acc pl*) of **ши́·л·о**

ши́мми (*indecl*) (M/T, a/c) shimmy, wheel wobble

шимпанзе́ (*m indecl*) (zool) chimpanzee, *Pan troglodytes*

ши́н·а, **-ы**, (*f*) tire, tyre; (elec) busbar, bar, bus; rubber gland (of pneumatic clutch); battening iron; (med) splint, bar

~, а́роч·н·ая ribbed tire/tyre

~, бес·ка́мер·н·ая tubeless tire

~ вы́·да·ч·и су́мм·ы sum output bus (computers)

~, груз·ов·а́я lorry/truck tire

~, легк·ов·а́я automobile tire

~, масси́в·н·ая solid tire

~, полу·масси́в·н·ая air-core tire

~, резино·бо́рт·н·ая beaded tire

~, с·бо́р·н·ая (elec) collecting bar; busbar

~, снего·хо́д·н·ая snow/winter tire

ши́нгл·, **-а**, (*m*) (build) shingle, shide

ши́нгл·ов·ый стан·о́к· (*m*) shingle/shingling saw

шине́л·ь, **-и**, (*f*) greatcoat, overcoat

шине́ль·н·ый, **-ая**, **-ое**, (*a*) of **шине́л·ь**

~ лист· (*m*) wrapper plate (boiler)

шинк·ова́ть, **-у́ют**, (*imp*) shred

шинк·о́вк·а, **-и**, (*g pl*) **-вок·**, (*f*) shredding; shredder

шинколобви́т·, **-а**, (*m*) (min) chinkolobwite

ши́н·ник·, **-а**, (*m*) tire worker, tire-man

ши́н·н·ый, **-ая**, **-ое**, (*a*) of **ши́н·а**

шино·держ·а́тел·ь, **-я**, (*m*) (elec) bus support

шино·пневмат·и́ческ·ий, **-ая**, **-ое**, (*a*) pneumatic

шино·ремо́нт·, **-а**, (*m*) tire repair

шиншилл·а, **-ы**, (*f*) (zool) chinchilla, *Chinchilla*

шип·, -а́, (*m*) (bot) prickle, thorn, acanthus; tongue, tenon; pivot pin, neck, journal (of rolls etc.); cramp-iron, spike (of shoes etc.); thorn needle (for sound playback); (fish) sturgeon

шип·е́ть, -я́т, (*imp*) hiss, fizz, sizzle

шип·ик·, -а, (*m*) (*dim*) of **шип·**; (biol) spinule

шип·ова́т·ый, -ая, -ое, (*a*) prickly, acanthaceous, spinous

шипо·ви́д·н·ый, -ая, -ое, (*a*) spiniform

шип·о́вник·, -а, (*m*) (bot) rose, dog rose

∼, соба́ч·ий dog rose, *Rosa canina*

шип·ов·о́й, -а́я, -о́е, (*a*) tongue, tenon; mortise, mortising

∼ долот·о́ (*n*) (wood) mortise chisel

∼ со·един·е́ни·е (*n*) mortise-and-tenon joint

шипо·но́с·к·и (*pl*) (ent) tumbling flower beetles, *Mordellidae*

шипо·ре́з·, -а, (*m*) (wood) tenoner

∼ стан·о́к· (*m*) (wood) tenon-making machine

шип·у́ч·ий, -ая, -ее, (*a*) hissing, fizzing, sizzling; sparkling, fizzy, carbonated, aerated (of beverages)

шир·е (*comp*) wider, broader

шир·е́ни·е, -я, (*n*) stretching; (text) tentering, stentering

шир·и́льк·а, -и, (*g pl*) **-лек·,** (*f*) stretcher

∼, цеп·н·а́я (text) clip stretcher

шир·и́льн·ый, -ая, -ое, (*a*) (text) tenter, stenter

∼ маши́н·а (*f*) (text) tenter, stenter

шир·ин·а́, -ы́, (*f*) width, breadth; face width (of gear wheel)

∼, габари́т·н·ая maximum/extreme breadth

∼ кры·л·а́ (a/c) wing depth

∼ между·ря́д·ь·я row spacing

∼ полови́н·ой мо́щ·ност·и (elec) half-power width

∼ полос·ы́ (rad) bandwidth

∼, рабо́ч·ая active face width (of gear wheel)

∼, реги́стр·ов·ая (shipb) registered breadth

∼ с·бро́с·а (geol) throw (vertical), heave (horizontal)

ши́р·ить, -ят, (*imp*) widen, broaden; **-ся** become wider/broader, spread

ши́рм·а, -ы, (*f*) screen; (ICE) shroud

шир·о́к·ий, -ая, -ое, (*a*) wide, broad

широко·вещ·а́ни·е, -я, (*n*) (rad) broadcasting

широко·вещ·а́тельн·ый, -ая, -ое, (*a*) broadcasting

широко·голо́в·ый, -ая, -ое, (*a*) (zool) platycephalous; chamaecephalic

широко·го́рл·ый, -ая, -ое, (*a*) wide-necked (bottles etc.)

широко·гру́д·н·ый, -ая, -ое, (*a*) broad-chested

широко·диапазо́н·н·ый, -ая, -ое, (*a*) wide-range/band

широко·коле́й·н·ый, -ая, -ое, (*a*) (rail) broad-gauge/gage

широко·ко́ст·н·ый, -ая, -ое, (*a*) big-boned

широко·кры́·л·ый, -ая, -ое, (*a*) broad-winged

широко·ли́ств·енн·ый, -ая, -ое, (*a*) broad-leaved, platyphyllous

широко·но́с·ост·ь, -и, (*f*) (zool) platyrrhiny

широко·но́с·ый, -ая, -ое, (*a*) broad-nosed, (zool) platyrrhine, platyrrhinian

∼ обезья́н·ы (*pl*) (zool) *Platyrrhina*

широко·пере·горо́д·чат·ый, -ая, -ое, (*a*) latimurate

широко·поло́с·н·ый, -ая, -ое, (*a*) (met) wide-strip, sheet; (rad) wide-band, broad-band

∼ анте́нн·а (*f*) (rad) broadband antenna

∼ стан· (*m*) (met roll) sheet mill (1·5mm–10mm × 600mm–2500mm)

∼ универса́ль·н·ая ста́л·ь (*f*) universal plates (steel)

широко·по́л·очн·ый, -ая, -ое, (*a*) (met) wide-flange

широко·по́л·ый, -ая, -ое, (*a*) wide-brimmed; full-skirted

широко·рас·кры́в·н·ый, -ая, -ое, (*a*) wide-angle (of radiations)

широко·ро́т·ы, -ов, (*m pl*) (fish) *Saccopharyngiformes*

широко·сифо́н·н·ый, -ая, -ое, (*a*) eurysiphonate

широко·сло́й·н·ый, -ая, -ое, (*a*) (wood) coarse-grained

широко·уго́ль·н·ый, -ая, -ое, (*a*) wide-angle

широко·ход·ов·о́й, -а́я, -о́е, (*a*) (mech eng) loose-running, loose

шир·от·а́, -ы́, (*f*) width, breadth; (geog) latitude; range

∼ вы́·бор·к·и (math) range of sample

∼, выс·о́к·ие (*pl*) high latitude(s)

шир·от·а́

~, геогра́ф·и́ческ·ая   (geog) terrestrial latitude

~, геоцентр·и́ческ·ая   geocentric latitude

~, ис·пра́вл·енн·ая   observed latitude

~, магни́т·н·ая   magnetic latitude

~, мг·нове́нн·ая   instantaneous latitude

~, неб·е́сн·ая   (astron) celestial latitude

~, обсерв·ова́нн·ая   latitude by observation

~, от·ше́д·ш·ая   latitude outwards

~, при·ше́д·ш·ая   latitude inwards

~, у·ме́р·енн·ые   (pl) temperate latitudes

~, фотогра́ф·и́ческ·ая   (phot) latitude of exposure

шир·о́тн·ый, -ая, -ое,   (a) (geog) latitude, latitudinal

~ ста́нци·я   (f) latitude observatory

шир·по·тре́б·, -а,   (m) consumer goods

ширстре́к·, -а,   (m) (shipb) sheer strake

шир·ь, -и,   (f) expanse, width, sweep (of view)

шисто·зо́м·а, -ы,   (f) (zool) schistosome, *Schistosoma*, (pl) schistosomes, *Schistosoma*

ш исто·зоматóз·, -а,   (m) (med) schistosomiasis

шисто·це́рк·а, -и,   (f) (ent) desert locust, *Schistocerca*

ши́·те   (pl imper) of **ши·ть**

ши́·т·ый, -ая, -ое,   (a) (past part pass) of **ши·ть;** sewn

ши·ть,   (pres 3rd pl) **шь·ют,**   (imp) sew, stitch; embroider

шит·ь·ё, -я́,   (n) sewing; needlework; embroidering, embroidery

ши́фер·, -а,   (m) roofing slate/tile; (min) slate

~, асбе́ст·ов·ый   (build) asbestos-cement slate/tile

~, сла́н·цев·ый   slate, natural slate tile

ши́фер·н·ый, -ая, -ое,   (a) of **ши́фер·**

шифóн·, -а,   (m) (text) chiffon

ши́фр·, -а,   (m) cipher, code; code number, mark, tally; classification number, call number (library); monogram, initials

шифр·а́т·, -а,   (m) cypher/coded text

шифр·а́тор·, -а,   (m) encoder; (autom, telecom) coder, quantizer

~, време́н·н·о́й   clock coder

~, и́мпульс·н·ый   pulse coder

шифр·а́ци·я, -и,   (f) = **шифр·о́вк·а**

шифр·ова́льн·ый, -ая, -ое,   (a) cipher, code

шифр·ова́льщик·, -а,   (m) cipherer, cipher clerk/officer

шифр·о́ванн·ый, -ая, -ое,   (a) (past part pass) of **шифр·ова́ть;** cipher

шифр·ова́ть, -у́ют,   (imp) cipher, code, encipher, encode

шифр·о́вк·а, -и,   (f) (v n) of **шифр·ова́ть;** cipher/code message/text

ши́фтинг·, -а,   (m) shifting board (in ship's hold)

Ши́фф·а, реа́кци·я   (f) (chem) Schiff test reaction

ши́фф·ов·ы основ·а́ни·я   (pl) (chem) Schiff bases

ши́хт·а, -ы,   (f) charge (for processing), burden (blast furnace)

шихт·ова́ть, -у́ют,   (imp) make up the burden/charge, burden; arrange in layers; (elec) arrange laminations

ши́хт·о́вк·а, -и,   (g pl) -вок·,   (f) (v n) of **шихт·ова́ть**

ши́хт·ов·ый, -ая, -ое,   (a) of **ши́хт·а**

ши́ш·ек·   (g pl) of **ши́ш·к·а**

ши́ш·ельник·, -а,   (m) (met cast) core-maker

ши́ш·ечк·а, -и,   (g pl) -чек·,   (f) (dim) of **ши́ш·к·а**

ши́ш·к·а, -и,   (g pl) -ш·ек·,   (f) lump, knob, bump; (met cast) core (of mould); hammer; (bot) cone

~, бо·ев·а́я   stone-breaking hammer

~, ворс·ова́льн·ая   (text) teasel

~, ел·о́в·ая   fir cone

~, сосн·о́в·ая   pine cone

~, споро·но́с·н·ая   (bot) strobile, strobilus

шиш·кова́т·ый, -ая, -ое,   (a) lumpy, knobby, bumpy, tuberculate

шишко·ви́д·н·ый, -ая, -ое,   (a) pineal, cone-like, cone-shaped, pine-cone-like

~ желез·а́   (f) (zool) epiphysis cerebri, pineal gland

шишко·но́с·н·ый, -ая, -ое,   (a) (bot) coniferous

шишко·обра́з·н·ый, -ая, -ое,   (a) cone-shaped

шишко·суш·и́лк·а, -и,   (f) core drier (foundry)

ши·я́   (pres gerund) of **ши·ть**

шкал·а́, -ы́,   (nom pl) **шка́л·ы,** (g pl) **шкал·,** (f) (see also under associated words); scale

~, без·нул·ев·а́я   (instr) suppressed-zero scale, setup scale

шкал·а́
~ времён·и time base
~ времён·и, по·движ·н·ая triggered time base
~, груз·ов·а́я (shipb) vertical displacement and deadweight scales
~, кру́гл·ая dial, circular scale
~, но́ниус·н·ая vernier scale
~, пере·во́д·н·ая conversion table/scale
~, равно·ме́р·н·ая uniform scale
~, рас·тя́·нут·ая expanded/extended/spread scale
~, с·тесн·ённ·ая compressed scale
~, сто·гра́дус·н·ая Celsius/centigrade scale
~, то́ч·н·ая microdial
шка́ль·н·ый, -ая, -ое, (a) of шкал·а́
~ ла́мп·а (f) (instr) dial lamp/light
шкап·, -а, (m) = шкаф·
шкато́рин·а, -ы, (f) (naut) bolt rope, leech rope
шкату́л·к·а, -и, (g pl) -л·ок·, (f) box, casket
шкаф·, -а, (nom pl) -ы́, (g pl) -о́в, (m) cupboard, cabinet, locker
~, вы·тяж·н·о́й (chem) fume cupboard/cabinet
~ за·движ·н·ы́х `воро́т· caisson chamber (dock)
~, за·щи́т·н·ый protective cabinet
~, кни́ж·н·ый bookcase
~, суш·и́льн·ый drier, desiccator
шкафу́т·, -а, (m) (naut) waist
шквал·, -а, (m) (meteor) squall
шква́л·ист·ый, -ая, -ое, (a) (meteor) squally
шква́р·к·а, -и, (f) (food) crackling, crisp fried fat
шквор·ен·ь, -р·н·я, (m) (mech) pin, pintle
~, по·воро́т·н·ый cotter pin
шквор·нев·о́й, -а́я, -о́е, (a) of шкво́р·ен·ь
шке́нтел·ь, -я, (m) (naut) short wire pendant
~ -гак·, -а, (m) cargo hook
~, груз·ов·о́й whip (of derrick)
шке́рт·, -а, (m) lanyard
шкив·, -а, (nom pl) -ы, (g pl) -о́в, (m) pulley, sheave
~, копр·о́в·ый (min) headgear sheave
~, на·тяж·н·о́й tension/bearing-up pulley
~, при·вод·н·о́й belt pulley
~ -эпици́кл·, -а, (m) epicyclic pulley
шки́пер·, -а, (m) skipper; (naut) boatswain's storekeeper

шки́пер·ск·ий, -ая, -ое, (a) of шки́пер·; (f as noun) boatswain's store
~ по·каз·а́ни·е (n) (naut) manifest
шко́л·а, -ы, (f) school
~, ча́ст·н·ая private school, public school (in Britain)
шко́ль·ник·, -а, (m) pupil, schoolboy
шко́ль·н·ый, -ая, -ое, (a) school
шко́т·, -а, (m) (naut) sheet (of sail)
шку́р·а, -ы, (f) hide, skin, pelt
~, пар·н·а́я raw hide
шкур·и́льн·ый стан·о́к· sand-paper machine
шку́р·ить, -ят, (imp) sand (wood)
шку́р·к·а, -и, (g pl) -р·ок·, (f) (dim) of шку́р·а; abrasive paper/sheet; band, sheet (of rubber)
~, нажда́ч·н·ая emery paper
~, стекл·я́нн·ая glass paper
~, шлиф·ова́льн·ая abrasive paper/sheet, sandpaper
шкурочно·шлиф·ова́льн·ый стан·о́к· (m) sander, sanding machine (woodworking)
шла (past fem sing) of ид·ти́
шлаг·, -а, (m) (naut) round turn (with rope); turn (of steering wheel)
шлагба́ум·, -а, (m) barrier, lifting/tipping barrier, tollgate
шлак·, -а, (m) slag, dross, cinder, clinker, scoria; poison (in nucl. reactor)
~, вс·пу́ч·енн·ый (build) foamed slag
~, до́мен·н·ый furnace slag, blast-furnace slag
~, жи́дк·ий fluid slag
~, жидко·тек·у́ч·ий (weld) fusible slag
~, марте́н·овск·ий (met) open-hearth slag
образ·ова́ни·е шлак·а (n) slagging
~, само·от·дел·я́ющ·ийся (weld) self-releasing/-peeling slag
~, тома́с·ов·ый (met) basic slag
шлако·бето́н·, -а, (m) slag concrete
шлак·ова́ни·е, -я, (n) slag formation, slagging, scorification; poisoning (of nuclear reactor)
шлако·ва́т·а, -ы, (f) slag wool
шлак·ова́ть, -у́ют, (imp) slag, form slag, scorify
шлак·ови́к·, -а, (m) dust catcher (furnace); slag chamber (open-hearth furnace)
шлак·ови́н·а, -ы, (f) (met) slag inclusion
шлако·во́з·, -а, (m) (met) cinder car

шла́к·ов·ый, -ая, -ое, (*a*) of шлак· scoriaceous

~ ва́т·а (*f*) slag wool

~ лёт·к·а (*f*) slag/cinder notch (blast furnace), slag hole

~ сто́пор· (*m*) cinder-notch stopper

шлако·образ·у́ющ·ий, -ая, -ее, (*a*) slag-forming

шлако·от·сто́й·ник·, -а, (*m*) (weld) slag basin

шлако·у·дал·е́ни·е, -я, (*n*) ash removal (from furnace); slagging (from steel bath)

шлако·портланд·цеме́нт·, -а, (*m*) Portland blast-furnace cement

шлако·у·лов·и́тел·ь, -я, (*m*) (met cast) dirt/slag trap

шла́м·, -а, (*m*) slime, sludge; slurry

~, бур·ов·о́й drilling mud

~, ма́рган·цев·ый manganese slime/ residue (fertilizer)

~, цеме́нт·н·ый cement slurry

шла́м·ов·ый, -ая, -ое, (*a*) of шлам·

шламо·образ·ова́ни·е, -я, (*n*) sludging

шламо·у·лов·и́тел·ь, -я, (*m*) sludge/ slime extractor

шла́нг·, -а, (*m*) hose, hosepipe, flexible tubing

~, водо·ла́з·н·ый diver's air pipe

~, гермет·и́чн·ый rubber sealing tube; inflatable rubber sealing tube (for doors, windows etc.)

~, ги́б·к·ий (elec) flexible conduit/ hose/tube

~, за·пра́в·очн·ый refuelling hose

~, от·лив·н·о́й discharge hose

~, по·жа́р·н·ый fire hose

~, при·ём·н·ый suction hose

~ -сигна́л·, -а, (*m*) breastrope (diver's)

шла́нг·ов·ый, -ая, -ое, (*a*) of шланг·; flexible (of elec. cable)

шле́й·к·а, -и, (*g pl*) -е·ек·, (*f*) breech/ breast-band (of harness)

шле́йф·, -а, (*m*) train (of dress etc.), trail; (elec) loop; (rad) stub; (agr) harrow

~ -анте́нн·а, -ы, (*f*) folded-dipole antenna

~ -борон·а́, -ы́, (*f*) (agr) float harrow

~, вал·у́нн·ый (geol) boulder train

~, ём·кости·ый capacitative stub

~, за́·мк·нут·ый closed loop

~, коротко·за́·мк·нут·ый closed stub

~, креп·я́щ·ий supporting stub

~, со·глас·у́ющ·ий matching stub

~, четверть·во́лн·ов·ый quarter wavelength stub

шле́йф·н·ый, -ая, -ое, (*a*) of шлейф·

шле́йф·ов·ый, -ая, -ое, (*a*) of шлейф·

~ осцилло́·граф· (*m*) loop oscillograph

шле́м·, -а, (*m*) helmet; dome (of still); (bot) galea

~, лёт·н·ый flying helmet

шл·ём (*pres 1st pl*) of сл·а́ть send, dispatch

шле́м·ник·, -а, (*m*) (bot) skullcap, *Scutellaria*

шлемо·ви́д·н·ый, -ая, -ое, (*a*) (biol) galeate, galeiform

шлемо·фо́н·, -а, (*m*) (rad) combined headset and laryngophone

шлёп·а·ть, -ют, slap, smack

шлёп·н·уть, -ут, (*perf*) see шлёп·а·ть

шлеп·о́к·, -п·к·а́, (*m*) slap, smack

шле́ппер·, -а, (*m*) transfer skid

~, кана́т·н·ый cable transfer skid

~, цеп·н·о́й chain transfer skid

шл·ёт, (*pres 3rd pl*) of сл·а́ть sends, dispatches

шл·ёте (*pres 2nd pl*) of сл·ать send, dispatch

шле·я́, -й, (*f*) breech-band (harness)

шли (*past pl*) of ид·ти́

шли́кер·, -а, (*m*) dross

~, во́д·н·ый (ceram) slip

шли́р·, -а, (*m*) drop (glass); (*pl*) streaks, schlieren, striae

~, интрузи́в·н·ые (*pl*) (geol) intrusive schlieren

шли́рен-фотогра́ф·и·я, -и, (*f*) schlieren photography

шл·и́те (*pl imper*) of сл·ать send, dispatch

шлиф·, -а, (*m*) microsection, section; metallographic specimen; sealed ground joint (glassware etc.), ground-in joint

~, плоск·остн·о́й planar section

~, полир·о́ванн·ый (met) crystallographic specimen; (min) polished section (for microscopic examination)

~, про·зр·а́чн·ый thin section

шлиф·ова́льн·о-полир·ова́льн·ый стан·о́к· (*m*) grinding-and-polishing machine

шлиф·ова́льн·ый, -ая, -ое, (*a*) grinding, polishing

~ круг· (*m*) grindstone

~ стан·о́к· (*m*) grinder

~ шку́р·к·а (*f*) abrasive sheet

шлиф·ова́ни·е, -я, (*n*) grinding, polishing

шлиф·ова́ть, -у́ют, (*imp*) grind, polish
~ в·сух·у́ю grind dry
~ на·чёрн·о grind rough/first
шлиф·о́вк·а, -и, (*f*) grinding, polishing
шлиф·о́вщик·, -а, (*m*) grinder
шли́х·, -а, (*m*) (min) heavy concentrate
шли́хт·а, -ы, (*f*) (text) size
шлихт·ова́льн·ая маши́н·а (*f*) (text) sizer, slasher
шлихт·ова́ть, -у́ют, (*imp*) (text) size
шли́ц·, -а, (*m*) spline
~ -шарни́р·, -а, (*m*) torsion link
шлиц·ева́ть (*imp*) spline, nick
шли́ц·ев·ый, -ая, -ое, (*a*) splined
шло (*past neut sing*) of **ид·ти́**
шлю́з·, -а, (*m*) (civ eng) lock; sluice
~, амальгамацио́н·н·ый (min) plate sluice
~, воз·ду́ш·н·ый (civ eng) air-lock
~, пере·да́т·очн·ый (nucl) transfer drawer
~ -регул·я́тор·, -а, (*m*) irrigation channel lock
шлю́з·н·ый, -ая, -ое, (*a*) = **шлюз·о́в·о́й**
шлюз·ова́ни·е, -я, (*n*) (*v n*) of **шлю́з·ова́ть**
шлюз·ова́ть, -у́ют, (*imp*) build navigation locks on; lock through (ships); (min) sluice
шлюз·ов·о́й, -а́я, -о́е, (*a*) (civ eng) lock; sluice
~ аппара́т· (*m*) air lock (to underwater working chamber)
~ ка́мер·а (*f*) (civ eng) lock chamber; air lock (to underwater working chamber); (met) pouring gate (continuous casting)
~ част·ь (*f*) dry dock entrance
шлюп·, -а, (*m*) (naut) sloop
шлюп·ба́л·к·а, -и, (*g pl*) **-л·ок·,** (*f*) (naut) boat's davit
~ за·ва́л·ивающ·аяся (naut) quadrant davit, luffing davit
шлюп·к·а, -и, (*g pl*) **-п·ок·,** (*f*) ship's boat, boat
~, греб·н·а́я rowing/pulling boat
~, рези́н·ов·ая rubber dinghy
~, с·пас·а́тельн·ая lifeboat; sea boat
шлюп·очн·ый, -ая, -ое, (*a*) boat, boat's
шл·ют (*pres 3rd pl*) of **сл·ать** send, dispatch
шл·ючи́ (*pres gerund*) of **сл·ать** sending, dispatching, while sending/dispatching

шл·ю́щ·ий, -ая, -ее, (*pres part act*) of **сл·ать** sending, dispatching, who/which is sending/dispatching
шля́мбур·, -а, (*m*) jumper, jumper borer
шля́п·а, -ы, (*f*) hat (garment); (tech) hat, cap
~, желе́з·н·ая (geol) gossan, iron hat
шля́п·к·а, -и, (*g pl*) **-п·ок·,** (*f*) hat (woman's); head (of nail etc.); (bot) cap
шля́п·н·ый, -ая, -ое, (*a*) of **шля́п·а**
~ магази́н· (*m*) hat shop (for men); milliner's (for women)
~ от·де́л·к·а (*f*) millinery
шмальти́н-хлоанти́т·, -а, (*m*) (min) smaltite, smaltine, smaltite-chloanthite
шмел·ь, -я́, (*m*) (ent) bumblebee, *Bombus*
шмуцти́тул·, -а, (*m*) (print) bastard/false title
шне́к·, -а, (*m*) screw (of extruder etc.); screw conveyor
шне́к·ов·ый, -ая, -ое, (*a*) of **шнек·**
~ пресс· (*m*) (food) continuous press
шни́тт·лу́к·, -а, (*m*) (bot) chives, *Allium schoenoprosum*
шно́ркел·ь, -я, (*m*) Schnorkel, snort (submarines)
шнур·, -а́, (*m*) cord, string, twine, twist, lacing; (expl) safety fuse, blasting fuse
~, амортизацио́н·н·ый shock cord
~, асбе́ст·ов·ый asbestos rope/cord
~, бикфо́рд·ов (expl) safety/miner's/Bickford fuse
~, быстро·гор·я́щ·ий instantaneous fuse
~, ги́б·к·ий (elec) flexible cord, flex
~, детон·и́рующ·ий detonating fuse
~, за·жиг·а́тельн·ый burning fuse
~, огне·про́·вод·н·ый blasting/safety/Bickford/miner's fuse
~, пере·но́с·н·ый patch cord (telephones)
~ пла́зм·ы (phys) plasma column, pinch
~, с·пуск·н·о́й (elec) drop cord; (gunn) firing lanyard
~, тэ́н·ов·ый primacord
шнур·ко́в·ый, -ая, -ое, (*a*) of **шнур·о́к·**
шнур·ова́ть, -у́ют, (*imp*) lace up, string
шнур·ов·о́й, -а́я, -о́е, (*a*) of **шнур·**
~ амортиз·а́тор· (*m*) (a/c) shock cord

шнур·ов·о́й, -а́я, -о́е
~ **блок·** (*m*) cord weight (telephones)
~ **со·един·е́ни·е** (*n*) patchcord (telephones)
шнур·о́к·, -р·к·а́, (*m*) (*dim*) of **шнур·**; lace, lacing, string
шов·, (*g*) **шв·а́,** (*m*) seam, joint; weld; (anat) suture, sutura, commisure
~, **бес·с·ко́с·н·ый** square joint
~, **вен·е́чн·ый** (anat) coronal suture
~, **во́·гн·ут·ый** concave weld
~ **в·при·ты́к·** (text) flat-butted seam
~ **в·ро́вен·ь** (text) butted seam
~, **вы́·пукл·ый** convex weld
~, **за·клёп·очн·ый** rivet seam; rivet weld
~, **за·чека́н·енн·ый** caulked seam
~, **ка́честв·енн·ый** sound weld
~, **кетл·ёвочн·ый** (text) looping/linking seam
~, **К-обра́з·н·ый одно·сторо́н·н·ий** double bevel joint
~, **кос·о́й** diagonal fillet weld
~, **кру́гл·ый** round seam (canvas etc.)
~, **лоб·н·ый** sutura frontalis
~, **лоб·ов·о́й** transverse fillet weld
~, **не·пре·ры́в·н·ый** continuous weld
~, **не·про·ва́р·енн·ый** shallow weld
~, **не·про́ч·н·ый** weak weld
~, **ни́ж·н·ий** flat weld
~, **норма́ль·н·ый** theoretical weld
~, **об·легч·ённ·ый** concave weld
~, **об·тач·н·о́й** (text) bound seam
~, **одно·про·хо́д·н·ый** single-pass weld
~, **о·са́д·очн·ый** keyed expansion joint
~, **па́·ян·ый** (met) brazing seam
~, **пло́ск·ий** flat seam
~, **под·ва́р·очн·ый** back-up weld
~, **по́р·ист·ый** porous/fibrous weld
~, **пре·ры́в·ист·ый** intermittent weld
~, **про́ч·н·о-пло́т·н·ый** composite weld
~, **ро́л·иков·ый** seam/line weld
~, **с·ва́р·н·о́й** weld, joint weld
~, **со·един·и́тельн·ый** commissure
~, **стрело·ви́д·н·ый** (anat) sagittal suture
~, **с·ты́к·ов·о́й** butt weld
~, **температу́р·н·ый** expansion joint
~, **тор·цев·о́й** edge weld
~, **угл·ов·о́й** fillet weld
~, **у·са́д·очн·ый** keyed expansion joint
~, **у·сил·ённ·ый** reinforced/convex weld
~, **у·си́л·ивающ·ий** strap weld
~, **фла́нг·ов·ый** side fillet weld
~, **цеп·н·о́й за·клёп·очн·ый** chain rivet seam

шов
~, **через·кра́й·н·ый** (text) oversewn seam
~, **шахма́т·н·ый за·клёп·очн·ый** zigzag rivet seam, reeled rivet seam
шовинист·и́ческ·ий, -ая, -ое, (*a*) chauvinistic
шо́в·н·ый, -ая, -ое, (*a*) of **шов·** (q.v.); sutural, commisural
шо́к·, -а, (*m*) (med) shock
шок·и́р·овать, -уют, (*imp*) shock, embarrass
шокола́д·, -а, (*m*) chocolate
шокола́д·ник·, -а, (*m*) (bot) cocoa tree, cacao tree, *Theobroma cacao*
шокола́д·н·ый, -ая, -ое, (*a*) chocolate
~ **де́рев·о** (*n*) cacao/cocoa tree, *Theobroma cacao*
шо́мпол·, -а, (*nom pl*) **-а́,** (*g pl*) **-о́в,** (*m*) draw-in rod, cleaning rod, ramrod; swab (oil well exploitation); stock indicator (of blast furnace)
шо́мпол·овани·е, -я, (*n*) (oil) swabbing
шо́мполь·н·ый, -ая, -ое, (*a*) of **шо́мпол·**
~ **руж·ь·ё** (*n*) (gunn) muzzle loader, musket
~ **эксплуат·а́ци·я** (*f*) = **шо́мпол·о·вани·е**
шооп·и́ровани·е, -я, (*n*) Schoop process/processing (metal spray coating)
шо́пот·, -а, (*m*) whisper; (telecom) unvoiced speech
шо́ппер·, -а, (*m*) (rubber) Schopper tester
шор·, -а, (*m*) (geol) shor, salina
Шо́р·а, монотро́н· (*m*) (instr) Shore monotron
~, **склеро·ско́п·** (*m*) (instr) Shore scleroscope
твёрд·ост·ь по шо́р·у (*f*) Shore hardness
шо́р·ник·, -а, (*m*) saddle-maker, saddler
шо́р·н·ый, -ая, -ое, (*a*) of **шо́р·ы;** saddler's
~ **де́л·о** (*n*) saddlery, saddler's trade
~ **из·де́л·и·я** (*pl*) saddlery, saddler's wares
шо́рох·, -а, (*m*) rustle, rustling noise; faint scraping noise
шо́ртгорн·, -а, (*m*) (agr) shorthorn
шо́р·ы (*f pl*) blinkers, blinders (for horses etc.); harness
шоссе́ (*n indecl*) highway, surfaced/macadamed/metalled road

шоссе́й·н·ый, -ая, -ое, (*a*) of шоссе́

шосс·и́р·овать, -уют, (*imp and perf*) build a road/highway; surface a dirt road

Шо́тки, дефе́кт· по (*m*) (cryst) Schottky defect

~, явл·е́ни·е (*n*) (phys) Schottky effect

шотла́нд·к·а, -и, (*g pl*) -д·ок·, (*f*) Scotchwoman, Scotswoman; (text) check fabric

шотла́нд·ск·ий, -ая, -ое, (*a*) Scottish, Scotch, Scots

~ котёл· (*m*) Scotch boiler, marine return-tube cylindrical boiler

Шо́ттки *see* Шо́тки

шофёр·, -а, (*m*) (M/T) driver; chauffeur

шпа́г·а, -и, (*f*) sword, rapier

шпага́т·, -а, (*m*) string, cord, twine

~, на·тяж·н·о́й trip wire

шпа́ж·ник·, -а, (*m*) (bot) gladiolus, *Gladiolus*

шпакл·ева́ть, -юют, (*imp*) putty

шпакл·ёвк·а, -и, (*f*) (build) putty, filler; puttying

шпа́л·а, -ы, (*f*) (rail) sleeper, tie, crosstie

шпале́р·а, -ы, (*f*) (hortic) espalier

шпало·за·ру́б·очн·ая маши́н·а (*f*) (rail) sleeper/tie cutter

шпало·под·бо́й·к·а, -и, (*g pl*) -о́·ек·, (*f*) (rail) ballast packer

шпало·про·пи́т·очн·ое ма́сл·о (*n*) anthracene oil (as wood preservative)

шпа́ль·н·ый, -ая, -ое, (*a*) (rail) sleeper, tie

~ я́щик· (*m*) tie crib

шпан·бло́к·, -а, (*m*) (naut) span block

шпангоу́т·, -а, (*m*) (shipb) frame (metal), rib (wood); (a/c) frame, ring

~, борт·ов·о́й (shipb) side frame

~, гла́в·н·ый (a/c) main frame

~, дн·и́щев·о́й ра́м·н·ый (shipb) double-bottom transverse frame

~, по·воро́т·н·ый (shipb) cant frame

~, ра́м·н·ый (shipb) web

~, сил·ов·о́й (a/c) thrust ring

~, скеле́т·н·ый (shipb) skeleton floor, open floor

~, терет·и́ческ·ий (shipb) ordinate station, transverse section (in lines drawing)

~, фальши́в·ый (shipb) false frame

~, це́л·ый (a/c) bulkhead

шпангоу́т·н·ый, -ая, -ое, (*a*) of шпангоу́т·

~ ра́м·к·а (*f*) (shipb) frame

шпа́р·ить, -ят, (*imp*) scald

шпару́т·к·а, -и, (*g pl*) -т·ок·, (*f*) (text) temple

шпару́т·очн·ый, -ая, -ое, (*a*) (text) temple

~ но́ж·ниц·ы (*pl*) (text) temple shears

шпа́т·, -а, (*m*) (min) spar (*and see under associated adjectives*); (vet) spath

~, ко́ст·н·ый (vet) spavin

~, пла́в·иков·ый (min) fluorspar, fluorite

~, тяжёл·ый (min) heavy spar, barytes

шпа́тел·ь, -я, (*m*) putty knife, stopping knife; palette knife; (med) spatula; brick set (for cutting bricks)

шпатл·ева́ть, -юют, (*imp*) *see* шпакл·ева́ть

шпатл·ёвк·а, -и, (*f*) *see* шпакл·ёвк·а

шпато·ви́д·н·ый, -ая, -ое, (*a*) (geol) sparry, spathic

шпа́т·ов·ый, -ая, -ое, (*a*) of шпа́т·

шпа́утер·, -а, (*m*) (met) spelter, commercial zinc

шпа́ци·я, -и, (*f*) (print) space; (shipb) frame-spacing

шпе́йз·а, -ы, (*f*) (met) speise, speiss

шпен·ёк·, -нь·к·а́, (*m*) pin, peg, prong

шперак·, -а, (*m*) (met forge) beak iron, beck iron, bickern

шпиг·, -а, (*m*) fat bacon, lard

шпига́т·, -а, (*m*) (naut) scupper

~, шторм·ов·о́й freezing port, wash port

шпи́гел·ь, -я, (*m*) (met) spiegel iron; (min) specular iron ore

шпиг·о́ванн·ый, -ая, -ое, (*past part pass*) of шпиг·ова́ть; (food) larded; (naut) thrummed

~ мат· (*m*) thrummed mat

~ пла́стыр·ь (*m*) canvas collision mat

шпиг·ова́ть, -у́ют, (*imp*) lard; thrum (mats etc.)

шпи́к·, -а, (*m*) = шпиг·; шпик·, -а́, spy, secret agent

шпил·ев·о́й, -а́я, -о́е, (*a*) capstan

~ маши́н·а (*f*) capstan engine

шпи́л·ечн·ый, -ая, -ое, (*a*) of шпи́ль·к·а

шпи́л·ь, -я, (*m*) (naut) capstan; (arch) spire, steeple

шпи́ль·к·а, -и, (*g pl*) -л·ек·, (*f*) stud bolt, stud; tack; hair-pin

~ вы·тяж·н·о́го тро́с·а rip-cord locking-pin (parachute)

~, ко́мпас·н·ая compass pivot

шпина́т·, -а, (*m*) (bot) spinach, *Spinacia*

шпингалéт·, -а,   (*m*) espagnolette/ upright bolt

шпи́ндел·ь, -я, (*m*) spindle (of machine tool); (horol) arbor; spindle, rod, shaft, pivot

~, вращ·áющ·ийся   live spindle (of machine tool)

~ зáд·н·ей бáб·к·и   tail spindle (of machine tool)

~, след·я́щ·ий   copy spindle (of machine tool)

~, трéф·ов·ый   (met roll) wobbler spindle

шпи́ндель·н·ый, -ая, -ое,   (*a*) of **шпи́ндел·ь**

~ аппарáт· (*m*) (agr) cotton-spindle picker

~ бáб·к·а (*f*) headstock (of machine tool)

шпинéл·ев·ый, -ая, -ое, (*a*) of **шпинéл·ь**

шпинел·и́д·, -а, (*m*) (min) spinellid

шпинелл·и́д·, -а, (*m*) (min) spinellid

шпинéл·ь, -и, (*f*) (min) spinel

шпиóн·, -а, (*m*) spy

шпион·áж·, -а, (*m*) espionage

шпирон·, -а, (*m*) prow, ram

шпиц·, -а, (*i*) **-ем,** (*m*) (arch) spire

шпиц·евáни·е, -я, (*n*) pointing (a wire)

шпицкáстен·, -а, (*m*) (min) spitzkasten

шплúнт·, -а, (*m*) split/cotter pin

~, раз·вóд·н·ый split/cotter pin

шпóн·, -а, (*m*) single ply (wood); (print) lead

шпóн·а, -ы, (*f*) = **шпон·**

шпóн·к·а, -и, (*g pl*) **-н·ок·,** (*f*) (mech eng) key (woodworking); keyed expansion joint (concrete)

~, за·бив·н·áя plain taper key

~, за·тяж·н·áя plain taper key

~, клин·ов·áя taper key

~ на лыск·е flat key

~, сегмéнт·н·ая Woodruff key

~, скольз·я́щ·ая (mech eng) feather

~, тангенциáль·н·ая Kennedy key

~, фрикциóн·н·ая hollow saddle key

шпóн·очн·ый, -ая, -ое, (*a*) of **шпóн·-к·а**

~ канáв·к·а (*f*) keyway

~ со·един·éни·е (*n*) keying, keyed joint; dowelled joint (wood)

~ стан·óк· (*m*) keyway cutting machine, keyway seating machine

шпóр·, -а, (*m*) (tech) heel

шпóр·а, -ы, (*f*) spur; (mech) groyne, spur; grouser (of caterpillar track); (orn) spur, calcar

шпóр·н·ый, -ая, -ое, (*a*) calcarine

шрéдер·, -а, (*m*) spreader

шпрединг-машúн·а, -ы, (*f*) (rubber) spreading machine

шпрéнгел·ь, -я, (*m*) (civ eng) strut frame

шпрéнгель·н·ый, -ая, -ое, (*a*) strut-frame, strut-framed

шпрúнг·, -а, (*m*) (naut) spring (berthing hawser)

шприц·, -а, (*i*) **-ем,** (*m*) syringe; grease gun; (ICE) primer; sausage-filling machine, extruder

~, за·лив·н·óй (ICE) primer

~ -машúн·а, -ы, (*f*) extruder

~, тавóт·н·ый grease gun

шприц·евáни·е, -я, (*n*) (*v n*) of **шприц·евáть**; extrusion

шприц·евáть, -ýют, (*imp and perf*) extrude; syringe

шприц·óванн·ый, -ая, -ое, (*a*) (*past part pass*) of **шприц·евáть**

шпрóт·, -а, (*m*) (fish) sprat, *Spratus*

шпул·ёвк·а, -и, (*f*) (text) bobbin winding

шпýль·к·а, -и, (*g pl*) **-л·ек·,** (*f*) (text) spool, bobbin

~, челн·óчн·ая shuttle, bobbin (of sewing machine)

шпýль·н·ый, -ая, -ое, (*a*) of **шпýл·я**

~ машúн·а (*f*) bobbin winder

шпýл·я, -и, (*f*) (text) spool, bobbin

шпул·я́рник·, -а, (*m*) (text) bank

шпунт·, -á, (*m*) tongue (carpentry); (civ eng) sheet piling

~ Лáрсен·а (civ eng) Larssen sheet pile

шпунт·гéбел·ь, -я, (*m*) grooving/ rabbet/tongue plane

шпунт·и́н·а, -ы, (*f*) (civ eng) sheet pile

шпунт·овáльн·ая машúн·а (*f*) tonguing-and-grooving machine

шпунт·овáть, -ýют, (*imp*) groove, rabbet, tongue

шпунт·ови́к·, -á, (*m*) hammer-head chisel; tonguing-and-grooving plane

шпунт·ов·óй, -áя, -óе, (*a*) of **шпунт·**

~ о·гражд·éни·е (*n*) (civ eng) sheet piling wall

~ пере·мы́ч·к·а (*f*) (civ eng) coffer-dam piling

~ пóяс· (*m*) (shipb) garboard strake

~ ряд· (*m*) (civ eng) piling, sheet piling

~ со·един·éни·е (*n*) tongue-and-groove joint

шпунт·ýбел·ь, -я, (*m*) *see* **шпунт·-гéбел·ь**

шпу́р·, -а, *(m)* (expl) shothole, blast-hole; (math) spur, trace (of a matrix); (m et) tapping hole, tap hole (of furnace)

∼, в·ру́б·ов·ый (min) cuthole, sumping hole, snubber

∼, вс·по·мог·а́тельн·ый (min) reliever hole

∼, котл·о́в·ый (expl) sprung hole

∼, от·бо́й·н·ый (expl) third-row hole, outside hole

∼, по до́·шв·енн·ый (expl) bench/foot/lifter hole

∼ прям·о́го в·ру́б·а (expl) burn-cut hole

шпур·ов·о́й, -а́я, -о́е, *(a)* of **шпур·**

Шра́ге, дви́г·ател·ь *(m)* (elec) Schrage motor

шра́м·, -а, *(m)* scar

∼, лист·ов·о́й (bot) cicatrix

шрапне́л·ь, -и, *(f)* (gunn) shrapnel shell

∼, пул·ев·а́я shrapnel

шра́тт·, -а, *(m)* (geol) crack

шре́ддер·, -а, *(m)* shredder, shredding machine

Шре́дер·а, у·равн·е́ни·е *(n)* (chem) Le Chatelier equation

Шре́дингер·а, у·равн·е́ни·е *(n)* (phys) Schrödinger's wave equation

Шре́йнемакерс·а, ме́тод· *(m)* (chem) Schreinemaker's method

шрекингери́т·, -а, *(m)* (min) schroek-ingerite

шре́ф·, -а, *(m)* (met roll) roll shoulder

шрифт·, -а, *(nom pl)* **-ы́,** *(g pl)* **-о́в,** *(m)* (print) type, print

∼, вы·дел·и́тельн·ый display type

∼, жи́р·н·ый bold/black/heavy type

∼, -ка́сс·а, -ы, *(f)* (print) type case

∼, курси́в·н·ый italics

ли́н·и·я шри́фт·а *(f)* type line, body line

∼, машинно·пи́с·н·ый typewriter face

∼, от·те́н·очн·ый shadow type

∼, полу·жи́р·н·ый half-dark type

∼, прям·о́й upright type

∼, руко·пи́с·н·ый script type

∼, с·би́т·ый batter

∼, све́тл·ый light type

∼, с·ме́ш·анн·ый bastard type

∼, у́з·к·ий lean/condensed type

шрифт·ов·о́й, -а́я, -о́е, *(a)* of **шрифт·**

шрифто·лит·е́йн·ый, -ая, -ое, *(a)* type-casting

шро́т·, -а, *(m)* (agr) oil-seed meal, extracted meal

Шрот-эффе́кт·, -а, *(m)* (elec) Schrott effect; shot noise

шта́б·, -а, *(nom pl)* **-ы́,** *(g pl)* **-о́в,** *(m)* staff, headquarters

штабел·ёр·, -а, *(m)* stacker

штабеле·у·кла́д·чик·, -а, *(m)* stacker

штабел·и́ровани·е, -я, *(n)* *(v n)* of **штабел·и́р·овать**

штабел·и́р·овать, -уют, *(imp)* pile, stack

шта́бел·ь, -я, *(nom pl)* **-я́,** *(g pl)* **-е́й,** *(m)* stack, pile

шта́б·ик·, -а, *(m)* *see* **шта́п·ик·**

штаб·н·о́й, -а́я, -о́е, *(a)* of **штаб·**

шта́г·, -а, *(m)* (naut) stay

шта́г·ов·ый, -ая, -ое, *(a)* of **штаг·**

∼ ого́н·ь *(m)* (naut) riding light

шта́мб·, -а, *(m)* trunk, stem, bole (of tree)

шта́мб·ов·ый, -ая, -ое, *(a)* of **штамб·**

шта́мм·, -а, *(m)* (microbiol) strain

шта́мп·, -а, *(m)* die, press tool, punch, stamp; title block (of a drawing)

∼, вы·ру́б·н·о́й blanking tool/die/punch

∼, вы·тя́ж·н·о́й deep-drawing die, cupping die/tool

∼, ги́б·очн·ый bending tool/die

∼, калибр·о́вочн·ый sizing die

∼, клейм·и́льн·ый lettering die

∼, ко́в·очн·ый drop/forging die

∼, кромко·за·ги́б·очн·ый beading die

∼, ма́стер- master die

∼, много·руч·ьев·о́й multi-impression die

∼, ни́ж·н·ий bed die

∼, об·ре́з·н·о́й trimming die

∼, об·ру́б·н·о́й cropping die

∼, от·ре́з·н·о́й parting die/tool

∼, пере·да́т·очн·ый offset die

∼, револьве́р·н·ый turret punch

∼, рихт·ова́льн·ый restriking die

∼, со·в·мещ·ённ·ый combination press tool

∼, фальц·о́вочн·ый horning tool

∼, чека́н·очн·ый coining/embossing die/punch

штамп·ова́льн·ый, -ая, -ое, *(a)* stamping, punching, pressing

штамп·ова́ни·е, -я, *(n)* = **штамп·о́вк·а**

штамп·о́ванн·ый, -ая, -ое, *(past part pass)* of **штамп·ова́ть**

штамп·ова́ть, -у́ют, *(imp)* stamp, punch, press, drop-forge, forge

штамп·о́вк·а, -и, (*g pl*) **-вок·,** (*f*) (*v n*) of **штамп·ова́ть**

~, гор·я́ч·ая hot/die forging

~, объ·ём·н·ая drop forging, closed-die forging

штамп·о́вочн·ый, -ая, -ое, (*a*) of **штамп·о́вк·а;** drop

~ мо́лот· (*m*) drop hammer

шта́мп·ов·ый, -ая, -ое, (*a*) of **шта́мп·**

~ стал·ь (*f*) die steel

шта́нг·а, -и, (*f*) rod, bar; weight (weight-lifting sport)

~, атлет·и́ческ·ая weight-lifter's bar

~, квадра́т·н·ая (oil) kelly

~, на·со́с·н·ая (oil) sucker rod

~, рас·по́р·н·ая spreader, spreader bar

~, с·круч·ивающ·ая (M/T) torsion bar

~, уда́р·н·ая sinker bar, drill stem (percussion drilling)

штанген- (*int component*) (instr) Vernier

штанген·глубино·ме́р·, -а, (*m*) (instr) height-and-depth gauge/gage, Vernier height-and-depth gauge/gage

штанген·микро́·метр·, -а, (*m*) Vernier micrometer

штанген·ре́йсмас·, -а, (*m*) Vernier height gauge/gage

штанген·ци́ркул·ь, -я, (*m*) (instr) Vernier caliper

~ -зубо·ме́р·, -а, (*m*) gear tooth Vernier

~, прост·о́й slide/sliding caliper

~, раз·мёт·очн·ый beam compasses, trammels (drawing instr.)

шта́нг·ов·ый, -ая, -ое, (*a*) rod, bar

штандо́л·ь, -я, (*m*) stand oil, lithographic varnish (paintmaking)

шта́нц·евать, -уют, (*imp*) cut/stamp/ blank out

штан·ы́ -о́в, (*m pl*) trousers, breeches

шта́пел·ь, -я, (*m*) (text) staple

шта́пель·н·ый, -ая, -ое, (*a*) (text) staple

~ волокн·о́ (*n*) (text) staple

шта́п·ик·, -а, (*m*) moulding, beading (wood)

Шта́рк·а, явл·е́ни·е (*n*) (phys) Stark effect

шта́рк·овск·ое рас·щепл·е́ни·е (*n*) (phys) Stark effect/splitting

шта́т·, -а, (*m*) staff, establishment; state

штати́в·, -а, (*m*) support, tripod

шта́т·н·ый, -ая, -ое, (*a*) staff, regular, permanent (of employment)

~ континге́нт· (*m*) establishment

шта́т·ск·ий, -ая, -ое, (*a*) civil; (*as noun*) civilian

шта́уфер·, -а, (*m*) grease cup

масл·ёнк·а шта́уфер·а (*f*) grease cup

штафи́р·к·а, -и, (*g pl*) **-р·ок·,** (*f*) trimming (shoemaking)

штев·ен·ь, -в·н·я, (*m*) (shipb) stem, stern post; (*pl*) stem and stern post

штейн·, -а, (*m*) (met) matte

~, бе́л·ый white metal

штеккер·, -а, (*m*) (telecom) manual switchboard plug

штемпел·ева́ни·е, -я, (*n*) (*v n*) of **штемпел·ева́ть**

штемпел·ева́льн·ый, -ая, -ое, (*a*) stamping, cancelling (by stamping)

~ маши́н·а (*f*) stamping/cancelling machine

штемпел·ева́ть, -юют, (*imp*) stamp, rubber-stamp, cancel (postage stamps etc.)

ште́мпел·ь, -я, (*nom pl*) **-я́,** (*g pl*) **-е́й,** (*m*) stamp, rubber stamp; stamp, stamp mark

~, почт·о́в·ый postmark, cancellation mark

ште́мпель·н·ый, -ая, -ое, (*a*) of **ште́мпел·ь**

~ поду́ш·к·а (*f*) stamp pad

ште́нгел·ь, -я, (*m*) (elec) exhaust tube

ште́псел·ь, -я, (*nom pl*) **-я́,** (*g pl*) **-е́й,** (*m*) (elec) plug

~ бес·кон·е́чн·ого со·противл·е́ни·я (elec) infinity plug

ште́псель·н·ый, -ая, -ое, (*a*) plug

~ ви́л·к·а (*f*) plug

~ доск·а́ (*f*) plug board; problem board (computers)

~ коло́д·к·а (*f*) (elec) plug

~ розе́т·к·а (*f*) (elec) socket outlet

~ со·един·е́ни·е (*n*) plug-and-socket connection

Ште́рн·а-Ге́рлах·а, о́·пыт· (*m*) (phys) Stern–Gerlach experiment

ште́рт·, -а, (*m*) lanyard ·

штив·а́·ть, -ют, (*imp*) (naut) even out, stow solidly (cargo)

шти́в·к·а, -и, (*f*) stowing, stowage

штил·ева́ть, -юют, (*imp*) be/become becalmed

штил·ев·о́й, -а́я, -о́е, (*a*) (meteor) calm

~ полос·а́ (*f*) calm belt, doldrums

шти́л·ь, -я, (*m*) (meteor) calm (0: Beaufort scale)

~, мёртв·ый dead calm

штиль
~, **польный**　calm (0: Beaufort scale)
Штифеля, агрегат (*m*) (met roll) Stiefel mill
штифт, -á, (*m*) (mech) pin, peg
~, **копирный**　former/follower/guide pin, tracer
~, **упорный**　locking pin
~ **форсунки**　spray/injector nozzle pintle
штифтовой, -áя, -óе, (*a*) of **штифт**
~ **клёпка** (*f*) rivet pin
штихель, -я, (*m*) (print) burin, graver
штихмасс, -a, (*m*) (instr) inside micrometer
шток, -a, (*m*) stem, rod (of valve, piston etc.); stock (of anchor); (geol) stock
штокверк, -a, (*m*) (min) stockwork
штольня, -и, (*g pl*) **-лен,** (*f*) (min) adit
~, **дренáжная** (min) drainage/efflux/water gallery/adit
штольцит, -a, (*m*) (min) stolzite
штопáльный, -ая, -ое, (*a*) darning
штопать, -ют, (*imp*) darn
штопор, -a, (*m*) corkscrew; (air) spin
~, **неустановившийся** (air) inadvertent spin
~, **перекрещивающий** cross spin
~, **плоский** flat spin
　прáвило штопора (*n*) (elec) corkscrew/screwdriver rule
штопорить, -ят, (*imp*) (air) spin
штопором (*adv*) (air) in a spin
штóра, -ы, (*f*) blind, window blind
штóрка, -и, (*g pl*) **-рок,** (*f*) (*dim*) of **штóра**; (phot) blind; (a/c) blind, face curtain (of ejection seat)
шторм, -a, (*m*) gale; (meteor) fresh gale (8: Beaufort scale)
~, **жестóкий** (meteor) storm (11: Beaufort scale)
~, **крéпкий** whole gale (10: Beaufort scale)
~, **равноденственный** equinoctial gale
~, **сильный** strong gale (9: Beaufort scale), whole gale (10: Beaufort scale), 42-48 knot wind
~, **урагáнный** hurricane (12: Beaufort scale)
штормовáть, -ýет, (*imp*) weather a storm, ride out a gale
штормовóй, -áя, -óе, (*a*) of **шторм**
~ **крышка** (*f*) (naut) deadlight
~ **лéер** (*m*) life line

штормовóй, -áя, -óе
~ **предостережéние** (*n*) (meteor) gale warning
штормтрáп, -a, (*m*) (naut) Jacob's/rope/jumping ladder
штóрно-щелевóй затвóр (*m*) (phot) focal-plane shutter
штóрный, -ая, -ое, (*a*) of **штóра**
штрангпресс, -a, (*m*) (plast) extruder
штрáнд, -a, (*m*) (geol) strand
штрáф, -a, (*m*) fine, penalty
штрафнóй, -áя, -óе, (*a*) of **штраф**
штрафовáть, -ýют, (*imp*) fine, impose penalty on
штрейкбрéхер, -a, (*m*) strike-breaker, blackleg, scab
штрек, -a, (*m*) (min) drift, gallery
~, **кореннóй** (min) main entry
~, **подэтáжный** (min) sublevel drift
~, **промежýточный** (min) intermediate level drift
~, **слоевóй** stoping drift
штрипка, -и, (*g pl*) **-пок,** (*f*) footstrap
штрипс, -a, (*m*) (met) skelp
штрипсовый, -ая, -ое, (*a*) skelp
штрих, -á, (*m*) stroke; (instr) graduation, blaze (of diffraction grating); (*pl*) hatching, hachure
~, **волоснóй** hair line, thin stroke
~, **лéдниковый** (geol) glacial score/scratch/scar
~, **основнóй** (print) main stroke, face stroke, stem
штрихóванный, -ая, -ое, (*past part pass*) of **штрихóвать**; (print) hatched, shaded
штрихóватость, -и, (*f*) striation
штрихóватый, -ая, -ое, (*a*) striated
штрихóвать, -ýют, (*imp*) make strokes, cover with strokes, hatch, striate
штрихóвка, -и, (*f*) striation, (print etc.) hatching
~, **лéдниковая** (geol) striae, glacial striation
штрихóвой, -áя, -óе, (*a*) of **штрих**
~ **линéйка** (*f*) graduated rule
~ **лúния** (*f*) dashed line
~ **рисýнок** (*m*) line drawing
штрихпунктúрный, -ая, -ое, (*a*) dot-and-dash
штудúровать, -уют, (*imp*) study
штýка, -и, (*f*) piece, thing, item
　пять штук five (objects)
штукатýр, -a, (*m*) (build) plasterer

штукату́р·ить, -ят,   (*imp*) (build) plaster, parget

штукату́р·к·а, -и,   (*g pl*) **-р·ок·,** (*f*) (build) plaster, plastering (inside); facing, rendering (outside)

~, бари́т·ов·ая barium plaster

~, звуко·по·глощ·а́ющ·ая acoustic plaster

~, орна́мент·н·ая (build) pargeting

штук·о́ванн·ый, -ая, -ое, (*past part pass*) of **штук·ова́ть**

штук·ова́ть, -у́ют, (*imp*) mend, darn, make an invisible mend

штук·о́вк·а, -и, (*f*) mending, darning; invisible mend, mend, darn

штурва́л·, -а,   (*m*) wheel, control/adjusting wheel; (M/T) steering wheel; (air) control wheel

~ у·правл·е́ни·я винт·о́м (a/c) propeller control wheel

~ у·правл·е́ни·я стабилиз·а́тор·ом (a/c) stabilizer adjusting wheel

штурва́ль·н·ый, -ая, -ое, (*a*) of **штурва́л·;** (*m decl as adj*) (naut) helmsman

~ колес·о́ (*n*) (a/c) control wheel

~ коло́н·к·а (*f*) (a/c) control column

шту́рм·, -а, (*m*) (mil) assault

Шту́рм·а, теоре́м·а (*a*) (math) Sturm's theorem

шту́рман·, -а, (*nom pl*) **-ы,** or **-а́,** (*m*) navigator, navigating officer

~ -на·во́д·чик·, -а, (*m*) (air) fighter controller

шту́рман·ск·ий, -ая, -ое, (*a*) navigator's, navigation

~ де́л·о (*n*) navigation

штурм·ова́ть, -у́ют, (*imp*) (mil) assault, storm

штурм·ови́к·, -а́, (*m*) ground attack aircraft, Stormovik

штурм·ов·о́й, -а́я, -о́е, (*a*) (mil) assault; (air) ground-attack

штур·тро́с·, -а, (*m*) (naut) tiller rope

~, цеп·н·о́й rudder chain

штур·тро́с·ик·, -а, (*m*) (naut) rudder line, yoke line

шту́ф·, -а, (*m*) (min) lump

штуф·н·о́й, -а́я, -о́е, (*a*) lump, in lumps

шту́цер·, -а, (*nom pl*) **-а́,** (*g pl*) **-о́в,** (*m*) union, nozzle (pipes)

~, в·са́с·ывающ·ий suction nozzle

~, за·ря́д·н·ый charging point (gas cylinders etc.)

~, на·гнет·а́тельн·ый discharge nozzle

~, на·по́р·н·ый discharge nozzle

шту́цер

~ от·со́с·а воз·ду́х·а hollow air bolt (underwater work)

~ фи́льтр·а filter cap

шту́ч·н·ый, -ая, -ое, (*a*) of **штук·а;** piece

~ о·пла́т·а (*f*) piecework payment, payment by the piece

~ рабо́т·а (*f*) piece work

~ това́р· (*m*) piece goods

штыб·, -а, (*m*) (min) gum, slack, duff; (geol) rubble

штык·, -а́, (*m*) bayonet; bend, hitch (tied); (agr) spade's depth, spit, surface layer

штык·ов·о́й, -а́я, -о́е, (*a*) of **штык·**

~ со·един·е́ни·е (*n*) bayonet joint

~ у·каз·а́тел·ь (*m*) dipstick

штыр·, -а, (*m*) *see* **штыр·ь**

штыр·ев·о́й, -а́я, -о́е, (*a*) of **штыр·ь**

~ анте́нн·а (*f*) (rad) whip antenna/aerial

~ изол·я́тор· (*m*) (elec) pin-type insulator

~ основ·а́ни·е (*n*) central pivot (of gun)

штыр·ь, -я́, (*m*) pintle, pin; cramp iron; (gunn) saddle pivot

штырь-крюк·, -а́, (*m*) (elec) J-type pin insulator

~ с·вя́з·и (rad) coupling probe

~, цо́коль·н·ый base post (vacuum tubes)

шу́б·а, -ы, (*f*) fur-coat

шу́б·н·ый, -ая, -ое, (*a*) of **шу́б·а**

шуг·, -а, (*m*) = **шуг·а́**

шуг·а́, -и́, (*f*) brash ice

шуго·с·бро́с·, -а, (*m*) (civ eng) ice chute/shoot

шультени́т·, -а, (*m*) (min) schultenite

шульци́т·, -а, (*m*) (min) schulzite

шу́м·, -а, (*m*) noise, sound, (med) murmur

~, бес·по·ря́д·очн·ый random noise

коэффицие́нт· шу́м·ов (*m*) noise factor

~, плот·ов·о́й (telecom) thump

~, по·сторо́н·н·ие (*pl*) ground/ambient noise

~, хаот·и́ческ·ий random noise

шу́манн·овск·ая фото·пласт·и́нк·а (*f*) (phot) Schumann plate

шум·ёр·, -а, (*m*) (theat) sound-effects man

шум·е́ть, -я́т, (*imp*) make a noise, be noisy

шум·ли́в·ый, -ая, -ое, (*a*) noisy

шу́м·н·ый, -ая, -ое, (*a*) noisy, loud

шум·овйк·, -á, (*m*) (theat) sound-effects man

шум·ов·ой, -áя, -óе, (*a*) noise

~ **оркéстр·** (*m*) percussion band

~ **эффéкт·ы** (*pl*) (theat) sound effects

шумо·мéр·, -а, (*m*) audio-noise meter, phonometer

шумо·пеленг·áтор·, -а, (*m*) (naut) hydrophone

шум·фáктор·, -а, (*m*) (rad) noise factor

шунгúт·, -а, (*m*) (geol) schungite

шýнт·, -а, (*m*) (elec) shunt

~ **гальванóметр·а** galvanometer shunt **магазин· шýнт·ов** (*m*) shunt box

шунт·úр·овать, -уют, (*imp*) (elec) shunt

шунт·úрующ·ий, -ая, -ее, (*pres part act*) of **шунт·úр·овать;** shunting, shunt

~ **ём·кост·ь** (*f*) shunt capacitance

шунт·ов·ой, -áя, -óе, (*a*) shunt, shunt-wound

~ **двйг·ател·ь** (*m*) shunt-wound motor, shunt motor

шур·а, -ы, (*f*) = **шуг·á**

шур·овáть, -ýют, (*imp*) rabble, turn, mix, poke (fuel in furnace)

шур·óвк·а, -и, (*g pl*) **-вок·,** (*f*) (*v n*) of **шур·овáть**

шур·óвочн·ый, -ая, -ое, (*a*) rabbling, stirring, poking

шурýп·, -а, (*m*) wood screw

шýрф·, -а, (*m*) (min) test pit, bore pit

шурф·овáни·е, -я, (*n*) sinking/making a bore pit, pitting

шурф·овáть, -ýют, (*imp*) dig a pit

шурш·áть, -áт, (*imp*) rustle

шут·úть, -ят, (*imp*) joke, make a joke/jest

шýт·к·а, -и, (*g pl*) **-т·ок·,** (*f*) joke

шýт·очн·ый, -ая, -ое, (*a*) comic, humorous

шуч·ý, (*pres 1st sing*) of **шут·úть**

шхéр·а, -ы, (*f*) (geog) skerry

шхúпер·, -а, (*m*) = **шкúпер·**

шхóт·, -а, (*m*) = **шкóт·**

шхýн·а, -ы, (*f*) (naut) schooner

~ **-бáрк·, -а,** (*m*) (naut) barquentine

~ **-брúг·, -а,** (*m*) (naut) brigantine

~**, мáрсель·н·ая** topsail schooner

шь·ёт (*pres 3rd sing*) of **ши·ть**

шь·ющ·ий, -ая, -ее, (*pres part act*) of **ши·ть**

Щ

щавелево·кисл·ый, -ая, -ое,　　(a)
(chem) oxalate (of)
~ ка́ли·й　(m) potassium oxalate
щавелево·у́ксус·н·ый, -ая, -ое,　　(a)
(chem) oxalacetic, oxalacetate
~ кисл·от·а́　(f) oxal acetic acid,
oxaloacetic acid
щаве́л·ев·ый, -ая, -ое,　(a) (chem)
oxalic; (bot) sorrel, dock
~ кисл·от·а́　(f) (chem) oxalic acid
щавел·ёк·, -ль·к·а́,　(m) (bot) sheep
sorrel, *Rumex acetosella*
щаве́л·ь, -я́, (m) (bot) sorrel, dock, *Rumex*
~, ки́сл·ый　common sorrel, *Rumex
acetosa*
щад·и́ть, -я́т,　(imp) spare, have care/
regard for
щеб·ёнк·а, -и,　(g pl) -нок·, (f) =
ще́б·ен·ь
щеб·ёночн·ый, -ая, -ое, (a) metalled
(of roads)
~ по·кры́т·и·е　(n) road metalling
ще́б·ен·ь, -б·н·я,　(m) rubble, chippings, crushed rock, ballast; road
metal, metalling; (rail) ballast; (geol)
debris, chippings
щебне·о·чист·и́тельн·ая маши́н·а (f)
(rail) ballast clearing machine
щего́л·, -гл·а́,　(m) (orn) goldfinch,
Carduelis carduelis
щёгол·ь, -я,　(m) dandy; (orn) black
redshank, *Tringa erythropus*
ще́др·ый, -ая, -ое, (a) generous, liberal
щек·а́, -и́, (acc sing) щёк·у, (nom pl)
щёк·и, (g pl) щёк·, (f) cheek; (mech
eng) crank arm/web; jaw (of vice etc.)
щек·ов·о́й, -а́я, -о́е, (a) of щек·а́
щеко·дроб·и́лк·а, -и,　(g pl) -лок·,
(f) (mech) jaw-crusher
щеко́лд·а, -ы,　(f) pawl, catch, trip
gear, trigger; locking bar
щеко·обра́з·н·ый, -ая, -ое,　　(a)
jaw-shaped
щекот·а́ть,　(pres 3rd pl) щеко́ч·ут,
(imp) tickle
щекот·ли́в·ый, -ая, -ое, (a) ticklish
щел·ева́т·ый, -ая, -ое, (a) fissured,
cracked
щеле·ви́д·н·ый, -ая, -ое,　(a) slit-
shaped, rimiform
щел·ев·о́й, -а́я, -о́е,　(a) chink, slit,
slot, slotted; fissured, cracked, rimate,
rimose; (hydr eng) notch, vee; (ling)
fricative

щел·ев·о́й, -а́я, -о́е
~ анте́нн·а　(f) (rad) slot antenna
~ водо·с·ли́в·　(m) (civ eng) vee-
notch weir
~ голо́в·к·а　(f) (plast) sheet die
~ спектр·　(m) slit spectrum
щелег·а, -и,　(f) underwater rock
щел·и́нн·ый, -ая, -ое,　(a) (ling)
fricative
щёлк·а·ть, -ют,　(imp) click, snap, pop,
crack
щёлк·н·уть, -ут,　(perf) give a click/
snap/pop/crack
щёлок·, -а, (m) (chem) lye, liquor
~, е́д·к·ий　caustic/alkali lye
~, сер·ни́ст·ый　(chem) sulphite liquor
щёлоче·сто́й·к·ий, -ая, -ое,　(a)
alkali resistant
щёлоче·у·по́р·ност·ь, -и,　(f) alkali
resistance
щелоч·и́ть, -а́т,　(imp) alkalize
щёлочно·земе́ль·н·ый, -ая, -ое,　(a)
alkali-earth
~ мета́лл·　(m) alkali-earth metal
щелоч·н·о́й, -а́я, -о́е,　(a) (chem)
alkali, alkaline; caustic
~ зе́мл·и　(pl) (min) alkali earths
~ мета́лл·　(m) alkali metal
~ рас·тво́р·　(m) alkaline solution
~ хру́п·кост·ь　(f) (met) caustic
embrittlement
щёлоч·ност·ь, -и,　(f) alkalinity
щёлочно·у·по́р·н·ый, -ая, -ое,　(a)
alkali-resistant
щёлоч·ь, -и, (nom pl) -и, (g pl) -е́й,
(f) (chem) alkali
~, е́д·к·ая　caustic alkali
щелч·о́к·, -чк·а́,　(f) click, snap, pop,
crack
щел·ь, -и, (nom pl) -и, (g pl) -е́й, (f)
chink, crack, slit, slot, cleft, (phys)
gap, (anat) fissure, fissura, rima,
(mil) slit trench
~, глаз·ни́чн·ая　(anat) orbital fissure
~, звук·ов·а́я　(cinema) sound gate
~ ис·то́ч·ник·а　(nucl) source slit
~, о·грани́ч·ивающ·ая　collimating
slit
~, пере·кры́·т·ая　shell slit
~, смотр·ов·а́я　visor
~, энергет·и́ческ·ая　(phys) energy gap
щел·ь·е, -я,　(n) smooth rock shore
щен·и́ться, -я́тся,　(imp) (zool) whelp,
cub

щен·о́к·, -н·к·а́, (*m*) puppy, cub, whelp; bergy bit (ice)

щен·я́т·а (*nom pl*) of **щен·о́к·**

щеп·а́, -ы́, (*nom pl*) **щéп·ы,** (*d pl*) **щеп·а́м,** (*f*) = **щéп·к·а**

щеп·а́ни·е, -я, (*n*) (*v n*) *see* **щеп·а́ть**

щеп·а́ть, ⸗пл·ю́т, (*imp*) chip, splinter

щéп·к·а, -и, (*g pl*) **-п·ок·,** (*f*) splinter, sliver, shaving, chip (of wood)

щепко·ло́в·к·а, -и, (*g pl*) **-в·ок·,** (*f*) splinter-screen

щепо́т·к·а, -и, (*g pl*) **-т·ок·,** (*f*) = **щепо́т·ь**

щепо́т·ь, -и, (*f*) pinch (quantity)

щерб·а́т·ый, -ая, -ое, (*a*) pitted; pock-marked

щерб·и́н·а, -ы, (*f*) pitting; gap (in teeth)

щет·и́н·а, -ы, (*f*) bristle, seta, chaeta

щет·и́нистост·ь, -и, (*f*) bristliness, hispidity

щет·и́нист·ый, -ая, -ое, (*a*) bristly, setaceous, hispid

щет·и́нк·а, -и, (*g pl*) **-нок·,** (*f*) bristle, seta

щетинко·ви́д·н·ый, -ая, -ое, (*a*) bristle-shaped, setiform

щет·и́нков·ый, -ая, -ое, (*a*) (zool) setal, setaceous; (zool) chaetopodan, chaetopodous, chaetophorous, (*pl as noun*) chaetopods, chetopods, *Chaetopoda*

щетинко·но́г·ий, -ая, -ое, (*a*) (zool) chaetopodan, chaetopodous, (*pl as noun*) chaetopods, chetopods, *Chaetopoda*

щетинко·но́с·н·ый, -ая, -ое, (*a*) setigerous, chaetiferous, chaetophorous

щетинко·челюст·н·о́й, -а́я, -о́е, (*a*) (zool) chaetognathous, chaetognathan, (*pl as noun*) chaetognaths, *Chaetognatha*

щет·и́нник·, -а, (*m*) (bot) millet, *Setaria*

щетино·ви́д·н·ый, -ая, -ое, (*a*) bristle-shaped, setiform

щетино·обра́з·н·ый, -ая, -ое, (*a*) setiform, setaceous

щет·и́ночк·а, -и, (*g pl*) **-чек·,** (*f*) fine bristle

щёт·к·а, -и, (*g pl*) **-т·ок·,** (*f*) brush; (cryst) druse

~, коллéктор·н·ая (elec) commutator brush

~, сло·и́ст·ая (elec) compound brush

щёт·к·а

~, со·бир·а́тельн·ая (elec) current-collecting brush

~, у́голь·н·ая (elec) carbon brush

щётко·держ·а́тел·ь, -я, (*m*) brush-holder

щёт·очн·ый, -ая, -ое, (*a*) brush

щёчно·губ·н·о́й, -а́я, -о́е, (*a*) (anat) buccolabial

щёчно·язы́ч·н·ый, -ая, -ое, (*a*) (anat) buccolingual

щёчно·шéй·н·ый, -ая, -ое, (*a*) (anat) buccocervical

щёч·н·ый, -ая, -ое, (*a*) cheek, buccal, malar

щи (*indecl*) letter Shch; **щ·и,** (*g pl*) **щ·ей,** (*d pl*) **щ·ам,** cabbage soup

щи́кол·к·а, -и, (*g pl*) **-л·ок·,** (*f*) ankle

щи́кол·отк·а, -и, (*g pl*) **-ток·,** (*f*) ankle

щип·а́льн·ый, -ая, -ое, (*a*) plucking, picking

щип·а́ть, ⸗л·ю́т, (*imp*) pinch, nip, tweak; pluck, pick; nibble, browse

щип·н·у́ть, -у́т, (*perf, single action*) pinch, nip, tweak; pluck, pick

щип·éц·, -п·ц·а́, (*m*) (*see also* **щип·-ц·ы́**) (arch) gable

щип·о́к·, -п·к·а́, (*m*) pinch, nip, tweak

щип·ц·ы́, -о́в, (*m pl*) tongs; (med) forceps

щи́п·чик·и, -ов, (*m pl*) pincers, tweezers

щири́ц·а, -ы, (*f*) (bot) amaranth, *Amaranthus*

щит·, -а́, (*m*) shield, screen; panel, board; baffle plate; (civ eng) gate, sluice gate; (gunn) target, practice target; (astron) the Shield, Scutum

~, водо·раз·дéль·н·ый circulation augmenter (boiler)

~, генера́тор·н·ый (elec) generator switchboard

~, голов·н·о́й (civ eng) top/head gate

~, групп·ов·о́й (elec) distribution board

~, диспéтчер·ск·ий control board (at power station)

~, контро́ль·н·ый (elec) supervisory control swichboard

~, мнемон·и́ческ·ий colorgraphic control panel

~, оруд·и́йн·ый gun shield

~, от·кид·н·о́й flap plate

~, про·хо́д·ческ·ий (civ eng) tunneling shield

~, рас·пре·дел·и́тельн·ый (elec) switchboard

~ у·правл·éни·я control board

щи́т·ен·ь, -т·н·я,   (*m*) (*pl* щи́т·н·и) (q.v.) (zool) *Lepidurus*; *Apus, Triops*; *Proterotriops*

~, обык·нове́нн·ый   *Apus cancriformis*

щитко·ви́д·н·ый, -ая, -ое,   (*a*) (bot) corymbose, corymbous

щит·ко́в·ые, -ых,   (*pl decl as adj*) (zool) ostracoderms, *Ostradermi*

щи́т·н·и, -ей,   (*m pl*) (*sing* щи́т·ен·ь) (zool) *Apodidae, Triopidae*

щи́т·ник·и, -ов,   (*m pl*) (ent) *Pentatomidae*

щитовидно·гло́т·очн·ый, -ая, -ое,   (*a*) (anat) thyropharyngeal

щитовидно·над·горта́н·н·ый, -ая, -ое,   (*a*) (anat) thyroepiglottic

щито·ви́д·н·ый, -ая, -ое,   (*a*) (anat) thyroid; (bot) peltate, scutate

щит·о́вк·а, -и,   (*g pl*) -в·ок·, (*f*) (ent) scale, *Coccodea subordo*

щит·ов·о́й, -а́я, -о́е,   (*a*) of щит·; (build) frame-end-panel

щитово·с·бо́р·н·ый, -ая, -ое,   (*a*) (build) frame-end-panel

щито·гру́д·ые, -ых,   (*pl decl as adj*) (zool) thoracostracans, *Thoracostraca*

щит·о́к·, -т·к·а́,   (*m*) (*dim*) of щит·; (a/c) split flap; (ent) corselet; (bot) corymb

~, аккумул·я́торн·ый   accumulator baffle plate

~ -за·кры́л·ок·, -л·к·а,   (*m*) (a/c) flap

~, по·са́д·очн·ый   (a/c) landing flap

~, пред·о·хран·и́тельн·ый   (elec) fuse board

~, прост·о́й   (a/c) plain split flap

~, с·ва́р·очн·ый   (weld) hand shield

~, тормоз·н·о́й   (a/c) air brake

~, фи́рм·енн·ый   maker's name-plate

щито·ли́ст·н·ый, -ая, -ое,   (*a*) (bot) scutifolious

щито·над·горта́н·н·ый, -ая, -ое,   (*a*) (anat) thyroepiglottic

щито·обра́з·н·ый, -ая, -ое,   (*a*) shield-shaped, scutiform

щито·раз·де́ль·н·ый, -ая, -ое,   (*a*) (bot) peltatipartite

щито·ше́й·н·ый, -ая, -ое,   (*a*) (anat) thyrocervical

щу́к·а, -и,   (*f*) (fish) pike, *Esox*

щу́к·ов·ые, -ых,   (*pl decl as adj*) (fish) pike, piskerel, *Esocidae*

щуко·обра́з·н·ые, -ых,   (*pl decl as adj*) (fish) *Esociformes*

щуп·, -а,   (*m*) про́бе; feeler gauge/gage; follower (cam-type computer)

~, масло·у·каз·а́тельн·ый   oil dipstick, sounding rod, sounding tube

~, электро·магни́т·н·ый   electro-magnetic gauge head (of machine tools)

щу́п·альц·е, -а,   (*g pl*) -л·ец·, (*n*) feeler, tentacle, palpus, (ent) antenna

щу́п·альцев·ые, -ых,   (*pl decl as adj*) (zool) *Tentaculata*

щупальце·но́с·ец·, -с·ц·а,   (*m*) (zool) lophophore

щупальце·но́с·н·ый, -ая, -ое,   (*a*) tentacular, tentacled, tentaculiferous

щу́п·а·ть, -ют,   (*imp*) feel, touch, probe, palpate

щу́п·ик·, -а,   (*m*) (zool) palpus

щу́р·ить, -ят,   (*imp*) half close (eyes)

щу́р·к·а, -и,   (*g pl*) -р·ок·, (*f*) (orn) bee eater, *Merops*

щу́чч·ий, -ья, -ье,   (*a*) of щу́к·а

щу́чк·а, -и,   (*g pl*) -ч·ек·, (*f*) (bot) hair grass, *Deschampsia*

Э

э (*abbr*) = **эрстед·** oe, oersted; = **эрг** erg

эбе́н·ов·ый, -ая, -ое, (*a*) ebony; (*pl*) *Ebenaceae*

Э́бергард·а, явл·е́ни·е (*n*) (phot) Eberhard effect

эбиги́т·, -а, (*m*) (min) ebigite

эбони́т·, -а, (*m*) ebonite, hard rubber

эбони́т·ов·ый, -ая, -ое, (*a*) ebonite

эбо́н·ов·ый, -ая, -ое, (*a*) = **эбе́н·о-в·ый**

эбу́лио- (*int component*) *see also* **эбуллио-**

эбулио́·метр·, -а, (*m*) (chem) ebulliometer

эбулио·метр·и́ческ·ий, -ая, -ое, (*a*) ebulliometer

~ ме́тод· (*m*) dynamic method (of measuring vapour pressure)

эбулио·ме́тр·и·я, -и, (*f*) ebulliometry

эбуллио- (*int component*) = **эбулио-**

эбуллио·ско́п·, -а, (*m*) ebullioscope

эбуллио·скоп·и́·я, -и, (*f*) ebullioscopy

эв (*abbr*) = **элекро́н-вольт·** (phys) ev, electron-volt

эв- (*int component*) eu- (good) *see also* **эй-**

эвакуацио́н·н·ый, -ая, -ое, (*a*) evacuation

эваку·а́ци·я, -и, (*f*) evacuation

эваку·и́рованн·ый, -ая, -ое, (*past part pass*) of **эваку·и́р·овать**

эваку·и́р·овать, -уют, (*imp and perf*) evacuate

эванси́т·, -а, (*m*) (min) evansite

эвапор·а́тор·, -а, (*m*) vaporizer

эвапор·и́т·, -а, (*m*) (geol) evaporite

эвапоро·гра́мм·а, -ы, (*f*) (meteor) evaporogram

эвапоро́·метр·, -а, (*m*) (instr) evaporimeter

эвапо·транспир·а́ци·я, -и, (*f*) evapotranspiration

эв·гедра́ль·н·ый, -ая, -ое, (*a*) (petr) euhedral, idiomorphic

эв·нег·о́л·, -а, (*m*) (biochem) eugenol

эв·гео·синклина́л·ь, -и, (*f*) (geol) eugeosyncline

эвгле́н·а, -ы, (*f*) (zool) Euglena

эвгле́н·ов·ые, -ых, (*pl decl as adj*) (zool) Euglenoidina

эв·диали́т·, -а, (*m*) (min) eudialyte

эв·дидими́т·, -а, (*m*) (min) eudidymite

эвдио́·метр·, -а, (*m*) (chem) eudiometer

эвдио·ме́тр·и·я, -и, (*f*) (chem) eudiometry

эвентуа́ль·н·ый, -ая, -ое, (*a*) eventual, provisionally possible

эверду́р·, -а, (*m*) (met) Everdur

эвер-дьюпойс·, -а, (*m*) avoirdupois

эвипа́н·, -а, (*m*) (pharm) evipan, hexabarbitone sodium

эвито́н·, -а, (*m*) (biol) E-viton

эвкази́н·, -а, (*m*) (pharm) eucasin

эвкайри́т·, -а, (*m*) (min) eucairite, eukairite

эвкали́пт·, -а, (*m*) (bot) eucalyptus, *Eucalyptus*

~, буго́р·чат·ый blue gum tree, *Eucalyptus globulus*

~, голуб·о́й blue gum tree, *Eucalyptus globulus*

эвкалипт·е́н·, -а, (*m*) (chem) eucalyptene

эвкали́пт·ов·ый, -ая, -ое, (*a*) eucalyptus

~ ма́сл·о (*n*) eucalyptus oil

эвкалопт·о́л·, -а, (*m*) (chem) eucalyptole, cineole

эвкалипт·оле́н·, -а, (*m*) (chem) eucalyptolene

эв·кла́з·, -а, (*m*) (min) euclase

Эвкли́д·а, алгори́тм· (*m*) (math) Euclid's algorithm

эвкли́д·ов·, -а, -о, (*a*) (math) Euclidean

~ геоме́тр·и·я (*f*) Euclidean geometry

~ про·стра́н·ств·о (*n*) Euclidean space

эвкода́л·, -а, (*m*) (pharm) eucodal, dyhydroxycodeinone hydrochloride

эвколи́т·, -а, (*m*) (min) eucolite, eukolite, eukolyte

эвколло́ид·, -а, (*m*) (chem) eucolloid

эв·крипти́т·, -а, (*m*) (min) eucryptite

эв·ксени́т·, -а, (*m*) (min) euxenite

эвлити́н·, -а, (*m*) (min) eulytine

эвлити́т·, -а, (*m*) (min) eulytite

эвольве́нт·а, -ы, (*f*) (math) involute

эвольве́нт·н·ый, -ая, -ое, (*a*) (math) involute

~ за·цепл·е́ни·е (*n*) (mech) involute gear

~ шли́ц·ев·ое со·един·е́ни·е (*n*) involute spline

эвольвенто·ме́р·, -а, (*m*) odontometer, involute gear gauge/gage

эволю́т·а, -ы, (*f*) (math) evolute

эволюти́в·н·ая ско́р·ост·ь (*f*) (air) safety speed

эволю́т·н·ый, -ая, -ое, (*a*) evolute

эволюцион·и́р·овать, -уют, (*imp and perf*) evolve, develop by evolution

эволюцио́н·н·ый, -ая, -ое, (*a*) evolutionary, evolution

эволю́ци·я, -и, (*f*) (biol) evolution; manoeuvre, evolution

~, скачко·обра́з·н·ая saltatory evolution

эвпари́н·, -а, (*m*) (biochem) euparin

эври·ба́т·н·ый, -ая, -ое, (*a*) (zool) eurybathic

эври·гали́н·н·ый, -ая, -ое, (*a*) (biol) euryhaline

эври·гали́н·ов·ый, -ая, -ое, (*a*) (biol) euryhaline

эври·птер·и́д·ы (*pl*) (pal) eurypterids, *Eurypterida*

эври·птер·о́ид·ы (*pl*) = эври·птер·и́д·ы

эврист·и́ческ·ий, -ая, -ое, (*a*) (math) heuristic

эври·терм·и́ческ·ий, -ая, -ое, (*a*) eurythermic, eurythermal, eurythermous

эври·те́рм·н·ый, -ая, -ое, (*a*) eurythermal, eurythermous

эври·то́п·н·ый, -ая, -ое, (*a*) (biol) eurytopic

эвро́п·и·й, -я, (*m*) europium, Eu

эв·синхи́т·, -а, (*m*) (min) eusynchite, descloizite

эв·ста́з·и·я, -и, (*f*) (ocean) eustacy

эв·стат·и́ческ·ий, -ая, -ое, (*a*) (ocean) eustatic

~ колеб·а́ни·я (*pl*) (ocean) eustatic movements

эв·сте́л·а, -ы, (*f*) (bot) eustele

эв·те́кт·ик·а, -и, (*f*) (phys) eutectic, eutectic point; eutectic ice block

~, жи́дк·ая eutectic solution

~, тверд·ая (met) eutectic alloy

эв·тект·и́ческ·ий, -ая, -ое, (*a*) eutectic

~ то́ч·к·а (*f*) (met) eutectic point

эв·тект·о́ид·, -а, (*m*) (met) eutectoid

эвтект·о́ид·н·ый, -ая, -ое, (*a*) eutectoid

~ ста́л·ь (*f*) eutectoid steel

эв·троп·и́ческ·ий, -ая, -ое, (*a*) (phys) eutropic

эв·тро́п·и·я, -и, (*f*) (phys) eutropy

эв·троф·ика́ци·я, -и, (*f*) (biol) eutrophication

эв·троф·и́·я, -и, (*f*) (biol) eutrophy

эв·тро́ф·н·ый, -ая, -ое, (*a*) (biol) eutrophic

эв·фемист·и́ческ·ий, -ая, -ое, (*a*) euphemistic

эв·фон·и́ческ·ий, -ая, -ое, (*a*) euphonic

эв·хрои́т·, -а, (*m*) (min) euchroite

эги́д·а, -ы, (*f*) aegis

эгири́н·, -а, (*a*) (min) aegirine, aegirite

~ -авги́т·, -а, (*m*) (min) aegirine-augite

эгито·гна́т·ый, -ая, -ое, (*a*) (zool) aegithognathous

эго·исти́ческ·ий, -ая, -ое, (*a*) egoistic, selfish

эго·центр·и́ческ·ий, -ая, -ое, (*a*) egocentric

эгу́тер·, -а, (*m*) (paper) dandy roll

эдаф·и́ческ·ий, -ая, -ое, (*a*) (biol) edaphic

эдафо·за́вр·ы, -ов, (*m pl*) (pal) *Edaphosauria*

эдафо·ло́г·и·я, -и, (*f*) edaphology, soil science

эдаф·о́н·, -а, (*m*) (biol) edaphon

Эделеан·у, о·чи́ст·к·а по (oil) Edeleanu refining

эдельве́йс·, -а, (*m*) (bot) edelweiss, *Leontopodium*

эдест·и́н·, -а, (*m*) (biochem) edestin

э́джер·, -а, (*m*) (met roll) edger, vertical stand

эдингтони́т·, -а, (*m*) (min) edingtonite

Э́дкок·а, анте́нн·а (*f*) (rad) Adcock antenna

э.д.с. (*abbr*) = электро·дви́ж·ущ·ая си́л·а e.m.f., electromotive force

эдт (*abbr*) (chem, rad) EDT, ethylene diamine tartrate

эжект·и́ровани·е, -я, (*n*) ejection

эжéктор·, -а, (*m*) ejector

~, водо·стру́й·н·ый water-jet ejector

~, паро·стру́й·н·ый steam-jet ejector

эжéктор·ный, -ая, ое, (*a*) ejector

эжéкци·я, -и, (*f*) ejection

эзельго́фт·, -а, (*m*) mast cap

эзери́н·, -а, (*m*) (chem) eserine, physostygmine

эзофаго·маля́ци·я, -и, (*f*) (med) oesophagomalacia, esophagomalacia

эзофаго·ско́п·, -а, (*m*) (med) oesophagoscope, esophagoscope

эзофаго·сто́м·а, -ы, (*f*) (zool) oesophagostome, esophagostome

эзо·филакси·я, -и, (*f*) (med) esophylaxis

эзо·фо́ри·я, -и, (*f*) (med) esophoria

эй (*int component*) eu- (good) *see also* эв-

эйген·валю́ (*indecl*) (math) eigenvalue

эйген·ве́ктор·, -а, (*m*) (math) eigenvector

эйген·велич·ин·а́, -ю́, (*f*) (math) eigenvalue

эйгенве́рт·, -а, (*m*) (math) eigenvalue

эйген·фу́нкци·я, -и, (*f*) (math) eigenfunction

эй·ка́ин·, -а, (*m*) (biochem) eucaine

эйко́ми·я, -и, (*f*) gutta percha tree

эйкона́л·, -а, (*m*) (opt) characteristic function

эйкоза́н·, -а, (*m*) (chem) eicosane

эйкоза́н·ов·ая кисл·от·а́ (*f*) eicosanic/ eicosanoic acid

эйкозе́н·, -а, (*m*) eicosene

эйксени́т·, -а, (*m*) (min) euxenite

эйкупино·токси́н·, -а, (*m*) (biochem) eucupinotoxin

Э́йлер·, -а, (*m*) (math etc.) Euler

Э́йлер·а, по·сто·я́нн·ая (*f*) (math) Euler's constant

~, у·равн·е́ни·е, -я, (*n*) Euler's equation (calculus of variables)

~, фо́рмул·а (*f*) Euler's formula; equation of Euler (differential geometry)

~, φ-фу́нкци·я (*f*) (math) indicator of an integer, Euler's φ-function

~ -Макло́рен·а, фо́рмул·а сумм·и́-ровани·я (math) Euler–Maclaurin sum formula

э́йлер·ов·, -а, (*a*) Eulerian

~ характер·и́стик·а (*f*) Euler characteristic

~ угл·ы́ (*pl*) (math) Eulerian/Euler's angles

эйнште́йн·, -а, (*m*) einstein (unit of light)

Эйнште́йн·а, эффе́кт· (*m*) (astron) Einstein shift

Эйнште́йн·а–де Ха́аз·а, эффект· (*m*) (phys) Einstein–de Haas effect

эйнште́йн·и·й, -я, (*m*) einsteinium, Es

эйри·зо́м·и·я, -и, (*f*) (biol) eurysomy

эйри·зо́м·н·ый, -ая, -ое, (*a*) (biol) eurysome

эй·ри́тм·и·я, -и, (*f*) (physiol) eurythmy, eurhythmy, eurythmia

эйро·фи́т·, -а, (*m*) (bot) aerophyte

эйфо́рб·и·й, -я, (*m*) (pharm) euphorbium

эй·фори́·я, -и, (*f*) (med) euphoria

эй·хини́н·, -а, (*m*) (pharm) euquinine

эйхо́рн·и·я, -и, (*f*) (bot) water hyacinth, *Eichhornia*

эка- (*int component*) (chem) eka-

~ -це́з·и·й, -я, (*m*) eka-cesium

эка́д·а, -ы, (*f*) (gen) orcad, ecad

экв·а́тор·, -а, (*m*) equator

экв·а́тор

~, зем·н·о́й terrestrial equator

~, магни́т·н·ый magnetic equator

~, неб·е́сн·ый (astron) equinoctial, celestial equator

~, основ·н·о́й fundamental equator

~, тепл·ов·о́й heat/thermal equator

~, терм·и́ческ·ий heat equator, thermal equator

экваториа́л·, -а, (*m*) equatorial (telescope)

экваториа́ль·н·ый, -ая, -ое, (*a*) equatorial

~ электро́д· (*m*) equatorial ring (of gyro compass)

экви·вале́нт·, -а, (*m*) equivalent; dummy, artificial device

~ анте́нн·ы (rad) artificial aerial

~, во́д·н·ый (phys) water equivalent

за·ко́н· эквивале́нт·ов (*m*) Einstein's law of photochemical equivalence

~, механ·и́ческ·ий (phys) mechanical equivalent (of heat, light)

~ на·гру́з·к·и artificial load

~, троти́л·ов·ый (expl) TNT equivalent

экви·вале́нт·ност·ь, -и, (*f*) equivalence

экви·вале́нт·н·ый, -ая, -ое, (*a*) equivalent

~ брус· (*m*) (mech) equivalent girder

~ схе́м·а (*f*) (elec) equivalent circuit

~ температу́р·а (*f*) (meteor) equivalent temperature

экви·грануля́р·н·ый, -ая, -ое, (*a*) equigranular

экви·либр·а́тор·, -а, (*m*) balancing stand

экви·либр·и́ст·, -а, (*m*) tightrope-walker, equilibrist

эквил·и́н·, -а, (*m*) (biochem) equiline

экви·молекуля́р·н·ый, -ая, -ое, (*a*) equimolecular

экви·моля́р·н·ый, -ая, -ое, (*a*) equimolar

экви·потенциа́л·, -а, (*m*) equipotential, equipotential line

экви·потенциа́ль·н·ый, -ая, -ое, (*a*) (phys) equipotential

экви·с·вя́з·ь, -и, (*m*) (cryst) equicohesion

э́кер·, -а, (*m*) *see* о́ккер·

экз·альт·а́ци·я, -и, (*f*) (med, opt) exaltation

экзальто́н·, -а, (*m*) exalton, cyclopentadecanone (used in perfumes)

экза́мен·, -а, (*m*) examination, exam

экзамен·ова́ть, -у́ют, (*imp*) examine (test knowledge)

экзамен·а́тор·, -а, (*m*) examiner

~ у́·ровн·ей (*m*) (surv) level trier

экзаменацио́н·н·ый, -ая, -ое, (*a*) of экза́мен·

экз·анте́м·а, -ы, (*f*) (med) exanthem, exanthema

экзара́ци·я, -и, (*f*) exaration

экз·артикул·я́ци·я -и, (*f*) exarticulation

эк·зе́м·а, -ы, (*f*) (med) eczema

эк·земато́з·н·ый, -ая, -ое, (*a*) (med) eczematous

экземпля́р·, -а, (*m*) copy, specimen

экзи́н·а, -ы, (*f*) (bot) exine

экзини́т·, -а, (*m*) (pal, bot) exinite

экзоа́ск·ов·ые гриб·ы́ (*pl*) (bot) *Exoascaceae*

экзо·гаме́т·а, -ы, (*f*) (gen) exogamete

экзо·га́м·и·я, -и, (*f*) (gen) exogamy

экзо·ге́н·н·ый, -ая, -ое, (*a*) exogenous

экзо·гео·синклина́л·ь, -и, (*f*) (geol) exogeosyncline

экзо·де́рм·а, -ы, (*f*) (biol) exoderm

экзо·ка́рп·и·й, -я, (*m*) (bot) exocarp

экзо·кри́н·н·ый, -ая, -ое, (*a*) (zool) exocrine

эк·золю́ци·я *see* экс·солю́ци·я

экзо·мо́рф·н·ый, -ая, -ое, (*a*) (geol) exomorphic

экзо·пла́зм·а, -ы, (*f*) (gen) exoplasm, ectoplasm

экзо·поди́т·, -а, (*m*) (zool) exopodite

экз·о́смос·, -а, (*m*) (chem) exosmosis

экзо·спо́р·и·й, -я, (*m*) (bot) exosporium, exine

экз·ост·о́з·, -а, (*m*) (med) exostosis

экзо·сфе́р·а, -ы, (*f*) (meteor) exosphere

экзо́т·, -а, (*m*) (biol) exotic

экзо·те́рм·и́ческ·ий, -ая, -ое, (*a*) (phys) exothermic, exothermal

~ реа́кци·я (*f*) (chem) exothermic reaction

экзо·те́ц·и·й, -я, (*m*) (bot) exothecium

экзот·и́ческ·ий, -ая, -ое, (*a*) exotic

экзо·токси́н·, -а, (*m*) (biochem) exotoxin

экзо·тро́п·изм·, -а, (*m*) (bot) exotropism, ectotropism

экзо·тро́п·н·ый, -ая, -ое, (*a*) exotropic, ectotropic

экзо·тро́ф·н·ый, -ая, -ое, (*a*) = экзо·тро́п·н·ый

экз·офта́льм·, -а, (*m*) (med) exophthalmos, exophthalmus

экзо·энерге́т·и́ческ·ий, -ая, -ое, (*a*) exoergic, exoenergic

экзу́ви·й, -я, (*m*) (zool) exuvia, (*pl*) exuviae

экипа́ж·, -а, (*m*) crew; ship's company (of major warships); carriage (horse-drawn)

~, с·ме́н·н·ый relief crew, (air) slip crew

~, флот·ск·ий naval depot, fleet manning pool

экип·иро́вк·а, -и, (*g pl*) -вок·, (*f*) (*v n*) of экип·ир·ова́ть; equipping, equipment, fitting out, outfit

экип·ир·ова́ть, -у́ют, (*imp and perf*) equip, fit out

э́ккер·, -а, (*m*) (surv) right-angle mirror

~, зеркал·ь·н·ый (surv) mirror/optical square

~, кресто·обра́з·н·ый (surv) cross-staff

Э́ккерт·а, про́б·а (*f*) (chem) Ekkert test reaction

эклект·и́ческ·ий, -ая, -ое, (*a*) eclectic

экли́·метр·, -а, (*m*) (surv) elevation meter

экли́пт·ик·а, -и, (*f*) (astron) ecliptic

~, основ·н·а́я fundamental ecliptic

эклоги́т·, -а, (*m*) (min) eclogite

Э́кман·а, спира́л·ь (*m*) (meteor, ocean) Ekman spiral

эко·гене́з·, -а, (*m*) (bot) ectogenesis

эко·лог·и́ческ·ий, -ая, -ое, (*a*) (biol) ecological

эко·ло́г·и·я, -и, (*f*) (biol) ecology

эконом·а́йзер·, -а, (*m*) (tech) economizer

~, вод·ян·о́й (st eng) economizer, feed-water heater

эконом·иза́тор·, -а, (*m*) welding economizer

эконо́м·ик·а, -и, (*f*) economics

эконом·и́ст·, -а, (*m*) economist; control clerk (on factory staff)

эконо́м·ить, -ят, (*imp*) economize, effect a saving in

эконом·и́ческ·ий, -ая, -ое, (*a*) economic, commercial; economical, economy, economizing

~ геогра́ф·и·я (*f*) economic geography

~ гор·е́лк·а (*f*) pilot burner

~ поли́т·ик·а (*f*) political economics

~ ско́р·ост·ь (*f*) (air) best speed for endurance; (naut) economical speed

эконом·и́чност·ь, -и, (*f*) economy, economic running, commercial efficiency/practicability

эконо́м·и́чн·ый, -ая, -ое, (*a*) economic, commercially efficient/practicable

эконо́м·и·я, -и, (*f*) economy, saving

эконо́м·нича·ть, -ют, (*imp*) economize, scrape and save, be mean

эконо́м·н·ый, -ая, -ое, (*a*) economical; sparing, mean

эконо́м·со·ве́т·, -а, (*m*) economic council

эко·систе́м·а, -ы, (*f*) (biol) ecosystem

эко·ти́п·, -а, (*m*) (biol) ecotype

эко·топо·ло́г·и·я, -и, (*f*) (biol) ecotopology

экра́н·, -а, (*m*) screen, shield, shade, baffle; shroud (of masked valve); (M/T) apron plate; apron (of dam)

~, водо·со·держ·а́щ·ий (nucl) hydrogenous shield

~, за·по·мин·а́ющ·ий storage surface (computers)

~, магни́т·н·ый magnetic screen

~, на·правл·я́ющ·ий baffle

~, от·раж·а́тельн·ый (rad) sheet reflector

~, про·све́ч·ивающ·ий fluoroscopic screen

~, свет·я́щ·ийся fluorescent screen

~, тепл·ов·о́й (nucl) heat shield; water wall (furnaces)

~, то́п·очн·ый water wall (boiler)

~, тру́б·чат·ый (elec) fluted screen

~, флюоресц·и́рующ·ий fluorescent screen

~, электро·стат·и́ческ·ий electrostatic shield, Faraday cage/screen

экраниз·а́ци·я, -и, (*f*) (*v n*) *see* **экраниз·и́р·овать**

экраниз·и́рованн·ый, -ая, -ое, (*a*) (*past part pass*) of **экраниз·и́р·о·вать**

экраниз·и́р·овать, -уют, (*imp and perf*) (cinema) screen, make a film of, film

экран·и́ровани·е, -я, (*n*) (*v n*) *see* **экран·и́р·овать**

~, тепл·ов·о́е water wall (of boiler or furnace)

~, электро·стат·и́ческ·ое electrostatic shield, electrostatic protection

эффе́кт· экран·и́ровани·я (*m*) (elec) screening effect

экран·и́рованн·ый, -ая, -ое, (*past part pass*) of **экран·и́р·овать;** screen-grid (of vacuum tubes); water-cooled (of furnaces)

~ ла́мп·а (*f*) (elec) screen-grid valve

экран·и́р·овать, -уют, (*imp and perf*) screen, shield, shade

экран·иро́вк·а, -и, (*f*) (*v n*) *see* **экран·и́р·овать**

конста́нт· экран·иро́вк·и (*m*) (nucl) screening constant

экран·и́рующ·ий, -ая, -ее, (*pres part act*) of **экран·и́р·овать**

~ се́т·к·а (*f*) screen grid, screen (of vacuum tube)

экра́н·н·ый, -ая, -ое, (*a*) of **экран·**

экрано·ско́п·, -а, (*m*) (cinema) screenoscope

экс- (*int component*) ex-

эксга́устер·, -а, (*m*) exhauster, exhaust fan

эксик·а́тор·, -а, (*m*) exsiccator, drier

эксик·а́ци·я, -и, (*f*) exsiccation

эксито́н·, -а, (*m*) (phys) exciton

эксито́н·н·ый, -ая, -ое, (*a*) exciton

~ со·сто·я́ни·е (*n*) (phys) exciton state

экситро́н·, -а, (*m*) (elec) excitron

экскав·а́тор·, -а, (*m*) (civ eng) excavator

~, ба́шен·н·ый tower cable excavator

~, гре́йфер·н·ый clamshell/grab excavator

~, гре́йфер·н·ый плов·у́ч·ий grab dredger

~, много·черп·а́ков·ый multi-/endless-bucket excavator

~, одно·черп·а́ков·ый плов·у́ч·ий dipper dredger

~, плов·у́ч·ий dredger

~, полно·по·воро́т·н·ый swing excavator

~, рото́р·н·ый bucket wheel excavator

~, траншей·н·ый trencher, trench excavator

~, шаг·а́ющ·ий walking excavator

экскав·а́ци·я, -и, (*f*) excavation

эксори·а́ци·я, -и, (*f*) (med) excoriation

экскре́ци·я, -и, (*f*) excretion

э́кскурс·, -а, (*m*) digression

экскурс·а́нт·, -а, (*m*) member of an excursion/trip/outing

экскурсио́н·н·ый, -ая, -ое, (*a*) excursion, trip, outing

экску́рс·и·я, -и, (*f*) excursion, trip, outing

эксогра́ф·, -а (*m*) shadowgraph

экспа́ндер·, -а, (*m*) (telecom) expandor; (eng) expansion engine; tube expander, expanding device

экспа́нс·и·я, -и, (*f*) expansion

экспатри·а́нт·, -а, (*m*) expatriate

экспатри·а́ци·я, -и, (*f*) expatriation

экспед·и́р·овать, -уют, (*imp and perf*) send off, despatch, forward, expedite

экспеди́тор·, -а, (*m*) (com) despatcher, despatch clerk; forwarding agent

экспедицио́н·н·ый, -ая, -ое, (*a*) expedition, expeditionary

~ су́д·н·о (*n*) scientific expedition ship; research ship

экспеди́ци·я, -и, (*f*) expedition; despatch, forwarding

эксперимéнт·, -а, (*m*) experiment

эксперимента́ль·н·ый, -ая, -ое, (*a*) experimental

эксперимент·а́тор·, -а, (*m*) experimenter; experimentalist

эксперимент·и́ровани·е, -я, (*n*) (*v n*) of эксперимент·и́р·овать; experimentation

эксперимент·и́р·овать, -уют, (*imp*) experiment

экспéрт·, -а, (*m*) expert, specialist

эксперти́з·а, -ы, (*f*) examination, inspection, expert examination/inspection; opinion, expert opinion; expert commission

экспéрт·н·ый, -ая, -ое, (*a*) of экспéрт·

экспир·а́торн·ый, -ая, -ое, (*a*) expiratory

экспир·а́ци·я, -и, (*f*) (physiol) expiration

эксплант·а́ци·я, -и, (*f*) (biol) tissue culture, explantation

эксплик·а́ци·я, -и, (*f*) explication, key (to symbols etc.)

эксплоат- (*int component*) *see* эксплуат-

эксплуата́тор·ск·ий, -ая, -ое, (*a*) exploiting (in derogatory sense)

эксплуатацио́н·н·ый, -ая, -ое, (*a*) operating, operation, operational

~ ка́честв·о (*n*) performance (of machinery)

~ контро́л·ь (*m*) (telecom) service checking

~ оборо́т· (*m*) service period (between checks)

~ срок· слу́ж·б·ы (*m*) service life

~ у·сло́в·и·я (*pl*) operating conditions (of a machine)

эксплуат·а́ци·я, -и, (*f*) exploitation; operation (of machine etc.), use в·вес·ти́ в эксплуат·а́ци·и put into operation

эксплуат·и́р·овать, -уют, (*imp*) exploit; operate, run, use (machinery)

экспозé (*n indecl*) resumé, exposé

экспози́·метр·, -а, (*m*) (phot) exposure meter

экспоз·и́ци·я, -и, (*f*) introductory material/passages; layout/arrangement of exhibits; (phot) exposure

~, двой·н·а́я (phot) double exposure

экспон·а́т·, -а, (*m*) exhibit

экспон·éнт·, -а, (*m*) exhibitor; (math) exponent

экспон·éнт·а, -ы, (*f*) (math) exponential curve

~, за·тух·а́ющ·ая (math) damped exponential

экспоненциа́ль·н·ый, -ая, -ое, (*a*) exponential потенциа́л· экспоненциа́ль·н·ой фо́рм·ы (*m*) (phys) exponential well

экспон·и́ровани·е, -я, (*n*) (*v n*) of экспон·и́р·овать; exposure

экспон·и́р·овать, -уют, (*imp and perf*) exhibit; (phot) expose

экспоно́·метр·, -а, (*m*) = экспози́·метр·; intensitometer (X-ray)

э́кспорт·, -а, (*m*) export; exports

~, бро́с·ов·ый (com) dumping

~, не·ви́д·им·ый (econ) invisible export(s)

экспорт·ёр·, -а, (*m*) exporter

экспорт·и́р·овать, -уют, (*imp and perf*) export

э́кспорт·н·ый, -ая, -ое, (*a*) export

экспрéсс·, -а, (*m*) express

~ -ана́лиз·, -а, (*m*) quick/approximate analysis

~ -лаборато́р·и·я, -и, (*f*) quick-analysis laboratory

экспрéсс·ност·ь, -и, (*f*) rapidity

экспрéсс·н·ый, -ая, -ое, (*a*) express, quick, rapid

экспропри·а́ци·я, -и, (*f*) expropriation

экспропри·и́р·овать, -уют, (*imp and perf*) expropriate

экс·солю́ци·я, -и, (*f*) (cryst) exsolution

экссуда́т·, -а, (*m*) (biol) exudate

экстензо́·метр·, -а, (*m*) (instr) extensometer

экстенси́в·н·ый, -ая, -ое, (*a*) extensive

экстéрн·, -а, (*m*) extramural student

экстерн·а́т·, -а, (*m*) extramural study/course

экстеро·рецéптор·, -а, (*m*) (physiol) exteroceptor

экс·территориа́ль·ност·ь, -и, (*f*) (dipl) extraterritoriality экстерьéр·, -а, (*m*) appearance, points (of livestock)

экстúнкци·я, -и, (*f*) extinction

экстирп·áци·я, -и, (*f*) extirpation

экстра·ваз·áци·я, -и, (*f*) (med) extravasation

экстраг·éнт·, -а, (*m*) extractant

экстраг·úровани·е, -я, (*n*) (*v n*) of экстраг·úр·овать; extraction

экстраг·úр·овать, -ую́т, (*imp and perf*) extract

~ со·в·мéст·н·о co-extract

экстраг·úруемост·ь, -и, (*f*) extractability

экстраг·úруем·ый, -ая, -ое, (*pres part pass*) of экстраг·úр·овать; extractable

экстраг·úрующ·ий, -ая, -ее, (*pres part act*) of экстраг·úр·овать; extracting

~ рас·твор·úтел·ь (*m*) extractant

экстрáкт·, -а, (*m*) extract

экстрактúв·н·ый, -ая, -ое, (*a*) extractive

~ пере·гóн·к·а (*f*) (chem) extractive distillation

экстрáктор·, -а, (*m*) extractor

экстракциóн·н·ый, -ая, -ое, (*a*) extraction, extractive

~ кóлб·а (*f*) (chem) extraction flask

экстрáкци·я, -и, (*f*) extraction

~, втор·úчн·ая (nucl) stripping, re-extraction

~, дву·жúдк·остн·ая liquid–liquid extraction

~, обрáт·н·ая back extraction

экстра·ординáр·н·ый, -ая, -ое, (*a*) extraordinary, unusual; supernumary

~ профéссор· (*m*) professor extra-ordinary

экстрапол·úровани·е, -я, (*n*) (math) extrapolation

экстрапол·úр·овать, -ую́т, (*imp and perf*) (math) extrapolate

экстраполяциóн·н·ый, -ая, -ое, (*a*) (math) extrapolated, extrapolation

экстрапол·яци·я, -и, (*f*) (math) extrapolation

экстра·систолú·я, -и, (*f*) (med) extrasystole

экстра·тóк·, -а, (*m*) (elec) extracurrent

экстремáль·н·ый, -ая, -ое, (*a*) extremal, extreme, extremum

~ регул·úровани·е (*n*) (autom) extremum adaptive control

экстрéмум·, -а, (*m*) (math) extremum, extreme value, extreme

э́кстрен·н·о (*adv*) urgently; (telecom) top precedence

э́кстрен·н·ый, -ая, -ое, (*a*) special, extraordinary; urgent, emergency; extra

~ вы́·пуск· (*m*) special edition

~ пó·езд· (*m*) special train

~ рас·хóд·ы (*pl*) extra charges

экстру́д·ер·, -а, (*m*) extruder

экстру́динг·, -а, (*m*) extruding, extrusion

экструд·úр·овать, -ую́т, (*imp and perf*) extrude

экструзúв·н·ый, -ая, -ое, (*a*) extrusive

~ гóр·н·ые по·рóд·ы (*pl*) (geol) extrusive rocks

экструзиóн·н·ый, -ая, -ое, (*a*) extrusion, extruding

~ машúн·а (*f*) extruder

экстру́зи·я, -и, (*f*) extrusion

экстрю́динг·, -а, (*m*) = экстру́динг·

эксудáт·, -а, (*m*) (biol) exudate

эксудациóн·н·ый, -ая, -ое, (*a*) exudative

эксудáци·я, -и, (*f*) exudation

эксфолиатúв·н·ый, -ая, -ое, (*a*) exfoliative

эксфоли·áци·я, -и, (*f*) (bot) exfoliation

эксцéдр·а, -ы, (*f*) (arch) niche

экс·цéнтр·ик·, -а, (*m*) (mech) eccentric, cam

экс·цéнтр·иков·ый, -ая, -ое, (*a*) of экс·цéнтр·ик·

~ бýгель (*f*) eccentric strap

~ диск· (*m*) (eng) eccentric sheave

экс·центрис·итéт·, -а, (*m*) (mech) eccentricity, eccentric position

экс·центр·úчност·ь, -и, (*f*) eccentricity, eccentric position/state

экс·центр·úчн·ый, -ая, -ое, (*a*) eccentric

эксцéсс·, -а, (*m*) excess

э́ктим·а, -ы, (*f*) (med) ecthyma

экто·генéз·, -а, (*m*) (biol) ectogenesis

экто·гéн·н·ый, -ая, -ое, (*a*) ectogenous

экто·дéрм·а, -ы, (*f*) (zool) ectoderm

экто·кáрп·ов·ый, -ая, -ое, (*a*) (biol) ectocarpous

экто·лóф·, -а, (*m*) (zool) ectoloph

экто·паразúт·, -а, (*m*) (microbiol) ectoparasite

эк·топ·úрованн·ый, -ая, -ое, (*a*) (med) ectopic

эк·топ·ú·я, -и, (*f*) (med) ectopia, ectopy

экто·плáзм·а, -ы, (*f*) (cyt) ectoplasm

экто·фúт·, -а, (*m*) (bot) ectophyte

эк·тропион·, -а, *(m)* (med) ectropion

экцитóн·, -а, *(m)* = экситóн·

элайó·метр·, -а, *(m)* elaiometer

элайо·плáст·, -а, *(m)* (bot) elaioplast

эласмо·брáнх·и·и *(pl)* (zool) elasmo-branchs, *Elasmobranchii*

эласмо·тéр·и·и, -ев, *(m pl)* (pal) elasmotheres, *Elasmotheriidae*

элáст·ик·, -а, *(m)* (plast) elastomer

эласт·úн·, -а, *(m)* (biochem) elastin

эласт·úческ·ий, -ая, -ое, *(a)* elastic

~ волóкн·а *(pl)* (biol) elastic/yellow fibres

эласт·úчност·ь, -и, *(f)* (*see also* упрý-гость); elasticity, resilience

~ на от·скóк· rebound resilience

~ по от·скóк·у rebound resilience

~ по Шóб·у Schob elasticity

эласт·úчн·ый, -ая, -ое, *(a)* elastic, resilient, flexible

~ мýфт·а *(f)* flexible coupling

эласто·вискозú·метр·, -а, *(m)* elasto-viscometer

эласто·мéр·, -а, *(m)* (plast) elastomer

эластó·метр·, -а, *(m)* (instr) elastometer

элáтер·а, -ы, *(f)* (bot) elater

элатер·úн·, -а, *(m)* (biochem) elaterin

элатер·úт·, -а, *(m)* (min) elaterite, elastic bitumen, mineral caoutchouc

элатеро·фóр·, -а, *(m)* (bot) elaterophore

элатúн·, -а, *(m)* (pharm) elatin

элевáтор·, -а, *(m)* elevator (storage); hoist, elevator, conveyor

~, америкáн·ск·ий (oil) single-link elevator

~, двух·штрóп·н·ый (oil) double-link elevator

~, зерн·ов·óй grain elevator

~, одно·штрóп·н·ый (oil) single-link elevator

~, пóл·очн·ый wing-tray elevator

~, ступ·éнчат·ый escalator

элевáтор·н·ый, -ая, -ое, *(a)* of элевáтор·

элевон·, -а, *(m)* (a/c) elevon

элевтеро·зó·и, -ев, *(m pl)* (zool) eleutherozoans, *Eleutherozoa*

элейтеро·мóрф·н·ый, -ая, -ое, *(a)* eleutheromorphic

электрéт·, -а, *(m)* electret

электриз·áци·я, -и, *(f)* electric/electro-static charging; (med) electrization

электриз·овáть, -ýют, (*imp and perf*) charge electrically/electrostatically; (med) electrize

элéктр·ик·, -а, *(m)* electrician, electrical engineer; grey-blue

электрúт·, -а, *(m)* electret

электрифик·áци·я, -и, *(f)* electrification

электрифиц·úр·овать, -уют, (*imp and perf*) electrify

электр·úческ·ий, -ая, -ое, *(a)* electric, electrical

~ прóч·ност·ь *(f)* electrical strength

~ стáнци·я *(f)* (elec) power station; generating set

~ схéм·а *(f)* (elec) circuit diagram; circuit

электр·úчеств·о, -а, *(n)* electricity

~, с·вя́з·анн·ое latent/disguised electricity

~ тр·éни·я frictional electricity

электро- (*int component*) electric, electro-; electrolytic

электро·абразúв·н·ая об·рабóт·к·а *(f)* (eng) electroabrasion

электро·акýст·ик·а, -и, *(f)* (phys) electroacoustics

электро·анáлиз·, -а, *(m)* (chem) electroanalysis

электро·блок·ирóвк·а, -и, *(f)* electric interlock

электро·блок·ирóвочн·ый, -ая, -ое, *(a)* electric-interlock

электро·бýр·, -а, *(m)* (oil) electrodrill

электро·бýс·, -а, *(m)* (min) electric trolley-and-battery car

электро·вáкуум·н·ый, -ая, -ое, *(a)* electronic

~ при·бóр· *(m)* electronic device

электро·валéнт·ност·ь, -и, *(f)* (chem) electrovalence, electrovalency

электро·валéнт·н·ый, -ая, -ое, *(a)* electrovalence

электро·вéтер·, -тр·а, *(m)* electric wind; (med) static breeze

электро·вибрациóн·н·ый грóхот· *(m)* electromagnetic vibrating screen (ore dressing)

электро·водо·по·ниж·éни·е, -я, *(n)* electroosmotic dewatering

электро·вóз·, -а, *(m)* electric locomotive

электро·вóз·н·ый, -ая, -ое, *(a)* of электро·вóз·; electric traction

электровозо·стро·éни·е, -я, *(n)* electric locomotive engineering

электро·во·оруж·ённост·ь, -и, *(f)* electric power/labour ratio (industrial statistics)

электро·вос·пламен·úтел·ь, -я, *(m)* (expl) fusehead (of detonator); electric squib

электро·генера́тор·, -а, (*m*) electric generator

электро·грав·ирова́льн·ая маши́н·а (*f*) electrogravure machine

электро́·граф·, -а, (*m*) (instr) electrograph; electric etcher

электро·гра́ф·и·я, -и, (*f*) electrography; electric etching

электро·гре́л·к·а, -и, (*g pl*) -л·ок·, (*f*) electric space heater

электро́д·, -а, (*m*) (elec) electrode

~, арм·и́рованн·ый (weld) sheathed electrode

~, водо·ро́д·н·ый (chem) hydrogen electrode

~ втор·о́го ро́д·а half-cell

~, га́з·ов·ый (elec) gas cell

~, газо·обра́з·у́ющ·ий (weld) shielded-arc electrode

~, го́л·ый (weld) bare electrode

~ -да́т·чик·, -а, (*m*) pick-up electrode

~, двух·по́люс·н·ый (elec chem) bipolar electrode

~, за·пир·а́ющ·ий (TV) pick-off electrode

~, ка́ломель·н·ый calomel electrode, calomel half-cell

~, кон·и́ческ·ий conical electrode

~, корон·и́рующ·ий corona-discharge electrode

~, модул·и́рующ·ий control electrode, control grid, shield (of C.R.T.)

~, не·пла́в·ящ·ийся (weld) non-consumable electrode

~, не·рас·хо́д·уем·ый non-consumable electrode

~, норма́ль·н·ый standard electrode

~, об·ма́з·анн·ый (weld) coated/covered electrode

~, о·сад·и́тельн·ый receiving electrode (of filter)

~, от·клон·я́ющ·ий deflector, deflecting electrode (of electron-beam tubes)

~, о·хра́н·н·ый (phys) guard electrode

~, о·чищ·а́ющ·ий sweeping electrode

~, пла́в·ающ·ий floating electrode

~, плоск·остн·о́й base electrode (of transistor)

~, под·жиг·а́ющ·ий (elec) keep-alive electrode (in TR tube)

~, под·све́ч·ивающ·ий intensifier electrode

~, по·кры́·т·ый (weld) covered/coated electrode

~, после·у·скор·я́ющ·ий intensifier electrode

электро́д

~, рас·хо́д·уем·ый consumable .electrode

~ с газо·обра́з·у́ющ·им по·кры́·т·и·ем (weld) shielded-arc electrode

~, сигна́ль·н·ый signal electrode

~, с·вод·я́щ·ий (elec) convergence electrode

~, со·бир·а́ющ·ий collector electrode, collector

~, станда́рт·н·ый (electrochem) standard cell

~, то́лсто·об·ма́з·анн·ый (weld) heavy coated electrode

~, тонко·об·маз·анн·ый (weld) dipped electrode, light covered electrode

~, тормоз·я́щ·ий retarding/decelerating electrode (electron-beam tubes)

~, у́голь·н·ый carbon electrode

~, у·правл·я́ющ·ий control electrode, control grid, shield (of C.R.T.)

~, у·скор·я́ющ·ий accelerator, accelerating electrode/anode (of electron-beam tubes)

~, фити́ль·н·ый (weld) cored electrode

~, форм·и·ру́ющ·ий beam-forming electrode (of vacuum tubes)

~, шлако·обра́з·у́ющ·ий (weld) heavy-coated electrode

электро·дви́г·ател·ь, -я, (*m*) electric motor, motor

~, асинхро́н·н·ый (elec) asynchronous motor

~, буше́ро (elec) boucherot motor, double squirrel-cage motor

~, колле́ктор·н·ый (elec) a.c. commutator motor

~, компа́унд·н·ый (elec) compound wound motor

~, конденса́тор·н·ый capacitor motor

~, репульсио́н·н·ый (elec) repulsion motor

~, рул·ев·о́й (shipb) electric steering motor

~, с конта́кт·н·ыми ко́льц·ами (elec) slip-ring motor, wound-rotor motor

~, с фа́з·ов·ым ро́тор·ом (elec) slip-ring motor, wound-rotor motor

~, се́риес·н·ый (elec) series-characteristic motor

~, синхро́н·н·ый (elec) synchronous motor

~, фла́н·цев·ый (elec) flange-mounted motor

~, шунт·ов·о́й (elec) shunt motor, shunt-characteristic motor электро·двиг·ательн·ый, -ая, -ое, (*a*) of электро·двиг·ател·ь; electro-motive

электро·движ·éни·е, -я, (*n*) electric propulsion

электро·движ·ущ·ий, -ая, -ее, (*a*) electromotive

~ сил·а (*f*) electromotive force, e.m.f.

электро·детонáтор·, -а, (*m*) (expl) electric detonator, detonator

~ за·мéдл·енн·ого дéй·стви·я long-period delay detonator

~ коротко·за·мéдл·енн·ого дéй·-стви·я short-period delay detonator

электро·держ·áтел·ь, -я, (*m*) (weld) electrode holder

электро·диагнóз·, -а, (*m*) (med) electrodiagnosis

электро·диáлиз·, -а, (*m*) (chem) electrodialysis

электро·динáм·ик·а, -и, (*f*) electro-dynamics

электро·динам·и́ческ·ий, -ая, -ое, (*a*) electrodynamic, moving-coil

~ машин·а (*f*) electrodynamic machine

~ об·рабóт·к·а (*f*) spark machining, electrosparking

~ при·бóр· (*m*) electrodynamic/moving-coil instrument

электро·динамó·метр·, -а, (*m*) electrodynamometer

электрóд·н·ый, -ая, -ое, (*a*) electrode

~ потенциáл· (*m*) (electrochem) electrode/electrolytic potential

~ прó·волок·а (*f*) continuous electrode

электро·держ·áтел·ь, -я, (*m*) (weld) electrode holder

электро·дóй·к·а, -и, (*g pl*) -ó·ек·, (*f*) (agr) electrical milking

электро·дóмен·н·ая печ·ь (*f*) electrical blast furnace

электро·дрен·áж·, -а, (*m*) (civ eng) electroosmotic dewatering

электро·дуг·á, -и́, (*nom pl*) -у́г·и (*f*) electric arc

электро·дуг·ов·óй, -áя, -óе, (*a*) electric-arc

~ с·вáр·к·а (*f*) electric-arc welding

электро·жезл·ов·óй аппарáт· (*m*) (rail) token instrument

электро·за·жиг·áтел·ь, -я, (*m*) (expl) electric powder fuse

электро·за·клёп·к·а, -и, (*f*) rivet weld

~, не·сквоз·н·áя plug weld

~, сквоз·н·áя rivet weld

электро·за·пáл·, -а, (*m*) (expl) fuse-head (of detonator); electric primer

электро·зá·пис·ь, -и, (*f*) electric recording, electric transcription (of radio transmission)

электро·за·щёлк·а, -и, (*f*) (rail) electric lock

электро·за·щи́т·а, -ы, (*f*) cathodic/sacrificial protection (against corrosion)

электро·звук·ов·óй, -áя, -óе, (*a*) electroacoustic

электро·зóл·ь, -я, (*m*) (chem) electrosol

электро·и́з·город·ь, -и, (*f*) electric fence

электро·из·мер·и́тельн·ый при·бóр· (*m*) electric measuring instrument

электро·изоляцóн·н·ый, -ая, -ое, (*a*) electrical-insulation

электро·инструмéнт·, -а, (*m*) power tool

электро·интегрáтор·, -а, (*m*) (autom) differential analyser

~ -сéт·к·а, -и, (*f*) network analyser

электро·искр·ов·áя об·рабóт·к·а (*f*) shark machining, electrosparking

электро·капилля́р·ност·ь, -и, (*f*) electrocapillarity

электро·кáр·, -а, (*m*) electric trolley

электро·кáр·а, -ы, (*m*) = электро·-кáр·

электро·кардио·грáмм·а, -ы, (*f*) (med) electrocardiogram

электро·кардиó·граф·, -а, (*m*) electro-cardiograph

электро·каротáж·, -а, (*m*) (*see also* каротáж·) (geophys) electrical well logging

электро·картóн·, -а, (*m*) electrical insulation cardboard

электро·кинет·и́ческ·ий, -ая, -ое, (*a*) (chem) electrokinetic

электро·контáкт·, -а, (*m*) electric contact

электро·контáкт·н·ый, -ая, -ое, (*a*) electrocontact

электро·контрóль·ник·, -а, (*m*) electric verifier (of punched cards)

электро·кору́нд·, -а, (*m*) electro-corundum, alundum (abrasive and refractory)

электро·ко·тёл·, -тл·á, (*m*) electric boiler

электро·культу́р·а, -ы, (*f*) (hortic) electro-culture; electric plant growing

электро·лáмп·ов·ый, -ая, -ое, (*a*) electric-lamp

электро·леч·éни·е, -я,   (n) (med) electrotherapy, electrotherapeutics, electropathy

электро́·лиз·, -а, (m) (phys) electrolysis

электро·лиз·ёр·, -а,   (m) electrolytic/ electrolysis cell/bath

электро·лит·, -а, (m) (chem) electrolyte

~, газо·обрáз·н·ый   dry electrolyte

~, за·гущ·ённ·ый   electrolytic paste

~, рас·плáвл·енн·ый   fused electrolyte

электро·лит·ик·а, -и,   (f) electrolytic capacitor

электро·лит·и́ческ·ий, -ая, -ое,   (a) electrolytic

~ о·сажд·éни·е   (n) electrodepositing, electrodeposition

~ о·чи́ст·к·а   (f) electrolytic cleaning

~ по·кры́т·и·е   (n) (met) electrolytic coating, electroplating

~ по·луч·éни·е   (n) (met) electroextraction, electrowinning

~ рас·твор·éни·е   (n) (chem) electrodissolution

электро·ло́г·и·я, -и, (f) (med) electrology, electrotherapy

электро·ло́т·, -а,   (m) (ocean) electrosounder

электро·лю́мин·есцéнт·н·ый,   -ая, -ое, (a) electroluminescent

электро·лю́мин·есцéнци·я, -и,   (f) electroluminescence

электро·магнет·и́зм·, -а, (m) electromagnetism

электро·магни́т·, -а,   (m) electromagnet

~ от·клю́ч·éни·я   (elec) tripping magnet

~, тормоз·н·о́й   electromagnetiс brake

~, у·дéрж·ивающ·ий   holding magnet

электро·магни́т·н·ый, -ая, -ое,   (a) electromagnetic

~ дренáж·   electrical drainage, cathode protection

~ дут·ь·ё   (n) magnetic blast

~ му́фт·а   (f) magnetic clutch; electric/ electromagnetic coupling

~ при·бо́р·   (m) electromagnetic device/ instrument; moving-iron instrument (specifically)

~ при·во́д·   (m) electromagnetic operating gear (of circuit breaker)

электро·маши́н·а, -ы,   (f) electrical machine/machinery

электро·маши́н·н·ый, -ая, -ое,   (a) (elec) dynamoelectric; electrical (used loosely)

электро·маши́н·н·ый, -ая, -ое

~ генерáтор·   (m) dynamoelectric generator

~ дáт·чик·   (m) (autom) dynamoelectric transmitter

~ регул·я́тор· (m) metadyne generator

~ регул·я́тор· по·перéч·н·ого по́л·я   amplidyne

~ регул·я́тор· про·до́ль·н·ого по́л·я   rototrol

~ у·сил·и́тел·ь   (m) = э· регуля́тор

электро·машино·стро·éни·е, -я, (n) heavy electrical engineering, electric machine-building

электро·мегафо́н·, -а, (m) loud-hailer, bullhorn

электро·металл·у́рг·и·я,   -и,   (f) electrometallurgy

электро́·метр·, -а,   (m) electrometer

~, бинант·н·ый   Hoffman electrometer

~, вибрацио́н·н·ый   vibrating-reed electrometer

~, дву·ни́т·н·ый   Wulf-string electrometer

~, капилля́р·н·ый   capillary electrometer

~, квадрáнт·н·ый   quadrant electrometer

~, одно·ни́т·очн·ый   string electrometer

электро·метр·и́ческ·ий, -ая, -ое, (a) electrometer

электро·мéтр·и·я, -и,   (f) electrical measurement, measurement of electrical values

электро·механ·и́ческ·ий, -ая, -ое, (a) electromechanical, electromechanical-engineering

~ звуко·зá·пис·ь   (f) electromechanical recording

~ об·рабо́т·к·а   (f) (mech eng) electromechanical machining, electromachining

~ трансми́ттер·   (m) (autom) mechanical reader

~ цех·   (m) electromechanical engineering shop

электро·мигрáци·я, -и,   (f) (phys) electromigration

электро·мио·грáмм·а, -ы,   (f) (med) electromyogram

электро·мио·грáф·и·я, -и,   (f) electromyography

электро·моби́л·ь, -я, (m) electromobile, battery-driven vehicle

электро·модел·и́рующ·ий, -ая, -ее, (a) (autom) analogue, analog

электро·модёл·ь, -и, (*f*) electronic simulator

~, сёт·очн·ая network analyser

электро·монт·а́ж·, -а, (*m*) electrical installation; electrical assembly (of small devices)

электро·монта́ж·н·ый по́·езд· (*m*) (rail) electrical installation and maintenance train

электро·монт·ёр·, -а, (*m*) electrician, wiring man, (cinema) rigger; maintenance engineer (power station)

электро·мото́р·, -а, (*m*) *see* **электро·-двиг·атель·ь**

электро·мото́р·н·ый, -ая, -ое, (*a*) of **электро·мото́р·**

~ от·сёк· (*m*) motor room (of submarine)

электро́н·, -а, (*m*) electron; (met) Elektron

~, бы́стр·ый fast electron

~, валёнт·н·ый valence electron

~, внёш·н·ий outer electron

~, вне·я́дер·н·ый extranuclear electron

~, вну́тр·енн·ий inner electron

~ -вольт·, -а, (*m*) electron-volt, eV, ev

~, вращ·а́ющ·ийся spinning electron

~, жёстк·ий hard electron

~, мёдл·енн·ый slow electron

~, мя́гк·ий soft electron

~ от·да́·ч·и recoil electron

~ про·вод·и́мост·и conduction electron

~, рас·сё·ивающ·ий scattering electron

~, с·вя́з·анн·ый bound electron

~ с·вя́з·и bonding electron

~, у·бег·а́ющ·ий runaway electron

электро·навигацио́н·н·ый, -ая, -ое, (*a*) electrical navigational, electro-navigation

электро·на·грё·в·, -а, (*m*) electric heating

~, градиёнт·н·ый electric heating in a gradient

электро́н·ик·а, -и, (*f*) electronics

электро·нейтра́ль·ност·ь, -и, (*f*) electric neutrality

электрон·микроскоп·, -а, (*m*) *see* **электро́н·н·ый микро·скоп·**

электро́н·н·о- (*component*) (*see also* **электрон·о-**) electron, electronic, tube/valve-type

электронно·ды́р·очн·ый, -ая, -ое, (*a*) electron-hole, *p-n* (semiconductors)

электронно·ды́р·очн·ый, -ая, -ое

~ па́р·а (*f*) electron-hole pair, *p-n* pair

~ пере·хо́д· (*m*) *p-n* junction, *p-n* semiconductor junction

электронно·ионизацио́н·н·ый, -ая, -ое, (*a*) ionization

электронно·луч·ев·о́й, -а́я, -о́е, (*a*) electron-beam; cathode-ray

~ ла́мп·а (*f*) electron-ray/-beam/-indicator tube/valve

~ при·бо́р· (*m*) electron-beam/cathode-ray instrument

~ тру́б·к·а (*f*) cathode-ray tube

электронно·механ·и́ческ·ий, -ая, -ое, (*a*) mechano-electronic

электронно·опт·и́ческ·ий, -ая, -ое, (*a*) optico-electronic

~ пре·образ·ова́тел·ь (*m*) (TV) image convertor

электронно·пневмат·и́ческ·ий, -ая, -ое, (*a*) pneumo-electronic, pneutronic

электронно·по·то́ч·н·ый, -ая, -ое, (*a*) electron-stream

электро́н·н·ый, -ая, -ое, (*a*) electron; electronic; tube-/valve-type

~ генера́тор· (*m*) (elec) valve oscillator

~ за·хва́т· (*m*) (phys) electron capture

~ из·обра́ж·ёни·е (*n*) (TV) electron image

~ ла́мп·а (*f*) (elec) valve, thermionic valve, electron/vacuum tube

~ микро·ско́п· (*m*) electron microscope

~ микро·ско́п·, зерка́ль·н·ый (*m*) mirror-type electron microscope

~ микро·ско́п·, от·раж·а́тельн·ый (*m*) mirror-type electron microscope

~ микро·ско́п·, про·свёч·ивающ·ий (*m*) emission-type electron microscope

~ при·бо́р· (*m*) electronic device/instrument, valve-/tube-type instrument

~ проёктор· (*m*) (TV) opticoelectronic enlarger

~ прожёктор· (*m*) electron gun (vacuum tubes), space-charge limited emission electron gun

~ пу́ш·к·а (*f*) high-current electron gun (vacuum tubes)

~ регул·я́тор· (*m*) electronic controller

~ с·род·ств·о́ (*n*) (phys) electron affinity

~ тео́р·и·я (*f*) (phys) electron theory

~ тёхн·ик·а (*f*) electronic engineering

~ у·сил·и́тел·ь (*m*) vacuum-tube amplifier

электро́н·н·ый, -ая, -ое
~ **эми́сси·я** (*f*) electron emission
электроно- (*int component*) (*see also* **электронно-**) electron
электроно·акце́птор·, -а, (*m*) (phys) electron acceptor
электроно·акце́птор·н·ый, -ая, -ое, (*a*) (phys) electron-seeking
электроно·гра́мм·а, -ы, (*f*) electron diffraction pattern
электроно·гра́ф·, -а, (*m*) electron diffraction instrument/camera
электроно·гра́ф·и·я, -и, (*f*) electron diffraction study
электроно·до́нор·, -а, (*m*) (phys) electron donor
электроно·до́нор·н·ый, -ая, -ое, (*a*) (phys) electron-donating, electron-donor
электроно·с·род·ств·о́, -а́, (*n*) (phys) electron affinity
электро·обо·гре́·в·, -а, (*m*) electrical heating
электро·обо·грев·а́ем·ый, -ая, -ое, (*a*) electrically heated
электро·об·ору́д·овани·е, -я, (*n*) electrical equipment
электро·о́пт·ик·а, -и, (*f*) (phys) electrooptics
электро·о·сажд·е́ни·е, -я, (*n*) electro-deposition
электро·о́смос·, -а, (*m*) (phys) electro-osmosis
электро·о·стано́в·, -а, (*m*) electrical stop mechanism
электро·от·риц·а́тельност·ь, -и, (*f*) electronegativity
электро·от·риц·а́тельн·ый, -ая, -ое, (*a*) electronegative
электро·о·чи́ст·к·а, -и, (*f*) electrostatic purification (of gases); (oil) electrorefining
электро·па́й·к·а, -и, (*g pl*) **-па́·ёк·** (*f*) electric soldering
электро·паст·у́х·, -а́, (*m*) electric fence
электро·пере·да́·ч·а, -и, (*f*) power transmission
    ли́н·и·я электро·пере·да́·ч·и (*f*) (elec) power transmission line
электро·пере·но́с·, -а, (*m*) (phys) electron transfer
~ **ио́н·ов** ion migration
электро·перфор·а́тор·, -а, (*m*) (oil) electric perforator
~, **стрел·я́ющ·ий** electric gun perforator

электро·пит·а́ни·е, -я, (*n*) power supply
электро·плав·и́льн·ая печ·ь (*f*) electric furnace
электро·плит·а́, -ы́, (*f*) electric hot plate, hotplate
электро·пневмат·и́ческ·ий, -ая, -ое, (*a*) electropneumatic
электро·поду́ш·к·а, -и, (*g pl*) **-ш·ек·,** (*f*) (med) hot pad
электро·полир·о́вк·а, -и, (*f*) (met) electropolishing, electrolytic polishing, electrobrightening
электро·по·лож·и́тельн·ый, -ая, -ое, (*a*) electropositive
электро·по·су́д·а, -ы, (*f*) (food) electric boilers
электро·по́чт·а, -ы, (*f*) electrical tube postal system
электро·при·во́д·, -а, (*m*) electric drive
электро·про́·вод·, -а, (*m*) electric conductor
электро·про·вод·и́мост·ь, -и, (*f*) electrical conductivity
электро·про·во́д·ност·ь, -и, (*f*) electrical conductivity, electrical conduction, conductance
электро·про·мы́шл·енност·ь, -и, (*f*) electrical industry
электро·про·и́гр·ыватель·, -я, (*m*) record player
электро·пыле·со́с·, -а, (*m*) vacuum cleaner
электро·раз·ве́д·к·а ме́тод·ом со·-против·ле́ни·я (geophys) resistivity prospecting
электро·раз·ря́д·, -а, (*m*) electric discharge; (*pl*) (cinema) static marks
электро·раз·ря́д·н·ый, -ая, -ое, (*a*) (elec) discharge
электро·рас·сек·а́тел·ь, -я, (*m*) electrodesiccator
электро·рас·щепл·е́ни·е, -я, (*n*) electrodisintegration
электро·рафинор·ова́ни·е, -я, (*n*) (met) electrorefining, electrolytic refining
электро·ремо́нт·н·ый, -ая, -ое, (*a*) electrical-repair
электро·рео·форе́з·, -а, (*m*) (chem) electrorheophoresis
электро·с·ва́р·к·а, -и, (*g pl*) **-р·ок·,** (*f*) electric welding
электро·с·ва́р·очн·ый, -ая, -ое, (*a*) electric-welding
~ **дро́ссел·ь** (*m*) (weld) transformer regulator

электро·сверл·о́, -а́,   (*n*) electric drill

электро·свеч·е́ни·е, -я,   (*n*) electro-luminescence

электро·с·вя́з·ь, -и,   (*f*) telecommunications, electrical communications

электро·секундо·ме́р·, -а,   (*m*) electric second timer

электро·сепар·а́тор·, -а,   (*m*) (*and see* **электро·фи́льтр·**) (min) electromagnetic separator

электро·сепар·а́ци·я, -и,   (*f*) (min) electromagnetic separation

электро·се́т·ь, -и,   (*f*) (elec) mains, power-supply system

электро·си́л·а, -ы,   (*f*) electric power, power

электро·сил·ов·о́й, -а́я, -о́е,   (*a*) (elec) power, electric-power

электро·си́нтез·, -а,   (*m*) electro-synthesis

электро·сковород·а́, -ы́,   (*f*) electric frying pan, electric fryer

электро·скоп·, -а,   (*m*) (phys) electro-scope

~, дозиметр·и́ческ·ий   (nucl) radio-scope

~, интегр·и́рующ·ий   capacitor-type electroscope

электро·скоп·и́ческ·ий, -ая, -ое,   (*a*) electroscope; electroscopic

электро·слесар·ь, -я,   (*m*) electrical fitter

электро·со·противл·е́ни·е, -я,   (*n*) electrical resistance (of a circuit); electrical resistivity (of a material)

электро·стале·плав·и́льн·ый, -ая, -ое,   (*a*) electrical steelmaking

электро·ста́л·ь, -и,   (*f*) electric steel (steel made in electric furnace)

электро·ста́нци·я, -и,   (*f*) (elec) power station; generating set

~, а́том·н·ая   atomic power station

~, ба́з·ов·ая   base-load power station

~, пи́к·ов·ая   peak-load power station

~, тепл·ов·а́я   thermal power station

~, тепло·фикацио́н·н·ая   heat-and-power station, district-heating power station

электро·ста́ртер·, -а,   (*m*) electric starter

электро·ста́т·ик·а, -и,   (*f*) electro-statics

электро·стат·и́ческ·ий, -ая, -ое,   (*a*) electrostatic

~ маши́н·а   (*f*) influence/induction/static/frictional machine. electrostatic generator

электро·стено́лиз·, -а,   (*m*) (chem) electrostenolysis

электро·стри́ж·к·а, -и,   (*f*) (agr) electrical sheep-shearing

электро·стри́кци·я, -и,   (*f*) electro-striction

электро·су́дорож·н·ая терап·и́·я   (*f*) (med) electric-shock treatment

электро·суш·и́лк·а, -и,   (*g pl*) **-лок·,** (*f*) electric dryer

электро·та́л·ь, -и,   (*f*) electric hoist

электро·теле́ж·к·а, -и,   (*f*) = **элек-тро·ка́р·**

электро·те́льфер·, -а,   (*m*) electric telpher

электро·тензо́·метр·, -а,   (*m*) strain gauge/gage

~ со·противл·е́ни·я   resistance-wire strain-gauge

электро·тепл·ов·а́я противо·об·лед·-ени́тельн·ая систе́м·а   (*f*) electrical de-icing system

электро·терм·и́ческ·ий, -ая, -ое,   (*a*) electrothermic

электро·терм·и́·я -и,   (*f*) electro-thermics

электро·те́хн·ик·, -а,   (*m*) electrician

электро·те́хн·ик·а, -и,   (*f*) electrical engineering/technology, electro-technics

электро·техн·и́ческ·ий, -ая, -ое,   (*a*) electrical-engineering, electrical

~ карто́н·   (*m*) (elec) insulating card-board

~ стал·ь   (*f*) (met) electrical steel, electrical-engineering steel

~ у́гол·ь   (*m*) electrical carbon

электро·ти́п·и·я, -и,   (*f*) (met) electro-forming; (print) electrotype

электро·то́к·, -а,   (*m*) electric current

электро·том·и́·я, -и,   (*f*) (med) electro-desiccation

электро·то́н·, -а,   (*m*) (anat) electro-tonus

электро·тра́вм·а, -ы,   (*f*) (med) injury due to electric current; electric shock

электро·троп·и́зм·, -а,   (*m*) (physiol) electrotropism

электро·тя́г·а, -и,   (*f*) electric traction

электро·тя́г·ов·ый, -ая, -ое,   (*a*) electric-traction

электро·у́гол·ь -гл·я,   (*m*) electrical carbon

электро·у·правл·я́ем·ый, -ая, -ое,   (*a*) electrically controlled

электро·фи́ль·н·ый, -ая, -ое,   (*a*) electrophylic

электро·фи́льтр·, -а, (*m*) electrostatic precipitator, electrostatic air filter

~, коро́н·н·ый corona-discharge precipitator

~, пласт·и́нчат·ый plate-type electrostatic precipitator

~, тру́б·чат·ый tube-type electrostatic precipitator

электро·фо́р·, -а, (*m*) electrophorus

электро·форе́з·, -а, (*m*) electrophoresis

электро·форе́з·ер·, -а, (*m*) vessel for electrophoresis

электро·форе́з·н·ый, -ая, -ое, (*a*) electrophoretic

электро·форм·о́вк·а, -и, (*f*) (elec) electroforming

электро·хим·и́ческ·ий, -ая, -ое, (*a*) electrochemical

~ ряд·на·пряж·е́ни·я electrochemical/electromotive series

электро·хи́м·и·я, -и, (*f*) electrochemistry

электро·хо́д·, -а, (*m*) motor ship (with electric propulsion)

электро·центра́л·ь, -и, (*f*) power house/station

электро·час·ы́, -о́в, (*pl*) electric clock

~, и́мпульс·н·ый impulse-driven clock

электро·шла́к·ов·ая с·ва́р·к·а (*f*) submerged-arc welding

электро·шлиф·ова́лк·а ле́нт·очн·ая (*f*) belt sander

электро·щу́п·, -а, (*m*) (text) electric feeler

электро·эндо́смос·, -а, (*m*) (chem) electro-endosmosis

электро·энцефело́·граф·, -а, (*m*) (med) electroencephalograph

электро·эрозио́н·н·ая об·рабо́т·к·а (*f*) (mech eng) electromachining, electroerosion

эле́ктрум·, -а, (*m*) (met) electrum

элеме́нт·, -а, (*m*) (*and see under associated adjectives*) element; unit, component; (elec) cell, battery cell, primary cell

~ авто·ши́н·ы tire component

~, аккумуля́тор·н·ый secondary cell

~, водо·на·лив·н·о́й inert cell

~ во́льт·а voltaic cell

~, вос·при·ним·а́ющ·ий sensing element, (autom) input element

~, в·ход·н·о́й (autom) input element

~, вы·ход·н·о́й (autom) final control element (U.S.), regulating unit (Brit)

~, газо·на·лив·н·о́й gas cell

элеме́нт

~, гальван·и́ческ·ий galvanic cell; localized galvanic cell (in metal corrosion)

~, гравитацио́н·н·ый gravity cell

~, до·ба́в·очн·ый end/regulator/emergency cell

~, доче́р·н·ий (nucl) daughter element

~, за·да·ю́щ·ий (autom) control point setting mechanism

~, ис·полн·и́тельн·ый (autom) final control element (U.S.), regulating unit (Brit)

~, ка́дм·иев·ый cadmium electrode

~ констру́кци·и structural element

~, концентрацио́н·н·ый concentration cell

~, коррозио́н·н·ый гальван·и́ческ·ий (met) local/localised galvanic cell

~ Лекланше́ Leclanché cell

~, мо́кр·ый wet cell

~, нейтра́ль·н·ый (math) unit/identity element

~, на·грев·а́тельн·ый heating element

~, нес·у́щ·ий (phys) carrier; (mech) load-bearing element/unit

~, обра́т·н·ый (math) inverse element, inverse

~ па́мят·и (autom) storage element

~, пере·хо́д·н·ый (chem) transition element

~, пит·а́ющ·ий power supply element

~, пре·образ·у́ющ·ий transducer, conversion/transducing element

~, пуск·ов·о́й (autom) operating element

~, рас·се́·ян·ый (geol) trace element

~, реш·а́ющ·ий computing element (of computer)

~, стака́н·чиков·ый dry cell, Leclanché cell

~, сух·о́й (elec) dry cell

~, сухо·на·лив·н·о́й inert cell

~, тепло·вы·дел·я́ющ·ий (nucl) fuel element

~, у·пру́г·ий expansible element

~, фильтр·у́ющ·ий filter element

~, цикло́н·н·ый cyclone unit

~, чувств·и́тельн·ый sensing/sensitive element, (autom) input element

~, электро·хим·и́ческ·ий electrolytic cell

элемента́р·ност·ь, -и, (*f*) elementary nature/character

элемента́р·н·ый, -ая, -ое, (*a*) elementary

элемéнт·н·ый, -ая, -ое, (*a*) of элемéнт·

~ коммутáтор· (*m*) battery-regulating switchboard

элéми (*indecl*) elemi gum

элео·лúт·, -а, (*m*) (min) elaeolite, eleolite

элербн·, -а, (*m*) (a/c) aileron

элефантиáз·, -а, (*m*) (med) elephantiasis

элиазúт·, -а, (*m*) (min) eliasite, gummite

элимин·áтор·, -а, (*m*) drop eliminator (air conditioning systems)

элимин·áци·я, -и, (*f*) elimination

элимин·úр·овать, -уют, (*imp and perf*) eliminate

элúмус·, -а, (*m*) (bot) wild rye, *Elymus*

элинвáр·, -а, (*m*) (met) Elinvar

элúт·а, -ы, (*f*) (agr) seed stock

элúтр·а, -ы, (*f*) (ent) elytron, elytrum, (*pl*) elytra

элитр·úт·, -а, (*m*) (med) elytritis

элитро·ррáф·и·я, -и, (*f*) (med) elytrorrhaphy

элитро·том·ú·я, -и, (*f*) (med) elytrotomy

элладо·тéр·и·й, -я, (*m*) (pal) helladotherium

эллестадúт·, -а, (*m*) (min) ellestadite

бллинг·, -а, (*m*) ship-repair/shipbuilding shed; boat shed; airship hangar

~, судо·ремóнт·н·ый ship repair shed

~, судо·стро·úтельн·ый (shipb) building shed, covered building berth

эллинист·úческ·ий, -ая, -ое, (*a*) Hellenistic, Hellenic

бллипс·, -а, (*m*) (math) ellipse

~ рас·сé·ивани·я zone of dispersion, beaten zone

бллипсис·, -а, (*m*) = бллипс·

эллипс·óид·, -а, (*m*) (math) ellipsoid

~ вращ·éни·я (math) ellipsoid of revolution, spheroid

~ на·пряж·éни·я (mech) stress ellipsoid

~, рас·тá·нут·ый prolate ellipsoid

~, с·плющ·енн·ый oblate ellipsoid

эллипсидáль·н·ый, -ая, -ое, (*a*) ellipsoidal

~ координáт·ы (*pl*) (math) ellipsoidal coordinate

эллипс·óид·н·ый, -ая, -ое, (*a*) ellipsoidal

эллипт·úческ·ий, -ая, -ое, (*a*) elliptic(al)

эллипт·úчност·ь, -и, (*f*) ellipticity, ellipticalness

элонг·áци·я, -и, (*f*) (astron) elongation

эльзáс·ск·ий, -ая, -ое, (*a*) Alsatian

Эльмор·а, мáсл·ян·ая флот·áци·я (*f*) Elmore flotation (ore dressing)

эльсвортúт·, -а, (*m*) (min) ellsworthite

элю·áт·, -а, (*m*) (chem) eluate

элю·áци·я, -и, (*f*) (chem) eluation

элю·úровани·е, -я, (*n*) elution

элювиáль·н·ый, -ая, -ое, (*a*) (geol) eluvial

элюв·и·й, -я, (*m*) (geol) eluvium

элюци·я, -и, (*f*) (chem) elution

эмáл·ев·ый, -ая, -ое, (*a*) enamel, enamelled

эмал·ир·овáть, -ýют, (*imp*) enamel

эмал·ирóвк·а, -и, (*g pl*) -вок·, (*f*) enamelling, enamel

эмал·ирóвочн·ый, -ая, -ое, (*a*) enamel, enamelling

эмáл·ь, -и, (*f*) enamel

~, пента·фтáл·ев·ая pentaerithritol phthalate enamel

эман·áци·я, -и, (*f*) emanation, emanon, Em

~ актúн·и·я (nucl) actinon, An

~ рáд·и·я (nucl) radon, Rn

~ тóр·и·я thoron

эман·úр·овать, -уют, (*imp and perf*) emanate

эманó·метр·, -а, (*m*) emanometer

эмаскул·áци·я, -и, (*f*) (gen) emasculation

эмбáрго (*n indecl*) embargo

эмбитóл·, -а, (*m*) (pharm) embitol

эмбихúн·, -а, (*m*) (pharm) embichine

эмблéм·а, -ы, (*f*) emblem

эмбол·, -а, (*m*) (med) embolus

эмбол·úт·, -а, (*m*) (min) embolite

эмбол·ú·я, -и, (*f*) (med) embolism, (zool) embole, emboly

эмболо·тéр·и·й, -я, (*m*) (pal) Embolotherium

эмбрио·генéз·, -а, (*m*) embryogenesis, embryogeny

эмбрио·лóг·и·я, -и, (*f*) (biol) embryology

эмбриóн·, -а, (*m*) (biol) embryo

эмбрион·áль·н·ый, -ая, -ое, (*a*) embryonic, foetal

эмбрио·том·ú·я, -и, (*f*) (med) embryotomy

эмбрит·úт·, -а, (*m*) (min) embrithite

эмеральдúн·, -а, (*m*) emeraldine (dye)

эмергéн·ец·, -н·ц·а, (*m*) (biol) emergence

эмет·амúн·, -а, (*m*) (biochem) emetamine

эмет·и́н·, -а,   (*m*) (pharm) emetine, emetir

эмигр·а́нт·, -а,   (*m*) emigrant, émigré

эмигр·а́ци·я, -и,   (*f*) emigration

эмигр·и́р·овать, -уют,   (*imp and perf*) emigrate

эмиссио́н·н·ый, -ая, -ое,   (*a*) of **эми́сси·я**

~ курс·   (*m*) (fin) issuing price

~ фото·элеме́нт·   (*m*) (opt) photoemissive cell

эми́сси·я, -и,   (*f*) emission; (fin) issue

~, а́вто·электро́н·н·ая   (elec) autoemission, cold emission

~, ио́н·н·ая   (phys) ion emission

~, втор·и́чн·ая электро́н·н·ая   secondary electron emission

~, обра́т·н·ая   (nucl) back emission

~, термо·электро́н·н·ая   thermionic emission

~, у·де́ль·н·ая   emissivity

~, электро́н·н·ая   electron emission, thermionic emission

эми́тер·, -а,   (*m*) = **эми́ттер·**

эмит·и́р·овать, -уют,   (*imp and perf*) emit; (fin) issue

эмит·и́рованн·ый, -ая, -ое,   (*past part pass*) of **эмит·и́р·овать**

эми́ттер·, -а,   (*m*) emitter

~, термо·электро́н·н·ый   cathode emitter

эми́ттер·н·ый, -ая, -ое,   (*a*) of **эми́ттер·**

эмитт·и́р·овать, -уют,   (*imp and perf*) see **эмит·и́р·овать**

эмка́р·, -а,   (*m*) (vet) emphysematous anthrax

эммен·и́н·, -а,   (*m*) (gen) emmenin, emmenine

э́ммер·, -а,   (*m*) (bot) emmer, *Triticum dicoccum*

эмметр·опи́·я, -и,   (*f*) (physiol) emmetropia

эммонси́т·, -а,   (*m*) (min) emmonsite

эмо́ци·я, -и,   (*f*) (med) emotion

эмпие́м·а, -ы,   (*f*) (med) empyema

эмпиревмат·и́ческ·ий, -ая, -ое,   (*a*) empyrheumatic

эмпи́р·ик·, -а,   (*m*) empiricist

эмпир·и́ческ·ий, -ая, -ое,   (*a*) empirical

эмплекти́т·, -а,   (*m*) (min) emplectite

эм·по́д·и·й, -я,   (*m*) (zool) empodium

эмпресси́т·, -а,   (*m*) (min) empressite

э́му   (*m indecl*) (orn) emu, *Dromiceius,* (*pl*) *Dromiceidae*

эмульг·а́тор·, -а,   (*m*) emulsifier

эмульг·а́ци·я, -и,   (*f*) emulsification

эмульс·а́тор·, -а,   (*m*) emulsifier

эмульс·и́н·, -а,   (*m*) (biochem) emulsin

эмульсио́н·н·ый, -ая, -ое,   (*a*) emulsion

~ кра́с·к·а   (*f*) emulsion paint

эмульс·и́ровани·е, -я,   (*m*) (*v n*) of **эмульс·и́р·овать;** emulsification

эмульс·и́р·овать, -уют,   (*imp and perf*) emulsify

эму́льс·и·я, -и,   (*f*) (chem) emulsion

эмульс·о́ид·, -а,   (*m*) emulsoid

эмульс·о́л·, -а,   (*m*) (mech eng) cutting fluid, coolant

эмфат·и́ческ·ий, -ая, -ое,   (*a*) (ling) emphatic

эм·физе́м·а, -ы,   (*f*) (med) emphysema

эм·физемато́з·н·ый, -ая, -ое,   (*a*) (med) emphysematous

э́мшер·ск·ий коло́д·ец·   (*m*) Imhoff tank (sewage)

энали́т·, -а,   (*m*) (min) enalite

эн·анте́м·а, -ы,   (*f*) (med) enanthema, enanthem

энантио·мо́рф·н·ый, -ая, -ое,   (*a*) (cryst) enantiomorphic, enantiomorphous

энанти·о·тро́п·н·ый, -ая, -ое,   (*a*) enantiotropic

энантово·ки́сл·ый, -ая, -ое,   (*a*) (chem) enanthate (of)

эна́нт·ов·ый, -ая, -ое,   (*a*) (chem) enanthic

~ альдеги́д·   (*m*) enanthaldehyde

~ кисл·от·а́   (*f*) enanthic acid

энарг·и́т·, -а,   (*m*) (min) enargite

энгадини́т·, -а,   (*m*) (min) engadinite

энгана́ж·, -а,   (*m*) (horol) depth

энгидри́т·, -а,   (*m*) (geol) enhydrite

Э́нглер·у, вя́з·кость·ь по   (oil) Engler viscosity

энглиши́т·, -а,   (*m*) (min) englishite

энд·артери́т·, -а,   (*m*) (med) endarteritis

энделлиони́т·, -а,   (*m*) (min) endellionite

энделли́т·, -а,   (*m*) (min) endellite

энде́м·ик·, -а,   (*m*) (biol) endemic form

эндем·и́ческ·ий, -ая, -ое,   (*a*) endemic

~ фо́рм·а   (*f*) (biol) endemic form

эндем·и́чн·ый, -ая, -ое,   (*a*) = **эндем·и́ческ·ий**

эндем·и́·я, -и,   (*f*) (med) endemy, endemia

эндлихи́т·, -а,   (*m*) (min) endlichite

эндо·вибр·а́тор·, -а,   (*m*) (phys) cavity resonator

эндо·га́м·и·я, -и,   (*f*) (biol) endogamy

эндо·генет·и́ческ·ий, -ая, -ое, (*a*) (geol) endogenetic

эндо·ге́н·н·ый, -ая, -ое, (*a*) endogenous

эндо·де́рм·а, -ы, (*f*) (zool) endoderm

эндо·де́рмис·, -а, (*m*) (bot) endodermis

эндо·зоохо́р·а, -ы, (*f*) (biol) endozoochore

эндо·ка́рд·, -а, (*m*) (anat) endocardium

эндо·кард·и́т·, -а, (*m*) (med) endocarditis

эндо·ка́рп·, -а, (*m*) = **эндо·ка́рп·и·й**

эндо·ка́рп·и·й, -я, (*m*) (bot) endocarp

эндо·ко́н·, -а, (*m*) (zool) endocone

эндо·кри́н·н·ый, -ая, -ое, (*a*) (physiol) endocrine

~ **же́лез·ы** (*pl*) (physiol) endocrine glands

эндо·крино·ло́г·и·я, -и, (*f*) (zool) endocrinology

эндо·кри́пт·и·я, -и, (*f*) (cryst) endocryptic process

эндо·лизи́н·, -а, (*m*) (bact) endolysin

эндо·ли́мф·а, -ы, (*f*) (zool) endolymph

эндо·ме́тр·и·й, -я, (*m*) (anat) endometrium

эндо·метри́т·, -а, (*m*) (med) endometritis

эндо·ми́ксис·, -а, (*m*) (zool) endomixis

эндо·мо́рф·, -а, (*m*) (cryst) endomorph

эндо·морф·и́зм·, -а, (*m*) (geol) endomorphism

эндо·парази́т·, -а, (*m*) (biol) endoparasite

эндо·пла́зм·а, -ы, (*f*) (cyt) endoplasm

эндо·поди́т·, -а, (*m*) (zool) endopodite

эндо·полиплоид·и́·я, -и, (*f*) (gen) endopolyploidy

эндо·се́пт·а, -ы, (*f*) (biol) endoseptum

эндо·скеле́т·, -а, (*m*) (ent) endoskeleton

эндо·ско́п·, -а, (*m*) (med) endoscope

эндо·скоп·и́·я, -и, (*f*) endoscopy

энд·о́смос·, -а, (*m*) endosmosis

эндо·спе́рм·а, -ы, (*f*) (bot) endosperm

эндо·спо́р·а, -ы, (*f*) (bot) endospore

эндо·спо́р·и·й, -я, (*m*) (bot) endosporium, (*pl*) endosporia

энд·о́ст·, -а, (*m*) (zool) endosteum

эндо·сти́л·ь, -я, (*m*) (zool) endostyle

эндо·те́ли·й, -я, (*m*) (biol) endothelium

эндо·тели·о́м·а, -ы, (*f*) (biol) endothelioma

эндо·телио·ци́т·, -а, (*m*) (biol) endotheliocyte

эндо·терм·и́ческ·ий, -ая, -ое, (*a*) endothermic

эндо·терм·и́ческ·ий, -ая, -ое

~ **реа́кци·я** (*f*) (chem) endothermic reaction

эндо·те́ци·й, -я, (*m*) (bot) endothecium

эндо·токси́н·, -а, (*m*) (bact) endotoxin

эндо·тро́ф·н·ый, -ая, -ое, (*a*) (biol) endotrophic

эндо·фи́т·, -а, (*m*) (bot) endophyte

эндо·энзи́м·, -а, (*m*) (chem) endoenzyme

энео·ли́т·, -а, (*m*) (geol) Aeneolithic, Eneolithic

энерге́т·ик·, -а, (*m*) power engineer

энерге́т·ик·а, -и, (*f*) power engineering; energetics

энергет·и́ческ·ий, -ая, -ое, (*a*) (phys) energy; (eng) power

~ **интерва́л·** (*m*) (phys) energy band

~ **реа́ктор·** (*m*) (nucl) power reactor

~ **со·сто·я́ни·е, основ·н·о́е** (*n*) (phys) ground state

~ **спектр·** (*m*) (cryst) energy distribution

энерг·и́чн·о (*adv*) energetically

энерг·и́чн·ый, -ая, -ое, (*a*) energetic, vigorous

~ **кип·е́ни·е** (*n*) vigorous boiling

эне́рг·и·я, -и, (*f*) energy, power

~ **актив·а́ци·и** activation energy

~, **а́том·н·ая** atomic energy

~ **а́том·н·ого ядр·а́** nuclear energy

~ **взаимо·де́й·стви·я** binding energy

~, **вну́тр·енн·яя** intrinsic energy

~, **внутри·а́том·н·ая** nuclear energy

~, **вод·ян·а́я** water power

~ **вращ·е́ни·я** rotational energy

~, **вы́·свобожд·енн·ая** released energy

~, **Зеема́н·овск·ая** Zeeman energy

~, **ка́ж·ущ·аяся** apparent power

~, **кинет·и́ческ·ая** kinetic energy

~ **колеб·а́ни·й** vibration energy

~, **луч·и́ст·ая** radiant energy

~ **магни́т·н·ой　　анизотро́п·и·и** anisotropy energy, magnetocrystalline energy

~ **на в·хо́д·е, по́лн·ая** (a/c) intake energy

~, **на·ча́ль·н·ая** zero energy

~, **нул·ев·а́я** zero energy

~ **образ·ова́ни·я ядр·а́** binding energy

~, **по́лн·ая** (aerodynam) total energy

~ **по·ступ·а́тельн·ого движ·е́ни·я** translation energy

~, **реакти́в·н·ая** reactive power, (elec) reactive voltamperes

~ **решёт·к·и** lattice energy

энéрг·и·я

~, свобóд·н·ая   free energy

~, с·вя́з·анн·ая   bound energy

~ с·вя́з·и   (phys) binding energy; (chem) bonding energy

~, скор·остн·áя   kinetic energy

~, сóб·ственн·ая   intrinsic/proper energy

~, электр·и́ческ·ая   electric power

энерго·во·оруж·ённост·ь, -и,   (f) power supply per production unit; (shipb) power/displacement ratio

энерго·ём·кост·ь, -и,   (f) energy content; power/energy capacity

энерго·за·трáт·а, -ы,   (f) (physiol) energy expenditure

энерго·пó·езд·, -а,   (m) power-generating train, (rail) mobile power unit

энерго·систéм·а, -ы,   (f) electric power system

энерго·снабж·éни·е, -я,   (n) power supply

энерго·цéх·, -а,   (m) power engineering shop

эн·зи́м·, -а,   (m) (biochem) enzyme

~, редуц·и́рующ·ий   (biochem) reductase

эн·зимат·и́ческ·ий, -ая, -ое,   (a) (biochem) enzymatic

эн·зи́м·н·ый, -ая, -ое,   (a) of **эн·зи́м·**

эн·зоóти·я, -и,   (f) (zool) enzootic, enzooty

энигмати́т·, -а,   (m) (min) enigmatite, aenigmatite

энкрини́т·, -а,   (m) (pal) encrinite

óн·н·ый, -ая, -ое,   (a) nth power/degree, any

эно·тéр·а, -ы,   (f) (bot) evening primrose, *Oenothera*

эно·циани́н·, -а,   (m) (chem) oenocyanine

óн·ск·ий, -ая, -ое,   (a) X (nameless for security purposes)

энстати́т·, -а,   (m) (min) enstatite

энтальпи́·я, -и,   (f) (phys) enthalpy

энтел·одóнт·, -а,   (m) (pal) entelodont, (pl) *Entelodontidae*

энтер·и́т·, -а,   (m) (med) enteritis

энтеро·анастомóз·, -а,   (m) (med) enteroanastomosis

энтеро·гепати́т·, -а,   (m) (med) enterohepatitis

энтеро·кинáз·а, -ы,   (f) (med) enterokinase

энтеро·кóкк·, -а,   (m) (microbiol) enterococcus

энтеро·коли́т·, -а,   (m) (med) enterocolitis

энтеро·ли́т·, -а,   (m) (med) enterolith

энтеро·птóз·, -а,   (m) (med) enteroptosis

энтеро·токси́н·, -а,   (m) (biochem) enterotoxin

энтеро·цéль·н·ый, -ая, -ое,   (a) (zool) enterocoelous

энто-   (int component) see also **эндо-**

энто·блáст·, -а,   (m) (cyt) entoblast, endoblast, hypoblast

энто·дéрм·а, -ы,   (f) (zool) entoderm, endoderm

энто·кони́д·, -а,   (m) (zool) entoconid

энтомо·лóг·и·я, -и,   (f) (zool) entomology

энтомо·стрáк·и   (pl) (zool) entomostracans, *Entomostraca*

энтомо·фил·и́·я, -и,   (f) (biol) entomophily, entomophilia

энтомо·фи́ль·н·ый, -ая, -ое,   (a) (bot) entomophilous

энто·парази́т·, -а,   (m) (biol) entoparasite, endoparasite

энто·фи́т·, -а,   (m) (bot) entophyte, endophyte

эн·троп·и́йн·ый, -ая, -ое,   (a) entropy

эн·троп·и́·я, -и,   (f) (phys etc.) entropy

энтузиаст·и́ческ·ий, -ая, -ое,   (a) enthusiastic

э·нукле·áци·я, -и,   (f) enucleation

энцефал·и́т·, -а,   (m) (med) encephalitis

энцефалó·граф·, -а,   (m) (med) encephalograph

энцефало·грáф·и·я, -и,   (f) encephalography

энцефало·маля́ци·я, -и,   (f) (med) encephalomalacia

энцефало·миокарди́т·, -а,   (m) (med) encephalomyocarditis

энцефало·миэли́т·, -а,   (m) (med) encephalomyelitis

энциклопед·и́ческ·ий, -ая, -ое,   (a) encyclopaedic, encyclopedic

энциклопéд·и·я, -и,   (f) encyclopaedia, encyclopedia

эо·ги́ппус·, -а,   (m) (pal) Eohippus

эоз·и́н·, -а,   (m) eosin (dye)

эозино·фи́л·, -а,   (m) (cyt) eosonophil

эозино·фил·и́·я, -и,   (f) (med) eosinophilia

эо·зóй·ск·ий, -ая, -ое,   (a) (geol) Eozoic

эо·зóон·, -а,   (m) (geol) eozoon

эо·кéмбр·и·й, -я,   (m) (geol) Eocambrian

эо·ли́т·, -а,   (m) (geol) eolith

эо́л·ов·ый, -ая, -ое, (*a*) aeolian, eolian, wind-borne/shaped, atmogenic

~ а́рф·а (*f*) (meteor) aeolian tones

эо́н·, -а, (*m*) (geol) eon, aeon

эо·палео·зо́·й, -я, (*m*) (geol) Eopalaeozoic, Eopaleozoic

эо·це́н·, -а, (*m*) (geol) Eocene

эпейр·и́ческ·ий, -ая, -ое, (*a*) (geog) epeiric

эпейро·ген·и́ческ·ий, -ая, -ое, (*a*) (geol) epeirogenic

эпейро·ген·и́·я, -и, (*f*) (geol) epeirogeny, epirogeny

эп·е́ндим·а, -ы, (*f*) (anat) ependyma

эпи·бе́нтос·, -а, (*m*) (ocean) epibenthos

эпи·бла́ст·, -а, (*m*) (bot) epiblast

эпи·бле́м·а, -ы, (*f*) (bot) epiblema, epiblem

эпи·бол·и́·я, -и, (*f*) (zool) epiboly

эпи·гене́з·, -а, (*m*) (gen) epigenesis

эпи·генет·и́ческ·ий, -ая, -ое, (*a*) epigenetic

эпи·ги́н·ов·ый, -ая, -ое, (*a*) (bot) epigynous

эпи·грамм·а, -ы, (*f*) diagram, curve, graph

эпи·демио·ло́г·и·я, -и, (*f*) (med) epidemiology

эпи·дем·и́ческ·ий, -ая, -ое, (*a*) (med) epidemic

~ желт·у́х·а (*f*) (med) infectious jaundice

эпи·де́м·и·я, -и, (*f*) epidemic

эпи·де́рм·а, -ы, (*f*) (biol) epiderm

эпи·де́рмис·, -а, (*m*) (biol) epidermis

эпи·дерм·и́ческ·ий, -ая, -ое, (*a*) epidermal, epidermic

эпи·диа·ско́п·, -а, (*m*) (opt) epidiascope

эпиди́димис·, -а, (*m*) (zool) epididymis

эпидидими́т·, -а, (*m*) (med) epididymitis

эпи·диори́т·, -а, (*m*) (geol) epidiorite

эпи·до́т·, -а, (*m*) (min) epidote

эпизо́д·, -а, (*m*) episode, occasion

эпизод·и́ческ·ий, -ая, -ое, (*a*) occasional

эпи·зоот·и́ческ·ий, -ая, -ое, (*a*) (vet) epizootic

эпи·зоо́т·и·я, -и, (*f*) (vet) epizootic

эпи·зоото·ло́г·и·я, -и, (*f*) (vet) epizootology

эпи·кард·, -а, (*m*) (anat) epicardium

эпи·ко́тил·ь, -я, (*m*) (bot) epicotyl

эпи·крит·и́ческ·ий, -ая, -ое, (*a*) (zool) epicritic

эпи·ле́пси·я, -и, (*f*) (med) epilepsy

эпи·лепт·и́ческ·ий, -ая, -ое, (*a*) (med) epileptic

эпи·лимнио́н·, -а, (*m*) epilimnion, (*pl*) epilimnia (layer of lake)

эпил·и́ровани·е, -я, (*n*) see эпил·я́ци·я

эпил·иро́вк·а, -и, (*f*) see эпил·я́ци·я

эпил·иро́вочн·ый, -ая, -ое, (*a*) of эпил·иро́вк·а

эпи·ло́г·, -а, (*m*) epilogue

эпил·я́ци·я, -и, (*f*) depilation, hair removal

эпи·морф·, -а, (*m*) (cryst) epimorph

эпи·на́ст·и·я, -и, (*f*) (bot) epinasty

эпи·ско́п·, -а, (*m*) (opt) episcope

эписпади́·я, -и, (*f*) (med) epispadias

эпи·ста́з·, -а, (*m*) (gen) epistasis

эпистемо·ло́г·и·я, -и, (*f*) epistemology, theory of knowledge

эпи·стильби́т·, -а, (*m*) (min) epistilbite

эпи·строфе́·й, -я, (*m*) (zool) epistropheus

эпи·такс·и́ческ·ий, -ая, -ое, (*a*) epitaxic

~ с·раст·а́ни·е (*n*) (met) epitaxis

эпи·та́кс·и·я, -и, (*f*) (cryst) epitaxy

эпи·та́фи·я, -и, (*f*) epitaph

эпи·те́к·а, -и, (*f*) (zool) epitheca

эпи·тели·а́льн·ый, -ая, -ое, (*a*) (biol) epithelial

эпи·те́ли·й, -я, (*m*) (biol) epithelium

эпитём·а, -ы, (*f*) (bot) epithem

эпи·терма́ль·н·ый, -ая, -ое, (*a*) (geol) epithermal

эпи·фи́з·, -а, (*m*) (anat) epiphysis

эпи·физ·арн·ый, -ая, -ое, (*a*) (anat) epiphyseal, epiphysial

эпи·фи́т·, -а, (*m*) (bot) epiphyte

эпи·фи́т·н·ый, -ая, -ое, (*a*) (bot) epiphytic, epiphyllous

эпи́·фор·а, -ы, (*f*) (med) epiphora

эпи·фра́гм·а, -ы, (*f*) (zool) epiphragm

эпи·хлор·гидри́н·, -а, (*m*) epichlorhydrin

эпи·це́нтр·, -а, (*m*) (seismol) epicentre, epicenter

~ вз·ры́в·а ground zero (of weapon explosion)

эпи·центра́ль·н·ый, -ая, -ое, (*a*) epicentral

эпи·ци́кл·, -а, (*m*) (math) epicycle

эпи·цикл·о́ид·а, -ы, (*f*) (math) epicycloid

эп·и́ческ·ий, -ая, -ое, (*a*) epic

эпокси- (*int component*) (chem) epoxy

эпокси́д·, -а, (*m*) (chem) epoxy

эпокси́д·н·ый, -ая, -ое, *(a)* epoxy
~ смол·а́ *(f)* (chem) epoxy resin
эпоксидо·образ·ова́ни·е, -я, *(n)* epoxidation
эпокси·смол·а́, -ы́, *(f)* epoxy resin
эпокси·со·един·е́ни·е, -я, *(n)* epoxy compound
эпо́х·а, -и, *(f)* epoch
э́псилон *(indecl)* epsilon, ε (Greek)
эпсоми́т·, -а, *(m)* (min) epsomite, (pharm) Epsom salts
эпю́р·, -а, *(m)* = **эпю́р·а**
эпю́р·а, -ы, *(f)* projection diagram; curve
~ из·ги́б·а́ющ·их моме́нт·ов curve of bending moments
~ на·гру́з·к·и loading diagram
~ на·пряж·е́ни·я stress sheet/envelope
~ пере·вя́з·ывающ·их сил· curve of shearing forces
~ рас·пре·дел·е́ни·я скор·ост·е́й velocity diagram
~ сил· curve/diagram of forces
~ сил· ве́с·а (shipb) curve of weights
~ сил· на·гру́з·к·и (shipb) curve of load
~ сил· под·держ·а́ни·я (shipb) curve of buoyancy
~ с·ре́з·ывающ·их сил· (mech) shear line
эпюрацио́н·н·ая коло́нн·а *(f)* (chem) fractionating column/tower
э́р·а, -ы, *(f)* era
   до на́ш·ей э́р·ы B.C. (of dates)
   на́ш·ей э́р·ы *(g)* A.D. (of dates)
э́рб·иев·ый, -ая, -ое, *(a)* of **э́рб·и·й**
э́рб·и·й, -я, *(m)* (chem) erbium, Er
э́рг·, -а, *(m)* erg (unit of work)
эргаст·и́ческ·ий, -ая, -ое, *(a)* (biol) ergastic
эрго·гра́ф·, -а, *(m)* (physiol) ergograph
эргод·и́ческ·ий, -ая, -ое, *(a)* (math) ergodic
~ тео́р·и·я *(f)* (dynam) ergodic theory
эргод·и́чност·ь, -и, *(f)* (math) ergodicity
эргод·н·ый, -ая, -ое, *(a)* (math) ergodic
эрго́·метр·, -а, *(m)* (physiol) ergometer
эрго·метри́н·, -а, *(m)* (biochem) ergometrine, ergotocine, ergobasine
эрго·но́м·и·я, -и, *(f)* (physiol) ergonomics
эрго·стери́н·, -а, *(m)* (biochem) ergosterol

эрго·стеро́л·, -а, *(m)* (biochem) ergosterol
эрготами́н·, -а, *(m)* (pharm) ergotamine
эргот·и́зм·, -а, *(m)* (med) ergotism
эргот·и́н·, -а, *(m)* (pharm) ergotin, ergotine
эрго·токси́н·, -а, *(m)* (chem) ergotoxine
эрдмани́т·, -а, *(m)* (min) erdmannite
эре́ктор·, -а, *(m)* erector, erector arm (tunnel driving)
эре́кци·я, -и, *(f)* (biol) erection
эремуру́с·, -а, *(m)* (bot) desert candle, *Eremurus*
эреп·си́н·, -а, *(m)* (biochem) erepsin
э́рик·а, -и, *(f)* (bot) heather, *Erica*
эрини́т·, -а, *(m)* (min) erinite
эринс-ба́кштаг·, -а, *(m)* (naut) vang
эрио́·метр·, -а, *(m)* (text) eriometer
эрио·хальци́т·, -а, *(m)* (min) eriochalcite
эрит·е́м·а, -ы, *(f)* (med) erythema
эрите́метр·, -а, *(m)* (med) erythemeter
эритра́зм·а, -ы, *(f)* (med) erythrasma
эритр·еми́·я, -и, *(f)* (med) erythraemia, erythremia
эритр·и́н·, -а, *(m)* (min) erythrite, cobalt bloom
эритр·и́т·, -а, *(m)* (chem) erythrite, erythritol
эритро·бла́ст·, -а, *(m)* (biol) erthroblast
эритр·ози́н·, -а, *(m)* (chem) erythrosine
эритро·сидери́т·, -а, *(m)* (min) erythrosiderite
эритро·фи́лл·, -а, *(m)* (bot) erythrophyll
эритро·ци́т·, -а, *(m)* (zool) erythrocyte
Э́рленмейер·а, ко́лб·а *(f)* (chem) Erlenmeyer flask
э́рлифт·, -а, *(m)* air-lift (pump)
Эрми́т·а, много·чле́н· *(m)* (math) Hermite polynomial
эрми́т·ов· опер·а́тор· *(m)* (math) Hermitian operator
эрод·и́рованн·ый, -ая, -ое, *(past part pass)* of **эрод·и́р·овать;** eroded
эрод·и́р·овать, -ую́т, *(imp)* erode
эрод·и́руемост·ь, -и, *(f)* erodibility
эрози́в·н·ый, -ая, -ое, *(a)* erosive
эрозио́н·ност·ь, -и, *(f)* erodibility
эрозио́н·н·ый, -ая, -ое, *(a)* erosive; erosional
эро́зи·я, -и, *(f)* erosion
~, под·зе́м·н·ая soil piping/tunneling
э́рос·, -а, *(m)* = **эро́т·**
эро́т·, -а, *(m)* (astron) Eros

эрот·и́ческ·ий, -ая, -ое, (*a*) erotic

эррат·и́ческ·ий, -ая, -ое, (*a*) erratic

эрсте́д·, -а, (*m*) oersted (unit of magnetic field strength)

эрстед·ме́тр·, -а, (*m*) (elec) oersted meter

эруди́ци·я, -и, (*f*) erudition

эру́к·ов·ая кисл·от·а́ (*f*) (chem) erucic acid

эрупти́в·н·ый, -ая, -ое, (*a*) eruptive

эру́пци·я, -и, (*f*) eruption; (astron) solar eruption

эска́др·а, -ы, (*f*) (naut) squadron

эска́др·енн·ый мино·но́с·ец· (*m*) (naut) destroyer

эскадри́л·ь·я, -и, (*g pl*) -л·и·й, (*f*) (air) squadron

эскадро́н·, -а, (*m*) (mil) squadron

эскала́тор·, -а, (*m*) escalator, moving stairs

эска́рп·, -а, (*m*) (geol) scarp

э́скер·, -а, (*m*) (geol) esker

эски́з·, -а, (*m*) sketch, freehand/rough drawing/sketch; draft, outline

эски́з·н·ый, -ая, -ое, (*a*) sketch, sketched; outline, rough, draft

~ прое́кт· (*m*) draft project

эскимо́ (*n indecl*) chocolate-coated ice-cream

эскимо́с·ск·ий, -ая, -ое, (*a*) Eskimo

эско́рт·, -а, (*m*) escort

эскорт·и́р·овать, -ую, (*imp*) escort

эско́рт·н·ый, -ая, -ое, (*a*) escort

эсми́н·ец·, -н·ц·а, (*i*) -н·ц·ем, (*m*) = эска́др·енн·ый мино·но́с·ец·

эспа́рто (*n indecl*) (bot) esparto grass, *Stipa tenacissima*

эспарце́т·, -а, (*m*) (bot) sainfoin, *Onobrychis*

эссе́нц·и·я, -и, (*f*) essence

~, мирба́н·ов·ая (chem) oil of mirbane, nitrobenzene

эстака́д·а, -ы, (*f*) pier, trestle, trestle bridge, scaffold bridge; service rack (vehicle inspection)

~, бу́нкер·н·ая hopper gantry (furnace charging)

~, на́·береж·н·ая piled wharf

~, на·лив·н·а́я (oil) loading rack/dock

эстака́д·н·ый, -ая, -ое, (*a*) of эстака́д·а

~ желе́з·н·ая доро́г·а (*f*) elevated/overhead railway/railroad

эста́мп·, -а, (*m*) block print

эстафе́т·а, -ы, (*f*) relay-race

эстафе́т·н·ый, -ая, -ое, (*a*) of эстафе́т·а; sequential, successive

эстезио́·метр·, -а, (*m*) (physiol) aesthesiometer, esthesiometer

эстерифик·а́ци·я, -и, (*f*) (chem) esterification

эстет·и́ческ·ий, -ая, -ое, (*a*) aesthetic, esthetic

эстив·а́ци·я, -и, (*f*) (biol) aestivation, estivation

эсто́н·ск·ий, -ая, -ое, (*a*) Estonian

эстра́д·а, -ы, (*f*) stage; (theat) music-hall, vaudeville

эстр·и́н·, -а, (*m*) (biochem) oestrin, estrin

эстро·ге́н·, -а, (*m*) (biochem) oestrogen, estrogen

эстр·о́н·, -а, (*m*) (biochem) oestrone, estrone

эстуа́р·иев·ый, -ая, -ое, (*a*) (geog) estuary, estuarine

эстуа́р·и·й, -я, (*m*) (geog) estuary

э́та (*pron*) (*nom sing f*) of э́тот this; eta, η (Greek)

эта́ж·, -а́, (*i*) -о́м, (*m*) (arch) storey, floor; (min) level; platen (of a press)

~, мансА́рд·н·ый (build) mezzanine floor

~, ни́ж·н·ий ground floor

~, пе́рв·ый (build) first floor/storey

~, про·меж·у́точн·ый (min) intermediate level

~, цо́коль·н·ый (build) semi-basement

этаж·е́рк·а, -и, (*g pl*) -р·ок·, (*f*) set of shelves, shelves, bookcase

эта́ж·ност·ь, -и, (*f*) number of storeys

эта́ж·н·ый, -ая, -ое, (*a*) of эта́ж·

этало́н·, -а, (*m*) standard (of measurement); standard measuring instrument

~, втор·и́чн·ый reference standard

~ -ко́п·и·я, -и, (*f*) reference standard

~ между·на·ро́д·н·ой свеч·и́ international candle

~ ме́тр·а the standard metre

~, на·сып·н·о́й powder standard

~, перв·и́чн·ый primary standard

~ част·от·ы́ перв·и́чн·ый primary standard of frequency

этало́н·и́ровани·е, -я, (*n*) = этало́н·и́ровк·а

этало́н·и́рованн·ый, -ая, -ое, (*past part pass*) calibrated

этало́н·и́ровк·а́, -и, (*f*) calibration

этало́н·н·ый, -ая, -ое, (*a*) calibration, standard, reference

ли́н·и·я этало́н·н·ой за·де́рж·к·и (*f*) (elec) standard lag line

~ то́пл·ив·о (*n*) reference fuel

эталóн·н·ый, -ая, -ое
~ **част·от·á** (*f*) (elec) standard frequency
эт·áн·, -а, (*m*) (chem) ethane
этанóл·, -а, (*m*) ethanol, ethyl alcohol
этанол·амúн·, -а, (*m*) (chem) ethanolamine
этáп·, -а, (*m*) stage
~ **бо·ев·óго пут·ú, контрóль·н·ый** pre-bombing run
~ **по·лёт·а** (rocket) phase
~, **техно·лог·úческ·ий** production/construction stage
этéн·, -а, (*m*) = **этилéн·**
этерифик·áци·я, -и, (*f*) (chem) esterification, etherification
этернúт·, -а, (*m*) (build) eternit
éти (*pron*) (*nom pl, acc pl inanim*) of **éтот** these
этизúн·, -а, (*m*) (pharm) ethisine
éт·ик·а, -и, (*f*) ethics
этикéт·к·а, -и, (*g pl*) **-т·ок·,** (*f*) label, tally plate, name plate, tag
этикето·нос·úтел·ь, -я, (*m*) label-carrier
эт·úл·, -а, (*m*) (chem) ethyl
~, **брóм·ист·ый** ethyl bromide, monobromoethane
~, **хлóр·ист·ый** ethyl chloride, monochloroethane
этилáл·ь, -я, (*m*) ethylal, diethylformal
этил·амúн·, -а, (*m*) ethylamine
этил·ацетáт·, -а, (*m*) ethyl acetate
этил·бензóл·, -а, (*m*) ethylbenzene
этил·борáт·, -а, (*m*) borate ester
этил·дихлорарсúн·, -а, (*m*) ethyldichloroarsine
этил·éн·, -а, (*m*) ethylene
~, **хлóр·ист·ый** ethylene chloride, dichloroethane
этилен·гликóл·ь, -я, (*m*) ethylene glycol
этилен·диамúн·, -а, (*m*) ethylenediamine
этилен·диамин·тетра·ацетáт· нá·три·я EDTA, versene, sequestrene, tetrasodium salt of ethylene diamine tetraacetic acid
этилен·ди·амин·тетра·úксус·н·ая кисл·от·á (*f*) ethylenediaminetetraacetic acid
этилен·изáци·я, -и, (*f*) (food) artificial ripening with ethylene treatment
этилéн·ов·ый, -ая, -ое, (*a*) ethylene, ethylenic
этилен·оксúд·, -а, (*m*) ethylene oxide
этилен·хлор·гидрúн·, -а, (*m*) ethylene chlorohydrin

этил·идéн·, -а, (*m*) (chem) ethylidene
~, **хлóр·ист·ый** ethylidene chloride, 1,1 dichlorethane
этилиминúн·, -а, (*m*) (chem) ethyleneimine, azocyclopropane
этил·úровани·е, -я, (*n*) ethylation
этил·úрованн·ый, -ая, -ое, (*a*) ethylated
этил·моч·евúн·а, -ы, (*f*) ethyl urea
этúл·ов·ый, -ая, -ое, (*a*) (chem) ethyl
~ **жúдк·ост·ь** (*f*) ethyl fluid, tetraethyl lead (gasoline additive)
~ **спирт·** (*m*) ethyl alcohol
~ **эфúр·** (*m*) ethyl ether; ethyl ester
~ **эфúр· азóт·н·ой кисл·от·ы́** ethyl nitrate
~ **эфúр· муравь·úн·ой кисл·от·ы́** ethyl formate
~ **эфúр· úксус·н·ой кисл·от·ы́** ethyl acetate
этил·сéр·н·ая кисл·от·á (*f*) ethyl sulphuric/sulfuric acid
этил·целлюлóз·а, -ы, (*f*) (plast) ethyl cellulose
éт·им (*pron, instr sing, dat pl*) of **éтот**
éт·ими (*pron, instr pl*) of **éтот**
этимо·лог·úческ·ий, -ая, -ое, (*a*) etymological
этимо·лóг·и·я, -и, (*f*) (ling) etymology
этиол·úрованн·ый, -ая, -ое, (*a*) (agr etc.) etiolated
этио·лóг·и·я, -и, (*f*) (med etc.) aetiology, etiology
этиóн·ов·ая кисл·от·á (*f*) (chem) ethionic acid
этио·порфирúн·, -а, (*m*) (chem) etioporphyrin
этио·трóп·н·ый, -ая, -ое, (*a*) (med) aetiotropic, etiotropic
этио·филлúн·, -а, (*m*) (biochem) etiophyllin
эт·úт·, -а, (*m*) (min) aetite, eaglestone
éт·их (*pron, gen/prep/acc anim pl*) of **éтот** these
эт·úческ·ий, -ая, -ое, (*a*) ethical
этм·оид·áльн·ый, -ая, -ое, (*a*) (anat) ethmoid
этм·оид·úт·, -а, (*m*) (med) ethmoiditis
этмо·лúт·, -а, (*m*) (geol) ethmolith
этн·úческ·ий, -ая, -ое, (*a*) ethnic
этно·граф·úческ·ий, -ая, -ое, (*a*) ethnographic
этно·лóг·и·я, -и, (*f*) ethnology
éт·о (*pron, nom/acc sing n*) of **éтот**
éт·ой (*pron, gen/dat/instr/prep f sing*) of **éтот** this
этоксúл·, -а, (*m*) (chem) ethoxyl

это·ло́г·и·я, -и, (*f*) (zool) ethology
э́т·ом (*pron, prep sing m, f*) of **э́тот**
э́т·ому (*pron, dat sing m, n*) of **э́тот**
этро́л·, -а, (*m*) (plast) Etrol
э́тот (*pron*) this
э́т·у (*pron, acc sing f*) of **э́тот**
эу- (*int component*) *see also* **эв-, ей-**
эу·литора́л·ь, -и, (*f*) eulittoral (of body of water)
эури- (*int component*) *see* **эври-, ейри-**
эу·сте́л·а, -ы, (*f*) (bot) eustele
эу·тро́ф·н·ый, -ая, -ое, (*a*) (biol) eutrophic
эу·филли́н·, -а, (*m*) (pharm) euphyllin(e)
эфе́др·а, -ы, (*f*) (bot) ephedra, *Ephedra*
эфедр·и́н·, -а, (*m*) (biochem) ephedrine, ephedrin
эфеме́р·, -а, (*m*) (bot) ephemer
эфемер·и́д·а, -ы, (*f*) (astron) ephemeris
эфемер·и́д·н·ый, -ая, -ое, (*a*) (astron) ephemeris
~ вре́м·я (*n*) (astron) ephemeris time
эфеме́р·н·ый, -ая, -ое, (*a*) ephemeral
эфемер·о́ид·, -а, (*m*) (bot) ephemeroid
эфио́п·ск·ий, -ая, -ое, (*a*) Ethiopian
~ о́бласт·ь (*f*) (zool) Ethiopian region
эфи́р·, -а, (*m*) (chem) ether; ester
~ адипи́н·ов·ой кислот·ы, диме-тил·ов·ый dimethyl adipinate
~ алкокси·кисл·от·ы́ ether ester
~, бензо́й·н·о-эти́л·ов·ый methyl benzoate
~, бензо́й·н·ый benzoate of, benzoic acid ester
~, винно·эти́л·ов·ый ethyl tartrate
~ глицери́н·а, диметакри́л·ов·ый glyceryl dimethacrylate
~, ди·эти́л·ов·ый diethyl ether, ether
~, карбамино́во·эти́л·ов·ый ethyl carbamate, urethane
~, ки́сл·ый acid ester, ether acid
~, мети́л·ов·ый methyl ether
~, метил·фени́л·ов·ый methyl phenyl ether
~ се́р·н·ой кисл·от·ы́, мети́л·ов·ый methyl sulphate/sulfate
~, с·ло́ж·н·ый ester, compound ether, ethereal salt
~ тио·кисл·от·ы́ thioester
~ у́голь·н·ой кисл·от·ы́ alkyl carbamate
~ у́ксус·н·ой кисл·от·ы́ acetic acid ester, acetate
~ у́ксус·н·ой кисл·от·ы́, эти́л·ов·ый ethyl acetate

эфи́р
~, уксусно·мети́л·ов·ый methyl acetate
~ фено́л·а, мети́л·ов·ый methyl phenoxide
~ фосфо́р·ист·ой кислот·ы́ phosphite
~ эфиро·кисл·от·ы́ ether acid ester
~ янта́р·н·ой кисл·от·ы́, глико́л·е·в·ый glycol succinate
~ окси·кето́н·а ether ketone
~ окси·кисл·от·ы́ hydroxy ether; hydroxy ester
~, по́лн·ый complete ether
~, прост·о́й ether
~ себацио́н·ов·ой кисл·от·ы́ sebacic acid ester, sebacate, sebate
эфи́р·а, -ы, (*f*) (zool) ephyra
эфи́р·н·ый, -ая, -ое, (*a*) ether; (chem) ethereal; ester
~ ма́сл·о (*n*) ethereal oil, (biochem) essential/volatile oil
эфиро·кисл·от·а́, -ы́, (*f*) (chem) ether acid
эфиро·но́с·, -а, (*m*) (bot) essential-oil plant
эффе́кт·, -а, (*m*) (*and see under associated words*) effect
~, бли́з·ост·н·ый proximity effect
~, вз·ры́в·н·о́й blast effect
~, до́н·н·ый (dynam) afterbody/bottom effect
~, кон·цев·о́й end/tip effect
~, масшта́б·н·ый scale effect
~, ноч·н·о́й night effect
~, про·стра́н·ствени·ый three-dimensional effect
эффекти́в·ност·ь, -и, (*f*) efficiency, efficacy, effectiveness
~ воздухо·за·бо́р·ник·а ram efficiency (of ram jet)
~ за·хва́т·а (nucl) capture efficiency
пре·де́л· эффекти́в·ност·и (*m*) effective range
эффекти́в·н·ый, -ая, -ое, (*a*) effective, efficacious, efficient
~ давл·е́ни·е, сре́д·н·ее (*n*) brake mean effective pressure
~ знач·е́ни·е (*n*) (elec) r.m.s. value, root mean square value (of an alternating current or voltage)
~ коэффицие́нт· поле́з·н·ого де́й-·стви·я (*m*) (mech) overall efficiency
~ мо́щ·ност·ь (*f*) brake horsepower, effective horsepower, shaft horsepower
~ на·пряж·е́ни·е (*n*) (mech) true stress

эффе·ктив·н·ый, -ая, -ое
~ у·дель·н·ый рас·ход· топл·ив·а (*m*) specific fuel consumption
эффект·н·ый, -ая, -ое, (*a*) impressive, striking, effective
эфферент·н·ый, -ая, -ое, (*a*) (zool) efferent
эффлюент·, -а, (*m*) effluent
эффузив·, -а, (*m*) (geol) effusion, effusive rock
эффузив·н·ый, -ая, -ое, (*a*) (geol) effusive
~ по·род·а (*f*) (geol) effusive rock
эффузио·метр·, -а, (*m*) (chem) effusiometer
эффузион·н·ый, -ая, -ое, (*a*) effusion
~ метод· (*m*) effusion method (of measuring vapour pressure)
эффуз·и·я, -и, (*f*) (phys) effusion, gaseous effusion; (geol) effusion
эффузор·, -а, (*m*) effuser
эхин·, -а, (*m*) (arch) echinus
эхино·кактус·, -а, (*m*) (bot) echinocactus
эхино·кокк·оз·, -а, (*m*) (med) echinococcosis
эхино·плутеус·, -а, (*m*) (zool) echinopluteus
эх·о, -а, (*n*) echo
~, много·крат·н·ое (rad) flutter echo
эхо·грамм·а, -ы, (*f*) (instr) echogram
эхо·граф·, -а, (*m*) recorder (of echo sounder)
эхо·зонд·, -а, (*m*) (instr) echometer

эхо·лот·, -а, (*m*) (ocean) echo sounder
~, звук·ов·ой sonic echo-sounder
~ низ·к·ой част·от·ы́ sonic echo-sounder, sonic depth finder
эхо·лот·ѝровани·е, -я, (*n*) echo-sounding
эхо·метр·ѝровани·е, -я, (*n*) (oil) echo depth-measuring
эхо·резонатор·, -а, (*m*) (rad) echo box
эцѝди·й, -я, (*m*) (bot) aecidium
эцидио·спор·а, -ы, (*f*) (bot) aecidiospore
эшелетт·, -а, (*m*) (instr) échelette; (*pl*) echelon grating
эшелон·, -а, (*m*) echelon; (naut) column on line of bearing; wave (of assault landing); military/troop/supply train
~ выс·от·а́ по·лёт·а (air) flight altitude
~ Майкельсон·а (opt) Michelson interferometer
эшелон·ѝровани·е, -я, (*n*) (*v n*) of эшелон·ѝр·овать; separation (of aircraft in the air)
эшелон·ѝр·овать, -уют, (*imp and perf*) put/form in echelon; stack (aircraft waiting to land)
эшелон·н·ый, -ая, -ое, (*a*) of эшелон·
~ рас·по·лож·ени·е (*n*) (mar eng) unit system
эшинѝт·, -а, (*m*) (min) aeschynite
эшлёпс·, -а, (*m*) (bot) goatgrass, *Aegilops*
эякул·ят·, -а, (*m*) (zool) ejaculate
эякул·яци·я, -и, (*f*) (zool) ejaculation

Ю

Ю. = **юг·** S, south
юбе́·я, -и, (*f*) (bot) coquito palm, *Jubaea*
юбиле́·й, -я, (*m*) anniversary, jubilee
юбил·я́р·, -а, (*m*) person/town etc. celebrating anniversary
юб·к·а, -и, (*g pl*) **юб·ок·,** (*f*) skirt; shed, petticoat (of electric insulator)
~, ни́ж·н·яя petticoat
~ по́рш·н·я piston skirt
~ цили́ндр·а (ICE) cylinder skirt
юб·очк·а, -и, (*g pl*) **-чек·,** (*f*) (*dim*) of **юб·к·а**
ювели́р·, -а, (*m*) jeweller
ювели́р·н·ый, -ая, -ое, (*a*) jewellery, jewelry, jeweller's
~ из·де́л·и·я (*pl*) jewellery, jewelry
ювени́ль·н·ый, -ая, -ое, (*a*) (bot, geol) juvenile
~ вод·а́ (*f*) (geol) juvenile water
юг·, -а, (*m*) south
юга́ль·н·ый, -ая, -ое, (*a*) (anat) jugal
югланди́н·, -а, (*m*) (pharm) juglandine
юг·о-вос·то́к·, -а, (*m*) south-east
юг·о-вос·то́ч·н·ый, -ая, -ое, (*a*) south-east/eastern/easterly
юг·о-за́·пад·, -а, (*m*) south-west
юг·о-за́·пад·н·ый, -ая, -ое, (*a*) south-west/western/westerly
юго·сла́в·ск·ий, -ая, -ое, (*a*) Yugoslav, Yugoslavian
югум·, -а, (*m*) (biol) jugum
юж·а́нин·, -а, (*nom pl*) **-а́н·е,** (*g pl*) **-а́н·,** (*m*) southerner
ю́ж·н·ее (*adv*) to the south of, further south
южно·африка́н·ск·ий, -ая, -ое, (*a* South African, South Africa
южно·кита́й·ск·ий, -ая, -ое, (*a*) South Chinese, South China
ю́ж·н·ый, -ая, -ое, (*a*) south, southern, southerly
~ крест· (*m*) (astron) Southern Cross, Crux
~ по́люс· (*m*) (geog) South Pole
Юз·а, телегра́ф·н·ый аппара́т· (*m*) Hughes printing telegraph
ю́зин·ь, -я, (*m*) (naut) houseline, strong spunyarn
ю́з·ом (*adv*) skidding; by dragging
ю́кк·а, -и, (*f*) (bot) yucca, jucca, *Yucca*
юкони́т·, -а, (*m*) (min) yukonite
юл·а́, -ы́, (*f*) (orn) woodlark, *Lullula arborea*

юлиа́н·ск·ий, -ая, -ое, (*a*) Julian
Юм-Ро́зери, пра́в·ил·о (*n*) (phys) Hume–Rothery law
юми́т·, -а, (*m*) (geol) humite
ю́мор·, -а, (*m*) humour, humor
юмор·исти́ческ·ий, -ая, -ое, (*a*) humourous, humorous
юмохо́йт·, -а, (*m*) (min) umohoite
ю́нг·а, -и, (*m*) (naut) apprentice
Ю́нг·а, мо́дул·ь (*m*) (mech) Young's modulus, Young's modulus of elasticity
ЮНЕ́СКО (*indecl*) UNESCO
ю́н·ост·ь, -и, (*f*) youth (period), adolescence
ю́нош·а, -и, (*g pl*) **-ей,** (*m*) youth, lad
ю́нош·еск·ий, -ая, -ое, (*a*) youth, young people's, juvenile, adolescent; youthful
ю́нош·еств·о, -а, (*n*) young people, juvenile, teenagers, adolescents; youth, adolescence
ю́н·ый, -ая, -ое, (*a*) young, youthful, adolescent
Юпи́тер·, -а, (*m*) (astron) Jupiter
юр·а, -ы, (*f*) (geol) Jurassic
юрид·и́ческ·и (*adv*) legally, in law
юрид·и́ческ·ий, -ая, -ое, (*a*) legal, law; juridicial
~ част·ь (*f*) legal department
юрисди́кци·я, -и, (*f*) (law) jurisdiction
юриско́нсульт·, -а, (*m*) legal consultant/adviser
юриспруде́нци·я, -и, (*f*) (law) jurisprudence
юри́ст·, -а, (*m*) lawyer
~, уч·ён·ый- jurist
ю́рк·ий, -ая, -ое, (*a*) nimble, brisk
ю́р·ск·ий, -ая, -ое, (*a*) (geol) Jurassic
юст·и́ровани·е, -я, (*n*) (*v n*) of **юст·и·р·овать;** *see* **юст·иро́вк·а**
юст·ир·ова́ть, -у́ют, (*imp and perf*) regulate, adjust; correct, align, position (optical systems); (rad) orientate (PPI display)
юст·иро́вк·а, -и, (*g pl*) **-вок·,** (*f*) (*v n*) *see* **юст·ир·ова́ть;** regulation, adjustment; (rad) orientation (of PPI display); positioning (a beam), (opt) alignment, aligning
юст·иро́вочн·ый, -ая, -ое, (*a*) of **юст·иро́вк·а**
~ га́й·к·а (*f*) adjusting nut

юст·иро́вочн·ый, -ая, -ое
~ **клин·** (*m*) (opt) correction wedge
юсти́ци·я, -и, (*f*) justice; law/legal
authorities
ю́т·, -а, (*m*) (shipb) after superstructure

ют·и́ться, -я́тся, (*imp*) huddle together
ю́ферс·, -а, (*m*) (naut) dead eye
ю́фт·ев·ый, -ая, -ое, (*a*) of **ю́фт·ь**
ю́фт·ь, -и, (*f*) Russian leather

Я

я (*pron*) I

я́блок·о, -а, (*n*) apple; bull's eye (of target); (shipb) boss (of sternpost)

~, глаз·н·о́е (anat) eyeball, bulbous oculi

~, чёрн·ое (instr) centre circle

я́блон·ев·ый, -ая, -ое, (*a*) of **я́блон·я**

я́блон·я, -и, (*f*) (bot) apple tree, *Malus*

~, дом·а́шн·яя apple, *Malus domestica*

я́блоч·к·о, -а, (*g pl*) **-чек·,** (*n*) (*dim*) of **я́блок·о**

Я́блочков·а свеч·а́ (*f*) (opt) Jablochkoff candle

яблочно·ки́сл·ый, -ая, -ое, (*a*) (chem) malate (of), malic acid

я́блоч·н·ый, -ая, -ое, (*a*) apple, pomaceous, malic

~ кисл·от·а́ (*f*) malic acid

яв·а́нтроп·, -а, (*m*) (pal) Javanthropus

яв·и́ть, -ят, (*perf*) *see* **явл·я́·ть**

я́в·к·а, -и, (*g pl*) **я́в·ок·,** (*f*) appearance, presence, attendance; showing, presentation (of documents etc.); secret meeting-place; secret meeting

явл·е́ни·е, -я, (*n*) (*and see under associated words*); phenomenon, feature, effect, occurrence, event; appearance

~, законо·ме́р·н·ое natural phenomenon

явл·я́·ть, -ют, (*imp*) show; **-ся** (+ *instr*) be, appear, present oneself; report, register

явно·бра́ч·н·ый, -ая, -ое, (*a*) (bot) phanerogamous; (*as noun*) phanerogam

я́в·н·о (*adv*) clearly, distinctly, obviously

явно·вы·раж·ённ·ый, -ая, -ое, (*a*) clearly defined/expressed, distinct

явно·кристалл·и́ческ·ий, -ая, -ое, (*a*) phanerocrystalline

явно·по́люс·н·ый, -ая, -ое, (*a*) (elec) salient-pole

я́в·н·ый, -ая, -ое, (*a*) evident, obvious, clear, plain, distinct, (math) explicit

я́вор·, -а, (*m*) (bot) sycamore, *Acer pseudoplatanus*

я́вор·ов·ый, -ая, -ое, (*a*) of **я́вор·**

я́в·очн·ый, -ая, -ое, (*a*) secret-meeting

я́в·ственн·ый, -ая, -ое, (*a*) distinctive, distinct

я́в·ств·овать, -уют, (*imp*) be obvious/clear

я́в·ь, -и, (*f*) reality

я́гел·ь, -я, (*m*) (bot) lichen

ягн·ёнок·, -нк·а, (*nom pl*) **н·я́т·а,** (*g pl*) **-н·я́т·,** (*m*) lamb

ягн·и́ться, -я́тся, (*imp*) lamb

ягн·я́т·а (*nom pl*) of **ягн·ёнок·**

я́год·а, -ы, (*f*) (bot) berry, bacca

~, ви́н·н·ая fig, dried fig

я́год·иц·а, -ы, (*i*) **-ей,** (*f*) (anat) buttock

я́год·ичн·ый, -ая, -ое, (*a*) of **я́год·иц·а**

я́год·ков·ые, -ых, (*pl decl as adj*) (bot) *Thymelaeaceae*

я́год·ник·, -а, (*m*) fruit bush plantation/patch; fruit bush

я́год·н·ый, -ая, -ое, (*a*) of **я́год·а;** berry, fruit, baccate

ягодо·ви́д·н·ый, -ая, -ое, (*a*) berry-shaped/like, bacciform

ягодо·но́с·н·ый, -ая, -ое, (*a*) (bot) berry-producing/bearing, bacciferous

ягодо·обра́з·н·ый, -ая, -ое, (*a*) berry-shaped, bacciform; baccate

ягуа́р·, -а, (*m*) (zool) jaguar, *Felis onca*

яд·, -а, (*m*) poison, toxin; (zool) venom

~, змей·н·ый snake venom

~, тру́п·н·ый ptomaine

ядерно·за·мещ·ённ·ый, -ая, -ое, (*a*) (chem) ring-substituted

ядерно·резона́нс·н·ый, -ая, -ое, (*a*) nuclear-resonance

ядерно·чи́ст·ый, -ая, -ое, (*a*) nuclear-pure

я́дер·н·ый, -ая, -ое, (*a*) of **ядр·о́**

~ реа́ктор· (*m*) nuclear reactor

~ реа́кци·я (*f*) nuclear reaction

~ те́хн·ик·а (*f*) nuclear engineering

~ фи́з·ик·а (*f*) nuclear physics

я́дер·щик·, -а, (*m*) nuclear engineer

ядовито·ды́м·н·ый, -ая, -ое, (*a*) toxic-smoke

яд·ови́тост·ь, -и, (*f*) toxicity

яд·ови́т·ый, -ая, -ое, (*a*) poisonous, toxic; (zool) venomous

ядо·но́с·н·ый, -ая, -ое, (*a*) poisonous, toxic, noxious, venomous

ядо·химика́т·, -а, (*m*) toxic/poisonous chemical

ядр·ён·ый, -ая, -ое, (*a*) vigorous, healthy; fresh, juicy (of fruit)

ядр·и́ц·а, -ы, (*i*) **-ей,** (*f*) hulled buckwheat

ядр·о́, -а́, (*nom pl*) **я́др·а,** (*g pl*) **я́дер·,** (*n*) nucleus, kernel, core; (chem) ring; heart, heartwood (timber); (meteor) centre, center; (math) kernel; shot (sport); gist, substance (e.g. of article)

азот· ядр·а́ (chem) ring nitrogen

ядр·о́

~, **аромат·и́ческ·ое** (chem) aromatic ring

~, **а́том·н·ое** atomic nucleus

~, **бензо́ль·н·ое** (chem) benzene nucleus/ring

~ **ге́ли·я** (nucl) helion

~ **дейте́р·и·я** (nucl) deuton, deuteron

~ **за·медл·е́ни·я** (math) slowing-down kernel

за·мещ·ённ·ый в ядр·е́ (chem) ring substituted

~, **итер·и́рованн·ое** (math) iterated kernel

~ **кристаллиз·а́ци·и** centre/center of crystallization

~, **от·кры́·т·ое** (math) interior (of a set)

~ **пла́зм·ы** plasma core

~, **пло́ск·ое** (math) plane kernel

~ **по·вы́ш·енн·ого давл·е́ни·я** (meteor) centre of high pressure

~ **по·то́к·а** (dynam) flow core

~, **про·меж·у́точн·ое** (phys) compound nucleus

~ **про·тык·а́ни·я** (geol) piercement/ diapiric core

~, **раз·реш·а́ющ·ее** (math) resolvent kernel

~ **с·мещ·е́ни·я** (math) displacement kernel

~, **тра́нспорт·н·ое** (math) transport kernel

ядро·по·до́б·н·ый, -ая, -ое, (a) nucleiform

я́др·ышк·о, -а, (g pl) **-шек·,** (n) (biol) nucleolus

я́зв·а, -ы, (f) (med) ulcer, sore; pitting (corrosion)

~, **коррозио́н·н·ая** pitting

~, **кра́с·н·ая** coral spot

~, **сиби́р·ск·ая** anthrax

я́зв·енник·, -а, (bot) anthyllis, *Anthyllis*

я́зв·енн·ый, -ая, -ое, (a) ulcerous

~ **бол·е́зн·ь** (f) (med) stomach ulcers

я́зв·ин·а, -ы, (f) pitting (corrosion); ulcer

язв·и́тельн·ый, -ая, -ое, (a) mocking, biting, caustic

язы́к·, -а́, (m) tongue, glossa; language; bolt (of lock); clapper, tongue (of bell)

~, **лед·нико́в·ый** glacier tongue

~, **маши́н·н·ый** machine language

~, **меа́ндр·ов·ый** (geol) meander lobe

~, **мор·ск·о́й** (fish) sole, (pl) *Soleidae*

~ **-по·сре́д·ник·, -а,** (m) intermediary language

~, **холо́д·н·ый** (meteor) cold tongue

языко·ве́д·, -а, (m) philologist, linguist

языко·ве́д·ени·е, -я, (n) philology, linguistics

языко·ве́д·ческ·ий, -ая, -ое, (a) philological, linguistic

язык·ов·о́й, -а́я, -о́е, (a) linguistic

язык·о́в·ый, -ая, -ое, (a) tongue, lingual

языко·гло́т·очн·ый, -ая, -ое, (a) (anat) glossopharyngeal

языко·зна́·ни·е, -я, (n) (see also **лингви́ст·ик·а**); philology, linguistics

языко·тру́б·, -а, (m) (bot) salpiglossis, *Salpiglossis*

язы́ч·еск·ий, -ая, -ое, (a) heathen, pagan

языч·ко́в·ый, -ая, -ое, (a) of **язы·чо́к**; (anat) uvular, staphyline, (bot) ligulate, (gram) glossal; (pl decl as adj) *Pentastomida*

~ **игл·а́** (f) (text) latch needle

язычно·нёб·н·ый, -ая, -ое, (a) (anat) glossopalatine

язычно·языч·ко́в·ый, -ая, -ое, (a) (anat) glossostaphyline

язы́ч·н·ый, -ая, -ое, (a) tongue, lingual

языч·о́к·, -ч·к·а́, (m) (dim) of **язы́к·**; (anat) uvula; (bot) ligule; lug, tag, catch; reed; (text) latch (of latch needle)

~ **за·мк·а́** lock catch

~, **пружи́н·ящ·ий** reed

яз·ь, -я́, (f) (fish) *Leuciscus idus*

яйч·к·о, -а, (g pl) **-ч·ек·,** (n) (dim) of **яйц·о́**; (anat) testicle, testis; (bot) ovule

яйч·ков·ый, -ая, -ое, (a) (anat) testicular

яйч·ник·, -а, (m) (anat) ovary

яйч·ников·ый, -ая, -ое, (a) (anat) ovarian

яйч·ница, -ы, (i)**-ей,** (f) (food) egg dish

~ **-болт·у́нь·я, -в,** (g pl) **-ни·й,** (f) scrambled eggs

яйч·н·ый, -ая, -ое, (a) egg

~ **бел·о́к·** (m) egg white, white of egg

~ **желт·о́к·** (m) egg yoke

~ **порош·о́к·** (m) egg powder

яйл·а́, -ы, (f) (geol) yaila

яйц·е- (root) egg, ovi-

яйце·ви́д·н·ый, -ая, -ое, (a) ovate, egg-shaped

яйце·во́д·, -а, (m) (zool) oviduct

яйце·живо·рожд·е́ни·е, -я, (n) (zool) ovoviviparity

яйце·кла́д·, -а, (*m*) (zool) ovipositor

яйце·клад·у́щ·ие, -их, (*pl decl as adj*) (zool) monotremes, *Monotremata*

яйце·клет·к·а, -и, (*g pl*) **-т·ок·,** (*f*) (zool) ovicell

яйце·но́с·кост·ь, -и, (*f*) egg production

яйце·но́с·н·ый, -ая, -ое, (*a*) (zool) oviferous

яйце·обра́з·н·ый, -ая, -ое, (*a*) egg-shaped, ovoid, oviform

~ ампут·а́ци·я (*f*) elliptical amputation

яйце·ро́д·н·ый, -ая, -ое, (*a*) (zool) oviparous, oötocous

яйце·род·я́щ·ий, -ая, -ее, (*a*) (zool) oviparous

яйце·рожд·е́ни·е, -я, (*n*) (zool) oviparity

яйц·о́, -а́, (*nom pl*) **я́йц·а,** (*n*) egg; ovum

~ с·пуск·ов·о́й стрел·ы́ (shipb) dog shore trigger

як·, -а, (*m*) (zool) yak, *Bos gruniens*

Яко́би, много·чле́н·ы (*pl*) (math) Jacobi polynomials

~, си́мбол· (*m*) (math) Jacobi symbol

якобиа́н·, -а, (*m*) (math) Jacobian, functional determinant

якобси́т·, -а, (*m*) (min) jacobsite

якобсо́н·ов· о́рган· (*m*) (zool) Jacobson's organ

я́кобы (*particle*) as though, as if, allegedly

якор·ёк·, -р·к·а, (*m*) (*dim*) of **я́кор·ь;** (acous) reed

~, пружи́н·ящ·ий reed

якор·н·ый, -ая, -ое, (*a*) of **я́кор·ь**

~ бо́ч·к·а (*f*) mooring buoy

~ земле·со́с· (*m*) moored suction dredger

~ ого́н·ь (*m*) (naut) riding/anchor light

~ с·бор· (*m*) anchorage/anchor dues, groundage

~ сте́рж·н·и (*pl*) (elec) armature bars

~ сто·я́нк·а (*f*) anchorage

я́кор·ь, -я, (*nom pl*) **-я́,** (*g pl*) **-е́й,** (*m*) anchor; (elec) armature; sinker (of a sea mine)

~, адмиралте́й·ск·ий admiralty pattern anchor

~, винт·ов·о́й mooring screw

~, га́з·ов·ый (oil) gas anchor

~, грибо·ви́д·н·ый mushroom anchor

~ Де́нн·а stockless anchor

~, кольц·ев·о́й (elec) ring armature

~, лед·о́в·ый ice anchor

я́кор·ь

~, мёртв·ый mooring anchor

~, песо́ч·н·ый (oil) sand anchor

~, плав·у́ч·ий sea anchor, drift anchor

~, по·воро́т·н·ый (elec) hinged armature

~ -реле́ (*indecl*) (elec) relay armature

~, ры́ск·ов·ый hedge anchor

~, станов·о́й bower anchor

~, четырёх·ла́п·н·ый grapnel

якути́нг·а, -и, (*f*) (geol) jacutinga

ял·, -а, (*m*) rowing boat

я́л·ик·, -а, (*m*) (naut) skiff

я́л·ов·ый, -ая, -ое, (*a*) barren, sterile (of animals)

ялпаи́т·, -а, (*m*) (min) jalpaite

я́м·а, -ы, (*f*) pit, hole, depression, well; (ocean) cauldron; (anat) fossa

~, аккумуля́тор·н·ая battery tank (of submarine)

~, вис·о́чн·ая (anat) temporal fossa

~, воз·ду́ш·н·ая air pocket

~, во́лч·ья concealed pit

~, потенциа́ль·н·ая (elec) potential well

~, сол·ян·а́я salt pan

~, у́голь·н·ая coal bunker

я́м·к·а, -и, (*g pl*) **-м·ок·,** (*f*) (*dim*) of **я́м·а;** (anat) fossa; skin hole (metal defect)

я́м·н·ый, -ая, -ое, (*a*) of **я́м·а**

я́м·очк·а, -и, (*g pl*) **-чек·,** (*f*) (*dim*) of **я́м·к·а;** dimple

я́м·чатост·ь, -и, (*f*) pit (plant disease)

я́м·чат·ый, -ая, -ое, (*a*) pitted

янва́р·ск·ий, -ая, -ое, (*a*) January

янва́р·ь, -я́, (*m*) January

яндов·а, -ы, (*f*) (build) valley

янтарно·ки́сл·ый, -ая, -ое, (*a*) (chem) succinate (of)

янта́р·н·ый, -ая, -ое, (*a*) amber; amber-coloured/colored

~ альдеги́д· (*m*) succinaldehyde

~ кисл·от·а́ (*f*) succinic acid

янта́р·ь, -я́, (*m*) (min) amber, succinite

янтини́т·, -а, (*m*) (min) ianthinite

япо́н·ск·ий, -ая, -ое, (*a*) Japanese

яр·, -а, (*m*) steep sandy bank

я́рд·, -а, (*m*) yard (0·914 m)

яре́м·н·ый, -ая, -ое, (*a*) (anat) jugular

я́р·к·ий, -ая, -ое, (*a*) bright; **vivid**, striking

ярко·вы́·раж·енн·ый, -ая, -ое, (*a*) pronounced

я́р·к·о-зелён·ый, -ая, -ое, (*a*) bright green

яр·к·о-кра́сн·ый, -ая, -ое, (*a*) bright red

ярко·ме́р·, -а, (*m*) brightness meter, nitometer

я́р·кост·ь, -и, (*f*) brightness; vividness

~, про·стра́н·ственн·ая spatial brightness

ярли́т·, -а, (*m*) (min) jarlite

ярлы́к·, -а́, (*m*) label, tag

я́рмар·к·а, -и, (*g pl*) **-р·ок·,** (*f*) fair, annual fair

ярм·о́, -а́, (*nom pl*) **я́рм·а,** (*g pl*) **ярм·,** (*n*) yoke

ярмо·зу́б·ые, -ых, (*pl decl as adj*) (pal) zeuglodonts, *Zeuglodontia*

яр·овиза́ци·я, -и, (*f*) vernalization (of seeds etc.)

яр·ов·о́й, -а́я, -о́е, (*a*) (agr) spring, vernal

~ се·в· (*m*) (agr) spring sowing

ярози́т·, -а, (*m*) (min) jarosite

я́р·остн·ый, -ая, -ое, (*a*) ardent, furious, violent

я́р·ост·ь, -и, (*f*) ardour, fury, rage; frenzy, madness

я́р·очк·а, -и, (*g pl*) **-чек·,** (*f*) ewe lamb

ярун·о́к·, -н·к·а́, (*m*) bevel (carpenter's tool)

я́рус·, -а, (*m*) tier; (geol) layer, stage; (theat) circle

~ облак·о́в (meteor) cloud level

я́рус·н·ый, -ая, -ое, (*a*) layered, tiered; stepped, staged, decked

яру́т·к·а, -и, (*g pl*) **-т·ок·,** (*f*) (bot) pennycress, *Thlaspi*

я́р·ый, -ая, -ое, (*a*) ardent

я́р·ь, -и, (*f*) (agr) spring corn; fury, rage; verdigris

яр·ь-мед·я́нк·а, -и, (*f*) verdigris

ясен·е́ц·, -н·ц·а́, (*m*) (bot) dittany, *Dictamnus*

ясен·ь, -я, (*m*) (bot) ash, *Fraxinus*

~, америка́н·ск·ий white ash, *Fraxinus americana*

~, обык·нове́нн·ый European ash, *Fraxinus excelsior*

~, чёрн·ый black ash, *Fraxinus nigra*

ясл·и, -ей, (*pl*) crèche (for infants); (agr) feeding rack

ясме́н·ник·, -а, (*m*) (bot) woodruff, *Asperula*

ясн·о- (*component*) clearly

ясн·ост·ь, -и, (*f*) clarity, lucidity, distinctness

ясно́тк·а, -и, (*f*) (bot) deadnettle, *Lamium*

я́сн·ый, -ая, -ое, (*a*) clear, lucid, distinct

я́стреб·, -а, (*nom pl*) **-а́,** (*m*) hawk, buzzard, (*pl*) *Falconiformes*

ястреб·и́нк·а, -и, (*g pl*) **-нок·,** (*f*) (bot) hawkweed, *Hieracium*

ястреб·и́н·ый, -ая, -ое, (*a*) of **я́стреб·**

яс·ь, -и, (*f*) jar (percussion drilling tool)

ятре́н·, -а, (*m*) (pharm) yatren, jatren

ятры́ш·ник·, -а, (*m*) (bot) orchis, *Orchis*

я́хт·а, -ы, (*f*) yacht

~, го́н·очн·ая racing yacht

~ -моно·ти́п·, -а, (*m*) one-design yacht

~, мото́р·н·ая motor yacht

я́хт·енн·ый, -ая, -ое, (*a*) of **я́хт·а**

яче́·ист·ый, -ая, -ое, (*a*) cellular; mesh (of net)

~ бето́н· (*m*) aerated/cellular concrete

яче́й·к·а, -и, (*g pl*) **-е·ек·,** (*f*) cell; compartment, cubicle; alveolus, pit; mesh (of net); pigeon-hole; (mil) foxhole

~, бина́р·н·ая binary cell (of electronic storage system)

~, за·па́ль·н·ая (nucl) seed unit

~, за·по·мин·а́ющ·ая storage cell (of computer)

~, зуб·н·а́я (anat) dental alveolus

~, кондукто·метр·и́ческ·ая (elec instr) conductivity cell

~, от·де́ль·н·ая discrete cell (computers)

~ па́мят·и storage cell/location, location (computers)

~ па́мят·и, не·чёт·н·ая odd location (computers)

~ рас·пре·дел·и́тельн·ого у·стро́й·-ств·а (elec) switch cell

~, стрел·ко́в·ая (mil) foxhole

~, три́ггер·н·ая (elec) flip-flop

~, элемента́р·н·ая (cryst) unit cell

ячейко·обра́з·н·ый, -ая, -ое, (*a*) cellular, compartmented

яче·я́, -й, (*f*) cell; mesh (of nets etc.); bay (for equipment)

ячме́н·н·ый, -ая, -ое, (*a*) barley

ячме́н·ь, -я́, (*m*) (bot) barley, *Hordeum*

я́шм·а, -ы, (*f*) (min) jasper

я́щер·, -а, (*m*) (*see also* **я́щер·ы**) (zool) pangolin, *Manis*

я́щер·иц·а, -ы, (*i*) **-ей,** (*f*) (zool) lizard, (*pl*) *Lacertilia*

я́щеро·га́д·ы, -ов, (*m pl*) (zool) rhynchocephalians, *Rhynchocephalia*

ящеро·но́г·ий, -ая, -ое, (*a*) (pal) sauropodous, sauropod, (*pl as noun*) sauropods, *Sauropoda*

ящеро·та́з·ов·ые, -ых, (*pl decl as adj*) (pal) saurischians, *Saurischia*

ящеро·хво́ст·ые, -ых, (*pl decl as adj*) (pal) saururans, *Saururae*

я́щер·ы, -ов, (*m pl*) (pal) saurians, *Sauria*; (zool) pangolins

я́щик·, -а, (*m*) box, chest, locker; drawer

~, батере́й·н·ый accumulator crate

~, воз·ду́ш·н·ый airspace (boiler); buoyancy tank (boat)

~ -гига́нт·, -а, (*m*) concrete pontoon

я́щик

~, диафра́гм·енн·ый (elec, met) cathode box

~, дым·ов·о́й smoke box (boiler)

~, като́д·н·ый (elec, met) cathode box

~, му́сор·н·ый dustbin

~ -понто́н·, -а, (*m*) concrete pontoon

~, почт·о́в·ый post box, letter box

~, сигна́ль·н·ый (naut) flag locker

~, у́голь·н·ый coal bunker

~, цеп·н·о́й (naut) cable locker

я́щич·н·ый, -ая, -ое, (*a*) of **я́щик·**

я́щур·, -а, (*m*) (vet) foot-and-mouth disease, *Aphtae episooticae*

SUPPLEMENTARY DICTIONARY OF RUSSIAN WORD ENDINGS

Ending	Grammar	Remarks
··· zero	1. *m, nom/acc sing*	
	2. *f, g pl*	See nom si = ···**а**
	3. *n, g pl*	See nom si = ···**о**
	4. *a, sf, m si*	Sentences with no verb (is/are is understood)
	5. *adv*	Invariable. See Dictionary
	6. *imp/perf, past, si*	Often from infin ending **-нуть**
···**а**	1. *m, nom si*	Rare
	2. *m, acc anim si*	Nom si = ··· zero. Uncommon
	3. *m, g si*	Nom si = ··· zero. Common
	4. *m, nom pl*	Often plurals of professional titles, nom si = ··· zero
	5. *m, acc pl*	Rare
	6. *f, nom si*	Common
	7. *n, g si*	Nom si = ···**о**. Common Nom si = ···**е**. Less common
	8. *n, nom/acc pl*	Nom si = ···**о**. Common Nom si = ··**е**. Less common
	9. *a f, sf, si*	Sentences with no verb (is/are is understood). Long form ···**ый**, ···**ой** or ···**ий**
	10. *ger, pres*	Uncommon
	11. *pron f, nom si*	See Dictionary
	12. *adv*	See Dictionary
··**аз·а**	*f, nom si*	English ending -ase
··**оз·а**	*f, nom si*	English ending -ose
··**к·а**	*f, nom si*	Often noun derived from verb
··**овк·а**	*f, nom si*	Often noun derived from verb
··**ировк·а**	*f, nom si*	Often noun derived from verb
··**ик·а**	*f, nom si*	Names of disciplines etc., cf. ending ··**ик-**
··**очк·а**	*m, g si*	Diminutive form
	f, nom si	Diminutive form
···**ла**	*imp/perf, past, f si*	Often from infin ending **-нуть**
··**а·ла**	*imp/perf, past, f si*	Infin ending **-а·ть**
··**е·ла**	*imp/perf, past, f si*	Infin ending **-е·ть**
···**ила**	*imp/perf, past, f si*	Infin ending ···**ить**
··**я·ла**	*imp/perf, past, f si*	Infin ending ··**я·ть**
··**аем·а**	*pres part pass, sf, f si*	= which is -ed
··**им·а**	*pres part pass, sf, f si*	= which is -ed
··**н·а**	*a f, sf, si*	Sentence with no verb (is/are is understood)
··**ан·а**	*past part pass, sf, f si*	= which was -ed from infin **-ать**
··**ен·а**	*past part pass, sf, f si*	= which was -ed from infin **-ить**
-**м·ен·а**	*n, nom/acc pl*	See nom si = **-м··я**
··**изн·а**	*f, nom si*	Qualities
··**ур·а**	*f, nom si*	English ending -ure
··**ет·а**	*f, nom si*	Often derived from adj ···**ый** or ···**ий**
··**от·а**	*f, nom si*	Often derived from adj ···**ый** or ···**ий**
··**ят·а**	*m, nom pl*	Often denotes young creatures. Nom si ··**енок·**

Ending	Grammar	Remarks
–б·	See ··· zero at beginning	
–в·	1. See ··· zero at beginning	
···в	2. *ger, past*	Infin ···**ть**
··а·в	*ger, past*	Infin ··**а·ть**
··ирова·в	*ger, past*	Infin ··**ирова·ть**
··е·в	1. *ger, past*	Uncommon
···ев	2. *m, acc anim pl*	Nom si ··· zero. Uncommon
	3. *m, g pl*	Nom si ··· zero. Common
···ив	*ger, past*	Infin ···**ить**
··лив·	*a m, sf, si*	Sentence with no verb (is/are is understood). Long form ··**лив·ый**
···ов	1. *m, acc anim pl*	Nom si ··· zero. Uncommon
	2. *m, g pl*	Nom si ··· zero. Common
··ну·в	*ger, past*	Infin ··**ну·ть**
–г·	See ··· zero at beginning	
–лог··	*m, nom si*	English ending -logist
–д··	See ··· zero at beginning	
–оид··	*m, nom/acc si*	English ending -oid
···е	1. *m, p si*	Nom si = ··· zero, ···**ь** or ···**й**. Less common
	2. *f, d si*	Nom si = ···**а** or ···**я**. Common
	3. *f, p si*	Nom si = ···**а** or ···**я**. Less common
	4. *n, nom/acc si*	Very common
	5. *n, p si*	Nom si = ···**о** or ···**е**. Less common
	6. *a n, sf, si*	From soft adj in **-ий**. Sentence with no verb
	7. *comp adj*	Invariable, see Dictionary
···ее	1. *a n, nom/acc si*	For suffixes see **-ий**. Very common
	2. *comp adj*	Find comparative or ordinary form of adj in Dictionary. Less common
	3. *adv*	Find comparative or ordinary form of adj in Dictionary. Less common
–ж··е	*comp adj*	Attributive forms of adj have **г, д, дк, зк** instead of **ж**
···ие	1. *a, nom/acc pl*	For suffixes see **-ий**
··и·е	2. *n, nom/acc si*	See below for suffixes
··стви·е	*n, nom/acc si*	
··ани·е	*n, nom/acc si*	Often verbal nouns -ation, -ing from infin **-ать**
··ивани·е	*n, nom/acc si*	Often verbal nouns -ation, -ing from infin **-ать**
··овани·е	*n, nom/acc si*	Often verbal nouns -ation, -ing from infin **-ать**
–из·овани·е	*n, nom/acc si*	English endings -ization, -izating, -izing
··ировани·е	*n, nom/acc si*	English endings -ation, -ing. Russian words often adopted from foreign languages
–из·ировани·е	*n, nom/acc si*	Russian words often adopted from foreign languages, English endings -ization, -izating, -izing
··ени·е	*n, nom/acc si*	Verbal nouns often from infin **-ить**
··яни·е	*n, nom/acc si*	Verbal nouns often from infin **-ять**

Ending	Grammar	Remarks
··ан·е	1. *m, p si*	See nom si = ··ан·
	2. *m, nom pl*	Nom si = ··анин·. Persons of given nationality or type
··ян·е	1. *m, p si*	Nom si = ··янин·. Persons of given nationality or type
	2. *m, nom pl*	
···ое	1. *a n, nom/acc si*	For suffixes see **-ый**
	2. *num*	Uncommon, see Dictionary
	3. *adv*	Uncommon, see Dictionary
-о·е	4. *m, p si*	Nom si = -о·й
···те	1. *imp/perf, imper, pl*	
-т·е	2. *m, p si*	Nom si = -т·
	3. *f, d/p si*	Nom si = ·от·а, ·ет·а
···ете	1. *imp, pres, 2 pl*	Infin -а·ть, -е·ть, -овать. For suffixes see below
	2. *perf, fut, 2 pl*	Infin -а·ть, -е·ть, -овать. For suffixes see below
··а·ете	1. *imp, pres, 2 pl*	Infin ··а·ть. See ···ете above
	2. *perf, fut, 2 pl*	Infin ···а·ть. See ···ете above
··ива·ете	*imp, pres, 2 pl*	Suffix ·ива· in imperfective verbs
··ыва·ете	*imp, pres, 2 pl*	Suffix ·ыва· in imperfective verbs
··е·ете	1. *imp, pres, 2 pl*	See infin ··е·ть
	2. *perf, fut, 2 pl*	See infin ··е·ть
··н·ете	1. *imp, pres, 2 pl*	See infin ··н·уть. Generally perfective
	2. *perf, fut, 2 pl*	
··у·ете	1. *imp, pres, 2 pl*	See infin ···овать
	2. *perf, fut, 2 pl*	
··ю·ете	1. *imp, pres, 2 pl*	See infin ···евать
	2. *perf, fut, 2 pl*	
··я·ете	1. *imp, pres, 2 pl*	See infin -я·ть. Generally imperfective
	2. *perf, **fut**, 2 pl*	
···ите	1. *imp, pres, 2 pl*	See infin ···ить
	2. *perf, fut, 2 pl*	See infin ···ить
-ит·е	3. *m, p si*	Nom si = -ит·
	4. *f, d/p si*	Nom si = -ит·а. Uncommon
···йте	*imp/perf, imper, pl*	The й is always preceded by a vowel
··ива·йте	*imp, imper, pl*	Infin ··ива·ть. Always imperfective
··ыва·йте	*imp, imper, pl*	Infin ··ыва·ть. Always imperfective
···ьте	*imp/perf, imper, pl*	
··ц·е	1. *m, p si*	See nom si = ··ец·
	2. *f, d/p si*	Nom si = ··ц·а. Uncommon
	3. *n, nom/acc si*	Sometimes a diminutive form
	4. *adv*	Invariable, see Dictionary
··айш·е	*superl adv*	Invariable, from superl adj ··айш·ий
··ейш·е	*superl adv*	Invariable, from superl adj ··ейш·ий
··ьш·е	*comp adj/adv*	Invariable, see Dictionary
···ые	*a, nom/acc pl*	For suffixes see **-ый**
·ев·ые	*a, nom/acc pl*	In biology used for taxonomic categories
··ов·ые	*a, nom/acc pl*	In biology used for taxonomic categories
··н·ые	*a, nom/acc pl*	In biology used for taxonomic categories

Ending	Grammar	Remarks
-ж·	See ··· zero at beginning	
-аж·	*m, nom/acc si*	Often words derived from French ending in -age
··еж·	*m, nom/acc si*	
-з	See ··· zero at beginning	
-лиз··	*m, nom/acc si*	English ending -lysis
-оз··	*m, nom/acc si*	English ending -osis
···и	1. *m, p si*	Rare, nom si = ·и·й
	2. *m, nom/acc pl*	Nom si = ···ь, ···й or -г·, -к·, -х·, -ж·, -ч·, -ш·, -щ·
	3. *f, g si*	See nom si ···я, ··и·я, ···ь
	4. *f, d si*	See nom si ··и·я, ···ь
	5. *f, p si*	See nom si ··и·я, ···ь
	6. *f, nom/acc pl*	See nom si ···я, ··и·я, ···ь
	7. *n, p si*	See nom si ··и·е, -м··я
	8. *a, sf, pl*	Sentence contains no verb (is/are is understood). Derived from long form, see ···ий
	9. *pron, nom/acc pl*	See Dictionary
	10. *imp/per, imper, si*	Infin ···ить
··и·и	1. *m, nom/acc pl*	Nom si = ··и·й
	2. *f, g si*	Nom si = ···я, ··и·я, ···ь
	3. *f, d si*	Nom si = ··и·я, ···ь
	4. *f, p si*	Nom si = ··и·я, ···ь
	5. *f, nom/acc pl*	Nom si = ···я, ··и·я, ···ь
	6. *n, p si*	Nom si = ··и·е
··аци·и	1. *f, g/d/p si*	See nom si ·аци·я
	2. *f, nom/acc pl*	See nom si ·аци·я
··яци·и	1. *f, g/d/p si*	See nom si ··яци·я
	2. *f, nom/acc pl*	
··ск·и	*adv*	From adj ··ск·ий
··еск·и	*adv*	From adj ··еск·ий
··ческ·и	*adv*	From adj ··ческ·ий
··ическ·и	*adv*	From adj ··ическ·ий English ending -ically
···ли	1. *imp/perf, past, pl*	
	2. *m, nom/acc pl*	See nom si -л··ь, -л·ь
	3. *f, g si*	See nom si -л··я, -л··ь
	4. *f, d si*	See nom si -л··ь
	5. *f, p si*	See nom si -л··ь
	6. *f, nom/acc pl*	Nom si -л··я, -л··ь
··а·ли	1. *imp/perf, past, pl*	Infin -а·ть, noun endings ал·ь, ал·я
	2. *m, nom/acc pl*	
	3. *f, g si*	
	4. *f, d si*	
	5. *f, p si*	
	6. *f, nom/acc pl*	
··ева·ли	*imp/perf, past, pl*	Infin ева·ть
··ива·ли	*imp, past, pl*	Infin ·ива·ть. Imperfective
·ова·ли	*imp/perf, pl*	Infin ·овать
·ыва·ли	*imp, past, pl*	Infin ·ыва·ть. Imperfective
··е·ли	1. *imp/perf, past, pl*	Infin -еть
	2. See ···ли	

Ending	Grammar	Remarks
···или	1. *imp/perf, past, pl* 2. See ···ли	Infin **-ить**
··о·ли	1. *imp/perf, past, pl* 2. See ···ли	Infin **-оть**
··я·ли	1. *imp/perf, past, pl* 2. See ···ли	Infin **-ять**
···ами	1. *m, i pl* 2. *f, i pl* 3. *n, i pl*	Common Common Common
···ими	*a, i pl*	Common
···ыми	*a, i pl*	Common
···ьми	1. *m, i pl* 2. *f, i pl* 3. *n, i pl*	Rare Rare Rare
···ями	1. *m, i pl* 2. *f, i pl* 3. *n, i pl*	Common Common Common
-м·ен·и	*n, g/d/p si*	See nom si **-м··я**
···ти	1. *imp/perf, infin* 2. *f, g/d/p si* 3. *f, nom/acc pl*	 For suffixes see nom si ···**ть** For suffixes see nom si ···**ть**
···ши	*ger, past*	
···вши	*ger, past*	
··а·вши	*ger, past*	Infin ·**а·ть**
··ирова·вши	*ger, past*	Infin ·**ирова·ть**
··ив·ши	*ger, past*	Infin **ить**
··нув·ши	*ger, past*	Infin **ну·ть**
··я·вши	*ger, past*	Infin **я·ть**
···ьи	1. *m, nom/acc pl* 2. *f, g si* 3. *f, nom/acc pl* 4. *a, nom/acc pl*	Nom si = ··**е·й** Nom si = ··**ь·я** Nom si = ··**ь·я** Declension as **трет··ий** = third
···й	1. *m, nom/acc si* 2. *f, acc anim pl* 3. *f, g pl* 4. *n, g pl* 5. *imp/perf, imper, si*	**-й** preceded by vowel Nom si ··**и·я**. Uncommon Nom si ··**и·я** Nom si ··**и·е**, ··**ь·е** **-й** preceded by vowel
··ива·й	*imp, imper, si*	Infin ··**ива·ть**. Imperfective
··ыва·й	*imp, imper, si*	Infin ··**ыва·ть**. Imperfective
···ей	1. *m, acc anim pl* 2. *m, g pl* 3. *f, i si* 4. *f, acc anim pl* 5. *f, g pl* 6. *n, g pl* 7. *a f, g/d/i/p si* 8. *pron f, g/d/i/p si* 9. *comp adj/adv*	For suffixes see nom si ···**ь**, **-ж·**, **-ч·**, **-ш·**, **-щ·**, **-ц** For suffixes see nom si ···**ь**, **-ж·**, **-ч·**, **-ш·**, **-щ·**, **-ц·** For suffixes see nom si ···**я**, ··**и·я**, ···**а** For suffixes see nom si ···**я**, ··**ь**, ···**а** For suffixes see nom si ···**я**, ··**ь**, ···**а** For suffixes see nom si ···**е** Masc nom si **-ий -ый** Masc nom si **-ий -ый** See Dictionary. If not given, deduce from adj ending **-ый**, **-ий**

Ending	Grammar	Remarks
-е·й	10. *m, nom/acc si*	See ···**й** above
	11. *f, acc anim pl*	See ···**й** above
	12. *f, g pl*	See ···**й** above
	13. *imp/perf, imper, si*	See ···**й** above
···**ий**	1. *a, nom/acc sing*	For suffixes see below
··**и·й**	2. *m, nom/acc si*	Often English words ending in -ium
	3. *f, g pl*	Nom si ··**и·я**
	4. *n, g*	Nom si ··**и·е**, ··**ь·е**
··**к·ий**	*a, nom/acc sing*	
··**ск·ий**	*a, nom/acc sing*	
евск·ий	*a, nom/acc sing*	
овск·ий	*a, nom/acc sing*	
еск·ий	*a, nom/acc sing*	
ческ·ий	*a, nom/acc sing*	
ическ·ий	*a, nom/acc sing*	
истск·ий	*a, nom/acc sing*	
тельск·ий	*a, nom/acc sing*	**т** preceded by vowel
ацк·ий	*a, nom/acc sing*	
ецк·ий	*a, nom/acc sing*	
еньк·ий	*a, nom/acc sing*	
оньк·ий	*a, nom/acc sing*	
н·ий	*a, nom/acc sing*	
ч·ий	*a, nom/acc sing*	**ч** preceded by vowel
··**вш·ий**	*past part act m, nom/acc si*	
··**айши·ий**	*superl a m, nom/acc si*	
··**чайш·ий**	*superl a m, nom/acc si*	
··**ейш·ий**	*superl a m, nom/acc si*	
··**ащ·ий**	*pres part act m, nom/acc si*	
··**ущ·ий**	*pres part act m, nom/acc si*	
··**ющ·ий**	*pres part act m, nom/acc si*	
··**ующ·ий**	*pres part act m, nom/acc si*	Infin **-овать**
··**ююющ·ий**	*pres part act m, nom/acc si*	Infin **-евать**
··**яюющ·ий**	*pres part act m, nom/acc si*	Infin **-я·ть**
··**ящ·ий**	*pres part act m, nom/acc si*	
···**ой**	1. *f, i si*	Nom si ···**а**
	2. *a m, nom/acc si*	Always stressed
	3. *a f, g/d/i/p*	Masc nom si **-ый**, **-ий**
-о·й	4. *m, nom/acc si*	
	5. *imp/perf, imper, si*	
···**ый**	*a m, nom/acc si*	For suffixes see below
··**ав·ый**	*a m, nom/acc si*	
··**ев·ый**	*a m, nom/acc si*	
··**ив·ый**	*a m, nom/acc si*	
··**ов·ый**	*a m, nom/acc si*	
··**яв·ый**	*a m, nom/acc si*	
··**л··ый**	*a m, nom/acc si*	
··**ал··ый**	*a m, nom/acc si*	
··**ел··ый**	*a m, nom/acc si*	
··**аем·ый**	*pres part pass m, nom/acc si*	Infin ··**а·ть**
··**еем·ый**	*pres part pass m, nom/acc si*	Infin ··**е·ть**

Ending	Grammar	Remarks
··уем·ый	*pres part pass m, nom/acc si*	Infin **-овать**
··юем·ый	*pres part pass m, nom/acc si*	Infin **-евать**
·яем·ый	*pres part pass m, nom/acc si*	Infin ··**я·ть**
··им·ый	*pres part pass m, nom/acc si*	Infin ···**ить**
··ом·ый	*pres part pass m, nom/acc si*	
··н·ый	*a m, nom/acc si*	-ic (chem)
··ан·ый	*a m, nom/acc si*	
··ебн·ый	*a m, nom/acc si*	
··обн·ый	*a m, nom/acc si*	
··авн·ый	*a m, nom/acc si*	
··ивн·ый	*a m, nom/acc si*	
··евн·ый	*a m, nom/acc si*	
··овн·ый	*a m, nom/acc si*	
··ен·ый	*a m, nom/acc si*	
··езн·ый	*a m, nom/acc si*	
··озн·ый	*a m, nom/acc si*	
··ин·ый	*a m, nom/acc si*	
··йн·ый	*a m, nom/acc si*	**й** preceded by vowel
··нн·ый	*past part pass m, nom/acc si* / *a m, nom/acc si*	
··анн·ый	*past part pass m, nom/acc si* / *a m, nom/acc si*	Infin ··**а·ть** / Infin ··**а·ть**
··енн·ый	*past part pass m, nom/acc si* / *a m, nom/acc si*	Infin ···**ить** / Infin ···**ить**
··инн·ый	*a m, nom/acc si*	
··онн·ый	*a m, nom/acc si*	
··унн·ый	*a m, nom/acc si*	
··ьнн·ый	*a m, nom/acc si*	
··янн·ый	*past part pass m, nom/acc si* / *a m, nom/acc si*	Infin **я·ть**
-рн·ый	*a m, nom/acc si*	**р** preceded by vowel
··есн·ый	*a m, nom/acc si*	
··отн·ый	*a m, nom/acc si*	
··остн·ый	*a m, nom/acc si*	
-чн·ый	*a m, nom/acc si*	**ч** preceded by vowel
··ушн·ый	*a m, nom/acc si*	
··юшп·ый	*a m, nom/acc si*	
··ищн·ый	*a m, nom/acc si*	
-льн·ый	*a m, nom/acc si*	**л** preceded by vowel
··ян·ый	*a m, nom/acc si*	
··т·ый	*past part pass m, nom/acc si* / *a m, nom/acc si*	

Ending	Grammar	Remarks
··ат·ый	*a m, nom/acc si*	**a** always stressed
·новат·ый	*a m, nom/acc si*	-ic (chem)
··чат·ый	*a m, nom/acc si*	
··ет·ый	*past part pass m, nom/acc si*	
	a m, nom/acc si	
··ит·ый	*a m, nom/acc si*	
··от·ый	*a m, nom/acc si*	
··аст·ый	*a m, nom/acc si*	
··ист·ый	*a m, nom/acc si*	-ous (chem)
··новатист·ый	*a m, nom/acc si*	hypo- (chem)
··нут·ый	*past part pass m, nom/acc si*	
	a m, nom/acc si	
··ят·ый	*past part pass m, nom/acc si*	
	a m, nom/acc si	
··ю·й	*imp/perf, imper, si*	
··я·й	1. *imp/perf, imper, si*	Infin ·ять
	2. *m, nom/acc si*	
–к·	See ··· zero at beginning	
··ак·	*m, nom/acc si*	
··ек·	1. *m, nom/acc si*	
	2. *f, g pl*	Nom si ··к·а
	3. *a m, sf, si*	Long form ··к·ий
·ик·	*m, nom/acc si*	
··ник·	*m, nom/acc si*	
··чик·	*m, nom/acc si*	English ending often -er
··щик·	*m, nom/acc si*	English ending often -er
··ок·	1. *m, nom/acc si*	
	2. *f, g pl*	Nom si ··к·а
	3. *a m, sf, si*	Long form ··к·ий
··енок·	*m, nom/acc si*	Often denotes young creature. Nom pl ··ят·а, ··ат·а
··онок·	*m, nom/acc si*	Often denotes young creature. Nom pl ··ят·а, ··ят·а
··як·	*m, nom/acc si*	
····л	1. *imp/perf, past m si*	Very common
	2. *m, nom/acc si*	Uncommon
	3. *f, acc anim pl*	Nom si –л·а. Uncommon
	4. *f, g pl*	Nom si –л·а. Uncommon
	5. *n, g pl*	Nom si –л·о. Uncommon
	6. *a m, sf, si*	Long form –л·ый. Uncommon
··ива·л	1. *imp, past, m si*	Infin ··ива·ть. Imperfective
	2. See ···л above	
··ыва·л	1. *imp, past, m si*	Infin ··ыфа·ть. Imperfective
	2. See ···л above	
–м·	See ··· zero at beginning	
····ам	1. *m, d pl*	
	2. *f, d pl*	
	3. *n, d pl*	

Ending	Grammar	Remarks
···ем	1. *m, i si*	Nom si ··· zero, ···ь, ···й
	2. *n, i si*	Nom si ···е, ··и·е
	3. *a m, p si*	For suffixes see masc nom si ···ий, ···ый
	4. *a n, p si*	For suffixes see masc nom si ···ий, ···ый
	5. *imp, pres, 1 pl*	
	6. *perf, fut, 1 pl*	
-ем·	7. *a m, sf, si*	Long form -ем·ый
-ем··	8. *f, g pl*	Nom si -м··а
··ива·ем	*imp, pres 1 pl*	Infin ива·ть. Imperfective
··ыва·ем	*imp, pres 1 pl*	Infin ыва·ть. Imperfective
··и·ем	1. *m, i si*	Nom si ··и·й
	2. *n, i si*	Nom si ··и·е
-м·ен·ем	*n, i si*	Nom si -м·я
-изм··	*m, nom/acc si*	English ending -ism
···им	1. *a m, i si*	For suffixes see ···ий, ···ый
	2. *a n, i si*	For suffixes see ···ий, ···ый
	3. *a, d pl*	For suffixes see ···ий, ···ый
	4. *imp, pres, 1 pl*	Infin of regular paradigms -ить
	5. *perf, fut, 1 pl*	Infin of regular paradigms -ить
-им··	6. *m, nom si*	Rare
··им·	7. *a m, sf, si*	Long form -м·ый
···ом	1. *m, i si*	Nom si ··· zero
	2. *n, i si*	Nom si ···о
	3. *a m, p si*	For suffixes see ···ый, ии
	4. *a n, p si*	For suffixes see ···ый, ии
	5. *adv*	Instrumental case derived from noun
-ом··	6. *m, nom/acc si*	
···ым	1. *a m, i si*	For suffixes see ···ый
	2. *a n, i si*	For suffixes see ···ый
	3. *a, d pl*	For suffixes see ···ый
···ям	1. *m, d pl*	
	2. *f, d pl*	
	3. *n, d pl*	
-н·	See ··· zero at beginning	
··ан·	1. See ··· zero at beginning	
	2. *Past part pass m, sf, si*	Regular infin -ать
··чан	*m, g pl*	Nationalities and types of people
··ен·	1. See ··· zero at beginning	
	2. *Past part pass m, sf, si*	Regular infin -ить
-м·ен	1. See ··· zero at beginning	
	2. *n, g pl*	Nom si -м·я
··ин·	1. See ··· zero at beginning	
	2. *m, nom/acc si*	
	3. *possess adj m, nom si*	
··анин·	*m, nom si*	Nationalities and types of people
··чанин·	*m, nom si*	Nationalities and types of people
··янин·	*m, nom si*	Nationalities and types of people
···о	1. *n, nom/acc si*	
	2. *a n, sf, si*	For suffixes see ···ый, ···ий
	3. *adv*	Derived from adj ···ый, ···ий, ···ой

Ending	Grammar	Remarks
··ств·о	*n, nom/acc si*	
···его	1. *a m, acc anim si*	For suffixes see ···**ий**, ···**ый** (rare)
	2. *a m, g si*	For suffixes see ···**ий**, ···**ый** (rare)
	3. *a n, g si*	For suffixes see ···**ий**, ···**ый** (rare)
···ого	1. *a m, acc anim si*	For suffixes see ···**ый**, ···**ий**
	2. *a m, g si*	For suffixes see ···**ый**, ···**ий**
	3. *a n, g si*	For suffixes see ···**ый**, ···**ий**
···ло	1. *imp/perf, past, n si*	Common
-л·о	2. *n, nom/acc si*	Uncommon
	3. *a n, sf, si*	Long form **-л·ый**. Uncommon
	4. *adv*	Derived from adj **-л·ый**. Uncommon
···ива·ло	*imp, past, n si*	Infin ··**ива·ть**. Imperfective
···ыва·ло	*imp, past, n si*	Infin ··**ыва·ть**. Imperfective
-п·	See ··· zero at beginning	
-р·	See ··· zero at beginning	
-атор·	*m, nom/acc si*	English ending often -er
-ятор·	*m, nom/acc si*	English ending often -er
-с·	See ··· zero at beginning	
-т·	See ··· zero at beginning	
···ат	1. *imp, pres, 3 pl*	
	2. *perf, fut, 3 pl*	
··ат·	3. *m, nom/acc si*	
	4. *f, g pl*	Nom si **-ат·а**
	5. *a m, sf, si*	Long form **-ат·ый**
···ет	1. *imp, pres, 3 si*	
	2. *perf, fut, 3 si*	
··ет·	3. *m, nom/acc si*	
	4. *f, g pl*	Nom si **-ет·а**
··а·ет	1. *imp, pres, 3 si*	Regular infin **а·ть**
	2. *perf, fut, 3 si*	Regular infin **а·ть**
··ива·ет	*imp, pres, 3 si*	Infin ··**ива·ть**. Imperfective
··ыва·ет	*imp, pres, 3 si*	Infin ··**ыва·ть**. Imperfective
··е·ет	1. *imp pres, 3 si*	Regular infin **-е·ть**
	2. *perf, fut, 3 si*	Regular infin **-е·ть**
··итет·	*m, nom/acc si*	English ending -ity, -ty
··у·ет	1. *imp, pres, 3 si*	Infin **-овать**
	2. *perf, fut, 3 si*	Infin **-овать**
··ю·ет	1. *imp, pres, 3 si*	Infin **-евать**
	2. *perf, fut, 3 si*	
··я·ет	1. *imp, pres, 3 si*	Infin **-я·ть**
	2. *perf, fut, 3 si*	
···ит	1. *imp, pres, 3 si*	Regular infin **ить**. Common
	2. *perf, fut, 3 si*	Regular infin **ить**. Common
-ит·	3. *m, nom/acc si*	English -itis (med), -ite (geol)
	4. *a m, sf, si*	Long form **ит·ый**
	5. *past part pass m, sf, si*	Long form **ит·ый**
··ист·	*m, nom/acc si*	English -ist
···ут	1. *imp, pres, 3 pl*	Common
	2. *perf, fut, 3 pl*	Common
-ут·	3. *m, nom/acc si*	Uncommon
···ют	1. *imp, pres, 3 pl*	
	2. *perf, fut, 3 pl*	

Ending	Grammar	Remarks
•а·ют	1. *imp, pres, 3 pl*	Regular infin **а·ть**
	2. *imp, pres, 3 pl*	Regular infin **а·ть**
··ива·ют	*imp, pres, 3 pl*	Infin ··**ива·ть**. Imperfective
··ыва·ют	*imp, pres, 3 pl*	Infin ··**ыва·ть**. Imperfective
··е·ют	1. *imp, pres, 3 pl*	Regular infin **еть**
	2. *perf, fut, 3 pl*	Regular infin **еть**
··у·ют	1. *imp, pres, 3 pl*	Infin **-овать**
	2. *perf, fut, 3 pl*	Infin **-овать**
··ю·ют	1. *imp, pres, 3 pl*	Infin **-евать**
	2. *perf, fut, 3 pl*	Infin **-евать**
··я·ют	1. *imp, pres, 3 pl*	Regular infin **я·ть**
	2. *perf, fut, 3 pl*	Regular infin **я·ть**
···ят	1. *imp, pres, 3 pl*	Regular infin **-ить**
	2. *perf, fut, 3 pl*	Regular infin **-ить**
	3. *m, acc/g pl*	Young creatures
-ят·	4. *numeral*	
	5. *a m, sf, si*	Long form **-ят·ый**
···у	1. *m, d si*	Nom si ··· zero
	2. *f, acc si*	Nom si ···**а**
	3. *n, d si*	Nom si ···**о**
	4. *possess adj m, d si*	Masc nom si ··**ов·**, ··**ев·**, ··**ин·**
	5. *possess adj f, acc si*	Masc nom si ··**ов·**, ··**ев·**, ··**ин·**
	6. *imp, pres, 1 si*	
	7. *perf, fut, 1 si*	
	8. *adv*	
···ему	1. *a m, d si*	For suffixes see ···**ий**, ···**ый**
	2. *a n, d si*	For suffixes see ···**ий**, ···**ый**
··ем·у	3. *f, acc si*	Nom si ··**ем·а**
···ому	1. *a m, d si*	For suffixes see ···**ый**, ···**ий**. Common
	2. *a n, d si*	For suffixes see ···**ый**, ···**ий**. Common
··ом·у	3. *m, d si*	Nom si **-ом·**. Uncommon
	4. *f, acc si*	Nom si **-ом·а**. Uncommon
-ф·	See ··· zero at beginning	
-х·	See ··· zero at beginning	
···ах	1. *m, p pl*	
	2. *f, p pl*	
	3. *n, p pl*	
··ах·	4. *m, nom/acc si*	Uncommon
-м·ен·ах	*n, p pl*	Nom si **-м·я**
···ех	1. *pron, acc anim pl*	
	2. *pron, g pl*	
	3. *pron, p pl*	
	4. *num, g pl*	Three or four
	5. *num, p pl*	Three or four
··ех·	6. *m, nom/acc sing*	Rare
···их	1. *a, acc anim pl*	For suffixes see ···**ий**, ···**ый**
	2. *a, g pl*	For suffixes see ···**ий**, ···**ый**
	3. *a, p pl*	For suffixes see ···**ий**, ···**ый**
···ух	1. *num, g/p*	Two
··ух·	2. *m, nom/acc si*	Rare
···ых	1. *a, acc anim pl*	For suffixes see ···**ый**, ···**ий**
	2. *a, g pl*	For suffixes see ···**ый**, ···**ий**
	3. *a, p pl*	For suffixes see ···**ый**, ···**ий**

Ending	Grammar	Remarks
···ях	1. *m, p pl* 2. *f, p pl* 3. *n, p pl*	
-ц·	See ··· zero at beginning	
-ч·	See ··· zero at beginning	
-ш·	See ··· zero at beginning	
-щ·	See ··· zero at beginning	
···ы	1. *m, nom/acc pl* 2. *f, g si* 3. *f, nom/acc pl* 4. *a, sf, pl* 5. *possess adj, nom/acc pl*	Nom si ··· zero Nom si ···**а** Nom si ···**а** For suffixes see long form nom si ···**ый**, ···**ой** Masc nom si ··**ов·**, ··**ев·**, ··**ин·**
···ь	1. *m, nom/acc si* 2. *f, nom/acc si* 3. *f, g pl* 4. *a m, sf, si* 5. *imp/perf, imper, si* 6. *imp/perf, infin* 7. *adv*	See below for suffixes See below for suffixes Nom si ···**я** Long form ···**ий**. Rare Uncommon **ь** preceded by **ч**
-тел·ь	*m, nom/acc si*	**т** preceded by vowel often signifies: agent, operator, doer
··ен·ь	1. *m, nom/acc si* 2. *f, g pl*	G si ··**н·я** Nom si ··**н·я**
·зн·ь	*f, nom/acc si*	
··ар·ь	*m, nom/acc si*	
-сь	1. *reflexive particle* (for meanings see -**ся**)	Added to the end of inflections. Look up letters before -**сь**. Very common
-с·ь	2. *f, nom/acc si*	Nearly all end -**вес·ь**, -**мес·ь**, -**кис·ь**, -**пис·ь**. Uncommon
	3. See remarks	Only common words ending in -**сь** are: **весь** = all; **здесь** = here
···ть	1. *imp, infin* 2. *perf, infin*	For suffixes see below For suffixes see below
-т·ь	3. *f, nom/acc si*	For suffixes see below
··а·ть	1. *imp/perf infin* 2. *f, nom/acc si*	Rare
··ева·ть	*imp/perf, infin*	
··ива·ть	*imp, infin*	
··ова·ть	*imp/perf infin*	Exception **кров·ат·ь** (*f*) = bed
-из·ов·ать	*imp/perf infin*	English often -ize
-ир·ова·ть	*imp/perf infin*	Often Russianized foreign words
-изир·ова·ть	*imp/perf infin*	English often -ize
··ыва·ть	*imp, infin*	
··е·ть	1. *imp/perf, infin*	
-ет·ь	2. *f, nom/acc si*	Rare
··ить	1. *imp/perf, infin*	
	2. *f, nom/acc si*	Rare
-от·ь	1. *m, nom/acc sing*	
··от·ь	2. *f, nom/acc sing*	
··о·ть	3. *imp/perf, infin*	

Ending	Grammar	Remarks
···сть	1. *imp/perf, infin*	Irregular verbs
-ст·ь	2. *f, nom/acc si*	For suffixes see below
··ест·ь	*f, nom/acc si*	Often signifies a quality or property. Derived from adj **-ий**, **-ый**, **-ой**
··учест·ь	*f, nom/acc si*	Often signifies a quality or property. Derived from adj **-ий**, **-ый**, **-ой**
···ост·ь	*f, nom/acc si*	Often signifies a quality or property. Derived from adj **-ий**, **-ый**, **-ой**
··ивост·ь	*f, nom/acc si*	Often signifies a quality or property. Derived from adj **-ий**, **-ый**, **-ой**
··овост·ь	*f, nom/acc si*	Often signifies a quality or property. Derived from adj **-ий**, **-ый**, **-ой**
··лост·ь	*f, nom/acc si*	Often signifies a quality or property. Derived from adj **-ий**, **-ый**, **-ой**
··аемост·ь	*f, nom/acc si*	Derived from pres part pass of verb. Infin **·а·ть**
··уемост·ь	*f, nom/acc si*	Infin **-овать**
··яемост·ь	*f, nom/acc si*	Infin **··я·ть**
··имост·ь	*f, nom/acc si*	Infin **-ить**
··ност·ь	*f, nom/acc si*	Derived from adj **··н·ый**
-тост·ь	*f, nom/acc si*	Derived from adj **-т·ый**
у··ть	*imp/perf infin*	
··н·уть	·*imp/perf infin*	
·я·ть	1. *imp/perf infin*	
	2. See also **··т·ь**	
···чь	1. *imp/perf, infin*	Irregular
··ч··ь	2. *f, nom/acc si*	
···шь	1. *imp, pres, 2 si*	
	2. *perf, fut, 2 si*	
···ешь	1. *imp, pres, 2 si*	Regular infin **··а·ть**
	2. *perf, fut, 2 si*	Regular infin **··а·ть**
··ива·ешь	*imp, pres, 2 si*	Infin **··ива·ть**
··ыва·ешь	*imp, pres, 2 si*	Infin **··ыва·ть**
··у·ешь	1. *imp, pres, 2 si*	Infin **овать**
	2. *perf, fut, 2 si*	
··ю·ешь	1. *imp, pres, 2 si*	Infin **евать**
	2. *perf, fut, 2 si*	
···ишь	1. *imp, pres, 2 si*	Regular infin **···ить**
	2. *perf, fut, 2 si*	Regular infin **···ить**
-э	Indeclinable nouns	
···ю	1. *m, d si*	Nom si **···ь**, **···й**
	2. *f, acc si*	Nom si **···я**, **··и·я**
	3. *n, d si*	Nom si **···е**, **··и·е**
	4. *imp, pres, 1 si*	
	5. *perf, fut, 1 si*	
··ива·ю	*imp, pres, 1 si*	Infin **··ива·ть**
··ыва·ю	*imp, pres, 1 si*	Infin **··ыва·ть**
··е·ю	1. *imp, pres, 1 si*	Regular infin **-е·ть**
	2. *perf, fut, 1 si*	Regular infin **-е·ть**
··ею	3. *f, i si*	Rare
	4. *a/pron f, i si*	Rare

Ending	Grammar	Remarks
··и·ю	1. *m, d si*	Nom si ··и·й
	2. *f, acc si*	Nom si ··и·я
	3. *n, d si*	Nom si ··и·е
-о·ю	1. *f, i si*	Rare
	2. *a/pron f, i si*	Rare. For suffixes see **-ый**, **-ой**
	3. See also ···ю	
···ую	1. *a f, acc si*	For suffixes see **-ый**, **-ий**
	2. *imp, pres, 1 si*	Infin **-овать**
	3. *perf, fut, 1 si*	Infin **-овать**
	4. *adv*	
-у··ю	5. *m, d si*	Rare
···ью	1. *f, i si*	Nom si ···ь
	2. *possess adj f, acc si*	Masc nom si **-ий**
	3. *adv*	Uncommon
-ь·ю	4. *n, d si*	Nom si ··ь·е. Common
···юю	1. *a f, acc si*	For suffixes see masc nom si ···ий
	2. *imp, pres, 1 si*	Infin **-евать**
	3. *perf, fut, 1 si*	Infin **-евать**
··я·ю	1. *imp, pres, 1 si*	Regular infin ··я·ть
	2. *perf, fut, 1 si*	Regular infin ··я·ть
···я	1. *m, acc anim si*	Nom si ···ь, ···й. Uncommon
	2. *m, g si*	Nom si ···ь, ···й. Common
	3. *f, nom si*	Nom si ···ь, ···й. Common
	4. *n, g si*	Nom si ···е, ··и·е. Very common
	5. *n, nom/acc pl*	Nom si ···е, ··и·е. Very common
	6. *a f, sf, si*	For suffixes see **-ий**
	7. *ger, pres*	Regular infin **-ить**
···ая	1. *a f, nom si*	For suffixes see **-ый**, **-ий**
	2. *f, nom si*	Declined like adj. Uncommon
	3. *m, g si*	Often names of rooms. Nom si **-а·й**. Uncommon
··а·я	4. *ger, pres*	Regular infin **-а·ть**
··и·я	1. *m, acc anim si*	Nom si ···ь, ···й. Uncommon
	2. *m, g si*	Nom si ···ь, ···й. Common
	3. *f, nom si*	Common
	4. *n, g si*	Nom si ···е, ··и·е. Very common
	5. *n, nom/acc pl*	Nom si ···е, ··и·е. Very common
··аци·я	*f, nom si*	English ending often -ation
··изаци·я	*f, nom si*	English ending often -ization
··есценци·я	*f, nom si*	English ending often -escence
··яци·я	*f, nom si*	English ending often -ation
-м··я	*n, nom/acc si*	G si **-м·ен·и**
		Nom pl **-м·ен·а**
		Uncommon declension with a few common examples, e.g. **время** = time; **имя** = name
		Note **тремя** = instr of **три** = three **двумя** = instr of **два** = two **четырьмя** = instr of **четыре** = four
·н·я	1. *f, nom si*	Often signifies a room or workshop
	2. See ···я	

Ending	Grammar	Remarks
··**альн·я**	1. *f, nom si* 2. See ···**я**	Often signifies a room or workshop
··**ильн·я**	1. *f, nom si* 2. See ···**я**	Often signifies a room or workshop
-**ся**	Reflexive particle. This appears on the end of the inflections of some verbs and the participles (verbal adjectives) and gerunds (verbal adverbs) derived from them. The inflections can be found elsewhere in this supplementary dictionary of word endings. (Simply look up words as if letters **-ся** were not present.) The meanings of the reflexive may be any of the following: 1. Reflexive action (e.g. it moves itself) 2. Reciprocal action (e.g. they move each other) 3. Passive (e.g. it is moved) 4. Intransitive (e.g. it moves) 5. No different from the ordinary non-reflexive form. The only common word which can be confused with the ending **-ся** is **вся** = all (*fem nom si*)	
··**у·я** -**у··я**	1. *ger, pres* 2. *f, nom si* 3. *m, g si*	Infin -**овать** Nom si **-у··й**

··**ь·я**	1. *m, acc anim si*
	2. *m, g si*
	3. *f, nom si*
	4. *n, g si*
	5. *n, nom/acc pl*
	6. Some irregular nom/acc plurals, e.g. nom si
	   сын (*m*)
	   стул (*m*)
	   лист (*m*)
	   крыло (*n*)
	   дерево (*n*)
	   перо (*n*)
	7. *a f, sf, si*
		Nom si **-е·й**
		Nom si **-е·й**
		
		Nom si **-ь·я**
		Nom si **-ь·я**

··**ю·я**	*ger, pres*	Infin -**евать**
···**яя**	1. *a f, nom si* 2. *ger, pres*	Masc nom si **-ий** Regular infin ··**я·ть**

ABBREVIATIONS

a	adjective
abbr	abbreviation, etc.
a/c	aircraft, aircraft industry
acc	accusative case
acous	acoustics
act	active
adj	adjective
adv	adverb
aerodynam	aerodynamics
agr	agriculture
air	aeronautics, air transport
anat	anatomy
arch	architecture
astron	astronomy
attrib	attributive
autom	automation, automatics, control engineering
bact	bacteriology
biol	biology
bot	botany
Brit	British
build	building
cast	casting, foundry work
ceram	ceramics
chem	chemistry
cinema	cinematography
civ eng	civil engineering
coll	colloquial
com	commerce
comp	comparative
component	word component, complex combining form
conj	conjunction
cryst	crystallography
cyt	cytology
d	dative case
deter	determinate
dim	diminutive
dipl	diplomatic
dynam	dynamics
econ	economics, economy
elec	electrical, electronics
eng	engineering
ent	entomology
esp	especially
expl	explosives
f	feminine noun
fem	feminine, feminine gender
fin	finance, financial
fish	fish, fisheries, fishing ichthyology
food	food, food technology
g	genitive case
g as adj	used in genitive case as an adjective
gen	genetics
geog	geography, geographical
geol	geology
ger	gerund
gram	grammar, grammatical
gunn	gunnery, guns, ordnance, ammunition
GW	guided weapons
horol.	horology
hortic.	horticulture
hydr.	hydraulics hydrology
hydr eng	hydraulic engineering
hydrog	hydrography
i	instrumental case
ICE	internal combustion engine
imp	imperfective verb
imper	imperative
indecl	indeclinable
indet	indeterminate
ins	insurance
instr.	instrumentation, experimental equipment
int	international
intrans	intransitive
law	law, legal
ling	linguistics
m	masculine noun
mar	marine
mar eng	marine engineering
mar ins	marine insurance
masc	masculine, masculine gender
math	mathematics, mathematical
mech.	mechanical, mechanics